KENT'S
MECHANICAL ENGINEERS' HANDBOOK
Design and Production Volume

WILEY ENGINEERING HANDBOOK SERIES

HANDBOOK OF ENGINEERING MATERIALS. Edited by DOUGLAS F. MINER and JOHN B. SEASTONE.

HANDBOOK OF ENGINEERING FUNDAMENTALS. Second Edition. Edited by OVID W. ESHBACH.

KENT'S MECHANICAL ENGINEERS' HANDBOOK. Twelfth Edition.

DESIGN AND PRODUCTION. Edited by COLIN CARMICHAEL.

POWER. Edited by J. K. SALISBURY.

ELECTRICAL ENGINEERS' HANDBOOK. Fourth Edition.

ELECTRIC POWER. Edited by HAROLD PENDER and WILLIAM A. DEL MAR.

ELECTRIC COMMUNICATION AND ELECTRONICS. Edited by HAROLD PENDER and KNOX MCILWAIN.

MINING ENGINEERS' HANDBOOK. Third Edition. Edited by the late ROBERT PEELE.

HANDBOOK OF MINERAL DRESSING.

ORES AND INDUSTRIAL MINERALS. Edited by ARTHUR F. TAGGART.

ELEMENTS OF ORE DRESSING. ARTHUR F. TAGGART.

KENT'S MECHANICAL ENGINEERS' HANDBOOK

IN TWO VOLUMES

Design and Production Volume

Prepared by a Staff of Specialists

COLIN CARMICHAEL

Editor

TWELFTH EDITION

WILEY ENGINEERING
HANDBOOK SERIES

JOHN WILEY & SONS, INC.
NEW YORK · LONDON · SYDNEY

TWELFTH EDITION
30 29 28 27 26 25 24 23 22

ISBN 0 471 46959 9

PRINTED IN THE UNITED STATES OF AMERICA

CONTRIBUTORS

J. O. Almen, Research Laboratories Division, General Motors Corp. ***Shot Peening.***

E. A. Anderson, Research Division, The New Jersey Zinc Co. (of Pa.). ***Zinc and Its*** *Alloys.*

E. Stanley Ault, Professor of Machine Design, Purdue University. ***Straight, Needle, and*** *Tapered Roller Bearings.*

Chester R. Austin, Supervisor, Ceramic Division, Battelle Memorial Institute. ***Ceramics.***

F. S. Badger, Haynes Stellite Co. *Precision Investment Casting.*

Claire C. Balke, Director of Research, Roberts and Mander Corp. ***Powder Metallurgy.***

A. M. Baltzer, Assistant Manager, Industrial Department, National Safety Council. *Safety Engineering.*

A. G. Baumgartner, The Cincinnati Shaper Co. ***Metal-working Shapers.***

Roger W. Bolz, Mechanical Engineer, Cleveland, and Associate Editor, ***Machine Design.*** *Barrel Finishing; Broaching.*

James A. Broadston, Aerophysics Field Laboratory, North American Aviation, Inc., and Chief Engineer, Surface Checking Gage Co. ***Measurement and Designation of Surface Quality.***

Allan H. Candee, Gleason Works. *Bevel-gear Cutting.*

Colin Carmichael, Editor, *Machine Design.* *Cone Pulleys; Fabricated Materials; Fasteners; Hardness, Density, and Compressibility; Hydraulic Torque Converters; Shafts, Keys, and Splines.*

George S. Cherniak, Vice President, Lessells and Associates, Inc., Consulting Engineers. *Strength of Materials.*

R. M. Cherry, Industrial Heating Division, General Electric Co. ***Industrial Electric Heating.***

Eugene C. Clarke, President, Chambersburg Engineering Co. ***Forging.***

Carter S. Cole, American Society for Testing Materials. ***Sand-cast Copper-base Alloys.***

Lester V. Colwell, Associate Professor, Department of Metal Processing, University of Michigan. *Stamping Processes.*

Theodore Crane, Consulting Engineer, Professor of Architectural Engineering, Yale University. *Building Construction.*

S. Leroy Crawshaw, Manager, Engineering and Sales, Western Gear Works. ***Gearing.***

Samuel Damon, Westinghouse Electric Corp. *Brazing.*

Robert C. Deale, The Sylvester Co. *Single-point Metal-cutting Tools.*

Charles K. Donoho, Chief Metallurgist, American Cast Iron Pipe Co. ***Iron and Steel*** *Casting.*

V. L. Doughtie, Professor of Mechanical Engineering, University of Texas. ***Mechanism,*** *Mechanics, and Machine Elements.*

W. J. Eckert, Director of Pure Science, International Business Machines Corp. *Calculating Devices.*

Hans Ernst, Director of Research, The Cincinnati Milling Machine Co. ***Milling Machine*** *Practice.*

C. T. Evans, Jr., Chief Metallurgist, Elliott Co. ***High-temperature-resistant Alloys.***

John L. Everhart, Battelle Memorial Institute. *The Less Common Metals.*

Paul G. Exline, Exline Engineering Co., Tulsa, Okla. *Lubricants.*

R. D. Faber, Development and Research Division, The International Nickel Co. *Nickel and Its Alloys.*

J. E. Faloon, Plastics Division, Chemical Department, General Electric Co., Pittsfield, Mass. *Plastics Molding.*

Robert O. Fehr, General Engineering and Consulting Laboratory, General Electric Co. *Vibration and Noise Control.*

Charles D. Ferris, Bigelow-Sanford Carpet Co., Thompsonville, Conn. ***Statistical Quality*** *Control.*

R. B. Foley, Sales Engineer, American Blower Corp. ***Dust Collection.***

J. C. Fox, Chief Metallurgist, Doehler-Jarvis Corp. ***Die-casting Alloys.***

A. F. Gagne, Jr., Development Engineering Division, E. I. du Pont de Nemours and Co. *Couplings.*

H. W. Gillett, Chief Technical Adviser, Battelle Memorial Institute. ***The Application of Data on Strength of Metallic Materials; Iron and Steel.***

W. T. Gloor, The Motch and Merryweather Machinery Co. ***Metal Sawing.***

W. P. Goepfert, Aluminum Company of America. *Aluminum-alloy Casting; Aluminum and Its Alloys.*

Bruce W. Gonser, Battelle Memorial Institute. *The Less Common Metals; Solders.*

William H. Gross, Magnesium Division, The Dow Chemical Co. *Magnesium-alloy Casting; Magnesium and Its Alloys.*

M. N. Halberg, Industrial Power Division, General Electric Co. *Electric Motors and Their Control.*

Herbert Harris, Jr., Sperry Gyroscope Co., Inc. *Servomechanisms.*

R. E. W. Harrison, Consultant, Management Engineering, Chambersburg, Pa. *Abrasive Processes.*

E. L. Hemingway, Research Engineer, Gisholt Machine Co. *Superfinishing.*

L. W. Herchenroeder, Westinghouse Electric Corp. *Industrial Electronic Control.*

W. R. Hibbard, Assistant Metallurgist, The American Brass Co. *Copper; Wrought Copper-base Alloys.*

Samuel G. Hibben, Director of Applied Lighting, Westinghouse Electric Corp. *Illumination.*

George S. Hoell, Division Director, Mechanical and Civil Engineering Division, The Franklin Institute. *Limitations on Planetary Gear Ratios.*

H. A. Stevens Howarth, Consulting Engineer on Oil Film Bearings, Philadelphia, Pa., and Research Engineer, Franklin Institute Laboratories for Research and Development. *Classification of Bearings; The Loading and Friction of Bearings on Films of Lubricant.*

James H. Lansing, Technical and Research Director, Malleable Founders' Society. *Malleable Iron Castings.*

E. T. Larson, Norton Co. *Abrasive Wheels; Factors Affecting Grinding-wheel Selection.*

William H. Lehmberg, Director of Engineering, American Felt Co. *Felt.*

John M. Lessells, President, Lessells and Associates, Inc., Consulting Engineers, and Associate Professor of Mechanical Engineering, Massachusetts Institute of Technology. *Strength of Materials.*

L. S. Linderoth, Jr., Professor of Mechanical Engineering, Iowa State College of Agriculture and Mechanic Arts. *Machine Hydraulics; Pressure Seals.*

J. Spotts McDowell, Harbison-Walker Refractories Co. *Refractories.*

L. J. Markwardt, Assistant Director, U. S. Forest Products Laboratory. *Wood.*

Mario Martellotti, Research Engineer, The Cincinnati Milling Machine Co. *Milling Machine Practice.*

John J. Martin, Bell Telephone Laboratories, Inc. *Plastics.*

E. L. Murray, The Warner and Swasey Co. *Lathe Work.*

R. M. Nardone, Chief Engineer, Victory Engineering Corp. *Brakes; Clutches.*

Waldemar Naujoks, Special Projects Engineer, Ladish Co., Cudahy, Wis. *Forging.*

P. E. Nokes, United Shoe Machinery Corp. *Machine-performance Testing.*

C. W. Obert, Consulting Engineer, New York. *Flame Cutting; Welding of Materials.*

Carl J. Oxford, Chief Engineer, National Twist Drill and Tool Co. *Drills and Drilling.*

Gunnar Palmgren, Chief Engineer, SKF Industries, Inc. *Rolling Bearings.*

Thomas D. Perry, Engineer in Wood Work, Moorestown, N. J. *Adhesives and Glues; Woodworking.*

Norman B. Pilling, Director, Research Laboratory, The International Nickel Co., Inc. *General Properties of Metals and Alloys.*

Matthew W. Potts, Materials Handling Consultant, New Rochelle, N. Y. *Materials Handling.*

M. L. Price, Professor of Machine Design, Worcester Polytechnic Institute. *Belt Drives.*

Philip G. Rhoads, J. E. Rhoads and Sons. *Short-center Belt Drives.*

E. F. Riesing, Manager, Automotive Engineering and Sales, Firestone Industrial Products Division of the Firestone Tire and Rubber Co. *Rubber and Rubber-like Resilient Materials.*

Bryce Ruley, Senior Engineer, SKF Industries, Inc. *Rolling Bearings.*

Harry M. St. John, Crane Co. *Copper-base Alloy Casting.*

E. L. Schwarz-Kast, University of Chicago. *Variable-speed Drives.*

G. W. Seagren, Senior Fellow, Mellon Institute. *Organic Coatings.*

Harold S. Sizer, Assistant Director of Design, Brown and Sharpe Manufacturing Co. *Specification of Dimensions and Tolerances.*

Ed S. Smith, Ballistician, Aberdeen Proving Ground. *Automatic Regulators.*

Robert B. Sosman, Professor of Ceramics, Rutgers University. *Thermometers and Pyrometers.*

Frank N. Speller, Metallurgical Consultant, Pittsburgh, Pa. *Corrosion of Metals.*

W. O. Sweeny, Haynes Stellite Co. *Precision Investment Casting.*

W. Trinks, Professor Emeritus of Mechanical Engineering, Carnegie Institute of Technology. *Industrial Furnaces.*

H. H. Uhlig, in Charge of Corrosion Laboratory, Massachusetts Institute of Technology. *Corrosion Resistance of Metals and Alloys.*

Arthur F. Underwood, Research Laboratories Division, General Motors Corp. *Bearing Metals.*

Harold G. Vogt, Physicist, Corning Glass Works. *Glass.*

A. M. Wahl, Research Laboratories, Westinghouse Electric Corporation. *Mechanical Springs.*

John M. Walter, Vice President in Charge of Engineering, The G. A. Gray Co. *Planer Work.*

William G. Waltermire, Chief Products Engineer, Lamson and Sessions Co. *Strength of Screw Fasteners.*

Ermand L. Watelet, Charge of Design, Small Tool Division, Brown and Sharpe Manufacturing Co. *Dimensional Measurements.*

Arthur M. Weinberg, Chemical Department, General Electric Co. *Silicones.*

G. L. Werley, Development Engineering Division, New Jersey Zinc Co. (of Pa.). *Zinc-alloy Casting.*

Sidney J. Williams, Assistant to the President, National Safety Council. *Safety Engineering.*

W. A. Williams, Chief Engineer, The American Pulley Co. *Pulleys.*

H. A. Wilson, Allis-Chalmers Manufacturing Co. *V-belt Drives.*

PREFACE

Much of the preliminary planning for this new edition of *Kent's Mechanical Engineers' Handbook* was done by Robert Kent prior to his untimely death in 1947. More than half of the contributors had been given their assignments and were already at work on their sections. The present editor has kept in constant view the objectives that guided both William Kent and Robert Kent in preparing the previous eleven editions of this Handbook. However, advantage has been taken of the latitude permitted in the selection of the remaining subjects and their authors to effect a slight change in emphasis, which is symbolized by the adoption of the subtitle *Design and Production*, in place of the previous *Design, Shop Practice*, for this volume of the Handbook.

The earlier title recognized the tie-in between the design engineering department and the shop. The new one underscores the fact that design and production are becoming recognized as simply two phases of a single activity—the engineering of a product from the inception of an idea to the completion of the finished article at an economical cost.

This volume is directed primarily toward the engineers who *design* and *manufacture* machinery, appliances, mechanical equipment, and other engineered products. The twenty-eight sections into which the book has been rearranged may be classified broadly as: Selection of Materials (Sections 1 to 6); Design Principles (Sections 7 to 9); Design and Selection of Machine Components (Sections 10 to 18); Production Processes (Sections 19 to 25); Production Plant Equipment (Sections 26 and 27); and Mathematical Tables (Section 28).

The presentation of each subject attempts to give the engineer a quick grasp of the essentials of the field, together with pertinent technical data in condensed form. These data consist essentially of: basic principles; working formulas, charts, and tables; standard dimensions, proportions, and specifications; and illustrated discussions of typical equipment. References included in the text and at the ends of many of the chapters guide the reader to sources of more detailed information.

With minor exceptions, the book has been almost completely rewritten. New material, required by progress in many fields, has outweighed the obsolete material that has been discarded, resulting in a book more than three hundred pages longer than the previous edition. The ninety-two contributors have prepared material ranging in length from a few paragraphs to entire sections. In addition, many industrial concerns have furnished data and illustrations, either to the individual authors or to the editor directly, the sources being indicated in all cases.

The publishers and the editor take this occasion to acknowledge with appreciation the cooperation of the many individuals and organizations whose interest and help have made this volume possible.

Colin Carmichael

March 1950

ABSTRACT FROM PREFACE TO THE FIRST EDITION, 1895

More than twenty years ago the author began to follow the advice given by Nystrom: "Every engineer should make his own pocket-book, as he proceeds in study and practice, to suit his particular business." The manuscript pocket-book thus begun, however, soon gave place to more modern means for disposing of the accumulation of engineering facts and figures, viz., the index rerum, the scrap-book, the collection of indexed envelopes, portfolios and boxes, the card catalogue, etc. Four years ago, at the request of the publishers, the labor was begun of selecting from this accumulated mass such matter as pertained to mechanical engineering, and of condensing, digesting, and arranging it in form for publication. In addition to this, a careful examination was made of the transactions of engineering societies, and of the most important recent works on mechanical engineering in order to fill gaps that might be left in the original collection, and insure that no important facts had been overlooked.

Some ideas have been kept in mind during the preparation of the pocket-book that will, it is believed, cause it to differ from other works of its class. In the first place it was considered that the field of mechanical engineering was so great, and the literature of the subject so vast, that as little space as possible should be given to subjects which especially belong to civil engineering.

Another idea prominently kept in view by the author has been that he would not assume the position of an "authority" in giving rules and formulae for designing, but only that of compiler, giving not only the name of the originator of the rule, where it was known, but also the volume and page from which it was taken, so that its derivation may be traced when desired. When different formulae for the same problem have been found they have been given in contrast, and in many cases examples have been calculated by each to show the difference between them. In some cases these differences are quite remarkable. Occasionally the study of these differences has led to the author's devising a new formula, in which case the derivation of the formula is given.

Much attention has been paid to the abstracting of data of experiments from recent periodical literature, and numerous references to other data are given. In this respect the present work will be found to differ from other pocket-books.

William Kent

CONTENTS

(Detailed tables of contents are given at the beginning of each section. An alphabetical index appears after Section 28.)

SECTION 1

GENERAL PROPERTIES OF MATERIALS

BY

NORMAN B. PILLING, *Director, Research Laboratory, The International Nickel Co., Inc.*

H. W. GILLETT, *Chief Technical Adviser, Battelle Memorial Institute*

GENERAL PROPERTIES OF METALS AND ALLOYS

By Norman B. Pilling

1. THE METALS

CLASSIFICATION. There are more than 90 chemical elements (Table 1), more than 70 of which are classified as metallic, 4 as metalloids, and 14 as nonmetallic. Of the metallic elements, some 20 or more lie in the groups of rare-earth and radioactive metals, which, because they are very scarce and difficult to separate, are chiefly of scientific interest only at the present time. The 4 metalloids are boron, carbon, phosphorus, and silicon. Although their chemical properties relate them to the nonmetallic elements, the metalloids form true alloys with many metals and are among the most useful metallurgical elements.

PROPERTIES. The metals owe their engineering importance to properties, especially combinations of properties, which they display either uniquely or in far greater intensity than other classes of materials. Among the distinctive properties of metals are their high conductivity for electricity and the facility with which they may be mixed, or alloyed, with each other to enhance some properties, or even to create new qualities not displayed by the component metals. The conservative employment of metals attaches great importance to the improvement and control of their properties.

The technologically interesting metallic and metalloidal elements, with the chief properties which have metallurgical significance, are listed in Table 3.

METALLIC ELEMENTS, METALS AND ALLOYS. Few metals indeed are ever seen in a state approaching complete purity. The term *metallic element* assumes such purity; the physical constants of the elements are expressed with such an implication or in relation to some stated degree of purity. Although all the metals that are sufficiently tractable to have importance can be prepared to a high degree of chemical purity provided sufficient effort is expended, the purity to which the metals are commonly refined is set by the requirements of their principal end uses, and to this their cost is to a considerable extent related. (See Table 2.)

By general usage, the term *metal* is employed in a generic sense, often referring both to simple metals as well as to metallic alloys. There can be no distinction between alloys and simple metals based upon order of purity; the word *alloy* implies a purposeful mixture.

2. STRUCTURE OF METALS AND ALLOYS

All substances whether solid, liquid, or gaseous, elemental or complex, are composed of atoms of the ninety-odd elements. The atom is not the most elementary particle of nature but has a complex internal structure which is frequently diagrammed as consisting of a central, electrically charged nuclear body surrounded by an atmosphere of oppositely charged particles of a different nature, the electrons. The arrangement of the elements in the periodic table parallels a systematic progression in the nucleus/electron pattern. The several distinctive properties of metals originate at different depths within the atomic structure. This is the field of atomic physics.

SIMPLE METALS. In general, metals in the gaseous and liquid states are characterized by absence of crystallinity. In the solid state, metals are almost completely crystalline, although little in their outward appearance suggests this. The evidence of crystallinity is complete, however, and is revealed by x-ray diffractive powers and by light reflective behavior of finely etched surfaces when suitably examined with a microscope. Crystalline texture is not necessarily combined with geometrically regular external form. This is because it generally happens that as metals solidify from the molten condition, a great multitude of crystals starts to form; these soon grow into each other and interfere with the further development of regular form. Metal crystals grow during solidification in skeleton shapes, as a series of geometrically related arms, branches, subbranches, etc., which interlace the liquid and finally trap the last drops to solidify in the interstices of this branching

Table 1. The Chemical Elements

(Data from *The Metals Handbook*, American Society for Metals, 1948. By permission.)

Element	Symbol	Atomic Number	Atomic Weight	Element	Symbol	Atomic Number	Atomic Weight
Actinium	Ac	89	227.05	Mercury	Hg	80	200.61
Aluminum	Al	13	26.97	Molybdenum	Mo	42	95.95
Americium	Am	95	241	Neodymium	Nd	60	144.27
Antimony	Sb	51	121.76	Neon	Ne	10	20.183
Argon	A	18	39.944	Neptunium	Np	93	237
Arsenic	As	33	74.91	Nickel	Ni	28	58.69
Astatine	At	85	211	Nitrogen	N	7	14.008
Barium	Ba	56	137.36	Osmium	Os	76	190.2
Beryllium	Be	4	9.02	Oxygen	O	8	16.0000
Bismuth	Bi	83	209.00	Palladium	Pd	46	106.7
Boron	B	5	10.82	Phosphorus	P	15	30.98
Bromine	Br	35	79.916	Platinum	Pt	78	195.23
Cadmium	Cd	48	112.41	Plutonium	Pu	94	239
Calcium	Ca	20	40.08	Polonium	Po	84	210
Carbon	C	6	12.010	Potassium	K	19	39.096
Cerium	Ce	58	140.13	Praseodymium	Pr	59	140.92
Cesium	Cs	55	132.91	Protoactinium	Pa	91	231
Chlorine	Cl	17	35.457	Radium	Ra	88	226.05
Chromium	Cr	24	52.01	Radon	Rn	86	222
Cobalt	Co	27	58.94	Rhenium	Re	75	186.31
Columbium	Cb	41	92.91	Rhodium	Rh	45	102.91
Copper	Cu	29	63.54	Rubidium	Rb	37	85.48
Curium	Cm	96	242	Ruthenium	Ru	44	101.7
Dysprosium	Dy	66	162.46	Samarium	Sm	62	150.43
Erbium	Er	68	167.2	Scandium	Sc	21	45.10
Europium	Eu	63	152.0	Selenium	Se	34	78.96
Fluorine	F	9	19.00	Silicon	Si	14	28.06
Francium	Fa	87	223	Silver	Ag	47	107.880
Gadolinium	Gd	64	156.9	Sodium	Na	11	22.997
Gallium	Ga	31	69.72	Strontium	Sr	38	87.63
Germanium	Ge	32	72.60	Sulfur	S	16	32.066
Gold	Au	79	197.2	Tantalum	Ta	73	180.88
Hafnium	Hf	72	178.6	Technetium	Tc	43	99
Helium	He	2	4.003	Tellurium	Te	52	127.61
Holmium	Ho	67	164.94	Terbium	Tb	65	159.2
Hydrogen	H	1	1.0080	Thallium	Tl	81	204.39
Illinium	Il	61	147	Thorium	Th	90	232.12
Indium	In	49	114.76	Thulium	Tm	69	169.4
Iodine	I	53	126.92	Tin	Sn	50	118.70
Iridium	Ir	77	193.1	Titanium	Ti	22	47.90
Iron	Fe	26	55.85	Tungsten	W	74	183.92
Krypton	Kr	36	83.7	Uranium	U	92	238.07
Lanthanum	La	57	138.92	Vanadium	V	23	50.95
Lead	Pb	82	207.21	Xenon	Xe	54	131.3
Lithium	Li	3	6.940	Ytterbium	Yb	70	173.04
Lutecium	Lu	71	174.99	Yttrium	Y	39	88.92
Magnesium	Mg	12	24.32	Zinc	Zn	30	65.38
Manganese	Mn	25	54.93	Zirconium	Zr	40	91.22

Table 2. Purity of the Principal Commercial Metals

Order of Total Impurity Content, %	Metals
Less than 0.01	Gold, silver
0.01 to 0.1	Copper, lead, molybdenum, platinum, tin, tungsten, zinc
0.1 to 1.0	Aluminum, iron (steel), magnesium, mercury, nickel
1.0 to 10	Chromium, cobalt, iron (cast iron), manganese, silicon

Table 3. The Metallic and Metalloidal Elements

Element	Symbol	Periodic Classification	Melting Point, °F	Boiling Point, °F	Density, grams/cm³	Hardness No.	Elastic Modulus, Tension, million psi	Thermal Expansion Coefficient, millionths per °F	Electrical Resistivity, microhm-cm	Thermoelectric Power, microvolts per °F	Crystalline Lattice Form	Lattice Constants a, Å	Lattice Constants c, Å
Notes		1				2	3	4		5	6	7	
Aluminum	Al	3B	1220		2.70	16	10	13.3	2.66	−0.38	FC Cubic	4.041	
Antimony	Sb	5B	1167	2980	6.62	42	11.3	6.3	39		Rhomb. Hex.	4.497	
Arsenic	As	5B	1505		5.73			2.14	35	−7.6	Rhomb. Hex.	4.135	
Barium	Ba	2A	1310	2700±	3.5	42					BC Cubic	5.015	
Beryllium	Be	2B	2400±		1.85	98	42.7	6.8	18.5		CP Hexagonal	2.281	
Bismuth	Bi	5B	520	2840	9.80	4.7	4.6	7.5	115	−49.5	Rhomb. Hex.	4.736	
Boron	B	3C	4200±		2.3	>1000		1.1	Over 1,000,000		Hexagonal		
Cadmium	Cd	2B	577	1412	8.65	35		16.6	7.59	+1.94	CP Hexagonal	2.973	5.606
Calcium	Ca	2A	1560		1.55	13		13.9	4.6		FC Cubic	5.560	
Carbon-Gr.	C	4C	6700±		2.22		0.7	0.67	1,000		Hexagonal	2.46	6.69
Cerium	Ce	3A	1500		6.9	21			78		FC Cubic	5.143	
Cesium	Cs	1A	86	1290	1.9			54	20		BC Cubic	6.2	
Chromium	Cr	6A	3490		7.14	70		4.5	13.1		BC Cubic	2.878	
Cobalt	Co	8A	2692		8.9	75±		6.7	9.7	−12	CP Hexagonal	2.507	4.072
Columbium	Cb	5A	3600		8.57	250±		4	20		BC Cubic	3.294	
Copper	Cu	1B	1981		8.94	30	16	9.1	1.68	+0.848	FC Cubic	3.608	
Gallium	Ga	3B	86	4100±	5.91	2.5		10.2	57		Orthorhombic	4.506	7.642
Germanium	Ge	4B	1757		5.36	4					D Cubic	5.647	
Gold	Au	1B	1945	5300±	19.3	25	11.3	8.0	2.42	+0.67	FC Cubic	4.070	
Indium	In	3B	313	2600±	7.31	1.0		18.3	9		FC Cubic	4.585	
Iridium	Ir	8B	4450		22.4			3.5	6.08		FC Cubic	3.831	
Iron	Fe	8A	2795	4900±	7.87	45	30	6.6	9.8	+9.0	BC Cubic	2.861	
Lanthanum	La	3A	1519		6.15	36			59		CP Hexagonal	3.75	6.06
Lead	Pb	4B	621	3185	11.34	2.9	2.56	16.4	20.6	0	FC Cubic	4.939	
Lithium	Li	1A	356	2420	0.53			31	8.5		BC Cubic	3.51	
Magnesium	Mg	2B	1202	2020	1.74		6.25	14.3	4.46	+1.13	CP Hexagonal	3.202	5.199
Manganese	Mn	7A	2268	3760	7.44			12.8			Complex Cubic	8.894	

Mercury	Hg	2B	−38	678	13.55				95.8	−0.23	FC Rhomb.	4.581	
Molybdenum	Mo	6A	4750		10.2	160	50.2	3.05	4.77		BC Cubic	3.140	
Nickel	Ni	8A	2651	5500±	8.9	70	30	7.6	6.9	−12.7	FC Cubic	3.517	
Osmium	Os	8B	4600±		22.5			3.2	9		CP Hexagonal	2.730	4.310
Palladium	Pd	8B	2830		12.0	50	17.0	6.4	10	−3.8	FC Cubic	3.882	
Phosphorus	P	5C	111	536	1.82			69	Over 10,000,000		FC Rhomb	3.31	
Platinum	Pt	8B	3223		21.45	42	21.4	4.3	9.83	+0.50	FC Cubic	3.916	
Potassium	K	1A	144	1400	0.86	0.04		46	7.0		BC Cubic	5.333	
Rhodium	Rh	8B	3571		12.44	110	42.5	4.5	4.93		FC Cubic	3.796	
Rubidium	Rb	1A	102	1300	1.53			50	12.5		BC Cubic	5.7	
Ruthenium	Ru	8B	3600±		12.2	220		4.7	10		CP Hexagonal	2.699	4.274
Selenium	Se	6B	428		4.81		16	20.6	12	+450	Hexagonal	4.337	4.944
Silicon	Si	4C	2620	4760±	2.4		10.3		85,000		D Cubic	5.417	
Silver	Ag	1B	1762	3940	10.5	27	11.2	10.5	1.62	+1.34	FC Cubic	4.077	
Sodium	Na	1A	208	1615	0.97	0.07		39.5	4.6		BC Cubic	4.282	
Strontium	Sr	2A	1420		2.6				22.7		FC Cubic	6.075	
Tantalum	Ta	5A	5400±		16.6	30	27.0	3.6	15.5		BC Cubic	3.296	
Tellurium	Te	6B	846	2530±	6.24			9.3		+280	Hexagonal	4.445	5.912
Thallium	Tl	3B	578	2645	11.85	3		15.6	18.1		CP Hexagonal	3.450	5.520
Thorium	Th	4A	3320		11.5	78		6.84	18		FC Cubic	5.077	
Tin	Sn	4B	450	4300±	7.30	14	7.4	12.7	11.5	−0.18	Tetragonal	5.319	3.175
Titanium	Ti	4A	3140		4.5	160		3.9			CP Hexagonal	2.953	4.729
Tungsten	W	6A	6100±		19.3	230	60	2.2	5.48		BC Cubic	3.159	
Uranium	U	6A	3070±		18.7	92			60		Orthorhombic	2.852	4.945
Vanadium	V	5A	3100±		5.68	250			26		BC Cubic	3.033	
Zinc	Zn	2B	787	1663	7.14	35	11.5±	16.5	5.8	+2.1	CP Hexagonal	2.659	4.936
Zirconium	Zr	4A	4000±		6.4	80		3.5	41		CP Hexagonal	3.223	5.123

1. Designation based on the periodic table; elements with like designations are chemically similar and have other general resemblances.
2. Brinell hardness number, soft condition.
3. psi = pounds per square inch.
4. At 70 F±.
5. Thermoelectric power relative to lead.
6. FC = face centered; BC = body centered; CP = close packed; D = diamond symmetry.
7. Å = Ångstrom unit = 10^{-8} cm.

system. Many of the qualities of metals and alloys are inherited from this mode of freezing. Evidences of their tree-like or dendritic texture are commonly seen in the microstructure of cast alloys. Geometrically shaped crystals are sometimes seen in metals condensed from vapor, and in the interior of large ingots from which the liquid contents have been suddenly emptied after partial freezing.

Crystalline Textures. The essence of crystalline texture is the arrangement of atoms on a precisely recurrent, regular three-dimensional pattern. For most metals, and the alloys derived from them, these patterns are rather simple. About 60% of the common metals crystallize in some cubic pattern, and 30% in some hexagonal pattern (Table 3.)

The three forms of crystalline pattern, or *space lattice*, most common to the metals are shown in Fig. 1. These three lattices develop directly from the geometry of layers of uniformly sized spheres arranged to achieve efficient packing.

The basic dimension of the lattice is the length of the sides of the unit cell, or *lattice constant*, which is expressed in Ångstrom units (1 Å = 10^{-8} cm). Because atoms differ in size this dimension varies from one metal to another and ranges from about 2.9 to 6.2 Å. Lattices other than cubic have more than one lattice constant, depending on the symmetry of the unit cell.

The significant systems of crystallographic planes are those on which the atoms lie in greatest numbers. These planes play a part in plastic straining and brittle cleavage and

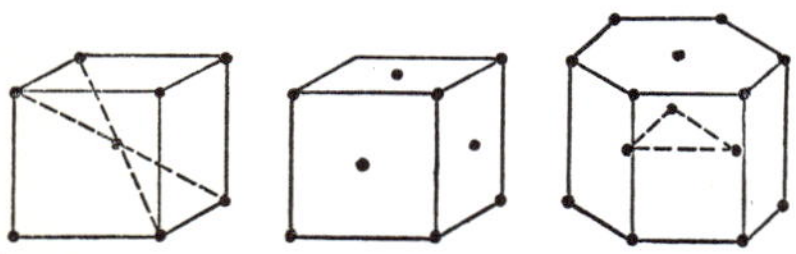

(*a*) Body-centered cubic (*b*) Face-centered cubic (*c*) Close-packed hexagonal

FIG. 1. Space lattices most common to the metals.

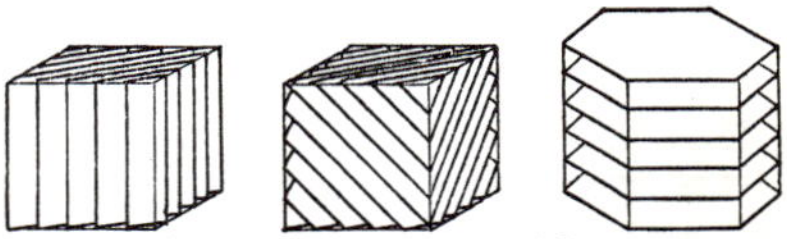

(*a*) BC cubic (*b*) FC cubic (*c*) CP hexagonal

FIG. 2. Significant crystalline planes.

often fix the pattern upon which precipitates form. The more important planes are shown upon which the slip mechanism operates in the three principal lattice types. (See Fig. 2.)

Grains and Grain Boundaries. It is unusual to encounter metals composed of only a single crystal; most metal parts are aggregates of crystals. In castings, i.e., metals that have merely solidified from the liquid state, the shape of the crystals, or *grains* as they are usually called, is irregular and almost never geometrically symmetrical. The grains differ randomly in relative orientation and for this reason can be separately distinguished. The grain size of castings is usually much larger than that of metals which have been annealed and recrystallized, and is determined by the number of points at which solidification independently starts. The formation of large grains in freezing is favored by slow, undisturbed cooling; on the other hand, turbulent motion and rapid extraction of heat at temperatures close to the freezing point diminish the grain size. Column-shaped grains result from unsymmetrical heat extraction, as against the walls of chill molds. Large ingots characteristically have a shell of columnar grains enclosing a central core of smaller, equiaxed grains.

Hot working, or annealing after cold working, results in recrystallization, whereby a new generation of grains comes into being. This process starts from many centers. The new grains enlarge by advancement of their boundaries in detail across old grains. The first result is a refinement in grain size, but prolonged overheating may lead to great coarseness in grain size.

Grain boundaries represent the surface along which the differently directed lattices of similar grains meet. In pure metals it is a tenuous zone in which the attractive forces of the adjacent crystals compete. The atoms in this zone are thought to be noncrystalline, and some qualities of polycrystalline metals harmonize with the supposition. At any rate, at ordinary temperatures grain boundaries are sources of strength, not weakness.

The grain structures of metals and alloys are revealed for study or inspection by the preparation of polished, cut sections which are lightly etched with slow-acting acid solutions. Many features become visible to the unaided eye and much is revealed by microscopic examination up to 1500 diameters enlargement. Crystalline textures are revealed and analyzed only through the diffraction of x-rays or other short radiations.

Allotropy. Some metals exist in more than one crystalline form, with each type of lattice stable over a definite temperature range (Table 4). These differing forms sometimes have widely different properties which may have great practical importance. For example, the two allotropic forms of iron have greatly different solubilities for carbon, making possible the development of hardening and heat treatment of steel. At the temperature of transformation from one allotropic phase to another, density, electrical, thermal, and other property changes may occur. The allotropes are designated by Greek letters.

Table 4. Allotropic Forms of the Metals

Metal	Allotropic Phase	Upper Temperature Limit of Stability, °F	Crystalline Form
Cobalt	Alpha	880	CP Hexagonal
	Beta	Melting point	FC Cubic
Iron	Alpha	1663	BC Cubic
	Gamma	2557	FC Cubic
	Delta	Melting point	BC Cubic
Manganese	Alpha	1368	Complex cubic
	Beta	2176	Complex cubic
	Gamma	Melting point	FC Tetragonal
Thallium	Alpha	450	CP Hexagonal
	Beta	Melting point	FC Cubic
Tin	Alpha	64	Tetrahedral cubic
	Beta	322	BC Tetragonal
	Gamma	Melting point	Hexagonal
Zirconium	Alpha	1584	CP Hexagonal
	Beta	Melting point	BC Cubic

ALLOYS. Two metals alloy if they melt together to form a homogeneous liquid. The solid alloy which freezes from it is crystalline and, like simple metals, consists usually of an aggregate of grains. These grains may be of several kinds, depending on the metal pair. (1) The grains may be homogeneous, the atoms of the two metals forming a single lattice similar to that of one of the metals, i.e., a *solid solution.* (2) The two metals may combine to form an *intermetallic compound* with an entirely new, and frequently complex, lattice structure. (3) The grains may be *composite*, as a physical mixture of two kinds of solid solution grains, a solid solution and an intermetallic compound, or two intermetallic compounds.

Alloying may also be accomplished at temperatures below the melting point by diffusion between contacting metals or reactive gases. Examples of these processes, which are slow, are carburization, cementation, and the bonding of electroplated layers. These methods of alloying are resorted to when fusion processes are not practical.

Two metals may show complete liquid insolubility when melted together, similar to oil and water. A limit of liquid solubility may be shown that is the maximum concentration dissolved before liquid separation occurs.

Solid Solubility. In the solid solution lattice, the atoms present in lesser proportion conform in position on the parent lattice but distort it to a degree depending on their relative lattice constants. Such accommodation is easiest when the lattice types are similar and when the size difference is less than 14%; greater distortions result in the formation or precipitation of other kinds of alloy crystal. Solid solubility is this limiting concentration.

The solid solubility of two metals may be complete, extending without break from one metal through all percentage ranges to the other. An example is nickel and copper. As nickel is increasingly added to copper, the lattice constant decreases continuously to that of nickel, and other properties such as electrical resistivity change continuously although not usually linearly. More frequently solid solubility is limited; the structure of a binary alloy of two metals, at a concentration which exceeds the solubility of either in the other, consists in a mixture of the two saturated solid solution phases, the proportions occurring being fixed by the respective solubilities.

Intermetallic compounds are the result of a more or less strong chemical union between two metals which differ considerably in position in the periodic table. They are generally hard and brittle, have complex lattice structures, and do not conform in combining proportions to the simple rules of chemical valence. Many dispersives in precipitation-hardened alloys are intermetallic compounds.

Equilibrium Diagrams. The complex relations of the structures of a two-component alloy series are described graphically and in quantitative detail by the equilibrium diagram of the series. This is a temperature-composition graph based on observed facts concordant with the phase rule of physical chemistry. Figures 3*a* and 3*b* are two of the simpler diagrams illustrating the types of information available for many binary alloy series. From such diagrams the melting temperatures and freezing ranges of the entire series may be read off; the course of freezing may be followed, and the related degrees of microsegregation may be estimated; eutectic ranges may be observed, and the existence of intermetallic compounds recognized; the extent of solid solubility and the way it changes with temperature may be noted.

Figure 3*a* shows two metals that form solid solutions in all proportions. The *liquidus* line describes the temperatures at which solidification normally begins upon cooling of the melted alloy. The *solidus* line represents the temperatures at which melting normally begins upon heating the solid alloy. These two curves are conjugate, and pairs of points on the two curves connected by a horizontal temperature line represent respectively the compositions of the liquid phase and the solid crystal which are in equilibrium with each other. In general, the composition of the first crystal to separate from the melt is not the same as the average composition of the melted alloy; it is richer in the metal of higher melting point. The vertical line *AB* illustrates the course of freezing of a molten 35%

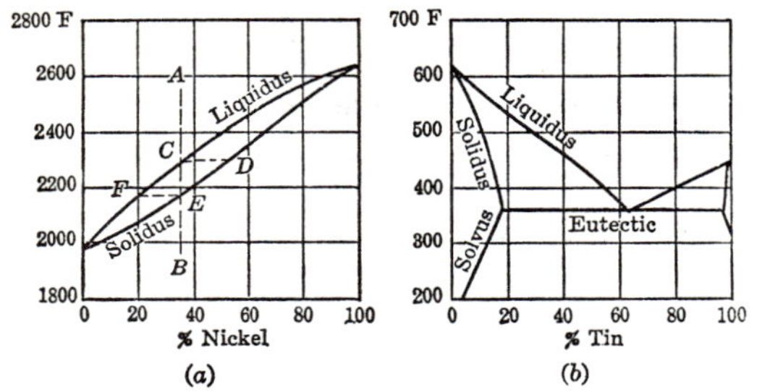

Fig. 3. Equilibrium diagrams of simple two-component alloy series: (*a*) copper-nickel and (*b*) lead-tin series.

nickel alloy: *C*, initial freezing temperature; *D*, composition at heart of first dendritic crystal; *E*, temperature of completion of freezing; *F*, composition of last drops of liquid to freeze within the dendritic network. The temperature spread *CE* is the freezing range. The extremes of heterogeneity thus indicated would be found within alloy grains were it not that diffusion at high temperatures acts in an opposite manner and sets in as soon as liquid no longer is in contact. After the completion of freezing, prolonged heating at temperatures close to the freezing range may succeed in obtaining equalization of composition throughout the grains, or *homogenization.*

Figure 3*b* shows two metals that have only limited solid solubility for each other. It will be noted that the melting point of each metal is depressed by the other and the two liquidus curves ultimately intersect. The alloy (32% tin) corresponding in composition to point of intersection is the *eutectic alloy* and has the lowest melting point of the series; its melting point is the *eutectic temperature.* By following the course of freezing of an alloy richer in lead than the eutectic it will be seen that the liquid last to freeze is always of eutectic composition. The amount of eutectic thus present is in proportion to composition limits, e.g., 80% tin yields 50% eutectic. The eutectic alloy has a characteristic structure of finely laminated alternating plates of the two phases, suggesting a seesawing mode of freezing. The *solvus* line traces the changes of solid solubility with temperature.

It must be emphasized that the data of the equilibrium diagram are for conditions of complete structural stability, a matter seldom occurring in practical circumstances. Many transient or unstable structures of real importance can be systematized only in empirical *structural diagrams* for particular conditions of treatment.

Three-component, or ternary, alloys can be dealt with by three-dimensional models and by plane diagrams corresponding to sections through, or projections from, such models, but alloys containing more than three components are not readily diagrammed.

3. ALLOYING POWERS OF THE METALS

Although there are more than 70 metallic elements, most of the industrially important alloys are based on the 8 relatively cheap and abundant ones: iron, lead, copper, zinc, aluminum, nickel, tin, and magnesium. The more basic alloying properties of some of the metals with this group, and with the two principal precious metals which have industrial use, are in Table 5. If the data of Table 3 and Table 5 are combined, it is possible to sketch out the principal features of these binary alloys.

Table 5. Alloying Properties of the Principal Metals

Basis Metal	Alloy Metal	Solid Solubility		Eutectic		Inter-metallic Cpd.,* %	Limit of Liquid Solubility, %
		Min, %	Max, %	Conc., %	Temp., °F		
Aluminum	Antimony	...	<0.01	1.1	1215	82	...
	Beryllium	...	0.05	1.4	1190	...	...
	Chromium	0.2	0.8	0.6	1220	24	...
	Cobalt	...	<0.1	1.0	1215	36	...
	Copper	0.5	5.7	33	1020	54	...
	Iron	...	<0.06	1.9	1210	40	...
	Lead	...	...	...	...	...	1
	Magnesium	2.9	15.4	34.5	845	38	...
	Manganese	0.1	0.7	2.2	1200	23	...
	Nickel	<0.001	0.05	5.7	1185	42	...
	Silicon	0.05	1.6	11.7	1070	...	...
	Zinc	2	69	95	715	...	...
Copper	Aluminum	7.5	9.5	10.5	1905	12.5	...
	Antimony	9	9	31	1195	49	...
	Beryllium	0.2	2.1	4.3	1590	12	...
	Cadmium	0.5	3	60	1020	47	...
	Chromium	0.05	1	1.5	1970	...	37
	Iron	0.2	3.8	...	...	...	20±
	Lead	<0.05	...	36	1750	...	38
	Magnesium	1	2.8	9.7	1330	16	...
	Manganese	20	34	35	1560	...	...
	Molybdenum	...	...	...	...	...	0
	Nickel	100	...	...	...	...	...
	Phosphorus	0.5	1.15	8.2	1305	...	14
	Silicon	4.1	6.8	10	1490	8	...
	Silver	0.4	8	71.5	1435	...	...
	Tin	14	16	99.2	440	32	...
	Zinc	32.5	39			38	...
Iron (Alpha)	Aluminum	33	36	51	2135	49	...
	Beryllium	1.2	6	9	2110	...	...
	Carbon	<0.01	1.7	4.3	2095	6.7	...
	Chromium	35	...	...	...	49	...
	Cobalt	72	78	...	...	...	...
	Copper	0.4	8.5	...	...	...	20±
	Lead	...	...	...	...	...	0
	Manganese	0	12.7	...	...	...	...
	Molybdenum	7	34	38	2625	53	...
	Nickel	0	7	...	...	...	...
	Phosphorus	1.0	2.8	10.5	1920	16	...
	Silicon	15	18	20	2185	33	...
	Silver	...	...	...	...	...	0
	Tin	9	18	51	2030	68	...
	Titanium	2	6	13.2	2370	23	...
	Tungsten	8	33	35	2775	69	...
Lead	Aluminum	...	...	...	...	...	0
	Antimony	0.2	2.4	13	477	...	...
	Bismuth	17	35	56.5	257	...	...
	Cadmium	...	4±	17.5	478	...	...
	Calcium	0.01	0.10	39	1290	72	...
	Chromium	0	0	72	2680	...	15±
	Copper	...	<0.1	...	...	...	7.5
	Iron	...	<0.001	...	...	...	0
	Nickel	<0.03	0.2	...	...	...	7
	Silicon	...	<0.01	...	...	...	0.2
	Silver	0.05	0.10	2.5	580	...	...
	Tin	2.5	19.5	61.9	362	...	...
	Zinc	...	<0.02	0.5	605	...	2
Magnesium	Aluminum	2.4	12.1	32	820	42	...
	Antimony	...	<1	35.7	1165	77	...
	Beryllium	...	...	...	...	...	0
	Copper	0.02	0.03	30.7	905	43	...
	Lead	15	26	68	875	81	...
	Manganese	<0.1	3.3	...	...	...	...
	Nickel	...	<0.1	23.5	945	55	...
	Silicon	...	<0.1	1.4	1195	37	...
	Tin	<2.3	14.8	36.4	1040	71	...
	Zinc	<1.8	8.4	53.5	645	73	...
Nickel	Aluminum	5.7	15	12	2500	32	...
	Beryllium	<1.4	2.8	5.2	2110	13	...
	Chromium	30	46	51	2430	...	...
	Cobalt	100	...	...	...	...	...
	Copper	100	...	...	...	...	...

* Intermetallic compound lowest in content of alloy metal.

(*Table continued on p. 1-10*)

Table 5. Alloying Properties of the Principal Metals—*Continued*

Basis Metal	Alloy Metal	Solid Solubility		Eutectic		Intermetallic Cpd.,* %	Limit of Liquid Solubility, %
		Min, %	Max, %	Conc., %	Temp., °F		
Nickel (continued)	Iron	80±	100	...	...	...	...
	Lead	<0.1	3	34	2445	...	35
	Magnesium	<0.1	<0.1	11	1980	17	...
	Manganese	23	44	59	1840	...	...
	Molybdenum	30±	33	37	2380	62	...
	Phosphorus	...	...	11	1615	15	...
	Silicon	5.5	8	10.5	2100	14	...
	Silver	<0.5	4	5	2595	...	5±
	Tin	13	15	33	2090	40	...
	Tungsten	<20	47	52	2750	34	...
	Zinc	29	43	...	...	52	...
Platinum	Copper	100	...	...	...	51	...
	Gold	5	100	...	...	...	...
	Iron	40	100	...	...	22	...
	Lead	...	10	95	555	51	...
	Palladium	100	...	...	...	...	...
	Nickel	100	...	...	...	...	...
	Rhodium	100	...	...	...	...	...
	Silver	2	12	...	...	...	...
	Tin	...	5	29	1960	17	...
Silver	Aluminum	5	6	28	1035	8	...
	Antimony	7	8.5	45	905	...	...
	Copper	0.4	8.8	28.5	1415	...	...
	Gold	100	...	...	...	...	...
	Iron	...	<0.001	...	...	...	0
	Lead	...	4	97.5	579	...	...
	Nickel	0.01	0.1	...	...	...	3
	Palladium	100	...	...	...	...	...
	Phosphorus	...	<0.03	1.0	1610	...	...
	Platinum	32	55	...	...	...	...
	Rhodium	...	...	...	...	...	0
	Tin	10.5	13.3	96.5	430	27	...
	Zinc	22	27	...	...	51	...
Tin	Aluminum	...	<0.1	0.5	445	...	...
	Antimony	7.5	10	...	...	42	...
	Bismuth	<1	19	42	280	...	...
	Cadmium	1.2	10	32.3	350	...	...
	Copper	<0.05	0.2	0.7	440	39	...
	Iron	...	<0.1	...	...	19	...
	Lead	<0.1	2.6	38.1	362	...	...
	Magnesium	...	<0.1	2.0	390	29	...
	Nickel	...	<0.005	0.1	448	33	...
	Silver	...	<0.1	3.5	430	73	...
	Zinc	...	<2	9.0	390	...	...
Zinc	Aluminum	0.2	1.0	5.0	715	...	...
	Bismuth	...	<0.1	1.9	780	...	2
	Cadmium	<0.1	2.6	82.6	510	...	...
	Copper	<0.3	2.7	...	...	14	...
	Iron	...	0.01	...	...	11	...
	Lead	...	<0.03	0.5	785	...	0.5
	Magnesium	...	<0.06	3.1	685	7	...
	Manganese	0.1	0.3	0.8	775	8	...
	Nickel	...	<0.1	0.2	785	11	...
	Tin	...	<0.1	91	390	...	...

4. METALLIC PROPERTIES

Physical and mechanical properties are primarily established by the chemical composition and the structure; the latter is in turn a product of the working and thermal treatments which have been given. These properties may accordingly be controlled or varied within narrow or wide limits by suitable attention to these factors.

ELASTIC AND PLASTIC PROPERTIES. Of the many qualities of metals, perhaps the most significant to the engineer are the related properties of elasticity and plasticity. Upon plasticity depends the ability to shape metals into useful forms; elasticity governs their safe and economical use as load-bearing members.

Elasticity. A body is said to be elastic if, after change in shape when externally loaded, it returns to its original shape with removal of the load. The elastic limit is the maximum stress at which the body behaves elastically. The proportional limit is the stress at which strain ceases to be proportional to applied stress; it is practically equivalent to elastic limit. The modulus of elasticity, or stiffness, is the stress which would be required to produce 100% of strain if there were no limit to elasticity, e.g., to double the length of a bar if loaded by axial tension.

Plasticity, as applied to metals, is a term used to express an ability to be deformed beyond the range of elasticity without fracture, resulting in permanent change in shape. A diagrammatic stress-strain curve for a test bar in tension is shown in Fig. 4, in which the elastic and plastic ranges are distinguished. Characteristically, the ratio of plastic to elastic deformability in the useful metals is high, of the order of 100 : 1 or 1000 : 1. Interrupting the course of plastic straining, as shown by the dotted lines, results in an elastic recovery and an increase in the proportional limit on resumption of straining; this is an evidence of strain hardening. A fuller development of the external aspects of elastic and plastic behavior will be found in Section 8, Strength of Materials.

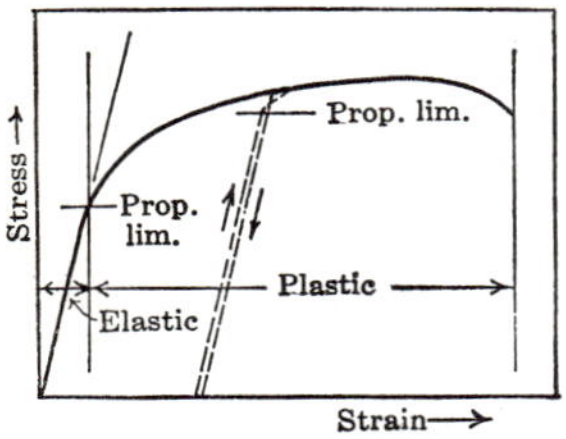

FIG. 4. Stress-strain curve for a test bar in tension, showing elastic and plastic ranges.

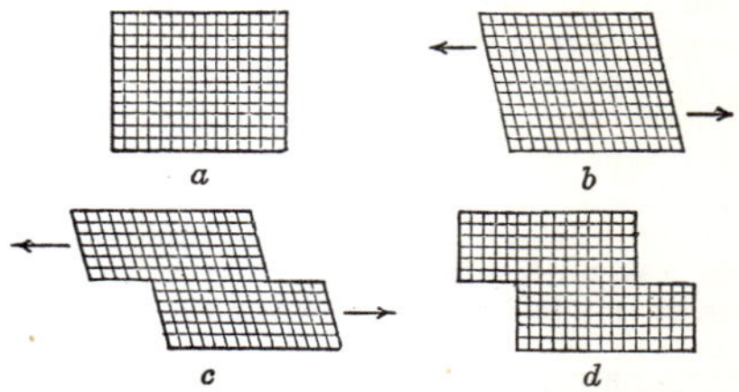

FIG. 5. Deformation of a cubic crystal lattice: (*a*) unstrained condition; (*b*) elastic deformation; (*c*) plastic deformation; (*d*) permanent set as result of slip.

Deformation in Single Metallic Crystals. A simplified two-dimensional representation of a cubic crystal lattice in an unstrained condition is represented by *a* in Fig. 5. If a shearing force is applied which is within the elastic range, the lattice is uniformly distorted as in *b*, the extent of the distortion being proportional to the applied force. When the force is removed, the lattice springs back to its original shape (*a*). However, when the force exceeds the elastic (or proportional) limit, a sudden change in the mode of deformation occurs; without further increase in the amount of elastic strain, the lattice shears along a crystallographic plane, or *slip plane;* one block of the lattice makes a long glide past the other and stops (*c*). Upon release of load, the lattice in the two displaced blocks resumes its original shape (*d*). If the applied force is continued, slip does not continue indefinitely along the original slip plane, which appears on the contrary to acquire resistance to further motion, but some parallel plane comes into action. Both the extent of slip per plane and the distance between active slip planes are large in comparison to the unit lattice dimensions. As slip shifts from one slip plane to another, progressively higher forces are required to accomplish it, i.e., the metal has been work hardened. At some stage, resistance to further slip along the primitive set of planes exceeds the resistance offered by some other set of differently directed slip planes, which then come into action. This process elaborates as plastic deformation progresses.

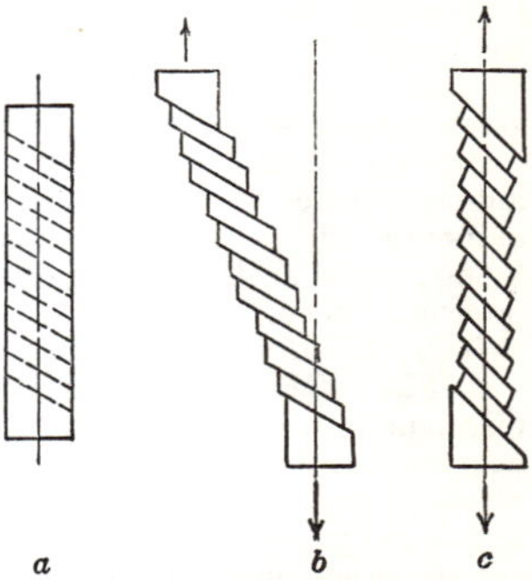

FIG. 6. Plastic deformation of a single crystal: (*a*) unstrained; (*b*) sheared along slip planes; (*c*) rotated to maintain alignment of applied load.

The slip mechanism outlined for a sample of lattice applies to the deformation behavior of a test bar made of a single grain. The simplest case is that of a hexagonally crystalline metal like zinc, in which there is but one principal set of slip planes. The basic effects are shown in Fig. 6. As the tensile force is exerted, blocks of the grain tend to shear, as in *b*, like the displacement of a stack of coins, at a stress which depends upon the angle between the slip planes and the direction of pull. Since external restraint compels the axis of the test bar to keep in line with the direction of pull, this accommodation occurs by an angular movement of the slip planes that brings them more nearly parallel to it (*c*). As this action occurs regardless of grain orientation relative to the axis of pull, it follows that an end result of severe plastic straining is the development of a preferred crystalline orientation, or fiber. As the single-grain test bar is further strained, its anisotropic qualities become apparent. An originally circular cross section becomes elliptical, and it finally draws down and breaks to a chisel-shaped edge.

These interesting effects are not always desirable; metals and alloys are generally in their most useful form if finely polycrystalline.

Deformation of Crystal Aggregates. Metal parts are normally an aggregate of a multitude of tiny, randomly positioned grains. The elastic and plastic behaviors of such an aggregate differ from that of a single grain for several reasons. (1) *Mutual reinforcement*, whereby those grains weakly positioned for slip are buttressed by more strongly positioned neighboring grains. (2) *Uniformity of properties*, since the strong directional qualities of the separate grains cancel out with their random placements. (3) *Grain boundary strength.* If insoluble components are not present, the grain boundary is an extremely thin zone in which the atoms are subject to the conflicting attractions for differently positioned lattices. At ordinary temperatures this zone is stronger than the rest of the grain; the path of fracture usually avoids grain boundaries. Increase in the intricacy of grain boundaries results in increase of hardness and related properties. (4) *Grain boundary segregations.* Many impurities or alloy components, if present as separate phases, show a tendency to collect along grain boundaries, intensifying the effects of their individual qualities.

Grain aggregates, when heavily cold worked, develop, for reasons explained in the preceding heading, a crystalline fiber or *preferred orientation* related to the principal direction of straining. Under some conditions of working and annealing this cold-worked fiber is transmitted to the recrystallized structure, forming an aggregate of grains regimented as to a lattice direction and showing some of the more annoying qualities of single grains.

INHERENT STRENGTH OF METALS. The stress required to rupture a metallic crystal in pure tension is not readily measurable, but calculations based on energetics disclose that the strength of metals as ordinarily measured is but a small fraction of their inherent strength. For copper, some significant comparisons are these:

Calculated tensile rupture stress	1,300,000 psi
Ultimate breaking strength	62,000 psi
Ultimate tensile strength	32,000 psi
Critical slip stress	150 psi

Similar relations are observed for other pure metals.

Inability to develop usefully the inherent strength of ductile metals results from (1) the relatively low slip shear strength; (2) the usual occurrence of fracture by degrees, starting at some internal focus and spreading. The local stress intensity at an internal discontinuity or at a focus of fracture may be many times greater than the average stress over the section.

Imperfections of many kinds may serve to localize and intensify stresses, such as flaws in the regularity of the crystal lattice, micro-cracks within a grain, shrinkage voids, nonmetallic inclusions, rough surfaces, and notches of all kinds. Many impurities owe their potency to a high degree of insolubility in the solid matrix coupled with high solubility in the fusion; this permits their freezing out relatively late in the solidification process, as concentrates or films between the grains, thus serving as effective internal notches. The great weakening effect of the graphite flakes in cast iron is an illustration.

Notches act not only as stress raisers but also as stress complicators, and frequently induce a degree of stress tri-axiality. The deeper the notch and the sharper the acuteness of its root, the more effective it is in this respect. A purely tri-axial stress could be relieved by no process of slip or shear, would result in fracture without deformation, and may occur internally in highly stressed parts. Notches are the great weakeners, and engineers do well to recognize their prevalence under many disguises.

EFFECT OF TEMPERATURE. Two effects of temperature on metals and alloys are distinguished: (1) temporary, in which properties characteristic of the temperature are acquired; and (2) permanent, in which changes occur which persist upon return of the metal to normal temperatures. At low, subatmospheric temperatures many steels and iron alloys become quite brittle, because of a change in the mode of deformation (cleavage). On the other hand, most nonferrous metals having crystal lattices of the face-centered cubic type, and their alloys as well as austenitic steels, remain tough at temperatures as low as that of boiling nitrogen (−320 F). At high temperatures, not only are metals subject to direct attack by atmospheric oxygen and by sulfurous and other combustion gases (scaling, sulfidation, and carburization) but they also take on temporary characteristics not readily disclosed by examination in the cold or by brief tests at temperature.

One common change in property is the reversal in strength at grain boundaries. At temperatures higher than the recrystallization temperature, the crystalline interior of the grain becomes stronger than the boundary, and metals broken at such temperatures show *intercrystalline fracture.* Metals and alloys intended for load-bearing service at such higher temperatures for this reason are preferred in a coarse-grained condition.

Flow under sustained loading, which is absent in elastic materials at ordinary temperatures, becomes the dominant characteristic at temperatures considerably below the recrystallization range. The evaluation of alloys for use in highly stressed parts operating at high temperatures is a difficult, lengthy, and expensive procedure, involving rates of creep, time to fracture, microstructural stability, and corrosion resistance, all as interrelated factors. Within or below the softening range, creep diminishes with time due to the accumulation of work-hardening effects, and this stage may persist for years. At higher temperatures or with higher stressing this stage may terminate with an unstable acceleration of creep, leading rapidly to fracture at stresses which may be as little as one-tenth the breaking stress in ordinarily rapid tension tests. As this termination with its serious consequences may be much delayed, creep tests are frequently required to approximate in length the expected life of the proposed equipment, such as several hundred hours for some types of aircraft engine or several years for heavy prime movers.

HARDENING OF METALS. All the effective means of hardening the metals accomplish it by greatly increasing the slip shear strength of the grain. These means are (1) work or deformation hardening, discussed under Article 5; (2) solid solution alloying; (3) formation of dispersed precipitates.

Solid Solution Hardening. Hardening by solid solution is accomplished by distortion of the crystalline lattice of the solvent metal by the solute atoms, with its resulting effect on the forces required to accomplish slip on warped or dimpled planes. The Rosenhain rule states that the specific hardening increment of a dissolved element is greater, the lower its maximum solubility in the matrix metal.

Precipitation Hardening. The hardening of metals by dispersed precipitates is widely practiced in current metallurgy. In principle it is applicable to any metal or kind of alloy and depends upon the development of a structure comprising, in the simplest case, grains of a single type in which are dispersed a multitude of extremely fine particles of some other metallic substance. The dispersed phase does not necessarily need to have a high intrinsic hardness. The hardness of its matrix results from a sort of mechanical interference with slip, the tiny particles in effect acting as "keys" against slip displacement. There is a most favorable size for dispersed particles to act in this manner, of the order of 10 to 100 times the lattice constant. Smaller or larger particles may have little hardening effect.

The basic requirements for obtaining such dispersed precipitates are (1) existence of a substance, usually an intermetallic compound, for which the matrix has a limited solid solubility, but which increases substantially in solubility with temperature; (2) provision of this substance in the alloy in an amount exceeding its lower limit of solubility; (3) establishment of a supersaturated solution by heating above the solubility-temperature limit, followed by rapid cooling; (4) reheating to a lower temperature at which the substance slowly separates from the matrix. If the temperature of step 4 is too high or the time allowed is too long, oversized and ineffective precipitates result. Details of the nature and necessary amount of precipitant and of heat treatment differ widely for various metal combinations.

Precipitation hardening has the advantage of being a reversible process; an alloy having this power may at will be put into a soft (solution treated) condition in which machining, forming, and other mechanical operations may easily be done. It may then be converted to the hard (precipitate) state for use, and the cycle repeated if desired.

5. TREATMENT OF METALS

The changes that can be evoked by mechanical work at various temperatures, by heat treatment for various purposes, and by altering the sequence with which a succession of these treatments is done afford ample opportunity for the skillful regulation and combination of properties.

HOT AND COLD WORKING. The deformation of metals or alloys in the cold, as by rolling, stamping, swaging, or drawing, results in an increase in hardness and tensile strength and a decrease in ductility. These changes are roughly proportional to the extent of deformation, usually expressed as the reduction in cross-sectional area or in thickness. The hardness and tensile strength of ductile alloys may in this way be increased two- or three-fold with, however, relatively much greater loss in ductility, to $^1/_{30}$ or $^1/_{50}$. Very heavy working develops fiber both as to included particles of impurities and as to grain structure, and a resulting directional difference in properties, as longitudinal versus transverse. Overworking results in edge cracking, internal bursts, and other permanent injuries. In general, a metal is cold worked at temperatures below the recrystallization range; it is hot worked at temperatures above it. Hot working, such as forging, hot rolling, and piercing, does not alter the mechanical properties as radically as cold working, since it is accompanied by simultaneous recrystallization and softening.

HEAT TREATMENT. Stress Relief. As a result of permanent deformation by cold working, internal elastic stresses remain which may cause distortions after cutting operations, lead to premature yielding, or even to fracture under corrosive attack. These stresses may be dissipated by heating to temperatures considerably below the softening range. The metal remains hard, but its elastic strength is improved. Residual tensile stresses are usually more detrimental than compressive stresses. Residual compressive stresses may be imparted to surfaces by an operation of blasting the surface, or peening, with fine steel shot; this operation may lead to improved performance of parts which fail by tensile fatigue, as in vibrational bending. (See Articles 10 to 12, Section 20.)

Annealing and Recrystallization. When metals which have been cold worked are heated, the first evidence of permanent change is the relief of internal stresses and a slow softening. Both occur before changes in microstructure are visible. The rate of softening of cold-worked copper (Fig. 7) depends upon the initial degree of cold work, and follows a logarithmic law that indicates steady but slow softening to ordinary temperatures. Other metals

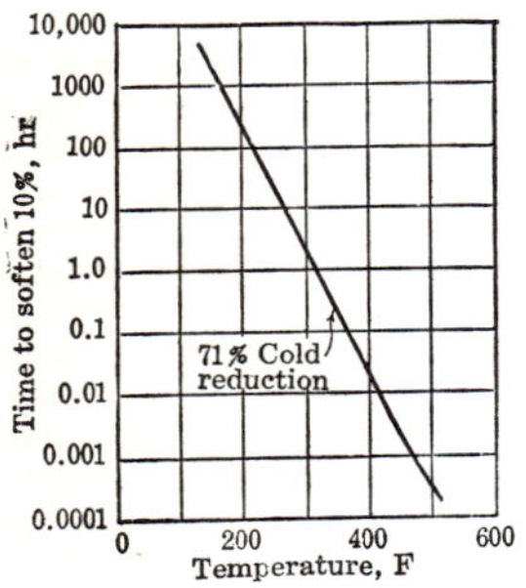

Fig. 7. Softening of hard rolled copper.

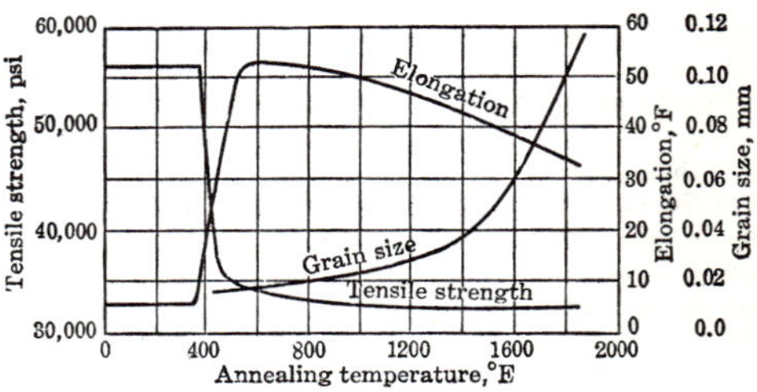

Fig. 8. Recrystallization of hard rolled copper.

and alloys presumably behave similarly at different rates. This kind of softening must be taken into account in applications which rely upon the maintenance of properties over long periods of time.

At somewhat higher temperatures the distorted, elongated, worked grains may be seen to break down with the formation of a relatively few, tiny equi-axed grains. As time is further allowed, or temperature is increased, this recrystallization spreads throughout, resulting in a uniformly fine-grained structure, with the restoration of the characteristic softness and ductility of the unworked metal. The average grain size developed is principally related to the annealing temperature, although tending to be diminished by increasing degrees of cold work and by the presence of insoluble impurities. With increase in annealing temperature, grain size increases rapidly. An example of softening and recrystallization is given in Fig. 8. Recrystallization temperatures are subject to variation but, in the case of simple metals, they increase with rise in melting point; solid solution alloys have higher recrystallization temperatures than the parent metals; increase in amount of cold-work decreases moderately the temperature of recrystallization. (See Table 6.)

Table 6. Lowest Recrystallization Temperatures After Severe Cold Work of Metals of Commercial Purity

Metal	Temperature, °F	Metal	Temperature, °F	Metal	Temperature, °F
Tungsten	2200	Platinum	750	Magnesium	300
Molybdenum	1600	Copper	400	Zinc	70
Nickel	1100	Silver	400	Tin	64 *
Iron	850	Aluminum	300	Lead	Subatmospheric

* Transformation temperature to Alpha tin.

Thermal Hardening. The operation of heat-treating an alloy which is responsive to precipitate hardening is designed to accomplish the two steps of putting the precipitant into solution and then throwing it out of solution under critical temperature conditions.

Solution temperatures may vary widely, depending both on the basis metal and the nature and solubility of the precipitant (Table 7). In some cases, notably alloys preserving

Table 7. Some Precipitate-hardening Alloy Systems

Predominant Metal in Matrix	Addition Elements	Precipitant Phase	Solution Temp., °F	Precipitating Temp., °F	Typical Hardness Range, BHN Soft	Hard
Aluminum	Copper	$CuAl_2$	1000	300	62	105
	Magnesium, silicon	Mg_2Si	970	310	26	80
Copper	Beryllium	CuBe	1450	650	110	360
	Nickel, silicon	Ni_2Si	1500	500	90	170
	Nickel, tin	Ni_3Sn	1400	550	75	180
Iron	Molybdenum	Fe_3Mo_2	2300	1150	160	430
Lead	Calcium	Pb_3Ca	600	70	7	19
Nickel	Beryllium	NiBe	1830	840	170	525
Nickel-copper	Aluminum	....	1700	1200	150	300
Nickel-iron	Titanium	Ni_2Ti	1750	1200	140	360
Silver	Copper	Cu	1450	600	45	110

load-carrying ability at high temperatures, the strengthening precipitate not only must be stable at the temperature of use, but also must have its solubility limit at substantially higher temperatures. This may lead to solution temperatures approaching the melting range.

The speed of cooling necessary to preserve a supersaturated solution also varies considerably. Aluminum- and copper-base alloys may require drastic quenching, but for many of the nickel-rich alloys air cooling even in sections up to 1 1/2 in. may suffice.

The reheat for precipitation is critical, both as to temperature and time. The operation frequently is a lengthy one, and overdoing it, such as to develop too coarse a precipitate, can be rectified only by repeating the whole cycle of heat treatment. If the precipitation reaction proceeds spontaneously at room temperature it is usually referred to as *aging*.

6. ENGINEERING ALLOYS

The alloys of engineering and commercial importance are few compared with the almost limitless number of possible mixtures. The tendency is toward fewer alloys and greater standardization to serve the ordinary requirements of industry. The great cost of development in bringing a meritorious novel alloy to the point of availability in all commonly used shapes, sizes, and quantities precludes its trade introduction unless tonnage uses are in sight. On the other hand, the engineer often is in the dilemma of urgently requiring alloys with highly specialized properties in quantities too limited to attract the attention of established sources of supply. The larger producers of metals frequently maintain development departments which have available many unpublished data and can assist in such problems.

STANDARDIZATION. For many alloys and simple metals standard compositions have been prescribed and specifications for their control have been prepared, based on the common experience of producers, users, and engineers. The principal standardizing agencies in the United States for metals are the American Society for Testing Materials, in Philadelphia, the Society of Automotive Engineers, in New York, and the Bureau of Standards, in Washington. In addition, many large industrial organizations issue their own purchase specifications, available on request.

A comprehensive and useful guide to the standard specifications of many industrial, government, and engineering society organizations is the *Cross Index of Chemically Equivalent Specifications and Identification Code*, published by General Motors Corp., Detroit (1945). See also *Standards and Specifications for Metals and Metal Products*, Miscellaneous Publication 120, U. S. National Bureau of Standards, Washington.

TRADE NAMES. In addition to numerically designated standard alloys, there are many proprietary alloys generally bearing names that do not suggest their composition, but similar to standard alloys in properties. There are also many obsolete, named alloys which have been described and have become established in handbooks but which no longer have distinctive properties. There are special lists for the identification of such materials: *List of Alloys*, by W. Campbell, ASTM (1930); *Engineering Alloys*, by N. E. Woldman and R. J. Metzler, ASM (1945); sections in *Metals Handbook*, ASM (1948).

GENERAL REFERENCES

The following sources are suitable for engineering readers requiring more complete metallurgical treatment.

GENERAL

C. S. Barrett, *Structure of Metals*, McGraw-Hill Book Co., New York, 1943.
Sir H. Carpenter and J. M. Robertson, *Metals*, Vols. 1 and 2, Oxford University Press, London, 1939.

APPLIED METALLURGY

Metals Handbook, American Society for Metals, Cleveland, 1948.
F. T. Sisco, *Modern Metallurgy for Engineers*, 2nd ed., Pitman Publishing Corp., New York, 1948.

TABULATED DATA

S. L. Hoyt, *Metals and Alloys Data Book*, Reinhold Publishing Corp., New York, 1943.
Mechanical properties of metals and alloys, *Natl. Bur. Standards Circ.* C447, 1943.

EQUILIBRIUM DIAGRAMS

M. Hansen, *Structure of Binary Alloys*, Edwards Brothers, Ann Arbor, Mich., 1943.

JOURNALS (CURRENT ISSUES)

J. Inst. Metals.
Proc. Am. Soc. Testing Materials.
Trans. Am. Foundrymen's Assoc.
Trans. Am. Inst. Mining Met. Engrs.

THE APPLICATION OF DATA ON STRENGTH OF METALLIC MATERIALS

By H. W. Gillett

The designer selects materials, chiefly metals if he is a mechanical engineer, to provide the properties essential to the object being designed and decides the geometrical shape and size which, in the material chosen, will allow the object to function under the conditions of service.

Correctness of combination of material and of its shape and size is established by the experience of service. When the previously used material becomes unavailable or too expensive, a substitute is sought; or a new material of improved properties opens the way to design of a better or cheaper object. If the experience of service is lacking, prediction of the behavior of the substitute or the new material must be made on some other basis. The object of a handbook is to present concise and correct information on properties and behavior to allow proper proportioning and shaping of old, substitute, and new materials.

TEST SPECIMENS. The properties of materials are determined by certain conventional tests upon "specimens," often chosen for ease in testing and reproducibility of results. These tests may or may not, and often do not, reproduce the conditions of service. The specimens may or may not, and often do not, exactly represent the quality of the material in, or the size or shape of, the actual object. Hence there are elements of uncertainty in the service behavior which can be fully resolved only by service experience or a carefully planned and controlled simulated service test.

To minimize the uncertainty, there must be added to the quantitative numerical data on conventional properties a qualitative understanding of the unmeasured properties essential in fabrication and in use, and of the peculiarities of the material.

The numerical data cited for a given material, prepared in a particular way, are usually for flawless material, and as a rule are obtained under test conditions which avoid stress concentration. The data are usually obtained by stressing "with the grain" and very often on specimens much smaller than the object to be made. The influences of stress concentration, directional differences, and size effects are not reflected in the conventional data on mechanical properties.

EFFECTS OF PROCESSING. The same material, chemically prepared and fabricated in different ways, may vary not only in the conventional properties on flawless material, but the preparation may affect the prevalence of flaws. The recorded data for the prototype material may not be representative of the actual material, whereupon testing of samples of the actual material, either by the manufacturer in the exercise of quality control or by the user prior to acceptance, becomes necessary. In either case, great care needs to be exercised in sampling so that the samples tested truly represent the material for use; a longitudinal specimen on a 1-in. rolled bar of a particular steel is not a correct specimen for appraising the properties of a 10-in. forging in the radial direction.

No compilation of properties within a handbook can be so comprehensive as to cover every variation of size and prior processing for all materials of engineering. Moreover, manufacturers' data and the official specifications are based on normal expectancy. Even with all details of processing under good control, supposedly identical materials show a range of properties, lying on a hairpin-shaped distribution curve, at either end of which come a few low and a few high values.

Testing and inspection will vary in rigorousness, depending on the presence or absence of hazards to life and limb through failure in service or, where such hazards are absent, on the cost of a few failures.

The likelihood of incidence of low values, the behavior in uses analogous to the one contemplated, and the limitations, idiosyncrasies, and peculiarities of a material are generally known to, and information on them available from, the engineering staffs of reliable producers, even though the advertising staffs do not stress the limitations of the material in their sales literature.

SELECTION OF MATERIAL. Knowledge of the peculiarities of a material is essential to intelligent selection by the designer. Use of handbook data should be accompanied by a realization of those behaviors of a material which may render such data inapplicable to a case in hand. That a handbook gave a certain figure for the normal expectancy of a material, normally processed in a certain section, is no alibi for a failure resulting from lower properties inherent with some other processing on some other section. A handbook is a supplement to, not a substitute for, intelligent thinking on the part of the engineer.

Hence special stress has been put in the articles that follow upon the fundamentals underlying the behavior of materials, in an effort to aid the designer in estimating the qualitative influence of factors that cannot be quantitatively expressed or tabulated.

Before recorded data on strength of materials can be correctly applied by the designer, a searching examination of several factors must be made.

7. STRESS RAISERS

APPLIED STRESSES. First the stresses to be applied in service must be evaluated, and this means not only the average stress throughout the whole section of the part but also, and most important, the highest local stress at the worst point of stress concentration, as at notches, nicks, scratches, tool marks, sharp fillets, irregularities at welds, cracks, keyways, and poor casting design. Such a stress raiser is the Achilles' heel. They occur most frequently at the surface rather than in the interior of a part. Design to avoid their pre-existence and protection to prevent their accidental production in service can do more to avoid failure than shifting from one material of construction to another.

Almen has said (Ref. 1):

> Industry has paid premium prices for alloy steels because of their fancied advantages when used as gears and in other parts. Fatigue tests on actual machine parts correlated with service records have shown that there is no detectable difference between the high-priced alloy steels and many of the low-priced alloy steels when used in many machine elements.

He goes on to say that any real differences in fatigue characteristics are

> . . . so small in comparison with the mechanical fatigue hazards introduced by the design and fabrication of the machine part as to be negligible.
>
> Furthermore [Ref. 2] . . . Fully 90 per cent of all fatigue failures occurring in service or during laboratory and road tests are traceable to design and production defects and only the remaining 10 per cent are primarily the responsibility of the metallurgist as defects in material, material specification, or heat treatment.
>
> Study of fatigue of materials is the joint duty of the metallurgical, engineering, and production departments. There is no definite line between mechanical and metallurgical factors that contribute to fatigue. This overlapping of responsibility is not sufficiently understood.
>
> Hence the engineers are constantly demanding new metallurgical miracles instead of correcting their own faults. Until metallurgists are less willing to look for metallurgical causes of fatigue and insist that equally competent examination for mechanical causes be made, we cannot hope to make full use of our engineering materials.

SERVICE CONDITIONS. Second, attention must be given to the conditions of service. If static loading only is concerned, very slight plastic flow may be able to redistribute the stress without damage. If the stress is repeated, however, the possibility of fatigue failure starting from a stress raiser is imminent, as is brought out in Fig. 1 by Roš and Eichinger (Ref. 3).

The location and some idea of the intensity of stress at a stress raiser can be found by use of short-gage-length electrical strain gages, brittle lacquers, or examination of a transparent model in polarized light. Merely giving thought to the problem of notches and fillets will often bring out the necessity of redesign because the presence of serious stress raisers will be obvious.

A good introduction to methods of location of stress raisers and of reducing their intensity by design is available in Ref. 4. Theoretical stress intensification factors for notches of varying severity are listed by Roark (Ref. 5). The relation of notches to fatigue behavior is discussed in detail in Ref. 6. Increasing the static strength of steel tends to increase fatigue resistance in the absence of stress raisers or internal stress. Very hard steels are more notch sensitive in fatigue than softer ones, however, and are likewise more prone to have internal stress. Corrosion combined with repeated stress vastly reduced the endur-

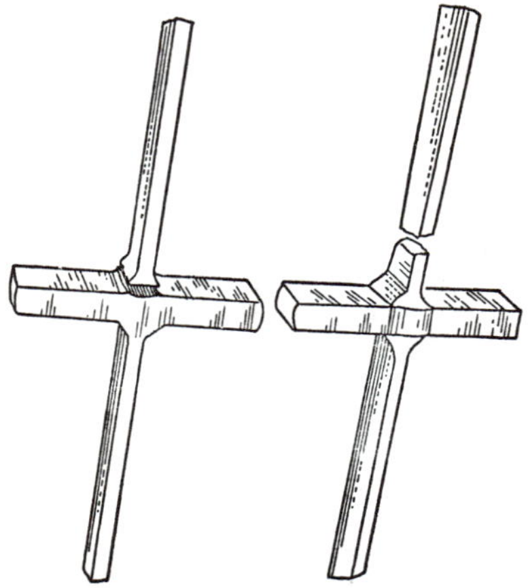

FIG. 1. Welded specimen at right failed under single static load of 33,000 psi. Specimen at left failed in fatigue after 1,194,400 pulsations of load, 0 to 9000 psi. (Illustration after Roš and Eichinger, reproduced by permission from *Prevention of the Failure of Metals under Repeated Stress*, John Wiley and Sons, Ref. 6)

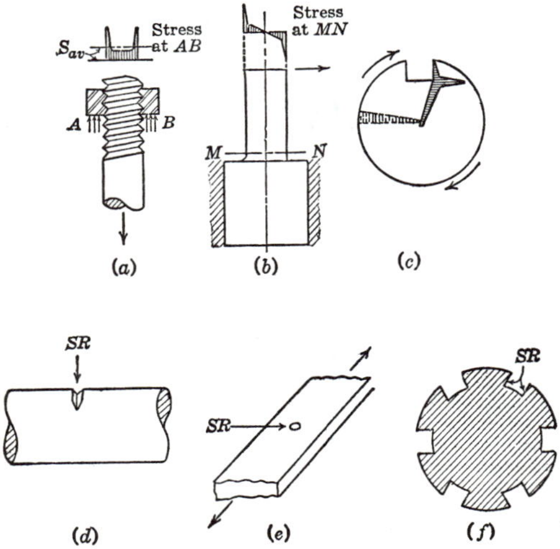

FIG. 2. Stress raisers that lead to fatigue failure under over-stress: (*a*) screw threads; (*b*) sharp fillets; (*c*) keyways; (*d*) nicks; (*e*) holes; and (*f*) sharp corners. Cracks, surface seams, and large sharp-cornered nonmetallic inclusions have like effects. (Reproduced by permission from Bullens, *Steel and Its Heat Treatment*, 5th Ed., Vol. 1, John Wiley and Sons. Ref. 7)

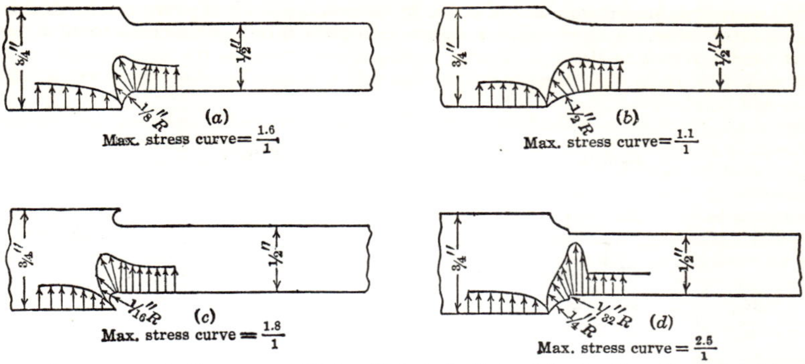

FIG. 3. Stress concentrations around fillets. Note bad effects of jog in (*d*). (Reproduced by permission from *Prevention of the Failure of Metals under Repeated Stress*, John Wiley and Sons, Ref. 6)

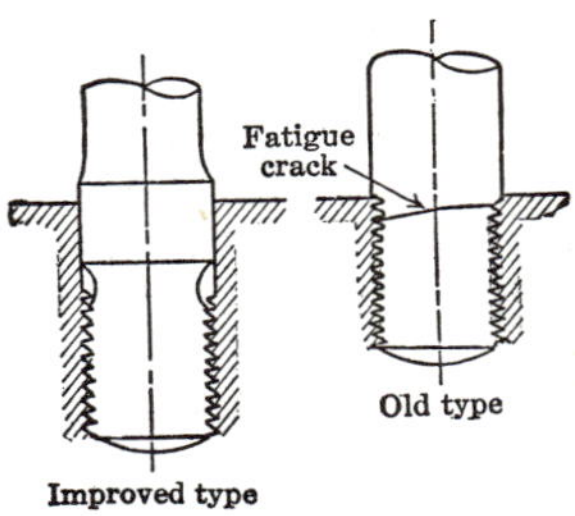

FIG. 4. Stud with stress-relieving groove, left, avoids fatigue failure which results with plain type, right. The fillet should extend beyond the first female thread. (MacGregor, Burn, and Bacon, *Trans. North East Coast Inst. Engrs. & Shipbuilders*, Ref. 8)

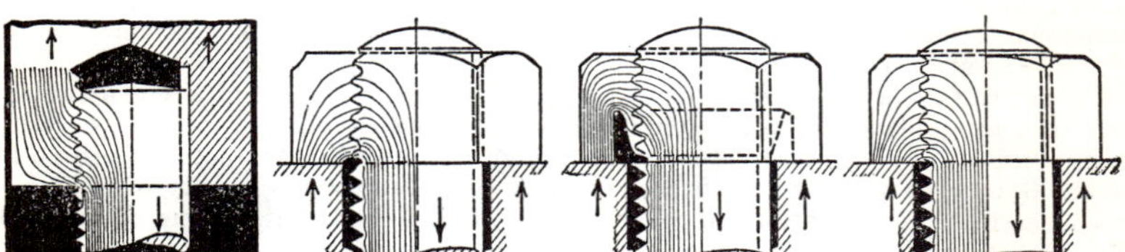

FIG. 5. Diagram showing the distribution of stress between stud and part with various bolt and nut designs. (MacGregor, Burn, and Bacon, *loc. cit.*)

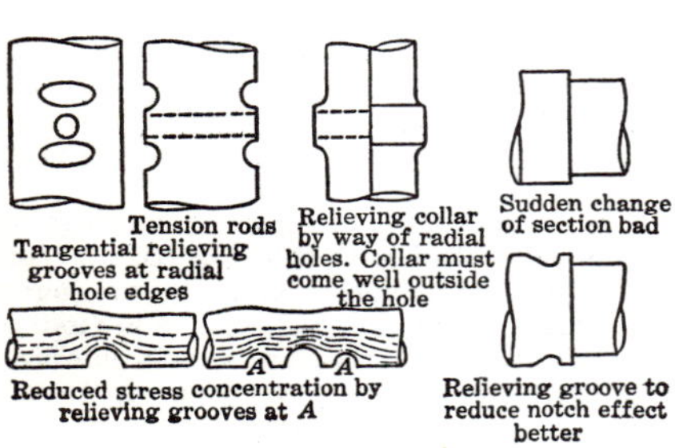

FIG. 6. Devices for relieving stress concentration. (MacGregor, Burn, and Bacon, *loc. cit.*)

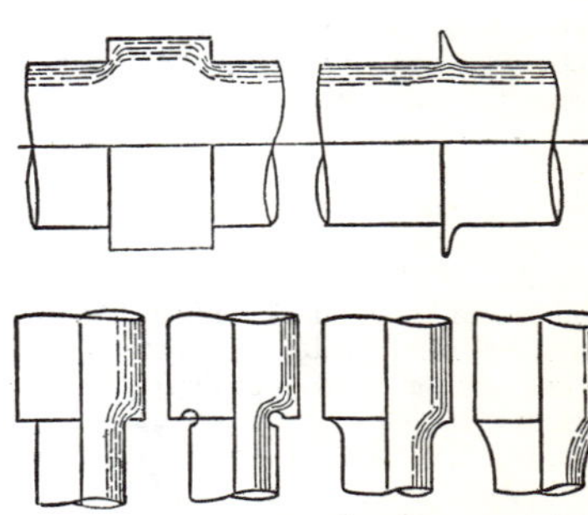

FIG. 7. Comparison of stress concentrations with wide and narrow collars (upper views). Effect of abrupt versus gradual change of section on stress concentration (lower views). (MacGregor, Burn, and Bacon, *loc. cit.*)

ance of corrodible steels, as indicated in Fig. 2, Section 2, p. 2-09. Stress raisers and designs for mitigation of stress concentration, taken from Refs. 6 to 9, which go into more detail, are shown in Figs. 2 to 10.

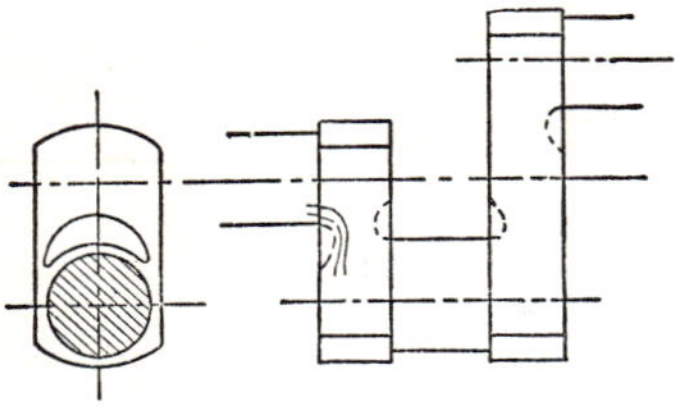

FIG. 8. Localized eccentric internal fillets used in crankshaft design to give reduced stress concentrations. (MacGregor, Burn, and Bacon, *loc. cit.*)

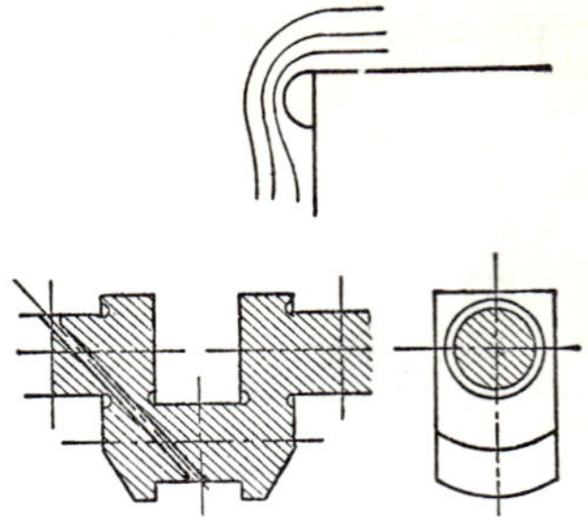

FIG. 9. Circular internal fillets, as used to allow increased bearing length. (MacGregor, Burn, and Bacon, *loc. cit.*)

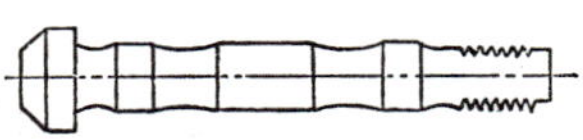

FIG. 10. Body of bolt turned down at nonbearing portions of length to increase resilience and reduce fatigue stresses. (MacGregor, Burn, and Bacon, *loc. cit.*)

8. LIMITATIONS OF PUBLISHED DATA

Data from conventional mechanical tests of materials of construction are usually too favorable, may be a long way from expressing the complete situation, and may be a source of serious misapprehension, from several causes.

SCATTER. There is a variation in properties among even consecutive heats of steels or other alloys, planned to be alike. Certain numbers, symbols, or trade names are used as shorthand to separate the alloys into classes, usually on the basis of chemical composition. Necessarily, there is a leeway in composition. Consequently, when the properties of a class are tabulated, a range of values should be given instead of a single value; and when they are plotted, the plot should not be a line but should be a band to express the normal scatter. Moreover, in steel, even though the chemical composition of two heats may be practically identical, each heat is likely to show an individuality of its own, primarily ascribable to variations in practice in "finishing" the heat. Some of these variations are controllable, others still elude precise control.

Hence, the designer does not know, without specific tests on each particular lot, whether that lot is at the top, the average, or the bottom of the scatter band. Statistical methods are necessary to indicate what risk the designer takes in assuming that the lot in question has average properties. Few data are available on the statistical basis; most published data relate to something fairly close to the average. They ignore the fact that poorer properties may be met or that, by "quality control" and adequate testing, lots with better properties may be selected. Quality control and statistical methods are familiar to the engineer in respect to staying within prescribed tolerances of dimensions in machining. (See Section 24.) Similar principles are applicable to evaluation and control of mechanical properties of materials.

SIZE EFFECT. Published data usually refer to specimens of relatively small section and fail to bring out the effect of size. In hot- or cold-rolled steel, the properties of 1/4-in., 1-in., and 2-in. plate, for example, fall off as the section increases. Similarly, the conventional test bar used to evaluate the quality of metal poured into a casting does not indicate whether the casting has the same properties as the test bar—they are usually inferior to those in the test bar—nor does it indicate whether the casting itself is sound, a matter ordinarily far more important than the mechanical properties of a sound bar.

The size effect is especially misleading in the case of quenched and tempered alloy steels, data for which are generally given from tests of bars not over 3/4 in. in diameter. There is a serious deficiency in yield strength of slack-quenched and tempered steels; that is, those quenched in so large a section that the center is not completely hardened. A relatively shallow hardening steel tested in 3/4-in. diameter may show up excellently, but in

1-in. diameter or larger, its center may have very mediocre properties. The properties of different steels, each quenched in such section that each hardens fully, when tempered back to the same hardness are practically indistinguishable. Tables and property charts, showing properties versus tempering temperature, really show only the response to tempering, a matter with which the heat treater is concerned but the designer is not, since he calls for a steel tempered to the particular strength he needs. not for one tempered at a given temperature for a given time.

DIRECTIONAL EFFECTS. In rolled or forged material there are directional effects resulting from stretching or compressing the material more in one direction than in another. In making gears from forged blanks, great care is exerted to work the blank so that the "grain never cuts across the teeth." Directional properties are more pronounced in alloys with more than one metallographic constituent and, especially, when nonmetallic inclusions are present. Yet, even in pure metal, directional properties may result from orientation of the constituent crystals in definite relation to the direction of rolling. For example, cold-rolled pure copper, pressed into cups, will show four "ears," representing differences in deformability of the copper crystal in different directions.

Directional effects are particularly prone to appear in steel because all steel contains certain nonmetallic inclusions, notably manganese sulfide, which is plastic at steel-working temperatures. The manganese sulfide particles, globular in the ingot, are rolled out into flattened plates or leaves, or into stringers extending in the direction of rolling. Brittle inclusions are strung out into "stepping stones." Cross rolling of plate makes the inclusions shorter and wider and with less effect on properties, but the laminations still exist in the through-plate direction.

Formability is reduced when the inclusions lie across the direction of bending. Static strength is not vastly reduced, but cracking starts at a smaller angle of bend. The inclusions decrease the transverse fatigue strength and the notched-bar impact values of transverse specimens.

How striking the directional differences may be is brought out in the tables on p. 2-05, Section 2. The cited values in handbooks and manufacturers' tables are almost invariably for specimens taken longitudinally; unless the designer is entirely certain that stresses in service will only be applied longitudinally, such specimens give a falsely rosy picture. Directional differences, usually of less vital nature, also result from micro or macro segregation in the ingot, causing a banded structure upon rolling, in which carbide-rich, strong, and possibly brittle bands are sandwiched between weak, carbide-poor bands. This structure often looks worse than it really is, but in extreme cases it may lead to inferior directional properties.

Castings which receive no mechanical work are almost free from directional properties. Specimens from the castings may not appear as good as specimens of comparable wrought material taken in the direction of working, but when transverse specimens of the wrought material are considered the casting might often prove more reliable.

The highest degree of directional difference is naturally met in screw stock that is exceptionally high in sulfur and hence in manganese sulfide. The foreign particles are intentionally introduced to favor machinability, this advantage outweighing, for many lightly loaded uses, the depreciation in transverse properties. For severe service, such steels are not considered safe.

Machinability is enhanced without marked depreciation of transverse properties by the use of 0.10 to 0.20% lead (Ref. 10) instead of large amounts of manganese sulfide. The lead particles appear to be in an extremely fine state of subdivision, not readily detected by the microscope even at high magnification.

SURFACE CONDITIONS. A further drawback to published values is that they are universally given from tests on specimens that have had the surface rather deeply machined off. Mill scale and surface roughness, which, obviously, will act as stress raisers, can be seen with the naked eye. When such material is to be used with the mill surface, test data should clearly be obtained without removing the surface. A more insidious surface defect is the presence of a decarburized "bark," which is not visible to the naked eye but requires metallographic examination for detection. The bark may be thin and not made clearly evident by hardness determinations.

These surface defects may or may not be serious in static loading, but they are highly serious under repeated stress. Fatigue strength decreases as carbon content decreases, and a fatigue crack, started at the weak surface, even though that surface is almost infinitesimally thin, can readily propagate through the interior. Moreover, when shot peening is applied in order to strengthen a surface against fatigue, the decarburized skin is not strengthened to the proper degree.

NEED FOR SIMULATED SERVICE TESTING. For all these reasons, ordinary tabular handbook or manufacturers' data should be examined with a critical eye, to learn whether they represent the size, direction, and surface condition characteristic of the actual service of the part to be made from that alloy or lot. Since there is seldom sufficient information to be had from published data, much less reliance is placed on supposedly representative data from conventional mechanical tests and much more on simulated service testing of the actual, full-sized part, stressed as it will be stressed in service.

THE DUCTILITY PROBLEM. In ordinary static tensile loading, ability to flow a trifle without cracking, thus to redistribute and even out a localized applied stress, as around a rivet hole, has utility; but the usable deformation is very low, generally taken to be of the order of 1%. Ability for such small deformation separates materials into (1) those which are glass brittle and will stand no misalignment whatever; (2) those like cast iron, ordinarily classed as brittle, but having some degree of toughness; and (3) ordinary metals and alloys with tensile ductility ranging from a very small percentage up to taffy-like behavior.

Glass itself is a useful material of construction, in its proper place. The cast-iron class also has its obvious uses. An outstanding example of a material, in this class as to brittleness, that is serving admirably where external toughness used to be called for is the Ford cast crankshaft alloy. With only a trace of tensile ductility or notched-bar impact resistance, these cast crankshafts are used instead of steel forgings in which very high toughness used to be specified. In crankshaft service, rigidity and fatigue resistance are called for; whereas any measurable plastic deformation makes the crankshaft valueless. The ductility of the forging cannot be used. An automobile fender, however, properly calls for material capable of plastic deformation.

In the third class is included cold-drawn bridge wire, classed as one of our most dependable materials of construction, yet having only some 2 to 6% tensile elongation. Big guns with 5% elongation have stood up all right in service.

The numerical value of a figure for elongation means nothing in itself; it has meaning only in relation to the level of ductility expected in normally processed material of that composition. Reduction of area is in the same category. Various useless "merit indices" have been suggested for heat-treated steels, in which great weight is given to the reduction of area determined on longitudinal specimens. Reduction of area is highly sensitive to direction of working; a steel showing 60% in the longitudinal direction will often show 35% on a specimen taken at a 30-degree angle to the direction of working.

High-strength aluminum alloys with 12% elongation and less were tried out with trepidation in place of older, weaker alloys of 18% elongation, but proved quite as dependable. It has been shown (Ref. 11) that an airplane wing would be useless were it to undergo 1 1/2% plastic deformation. The ability for larger plastic deformation in the alloy of the wing is unusable. Nowadays it is recognized that each alloy has a characteristic level of ductility representative of its own goodness, but the goodness is not regulated by the level itself.

Failure to show the expected level, whatever that is, is a sign to the metallurgist of some unexpected happening in the preparation and processing. The happening might be bad in respect to some attribute not measured by the conventional test. This works both ways. If a steel casting supposedly of given composition and heat treatment showed the ductility ordinarily met in wrought stock of that composition and treatment, when tested on a *longitudinal* wrought specimen, it would be suspected that something had gone wrong somewhere, and it would ordinarily turn out that the steel casting had low yield strength.

The need for toughness depends entirely upon the type of service to be met; and when toughness is required, specific proof that it exists, in the actual part, in the actual service is needed. The indirect evidence of conventional tensile and impact tests is insufficient. The significance of the tension test has been fully discussed in Ref. 12.

Data from conventional tests that are recorded in percentage or other values instead of pounds per square inch, such as elongation and reduction of area in the tensile test, do not apply directly in design. They are indirect measures of cold formability in fabrication and show whether the material has the usual characteristics expected upon correct processing.

Less readily evaluated features, such as machinability and weldability, often play a larger part in selection than the more measurable qualities. It is particularly to be noted that notched-bar impact data, reported in foot-pounds energy absorbed by a certain size of specimen with a specified notch, cannot be correctly expressed in foot-pounds per square inch and converted to give data for any other size or notch. Moreover, the figure for energy absorbed in a notched-bar impact test bears no relation whatever to behavior in notched fatigue.

9. ALTERNATE MATERIALS AND PRODUCTION METHODS

Many parts are so lightly loaded that neither static nor fatigue failure is at all likely, nor is wear or corrosion resistance a major factor. Where the piece must have definite contour and dimensions, but few other requirements, a wide variety of materials and methods present themselves for choice, the choice resting on cost of the finished piece. Because this piece or an analogous one was formerly made in a given way from a given material is no reason for clinging to that material and method in the next case, especially where the quantity to be produced is large enough to amortize the equipment required for a different process. Changes in raw-material cost and availability, and technical advances in processing, continually shift the economic picture.

Often the real criterion is how close to final dimensions the piece can be made so as to require the least and the easiest and fastest machining. A small, lightly loaded gear, for example, could be stamped from brass sheet, requiring a die and some finishing, or be made of an aluminum alloy extrusion, again requiring a die, the gear being sliced off by sawing the extrusion. It could be sand cast in gray iron, malleable iron, or secondary aluminum alloy, allowing appreciable finish when machined; it could be die cast very closely to dimensions in a zinc-base die casting, alternately in aluminum; or it could be made from iron powder by die pressing, sintering, and repressing in a coining die.

Other objects might be made on automatic machines from steel, brass, or aluminum free-machining stocks. Brass casting in plaster molds might also be considered. Offhand, the aluminum extrusion and the zinc-base die casting might be given primary consideration for long runs. A meter case with bosses and attachment holes could be made as a sand casting of several alternate materials, as a die casting or pressed from sheet steel with the bosses, which might be made by a screw machine, attached by copper brazing, or by a variety of welding processes. Silver brazing also finds increasing use.

In particular cases, use of seamless or welded tubing or centrifugally cast tubes would offer a choice, affected by the tube dimensions. Centrifugal casting is a logical choice for tubular-type objects, but it is also applied to many patterns normally cast by static pouring. Smaller gates and risers, hence less scrap to remelt, are required than in static casting. Permanent and semipermanent molds are used to secure closer dimensions than in sand casting.

Precision casting, by the lost wax process, long used in dentistry, is applicable to some alloys for high-temperature use that could not otherwise be made to the required dimensional tolerances. (See Section 19, Articles 22 to 24.)

Where a wear-resistant surface, or one with greater strength than the body of the metal, is wanted, carburized, cyanided, or nitrided cases may be applied. Since it is the surface that counts, a very thin case, backed up by a rather strong core, is increasingly used in place of heavy carburized cases on very soft cores. This change is made in order to save processing time and because once a crack is formed in a brittle case it propagates readily through a core, making the core act brittle despite great toughness of the core with the case removed.

Electroplated coatings, for example, chromium, metal-spray coatings, and puddled-on special wear-resistant surface coatings, are used for special cases where wear resistance is needed.

Of great importance in this connection are the induction and flame-hardening processes, carried out with great rapidity, fitting into production lines and providing a hard, strong surface as well as introducing useful compressive stress at the surface. Shot peening, similarly, both hardens the surface and introduces compressive stress. Cheap carbon and low-alloy steels are suitable for these surface treatments. Many alternates are thus available for uses for which heavily carburized, high-alloy steels used to be specified.

Appreciation of the limitations of conventional test data on metals, of the vulnerability of metals to notches and unsoundness which can be avoided or mitigated by thoughtful design, and of the many possibilities in alternate steels and alternate methods of fabrication and processing will save the designer from many pitfalls.

REFERENCES

1. J. O. Almen, *Automotive and Aviation Inds.*, Vol. 88, Feb. 1, 1943, pp. 28–32.
2. J. O. Almen, *SAE J.*, Vol. 51, May 1943, pp. 28–29.
3. M. Roš and A. Eichinger, Symposium on the welding of iron and steel, *Iron Steel Inst.*, Vol. 2, 1935, pp. 843–866.
4. Symposium, Experimental stress analysis, *Product Eng.*, Vol. 17, Dec. 1946.
5. R. J. Roark, *Formulas for Stress and Strain*, pp. 298–309, McGraw-Hill Book Co., New York, 1938

6. Metallurgical Staff of Battelle Memorial Institute, *Prevention of the Failure of Metals under Repeated Stress*, John Wiley and Sons, New York, 1941.
7. D. K. Bullens and the Metallurgical Staff of Battelle Memorial Institute, *Steel and Its Heat Treatment*, Vols. 1 and 2, 5th ed., John Wiley and Sons, New York, 1948.
8. R. H. MacGregor, W. S. Burn, and L. Bacon, *Trans. North East Coast Inst. Engrs. & Shipbuilders*, Vol. 51, 1935, pp. 161–288.
9. H. F. Moore, *Trans. Am. Inst. Mining Met. Engrs.*, Vol. 120, 1936, pp. 13–35.
10. T. Swinden, *J. Iron Steel Inst.* (London), Vol. 148, 1943, pp. 441–460. F. J. Robbins, *Iron Age*, Vol. 142, Nov. 17, 1938, p. 2233. Staff article, *Engineering* (London), Vol. 152, Sept. 2, 1941, p. 205.
11. R. K. Koegler and A. Schmitt, *J. Aeronaut. Soc.*, Vol. 10, 1943, pp. 273–284.
12. Symposium, *Proc. Am. Soc. Testing Materials*, Vol. 40, 1940, pp. 501–609.

HARDNESS, DENSITY, AND COMPRESSIBILITY

10. HARDNESS TESTING OF MATERIALS

Hardness of a substance may be defined as its resistance to penetration by another substance. Several methods of determining hardness, in commercial use, are described in the following paragraphs.

BRINELL HARDNESS NUMBER (abbreviated to BHN, or simply Brinell) is calculated from the diameter of impression left after a hardened steel ball, 10 mm in diameter, is forced with a pressure of 3000 kg (500 kg for soft metals) into a flat surface on the sample to be tested. The procedure provides for a definite time of load application. The hardness number is defined as the load divided by the *spherical* area of the indentation, using the equation

$$\text{BHN} = \frac{P}{(\pi D/2)(D - \sqrt{D^2 - d^2})}$$

where P = load (kg); D = diameter of ball (= 10 mm); d = diameter of indentation (mm). Tables of hardness numbers corresponding to different values of d, calculated from the foregoing formula, are available.

VICKERS PYRAMID HARDNESS (VPH) or diamond pyramid hardness (DPH) is determined from a test similar to the Brinell. The penetrator, however, is a small diamond pyramid (136°) with a square base, and the load is much lighter, permitting its use on thinner specimens than is possible with the Brinell test. Depth of the indentation can be as little as 0.002-in. The formula for calculation of hardness is

$$\text{VPH} = \frac{1.8544 \times (\text{Load})}{(\text{Diagonal of indentation})^2}$$

By varying the load, hardness determinations may be made on a wide range of alloys, from the hardest to the softest, without the necessity of shifting from one scale to another.

KNOOP HARDNESS NUMBER (KHN) is determined from the use of a diamond indenter in the form of a pyramid, with a rectangular base, and having 172° 30′ longitudinal angle and 130° 0′ transverse angle. The impression is shallow and about three times as long as that produced by the Vickers pyramid at the same load. The long diagonal does not change dimension when the load is removed, but the short diagonal shortens because of the elastic recovery of the material. The hardness is obtained by dividing the load in kilograms by the *unrecovered* projected area of indentation in square millimeters. The formula is

$$\text{KHN} = \frac{L}{l^2 Cp}$$

where L = load (kg); l = measured length of long diagonal of indentation (mm); and Cp = constant relating l to projected area = 7.028×10^{-2} for the indenter described in the foregoing. The Knoop indenter is generally used in the Tukon tester, which applies a known weight through a balance beam, the load being varied according to the hardness of the material to be tested. With loads of 200 grams or over, the Knoop hardness figures do not vary much with load and coincide with those of the Vickers for the usual heat-treated steels. At low loads the values depend on load and should be stated in reporting test results.

ROCKWELL HARDNESS is determined by a penetration test similar to the foregoing methods. However, instead of the operator's having to measure the breadth of the im-

Table 1. Approximate Hardness Conversion Numbers for Steel, Based on DPH (Vickers)

(Reprinted by permission from the *SAE Handbook*, 1949 edition)

Diamond Pyramid Hardness Number, Vickers	Brinell Hardness Number, 10-mm Ball, 3000-kg Load			Rockwell Hardness Number				Rockwell Superficial Hardness Number Superficial Brale Penetrator			Shore Scleroscope Hardness Number	Tensile Strength (Approx.) 1000 psi	Diamond Pyramid Hardness Number, 50-kg Load
	Standard Ball	Hultgren Ball	Tungsten Carbide Ball	A-Scale 60-kg Load Brale Penetrator	B-Scale, 100-kg Load 1/16 in. Diam. Ball	C-Scale 150-kg Load Brale Penetrator	D-Scale 100-kg Load Brale Penetrator	15-N Scale 15-kg Load	30-N Scale 30-kg Load	45-N Scale 45-kg Load			
1	2	3	4	5	6	7	8	9	10	11	12	13	14
940	....	....	....	**85.6**	....	**68.0**	**76.9**	**93.2**	**84.4**	**75.4**	97	....	**940**
920	....	....	....	**85.3**	....	**67.5**	**76.5**	**93.0**	**84.0**	**74.3**	96	....	**920**
900	....	....	....	**85.0**	....	**67.0**	**76.1**	**92.9**	**83.6**	**74.2**	95	....	**900**
880	....	....	767	**84.7**	....	**66.4**	**75.7**	**92.7**	**83.1**	**73.6**	93	....	**880**
860	....	....	757	**84.4**	....	**65.9**	**75.3**	**92.5**	**82.7**	**73.1**	92	....	**860**
840	....	....	**745**	**84.1**	....	**65.3**	**74.8**	**92.3**	**82.2**	**72.2**	91	....	**840**
820	....	....	**733**	**83.8**	....	**64.7**	**74.3**	**92.1**	**81.7**	**71.8**	90	....	**820**
800	....	....	**722**	**83.4**	....	**64.0**	**73.8**	**91.8**	**81.1**	**71.0**	88	....	**800**
780	....	....	**710**	**83.0**	....	**63.3**	**73.3**	**91.5**	**80.4**	**70.2**	87	....	**780**
760	....	....	**698**	**82.6**	....	**62.5**	**72.6**	**91.2**	**79.7**	**69.4**	86	....	**760**
740	....	....	**684**	**82.2**	....	**61.8**	**72.1**	**91.0**	**79.1**	**68.6**	84	....	**740**
720	....	....	**670**	**81.8**	....	**61.0**	**71.5**	**90.7**	**78.4**	**67.7**	83	....	**720**
700	....	**615**	**656**	**81.3**	....	**60.1**	**70.8**	**90.3**	**77.6**	**66.7**	81	....	**700**
690	....	**610**	**647**	**81.1**	....	**59.7**	**70.5**	**90.1**	**77.2**	**66.2**	....	....	**690**
680	....	**603**	**638**	**80.8**	....	**59.2**	**70.1**	**89.8**	**76.8**	**65.7**	80	329	**680**
670	....	**597**	**630**	**80.6**	....	**58.8**	**69.8**	**89.7**	**76.4**	**65.3**	....	324	**670**
660	....	**590**	**620**	**80.3**	....	**58.3**	**69.4**	**89.5**	**75.9**	**64.7**	79	319	**660**
650	....	**585**	**611**	**80.0**	....	**57.8**	**69.0**	**89.2**	**75.5**	**64.1**	....	314	**650**
640	....	**578**	**601**	**79.8**	....	**57.3**	**68.7**	**89.0**	**75.1**	**63.5**	77	309	**640**
630	....	**571**	**591**	**79.5**	....	**56.8**	**68.3**	**88.8**	**74.6**	**63.0**	....	304	**630**
620	....	**564**	**582**	**79.2**	....	**56.3**	**67.9**	**88.5**	**74.2**	**62.4**	75	299	**620**
610	....	**557**	**573**	**78.9**	....	**55.7**	**67.5**	**88.2**	**73.6**	**61.7**	....	294	**610**
600	....	**550**	**564**	**78.6**	....	**55.2**	**67.0**	**88.0**	**73.2**	**61.2**	74	289	**600**
590	....	**542**	**554**	**78.4**	....	**54.7**	**66.7**	**87.8**	**72.7**	**60.5**	....	284	**590**
580	....	**535**	**545**	**78.0**	....	**54.1**	**66.2**	**87.5**	**72.1**	**59.9**	72	279	**580**
570	....	**527**	**535**	**77.8**	....	**53.6**	**65.8**	**87.2**	**71.7**	**59.3**	....	274	**570**
560	....	**519**	**525**	**77.4**	....	**53.0**	**65.4**	**86.9**	**71.2**	**58.6**	71	269	**560**
550	**505**	**512**	**517**	**77.0**	....	**52.3**	**64.8**	**86.6**	**70.5**	**57.8**	....	264	**550**
540	**496**	**503**	**507**	**76.7**	....	**51.7**	**64.4**	**86.3**	**70.0**	**57.0**	69	260	**540**
530	**488**	**495**	**497**	**76.4**	....	**51.1**	**63.9**	**86.0**	**69.5**	**56.2**	....	254	**530**
520	**480**	**487**	**488**	**76.1**	....	**50.5**	**63.5**	**85.7**	**69.0**	**55.6**	67	250	**520**
510	**473**	**479**	**479**	**75.7**	....	**49.8**	**62.9**	**35.4**	**68.3**	**54.7**	....	244	**510**
500	**465**	**471**	**471**	**75.3**	....	**49.1**	**62.2**	**85.0**	**67.7**	**53.9**	66	240	**500**
490	**456**	**460**	**460**	**74.9**	....	**43.4**	**61.6**	**84.7**	**67.1**	**53.1**	....	234	**490**
480	**448**	**452**	**452**	**74.5**	....	**47.7**	**61.3**	**84.3**	**66.4**	**52.2**	64	230	**480**
470	**441**	**442**	**442**	**74.1**	....	**46.9**	**60.7**	**83.9**	**65.7**	**51.3**	....	224	**470**
460	**433**	**433**	**433**	**73.6**	....	**46.1**	**60.1**	**83.6**	**64.9**	**50.4**	62	220	**460**
450	**425**	**425**	**425**	**73.3**	....	**45.3**	**59.4**	**83.2**	**64.3**	**49.4**	....	214	**450**
440	**415**	**415**	**415**	**72.8**	....	**44.5**	**58.8**	**82.8**	**63.5**	**48.4**	59	210	**440**
430	**405**	**405**	**405**	**72.3**	....	**43.6**	**58.2**	**82.3**	**62.7**	**47.4**	....	204	**430**
420	**397**	**397**	**397**	**71.8**	....	**42.7**	**57.5**	**81.8**	**61.9**	**46.4**	57	200	**420**
410	**388**	**388**	**388**	**71.4**	....	**41.8**	**56.8**	**81.4**	**61.1**	**45.3**	....	195	**410**
400	**379**	**379**	**379**	**70.8**	....	**40.8**	**56.0**	**81.0**	**60.2**	**44.1**	55	190	**400**
390	**369**	**369**	**369**	**70.3**	....	**39.8**	**55.2**	**80.3**	**59.3**	**42.9**	....	185	**390**
380	**360**	**360**	**360**	**69.8**	(110.0)	**38.8**	**54.4**	**79.8**	**58.4**	**41.7**	52	180	**380**

(*Table continued on p. 1-26*)

Table 1. Approximate Hardness Conversion Numbers for Steel, Based on DPH (Vickers)—*Continued*

(Reprinted by permission from the *SAE Handbook*, 1949 edition)

Diamond Pyramid Hardness Number, Vickers	Brinell Hardness Number, 10-mm Ball, 3000-kg Load: Standard Ball	Brinell: Hultgren Ball	Brinell: Tungsten Carbide Ball	Rockwell Hardness Number: A-Scale 60-kg Load Brale Penetrator	Rockwell: B-Scale, 100-kg Load 1/16 in. Diam. Ball	Rockwell: C-Scale 150-kg Load Brale Penetrator	Rockwell: D-Scale 100-kg Load Brale Penetrator	Rockwell Superficial Hardness Number Superficial Brale Penetrator: 15-N Scale 15-kg Load	Superficial: 30-N Scale 30-kg Load	Superficial: 45-N Scale 45-kg Load	Shore Scleroscope Hardness Number	Tensile Strength (Approx.) 1000 psi	Diamond Pyramid Hardness Number, 50-kg Load
1	2	3	4	5	6	7	8	9	10	11	12	13	14
370	350	350	350	69.2	....	37.7	53.6	79.2	57.4	40.4	....	175	370
360	341	341	341	68.7	(109.0)	36.6	52.8	78.6	56.4	39.1	50	170	360
350	331	331	331	68.1	....	35.5	51.9	78.0	55.4	37.8	....	166	350
340	322	322	322	67.6	(108.0)	34.4	51.1	77.4	54.4	36.5	47	161	340
330	313	313	313	67.0	....	33.3	50.2	76.8	53.6	35.2	....	156	330
320	303	303	303	66.4	(107.0)	32.2	49.4	76.2	52.3	33.9	45	151	320
310	294	294	294	65.8	....	31.0	48.4	75.6	51.3	32.5	....	146	310
300	284	284	284	65.2	(105.5)	29.8	47.5	74.9	50.2	31.1	42	141	300
295	280	280	280	64.8	....	29.2	47.1	74.6	49.7	30.4	....	139	295
290	275	275	275	64.5	(104.5)	28.5	46.5	74.2	49.0	29.5	41	136	290
285	270	270	270	64.2	....	27.8	46.0	73.8	48.4	28.7	....	134	285
280	265	265	265	63.8	(103.5)	27.1	45.3	73.4	47.8	27.9	40	131	280
275	261	261	261	63.5	....	26.4	44.9	73.0	47.2	27.1	....	129	275
270	256	256	256	63.1	(102.0)	25.6	44.3	72.6	46.4	26.2	38	126	270
265	252	252	252	62.7	....	24.8	43.7	72.1	45.7	25.2	....	124	265
260	247	247	247	62.4	(101.0)	24.0	43.1	71.6	45.0	24.3	37	121	260
255	243	243	243	62.0	....	23.1	42.2	71.1	44.2	23.2	....	119	255
250	238	238	238	61.6	99.5	22.2	41.7	70.6	43.4	22.2	36	116	250
245	233	233	233	61.2	....	21.3	41.1	70.1	42.5	21.1	....	114	245
240	228	228	228	60.7	98.1	20.3	40.3	69.6	41.7	19.9	34	111	240
230	219	219	219	....	96.7	(18.0)	......	......	......	......	33	106	230
220	209	209	209	....	95.0	(15.7)	......	......	......	......	32	101	220
210	200	200	200	....	93.4	(13.4)	......	......	......	......	30	97	210
200	190	190	190	....	91.5	(11.0)	......	......	......	......	29	92	200
190	181	181	181	....	89.5	(8.5)	......	......	......	......	28	88	190
180	171	171	171	....	87.1	(6.0)	......	......	......	......	26	84	180
170	162	162	162	....	85.0	(3.0)	......	......	......	......	25	79	170
160	152	152	152	....	81.7	(0.0)	......	......	......	......	24	75	160
150	143	143	143	....	78.7	....	......	......	......	......	22	71	150
140	133	133	133	....	75.0	....	......	......	......	......	21	66	140
130	124	124	124	....	71.2	....	......	......	......	......	20	62	130
120	114	114	114	....	66.7	....	......	......	......	......	....	57	120
110	105	105	105	....	62.3	....	......	......	......	......	....	....	110
100	95	95	95	....	56.2	....	......	......	......	......	....	....	100
95	90	90	90	....	52.0	....	......	......	......	......	....	....	95
90	86	86	86	....	48.0	....	......	......	......	......	....	....	90
85	81	81	81	....	41.0	....	......	......	......	......	....	....	85

Note. The values in this table shown in bold faced type correspond to the values shown in the corresponding joint SAE—ASM—ASTM Committee on Hardness Conversions as printed in ASTM E48-43T, Table 1.

Values in ( ) are beyond normal range; given for information only.

pression, the depth of penetration is measured by the testing instrument and indicated as hardness number on a dial indicator. A variety of penetrators and standard loads can be used, giving a series of different scales identified by letters. In reporting tests the scale must always be stated. The regular penetrator is a diamond cone (120° with a slightly rounded cone), and the regular load is 150 kg, giving the Rockwell C scale, sometimes abbreviated to Rc, which is used for the harder metals. For the less hard metals the Rockwell B scale is used, employing a 1/16-in. steel ball penetrator and a 100-kg load. The Rockwell method is much faster than the other penetration methods, and produces an indentation on the order of 0.005-in. It is, therefore, suitable for testing thin specimens. For very thin hard surfaces, such as those on nitrided steel, the Rockwell superficial-hardness tester has been devised, the depth of penetration with which is only about 0.002-in.

THE SHORE SCLEROSCOPE employs a diamond-tipped hammer which is dropped from a definite height upon the surface to be tested. The hardness is measured by the height of rebound, the scale being graduated in arbitrary units such that quenched plain high-carbon steels indicate about 100. Inasmuch as the rebound is affected more by the elastic limit of the specimen than by its tensile strength, the Scleroscope does not measure exactly the same type of hardness as the indentation methods.

Because depth of penetration is very small, the instrument can be used on case-hardened surfaces. Being portable it is especially useful for testing on parts too large to be inserted in one of the other types of instruments. It is also popular when a rapid check on production is needed. For general testing work, however, the Scleroscope is not widely used.

HARDNESS CONVERSION TABLES. It is not possible to translate from one arbitrary hardness scale to another with any degree of precision. However, one scale can be calibrated in terms of another for a particular material or type of material. Conversion tables are given in ASTM E48-43T, *Tentative Hardness Conversion Tables for Steel,* and in ASTM E33-42, *Standard Hardness Conversion Table for Cartridge Brass.*

Table 1 provides a general comparison between hardness scales, but it may be in considerable error for some materials, particularly those of high hardness.

MOHS' SCALE is a scratch hardness scale, consisting of a series of minerals each of which can be scratched by the one next above it in the scale. The differences are not uniform, and the intervals between successive numbers increase rapidly as the hardness number increases. The step from 9 to 10 probably is as great as from 1 to 9. This is the Mohs' scale:

Talc	1	Orthoclase	6
Gypsum	2	Quartz	7
Calc spar	3	Topaz	8
Fluor spar	4	Sapphire	9
Apatite	5	Diamond	10

Table 2 gives the relative hardness of various materials according to the Mohs' scale.

Table 2. Hardness of Various Materials by Mohs' Scale

(Data reprinted by permission from the Smithsonian Tables)

Material	Hardness	Material	Hardness	Material	Hardness	Material	Hardness
Agate	7.0	Bismuth	2.5	Gypsum	1.6–2.0	Plat-iridium	6.5
Alabaster	1.7	Boric acid	3.0	Hematite	6.0	Pyrite	6.3
Alum	2.0–2.5	Brass	3.0–4.0	Hornblende	5.5	Quartz	7.0
Aluminum	2.0	Calamine	5.0	Iridium	6.0	Rock-salt	2.0
Amber	2.0–2.5	Calcite	3.0	Iridosmium	7.0	Ross metal	2.5–3.0
Andalusite	7.5	Copper	2.5–3.0	Iron	4.0–5.0	Silver chloride	1.3
Anthracite	2.2	Corundum	9.0	Kaolin	1.0	Sulfur	1.5–2.5
Antimony	3.3	Diamond	10.0	Loess (0 C)	0.3	Stibnite	2.0
Apatite	5.0	Dolomite	3.5–4.0	Magnetite	6.0	Serpentine	3.0–4.0
Aragonite	3.5	Feldspar	6.0	Marble	3.0–4.0	Silver	2.5–3.0
Arsenic	3.5	Flint	7.0	Meerschaum	2.0–3.0	Steel	5.0–8.5
Asbestos	5.0	Fluorite	4.0	Mica	2.8	Talc	1.0
Asphalt	1.0–2.0	Galena	2.5	Opal	4.0–6.0	Tin	1.5
Augite	6.0	Garnet	7.0	Orthoclase	6.0	Topaz	8.0
Barite	3.3	Glass	4.5–6.5	Palladium	4.8	Tourmaline	7.3
Beryl	7.8	Gold	2.5–3.0	Phosphor bronze	4.0	Wax (0 C)	0.2
Bell metal	4.0	Graphite	0.5–1.0	Platinum	4.3	Wood's metal	3.0

11. SPECIFIC GRAVITY

THE SPECIFIC GRAVITY of a solid or liquid substance is the ratio of the mass of a given volume of the substance to the mass of an equal volume of water at some standard temperature. The temperature commonly used in engineering work is 60 F. The temperature used by physicists is 4 C. In the metric system the specific gravity is equal to the density, i.e., to the weight in grams of one cubic centimeter.

The specific gravity of gases is taken as the ratio of the masses of equal volumes of the given gas and of air or hydrogen, usually at 32 F and 29.921 in. of mercury barometric pressure (0 C and 760 mm pressure). It is necessary to state which basis is used, i.e., air = 1 or hydrogen = 1.

Determination of Specific Gravity of Solids. Determine the weight of the body in air; then submerge it in water and determine its weight. If W = weight in air and w = weight when submerged,

$$\text{Specific gravity} = \frac{W}{W - w}$$

If the body is lighter than water, attach to it a weight, w_1, sufficient to sink it. Then

$$\text{Specific gravity} = \frac{W - w_1}{(W - w_1) - w}$$

Given the specific gravity, referred to water at 60 F,

$$\text{Weight per cu ft} = \text{specific gravity} \times 62.36.$$
$$\text{Weight per cu in} = \text{specific gravity} \times 0.036088.$$
$$\text{Specific gravity} = \text{weight per cu ft} \times 0.016036.$$

THE HYDROMETER. The hydrometer is an instrument for determining the density of liquids. It usually is made of glass, and consists of three parts: (1) the upper part, a graduated stem or fine tube of uniform diameter; (2) a bulb, or enlargement of the tube, containing air; and (3) a small bulb at the bottom, containing shot or mercury which causes the instrument to float in a vertical position. The graduations are figures representing either specific gravities or the numbers of an arbitrary scale, as the Baumé, API, Twaddell, or Beck.

The Baumé hydrometer scale is based on the following formulas:

For liquids lighter than water:

$$\text{Specific gravity} = \frac{140}{130 + \text{degrees Baumé}}$$

For liquids heavier than water:

$$\text{Specific gravity} = \frac{145}{145 - \text{degrees Baumé}}$$

The API hydrometer scale, adopted to avoid the confusion incident to two Baumé scales, is recommended by the American Petroleum Institute, the U. S. Bureau of Mines and the U. S. Bureau of Standards for use in the petroleum and oil industries. It is based on the formula

$$\text{Specific gravity 60/60 F} = \frac{141.5}{131.5 + \text{degrees API}}$$

Beck Hydrometer Scale. The relation between specific gravity and the Beck hydrometer scale is expressed by the formulas:

For liquids heavier than water:

$$\text{Specific gravity} = \frac{170}{170 - \text{Beck degrees}}$$

For liquids lighter than water:

$$\text{Specific gravity} = \frac{170}{170 + \text{Beck degrees}}$$

Twaddell Hydrometer Scale. The relation between specific gravity and the Twaddell hydrometer scale, used only for liquids heavier than water, is expressed by the formula

$$\text{Specific gravity} = \frac{5 \times \text{Twaddell degrees} + 1000}{1000}$$

Table 3. Specific Gravity and Weight of Solid Substances, Except Metals and Wood

Substance	Specific Gravity	Average Weight, lb per cu ft	Substance	Specific Gravity	Average Weight, lb per cu ft
Asbestos	2.1–2.8	153	Gypsum	2.08–2.4	140
Ashes	....	43	Hornblende	3.2–3.52	210
Asphaltum	1.39	87	Ice	0.88–0.92	55–57
Barytes	4.50	281	Leather	0.86–1.02	59
Basalt	2.7–3.2	184	Lime, quick, in bulk	0.8–0.96	55
Bauxite	2.55	159	Limestone	2.3–2.9	160
Bluestone	2.2–2.5	147	Magnesia, carbonate	2.4	150
Brick, soft	1.6	100	Magnesite	3.0	187
common	1.79	112	Marble	2.56–2.88	170
hard	2.0	125	Masonry, dry, rubble	2.24–2.56	150
pressed	2.16	135	dressed	2.24–2.88	160
fire	2.24–2.4	145	Mica	2.80	175
sand-lime	2.18	136	Mortar	1.44–1.6	95
Brickwork, mortar	1.6	100	Mud	1.7–1.8	111
cement	1.79	112	Paper	0.70–1.15	58
Borax	1.7–1.8	109	Paraffin	0.87–0.91	56
Cement, American, natural	2.8–3.2	187	Peat	0.65–0.85	47
portland	3.05–3.15	190	Phosphate rock	3.2	200
loose	1.44	92	Pitch	1.15	72
barreled	1.84	115	Plaster-of-Paris	1.5–1.8	103
slag	1.9–2.3	130	Porcelain	2.3–2.5	156
Chalk	1.8–2.6	137	Porphyry	2.6–2.9	172
Clay	1.92–2.4	137	Pumice	0.37–0.90	40
Coal, anthracite	1.4–1.8	97	Riprap, limestone	1.3–1.4	80–85
bituminous	1.2–1.5	84	sandstone	1.4	90
lignite	1.1–1.4	78	shale	1.7	105
charcoal	0.27–0.58	18	Rubber, caoutchouc	0.92–0.96	59
Coke	0.27	30.8	manufactured	1.0–2.0	95
Concrete	1.92–2.48	144	Salt	0.78–1.25	49–78
Cork	0.22–0.26	15	Saltpeter	1.07	67
Dolomite	2.9	181	Sand	1.44–1.76	100
Earth, dry, loose	1.2	75	wet	1.89–2.07	125
packed	1.5	93	Sandstone	2.24–2.4	145
moist, loose	1.3	81	Serpentine	2.4–2.7	165
packed	1.6	100	Shale	2.6–2.9	172
mud, flowing	1.7	106	Slag, bank	1.1–1.2	69
packed	1.8	112	screenings	1.5–1.9	107
Emery	4.0	250	machine	1.5	96
Feldspar	2.5–2.6	159	sand	0.8–0.9	53
Glass, common	2.5–2.75	164	Slate	2.72–2.88	175
crystal	2.90–3.00	184	Soapstone	2.65–2.8	170
flint	3.2–4.7	188	Starch	1.53	96
plate	2.45–2.72	161	Stone, various	2.16–3.4	135–200
Gneiss	2.4–2.7	165	crushed	1.6	100
Granite	2.5–3.1	179	Sulfur	1.93–2.07	125
Graphite	1.9–2.3	126	Talc	2.6–2.8	169
Gravel, dry, loose	1.4–1.7	90–105	Tar, bituminous	1.20	75
packed	1.6–1.9	100–120	Terra-cotta	1.9	119
wet	1.9	120	Tile	1.76–1.92	115
Greenstone	2.8–3.2	187	Trap rock	2.72–3.4	185
			Wool	1.32	82

Table 4. Specific Gravity of Liquids

Water at 39.1 F (sp. gr. = 1.000) weighs 62.43 lb per cu ft

Liquid	Specific Gravity	Liquid	Specific Gravity	Liquid	Specific Gravity
Acid, acetic	1.06	Carbon disulfide	1.293	Oil, neat's-foot	0.913–0.917
carbolic	0.96	Chloroform	1.480	olive	0.918
fluoric	1.50	Ether	0.736	palm	0.905
muriatic	1.20	Gasoline	0.66–0.75	petroleum	0.78–0.88
nitric	1.54	Glycerol	1.260	rape	0.92
phosphoric	1.56	Kerosene	0.78–0.82	resin	0.955
sulfuric, 87%	1.80	Lye, soda, 66%	1.70	sperm	0.88
Alcohol, ethyl, 100%	0.807	Naphtha	0.76	turpentine	0.873
methyl, 100%	0.810	Oil, cottonseed	0.93	whale	0.92
Ammonia, 27.9%	0.891	lard	0.92	Water, 39.1 F	1.00
Benzine	0.899	linseed, boiled	0.942	212 F	0.958
Bromine	3.187	mineral	0.92	sea	1.02–1.03

The Salinometer is a hydrometer used for determining the density of brine solutions. The indications are four times as large as those on the corresponding Baumé scale.

SPECIFIC GRAVITY OF VARIOUS SUBSTANCES. The specific gravity of the useful metals is given in Table 3, p. 1-04. Table 3 on p. 1-29 gives the specific gravity of various solids, except the metals and woods, and Table 4 gives the specific gravity of liquids. For densities of the various woods, see p. 5-73ff.

12. COMPRESSIBILITY

COMPRESSIBILITY OF LIQUIDS. The coefficient of compression of a liquid is $\beta = (1/V_1) \times (V_1 - V_2)/(p_2 - p_1)$, where V_1 and V_2 are the volumes at the same temperature under pressure p_1 and p_2, respectively. In general, as p increases β decreases, rapidly at first and then slowly. The change of β with temperature is large at low pressures but very small at pressures above 1000 to 2000 atm. Values of $\beta \times 10^6$ are given in Table 5, the pressures being measured in megabaryes (1 megabarye = 10^6 dynes per sq cm = 0.987 atm = 14.50 psi). The units of β are megabaryes^{-1}. To convert to square inches per pound, multiply the value of β by 0.06895. *Example:* For glycerol the value of β is 22×10^{-6} megabaryes^{-1}, which is $22 \times 10^{-6} \times 0.06895 = 1.52 \times 10^{-6}$ sq in. per lb.

Table 5. Compressibility of Liquids

(From Smithsonian Tables, by permission)

Liquid	Temp., °C	Pressure, megabaryes	Compressibility, $\beta \times 10^6$
Benzene	17	5	89
	20	200	77
	20	400	67
Carbon tetrachloride	20	200	86
	20	400	73
Chloroform	20	200	83
	20	400	70
Ethyl alcohol	14	23	100
	20	500	63
	20	1,000	54
	20	12,000	8
Ethyl ether	25	23	188
	20	500	84
	20	1,000	61
	20	12,000	10
Glycerol	15	5	22
Kerosene	20	500	55
	20	1,000	45
	20	12,000	8
Mercury	20	300	3.95
	22	500	3.97
	22	1,000	3.91
	22	12,000	2.37
Methyl alcohol	15	23	103
	20	200	95
	20	400	80
	20	500	65
	20	1,000	54
	20	12,000	8
Nitric acid	0	17	32
Oils: almond	15	5	53
castor	15	5	46
linseed	15	5	51
olive	15	5	55
rape-seed	20		59
Toluene	20	200	74
	20	400	64
Turpentine	20		74
Water	20	13	49
	20	200	43
	20	400	41
	20	500	39
	40	500	38
	40	1,000	33
	40	12,000	9

COMPRESSIBILITY OF SOLIDS. Table 6 gives values of $10^{-6}/\beta$, where β is defined as in the previous paragraph and is in megabaryes^{-1}. Values in the table therefore represent the bulk modulus in megabaryes. To convert to pounds per square inch, multiply the value of $1/\beta$ by 14.5. *Example:* For aluminum $1/\beta = 0.75 \times 10^6$ megabaryes, which is $0.75 \times 10^6 \times 14.5 = 10.9 \times 10^6$ psi.

Table 6. Compressibility of Solids

(From Smithsonian Tables, by permission)

Material	$10^{-6}/\beta$	Material	$10^{-6}/\beta$	Material	$10^{-6}/\beta$
Tungsten	3.7	Silver	1.01	Sodium chloride	.24
Boron	3.0	Aluminum	0.75	Arsenic	.22
Silicon	3.1	Germanium	0.71	Calcium	.175
Platinum	2.6	Zinc	0.57	Potassium chloride	.135
Nickel	2.3	Tin	0.53	Strontium	.120
Molybdenum	2.2	Gallium	0.48	Lithium	.111
Tantalum	1.9	Cadmium	0.46	Phosphorus (red)	.109
Palladium	1.9	Plate glass	0.45	Selenium	.083
Cobalt	1.82	Lead	0.44	Ice	.083
Nichrome	1.79	Thallium	0.43	Sulfur	.078
Iron	1.67	Antimony	0.42	Iodine	.077
Gold	1.67	Quartz	0.37	Sodium	.064
Pyrite	1.4	Magnesium	0.34	Hard rubber	.051
Copper	1.33	Bismuth	0.33	Phosphorus (white)	.049
Manganese	1.19	Graphite	0.33	Potassium	.032
Brass	1.12	Silica glass	0.32	Rubidium	.025
Chromium	1.12	Cerium	0.27	Caesium	.016

COMPRESSIBILITY OF GASES. At moderate temperatures and pressures, the compressibility of a gas is equal to the reciprocal of the pressure, the bulk modulus being equal to the pressure. Where pressures are high or temperatures low, or the gas is not far removed from its boiling point, a more exact relationship is needed. Table 7 gives the relative change of volume of several gases under various pressures and temperatures. For additional data, see *Power* Volume of Kent's *Mechanical Engineers' Handbook*, Sections 1, 4, and 11.

Table 7. Compressibility of Gases

Relative volume at various temperatures and pressures. Volume at 0 C and 1 atm pressure = 1,000,000.

(Data from the Smithsonian Tables)

Pressure, atmospheres	Oxygen			Air			Nitrogen		
	0 C	99.5 C	199.5 C	0 C	99.4 C	200.4 C	0 C	99.5 C	199.6 C
100	9,265	....	....	9,730	....	....	9,910	....	....
200	4,570	7,000	9,095	5,050	7,360	9,430	5,195	7,445	9,532
300	3,208	4,843	6,283	3,658	5,170	6,622	3,786	5,301	6,715
500	2,312	3,244	4,100	2,680	3,565	4,422	2,780	3,655	4,515
1,000	1,735	2,151	....	1,992	2,415	2,828	2,068	....	....
	Hydrogen			**Ethylene**			**Carbon Dioxide**		
	0 C	99.3 C	200.5 C	0 C	100 C	198.5 C	0 C	100 C	198 C
50	....	....	....	3,520	23,840	33,040	2,100	24,120	....
100	....	....	....	3,100	10,050	15,800	2,020	10,300	12,582
200	5,690	7,567	9,420	2,875	4,730	7,685	....	....	....
300	4,030	5,286	6,520	2,686	3,770	5,426	1,863	2,966	4,976
500	2,713	3,462	4,210	2,512	3,156	3,970	1,782	2,402	3,356
1,000	1,720	2,093	....	2,289	2,643	....	1,656	1,999	....

SECTION 2

IRON AND STEEL

By

H. W. GILLETT, *Chief Technical Adviser, Battelle Memorial Institute*

IRON AND STEEL

Iron is a chemical element which, in its pure form, has almost no commercial use. The addition of other elements to iron radically changes its properties, resulting in countless varieties of cast and wrought irons and steels.

Iron is smelted in blast furnaces from *ore* (iron oxide plus impurities) which is reduced by combination of the oxygen with the carbon in *coke*, the impurities being removed in the slag formed by fusion with *limestone* flux. The product of the blast furnace is pig iron containing 3 to 4% carbon, 1 to 3% silicon, 0.3 to 1.5% phosphorus, 0.1 to 1.0% manganese, and under 1% sulfur. Pig iron partly refined in a *cupola* produces the various grades of *cast iron*. By puddling or shotting processes *wrought iron* also is produced from pig iron. Pig iron plus scrap melted in an *open-hearth furnace* with a flux produces *steel*, part of the carbon being burned out and the impurities further removed. Steel is also produced in *Bessemer converters*, and high-grade steel in *electric furnaces*. Alloying elements may be added at appropriate stages in any of these processes.

Cast steel results from pouring into molds of appropriate shape. To produce the infinite varieties of *wrought steel*, the molten metal is cast into *ingots* of suitable size and shape for hot rolling into *blooms* (cross section more than 6 in.), *billets* (less than 6 in.), or *slabs* (wide sections). Blooms and billets are subsequently hot rolled into bars or structural-steel shapes, or formed into forging blanks. Slabs are hot rolled into plate or sheet form. Further hot working produces finished forgings, tubing, rods, etc. Subsequent cold-working operations include cold rolling, wire drawing, stamping and drawing, and cold heading. All the foregoing processes have profound effects on the ultimate properties of the material in the finished part and must be given due consideration in choosing the particular metal or alloy.

1. LIMITATIONS OF MECHANICAL PROPERTY DATA FOR DESIGN

Selection of engineering materials for machine design aims at a final product with the properties demanded by the particular service, at the lowest final cost, which includes both cost of raw material and the cost of fabricating it. The final properties required are frequently not easily measured, nor are they necessarily implied by the usually listed mechanical properties determined by conventional tests and included in usual specifications.

Fitness for fabrication is even less readily measured in truly quantitative terms. *Machinability*, the quality required for fabrication by machining, grinding, honing, and the like, varies with the particular type of machining operation employed and with the criterion imposed, which may be tool life, energy required to cut, or surface finish produced. What may be unmachinable with one type of tool may become readily machinable with another.

Weldability, a property more often required in modern construction than in earlier days, depends on the welding process, on the preheating and postheating that can be applied, and on the atmosphere surrounding the weld as it is being made. What may be unweldable with one technique may be readily weldable with another.

Formability by plastic deformation, hot or cold, likewise depends on the deformation, bend or draw, and on the technique used.

Castability, suitability to be cast into an approximation to finished size without too much stock to be removed by machining, depends not only on the material itself but also on the foundry technique and, importantly, on design, which may facilitate or prevent production of a sound casting.

None of these abilities for fabrication are wholly inherent in the material, independent of the particular process. The economical process among several possible ones depends on the number of pieces to be made. All that can ordinarily be recorded on ability for fabrication consists of qualitative statements on general order of excellence.

Metallurgy. Many problems arise during the processing of materials to the semi-finished stage in which they come to the plant of the machine designer. In processing iron and steel, metallurgical problems, of some concern to the designer, are involved. Discussion of metallurgical principles and of iron and steel materials herein will be confined to aspects directly concerning the designer: the specifications for the semifinished material, its amenability for fabrication within the plant that makes machines, and the properties of the finished parts.

Service Conditions. The ultimate requirement of a finished part is that it shall last for the desired life of the whole structure, or, if designed for replacement, that it shall have a predetermined life. The life can be ended by wear or corrosion just as certainly as by actual fracture or by being bent or battered out of shape. The environment—the actual conditions of service—controls wear and corrosion quite as much as do the properties of the material itself. Hence, no quantitative values can be ascribed to the wear and corrosion resistance of materials by themselves, without relation to service.

Design must therefore consider all the conditions of service, not merely those relating to fracture or to unbearable permanent deformation. Only these features—deformation and fracture—are appraised by conventional mechanical tests and quantitatively appraised by those tests expressible in pounds per square inch.

When, and only when, the designer can appraise the conditions of service in pounds per square inch can he apply these test results to design. For resistance to a single static stress, yield strength is applicable to design where failure would come by permanent deformation. For resistance to repeated loading, endurance strength is applicable when failure would come from fatigue fracture, but this requires that the concentrated rather than the average applied stress be known, and that the endurance strength be determined at the same stress concentration. The modulus of elasticity, another pounds-per-square-inch value, is directly applicable for calculation of stiffness.

Other numerical values among those obtained by conventional mechanical testing, such as elongation and reduction of area in the tensile test and energy absorbed by a notched bar in a static bend or an impact test, do not have direct applicability to design; what bearing they have is a matter of implication. That they are matters of implication is illustrated by many familiar examples.

TENSILE DUCTILITY. Ductility in a tensile test is often thought of as insurance against brittle, shattering failure. But much commercial steel, of high tensile ductility, can fail with a brittle fracture, propagated with explosive rapidity, as was demonstrated by wartime failures of welded ships fabricated from steel with a high level of tensile ductility.

Cast iron lacks tensile ductility and may fracture under shock, as happens when cast-iron fittings in a ship are subjected to shock resulting from impact of a projectile elsewhere upon the ship, or to one transmitted to the ship through the water, say from a "near miss" of an explosive charge. But when conditions of service permit no such shock, as in the base of a machine tool, this brittleness is no drawback. Automobile crankshafts used to be specified of high tensile ductility and notched-bar energy absorption on the assumption that a shattering failure would ruin the whole engine. Finally, it was realized that a bent crankshaft was equally ruinous, and that stiffness was a prime necessity. The Ford cast crankshaft, designed for stiffness, and made from a metal with almost zero tensile ductility and notched-bar energy absorption, has been proved adequate by many millions in actual operation.

Local Ductility. In a riveted bridge, various members fail to match exactly, and the initial stress distribution is not the average called for by design but is concentrated at local spots. When this local stress concentration exceeds the yield strength, the surrounding metal stretches locally until the stress is distributed and the other members take up their share. That much ductility is essential, but the necessary amount of permanent deformation is very small. It is local ductility that is needed rather than the general extension measured in the tensile test. Similarly, ability of a material to give locally at the base of a notch and thus relieve stress concentration in fatigue, rather than to extend the notch into a crack, is an asset. But this "crackless plasticity" is not quantitatively measured by the ductility of the tensile test.

Ability to be cold formed without cracking, as in the deep drawing of a complicated automobile fender, does not go hand in hand with conventional tensile ductility, for tensile ductility alone will not unfailingly sort out materials of good and poor behavior. The applied stresses are not purely tensile. A satisfactory test must more closely duplicate the conditions of actual forming.

Ductility, Toughness, and Notch Sensitivity. With very few exceptions, ferrous-base materials, so alloyed, so cold worked, or so heat treated as to be strengthened, have their ductility, notched-bar toughness, and resistance to notch sensitivity in fatigue decreased. But none of these three quantitatively measures either of the other two. Though all three go in the same direction, no two go at the same rate. By experience and analogy it is found that, in material of the same composition, processed in the same manner, to a given yield strength, a certain ductility connotes a certain notch toughness and a certain notch sensitivity in fatigue, but any "conversion factors" so obtained do not hold with exactness for material of another composition or processing. As to notch toughness, supposedly identical heats may act individually, displaying widely different behaviors.

Excessive Ductility. Ductility values may reveal metallurgical differences and thus be a clue to differences in processing, or to internal unsoundness which may not measurably affect yield strength. Thus they are useful indicators to the metallurgist striving to obtain complete uniformity in a product. Many metallic materials have far more ductility than the designer can allow to be displayed in service, and the excess, in itself, is then of no value in the material. Some other property that happens to go along with, but is not measured by, ductility, may be of value. If so, that other property is the one that should be measured.

Big guns, centrifugally cast, autofrettaged, and heat treated to 90,000 to 95,000 psi yield, have been proof and service tested, with satisfactory performance, when made from steel with 5 to 11% elongation and 36 to 48% reduction of area, although it is feasible to get 17 and 60% respectively with proper choice of steel and processing method (Ref. 1). Insistence on greater elongation as "insurance" for steel in less severe service seems scarcely justified.

Each material has its own level of ductility. Ordinary structural bridge steel has some 30% elongation at around 35,000 psi yield strength; cold-drawn bridge wire has some 2 to 6% elongation at around 180,000 psi yield strength. Each needs some ability for permanent deformation when stressed above the yield, in order to redistribute stress, and each has it. Obviously, the structural steel does not need any more ductility than the wire to accomplish this purpose. That it has 30 instead of 2% does not make it any better in service, but if a 35,000-yield-strength structural steel displayed only 6%, it would be obvious that something had gone wrong in its manufacture and it did not represent normal material.

Qualitative information is obtainable from ductility figures. In cast steel, incorrect deoxidation practice makes the ever-present manganese sulfide, which is brittle at room temperature, collect around grain boundaries in films; correct practice forms them as widely separated globules, practically innocuous. If two grains separated by a brittle film are so oriented at the surface of a piece that the boundary is actually a notch, filled by brittle material, the behavior in fatigue will be inferior. Even in the worst cases, the films are not continuous enough to affect the yield strength appreciably, and ample ductility for static loading still remains, but the tensile ductility drops sufficiently to serve as a warning against the use of poorly deoxidized steel for fatigue service. Microscopic examination would give equal information, and the per cent elongation is no quantitative measure of the reduction in endurance.

Low ductility values, perhaps accompanied by low yield strength, may be met in "flaky steel." These "flakes" are unwelded voids, that is, small cracks, generally accepted as resulting from hydrogen in the steel which was first in solid solution but came out of solution at some microscopic imperfection in the steel crystal, the gaseous hydrogen exerting enough pressure to rupture the steel locally. Flakes may appear in some highly alloyed steels, and an analogous defect, shatter cracks, in rail steel. Less obvious analogs to flakes may occur from hydrogen in ordinary cast steel. In rails and small steel castings, controlled cooling, which gives time for hydrogen to diffuse out without collecting as gas within the steel, helps a good deal. Ridding the stiffer high-alloy steels of the effects of hydrogen is more difficult.

Flakes and shatter cracks are nuclei for fatigue failures. Their presence is more clearly shown by metallographic examination or by examination of a fracture than by lowering of tensile ductility or other conventional tests, but lower-than-normal ductility helps to point out, merely in qualitative fashion, the need for specific examination.

Thus, there is some advantage in knowing what ductility normally accompanies a given yield strength, though the ductility values do not provide quantitative design information. For some applications, steel with subnormal ductility might serve quite as well as normal material.

DIRECTIONAL PROPERTIES. The ductility of wrought steel can, and usually does, vary immensely according to the angle at which the test piece is taken in respect to the direction of working. All steel contains nonmetallic inclusions, notably manganese sulfide. This sulfide is plastic at hot-working temperatures and is flattened out into leaflike plates or stringers which interrupt the continuity of the metal matrix in the direction normal to the direction of working. Metallic constituents themselves may give the matrix a banded structure, resulting from an original coarse structure of the two metallographic constituents of steel of usual composition in large, slowly cooled ingots, and the rolling down of these two sorts of crystals, one harder and one softer, into a "banded" structure. The mechanical effects of this are not so marked as those of nonmetallic inclusions, but any wrought steel, no matter how clean, has directional properties if it has received more hot deformation in one direction than in another. The yield strength is not greatly affected by the direction in which the specimen is taken unless the amount of nonmetallic inclusions is very high. Fatigue strength may be reduced. Data have been cited (Ref. 2) for a certain lot of steel, heat treated to 115,000 psi yield, which gave 78,000 psi endurance limit, 59 ft-lb Izod on longitudinal specimens, but only 58,000 psi and 36 ft-lb on transverse specimens. The endurance tests were on unnotched, polished specimens. Gear makers take great care to forge gear blanks so that the "grain" is not transverse to the applied stress at any point.

Lest the citing of endurance and Izod values together should give an erroneous impression, data may be given (Ref. 3) for a nickel-chromium steel, originally with 38 ft-lb notched-bar impact and an endurance limit of 23,500 psi on a threaded specimen. Long-continued use at high temperature dropped the notched-bar value to 4 ft-lb, but the threaded specimen from this embrittled steel still showed 23,500 psi endurance limit. Static notched-bar and notched endurance values have no relation.

Directional differences are evidenced both by ductility and by notched-bar values. The following data on low-alloy, as-rolled plate steel, are illustrative:

Specimen	Tensile Strength, psi	Yield Strength, psi	Elongation, %	Reduction of Area, %
Longitudinal	65,000–75,000	45,000–50,000	30	65
Transverse	65,000–75,000	45,000–50,000	20	55
Normal to plate	65,000–75,000	45,000–50,000	2–4	5–15

Another example is hot-rolled 0.10% C, 0.75% Mn, 0.20% S free-machining bar stock, with:

Specimen	Tensile Strength, psi	Yield Strength, psi	Elongation, %	Charpy Keyhole Impact, ft-lb	Reduction of Area, %
Longitudinal	67,000	40,000	37	16–19 1/2	60
Transverse	62,300	36,000	22 1/2	8– 9 1/2	24 1/2

Both tensile and notched-bar data are ordinarily reported only as obtained on longitudinal specimens. The designer is seldom informed that directional differences exist. That so much difference does exist in material that has been much used, with satisfactory results, indicates that, in the applications where it has been found satisfactory, there is little actual reason for calling for extremely high longitudinal ductility and toughness, since it would be rare that service stresses would come solely in the longitudinal direction.

High longitudinal ductility is regularly accompanied by lower transverse ductility. Cross rolling of plate tends to equalize the two but does not affect the ductility in the through-plate direction.

Castings are sometimes considered inferior to rolled or forged material of equal yield strength on the basis of lower ductility than that displayed on longitudinal specimens from the wrought material, whereas the lack of directional properties in the casting would reverse the comparison were transverse ductility of the wrought product considered the criterion.

SIZE EFFECTS. Another fallacy is the assumption that reported values for mechanical properties determined on a light rolled or cast section represent the properties of a heavy section. This is true neither in wrought steel as rolled or normalized nor in cast steel. (It is likely to be even farther from the truth in heat-treated steels, as will be discussed later.) The yield strength and ductility of 1/2-in. plate are appreciably higher than those of 2-in. plate, for example.

The properties of castings vary materially with the size of section—this is particularly true of cast iron and noteworthy in cast steel—the larger the section, the lower the strength. The mechanical properties are usually reported from test bars cast separately rather than from sections cut from an actual casting. Good practice for cast iron uses a test bar of section equal to that of the casting, but tabulated values may not take section into account. The test bar for cast steel is ordinarily 1 in. square in section and cast with a very large riser to insure soundness. The casting itself may be much heavier, with resulting lower yield strength, and it is seldom as adequately fed as the test bar is. The test bar results merely tell something about the quality of the molten steel in the ladle. Design of the casting and its gating govern whether good steel makes a sound or unsound casting. For reliability, it is far more important that a casting be sound than that a separate piece of metal have high mechanical properties.

These matters can be summed up in the statement that true sampling is required and that the properties on which design is based must be determined on specimens from the section called for in design. Tabulated values from conventional tests on small specimens are open to question.

NOTCHED-BAR TESTS. Particularly misleading on account of size effect are conventional notched-bar impact tests such as the Charpy and Izod, which are made in the hope that they will tell something about toughness. As already mentioned, ductility values from a tensile test are no sure indication of toughness, brittle fractures occurring in many commercial structural steels of great tensile ductility.

Restraint to Deformation. When steel starts to fracture, it tries to pull out and neck down by plastic deformation. If not restrained, it does move; the fracture is tough and occurs relatively slowly, the tensile specimen exerting little restraint. Steels very high in carbon, unless specially heat treated, or those with medium carbon and very high phosphorus, do not deform much, and they break with a flat fracture even in the tensile test. Similar "stiffness" appears in some steels at extremely low temperatures, and flat fractures can be produced even in tensile tests of such steels at such temperatures even though they are highly ductile at room temperature. Lowering the temperature naturally makes a metal stronger and less plastic, just as raising the temperature tends to make it weaker and more plastic; thus the tensile and yield strengths rise as temperature drops. Ductility in the tensile test, however, is reduced less than the strength is raised. Thus the metal itself shifts in properties and behavior.

However, this behavior is quite different when the metal that is trying to let itself be displaced by plastic flow has no place to go. If the applied stress is localized, as at the base of a notch, the notch has to open its lips and its edges rise above their original position while the base of the notch is rounding off and distributing the stress. If this motion is restrained by adjacent metal instead of being allowed by a free surface, the motion is restricted and, instead of the base of the notch rounding off, a crack starts which may propagate with explosive rapidity (about a mile a second). The behavior thus is specifically related to the geometrical restraint and is a primary function of the size and shape of the piece. Naturally, it depends considerably on the sharpness of the notch, i.e., the degree of stress concentration, as well as on the possibility of relieving the stress concentration. The ability of the metal to move is affected by its own make-up. Coarse grains appear less able to deform than fine ones. The grain boundaries are less restrained than the body of the grain, and, just as on the macro scale a larger metal mass discourages movement, so on the micro scale the larger mass of metal in a coarse grain also discourages it. Lowering the temperature, as mentioned above, makes the metal less able to move under a given stress since its yield strength is increased.

Thus the behavior in a notched-bar test is governed only partly by the nature of the metal but very greatly by the geometry of the piece being fractured. Notched-bar values therefore represent behavior in a particular piece, of given dimensions and given notch sharpness, and can be expressed in foot-pounds for that piece, but not in foot-pounds per square inch for any other piece. Even the Izod and Charpy values cannot be accurately converted one to the other. Thus, such figures do not define the material by itself and are of no direct value in design. The *only* figure of any value in design is that obtained on a *full-sized* piece, with the *exact* stress concentration of service, tested *at* the temperature of service. The conventional notched-bar Charpy specimen with a breaking section of about 0.2 in. by 0.4 in. is extremely tiny compared to most actual structures and exerts small geometrical restraint.

Temperature and Size Effects. In the steels that display brittle behavior in large sections and at lowered temperature, but ductile behavior in small sections at room temperature, the transition from ductile to brittle behavior can occur with suddenness as soon as a limiting size is overstepped or the piece cooled to a limiting temperature. This transition is quite as clearly marked by the type of fracture as by the energy absorbed. The fracture may be ductile, by tearing, around the edges, but flat, by cleavage, in a spot in the center, when restraint becomes too much for plastic flow to continue. With increasing size of piece or decreasing temperature the flat spot increases in size. The actual brittle fractures met in service, as in the failures of welded ships, start with deformation and thinning down of the plate edges at the initial fracture, but as the fracture progresses away from a free surface it shifts to the cleavage type.

Observation of the outline of the fractured surface of a notched-bar specimen gives a picture of the degree of deformation that has occurred. The figure for energy absorbed may be quite misleading, for it may still be high when the periphery is deformed but the center is cleaving without deformation and the piece is actually on the verge of shifting to a preponderantly brittle fracture.

Under transition conditions very slight differences in internal structure or in sharpness of notch will affect the behavior so that some "duplicate" specimens may show high energy absorbed, others very low. If only two or three duplicates are tested it may happen that only high values are obtained and the imminence of transition is not shown. See Fig. 1.

Obviously, if room temperature and a given size and shape of specimen are used, differences in the behavior of different steels will not be shown up unless one steel happens to have its transition temperature above, and the other below, room temperature. Progressively increasing the size of the specimen, as by use of double or triple widths, is more likely to bracket the transition, but the easiest way to bracket it is by using the ordinary specimen at a decreasing range of temperature. Such a series evaluates different steels, and different heats of the same steel, as to their relative dislike for conditions of restraint. The evaluation should be based not merely on energy absorbed, but equally on type of fracture and outline of the fractured piece.

Design. This evaluation by conventional specimens gives qualitative information to the metallurgist in his study of composition, grain size, and finishing practice. But it gives the designer no basis for safe design. Specifications based on energy absorbed by a tiny specimen give the appearance of safe selection of material without the reality. The alloy-steel lining of a liquefied-gas storage tank at Cleveland met the specification of 20 ft-lb Charpy at −260 F on the conventional specimen (Ref. 4). Nevertheless, the tank exploded with disastrous results. The restraint upon deformation in the full-size lining is obviously much greater than in the tiny test specimen.

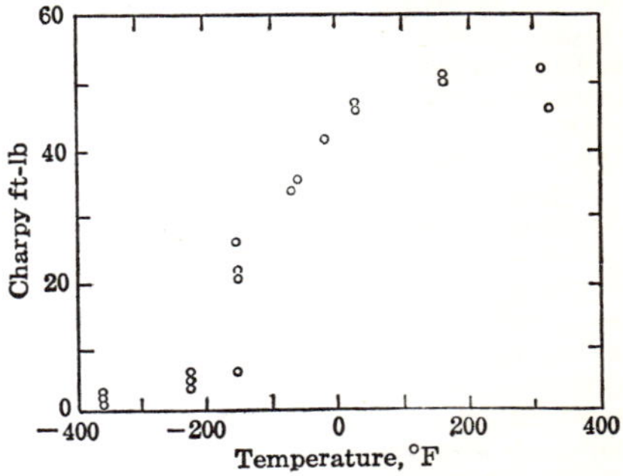

FIG. 1. Charpy impact data for a heat of 1030 steel, normalized. Note the variable results at −150 F.

Designing in layer construction, with thin layers, utilizes the superior geometric toughness of small sections. Pressure vessels, built up by wrapping thin strips upon themselves, have been made to fail only in ductile fashion (Ref. 5) when heavy walled ones might fail in brittle fashion. Large crane hooks built up of a pack of thin plates side by side find safe use where solid ones are likely to fail in brittle fashion.

The low-temperature and size-effect toughness problem is acute only in certain ferrous materials. Commercial nonferrous alloys in general do not display the tough-to-brittle transition, nor do wholly austenitic steels. These materials, if tough at room temperature, remain tough at the lowest temperatures met commercially. If relatively brittle at room temperature, they remain so, but not appreciably more so at low temperature.

Economic requirements do not often permit use of materials not subject to transition of behavior, and steel has to be used. Some unkilled structural steels have the transition at or even above room temperature if the section is large, but this does not prevent their use in small sections or where stress concentration as well as geometrical restraint is absent. Other things being equal, killed structural steels maintain tough behavior to lower temperature than rimmed steels, and, the finer the grain, the lower the temperature.

Fine-grained killed steel, extremely low in carbon and with some of the ferrite-strengthening elements added to make up for deficiency in static strength, shows still lower transition temperature, and heat-treatable steel, quenched to a fully martensitic structure, then tempered, also shows a low transition temperature. (See Fig. 7.) But any of these, under sufficient stress concentration and sufficient restraint upon deformation arising from section and size, can display the transition to brittle behavior at a temperature far above that displayed in the conventional notched-bar test. Therefore, sound design can be based only on full-scale tests with the full stress concentration, restraint, and temperature occurring in service. Nor should it be forgotten that individual heats of steel, supposedly alike, vary in transition behavior according to minor variations in the way they are "killed," so that immunity to brittle behavior shown by one lot on the full-scale test does not guarantee that another lot will have like immunity.

In such connections, the conventional notched-bar impact values stand as a snare and a delusion to the designer. High values may come far from truly indicating utility.

Temper Brittleness. Notched-bar tests reveal a condition, not evidenced by other conventional tests, known as "temper brittleness." Some highly alloyed, quenched steels, tempered at rather high temperatures, show relatively low energy absorbed in the notched-bar test if they have been slowly cooled from the tempering temperature, higher energy absorption if rapidly cooled. Some are brittle if tempered in the range 800–1100 F, even if rapidly cooled. This behavior appears to be due to separation at grain boundaries of some material not completely identified. The effect of this condition on static or fatigue behavior is probably slight, being marked only in single-blow notch tests. This condition is taken seriously by makers of armor, some of whom feel that ballistic and notched-bar behaviors are invariably closely related. Ballistic records, however, show cases of poor ballistic performance with high notch toughness, and vice versa, depending somewhat on the relative sizes of the projectile and the armor. The temper-brittle condition in such steels is alleviated if not entirely removed by the addition of sufficient molybdenum, though it is possible to give an overdose.

Low values from the conventional notched-bar test may lead to avoidance of certain materials because of their indicated "brittleness" when their actual performance belies the term. Regular malleable cast iron has a room-temperature Charpy notched-bar impact value up to 6 or 7 ft-lb, yet it finds use in tank track guides, brackets for 3-in. guns, and a multitude of other applications in which toughness would seem important. Frye (Ref. 6) illustrates some applications to severe service. Makers of malleable iron use a "wedge curl" test rather than a notched-bar test for evaluation of toughness. This test and the room- and low-temperature behavior of various lots of malleable are discussed by Simmons, Rosenthal, and Lorig (Ref. 7).

The metallurgist can get qualitative information from notched-bar values and modes of fracture, but, on the whole, the *designer* can place *no* reliance on conventional notched-bar figures for evaluation of materials. This is a field in which no sample short of the actual piece itself is a true sample. The values are true for the "specimen," fictitious for anything else.

ENDURANCE. Most failures of machine parts are fatigue failures, resulting from repeated stress and starting from some location of stress concentration as a nucleus. Although, under a single static load reaching the yield stress, a metal may deform, distribute the load, and relieve the local stress, stresses below the yield stress do not readily bring this about and the stress concentration remains at a high level. As mentioned earlier, single-blow notched-bar behavior, taking the material to fracture as a single application of stress, is no clue whatever to a metal's behavior in notched fatigue. Tensile ductility values likewise give no direct measure. Notch sensitivity in fatigue is appraisable only by fatigue tests.

Design that is entirely safe for static loading may easily be unsafe for repeated stress because, under repeated stress never reaching the yield stress, fatigue failure results in any but the softest steels, and the fatigue notch sensitivity increases as the static strength increases.

Conventional fatigue tests are made on carefully polished specimens, as free as possible from any notch or other stress raiser. These values are in pounds per square inch and are applicable to design, *if* the actual part is equally well finished, completely free from stress raisers, and always maintained in that condition. Thus, a nicely contoured and polished connecting rod not subject to scratching in use could reasonably be designed on the basis of conventional fatigue data, with due allowance for what stress concentration does exist and for inevitable deviation from perfect axial loading, which will raise the true stress at some location on the surface above the nominal design stress.

Few parts have this ideal initial and continued freedom from stress raisers. An automobile valve spring may remain in good condition and serve well though stressed close to its polished-bar endurance limit. An exposed automobile spring leaf may become roughened by chafing against its neighbors, scratched by grit from the road, and pitted by rusting. Then stress-raising factors soon so depreciate the endurable repeated load that design on the basis of polished-bar values would be unsafe. A polished-bar specimen, unless it faithfully represented any decarburization existing in the actual spring, would not correctly depict even the initial low ability of that spring to carry repeated loads.

Surface Effects. Fatigue fractures start from the surface or, very rarely, from internal stress raisers close to the surface. The surface condition is therefore paramount. Unless a perfect surface is obtained, the notch sensitivity in fatigue comes into play, and this increases with the hardness of the steel. If notch sensitivity were not to be taken into account, fatigue behavior would improve as hardness increases. Since failure starts at the surface, making the surface harder, as by carburizing or nitriding, by induction or flame hardening, or by cold working it, acts to improve endurance of a part; softening it, as by decarburization of the surface, lowers it. Conventional fatigue specimens from which decarburization has been removed do not correctly measure the behavior of parts which do have a decarburized surface. The depth of decarburization beneath the surface makes little difference; it is the extreme surface that is the Achilles' heel, so "only a trace of decarb" may easily be fatal. As for any test, a steel sample for a fatigue test should be a true sample.

Polished-bar fatigue resistance is more closely related to tensile strength and hence to hardness than to any other of the conventional mechanical properties. The endurance ratio (polished endurance limit divided by tensile strength) approximates to a constant 50%, speaking very broadly. Higher or lower ratios, however, are frequent, sometimes being unexplainable, sometimes ascribable to definite causes such as large, sharp-edged inclusions so oriented as to act as stress raisers or to varying degrees of relief of internal stress. The inclusions commercially met are relatively harmless in very soft steels, increasingly harmful as the hardness increases. Since the inclusions at the surface act as stress raisers, this ties in with the notched endurance sensitivity, demonstrated when known notches are created on the surface. If steel is roughened by corrosion, the corrosion pits are stress raisers. Moreover, if corrosion occurs while the steel is being repeatedly stressed, fatigue failure is vastly accelerated, so much so that no endurance limit can be stated for a corrodible steel subjected to corrosion fatigue.

A schematic picture of the relation of polished-bar, notched-bar, and corrosion fatigue behavior in relation to hardness is shown in Fig. 2, which brings out the great penalty paid for design that allows notches or other stress raisers. It shows that, when stress raisers are allowed to be present, little is gained in fatigue resistance by increasing the tensile strength above 125,000 psi, whereas much is gained, up to some 220,000 psi, if there is no stress raiser. The gap between the bands for polished and notched specimens indicates the rising notched endurance sensitivity as hardness increases.

Avoidance, in actual design, of extremely hard steels which, on the basis of yield strength, would appear the better materials of construction is largely based on sad experience with fatigue failures when stress raisers were present. The width of the bands in Fig. 2 indicates that such wide variation in behavior demands more information on fatigue behavior of a given lot of steel than can be secured by analogy with hardness or tensile strength.

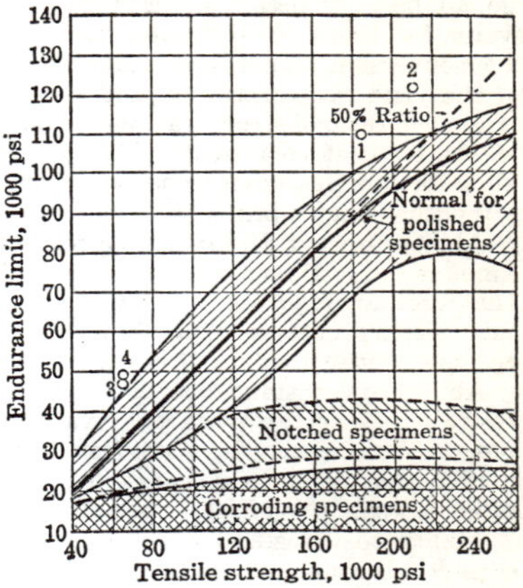

FIG. 2. Relation between tensile strength and endurance limit of steel. Points 1 and 2 are for specimens tempered for extremely long periods of time; points 3 and 4 are for high-phosphorus, mild-alloy steels. (Reproduced by permission from Bullens, *Steel and Its Heat Treatment*, 5th ed., Vol. I, Wiley, 1948)

Design Data. But conventional polished-bar tests, though they evaluate the properties of the steel in a fashion, tell nothing about how it will act in an actual part, with its actual notches or stress concentration. Moreover, there is a size effect in fatigue, though not a very important one. Hence for useful design values the full-sized part, with its actual stress raisers, made from the steel to be studied, must be tested under the actual repeated stress conditions of service. *If* these service stress conditions are known, and design is to be on the basis of indefinite life, with the applied stresses kept below the endurance limit, such tests give design figures.

But the complication generally arises, either that the number and intensity of applied stresses are unknown, or that a limited life will do rather than an infinite life, or both. Design for limited life is based on life at stresses above the endurance limit, where the steel will fail if there are sufficient repetitions of stress. In this range of stresses duplicate tests often show wide scatter, and a great number of tests is required to establish what is the shortest life that may be expected.

Further complication ensues when the repeated stresses are sometimes above and sometimes below the endurance limit. Repeated stresses below the endurance limit strengthen by understressing, and above the limit they damage by overstressing, so that the future behavior in fatigue depends on what number of stresses at what levels have previously been applied. Thus, the practical fatigue problem removes itself from the realm of conventional testing and resolves itself into that of special testing, and special testing of *design* rather than *materials*. That is, notched endurance sensitivity is so predominating that study of the part for locations and levels of stress concentration, by photoelastic study of models, by brittle-lacquer studies, by short gage length electrical strain gages and like methods, followed by redesign to minimize stress concentration, precedes the choice of materials.

If the design cannot be modified to avoid large stress intensification, a relatively soft steel is needed. To the degree that it can be so modified, a harder steel can be applied.

No final evaluation of a steel or treatment can be made that is useful for anti-fatigue design without close simulation of the actual design, size, and stress application. Even simulated tests are not easy to plan.

Hence, the designer ordinarily has to fall back on a lively appreciation of the effect of stress raisers in fatigue and a general qualitative understanding of the trends of notched fatigue sensitivity. No tabulated values of notched fatigue resistance can be assembled, and polished-bar values, serving only as a practically unattainable par, would have to be given with plus and minus deviations of the bands in Fig. 2. Some qualitative statements about fatigue resistance of various classes of steel and various microstructures can be made, but such statements cannot be put in "handbook values."

SUMMARY—CONVENTIONAL TESTS OF STEEL. The modulus of elasticity is a sound basis of design for stiffness. Yield strength is a sound, pounds-per-square-inch value for design for static loading. Tensile strength (which can be closely approximated by conversion of hardness values) formerly was used in static design and is applicable if the yield ratio (ratio of yield to tensile strength) of the steels considered is constant. The yield ratio of structural steels is materially increased by alloying and, in heat-treatable steels, by heat treatment. Moreover, design for lack of permanent deformation rather than for complete fracture is customary, and so behavior between the yield and the tensile strengths is not a primary factor.

Ability to yield *slightly* is required in many cases for stress distribution but the required amount of plastic action between the yield strength and the tensile strength is far below that ordinarily present. Ability to yield greatly without fracture is required in some fabrication processes. This ability roughly goes along with, but is not adequately measured by, the ductility values of the tensile test.

Ability to withstand a concentrated applied stress under conditions of severe geometrical restraint, especially when aggravated by low temperature, may be required in service, but tensile ductility is notoriously inadequate in measuring that ability. Conventional notched-bar tests, sometimes alleged to measure this ability, do not measure it, because of an extreme size effect. Special rather than conventional tests are required.

Ability to withstand repeated stress, partly a function of the steel and of its tensile strength, is much more a function of surface condition. Sensitivity to stress raisers is not precisely measured by tensile ductility values. Again, special rather than conventional tests are required for design values.

Directional differences and size effects are shown up by conventional tests only when honest sampling is employed. There remain very few handbook values that can be used, in cookbook recipe fashion, for design without exercise of engineering judgment.

The conventional mechanical tests give, except for yield strength, a generally pseudo-quantitative terminology which really has only qualitative meaning. They indicate conformity to, or deviation from, expected values for steels and treatments, and thus are signposts to the metallurgist of differences, not measured by them, which may in one way or another affect performance, but they do not constitute sound design values.

Within a given class of steels, agreement shown by one steel, or one heat of steel, with the conventional values characteristic of other steels or heats that have behaved well in service may correctly indicate to the designer that the new lot will also behave well, or it may entirely fail to indicate vital differences. Incorrect indications from the conventional values are especially to be feared when different classes of steels, treatments, or structures are compared. To differentiate among classes where conventional values may or may not be useful signposts, the designer must be aware of certain metallurgical facts in making an intelligent exercise of engineering judgment.

2. METALLURGY OF IRON AND STEEL FROM THE DESIGNER'S POINT OF VIEW

Primary factors affecting the structure and properties of iron and steel are *carbon content* and *rate of cooling*.

Carbonless iron is soft and weak. Pure iron is not commercially available in quantity. Both the usual electrolytic product and commercial ingot iron contain appreciable amounts of iron oxide. A nearly all-iron matrix can be produced by binding the carbon of a low-carbon steel to an element that forms an extremely stable carbide. Wrought iron is a very low-carbon material, usually rather high in phosphorus, and invariably containing large amounts of slag, i.e., nonmetallic oxides.

Inclusions. All ferrous products contain small amounts of sulfur. In the presence of iron without manganese, ferrous sulfide exists, tending to embrittle the product. Addition of sufficient manganese causes the formation of a manganese (or manganese-iron) sulfide, with less embrittling tendency, which is plastic at rolling temperatures. All steel contains small amounts of manganese sulfide inclusions, the amount being intentionally increased in free-machining steels. All steel contains at least tiny amounts of other non-metallic inclusions. All steel contains at least traces of phosphorus. In high-carbon steels, high phosphorus leads to embrittlement; in low-carbon steels, it acts as a strengthening element.

Carbon. Beside these minor amounts of phosphorus, manganese sulfide, and other inclusions, steel contains carbon, as a definite chemical compound, or at least a definite crystal aggregate that amounts to much the same thing as a true chemical compound, called *cementite*. This is iron carbide in which three atoms of iron combine with one of carbon (Fe_3C), or that compound in which part of the iron is replaced by some other carbide-forming element such as manganese, chromium, or molybdenum. Special high-alloy steels may contain other carbides, differing from cementite in chemical composition, crystal structure, and properties. Cementite is hard. The special carbides are even harder.

Steel contains cementite particles distributed through the iron matrix, which matrix is called *ferrite*. The properties of steel are overwhelmingly governed by the amount, size, and distribution of cementite particles within the ferrite matrix. The more the paths through the soft matrix are interrupted by the hard particles, the stronger is the steel. Control of the amount of carbide and its distribution provides primary control over steel properties. The effects of alloys in themselves are minor except as they allow increased control over the distribution of carbide, though they somewhat strengthen the ferrite matrix and, in suitable amount, alter the nature of the carbide.

Slow Cooling. The amount of carbon in steel affects its response to different rates of cooling. In ordinary steels, at forging or rolling temperatures or at temperatures used in heating for quenching, neither of the two separate constituents of cold steel, ferrite and cementite, is present. Ordinary steels, at the high temperature, have only one constituent, called *austenite*, which is iron holding carbon in solid solution. As the austenite cools and passes slowly enough through the "critical temperature" (which depends on the carbon content), the carbon is thrown out of solution, and the iron of the austenite partly transforms to ferrite and partly forms iron carbide. On slow cooling, the iron carbide associates itself with a definite amount of ferrite to form an aggregate called *pearlite*. If the austenite contains 0.83% carbon, all the ferrite is tied up in this aggregate, which is made up of alternate layers of ferrite and cementite. The width of the layers depends on the rate of cooling, and upon this and on the grain-coarsening characteristics of the austenite depends also the size of the individual grains of the pearlite aggregate. Rail steel is nearly all pearlite.

If carbon is less than 0.83%, more ferrite is formed in the transformation than can be held in the pearlite. This excess ferrite is formed first, at a higher temperature than cementite, so that it forms a matrix in which the pearlite patches are embedded. Slow cooling, as in full annealing, produces large pearlite patches or islands in the sea of ferrite; faster cooling, as in air cooling of ordinary sections, gives smaller islands, closer together. Most commercial steels are under 0.83% carbon. If there is more than 0.83% carbon, the excess carbon first separates as cementite, and the matrix is pearlite. Drill rod has such a structure. The presence of alloys usually cuts the 0.83% carbon of pearlite to a slightly lower content, but the relative behavior is the same.

Transformation Temperature. The size and distribution of pearlite patches in the ferrite matrix are governed by the temperature at which the transformation occurs. It takes some time for the carbon atoms in austenite to elbow away the surrounding iron atoms with which they are not going to unite, and then to associate themselves with the ferrite that is going to form pearlite. In the process of cooling, ferrite is being liberated and the pearlite aggregate is collecting into patches. If the austenite is cooled fairly rapidly, these things do not occur at as high a temperature as they would on very slow cooling.

If cooling is very slow, these changes take place at the highest temperature at which they can occur. The resulting structure is coarse, the pearlite patches large, the paths through the ferrite broad and unobstructed; hence the steel, after this "full anneal," is in the softest condition it can attain short of a "spheroidizing anneal" which breaks up the pearlite patches. Faster cooling prevents this aggregation, for the steel is becoming stiffer. It is also possible, by alloying, to make austenite more sluggish, hence to accomplish the same result at a less rapid rate of cooling.

If the transformation can be held back to a lower temperature, the pearlite patches are smaller when deposited, and, once deposited, they stay put. It takes very long heating of the solid ferrite and pearlite to cause the pearlite to break up, free its ferrite and add it to the matrix, and change the cementite layers into balls, though just this can be done by a long "spheroidizing anneal." This is because the solubility of carbon in ferrite and its rate of diffusion in ferrite are very low.

Austenitic Grain Size. On the other hand, the solubility of carbon in austenite is high, and diffusion of carbon in it at elevated temperature is quite rapid. Austenite is solid and exists in crystal grains. If austenite is held at higher temperatures and for longer times, its grains grow. Rejection of excess ferrite begins at the boundaries of the austenite grains, as they start to transform on cooling. Coarse austenite grains therefore favor a more connected ferrite matrix in the ferrite-pearlite structure and lead to a less uniform distribution of pearlite, and it is the uniformity of distribution on which toughness and strength depend. Austenitic grain size is therefore a factor in the properties of the transformed steel when it gets down to room temperature.

Hence, the properties of as-rolled steel, or of steel reheated to the temperature at which austenite exists (the transformation is reversible; the ferrite-pearlite structure becomes austenite on heating), depend not only on the carbon content and on the rate of cooling but also on the grain size (and homogeneity) of the austenite. The rate of cooling depends on the temperature at which the cooling of the austenite is started, on the surrounding medium (still air, air blast, oil, water, brine, etc.) used to abstract the heat from the austenite, and importantly on the size of the piece being cooled. No matter how fast the outside is cooled, heat is not instantaneously abstracted from the center—it can flow only as fast as the low thermal conductivity of steel will allow.

Control of Transformation. Thus the ability to cool austenite fast enough, to make it wait for transformation until it has reached the temperature at which transformation will give the desired structure, is strictly limited by the size of the piece. In a small piece considerable control is possible, but in a very huge piece there is almost none at all upon the center because the center takes its own time to cool even though the skin has been cooled rapidly. The only recourse for control in large sections is to alloy the austenite enough to make it sufficiently sluggish to act as desired under the cooling conditions created. Thus there is a size effect in the heat treatment of steel.

In small enough sections or with a sufficiently sluggish austenite considerable control can be exerted producing, in air cooling, a desirable distribution of ferrite and pearlite. Further control is had by controlling the austenite grain size, as by reheating only slightly into the range of austenite temperatures before cooling, thus giving the austenite small opportunity to grow (but for a long enough time so that it becomes homogeneous), hence getting a fine distribution of pearlite in ferrite on this cooling. (Such reheating and cooling is called normalizing because the structure is considered normal for an air-cooled steel.) Similar control can be applied by introducing tiny amounts of *grain-control elements* which prevent austenite grains from growing until a very high temperature is reached. Fine grain diminishes the tendency toward air hardening, which is to be avoided in normalizing.

Bainite. Indeed, if the section is small enough, and the sluggishness and cooling rate great enough, the transformation can be delayed so far that the regulation pearlite aggregate is not produced. Instead, there results one consisting of patches or needles with less carbon than normal pearlite and with an even finer distribution. This structure is called *bainite.* It ranges from not much stronger than a fine ferrite-pearlite structure to one very much stronger, according to how much the transformation temperature is decreased. However, the structure and the degree of strength produced are highly sensitive to the cooling rate, i.e., to the size effect; hence uniform properties from surface to center are produced only in thin sections whose cooling is arrested at the desired temperature and held there for a more or less extended period.

Martensite. If the transformation temperature can be lowered still further, again by having the section small enough, the rate of cooling high enough (as by quenching instead of air cooling), and the austenite sluggish enough, it is possible to avoid the transformation to ferrite and pearlite or to bainite and to allow austenite to transform to a quite different product, *martensite.* The upper transformations will go on at constant temperature if the steel is held at that temperature. Martensite, however, forms only on a falling temperature.

Martensite, as quenched, is brittle, strong, and very hard, its hardness and brittleness increasing with the carbon content. It consists of ferrite and cementite (or if, as usually considered, of ferrite and carbon, the carbon combines to form cementite on slight heating) with the cementite in a state of subdivision far finer than can be had in the ferrite and pearlite structure or in bainite. Its hardness gives it wear resistance, but its brittleness prevents its use save in certain very special applications. But this brittle material can be toughened, and its static notch sensitivity and notched endurance sensitivity decreased, by heating (*tempering*) through agglomeration of the carbide and thus removing some of the interruption of the paths through the ferrite matrix.

Tempering. The properties of tempered martensite can be varied within wide limits by tempering back to a particular hardness. A high yield strength and high yield ratio are characteristic of tempered martensite. Quenching and tempering provide a range of strengths combined with toughness not otherwise attainable.

Bainite forms both above, and in, the temperature range in which martensite forms. That produced at a very low temperature is nearly as hard as the martensite of the same steel, but the bainite, tempered back to the same hardness as the tempered martensite, h . lower yield strength. The ultimate goal in quenching is therefore to produce all martensite without bainite. The goal is not always attainable, nor does it always need to be attained, but the fact remains that a tempered, full-quenched steel, initially all martensite, has higher core strength than a slack-quenched steel whose core is a mixture of tempered martensite and bainite, or some range of bainites, or even so slack quenched as to be ferrite and pearlite. A quenched and tempered spring with a bainite core takes a set at lower load than one that was all martensite, and armor plate whose interior was bainite is not as good armor as one that was all martensite.

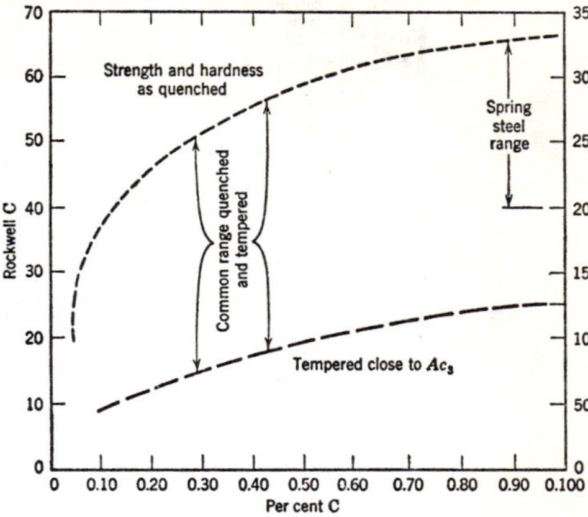

FIG. 3. Range of strength obtainable in steel by varying carbon content and heat treatment. (Reproduced by permission from Bullens, *op. cit.*)

Retained Austenite. By heavy alloying, even the production of martensite may be made incomplete or entirely absent, and the steel may retain some or all austenite at room temperature. This seldom occurs in the steels used in everyday design, though it is intentional in certain steels, notably the 18% chromium 8% nickel type of stainless steel. It occurs in high-speed tool steels and may occur in carburized cases of highly alloyed carburizing steels. Consideration of retained austenite is beyond the scope of the present discussion.

Control of Properties. The range of strength attainable in steel by variation in carbon and heat treatment is shown in Fig. 3. The effect of carbon content in annealed carbon steels, the inverse relation between strength and ductility, and the effect of the ratio of ferrite to pearlite, a result of the carbon content, are shown in Fig. 4.

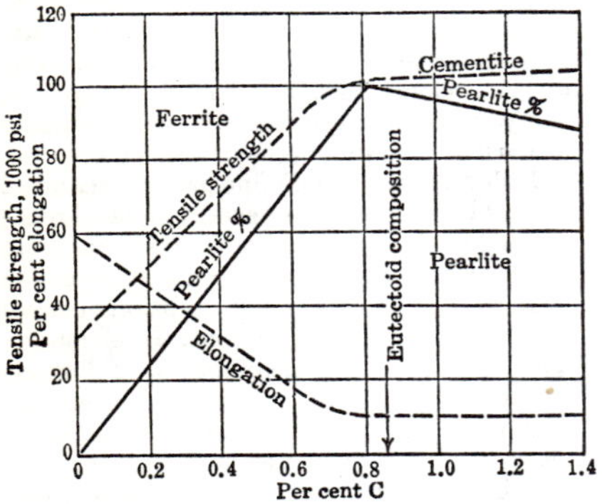

FIG. 4. Ferrite-pearlite-cementite composition (solid lines), compared with tensile strength and elongation of annealed steels. (Reproduced by permission from Bullens, *op. cit.*)

The influence of carbon on the hardness of quenched martensite, before tempering, and in normalized (pearlitic) and drastically annealed (spheroidized) plain carbon steels is shown in Fig. 5*a*. The hardness of martensite is not appreciably influenced by alloy content. The curve is dashed below 0.15% C to indicate that such low-carbon steels, unless accompanied by a huge amount of alloy, do not have the structure of martensite, even when quenched, but bainite or ferrite and pearlite, i.e., a normalized structure.

In ordinary heat treatment of steels with more than 0.60% carbon, the quenched hardness does not increase much above that at 0.60%, because the heating for production of austenite is not high enough to take all the carbide into solution. Therefore, as the lower boundary of the band in Fig. 5*b* shows, the quenched hardness remains at 65 to 67 Rockwell C in spite of the higher carbon; but if all the carbon is taken into solution by use of very high temperature the hardness can be increased, as shown by the upper boundary. This high-temperature treatment is ordinarily avoided because of the danger of excessive grain growth, of oxidation or of decarburization, and other disadvantages.

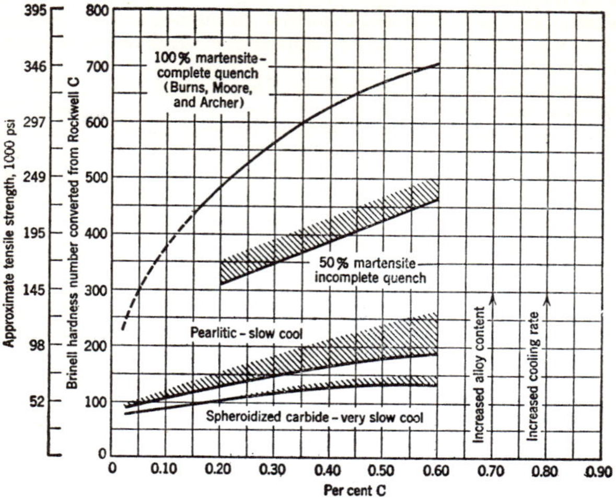

FIG. 5a. Effect of carbon content, method of cooling, and slack quenching on strength and hardness of steel. (Reproduced by permission from Bullens, *op. cit.*)

Carbon Content. Steels for quenching and tempering ordinarily run between 0.25 and 0.50% carbon and still more commonly in the range of 0.30 to 0.40%. Steels for foolproof welding generally run below 0.20% carbon, and below 0.15% if alloyed. Those for deep drawing come closer to 0.10%. Steels to be normalized may run from around 0.10% to, say, 0.40%, depending on section and on alloy content. Cast steels hover closely around 0.25 to 0.30%.

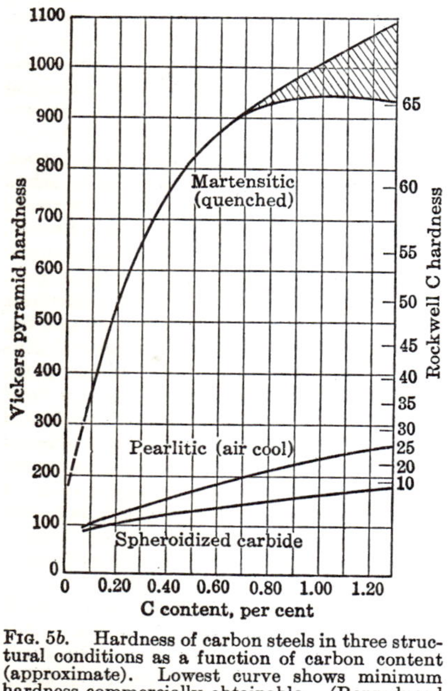

FIG. 5b. Hardness of carbon steels in three structural conditions as a function of carbon content (approximate). Lowest curve shows minimum hardness commercially obtainable. (Reproduced by permission from Bullens, *op. cit.*)

Melting Practice. The carbon level and the degree of need for grain-size control influence the choice of method of melting the steel and the type of deoxidation practice. Steel may be melted in the *open-hearth furnace*, the *converter*, or the *electric furnace*. Any steel save those high in oxidizable alloying elements may be made by any of the processes; those high in such elements are usually made in the electric furnace.

Killed Steel. When fine grain is required it may be produced, especially in the lower-carbon steels, by a normalizing process from not far above the critical temperature of the change to austenite, or by addition of small amounts of grain-size-control elements. These elements are deoxidizers, and steels containing them in amount sufficient for high restraint upon grain growth are *killed*. Other combinations of elements can be used to kill without restraining grain growth. Relatively low-carbon steels may be produced either killed or unkilled. Alloy steels are almost invariably killed. The term arises from the fact that unkilled steel boils and spits in the ladle; killed steels lie "dead."

A fully killed steel develops no gas while freezing and, because of the decrease in volume between liquid and solid steel, *pipes* in the ingot, i.e., has a cavity which is at the top and is cut off (*cropped*) from the ingot before rolling if the steel is cast in a *big-end-up* ingot. If the steel is cast in a *big-end-down* ingot, the cavity or cavities are left in, in the hope that they will weld up on rolling.

Unkilled Steel. An unkilled steel in the hot, molten condition contains iron oxide which reacts with the carbon of the steel to form carbon monoxide as the steel cools to the freezing point. The bubbles of gas are trapped in the ingot as it freezes. The conditions are so adjusted by the steelmaker that the ingot does not pipe, the volume shrinkage being balanced by the volume of gas bubbles. The ingot is not cropped, and this absence of discard makes unkilled steel cheaper than killed. Unkilled steel first separates a solid rim of relatively pure ferrite without gas holes. This gives it the name *rimmed* steel. It is also called *open* steel. The core, freezing a bit later, contains more carbon and has the holes. The clean, soft skin gives a nice surface to the final rolled product and to stampings made therefrom. (Unkilled steels are not always rimmed; the holes may come close to the surface.)

Unless the holes fully weld in rolling, the core has some degree of unsoundness which may or may not matter, depending on what fabrication processes and final uses are contemplated. The unsound core is usually avoided in high-strength steels, no matter how that strength is to be achieved. All steel castings, all steels for quenching and tempering, and most steels for normalizing are killed.

Effects of Steel-making Processes. The best quality in high-strength steels is obtained from big-end-up, cropped ingots. The internal pipe in big-end-down ingots results, in the material rolled from the ingot, in good sound material rolled out from those places in the ingot where there happened to be no pipe, but of questionable material from the piped locations. Random tests do not necessarily hit the stock coming from the piped locations.

The cleanliness of the steel—its relative freedom from nonmetallic inclusions—is not related primarily to the method of melting but more to the time and the temperature at which the molten metal is held, though it is considered easier to produce cleaner "aircraft-quality" steel from the electric. Converter steel, of like chemical composition to open-hearth or electric and with like deoxidation practice, has been barred by some specifications, but without reason.

A steel of the same chemical composition can be produced by any of these melting processes or any combination thereof; it may be rimmed or killed; if killed it may be grain size controlled or not, and may be cast in a big-end-up or big-end-down ingot; and it may contain many or few nonmetallic inclusions. Trying to state properties and utility for a given steel composition without mentioning these factors is obviously a too-sweeping effort.

Specifications. All these factors affect the cost of steel, and the mills charge "extras" for applying such variations in practice as increase their production cost of finished steel. If the designer does not need some of these quality features of mill practice to insure satisfactory service of his product, he can save money by not specifying them. If he needs them, it is economy to call for them. The mills are anxious to provide useful material, and their appraisal of fitness for an intended use, when that use is fully described to them, is worth having. Such appraisal may well result in mutually acceptable purchase specifications for chemical composition limits, narrow or wide as the case may justify, grain size, degree of cleanliness, and other factors that make the specification have much more bearing than specifications based on conventional mechanical properties, which notoriously fail to insure the particular deep-drawing properties required for a particular part like an automobile fender.

Reaching a strength level much above that of everyday structural steel requires a normalizing heat treatment at least and, for really high strength, quenching and tempering. Carbon steels have size limitations, under these treatments, which are extended by the introduction of alloying elements. Sales propaganda has tended to obscure some of the vital metallurgical facts affecting the choice of alloy steels. The principles involved are much simpler than they used to be made to appear.

3. METALLURGICAL PRINCIPLES INVOLVED IN STRENGTHENING STEEL

FERRITE STRENGTHENING. A family of "low-alloy high-yield-strength" or, more concisely, "mild-alloy" steels has been built up to meet designers' needs for steel of 50,000 psi yield strength obtained with no heat treatment or the simplest possible heat treatment—mere normalizing—but combining, with this yield strength, high cold formability and foolproof weldability. Raising the carbon content of plain carbon steel to provide this strength depreciates formability and weldability. Low carbon is therefore essential, and, with low carbon, strength has to be supplied in some other way. It is supplied in two ways. Alloying elements make austenite sluggish. If it is thus made sluggish enough to delay its transformation to ferrite and pearlite until a lower temperature is reached than with low and medium straight carbon steels, the pearlite patches are finer and more uniformly distributed. Concurrently, the ferrite matrix is itself made stronger by entrance of alloying elements. This is the second way of strengthening. By what combination of elements these two results are secured is relatively immaterial. Naturally, the elements that are cheap and fairly powerful, or very powerful, even if expensive, have been chosen.

One attribute extraneous to strength has entered into this choice. Long before the days of these steels, it was known that structural steel containing to 0.20% or above of copper (especially if the phosphorus content is somewhat above normal), i.e., so-called copper-bearing steel, acquires a more adherent type of rust when exposed, bare, to the atmosphere, than steels very low in copper and phosphorus and hence (for atmospheric exposure only) subsequently rust more slowly and have a longer life. It *starts* to rust just as fast as the other steels. This ultimate rust resistance was considered a major asset, which it may or may not be, depending on how the steel is actually going to be used and whether it will be kept painted or not.

Both copper and phosphorus are effective in making austenite sluggish and strengthening ferrite. Phosphorus embrittles medium- and high-carbon steels. Roughly, 0.25% as the sum of carbon and phosphorus is the permissible limit. For tough low-temperature behavior, the limit may be lower.

Copper is universally used in this class of steels. Phosphorus appears in a large proportion. Manganese is often made use of. Chromium, nickel, and molybdenum appear in several. Grain-size control plus some strengthening is achieved by vanadium, titanium, or aluminum, alone or in combination.

Table 6, p. 2-43, lists some of the many variations in chemical content displayed by mild alloy steels of closely equivalent mechanical properties, cold formability, and weldability. Broadly speaking, any one can be substituted for any other.

When formability and especially weldability are less important, as in some riveted structures, for example, the carbon content can be raised and a somewhat higher yield strength can be reached, as discussed on p. 2-44.

The properties given are for longitudinal tests on plate $^1/_2$ to $^3/_4$ in. thick. Transverse toughness, of course, will be less. Heavier plate will drop slightly in yield strength and ductility, but yield strength can be raised by increasing alloy content, molybdenum often being used for this purpose.

PRECIPITATION HARDENING. The next upward step for strength in simply treated steels utilizes the precipitation-hardening phenomenon which can be produced in low- and medium-carbon steels containing copper over about 0.60%, most notably with about 1 to 1 $^1/_4$%. Up to around 0.60% the copper is held in solution in the ferrite, strengthening it and conferring the atmospheric corrosion resistance previously mentioned. The excess over about 0.60% is held in apparent solution, even on quite slow cooling, but on reheating to around 1050 F for about an hour this excess is thrown out of solution, in extremely finely divided form, peppered throughout the grains. Just as a fine and uniform distribution of cementite in ferrite raises the strength, so does this distributed material, in spite of the softness of massive copper itself.

This reheating raises the tensile and yield strengths of low- and medium-carbon steels, irrespective of what other alloys are present, by 10,000 to 20,000 psi. Straight carbon, 1% copper steels show such an increase, and, applied to the 50,000 psi yield mild alloy steels containing the proper amount of copper, the reheating brings the yield to around 70,000 psi. Some toughness and ductility are, of course, sacrificed concurrently with the strengthening, but ample remains for most uses. However, in the "solution-treated" state, i.e., before reheating and precipitation, there is high ability for cold forming. The formed article is then reheated for precipitation and strengthening.

This copper-precipitation strengthening is not markedly subject to size effect; the apparent or "forced" solution of the excess copper persists even under remarkably slow rates of cooling, so that even the center of quite large sections retains it until reheated, and uniform strengthening occurs over the whole section. This class of material and type of treatment afford a simple and effective method of getting yield strength well above that of the regular mild-alloy-steel class and bridge a gap between those steels and quenched and tempered steels.

Precipitation hardening occurs in many special alloys especially concocted to utilize the phenomenon, which do not fall within the limits of the present discussion. It occurs as an adventitious, sometimes detrimental, phenomenon in some steels used by the designer. When the precipitation occurs within the grains, useful strengthening tends to result; but when it is confined to grain boundaries, and especially when the precipitated material is itself brittle, it tends toward brittleness. Various compounds are suspected of such precipitation, and it is not always clear just which elements among those that might so act are really responsible. Temper brittleness is suspected as such a phenomenon, and the beneficial effect of molybdenum is suspected to be exerted by causing precipitation in, instead of around, the grains.

Very low-carbon unkilled deep-drawing steels are prone to "aging." When fresh from even slight cold rolling they form without cracking or "stretcher straining," and their stress-strain curves are smooth, without a yield-point jog. A month at room temperature or half an hour in boiling water produces the jog and poorer deep-drawing behavior. Properly killing the steel stabilizes it against such behavior, perhaps by rendering insoluble some minor constituents that otherwise go into solution and then precipitate, or by preventing a sort of annealing of the distorted cold-worked crystals that can go on even at room temperature.

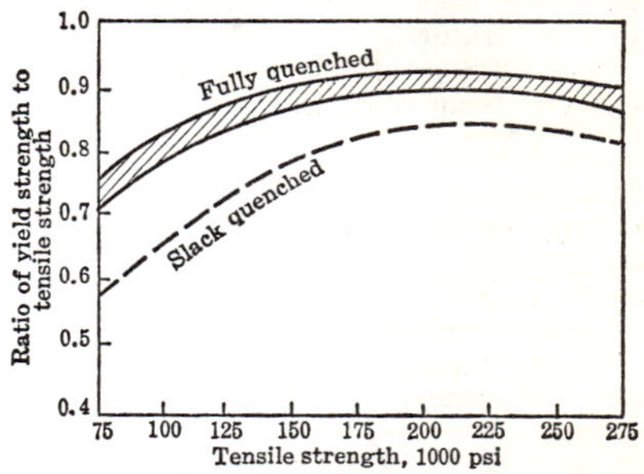

FIG. 6. Loss in yield ratio caused by slack quenching.

Austenitic steels, such as 18-8 stainless and Hadfield manganese steel, whose austenite contains carbon in solution, can precipitate carbide if held in a certain temperature zone, the brittle carbide being concentrated at the grain boundaries. Some of the properties are thus badly deteriorated.

QUENCHED AND TEMPERED STEELS. In the heat treatment of quenching and tempering necessary for production of higher strength levels, two facts stand out.

First, the yield strength, at given hardness and tensile strength, of any tempered-steel structure that was not initially martensite, before tempering, is inferior to that of a steel initially wholly martensite. That is, the slack-quenched interior of a section that does not harden is inferior when the whole piece has been tempered to the degree desired for the exterior which has hardened to martensite. The degree of inferiority in yield strength and other properties is the greater, the higher the temperature at which the core has transformed to other structures. The effects of slack quenching are brought out in Figs. 5, 6, and 7.

The intended use may not require high core properties. In fatigue the surface is usually the controlling factor. Even under tensile stress where it is intended that the whole section bear its share of the load, loading may easily be so eccentric that the exterior only is of major importance. But some uses do demand equal strength in the interior, and, if the heat-treated piece is ground or machined so that stress concentration, or an actual notch, extends into the weaker material, the piece acts in accordance with the lower strength of that material. Considerable engineering judgment is required in determining what degree of depth hardening is needed in a heat-treated part. Although data are lacking for tabulation of all shades of difference in mechanical properties as one goes from outside to inside of a piece of shallow-hardening, or slack-quenched steel, a hardness survey through the cross section, especially when coupled with metallographic examination of the changing structure through the section, will give qualitative information.

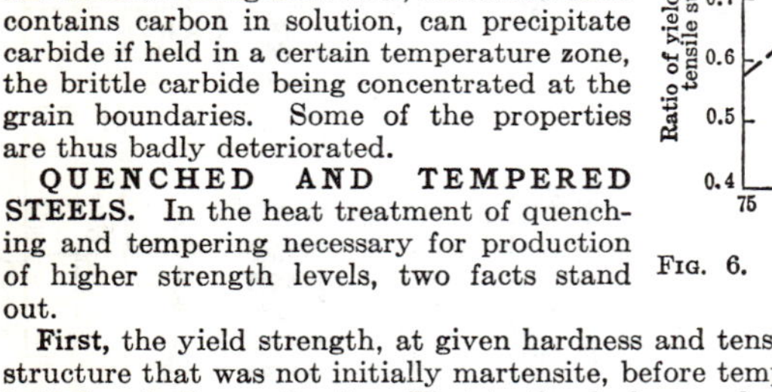

FIG. 7. Notched-bar behavior of 1045 steel heat treated in various ways. In the size used, oil quenching was "slack." (Reproduced by permission from Bullens, *op. cit.*)

As a first approximation, if two steels of the same carbon content are quenched and tempered to the same surface hardness, and their hardness gradients from surface to center are identical, quite similar properties at corresponding points and quite similar service can be expected. This is not necessarily true in every minute detail, but such a situation creates a high probability of close similarity.

The second vital fact about quenched and tempered steels, an even more universally true observation, is that, if two steels of corresponding carbon levels are quenched throughout to all-martensite, and then tempered back to the same hardness, their properties are alike, practically as alike as would be found in duplicate tests of the same steel. Similarity of grain size is also a factor in similarity of properties.

This likeness in other properties at the same hardness level among different steels, without regard to alloy composition, gives a quite different concept from that of earlier days. Then each alloying element or combination thereof was supposed to possess particular advantages or disadvantages relative to the others, depending on which element its proponent was trying to sell. What appeared to be radical differences were displayed in "property charts" of curves for tensile and yield strengths, elongation and reduction of area, perhaps hardness and notched-bar figures, all plotted against tempering temperature for fully quenched, initially martensitic, small specimens. Similarly, the data from which the curves were plotted were given in long tables of properties, with a line for each tempering temperature.

The information in these plots was really temperability, i.e., what temperature, at ordinary tempering times, is required to soften martensite back to a required yield strength.

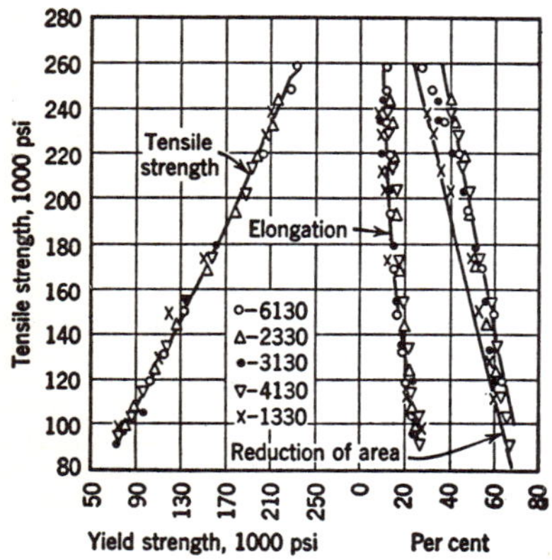

FIG. 8. Properties of various steels of 0.30% carbon content water quenched to martensite and tempered to various strength levels. (Reproduced by permission from Bullens, *op. cit.*)

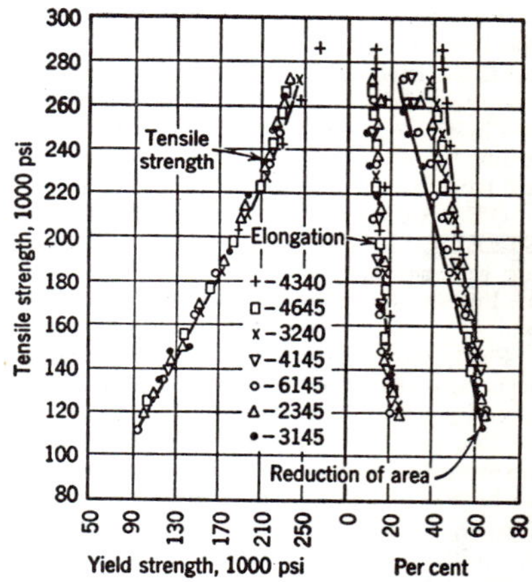

FIG. 9. Properties of various steels of 0.40 to 0.45% carbon content, oil quenched to martensite and tempered to various strength levels. (Reproduced by permission from Bullens, *op. cit.*)

The metallurgist and heat treater are concerned with temperability, but the designer is not; he uses the material in parts that the heat treater has tempered for him.

Quenched and Tempered Steels, Basic Data. When the data are reduced to a tensile strength, yield strength, or hardness basis as has been done by Janitsky and Baeyertz (Ref. 8), Patton (Ref. 9), Herres and Jones (Ref. 10), and many others, plots like Figs. 8 and 9 result, which show that for each strength level up to at least 200,000 psi tensile strength there is a closely defined corresponding yield strength, tensile strength, and elongation and not very much scatter in reduction of area. Even the Izod notched-bar impact values fall pretty close to the same expectation, as Herres and Jones (Fig. 10), also Crafts and Lamont (Ref. 11) show.

For the slack-quenched interior, initially softer than martensite, and further softened by tempering, Fig. 11 shows the relations. At the right of this figure, R_M denotes Rockwell C hardness of martensite, R_Q Rockwell C hardness for that steel as actually quenched. The difference, $R_M - R_Q$, is a measure of the degree of slack quenching, i.e., presence of constituents other than martensite. As the tempering temperature is increased the impact value increases.

Figure 11 was plotted from data on carbon, low-alloy, and high-alloy steels, including the usual SAE and NE steels. Crafts and Lamont say that "the effects of alloying elements are non-specific, so, in general, alloys may be used interchangeably without sacrifice of serviceability." It is hardenability that is conferred by alloys; at the same grain size and with full hardening in quenching, the maximum combination of tensile and impact strength is achieved in the tempered product. The plotted data are for fine-grained steels. Coarse-grained steels have lower toughness than fine-grained steels anyway, but Crafts and Lamont state that, where this is suitably taken into account, the coarse-grained steels show even more reduction in notched-bar values as a result of slack quenching than the fine-grained ones.

Impact strength of tempered steel can be used as a measure of original slack quenching. Thus, applied to the centers of bars of different diameters, it is a measure of hardenability, and Crafts and Lamont give a hardenability table on this basis which corresponds to similar tables based on the more direct Jominy data. Indeed, the end-quenched and tempered Jominy bar can be made into impact specimens with the notch located at the position corresponding to that at the center or at a selected distance from the center of a large bar to give a rough idea of the impact behavior of the heat-treated large bar, were it sampled at the selected location.

At strengths above 200,000 psi tensile, the values for reduction of area and notched-bar values scatter more, because different degrees of stress relief may be obtained at such a level of hardness without affecting hardness and because retained austenite in different amounts will be present in some of the highly alloyed steels, producing such strengths, as

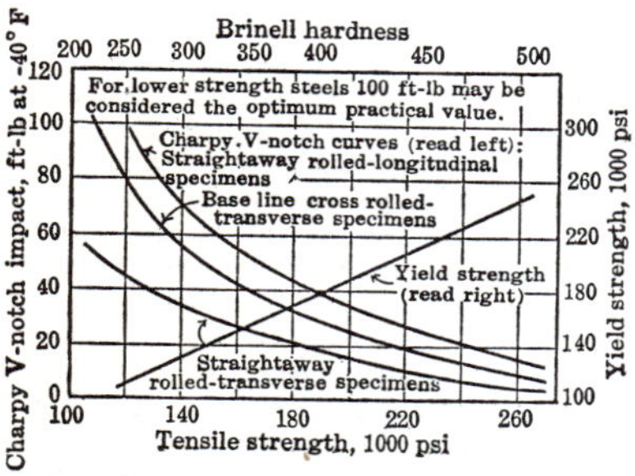

FIG. 10. Relationship of yield strength, Charpy V-notch impact resistance, hardness, and tensile strength of properly heat-treated and fully quenched reasonably clean steel. (Reproduced by permission from A Method for Evaluating Toughness of Steel, by Herres and Jones in *Metal Progress*, Sept. 1946)

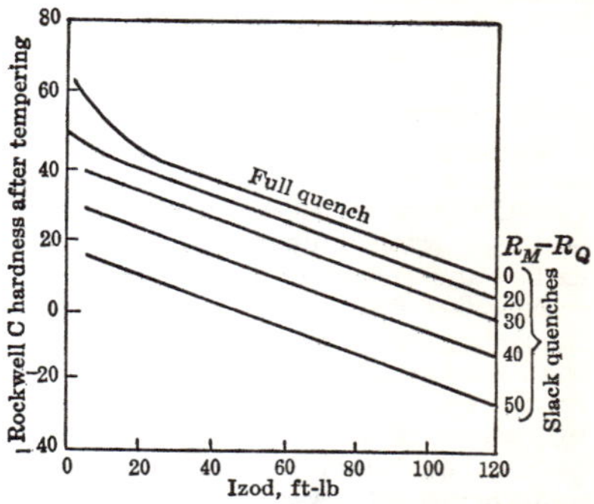

FIG. 11. Loss in notched-bar impact values because of slack quenching. (Adapted from Crafts and Lamont, The Izod Impact Strength of Heat-treated Alloy Steel, *AIMME Tech. Pub.* 2134, *Metals Technol.*, Vol. 14, Feb. 1947)

well as martensite. Unless the designer has a use to which steel of over 200,000 psi tensile or 180,000 psi yield strength is adapted, these data give no reason for selecting one fully hardened, then tempered, alloy steel over another. It is, indeed, open to question whether the scatter in values at still higher strength levels is not so likely to occur between duplicate heats of the same steel as between different steels. Temper-brittleness effects, which are notably variable from heat to heat, may come into the notched-bar behavior, but if these are eliminated, Herres and Jones extend their smooth curve of expectancy of notched-bar behavior of quenched and tempered steels clear up to 275,000 psi tensile, just as Crafts and Lamont carry their curves to corresponding hardnesses.

It is not proved that there is appreciable reliable difference among very hard-quenched and tempered steels of commercial compositions used for constructional purposes, but it is proved that, at least up to 200,000 psi tensile, there is no appreciable difference. Fatigue behavior, to say nothing of machinability, usually sets even a lower limit for the great bulk of design.

This similarity holds only for those steels whose carbide is of the Fe_3C type. When different carbides are produced in high-alloy steels, the behavior can differ. The composition limits at which different carbides are produced are rather well known to all producers of constructional steels, and commercially obtainable steels for constructional purposes seldom fall outside those limits. Special-purpose steels are another story.

Hence, properties of many compositions of steels, instead of being tabulated separately, can be as adequately shown in the master curves of Figs. 8, 9, and 10 for every purpose of the designer. It should be remembered that all these are for longitudinal specimens except as noted in Fig. 10.

Heat Treatment. From the heat treater's point of view, composition has an influence upon proclivity toward or away from cracking on quenching, especially in parts of unbalanced section. If the temperature at which martensite starts to form is too low, quench cracking is likely to occur upon quenching in water or brine, and the more expensive, less effective, oil quench must be used. The martensite-formation temperature is most rapidly depressed by carbon itself, next most rapidly by manganese. Hence, for reasons of processing rather than properties of successfully processed material, the heat treater prefers that both carbon and manganese be limited and that the desired sluggishness of austenite be secured by the addition of other elements less active in depressing the martensite-formation temperature. He also prefers that the temperature of completion of the martensite transformation be above room temperature, else he is confronted with the retention of austenite and the necessity of using cold treatment or repeated temperings to destroy the austenite so that it will not transform during service of the part to brittle, untempered martensite, if the part heats up and then cools off. Within these limits, the heat treater does not care what the composition is. His concern is in the satisfactory production of an all-martensite structure upon quenching.

Were all quenched and tempered sections as thin as watch springs, there would be no demand for alloy constructional steels, only for the special steels. Using alloy constructional steels instead of plain carbon steel in watch springs does not alter their mechanical properties or performance, because the carbon steel quenches to all-martensite. (Securing these properties plus corrosion resistance in watch springs is another matter, requiring alloys that can hardly be called steels.)

Using an alloy steel that will harden throughout, however, would materially improve the properties and performance of a railway car spring, with its large cross section.

Chemical Composition Specification. Alloy steels used to be defined and discussed in terms of chemical composition; to facilitate definition and discussion, the SAE and later the AISI introduced a shorthand numbering system which designates the alloy group and the carbon content and, by establishment of permissible ranges for the variation in carbon and each other element from heat to heat, also conveys a crude idea of the way each so-classified steel will harden on quenching. This depth-hardening behavior was merely implied and not directly expressed. As a matter of fact, within these "permissible" ranges the hardenability could vary widely. In order to insure that the hardenability needed for the section to be quenched would be obtained when a heat on the low-hardenability side was encountered, it was practically necessary to specify a steel of a second, higher-hardenability class than would have served had the hardenability of the first class been uniform instead of variable. This meant that the higher-hardenability heats of the second class might have the martensite-formation temperature so low that quench cracking appeared in such heats under the quenching practice necessary for the low-hardenability heats. Moreover, the higher alloy content of the second class of steels made it more expensive. To solve this dilemma, users began to narrow the chemical composition ranges so as to secure more uniform hardenability, and this made mill control far more difficult, calling for price extras that increase the cost of the steel.

Hardenability Specification. That hardenability and uniform hardenability were prime desiderata in steels for quenching has long been recognized, and it is beginning to be recognized that a more logical system of classification would be on the basis of hardenability: for instance, all steels that harden to martensite at the center when commercially water quenched in 1/2 in. diameter but not in 3/4 in. would be in one classification; those hardening in 3/4 in. but not 1 in. in another, and so on; likewise those that harden on oil quenching in such section ranges. To test a heat for such behavior by quenching and sectioning for examination all sizes involved, say from 1/2 to 2 in. or 1 1/2 to 4 in., as the case might be, would entail much trouble and expense.

However, a quick and simple test has been devised which gives equivalent information on a single specimen. It has become a conventional test for ability to be processed. It does not directly give design data, but it does evaluate uniformity in hardenability, and uniform hardenability plus proper heat treatment insures that the mechanical properties of the quenched and tempered piece will be so uniform and reproducible that mere hardness testing is ordinarily adequate for appraisal of the properties wanted by the designer, without necessity for expensive tensile testing of the finished product.

Chemical appraisal thus gives way to appraisal by hardenability, although we are still in the transition stage where alloy steels are referred to by AISI numbers, and hence by alloy composition, rather than merely by carbon content and hardenable diameter.

END QUENCH HARDENABILITY TESTING. The conventional test for hardenability consists in heating a specimen 1 in. in diameter by 4 to 6 in. long, to the temperature (and for the time) to be used in actual heating of the steel for quenching, and directing a stream of water upon one end face only, the length and the other end contacting air but no water. The piece thus has all gradations from the fully quenched end through a slack-quenched, down to a merely air-cooled, normalized structure.

Jominy Curves. A small flat is carefully ground along the length of the bar, and the hardness gradient is determined by hardness tests, usually Rockwell C, at spaced intervals. When radically different steels are being compared, metallographic examination is made along the bar length to show how far pure martensite extends and what the gradation through bainite down to ferrite and pearlite is. The data are then plotted as hardness against distance from the quenched end, also in terms of per cent martensite, bainite, and ferrite, plus pearlite at the various distances. From the name of one of its originators, the plot of hardness against distance is usually called a "Jominy curve." A similar plot can also be made after tempering the end-quenched bar to evaluate temperability.

By means of suitable conversion factors, based on experiment, the Jominy distances determined on the end-quenched bar can be converted to those that would be obtained were the specimen end-quenched by oil instead of water. Either set of distances can be

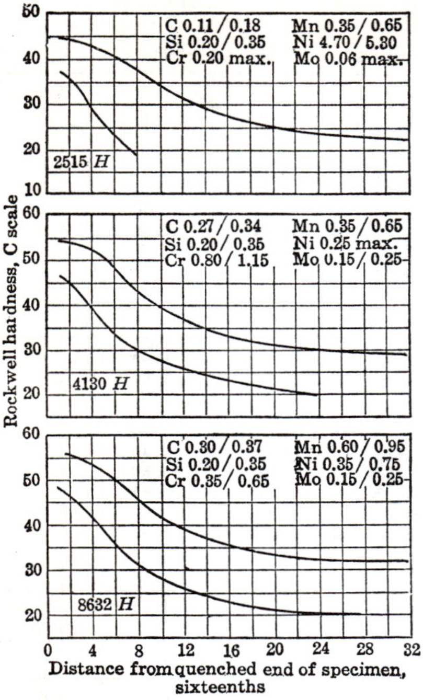

FIG. 12. End-quench test (Jominy) curves and bands. (Reproduced by permission from Bullens, *op. cit.*)

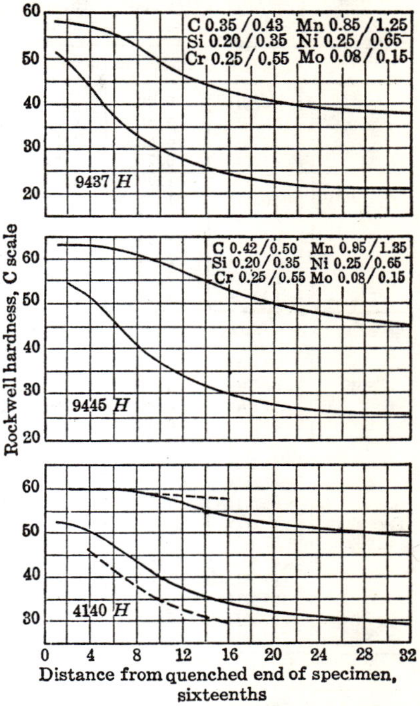

FIG. 13. End-quench test (Jominy) curves and bands. On bottom chart solid line is for 4140H, dotted line for standard chemical composition. (Adapted by permission from Bullens, *op. cit.*)

converted into the diameter of a bar which, so quenched, will harden at its center to all-martensite or to any chosen percentage of martensite mixed with bainite. This size information is what is wanted. Thus, by the Jominy test the depth of full quenching and of slack quenching of the particular lot tested is experimentally established. A modification of the bar is used when extremely shallow-hardening steels are evaluated, but the principle is the same.

For any one heat of steel, with the same time and temperature of heating before quenching, there is a characteristic Jominy curve, reasonably reproducible in duplicate tests. Different heats of the same composition, ostensibly processed alike in melting and "finishing," often show differences, and so for one composition the curve is a (narrow) band. For compositions ranging from everything on the high side to everything on the low side in chemistry, within the range permitted by the definitions of steel classes designated by SAE or AISI numbers, the band is extremely wide. Figures 12 and 13 are examples of Jominy curves and bands. As a result of this method of approach, hardenability information can be expressed in extremely comprehensive form, as in Fig. 14.

H Steels. Steels approximately conforming to chemical limits expressed by AISI numbers but falling within more restricted hardenability limits are commercially obtainable as H, i.e., controlled hardenability, steels without a price extra for observing the hardenability limits. Hardenability bands for such steels have been published (Refs. 12,

(Continued on p. 2-24)

Table 1. Maximum Sizes of Bars of Required As-quenched Center Hardness (H Steels)

Reprinted from data sheet by A. L. Boegehold, *Metal Progress*, Feb. 1947, by permission

Martensite in Microstructure at Center	Steel	After Oil Quench (Present SAE Correlation)		After Water Quench			Maximum Hardenability
				Minimum Hardenability according to			
		Minimum Hardenability	Maximum Hardenability	Revised Correlation of Cooling Rates	Present SAE Correlation	Revised Correlation of Cooling Times	
Max. Diameter of Bar That Will Harden at Center to Rockwell C-25							
Less than 20%	4620H	0.40	2.50	0.75	1.00	1.00	3.50
	8620H	0.50	2.70	0.90	1.10	1.10	3.40
	8720H	0.60	3.15	1.00	1.20	1.20	4.00
	8622H	0.75	3.50	1.30	1.40	1.60	4.00+
	8722H	1.00	4.00+	1.55	1.60	1.85	4.00+
	8625H	1.20	4.00+	1.75	1.85	2.15	4.00+
	8725H	1.40	4.00+	2.00	2.10	2.40	4.00+
	8727H	1.60	4.00+	2.20	2.30	2.60	4.00+
	8730H	2.15	4.00+	2.85	2.85	3.20	4.00+
	8732H	2.65	4.00+	3.45	3.40	3.70	4.00+
	8735H	3.15	4.00+	4.00	4.00	4.00+	4.00+
Max. Diameter of Bar That Will Harden at Center to Rockwell C-30							
20%	4620H	0.25	1.25	0.45	0.65	0.65	1.90
	8620H	0.35	1.50	0.60	0.80	0.80	2.25
	8720H	0.40	1.80	0.75	1.00	1.00	2.50
Less than 20%	8622H	0.40	1.95	0.75	1.00	1.00	2.70
	8722H	0.50	2.35	0.90	1.10	1.10	3.05
	8625H	0.70	2.50	1.15	1.30	1.40	3.25
	8725H	0.75	3.15	1.30	1.40	1.60	4.00
	8627H	1.00	3.15	1.55	1.60	1.85	4.00
	8727H	1.00	4.00	1.55	1.60	1.85	4.00+
	8730H	1.40	4.00+	2.00	2.10	2.40	4.00+
	8732H	1.80	4.00+	2.40	2.50	2.85	4.00+
	8735H	2.05	4.00+	2.75	2.80	3.10	4.00+
	8737H	2.35	4.00+	3.05	3.05	3.40	4.00+
	8740H	2.65	4.00+	3.45	3.40	3.70	4.00+
	8742H	3.25	4.00+	4.10	4.00	4.25	4.00+
Max. Diameter of Bar That Will Harden at Center to Rockwell C-35							
65%	8622H	0.25	1.15	0.45	0.65	0.65	1.85
	8722H	0.35	1.40	0.90	1.10	1.15	2.10
50%	8625H	0.40	1.60	0.75	1.00	1.00	2.30
	8725H	0.45	2.00	0.75	1.00	1.00	2.70
40%	8627H	0.60	2.00	1.00	1.20	1.20	2.70
	8727H	0.60	2.35	1.00	1.20	1.20	3.05
35%	8630H	0.75	2.35	1.30	1.40	1.60	3.05
	8730H	0.85	3.00	1.40	1.50	1.75	3.75
25%	8732H	1.20	3.85	1.75	1.85	2.15	4.00+
	8735H	1.50	4.00+	2.00	2.20	2.50	4.00+
	8737H	1.65	4.00+	2.30	2.35	2.75	4.00+
	8740H	1.90	4.00+	2.55	2.60	2.90	4.00+
Less than 20%	8742H	2.35	4.00+	3.05	3.05	3.40	4.00+
	8745H	2.65	4.00+	3.45	3.40	3.70	4.00+
	8747H	3.00	4.00+	3.60	3.80	4.00	4.00+
	8750H	3.40	4.00+	4.00+	4.00+	4.00+	4.00+
Max. Diameter of Bar That Will Harden at Center to Rockwell C-40							
75%	8627H	0.30	1.40	0.60	0.80	0.80	2.10
	8727H	0.40	1.60	0.75	1.00	1.00	2.30
70%	4130H	0.50	1.40	0.90	1.10	1.10	2.10
60%	4132H	0.60	2.25	1.00	1.20	1.20	3.00
70%	8630H	0.60	1.60	1.00	1.20	1.20	2.30
	8730H	0.60	1.95	1.00	1.20	1.20	2.70
60%	8632H	0.75	1.95	1.30	1.40	1.60	2.70
	8732H	0.75	2.45	1.30	1.40	1.60	3.15
50%	8735H	1.00	3.00	1.65	1.60	2.05	3.80

Table 1. Maximum Sizes of Bars of Required As-quenched Center Hardness (H Steels)—*Continued*

Martensite in Micro-structure at Center	Steel	After Oil Quench (Present SAE Correlation): Minimum Harden-ability	After Oil Quench (Present SAE Correlation): Maximum Harden-ability	After Water Quench, Minimum Hardenability according to: Revised Correlation of Cooling Rates	After Water Quench, Minimum Hardenability according to: Present SAE Correlation	After Water Quench, Minimum Hardenability according to: Revised Correlation of Cooling Times	Maximum Harden-ability
45%	8737H	1.20	3.65	1.75	1.85	2.15	4.00+
35%	8740H	1.40	4.00+	2.00	2.10	2.40	4.00+
25%	8742H	1.60	4.00+	2.30	2.30	2.75	4.00+
20%	8745H	1.95	4.00+	2.65	2.70	3.05	4.00+
Less than 20%	8747H	2.35	4.00+	3.05	3.05	3.40	4.00+
	8750H	2.65	4.00+	3.45	3.40	3.70	4.00+
Max. Diameter of Bar That Will Harden at Center to Rockwell C-45							
85%	4130H	0.25	1.15	0.45	0.65	0.85	1.85
	8730H	0.25	1.40	0.45	0.65	0.65	2.10
	4130H	0.30	1.60	0.60	0.80	0.80	2.30
87%	8632H	0.40	1.40	0.75	1.00	1.00	2.10
	8732H	0.40	1.60	0.75	1.00	1.00	2.30
	8635H	0.60	1.65	1.00	1.20	1.20	2.40
82%	8735H	0.60	1.95	1.00	1.20	1.20	2.70
	4135H	0.70	2.60	1.15	1.30	1.40	3.30
80%	8637H	0.75	1.95	1.30	1.40	1.60	2.70
	8737H	0.75	2.50	1.30	1.40	1.60	3.25
	4137H	0.90	3.85	1.40	1.50	1.75	4.00+
70%	8640H	0.90	2.35	1.40	1.50	1.75	3.05
	8740H	1.00	3.00	1.55	1.60	1.85	3.80
60%	8642H	1.00	3.00	1.55	1.60	1.80	3.80
70%	4140H	1.15	4.25	1.75	1.85	2.15	4.00+
60%	8742H	1.20	4.00	1.75	1.85	2.15	4.00+
45%	8645H	1.20	3.50	1.75	1.85	2.15	4.00+
60%	4142H	1.40	4.00+	2.00	2.10	2.40	4.00+
45%	8745H	1.40	4.00+	2.00	2.10	2.40	4.00+
50%	8647H	1.40	4.00+	2.00	2.10	2.40	4.00+
45%	4145H	1.75	4.00+	2.45	2.50	2.85	4.00+
50%	8747H	1.60	4.00+	2.20	2.30	2.60	4.00+
30%	8650H	1.70	4.00+	2.30	2.40	2.75	4.00+
50%	4147H	2.10	4.00+	2.75	2.80	3.15	4.00+
30%	8750H	1.95	4.00+	2.65	2.70	3.05	4.00+
30%	4150H	2.35	4.00+	3.05	3.05	3.40	4.00+
70%	4340H	4.00	4.00+	4.00+	4.00+	4.00+	4.00+
Max. Diameter of Bar That Will Harden at Center to Rockwell C-50							
92%	8637H	0.25	1.40	0.50	0.65	0.70	2.10
	4137H	0.25	2.35	0.45	0.70	0.65	3.05
	8737H	0.40	1.80	0.75	1.00	1.00	2.50
90%	8640H	0.40	1.70	0.75	1.00	1.00	2.40
	8740H	0.40	2.15	0.90	1.00	1.15	2.85
85%	8642H	0.60	2.15	1.00	1.20	1.30	2.85
90%	4140H	0.60	4.00	1.00	1.20	1.30	4.00+
85%	8742H	0.75	2.65	1.25	1.40	1.60	3.40
80%	8645H	0.75	2.50	1.30	1.40	1.60	3.25
85%	4142H	0.80	4.25	1.40	1.40	1.75	4.00+
80%	8745H	1.00	3.40	1.50	1.60	1.90	4.00+
75%	8647H	1.00	3.00	1.50	1.60	1.90	3.80
80%	4145H	1.15	4.25	1.75	1.85	2.15	4.00+
75%	8747H	1.20	4.00+	1.75	1.85	2.15	4.00+
70%	8650H	1.20	3.50	1.75	1.85	2.15	4.00+
75%	4147H	1.40	4.00+	2.00	2.10	2.40	4.00+
70%	8750H	1.40	4.00+	2.10	2.10	2.50	4.00+
70%	4150H	1.75	4.00+	2.40	2.50	2.85	4.00+
90%	4340H	2.85	4.00+	3.65	3.60	3.85	4.00+
Max. Diameter of Bar That Will Harden at Center to Rockwell C-55							
95%	4147H	0.60	4.00+	1.00	1.20	1.30	4.00+
90%	4150H	1.00	4.00+	1.50	1.60	1.90	4.00+

13, 14), and the average expectancy for center hardness and per cent martensite, grouped according to carbon levels, is given in Table 1. Chemical composition ranges for the AISI steels and their old SAE equivalents are given in Tables 2 to 5. Some NE steels were used for a time, then abandoned. The compositions of the obsolete ones are on record (Ref. 15).

The degree of conformity with the average hardenability expectations of any particular heat is ascertainable by an actual Jominy test, when more precise information is wanted than is provided by the fact that the heat has been supplied as an H steel, so that its Jominy curve is known to be within the band. Heats unpedigreed as to hardenability can likewise be put into their true category by actual Jominy test.

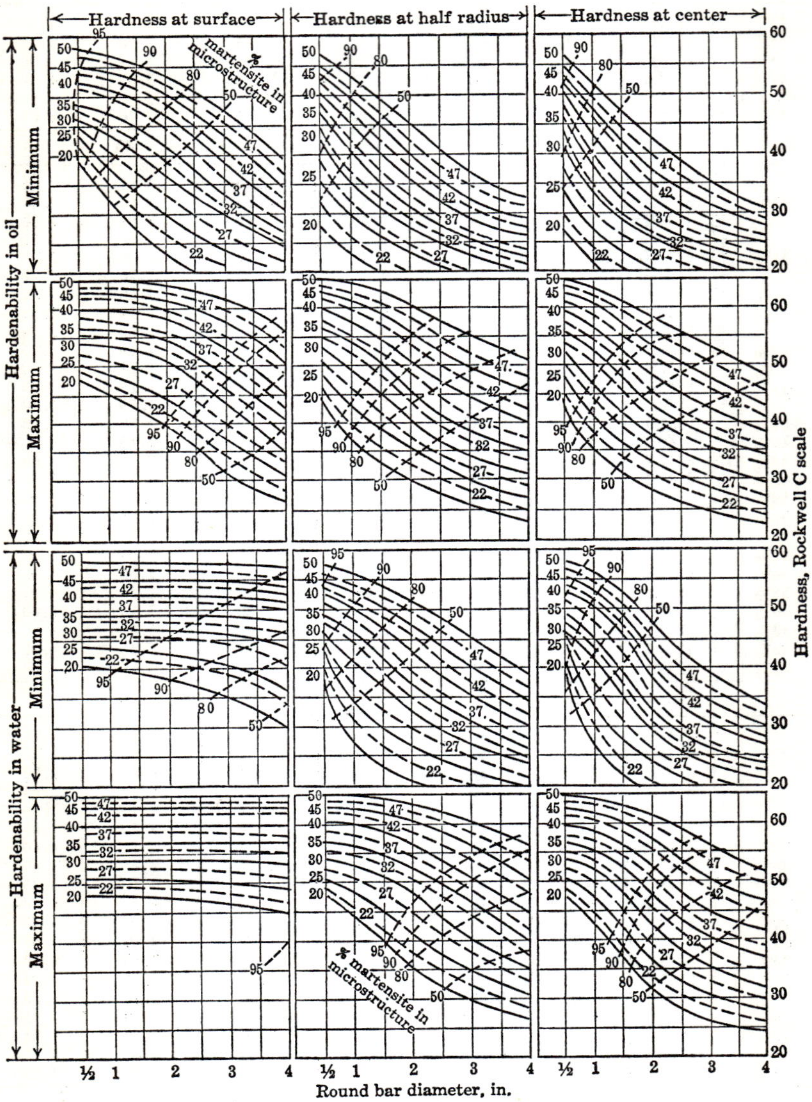

FIG. 14. Hardenability of NE 8600H steels. Mean carbon content is shown on each curve. (Reproduced by permission from a data sheet by A. L. Boegehold in *Metal Progress*, Dec. 1946)

Table 2. Standard Carbon and Free-cutting Steels

Revised July 1, 1946

Basic Open-hearth and Acid Bessemer Carbon Steels

AISI Number *	Semi-finish	Hot-rolled Bars	Cold-finished Bars	Rods	Wire	C	Mn	P †	S †	1946 SAE Number
C 1006	......	......	......	✓	✓	0.08 max	0.25–0.40	0.040	0.050	1006
C 1008	✓	✓	✓	✓	✓	0.10 max	0.30–0.50	0.040	0.050	1008
C 1010	✓	✓	✓	✓	✓	0.08–0.13	0.30–0.60	0.040	0.050	1010
C 1012	✓	✓	......	✓	✓	0.10–0.15	0.30–0.60	0.040	0.050	......
C 1013	......	......	......	✓	✓	0.11–0.16	0.60–0.90	0.040	0.050	......
C 1015	✓	✓	✓	✓	✓	0.13–0.18	0.30–0.60	0.040	0.050	1015
C 1016	✓	✓	✓	✓	✓	0.13–0.18	0.60–0.90	0.040	0.050	X1015 1016
C 1017	✓	✓	......	✓	✓	0.15–0.20	0.30–0.60	0.040	0.050	1017
‡	✓	✓	✓	✓	......	0.15–0.20	0.60–0.90	0.040	0.050	1018
C 1019	✓	✓	✓	✓	✓	0.15–0.20	0.70–1.00	0.040	0.050	1019
C 1020	✓	✓	✓	✓	✓	0.18–0.23	0.30–0.60	0.040	0.050	1020
C 1022	✓	✓	✓	✓	✓	0.18–0.23	0.70–1.00	0.040	0.050	X1020 1022
C 1023	✓	✓	......	......	......	0.20–0.25	0.30–0.60	0.040	0.050	......
C 1025	✓	✓	✓	✓	✓	0.22–0.28	0.30–0.60	0.040	0.050	1025
‡	✓	✓	✓	✓	......	0.22–0.28	0.40–0.60	0.040	0.050	1026
C 1030	✓	✓	✓	✓	✓	0.28–0.34	0.60–0.90	0.040	0.050	1030
‡	✓	✓	✓	✓	......	0.30–0.36	0.60–0.90	0.040	0.050	1033
C 1034	......	......	......	✓	✓	0.32–0.38	0.50–0.80	0.040	0.050	......
C 1035	✓	✓	✓	......	......	0.32–0.38	0.60–0.90	0.040	0.050	1035
C 1038	......	......	......	✓	✓	0.35–0.42	0.60–0.90	0.040	0.050	......
C 1040	✓	✓	✓	✓	✓	0.37–0.44	0.60–0.90	0.040	0.050	1040
C 1041	......	......	......	✓	✓	0.36–0.44	1.35–1.65	0.040	0.050	......
C 1043	✓	✓	......	......	......	0.40–0.47	0.70–1.00	0.040	0.050	......
C 1045	✓	✓	✓	......	......	0.43–0.50	0.60–0.90	0.040	0.050	1045
D 1049	......	......	......	✓	✓	0.43–0.50	0.50–0.80	0.050	0.050	......
C 1050	✓	✓	✓	......	......	0.48–0.55	0.60–0.90	0.040	0.050	1050
C 1051	......	......	......	✓	......	0.45–0.56	0.85–1.15	0.040	0.050	......
‡	✓	✓	✓	✓	......	0.47–0.55	1.20–1.50	0.040	0.050	1052
C 1054	......	......	......	✓	......	0.50–0.60	0.50–0.80	0.040	0.050	......
D 1054	......	......	......	✓	✓	0.50–0.60	0.50–0.80	0.050	0.050	......
C 1055	✓	✓	......	......	......	0.50–0.60	0.60–0.90	0.040	0.050	1055
C 1057	......	......	......	✓	......	0.50–0.61	0.85–1.15	0.040	0.050	......
C 1059	......	......	......	✓	......	0.55–0.65	0.50–0.80	0.040	0.050	......
D 1059	......	......	......	✓	✓	0.55–0.65	0.50–0.80	0.050	0.050	......
C 1060	......	✓	......	✓	......	0.55–0.65	0.60–0.90	0.040	0.050	1060
C 1062	......	......	......	✓	......	0.54–0.65	0.85–1.15	0.040	0.050	1062
C 1064	......	......	......	✓	......	0.60–0.70	0.50–0.80	0.040	0.050	......
D 1064	......	......	......	✓	✓	0.60–0.70	0.50–0.80	0.050	0.050	1064
C 1065	......	✓	......	......	......	0.60–0.70	0.60–0.90	0.040	0.050	......
C 1066	......	......	......	✓	......	0.60–0.71	0.85–1.15	0.040	0.050	X1065 1066
C 1069	......	......	......	✓	......	0.65–0.75	0.40–0.70	0.040	0.050	......
D 1069	......	......	......	✓	✓	0.65–0.75	0.40–0.70	0.050	0.050	......
C 1070	......	✓	......	......	......	0.65–0.75	0.60–0.90	0.040	0.050	1070
‡	✓	✓	✓	✓	......	0.70–0.80	0.50–0.70	0.040	0.050	1074
D 1075	......	......	......	✓	✓	0.70–0.80	0.40–0.70	0.050	0.050	......
C 1075	......	......	......	✓	......	0.70–0.80	0.40–0.70	0.040	0.050	......
C 1078	......	✓	......	✓	......	0.72–0.85	0.30–0.60	0.040	0.050	1078
C 1080	......	✓	......	......	......	0.75–0.88	0.60–0.90	0.040	0.050	1080
C 1085	✓	✓	......	......	......	0.80–0.93	0.70–1.00	0.040	0.050	1085
‡	✓	✓	✓	✓	......	0.82–0.95	0.30–0.50	0.040	0.050	1086
‡	✓	✓	✓	✓	......	0.85–1.00	0.60–0.90	0.040	0.050	1090
C 1095	......	✓	✓	✓	......	0.90–1.05	0.30–0.50	0.040	0.050	1095

(Table continued on p. 2-26)

Table 2. Standard Carbon and Free-cutting Steels—*Continued*

Revised July 1, 1946

Free-cutting Steels

AISI Number *	Semi-finished Products	Bars §	Rods	C	Mn	P †	S †	1946 SAE Number
B 1006	....	......	‖	0.08 max	0.45 max	0.07–0.12	0.060 max	....
B 1010	....	✓	‖	0.13 max	0.30–0.60	0.07–0.12	0.060 max	....
C 1108	....	......	✓	0.08–0.13	0.50–0.80	0.045 max	0.07–0.12	....
C 1109	✓	✓	✓	0.08–0.13	0.60–0.90	0.045 max	0.08–0.13	....
B 1111	....	✓	✓	0.08–0.13	0.70–1.00	0.07–0.12	0.10–0.15	1111
C 1111	....	......	✓	0.08–0.13	0.60–0.90	0.045 max	0.16–0.23	....
B 1112	....	✓	✓	0.08–0.13	0.70–1.00	0.07–0.12	0.16–0.23	1112
C 1112	....	✓	✓	0.10–0.16	1.00–1.30	0.045 max	0.08–0.13	....
B 1113	....	✓	✓	0.08–0.13	0.70–1.00	0.07–0.12	0.24–0.33	X1112 1113
C 1115	✓	✓	✓	0.13–0.18	0.70–1.00	0.045 max	0.08–0.13	1115
C 1116	....	✓	......	0.14–0.20	1.10–1.40	0.045 max	0.16–0.23	....
C 1117	....	✓	✓	0.14–0.20	1.00–1.30	0.045 max	0.08–0.13	X1314 1117
C 1118	✓	✓	✓	0.14–0.20	1.30–1.60	0.045 max	0.08–0.13	X1315 1118
C 1120	✓	✓	✓	0.18–0.23	0.70–1.00	0.045 max	0.08–0.13	....
‡	✓	✓	✓	0.27–0.34	1.35–1.65	0.045 max	0.08–0.13	X1330 1132
C 1137	✓	✓	✓	0.32–0.39	1.35–1.65	0.045 max	0.08–0.13	X1335 1137
C 1141	....	✓	✓	0.37–0.45	1.35–1.65	0.045 max	0.08–0.13	X1340 1141
C 1144	....	✓	......	0.40–0.48	1.35–1.65	0.045 max	0.24–0.33	....
C 1145	✓	✓	......	0.42–0.49	0.70–1.00	0.045 max	0.04–0.07	1145
C 1151	✓	✓	......	0.48–0.55	0.70–1.00	0.045 max	0.08–0.13	1151

Note on silicon content: When silicon is specified in standard basic open-hearth steels, silicon may be ordered only as 0.10% max; 0.10 to 0.20%, or 0.15 to 0.30%. In the case of many grades of basic open-hearth steel, special practice is necessary in order to comply with a specification including silicon. Acid Bessemer steel is not furnished with specified silicon content.

* Code for prefix letters: B is acid Bessemer carbon steel. C is basic open-hearth carbon steel. D is acid open-hearth carbon steel.

† Phosphorus and sulfur are maximum for basic open-hearth and acid Bessemer steels; sulfurized or phosphorized steels are not subject to check analysis for sulfur or phosphorus.

‡ This steel was dropped from the AISI lists of standard steels dated July 1, 1946.

§ Specifications are for both hot-rolled bars and cold-finished bars.

‖ B 1006 and B 1010 are standard basic open-hearth and acid Bessemer steels for rods and wire.

It might be asked why hardenability, since it is related to chemical composition, is not computed from the chemical analyses. As a rough first approximation, it can be, but in the absence of a very complete chemical analysis and full knowledge of grain size and of all details of finishing the heat, including the use of hardenability-conferring agents, some of which are effective when present in mere traces, the method is likely to be woefully inaccurate.

Despite much effort, and many publications, on hardenability calculations they can be dismissed as inadequate, often highly misleading, and not to be compared in value with the direct information from an actual Jominy test. This situation results from the fact that the effect of a given amount of a given alloy is vastly modified by what amounts of what other alloys are present, so that the straight-line relations necessary for mathematical computation do not hold for many commercial compositions.

Low-alloy Steels. The advent of hardenability testing as an inspection tool and of hardenability limits as a means of purchase specification has resulted in the rise of steels of adequate hardenability for a wide variety of uses that can be made more cheaply than former steels of like hardenability. These arose as a wartime expedient for conservation of scarce alloying elements and maximum utilization of the alloying elements in scrap. At that time they were known as the National Emergency or NE steels (Table 4). Their cheapness, utility, and wide commercial acceptance have since put them into the category of AISI numbered steels, and they are no longer thought of as temporary expedients. The 8600 series has become the most-used series among the alloy steels.

The use of these steels has emphasized the importance, in respect both to cost and to behavior, of exactly matching the hardenability of the steel with the size of the section to be hardened. Deficiency in hardenability sacrifices yield strength in the interior of the slack-quenched steel, and excess hardenability invites quench cracking.

For the designer, this means that a part formerly made from a high-alloy steel, different lots of which had different hardenability, may to advantage be made instead from a lower-alloy steel of controlled hardenability. "What used to be used" is not the modern engineering criterion for selection of steels.

Table 3. AISI Standard Alloy Steel Compositions

Revised July 1, 1946

Open-hearth and Electric Furnace Alloy Steels; Blooms, Billets, Slabs, Bars, and Bar-strip

AISI Number *	C	Mn	P Max †	S Max †	Si †	Ni	Cr	Mo	SAE Number
A 1320	0.18–0.23	1.60–1.90	0.040	0.040	0.20–0.35	....	....	....	1320
A 1330	0.28–0.33	1.60–1.90	0.040	0.040	0.20–0.35	....	....	....	1330
A 1335	0.33–0.38	1.60–1.90	0.040	0.040	0.20–0.35	....	....	....	1335
A 1340	0.38–0.43	1.60–1.90	0.040	0.040	0.20–0.35	....	....	....	1340
A 2317	0.15–0.20	0.40–0.60	0.040	0.040	0.20–0.35	3.25–3.75	....	....	{2315 {2317
A 2330	0.28–0.33	0.60–0.80	0.040	0.040	0.20–0.35	3.25–3.75	....	....	2330
A 2335	0.33–0.38	0.60–0.80	0.040	0.040	0.20–0.35	3.25–3.75	....	....	....
A 2340	0.38–0.43	0.70–0.90	0.040	0.040	0.20–0.35	3.25–3.75	....	....	2340
A 2345	0.43–0.48	0.70–0.90	0.040	0.040	0.20–0.35	3.25–3.75	....	....	2345
E 2512	0.09–0.14	0.45–0.60	0.025	0.025	0.20–0.35	4.75–5.25	....	....	....
A 2515	0.12–0.17	0.40–0.60	0.040	0.040	0.20–0.35	4.75–5.25	....	....	2515
E 2517	0.15–0.20	0.45–0.60	0.025	0.025	0.20–0.35	4.75–5.25	....	....	....
A 3115	0.13–0.18	0.40–0.60	0.040	0.040	0.20–0.35	1.10–1.40	0.55–0.75	....	3115
A 3120	0.17–0.22	0.60–0.80	0.040	0.040	0.20–0.35	1.10–1.40	0.55–0.75	....	3120
A 3130	0.28–0.33	0.60–0.80	0.040	0.040	0.20–0.35	1.10–1.40	0.55–0.75	....	3130
A 3135	0.33–0.38	0.60–0.80	0.040	0.040	0.20–0.35	1.10–1.40	0.55–0.75	....	3135
A 3140	0.38–0.43	0.70–0.90	0.040	0.040	0.20–0.35	1.10–1.40	0.55–0.75	....	3140
A 3141	0.38–0.43	0.70–0.90	0.040	0.040	0.20–0.35	1.10–1.40	0.70–0.90	....	{X3140 { 3141
A 3145	0.43–0.48	0.70–0.90	0.040	0.040	0.20–0.35	1.10–1.40	0.70–0.90	....	3145
A 3150	0.48–0.53	0.70–0.90	0.040	0.040	0.20–0.35	1.10–1.40	0.70–0.90	....	3150
(*a*)	0.38–0.45	0.40–0.60	0.040	0.040	0.20–0.35	1.65–2.00	0.90–1.20	....	3240
E 3310	0.08–0.13	0.45–0.60	0.025	0.025	0.20–0.35	3.25–3.75	1.40–1.75	....	{3310 {3312
E 3316	0.14–0.19	0.45–0.60	0.025	0.025	0.20–0.35	3.25–3.75	1.40–1.75	....	....
A 4023	0.20–0.25	0.70–0.90	0.040	0.040	0.20–0.35	....	....	0.20–0.30	4023
A 4024	0.20–0.25	0.70–0.90	0.040	(*b*)	0.20–0.35	....	....	0.20–0.30	....
A 4027	0.25–0.30	0.70–0.90	0.040	0.040	0.20–0.35	....	....	0.20–0.30	4027
A 4028	0.25–0.30	0.70–0.90	0.040	(*b*)	0.20–0.35	....	....	0.20–0.30	....
A 4032	0.30–0.35	0.70–0.90	0.040	0.040	0.20–0.35	....	....	0.20–0.30	4032
A 4037	0.35–0.40	0.70–0.90	0.040	0.040	0.20–0.35	....	....	0.20–0.30	4037
A 4042	0.40–0.45	0.70–0.90	0.040	0.040	0.20–0.35	....	....	0.20–0.30	4042
A 4047	0.45–0.50	0.70–0.90	0.040	0.040	0.20–0.35	....	....	0.20–0.30	4047
A 4063	0.60–0.67	0.75–1.00	0.040	0.040	0.20–0.35	....	....	0.20–0.30	4063
A 4068	0.64–0.72	0.75–1.00	0.040	0.040	0.20–0.35	....	....	0.20–0.30	4068
(*a*)	0.17–0.22	0.70–0.90	0.040	0.040	0.20–0.35	....	0.40–0.60	0.20–0.30	4119
(*a*)	0.23–0.28	0.70–0.90	0.040	0.040	0.20–0.35	....	0.40–0.60	0.20–0.30	4125
A 4130	0.28–0.33	0.40–0.60	0.040	0.040	0.20–0.35	....	0.80–1.10	0.15–0.25	{X4130 { 4130
E 4132	0.30–0.35	0.40–0.60	0.025	0.025	0.20–0.35	....	0.80–1.10	0.18–0.25	....
E 4135	0.33–0.38	0.70–0.90	0.025	0.025	0.20–0.35	....	0.80–1.10	0.18–0.25	....
A 4137	0.35–0.40	0.70–0.90	0.040	0.040	0.20–0.35	....	0.80–1.10	0.15–0.25	4137
E 4137	0.35–0.40	0.70–0.90	0.025	0.025	0.20–0.35	....	0.80–1.10	0.18–0.25	....
A 4140	0.38–0.43	0.75–1.00	0.040	0.040	0.20–0.35	....	0.80–1.10	0.15–0.25	4140

(*Table continued on p. 2-28*)

Table 3. AISI Standard Alloy Steel Compositions—*Continued*

Revised July 1, 1946

Open-hearth and Electric Furnace Alloy Steels; Blooms, Billets, Slabs, Bars, and Bar-strip

AISI Number *	C	Mn	P Max †	S Max †	Si †	Ni	Cr	Mo	SAE Number
A 4142	0.40–0.45	0.75–1.00	0.040	0.040	0.20–0.35	....	0.80–1.10	0.15–0.25	....
A 4145	0.43–0.48	0.75–1.00	0.040	0.040	0.20–0.35	....	0.80–1.10	0.15–0.25	4145
A 4147	0.45–0.50	0.75–1.00	0.040	0.040	0.20–0.35	....	0.80–1.10	0.15–0.25	....
A 4150	0.48–0.53	0.75–1.00	0.040	0.040	0.20–0.35	....	0.80–1.10	0.15–0.25	4150
A 4317	0.15–0.20	0.45–0.65	0.040	0.040	0.20–0.35	1.65–2.00	0.40–0.60	0.20–0.30	....
A 4320	0.17–0.22	0.45–0.65	0.040	0.040	0.20–0.35	1.65–2.00	0.40–0.60	0.20–0.30	4320
E 4337	0.35–0.40	0.60–0.80	0.025	0.025	0.20–0.35	1.65–2.00	0.70–0.90	0.20–0.30	....
A 4340	0.38–0.43	0.60–0.80	0.040	0.040	0.20–0.35	1.65–2.00	0.70–0.90	0.20–0.30	{ X4340 4340
E 4340	0.38–0.43	0.60–0.80	0.025	0.025	0.20–0.35	1.65–2.00	0.70–0.90	0.20–0.30	....
A 4608	0.06–0.11	0.40 max	0.040	0.040	0.25 max	1.40–1.75	....	0.15–0.25	....
A 4615	0.13–0.18	0.45–0.65	0.040	0.040	0.20–0.35	1.65–2.00	....	0.20–0.30	4615
E 4617	0.15–0.20	0.45–0.65	0.025	0.025	0.20–0.35	1.65–2.00	....	0.20–0.27	....
A 4620	0.17–0.22	0.45–0.65	0.040	0.040	0.20–0.35	1.65–2.00	....	0.20–0.30	4620
E 4620	0.17–0.22	0.45–0.65	0.025	0.025	0.20–0.35	1.65–2.00	....	0.20–0.27	....
A 4621	0.18–0.23	0.70–0.90	0.040	0.040	0.20–0.35	1.65–2.00	....	0.20–0.30	....
A 4640	0.38–0.43	0.60–0.80	0.040	0.040	0.20–0.35	1.65–2.00	....	0.20–0.30	4640
E 4640	0.38–0.43	0.60–0.80	0.025	0.025	0.20–0.35	1.65–2.00	....	0.20–0.27	....
A 4812	0.10–0.15	0.40–0.60	0.040	0.040	0.20–0.35	3.25–3.75	....	0.20–0.30	....
A 4815	0.13–0.18	0.40–0.60	0.040	0.040	0.20–0.35	3.25–3.75	....	0.20–0.30	4815
A 4817	0.15–0.20	0.40–0.60	0.040	0.040	0.20–0.35	3.25–3.75	....	0.20–0.30	....
A 4820	0.18–0.23	0.50–0.70	0.040	0.040	0.20–0.35	3.25–3.75	....	0.20–0.30	4820
A 5120	0.17–0.22	0.70–0.90	0.040	0.040	0.20–0.35	....	0.70–0.90	....	5120
A 5130	0.28–0.33	0.70–0.90	0.040	0.040	0.20–0.35	....	0.80–1.10	....	....
A 5140	0.38–0.43	0.70–0.90	0.040	0.040	0.20–0.35	....	0.70–0.90	....	5140
A 5145	0.43–0.48	0.70–0.90	0.040	0.040	0.20–0.35	....	0.70–0.90	....	....
A 5150	0.48–0.53	0.70–0.90	0.040	0.040	0.20–0.35	....	0.70–0.90	....	5150
A 5152	0.48–0.55	0.70–0.90	0.040	0.040	0.20–0.35	....	0.90–1.20	....	....
E 52095	0.95–1.10	0.25–0.45	0.025	0.025	0.20–0.35	....	0.40–0.60	....	50100
E 52098	0.95–1.10	0.25–0.45	0.025	0.025	0.20–0.35	....	0.90–1.15	....	51100
E 52100	0.95–1.10	0.25–0.45	0.025	0.025	0.20–0.35	....	1.20–1.50	....	....
E 52101	0.95–1.10	0.25–0.45	0.025	0.025	0.20–0.35	....	1.30–1.60	....	52100
A 6120	0.17–0.22	0.70–0.90	0.040	0.040	0.20–0.35	....	0.70–0.90	0.10 min V	....
A 6145	0.43–0.48	0.70–0.90	0.040	0.040	0.20–0.35	....	0.80–1.10	0.15 min V	....
A 6150	0.48–0.53	0.70–0.90	0.040	0.040	0.20–0.35	....	0.80–1.10	0.15 min V	6150
E 6150	0.48–0.53	0.70–0.90	0.025	0.025	0.20–0.35	....	0.80–1.10	0.15 min V	....
A 6152	0.48–0.55	0.70–0.90	0.040	0.040	0.20–0.35	....	0.80–1.10	0.10 min V	....
A 8615 to A 8660; A 8720 to A 8750; A 9255 to A 9262, See Table 4									
E 9310	0.08–0.13	0.45–0.65	0.025	0.025	0.20–0.35	3.00–3.50	1.00–1.40	0.08–0.15	....
E 9315	0.13–0.18	0.45–0.65	0.025	0.025	0.20–0.35	3.00–3.50	1.00–1.40	0.08–0.15	....
E 9317	0.15–0.20	0.45–0.65	0.025	0.025	0.20–0.35	3.00–3.50	1.00–1.40	0.08–0.15	....
A 9437 to A 9445; A 9747 and A 9763; A 9840 to A 9850, See Table 4									

NOTES: (*a*) Eliminated from AISI Standards, as revised July 1, 1946. (*b*) Sulfur range: 0.035 to 0.050%.

* Prefix A means basic open-hearth alloy steel; E means electric furnace steel.

† Lowest standard maximum phosphorus or sulfur content for acid open-hearth or acid electric furnace alloy steel is 0.05% each; silicon is 0.15% min.

Table 4. National Emergency Alloy Steels *

Open-hearth Alloy and Electric Furnace Alloy Steels; Blooms, Billets, Slabs, Bars, and Hot-rolled Strip

Designation	C	Mn	Si †	Ni	Cr	Mo
NE8615	0.13–0.18	0.70–0.90	0.20–0.35	0.40–0.70	0.40–0.60	0.15–0.25
NE8617	0.15–0.20	0.70–0.90	0.20–0.35	0.40–0.70	0.40–0.60	0.15–0.25
NE8620	0.18–0.23	0.70–0.90	0.20–0.35	0.40–0.70	0.40–0.60	0.15–0.25
NE8622	0.20–0.25	0.70–0.90	0.20–0.35	0.40–0.70	0.40–0.60	0.15–0.25
NE8625	0.23–0.28	0.70–0.90	0.20–0.35	0.40–0.70	0.40–0.60	0.15–0.25
NE8627	0.25–0.30	0.70–0.90	0.20–0.35	0.40–0.70	0.40–0.60	0.15–0.25
NE8630	0.28–0.33	0.70–0.90	0.20–0.35	0.40–0.70	0.40–0.60	0.15–0.25
NE8632	0.30–0.35	0.70–0.90	0.20–0.35	0.40–0.70	0.40–0.60	0.15–0.25
NE8635	0.33–0.38	0.75–1.00	0.20–0.35	0.40–0.70	0.40–0.60	0.15–0.25
NE8637	0.35–0.40	0.75–1.00	0.20–0.35	0.40–0.70	0.40–0.60	0.15–0.25
NE8640	0.38–0.43	0.75–1.00	0.20–0.35	0.40–0.70	0.40–0.60	0.15–0.25
NE8641 ‡	0.38–0.43	0.75–1.00	0.20–0.35	0.40–0.70	0.40–0.60	0.15–0.25
NE8642	0.40–0.45	0.75–1.00	0.20–0.35	0.40–0.70	0.40–0.60	0.15–0.25
NE8645	0.43–0.48	0.75–1.00	0.20–0.35	0.40–0.70	0.40–0.60	0.15–0.25
NE8647	0.45–0.50	0.75–1.00	0.20–0.35	0.40–0.70	0.40–0.60	0.15–0.25
NE8650	0.48–0.53	0.75–1.00	0.20–0.35	0.40–0.70	0.40–0.60	0.15–0.25
NE8655	0.50–0.60	0.75–1.00	0.20–0.35	0.40–0.70	0.40–0.60	0.15–0.25
NE8660	0.55–0.65	0.75–1.00	0.20–0.35	0.40–0.70	0.40–0.60	0.15–0.25
NE8740	0.38–0.43	0.75–1.00	0.20–0.35	0.40–0.70	0.40–0.60	0.20–0.30
NE8742	0.40–0.45	0.75–1.00	0.20–0.35	0.40–0.70	0.40–0.60	0.20–0.30
NE8745	0.43–0.48	0.75–1.00	0.20–0.35	0.40–0.70	0.40–0.60	0.20–0.30
NE8747	0.45–0.50	0.75–1.00	0.20–0.35	0.40–0.70	0.40–0.60	0.20–0.30
NE9255	0.50–0.60	0.70–0.95	1.80–2.20	§	‖	¶
NE9260	0.55–0.65	0.70–1.00	1.80–2.20	§	‖	¶
NE9261	0.55–0.65	0.75–1.00	1.80–2.20	§	0.10–0.25	¶
NE9262	0.55–0.65	0.75–1.00	1.80–2.20	§	0.25–0.40	¶
NE9437 **	0.35–0.40	0.90–1.20	0.20–0.35	0.30–0.60	0.30–0.50	0.08–0.15
NE9440 **	0.38–0.43	0.90–1.20	0.20–0.35	0.30–0.60	0.30–0.50	0.08–0.15
NE9442 **	0.40–0.45	1.00–1.30	0.20–0.35	0.30–0.60	0.30–0.50	0.08–0.15
NE9445 **	0.43–0.48	1.00–1.30	0.20–0.35	0.30–0.60	0.30–0.50	0.08–0.15
NE9747	0.45–0.50	0.50–0.80	0.20–0.35	0.40–0.70	0.10–0.25	0.15–0.25
NE9840	0.38–0.43	0.70–0.90	0.20–0.35	0.85–1.15	0.70–0.90	0.20–0.30
NE9845	0.43–0.48	0.70–0.90	0.20–0.35	0.85–1.15	0.70–0.90	0.20–0.30
NE9850	0.48–0.53	0.70–0.90	0.20–0.35	0.85–1.15	0.70–0.90	0.20–0.30

* All National Emergency steels are subject to the conditions concerning chemical ranges and phosphorus and sulfur maxima as outlined for standard alloy steels on pages 19 and 20 inclusive in the latest revision of American Iron & Steel Institute's *Steel Product Manual* 10, entitled "Alloy Steels" and to the standard permissible variations from specified ranges as listed on page 27 of that *Manual.* Large sections are subject to modifications of carbon content only to a range of 0.10 (10 points); see page 24 of *Manual* 10.

Phosphorus and Sulfur Limits. When basic electric furnace steel is specified, phosphorus and sulfur limits are to be 0.025% max each; for basic open-hearth steel 0.040% max each; for acid open-hearth or electric steel 0.050% max each. (These elements not used in computing hardenability.)

† For basic steel. Lowest minimum silicon limit to be specified for acid steel is 0.15%.

‡ Sulfur 0.040 to 0.060%.

§ Supposedly none. Residual assume 0.10–0.25.

‖ Ditto. Assumed 0.05–0.10.

¶ Ditto. Assumed 0.00–0.08.

** NE9400 series, as established Aug. 17, 1942, had high silicon (0.40–0.60%), and low nickel and chromium (0.20 to 0.40% each). Silicon, nickel, and chromium were dropped to their present range on Aug. 15, 1943. Also take prefix A.

Table 5. H Steels; Chemical Composition Ranges

Electric Furnace or Open-hearth Bars, Billets, or Blooms

SAE or AISI Designation	Chemical Composition *					
	C	Mn	Si	Ni	Cr	Mo
2512H	0.08–0.15	0.35–0.65	0.20–0.35	4.70–5.30	....	....
2515H	0.11–0.18	0.35–0.65	0.20–0.35	4.70–5.30	....	....
2517H	0.16–0.21	0.35–0.65	0.20–0.35	4.70–5.30	....	....
3310H	0.07–0.14	0.35–0.65	0.20–0.35	3.20–3.80	1.35–1.75	....
3316H	0.13–0.20	0.35–0.65	0.20–0.35	3.20–3.80	1.35–1.75	....
4130H	0.27–0.34	0.35–0.65	0.20–0.35	....	0.80–1.15	0.15–0.25
4132H	0.30–0.37	0.35–0.65	0.20–0.35	....	0.80–1.15	0.15–0.25
4135H	0.32–0.39	0.60–0.95	0.20–0.35	....	0.80–1.15	0.15–0.25
4137H	0.35–0.43	0.60–0.95	0.20–0.35	....	0.80–1.15	0.15–0.25
4140H	0.37–0.45	0.70–1.05	0.20–0.35	....	0.80–1.15	0.15–0.25
4142H	0.40–0.48	0.70–1.05	0.20–0.35	....	0.80–1.15	0.15–0.25
4145H	0.42–0.50	0.70–1.05	0.20–0.35	....	0.80–1.15	0.15–0.25
4147H	0.44–0.52	0.70–1.05	0.20–0.35	....	0.80–1.15	0.15–0.25
4150H	0.46–0.54	0.70–1.05	0.20–0.35	....	0.80–1.15	0.15–0.25
4317H	0.14–0.21	0.40–0.70	0.20–0.35	1.50–2.00	0.35–0.65	0.20–0.30
4320H	0.16–0.23	0.40–0.70	0.20–0.35	1.50–2.00	0.35–0.65	0.20–0.30
4340H	0.37–0.45	0.60–0.95	0.20–0.35	1.50–2.00	0.65–0.95	0.20–0.30
4620H	0.17–0.24	0.40–0.70	0.20–0.35	1.50–2.00	....	0.20–0.30
4640H	0.37–0.45	0.55–0.85	0.20–0.35	1.50–2.00	....	0.20–0.30
4815H	0.12–0.19	0.35–0.65	0.20–0.35	3.20–3.80	....	0.20–0.30
4820H	0.17–0.24	0.45–0.75	0.20–0.35	3.20–3.80	....	0.20–0.30
8620H	0.17–0.24	0.60–0.95	0.20–0.35	0.35–0.75	0.35–0.65	0.15–0.25
8622H	0.20–0.27	0.60–0.95	0.20–0.35	0.35–0.75	0.35–0.65	0.15–0.25
8625H	0.22–0.29	0.60–0.95	0.20–0.35	0.35–0.75	0.35–0.65	0.15–0.25
8627H	0.25–0.32	0.60–0.95	0.20–0.35	0.35–0.75	0.35–0.65	0.15–0.25
8630H	0.27–0.34	0.60–0.95	0.20–0.35	0.35–0.75	0.35–0.65	0.15–0.25
8632H	0.30–0.37	0.60–0.95	0.20–0.35	0.35–0.75	0.35–0.65	0.15–0.25
8635H	0.32–0.39	0.70–1.05	0.20–0.35	0.35–0.75	0.35–0.65	0.15–0.25
8637H	0.35–0.43	0.70–1.05	0.20–0.35	0.35–0.75	0.35–0.65	0.15–0.25
8640H	0.37–0.45	0.70–1.05	0.20–0.35	0.35–0.75	0.35–0.65	0.15–0.25
8642H	0.40–0.48	0.70–1.05	0.20–0.35	0.35–0.75	0.35–0.65	0.15–0.25
8645H	0.42–0.50	0.70–1.05	0.20–0.35	0.35–0.75	0.35–0.65	0.15–0.25
8647H	0.44–0.52	0.70–1.05	0.20–0.35	0.35–0.75	0.35–0.65	0.15–0.25
8650H	0.46–0.54	0.70–1.05	0.20–0.35	0.35–0.75	0.35–0.65	0.15–0.25
8720H	0.17–0.24	0.60–0.95	0.20–0.35	0.35–0.75	0.35–0.65	0.20–0.30
8722H	0.20–0.27	0.60–0.95	0.20–0.35	0.35–0.75	0.35–0.65	0.20–0.30
8725H	0.22–0.29	0.60–0.95	0.20–0.35	0.35–0.75	0.35–0.65	0.20–0.30
8727H	0.25–0.32	0.60–0.95	0.20–0.35	0.35–0.75	0.35–0.65	0.20–0.30
8730H	0.27–0.34	0.60–0.95	0.20–0.35	0.35–0.75	0.35–0.65	0.20–0.30
8732H	0.30–0.37	0.60–0.95	0.20–0.35	0.35–0.75	0.35–0.65	0.20–0.30
8735H	0.32–0.39	0.70–1.05	0.20–0.35	0.35–0.75	0.35–0.65	0.20–0.30
8737H	0.35–0.43	0.70–1.05	0.20–0.35	0.35–0.75	0.35–0.65	0.20–0.30
8740H	0.37–0.45	0.70–1.05	0.20–0.35	0.35–0.75	0.35–0.65	0.20–0.30
8742H	0.40–0.48	0.70–1.05	0.20–0.35	0.35–0.75	0.35–0.65	0.20–0.30
8745H	0.42–0.50	0.70–1.05	0.20–0.35	0.35–0.75	0.35–0.65	0.20–0.30
8747H	0.44–0.52	0.70–1.05	0.20–0.35	0.35–0.75	0.35–0.65	0.20–0.30
8750H	0.46–0.54	0.70–1.05	0.20–0.35	0.35–0.75	0.35–0.65	0.20–0.30
NE9420H	0.17–0.24	0.80–1.15	0.20–0.35	0.25–0.65	0.25–0.55	0.08–0.15
NE9422H	0.20–0.27	0.80–1.15	0.20–0.35	0.25–0.65	0.25–0.55	0.08–0.15
NE9425H	0.22–0.29	0.80–1.15	0.20–0.35	0.25–0.65	0.25–0.55	0.08–0.15
NE9427H	0.25–0.32	0.80–1.15	0.20–0.35	0.25–0.65	0.25–0.55	0.08–0.15
NE9430H	0.27–0.34	0.85–1.25	0.20–0.35	0.25–0.65	0.25–0.55	0.08–0.15
NE9432H	0.30–0.37	0.85–1.25	0.20–0.35	0.25–0.65	0.25–0.55	0.08–0.15
NE9435H	0.32–0.39	0.85–1.25	0.20–0.35	0.25–0.65	0.25–0.55	0.08–0.15
NE9437H	0.35–0.43	0.85–1.25	0.20–0.35	0.25–0.65	0.25–0.55	0.08–0.15
NE9440H	0.37–0.45	0.85–1.25	0.20–0.35	0.25–0.65	0.25–0.55	0.08–0.15
NE9442H	0.40–0.48	0.95–1.35	0.20–0.35	0.25–0.65	0.25–0.55	0.08–0.15
NE9445H	0.42–0.50	0.95–1.35	0.20–0.35	0.25–0.65	0.25–0.55	0.08–0.15
NE9447H	0.44–0.52	1.15–1.55	0.20–0.35	0.25–0.65	0.25–0.55	0.08–0.15
NE9450H	0.46–0.54	1.15–1.55	0.20–0.35	0.25–0.65	0.25–0.55	0.08–0.15

* *Permissible Chemical Variations.* Phosphorus and sulfur in open-hearth steel to be 0.040% maximum each. Phosphorus and sulfur in electric furnace steel to be 0.025% maximum each.

Small quantities of certain elements may be found in alloy steel which are not specified or required. These elements are to be considered as incidental and acceptable to the following maximum amounts: copper 0.35%, nickel 0.25%, chromium 0.20%, molybdenum 0.06%.

The chemical ranges and limits shown are subject to the standard permissible variations for check analysis over the maximum limit or under the minimum limit. These permissible variations are: 0.01 for all ranges of carbon; 0.03 for manganese up to and including 0.90%, and 0.04 for over 0.90 to 1.55% inclusive; 0.005 for either phosphorus or sulfur; 0.02 for silicon; 0.03 for nickel up to and including 1.00%, 0.05 for nickel over 1.00 up to and including 2.00%, and 0.07 over 2.00 up to and including 5.30%; 0.03 for chromium up to 0.90% inclusive, and 0.05 for chromium over 0.90 to 1.75% inclusive; 0.01 for molybdenum up to 0.20% inclusive, 0.02 for molybdenum over 0.20 to 0.30% inclusive.

Quality Conditions. All conditions and quality features, except as detailed above, shall be in accordance with the regulation, shown in the American Iron and Steel Institute's *Steel Products Manual,* Section 10 on alloy steels.

NOTE. NE denotes National Emergency Standard Steel.

4. PRESENTATION OF DATA

LIMITATIONS. Appreciation of the limitations of conventional mechanical tests and of the importance of properties not, or not easily, measurable in quantitative figures, plus appreciation of the similarities among fully quenched, quenched, and tempered steels, leads to abandonment of the effort to express in figures properties not so expressible for direct use in design.

The indirect connotations implied, but not measured, by conventional mechanical tests still have some value if their indirectness and their frequent inaccuracy are kept in mind, and so some such values will still be given. But primary attention will be given to listing groups of ferrous materials available to the designer, with running comment on advantages and disadvantages, whether these can be put in figures or must be only qualitatively described.

CLASSES OF FERROUS MATERIALS FOR ENGINEERING DESIGN. Broad classifications of ferrous materials of construction are *cast* and *wrought*, with further subdivisions as to method of manufacture and intended processing and use; *duplex* materials; and the catch-all class of *special* materials.

5. CAST IRON AND MALLEABLE IRON

The selection of iron-base cast metals is not handicapped by many chemical requirements imposed by official specifications. Classification is on the basis of mechanical properties, the foundryman being left free to use his experience and judgment in adjusting composition. Cast materials in general are free from the directional differences in properties characteristic of most wrought materials. They are size-sensitive and demand suitable initial design to facilitate gating and feeding for soundness and experienced foundry technique to insure soundness. Consultation with patternmaker and foundryman before freezing a design tends to insure sounder castings, lower production costs, and less scrap both in foundry and machine shop.

GRAY CAST IRON (see also Section 19). This metal is easy to cast sound, very readily machinable, not ordinarily considered weldable, and incapable of cold forming. It is evaluated on tensile strength or transverse strength; it has no yield point but displays a curved stress-strain diagram. It is relatively brittle; its approach toward toughness is evaluated by special impact tests on unnotched bars. Hardness is not necessarily a direct reflection of strength.

Composition. Gray iron has a matrix of silicon-bearing iron in which are embedded graphite flakes. The silicon-ferrite matrix may be accompanied by pearlite particles, that is, by iron carbide in layer formation, as in steel, or the matrix may be all pearlite. Hardness rises and machinability decreases as the matrix structure approaches all pearlite. The softer irons have higher carbon, the range being about 3.25 to 3.75%, with silicon at about 1.50 to 2.50%, strong irons near 3% carbon, silicon lower, sometimes down to 1%. The balance among carbon, silicon, manganese, sulfur, phosphorus, and any alloying elements and the use of various inoculating ladle treatments for graphite control depend on the strength desired in the section to be cast, the type of casting, and the sort of matrix wanted. Graphite size and distribution play a very large part. Composition should not be specified but should be left for the foundryman to fit to the particular case in hand. The higher the silicon and carbon, the softer the iron, and the larger the graphite flakes. The foundryman balances the carbon and silicon to suit the casting to be made. Strength depends more on shape and distribution of graphite than on its amount. Composition is not the sole regulator of strength; melting practice and addition of "modifiers" are factors. Structure is altered by variations in rate of freezing. Rapidly chilled corners may be white, very hard, and brittle.

Tests. The strength of a given composition varies with the section cast. Transverse and tensile tests should be made on specimens of like section to that of the casting. Conventional sizes are 0.875, 1.2, and 2.0 in. diameter. Transverse tests are often used for foundry control; tensile, for design data.

Mechanical Properties. Interrelations among the various mechanical properties have been discussed in detail (Ref. 16). Modulus of elasticity varies with strength, weakest irons being about 12 million psi, strongest about 19 million. Damping capacity, ability to absorb energy without transmitting it, is high, decreasing with increased strength. Torsional strength runs 1.2 to 1.4 times tensile strength, compressive strength is about $3\,^1/_2$ to $4\,^1/_2$ times tensile strength, making gray iron highly useful against compressive loading. Machinability decreases as strength increases but remains relatively high.

Endurance limit, unnotched, runs 1/3 to 1/2 tensile strength. Notched endurance is equal to unnotched in weakest irons, only moderately reduced in strong irons. Even under corrosion fatigue, in fresh water at least, notched specimens have relatively long life for a ferrous material of such strength. (See also Section 19, Article 2, Iron and Steel Casting.)

Specifications. Commercial gray irons are classified according to minimum tensile strength, pounds per square inch. ASTM specifications A48-46 cover the following grades: 20,000, 25,000, 30,000, 35,000, 40,000, 50,000, and 60,000 tensile irons. (See Section 19, Article 2.) Specifications A159-44T for automotive castings cover 20,000, 30,000, 35,000, 40,000, and 45,000 tensile irons. (See Section 19, Article 2.)

Specific gravity rises from about 7 to 7.7 as strength increases. Irons up to and somewhat above 40,000 psi tensile are widely available commercially. Higher-strength irons are more difficult to produce, especially in large sizes, and less easily controlled for soundness. Irons for some special purposes can be supplied if the use is clearly described.

Temperature Effects. The properties of gray iron are not widely different at subnormal temperatures or at elevated temperatures, up to around 600 F, but above that ordinary cast iron is subject to oxidation "growth" and consequent deterioration. Slight chromium additions stabilize the carbide and avoid growth to temperatures where oxidation resistance is insufficient for service. The linear coefficient of thermal expansion of cast iron is lower than that of steel, varying from around 0.000005 to around 0.00001 per degree F with class of iron and temperature range. The thermal conductivity likewise varies but is of the same order as that of most steels.

Wear resistance of gray iron is a function of the type of service and of the structure of the iron.

Alloying. The 40,000 psi tensile irons may be, and higher-strength irons, especially for large sections, usually are, alloyed. The alloying elements are much the same as those used in steel, and, as in steel, various elements and various combinations may be used interchangeably. Most cast irons are made in the cupola; the very-high-strength and the alloyed irons, which normally carry less carbon, are often made in electric or air furnaces.

Heat treatment of cast iron is not carried out to improve properties as a structural material but may be useful in respect to wear resistance. Surface hardening by flame or induction can be applied to produce a hard, wear-resisting surface.

Stress Relief. As in any casting, internal stresses may be present which, if not relieved, may cause warpage on machining. In earlier days iron castings were allowed to "season" for long periods at atmospheric temperature, which provided some slow and slight stress relief. A stress relief anneal at 800 to 1050 F for a few hours, followed by slow cooling, relieves these stresses without affecting mechanical properties.

Casting. Most gray iron castings are made in sand. Pipe is often made by centrifugal casting, which may result in a thin outer skin of chilled iron, which is graphitized and softened by heat treatment. Gray iron is seldom cast to extremely precise dimensions, since it is so easily machined. However, small castings can be made to close dimensions by casting in permanent metal molds, using a soft iron, and coating the molds to avoid chilling the edges and surfaces.

NODULAR CAST IRON. A new material of construction which has interesting potentialities is a variety of cast iron with much the same properties as pearlitic malleable (p. 2-34). By treatment of the melt with *cerium*, as in the product developed in England by the British Cast Iron Research Association, or with *magnesium* (added as a rather low magnesium content alloy with copper or nickel), as in that developed in the United States, the iron can be made to freeze with its graphite in the nodular form familiar in malleable iron, without requiring the more or less lengthy heat treatment required in the various varieties of malleable.

The ductility and toughness obtained with the nodular rather than the flake graphite structure give the as-cast nodular iron a combination of properties such as 60,000 psi yield, 100,000 psi tensile, and around 5% elongation, the properties, of course, shifting with section size. By heating for one hour at 1650 F, then slowly cooling, the properties become of the order of 50,000 psi yield, 75,000 psi tensile, and 20% elongation (Refs. 50 and 51).

While it remains for simulated service tests and actual use to demonstrate just what such ductilities imply, as to service toughness, the structure and properties are so similar to some of the malleables as to imply suitability for uses heretofore open to malleable but closed to ordinary cast iron. A considerable amount of experience and know-how will be required to insure obtaining the correct structure, so some time may elapse before the material becomes widely available.

WHITE AND CHILLED IRON. Cast iron of suitable composition may be gray, i.e., with a gray fracture from the presence of much graphite, when cooled slowly during freezing, but may be white, with fracture practically free from graphite, when more rapidly cooled, as when cast against a metal chill which abstracts heat more rapidly than a sand mold. The white, chilled face is very hard and brittle but wear resistant. It contains internal stress which needs to be relieved by heat treatment.

Wearing portions of a gray cast-iron camshaft are often chilled on their working faces, but little structural use is made of castings that are white throughout, because they are too brittle. However, sand-blast nozzles and castings for a few analagous uses are exceptions.

White iron upon wearing portions, such as plow points and railway car wheels, supported by less brittle material, meets a need in services where the chilled layer is not subjected to tensile stress. Such castings, notably car wheels with chilled running faces, are subjected to a heat treatment reminiscent of a "full anneal" of steel, to alter structure somewhat and relieve internal stress.

Inasmuch as the foundry adjusts the chemical composition of castings with chilled surfaces to meet the particular conditions, the user should not specify composition.

Completely white iron, chilled throughout, is the starting point for the production of malleable cast iron.

MALLEABLE CAST IRON. By long heat treatment of white iron at sufficiently high temperature, graphite, usually called "temper carbon," is deposited in tiny, rounded clumps, or nodules, much smaller and more uniformly distributed than the graphite of gray iron, leaving the matrix silicon-containing ferrite (ordinarily around 1% silicon) almost carbon free. The carbon content of malleable ranges from about 1.75 up to around 3%, i.e., lower than in gray iron. The silicon hardens the ferrite and makes it less gummy on machining than low-silicon ferrite. The temper carbon nodules make the chips break up and prevent welding of the chip to the tool. Fully malleabilized iron is the most easily machined of all ferrous products. (See also Section 19, Articles 9, 10, 11, and 12.)

Specification. Malleable is an exception among ferrous products in that increased strength is usually accompanied by increased rather than decreased ductility and toughness. ASTM specification A47-33 covers two grades, and A197-39 covers cupola malleable.

ASTM No.	Tensile Strength, psi min	Yield Strength, psi min	Elongation, in 2 in., % min	Specific Gravity
32510	50,000	32,500	10	7.25–7.35
35018	53,000	35,000	18	7.20–7.45
Cupola	40,000	30,000	5	7.10–7.20

Higher strengths and elongations are frequently produced in all these. For example, some manufacturers will supply 33,000 psi minimum yield, 12 1/2% minimum elongation. The two higher-strength grades are produced in the open hearth, air furnace, or electric furnace, since to secure these properties the carbon content needs to be lower than is readily obtainable from the cupola.

The grade with 10% elongation has usually been considered adequate for automotive uses. The 5% elongation grade, cupola malleable, is widely used for pipe fittings. "Toughness" as measured by Charpy keyhole notch specimens runs about 7, 6 1/2, and 5 ft-lb, respectively, for 32510, 35018, and cupola grades, but wider differences and more toughness for the first two grades than the figures would ordinarily be taken to indicate are shown by the "wedge curl" test. A bend test also demonstrates toughness and the ability of malleable to be cold formed. As Rehder (Ref. 17) points out, elongation in the tensile test is no sure criterion of bending toughness.

Special-alloy malleable iron can be produced with 60,000 psi tensile, 40,000 psi yield, 18% elongation; 65,000, 54,000, and 14% have been reported for precipitation-hardened copper-alloyed malleable.

The endurance limit, determined on unnotched specimens, runs close to one-half the tensile strength. Scanty notched endurance data indicate rather better behavior for malleable than for cast steel of equal yield strength. The elastic modulus runs from 20 to 26 million psi.

Temperature Effects. Up to around 600 F the elevated temperature properties do not depart greatly from those at room temperature, and good behavior in creep up to 775 F has been reported. At low temperature, notched-bar and wedge-curl values may be slightly depreciated or may drop not far below room-temperature values. When information on low-temperature behavior is wanted, the particular lot to be evaluated should be tested.

Malleable iron is sometimes zinc coated by the hot-dip process, and the necessary temperature may or may not produce some embrittlement. Means are known for avoiding this, as by prior heat treatment. Cupola malleable is not prone to galvanizing embrittlement.

Malleable is not practically weldable, as the welding temperature changes the structure and a complete reheat treatment would be required. Low-temperature brazing may be employed for joining. It can be flame or induction hardened on the surface.

Malleable does not display the higher compressive than tensile strength displayed by gray cast iron. It does not have the tendency to "growth" at elevated temperature shown by some gray cast irons. Magnetic behavior is superior to that of gray iron.

PEARLITIC MALLEABLE. The heat treatment by which white iron is converted from its initial brittle structure to its final fully malleablized ductile structure is a long one, requiring some 3 to 6 days for completion. The cycle can be shortened, and various stages of approach toward full malleablization reached.

Incomplete malleablization produces more or less pearlite in the matrix accompanied by less or more deposition of temper carbon. A range of structures stronger and less ductile than regular malleable are thus obtained. (Alloying with elements that stabilize carbide can produce similar structures in the regular, long, malleablizing treatment.) These stronger structures machine less readily than regular malleable but nevertheless more readily than steel of like yield strength.

Specification. Neither chemical composition nor specific malleablizing heat treatment is specified. ASTM specification A220-44T gives the data in the accompanying tabulation.

Number	Minima: Tensile Strength, psi	Minima: Yield Strength, psi	Elongation in 2 in., %
43,010	60,000	43,000	10
50,007	65,000	50,000	7
60,005	75,000	60,000	5
70,003	90,000	70,000	3

Other gradations are commercially produced. It will be noted that, unlike regular malleable, but like most classes of ferrous alloys, pearlitic shows ductility decreasing with increase in strength. All these grades can be produced from the same metal by varying the thermal treatment. Like properties can be secured by starting with regular malleable, completely malleablized, and reheating it to take some of the carbon into solution in austenite, then controlling the graphite distribution and the matrix structure by adjusting the cooling rate.

Application. Pearlitic malleable is a material corresponding in many ways to the often-expressed desire for a "strong cast iron with toughness." Designers in general have not given this material the attention it deserves, probably being influenced by traditional fear of relatively low ductility. The automotive industry, however, has made large use of it, pistons and camshafts being some examples. As with other types of malleable, hard, wear-resistant surfaces can be produced by flame or induction hardening. One automotive producer calls its pearlitic malleable "graphite steel" or Armasteel (steel produced by arrested malleablization), but this does not seem good terminology since the accepted definition of steel calls for not over 1.80% carbon and specifies that it be malleable as cast, which is not true of this material.

High fatigue resistance and resistance to shock at subzero temperatures are attributed (Ref. 18) to pearlitic malleable. Laboratory tests (Ref. 19), however, indicate that not all pearlitic malleables display the latter characteristic.

Alloying elements, especially around 1% copper, with or without a small amount of molybdenum, are frequently included in pearlitic malleable as aids in securing the desired structure when a relatively rapid annealing cycle is used. The mechanical properties are controlled by the structure. Commercially available pearlitics are discussed by Lorig (Ref. 19). They usually contain around 2.35% carbon, 1% silicon.

FORD ALLOYS. So-called "copper silicon steel" alloys are cast alloys of lower carbon content (around 1 1/2%), about the same silicon (1%), and more copper (around 1 3/4%), cast white like regular and pearlitic malleable, but malleablized in still shorter cycles. Also misnamed "steel," they were developed by the Ford Motor Company. Such an alloy, which is used for brake drums, can be made to give 85,000 psi tensile, 70,000 psi yield, and 6% elongation. The addition of around 1/2% chromium and the use of a particular time-temperature annealing cycle provide properties of around 105,000 psi tensile, 92,000 psi yield, 1 3/4% elongation, and 2 1/4% reduction of area. This alloy has been used in millions of Ford crankshafts. Specimens taken from commercial crankshafts may run about 80,000 psi tensile with no measurable elongation or reduction of area, and only 1 to 2 ft-lb Charpy impact. The endurance limit on polished bars is as high as or higher than that of steel of equal tensile strength, and the notched endurance limit is higher (Ref. 20).

AVAILABLE RANGES OF STRENGTH. Gray and malleable iron castings are available as materials of construction to carry static tensile loads from 20,000 psi in soft gray iron to 70,000 psi in special malleable iron, with excellent fatigue and notched-fatigue values, and with appreciable (for even gray iron is not exactly "glass brittle") to considerable toughness, toughness which experience has proved adequate for many types of severe service. Their compositions and structures are matched to the section and the service, not laid down by arbitrary specifications. They are all relatively high in carbon and silicon.

6. CAST STEELS

Cast steel is, ordinarily, much lower in carbon, running from 0.20 to 0.35% with up to 0.50% silicon, approximately, and very often with manganese at quite a high level, up to 1% and more. Such a composition, and those with higher manganese or other alloy additions, are amenable to the simple heat treatment of normalizing, or to quenching and tempering. Raw, unnormalized castings of low carbon content and fully annealed castings of higher carbon content were employed in very early days but are now outmoded, except that for castings of very large section full annealing is often still applied. The great bulk of steel castings are normalized, with or without a tempering treatment after normalizing.

SPECIFICATION. Foundrymen control the chemical composition, including sulfur and phosphorus, for self-protection against high scrap losses, and, though specifications restrict these elements to 0.06% sulfur and 0.05% phosphorus, the stipulations are scarcely necessary. Cast steels are classified, and usually specified, on the primary basis of mechanical properties. Annealing may be substituted for normalizing if the mechanical properties are met.

ASTM specification A27-46T covers mild to medium strength (30,000 to 36,000 psi yield) carbon steel castings for general application, for those grades which are required to be weldable without special precautions. (See Section 19, Article 5, Table 10.)

By agreement between buyer and seller a limitation that not more than 1% total of copper, nickel, chromium, molybdenum, and tungsten (with individual limitations on each) may be imposed, to deal with a case of unusual amounts of residual metals coming from scrap. This is an indirect and awkward way of limiting excess hardenability which might lead to cracking unless special precautions are taken in welding. Such requirements could be more simply stated in terms of a Jominy test.

The steels may be made by any usual commercial melting process. The annealing, normalizing, or quenching and tempering heat treatment can be as the producer desires. For only one grade (0.25% carbon maximum) is some form of heat treatment not required. Other grades are normalized or otherwise heat treated. Chemical and physical test requirements of this specification are listed in Table 10 of Article 5, Section 19.

Ductility. The specified ductilities follow naturally in properly killed steels having compositions and treatments producing these yield strengths. It is that fact rather than any requirement of ordinary service for such large ductilities that has led to their inclusion. A Navy specification corresponding to Class 70-36 but permitting a somewhat lower substitution of manganese for carbon calls for the values tabulated. Higher properties are often secured such as 70,000 to 75,000 psi tensile, 40,000 to 45,000 psi yield, 30 to 35% elongation, 50 to 55% R.A.

Size	Maximum Carbon, %	Tensile, psi	Yield, psi	Elongation, %	R.A., %	Cold Bend, degrees
Castings weighing not over 5000 lb	.35	70,000	40,000	24	30	120
Those over 5000 lb	.35	70,000	35,000	20	30	90

A common terminology and specification for cast steel was "Grade B," which was supposed to contain not over 0.85% manganese, carbon not specified, and had, as specified minima, 70,000 psi tensile, 38,000 psi yield, 24% elongation, 36% R.A. The ASTM specification, which included Grade B, has been superseded by A27-46T.

Grain-size control and double normalizing treatments favor higher properties. The endurance limit runs a little under half the tensile strength on smooth specimens; notched fatigue values are around one-fourth. These plain carbon steel castings cover the yield strength range of 30,000 to 35,000 psi, possibly up to 40,000 psi, with rather foolproof weldability, acceptable machinability, and considerable ability for cold deformation.

Steel castings for constructional use are, however, more frequently in the low-alloy than in the plain carbon class. ASTM specification A148-46T *High-Strength Steel Castings* for structural purposes permits any usual melting method, requires some type of heat treatment which the producer may choose, sets no chemical limits beyond the conventional 0.06% sulfur, 0.05% phosphorus, and classifies the steels solely on mechanical properties, for which minima are specified as in Table 11, Article 5, Section 19.

The polished- and notched-bar fatigue behavior can be approximated from Fig. 2. Behavior of Grade B steel in single-blow notched-bar impact tests at room and lower temperatures has been described by Sims and Dahle (Ref. 21), who show that the deoxidation practice has a major influence. When such behavior is to be appraised, tests need to be made on each heat, unless it is known that proper deoxidation practice will always be applied.

Foundry Practice. The foundryman is free to adjust carbon content, grain-size control, and alloy content to the heat treatment chosen. He will ordinarily prefer so to alloy as to secure the specified properties by normalizing or normalizing and drawing, and for the 50,000 and 60,000 psi yield strength classes the foundryman usually chooses "intermediate manganese steels" with around 0.33% carbon and 1 1/2 and 1 3/4% manganese respectively. If other elements are present, from scrap, which help make austenite sluggish and thus allow normalizing to a fine, strong structure, he reduces the manganese accordingly. However, normal manganese and a wide variety of other elements, singly or with small amounts of several, give equivalent results.

Ductility. The 90-60 grade was earlier known as "Grade C," for which the specification, now replaced by H148-46T, required 90,000 psi tensile, 60,000 psi yield, 22% elongation, 45% R.A., instead of the 20% and 40% now required. That is, a steel with 21% and 42%, rejected in purchasing on the old specification becomes acceptable on the present one. From this example it is evident that ductility specifications are not really sacred. A good many heats of 60,000 psi yield will show 26% elongation and 60% R.A. It would take a lot of proving to demonstrate that such heats are any better for service than those with 20 and 40%, especially when it is noted that the ductility required for grades 80-40 and 105-85, admitted to be respectable constructional materials, is still lower. The ductilities stated are merely those that accompany ordinary compositions, heat treated in the ordinary way, to a given yield strength.

Yield Strength. The grades of higher yield strength are obtainable by quenching and tempering of low-alloy compositions, the amount of alloy increasing as higher yield strengths are desired. There are "hard" grades of carbon and of alloy cast steels in which carbon may go up, but rarely above 0.50%, when wear resistance is wanted and weldability, machinability, and toughness are relatively unimportant. The higher carbon content is ordinarily steered clear of, especially when the steel is to be quenched.

Cast carbon and alloy steels can be quenched and tempered to yield strengths closely corresponding to those obtainable in the same steels when wrought, but, since castings are often complex and unsymmetrical with thick and thin portions, it has been considered hazardous to quench them for fear of quenching cracks. However, experience with quenching of huge castings for Army tanks, which were alternatively used with parts fabricated from plate by riveting or welding, proves that, with proper design and skilled control of heat treatment, supposedly unquenchable castings behave nicely.

At equal yield strengths, castings will not show as high ductility as wrought products, rolled or forged, tested in the direction of working, but this is compensated for or more than compensated for by the casting having like ductilities in all directions.

Size Effects. It is necessary to remember that the mechanical properties of steel castings are conventionally determined on separately cast specimens 1 in. by 1 in. in cross section, very adequately fed by a huge riser. There is a strong size effect in steel castings; very thin sections will have higher properties, heavy ones, in their interiors, appreciably lower ones. Center unsoundness may be marked, especially in large, flat castings, and, of course, when quenching and tempering are applied, if hardenability is insufficient to avoid slack quenching, the tiny specimen, quenched and tempered, is as unrepresentative of the interior of the casting as a small wrought-steel specimen.

CASTING DESIGN. Honest sampling of a casting from the actual heat of steel, from the full section, with its actual heat treatment, is imperative for sound design. Even such sampling, unless it hits an unsound spot, does not insure that the casting, as a structure, may not be unreliable because of gas holes, shrinkage cavities, or other interior defects. These defects are avoidable, with decent foundry practice, if the design allows proper feeding and escape of gas. A heavy section cannot be fed through a light one that freezes first, nor can gas bubbles escape through the frozen portion. Incompletely fed heavy sections inevitably contain shrinkage cavities or spongy metal. Chills on heavy sections help to approach more uniform freezing but are seldom completely effective.

The designer must do his part by straining every nerve to avoid unequal sections, to provide good fillets, to avoid sharp angles at which cracks may start from cooling stresses, and to allow the patternmaker so to construct the pattern that the foundryman can use proper gates and risers. He should check with both of them in time to alter the design to meet their needs. If the requirements for making a sound casting are not met in a proposed design, and the design cannot be modified to meet them, then the structure should be subdivided into parts that can be cast.

These elementary precautions are needed in all the cast ferrous products, whether gray, malleable, or pearlitic malleable irons, or cast steel.

These precautions for getting sound castings and detailed discussion of many features of cast ferrous metals, as well as the classes of products for which each type has been successfully applied, are given in the *Cast Metals Handbook*, published by the American Foundrymen's Association, 222 West Adams Street, Chicago, Illinois.

One final feature about castings should be mentioned. If a small amount of a special tailor-made iron or steel, not readily available in the market, is needed for a particular, special use, it can easily be supplied as a casting, or as stock for forging, by foundries having suitable metallurgical know-how and small electric furnaces. Such foundries can render service that steel mills or foundries equipped only with large furnaces cannot.

A relatively unexplored possibility for saving is by use of a casting, made nearly to final size, as a forging blank, to reduce the number of expensive forging dies and operations.

7. WROUGHT IRONS AND PLAIN CARBON STEELS

IRONS AND STEELS NOT INTENDED FOR HEAT TREATMENT. Many objects may be fabricated by cold bending, deep drawing, spinning, or a variety of other ways of securing plastic deformation, into final shape, or into subshapes thereafter joined mechanically or by welding, brazing, soldering, etc. They may be painted, metal-coated, or coated with vitreous enamel. Formability is often the prime requisite, the final object being designed more for stiffness than for strength, and that stiffness being normally accompanied by ample strength for the conditions of service, even when made from the weakest of steels. Very thin stock can be more easily deformed than thick and, if properly protected against corrosion, often serves. When thin stock does not supply the required stiffness or strength, heavier stock, still of weak steel, will often be chosen, unless weight considerations are important.

Plastic deformation hardens and embrittles iron or steel, and at some stage in such deformation a condition may be reached where any further attempt at deformation would produce a crack. In such a case, deformation is stopped short of that stage, and the steel is returned to approximately its original soft and ductile condition either by a "process anneal," carried out at a temperature at which the steel matrix remains ferritic, or, more rarely, by a "full anneal" carried out by heating far enough for the steel to become austenitic. With stronger grades of steel or with sections of weak ones so heavy that they will stand too little cold deformation without cracking, the steel may have to be heated and deformed while still hot.

Processing. Some purposes require nothing but formability; others may superimpose requirements for weldability, suitability for vitreous enameling, or the like. Irons and steels very low in carbon meet formability, weldability, and various other requirements more readily than steel with higher carbon. Very-low-carbon steels are more cheaply produced rimmed or semi-killed instead of fully killed. The still lower carbon exterior of rimmed steels has the advantage of a better surface for deep drawing, but such steels, unless properly killed and their grain size controlled, are subject to "stretcher straining," to surface markings, and to a precipitation-hardening type of "aging embrittlement" by which a steel, easily formable when freshly rolled, cracks during forming after this embrittlement has progressed with time.

Attempts to specify utility in indirect terms of chemical composition and conventional mechanical properties fall far short of adequate evaluation of the thin stock under discussion. Such materials, however, are in wide demand and are in quantity production by all steel mills, which, consequently, have evolved processing methods resulting in suitability for specific types of further processing by the purchaser, though the degree of suitability may be unmeasurable save by putting the material through this actual further processing. Yet the supplier, fully informed as to the use to be made, can often correlate his own practice with good and bad behavior in the user's hands.

Unless the user has similar knowledge and can devise some simple test proved to be applicable to his own case, he does better to order in terms of the fully described use and final processing methods than to resort to published specifications, if such exist, or to try to build up a specification, assumed to be applicable, on the basis of conventional tests.

THIN STOCK. Thus there is no virtue, and would be considerable danger, in a handbook listing of properties of low-carbon steels in thin stock which might, by inference, be interpreted as suitable data on which to build a purchase specification. Purchase specifications exist for thin stock where mechanical properties come into consideration and structural purposes are intended, but they do not purport to indicate directly anything about suitability for processing, and specific statements are contained in ASTM Specifications to this effect.

Rolling Practice. In earlier practice, thin gages were hot rolled singly down to, say, 1/16 in. thick, then pack rolled hot, and finally box annealed for a long time at around 1200 F. More economical production first hot rolls as before. Further reduction, however, is by cold rolling in continuous mills, then either a long box annealing of coils or a short normalizing heat treatment at about 1700 F in a continuous furnace is employed for softening prior to difficult cold forming. The cold-rolled strip, hard from its cold rolling, may be process annealed for times and temperatures which only partly soften it, leaving it stronger but less amenable to cold forming by the user.

ASTM Specifications A246-44T (for material under 0.05 in. thick) and A245-44T (for 0.05 in. to 0.25 in.) cover open-hearth and acid Bessemer steels and set the following chemical maxima:

	Carbon, %	Phosphorus, %	Sulfur, %
Open-hearth	.25	.06	.05
Bessemer	.11	.11	.06

When "copper-bearing" steel is ordered, for resistance to atmospheric corrosion, 0.20% minimum copper is required.

For material under 0.05 in. the mechanical requirements are:

	Tensile	Yield	Elongation in 2 in., %	Bend, without Cracking
Commercial grade	*	*	*	Flat in any direction, then doubled flat in opposite direction
Grade A	45,000	25,000	25	Flat in any direction
Grade B	50,000	30,000	23	180° in any direction over its own thickness
Grade C	52,000	33,000	18 †	In longitudinal direction only, over 1 1/2 times its thickness

* Commercial grade is subject to bend test only.

† Sliding scale, lower as thickness decreases.

For the heavier stock, up to 1/4 in. thick, the requirements are:

	Tensile	Yield	Elongation in 2 in., %	Elongation in 8 in., %	Bend, without Cracking
Commercial grade	*	*	*	*	Requirements for each grade same as for thin stock, above
Grade A	48,000	25,000	23	29	
Grade B	52,000	30,000	20	25	
Grade C	55,000	33,000	18	22	

* Commercial grade is subject to bend test only.

Carbon will seldom be at the 0.25% maximum in commercially available material. The lower carbon in Bessemer reflects the strengthening effect of the phosphorus content of Bessemer, requiring that carbon be lower than in open-hearth in order to retain the desired formability. All tensile specimens for both thickness ranges are taken in the longitudinal direction.

HEAVIER STRUCTURAL STEEL (HOT-ROLLED). A blanket specification for ordinary hot-rolled plates, shapes, and bars is ASTM A7-46. They are to contain 0.20% copper when ordered copper-bearing. Material 7/16 in. and under may be Bessemer or open-hearth or electric, heavier stock only the last two.

Carbon limits are not specified, but carbon content is to be reported to the purchaser. In Bessemer, phosphorus is 0.10% maximum, no sulfur limit stated. Phosphorus in acid open-hearth or electric is 0.06%, in basic 0.04% maximum. Sulfur is 0.05% maximum. No limitation is stated as to killing; rimmed, semikilled, or killed are all permissible, nor are grain-size limitations set.

Strength requirements are: tensile 60,000 to 72,000 psi (upper limit set in interest of formability); yield, half the tensile strength but not below 33,000 psi; elongation in 2 in., 22% minimum; elongation in 8 in., 1 1/2 million divided by tensile strength, this requirement being reduced for material over 3/4 in. or under 5/16 in. thick.

Cold-bend requirements for plates, shapes, and bars are 180° without cracking over a pin whose diameter bears the following relation to thickness of material:

Thickness, in.	Ratio of Pin Diameter to Thickness
3/4 and under	1/2
3/4 to 1	1
Over 1 to 1 1/2	1 1/2
Over 1 1/2 to 2	2 1/2
Over 2	3

All this sums up to the fact that commercial hot-rolled carbon steel has low yield strength, which ranges from 25,000 to 36,000 psi. Ductility and formability, as indicated by elongation and bend, vary with thickness, the usual expectations as to what degree of these goes along with the yield strength being expressed by the specification figures.

INGOT AND ENAMELING IRONS. Implied in some ASTM specifications, a low-carbon rimmed material, usually under 0.04%, called ingot iron is commercially produced. A typical analysis of this material is 0.015% carbon, 0.025% manganese, 0.005% phosphorus, 0.025% sulfur, and trace of silicon.

Ingot iron * is a commercially pure iron made by the basic open-hearth process. Tensile strength ranges from 38,000 to 48,000 psi, and yield point from 18,000 to 32,000 psi, depending on the processing, with elongation as high as 48% in 2 in. for the dead soft material. Endurance limit in reversed bending is 26,000 psi for annealed rods or plates, impact is 90 ft-lb (longitudinal) and 56.3 ft-lb (transverse) on the Izod scale for normalized material, and Brinell hardness ranges from 67 to 220. Magnetic and electrical properties of ingot iron make it especially suitable for direct-current applications. If ingot iron for such uses is processed by any form of mechanical working the magnetic properties, except permeability at high inductions, are impaired and must be restored by suitable annealing. The relatively low electrical resistance of ingot iron as compared with steel has led to its use in telegraph wire and bond wire for rail signal circuits and as a third rail for electric railways. High magnetic permeability and low retentivity have stimulated its use for a wide range of magnetic applications. In sheet form ingot iron is extensively used for porcelain enameling.

Because of its high ductility ingot iron can be severely cold formed. In hot working there is a critical range from 1580 to 1920 F, due to low manganese content, which must be avoided. The high ductility and extreme toughness make it necessary to employ special tool shapes in cutting (see Section 23). Both plates and welding wire of ingot iron possess good characteristics for gas, arc, and resistance welding. Slightly higher heat is used than for steel because of the higher melting point of ingot iron. Rivets of ingot iron up to 3/8 in. in diameter may be driven cold.

ASTM specification A129-46, Open Hearth Iron Plates of Flange Quality, covers ingot iron under the designation Type A, for which tensile strength is given as 44,000 psi, yield point 27,500 psi, and elongation 25% in 8 in. ASTM specifications A178-46, Electric Resistance Welded Steel and Open Hearth Iron Boiler Tubes, and A83-46, Lap Welded and Seamless Steel and Lap Welded Iron Boiler Tubes, cover ingot iron under the designation Type B, Open Hearth Iron. (See also Refs. 53 and 54.)

Special enameling irons are also commercially available. One of them carries a high titanium content, sufficient to combine the already low carbon content as titanium carbide instead of iron carbide, leaving the matrix practically pure ferrite with inert titanium carbide inclusions, the idea being to avoid iron carbide which can react with oxides of the enamel during firing of the enamel to produce carbon monoxide gas and hence blisters.

The relatively pure iron matrix has a high critical temperature, in the absence of iron carbide, and remains in the ferritic condition at usual enamel firing temperatures, thus being stiffer and sagging less than when the austenitic condition is produced. Higher critical temperature and resistance to sagging are also met in low-carbon, high-phosphorus enameling irons.

WROUGHT IRON is a low-carbon, high-phosphorus material very high in intentional slag content. It used to be made by laborious hand processes, but is produced now by intermingling overblown, low-carbon Bessemer steel with molten slag to form a conglomerate which is worked down to the same structure as the older product.

The slag, rolled down into plates and stringers, produces highly directional properties and low transverse ductility, but when wrought iron is broken across the grain it fractures woodily, much as would a laminated pack of plates.

Wrought iron is readily weldable. Depending on section, its longitudinal tensile strength runs from 45,000 to 48,000 psi, yield strength 22,500 to 29,000 psi, longitudinal ductility

---

* Data furnished by Armco Steel Corp., Reid L. Kenyon, associate director, research laboratories.

high, transverse ductility very low. Several ASTM specifications exist for a variety of wrought-iron products. Claims for corrosion resistance of both ingot iron and wrought iron, in comparison with steel, have been made, not all of which have been borne out. All these low-carbon materials are lower in strength than ordinary structural steel.

COLD-WORKED CARBON STEEL, THIN SECTIONS. Yield strength can be increased by cold work—cold rolling for flat stock or cold drawing for wire, bar stock, and tubes.

Specification. Cold-rolled strip steel, open-hearth or electric, 1/4 in. or under in thickness, is covered by ASTM A109-38. Material not required to take more than slight bending may contain up to 0.25% carbon. Otherwise, the chemical limits are carbon 0.04 to 0.12, manganese 0.30 to 0.60, phosphorus 0.04 maximum, sulfur 0.05% maximum. As to mechanical properties, bend tests only are required.

Temper. The cold-rolled stock is described and ordered by the vague term "temper" in five grades: 1, hard; 2, half-hard; 3, quarter-hard; 4, soft; and 5, dead-soft tempers. No bend requirement is imposed for hard (No. 1) grade. No. 2 is to bend to a sharp right angle across the grain, No. 3 flat on itself across the grain, to a sharp right angle along the grain, No. 4 and No. 5, flat on themselves both ways of the grain.

Mechanical properties, expected but not specified, are:

	Rockwell B Hardness	Tensile, psi	Elongation* in 2 in., %
No. 1	84–96	68,000–92,000	1– 5
No. 2	75–85	56,000–72,000	4–14
No. 3	64–74	48,000–60,000	13–27
No. 4	52–64	43,000–53,000	24–36
No. 5	38–52	40,000–48,000	33–45

* Elongation values for material 0.05 in. thick. Elongation varies with thickness.

Yield strength, not included in the ASTM discussion, will vary according to thickness of stock and the way the tensile strength has been produced, i.e., whether by a very light reduction, or by a severe reduction followed by process annealing and how far the softening by process annealing has been allowed to progress.

With the elongation measured on stock 0.03 in. or thicker, one producer gives the following relations, longitudinal specimens:

Tensile, psi	Yield, psi	Elongation in 2 in., %, minimum
40,000	25,000	35
45,000	40,000	28
54,000	50,000	20
82,000	80,000	5

Transverse and longitudinal tensile properties cited by this producer do not vary noticeably. Compressive yield strength is quite close to tensile yield strength.

In the bending or forming of hot-rolled stock, or of soft-temper cold-rolled, very severely deformed portions acquire hardness and mechanical properties akin to those of hard cold-rolled stock; by suitable heat treatment those portions may be softened, just as in the soft tempers of cold-rolled stock.

High Strength. By starting with steel of higher carbon content and giving it severe cold reduction, much higher strength can be obtained. Spring wire, hard drawn, is covered by ASTM 227-41 and music spring wire by A228-41. They require killed steel. The other chemical requirements are:

	C, %	Mn, %	Si, %	P, %	S, %
227–41	0.45–0.75	0.60–1.20	0.10–0.30	0.045 max	0.05 max
228–41	0.70–1.00	0.20–0.60	0.12–0.30	0.03 max	0.03 max

Tensile strength runs as follows:

ASTM	Diameter, in.	Tensile Strength, psi
227-41	.032	275,000–310,000
	.500	165,000–195,000
228-41	.004	390,000–440,000
	.156	255,000–285,000

Intermediate sizes have intermediate strengths. The stress-strain diagrams for heavily cold-worked steels are curved because of internal stress. Tempering at around 400 F straightens out the diagram and gives a yield strength around 90 to 95% of the tensile strength. Strength can be lowered and ductility increased by regulated process anneals.

Wire of around 0.80% carbon previously heat-treated to give it a structure suitable for cold drawing, then cold-drawn, is used in bridge suspension cables. In wire of 0.192 in. diameter, this material shows around 220,000 psi tensile, 183,000 yield, around 6% elongation in 10 in., and about 40% reduction of area.

HOT-ROLLED CARBON-STEEL BARS. In hot-rolled carbon steels, strength depends primarily upon the amount of pearlite, that is, upon carbon content. Increase in manganese content, within the limits accepted by the definition of carbon steels, also increases strength. The amount of hot reduction, the finishing temperature at which rolling is stopped, the grain size, and the section size (which controls the rate of cooling after hot rolling) also exert influence.

AISI 1000 Series. Readily available commercial hot-rolled carbon steels are those included in the AISI 1000 series. The median carbon content, from which deviation of a few points is permitted, varying from a total range of 0.05% carbon in the lower carbon steels, up to 0.15% in the highest, is denoted by the last two digits of the assigned number. Thus, for example, 1040 covers 0.37 to 0.44% carbon. With a few exceptions the lower carbon steels, through 1025, carry 0.30 to 0.60% manganese (1016 carries 0.60 to 0.90% and 1019 and 1022 carry 0.70 to 1.00%). Steels 1030 through 1080 carry 0.60 to 0.90% manganese, whereas 1095 carries 0.30 to 0.50%. Steels 1008 through 1020 are rimmed or semi-killed, whereas 1025 to 1095 are killed, with silicon running 0.10 to 0.30%. Phosphorus is 0.04 and sulfur 0.05% maximum.

The steels may be had as hot-rolled, or "cold-finished," or cold-drawn. The properties of the cold-drawn stock will depend on the degree of cold reduction and the stress-relief heating that is applied. This treatment is ordinarily given by the producer, rather than the user. "Typical" longitudinal tensile values for cold-finished bars of about 1 in. diameter are cited in ASTM Specification A108-46, for information only. Typical values, also for information only, for hot-rolled bars about 1 in. in diameter are given in ASTM A107-46. Combining these:

AISI	Hot-rolled 1-in. Bars			Cold-finished 1-in. Bars		
	Tensile, 1000 psi	Yield, 1000 psi	Elongation in 2 in., %	Tensile, 1000 psi	Yield, 1000 psi	Elongation in 2 in., %
1008	50	26	40	58	48	18
1010	54	27		60	50	16
1015	58	29		65	55	15
1016	64	36		68	58	14
1019	....	....		70	60	14
1020	62	34	35	68	58	14
1022	69	41		72	62	13
1025	65	37		72	62	13
1030	78	42		80	70	12
1035	83	45		85	75	12
1040	89	48	25	90	80	12
1045	95	52		95	85	10
1050	100	56		100	90	8
1055	104	59	15	....	....	....
1060	108	62		....	....	....
1070	120	68		....	....	....
1080	125	72	10	....	....	....
1095	135	78	6	130	115	4

NOTE: Longitudinal elongation values for hot-rolled are not included in ASTM table. Values have been spotted in from figures given by Sisco (Ref. 22). Intermediate compositions have intermediate values. As Sisco shows, transverse elongation values on comparable plate are lower, sometimes only half as large as the longitudinal. Directional differences are often even more marked in the cold-finished steels.

A plain carbon Bessemer steel, B1010, in which carbon is held to 0.13%, phosphorus 0.07 to 0.12%, sulfur 0.06% maximum, and manganese 0.30 to 0.60% shows: Hot-rolled, 63,000 psi tensile, 36,000 psi yield; cold-finished, 65,000 psi tensile, 55,000 psi yield, and 12% elongation.

Yield strengths ranging from around 25,000 to around 75,000 psi for the range of compositions in hot-rolled are increased from around 50,000 to 100,000 psi, with appreciable ductility still retained, by cold finishing. Cold-finished flats up to 6 in. in width and rounds, squares, and hexagons are available. Shafting is cold finished up to 8 in. in diameter.

Free-machining steels and steels higher in manganese than the above are supplied both hot-rolled and cold-drawn. Some of them are included in ASTM A107-46 and A108-46. They will be mentioned later under alloy steels.

Flange Quality. Hot-rolled structural iron or carbon steel plates are supplied for uses requiring a higher degree of formability than that of everyday structural steel, for example, "flange quality," ASTM A129-46, which calls for 44,000 psi tensile, 27,500 psi yield, 25% elongation in 8 in. longitudinal values for stock $^3/_4$ in. thick.

Flange and firebox plates A285-46 come in three grades, with 0.80% maximum manganese, and carbon ranging from 0.15 to 0.30%, depending upon grade and thickness of plate. The three grades have 50,000, 55,000, and 60,000 psi tensile, each one ±5000 psi; yield strength half the tensile, with 24,000, 27,000, and 30,000 psi respectively as minima; and elongation in $^3/_4$-in. plate equal to 1,550,000 divided by tensile strength, with usual modifications for other thicknesses. These grades are for welding, and need not be killed. They must withstand a more severe bend test than that of ordinary structural A7-46.

Steel for locomotives and cars A113-46 includes much the same range of properties displayed by the range covered by the three grades of steel just mentioned. Again, no silicon content is specified, the steel does not have to be killed, only to meet the mechanical properties.

Higher mechanical properties are produced in steels which have been given the misnomer "carbon silicon" steel. The strength is not due to silicon, but rather to high carbon and manganese. One of these, "structural silicon steel," which has been used in riveted structures but in which the maximum carbon (0.40%) in ASTM A94-46 is too high for easy welding, has 80,000 to 95,000 psi tensile and 45,000 psi minimum yield and is required to stand the usual ductility and bend test requirements. Grades intended for fusion welding, restricted to lower carbon and 0.80% maximum manganese (A201-46), have 55,000 to 65,000 psi and 60,000 to 70,000 psi tensile; yield, half the tensile; and the usual other requirements.

Stronger plates, up to 4 $^1/_2$ in. thick, with higher carbon and up to 0.90% manganese permitted (A212-46) give 65,000 to 77,000 psi and 70,000 to 82,000 psi tensile ranges, yield half the tensile. Another specification (A284-46T) for plates over 2 in. thick, to be welded, calls for 0.80% manganese maximum, 0.31 and 0.35% carbon maximum, in two grades having 55,000 and 60,000 psi tensile, 27,500 and 30,000 psi yield minimum, respectively.

In all these silicon is required, a range of 0.15 to 0.30% being set except in A94-46, where 0.20% minimum is called for. That is, these are killed steels.

From the design point of view, the steels in each of the different grades, when of equal yield strength, behave pretty much alike, save that differences in manner of killing will be detectable in low-temperature behavior, and differences in grain size in high-temperature behavior, which the conventional acceptance tests do not cover. The gradations relate to formability and weldability—to processing—not to serviceability in final use. Steel mills charge different "extras," according to the formability required. They appraise the formability more on the basis of their knowledge of how a heat was processed than on conventional mechanical tests.

Ship steel takes one of the highest "extras." A steel in which no silicon requirement is stated, and carbon and manganese are unrestricted, but a maximum tensile is set in the interest of formability is "structural steel for ships," running 60,000 to 70,000 psi tensile, yield half the tensile but not less than 33,000 psi, elongation varying with thickness but not less than 22%. Since this steel does not have to be killed, its low-temperature behavior depends on degree of killing actually employed. Failures in welded ships made from steel complying with this specification testify that meeting a specification drawn in terms of conventional tensile and bend tests is no complete assurance that it is exactly fitted for a particular use. Although revised in 1946, the ASTM specification makes no mention of ship failures.

In all the grades of steel mentioned above, the degrees of weldability and formability required are accompanied by low yield strength. Designers often require a considerable degree of each of these in steel of distinctly higher yield strength. In plain carbon steel, raising the carbon enough to reach this higher yield strength level sacrifices too much in these properties. Hence, alloying of low carbon steels is resorted to, when a combination is needed. Of course, when strength only is called for, high carbon steel can be used, as in rails, ASTM Specification A1-46 and A2-27, where different carbon ranges within 0.55 to 0.90%, with relatively high manganese and silicon sufficient to kill, are used. Drop tests are imposed in A1-46 and a "Giant Brinell" test in A2-27. For use as concrete reinforcement, steels of 33,000 to 50,000 psi yield are covered by specifications A15-39, A16-35, A160-39, A184-37 and cold-drawn wire of 64,000 psi yield by A82-34.

Rerolled rails, analogous to A16-35, expected to show 50,000 psi yield are commercially available in shapes small enough to be rolled from sections obtained from used rails by cutting the head and base from the web. When formability, toughness, and weldability are not required, such material is available to the designer.

8. MILD-ALLOY OR HIGH-YIELD-STRENGTH LOW-ALLOY STEELS

LOW-ALLOY STRUCTURAL STEEL. A blanket specification, A242-46, covering material 3/16 to 2 in. thick, sets a carbon maximum of 0.20% and a manganese maximum of 1.25%, with no restriction on phosphorus or on alloying elements other than manganese; it is intended, however, that alloying elements be used that increase resistance to atmospheric corrosion. The size effect is recognized by the following gradations:

Size, in.	Tensile, psi	Yield, psi	Elongation in 8 in., %
3/16– 3/4	70,000	50,000	1,500,000 divided by tensile,
3/4 –1 1/2	66,000	45,000	with deductions for material
1 1/2 –2	63,000	40,000	under 5/16 in. or over 3/4 in.

Cold-bend requirements: Bend 180° over a pin whose diameter is given by the following:

Plate thickness	3/16–3/4 in.	3/4–1 in.	1–1 1/2 in.	1 1/2–2 in.
Ratio $\frac{\text{pin diameter}}{\text{plate thickness}}$	1	1 1/2	2	2 1/2

The steel is intended to be suitable for welding. It does not have to be killed, but usually is.

BOILER PLATE. Certain steels with about 40,000 psi minimum yield up to nearly 50,000 maximum (37,000 to 43,000 psi minimum, in the case of A202-46, up to 55,000 maximum) for plates for boilers and other pressure vessels are listed by ASTM with chemical limits superimposed on mechanical requirements under A202-46 (chromium, manganese, silicon); A203-46 (nickel steel); A204-46 (molybdenum steels); A225-46 (manganese vanadium).

Wider use is made of steels of more complex composition designed to give, in the welding grade, 50,000 psi yield in 1/4 in. thickness, and 55,000 or even up to 60,000 to 70,000 psi in the grade for riveting. Around 25 to 35% elongation in 2 in. accompanies the 50,000 psi yield strength. In each grade it is usually feasible to achieve the 50,000 or 55,000 psi yield in stock considerably thicker than 1/4 in. by suitably adjusting the alloy content. Grain-size control and finishing practice are influential in maintaining the desired yield strength.

"VANITY" STEEL, low-carbon, high-manganese, with tiny additions of either vanadium or titanium, usually both, whence the name, has received much attention for high-strength ship plate (Ref. 23). Copper, for atmospheric corrosion resistance as well as for strength, is usually present in rather generous amount. Phosphorus also confers corrosion resistance, and is a potent strengthener. If the sum of carbon plus phosphorus is too high (0.25% total is around the practical limit), toughness is reduced. But phosphorus does not injure weldability, as too high carbon does.

Killed Bessemer steel, which has been much used for pipe, of around 0.15% carbon, 0.10% phosphorus, 0.50% manganese, 0.20% silicon, aluminum-treated, has 45,000 to 55,000 psi yield strength, 25 to 35% elongation in 2 in.

Hence, two types of this class of steel, with indistinguishable room temperature mechanical properties, are made, one with medium carbon and low phosphorus, the other with very low carbon and relatively high phosphorus.

Characteristic compositions (sulfur, ordinarily 0.02 to 0.03%) are given in Table 6.

Table 6

	C	P	Mn	Si	Cu	Ni	Cr	Mo	V	Ti	Al	Zn
Vanity	.16	.02	1.20	.25	.20 *	.15 *	.03 *	.02 *	.05	.015	.015	...
	.09	.10	1.20	.55	.45	....	...	.25	...	....	....	...
	.10	.10	.60	...	.60	.60	...	.20	...	....	....	...
	.10	.08	.60	.10	.60	.85	...	...	...	.06	....	...
	.13	.02	.70	.80	.20	.20	.60	...	...	....	....	.12
	.11	.02	.75	...	.80	1.00	...	.10	...	....	....	...
	.08	.02	.50	.25	1.00	1.60	...	...	...	....	....	...
	.10	.12	1.10	.05	.35	....	...	...	...	...	.....	...
	.10	.10	.60	.15	1.10	.55	...	...	...	....	....	...
	.10	.10	.25	.75	.40	.50	.75	...	...	....	....	...
	.10	.10	.75	.30	.50	.50	.75	...	...	....	....	...

* Residuals.

For the present at least, all such steels carry the same price extra. Each of the major steel producers makes a steel of this class. Each maker, through experience with his pet composition, may be able to produce that steel in better quality than one he is not used to.

From the user point of view, the steels appear interchangeable. Possibly the low-temperature behavior of the high phosphorus may not be up to that of the low phosphorus, although properly deoxidized heats of the phosphorus class show excellent behavior.

RIVETING GRADE 55,000 TO 65,000 PSI YIELD STRENGTH. When weldability is not required and formability is less important, yield strength of as-rolled mild alloy steels can be raised, either by raising the carbon or raising the alloy. Such steels can be welded if proper welding technique, usually including preheating and postheating (stress-relief anneal) can be applied, but they are not as foolproof as the "welding grade." In the high phosphorus group the carbon is not raised, but the alloy content is, a dash of molybdenum being especially effective.

One of the earliest steels in this yield strength range was a killed 0.30% carbon, 1.50% manganese, 0.20% copper steel. A steel of similar composition but with carbon lower, around 0.20 to 0.25%, and with about 0.15 to 0.17% vanadium added, is suitable for normalizing, even in decidedly large sections. To secure the highest yield strength, normalizing from a temperature at which grain growth does not occur is applied rather than leaving the steels of this group in the as-rolled condition, since heating to rolling temperature may produce too coarse a grain.

In the high phosphorus group there is a steel of 0.17% carbon, 0.09% phosphorus, 1.00% manganese, 1.00% copper plus nickel, and 0.20% molybdenum.

Steel of around 0.25% carbon or a bit under, with manganese at around 0.90%, plus silicon and copper each at about 0.50%, with nickel and chromium each approaching 0.25%, also falls in this yield strength category, an example of using small amounts of several elements.

Other examples of cutting manganese down a bit from the 1 1/2 level and using other alloys to add the desired strength are:

Carbon, %	Manganese, %	Silicon, %	Copper, %	Chromium, %	Nickel, %	Molybdenum, %
0.20	1.25	0.75	0.20	0.50	....	....
0.25 *	1.00	0.25	1.40	....	0.90	0.25
0.22	0.80	0.25	0.95	....	1.90	....

* Yield strength at this carbon level about 70,000 psi.

At 55,000 yield, the riveting class will have about 25 to 30% elongation in 2 in., dropping to around 15% at 70,000 psi.

FORMABLE 75,000 PSI YIELD STRENGTH STEEL. As has been previously stated, by using a high copper content, forming prior to reheating, then reheating for precipitation hardening, mild alloy steels of fairly low carbon can be raised to 75,000 psi yield.

The factor that gives mild alloy steels their high yield strength is the sluggishness conferred by the alloy content upon austenite, by virtue of which its transformation to ferrite and pearlite is held back to occur, upon air cooling, at a lower temperature and with resultant finer structure than in steel with negligible alloy content.

NE NORMALIZING STEELS. By increasing the cooling rate, as by quenching, this transformation can be avoided and quench-hardening induced, if the austenite is sufficiently sluggish. Hence this class merges into that of the "shallow-hardening" steels, which could equally well be classed as "deep-normalizing" steels. Since their air-cooled structure will contain bainite and they contain internal stress, the deep-normalizing steels need tempering.

Properties. Stewart and Wiley (Ref. 24) studied a large group of the NE steels, those containing enough silicon, 0.20 to 0.35%, for killing, and alloying amounts of manganese (0.85 to 1.40%), chromium, nickel, each in the neighborhood of 0.50%, and molybdenum at about 0.25%, at 0.30 to 0.55% carbon normalized in 1 in. and 8 in. round, and tempered at various temperatures.

In 1-in. section, yield strengths ranging from 65,000 psi at the 0.30% carbon end tempered at 1200 F to 100,000 psi at 0.45% carbon tempered somewhat lower, with corresponding elongation of 30 and 20%, were obtained. The yield strength is, of course, lower for a given hardness and tensile strength than would be obtained by martensitic quenching and tempering back to that hardness.

In 8-in. section, Stewart and Wiley determined both longitudinal and transverse tensile, Charpy notched-bar, and endurance limit on longitudinal smooth unnotched specimens. Even in this large section, yield strengths of 50,000 to 55,000 psi were obtained at the 0.30% carbon level and a high tempering temperature and 60,000 to 80,000 psi at the 0.45% level, depending upon alloy content and tempering temperature. The transverse yield strength equaled the longitudinal, but the usual directional differences were met in ductility and notched-bar behavior.

Application. These molybdenum-containing, deep-normalizing steels, although designed for quenching and tempering in relatively small sizes, offer interesting possibilities as engineering materials for large normalized sections. While not all the NE steels studied by Stewart and Wiley remain in regular production, the steels initially classed as NE and now included as regular AISI steels are readily available and, with adjustment of carbon content and tempering temperature, will cover the range of yield strengths shown by the group studied.

The variety of alloy compositions in the mild alloy and NE low-alloy groups, applicable for strengthening a base steel of a given carbon content and with mechanical properties indistinguishable at the same carbon content and same hardness, is evidence that the combination of alloying elements used to make austenite sluggish and produce the proper structure is immaterial. The structure and properties result almost exclusively from the sluggishness, not from the alloy combination, even though, because of its potency, molybdenum is used in the NE steels as cap stone of the alloy structure. Molybdenum-free compositions could readily be substituted.

Hence, instead of considering that he should specify some particular alloy composition, the designer might well ask for normalizing steel on the basis of the carbon level and the yield strength needed in a given section, and let the alloy chips fall where they may.

9. FREE-MACHINING STEELS

The free-machining behavior of steel results from its structure. The structure may be somewhat controlled by heat treatment further altered by introducing "foreign material" which forms a new structural constituent in the matrix, and by modification of the matrix itself.

The idea that softness makes for easy machining and hardness for difficult machining is only partly true. Pure ferrite or structures with much ferrite uninterrupted by harder constituents are gummy and stick to the tool, producing a rough, torn finish. After the chip is separated from the work piece, the chip has to be disposed of and kept from clogging the spaces between teeth of a multi-tipped tool. Turning is therefore one thing; broaching, sawing, etc., another.

Machinability is variously evaluated, sometimes primarily on the surface left, especially in finishing cuts, sometimes on energy required to remove a cubic inch of metal, especially in hogging cuts, and frequently on the tool wear life between grinds. What the tool will stand, i.e., whether it is carbon, high-speed or carbide, its angles, the rigidity of the machine setup, etc., have as much to do with the dividing line between machinable and unmachinable as the material cut. Hence no blanket definition of, and no single conventional test for, machinability can be drawn up.

Qualitatively, machinability is enhanced by having enough carbon, say 0.20%, to supply pearlite patches sufficient to break up the continuity of the ferrite matrix. With lower carbon content, raising phosphorus (as in the mild alloy steels) stiffens the matrix and makes it less gummy. For hogging cuts coarse grain size facilitates machining, but it leaves a rougher surface under a finishing cut than fine grain.

Heat Treatment. In medium-carbon steels, heat treatment leaving a lamellar pearlite structure is usually advised for milling, drilling, and broaching, and a spheroidized structure for turning, though in the higher carbon steels the latter may not be the better. In very-high-carbon steels, the spheroidized structure may be required for most types of machining.

Since the structure and grain size can be controlled by heat treatment, it is often advisable to give a preliminary treatment to produce the structure needed for the particular type of machining to be done, then machine, leaving stock for a light finishing cut, then reheat-treat to produce the structure wanted for yield strength, and take the finishing cut.

Thus the designer can rely on the heat treater to mitigate relatively poor machining properties of steels for uses where mechanical properties are paramount and cost of machining a minor factor.

Sulfide Inclusions. When yield strength is the chief design need and other mechanical properties, especially toughness and fatigue resistance, are not material, machinability can be greatly enhanced by raising the sulfur content, and the manganese content along with it, to supply manganese sulfide inclusions to break up the continuity of the ferrite matrix and produce short chips. Such additions may be accompanied by high phosphorus to stiffen the ferrite matrix. Selenium can be used instead of sulfur; it is more common in austenitic stainless steels than in ordinary constructional steels. Bismuth or lead may also be used in the austenitics.

These sulfide inclusions lead to great directional differences in toughness and fatigue behavior as has been discussed on p. 2-04.

In steels for quenching and tempering, where an extreme of these directional effects would be hazardous at the strength level desired, a compromise is often made by just raising the sulfur content a point or so above the "maximum" sulfur restriction of the conventional specification.

Lead. One method of materially raising machinability without introducing appreciable directional effects or loss of toughness, even in quenched and tempered steels at high hardness levels, is by introduction of lead, analogous to making brass more free machining by microscopically visible lead globules.

In steel, 0.05 to 0.20% lead greatly improves machinability and, probably because the lead particles in steel are too small to be found microscopically, save perhaps by very special technique, the mechanical properties are practically unaltered in any direction. In England, large amounts of lead-alloyed steels were used during World War II, even in projectiles at high hardness where no loss in toughness could be tolerated. The leaded steels are produced only on special order in the United States at present.

Silver is insoluble in solid steel, just as lead is, and small silver additions are sometimes made to die-block steels where the final tooling is at high hardness. Machinability is improved without impairing mechanical properties. The silver-treated steels, also, are specialty products.

Molybdenum. The statement is often made that quenched and tempered steels high in molybdenum machine appreciably better at rather high hardness than comparable steels without it.

SPECIFICATION. The conventional free-machining carbon steel bar stocks, for use where decreased toughness and fatigue resistance can be borne, are described in ASTM A107-46 for hot-rolled and A108-46 for cold-drawn. Cold-drawn is commonly used in automatic screw machines because of closer dimensional tolerance achieved by cold drawing.

Bessemer. Of these the Bessemer (high-phosphorus) steels with high sulfur are considered the easiest machining, B1113 standing at the top:

					Longitudinal Specimens				
					Hot-rolled		Cold-drawn		
AISI No.	C, %	Mn, %	P, %	S, %	Yield, min	Tensile, ave	Yield, ave	Tensile, ave	Elongation in 2 in., %
B1010	0.15 max	0.30–0.60	0.07–0.12	0.06 max	36,000	63,000	55,000	65,000	12
B1011	0.08–0.13	0.70–1.00	0.07–0.12	0.10–0.15	40,000	65,000	55,000	75,000	10
B1112	0.08–0.13	0.70–1.00	0.07–0.12	0.16–0.23	40,000	65,000	55,000	75,000	10
B1113	0.08–0.13	0.70–1.00	0.07–0.12	0.24–0.33	40,000	65,000	55,000	75,000	10

Open-hearth steels, all with phosphorus 0.045% maximum, are:

				Longitudinal Specimens				
				Hot-rolled		Cold-drawn		
AISI No.	C, %	Mn, %	S, %	Yield, psi	Tensile, psi	Yield, psi	Tensile, psi	Elongation in 2 in., %
1115	0.13–0.18	0.70–1.00	0.10–0.15	35,000	62,000	55,000	65,000	16
1117	0.14–0.20	1.00–1.30	0.08–0.13	41,000	68,000	60,000	70,000	15
1118	0.14–0.20	1.30–1.60	0.08–0.13	43,000	70,000	65,000	75,000	14
1120	0.18–0.23	0.70–1.00	0.08–0.13	36,000	66,000	......	....	..
1137	0.32–0.39	1.35–1.65	0.08–0.13	55,000	95,000	80,000	90,000	10
1141	0.37–0.45	1.35–1.65	0.08–0.13	60,000	100,000	85,000	95,000	10
1151	0.48–0.55	0.70–1.00	0.08–0.13	65,000	110,000	90,000	105,000	8

Numbers 1115 to 1120 are usually unkilled, and are mainly used for carburizing; 1137 to 1151 usually carry silicon for killing. The 1137 to 1151 steels are amenable to quenching and tempering.

Cold-drawn 1337, stress-relieved, ranges on longitudinal tests from 100,000 psi tensile, 80,000 psi yield, 15% elongation, 45% reduction in $^1/_2$ in. diameter, to 90,000 psi, 70,000 psi, 15%, 45% in 2 in. and 85,000 psi, 65,000 psi, 12%, 35% in 3 in. If especially heavy drafts are taken in cold drawing, as Landis (Ref. 25) explains, these values are somewhat improved. It is to be remembered that, unlike the center of a quenched and tempered shallow-hardening steel, the centers of the cold-drawn bars have properties closely similar to those of the periphery.

Many analogies will be seen between this group and the mild alloy and low alloy previously discussed. In the Bessemer group, the usual carbon plus phosphorus limit is not exceeded. In the open-hearth group, the as-rolled strength increases as either carbon or manganese rises, and at the highest carbon level, in 1151, manganese drops in order to retain some toughness.

Open-hearth steels with higher than the 0.13% maximum sulfur can also be obtained, for example, modified 1144 with 0.20 to 0.32% sulfur.

It will be noted that yield strength is a function of carbon, phosphorus, and manganese, and is not altered by raising sulfur even to around 0.30%.

Not mentioned in the ASTM specifications cited is 8640, with sulfur at 0.07% maximum, for a combination of ability to secure the usual mechanical properties without great directionality, yet with improved machinability, as is pointed out by Boyer (Ref. 26), who rates it, cold drawn, as of excellent machinability. His argument for a specific method of adding sulfur to open-hearth steel probably finds little general agreement, but his listing of relative machinability of quite a group of steels probably would find general agreement. He also lists the "extras" obtaining at the date of the article.

It is generally considered that cold drawing adds materially to machinability, but for some types of machining, at least, this belief has been challenged.

GRAPHITIC STEELS. Another free-machining steel starts its career in much the same "white-iron" condition as pearlitic malleable, but its composition is so balanced that the material can be hot rolled, or forged from the ingot without appreciable graphitization. Upon slow cooling, however, a process applied by the manufacturer, graphite or temper carbon separates out. Such steels contain around 1 to 1 1/2% carbon and up to around 1% silicon, with less silicon when another graphitizing element such as aluminum is present. The slow cooling causes something like half the carbon to separate as tiny graphite nodules, leaving a matrix of pearlite and silico-ferrite which is heat treatable. Depending upon the temperature and time of austenitization, very little or quite a proportion of the separated graphite can be taken back into solution, so the effective carbon content of the matrix can be controlled.

Hardenability. The austenitized matrix, carrying graphite "inclusions," can be quenched and tempered. The matrix of one grade has about the same composition as one of the tough, oil-hardening tool steels. The matrix can be made shallow hardening even in water, by omitting elements that promote hardenability, keeping the silicon low, and adding about 0.15% aluminum. With higher silicon a more deeply hardening, water-hardening steel is produced. With a little molybdenum on top of higher silicon, a still more deeply hardening, oil-hardening steel results. Tungsten, which produces harder carbide, is added in special wear-resistant uses. Piling in silicon, manganese, nickel, molybdenum, and chromium produces an air-hardening steel.

In all the modifications the separated graphite nodules provide remarkably easy machining at relative high hardness. Like all high-silicon steels, these are prone to surface decarburization and need care in heat treatment to avoid it.

The tensile and yield strengths of the matrix, if fully quenched to martensite and then tempered, bear the same relation to each other as holds in all fully quenched and tempered steels; they follow the master plots of Figs. 8 and 9. Ductility is, of course, reduced by the graphite particles, but since these particles are produced after rolling, and uniformly distributed, directional differences are less than in steels made free machining by manganese sulfide.

Lubricant. The graphite particles appear to serve as receptacles to retain oil lubricant, as well as to supply lubricating effects from graphite itself. These effects result in long-wearing punches and dies.

Yield strengths in the quenched and tempered condition can be varied, by adjustment of composition and tempering temperature from 100,000 to 175,000 psi or, for wear-resistant purposes, to the order of 225,000 psi.

Application. The graphitic steels are commercially available. Although they are used primarily as wear-resistant punches and the like, cams and slide ways are applications in ordinary design. For a complex machining job that needs to be done at unusually high hardness, even where only structural strength is required, without wear resistance, they are worth keeping in mind.

10. ALLOY CONSTRUCTIONAL STEELS FOR QUENCHING AND TEMPERING

In earlier handbooks, alloy constructional steels were viewed as distinct entities, plots of tensile properties against tempering temperature, or analogous tables, being given for each of the steels then designated by SAE numbers, and lists of products for which each steel was frequently used were often included. Without statement of the section size of the particular product, such lists are meaningless.

Properties. As discussed earlier, property charts, based on sections which harden fully to martensite, can be condensed into the master plots of Figs. 8 and 9. The properties of the interior of a section which does not harden to martensite throughout, after tempering of the nonmartensitic core along with the martensitic outer zone, depend upon the size of the section, which controls the rate of cooling, as well as upon the sluggishness of the austenite. Yield strength and other properties grade down, in a sufficiently large section, from those moderately inferior to the surface properties to the distinctly inferior properties of a normalized steel, as the center is approached.

Design. The designer may generalize the problem by determining to what depth a fully martensitic quenched outer layer is needed, to what yield strength and accompanying toughness this layer should be tempered back, and what strength is needed in the underlying material to support this outer layer.

Quenched hardness of the initially martensitic outer layer is determined by its carbon content. The lower the carbon, the less the danger of cracking an unbalanced section upon quenching, but the more alloy is required to produce deep hardening.

Past experience or analogy with other examples indicates what depth of hardening and what carbon level are suitable for the case in hand, and similar experience tells whether water can be used or whether oil quenching is required to avoid quench cracking. The steel that has fitted these requirements in the past has probably been chosen with considerable excess average hardenability, secured at the expense of costly total alloy content to insure this hardenability at the lowest carbon and alloy levels permitted by the earlier specification for that steel, based on chemical composition.

Hardenability Specification. If the Jominy hardenability curve for that steel at this low chemical level is determined, that curve can be matched with the lower boundary of the Jominy hardenability band of one or more of the cheaper, less heavily alloyed, steels now available as "H," i.e., controlled hardenability steels. This procedure automatically matches the carbon levels but does not imply the same alloy content. Parker (Ref. 14) shows the closer hardenability limits of some of the 8600, 4100, and 4600 series of steels when ordered as "H" steels instead of by composition, and discusses the matching of the Jominy values to the section to be treated.

The economics of selection on a hardenability basis and of considering price in choosing among steels of like hardenability has been interestingly discussed by Brown (Ref. 27). Any such steel, tempered back to the same surface hardness as was used for the satisfactory more highly alloyed steel, has a high degree of probability for showing identical service. Thus the 8600, 8700, and 9400 series, with controlled hardenability, are rapidly replacing the 2300, 3100, 4100, 4600, and 6100, even the 4300 series. A few of the many published statements attesting to such replacements are listed in References 26 to 33.

Compositions. Tables 2 to 5 summarize the compositions of the steels in the AISI list. In order to secure optimum yield strengths, the steels, especially above 0.20% carbon, are usually specified to be fine grained.

The wide hardenability bands for the various AISI numbered steels when ordered merely to the chemical composition ranges for those numbers and the narrower ones for the controlled hardenability "H" steels are available in pamphlets published by the AISI (Ref. 12). A few examples are given in Figs. 12 and 13.

Carbon content for steels to be quenched ranges from 0.20 to 0.55% and total alloy content from not much over 1% up to around 5%. With low-carbon and low-alloy content the steels are shallow hardening and can be water quenched; with both carbon and alloy content high, they are deep hardening and require oil quenching. Steels high in carbon and so highly alloyed as to be hardenable on air cooling in small or medium sections are not ordinarily applied to constructional uses.

Condensed information on sizes hardening in oil or water with carbon from 0.20 to 0.50% for the 8600 series to whatever hardness is selected, and the relation of the hardness gradients to the conventional Jominy curve, are contained in Boegehold's plot (Fig. 14).

Steels with higher than 0.55% carbon occur in the plain carbon 1000 series; they are used for cutting tools not expected to operate at high temperature, for wear-resistant purposes, in railway rails, and in small springs.

Steels of the 52100 series, with carbon at around 1% and chromium up to 1 1/2%, are used for ball and roller bearings and races; 6195, with about 1% each of carbon and chromium plus about 0.20% vanadium, and steels with 0.50 to 0.65% carbon, about 3/4% manganese and about 2% silicon, as well as 8650, are used for springs of fairly heavy sections.

11. STEELS FOR CASE HARDENING

CARBURIZING. Two general classes of steels are used for carburizing: first, those of 0.10 to 0.20%, second, those of around 0.30 to 0.35% carbon, each with or without alloying elements.

Case Depth. Low-carbon steels are ordinarily given a thick case, thick enough to withstand crushing by virtue of its own strength in spite of being underlain by a very soft core. Higher carbon steels may be given a relatively thin case since it is adequately supported by the harder core.

The hard, brittle, high-carbon case is applied for wear resistance. In the vast majority of uses the thin case is thick enough to provide for all permissible removal of metal by wear, and the chief reason for a thick case is to allow grinding tolerance upon a warped piece, so that a sufficient case thickness still remains. It would be better to avoid warping and save the time and expense of applying the thick case.

Thick cases are usually applied by heating the pieces for a long time along with a solid carburizing agent, thin ones more commonly for a much shorter time in a gaseous carburizing medium. In either case, heat treatment is applied after carburizing, the heat treatment involving at least a quench to produce martensite in the case, followed by a low-temperature tempering to "toughen the case" without materially softening it, leaving it file-hard. In spite of this "toughening," the case is always brittle. Case toughness is sometimes demanded to allow straightening a warped piece without cracking the case. It is very doubtful if much straightening can really be done without cracking the case, although the crack may not be visible and the case, though cracked, may still adhere to the core.

The case may have only about 0.80% carbon, or it may run up to well over 1%. Alloy steels containing strong carbide-forming elements such as chromium or molybdenum may be used to provide more wear-resistant carbides, but they tend to build up a higher carbon content on the surface than is desired, and a steeper carbon gradient. Even when the surface carbon is not very high, whatever the steel composition, appreciable time must be allowed for diffusion of carbon into the core so as not to have a sharp line of demarcation between high carbon case and lower carbon core, but instead a gradual gradient, in order to avoid spalling of the case.

This diffusion time means that, despite differences in the rate at which the extreme surface picks up carbon in different alloy steels, the time required for producing a given depth of a proper type of case and diffusion zone does not vary greatly, though the presence of high manganese appears somewhat beneficial.

Core properties have received much misapplied attention. Because the case is brittle anyway, it has been supposed that the core should have extreme toughness; hence the very low carbon of the steels that were first used, in spite of their gumminess and poor machinability. The toughness of the core used to be evaluated on "pseudo-carburized" stock that was put through the whole heat treatment cycle, but without carburizing agents, or on the core after the case had been removed. But when engineers became curious enough to study the carburized piece as a whole (Refs. 34, 35), as it is used in service, it turned out that impact values, either notched or unnotched, are highly variable, but low. That is, a crack once formed in the brittle case propagates readily even through a core that behaves tough when the case is removed.

The behavior of the case can be studied by making a bend test of a carburized specimen, noting the deflection at which a "ping," from the development of a crack, is heard. The case behavior, which depends upon the carbon content, the carbon gradient, and the thickness of the case, is determined by the carburizing technique but very little affected by the steel composition. Few uses of carburized parts admit enough deflection to crack the case, and for gears, Almen and Boegehold concluded that impact test figures did not correlate with service behavior.

Automobile tire chains are in an unusual category in that the case serves only to delay initial wear, the core being exposed for a large part of the life.

As long as the toughness of the core was regarded as having meaning in respect to service, great pains were taken to alloy and heat-treat the steel to obtain the aimed-at core properties. Since the low-carbon core becomes austenitic only at high temperature whereas the high-carbon case becomes so at much lower temperature, a double quench was used. A high-temperature quench was used to refine the core, followed by reheating to a lower temperature, at which the case becomes austenitic but the core does not, then again quenching to harden the case.

The core refinement was applied because, in the long sojourn at carburizing temperature, excessive grain growth occurs in steels that coarsen easily. Use of grain-control agents to produce fine-grained steels avoids this trouble.

Steels Used. There is danger of warpage, even of cracking the case, every time the part is quenched, to say nothing of the cost of the extra operation. For this reason, the use of fine-grained steels and a single quench for case properties, with less regard for the alleged need of core toughness, is the general rule. Low-alloy steels serve, and very little use is made of the very highly alloyed ones, such as the 1 1/2% chromium, 4% nickel steels that were favored for maximum core toughness. Boyer (Ref. 26) remarks that

8617 and 8620 are almost the only alloy steels now in wide use for carburizing, 1019 being preferable among plain carbon steels for deep carburizing.

A growing tendency appears for the use of higher carbon steels, 0.30 to 0.35% and a thin, gas-carburized case upon this stronger core. A still thinner hard case, containing both carbide and nitride, can be applied by use of molten salt baths containing cyanide. Here the core is often made still harder, for adequate support, and steels of 0.35 to 0.45% carbon are utilized. With heat treatment applied only for case properties, the core properties approach those of a normalized steel as previously described.

SURFACE HARDENING. When a very thin hard surface will serve and minimum distortion is needed, carburizing is omitted, and a still higher carbon steel chosen for support. This is first heat treated to secure the desired strength, then the skin only heated and quenched. Automatic devices for flame hardening and for induction hardening have come into wide use. The steel is usually plain carbon of about 0.45% carbon, or a very mildly alloyed steel, since no ability for depth hardening is required. The processes are very rapid, the equipment fits readily into a production line, and there is no need for expensive steels.

NITRIDING. A supremely hard and wear-resistant, but very thin, case is produced by a long, low-temperature treatment (below the austenitic range), in ammonia, of steels alloyed with elements that readily form nitrides, notably aluminum, chromium, molybdenum, and vanadium. The steel, usually of 0.25 to 0.40% carbon, must first be quenched and tempered, machined or ground free from any decarburized surface, nitrided and the nitrided surface preferably lightly ground to remove a very small part of the thin nitrided layer, since the outer film is not so hard as the underlying material and more liable to spall. Moreover, this can be made to compensate for the very tiny increase in dimensions.

The nitrided layer is brittle and nitrided pieces cannot be straightened, but warpage is at a minimum. Half a dozen compositions of nitriding steel are commercially available. Their price is high, since they contain 2 or 3% of alloys, and the process is long and costly. The hardness is retained at high temperatures that would materially soften a quenched, carburized case.

The nitriding steels quench and temper to much the usual set of properties shown in the master curves of Figs. 8 and 9. The elongations hug the lower boundary of the band, and the reduction of area values may even be a bit below it, but the variation is not large. From their high alloy content, most of the compositions are deep hardening.

Since fatigue resistance is ordinarily governed by the strength of the surface, the polished-bar fatigue behavior is exceptionally good, and the very hard skin resists scratching. Johnson and Oberg (Ref 36) obtained over 80,000 psi endurance limit, even when the nitrided layer was scratched. Although nitrided steels are ordinarily employed for wear resistance, they are worth considering for uses where fatigue resistance is paramount.

12. DUPLEX MATERIALS

In a sense, the nitrided, cyanided, or carburized steels are duplex, since their surfaces are of different composition and properties from the interior. Induction or flame-hardened pieces are akin, in having different skin properties, though skin and core have the same composition. Cold working the surface, as by shot-peening, is often highly effective in improving fatigue resistance. Highly undesirable for fatigue resistance is a steel made "duplex" by decarburization of its surface.

CORROSION-RESISTANT COATINGS. Since ordinary steels rust, corrosion-resistant coatings are applied, familiar in tinplate and hot-galvanized or electroplated zinc-coated stock. Steel sheet and strip plated with many other metals are now commercially available, as is hot-dipped aluminum-coated stock. Steel clad with a heavier surface layer, on one or both sides, of copper, nickel, stainless steel, etc., is available in both thin gages and in heavy plate. They are produced by casting the desired coating metal about a steel core and rolling the whole, or by hot rolling the layers together so that they weld.

These materials deserve careful consideration by the designer since they have been developed to the point where quantity production has brought the cost of most of them well below that of a solid piece of the rust- or corrosion-resistant coating metal. The thickness of the coating can be varied according to the intended use and the design strength provided by the steel backing.

"Soft center plow steel," made similarly to clad steel, with a low carbon core and high carbon exterior, thus much like an extremely heavily cased carburized steel, is also procurable. "Sandwich" construction, in which a light, weak, often a porous, nonmetallic backing is faced with strong metal, is very much to the fore in present-day design. The clad steels fit nicely into sandwich assembly. (See Ref. 52.)

SPECIFICATIONS. ASTM Specification A263-44T covers steel plate, sheet and strip clad with corrosion-resisting chromium steel of low carbon content and in different grades from 11 1/2 to 18% chromium, ferritic steel. On sheet and strip a 180-degree cold-bend test is required.

A264-44T covers a similar clad product with the cladding from the group of austenitic corrosion-resistant chromium-nickel alloy steels, and A265-43T, those with the cladding of nickel, nickel-copper, or nickel-chromium-iron alloy.

13. SPECIAL STEELS

The steels previously discussed have been those whose properties depend primarily on the amount and distribution of ferrite and "everyday" carbide and the behavior of ferrite and carbide on heat treatment. Save as the alloy composition serves to make austenite sluggish and thus to control its transformation to ferrite and carbide in the desired state of distribution, the particular alloys in the alloy content have been of very secondary importance except where copper is used for precipitation hardening. The heat treatment of these, except with copper, follows the same general pattern whatever the alloy.

A few steels, mostly high carbon and often highly alloyed, have somewhat special properties because the everyday carbide contains enough of one or more of the carbide-forming elements to modify the carbide somewhat, especially to make it harder. Many tool and die steels come in this category. Since this discussion deals with constructional steels, that class will not be discussed in detail. The carbide in some of the more highly alloyed constructional steels may change in content of the carbide-forming element that replaces iron in cementite, according to the tempering temperature, as discussed by Austin (Ref. 37), but the effect is probably small in the commercial compositions previously discussed.

From the constructional point of view, thermal stability of the carbide allows use of thermally stable carbide-containing steels at elevated temperature. As service temperature goes up, resistance to oxidation is increasingly needed; it is provided by large amounts of chromium, with smaller amounts of aluminum and silicon helping somewhat. Resistance to corrosion, especially by chemicals of an oxidizing nature, is also a characteristic of high-chromium steels.

When the carbon content and the content of carbide-forming elements rise sufficiently, everyday carbide is no longer formed. Instead, a special carbide of different crystallographic form and often of very high hardness appears. Wear resistance and cutting ability may be greatly enhanced by such carbides. The heat treatment required may vary from that for the usual constructional steels.

Large amounts of certain elements and certain combinations of elements make austenite so sluggish that it does not transform, even on drastic quenching, and a class of steels, austenitic at room temperature, results. These steels are necessarily described by chemical composition and must largely be classified in terms of intended use.

HIGH-MANGANESE WEAR-RESISTANT AUSTENITIC (HADFIELD) STEEL. For uses in which heavy battering wear (but not light impact or pure abrasion, as by sand) is involved, steel with 1.0 to 1.4% carbon and 11 to 12% manganese, with or without smaller amounts of other alloying elements, is applied. It is invariably water quenched from 1830 F or higher in order to retain the carbide in solid solution. Regular annealing produces a weak, brittle structure, as does reheating to 650 F or over. The heating and quenching treatment is called a "quench anneal" since this treatment toughens. In the tough condition, the steel starts to yield slightly at rather low loads, but as it deforms, it work-hardens very rapidly, the skin becoming file hard. In wrought stock, in the tough condition, it has around 100,000 to 140,000 psi tensile, 40,000 to 55,000 psi yield with anywhere from 30 to 80% elongation. Machining is extremely difficult because of the work-hardening propensities.

Austenitic steels in general retain toughness at extremely low temperatures, but elevated temperature service of the manganese steel is limited to below 650 F. Hadfield steel is not useful where no change in dimensions of the piece is permissible.

Castings are covered by ASTM A128-33. The only mechanical tests mentioned are a conventional bend test, or the practical one of deforming a casting by pressure or by blows to demonstrate that it is tough and ductile. The cast steel, toughened, has 80,000 to 120,000 psi tensile, 40,000 to 45,000 yield, and 15 to 40% elongation.

A similar steel with lower carbon, the usual manganese plus about 4% nickel, is reasonably tough on ordinary normalizing, without water quenching, and is serviceable at higher temperatures.

HIGH-CARBON, HIGH-CHROMIUM, WEAR-RESISTANT STEEL. For those uses in which wear is combatted by high hardness, steels of about 12% chromium, 1 1/2 to 2 1/4% carbon, often with around 1% molybdenum and 1/4% vanadium, air or oil quenched, find a place. Lathe centers are an example. Such steels are very deep hardening, have low thermal conductivity, require slow heating, careful heat treatment, and avoidance of decarburization. The special carbide present makes them very resistant to tempering, around 58-59 Rockwell C being retained after tempering at 600 F.

PEARLITIC STEELS FOR HIGH-TEMPERATURE SERVICE. Ordinary carbon steels, killed to avoid complications in behavior on elevated temperature aging, drop slowly in yield strength and long-time load-carrying capacity at 650 to 750 F, thereafter more rapidly. For long service at higher temperatures, up to those at which ordinary steels oxidize too rapidly, molybdenum, usually around 0.50%, and a coarse- rather than fine-grained structure are specified. At 1000 F this steel has more than double the load-carrying capacity of plain carbon steel. This is in line with the resistance to tempering of normalized or quenched molybdenum steels; they require higher temperatures for equivalent softening than most other constructional steels. Vanadium also improves high-temperature load-carrying ability in pearlitic steels.

Use of the molybdenum steel, aluminum-killed, in the range of 850 to about 1100 F for several years, resulted in some cases of graphitization and embrittlement. Addition of 1/2 to 1 1/2% chromium materially stabilizes the carbide and prevents or minimizes graphitization. Aluminum killing is avoided when possible; but where it must be used, the chromium is raised from the lower level.

Still higher chromium contents, up to around 9%, are used to add a slowly increasing degree of oxidation and corrosion resistance. Smaller amounts of aluminum and silicon are also employed. A 5% chromium, 0.50% molybdenum composition is in large production.

All these steels ordinarily run in the low carbon range, 0.10 to 0.15%, and are generally used in the normalized condition, for high-temperature steam lines and in the petroleum and chemical industries.

Bolting stock for high temperature use is varied in composition and properties, depending upon the yield-strength requirements and the temperature of service. Since the bolts are usually at a lower temperature than the pipe, heat-treated steels may be used if the operating temperature does not temper the bolt and reduce its initial yield strength too far.

A large variety of ferritic steels with carbon content suitable for quench hardening is in use for bolting stock. The real requirement is a limited stretch or "relaxation" under operating conditions. Relaxation tests are not easy to make, but in these steels the service behavior for the different compositions and treatments are fairly well known by the producer so that the room temperature properties are used for specification purposes. In sizes 2 1/2 in. and under, and tempering at 1000, 1100, or 1200 F, which tempering temperatures are always to be very considerably above the temperature of use, the compositions are chosen so that most of the steels, each after its particular tempering, give 125,000 psi tensile, 105,000 psi yield, 16% elongation in 2 in., 50% reduction of area. Yield strength after tempering at high temperatures is lower in larger sizes.

Pipe. Carbon, carbon-molybdenum, and chromium-molybdenum pipe for high-temperature service are respectively covered by ASTM Specifications A106-46T, A206-46T, and A280-46aT. Some of the steels in the chromium-molybdenum and like alloy steel series for seamless pipe are covered by A158-46T, the bolting materials by A193-46T, and nuts by A194-46. A213-46 covers similar steels for boiler and superheater tubes.

Seamless low-carbon and carbon-molybdenum still tubes come under ASTM A161-46, alloy-steel tubes of these and various grades with chromium under A200-46 and A199-46.

Specifications for castings for fusion welding for high-temperature service are carbon steel, A216-44T, alloy steel, A217-46T. Wrought pipe welding fittings of carbon and carbon-molybdenum steel are covered by A234-44, and carbon-steel wrought pipe flanges and valves by A105-46. They are for high-temperature use.

Castings. Chromium alloy-steel castings, of the ferritic type, are covered by ASTM Specification A221-39 in three classes:

	C, %	Cr, %	Ni, %	Cu, %	Tensile, psi	Yield, psi	Elongation in 2 in., %	R.A., %
A	0.15 max	11.5–14	0.80 max	....	80,000	50,000	20	30
B	0.30 max	17–21	1.00	0.90–1.20	90,000	55,000	5	5
C	0.25–0.50	26–30	3.00	....	55,000	30,000	..	..

Castings with a still wider range of composition, with chromium up to 30%, with different nickel maxima are covered by ASTM A296-46T.

Tubing in the range 11.5 to 28% chromium comes under A268-44T.

Bars. Hot-rolled, cold-finished bars of 11.5 to 18% chromium, with 70,000 tensile, 40,000 yield, 16% elongation, 45% reduction of area are included in A276-44T.

Corrosion Resistance. Some of these steels are discussed and specifications given in the ASME Boiler Construction Code. For general high-temperature service the conditions of oxidation or corrosion have to be carefully considered. These conditions are combatted by chromium, as a mainstay. Other alloying elements are introduced for added resistance to specific types of corrosion, or to permit raising the chromium to the desired level without sacrifice of, or with enhancement of, mechanical properties and workability. For resistance to atmospheric rusting, i.e., to produce a "rustless" or "stainless" steel at least 11 to 12% chromium is required. As the chromium content is raised to around 30%, resistance to severe oxidizing corrosion conditions is increased and resistance to oxidation in air is carried to higher and higher temperatures. The carbon content and the content of other alloys are adjusted to make a steel in which the desired chromium level can be held and the necessary mechanical properties for fabrication and for the particular service contemplated can still be obtained.

Formability. There are limitations to the formability of plain carbon-chromium steels. They are very sensitive to changes in carbon content. Although a carbon steel to be quench hardened contains about 0.25% or more carbon and is all pearlite in the normalized condition, at 0.80% carbon, a steel with 13% chromium quench-hardens (even in air in thin sections) with 0.09% and is all pearlite at 0.30% carbon.

In the "rustless" class, which implies about 13% chromium instead of 18% or more in the "stainless" class, at 0.08 to 0.12% carbon, oil quenched and tempered at 1200 F, a tough product with 100,000 to 120,000 psi tensile, 80,000 to 100,000 psi yield, 20 to 25% elongation, and 60 to 65% reduction of area is obtained and serves for steam turbine buckets and the like. Tempered at 600 F, it runs about 190,000 psi, 180,000 psi, 20%, 60%. For decorative purposes involving cold formability still lower carbon is preferred.

Commercial grades, the 400 AISI series, in the soft, workable condition have about 70,000 psi tensile, 35,000 yield, 25% elongation in 2 in., and can be had ranging from 11 1/2 to 18% chromium. ASTM A176-44 and 240-44 cover solid, A263-44T covers clad plate, sheet and strip in such gradations, designated 403, 410, and 430.

In solid plate sheet and strip higher carbon materials, 442 and 446, with about 0.35% carbon and 18 to 27% chromium, having 75,000 psi tensile, 40,000 psi yield, 20% elongation, are also covered by A176-44. The first fall into the class of "cutlery stainless," which may contain as low as 13% chromium, or as high as 17%, the latter with around 0.70% carbon. After oil quenching, they can, by varying the tempering temperature, have the properties adjusted from 250,000 psi tensile, 200,000 psi yield with low ductility but in the most corrosion-resistant condition, down to around 150,000 psi, 100,000 psi, 10%, 40% in the annealed state. These steels work-harden readily and are difficult to machine, but not so much so as the austenitic types.

At around 16 to 30% chromium, at low carbon content, the chromium steels are ferritic, they do not convert to austenite on heating, and hence are not quench hardenable on cooling. In the lower chromium range, increasing the carbon allows production of austenite, hence quench hardenability. Around 27% chromium the oxidation and corrosion resistances are so excellent that considerable commercial use has been made of such alloys in spite of the fact that if held long in the 750 to 1025 F range, they become excessively brittle upon cooling to room temperature although tough at elevated temperatures. In the annealed state, they are formable unless the previous processing has been such as to create large grain size.

AUSTENITIC STAINLESS STEELS. The addition of carbon, copper, manganese, nickel, and nitrogen to the high-chromium steels tends to make them austenitic. For stainlessness under atmospheric and many corrosive conditions, at least 17% chromium, preferably more, is used, along with varying amounts of the elements just mentioned with a minimum of 7% nickel. The typical composition is 18% Cr, 8% Ni. Some of the nickel can be replaced by manganese, but with complete replacement the chromium cannot be brought up to the desired level and have the steel fully austenitic. Copper is added to improve resistance to certain types of corrosion. Molybdenum is also used for this purpose, and improves high-temperature load-carrying ability as well.

Carbon is held to a low limit in wrought products, especially those to be welded, since a deleterious precipitation effect occurs if the steel is held in the middle of the range 750 to 1500 F or cooled too slowly through it. Carbide precipitates at the grain boundaries which are made more readily subject to chemical attack or to oxidation. With carbon at 0.02 to 0.03%, this does not occur so readily. Such material is now available commercially, avoiding the welding difficulty, but not that of long sojourn in the precipitation range and subsequent exposure to the very active corrodents sometimes used in conventional testing. Such tests may be very misleading. The actual corrodent should be used.

Usually the carbon is kept as low as is economically feasible, usually in the range 0.08 to 0.15%, and carbide stabilizers, titanium or columbium, are added when grain boundary precipitation would be met.

Like the austenitic manganese steel, austenitic stainless is "quench-annealed" by rapid cooling from a high temperature and, like it, work-hardens readily and is difficult to machine.

Strength. Since the austenitic steels cannot be hardened by quenching and tempering, strength is imparted by cold working. Properties ranging from those in the soft, formable quench-annealed state of 80,000 to 95,000 psi tensile, 35,000 to 45,000 psi yield, 55 to 60% elongation, 55 to 65% reduction of area can be shifted by cold work to anything between 100,000 psi, 60,000 psi, 50%, 65% up to 300,000 psi, 250,000 psi, 2%, 30%. Even when cold rolled to high yield strength determined by the offset method, there is still a tiny amount of plastic action which makes the stress-strain curve depart a trifle from a straight line even at as low as 24,000 to 30,000 psi, though very much higher stresses are withstood, because of the rapid rate of work hardening, without exceeding a permanent set permissible in most design.

Toughness. The austenitic material is tough, from all points of view, until work hardening has been carried to a great extreme. It is tough in static loading, tough in notched-bar behavior even in large section and at extremely low temperature, and it is outstanding in resistance to notched fatigue. These features make it a useful, though very expensive, material for general construction even when its special properties of resistance to oxidation and corrosion are not required. Its main drawback, other than price, is difficult machinability. Selenium or other additions can be made to improve machinability without greatly depreciating corrosion resistance or mechanical properties.

Appearance. Where decorative appearance is important, as on many parts formed from sheet, a wide variety of finishes ranging from matte to mirror-bright are commercially produced in the sheet itself, the price varying with the finish.

Corrosion Resistance. For increased resistance to certain types of corrosion, addition of up to around 3% molybdenum is made. Digesters, valves, and fittings for the pulp and paper industry make much use of this type. Tabulations of the corrosive conditions combatted by various grades of stainless are available (Ref. 38).

Specification. ASTM A240-44 covers grades of stainless sheet containing molybdenum, columbium, or titanium all with 75,000 psi tensile, 30,000 yield, 30% elongation in 2 in. A167-44 for plate is similar, but requires 40% elongation. A264-44T is for clad steel, and includes coatings of some of these grades.

One specification for 18-8 stainless tubing is A270-44T; that for similar castings is A198-39.

Tubes with 17 to 20% chromium, with various amounts of nickel, and with titanium, columbium, or molybdenum come under ASTM 269-44T.

Cold-worked 18-8 sheet covered by ASTM A177-44 comes in four commercial tempers:

	Tensile, psi	Yield, psi	Elongation in 2 in., %
Quarter hard	125,000	75,000	25
Half hard	150,000	110,000	9 to 10, depending on thickness
Three quarters hard	175,000	135,000	3 to 7, depending on thickness
Full hard	185,000	140,000	3 to 5, depending on thickness

Hot-rolled, cold-finished bars, low carbon (0.08 to 0.10%), of the 18% chromium, 8% nickel type, covered by A276-44T, show 75,000 to 90,000 psi tensile, 30,000 to 45,000 psi yield, 25 to 35% elongation, 45 to 50% R.A. according to size. With 0.15% carbon maximum, the figures are 120,000 to 125,000 psi, 65,000 to 85,000 psi, 12 to 15% and 35% according to size.

Castings of two grades, 0.07 and 0.20% carbon maximum, with 18% or more chromium and 8% or more nickel, both requiring 70,000 psi tensile, 30,000 psi yield, 35% elongation, 40% R.A., are covered by ASTM A198-50. Carbide stabilizers, molybdenum, tungsten, titanium, columbium, or vanadium, and free-machining additions, selenium plus phosphorus or molybdenum plus sulfur, may be made as needed.

ASTM A222-39 deals with castings of higher chromium content:

Class	C, % max	Cr, %	Ni, %	Tensile, psi	Yield, psi	Elongation, %	R.A., %
A	.25 *	22–26	11–13	70,000	45% of tensile	30	30
B	.30	23–27	19–22	72,500	45% of tensile	25	25
C	.30	26–30	8–10 †	85,000	45% of tensile	20	20

* Up to 0.40% for high-temperature use.

† Up to 12% when desired.

Stabilizing elements may be introduced.
ASTM A223-39 covers the high-nickel type:

Class	C, % max	Cr, %	Ni, %	Tensile, psi	Yield, psi	Elongation, %	R.A., %
A	.20 *	8–10	19–22	65,000	30,000	30	35
B	.60	14–17	34–37	60,000	45% of tensile	..	..

* Up to 0.35% for high-temperature use.

A casting alloy of 0.30 to 0.35% carbon around 12% chromium, around 8% nickel, 5% manganese, 2% silicon was introduced as a war substitute (Ref. 39).

Properties. The room-temperature properties merely strive to measure the similarity of a heat to another previous heat whose high-temperature behavior has been studied, and they form no basis for high-temperature design, nor for design for corrosive conditions. The high-temperature behavior of the austenitic high-chromium alloys has been exhaustively studied for load-carrying ability, oxidation resistance in air, and resistance to attack in other atmospheres. One partial study (Ref. 40) of one alloy covers 64 pages. The literature of corrosion resistance of austenitic and ferritic alloy steels is also very extensive.

The way to procure such alloys is to consult the producers, giving all details of the required forming, welding, etc., and complete information on the conditions of service, not only as to temperature of use and its fluctuations, necessary load-carrying ability, permissible deformation and the like but also as to the surrounding atmosphere or, for corrosion-resistant service, like details as to the corrodent. Because design to facilitate production of sound, properly fed castings is highly important, consultation should also be had on this topic.

High-temperature load-carrying ability of the heat-resistant austenitic alloys is much higher than that of the ferritic chromium steels of like chromium level. Alloys for extremely severe high-temperature service are built with sufficient chromium for resistance to oxidation, made austenitic by suitable alloys, usually nickel, and given much increased load-carrying ability by adding several per cent of molybdenum and/or columbium and/or tungsten. The carbon content varies widely. The composition range is sometimes narrow, since a brittle iron-chromium structure, metallographically termed sigma, often tends to appear as chromium is increased above a certain level, a level which varies with the other alloys present.

Still higher load-carrying ability is conferred by large amounts of cobalt. Iron is often reduced so far that the alloys are far from the category of steels.

Blades and rotors for gas turbines, combustion chambers and nozzles for jet propulsion, and like specialties are calling for super-alloys, for discussion of which space is lacking here. The alloys are in rapid state of development, and what is "standard" today becomes obsolete tomorrow. Evaluation of such alloys is made, for purposes of elimination and for load-carrying design values by high-temperature testing, according to the desired life, which may range from years to seconds. No one type of test puts the alloys in correct order for both long and short lives, and adequate design values are not obtainable short of service experience or simulated service tests that include the true conditions or service.

PRECIPITATION-HARDENED FERRITIC 17-7 STEEL. A relatively new material of construction, in which strength is of the order of that of cold-worked 18-8 austenitic stainless, is produced by precipitation hardening induced by titanium, in the presence of some aluminum. The corrosion resistance is similar to that of fully austenitic 18-8.

In the soft, annealed condition the properties approximate 120,000 psi tensile, 75,000 psi yield, with 3% (thin sheet) to 10% (bars over 1/2 in.) elongation. Application of precipitation hardening heat treatment produces from about 210,000 psi tensile, 195,000 psi yield with the same elongation as in the annealed condition, down to about 150,000 psi tensile, 125,000 psi yield, with 5% (thin sheet) to 12% (bars over 1/2 in.) elongation, according to the temperature (925 to 1050 F) and time for precipitation. These precipitation-hardened properties hold for large sections, no appreciable differences appearing between 1 in. diameter and 5 in. diameter bars. Castings as well as wrought shapes can be precipitation hardened.

On material of 170,000 psi tensile, a polished-bar endurance limit of 85,000 psi and a notched-bar one of 38,000 psi have been recorded. It is reported to hold its strength rather well up to about 600 F. Low-temperature behavior is like that of ferrite. The material can be welded and nearly full strength obtained if the joint is precipitation treated after welding. The range of properties obtained has been quite fully described (Ref. 41).

14. SUMMARY—GENERAL FERROUS MATERIALS OF CONSTRUCTION

The major commercially available ferrous materials of construction have been listed in the foregoing, with their ranges of room-temperature mechanical properties insofar as these are revealed by conventional tests and described in current purchase specifications. Seldom do such tests and specifications go to the root of the problem, suitability for intended processing and final use.

However, experience has shown the range of applicability of the different commercial materials, and the tests and specifications are designed to insure getting more material of like applicability. They do not imply that other materials may not be equally applicable, better, or cheaper, nor do they prove that all materials passing the tests and specifications are suitable. There are many minute details in the processing of irons and steels that affect serviceability. The effects of some of them can sometimes be differentiated by special tests that the user could apply; sometimes they are detectable only by actual or simulated service. Experienced producers, fully notified of the conditions under which the purchaser will process and finally use the material, will aid in the selection of proper and the avoidance of improper material. The more of a commercial material they have produced, the larger their background of experience as to its behavior. This fact, as well as the item of cost, is a reason for designing on the basis of commercial material that is in large production.

This is not invariably the case. Materials of well-known properties not now being produced in large volume may offer advantages not so obvious to the potential producer as to the consumer. It may require consumer pressure to make such material available. On the basis of British war experience, reported by Swinden (Ref. 42), leaded steels might come into this category.

The producers also know the replaceability of one material by another. If they are invited to suggest a replacement, rather than being hampered by imposition of strict chemical composition requirements as well as mechanical ones, they can often suggest a more available and more economical material, equally applicable, instead of the one the designer formerly used. When the designer specifies without consultation, the producer's duty is done when he meets the specifications. When he is consulted and fully informed on the way the material will be used, he tries to supply the material that is most suitable; and neither the conventional tests nor the current specifications may measure suitability. Previous comment on directional differences gives point to this statement.

Some of the idiosyncrasies of irons and steels that affect suitability, but for a discussion of which space is lacking here, are brought out in recent metallurgical publications (*ASM Metals Handbook*, 1948 edition, Cleveland, Ohio, and Bullens, *Steel and Its Heat Treatment*, fifth edition, in 3 volumes, John Wiley and Sons, 1948). Of the factors in design which, quite as much as the iron or steel selected, govern the serviceability of a part, castings are treated in the *AFA Cast Metals Handbook*. Those affecting life under repeated stress are treated in reference books (Refs. 43, 44).

15. SPECIAL-PURPOSE MATERIALS

Because they enter only highly specialized designs, some ferrous materials call for mere mention here, rather than description.

TOOL AND DIE STEELS. Tools and dies are of huge importance in fabrication, but, save for some wear-resistant parts, the steels used for them do not enter ordinary design. For information in this specialized field, the book by Gill and co-authors (Ref. 45) should be consulted.

The most important of this group is high-speed steel, originally high in tungsten. Prior to and during World War II, the alternative use of molybdenum and finally of compositions utilizing both, grew up. Many controversial discussions of the relative value of the various types have been published. The facts are summed up by Stotz (Ref. 46): ". . . during the height of the war production, no tools made of properly heat-treated molybdenum high-speed steel failed to hold their own with the equivalent steels in the tungsten family, and in many cases they did better." This is further evidence that chemical composition is of secondary and structure of primary importance, for the structures of the tungsten and molybdenum high speeds are indistinguishable.

SPECIAL CORROSION-RESISTANT FERROUS MATERIALS. For resistance to most acids, except to hydrofluoric and hot hydrochloric, a cast iron with about 15%

silicon is commercially available. It is weak, about 17,000 psi tensile, excessively brittle, and requires grinding instead of machining. Its use is an interesting case of protecting an installation against conditions where its brittleness would invite failure, in order to utilize an outstandingly good characteristic.

Austenitic cast irons, high in nickel or nickel and copper, are resistant to a variety of severe corrosive conditions and are machinable. Details are discussed in the *AFA Cast Metals Handbook*.

MAGNETS AND FERROUS MATERIALS USED FOR MAGNETIC PROPERTIES. For discussion of these special materials, see *ASM Metals Handbook and* Mechanical Properties of Metals and Alloys, *Circular* 447, National Bureau of Standards, 1943.

VALVE STEELS—INTAKE AND EXHAUST. For these, see Heron *et al.* (Ref. 47) and Hoyt (Ref. 48).

POROUS PRODUCTS. Porous shapes can be made by compressing iron or low-carbon steel powder in a die, sintering and shrinking the compact so made at elevated temperature, then repressing in a coining die, or machining to size when close dimensions are required. Large objects are difficult to produce with interior properties similar to the exterior because the powder does not effectively transmit pressure very far. Addition of carbon or graphite, with time for diffusion during sintering, as in carburizing, or carburizing by gaseous media, produces a steel-like matrix around the holes.

Because of the porosity, the powder iron objects rust easily and must be impregnated with oil. The product is used for very lightly loaded bearings where its oil retention is utilized. The strength is low, of the order of weak cast iron, and the brittleness is even greater. Attempts to produce objects quite free from porosity by use of extremely high pressures, to fill the pores with some other metal such as copper, and to improve properties by heat treatment of carburized objects have not been notably successful. Strength is inferior to that of solid cast iron or steel produced by conventional methods, and extreme brittleness is seldom avoided.

For production of small objects that are costly to machine from solid stock, in sufficient quantity to amortize the cost of presses, pressing dies, and coining dies, when the objects require only very mediocre strength and are protected against rusting and corroding, the process and products deserve consideration. However, they come more nearly at the bottom of the list of ferrous materials of construction than some overenthusiastic appraisals would indicate. Their best applications are where the porosity is utilized.

PROPERTIES OTHER THAN STRENGTH. Certain properties other than strength often enter into design. Ordinarily, however, they are not so crucial but that a knowledge of the general order of magnitude will suffice. When highly precise information must be had, specific tests of the exact material are needed.

POISSON'S RATIO. This approximates 0.3 for ferrous materials.

MODULUS OF ELASTICITY IN TENSION. The elastic modulus of steel is little affected by composition or structure. The usual design value is 29,000,000 psi. Values for ordinary steels ranging from 26,000,000 to 33,000,000 psi have been reported, but major deviations from 29,000,000 $\pm$ 1,000,000 are unlikely to be substantiated. The modulus increases with decrease in temperature and decreases practically linearly to about 25,000,000 or 26,000,000 psi at 700 F. Thereafter, it may in steels with high-temperature strength continue on the straight line to 1000 F or more, but ultimately drops rapidly.

Austenitic manganese and stainless steels, especially when strain hardened and not stress relieved, have slightly lower but about the same modulus over the lower part of the stress-strain diagram as other steels. However, the effective modulus up to the yield strength is lower because of the shape of the diagram. Values of 29,000,000 psi for quench-annealed and 27,000,000 psi for strain-hardened austenitic stainless are commonly used. Less decrease in modulus with increased temperature is to be expected than with ordinary steels.

Malleable irons show around 25,000,000 psi. The ordinary commercial gray cast irons range from about 12,000,000 psi for very soft, up to 23,000,000 for *very strong irons*, with 18,000,000 as a common value for average iron.

SPECIFIC GRAVITY. Specific gravity of carbon steel as rolled ranges from about 7.85 for low carbon to around 7.81 for over 1% carbon. Quenched to martensite it ranges from 7.83 to 7.76. Tempering at increasing temperatures raises the specific gravity to about the as-rolled figures. Cast steel is usually taken at about 7.85.

Ordinary constructional alloy steels run about the same as carbon steels, since the densities of most of the alloying elements do not differ sufficiently from that of iron for their small content to have much effect. Carbon and silicon are much less dense than iron, so that gray cast iron ranges from 7.0 for very soft iron to 7.3 for the stronger irons.

Malleable iron runs from 7.2 to 7.45, but the usual value taken is 7.30 to 7.40 with pearlitic malleable about the same.

The wrought austenitic stainless steels, low in carbon, run from around 7.93 for 18-8 to 7.83 for 25-12.

THERMAL EXPANSION. Thermal expansion values are ordinarily given in terms of the linear coefficient per degree centigrade times 10^{-6}. To transform to coefficient per degree Fahrenheit, multiply by 5/9.

The coefficient increases with increasing temperature, i.e., 11 to 12 from zero to 212 F, 13 to 15 from zero to 750 F for carbon and constructional alloy steels. Cast steels usually show 12 and 15. 18% chromium, 8% nickel stainless gives around 16 1/2 to 17 and 17 to 19 over these ranges. The 18-8 austenitic steel expands nearly half as much again as ordinary steels, the 25% chromium, 12% nickel about a quarter as much again. Plain chromium ferritic stainless steels expand less, about 90% as much, as ordinary steel.

Gray cast iron gives 10 to 8 1/2 and 12 to 13, malleable about 12 and 14, over these ranges. Detailed data are given by Mochel (Ref. 49). For cast materials, see the *AFA Cast Metals Handbook.*

THERMAL AND ELECTRICAL CONDUCTIVITY. In the ordinary ferrous materials conductivity decreases greatly with increase in carbon, in alloy content, and in temperature. Ordinary steels and cast irons have around one-eighth to one-ninth the thermal conductivity of copper at room temperature. In the quenched, martensitic condition, the conductivity is still low.

The austenitic steels have much lower conductivity, 18% chromium, 8% nickel, about one-twenty-fifth the thermal conductivity of copper at room temperature, and the Hadfield manganese steel of the same order. However, the conductivity of the austenitics rises slightly with temperature but is still low, less than half at 750 F of the value for ordinary steels and cast irons at 750 F.

Electrical resistivity in microhm-centimeters puts the ferrous materials in the following order as compared with copper, all the figures for ferrous alloys being approximate:

Copper	1.682
Low-carbon steel	11
Cast steel	15
High-carbon steel	20
Malleable	30– 35
Gray cast iron	50–100
Austenitic 18-8	90–130

Both the thermal and electrical conductivity figures reported by various observers differ greatly and cannot be depended upon for precise design.

REFERENCES

1. T. C. Dickson, Casting guns centrifugally, *Trans. Am. Soc. Steel Treating,* Vol. 18, pp. 21 ff., 1930.
2. J. B. Johnson, Aircraft engine materials, *S.A.E. J.,* Vol. 40, pp. 153–162, March 1937.
3. N. P. Inglis, Relationship between mechanical tests of materials and their suitability for specific working conditions, *Trans. North East Coast Inst. Engrs. & Shipbuilders,* Vol. 53, pp. 101–136, 1936–1937.
4. J. O. Jackson, Welded liquefied natural gas storage tanks, *Welding J. (N. Y.),* Vol. 20, pp. 833–836, December 1941.
5. T. M. Jasper, Multi-layer construction of thick-wall pressure vessels, *Proc. Am. Petroleum Inst.,* Vol. 20, pp. 7–16, 1939.
6. J. H. Frye, Malleable iron castings in ordnance, *Foundry,* Vol. 71, pp. 102–104, 162–163, September 1943.
7. O. W. Simmons, P. C. Rosenthal, and C. H. Lorig, Malleable castings for military applications, *Foundry,* Vol. 71, pp. 102–104, 190, October 1943; pp. 122–125, 190–192, November 1943; pp. 107–108, 202–204, December 1943.
8. E. J. Janitsky and M. Baeyertz, The marked similarity in tensile properties of several heat treated S.A.E. steels, *ASM Metals Handbook,* American Society for Metals, 1939 ed., pp. 515–518.
9. W. G. Patton, Mechanical properties of N.E., S.A.E., and other hardened steels, *Metal Progress,* Vol. 43, pp. 726–733, May 1943.
10. S. A. Herres and A. F. Jones, A method for evaluating toughness of steel, *Metal Progress,* Vol. 50, pp. 462–469, September 1946.
11. W. Crafts and J. L. Lamont, The Izod impact strength of heat treated alloy steel, *Am. Inst. Mining Met. Engrs. Tech. Pub.* 2134, *Metals Technol.,* Vol. 14, February 1947, 15 pp.
12. *Contributions to the Metallurgy of Steel,* American Iron and Steel Institute, No. 9, Report on special alloy addition agents, November 1942, 30 pp; No. 11, Tentative hardenability bands, July 1944, 50 pp.
13. L. E. Eckholm, "H" steels and their specifications, *Metal Progress,* Vol. 48, pp. 673–683, October 1945.
14. C. M. Parker, Selecting steels by hardenability bands, *Materials & Methods,* Vol. 25, pp. 68–72, March 1947.
15. Data sheet, *Metal Progress,* Vol. 50, p. 832, November 1946.
16. V. H. Schnee, Gray irons vs. gray iron, *Iron Age,* Vol. 143, pp. 19–23, 48, April 20, 1939.
17. J. E. Rehder, Malleable iron test bars, *Canadian Metals Met. Ind.,* Vol. 10, pp. 14–17, February 1947.

18. C. F. Joseph, Armasteel, its manufacture, control and applications, *Foundry Trade J.*, Vol. 67, pp. 65–69, 385–389, 397, 1942; Vol. 68, pp. 11–15, 1943.
19. C. H. Lorig, Properties of commercial pearlitic malleable irons, *ASTM Bull.*, pp. 29–34, August 1940.
20. H. W. Russell, Resistance to damage by overstress of precipitation hardened copper steels and copper malleable, *Metals & Alloys*, Vol. 7, pp. 321–324, December 1936.
21. C. E. Sims and F. B. Dahle, Quality of converter cast steel, *Proc. Am. Soc. Testing Materials*, Vol. 42, pp. 532–555, 1942.
22. F. T. Sisco, *Alloys of Iron and Carbon*, Vol. II, *Properties*, McGraw-Hill Book Co., New York, 1937, 777 pp.
23. C. M. Motley, The application of statistical methods to the development and quality control of high tensile steel, *J. Am. Soc. Naval Engrs.*, Vol. 57, pp. 21–55, February 1945.
24. W. C. Stewart and R. E. Wiley, Chemical and mechanical properties of some of the N.E. steels *J. Am. Soc. Naval Engrs.*, Vol. 56, pp. 396–411, August 1944.
25. M. N. Landis, Cold finished bars to physical property specifications, *Metal Progress*, Vol. 48, pp. 769–777, October 1945.
26. G. D. Boyer, Popular types of steels, *Product Eng.*, Vol. 18, pp. 81–85, January 1947.
27. D. I. Brown, An economic approach to hardenability, *Iron Age*, Vol. 159, pp. 42–47, March 27, 1947; pp. 53–55, 160, April 3, 1947
28. J. A. Rosa, Alloy steels for maintenance, *Iron Steel Engr.*, Vol. 22, p. 37, September 1945.
29. W. Lehrer, Bright gas quenching of SAE X4130 and NE 8630 welded aircraft tubes, *Trans. Am. Soc. Metals*, Vol. 33, pp. 290–319, 1944.
30. C. Carmichael, Primary factors in choosing aircraft materials, *Machine Design*, Vol. 17, p. 113, October 1945.
31. C. M. Schwitter, Triple alloy structural tubing, *Iron Age*, Vol. 155, p. 46, Aug. 22, 1946.
32. J. H. Clark, J. D. Walker, and A. S. Jameson, N. E. steels for steering knuckles and axle drive shafts, *Metal Progress*, Vol. 48, pp. 280–287, August 1945.
33. H. C. Keysor, Carbon and alloy steel materials for hot-formed springs, *Product Eng.*, Vol. 27, pp. 86–89, November 1946.
34. O. W. McMullan, Physical properties of case hardened steel, *Trans. Am. Soc. Metals*, Vol. 23, pp. 319–381, 1935.
35. J. O. Almen and A. L. Boegehold, Rear axle gears, Factors which influence their life, *Proc. Am. Soc. Testing Materials*, Vol. 35, Pt. 2, pp. 99–146, 1935.
36. J. B. Johnson and T. T. Oberg, Effect of notches on nitrided steel, *Metals & Alloys*, Vol. 5, pp. 129–130, 1934.
37. J. B. Austin, Effect of changes in carbides on the properties of steel, *Metal Progress*, Vol. 50, pp. 1191–1195, December 1946.
38. C. A. Zapffe, Corrosion resistance of the stainless steels, *Metal Progress*, Vol. 48, pp. 693–707, October 1945.
39. O. E. Harder and J. T. Gow, An emergency heat resistant alloy, *Trans. Am. Soc. Metals*, Vol. 32, pp. 408–475, 1944
40. O. E. Harder and J. T. Gow, Balancing the composition of cast 25 Cr 12 Ni type alloys, *Trans. Am. Soc. Metals*, Vol. 30, pp. 855–919, 1942.
41. R. Smith, E. H. Wyche, and W. W. Gorr, A precipitation-hardening stainless steel of the 18 Cr 8 Ni type, *Am. Inst. Mining Met. Engrs. Tech. Pub.* 2006, *Metals Technol.*, June 1946.
42. T. Swinden, Leaded manganese molybdenum steel, Paper 13–1943, Alloy Steels Research Committee, British Iron & Steel Institute. Abstract *Metal Progress*, Vol. 47, pp. 406–407, February 1945.
43. Battelle Memorial Institute, *Prevention of the Failure of Metals under Repeated Stress*, John Wiley and Sons, New York, 1941 (1946 printing), 295 pp. A book on design of structures to resist fatigue, same author and publisher, is in preparation.
44. University of Melbourne Symposium, *The Failure of Metals by Fatigue*, University of Melbourne Press, 1947, 505 pp.
45. J. P. Gill, R. S. Rose, G. A. Roberts, H. G. Johnstin, R. B. George, *Tool Steels*, American Society for Metals, Cleveland, Ohio, 1944, 577 pp.
46. N. Stotz, quoted in *Steel*, Vol. 120, p. 37, March 31, 1947.
47. S. D. Heron, O. E. Harder, and M. R. Nestor, Exhaust valve materials for internal combustion engines, *Proc. Am. Soc. Testing Materials*, Vol. 40, pp. 1–26, 1940.
48. S. L. Hoyt, *Metals and Alloys Data Book*, Reinhold Publishing Corp., New York, 1943, 334 pp.
49. N. L. Mochel, Thermal expansion of metals, *Symposium on Effect of Temperature on Metals*, pp. 683–718, American Society for Testing Materials and American Society of Mechanical Engineers, 1931.
50. C. K. Donoho, Producing nodular graphite with magnesium, *Am. Foundryman*, Vol. 15, pp. 30–37, Feb. 1949.
51. A. P. Gagnebin, K. D. Mills, and N. B. Pilling, Ductile cast iron, a new engineering material, *Iron Age*, Vol. 163, pp. 77–84, Feb. 17, 1949.
52. A. G. H. Dietz (Ed.), *Engineering Laminates*, John Wiley & Sons, New York, 1949.
53. Cleaves and Thompson, *The Metal-Iron*, McGraw-Hill Book Co., New York, 1935.
54. Reid L. Kenyon, Physical and mechanical properties of iron, *Metals Handbook*, American Society for Metals, Cleveland, p. 423, 1948.

SECTION 3

CORROSION AND CORROSION RESISTANCE

By

FRANK N. SPELLER, *Metallurgical Consultant, Pittsburgh, Pa.*

H. H. UHLIG, *in Charge of Corrosion Laboratory, Massachusetts Institute of Technology*

CORROSION OF METALS

By Frank N. Speller

CORROSION RESISTANCE OF METALS AND ALLOYS

By H. H. Uhlig

CORROSION OF METALS

By Frank N. Speller

1. CORROSION IN GENERAL

Corrosion may be defined broadly as the deterioration or destruction of a metal by electrochemical or chemical action of its environment. Under most ordinary conditions of exposure the corrosion products comprise mainly oxides (more or less hydrated), carbonates, and sulfides. At high temperatures the products consist mainly of oxides, though indications are that corrosion of this type also progresses more rapidly when some moisture is present. Corrosion may be entirely superficial or may attack the intercrystalline boundaries and cause disintegration in the metal, as dezincification of brass or so-called caustic embrittlement in boiler steel.

In nearly all cases of corrosion at ordinary, or elevated, temperatures, comprising those most important in practice, the driving force of the corrosion reaction between metal and environment is electrochemical. This discussion refers mainly to iron and steel.

SOME ESTABLISHED FACTS REGARDING CORROSION. (*a*) In most cases both moisture and oxygen are necessary for corrosion. (*b*) The initial rate of corrosion usually is comparatively rapid, slowing as a protective film forms. (*c*) Surface films are important in controlling the rate and distribution of corrosion. (*d*) Increased rate of motion increases corrosion in water up to a certain velocity; the rate then may drop where strong protective films are formed. (*e*) Dissimilar metals in electrical contact accelerate corrosion of the ones that happen to be anodic. Galvanic action is a most active agent of corrosion. It occurs when two metals, one electronegative to the other, are placed in contact and exposed to moisture. (*f*) The composition of ordinary iron and steel has little or no effect on their relative rates of corrosion under water or under ground. Under these conditions the particular ferrous metal usually is not so important as environment. However, the chemical composition of wrought iron and carbon or low-alloy steel has considerable influence in corrosive atmospheres or acids. (*g*) Variation in the concentration or composition of a solution in contact with a metal tends to localize corrosion. Where this is due to oxygen in solution, the surface to which the oxygen diffuses most rapidly is cathodic to the area that is protected from diffusion of oxygen. The smaller the anodic areas in relation to cathodic areas the greater is the rate of penetration at anodic points. Corrosion, like other chemical reactions, increases with temperature, other factors being the same. Anaerobic bacteria in absence of oxygen may cause serious corrosion by breaking down sulfates in water or soil and indirectly accelerate corrosion by depolarizing cathodic areas as they remove the hydrogen.

The above, and other facts, are the basis of the electrochemical mechanism which now is generally accepted to explain corrosion in the presence of water (Refs. 1, 2, and 3).

2. MECHANISM OF CORROSION

CORROSION OF IRON OR STEEL IN GENERAL. While there is a vast difference in the amount of corrosion of iron and steel under different conditions, and while the relative corrosion varies with surrounding condition, certain fundamental causes underlie all corrosion.

Every metal when placed in water is subjected to a fixed tendency to go into solution. This is wholly a matter of electrochemical activity and varies to a definite extent with each metal. The initial reaction in the process of corrosion is analogous to solution in acid. The acidity of the water determines the initial speed of attack. As in all electrochemical reactions, however, the initial speed of solution soon slows down as the electrochemical couples which form over the surface of the metal become "polarized" due to the accumulation of hydrogen on their cathodic surfaces. Thus corrosion may be practically stopped. In the case of iron in aqueous media, ferrous hydrate film forms which also retards the rate of solution of the metal. In other words, iron, like all other metals, when in contact with water or a solution has a definite inherent tendency to go into solution in the form of electrically charged particles (iron ions). The solution must remain electrically neutral, and these positive ions can enter it only if an equivalent number of positive ions of some other elements are displaced. Thus, iron immersed in a solution of copper sulfate begins to go into solution, but, simultaneously, an equivalent amount of copper is plated out on the surface of the iron and protects it somewhat from further action. Similarly when iron is immersed in water, hydrogen is plated out and forms a thin invisible film on the iron surface.

This first stage of corrosion of iron, while essential, is not serious if the reaction between water and iron can be stopped at this point. In fact, it cannot proceed unless, in some manner, the hydrogen film protecting the metal is removed. This is usually accomplished by some depolarizing agent, generally free oxygen, which also combines with ferrous hydrate, forming insoluble ferric hydrate, commonly known as rust.

In solutions that are nearly neutral such as domestic water the amount of hydrogen evolved from cathodic surfaces in the gaseous form is small, equivalent only to about 3% of the overall corrosion. The amount of hydrogen released as gas increases with the acidity of the solution. The rate of corrosion is not materially influenced by the *p*H of the solution between *p*H 4 and 9. Low *p*H generally connotes increased hydrogen evolution while high *p*H (over 9) is usually accompanied by a lower rate of overall corrosion. At an intermediate *p*H (around 8) corrosion is localized, and the penetration rate may be serious.

This second stage of corrosion is thus caused by the oxidation of hydrogen and ferrous hydrate in solution by free oxygen from the atmosphere. Oxygen is soluble in water at normal temperature and pressure to the extent of about eight parts by weight per million, and as water readily dissolves oxygen, the reaction will continue until the metal has been destroyed, or the reaction may be stopped by exhaustion of the available supply of oxygen. The latter case is illustrated in any closed water system, such as a metallic water pipe, where the free oxygen will decrease as the water travels from the inlet. If, then, the pipe is long enough the corrosion will finally be reduced to a negligible amount.

FACTORS CONTROLLING CORROSION. Primary factors, which are those having to do with the initial tendency toward solution, are associated mainly with the metal itself. Secondary factors, those that influence the subsequent rate of corrosion, usually are much more important than primary factors. They depend mainly on the environment of the metal. The *main* factors influencing corrosion are listed below, regardless of their relative importance, which changes materially with environment.

Primary Factors Having to Do Mainly with the Metal. (1) Effective electrode potential of the metal in solution. (2) Chemical and physical homogeneity and texture of its surface. (3) Its inherent ability to form an insoluble protective film. (4) Overvoltage of hydrogen on the metal surface.

Factors Which Vary with Environment. (5) Hydrogen-ion activity (*p*H) in the solution. (6) Effective supply of oxygen in the solution adjacent to the metal. (7) Distribution of available oxygen on the metal surface. (8) Specific nature, concentration, and distribution of other ions in the solution. (9) Effective rate of flow of the liquid past the metal surface. (10) Presence of solid particles (dirt), or a coating of any kind, as mill scale, on the metal surface, or contact with other conducting material in the presence of an electrolyte. (11) Temperature. (12) Whether the metal is under static stress or cycles of alternating stress (the so-called corrosion fatigue). (13) The ability of environment to build up protective films on the metal. (14) Biological influences. (15) The corrosion reaction may be restrained or prevented by inhibitors, such as chromates, silicates, and carbonates in water.

While all the above factors are important under some conditions, the following deserve special attention and should be more clearly recognized. One or more factors usually dominate and therefore control the rate of attack in any specific environment; moisture in air and dissolved oxygen in water are common examples of "controlling factors."

Oxygen. In most cases, oxygen concentration and the rate at which oxygen diffuses to the metal are controlling factors when the metal is immersed in water or buried in soil. The part to which the oxygen diffuses most readily is cathodic to the adjacent areas that are shielded by protective coatings or other material. That is, the areas that are less accessible to oxygen corrode more rapidly than areas to which the oxygen has freer access. This and the galvanic effect from contact between dissimilar materials generally cause local corrosion or pitting.

Surface Films. Clean metallic surfaces exposed to air quickly acquire a film of gas or oxide, which at ordinary temperature is thin and invisible. At higher temperatures, the oxide is thick enough to give well-known characteristic colors, varying with the temperature. The nature and properties of the film depend on the composition of the metal itself and its environment. Aluminum and stainless steels, for instance, owe their high durability in certain environments to the formation of a continuous and permanent film that is stable under the conditions of exposure. In other environments the film may not form or is quickly dissolved. The metal, therefore, has short life under such conditions. The addition of 0.25% copper to ordinary steel increases the life of the steel in industrial atmospheres, because of the formation of a dense adherent layer apparently consisting of iron oxide with a little copper oxide on the surface of the metal. When copper steel is immersed in water or buried in soil, this protective layer does not form. Under such conditions, copper steel is no more durable than ordinary steel without copper.

If the metal has no inherent power to form a protective film in solution, a film may be built up by adding film-forming reagents, as the chromates or alkalies, to the water. When removed from the solution, however, these films are only temporary. To secure permanent protection it is necessary to maintain in the water the proper concentration of chromate, or other material, used for this purpose. Films, when formed, should be self-healing, so that if injured or removed in any way a new film will form immediately. Chromium in iron over about 2% contributes to formation of a resistant invisible film under oxidizing conditions, but in dilute hydrochloric acid, where such films cannot exist, chrome steels dissolve even more rapidly than pure iron. Metals which acquire a stable protective film in a passivating environment become passive. In this condition they are at least temporarily resistant to corrosion when removed from the film-forming solution. The presence of thin invisible oxide films on passive material has been determined by the change in potential of the metal and by the removal of the films from the surface by selective dissolution of the underlying metal. (Refs. 4 and 5.) Where the corrosion products form spongy layers of irregular thickness, they do not protect, and may even accelerate, the normal action by interfering with the rate of diffusion of oxygen to the surface of the metal in water, and by holding moisture on the metal in the atmosphere.

Corrosion Fatigue. Metals subjected to repeated application of stress fail at considerably below their static ultimate tensile strength. However, the metal may be cyclically stressed indefinitely without apparent injury, below the so-called air fatigue or endurance limit. In absence of corrosion, carbon steels may have an endurance limit about one-half their tensile strength, but even a very small amount of corrosion present with cyclic stress will considerably reduce this endurance limit. Concentrated stress tends to accelerate the rate of corrosion. A notch is soon formed at the point of maximum stress, and, after this reaches a certain depth, the metal under cyclic stress fails by ordinary fatigue. The combined action of cyclic or repeated stress and corrosion is termed *corrosion fatigue*. The amount of cyclic stress that a metal will stand under these conditions depends mainly on its corrodibility and notch sensitivity and on the nature of the protective films formed by the metal or its environment. Hence, there is no fixed corrosion fatigue limit for any metal under all conditions of exposure. Corrosion factors predominate in corrosion fatigue and corrosion fatigue is usually independent of the tensile strength. Heat treatment of steel raises the tensile strength and the air endurance limit, but if it does not increase resistance to corrosion it will have no material influence on the resistance to corrosion fatigue. Even 18% chromium steel and 18-8% chromium-nickel steels show a considerable reduction in endurance limit in corrosive environment. Carbon steels in the as-rolled condition may fail at 20 to 75% of their air-endurance limit when kept continuously wet, depending on the composition of the water and the amount of oxygen present. (See also Ref. 3.)

An exceedingly small amount of corrosion is sufficient to depress the endurance limit of the metal under stress. In practice, therefore, it is important to prevent contact between stressed portions of metal and corrosive solutions. For instance, the endurance limit (based on 10^7 stress cycles at 1750 cycles per min) of SAE 1035 steel of 85,000 tensile strength showed:

In air	40,600 psi
In brine	24,600 psi
In brine containing H_2S	10,600 psi

Corrosion fatigue limits are higher in fresh water than in salt water—both aerated. They are not materially affected by changes in composition or heat treatment unless the steel is made thereby more resistant to corrosion. Chromium is the most effective alloying element for increasing the corrosion fatigue resistance in oil-well brines in absence of H_2S, but nickel is most effective when H_2S is present. A 3% Ni–Mo iron had a fatigue limit of 21,900 psi in the sulfide brine. Corrosion fatigue is manifested by cracks transverse to the direction of the maximum stress originating from corrosion pits. The most effective means for preventing corrosion fatigue is by reducing operating stress; the next most effective is by control of corrosion by corrosion-resistant alloys, protective coatings or inhibitors. Strong inhibitors like sodium chromate tend to prevent corrosion fatigue, except in high chloride waters.

An *inhibitor* is a chemical substance or mixture which, when added to an environment, usually in small concentration, effectively decreases corrosion. A *passivator* is a special form of inhibitor which forms a thin and relatively inert film of protective matter on the metal surface. This film is thought to consist of a very thin adsorbed monolayer of oxygen or passivating ions such as CrO_4, sometimes reinforced by a thin layer of reaction products. Metals thus passivated are less active and usually have a more noble potential than the active metal. (See also Refs. 6 and 7.)

3. CORROSION TESTS

CORROSION TESTING may be carried out in the laboratory or in the field under service conditions. Laboratory tests permit close control of dominant factors. Study of fundamentals, such as influence of oxygen concentration and velocity of motion and electrical potential differences, are examples. *Field tests* should be made with all variations in temperature, light, rate of motion, oxygen concentration, humidity, etc., that usually occur in nature. Laboratory tests are well adapted to the study of fundamentals and preliminary study of problems in the chemical industry, but tests of metals or protective measures are best made in the field or plant or, if time permits, in actual service. In no case should a factor such as an acid be introduced that does not occur in service. This has been done to accelerate the rate and shorten the time required but usually will give misleading results.

RELATIVE CORROSION. Difference of opinion regarding the relative corrosion of iron and steel and other metals apparently is due to failure to appreciate that different factors control in different environments. Factors external to the metal usually determine the corrosion rate, and these vary considerably with environment. The relative rate, therefore, may vary under different conditions of exposure. This fact should be the basic principle on which all corrosion tests (laboratory or service) are designed. Thus, in a study of metals, under atmospheric condition, the dominant factor should be an excess of oxygen. If the endurance limit of metal under water is in question, the test should be made in that environment alone. It is a mistake to take the results of tests made under one type of corrosion, i.e., in the atmosphere or in acids, and to apply the conclusions to the conditions in which entirely different factors control. An extreme instance is the well-known acid test made by exposing the metals to dilute sulfuric acid, in regard to which the ASTM has issued the following statement:

> The continued functioning during the past 25 years of the Committees on Corrosion of Iron and Steel of the A.S.T.M. has developed no further information as to the relation of the accelerated acid test to corrosion by the elements, or the trustworthiness of the acid test as a measure of natural corrosion.
>
> Studies were made to determine whether consistent results can be obtained in testing iron and steel sheets in 20% sulphuric acid, and if the results in acid compare favorably with the results of corrosion tests in the atmosphere. See Report on Total Immersion Tests, *Proc. A.S.T.M.*, Vol. 31, Part I, p. 176, 1931.
>
> A representative average set of samples from two sheets each of ten materials of iron and steel, both copper bearing and non-copper bearing, part of each sheet having been exposed to the atmosphere, was tested by five co-operating laboratories following an exact accelerated acid test procedure. The life in acid of similar sheets in atmospheric corrosion tests at Pittsburgh, Pa., and Key West, Fla., indicated that an acid test conducted by the procedure followed by the committee is not capable of consistent repetition. Further, the data from such an acid test should not be used to forecast the relative life of ordinary ferrous materials in the atmosphere.

Clearly, the subject of relative corrosion should be studied on the basis of type of service. The service test is the final criterion of the value of the metal, although judiciously made laboratory tests, with proper consideration given to the factors involved, will show under selected conditions a fair approximation of results in service. Three common types of service to be considered are atmospheric, water, and soil corrosion.

ATMOSPHERIC CORROSION. In atmospheric corrosion, oxygen is the dominant factor. In general, results of tests indicate little difference between the various kinds of iron and steel except where copper is present in excess of 0.07%. With copper content of 0.2 to 0.5%, the durability of steel in air is two to three times that of steel without copper in industrial or marine atmospheres. Figure 1 shows the influence of copper on corrosion. The greatest contrast between copper steel and noncopper steel is found in industrial atmospheres. Wrought iron without copper shows slightly longer life than noncopper steel under outdoor conditions. This probably is due to the better adherence of the products of corrosion to the rough surface of the metal. Sulfur in excess of 0.06% tends to increase the rate of corrosion in industrial air. For results of authoritative work on this type of corrosion, see reports of Committee A-5 in *Proc. ASTM* dating from 1915 to the present time.

A critical study of ASTM tests (see Ref. 8) is summarized as follows:

A study of the Pittsburgh and Fort Sheridan results confirms previous conclusions as to the effect of copper. At the close of eleven years of observations of these tests, the sub-committee would direct attention to the fact that failures at the Fort Sheridan location confirm the findings at the Pittsburgh location, that copper-bearing metal shows marked superiority in rust-resisting properties as compared to non-copper bearing metal of substantially the same general composition under atmospheric exposure. (Report of Committee A-5 on Corrosion of Iron and Steel, *Proc. A.S.T.M.*, Vol. 28, Pt. I, p. 156, 1928.) This paper also points out that at the Pittsburgh location, phosphorus combined with copper has beneficial effects from an atmospheric corrosion standpoint. Manganese and sulphur, within the limits of composition involved, seem to have little, if any, effect either way. The Fort Sheridan data, although limited on account of the incompleteness of the test, show similar beneficial results for high-copper material. The A.S.T.M. total immersion tests have been thoroughly studied, and, contrary to the atmospheric tests, the results indicate that the variation of the common elements, either singly or in combination, within range of composition and under the conditions of these tests, has no material influence on corrosion resistance and is entirely secondary to external conditions.

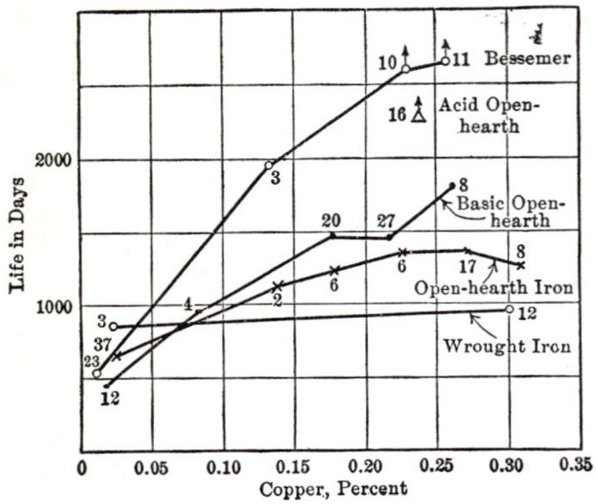

FIG. 1. Influence of copper on corrosion.

Tests of vent lines on the roofs of buildings were carried out by the National Tube Co., with wrought-iron, steel, and copper-steel pipe. Results based on the decrease in wall thickness indicated the relative loss of these metals to be as follows, taking steel as 100: wrought iron, 92%, copper steel, 36%.

Tests on hopper and gondola cars, constructed one-half of copper steel and the other half of plain steel, showed after 13 years that the copper steel had about 200% longer life in the hopper cars (subjected to some abrasion with corrosion) and 300% longer life for the side sheets of the gondolas (subject to corrosion only) in comparison with plain steel of the same composition. The number of cars tested was 100. J. S. Unger (Ref. 9) summarizes the experience with these cars as follows:

(1) Paint adheres much better to copper-bearing steel in a car body than to plain open-hearth steel. (2) The saving in repainting cars, due to better adherence of paint to copper-bearing steel, would be sufficient to justify the use of copper-bearing steel in the bodies of steel freight cars. (3) Where mechanical abrasion has not been a serious factor, as in the side sheets of gondolas, the loss in thickness for the copper-bearing steel was only one-third as great as for the plain open-hearth steel. (4) Where the steel was subjected to severe mechanical abrasion as well as corrosion, the loss in thickness for the copper-bearing steel was approximately 60% as great as for the plain open-hearth steel. (5) From the results of this investigation, it can be conservatively stated that the use of all copper-bearing steel in the body of the cars would increase the life of the car body from 33 1/3 to 50%.

Service tests on copper steel tie plates, containing 0.25% copper, over a period of 11 years, show 5% loss in weight for copper steel and 20% loss for plain steel. Similar results have been obtained on cut spikes. (See Ref. 10.) Tests of 730 plain and a similar number of copper steel ties, in track for 9 years, indicate at least double the life for copper steel ties with a subgrade of broken slag and ashes for ballast. (See Ref. 11.)

Low-alloy steels containing small amounts of chromium, silicon, and sometimes up to 0.15% phosphorus last about twice as long as copper steels in industrial air. One of the best commercial low-alloy steels of this type has the following analysis: C 0.09, Mn 0.24, P 0.154, S 0.024, Si 0.8, Cu 0.43, Ni 0.05, Cr 1.07% (Cor-ten).

WATER CORROSION. When material is in continuous contact with water, the available oxygen is limited to its solubility in water. This affects the nature of the film formed by corrosion products, and therefore the rate of corrosion. Immersion tests were made by ASTM on the materials referred to above, in their atmospheric tests. The results show no material difference between the copper steels and similar steels without copper. Wrought-iron, steel, and ferrous alloys commonly used in the manufacture of pipe, tested in service in warm water and return steam pipe lines, show practically the same rate of corrosion and pitting as in Table 1. (See Ref. 1.)

Table 1. Summary of Grand Averages of Complete Record of Comparative Service Corrosion Tests of Pipe

Tests conducted in cooperation with outside interests by National Tube Co., Pittsburgh, Pa. Percentages based on steel = 100; and for galvanized materials, galvanized steel = 100.

Character of Service *	No. of Tests	Steel	Wrought Iron		Galvanized Steel		Galvanized Wrought Iron		Copper Steel		Scale-free Steel	
		Depth, in.	Depth, in.	Per Cent	Depth, in.	Per Cent	Depth, in.	Per Cent	Depth, in.	Per Cent	Depth, in.	Per Cent
		Average of Deepest Pits (One pit, the deepest, from each pipe in test)										
HWS	38	.076	.084	111	......	.....	......	.....	......	.....	......	.....
HWS	7	......	......	.....	.084	100	.082	98	......	.....	......	.....
HWS	11	.084	......	.....	......	.....	......	.....	.080	95	......	.....
HWS	4	.069	......	.....	......	.....	......	.....	......	.....	.047	68
SR	9	.052	.063	121	......	.....	......	.....	......	.....	......	.....
SR	5	.043	......	.....	......	.....	......	.....	.027	63	......	.....
MW	1	.031	.070	226	......	.....	......	.....	......	.....	......	.....
HSW	1	.086	.135	157	......	.....	......	.....	......	.....	......	.....
HSW	1	......	......	.....	.061	100	.083	136	......	.....	......	.....
RBA	1	.095	.072	76	......	.....	......	.....	......	.....	......	.....

* Character of service: HWS = Hot water; SR = Steam return; MW = Mine water; HSW = Hot salt water; RBA = Refrigerating brine and atmosphere.

Examination of test samples from various parts of the country indicates no fundamental difference in the rate of corrosion of any of these materials *in water*, except that the copper steels are somewhat more durable for return steam lines, where corrosion at times is of the wet and dry type as in air.

Low-alloy steels suffer about the same loss in weight as carbon steels or copper steels in industrial or sea water, but usually show about one-half the penetration by pitting. Steels having 2 1/2% chromium are exceptional in that they suffer only about one-half the penetration by pitting of other low-alloy steels or copper steel in water or air. This has been confirmed by the 10-year sea water test at Kure Beach, N. C. For further information and practical experience with steel pipe for water service, see Ref. 12.

SOIL CORROSION. Under ground, the dominant factor is local electrolytic action, resulting in pitting. The rate of penetration varies with moisture content, electrical resistivity, physical structure and chemical composition of the soil and ground water, and stray electric currents leaving the pipe (electrolysis).

The National Bureau of Standards, in co-operation with API, AGA, and a number of the leading pipe coating manufacturers, has conducted an extensive investigation of soil corrosion in forty-seven typical soils in the United States. This comprehensive test in-

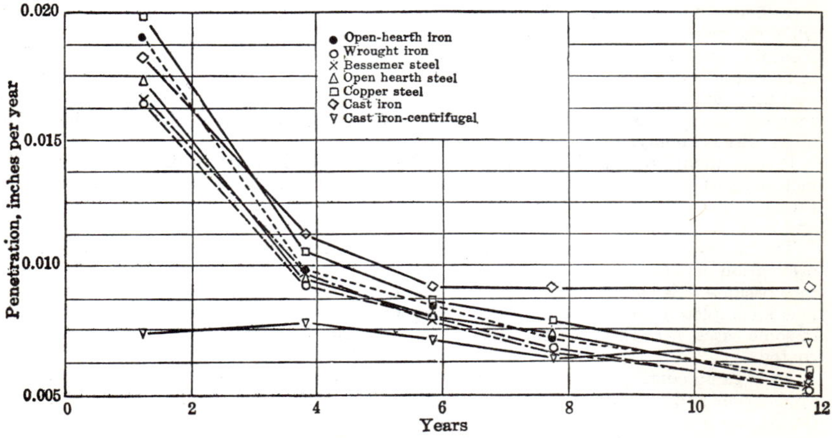

FIG. 2. Results of Bureau of Standards soil-corrosion study.

cluded some 15,000 samples. For detailed records of the tests, see Ref. 13. In Fig. 2 are summarized the weighted average results, based on maximum pits on ferrous metals from twenty-six corrosive soils included in these tests over a period of twelve years.

CHEMICAL CORROSION includes the action of acids and other chemicals. It is possible only to determine the relative durability of ferrous materials under such variable conditions by actual tests, in each case, in the laboratory or preferably in service. Before tests are made in the chemical industry the problem should be carefully appraised and all important factors determined and their relation to one another established.

4. PREVENTION OF CORROSION

Deterioration of metals can generally be prevented or reduced in three ways: (1) By using a more resistant metal. (2) By application of a suitable protective coating. (3) By making the environment less corrosive or (4) by cathodic protection (which might be included in the third group). The amount of protection specifically depends on the corrosive nature of the environment and must be chosen to suit the conditions and with a view to economy. In most cases involving iron or low-alloy steels the environment has much more influence on the rate of corrosion and pitting than the composition of the metal. This of course does not apply so generally to the more expensive types of corrosion-resisting ferrous or nonferrous metals.

5. PROTECTION AGAINST ATMOSPHERIC CORROSION

ORGANIC COATINGS.* Organic coatings are a class of materials, used mainly to protect metals from corrosion, which are applied as liquids and act chiefly as barriers between environment and the metals to be protected. Included in this class are varnishes, paints, lacquers, and enamels. Generally, their components are of two kinds, volatile and nonvolatile. The volatile component, called a thinner, consists of solvents and diluents. The nonvolatile component, i.e., the film-forming material, such as resin, oil, or wax, may also include both pigments and plasticizers.

The film-forming material is the barrier substance; but, since no organic film-forming materials completely exclude moisture from the metal surface, pigments may be added, either to reduce the rate of moisture penetration or to reduce the corrosiveness of the penetrating moisture. Also, pigments are added to reduce the harmful effects of sunlight upon the organic film, to give the vehicle more body so that it can be more readily applied, and to color the organic coating for decorative purposes. Some organic film-forming material may be too brittle initially or it may become too brittle during service. This property of a coated film can be controlled by means of plasticizers which increase the flexibility of the organic coating.

The proper application of an organic coating is one of the most important means of assuring the satisfactory performance of the coating film. The kind and quality of the ingredients of a coating are frequently blamed for the premature failure of an applied film, when the true cause may be improper preparation of the surface to be coated or the detrimental contamination of the surface prior to painting. For best results, the metal surface to be coated should be free of mill scale, corrosion products, oil, grease, and dirt before any coating is applied. The ultimate in surface cleanliness is not always attained because of economic factors; but when effective protection by organic films is the goal, this is too often false economy.

Pretreatments of metal surface are used extensively in shop coating to convert the metallic surface to a less active compound of the metal, such as metallic phosphate. If properly done, this results in cleaner surfaces and in better adhesion of the organic coating and reduces the spread of corrosion from holes in the organic coating.

The factors which cause corrosion should be understood before organic coatings for prevention of corrosion are applied. For example, if a metal is corroding at an accelerated rate because it is coupled with a more noble metal, coating the potentially corrodible surface incompletely results in more intense attack at discontinuities and in deep pitting or, in some cases, perforation of the metal. In such cases, both the coupled metals should be coated.

An organic coating is essentially a barrier. However, unless great care is exercised in application and in maintenance, discontinuities are sure to be present either initially or during the service life of the coating.

* By G. W. Seagren, senior fellow, Mellon Institute.

Definitions. The following definitions are recommended by the American Society for Testing Materials (ASTM D-16-47):

Varnish: A liquid composition which is converted to a transparent or translucent solid film after application as a thin layer.

Bituminous Varnish: A dark-colored varnish containing bituminous ingredients, either oil- or spirit-type varnish.

Oil Varnish: A varnish which contains resin and drying oil as the basic film-forming ingredients and is converted to a solid film by chemical reaction.

Spirit Varnish: A varnish which is converted to a solid film primarily by solvent evaporation.

Paint: A pigmented liquid composition which is converted to an opaque solid film after application as a thin layer.

Oil Paint: A paint that contains drying oil or oil varnish as the basic vehicle ingredient.

Water Paint: A paint that contains a water emulsion or dispersion as the vehicle.

Enamel: A paint which is characterized by an ability to form an especially smooth film.

Lacquer: A liquid coating composition containing, as the basic film-forming ingredients, cellulose esters or ethers and plasticizers, with or without resin.

References (Organic Coatings)

Protective and Decorative Coatings, edited by J. J. Mattiello, John Wiley and Sons, New York, 1941–1946. 5 volumes.

National Paint Dictionary, by J. R. Stewart. Washington, D. C., Stewart Research Laboratory, 1948.

Physical and Chemical Examination of Paints, Varnishes, Lacquers, Colors, by H. A. Gardner and G. G. Sward, Bethesda, Md., distributed by A. Gardner Laboratory, Inc., 1946.

Protective Coatings for Metals, by R. M. Burns and A. E. Schuh, Reinhold Publishing Corporation, New York, 1939. (*ACS Monograph Series*, No. 79.)

PROTECTION OF PIPES. Both cast-iron and steel water mains corrode more or less and lose carrying capacity rapidly by tuberculation unless protected from corrosive waters. Interior protection is now usually obtained by the hot application of bituminous enamel about $^1/_{16}$ in. thick by the centrifugal process usually under AWWA Standard Specifications 7A.5 and 7A.6–1940.

Air-drying paints have been materially improved for service in water, but except perhaps as a priming coat for bituminous enamel they lack the desired durability for protection of water lines. Certain baking-type phenolic-resin formulations are, however, much more desirable than air-drying paints and promise to prolong considerably the life of steel pipe and tanks in hot- and cold-water and steam-heating systems.

Paints for underwater service at normal temperatures have been greatly improved during recent years by the use of phenolic-chlorinated rubber and other synthetic resin bases. Substitution of tung oil for all or 60% of the linseed oil previously used and the addition of inhibitive pigments have also played an important part in prolonging the useful life of paints in water (Ref. 14). Several of these paints are now being made by leading manufacturers.

Portland cement coatings have given long service as interior coatings for large water mains and can be readily applied to old tuberculated mains after cleaning (Ref. 1, pp. 347–355). The exterior of water mains are usually protected from soil corrosion by reinforced bituminous coatings with cathodic protection.

Preparation of Metal Surface for Painting. Sand or steel grit blasting and acid pickling are now recognized as the best methods of preparing the metal for painting. All mill scale and rust should be completely removed and the paint applied when the surface is dry and at a temperature above 10 C (50 F). Mill scale is cathodic to iron and therefore promotes pitting.

Aqueous inhibitive solutions containing phosphoric acid applied to the clean metal before painting have been found to reduce but not entirely prevent blistering of paints *in water*, but their effect is markedly different with different paints. These inhibitors are usually most effective with the better-class paints. Under *atmospheric* exposure the beneficial influence of these inhibitive washes is much less noticeable or entirely lacking (Ref. 15).

CEMENT. Ironwork often is bedded in concrete which, if free from cracks and voids, is an efficient protection. The metal should be cleaned and then preferably washed with neat cement before embedding. Rich Portland cement mixtures are used for lining of steel and cast-iron water pipe.

BOWER-BARFF PROCESS. Dr. Barff's method was to heat articles to be coated to about 1800 F and inject steam heated to 1000 F into the muffle. George W. Maynard and A. S. Bower used air instead of steam, then carbon monoxide (producer gas) to reduce the red oxide. (See also Ref. 16.)

RUSSIA AND PLANISHED IRON. Russia iron is made by cementation and slight oxidation. W. Dewees Wood (U. S. Patent 252,166 of 1882) treated planished sheets with hydrocarbon vapors or gas and superheated steam within an airtight and heated chamber.

NITER PROCESS is an old process improved by Col. A. R. Buffington in 1884. The articles are stirred about in a half-and-half mixture of fused sodium and potassium nitrate (saltpeter) to which is added manganese dioxide in the ratio of 1 part oxide to 50 parts niter by volume. They then are suspended in the vapors, dipped and washed in boiling water, and finally immersed in hot oil or wax for protection against corrosion. Pure chemicals are essential. This process is used for small arms and pieces which cannot stand the high heat of other processes. The temperature of the bath should be 600 to 650 F to yield a temper blue, and 100 F to yield the dark gun-metal finish used for rifle parts.

PHOSPHATE COATINGS (Ref. 17). Iron and steel articles, properly cleaned and immersed in a dilute phosphoric acid solution saturated with one or more of the metals iron, manganese, or zinc, become coated at the point of solution with crystalline phosphate. The reaction of the acid with the metal takes place with the evolution of hydrogen gas until the metallic surface is converted to a protective phosphate coating. The protective value of such coatings depends upon their continuity and for this reason thickness and density of coating are important factors. Coatings having a wide range of thickness can be produced depending upon the method of cleaning prior to phosphating, solution concentration, and time of processing. Because no electric current is used and formation of the coating is solely dependent upon the phosphating solution contacting the metal surface, uniform coatings are produced on irregularly shaped articles, in recessed areas, and on threaded surfaces with facility equal to that on flat surfaces.

Coslett Process. The first commercial application of phosphate coatings followed the proposal of Coslett in 1907 to treat iron and steel articles in a hot dilute phosphoric acid solution with a control agent such as iron filings. Iron and steel articles immersed in this solution for 3 to 4 hours at boiling temperature become coated with a rust-resisting ferrous phosphate coating. The difficulty in obtaining uniform coatings, the great amount of suspended matter adhering to the work after rinsing, and the nonuniform appearance made the adoption of the process slow.

Parkerizing. Parkerizing utilizes a phosphating bath so prepared and replenished that the solution remains in balance at all times and is easy to operate and control. In operation the properly cleaned articles are immersed in the hot Parkerizing bath for 15 to 45 minutes where a thin layer of the metal being treated is dissolved, resulting in the formation of the Parkerized coating chemically bonded to the base metal. Immediately following the Parkerizing treatment, processed articles are rinsed in clean hot water and dried. The affinity of Parkerized coatings for oil, wax, or color is utilized in practice to enhance further their corrosion resistance. Frequently such coated articles are finished by dipping in nondrying or drying type of oils, followed by centrifuging to remove the excess. The finish has a black matte appearance.

Bonderizing. Bonderizing produces on iron, steel, zinc, cadmium, and aluminum surfaces a corrosion-inhibiting phosphate coating for increasing the adhesion and resultant durability of paint finishes. Since its introduction in 1929 it has become recognized as a useful method of preparing surfaces for painting wherever the maximum durability of the paint finish is required. Usually either the article is immersed in the phosphating solution or the solution is sprayed onto the articles to be coated. In general, small parts which can be more conveniently handled as a unit load are treated by immersion application whereas larger parts such as refrigerator cabinets, automobile bodies, and fenders are treated by spray application. The immersion process requires 2 to 5 min for application of the coating, and the spray process 45 sec to 1 min.

GALVANIZING is a method of coating steel or iron articles with zinc, the product being known as galvanized steel or iron. Galvanized coatings usually resist, for a reasonable time, such ordinary corrosive conditions as fairly pure air, moderately hard domestic waters, sea water, and brines. The zinc, particularly under atmospheric conditions, often becomes covered with a protective layer or film believed to be a zinc-hydroxide-carbonate mixture. Zinc is anodic to iron and therefore protects the base metal, where exposed, until the adjacent outer layer of zinc is dissolved. Galvanized coatings are unsuitable where the corroding agent is very acid, as in certain mine waters and industrial solutions, or where the solutions are excessively alkaline, e.g., calcium brines with an excess of caustic alkalinity. Favorable water *p*H values are between 6 and 8.5 and temperatures below 130 F. In brines the *p*H should not exceed 8. Very hot soft water (over 160 F) often attacks galvanizing rapidly. The potential of zinc to iron above 140 C is reversed in some water so that the pitting rate is increased.

Galvanized coatings are commercially applied: (*a*) by dipping the article in molten zinc; (*b*) by electroplating; (*c*) by heating in contact with zinc powder.

Hot-dip Galvanizing. Articles to be hot-dipped-galvanized usually are cleaned by acid pickling, washed, and fluxed by immersion in a bath of zinc ammonium chloride or hydrochloric acid. They then are dipped in molten zinc maintained at 830 to 920 F. When the steel has reached approximately the zinc temperature, requiring a few seconds in the case of wire and several minutes for heavier sections, it is drawn slowly from the bath, allowing time for draining. In the case of wire, excess zinc often is removed and the coating smoothed mechanically by asbestos or charcoal wipers. Long, slightly corrugated rolls are similarly used for sheets. The article may cool in air, producing a spangled (crystalline) zinc surface, or it may cool rapidly in water, producing a matte surface.

Adherence of zinc coatings depends on the formation of suitable zinc-iron alloys. Some grades of wire and sheets, conduit pipe, etc., requiring maximum bending qualities are made with light coatings consisting largely of zinc-iron alloy formed by wiping after dipping. Where bending requirements are secondary and maximum corrosion resistance is desired, as for other grades of iron and sheets, galvanized water pipe, galvanized wire, etc., the galvanizing should be considerably heavier, and with more pure zinc in the coating. The average weight of coating applied varies from $1/2$ to 2 $1/2$ oz per sq ft of covered surface, depending on the thickness of base metal, temperature, and time in bath, and amount of wiping. The protective value varies directly as the thickness of the zinc.

In the process of galvanizing, tin is at times added to the zinc bath to improve the spangle, whereas aluminum may be added to improve the fluidity of the bath and brightness and smoothness of the coating. The vapor of burning sulfur or ammonium chloride dust on the zinc surface as it comes from the bath has been used to improve brightness, but the sulfur may impair the corrosion resistance of the zinc.

Electrogalvanizing. Electrogalvanized zinc coatings are applied to conduit pipe, wire, and intricate shapes. Electrogalvanized coatings on the threads of pipe couplings prevent sticking or galling of the threaded joints, and generally are used for tubular goods in petroleum production. Articles to be electrogalvanized may be cleaned by pickling with or without the aid of electric current, or machined surfaces may be cleaned with alkaline grease removers. The zinc is deposited from sulfate or cyanide baths, using low-voltage plating current (see ASTM Specification A164-40 T, Electrodeposited Zinc on Steel).

GALVANIZING BY CEMENTATION. SHERARDIZING. Sherard Cowper-Cowles (England) refined and perfected the process of cementation with zinc, and British and United States patents were granted to him in 1902 for the method known as sherardizing. It consists in packing the articles to be sherardized, usually steel, in a retort with zinc powder known as *blue dust*. The retort or drum is tightly sealed, and by suitable means so heated that its contents attain a temperature approaching the melting point of zinc. The melting point of zinc is 786 F, and sherardizing usually is carried on at 600 to 700 F, although it can be done at temperatures above or below this range. Better results are obtained if the drums are slowly rotated, and this generally is done in order to bring a fresh supply of dust into intimate contact with the articles being treated. The time required is 2 to 4 hours. After drums are removed from the heating oven, they are allowed to cool before being opened. Prolonged heating and slow cooling of steel articles embedded in zinc dust, which is a poor conductor of heat, tends to soften them, which is highly desirable for objects which must be cut and bent.

Blue dust is a natural by-product of smelting zinc, but, when especially made of greater purity and uniformity for the purpose of sherardizing, is known as sherardizing zinc dust. It consists of minute particles of zinc metal, each encased in a thin skin, shell, or pellicle of zinc oxide. Commercial sherardizing zinc contains about 95 to 97% metallic zinc. The remainder is chiefly zinc oxide, with traces of impurities such as iron and lead. The resultant coating is a true zinc-iron alloy, ranging from $FeZn_3$ at the base metal line to $FeZn_{10}$ near the surface, and a layer of relatively pure zinc on the exterior; exact analysis naturally is difficult.

Characteristics of a sherardized coating are *color*, a rich gray; *texture*, which is matte, caused by innumerable minute fissures not visible to the unaided eye, and has a velvety feel; *ability to receive and retain paints*, varnishes, enamels, etc., due to the bonding effect in the minute fissures; *uniform distribution and penetration*, making possible the effective coating of irregularly shaped and hollow articles, screw threads, holes, etc., without building up on corners or actually filling up depressions, although allowance must be made for thickness on threads and other closely fitted parts; *ductility and tenacity*, a reasonable amount for bending, as in electric conduit; *ability to take a high polish*, the surface being relatively hard.

Corrosion Resistance of Sherardizing. Concerning exposure to ordinary attacks indoors, and when exposed to weather and the zinc is oxidized, Rawdon says (Ref. 18):

> The coating gradually assumes a dark appearance as the iron likewise is oxidized, finally becoming a very dark gray and almost black if the iron content is not too high. Coatings high in iron, and hence containing the iron-rich compound, usually show a red dusty surface on weathering from the ferric

oxide produced. A wrong impression often is formed concerning such coatings, that is, that the coating has been entirely corroded away and that the red dust indicates an exposure of the base metal. It can generally be shown that such is not the case.

When this color is not due to rusting of the base metal, the red or yellowish formation usually disappears in a short time, and the surface darkens. The black coating is believed to be the magnetic oxide of iron. When sherardizing acquires this surface formation it becomes highly resistant to atmospheric corrosion. The relation of coating thickness to durability is important, since a heavy coating is, for mechanical reasons, objectionable on some articles, particularly those subject to bending and flexing. For Comparative Life of Different Types of Zinc Finishes, see Ref. 19, p. 95.

TESTS AND SPECIFICATIONS FOR GALVANIZED COATINGS. Galvanized coatings usually are specified as a given weight of zinc per unit area. If bending or forming is anticipated, bend tests may be specified. Coatings may be expressed as ounces of zinc per square foot of surface, and determined by weighing a suitable sample of the galvanized article, removing the zinc, and reweighing. The usual means of removing the zinc is with hydrochloric acid containing sufficient antimony chloride to prevent attack on the base metal. The minimum thickness of galvanized coatings, particularly on wire, also is expressed in terms of the Preece test, which consists of dipping the zinc-coated article in a neutral solution of copper sulfate (sp. gr. 1.186 at 18 C) for periods of one minute, washing and wiping between each dip until the zinc is penetrated, as shown by the formation of adherent copper on the exposed surface of the steel. Coating weight is expressed as number of dips to cause failure. For details of both methods, see Standard Methods of Determining Weight of Coating on Zinc Coated Articles, ASTM Specification, Designation A90-39 and A239-41. (See ASTM book of standards.)

Specifications for most commercial products covering weight of coating and bend tests when required are available in 1946 and subsequent ASTM Standards, Part 1A, Metals.

On sheets, coating weights are given as weight per square foot of sheet, i.e., 2 sq ft of coated surface. Ordinary coatings on sheets range from 1.25 to 2.75 oz per sq ft of sheet coated on both sides. The weight of zinc per single covered surface would be one-half this amount. Zinc coatings on pipe, structural shapes, etc., generally range between 1.5 and 3.0 oz per sq ft of covered surface. The weight of coating on wire varies with the size of wire and process of coating and, except in fine sizes, will range from 0.15 to 1.25 oz per sq ft of covered surface.

LEAD COATINGS. Lead is very resistant to certain types of corrosion, and lead coatings on steel have been used to some extent, where sulfate and other insoluble lead compounds are formed. A thin coating of lead, however, is difficult to apply, either by electroplating or hot dipping, so as to be sufficiently free from pinholes to be effective when submerged in corrosive water. The atmospheric resistance of such coatings usually is quite good. Hot-dipped lead-tin alloy sheets, known as terne plate, are extensively used. Wrought steel or iron pipes protected on the interior by thick lead tubes mechanically applied, when carefully installed with suitable fittings and valves, are effective for carrying corrosive acids and similar liquids.

FLAME SPRAYING (metallizing) is the process by which zinc, aluminum, and other metals in the form of wire or powder are passed through an oxyacetylene flame by air pressure and thus atomized and plated on the material to be coated. Steel should be well cleaned by grit blasting. Plastics that can be powdered can be applied by this process. Sprayed zinc coatings have a porous outer surface and afford a good base for an outer coat of paint. Sprayed zinc, 0.010 in. thick, has protected steel dams in fresh water for 10 years.

CADMIUM COATINGS (electroplated) have a pleasing appearance and afford satisfactory protection, indoors, against rusting. Cadmium like zinc is usually anodic to iron and therefore protects by sacrifice of the coating. It has the advantage of not developing bulky and unsightly corrosion products, in contrast to zinc. However, the increasing cost of cadmium and its inferiority to zinc in marine and industrial atmospheres have limited its field of use. Practically all cadmium coatings are electroplated from cyanide baths.

A joint atmospheric corrosion test on electroplated coatings being conducted by the American Electroplater's Society, the ASTM, and the National Bureau of Standards includes cadmium coatings. After 180 weeks in several types of atmospheres, all the quality ratings of the cadmium coating are lower than for zinc of similar thickness (Ref. 19, p. 127). The cadmium coatings are somewhat closer to zinc in marine locations. The salt spray test indicated cadmium coatings superior to zinc, but this has not been borne out in subsequent field tests in marine air.

RUST-RESISTING METALS. Iron. Various modifications of plain iron and steel have been introduced to give better service under atmospheric conditions. These grades of iron and steel are included in the ASTM corrosion tests referred to on p. 3-06.

Copper steel has found a wide commercial application on account of its slow rusting properties under industrial and marine atmospheric conditions of exposure. (See Ref. 20.) Addition of 0.15 to 0.25% copper at an increased price of about $3.00 per ton usually will prolong the life of the metal two times under such conditions. (See also p. 3-06.)

Corrosion-resistant alloys, containing high percentages of chromium and nickel, are widely used in industrial and marine atmospheres.

6. PROTECTION AGAINST UNDERWATER CORROSION

The main factors controlling rate of corrosion in water are dissolved oxygen and carbonic acid, temperature, rate of motion, and soluble salts in the water. The chemical composition of ferrous metals, within rather wide limits, does not seem to have a marked influence on the character or the amount of corrosion. The high-chromium series of steels are an exception to this. With cold water the rate of corrosion is usually slow, but in hot water the corrosive action is accelerated according to the temperature, the rate being about doubled for every 25 F increase in temperature.

Localized corrosion or pitting usually is much deeper near mill scale, especially where the scale is thick and firmly adherent, due to the cathodic character of the scale with respect to iron. Old rust acts somewhat similarly but is less harmful. Removal of mill scale and rust reduces the tendency to pitting, provided the metal itself is fairly homogeneous and free from strain. Surface finish is a greater controlling factor in corrosion of iron and steel than the variations in the composition of the metal itself.

Within certain limits, corrosion varies directly with the temperature and the rate of flow and also with the composition of the water. Corrosion is likely to be more rapid on unprotected metal with increasing velocity of the water. The amount of corrosion is almost directly proportional to the amount of oxygen in the solution in the water. Hot-water heating systems show practically no corrosion after 50 years of use, whereas hot-water supply systems, operating at the same temperatures, have failed in 6 to 8 years. The only difference between the two is that in the heating system free oxygen is excluded, while the water in the supply system usually is saturated with that gas.

DEACTIVATION AND DEAERATION OF WATER. In some cases it is more economical to prevent internal corrosion by removing dissolved oxygen from the water, especially with hot water. Oxygen removal may be accomplished: (1) by mechanical deaeration for which there are two or three satisfactory designs of apparatus in use; (2) by the chemical combination of the oxygen with some material such as scrap and sheet iron or sodium sulfite. The two methods may be combined by using mechanical deaeration to remove most of the oxygen and chemical treatment for the remainder.

Dissolved oxygen is usually the controlling factor in corrosion of iron in water. In the return condensate lines of heating systems, free CO_2 is also of prime importance. These facts have been established by laboratory experiments and by experience in service. Means for removal of free oxygen and CO_2 from water have been considerably improved in recent years. (See Speller's chapter in Ref. 2; also Ref. 27.)

CHEMICAL TREATMENT OF WATER. Hot-water Heating Systems. The water is usually obtained from domestic supplies and is either soft or of medium hard quality. These systems are closed except for a small air vent so that the free oxygen is soon exhausted, and corrosion is therefore negligible. A small amount of corrosion occurs in water which has been deactivated in this way, with the evolution of hydrogen which may be drawn off through petcocks near the tops of radiators. By adding sufficient caustic soda to bring the pH of the water up to 8 or 8.5, the generation of hydrogen can be prevented and the gas nuisance eliminated.

Cooling water that is recirculated in contact with the atmosphere, as in spray ponds or secondary heat exchangers (as for stationary diesel or gas engines), can generally be inhibited to prevent serious corrosion of steel by the addition of about 500 ppm of sodium chromate, which may be reduced materially after the protective film has formed. Aeration does not impair the efficiency of chromates. The optimum amount required, which depends on the composition of the water, may be determined approximately by laboratory tests. In closed cooling systems (recirculating) it is better to keep the concentration above 1000 ppm, or even higher, where much chloride is present (Ref. 21). The manufacturers who specialize in the production of chromate for such service are usually able to give useful information as to the amount required for specific service (Ref. 22).

Sodium nitrite gives about the same protection as the chromate. Water-soluble organic inhibitors (such as the glucosates or tannin compounds) may also be used as inhibitors in cooling systems. (While these are usually somewhat more expensive than the chromates, they have no toxic effect in the concentrations employed.) Advice on their application and concentrations required for specific conditions may be obtained from manufacturers. (See also Ref. 22.)

Domestic Water Systems. Natural waters are selected and often treated for domestic use. They are used for boiler feed and cooling, with or without local treatment.

Water treatment for very large systems is usually limited by the cost to *p*H control and deposition of $CaCO_3$ scale. The carbonate balance system of treatment aims at having an excess of $CaCO_3$ present so that a continuous protective layer of carbonate or lime is deposited on the inside of pipe and containers. This is accomplished by adjusting the *p*H with respect to the carbonate, bicarbonate, and free CO_2.

Silicate of soda has been used since 1920 to protect iron, lead, and brass water pipe. For most water a solution of $Na_2O : 3\ SiO_2$ is recommended and may be fed by any convenient type of proportional feed apparatus.

Hexametaphosphate and other polyphosphates of soda have proved to be highly useful in preventing the formation of scale deposits due to excess of calcium and magnesium salts in water; 2 to 5 ppm is usually sufficient for this purpose. These compounds also prevent discoloration from iron when added in the proportion of 2 to 1 ppm of iron in the water. For this purpose the phosphate must be added before the water is chlorinated or exposed to air.

When the water at normal temperature is in turbulent flow (about one foot per second or higher in pipe under one inch in diameter) 10 ppm hexametaphosphate has been shown to retard corrosion in hot soft waters, apparently by formation of an insoluble film on the metal (Ref. 23).

Pipe Materials for Water Supply. The choice lies mainly between galvanized ferrous pipe, copper, and two or three grades of brass. Each is available in three thicknesses and standard inside nominal diameters with threads and couplings. Copper is available in thinner wall pipe provided with solderable streamline copper fittings.

Soft waters (nearly neutral) are much more reactive with metals than those of higher primary alkalinity or when supersaturated with calcium carbonate. Under the latter, galvanized pipe in cold-water lines has given satisfactory results.

For monumental buildings, copper or 85(Cu)–15(Zn) brass of the standard pipe gage is advisable unless the water is deactivated by removal of the dissolved oxygen or by other adequate treatment.

Brasses with 60 to 67% copper are dezincified in some corrosive waters, and in certain localities they are not much more serviceable than galvanized iron or steel pipe. The zinc may be leached out locally, leaving a plug of porous copper, or the dezincing effect may be spread out, resulting in a general weakening of the pipe, which is especially noticeable under the threads. Dezincification is retarded by the use of silicate of soda (8 ppm added silicate). Brasses with 85% or more copper do not show this (Ref. 24). Evidently no kind of pipe, ferrous or nonferrous, should be used indiscriminately without regard to the corrosive character of the water.

Promising synthetic resin coatings of the baking type have been developed that are proving useful for protection of hot-water tanks and piping.

In salt or fresh water there is no material difference in rate of pitting of wrought iron, steel, low metalloid steels, or copper-bearing steels. This is contrary to relative performance of these metals in atmosphere. Low-alloy steels containing nickel and chromium are not enough better in this respect to warrant their extra cost in form of pipe for water-supply services in general.

ANAEROBIC BACTERIA cause serious damage from corrosion of cast iron and steel pipe in certain soils or water. They thrive when sulfates and a little organic matter occur, provided the soil or water is low in free oxygen. These bacteria take up hydrogen and reduce sulfates with liberation of free sulfur or hydrogen sulfide. When they operate on the metal surface they evidently obtain most of their hydrogen from cathodic surfaces and thus keep the corrosion process active as in ordinary corrosion by the familiar electrochemical process of depolarization. Anaerobic bacteria may operate in water, as in the water seals of gas holders, within a *p*H of 5.5 to 8.5. Protection may be obtained by reinforced coal tar coatings $^1/_{16}$ in. or more thick and by cathodic protection, or by keeping the *p*H of the water over 9. (See Ref. 2.)

CORROSION IN STEAM BOILERS. Internal corrosion may be caused by the use of water containing free acid, by water containing chloride of magnesium, which decomposes when heated, liberating acid, or by water containing air or carbonic acid in solution. External corrosion rarely takes place when a boiler is kept hot, but when cold it is likely to corrode rapidly in those portions where it joins the brickwork or where it is covered by dust or ashes, or wherever water may lodge (Ref. 2, p. 520).

STEAM CONDENSATE lines suffer from corrosion due to free oxygen and carbon dioxide in the condensate. Oxygen is a more important factor than CO_2, but the combination is worse than either separately. The noncondensable gases entrained with the steam accumulate so that the amount present in the vapor phase may be much higher than in the incoming steam. As a result, the amount dissolved in the condensate may equal, or even exceed for short periods, the amount entrained by the steam (see Ref. 25). There are four means available to minimize corrosion in steam condensate systems: (1) Treatment of boiler feed water to eliminate free O_2 and CO_2 in particular. (2) Design to minimize re-solution of deleterious gases in the condensate. (3) Chemical treatment of the condensate, as by addition of an amine of suitable composition. (These are volatile alkalies like ammonia.) (4) Use of more resistant metals. Highly resistant alloys like 18-8 steel, for economic reasons, can be used only on limited sections. (See also Refs. 1 and 26.)

CORROSION OF NONFERROUS PIPES. The corrosion of brass, zinc, and other metals is greatly retarded or entirely stopped by removing oxygen from the water in contact with these metals. Brass and copper piping are sometimes seriously attacked by heated water under pressure. Water heated in a domestic copper boiler of the usual type may lose nearly one-half of its free oxygen in passing through the heater. Brass pipes after 10 or 12 years frequently have lost most of the zinc content by "dezincification" (Ref. 2, p. 69) in some waters. The same has been observed in brass condenser tubes. It is now generally recognized that no one metal will meet all requirements of service. Ordinary brass pipe (35 Zn, 65 Cu) is adequate where the water is fairly pure and low in chlorides; red brass (15 Zn, 85 Cu) is much more durable where the water carries considerable amounts of soluble salts. It is much less likely to suffer dezincification. This grade is preferable for sea water. Copper pipe usually is more durable than brass for practically all domestic water service. (See Refs. 24 and 28.)

7. PROTECTION AGAINST SOIL CORROSION

SOIL CORROSION is usually highly localized and is caused by electrochemical action resulting from variations in the soil and solutions therein. Low-alloy steels, i.e., steels with less than 10% chromium, do not show enough advantage, from an economic standpoint, under severe soil corrosion to warrant their use, and the solution of the problem of longer life of underground pipe is practically limited to a selection of suitable protective coatings and electrical protection (p. 3-16).

Soil corrosion varies considerably from point to point on pipe lines. On long lines serious damage is confined to 10% or less of the line as a rule. It is, therefore, advantageous to have a soil survey of the territory through which the pipe line will run, before deciding on the type of protective coating, if any, to be used. Much progress has been made in the study of soil factors at the National Bureau of Standards, so that a careful survey, judiciously interpreted, with regard to all the factors involved, should give an approximate idea of the corrosivity of the soil. (See National Bureau of Standards Research Papers 95, 298, 329, 359, 365, and Bulletin C450.) A soil survey, with experience on pipe lines situated in similar soils, affords a fairly sound basis for selecting the most economical coating, with or without cathodic protection, for a particular location.

BITUMINOUS COATINGS have been used mostly for underground protection. They consist of the well-known coal tar and asphalt dip coatings, preferably applied, after removing all grease, dirt, and loose mill scale, by vertically dipping the pipe in a bituminous bath at about 350 F. The melting point of the asphalt mixture should be varied between 140 and 190 F to suit climatic conditions. Coatings of this type, properly applied at the mill, are suitable only for mild corrosive conditions, but afford fair protection on the inside of water mains. In wet, marshy places, or in alkaline or acid soils, more substantial protection often is required. A thicker coating may be applied at the trench, but the pipe should receive either an asphalt or coal tar dip coating, or a coating of a suitable bituminous primer applied cold at the plant before shipment to prevent rusting in transit. Bituminous coatings usually are reinforced with saturated asbestos felt wound spirally on the pipe over the dip coat. A Fiberglas fabric is now available for this purpose. In certain soils, unless these bituminous coatings have a high melting point and are unusually rigid, they are subject to soil stress; that is, some soils have a tendency to cling to the coating and pull it away from the pipe when the soil shrinks by drying, or irregular pressure in the trench may distort the bitumen and cause thin spots in the coating. In such soils a shielding material should be used to prevent damage to the coating by soil stress. Experience shows that many bituminous coatings that otherwise would have afforded adequate protection have been destroyed or rendered ineffective in this way. Thin, stiff, rolled steel sheet (26 gage) or copper foil may be used as a shield. Asphalt mastics consisting of 30% bitumen and 70% of a graded sand have been used with good results, especially for protection of pipe conveying hot oil through very corrosive soils. These mixtures are applied about $^3/_8$ in. thick, on 8-in. pipe, and, while expensive, this coating has proved superior to other bituminous coatings.

PORTLAND CEMENT CONCRETE has been used for over 40 years for protecting pipe in highly corrosive soil. It has a record as one of the most practical means of affording permanent protection. The minimum thickness usually is about 1 in.; the mixture may be applied in a wooden box surrounding the pipe. A portable steel form has been developed so that a concrete mixture consisting of 3 sand and 1 Portland cement can be poured around the pipe with a minimum thickness of $^5/_8$ in. Portions of the Hetch Hetchy steel conduits have been protected with such a mixture applied $^1/_2$ in. thick, wrapped with fabric.

Portland cement coatings are applied in such thicknesses at about the same cost as fabric-wrapped bituminous coatings on 8-in. to 12-in. pipe lines. Concrete coatings offer some resistance to stray electric currents but should not be relied upon to protect pipe against electrolysis. Where pipe lines pass under electric railways, or in situations where complete electrical insulation is necessary, the line has been effectively protected by boxing and coating with hot asphalt with a minimum thickness of about 1 in. for a distance of 100 ft on each side of the track.

A comprehensive field test of pipe coatings has been conducted under the auspices of the National Bureau of Standards with the cooperation of committees of the American Gas Association and API. These tests were made on short lengths of pipe, and also on operating oil lines, and give useful information concerning the durability of all materials now available for such protection. (See Refs. 13 and 29.)

CATHODIC PROTECTION. Cathodic protection is obtained by rendering the structure that is corroding slightly cathodic to the surrounding water or soil by means of a controlled difference of potential. This method, which has proved satisfactory and economical on a number of gas and oil pipe lines underground, has also been applied successfully to the protection of the inside of water-storage tanks and other structures that are in contact continuously with water. Protective coatings that insulate a large portion of the metal surface will very materially reduce the total amount of protective current that it is necessary to impress on bare anodic areas to arrest corrosion. This treatment is usually too costly to use without a fairly good coating. The protective current may be obtained from zinc, aluminum, or magnesium anodes or it may be supplied from external sources and applied through large anodes of steel or iron scrap, surrounded underground with coke breeze.

For each structure it is necessary to determine or estimate the minimum current density required, and to place the anodes so that the necessary protection of all parts in the corrosive area can be obtained most economically. In water having relatively high electrical conductivity, such as sea water, this is easy as compared with fresh-water systems. In the latter the composition of the water is a major factor. It is therefore desirable to obtain experienced engineering advice in the analysis of these problems in order to obtain an accurate estimate of the minimum current density required. The protective current is controlled by the potential between the anode and the structure to be protected. Usually at least 0.25 volt difference in potential between the buried structure and soil is required to insure sufficient current density on the structure as cathode to prevent corrosion.

Rectifiers have generally proved to be the most practical means for supplying the necessary current for protection of surfaces in contact with neutral waters (Ref. 30). Zinc, magnesium and aluminum anodes (in suitable back-fill) have given good protection in medium- and low-resistance soils, as for river crossings on pipe lines. (See Ref. 2.)

8. SOURCES OF INFORMATION

The economic importance of corrosion control, now generally recognized, is evidenced by the numerous committees of national engineering societies working on various phases of the problem. The ASTM has conducted extensive field tests and has formulated specifications.

The National Association of Corrosion Engineers, organized in 1944, is the center of discussions on corrosion prevention (headquarters, Southern Standard Building, Houston 1, Texas). The NACE publishes a monthly journal that contains a bibliography of current literature. Another bibliography, dealing particularly with the metallurgical side of the problem, is published monthly in *Metal Review* by the American Society for Metals, and *Chemical Abstracts* (American Chemical Society) covers papers of interest to the chemical engineer.

The Electrochemical Society has a corrosion division that specializes on the fundamental phases of the subject, but includes in its published transactions many papers on applied aspects. A *Corrosion Handbook*, including about 100 articles by specialists covering scientific data and its engineering application, has been prepared and sponsored by the Society (Ref. 2).

The National Bureau of Standards has carried on authoritative research and tests on soil corrosion, as has the American Petroleum Institute and the American Gas Association in their respective fields.

REFERENCES

1. F. N. Speller, *Corrosion, Causes and Prevention*, McGraw-Hill Book Co., New York, 1935.
2. H. H. Uhlig (Editor), *Corrosion Handbook*, John Wiley and Sons, New York, 1948.
3. B. B. Westcott, "Corrosion Fatigue," chapter in Ref. 2.
4. U. R. Evans, Thin films in relation to corrosion problems, *J. Inst. Metals*, Vol. 46, pp. 7–23, 1931.
5. *Trans. Electrochem. Soc.*, p. 1, Jan. 1948.
6. D. J. McAdam, Stress-strain cycle relationship and corrosion fatigue of metals, *Proc. Am. Soc. Testing Materials*, Vol. 26, Pt. II, pp. 224–254, 1926; Corrosion fatigue of non-ferrous metals, *Proc. Am. Soc. Testing Materials*, Vol. 27, Pt. II, p. 102, 1927; Some factors involved in corrosion and corrosion-fatigue of metals, *Proc. Am. Soc. Testing Materials*, Vol. 28, Pt. II, pp. 117–158, 1928.
7. F. N. Speller, I. B. McCorkle, and P. F. Mumma, Influence of corrosion accelerators and inhibitors on fatigue of ferrous metals, *Proc. Am. Soc. Testing Materials*, Vol. 28, Pt.II, pp. 159–167, 1928: Vol. 29, Pt. II, p. 238, 1929.
8. V. V. Kendall and E. S. Taylerson, A critical study of the ASTM corrosion data on uncoated commercial iron and steel sheets, *Proc. Am. Soc. Testing Materials*, Vol. 29, Pt. II, pp. 204–219, 1929.
9. J. S. Unger, Results of using copper bearing steel in freight cars, *Ry. Age*, Vol. 84, p. 1062, May 1928.
10. Report of Committee on Iron and Steel Structures, American Railway Engineering Association, *Ry. Age*, Vol. 86, No. 562, pp. 166–167, 1929.
11. J. S. Unger, Comparison of corrosion resistance of copper steel and plain steel railroad ties, *Steel Maker*, Vol. 2, No. 1, pp. 93–95, Summer 1931.
12. F. N. Speller, Corrosion of pipe in buildings, *Blast Furnace Steel Plant*, Sept. 1940
13. *Bulletin* 450C, Research Papers 95 and 329, National Bureau of Standards, 1945.
14. F. N. Speller, Third Water Conference, *Proc. Engrs.' Soc. West. Penna.*, 1942.
15. Report of Subcommittee XXIX of Committee D-1, American Society for Testing Materials, 1949.
16. George W. Maynard, *Trans. Am. Soc. Mech. Engrs.*, Vol. 4, p. 315, 1883.
17. Van M. Darsey and Walter R. Cavanagh, Formation and application of phosphate coatings, *Trans. Electrochem. Soc.*, Vol. 91, p. 351–365, 1947.
18. Rawdon, *Protective Metallic Coatings*, p. 112, Chemical Catalog Co., 1928.
19. R. M. Burns and A. E. Schuh, *Protective Coatings for Metals*, Reinhold Publishing Corp., New York, 1939.
20. D. M. Buck, A review of the development of copper steel, *Yearbook Am. Iron Steel Inst.*, pp. 373–379, 1920.
21. Marc Darrin, *Ind. Eng. Chem.*, p. 368, April 1946.
22. ACS symposium on inhibitors for aqueous corrosion, *Ind. Eng. Chem.*, Aug., Sept. 1945.
23. O. Rice and G. Hatch, Corrosion control with threshold treatment, *Ind. Eng. Chem.*, Vol. 32, p. 1572, 1940.
24. E. P. Polushkin and H. L. Shuldener, *Metal Technology* (AIME), Oct. 1944.
25. L. F. Collins and E. L. Henderson, in *Heating, Piping and Air Conditioning*, Sept., Oct., Nov., Dec. 1939; also Jan. 1945.
26. *Heating, Ventilating, Air Conditioning Guide*, 1947, ASHVE, New York.
27. J. R. McDermot, Separation of dissolved gases from water, *Trans. Am. Soc. Mech. Engrs.*, Vol. 42, p. 267, 1920.
28. F. N. Speller, W. H. Walker, and R. G. Knowland, *J. Am. Soc. Heating Ventilating Engrs.*, 1918–1920.
29. *Proc. Am. Petroleum Inst.*, Sec. A, Production, 1940.
30. L. P. Sudrabin, Cathodic protection of steel equipment submerged in water, Fifth Water Conference, *Proc. Engrs.' Soc. West. Penna.*, October 1944.

CORROSION RESISTANCE OF METALS AND ALLOYS

By H. H. Uhlig

9. FORMS OF CORROSION

Metals corrode by uniform surface reaction as in the rusting of iron or tarnishing of copper, but in addition they may exhibit attack in other forms. These forms are particularly important in some of the corrosion-resistant alloys for which uniform corrosion is often negligible, but corrosion along localized paths may be correspondingly pronounced. In the application of these alloys, therefore, proper regard should be paid to factors of alloy treatment and environment that are known to influence localized attack.

Precautions are included in the following discussion pointing out some of these factors. For example, with some alloys proper quench from high temperatures and, in others, stress relief anneal are necessary conditions for satisfactory corrosion resistance in even mild environments.

The various types of corrosion can be summarized as follows.

UNIFORM CORROSION. This is corrosion represented by rusting of iron, tarnishing of silver, fogging of nickel, or high-temperature oxidation of iron or stainless steels.

PITTING. This is localized attack at random on specific areas, in the form of either shallow, saucer-like indentations or deep, narrow channels appearing at times as if a worm had bored through the structure.

DEZINCIFICATION AND PARTING. Dezincification is a type of corrosion occurring with zinc alloys in which there is preferential loss of zinc. Yellow brass, for example, often corrodes in water or aqueous solutions in this fashion. A porous, mechanically weak, spongy copper remains, having approximately the same shape as the original brass. If dezincification occurs only in relatively small areas, it is known as *plug-type* dezincification. If the attack is general, it is called *layer-type* dezincification.

Parting is similar to dezincification in that one component of the alloy is preferentially corroded. It is known primarily in the noble metal alloys. An alloy of 25% silver and 75% gold, for example, corrodes preferentially in nitric acid, leaving as residue a porous mass of pure gold.

INTERGRANULAR CORROSION occurs locally along narrow regions of metal separating the various metal crystals or grains one from the other. This type of attack occurs, for example, when Duralumin (aluminum plus approximately 4% copper) or 18-8 stainless steel (18% Cr, 8% Ni, bal. Fe) is improperly heat treated and later exposed to a corrosive environment. Heat treatment in specific temperature zones, either intentional or accidental, precipitates compounds in the grain boundary regions, making the boundary a preferred path for electrochemical action. Mechanical failure of an appreciable cross section of metal can be rapid when this type of corrosion is the cause.

CRACKING. Corrosion in the form of cracking has either one of two names. If a metal exposed to a corrosive environment contains residual stresses, or is constantly stressed externally, the failure is called *stress-corrosion cracking.* If the metal is repeatedly or alternately stressed, such as in vibration, the failure is called *corrosion fatigue.*

Stress-corrosion Cracking. Practically speaking, any of the commercial metals or alloys are subject to stress-corrosion cracking if the conditions of combined stress and environment are critical. Unless the corrosive environment is of the right kind and concentration, however, no cracking occurs. For this reason, stress corrosion cracking is not commonly observed in practice. If the environment happens to be the critical one, however, trouble is encountered whenever the metal is subject simultaneously to the critical stress. For example, stressed brasses crack in ammonia or amine plus oxygen atmospheres, but not in other atmospheres; stressed stainless steels (18-8) crack specifically in slightly acid chloride solutions; some stressed aluminum or magnesium alloys fail similarly in neutral chlorides; and steel stressed beyond the elastic limit cracks in hot concentrated nitrates or caustic soda.

Corrosion fatigue, on the other hand, is an accelerated form of fatigue cracking owing to the presence of a corrosive environment. The environment is not so specific as with stress-corrosion cracking. The metal, furthermore, subject to the same alternate stresses, may or may not fail in the absence of such an environment, depending on whether the stresses are above or below the endurance limit (the endurance limit is the maximum stress that may be applied to a metal without failure after an indefinitely large number of stress cycles). In general, the endurance limit is reduced, and in some cases may not exist at all (reduced to zero stress) if the metal is corroded at the same time that it is subject to alternate stresses. Air alone is sufficiently corrosive to some metals alternately stressed, e.g., copper and brass, so that their endurance limits are reduced as much as 15 to 25% below corresponding values in vacuum. A similar effect is found also for steel, but the reduction is not so large.

EMF AND GALVANIC SERIES. Dissimilar metals coupled together and exposed to an electrolyte form a short-circuited galvanic cell, the operation of which usually accelerates corrosion of one metal in the couple. This accelerated corrosion is called galvanic corrosion. The extent of galvanic corrosion depends on conductivity of the electrolyte, polarization characteristics of the electrodes, and operating potential of the cell. In general, the EMF series (Table 1) has been used to indicate the approximate potential difference one may expect when two metals are coupled together. Metals near one another in the series show relatively less potential difference than metals far apart. For this reason, the statement is made that it is inadvisable to couple together metals far apart in the series, since the increased potential difference induces larger currents, causing correspondingly greater galvanic corrosion.

There are limitations to the use of the EMF series, however, that deserve consideration. The EMF series essentially arranges metals in order of their standard reversible potentials, that is, metal potentials in solutions of their respective ions having thermodynamic concentration (activity) equal to one molal. In practice, however, concentrations of this

Table 1. Electromotive Force Series

Standard Electrode Potential, E_0, 25 C

Electrode Reaction	Volts
Active Metals	
$K = K^+ + e^-$	−2.922
$Ca = Ca^{++} + 2e^-$	−2.87
$Na = Na^+ + e^-$	−2.712
$Mg = Mg^{++} + 2e^-$	−2.34
$Be = Be^{++} + 2e^-$	−1.70
$Al = Al^{+++} + 3e^-$	−1.67
$Mn = Mn^{++} + 2e^-$	−1.05
$Zn = Zn^{++} + 2e^-$	−0.762
$Cr = Cr^{+++} + 3e^-$	−0.71
$Ga = Ga^{+++} + 3e^-$	−0.52
$Fe = Fe^{++} + 2e^-$	−0.440
$Cd = Cd^{++} + 2e^-$	−0.402
$In = In^{+++} + 3e^-$	−0.340
$Tl = Tl^+ + e^-$	−0.336
$Co = Co^{++} + 2e^-$	−0.277
$Ni = Ni^{++} + 2e^-$	−0.250
$Sn = Sn^{++} + 2e^-$	−0.136
$Pb = Pb^{++} + 2e^-$	−0.126
$H_2 = 2H^+ + 2e^-$	0.000
$Cu = Cu^{++} + 2e^-$	0.345
$Cu = Cu^+ + e^-$	0.522
$2Hg = Hg_2^{++} + 2e^-$	0.799
$Ag = Ag^+ + e^-$	0.800
$Pd = Pd^{++} + 2e^-$	0.83
$Hg = Hg^{++} + 2e^-$	0.854
$Pt = Pt^{++} + 2e^-$	ca 1.2
$Au = Au^{+++} + 3e^-$	1.42
$Au = Au^+ + e^-$	1.68
Noble Metals	

kind are seldom, if ever, met with so that actual potentials in more dilute solutions may actually reverse the potential relations predicted by the EMF series. For example, metal complex ions sometimes form which reduce the thermodynamic activity of metal ions to very small values and thereby increase tendency of the metal to corrode. Complex formation of this kind accounts for the fact that copper, although normally noble to iron, is less noble than iron in cyanide plating baths, and, for the same reason, lead, although normally cathodic to iron or steel, reverses potential relations in alkaline waters.

Another limitation of the series arises because of the phenomenon of passivity. As practically employed, many metals such as aluminum, chromium, and nickel are passive, that is, they behave electrochemically as more noble metals. Since the EMF series deals only with metals in their nonpassive or active state, no information is given relating to

Table 2. Galvanic Series in Sea Water

(Based on galvanic series in sea water by F. L. LaQue and G. L. Cox, *Proc. ASTM*, Vol. 40, p. 670, 1940. Places for chromium, determined from parallel data by other investigators, are approximate.)

Active or Anodic Metals

Magnesium
Magnesium alloys

Zinc
Galvanized steel or galvanized wrought iron

Aluminum 52SH
Aluminum 4S
Aluminum 3S
Aluminum 2S
Aluminum 53S-T
Alclad

Cadmium

Aluminum A17S-T
Aluminum 17S-T
Aluminum 24S-T

Mild steel
Wrought iron
Cast iron

Ni-resist

13% chromium stainless steel, type 410 (*active*)

50-50 lead tin solder

18-8 stainless steel, type 304 (*active*)
18-8, 3% Mo stainless steel, type 316 (*active*)
Chromium (*active*)

Lead
Tin

Muntz metal
Manganese bronze
Naval brass

Nickel (*active*)
78% Ni–13.5% Cr–6% Fe (Inconel) (*active*)
Yellow brass
Admiralty brass
Aluminum bronze
Red brass
Copper
Silicon bronze
5% Zn–20% Ni, Bal. Cu (Ambrac)
70% Cu–30% Ni
88% Cu–2% Zn–10% Sn (Composition G-bronze)
88% Cu–3% Zn–6.5% Sn–1.5% Pb (Composition M-bronze)

Nickel (*passive*)
78% Ni–13.5% Cr–6% Fe (Inconel) (*passive*)

70% Ni–30% Cu (Monel)

18-8 stainless steel, type 304 (*passive*)
18-8, 3% Mo stainless steel, type 316 (*passive*)
Chromium (*passive*)

Passive or Cathodic Metals

the passive metals as employed in practice. Furthermore, the EMF series as defined is unable to accommodate the many alloys of commerce.

To take care of these factors adequately, the so-called galvanic series have been proposed. These series arrange metals, and also alloys, in order of their actual potentials in some stated environment. The potentials are not necessarily thermodynamically reversible, as are the potentials in the EMF series, but may represent steady-state potentials and, correspondingly, the actual potentials that operate when dissimilar metals are coupled in the given environment. Table 2 lists the galvanic series for sea water. In general, coupling metals far apart in this series leads to more serious galvanic corrosion than coupling metals adjoining one another. The series may be similar or may vary as one chooses environments other than sea water.

10. THE STAINLESS STEELS

DEFINITION. Iron-base alloys containing 12% chromium or more are known as stainless steels. They are so called because of their property of passivity at and above this critical percentage of chromium. Passivity is related to thin films, possibly monolayers of adsorbed substance (gases, solute ions, etc.) on the surface, and a specific electronic structure of the alloy which combine to impart considerable corrosion resistance. Sometimes a thicker corrosion-product layer, forming later, supplements the corrosion resistance accounted for by the primary film.

At elevated temperatures, the stainless steels are oxidation resistant, attributed again to their chromium content. The chromium content, however, is not critical as in passivity, the oxidation resistance increasing with percentage of chromium. The mechanism of resistance at high temperatures lies in physical protection of relatively thick layers of metal oxides on the surface.

Table 3. Stainless Steels

(All AISI Types)

MARTENSITIC

Type 403: 0.15 max C, 11.5—13.0 Cr. **Thermal treatment:** Full annealing, cool slowly from 1500–1650 F; low annealing, cool slowly from 1200–1400 F; hardening, cool rapidly from 1700–1850 F; tempering, 400–1400 F.

Used for steam turbine blades.

Type 410: 0.15 max C, 11.50–13.50 Cr. **Thermal treatment:** Same as type 403 (*above*).

Type 414: 0.15 max C, 11.50–13.00 Cr, 1.25–2.50 Ni. **Thermal treatment:** Full annealing, not practical; low annealing, 1200–1300 F; hardening, cool rapidly from 1800–1900 F; tempering, 400–1300 F.

Somewhat better corrosion resistance than type 410.

Type 416: 0.15 max C, 12.00–14.00 Cr, 0.60 Mo or Zr, 0.07 min P, S, or Se. **Thermal treatment:** Full annealing, cool slowly from 1500–1650 F; low annealing, 1200–1400 F; hardening, cool rapidly from 1700–1850 F; tempering, 400–1400 F.

Free machining and nonseizing. Suitable for automatic screw machine.

Type 420: C $>$0.15, 12.00–14.00 Cr. **Thermal treatment:** Annealing, cool slowly from 1550–1650 F; low annealing, 1350–1450 F; hardening, cool rapidly from 1800–1900 F; tempering, 300–700 F.

Can be hardened to 500 Brinell. Maximum corrosion resistance only in fully hardened condition. Used for cutlery, valves, surgical instruments.

Type 431: 0.20 max C, 15.00–17.00 Cr, 1.25–2.50 Ni. **Thermal treatment:** Full annealing, impractical; low annealing, 1150–1225 F; hardening, cool rapidly from 1800–1950 F; tempering, 400–1200 F.

Called 16-2 stainless steel. Better corrosion resistance than types 410, 420, 430, and 440.

Type 440A: 0.60–0.75 C, 16.00–18.00 Cr, 0.75 max Mo. **Thermal treatment:** Full annealing, cool slowly from 1550–1650 F; low annealing, 1350–1450 F; hardening, cool rapidly from 1850–1950 F; tempering, 300–800 F.

Greater hardness than type 420. Greater toughness than types 440B and C. Types 440A, B, C are used for cutlery, valves, instruments.

Type 440B: 0.75–0.95 C, 16.00–18.00 Cr, 0.75 max Mo. **Thermal treatment:** Full annealing, cool slowly from 1550–1650 F; low annealing, 1350–1450 F; hardening, cool rapidly from 1850–1950 F; tempering, 300–800 F.

Corrosion resistant only in fully hardened condition. Greater hardness than type 440A and greater toughness than type 440C.

Type 440C: 0.95–1.20 C, 16.00–18.00 Cr, 0.75 max Mo. **Thermal treatment:** Full annealing, cool slowly from 1550–1650 F; low annealing, 1350–1450 F; hardening, cool rapidly from 1850–1950 F; tempering, 300–800 F.

Acquires highest hardness of a corrosion-resistant steel. Corrosion resistant only in fully hardened condition.

Table 3. Stainless Steels—*Continued*

(All AISI Types)

Ferritic

Type 405: 0.08 max C, 11.50–13.50 Cr, 0.10–0.30 Al. **Thermal treatment:** Low annealing, 1350–1500 F; hardening, not appreciably hardenable.

Can be cooled from elevated temperatures without marked hardening.

Type 406: 0.15 max C, 12.00–14.00 Cr, 3.50–4.50 Al. **Thermal treatment:** Low annealing, 1350–1500 F; hardening, not appreciably hardenable.

Used for electrical resistor elements. Low thermal coefficient of electrical resistance up to 1000 F.

Type 430: 0.12 max C, 14.00–18.00 Cr. **Thermal treatment:** Low annealing, 1400–1500 F; hardening, not appreciably hardenable.

Better corrosion resistance than type 410. Used for automobile trim, nitric acid equipment, etc.

Type 430F: 0.12 max C, 14.00–18.00 Cr, 0.60 max Mo or Zr, 0.07 min P, S, or Se. **Thermal treatment:** Low annealing, 1250–1400 F; hardening, not appreciably hardenable.

Free-machining variation of type 430. Suitable for automatic screw machines.

Type 442: 0.25 max C, 18.00–23.00 Cr. **Thermal treatment:** Low annealing, 1400–1500 F; hardening, not hardenable by thermal treatment.

Slightly better corrosion and heat resistance than type 430. Resists destructive scaling up to 1600 F.

Type 443: 0.20 max C, 18.00–23.00 Cr, 0.90–1.25 Cu. **Thermal treatment:** Low annealing, 1400–1500 F; hardening, not hardenable by thermal treatment.

Added corrosion resistance to some types of chemicals because of Cu content.

Type 446: 0.35 max C, 23.00–27.00 Cr, 0.25 max N. **Thermal treatment:** Low annealing, 1450–1600 F; hardening, not hardenable by thermal treatment.

Contains maximum chromium consistent with commercial malleability. Used principally for high-temperature oxidation resistance. Resists destructive scaling up to 2000 F. Embrittles (only at room temperature) on prolonged heat treatment at 750–1025 F.

Austenitic

Type 301: 0.08–0.15 C, 16.00–18.00 Cr, 6.00–8.00 Ni. **Thermal treatment:** Annealing, cool rapidly from 1850–2050 F; hardening, hardenable by cold work only; stress relieving, 400–750 F.

Attains high tensile strength and ductility by cold working. Used largely as cold-rolled sheet, strip, and wire.

Type 302: 0.08–0.15 C, 17.00–19.00 Cr, 8.00–10.00 Ni. **Thermal treatment:** Same as type 301 (*above*).

This is the general-purpose austenitic stainless steel. Corrosion resistance better than type 301. Used largely in annealed condition. Can be cold worked to high tensile strength, but with slightly lower ductility than type 301.

Type 302B: 0.08–0.15 C, 17.00–19.00 Cr, 8.00–10.00 Ni, 2.00–3.00 Si. **Thermal treatment:** Same as type 301 (*above*).

Resistance to scaling better than type 302. Used for high-temperature applications.

Type 303: 0.15 max C, 17.00–19.00 Cr, 8.00–10.00 Ni, 0.60 max Mo or Zr, 0.07 min P, S, or Se. **Thermal treatment:** Same as type 301 (*above*).

Free machining and nonseizing 18-8. Suitable for automatic screw machines.

Type 304: 0.08 max C, 18.00—20.00 Cr, 8.00–11.00 Ni. **Thermal treatment:** Same as type 301 (*above*).

Low carbon 18-8; somewhat better corrosion resistance than type 302. It is less readily sensitized with respect to intergranular corrosion in welding.

Type 308: 0.08 max C, 19.00–21.00 Cr, 10.00–12.00 Ni. **Thermal treatment:** Same as type 301 (*above*).

Somewhat better corrosion and heat resistance than type 304.

Type 309: 0.20 max C, 22.00–24.00 Cr, 12.00–15.00 Ni. **Thermal treatment:** Annealing, cool rapidly from 1900–2050 F; hardening, hardenable by cold work only; stress relieving, 400–750 F.

Good heat-resistant qualities.

Type 310: 0.25 max C, 24.00–26.00 Cr, 19.00–22.00 Ni. **Thermal treatment:** Annealing, cool rapidly from 1900–2100 F; hardening, hardenable by cold work only; stress relieving, 400–750 F.

Still better heat-resistant qualities than type 309.

Type 316: 0.10 max C, 16.00–18.00 Cr, 10.00–14.00 Ni, 1.75–2.75 Mo. **Thermal treatment:** Annealing, cool rapidly from 1850–2050 F; hardening, hardenable by cold work only; stress relieving, 400–750 F.

Superior corrosion resistance, especially to chlorides, nonoxidizing acids, and the atmosphere. Superior creep strength at elevated temperatures. Susceptible to intergranular corrosion if heated in range 800–1650 F, but less so than types 302 or 304.

Type 321: 0.10 max C, 17.00–19.00 Cr, 8.00–11.00 Ni, Ti 5 × carbon, min. **Thermal treatment:** Annealing, cool rapidly from 1750–1950 F; hardening, hardenable by cold work only; stress relieving: 400–750 F.

"Stabilized" 18-8. Recommended for parts fabricated by welding which cannot be annealed subsequently. (For spot welding, types 304 and 316 are also satisfactory without subsequent annealing.) Also recommended for parts to be used between 800 and 1650 F.

Type 347: 0.10 max C, 17.00–19.00 Cr, 9.00–12.00 Ni, Cb 10 × carbon, min. **Thermal treatment:** Annealing, cool rapidly from 1850–2050 F; hardening, hardenable by cold work only; stress relieving, 400–750 F.

"Stabilized" 18-8. Recommended for parts fabricated by welding which cannot be annealed subsequently. (For spot welding, types 304 and 316 are also satisfactory without subsequent annealing.) Also recommended for parts to be used between 800 and 1650 F.

TYPES. There are three classifications of the stainless steels. Their compositions, thermal treatment, and miscellaneous characteristics are listed in Table 3.

Martensitic stainless steels are magnetic, and can be hardened by heat treatment.

Ferritic stainless steels are magnetic, and are not appreciably hardenable. Above approximately 20% chromium they embrittle if heated for long periods in the temperature range of 750 to 1025 F. This embrittlement, however, is observed only at room temperature, the alloys having normal ductility at operating temperature.

Austenitic stainless steels are nonmagnetic and are not hardenable by heat treatment. They are annealed by rapid cooling from high temperatures, acquiring maximum softness and ductility. They can be hardened considerably by cold work. In this process, some compositions become not only hard and acquire high strength, but magnetic as well. Types 308 and 309, because of their higher nickel content, become less magnetic than some other types; whereas type 310 remains essentially nonmagnetic on cold working. The cold-worked materials are also corrosion resistant.

The austenitic alloys corrode *intergranularly* if heated for appreciable lengths of time within the range of 800 to 1650 F, or are slowly cooled in this range. The minimum time for sensitization to intergranular attack amounts to a few minutes in the higher temperature range, and many hours in the lower range of temperature. In welding operations, sensitization usually occurs to some extent at areas away from the weld. The higher the carbon content, the more readily are the alloys sensitized. Alloys so sensitized must be quenched from 1850 to 2050 F to restore corrosion resistance.

The "stabilized" steels (types 321 and 347) are immune to this type of attack, regardless of heat treatment or welding operation. The type 316 alloy (Mo 18-8), although attacked less rapidly than type 302 or 304, will also corrode intergranularly if heated in the temperature range cited above.

Recently, special low-carbon grades of stainless steels have been produced which are less susceptible to intergranular corrosion when heated in the critical temperature zone than are the usual unstabilized types.

GENERAL CORROSION CHARACTERISTICS. Stainless steels, in general, have maximum corrosion resistance when quenched from high temperatures. Cold work tends to decrease only slightly their resistance to some media; tempering decreases resistance somewhat more.

They possess maximum passivity and, hence, maximum corrosion resistance in fully aerated or oxidizing solutions. For example, they show outstanding corrosion resistance to *nitric acid*. They are not satisfactorily resistant to *sulfuric acid* except in aerated dilute concentrations, or in acid properly inhibited with Cu^{++}, Fe^{+++}, or NO_3^- additions. The type 316 alloy is satisfactorily resistant to *sulfurous acid*. None of the stainless steels withstands *hydrochloric acid*.

All the alloys resist attack by *alkalies* at ordinary temperatures, but corrode at elevated temperatures, particularly in caustic alkalies that are of medium or high concentration. The alloys can be used with *ammonia* or ammonium hydroxide at room temperature without attack. They are nitrided in contact with ammonia at elevated temperatures. Stainless steels hardened by nitriding or carburizing suffer in some degree loss of passivity and corrosion resistance.

The halogen ions break down passivity, and stainless steels, therefore, tend to corrode in such solutions. In hydrochloric acid or in nonoxidizing acid halide salts, e.g., $AlCl_3$, corrosion is uniform. However, in neutral or near-neutral halide salts, corrosion usually occurs by pitting. In *sea water*, for example, serious pitting may be observed after a few months. Type 316 (Mo 18-8) is more resistant, but pitting of this alloy in sea water eventually occurs (usually within 2 1/2 years or less). All the alloys pit badly in heavy metal chloride solutions, such as ferric chloride, cupric chloride, and mercuric chloride. Buried underground, their corrosion resistance is good. They tend to fail by pitting in some types of *soil*, but this tendency is much less pronounced than in sea water.

The stainless steels, particularly those of high alloy content, remain bright for many years in rural *atmospheres*. A slight tarnish may form in industrial or marine atmospheres. Usually, attack can be avoided or minimized by frequent cleaning of the surface so as to remove grime and dust. Type 316 18-8 is the most resistant of the common types of stainless steels.

Stainless-steel surfaces in any application should always be kept clean and bright. At a contacting surface in presence of moisture, it is common for an electrolytic cell to establish itself so that a pit forms underneath the area of contact (contact or crevice corrosion). If the surface is clean, however, such cells do not set up and the surface remains bright. For maximum corrosion resistance, therefore, crevices, cracks, or overlapping joints should be avoided, particularly if such areas permit the accumulation of dirt or act as reservoirs for stagnant solutions.

PITTING. Since the stainless steels, properly heat treated, corrode, if at all, usually by pitting, the following precautions help avoid attack of this kind:

1. Avoid concentration of halogen ions.
2. Insure uniform concentrations of dissolved oxygen or oxidizing solutions. Solutions should be agitated, and stagnant pools avoided.
3. Increase *p*H. Appreciably alkaline chloride solutions cause less pitting or none at all compared to neutral or acid chlorides.
4. Operate at lowest temperatures possible.
5. Apply cathodic protection. Stainless steels, galvanically coupled to steel, aluminum, or zinc, for example, apparently do not pit in sea water.
6. Cleanse or burnish surfaces periodically, thus helping to avoid contact (crevice) corrosion. Alkaline cleaners and stainless steel wool are satisfactory.

STRESS-CORROSION CRACKING. Stainless steels are subject to stress-corrosion cracking in specific media, particularly acid chlorides or dilute hydrochloric acid. The failure is usually transgranular, distinguishing stress-corrosion cracking of these alloys from intergranular corrosion, discussed above, not requiring stress. Failure from this source does not depend on precipitated carbides; hence the columbium and titanium stabilized stainless steels are also susceptible. Because conditions for cracking require both a minimum stress and a highly specific environment, failures of this kind are only rarely observed in service.

Boiling 42% magnesium chloride is extremely active in producing stress-corrosion cracking of the stainless alloys and, hence, is used in accelerated tests. In laboratory tests, cracking has been observed with boiling calcium chloride, cobalt chloride, zinc chloride, lithium chloride, and ammonium chloride solutions. Ethyl chloride containing water can also cause cracking.

The stress necessary to cause transgranular cracking in specific media can be residual or applied. Stencil-stamped specimens and deep-drawn stainless steel sheets will crack in some of the media listed above. Stresses resulting from cold straightening, quenching, grinding, or differential thermal expansion are adequate to cause stress-corrosion cracking. The minimum critical stress for stress-corrosion cracking of 18-8 in boiling 42% magnesium chloride has been found to be less than 20,000 psi.

The ferritic stainless steels are less susceptible than the austenitic alloys. With the austenitic alloys, it is reported that resistance to cracking increases with nickel content. Furthermore, maximum sensitivity of 18-8 to cracking in a hot calcium chloride solution containing $^1/_{10}$ to 1% mercuric chloride increases with carbon content up to 0.045%, then decreases. For similar alloys containing higher nickel, the corresponding carbon content for greatest sensitivity is said to be higher.

GALVANIC COUPLING. The stainless steels when passive have approximately the same galvanic potential as silver. Hence, they are cathodic to and somewhat accelerate corrosion of, e.g., steels, zinc, and cadmium with which they may be coupled. When they lose passivity, as in dilute acid chlorides, they have approximately the same potential as iron. Since they are applied usually as passive materials, it is commonly satisfactory to couple one type of stainless steel with another type, or with metals such as silver, copper, aluminum and aluminum alloys, nickel, Inconel, and Monel.

REMOVING OXIDES FROM THE SURFACE. A stainless-steel surface oxidized by heat treatment or welding operations can be cleaned either by sand blasting, grinding, or pickling. Sand blasting is particularly useful for general removal of scale when a heavy layer is formed. It is not recommended if surfaces are to be polished.

Pickling is accomplished using the following solution:

Sulfuric acid (specific gravity 1.84), 10 parts by volume.
Water, 100 parts by volume.
Rock salt, $^1/_4$ lb per gal of solution.
Temperature, 160 to 180 F.
Time, 10 to 45 min.

This solution is used only to loosen the scale. It can be made more active by addition of hydrochloric acid. Attack of the metal is reduced by addition of a corrosion inhibitor. The work after pickling should be rinsed in hot water and scrubbed.

If further treatment is required, the following solution should be used. This solution is also useful alone for removing light oxide or tarnish films not requiring the pretreatment above.

Nitric acid (specific gravity 1.42), 10 to 20 parts by volume.
Hydrofluoric acid, 1 to 4 parts by volume.
Water, 100 parts by volume.
Temperature, 130 to 150 F.

The latter should be used only under conditions of adequate ventilation since the fumes are toxic. The work, as before, should be rinsed and scrubbed thoroughly in hot water.

Sometimes the stainless steels are "passivated" after fabrication and heat treatment to impart maximum corrosion resistance. This operation, however, is not necessary after the pickling treatments outlined above, since no permanent additional corrosion resistance is imparted. The process is carried out in this solution:

Nitric acid (specific gravity 1.42), 5 to 20% by volume.
Temperature, 130 to 140 F.
Time, 10 to 20 min.

The work finally should be scrubbed and rinsed in hot water.

MATERIALS FOR PICKLING TANKS. Tanks for any of the pickling or "passivating" solutions are most conveniently constructed of wood. Although they will not give long life, they are easily constructed and inexpensively replaced.

Wooden tanks lasting somewhat longer than more ordinary tanks are constructed with an outer box of cypress coated with bakelite base varnish. This box is lined with a bituminous paint, then with fir also coated with bakelite finish.

Acidproof brick can also be used. Such tanks consist of an outer shell of steel or concrete lined with heavy asphaltum. Acidproof brick is "set in" and bonded with acidproof cement.

Tanks for the nitric acid "passivation" bath can be made from welded type 304 low-carbon 18-8.

11. SILICON-IRON ALLOYS

Silicon-iron alloys containing 14–15% silicon are known under various trade names, some of which are Duriron, Tantiron, and Corrosiron. They are generally corrosion resistant to a wide variety of media, but are used most generally for handling acids or acid salts. The alloys are not machinable, hence must be cast or ground to shape. They have high hardness and brittleness, and possess poor resistance to thermal shock.

The alloys are generally satisfactory in resisting sulfuric, nitric, acetic, lactic, and many other commercial acids in any concentration and at elevated temperatures. The corrosion rate in 10% sulfuric acid at 175 F is about 0.01 ipy (inches per year). They also satisfactorily resist sea water and mine water. They can be used to handle caustic solutions under moderately corrosive conditions, although some other common metals usually offer better advantages.

The alloy should not be used with halogen gases or their aqueous solutions, fused alkalies, hydrofluoric acid, crude phosphoric acid, concentrated hydrofluosilicic acid, sulfurous acid, hot hydrochloric acid, aqua regia, and solutions of heavy metal chlorides.

THE 14.5% Si–3% Mo–Fe ALLOY. Resistance of the Si-Fe alloy to hot concentrated HCl is accomplished by the metallurgical addition of 3% Mo. This alloy is known as Durichlor. The corrosion rate in 30% HCl at 85 F is 0.01 to 0.02 ipy, at 150 F it is 0.02 to 0.035 ipy, and at 175 F it is 0.06 to 0.1 ipy. This alloy is not recommended for use with hot, heavy metal chloride solutions in concentrated acids, with mixtures of chlorine and steam, or with strong sulfuric or dilute nitric acids alternated with hydrochloric acid.

12. NICKEL AND NICKEL ALLOYS

NICKEL (see also Table 4) is more resistant under reducing conditions than under oxidizing conditions. For that reason, it can be used in service with sulfuric acid with only moderate attack, whereas it is not useful with nitric acid. It is outstandingly resistant to alkalies (except ammonium hydroxide) at any concentration or temperature.

Nickel is resistant to *fresh waters*, whether soft or hard. Corrosion in *sea water* is accompanied by pitting (0.07 to 0.09 in. maximum penetration in 3- to 4-year tests). Dissolved hydrogen sulfide accelerates the corrosion rate; resistance of nickel, nevertheless, is adequate in brine and natural waters containing this gas.

The corrosion rate of nickel in *sulfuric acid* is less in deaerated acids (0.003 ipy in hydrogen-saturated 5% sulfuric acid, 85 F) than in aerated acid (0.04 ipy in air-saturated 5% sulfuric acid, 85 F). The same is true in *hydrochloric acid*, the rates in 5% HCl being about twice those in 5% H_2SO_4. Nickel is not satisfactory for use with *crude phosphoric acid* or with *sulfurous acid*. In pure dilute *phosphoric acid* rate of attack is moderate, but in hot concentrated acid, corrosion is severe.

Organic acids at room temperature are only moderately corrosive in presence of nickel. At elevated temperatures, in highly aerated acids such as formic or acetic, corrosion may

Table 4. Nickel and Nickel Alloys

Metal	Best Resistance to Media That Are:	Acids: Sulfuric (Aerated)	Acids: Hydrochloric	Acids: Nitric	Acids: Phosphoric	Acids: Sulfurous	Acids: Acetic (Aerated)	H_2S	Ferric Chloride	Fatty Acids	Caustic Soda
Nickel	Reducing.	Less resistant than Cu-Ni, Pb, or Si-Fe alloys. If acid is boiling, use conc. under 10%.	Dilute only (<10%).	N.G.	N.G.	N.G.	0.01 ipy for 6% acid, room temperature.	O.K. for brines, and natural waters containing H_2S.	N.G.	O.K. (0.004 ipy at 440 F).	Excellent. For fused caustic, sulfur or SO_4 plus organic matter must be absent.
Monel 67% Ni, 30% Cu, 1.4% Fe, 1% Mn.	More resistant than Ni in reducing media. Better resistant than Cu in oxidizing media.	Up to 80% acid, <0.04 ipy.	Dilute only.	N.G.	N.G.	N.G.	<0.02 ipy at any concentration, room temperature.	Same as nickel	N.G.	O.K. (0.004 ipy at 500 F).	Good. (Nickel is better.)
Inconel 79.5% Ni, 13% Cr, 6.5% Fe.	Oxidizing.	Not as resistant as Monel. 0.06 ipy in 5% acid.	In dilute acid, same rate as Ni. In concentration > 10%, rate higher than for Ni.	O.K. in conc. above 20%, room temperature. Not as good as stainless steels.	Useful for all concentrations at room temperature.	N.G.	0.0008 ipy for 10% acid, room temperature.	O.K.	N.G.	O.K.	Good. (Nickel is better.)

be severe. Nickel is usefully resistant to *fatty acids* such as stearic and oleic at ordinary or elevated temperatures.

Nickel, which shows excellent resistance to *alkalies*, is used widely to handle caustic solutions where contamination and corrosion must be a minimum. It is second only to silver in this resistance. In fused alkalies it is essential that sulfides, or sulfates plus organic materials, be absent; otherwise intergranular penetration and weakening of nickel result. In handling 75 to 98% NaOH at high temperatures, nickel, if highly stressed, is subject to stress-corrosion cracking. Annealing at 1600 F for 3 to 5 min avoids any difficulty. "L" (low-carbon) nickel is better resistant than ordinary grades in fused caustic soda, particularly at welds.

Nickel is not resistant to aqueous ammonia.

Atmospheric Corrosion. Nickel remains reasonably bright and free from tarnish indoors, being superior to silver, copper, and brass in this respect. The metal becomes dull ("fogs") when exposed outdoors, a condition that can be avoided by a thin covering electro-deposit of chromium.

Electrodeposited on iron, a thickness of nickel equal to 0.5 mil is usually considered a satisfactory coating for indoor exposures, and 1.5 mil for outdoor exposures.

MONEL (NICKEL-COPPER) (see also Table 4). The composition of Monel is given nominally as 67% Ni, 30% Cu, 1.4% Fe, 1% Mn, 0.1% Si, 0.15% C, and 0.01% S. There are other Monels of the same basic composition to which are added aluminum, for example, to impart age-hardening characteristics, or sulfur for free machining. There are also casting compositions containing silicon added to increase hardness and impart resistance to galling. The following description applies only to Monel without additional elements, although in many environments one expects that the special Monels will behave comparably.

Monel is better resistant under reducing conditions than oxidizing conditions. It is highly resistant to corrosion by *fresh waters*, whether hot or cold (hence Monel hot-water boilers) and very resistant to *steam condensates*. The alloy is useful in *sea water*, especially under conditions of high velocity. In stagnant exposures, it tends to pit (0.005- to 0.016-in. maximum penetration in 3- to 4-year tests at Kure Beach, North Carolina).

In aerated *sulfuric acid* at 85 F, Monel is usefully resistant in concentrations of acid below 85%. The maximum corrosion rate of 0.04 ipy in the useful concentration range comes at 5% acid. Corrosion rates in hot *hydrochloric acid*, hot *phosphoric acid*, and hot *sulfurous acid* are high, as they are for nickel. Monel is resistant to unaerated *hydrofluoric acid* solutions including the anhydrous acid at low or high temperatures. It is usefully resistant to *acetic acid* at any concentration (less than 0.02 ipy at room temperature) and most common organic acids, including *fatty acids*.

The alloy is severely attacked by *nitric* or *nitrous acids*.

Monel, although attacked by concentrated *hypochlorite solutions*, is said to be satisfactory on continuous exposure to solutions containing less than 500 ppm of available chlorine. Resistance is increased by addition of sodium silicate or sodium phosphate (0.05 to 0.2%) to the hypochlorite solution.

Monel is resistant to *alkalies*, although inferior to nickel, especially in concentrated caustic soda at high temperatures. The alloy corrodes in aqueous *ammonia*.

Atmospheric Exposure. Monel remains reasonably bright indoors, but tarnishes outdoors, especially in industrial atmospheres. The corrosion rate is negligible over a long period of years.

INCONEL (Ni–Cr–Fe ALLOY) (see also Table 4). The nominal composition of Inconel is 79.5% Ni, 13% Cr, and 6.5% Fe. The alloy, in contrast to nickel and Monel, is better resistant to oxidizing than reducing media.

Inconel satisfactorily resists *fresh waters* and *steam condensates*. It is also resistant to *mine waters* containing ferric sulfate. In *sea water*, corrosion by pitting is likely to be pronounced, the tendency in this direction being greater than with Monel. In marine atmospheres, however, it is superior to Monel.

Resistance to *sulfuric acid* is not so good as with Monel. The corrosion rate increases with degree of aeration of the acid.

Although resistance to *phosphoric acid*, pure or crude, at room temperature is good, appreciable corrosion occurs in hot concentrated acid.

Inconel is resistant to *nitric acid* at room temperature above 20% concentration. However, the alloy is not so usefully resistant to this acid as are the stainless steels.

Resistance to many *organic acids* is satisfactory (although not in boiling concentrated acetic or formic acids). Resistance to *fatty acids* even at distillation temperatures (400 to 600 F) is excellent.

Inconel is resistant to *alkalies*, including concentrated NaOH at high temperatures The corrosion rate in hot 75% NaOH at 275 F is reported to be 0.0013 ipy.

In hot alkaline solutions, *intergranular corrosion* has been reported. Corrosion of this kind presumably can be avoided by heat treatment at 1550 to 1600 F for approximately 10 hours or longer.

Inconel is better resistant to *sulfur* and *sulfides* at elevated temperatures than are Monel and nickel. It is also different in that it resists aqueous *ammonia.* Pitting occurs in *sodium hypochlorite* solutions, although the alloy demonstrates satisfactory resistance in concentrations used for textile bleaching or for sterilization.

Atmospheric Exposure. On indoor exposure, Inconel remains bright. Outdoors, tarnish occurs eventually when exposed to industrial atmospheres. The corrosion rate is extremely small.

HASTELLOYS. The Hastelloys are nickel-base alloys containing molybdenum and other alloying elements. Compositions of four Hastelloys are given in the accompanying table.

	Per Cent				
	Ni	Mo	Fe		Available Forms
Hastelloy *A*	60	20	20		Wrought and cast
B	65	30	5		Wrought and cast
C	59	17	5	14% Cr, 5% W	Hot-rolled plate, and cast
D	85			10% Si, 3% Cu, 2% Al	Cast

They are alloys essentially designed for resistance to heavy chemicals. Grades *A*, *B*, and *C* are resistant, and *D* moderately resistant, to *hydrochloric acid* in all concentrations at ordinary temperatures. Resistance drops off as temperature increases. For example, Grade *A* corrodes at the high rate of 0.26 ipy in 10% hydrochloric acid at the boiling point. At 155 F, however, the rate is 0.08 ipy, which is said to mark approximately the highest useful temperature for this grade used with hydrochloric acid.

Grade *B* is more resistant to hydrochloric acid, the rate being 0.034 ipy at 155 F and 0.012 ipy at the boiling point. Grades *C* and *D* in HCl corrode at higher rates than either *A* or *B*.

In *sulfuric acid*, Hastelloy *A* is suitable for handling acid in concentrations up to 50% and at temperatures including the boiling point. In concentrations over 50%, it is resistant up to about 155 F. Grade *B* is more resistant to boiling sulfuric acid than Hastelloy *A* in concentrations up to 60%. Grade *D* is economically resistant to all concentrations at temperatures including the boiling point. The alloy can be used for concentrating sulfuric acid by evaporation (corrosion rate = approximately 0.05 ipy).

Grade *C* resists oxidizing agents such as *nitric acid* (any concentration at room temperature, dilute only up to 150 F) and *chromic acid*, or mixtures with sulfuric acid. It is less resistant to most nonoxidizing acids than *A* or *B*.

The alloys resist *alkalies*, although there are less expensive metals available that are equally resistant. Hastelloy *C* is usefully resistant to *ferric chloride* and *cupric chloride* at room temperatures, but pitting eventually occurs, particularly at above room temperatures.

Grade *C* has excellent resistance to *sea water*. In a 6-year test at Kure Beach, North Carolina, the corrosion rate was only 0.00005 ipy, and no pitting occurred. The latter grade is also resistant to wet *chlorine* at room temperature and to *sulfur dioxide* up to about 155 F.

13. COPPER ALLOYS

The important copper alloys are the copper-zinc alloys, called *brasses*, and the copper-tin alloys, called *bronzes*. Copper and some copper alloys are especially useful in resisting natural waters, including sea water. The corrosion rate of copper in sea water is approximately 0.001 ipy. In common with high-copper alloys, copper is subject to little or no fouling by marine organisms in contrast to most metals and alloys. In soft waters and waters high in dissolved carbon dioxide, the corrosion rate is higher than in hard waters. The alloys resist *nonoxidizing* acids satisfactorily if dilute and, when dissolved oxygen or oxidizing salts are absent, in higher concentrations up to boiling temperatures. They corrode rapidly in *nitric acid*, or nonoxidizing acids containing heavy metal ions such as ferric or cupric.

The alloys have excellent resistance to *atmospheric attack*. In industrial atmospheres or near the ocean they form a characteristically protective green-colored patina after long periods of exposure.

Copper and copper alloys are satisfactory in many types of soils. In soils high in organic matter, or subject to seepage by waters high in dissolved carbon dioxide and organic acids, the corrosion rate may exclude copper as an economical material.

BRASSES. Dezincification. Brasses often corrode by dezincification. The tendency increases with zinc content of the alloy, and is accelerated in stagnant solutions. *Red brass* (85% Cu, 15% Zn) and alloys containing more than 85% Cu are largely immune to dezincification. This composition range including copper, however, is more susceptible to corrosion by solutions in motion (*impingement attack*). *Yellow brass* (70% Cu, 30% Zn) if alloyed with 1% tin (*Admiralty metal*) has increased resistance to dezincification. Furthermore, if a few tenths per cent of arsenic, antimony, or phosphorus is added, the alloy, called inhibited Admiralty metal, still further resists dezincification. This alloy is used widely for condenser tubes in sea water and fresh water service.

Aluminum brass contains approximately 2% aluminum, 76% copper, 0.1% arsenic, antimony or phosphorus, and the remainder zinc. This alloy, which is particularly resistant to high-velocity media, is used where impingement attack of the usual brasses is a factor. In stagnant chloride solutions, the alloy tends to corrode by pitting.

Galvanic Coupling. Copper and the brasses can be coupled to iron in cold fresh waters without serious *galvanic corrosion*. In hot water, the action is more rapid, and in sea water still more rapid, causing accelerated failure of the iron near the point of coupling. Coupling with stainless steels is usually acceptable, but coupling with aluminum or magnesium leads to serious attack of the light metals.

Season Cracking. If residual stresses are present in brasses after fabrication, the alloys corrode by cracking when exposed to an atmosphere containing dilute ammonia or amines. This type of failure is called *season cracking*. The presence of ammonia or amines appears to be necessary to failure of this kind.

Stresses sufficient to cause season cracking can be checked by the so-called mercury test in which a solution of mercurous nitrate in dilute nitric acid causes cracking within a few minutes if the alloy has not been properly annealed.*

Alloys containing 85% copper or more are more resistant to season cracking than alloys of lower copper content such as yellow brass or Muntz metal.

COPPER-TIN ALLOYS are called bronzes or tin bronzes, and when deoxidized with phosphorus are also known as *Phosphor Bronze*. The alloys are especially resistant to impingement attack. For high-velocity waters, it is desirable that the alloy contain at least 5% tin. Alloys containing more than 8% tin are usually used in the form of castings. The corrosion rate of the 5 to 8% tin alloy in sea water is approximately 0.001 ipy. The alloys have good resistance to corrosion fatigue and stress-corrosion cracking.

COPPER-SILICON ALLOYS (Everdur) contain 1 1/2 to 4% silicon. These are high-strength alloys behaving in corrosive media approximately like copper. They are more resistant to stress-corrosion cracking than the brasses.

COPPER-BERYLLIUM ALLOYS are high-strength, heat-treatable alloys also having corrosion resistance comparable to that of copper. They contain up to 2.75% beryllium and often 0.35% nickel. They have good resistance to corrosion fatigue and intergranular corrosion. The highly stressed alloys have been known to stress-corrosion-crack in ammonia atmospheres.

COPPER-NICKEL (CUPRO-NICKEL) ALLOYS contain 70% copper and 30% nickel, and sometimes 80% copper and 20% nickel. They are used for condenser tubes when corrosive conditions are especially severe, and when strength above atmospheric temperatures must be maintained. They resist impingement attack and are not subject to season or stress-corrosion cracking in normal applications.

14. ALUMINUM AND ALUMINUM ALLOYS

ALUMINUM is one of the metals exhibiting passivity, hence is better corrosion resistant in fully aerated or oxidizing media than in reducing or partially aerated environments. Chlorides and other halogen ions tend to break down passivity locally; hence aluminum in their presence may corrode by pitting. As with other passive metals, crevices and partially shielded areas must be avoided to avoid pitting at such areas. Of outstanding importance is serious lack of resistance to alkalies. Even in dilute alkalies, such as are leached from fresh Portland cement, aluminum is attacked.

Pure aluminum is satisfactory for handling distilled water, and resists action of steam up to 350 to 500 F. It corrodes in sea water with formation of pits.

ALUMINUM ALLOYS, in general, are less resistant than pure aluminum (2S). Pure aluminum, accordingly, has better resistance to saline atmospheres and solutions, as well as to dilute nonoxidizing acids such as acetic acid.

* The ASTM requirements for the mercurous nitrate test specify immersion for 15 min without cracking in a solution containing 100 grams per liter of mercurous nitrate and 13 ml of nitric acid (specific gravity 1.42).

Aluminum is satisfactorily resistant to acetic, lactic, tartaric, citric, and boric acids at room temperature. Formic and oxalic acids are definitely corrosive, as are hydrochloric, phosphoric, and perchloric acids. It is moderately resistant to sulfuric acid below 10% concentration, or in the concentration range of 95 to 100%.

Aluminum withstands oils, gasoline, and chlorinated solvents particularly *if dry.* Halogenated solvents in presence of moisture are corrosive.

Aluminum and its alloys should not be exposed to any alkali. They should be protected against concrete or lime, and any leachings from these materials.

Mercury and mercury salts must be rigidly excluded from contact with aluminum. They can cause perforation of aluminum sheet in very short time.

CLAD ALUMINUM ALLOYS. The better corrosion resistance of pure aluminum is made use of in *cladding* to protect stronger but inherently less corrosion-resistant alloys. They are called Alcald or Pureclad. If a pit penetrates the outer layer, perforation of the sheet is retarded by galvanic currents set up between outer and inner layers. The inner layer composed of, for example, a Duralumin-type alloy (17S or 24S) is noble with respect to the cladding, so that the core is cathodically protected by the cladding. Only after considerable portions of the cladding are corroded, which takes appreciable time, is the Duralumin core attacked. The protection is similar in mechanism to the protection of iron by a galvanized coating.

DURALUMIN. The Duralumin-type alloy if slowly cooled from the solution heat-treatment temperature of 920 F, or heated for a critical time above about 250 F, *corrodes intergranularly.* If the material is solution-heat-treated and quenched, this type of corrosion does not occur.

STRESS-CORROSION CRACKING. Some highly stressed aluminum alloys are subject to *stress-corrosion cracking* in certain mildly corrosive environments, such as sodium chloride solutions. Their relative susceptibility is often dependent on heat treatment. Commercially pure aluminum (2S), on the other hand, is resistant.

GALVANIC COUPLING. In the atmosphere and in many fresh waters, aluminum is cathodic to iron. In sea water, however, aluminum becomes more anodic and, therefore, cathodically protects iron by sacrificial corrosion. Cadmium and aluminum have similar potentials in many aqueous media, so that these two metals may be coupled together. Zinc is anodic to aluminum in most neutral or acid media, but the potential reverses in alkaline media, causing accelerated attack of aluminum.

In general, it is inadvisable to couple aluminum with copper or brass without due regard to accelerated corrosion that may result. Certain fresh waters, particularly soft waters, passing through copper pipe may pick up enough copper to damage aluminum subsequently.

BEHAVIOR IN SOILS. Aluminum buried underground in some types of soil may perforate by pitting. Alkaline or acid soils may cause rapid general attack. For long life, therefore, the metal should usually be protected by a suitable coating.

ATMOSPHERIC EXPOSURE. Of aluminum-base materials, 2S aluminum is among compositions showing lowest rate of attack in outdoor exposures. Corrosion rates, in general, are low. The use of clad alloys is recommended where perforation of the sheet must be avoided over as long a time as possible.

15. MAGNESIUM AND MAGNESIUM ALLOYS

Magnesium is active in the EMF series, but exhibits, nevertheless, good corrosion resistance in many environments. Degree of aeration of the environment is not a factor in the corrosion rate as with iron, corrosion proceeding largely by hydrogen evolution. Unlike aluminum, it satisfactorily resists attack by alkaline solutions.

Pure magnesium is considerably more resistant than commercial magnesium or its alloys. Pure magnesium, for example, corrodes in sea water at a rate of 0.012 ipy, whereas commercial magnesium corrodes at a rate 100 to 500 times higher.

Magnesium satisfactorily withstands hydrofluoric acid in concentrations above 2% and can be used to handle this acid. Some pitting may develop at the water line. Magnesium is also moderately resistant to pure chromic acid. It is not resistant to any of the mineral acids other than hydrofluoric and is attacked by most organic acids.

In alkaline solutions, magnesium corrodes at only a low rate. In dilute alkalies, attack is slight up to temperatures including the boiling point, and in concentrated alkalies, to somewhat lower temperatures. For example, the rate for 10% Al–0.2% Mn–Mg alloy in 50% NaOH at 140 F (21-day test) is only 0.001 ipy but this value increases above 140 F. Chlorides also increase the rate.

Magnesium satisfactorily resists many fresh waters, whether soft or hard. Traces of heavy metal salts such as cupric or ferric cause increased attack. The metal is resistant to ethylene glycol solutions. When magnesium is coupled to iron in ethylene glycol solutions at elevated temperatures, corrosion of magnesium increases, but can be inhibited by the addition of 0.5% NaF to the solution.

Magnesium satisfactorily withstands oils and gasoline, particularly *if dry*. If gasolines are displaced by water, those containing tetraethyl lead and ethylene dibromide cause increased tendency to attack especially at the water line. Organic acids in lubricating oils cause some corrosion.

Although denatured alcohol has little effect on magnesium, anhydrous methyl alcohol (methanol) causes serious attack. The reaction can be partially inhibited by addition of water, even $^{1}/_{10}$% greatly decreasing rate of attack. Phenol can be stored in magnesium-alloy drums. Fluorinated or chlorinated solvents do not produce corrosion except in the presence of water.

GALVANIC COUPLING. In a conducting solution such as 3% sodium chloride, all common metals in contact with magnesium act as cathode and, hence, accelerate attack of magnesium. The 5% Mg–Al alloy causes much less galvanic corrosion in comparison with other readily available metals.

Cadmium or zinc plating of the more cathodic metals such as iron or steel reduces galvanic corrosion to one-tenth the rate. A reduction in conductivity, for example a change from 3% sodium chloride to tap water, effects an even greater reduction.

Under conditions where the cathodic corrosion product is not continuously removed, or where the environment may become strongly alkaline, both magnesium and an amphoteric metal to which it is coupled, such as aluminum, may be attacked. Aluminum alloys containing appreciable magnesium such as 52S, 53S, or 56S are attacked least in this manner by chloride media.

One means of materially decreasing galvanic attack is to increase the distance between dissimilar metals by use of inset caulking compounds applied in a groove, machined either in the insert or in the magnesium alloy immediately adjacent. Nonconducting washers between dissimilar metals galvanically coupled will produce the same result.

Galvanic corrosion in the atmosphere is very much less than in aqueous media.

STRESS-CORROSION CRACKING. Magnesium alloys, if highly stressed, are subject to stress-corrosion cracking in water or the atmosphere. Tendency to cracking is especially severe in potassium chromate–sodium chloride solutions. The alloys are resistant to stress-corrosion cracking in solutions above *p*H 10.2, and also in fluoride solutions of any *p*H. Stress relief of fabricated parts reduces or eliminates tendency to failure of this kind.

The 1.5% Mn–Mg alloy, as well as the commercial casting alloys, are resistant to stress-corrosion cracking even though stressed up to the yield point. The magnesium alloys containing aluminum, zinc, and manganese, on the other hand, are sensitive below the yield point.

16. LEAD

In media that corrode lead more than slightly, the rate is usually proportional to the concentration of dissolved oxygen. Lead is outstandingly resistant to atmospheric attack and to attack by sea water. It withstands the action of many mineral acids such as sulfurous, phosphoric, chromic, and hydrofluoric (cold). Lead is particularly useful in handling sulfuric acid in any concentration and at any temperature. The rate of corrosion in 20% sulfuric acid at 75 F is approximately 0.007 ipy. Corrosion rates are higher when velocity of the acid succeeds in eroding the protective sulfate film.

Aerated acetic acid is very corrosive, even at room temperature. Acids from wood, especially green woods, can cause damage if lead is in contact with them or their vapors. Permissible woods for contact with lead are cedar or hemlock; otherwise lead should be protected by tar, asphaltum, paint, or lacquer.

Lead resists the action of alkalies only poorly. The rates are high enough so that use of other materials such as iron is more economical. The corrosion rate in dilute alkalies such as calcium hydroxide is sufficiently high, so that lead should not be brought into contact with fresh concrete or with water leaching from concrete.

GALVANIC CORROSION. Corrosion of iron is usually accelerated when coupled to lead in acid solutions, but in alkaline solutions the reverse is true and corrosion of lead is accelerated by iron. Corrosion of copper is accelerated by lead in strong acid solutions, but the reverse is true in alkaline solutions. It is usually satisfactory to couple the various commercial lead alloys in handling sulfuric acid and sulfates.

Galvanic corrosion of copper may occur, as for example with copper roofing, along a soldered seam if solder (Pb–Sn) is applied at such a temperature that a noble copper-tin intermetallic compound forms. If excessive heat is not used, corrosion of copper from this source is minimized.

TOXIC PROPERTIES. Lead, although a durable metal for handling fresh waters, may corrode sufficiently to render water questionable for drinking purposes because its salts are cumulatively toxic in small quantities. Distilled, soft, and carbonated waters are particularly aggressive in this respect. Likewise, alcoholic beverages and foodstuffs may accumulate dangerous quantities of lead if brought into contact with lead or its alloys.

GENERAL REFERENCES

H. H. Uhlig (Editor), *Corrosion Handbook*, John Wiley and Sons, New York, 1948.
F. N. Speller, *Corrosion, Causes and Prevention*, McGraw-Hill Book Co., New York, 1935.
R. J. McKay and R. Worthington, *Corrosion Resistance of Metals and Alloys*, Reinhold Publishing Corp., New York, 1936.
U. R. Evans, *Metallic Corrosion, Passivity and Protection*, Edward Arnold and Co., London, 1946.
Steel Products Manual (stainless and heat-resisting steels), American Iron and Steel Institute, New York, July 1947.

SECTION 4

NONFERROUS METALS AND ALLOYS

BY

W. P. GOEPFERT, *Aluminum Company of America*

W. R. HIBBARD, *Assistant Metallurgist, The American Brass Co.*

CARTER S. COLE, *American Society for Testing Materials*

JOHN L. EVERHART, *Battelle Memorial Institute*

WILLIAM H. GROSS, *Magnesium Division, The Dow Chemical Co.*

R. D. FABER, *Development and Research Division, The International Nickel Co.*

E. A. ANDERSON, *Research Division, The New Jersey Zinc Co. (of Pa.)*

BRUCE W. GONSER, *Battelle Memorial Institute*

ARTHUR F. UNDERWOOD, *Research Laboratories Division, General Motors Corp.*

J. C. FOX, *Chief Metallurgist, Doehler-Jarvis Corp.*

C. T. EVANS, JR., *Chief Metallurgist, Elliott Co.*

ALUMINUM AND ITS ALLOYS

By W. P. Goepfert *

1. PROPERTIES OF ALUMINUM

Aluminum is one of the lightest of the metals used for structural purposes. The principal source is bauxite, a hydrated oxide of aluminum. In the production of aluminum, pure alumina (Al_2O_3), obtained by the treatment of bauxite ore with caustic soda, is reduced electrolytically to produce the metal. Silica and iron oxide impurities in the alumina will appear in the metal as silicon and iron. Purities as high as 99.95% have been produced for commercial application although 99.0 to 99.8% purity is more common for commercial usage.

RESISTANCE TO CORROSION. Aluminum possesses good resistance to corrosion under most atmospheric conditions. This characteristic is demonstrated by the successful use of high-purity aluminum (99.5% approx.) for transmission lines and outside bus bar installations. It is resistant to distilled water, many food products, fruit juices, milk and beer, sulfur dioxide, ammonia, and many other commercial items. It is readily dissolved by hydrochloric and hydrofluoric acids and by caustic solutions. The high resistance to corrosion of aluminum is due to the protective oxide film which is formed on its surface. In general, aluminum of the highest purity has the highest resistance to corrosion. Except for a few special cases, the presence of any other metals, either as impurities or alloying elements, tends to lower the resistance to corrosion.

PHYSICAL PROPERTIES. Some of the more important physical characteristics of high-purity aluminum are given in Table 1.

Table 1. Properties of High-purity Aluminum

Atomic weight	26.97
Density at 68 F	168.56 lb per cu ft
Density at melting point (solid)	159.20 lb per cu ft
Density at melting point (liquid)	148.59 lb per cu ft
Melting point	1220 F
Boiling point	3272 F
Coefficient of expansion per °F, 68–212 F	0.000 0132
Coefficient of expansion per °F, 68–572 F	0.000 0143
Coefficient of expansion per °F, 68–932 F	0.000 0154
Heat of vaporization	3510–3600 Btu per lb
Latent heat of fusion	167.36
Specific heat, 32–212 F	0.226
Thermal conductivity, Btu per hr per sq ft per °F per 1 in. thickness	1509
Electrical Properties	
Electrical resistivity at 20 C	2.688 microhms per cu cm
Electrical resistivity at 20 C (commercial conductors)	2.83 microhms per cu cm
Temperature coefficient of resistivity at 20 C	0.00403
Mechanical Properties	
Tensile strength	9000 psi
Yield strength (set = 0.2%)	3000 psi
Elongation in 2 in., %	60
Brinell hardness, 10-mm ball, 500-kg load	15
Modulus of elasticity	10,000,000
Modulus of rigidity (torsion)	3,800,000
Poisson's ratio	0.32
Miscellaneous Properties	
Crystallization shrinkage	6.6%
Reflectivity for white light	90%
Magnetic susceptibility (cgs units)	0.000 000 58
Compressibility (change in volume per megabar pressure)	0.000 0014

* Revision of material originally prepared by Zay Jeffries.

Note. Some of the processes and products described in this chapter on aluminum are covered by U. S. patents.

Table 2. Composition and Typical Properties of Sand-cast Aluminum Alloys *

Alloy ¶		Nominal Chemical Composition	Tension †			Compression ‡	Hardness	Shear §	Fatigue ‖	Density
			Yield ** Strength, psi	Ultimate Strength, psi	Elongation in 2 in., %	Yield ** Strength, psi	Brinell, 10-mm ball, 500-kg load	Psi	Endurance Limit, psi	Lb per cu in.
43	ASTM-S1, SAE 35	Si-5%	9,000	19,000	6.0	10,000	40	14,000	6,500	.095
108		Cu-4%, Si-3%	14,000	21,000	2.5	14,000	55	20,000	8,000	.099
113	ASTM-CS22, SAE 33	Cu-7%, Si-2%, Zn-1.7%	15,000	24,000	1.5	17,000	70	20,000	9,000	.104
142-T21	ASTM-CN21, SAE 39	Cu-4%, Mg-1.5%, Ni-2%	18,000	27,000	1.0	18,000	70	21,000	6,500	.100
142-T571			28,000	32,000	0.5	34,000	85	27,000	8,000	.100
142-T77			25,000	28,000	2.0	25,000	75	24,000	9,500	.100
195-T4	ASTM-C1, SAE 38	Cu-4.5%, Si-0.8%	16,000	32,000	8.5	16,000	60	24,000	6,000	.100
195-T6			24,000	36,000	5.0	25,000	75	30,000	6,500	.100
195-T62			30,000	40,000	2.0	38,000	95	31,000	7,000	.100
212		Cu-8%, Si-1.2%	14,000	23,000	2.0	14,000	65	20,000	8,000	.102
214	ASTM-G1, SAE 320	Mg-3.8%	12,000	25,000	9.0	12,000	50	20,000	5,500	.095
B214		Mg-3.8%, Si-1.8%	13,000	20,000	2.0	15,000	50	17,000	....	.094
F214		Mg-3.8%, Si-0.5%	12,000	20,000	3.0	12,000	50	....	....	.094
220-T4	SAE 324	Mg-10%	25,000	46,000	14.0	26,000	75	33,000	7,000	.091
319		Si-6.3%, Cu-3.5%	18,000	27,000	2.0	18,000	70	....	10,000	.099
319-T6			24,000	36,000	2.0	24,000	80	....	10,000	.099
355-T51	ASTM-SC-21, SAE 322	Si-5%, Cu-1.3%, Mg-0.5%	23,000	28,000	1.5	24,000	65	22,000	7,000	.096
355-T6			25,000	35,000	2.5	29,000	80	30,000	8,500	.096
355-T61			35,000	39,000	1.0	37,000	90	32,000	....	.096
355-T7			36,000	38,000	0.5	36,000	85	26,000	8,500	.096
355-T71			29,000	35,000	1.5	29,000	75	....	10,000	.096
356-T51	ASTM-SG1, SAE 323	Si-7%, Mg-0.3%	20,000	25,000	2.0	22,000	60	18,000	7,500	.095
356-T6			24,000	33,000	4.0	24,000	70	27,000	8,000	.095
356-T7			30,000	34,000	2.0	30,000	75	18,000	....	.095
356-T71			21,000	28,000	4.5	21,000	60	16,000	....	.095
406		Mn-2%	9,000	19,000	12.0	9,000	35	14,000	5,500	.099

* In absence of more precise data, average modulus of elasticity in tension and compression may be taken as 10,300,000 psi for all these alloys. Exact moduli will vary with alloy compositions. Tension, compression, hardness, and shear tests made on separately cast test bars without machining the surface.

† Tension values determined on standard $1/2$-in. diameter test bars.

‡ Compression values determined on specimens having an l/r ratio of 12.

§ Single-shear strength values determined by double-shear tests.

‖ Endurance limits are based on 500,000,000 cycles of completely reversed stress using the R. R. Moore type of machine and specimen.

¶ Alloy numbers, except ASTM and SAE, refer to designations used by Aluminum Company of America. Suffix indicates the heat treatment applicable. Where no suffix appears after the alloy number, properties are for "As cast" material.

** Yield-strength values determined from stress-strain curve as the point where the curve departs 0.2% from the initial modulus line produced.

Table 3. Composition and Typical Properties of Permanent Mold-cast Aluminum Alloys *

Alloy ¶		Nominal Chemical Composition	Tension †			Compression ‡	Hardness	Shear §	Fatigue ‖	Density
			Yield ** Strength, psi	Ultimate Strength, psi	Elongation in 2 in., %	Yield ** Strength, psi	Brinell, 10-mm ball, 500-kg load	Psi	Endurance Limit, psi	Lb per cu in.
43	ASTM-S1, SAE 35	Si-5%	9,000	24,000	9.0	9,000	45	18,000	....	.095
A108	ASTM-SC1	Si-5.5%, Cu-4.5%	16,000	28,000	2.0	16,000	70	25,000	....	.100
113	ASTM-C3, SAE 33	Cu-7%, Si-2%, Zn-1.7%	19,000	28,000	2.0	19,000	70	23,000	....	.104
C113		Cu-7%, Si-3.5%	24,000	30,000	1.0	24,000	80	24,000	9,500	.104
A132-T551	ASTM-SN41, SAE 321	Si-12%, Ni-2.5%, Mg-1.2%, Cu-0.8%	28,000	36,000	0.5	30,000	105	24,000	....	.097
A132-T65			43,000	47,000	0.5	43,000	125	27,000	....	.097
138		Cu-10%, Si-4%, Mg-0.3%	24,000	32,000	1.5	32,000	100	22,000	....	.105
142-T571	ASTM-CN21, SAE 39	Cu-4%, Ni-2%, Mg-1.5%	34,000	40,000	1.0	34,000	105	26,000	10,500	.100
142-T61			42,000	47,000	0.5	46,000	110	31,000	9,500	.100
B195-T4	ASTM-CS4, SAE 380	Cu-4.5%, Si-2.5%	19,000	37,000	9.0	22,000	75	30,000	9,500	.101
B195-T6			26,000	40,000	5.0	26,000	90	32,000	10,000	.101
B195-T7			20,000	39,000	4.5	20,000	80	....	....	.101
A214		Mg-3.8%, Zn-1.8%	16,000	27,000	7.0	17,000	60	22,000	....	.096
319		Si-6.3%, Cu-3.5%	19,000	34,000	2.5	19,000	85	24,000	....	.099
319-T6			27,000	40,000	3.0	27,000	95	....	....	.099
355-T51	ASTM-SC21, SAE 322	Si-5%, Cu-1.3%, Mg-0.5%	24,000	30,000	2.0	24,000	75	24,000	....	.096
355-T6			27,000	43,000	4.0	27,000	90	30,000	9,000	.096
355-T62			40,000	45,000	1.5	40,000	105	36,000	....	.096
355-T7			30,000	40,000	2.0	30,000	85	30,000	....	.096
355-T71			31,000	36,000	3.0	31,000	85	26,000	....	.096
356-T6	ASTM-SG1, SAE 323	Si-7%, Mg-0.3%	27,000	40,000	5.0	27,000	90	....	....	.095
356-T7			24,000	33,000	5.0	24,000	70	....	....	.095
750-T5		Sn-6.5%, Cu-1%, Ni-1%	8,500	20,000	10.0	8,500	45	14,000	9,000	.102

* In absence of more precise data, average modulus of elasticity in tension and compression may be taken as 10,300,000 psi for all these alloys. Exact moduli will vary with alloy compositions. Tension, compression, hardness, and shear tests made on separately cast test bars without machining the surface.

† Tension values determined on standard 1/2-in. diameter test bars.

‡ Compression values determined on specimens having an l/r ratio of 12.

§ Single-shear strength values determined by double-shear tests.

‖ Endurance limits are based on 500,000,000 cycles of completely reversed stress using the R.R. Moore type of machine and specimen.

¶ Alloy numbers, except ASTM and SAE, refer to designations used by Aluminum Company of America. Suffix indicates the heat treatment applicable. Where no suffix appears after the alloy number, properties are for "As cast" material.

** Yield-strength values determined from stress-strain curve as the point where the curve departs 0.2% from the initial modulus line produced.

2. CAST ALUMINUM ALLOYS

Pure aluminum does not cast easily, nor does such a casting have attractive commercial properties. Alloys of aluminum, however, have been formulated to produce the desired commercial properties and to be compatible with the particular casting method for which they are intended. Although there has been some small production of aluminum castings by such methods as plaster mold casting, precision investment casting, and centrifugal casting, the three most significant methods are sand casting, permanent mold casting, and die casting. They are discussed in Aluminum Alloy Casting, Section 19, Article 18.

CHEMICAL COMPOSITION AND PHYSICAL PROPERTIES. The nominal chemical compositions, densities, and typical mechanical properties of the majority of the more widely used sand-casting alloys and tempers are contained in Table 2. Permanent mold-casting alloys are listed in Table 3, and die-casting alloys in Table 6, p. 4-81.

HEAT-TREATMENT DESIGNATIONS. Aluminum Company of America's basic heat treatment designations for the conditions of the casting alloys discussed herein are T2, T4, T5, T6, and T7. Suffix numbers affixed to these indicate certain variations within the basic type of treatment to obtain desired characteristics. The basic T2 treatment is that generally classified as an annealing treatment. It consists of heating at an intermediate temperature to relieve internal stresses and to stabilize dimensions. In order to obtain the potential high properties of the heat-treatable alloys, the castings must first be given a

Table 4. Typical Properties of Cast Aluminum Alloys at Elevated Temperatures

(Lowest strength during 10,000 hours of heating at testing temperature)*

Alloy	Method of Casting	Tension †	Temperature, °F				
			75	300	400	500	600
A132–T551 ASTM-SN41 SAE-321	Permanent mold cast	Yield strength Ultimate strength Elongation	28,000 36,000 0.5	22,000 31,000 1.0	13,500 23,000 2.0	9,500 17,500 2.0	5,000 11,000 8.0
142-T21 ASTM-CN21 SAE-39	Sand cast	Yield strength Ultimate strength Elongation	18,000 27,000 1.0	18,000 27,000 1.0	14,000 21,000 1.5	8,000 16,000 3.0	4,000 8,000 9.0
142-T571 ASTM-CN21 SAE-39	Permanent mold cast	Yield strength Ultimate strength Elongation	34,000 40,000 1.0	33,000 37,000 1.0	22,000 28,000 2.0	9,000 15,000 10.0	5,000 9,000 30.0
195-T4 ‡ ASTM-C1 SAE-38	Sand cast	Yield strength Ultimate strength Elongation	16,000 32,000 8.5	13,000 24,000 9.0	9,000 15,000 20.0	6,000 9,500 25.0	3,000 4,000 80.0
B195-T6 ASTM-CS4 SAE-380	Permanent mold cast	Yield strength Ultimate strength Elongation	26,000 40,000 5.0	21,000 33,000 4.0	9,000 17,000 10.0	5,000 8,000 25.0	2,500 4,000 65.0
214 ASTM-G1 SAE-320	Sand cast	Yield strength Ultimate strength Elongation	12,000 25,000 9.0	12,000 23,000 7.0	12,000 18,500 9.0	8,000 13,500 12.0	4,000 9,000 17.0
355-T6 ASTM-SC21 SAE-322	Sand cast	Yield strength Ultimate strength Elongation	25,000 35,000 2.5	25,000 30,000 1.5	9,000 13,000 12.0	5,000 8,000 22.0	3,500 6,000 30.0
355-T51 ASTM-SC21 SAE-322	Sand cast	Yield strength Ultimate strength Elongation	23,000 28,000 1.5	17,000 22,000 2.5	9,000 13,000 12.0	5,000 8,000 22.0	3,500 6,000 30.0
355-T6 ASTM-SC21 SAE-322	Permanent mold cast	Yield strength Ultimate strength Elongation	27,000 43,000 4.0	24,000 32,000 2.0	9,000 12,000 20.0	6,000 8,000 25.0	3,000 4,500 50.0
356-T6 ASTM-SG1 SAE-323	Sand cast	Yield strength Ultimate strength Elongation	24,000 33,000 4.0	16,000 21,000 5.0	9,000 13,000 8.0	5,500 8,000 20.0	3,000 4,500 45.0

* Tests made on separately cast test bars without machining the surface.

† Yield and ultimate strengths are pounds per square inch and elongation is percentage in 2 in.

‡ The properties shown for temperatures of 300 F and higher also apply to the T6 condition.

solution heat treatment. This treatment, identified as T4, provides for heating the material at a sufficiently high temperature to obtain solution of the soluble constituents and for quenching to retain a large part of the solute in solution. A considerable increase in properties is obtained by this treatment. For most alloys, however, a subsequent aging treatment at a low temperature to effect precipitation of the dissolved constituents will further increase the properties. Solution treatment followed by such aging is designated as T6. Treatments consisting of aging, subsequent to solution treatment, under such conditions as to carry the hardness of the alloy beyond maximum and to control growth and residual stress bear the designation T7. Material in the as-cast condition when given an aging treatment without a previous solution treatment will experience an increase in properties above those in the as-cast condition. This treatment, which is known as T5, improves mechanical properties, reduces internal stress, and increases the dimensional stability at elevated temperature.

ALUMINUM-ALLOY CASTINGS AT HIGH TEMPERATURE. Strength and hardness of pure aluminum decreases progressively with rising temperature. Certain alloys, however, maintain their properties at elevated temperatures better than others and are therefore used for high-temperature applications. Properties of some of the casting alloys are given in Table 4. The values in this table are the lowest strength during 10,000 hours of heating at testing temperature. Room temperature properties after heating will be higher than the properties at an elevated temperature.

ALUMINUM-ALLOY CASTINGS AT LOW TEMPERATURES. In general, strength and hardness of aluminum alloys remain unchanged or improve at subzero temperatures.

3. WROUGHT ALUMINUM ALLOYS

CLASSIFICATION. Wrought aluminum alloys may be classified as: (1) nonheat-treatable or those in which the obtainable properties are produced by strain hardening; (2) heat-treatable or "strong alloys" in which the obtainable properties are produced by heat treatment alone or in combination with a certain amount of strain hardening. In general, higher properties are obtainable with alloys of Class 2. The Aluminum Company of America designates its wrought alloys by the alloy number followed by a temper or heat-treatment designation. The letter "S" used in terminating the alloy number merely indicates that the alloy is in wrought form as differentiated from cast form.

TEMPER DESIGNATIONS. With alloys of the nonheat-treatable type, material in a definite strain-hardened state is indicated by a temper designation representing a particular set of mechanical properties produced by a particular type of treatment for the pertinent alloy.

Formerly, these temper designations ranged with increasing amounts of strain hardening from "O" for annealed material to "H" for hard temper, intermediate tempers being designated as fractions of H. This system of nomenclature, however, has been found too inflexible to permit differentiation of certain processing variables, and the Aluminum Company of America has introduced a more complete system. In the new system O continues to designate annealed material for all wrought alloys. The strain-hardened conditions are indicated basically by H1 when the properties are the result of strain hardening only, H2 when strain hardened and partially annealed, and H3 when strain hardened and stabilized. These basic designations are followed by one or more digits to indicate the final degree of strain hardening remaining in the material. For example, 2S-$^1/_2$H now becomes 2S-H14 or 2S-H24, depending on whether it was produced by strain hardening alone or by strain hardening and partial annealing, and 52S-$^1/_2$H (strain hardened and stabilized) becomes 52S-H34.

Similarly, the old system of nomenclature for the heat-treatable alloys, where basically "W" indicated the solution heat-treated temper and "T" the solution heat-treated and aged temper, has been found inadequate to differentiate between all the pertinent processing differentials recognized in the industry today. The new system introduced by the Aluminum Company of America for these alloys is basically:

- -W Solution heat treated—unstable temper.
- -T3 Solution heat treated and then cold worked.
- -T4 Solution heat treated and naturally aged to a substantially stable condition.
- -T5 Artificially aged only.
- -T6 Solution heat treated and then artificially aged.
- -T8 Solution heat treated, cold worked, and then artificially aged.

Subsequent digits are used to indicate variations within the above basic classifications. For example, 24S-T becomes 24S-T3 or T4, depending on whether the material is cold worked after solution heat treatment or not.

(Continued on p. 4-10)

Table 5. Composition and Typical Properties of Nonheat-treatable Wrought Aluminum Alloys *

Alloy and Temper §		Old Temper Desig-nation	Nominal Chemical Composition	Tension				Modulus of Elasticity,† psi	Hardness, Brinell, 10-mm ball, 500-kg load	Shear, psi	Fatigue,‡ Endurance Limit, psi	Density, lb per cu in.
				Yield ‖ Strength, psi	Ultimate Strength, psi	Elongation in 2 in., %						
						Sheet Specimen, 1/16 in. thick	Round Specimen, 1/2 in. diam.					
2S-O	SAE 25	O	99.0% min Al	5,000	13,000	35	45	10,000,000	23	9,500	5,000	.098
2S-H12		1/4 H		14,000	15,500	12	25	10,000,000	28	10,000	6,000	.098
2S-H14		1/2 H		16,000	17,500	9	20	10,000,000	32	11,000	7,000	.098
2S-H16		3/4 H		18,000	20,000	6	17	10,000,000	38	12,000	8,500	.098
2S-H18		H		22,000	24,000	5	15	10,000,000	44	13,000	8,500	.098
3S-O	SAE 29	O	Mn-1.2%	6,000	16,000	30	40	10,000,000	28	11,000	7,000	.099
3S-H12		1/4 H		17,000	19,000	10	20	10,000,000	35	12,000	8,000	.099
3S-H14		1/2 H		19,000	21,500	8	16	10,000,000	40	14,000	9,000	.099
3S-H16		3/4 H		22,000	25,000	5	14	10,000,000	47	15,000	9,500	.099
3S-H18		H		26,000	29,000	4	10	10,000,000	55	16,000	10,000	.099
52S-O	SAE 201	O	Mg-2.5%, Cr-0.25%	12,000	27,000	25	30	10,200,000	45	18,000	17,000	.097
52S-H32		1/4 H		27,000	34,000	12	18	10,200,000	62	20,000	17,500	.097
52S-H34		1/2 H		31,000	37,000	10	14	10,200,000	67	21,000	18,000	.097
52S-H36		3/4 H		34,000	39,000	8	10	10,200,000	74	23,000	18,500	.097
52S-H38		H		36,000	41,000	7	8	10,200,000	85	24,000	19,000	.097

* The properties given in this table are generally applicable to all wrought products made in these alloys, although some variation with product and direction of testing will occur.

† Average modulus of elasticity in tension and compression. Tensile modulus will be approximately 1% lower and compressive modulus approximately 1% higher than this average value.

‡ Endurance limit values are based on 500,000,000 cycles of completely reversed stress using the R. R. Moore type of machine and specimen.

§ Alloy numbers, except SAE, refer to designations used by the Aluminum Company of America. Suffix indicates temper applicable.

‖ Yield-strength values determined from stress-strain curve as the point where the curve departs 0.2% from the initial modulus line produced. Compressive yield strength is the same as tensile yield strength for all practical purposes.

Table 6. Composition and Typical Properties of Heat-treatable Wrought Aluminum Alloys *

Alloy and Temper §		Old Temper Designation	Nominal Chemical Composition	Tension: Yield ‖ Strength, psi	Tension: Ultimate Strength, psi	Tension: Elongation in 2 in., %: Sheet Specimen, 1/16 in. thick	Tension: Elongation in 2 in., %: Round Specimen, 1/2 in. diam.	Modulus of Elasticity,† psi	Hardness, Brinell, 10-mm ball, 500-kg load	Shear, psi	Fatigue,‡ Endurance Limit, psi	Density, lb per cu in.
11S-T3		T3	Cu-5.5%, Bi-0.5%, Pb-0.5%	48,000	55,000	....	15	10,200,000	95	32,000	18,000	.102
11S-T8		T8		45,000	59,000	....	12	10,200,000	100	35,000	18,000	.102
14S-O	SAE 260	O	Cu-4.4%, Si-0.8%, Mn-0.8%, Mg-0.4%	14,000	27,000	....	18	10,600,000	45	18,000	13,000	.101
14S-T4		W		44,000	62,000	....	20	10,600,000	105	38,000	18,000	.101
14S-T6		T		60,000	70,000	....	13	10,600,000	135	42,000	18,000	.101
Alclad 14S-O		O	Core: 14S	10,000	25,000	21	....	10,500,000	....	18,000	....	.101
Alclad 14S-T3		W	Cladding: Mg-1.3%, Si-0.7%, Cr-0.25%	40,000	63,000	20	....	10,500,000	....	37,000	....	.101
Alclad 14S-T4		W		37,000	61,000	22	....	10,500,000	....	37,000	....	.101
Alclad 14S-T6		T		60,000	68,000	11	....	10,500,000	....	41,000	....	.101
17S-O	SAE 26	O	Cu-4%, Mn-0.5%, Mg-0.5%	10,000	26,000	....	22	10,500,000	45	18,000	13,000	.101
17S-T4		T		40,000	62,000	....	22	10,500,000	105	38,000	18,000	.101
18S-T61	SAE 270	T	Cu-4.0%, Ni-2.0%, Mg-0.6%	48,000	62,000	....	12	10,800,000	120	39,000	17,000	.101
24S-O	SAE 24	O	Cu-4.5%, Mg-1.5%, Mn-0.6%	11,000	27,000	19	22	10,600,000	47	18,000	13,000	.100
24S-T3		T		50,000	70,000	16	....	10,600,000	120	41,000	18,000	.100
24S-T4		T		48,000	68,000	20	19	10,600,000	120	41,000	18,000	.100
24S-T36		RT		57,000	73,000	13	....	10,600,000	130	42,000	18,000	.100
Alclad 24S-T3	SAE 240	T	Core: 24S	44,000	64,000	15	....	¶	....	40,000	....	.100
Alclad 24S-T4		T	Cladding: Al-99.3%	42,000	64,000	19	....	¶	....	40,000	....	.100
Alclad 24S-T36		RT		53,000	67,000	11	....	¶	....	41,000	....	.100
Alclad 24S-T81		T81		60,000	65,000	6	....	¶	....	....	....	.100
Alclad 24S-T86		T86		66,000	70,000	6	....	¶	....	....	....	.100
25S-T6	SAE 27	T	Cu-4.5%, Si-0.8%, Mn-0.8%	37,000	58,000	....	19	10,400,000	110	35,000	18,000	.101

32S-T6	SAE 290	T	Si-12.2%, Mg-1.1%, Cu-0.9%, Ni-0.9%	46,000	55,000	....	9	11,400,000	120	38,000	16,000	.097
A51S-T6	SAE 280	T	Si-1.0%, Mg-0.6%, Cr-0.25%	43,000	48,000	....	17	10,200,000	100	32,000	11,000	.097
53S-O	SAE 282	O	Mg-1.3%, Si-0.7%, Cr-0.25%	7,000	16,000	....	35	10,000,000	26	11,000	8,000	.097
53S-T4		W		20,000	33,000	....	30	10,000,000	65	20,000	13,000	.097
53S-T6		T		33,000	39,000	....	20	10,000,000	80	24,000	13,000	.097
61S-O	SAE 281	O	Mg-1.0%, Si-0.6%, Cu-0.25%, Cr-0.25%	8,000	18,000	22	30	10,000,000	30	12,500	9,000	.098
61S-T4		W		21,000	35,000	22	25	10,000,000	65	24,000	13,500	.098
61S-T6		T		40,000	45,000	12	17	10,000,000	95	30,000	13,500	.098
63S-F		As extruded	Mg-0.7%, Si-0.4%	13,000	22,000	20	....	10,000,000	42	14,000	....	.098
63S-T5		....		25,000	30,000	12	....	10,000,000	65	18,000	....	.098
63S-T6		....		30,000	35,000	12	....	10,000,000	73	22,000	....	.098
75S-O		O	Zn-5.6%, Mg-2.5%, Cu-1.6%, Cr-0.3%	15,000	33,000	17	16	10,400,000	60	22,000	....	.101
75S-T6		T		72,000	82,000	11	11	10,400,000	150	49,000	21,000	.101
Alclad 75S-O		O	Core: 75S	14,000	32,000	17	....	**	....	22,000	....	.101
Alclad 75S-T6		T	Cladding: Zn-1.0%	67,000	76,000	11	....	**	....	46,000	....	.101

* The properties given in this table are generally applicable to all wrought products made in these alloys, although some variation with product and direction of testing will occur.

† Average modulus of elasticity in tension and compression. Tensile modulus will be approximately 1% lower and compressive modulus approximately 1% higher than this average value.

‡ Endurance limit values are based on 500,000,000 cycles of completely reversed stress using the R. R. Moore type of machine and specimen.

§ Alloy numbers, except SAE, refer to designations used by the Aluminum Company of America. Suffix indicates heat treatment applicable.

‖ Yield-strength values determined from stress-strain curve as the point where the curve departs 0.2% from the initial modulus line produced. Compressive yield strength is the same as tensile yield strength for all practical purposes.

¶ Alclad 24S has a primary and secondary modulus as a result of the aluminum cladding. Primary modulus is 10,600,000 for all gages. Secondary modulus for gages 0.064 in. and greater is 10,100,000, for gages less than 0.064 in., 9,600,000. Primary modulus effective for approximately the first 10,000 psi.

** Alclad 75S has a primary and secondary modulus as a result of the aluminum alloy cladding. Primary modulus is 10,400,000 and the secondary modulus is 9,600,000 for all gages. Primary modulus effective for approximately the first 10,000 psi.

CHEMICAL COMPOSITION AND PHYSICAL PROPERTIES. The nominal chemical compositions, densities, and typical mechanical properties for the majority of the more widely used wrought alloys and tempers are given in Tables 5 and 6 for nonheat-treatable and heat-treatable alloys, respectively.

ALUMINUM ALLOY WROUGHT PRODUCTS AT HIGH TEMPERATURE. In general wrought-aluminum alloys will experience a decrease in strength when heated above a certain critical temperature. For certain alloys which precipitation-harden during heating, there will be a loss in strength in the very early stages of heating followed by an increase in strength which eventually will drop off to a stabilized condition with increasing heating times. Properties of some of the wrought alloys at elevated temperatures are given in Tables 7 and 8 for the nonheat-treatable and the heat-treatable alloys, respectively. The values in these tables are the lowest strengths that will be obtained during 10,000 hours of heating at the test temperature. Room-temperature properties after heating will in general be higher than the properties at an elevated temperature.

Table 7. Typical Properties of Nonheat-treatable Wrought Aluminum Alloys at Elevated Temperatures

(Lowest strength during 10,000 hours of heating at testing temperature)

Alloy and Temper	Old Temper Designation	Tension *	Temperature, °F 75	212	300	400	500	600	700
2S-O	O	Yield strength	5,000	4,500	4,000	3,000	2,000	1,500	1,000
SAE-25		Ultimate strength	13,000	10,500	8,500	6,000	3,500	2,500	1,500
		Elongation	45	55	65	70	85	90	95
2S-H14	1/2 H	Yield strength	16,000	13,000	9,000	5,000	2,000	1,500	1,000
SAE-25		Ultimate strength	17,500	15,000	13,000	9,500	3,500	2,500	1,500
		Elongation	20	21	22	25	85	90	95
2S-H18	H	Yield strength	22,000	16,500	13,000	3,000	2,000	1,500	1,000
SAE-25		Ultimate strength	24,000	21,500	17,500	6,000	3,500	2,500	1,500
		Elongation	15	15	16	70	85	90	95
3S-O	O	Yield strength	6,000	5,500	5,000	4,500	3,500	2,500	2,000
SAE-29		Ultimate strength	16,000	13,000	11,000	8,500	6,000	4,000	3,000
		Elongation	40	43	47	50	60	60	60
3S-H14	1/2 H	Yield strength	19,000	16,000	12,500	8,000	4,000	2,500	2,000
SAE-29		Ultimate strength	21,500	19,500	17,500	14,000	10,000	5,000	3,000
		Elongation	16	16	17	22	25	40	60
3S-H18	H	Yield strength	26,000	19,000	14,500	6,500	4,000	2,500	2,000
SAE-20		Ultimate strength	29,000	25,500	22,500	16,500	10,000	4,000	3,000
		Elongation	10	10	12	15	25	55	60
52S-O	O	Yield strength	12,000	11,000	10,000	9,000	7,000	4,500	2,500
SAE-201		Ultimate strength	27,000	26,000	20,000	15,000	11,000	7,500	5,000
		Elongation	30	40	55	65	100	105	120
52S-H36	3/4 H	Yield strength	34,000	32,000	27,000	11,000	8,000	4,500	2,500
SAE-201		Ultimate strength	39,000	37,000	32,000	24,000	12,000	8,000	5,000
		Elongation	10	12	16	35	80	100	120

* Yield and ultimate strengths are pounds per square inch and elongation is percentage in 2 in.

ALUMINUM-ALLOY WROUGHT PRODUCTS AT LOW TEMPERATURE. In general, strength and hardness of aluminum alloys remain unchanged or improve at subzero temperatures.

EFFECTS OF COLD WORKING. Many commercial wrought products of aluminum and its alloys are supplied in various tempers determined by the amounts of cold working given the products after final annealing. See R. L. Templin in *The Aluminum Industry*, Vol. II, Chapter xix, p. 396, and *Transactions* of the American Institute of Mining and Metallurgical Engineers, Institute of Metals Division, 1930, p. 466, for a discussion of the effect of cold working on the physical properties of aluminum and its alloys. The formulas and diagrams which follow are taken from these references.

Table 8. Typical Properties of Heat-treatable Wrought Aluminum Alloys at Elevated Temperatures

(Lowest strength during 10,000 hours of heating at testing temperature)

Alloy and Temper	Old Temper Designation	Tension *	Temperature, °F						
			75	212	300	400	500	600	700
11S-T3	T3	Yield strength	48,000	39,000	19,000	14,000	5,000	2,000	1,500
		Ultimate strength	55,000	49,000	28,000	20,000	9,000	5,000	3,000
		Elongation	15	16	25	35	45	90	125
14S-T6 SAE-260	T	Yield strength	60,000	56,000	40,000	12,000	8,500	5,000	3,500
		Ultimate strength	70,000	66,000	47,000	18,000	11,000	6,500	4,500
		Elongation	13	14	15	35	45	65	70
17S-T4 SAE-26	T	Yield strength	40,000	37,000	30,000	17,000	9,500	5,000	3,500
		Ultimate strength	62,000	56,000	40,000	22,000	12,000	6,500	4,500
		Elongation	22	18	16	28	45	95	100
A17S-T4	T	Yield strength	24,000	21,000	17,000	12,000	5,500	3,500	2,000
		Ultimate strength	43,000	36,000	30,000	16,000	7,500	4,500	3,000
		Elongation	27	16	20	35	55	80	110
18S-T61 SAE-270	T	Yield strength	48,000	46,000	44,000	17,000	7,000	4,000	2,500
		Ultimate strength	62,000	58,000	48,000	21,000	11,000	6,000	4,000
		Elongation	12	12	13	25	40	60	85
24S-T3 SAE-24	T	Yield strength	48,000	45,000	37,000	22,000	10,000	5,000	3,500
		Ultimate strength	68,000	61,000	43,000	26,000	14,000	7,000	5,000
		Elongation	19	17	17	22	45	75	100
25S-T6 SAE-27	T	Yield strength	37,000	34,000	29,000	14,000	6,500	4,500	3,000
		Ultimate strength	58,000	49,000	36,000	19,000	9,000	5,000	4,000
		Elongation	19	19	18	35	50	70	80
32S-T6 SAE-290	T	Yield strength	46,000	42,000	33,000	11,000	6,500	3,500	2,000
		Ultimate strength	55,000	48,000	38,000	16,000	8,500	6,000	3,500
		Elongation	9	9	9	30	40	50	80
A51S-T6 SAE-280	T	Yield strength	43,000	32,000	18,000	6,000	5,000	4,000	3,500
		Ultimate strength	48,000	36,000	21,000	7,500	6,000	4,500	4,000
		Elongation	17	20	25	58	59	60	65
53S-T6 SAE-282	T	Yield strength	33,000	29,000	25,000	12,000	4,000	2,500	2,000
		Ultimate strength	39,000	34,000	26,000	14,000	6,000	4,000	3,000
		Elongation	20	19	17	30	70	80	85
61S-T6 SAE-281	T	Yield strength	40,000	37,000	30,000	16,000	5,000	2,500	2,000
		Ultimate strength	45,000	41,000	32,000	19,000	7,000	4,000	3,000
		Elongation	17	18	18	25	55	90	105
75S-T6	T	Yield strength	72,000	60,000	21,000	12,000	8,500	6,500	4,500
		Ultimate strength	82,000	65,000	25,000	14,000	11,000	8,500	6,500
		Elongation	11	15	30	60	65	80	65

* Yield and ultimate strengths are pounds per square inch and elongation is percentage in 2 in.

Cold working is defined as the change in cross-sectional area of the metal resulting from rolling, forging, drawing, or extrusion at about room temperature, i.e., 60 to 80 F. Generally, sheet or plate may be considered to change in cross-sectional area solely from the reduction in thickness of the product, since there is little change in width during rolling.

If aluminum and its alloys are worked at temperatures above room temperature, the effects of cold working are less in proportion to the amount of work done than when they are worked at room temperature. Conversely, the work done at temperatures below room temperature is more effective than the same amount ot work at room temperature.

Cold working is expressed as the percentage change in the original cross-sectional area (commonly called reduction) as follows:

In general,

$$C = \frac{100(A - a)}{A}$$

where C = amount of cold working, %; A = initial area of material in properly annealed condition; a = area of the cold-worked material.

For sheet,

$$C = \frac{100(t_0 - t_f)}{t_0}$$

where t_0 = initial thickness of the material and t_f = final thickness.

The effect of cold working on the tensile strength of wrought aluminum and its alloys is shown by $TS = TS_0(1 + 0.9R)$, where TS = tensile strength, psi; TS_0 = tensile strength of the properly annealed material; R = reduction in area expressed as a decimal.

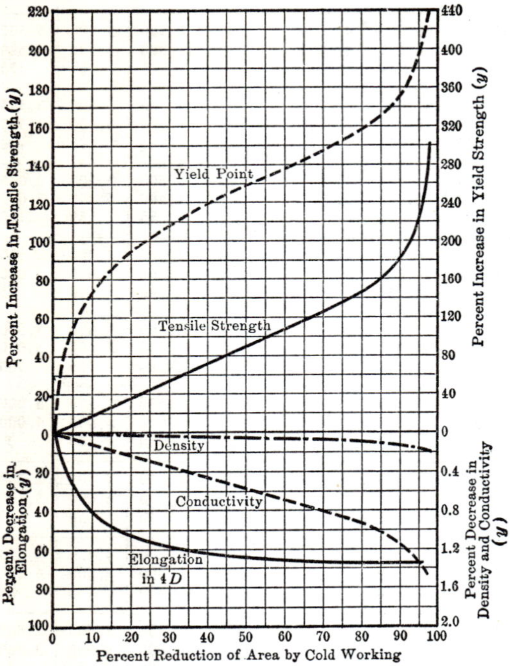

Fig. 1. Effects of cold working on the physical properties of aluminum.

The average tensile strengths of the annealed commercial alloys that are fabricated into products of various intermediate tempers by cold working are as given in Table 5.

The yield strength of aluminum and its alloys is defined as the stress corresponding to a limiting permanent set of 0.2%. In properly annealed wrought metal this value will be approximately one-third the ultimate tensile strength. For all the other tempers of a cold-worked product, the yield strength will vary from 75 to 95% of the ultimate tensile strength.

Elongation is affected considerably both by the size and shape of the test specimen, as well as by the ductility.

The relation between tensile strength and Brinell hardness of wrought aluminum and its alloys is approximately TS = 575 × Brinell hardness number.

Figure 1 shows graphically the effects of cold working on the physical properties of aluminum and its alloys.

4. DESIGN CONSIDERATIONS AND ALLOY SELECTION

WORKING STRESSES. Selection of the working stresses for a particular application is a function of the discretionary ability of the design engineer founded on experience in the particular field of endeavor. Small amount of reflection is necessary to see that it is not practicable to establish tables of working stresses for the various alloys and tempers or heat treatments that would be suitable for all applications and fields of design. Such values must be selected on the basis of as complete a knowledge of the operating conditions of the part being designed and of the properties of the material being used as it is practicable to obtain. The more complete and accurate this knowledge, the smaller can be the factor of safety reliably employed in establishing the working stress for the particular application.

For aircraft design data reference should be made to *ANC Bulletin ANC-5* (Strength of Aircraft Elements), issued by the Air Force-Navy-Civil Sub-committee on Aircraft Design Criteria; it can be purchased from the Superintendent of Documents, Government Printing Office, Washington 25, D.C. As a general comparison of the use of aluminum in ordinary structural applications, allowable unit tensile stresses of 16,000 psi for 61S-T6 and of 21,000 psi for 14S-T6 are used for applications where a stress of 18,000 psi is used for ordinary grade of carbon steel. To supplement the information given in the previous sections, additional information is presented here as an aid to the designer in his selection of material and allowable working stresses for a particular application. Table 9 contains typical values of rotating beam fatigue strength of some of the wrought alloys at five

Table 9. Rotating-beam Fatigue Data *

Alloy and Temper	Old Temper Designation	Approximate Maximum Stresses Which Material Will Withstand for Various Numbers of Cycles †				
		100,000 Cycles	1,000,000 Cycles	10,000,000 Cycles	100,000,000 Cycles	500,000,000 ‡ Cycles
3S-H14	1/2 H	17,000	12,000	10,000	9,000	9,000
3S-H16	3/4 H	17,500	12,500	10,500	9,500	9,500
14S-T6	T	40,000	32,000	25,000	20,000	18,000
17S-T4	T	40,000	32,000	25,000	20,000	18,000
24S-T3	T	40,000	32,000	25,000	20,000	18,000
A51S-T6	T	29,000	21,000	15,500	12,500	11,000
52S-O	O	20,000	19,000	18,000	17,500	17,000
52S-H32	1/4 H	23,000	20,000	18,500	17,500	17,500
52S-H34	1/2 H	26,000	20,500	19,000	18,000	18,000
52S-H36	3/4 H	28,000	21,000	19,000	18,500	18,500
53S-O	O	12,000	10,500	9,500	8,500	8,000
53S-T4	W	26,000	20,000	16,000	14,000	13,000
53S-T6	T	29,000	22,500	17,500	14,000	13,000
61S-O	O	16,000	13,000	11,000	9,500	9,000
61S-T4	W	27,500	22,500	17,500	15,000	13,500
61S-T6	T	31,000	22,500	17,500	15,000	13,500

* Values given were determined by testing 0.3-in. diameter machined specimens in R. R. Moore's rotating-beam fatigue machines. They represent extreme fiber stresses which such specimens will withstand in completely reversed flexure.

† All values of stress in pounds per square inch.

‡ Values given for 500,000,000 cycles are commonly known as endurance limits.

different cycle levels for complete reversal of stress. Table 10 contains typical values of direct tension-compression fatigue strength of some of the wrought alloys at five different cycle levels for different conditions of minimum stress. Table 11 contains column formulas for ultimate strength for various wrought-aluminum alloys.

ALLOY SELECTION. As for selecting working stresses, the design engineer's most reliable basis for selection of alloy is experience. The alloy that has been found very successful in a closely similar application is in most instances the one best suited to a particular application. This generalization, however, must be tempered to the extent of not interfering with the utilization of new and improved alloys designed to supplant some of the older more popular alloys. Some general characteristics of the various alloys and a few typical applications are presented here as an indication of where some experience has been acquired with the pertinent material. Composition and properties of the sand-casting alloys are given in Table 2, p. 4-03; of permanent mold casting alloys in Table 3, p. 4-04; of die-casting alloys in Table 6, p. 4-81; of wrought alloys in Tables 5 and 6, pp. 4-07 and 4-08.

Casting Alloys. 13 alloy is a die-casting alloy having excellent castability and resistance to corrosion. It finds application in castings having large thin-walled sections and intricate designs. Typewriter frames, outboard motor pistons, and instrument cases are typical applications.

43 alloy has limited application in die castings requiring above average ductility. The alloy has excellent resistance to corrosion and weldability. It is also used extensively in sand and permanent-mold castings where these characteristics are desired. Cooking utensils, food-handling equipment, marine and architectural fittings, and miscellaneous thin-sectioned castings are typical applications.

85 alloy is a general-purpose die-casting alloy having a good combination of strength and ductility. Brackets, levers, and frames having thick sections are typical applications.

108 and A108 alloys are used where good casting characteristics, pressure tightness, and weldability are required. Ornamental grilles, manifolds, and valve bodies are typical applications.

113, C113, and 212 alloys are used where good casting and pressure tightness characteristics and very good machinability are required. Of these alloys C113 is the most popular for permanent-mold casting.

Table 10. Direct Tension-compression Fatigue Data *

Minimum † Stress in Each Cycle		Approximate Maximum Stresses Which Material Will Withstand for Various Numbers of Cycles †				
		100,000 Cycles	1,000,000 Cycles	10,000,000 Cycles	100,000,000 Cycles	500,000,000 Cycles
14S-T6, 17S-T4, 24S-T3 Alloy	−25,000	+38,000	+26,000	+16,000	+10,000	+ 7,000
	−20,000	+41,000	+30,000	+20,000	+14,000	+11,000
	−15,000	+43,000	+34,000	+24,000	+18,000	+15,000
	−10,000	+46,000	+37,000	+28,000	+22,000	+19,000
	− 5,000	+48,000	+40,000	+32,000	+26,000	+24,000
	0	+51,000	+43,000	+35,000	+30,000	+27,000
	+ 5,000	+53,000	+46,000	+39,000	+34,000	+31,000
	+10,000	+55,000	+48,000	+42,000	+38,000	+35,000
	+15,000	+57,000	+51,000	+45,000	+41,000	+39,000
	+20,000	+58,000	+53,000	+48,000	+45,000	+43,000
	+25,000	+59,000	+55,000	+51,000	+48,000	+46,000
52S-H36 Alloy	−25,000	+19,000	+ 9,000	+ 6,000	+ 4,000	+ 3,000
	−20,000	+22,000	+14,000	+11,000	+10,000	+ 9,000
	−15,000	+26,000	+19,000	+17,000	+16,000	+15,000
	−10,000	+29,000	+24,000	+22,000	+21,000	+20,000
	− 5,000	+32,000	+28,000	+27,000	+26,000	+25,000
	0	+34,000	+32,000	+31,000	+30,000	+29,000
	+ 5,000	+36,000	+35,000	+34,000	+33,000	+32,000
53S-T6 Alloy	−20,000	+16,000	+10,000	+ 6,000	+ 3,000	+ 1,000
	−15,000	+20,000	+14,000	+10,000	+ 8,000	+ 6,000
	−10,000	+23,000	+18,000	+14,000	+12,000	+11,000
	− 5,000	+26,000	+22,000	+19,000	+16,000	+15,000
	0	+29,000	+25,000	+22,000	+20,000	+19,000
	+ 5,000	+31,000	+28,000	+26,000	+24,000	+23,000
	+10,000	+33,000	+31,000	+29,000	+28,000	+27,000
	+15,000	+34,000	+33,000	+32,000	+31,000	+30,000
61S-T6 Alloy	−20,000	+29,000	+20,000	+12,000	+ 6,000	+ 3,000
	−15,000	+32,000	+24,000	+16,000	+10,000	+ 7,000
	−10,000	+34,000	+27,000	+20,000	+15,000	+12,000
	− 5,000	+36,000	+30,000	+24,000	+19,000	+17,000
	0	+38,000	+33,000	+28,000	+24,000	+22,000
	+ 5,000	+40,000	+35,000	+31,000	+28,000	+27,000
75S-T6 Alloy	−30,000	+39,000	+27,000	+17,000	+10,000	+ 8,000
	−25,000	+42,000	+30,000	+20,000	+14,000	+12,000
	−20,000	+44,000	+34,000	+24,000	+18,000	+16,000
	−15,000	+47,000	+37,000	+27,000	+22,000	+19,000
	−10,000	+49,000	+40,000	+30,000	+25,000	+23,000
	− 5,000	+52,000	+43,000	+34,000	+29,000	+27,000
	0	+54,000	+45,000	+37,000	+32,000	+30,000
	+ 5,000	+56,000	+48,000	+40,000	+35,000	+34,000
	+10,000	+58,000	+51,000	+43,000	+39,000	+37,000

* Values given were determined by testing 0.2-in. diameter machined specimens in direct stress fatigue machines and represent uniformly distributed stresses which such specimens will withstand under repeated axial loads.

† All values of stress in pounds per square inch. Stresses considered algebraically: plus (+) means tension, minus (−) means compression.

Automotive cylinder heads, internal-combustion engine timing gears, washing machine agitators, and miscellaneous general-purpose castings are typical applications.

A132 alloy is used where good high-temperature strength, low coefficient of thermal expansion, and good wear resistance are required. Automotive and diesel pistons and pulleys and sheaves are typical applications.

138 alloy is used where high hardness at elevated temperatures and good polishing characteristics are desired. The major application of this alloy is sole plates for electric flat irons.

195 and B195 alloys are used where a good combination of mechanical properties, fair castability, and good machinability is required. Aircraft fittings, aircraft wheels, railroad car seat frames, mechanical equipment parts, and flywheel and gear housing are typical applications.

214, A214, B214, and F214 alloys are used where excellent resistance to corrosion and tarnish is required. Dairy and food-handling equipment, architectural applications, cooking utensils, and fittings for chemical and sewage use are typical applications.

218 alloy is a die-casting alloy used in applications requiring exceptional mechanical properties, resistance to corrosion and finishing characteristics. Aircraft brake shoes, cylinder blocks for outboard motors, fittings, and brackets for marine application are typical applications.

220 alloy is used where very high strength and elongation, excellent machinability, and resistance to corrosion are required. Aircraft fittings, railroad and truck frame and bolster parts, and miscellaneous castings requiring strength and shock resistance are typical applications.

319 alloy is used where good castability and a good combination of mechanical properties are desired. It is used today for many applications formerly using 112, 113, C113, and 212 alloys. Piano plates, internal-combustion engine crankcases, oil pans, gasoline and oil tanks, and miscellaneous pressure tightness applications are typical applications.

Table 11. Ultimate Strength Formulas for Axially Loaded Aluminum-alloy Columns

(Representative of material having the typical properties shown in Tables 5 and 6)

Alloy and Temper	Old Temper Designation	Typical Compressive Yield Strength, psi	For KL/r less than C	C	For KL/r greater than C
3S-O	O	6,000	$\frac{P}{A} = 6{,}200 - 18\frac{KL}{r}$	194	$\frac{P}{A} = \frac{102{,}000{,}000}{\left(\frac{KL}{r}\right)^2}$
3S-H12	1/4 H	15,000	$\frac{P}{A} = 16{,}100 - 78\frac{KL}{r}$	138	
3S-H14	1/2 H	18,000	$\frac{P}{A} = 19{,}600 - 105\frac{KL}{r}$	125	
3S-H16	3/4 H	21,000	$\frac{P}{A} = 23{,}200 - 135\frac{KL}{r}$	115	
14S-T6	T	58,000	$\frac{P}{A} = 74{,}800 - 780\frac{KL}{r}$	64	
17S-T4	T	37,000	$\frac{P}{A} = 43{,}800 - 350\frac{KL}{r}$	83	
24S-T3	T	46,000	$\frac{P}{A} = 56{,}600 - 510\frac{KL}{r}$	73	
52S-O	O	14,000	$\frac{P}{A} = 15{,}000 - 70\frac{KL}{r}$	143	
52S-H32	1/4 H	26,000	$\frac{P}{A} = 29{,}400 - 192\frac{KL}{r}$	102	
52S-H34	1/2 H	29,000	$\frac{P}{A} = 33{,}200 - 230\frac{KL}{r}$	96	
52S-H36	3/4 H	34,000	$\frac{P}{A} = 39{,}800 - 302\frac{KL}{r}$	88	
53S-T4	W	20,000	$\frac{P}{A} = 22{,}000 - 124\frac{KL}{r}$	118	
53S-T6	T	33,000	$\frac{P}{A} = 38{,}400 - 287\frac{KL}{r}$	90	
61S-T4	W	21,000	$\frac{P}{A} = 23{,}200 - 135\frac{KL}{r}$	115	
61S-T6	T	39,000	$\frac{P}{A} = 46{,}600 - 384\frac{KL}{r}$	81	

Notes. Members are assumed to be compact enough so that no local failure will occur. P/A = ultimate strength of column, pounds per square inch; L = unsupported length of column, inches; r = corresponding radius of gyration, inches; K = 0.5 for both ends fixed; K = 1.0 for both ends hinged.

355 alloy is used where good castability, weldability, and pressure tightness are required. Aircraft supercharger covers and impellers, fuel pump bodies, air compressor pistons, water-cooled cylinder heads, liquid-cooled engine crankcases, and blower housings are typical applications.

356 alloy is used where excellent castability and good weldability, pressure tightness, and resistance to corrosion are required. Aircraft pump parts, marine transmission cases, aircraft fittings and control parts, liquid-cooled engine crankcases and aircraft, truck and bus wheels are typical applications.

360 and A360 are cold-chamber die-casting alloys used in large thin-walled applications and intricate designs where good properties and resistance to corrosion are desired. Cover plates and instrument cases are typical applications.

380 and A380 are general-purpose die-casting alloys having better castability than 85 alloy.

406 alloy has excellent resistance to corrosion but poor castability. It has found some application in pipe and tank fittings.

750 is a bearing alloy that has found use in miscellaneous internal-combustion engine-bearing and bushing applications.

Wrought Alloys. 2S and 3S alloys are used where good formability, high resistance to corrosion, and good weldability are governing factors. 3S is preferred where higher strength is desired. Gasoline and oil tanks, trim in transportation equipment, cooking utensils, food and chemical handling, and storage equipment are typical applications.

Alloy 11S is used where excellent machinability and good strength are required. Screw machine parts and miscellaneous forgings requiring considerable machining are typical applications.

Alloy 14S is used where high strength, relatively good formability, and high hardness are required. Heavy-duty forgings, power shovel bails, aircraft fittings, and structural applications are typical.

Alloy 17S is currently not used other than for rivets and screw machine parts. It has been largely superseded by 14S and 24S alloys.

Alloy 18S is used for forgings requiring good strength at elevated temperatures. Cylinder heads and pistons are typical applications.

Alloy 24S is used in applications where higher strengths than 17S are desired. Aircraft structures, nails, rivets and hardware, and miscellaneous structural applications are typical.

Alloy 25S is used for forgings where good forgeability and fairly high strength are required. Airplane propellers, link rods, supercharger impellers, and jet engine compressor wheels are typical applications.

Alloy 32S is used for forgings requiring good forgeability and low coefficient of thermal expansion. Forged pistons are the major application.

Alloy A51S is used for forgings where very good forgeability, good strength, and good resistance to corrosion are required. Crankcases, fuse parts, machine and automotive parts are typical applications.

Alloy 52S is used in applications where good workability, very good resistance to corrosion, high fatigue strength, and moderate static strengths are desired. Aircraft fuel and oil lines, fuel tanks, miscellaneous marine and transportation applications are typical.

Alloy 53S is used in applications where good workability, very good resistance to corrosion, and moderate strength are desired. Petroleum-refining equipment, beer barrels, miscellaneous naval construction, and architectural applications are typical.

Alloy 61S is used in applications where good strength, good workability, and very good resistance to corrosion are required. Aircraft landing mats, pontoon boats, canoes, furniture, vacuum-cleaner tubes, and transportation equipment are typical applications.

Alloy 63S has its major application in architecture, where a bright silvery anodized finish is desired.

Alloy 75S is used where very good strength and good resistance to corrosion are required. Aircraft structural parts are currently the major field of application for the alloy.

FABRICATED PRODUCTS. Although aluminum and its alloys are produced in the wrought form as sheet and plate, foil, bar, rod, shapes, wire, tubing, and forgings, not all the wrought alloys are commercially produced in all these products. Availability of the alloy desired in the form and size contemplated should be determined before proceeding too far with design. Plate, sheet, wire, and tubing are commercially available with a clad coating of higher resistance to corrosion and of higher solution potential than the core material. The clad coating is integrally bonded to the core and is of such thickness as to retain maximum physical properties consistent with adequate protection of the core.

RESISTANCE TO CORROSION OF ALUMINUM ALLOYS. Resistance to corrosion of a material is a characteristic of major importance to the designer. Because of the inability of expressing this characteristic in measurable terms of general applicability, there is no substitute for experience in uses under similar operating environments, and the advice of specialists in the field should be sought in selection of alloys. As a guide to the relative merits of the alloy systems composing the more commonly used aluminum alloys, the following brief analysis is presented.

Aluminum-manganese Alloys. The wrought aluminum-manganese alloy 3 S, because of its excellent resistance to corrosion, has been used for many years in a wide variety of applications. Casting alloy 406, although not used extensively, has a comparable resistance to corrosion.

Aluminum-magnesium Alloys. Wrought alloys and cast alloys having maximum magnesium contents of about 5 and 10% respectively, have excellent resistance to corrosion, and for this reason the wrought alloys 52S and 56S and the cast alloys 214, 218, and 220 have been used under relatively severe conditions of exposure without protection. The cast alloy A214, which contains additions of zinc, as well as alloys B214 and F214, which contain additions of silicon, have a resistance to corrosion very closely approaching that of the foregoing alloys.

Aluminum-magnesium Silicide Alloys. Alloys of this system having magnesium and silicon contents in the proportion required to produce Mg_2Si have excellent resistance to corrosion, comparable to that of 2S, 3S, and 52S. Wrought alloys 53S, 61S, and 63S fall within this category. Alloy A51S has an excess of silicon, which results in a slight reduction in resistance to corrosion in the heat-treated and aged conditions.

Aluminum-silicon Alloys. The binary alloys 13 and 43, and those of this group such as 356, 360, and A360 which contain magnesium additions, possess excellent resistance to corrosion. Alloy 355, which contains additions of copper, and alloy A132, which contains additions of nickel and copper, have resistance to corrosion closely approaching that of the other alloys included in this system.

Aluminum-copper Alloys. With proper heat treatment, the commercially available wrought alloys of this system such as 11S, 14S, 17S, 24S, and casting alloys 113, C113, 142, 195, and 212 have good resistance to corrosion but should be given a protective treatment when used outside. Wrought alloys 18S and 25S have lower resistance to corrosion than 17S and 24S, but considerable quantities have been used in aircraft engine and propeller applications, respectively, with satisfactory results.

Aluminum-copper-silicon Alloys. Some of the casting alloys contain both copper and silicon in relative percentages to justify grouping in another system. Of this system, alloys 85, 108, A108, 138, B195, 380, and A380 can be classed with the aluminum-copper alloys with regard to resistance to corrosion.

Aluminum-zinc-magnesium Alloys. 75S alloy, which is of this system, has a resistance to corrosion comparable to that of other high-strength aluminum alloys such as 24S.

Clad Alloys. For applications where the mechanical properties of some of the wrought alloys are desired but better resistance to corrosion is needed, or where good resistance to perforation is desired, clad products have been developed. Clad 3S, 14S, 24S, and 75S are examples of such products. These products have excellent resistance to corrosion, the cladding also offering protection against corrosion at the cut edges or small areas of exposed core material by electrochemical sacrifice. Sprayed aluminum coatings are sometimes used to protect castings and forgings by the same principle.

5. HEAT TREATMENT OF ALUMINUM ALLOYS

SOLUTION HEAT TREATMENT. A change in properties as a result of heat treatment of aluminum alloys usually is associated with a change in solubility of one or more of the alloying ingredients with change in temperature. Thus, over 5% copper is soluble in solid aluminum just below the melting point of the eutectic; at room temperature the solubility is well below 1%. If aluminum alloys containing about 4% copper are heated to a temperature slightly below the melting point of the eutectic, the copper will dissolve in the aluminum at that temperature. If the alloy then is cooled quickly to a low temperature, it becomes supersaturated with copper at the lower temperature. This treatment is known as a solution heat treatment. In certain alloys, like 17S or 24S, the supersaturated solid solution breaks down at ordinary temperatures, precipitating minute particles of the solute. This precipitation, known as aging, changes the physical properties.

In carrying out the solution heat treatment, temperature must be controlled closely. Thin pieces of wrought metal may be heated to the proper temperature and held there only a short time before cooling, with good results. The operation may be carried on in a hearth or pit furnace or in a bath of fused niter. Castings and large wrought pieces, however, must be held at the heat-treating temperature for considerable periods of time, and automatic temperature control is practically necessary. Quenching after the proper heating cycle should be as rapid as practicable. Wrought products are usually quenched in cold water whereas castings and complicated forgings are more commonly quenched in hot water to minimize internal stress and avoid distortion.

PRECIPITATION HEAT TREATMENT. In certain alloys, the precipitation changes produced in the solution heat-treated material at ordinary temperatures are insufficient for the requirements, and the desired results are effected by treatment at elevated temperatures. Such a treatment is known as precipitation heat treatment or artificial aging. In general, room-temperature aging will increase hardness and strength of solution heat-treated aluminum alloys. The same is true of the precipitation heat treatment, if it is not carried too far. The latter is capable of producing a wide range of properties without change in the allcy composition. It is sometimes used on material in the "as fabricated" condition, without intermediate solution heat treatment, taking advantage of the partial solution of the alloying constituent in this condition.

The precipitation heat treatment may be effected by heating in pressure vessels with steam or by heating in baths, such as oil, or in furnaces heated by oil, gas, or electricity. The temperature should be well controlled for best results, and, if an air furnace is used, circulation is desirable. The rate of cooling after the precipitation heating usually is not an important factor.

Other effects of heat treatment of aluminum alloys are the removal of internal strains, stabilization of dimensions of parts used or processed at elevated temperatures, and changes in particle size of constituents such as silicon, which may be present in amounts greater than can be dissolved in solid aluminum even at the eutectic temperature.

6. MACHINING OF ALUMINUM

Aluminum and its alloys may be machined readily. In many instances, satisfactory results may be obtained with tools ordinarily used for steel, provided they are sharp and in good condition. Because of the inherent machining characteristics of aluminum, better results may be obtained, however, when certain changes in tool shape and machining practice are observed. Generally speaking, aluminum and most of its alloys have similar machining characteristics.

Cutting tools for aluminum resemble those for cutting hard wood in many respects. It should be emphasized that the cutting edges must be ground to a sharp included angle, must be keen, and should be provided with smoothly finished surfaces. The tools generally are ground with fine abrasive wheels, supplemented by hand stoning with an oilstone to give the desired finish of the surfaces and edges.

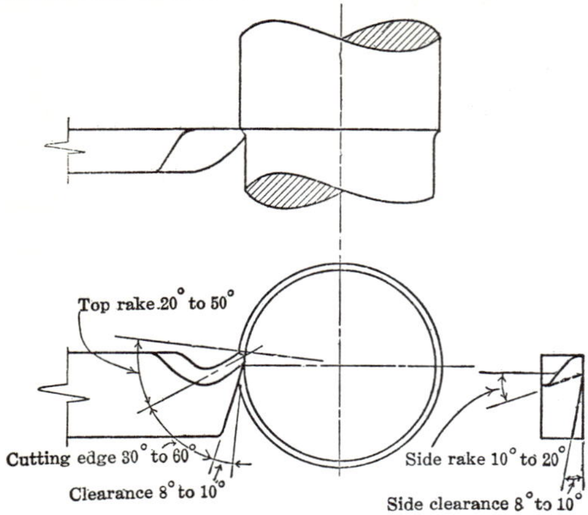

FIG. 2. Lathe tool for turning aluminum.

LATHE, SHAPER, AND PLANER TOOLS. Figures 2 to 4 illustrate typical lathe and planer tools for machining aluminum. Attention is called to the unusually large amount of top and side rake.

MILLING CUTTERS for aluminum should be of the coarse-tooth spiral type, with a considerable amount of top rake on the cutting edges. The more recent helical milling cutters, designed primarily for machining steel, work especially well with aluminum and its alloys, if the cutting edges are provided with suitable top rake.

DRILLS. Ordinary twist drills work fairly satisfactorily in aluminum if properly ground with keen edges, but a better type of drill is one with more twist to the flutes. Both types are illustrated in Fig. 5. Fairly high speeds may be employed if ample cutting compound is used. (See also Section 23, Articles 23 and 24.)

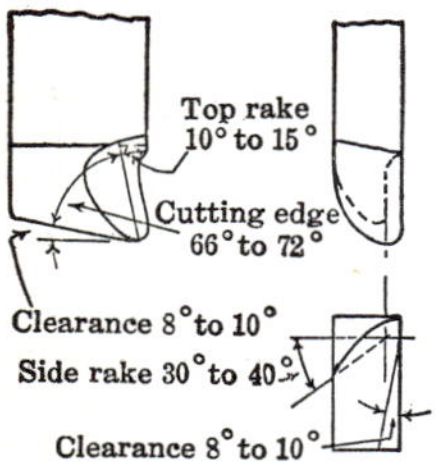

FIG. 3. Planer tool for roughing aluminum.

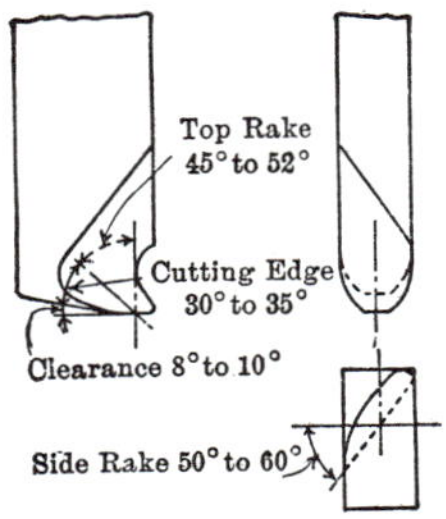

FIG. 4. Planer tool—finishing.

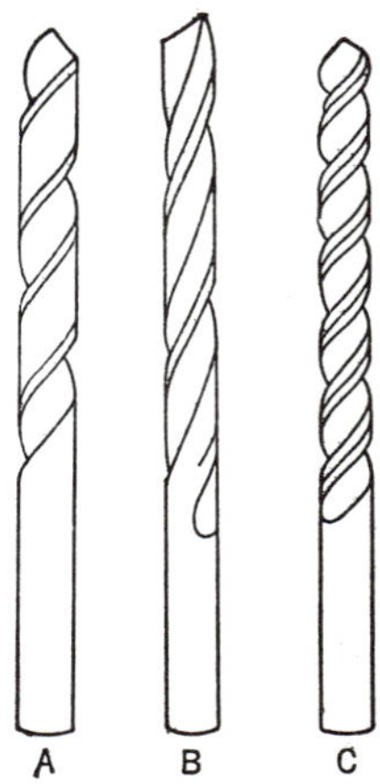

FIG. 5. Three types of twist drills, 3/8 in. diam. *A*, Single flute twist-drill, 40° spiral angle. *B*, Double flute or standard twist-drill, 24° spiral angle. *C*, Special double fluted twist drill, 47° spiral angle.

REAMERS of the spiral fluted type produce the best results for reaming holes in aluminum.

TAPS, generally designated "gun" taps or spiral pointed taps (Fig. 6), which have a short spiral ground at the nose, work very satisfactorily in aluminum, especially for through holes where the cuttings can be pushed ahead. Better results may be obtained with spiral fluted taps (Fig. 7), now offered by a number of manufacturers of small tools.

SAWING. The shapes of saw teeth should be patterned after those of the cutting tools described in the foregoing, so far as practicable. Teeth usually should have some top rake or hook, the amount depending on the type of saw, although very little can be used with band saws or circular saws where the work is fed by hand. The teeth should be comparatively coarse, with curved gullets, free from sharp corners and burrs or scratches which are likely to pick up cuttings.

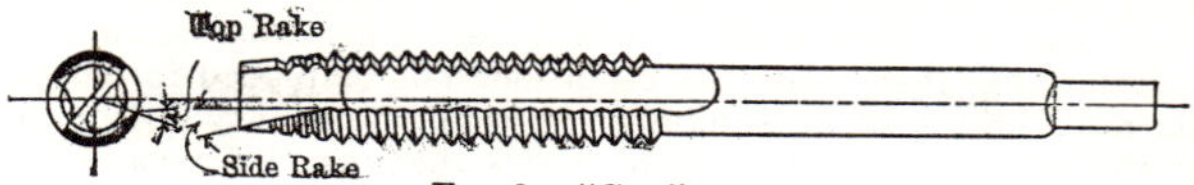

FIG. 6. "Gun" tap.

Circular saws for aluminum may be operated at peripheral speeds of about 10,000 ft per min. Speeds of about 5000 ft per min are recommended for band saws.

A cutting lubricant or coolant is desirable, and for heavy work is necessary. Mineral-base lubricating oils thinned with kerosene and, in some instances, soluble-oil cutting compounds have been found satisfactory.

Hand hack-saw blades of the "wavy-set" type work especially well with aluminum.

FILES. Some of the coarse-tooth files with top and side rake to the cutting teeth work well on aluminum.

CUTTING SPEEDS AND FEEDS. A wide range of cutting speeds and feeds may be used in machining aluminum and its alloys. Roughing cuts may be quite heavy. For highly finished surfaces, cuts only a few thousandths of an inch deep may be used, provided the cutting tools are properly shaped and finished. Speeds necessarily depend on such factors as the kind of machining operation, type and condition of machine tool, and kind of cutting material. Conventional machine tools can generally be operated at their maximum speeds in machining aluminum and in certain special-purpose machines; cutting speeds as high as 20,000 ft per min have been used to advantage.

CUTTING LUBRICANTS. Some of the alloys of aluminum have been machined successfully without any lubricant or coolant, with tools of the form described, but the use of some form of compound is desirable for best results. The soluble cutting oils usually are satisfactory for milling, sawing, and drilling. For such operations as turning, shaping, and tapping, a mixture of equal parts of lard oil and kerosene generally is recommended. Cutting compounds of paraffin oil, such as are used in machining brass, are not satisfactory for aluminum.

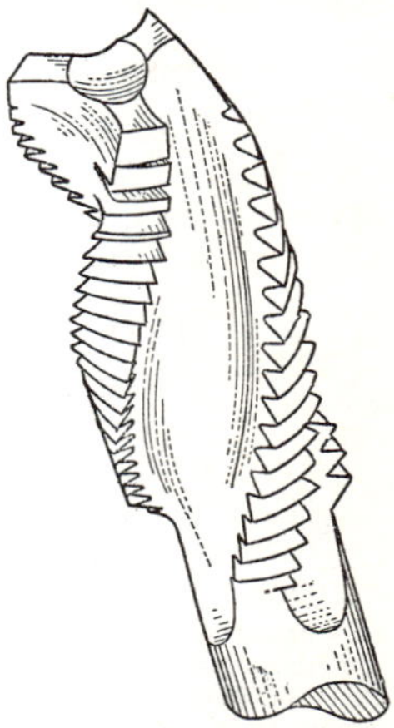
FIG. 7. Spiral fluted tap.

TOOL MATERIALS. Tools for machining aluminum may be either of high-carbon or high-speed steel. Cemented tungsten-carbide tools have been found indispensable, especially for machining aluminum alloys with high silicon content; they also are excellent for high production.

7. JOINING OF ALUMINUM

WELDING. Aluminum alloys may be joined by welding by the torch, electric arc, or resistance processes. In fusion welding, the equipment is the same as that used for welding steel, but in electrical resistance welding, machines of slightly different design and greater electrical capacity are required. The technique varies slightly from that required for steel. The metal arc generally will make stronger welds than can be made with the torch; however, $^3/_{32}$ in. is the practical minimum thickness of metal that can be so welded.

For metal arc electrodes and for torch welding filler rods for joining strong alloys, alloy 43S is most commonly used. A flux is used in fusion welding. In the metal arc process, the flux is best applied as a heavy coating on the electrode. When inert atmosphere is used, such as in argon, shielded-arc welding flux is not required.

Because of the effect of welding heat on the mechanical properties and resistance to corrosion not all the aluminum alloys are suitable for this method of joining. Wrought alloys 2S, 3S, 52S, 53S, 61S, and 63S find general application in components welded by gas or arc methods. Some use has been made of arc welding for 14S alloy. Resistance welding has been used for all the wrought and cast alloys. Arc welding has been employed for welding of all permanent-mold and sand-casting alloys except 220 and 750. Gas welding is applicable to some permanent-mold and sand-casting alloys such as 43, 195, B195, 214, 355, and 356. Die castings are normally neither gas nor arc welded.

SOLDERING. There are available a variety of solders and fluxes adapted to nearly all soldering requirements. The use of proper solder and flux enable aluminum alloy parts to be joined in much the same way and for about the same type of service as solder joints of other metals. Suitable directions can be obtained from the supplier of solder and flux. Of the wrought alloys, 2S, 3S, 53S, 61S, and 63S are used in soldered applications. All permanent-mold and sand-casting alloys can be soldered. Die castings are soldered by first copper plating and then using the usual methods employed for copper base alloys. The dissimilar metal in the soldered joint presents the possibility of electrolytic corrosion so that adequate protection is necessary for joints that may be exposed to corrosive conditions.

BRAZING. Aluminum alloys are brazed by furnace, torch, or dip methods. Not all the alloys can be brazed because of relationship of the melting points of the available brazing alloy and the various aluminum alloys. Of the wrought alloys, 2S, 3S, 53S, 61S, and 63S can be brazed. Of the permanent-mold and sand-casting alloys, 43, 355, 356, and 406 are brazed. Die-casting alloys are not brazed.

RIVETING. The joining of aluminum-alloy components by riveting is a common fabricating procedure. The particular alloy rivet selected depends on the strength requirement of the application. Table 12 lists typical shear strengths of the driven rivets for the

Table 12. Average Ultimate Shear Strengths of Driven Rivets *

Rivet	Driving Procedure	Shear Strength, psi
2S	Cold, as-received	11,000
A17S-T3	Cold, as-received	33,000
17S-T3	Cold, as-received	39,000
17S-T31	Cold, immediately after quenching	34,000 †
24S-T31	Cold, immediately after quenching	42,000 †
53S-T61	Cold, as-received	23,000
53S-T6	Cold, as-received	26,000
17S-T41	Hot, 930 to 950 F	33,000 †
53S-T41	Hot, 960 to 1050 F	18,000 †,‡
Steel	Hot, 1700 to 1900 F	45,000

* Values for rivets driven with cone-point heads. Rivets driven with heads requiring more pressure may be expected to develop slightly higher strengths.

† Immediately after driving, the shear strengths of these rivets is about 75% of the value shown. On standing at ordinary temperatures, they age-harden to develop their full strength, this action being completed in about four days for 17S and 24S rivets, and in about two weeks for 53S rivets.

‡ Shear strength for rivets driven at temperatures of 960 to 980 F. The shear strength increases about 1000 psi for each increase of 12 F in driving temperature.

alloys and condition of the aluminum-alloy rivets which have found most frequent application. When greater strength is required, zinc-coated steel rivets are sometimes used. It will be noted in Table 12 that properties are given for 17S and 24S alloys driven immediately after quenching (T31 temper). For most applications, particularly with rivets of larger size, it is not possible to do a satisfactory job of riveting with these alloys in the T4 condition, the alloy being too hard to drive without excessive cracking. It is common practice, therefore, with these alloys either to drive the rivets within 2 hours after solution heat treatment or as a convenient method to solution heat-treat the rivets and to store them at temperatures of 32 F or below until it is time to drive. By these methods the rivets are driven before precipitation hardening has progressed to such a degree as to make the rivets too hard for driving. At 32 F the solution heat-treated rivets can be maintained in a condition suitable for driving for periods up to about 36 hours. At lower temperatures, longer storing times are possible.

Aluminum-alloy rivets are sometimes driven hot to facilitate driving. In order to obtain desirable strength and resistance to corrosion properties the rivets are heated to the solution heat-treatment temperature and driven. The contact of the hot rivet with the cold material around the rivet hole constitutes a sufficient quench to produce the desired properties.

In die casting, small studs are sometimes cast as an integral part of the casting. These studs are used instead of rivets in assembling the mating parts, the studs being pressed after assembly to form a "riveted" union.

Rivet holes may be either drilled, punched, subdrilled or subpunched, and reamed. The best clearance, which is the minimum that will permit expeditious insertion of the rivet in the hole, is usually about 2 or 3% of the rivet shank diameter. Hot riveting will require slightly greater clearance than cold riveting to facilitate insertion. Too great a clearance will result in poor upsetting during the driving operation. Taper bridge reamers of the spiral fluted type are recommended for reaming subdrilled or subpunched holes.

8. FINISHING OF ALUMINUM

FROSTED FINISH. This finish is used primarily for its appearance. It is common practice to anodize the "frosted" article to maintain the desired finish in service. Frosted finish usually is applied by etching in a hot solution of sodium hydroxide (caustic soda), followed by treatment with strong nitric acid to neutralize any remaining alkali and to remove any remaining smudge. For ordinary operations, the solution contains about 5% of caustic soda, although solution concentrations of 2 to 25% are sometimes used. The temperature of the solution is commonly kept at about 170 F. The time of immersion is determined by the temperature of the solution, its concentration, and the depth of etch desired. Usually treatments of about one minute are sufficient.

After removal from the caustic solution, the part is rinsed in clear water and immersed in a 1 to 1 nitric acid and water solution. Occasionally a weaker solution, for example, 10 to 15% nitric acid, is used.

MECHANICAL FINISHES. Aluminum alloys can be given a very pleasing variety of mechanical finishes by grinding, polishing, buffing, coloring, scratch brushing, hammering, fluting, sandblasting, and burnishing. These finishes can be applied in such a manner as to produce attractive designs on the article for decorative purposes or to produce a highly reflective or low reflective surface, as desired.

ELECTROPLATING. Aluminum may be electroplated with other metals to produce an attractive finish, to increase the wear resistance, to change the frictional characteristics, and for other more specialized purposes. A few examples of plating uses are: *silver* for increasing surface electrical conductivity; *brass* to facilitate vulcanization of rubber; *chromium* to increase wear resistance; *tin* to reduce friction; *copper* to permit soft soldering; and *zinc* for improving anti-seize characteristics of threaded parts.

Prior to electroplating, the part is given a pretreatment which provides adequate anchorage for the plated metal. The most commonly used method of pretreatment is the zinc immersion process, in which a chemically deposited zinc coating is applied. Anodizing and other etching methods are also used. After this pretreatment the desired metal is plated on in accordance with common plating procedures. Plating thickness will vary but should be sufficient to provide an impervious coating as a protection against undesirable galvanic action.

OXIDE COATINGS are one of the most important finishes for aluminum. These coatings are used to improve the corrosion resistance and appearance of the article and to provide a good base for subsequent paint coatings. In addition the anodic oxide coatings applied by the Alumilite process are sometimes used because of their hard abrasion-resistant characteristic. The anodic treatments consist of making the article to be coated the anode in a suitable electrolyte and passing the current necessary for producing the desired film characteristics. The conditions of coating will vary with alloy, electrolyte, and nature of coating desired. The coatings as produced are submicroscopically porous and range from a clear transparent film to one that is translucent or opaque, the color being governed by the alloy being coated and the process employed. The porous nature of the films is used to advantage in subsequent sealing treatments to cause corrosion inhibitors or dyes to be absorbed by the film. A great variety of colored oxide coatings can be produced which have desired eye appeal and resistance to corrosion. The characteristic of the oxide coating that reproduces the texture of the surface on which it is formed is made use of to preserve applied etches and mechanical finishes during service.

Oxide coatings are produced by chemical-treatment methods as well as by anodic oxidation. A series of patented chemical dip treatments, developed by the Aluminum Company of America and known as "Alrok" treatments, have found wide commercial use.

PAINTING AND LACQUERING. Aluminum may be coated with paint or lacquer by much the same technique as followed for other metals. Surface preparation is important. Frequently, all that is required is a cleansing from grease and dirt by wiping the surface with a cloth wet with mineral spirits or other solvent. Chromic acid solution and certain chemical cleansing agents, such as a phosphoric acid-alcohol mixture, sold under various trade names, give greatly improved paint adhesion. With castings, a light sand blast improves adhesion of paint. An anodic or chemical oxide coating provides exceptionally good adhesion, especially in painting for protection against salt-water corrosion.

A priming coat containing zinc chromate pigment is effective on aluminum. Aluminum paint is made by mixing aluminum powder or paste with a suitable vehicle. Aluminum paint, in the proportion of two pounds of powder to a gallon of vehicle, forms a coating of high moisture resistance and superior protective qualities. By proper choice of vehicle it can be used on metal, wood, plaster, or concrete and is particularly recommended for the painting of aluminum and its alloys.

Transparent varnishes or lacquers may be used to preserve polished surfaces. Many of them are sufficiently durable to be applied to exterior ornamental work. A varnish of light color and relatively insensitive to light exposure should be selected. Synthetic resins have increased durability; no clear finish, however, will last as long as pigmented finishes.

VITREOUS ENAMEL. Some application has been made of coatings of this type. Because of the relatively low melting point of aluminum alloys, however, the process is restricted to the use of a low temperature frit. They are generally of low resistance to corrosion and weathering.

9. ELECTRICAL USES OF ALUMINUM

Aluminum was first used as a conductor in high-tension wires in the United States in 1898. The first conductors were made entirely of aluminum. Later, however, a cable with an extra high strength steel reinforcing core for the aluminum strands was developed and has since been used with marked success.

CONDUCTIVITY. The AIEE standard of conductivity for hard-drawn aluminum wire is 60.97% of the International Annealed Copper Standard. On this scale the conductivity of commercial hard-drawn copper usually is taken at 97%.

The electrical conductivity of annealed high-purity (99.97%) aluminum is 64.6%, based on the International Annealed Copper Standard. The electrical conductivity is lowered by the addition of alloying elements, the reduction varying with the nature of the element and the amount added. Heat treatment and mechanical working also influence this property to a marked degree. The following are the approximate conductivities of various grades of aluminum and aluminum alloys:

Metal	99.97 Per Cent Al	Hard-drawn Conductors	2S-O	2S-H	3S-O	3S-H	14S-T	24S-T	61S-T
Conductivity, per cent of International Annealed Copper Standard	64.6	61	59	57	50	40	40	30	40

Conductivities of a number of other alloys are contained in AIMME, *Transactions of Metals Division*, Vol. 124, 1937, pp. 287–299.

CABLES AND CONDUCTORS. Aluminum cable, steel reinforced, is used for messenger cables of overhead catenary railway lines. In the smaller sizes, it also is used for signal circuits, and for telephone lines supported on the same structure with power lines, where its high strength permits the use of as long spans as the transmission conductors. Some spans as long as one mile have been used successfully. Because of its light weight, aluminum cable has made a welcomed contribution to the aircraft industry as an electrical conductor. It is also finding considerable application as building wire. Large all-aluminum cables have been used as railway feeders, where they carry heavy currents at relatively low voltages. Particularly in large conductors, the use of aluminum effects a great saving in weight. Conductors for high-voltage installations, such as outdoor substations, may advantageously be made of aluminum tubing. They are rigid and light, and permit the use of long spans which often are necessary to obtain electrical clearances, and show economy in supporting structures.

SWITCHBOARD CONSTRUCTION. Aluminum may be used to advantage in bus bar construction, both on switchboards and for general transmission of power. Aluminum bus bars are rolled and are available in the usual rectangular cross sections, as well as in channels and certain other structural shapes, which offer definite advantages in strength and current-carrying capacity. The current rating of aluminum bus bars may be conveniently taken as 84% of the accepted values for copper of equal cross section and similar shape.

ELECTRICAL MACHINERY. Aluminum is used for casting the rotor conductors of squirrel-cage induction motors and also for field windings in certain types of motors and high-speed turbo-generators.

COPPER AND COPPER-BASE ALLOYS

10. COPPER

By W. R. Hibbard *

COMPOSITION OF COMMERCIAL COPPER. Specifications for copper, generally accepted by industry, are the ASTM standard specifications. These cover lake copper, electrolytic copper, and fire-refined copper. Lake copper, which must originate on the upper peninsula of Michigan, may be low-resistance or high-resistance lake copper.

Low-resistance lake copper, used for electrical purposes, may be electrolytically or fire refined. It is required to have a content of (copper + silver) of not less than 99.900%. Maximum permissible resistivities in international ohms (meter, gram) are: copper wire bars, 0.15436; ingots and ingot bars, 0.15694.

High-resistance lake copper is lake copper of resistivity exceeding 0.15694 international ohm (meter, gram) at 20 C. Its (copper + silver + arsenic) content is not less than 99.900%.

Physical Properties of Copper

General Properties

Density	0.322 lb per cu in.	Specific heat at 77 F	0.0917
Melting point	1981 F	Electrical resistivity at 20 C	1.27 microhm-cm
Coefficient of expansion per °F	77–100 F 0.000 009 332; 77–572 F 0.000 009 888	Temperature coefficient of resistivity at 20 C	0.0039 per °C
Pattern shrinkage	1/4 in. per ft.	Magnetic properties	Diamagnetic
Thermal conductivity at 77 F, Btu per hr per sq ft per in. thickness per °F	2679	Optical properties	Selectively reflecting
		Young's modulus	17,300,000

Tensile Properties

	Annealed	Cold Rolled or Drawn	Cast
Tensile strength, psi	30,000–40,000	50,000–70,000	20,000–30,000
Elongation in 2 in.	40–60%	2–4%	25–45%
Reduction of area	40–60%	2–4%	....

Hardness

	Annealed	Cold Rolled or Drawn
Brinell (500 kg)	30–40	60–100
Scleroscope (Universal Hammer)	6–8	20– 25

Electrolytic copper specifications for purity and resistivity are the same as those of low-resistance lake copper.

Fire-refined copper, other than lake, is used for rolling into sheets and shapes. It is not acceptable for electrical uses. ASTM Specification B216-47T, Fire-refined Copper for Wrought Products and Alloys, calls for the following analysis: Cu + Ag, min, 99.88%; As, max, 0.012%; Sb, max, 0.003%; Se + Te, max, 0.025%; Ni, max, 0.05%; Bi, max, 0.003%; Pb, max, 0.004%.

Oxygen-free high-conductivity copper is a highly ductile material, made under conditions that prevent the entrance of oxygen and formation of copper oxide. It is utilized in deep-drawing, spinning, and edge-bending operations, and in welding, brazing, and other hot-working operations where embrittlement must be avoided. It has the same conductivity and tensile properties as touch pitch electrolytic copper.

HARDENED COPPER. There are only two methods for hardening copper: by cold working it or by alloying with it small amounts of other elements such as silicon, tin, zinc, and aluminum. Both these methods are in commercial use. When copper is hardened with tin, silicon, or aluminum it generally is called bronze; when hardened with zinc, brass.

CORROSION. Copper is resistant to the action of sea water, and to atmospheric corrosion. It is not resistant to the common acids, and is unsatisfactory in service with ammonia and with most compounds of sulfur. Manufacturers should be consulted in regard to its use under corrosive conditions.

FABRICATION. Copper may be hot or cold forged, hot or cold rolled, hot extruded, hot pierced, and drawn, stamped, or spun cold. It can be silver-soldered, brazed, and welded. For welding by the oxyacetylene torch or electric arc, the oxidized copper will give more satisfactory welds than electrolytic or lake copper. Copper cannot be cut by

* Revision of material originally prepared by the late W. H. Bassett.

the oxygen jet because of its high heat conductivity. Copper is annealed from 480 to 1400 F, depending on the properties desired, and in a slightly oxidizing atmosphere. Ordinary commercial annealing is done in the neighborhood of 1100 F. A reducing atmosphere will make the copper brittle and unfit for use. Copper may be electrodeposited from the alkaline cyanide solution, or from the acid sulfate solution.

STANDARDS. The following is a list of the specifications for copper for various purposes given in the 1947 ASTM Standards:

Designation B4-42, Lake Copper Wire Bars, Cakes, Slabs, Billets, Ingots, and Ingot Bars; B5-43, Electrolytic Copper Wire Bars, Cakes, Slabs, Billets, Ingots, and Ingot Bars; B72-47T, Fire-refined Casting Copper; B1-47, Hard-drawn Copper Wire; B2-47, Medium Hard-drawn Copper Wire; B3-45, Soft or Annealed Copper Wire; B33-46, Tinned Soft or Annealed Copper Wire for Electrical Purposes; B8-48, Concentric-lay-stranded Copper Conductors, Hard, Medium-hard, or soft; B11-47, Copper Plates for Locomotive Fireboxes; B12-47, Copper Rods for Locomotive Staybolts; B13-41, Seamless Copper Boiler Tubes; B42-47, Copper Pipe, Standard Sizes; B47-46, Copper Trolley Wire; B48-48T, Soft Rectangular and Square Bar Copper Wire for Electrical Conductors; B49-41, Hot-rolled Copper Rods for Electrical Purposes; B68-47, Seamless Copper Tubing, Bright Annealed; B75-48T, Seamless Copper Tubes; B88-48, Copper Water Tube; B116-46, Figure-9 Deep-section Grooved and Figure-8 Copper Trolley Wire for Industrial Haulage; B133-48T, Copper Rods, Bars and Shapes; B152-48T, Copper Sheet, Strip and Plate; B170-47, Oxygen-free Electrolytic Copper Wire Bars, Billets, and Cakes; B172-45T, Rope-lay-stranded Copper Conductors Having Bunch-stranded Members for Electrical Conductors; B173-47T, Rope-lay-stranded Copper Conductors Having Concentric-stranded Members for Electrical Conductors; B174-45T, Bunch-stranded Copper Conductors for Electrical Conductors; B187-48T, Copper Bus Bars, Rods, and Shapes; B188-47T, Copper Bus Pipes and Tubes; B189-48, Lead-coated and Lead-alloy-coated Soft Copper Wire for Electrical Purposes; B216-47T, Fire-refined Copper for Wrought Products and Alloys.

11. SAND-CAST COPPER-BASE ALLOYS

By Carter S. Cole *

The principal standard alloys used by engineers and architects in the form of sand castings of various copper-base alloys are given in Tables 1 and 2. A cross index of the approximately equivalent specifications for the various alloys is given in Table 3. Information on high-conductivity castings and other special alloys, such as high-tin bronzes and gear bronzes, is given later in this article.

SELECTION OF ALLOY. Table 1 is an outline of the various types of alloys generally used for certain purposes. In selecting a specific alloy for a given use the foundry should be consulted, and it should always be borne in mind that experience in the same or similar applications is the best criterion for selection. This is particularly important when corrosion resistance, or some other special property, is required. All the copper-base alloys are generally resistant to corrosion, but for specific environments, especially chemicals, the manufacturer of the alloy should be consulted as to its use under the proposed conditions, or tests made on samples to prove suitability.

The relative value of these alloys for mechanical purposes varies with temperature, and information on this point should be obtained before subjecting them to high-temperature service.

DESIGN OF CASTINGS. (See Section 19, Article 15.)

SPECIFICATION AND INSPECTION. (See Section 19, Article 16.)

FABRICATION. Nearly all the sand-cast copper-base alloys listed are readily machined. Those containing lead fall especially in this category. Some of the stronger alloys, such as aluminum-bronze and manganese-bronze, are tough and difficult to machine, but with proper tool angles and machine conditions they can be handled. The silicon-bronze, aluminum-bronze, and manganese-bronze alloys are the more readily weldable, but those containing lead do not lend themselves to welding technique. It should be noted that, for each of the alloys listed in the group of high-tensile brasses and bronzes in Table 2, a number of variants other than the "nominal" compositions shown in Table 2 conform to the specification requirements for these alloys shown in Table 3.

PHYSICAL PROPERTIES. The physical properties of the more important standard alloys are given in Table 2, along with remarks regarding the general characteristics of each alloy and typical reported uses. Alloy number designations used in this and the other tables are the standard designations for the alloys adopted by the American Society for Testing Materials and the Brass and Bronze Ingot Institute (formerly the Non-ferrous Ingot Metal Institute) from whose Handbook most of the data given in Table 2 have been taken. The rest of the data given in Table 2 are taken from other sources deemed

* Revision of material originally prepared by W. M. Corse.

reliable. Attainment of the properties indicated is, of course, dependent on proper foundry technique, proper melting, pouring, gating, venting, and feeding as well as conformance to the requirements of the specifications for the alloys shown in Table 3.

Table 1. Types of Cast-brass or Bronze Alloys Reported Used for Various Services

(Refer to Tables 2 and 3 for properties of and specifications covering the various alloys listed.)

Quality or Use	Type of Alloys	Alloy Number
Abrasion resisting	Gear bronze	*See* text
Acid resisting	*See* Corrosion resistance	
Air fittings	Semi-red brass	5B
	Yellow brass	6A-6B
Architectural trim	Nickel-silver	10A-11A
Bathroom fittings	Nickel-silver	10A-11A
Bearing backs	Red brass	2A-4A
Bearings	*See* p. 4-75	
Bell metal	High tin bronze	*See* text
Builders' hardware	Nickel-silver	10A-11A
	Yellow brass	6A-6B
Castings, chill	Aluminum bronze	9A-9B-9C
Common	Yellow brass	6A-6B
Light	Yellow brass	6A-6B
Strong and tough	Manganese bronze	7A-8A-8C
Corrosion resistance to:		
Acids	Aluminum-bronze	9A-9B-9C
	Nickel-silver	10A-11A
	Silicon-bronze	12A
Alkalies	Silicon-bronze	12A
Atmosphere	Semi-red brass	5B
Phenol	Silicon-bronze	12A
Sea water	Aluminum-bronze	9A-9B
	Silicon-bronze	12A
	Red brass	4A
Water	Red brass	4A-5A-5B
Electrical equipment overhead	Semi-red brass	4B
Electrical fittings	Conductivity castings	*See* text
Fittings		
Air	Semi-red brass	5A-5B
Chemical	*See* Corrosion resistance	
Electrical	Conductivity castings	*See* text
Gas	Semi-red brass	5B
Steam	Bronze and red brass	2A-4A
Valve	Bronze and red brass	2A-4A
Water	Semi-red brass	4A-5A-5B
Gears	Aluminum-bronze	9B-HT, 9C-HT
	Gear bronze	*See* text
Hardness	Aluminum-bronze	9B-HT, 9C-HT
	Silicon-brass	13B
	Manganese-bronze	8A-8C
	Nickel-bronze	*See* text
	Bell metal	*See* text

Quality or Use	Type of Alloys	Alloy Number
Machine parts	Bronze	1B-2B
Miscellaneous work	Yellow and naval brass	6A-6B-6C
Noncorrosive	*See* Corrosion resistance	
Ornamental work	Yellow brass	6B
	Nickel-silver, 12% and 20%	10A-11A
Patterns, metal	Red brass	4A
Pressure resisting	Aluminum-bronze	9A-9B-9C
	Bronze	1B-2A-4A
Propellers	Aluminum-bronze	9A-9B
	Manganese bronze	8A
Pipe fittings	*See* Fittings, air, etc.	
Pump bodies	Bronze and red brass	2A-4A
Pumps		
Lubricating oil	Yellow brass	6B
Water	Red brass and semi-red brass	4A-5A-5B
Radiator valves	Red brass	4A
Resistance to wear	Aluminum-bronze	9A-9B-9C
Resistance to corrosion	*See* Corrosion resistance	
Resistance to shock	Aluminum-bronze	9A-9B
Severe duty	Bronze	1B-2A
Steam fittings	Bronze or red brass	2A-4A
Steam metal	Bronze	2A
Strength	Aluminum-bronze	9A-9B-9C
	Manganese-bronze	8A-8C
	Naval brass	6C
Structural parts	Manganese-bronze	8A-8C
	Silicon bronze and brass	12A-13B
Toughness	Aluminum-bronze	9A-9B-9C
	Manganese-bronze	8A
Under-water fittings	Manganese-bronze	8A
Valve bodies		
Chemical	Aluminum-bronze	9A-9B
High-pressure	Bronze	1B
Valve stems	Manganese-bronze	7A
Valves	Bronze	1B-2B
	Red brass	2A-4A
Low-pressure	Red brass	4A
Radiator	Red brass	4A
Water Fittings	Red brass and semi-red brass	4A-5A-5B
Water tightness	Bronze	2A-4A
Wear resistance	Aluminum-bronze	9A-9B
Worm wheels	Aluminum-bronze	9A-9B
	Manganese-bronze	8A

SPECIAL ALLOYS. There are a number of other recognized standard alloys and many alloys used for special purposes that have not been listed in Table 2. They were omitted primarily because they are of relatively low tonnage use and also because, for most of the special-purpose alloys, standardization of specifications has not advanced as far as it has for the alloys listed in Tables 2 and 3. Among the more important of such special uses of cast copper-base alloys are the following.

High-conductivity castings.* Although it is possible to cast unalloyed copper, the foundry problems are such that many special deoxidants are used. Consequently, such castings have a conductivity of about 80 to 85%. The term *high conductivity* is, however, variously applied to castings with a conductivity in excess of 30%.

One of the commonly used alloys in this class contains 0.80% chromium and 0.08% silicon. This alloy is heat treatable, and a conductivity of 85%, a yield strength of 35,000 psi, a tensile of 52,000 psi, and an elongation in 2 in. of 13% can be attained. Another silicide precipitation alloy with 2.5% nickel and 0.60% silicon can be produced with a conductivity of 45%, a yield strength of 65,000 psi, a tensile strength of 75,000 psi, and a Brinell hardness of 140.

(*Continued on p. 4-29*)

* See High conductivity copper castings, by A. B. Kinzel, *Trans. Am. Foundrymen's Assoc.*, Vol. XLVII, No. 4, June 1940.

Table 2. Sand-cast Copper-base Alloys

Alloy No.	Commercial Designation	Nominal Composition, %					Yield Strength, 1000 psi	Tensile Strength, 1000 psi	Elongation in 2 in., %	Reduction of Area, %	Compressive Strength		Brinell Hardness, 10-mm ball, 500-kg load	Impact Strength (Izod), ft-lb	Remarks and Typical Reported Uses
		Cu	Sn	Pb	Zn	Other					0.001-in. set 1000 psi	0.1-in. set 1000 psi			
							BRONZES, BRASSES AND NICKEL-SILVERS								
1B	Navy "G"	88	8	......	4	....	18–25	40–50	20–35	15–40	12–15	40–48	65–80	9–13	High-grade alloy for structural uses, pressure castings, and corrosion resistance.
2B	Commercial "G"	87	8	1	4	....	16–24	33–43	18–30	15–30	9–11	32–38	60–75	12–15	Strong general-utility structural bronze. Steam-pressure castings and high-pressure valves.
2A	Navy "M"	88	6	1.5	4.5	....	16–21	36–48	25–40	16–33	12–15	30–38	60–72	11–16	High-grade steam and valve-bronze. Oil pumps. Gears. Backing for babbitt-lined bearings. Bushings and many other uses.
3A	High lead tin bronze 80-10-10	80	10	10	......	....	15–22	27–37	8–12	5–11	12–16	40–47	55–70	2–8	General-purpose bushing and bearing alloy, used for many types of machine bearings, including high speed and heavy pressure.
3E	High lead tin bronze 70-5-25	70	5	25	......	....	11–15	23–30	7–16	5–12	12–14	22–25	42–55	4–6	General bearing alloy for high speed under light or medium pressure. Used for locomotive driving boxes, etc.
4A	Ounce metal 85-5-5-5	85	5	5	5	....	17–24	33–46	20–35	17–32	10–12	30–33	55–65	6–12	High-grade red brass for general service, valves, fittings, bearings, small gears, etc. Very widely used alloy.
4B	Commercial red brass	83	4	6	7	....	12–17	30–38	15–27	12–25	11–12	28–30	50–60	....	Valves, fittings, pump bodies, hardware. General-purpose free-machining red brass alloy.
5A	Valve composition 81-3-7-9	81	3	7	9	....	13–17	29–39	18–30	15–27	....	....	50–60	6–10	Low-pressure valves and fittings, hardware, and ornamental castings. Plumbers' supplies and fittings.
5B	Semi-red brass	76	3	6	15	....	12–16	30–40	20–35	15–30	8–10	29–32	50–60	....	Plumbing fixtures. Air and gas fittings. Hardware and low-pressure valves and fittings.

6A	High copper yellow brass	71	1	3	25	...	12–14	35–40	25–40	20–40	8–10	28–32	40–55	...	General-purpose yellow brass. Valves and fittings, hardware and ornamental castings.
6B	Commercial yellow brass	67	1	3	29	...	11–15	30–38	20–35	15–30	8–10	28–32	40–60	...	General-purpose yellow brass. Hardware and trim. Ornamental castings. Valves and fittings.
6C	Naval brass	61	1	1	36.7	Al = 0.3	14–20	40–45	15–25	18–30	...	...	50–75	...	Hardware and ornamental castings. Ship trim. Plumbers' flanges, etc.
10A	12% Nickel-silver	57	2	9	20	Ni = 12	15–20	30–40	10–25	7–20	...	...	50–60	...	Architectural trim, bathroom and kitchen fittings, hardware, and ornamental castings.
11A	20% Nickel-silver	64	4	4	8	Ni = 20	17–30	40–60	15–25	11–22	...	...	76–120	...	Valves, fittings, hardware, building trim, and ornamental castings.
	High Tensile Brasses and Bronzes (Note 1)														
		Cu	Zn	Fe	Al	Mn	Other								
7A	Leaded manganese bronze	61	35.5	1	0.75	0.25	Sn = 0.75 Pb = 0.75	25–40	60–78	15–30	15–30	20–26	85–90	80–95	20–40	Free-machining manganese bronze. Valve stems. Pump bodies. Gears.
8A	No. 1 manganese bronze	58	39.25	1.25	1.25	0.25	...	28–40	70–88	20–35	20–40	22–26	85–90	90–120 115–150 *	20–40	Standard manganese bronze. Propeller hubs and blades. Engine frames and machinery parts requiring strength and toughness.
8C	High-strength manganese bronze	60–68	Rem.	2–4	3–7.5	2.5–5	...	65–90	110–120	12–18	5–18	55–65	90–120	170–225 190–235 *	7–12	Extra-heavy-duty high-strength alloy. Gears. Screw-down nuts. Cams. Bridge parts. Combines toughness and hardness.
9A	Aluminum bronze	88	...	3	9	...	...	25–30	70–87	22–38	20–36	27–31	65–70	90–125	35–50	These alloys combine high strength, toughness, hardness, and corrosion resistance. For valve seats, guides, stems; worm wheels, marine equipment, propellers, acid-resisting pump parts, bushings and bearings, gears, etc.
9B	Aluminum bronze	89	...	1	10	...	...	30–35	70–85	20–35	15–30	16–20	75–83	100–140	30–36	
9B-HT	Aluminum bronze (heat treated)	89	...	1	10	...	...	40–55	80–95	12–16	12–16	35–45	...	160–225 *	...	
9C	Aluminum bronze	85	...	4	11	...	...	30–41	75–95	12–20	12–20	...	...	150–185 *	7–22	
9C-HT	Aluminum bronze (heat treated)	85	...	4	11	...	...	45–52	90–110	6–15	7–18	...	...	190–235 *	...	
12A	Silicon bronze	90–95	Note 2	Note 2	Note 2	1	Si = 4	18–30	45–55	20–30	...	15–22	50–70	80–130	15–60	Combines high strength, toughness, and resistance to corrosion.
13B	Silicon brass	82	14	...	...	...	Si = 4	24–30	60–70	16–20	...	...	...	130–150	...	Combines high strength and hardness.

Note 1. For the alloys 7A to 13B, inclusive, there are variants within the specification limits other than the "nominal compositions" shown. *Note 2.* Zinc, iron, and/or aluminum, as well as tin, may be present. * 3000-kg load.

Most of the data in this table are from the Handbook of the Brass and Bronze Ingot Institute.

Table 3. Index of Approximately Equivalent Specifications (Note 1)

Alloy No.	Commercial Designation	Am. Soc. for Testing Materials		Federal		Navy		Soc. Automotive Engrs.
		Specification Number	Alloy No.	Specification Number	Alloy or Grade	Specification Number	Grade or Class	Specification Number
1B	Navy "G"	B143	1B	QQ-B-691b	5	46M6i	....	620
2A	Navy "M"	B143 B61	2A	QQ-B-691b	1	46B8i	....	622 ....
2B	Commercial "G"	B143	2B	QQ-B-691b	6 6x	46B5j	I and II	621
3A	High-leaded tin bronze 80-10-10	B144 B22	3A C	....	....	....	....	64 and 792 ....
3E	High-leaded tin bronze 70-5-25	B144 B66 B67	3E Soft bronze Backs	....	....	46B22g	V	794 ....
4A	Ounce metal or 85-5-5-5	B145 B62	4A	.... QQ-B-691b	 2	46B23e 46B5j	 III	 40
4B	Commercial red brass	B145	4B	....	....	....	....	....
5A	Valve composition 81-3-7-9	B145	5A	QQ-B-691b	4	46B24e 46B5j	 III	....
5B	Semi-red brass	B145	5B	....	....	....	....	....
6A	High copper yellow brass	B146	6A	QQ-B-621a	C	46B11h	II	....
6B	Commercial yellow brass	B146	6B	QQ-B-621a	B	46B11h	I	41
6C	Naval brass	B146	6C	QQ-B-621a	A	46B10h	....	....
7A	Leaded-manganese bronze	B147 B132	7A A	QQ-B-726c	D	....	....	....
8A	No. 1 manganese bronze	B147	8A	QQ-B-726c	A	49B3f	....	43
8C	High-strength manganese bronze	B147	8C	QQ-B-726c	C	46B29a	A	430B
9A	Aluminum bronze	B148	9A	QQ-B-671a	A	46B18d	No. 1	68A
9B	Aluminum bronze	B148	9B 9B-HT	QQ-B-671a	B	46B18d	No. 2	68B 68B (heat treated)
9C	Aluminum bronze	B148	9C 9C-HT	QQ-B-671a QQ-B-671a	C D	46B18d	No. 3	....
10A	12% Nickel-silver	B149	10A	....	....	....	....	....
11A	20% Nickel-silver	B149	11A	Note 2	....	....	....	....
12A	Silicon bronze	B198	12A	QQ-C-593	....	46B28a	b	....
13B	Silicon brass	B198	13B	....	....	46B28a	a	....

Note 1. Although the specifications for the several alloys as listed in this table are approximately equivalent in most of the essential details, there are some requirements that may differ. Therefore, if material is ordered to any one of these specifications, that particular specification should be consulted.
ASTM specifications may be obtained from the American Society for Testing Materials, 1916 Race St., Philadelphia, 3, Pa. Federal specifications may be obtained from the Superintendent of Documents, Washington, 25, D.C. Navy specifications may be obtained from the Bureau of Supplies and Accounts, Navy Department, Washington, 25, D.C. SAE specifications may be obtained from the Society of Automotive Engineers, 29 West 39th St., New York, 18, N.Y.

Note 2. Federal Specifications: FF-H-106, FF-H-111, FF-H-116b, FF-H-121a, FF-H-136, and WW-P-541a.

Beryllium-copper castings with about 2.0% beryllium, 0.50% cobalt plus iron, and 0.25% silicon have a minimum tensile of 78,000 psi and an elongation in 2 in. of 29% in the "as cast" condition and, when heat-treated, 155,000 psi tensile with 2% minimum elongation. This alloy has a Rockwell hardness of B76 as cast, and C35 when heat treated. Conductivities of some 50 to 65% can be realized.

There are also "conductivity" castings with over 90% copper made by using small amounts of tin and zinc—just sufficient to give good casting properties. However, none of these conductivity alloys is as yet covered by commercial specifications (except for the Navy Specification 46C11a for beryllium copper).

Bell Metal. The now obsolete Navy Specification 23B2 called for 78% copper, 22% tin. (The revised Navy Specification for Bells—23B2a—called for a silicon-bronze alloy.) Commercial bells are made with an 81-19 copper-tin alloy, or even lower tin.

Bridge Bearing Plates. Generally high tin bronzes are used. [See ASTM Specification B22—Class A (19% tin); Class B (16% tin)—both 1.0 maximum phosphorus.] The 80-10-10 alloy and 88-10-2 are also used where bearing loads are lighter.

Piston Rings. High tin bronzes similar to B22, Classes A and B, are generally used. May be centrifugally cast.

Rolling Mill Screw Down and Stripper Nuts. High tin bronzes of about 16% tin deoxidized with phosphorus are generally used with a Brinell hardness (500-kg load) of 83 to 93.

Gear Bronzes. High tin alloys—usually with about 11% tin (see SAE 65)—are quite generally used. One% or more nickel is often added and sometimes 0.5 to 1.5% lead. (See, for example, SAE 640.)

For all these special applications it is especially advisable that the foundry be consulted so that the most suitable alloy for the purpose in hand will be chosen.

GENERAL REFERENCES

BOOKS

ASTM Book of Standards, Part 2, American Society for Testing Materials, Philadelphia, 1949.
Cast Metals Handbook, American Foundrymen's Society, Chicago, 1944.
Corrosion Resistance of Metals and Alloys, R. J. McKay and R. Worthington, Reinhold Publishing Corp., New York, 1936.
Engineer's Handbook, Lumen Bearing Co.
Ingot Brass and Bronze, Brass and Bronze Ingot Institute, Chicago.
Metal Industry Handbook, London.
Recommended Practices for the Casting of Non-ferrous Alloys, American Foundrymen's Society, Chicago
SAE Handbook, Society of Automotive Engineers, New York.

PERIODICALS

American Foundryman, Chicago.
Chemical Engineering, New York.
The Foundry, Cleveland.
Materials and Methods, New York.
Metal Finishing, New York.
Metal Industry, London.

PROCEEDINGS AND TRANSACTIONS

Journal, Institute of Metals, London.
Proceedings, American Society for Testing Materials, Philadelphia.
Transactions, American Foundrymen's Society, Chicago.
Transactions, American Institute of Mining and Metallurgical Engineers, New York.

12. WROUGHT COPPER-BASE ALLOYS

By W. R. Hibbard *

The nomenclature used in the nonferrous metals trade is confusing to those who are not familiar with the business, but through long usage it has become so thoroughly rooted that it is practically impossible to change it. The trade names have an actual value to the producers of various materials, and they would, therefore, naturally resist changes.

All questions of properties of engineering material are tied up with factors of safety, factors of reliability, conditions of exposure (which are always local), and similar matters. The requirements of special alloys for specific purposes may be considered from the viewpoint of the physical requirements or resistance to corrosion or both. The physical requirements and properties may be studied in the laboratory. The question of corrosion is less easily solved, because of the difficulty of reproducing in the laboratory the conditions to which the metal may be exposed in service. (*Continued on p. 4-40*)

* Revision of material originally prepared by the late W. H. Bassett.

Table 4. Average Chemical and Physical Properties of Wrought Copper-base Alloys

Data from *Bulletin B-32*, American Brass Co.

(Variations must be expected in practice.)

	Material	Form *	Approximate Composition, %					Tensile Strength, psi		Elongation in 2 in., %		Yield Strength (c), psi		Modulus of Elasticity, psi × 10^{-6}, Hard †	Shearing Strength, psi	
			Copper	Zinc	Lead	Tin		Hard †	Soft	Hard †	Soft	Hard †	Soft		Hard †	Soft
B-1	Copper, electrolytic	S	99.90+					46,000	33,000	5	35	40,000	10,000	16.0		
	tough pitch	W	99.90+					66,000	35,000	1(b)	35(b)					
		R	99.90+					45,000	32,000	15	45	40,000	10,000		30,000	21,000
B-2	Deoxidized copper	T	99.90+				Phosphorus present	45,000	35,000	10	45	40,000	10,000	16.0		
		S	99.90+					46,000	35,000	5	35	40,000	10,000			
		R	99.90+					45,000	32,000	15	45	40,000	10,000			
		W	99.90+					66,000	35,000	1(b)	35(b)	40,000	10,000			
	Brasses															
B-3	Gilding metal	S	95.00	5.00				55,000	35,000	5	38	44,000	11,000	15.0		
B-4	Commercial bronze	S	90.00	10.00				62,000	37,000	6	40	47,000	12,000	15.0		
		W	90.00	10.00				80,000	38,000	1(b)	40(b)					25,000
B-5	Red brass 85%	S	85.00	15.00				69,000	40,000	7	45	55,000	15,000	15.0		
		W	85.00	15.00				88,000	42,000	1(b)	42(b)					
		T	85.00	15.00				69,000	40,000	10	50	55,000	15,000			
B-6	Low brass	S	80.00	20.00				73,000	43,000	8	50	60,000	16,000	15.0	43,000	
		W	80.00	20.00				100,000	47,000	1(b)	45(b)					31,000
B-7	Cartridge brass	S	70.00	30.00				76,000	47,000	10	65	60,000	16,000	14.0		
		W	70.00	30.00				110,000	50,000	1(b)	50(b)				45,000	35,000
		T	70.00	30.00				76,000	47,000	10	60	62,000	17,000			
B-8	Yellow brass	S	65.00	35.00				73,000	45,000	10	60	60,000	17,000	14.0		
B-9	Yellow brass	R	63.00	37.00				65,000	46,000	20	60	50,000	17,000	14.0		
		W	63.00	37.00				105,000	50,000	1(b)	50(b)					
B-10	Muntz metal	S	60.00	40.00					54,000		45	60,000	20,000	12.8		
		R	60.00	40.00					54,000		50		20,000			
	Leaded Brasses															
B-11	Leaded commercial bronze	S	89.60	10.00	0.40			62,000	37,000	6	40	47,000	12,000	15.0		
B-12	Leaded commercial bronze	R	88.50	10.00	1.50			54,000	37,000	15	40	45,000	12,000	15.0	35,000	
B-13	Leaded brass	R	69.00	29.50	1.50			65,000	46,000	15	40	45,000	17,000	14.0		
B-14	Yellow brass	T	66.50	33.00	0.50			73,000	45,000	10	55	60,000	17,000	14.0		
B-14	Butt brass	S	64.00	35.00	1.00			73,000	45,000	8	55	60,000	17,000	14.0		

Note. See table footnotes on p. 4–37.

(*Table continued on p. 4-32*)

	Material	Form*	Rockwell B Hardness No.		Bend Test, degree		Melting Point, °F	Specific Gravity	Density, lb per cu in.	Coefficient of Expansion $\times 10^7$	Electrical Properties at 20 C		Thermal Conductivity (g)	Uses—Remarks
			Hard †	Soft	Hard †	Soft					Resistivity ohms (mil ft)	Conductivity, % IACS		
B-1	Copper, electrolytic tough pitch	S W R	51	Too soft	180	180	1983	8.89	.3217	98	10.268	101.0	2678.02	For electrical parts requiring high conductivity, also for rods, sheets, and wire for commercial purposes.
B-2	Deoxidized copper	T S R W	50	Too soft	180	180	1983(*a*)	8.93(*a*)	.323	98				Copper pipe and tube. For fabrication by welding.
	Brasses													
B-3	Gilding metal	S	61	Too soft	180	180	1949	8.666	.320	101	18.52	56.0	1672.13	For jewelry trade and manufacturing when soft pliable metal is required.
B-4	Commercial bronze	S W	70	1	180	180	1913	8.804	.318	101	24.1	43.0	1294.74	For window screen wire, automobile radiators, bullet jackets, etc.
B-5	Red brass 85%	S W T	76	5	180	180	1877	8.745	.316	104	28.03	37.0	1103.14	For architectural work on account of its color and resistance to corroding agents, hardware, and pipe.
B-6	Low brass	S W	81	10	180	180	1832	8.667	.313	106	31.95	32.5	972.51	For its color and resistance to corroding agents. Fourdrinier wire.
B-7	Cartridge brass	S W	83	20		180	1750	8.528	.308	111	38.4	27.0	841.87	Deep-drawing brass, primers, cartridges, tubes, shot shell heads.
		T	83	20		180								
B-8	Yellow brass	S	80	15		180	1706	8.460	.306	112	39.8	26.0	827.36	Common brass used for a large variety of articles.
B-9	Yellow brass	R W	75	20		180	1697	8.437	.305	114	39.8	26.0	827.36	For rivets, screws, pins, and other heading operations.
B-10	Muntz metal	S		45		180	1661	8.396	.303	116	36.5	28.4	870.90	For sheathing, bolts, nuts, pin wire, etc.
		R		45		180								
	Leaded Brasses													
B-11	Leaded commercial bronze	S	70	1			1913	8.804	.318	101	24.63	42.1	1248.00	For electrical parts.
B-12	Leaded commercial bronze	R	58	1			1900.2	8.830	.319	102	25.5	41.0	1254.10	For hardware, jewelry, free cutting.
B-13	Leaded brass	R	75	15			1742	8.562	.309	111	38.4	27.0	840	For special shapes where high lead is detrimental.
B-14	Yellow brass	T	80	15			1715(*a*)	8.528	.308	113	39.9	26.0	816	For pipe and tube.
B-14	Butt brass	S	80	15			1692.8	8.460	.306	113	39.9	26.0	816	For hinges, for threading.

(Table continued on p. 4-33)

Table 4. Average Chemical and Physical Properties of Wrought Copper-base Alloys—*Continued*

(Variations must be expected in practice.)

	Material	Form *	Approximate Composition, %					Tensile Strength, psi		Elongation in 2 in., %		Yield Strength (*b*), psi		Modulus of Elasticity, psi × 10^{-6}, Hard †	Shearing Strength, psi	
			Copper	Zinc	Lead	Tin		Hard †	Soft	Soft	Hard †	Hard †	Soft		Hard †	Soft
B-15	Free-cutting yellow brass	R	62.00	35.00	3.00			58,000	47,000	18	60	42,000	18,000	15.0	35,000	
B-16	Clock brass	S	61.50	37.00	1.50			73,000	45,000	7	50	60,000	17,000	14.0		
B-17	Forging brass	R	60.00	38.00	2.00				54,000		45		20,000			
B-18	Extruded architectural bronze		56.00	41.50	2.50				60,000		25		20,000			
	Special Brasses															
B-19	High-strength commercial bronze	R	90.25	6.90	1.75		Nickel 1.00 Phosphorus 0.10	70,000		12		60,000				
B-20	Oreide	S	87.25	11.50		1.25		72,000	43,000	8	47	60,000	16,000			
B-21	Aluminum brass	T	76.00	21.95			Aluminum 2.00 Arsenic 0.05	85,000	52,000	10	65	60,000	20,000			
B-22	Admiralty	T	70.00	29.00		1.00		85,000	52,000	10	65		20,000			
		S	70.00	29.00		1.00			48,000		65		18,000			
B-23	Naval brass	S	60.00	39.25		0.75			56,000				22,000			
		R	60.00	39.25		0.75		63,000	56,000	30	40	35,000	22,000	15		
B-24	Tobin bronze	S	60.00	39.25		0.75		85,000	56,000	8	45	62,000	22,000			
		R	60.00	39.25		0.75		63,000	56,000	35	45	35,000	22,000	15	39,000	
B-25	Manganese bronze	R	59.00	39.00		0.70	Iron 0.80 Manganese 0.50	75,000	60,000	20	30	45,000	30,000			
	Phosphor Bronzes															
B-26	Phosphor bronze	S	98.75			1.25		65,000	40,000	6	48	50,000	14,000			
B-27	Signal bronze	W	98.25			1.75	Phosphorus 0.25 (B-27 to B-29)	105,000	45,000	1(*b*)	40(*b*)					
B-28	Phosphor bronze (Grade A)	W	95.75			4.00		110,000	52,000	1(*b*)	40(*b*)	65,000	20,000			
		R	95.75			4.00		65,000	48,000	30	50	55,000	20,000	15		40,000
B-29	Phosphor bronze (Grade A)	S	95.00			5.00		80,000	48,000	8	50	65,000	20,000	15		37,000
B-30	Leaded phosphor Bronze (Grade B)	R	94.00		1.00	5.00		65,000	48,000	25	40	55,000	20,000			

Note. See table footnotes on p. 4-37.

(*Table continued on p. 4-34*)

	Material	Form*	Rockwell B Hardness No.		Bend Test, degree		Melting Point, °F	Specific Gravity	Density, lb per cu in.	Coefficient of Expansion $\times 10^7$	Electrical Properties at 20 C		Thermal Conductivity (g)	Uses—Remarks
			Hard †	Soft	Hard †	Soft					Resistivity ohms (mil ft)	Conductivity, % IACS		
B-15	Free-cutting yellow brass	R	70	20	120	180	1653(*a*)	8.489	.307	114	41.5	25.0	748.97	For automatic machine work
B-16	Clock brass	S	80	15			1666.2	8.44	.305	113	39.9	26.0	816	For clock and meter parts, free machining.
B-17	Forging brass	R		45		120	1653(*a*)	8.44	.305	115	39.9	26.0	744	For hot forgings, machines easily.
B-18	Extruded architectural bronze			65			1623.6	8.432	.305	116	37.0	28.0	852	For architectural shapes.
	Special Brasses													
B-19	High-strength commercial bronze	R	80					8.666	.320		32.4	32.0	972	For line hardware, free machining.
B-20	Oreide	S	80	15		180	1886	8.791(*a*)	.317	104	34.3	30.2	972	For color.
B-21	Aluminum brass	T	85	30			1778(*a*)	8.336(*a*)	.301	108	45.9	22.6	696	Condenser tubes.
B-22	Admiralty	T		30			1715(*a*)	8.528	.308	112	42.1	24.7	768	Condenser tubes, tube sheet, head sheet.
		S		25										
B-23	Naval brass	S		50		180	1625(*a*)	8.404(*a*)	.304	119	40.3	25.8	804	Resists salt-water corrosion.
		R	65	50										
B-24	Tobin bronze	S	90	50		180	1625(*a*)	8.404(*a*)	.304	117	40.3	25.8	804	Shafting, piston rods, bolts, nuts, plates, welding rod.
		R	65	50										
B-25	Manganese bronze	R	85				1664.4(*a*)	8.370(*a*)	.302	119	45.0	23.0	696	For strength and resistance to corrosion.
	Phosphor Bronzes													
B-26	Phosphor bronze	S	75			180	1967	8.89	.321	99	24.1	43.0	898.92	For flexible metal hose.
B-27	Signal bronze	W				180	1958	8.89	.321	99	27.5	37.8	1016.05	For electrical purposes. Fourdrinier wire.
B-28	Phosphor bronze (Grade A)	W				180								For springs, electric switches, welding rod.
		R	75	28		180	1922	8.87(*a*)	.320	106	75.6	13.7	433	
B-29	Phosphor bronze (Grade A)	S	86	28			1922	8.87(*a*)	.320	99	56.46	18.4	566.09	For window weight chain, etc.
B-30	Leaded phosphor bronze (Grade B)	R	75	25		180	1922	8.929(*a*)	.322	99	56.5	18.4	577.70	Phosphor bronze with good machining properties.

(*Table continued on p. 4-35*)

Table 4. Average Chemical and Physical Properties of Wrought Copper-base Alloys—*Continued*

(Variations must be expected in practice.)

	Material	Form *	Approximate Composition, %					Tensile Strength, psi		Elongation in 2 in., %		Yield Strength (c), psi		Modulus of Elasticity, psi × 10^{-6}, Hard †	Shearing Strength, psi	
			Copper	Zinc	Lead	Tin		Hard †	Soft	Hard †	Soft	Hard †	Soft		Hard †	Soft
B-31	Phosphor bronze (Grade C)	S	92.00			8.00		93,000	60,000	10	65	68,000	24,000	14		
		R	92.00			8.00		80,000	60,000	30	65		24,000			
		W	92.00			8.00		130,000	62,000	1(*b*)	40(*b*)				80,000	
B-32	Phosphor bronze (Grade D)	S	89.50			10.50		102,000	66,000	12	65	70,000	28,000	14		
		R	89.50			10.50		85,000	65,000	25	65					
		W	89.50			10.50		145,000	68,000	1(*b*)	40(*b*)					
B-33	Special free-cutting phosphor bronze	R	88.00	4.00	4.00	4.00		60,000		20		45,000		15		
	Cupronickels															
B-34	30% Cupronickel	S	70.00				Nickel 30.00 Sometimes 0.40 iron	77,000	55,000	5	40	70,000	22,000	20		
		R	70.00					70,000	55,000	25	40	60,000	22,000		44,000	38,000
		W	70.00					95,000	58,000	1(*b*)	35(*b*)					
		T	70.00					70,000	55,000	10	45	60,000	22,000			
B-35	20% Cupronickel	S	80.00				Nickel 20.00	74,000	48,000	5	40	67,000	20,000	19		
		T	80.00						50,000		40		20,000			
	Nickel Silvers															
B-36	10%	S	65.00	25.00			Nickel 10.00	88,000	55,000	7	42	70,000	20,000	17.5		
B-37	18% Grade A	S	65.00	17.00			Nickel 18.00	85,000	58,000	4	40	70,000		18		
		R	65.00	17.00				70,000	58,000	20	45		22,000			
		W	65.00	17.00				110,000	60,000	1(*b*)	40(*b*)					
B-38	18% Grade B	S	55.00	27.00			Nickel 18.00	99,000	60,000	4	45	75,000	22,000			
		R	55.00	27.00				80,000	60,000	20	45	60,000	22,000		55,000	
		W	55.00	27.00				130,000	65,000	1(*b*)	40(*b*)					
B-39	Ambrac	S	75.00	5.00			Nickel 20.00	82,000	50,000	5	35	55,000	18,000	19		
		R	75.00	5.00				70,000	50,000	20	40	60,000	18,000			
		W	75.00	5.00				100,000	55,000	1(*b*)	35(*b*)					35,000
		T	75.00	5.00				75,000	50,000	10	45	55,000	18,000			
	Leaded Nickel Silvers															
B-40	Extruded leaded nickel silver 10%	R	46.50	40.75	2.75		Nickel 10.00	70,000		15		40,000			40,000	
B-41	Leaded nickel silver 12%	R	65.00	22.00	1.00		Nickel 12.00	68,000		15		60,000				

Note. See table footnotes on p. 4-37.

(*Table continued on p. 4-36*)

	Material	Form*	Rockwell B Hardness No.		Bend Test, degree		Melting Point, °F	Specific Gravity	Density, lb per cu in.	Coefficient of Expansion × 10^7	Electrical Properties at 20 C		Thermal Conductivity (g)	Uses—Remarks
			Hard †	Soft	Hard †	Soft					Resistivity ohms (mil ft)	Conductivity, % IACS		
B-31	Phosphor bronze (Grade C)	S	94	50		180	1877	8.815(a)	.318	101	76.2	13.6	435.45	For electric switches, contact fingers, radio parts, Fourdrinier wire.
		R		50										
		W				180								
B-32	Phosphor bronze (Grade D)	S	98	55		180	1832	8.78(a)	.317	102	94.2	11.0	351.26	For very stiff resilient springs, chain, welding rod.
		R												
		W				180								
B-33	Special free-cutting phosphor bronze	R	75				1832(a)	8.86(a)	.320	100	85	12.2	386.10	For good machining properties.
	Cupronickels													
B-34	30% Cupronickel	S	84	35		180	2237	8.950	.323	90	226	4.6	200.31	For condenser tubes, head plates, tube sheets, resistance to sea water.
		R												
		W												
		T	80	35										
B-35	20% Cupronickel	S	79			180	2192	8.93	.323	91	160.2	6.5	252.56	For turbine blades, tubes, resistance to corrosion.
		T												
	Nickel Silvers													
B-36	10%	S	87	30		180	1850	8.675	.313	91	125	8.4	319.33	For cheaper grades silver plated ware.
B-37	18% Grade A	S	85	40		180	2030	8.752(a)	.316	90	175	6.0	232.24	For silver plated table and hollow ware.
		R		40										
		W												
B-38	18% Grade B	S	93	45		180	1931	8.68(a)	.314	93	195.6	5.3	206.11	For ornamental purposes.
		R	85	45										
		W												
B-39	Ambrac	S	85	25		180	2102	8.860	.320	91	172	6.1	260.08	Resistance to corrosion, color.
		R	80	25										
		W												
		T	82	25										
	Leaded Nickel Silvers													
B-40	Extruded leaded nickel silver 10%	R	70				1715(a)	8.460	.306		122	8.5	324	For color and machining.
B-41	Leaded nickel silver 12%	R					1859(a)	8.680	.314	93	122	8.5	324	For color, watch parts and machining.

(*Table continued on p. 4-37*)

Table 4. Average Chemical and Physical Properties of Wrought Copper-base Alloys—*Continued*

(Variations must be expected in practice.)

	Material	Form *	Approximate Composition, %					Tensile Strength, psi		Elongation in 2 in., %		Yield Strength (*c*), psi		Modulus of Elasticity, psi × 10^{-6}, Hard †	Shearing Strength, psi	
			Copper	Zinc	Lead	Tin		Hard †	Soft	Hard †	Soft	Hard †	Soft		Hard †	Soft
	Aluminum Bronzes															
B-42	5%	S	95.00					92,000	55,000	7	65	65,000	22,000	15		
		R	95.00				Aluminum 5.00		55,000		65		22,000			
		T	95.00						60,000		60					
B-43	8%	S	92.00					105,000	65,000	7	60	65,000	25,000	15		
		R	92.00				Aluminum 8.00	80,000	65,000	30	65	50,000	25,000			
							Aluminum 9.25	95,000	80,000	16	22	55,000	40,000		50,000	
B-44	Avialite Type I	R	89.25		0.40		Nickel 0.50									
							Iron 0.60									
B-45	Ambraloy #917	P	82.00				Aluminum 9.50		90,000		12					
	Type II	R	82.00				Nickel 5.00	105,000		12		60,000				
							Iron 2.50									
	Cadmium Bronzes															
B-46	Hitenso BB Grade 85	S	99.00				Cadmium 1.00	55,000	37,000	6	50	48,000	12,000	15		
		W	99.00					90,000	40,000	1(*b*)	40(*b*)					28,000
B-47	Hitenso C Grade 65	W	98.60			0.60	Cadmium 0.80	95,000	42,000	1(*b*)	40(*b*)			15		
	Copper Silicon Alloys															
B-48	Everdur 1010	S	95.80					95,000	58,000	7	60	60,000	22,000	15	58,000	43,000
	Type A	R	95.80				Silicon 3.10	90,000	58,000	18	70	60,000	22,000			
		W	95.80				Manganese 1.10	145,000	60,000	1(*b*)	50(*b*)					
B-49	Everdur 1012	R	95.60		0.40		Silicon 3.00	90,000		18		60,000		15		
	Type D						Manganese 1.00									
B-50	Everdur 1015	S	98.25					65,000	40,000	8	46	50,000	15,000	15		
	Type B	R	98.25				Silicon 1.50	70,000	40,000	15	50	55,000	15,000		40,000	28,000
		W	98.25				Manganese 0.25	122,000	42,000	1(*b*)	40(*b*)					
		T	98.25					65,000	40,000	8	50	55,000	15,000			
B-51	Everdur 1014	R	91.00				Aluminum 7.00		95,000		25		53,000	15		
							Silicon 2.00									
	Special Alloys															
B-52	Calsun bronze Grade 13	W	95.50			2.00	Aluminum 2.50	135,000	52,000	1(*b*)	40(*b*)			15		
B-53	Chromium copper	R	99.05				Chromium 0.85	62,000	35,000	20	40	57,000	15,000	15		
							Silicon 0.10	72,000 (*e*)	63,000 (*f*)	25(*e*)	25(*f*)	61,000 (*e*)	45,000 (*f*)	16		

No.	Material	Form*	Rockwell B Hardness No. Hard †	Rockwell B Hardness No. Soft	Bend Test, degree Hard †	Bend Test, degree Soft	Melting Point, °F	Specific Gravity	Density, lb per cu in.	Coefficient of Expansion × 10^7	Electrical Properties at 20 C: Resistivity ohms (mil ft)	Electrical Properties at 20 C: Conductivity, % IACS	Thermal Conductivity (*g*)	Uses—Remarks
	Aluminum Bronzes													
B-42	5%	S	92	35			1940	8.176(*a*)	.295	99	61	17.0	575	For color, diaphragms.
		R		35										
		T												
B-43	8%	S	96	50			1904	7.80(*a*)	.281	99	70.08(*d*)	14.8(*d*)	502.22	For color, diaphragms.
		R												
B-44	Avialite Type I	R		50			1908(*a*)	7.585(*a*)	.274	94	82.2	12.8	418	For valve seats and guide bushings airplane engines.
B-45	Ambraloy #917 Type II	P					2111(*a*)	7.585(*a*)	.274		138.5	7.5	264	For shafting, tube sheets, and header sheets.
		R	105											
	Cadmium Bronzes													
B-46	Hitenso BB Grade 85	S	65	Too soft		180	1969	8.89(*a*)	.3212	98	11.67	89.0	2392	For high-strength electrical wire and cables.
		W												
B-47	Hitenso C Grade 65	W					1958(*a*)	8.89(*a*)	.3212	98	16.9	61.0	1614	For high-strength electrical wire and cable.
	Copper Silicon Alloys													
B-48	Everdur 1010 Type A	S	92	35		180	1866(*a*)	8.539(*a*)	.308	100	154.5	6.5	226.43	For strength and resistance to corrosion. Welding rod.
		R	90	35										
		W				180								
B-49	Everdur 1012 Type D	R	90				1866(*a*)	8.740(*a*)	.308	99	154.5	6.5	226.4	Good machining.
B-50	Everdur 1015 Type B	S	77	Too soft		180	1931(*a*)	8.740(*a*)	.316	99	94.3	11.0	374	Bolt stock, pipes, tubes.
		R	80	Too soft		180								
		W				180								
		T	75	Too soft		180								
B-51	Everdur 1014	R		95			1841(*a*)	7.685(*a*)	.278	100	112.7	9.2	312	High strength, free machining.
	Special Alloys													
B-52	Calsun bronze Grade 13	W				180	1929.6(*a*)	8.539(*a*)	.308	99	54.6	19.0	600	For electrical wire and cable, etc.
B-53	Chromium copper	R	70	Too soft		180	1967(*a*)	8.89(*a*)	.321		25.9	40.0	1188	For resistance welding electrodes and grid support rods
			77	65				8.89(*a*)	.321	98	12.8	80.0	2244	

* R = Rod. S = Sheet. T = Tube. W = Wire. P = Plate.

† For some alloys the figures are for a temper slightly different from that commonly known as "hard."

[a] Determination.

[b] Elongation of wire in 10 in., %.

[c] Yield point taken as the load producing an extension under stress of 0.50%.

[d] At 18.1 C.

[e] Cold worked and heat treated.

[f] Annealed, quenched, and heat treated.

[g] Btu per sq ft per hr per 1 in. thick per 1° F, at 68 F.

Table 5. Types of Wrought Copper-base Alloys for Various Services

Use or Quality	Alloy	Alloy No.
Architectural shapes	Extruded architectural bronze	B18
Architectural work	Red brass 85%	B5
Bolts	Muntz metal	B10
Bolts	Tobin bronze	B24
Bolt stock	Everdur 1015	B50
Bullet jackets	Commercial bronze	B4
Bushings, valve guide	Avialite	B44
Cartridges	Cartridge brass	B7
Chain	Phosphor bronze D	B32
Chain window weight	Phosphor bronze A	B29
Clock parts	Clock brass	B16
Color	Red brass 85%	B5
Color	Low brass	B6
Color	Oreide	B20
Color	Ambrac	B39
Color	10% Leaded nickel silver	B40
Color	12% Leaded nickel silver	B41
Color	5% Aluminum bronze	B42
Color	8% Aluminum bronze	B43
Condenser tubes	Aluminum brass	B21
Condenser tubes	Admiralty	B22
Contact fingers	Phosphor bronze C	B31
Corrosion resistance	Red brass 85%	B5
Corrosion resistance	Low brass	B6
Corrosion resistance	20% Cupronickel	B35
Corrosion resistance	Ambrac	B39
Corrosion resistance	Manganese bronze	B25
Corrosion resistance	Everdur 1010	B48
Diaphragms	5% Aluminum bronze	B42
Diaphragms	8% Aluminum bronze	B43
Drawing, deep	Cartridge brass	B7
Electrical parts	Leaded commercial bronze	B11
Electrical parts, high conductivity	Copper	B1
Electrical purposes	Signal bronze	B27
Electrical wire and cable	Hitenso BB	B46
Electrical wire and cable	Hitenso C	B47
Electrical wire and cable	Calsun bronze	B52
Forging, hot	Forging brass	B17
Grid support rods	Chromium copper	B53
Hardware	Red brass 85%	B5
Hardware	Leaded commercial bronze	B12
Hardware, line	High-strength commercial bronze	B19
Heading operations	Yellow brass	B9
Hinges	Butt brass	B14
Hose, flexible metal	Phosphor bronze	B26
Jewelry trade	Gilding metal	B3
Jewelry trade	Leaded commercial bronze	B12
Machine work, automatic	Free-cutting yellow brass	B15
Machining	Leaded commercial bronze	B12
Machining	Forging brass	B17
Machining	Leaded phosphor bronze B	B30
Machining	Clock brass	B16
Machining	Free-cutting phosphor bronze	B33
Machining	High-strength commercial bronze	B19
Machining	10% Leaded nickel silver	B40
Machining	12% Leaded nickel silver	B41

Use or Quality	Alloy	Alloy No.
Machining	Everdur 1012	B49
Machining	Everdur 1014	B51
Meter parts	Clock brass	B16
Nuts	Muntz metal	B10
Nuts	Tobin bronze	B24
Ornamental work	18% Nickel silver B	B38
Pins	Yellow brass	B9
Pipe	Deoxidized copper	B2
Pipe	Red brass 85%	B5
Pipe	Yellow brass	B14
Pipe	Everdur 1015	B50
Piston rods	Tobin bronze	B24
Plates	Tobin bronze	B24
Plates, head	30% Cupronickel	B34
Primers	Cartridge brass	B7
Radiators, automobile	Commercial bronze	B4
Radio parts	Phosphor bronze C	B31
Rivets	Yellow brass	B9
Rod, copper	Copper	B1
Rod, welding	See welding rod	
Screws	Yellow brass	B9
Sea water, resistance to	Naval brass	B23
Sea water, resistance to	30% Cupronickel	B34
Shafting	Tobin bronze	B24
Shafting	Ambraloy 917	B45
Shapes, special	Leaded brass	B13
Sheathing	Muntz metal	B10
Sheets	Copper	B1
Sheet, tube and head	Admiralty	B22
Sheet, tube and head	30% Cupronickel	B34
Sheet, tube and head	Ambraloy 917	B45
Shot shell heads	Cartridge brass	B7
Silver-plated ware	10% nickel silver	B36
Silver-plated ware	18% nickel silver A	B37
Springs	Phosphor bronze A	B28
Springs	Phosphor bronze D	B32
Strength	Manganese bronze	B25
Strength	Everdur 1010	B48
Strength	Everdur 1014	B51
Switches, electric	Phosphor bronze A	B28
Switches, electric	Phosphor bronze C	B31
Tableware	18% Nickel silver A	B37
Threading and similar operations	Butt brass	B14
Tubes	Cartridge brass	B7
Tubes	Deoxidized copper	B2
Tubes	Yellow brass	B14
Tubes	20% Cupronickel	B35
Tubes	Everdur 1015	B50
Tubes, condenser	Aluminum brass	B21
Tubes, condenser	Admiralty	B22
Tubes, condenser	30% Cupronickel	B34
Turbine blades	20% Cupronickel	B35
Valve seats (airplane engines)	Avialite	B44
Watch parts	12% Leaded nickel silver	B41
Welding electrodes, resistance	Chromium copper	B53
Welding fabrication	Deoxidized copper	B2
Welding rod	Tobin bronze	B24
Welding rod	Phosphor bronze A	B28
Welding rod	Phosphor bronze D	B32
Welding rod	Everdur 1010	B48
Wire	Copper	B1
Wire, Fourdrinier	Low brass	B6
Wire, Fourdrinier	Signal bronze	B27
Wire, Fourdrinier	Phosphor bronze C	B31
Wire, pin	Muntz metal	B10
Wire, window screen	Commercial bronze	B4

Table 6. ASTM Specifications for Wrought Copper-base Alloys

Alloy No. (See Table 4)	Strip, Sheet, and Plate	Rods and Bars	Wire	Tubes
B-1	B152-48T Type A	B133-48T B187-48T	B1-47 B2-47 B3-45 B48-48T	B188-47T
B-2	B152-47T Type B			B75-48T Type B B68-47 Type B B111-48 B88-48 B42-47
B-3	B36-48T Alloy No. 1		B134-48 Alloy No. 1	
B-4	B36-48T Alloy No. 2		B134-48 Alloy No. 2	
B-5	B36-48T Alloy No. 3		B134-48 Alloy No. 3	B135-47T Alloy No. 1 B111-48 B43-47
B-6	B36-48T Alloy No. 4		B134-47 Alloy No. 4	
B-7	B36-48T Alloy No. 6		B134-47 Alloy No. 6	B135-47T Alloy No. 2
B-8	B36-48T Alloy No. 8		B134-47 Alloy No. 7	
B-9			B134-47 Alloy No. 8	
B-10		B124-47 Alloy No. 1		B135-47T Alloy No. 5 B111-48
B-11	B121-47T Alloy No. 1			
B-12		B140-47 Type B		
B-13	....	....	....	....
B-14(t)	B121-47T Alloy No. 2			B135-47T Alloy No. 3
B-14(S)	B121-47T Alloy No. 3			
B-15		B16-47		
B-16	B121-47T Alloy No. 4			
B-17		B124-48 Alloy No. 2		
B-18	....	....	....	....
B-19	....	....	....	....
B-21				B111-47 Type B
B-22	B171-48			B111-47 Type B
B-23	B171-48	B21-48T Grade A B124-48 Alloy No. 3		
B-24	B171-48		B124-48 Alloy No. 3	
B-25		B21-48T Grade A B138-47 Type A B124-47 Alloy No. 4		
B-26	....	....	....	....
B-27			B105-47 Grade 30	
B-28	B103-47 Grade A	B139-48T Grade A	B159-47T Grade A	
B-29	B103-47 Grade A B100-47 Type A			
B-30		B139-48T Grade B1		
B-31	B103-47 Grade C	B139-48T Grade C	B159-47T Grade C	
B-32	B103-47 Grade D B100-47 Type A	B139-48 Grade D	B159-47T Grade D	
B-33		B139-48 Grade B2		
B-34	B122-48T Alloy No. 5 B171-48			B111-48
B-35	B122-48T Alloy No. 6			B111-48
B-36	B122-48T Alloy No. 3			
B-37	B122-48T Alloy No. 2	B151-48T Alloy A	B206-47T Alloy A	
B-38	B122-48T Alloy No. 4	B151-48T Alloy B	B206-47T Alloy B	
B-39	B122-48T Alloy No. 7			B111-48
B-40	....	....	....	....
B-41	....	....	....	....
B-42	B169-47T Alloy A	B124-48 Alloy No. 10		B111-48
B-43	B169-47T Alloy C	B150-48T Type I B124-48 Alloy No. 11		
B-44		B150-48T Type I B124-48 Alloy No. 11		
B-45	B171-48	B150-48T Type II		
B-46			B105-47 Grade 80	
B-47			B105-47 Grade 55	
B-48	B97-47 Type A B96-47 Type A B100-47 Type B	B98-47 Type A B124-48 Alloy No. 7,9	B99-47 Type A	
B-49		B98-47 Type D	B99-47 Type D	
B-50	B97-47 Type B	B98-47 Type B B124-48 Alloy No. 8	B99-47 Type B B105-47 Grade 8.5	
B-51		B150-48T Type I B124-48 Alloy No. 11		
B-52			G105-47 Grade 15	
B-53	....	....	....	....

The joint use of "dissimilar" alloys has caused much trouble. For instance, the use of one of the so-called stronger brasses (one of the brasses sometimes called bronzes which contains a considerable percentage of zinc together with tin or iron or a combination of these elements) in connection with a straight copper-tin bronze in an electrolyte such as sea water will result in the weakening of the alloy containing the larger proportion of zinc. Where brass strainers have been held in position with copper or gun-metal bolts and where brass screens have been placed in position with copper rivets, galvanic action has occurred, destroying the alloy containing the greater percentage of zinc.

Before proceeding with design, engineers not thoroughly familiar with the composition and properties of the alloys with which they propose to deal should consult reputable metallurgists, giving a thorough review of the conditions under which the material is to be used. The question of conditions of use is very important and should be carefully considered.

The so-called season cracking of brasses and bronzes has given much trouble in certain places. This phenomenon is due to the internal stress in the alloy brought about by cold working, such as rolling and drawing, and corrosive action. Copper and the copper-zinc alloys containing 20% or less of zinc are seldom subject to this trouble. The stronger copper-zinc alloys, as well as those containing tin and other elements, may be prepared so as to prevent this difficulty if the conditions of exposure are understood beforehand.

Table 4 gives an idea of what may be expected in the way of physical properties from some of the more common combinations of metals, but should be used as a general guide only. Table 5 is an index to the alloys of Table 4. Table 6 lists the ASTM specifications covering the alloys listed in Table 4.

Table 4 gives the more common wrought alloys with their ordinary physical properties, both "hard rolled" or "drawn" and "annealed" or "soft." The figures are averages that ordinarily may be expected in practice but may be modified by "extra hard," "half hard," or "quarter hard" rolling. In other words, almost any physical properties between the figures for "hard" and "soft" may be obtained by appropriate working, and higher figures for tensile strength, or hardness, may be obtained by a greater amount of working. The alloys have been chosen to represent the more commonly used copper-zinc, copper-zinc-nickel, copper-tin, and copper-aluminum mixtures. The alloys intermediate in composition between those chosen have physical properties lying between those given for the nearest compositions.

MAGNESIUM AND ITS ALLOYS

By William H. Gross

Magnesium, with a specific gravity of only 1.74, is the lightest metal available for engineering use. Chief source is sea water which contains about 0.13% magnesium. It is first mixed with lime produced by the roasting of oyster shells. Calcium in the lime replaces magnesium in the sea water, producing magnesium hydroxide, which is treated with hydrochloric acid prepared from natural gas and chlorine. Magnesium chloride produced by this reaction is electrolytically converted to metallic magnesium and chlorine gas, the magnesium being cast into pigs.

13. PROPERTIES AND AVAILABILITY

Like other pure metals, magnesium must be alloyed to give maximum strength and usefulness. Although lightness is its outstanding characteristic, it has other properties that are equally significant. Magnesium is extremely easy to machine and is adaptable to practically all the usual methods of metal working. It can be joined by gas, arc, and electric resistance welding and by riveting. Other significant properties include good stability to atmospheric exposure and resistance to attack by alkalies, chromic and hydrofluoric acids, and many organic chemicals such as hydrocarbons, aldehydes, alcohols, phenols, amines, esters, and most oils. The alloys of magnesium have relatively low electrical resistivity and high thermal conductivity. Magnesium is nontoxic and nonmagnetic.

The common commercial alloys of magnesium are listed in Table 1 under the designations of the ASTM and of the two largest producers of alloy ingot. Physical constants are given in Table 2.

AVAILABILITY. Pure magnesium is furnished in ingot for remelting and in the form of extruded sticks for alloying with other metals or for chemical uses. The alloys are produced in the form of ingots or billets; as sand, permanent mold, and die castings; forgings; extruded bars, round rods, tubing, and solid shapes; and as rolled sheets, strip, and plate.

CASTINGS. Sand castings of magnesium find widespread use in aircraft engines and aircraft landing wheels where light weight, high strength, and shock resistance are of prime importance. In high-speed rotating and reciprocating parts, their light weight is used to effect smoother operation and to reduce starting inertia. The printing and textile industries offer many examples of such uses. Portable tools and manually handled equipment of all kinds, such as vacuum cleaners, portable grinders, polishers, and foundry flasks have always presented a large market for cast-magnesium alloys. The excellent machining characteristics of magnesium sand castings provide reason for using magnesium in some instances.

Table 1. Magnesium Alloys in Common Use

ASTM Designation	Commercial Designations		Typical Composition, %				Uses
	The Dow Chemical Company	American Magnesium Corporation	Al	Zn	Other	Mg	
A10	Dowmetal G	AM240	10.0	.....	....	Remainder	Permanent mold castings
AZ31	Dowmetal FS	AM52S	3.0	1.0	....	"	Extrusions, sheet, and forgings
AZ61	Dowmetal J	AM57S	6.5	1.0	....	"	Extrusions and forgings
AZ63	Dowmetal H	AM265	6.0	3.0	....	"	Sand and permanent-mold castings
AZ80	Dowmetal O	AM58S	8.5	0.5	....	"	Extrusions and forgings
AZ91	Dowmetal R	AM263	9.0	0.6	....	"	Die castings
AZ92	Dowmetal C	AM260	9.0	2.0	....	"	Sand and permanent-mold castings
M1	Dowmetal M	AM3S	......	.....	1.2 Manganese	"	Extrusions, forgings, sheet, and special sand castings

Table 2. Physical Constants of Magnesium Alloys

ASTM Designation	Specific Gravity at 75 F	Density, lb/cu in.	Melting Point, °F	Coefficient of Thermal Expansion (65–750 F)	Thermal Conductivity, cgs units (100–300 C)	Electrical Resistivity, microhm-centimeters at 20 C (68 F)
A10	1.81	.066	1100	.000016	.17	15.0
AZ31	1.77	.064	1160	.000016	.23	10.0
AZ61	1.80	.065	1145	.000016	.19	12.5
AZ63	1.83	.066	1135	.000016	.18	11.5
AZ80	1.80	.065	1130	.000016	.18	14.5
AZ91	1.81	.063	1120	.000016	.17	17.0
AZ92	1.82	.066	1110	.000016	.16	16.0
M1	1.76	.064	1200	.000016	.30	5.0

Modulus of elasticity, 6,500,000 psi. Modulus of rigidity, 2,400,000 psi. Poisson's ratio, 0.35.

Permanent-mold castings, in addition, have closer tolerances on dimensions, better finish, and require less machining. Magnesium alloy permanent mold castings are being used for such parts as nose pieces on aircraft engines, fittings, wheel flanges, pistons, brackets, and various types of housings.

Die castings possess such advantages as extremely accurate dimensions, reduction of machining, excellent surface finish, and low cost. These characteristics have caused magnesium die castings to be widely used in such applications as small aircraft engine parts, instrument parts, various kinds of housings, aircraft landing wheels, vacuum cleaner parts, miscellaneous boxes and covers, and innumerable other parts where light weight is essential.

Table 3 lists typical mechanical properties of magnesium castings. Sand and permanent-mold castings have essentially the same properties, and some of the alloys are used interchangeably for castings of these types. Only one alloy composition is in common use for die casting. This alloy is sometimes sand cast for special purposes.

Table 3. Typical Mechanical Properties of Magnesium Sand, Die, and Permanent Mold Casting Alloys

Alloy	Condition	Tension			Compression	Hardness		Shear	Impact
		Tensile Strength, psi	Yield Strength,* psi	Elongation in 2 in., %	Yield Strength,* psi	Brinell, 500-kg load, 10-mm ball	Rockwell E	Ultimate Stress, psi	Izod, ft-lb
AZ92	As cast	24,000	14,000	2	16,000	65	77	18,000	1
AZ92	Heat treated	40,000	14,000	10	16,000	63	75	20,000	4
AZ92	Heat treated and aged	40,000	23,000	2	23,000	84	90	21,000	1
A10	As cast	22,000	13,000	2	13,000	54	65	17,000	2
A10	Heat treated	33,000	12,000	8	12,000	52	62	19,000	4
A10	Heat treated and aged	34,000	19,000	2	19,000	69	80	21,000	2
AZ63	As cast	29,000	14,000	6	14,000	50	59	18,000	3
AZ63	Heat treated	40,000	14,000	12	14,000	55	66	19,000	5
AZ63	Heat treated and aged	40,000	19,000	5	19,000	73	83	20,000	2
M1	As cast	14,000	4,500	5	4,500	33	3	11,000	9
AZ91	Die cast	33,000	22,000	3	22,000	60	72	20,000	....

* Yield strength is defined as the stress at which the stress-strain curve deviates 0.2% from the modulus line.

FORGINGS. Magnesium forgings, combining good strength, light weight, and pressure tightness, are suited for a wide variety of applications. Typical uses for alloy forgings include aircraft engine bearing caps, rocker-arm supports, door and aileron hinges, engine mountings, cylinder heads, valve and pump bodies, hydraulic cylinders, supercharger parts, horn disks, miscellaneous control levers, brackets, fittings, and similar applications. Table 4 contains the typical mechanical properties of magnesium forgings.

Table 4. Typical Mechanical Properties of Magnesium Forgings

Alloy	Condition	Tension			Compression	Hardness	
		Tensile Strength, psi	Yield Strength,* psi	Elongation in 2 in., %	Yield Strength,* psi	Brinell, 500-kg load, 10-mm ball	Rockwell E
AZ61	As forged	43,000	26,000	12	17,000	55	66
M1 †	As forged	36,000	23,000	7	....	47	54
AZ80	As forged	46,000	31,000	8	25,000	69	80
AZ80	Heat treated and aged	50,000	34,000	5	27,000	72	82

* Yield strength is defined as the stress at which the stress-strain curve deviates 0.2% from the modulus line.

† Properties for M1 alloy are based on both press and hammer forgings; those for AZ61 and AZ80 are for press forgings only.

EXTRUSIONS. Magnesium extrusions have been used extensively in the aircraft, textile, and transportation industries, and in portable equipment. Other specific uses of magnesium extrusions include such applications as floor beams, moldings, stiffener elements in aircraft structures, the framework of bus and airplane seats, bus and trailer body frames, and structural members in a wide variety of parts. In Table 5 are given the typical mechanical properties of extruded magnesium rods and bars, tubes, and solid shapes.

SHEET AND PLATE. Magnesium sheet can be arc, gas, and spot welded; formed; drawn; spun; and riveted. Magnesium sheet and plate have been used in such parts as aircraft wings, wing tips, ailerons, fairings, oil and fuel tanks, cases, floor plates, ducts, conveyors, seats, venetian blinds, bus and trailer bodies, and dock boards. Typical mechanical properties of magnesium sheet and plate are listed in Table 6.

NONSTRUCTURAL USES. The fields of usefulness for magnesium alloys can be divided into two types, chemical and structural. Chemical uses include magnesium as an

Table 5. Typical Mechanical Properties of Magnesium Extrusions

Alloy	Condition	Tension			Compression	Shear	Hardness	
		Tensile Strength, psi	Yield Strength,* psi	Elongation in 2 in., %	Yield Strength,* psi	Ultimate Stress, psi	Brinell, 500-kg load, 10-mm ball	Rockwell E
	Bars and Rods †							
AZ31	As extruded	40,000	30,000	15	17,000	19,000	49	57
AZ61	As extruded	45,000	32,000	15	20,000	19,000	58	70
M1	As extruded	38,000	26,000	10	14,000	16,000	42	41
AZ80	As extruded	49,000	33,000	11	22,000	20,000	55	77
AZ80	Heat treated and aged	54,000	41,000	8	30,000	23,000	80	87
	Shapes							
AZ31	As extruded	38,000	26,000	15	15,000	....	50	59
AZ61	As extruded	44,000	28,000	16	19,000	....	64	76
M1	As extruded	34,000	20,000	9	11,000	....	46	50
AZ80	As extruded	46,000	29,000	9	22,000	....	67	78
AZ80	Heat treated and aged	53,000	35,000	9	30,000	....	85	90
	Tubing							
AZ31	As extruded	36,000	21,000	16	15,000	....	46	51
AZ61	As extruded	40,000	21,000	14	15,000	....	50	60
M1	As extruded	33,000	21,000	9	10,000	....	42	41

* Yield strength is defined as the stress at which the stress-strain curve deviates 0.2% from the modulus line.

† The properties for bars and rods apply to sizes 1/4 to 1 1/2 in. in diameter or 2 1/2 sq in.

alloying element in aluminum and zinc alloys. Other nonferrous alloys also utilize magnesium as an alloying ingredient although to a lesser extent. Magnesium is also used metallurgically as an oxygen scavenger or deoxidizer in the manufacture of copper and nickel alloys. One of the best-known nonstructural applications for magnesium is as a catalyst in the well-known Grignard reaction of organic chemistry. When alloyed with 30% or more of aluminum and ground into a powder, magnesium finds use in pyrotechnics. The magnesium anode is used for the cathodic protection of other metals from corrosion.

Table 6. Typical Mechanical Properties of Magnesium Plate, Sheet, and Strip

Alloy	Condition	Tension			Compression	Shear	Hardness	
		Tensile Strength, psi	Yield Strength,* psi	Elongation in in., %	Yield Strength,* psi	Ultimate Stress, psi	Brinell, 500-kg load, 10-mm ball	Rockwell E
AZ31	Annealed	37,000	22,000	21	16,000	21,000	56	67
AZ31	Hard rolled	43,000	33,000	11	26,000	23,000	73	83
M1	Annealed	33,000	15,000	17	12,000	17,000	48	55
M1	Hard rolled	37,000	29,000	8	20,000	17,000	56	67

* Yield strength is defined as the stress at which the stress-strain curve deviates 0.2% from the modulus line.

Successful protection has been provided on underground metal structures like oil, brine, and gas lines, well casings, telephone and telegraph cables, and buried tanks. In order to protect such structures it is necessary only to bury magnesium anodes in the ground at intervals adjacent to the structure to be protected and then connect the two by means of a wire.

14. MACHINING OF MAGNESIUM ALLOYS

Magnesium can be worked by practically all methods known to the metal-working art, using the same tools employed on other metals. In some of these operations magnesium exhibits a pronounced superiority to other metals in common use. Particularly outstanding is its machinability.

High-speed steels are recommended as cutting-tool materials although carbide-tipped tools produce worthwhile economies on high-production jobs. Oil-type coolants are quite generally used, particularly for parts which must be held to close dimensions, on small parts, or when taking fine cuts. When a coolant is indicated, only oil type—never water-soluble oils or emulsions—should be used. This is important because magnesium chips cannot be easily ignited when generously wetted with oil. On the other hand, damp chips can be ignited by a chance spark.

TURNING, BORING. Figure 1 shows recommended angles for grinding single-point tools for rough turning and finishing operations on magnesium. Angles conform to the terminology advocated by the American Standards Association. Table 7 gives feeds, speeds, and position of cut for turning and boring magnesium.

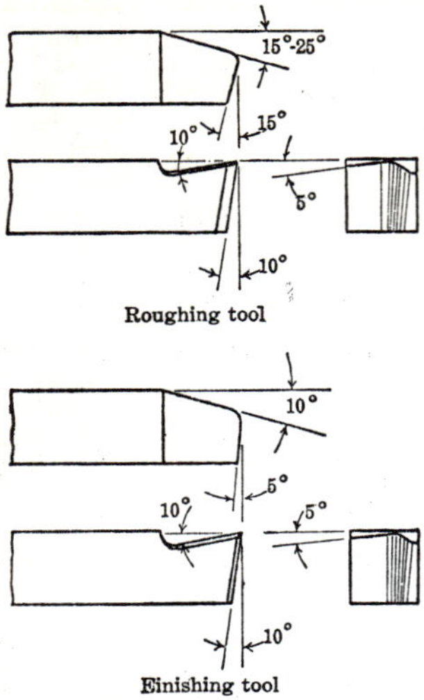

FIG. 1. Single-point tools for turning magnesium alloys.

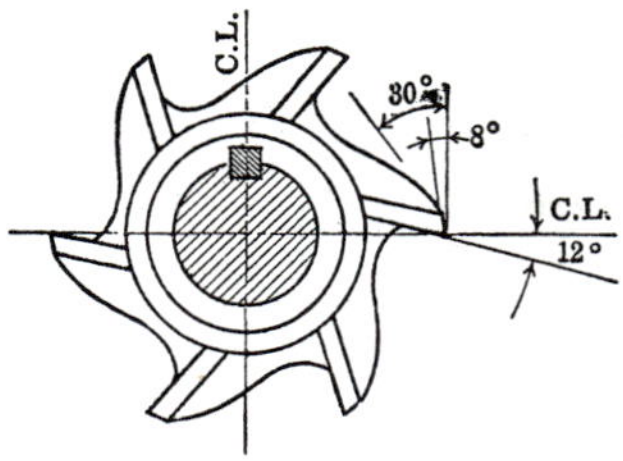

FIG. 2. Milling cutter suitable for magnesium alloys.

Table 7. Recommended Speeds, Feeds, and Depths of Cut for Turning and Boring Magnesium

Operation	Speed, ft per min	Feed, in. per revolution	Maximum Depth of Cut, in.
Roughing	300 to 600	0.030 to 0.100	0.500
	600 to 1000	0.020 to 0.080	0.400
	1000 to 1500	0.010 to 0.060	0.300
	1500 to 2000	0.010 to 0.040	0.200
	2000 to 5000	0.010 to 0.030	0.150
Finishing	300 to 600	0.005 to 0.025	0.100
	600 to 1000	0.005 to 0.020	0.080
	1000 to 1500	0.003 to 0.015	0.050
	1500 to 2000	0.003 to 0.015	0.050
	2000 to 5000	0.003 to 0.015	0.050

MILLING. Coarse-toothed milling cutters having one-half to one-third as many teeth as cutters commonly used on steel are best for magnesium. Figure 2 shows proper tool angles; Table 8 gives recommended operating details.

Table 8. Recommended Speeds, Feeds, and Depths of Cut for Milling Magnesium

Operation	Speed, ft per min	Feed		Depth of Cut, in.
		in./min	in./tooth	
Roughing	Up to 900	10 to 50	0.005 to 0.025	Up to 0.500
	900 to 1500	10 to 60	0.005 to 0.020	Up to 0.375
	1500 to 3000	15 to 75	0.005 to 0.010	Up to 0.200
Finishing	Up to 900	10 to 50	0.005 to 0.015	Up to 0.075
	1000 to 3000	10 to 70	0.004 to 0.008	0.005 to 0.050
	3000 to 5000	10 to 90	0.003 to 0.006	0.003 to 0.030
	5000 to 9000	10 to 120	0.002 to 0.005	0.003 to 0.030

DRILLING. Three types of drills (see Fig. 3) are in use for the various types of drilling requirements normally included in machine-shop work. They include a drill for deep holes, one for shallow holes, and a low-helix-angle drill recommended to prevent the work from "walking" up the drill when drilling thin sheet. Table 9 gives recommended drilling data. Reamers, like milling cutters, should have fewer flutes than usual. Typical reamer for use on magnesium is shown in Fig. 4.

Table 9. Recommended Speeds and Feeds for Drilling Magnesium

Drill Diameter, in.	Speed, ft per min	Feed, in. per revolution		
		Sheet	Shallow Holes	Deep Holes
1/4	300 to 2000	0.005 to 0.030	0.004 to 0.030	0.004 to 0.008
1/2	300 to 2000	0.010 to 0.030	0.015 to 0.040	0.012 to 0.020
1	300 to 2000	0.010 to 0.030	0.020 to 0.050	0.015 to 0.030

TAPPING. Ground-thread high-speed steel taps are advocated for magnesium.

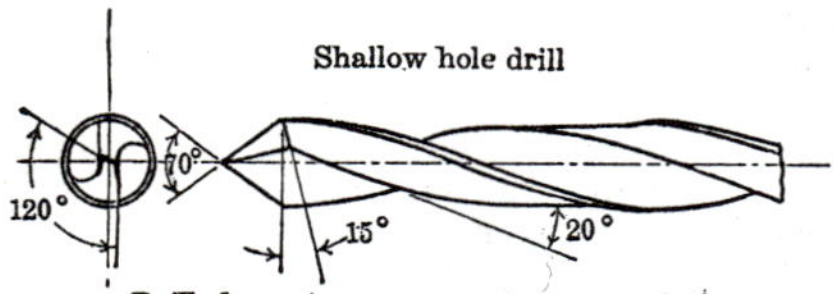

Drill element	Recommendation
Point angle	70 to 118°
Helix angle	10 to 30°
Chisel edge angle	120 to 135°
Relief angle	12 to 15°
Corners	Rounded
Flutes	Polished

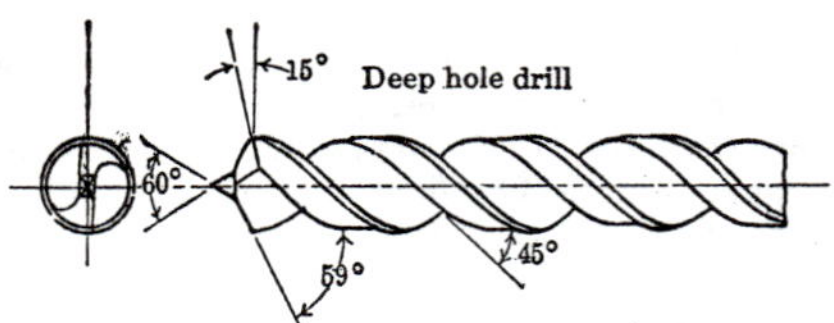

Drill element	Recommendation
Point angle	118°
Helix angle	40 to 45°
Chisel edge angle	135 to 150°
Web	Constant thickness
Flutes	Opened and polished
Point	118° plus spur

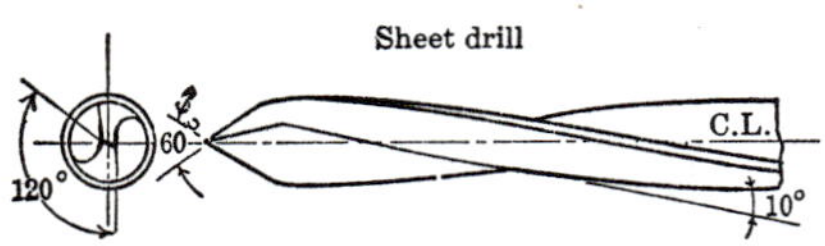

Drill element	Recommendation
Point angle	60°
Helix angle	10°
Chisel edge angle	120 to 135°
Web	Thinned at point
Corners	Rounded

FIG. 3. Three types of drills used on magnesium alloys.

SAWING AND FILING. Saws for use on magnesium should have fewer teeth than usual to take care of the extra dust produced by the heavier feeds of operation which the easy machinability of magne-

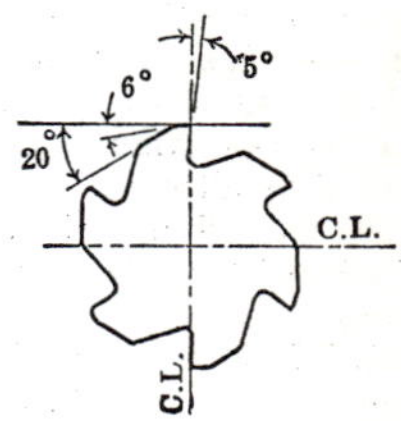

FIG. 4. Reamer for use on magnesium alloys.

sium makes possible. By the same token the best files for use on magnesium are single and double cut of the coarse bastard and single-cut types. Hardened, high-speed steel rotary files and burrs are best for power-operated filing operations.

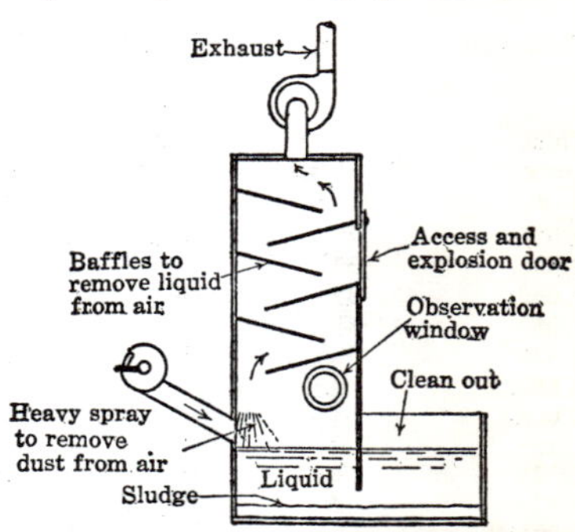

FIG. 5. Diagram of dust collection system considered safe for use with magnesium.

HANDLING. Good housekeeping is the answer to safe handling of magnesium in finely divided form. Heavy pieces of solid magnesium are extremely difficult to ignite since the material must be heated to about its melting point. Fines, particularly if moistened with water, might be ignited by a stray spark. For this reason chips, turnings, saw dust, and similar fines should be swept up frequently and placed in a covered metal container. If a magnesium fire should occur on a machine tool it can be extinguished with dry graphite powder or cast-iron borings. Liquid or gas-type fire extinguishers or water should not be used on such fires. Proper dust-collection equipment should be used on grinding. A safe unit is one in which dust is immediately drenched with water as soon as it leaves the grinding wheel. A diagram of a safe set-up is shown in Fig. 5. It has been found very desirable to have the power connected in such a way that the grinder, air blower, and water circulation will operate simultaneously.

15. JOINING OF MAGNESIUM ALLOYS

Magnesium structures can be joined by a number of the common methods, including riveting, arc welding, gas welding, spot welding, and, of course, by bolts and screws.

RIVETING. Aluminum-alloy rivets are universally used in magnesium joints. Aluminum 56S rivet alloy is recommended because of the minimum danger of galvanic corrosion. 17S and 24S compositions have good strength but should not be used for joints which will be exposed to the weather or corrosive atmosphere. 2S aluminum rivets are easy to drive and give a minimum of trouble from galvanic corrosion, but their use is limited to lightly stressed joints. Choice of rivet diameter is important and is governed by proper balance between the shear strength of the rivet and the bearing strength of the sheet. With 56S rivets in the quarter-hard condition (1/4 H) a good rule is to use a rivet diameter approximately three times the sheet thickness. The distance from the center line of the first row of rivets in a joint to the edge of the sheet, defined as "edge distance," should not be less than 2 1/2 times the rivet diameter. Spacing between rivets should be not less than three times the rivet diameter. Pneumatic hammers can be used for driving rivets into magnesium joints although hydraulic squeeze rivets are preferred when they can be used. Although 56S aluminum rivets are usually driven cold in the quarter-hard condition (1/4 H), rivets with diameters of 5/16 in. and greater are often heated to 650 F immediately before driving.

ARC WELDING of magnesium is done with a special torch equipped with a small cup surrounding the electrode through which an inert gas such as argon or helium flows during welding. The inert gas forms a blanket over the molten metal and prevents oxidation. This eliminates the need for a flux, which is required in gas welding. Either direct current with reversed polarity or alternating current with superimposed high frequency is used in the arc-welding process. Magnesium welding rod of approximately the same composition as the metal being welded is used during arc welding. Table 10 gives machine settings and other recommendations for arc-welding magnesium. The flow of inert gas is given in cubic feet per minute, and it is possible to obtain meters that will read directly in argon or helium flow in cubic feet per minute.

Table 10. Recommendations for Welding Magnesium with Helium or Argon Shielded Arc

Thickness, in.	Current, amp *	Filler Rod Diameter, in.	Helium Flow, cu ft/min	Argon Flow, cu ft/min	Diameter of Tungsten Electrode (d-c), in.	Diameter of Tungsten Electrode (a-c), in.	Cup Size ID, in.
0.030	20	1/16	.3	.12	3/32	1/16	7/16
0.040	30	1/16	.3	.12	3/32	1/16	7/16
0.050	35	3/32	.3	.12	1/8	3/32	7/16
0.060	45	3/32	.3	.12	1/8	3/32	7/16
0.070	55	1/8	.3	.12	3/16	3/32	7/16
0.080	60	1/8	.3	.12	3/16	3/32	7/16
0.090	65	1/8	.3	.12	3/16	3/32	7/16
0.100	70	1/8	.3	.12	3/16	3/32	7/16
0.125	75	1/8	.3	.12	3/16	1/8	7/16
0.150	80	5/32	.3	.16	3/16	1/8	7/16
0.200 †	90	5/32	.5	.16	1/4	1/8	5/8
0.250	100	5/32	.5	.16	1/4	1/8	5/8
0.500	115	5/32	.5	.16	1/4	1/8	5/8
1.000	130	5/32	.5	.16	1/4	1/8	5/8
Machine Welds (One Pass Each Side on Square Edges)							
3/8	300	HF a-c 3/16	..	.31	...	1/4	3/4
1/2	375	HF a-c 3/16	..	.31	...	1/4	3/4

* Currents shown are for all alloys except ASTM alloy M1, which requires 5 to 10 amp more current for material up to 0.050 in. and 15 to 30 amp more current for thicker material. Currents given are for welding speeds of 12 in. per min.

† Sheet thicker than 0.150 in. should be welded in more than one pass. A current of about 60 amp is used on the first pass, and the currents given in the table are used for subsequent passes.

GAS WELDING. Oxyacetylene, oxyhydrogen, and oxycarbohydrogen flames can be used for gas-welding magnesium. Oxycarbohydrogen is recommended in preference to the other two. It provides a cooler flame than oxyacetylene and a more easily seen and, therefore, a more easily adjusted flame than the oxyhydrogen. A flux must be used when gas-welding magnesium, and for this reason gas welds are limited to butt joints and to single-pass welds. Flux entrapped in the joint can result in corrosion difficulties later because the flux is hygroscopic and will take on moisture. Normally the filler rod is of the same composition as the material being welded. Fluxes are prepared for use by mixing two parts of the dry powder with one part of water, forming a paste which should be brushed onto both edges of the joint and on the rod. The choice of rod size and other necessary operating information is given in Table 11 for oxycarbohydrogen and oxyacetylene flames. Immediately after welding, the bead should be scrubbed with a stiff

Table 11. Recommendations for Gas-welding Magnesium

OXYACETYLENE

Metal Thickness, in.	Welding-rod Diameter, in.	Orifice Diameter, in.	Drill Number	Oxygen Pressure, psi	Acetylene Pressure, psi
.020	1/16 to 3/32	.035	65	1 to 4	1 to 3
.032	1/16 to 3/32	.035	65	3 to 5	1 to 3
.040	3/32 to 1/8	.035	65	3 to 5	1 to 3
.051	3/32 to 1/8	.035	65	3 to 5	1 to 3
.064	3/32 to 1/8	.040	60	3 to 5	3 to 5
.072	3/32 to 1/8	.040	60	3 to 5	3 to 5
.081	3/32 to 1/8	.046	56	3 to 5	3 to 5
.091	1/8 to 5/32	.046	56	3 to 5	3 to 5
.128	1/8 to 5/32	.059	53	3 to 5	3 to 5

OXYCARBOHYDROGEN

Metal Thickness, in.	Welding-Rod Diameter, in.	Orifice Diameter, in.	Drill Number	Oxygen Pressure, psi	Carbohydrogen Pressure, psi
.020	1/16 to 3/32	.040	60	1 to 4	1 to 4
.032	1/16 to 3/32	.046	56	3 to 5	3 to 5
.040	3/32 to 1/8	.059	53	3 to 5	3 to 5
.051	3/32 to 1/8	.059	53	3 to 5	3 to 5
.064	3/32 to 1/8	.059	53	3 to 5	3 to 5
.072	3/32 to 1/8	.059	53	3 to 5	3 to 5
.081	3/32 to 1/8	.059	53	3 to 5	3 to 5
.091	1/8 to 5/32	.067	51	3 to 5	3 to 5
.128	1/8 to 5/32	.081	46	3 to 5	3 to 5

brush dipped in hot water. Next the part should be dipped for one minute in a solution containing 1 1/2 lb sodium dichromate and 1 1/2 pints of concentrated nitric acid, and water to make a total of one gallon. This is the same solution used for applying the chrome-pickle treatment to provide a paint base on magnesium, described later. After one minute in this solution the welded part should be washed in running water and then boiled for 2 hours in a solution containing 1/2 lb of sodium dichromate and 1 gal of water. Upon removal from this solution the welded part should be rinsed thoroughly in cold running water and then dipped in hot water to facilitate drying.

STRESS RELIEF OF ARC AND GAS-WELDED JOINTS. Both arc and gas welding of magnesium set up residual stresses in the metal which must be relieved by suitable treatment. Table 12 gives times and temperatures for various alloys in cast and wrought forms.

Table 12. Recommended Stress-relieving Times and Temperatures for Magnesium Alloys

Magnesium Alloy	Form	Stress-relieving Treatment: Time	Stress-relieving Treatment: Temperature, °F
M1h *	Hard-rolled sheet	1 hour	400
AZ31h	Hard-rolled sheet	1 hour	275
AZ51h	Hard-rolled sheet	1 hour	375
A11	Annealed sheet and extrusions	15 min	500
A11	Sand castings	1 hour	500

* Alloy M1 is not subject to stress-corrosion cracking but stress relief of welded joints may be desirable to prevent distortion during machining.

SPOT WELDING. Electric resistance spot welding is an excellent method of joining wrought magnesium. Both d-c and a-c welding machines are satisfactory. Welding electrodes should have dome tips made from a hard copper alloy with high electrical conductivity. Actual machine settings can be obtained from a magnesium manufacturer, but to those already familiar with spot-welding procedures on other metals it may suffice to say that machine settings will be very similar to those used for aluminum. Before welding is begun, contacting surfaces should be thoroughly cleaned with a solvent to remove oil and grease. Next, the surface should be polished to a bright luster with a wire brush or steel wool. Stainless steel wire brushes and steel wool are recommended because stainless steel is not magnetic and, therefore, particles will not be attracted to the welding area to contaminate the surface. Electrode tips should be kept clean during welding by polishing with an emery cloth. Some alloys require that the electrode tips be cleaned after every ten or fifteen welds. This is true of pure magnesium and alloy M1. With other alloys, such as alloy AZ51, as many as one hundred and fifty welds can be made between cleanings.

16. FORMING OF MAGNESIUM ALLOYS

Most of the common methods used for converting metal sheet and extrusions into useful shapes are applicable to magnesium. Deep drawing, shallow drawing, pressing, stretch forming, spinning, and similar processes are in regular commercial use.

PRESS DRAWING is done on both hydraulic and mechanical presses although hydraulic presses usually are preferred because of better control of drawing speed. Most forming is done hot, which has advantages over cold working of magnesium. Because of its crystalline structure, the working of magnesium at elevated temperatures puts more planes of slip into play, and many intricate parts can be drawn in one operation without intermediate annealing steps and with only one set of dies being necessary. Hot working likewise results in parts made to exact size with no "spring back" when the finished piece is removed from the die. An entering radius on the draw ring five to seven times the sheet thickness gives maximum draw ability. The male die or punch is usually heated by means of gas burners or electric strip heaters inside the punch. The draw ring is also heated. Care should be taken when using a gas flame that it impinges only on the surface to be heated and not on the magnesium sheet. The magnesium sheet is usually preheated in an oven prior to forming. Shallow parts, flanges, and similar shapes not requiring deep drawing action can be rubber formed by the Guerin process of rubber forming.

DROP FORMING. The drop hammer has been successfully used as a means of forming magnesium sheet. Zinc alloys are satisfactory for drop hammer dies when the temperatures do not exceed 500 F. Lead-antimony alloy, which can be cheaply converted into a punch by casting in the zinc-alloy die, is satisfactory for working temperatures up to 400 F. As in deep drawing, both drop hammer and die should be heated. A ring burner may be used in the die.

STRETCH FORMING. The possession of a very high elongation at elevated temperatures makes magnesium ideally suited to stretch forming. Punches for hot-stretch forming are conveniently made of zinc alloy if temperatures do not go too far above 400 F. The punch can be heated by means of gas burners, and the sheet can be preheated in an oven and kept hot during forming by means of contact with the punch.

BENDING. Magnesium sheet and extrusions can be bent with the same type of tools customarily used for other metals. Bending rolls are used on sheet for such operations as the production of cylindrical tanks. Pipe-bending machines of standard design are suitable for bending magnesium tubing. Press brakes and folding-leaf-type brakes are being used for magnesium sheet. Other extrusions such as bars, rods, and shapes are bent by the same methods as those used for aluminum and other metals. When heating of the stock is necessary for making bends, the heat can be applied in any convenient manner. On small production runs simply passing a torch over the work may be adequate. On long runs, and where the equipment permits, electric strip heaters can be attached to bending brakes. The desirable working temperature for annealed magnesium sheet and extrusions is 500 to 650 F. Sheet in the hard-rolled condition should not be heated over 400 F lest annealing result. A contact pyrometer should be used to check the temperature of the material. In case such an instrument is not available, the crude but effective temperature indicators known to many metal workers can be substituted. Typical of them are: A smear of laundry soap on the metal turns dark brown at about 400 F; a mark made with blue carpenter's chalk will lose its color at approximately 600 F.

SPINNING. When the quantity of shaped magnesium parts does not justify the cost of die equipment for deep and shallow drawing, spinning can sometimes be substituted.

When heating seems desirable to improve the workability of the metal, an ordinary welding torch may be used.

17. SURFACE FINISHING

The finishing of magnesium is of two types—protective and decorative. Protective finishing includes a chemical treatment which passivates the surface, followed by the application of paints. Decorative finishing makes use of the many attractive paint finishes used for the purpose or it might be simply a chemical treatment followed by a clear lacquer to provide durability.

CLEANING. Proper surface cleaning is a very important consideration prior to either protective or decorative finishing. For this purpose solvent-type cleaners, such as vapor degreasers, or alkaline cleaners applied by soaking in hot solutions, can be used. The magnesium also may be simply dipped in solvents. The hot alkaline cleaners have the advantage that they also remove any previously applied chemical finish or old paint coatings. Solvent and alkaline cleaners function to remove oil and grease. The cleaning of the surface to remove any oxide coatings, surface contamination, and previously applied chemical finishes can be accomplished by acid pickling or mechanically by wire brushing, sanding, sand blasting, and the like.

PROTECTIVE CHEMICAL TREATMENTS. Two protective chemical treatments are most commonly used for magnesium. One is the chrome-pickle treatment which is applicable to practically all forms of magnesium in order to provide protection during shipment and storage. It also makes a satisfactory base for a subsequently applied paint coating. The treatment consists of about a one-minute dip in a solution containing 1 1/2 lb of sodium dichromate and 1 1/2 pints of concentrated nitric acid and water to make 1 gal of solution. The protective value of the finish produced in this solution can be improved by boiling parts after treatment in a water solution containing 1 1/2 lb of sodium dichromate and 3/4 oz of magnesium fluoride per gallon of solution. After treatment, parts should be rinsed in cold running water and then dipped in hot water to facilitate drying.

The other chemical treatment is known as the dichromate treatment. It is used also as a base for paint coatings. The treatment does not remove metal, as does the chrome pickle, and, therefore, can be used on parts which have been machined to close dimensions. The dichromate consists of two steps. Parts are placed for 5 min in a solution prepared by diluting one volume of 48 to 52% hydrofluoric acid with two volumes of water in a lead-lined tank. Hydrofluoric acid must be handled with care to avoid contact of the acid with the skin and to avoid inhalation of the fumes. The bath should be maintained at a temperature between 70 and 90 F during use. Upon removal from the hydrofluoric acid bath, parts should be thoroughly rinsed in cold running water and then boiled for 45 min in a dichromate bath, which is easily prepared by dissolving a pound of sodium dichromate in a gallon of water. After this, parts are thoroughly rinsed in cold running water, dipped in hot water, and allowed to dry. They are then ready for painting.

PAINTING. After the application of one of the two protective chemical treatments just described, magnesium parts are ready for painting after thorough rinsing and drying. The method of applying paints is not different from that used on other metals; in fact, the same materials can be used for most purposes. Coastal exposure or indoor locations subject to corrosive conditions require the use of inhibitive-type primer and additional top coats. Zinc yellow is the most effective pigment for use with magnesium.

The refinishing of magnesium does not require special consideration, unless the old paint has been removed by sand blasting or other methods which expose portions of the metal. Such areas of exposed metal should be chemically treated before painting is attempted. Small parts present no problem because they can be stripped of their paint and chemically treated by dipping. Objects too large to be dipped, however, require that a freshly prepared chrome-pickle solution be brushed onto areas from which the loose paint has been removed and underneath which the metal has been sanded until bright.

Assembly protection is an important consideration too, since contact between magnesium and dissimilar metals should be avoided. Such dissimilar metal contact can result in what is known as galvanic corrosion. Two or three coats of paint or plastic sealing and caulking compounds are recommended at the point of contact between magnesium and other metals.

DECORATIVE FINISHING. Decorative finishes for magnesium, as their name implies, are used to provide ornamental effects. When paints are used, and they should always be applied over the chrome-pickle or dichromate treatment, decorative finishes can provide protection as well as enhanced sales appeal. At other times, however, they may provide little or no protection and should be used only for mild service.

Mechanical Finishes. Scratch brushing, buffing, sanding, and similar mechanical methods impart a luster to magnesium surfaces prior to subsequent chemical finishing. The use of a clear lacquer directly over mechanically cleaned surfaces is not recommended for long service.

Chemical Finishes. A bright metallic luster which, when protected by clear lacquer, provides an attractive effect is produced by a chemical treatment known as bright pickle. It is applied by immersing the part in the following solution for 2 to 5 min at 70 to 80 F. It is necessary that the magnesium part be given a thorough alkaline cleaning before treatment.

Chromic acid	1 1/2 lb
Sodium nitrate	4 oz
Magnesium fluoride	1/3 oz
Water	to make one gallon

Upon removal from this solution, parts should be neutralized by being dipped momentarily into a water solution containing 6 oz per gal of sodium carbonate, after which they should be thoroughly rinsed in cold running water, dipped in hot water, and then allowed to dry. Clear lacquer should be applied immediately. Electrochemical methods produce the best decorative coating. One method is an anodic treatment which provides a coating that can be colored with organic dies. Methods of electroplating have also been worked out, and electrodeposited coatings of nickel, chromium, and other metals can be applied to magnesium.

NICKEL AND ITS ALLOYS

By R. D. Faber

Nickel is distributed widely in the earth's crust, but in only a few places are there economically workable ore bodies. Its major uses are as an alloying element in hardenable steels, stainless steels, corrosion-resistant and high-temperature alloys, cast-iron, and other alloys such as copper-nickel, "nickel silvers," and aluminum-nickel. The metal and many of its alloys are silvery-white, corrosion-resistant, malleable, ductile, and can be processed into forgings, rods, bars, tubes, plates, sheet, strip, and wire. To its more than 3000 alloys, nickel imparts strength, toughness, ductility, increased resistance to oxidation and to corrosion, and in a few cases special magnetic, magnetostrictive, and thermal-expansive characteristics.

CLASSIFICATION. Nickel and its alloys may be classified into the following groups on the basis of chemical composition (Ref. 1).

Nickel. (1) Pure nickel, electrolytic (99.95% Ni); (2) commercially pure wrought nickel (99.4% Ni); (3) nickel shot and ingot (99.6% Ni); (4) anodes (99% Ni); (5) coinage (99.4% Ni).

Nickel and Copper. (1) Low-nickel alloys (2 to 13% Ni); (2) cupro nickels (10 to 30% Ni); (3) coinage alloy (25% Ni); (4) electrical resistance alloy (45% Ni); (5) nonmagnetic alloys (up to 60% Ni); (6) high-nickel alloys, Monel * (over 60% Ni).

Nickel and Iron. (1) Wrought alloy steels (1/2 to 9% Ni); (2) cast alloy steels (1/2 to 9% Ni); (3) alloy cast irons (1 to 6 and 14 to 36% Ni); (4) magnetic alloys (20 to 90% Ni): (*a*) temperature-compensating alloys (29.5 to 32.5% Ni) and (*b*) high-permeability alloys (49 to 80% Ni); (5) nonmagnetic alloys (10 to 20% Ni); (6) clad steels (5 to 40% Ni); (7) thermal expansion alloys: (*a*) low expansion (36 to 50% Ni) and (*b*) selected expansion (22 to 50% Ni).

Nickel and Chromium. (1) Electrical resistance alloys (80% Ni); (2) electromotive alloy (90% Cr).

Nickel and Aluminum. (1) Cylinder head and piston alloys (2% Ni); (2) low expansion alloys (1 to 2.5% Ni); (3) bearing alloy (1% Ni).

Nickel, Chromium, and Iron. (1) Stainless steels (2 to 25% Ni); (2) heat-resisting alloys (12 to 85% Ni); (3) electrical resistance alloys (35 to 60% Ni).

Nickel, Copper, and Zinc. (1) "Nickel silvers" (5 to 30% Ni).

Nickel, Copper, and Tin. (1) Nickel brasses, and bronzes (1 to 15% Ni).

Other Alloys. (1) Age-hardenable alloys (0.5 to 98% Ni); (2) nickel-manganese alloys (94 to 98% Ni); (3) nickel-molybdenum alloys (55 to 80% Ni); (4) nickel-silicon alloys (95 to 98% Ni).

The following summary is concerned with commercially pure nickel, nonferrous alloys

* Registered U. S. Patent Office.

containing more than 50% nickel, and a few alloys containing less than 50% nickel, such as "nickel silvers" and cupro nickels.

Nominal chemical compositions and complete range of mechanical properties are given in Tables 1 and 2, respectively. ASTM Specifications covering certain nickel and nickel-alloy products are listed in Table 3. Physical properties of nickel, Monel, and Inconel are given in Table 4.

Table 1. Nominal Chemical Composition of Nickel Alloys, %

Material	Condition (C = Cast, W = Wrought)	Ni *	Cu	Fe	Mn	Si	Cr	C	Other Elements
Nickel	C	97	0.3	0.25	0.5	1.5		0.5	
Nickel	W	99.4	0.1	0.15	0.2	0.05		0.1	
"L" Nickel	W	99.4	0.05	0.1	0.2	0.15		0.2 max	
Duranickel	W	93.7				0.5		0.2	Al-4.4
"D" Nickel	W	95.2	0.05	0.15	4.5	0.05		0.1	
"E" Nickel	W	97.7	0.05	0.1	2.0	0.05		0.1	
Monel	C	67	29	1.5	0.9	1.25		0.3	
Monel	W	67	30	1.4	1.0	0.1		0.15	
"R" Monel	W	67	30	1.4	1.1	0.05		0.1	S-0.035
"K" Monel	W	66	29	0.9	0.75	0.5		0.15	Al-2.75, Ti-0.5
"KR" Monel		66	29	0.9	0.75	0.5		0.3	Al-2.75, Ti-0.5
"H" Monel	C	65	29.5	1.5	0.9	3		0.1	
"S" Monel	C	63	30	2	0.9	4		0.1	
Constantan	W	43	55	0.25	0.5–1.0				
Cupronickel, 70-30	W	30	70						
Nickel Silvers									
20% Nickel	C	20	65						Zn-6, Pb-5, Sn-4
13% Nickel	C	12.5	55						Zn-20, Sn-2, Pb-10.5
18% Nickel	W	18	64						18-Zn
10% Nickel	W	10	65						25-Zn
80-20 Nickel-chromium	W	76–78		0.5	0.2–2.0	1.25–1.4	20	0.1	
Inconel	C	77.5	0.25	6	0.8	2	13.5	0.2	
Inconel	W	77	0.2	7	0.25	0.25	15	0.08	
Inconel "X"	W	70	0.05	9	0.5	0.4	14–16	0.04	Al-0.9, Cb-1, Ti-2.5
60-15 Nickel-chromium-iron	C and W	68–60		Remainder	0.2–1.35	1.0–1.4	18–12	0.5–0.06	
78-22 Nickel-iron	W	78–80				0.15–0.5		0.1	Cu, Cr, or Mo
50-50 Nickel-iron	W	50		49	0.1	0.05		0.05	
42-58 Nickel-iron	W	42		58	0.1	0.05		0.05	
36-64 Nickel-iron	W	36		64	0.1	0.05		0.05	
Ni Span C	W	42		Remainder	0.4	0.5	5.5	0.06 max	Ti-2.4, Al-0.5
Hastelloy A	C and W	57		20	2.0	1		0.1	Mo-20
Hastelloy B	C and W	62		5	1.0	1		0.1	Mo-30
Hastelloy C	C and W	58		5	1.0	1	14	0.1	Mo-17, W-5
Hastelloy D	C	85	3		1.0	10			Al-1
Illium R	W	60	3	8	1.2	0.65	21	0.1	Mo-5, W-1
Illium G	C	58	6	6	1.2	0.6	22	0.2	Mo-6

* Includes a small amount of cobalt.

18. NICKEL

Nickel is used in cast as well as wrought forms. Mechanical properties of nickel in various forms are listed in Table 5. Commercial, wrought nickel has a nominal nickel content of 99.4%, this includes a small amount of cobalt, less than 1%, which is counted as nickel because its effect upon the significant properties of nickel is slight and not detrimental (Ref. 2). The mechanical properties (Table 5), resistance to corrosion, and thermal and electrical conductivities of nickel are good. Its modulus of elasticity and thermal expansion are comparable to those of steel.

Table 2. Mechanical Properties of Nickel Alloys

Material	Condition (C = Cast, W = Wrought)	Yield Strength, 0.2% offset 10^3 psi	Tensile Strength, 10^3 psi	Elongation in 2 in., %	Reduction in Area, %	Hardness	
						Brinell, 3000 kg	Rockwell
Nickel	C	20– 30	45– 60	30–15		80–125	40–70B
Nickel	W	10–155	50–165	65– 2	75–60	90–230	40–100B
"L" Nickel	W	15– 25	55– 75	55–35	75–50		25–55B
Duranickel	W	30–180	90–240	50– 2	65–15		75B–46C
"D" Nickel	W	35–125	75–140	50– 2	75–60	140–230	
"E" Nickel	W	35– 80	75–100	40–25	75–60	140–190	
Monel	C	32– 40	65– 90	45–25		125–150	
Monel	W	25–160	70–170	50– 2	75–50	110–250	60B–23C
"R" Monel	W	25–130	70–140	50– 4	70–50	110–230	60–100B
"K" Monel	W	40–175	90–200	45– 2	70–25	140–320	
"KR" Monel	W	40–175	90–200	45– 2	70–25	140–320	
"H" Monel	C	45– 75	90–115	20–10		175–250	
"S" Monel	C	70–115	90–145	4– 1		275–375	
Constantan	W	30–120	60–135	45– 0.5	70–25		50–93B
Cupronickel 70-30	W	20– 70	40– 80	55– 2	75–50		50–85B
Nickel silvers							
20% Nickel	C	23– 28	42– 48	16–12	18–14	90–100	
13% Nickel	C	20	35	15		15	
18% Nickel	W	25– 80	60–105	40– 3		70–200	
10% Nickel	W	20– 80	55–110	45– 4			
80-20 Nickel-chromium	W	50–100	100–190	35– 0.5	70–40	150–320	85B–30C
Inconel	C	30– 45	70– 95	30–10		160–190	
Inconel	W	25–175	80–185	50– 2	70–40	120–290	65B–35C
Inconel "X"	W	30–250	105–300	40– 2		140–480	77B–50C
60-15 Nickel-chromium	C	40– 60	60– 70	5– 2		170–210	
60-15 Nickel-chromium	W	40–160	80–175	35– 0.5	70–40	180–235	
78-22 Nickel-iron	W	25–130	80–140	60– 0.5	60–30		85–103B
50-50 Nickel-iron	W	25–140	70–150	35– 0.5	60–30		80–105B
Ni Span "C"	W	35–180	90–200	40– 6		145–395	78B–42C
Hastelloy "A"	C	43– 45	69– 78	12– 8		155–200	
Hastelloy "A"	W	47– 52	110–120	48–40	54–40	200–215	
Hastelloy "B"	C	55– 57	75– 82	9– 6		190–230	
Hastelloy "B"	W	60– 65	130–140	45–40	45–40	210–235	
Hastelloy "C"	C	45– 48	72– 80	15–10		175–215	
Hastelloy "C"	W	55– 65	115–128	50–25		160–210	
Hastelloy "D"	C	36– 41	36– 41	0			50–55C
Illium "R"	W	50–100	100–150	32–12		190–365	
Illium "G"	C	50– 64	60– 73	9– 4		160–210	

Table 3. ASTM Specifications * Covering Nickel and Nonferrous Nickel Alloys

Alloy	Rods, Bars	Cold-drawn Pipe, Tubing	Plate, Sheet Strip	Seamless Condenser Tubes, Ferrule Stock	Wire	Drawn Rolled
Nickel	B160	B161	B162	B163	B175	
Nickel-copper (70-30)	B164	B165	B127	B163		
Nickel-chromium (80-20)						B82
Nickel-chromium-iron (78-15-7)	B166	B167	B168	B163		
Nickel-chromium-iron (60-16-24)						B83
Copper-nickel (70-30)			{B122, B171}	B111		
Copper-nickel-zinc	B151		B122	B111	B206	

* See ASTM Standards for latest date of issue.

Table 4. Typical Physical Properties of Nickel, Monel,* and Inconel *

	"A" Nickel	Monel	Inconel
Density at 68 F, g/cc,	8.88	8.84	8.51
lb/cu in.	0.321	0.319	0.307
Liquidus temperature, °F	2635	2460	2600
Solidus temperature, °F	2615	2370	2540
Specific heat (32–212 F), cal/g	0.13	0.127 †	0.109 ‡
Latent heat of fusion, cal/g	73	68	
Thermal conductivity (32–212 F), cal/sq cm/cm/°C/sec,	0.16	0.06	0.038
Btu/sq ft/hr/°F/in.	464	174	110
Electrical resistivity at 68 F, microhm/cm,	10	48.2	100
ohms/cir mil ft	60	290	600
Temperature coefficient of electrical resistivity (68–212 F), per °F	0.0023	0.0011	0.00007 §
Initial permeability	100–200		
Maximum permeability	500–2000		1.005 ‖
Coercive force, oersteds	3–5 ¶		
Curie temperature, °F	680 **	110 to 140	Below −40
Saturation magnetization, gausses at 68 F	6100	1000 to 2000 ††	
Hot-working temperature range, °F	1200–2300	1200 to 2150	1600–2300
Annealing temperature, °F	1200–1400	1400 to 1800	1600–1900
Casting or pouring temperature range, °F	2750–2900	2700 to 2850	2800–2900
Tension modulus of elasticity, psi	30×10^6	26×10^6	31×10^6
Torsional modulus of elasticity, psi	11×10^6	9.5×10^6	11×10^6

* Registered U.S. Patent Office.
† Specific heat at 2300 F, cal/g.
‡ Specific heat at (77–212 F), cal/g.
§ Temperature range (68–932 F).
‖ Permeability of vacuum, 1.000.
¶ Annealed condition.
** Approximate.
†† At 200 oersteds.

Nickel has many uses in the chemical, soap, caustic, and allied industries for the construction of evaporators, tanks, jacketed kettles, heating coils, tubular condensers, and various other processing equipment. Nickel is noteworthy for its resistance to hot concentrated caustic soda, chlorine, and fluorine gases at temperatures up to 1000 F. It is useful for magnetostriction devices and has various applications in the incandescent lamp and radio industries. Nickel is also available in cast form and in the form of clad steel.

"L"* Nickel is wrought "carbon-free" nickel. It is softer than regular nickel, does not work-harden as rapidly, and finds special applications as base metal for high-quality rolled gold plate in the jewelry and optical industry, as containers for fused oxidizing salts, difficult deep-drawing and spinning applications, and crucibles and combustion boats for analytical laboratory use.

Duranickel * is a wrought, age-hardenable nickel alloy possessing high strength and hardness in addition to the resistance to corrosion characteristic of nickel. It is used for springs, diaphragms, radio antennae, extrusion dies in the plastic and textile fields, pump parts, and many other special applications requiring corrosion resistance together with high mechanical properties.

"E" * Nickel and "D" * Nickel contain 2.0 and 4.5% manganese, respectively; the manganese increases resistance to oxidation in sulfidizing atmospheres. "D" Nickel is used for ignition tubes, carbon arc electrode holders, and radio tube grids. "E" Nickel is used for furnace lead-in wires and support wires in incandescent lamps.

ELECTROPLATING. Approximately 10% of the total annual production of nickel is consumed by electroplating processes. Nickel can be electrodeposited to develop mechanical properties of the same order as wrought nickel; however, special plating baths are available which will yield nickel deposits possessing a hardness as high as 450 Vickers (425 BHN). Other types of plating baths in commercial use will yield bright, rather hard deposits which may require no further buffing. Adherent bonds, approaching the strength of the metal, can be developed between the nickel plate and ordinary steel, and very strong bonds may be secured to copper, brass, aluminum, zinc, and numerous other metals and alloys, including the stainless steels and Inconel. Nickel electroplate may be used to facilitate or enhance the silver brazing of various chromium-content alloys.

Very heavy nickel electrodeposits are employed for building up worn or overmachined parts. These parts can then be finished with a chromium plate or special hard nickel plate where high surface hardness is required. Special nickel shapes, difficult or virtually impossible to produce otherwise, are electroformed by depositing a heavy layer of nickel upon a suitable conducting matrix, which is subsequently removed. Electroformed nickel

* Registered U. S. Patent Office.

Table 5. Nominal Mechanical Properties of Nickel

Form and Condition	Tensile Properties			Hardness	
	Tensile Strength, 1000 psi	Yield Strength (0.20% offset), 1000 psi	Elongation in 2 in., %	Brinell, 3000 kg	Rockwell, scale as shown
Rod and bar:					
Annealed	55– 80	15– 30	60–40	90–120	45–68B
Hot rolled, as rolled	55– 85	15– 40	55–35	90–130	45–72B
Forged	60– 90	20– 60	55–30	100–170	54–86B
Cold drawn, as drawn	80–115	60– 90	35–15	130–200	72–93B
Wire, cold drawn:					
Annealed	50– 80	10– 30	50–30		
No. 1 temper	90–105	75– 90	20– 5		
Regular temper	105–135	85–130	15– 4		
Spring temper	125–165	125–155	5– 2		
Plate, hot rolled:					
Annealed	55– 80	15– 30	50–35	90–120	45–68B
Hot rolled, as rolled	60–100	20– 75	45–30	100–200	54–92B
Sheet: *					
Standard cold rolled					
Soft temper	55– 80	15– 30	55–35		55–70B
Hard temper	90–115	70–105	15– 2		92B min
Special cold rolled					
Soft, spinning, and deep drawing	55– 80	15– 25	55–35		64B max
1/4 Hard temper	60– 80	30– 50	45–30		71–79B
1/2 Hard temper	65– 85	50– 65	40–25		80–85B
3/4 Hard temper	70– 90	55– 70	25–15		86–91B
Strip: *					
Standard cold rolled					
Soft temper	55– 80	15– 30	55–35		64B max
Spring temper	90–130	70–115	15– 2		96B min
Special cold rolled					
Soft, spinning, and deep drawing	55– 80	15– 25	55–35		64B max
Skin hard temper	60– 80	20– 35	50–35		65–70B
1/4 Hard temper	65– 85	30– 50	45–30		71–79B
1/2 Hard temper	70– 90	50– 65	40–25		80–85B
3/4 Hard temper	75– 95	55– 70	25–15		86–91B
Hard temper	90–115	70–105	15– 2		92–95B
Tubing:					
Soft	55– 80	15– 30	60–40		40–60B
As drawn	70– 95	50– 80	35–15		80–100B

* Sheet and strip products are available on the basis of hardness for the various tempers shown.

dies for pressing in the sound tracks on phonograph records illustrate the fine detail which can be reproduced by the nickel electroforming process.

The most extensive use of nickel plate is for corrosion protection of iron and steel parts and zinc-base die castings used in the automotive field. For these applications, a layer of nickel, 0.0015 to 0.003 in. thick is used. This nickel plate is then finished or covered with a chromium plate consisting in thickness of about 1% of the underlying nickel plate thickness in order to maintain a brilliant, tarnish-free, hard exterior surface.

19. ALLOYS OF NICKEL

NICKEL-COPPER ALLOYS. **Monel** * is the most important nickel-copper alloy. It contains about two-thirds nickel and one-third copper and has high strength, excellent ductility, and good corrosion resistance to flowing salt water, dilute mineral acids, hydrofluoric, hydrochloric, sulfuric, phosphoric, and most organic acids and also strong caustic soda. Cold-worked, welded, and cast forms behave substantially the same in the above-

* Registered U. S. Patent Office.

Table 6. Nominal Mechanical Properties of Monel

Form and Condition	Tensile Properties				Hardness		Impact Strength, ft-lb	
	Tensile Strength, 1000 psi	Yield Strength (0.20% offset), 1000 psi	Elongation in 2 in., %	Reduction of Area, %	Brinell 3000 kg	Rockwell, scale as shown	Izod	Charpy U-notch
Rod and bar								
Annealed	70– 85	25– 40	50–35	75–60	110–140	60–75B	120– 90	215
Hot rolled, as rolled	80– 95	40– 65	45–30	75–60	140–185	75–90B	120–100	220
Forged	75–110	40– 85	40–20	70–60	130–220	75–98B	115– 75	
Cold drawn, as drawn	85–125	55–120	35–15	70–50	160–250	85B–23C	115– 75	150
Cold drawn, No. 1 temper (under 1/2 in. diam.)	95–110	50– 85	20– 5					
Wire, cold drawn								
Annealed	70– 95	25– 40	50–30					
No. 1 temper	95–110	80–100	20– 5					
Regular temper	110–140	100–130	15– 4					
Spring temper (under 5/16 in. diam.)	140–170	130–160	10– 2					
Plate, hot rolled								
Annealed	70– 85	30– 45	50–30		108–140	60–75B		
As rolled	75–110	40– 90	45–20		150–228	80–98B		
Sheet *								
Standard cold rolled								
Soft temper	70– 85	25– 45	50–30			61–73B		
Hard temper	100–120	90–110	15– 2			94B min		
Special cold rolled								
"35" Monel (0.021 to 0.093 in.)	78– 85	45– 65	40–20			74–89B		
"35" Monel (over 0.093 and under 0.021 in.)	70– 85	25– 45	50–30			61–73B		
Soft, spinning, and deep drawing	70– 85	25– 40	50–30			68B max		
1/4 Hard temper	70– 85	35– 60	45–25			74–82B		
1/2 Hard temper	75– 90	50– 65	30–20			83–89B		
3/4 Hard temper	85–100	70– 85	20–10			90–93B		
Strip *								
Standard cold rolled								
Soft temper	70– 85	25– 45	50–30			68B max		
Spring temper	100–140	90–130	10– 2			98B max		
Special cold rolled								
Soft, spinning, and deep drawing	70– 85	25– 40	50–30			68B max		
Skin hard temper	70– 85	30– 45	50–30			69–73B		
1/4 Hard temper	70– 85	35– 60	45–25			74–82B		
1/2 Hard temper	75– 90	50– 65	30–20			83–89B		
3/4 Hard temper	85–100	70– 85	20–10			90–93B		
Hard temper	100–120	90–110	15– 2			94–97B		
Tubing								
Soft	70– 85	25– 45	50–35			60–73B		
As drawn	85–125	60–120	30–10			85B–24C		

* Sheet and strip products are available on the basis of hardness for the various tempers shown.

mentioned corrosive media. The alloy is not resistant to corrosion by strongly oxidizing solutions such as nitric acid and ferric chloride. Monel is available in all wrought and cast forms and is used widely in the chemical and marine industries. It is also available in the form of Monel-clad steel plates. Mechanical properties of Monel in various forms are listed in Table 6.

"R" * Monel contains approximately 0.035% of sulfur, which is added to secure improved machinability, particularly in automatic screw-machine operations. It is produced only in wrought forms.

* Registered U. S. Patent Office.

"K"* Monel is an age-hardenable wrought alloy which has mechanical properties comparable to those of heat-treated alloy steel of the same order of hardness, and resistance to corrosion similar to that of Monel. It is nonmagnetic down to −150 F and is used for aircraft parts, turbo-pump shafts, marine-equipment parts, steam and hydraulic valve disks and seats, sleeve bearings, diaphragms, and spring parts.

"KR"* Monel is a wrought alloy which contains a higher carbon content than "K" Monel. The higher carbon results in improved machinability.

"H"* Monel and **"S"* Monel** are two cast age-hardenable alloys containing 3 and 4% silicon, respectively. They possess the general corrosion resistance of this copper-nickel group of alloys together with improved resistance to galling and erosion for applications which must withstand severe abrasion as well as corrosion. Such applications are valve seats and plugs, bushings, pump liners, and sleeve bearings, and other sliding or moving elements.

Constantan and the cupro nickels and their modifications, marketed under various trade names (e.g., Advance, Copel, Cupron, Excelsior, Jelliff Alloy 45, Chase Cupro-nickel 70-30, Cupro-nickel Alloy No. 510, Super Nickel-701), are best known for their electrical-resistance and corrosion-resisting characteristics. The 70-30 cupro-nickel alloy, and to a lesser extent the 80-20 alloy, resists impingement corrosion in sea-water condensers. Their applications include marine-turbine condensers and power-plant condensers, feed-water heaters, salt-water lines, and salt evaporators in the chemical industry. The 75-25 cupro nickel is extensively used as a coinage alloy. Constantan, containing 55% copper, possesses a coefficient of electrical resistivity which is near zero and has a high thermoelectric potential against copper, iron, and nickel-chromium alloys. It is used principally for electrical resistors, thermocouples, and thermocouple lead-in wires.

"NICKEL-SILVER" ALLOYS, formerly called German silvers, are ternary copper-nickel-zinc alloys which contain no silver. More than 200 compositions are known by a wide variety of trade names (e.g. Adnic, Ambrac 850). They contain copper 52 to 80%, nickel 5 to 30%, and zinc 5 to 35%. Nickel imparts a silver-white color to these copper-base alloys and improves their strength and resistance to corrosion. The whiter alloys of this group contain about 18% or more of nickel; other alloys contain approximately 12% of nickel.

The nickel silvers are used largely for silver-plated tableware, jewelry, springs, key stock, optical frames, watch parts, marine fixtures, plumbing fixtures, and architectural trim. In some cases they are employed as electrical resistors.

NICKEL-SILICON ALLOYS. The best-known commercial alloy in this group is Hastelloy D.* It contains approximately 10% silicon and is available only in the cast condition. This alloy is strong, tough, and extremely hard; it is not recommended for any machining operations and must be finished by grinding. Its outstanding characteristic is excellent corrosion resistance to hot and cold sulfuric acid of all concentrations; it also possesses good corrosion resistance to other acids such as acetic, formic, and phosphoric.

NICKEL-CHROMIUM ALLOYS. The principal material in this group is an 80-20 nickel-chromium alloy, known by various trade names (e.g., Alray A, Chromel A, Nichrome V, Tophet A, Evanohm), which was developed to give maximum life as an electrical heating element at temperatures up to 2100 F and for electrical resistors at moderate temperatures. The 10% chromium alloy, Chromel P,* is widely used as one component of the Chromel-alumel thermocouple.

NICKEL-IRON ALLOYS. Nickel and iron are completely soluble in the liquid or solid states. Alloys of these elements are ferritic or austenitic, depending principally upon their nickel content, and are known and used chiefly for their magnetic, thermal-expansion, and thermoelastic characteristics.

The alloys containing 70 to 80% nickel (e.g., 4-79 Moly Permalloy, Hy Mu 80, Mumetal) possess extremely high initial permeability (about 20,000), high maximum permeability, and low hysteresis loss when annealed in exceptionally pure dry hydrogen at about 2060 F. followed by slow cooling. Saturation magnetization for this alloy is about 8500 gauss, They are used for cores in loading coils, filters, magnetic shields, pulse and high-frequency transformers where high initial permeability is required. The 50-50 nickel-iron alloys (e.g., 45 Permalloy, 52 Alloy, 4750 Metal, High Permeability 49, Hipernik, Nicaloi) have quite high permeability and can be used at higher flux densities than the 70 to 80% nickel-iron alloys. Because their saturation value is about 15,500 gauss they are employed usually as cores in audio and current transformers, and magnetic amplifiers.

The alloys with 36 to 50% nickel (e.g., Bethlehem No. 6, Gamma Nickel Steel, Invar 36, Nilvar) comprise those special-purpose alloys with low or controlled thermal expansion and thermoelastic characteristics. In this group are the Invars (36% Ni) and the alloys containing 42% Ni (e.g., Allegheny 42, Bethlehem No. 7, Low Expansion 42). Modifica-

* Registered U. S. Patent Office.

Table 7. Nominal Mechanical Properties of Inconel

Form and Condition	Tensile Properties				Hardness		Impact Strength, ft-lb	
	Tensile Strength, 1000 psi	Yield Strength (0.20% offset), 1000 psi	Elongation in 2 in., %	Reduction of Area, %	Brinell, 3000 kg	Rockwell, scale as shown	Izod	Charpy U-notch
Rod and bar								
Cold drawn								
Annealed	80–100	25–50	55–35	70–60	120–170	65–85B	120+	230
As drawn	95–150	70–125	30–10	60–40	180–290	8–30C	100–70	151
Hot rolled								
Annealed	80–100	25–50	55–35	70–60	120–170	65–85B	120+	230
As rolled	85–120	35–90	50–30	65–50	140–210	75–95B	120–100	191
Forged	85–120	35–90	45–25	65–50	140–210	75–95B	120–100	
Wire, cold drawn								
Annealed	80–115	25–50	50–25					
Regular temper	130–165	115–165	12–3					
Spring	165–185	150–175	10–2					
Plate, hot rolled								
Annealed	80–110	30–60	50–35		120–170	65–85B		
As rolled	90–140	45–95	40–20		160–240	85–100B		
Sheet *								
Standard cold rolled								
Soft	80–100	30–45	55–35			76–88B		
Hard	125–150	90–125	15–2			21–26C		
Special cold rolled								
Soft, spinning, and deep drawing	80–100	25–45	55–35			80B max		
1/4 Hard	100–120	50–70	50–30			89–93B		
1/2 Hard	105–130	65–80	30–15			94–97B		
3/4 Hard	110–135	80–100	20–10			98–101		
Strip *								
Standard cold rolled								
Soft	80–100	30–45	55–35			84B max		
Spring temper	140–170	100–160	10–2			30C min		
Special cold rolled								
Soft, spinning, and deep drawing	80–100	25–45	55–35			80B max		
Skin hard	85–105	35–55	55–35			85–88B		
1/4 Hard	100–120	50–70	50–30			89–93B		
1/2 Hard	105–130	65–80	30–15			94–97B		
3/4 Hard	110–135	80–100	20–10			98–101B		
Hard	125–150	90–125	15–2			25–29C		
Tubing								
Cold drawn								
Annealed	80–100	30–50	55–35			65–85B		
As drawn	110–160	70–140	25–2			8–35C		

* Sheet and strip products are available on the basis of hardness for the various tempers shown.

tions of the latter, containing about 5% chromium, are used to seal to soft glass. In hard glass, Kovar * (about 29% Ni, 17 Co, remainder Fe) is used (e.g., 14 Alloy, 426 Alloy, Alloy No. 4, HC4 Alloy, Kovar, Fernico). Two of the alloys in this group are age hardenable, and their thermoelastic characteristics can be controlled closely by thermal treatment; Ni Span C,* which has a constant modulus of elasticity over a moderate range of temperature (0 to 200 F) is used for hair springs in chronometers and watches.

The magnetic temperature-compensating alloys (e.g., Carpenter Temperature Compensators 30, Super Nilvar) contain 29 to 33% nickel and are used as magnetic shunts in automotive voltage regulators, aircraft tachometers, and watthour meters.

All the nickel-iron alloys may contain one or more additional elements, such as chromium, copper, molybdenum, and cobalt in order to enhance or modify the aforementioned characteristics.

* Registered U. S. Patent Office.

NICKEL-CHROMIUM-IRON ALLOYS. The many alloys in this group contain 78.5 to 60.0% nickel, 13.0 to 18.0% chromium, and the remainder principally iron.

Inconel * contains 77% nickel and 15% chromium, has high mechanical properties, excellent resistance to many corrosive media and to oxidation at temperatures up to 2200 F, and is used extensively both in oxidizing and carburizing atmospheres at elevated temperatures. Mechanical properties are listed in Table 7. An age-hardenable modification, Inconel "X",* is suitable for gas-turbine, supercharger, jet-propulsion, and other applications in the temperature range of 1200 to 1500 F. This precipitation-hardening alloy will maintain most of its room temperature strength properties up to 1000 F.

Alloys containing essentially 60 to 65% nickel and 15 to 20% chromium are available under a variety of trade names such as Nichrome,* Chromel C,* Tophet C,* and others; † they are used chiefly for heat-resisting purposes.

NICKEL-MOLYBDENUM-IRON ALLOYS. Hastelloy A * and Hastelloy B,* available in both wrought and cast forms, possess high mechanical properties and resistance to corrosion. Hastelloy A,* and others ‡ containing approximately 20% molybdenum, resists hydrochloric acid at moderate temperatures, and dilute sulfuric acid. Hastelloy B,* and others § containing about 30% molybdenum, has excellent resistance to boiling hydrochloric acid and wet hydrogen chloride gas, and resists boiling concentrated phosphoric acid. Optimum corrosion resistance is developed by a high-temperature solution anneal (2125 F), followed by quenching. The alloys are used in chemical and allied industries for equipment to handle, transport, and store acids and other corrosive materials.

NICKEL-CHROMIUM-MOLYBDENUM-IRON ALLOYS have outstanding resistance to strong oxidizing acids and salts. Hastelloy C is used to resist corrosion by dilute nitric acid up to 150 F and by cold solutions of nitric acid containing various amounts of ferric and cupric salts, hydrochloric, and/or sulfuric acids. Illium R and Illium G resist both sulfuric and nitric acids over a wide range of concentration and temperature conditions.

20. FABRICATION OF NICKEL ALLOYS

The working methods for the nickel alloys are determined by the chemical composition, mechanical properties, and physical constants of the materials, particularly strength, hardness, toughness, and thermal conductivity.

The high-nickel materials are stronger, harder, tougher, and work-harden more rapidly than low-carbon steel, copper, brass, aluminum, and aluminum-base alloys. In general, fabrication methods are more similar to those used for steel than for the softer nonferrous materials. Heavier equipment is required for forging, deep-drawing, and spinning, and the materials must be annealed more frequently during cold-working operations.

In hot-working operations, the high melting points of the nickel materials require temperatures that are practically identical with those used for steel; in addition, heating atmospheres which are sulfur-free are required because of the embrittling effect of this element upon nickel. Annealing temperatures for all the high-nickel materials except pure nickel are considerably higher than those required for the common steels. Bright annealing of the chromium-free alloys is easier than with steel.

The procedure for age-hardening is essentially the opposite of that for steel; the four nickel age-hardenable alloys are soft when quenched from high temperatures and are hard when held at and cooled from intermediate temperatures. The other commercial high-nickel alloys have the same hardness whether quenched or slowly cooled.

The different physical and mechanical characteristics of the nickel alloys require slight modifications for welding procedures from those employed for steel. In general, the welding of the nickel alloys is characterized, in contrast to the welding of steel, by less penetration and the need for attention to cleanliness and more skill in manipulation. Resistance welding requires less current because of the high electrical resistivity which is characteristic of most high-nickel alloys, particularly those containing chromium, molybdenum, aluminum, and titanium.

The high-nickel materials are not difficult to machine, although the combinations of toolings, speeds, feeds, and cutting compounds required are different from those used with steel or brass. The methods for deep-drawing, spinning, shearing, perforating, and

---

* Registered U. S. Patent Office.

† For example, Amsco F5, Amsco F6, B and W's 700 and 701, BTG, Calite N, Empire 60–20, Fahralloy's No. 5 and 6, Fahrite, Fernite 1, Firarmor's A and B, Misco HN, NIC, Nichroloy, Par 8, Premier C, Q Alloy A and B, SSS, Standard Alloy HR4, Sweetalloy, Thermalloy A, Wehralloy No. 9, Westeeco No. 5.

‡ Chlorimet 3, Chemalloy H1.

§ Chlorimet 2, Chemalloy H2.

tube and pipe fabrication require little modification of the standard technique for steel. Pickling is quite different because the high-nickel materials are strongly resistant to many acids, particularly those used for steel and the high-copper alloys. Further fabrication details will be found in Refs. 3, 4, and 5.

21. LOW- AND HIGH-TEMPERATURE PROPERTIES OF NICKEL ALLOYS

At subzero temperatures, the nickel alloys have strength and hardness without sacrifice of ductility and toughness. Current research in the field has developed the low-temperature mechanical values for the high-nickel alloys shown in Table 8.

Table 8. Mechanical Properties at Low Temperatures

Material	Conditions	Temp., °F	Yield Strength (0.2% Offset), psi	Tensile Strength, psi	Elongation in 2 in., %	Reduction of Area, %	Charpy Impact Strength, ft-lb
Monel	Forged	70	67,000	92,000	31.0	72.7	216
		−297	91,500	128,250	44.5	71.8	216
		−423	96,400	142,000	38.5	61.0	......
	Annealed	70	31,300	78,650	51.5	75.0	216
		−297	49,500	115,250	49.5	73.9	212
Nickel	Hot rolled	70	24,600	65,600	....	....	195
		−112	27,500	76,400	....	....	236
		−292	28,000	98,000	....	....	......
		−310	....	103,000	51.0	....	227
Inconel	Hot rolled	70	....	87,000	42.5	71.9	213
		−315	....	116,750	51.0	72.7	169
	Cold drawn, 50% reduction	70	....	145,000	10.0	35.1	69
		−315	....	182,000	10.0	49.5	61
"K" Monel	Cold drawn, age harden	70	125,900	157,300	15.5	37.4	......
		−110	134,600	171,550	17.3	41.1	......

Table 9. Mechanical Properties * at Elevated Temperatures

Temperature, °F	Room	600	700	800	900	1000	1100	1200	1400	1600	1800
MONEL, HOT ROLLED, 10^3 PSI											
Yield strength (0.2% offset)	30	21.5	21.5	21	21	20	17.5	14.5	11	6.5	2.5
Tensile strength, psi	79	75	70	63.5	55	45.5	36	26.5	17.5	9	5
Elongation in 2 in., %	48	50	51	50	37	26	28	36	44	52	60
Tensile modulus, 10^6 psi	26	25	23.5	21	18.5	16	14.5	13	10.5	8	5
Creep Strength, psi											
Stress to produce a creep rate of											
0.1% in 10,000 hr (0.1 Cru)	....	....	30	14.5	4	....	....	....	....	.....	......
1.0% in 10,000 hr (1 Cru)			40	25.5	13.5	3	....	....	....	.....	......
NICKEL, HOT ROLLED, 10^3 PSI											
Yield strength (0.2% offset)	21	19.5	18	16.5	15.5	13.5	12	10	7.5	.....	......
Tensile strength, psi	63	61.5	54	43.5	37	31.5	26.5	21.5	14	.....	......
Elongation in 2 in., %	51	54	61	64	66	69	72	76	89	.....	......
Tensile modulus, 10^6 psi	30	29	27.5	26	24	22	21	18	15.5	.....	......
Creep Strength, psi											
Stress to produce a creep rate of											
0.1% in 10,000 hr (0.1 Cru)	....	23.5	8.8	3.7	....	....	....	....	....	.....	......
1.0% in 10,000 hr (1 Cru)	....	52.0	19.0	7.8	3.6	3.1	2.6	2.4	....	.....	......
INCONEL, HOT ROLLED, 10^3 PSI											
Yield strength (0.2% offset)	36.5	31	30	29.5	29	28.5	27.5	26.5	17	9	4
Tensile strength, psi	90.5	90.5	90	88.5	86.5	84	79	65	27.5	15	7.5
Elongation in 2 in., %	47	46	48	49	49	47	41	39	46	80	118
Tensile modulus, 10^6 psi	31	29.5	28.5	28	27	25	22.5	20	17	14	11.5
Creep Strength, psi											
Stress to produce a creep rate of											
0.1% in 10,000 hr (0.1 Cru)	....	....	....	46	30	14.5	7.2	5.5	1.7	0.46	0.32
1.0% in 10,000 hr (1 Cru)	....	....	....	57	42	25	12	8.3	3.5	0.64	0.50

* Values in this table are not to be considered ultimate design values. These data may be considered a basis for good design practice where the safety factors accepted today are applied.

High-temperature properties of materials have become of extreme importance during the recent years of inventive development and design. Table 9 gives the short-time tensile as well as creep data for the nickel alloys shown. Although some engineers still use short-time tensile data, such data are now considered unreliable design criteria for applications at elevated temperature; design values based on long-time creep tests are to be preferred.

REFERENCES

1. *The Romance of Nickel*, The International Nickel Co., 1947.
2. W. A. Wissler, Cobalt alloys, *Corrosion Handbook* (H. H. Uhlig, Editor), p. 57, John Wiley and Sons, New York, 1948.
3. N. E. Woldman, Nickel and high-nickel alloys, *Materials and Methods*, Vol. 24, No. 6, 1946, pp. 1484–1490.
4. W. A. Mudge, The fabrication and treatment of nickel and high-nickel alloys, *Trans. Can. Inst. Min. Met.*, Vol. 46, 1943, pp. 506–534.
5. The working of high-nickel alloys, *American Machinist*, Vol. 86, No. 17, 1942, pp. 885–904.

ZINC AND ITS ALLOYS

By E. A. Anderson

Zinc is produced commercially as slabs, anodes, strip, sheet, wire, powder, and rod. It is consumed in the form of stampings, die castings, and other fabricated shapes. It is also used in the form of anodes for electrogalvanizing and as coatings on steel.

22. SLAB ZINC AND ITS USES

ASTM Standard Specification B6-46 for slab zinc (Spelter) gives the values of Table 1 for the various grades of slab zinc.

Table 1. Standard Specification of Slab Zinc

Grade	Maximum Allowable Percentages of Lead	Cadmium	Iron	Aluminum	Lead + Cadmium + Iron Not to Exceed
1a Special High Grade	0.006	0.004	0.005	None	0.010
1 High Grade	0.07	0.07	0.02	None	0.10
2 Intermediate	0.20	0.50	0.03	None	0.50
3 Brass Special	0.60	0.50	0.03	None	1.0
4 Selected	0.80	0.75	0.04	None	1.25
5 Prime Western	1.60	.....	0.08	.....	.....

GALVANIZING is the process of coating iron and steel with zinc as a protection against corrosion. The product is known as galvanized or zinc-coated iron or steel. Zinc coatings are desirable not only because the zinc corrodes more slowly than iron but also because at cut edges or similar discontinuities in the coating the zinc, being anodic to the iron, affords galvanic protection and the iron is not corroded until the zinc has been removed completely for some distance by corrosion. The distance over which zinc will protect iron galvanically depends on the conductivity of the moisture in contact with the couple and is greatest in moisture having the highest conductivity.

The overall protection afforded by zinc depends largely on the thickness of the coating. For protection indoors, a few tenths of an ounce per square foot of surface is sufficient. For outdoor service, a coating of at least $^3/_4$ oz per sq ft of surface and preferably 1 oz or more is necessary.

The less pure grades of zinc commonly are used for galvanizing. Unless special coating procedures are used, however, the coating so produced will not withstand much deformation without cracking and flaking. High Grade and Special High Grade zinc, nearly free from cadmium, are used for galvanizing bridge wire, telegraph wire, and other products that are subject to major deformation such as twisting. Regardless of the method of application, the life of zinc coatings has been shown to be directly proportional to the thickness or weight of the coating. The proportion of zinc to zinc-iron alloys in the coatings has no demonstrable effect.

Specifications and Tests for Galvanized Coatings. The amount of zinc on galvanized articles is usually expressed in terms of ounces per square foot of surface excepting for sheet, where the figures are given as ounces per square foot of sheet, that is per 2 sq ft of surface. One ounce per square foot of surface corresponds to a thickness of 0.00174 in. For details of methods of determining the weight of coating, see Standard Methods for Determining Weights of Coating on Zinc-coated Articles, ASTM Specification A90-39.

Specifications for most commercial products covering weight of coating and bend tests, when required, are available in ferrous metals, ASTM Standards, 1946, Part 1A.

Ordinary coatings on sheets range from 0.75 to 2.75 oz per sq ft of sheet. The weight of zinc per single covered surface, as for wire, for example, would be one-half this amount. Zinc coatings on pipe, structural shapes, hardware, etc., generally range between 1 and 4 oz per sq ft of covered surface. The weight of coating on wire varies with the gage of the wire and the process of covering and, except in fine sizes, will range between 0.15 and 1.25 oz per sq ft of covered surface.

METHODS OF APPLICATION. Zinc coatings are applied (1) by dipping the article in molten zinc, (2) by electrodeposition, (3) by heating in contact with zinc dust, and (4) by metal spraying.

Hot-dip Galvanizing. All adhering scale or rust is removed from the wire or steel by pickling, usually in hot sulfuric acid. A rinse and a dip in cold hydrochloric acid usually follow, but in better practice iron salts are removed from the pickled material by vigorous rinsing and scrubbing, followed, if necessary, by dipping in slightly acidulated 25% zinc chloride. After partial drying, the material enters the bath of molten zinc sometimes through a flux of zinc ammonium chloride. Interaction with the molten zinc causes the rapid formation of an adherent layer of iron-zinc alloys on the iron surface. The longer the immersion, the thicker is the alloy layer. The material is withdrawn from the zinc bath through rolls kept clean by zinc ammonium chloride flux for sheet or through charcoal, sand, or asbestos wipes for wire. A normal coating will consist of an inner layer of iron-zinc alloy reaction products and an outer layer of zinc from the galvanizing bath which has adhered to the alloy layer. Speed of withdrawal and nature of wipe influence the thickness of the layer of molten zinc adhering to the work. Tin may be added to the zinc bath to produce coatings with bright spangles. Aluminum may be added in the hand-dip galvanizing of pipes, shapes, and hardware to produce a smooth coating. The vapor of burning sulfur or ammonium chloride dust sometimes is blown over the surface of the sheets to improve coating brightness. There are a number of patented hot-dip galvanizing processes which depart in one or more ways from the above, usually in the preparation of the steel for dipping. One such process also provides for the addition of aluminum to the zinc.

Electrogalvanizing. Zinc coatings may be applied to wire, strip, conduit, hardware, and fabricated shapes by electroplating. Small articles intended for indoor use usually are plated in proprietary cyanide-type solutions capable of producing bright deposits. Sulfate and chloride baths are used in the continuous production of electrogalvanized strip, wire, etc. Special High Grade zinc anodes are required except in the patented Tainton process, which uses insoluble anodes. Electrogalvanizing avoids changes in dimensions and properties due to heat and does not produce excessive deposits in recesses, such as the base of threads, which would affect dimensional tolerances. The thicknesses of coating required for different types of service are the same as those needed in hot-dip galvanized coatings and are expressed as thickness in thousandths of an inch. ASTM Specification A164-40T covers three classes of electrodeposited zinc coating having minimum thicknesses of 0.00015, 0.0005, and 0.001 in. ASTM Specification A219-45T defines methods for measuring the thickness of electrogalvanized coatings.

Sherardizing. Zinc-rich coatings are produced on iron and steel by heating the material for several hours at 600 to 700 F in contact with zinc dust in a closed, usually rotated, vessel. The resulting coating, which is quite uniform, consists almost entirely of iron-zinc alloys, richer in iron near the base and containing some entrained zinc dust at the surface. (See also p. 3-11.)

Sherardizing may result in some heat treatment of the steel. The coatings tend to dust rather than flake when bent. Paint films adhere well. Sherardized coatings will perform equally as well in corrosive service as hot-dip or electrogalvanized coatings of the same thickness.

Metal Spraying. Sprayed zinc coatings are applied by exposing the iron or steel surface to a forcible spray of zinc particles produced in a suitable gun from zinc wire or zinc powder. Grit blasting of the steel and care in the operation of the gun are essential to good coating adhesion. By going over the same area a number of times, very heavy coatings may be built up. Care must be taken, however, to avoid thin spots. Sprayed zinc coatings tend to corrode at a somewhat faster rate than other types of zinc coatings, but this is normally compensated for by the heavier coatings applied.

BRASS. The second largest use of zinc is in the manufacture of brass. Brass alloys are discussed in detail on pp. 4-24 to 4-40. Since lead is objectionable in alpha brass for hot rolling and in brass used for drawing, grades of zinc low in lead are largely used in brass manufacture.

DIE CASTING. Suitable alloys for die casting require zinc of Special High Grade purity. Comparative freedom from lead, cadmium, and tin is essential. By alloying with about 4% aluminum, a few hundredths of 1% magnesium, and 0 to 3% copper, alloys amenable to die casting and having excellent mechanical properties in the die-cast state are obtained. (See also p. 4-79.) Alloys of this sort have also been used in the sand-cast condition as dies of large size for the fabrication of thin sheet metal, such as aluminum airplane parts.

SLUSH CASTING. Alloyed Special High Grade zinc containing 4 to 5.5% of aluminum and sometimes small additions of copper is used in the production of slush castings largely for use in lighting fixtures and other ornamental parts. Alloys of this sort are less stable than the die-casting alloys which, in turn, cannot be used satisfactorily in the slush-casting process. (See also Section 19, Article 21.)

23. ROLLED ZINC

COMPOSITIONS. ASTM Specification B69-39 lists the rolled zinc compositions in Table 2 as typical.

Table 2. Typical Rolled Zinc Compositions

Lead, %	Iron Maximum, %	Cadmium, %	Copper, %	Magnesium, %	Zinc, %
0.05 max	0.010	0.005 max	0.001 max	....	Remainder
0.05 to 0.12	0.012	0.005 max	0.001 max	....	Remainder
0.30 to 0.65	0.020	0.20 to 0.35	0.005 max	....	Remainder
0.05 to 0.12	0.012	0.005 max	0.65 to 1.25	....	Remainder
0.05 to 0.12	0.015	0.005 max	0.75 to 1.25	0.007 to 0.02	Remainder

Other compositions of a generally similar type also are available and, where consumption is large, special compositions usually can be arranged for.

CLASSIFICATION OF ROLLED ZINC. Rolled zinc is classified as follows (see ASTM Specifications B-69-39, for rolled zinc):

Type I. Coils or sheets cut from strip (ribbon) rolled zinc.

Type II. Pack rolled zinc.

Type III. Zinc plates such as boiler and hull plates produced by any rolling method.

Strip zinc may be obtained in coils or flat lengths in thicknesses of 0.004 to 0.125 in. Depending on the gage, maximum width varies from 12 to 20 in., but widths less than 12 in. are obtainable in any gage. Lengths of coil may vary from 100 to 250 ft in heavy and medium gages and up to 2000 ft and over in the lightest gage. Thicknesses below 0.004 in. are obtainable as foil. Dimensions of sheets range as follows: thickness, 0.006 to 0.10 in.; width, 18 to 60 in.; length, 72 to 144 in.

BOILER PLATES. Plates of rolled zinc, usually about 1/2 to 1 in. thick, termed boiler plates or hull plates, are attached to the inside of boilers or to the outside of steel hulls to give galvanic protection against corrosion. Cast- or rolled-zinc anodes, buried in the soil and connected to steel pipe lines by insulated wires, similarly protect the pipes. The process, frequently called cathodic protection, results in protection of the steel at the expense of the zinc which dissolves. Best results have been obtained with zinc of good purity.

ZINC GAGE. Table 3 lists the standard zinc gages together with the corresponding thicknesses and weights per square foot. The use of this special gage is discouraged, and to avoid error, the actual thickness, expressed in thousandths of an inch, is preferable. For comparison of the zinc gage with other standard gages, see p. 6-06.

ASTM Specifications B69-39 permit the following thickness tolerances:

	Thickness, in.	Permissible Variation in Thickness
Type I.	Up to and including 0.020	±0.001
	Over 0.020	±5%
Type II.	Up to and including 0.032	±0.002
	Over 0.032	±6%
Type III.	All thicknesses	±6%

Established limits for variations in length and width and for amount of curvature and bend will be found in the ASTM specification.

Table 3. Zinc Gage

	Rolled Zinc Gage			Rolled Zinc Gage	
No.	Weight of 1 sq ft, lb	Thickness, in.	No.	Weight of 1 sq ft, lb	Thickness, in.
3	0.22	0.006	16	1.68	0.045
4	0.30	0.008	17	1.87	0.050
5	0.37	0.010	18	2.06	0.055
6	0.45	0.012	19	2.25	0.060
7	0.52	0.014	20	2.62	0.070
8	0.60	0.016	21	3.00	0.080
9	0.67	0.018	22	3.37	0.090
10	0.75	0.020	23	3.75	0.100
11	0.90	0.024	24	4.70	0.125
12	1.05	0.028	25	9.40	0.250
13	1.20	0.032	26	14.00	0.375
14	1.35	0.036	27	18.75	0.500
15	1.50	0.040	28	37.50	1.000

MECHANICAL PROPERTIES AND TEST METHODS. Ordinary methods of mechanical testing have only limited usefulness in testing zinc. The strong dependence of properties on the rate and duration of loading renders the usual tensile property determinations of little value, since normal testing speeds are too low to reproduce the rates of deformation in fabricating operations and far too high to apply in cases of long-continued stressing in service. For this reason, no figure for the modulus of elasticity can be given for zinc. The best basis for designing zinc parts is the informed use of creep data (see *ASM Metals Handbook*, 1948 edition). Consultation with a rolled zinc producer on specific problems is recommended.

Since the best test for fabricating properties is behavior in actual fabrication, most rolled zinc is custom rolled to suit individual requirements. ASTM Specifications B69-39 give information on two tests, temper and dynamic ductility, which are commonly used as mill control tests and to a limited degree in purchase specifications.

Temper is measured by a testing machine that wraps a specimen tightly around a mandrel in which one end of the specimen is clamped. The degree to which the free end springs away from the mandrel is measured on an arbitrary scale ranging from zero, representing no spring whatever, to 100, representing material that returns to its original straight condition.

The three types of rolled zinc are classified according to temper: Type I, dead soft, soft, half hard, and hard; Type II, commercial; Type III, no classification. It should be noted that the terms soft, dead soft, etc., serve to designate certain specific ranges of properties regardless of the combination of composition and rolling treatment by which the properties are obtained. Table 4 gives the temper requirements of Type I and Type II rolled zinc.

Dynamic ductility is determined by forming a series of cups by forcing a spherical-ended plunger into a test strip, the plunger being adjustably mounted in a drawing press to permit varying the depth of the cup. The maximum depth of the cup in inches which can be formed where "necking down" is just visible is the dynamic ductility. See Table 4 for dynamic ductility characteristics of the various types of rolled zinc. Tensile strength is of secondary importance in the application of rolled zinc. Where such tests are required, the ASTM standard tensile specimen for thin sheet metal is entirely suitable. A free head speed of 0.25 in. per min should be used. Under these conditions, the tensile strength of rolled zinc in the direction of rolling will vary from about 18,000 psi for soft temper High Grade zinc to 35,000 psi for hard temper Prime Western zinc. Alloy zincs can be raised to higher strength levels.

The elongation will vary from 5% for the hardest to at least 65% for the softest metal. The tensile strength at right angles to the rolling direction will be higher and the elongation somewhat lower. Creep testing is of much greater importance than ordinary tensile testing in the evaluation of rolled zinc for specific applications. Much work along these lines has been carried out by rolled-zinc producers. The results have been correlated with actual loading tests on corrugated zinc roofing sheets.

FABRICATION. The ductility of zinc is high at and above room temperature but falls off rapidly at temperatures below 70 F. All drawing and bending operations should be performed at temperatures of 70 F or above. Frequently, an operation which is just too severe to be successful at temperatures near the lower limit may be successfully accomplished by warming the zinc and tools, for example, to 100 F. In any type of extrusion, it is necessary to use an elevated temperature in order to operate with reasonable pressures.

Table 4. Dynamic Ductility and Temper, Types I and II Rolled Zinc

American Zinc Gage, No.	Thickness, in.	Type I								Type II	
		Dead Soft		Soft		Half Hard		Hard			
		Dynamic Ductility, min, in.	Temper, max, %	Dynamic Ductility, min, in.	Temper, max, %	Dynamic Ductility, min, in.	Temper, max, %	Dynamic Ductility, min, in.	Temper, max, %	Dynamic Ductility, min., in.	Temper, max, %
3	.006	.275	70	.255	76	.215	80	.165	86	.165	83
4	.008	.280	65	.260	72	.220	75	.170	83	.170	77
5	.010	.285	60	.265	66	.225	70	.175	80	.175	73
6	.012	.290	55	.270	62	.230	65	.180	77	.180	69
7	.014	.295	52	.275	58	.235	61	.185	74	.185	65
8	.016	.300	48	.280	54	.240	58	.190	72	.190	62
9	.018	.305	46	.285	51	.245	54	.195	69	.195	58
10	.020	.310	43	.290	48	.250	51	.200	68	.200	57
....	.022	.315	41	.295	45	.255	48	.205	65	.205	55
11	.024	.320	40	.300	43	.260	45	.210	63	.210	54
....	.026	.325	39	.305	40	.265	43	.215	61	.215	52
12	.028	.330	37	.310	37	.270	42	.220	59	.220	51
....	.030	.335	37	.315	37	.275	42	.225	58	.225	49
13	.032	.340	36	.320	36	.280	41	.230	57	.230	48
....	.034	.345	36	.325	36	.285	40	.235	56	.235	47
14	.036	.345	......	.325	......	.290	......	.240	......	.240	......
....	.038	.350	......	.330	......	.295	......	.245	......	.245	......
15	.040	.350	......	.330	......	.300	......	.250	......	.250	......
....	.042	.350	......	.330	......	.300	......	.250	......	.250	......
....	.044	.355	......	.330	......	.305	......	.255	......	.255	......
....	.046	.355	......	.330	......	.305	......	.255	......	.255	......
....	.048	.355	......	.335	......	.310	......	.260	......	.260	......
17	.050	.355	......	.335	......	.310	......	.260	......	.260	......

For certain ornamental sheet metal work, heavy gage zinc is hand formed at temperatures as high as 480 F.

Lubrication. In all operations where a lubricant is required, excepting those involving temperatures well above 70 F, soapy water is the cheapest lubricant and is quite satisfactory. A neutral soap should be used to avoid staining. If finish is important, it may be necessary, in order to avoid spotting, to wash the work and dry it in sawdust. Loading or building up of zinc deposits on the tools sometimes occurs. This may be minimized by including a small amount of a very fine abrasive such as French chalk in the drawing lubricant.

Bending usually is performed either on an ordinary brake or in forming rolls. In a brake, the edges of the bending and clamping jaws should be rounded as much as possible, at least to the extent of removing the sharp edge. If possible, bends in rolled zinc should be at right angles to the grain or rolling direction. It then is easy to obtain proper bending properties without the sacrifice of hardness and stiffness.

Bends frequently must be made with the bend axis in the rolling direction because of width limitations in strip zinc or to take advantage of automatic machinery using coils of ribbon zinc. Zinc manufacturers should be consulted in regard to the best combination of high bending properties and high stiffness for the particular conditions involved.

Temperature is most important in bending zinc as in other forming operations. A slight warming of the metal sometimes makes possible an otherwise impossible bending operation.

Drawing. Operations in which cups are formed from flat blanks without any general reduction in gage are called *drawing*. Shop methods in drawing zinc vary considerably from one plant to another, but in general brass practice is followed with departures in detail, depending on the particular operation. It is unnecessary to anneal pure zinc between operations, and the work-hardening alloys require fewer annealings than brass. Reduction in diameter from blank to the first cup should not exceed 40% in commercial operation. A somewhat lower reduction usually is preferable. It is possible to reduce the wall thickness about 10% in redrawing operations in which the cup diameter is reduced 20% or less.

Extrusion. Zinc is successfully extruded into any of the forms commonly produced by this process. Tubing and rod have been produced, but in the United States development has been limited principally to the impact extrusion of cups, buttons, etc. Blanks for the extrusion of small articles usually are punched from rolled metal, although cast billets have been used for rods. Because of the directional properties of rolled zinc, some trouble due to nonuniform flow is encountered in some classes of work. A specially rolled zinc can be supplied to minimize this difficulty.

Welding of zinc requires some care because of the low melting point and the tenacious nature of the oxide. Zinc cannot be welded without seriously weakening areas adjacent to the weld by coarse grain growth. If strength is important, some cold-working operation to refine the grain should follow welding.

Soldering. Zinc is easily soldered. Low soldering temperature is necessary to avoid excessive grain growth in areas adjacent to the joint. This is best obtained by using a large iron hot enough to accomplish the soldering with a single pass. Half-and-half solder generally is used, although solder with a lower melting point is of advantage. A solder composed of Pb 55%, Sn 30%, and Cd 15% is suitable for zinc. It has a low melting point combined with high strength. Acidulated zinc chloride or killed hydrochloric acid makes an excellent flux, although much zinc is soldered without a flux. Soldered zinc should not be remelted because the resulting tin content causes hot shortness.

Hot Stamping. Certain complicated and deep draws such as occur in ornamental sheet metal work are made in heavy gage zinc by preheating the metal to about 480 F before stamping. This work requires knowledge of the working qualities of zinc at this temperature and should be done by skilled workmen. For some uses, as in the manufacture of organ pipe, the rolled zinc is given a thorough anneal prior to working. A 30-min anneal at 480 F, while removing much of the desirable bending and drawing qualities, will reduce the metal to the very soft, nonresonant condition required for this use.

Machining. In machining zinc, it is common practice to use tools with a greater rake than is normally used and to use soapy water as a lubricant.

24. PROPERTIES OF ZINC

PHYSICAL PROPERTIES. The more important physical constants of zinc are given in Table 5.

Table 5. Physical Constants of Pure Zinc

Atomic weight	65.38
Density (cast), 77 F	7.133 g/cm^3
	0.258 lb/in.^{3}
Density (rolled), 77 F	7.142 g/cm^2
Melting point	787 F
Boiling point	1663 F
Specific heat, 68 F	0.0915
Latent heat of fusion	43.4 Btu/lb
Latent heat of vaporization	767.3 Btu/lb
Coefficient of linear expansion	
Cast	21.7×10^{-6}/°F
Rolled, with grain	17.8×10^{-6}/°F
Rolled, across grain	12.8×10^{-6}/°F
Thermal conductivity	786.5 Btu/sq ft/in./hr/°F
Electrical resistance *	2.33 microhm-in.
Temperature coefficient of electrical resistance, 32–212 F	0.00232 per °F
Electrochemical equivalent	0.3388 mg/coulomb
Electrolytic solution potential vs. hydrogen electrode	−0.762 volt
Magnetic susceptibility	-0.15×10^{-6} cgs
Velocity of sound (room temperature)	12,041 ft/sec.
Crystal structure	Hexagonal, close packed
a_0	2.6594KX
c_0	4.9370KX
c/a	1.8564

* Based on soft rolled zinc strip 99.99% pure. Resistivity of strip zinc may vary 3%, depending on rolling treatment, and in commercial grades about the amount, depending on purity.

METALLOGRAPHY. The common impurities in zinc are lead, cadmium, and iron. Their importance varies individually with the use to which the metal is put. These elements as well as copper, magnesium, and aluminum, which are the ones most commonly added in zinc alloys, are discussed below.

Lead, which enters into extremely limited solid solution, has no great effect on the mechanical properties of unalloyed zinc. In zinc-alloy die castings it has the very undesirable effect of stimulating intercrystalline corrosion.

Cadmium, in the amounts present in commercial zinc, is completely in solid solution and exerts an important hardening effect. It also raises the recrystallization temperature.

Whereas pure zinc recrystallizes at about room temperature and cannot be appreciably work hardened, zinc containing cadmium can absorb considerable amounts of work hardness during cold rolling or drawing. Cadmium, like lead, stimulates intercrystalline corrosion in zinc die-casting alloys which contain aluminum.

Iron has a limited solid solubility of the order of 0.001 to 0.003%. Attention must be given to the iron content of zinc for rolling. Some iron is desirable, but excessive amounts are harmful to properties.

Copper is the alloying element most frequently added to zinc. It enters into solid solution in zinc up to approximately 1% and with about the same effect as cadmium, except that the copper-zinc alloys are more ductile and easier to roll.

Magnesium has a solid solubility ranging from 0.002 to 0.1%, but even these amounts exert a strong influence on its creep properties. When magnesium is added in the presence of copper, a striking increase in creep resistance is observed.

Aluminum. Except in carefully controlled compositions comparable to those used in die casting, zinc alloys containing aluminum are unstable when rolled. Alloys of the die-casting type held within the required limits for composition are essentially stable in either the cast or the rolled condition.

CHEMICAL PROPERTIES. Atmospheric Corrosion. Rolled zinc corrodes smoothly and uniformly in the outdoor atmosphere. The absence of pitting permits estimation of the corrosion rate from weight change data. The data in Table 6 are typical. The corrosion rate of zinc is greatest in industrial atmospheres and least in rural and desert atmospheres. Although there is a tendency for the purest zincs to have lower corrosion rates than the impure zincs, particularly in southern exposures, it is generally true that the corrosion rate of rolled zinc is substantially independent of the composition.

Table 6. Corrosion of Rolled Zinc in Typical Outdoor Atmospheres

(Data from published reports of Committee B-3 of ASTM)

Type Zinc	Corrosion Rate													
	Altoona, Pa.*		New York City *		Sandy Hook, N.J.†		Key West, Fla.†		LaJolla, Calif.†		State College, Pa.‡		Phoenix, Ariz.‡	
	mdd	ipy	mdd	ipy	mdd	ipy	mdd	ipy	mdd	ipy	mdd	ipy	mdd	ipy
Special High Grade	0.94	.00019	0.99	.00020	.29	.000058	.080	.000016	.26	.000053	.19	.000039	.025	.0000051
High Grade	1.0	.00021	1.1	.00023	.29	.000059	.094	.000019	.34	.000069	.18	.000037	.038	.0000077
Prime Western	1.0	.00021	1.0	.00021	.29	.000059	.11	.000022	.37	.000075	.21	.000043	.055	.000011

mdd = milligrams per square decimeter per day.
ipy = inches penetration per year.

* Industrial.
† Seacoast.
‡ Rural and desert.

Aqueous Corrosion. Zinc coatings are widely used in the handling of water supplies with generally satisfactory results. Rolled zinc and zinc die castings are less commonly used in contact with water. The rate of corrosion in water is, in general, much lower than that of iron. Temperature, *p*H, and concentration of dissolved oxygen all influence the corrosion rate. The corrosion of zinc is high in acid and strongly alkaline solutions but becomes relatively low in the *p*H range 7 to 12.5. The rate of corrosion is stimulated by temperature, the effect being small for temperatures up to about 100 F and increasing rapidly at higher temperatures. Increasing concentrations of oxygen increase the corrosion rate, but very low concentrations, such as exist in stagnant water, result in rapid pitting. Carbon dioxide increases the rate of corrosion of zinc in aerated water.

Corrosion Inhibitors. The rate of corrosion of zinc in water can be reduced substantially by small additions of sodium dichromate, borax, or sodium silicate to the water. The rapid pitting type of attack which develops in contact with stagnant water can be prevented by the application to the surface of a chromate film. Such films are produced by reaction between the surface and a solution containing chromic acid or sodium dichromate suitably activated with sulfates.

Organic Substances. Zinc is inert to such common organic substances as gasoline and kerosene in the absence of water. Soaps do not attack it, although in dilute soap solutions there is a tendency to form protective films of zinc soaps. Zinc is entirely inert toward anhydrous alcohol but is corroded by mixtures of alcohol and water.

Electrochemical Properties. Zinc is a chemically active metal. It has a high electronegative potential, which is used in dry cells and in the galvanic protection of steel. Zinc anodes are used to protect steel hulls of ships and steel pipe lines buried in the ground. In contact with acids, zinc dissolves, releasing hydrogen. It forms both basic and acid radicals and is soluble both in acids and in strong alkalies.

Toxicity. Except for the salts of strong acids, which are corrosive or zinc compounds containing poisonous elements or radicals, the compounds of zinc are not toxic. Zinc is a normal constituent of many foods, and considerable excesses of zinc taken into the body are readily eliminated without ill effects. Excess zinc content in acid foods or beverages may cause unpleasant symptoms. The amounts necessary to cause such symptoms, however, are far beyond those normally encountered. The maximum safe zinc content for drinking water has been set by the U. S. Bureau of Public Health at 15 parts per million. Larger contents may be ingested safely, but the taste is unpleasantly metallic.

Plating. Articles produced from both die-cast and wrought zinc may be plated readily with any of the common plated finishes. It is customary to apply a primary copper coating to a thickness of at least 0.0002 in. before depositing other metals. For decorative purposes in use outdoors, minimum total coating thicknesses of all coats of 0.0010 to 0.0015 in. are required. Thinner coatings are suitable for indoor use.

Painting. Organic finishes do not adhere well to untreated zinc surfaces. A phosphate pretreatment such as Bonderite insures good initial adhesion and retention of adhesion. Baking enamels to be used on zinc should require relatively low baking temperatures to avoid alteration of the properties of the zinc.

GENERAL REFERENCES

The Corrosion Handbook (H. H. Uhlig, Editor, and sponsored by the Electrochemical Society), John Wiley and Sons, New York, 1948.

Metals Handbook, American Society for Metals, Cleveland, 1948.

Zinc and its alloys, *Natl. Bur. Standards Circ.* 395, 1931. This circular has a complete bibliography.

Valuable bulletins on various subjects relating to zinc and its uses may be obtained from the principal producers of zinc and from the American Zinc Institute.

THE LESS COMMON METALS

By Bruce W. Gonser and John L. Everhart

Some of the industrial applications of the less common metallic elements are summarized briefly in the following paragraphs. For more detailed information, the publications listed at the end of this chapter (p. 4-72) may be consulted. See also Table 3, p. 1-04.

ANTIMONY. This bluish-white metal is harder than copper and so brittle that it may be pulverized readily. The metal, which is a poor conductor of heat and electricity, is used almost exclusively as an alloying addition. With tin and copper it is used in bearing metals (babbitt); with tin in Britannia metal; and has been suggested for alloying with copper and nickel as an antimony worm gear bronze.

Antimony is used extensively for hardening lead. Such lead-base alloys containing antimony are applied as bearing metals and type metals (with varying content of antimony and other metals), battery plates (4 to 12% antimony), solders (up to 3% antimony), cable sheath and lead foil (1% antimony), collapsible tubes (2 to 3% antimony), and chemical equipment (12% antimony).

ARSENIC. Metallic arsenic is a brittle gray material which sublimes at 842 F without melting. The metal is used to increase the recrystallization temperature of copper (arsenical copper), in lead shot to increase the hardness and surface tension, and in insoluble lead-antimony anodes (up to 0.8% arsenic).

BARIUM, commercially available with a purity of 99.95%, is soft and malleable with a hardness similar to sodium. Barium is used as a "getter" to remove the last traces of gas in vacuum tubes and as a lead-barium-calcium bearing metal. A high-nickel alloy containing barium finds application in automotive ignition equipment.

BERYLLIUM, with a specific gravity of 1.82, is two-thirds as heavy as aluminum and is the lightest of all metals with the exception of magnesium and the alkalies. It melts at about 2350 F. The massive metal is not attacked by air at ordinary temperatures and does not tarnish in the presence of sulfur. It has a great affinity for oxygen, however, and beryllium powder, if ignited in air, burns brilliantly.

Beryllium is used for windows in x-ray tubes, electrodes in Neon tubes, in cyclotrons.

and in connection with atomic energy development. However, it is applied most widely in copper-beryllium alloys. These alloys, containing usually 2 to 2.5% beryllium, are hardenable by heat treatment. The increase in strength in the copper-beryllium alloys is not accompanied by a great loss in electrical conductivity, compared to the loss obtained with many other elements which harden copper, such as silicon or tin. Beryllium-copper is used for springs in electrical applications, for nonsparking tools, and for various wear-resisting parts. The alloys are available in the form of strip, rod, wire, forgings, and castings.

Several modified alloys are available commercially also. One which contains 2.6% cobalt and 0.5% beryllium has excellent electrical conductivity and heat resistance after heat treatment and may be used for electrical contact springs that are subject to overheating. Nickel-beryllium alloys are also heat treatable to high strengths, as are several complex iron-chromium-nickel-beryllium alloys. Beryllium oxide resists thermal shock and retains its high electrical resistance at high temperatures. With a melting point of about 4660 F, it is a useful refractory for high-temperature service.

BISMUTH. This very brittle material is white with a distinct pink tinge. It expands 3.3% during solidification. The thermal conductivity of solid bismuth, 4.85 Btu per hr per ft per °F, is less than that of any other metal except tellurium and mercury.

Bismuth is used most widely in fusible alloys and low-melting-point solders. A bismuth-lead-tin-cadmium alloy is employed for tube bending. The bismuth-lead eutectic alloy, melting at 254 F, neither expands nor contracts when cast into sand molds and is used for pattern duplication and the production of master patterns in foundry operations. A nonshrinking bismuth-lead-tin-antimony alloy, known as Matrix alloy, has a pouring range of 480 to 660 F. The alloy will sustain continuous loads of 5 tons per sq in. and short period loads of 8 tons per sq in. It is used for short-run dies, forming dies for sheet metal, to mount parts of complex punches and dies, and in repairing broken dies. Other applications of the high-bismuth alloys include oiltight, airtight, and watertight seals between glass and brass or copper tubing. A bismuth-tin-lead low-melting alloy is used effectively as a die by casting to shape, dipping in liquid nitrogen, and frequently redipping to maintain the high compressive strength that it reaches at low temperatures.

BORON is a light metal (sp. gr. 2.3) that is frequently classed as a nonmetal. There are only insignificant uses for the pure metal and only a few applications of boron as an alloying constituent. Ferroboron is used to add very small amounts of boron to steel (as 0.002%) since it increases the capacity of steels to harden upon quenching.

Boron carbide and boron nitride are the hardest materials known, next to the diamond. As such, they are used where erosion is extreme, as for sandblasting nozzles. Boron is used as a constituent in some hard-facing alloys.

CADMIUM is a soft ductile metal with a bluish-white color, having a melting point of 610 F. It is used most extensively as a protective coating for steel. It is a constituent of a number of low-melting alloys employed as solders or as fusible plugs in sprinkler systems, also of a number of silver (hard) solders. Cadmium-silver and cadmium-nickel alloys have been used as bearing metals. To improve the strength and wear resistance without seriously reducing the electrical conductivity, 1% cadmium is frequently added to copper for use in trolley wires.

CALCIUM, a silvery white metal soft enough to be cut with a knife, oxidizes rather rapidly on exposure to moist air. The addition of 0.04% calcium to lead improves the strength and resistance to fatigue, and the alloy has been used as sheathing for telephone cables. Several lead-base alloys containing calcium are employed as bearing metals, particularly in Germany. Calcium together with lithium is used to deoxidize copper because the quantities that remain in the solid copper have very little effect on the electrical conductivity.

CERIUM is a steel-gray, malleable, and ductile metal. It is used as an alloying element, particularly in magnesium to enhance high-temperature properties, and (in very small percentages) in such alloys as 80 nickel–20 chromium electrical resistance elements where it is claimed to give longer life at high temperatures.

Misch metal, the most used commercial alloy (containing cerium and other rare earths with varying quantities of iron), is brittle and pyrophoric. It is used in the manufacture of lighter flints, tracer bullets, and other ignition devices, as well as an alloying addition to the light metals and as a less expensive substitute for cerium alone.

CHROMIUM is a highly lustrous white metal with a distinctly bluish tinge, having a specific gravity of 7.1 and a melting point of about 3380 F. High-purity chromium is sufficiently ductile to be drawn into wire. Because of its excellent resistance to atmospheric corrosion and tarnishing, chromium is used extensively in plating other metals for decorative purposes, particularly in the automotive industry. Being very hard and wear resistant, chromium plating is applied also to increase the life of machine and engine parts, punches, dies, and gages. The resistance of chromium to corrosion makes the plate valuable in the chemical industries.

Chromium is a constituent of the most widely used stainless and heat-resisting steels and is applied broadly in the low-alloy steel field. Nickel-chromium alloys are used for electrical resistance elements. Alloys consisting primarily of cobalt, chromium, and tungsten form a class of high-speed cutting tools. Chromium (up to 1%) is added to copper to increase the strength without greatly reducing the electrical conductivity. A number of chromium-base alloys are under development for severe service at high temperature.

COBALT is a fairly hard metal resembling nickel in many of its properties. It has a specific gravity of 8.9 and a melting point of about 2695 F. Cobalt is highly magnetic, being second only to iron in this property.

Many of the alloys developed in modern times for cutting tools or for wear resistance contain cobalt. They include high-speed steels, various tool and die alloys, stellite, and the cemented carbides. Cobalt is used extensively in permanent-magnet alloys. A cobalt-chromium-aluminum alloy is one of the principal electrical-resistance materials. Potentially, the most important application of cobalt lies in cobalt-base alloys for severe service at high temperatures. These alloys, usually precision cast to shape, are characterized by their resistance to wear, heat, corrosion, and erosion. During the war, supercharger blades for operation at 1500 F were cast from an alloy containing about 60 to 65% cobalt.

COLUMBIUM. This light-gray ductile material is approximately as hard as wrought iron. It can be hardened locally or throughout by heating in nitrogen, hydrogen, carbon-bearing gases, or in air. Although resistant to most chemical reagents, it is attacked by concentrated sulfuric acid and fused alkalies. Similar to tantalum in most respects, it is only about half as heavy and is not quite so resistant to corrosion by some acids.

Columbium is used as ferrocolumbium to prevent the intergranular corrosion of stainless steels at elevated temperatures.

GALLIUM. This grayish metal, of low malleability, has a specific gravity of 5.9, a melting point of 85 F, and a boiling point of 4170 F. The wide range between melting and boiling points makes gallium useful for high-temperature quartz thermometers ranging up to 1830 F. The metal, which has the property of forming exceedingly white mirrors on glass, may find application in optical service.

GERMANIUM is one of the unusual metals which only recently has become available in commercial quantities. Like bismuth, it expands slightly on freezing, and this property persists even when alloyed with as much as 92% gold. The very high electrical resistance of germanium has suggested its application as resistors, particularly as thin films of metal deposited on glass. Although germanium melts at about 1760 F, when alloyed with 88% gold, the melting point drops to 672 F. Such low-melting gold alloys make good solders for gold, and good dental and jewelry castings. Germanium is chiefly used, however, as an electric current rectifier, as in rectification of microwaves for radar applications.

GOLD. This most ductile and malleable of all metals has a specific gravity of 19.3 and a melting point of 1945 F. It is a good conductor of heat and electricity and is highly resistant to the action of most chemicals. Coinage and jewelry applications have obscured the fact that its properties make it useful for corrosion-resistant apparatus. Considerable quantities are used for this purpose.

INDIUM. This soft and ductile white metal resembles tin in many of its properties. The metal has been electroplated on silver alloys with the aim of improving the tarnish resistance. Silver bearings are often plated with lead to avoid seizure, and indium plated over the lead to prevent corrosion of the lead by corrosive lubricating oil. The addition of 18% indium to Wood's metal reduces the melting point to 117 F and makes the alloy valuable as a very low-melting solder.

IRIDIUM is one of the hardest, most refractory, and difficult to work of the metals, although it may be forged at high temperatures. It has a specific gravity of 22.4 and melts at 4450 F. It is the most corrosion-resistant element known, being unaffected by common acids and resistant to aqua regia and fuming sulfuric acid.

It is used to harden platinum for applications in electrical contacts, for voltage regulators, high-tension magnetos, relays, etc. Iridium-platinum alloys serve as insoluble anodes in electroplating.

LITHIUM. Lightest of the metallic elements, lithium has a specific gravity of 0.53. It is a silver-white metal that in commercial purity oxidizes readily in air and reacts with water, although not so actively as other alkali metals. It is used in a number of lead-base bearing metals, particularly in Germany. Copper, deoxidized with lithium and calcium, retains a high electrical conductivity. Lithium hydride is sometimes used as a convenient source of hydrogen.

MANGANESE. Electromanganese of high purity is silver white, so hard that it is not scratched by a file, and very brittle.

In ferrous metallurgy, manganese is universally employed as a scavenger to counteract the effect of sulfur. It is applied similarly in nickel alloys. In low-alloy steels, the metal is added frequently to increase the strength. One of the earliest austenitic steels, containing 10 to 14% manganese, has very good resistance to abrasion when the pressure is high, and is applied to the manufacture of railroad frogs and crossings, rock crushers, ball-mill linings, etc.

Manganese is used to harden aluminum, primarily because it is one of the few hardening agents that do not have an adverse effect on the corrosion resistance. In copper alloys it is a constituent of the silicon bronzes and the high-strength brasses. Manganin, a copper-manganese-nickel alloy has a low temperature coefficient of electrical resistance and is used for standard resistances. In the magnesium alloys, manganese is particularly valuable for improving the resistance of the alloys to salt-water corrosion and to increase the yield strength. An alloy containing 72% manganese, 18% copper, and 10% nickel has a very high coefficient of expansion and is used as the high-expansion side in a number of bimetallic thermoregulators.

MERCURY. The only metal which is liquid at room temperature, mercury has a specific gravity of 13.5, a melting point of −38 F, and a boiling point of 675 F. It is used in mercury-arc rectifiers, mercury-vapor lamps, small dry cells (as for hearing aids), automatic control devices in industrial instruments, and scientific instruments. The mercury-vapor boiler, which operates at high efficiency, low power cost, and in small space, is an important potential application of the metal. A mercury clutch for small motors is being developed, and mercury poured into a mold and frozen is considered to have some advantages as a pattern for precision casting.

MOLYBDENUM is a ductile white metal having a specific gravity of 10.2, and a melting point of 4750 F. It can be rolled and drawn, and is available commercially in wire diameters as small as 0.002 in., and in sheets down to 0.001-in. thickness. It is used for electrical contact points, spot-welding electrodes, targets in x-ray tubes, and filament supports and grids in vacuum tubes. In an atmosphere of hydrogen, molybdenum is employed as a heating element in electrical-resistance furnaces operating up to 3600 F. Heavy molybdenum sheet is used for electrodes in glass-melting furnaces.

Molybdenum is employed widely in the low-alloy steels. It acts as a mild stabilizer of the carbides in stainless steels (18% chromium–8% nickel) and increases the strength at elevated temperatures. It also increases the resistance to corrosion by sulfuric acid and, to a lesser extent, to dilute hydrochloric acid. Molybdenum has been substituted for tungsten in high-speed steels. It is used also to improve the strength and resist the growth in cast iron at elevated temperatures. Examples of its use as a constituent in a number of nonferrous alloys are Stellite, Illium, and molybdenum-permaloy.

OSMIUM. Heaviest of the metals, osmium has a specific gravity of 22.48 and melts at about 4900 F. It is harder than glass or quartz, but is scratched by topaz. Osmium has not been worked successfully, for it is brittle even at high temperatures. However, it can be sintered at 3600 F. The principal use is in hard corrosion-resistant alloys, as for tipping pen nibs or for phonograph needles. It is applied also in special hard electrical contact materials.

PALLADIUM resembles platinum in many of its properties, but is slightly harder and stronger. It is very ductile. Like gold, it can be drastically cold worked and has been beaten into sheets $^1/_{250,000}$ in. thick. The metal has excellent corrosion resistance. Palladium is used extensively for contacts in telephone relays, while base metals clad with palladium are applied in chemical equipment for the production of fine chemicals.

PLATINUM. A very ductile white metal, platinum may be worked hot or cold and has been drawn to 0.00005-in.-diameter wire. The metal has excellent corrosion resistance and retains high strength at elevated temperatures. Either massive or clad on base metals, platinum is applied in corrosion-resisting apparatus. It is resistant to the attack of molten glass and is used in furnace construction for making glass fibers. Platinum alloys are employed for resistance-furnace windings and electrical contacts.

RADIUM is a brilliantly white radioactive metal having a specific gravity of 5.0 and a melting point of about 1290 F. The metal forms a nitride on exposure to air, attacks glass and quartz, and decomposes water rapidly. It is used in therapy, as a substitute for x-rays in the examination of metals, and for making luminous dials in industrial instruments.

RHENIUM has a specific gravity of 20.5 and has the unusually high melting point of about 5730 F. Being similar to tungsten in massive form, it can be sintered at high temperatures and rolled or forged above 1470 F. It is attacked by nitric and sulfuric acids, but resists hydrochloric acid.

Electrodeposited rhenium is very hard and has been suggested for lining vessels for the transportation of hydrochloric acid. Added to platinum metals, rhenium is said to improve their ductility. It has been used in pen-point alloys.

RHODIUM. A white metal which hardly can be distinguished from silver, rhodium can be forged or swaged at room temperature and is very malleable at high temperatures. The metal is harder than silver and is resistant to oxidation and to chemical and industrial fumes.

The most important application is in plating searchlight reflectors. Plated rhodium accurately reproduces the base-metal surface; that is, a polished surface remains bright while a matte surface retains its matte finish. Rhodium alloys are used for electrical contacts, spinnerets in the rayon industry, resistance furnace windings, and high-temperature thermocouples.

RUTHENIUM, another metal of the platinum group, is hard and difficult to work, but can be forged at 2700 F. Its corrosion resistance is similar to that of iridium. It is used to harden and increase the corrosion resistance of platinum and palladium.

SELENIUM exists in four modifications; the metallic form is gray, with a specific gravity of 4.81, and a melting point of 428 F. Because of its change in electrical resistance with the intensity of light, it has been used in light-sensitive cells. Other uses are in rectifiers to convert alternating current to direct current and in free-machining copper alloys.

SILICON is the most abundant element in the earth's crust, and one which can be cheaply produced. Although possessing excellent resistance to atmospheric corrosion, even at elevated temperatures, and resistance to acids (except hydrofluoric), the use of the pure metal is negligible because of its brittleness and lack of formability. As ferrosilicon, it is a common addition to special steels, as for transformer sheet, and as a deoxidizer in ordinary steels. Silicon is also much used as an alloying constituent with copper in silicon brasses and bronzes, and with aluminum, magnesium, and their alloys.

SILVER. The whitest of all metals, silver is exceeded only by gold in malleability and ductility. In the annealed condition, silver is the best conductor of heat and electricity of the elements. It has a high resistance to attack by alkalies, organic acids, and many other elements.

The metal is used either solid or clad on base metals for corrosion-resistant equipment. Alloys are applied in electrical contacts, bearings (silver-cadmium), soft solders (silver-lead), and hard solders (silver-copper-zinc base).

SODIUM. This light, very soft metal has a specific gravity of 0.97, and a melting point of 216 F. It readily oxidizes in air and decomposes water violently. Although nearly as inexpensive as pig iron on a volume basis, there are few uses because of its great reactivity. It is used as a cooling medium in aircraft engine valves.

TANTALUM. The melting point of tantalum (about 5425 F) is higher than that of any other metal, excepting rhenium and tungsten. The metal may be worked at room temperatures, like tungsten and molybdenum, but absorbs gases readily at elevated temperatures. It can be drawn to fine wire (0.001-in. diameter) and rolled to thin sheet (0.002-in. thick). Tantalum has a chemical resistance similar to glass. It resists attack by the common acids, including hydrochloric, but is attacked by fused alkalies.

Alone or clad on base metals, tantalum is used in corrosion-resistant equipment. It is employed in the rayon industry for spinnerets and dies. Tantalum and tungsten carbides are used for tools and dies for severe service, including the swaging of molybdenum and tungsten at elevated temperatures. The solid solution of 20 atomic per cent of hafnium carbide in tantalum carbide has the highest melting point (over 7200 F) of any known material.

TELLURIUM is practically a nonconductor of electricity and, unlike selenium, shows no increase in conductivity with light intensity. Small quantities of tellurium (up to 0.1%) are added to lead to improve the strength and fatigue resistance. Even smaller additions, as 0.1 lb per ton, are remarkably effective in increasing the chill of cast iron. The metal has been used also to improve the machining properties of copper alloys and of stainless steel.

THALLIUM. A malleable metal, comparatively soft and with a low melting point (572 F), thallium resembles lead in many of its properties and is used largely in lead alloys. It tarnishes rapidly on exposure to air, yet is added to lead for better corrosion resistance. Thallium has been suggested for use in lead-base bearing metals to improve the resistance to deformation. A thallium-lead alloy is employed in electrical fuses.

THORIUM is a grayish white metal having a specific gravity of 11.5 and a melting point of 3350 F. It is readily cold worked and may be easily rolled to thin sheet. Although ductile and soft, the mechanical strength is low. When ignited, it burns brightly in air. The metal is radioactive and is subject to nuclear fission when bombarded with slow neutrons.

It is used in photoelectric cells, x-ray targets, and glow tube electrodes. Small additions of thorium to nickel-chromium resistor alloys and to platinum restrict grain growth and prolong the life at high temperature. The oxide is used to restrict grain growth in tungsten filaments. Having the highest melting point of all refractory oxides, thorium oxide is used for high-temperature furnace linings.

TITANIUM. A white metal, having a specific gravity of 4.50 and a melting point of 3275 F, pure titanium can be drastically cold rolled into strip or swaged into rod form. It can be drawn into fine wire, although its tendency to seize complicates the operation. The metal is very active and, at elevated temperatures, readily absorbs oxygen, hydrogen, nitrogen, or carbon from carbon-bearing gases. Its general resistance to corrosion is similar to that of stainless steel.

Titanium is used in stainless steels to stabilize the carbides. A nickel-cobalt-iron-titanium alloy, Konel, has high strength at elevated temperatures. The carbide, usually in combination with other carbides, is applied in cutting tools. Titanium is potentially important because of an excellent strength-weight ratio, good mechanical and corrosion-resistant properties, and abundance of ores. Many alloys, notably those with a copper base, are made precipitation-hardenable upon addition of a small percentage of titanium.

TUNGSTEN has the highest melting point (6115 F) of all the elements. It is obtained in massive form by powder metallurgical methods. After sintering, it can be hot swaged and drawn into wire. The essential practice is to work the metal from a high temperature to a low temperature to get a fibrous structure for high strength and some ductility. It is very resistant to corrosion by acids but is attacked by alkalies. The metal oxidizes rapidly at high temperatures.

Tungsten is used for electrical contacts, lamp filaments, and x-ray targets. As an alloying element, it is the principal constituent of the high-speed steels and complex nonferrous alloys for high-temperature service. Cobalt-chromium-tungsten alloys and tungsten carbide are widely used as cutting tools and dies. A sintered tungsten alloy containing copper and nickel has a specific gravity of 16.5 and is used for balancing crankshafts and for protection from x-rays. Tungsten is very resistant to attack by liquid zinc and has been used for pump parts in handling this material.

URANIUM is a silvery white metal having a specific gravity of 18.7 and melting at approximately 3075 F. The powder can be sintered and drawn into fine wire, although the fine powder is pyrophoric. It forms alloys readily with many metals and some (as 66 uranium, 34 nickel, or up to 20% uranium in copper) are resistant to corrosion, although the resistance of uranium alone to corrosion is very poor. A radioactive metal, it has been used in glow-discharge tubes and x-ray tubes, but its greatest application is in the production of atomic energy by nuclear fission.

VANADIUM. Pure vanadium is a white metal sufficiently ductile to be drawn into fine wire. Metal of this purity is not produced commercially, however, and nearly all vanadium used in metallurgy is in the form of ferrovanadium. The principal application is in steels, in which it is a carbide-forming element.

ZIRCONIUM. Zirconium is a ductile metal having a specific gravity of 6.4 and a melting point of 3370 F. It is very resistant to attack by common acids, including hydrochloric and nitric, but is attacked by hydrofluoric. Its resistance to attack by alkalies is excellent. The metal burns in air at high temperatures and, like titanium, is so reactive to all but the inert gases that it can be used effectively as a "getter" in vacuum-tube manufacture. Ductile zirconium has excellent mechanical properties, which, with its resistance to corrosion, makes it potentially a valuable material of construction.

Its uses include igniters in photoflash bulbs, primers in ammunition, electric-light filaments, filters in x-ray tubes having molybdenum targets, and in electronic applications. The excellent corrosion properties of zirconium indicate that it will be applied increasingly in corrosion-resisting equipment. It is used as ferrozirconium to a limited extent in steel metallurgy.

The oxide, zirconia, has been used as a high-temperature refractory furnace lining, as for melting chromium and platinum.

GENERAL REFERENCES

J. DeMent and H. C. Dake, *Rarer Metals*, Chemical Publishing Co., New York, 1946.

B. S. Hopkins, *Chapters in the Chemistry of the Less Familiar Metals*, Stipes Publishing Co., Champaign, Ill., 1938–1939.

Mechanical properties of metals and alloys, *Natl. Bur. Standards Circ.* C447, 1943.

Modern Uses of Nonferrous Metals, American Institute of Mining and Metallurgical Engineers, 1935.

The Platinum Metals and Their Alloys, The International Nickel Co., 1941.

D. M. Yost, H. Russell, Jr., and C. S. Garner, *The Rare-Earth Elements and Their Compounds*, John Wiley and Sons, New York, 1947.

BEARING METALS

By Arthur F. Underwood

25. SELECTION OF BEARING METAL

Bearings are subjected to such varied influences that no one metal is suitable for all applications. Only by appreciating these influences and selecting an appropriate metal having the requisite characteristics can the best design be made. The individually different requirements of a bearing are (1) score resistance, (2) compressive strength, (3) deformability, including conformability and embedability, (4) fatigue strength, and (5) corrosion resistance. No order of relative importance is suggested by the listing. For one application, fatigue strength may be the predominant requirement, whereas another application can require the very best score resistance with maximum corrosion resistance. The selection of a bearing metal is usually a compromise, and the best choice is made by first recognizing what the bearing requirements are, and then selecting a commercial alloy having these qualities.

SCORE RESISTANCE. The score resistance, or anti-seizure characteristic, of metals used for bearings is often the prime requirement. Most bearings are designed to run on oil films. But metal-to-metal contact between the journal and bearing is caused by starting, stopping, deflection of the components, or heavy loads. On the other hand, many bearings operate continuously under partial or "boundary" lubrication conditions which produce rubbing contact of the surfaces. If the score resistance is not sufficiently high for the surface speed, the bearing metal will score, or ultimately seize if severe enough. At low surface speed, the heat generated at the contacts may be dissipated through the metal and heavy loads can be carried.

Experience has provided a guide to the many available metals, which is given in Table 1. Low loads are under 1000 psi, moderate loads are up to 2000 psi, and over 3000 psi is a heavily loaded bearing. Surface speeds can be considered low up to 5 to 10 ft per sec, moderate up to 15 to 20 ft per sec, and high when above 30 ft per sec. For each class a typical analysis (T) is given. It should be noted that such a general classification does not allow the inclusion of special bearing metals.

Table 1. Score Resistance and Compressive Strength of Bearing Metals

Composition, %				Order of Score Resistance	Order of Compressive Strength	Suggested Application
Cu	Zn	Sn	Pb			
85–90 (T) 86	(Fe: 2) (Fe: 2)	(Al: 4–10) (Al: 4)	.....	8	1	Very poor anti-score quality. Used in minor locations.
60–90 (T) 90	10–40 9	....	0–2 1/2	7	2	Same as above. Used with hypoid gear oils.
85–90 (T) 88	1/2–3 2	7–10 10	0–5	6	3	For heavily loaded wrist pins and similar motions where surface speed is very low.
80–90 (T) 88 (T) 80	0–5 4	3–10 4 10	5–10 4 10	5	4	Average bushing materials for low-speed, small-diameter shafts.
70–75 (T) 72	....	4–10 4	20–25 20	4	5	For moderate-speed bushings; and where lubrication is not well maintained.
Aluminum alloys				3	6	For heavily loaded moderate-speed bearings.
Copper-leads				2	7	For heavily loaded high-speed bearings.
White metal alloys				1	8	Low to moderately loaded, low- to high-speed bearings, where dirt and deflections are present.

An increase in lead content and a decrease in zinc and tin improve the score resistance of brass and bronze. In some instances the high-score quality of white metals is utilized by plating a very thin layer of it over an intermediate coating of silver or appropriate bronze which imparts increased strength.

COMPRESSIVE STRENGTH. Especially under steady, unidirectional loads, the bearing must have sufficient compressive strength not to extrude or crack. It should be recognized, however, that fatigue strength needed by a bearing to withstand an alternating load is a separate property. In Table 1 the metals are shown in order of decreasing compressive strength. In general, decreasing strength is associated with increasing score resistance. As the inherent compressive strength becomes less, thinner layers are used to improve the effective strength. This bearing metal requirement is one which can be directly related to the physical properties found in handbooks and suppliers' catalogs. It is important to consider the effect of temperature, especially when choosing a white metal. The data for yield point and ultimate strength in Table 4 demonstrate the softening of certain commercial babbitts.

FATIGUE STRENGTH. Fatigue resistance is important for bearings subjected to alternating or rotating loads. Well-known examples are the connecting rod and main bearings of internal-combustion engines. Failure starts as small cracks at the surface of the bearing; these work down to the bonding material where they progress parallel to the backing. "Tiles" of metal are thus loosened and may be eroded away by the lubricant action. Such failures should not be confused with those resulting from poor workmanship giving bad bond, when the difficulty shows up very early and can usually be identified by no bonding material on the backing metal.

The operating conditions (especially time and temperature) and the design of the bearing housings have so much influence that it is difficult to state specific loads. In Table 2, however, a guide to the maximum load is given.

Table 2. Fatigue Strength and Deformability of Bearing Metals

	Order of Fatigue Strength	Approximate Maximum Load, psi	Order of Deformability
Bronzes	1	10,000	7
Copper-lead with tin or silver	2	3000–4000	6
Thin babbitt overlays (0.003 in. or less)	3	2000–4000	5
Aluminum alloys	4	2000–3000	4
Copper-lead	5	1500–2500	3
Cadmium alloys	6	1200–1500	2
Lead and tin base babbitts	7	800–1500	1

The method and techniques of manufacturing these metals affect their strength and fatigue life. The performance of white metals has been improved by precision manufacturing methods to obtain thin layers. Figure 1 shows the data on babbitts. The fatigue strength of aluminum alloys reaches its highest value when securely bonded to a steel back. However, long life is found with solid aluminum alloy * bearings when the shells can be made quite thick and the temperature rise in the steel housing is not sufficient to cause permanent set.

Silver is a specialized bearing material used in aircraft engines because of its high fatigue strength.

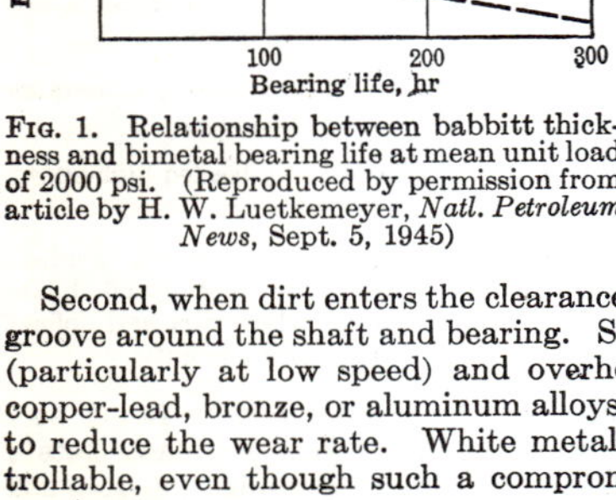

FIG. 1. Relationship between babbitt thickness and bimetal bearing life at mean unit load of 2000 psi. (Reproduced by permission from article by H. W. Luetkemeyer, *Natl. Petroleum News*, Sept. 5, 1945)

DEFORMABILITY. When loads are applied to a structure the bearings or journals deflect, often producing edge contact. If a conformable bearing metal is selected for the design, an adjustment is made without producing high temperatures that lead to a burned-out or fatigued bearing.

Second, when dirt enters the clearance space, it must embed in the bearing metal or cut a groove around the shaft and bearing. Silica (dust, core sand) or metal particles cause wear (particularly at low speed) and overheating (at high speed with hard metals). When copper-lead, bronze, or aluminum alloys are used, it is good practice to harden the journals to reduce the wear rate. White metals are preferred where dirty conditions are uncontrollable, even though such a compromise may involve a lowering of another desirable requirement.

Frequently the ability to embed when dirt is present and to conform when distortion is

* H. Y. Hunsicker and L. W. Kempf, *SAE Quarterly Trans.*, Vol. 1, No. 1, 1947.

ment are treated as separate characteristics. Because in practice a material responds about equally to each of these influences Table 2 shows this characteristic as deformability.

CORROSION RESISTANCE. Bearing metals containing unalloyed percentages of silver, copper, lead, and cadmium are subject to the corrosive attack of oils containing acids. As pointed out in Table 1, a leaded brass greatly decreases corrosion of bronze by sulfur and is to be used in designs calling for EP gear oils. Lead base alloys containing over 3% tin or suitable amounts of antimony, arsenic, or indium are protected from acidic lubricants. Cadmium bearings can be protected by small amounts of indium.

In Table 3, bearing metals are classified for their resistance to corrosion. The table should be used with discretion as the "intermediate" and "corrodible" alloys may reverse

Table 3. Corrosion Resistance of Bearing Metals

Noncorrodible	Intermediate	Corrodible
Aluminum alloys	Bronzes (high lead)	Cadmium alloys
Tin-base babbitt	Copper-lead	
Lead-base babbitt (properly alloyed)	Alkali-hardened lead	
	Silver *	
Cadmium-indium alloys		
Bronzes (low lead)		

* Attacked by sulfur under certain conditions.

their activity in the presence of certain lubricants. By proper additives to lubricants, the possibility of bearing corrosion is greatly diminished, and the designer may specify the correct oil to be used. Lubricant specialists should be consulted.

26. COMMERCIAL BEARING METALS

BABBITT METALS. The tin-base bearing materials, often referred to as "babbitts," are among the best metal alloys as they have excellent characteristics, except the ability to carry more than low loads. Their cost is low compared to most other metals, and in relatively unimportant locations lead can be added in amounts over 18% to further reduce the expense. Specifications of ASTM grade 1 through 5 (Table 4) and SAE 10 through 12 (Table 5) cover most of the standard tin-base babbitts.

ASTM grade 1 (SAE 10) is the softest alloy and is most liable to extrusion; ASTM grade 3 is the hardest and is a good choice for high compressive loads. ASTM grade 2 (SAE 12) is recommended as a general selection, especially for designs involving fatigue loading. ASTM grades 4 and 5 contain high percentages of lead which allows a mushy eutectic to form at low temperature (370 F) and, therefore, restricts their use to very light-duty applications.

Lead-base bearings have the advantage of lowest cost but do not support as high loads as tin-base babbitt when thick linings are used. However, when the thickness is less than about 0.010 in., no appreciable difference is to be found in load-carrying ability. As with tin-base babbitts, these alloys have excellent characteristics except the loading factor. Tables 4 and 5 include the ASTM and SAE specifications. Many of these are variants of a general class. ASTM grade 7 (SAE 14) is typical of the lead-tin-antimony alloys, which are widely used for all bearing applications. ASTM grades 10, 11, and 12 are the cheapest babbitts and are serviceable for light loads. Typical of arsenical leads is ASTM grade 15 (SAE 15), which is widely used in automotive bearings where thin layers (0.002 to 0.006 in.) allow moderate fatigue loads to be carried. Alloy grade 19 is similar to the chemical composition of lining specified for car and tender journal bearings (ASTM B67).

COPPER-LEAD ALLOYS. Copper-lead bearings are backed by steel to give strength and rigidity. They are widely used where loads are higher than can be carried on babbitts. Table 6 gives the SAE specifications of standard commercial analyses. The higher lead content gives a higher score quality but at the expense of strength. Because SAE 480 has a broad lead range to cover the numerous commercial compositions the designer should specify closer tolerance for purchasing.

To obtain higher fatigue strength, tin up to 3% and silver up to 5% are added, at the sacrifice of score quality. This result can be offset by overplating a thin coating of lead alloy.

BRONZES. Bronzes are especially valuable because of their low cost and wide range of bearing characteristics as determined by the many various compositions. They are especially resistant to wear when used with hard shafts.

The number of commercial bronze alloys is very large. The SAE specifications in Table 7 can be used as a guide in conjunction with Table 1, which gives typical classifica-

Table 4. Composition and Physical Properties* of White Metal Bearing Alloys

(ASTM Standard B23-48T)

ASTM Alloy Grade †	Specified Nominal Composition of Alloys				Specific Gravity ‡	Compositions of Alloys Tested				Yield Point, psi ‡		Johnson's Apparent Elastic Limit, psi §		Ultimate Strength in Compression, psi ‖		Brinell Hardness ¶		Melting Point		Temperature of Complete Liquefaction		Proper Pouring Temperature	
	Tin, %	Antimony, %	Lead, %	Copper, %		Tin, %	Antimony, %	Lead, %	Copper, %	20 C	100 C	20 C	100 C	20 C	100 C	20 C	100 C	°F	°C	°F	°C	°F	°C
1	91.0	4.5	....	4.5	7.34	90.9	4.52	None	4.56	4400	2650	2450	1050	12,850	6950	17.0	8.0	433	223	700	371	825	441
2	89.0	7.5	....	3.5	7.39	89.2	7.4	0.03	3.1	6100	3000	3350	1100	14,900	8700	24.5	12.0	466	241	669	354	795	424
3	83.33	8.33	....	8.33	7.46	83.4	8.2	0.03	8.3	6600	3150	5350	1300	17,600	9900	27.0	14.5	464	240	792	422	915	491
4	75.0	12.0	10.0	3.0	7.52	75.0	11.6	10.2	3.0	5550	2150	3200	1550	16,150	6900	24.5	12.0	363	184	583	306	710	377
5	65.0	15.0	18.0	2.0	7.75	65.5	14.1	18.2	2.0	5050	2150	3750	1500	15,050	6750	22.5	10.0	358	181	565	296	690	366
6	20.0	15.0	63.5	1.5	9.33	19.8	14.6	63.7	1.5	3800	2050	3550	1800	14,550	8050	21.0	10.5	358	181	531	277	655	346
7	10.0	15.0	75.0	....	9.73	10.0	14.5	75.0	0.11	3550	1600	2500	1350	15,650	6150	22.5	10.5	464	240	514	268	640	338
8	5.0	15.0	80.0	....	10.04	5.2	14.9	79.4	0.14	3400	1750	2650	1200	15,600	6150	20.0	9.5	459	237	522	272	645	341
10	2.0	15.0	83.0	....	10.07	2.05	15.7	82.0	0.12	3350	1850	2250	1200	15,450	5750	17.5	9.0	468	242	507	264	630	332
11	....	15.0	85.0	....	10.28	0.09	14.8	84.7	0.19	3050	1400	2750	1100	12,800	5100	15.0	7.0	471	244	504	262	630	332
12	....	10.0	90.0	....	10.67	0.11	9.9	89.4	0.12	2800	1250	2250	950	12,900	5100	14.5	6.5	473	245	498	259	625	329
15 ††	1.0	15.0	82.5	0.5	10.05	....	....	....	....	......	......	......	......	....	......	21.0	13.0	479	248	538	281	662	350
16	10.0	12.5	77.0	0.5	9.88	....	....	....	....	......	......	......	......	....	......	27.5	13.6	471	244	495	257	620	327
19	5.0	9.0	86.0	....	10.50	....	....	....	....	......	......	......	......	15,600	6100	17.7	8.0	462	239	495	257	620	327

* The compression test specimens were cylinders 1.5 in. in length and 0.5 in. in diameter, machined from chill castings 2 in. in length and 0.75 in. in diameter. The Brinell tests were made on the bottom face of parallel machined specimens cast in a steel mold, at room temperature, 2 in. in diameter and 0.625 in. deep.

† Alloy grade No. 9 was discontinued in 1946.

‡ The values for yield point were taken from stress-strain curves at a deformation of 0.125% reduction of gage length.

§ Johnson's apparent elastic limit is taken as the unit stress at the point where the slope of the tangent to the curve is 1 1/2 times its slope at the origin.

‖ The ultimate strength values were taken as the unit load necessary to produce a deformation of 25% of the length of the specimen.

¶ These values are the average Brinell number of three impressions on each alloy using a 10-mm ball and a 500-kg load applied for 30 sec.

** The specific gravity multiplied by 0.0361 equals the density in pounds per cubic inch.

†† Also nominal arsenic 1.0%.

tions and their use. Numerous variations of these bronzes and special-purpose bronze alloys are to be found in the ASTM Standards, Part 1-B, and the *SAE Handbook.* SAE 40, 62, 64, and 68A correspond to ASTM Designations B145-4A, B143-1A, B144-3A, and B148-9A respectively.

Table 5. SAE Standard Babbitt Metals

(*SAE Handbook*)

	Compositions, %					
	SAE 10	SAE 11	SAE 12	SAE 13	SAE 14	SAE 15
Tin	90.0	86.0	88.25	5.0–7.0	9.25–10.75	0.9 –1.25
Antimony	4.0–5.0	6.0–7.5	7.0–8.5	9.0–11.0	14.0–16.0	14.50–15.50
Lead	0.5	0.5	0.5	Remainder	Remainder	Remainder
Copper	4.0–5.0	5.0–6.5	2.25–3.75	0.50	0.50	0.60
Arsenic	0.10	0.10	0.10	0.25	0.60	0.80–1.10
Iron and bismuth, each	0.08	0.08	0.08	....	....	....
Other elements *	0.20	0.20	0.20	0.20	0.20	0.20

* Zinc and aluminum maximum 0.005% in a 10-gram sample for each of all specifications.

Table 6. SAE Standard Copper-lead Bearing Alloys

(*SAE Handbook*)

	Composition, %		
	SAE 48	SAE 480	SAE 49 *
Copper	67.0–74.0	60.0–70.0	73.0–79.0
Lead	25.0–32.0	30.0–40.0	21.0–27.0
Silver	1.5	1.5	....
Zinc	0.1	....	....
Phosphorus	0.025	....	....
Iron	0.35	0.35	0.35
Tin	0.25	0.50	0.50
Other elements	0.15	0.3	0.45

* This alloy to be used in combination with a corrosion-resistant overlay.

Table 7. SAE Standard Brass and Bronze Bearing Alloys

(*SAE Handbook*)

	Composition, %								
	SAE 40 Cast Lining	SAE 62 Cast Lining	SAE 64 Cast Lining	SAE 68A Cast Lining	SAE 791 Rolled or Wrought	SAE 792 Cast Lining	SAE 793 Cast Lining	SAE 794 Cast Lining	SAE 795 Rolled or Wrought
Copper	84.0–86.0	86.0–89.0	78.0–82.0	86.0–89.0	remainder	remainder	83.0 min	68.5–75.5	88.0–92.0
Tin	4.0–6.0	9.0–11.0	9.0–11.0	0.50	3.5–4.5	9.0–11.0	3.5–4.5	3.0–4.0	0.25–0.75
Lead	4.0–6.0	0.30	8.0–11.0	....	3.5–4.5	9.0–11.0	7.0–9.0	21.0–25.0	....
Zinc	4.0–6.0	1.0–3.0	0.75	....	1.5–4.0	0.5	4.0	3.0	remainder
Antimony	....	....	0.50	....	......	0.5	0.5	0.5	....
Nickel	1.0	1.0	0.50	....	......	0.5	0.5	0.5	....
Iron	0.30	0.15	0.15	2.5–4.0	0.1	0.35	0.35	0.35	0.1
Aluminum	0.005	0.005	0.005	8.5–9.5	......	....	......	....	....
Phosphorus	....	....	0.25	....	......	....	......	....	....
Silicon	0.005	....	0.005	....	......	....	......	....	....
Other elements	....	....	....	....	......	0.2	0.3	0.4	0.2

ALUMINUM ALLOYS. Commercially two aluminum bearing alloys are available as shown in Table 8. Bearing manufacturers should be consulted for specific information on the best applications for these alloys.

Table 8. Aluminum Bearing Alloys

	Composition, %	
	750	A-750
Aluminum	Remainder	Remainder
Tin	6.5	6.5
Nickel	1	0.5
Copper	1	1
Silicon		2.5

CADMIUM ALLOYS. These alloys are used in very special designs where their added cost is justified by the increased fatigue life. Table 9 shows the SAE specifications for the two alloys which are essentially the same in practical use. These alloys are attacked by corrosive oils unless treated with indium.

Table 9. SAE Standard Cadmium Bearing Alloys

	Composition, %	
	SAE 18	SAE 180
Cadmium, min	98.4	98.25
Nickel	1.0–1.6	0.25
Silver	0.01	0.5–1.0
Copper	0.20	0.4–0.75
Tin	0.02	0.02
Lead	0.05	0.05
Zinc	0.15	0.15
Bismuth	0.02	0.02

MISCELLANEOUS ALLOYS. Oil-less bearings are important in designs where infrequent lubrication is expected. Specially prepared bronze is treated with oil or graphite. In general, low load and low speed are required for successful design.

Certain zinc-base alloys are used where the imposed conditions are low and cheapness is desired.

DIE-CASTING ALLOYS

By J. C. Fox

The necessary requirements of a suitable die-casting alloy are: (1) Physical and mechanical properties—strength, hardness, etc. (2) Permanence of properties and dimensions in use. (3) Fluidity. (4) Strength at temperatures of solidification. (5) Solidification range. (6) Low shrinkage. (7) Machinability. (8) Polishing, plating, and finishing properties. (9) Corrosion resistance. (10) Weight and cost.

The alloys used in modern die-casting practice may be classified into six main groups: lead, tin, zinc, aluminum, magnesium, and copper base alloys. The general characteristics, properties, and applications of each of these groups of alloys are given in the following pages.

27. LEAD- AND TIN-BASE ALLOYS

Table 1 shows the chemical requirements of lead and tin-base die-casting alloys under ASTM B102-44.

Table 1. Lead- and Tin-base Die-casting Alloys

(ASTM B102-44)

Alloy No.	Tin, %			Antimony, %			Lead, %			Copper, %			Other Elements, max %			
	Min	Desired	Max	Min	Desired	Max	Min	Desired	Max	Min	Desired	Max	Iron	Arsenic	Zinc	Aluminum
1 *	90	91	92	4	4.5	5	.....	.....	0.35	4	4.5	5	0.08	0.08	0.01	0.01
2	80	82	84	12	13	14	.....	.....	0.35	4	5	6	0.08	0.08	0.01	0.01
3 *	64	65	66	14	15	16	17	18	19	1.5	2.0	2.5	0.08	0.15	0.01	0.01
4 *	4	5	6	14	15	16	79	80	81	.....	.....	0.50	......	0.15	0.01	0.01
5 *	.....	.....	.....	9.25	10.0	10.75	89	90	91	.....	.....	0.50	......	0.15	0.01	0.01

* These alloys are similar to grades 1, 5, 8, and 12, respectively, of the Tentative Specifications for White Metal Bearing Alloys (known commercially as Babbitt Metal, ASTM B23). See Article 26 in this section.

LEAD BASE alloys are usually employed where a low-cost noncorrosive metal is required and where strength, hardness, and other mechanical properties are unimportant. Parts which must withstand the action of strong mineral acids, as in fire-extinguisher parts, batteries, and other similar chemical apparatus, are produced as lead die castings. X-ray equipment employs some lead die castings because of the resistance lead shows to the passage of x-rays. The high specific gravity of lead is taken advantage of in the use of lead die castings for governor weights and similar uses.

TIN-BASE die castings have been extensively used in the past for bearings, especially for automobiles. This use, however, has greatly dwindled because of the substitution of other bearing materials. Tin-base die castings are used particularly for their corrosion resistance for such parts as soda-fountain equipment, milking machines, syrup pumps, dental appliances, and surgical instruments.

The mechanical properties of both lead- and tin-base alloys are very low, and die castings of them therefore represent only a very small percentage of the total die-casting consumption. For some specific applications, such as those given above, these are quite indispensable.

28. ZINC-BASE ALLOYS

Table 2 shows the chemical composition of the three standard zinc-base die-casting alloys, covered under ASTM Specification B86-46, and also by the SAE.

Table 2. Zinc-base Die-casting Alloys

ASTM No. SAE No.	Alloy XXI 921	Alloy XXIII 903	Alloy XXV 925
Copper, %	2.5–3.5	0.10 max	0.75–1.25
Aluminum, %	3.5–4.5	3.5–4.3	3.5–4.3
Magnesium, %	0.02–0.10	0.03–0.08	0.03–0.08
Iron, %	0.100 max	0.100 max	0.100 max
Lead, %	0.007 max	0.007 max	0.007 max
Cadmium, %	0.005 max	0.005 max	0.005 max
Tin, %	0.005 max	0.005 max	0.005 max
Zinc, %	Remainder	Remainder	Remainder

It should be noted that these three alloys, which have been adopted by all standards groups, are based on the use of 99.99+% grade of zinc, which is mandatory. They all contain approximately 4% aluminum and vary only in the copper content. Alloy XXI, having 2.75% copper, is gradually being obsoleted in favor of the other two alloys, XXIII and XXV, because the XXI alloy shows poorer aging characteristics than the others. There is little to choose between the XXIII and XXV alloys. The XXIII alloy possesses somewhat better aging properties over a long period of time, although having initially lower tensile strength and hardness than the XXV alloy. The latter alloy possesses slightly better surface corrosion resistance when used without surface protection.

The limits of each constituent of these alloys have been carefully established after many years of experience and thorough testing. Although each constituent has a minimum and maximum content specified, there is an optimum content for each that produces the best all-round results.

Aluminum. Although the aluminum content is specified from 3.5 to 4.3, the best content is considered to be about 4.0%. Aluminum appreciably above 4.30% affects impact strength. At 5.0% the alloy is quite brittle. On the low side, aluminum below the minimum of 3.50% results in poorer castability and lower physical properties

Copper increases tensile strength, hardness and surface corrosion resistance. However, copper adversely affects the ductility, impact strength, and dimensional stability on aging.

The XXIII alloy can stand a copper content up to 0.25% without loss of properties due to aging. Higher copper contents than this quantity causes the alloy to be more susceptible to the effects of tin.

The XXV alloy is not seriously affected by copper contents outside the specification limits. Copper contents appreciably above the maximum will increase surface corrosion resistance, hardness, and strength, but will show aging effects with subsequent loss of impact strength and more dimensional growth than if the copper were held at the mean amount.

Magnesium is a necessary addition to the zinc alloys, primarily to act as an inhibitor of subsurface corrosion which may occur without it. The optimum content is 0.04%, with 0.03% as the very minimum to maintain permanence of properties and dimensions. Magnesium, above 0.05% and especially approaching the 0.08% allowed in the specifications, is conducive to "hot-shortness" in casting and to lowering the impact strength.

Lead should be kept under the 0.007% maximum specified to prevent occurrence of intercrystalline corrosion. Care should be exercised in selection and use of the quality of aluminum in alloying with the zinc. No difficulty is found in holding to this maximum lead content of die castings. The aluminum used must not contain more than 0.02% lead. Contamination of lead-containing materials, such as solder, foil, cigarette papers, lead paint, must, of course, be guarded against.

Cadmium. Unless contaminated by material containing cadmium, there is no difficulty in holding to the 0.005% maximum set forth in these specifications.

Tin, like lead, promotes subsurface corrosion and it is therefore necessary to hold to a maximum limit for this element. The results of extensive experimental work on the effect of tin, in this type of alloy, establishes the maximum safe limit of 0.005%.

Iron. About 0.02% iron is soluble in the liquid zinc alloy. Iron present above this amount is combined with aluminum, which tends to rise to the surface and is periodically removed as dross. Iron up to 0.10% maximum will have no effect on the permanence of properties and dimensions. If any of the high-iron aluminum compounds should inadvertently get into the casting, they will show up as hard spots and affect machinability. This, however, can happen only by carelessness.

Contamination. In addition to the care necessary in the selection of the proper grades of raw metal used in the preparation of the zinc die-casting alloy, extreme precaution must be exercised to insure against accidental contamination by injurious metals and alloys, such as lead, tin, cadmium, bismuth, and some others.

Table 3. Physical Properties of Zinc-base Alloys

	Alloy XXI	Alloy XXIII	Alloy XXV
Impact strength, ft-lb	19	20	17
Tensile strength, psi	47,900	40,300	44,500
Tensile elongation, in 2 in., %	5.1	4.7	3.0
Brinell hardness	83	74	76
Compressive strength, psi	93,100	60,500	87,300
Shearing strength, psi	45,800	30,900	38,400
Transverse deflection, in.	0.22	0.27	0.16
Electrical conductivity, ohms/cmr 25 C	144,000	155,000	150,000
Melting point, °F	715.1	717.6	717.1
Specific gravity	6.7	6.6	6.7
Specific heat, cal/g/°C	0.10	0.10	0.10
Thermal conductivity, cal/sec/cm^3	0.25	0.27	0.26
Thermal expansion, per °F	15.4×10^{-6}	15.2×10^{-6}	15.2×10^{-6}

Table 4. Effect of Normal and Accelerated Aging on Properties of Zinc-base Alloys

Alloy	Condition	Impact Strength *	Tensile Strength	Tensile Elongation, in 2 in., %	Brinell Hardness
XXIII	As cast	43	41,000	10	82
	Accelerated aging †	36	31,000	17	..
	Normal aging ‡	41	33,800	16	..
XXV	As cast	48	47,600	7	91
	Accelerated aging †	7	35,200	13	..
	Normal aging ‡	42	38,300	15	..

* Charpy, ft-lb on 1/4 in. by 1/4 in. bars.
† Two years at 95 C in dry air.
‡ Seven years at room temperature.

Table 5. Mechanical Properties of Zinc-base Die-casting Alloys at Subnormal and Elevated Temperatures

Temperature	−40 C −40 F	−20 C −4 F	0 C 32 F	21 C 70 F	40 C 104 F	95 C 203 F
XXIII Alloy						
Tensile strength, psi	44,800	43,700	41,300	41,000	35,500	28,300
Impact strength, ft-lb 1/4 in. x 1/4 in. bars	2	4	23	43	42	40
Brinell hardness	91	87	82	82	68	43
Elongation, in 2 in., %	3	4	6	10	16	30
XXV Alloy						
Tensile strength, psi	48,900	49,400	48,300	47,600	42,900	35,100
Impact strength, ft-lb 1/4 in. x 1/4 in. bars	2	4	41	48	46	43
Brinell hardness	107	194	99	91	89	62
Elongation, in 2 in., %	2	3	6	7	13	23

Zinc die casting must be segregated from the casting of other alloys, particularly the casting of tin- and lead-base alloys. The presence of lead and tin materials, such as lead hammers, counterweights, matrix metal, solder, foil, tobacco wrappings, and other like articles must be prohibited to prevent possible contamination. Likewise, cadmium materials such as plated inserts and castings must be prevented from coming into contact with the molten zinc die-casting alloy.

PHYSICAL AND MECHANICAL PROPERTIES. Table 3 shows physical and mechanical properties of the XXI, XXIII, and XXV alloys, as cast. Table 4 shows average mechanical properties of the XXIII and XXV alloys from die-cast test bars cast under the best conditions, before and after normal and accelerated aging. Table 5 shows the effect of temperature on mechanical properties.

DIE CASTING OF ZINC ALLOYS. See Section 19, Article 21.

29. ALUMINUM-BASE ALLOYS

The outstanding characteristics of aluminum die castings are: (1) Low specific gravity. (2) Corrosion resistance. (3) Retention of polished luster for long periods of time. (4) Freedom from dimensional changes. (5) Retention of properties at subnormal temperatures. (6) High thermal and electrical conductivity.

This combination of properties has resulted in the wide use that aluminum die castings have reached. The average specific gravity of aluminum die castings is 2.80, which is compared with 1.80 for magnesium alloys. For many applications, especially small parts, this weight differential between these two metals becomes negligible.

The ability of a part to take and to hold a polished luster is important, and advantage is taken of this property in the finishing of many aluminum die castings, especially household utility parts, by simply polishing the casting—resulting in a final finish highly decorative and of low cost. The binary magnesium-aluminum type alloys containing 4 to 10% magnesium are superior to other types of aluminum-base alloys in this property since they are capable of producing a whiter and highly lustrous finish.

Aluminum die castings have excellent corrosion resistance in the as cast condition. When it is necessary to increase resistance to attack by corrosion media, resort can be had to the anodic oxidation treatment, which greatly enhances corrosion resistance even in salt atmospheres. Here, again, the aluminum-magnesium excel over other alloys, particularly where decorative effects through anodizing are desired.

Aluminum die castings are remarkably stable and are not subject to changes of dimensions or properties on aging. Whenever it is required to hold to extremely close machined dimensions, however, say of the order of 0.0001 in., in some applications, it may be expedient to anneal the die castings prior to machining, to relieve any residual casting stresses set up as a result of the rapid chilling of the liquid metal. A simple anneal of about 400 F for about one hour is usually sufficient to accomplish complete stability.

Aluminum is one material capable of withstanding extreme low temperatures without affecting original ductility. For this reason aluminum has found extensive use in aircraft, refrigerators, and similar equipment where temperatures as low as −60 F may be met.

Table 6 lists the compositions of the aluminum-base die-casting alloys which have been standardized by ASTM under specification designation B85-48T, together with the equivalent SAE and Aluminum Company of America numbers.

Table 6. Aluminum-base Die-casting Alloys

Designation *			Chemical Composition, %								
ASTM No.	SAE No.	Alcoa No.	Copper	Fe, max	Silicon	Mn, max	Magnesium	Zinc, max	Ni, max	Tin, max	Other max
S4	304	43	0.6 max	2.0	4.5– 6.0	0.3	0.1 max	0.5	0.5	0.1	0.2
S5	305	13	0.6 max	2.0	11.0–13.0	0.3	0.1 max	0.5	0.5	0.1	0.2
S9	.....	A13	0.6 max	1.3	11.0–13.0	0.3	0.1 max	0.5	0.5	0.1	0.2
SC2	307	85	3.0–4.0	2.0	4.5– 5.5	0.5	0.1 max	1.0	0.5	0.3	0.5
SC5	.....	A85	3.0–4.0	1.3	4.5– 5.5	0.5	0.1 max	0.6	0.5	0.3	0.5
SC6	306	A380	3.0–4.0	1.3	7.5– 9.5	0.5	0.1 max	0.6	0.5	0.3	0.5
SC7	308	380	3.0–4.0	2.0	7.5– 9.5	0.5	0.1 max	1.0	0.5	0.3	0.5
SG2	.....	A360	0.6 max	1.3	9.0–10.0	0.3	0.4–0.6	0.5	0.5	0.1	0.2
SG3	.....	360	0.6 max	2.0	9.0–10.0	0.3	0.4–0.6	0.5	0.5	0.1	0.2
G2	.....	218	0.2 max	1.8	0.3 max	0.3	7.5–8.5	0.1	0.1	0.1	0.2

* Detail composition limits are ASTM specifications. SAE and Alcoa numbers denote equivalent alloys.

Table 7 shows the typical physical properties that may be expected of tension and impact test specimens when made in a die in regular production routine and conforming to the chemical composition specified. It is to be understood that these figures represent die-cast test specimens and not specimens cut from commercial die-cast parts.

Table 7. Typical Mechanical Properties of Aluminum-base Die-casting Alloys

	Tension Tests on Round Specimens			Charpy Impact
Designation	Ultimate Strength, psi	Yield Strength,* psi	Elong. in 2 in., %	on Square Specimens, ft-lb
S4	30,000	16,000	5.0	4.5
S5	37,000	20,000	1.8	2.0
S9	35,000	......	3.5	...
SC2	38,000	23,000	2.5	2.5
SC5	35,000	......	3.0	3.0
SC6	41,000	22,000	3.0	3.5
SC7	43,000	25,000	2.0	3.2
SG2	43,000	24,000	5.0	4.2
SG3	43,000	26,000	3.0	2.7
G2	41,000	25,000	7.5	8.3

* Point where stress-strain curve departs 0.2% from the initial modulus line produced.

APPLICATIONS. Aluminum-base alloy die castings represent about 28% of the total die-casting consumption, extending to practically all industries—household utilities, such as vacuum sweepers, washing machines, ironers, food-mixing devices, slicing machines, cooking utensils, and others. The automotive industry uses aluminum die castings for transmission parts, carburetors, hydraulic brake parts, and trim. The electrical industry uses them for motor housings, instrument cases and frames, and radio equipment. The optical uses extend to cameras, projectors, binoculars, and the like. There are many other industrial uses of aluminum die castings, too numerous to list here.

30. MAGNESIUM-BASE ALLOYS

The standard magnesium-base die-casting alloys shown in Table 8 are covered by ASTM B94-47T.

Table 8. Magnesium-base Die-casting Alloys

	Alloy No. AS100	Alloy No. AZ91	Alloy No. AZ91X
Aluminum, %	9.0–11.0	8.3–9.7	8.3–9.7
Manganese, min %	0.10	0.13	0.13
Zinc, %	0.3 max	0.4–1.0	0.4–1.0
Silicon, max %	1.0	0.5	0.3
Copper, max %	0.05	0.10	0.05
Nickel, max %	0.03	0.03	0.002
Iron, max %	....	....	0.004
Magnesium, %	Remainder	Remainder	Remainder

Typical physical properties which may be expected of these alloys in the die-cast state are given by Table 9.

Table 9. Typical Mechanical Properties of Magnesium-base Die-casting Alloys

Alloy	Tensile Strength, psi	Elongation in 2 in., %	Charpy Impact, ft-lb	Remarks
AS100	29,000–34,000	1–3	0.5–2.0	Good casting and good physical properties
AZ91	29,000–34,000	2–5	1 –3.0	Good casting and better physical properties
AZ91X	29,000–34,000	2–5	1 –3.0	Good casting, good resistance to salt-water corrosion

APPLICATIONS. Magnesium die castings are used mostly for applications where lightness, the chief characteristic of the metal (specific gravity of the foregoing alloys is about 1.81), is the main requirement, such as portable typewriters, stenotype and other business machine cases and housings, portable tools, and other portable equipment. Magnesium die castings are also used in reciprocating and moving parts for textile, conveying, and packaging machinery.

Other outstanding properties of magnesium alloys, such as ease of machinability, rigidity, vibration, and sound damping, are taken advantage of in a variety of applications.

31. COPPER-BASE (BRASS) ALLOYS

Present ASTM B176-47T covers two brass alloys for pressure die casting. These alloys have the chemical requirements indicated in Table 10.

Table 10. Copper-base (Brass) Die-casting Alloys

	Alloy A	Alloy B
Copper, %	57.0 min	63.0–67.0
Silicon, %	0.25 max	0.75–1.25
Lead, max %	1.50	0.25
Tin, max %	1.50	0.25
Manganese, max %	0.25	0.15
Aluminum, max %	0.25	0.15
Iron, max %	0.25	0.15
Other elements,* max %	0.50	0.50
Zinc, %	30.0 min	Remainder

* Of the maximum other elements specified, the amount of arsenic, antimony, and sulfur shall not exceed 0.05% for each, and phosphorus shall not exceed 0.01%.

Alloy A represents a general-purpose low-cost brass alloy with wide limits on impurities. Alloy B is a high-purity alloy capable of excellent castability and physical properties in the die-cast state.

The *minimum* tensile properties of these alloys shown in the foregoing ASTM specifications are:

	Alloy A	Alloy B
Ultimate tensile strength, psi	45,000	58,000
Yield strength, psi	25,000	30,000
Elongation, in 2 in., %	10	15

Die-cast test bars of these alloys, however, actually show far greater properties than these minimum figures. For instance, results from three laboratories on die-cast test bars of these alloys show an average ultimate tensile strength of over 60,000 psi for alloy A and over 80,000 psi for alloy B. The average yield strength for alloy A is 37,000 psi and over 50,000 psi for alloy B.

In addition to these alloys, another alloy is being extensively used, particularly for its high strength, hardness, wear, and corrosion resistance. This alloy has the following nominal composition and mechanical properties:

Copper, %	80–81.5
Silicon, %	3.5–4.5
Zinc, %	Remainder
Ultimate tensile strength, psi	85,000–95,000
Yield strength, psi	45,000–50,000
Elongation, in 2 in., %	12–20
Brinell hardness	160–180

This alloy is finding considerable use where resistance to wear is an important requirement. For this particular use, die castings of this alloy have been found to outwear other hard materials, and in a number of applications the alloy has outworn and replaced steel.

Die castings can be produced from other copper-base alloys, but present practice dictates the use of only those alloys having melting points under 1650 F. Aluminum bronze, tin bronze, low brasses, and other high-copper alloys can be readily die-cast, but the melting points and casting temperatures are too high to be practical or economical with the die materials at present available.

APPLICATIONS. The salient properties of high strength, toughness, corrosion, and wear resistance of brass-alloy die castings make them suitable for many varied uses and open to industry a source for engineering and structural parts having the accuracy, intricacy, stability, and economy afforded by the die-casting process. Automotive lock parts, synchronizer rings, brush holders, bearing caps, hydraulic brake pistons, transmission and clutch parts, electrical switchboard parts, contactors, refrigerator parts, fittings, bearings, oil-burner parts, general engineering fittings, and household hardware are a few of the applications of brass die castings.

SOLDERS

By Bruce W. Gonser

The two types of solder most generally used are lead-and-tin solders and silver solders. The latter are used in connection with high operating temperatures and for high strength. Brazing alloys of phosphorus-copper are used at temperatures between the melting point of copper and the softening range of the comparatively low-melting silver solders.

LEAD-TIN SOLDERS. The composition and melting range of lead-tin alloys used as solders and commercially known as soft solders, are given in ASTM standard specification, Designation B32-46T. Grade A solders should be used on zinc and galvanized iron, since antimony in the solder is harmful when soldering zinc.

Table 1 shows the composition and melting range of these solders. The grade of solder

Table 1. Composition and Melting Range of Lead-tin Solders

Alloy Grade,* †	Tin, %	Lead, %	Antimony, % (Max.)	Melting Ranges			
				Solidus		Liquidus	
				°C	°F	°C	°F
70A	70	30	0.12	183	361	192	378
70B	70	30	0.50	183	361	192	378
60A	60	40	0.12	183	361	190	374
60B	60	40	0.50	183	361	190	374
50A	50	50	0.12	183	361	216	421
50B	50	50	0.50	183	361	216	421
45A	45	55	0.12	183	361	227	441
45B	45	55	0.50	183	361	227	441
40A	40	60	0.12	183	361	238	460
40B	40	60	0.50	183	361	238	460
40C	40	58	2.4	185	365	231	448
35A	35	65	0.25	183	361	247	477
35B	35	65	0.50	183	361	247	477
35C	35	63.2	2.0	185	365	243	470
30A	30	70	0.25	183	361	255	491
30B	30	70	0.50	183	361	255	491
30C	30	68.4	1.8	185	365	250	482
25A	25	75	0.25	183	361	266	511
25B	25	75	0.50	183	361	266	511
25C	25	73.7	1.5	184	364	263	504
20B	20	80	0.50	183	361	277	531
20C	20	79	1.2	184	364	270	517
15B	15	85	0.50	183	361	288	550
10B	10	90	0.50	183	361	299	570
5B	5	95	0.50	270	518	312	594

* For elements other than those mentioned in the table, the maximum content in the alloy shall be:

Bismuth	0.25%
Copper alloy grades 70A to 5B inclusive	0.08%
Iron	0.02%
Aluminum, Zinc } each shall not exceed	0.005%

† The chemical requirements of SAE Specifications 1A, 2A, 2B, 3A, 3B, 4A, 4B, 5A, 5B, and 6A conform substantially to the requirements for alloy grades 45B, 40B, 40C, 30B, 30C, 25B, 25C, 20B, 20C, and 15B, respectively.

metal should be selected which contains the least amount of tin required to give suitable flowing and adhesive qualities for the work on hand. The alloys are completely solid below the lower point given, designated "solidus," and completely liquid above the higher point given, designated "liquidus." Between these two points the alloys are partly solid and partly liquid. Grades 30 and 35 are particularly suitable for making wiped joints because of the large pasty range.

Because of wartime restrictions on the use of tin, substitute solders have been used for many applications. The most common substitutes are based on 15 to 20% tin, remainder lead, with and without the addition of various amounts of silver, antimony, bismuth, cadmium, and zinc. Substitute solders are reviewed by Rhines and Anderson* and Russell and Mack.† For outside seams and dry packs, the canning industry uses a lead-tin solder containing as low as 2% tin. Low-tin solders are more difficult to use because of the higher melting points, but industry has overcome the initial difficulties and little trouble is now experienced in their use.

SILVER SOLDERS, which contain a greater or less amount of silver together with varying quantities of copper and zinc, are used for brazing. Table 2 gives the composition and characteristics of silver solders covered by ASTM standard specifications, Designation B73-29. Grades 1 to 3 are suitable for brazing where a freer flowing solder than the ordinary copper-zinc brazing solders is required. Grades 4 and 5 will flow more freely at lower temperatures than grades 1 to 3. They are used where strong joints are required, and where the temperature necessary to flow grades 1 to 3 would injure the metal to which the solder is applied. Grades 6 to 8 are used where a high degree of malleability or ductility is required. (See also Section 21, Articles 6 to 9.)

Table 2. Composition and Properties of Silver Solders

Grade	1	2	3	4	5	6	7	8
Ag, %	10	20	20	45	50	65	70	80
Cu, %	52	45	45	30	34	20	20	16
Zn, %	38	35	30	25	16	15	10	4
Cd, %	†	†	5	Nil	Nil	Nil	Nil	Nil
Impurities, % *	0.15	0.15	0.15	0.15	0.15	0.15	0.15	0.15
Melts at, °F	1510	1430	1430	1250	1280	1280	1335	1360
Flows at, °F	1600	1500	1500	1370	1425	1325	1390	1460
Color	Yellow	Yellow	Yellow	Nearly white	Nearly white	White	White	White

* Maximum. † The addition of not over 0.50% cadmium to assist fabrication shall not be considered a harmful impurity.

SOLDERING FLUXES. Neutral fluxes such as rosin or rosin in alcohol are noncorrosive. They are used where corrosive action must be avoided and where removal of the flux residue is impractical. Their cleaning action is slight, and they are not suitable if the surface is oxidized. Activated rosin fluxes, containing a small amount of an organic compound, are now available. These activated rosin fluxes are much more efficient than rosin when soldering oxidized surfaces and are relatively noncorrosive. More active fluxes containing zinc chloride or ammonium chloride, or both, dissolved in water or mixed with petrolatum jelly, are manufactured and sold by several companies. The zinc chloride fluxes are corrosive, and the joints should be washed with weak muriatic acid (5 : 95) and then with water after soldering to remove the insoluble and hygroscopic residue of zinc oxychloride. A very active but corrosive flux suitable for soldering stainless steel contains zinc chloride, ammonium chloride, and stannic chloride with a small amount of hydrochloric acid. For silver solder and brazing alloys, powdered borax or the proprietary brazing fluxes are used.

HIGH-TEMPERATURE-RESISTANT ALLOYS

By C. T. Evans, Jr.

Materials used in high-temperature applications, particularly gas turbines, are currently the subject of intensive research and development. Table 1 lists the chemical compositions of selected old and new alloys employed for high-temperature applications in 1948. Other alloys still to be developed or whose compositions have not been published may ultimately replace some of these materials. The superalloys in the table are classified under three categories: class A alloys, normally not heat treated; class B alloys, normally heat treated; class C alloys, fabricated by casting.

Two criteria of suitability for high-temperature high-stress applications are creep and stress rupture. The stress to produce a secondary creep rate of 1% in 10,000 hours is plotted in Fig. 1 against temperature. Ranges covered by superalloys of the three classes listed in Table 1 are compared with each other and with four typical older materials. Corresponding data for the stress to produce rupture in 1000 hours are plotted in Fig. 2. Stresses to produce creep and rupture in 100,000 hours are given in Figs. 3 and 4.

* Rhines and Anderson, *Metals and Alloys*, Vol. 14, 1941, pp. 704–711.
† Russell and Mack, *AIME Metals Technology*, T.P. 1770, 1944.

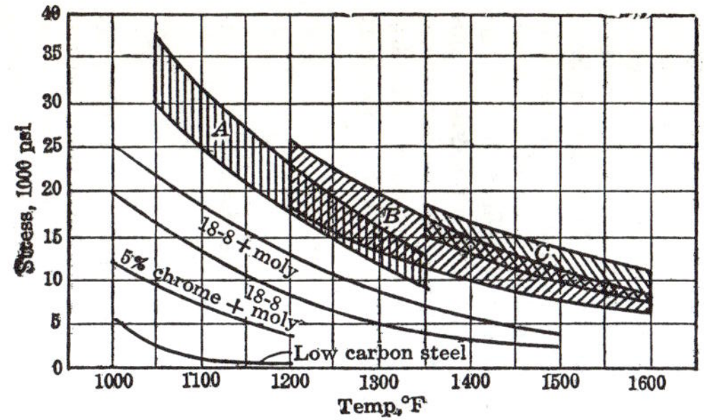

FIG. 1. Stress to produce secondary creep rate of 1% in 10,000 hours.

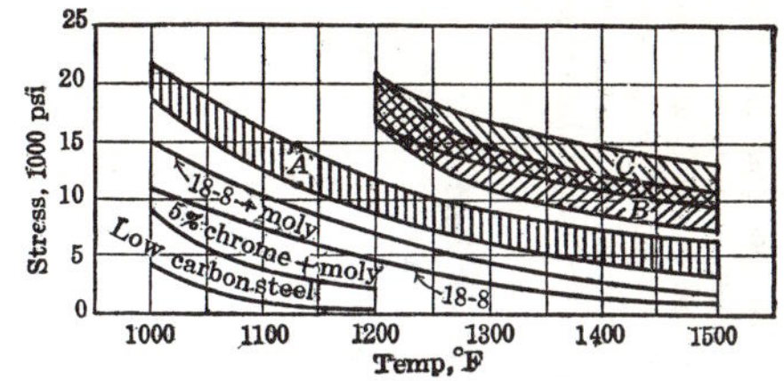

FIG. 3. Stress to produce secondary creep rate of 1% in 100,000 hours.

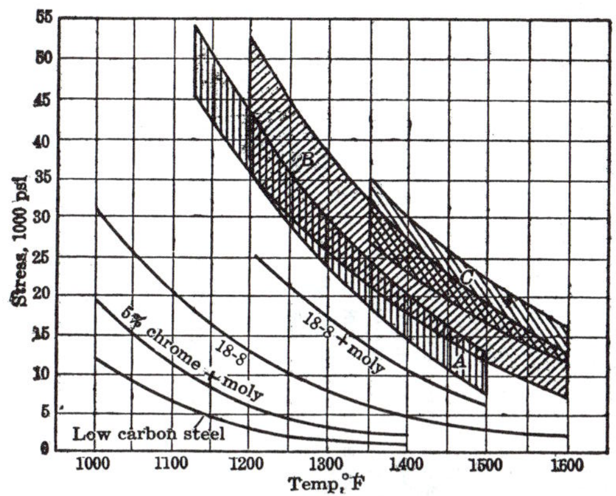

FIG. 2. Stress to produce rupture in 1000 hours.

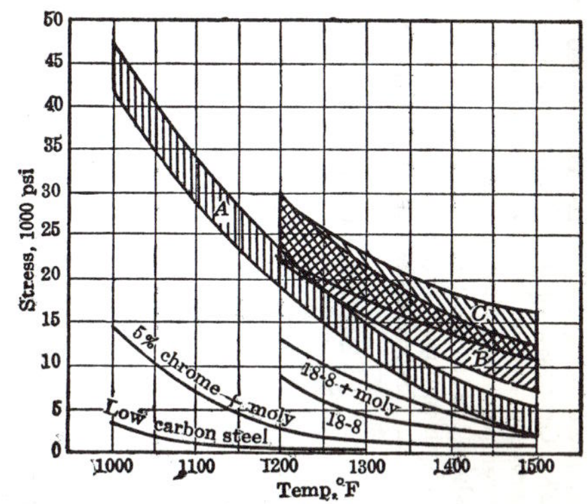

FIG. 4. Stress to produce rupture in 100,000 hours, extrapolated.

Table 1. Chemical Analyses of Heat-resisting Alloys

(Data from the Elliott Co., Jeannette, Pa)

	Old Alloys			Inconel	New (Super) Alloys									
					Class A		Class B						Class C	
	18-8 Columbium	18-8 Moly	25-20 (Cast)		19-9DL	Timken 16-25-6	N-155	S-588	S-590	S-816	Inconel X	Refractaloy 26	Vitallium (Cast)	X-40 (Cast)
Carbon	0.07	0.07	0.35	0.09	0.30	0.08	0.15	0.40	0.40	0.40	0.04	0.03	0.30	0.50
Manganese	1.00	1.00	0.75	0.75	0.75	1.50	1.00	0.75	0.75	0.75	0.75	0.70	0.30	
Silicon	0.50	0.50	1.00	0.30	0.50	0.80	0.50	0.65	0.65	0.65	0.50	0.65	0.25	
Chromium	18.50	17.00	25.00	13.50	19.00	16.50	21.00	19.00	20.00	20.00	15.00	18.00	28.50	25.00
Nickel	11.00	12.00	20.50	78.00	9.00	25.00	20.00	20.00	20.00	20.00	73.00	37.00	2.00	10.00
Cobalt	......	......	......	......	......	......	20.00	......	20.00	44.00	......	20.00	62.00	55.00
Molybdenum	......	2.50	......	......	1.25	6.25	3.00	4.00	4.00	4.00	......	3.00	5.50	......
Tungsten	......	......	......	......	1.25	......	2.00	4.00	4.00	4.00	......	......	......	7.00
Columbium	10 × C min.	......	......	......	0.40	......	1.00	4.00	4.00	4.00	1.00	......	......	......
Titanium	......	......	......	......	0.35	......	......	......	......	......	2.50	3.00	......	......
Nitrogen	......	......	......	......	......	0.10	0.12	......	......	......	......	......	......	......
Iron	Rem.	Rem.	Rem.	Rem.	Rem.	Rem.	Rem.	Rem.	Rem.	3.00	7.00	Rem.	1.00	0.60
Aluminum	......	......	......	......	......	......	......	......	......	......	0.60	0.30	......	......

SECTION 5

NONMETALLIC MATERIALS

BY

CHESTER R. AUSTIN, *Supervisor, Ceramic Division, Battelle Memorial Institute*

WILLIAM H. LEHMBERG, *Director of Engineering, American Felt Co.*

HAROLD G. VOGT, *Physicist, Corning Glass Works*

JOHN J. MARTIN, *Bell Telephone Laboratories, Inc.*

J. SPOTTS McDOWELL, *Harbison-Walker Refractories Co.*

E. F. RIESING, *Manager, Automotive Engineering and Sales, Firestone Industrial Products Division of the Firestone Tire and Rubber Co.*

ARTHUR M. WEINBERG, *Chemical Department, General Electric Co., Pittsfield, Mass.*

L. J. MARKWARDT, *Assistant Director, U. S. Forest Products Laboratory*

CARBON AND GRAPHITE

(Data furnished by National Carbon Co., Inc.)

Elemental carbon occurs in widely varied forms, from the transparent diamond with its extreme hardness to graphite which is gray and soft. Lampblack, charcoal, petroleum coke, and coal from which the volatile matter has been removed are nearly amorphous forms of carbon. Graphite is a crystalline form of carbon in which the atoms form a hexagonal pattern in a layer-like structure. Deposits of natural graphite are widely distributed. Large quantities have been found in Madagascar, Ceylon, and Mexico. Artificial graphite is prepared by heating certain forms of carbon, such as petroleum coke, to nearly 3000 C. In this temperature range, carbon assumes an orderly hexagonal crystal pattern similar to that of natural graphite. The transformation from carbon to graphite does not occur at a specific temperature but is a function of both time and temperature. Some forms of carbon, such as charcoal or lampblack, cannot be graphitized to the same degree as petroleum coke.

1. PHYSICAL PROPERTIES AND AVAILABLE FORMS

Carbon is chemically inactive, resisting the action of most acids (notably hydrofluoric), alkalies, and other chemicals, which are not highly oxidizing in their behavior. At incandescent temperatures, it does not melt, and the rate of oxidation is relatively slow. The thermal conductivity of carbon is high in comparison with most nonmetallic solid materials and, in graphitic form, exceeds that of many metals. This property, in combination with low coefficient of thermal expansion, gives to carbon products a very high resistance to thermal shock. Although much higher in electrical resistance than the metals, carbon is a good conductor of electricity and, for this reason, finds many applications in the electrical field and in allied industries. Physical properties of carbon are listed in Table 1, and specific heat is given in Fig. 1.

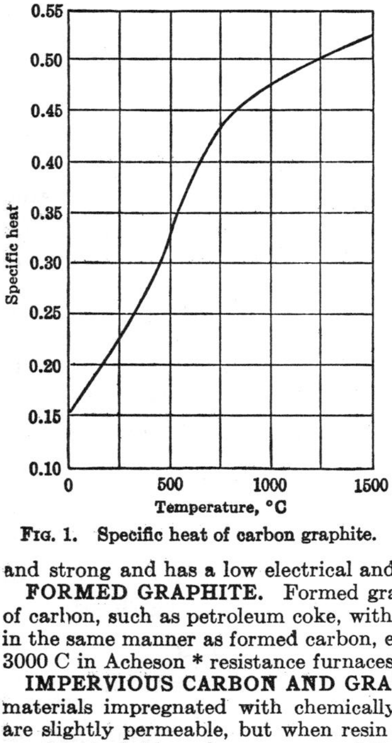

FIG. 1. Specific heat of carbon graphite.

FORMED CARBON. Formed carbon is usually prepared by mixing petroleum coke and/or coal or lampblack with pitch or tar and extruding or molding the mixture under pressure in the form of cylinders or blocks of various sizes. These shapes are baked on slow schedules to 700 to 1200 C to carbonize the binder and remove the volatile matter. The resultant product is a nearly amorphous material which is hard and strong and has a low electrical and thermal conductivity. (See Table 2.)

FORMED GRAPHITE. Formed graphite is prepared by mixing a graphitizable form of carbon, such as petroleum coke, with tar or pitch. This mixture is extruded or molded in the same manner as formed carbon, except that it is finally baked to temperatures near 3000 C in Acheson * resistance furnaces to graphitize the carbon. (See Table 2.)

IMPERVIOUS CARBON AND GRAPHITE (Karbate *) are carbon and graphite base materials impregnated with chemically resistant synthetic resins. The base materials are slightly permeable, but when resin is deposited in their pores they are rendered impervious to fluids under pressure and their strength is increased materially. Impregnation does not change the thermal conductivity of the base materials. By selection of the proper

* The terms National, Karbate, and Acheson appearing in this chapter are registered trademarks distinguishing products of National Carbon Co., Inc.

Table 1. Physical Properties of Carbon

Chemical Symbol, C Atomic Weight, 12.010

Volatilization Point, 3925–3970 K, 3652–3697 C, 6606–6687 F

Melting Point. Carbon, in all forms, volatilizes without melting at ordinary pressures.

Specific Gravity of Carbon Materials Used in the Production of Carbon Products

Material	Specific gravity	Material	Specific gravity
Gas calcined petroleum coke carbon	1.98–2.10 *	Electrically calcined anthracite	1.90–1.97 *
Gas calcined lampblack carbon	1.80–1.85 *	Graphitized petroleum coke carbon	2.20–2.24 *
Gas calcined anthracite	1.79 *	Graphitized lampblack	1.98–2.08 *

Electrical Resistivity of Formed Carbon or Graphite Products, at 20 C

	Ohm–inch †	Ohm-centimeter †
Carbon, petroleum coke base	.0014–.0018	.0035–.0046
Graphite, petroleum coke base	.0003–.0005	.0008–.0013
Carbon, anthracite base	.0013–.0026	.0033–.0066
Carbon, lampblack base, gas baked	.0023–.0032	.0058–.0081
Carbon, lampblack base, graphitized	.0018–.0026	.0046–.0066

Specific Heat

	Temperature, °C					
	26–76	26–282	26–538	36–902	47–1193	56–1450
Carbon, anthracite or petroleum coke base	.168	.200	.234	.315	.352	.387
Graphite, petroleum coke base	.165	.195	.234	.324	.350	.390

* Dependent on source and degree of calcination.
† This property is also expressed as ohms/in.^{3} or ohms/cm.^{3}

type of resin impregnant, Karbate brand materials may be utilized in practically all chemical conditions in which the base materials are resistant. (See Table 2.)

LOW-TEMPERATURE OXIDATION OF CARBON AND GRAPHITE. Carbon and graphite are thermally stable and in reducing or nonoxidizing atmospheres will withstand temperatures up to 3600 C. In air the threshold oxidation temperature * is 400 C for commercial graphite and 350 C for carbon.

BRICK. Carbon and graphite corrosion-resistant brick are recommended for many types of construction where corrosive conditions or thermal shock are encountered. They resist the corrosive action of all acids, alkalies, and solvents except those of highly oxidizing character, adhere firmly to bonding materials, and are mechanically strong. (See Table 3.)

Graphite brick, although somewhat softer than carbon brick, possess the advantages of carbon brick plus a much higher thermal conductivity and greater resistance to oxidation.

Carbon and graphite brick and Karbate impervious brick are manufactured in practically all standard shapes, including scored and circle brick. Special shapes also can be provided. Carbon brick construction weighs approximately 100 lb per cu ft.

BEAMS, BLOCKS, PLATES, TILE, AND RODS. Carbon and graphite and impervious carbon and graphite are used in a variety of square, rectangular, round, and special sections, including beams, blocks, plates, tile, and rods. (See Table 4.)

Machined carbon beams are used as a material of construction where corrosive chemicals and severe thermal conditions are encountered. Installation of large beams is accomplished in a small fraction of the time required for standard brick construction. Premachining of parts also greatly simplifies assembly, with resultant savings in time and money.

Important uses of machined carbon beams are in electrostatic precipitators and supporting members in tanks, or vessels in chemical plants and paper mills.

TOWER PACKING. Carbon Raschig rings are suitable packing material for absorption, scrubbing, fractionation, and distillation towers. They are unaffected by acid, alkali, or organic conditions except those involving active oxidation. Because of their light weight and low thermal expansion, they are less susceptible to crushing, cracking, and spalling. (See Table 5.)

Other forms of carbon tower packing, such as tile, checker work, grills, and perforated plates, are also employed. Graphite and Karbate impervious graphite plates and bubble-caps are suitable for processes corrosive to the more common metals.

---

* The threshold oxidation temperature is defined as the temperature at which carbon or graphite undergoes a weight loss of less than 1% in 24 hours when exposed to air.

Table 2. Physical Characteristics of Structural Carbon and Graphite Products *

Product and Trademark	Apparent Density	Weight, lb/cu ft	Strength, psi: Tensile	Strength, psi: Compressive	Strength, psi: Flexural Stress	Elastic Modulus, psi (Multiply by 10^5)	Electrical Resistivity, ohm-in.	Thermal Expansion (see Note) a Values	Thermal Conductivity Btu/(hr)(sq ft)(°F/ft)
Carbon—National †									
Cylinders									
8 in. diameter	1.57	98	400	2,400	1100	12	.0011	12	8.7
10–14 in. diameter, incl.	1.57	98	290	2,000	800	9.5	.0014	12	8.7
17–40 in. diameter, incl.	1.57	98	150	1,700	400	7	.0020	12	8.7
Squares and Slabs									
4 x 4 in. to 6 x 6 in., incl.	1.57	98	400	2,400	1100	12	.0011	12	8.7
6 x 6 in. to 20 x 20 in., incl.	1.57	98	300	2,140	800	9.5	.0016	12	8.7
15 x 30 in., 24 x 30 in., and 24 in. square	1.57	98	200	2,300	550	7.5	.0020	12	8.7
Pipe and Tubes									
1–6 in. ID, incl.	1.55	97	900	9,000	2600	19	.0015	12	3.0
Brick (See Table 3)									
Graphite—National † *and Acheson* †									
Cylinders									
To 5 1/8 in. diameter, incl.	1.56	97.5	900	3,100	1650	13.5	.00036	9	81
6–12 in. diameter, incl.	1.55	97	600	2,400	1100	10	.00039	9	75
14–20 in. diameter	1.51	94.5	440	1,700	800	8	.00041	9	71
Squares and Slabs									
To 5 in. thick, incl.	1.56	97.5	900	3,100	1650	13.5	.00036	9	81
6 in. thick to 144 sq in.	1.55	97	600	2,400	1100	10	.00039	9	75
Over 144 sq in., section	1.51	94.5	440	1,700	800	8	.00041	9	71
Pipes and Tubes									
1–6 in. ID, incl.	1.67	104	800	5,000	2800	14	.0003	9	97
Brick (See Table 3)									
Impervious Carbon—Karbate †									
Pipes and Tubes									
1–6 in. ID, incl.	1.77	120	1800	10,000	4400	28	.0016	28	3.0
Impervious Graphite—Karbate †									
Pipes and Tubes									
1–6 in. ID, incl.	1.87	127	2500	9,000	4700	22	.00034	23	86

Note: Lineal expansion in percentage for temperature rise from

$$70\text{ F to temp } t\ (°\text{F}) = \frac{a(t-70) + 0.00555(t-70)^2}{100{,}000}\ \%$$

$$21\text{ C to temp } t\ (°\text{C}) = \frac{1.8a(t-21) + 0.018(t-21)^2}{100{,}000}\ \%$$

* The physical characteristics in this table are for the most commonly used stock. Materials having other characteristics are available.

† The terms National, Karbate, and Acheson are registered trademarks distinguishing products of National Carbon Co., Inc.

Table 3. Average Physical Properties of Carbon and Graphite Brick

Brick	Strength, psi: Tensile	Strength, psi: Crushing	Strength, psi: Flexural Stress	Electrical Resistivity, ohm-in.	Thermal Conductivity, Btu/(hr)(sq ft)(°F/ft)	Thermal Expansion (see Note) a Values	Apparent Density	Approx. Weight, lb/cu ft
Carbon (CB3)	1530	8,320	3070	.0016	3	12	1.55	100
Graphite	880	3,050	1750	.00036	81	9	1.56	100
Karbate impervious carbon	2000	10,500	4650	.0016	3	28	1.76	110
Karbate impervious graphite	2350	10,500	5000	.00033	87	22	1.91	120

Note: Lineal expansion in percentage for temperature rise from

$$70\text{ F to temp } t\ (°\text{F}) = \frac{a(t-70) + 0.00555(t-70)^2}{100{,}000}\ \%$$

$$21\text{ C to temp } t\ (°\text{C}) = \frac{1.8a(t-21) + 0.018(t-21)^2}{100{,}000}\ \%$$

Table 4. Maximum Sizes for Blocks and Cylinders Normally Available *

Section	Carbon	Graphite
Rectangular	24 x 30 x 180 in.	20 x 20 x 72 in.
Cylindrical	40 in. diameter x 110 in.	20 in. diameter x 72 in.
		24 in. diameter x 66 in.
		30 in. diameter x 72 in.

* Sizes below limits shown in table can be supplied.

Table 5. Stock Sizes and Physical Data—Carbon Raschig Rings

Specifications	Size Number							
	0	1	2	3	4	5	6	7
Outside diameter, in.	1/4	1/2	3/4	1	1 1/4	1 1/2	2	3
Inside diameter, in.	1/8	3/8	1/2	3/4	7/8	1	1 1/2	2 3/8
Length, in.	1/4	1/2	3/4	1	1 1/4	1 1/2	2	3
Approximate number of rings per cubic foot	85,000	10,600	3140	1325	678	392	166	49
Square feet of absorption surface per 1000 rings	2.50	10.75	24	43	67	96	172	388
Square feet of absorption surface per cubic foot of tower space	212	114	75	57	45	37.5	28.5	19
Pounds of rings per cubic foot of tower space	46	27	34	27	31	34	27	23
Minimum crushing strength across diameter, lb	17	15	47	44	196	194	184	170
Load on each bottom ring (dry rings) per 10 ft of tower height, lb	0.25	0.56	1.6	2.25	4.0	6.4	9.0	17.5
Percentage of free gas space	55	74	67	74	69	67	74	78
Square feet of absorption surface per cubic foot of free gas space ($F \div J$)	387	155	113	77	65	56	39	25
Relative scrubbing capacity ($F \times J$)	116	84	50	42	31	25	21	15

POROUS CARBON AND GRAPHITE. (See Tables 6, 7, 8, and 9.) Porous carbon is essentially a mass of carefully sized carbon particles bonded with carbon to form a rigid, uniformly porous material.

Table 6. Standard Porous Carbon Tubes

OD in.	ID in.	Length in.	OD in.	ID in.	Length, in.
1 3/4	5/8	4	4	2	9
2	3/4	12	4	2	12
2 1/2	1	4	4	2	24
2 1/2	1	6	4 1/2	2	36
2 1/2	1	12	6 5/8	3 3/4	24
2 1/2	1	24	6 5/8	3 3/4	36

Table 7. Standard Porous Carbon Plates

12 x 12 x 1 1/2 in., all grades
12 x 12 x 1 in., grades 50 and 60

Table 8. Permeability of Porous Carbon and Graphite (Average Values)

Grade	Average Water Permeability,* gal/sq ft/min at 5 psi Pressure	Average Air Permeability,† cu ft/sq ft/min at 2 in. Water Pressure
60	14.0	....
50	30.0	....
40	45.0	4.0
30	80.0	8.5
20	120.0	17.0
10	175.0	33.0

* Water at 70 F, 1-in. thick sample.
† Expressed at 70 F and 760 mm Hg pressure, 15% relative humidity, 1-in. thick sample.

Table 9. Physical Properties of Porous Carbon and Graphite (Average Values)

Grade	Material	Weight, lb/cu ft	Effective Porosity, %	Thermal Expansion (see Note) a Values	Electrical Resistivity, ohm-in.	Minimum Strength			Average Pore Diameter		Filter Action, Minimum Diameter Particle Retained, (in.)
						Tensile, psi	Flexural Stress	Compressive, psi	In.	Microns	
60	Carbon	69	48	25	.0070	190	600	850	.0013	33	.00047
60	Graphite	69	48	19	.0012	110	250	500	.0013	33	.00047
50	Carbon	69	48	25	.0070	180	500	830	.0019	48	.00079
50	Graphite	69	48	19	.0012	110	250	500	.0019	48	.00079
40	Carbon	69	48	25	.0057	120	320	900	.0027	69	.00098
40	Graphite	69	48	19	.0013	100	190	500	.0027	69	.00098
30	Carbon	69	48	25	.0070	100	250	770	.0039	99	.00173
30	Graphite	69	48	19	.0017	80	200	520	.0039	99	.00173
20	Carbon	68	48	25	.0070	90	240	700	.0055	140	.0030
20	Graphite	68	48	19	.0020	60	140	310	.0055	140	.0030
10	Carbon	68	48	24	.0080	80	160	300	.0075	190	.0059
10	Graphite	68	48	19	.0020	50	140	270	.0075	190	.0059

Note: Lineal expansion in percentage for temperature rise from

$$70\text{ F to temp } t\ (^\circ\text{F}) = \frac{a(t-70) + 0.00555(t-70)^2}{100{,}000}\ \%$$

$$21\text{ C to temp } t\ (^\circ\text{C}) = \frac{1.8a(t-21) + 0.018(t-21)^2}{100{,}000}\ \%$$

Porous carbon and porous graphite are chemically inert in practically all acids, alkalies, and organic compounds except those of strong oxidizing character. They are immune to thermal shock and are stable at all commercial temperatures in reducing atmospheres. Porous carbon starts to show significant oxidation in air at 660 F and porous graphite at 840 F.

Plates and tubes are particularly adaptable to the filtration of corrosive fluids and molten materials. A variety of forms and types of filters may be constructed. Porous carbon diffuser elements of several sizes and designs are used extensively in many chemical, physical, and biological reactions depending on the dispersion of a gas into a liquid.

2. APPLICATIONS OF CARBON AND GRAPHITE

HEAT EXCHANGERS. Shell- and tube-type exchangers, with heat transfer surfaces up to 700 sq ft per unit. The most common construction mounts Karbate (see p. 5-02) tube bundles in steel shells with metal baffle assemblies. Karbate head covers with standard connections are usually employed, but the covers may be made of other suitable materials. Where operating conditions require corrosion-resistant shells as well as tube bundles, all parts of the exchanger in contact with the corrosive fluids may be made of Karbate materials.

Plate-type heat exchangers, durable, compact units, resistant to thermal and mechanical shock, for heating or cooling pickling and plating solutions and other corrosive fluids; available in various styles ranging from 4 to 56 sq ft of heat transfer surface per unit.

Bayonet-type exchangers, conventional construction.

Sectional cascade coolers, constructed from prefabricated sections, 9 ft long, by sections stacked to form a series flow, vertical bank. Maximum cooling surfaces of approximately 360 sq ft are obtainable with any of the five pipe sizes—1, 1 1/2, 2, 3, and 4 in.

Jacketed pipe exchangers, with either metal or Karbate pipe jackets.

CONVEYING EQUIPMENT. Pipe and fittings, in twelve sizes from 1/2 in. through 10 in. Threaded, flanged, and flexible joints are available.

Globe valves, light in weight with short face-to-face dimensions. Corrosive fluids contact Karbate materials only.

Centrifugal pumps, open impeller type. All wet end parts are of corrosion-resistant Karbate material.

Jet pumps, Karbate impervious graphite parts for steam jet ejectors and injectors.

TOWERS. Tower sections and auxiliary parts such as support grills and distributor sections are available in sizes ranging from 8 in. ID to 33 in. ID. Complete towers are readily assembled from standard parts.

HYDROGEN CHLORIDE ABSORBERS. Standard absorbers complete with supporting structure and control equipment are available for gas concentrations ranging from 40 to 100% hydrogen chloride. Other types of absorbers are easily assembled from the standard line of Karbate equipment.

METALLURGICAL APPLICATIONS. Carbon (petroleum coke, anthracite coal, or lampblack base) and electric furnace graphite find many uses in the metallurgical field where materials are subjected to high and rapid changes in temperatures. The significant properties of carbon and graphite are given in Table 1. The materials are not attacked by, nor do they fuse with, slag. They are not "wet" by most metals, they do not warp at high temperature, and they have excellent resistance to heat spalling. Table 10 lists typical metallurgical applications.

Table 10. Metallurgical Applications for Carbon and Graphite

Application	Carbon	Graphite
Blast furnaces	Furnace linings Run-out trough linings Cinder notch cooler linings Cinder notch plugs	
Steel mills	Mold plugs	Mold stool inserts Molds for Jominy test bars
Steel foundries		Riser rods
Iron foundries	Core shapes Chill plates	Core shapes Chill plates
Aluminum mills and foundries	Core shapes Chill plates	Fluxing tubes Thermocouple sheaths Stirring rods or puddlers Core and chill plates
Brass-bronze mills and foundries	Core shapes Chill plates	Pig ingot molds Crucible skimmer floats Molds for centrifugally cast bushings Core or chill plates Mold wash powders
Powder metallurgy	Powders	Die assemblies Sintering boats Powders
Miscellaneous	Resistor tubes	High purity induction heating crucibles Continuous casting dies

Table 11. Approximate Current-carrying Capacities of Carbon Electrodes

Nominal Diameter, in.	Area, sq in.	Current-carrying Capacity, amp	Current Density, amp/sq in.
Round Electrodes			
8	50	2000– 3000	40–60
10	79	3000– 4800	40–60
12	113	4500– 6800	40–60
14	154	5400– 8500	35–55
17	227	7900–12500	35–55
20	314	11000–17300	35–55
24	452	15800–24800	35–55
30	707	24700–35300	35–50
35	962	28800–38400	30–40
40	1257	37700–50200	30–40
Square or Rectangular Electrodes			
Side, in.			
8	64	2500– 3800	40–60
10	100	4000– 6000	40–60
12	144	5700– 8600	40–60
14	196	6800–10800	35–55
16	256	9000–14000	35–55
20	400	14000–22000	35–55
24	576	20200–25900	35–45
24 x 30	720	21600–28800	30–40

Variations in Dimensions of Carbon Electrodes. Dimensions are nominal. Variations permitted are: diameter or side, ±2%; length, ±2 in.; curvature not to exceed 1/2% of length chord to arc. Ash: cylindrical, not more than 8 1/2%; square, 12%.

ELECTRODES. (See Tables 11 and 12.) In electrothermic applications, experience indicates that carbon electrodes operate to best advantage and give lowest costs in the production of carbide, abrasives, ferro alloys, and phosphorus. In these industries, carbon electrodes are used exclusively except in the production of some special grades of abrasives, ferro vanadium, and the very low carbon ferro alloys, where graphite electrodes are more satisfactory.

In the electric-furnace production of iron and steel, both types of electrodes are used, and the choice is based primarily on overall cost. Graphite electrodes cost about twice as much per pound as carbon, but the electrical conductivity of the former is approximately four times that of carbon of the same cross section. Consequently, a graphite electrode of about one-half the diameter of the carbon will carry the same current. In making a choice, however, the electrode cost is not always the governing factor, as power consumption, refractory costs, metallurgical conditions, freight, and many other items influence the final cost of the metal.

Table 12. Approximate Current-carrying Capacities of Graphite Electrodes

ROUND ELECTRODES

Nominal Diameter, in.	Area, sq in.	Current-carrying Capacity, amp	Current Density, amp/sq in.
2	3.1416	600– 1000	200–320
2 1/2	4.9087	800– 1500	160–310
3	7.0686	1200– 2100	170–300
4	12.566	1800– 3000	140–240
5 1/8	20.629	2300– 4100	110–200
6	28.274	3100– 5400	110–190
7	38.485	4200– 6900	110–180
8	50.265	5500– 9000	110–180
9	63.617	6400–10800	100–170
10	78.540	7800–12500	100–160
12	113.10	11300–17000	100–150
14	153.94	15400–21500	100–140
16	201.06	20100–26100	100–130
17	226.98	22700–28400	100–125
18	254.47	25500–30500	100–120
20	314.16	28300–34600	90–110
24	452	38000–41000	85– 90
30	707	56000–60000	80– 85

SQUARE ELECTRODES

Side, In			
2	4	650– 1000	160–250
4	16	1900– 3000	120–180
6	36	3600– 4900	100–140
8	64	6400– 8300	100–130
10	100	9000–13000	90–130
12	144	13000–17300	90–120
16.35	267.32	21400–26700	80–100

Variations in Dimensions of Graphite Materials. Dimensions are nominal. Average tolerances of unmachined materials are as follows:

Round: Under 2 1/2 in., ±2% of nominal diameter.
2 1/2 in., −1/16 in., +3/32 in.
Over 2 1/2 in., up to 6 in., ± 1/8 in.
6 in. and over, − 1/8 in., + 1/4 in.
Rectangular: ±2% of width and thickness.
Length: Less than 24 in., −1 1/2%, +2% of nominal length.
Over 24 in., −1/2%, +2%.
Curvature lengthwise, measured chord to arc, 1/2% of length.
Ash: Not more than 1 1/2%.

In electrolytic cells, anodes of graphite are preferable to carbon with but very few exceptions. The density, oxidation resistance, high purity, high electrical conductivity, and ease of machining and assembly are the characteristics which make graphite more useful and economical than carbon in this field.

In the electrolysis of aqueous or fused halogen salts, Acheson graphite anodes have met successfully the basic requirements of the ideal insoluble electrode as defined by high purity, good electrical conductivity, low halogen discharge potential and over-voltage, high mechanical strength, ease of machining and assembly, high resistance to anodic corrosion, and low initial cost.

Except for attack by nascent oxygen, at moderate temperatures graphite anodes are practically inert to all other chemical action. Consumption rate depends primarily on the degree of anodic oxygen generation; therefore, the extent to which oxidation can be controlled usually determines the relative cost for anodes in a particular application. Electrolytes containing sulfate, nitrate, or high concentrations of carbonate ions result in anodic discharge of oxygen, in turn leading to comparatively rapid anode disintegration. The limiting factor, therefore, in the economy of graphite as the anode material is the oxidation reaction rate and other related operating costs. When employed as cathode, graphite is not subject to oxidation and for this reason will show little or no deterioration in such service.

For electrochemical processes, Acheson graphite anodes are supplied in three grades: (1) regular quality, (2) "X" quality, and (3) special treated. The choice of these grades depends upon a number of different operating factors which vary with the type of electrolyzing equipment and the prevailing cost of power. Under what can ordinarily be called normal operating conditions, "X" quality will consume at a rate 15 to 20% less than regular quality and consequently is the choice in most processes where untreated anodes are employed. By reason of a higher apparent density, the "X" quality has approximately a 6% greater weight per piece than regular quality, but with the cost per pound the same the former has an operating cost advantage. Acheson special treated anodes under most comparable operating conditions will show consumption rates well below those for "X" quality; however, a slightly higher average cell voltage is usually encountered and must be considered in balancing increased power consumed against a saving in graphite cost.

CERAMICS

By Chester R. Austin

3. CHEMICAL STONEWARE

Chemical stoneware is made from a clay, or mixtures of clays, usually with a flux added. Proper proportioning of the raw materials permits ware of various properties, as desired, to be made. Chemical stoneware covers a range of ceramic products which at one extreme are as coarse in texture and as porous as firebrick and at the other extreme are as fine and dense as porcelain. The more porous bodies are more resistant to temperature changes; the denser ones are more resistant to corrosion. Chemical stoneware is used in places where resistance to acids, alkalies, and other chemicals is wanted. Concentrated and dilute materials of this nature, hot or cold, can be handled in chemical stoneware, with the

Table 1. Typical Properties of Chemical Stoneware

(Data from General Ceramics and Steatite Corporation, Keasbey, New Jersey, and Maurice A. Knight, Akron, Ohio.)

	General Range of Properties of Chemical Stoneware for Various Applications	Specific Data on Material Where No Exceptional Conditions Are to Be Met *
Bulk specific gravity	2.2	2.2
Hardness, Mohs' scale	....	8.5
Scleroscope number	100	....
Ultimate tensile strength, psi	2,000–3,000	2,500
Ultimate compressive strength, psi	80,000–100,000	80,000
Modulus of rupture, psi	6,000–13,000	6,500
Modulus of elasticity, psi	7,000,000–10,000,000	10,000,000
Specific heat, cal/gram/°C, Btu/lb °F	0.2	0.2
Pore volume, %	....	1.5
Water absorption, %	0–2	....
Thermal conductivity, Btu hr^{-1} ft^{-2} in. $°F^{-1}$, 68–212 F	8–24	8
Linear coefficient of thermal expansion, per °F, 68–212 F	2.0×10^{-6}	2.7×10^{-6}

* Such as used for acid jars and other containers, electroplating tanks, and similar equipment not subject to appreciable mechanical or thermal stress.

Table 2. Typical Properties of Steatite

(Data from American Lava Corporation, Chattanooga, Tennessee, and General Ceramics and Steatite Corporation, Keasbey, New Jersey. General reference: J. S. McDowell, Specialized insulator bodies for the electronic field, *Ceram. Age*, Vol. 50, 1947, pp. 32–36.)

	General Range of Properties of Steatite	Specific Properties of General Grade *	Specific Properties of High-frequency Grade *
Specific gravity	2.5–2.7	2.7	2.8
Density, lb/cu in.	0.090–0.098	0.098	0.102
Volume, cu in./lb	10.26–11.11	....	....
Water absorption, %, ASTM D116-42(a)	0–1	0–0.05	0–0.05
Softening temperature, ASTM C24-35	1400–1450 C	....	....
	2552–2642 F	....	....
Resistance to heat (safe limit for constant temperature)	1000 C	....	....
	1832 F	....	....
Hardness, Mohs' scale	7.5	7.5	8.0
Linear coefficient of thermal expansion per °C			
25–100 C	$6.9–7.7 \times 10^{-6}$	....	....
25–700 C	$8.6–10.2 \times 10^{-6}$	....	....
20–100 C		6.5×10^{-6}	6.0×10^{-6}
20–400 C		7.8×10^{-6}	7.3×10^{-6}
Tensile strength, psi, ASTM D116-42	7,500–10,000	9,500	9,000
Compressive strength, psi, ASTM D667-42T	65,000–85,000	81,500	80,000
Flexural strength, psi, ASTM D667-42T	18,000–20,000	19,000	20,000
Resistance to impact (1/2-in. rod), in.-lb, Charpy D667-42T	4.5–5.0	2.30	1.90
Thermal conductivity approx., $\frac{\text{g-cal} \times \text{cm thick}}{\text{cm}^2 \times \text{sec} \times °\text{C}}$	0.006	0.006	0.006
Dielectric strength, volts/mil, step 60 cycles, test disks 1/4 in. thick, ASTM D667-42T	210–240	240	200
Volume resistivity at various temperatures, ohms/cm cube			
°C °F			
20 68	....	....	10^{14}–....
25 77	$>10^{14}$–....	....	....
100 212	$2.1 \times 10^{12}–>10^{14}$	....	....
200 392	....	....	2.26×10^{13}–..
300 572	$6 \times 10^{7}–9 \times 10^{12}$	1×10^{13}–....	3.10×10^{12}–..
500 932	$3.2 \times 10^{5}–3.5 \times 10^{10}$	2×10^{9}–....	1.81×10^{10}–..
700 1292	$2.3 \times 10^{14}–4.8 \times 10^{8}$	1.7×10^{7}–....	1.64×10^{9}–..
900 1652	$7.0 \times 10^{3}–2.5 \times 10^{7}$	....	....
Te value †			
°C	400–>1000	700	700
°F	824–>1832	....	....
Dielectric constant, ASTM D667-42T			
60 cycles	5.9–6.3	....	....
1 megacycle	5.8–6.0	5.85	6.3
10 megacycles	5.7–5.8	....	....
100 megacycles	5.6–5.7	....	....
Power factor, ASTM D667-42T			
60 cycles	0.015 ‡ –0.002	....	....
1 megacycle	0.0035 ‡–0.0004	0.0012	0.00065
10 megacycles	0.003 ‡ –0.0003	....	....
100 megacycles	0.0028 ‡–0.0003	....	....
Loss factor, ASTM D667-42T			
60 cycles	0.09 –0.013	....	....
1 megacycle	0.021 –0.0023	0.007	0.004
10 megacycles	0.017 –0.0017	....	....
100 megacycles	0.016 –0.0017	....	....
Capacity change per °C, ppm	+120 – +160	....	....

* Values obtained by using similar, but not necessarily identical, techniques as were employed in determining the general range of properties.

† Temperature at which a 1-cm cube has a resistance to 1 megohm.

‡ Measured wet, after immersion in water for 48 hours.

exceptions of hydrofluoric acid and strong caustic solutions. Severe temperature changes should not be permitted when chemical stoneware is employed.

In manufacturing chemical stoneware, selected raw clays and, usually, fluxes are made into a water suspension by blunging, screened, passed through a magnetic separator, and filtered to remove excess water. The filter press cakes are mixed, with any nonplastic ingredients to be added, in a wet pan. The mixes then are put through a pug mill or vacuum machine for final tempering. The mix is aged either after the mixing in the wet pan or after pugging to insure uniform moisture distribution and to improve the working properties of the mass. Machines are used to form brick, tower packings, stoppers, tile, and pipe. Piping is finished by hand. Other equipment is built entirely by hand by skilled workmen. Plaster molds are used to assist in forming certain pieces; others are made entirely without molds. Some pieces are made by casting a slip into plaster molds. After drying, the pieces are set in the kiln and fired to impart the desired properties. A salt glaze, formed by the reaction of salt, volatilized in the fireboxes, can be applied during the firing operation. Such a glaze improves the finish and appearance of certain ware. The fired ware frequently is machined with high-speed silicon carbide grinding wheels. Lapping and drilling with silicon carbide grain and cutting with diamond-impregnated blades also are employed.

Data on representative chemical stoneware bodies are given in Table 1.

4. STEATITE

Steatite is a generic term covering a class of ceramic products in which a preponderance of talc is used as a raw material. Oxides or compounds of the alkaline earths are added as fluxes. Clay is either entirely absent or present in small quantities, and an organic gum or plasticizer is added. Barium carbonate or other salt of barium is the most common flux used throughout the industry for production of a vitrified steatite body. Zircon or compounds of zircon are sometimes added to increase the thermal conductivity and decrease the thermal expansion. The ware is formed by machines and fired in ceramic kilns. The fired product is chiefly clinoenstatite ($MgO \cdot SiO_2$) crystals in a glassy matrix. Electrical loss properties of these bodies are controlled largely by the glass phase, whereas the other properties are controlled primarily by the crystal phases which are more or less constant for the entire family. A series of steatite bodies with different fluxes will not vary over 10 to 15% in their physical properties, but they may vary 3500% in volume resistivity at elevated temperatures.

Steatite bodies generally are prepared for fabrication by dry-process techniques in which the raw materials are blended thoroughly in a muller-type of mixer. In a few instances, the bodies are prepared by blunging or by kneading.

Steatite is formed by wet pressing, dry pressing, extruding, and casting. The method of fabrication depends on the shape, size, and purpose of the product. In the production of some items, secondary fabrication of the unburned ware is necessary, such as drilling, tapping, cutting, and machining; in other instances, the ware is ground or cut to size after burning.

Data on the properties of steatite are given in Table 2.

5. ELECTRICAL PORCELAIN

Electrical porcelain is a vitrified ceramic product produced from a mixture of clays, silica, and a flux or fluxes. Feldspar is the most common flux.

The general manufacturing procedures can be classified as wet process and dry process. The wet process is used for high-voltage purposes, and the dry process is employed for low-voltage insulation. The wet process can be subdivided into (1) plastic and (2) casting methods.

In wet-process porcelain, the raw materials are mixed thoroughly with water, usually in a blunger, to form a suspension, passed over screens and through a magnetic separator, and then filter pressed. For the cast method of wet process, the filter-press cakes are reblunged with water and electrolytes, and the resultant suspension, of thick-cream consistency, is ready for use. In the plastic wet-process method, the filter-press cakes usually are put through a de-airing machine before final forming.

In the dry process, the filter-press cakes may be dewatered by storage; the material is granulated in some type of pulverizer and screened, and then it is ready for final forming. An alternative method (dry mixing) is to start with a blending mixer, employ only fine

(Continued on p. 5-14)

Table 3. Properties of Wet-process and Dry-process Porcelains *

(Data from (1) General Electric Company, Schenectady, New York; (2) *Ceramic Whiteware,* Rexford Newcomb, Jr., Pitman Publishing Company, New York; (3) Ralston Russell, Jr., Twenty-five years of progress: electrical porcelain, *Ceram. Age,* Vol. 50, 1947, pp. 19–27.)

Property	Wet-process Porcelain		Dry-process Porcelain
	Plastic	Cast	
Dielectric strength (thickness = 0.5 in.) V/mil (Note 3)	135	135	40
Dielectric strength vs thickness	See Fig. 1	See Fig. 1	...
Dielectric strength vs. temperature	See Fig. 2	See Fig. 2	...
Resistivity (ohm-cm)			
At 25 C	1×10^{14}	1×10^{14}	...
At 75 C	3.0×10^{11}	...	...
At 100 C	2.3×10^{10}	...	...
At 150 C	3.4×10^{9}	...	...
At 200 C	7.0×10^{8}	...	...
At 250 C	1.6×10^{8}	...	...
At 300 C	4.0×10^{7}	...	...
Power factor — *Frequency, cycles*; *Temp., °C*; *Condition*			
60; 25; As received	0.025–0.030	...	...
1,000,000; 25; As received	0.006–0.008	0.006–0.008	0.0065–0.008
1,000,000; 25; After 48 hr in water	0.007–0.020	...	>0.10
Dielectric constant — *Frequency, cycles*; *Temp., °C*; *Condition*			
60; 25; As received	6.0–7.0	...	...
1,000,000; 25; As received	6.0–7.0	6.0–7.0	6.0–7.0
1,000,000; 25; After 48 hr in water	6.0–7.0	...	...
Tensile strength at 25 C, psi (Note 2); specimen = 1 1/8 in. diameter	6,000	5,000	...
Compressive strength at 25 C, psi; specimen 1 1/8 in. diameter, 1 1/8 in. high	55,000	50,000	30,000
Flexural strength at 25 C, psi; specimen 1 1/8 in. diameter, 6 in. long; on 5 in. supports	11,000	10,000	5,000
Resistance to impact, ft-lb, Charpy; specimen 1 1/8 in. diameter, 6 in. long	1.5	...	1.2
Porosity			
Pore volume, %	Less than 0.3	Less than 0.7	Less than 4.0
Water absorption, %	Nil	Nil	...
Dye penetration under approx. 7000 psi	Nil	Nil	...
Coefficient of expansion per Deg. C. (temperature range, 0–200 C)	5.7×10^{-6}	5.7×10^{-6}	...
Thermal conductivity, watts/sq cm/°C/cm	0.0144	0.0144	...
Specific heat			
Temperature range, 0–200 C	0.202	0.202	...
Temperature range, 200–400 C	0.221	0.221	...
Resistance to thermal changes	See Note 4	See Note 4	See Note 4
Specific gravity	2.442	2.442	2.366

* These are average values obtained on unglazed porcelain in accordance with ASTM standard methods of testing, serial D116-42. See Notes 1, 2, and 3.

Note 1. The dielectric and mechanical strength values obtained from testing standard specimens are for general guidance only. The shape of specific designs, the nature of assembled metal parts, mechanical stresses, and other factors may have appreciable effect upon the strength of insulators. Design tests should, therefore, always be made definitely to establish breakdown values.

Note 2. Tensile strength is determined on a specimen designed to eliminate other forces, such as shearing. An increase in diameter of the test specimen results in a decrease in strength per unit area. This relationship is neither linear nor a simple curve. The test specimens used in obtaining the above average values for tensile strength were relatively large (1 1/8 in. diameter).

Note 3. Dielectric strength obtained on standard test specimens at 60 cycles under oil using 3/4-in. mercury sealed-in electrodes (ASTM method D116-42). Short-time tests: Voltage raised at rate of 0.5 kv/sec.

Note 4. In the assembly of porcelain bushings, studs, shells, etc., molten alloys and insulating compounds are often used. This work must be carried out very carefully to avoid cracking the porcelain. It is well to warm the porcelain to about 100 F before the operation and to avoid having the alloy or compound hotter than required for the proper fluidity for filling the joints.

Even slowly applied heat, if carried as high as 500 C, is very apt to leave the porcelain with substantially reduced mechanical strength.

Table 4. Properties of Electrical Porcelain Bodies

(Data from (1) General Electric Company, Schenectady, New York; (2) *Ceramic Whiteware*, Rexford Newcomb, Jr., Pitman Publishing Company, New York; (3) Ralston Russell Jr., Twenty-five years of progress: electrical porcelain, *Ceram. Age*, Vol. 50, 1947, pp. 19–27.)

Properties	Conventional Minimum	Conventional Maximum	Low-Voltage Porcelains	High-Voltage Porcelains	Mullite Porcelains	Cordierite Porcelains	Steatite Porcelains	Zircon Porcelains	Alumina (75–100%) Porcelains	Titania and Titanate Porcelains
Specific gravity	1.90	2.48	2.3–2.4	2.3–2.5	2.4–3.1	2.1–2.2	2.5–2.8	3.0–3.8	3.1–3.9	3.5–4.0
Compressive strength, psi	30,000	65,000	25,000–50,000	40,000–80,000	50,000–120,000	30,000–70,000	65,000–130,000	80,000–150,000	80,000–410,000	40,000–125,000
Tensile strength, psi	1,500	6,000	1,500– 2,500	3,400– 8,000	5,000–10,000	1,500– 5,000	6,200– 13,000	10,000– 15,000	8,000– 50,000	4,000– 11,000
Modulus of rupture, psi	3,000	8,500	3,500– 6,000	9,000–15,000	13,000– 21,000	4,000–12,000	14,000– 24,000	20,000– 35,000	18,000– 65,000	10,000– 22,000
Water absorption, %	0.2	2.0 *	0.5–2.0	0.0	0.0	0.0–8.0	0.0–1.0	0.0	0.0	0.0–0.1
Impact resistance, ft-lb/sq in.	0.50	0.90	…	…	…	…	…	…	…	…
Thermal conductivity, cal/sec/sq in./in.°C	0.0020	0.0039	0.0040–0.0045	0.0025–0.0050	0.0037–0.0073	0.0025–0.0061	0.0053–0.0061	0.010–0.015	0.0073–0.0560	…
Coefficient of linear thermal expansion (per °C × 10^{-6}) (20–700 C)	2.5	5.5	5.5–6.5	5.0–6.8	4.3–5.0	2.0–2.8	7.8–10.6	3.5–5.5	5.5–8.1	7.0–10.0
Thermal shock differential, °F	650	1,500	…	…	…	…	…	…	…	…
Thermal shock resistance	…	…	Moderate	Moderate	Good	Excellent	Poor	Good	Excellent	Poor
Modulus of elasticity, psi	…	…	7–10 × 10^6	7–14 × 10^6	…	7–13 × 10^6	13–15 × 10^6	20–30 × 10^6	15–52 × 10^6	…
Mohs' hardness	…	…	7.0	7.0	7.5	6.0–7.5	7.0–7.5	7.5–8.5	8.5–9.0	8.0
Safe operation temperature, °C	…	…	1,000	1,000	1,250	1,200	1,000–1,100	1,000–1,300	1,350–1,600	1,000
Power factor at 1 megacycle †	0.60	0.95	0.010–0.020	0.006–0.010	0.004–0.005	0.003–0.007	0.0002–0.0035	0.0002–0.0020	0.0010–0.0020	0.0002–0.50
Dielectric constant	6.0	7.5	6.0–7.5	6.0–7.5	6.2–6.8	4.1–5.4	5.5–7.5	8.0–10.5	7.3–11.0	18.0–20.000
Loss factor	…	…	0.060–0.150	0.036–0.075	0.025–0.034	0.012–0.038	0.0011–0.026	0.0016–0.0210	0.007–0.022	…
Dielectric strength, volts/mil, specimen 1/4 in. thick	90	200	40–100	250–400	250–400	100–250	200–350	250–350	400–1,100	50–300
Resistivity (ohm-cm)	1.18 × 10^{11}	1.18 × 10^{12}	10^{12}–10^{13}	10^{12}–10^{14}	10^{14}–10^{15}	10^{11}–10^{15}	10^{13}–10^{15}	10^{13}–10^{15}	10^{14}–10^{15}	10^{8}–10^{15}
Te value, °C ‡	…	…	250–300	300–500	500–650	500–700	440–>700	>700	>700	…

* Porous refractory insulator bodies may have absorption of 10% or higher They are special bodies to resist very rapid changes of temperature, as in electric ranges, rheostats, immersion heaters, and similar equipment.

† Power factor and dissipation factor are synonymous for values in the ranges cited.

‡ "Te" value is the temperature in °C at which the resistivity is 1 megohm per cm cube.

clays in the usual batch, and add the necessary amount of water. When discharged, the material is ready for final forming.

The wet-process plastic porcelain is shaped by being formed into approximately the desired shapes (blanks) and then dried to a certain extent. Final design is imparted by hand with spiral tools while the blank is revolved mechanically.

The wet-process cast porcelain is shaped by pouring the mix of thick-cream consistency into gypsum molds, which impart the desired shape; the water is absorbed sufficiently by the mold to enable the piece to be handled, and it then is removed from the mold. It may undergo further processes, such as trimming or grooving, before it is fired.

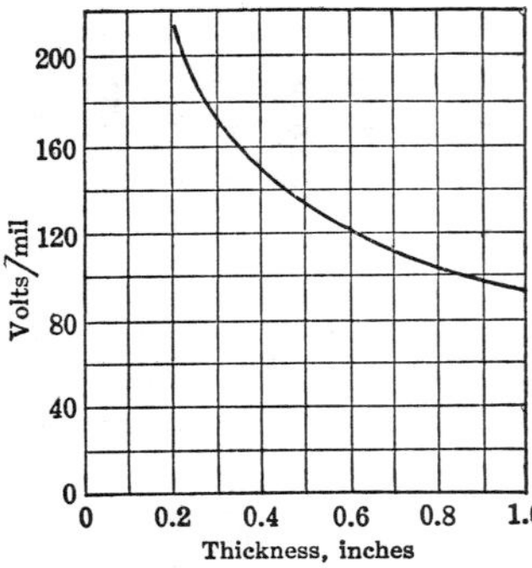

FIG. 1. Dielectric strength vs. thickness of unglazed, wet-process porcelain.

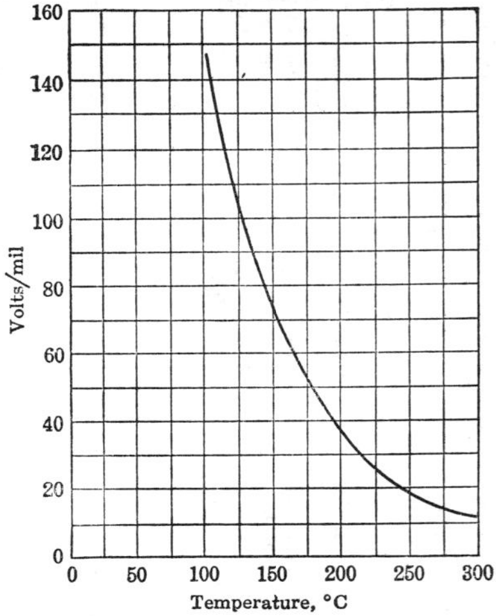

FIG. 2. Dielectric strength vs. temperature of unglazed wet-process porcelain. Tests made on a cylinder (wall thickness 0.136 in.) at 60 cycles in air, one minute step-by-step, using 1-in. lead-foil electrodes.

Dry-press porcelain is formed in steel dies by compressing the damp, coarse powder prepared by one of the two general methods described previously. Parts having lower porosity and higher dielectric strength can be made by pressing the mixture under vacuum.

After drying, pieces which require glazing are dipped or sprayed and are burned in ceramic kilns, usually at a temperature of about 2300 F.

Data on some specific electrical porcelain bodies and general ranges of properties are given in Tables 3 and 4.

6. BRICK AND TILE

BUILDING BRICK considered in this section are made from clay or shale and are fired after appropriate forming methods. The characteristics of the brick vary, depending on the particular clay and/or shale used, grinding, forming method, and burning treatment. In general, such brick are classified as face or common, depending on the intended application. Face brick are intended to be pleasing when exposed to view, and common brick most commonly are used for back-up purposes.

Building brick are made by grinding the raw clay and/or shale, tempering with water, in a pug mill or wet pan, and forming by machine. Extrusion is the most commonly employed method in which a solid column, extruded from an auger machine, is cut into appropriate sizes. Cores are used frequently in the extrusion machine to produce holes in the brick. Presses are also used. A few hand-molded brick are made. The brick then are dried and fired in ceramic kilns.

ASTM recognizes three classes of brick (ASTM C62-44):

SW—Brick intended for use where a high degree of resistance to frost action is desired and the exposure is such that the brick may be frozen when permeated with water. Brick used for foundation courses and retaining walls should conform to this grade. This grade is recommended for use where a high and uniform degree of resistance to weathering is desired.

MW—Brick intended for use where exposed to temperatures below freezing but unlikely to be permeated with water, or where a moderate and somewhat nonuniform resistance to frost is permissible. Brick used in the face of a wall above ground should conform to this grade. It should be emphasized that disintegration is not necessarily a characteristic of brick in this grade. They may, however, be a mixture of durable and nondurable brick. Examination of brick in this grade in the user's own territory after 10 to 20 years' service is the best method of making a selection.

NW—Brick intended for use as back-up or interior masonry, or, if exposed, for use where no frost action occurs; or if frost action occurs where the average annual precipitation is less than 20 in.

The physical requirements for brick to be classified in these grades are covered in ASTM specifications, the tests on which they are based being described in ASTM C67.

ASTM requires that the net cross-sectional area of the brick shall be at least 75% of the gross cross-sectional area (including holes) when measured in a plane parallel with the bearing surface of the brick. This limits the "coring" that may be used in fabrication.

In its report, *Recommended Minimum Requirements for Masonry Wall Construction* (1931), the Building Code Committee of the U. S. Department of Commerce recognizes four grades of brick according to compressive strength, tested flatwise: 1500–2500 psi; 2500–4500 psi; 4500–8000 psi; and over 8000 psi.

A standard size of brick is 2 1/4 by 3 3/4 by 8 in. Brick of this size, however, are not produced in all sections of United States.

STRUCTURAL CLAY TILE (ASTM C43-36) are hollow burned-clay masonry building units with parallel cells. They are made from fire clay, surface clay, shale, or mixtures of these materials. They are made by the same general procedure as building brick (clay or shale), but the forming is done by extrusion.

The American Society for Testing Materials (ASTM C34-41, C56-41, and C57-39) and the American Standards Association (ASA A74.1-1942, A76.1-1942, and A77.1-1942) recognize three general classifications of these products:

1. Structural Clay Load-bearing Wall Tile. *Grade LBX.* Suitable for general use in masonry construction and adapted for use in masonry exposed to weathering, provided they are burned to the normal maturity of the clay. They may also be considered suitable for the direct application of stucco.

Grade LB. Suitable for general use in masonry where not exposed to frost action, or for use in exposed masonry where protected with a facing of 3 in. or more of stone, brick, terra cotta, or other masonry.

The physical requirements of these tile and the requirements for the minimum number of cells in the direction of wall thickness and average weight are given in the standards referred to.

2. Structural Clay Non-load-bearing Tile. Partition, fireproofing, and furring tile are included in this group. Only one grade, NB, is recognized. The absorption requirements are given in the standards. Fireproofing tile intended for use in load-bearing masonry must have the same properties as structural clay load-bearing wall tile.

Partition and split furring tile requirements in regard to total number of cells, number of cells in the direction of wall thickness, and dry weights per square foot of tile are given in the standards. Requirements for minimum weights of structural clay non-load-bearing tile shall be waived if the overall thickness of the shells, measured between the inner and extreme outer surfaces, is not less than 5/8 in. and the thickness of the webs is not less than 1/2 in.

3. Structural Clay Floor Tile. [These tile are used as structural units in floor and roof construction.] Two grades are covered, designated Grade FT1 and Grade FT2. Both grades are suitable for use in flat or segmental arches or in combination tile and concrete ribbed-slab construction. Flat and segmental arch tile must meet the requirements for number of cells in the depth of the arch and average dry weights per square foot of tile. Tile designed to resist stresses in combination tile and concrete ribbed-slab construction should conform to the prescribed requirements as to number of cells in the direction of depth of slab, and average dry weights per square foot of tile. The requirements for minimum weights of structural clay floor tile may be waived if the overall thickness of the shells, measured between the inner and extreme outer surfaces, is not less than 5/8 in. and the thickness of the webs is not less than 1/2 in.

The standard methods for determining the physical requirements are given in ASTM C112-36 and ASA 83.1-1942.

Salt-glazed tile have a vitreous glaze on the surface produced by volatilizing salt in the kiln at the temperature used in finishing the burning.

In porous structural clay tile, the natural porosity of the clay or shale has been increased by the admixture of other ingredients.

DRAIN TILE (clay) are made from shale, fire clay, surface clay, or mixtures of these. The general procedure of manufacturing is about the same as for manufacturing structural clay tile. The extrusion machine, of course, always has a core to insure the proper wall thickness.

Three classes of drain tile are recognized by the American Society of Testing Materials (ASTM C4-24) and the American Standards Association (ASA A6-1925). These are:

1. **Farm drain tile,** for ordinary private drainage work on farms, for moderate sizes and depths.
2. **Standard drain tile,** for ordinary district land drainage at moderate depths.

3. **Extra-quality drain tile,** intended for district land drainage, for considerable depths and where an extra quality is desired.

In certain instances, freezing and thawing tests or special chemical requirements as to the resistance of the tile to chemical action in specific environments may be desirable. Visible grains or masses of caustic lime, iron pyrites, or any other minerals which are known to cause slaking or disintegration of the tile may be a valid ground for rejection unless satisfactory proof is afforded that the tile are permanent and durable.

CLAY SEWER PIPE * are made from surface clay, shale, fire clay, or a combination of these materials. The materials are formed into the desired shape and subjected to a suitable temperature for developing strength and density. The finished product is a composite of crystalline mineral grains bonded together with a glassy matrix. Because of its composition, it is dense, strong, and resistant to chemical attack. Because of the high degree of vitrification developed in the fired bodies, they commonly are referred to as vitrified clay pipe. They are used for domestic and municipal sewage disposal and for the handling of industrial waste and corrosive liquids.

In the manufacture of vitrified clay pipe, the raw material after mining is ground, screened, and mixed with the proper amount of water for forming. The prepared body is extruded through an annular die of the desired size and shape by one of several methods. After removal of the plasticizing water, the ware is fired in either periodic or continuous car tunnel kilns to a temperature suitable for developing the strength and absorption values desired. These temperatures vary between 1900 and 2100 F, depending on the refractoriness of the clays or shales.

Vitrified clay pipe usually are coated with a salt glaze which is applied during firing, or with a ceramic glaze which is applied before firing and is matured during the firing process. Some ware now are being made without any glaze coating.

Most pipe are made with the conventional socket joint which usually is filled with cement, asphalt, or coal-tar compounds. There are several prefabricated joints available for socket pipe. Ring pipe, cylindrical tubes without sockets, are produced principally in California where they are used widely with mechanical joints, which combine strength with some flexibility. A number of such joints are available.

Vitrified clay pipe are manufactured in two strength classifications identified as Standard and Extra Strength. They vary primarily in wall thickness and in the load which they are required to support. ASTM requirements are listed in ASTM C13-44T and C200-44T.

CORK

(Data furnished by Armstrong Cork Co.)

Cork is the spongy outer bark of the cork oak tree, which grows principally in Spain, Portugal, and northern Africa. It is of cellular structure, each cell comprising a strong, impenetrable skin, enclosing air.

NATURAL CORK. The uses of natural cork are limited by the variability of the raw material, by structural flaws, which are uncontrollable in a natural material, and by the irregularity in size, thickness, and shape. These limitations can be overcome by laminating the corkwood to produce a material of greater uniformity and structural strength.

CORK COMPOSITIONS. Finely ground cork combined with various binders form materials having all the desirable basic qualities of natural cork. Composition can be made to provide for expansion or contraction, and to be fabricated into sheets, disks, tubes, blocks, rings, strips, balls, wheels, rods, and irregularly shaped articles. Composition is laminated with cloth, vulcanized fiber, or other material to increase the tensile strength.

Cork and rubber compositions are employed for applications where certain properties of the various synthetic or natural rubbers are required but where their normal side flow under compression is objectionable. By combining compressible cork particles with non-compressible rubbers, there results a group of compositions in which side flow is either reduced to a specified value or virtually eliminated.

PROPERTIES OF CORK. Resistance to Liquid Penetration. Cork is highly resistant to penetration by liquids. Since it is not fibrous, liquids are not drawn through it by capillary action.

---

* Data from Dr. J. O. Everhart, Engineering Experiment Station, Ohio State University, Columbus, Ohio.

Compressibility. Cork is truly compressible, undergoing actual reduction of volume when compressed. This property is often referred to as compressibility without side flow, inasmuch as Poisson's ratio is virtually zero.

Resilience. Long subjection to pressure does not destroy the structure or resilience of cork. It will not harden nor deteriorate under normal conditions.

Buoyancy. Natural cork, after removal of the outer layer of "hardback," has a specific gravity of 0.15 to 0.20 and in this form is used for floats or other devices requiring buoyancy.

Thermal Conductivity. The air-cell structure of cork forms an excellent insulator. It is highly resistant to charring and ignition. At about 250 F, slow distillation begins and the cork discolors slightly. It does not ignite except in contact with flame, and it burns only as long as flame is applied from an external source. Combustion produces a carbonized coating which retards further burning.

Chemical Inertness. Cork resists chemical action better than most natural materials and can be safely used in contact with most chemicals. In industrial processes it does not corrode, discolor, or deteriorate and is not affected by oil, grease, gasoline, or other petroleum products. Alkalis, certain acids, and a few organic solvents have destructive action; as a rule, cork should not be used in contact with them. Iron solutions sometimes react with the natural tannin in the cork to cause the surface discoloration, which does not affect the properties of the cork.

Stability. Cork retains its initial properties practically unimpaired under widely varying conditions of temperature, humidity, or other atmospheric conditions. Its compressibility and resilience enable it to withstand abrasive wear. It is not subject to progressive deterioration with age.

Friction Properties. Owing apparently to the fact that the cell structure presents a myriad of minute vacuum cups, the coefficient of friction of natural cork and compositions is higher than that of most materials used for friction. This property is not materially decreased by the presence of moisture or oil. Operating temperatures are the principal limiting factor, as cork does not stand sustained high temperatures in excess of 250 F, although it may successfully withstand surges to 300 or even 350 F. This factor tends to limit its use to light-duty applications or to heavy-duty designs involving little slippage. However, if special means of heat dissipation are provided, clutches involving a comparatively high amount of intentional slippage may successfully use cork facings. Thus, designers of oil-filled clutches and torque converters for automobiles have taken advantage of cork's ability to retain a high coefficient of friction in oil, while the presence of the oil avoids the build-up of frictional heat which would be damaging to the cork.

Compositions are used more extensively than natural cork for reasons of uniformity and control of compressibility and other properties. In practice, laboratory data have defied translation into actual performance data, and the "cut-and-try" method of evaluation is strongly recommended. As a starting point in design calculations, a coefficient of friction of 0.5 is suggested for dry assemblies involving quick starts or stops; 0.25 is suggested for oil-filled devices.

USES. Following are the principal applications of cork and cork products:

Gaskets and Packings. Cork gaskets are widely used in many types of machines and devices for sealing liquids such as water, gasoline, oil, etc., and for cushioning between adjoining surfaces of metal and glass or similar combinations. The compressibility and resilience of cork permit a seal even when the adjoining surfaces are comparatively rough and irregular.

Insulation (Thermal). Cork and cork compositions find wide application as insulation for refrigerating equipment and in cold-storage plants, ships, cars, trucks, etc. See Heat Insulation in *Power* Volume, Kent's *Mechanical Engineers' Handbook.* Special parts also are made in the form of handles for industrial equipment where heat insulation is required.

Insulation (Noise and Vibration). Cork and cork compositions are used as resilient cushions under machines to isolate them from their surroundings. They are also used in the form of panels for sound deadening of walls, to reduce reflection of sound waves back into a room. See Section 9, Vibration and Noise Control.

Friction. Applications include clutch facings, friction wheels, feed rolls, high-friction grips for fishing rods, tools, and instruments.

Floats. Applications include gage mechanisms, carburetors, dispensing pumps for oil and gasoline, and life preservers.

FELT

By William H. Lehmberg

Felt is one of the most versatile nonmetallic materials employed for mechanical purposes. Its adaptable range of densities from soft to hard and the fact that it is resilient, durable, and will not ravel when cut are among the properties responsible for its extensive use.

DEFINITION. Felt is a fabric built up by the interlocking of fibers by a suitable combination of mechanical work, chemical action, moisture, and heat, without spinning, weaving, or knitting. It may consist of one or more classes of fibers; wool, reprocessed wool, and/or reused wool; with or without admixture with animal, vegetable, and synthetic fibers.

MANUFACTURE. Processes in the manufacture of felt follow the general pattern of wool-textile manufacture except for two distinct operations, hardening and fulling, which replace conventional spinning and weaving. Processes in order of sequence include: fiber blending and picking, carding, hardening, fulling, scouring and neutralizing, dyeing, drying, sanding and/or shearing, and press finishing. Cross laying of carded webs is employed to develop balanced strength in lengthwise and widthwise directions. The amount of felting (hardening and fulling), in combination with selective fiber blending, governs felt density and the performance characteristics of the finished felt.

MECHANICAL FELT STANDARDS AND SPECIFICATIONS. Mechanical felt standards and specifications apply to: (1) roll felts, which, depending on type and density, are customarily fabricated in 15- to 30-yd lengths, $^3/_{64}$ to 1 in. thick, 60 or 72 in. wide; and (2) sheet felts, which are customarily made in 36-in. squares, $^1/_8$ to 3 in. thick. Specification conformance tests are governed by ASTM D461, *Standard Methods of Testing Felt.*

Roll Felt Specifications. Specifications covering standard types of roll felt for automotive, aircraft, and other mechanical uses are given in Table 5. SAE, ASTM, and Felt Association type number designations are listed.

Sheet Felt Specifications. Standard types of sheet felt including weight and thickness specifications were established by the Felt Association in September 1948. (See Table 6.) These specifications cover all standard sheet felt types, thicknesses, and densities for mechanical use and apply only to those felts which are manufactured in individual sheets rather than in roll or yardage form.

PROPERTIES OF FELT. Depending on the purposes they are designed to serve, felts differ in density, fiber blend, and construction. Distinctions between these characteristics largely distinguish the performance of one type or grade from another. Engineering data typifying the general behavior of mechanical felts in several fields of application and under various working conditions are given in the following paragraphs. These data have been established from test results made available by the Felt Association, and, although they do not constitute specification material, they provide a general guide to the selection of standard grades for mechanical purposes.

Formability. Felt is particularly suited to form shaping by stretching or molding. In tension, it develops a progressive reaction to increased loading which results in a high rate of elongation before rupture.

Compressibility and Resilience. Load deflection and compression set values for standard mechanical roll felts are given in Tables 1 and 2. Wool felts are characteristically resilient, will not crumble or flow, and do not deteriorate with age. Load deflections up to 30% produce less than 5% permanent deformation. The high degree of compressibility and resilience of the lower-density felts makes them ideally adaptable to cushioning, form sealing, packing, padding, and damping.

Table 1. Load Deflection of SAE Felts *

	Deformation, %					
Load, psi	F-1	F-3	F-5	F-10	F-11	F-15
10	5	8	15	30	25	30
20	9	14	22	39	33	37
40	17	21	30	46	42	45
60	23	25	35	50	47	49
80	26	29	39	54	51	53
100	29	31	41	56	54	55
200	37	39	49	62	61	62
400	47	49	58	71	69	70

* Deformation reported is the percentage of decrease in original measured thickness 1 min after loading to the nominal compressive stress.

Table 2. Compression Set of SAE Felts *

Load Deflection, %	F-1		F-3		F-5		F-10		F-11		F-15	
	Load, psi	Comp. Set, %	Load, psi	Comp. Set, %	Load, psi	Comp. Set, %	Load, psi	Comp. Set, %	Load, psi	Comp. Set, %	Load, psi	Comp. Set, %
10	21	1	13	1	6	1	4	1	4	1	3	1
20	45	2	35	3	17	2	7	2	7	2	6	2
30	108	4	93	6	41	4	9	3	15	4	10	4
40	246	7	210	9	91	6	24	6	34	6	26	6

* Compression set reported is the percentage of difference between original and final felt thickness as determined by loading for 1 min to the stated deformation and measuring final thickness 1 min after unloading for recovery.

Frictional Properties. The coefficient of friction of felt is largely independent of density and type and averages approximately 0.22 for dry felt against steel. The frictional force developed by felt, however, depends on the texture and smoothness of adjacent surfaces as well as on the viscosity and character of associated lubricating oils and greases or other liquids. Coefficients of static friction of felt against smooth, clean surfaces of wood and glass average 0.37 and 0.26, respectively. Against steel lightly lubricated with SAE 30 aviation oil, the coefficient of friction of felt is 0.18.

Wear Resistance. The wear resistance of felt varies in direct proportion to density and fiber quality and in inverse proportion to applied load. The excellent wear resistance of high-density roll and sheet felts is one of their most important properties. Fabricated into laps, wheels, and bobs, these felts are used extensively for polishing optical and ophthalmic lenses, plate glass, plastics, and metals. Others are similarly employed for the continuous wiping of brass and steel in sheet and strip mills.

Sound Absorption and Thermal Insulation. Owing to its structure, felt is a highly efficient sound-absorbing and thermal-insulating material. It is easy to handle, will not disintegrate from vibration, and will not collapse when wet. Thermal conductivity decreases with decreased felt density, and many of the lighter weight mechanical grades have *K* factors of less than 0.25. See Table 3. Combinations of kapok fiber with wool produce felts of 0.21 *K* factor and of 0.80 sound absorption coefficient at frequencies above 512 cps. These felts weigh approximately 3.2 lb per sq yd for 1 in. thickness and are designed specially for military and commercial aircraft insulation. Kapok felts as well as standard mechanical types are applied industrially at temperatures up to 200 F to insulate process equipment and instruments, reduce general noise level, soundproof test rooms and ventilating ducts, and dampen noise generated by machine vibration. For most applications, sound and thermal insulating felts are proofed for resistance to flame, fungi, mildew, moth, and vermin.

Table 3. Thermal Conductivity of SAE Felts *

	Btu/Hr/Sq Ft/°F				
	F-1	F-3	F-5	F-11	F-15
K Factor, 1-in. thickness	0.36	0.35	0.30	0.23	0.24

* Thermal insulation felts blended with kapok fiber have tested values of 0.21 *K* factor.

Absorption Capacity and Capillarity. Foremost among the important functional properties of felt is its ability to absorb, store, or feed oil and grease in mechanical sealing,

(*Continued on p. 5-24*)

Table 4. Percentage of Oil Absorption Capacity of SAE Felts *

	Viscosity, Saybolt Universal Seconds at 70 F					
	92	212	279	575	940	1400
SAE No.	Specific Gravity, 60/60 F					
	0.845	0.853	0.865	0.890	0.898	0.925
F-1	165%	157%	170%	161%	173%	179%
F-3	203	183	207	212	200	201
F-5	238	225	244	240	241	244
F-7	232	206	221	252	241	233
F-10	427	450	464	456	476	470
F-11	387	382	352	341	367	461
F-13	356	359	378	378	394	384
F-15	334	331	334	349	353	348

* Values given are percentage of weight increase of oil-saturated felt as determined by measuring the volumetric displacement of oil of known gravity at 70 F.

Table 5. Standard Mechanical Roll Felt Specifications

SAE Specification No.	ASTM and Felt Association Type No.	Thickness, in.		Weight, lb/sq yd		Wool Content,‖ % (min)	Carbon Tetrachloride Soluble, % (max)	Water Soluble, % (max)	Combined Carbon Tetrachloride and Water, % (max)	Ash, % (max)	Tensile Strength, psi (min)	Splitting * Resistance, lb/2-in. Width (min)	Trade Designation	Color	Standard Width, in.
		Nominal	Limits	Nominal	Limits										
....	18R1	1/8	0.113–0.137	2.25	2.15 – 2.35	95	2.5	2.5	3.0	1.5	600	35	Laundry	White	60
		3/16	0.175–0.201	3.38	3.23 – 3.53										
		1/4	0.236–0.264	4.50	4.30 – 4.70										
		5/16	0.298–0.328	5.63	5.38 – 5.88										
		3/8	0.359–0.391	6.75	6.45 – 7.05										
		1/2	0.481–0.519	9.00	8.60 – 9.40										
F-1 †	16R1	1/8	0.113–0.137	2.0	1.90 – 2.10	95	2.5	2.5	3.0	1.5	500	33	Back check	White	60
		3/16	0.175–0.201	3.0	2.85 – 3.15										
		1/4	0.236–0.264	4.0	3.80 – 4.20										
		5/16	0.298–0.328	5.0	4.75 – 5.25										
		3/8	0.359–0.391	6.0	5.70 – 6.30										
		1/2	0.481–0.519	8.0	7.60 – 8.40										
		5/8	0.603–0.647	10.0	9.50 –10.50										
		3/4	0.725–0.775	12.0	11.40 –12.60										
		7/8	0.847–0.903	14.0	13.30 –14.70										
		1	0.969–1.031	16.0	15.20 –16.80										
F-2	16R2	Same thicknesses and weights as F-1				95	2.5	2.5	4.0	2.0	500	28	Back check	Any color except gray or black	60
F-3 ‡	16R3	1/8	0.113–0.137	1.95	1.80 – 2.10	90	2.5	3.0	4.5	2.5	400	22	Back check	Gray	60
		3/16	0.175–0.201	2.93	2.71 – 3.15										
		1/4	0.236–0.264	3.90	3.61 – 4.19										
		5/16	0.298–0.328	4.87	4.50 – 5.24										
		3/8	0.359–0.391	5.85	5.41 – 6.29										
		1/2	0.481–0.519	7.80	7.21 – 8.39										
		5/8	0.603–0.647	9.75	9.01 –10.49										
		3/4	0.725–0.775	11.70	10.81 –12.59										

		7/8	0.847–0.903	13.65	12.61 –14.69									
		1	0.969–1.031	15.60	14.41 –16.79									
F-5	12R1	1/8	0.111–0.139	1.53	1.45 – 1.61	95	2.5	2.5	3.0	2.0	400	18	Extra firm pad	White	60
		3/16	0.172–0.204	2.29	2.17 – 2.41									
		1/4	0.232–0.268	3.06	2.90 – 3.22									
		5/16	0.293–0.333	3.82	3.62 – 4.02									
		3/8	0.353–0.397	4.59	4.35 – 4.83									
		1/2	0.474–0.526	6.12	5.80 – 6.44									
		5/8	0.595–0.655	7.65	7.25 – 8.05									
		3/4	0.716–0.784	9.18	8.70 – 9.66									
		7/8	0.837–0.913	10.71	10.15 –11.27									
		1	0.958–1.042	12.24	11.60 –12.88									
F-6	12R2	Same thicknesses and weights as F-5				92	2.5	2.5	4.5	2.5	275	16	Extra firm pad	Gray	60 or 72
F-7 §	12R3	Same thicknesses and weights as F-5				80	3.0	4.0	7.0	3.0	250	12	Extra firm pad	Gray	72
F-10	9R1	1/8	0.105–0.145	1.06	0.98 – 1.14	95	2.5	2.5	3.0	2.5	225	8	Firm pad	White	72
		3/16	0.165–0.211	1.59	1.47 – 1.71									
		1/4	0.224–0.276	2.12	1.96 – 2.28									
		5/16	0.284–0.342	2.65	2.45 – 2.85									
		3/8	0.343–0.407	3.18	2.94 – 3.42									
		1/2	0.462–0.538	4.24	3.92 – 4.56									
		5/8	0.581–0.669	5.30	4.90 – 5.70									
		3/4	0.700–0.800	6.36	5.88 – 6.84									
		7/8	0.819–0.931	7.42	6.86 – 7.98									
		1	0.938–1.062	8.48	7.84 – 9.12									
F-11	9R2	Same thicknesses and weights as F-10				92	3.0	2.5	4.5	3.0	200	6	Firm pad	Gray	72
F-12	9R3	Same thicknesses and weights as F-10				85	4.0	2.5	6.5	3.5	100	3	Firm pad	Gray	72
F-13	9R4	Same thicknesses and weights as F-10				75	4.0	4.0	8.0	3.5	75	2	Firm pad	Gray	72
F-15	9R5	Same thicknesses and weights as F-10				55	4.0	5.0	9.0	4.0	75	2	Firm pad	Gray	72
F-26	8R5	1/8	0.085–0.165	0.90	0.81 – 0.99	45	8.0	6.0	14.0	5.0	....	....	Soft pad	Gray	72
		1/4	0.194–0.306	1.80	1.62 – 1.98									
		3/8	0.303–0.447	2.70	2.43 – 2.97									
		1/2	0.412–0.588	3.60	3.24 – 3.96									
		3/4	0.630–0.870	5.40	4.86 – 5.94									
		1	0.848–1.152	7.20	6.48 – 7.92									

(Table continued on p. 5-22)

Table 5. Standard Mechanical Roll Felt Specifications—*Continued*

SAE Specification No.	ASTM and Felt Association Type No.	Thickness, in.		Weight, lb/sq yd		Wool Content,‖ % (min)	Carbon Tetrachloride Soluble, % (max)	Water Soluble, % (max)	Combined Carbon Tetrachloride and Water, % (max)	Ash, % (max)	Tensile Strength, psi (min)	Splitting * Resistance, lb/2-in. Width (min)	Trade Designation	Color	Standard Width, in.
		Nominal	Limits	Nominal	Limits										
F-50	16R1X	3/64	0.040–0.054	0.750	0.712–0.788	95	2.5	2.5	3.0	1.5	500	....	Ball-bearing felt	White	60 or 72
		1/16	0.056–0.070	0.975	0.937–1.013										
		5/64	0.071–0.085	1.200	1.162–1.238										
		3/32	0.087–0.101	1.425	1.387–1.463										
F-51	16R3X	Same thicknesses and weights as F-50				92	2.5	2.5	4.5	2.5	300	....	Ball-bearing felt	Gray	60 or 72
F-55	12R3X	1/16	0.056–0.070	0.750	0.712–0.788	75	4.0	4.0	8.0	3.0	200	....	Lining	Gray or black	60 or 72
		3/32	0.087–0.101	1.125	1.087–1.163										

* Splitting resistance is not applicable to felts where the thickness is less than 3/16 in. For materials less than 3/16 in. in thickness, breaking strength only is recommended as an indicative test.

† For thicknesses under 1/8 in., see F-50.

‡ For thicknesses under 1/8 in., see F-51.

§ For thicknesses under 1/8 in., see F-55.

‖ The wool content indicates the percentage of wool by chemical analysis, but does not differentiate between wool, reprocessed wool, and reused wool, and is exclusive of traces of other fibers and impurities present in the wool used in fabricating the several types of felt. For example, Type 16R1, fabricated from 100% wool, may contain incidental traces of other fibers, residual wool fats, and oils or soaps used in processing which may reduce the actual wool fiber content on analysis to a minimum of 95%.

Table 6. Standard Sheet Felt Weight and Thickness Tolerances *

(Standard sheet felt types and tolerances are representative of industry practice as adopted by the Felt Association, September 17, 1948.)

Density		12S (Extra Soft)					16S (Soft)					20S (Medium)					26S (Hard)					32S (Rock Hard)				
Type		12S1	12S2		12S3	12S4	16S1	16S2		16S3	16S4	20S1	20S2		20S3	20S4	26S1	26S2		26S3	26S4	32S1	32S2		32S3	32S4
Trade Designation		Fine Spanish	Spanish		Mexican	Coarse Mexican	Fine Spanish	Spanish		Mexican	Coarse Mexican	Fine Spanish	Spanish		Mexican	Coarse Mexican	Fine Spanish	Spanish		Mexican	Coarse Mexican	Fine Spanish	Spanish		Mexican	Coarse Mexican
Nominal Thickness,† in.		Weight, lb/sq yd			Thickness, in.		Weight, lb/sq yd			Thickness, in.		Weight, lb/sq yd			Thickness, in.		Weight, lb/sq yd			Thickness, in.		Weight, lb/sq yd			Thickness, in.	
		Nominal	Limits		Limits		Nominal	Limits		Limits		Nominal	Limits		Limits		Nominal	Limits		Limits		Nominal	Limits		Limits	
			Min	Max	Min	Max		Min	Max	Min	Max		Min	Max	Min	Max		Min	Max	Min	Max		Min	Max	Min	Max
1/4	0.250	3.00	2.80	3.20	0.230	0.270	4.00	3.70	4.30	0.234	0.266	5.00	4.60	5.40	0.236	0.264	6.50	5.90	7.10	0.239	0.261	8.00	7.00	9.00	0.243	0.257
3/8	0.375	4.50	4.20	4.80	0.351	0.399	6.00	5.60	6.40	0.356	0.394	7.50	7.00	8.00	0.359	0.391	9.75	8.95	10.55	0.362	0.388	12.00	10.90	13.10	0.366	0.384
1/2	0.500	6.00	5.60	6.40	0.470	0.530	8.00	7.50	8.50	0.478	0.522	10.00	9.40	10.60	0.481	0.519	13.00	12.00	14.00	0.484	0.516	16.00	14.80	17.20	0.489	0.511
5/8	0.625	7.50	7.05	7.95	0.590	0.660	10.00	9.40	10.60	0.599	0.651	12.50	11.80	13.20	0.603	0.647	16.25	15.15	17.35	0.607	0.643	20.00	18.70	21.30	0.612	0.638
3/4	0.750	9.00	8.50	9.50	0.710	0.790	12.00	11.30	12.70	0.721	0.779	15.00	14.20	15.80	0.725	0.775	19.50	18.30	20.70	0.730	0.770	24.00	22.60	25.40	0.735	0.765
7/8	0.875	10.50	9.95	11.05	0.829	0.921	14.00	13.25	14.75	0.843	0.907	17.50	16.60	18.40	0.848	0.902	22.75	21.45	24.05	0.853	0.897	28.00	26.50	29.50	0.858	0.892
1	1.000	12.00	11.40	12.60	0.949	1.051	16.00	15.20	16.80	0.964	1.036	20.00	19.00	21.00	0.970	1.030	26.00	24.60	27.40	0.975	1.025	32.00	30.40	33.60	0.981	1.019
1 1/4	1.250	15.00	14.30	15.70	1.188	1.312	20.00	19.10	20.90	1.208	1.292	25.00	23.90	26.10	1.214	1.286	30.75	29.25	32.25	1.221	1.279	40.00	38.30	41.70	1.228	1.272
1 1/2	1.500	18.00	17.20	18.80	1.428	1.572	24.00	23.00	25.00	1.451	1.549	30.00	28.80	31.20	1.459	1.541	35.50	33.90	37.10	1.467	1.533	48.00	46.20	49.80	1.474	1.526
1 3/4	1.750	21.00	20.10	21.90	1.667	1.833	28.00	26.90	29.10	1.694	1.806	35.00	33.70	36.30	1.703	1.797	40.25	38.55	41.95	1.712	1.788	......	......	......	......	......
2	2.000	24.00	23.00	25.00	1.906	2.094	32.00	30.80	33.20	1.937	2.063	40.00	38.60	41.40	1.947	2.053	45.00	43.20	46.80	1.958	2.042	......	......	......	......	......
2 1/2	2.500	30.00	28.95	31.05	2.385	2.615	40.00	38.75	41.25	2.424	2.576	50.00	48.55	51.45	2.436	2.564	54.50	52.60	56.40	2.449	2.551	......	......	......	......	......
3	3.000	36.00	34.90	37.10	2.864	3.136	48.00	46.70	49.30	2.910	3.090	60.00	58.50	61.50	2.925	3.075	64.00	62.00	66.00	2.940	3.060	......	......	......	......	......

* Made in sheets 36 by 36 in.

† Sheets of 1/8-in. thickness are manufactured on order in Nos. 20-S, 26-S, and 32-S.

lubricating, and wicking applications. Absorption or storage capacity is principally a function of felt density, increasing in direct inverse proportion to density as shown in Table 4. The capillarity of felt, or distance to which felt will wick liquids, increases with felt density but varies in inverse proportion to fluid viscosity. For design purposes, delivery rates can be estimated from the following formula derived from experimental test data:

$$Q = \frac{0.8 \times 60DCAR}{G}$$

where Q = delivery rate, cubic centimeters per hour.
0.8 = average coefficient of capillarity.
D = density of felt, grams per cubic inch
$= \dfrac{\text{per cent specific gravity}}{100} \times 16.4.$
C = absorption capacity of felt, percentage.
A = cross-sectional area of wick, square inches.
R = wick-rise rate, inches per minute at specified wicking distance.
G = specific gravity of fluid.
60 = conversion from cubic centimeters per minute to cubic centimeters per hour.

Chemical Inertness. The effects of distilled, tap, chlorinated, or sea water on the physical properties of felt after extended immersion at 70 F and 212 F are moderate thickness swell, slight dimensional shrinkage in length and width, increased rate of elongation under tensile stress, and decreased wet breaking strength to the extent of about 25%. Felts may be treated for water repellency and when treated will demonstrate improved stability.

Extended exposure of felt to lubricating oils, greases, and organic solvents has no appreciable effect on the dimensional stability or physical properties of felt. Strong inorganic acids are noninjurious in concentrations up to 14 Bé. Nitric acid in 14 Bé concentration, however, reduces the breaking strength of wool felt by 30 to 40% after 5-day immersion. Strong bases or alkalis are injurious.

Felt, unlike many organic substances, may be considered a nonaging material. The natural destructive forces of atmospheric temperature, ozone, and ultraviolet radiation are essentially nonreactive. Felt is unchanged by increased temperatures up to 212 F. Above 212 F, wool loses its hygroscopic moisture and becomes harsh to the touch, and the physical properties of breaking strength, elongation, and splitting resistance are adversely affected. At low temperatures, felt is the same at −60 F, in the absence of condensate, as at normal room temperatures. At −80 F, its properties are only slightly different.

Mechanical Felt Applications. There are, as a minimum, fifteen important mechanical functions fulfilled by felt, i.e., sealing, wicking, wiping, polishing, vibration isolation, filtration, sound absorption, thermal insulation, shock absorption, cushioning, dampening, padding, packaging, surfacing, and frictional applications.

Precision-cut washers, fabricated from felt or felt laminated with oil resistant synthetic elastomers, seal against loss of lubricant and protect ball and antifriction bearings from dirt, grit, and moisture. Felt filters for industrial respirators, air-conditioning units, process and product filters, including pigments, paint, lacquer, photographic emulsions, and food syrups, are highly retentive, permeable, and characteristically slow plugging. Vibration mountings for machines, weather stripping, automotive window channels, antisqueak body spacers, polishing wheels, and laps are other typical felt applications, as are wicks for motor-bearing lubrication, sound insulation for commercial and military aircraft, and nonskid scratch protection for fan bases, instruments, and furniture.

TYPICAL MECHANICAL ROLL FELT APPLICATIONS. Type 18R1 is recommended where unusual strength and hardness are required. Typical uses include: small-diameter wicks, polishing wheels, blocks and pads, bumpers, grommets, and precision washers.

Type 16R1 (SAE F-1) is suitable for oil retention in installations where the felt is not compressed or for feeding low viscosity or light oil. Typical uses are: washers; bushings; wicks; door bumpers; polishing blocks, wheels, and pads; grommets; window channels; resilient mountings, antivibration and dampening pads; and parts where wear and resistance to abrasion are required.

Types 16R2 (SAE F-2) and 16R3 (SAE F-3) are recommended for vibration mountings and oil and grease shields, and for the same general purposes as SAE F-1 but where a felt of slightly lower tenacity is satisfactory.

Types 12R1 (SAE F-5), 12R2 (SAE F-6), and 12R3 (SAE F-7) are recommended for dust shields, wipers, grease retainer washers, wicks, vibration mountings, and where a resilient felt is required.

Types 9R1 (SAE F-10), 9R2 (SAE F-11), and 9R3 (SAE F-12) are recommended for grease and oil retention where the felt is confined and compressed in assembly. They are also recommended for dust shields under less severe operating conditions, where 12R1, 12R2, and 12R3 are not required.

Types 9R4 (SAE F-13) and 9R5 (SAE F-15) are recommended for sound deadening, chassis strips, spacers, dust shields, pedal pads, and dash liners, and for mechanical purposes where abrasion and wear are not important factors.

Type 8R5 (SAE F-26) is suitable for packing or padding when held in place between other materials. This grade should not be used for mechanical purposes.

Type 16RX1 (SAE F-50) is recommended for ball- and roller-bearing oil retainer washers, and small dust-excluding washers, and for mechanical purposes where an accurate, thin, smooth high grade felt is required.

Type 16RX3 (SAE F-51) is recommended for the same general uses as F-50 but in installations where tolerances and length of life are not so important. It is also used for thin cut parts such as gaskets and liners.

Type 12RX3 (SAE F-55) is recommended for antisqueak strips and for lining when cemented to fiber board or metal panels.

TYPICAL SHEET FELT APPLICATIONS. Types 12S1 to 12S4, depending on fiber fineness required, are suitable as absorbent pads, vibration and shock mountings, and dampeners, and for fine and soft metal polishing, buffs, and the like.

Types 16S1 to 16S4 are suitable for the same general uses as 12S1 to 12S4 where a slightly denser felt is required, as well as for oil and fluid wicks, washers, plugs, bushings, channels, bumpers, ink and fluid transfer rolls, and general-purpose polishing blocks.

Types 20S1 to 20S4, depending on fineness requirements, are recommended for mirror and glass polishing, lens polishing, cash carrier heads, bumpers, ink rolls, rubber stamp pads, shoe insoles, scratch wheels, wood polishing, furniture rubbing, and punched wicks.

Types 26S1 to 26S4, depending on fineness requirements, are recommended for glass tumbler polishing, marble and granite polishing, artificial limbs, furniture casters, shoe and boot soles, block cutters, and for the same general uses as 20S1 to 20S4 where a denser felt is required.

Types 32S1 to 32S4 are used for heavy-duty grinding and polishing wheels as well as for the same general purposes as 26S1 to 26S4 where a dense quality is needed.

GLASS

By Harold G. Vogt

7. DEFINITION AND CHEMICAL NATURE OF GLASS

Glass is an inorganic product of fusion which has been cooled to a rigid condition without crystallizing (ASTM C162-45T). It is typically hard and breaks with conchoidal fracture. Although basically homogeneous and transparent, masses or bodies of glass may be made colored, translucent, or opaque by the inclusion of dissolved, amorphous, or crystalline materials.

Glass is made by heating a mixture of dry, powdered, or granular ingredients at about 2800 F in a refractory container until a viscous, homogeneous liquid is formed. The batch consists of the following raw materials in the form of oxides or of compounds which yield such oxides: (1) almost always silica (sand) as the glass former, (2) usually some lime and soda ash for fluxing, and (3) often one or more of the following for modifying purposes: potash, lead oxide, boron oxide, alumina, magnesia, barium oxide, and zinc oxide. Small pieces of scrap glass, "cullet," from a previous melt are usually added and, depending upon many special circumstances, the batch may include small amounts of one or more oxidizing, decolorizing, coloring, or opacifying agents.

The proportions of the raw materials in the batch very seldom correspond to those of chemical compounds. The proportions may be varied continuously over relatively wide ranges; some of the physical and chemical properties of the resulting glasses vary continuously over wide ranges, some remain practically constant, and others vary over small ranges. Also, there is considerable choice as to which and how many ingredients are to be used. Through a study of the properties of some 50,000 glasses, it has been possible to select the formulas that are best fitted for specific applications. About 400 different glasses, which differ considerably in some of their properties, are melted commercially each year. Of these glasses, a relatively small number have such general usefulness that they are melted continuously and in great volume.

TYPES OF GLASSES. The great bulk of commercial production is commonly classified in the following chemical types of silicate glass, listed in order of decreasing costs, decreasing softening temperatures, decreasing thermal shock resistances, and increasing thermal expansion coefficients. (1) 99.8% silica glass, i.e. fused quartz or fused silica, (2) 96% silica glasses, (3) borosilicate glasses, (4) lead glasses, and (5) soda-lime glasses. The commercial ranges in compositions of these types, as determined by analysis and calculation of oxide equivalents of the finished glasses, are given in Table 1, and the physical properties of some typical representatives of each type are given in Table 2.

Table 1. Composition Ranges of Commercial Glasses

Oxide	Soda Lime	Lead	Borosilicate	96% Silica	Fused Quartz
SiO_2	70–75%	53–67%	73–82%	96%	99.8%
Na_2O	12–18	5–10	3–10	....	......
K_2O	0–1	1–10	0.4–1	....	......
CaO	5–14	0–6	0–1	....	......
PbO	....	20–40	0–10	....	......
B_2O_3	....	....	5–20	3+	.....
Al_2O_3	0.5–1.5	....	2–3	....	......
MgO	0–4	....	....	....	......

In other types of glasses, such as colored, opal, and optical, the chemical composition is not the distinguishing characteristic. Such glasses are usually made from the foregoing types or "hybrids" thereof. Colored glasses are made by adding traces generally of metallic oxides to the batch, plus proper heat treatment of the finished glass where needed. Opal glasses most often are made by adding fluorides or phosphates to the batch, and the finished glass is treated so that small particles appear in the body of otherwise transparent glass and cause the scattering light that makes opal glasses appear "milky." Optical glasses are made from a variety of compositions to get desired combinations of index of refraction and dispersive constant. Their distinguishing characteristic lies in the stringent degree of freedom from physical imperfections required in the finished glasses.

There are some nonsilicate glasses, e.g., borate and phosphate glasses, made in small quantities for special uses.

8. PHYSICAL AND CHEMICAL PROPERTIES

During recent decades, industrial users of glass have adopted much more of an engineering approach toward its application, and glassmakers, through intensive research, have acquired scientific understanding and technical control of the material. New, superior, chemical types have been developed, particularly the low-expansion borosilicate glasses and 96% silica glasses. Technological improvements have been made in the old soda-lime and lead types, and designing and finishing treatments have advanced greatly. The resulting improvements in mechanical strength, thermal endurance, chemical durability, and electrical properties have been so marked that glass, particularly of the low-expansion types, has gained wide recognition as a versatile engineering material.

Despite the fact that relatively low tensile strength and limitations on cold working are disadvantages for many engineering uses, the toughness of most glass products today is such that the lay term, "brittleness," associated with the products of a few decades ago, is inappropriate. Modern glass is turning out to be strong enough for a great number and variety of new industrial uses. Its unique and excellent combinations of physical and chemical properties are great advantages. Some of its fine characteristics are:

1. Transparency—very high and permanent. Translucency or opaqueness can be obtained as desired, also versatile control over differential transmission or blocking of infrared, visible, or ultraviolet light.
2. Smoothness of molded surfaces or supersmoothness of ground and polished surfaces.
3. Great resistance to the attack of chemicals except hydrofluoric acid, i.e., superlative corrosion resistance. Correlated with this is the freedom from contamination of materials in contact with it.
4. Good hardness and resistance to scratching.
5. Dimensional stability or permanence of shape better than that of concrete or steel (hence the use of glass for the 20-ton, 200-in. mirror disk for the Mount Palomar Observatory telescope). Glass is free from creep, from elastic hysteresis, and from liability to permanent deformation under thermal or mechanical stresses that do not exceed breaking values.
6. Lightness or low specific gravity.
7. Low heat conductivity.
8. Good-to-excellent resistance to heat.
9. Very low expansion coefficients available in borosilicates or 96% silica glasses.
10. Good resistance to thermal stresses of the low-expansion types.
11. Small coefficient of reflection for radiant heat. Baking dishes of glass, despite their lower heat conductivity, bake their contents faster than do shiny metal dishes.
12. Excellent electrical insulation and dielectric properties.
13. Ease of hot-forming intricate shapes.
14. Wide range of fine, permanent colors.

Table 2. Physical Properties of Some Typical Glasses *

	Specific Gravity	Tensile Strength (annealed samples), 10^3 psi	Modulus of Elasticity, 10^6 psi	Linear Thermal Expansion Coefficient, 10^{-5} cm/cm/°C	"Softening" Point °F	Annealing Point, °F	Strain Point, °F	Thermal Conductivity, 10^{-4} g-cal/sec/cm²/(°C/cm)	Specific Heat, g-cal/g/°C	Volume Resistivity (annealed samples), ohm-cm	Dielectric Constant, at 1 megacycle	Dielectric Power Factor, %, at 1 megacycle	Dielectric Strength, volts/mil at 60 cps	Refractive Index, for sodium D line
Typical soda-lime glass, Corning's #0080	2.47	4–10 †	9.8	.92	1280	950	890	22 ‡	.20 ‡	$10^{12.5}$ at 20 C $10^{6.4}$ at 250 C $10^{5.2}$ at 350 C	7.2	.9	>1000 *	1.512
Typical lead glass, Corning's #0010	2.85	4–10 †	9.0	.91	1160	800	745	22 ‡	.20 ‡	$10^{9.1}$ at 250 C $10^{7.1}$ at 350 C	6.6	.16	>1000 *	1.542
Typical borosilicate glass, Corning's #7740	2.23	4–10 †	9.8	.325	1510	1030	960	28	.20 ‡	$10^{14.5}$ at 20 C $10^{8.2}$ at 250 C $10^{6.7}$ at 350 C	4.6	.46	>1000 *	1.474
Typical 96% silica glass, Corning's #7900	2.18	4–10 †	9.7	.08	2730	1670	1510	33 ‡	.20 ‡	$10^{10.2}$ at 250 C $10^{8.5}$ at 350 C	3.8	.05	>1000 *	1.458
Fused quartz, 99.8% silica	2.20	4–10 †	12.0	.055	3000	2080	1960	33	.20 ‡	10^{18} at 20 C $10^{11.7}$ at 250 C $10^{9.5}$ at 350 C	3.8	.02	>1000 *	1.459

* Use in conjunction with general remarks in text on properties.
† For safe design practice, tensile stresses should not exceed 1000 psi for annealed glass, 2000 to 4000 psi for tempered glass, depending on piece.
‡ Approximate values.

In applying the specific values for various properties of glass listed in Table 2, it is necessary to keep in mind the following general comments because the physical nature of glass is so different from that of other engineering materials.

MECHANICAL PROPERTIES. Glass is a supercooled liquid and is not crystalline or endowed with any directional properties as crystals are. It is nonductile and perfectly elastic, i.e., it shows no cold flow, up to the point of stress necessary to break it. It seems to be broken only by tensile stresses, the fracture starting in almost all cases from imperfections in the surface. Glass does not deform or decompose with aging or exposure to moisture, and it suffers no permanent deformation when subjected to mechanical or thermal stresses, even at very low or quite high temperatures. Having no porosity, it does not swell or shrink with changes of humidity or immersion.

Strength. For safe design practice, tensile stresses should be limited to 1000 psi maximum for annealed articles and to 2000 psi for tempered articles of glass. This is true since, although annealed articles have tensile strengths ranging from 4000 to 10,000 psi and although many products can be tempered to have two to five times the strengths of annealed products, the evaluation of safety factors is rather difficult and depends on the type of application and the condition of the surface.

The nature of the surface has a great effect on the tensile strength of glass. Not only are loading stresses considerably magnified in regions of surface imperfections, e.g., scratches, but also the stress concentrations cannot be relieved by any yielding or cold flow since glass is nonductile. Consequently, the strength of most glass articles is markedly less than the 500,000 psi and upwards found with some fibers of glass. There are also noticeable variations in the measured strengths. Tempering puts a permanent compressive stress in the surface and thereby strengthens a glass article and makes the glass more resistant to abrasion and scratching.

Glass tested in compression or shear seems always to be broken by the tension components of stress, and the terms compressive strength and shear strength are generally considered to have little meaning for glass. However, the apparent or effective compression strength is about ten times as great as the normally measured tensile strength, and consequently it is a decided advantage, when glass is used in conjunction with metals or other materials, to plan the design so that forces on the glass are compressive rather than tensile. This is particularly true when temperature changes may introduce considerable stress in the glass because of differences in expansion coefficients.

Modulus of Elasticity. Young's modulus ranges from 6.5×10^6 to 12×10^6 psi for the various glasses. Poisson's ratio ranges from 0.18 to 0.30, but 0.22 may safely be used in calculations.

Hardness. The nonductility of glass makes it difficult to assess the significance of indentation tests, and glass manufacturers use tests different from those used for metals. On the Moh scale of resistance to scratching, where diamond is arbitrarily assigned a figure of 10, glasses range from 5 to 7.

The terms hard and soft as used in the glass trade seldom refer to the mechanical hardness but rather to the question of whether a glass has a high or low softening temperature.

THERMAL PROPERTIES. Reference Temperatures and Maximum Operating Temperature. Consonant with the fact that glass is a supercooled liquid, it has no critical temperatures. It has no freezing or melting point, and the viscosity changes continuously with change of temperature. The "melting temperature" of glass is merely a high temperature at which the batch of materials will fuse together and give a homogeneous, viscous liquid in a reasonable length of time. The viscosities of glasses are about 100 poises, even at the melting temperatures. For practical control of shaping and annealing operations, certain useful reference temperatures have been defined. The softening, annealing, and strain points are the temperatures at which glasses have more or less closely the viscosities of 4.5×10^7, 2.5×10^{13}, and 4.0×10^{14} poises, respectively. At the annealing point, the upper end of the annealing temperature range, internal stresses are substantially relieved in 15 min, whereas at the strain point temperature, the lower end of the annealing range, it takes 16 hr. In cooling through the strain point temperature, the properties have already begun to lag appreciably behind the temperature, and thus the viscosity reached in annealing practice may be appreciably greater or less than 4.0×10^{14} poises, depending on the speed of cooling. It is for these reasons that the *maximum operating temperature* for annealed ware, if undue thermal shock is avoided, may approach the strain point with no danger of introducing appreciable permanent strain, but for tempered ware it must be lower in order not to decrease the greater strength achieved by tempering.

Linear Thermal Expansion Coefficients. Coefficients of glasses range from values equal to those of metals to values one-thirtieth those of metals. The values tabulated are averages for the temperature range 0 to 300 C, but they differ little from the values for the range 20 to 80 C.

Thermal conductivity at room temperature ranges from 0.0018 to 0.0033 g-cal/sec/cm²/(°C/cm) for different glasses. For a typical borosilicate glass, the value at 250 C is 20 to 25% higher than at room temperature.

Thermal endurance is the ability to withstand, without cracking or breaking, rapid or uneven changes of temperature of the glass. Since thermal endurance depends on all those complicated factors controlling tensile strength and on other factors, no general formula has been found for glass articles. There are, however, several useful generalizations. (1) The smaller the coefficient of expansion, the greater the thermal endurance. (2) The thicker the glass, the more apt it is to break from thermal shock. (3) Glass will stand a much greater total immersion thermal shock when heated than when cooled since in heating the surface is put into compression whereas in cooling the surface is put into tension. Ordinary bottles are made of soda-lime glass and will usually break if plunged from boiling water into ice water, but articles of the low-expansion borosilicate glasses usually will not crack even with a wall thickness up to 1/2 in. The 96% silica glasses do not break even if plunged into ice water while they are at a glowing red heat.

ELECTRICAL PROPERTIES. Resistivity. The very high values of resistivity of most glasses make it difficult to obtain reliable data at room temperature. Volume resistivities of annealed glasses at 20 C range from 10^8 to 10^{18} ohm-cm, but generally they are 10^{12} ohm-cm or higher.

Dielectric Strength. The dielectric strengths of glasses in air range from 1000 to 8000 volts per mil of thickness and are so high that measurements often are made under dry oil to prevent "flashover." Under these conditions, the oil generally fails first and sets up point electric stresses of extreme magnitude, with the result that only a fraction of the true dielectric strength of glass is measured. Consequently, data taken under oil should be used only when the insulator is intended for service under oil.

CORROSION RESISTANCE AND CHEMICAL DURABILITY. One of glass's most outstanding properties is its great stability or chemical inertness. The resistance which glass offers to the attack of water, to weathering which is due primarily to water and carbon dioxide, and to the attack of aqueous solutions of acids (except hydrofluoric), bases, and salts is the main reason why glass is used so widely for the storage of foods, drugs, and chemicals, for chemical laboratory apparatus, and for chemical plant equipment such as containers, pipeline, and pump parts. The *chemical durability* of glass is the lasting quality (both visible and chemical) of its surface. It is frequently evaluated, after prolonged weathering or storing, in terms of chemical and physical changes in the glass surface, or in terms of changes in the contents of a vessel.

No accelerated laboratory test or practical series of tests has been devised to give a general numerical value for chemical durability, and results of the various tests must be used in combination with a background knowledge of actual service experience.

Certain commercial borosilicate glasses have good all-round corrosion resistance; 96% silica glass has very superior chemical durability and in this respect, as well as in others, compares very favorably with fused quartz. The 96% silica glasses, moreover, have an advantage over fused quartz in that they can be fabricated more feasibly into larger and more intricate articles.

9. FABRICATION OF INDUSTRIAL GLASSWARE

Cold glass cannot be turned in a lathe, surfaced in a shaper, or drilled with a machine drill. In other words, it cannot be worked with a cutting tool bit. The drilling of small holes, 1/4 in. in diameter or less, by a triangular piece of hard steel or Carboloy, fed with turpentine and operated in a drill press, might seem to be an exception. Larger holes can be made with a metal cylinder having a serrated end, operated in a drill press, and fed with a mixture of water and abrasive particles.

On the other hand, glass can be ground very much as metals except for slower speeds and extra mounting precautions. It can be polished to a much smoother surface. Relatively large cylinders can be cut by a variety of simple "crack-off" methods. With small sizes of rod and tubing, a scratch with a diamond or carbide wheel glass cutter allows the glass to be broken off squarely by putting the scratched region into tension. A more effective way of cutting thicker-walled tubing is to place an electrically heated Nichrome wire in a file scratch around the tubing and then put cold water on the scratched region. Heavy pieces can be sawed with abrasive or diamond wheels.

Crews can be quickly trained to do fitting of glass piping in the field with the hot-wire crack-off and high-frequency electric sealing methods.

Yellow hot glass is quite moldable and can be blown or pressed into complex shapes. A variety of mass-production methods exists for shaping hot glass as it emerges from the melting tank. Also, cold glass can be reheated and then reshaped and welded.

Glass is being applied in a great and growing diversity of ways in industry. Glassware, produced for industrial purposes in great volume, is made by a variety of mass-production methods of automatic fabrication.

Three principal methods—blowing, pressing, and drawing—are used for forming the ware from glass in the hot, moldable state. Each method is suitable for certain types and shapes of ware; and each method, as well as the nature of glass, imposes some restrictions on design and weight.

Sheet glass is made by drawing molten glass continuously and directly from continuous-melting tank furnaces and passing it right on through annealing ovens. Plate glass has more or less the same soda-lime composition as sheet glass and is fabricated by two processes. In the intermittent method, molten glass is poured from pots in front of iron forming rolls and flattened. After annealing, the sheets are ground to a greater flatness and polished. In the continuous process, a stream of molten glass flows from the tank to water-cooled rollers which shape an endless flat ribbon of uniform width and thickness. After annealing, or lehring, the ribbon is cut, inspected, and passed through continuous grinding and polishing machines.

Flat forms of special compositions, such as for heat-resistant windows, color filters, or electrical insulators, are generally made by pressing molten glass in iron molds. For the accurate control of the light from modern high-intensity sources, Fresnel prisms are pressed with precision in the surfaces of glass panels.

Cylindrical tubing and rods are drawn automatically and continuously from the molten glass in the tank. The tubing ranges in diameter from about 1/8-in. up to 4 1/2-in. OD, with various wall thickness ranging up to 1/4 in. for the large OD sizes. The maximum length of pieces of flanged piping is 20 ft. The rods range up to 2 in. in diameter and 15 ft in length.

Blown glassware is made by inserting an elongated, thick-walled bubble of yellow hot glass, which is attached to a hollow blowing iron part, into an iron mold and blowing the glass out against the mold. This gives an article which is correct in shape and size in one operation except for cutting off the excess part which is attached to the blowing mechanism. With cylindrical ware that permits rotation in the mold during blowing, the iron mold is lined with a wet carbonaceous paste that generates a steam cushion for the glass to ride on. This results in an article with a surface that is glossier and smoother than that obtained with pressed ware or with ware blown in a "hot iron" mold where there is no lining and, of course, no rotation. A "hot iron" mold has to be used whenever the blown ware is nonsymmetrical about an axis or has raised lettering or other projections.

The automatic manufacturing of bottles is an example of "press-blow" fabrication which results in an article requiring no subsequent shaping operation. A blank of hot glass is pressed by forming dies which shape the neck complete with a screw thread and which also partially form the cavity part of the bottle. Then, while the glass is still hot enough for further shaping, it is inserted automatically into a blow mold, and the shaping is completed by blowing.

For pressed ware, a lump of hot glass of a definite weight is fed into an iron mold, and an iron plunger is moved down to close the opening in the retainer ring. Then with relatively low pressure the plunger is forced down until the glass fills the cavity formed between the plunger and the mold and hardens. A few degrees of taper must be allowed on the plunger so that it can be withdrawn. Hinged molds often are used for more complicated shapes; and for special articles the plunger extrudes the glass into a multiplicity of mold cavities in a "font" mold.

Some articles are made by fusion-sealing together two or more parts that have been made by the above processes, for example, lightning arrestor bodies for power-line use and large television tube envelopes. Flanged pipe has pressed flanges sealed to drawn tubing.

Multiform glass products are made by cold pressing or cold casting powdered glass, followed by a firing to give perfectly vitreous articles that may have holes, grooves, and screw threads and that may be accurate to a few thousandths of an inch in some dimensions without requiring any finish shaping.

After shaping, glassware is usually annealed and then may be treated by one or more of the following finishing or related operations: inspection, tempering, removal of excess glass, grinding and polishing, cleaning in special cases, precision finishing, calibration, attachment of metal parts, metalization for solder sealing to metal parts, making of holes, reworking and welding by flame heating or high frequency electric heating, and frosting.

PRINCIPAL INDUSTRIAL APPLICATIONS. Chemical (for food- and drug-processing industries and for chemical industries; for the most part of low expansion chemical- and heat-resistant glasses): containers, tanks, jars, piping, centrifugal pump parts, laboratory ware, tubing for heat exchangers, cylinders for fractionating columns, sight and gage glasses, and Godet wheels for pulling rayon thread out of acid baths.

Electrical and Electronic: insulators, lightning arrestor bodies for power-line use, ribbon glass condensers, fuse plug bodies, x-ray tube envelopes and x-ray absorbing sheet or molded shields, electronic tube envelopes and tubing, mercury switches, metalized glass for hermetically sealed containers, metalized glass electronic components, and special glasses for matched expansion glass-metal fusion seals.

Lighting: incandescent lamp envelopes, fluorescent lamp tubing, sign light tubing, projection lamp envelopes and cover lenses, diffusing glassware, and lens-action prismatic light-controlling glassware.

Light Filters and Colored Signalware: ultraviolet transmitting: germicidal lamp tubing, sun-lamp bulbs, blacklight vision tubing and bulbs; heat or infrared transmitting: heat-lamp bulbs, blacklight vision filters; and colored glass for filters and for signalware that often has accurate prismatic surfaces on cylindrical, flat, or lens shapes.

Thermal: heat exchanger tubing, furnace windows, induction furnace crucibles, calcining jars or trays, thermocouple protection tubes, heat-absorbing sheet glass, furnace cores or linings, Thermos and Dewar flasks, thermometer tubing, heat-insulation Foamglas, heat-insulation Fiberglas products, and heat-insulation double-glazed (air-space) windows.

Optical: instrument lenses, prisms, mirrors, eyeglasses, and goggles.

Mechanical: jewel bearings for precision instruments, optically flat surface plates, machinists' gages, torsion fibers, paraboloidal reflector shells for sealed-beam headlamps (made by pressing), accurate bore tubing for gas and liquid flowmeters, blueprint cylinders and sheets, molds for rubber and plastics, vacuum pumps (diffusion types), lubricator and oil cups, rollers for yarns and thread, Fiberglas filters, slide valve disks for gas-meter valves, rotary valve seats, Fiberglas textiles, humidifier floats, multiform glass pipeline filters, centrifuge containers, and level vials.

Acoustical: Fiberglas wool for sound absorption.

Structural: laminated safety-glass windshields; tempered sheet for doors, machine windows, fire screens, and pickling tanks; building blocks; Fiberglas-plastic board of very high strength-to-weight ratio; cases and covers; ultra-low expansion spacers; mirror disks for ultra-precision reflecting telescopes; and Foamglas floats.

PLASTICS

By John J. Martin

Prior to the development of modern synthetic resins, the most important components of plastics were natural resins such as copal, rosin, pitch, amber, and shellac. Depending on the purpose, these resins were mixed with various types of fillers. In general, these materials were limited in strength and resistance to heat; consequently their use as structural materials was severely restricted. Most of these resins are now used in large quantities in varnishes. Shellac also is still employed as a molding composition for phonograph records and in laminated electrical insulations.

The modern plastics industry in this country originated with the work on Celluloid (cellulose nitrate) in 1869 and received its greatest stimulus in 1909 with the introduction of the first commercial synthetic resin, Bakelite, which was produced by certain reactions between phenol and formaldehyde. A chemical classification of plastics and synthetic elastomers published by the government during 1943 lists approximately 100 items.

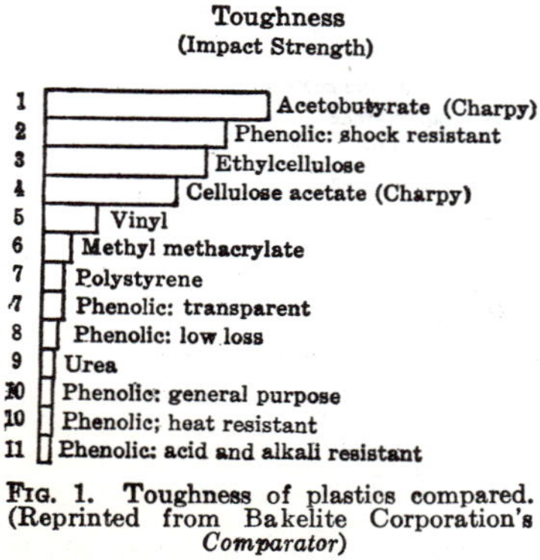

FIG. 1. Toughness of plastics compared. (Reprinted from Bakelite Corporation's *Comparator*)

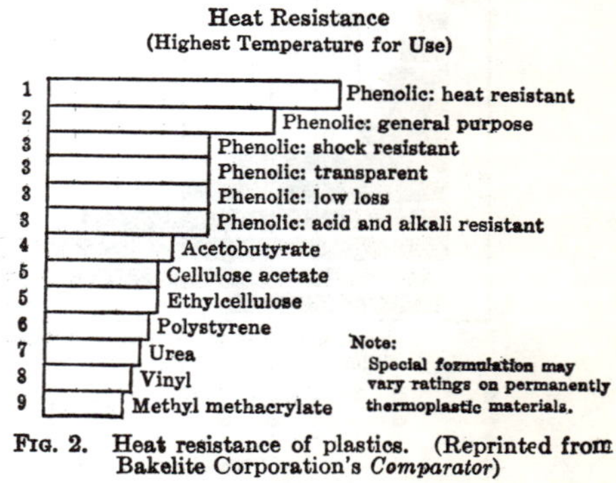

FIG. 2. Heat resistance of plastics. (Reprinted from Bakelite Corporation's *Comparator*)

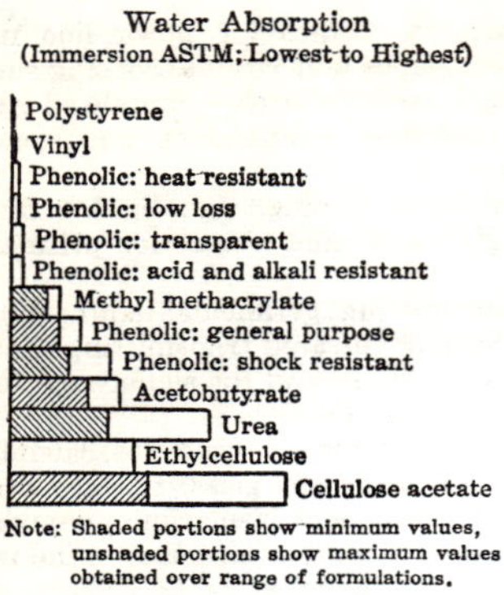

Fig. 3. Water absorption of plastics. (Reprinted from Bakelite Corporation's *Comparator*)

These materials are as unlike among themselves as the different metals which the design engineer knows have varying properties, depending on the additive element used to form the desired alloy. For example, specifications for one of the commonly used plastics (cellulose acetate) include four types and thirteen grades. Even with all these variations to meet the specifications, special materials have to be compounded for particular applications. For this reason, engineers should consult the manufacturers before making a choice of material in order that they may avoid wrong applications which not only result in financial loss but also give an entirely erroneous impression of the usefulness of plastics. A list of material manufacturers is contained in the *Encyclopedia of Plastics* published by Plastics Catalogue Corporation, 122 East 42 Street, New York, N. Y.

The noun "plastic" has been defined by the American Society for Testing Materials as follows:

"A plastic is a material which contains as an essential ingredient an organic substance of large molecular weight, is solid in its finished state, and, at some stage in its manufacture or in its processing into finished articles, can be shaped by flow."

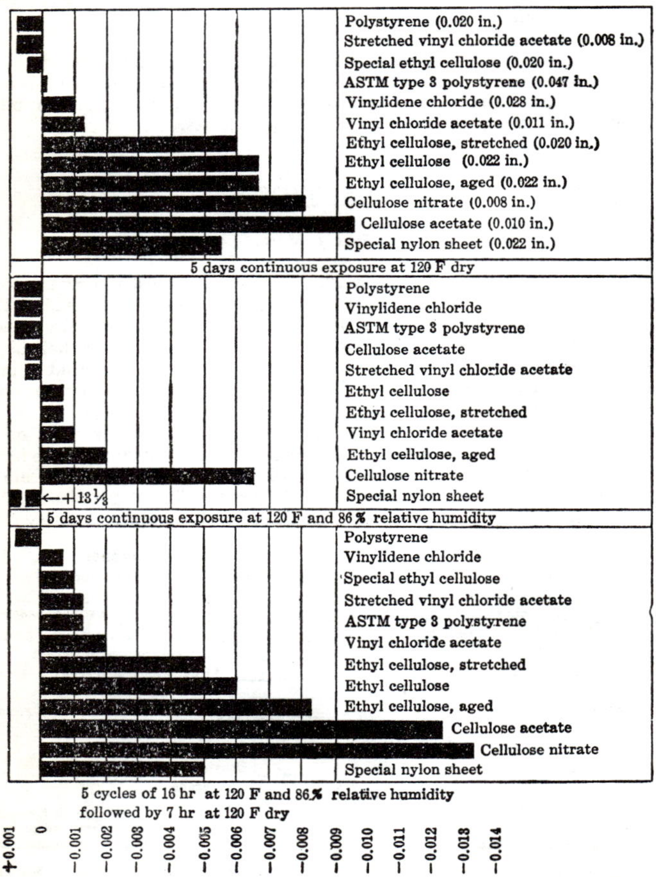

Fig. 4. Dimensional stability of plastic sheeting. (Reprinted by permission from Dimensional Stability of Plastics, by Robert Burns, *ASTM Bulletin*, May, 1945)

THERMOPLASTIC MATERIALS.

"The adjective *thermoplastic* means capable of being repeatedly softened by increase of temperature and hardened by lowering of temperature. *Thermoplastic* applies to those materials whose change upon heating is substantially physical." (ASTM definition.)

THERMOSETTING MATERIALS.

"The adjective *thermosetting* carries the meaning as stated of changing into a substantially infusible or insoluble product when cured under application of heat or chemical means." (ASTM designation.)

PLASTICIZERS. Most thermoplastic resins, particularly the cellulose derivatives, are not fusible enough to be molded or softened without thermal decomposition. In such cases, it is necessary to add a softening agent or plasticizer to develop such properties as resiliency, elasticity, and flexibility. These plasticizers are not capable of forming resinous products in themselves and should not be confused with plastics. They are liquid, semiliquid, or solid. Examples are dibutyl phthalate, diethyl phthalate, tricresyl phosphate, diethoxy ethyl phthalate, and dibutyl sebacate. The nature and amount of plasticizer

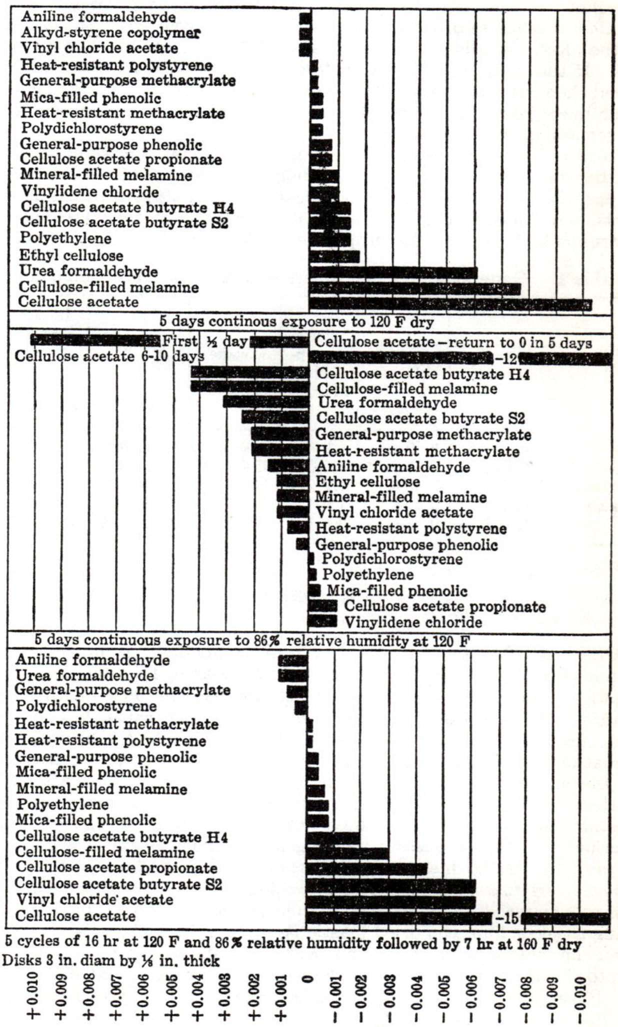

FIG. 5. Dimensional stability of molded plastics. (Reprinted by permission from Robert Burns, *loc cit.*)

has an important bearing on the physical properties of the resulting plastic. A gain in one property generally is obtained at a sacrifice of another, which accounts for the wide spread in the physical properties given in the various tables of this section.

COMPARATIVE PROPERTIES. As a general guide to the properties of various types of plastics, their toughness, heat resistance, and water-absorption characteristics are compared in Figs. 1, 2, and 3. Dimensional stability of sheeting and of molded plastics are compared in Figs. 4 and 5.

10. THERMOPLASTICS

CELLULOSE NITRATE. Cellulose nitrate compounds, commercially known as Celluloid, Pyralin, Nitron, and Nixon CN, are plastics formed by treating cotton or purified wood cellulose with sulfuric and nitric acids. After nitration, the spent acid is removed, and the cellulose nitrate is thoroughly washed and "wet" with alcohol, as it is not safe to store or ship it when dry. Camphor, a commonly used plasticizer, is mixed with the alcohol-wet cellulose nitrate mixture which is then strained. The remaining mass of plasticized cellulose nitrate is kneaded on heated rolls while the desired colors are added and the alcohol is allowed to evaporate. It is then ground into proper grain size for molding. If the material is to be made into sheets, slabs from the rolls are pressed into cakes and then cut into sheets. Tubes or rods, or continuous sheet, can be formed by extruding the mass from the rollers. For these, the material must be allowed to age to remove the remaining solvent.

Cellulose nitrate has good chemical resistance to solvents other than alcohols, ketones, and esters. Because this plastic is highly inflammable and tends to become yellow and brittle with age, the material has limited applications. Cellulose nitrate is used for handles for small tools, mallet heads, spectacle frames, dials and crystals for clocks and instruments, drawing instruments, toys, etc. Its properties are listed in Table 1.

Table 1. General Range of Properties of the Cellulose Compounds

Material	Specific Gravity	Cost, dollars/lb	Water Absorption in 24 hr, %	Tensile Strength, psi	Flexural Strength, psi	Impact Strength (Izod) at 77 F, ft-lb/in. of notch	Linear Thermal Expansion, in./in./°C
Cellulose nitrate	1.35–1.57	0.50–1	1.0–4.0	3000–8000	4000–12000	1.0–6.0	$9–16 \times 10^{-5}$
Cellulose acetate	1.27–1.34	0.35–0.54	2.0–5.0	1500–8000	3000–12000	0.5–6.0	$8–16 \times 10^{-5}$
Cellulose acetate butyrate	1.18–1.25	0.47–0.61	1.7–2.4	1400–7000	2000–12000	0.6–9.0	$11–17 \times 10^{-5}$
Cellulose propionate	1.17–1.22	0.61–0.68	1.0–1.7	2000–6500	3000–10500	0.5–11.0	$13–16 \times 10^{-5}$
Ethyl cellulose	1.09–1.16	0.58–0.65	0.8–1.8	2000–8000	4000–12000	2.0–8.0	$10–20 \times 10^{-5}$
ASTM Test Method	D792		D570	D638	D790	D256	D696

Note: ASTM procurement specifications for cellulosic molding compounds: cellulose nitrate, D701; cellulose acetate, D706; cellulose acetate butyrate, D707; ethyl cellulose, D787.

CELLULOSE ACETATE. Cellulose acetate compounds commercially known as Tenite I, Lumarith, Plastacele, Bakelite Cellulose Acetate, Fibestos, and Chemaco CA, etc., are plastics made from cotton linters which are dissolved in a mixture of acetic acid and acetic anhydride, with sulfuric acid or other catalyst. A calculated amount of water, sufficient to cause hydrolysis but not precipitation, is added to the heavy syrup resulting from this treatment, and the syrup is allowed to "ripen" or stand until the original triacetate has hydrolyzed to an acetate of the desired acetyl content. After this action is completed, the cellulose acetate solution is plunged into cold water which precipitates the cellulose acetate into small flakes. These flakes are thoroughly washed and dried. In the manufacture of cellulose acetate molding compounds, plasticizers and related compounds are added either by means of solvents or hot-mix methods to bring the cellulose into a homogeneous mass. The choice and blending of the plasticizers influence the properties of this material. The material is then cooled and ground into molding powder.

For making sheet or film, the dried plastic is mixed with a plasticizer, color is added if required, and the mass is worked on a hot roll. The slabs from the roll are pressed into cakes, cut into sheets, and seasoned. Rods and tubes can be extruded to the desired size and shape. Sheet material especially 30 mils or over in thickness is also produced by extruding. The films are cast on a drum.

Cellulose acetate compounds are similar to cellulose nitrate compounds but show considerable improvement in many respects. They are slow burning and are easily molded by conventional methods. Strong acids, alkalies, and acetone dissolve the cellulose acetate compounds. Their chief disadvantage is their comparatively poor dimensional stability under adverse atmospheric conditions.

Typical commercial applications of these compounds are name plates, flashlight cases and lenses, tool handles, terminal blocks, interleaving in relay coils and other wound coils, backing for electrical grades of pressure sensitive tapes, etc. The properties of cellulose acetate are listed in Table 1.

CELLULOSE ACETATE BUTYRATE. Cellulose acetate butyrate compounds, commercially known as Tenite II, etc., are similar in preparation to cellulose acetate compounds except that butyric acid is used in addition to the acetic acid and acetic anhydride. This facilitates the production of molding materials that have better moisture resistance and improved dimensional stability. A disadvantage of the cellulose acetate butyrates is that they may have an unpleasant odor. Molded materials have been used for producing such items as gun stocks, auto steering wheels, combined telephone-set housings, building hardware, etc. For properties of this material, see Table 1.

CELLULOSE PROPIONATE. Cellulose propionate compound is produced through the reaction of propionic acid and propionic anhydride with cellulose in the presence of a catalyst, the flake being mixed in the usual manner with plasticizers, color, etc. It is an unusually stable cellulosic compound comparing favorably with cellulose acetate butyrate. The material is sold under the trade name of Forticel. Its properties are listed in Table 1.

ETHYL CELLULOSE. Ethyl cellulose is produced by the action of a strong alkali such as sodium hydroxide upon cellulose and by treatment of the resulting product with ethyl chloride or ethyl sulfate. In common with the other cellulosics, it is compounded with plasticizers, colors, etc. The chief improvement of this material over other cellulosics is its low-temperature flexibility, high-temperature stability, and high dielectric properties. It has the disadvantage, however, of softening with carbon tetrachloride and alcohol, both of which are common in many cleaning processes. The material is known commercially as Ethocel, Lumarith E.C., Chemaco E.C., and Nixon Ethyl Cellulose. It is used for containers, flashlight cases, etc.

POLYVINYL CHLORIDE. Monomeric vinyl chloride, with a boiling point of −13 C, is a gas at room temperature. Under suitable conditions of heat and pressure and in the presence of catalysts, vinyl chloride is transformed to a white fluffy polymer. The polymer, when plasticized, can be molded under heat and pressure to give nonrigid materials characterized by low moisture absorption and inertness to a variety of chemical solvents, as well as by resistance to flame. Since plasticized polyvinyl chloride is too soft to serve as a rigid molding material, it is generally copolymerized with vinyl acetate. This process will be discussed later.

The polyvinyl chloride resins known as Koroseal, Geon, etc., are used for producing films, sheets, tubing, wire and cable coatings, belts, watch straps, chemical and water-resistant fabric coatings, etc. Their properties are shown in Table 2.

Table 2. Approximate Physical Properties of Polyvinyl Chloride and Vinyl Chloride Acetate Compounds

Properties	Rigid Vinyl Chloride Acetate Resins	Nonrigid Polyvinyl Chloride Resins
Specific gravity	1.35–1.45	1.18–1.65
Tensile strength, psi	4,500– 8,000	1000–3000
Impact strength (Izod), at 77 F, ft-lb/in. of notch	0.25–0.75	....
Flexural strength, psi	7,500–14,500	....
Linear thermal expansion, in./in./°C, below 55–65 C	6.9×10^{-5}	....
Water absorption in 24 hr, %	0.05–0.15	0.10–2.25
Cost, dollars/lb	0.59–0.81	....

POLYVINYL ACETATE. Vinyl acetate is a volatile liquid which boils at 73 C. The vinyl acetate monomer is obtained by reaction of acetylene with acetic acid in the presence of suitable catalysts. The monomeric acetate can be polymerized to form resins. Polyvinyl acetate resins occupy a unique position in the vinyl resin field in that they are also used as starting materials for the production of other resins. However, the acetate is not used by itself for molded application because of its low softening temperature and high cold flow.

The polyvinyl acetate resins are soluble in a wide range of organic liquids. They have exceptionally good resistance to greases and oils and are used for coating grease-proof paper and containers.

POLYVINYL CHLORIDE ACETATE. Vinyl chloride acetate molding compounds, commercially known as Vinylite, Tygon, etc., are copolymers of vinyl acetate and vinyl chloride formed by the reaction of these materials in the presence of a solvent and a catalyst. The proportional amounts of the two esters can be varied over a wide range for specific purposes. The most useful vinyl chloride acetate copolymers contain 85% or more of vinyl chloride.

These molding compounds exhibit high resistance to humidity and chemicals and are difficult to ignite. However, they have a low temperature limit precluding their use for applications where the molded parts may operate at elevated temperatures. The compounds are used commercially for phonograph records, watch crystals, safety goggle frames, straps, belts, etc. The properties are given in Table 2.

POLYVINYL ALCOHOL. Polyvinyl alcohol is prepared by hydrolysis of polyvinyl acetate. The conditions under which hydrolysis takes place have an important bearing on the properties of the finished material. When completely hydrolyzed the alcohol is soluble in hot water, and when suitably compounded it is used to coat or impregnate textiles, paper, etc.

The compounds are highly resistant to hydrocarbons and to most organic solvents of the ester, ether, alcohol, and chlorinated hydrocarbon classes, as well as carbon disulfide. They are completely unaffected by oils, fats, or greases. Films prepared from polyvinyl alcohol are tough and highly impermeable to most gases.

Polyvinyl alcohol compounds can be extruded as seamless, flexible tubing, either plain or reinforced with braided fabric or wire. Because of the water solubility of polyvinyl alcohol, these products cannot be employed for handling water or aqueous solutions. Such materials are sold in the trade under the names of PVA and Resistoflex.

POLYVINYL BUTYRAL. Polyvinyl butyral resins were developed specifically for safety-glass sheeting. This plastic, commonly known as Butacite, etc., is outstanding for its toughness, adhesive characteristics, clarity, and resistance to shock, even at low temperatures. The resin may be manufactured by the reaction of butyraldehyde with polyvinyl alcohol.

Safety-glass sheeting is made either by calendering, or by extruding as a thick solution and evaporating the solvent. To make the laminated safety glass, the plastic sheeting is inserted between two clean glass plates which are bonded together by means of heat and pressure. It is also used between sheets of methyl methacrylate to protect enclosures of pressurized cabins of airplanes against shattering.

POLYVINYLIDENE CHLORIDE. The basic raw materials of the vinylidene chloride resins are petroleum and brine. Ethylene from petroleum and chlorine from brine combine to form monomeric vinylidene chloride, a clear colorless liquid with a boiling point at about 89 F.

Vinylidene chloride is polymerized by a number of catalysts such as peroxides. The most useful polymers are not produced from vinylidene chloride alone. They are chiefly copolymers of vinylidene chloride with other unsaturated compounds. Vinyl chloride is generally used. The copolymers of vinyl chloride and vinylidene chloride are sold in the trade under the name of Saran. The material is quite resistant to water, acids, alkalies, and organic solvents.

Vinylidene chloride is utilized in the fabrication of gaskets and valve seats. Tubing is used for handling certain fluids in refrigeration and air-conditioning equipment. Molded couplings of polyvinylidene chloride are available for both tubing and pipe. Cold-stretched monofilament, in diameters of 0.007 to 0.100 in., is employed in woven textiles for automobile and airplane upholstery, fishing leaders, and corrosion-resistant, partially transparent screening. It is claimed that the screening will outlast any metal screening; it is not affected by fungus and will not burn. Vinylidene chloride is also available in film form for packaging.

METHACRYLATE. Methacrylate molding compounds, commercially known as Lucite and Plexiglas, are produced by treating an acetone compound with methanol in the presence of sulfuric acid, and polymerizing the resulting methyl methacrylate. Heat, light, and oxygen serve as polymerizing accelerators.

These materials are particularly valuable for their crystal clarity and for their outstanding optical properties. Molded parts made from methyl methacrylate resin are brittle compared to the cellulosic plastics, but they have reasonably good scratch-resisting properties. They do not shatter like glass when broken.

Methacrylate resin is also used in large quantities in the form of cast sheets, rods, and tubes. The sheets are available in thicknesses from 0.060 to 2.50 in. Rods are furnished up to 2 in. in diameter. The tubes are manufactured in various wall thicknesses and diameters.

The properties of cast products are similar to those of molded resins. They are resistant to moisture or aqueous solutions of mineral salts and alkalies. Solvents include acetone, toluene, chloroform, and ethylene dichloride. These products are attacked by a number of alcohols, esters, ketones, aromatic hydrocarbons and chlorinated solvents.

If subjected to excessive fiber stress, methacrylates will craze, i.e. develop tiny cracks, and thereby destroy their desirable optical properties.

Molded and cast materials are used for aircraft and marine windows, lenses, dials, handles, dentures, watch crystals, radiator ornaments, draft gages, edge-lighted instrument panels, transparent models, traffic directors, miscellaneous novelties, and costume jewelry. Properties are given in Table 3.

Table 3. Approximate Physical and Electrical Characteristics of Acrylic, Polystyrene, and Polyethylene Resins

Property	Methacrylate Molding Compounds	Polystyrene Molding Compounds	Polyethylene Molding Compounds
Specific gravity	1.2	1.05–1.07	0.92
Tensile strength, 1000 psi	6–10	5.5–9.0	1.7–3.0
Flexural strength, 1000 psi	12–17	8.0–19.0	1.6
Impact strength, (Izod) at 77 F, ft-lb/in. notch	0.2–0.5	0.25–0.40	>4
Thermal expansion in./in./°C	9×10^{-5}	$6\text{–}8 \times 10^{-5}$	$16\text{–}18 \times 10^{-5}$
Water absorption in 24 hr, %	0.5	0.05	<0.01
Dielectric constant, 10^6 cycles	2.7–3.2	2.6	2.3
Power factor, 10^6 cycles	0.02–0.03	0.0001–0.0005	0.0003
Refractive index, n_D	1.5	1.6	....
Cost, dollars/lb	0.65–0.80	0.27–0.42	0.49–0.56
ASTM procurement specification	D788	D703	....

POLYSTYRENE. Polystyrene molding compounds, commercially known as Styron, Lustron, Loaline, Bakelite Polystyrene, etc., are made by forcing ethylene, made from natural gas or petroleum, through benzene, a by-product from coke ovens, in the presence of a suitable catalyst, such as ethyl chloride, to produce ethyl benzene. This material is then decomposed under heat, hydrogen splitting away from the rest of the molecule to form the monomeric styrene. Polystyrene is prepared by polymerization.

Polystyrene molding compounds have excellent electrical properties, especially at the higher frequencies. Resistance to alkalies is excellent, but some strong acids attack the resin. Polystyrene has low specific gravity, excellent dimensional stability, and remarkably low moisture absorption. Although it has reasonably good tensile strength, it is somewhat brittle. It is another of the few plastics suitable for making optical lenses. Perhaps the greatest disadvantage of polystyrene compounds is their tendency to craze and crack. This can be alleviated to some extent by heat treating. Polystyrene is also furnished in film form and is used in the construction of condensers and coils.

A great amount of effort has been expended toward developing high-temperature-resistant polystyrenes. Polystyrene P-8 produced by the Chemical Division of the Koppers Company and Cerex X-250 manufactured by the Monsanto Chemical Company are examples of such materials.

Polystyrene is unique among plastics in that it gives a somewhat metallic ring when dropped on a hard surface or struck sharply. This property can often be used as a simple test to distinguish it from other plastics.

Polystyrene molding compounds find their greatest application in electrical insulators. In certain optical applications they are used similarly to methacrylate molding compounds. Polystyrene is also being used for making a number of parts for the refrigerator industries.

Applications of polystyrene involving chemical resistance include articles such as closures for chemical bottles, photographic equipment, battery cases, etc. The properties of typical materials are given in Table 3.

STYRENE COPOLYMERS. Styrene was one of the materials used in the wartime synthetic rubber program and for this reason was in good supply after the war.

Little information is available pertaining to the formulation or preparative methods of many styrene materials. When copolymerized with other materials, styrene yields heat-resistant copolymers. Included among these materials are Textolite 1421 and 1422 cast plastics which are available in sheets and rods. These materials cost $1.40 to $10.50 per lb, depending on their shape. Electrical properties of Textolite 1421 and 1422 are comparable to those of polystyrene, but the materials have a much higher heat distortion point. The materials can be machined to close tolerances, and they retain their shape and strength at elevated temperatures. Textolite 1422 is dimensionally stable in the temperature range of 250 to 300 F. It is not moldable, but castings can be fabricated into structural shapes for equipment requiring a low-loss material and operating at a temperature unsafe for polystyrene.

A styrene copolymer known as Plexene M has been developed to combine high heat resistance, greater strength, improved machining qualities, and better weathering characteristics with many of the excellent properties of the styrene polymers. Moldings made of Plexene M do not show the marked tendency to crazing, characteristic of many unmodified styrene compounds. Although the dielectric properties are not so desirable as those of polystyrene, physical strength characteristics are exceptionally high in relation to the low specific gravity (1.08) of the material. The material is resistant to acids, alkalies, dilute alcohol, and gasoline. Plexene M is available at about 65 cents per pound.

Another group of styrene copolymers which possess qualities of both rubberlike materials and rigid thermoplastics are sold under the trademark Styraloy. These resins, having a specific gravity of 0.95 to 0.97, vary from flexible materials in thin sections to fairly rigid products in thick sections. Styraloy 22 was developed during the war primarily for a one-piece cable sheathing with a low power loss at high frequencies, a low temperature flexibility, and minimum cold flow at elevated temperatures. Styraloys, like natural rubber, are attacked by common solvents, but to a lesser degree. They have good resistance to strong mineral acids and strong alkalies. They are now finding many applications in the aviation, automotive, and chemical industries.

UNSATURATED POLYESTER CAST PLASTICS. The industrial applications of these materials, sometimes referred to under the generic term cross-linked addition polymers, are rapidly expanding. The polyesters consist essentially of plastics in liquid form, with or without fillers, which are poured into suitable molds and polymerized by the use of a catalytic agent and the application of heat. Room temperature cures are frequently feasible. The resultant plastic is hard, dense, and thermosetting in nature. Most commercial materials are mixtures of styrene and polyester monomers, the mixture undergoing a volume shrinkage of about 12% when cured. Although the relatively large degree of shrinkage frequently causes cracking around inserts, this difficulty is sharply reduced by the judicious use of fillers. These polymers are in general satisfactorily resistant to organic solvents, acids, alkalies, and oxidizing agents.

A reduced degree of cross-linking is possible in this class of polymers, producing a rubbery-type mass. Such materials are promising for potting and filling operations where resistance to shock as well as to moisture is essential.

Cross-linked addition polymers are commercially available as Laminac, Selectron, Vibrin, Paraplex, MR-Resins, etc.

Properties of the rigid castings are shown in Table 4.

Table 4. Approximate Properties of Polyester Cast Plastics

Specific gravity	1.20
Tensile strength, psi	6,000
Flexural strength, psi	15,000
Modulus of elasticity, flexure	$3\text{–}8 \times 10^5$
Impact at 77 F, Izod ft-lb/in. notch	0.3–0.4
Hardness	M90
Water absorption in 24 hr, %	0.50
Power factor, 10^6 cycles	0.03
Dielectric constant, 10^6 cycles	3.5
Dielectric strength, v/mil (short time)	450

ALLYL CASTING RESINS are available from the manufacturer in the form of liquid monomers. Transparent cast sheets are also furnished by the suppliers. The allyl monomers, like the polyesters, can be cast polymerized by the fabricator in simple molds. Iron and steel, aluminum, low-melting alloys, glass, and glazed ceramic molds may be used to form thermosetting parts. The allyl resin can be cured at moderate temperature and need not be pressurized during the cure.

Castings made of allyl resin are resistant to chemicals and solvents. They also have good resistance to abrasion and marring. The mechanical properties of these materials are quite similar to those of the polyesters.

They may be employed for such parts as dial, indicator, and gage covers; aircraft enclosures; heat-resistant electric insulating components; decorative novelties; television lenses, etc.

Allyl casting resins, in the liquid monomer form, are commercially available as Laminac, Kriston, and Allyl Diglycol Carbonate.

POLYETHYLENE. Polyethylene molding compounds, known commercially as Polythene and Bakelite Polyethylene, are made by the high-temperature, high-pressure polymerization of ethylene. A commonly used catalyst is oxygen in very small quantities.

The compounds are resistant to both alkalies and acids and to most of the organic solvents. They are extremely flexible and tough over a wide range of temperatures. Like polystyrene, they have excellent electrical properties and a low specific gravity. Their resistance to water and vapor is exceptionally good. The materials are used for wire and cable insulation, acid-resistant clothing in the chemical industry, shower curtains, etc. Their properties are shown in Table 3.

SYNTHETIC POLYMERIC AMIDES. Nylon is a generic term that was assigned by E. I. du Pont de Nemours and Co. to a whole group of long-chain synthetic polymeric amides which have recurring amide groups as an integral part of the main polymer chain and are capable of being formed into a filament in which the structural elements are oriented in the direction of the axis.

Although the greatest volume of the material is used as yarn for making stockings and other textile items, molded articles from nylon powders are steadily increasing in use. These include coil forms, slide fasteners, bearings, tableware, etc. The molding powders are quite costly, the price ranging from $1.60 to $4.50 per lb. Nylon is also available in sheet, tube, and rod form.

The physical and electrical properties of several nylon molding compounds are given in Table 5.

Table 5. Properties of a Few of the Nylon Molding Compounds

(The material included in this table appeared in E. I. du Pont de Nemours & Company, Catalog issued by their Plastics Department, 1948. Reprinted by permission.)

Properties	ASTM Test Method	FM-10001	FM-3001
Specific gravity	D-792	1.14	1.09
Tensile strength, 77 F, psi	D-638	10,900	7,000
Elongation, 77 F	D-638	50	135
Modulus of elasticity, 77 F, psi	D-638	400,000	260,000
Flexural strength, 77 F, psi	D-790	14,600	9,500
Impact strength, 77 F, ft-lb/in.	D-256	1.5	2.7
Shear strength, psi	D-732	9,600	8,400
Hardness, Rockwell	D-785	R118	R111
Coefficient of linear thermal expansion per °F	D-696	5.5×10^{-5}	8.2×10^{-5}
Dielectric strength, short-time, volt/mil	D-149	385	470
Dielectric constant, 10^6 cycles	D-150	3.4	3.5
Power factor, 10^6 cycles	D-150	0.04	0.03
Water absorption, %	D-570	1.5	0.4

POLYTETRAFLUORETHYLENE. This plastic, known as Teflon, is most oustanding from standpoints of high temperature and chemical resistance. It is produced by polymerizing gaseous tetrafluorethylene to yield a solid granular polymer. It is claimed that the material does not soften appreciably until temperatures of about 500 F have been reached and will resist attack by all acids including boiling aqua regia. It is also claimed that no substance has been found that is capable of dissolving or swelling polytetrafluorethylene. It is suggested for use as a gasket and packing material, as tubing for chemical process work, and for electrical applications where low dielectric losses and high heat resistance are essential. (See Table 6.) It can be extruded in the form of rods and tubes or as wire coating, and compression moldings of simple articles can be made by special techniques.

TRIFLUOROCHLOROETHYLENE (Kel-F), like Teflon, is another high-temperature compound. The material is produced by M. W. Kellogg Company, of Jersey City, New Jersey. Made up of fluorine and chlorine to the extent of four-fifths of its weight, it is related to the new family of organic compounds, the fluorocarbons, which reached practical development during the war. It is understood that this material is extremely resistant to concentrated sulfuric, hydrofluoric, and hydrochloric acids. The material is reported to be suitable for use at temperatures as high as 400 F and also at very low temperatures. It absorbs no water and has a very low water-vapor transmission.

The electrical properties of Kel-F are excellent, and the material should provide good electrical insulation for parts exposed to extremes of temperature and to corrosive or humid atmospheres.

CASEIN (Protein). The raw material for casein plastics is skimmed milk, which is treated with rennet under controlled conditions to give a curd consisting of a loose complex of the calcium phosphate of the protein components of the milk. After being washed and dried, the curd is obtained as a coarse powder, known as rennet casein. In the manufacture of plastics material, the rennet casein is ground to a fine powder and mechanically mixed with about 30 to 40% of its weight of water until the water is absorbed to give individual particles of casein gel which is extruded in rod form from screw-type extruders.

Table 6. Typical Properties of Teflon

(Data from du Pont's *Technical Service Bulletin* 13, April 1, 1949. Reprinted by permission.)

Data given represent standard commercial grades of material and standard methods of testing except where otherwise noted.

Property	Value		Method
Specific gravity	2.1–2.3		D792-44T
Tensile strength at 77 F, psi	1,500–2,500	(1)	D638-44T
Elongation at 77 F, %	100–200		D638-44T
Flexural strength at 77 F, psi	2,000	(2)	D650-42T
Stiffness at 77 F, psi (0.125 in.)	60,000		D747-43T
Impact strength, Izod, −70, 77, 170 F, ft-lb/in.	2.0 4.0 6.0		D256-41T
Hardness, "D," Durometer	55–70		....
Compressive strength, psi, at 0.1% deformation	1,700		D695-42T
Deformation under load at 50 C under 1200 psi, % in 85 hr	4–8		D621-43
Heat-distortion temperature, low-load, °F	270		D648 (B)-44T
Specific heat, Btu/lb°F	0.25	(3)	
Coefficient of linear expansion per °F (77–140 F)	5.5×10^{-5}		D696-42T
Thermal conductivity, Btu/hr/ft²/°F in. (0.18 in.)	1.7	(4)	Arl. P-32
Dielectric strength, short-time, volt/mil (0.080 in.)	400–500	(5)	D149-40T
Surface arc-resistance, seconds	700	(6)	D495-42
Volume resistivity, ohm-cm	10^{16}		D257-38
Surface resistivity at 100% rel. humidity, megohms	3.6×10^{6}		D257-38
Dielectric constant at 60, 10^3, 10^6, 10^8 cycles	2.0–2.05		D150-40T
Power factor at 60, 10^3, 10^6, 10^8 cycles	0.0005		D150-40T
Water absorption, %	0.00	(7)	D570-42
Static coefficient of friction against polished steel	0.09–0.12	(8)	
Vapor pressure, mm Hg at 25 C	10^{-6}		
Outdoor weathering	No detectable change in 3 years		

(1) Tensile strength in oriented film may be as high as 15,000 psi. (2) Specimens do not break. (3) Method of mixtures. (4) Cenco-Fitch apparatus. (5) 1000–2000 volts/mil in thicknesses 5 to 12 mils. (6) Does not track. (7) Not wet by water unless a surface-active agent is added. (8) Inclined-plane method.

If the final product is to be rods, the extruded rods are hardened by immersion in dilute formaldehyde. If sheets are desired, the extruded rods are hot pressed in hydraulic presses, and the sheets so formed are hardened in formaldehyde. After they have hardened, the rods or sheets are carefully dried to a controlled moisture content, which may require weeks to months, depending on their size, diameter, or thickness.

This material is nonburning, has good resistance to organic solvents, and may be had in a wide range of colors. It is hygroscopic and affected by atmospheric moisture changes.

Casein plastics commercially known as Ameroid, Galorn, etc., are used in the manufacture of buttons, handles, small novelties, and trimming accessories.

ANILINE FORMALDEHYDE. This resin, known in the trade as Dilectene or Cibanite, is not to be classified as either a thermosetting resin or a thermoplastic resin. It will soften somewhat when heated, but it will not flow as do resins of the thermoplastic type. Under pressure, however, it may be caused to assume various shapes and forms when hot. The chemical stability of aniline formaldehyde resin is very good. Although it is attacked by strong acids, it is unaffected by alkalies and is not soluble in any of the common organic solvents; alcohol and aromatic hydrocarbons particularly have no effect on the material.

Table 7. Properties of Unfilled Aniline Formaldehyde Resin

Specific gravity	1.21
Tensile strength, psi	10,500
Flexural strength, psi	15,000
Modulus of elasticity in flexure, psi	510,000
Compression strength, psi	20,000
Impact strength (Izod), 1/2 × 1/2-in. specimen, parallel to molding pressure, ft-lb/in. of notch	0.33
Thermal expansion, in./in./°C	3×10^{-5}
Water absorption in 24 hr, 3 × 1 × 0.075-in. specimen, %	0.08
Dielectric constant, 10^6 cycles	3.5
Power factor, 10^6 cycles	0.006
Cost, dollars/lb	2.00

The material is of a reddish brown color and can be dyed in dark colors only. It has found considerable use in high-frequency applications such as coil forms, vacuum tube sockets, and antenna housings. The resin, having good mechanical strength and a high order of insulation resistance, which is maintained under the influence of moisture and atmospheric conditions, is also employed for terminal boards, strips, and blocks. The properties of the material are given in Table 7.

11. THERMOSETTING PLASTICS

PHENOL FORMALDEHYDE MOLDING COMPOUNDS. These compounds, commercially known as Bakelite, Durez, Resinox, etc., are made when phenols react with formaldehyde in the presence of a catalyst to form the phenolic resin. The type of catalyst has a considerable influence on the properties of the resultant plastic. The base resin is ground and mixed with a lubricating agent and other materials to form a honey-colored resin compound. This resin is rather hard and brittle, and, in order to obtain satisfactory impact strength, fillers and pigments are added to achieve the desired properties and coloration.

The fillers influence the particular properties of the finished mixture. Listed below are a few of the common fillers and the type of application for which each is used. The materials, depending on the type of filler, have specific gravities ranging from 1.27 to 2.00, and the cost per pound varies from \$0.15 to \$0.61.

Filler	Application
Wood flour	General purpose
Cotton flock	Shock resistant
Shredded or chopped fabric	Shock resistant
Chopped canvas or twisted cord	Heavy impact
Inorganic fillers (Asbestos, silica, mica)	Low water absorption, good heat resistance, improved dimensional stability, and low thermal coefficient of expansion

Mica-filled molding compounds also exhibit excellent electrical properties, especially with regard to low dielectric loss.

Molded phenolics have good resistance to most solvents and to acid attack but are decomposed by strong alkalies. Because their coefficient of thermal expansion is greater than that of metals, inserts are held securely in molded parts as the material shrinks when cooled from molding temperature. Molded phenolics are used for such parts as clock cases, telephone equipment, radio cabinets, handles, electrical insulators, and vacuum tube sockets. Their properties are listed in Table 8.

Table 8. Approximate Physical and Electrical Properties of Molded Thermosetting Plastic Materials

Type	Tensile Strength, 1000 psi	Flexural Strength, 1000 psi	Impact at 77 F (Izod), ft-lb/in. of notch	Coefficient of Thermal Expansion, in./in./°C	Heat Resistance, continuous, °F	Dielectric Constant, 10^6 cycles	Power Factor, 10^6 cycles
General-purpose phenolic	5.5– 8.5	9.0–12.0	0.24–0.44	$3.0–4.5 \times 10^{-5}$	300	4.5–6.0	0.03 –0.06
High-impact phenolic	4.0– 9.0	8.0–15.0	2.0–8.0	$2.9–4.4 \times 10^{-5}$	240	4.5–6.0	0.03 –0.06
Low-loss phenolic	4.0– 7.0	8.0–12.0	0.30–0.38	$1.8–1.9 \times 10^{-5}$	275	4.0–5.0	0.005–0.03
General-purpose urea formaldehyde	6.0–13.0	10.0–16.0	0.20–0.35	$2.5–3.0 \times 10^{-5}$	170	6.6–7.7	0.027–0.035
General-purpose melamine formaldehyde	7.5–13.0	9.0–16.0	0.20–0.35	$2.0–4.5 \times 10^{-5}$	210	7.2–8.0	0.025–0.45
Fabric-filled melamine formaldehyde	8.0–11.0	11.0–17.0	0.6–1.4	$2.0–4.5 \times 10^{-5}$	250	6.1–6.9	0.040–0.065
Asbestos-filled melamine formaldehyde	5.5– 7.0	7.5–11.0	0.25–0.40	$2.0–4.5 \times 10^{-5}$	>250	6.7	0.041

Note: Procurement specifications: phenolic molding compounds—ASTM. D700; melamine formaldehyde molding compounds—ASTM. D704; urea formaldehyde molding compounds—ASTM D705.

UREA FORMALDEHYDE MOLDING COMPOUNDS. Urea formaldehyde molding compounds, commercially available as Beetle, Bakelite Urea, Plaskon, etc.. are made when urea is compounded with formaldehyde in the presence of an alkaline catalyst such as sodium hydroxide. The filtered solution of urea formaldehyde in water is mixed with purified wood cellulose as a filler and dried. Acid or acid-producing catalysts, such as oxalic acid, are added to assist in the final polymerization.

The urea formaldehyde molding compounds are extremely color fast and readily take dyes or pigments. Except for dimensional stability, which is somewhat poorer than that of general-purpose phenolics, the physical and chemical properties of these materials are similar to those of phenol formaldehyde molding compounds having similar fillers. However, they exhibit good arc resistance and, unlike the colored phenolic compounds, can be made with a considerable degree of translucency. The specific gravity of the urea formaldehyde plastics varies from 1.45 to 1.55. The compounds cost from $0.22 to $0.65 per pound.

Urea formaldehyde molding compounds find typical applications in reflectors, clock and radio housings, instrument dials, closures, electrical parts, etc. The physical and electrical properties are given in Table 8.

MELAMINE FORMALDEHYDE MOLDING COMPOUNDS. Melamine formaldehyde molding compounds, commercially available as Melmac, Plaskon, Catalin, Resimene, etc., are prepared by compounding melamine, a complex organic material fabricated from calcium carbide, nitrogen, and formaldehyde. The melamine formaldehyde resin resulting is then dried and is filled with either purified wood cellulose or fabric fillers. The materials have a specific gravity of 1.45 to 1.55 and cost from $0.38 to $0.55 per pound.

These compounds, like the urea compounds, have high arc resistance but excel the ureas in hardness, are more heat resistant, show greater resistance to dilute acids and alkalies, and are less affected by boiling water.

Typical commercial applications of the melamine formaldehyde molding compounds are in the manufacture of terminal blocks, aircraft ignition parts, circuit breakers, tableware, etc. Their properties are given in Table 8.

CAST PHENOLICS. The bulk of cast phenolics is produced in the form of sheets, rods, and tubes. Trade names are Bakelite cast resinoid, Catalin, Marblette, Crystle, Prystal, etc. The finished articles are fabricated by machining and polishing. They are available in an unlimited range of color, both opaque and transparent. In addition to the stock forms, custom-made castings are produced in almost any shape desired. These are cast by hand-pouring the specially prepared, partially resinified mass into a mold and allowing it to stand for a few hours to several days, depending on the temperature and catalyst. With large castings, longer curing times are required. Usually the dimensional stability of cast phenolics is not so good as that of molded products, for resinification continues within the plastic over a considerable period of time, resulting in volume changes.

Typical applications of cast phenolics are personal accessories, hardware, display fixtures, toys, etc. Special castings are made in the production of small radio cabinets.

12. LAMINATED PLASTICS

HIGH-PRESSURE LAMINATES. The fundamental difference between thermosetting laminated plastics and molded plastics is in the nature of the filler which, to a great extent, determines the properties of the laminate. The fibrous sheets principally employed for these purposes are of paper, cotton fabric, canvas duck, asbestos paper and fabric, glass cloth, and glass fibers. In general, synthetic resins of the phenol formaldehyde type are most widely used, although urea formaldehyde and melamine formaldehyde resins are also employed. In some cases, in order to produce materials for certain applications, combinations of the three resins are required. The resin is used in the form of a varnish in solvents or in water.

In the production of the laminates, the fibrous sheets are passed through a bath of varnish, the excess resin is squeezed out either by rolls or by a doctor blade, and the sheets are then passed through a heated oven in which the solvents are evaporated and the resins are partially cured. The viscosity of the resin solution and the setting of the squeeze rolls or doctor blades are carefully controlled to give the desired resin content in the impregnated filler sheets. In the drying operation, the volatile content of the treated filler sheets must also be closely controlled so that the resin solubility and residual flow is sufficient to obtain a good bond in the subsequent pressing operation.

The impregnated stock thus produced is manufactured into sheets, tubes, rods, and special shapes. When sheets are made, the impregnated material is cut into sheets of the desired length, placed in stacks between polished or otherwise treated plates to give varied surface textures, and subjected to heat and pressure in a hydraulic press.

To produce tubes, the impregnated material is wound over a mandrel the size of the inside diameter of the finished tube and either baked in an oven to produce a rolled tube or formed in a mold under heat and pressure if a molded tube is desired. The manufacture of molded rods also starts by winding the impregnated material on a mandrel with a very small diameter. The mandrel is withdrawn before the curing operation is performed in a closed mold. Rods are also molded by placing strips of the impregnated stock in a closed mold and applying pressure edgewise, causing the laminations to crinkle. Rods may also be turned from sheet stock.

The laminated thermosetting plastics, which are known commercially as Formica, Micarta, Synthane, Textolite, Dilecto, Panelyte, Spauldite, Phenolite, etc., are essentially used as structural members or electrically insulating parts. They are employed for such applications as automotive timing gears, insulation in keys, jacks, relays, and other electrical mechanisms, diaphragms for loudspeakers, bearings, helmets, panels, refrigerator door parts, nameplates and dials, counter tops and paneling, coil forms, and terminal strips.

Generally speaking, 225 F is the maximum safe continuous operating temperature for any of the paper- and fabric-base materials. The asbestos-base grades may be used at temperatures of the order of 250 to 300 F. Table 9 shows the physical properties of the various laminated phenolic materials.

Table 9. Physical Properties of the Various Laminated Phenolic Materials

Filler	Tensile Strength, 1000 psi	Flexural Strength, 1000 psi	Compressive Strength, 1000 psi	Water Absorption in 24 hr (max), 3 x 1 x 1/16 in. sample, %	Edgewise Impact Strength (Izod), at 77 F, ft-lb/in. of notch	Specific Gravity	Linear Thermal Expansion 10^{-5} in./in./°C
Paper base	5–10	11–20	25–35	1.2–6.0	0.3–0.5	1.30–1.36	1.7–2.5
Cotton fabric base							
Fine weave	6.5–7.0	15	30–33	1.8–2.5	1.0–1.2	1.30–1.36	1.7–3.0
Canvas base	6.5–7.5	13–16	35	4.4	2.0	1.30–1.36	1.7–3.0
Asbestos paper	5.0	10	30	1.5	0.80	1.70	1.0–1.6
Asbestos fabric	8.0	16	35	1.9	3.0	1.6–1.8	1.7–2.5

Note: Procurement specifications—ASTM D709.

LOW-PRESSURE AND NO-PRESSURE LAMINATES. Low-pressure laminating is chiefly a wartime development brought about by the need for large sections of plastic laminate, of high strength, molded in shapes too complicated for high-pressure techniques. The low-pressure laminates can be cured by the application of heat at pressures generally below 250 psi, using inexpensive molds made from plaster of Paris, wood, lightweight sheet metal, etc.

A great deal of interest has also been shown in the construction of laminates prepared by the application of slight or no pressure, with or without heat. The pressure in most cases need only be sufficient to exclude air pockets. The resins used to manufacture these laminates are called contact resins and are usually furnished in the form of viscous, partially condensed, unsaturated fluids.

In the production of some of the low-pressure and no-pressure laminates, resins such as unsaturated polyesters, phenolics, modified melamine formaldehyde, allyls, cellulose acetate and ethyl cellulose are used.

The type of filler used in the construction of these laminates is also of great importance in obtaining adequate strength properties. Such fillers may be veneers or thin plies of wood, high-strength paper, woven fibers of cotton, or glass filaments used either in felted form or woven into fabric. Laminates made with glass fabrics possess an exceptionally high strength-weight ratio.

The foregoing laminates are employed for such items as aircraft rudders, doors, gasoline tanks, wall panels, instrument cases, luggage, buckets, ducts, boats, and decorative applications.

POST-FORMING LAMINATED PHENOLIC PLASTICS. A special laminated forming sheet has been developed by several manufacturers of laminated plastic sheet. The material generally used has been a special grade of phenolic resin-laminated canvas. Normally the use of laminated phenolics has been limited to shapes that can be derived from simple forms, such as plates, rods, tubes, angles, and channels. Forms more intricate than these have always required expensive steel molds. With the new material, forming is accomplished by heating the sheet rapidly to a high temperature (below the blistering point), quickly bending or drawing it into the new contour, and holding it in this shape until it has cooled. Common heating methods are (1) parallel hot plates, (2) infrared radiant heat, (3) hot-air ovens, and (4) dielectric heating.

The molds for this type of laminate may be made from wood, high-pressure laminates, Masonite die stock, or concrete. In large production jobs, where the number of pieces

Table 10. Plastics and Synthetic Resin Production, in Pounds

(Figures from U. S. Tariff Commission. Table reproduced from *Modern Plastics Encyclopedia*, 1949 edition, by permission)

	1947	1948
Cellulose acetate and mixed ester plastics: *		
Continuous sheets:		
Under 0.003 gage	7,227,370	7,822,099
0.003 gage and up	7,843,613	6,859,452
All other sheets, rods, and tubes	3,851,354	3,091,853
Molding and extrusion materials	58,544,379	48,531,060
Total	77,466,716	66,304,464
Cellulose nitrate plastics *		
Sheets	9,216,859	7,227,868
Rods and tubes	3,668,652	3,147,991
Total	12,885,511	10,375,859
Other cellulose plastics *, †, ‡	4,000,000 §	8,888,625
Phenolic and other tar acid resins:		
Laminating (dry basis)	40,546,186	43,061,642
Adhesive (dry basis)	21,069,109	20,723,400
Molding materials *	194,235,872	171,116,725
All other, including casting (dry basis *, ‡)	63,525,691	61,415,818
Total	319,376,858	296,317,585
Urea and melamine resins:		
Adhesives (dry basis)	48,472,384	52,215,457
Textile and paper treating (dry basis)	16,862,315	18,858,681
All other, including laminating (dry basis) ‡, ‖	7,785,087	2,788,896 ¶
Total	73,119,786	73,863,034
Polystyrene ‡, **	94,985,618	145,299,338
Vinyl resins:		
Sheeting and film, incl. safety glass sheeting *	66,425,689	98,020,686
Textile and paper coating resins (resin content)	17,235,802	31,086,486
Molding and extrusion materials (resin content)	73,113,605	73,635,165
All other, including adhesives, but excluding protective coatings (resin content)	27,062,369	21,147,802
Total	183,837,465	223,890,139
Miscellaneous resins:		
Molding materials *, ††	56,148,925	69,999,064
All other (dry basis) ‡, ‡‡	29,558,704	33,729,857
Total	85,707,629	103,728,921
Grand Total	851,379,583	928,667,965

* Includes filler, plasticizers, and extenders.
† Includes ethylcellulose and related plastics.
‡ Excludes data for protective coating resins.
§ Figure for 1947 estimated.
‖ Excludes urea and melamine molding materials; see footnote ††.
¶ Jan. to May, incl., only. Figures for remainder of year included with "Miscellaneous resins, All other."
** Dry basis, including necessary coloring material.
†† Includes data for urea and melamine, acrylic acid and miscellaneous molding materials.
‡‡ Includes data for petroleum resins, acrylic acid ester resins, mixtures, and miscellaneous synthetic resin materials.

may run into thousands, blanking dies of steel or Kirksite (cast zinc alloy) are sometimes provided.

The post-forming material used during World War II to manufacture numerous aircraft components is now being adapted to the production of many commercial items.

SANDWICH CONSTRUCTIONS. These materials consist of thin-gage, high-strength surface sheets, bonded to low-density cores by means of waterproof, high-strength synthetic adhesives. They are finding application as structural and nonstructural members of boats, aircraft, furniture, refrigerators, and other items requiring high-strength, lightweight materials. The surface sheets or skins furnish rigidity and strength to the structure and are made of such materials as stainless steel, aluminum, magnesium, and high- and low-pressure plastic laminates using glass fiber, paper, duck, or wood as the filler sheets. Balsa wood was first used as a core material, but expanded (porous) plastics are now used for this purpose. Cellular cellulose acetate Strux CCA, expanded polystyrene Styrofoam, expanded phenolic resin Textolite Foam, expanded rubber Flotofoam, Cell-tite, and expanded polyvinyl formal Plastazote are some of the most widely used plastics for core materials. The rigid plastic foams weigh from 1.5 to 30 lb per cu ft.

In addition to the expanded plastic cores, a honeycomb core is sometimes used in sandwich constructions. The core is made from resin-impregnated sheet material which has been cured and assembled into the form of nested hexagonal-shaped cells resembling bees' honeycomb.

Scientists are endeavoring to develop the stiffest and lightest material to fill the wings on superspeed planes (over 750 mph).

PLASTICS PRODUCTION. Relative output of the various types of plastics and synthetic resins is indicated in Table 10.

TRADE NAMES. A partial list of trade names of plastics materials appears in Table 11. For a complete directory of trade names of plastics materials see *Modern Plastics Encyclopedia*, published by Plastics Catalogue Corporation, 122 East 42 St., New York 17, N. Y.

Table 11. Trade Names of Plastic Materials

Type of Composition	Trade Name
Phenol formaldehyde	Bakelite, Catalin, Co-ro-lite, Crystle, Durez, Durite, Dilecto, Formica, Insurok, Lamicoid, Micarta, Marblette, Makalot, Phenolite, Panelyte, Resinox, Spauldite, Synthane, Textolite
Urea formaldehyde	Bakelite Urea, Beetle, Plaskon, Uformite
Melamine formaldehyde	Catalin, Melmac, Plaskon, Resimene
Analine formaldehyde	Cibanite, Dilectene
Cellulose nitrate	Celluloid, Nixon CN, Nitron, Pyralin
Cellulose acetate	Bakelite C.A., Chemaco C.A., Fibestos, Lumarith, Nixon CA, Plastacele, Tenite I
Cellulose acetate butyrate	Tenite II
Cellulose propionate	Forticel
Ethyl cellulose	Chemaco E.C., Ethocel, Lumarith E.C., Nixon E.C.
Polyvinyl chloride	Geon, Koroseal
Polyvinyl alcohol	Resistoflex, PVA
Polyvinyl chloride acetate	Vinylite, Tygon
Polyvinyl butyral	Butacite
Polyvinylidene chloride	Saran
Methyl methacrylates	Lucite, Plexiglas
Polymeric amides	Nylon
Polystyrene	Styron, Loaline, Lustron, Bakelite Polystyrene
Styrene copolymers	Textolite 1421, Textolite 1422, Plexene M, Styraloy 22
Unsaturated polyesters (cross-linked addition polymers)	Laminac, Selectron, Thalid, Paraplex
Polyethylene	Polythene, Bakelite Polyethylene
Polytetrafluorethylene	Teflon
Trifluorochloroethylene	Kel-F
Casein	Ameroid, Galorn

GENERAL REFERENCES

ASTM Standards on Plastics, American Society for Testing Materials, Philadelphia, Sept. 1948.
Data sheets on Dilectene, Continental Diamond Fibre Company, Newark, Delaware, 1941.
E. I. du Pont de Nemours and Co., catalog covering polyethylene, April 1946.
J. S. Hicks and R. J. Francis, *Low-pressure Laminating of Plastics*, Reinhold Publishing Corp., New York, 1947.
G. M. Kline, The chemists' wonderland plastics through the looking glass, *Chem. Eng. News*, June 10, 1944.

Modern Plastics Encyclopedia, Plastics Catalogue Corp., New York, 1947.
Plastics Properties Chart, Plastics Catalogue Corp., New York, 1946.
Dr. C. A. Refarn, Charts on plastics, *Brit. Plastics*, May 1943.
H. M. Richardson and J. W. Wilson, *Fundamentals of Plastics*, McGraw-Hill Book Co., New York, 1946.
K. Rose, Plastic laminates, *Materials & Methods*, Sept. 1946.
K. Rose, A plastics primer for engineers, *Materials & Methods*, April 1947.
Dr. J. W. Shackleton, Polyethylene, *Modern Plastics*, Feb. 1944.
Standard Commodity Classification, *Technical Paper* 26, Washington, D. C., May 1943.
F. Sullivan, Post-formed laminated phenolic plastic, *Metals & Alloys*, Feb. 1945, p. 392.
Technical Data on Plastics, Plastic Materials Manufacturers' Association, Washington, D. C., May 1948.
L. Wakeman, *The Chemistry of Commercial Plastics*, Reinhold Publishing Corp., New York, 1947.
A. J. Warner, Plastics—a brief review of their physical and electrical properties, *Elec. Commun.*, Vol. 22, No. 1, 1944.
New plastic insulator, *Science News Letter*, Dec. 11, 1948, p. 374.
F. B. Stanley, Casting techniques for polyesters, *Modern Plastics*, Aug. 1948, p. 103.

REFRACTORIES

By J. Spotts McDowell

Refractories are structural materials used at high temperatures in industrial furnaces. They are supplied mainly in the form of brick of standard sizes and of special shapes. Refractories also include refractory cements, bonding mortars, plastic firebrick, castables, ramming mixtures, and other bulk materials such as dead-burned grain magnesite, chrome ore, ground ganister and special clays.

The types of refractory brick most generally used are fireclay, superduty fireclay, high alumina, silica, magnesite, magnesite chrome, chrome, chrome magnesite, and forsterite brick and insulating firebrick of various classes. Other refractory products, such as silicon carbide, sillimanite, and zirconia, usually referred to as special refractories, are marketed in relatively smaller tonnages. With the exception of silicon carbide, they are not discussed in this chapter.

The life of refractories is affected by the high temperatures to which they are subjected and usually by one or more of the following influences: abrasion, corrosion through chemical action of slags and fluxes, erosion by mechanical action of molten metal or slag or by dust-laden gases moving at high velocity, spalling of various types, or the chemical or physical effect of various gases or fumes.

13. TYPES OF REFRACTORIES

FIRECLAY BRICK. Of all the various types of refractories, the most widely used are fireclay brick, made from flint and plastic fireclays. Hydrous aluminum silicate minerals, of which kaolinite is the most common, are the essential components of fireclay. The theoretical analysis of kaolinite, free from accessory components, is 39.5% alumina, 46.5% silica, and 14% water. At high temperature the water is driven off, and the resulting calcined material theoretically contains 45.9% alumina and 54.1% silica. However, chemical analyses of even the purest clays disclose the presence of small amounts of other compounds, such as those of iron, calcium, magnesium, and titanium, as well as alkalies, and usually free silica.

Flint clays, known also as hard clays, are hard and dense and break with a conchoidal fracture. After they are ground and mixed with water, their plasticity and drying shrinkage are very low; their firing shrinkage is somewhat variable and moderately low. Flint clays possess high refractoriness. Deposits of flint clay of considerable commercial importance occur in rather limited areas in central Pennsylvania, Maryland, eastern Kentucky, southern Ohio, and eastern Missouri.

Plastic and semiplastic clays cover a wide range in refractoriness, plasticity, and bonding strength. Their shrinkage during drying and firing is fairly high. Many of these clays have high plasticity and excellent bonding power.

White-burning clays are generally known as kaolins. Pure kaolins usually are only moderately plastic, but their drying and firing shrinkages are high. However, kaolins that contain appreciable amounts of free silica have low drying and firing shrinkages. Some of the largest deposits of kaolin in America are found in Georgia. Kaolins suitable for the manufacture of refractories are found also in various other states.

Most standard sizes of fireclay brick are made by machine, by either the power-press or the extrusion process. The power-press process comprises (1) grinding the clays in a dry pan, (2) screening them to desired sizes, (3) blending and thoroughly mixing them with a small but definitely controlled amount of water, and (4) forming the brick under high pressure by means of a mechanically operated press.

The steps in the manufacture of brick by the extrusion process are (1) grinding the clay or clays in a dry pan, (2) screening and blending them and tempering them with a controlled amount of water in a mixer, (3) extruding the tempered clay through a die as a column, (4) cutting the column of clay into desired sizes by means of wires, and (5) press ing the sizes mechanically to form brick with sharp corners and edges and smooth surfaces.

Special shapes are generally made by hand. The process consists in (1) grinding a single clay or a mixture of clays with water in a wet pan and (2) molding the brick by hand. A relatively small proportion of standard sizes is made by hand. After being molded and partially dried, these brick are re-pressed in a hand-operated press.

The brick formed by any of the processes described are dried either on hot floors or in tunnel driers, after which they are fired in periodic downdraft kilns or in continuous car-type tunnel kilns.

To meet certain requirements, brick are manufactured by a method in which a high vacuum is applied during the forming operation. In the extrusion process, the method is called de-airing. As applied to the power-press process, it is known as vacuum pressing. Brick so made are harder, stronger, less porous, and slightly heavier than brick made otherwise.

Fireclay Brick have been classified as follows by the American Society for Testing Materials (ASTM C27-41):

Superduty Fireclay Brick. A fireclay brick having a pyrometric cone equivalent (see p. 5-53) not lower than cone No. 33 on the fired product, not more than 1% linear shrinkage in the linear change test, schedule C (2910 F, 1600 C), and not more than 4% loss in the panel spalling test (preheated at 3000 F, 1650 C).

High-heat Duty Fireclay Brick. A fireclay brick conforming to one of the following requirements: pyrometric cone equivalent not lower than cone No. 31-32, or not more than 1.5% deformation in the 2460 F (1350 C) load test.

Intermediate-heat Duty Fireclay Brick. A fireclay brick conforming to one of the following requirements: pyrometric cone equivalent not lower than cone No. 29, or not more than 3% deformation in the 2460 F (1350 C) load test.

Low-heat Duty Fireclay Brick. A fireclay brick having a pyrometric cone equivalent not lower than cone No. 19.

HIGH-ALUMINA BRICK. Brick of the alumina-silica group of refractories containing more than 47.5% alumina are known as high-alumina refractories. The raw materials used in their manufacture include diaspore, diaspore clays, bauxite, kyanite, andalusite, sillimanite, and crystalline alumina or corundum. Some of these materials are used singly; others are combined or blended with clays or other bonding agents, depending on the properties and the percentage of alumina desired in the finished product. High-alumina brick are made by processes similar to those used in the manufacture of fireclay brick.

High-alumina brick serve advantageously in place of superduty and high-heat duty fireclay brick under service conditions beyond the safe working limits of brick of those classes.

Various grades of high-alumina brick are available in the commercial classes containing approximately 50, 60, 70, 80, 90, and 99% alumina. Brick made mainly from diaspore clay have been classified as follows by the American Society for Testing Materials (ASTM C27-41):

50% Alumina-diaspore Fireclay Brick. Refractory brick having diaspore or nodular fireclay as an essential original ingredient, an alumina content of 50 $\pm$ 2.5%, and a pyrometric cone equivalent not lower than cone No. 34.

60% Alumina-diaspore Fireclay Brick. Refractory brick having diaspore or nodular fireclay as an essential original ingredient, an alumina content of 60 $\pm$ 2.5% and a pyrometric cone equivalent not lower than cone No. 35.

70% Alumina-diaspore Fireclay Brick. Refractory brick having diaspore as an essential original ingredient, an alumina content of 70 $\pm$ 2.5%, and a pyrometric cone equivalent not lower than cone No. 36.

SILICA BRICK. Quartzite rock, commonly referred to as ganister, is the raw material used in the manufacture of most silica brick. It is a hard, fine-grained rock, containing 97 to 98% silica. The steps in the manufacture of silica brick are (1) crushing and grinding the ganister in a unit known as a wet pan with the addition of water and approximately 1.75 to 3% lime, (2) molding in a power press or by hand; (3) drying in tunnel driers; and (4) firing, either in periodic downdraft kilns or in continuous car-type tunnel kilns. In a modified process, the ganister is ground in a dry pan, screened to definite sizes, and stored in bins. The separate sizes are later blended in desired proportions and mixed with water and approximately 1.75% to 3% lime. Molding, drying, and firing follow in order, as already outlined.

The lime added to the charge serves as a bond. During firing, it reacts with a portion of the quartzite and gives strength to the finished product.

During the firing operation, a permanent expansion of the brick takes place. Silica brick show a slight permanent expansion in high-temperature service, since it is not practicable to carry the permanent expansion to completion during the firing. The linear thermal expansion of well-burned silica brick amounts to 1.2 to 1.6%, the greater part of which occurs below 1200 F.

When silica brick is laid, it is important to provide for the reversible thermal expansion of about $^3/_{16}$ in. per ft. Since the joints take up some expansion, it is customary to allow for expansion at the rate of $^1/_8$ in. per ft. Expansion joints should be kept free from mortar by packing them with a filler which will burn out.

Silica brick are sensitive to rapid changes in temperature below 1200 F and should therefore be heated and cooled slowly between 1200 F and atmospheric temperature.

Three classes of silica brick are available: (1) superduty silica brick, containing not more than 0.5% alumina, titania, and alkalies; (2) brick which might well be called high-heat duty silica brick, containing a total of about 0.9 to 1.25% alumina, titania, and alkalies; and (3) sandrock silica brick, containing a total of approximately 1 $^1/_2$ to 2 $^1/_2$% alumina, titania, and alkalies.

MAGNESITE BRICK. The outstanding characteristics of magnesite brick are their strongly basic chemical behavior and their extremely high refractoriness. Prior to 1939, practically all magnesite refractories were made from a rock which consists almost entirely of the mineral magnesite (magnesium carbonate). The large deposits of crystalline magnesite in Europe, in Manchuria, and in the state of Washington supplied by far the larger portion of all the magnesite used for refractory purposes. Natural magnesite is still used for the production of the major portion of the refractory magnesia employed in industry. The rock consisting of the mineral brucite (magnesium hydroxide), mined in Nevada, is used to a minor extent. A substantial amount of magnesium hydroxide is recovered from natural brines, such as seawater, bittern from salt works, and brine from deep wells.

The first step in the manufacture of magnesite brick is the firing of the raw material at high temperature in a rotary kiln. In the firing process, the carbon dioxide or the water of combination, as the case may be, is expelled, and practically complete shrinkage of the magnesia is effected. The product, known as dead-burned magnesite, is inert to atmospheric moisture and carbon dioxide gas.

For refractory purposes, dead-burned magnesite is supplied in the form of brick, of finely ground material for brick laying, and of granular material (dead-burned grain magnesite or magnesite ramming mixtures) for making and repairing furnace bottoms.

To make magnesite brick, the dead-burned magnesite is ground, mixed with a controlled amount of water, and molded at high pressure in a power press. Special shapes are molded by hand. The molded brick are thoroughly dried in tunnel driers and fired at extremely high temperature under a controlled time-temperature schedule in downdraft periodic kilns or in continuous car-type tunnel kilns.

Grain magnesite, which is widely used for the construction of monolithic furnace hearths, consists of a mixture of granules, varying in size from about $^5/_8$ in. in diameter to particles of extreme fineness. It is fritted into place in thin layers by the application of heat. Magnesite ramming mixtures having a maximum particle size of about $^1/_4$ in. in diameter are rammed into place with air hammers and take a strong set merely upon drying.

CHROME BRICK. The raw material in the manufacture of chrome brick is a rock known as chrome ore. The composition of ores suitable for the manufacture of refractories ranges from 30 to 50% chromic oxide (Cr_2O_3), 12 to 18% ferrous oxide (FeO), 13 to 18% magnesia (MgO), 13 to 32% alumina (Al_2O_3), and 4 to 8% silica (SiO_2). In the manufacture of chrome brick the ore is ground, properly sized, mixed with a small amount of water, and molded under heavy pressure. The molded brick are dried and then fired at high temperature.

MAGNESITE-CHROME AND CHROME-MAGNESITE BRICK. It has been demonstrated that brick with properties superior for various purposes to those made from magnesite alone or from chrome ore alone can be manufactured from compositions consisting of blends of chrome ore and dead-burned magnesite in suitable proportions. The two-component brick made from such blends are known as magnesite-chrome brick and chrome-magnesite brick; in each case, the principal component is named first. In general, these two-component brick have superior resistance to spalling and superior rigidity at high temperature under load. Such brick are usually bonded chemically instead of by firing.

However, one type of chrome-magnesite brick is bonded by firing at extremely high temperature.

METAL-ENCASED BRICK. Thin sheets of steel in the joints between certain types of basic brick add materially to their life by decreasing their tendency to spall and by increasing their rigidity at furnace operating temperatures.

With straight, arch, wedge, and key brick and with brick of certain special shapes, steel is applied in the form of a casing which covers three or four faces of the brick. Other metal-encased brick consist of basic refractory material packed solidly in steel tubes 3 in. in diameter and 9, 13 1/2, or 18 in. long. The materials most generally used with steel casings are fired magnesite brick and chemically bonded magnesite and magnesite-chrome brick.

FORSTERITE BRICK. The mineral forsterite is a silicate of magnesium, having the theoretical chemical formula $2MgO \cdot SiO_2$, corresponding to a composition of 51.3% magnesia and 42.7% silica. The melting point of pure forsterite is 3470 F. In the United States, forsterite brick were first manufactured commercially in 1933.

In nature, forsterite rarely occurs in pure form but is usually associated with a silicate of iron known as fayalite ($2FeO \cdot SiO_2$), in the mineral olivine. A low-iron olivine rock known as dunite is the raw material from which forsterite brick are made.

Accessory minerals which occur with the olivine are converted into additional forsterite through reaction with magnesia added during the process of manufacture. The brick are molded principally by the power-press process under extremely high pressure. The required firing temperature is higher than that necessary for most other refractory materials.

INSULATING FIREBRICK. The primary function of furnaces built of refractory materials is to confine and to utilize heat energy. The quantity of heat which passes through furnace walls, to be dissipated into the surrounding air, often constitutes a considerable proportion of the total heat input. Such heat losses may be reduced greatly by the application of insulating firebrick to the outer faces of the refractory walls.

In intermittently operated furnaces, the heat required to bring the furnace to operating temperature is lost when the furnace is permitted to cool. This heat loss is often of greater significance in the heat balance than is the amount of heat necessary to maintain the required temperature for the few consecutive hours of operating time. Because of their light weight, insulating firebrick have low heat capacities. Consequently their use in intermittently operated furnaces reduces not only the amount of heat lost by conduction and radiation, but also the amount of heat stored in the brickwork, which is lost when the furnace is cooled to atmospheric temperature.

Insulating firebrick have been classified by the American Society for Testing Materials (ASTM C155-47) as indicated in Table 1. The grouping of insulating firebrick in accord-

Table 1. Grouping of Insulating Firebrick

Group Identification	Permanent Linear Change * Not More than 2% When Tested at	Bulk Density Not Greater Than
16	1550 F (845 C)	34 lb/cu ft
20	1950 F (1065 C)	40 lb/cu ft
23	2250 F (1230 C)	48 lb/cu ft
26	2550 F (1400 C)	52 lb/cu ft
28	2750 F (1510 C)	60 lb/cu ft

* Maximum temperature held 24 hr in accordance with Standard Method of Reheat Change of Insulating Fire Brick, ASTM C210-46.

ance with Table 1 is based on bulk density (weight per cubic foot) and the behavior in the reheat change test conducted at the specified temperature.

MORTAR MATERIALS. The two classes of mortar materials for refractory brick masonry are (1) hot-setting bonding mortars shipped in dry form and (2) cold-setting bonding mortars shipped in either dry or wet form. Each class includes materials of various compositions, each of which has a definite field of application. The bonding strength of hot-setting bonding mortars is developed only at high temperatures, with the formation of what is known as a ceramic bond. These mortars include such materials as finely ground fireclay, silica, high-alumina, magnesite and chrome ore, and forsterite, as well as a great variety of blends of the different commercially available refractory materials. Cold-setting bonding mortars are distinguished by their ability to take a rigid cold set and to form a strong bond at atmospheric temperature. These mortars are similar in composition to those of the hot-setting class, but in their preparation chemical bonding agents are added which impart cold-setting properties.

The type of refractory brick to be bonded and the service conditions to which they will be exposed govern the selection of the mortar material best suited to meet the requirements.

PLASTIC FIREBRICK. Plastic firebrick is made from blended raw materials of various compositions. It is supplied in standard and superduty qualities, ready mixed, with just a sufficient amount of water to render it suitable to be rammed into place. It serves admirably for emergency repairs and under certain conditions may be used for forming complete one-piece furnace linings.

CASTABLES. Castables are refractory materials of various compositions, supplied in dry form, which require thorough mixing with water to be suitable for casting. The ease with which castables can be applied and the fact that they quickly develop a strong bond make them useful where time is important for such applications as linings of irregular dimensions in small furnaces, patching, and the production of intricate special shapes. Castables are adapted for the construction of baffles, stoker installations, oil burner blocks and ports, door linings, ash-pit linings, and domestic oil-fired furnace linings. The three available classes of castables are standard, superduty, and lightweight.

RAMMING MIXTURES. Ramming mixtures consist essentially of ground refractory materials of graded particle size, mixed with small amounts of other ingredients to give good working qualities and a cold set. Various of these mixtures have a chrome or a magnesite base. They are available in both dry and plastic forms.

Upon air-drying, ramming mixtures set to form a strong, dense, well-bonded refractory. The strength and density may increase at high temperatures. Their principal application is in the bottoms and hearths of furnaces, lining of ladles and converters, roofs and linings of small electric furnaces, burner blocks, and various other constructions.

SILICON CARBIDE REFRACTORIES (data supplied by The Carborundum Company, Refractories Division, Perth Amboy, N. J.) Silicon carbide is composed of carbon and silicon, the chemical formula being SiC. It is made in the electric resistance furnace. The crystalline product is generally considered to form at approximately 3350 F or higher and to dissociate at approximately 4050 F, below which temperature no softening or fusion occurs. Its hardness is 9.6 on the Moh scale, giving a remarkable resistance to abrasion. It is inert to most acids, liquid or gaseous, but may be decomposed by alkali when subjected to high temperature. Silicon carbide reacts with basic slags and molten iron and steel. It is made into refractory brick and shapes varying widely in size, weight, design, etc.

Silicon carbide refractories comprise silicon carbide crystals bonded with small percentages of permanent binders of which there are many, depending on the type of service. The standard 9 by 4 1/2 by 2 1/2-in. brick weighs approximately 9.25 lb. Many different special shapes are made for boilers, heat-treating, malleable-iron-annealing, forge, and melting furnaces, soaking pits, regenerators, electric furnaces, zinc retorts, coking and calcining retorts, gas producers, water gas generators, porcelain enameling furnaces, periodic and tunnel kilns, glass lehrs, rotary kilns, roasting furnaces, etc. Silicon carbide refractories are especially resistant to the destructive action of clinker formed from coal ash. Thermal conductivity of the higher grades of silicon carbide refractories averages 108 Btu per sq ft per inch of thickness per °F per hr in the temperature range 1200 to 2800 F, making them suitable for indirect heating, as in muffles, retorts, recuperators, regenerators, etc. Other physical properties are: modulus of rupture at 2462 F, 900 to 6250 psi; crushing strength at 2462 F 10,000 to 15,000 psi; at 68 F, 15,000 to 20,000 psi. The mean specific heat in the range 0 to 2552 F is 0.285. Silicon carbide refractories are characterized by extremely high spalling resistance.

14. PROPERTIES OF REFRACTORY BRICK

ABRASION. As applied to refractory brick, abrasion is defined as "wearing away of refractory surfaces by the scouring action of moving solids." (*The Standard Definitions of Terms Relating to Refractories*, ASTM C71-47.)

In certain types of furnaces, as, for example, in shaft lime kilns, the refractory linings are subjected to more or less severe abrasion by moving charges. The sandblast action of hot dust-laden gases impinging upon refractories (furnace walls) is another example of abrasive action.

Tests have been developed which indicate clearly the relative resistance of cold brick to abrasion. These tests, however, do not measure the resistance to abrasion of brick at furnace operating temperatures. No entirely satisfactory method has been devised for evaluating the resistance of hot brick to abrasion. However, brick which are mechanically strong at a given operating temperature will usually be resistant to abrasion.

CHEMICAL COMPOSITION. (See Tables 2 and 3.) The compositions of the different types of refractories vary considerably. For Chemical Analysis of Refractory Materials, see ASTM C18-45, *Manual of American Society for Testing Materials, Standards on Refractory Materials*, 1948.

Table 2. Typical Chemical Compositions of Alumina-silica Refractories

Type and Class of Brick	Composition, % SiO_2	Al_2O_3	TiO_2	Accessory Oxides
Fireclay				
Superduty	49–53	40–44	2.1–2.5	3–4
High-heat duty (aluminous)	52–60	35–40	1.8–3.3	2–6
High-heat duty (siliceous)	65–75	20–30	1.0–2.0	2–4
Intermediate heat duty	60–65	30–36	1.8–2.1	4–7
High-alumina				
50% alumina class	41–47	47.5–52.5	2.2–2.9	3–4
60% alumina class	31–37	57.5–62.5	2.2–3.5	3–4
70% alumina class	20–26	67.5–72.5	3.2–4.0	3–4
80% alumina class	10–16	77.5–82.5	3.0–4.0	3–4
90% alumina class	8–9	89–91	0.5–0.8	0.5–1.0

Table 3. Typical Chemical Compositions of Silica and Basic Refractories

Type of Brick	Composition, % Al_2O_3	SiO_2	CaO	MgO	FeO	Fe_2O_3	Cr_2O_3	Accessory Oxides
Silica		95–97	1.8–3.0					0.5–2.5
Magnesite (fired)				83–93		2–7		5–10
Chrome (fired)	15–33	3–6		14–18	11–17		30–45	
Forsterite		33–40		47–53		9–11		3–4

COLD CRUSHING STRENGTH. (See Table 4.) This property provides little or no indication of the strength at furnace operating temperatures. However, if the cold strength is extremely low, the brick may not withstand handling and shipping without excess scuffing and breakage. For details, see *Standard Method of Test*, ASTM C133-39.

Table 4. Cold Crushing Strength of Refractories

Type of Refractory	Cold Crushing Strength, 9-in. brick set on end, psi
Fireclay	
Superduty	900–4000
High-heat duty (handmade)	700–2500
High-heat duty (machine-made)	1000–6000
Intermediate-heat duty	1200–6000
High alumina	
50% alumina class	2500–5000
60% alumina class	1800–6000
70% alumina class	4000–6000
80% alumina class	5000–8000
90% alumina class	4000–7000
Silica	2000–4000
Magnesite	2000–10000
Magnesite-chrome (chemically bonded)	2000–4000
Chrome	2000–6000
Chrome-magnesite (chemically bonded)	2000–4000
Forsterite	1500–4000

THERMAL CONDUCTIVITY. Table 5 gives the approximate thermal conductivities of refractories over a wide range of temperatures. For details of *Standard Method of Test*, see ASTM C182-47, also C201-47 and C202-47.

Table 5. Approximate Thermal Conductivities of Refractories

(Btu/hr/sq ft of wall area, per °F temperature difference, for 1-in. thickness)

(From *Modern Refractory Practice*, 2nd ed., Harbison-Walker Refractories Co., Pittsburgh, Pa., 1937 by permission.)

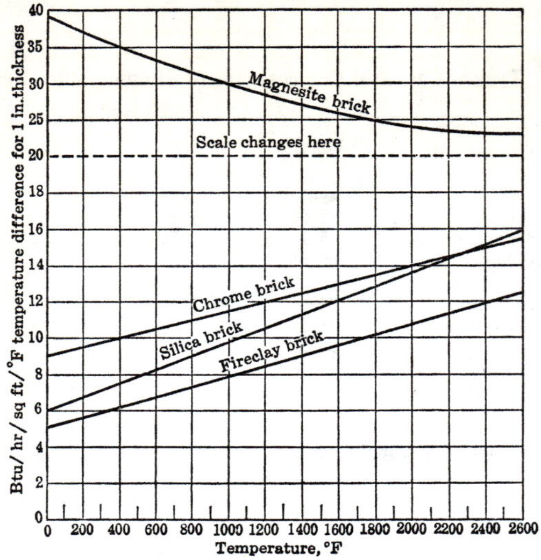

Kind of Brick	Temperature, °F													
	0	200	400	600	800	1000	1200	1400	1600	1800	2000	2200	2400	2600
Magnesite brick	39.5	37.2	35.4	33.3	31.5	30.0	28.3	26.9	25.9	25.0	24.0	23.5	23.2	23.0
Chrome brick	9.0	9.5	10.0	10.5	11.0	11.5	12.0	12.5	13.0	13.5	14.0	14.5	15.0	15.5
Silica brick	6.0	6.8	7.5	8.3	9.0	9.8	10.5	11.3	12.0	12.8	13.5	14.3	15.0	15.8
Fireclay brick	5.0	5.6	6.1	6.7	7.3	7.8	8.4	9.0	9.6	10.1	10.7	11.3	11.8	12.4

REVERSIBLE THERMAL EXPANSION. The reversible linear expansion of various types of refractory brick is given in Fig. 1. The curves represent average values from atmospheric temperatures to the temperature indicated.

LOAD RESISTANCE AT HIGH TEMPERATURE. In high-temperature service, refractories may be subjected to heavy pressure due to arch stresses or to the weight of overlying courses of brickwork. Fireclay refractories are least resistant to load at high temperature. They are strong and rigid up to 2000 F. However, at 2250 F and under a load of 25 psi, high-heat duty fireclay brick exhibit slow plastic deformation which increases with rising temperature. Silica brick remain strong and rigid almost up to their temperature of failure. High-alumina brick up to those of the 70% alumina class show appreciable subsidence in the standard load test at 2462 F, as can be seen from Table 6. Basic brick

Table 6. Behavior of Refractory Brick under Load at High Temperature

Type of Brick	Linear Subsidence at 25 psi and 2460 F,* %	Temperature of Failure at 25 psi, °F
Fireclay		
Superduty	2–6	
High-heat duty (aluminous)	3–10	
High-heat duty (siliceous)	2–4	
High alumina (70% alumina class)	0.5–4	
Silica		2900–3100
Magnesite		2450–2800
Magnesite-chrome		2750–3000
Chrome		2450–2800
Chrome-magnesite		2750–3000
Forsterite		2700–3100

* See ASTM C16-41.

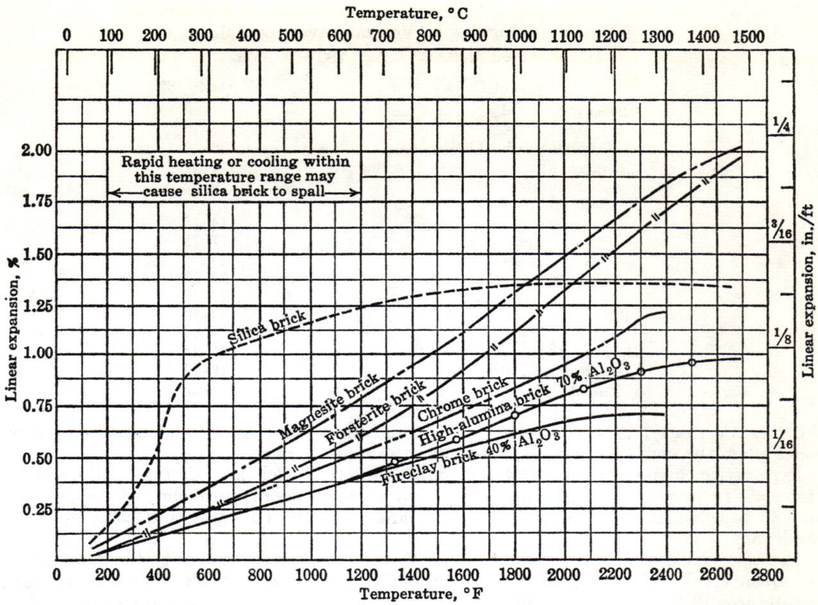

Fig. 1. Approximate reversible thermal expansion of refractories. To convert linear percentage of expansion into thirty-seconds of an inch per foot, multiply by 3.8. (Reproduced by permission from *Modern Refractory Practice*, 2nd ed., Harbison-Walker Refractories Co., Pittsburgh, 1937)

of various brands show wide variation in their behavior under load at high temperature. In general, they do not show plastic flow under load but fail abruptly.

MODULUS OF RUPTURE (see Table 7). The modulus of rupture of refractory brick

Table 7. Modulus of Rupture at Atmospheric Temperature

Type of Brick	Modulus of Rupture, psi
Fireclay	
Superduty	600–1500
High-heat duty (handmade)	400–1500
High-heat duty (machine-made)	400–2000
Intermediate-heat duty	600–2200
High alumina	
50% alumina class	1000–1500
60% alumina class	600–1600
70% alumina class	1200–1500
80% alumina class	1500–2000
90% alumina class	1200–1400
Silica	400–1200
Magnesite	750–3000
Chrome	900–2000
Forsterite	450– 800

is indicative only of their strength at atmospheric temperature and provides little or no indication of their strength at furnace operating temperatures. It is determined in accordance with *Standard Method of Testing*, ASTM C133-39.

POROSITY (see Table 8). The porosity of refractory brick is related to their permeability, rigidity under load at high temperature, thermal conductivity, resistance to abrasion, and resistance to slag action. Low porosity is extremely important for brick used in contact with molten slag.

PYROMETRIC CONE EQUIVALENT. Pyrometric cone equivalent, as applied to refractories, is defined as the number of that standard cone whose tip would touch the supporting plaque simultaneously with that of a cone of the material being investigated, when tested in accordance with the standard method of test for pyrometric cone equivalent (PCE) of refractory materials. The determination of PCE consists in testing cones of

Table 8. Porosities by Volume, Average Values

(See *Standard Methods of Test*, ASTM C20-46.)

Type of Brick	Porosity, %
Fireclay	
Superduty	12–26
High-heat duty (handmade)	20–30
High-heat duty (machine-made)	12–27
Intermediate-heat duty	10–30
High alumina	
50% alumina class	13–25
60% alumina class	17–25
70% alumina class	23–33
80% alumina class	20–27
90% alumina class	21–26
Silica	22–30
Magnesite	18–24
Chrome	19–24
Chrome-magnesite	18–22
Forsterite	18–27

the ceramic material by comparing their behavior with that of standard pyrometric cones when the cones are heated together in the same furnace.

Pyrometric cones, when used for determining the relative heat-resisting qualities of the materials for which they are applicable, give an accurate measurement of temperature only in tests which follow exactly the standard schedule of heating and other standard conditions.

The temperatures associated with the particular cones do not apply to the slower rates of heating common in commercial firing and use of refractory materials. When it is desired to interpret PCE values approximately in terms of temperature, Table 9 may be used.

Pyrometric cone equivalents for various types of refractory brick are given in Table 10. For complete details of the test, see ASTM C24-46.

Table 9. Pyrometric Cones Used in Connection with Refractories

(Based on the work of Fairchild and Peters as published in the *J. Am. Ceram. Soc.* Vol. 9, No. 11, Nov. 1926, p. 700.)

Cone No.	End Point, °F	Cone No.	End Point, °F	Cone No.	End Point, °F	Cone No.	End Point, °F
12	2435	18	2714	28	2939	33	3173
13	2462	19	2768	29	2984	34	3200
14	2552	20	2786	30	3002	35	3245
15	2615	23	2876	31	3056	36	3290
16	2669	26	2903	32	3092	37	3308
17	2687	27	2921	32 1/2 *	3131	38	3335

Note: The temperatures given are approximate.

* Not included in the tests of Fairchild and Peters.

Table 10. Pyrometric Cone Equivalents (PCE) for Various Types of Refractory Brick

Type	PCE	Temperature °F	Temperature °C
Fireclay			
Superduty	33 to 34	3173–3200	1745–1760
High-heat duty (aluminous)	31–32 to 32 1/2	3074–3137	1690–1725
High-heat duty (siliceous)	28 to 31	2939–3056	1615–1680
Intermediate-heat duty	29 to 31	2984–3056	1640–1680
Low-heat duty	19 to 26	2768–2903	1520–1595
High alumina			
50% alumina class	34 to 35	3200–3245	1760–1785
60% alumina class	35 to 36–37	3245–3299	1785–1815
70% alumina class	36 to 38	3290–3335	1810–1835
80% alumina class	38 to 39	3335–3389	1835–1865
90% alumina class	39 to 40–41	3389–3425	1865–1885

CHEMICAL REACTIONS BETWEEN REFRACTORIES. Dissimilar types of refractory brick (for example, magnesite and fireclay brick) will react chemically with each other at high temperatures. (See Table 11.) It is common practice to separate such refractories

Table 11. Reaction Temperatures between Different Refractories, °F

Type of Refractory Brick	Silica A	Silica B	High-heat Duty Fireclay A	High-heat Duty Fireclay B	High Alumina (70% Al_2O_3) A	High Alumina (70% Al_2O_3) B	Chrome A	Chrome B	Magnesite A	Magnesite B	Forsterite A	Forsterite B
Silica	....	....	2730	*	2910	*	No reaction		2730	2910	3100	3100
High-heat duty fireclay	2730	*	....	....	No reaction		2910	*	2550	2730	2730	2910
High alumina (70% Al_2O_3)	2910	*	No reaction		....		2910	2910	2730	*	3100	3100
Chrome	No reaction		2910	*	2910	2910	.....	.....	No reaction		No reaction	
Magnesite	2730	2910	2550	2730	2730	*	No reaction		....		No reaction	
Forsterite	3100	3100	2730	2910	3100	3100	No reaction		No reaction		....	

A—reaction first observed. *B*—reaction first became damaging.
* Not damaging at 3090 F.

by interposing a course of neutral material, usually chrome brick, which will not react with either.

REFRACTORINESS (see Pyrometric Cone Equivalent). The fusion point, softening point, and melting point are indefinite when applied to ceramic materials, which soften over a wide temperature range. In general, no definite temperature marks the transition from the distinctly solid to the distinctly liquid state. Instead, the pyrometric cone equivalent (PCE) is generally used as an index of the heat-resisting qualities or refractoriness of fireclay and high-alumina brick.

The ability of a refractory brick to support load at high temperatures in service is frequently of major importance. Consequently, the standard load test is of considerable significance in evaluating the refractoriness of materials for which the PCE test is not applicable.

SPALLING RESISTANCE. Spalling may be defined as the loss of fragments or spalls from the face of a refractory structure through superficial cracking and rupture, with exposure of inner portions of the original refractory mass.

Of the various types of refractory brick most widely used, superduty fireclay brick have the greatest resistance to thermal spalling. Both superduty fireclay brick and high-alumina brick may be considerably more resistant to spalling than other refractory brick in the presence of certain types of slag, owing to their high refractoriness and chemical composition which enable them to resist change in texture due to high temperature or slag action. Silica brick are extremely sensitive to temperature change below 1200 F. Magnesite and chrome brick are also sensitive to temperature change. Magnesite-chrome and chrome-magnesite brick have higher resistance to spalling than do magnesite or chrome brick.

No single method for determining resistance to spalling can be applied to all types of refractory brick. The most informative spalling test is the panel spalling test developed at the laboratory of the Refractories Fellowship of the Mellon Institute of Industrial Research, Pittsburgh, Pennsylvania. For *Basic Procedure in Panel Spalling Test,* see ASTM C38-47; see also ASTM C71-47, C107-47, C122-47, and C180-47.

Table 12. Specific Heats of Refractory Materials

t F	Mean Specific Heats between 32 and *t* F: Fireclay Brick	Silica Brick	Magnesite Brick	Chrome Brick
32	.193	.169	.208	.170
200	.199—	.188	.219	.175
400	.205	.211	.232	.182
600	.212	.229	.243	.188
800	.220	.238	.250	.194
1000	.227	.246	.257	.199
1200	.234	.252	.263	.204
1400	.242	.256	.267	.208
1600	.248	.260	.273	.212
1800	.253	.264	.277	.216
2000	.258	.269	.282	.220
2200	.261	.272	.287	.222
2400	.266	.276	.293	.224
2600	.269	.279	.297	.226

SPECIFIC HEAT. The heat content of a refractory brick at any temperature is a function of its specific heat and of its weight. The rate at which it will absorb or emit heat depends upon these factors and upon its thermal conductivity and emissivity.

The figures in Table 12 appear to be the most probable values for the specific heats of refractory materials. They are based on the work of various observers, whose data do not agree with exactness.

Table 13. Weights of Refractory and Insulating Firebrick

Kind of Brick	Average Weights: Pounds per cubic foot	Average Weights: Pounds per 9-in. straight 9 × 4 1/2 × 2 1/2 in.
Chrome and chrome-magnesite	190–205	11.2–12.1
Fireclay (except superduty)	115–140	6.8– 8.2
Fireclay (superduty)	135–155	7.9– 9.1
Forsterite	150–165	8.8– 9.7
High alumina		
50–70% alumina classes	130–150	7.6– 8.8
80% alumina class	155–165	9.1– 9.7
90% alumina class	165–175	9.7–10.3
Insulating firebrick		
1600 F class	30– 34	1.8– 2.0
2000 F class	36– 40	2.1– 2.35
2300 F class	44– 48	2.6– 2.8
2600 F class	48– 54	2.8– 3.2
2800 F class	57– 60	3.4– 3.5
Magnesite and magnesite-chrome	170–195	10.0–11.4
Silica	102–115	6.0– 6.8

Table 14. Weights of Bulk Refractory Materials

Material	Weight per cubic foot for estimating, lb
Fine ground fireclay	100
Fine ground high-alumina clay	110
Plastic firebrick	
In place, undried	130
Super-plastic firebrick	
In place, undried	135
Calcined fireclay	
Lumps of approximately 1-in. diameter, including fines	95
Fine ground	95
Standard castable	
In place, dried	110
Super castable	
In place, dried	120
Chrome ore	
Lumps	170
Fine ground	160
Magnesite	
Dead-burned grains (burned-in)	170
Fine ground	155
Silica fireclay	100

RUBBER AND RUBBER-LIKE RESILIENT MATERIALS

By E. F. Riesing

When properly compounded, natural rubber and recently developed synthetic elastomers that are rubber-like in behavior provide the designer with engineering materials having specific properties to fill many exacting requirements. These materials will endure where no other material can qualify under the abusive action of repeated impact, continuous flexing, vibration, abrasion, and the destructive action of chemicals, oils, weather, ozone, etc.

15. ELASTOMERS AND COMPOUNDS

Various synthetic elastomers can be polymerized to produce tailor-made molecular structures. These polymer structures can be blended with other elastomers, including natural rubber, in varying percentages, making an almost infinite number of combinations of vulcanizates. All these vulcanizates can be sheeted, compression or injection molded, or extruded to practically any shape. They can be chemically or mechanically bonded or adhered to metallic or nonmetallic surfaces. They can be frictioned, skimmed, calendered, or extruded and adhered to textile or metallic fabrics or cords. They can be extended and reinforced with fillers and flocks. They can be permanently lubricated by the addition of various waxes, graphite, etc. Some elastomers can be pigmented to produce a variety of colors, including white. All elastomers have electrical insulating qualities, but some may be specially compounded to be conducting to various degrees, depending on type and quantity of carbon black filler. Further, the majority of the elastomers can also be processed into sponge structures.

From the foregoing, it is apparent that to provide technical data on all the infinite variations of structures with the different elastomers would require volumes. The data selected for presentation here therefore cover only a small practical range group that will be helpful to the design engineer for such average applications as bumpers, vibration insulators, engine and motor supports, dampers, suspension bushings, oil and grease seals, etc.

BASIC ELASTOMERS. The range group covered here includes three different Shore Durometer hardnesses (40 Durometer is a soft grade, 50 Durometer is a medium hardness, and 60 Durometer is a firm hardness) of five different basic elastomers. General properties of these, plus Thiokol, are listed in Table 1. These five basic elastomers are:

Natural rubber, a natural high polymer of isoprene coagulated from the latex of the *Hevea* tree. Natural rubber is not oil resistant, but it is the most outstanding elastomer of this entire group for elements where a minimum of energy absorption is desired and long flexing life is required.

Neoprene, a polymer of chloroprene of the monovinyl acetylene type. Neoprene compounds have the desirable characteristic of petroleum-base oil resistance in varying degrees, plus excellent resistance to acids, ozone, and flame. Neoprene compounds are used for oil-resistant hose, belting, gaskets, sheet packings, tubes, tapes, printing rollers, covering for electric wires and cables, pipe lines, etc.

Butadiene styrene, a copolymer of butadiene and styrene. Butadiene styrene copolymers have practically the same lack of resistance to oils and gasoline as natural rubber and, therefore, are definitely non-oil-resisting synthetics. Compounds of this elastomer are used extensively as a substitute for natural rubber in seals, non-oil-resisting gaskets, electric cable insulation, etc.

Butadiene acrylonitrile, a copolymer of butadiene and acrylonitrile. The butadiene acrylonitrile copolymers provide excellent resistance to petroleum-base oils or other organic fluids. Butadiene acrylonitrile compounds are widely used for oil and gasoline hose, tank linings, packings, gaskets, hydraulic piston seals, bullet sealing, gasoline tanks and hose, etc.

Butyl, a copolymer of isobutylene and small amounts of unsaturated hydrocarbons such as isoprene. Although butyl is essentially a non-oil-resistant elastomer, it does provide good resistance to some vegetable oils. However, it is most outstanding for its low rate of diffusion of gases and excellent resistance to acids and ozone. Compounds of this elastomer are recommended for application where maximum resistance to ozone and acids is required. It is extensively used for automobile-tire inner tubes because of its greater resistance to air permeability and also because of its excellent tear resistance.

PHYSICAL PROPERTIES. Table 2 gives the conventional physical properties of these elastomeric polymers and copolymers, as well as the SAE–ASTM grade numbers of these five compounded elastomers in three different Shore Durometer hardnesses.

(*Continued on p. 5-60*)

Table 1. General Properties of Commercial Synthetic Rubbers Produced in the United States

Classification	Oil-resistant Type			Non-oil-resistant Type		Natural Rubber Reference Standard
Elastomeric polymer	Butadiene acrylonitrile	Neoprene	Thiokol	Butadiene styrene	Butyl	Natural rubber
Rubber Reserve Co. designation	GR-A	GR-M	GR-P	GR-S	GR-I	
Government color designation	Orange	Yellow	Green	Red	Blue	White
Physical and Chemical Properties						
Specific gravity (pure gum)	1.00	1.25	1.35	0.94	0.91	0.93
Vulcanization method	Heat and sulfur	Heat alone *	Heat alone *	Heat and sulfur	Heat and sulfur	Heat and sulfur
Odor	Poor	Poor	Bad	Fair	Excellent	Good
Tensile strength, psi						
Pure gum	600	3500	300	400	3000	3000
Black reinforced	3500	3500	1500	3000	3000	4500
Tension modulus at 400% elongation						
Pure gum	200	400	450	200	300	400
Black reinforced	2500	2500	1500	2000	1000	2500
Elongation, %						
Pure gum		800	300		1000	700
Black reinforced	600	600	500	500	400	600
Hardness, durometer	Any range	25–90	35–90	Any range	28–85	Any range
Efficiency						
Rebound, hot	Good	Good	Poor	Good	Good	Excellent
Rebound, cold	Fair to good	Fair to good	Poor	Fair to good	Very poor	Excellent
Speed of recovery	Poor	Fair	Very poor	Good	Very poor	Excellent
Flexometer, (heat build-up)	Medium	Medium	High	High	High	Low
Tear resistance	Fair	Fair to good	Fair	Poor to fair	Good	Good
Abrasion resistance	Good	Excellent	Poor	Good	Good	Excellent
Aging resistance in:						
Sunlight (ozone)	Poor	Excellent	Good	Poor	Excellent	Poor
Oxidation (oxygen)	Fair	Good	Good	Good	Good	Good
Heat	Good	Very good	Poor	Fair	Good	Fair
Static (shelf aging)	Good	Good	Good	Good	Good	Fair

Resistance to flex cracking						
Extension flexing	Fair	Good	Poor	Poor to fair †	Good	Good
Bending flexing	Good	Good	Poor	Good	Very good	Good
Resistance to compression set (or permanent set)	High	Medium	Very low	Medium	Medium	High
Solvent resistance						
Hydrocarbon	Good	Fair to good, except aromatics	Excellent	Poor	Poor	Poor
Oxygenated solvents	Poor	Poor	Good	Good	Good	Good
Oil and gasoline resistance	Excellent	Good	Excellent	Poor to fair	Poor	Poor
Acid resistance	Good	Excellent	Fair	Good	Excellent	Excellent
Flame resistance	Poor	Good	Poor	Poor	Poor	Poor
Cold resistance (freezing or hardening upon cooling)	Fair	Poor	Poor	Fair to good	Fair	Good
Permeability to gases	Medium	Low	Low	Medium	Extra low	Medium
Volume resistivity ohm/cu cm at 25 C	10^{10}	10^{11}	10^{13}	10^{15}	10^{17}	10^{17}–10^{9}
Dielectric constant at 1000 cycles	13.0	9.0	6.0	2.9	2.4	2.7
Power factor at 1000 cycles	5.5	3.0	....	0.32	0.3	0.23
Resistance to hot-water swelling	Excellent	Good	Excellent	Excellent	Excellent	Poor to fair
Adhesion to woven fabric	Fair	Excellent	Fair	Fair	Good	Good
Reclaimability	Yes	Yes	Questionable	Yes	Yes	Yes
PROCESSING CHARACTERISTICS						
Mixing efficiency	50% of natural rubber	Equal to natural rubber	75% of natural rubber	85% of natural rubber	90% of natural rubber	100%
Extruding	50% of natural rubber	75% of natural rubber	75% of natural rubber	90% of natural rubber	90% of natural rubber	100%
Calendering	50%	90%	50%	90%	100%	100%
Solvents	Methyl ethyl ketone, ethylene dichloride	Aromatic or petroleum hydrocarbons	Trichloroethylene	Petroleum hydrocarbons	Petroleum hydrocarbons	Petroleum hydrocarbons
Solutioning	Slow in special solvents	Fair to good in special solvents	Slow in special solvents	Good	Good	Good
Spreading	Fair	Excellent	Fair	Excellent	Excellent but hard to dry	Excellent
Cohesive properties	Poor	Good	Poor	Fair	Good	Excellent
Molding characteristics	Fair to excellent, depending on type	Good	Good	Fair	Good	Very good

* Metallic oxides required for commercial cures.
† Owing to poor tear resistance.

Table 2. Conventional Physical Properties of Five Elastomers

Elastomer	Natural Rubber	Neoprene	Butadiene Styrene	Butadiene Acrylonitrile	Butyl
40 SHORE DUROMETER GROUP					
SAE–ASTM grade number	RN-430-ABF	SC-425-ABEF	RS-415-ABF	SB-410-ABEF	RS-420-AB
Ultimate tensile at 70 F, psi	3075	2800	1575	1325	2400
Elongation at rupture at 70 F, %	690	880	800	460	660
Shore Durometer hardness at 70 F	41	41	40	39	40
Compression set,* "A" (constant load)	15.1	14.6	12.5	9.0	12.7
Compression set,* "B" (constant deflection)	22.6	22.1	18.5	27.4	14.8
Specific gravity	1.05	1.33	1.09	1.13	1.04
Vulcanization shrinkage, %	2.44	1.99	2.30	2.45	1.85
50 SHORE DUROMETER GROUP					
SAE–ASTM grade number	RN-530-ABF	SC-520-ABEF	RS-518-ABF	SB-520-ABEF	RS-525-AB
Ultimate tensile at 70 F, psi	3650	2175	2050	2200	2975
Elongation at rupture at 70 F, %	640	740	710	510	710
Shore Durometer hardness at 70 F	52	49	50	49	51
Compression set,* "A" (constant load)	12.0	20.2	14.6	7.2	13.5
Compression set,* "B" (constant deflection)	19.8	30.6	16.9	26.9	18.0
Specific gravity	1.11	1.37	1.10	1.21	1.085
Vulcanization shrinkage, %	2.16	1.61	2.13	1.91	1.66
60 SHORE DUROMETER GROUP					
SAE–ASTM grade number	RN-630-ABF	SC-630-ABEF	RS-625-ABF	SB-617-ABEF	RS-625-AB
Ultimate tensile at 70 F, psi	3450	3275	2850	2250	2725
Elongation at rupture at 70 F, %	610	690	720	410	610
Shore Durometer hardness at 70 F	61	60	59	59	58
Compression set,* "A" (constant load)	10.1	11.2	13.7	21.8	14.2
Compression set,* "B" (constant deflection)	20.0	24.6	20.0	24.0	19.3
Specific gravity	1.21	1.34	1.18	1.24	1.19
Vulcanization shrinkage, %	2.01	1.72	1.91	1.71	1.04

* Compression set is the residual decrease in thickness, expressed as a percentage of the original thickness, of a test specimen measured 30 min after removal from a loading device in which it had been subjected for a definite time to compressive deformation under specified conditions of load application and temperature. (See ASTM D395-47T.)

SAE–ASTM CLASSIFICATION grade number has been used to identify each of the 15 compounds in Table 2. The key to the SAE–ASTM grade number system given in Table 3 explains prefix letters, hardness range number, tensile numbers, and suffix letters. The SAE–ASTM classification simplifies the writing of specifications for a wide range of compounds and conditions. Table 2 illustrates three elastomers in the "Type R" non-oil-resistant solid rubber classification, and two elastomers in the Type S oil-resistant solid rubber classification. For example, the designation **RN-430-ABF** indicates (Table 3) a rubber of grade 40 Durometer and 3000 psi tensile requirement, with ovenaging requirement, low compression set, and flexibility down to −40 F.

Space does not permit presenting properties of sponge rubbers referred to in the classification chart, nor has any attempt been made to tabulate properties of either the large group of rubber-like flexible plastics or the "Type SA" group of maximum oil-resistant synthetics, such as Thiokol. (Thiokol is an organic polysulfide polymer available in a variety of different types. Its general properties are given in Table 1.) Both these groups lack the resilience characteristics needed in parts under appreciable mechanical action. Some of the materials in these groups are particularly resistant to aromatic fuels; however, as a class they are subject to high permanent compression set and will not resist elevated operating temperatures.

The conventional physical property data listed in Table 2, and the SAE–ASTM classification in Table 3 may provide enough information to enable the engineer to make his selection of the proper resilient material for simple applications, working at room temperature. However, for special design work it is frequently required to know more about such

Table 3. Key to SAE–ASTM Classifications and Grade Numbers

CLASSIFICATION *

Type R (non-oil-resistant)
- **RN**—Rubber or reclaim
- **RS**—Synthetic rubber

Type S (oil-resistant synthetic)
- **SA**—Maximum oil resistance
- **SB**—Good oil resistance
- **SC**—Medium oil resistance

GRADES

Solid Rubber

Grade	Durometer	Grade	Tensile
3	30 ± 5	**00**	Not req'd
4	40 ± 5	**02**	200 psi
5	50 ± 5	**03**	300 psi
6	60 ± 5	**04**	400 psi
7	70 ± 5	**05**	500 psi
8	80 ± 5	**06**	600 psi
9	90 ± 5	**07**	700 psi
		08	800 psi
		09	900 psi
		10	1000 psi
		14	1400 psi
		15	1500 psi
		15–6 †	1500 psi
		17	1700 psi
		18	1800 psi
		20	2000 psi
		25	2500 psi
		30	3000 psi
		35	3500 psi

Sponge Rubber

	Grade	Psi ‡
Chemically blown	**11**	3 1/2 ± 1 1/2
	12	7 ± 2
	13	11 ± 2
	14	15 ± 2
	15	20 1/2 ± 3 1/2
Latex foam, cored	**21**	0.19 ± 0.09
	22	0.35 ± 0.09
	23	0.57 ± 0.13
	24	0.85 ± 0.15
	25	1.17 ± 0.17
	26	1.53 ± 0.19
	27	1.93 ± 0.21
	28	2.37 ± 0.23
Latex foam, uncored	**31**	0.50 ± 0.30
	32	0.80 ± 0.40
	33	1.45 ± 0.75
	34	2.50 ± 1.10
	35	3.50 ± 1.30
	36	4.60 ± 1.40
Closed cell	**41**	2 to 5
	42	5 to 9
	43	9 to 13

SUFFIX LETTERS

- **A** —Oven-aging requirement
- **B** —Low compression set requirement
- **C** —Weather-resistance requirement
- **D** —Load-deflection requirement
- **E** —Oil-aging requirement
- **F** —Flexible at −40 F
- **FF**—Flexible at −70 F
- **G** —Maximum resistance to tear
- **H** —Maximum resistance to severe flexing
- **J** —Superior abrasion resistance
- **K** —Adhesion requirement
- **L** —Maximum resistance to moisture absorption
- **M**—Maximum inflammability resistance
- **N**—Maximum resistance to repeated impact
- **P**—Minimum staining effect of enamel and lacquer
- **Z**—An additional laboratory test over and above those designated by the suffix letters listed above, or life or performance requirements

* Classification of the compound is determined by the predominating basic elastomer therein.
† A suffix number following the grade designates a special elongation figure.
‡ Pounds per square inch required to compress original thickness 25%, based on 50 sq in. Scott indenter foot.

additional properties as moduli of elasticity of the compound in compression, tension, or shear; also the effect that high and low temperatures have in influencing these moduli values. It must be emphasized that the moduli values of a particular rubber manufacturer's compound that meets the SAE–ASTM conventional physical properties for a given grade and classification may, and in all probability will, vary widely from those of some other rubber manufacturer because different reinforcing materials, organic accelerators, etc., may have been used.

Therefore, in the limited space available, an attempt will be made to explain some of the properties beyond the conventional physicals, and some of the fundamentals of rubber design.

16. DESIGN OF RUBBER PARTS

COMPRESSION DESIGN. For the majority of designs for insulating impact shock, or for isolating vibration and reducing the transmission of noise, rubber-like resilient materials are utilized predominantly in compression. Most desirable results and longest life are obtained by bonding the rubber to plates, thereby preventing slippage and destructive abrasion of the load-carrying surfaces.

Designs employing rubber in compression should be calculated so as to limit the deflection under most working loads to 15 or 20% of the free unloaded height of rubber thickness;

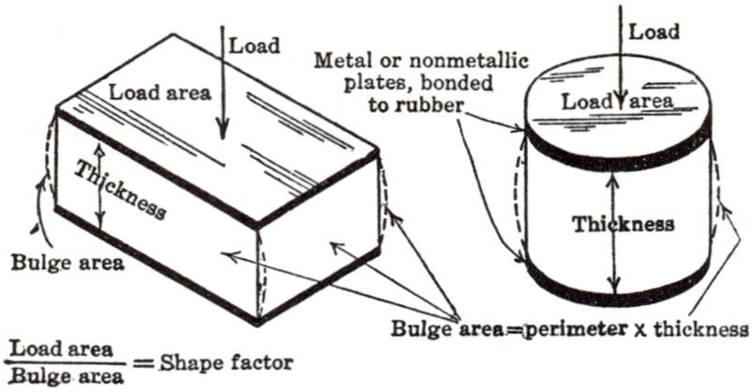

FIG. 1. Load area and bulge area of compression mounting.

10% is recommended. Constant high loadings causing deflections greater than 20% induce excessive permanent set and drift in the compound over long periods of time. It should also be noted that, for practical purposes, all resilient elastomers may be considered to have no volume compressibility. Actually, they are less compressible than water. To

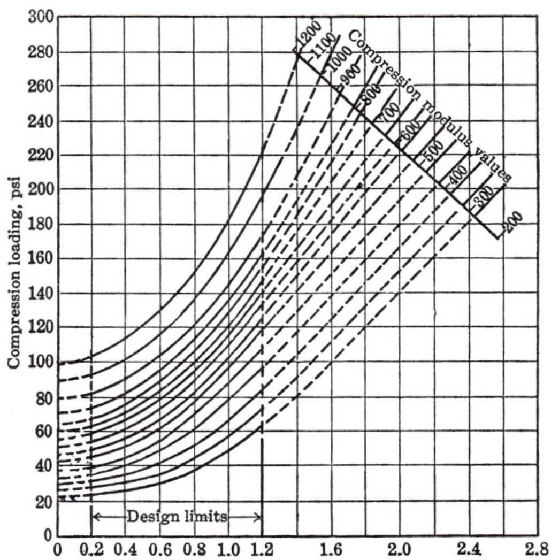

FIG. 2. Chart for determining compression modulus of elastomers for bonded compression designs of uniform thickness, for 10% deflections.

obtain resilience, therefore, the material must not be confined but must be allowed to bulge at the surfaces normal to the load-carrying area.

Since the compression modulus of various elastomers is so dependent on the shape of the sample tested, it is imperative that the designer keep the dimensions of the compression pad or spring under consideration to within a "shape factor" of 0.2 to 1.2. This shape factor is the ratio of *load area* to *bulge area*, Fig. 1. To facilitate calculations, Figs. 2 and 3 have been plotted for bonded compression designs for 10% and 20% deflections, respectively. These charts can be used as follows:

With the load to be supported known, the load-carrying area of the design may be determined by assuming a loading within the range of 50 to 200 psi. Thus the load to be supported, in pounds, divided by the compression loading, in pounds per square inch, equals the load area in square inches. A thickness must then be assumed. From the selected thickness and with the load area determined, the shape factor can readily be established (see Fig. 1). The intersection of the shape factor and the assumed compression loading on Fig. 2 or 3 establishes the compression modulus value of the compound required.

Compression moduli values are given in Table 4 for the 15 different compounds listed in Table 2. These values apply for deflections up to 20% compression and for static loadings as well as for dynamic loadings up to 150 cycles per minute. Compression moduli values for dynamic loadings above 150 cycles per minute should be obtained by contacting the technical division of any large rubber manufacturer. The effect of temperature on compression modulus should be noted particularly in Table 4.

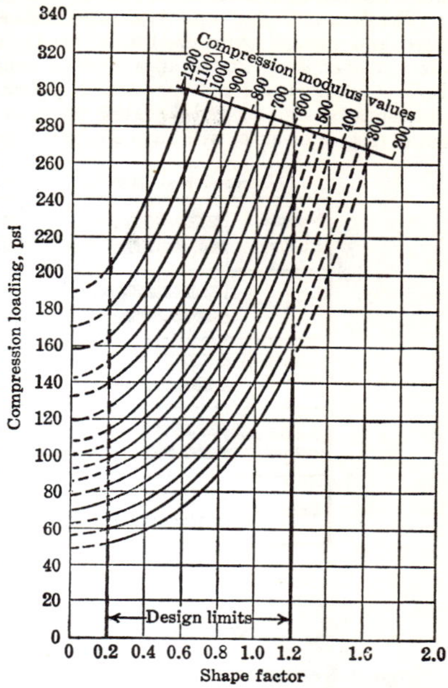

Fig. 3. Chart for determining compression modulus of elastomers for bonded compression designs of uniform thickness, for 20% deflections.

Example. It is required to design a bonded compression pad to support a load of 1200 lb. Location of pad allows a thickness of 1 in., and a deflection of 20% is desired. Assuming a compression loading of 150 psi, the load area becomes 1200/150 = 8 sq in.

For a *circular disk* pad, the diameter = $\sqrt{8 \times 4/\pi}$ = 3.2 in.; hence the bulge area = $\pi \times 3.2 \times 1$ = 10 sq in. The shape factor is therefore 8/10 = 0.8, and, from Fig. 3, the elastomer must have a compression modulus of 510 psi.

For a *square* pad, the side dimension = $\sqrt{8}$ = 2.83 in.; hence the bulge area = $2.83 \times 4 \times 1$ = 11.3 sq in. The shape factor is therefore 8/11.3 = 0.71, and, from Fig. 3, the elastomer must have a compression modulus of 550 psi.

For a *rectangular* pad where the length of the pad is twice the width, the dimensions should be 2×8 = 8; hence the bulge area = $2 \times (2 + 4) \times 1$ = 12 sq in. The shape factor is 0.667, and, from Fig. 3, the elastomer should have a compression modulus of 590 psi.

It should be noted that, as the perimeter of the pad keeps increasing, providing a larger bulge area, the necessary modulus increases if the same deflection is to be maintained.

Table 4. Compression Modulus, psi

(Up to 20% compression)

Temperature	Natural Rubber	Neoprene	Butadiene Styrene	Butadiene Acrylonitrile	Butyl
		40 Durometer			
158 F	400	470	360	212	340
70 F	405	475	415	270	360
0 F	410	480	810	2100	420
		50 Durometer			
158 F	540	690	480	340	410
70 F	635	700	620	500	460
0 F	690	900	880	3250	660
		60 Durometer			
158 F	990	900	660	460	540
70 F	1050	915	800	685	620
0 F	1250	940	1350	4300	900

The charts may be employed in reverse process also; that is, with the compression modulus known for a definite compound, the desirable compression loading in pounds per square inch can be established. Interpolations can also be made on the charts. Results should be accurate within 10 to 15%.

The outstanding feature of Figs. 2 and 3 is that the compression modulus values obtained will apply for any of the natural rubber and synthetic rubber compounds listed in the SAE–ASTM classifications of RN, RS, SC, and SB. The compression modulus value obtained by the foregoing method, along with information as to the type of service to which the compression pad or spring will be placed, the service temperature, whether oil or non-oil conditions exist, whether the design is subject to impact or occasional overloads, etc., will make it possible to procure the properly compounded elastomer.

The data in Figs. 2 and 3 resulted from a study of a large number of rubber compression pads and springs vulcanized of a variety of compounds of different elastomers, in various sizes, thicknesses, and shapes such as square, rectangular, and cubical, and in solid and hollow cylindrical forms bonded to plates. Tests have proved that similar compression elements bonded to metal plates, made of the identical compound and vulcanized to the same degree of cure, will have the same stress-strain curves up to 20% strain.

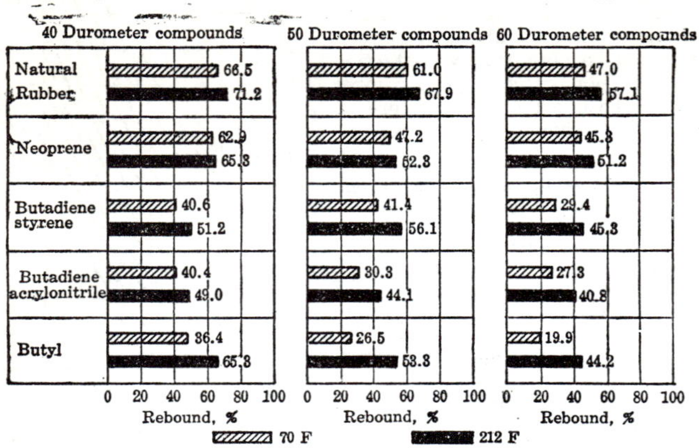

FIG. 4. Impact resilience of rubber and synthetic compounds.

Impact Resilience. Bumper designs or compression pads subject to periodic impact warrant the investigation by the designer of the impact resilience values. Resilience can be determined by several methods. The ball rebound device is an excellent, easy way to draw comparisons between various compounded elastomers and can be employed as an inexpensive production method of inspection on some shapes of the finished molded or extruded vulcanizate. Simple energy considerations based on the theory of free vibration show that the ball rebound device permits calculation of a factor which can be called impact resilience. This value is expressed by dividing the first rebound height H_1 by the distance of the free fall H_0.

$$\frac{H_1}{H_0} \times 100 = \text{Impact resilience, \%}$$

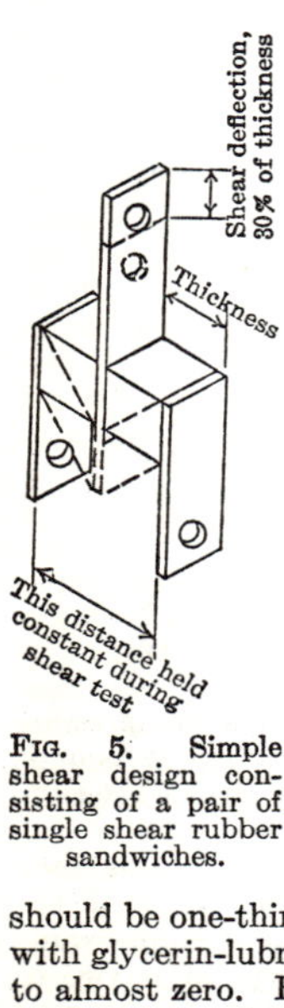

FIG. 5. Simple shear design consisting of a pair of single shear rubber sandwiches.

Figure 4 shows, in bar graph form, the impact resilience values at 70 F room temperature and the change in these values caused by the elevation in temperature to 212 F for the same 15 compounds listed in Table 2.

The hysteresis or internal friction of a compounded elastomer is an important factor when considering load-carrying insulators that are subject to vibrations of variable impressed frequencies that may force the resilient load-carrying insulator through a short-interval resonance period. Hysteresis is a function of the energy absorbed by the resilient compounded elastomer when forces are applied to it. It may be measured by the area of the "hysteresis loop" of the stress-strain extension-retraction diagram, or by the decay of free vibrations. The hysteresis value varies with temperature and frequency.

SHEAR DESIGNS. Designs requiring large deflections and low natural frequency oscillations cannot be satisfied by stressing the elastomers in compression. Designs of this type should utilize the resilient material by stressing it in simple shear when bonded to metal plates. Rubber is most compliant when used in simple shear, since the modulus of elasticity in shear is approximately one-sixth the compression modulus when bonded to plates. Theoretically the shear modulus should be one-third compression modulus for small deformations. This is based on studies with glycerin-lubricated compression surfaces on which the coefficient of friction is reduced to almost zero. However, experimental data obtained under conditions of 20% compression deflection between bonded plates have shown that this ratio is close to 1/6.

The basic simple form for this design is a pair of single shear sandwiches, such as pictured in Fig. 5. If greater deflections are necessary than can be obtained by a pair of single shear sandwiches, the designer may resort to a double pair or a quadruple pair. For a pair of single shear sandwiches, the following formula applies:

$$d = \frac{WT}{AG_s}$$

Also

$$\text{Shear rate per inch} = \frac{W}{d} = \frac{AG_s}{T}$$

where d = shear deflection, inches; W = load, pounds; T = free molded thickness, inches (not laterally compressed thickness); A = twice the area, square inches, at smallest cross section; and G_s = shear modulus, pounds per square inch (see Table 5).

Note: Correction must be made in the shear modulus value where the ratio d/T is more than 0.3. The value of G_s must be multiplied by 0.9 where d/T is 0.35 to 0.8; and it must be multiplied by 0.833 if the value of d/T is 0.8 to 1.2.

Table 5. Shear Modulus at 30% Deflection

(See Fig. 5)

Temperature	Natural Rubber	Neoprene	Butadiene Styrene	Butadiene Acrylonitrile	Butyl
		40 Durometer			
158 F	58	64	42	30	58
70 F	59	65	49	37	59
0 F	61	69	65	100	60
		50 Durometer			
158 F	82	97	80	44	70
70 F	86	107	104	71	71
0 F	90	145	139	183	78
		60 Durometer			
158 F	132	120	98	69	86
70 F	134	127	126	110	90
0 F	147	153	183	257	93

Where square, rectangular, or circular shear sandwich constructions are used, the static shear stress limit should be 20 to 30 psi of the smallest bonded area. This applies to engine mountings, motor supports, resilient wheels, and rubber springs of various types where permanent set in the rubber is an important consideration. In designs of this type, stress relief at the edges of the rubber should always be worked into the design by the incorporation of fillets in the rubber or radii on the edge of the metal part or both, inasmuch as most fatigue failures start at the edge of the bond. It should be noted here that lateral compression of the rubber in the shear sandwich is frequently resorted to by the designer in order to facilitate anchoring of the metal elements to which the rubber in shear is bonded. This lateral compression changes the shear characteristics of the design only slightly, for lateral compression loadings ranging from 10 to 50 psi, with shear calculations based on the original free thickness between the bonded plates. Table 5 gives moduli of elasticity values in shear for 30% deflection when load is applied parallel to the bonded surfaces, for the 15 compounds referred to in Table 2. Here, again, the effect of temperature is shown in influencing the shear rate or modulus.

The effect of high and low temperatures on the stiffness or modulus of natural and synthetic rubbers cannot be minimized. This must be fully recognized by the design engineer for resilient applications operating at temperatures above and below room temperatures. At temperatures of 0 F and lower, some of the elastomers rapidly begin to lose their elasticity, some actually crystallize, and others harden to a point where the material is not serviceable. At temperatures of 158 F and above, aging effects take place rapidly and shorten the service life. Both the tension modulus and the Young's modulus values illustrate outstandingly this temperature influence. Table 6 shows the change of ultimate tensile strength and the change in elongation at rupture for temperatures of 70, 158, and 212 F for the 15 compounds.

The reaction of these elastomers to temperatures above and below room temperatures as indicated in the tabulation is not to be confused with the "Joule effect." The Joule effect phenomenon occurs while the elastomer is under strain; an increase in temperature tends to contract or tighten up the molecular structure, and lowering the temperature causes an opposite condition of relaxing the structure.

Table 6. Effect of Temperature Change on Tensile Strength and Elongation at Rupture

Temperature	Physicals at Break	40 Durometer					50 Durometer					60 Durometer				
		Natural Rubber	Neoprene	Butadiene Styrene	Butadiene Acrylonitrile	Butyl	Natural Rubber	Neoprene	Butadiene Styrene	Butadiene Acrylonitrile	Butyl	Natural Rubber	Neoprene	Butadiene Styrene	Butadiene Acrylonitrile	Butyl
+212 F	Tensile, psi	1500	125	275	475	250	2125	475	275	400	675	2225	650	800	1025	1075
	Elong., %	720	200	275	297	250	730	430	300	230	580	640	320	390	340	595
+158 F	Tensile, psi	2400	275	550	625	450	2850	800	525	575	525	2350	1325	1400	1475	1750
	Elong., %	730	330	400	310	410	700	520	420	280	725	540	430	520	370	700
+70 F	Tensile, psi	3075	2800	1575	1325	2400	3650	2175	2050	2200	2975	3450	3275	2850	2250	2725
	Elong., %	690	870	800	460	660	640	740	710	510	715	610	690	725	410	610

The effect of low temperature on the 15 compounds in Table 2 can best be illustrated by showing the stiffening effect on the Young's modulus of elasticity, see Figs. 6, 7, and 8. Young's modulus is calculated by means of the beam bending formula:

$$E = \frac{Wl^3}{4dbh^3}$$

where E = Young's modulus, pounds per square inch; W = load producing a deflection, pounds; d = deflection produced by load W, inches; l = distance between center lines of

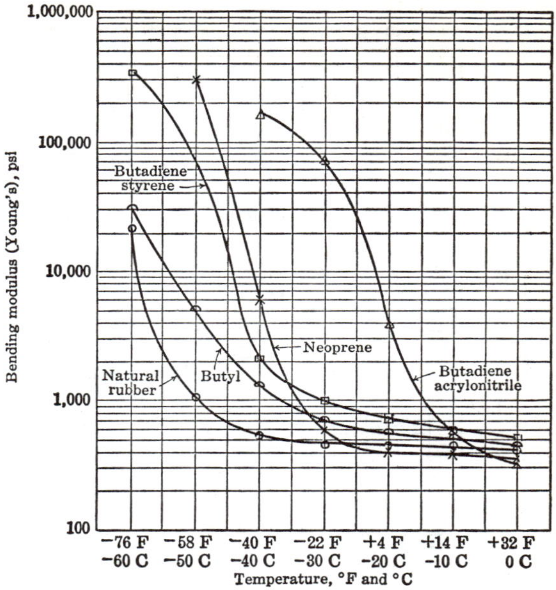

FIG. 6. Effect of temperature on elastic modulus for compounds of 40 durometer hardness.

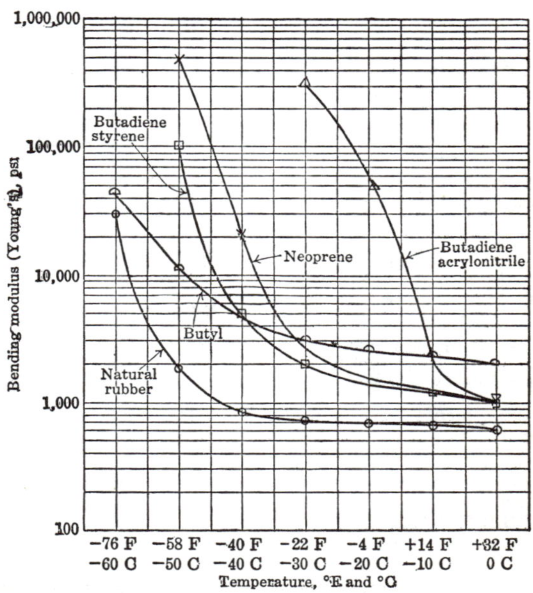

FIG. 7. Effect of temperature on elastic modulus for compounds of 50 durometer hardness.

sample supports, inches; b = width of sample, inches; and h = thickness of sample, inches.

A cylindrical rubber spring for loading parallel to the longitudinal centerline is illustrated in Fig. 9. Deflection of this type of spring may be calculated from the following formula:

$$d = \frac{W}{2\pi G_s}\left(\frac{r_2 - r_1}{h_2 r_1 - h_1 r_2}\right)(\log_e h_2 r_1 - \log_e h_1 r_2)$$

where d = deflection, inches; W = load, pounds; h_1 = height of bonded area on inside tube, inches; h_2 = height of bonded area on outside tube, inches; r_1 = radius of bonded

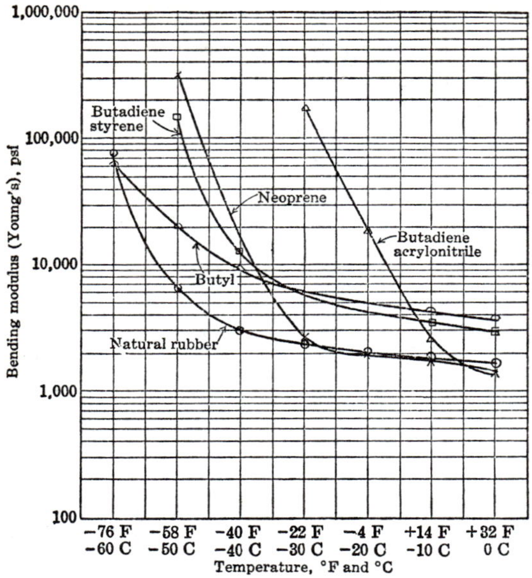

Fig. 8. Effect of temperature on elastic modulus for compounds of 60 durometer hardness.

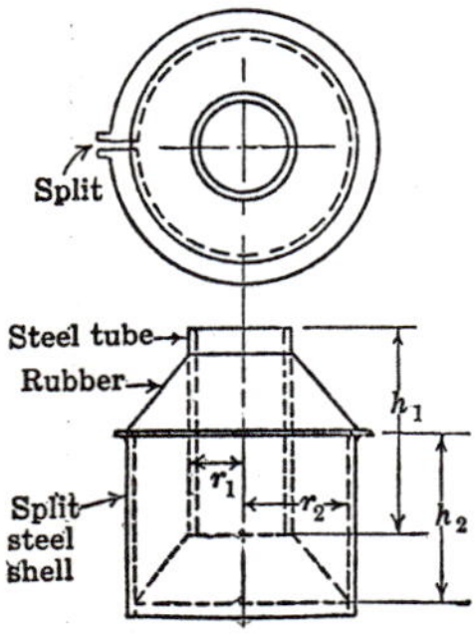

Fig. 9. Cylindrical rubber spring for loading parallel to longitudinal center line.

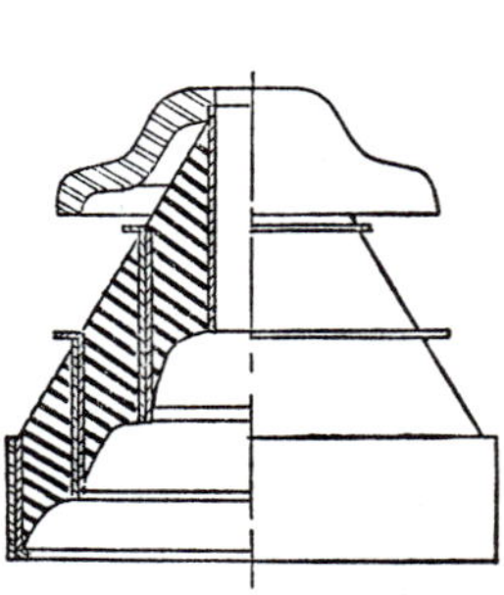

Fig. 10. Three-stage cylindrical rubber spring.

area on inside tube, inches; r_2 = radius of bonded area on outside tube, inches; and G_s = modulus of elasticity of rubber in shear, pounds per square inch.

A three-stage cylindrical rubber spring is shown in Fig. 10. Deflection of the separate stages may be calculated by the foregoing formula, the total deflection being the sum of the separate deflections.

TORSION RUBBER BONDED TYPE BUSHING. In this design, Fig. 11, rubber is bonded to inner and outer tubular concentric metals. The ends of the rubber in these designs should be contoured to relieve concentration of stress at the smallest bonded area, the length of the rubber being greater on the small tube than on the outer. The rubber should preferably be under radial compression in the final assembly, even if only to 3% radial compression.

The following formula can be used for design of rubber torsion bushings or springs for ratios of thickness of rubber to shaft diameter of 0.3 to 0.7 (see Fig. 11):

$$\theta = \frac{Wl}{4\pi BG_s}\left(\frac{1}{r_1^2} - \frac{1}{r_2^2}\right)$$

where Wl = torque, inch-pounds; θ = angular deflection, radians; B = effective length of rubber, inches; r_1 = radius of inside bonded area of rubber, inches; r_2 = radius of outside bonded area of rubber, inches; and G_s = modulus of elasticity in shear, pounds per square inch.

Two factors that must be considered in the design of rubber torsion springs are the stress on the bonded area and the strain in the rubber. Referring to Fig. 11, for a given

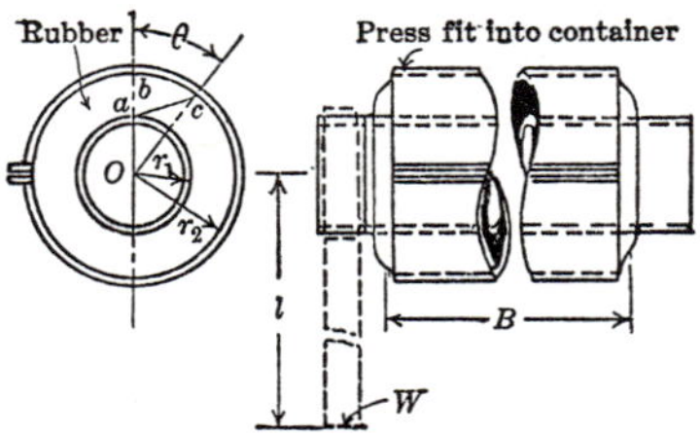

FIG. 11. Bonded rubber torsion spring.

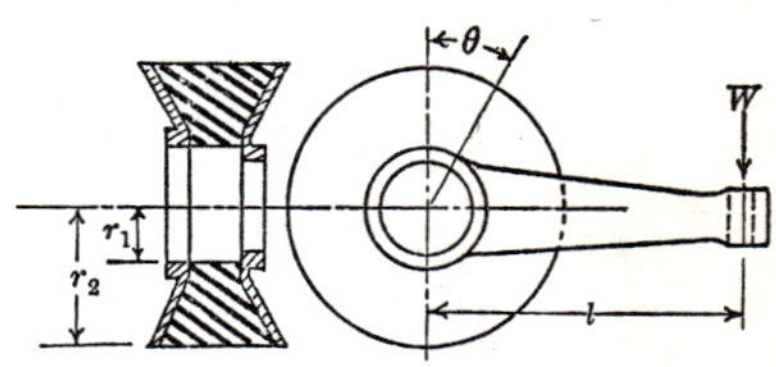

FIG. 12. Modified torsion spring.

angular movement θ of the torsion bushing, the strand of rubber ab becomes approximately ac, and the ratio of stretch to initial length is $(ac - ab)/ab$. The length ac, from the geometry of the figure, is equal to

$$ac = \sqrt{r_1^2 + r_2^2 - 2r_1r_2 \cos \theta}$$

If ab is small in proportion to r_1, the strain in the rubber for an angular movement θ will be excessive, even though the stress on the bonded area of the inner metal shell is low. Conversely, if ab is large in proportion to r_1, the strain in the rubber will be low, even though the stress on the bonded area is high. Hence, in the design of these bushings, it is best to limit the strain in the rubber to approximately 80% and the bond to approximately 110 psi of the area of the inner shaft under normal static loading.

A modified torsion spring is illustrated in Fig. 12. The formula for angular deflection of this type is

$$\theta = \frac{2WlT}{G_s(r_2^4 - r_1^4)}$$

where θ = angular deflection, radians; Wl = torque, inch-pounds; T = mean thickness of rubber, inches; r_1 = inside radius of rubber, inches; r_2 = outside radius of rubber, inches; and G_s = modulus of elasticity of rubber in shear, pounds per square inch.

SILICONES

By Arthur M. Weinberg

The silicones are a group of versatile synthetics developed through research in the field of organo-silicon chemistry. They are semiorganics combining the most desirable virtues of organic and inorganic substances. They possess, to a high degree, immunity to heat, cold, chemicals, and weathering.

Silicone products now being produced are classed as *water repellents*, *resins*, *oils and greases*, and *rubber*. These compounds differ chemically only in the number and arrangement of the organic groups attached to the —Si—O—Si— backbone.

WATER REPELLENTS. The water repellents are a group of liquids which enable surfaces of materials such as ceramics, paper, fabrics, plastics, glass, quartz, and leather to shed water. On many surfaces they produce a far higher degree of water repellency than can be obtained with any other materials. A typical repellent has the following advantages in comparison to wax coatings for treating steatite and porcelain insulators: (1) freedom from plugging small holes; (2) resistance to solvents, permitting cleaning after assembly operations; (3) cleaning of soldering fluxes possible without damaging coating; (4) superior electrical properties under long humidity cycles.

RESINS. Silicone resins resemble closely conventional resins in appearance but combine high thermal stability at machinery operating temperatures, improved dielectric properties, and improved resistance to moisture and severe chemical conditions. Their greater thermal stability makes possible a reduction in size and weight of equipment without a corresponding reduction in service rating; an increased motor output where operating temperatures can be raised; longer life at normal operating temperatures; and a much greater safety factor where severe overloads sometimes occur.

Silicone resins are baked at slightly higher temperatures than ordinary resins and are furnished with ordinary commercial solvents. They are particularly valuable as coatings, impregnants, and bonding agents where extremely high (482 F) or low (−58 F) temperatures are encountered. Table 1 shows characteristics of a typical silicone resin, and Table 2 shows chemical resistance of three silicone resins.

Table 1. Typical Silicone Resin

(Data from General Electric Co.)

Base, %	58–60
Color	Light straw
Drying time, 3-mil film on aluminum panel	
Room temperature	10 min
212 F	5 min
302 F	5 min
392 F	5 min
Rocker hardness (bake 2 hr, 302 F)	34
Gloss	Very good
Craze life, 482 F, glass tape	720 hr
Flexibility, 482 F, 3-mil film, iron plate, 1/8-in. mandrel	96 days
Recommended cure * (thin films)	
347 F	24 hr
392 F	8–10 hr
482 F	2–4 hr
Tensile strength, room temperature, cure 48 hr at 482 F	1800 psi

* A precure of 1/2–1 hr at 257–302 F to eliminate bubbling is recommended.

Table 2. Typical Chemical and Solvent Resistance of Silicone Resins

(Data from General Electric Co.)

	Resin A	Resin B	Resin C
HCl (conc)	Unattacked	Unattacked	Unattacked
HNO_3 (conc)	Unattacked	Darkens after 2–6 hr	Darkens after 2–6 hr
H_2SO_4 (conc)	Attacked	Attacked	Attacked
NaOH (10%)	Unattacked	Unattacked	Unattacked
NH_4OH (conc)	Unattacked	Unattacked	Unattacked
CCl_4	Softens and swells	Softens	Softens and swells
Oil (25 C) (No. 10 transil)	Softens	Softens slightly	Unattacked
Oil (100 C) (No. 10 transil)	Softens	Softens slightly	Softens slightly

Silicone resin-based protective and decorative finishes are unusually weather resistant, durable under high temperatures, and generally impervious to chemicals. Pigments are dispersed in silicone resins with ball or roller mills. Metal salts such as cobalt, iron, and zinc are added to catalyze curing, and common aromatic and aliphatic solvents thin them for more convenient application. Coatings are applied by spraying, rolling, dipping, or knifing, and are baked or cured at 400 F from 30 to 60 min. A single silicone paint may resist both high and low temperatures (−60 F to +500 F), withstand exposure to both acids and alkalies, and be both hard and flexible.

OILS AND GREASES. Silicone oils are clear, water-white liquids made in viscosities ranging from light machine oils to oils of semisolid consistencies. They possess three outstanding properties: low flow point (−121 F to +130 F); high-temperature stability depending on conditions (300 to 500 F); minimum change in viscosity from low to high temperatures. (See Fig. 1, p. 12-62.) These properties make silicone oils of interest as hydraulic fluids for aircraft, machine tools, fluid couplings, hydraulic brakes, and clutches, as manometer fluids, as cooling or heating baths, and, with some limitations, as lubricants. High solid emulsions of silicone oil (85%) are available for mold release applications and are particularly suitable for use in the rubber industry for tires, heels and soles, and mechanical rubber goods.

The silicone oils are inert in contact with common metals and plastics up to 300 F and have excellent electrical properties which recommend them as dielectric fluids for transformers and capacitors. (See Table 3.) These oils are available in viscosities ranging

Table 3. Some Properties of Silicone Oils

(Data from General Electric Co.)

Surface tension, dynes/cm	20
Solubility of air in oil at 82 F, by volume, %	21
Specific heat at 80 F, Btu/lb	0.47
Thermal conductivity; Btu/hr/sq ft/°F/ft	0.13
Volatile at 150 C, NV grade, %	0.5
Viscosity-temperature coefficient	0.6
Refractive index at 68 F	1.40
Dielectric constant at 60 cycles	2.58
D-c volume resistivity, ohms/cu cm	7.9×10^{14}
A-c volume resistivity at 760 cycles, ohms/cu cm	6×10^{13}
Power factor at 60 cycles	0.0002
Sound velocity in 70 NV oil at 6.5 mc, meters/sec *	984
Compressibility in 70 NV oil at 6.5 mc, sq cm/dyne *	1.05×10^{-10}

* Data courtesy of Dr. D. L. Arenberg, *Report* 932, NDRC Radiation Laboratory, Massachusetts Institute of Technology, Cambridge, Mass.

from 4 to 1000 centistokes at 100 F. (See Fig. 1, p. 12-62.) Intermediate viscosities between the standards are obtained by blending the oils. Oxidation inhibitors can be added to the oil to extend its temperature range and life.

The silicone oils are remarkably stable to heat and oxidation and remain free from discoloration, acid formation, sludging, and similar phenomena in the presence of air at temperatures as high as 300 F. Copper, bronze, brass, aluminum, magnesium, steel, tin, cadmium, and chromium are not affected by the oil. Silicone oils are miscible in petroleum ether, naphthas, ether, benzene, toluene, xylene, chloroform, carbon tetrachloride, amyl alcohol, and ethyl acetate; partially miscible with methanol, ethanol, propanol, butanol, and acetone; and immiscible with water and petroleum oils. Silicone oils have been added to paints and petroleum oil in extremely small quantities to prevent foaming. They have also been useful in preventing floating, streaking, and flocculation in paints and enamels.

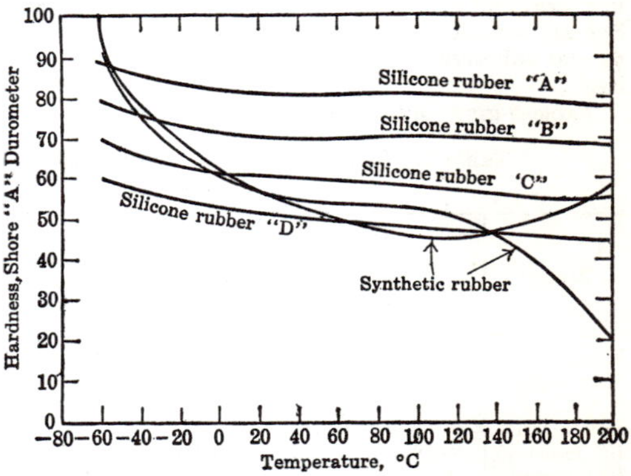

FIG. 1. Hardness vs. temperature change for typical silicone rubbers, compared with synthetic rubbers.

Silicone greases with semilubricating qualities have been developed for lubricating plug cocks and valves. The greases are stable over temperatures ranging from −70 to +400 F.

SILICONE RUBBER. Silicone rubber can be held continuously at temperatures of 300 to 350 F with little increase in surface hardness or decrease in flexibility and resilience. (See Fig. 1.) It does not soften and become sticky at 520 F and can be held at this temperature for hours (Table 4). With standard compression set procedure D395-37T, General Electric silicone rubber showed compression set values of 8 to 10%, which compares favorably with natural rubber stocks. Silicone rubber remains flexible and resilient down to −70 F, and no plasticizer is needed to improve low-temperature characteristics.

Polymeric silicone gum may be compounded with some fillers used in the rubber industry and can be processed on standard rubber equipment. It may be extruded in a standard continuous extrusion machine and molded and vulcanized in a steam-heated press. Vulcanization with sulfur is not possible, but other methods have been developed.

Silicone rubber may be made to any desired hardness over the range 40 to 85 as measured with a Shore "A" Durometer. The harder rubbers have slightly less tear resistance and a higher modulus with a relatively small elongation. The tensile strength of silicone rubber

Table 4. Typical Mechanical Properties of Silicone Rubbers Before and After Exposure to High Temperature

(Data from General Electric Co.)

	Rubber A	Rubber B	Rubber C	Rubber D
Original properties				
Hardness (Shore "A" Durometer)	75–85	65–75	55–65	45–55
Tensile strength, psi	500	650	450	200
Elongation, %	100	110	150	200
Tear resistance, lb/in.	35	45	45	35
After 500 hr at 200 C (392 F)				
Hardness change	+6–8	+8–10	+15–20	+15–20
Tensile change, %	±0	±0	+5	+20
Elongation change, %	−25	−25	−50	−50
Weight loss, %	2–3	2–3	2–3	2–3
Low-temperature resistance, flex test after 6 hr				
At −60 C At −76 F	No cracks or crazing	No cracks or crazing	No cracks or crazing	No cracks or crazing
Compression set				
Compression 30%, 22 hr at 70 C (158 F)	10%	10%	15%	10%
Compression 30%, 6 hr at 150 C (302 F)	25%	25%	40%	30%

is lower than that of most natural rubbers except at high temperatures, and the most probable value of tensile strength for a break is from 500 to 600 psi. Maximum elongation varies between 100 and 300%. Tear resistance of silicone rubber as compared to conventional rubbers is relatively poor, although inorganic fabrics can be added to improve tear resistance.

Silicone rubber in contact with metals has no corrosion action because of its chemical inertness and because it contains no added ingredients which might cause such corrosion. Special techniques have been developed for bonding silicone rubber to metals, and a special adhesive developed by the General Electric Co. will also bond silicone rubber to itself.

Table 5 gives data on the relative chemical resistances of four typical silicone rubbers.

Table 5. Chemical Resistance of Silicone Rubbers (ASTM D543-43)

(Data from General Electric Co.)

Reagent	Rubber A		Rubber B		Rubber C		Rubber D	
	Change of Weight, %	Change of Volume, %	Change of Weight, %	Change of Volume, %	Change of Weight, %	Change of Volume, %	Change of Weight, %	Change of Volume, %
Water	+0.59	0	+0.6	0	+0.6	+1	+2.4	+7
Water at 100 C	+3.0	0	+3.2	0	+6.8	+15	+7.1	+14
H_2SO_4 (conc)	Dissolved		Dissolved		Dissolved		Dissolved	
HCl (conc)	+2.0	+8	+2.4	+3	+24.5	+43	+11.7	+42
HNO_3 (conc)	+6.4	+8	+7.3	+8	+82.3	+118	Dissolved	
NH_4OH (conc)	+2.7	0	+2.1	0	+5.8	+11	+4.9	+13
KOH, 50%	−4.5	−5	−3.0	+1	Dissolved		Dissolved	
H_2SO_4, 10%	+0.5	0	+0.9	0	+18.0	+38	+16.0	+37
NH_4OH, 10%	+1.7	0	+1.7	0	+4.4	+10	+4.5	+13
NaCl (conc)	+0.1	+1	−0.1	−1	+0.1	0	+0.2	+8
Mineral oil								
At 25 C	+2.5	+2	+3.0	+4	+2.5	+3	+2.5	+2
At 168 C	....	+42	....	+38	....	+29	Dissolved	
Castor oil	+0.1	+1	0	+1	0	0	+3.0	0
Linseed oil	+0.4	+2	−0.4	+3	+0.3	−1	+3.8	+3
Glycerin	0	0	+0.3	+1	0	−1	+0.5	0
Alcohol	....	+5	....	+5	....	+5	....	+5
Acetone	....	+25	....	+25	....	+25	....	+25
Benzene	....	+85	....	+85	....	+150	....	+160
Toluene	....	+85	....	+85	....	+200	....	+200

WOOD

By L. J. Markwardt

17. PROPERTIES OF WOOD *

(Data from U. S. Forest Products Laboratory, Madison, Wis.)

CHEMICAL COMPOSITION. Although proportions of the elements comprising wood vary with species, with heartwood and sapwood, and among individual pieces of a species, the different woods nevertheless exhibit generally similar chemical characteristics. Average results of analyses of the sapwood and heartwood of larch, pine, spruce, oak, and beech are as follows: carbon, 50%; oxygen, 43 1/2%; hydrogen, 6%; nitrogen, 1/4%; and ash 1/4% (*F.P.L. Technical Note* 235, U. S. Department of Agriculture).

From the standpoint of basic components of primary importance in chemical conversion and utilization, the composition of wood is as follows: inorganic material, 1/2%; extractives, 5 1/2%; lignin, 22%; and holocellulose, 72%, the holocellulose being separable into hemicellulose (23%) and alpha cellulose (49%). Lignin forms the major portion of the complex material which serves to bind the cellulose fibers together; it is also distributed throughout the fiber wall.

It is impossible by chemical analysis to identify wood as to species. Physical structure provides the most satisfactory basis for identifying wood samples, and keys for identification by this means are available (*U. S. Dept. Agr. Misc. Cir.* 66; U.S.D.A. Forest Service, *Guidebook for the Identification of Woods Used for Ties and Timbers*).

Broad-leaved species (so-called hardwoods) differ from coniferous woods (so-called softwoods) in that they are about twice as high in pentosan content, give a higher yield of acetic acid on hydrolysis, and are usually higher in methoxyl content.

FUEL VALUE. The calorific value of dry wood is closely related to its density and is influenced further by the resin content. For information on the composition and heating value of wood see the *Power* Volume, Kent's *Mechanical Engineers' Handbook*, p. 2-40.

CHARACTERISTICS AND USES. Commercial woods are listed in the following pages, with notes on their characteristics and chief uses. (See also Table 2, p. 5-84.)

Alder, red (*Alnus rubra*) grows along the Pacific coast from Washington to California. Although red alder is mainly of local importance, it is the leading hardwood of the Pacific Northwest. The color varies from almost white to pale pinkish brown. The pores are comparatively small, fairly uniform in size, and uniformly distributed. The wood is moderately light (28 lb per cu ft, air dry), rather weak, low in shock resistance, moderately stiff, and shrinks little. It is used chiefly for furniture.

Ash (*Fraxinus* species) comprises some 18 species native to the United States. Commercial white ash, composed principally of white and green ash, is grown in the eastern United States. Commercial white ash ranks high in weight (41 lb per cu ft, air dry), strength, hardness, shock resistance, resistance to wear, and ability to retain shape. This combination of properties makes the wood especially desirable for such uses as handles, vehicle parts, and agricultural implements. In selecting ash, material from the swelled butts of swamp-grown trees, which is frequently low in weight, shock resistance, and stiffness, should be avoided.

Aspen (*Populus tremuloides*) grows in northern United States, east of the Rocky Mountains. It is one of the most widely distributed trees in the United States. The wood is soft, weak, moderately light in weight (26 lb per cu ft, air dry), uniform in texture, and easily worked; it seasons rapidly, although it is subject to warping and checking, and decays rapidly under damp conditions. The broad sapwood is creamy white, and the heartwood is light brown. The principal market for aspen logs is for pulp, although the species is also used for lumber, shipping containers, cooperage, excelsior, and matches.

Basswood (*Tilia glabra*) is distributed over the eastern half of the United States, the greater portion coming from the Great Lakes region. Basswood is light in weight (26 lb per cu ft, air dry), low in strength, soft in texture, easily workable, and straight grained. The heartwood is creamy brown, and the sapwood creamy white, changing more or less gradually into the heartwood. The sapwood, especially when of rapid growth, has a clean look that, with its other qualities, makes it particularly desirable for boxes, fruit baskets, and woodenware. Other important uses are for furniture and millwork.

Beech (*Fagus grandifolia*) grows throughout the eastern part of the United States. The wood of beech is heavy (45 lb per cu ft, air dry), strong, hard, of uniform texture, does not

* See also O. W. Eshbach, Editor, *Handbook of Engineering Fundamentals*, John Wiley and Sons, New York, 1936, p. 12-38.

impart taste or odor to food, and resists abrasion. It is low in resistance to decay. Beech is used for flooring, chairs, and other furniture, handles, woodenware, and laundry appliances. It is also highly suitable for food containers and when treated is used for railway ties.

Birch. Between 15 and 20 species of birch grow in the United States. Of these, yellow birch (*Betula lutea*) is by far the most abundant and most important commercially. Sweet birch (*B. lenta*) ranks next, followed by paper birch (*B. papyrifera*). Yellow birch, sweet birch, and paper birch grow principally in the northeastern part of the United States, in the Lake States, and in the Appalachian Mountains. Yellow and sweet birch are cut and sold without distinction simply as birch. Yellow birch has white sapwood and light reddish-brown heartwood. Sweet birch has light-brown sapwood and dark-brown heartwood tinged with red. The wood of both yellow birch and sweet birch is heavy (average 44 lb per cu ft, air dry), hard, stiff, and strong, and has good shock-resisting ability. It is low in natural resistance to decay. The wood is fine and uniform in texture and is capable of taking a beautiful natural finish. Interior finish and furniture take a large part of the birch cut. Other uses are flooring, musical instruments, treated railway ties, and woodenware.

Cedar, Alaska yellow, (*Chamaecyparis nootkatensis*) grows in the Pacific coast region of North America from southwestern Alaska to northern Oregon. The wood is of a fine, uniform texture and has very small shrinkage. It is moderately light (31 lb per cu ft, air dry) and moderately low in strength. It has excellent working and finishing properties and ranks high in resistance to decay. The heartwood is a bright, clear yellow, and the thin, scarcely distinguishable sapwood is a shade lighter. A large part of the cut is used locally for interior finish, furniture, small boat hulls, cabinet work, and novelties.

Cedar, Port Orford white, (*Chamaecyparis lawsoniana*) grows naturally only in a narrow belt bordering the coast of southern Oregon and northern California. The heartwood, which is highly resistant to decay, is light yellow to pale brown in color, and the usually thin sapwood is of similar appearance. The wood is moderately light in weight (29 lb per cu ft, air dry), of moderate strength and hardness, with a fine and unusually uniform texture and a spicy odor. It is used for battery separators, Venetian blinds, boats, interior finish, siding, and shingles. The lower grades are consumed in dock plank, ties, mine timbers, and general construction purposes.

Cedar, western red, (*Thuja plicata*) is an important tree in the humid regions along the North Pacific coast and extends inland from Washington to Montana. The heartwood is reddish brown, and the narrow sapwood is white. The wood is light in weight (23 lb per cu ft, air dry), easily worked, rather soft and weak and has very little shrinkage. The heartwood is highly resistant to decay. Western red cedar is used for shingles, siding, porch columns, and greenhouse construction, where durability and ease of working are important factors. It holds paint well. Large quantities are used for poles and posts. It is not used extensively for piling, since the wood is likely to crush under heavy driving.

Cedar, white, includes two species, northern white (*Thuja occidentalis*) and southern white cedar (*Chamaecyparis thyoides*). Northern white cedar grows in the Lake States and in the northeastern part of the United States. Southern white cedar grows in the eastern part of the United States, near the coast. Both woods are very light in weight (22 and 23 lb per cu ft, air dry, respectively), of low shrinkage, and soft, brittle, and weak. They split readily and have a characteristic aromatic odor. Their texture is fine and uniform. The heartwood, which is resistant to decay, is light brown with a reddish tinge. The sapwood is lighter in color than the heartwood, sometimes nearly white. Northern white cedar, which is a relatively small tree, is used principally for posts, ties, poles, and shingles. It is especially desirable for service in contact with the ground. The small quantity of lumber that is manufactured is used principally in small-boat construction. Southern white cedar, which is a medium-sized tree, is used for shingles, lumber, woodenware, posts, and poles. The lumber, which contains many knots, is made into siding, porch lumber, tanks, and boxes.

Cherry, black, (*Prunus serotina*) grows from Maine to eastern North Dakota and southward to central Florida and central Texas. It is, however, scattered sparsely throughout the forests. The scattered distribution over a large area insures a fairly steady though limited supply. The wood of black cherry is strong and moderately hard; it has good shock-resisting ability, a moderate degree of stiffness, and a very moderate shrinkage. Density is 35 lb per cu ft, air dry. The heartwood varies in color from light to dark reddish brown and has a beautiful and distinctive luster. The sapwood is narrow and nearly white. The annual rings show on a cross section but are not prominent. The wood has a fairly uniform texture and contains numerous scattered pores. Black cherry stays in place well after seasoning. It is exceptionally well suited for engraver's block and is also desirable for paneling and furniture.

Chestnut (*Castanea dentata*), which has long been one of our important commercial woods, has many uses. The blight that is now exterminating chestnut is only the bark disease, so that virtually all the wood of blight-killed chestnut trees is entirely satisfactory for lumber if the trees are cut and utilized before checking, insect attack, or decay begins. The principal commercial range of this species was in southern New England and the Appalachian Mountains. Most of the remaining chestnut timber is in the southern Appalachians. Chestnut is moderately low in weight (30 lb per cu ft, air dry) and strength, is straight grained, and stays in place well. The wood has large distinct pores, and the sapwood is narrow. The heartwood, which is grayish brown, is highly resistant to decay. Poles and fence posts are among the important uses for chestnut. Planing-mill products, core stock in veneer panels, caskets and coffins, furniture, and boxes and crates are other extensive uses.

Cottonwood includes 10 or 12 closely related species of trees that grow in the United States, of which eastern cottonwood (*Populus deltoides*) and northern black cottonwood (*P. trichocarpa hastata*) are the most important. Eastern cottonwood is widely distributed east of the Rocky Mountains, and black cottonwood in the Northwest. The wood is light in weight (28 and 24 lb per cu ft, air dry, respectively), weak, and has a tendency to warp in seasoning. It is odorless, has unusually uniform texture, works nicely, and does not split easily. The annual growth rings are scarcely noticeable. It decays quickly when in contact with the ground.

Uses include boxes and crates, vehicle parts, planing-mill products, agricultural implements, woodenware, laundry appliances, and trunks. In addition to the lumber cut, large amounts of cottonwood are cut into veneer.

Cypress, bald, (*Taxodium distichum*) is found in the low swamp lands along the southeastern coast of the United States and as far up the Mississippi River Valley as Missouri. When grown in swamps near salt water, lumber from this species is called, by the trade, red cypress or tidewater red cypress, and the inland or upland growth is called yellow or white cypress. The heartwood from cypress near salt water will vary in color from slightly reddish to almost black, whereas that found farther inland is only slightly reddish or yellowish brown. Bald cypress is moderately light (32 lb per cu ft, air dry) and moderately strong, and the heartwood is highly resistant to decay. The wood is largely used where such resistance is an important factor. It holds paint well. Doors, sash, porch material, siding, tanks, and greenhouse construction are among the important uses for southern cypress.

Douglas fir (*Pseudotsuga taxifolia*) grows from the Rocky Mountain region to the Pacific Northwest. The strongest wood of this species, as well as the greatest volume of production, comes from the coastal region of Washington, Oregon, and California. Some Douglas fir trees in the coast region are very large—up to 8 ft in diameter, breast high. On rare occasions, trees up to 14 ft are found. The quality of the wood in different portions of these trees varies greatly. Douglas fir in the form of lumber and timbers is one of the most desirable native woods for structural purposes and also is extensively used for poles, piling, and ties. Large quantities are cut into veneer for plywood and other purposes. It is strong, moderately hard and heavy (34 lb per cu ft, air dry), and the heartwood is moderately durable. In general, it has a tendency to check and split and does not hold paint well, although its paint retention can be materially improved. All virgin Douglas fir has little sapwood. The existing stand of merchantable Douglas fir is larger than that of any other species.

Elm. The three species constituting the greater part of the elm lumber produced in the United States are American (*Ulmus americana*), slippery (*U. fulva*), and rock elm (*U. racemosa*). The growth range of American and slippery elm (sold as commercial soft elm) includes most of the area east of the Rocky Mountains. Rock elm is confined largely to the North Central States. The principal supply of these three species comes from the Lake States and the Mississippi Valley. Rock elm is heavy (44 lb per cu ft, air dry), hard, and strong, and has a high capacity to resist shock. It is valuable for its strength and bending qualities. Soft elm is lighter (35 to 37 lb per cu ft) and has lower strength properties than rock elm, although very dense pieces have sufficient strength to be used with rock elm. Elm, in general, is used to a large extent for slack cooperage, some vehicle parts, and boxes and crates.

Fir, balsam, (*Abies balsamea*) grows principally in New England, southward to Pennsylvania, and in the Lake States. The wood is nearly white and has little figure. It is light in weight (25 lb per cu ft, air dry), soft, and low in strength. Much of it is used in rough form for buildings.

Fir, white, (*Abies concolor*) and **lowland white fir** (*A. grandis*) are both known commercially as white fir. Lowland white fir is cut mainly in Idaho and in California near the coast, and true white fir in the mountain region of California. Commercial white fir is light in weight (27 lb per cu ft, air dry) and moderately low in strength, moderately soft, straight grained, of medium and fairly uniform texture, nonresinous, and easily worked. Its resistance to decay is low. The heartwood is white with a reddish tinge, and the sapwood is not distinguishable from the heartwood. It has less color than any other commercial softwood. The chief consumption of this wood is in the construction of small houses, largely as dimension and common boards. Other uses are for general millwork, planing-mill products, and boxes.

Gum, sweet, (*Liquidambar styraciflua*) is a native of the swamp and bottom lands in the South that are dry for the greater part of the year. Although its tendency to warp badly in seasoning long kept it out of commercial use, research ultimately found the proper methods of drying and handling. Consequently, sweet gum lumber, as produced by responsible manufacturers, has for years been a satisfactory commercial material. The color of the heartwood, which is sold as red gum, ranges from a light to a deep reddish brown. The sapwood, which is commercially called sap gum and looked upon as a distinct kind of lumber, is nearly white. The wood is moderately heavy (34 lb per cu ft, air dry) and strong, of fine, uniform texture, and will take a beautiful finish. The heartwood is moderately durable. Sweet gum is used both in lumber and veneered form for finish and furniture. The lower grades are used for boxes and crates.

Hackberry (*Celtis occidentalis*) grows from New England to Virginia and westward to eastern North Dakota, Iowa, southwestern Missouri, and western Kansas. Sugarberry (*C. laevigata*) occurs from the coast of Virginia to southern Florida, and westward through the Gulf States to the valley of the Rio Grande and to eastern Texas, Arkansas, eastern Oklahoma, Missouri, eastern Kansas, southern Illinois, Indiana, Kentucky, and Tennessee. Both sugarberry and hackberry species are marketed as hackberry, though hackberry represents the far greater percentage of the lumber cut. Both species occur in stands associated with elm, walnut, hickory, ash, and other hardwood species. Although hackberry is widely distributed, its commercial range is chiefly the river bottoms and creek valleys of the Middle West. The annual reported cut is not large, but this is only a part of the total output, since it is often sold with the lower grades and for the less exacting uses of ash and also sometimes sold with elm. In color it is more yellowish than ash, but in its annual rings of growth it resembles that wood. The wood is moderately heavy (37 lb per cu ft, air dry), rather weak as a beam, is limber, good in shock resistance, and moderately hard. It is used mostly for furniture and vehicles. Some is used for refrigerators and kitchen cabinets, baskets, boxes, and crates.

Hemlock, eastern, (*Tsuga canadensis*) grows principally in the Lake States and in the mountains in the eastern part of the United States. The wood is pale buff in color, with a reddish tinge, is moderately light in weight (28 lb per cu ft, air dry) and moderately low in strength. It has a tendency to splinter, is subject to ring shake, and is not decay-resistant. The lumber is nonresinous and holds nails well, and the knots are characteristically small. Eastern hemlock is used largely for framing, sheathing, roofing, and subflooring. In the nominal 2-in. size the stock is cut extra thick, thus partly equalizing the greater strength of some other structural woods. Eastern hemlock lumber is carefully manufactured and graded and consequently can be purchased in quality well suited for general building. A small mount of tamarack (*Larix laricina*), a species heavier and stronger than eastern hemlock, is sawed into lumber and is often sold with eastern hemlock. Most tamarack, however, is utilized in the form of poles, posts, ties, and mine props, reaching the market under its own name.

Hemlock, western, (*Tsuga heterophylla*) grows along the Pacific coast from Alaska to northern California, penetrating farther inland just below the Canadian boundary than elsewhere. This species, which is sold commercially as west coast hemlock, is distinct from eastern hemlock. Western hemlock is a uniform, fine-textured wood and is comparatively free of ring shakes. It is moderately light in weight (29 lb per cu ft, air dry), moderately high in strength properties, usually straight grained, and nonresinous; its many small black knots are ordinarily fixed firmly in place. Its resistance to decay is low. Western hemlock in the common grades is mixed and sold with Douglas fir. It is used largely in house construction for framing, sheathing, and subfloors. Large quantities go also into siding, ceiling, flooring, and shiplap. Other uses are sash, doors, blinds, general millwork, and crossties.

Hickory, pecan, includes **bitternut** (*Hicoria cordiformis*), **nutmeg** (*H. myristicaeformis*), **pecan** (*H. pecan*), and **water hickory** (*H. aquatica*). It is found throughout much of the eastern half of the United States. Density is 45 lb per cu ft, air dry. Although these hickories rank relatively high in strength properties, the true hickories are somewhat superior, especially in shock resistance. Pecan is used for the same purposes as the true hickories, namely, for vehicle parts, tool and implement handles, and special products such as ladder rungs, doweling, and gymnasium apparatus. The bulk of the hickory consumed for these purposes, however, is true hickory. Like the true hickories, pecan is commonly classified as white, which is the sapwood, or red, which is the heartwood. As with the other hickories, color is no criterion of its strength.

Hickories, true, are found throughout most of the eastern half of the United States and are all similar in their properties. The species most important commercially are **bigleaf shagbark** (*H. laciniosa*), **mockernut** (*H. alba*), **pignut** (*H. glabra*), and **shagbark** (*H. ovata*).

Hickory is commonly classified as white, which is the sapwood, and red, which is the heartwood. Color, however, is no criterion of its strength because, weight for weight, sound hickory has the same strength and toughness, regardless of whether it is red, white, or mixed. True hickory is very tough, heavy (51 lb per cu ft, air dry), hard, and strong; the combination of high strength, stiffness, hardness, and shock resistance has not been found in any other native commercial wood. It is low in natural decay resistance and shrinks considerably in drying. Hickory having not more than 20 rings to the inch is generally the strongest, although slow-growth hickory, if heavy, also has high strength properties. True hickory, which is one of the prominent specialty woods, is used principally in the vehicle industry for spokes, rims, poles, and shafts, in the tool-handle industry for ax, pick, sledge, and hatchet handles, and for special products, such as golf-club shafts, ladder rungs, doweling, and gymnasium apparatus.

Larch, western, (*Larix occidentalis*) is native principally to the "Inland Empire." The heartwood is reddish brown, and the sapwood, which usually is not more than 1 in. thick, is yellowish white. In its natural color the wood finishes well, but does not hold paint well. It splits easily and is subject to ring shakes. Knots, although common, are usually small. Resistance to decay is moderate. It is rated as moderately heavy (36 lb per cu ft, air dry), strong, stiff, and moderately good in shock resistance. Western larch and Douglas fir are logged together and are frequently manufactured, graded, and sold together as larch-fir. Western larch is used largely in building construction as rough dimension, small timbers, planks, and boards. Planing-mill products, sash, doors, boxes, and crates take a small portion of it. It is a satisfactory wood for rough construction, and its use as an interior trim is increasing because of its pleasing appearance and ease of finishing.

Locust, black, (*Robinia pseudoacacia*) grows in the Appalachian Mountains from Pennsylvania to northern Georgia. The natural range also includes the southern part of Illinois and Indiana, and part of Arkansas, Oklahoma, and Tennessee. Black locust is one of the heaviest (48 lb per cu ft, air dry) and hardest native woods. It ranks very high in shock-resisting ability and strength. It has a moderately small shrinkage which for its density is very low. The heartwood is very durable. Black locust is subject to wormholes, cross grain, and bird pecks. Checking and splitting generally occur to only a moderate degree. Principal uses are for insulator pins, treenails, wagon hubs, fence posts, mine timbers, and poles.

Magnolia, evergreen, (*Magnolia grandiflora*) grows along the coast from North Carolina to Florida and westward in the Gulf Coast region to Texas and through western Louisiana to southern Arkansas. The range of **cucumber magnolia** (*M. acuminata*) extends from western New York to Alabama, following the Appalachian Mountains, and westward to Illinois and Mississippi, appearing west of the Mississippi River in Arkansas. It is largest and most abundant in the narrow valleys of eastern Tennessee and in the western parts of the Carolinas. The wood of the evergreen and cucumber magnolias resembles that of yellow poplar, but it usually is somewhat harder and heavier (35 and 33 lb per cu ft, air dry, respectively) and often has a light to dark purplish color. A large part of cucumber magnolia and some of the evergreen magnolia lumber are marketed with the lower grades of yellow poplar. During the years immediately preceding the economic depression, an average of about 27 million feet of evergreen magnolia was marketed annually as magnolia, and about one-half million feet of cucumber magnolia was marketed as cucumber. Important uses of these woods include furniture, planing-mill products, finish, siding, and boxes and crates.

Maple. Of the 13 species of maple that grow in the United States, **sugar maple** (*Acer saccharum*) and **black maple** (*A. nigrum*), jointly classed as hard maple, are by far the most important as well as the most abundant. Others of commercial importance are **silver maple** (*A. saccharinum*) and **red maple** (*A. rubrum*), which are commonly called soft maple, and **bigleaf** or **Oregon maple** (*A. macrophyllum*), which belongs in the soft-maple group on the basis of its properties. The commercial maples, as a group, grow throughout the eastern part of the United States, with the exception of bigleaf maple, which appears on the Pacific coast. Hard maple, which has a wide range, is found principally in the Lake States, the Northeast, and in the Appalachians. The heartwood of hard maple is light reddish brown, and its sapwood is white with a slight reddish-brown tinge. The white maple of commerce is the unstained sapwood of sugar maple. The wood of this species is heavy (44 lb per cu ft, air dry), hard, strong, and stiff, has good resistance to shock, and wears well under abrasion, excelling the other maples in these properties. It has moderate shrinkage and uniform texture. The grain is usually straight but sometimes curly, wavy, or birdseye grain occurs. Both silver maple and red maple are lighter in color and not so heavy, hard, or strong as sugar maple. Silver maple is lighter in weight (33 lb per cu ft, air dry) and weaker than red maple (38 lb per cu ft, air dry). The strength properties of bigleaf maple are intermediate between those of silver and red maple. Maple is used for flooring and other planing-mill products, interior trim, and furniture. Large quantities also go into agricultural implements, handles, and vehicle parts.

Oak. All the oaks of commercial importance, about 15 of the 60 or more native species, grow in the eastern half of the United States, especially in the Mississippi Valley and the South. In the lumber trade the oaks are commonly divided into two groups, red oak and white oak, and no attempt at a finer distinction is made in most cases. The common and botanical names of commercial white oak are: **White oak** (*Quercus alba*), **chestnut oak** (*Q. montana*), **swamp chestnut oak** (*Q. prinus*), **swamp white oak** (*Q. bicolor*), **post oak** (*Q. stellana*), **overcup oak** (*Q. lyrata*), and **bur oak** (*Q. macrocarpa*). The common and botanical names of commercial red oak are: **Red oak** (*Quercus borealis*), **swamp red oak** (*Q. rubra pagodaefolia*), **southern red oak** (*Q. rubra*), **pin oak** (*Q. palustris*), **black oak** (*Q. velutina*), **water oak** (*Q. nigra*), and **willow oak** (*Q. phellos*). White oak heartwood is more durable and of lighter color than red oak. Furthermore, the pores in the heartwood of white oak are for the most part plugged with a growth called tyloses, which makes the wood much less permeable to liquids than that of red oak. Oak is commonly sold both plain and quarter-sawed, with advantages in each, among which are more pronounced figure in plain-sawed material and less shrinkage in quarter-sawed material. Oak is heavy (44 lb per cu ft, red, to 47 lb, white, air dry), hard, stiff, and strong. Although there is no important difference in the properties of red and white oak, the oaks like all other species vary in their properties within a single species. Rapidly grown oak is generally stronger, harder, and tougher than slow-growth oak, which is finer grained, softer, and more easily worked. Flooring, interior trim, furniture, motor-vehicle parts, and implements consume a large part of the oak cut; in addition, it is used for cooperage, piling, crossties, and timbers.

Pine, lodgepole, (*Pinus contorta*) is widely distributed throughout the western part of the United States. The wood is light in color, with the heartwood only slightly darker than the sapwood. It is one of the lightest in weight of the yellow pines (29 lb per cu ft, air dry), is moderately low in strength, and has small knots. Lodgepole pine is not durable in contact with the ground, but the sapwood accepts preservative treatment readily. In mature stands, only about 20% of the material is large enough for saw timber; much of the cut is used for poles, mine props, and ties.

Pine, eastern white, (*Pinus strobus*), originally known just as "white pine," is found principally in the Lake States, the Northeastern States, and the Appalachian region. The sapwood is white, and the heartwood ranges from cream to light reddish brown. The wood is moderately light in weight (25 lb per cu ft, air dry), moderately low in strength, and usually straight grained. The soft, uniform texture of the virgin growth has won for it extensive use in building and in millwork. Since its dimensions change little with changes in moisture content and it is easily worked, white pine makes a desirable wood for patterns. The second-growth pine is usually of faster growth and more knotty than old growth. Second growth and the lower grades of virgin-growth pine are largely used for boxes and crates. Virgin growth is becoming scarce.

Pine, red, (*Pinus resinosa*) of the Lake States, although belonging to the yellow pine group, is sold with northern white pine in the lower common grades. It is somewhat coarser in grain and texture, however, has more strongly marked annual rings, is heavier (34 lb per cu ft, air dry) and stronger, and is somewhat more resinous.

Pine, ponderosa, (*Pinus ponderosa*) grows from Washington to the Black Hills and southward in the Rocky Mountain and Pacific coast regions. The wood is moderately light in weight (28 lb per cu ft, air dry), moderately low in strength, and ranks relatively low in shrinkage. It is easy to work, stays in place well, and has a uniform texture, being somewhat comparable in these respects with the white pines. The principal uses of ponderosa pine are for sash and frames, doors, general millwork, building construction, and boxes and crates. Considerable quantities are treated and used for railway ties.

Pine, southern yellow, is a general name for a number of closely related species, chiefly longleaf (*Pinus palustris*), shortleaf (*P. echinata*), loblolly (*P. taeda*), slash (*P. caribaea*), and pond pine (*P. rigida serotina*) which grow in the southern and eastern United States. Except for the fact that the pith often distinguishes longleaf from loblolly, shortleaf, and slash pines, the species cannot be positively distinguished by the wood alone. Southern yellow pine exhibits a considerable range in density and provides material for a wide variety of uses. The denser material (41 lb per cu ft, air dry) provides excellent structural material, whereas the less dense stock (36 lb per cu ft), which is more easily worked, splits less in nailing, often holds a coating of finish better, is superior in thermal insulation, and because of lighter weight may be handled and shipped at lower cost, is used extensively for light buildings, temporary construction, millwork, boxes, and crates. Southern yellow pine is an excellent general utility wood. In general, it has a tendency to check and split, and it does not hold paint well. The sapwood of all the southern pines is easily treated for preservation, and pieces containing a large proportion of sapwood are readily obtained.

Pine, sugar, (*Pinus lambertiana*) is native to California and southern Oregon. It is similar in appearance and properties to eastern white pine but can usually be distinguished by its more conspicuous resin ducts. Density is 25 lb per cu ft, air dry. Sugar pine is the largest of the species of the white pine group, and the bulk of the exceptionally wide, thick stock of soft pine lumber is cut from this species. A large part goes into the manufacture of planing-mill products, such as sash, blinds, and drain boards, and into millwork. It is used interchangeably with eastern white and western white pine for many purposes.

Pine, western white, (*Pinus monticola*), sold commercially as Idaho white pine or Idaho pine, grows principally in northern Idaho, eastern Washington, and western Montana. Western white pine resembles eastern white pine so closely that the clear wood of the two is indistinguishable. Patternmakers state that western white pine is somewhat more difficult to work, although both are regarded highly for their workability. Western white pine is light in color, moderately light in weight (27 lb per cu ft, air dry), and straight grained. It swells and shrinks a little more with changes in moisture content but on the whole compares favorably with eastern white pine and is largely used for the same purposes.

Poplar, yellow, (*Liriodendron tulipifera*) is one of the largest native trees supplying lumber; under favorable conditions it reaches a diameter of 6 to 8 ft and a height of 100 to 160 ft. It grows throughout the eastern part of the United States, with the largest stands in the southern Appalachian Mountains. The sapwood, frequently several inches in thickness, is white. The heartwood is yellowish brown with a greenish tinge. The wood is moderately light in weight (28 lb per cu ft, air dry), straight grained, moderately soft, moderately weak, comparatively uniform in texture, easy to nail because it does not split readily, easy to glue, stays in place well, holds paint and enamel well, is easily worked, and finishes smoothly. Yellow poplar is especially suitable for finish, siding, furniture, and other products that are to be painted or enameled. It is largely used for planing-mill products, furniture, veneer, and panels. The lower grades are used for boxes and crates. Other less important uses are vehicle parts, musical instruments, and car construction. **Cucumber magnolia** (*Magnolia acuminata*) and **evergreen magnolia** (*M. grandiflora*) are sometimes sold with the lower grades of yellow poplar, though they can be purchased under their own names. The wood of these species resembles that of yellow poplar, but it usually is somewhat harder and heavier and often has a light to dark purplish color.

Redwood. The commercially important stands of redwood (*Sequoia sempervirens*) are chiefly confined to a narrow belt, 10 to 30 miles wide, that extends from southern Oregon to within a few miles of San Francisco. The redwood forest in a small area south of San Francisco is also of commercial importance. The heartwood of redwood varies in color from cherry to dark mahogany, and the sapwood, which is very narrow, is almost white. Many of the redwoods are extremely large, individual trees commonly reaching diameters of 5 to 10 ft. Exceptional trees attain a diameter of 20 ft or more. Such size permits a high percentage of clear lumber, and unusually wide clear boards are easily obtained. Redwood is moderately light in weight (28 lb per cu ft, air dry) and moderately strong. Because of its high durability, the heartwood is especially valuable where resistance to decay is important. It has a low shrinkage, stays in place well, and is not difficult to work. It is used for sash, doors, frames, siding, and interior and exterior finish, and is valuable also for posts, crossties, tanks, pipes, flumes, greenhouses, and other service in which the structural material is subject to conditions that favor decay. It holds paint well. Redwood is also used as structural timber for highway bridges, trestles, mill roofs, and similar structures.

Spruce, "eastern," is a term which includes three species: **red** (*Picea rubra*), **white** (*P. glauca*), and **black spruce** (*P. mariana*). White spruce and black spruce grow principally in the Lake States and New England, and red spruce in New England and the Appalachian Mountains. All three species have about the same properties, and in commerce no distinction is made between them. The wood dries easily, stays in place well, is moderately light in weight (28 lb per cu ft, air dry) and easily worked, and has a moderate shrinkage and a moderate degree of strength, stiffness, toughness, and hardness. Eastern spruce is not resistant to decay. The wood is light in color. and there is little difference between the heartwood and sapwood. Eastern spruce is used principally for framing material and general millwork. Large quantities are also used in boxes and crates.

Spruce, Sitka, (*Picea sitchensis*) grows in a narrow strip, seldom more than 40 miles wide, along the pacific coast from Alaska to Northern California. The wood is light in weight (28 lb per cu ft, air dry) and has a moderate degree of hardness, stiffness, toughness, and strength. It ranks high in strength for its weight. On this account and also because it can be obtained in clear, straight-grained pieces of large size and uniform texture, Sitka spruce is the favored wood for airplanes. The wood dries easily, stays in place well, has a moderate shrinkage, and is easily worked. It is not resistant to decay. It is used for sash, doors, siding, ceiling, flooring, exterior and interior finish, and aircraft construction. Large quantities go into boxes, crates, and ladders.

Sycamore (*Platanus occidentalis*) grows widely in the eastern half of the United States, but the greater portion comes from the bottom lands along the Mississippi River and its tributaries. Sycamore is moderately heavy (34 lb per cu ft, air dry) and moderately strong. The heartwood is reddish brown, and the sapwood is somewhat lighter in color. In quarter-sawed sycamore the rays are conspicuous, but the plain-sawed lumber has little figure. Seasoning the wood without warping is difficult, especially when it is plain-sawed. Sycamore is used principally for interior trim and fancy paneling. Other uses are baskets, boxes and crates, and furniture.

Tupelo, water, (*Nyssa aquatica*) is found principally in the swampy regions of the South. **Black gum** (*N. sylvatica*) is native to all states east of the Mississippi River. In the South, it grows as far west as Texas. Water tupelo and black gum are sold both under their respective names and in mixture as one or the other, because there is no way of positively identifying the wood of the two species. These species were not cut for lumber to any great extent until about 25 years ago, at which time better means of handling and utilization were developed. The lumber of water tupelo and black gum is largely sapwood, which is grayish white when dry, soft, of fine texture, moderately heavy (35 lb per cu ft, air dry), and moderately strong, but is cross-grained, tends to warp, and is difficult to split. It is used largely for boxes and crates, flooring, inside finish, furniture, and vehicle parts. For factory floors, bridge flooring, and platforms subjected to heavy wear it has proved very satisfactory, but it requires preservative treatment when used under conditions that favor decay. The wood cut from the true black gum tree is, on the average, harder and stronger than that of the true water tupelo. Where a pretense is made commercially of differentiating between black gum and water tupelo lumber, it is usually on the basis that the harder material passes as black gum and the softer as water tupelo, even though actual species lines are disregarded in segregation.

Walnut, black, (*Juglans nigra*), one of the most beautiful native woods, is found in commercial quantities chiefly in Missouri, Iowa, Illinois, Indiana, Ohio, Kentucky, and Tennessee, although it grows to some extent from New England to Texas. The heartwood, which is brown, varies from light to dark in color. The sapwood is nearly white. The wood is heavy (38 lb per cu ft, air dry), hard, strong, and stiff. It stays in place well, takes stain and other finishes exceedingly well, and will also take a good polish. Besides this, it is easily worked and glues satisfactorily. Black walnut is used largely for cabinet work, furniture, interior finish, and gunstocks. Figured and higher-grade black walnut is cut into veneer and used as faces in plywood and panel construction. **Butternut** (*Juglans cinerea*) is related to walnut botanically and is sometimes actually called white walnut, but the wood is radically different in most respects. Density is 27 lb per cu ft, air dry. Butternut is soft and has low strength properties, but the grain and color are attractive and the wood is used to a limited extent for decorative purposes, particularly for veneer.

IMPORTED WOODS. **Balsa** (*Ochroma lagopus*) is native to tropical America. It is one of the lightest commercial woods but has a wide range in density. The specific gravity, based on weight when oven-dry, may vary from about 0.05 to 0.40 (3 to 25 lb per cu ft, air dry), so that for some uses the wood is selected or classified for density. The wood is pale gray or slightly brownish in color. Uses include thermal insulation, cushioning supports for machinery, and, when waterproofed, life preservers, rafts, and floats. It also provides an excellent core for sandwich construction.

Brazilian araucaria (*Araucaria angustifolia*) or Parana pine is native to southern Brazil and adjacent Paraguay and Argentina. The sapwood is yellowish and the heartwood light brown, resembling medium-colored bald cypress. The heartwood may contain bright red streaks. The average weight when the wood is air dry is about 33 lb per cu ft. Uses include construction lumber, matches, cooperage, turned products, paper pulp, shingles, shipping containers, furniture, interior trim, flooring, and veneer.

Crabwood or *Cedro macho* (*Carapa* species, chiefly *C. guianensis*) is native to tropical America. It should not be confused with Spanish cedar, which is also sometimes called

Cedro macho. The heartwood varies from pinkish to dark reddish brown, somewhat resembling walnut in appearance. The average weight per cubic foot is about 42 lb. The wood is used for cabinet work, furniture, and interior fittings.

Ebony (*Diospyros* species) has a naturally dark or black heartwood. The most valued species are native to India and Ceylon, but the material is also found elsewhere in the tropics. The wood weighs about 70 lb per cu ft, air dry. The dark wood is valued for its color and because it turns and polishes well, making it suitable for cabinet work, musical instruments, inlays, handles, and fancy articles.

Greenheart (*Ocotea rodiaei*) is found chiefly in British Guiana. The color of the heartwood varies from light to dark olive or nearly black. The wood weighs about 65 lb per cu ft, air dry. The heartwood is resistant to decay and termites and is especially valued for ship and marine construction.

Khaya (*Khaya* species) or so-called African mahogany is native to tropical West Africa. The heartwood is pale red to dark reddish brown, often purplish. The average weight is about 31 lb per cu ft, air dry. The principal uses are furniture, fixtures, and interior finish, much in the form of veneer.

Lignum vitae (usually *Guaiacum officinale* but sometimes also *G. sanctum*) is native to tropical America. The heartwood is olive-brown or bluish to dark brown or nearly black. It weighs about 78 lb per cu ft and is one of the hardest, heaviest, and most durable woods. Lignum vitae has an interlocked grain and an oily resinous content which acts as a natural lubricant. Important applications are water-lubricated bearings for ship propeller shafts and pumps, bearings, pulley sheaves, small turnings, mallet heads, furniture casters, tool handles, stencil blocks, and band-saw guides.

Mahogany (*Swietenia* species) is native to the West Indies and tropical America. The heartwood is reddish brown. Unlike most species, the radial and the tangential shrinkages are about equal. The relatively small shrinkage and easy working qualities contribute to making mahogany an excellent pattern wood. It weighs about 32 lb per cu ft, air dry. Mahogany is highly valued for cabinet work, interior finish, furniture, veneer, boats, and patternmaking.

Okoume (*Aucoumea Klaineana*) or Gaboon is native to West Africa. The pink or pinkish-brown heartwood has an average weight of about 29 lb per cu ft, air dry. Principal uses are veneer, interior woodwork, furniture, and boats.

Lauan or so-called Philippine mahogany comprises woods chiefly from the genera *Shorea Parashorea*, *Pentacme*, and *Tarrietia* of the Philippine Islands. The color of heartwood varies from dark red or reddish brown to a lighter reddish gray. The average weight of these woods is about 35 lb per cu ft. The principal uses in the United States are for cabinet work, furniture, boats, and interior trim.

Primavera (*Cybistax donnellsmithii*) is native to tropical America. The heartwood ranges in color from creamy or yellowish white to brownish with occasional rose-pink streaks. The average weight is about 30 lb per cu ft, air dry. The chief uses are for "blonde" furniture, cabinet work, and interior trim.

Sapele (*Entandrophragma* species), also called Sapele mahogany, is native to Central and West Africa. The heartwood of these species is dark or brownish red and often more purplish than the wood of *Khaya*. Sapele weighs about 40 lb per cu ft, air dry. The principal uses are for veneer, cabinet work, interior trim, and boats.

Teak (*Tectona grandis*) is found naturally in commercial quantities in India, Burma, Siam, French Indo-China, and Java, but plantations have also been established in many parts of the world. The yellowish or grayish brown heartwood is highly resistant to decay, termites, and marine borers. Teak from Burma averages about 43 lb per cu ft, air dry, and has relatively low shrinkage. It is especially valued for ship decking, furniture, and interior finish.

DECAY RESISTANCE. The development of fungi which cause decay in wood depends on suitable temperature, moisture conditions, and air supply. When any of these conditions is lacking, wood will last indefinitely. Hence, wood kept either constantly dry or continually submerged in water does not decay, regardless of species of wood or presence of sapwood.

Under conditions favorable to decay, species with natural resistance to decay should be chosen, or preservative treatment should be resorted to, or both precautions should be taken. For information on preservative treatment, see Section 25, Article 10. Different woods vary greatly in their natural decay resistance, but even in durable woods the decay resistance lies in the heartwood. When untreated, the sapwood of practically all species has low resistance to decay and usually has a short life under decay-producing conditions. Individual pieces of wood of a species may also vary from others of the same species in durability. The estimated classification of a number of species with respect to decay resistance follows:

1. Heartwood very durable even when used under conditions favoring decay: Cedar (Alaska yellow, eastern red, northern white, Port Orford white, southern white, western red); chestnut; bald cypress; black locust; Osage orange; redwood; black walnut; Pacific yew.

2. Heartwood nearly as durable as that of species listed as very durable: Douglas fir (dense); honey locust; white oak; southern yellow pine (dense).

3. Heartwood of intermediate durability: Douglas fir (unselected); sweet gum; western larch; southern yellow pine (unselected); tamarack.

4. Heartwood less durable than intermediate group: Commercial white ash; beech; birch (sweet, yellow); hemlock (eastern, western); hickory; sugar maple; red oak; spruce (black, Engelmann, red, Sitka, white).

5. Heartwood low in durability when used under conditions favorable to decay: Aspen; basswood; cottonwood; commercial white fir; black willow.

MODIFIED WOOD AND WOOD-BASE MATERIALS. There have been developed various modified forms of wood by means of exposure to high temperatures and pressures and/or resin treatment. A brief description of these products, which have greatly modified properties and many special uses, follows:

Impreg is the name given resin-treated wood that has not been compressed. A moderate increase in weight results due to the resin, accompanied by some improvement in compressive strength, but more significantly by greatly increased dimensional stability (reduced shrinking and swelling), decay resistance, electrical resistance, and acid resistance. These improved properties are accompanied by decreased toughness of the wood. A vital point in the improved process is impregnation with a phenolic resin-forming material that has an affinity for wood and penetrates the finer cell structure prior to polymerization.

Compreg is the general class of densified product comprising resin-treated and compressed wood, usually in the form of a laminated material. With resin impregnation of about 20% (based on the combined weight) and full compression, the density is about 1.35 to 1.40. The modulus of rupture and tensile strength parallel to the grain are on the order of 30,000 to 40,000 psi, with modulus of elasticity in the range of 3,000,000 to 4,000,000 psi. Marked antishrink efficiency is obtained by impregnating the wood with stabilizing phenolic resins as in the case of impreg.

Staypak is the densified product resulting when wood is compressed under conditions of moisture content and temperature that cause sufficient flow of lignin to relieve internal stresses and thus avoid the springback under the moisture exposure condition inherent in ordinary compressed wood. No resin is employed when staypak is made from solid wood. The density is about 1.35 to 1.40, approaching that of wood substance. Although not dimensionally stabilized like compreg, staypak is about twice as tough and has somewhat higher values in tension, flexure, and modulus of elasticity.

Papreg is the name given to resin-treated, compressed, laminated paper. Its density when fully compressed and with a resin content of 35% is about 1.40. Significant improvements in tensile and compressive strength and stiffness have been attained by improvements in the paper base used. Papreg has a tensile strength of 35,000 to 40,000 psi for parallel-laminated constructions and can be processed with varying properties to meet special requirements. It can be moderately post formed. Research has shown the feasibility of laminating with pressures as low as 50 psi, greatly facilitating molding of special shapes.

PLYWOOD (see also Section 25, Article 8) consists of a combination of three or more layers of veneer, usually glued together with the grain of alternate plies at right angles. Thus, although special properties result from the method of construction of plywood, the mechanical properties are complicated not only by the orthotropic structure of the wood itself but also by the arrangement of the plies. Compared with solid wood, the chief advantages of plywood are an equalization of strength properties along the length and width of the panel, greater resistance to checking and splitting, and greater dimensional stability with changes in moisture content.

Because plywood can be made with any desired number of plies from veneer of any desired thickness and species, or with any combination of thicknesses and species, it presents a versatility that affords almost unlimited choice in design.

18. DESIGN OF WOOD STRUCTURES

Natural and dressed sizes of American standard yard lumber and timber are listed in Table 1.

BASIC UNIT STRESSES for **clear timber** in bending, tension parallel to grain, compression, and shear are given in Table 2. For design with other than clear material, cer-

Table 1. American Standard Yard Lumber and Timber Sizes

Nominal Size, in.	Dressed Size, in.	Nominal Size, in.	Dressed Size, in.	Nominal Size, in.	Dressed Size, in.
2 × 4	1 5/8 × 3 5/8	6 × 18	5 1/2 × 17 1/2	14 × 14	13 1/2 × 13 1/2
2 × 6	1 5/8 × 5 5/8	6 × 20	5 1/2 × 19 1/2	14 × 16	13 1/2 × 15 1/2
2 × 8	1 5/8 × 7 1/2	6 × 22	5 1/2 × 21 1/2	14 × 18	13 1/2 × 17 1/2
2 × 10	1 5/8 × 9 1/2	8 × 8	7 1/2 × 7 1/2	14 × 20	13 1/2 × 19 1/2
2 × 12	1 5/8 × 11 1/2	8 × 10	7 1/2 × 9 1/2	14 × 22	13 1/2 × 21 1/2
2 × 14	1 5/8 × 13 1/2	8 × 12	7 1/2 × 11 1/2	14 × 24	13 1/2 × 23 1/2
2 × 16	1 5/8 × 15 1/2	8 × 14	7 1/2 × 13 1/2	14 × 26	13 1/2 × 25 1/2
2 × 18	1 5/8 × 17 1/2	8 × 16	7 1/2 × 15 1/2	14 × 28	13 1/2 × 27 1/2
2 × 20	1 5/8 × 19 1/2	8 × 18	7 1/2 × 17 1/2	14 × 30	13 1/2 × 29 1/2
3 × 4	2 5/8 × 3 5/8	8 × 20	7 1/2 × 19 1/2	16 × 16	15 1/2 × 15 1/2
3 × 6	2 5/8 × 5 5/8	8 × 22	7 1/2 × 21 1/2	16 × 18	15 1/2 × 17 1/2
3 × 8	2 5/8 × 7 1/2	8 × 24	7 1/2 × 23 1/2	16 × 20	15 1/2 × 19 1/2
3 × 10	2 5/8 × 9 1/2	10 × 10	9 1/2 × 9 1/2	16 × 22	15 1/2 × 21 1/2
3 × 12	2 5/8 × 11 1/2	10 × 12	9 1/2 × 11 1/2	16 × 24	15 1/2 × 23 1/2
3 × 14	2 5/8 × 13 1/2	10 × 14	9 1/2 × 13 1/2	16 × 26	15 1/2 × 25 1/2
3 × 16	2 5/8 × 15 1/2	10 × 16	9 1/2 × 15 1/2	16 × 28	15 1/2 × 27 1/2
3 × 18	2 5/8 × 17 1/2	10 × 18	9 1/2 × 17 1/2	16 × 30	15 1/2 × 29 1/2
3 × 20	2 5/8 × 19 1/2	10 × 20	9 1/2 × 19 1/2	18 × 18	17 1/2 × 17 1/2
4 × 4	3 5/8 × 3 5/8	10 × 22	9 1/2 × 21 1/2	18 × 20	17 1/2 × 19 1/2
4 × 6	3 5/8 × 5 5/8	10 × 24	9 1/2 × 23 1/2	18 × 22	17 1/2 × 21 1/2
4 × 8	3 5/8 × 7 1/2	10 × 26	9 1/2 × 25 1/2	18 × 24	17 1/2 × 23 1/2
4 × 10	3 5/8 × 9 1/2	10 × 28	9 1/2 × 27 1/2	18 × 26	17 1/2 × 25 1/2
4 × 12	3 5/8 × 11 1/2	10 × 30	9 1/2 × 29 1/2	18 × 28	17 1/2 × 27 1/2
4 × 14	3 5/8 × 13 1/2	12 × 12	11 1/2 × 11 1/2	18 × 30	17 1/2 × 29 1/2
4 × 16	3 5/8 × 15 1/2	12 × 14	11 1/2 × 13 1/2	20 × 20	19 1/2 × 19 1/2
4 × 18	3 5/8 × 17 1/2	12 × 16	11 1/2 × 15 1/2	20 × 22	19 1/2 × 21 1/2
4 × 20	3 5/8 × 19 1/2	12 × 18	11 1/2 × 17 1/2	20 × 24	19 1/2 × 23 1/2
6 × 6	5 1/2 × 5 1/2	12 × 20	11 1/2 × 19 1/2	20 × 26	19 1/2 × 25 1/2
6 × 8	5 1/2 × 7 1/2	12 × 22	11 1/2 × 21 1/2	20 × 28	19 1/2 × 27 1/2
6 × 10	5 1/2 × 9 1/2	12 × 24	11 1/2 × 23 1/2	20 × 30	19 1/2 × 29 1/2
6 × 12	5 1/2 × 11 1/2	12 × 26	11 1/2 × 25 1/2	24 × 24	23 1/2 × 23 1/2
6 × 14	5 1/2 × 13 1/2	12 × 28	11 1/2 × 27 1/2	24 × 26	23 1/2 × 25 1/2
6 × 16	5 1/2 × 15 1/2	12 × 30	11 1/2 × 29 1/2	24 × 28	23 1/2 × 27 1/2
				24 × 30	23 1/2 × 29 1/2

tain of these basic values must be adjusted downward, and for seasoned material used under dry conditions of service certain of these values may be increased. The amount of the reduction necessary, or of the increase permissible, depends on the property, on the size of timber, and on the use.

Structural grade and use factors as they relate to the modification of the basic unit stresses of Table 2 are presented in Table 3. As an example, the allowable stress for extreme fiber in bending for a grade of beams and stringers having defects that reduce the strength by one-fourth would be three-fourths of the value in Column 2, Table 2, for the species in question. For detailed information on methods of evaluating the **grade-strength ratio** timbers and for modifying basic stresses, see *Tentative Methods for Establishing Structural Grades of Lumber*, ASTM Designation D245-49T. The principal lumber associations have adopted stress grades applicable to the species used for structural purposes that they produce, and stress-graded lumber is commercially available.

BEAM DESIGN. To apply the bending formula to wood beams, it is necessary to consider the form factor.

Form Factors. When the height of a beam of rectangular cross section is increased, the modulus of rupture decreases slightly. This decrease should be considered in accurate comparisons of beams of different heights. The fractional decrease is the difference between unity and the form factor of the beam. A form factor is the ratio of the unit strength value of a beam of structural size to the corresponding unit value for a beam 2 by

Table 2. Basic Stresses for Clear Material *

(For use in determining working stresses according to the grade of timber and other applicable factors All values are in pounds per square inch. U. S. Forest Products Laboratory.)

Species	Extreme Fiber in Bending or Tension Parallel to Grain	Maximum Horizontal Shear	Compression Perpendicular to Grain	Compression Parallel to Grain L/d = 11 or Less	Modulus of Elasticity in Bending
Softwoods					
Baldcypress (Southern cypress)	1900	150	300	1450	1,200,000
Cedars					
Redcedar, Western	1300	120	200	950	1,000,000
White-cedar, Atlantic (Southern white-cedar) and northern	1100	100	180	750	800,000
White-cedar, Port Orford	1600	130	250	1200	1,500,000
Yellow-cedar, Alaska (Alaska cedar)	1600	130	250	1050	1,200,000
Douglas-fir, coast region	2200	130	320	1450	1,600,000
Douglas-fir, coast region, close-grained	2350	130	340	1550	1,600,000
Douglas-fir, Rocky Mountain region	1600	120	280	1050	1,200,000
Douglas-fir, dense, all regions	2550	150	380	1700	1,600,000
Fir, California red, grand, noble, and white	1600	100	300	950	1,100,000
Fir, balsam	1300	100	150	950	1,000,000
Hemlock, Eastern	1600	100	300	950	1,100,000
Hemlock, Western (West Coast hemlock)	1900	110	300	1200	1,400,000
Larch, Western	2200	130	320	1450	1,500,000
Pine, Eastern white (Northern white), ponderosa, sugar, and Western white (Idaho white)	1300	120	250	1000	1,000,000
Pine, jack	1600	120	220	1050	1,100,000
Pine, lodgepole	1300	90	220	950	1,000,000
Pine, red (Norway pine)	1600	120	220	1050	1,200,000
Pine, southern yellow	2200	160	320	1450	1,600,000
Pine, southern yellow, dense	2550	190	380	1700	1,600,000
Redwood	1750	100	250	1350	1,200,000
Redwood, close-grained	1900	100	270	1450	1,200,000
Spruce, Engelmann	1100	100	180	800	800,000
Spruce, red, white, and Sitka	1600	120	250	1050	1,200,000
Tamarack	1750	140	300	1350	1,300,000
Hardwoods					
Ash, black	1450	130	300	850	1,100,000
Ash, commercial white	2050	185	500	1450	1,500,000
Beech, American	2200	185	500	1600	1,600,000
Birch, sweet and yellow	2200	185	500	1600	1,600,000
Cottonwood, Eastern	1100	90	150	800	1,000,000
Elm, American and slippery (white or soft elm)	1600	150	250	1050	1,200,000
Elm, rock	2200	185	500	1600	1,300,000
Gums, blackgum, sweetgum (red or sap gum)	1600	150	300	1050	1,200,000
Hickory, true and pecan	2800	205	600	2000	1,800,000
Maple, black and sugar (hard maple)	2200	185	500	1600	1,600,000
Oak, commercial red and white	2050	185	500	1350	1,500,000
Tupelo	1600	150	300	1050	1,200,000
Yellow poplar	1300	120	220	950	1,100,000

* These stresses are applicable with certain adjustments to material of any degree of seasoning.

2 in. in cross section. The form factor is already included in the safe stresses in bending and for ordinary purposes need not be considered further.

The form factor, F, is inserted in the usual beam formula as follows:

$$M = \frac{FsI}{c}$$

in which M represents the bending moment; s, the unit stress due to the bending moment upon the fiber most remote from the neutral axis; I, the moment of inertia of the cross section about the neutral axis; and c, the normal distance from the neutral axis to the outermost fiber. For a rectangular beam, this formula reduces to

$$M = \frac{Fsbd^2}{6}$$

in which b represents the breadth and d the depth of the beam. The value of F is calculated by means of the formula

$$F = 1 - 0.07\left(\sqrt{\frac{d}{2}} - 1\right)$$

in which d is the depth of the beam. The value of F becomes unity when d is equal to 2. All standard bending tests on clear wood are made on beams 2 by 2 in. in cross section, and the formula for F was accordingly based on the value of the factor, unity, for a 2-in. height.

In I and box beams it is important to consider the form factor, since the safe extreme fiber stresses for wood beams apply directly only to beams of rectangular section. The reduction factor for the modulus of rupture may be obtained by substituting in the following empirical formula:

$$F = 0.50 + 0.50\left(\frac{K(t_2 - t_1)}{t_2} + \frac{t_1}{t_2}\right)$$

in which F represents the form factor for modulus of rupture; K, a coefficient depending upon the ratio of flange depth to total depth of beam (see Table 4); t_1, the thickness of

Table 3. Table Showing Which Working Stresses Are and Which Are Not Modified by Each of the Grade and Use Factors *

(U. S. Forest Products Laboratory)

Kind of Working Stress	Size Classification	Grade or Use Factor					
		Grade	Rate of Growth	Density	Seasoning	Conditions of Exposure	Duration of Load
Extreme fiber in bending	Joists and planks	Yes	Yes	Yes	Yes †	Yes ‡	Yes
	Beams and stringers	Yes	Yes	Yes	No	Yes ‡	Yes
Horizontal shear	Joists and planks	Yes	No	Yes	Yes †	No	Yes
	Beams and stringers	Yes	No	Yes	Yes	No	Yes
Compression perpendicular to grain	Joists and planks	No	Yes	Yes	Yes	Yes ‡	Yes
	Beams and stringers	No	Yes	Yes	Yes	Yes ‡	Yes
	Posts and timbers	No	Yes	Yes	Yes	Yes ‡	Yes
Compression parallel to grain	Joists and planks	Yes	Yes	Yes	Yes †	Yes ‡	Yes
	Posts and timbers	Yes	Yes	Yes	No	Yes ‡	Yes
Modulus of elasticity	Joists and planks	No	No	No	No	No	No
	Beams and stringers	No	No	No	No	No	No
	Posts and timbers	No	No	No	No	No	No

* Modification for grade (Col. 3) is accomplished by application of the grade strength ratio. Modifications for rate of growth and density (Cols. 4 and 5) are shown under the appropriate species in the basic stresses of Table 2. Modifications for seasoning, conditions of exposure, and duration of load (Cols. 6, 7, and 8) are to be made by the designer to fit the particular conditions for which the design is made.

† Modification for seasoning in current commercial practice is accomplished by liberalizing defects in grade rather than increasing working stress.

‡ It is recognized that timber design in general should be on a more conservative basis when decay hazard is present. However, the specific amount of stress reduction for the decay hazard is left for the designer to determine on the basis of the conditions applicable in each particular instance.

the web of an I beam or the combined thickness of the two webs of a box beam; and t_2 the overall width of beam.

The form factor for fiber stress at proportional limit is not the same as the modulus of rupture form factor. It is obtained by substituting in the formula

$$F_1 = 0.58 + 0.42\left(\frac{K(t_2 - t_1)}{t_2} + \frac{t_1}{t_2}\right)$$

in which K has the values given in Table 4.

PLYWOOD DESIGN. The following recommended methods for calculating the strength and stiffness of plywood (see Table 5) are an abridgment and simplification of more elaborate mathematical formulas that have been developed. Although they should be regarded as approximate, they may be considered sufficiently accurate for most structural design purposes where *buckling of the plywood does not occur.* If these approximate methods are applied to structures or structural elements so proportioned that the plywood will buckle, the resulting values will be higher than those which can be expected of the structure.

Appropriate unit stresses for Table 5 may be obtained from Table 2 as follows:

Clear Plywood without Defects: Use values from Table 2 with or without modification as directed in the last column of Table 5.

Plywood with Defects. When defects are present, reduce the foregoing stresses according to the structural grade ratio. If the defects reduce the strength by one-fourth, multiply the foregoing stresses for clear plywood by $^3/_4$, etc., except compression perpendicular to grain and modulus of elasticity, which are not affected by grade.

Plywood Used in Dry Locations. The unit stresses for extreme fiber in bending, compression parallel to grain, and compression perpendicular to grain may be increased by 25%. No increase for seasoning is provided for horizontal shear or modulus of elasticity.

BASIC FACTS. Durability. Wood, when not subjected to such deteriorating agencies as decay, will last indefinitely. In Europe, many wood structures and structural units that are centuries old still exist.

Decay of wood is caused by fungi, of which there are many varieties. Aside from food, three essentials to the development of fungi are air, suitable temperature, and favorable moisture content. Wood that is completely submerged in water does not decay because the necessary air is lacking. Wood with a moisture content constantly below about 20% does not decay because insufficient moisture is available for decay-producing organisms. The so-called "dry-rot" develops in timber that is apparently below such a moisture content because this fungus is capable of conducting the needed moisture from sources outside the timber itself.

The natural heartwood of some species is very resistant to decay, whereas the sapwood of all species is nondurable when exposed to conditions favoring decay. Examples of the most durable species are redwood, Western red cedar, and bald cypress.

Decay resistance of wood can be greatly increased by impregnation with wood-preserving chemicals, such as creosote. (See Section 25, Article 10.)

The heartwood of most species of wood is more resistant to impregnation with preservatives than sapwood. Hence, sapwood is generally to be preferred to heartwood for the outer portion of timbers that are to be treated.

Creosote in itself does not affect the strength properties of wood, but severe methods of treatment to obtain desired impregnation and penetration may seriously reduce the strength.

Untreated timber should not be completely imbedded in masonry; ventilation space should be provided at the ends of joists or girders supported in masonry walls to avoid decay conditions.

Kiln drying affords more rapid seasoning than air drying; properly kiln-dried wood is fully equal in strength to air-dried material. (See Section 25, Article 6.)

Appearance of the wood is not a reliable criterion of the effect of drying or treating process upon its strength.

Specific gravity of wood substance itself is practically constant for all species, being about 1.53. Differences in the bulk density of different species are hence due to differences in the amount of wood substance present.

Density and Strength. There is a general relation between the density of wood and its strength properties, the strength increasing with density. Some properties, such as compressive strength, increase as the first power of density, and others, such as hardness, vary as the 2 $^1/_4$ power among species.

Summer wood is in general appreciably denser than **spring wood,** and hence in some species, such as southern yellow pine and Douglas fir, the proportion of summer wood may

Table 4. Values of K in the Form Factor Formula

Ratio of Depth of Compression Flange to Depth of Beam	Values of K	Ratio of Depth of Compression Flange to Depth of Beam	Values of K
0.10	0.085	0.60	0.875
0.15	0.155	0.65	0.920
0.20	0.230	0.70	0.950
0.25	0.315	0.75	0.970
0.30	0.400	0.80	0.985
0.35	0.490	0.85	0.995
0.40	0.575	0.90	0.998
0.45	0.660	0.95	1.000
0.50	0.740	1.00	1.000
0.55	0.810		

Table 5. Design Method and Allowable Stresses for Plywood *

(U. S. Forest Products Laboratory)

Property	Direction of Stress with Respect to Direction of Face Grain	Area to Be Considered	Unit Stress to Be Used
Tension	Parallel or perpendicular	Parallel plies only †	Unit stress for extreme fiber in bending
	±45°	Full cross-sectional area	One-sixth unit stress for extreme fiber in bending
Compression	Parallel or perpendicular	Parallel plies only †	Unit stress in compression parallel to grain
	±45°	Full cross-sectional area	One-third unit stress in compression parallel to grain
Bearing at right angles to plane of plywood		Loaded area	Unit stress in compression perpendicular to grain
Load in bending	Parallel or perpendicular	Bending moment $M = KSI/c$, where S = unit stress for extreme fiber in bending; I = moment of inertia computed on basis of parallel plies only; c = distance from neutral axis to outer fiber of outermost ply having its grain in the direction of the span; $K = 1.50$ for three-ply plywood having the grain of the outer plies perpendicular to the span; $K = 0.85$ for all other plywood.	Unit stress for extreme fiber in bending
Deflection in bending	Parallel or perpendicular	Deflection may be calculated by the usual formulas, taking as the moment of inertia that of the parallel plies plus one-twentieth that of the perpendicular plies. (When face plies are parallel, the calculation may be simplified, with but little error, by taking the moment of inertia as that of the parallel plies only.)	Unit value for modulus of elasticity
Deformation in tension or compression	Parallel or perpendicular	Parallel plies only †	Unit value for modulus of elasticity
Shear through thickness	Parallel or perpendicular	Full cross-sectional area	Double unit stress in horizontal shear
	±45°	Full cross-sectional area	Four times unit stress in horizontal shear

* The suggested simplified methods of calculation apply reasonably well with usual plywood types under ordinary conditions of service. It is recognized, however, that they are not entirely valid for all types of plywood and plywood constructions, or for all spans and span-depth ratios. Also, the methods given are not applicable to structures so proportioned that the plywood is in the buckling range, in which event the results will be too high.

† By "parallel plies" is meant those plies whose grain direction is parallel to the direction of principal stress.

(Table continued on p. 5-88)

Table 5. Design Method and Allowable Stresses for Plywood *—*Continued*

(U. S. Forest Products Laboratory)

Property	Direction of Stress with Respect to Direction of Face Grain	Area to Be Considered	Unit Stress to Be Used
Shear in plane of plies	Parallel or perpendicular	Full shear area. 1. Plywood beams. Horizontal shear: 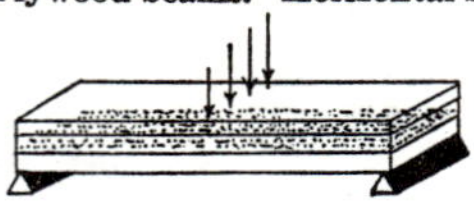	1. Three-fourths unit stress in horizontal shear
		Area of contact between plywood and flange or framing member. 2. I or box beams with plywood webs. Shear between plies of web or between web and flange: 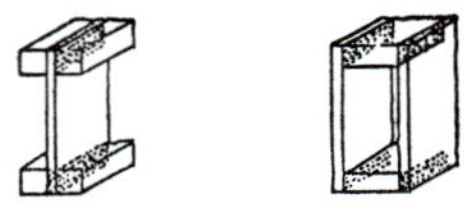	2. Three-eighths unit stress in horizontal shear
		3. Panels having plywood covers stressed in compression or tension, or both. Shear between plies or between cover and framing members when depth of member exceeds twice its width and end headers are used or when depth is not more than twice the width and no headers are used. A. Interior framing members: 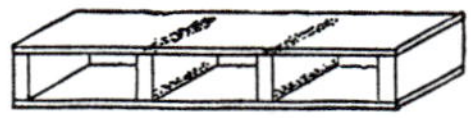	3. A. Three-fourths unit stress in horizontal shear
		B. Framing members at edge of panel: 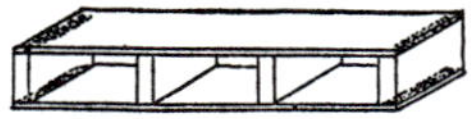	B. Three-eighths unit stress in horizontal shear
	±45°	Area of contact between plywood and flange or framing member. 1. I or box beams with plywood webs. Shear between plies of web or between web and flange: 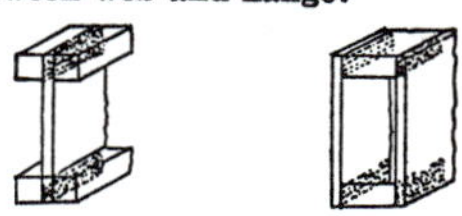	1. One-half unit stress in horizontal shear
		2. Panels having plywood covers stressed in compression, or tension, or both. Shear between plies or between cover and framing members when depth of member exceeds twice its width and end headers are used or when depth is not more than twice the width and no headers are used. A. Interior framing members: 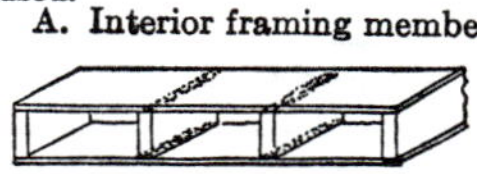	2. A. Unit stress in horizontal shear
		B. Framing members at edge of panel: 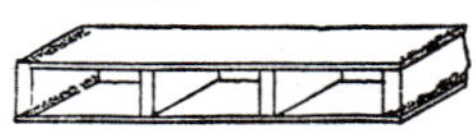	B. One-half unit stress in horizontal shear

be used as a criterion of strength. The so-called "density rule" in the structural grading of these species is based on this principle and allows a $^1/_6$ increase in working stress.

Density Variation. There is considerable variation in the density and strength of individual pieces of wood from a single tree, and also variation among different trees of a species. The estimated probable variation in bulk density of an individual piece from the average for the species is about 8%.

Moisture in wood in the green state is in part adsorbed by the cell walls and in part held as free water in the cell cavities. As wood dries, the cell walls do not give off moisture until the adjacent cell cavities are empty. The condition in which the cell walls are fully saturated and the cell cavities are empty is known as the "fiber-saturation point." It varies from 25 to 35% moisture content but for general purposes is accepted as 30%.

Wood is hygroscopic and, when left exposed a sufficient time to any given temperature and humidity conditions, will come to constant weight. The amount of moisture held in adsorption and as free water is proportional to the amount of wood substance present and is expressed as a percentage of the weight of the oven-dry wood, as "the moisture content." There is a definite relationship between the moisture content of wood and relative humidity at any given temperature. Wood held in a room controlled to a temperature of 75 F and a relative humidity of 64% will come to equilibrium at 12% moisture content.

Green wood begins to shrink across the grain when the moisture content is lowered below the fiber-saturation point. Wood shrinks and swells across the grain with the lowering or raising of the moisture content, for moisture content values below fiber saturation. The longitudinal shrinkage of normal wood is negligible for most uses.

Increase in strength of wood in small sizes begins when the wood is dried below the fiber-saturation point. For the properties affected, the change in strength is related to the change in moisture content, the amount of change varying with the property. For timbers in structural sizes, the increase in strength is partially or largely offset by seasoning defects, depending on the size and grade.

Grade. Some dimension grades of lumber are based on appearance characteristics which are not definitely related to strength. Structural timber grades are based on classification with respect to size and location of defects, for the express purpose of permitting assignment of working stresses. When a structural timber is resawn, the grade of the smaller stock produced bears no relation to the grade of the original structural timber.

Rate of growth as indicated by the width of the annual rings, although of some assistance in appraising quality, is not a reliable criterion of strength. Density is a much better guide to strength.

Change from sapwood to heartwood in a tree does not affect the strength properties. The conditions of growth that prevail when the wood is formed determine the density and the strength. Hence, specifying "heartwood" or "sapwood" does not in itself assure superior strength but may be desirable for other reasons, such as "heartwood" for decay resistance, or "sapwood" for ease of preservative treatment.

Dead Trees. Sound wood from trees killed by fire, insects, or fungi is suitable for structural uses if not unduly checked.

Grain. Wood may be treated mathematically as an orthotropic material with three principal axes of symmetry, longitudinal, radial, and tangential. Consequently, the "grain" directions play an important part in design and make wood more complicated than isotropic materials. From the standpoint of close design, wood has three moduli of elasticity, three moduli of rigidity, and six Poisson's ratios.

Camber. Good design practice suggests consideration of adequate camber for roof trusses and like structures where sag is undesirable.

Safety Factor. Because the ratio of the ultimate strength of small clear pieces of wood in bending to the recommended fiber stress for design of structural timbers is on the order of 4, or more, it cannot be assumed that this is the true factor of safety under continued loading. The true factor of safety for the poorer pieces is less than 2, when allowances are made for variability, duration of stress, size factor, and grade.

Thermal expansion of wood is so small as to be unimportant in ordinary usage.

SECTION 6

FABRICATED MATERIALS

FABRICATED MATERIALS

1. PLATES, SHEETS, BARS, AND RODS

WEIGHT OF PLATE AND SHEET. Table 1 gives the specific gravity and density of the common plate and sheet materials. Column 6 gives the weight in pounds per square foot of plate 1 in. thick, from which the weight of any other thickness is found by multiplying by the thickness t. A conveniently remembered rule is that an iron or steel plate $^1/_4$ in. thick weighs approximately 10 lb per sq ft.

Table 1. Weight of Plates, Bars, Rods, and Spheres

1	2	3	4	5	6	7	8	9	10
Material	Specific Gravity	Density		Relative Weight to Steel	Plate, lb/sq ft, $t\times$	Bar and Rod, lb/ft			Spheres, lb, $D^3\times$
		Lb/cu ft	Lb/cu in.			Square or Flat, s^2 or $bt\times$	Hexagonal, $f^2\times$	Round, $D^2\times$	
Magnesium (Dowmetal M)	1.76	110	.064	0.224	9.2	0.76	0.66	0.60	.0333
Aluminum (24S alloy)	2.77	173	.100	0.353	14.4	1.20	1.14	0.94	.0524
Zinc	7.14	446	.258	0.910	37.2	3.10	2.68	2.43	.135
Cast iron	7.22	450	.260	0.918	37.5	3.12	2.71	2.45	.136
Wrought iron	7.70	480	.278	0.980	40.0	3.33	2.89	2.62	.145
Steel	7.85	490	.283	1.000	40.8	3.40	2.95	2.67	.148
Yellow brass	8.47	529	.306	1.079	44.1	3.67	3.18	2.88	.162
Copper, nickel, bronze	8.92	556	.322	1.136	46.4	3.86	3.35	3.03	.169
Lead	11.35	708	.410	1.445	59.0	4.92	4.26	3.86	.215

(s = side of square, inches; b = width of flat, inches; t = thickness of flat, inches; f = width across flats of hexagon, inches; D = diameter, inches.)

In calculating areas of square and round plates and sheets, the tables on pp. 28-26 to 28-42, and p. 28-46, will be found helpful.

WEIGHT OF BARS AND RODS. Columns 7, 8, and 9 of Table 1 list coefficients from which the weights of flat, hexagon, and round bars may be calculated by multiplying by the dimensions at the heads of the columns. Actual weights of steel bars and rods are given in Tables 2, 3, and 4; for other materials, the values in Tables 2, 3, and 4 should be multiplied by the constants in column 5 of Table 1. Octagon bars weigh 17.16% less than square bars of the same width across flats.

WEIGHT OF TUBING. To find the weight per foot of tubing, the coefficients in column 9 of Table 1 should be multiplied by $(D^2 - d^2)$ or $4t(d + t)$ or $4t(D - t)$, where d is the bore of the tube, D is the outside diameter, and t is the wall thickness, all in inches.

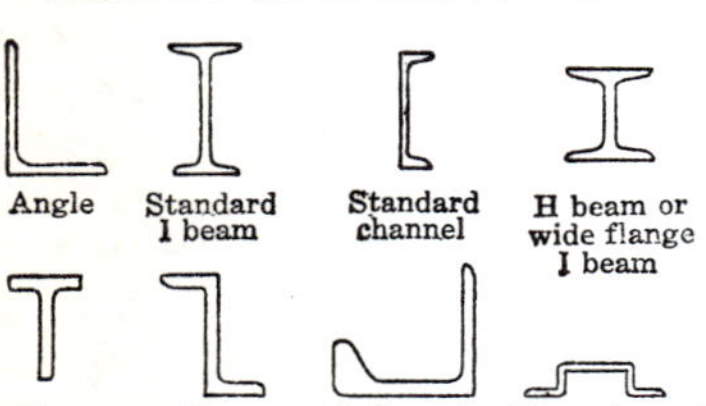

FIG. 1. Standard structural shapes.

STRUCTURAL SHAPES illustrated in Fig. 1 are available in a wide range of sizes in steel and aluminum; many are also available as magnesium extrusions. For shapes and sizes available as well as detail dimensions and section properties, see Eshbach's *Handbook of Engineering undamentals*, John Wiley and Sons, 1936, and the handbooks issued by the Bethlehem Steel Co., Carnegie Illinois Steel Corp., Aluminum Company of America, Dow Chemical Co., and other manufacturers.

Table 2. Weights of Rectangular Sections—Steel

(Pounds per lineal foot. For other materials, multiply by coefficients in column 5 of Table 1.)

Width, in.	Thickness, in.															
	1/16	1/8	3/16	1/4	5/16	3/8	7/16	1/2	9/16	5/8	11/16	3/4	13/16	7/8	15/16	1
1/4	0.053	0.106	0.159	0.213	0.266	0.319	0.372	0.425	0.478	0.531	0.584	0.638	0.691	0.744	0.797	0.850
1/2	.106	.213	.319	.425	.531	.638	.744	.850	.956	1.063	1.169	1.275	1.381	1.488	1.594	1.700
3/4	.159	.319	.478	.638	.797	.956	1.116	1.275	1.434	1.594	1.753	1.913	2.072	2.231	2.391	2.550
1	.213	.425	.638	.850	1.063	1.275	1.488	1.700	1.913	2.125	2.338	2.550	2.763	2.975	3.188	3.400
1 1/4	.266	.531	.797	1.063	1.328	1.594	1.859	2.125	2.391	2.656	2.922	3.188	3.453	3.719	3.984	4.250
1 1/2	.319	.638	.956	1.275	1.594	1.913	2.231	2.550	2.869	3.188	3.506	3.825	4.144	4.463	4.781	5.100
1 3/4	.372	.744	1.116	1.488	1.859	2.231	2.603	2.975	3.347	3.719	4.091	4.463	4.834	5.206	5.578	5.950
2	.425	.850	1.275	1.700	2.125	2.550	2.975	3.400	3.825	4.250	4.675	5.100	5.525	5.950	6.375	6.800
2 1/4	.478	.956	1.434	1.913	2.391	2.869	3.347	3.825	4.303	4.781	5.259	5.738	6.216	6.694	7.172	7.650
2 1/2	.531	1.063	1.594	2.125	2.656	3.188	3.719	4.250	4.781	5.313	5.844	6.375	6.906	7.438	7.969	8.500
2 3/4	.584	1.169	1.753	2.338	2.922	3.506	4.091	4.675	5.259	5.844	6.428	7.013	7.597	8.181	8.766	9.350
3	.638	1.275	1.913	2.550	3.188	3.825	4.463	5.100	5.738	6.375	7.013	7.650	8.288	8.925	9.563	10.20
3 1/4	.691	1.381	2.072	2.763	3.453	4.144	4.834	5.525	6.216	6.906	7.597	8.288	8.978	9.669	10.36	11.05
3 1/2	.744	1.488	2.231	2.975	3.719	4.463	5.206	5.950	6.694	7.438	8.181	8.925	9.669	10.41	11.16	11.90
3 3/4	.797	1.594	2.391	3.188	3.984	4.781	5.578	6.375	7.172	7.969	8.766	9.563	10.36	11.16	11.95	12.75
4	.850	1.700	2.550	3.400	4.250	5.100	5.950	6.800	7.650	8.500	9.350	10.20	11.05	11.90	12.75	13.60
4 1/4	.903	1.806	2.709	3.613	4.516	5.419	6.322	7.225	8.128	9.031	9.934	10.84	11.74	12.64	13.55	14.45
4 1/2	.956	1.913	2.869	3.825	4.781	5.738	6.694	7.650	8.606	9.563	10.52	11.48	12.43	13.39	14.34	15.30
4 3/4	1.000	2.019	3.028	4.038	5.047	6.056	7.066	8.075	9.084	10.09	11.10	12.11	13.12	14.13	15.14	16.15
5	1.063	2.125	3.188	4.250	5.313	6.375	7.438	8.500	9.563	10.63	11.69	12.75	13 81	14.88	15.94	17.00
5 1/4	1.116	2.231	3.347	4.463	5.578	6.694	7.809	8.925	10.04	11.16	12.27	13.39	14.50	15.62	16.73	17.85
5 1/2	1.169	2.338	3.506	4.675	5.844	7.013	8.181	9.350	10.52	11.69	12.86	14.03	15.19	16.36	17.53	18.70
5 3/4	1.222	2.444	3.666	4.888	6.109	7.331	8.553	9.775	11.00	12.22	13.44	14.66	15.88	17.11	18.33	19.55
6	1.275	2.550	3.825	5.100	6.375	7.650	8.925	10.20	11.48	12.75	14.03	15.30	16.58	17.85	19.13	20.40
6 1/4	1.328	2.656	3.984	5.313	6.641	7.969	9.297	10.63	11.95	13.28	14.61	15.94	17.27	18.59	19.92	21.25
6 1/2	1.381	2.763	4.144	5.525	6.906	8.288	9.669	11.05	12.43	13.81	15.19	16.58	17.96	19.34	20.72	22.10
6 3/4	1.434	2.869	4.303	5.738	7.172	8.606	10.04	11.48	12.91	14.34	15.78	17.21	18.65	20.08	21.52	22.95
7	1.488	2.975	4.463	5.950	7.438	8.925	10.41	11.90	13.39	14.88	16.36	17.85	19.34	20.83	22.31	23.80
7 1/4	1.541	3.081	4.622	6.163	7.703	9.244	10.78	12.33	13.87	15.41	16.95	18.49	20.03	21.57	23.11	24.65
7 1/2	1.594	3.188	4.781	6.375	7.969	9.563	11.16	12.75	14.34	15.94	17.53	19.13	20.72	22.31	23.91	25.50
7 3/4	1.647	3.294	4.941	6.588	8.234	9.881	11.53	13.18	14.82	16.47	18.12	19.76	21.41	23.06	24.70	26.35
8	1.700	3.400	5.100	6.800	8.500	10.20	11.90	13.60	15.30	17.00	18.70	20.40	22.10	23.80	25.50	27.20
8 1/4	1.753	3.506	5.259	7.013	8.766	10.52	12.27	14.03	15.78	17.53	19.28	21.04	22.79	24.54	26.30	28.05
8 1/2	1.806	3.613	5.419	7.225	9.031	10.84	12.64	14.45	16.26	18.06	19.87	21.68	23.48	25.29	27.09	28.90
8 3/4	1.859	3.719	5.578	7.438	9.297	11.16	13.02	14.88	16.73	18.59	20.45	22.31	24.17	26.03	27.89	29.75
9	1.913	3.825	5.738	7.650	9.563	11.48	13.39	15.30	17.21	19.13	21.04	22.95	24.86	26.78	28.69	30.60
9 1/4	1.966	3.931	5.897	7.863	9.828	11.79	13.76	15.73	17.69	19.66	21.62	23.59	25.55	27.52	29.48	31.45
9 1/2	2.019	4.038	6.056	8.075	10.09	12.11	14.13	16.15	18.17	20.19	22.21	24.23	26.24	28.26	30.28	32.30
9 3/4	2.072	4.144	6.216	8.288	10.36	12.43	14.50	16.58	18.65	20.72	22.79	24.86	26.93	29.01	31.08	33.15
10	2.125	4.250	6.375	8.500	10.63	12.75	14.88	17.00	19.13	21.25	23.38	25.50	27.63	29.75	31.88	34.00
10 1/2	2.231	4.463	6.694	8.925	11.16	13.39	15.62	17.85	20.08	22.31	24.54	26.78	29.01	31.24	33.47	35.70
11	2.338	4.675	7.013	9.350	11.69	14.03	16.36	18.70	21.04	23.38	25.71	28.05	30.39	32.73	35.06	37.40
11 1/2	2.444	4.888	7.331	9.775	12.22	14.66	17.11	19.55	21.99	24.44	26.88	29.33	31.77	34.21	36.66	39.10
12	2.550	5.100	7.650	10.20	12.75	15.30	17.85	20.40	22.95	25.50	28.05	30.60	33.15	35.70	38.25	40.80
12 1/2	2.660	5.310	7.970	10.63	13.28	15.94	18.59	21.25	23.91	26.56	29.20	31.90	34.50	37.20	39.80	42.50
13	2.760	5.530	8.290	11.05	13.81	16.58	19.34	22.10	24.86	27.63	30.40	33.20	35.90	38.70	41.40	44.20
13 1/2	2.870	5.740	8.610	11.48	14.34	17.21	20.08	22.95	25.82	28.69	31.60	34.40	37.30	40.20	43.00	45.90
14	2.980	5.950	8.930	11.90	14.88	17.85	20.83	23.80	26.78	29.75	32.70	35.70	38.70	41.70	44.60	47.60
14 1/2	3.080	6.160	9.240	12.33	15.41	18.49	21.57	24.65	27.73	30.81	33.90	37.00	40.10	43.10	46.20	49.30
15	3.190	6.380	9.560	12.75	15.94	19.13	22.31	25.50	28.69	31.88	35.10	38.30	41.40	44.60	47.80	51.00
15 1/2	3.290	6.590	9.880	13.18	16.47	19.76	23.06	26.35	29.64	32.94	36.20	39.50	42.80	46.10	49.40	52.70
16	3.400	6.800	10.20	13.60	17.00	20.40	23.80	27.20	30.60	34.00	37.40	40.80	44.20	47.60	51.00	54.40
16 1/2	3.510	7.010	10.52	14.03	17.53	21.04	24.54	28.05	31.56	35.06	38.60	42.10	45.60	49.10	52.60	56.10
17	3.610	7.230	10.84	14.45	18.06	21.68	25.29	28.90	32.51	36.13	39.70	43.40	47.00	50.60	54.20	57.80
17 1/2	3.720	7.440	11.16	14.88	18.59	22.31	26.03	29.75	33.47	37.19	40.90	44.60	48.30	52.10	55.80	59.50
18	3.830	7.650	11.48	15.30	19.13	22.95	26.78	30.60	34.43	38.25	42.10	45.90	49.70	53.60	57.40	61.20
18 1/2	3.930	7.860	11.79	15.73	19.66	23.59	27.52	31.45	35.38	39.31	43.20	47.20	51.10	55.00	59.00	62.90
19	4.040	8.080	12.11	16.15	20.19	24.23	28.26	32.30	36.34	40.38	44.40	48.50	52.50	56.50	60.60	64.60
19 1/2	4.140	8.290	12.43	16.58	20.72	24.86	29.01	33.15	37.29	41.44	45.60	49.70	53.90	58.00	62.20	66.30
20	4.250	8.500	12.75	17.00	21.25	25.50	29.75	34.00	38.25	42.50	46.80	51.00	55.30	59.50	63.80	68.00

Table 3. Weights and Areas of Hexagon Steel

(For other materials, multiply values by coefficients in column 5 of Table 1.)

Short Diam., in.	Weight per ft., lb.	Area, Sq. in.	Short Diam., in.	Weight per ft., lb.	Area, Sq. in.	Short Diam., in.	Weight per ft., lb.	Area, Sq. in.
1/16	0.0115	0.0034	1 1/16	3.3316	0.9776	3 1/2	36.1528	10.6082
3/32	.0259	.0076	1 1/8	3.7351	1.0960	3 5/8	38.7828	11.3799
1/8	.0461	.0135	1 3/16	4.1616	1.2211	3 3/4	41.5034	12.1782
5/32	.0721	.0211	1 1/4	4.6114	1.3531	3 7/8	44.3141	13.0030
3/16	.1037	.0304	1 5/16	5.0841	1.4918	4	47.2188	13.8553
7/32	.1412	.0414	1 3/8	5.5797	1.6373	4 1/8	50.2174	14.7351
1/4	.1845	.0541	1 7/16	6.0987	1.7895	4 1/4	53.3062	15.6415
9/32	.2334	.0685	1 1/2	6.6401	1.9484	4 3/8	56.4889	16.5734
5/16	.2882	.0846	1 9/16	7.2053	2.1142	4 1/2	59.7618	17.5358
11/32	.3487	.1023	1 5/8	7.7930	2.2867	4 5/8	63.1287	18.5237
3/8	.4150	.1218	1 11/16	8.4040	2.4660	4 3/4	66.5895	19.5392
13/32	.4871	.1429	1 3/4	9.0383	2.6521	4 7/8	70.1367	20.5800
7/16	.5649	.1658	1 13/16	9.6955	2.8449	5	73.7817	21.6496
15/32	.6485	.1903	1 7/8	10.3757	3.0445	5 1/8	77.5168	22.7455
1/2	.7378	.2165	1 15/16	11.0787	3.2508	5 1/4	81.3458	23.8691
17/32	.8329	.2444	2	11.8051	3.4639	5 3/8	85.2651	25.0191
9/16	.9317	.2740	2 1/16	12.5544	3.6838	5 1/2	89.2745	26.1956
19/32	1.0404	.3053	2 1/8	13.3269	3.9105	5 5/8	93.3779	27.3996
5/8	1.1529	.3383	2 3/16	14.1224	4.1439	5 3/4	97.5752	28.6312
21/32	1.2710	.3729	2 1/4	14.9408	4.3841	5 7/8	101.866	29.8904
11/16	1.3949	.4093	2 5/16	15.7822	4.6309	6	106.244	31.1749
23/32	1.5246	.4474	2 3/8	16.6468	4.8846	6 1/4	115.285	33.8278
3/4	1.6601	.4871	2 7/16	17.5347	5.1452	6 1/2	124.690	36.5876
25/32	1.8013	.5286	2 1/2	18.4452	5.4123	6 3/4	134.468	39.4565
13/16	1.9483	.5717	2 5/8	20.3361	5.9672	7	144.614	42.4336
27/32	2.1011	.6165	2 3/4	22.3190	6.5490	7 1/2	166.010	48.7118
7/8	2.2596	.6630	2 7/8	24.3940	7.1579	8	188.879	55.4222
29/32	2.4238	.7112	3	26.5614	7.7938	9	239.051	70.1441
15/16	2.5939	.7611	3 1/8	28.8209	8.4568	10	295.127	86.5982
31/32	2.7697	.8127	3 1/4	31.1728	9.1470	11	357.102	104.7834
1	2.9513	.8660	3 3/8	33.6168	9.8641	12	424.991	124.7041

2. WIRE GAGES AND SHEET METAL GAGES

The two wire gages in most extensive use in the United States are the American Wire Gage, also known as the Brown & Sharpe Gage, and the United States Steel Wire Gage more commonly called simply Steel Wire Gage.

Three other wire gages have remained in use but are now almost obsolete. These are the Birmingham Wire Gage, the Stubs Steel Wire Gage, and the Old English Wire Gage. There are also certain special gages, as the drill and screw rod gages and the music wire gage. The most frequently used special gage is the Birmingham Gage, sometimes called the Stubs Wire Gage. Its use in this country has been practically limited to large telegraph and telephone companies and certain departments of the Government. The Treasury Department used this gage for collection of duties on imports of wire.

As there are many gages differing from each other, and since even the thicknesses of a certain specified gage, as the Birmingham, are not assumed the same by all manufacturers, *orders for sheets and wires should always state the weight per square foot, or the thickness in thousandths of an inch.*

THE AMERICAN WIRE GAGE (Brown & Sharpe Gage) is practically the only gage used for copper and aluminum wire and, in general, for all wire in electrical work. (See Table 5.) Sizes of stranded conductors larger than No. 0000 AWG are specified by the cross section in circular mils (see below). The American Wire Gage also is used for sheets of copper, aluminum and other nonferrous metals. Wire diameters of successive gage numbers are in geometric progression, the ratio being $92^{1/39} \approx 1.1229$.

CIRCULAR MIL. A circular mil is the area of a circle 1 mil (0.001 in.) in diameter; hence the area of any circle in circular mils is equal to the square of its diameter in mils.

THE STEEL WIRE GAGE is the same gage which formerly was known by the names of Washburn & Moen Gage and the American Steel & Wire Co. Gage. This gage, with certain of its sizes rounded off to even thousandths of an inch, is identical with the Roebling Gage. It is used for practically all steel wire in the United States. (See Table 5.)

Table 4. Weights and Areas of Square and Round Steel Bars

(For weight of bars of other materials, multiply values by coefficients in column 5 of Table 1.)

Size, in.	Weight, lb per ft		Area, sq in.		Size, in.	Weight, lb per ft		Area, sq in.		Size, in.	Weight, lb per ft		Area, sq in.	
	Square	Round	Square	Round		Square	Round	Square	Round		Square	Round	Square	Round
0	......	......	......	......	**3**	30.600	24.033	9.000	7.0686	**6**	122.40	96.13	36.000	28.274
1/16	0.013	0.010	0.0039	0.0031	3 1/16	31.89	25.05	9.379	7.366	6 1/8	127.55	100.18	37.516	29.465
1/8	.053	.042	.0156	.0123	3 1/8	33.20	26.08	9.766	7.670	6 1/4	132.81	104.31	39.063	30.680
3/16	.120	.094	.0352	.0276	3 3/16	34.54	27.13	10.160	7.980	6 3/8	138.18	108.53	40.641	31.919
1/4	.213	.167	.0625	.0491	3 1/4	35.91	28.21	10.563	8.296	6 1/2	143.65	112.82	42.250	33.183
5/16	.332	.261	.0977	.0767	3 5/16	37.31	29.30	10.973	8.618	6 5/8	149.23	117.20	43.891	34.472
3/8	.478	.376	.1406	.1105	3 3/8	38.73	30.42	11.391	8.946	6 3/4	154.91	121.67	45.563	35.785
7/16	.651	.511	.1914	.1503	3 7/16	40.18	31.55	11.816	9.281	6 7/8	160.70	126.22	47.266	37.122
1/2	.850	.668	.2500	.1963	3 1/2	41.65	32.71	12.250	9.621	**7**	166.60	130.85	49.000	38.485
9/16	1.076	.845	.3164	.2485	3 9/16	43.15	33.89	12.691	9.968	7 1/8	172.60	135.56	50.766	39.871
5/8	1.328	1.043	.3906	.3068	3 5/8	44.68	35.09	13.141	10.321	7 1/4	178.71	140.36	52.563	41.282
11/16	1.607	1.262	.4727	.3712	3 11/16	46.23	36.31	13.598	10.680	7 3/8	184.93	145.24	54.391	42.718
3/4	1.913	1.502	.5625	.4418	3 3/4	47.81	37.55	14.063	11.045	7 1/2	191.25	150.21	56.250	44.179
13/16	2.245	1.763	.6602	.5185	3 13/16	49.42	38.81	14.535	11.416	7 5/8	197.68	155.26	58.141	45.664
7/8	2.603	2.044	.7656	.6013	3 7/8	51.05	40.10	15.016	11.793	7 3/4	204.21	160.39	60.063	47.173
15/16	2.988	2.347	.8789	.6903	3 15/16	52.71	41.40	15.504	12.177	7 7/8	210.85	165.67	62.016	48.707
1	3.400	2.670	1.0000	.7854	**4**	54.40	42.73	16.000	12.566	**8**	217.60	170.90	64.000	50.265
1 1/16	3.838	3.015	1.1289	.8866	4 1/16	56.11	44.07	16.504	12.962	8 1/8	224.45	176.28	66.016	51.849
1 1/8	4.303	3.380	1.2656	.9940	4 1/8	57.85	45.44	17.016	13.364	8 1/4	231.41	181.75	68.063	53.456
1 3/16	4.795	3.766	1.4102	1.1075	4 3/16	59.62	46.83	17.535	13.772	8 3/8	238.49	187.30	70.141	55.088
1 1/4	5.313	4.172	1.5625	1.2272	4 1/4	61.41	48.23	18.063	14.186	8 1/2	245.65	192.93	72.250	56.745
1 5/16	5.857	4.600	1.7227	1.3530	4 5/16	63.23	49.66	18.598	14.607	8 5/8	252.93	198.65	74.391	58.426
1 3/8	6.428	5.049	1.8906	1.4849	4 3/8	65.08	51.11	19.141	15.033	8 3/4	260.31	204.45	76.563	60.132
1 7/16	7.026	5.518	2.0664	1.6230	4 7/16	66.95	52.58	19.691	15.466	8 7/8	267.80	210.33	78.766	61.862
1 1/2	7.650	6.008	2.2500	1.7671	4 1/2	68.85	54.07	20.250	15.904	**9**	275.40	216.30	81.000	63.617
1 9/16	8.301	6.519	2.4414	1.9175	4 9/16	70.78	55.59	20.816	16.349	9 1/8	283.10	222.35	83.266	65.397
1 5/8	8.978	7.051	2.6406	2.0739	4 5/8	72.73	57.12	21.391	16.800	9 1/4	290.91	228.48	85.563	67.201
1 11/16	9.682	7.604	2.8477	2.2365	4 11/16	74.71	58.67	21.973	17.257	9 3/8	298.83	234.70	87.891	69.029
1 3/4	10.41	8.178	3.0625	2.4053	4 3/4	76.71	60.25	22.563	17.721	9 1/2	306.85	241.00	90.250	70.882
1 13/16	11.17	8.773	3.2852	2.5802	4 13/16	78.74	61.85	23.160	18.190	9 5/8	314.98	247.38	92.641	72.760
1 7/8	11.95	9.388	3.5156	2.7612	4 7/8	80.80	63.46	23.766	18.665	9 3/4	323.21	253.85	95.063	74.662
1 15/16	12.76	10.02	3.7539	2.9483	4 15/16	82.89	65.10	24.379	19.147	9 7/8	331.55	260.40	97.516	76.589
2	13.60	10.68	4.0000	3.1416	**5**	85.00	66.76	25.000	19.635	**10**	340.00	267.04	100.000	78.540
2 1/16	14.46	11.36	4.2539	3.3410	5 1/16	87.14	68.44	25.629	20.129	10 1/8	348.55	273.75	102.52	80.516
2 1/8	15.35	12.06	4.5156	3.5466	5 1/8	89.30	70.14	26.266	20.629	10 1/4	357.21	280.55	105.06	82.516
2 3/16	16.27	12.78	4.7852	3.7583	5 3/16	91.49	71.86	26.910	21.135	10 3/8	365.98	287.44	107.64	84.541
2 1/4	17.21	13.52	5.0625	3.9761	5 1/4	93.71	73.60	27.563	21.648	10 1/2	374.85	294.41	110.25	86.590
2 5/16	18.18	14.28	5.3477	4.2000	5 5/16	95.96	75.36	28.223	22.166	10 5/8	383.83	301.46	112.89	88.664
2 3/8	19.18	15.06	5.6406	4.4301	5 3/8	98.23	77.15	28.891	22.691	10 3/4	392.91	308.59	115.56	90.763
2 7/16	20.20	15.87	5.9414	4.6664	5 7/16	100.53	78.95	29.566	23.221	10 7/8	402.10	315.81	118.27	92.886
2 1/2	21.25	16.69	6.2500	4.9087	5 1/2	102.85	80.78	30.250	23.758	**11**	411.40	323.11	121.00	95.033
2 9/16	22.33	17.53	6.5664	5.1572	5 9/16	105.20	82.62	30.941	24.301	11 1/8	420.80	330.50	123.77	97.205
2 5/8	23.43	18.40	6.8906	5.4119	5 5/8	107.58	84.49	31.641	24.850	11 1/4	430.31	337.97	126.56	99.402
2 11/16	24.56	19.29	7.2227	5.6727	5 11/16	109.98	86.38	32.348	25.406	11 3/8	439.93	345.51	129.39	101.62
2 3/4	25.71	20.20	7.5625	5.9396	5 3/4	112.41	88.29	33.063	25.967	11 1/2	449.65	353.16	132.25	103.87
2 13/16	26.90	21.12	7.9102	6.2126	5 13/16	114.87	90.22	33.785	26.535	11 5/8	459.48	360.88	135.14	106.14
2 7/8	28.10	22.07	8.2656	6.4918	5 7/8	117.35	92.17	34.516	27.109	11 3/4	469.41	368.68	138.06	108.43
2 15/16	29.34	23.04	8.6289	6.7771	5 15/16	119.86	94.14	35.254	27.688	11 7/8	479.45	376.55	141.02	110.75
										12	489.60	384.53	144.00	113.10

Although the American and Steel gages are practically standard and have official although not legal sanction, the trend of practice is toward the specifying of all sizes of wire and sheet material in decimals of an inch rather than by gage numbers.

STUBS STEEL WIRE GAGE has a somewhat limited use for tool steel wire and drill rods. It should not be confused with the Birmingham Wire Gage, which is sometimes known as Stubs Iron Wire Gage. (See Tables 5 and 6.)

THE STANDARD WIRE GAGE, also known as the New British Standard, the English Legal Standard, or the Imperial Wire Gage, is the legal standard of Great Britain for all wires. It was constructed by so modifying the Birmingham Wire Gage that the differences between successive diameters were the same for short lengths. (See Table 5 under British Standard Wire Gage.)

MUSIC WIRE GAGE. Several gages are in use for music wire. Two of the most extensively used are given in Table 5. For data on others, see *Machinery's Handbook*, the Industrial Press.

Table 5. Wire and Sheet Metal Gages

Gage No.	American Wire Gage, B. & S. in.	Steel Wire Gage (Washburn & Moen), in.	Birmingham Wire Gage (Stubs Iron), in.	Old English Wire Gage (London), in.	Stubs Steel Wire Gage,* in.	British Standard Wire Gage, in.	Music Wire Gage		U. S. Standard Sheet Metal Gage		
							American Steel Wire Co.	Roebling, Trenton Iron Co.	Thickness, in., Approx.	Weight, lb. per sq. ft. Iron	Steel
0 000 000	....	0.4900	....	....	....	0.500	....	....	0.5000	20.0000	20.400
000 000	....	.4615	....	....	....	.464	0.004	....	.4687	18.7500	19.125
00 000	....	.4305	....	....	....	.432	.005	....	.4375	17.5000	17.850
0 000	0.460	.3938	0.454	0.454	....	.400	.006	0.007	.4062	16.2500	16.575
000	.410	.3625	.425	.425	....	.372	.007	.0075	.3750	15.0000	15.300
00	.365	.3310	.380	.380	....	.348	.008	.0085	.3437	13.7500	14.025
0	.325	.3065	.340	.340	....	.324	.009	.009	.3125	12.5000	12.750
1	.289	.2830	.300	.300	.227	.300	.010	.010	.2812	11.2500	11.475
2	.258	.2625	.284	.284	.219	.276	.011	.011	.2656	10.6250	10.837
3	.229	.2437	.259	.259	.212	.252	.012	.012	.2500	10.0000	10.200
4	.204	.2253	.238	.238	.207	.232	.013	.013	.2344	9.3750	9.562
5	.182	.2070	.220	.220	.204	.212	.014	.014	.2187	8.7500	8.925
6	.162	.1920	.203	.203	.201	.192	.016	.016	.2035	8.1250	8.287
7	.144	.1770	.180	.180	.199	.176	.018	.018	.1875	7.5000	7.650
8	.128	.1620	.165	.165	.197	.160	.020	.020	.1719	6.8750	7.012
9	.114	.1483	.148	.148	.194	.144	.022	.022	.1562	6.2500	6.375
10	.102	.1350	.134	.134	.191	.128	.024	.024	.1406	5.6250	5.737
11	.091	.1205	.120	.120	.188	.116	.026	.026	.1250	5.0000	5.100
12	.081	.1055	.109	.109	.185	.104	.029	.028	.1094	4.3750	4.462
13	.072	.0915	.095	.095	.182	.092	.031	.030	.0937	3.7500	3.825
14	.064	.0800	.083	.083	.180	.080	.033	.032	.0781	3.1250	3.187
15	.057	.0720	.072	.072	.178	.072	.035	.034	.0703	2.8125	2.869
16	.051	.0625	.065	.065	.175	.064	.037	.036	.0625	2.5000	2.550
17	.045	.0540	.058	.058	.172	.056	.039	.038	.0562	2.2500	2.295
18	.040	.0475	.049	.049	.168	.048	.041	.040	.0500	2.0000	2.040
19	.036	.0410	.042	.040	.164	.040	.043	.042	.0437	1.7500	1.785
20	.032	.0348	.035	.035	.161	.036	.045	.044	.0375	1.5000	1.530
21	.0285	.0317	.032	.0315	.157	.032	.047	.046	.0344	1.3750	1.402
22	.0253	.0286	.028	.0295	.155	.028	.049	.048	.0312	1.2500	1.275
23	.0226	.0258	.025	.0270	.153	.024	.051	.051	.0281	1.1250	1.147
24	.0201	.0230	.022	.0250	.151	.022	.055	.055	.0250	1.0000	1.020
25	.0179	.0204	.020	.0230	.148	.020	.059	.059	.0219	0.8750	0.892
26	.0159	.0181	.018	.0205	.146	.018	.063	.063	.0187	.7500	.765
27	.0142	.0173	.016	.01875	.143	.0164	.067	.067	.0172	.6875	.701
28	.0126	.0162	.014	.01650	.139	.0148	.071	.071	.0156	.6250	.637
29	.0113	.0150	.013	.01550	.134	.0136	.075	.074	.0141	.5625	.574
30	.0100	.0140	.012	.01375	.127	.0124	.080	.078	.0125	.5000	.510
31	.0089	.0132	.010	.01225	.120	.0116	.085	.082	.0109	.4375	.446
32	.0080	.0128	.009	.01125	.115	.0108	.090	.086	.0101	.40625	.414
33	.0071	.0118	.008	.01025	.112	.0100	.095	.090	.0094	.3750	.382
34	.0063	.0104	.007	.00950	.110	.0092	.100	.095	.0086	.34375	.351
35	.0056	.0095	.005	.00900	.108	.0084	.106	.100	.0078	.3125	.319
36	.0050	.0090	.004	.00750	.106	.0076	.112	.105	.0070	.28125	.287
37	.0045	.0085	....	.00650	.103	.0068	.118	.110	.0066	.26562	.271
38	.0040	.0080	....	.00575	.101	.0060	.124	.115	.0062	.2500	.255
39	.0035	.0075	....	.00500	.099	.0052	.130	.120	....	....	....
40	.0031	.0070	....	.00450	.097	.0048	.138	.125	....	....	....
41	....	.0066	....	....	.095	.0044	.146	.130	....	....	....
42	....	.0062	....	....	.092	.0040	.154	....	....	....	....
43	....	.0060	....	....	.088	.0036	.162	....	....	....	....
44	....	.0058	....	....	.085	.0032	.170	....	....	....	....
45	....	.0055	....	....	.081	.0028	.180	....	....	....	....
46	....	.0052	....	....	.079	.0024	....	....	....	....	....
47	....	.0050	....	....	.077	.0020	....	....	....	....	....
48	....	.0048	....	....	.075	.0016	....	....	....	....	....
49	....	.0046	....	....	.072	.0012	....	....	....	....	....
50	....	.0044	....	....	.069	.0010	....	....	....	....	....

* See also Table 6.

Table 6. Stubs Steel Wire Gage
(For Nos. 1 to 50, see Table 5)

No.	Size, in.	No.	Size, in.	No.	Size, in.	No.	Size, in.	No.	Size, in.	No.	Size, in.
Z	.413	P	.323	F	.257	51	.066	61	.038	71	.026
Y	.404	O	.316	E	.250	52	.063	62	.037	72	.024
X	.397	N	.302	D	.246	53	.058	63	.036	73	.023
W	.386	M	.295	C	.242	54	.055	64	.035	74	.022
V	.377	L	.290	B	.238	55	.050	65	.033	75	.020
U	.368	K	.281	A	.234	56	.045	66	.032	76	.018
T	.358	J	.277	1	See	57	.042	67	.031	77	.016
S	.348	I	.272	to	Table	58	.041	68	.030	78	.015
R	.339	H	.266	50	5	59	.040	69	.029	79	.014
Q	.332	G	.261			60	.039	70	.027	80	.013

EDISON WIRE GAGE. The size of a wire on this gage equals its cross-sectional area in circular mils divided by 1000. This gage is now rarely used.

The U. S. STANDARD GAGE FOR SHEET AND PLATE IRON AND STEEL. An Act of Congress in 1893 established the Standard Gage for sheet iron and steel. It is based on the fact that a cubic foot of iron weighs 480 lb.

A sheet of iron 1 ft square and 1 in. thick weighs 40 lb, or 640 oz, and 1 oz in weight should be $1/640$ in. thick. The scale has been arranged so that each descriptive number represents a certain number of ounces in weight and an equal number of 640ths of an inch in thickness.

According to the law, on and after July 1, 1893, the new gage shall be used in determining duties and taxes levied on sheet and plate iron and steel; and in its application a variation of $2\,1/2\%$ either way may be allowed.

The U. S. Standard Gage for sheet and plate iron is a *weight* gage, not a thickness gage, and is based on the following assumptions:

	Iron	Steel
Specific gravity	7.7	7.854
Weight per cubic foot	480.0	489.6
Weight per cubic inch	0.2778	0.2833

It applies to iron only; for steel the thickness would vary as 480/489.6.

ZINC GAGE is a special gage applied to sheet zinc (see Table 3, p. 4-63). Its use is discouraged, actual thickness in thousandths of an inch being preferred.

METRIC WIRE GAGE number is ten times the diameter in millimeters.

TWIST DRILL GAGE numbers are listed in Table 7. Use of these gage numbers is discouraged in favor of the American Standard series of drill sizes, which consists of 116

Table 7. Twist Drill and Steel Wire Gage
(Manufacturers' standard)

No.	Size, in.	No.	Size, in.	No.	Size, in.	No.	Size, in.	No.	Size, in.	No.	Size, in.
1	.2280	14	.1820	27	.1440	40	.0980	53	.0595	67	.0320
2	.2210	15	.1800	28	.1405	41	.0960	54	.0550	68	.0310
3	.2130	16	.1770	29	.1360	42	.0935	55	.0520	69	.0292
4	.2090	17	.1730	30	.1285	43	.0890	56	.0465	70	.0280
5	.2055	18	.1695	31	.1200	44	.0860	57	.0430	71	.0260
6	.2040	19	.1660	32	.1160	45	.0820	58	.0420	72	.0250
7	.2010	20	.1610	33	.1130	46	.0810	59	.0410	73	.0240
8	.1990	21	.1590	34	.1110	47	.0785	60	.0400	74	.0225
9	.1960	22	.1570	35	.1100	48	.0760	61	.0390	75	.0210
10	.1935	23	.1540	36	.1065	59	.0730	62	.0380	76	.0200
11	.1910	24	.1520	37	.1040	50	.0700	63	.0370	77	.0180
12	.1890	25	.1495	38	.1015	51	.0670	64	.0360	78	.0160
13	.1850	26	.1470	39	.0995	52	.0635	65	.0350	79	.0145
								66	.0330	80	.0135

decimal-size drill diameters, ranging from 0.0156 to 0.500 in., inclusive. See ASA B5.12-1940, *Twist Drills*, also Table 1 in Article 23, Section 23, in this volume.

3. WIRE

STEEL WIRE specifications for engineering applications cover wire intended for springs and in wire rope.

Spring Wire. Table 8 gives the strength of selected sizes of various grades of steel spring wire. These are condensed from ASTM specifications, which should be consulted

Table 8. Ultimate Tensile Strength of Steel Spring Wire

(Condensed from ASTM Standards)

Wire Diameter, in.	Ultimate Tensile Strength, 1000 psi	Wire Diameter, in.	Ultimate Tensile Strength, 1000 psi	Wire Diameter, in.	Ultimate Tensile Strength, 1000 psi
Carbon Steel, Valve Spring Quality (A230-47)		Hard-drawn Steel (A227-47)		Chromium-nickel Corrosion-resisting Steel (A313-47)	
.093–.128	210–230	.028	271–311	.009 and under	325–355
.129–.162	205–225	.041	255–293	.015	310–340
.163–.192	200–220	.072	232–266	.020	300–330
.193–.250	195–215	.120	210–241	.062	255–285
		.177	195–225	.120	225–255
		.250	182–210	.250	175–205
		.375	167–193	.375	140–170
		.500	156–180		
		.625	147–170		
Music Spring Wire (A228-41)		Oil-tempered Steel (A229-41)		Chromium-vanadium Steel, Valve Spring Quality (A231-41)	
.004	439–485	.032	275–310	.032 and under	295–313
.010	387–428	.072	236–271	.041	285–305
.015	365–404	.120	215–250	.054	275–295
.026	337–373	.177	195–225	.062	265–285
.036	321–355	.250	180–210	.080	255–275
.051	303–335	.375	175–205	.105	245–265
.072	287–317	.500	165–195	.135	235–255
.100	271–300			.162	225–245
.121	263–290			.192	220–240
.140	256–283			.244	210–230
.156	251–277			.283	205–225
				.437	195–215

for details of composition, tolerances, etc. *Carbon steel valve spring quality* wire (ASTM A230-47) is a 0.60 to 0.70% carbon steel, hardened and tempered to the desired physical properties. *Music spring wire* (A228-41) is a 0.70 to 1.00% carbon steel, cold drawn to produce the desired physical properties. *Hard-drawn steel spring wire* (A227-47) is a 0.45 to 0.75% carbon steel, cold drawn to produce the desired physicals. *Oil-tempered steel spring wire* (A229-41) is a 0.55 to 0.75% carbon steel, hardened and tempered to produce the desired physicals. *Chromium-nickel corrosion-resisting steel spring wire* (A313-47) is an 18-8 stainless steel, cold drawn to produce the desired physicals. It will operate up to 350 F without noticeable loss in physical properties. Even at 600 F, the loss is small enough to permit its use for thermostat, stove, and other applications. *Chromium-vanadium* steel spring wire (A231-41) and valve spring quality wire (A232-47) are annealed and cold drawn or oil tempered to produce the desired physical properties. They possess high fatigue-resisting properties, especially at moderately elevated temperatures.

Wire for Rope. Table 9 gives the strength of uncoated and galvanized wire used in the manufacture of wire rope. These are part of the Federal specifications for wire rope.

Table 9. Ultimate Tensile Strength of Main Wires for Rope

(Data from Federal Standard Stock Catalog)

Wire Material and Diam, in.	Minimum Tensile Strength, 1000 psi Uncoated	Galvanized
Iron	70	55
Annealed steel	...	60
Highway-guard steel	...	90
Traction steel	160	.....
Mild plow steel		
.030 and under	184	165.6
.031–.060	180	162
.061–.100	175	157.5
.101–.140	170	153
.141–.190	165	148.5
Plow steel and bridge-rope steel		
.030 and under	212	190.8
.031–.060	206	185.4
.061–.100	200	180
.101–.140	195	175.5
.141–.190	190	171
Improved plow steel		
.030 and under	244	219.6
.031–.060	238	214.2
.061–.100	230	207
.101–.140	225	202.5
.141–.190	218	196.2
Phosphor bronze	90	.....

The strengths of various wire rope constructions are given in Article 4 of this section. Zinc-coated steel wire strand is also covered by ASTM A122-41 and A218-41.

COPPER WIRE specifications for engineering applications cover wire intended for use in electrical conductors. Strength requirements of hard-drawn copper wire (ASTM B1-47), medium-hard copper wire (B2-47), and soft or annealed copper wire (B3-45) are given in Table 10. Other ASTM specifications cover tinned soft or annealed copper wire for electrical purposes (B33-46), as well as copper trolley wire and various types of stranded conductors. Copper alloy wires are also covered by ASTM specifications, including hard-drawn copper alloy wires for electrical conductors (B105-47), beryllium-copper alloy wire (B197-47T), brass wire (B134–48), and bronze trolley wire. Availability of various copper alloys in wire form is indicated in Table 4 on pp. 4-30 to 4-37.

4. WIRE ROPE *

Wire rope is a machine element designed primarily to transmit motion in any plane, turn corners readily and easily, and operate in restricted space. Because of its high weight-strength ratio it has almost entirely superseded chain or hemp rope for hauling, hoisting, or supporting loads. The multiple-wire construction insures the maintenance of adequate safety factors and provides the means for estimating its condition during

* The assistance of John A. Roebling's Sons Co. in revising this article is acknowledged with appreciation.

Table 10. Properties of Copper Wire at 68 F

(Condensed from ASTM Specifications)

Diameter, in.	Area		Hard-drawn (B1-47)		Medium-hard (B2-47)			Soft (B3-45)	
	Cir. mils	Sq in.	Minimum Tensile Strength, psi	Minimum Elongation,* %	Tensile Strength, psi Min.	Tensile Strength, psi Max.	Minimum Elongation,* %	Minimum Tensile Strength, psi	Minimum Elongation in 10 in., %
.4600	211,600	.1662	49,000	3.75	42,000	49,000	3.75	36,000	35
.4096	167,800	.1318	51,000	3.25	43,000	50,000	3.60	36,000	35
.3648	133,100	.1045	52,800	2.80	44,000	51,000	3.25	36,000	35
.3249	105,500	.08289	54,500	2.40	45,000	52,000	3.00	36,000	35
.2893	83,690	.06573	56,100	2.17	46,000	53,000	2.75	37,000	30
.2576	66,370	.05213	57,600	1.98	47,000	54,000	2.50	37,000	30
.2294	52,640	.04134	59,000	1.79	48,000	55,000	2.25	37,000	30
.2043	41,740	.03278	60,100	1.24	48,330	55,330	1.25	37,000	30
.1819	33,100	.02600	61,200	1.18	48,660	55,660	1.20	37,000	30
.165 †	27,225	.02138	62,000	1.14					
.1620	26,250	.02062	62,100	1.14	49,000	56,000	1.15	37,000	30
.1443	20,820	.01635	63,000	1.09	49,330	56,330	1.11	37,000	30
.134 †	17,956	.01410	63,400	1.07					
.1285	16,510	.01297	63,700	1.06	49,660	56,660	1.08		
.1144	13,090	.01028	64,300	1.02	50,000	57,000	1.06	37,000	30
.104 †	10,816	.008495	64,800	1.00					
.1019	10,380	.008155	64,900	1.00	50,330	57,330	1.04	38,500	25
.092 †	8,464	.006648	65,400	0.97					
.09074	8,234	.006467	65,400	0.97	50,660	57,660	1.02	38,500	25
.08081	6,530	.005129	65,700	0.95	51,000	58,000	1.00	38,500	25
.080 †	6,400	.005027	65,700	0.94					
.07196	5,178	.004067	65,900	0.92	51,330	58,330	0.98	38,500	25
.065 †	4,225	.003318	66,200	0.91					
.06408	4,107	.003225	66,200	0.90	51,660	58,660	0.96	38,500	25
.05707	3,257	.002558	66,400	0.89	52,000	59,000	0.94	38,500	25
.05082	2,583	.002028	66,600	0.87	52,330	59,330	0.92	38,500	25
.04526	2,048	.001609	66,800	0.86	52,660	59,660	0.90	38,500	25
.04030	1,624	.001276	67,000	0.85	53,000	60,000	0.88	38,500	25

* In 10 in. for wire diameter 0.2294 in. and above; in 60 in. for wire diameter 0.2043 in. and below.

† Often employed for communication lines; not in the American Wire Gage (B & S) series.

operation. Like any other intricate equipment, however, it must be handled and maintained with care in order to provide safe and economical service.

Wire rope is made in four standard grades which, in ascending order of strength, are: iron, mild plow steel, plow steel, and improved plow steel. The most common constructions are made of six strands laid around a fiber or metallic core. The metallic core may be a wire strand or an independent wire rope. Numerous types and sizes of wire rope are engineered for specific purposes, the number, size, and arrangement of wires in each strand providing the different properties. For example, a rope with six strands of seven wires each is less flexible than one with 37 wires to the strand, but the 6 by 7 rope is more resistant to wear and abrasion.

Selection of the proper size, grade, and construction should be done by thorough consideration of all pertinent data, including all loads; acceleration; deceleration; rope speed; rope attachments; the number, size, and arrangement of all sheaves and drums; conditions causing corrosion and abrasion; length of rope used; economical rope life; and protection of life and property. The following information, therefore, is briefed in order to help in evaluating the foregoing factors.

BREAKING STRENGTH OF WIRE ROPE. The various manufacturers have adopted standard figures for the strength of all sizes and qualities of wire rope. Formerly it was the custom to test the individual wires and to consider their combined strength as the strength of the rope as a whole. These strengths were greater than the actual strength obtained by breaking the finished rope. The figures given in Tables 13 to 28 represent actual breaks of the various ropes and range from 95 to 80% or less of the combined strength of the single wires, depending on the construction.

LAY OF WIRE ROPE. Regular lay of wire rope comprises wires in the strands laid to the left, the strands being laid to the right. This is known as right-lay rope. If the wires are laid to the right and strands to the left, it is known as left-lay rope. Standard rope is made regular lay, right lay. In Lang lay rope, the wires in the strands and the strands themselves are laid in the rope in the same direction, either right or left.

Lang lay rope is somewhat more flexible than regular lay rope and, as the wires are laid more axially in the rope, longer surfaces are exposed to wear and the endurance thereby is increased. Owing to improvements in its manufacture, Lang lay rope now is used extensively in the United States. Standard Lang lay rope requires more care in handling and use than regular lay rope. By preforming, its tendency to untwist has been greatly reduced; also, the unlaying of the strands when the rope is cut has been eliminated. Improved handling, flexibility, drum winding, and ultimate rope life are also obtainable.

SHEAVES AND DRUMS. Drums and sheaves of the largest practicable diameter are recommended in all wire rope installations. See Table 11. If possible, drums should

Table 11. Recommended Minimum Diameter of Sheaves and Drums for Wire Rope

(Federal Specification RR-R-571a)

No. of Strands	Wires per Strand	Minimum Sheave or Drum Diameter
5	19	45
6	7	72
6	19	45
6	25	45
6	29	30
6	37	27
6	42	24
8	19	31

(Diameters are × the nominal diameter of the wire rope.)

Average recommended diameter is about 50% greater than the minimum.

be lagged, and, where feasible, a grooved drum on hoists is more desirable than a flat drum. The grooves should give ample clearance between successive windings; thus a drum for 3/4-in. rope should have the grooves at least 7/8 in. apart on centers. The grooves should be made smooth in order not to cut the rope, and they should be of slightly larger radius than the rope in order to avoid wedging or pinching it. See Table 12. Overwinding,

Table 12. Requirements for Groove Diameter

(Federal Specification RR-R-571a)

Nominal Rope Diameter, in.	Amount Groove Diameter Should Be Larger than Nominal Rope Diameter: For Minimum Conditions	For New or Regrooved Sheaves
1/4 to 5/16	1/64	1/32
3/8 to 3/4	1/32	1/16
13/16 to 1 1/8	3/64	3/32
1 3/16 to 1 1/2	1/16	1/8
1 9/16 to 2 1/4	3/32	3/16
2 5/16 and over	1/8	1/4

that is, the winding of the rope in more than one layer, is to be avoided, if possible, by making the drum large enough to take all the rope in a single layer. Overwinding will rapidly destroy the rope, and the extra cost of the larger drum will be more than compensated by the greater life of the rope. The best possible alignment of sheaves and drums should be made to avoid undue wear on the sides of the sheaves and the rope. In general, the lead sheaves over which the rope runs from the drum should be aligned with the center of the drum or, if the drum is not entirely filled, with the center of the portion on which the rope is wound. The distance between the drum and lead sheave should be such as to cause an angle not exceeding 1° 30′ between the line from the center of the sheave to the center of the drum, and the line from the center of the sheave to the outer side of the drum. When the sheaves become worn, they should be replaced or the grooves turned before they are used with a new wire rope; otherwise the rope will not work properly. When changing an old rope for a new one, the sheaves should be examined, and, if scored or cut from the old rope, they should be turned, relined or renewed, so that the new rope will not receive undue wear. For many purposes, particularly mine service, the grooves can advantageously be lined with well-seasoned, hardwood blocks set on end, which can be renewed when worn. Large sheaves, running at high velocity, should be lined with leather set on end, or with rubber. This is the usual practice for power transmission between distant points, where the rope frequently runs at a velocity of 4000 ft per min.

REVERSE BENDING. Reverse bending, that is, bending the wire rope first in one direction over sheaves and then in the opposite direction, is to be avoided wherever possible. This practice will wear out a rope more quickly than any other known method. A little care in design usually will eliminate all situations that call for reverse bending, and it is even desirable to change existing constructions if necessary to remove this condition. The expense of rope renewals will more than equal the cost of change, as a rule.

HANDLING WIRE ROPE. Wire rope must not be coiled or uncoiled like hemp rope. When received in a coil, it should be rolled on the ground like a hoop and straightened out before being put on the sheaves. If on a reel, it should be mounted on a spindle or a flat turntable and properly unwound. Kinking or untwisting must be avoided.

PROTECTION AND LUBRICATION OF WIRE ROPE. Wire rope should be protected by a suitable lubricant, both externally and internally, to prevent rust and to keep it pliable. The lubricant should be suited to the conditions under which the rope is used. It should not contain acids or other corrosive agents and should readily penetrate the strands. It should be self spreading, adhesive, and water repelling. The higher the speed of the rope, the lower should be the viscosity of the lubricant. For stationary ropes for low-speed work, as on shovels, cranes, or derricks, a heavy petroleum oil or grease, viscosity approximately 1000 sec at 210 F, applied hot, is advised. For heavy haulage, as slope and logging ropes, a lubricant of viscosity of 500 to 800 sec at 210 F should be used. For high-speed hoisting, as in mine hoists, a lubricant of viscosity of 400 to 500 sec at 100 F is recommended, with automatic lubrication. In all cases, the viscosity should be such that the lubricant will cover the rope, penetrate thoroughly, withstand pressure, but yet not be so thin as to run off the rope. Crude oils should not be used. New lubricants only should be applied. Lubricants should be applied hot, as the penetration thereby is increased, and when cooled the lubricant exists as a plastic filler to lubricate the wires and to exclude moisture. The lubricant is best applied by passing the rope through a bath of heated oil. Placing a drum of heated oil over the rope and allowing the oil to flow on the rope is wasteful of oil. When hand lubrication is necessary, a piece of sheepskin with the wool coat on the inside is recommended as a wiper, the wool being saturated with oil, and held around the rope as it passes through the wiper. While in storage, wire rope should be kept dry and protected from the weather, acid, and acid fumes. When a used rope is stored, it should be thoroughly cleaned before oiling and then treated as a new rope. For special cases, the manufacturers of wire rope should be consulted as to the type and method of lubrication.

EXPOSURE TO HEAT. Where wire rope is exposed to intense heat, as in foundry or steel mill service, a wire core is often substituted for the fiber core.

GALVANIZED WIRE ROPE is used for "standing" service, where no bending is required, as guys, ship's rigging, towing and mooring lines, etc. It is not adapted to hoisting service, as the galvanized coating is abraded and worn away by the sheaves, thus promoting corrosion. The more commonly used types of galvanized wire rope are described below.

Galvanized standing and guy rope is made in all grades. Iron guy rope, being soft, is more readily subject to damage than the harder, high-strength, steel grades and is not recommended. The higher grades have an elastic limit which is a greater percentage of their ultimate strength and hence are more adaptable to designs where the elastic limit is the design criterion. Galvanized rope in the higher-strength grade has often been used for guying radio towers, stacks, and derricks.

Guy rope is made with 6 strands of 7 or 19 wires per strand. A steel core is generally preferred, owing to the higher strength and decreased stretch which it affords. Galvanized iron rigging and guy rope (6 by 7 sizes) are listed in Table 13. Steel rope of 6 by 7 sizes

Table 13. Standard 6 by 7 Galvanized Iron Rigging and Guy Rope

(U. S. Department of Commerce, *Simplified Practice Recommendation R198-43*)

Diameter, in.	Approximate Weight, lb/ft	Breaking Strength in Tons of 2000 lb	Diameter, in.	Approximate Weight, lb/ft	Breaking Strength in Tons of 2000 lb	Diameter, in.	Approximate Weight, lb/ft	Breaking Strength in Tons of 2000 lb
1/4	0.094	0.918	9/16	0.48	4.51	1	1.50	13.8
5/16	0.15	1.42	5/8	0.59	5.54	1 1/16	1.70	15.5
3/8	0.21	2.04	3/4	0.84	7.90	1 1/8	1.90	17.3
7/16	0.29	2.76	13/16	0.99	9.23	1 3/16	2.12	19.2
1/2	0.38	3.58	7/8	1.15	10.7	1 1/4	2.34	21.2

is listed in Table 14 and 6 by 19 sizes in Table 15. It should be noted that galvanizing decreases the strengths 10% from those listed in Tables 14 and 15.

Table 14. Standard 6 by 7 Wire Rope *

(U. S. Department of Commerce, *Simplified Practice Recommendation R198-43*)

Diameter, in.	Approximate Weight, lb/ft	Breaking Strength in Tons of 2000 lb			Diameter, in.	Approximate Weight, lb/ft	Breaking Strength in Tons of 2000 lb		
		Improved Plow Steel	Plow Steel	Mild Plow Steel			Improved Plow Steel	Plow Steel	Mild Plow Steel
1/4	0.094	2.64	2.30	2.00	3/4	0.84	22.7	19.8	17.2
5/16	0.15	4.10	3.56	3.10	7/8	1.15	30.7	26.7	23.2
3/8	0.21	5.86	5.10	4.43	1	1.50	39.7	34.5	30.0
7/16	0.29	7.93	6.90	6.00	1 1/8	1.90	49.8	43.3	37.7
1/2	0.38	10.3	8.96	7.79	1 1/4	2.34	61.0	53.0	46.1
9/16	0.48	13.0	11.3	9.82	1 3/8	2.84	73.1	63.6	55.3
5/8	0.59	15.9	13.9	12.0	1 1/2	3.38	86.2	75.0	65.2

* For these ropes with steel cores, add 7 1/2% to the above strengths. For these ropes when galnized, deduct 10% from the above strengths.

Table 15. Standard 6 by 19 Wire Rope *

(U. S. Department of Commerce, *Simplified Practice Recommendation R198-43*)

Diameter, in.	Approximate Weight, lb/ft	Breaking Strength in Tons of 2000 lb			Diameter, in.	Approximate Weight lb/ft	Breaking Strength in Tons of 2000 lb		
		Improved Plow Steel	Plow Steel	Mild Plow Steel			Improved Plow Steel	Plow Steel	Mild Plow Steel
1/4	0.10	2.74	2.39	2.07	1 1/4	2.50	64.6	56.2	48.8
5/16	0.16	4.26	3.71	3.22	1 3/8	3.03	77.7	67.5	58.8
3/8	0.23	6.10	5.31	4.62	1 1/2	3.60	92.0	80.0	69.6
7/16	0.31	8.27	7.19	6.25	1 5/8	4.23	107.0	93.4	81.2
1/2	0.40	10.7	9.35	8.13	1 3/4	4.90	124.0	108.0	93.6
9/16	0.51	13.5	11.8	10.2	1 7/8	5.63	141.0	123.0	107.0
5/8	0.63	16.7	14.5	12.6	2	6.40	160.0	139.0	121.0
3/4	0.90	23.8	20.7	18.0	2 1/8	7.23	179.0	156.0	
7/8	1.23	32.2	28.0	24.3	2 1/4	8.10	200.0	174.0	
1	1.60	41.8	36.4	31.6	2 1/2	10.0	244.0	212.0	
1 1/8	2.03	52.6	45.7	39.8	2 3/4	12.10	292.0	254.0	

* For these ropes with steel cores, add 7 1/2% to the above strengths. For these ropes when galvanized, deduct 10% from the above strengths.

STEEL WIRE ROPE. Simplified Practice Recommendation R198-43 published by the U. S. Department of Commerce lists four constructions, standard sizes of which are given in Tables 14, 15, 16, and 17. Their characteristics and uses are as follows:

Table 16. Standard 8 by 19 Wire Rope *

(U. S. Department of Commerce, *Simplified Practice Recommendation R198-43*)

Diameter, in.	Approximate Weight, lb/ft	Breaking Strength in Tons of 2000 lb		Diameter, in.	Approximate Weight, lb/ft	Breaking Strength in Tons of 2000 lb	
		Improved Plow Steel	Plow Steel			Improved Plow Steel	Plow Steel
1/4	0.09	2.35	2.04	3/4	0.82	20.5	17.8
5/16	0.14	3.65	3.18	7/8	1.11	27.7	24.1
3/8	0.20	5.24	4.55	1	1.45	36.0	31.3
7/16	0.28	7.09	6.17	1 1/8	1.84	45.3	39.4
1/2	0.36	9.23	8.02	1 1/4	2.27	55.7	48.4
9/16	0.46	11.6	10.1	1 3/8	2.74	67.1	58.3
5/8	0.57	14.3	12.4	1 1/2	3.26	79.4	69.1

* For these ropes with steel cores, add 7 1/2% to the above strengths. For these ropes when galvanized, deduct 10% from the above strengths.

Table 17. Standard 6 by 37 Wire Rope *

(U. S. Department of Commerce, *Simplified Practice Recommendation R198-43*)

Diameter, in.	Approximate Weight, lb/ft	Breaking Strength in Tons of 2000 lb		Diameter, in.	Approximate Weight, lb/ft	Breaking Strength in Tons of 2000 lb	
		Improved Plow Steel	Plow Steel			Improved Plow Steel	Plow Steel
1/4	0.10	2.59	2.25	1 1/2	3.49	87.9	76.4
5/16	0.16	4.03	3.50	1 5/8	4.09	103.0	89.3
3/8	0.22	5.77	5.02	1 3/4	4.75	119.0	103.0
7/16	0.30	7.82	6.80	1 7/8	5.45	136.0	118.0
1/2	0.39	10.2	8.85	2	6.20	154.0	134.0
9/16	0.49	12.9	11.2	2 1/8	7.00	173.0	150.0
5/8	0.61	15.8	13.7	2 1/4	7.85	193.0	168.0
3/4	0.87	22.6	19.6	2 1/2	9.69	236.0	205.0
7/8	1.19	30.6	26.6	2 3/4	11.72	284.0	247.0
1	1.55	39.8	34.6	3	13.95	335.0	291.0
1 1/8	1.96	50.1	43.5	3 1/4	16.37	390.0	339.0
1 1/4	2.42	61.5	53.5	3 1/2	19.00	449.0	390.0
1 3/8	2.93	74.1	64.5				

* For these ropes with steel cores, add 7 1/2% to the above strengths. For these ropes when galvanized, deduct 10% from the above strengths.

Standard 6 by 7 wire rope with fiber core (Table 14) is made of large wires and is the least flexible of any standard rope. Mild plow-steel rope is particularly adapted to mine haulage work, including tail rope and endless haulage systems, gravity hoists, and coal and ore dock haulage, roads, operating small grip cars, etc. The sizes 3/8 to 5/8 in., inclusive, are used for sand lines in oil wells, and from 5/8 to 1 in. for oil-well drilling. In general, it can be used for severe service and where minimum flexibility is required. The plow-steel rope is advised for situations similar to those for which the mild plow-steel ropes are used but where it is necessary to secure increased strength without altering the working conditions. The wires are harder and capable of standing greater wear than any of the foregoing ropes. Improved plow-steel rope is the strongest and stiffest of all and is used for work demanding the greatest strength and lightest rope possible. Sheaves for this rope should, if possible, be somewhat larger than for other grades.

Standard 6 by 19 wire rope with fiber core (Table 15) is more flexible than the 6 by 7 rope, inasmuch as the wires are smaller for the same size rope. It will not stand as much abrasion as transmission rope. The fiber-center rope is very pliable and will wind on moderate-size drums and sheaves without injury. If possible, drums and sheaves should be larger than shown in Table 11. The mild plow-steel, plow-steel, and improved plow-steel ropes are adapted to mine hoisting, logging, quarrying, elevators, derricks, hay presses, dredges, cableways, inclined planes, coal hoists, conveyors, ballast unloaders, skip hoists, and similar applications.

Standard 8 by 19 wire rope with fiber core (Table 16) has even greater flexibility, permitting its use on smaller sheaves and drums, such as usually are found on derricks. It is not advisable to use it where there is much overwinding, as it will flatten much more quickly than the 6 by 19 standard rope. Its uses are the same as those of standard 6 by 19 hoisting rope.

Standard 6 by 37 wire rope with fiber core (Table 17) is extremely flexible and is especially adapted to high-speed service on cranes or where sheaves are small. It will not stand so much abrasion as the 6 by 19 rope, but it is particularly efficient, as over 50% of the wires are in the inner layers and are protected from abrasion. The improved plow-steel rope is largely used on dredges for both main and spud ropes as well as hoist and boom ropes on large power shovels and draglines.

FLATTENED STRAND ROPE has triangular or sector-shaped strands which allow greater metallic area than is possible with round strand ropes. The shape of the strands gives more wearing surface and increased resistance to crushing. However, this construction is not so flexible or so fatigue resistant as round strand rope of the same general classification.

Flattened strand ropes are made in three styles, B and G for hoisting and D for haulage (see Tables 18, 19, and 20).

NONROTATING ROPE (Table 21) is fabricated by laying 18 7-wire strands around a fiber core in two concentric layers, the inside and outside layers being laid in opposite directions. The tendency for twisting between the inside and outside layers is therefore opposite, which balances the rope and establishes the nonrotating characteristic.

Table 18. Standard 6 by 25 Type B Flattened Strand Wire Rope *

(U. S. Department of Commerce, *Simplified Practice Recommendation R198-43*)

Diameter, in.	Approximate Weight, lb/ft	Breaking Strength in Tons of 2000 lb		Diameter, in.	Approximate Weight, lb/ft	Breaking Strength in Tons of 2000 lb	
		Improved Plow Steel	Mild Plow Steel			Improved Plow Steel	Mild Plow Steel
3/8	0.25	6.71	....	1 3/8	3.40	85.5	....
1/2	0.45	11.8	8.94	1 1/2	4.05	101.0	....
9/16	0.57	14.9	11.2	1 5/8	4.75	118.0	....
5/8	0.70	18.3	13.9	1 3/4	5.51	136.0	....
3/4	1.01	26.2	19.8	2	7.20	176.0	....
7/8	1.39	35.4	26.8	2 1/4	9.10	220.0	....
1	1.80	46.0	34.8	2 1/2	11.20	269.0	....
1 1/8	2.28	57.9	43.8	2 3/4	13.60	321.0	....
1 1/4	2.81	71.0	53.7				

* For these ropes with steel cores, add 7 1/2% to above strengths.

Table 19. Standard 6 by 30 Type G Flattened Strand Wire Rope *

(U. S. Department of Commerce, *Simplified Practice Recommendation R198-43*)

Diameter, in.	Approximate Weight, lb/ft	Breaking Strength in Tons of 2000 lb		Diameter, in.	Approximate Weight, lb/ft	Breaking Strength in Tons of 2000 lb	
		Improved Plow Steel	Mild Plow Steel			Improved Plow Steel	Mild Plow Steel
5/8	0.70	18.3	13.9	1 1/2	4.05	101.0	....
3/4	1.01	26.2	19.8	1 5/8	4.75	118.0	....
7/8	1.39	35.4	26.8	1 3/4	5.51	136.0	....
1	1.80	46.0	34.8	2	7.20	176.0	....
1 1/8	2.28	57.9	43.8	2 1/4	9.10	220.0	....
1 1/4	2.81	71.0	53.7	2 1/2	11.20	269.0	....
1 3/8	3.40	85.5	....	2 3/4	13.60	321.0	....

* For these ropes with steel cores, add 7 1/2% to above strengths.

Table 20. Standard 6 by 8 Type D Flattened Strand Wire Rope

(U. S. Department of Commerce, *Simplified Practice Recommendation R198-43*)

Diameter, in.	Approximate Weight, lb/ft	Breaking Strength in Tons of 2000 lb		Diameter, in.	Approximate Weight, lb/ft	Breaking Strength in Tons of 2000 lb	
		Improved Plow Steel	Mild Plow Steel			Improved Plow Steel	Mild Plow Steel
1/2	0.45	11.1	8.37	1 1/8	2.28	53.5	40.5
5/8	0.70	17.1	12.9	1 1/4	2.81	65.5	49.5
3/4	1.01	24.4	18.5	1 3/8	3.40	78.6	59.4
7/8	1.39	33.0	24.9	1 1/2	4.05	92.7	70.1
1	1.80	42.7	32.3				

Table 21. Standard 18 by 7 Nonrotating Rope *

(U. S. Department of Commerce, *Simplified Practice Recommendation R198-43*)

Diameter, in.	Approximate Weight, lb/ft	Breaking Strength in Tons of 2000 lb		Diameter, in.	Approximate Weight, lb/ft	Breaking Strength in Tons of 2000 lb	
		Improved Plow Steel	Plow Steel			Improved Plow Steel	Plow Steel
3/8	0.24	5.59	4.86	1	1.73	38.3	33.3
7/16	0.33	7.58	6.59	1 1/8	2.19	48.2	41.9
1/2	0.43	9.85	8.57	1 1/4	2.70	59.2	51.5
9/16	0.55	12.4	10.8	1 3/8	3.27	71.3	62.0
5/8	0.68	15.3	13.3	1 1/2	3.89	84.4	73.4
3/4	0.97	21.8	19.0	1 5/8	4.57	98.4	85.6
7/8	1.32	29.5	25.7	1 3/4	5.30	114.0	98.8

* For these ropes when galvanized, deduct 10% from the above strengths.

Nonrotating rope is used for hoisting on equipment where rotation is to be restricted. However, such ropes have less strength than similar-size ropes of other construction.

ELEVATOR ROPE. Table 22 gives the characteristics of standard 6 by 19 and 8 by 19 elevator rope.

Table 22. Standard Elevator Rope

(U. S. Department of Commerce, *Simplified Practice Recommendation R198-43*)

Diameter, in.	6 by 19 Elevator Rope				8 by 19 Elevator Rope			
	Approximate Weight, lb/ft	Breaking Strength			Approximate Weight, lb/ft	Breaking Strength		
		Iron	Traction Steel	High-rise Traction Steel		Iron	Traction Steel	High-rise Traction Steel
3/16	0.06	1,300	....	....	0.05	1,000	....	....
1/4	0.10	2,200	3,600	....	0.09	1,800	3,600	....
5/16	0.16	3,200	5,600	....	0.14	2,900	5,600	....
3/8	0.23	5,000	8,200	....	0.20	4,200	8,200	....
7/16	0.31	6,400	11,000	....	0.28	5,600	11,000	....
1/2	0.40	8,400	14,500	....	0.36	7,200	14,500	....
9/16	0.51	10,600	18,500	....	0.46	9,200	18,500	....
5/8	0.63	12,800	23,000	....	0.57	11,200	23,000	....
11/16	0.76	....	27,000	30,000	0.69	....	27,000	30,000
3/4	0.90	18,200	32,000	....	0.82	16,000	32,000	....
13/16	1.06	....	37,000	46,000	0.96	....	37,000	46,000
7/8	1.23	24,800	42,000	....	1.11	21,400	42,000	....
15/16	1.41	....	48,000	60,000	1.27	....	48,000	60,000
1	1.60	32,000	54,000	....	1.45	28,000	54,000	....
1 1/16	1.81	....	61,000	....	1.64	....	61,000	....

MARLINE-CLAD ROPE is a round strand rope, each strand being given an external serving of tarred marline (see Table 23). It is particularly adapted for ship rigging, cargo hoists, power transmission, grain shovels, etc.

Table 23. Standard 5 by 19 Marline Clad Rope *

(U. S. Department of Commerce, *Simplified Practice Recommendation R198-43*)

Diameter, in.		Approximate Weight, lb/ft	Breaking Strength in Tons of 2000 lb		Diameter, in.		Approximate Weight, lb/ft	Breaking Strength in Tons of 2000 lb	
Before Serving	After Serving		Plow Steel	Mild Plow Steel	Before Serving	After Serving		Plow Steel	Mild Plow Steel
1/4	9/16	0.21	2.17	1.89	7/8	1 1/4	1.32	25.5	22.1
5/16	5/8	0.28	3.37	2.93	1	1 3/8	1.70	33.1	28.7
3/8	11/16	0.36	4.82	4.20	1 1/8	1 1/2	2.12	41.6	36.2
7/16	3/4	0.42	6.53	5.68	1 1/4	1 5/8	2.58	51.1	44.4
1/2	13/16	0.51	8.50	7.39	1 3/8	1 3/4	3.14	61.4	53.4
9/16	7/8	0.62	10.7	9.31	1 1/2	1 7/8	3.69	....	....
5/8	1	0.81	13.2	11.4	1 5/8	2	4.29	....	....
3/4	1 1/8	1.10	18.8	16.4	1 3/4	2 1/8	5.00	....	....

* For these ropes with steel cores, add 7 1/2% to the above strengths. For these ropes when galvanized, deduct 10% from the above strengths.

FLAT ROPE. Flat rope consists of a number of "flat-rope" strands, laid alternately right and left, then placed side by side and sewed together with soft steel wire, to form a flat rope of the desired width and thickness. The soft sewing wires wear much quicker than the rope wire and have to be replaced from time to time, at which time worn strands can also be renewed. Flat rope is used principally for hoisting heavy loads out of deep shafts. It requires a reel but little larger than the width of the rope, whereas round rope necessitates the use of a large drum. Its use is recommended where saving of machinery space is an object. It does not twist or spin in the shaft. It is also used for operating spouts on coal and ore docks, for raising and lowering emergency gates on canals, and for similar machinery. For details of methods of fastening it to drums, the manufacturers should be consulted. Drums and sheaves should be as large as possible. A rule for the diameter of the drum is $D = ct$, where D is diameter of drum at bottom in inches, t = thickness of rope in inches, and c = 100 for drums and 160 for sheaves. Sheaves should be crowned at the center and have deep flanges to guide the rope. (See Table 24.)

Table 24. Standard Flat Rope

(U. S. Department of Commerce, *Simplified Practice Recommendation R198-43*)

Width and Thickness, in.	Number of Ropes	Approximate Weight, lb/ft	Breaking Strength in Tons of 2000 lb		Width and Thickness, in.	Number of Ropes	Approximate Weight, lb/ft	Breaking Strength in Tons of 2000 lb	
			Plow Steel	Mild Plow Steel				Plow Steel	Mild Plow Steel
1/4 x 1 1/2	7	0.69	16.8	14.6	1/2 x 4	9	3.16	81.8	71.2
1/4 x 2	9	0.88	21.7	18.8	1/2 x 4 1/2	10	3.82	90.9	79.1
1/4 x 2 1/2	11	1.15	26.5	23.0	1/2 x 5	12	4.16	109.0	94.9
1/4 x 3	13	1.34	31.3	27.2	1/2 x 5 1/2	13	4.50	118.0	103.0
5/16 x 1 1/2	5	0.77	18.5	16.0	1/2 x 6	14	4.85	127.0	111.0
5/16 x 2	7	1.05	25.8	22.4	1/2 x 7	16	5.85	145.0	126.0
5/16 x 2 1/2	9	1.33	33.2	28.8	5/8 x 3 1/2	6	3.40	85.8	74.6
5/16 x 3	11	1.61	40.5	35.3	5/8 x 4	7	3.95	100.0	87.1
5/16 x 3 1/2	13	1.89	47.9	41.7	5/8 x 4 1/2	8	4.50	114.0	99.5
5/16 x 4	15	2.17	55.3	48.1	5/8 x 5	9	5.04	129.0	112.0
3/8 x 2	6	1.25	31.4	27.3	5/8 x 5 1/2	10	5.59	143.0	124.0
3/8 x 2 1/2	8	1.64	41.8	36.4	5/8 x 6	11	6.14	157.0	137.0
3/8 x 3	9	1.84	47.1	40.9	5/8 x 7	13	7.23	186.0	162.0
3/8 x 3 1/2	11	2.23	57.5	50.0	5/8 x 8	15	8.32	214.0	186.0
3/8 x 4	12	2.44	62.7	54.6	3/4 x 5	8	6.50	165.0	143.0
3/8 x 4 1/2	14	2.83	73.2	63.7	3/4 x 6	9	7.31	185.0	161.0
3/8 x 5	15	3.03	78.4	68.2	3/4 x 7	10	8.13	206.0	179.0
3/8 x 5 1/2	17	3.42	88.9	77.3	3/4 x 8	11	9.70	227.0	197.0
3/8 x 6	18	3.63	94.1	81.9	7/8 x 5	7	7.50	190.0	165.0
1/2 x 2 1/2	6	2.13	54.5	47.4	7/8 x 6	8	8.56	217.0	188.0
1/2 x 3	7	2.47	63.6	55.4	7/8 x 7	9	9.63	244.0	212.0
1/2 x 3 1/2	8	2.82	72.7	63.3	7/8 x 8	10	10.69	271.0	236.0

RUNNING ROPES AND HAWSERS are made of 6 12-, 24-, or 37-wire strands. (See Tables 25, 26, and 27.) The 6 x 12 and 6 x 24 constructions have fiber core in each strand for increased flexibility; the 6 x 37 hawsers have all steel strands. They are lighter than hemp hawsers of equal strength and are flexible enough to work around the same bitts as hemp rope of equivalent strength. The following is a comparison of hawsers of equal strength:

	Tarred Russian Hemp	White Manila Hawser	4-in. Galvanized Steel Hawser	Studded Link Chain
Weight per fathom, lb	39	18	13	68
Circumference, in.	13	10	4	1 1/8 diam.

Table 25. Standard 6 by 12 Galvanized Running Rope and Hawsers

(U. S. Department of Commerce, *Simplified Practice Recommendation R198-43*)

Diameter, in.	Approximate Weight, lb/ft	Breaking Strength in Tons of 2000 lb			Diameter, in.	Approximate Weight, lb/ft	Breaking Strength in Tons of 2000 lb		
		Galvanized Improved Plow Steel	Galvanized Plow Steel	Galvanized Iron			Galvanized Improved Plow Steel	Galvanized Plow Steel	Galvanized Iron
5/16	0.10	2.34	2.04	0.905	1 3/16	1.48	32.2	28.0	....
3/8	0.15	3.36	2.92	1.30	1 1/4	1.64	35.6	30.9	....
7/16	0.20	4.55	3.95	1.76	1 3/8	1.99	42.8	37.2	....
1/2	0.26	5.91	5.14	2.28	1 7/16	2.17	46.7	40.6	....
9/16	0.33	7.45	6.48	2.88	1 1/2	2.36	50.7	44.1	....
5/8	0.41	9.16	7.97	3.54	1 5/8	2.77	59.2	51.4	....
3/4	0.59	13.1	11.4	5.06	1 11/16	2.99	63.6	55.3	....
13/16	0.69	15.3	13.3	5.92	1 3/4	3.22	68.3	59.4	....
7/8	0.80	17.7	15.4	6.85	1 13/16	3.45	73.0	63.5	....
1	1.05	23.0	20.0	8.89	1 15/16	3.94	83.0	72.2	....
1 1/16	1.19	25.9	22.5	10.0	2	4.20	88.2	76.7	....
1 1/8	1.33	29.0	25.2	....	2 1/16	4.47	93.6	81.4	....

Table 26. Standard 6 by 24 Galvanized Steel Mooring Lines and Hawsers

(U. S. Department of Commerce, *Simplified Practice Recommendation R198-43*)

Diameter, in.	Approximate Weight, lb/ft	Breaking Strength in Tons of 2000 lb		Diameter, in.	Approximate Weight, lb/ft	Breaking Strength in Tons of 2000 lb	
		Galvanized Improved Plow Steel	Galvanized Plow Steel			Galvanized Improved Plow Steel	Galvanized Plow Steel
3/8	0.194	4.77	4.14	1 3/8	2.61	61.0	53.1
1/2	0.35	8.40	7.30	1 7/16	2.85	66.5	57.9
5/8	0.54	13.0	11.3	1 1/2	3.11	72.3	62.9
3/4	0.78	18.6	16.2	1 5/8	3.64	84.5	73.4
13/16	0.91	21.8	19.0	1 11/16	3.93	90.9	79.0
7/8	1.06	25.2	21.9	1 3/4	4.23	97.5	84.8
1	1.38	32.8	28.5	1 13/16	4.53	104.0	90.8
1 1/16	1.56	36.9	32.1	1 15/16	5.18	119.0	103.0
1 1/8	1.75	41.2	35.9	2	5.52	126.0	110.0
1 3/16	1.95	45.9	39.9	2 1/16	5.87	134.0	116.0
1 1/4	2.16	50.7	44.1				

Table 27. Standard 6 by 37 Galvanized Steel Hawsers

(U. S. Department of Commerce, *Simplified Practice Recommendation R198-43*)

Diameter, in.	Approximate Weight, lb/ft	Breaking Strength in Tons of 2000 lb		Diameter, in.	Approximate Weight, lb/ft	Breaking Strength in Tons of 2000 lb	
		Improved Plow Steel	Plow Steel			Improved Plow Steel	Plow Steel
3/4	0.87	21.0	18.2	1 5/8	4.09	95.3	82.9
13/16	1.02	24.5	21.3	1 11/16	4.41	103.0	89.2
7/8	1.19	28.4	24.7	1 3/4	4.75	110.0	95.7
1	1.55	36.9	32.1	1 13/16	5.09	118.0	102.0
1 1/16	1.75	41.6	36.1	1 15/16	5.82	134.0	117.0
1 1/8	1.96	46.5	40.4	2	6.20	143.0	124.0
1 3/16	2.19	51.7	44.9	2 1/16	6.59	151.0	132.0
1 1/4	2.42	57.1	49.7	2 1/8	7.00	160.0	139.0
1 3/8	2.93	68.8	59.8	2 1/4	7.85	179.0	156.0
1 7/16	3.20	75.0	65.3	2 5/16	8.29	189.0	164.0
1 1/2	3.49	81.5	70.9	2 3/8	8.74	199.0	173.0

AIRCRAFT CORD AND STRAND. Special types of wire rope have been developed for aircraft use, as multiple-wire strand and as multiple-strand cord (Tables 28 and 29). They are furnished in tinned or galvanized carbon steel or in stainless steel and meet Army–Navy specifications. Aircraft cord is also available with an aluminum alloy tube swaged over the outside, providing a smooth surface, less stretch, and a coefficient of thermal expansion close to that of the aluminum alloy airframe. Such clad cord can be used only on straight runs. It is covered by U. S. Patent 2,308,669.

WIRE ROPE FITTINGS. (See also page 26-05.) **Splices.** Two types of splices are used with wire rope—the long or endless, and the loop or thimble splice. The long splice is used to attach one rope to another or to make a rope endless by interlocking the strands. Ropes of similar construction can be so spliced, but a right lay rope cannot be spliced into a left lay rope. In making a splice, a certain amount of rope is used up so additional length must be allowed for it.

The loop splice is made by forming a loop in the end of the rope and then interlocking the strands. The loop often contains a thimble or other fitting.

The endless splice is made in the same manner as the end-to-end splice.

The thimble splice is the forming of a loop in the end of a rope and then interlocking the strands, the loop usually containing a thimble or other fitting (Fig. 2).

Shackles are used to connect ropes whose ends are equipped with thimbles, sockets, turnbuckles, etc. (See Fig. 2.)

Clips are fastenings usually used to form a thimble loop where the use of the rope is temporary. The U-bolt of the clip should rest on the short end of the rope so that the flat bears on the tension side of the loop. Otherwise the rope will be injured by the clip

Table 28. Aircraft Cord and Strand

(Data from John A. Roebling's Sons Co.)

Diameter, in.	19-Wire Strand			6 by 7 Cord		7 by 7 Cord			7 by 19 Cord		
	Approx. Weight, lb/100 ft	Breaking Strength, lb		Approx. Weight, lb/100 ft	Breaking Strength, lb	Approx. Weight, lb/100 ft	Breaking Strength, lb		Approx. Weight, lb/100 ft	Breaking Strength, lb	
		Carbon Steel, Galvanized	Stainless Steel				Tinned or Galvanized	Stainless Steel		Tinned or Galvanized	Stainless Steel
1/32	0.25 *	185 *	150 *	...	...	...	...	...	...	...	...
3/64	0.55 *	375 *	300 *	...	...	...	...	...	...	...	...
1/16	0.85	500	500	0.68	400	0.75	480	480	...	...	...
5/64	1.4	800	800	1.0	550	1.1	650	650	...	...	...
3/32	2.0	1,200	1,200	1.45	800	1.6	920	920	...	...	...
7/64	2.7	1,600	1,600	2.0	1,050	2.2	1,260	1,260	...	...	...
1/8	3.5	2,100	2,100	2.55	1,440	2.8	1,700	1,700	2.9	2,000	1,900
5/32	5.5	3,300	3,300	3.9	2,200	4.3	2,600	2,600	4.5	2,800	2,600
3/16	7.7	4,700	4,700	5.6	3,150	6.2	3,700	3,700	6.5	4,200	3,900
7/32	10.2	6,300	6,300	7.5	4,100	8.3	4,800	4,800	8.6	5,600	5,200
1/4	13.5	8,200	8,200	9.7	5,200	10.6	6,100	6,100	11.0	7,000	6,600
9/32	17.0	10,300	10,300	12.2	6,600	13.4	7,600	7,600	13.9	8,000	8,000
5/16	21.0	12,500	12,500	15.2	8,000	16.7	9,200	9,100	17.3	9,800	8,200
11/32	...	...	...	18.2	9,500	20.1	11,100	10,800	20.7	12,500	...
3/8	...	...	...	21.5	11,500	23.6	13,100	12,600	24.3	14,400	12,000

* 7 wires per strand.

Table 29. Aircraft 6 by 19 Cord with 7 by 7 Independent Wire Rope Core

(Data from John A. Roebling's Sons Co.)

Diameter, in.	Approximate Weight, lb/100 ft	Breaking Strength, lb		Diameter, in.	Approximate Weight, lb/100 ft	Breaking Strength, lb	
		Tinned or Galvanized	Stainless Steel			Tinned or Galvanized	Stainless Steel
7/16	35.60	17,600	16,000	1	187.00	85,400	85,400
1/2	45.80	22,800	22,800	1 1/8	240.00	106,400	106,400
9/16	59.00	28,500	28,500	1 1/4	290.00	129,400	129,400
5/8	71.50	35,000	35,000	1 3/8	330.00	153,600	153,600
3/4	105.20	49,600	49,600	1 1/2	420.00	180,500	180,500
7/8	143.00	66,500	66,500				

putting a crimp in the tension side. Not less than 2, and preferably 4 or 6, clips should be used, especially on the larger diameters of rope.

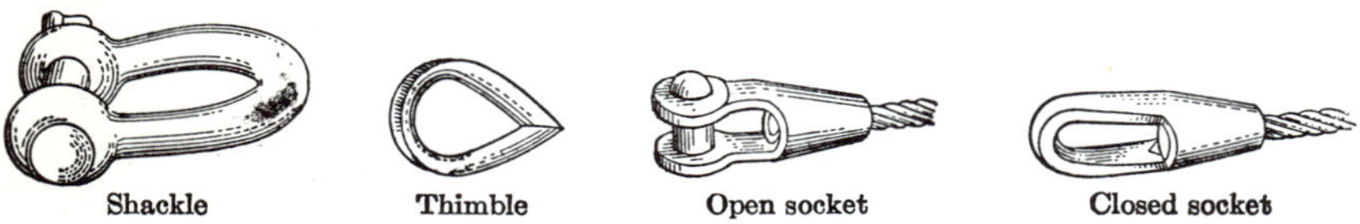

FIG. 2. Wire rope fastenings.

Clamps are temporary fastenings, usually used to form loops without thimbles. From 2 to 6 clamps should be used for each end fastening, and alternate clips and clamps are preferable.

Turnbuckles are necessary where slack in rope is to be taken up or a uniform tension maintained. The turnbuckles should always be equal in strength to that of the rope used with them. Where a greater take-up is required than that given in the table, two or more turnbuckles should be used. Turnbuckles are usually made with the styles of ends shown in Fig. 3.

Sockets are the strongest type of rope fastening. (See Fig. 2.) Bridge sockets of both open and closed type are also made for bridge cables and locked tramway and cableway strand. Sockets develop the full strength of the rope and are stronger than any rope that may be inserted in them.

Swaged Fittings. Where a light, compact, and highly efficient connection is desired, a swaged or compressed fitting may be used. When properly designed and attached, they develop the full strength of the cord or rope and thus have found many industrial applications. Swaged fittings for all types of aircraft control cords have superseded former methods of making attachments. The adaptability of this method has allowed the use of special fittings, such as shown in Fig. 4, which previously were unavailable.

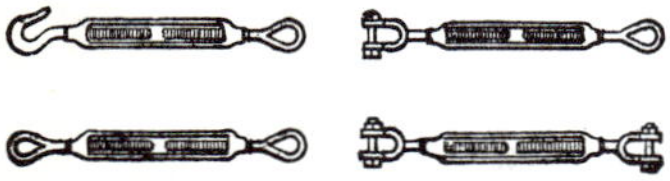

FIG. 3. Turnbuckles.

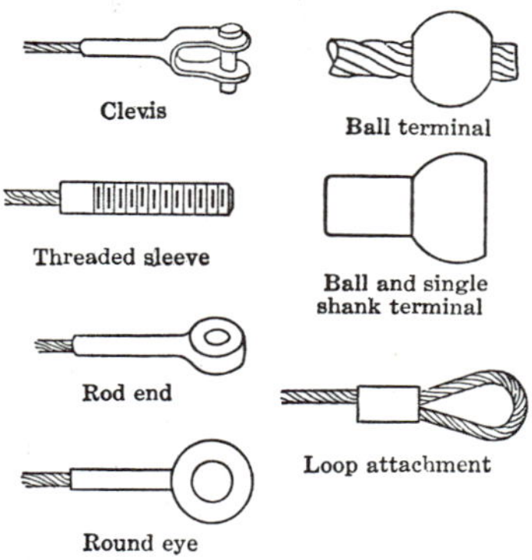

FIG. 4. Swaged fittings for wire rope.

5. ROPE

MANILA ROPE. Table 30 gives the weight and strength of manila rope in the sizes available. These figures correspond to the specifications of the U. S. Bureau of Standards.

STRESS DUE TO LENGTH OF VERTICAL ROPE. In a very long rope, the weight of the rope itself adds a considerable amount to the stress in the rope and decreases the

Table 30. Weight and Strength of Three-strand Manila Rope *

(Data from Columbian Rope Company, Auburn, N. Y.)

Size, in.		Approx. Weight, lb/ft	Length in 1 lb	Strength, lb		Size, in.		Approx. Weight, lb/ft	Length in 1 lb	Strength, lb	
Diameter	Circumference			Breaking	Working	Diameter	Circumference			Breaking	Working
3/16	5/8	0.015	66′ 7″	450	90	1 5/16	4	0.480	2′ 1″	15,000	3,000
1/4	3/4	0.020	50′ 0″	600	120	1 1/2	4 1/2	0.600	1′ 8″	18,500	3,700
5/16	1	0.029	34′ 6″	1,000	200	1 5/8	5	0.744	1′ 4″	22,500	4,500
3/8	1 1/8	0.041	24′ 5″	1,350	230	1 3/4	5 1/2	0.895	1′ 1″	26,500	5,300
7/16	1 1/4	0.0525	19′ 0″	1,750	330	2	6	1.080	11″	31,000	6,200
15/32	1 3/8	0.0625	16′ 0″	2,250	450	2 1/8	6 1/2	1.250	9 1/2″	36,000	7,200
1/2	1 1/2	0.0750	13′ 4″	2,650	530	2 1/4	7	1.460	8″	41,000	8,200
9/16	1 3/4	0.104	9′ 7″	3,450	690	2 1/2	7 1/2	1.670	7″	46,500	9,300
5/8	2	0.133	7′ 6″	4,400	880	2 5/8	8	1.910	6 1/4″	52,000	10,400
3/4	2 1/4	0.167	6′ 0″	5,400	1,080	2 7/8	8 1/2	2.150	5 5/8″	58,000	11,600
13/16	2 1/2	0.195	5′ 2″	6,500	1,300	3	9	2.420	5″	64,000	12,800
7/8	2 3/4	0.225	4′ 5″	7,700	1,540	3 1/8	9 1/2	2.690	4 1/2″	71,000	14,200
1	3	0.270	3′ 9″	9,000	1,800	3 1/4	10	2.990	4″	77,000	15,400
1 1/16	3 1/4	0.313	3′ 2″	10,500	2,100	3 1/2	11	3.670	3 1/4″	91,000	18,200
1 1/8	3 1/2	0.360	2′ 9″	12,000	2,400	3 3/4	12	4.360	2 3/4″	105,000	21,000
1 1/4	3 3/4	0.418	2′ 5″	13,500	2,700						

* Four-strand rope weighs about 5% more than 3-strand rope; the tensile strength is about the same as that of 3-strand rope.

load that the rope can safely carry. The weight of the rope, as a whole, can be diminished by making it up of a series of ropes of constantly decreasing diameters, the smallest rope being at the lower end. Table 31 shows the length of rope, hanging vertical, that will develop the maximum stress in the rope. The figures are based on the values of Table 30.

Table 31. Length of Vertical Rope to Develop Maximum Stress

Circumference, in.	Length to Develop Maximum Stress, ft		Circumference, in.	Length to Develop Maximum Stress, ft		Circumference, in.	Length to Develop Maximum Stress, ft	
	Breaking	Working		Breaking	Working		Breaking	Working
3/4	30,000	6,000	3 1/2	33,000	6,600	7	27,300	5,460
1	34,500	6,900	4	31,200	6,240	8	27,100	5,420
1 1/2	35,300	7,060	4 1/2	30,800	6,160	9	26,300	5,260
2	33,000	6,600	5	30,000	6,000	10	25,700	5,140
2 1/2	33,600	6,720	5 1/2	28,700	5,740	11	24,700	4,940
3	33,800	6,760	6	28,400	5,680	12	24,000	4,800

MANILA ROPE FOR HOISTING OR TRANSMISSION. Rope for hoisting or power transmission is subject to severe usage. Ordinary rope chafes and grinds to powder in the center, although the exterior may look little worn. In bending a rope over a sheave, the strands and the yarns of these strands slide a small distance upon each other, causing friction, and wearing the rope internally. This condition can be alleviated by proper lubrication of the rope. The selection of the lubricant depends on the use of the rope. For indoor use, the lubricant should be one that will not dry out the rope nor rub off easily. For outdoor use, a graphite lubricant of fine flake graphite should be worked into the yarns while the rope is being made, so that each fiber is impregnated.

The amount of work that the rope will do varies greatly. It depends not only on the quality of the fiber and the method of laying up the rope but also on weather when the rope is used, the blocks or sheaves over which it is run, and the strain in proportion to the strain put upon the rope. The principal wear comes in practice from defective or badly set sheaves, excess of load, and exposure to storms.

Loads put on the rope should not exceed those given in the tables, for the most economical wear. Indications of excessive load will be the twist coming out of the rope, or one of the strands slipping out of its proper position. A certain amount of twist comes out in use in the first day or two, but the rope then should remain substantially the same. If it does not, the load is too great for the durability of the rope. Wear of the rope on the outside, while it is good on the inside, indicates chafing in running over pulleys or sheaves. Very small blocks will increase the sliding of the strands and threads and result in a more rapid internal wear. For further information on rope for power transmission, see Article 26, Section 15.

TECHNICAL TERMS RELATING TO CORDAGE most frequently used are:

Yarn. Fibers twisted together. Yarns are laid up left-handed into *strands.*

Thread. Two or more *small yarns* twisted together.

String. The same as a thread but of a little larger *yarns.*

Strand. Two or more *large yarns* twisted together. Strands are laid up right-handed into rope.

Cord. Several threads twisted together.

Rope. Several *strands* twisted together.

Hawser. A rope of three *strands.* Hawsers are laid up left-handed into a cable.

Shroud laid. A rope of four *strands.*

Cable. Three hawsers twisted together.

A rope is:

Laid by twisting strands together in making the rope.

Spliced by joining to another rope by interweaving the strands.

Whipped by winding a string around the end to prevent untwisting.

Served when covered by winding a yarn continuously and tightly around it.

Parceled by wrapping with canvas.

Seized when two parts are bound together by yarn, thread, or string.

Payed when painted, tarred, or greased to resist wetting.

Hauled when pulled on.

Taut when drawn tight or strained.

SPLICING OF ROPE. The splice in a transmission rope is the weakest part of the rope and the first part to fail when the rope is worn out. If the rope is larger at the splice, the projecting part will wear on the pulleys and the rope will fail from the cutting off of the strands.

EFFICIENCY OF SPLICES AND KNOTS as a percentage of the full strength of the rope, and the factor of safety when used with the stresses given below, are as follows:

Kind of Knot	Efficiency, %	Factor of Safety
Eye splice over an iron thimble	90	6.3
Short splice in the rope	80	5.6
Timber hitch, round turn, half-hitch	65	4.5
Bowline slip knot, clove hitch	60	4.2
Square knot, weaver's knot, sheet bend	50	3.5
Flemish loop, overhand knot	45	3.1
Full strength of dry rope, average of four tests	100	7.0

Diameter of rope, in.	1	1 1/8	1 1/4	1 3/8	1 1/2	1 5/8	1 3/4
Working stress, lb	1000	1250	1500	1900	2200	2600	3000

6. CHAIN

CLASSES OF CHAIN. Two classes of chain are covered by ASTM A56-39; *crane chain* and *proof coil.* Strength, weight, and dimensions of these are given in Tables 32 and 33.

Crane chain is intended for slings, cranes, steam shovels, and marine uses where an all-iron chain is desired. It is made from all pig-puddled or processed wrought iron, free from any admixture of iron scrap or steel, and lap fire welded. Composition includes not more than 0.06% manganese.

Proof coil is intended for railroad cars, construction, and forestry work. It is made from open-hearth steel and may be either fire welded or electrically welded. Composition includes not more than 0.05% phosphorus and 0.05% sulfur.

Physical Properties and Tests. Chain subjected to the proof test load, Table 32, should show no defects. The break test load is applied to a test specimen not less than 2 ft in length cut from the finished chain. Elongation in 12 to 18 in. to the nearest link should be not less than 15%.

Standard weights and dimensions are given in Table 33. Permissible variation for length and weight of chain is ±4%. Length is determined by measuring to the inside of the end links, after the proof test, with a load not exceeding 10% of the proof test load to take up slack. Diameter of the weld should not exceed 25% of the bar diameter unless otherwise specified.

CAUSES OF CHAIN FAILURE. 1. Brittle failure is caused by cold-working of the metal surface. The metal fractures with no elongation, or with less elongation than in the ordinary tensile failure mentioned under (6). This type of fracture usually is caused by a brittle surface skin developed on the metal by impacts received in service. Brittle failures show a more or less crystalline fracture, especially where the brittleness is present in an extreme degree. In service, fractures caused by cold-working the surface occur most frequently in the sides or near the shoulder of the link. Where the surface hardening is extreme, the fracture extends directly across the metal of the link in an approximately plane surface. Brittle failures usually are found in chain which has failed under the impact of suddenly applied loads.

Table 32. Proof and Break Test Loads

(ASTM A56-39)

Nominal Size of Chain Bar, in.	Proof Test Load, lb		Break Test Load, lb		Safe Working Load, lb	
	Crane Chain	Proof Coil	Crane Chain	Proof Coil	Crane Chain	Proof Coil
1/4	1,345	1,700	3,535	3,400	1,060	850
5/16	2,100	2,650	5,520	5,300	1,655	1,325
3/8	3,020	3,850	7,950	7,700	2,385	1,925
7/16	4,115	5,250	10,830	10,500	3,250	2,625
1/2	5,375	6,850	14,145	13,700	4,240	3,425
9/16	6,800	8,650	17,895	17,300	5,370	4,325
5/8	8,395	10,700	22,095	21,400	6,630	5,350
3/4	12,080	15,350	31,800	30,700	9,540	7,675
7/8	16,430	20,900	43,245	41,800	12,960	10,450
1	21,490	27,350	56,550	54,700	16,950	13,675
1 1/8	25,380	....	66,800	....	20,040	....
1 1/4	31,350	....	82,500	....	24,750	....
1 3/8	37,920	....	99,800	....	29,910	....
1 1/2	45,100	....	118,700	....	35,600	....
1 5/8	53,000	....	139,500	....	41,800	....
1 3/4	61,400	....	161,600	....	48,450	....
1 7/8	70,500	....	185,500	....	55,300	....
2	80,200	....	211,100	....	63,300	....

Table 33. Nominal Weights and Dimensions of Chain

(ASTM A56-39)

Nominal Size of Chain Bar, in.	Actual Size of Material, in.	Nominal Length of 100 Links, in.		Nominal Weight per 100 ft, lb		Nominal Dimensions of Link, in.							
						Crane Chain				Proof Coil			
						Outside		Inside		Outside		Inside	
		Crane Chain	Proof Coil	Crane Chain	Proof Coil	Length	Width	Length	Width	Length	Width	Length	Width
1/4	9/32	86	100	78	70	1 27/64	1	55/64	7/16	1 9/16	1 1/16	1	1/2
5/16	11/32	100	111	115	105	1 11/16	1 3/16	1	1/2	1 51/64	1 3/16	1 7/64	1/2
3/8	13/32	109	123	166	158	1 29/32	1 7/16	1 3/32	5/8	2 3/64	1 7/16	1 15/64	5/8
7/16	15/32	122	138	220	210	2 5/32	1 5/8	1 7/32	11/16	2 5/16	1 11/16	1 3/8	3/4
1/2	17/32	134	150	275	265	2 13/32	1 13/16	1 11/32	3/4	2 9/16	1 7/8	1 1/2	13/16
9/16	19/32	156	175	350	335	2 3/4	1 15/16	1 9/16	3/4	2 15/16	2 1/16	1 3/4	7/8
5/8	21/32	169	188	430	410	3	2 3/16	1 11/16	7/8	3 3/16	2 5/16	1 7/8	1
3/4	25/32	188	213	615	580	3 7/16	2 9/16	1 7/8	1	3 11/16	2 11/16	2 1/8	1 1/8
7/8	29/32	225	250	820	780	4 1/16	3 1/16	2 1/4	1 1/4	4 5/16	3 3/16	2 1/2	1 3/8
1	1 1/32	256	275	1045	1000	4 5/8	3 7/16	2 9/16	1 3/8	4 13/16	3 9/16	2 3/4	1 1/2
1 1/8	1 5/32	288	......	1310	......	5 3/16	3 15/16	2 7/8	1 5/8	......	......	......	......
1 1/4	1 9/32	306	......	1600	......	5 5/8	4 5/16	3 1/16	1 3/4	......	......	......	......
1 3/8	1 13/32	363	......	1930	......	6 7/16	4 11/16	3 5/8	1 7/8	......	......	......	......
1 1/2	1 17/32	387	......	2335	......	6 15/16	5 1/16	3 7/8	2	......	......	......	......
1 5/8	1 21/32	425	......	2740	......	7 9/16	5 7/16	4 1/4	2 1/8	......	......	......	......
1 3/4	1 25/32	475	......	3180	......	8 5/16	5 13/16	4 3/4	2 1/4	......	......	......	......
1 7/8	1 29/32	525	......	3650	......	9 1/16	6 5/16	5 1/4	2 1/2	......	......	......	......
2	2 1/32	575	......	4100	......	9 13/16	6 11/16	5 3/4	2 5/8	......	......	......	......

2. Failure in the weld covers any failure involving partial or total separation of the welded surfaces. In old chain, welded surfaces often show signs of progressive deterioration and oxidation.

3. Failure due to repeated severe bending or deformation of the links is similar to that produced in a bar by repeated bending back and forth. This is common in long-link chains, as conveyor chains. The fractured surface usually is crystalline, both in wrought iron and steel.

4. Fatigue failure is produced by many repetitions of stress exceeding the fatigue limit of the metal. In fatigue tests on chain links, the fracture usually occurs at about the quarter point of the link. The fracture extends directly across the side of the link. Fatigue failure occurs in chain in machines or appliances used for repeated liftings of a given load, but it is not common in crane chains and sling chains. There is no elongation of the metal. The fractured surface is usually plane and crystalline.

5. Brittle failure caused by defects in the metal are comparatively rare. The proof test by the manufacturer detects brittleness from this cause, and the links are removed. Brittle fractures, due to overheating during welding, are rare.

6. Ordinary tensile failure, accompanied by full elongation of the metal of the links, usually occurs in new chain of good quality when the chain is required to carry a load greater than its ultimate tensile strength. The metal of the links elongates greatly and the chain "pulls stiff" before final failure. The fracture usually occurs in the sides of the links near the welded end. The metal "necks down" and pulls apart as in an ordinary tension test. The fractured surfaces are generally silky in steel, or fibrous in wrought iron.

USE OF CHAIN. The diameter of a sheave or winding drum used with a short link chain should be from 25 to 30 times the diameter of the stock used for the link, to minimize the bending stresses in the chain as it passes over the sheave or drum. Where the application requires a longer link, the chain manufacturer should be consulted as to the proper size of drum or sheave. Sheaves and drums used should be grooved to receive the chain. These grooves should be of such shape and size as to allow the chain to work smoothly without twisting. Chain should not be wound on sheaves, shafts, or drums without proper grooves.

Splicing of Broken Chain. A dangerous practice, which should be prohibited, is that of splicing a broken chain by inserting a bolt between two links, with the head of the bolt and the nut sustaining the load, or of passing one link through another and inserting a bolt or nail to hold it.

Kinked, Knotted, or Twisted Chain. A lift never should be made with a kink, knot, or twist in the chain. The chain should be perfectly straight. Twisting and dragging chain over pipe and other materials injures the chain.

Safe Loads. Some plants stamp on a metal tag attached to each chain the safe vertical load which may be lifted with that chain. A better way is to stamp the safe load or a reference number for the chain load on the ring or hook. Stamp marks should not be placed on chain links as they may form points of weakness in the surface of the links when they are in cold-worked condition, which may start a crack. Some chain users weld to the ring or connecting link of the chain a flat link on which the safe load or number may be marked. This flat link does not form a part of the chain to which the load is applied. A greater load can be lifted safely with a given size of chain or sling when the chain or all the legs of the sling are in a vertical position.

LUBRICATION. To lessen wear of the chain links when wound on a drum and passed over sheaves, they should be lubricated at frequent and regular intervals with grease or heavy oil mixed with graphite. When lubricating, the chain should be slack and the oil brushed well in between the ends of the links. Grease and oil should not be used in places such as foundries where grit and sand particles are present, resulting in severe abrasive action. Sling chains should not be lubricated, as the lubricant would be transferred to the bodies of men handling the chain, introducing a hazard. The lubricant may also increase the slipping hazard.

METHOD OF EXAMINATION. It is important to examine all hoisting chains at frequent and regular intervals. The examination should include inspection of the links for stretch, wear, gouge marks, open welds, and especially a careful inspection of the surface to ascertain the need of annealing or normalizing, as indicated by the peened, hammered, or polished surface of the links. The examination should be made by a person who is conversant with the defects and flaws liable to be found and who has the knowledge and authority to decide when the chain should be removed from service for repair, annealing, normalizing, or replacement.

BENT HOOKS. Hooks which have become bent owing to overloads, should not be straightened and put back into service. Once a hook is bent, the stress in the metal tends to weaken it so that its strength afterward is doubtful.

SECTION 7

MECHANISM, MECHANICS, AND MACHINE ELEMENTS

By

V. L. DOUGHTIE, *Professor of Mechanical Engineering, University of Texas*

MECHANISM

MECHANICS

MACHINE ELEMENTS

Revision of material originally prepared by William Kent and R. T. Kent.

MECHANISM

1. DEFINITIONS

A MACHINE is a combination of resistant bodies so arranged as to compel the forces of nature to produce work with determinate motions.

A MECHANISM is a combination of resistant bodies so arranged that the motion of one compels the motion of the others, according to a law depending on the nature of the combination. A machine modifies energy and a mechanism modifies motion.

A STRUCTURE is a combination of resistant bodies capable of transmitting forces or carrying loads but having no relative motion between parts.

THE FRAME of a machine is a structure which supports the moving parts and regulates the path, or kind of motion, of some of the parts.

THE DRIVER is that piece of a mechanism which causes motion.

THE FOLLOWER is that piece of a mechanism whose motion is effected.

A PAIR OF ELEMENTS is the geometrical form used to join two parts of a machine so as to obtain definite relative motion between the parts.

A LINK is a rigid piece or a nonelastic substance which serves to transmit force from one piece to another or to cause or control motion.

A LINKAGE consists of a number of pairs of elements connected by links.

A CRANK is a body rotating or oscillating about an axis. When two cranks on the same axis are rigidly connected to each other the combination is called *a lever.*

A CONNECTING ROD is the body joining the moving ends of the two cranks of a four-bar mechanism.

EFFICIENCY OF A MACHINE is the ratio of the useful work performed to the total work received. The efficiency is usually expressed as a percentage but used as a decimal quantity. The limit to the efficiency of a machine is unity, denoting the efficiency of a perfect machine in which no work is lost.

2. LINKAGES

A FOUR-BAR LINKAGE (Fig. 1) consists of two cranks, 2 and 4, a stationary piece, 1, or frame, called the line of centers and the connecting rod, 3. In general, any mechanism can be analyzed as a four-bar linkage or a combination of such linkages. The angular speeds of the two cranks of a four-bar linkage are inversely proportional to the lengths of the perpendiculars or any two parallel lines drawn from the fixed centers to the center line of the connecting rod. In Fig. 1

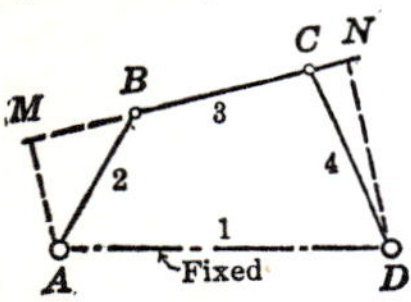

FIG. 1. Four-bar linkage.

$$\frac{\text{Angular velocity of crank 2}}{\text{Angular velocity of crank 4}} = \frac{DN}{AM} \tag{1}$$

CRANK AND ROCKER (Fig. 2). In this combination the link AD is fixed, the crank AB revolves in a complete circle about its fixed axis A and imparts an oscillating motion to the rocker CD about the fixed axis D through the connecting rod BC. In order for this linkage to operate, the following conditions must exist:

1. $AB + BC + DC > AD$
2. $AB + AD + DC > BC$
3. $AB + BC - DC < AD$
4. $BC - AB + DC > AD$

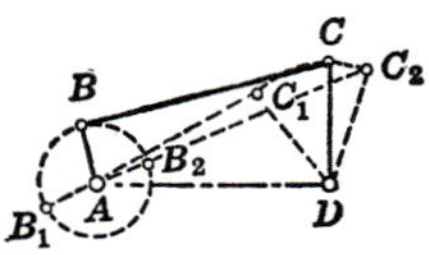

FIG. 2. Crank and rocker mechanism.

Either AB or DC may be the driving crank. If AB drives, the mechanism always will operate. If DC drives, it will not operate unless a flywheel or other aid carries the crank over the dead-center points.

DRAG LINK (Fig. 3). If the link AB in Fig. 2 is fixed and the proportions of the links are nearly the same the links BC and AD will rotate about the fixed points B and A, respectively, becoming cranks, and CD becomes the connecting rod. This mechanism

(Fig. 3) is called a drag link. In order for the cranks to make complete revolutions without any dead centers the following conditions must exist:

1. $CD > AB + AD - BC$ 2. $CD < AD + BC - AB$

PARALLEL CRANK FOUR-BAR LINKAGE. If in Fig. 1 the cranks AB and DC are of equal length and the connecting rod BC is equal in length to the line of centers AD, the two cranks always will turn at the same angular speed.

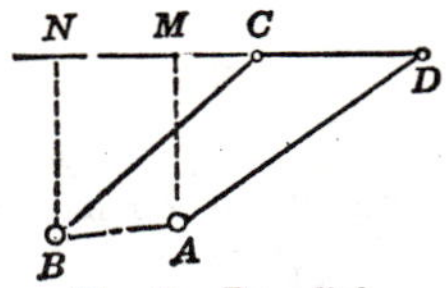

FIG. 3. Drag link.

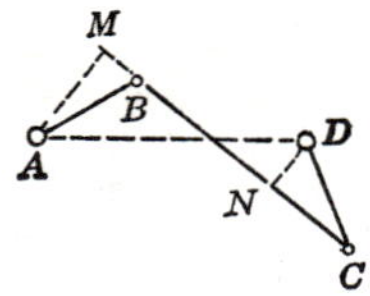

FIG. 4. Nonparallel equal crank linkage.

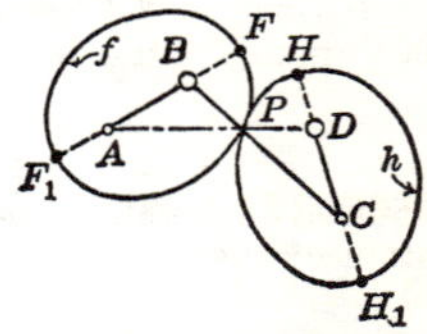

FIG. 5. Elliptic gears. Compare with Fig. 4.

NONPARALLEL EQUAL CRANK LINKAGE (Fig. 4). If the cranks AB and DC are made of equal length. but point in opposite directions from their centers, motion imparted to one crank will cause the other to rotate in the opposite direction. If crank AB is given a uniform angular motion, crank DC will have a varying angular speed. The opposite sense of rotation may be received by providing some means of causing the cranks to pass the dead centers in the proper sense. This may be accomplished by some device placed at the instantaneous axis of the connecting rod when in these positions. Linkages of this type may be replaced by gears (Fig. 5) which are cut on the circumferences of ellipses f and h whose foci are A and B and D and C, respectively, and whose major axes are AD and BC. The intersection of links BC and AD at all positions of the linkage is a point on the circumference of the two ellipses and, therefore, is the point of tangency.

SLOW-MOTION LINKWORK. Figure 6 shows a four-bar linkage proportioned so as to produce slow motion of the crank DC. As the crank AB is rotated clockwise it imparts motion to crank DC, whose angular velocity decreases to zero when the crank AB reaches its dead-center position with the connecting rod BC in line with the crank.

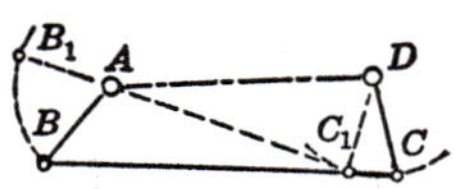

FIG. 6. Slow-motion linkage.

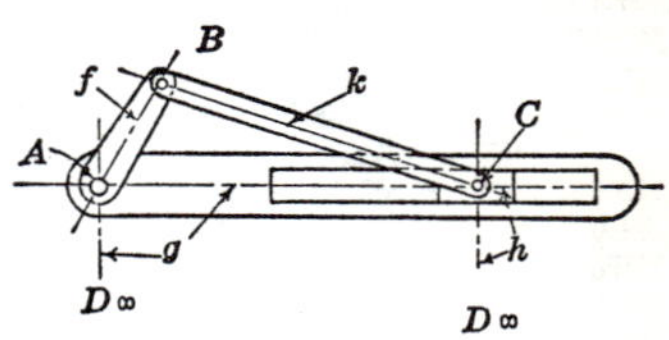

FIG. 7. Slider crank mechanism.

LINKAGE WITH A SLIDING PAIR. In Fig. 7, the four links of the linkage are AB, BC, $AD\infty$, and $CD\infty$, the lines $AD\infty$ and $CD\infty$ meeting at infinity, that is, being parallel, and perpendicular to the center line of the slot g. The line $AD\infty$ is on the piece g and the line $CD\infty$ is on the piece h. The piece g and the block h replace, and constitute the equivalent of, the two infinite links AD and CD. Two general classes of this combination exist: (1) one member of the sliding pair fixed; (2) one member of the sliding pair turning about a fixed axis. In group 1, the connecting rod BC can be longer, shorter, or equal to the length of the crank AB. In group 2, the crank can be longer, shorter, or equal to the line of centers.

Sliding-block Linkage. Figure 8 is a line drawing, notated as in Fig. 7, of the cylinder, piston, connecting rod, crank, and crankshaft of a gasoline engine. This linkage is commonly called a slider-crank mechanism. The cylinder g, though not absolutely fixed to the earth, is fixed to the frame and is considered as the fixed piece of the four-bar linkage. The crankshaft A turns in bearings rigidly attached to the cylinder. The line $AD\infty$ is the line of centers, f the finite crank, k the connecting rod, and the line $CD\infty$ the infinite crank. The piston h receives its energy from the exploding fuel and is the driving member. The same mechanism is used in steam engines and reciprocating pumps and compressors. In these machines the slider h, called the cross head, is connected rigidly to the piston by the piston rod. In the steam engine the cross head is the driver; in the pump and compressor the crank f is the driver. When the slider C is the driver, means must be provided to carry the crank over the dead-center positions, as by a

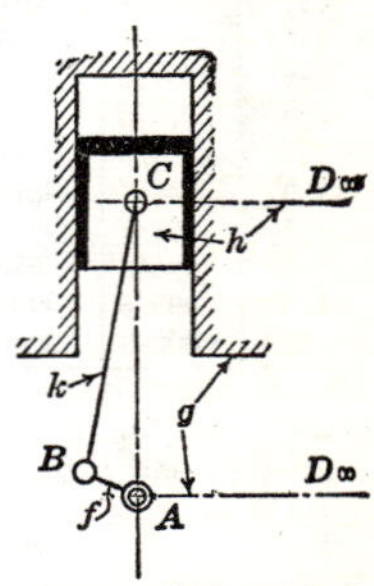

FIG. 8. Sliding-block linkage.

flywheel or by two such linkages, the cranks of which are keyed to the same shaft and set at right angles to each other. The displacement of the piston C from the head end dead center is

$$X = R(1 - \cos\theta) + L\left(1 - \sqrt{1 - \left(\frac{R}{L}\right)^2 \sin^2\theta}\right) \tag{2}$$

where X = displacement, inches; R = length of crank AB, inches; L = length of connecting rod BC, inches; and θ = crank angle (angle CAB in Fig. 8), degrees. An approximately correct formula for the displacement is

$$X = R(1 - \cos\theta) + \frac{R^2 \sin^2\theta}{2L} \tag{3}$$

The position of the piston for various crank angles is given in Table 1. If the center line of motion of the reciprocating parts is not in the same plane as the center line of the crankshaft but is at a perpendicular distance d away from it, the stroke is greater than $2R$, and the dead centers are not symmetrical. In this case the length of the stroke S is

Table 1. Piston Positions for Various Crank Angles

From beginning of stroke toward crankshaft. Distance of piston from beginning of stroke = tabular value × length of stroke.

Crank Angle, degrees	Ratio (Length of Connecting Rod/Length of Crank) 2	2 1/2	3	3 1/2	4	4 1/2	5	5 1/2	6	7	8
5	.00285	.00266	.00254	.00245	.00238	.00233	.00228	.00224	.00222	.00217	.00214
10	.01137	.01061	.01011	.00975	.00948	.00927	.00910	.00896	.00885	.00867	.00854
15	.02545	.02375	.02263	.02183	.02123	.02075	.02040	.02008	.01983	.01942	.01913
20	.04488	.04191	.03993	.03852	.03748	.03667	.03601	.03548	.03500	.03433	.03381
25	.06943	.06484	.06151	.05965	.05799	.05680	.05579	.05499	.05430	.05323	.05244
30	.09873	.09224	.08797	.08494	.08268	.08092	.07952	.07838	.07742	.07593	.07481
35	.13243	.12377	.11810	.11408	.11110	.10881	.10693	.10542	.10416	.10219	.10072
40	.17003	.15900	.15181	.14674	.14297	.14005	.13772	.13582	.13424	.13177	.12991
45	.21103	.19749	.18871	.18253	.17795	.17440	.17158	.16927	.16735	.16434	.16210
50	.25487	.23874	.22833	.22102	.21563	.21145	.20812	.20542	.20316	.19964	.19699
55	.30094	.28222	.26962	.26182	.25556	.25080	.24699	.24388	.24130	.23726	.23423
60	.34861	.32739	.31386	.30442	.29744	.29195	.28778	.28431	.28142	.27689	.27343
65	.39726	.37372	.35878	.34838	.34071	.33480	.33011	.32628	.32311	.31815	.31444
70	.44624	.42066	.40448	.39324	.38496	.37859	.37353	.36942	.36602	.36067	.35667
75	.49495	.46766	.45047	.43856	.42978	.42304	.41767	.41334	.40972	.40407	.39986
80	.54281	.51425	.49630	.48388	.47474	.46772	.46215	.45763	.45386	.44799	.44360
85	.58930	.55995	.54155	.52881	.51944	.51225	.50654	.50191	.49806	.49205	.48756
90	.63398	.60436	.58678	.57295	.56351	.55627	.55048	.54583	.54196	.53591	.53138
95	.67646	.64711	.62870	.61596	.60659	.59940	.59370	.58906	.58522	.57921	.57471
100	.71646	.68790	.66995	.65753	.64839	.64137	.63580	.63127	.62751	.62163	.61725
105	.75376	.72648	.70929	.69737	.68860	.68186	.67650	.67216	.66854	.66288	.65867
110	.78826	.76268	.74650	.73526	.72698	.72062	.71555	.71144	.70804	.70269	.69869
115	.81987	.79634	.78140	.77100	.76332	.75741	.75272	.74890	.74573	.74077	.73706
120	.84862	.82739	.81386	.80442	.79744	.79196	.78778	.78431	.78142	.77689	.77343
125	.87455	.85580	.84319	.83539	.82917	.82438	.82057	.81746	.81487	.81083	.80781
130	.89766	.88152	.87112	.86381	.85842	.85424	.85091	.84820	.84595	.84242	.83977
135	.91814	.90460	.89581	.88964	.88506	.88151	.87868	.87637	.87446	.87145	.86921
140	.93608	.92504	.91786	.91279	.90902	.90610	.90377	.90186	.90029	.89781	.89595
145	.95158	.94292	.93725	.93323	.93025	.92796	.92608	.92457	.92332	.92135	.91987
150	.96476	.95827	.95399	.95097	.94870	.94695	.94554	.94440	.94345	.94196	.94083
155	.97574	.97114	.96782	.96596	.96430	.96310	.96210	.96128	.96061	.95954	.95875
160	.98457	.98160	.97963	.97822	.97717	.97636	.97570	.97517	.97469	.97403	.97350
165	.99137	.98968	.98855	.98776	.98715	.98669	.98632	.98601	.98576	.98536	.98505
170	.99618	.99542	.99492	.99457	.99429	.99408	.99391	.99377	.99366	.99348	.99335
175	.99905	.99885	.99873	.99864	.99858	.99852	.99847	.99844	.99841	.99836	.99833
180	1.00000	1.00000	1.00000	1.00000	1.00000	1.00000	1.00000	1.00000	1.00000	1.00000	1.00000

$$S = \sqrt{(1 + R)^2 - d^2} - \sqrt{(1 - R)^2 - d^2} \tag{4}$$

The linear velocity of the piston in inches per second, V, is approximately

$$V = R\omega\left(\sin\theta + \frac{R}{2L}\sin 2\theta\right) \tag{5}$$

where ω = angular velocity of crank AB, radians per second. For velocities of piston at various crank angles and L/R ratios, see Table 1, p. 7-36.

The linear acceleration of the piston in inches per second per second, A, is approximately

$$A = R\omega^2\left(\cos\theta + \frac{R}{L}\cos 2\theta\right) \tag{6}$$

For acceleration factors at various crank angles and L/R ratios, see Table 2, p. 7-38.

Swinging-block Quick-return Mechanism. Figure 9 illustrates a case of the second class noted above wherein the line of centers BC is longer than the finite crank AB. The gear f, driven by the source of power, rotates about the fixed axis B. Attached to f is the pin whose axis is A. On this pin is the block g which slides in the slot in the swinging arm h. The arm h oscillates about the fixed axis C. The driving crank f, being a part of the gear, turns at a uniform angular speed. The ratio of the time of the working stroke to the time of the return stroke is equal to the ratio of the angles through which f turns during these respective motions.

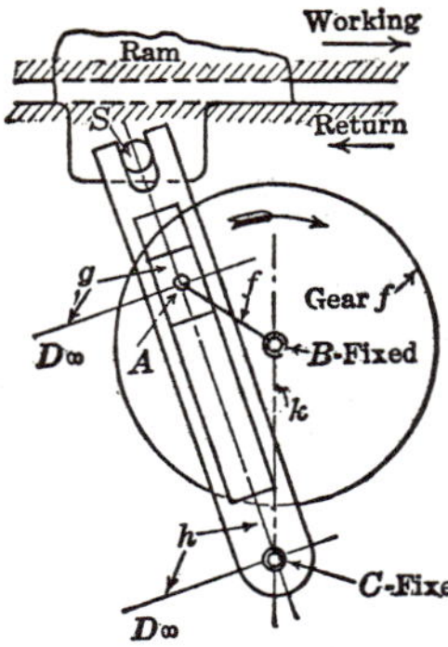

FIG. 9. Swinging-block quick-return mechanism.

Turning-block Linkage. Figure 10, a diagram of the Whitworth quick-return mechanism, illustrates another case of the second class noted above wherein the line of center AB is shorter than the finite crank BC. The action of this mechanism is similar to that of the swinging-block quick-return mechanism except that the crank g makes complete revolutions with variable angular speed but the arm h in Fig. 9 oscillates.

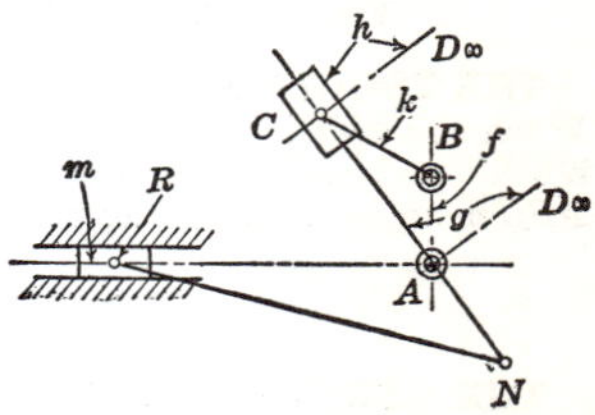

FIG. 10. Turning-block or Whitworth quick-return mechanism.

STRAIGHT-LINE MECHANISMS. If the connecting rod CB in Fig. 7 is made equal to the crank AB an isosceles sliding-block linkage results. If the connecting rod CB is extended to E making BE equal to CB (Fig. 11) the point E will travel in an exact straight line. This linkage may be modified by replacing the block h with a crank CD as in Fig. 12. This linkage is a form of the Scott Russell mechanism. The point E will trace an approximate straight line; the longer the crank CD is made, the less will be the deviation of E from the straight line yy. Figure 13 shows a form of the four-bar linkage as used in the Watt straight-line mechanism. The point P is so located as to

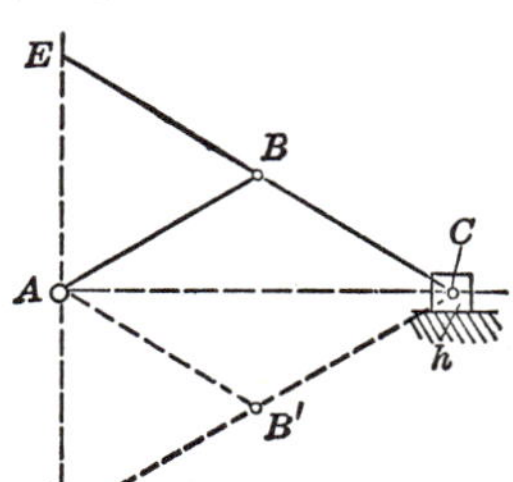

FIG. 11. Straight-line mechanism.

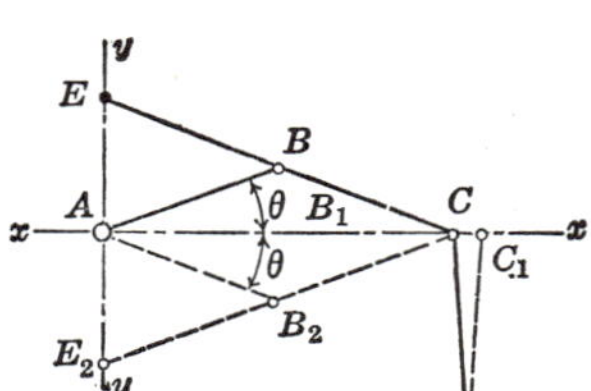

FIG. 12. Scott Russell straight-line mechanism.

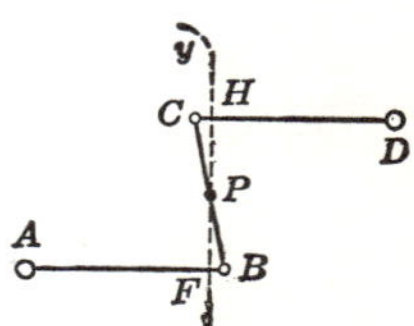

FIG. 13. Watt straight-line mechanism.

cause it to trace a complex path yy, a portion of which is an approximate straight line. If the location of the fixed axes A and D, the position of the line yy along which the point P is to be guided, and the desired length of the stroke S are known, AB, BC, DC, and the location of the point P are found as follows:

Draw AF and DH perpendicular to yy. Then

$$AB = AF + \frac{S^2}{16AF} \tag{7}$$

$$DC = DH + \frac{S^2}{16DH} \tag{8}$$

Join B and C; P will be the point where BC cuts yy.

THE PANTOGRAPH is a four-bar linkage with the opposite sides parallel and forming a parallelogram. In Fig. 14, the point E on link CD and the point P on BC extended will move parallel and similar to each other over any straight or curved path provided that E and P lie on the same straight line passing through the fixed point A. The displacement of P will equal the displacement of E times the ratio AP/AE.

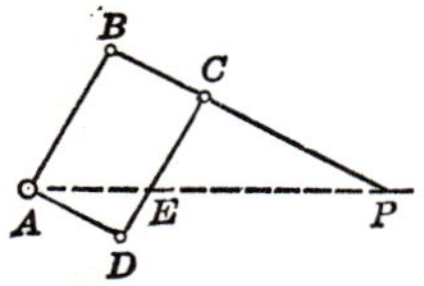

FIG. 14. Pantograph.

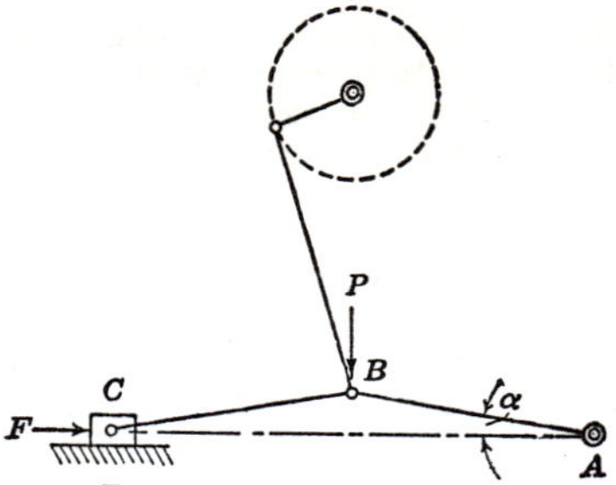

FIG. 15. Toggle mechanism.

THE TOGGLE MECHANISM (Fig. 15) is a combination of two four-bar mechanisms. When the block C approaches the end of its stroke, its velocity decreases, causing the block C to be capable of exerting a high force F with a low applied force P. If the linkage ABC is an isosceles sliding-block linkage,

$$\frac{P}{F} = \frac{2 \sin \alpha}{\cos \alpha} = 2 \tan \alpha \tag{9}$$

THE GENEVA WHEEL (Fig. 16) offers a means of obtaining intermittent motion. In this particular case the starwheel A makes one-sixth of a revolution for one turn of the driving wheel B; however, velocity ratios other than 6 to 1 may be used. Motion is imparted to A by the pin b working in the radial slots b'. The circular portion c of the driver, coming in contact with the corresponding similar parts c', causes A to remain in position when the pin b is out of action. The wheel B is cut away as shown to provide clearance for A in its motion. In laying out the mechanism the angle β is made equal to 360 degrees divided by the number of slots required. The angle θ is drawn so that its side ab is perpendicular to the center line of the slot bd. If another pin is placed opposite b on wheel B as shown by dotted lines and proper clearance is allowed, wheel A would move one-sixth turn each half turn of B. If one of the slots on A, say b', is closed, wheel B can rotate only a little over 5 1/2 revolutions in either direction before pin b strikes the closed slot. This variation of the mechanism is called the *Geneva stop.*

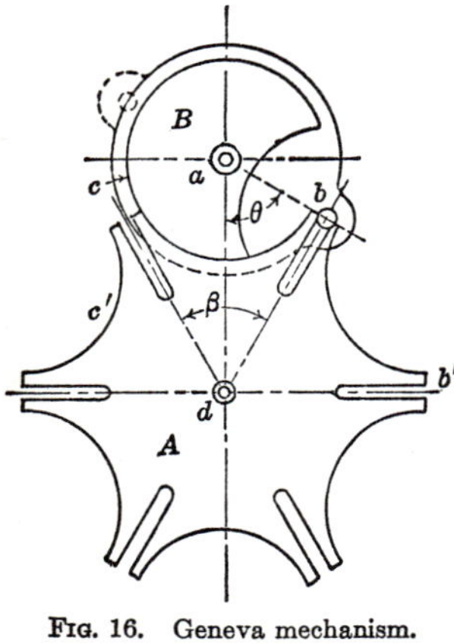

FIG. 16. Geneva mechanism.

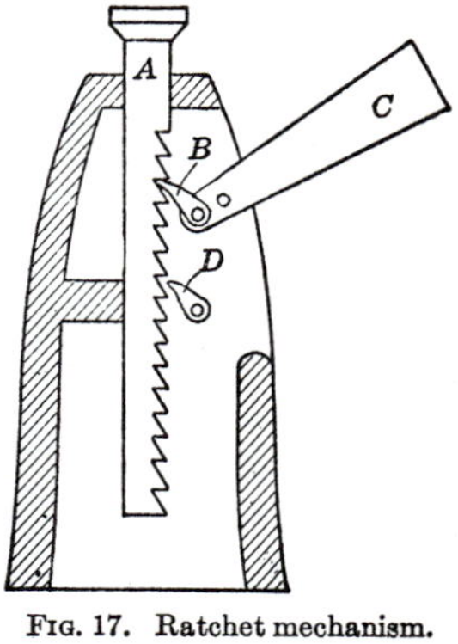

FIG. 17. Ratchet mechanism.

RATCHET MECHANISMS are used for obtaining intermittent rectilinear or rotary motion from another member which usually moves with reciprocating or oscillating motion. Figure 17 shows a ratchet mechanism used in a lifting jack. The *pawl, detent, click,* or *catch* B engages with the teeth on A, and by forcing the handle C downward, A is raised. The pawl D prevents A from moving downward when pawl B engages another tooth on A

Figure 18 illustrates a ratchet mechanism where rotary motion is imparted to the wheel A by the movement of the arm C causing pawl B, in mesh with a tooth b on A, to rotate A about the axis a. Wheel A may be prevented from moving backward on the return of the

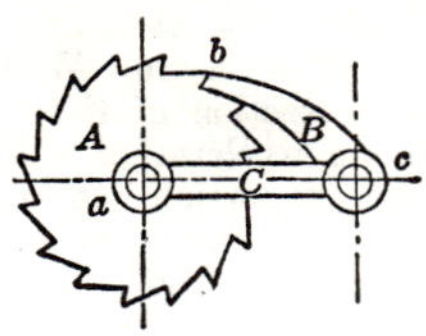

FIG. 18. Ratchet wheel.

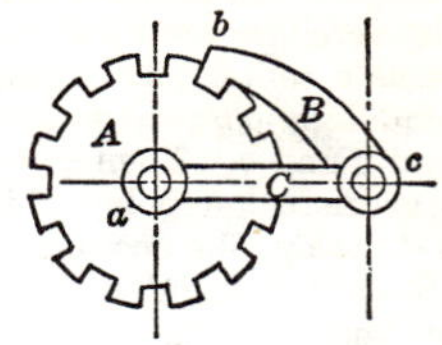

FIG. 19. Retaining pawl.

pawl B by using a fixed pawl similar to B, turning on a fixed pin. Figure 19 shows a retaining pawl which prevents rotation of the wheel A in either direction.

VELOCITIES IN LINKAGES may be readily determined by graphical methods. In Fig. 20 (drawn to scale) use is made of relative velocities in drawing a velocity polygon from which the linear velocities of points in the various links and the angular velocities of the links may be determined. If the angular speed N of the crank Q_2A is known, the linear velocity of the crankpin A may be calculated. qa represents this velocity, laid off to scale and in its correct sense. A line representing the velocity of B relative to A is drawn from a perpendicular to AB. Another line representing the velocity of b is drawn from q perpendicular to Q_4B. At the intersection of these lines locate b. Then ab represents, to the same scale as qa was drawn, the velocity of B relative to A, and qb represents the velocity of b. In like manner, bc perpendicular to BC is the velocity of C relative to B, and qc is the velocity of C. cd is perpendicular to CD, and qd, drawn parallel to the motion of D, is the velocity of the block D. The point P on link AB is located on ab by the proportion $ap = AP(ab/AB)$. Then the velocity of p is represented to scale by the line qp. The angular velocity of AB equals the linear velocity of B relative to A (ab to scale), divided by the length of AB on the machine in the same linear units as the linear units of the linear velocity of B relative to A. The angular velocity of the triangular-shaped link Q_4BC equals the linear velocity of B (qb to scale) divided by the length of Q_4B. The angular velocity of CD equals the linear velocity of C relative to D (cd to scale) divided by the length of CD.

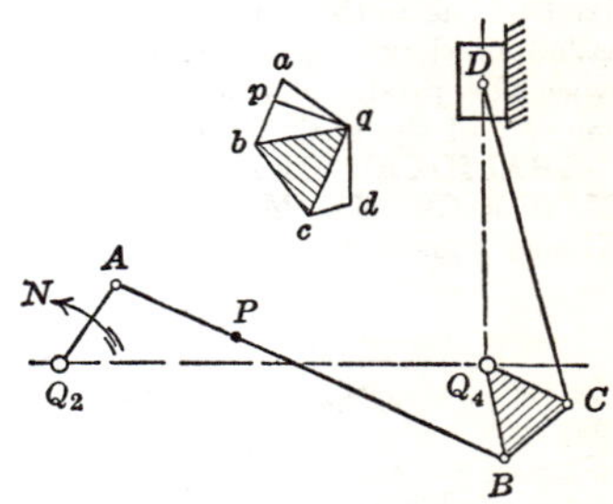

FIG. 20. Graphical determination of velocities in a linkage.

ACCELERATIONS IN LINKAGES may also be determined graphically. Figure 21 represents a crank and rocker mechanism drawn to scale. Choose convenient velocity and acceleration scales. Draw the velocity polygon. The acceleration polygon is drawn as follows: (1) Calculate the normal acceleration of A, denoted by $A_a{}^n$, from the relation $A_a{}^n = V_a{}^2/Q_2A$ and the tangential acceleration of A, denoted by $A_a{}^t$, from the relation $A_a{}^t = \alpha_2 \times Q_2A$. (2) Lay off $A_a{}^n$ (to scale) parallel to Q_2A and $A_a{}^t$ perpendicular to Q_2A, obtaining qa, the acceleration of A. (3) Calculate the normal acceleration of B relative to A ($A_{ba}{}^n$) from the relation $A_{ba}{}^n = V_{ba}{}^2/AB$ and the normal acceleration of B ($A_b{}^n$) from the relation $A_b{}^n = V_b{}^2/Q_4B$. (4) Lay off $A_{ba}{}^n$ (to scale) parallel to BA and draw a line perpendicular to AB representing the direction of the tangential accelera-

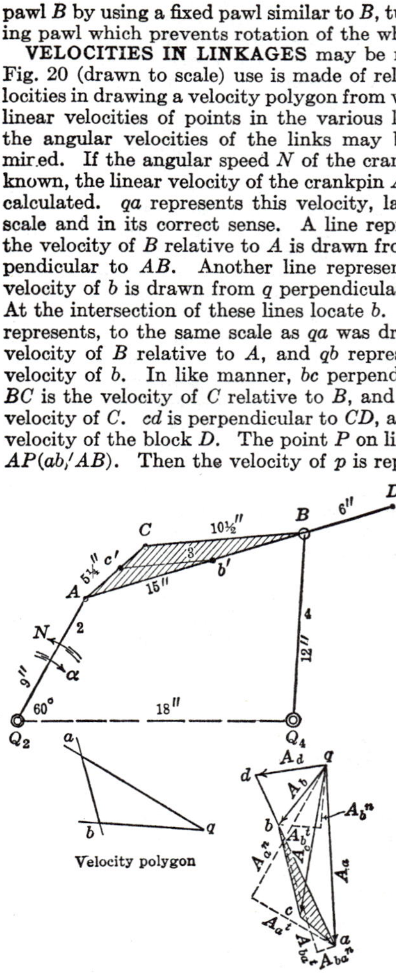

FIG. 21. Graphical method of determining accelerations in a linkage.

tion of B relative to A. (5) Lay off $A_b{}^n$ (to scale) parallel to Q_4B and draw a line perpendicular to Q_4B representing the direction of the tangential acceleration of B. At the intersection of these two lines representing the tangential acceleration of B relative to A and the tangential acceleration of B, locate b. Then qb represents to scale the linear acceleration of B (A_b). (6) Lay off $Ab' = ab$. Draw $b'c'$ parallel to BC. With a as a center and a radius equal to Ac', draw an arc. With b as a center and a radius equal to $b'c'$, draw an arc. At the intersection of these arcs locate c. Join a and c, and b and c. Then qc equals (to scale) the acceleration of C, A_c. (7) The point d in the acceleration polygon is located by the proportion $abd = ABD(ab/AB)$. Then qd equals (to scale) the acceleration of D, A_d. (8) The angular acceleration of link 3 is $\alpha_3 = A_{ba}{}^t/AB$ and the angular acceleration of link 4, $\alpha_4 = A_b{}^t/Q_4B$. AB and Q_4B must be the actual lengths of the links in the machine in the same linear units as the linear units in the tangential acceleration.

CORIOLIS' LAW states that when a particle is moving along a path which is also in motion the absolute linear acceleration of this particle is the vector sum of (1) the acceleration which the particle would have if the path were fixed and the particle moved only along the path, (2) the acceleration which the particle would have if the particle were fixed to the path and the path moved, and (3) a compound supplementary acceleration called Coriolis' acceleration. This compound supplementary acceleration is equal to twice the product of the velocity of the particle relative to the path and the angular velocity of the path. For a detailed discussion of this law see Ref. 1, pp. 391–395.

GRAPHICAL DETERMINATION OF VELOCITY AND ACCELERATION OF A SLIDER-CRANK MECHANISM. Figure 22 is a graphical construction, known as Klein's construction, for the determination of the velocity and the acceleration of the sliding member B. At Q erect a perpendicular to the path of travel of the slide. At the intersection of the perpendicular with the connecting rod AB (extended if necessary) locate W. With AW as a radius and A as a center, draw a circle; with a radius equal to one-half the length of the connecting rod and a center at M, the midpoint of the connecting rod, draw a circle. Join the intersections, T and J, of these two circles with the line TJ. Locate H at the intersection of TJ and a line through Q parallel to or coinciding with the path of travel of the slide. HQ is equal to the linear acceleration of the slide B to scale, and QW is equal to the linear velocity of the slide B to scale. For this construction to hold, the mechanism must be a slider-crank mechanism, with the guides stationary; the crank must rotate uniformly; the length of the crank QA on the drawing must equal to scale the linear velocity of the crankpin A; and the length of the crank QA on the drawing must equal to scale the normal acceleration of the crankpin A. The acceleration of the slide, represented by HQ, is to the same scale as the normal acceleration of A, and the velocity of the slide, represented by QW, is to the same scale as the linear velocity of the crank A. The mechanism may be drawn to any convenient scale.

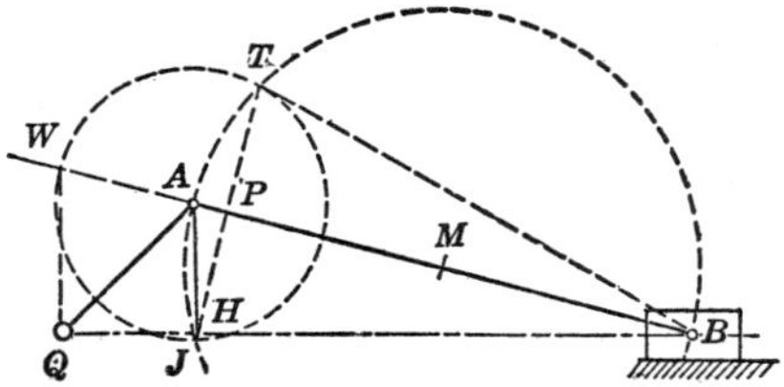

FIG. 22. Klein's construction for velocity and acceleration of slider block.

For a full discussion of linkages, see Ref. 2.

3. POWER SCREWS

THE SCREW is an inclined plane wrapped around a cylinder in such a way that the height of the plane is parallel to the axis of the cylinder. If the screw is formed on the internal surface of a hollow cylinder, it is called a nut. When force is applied to raise a weight or overcome a resistance by means of a screw and nut, either the screw or the nut may be fixed, the other being movable. The force generally is applied at the end of a wrench or lever arm, or at the circumference of a wheel. If r is the radius of the wheel or lever arm, inches, and p is the pitch of the screw, or distance between threads, that is, the height of the inclined plane for one revolution of the screw, inches, P is the applied force, pounds, and W is the resistance overcome, pounds, then, neglecting resistance due to friction,

$$2\pi rP = Wp \tag{10}$$

The ratio of P to W is thus independent of the diameter of the screw. In actual screws, much of the power transmitted is lost through friction.

Efficiency of a Screw. Let α be the helix angle or angle of the thread, that is, the angle whose tangent is the pitch of the screw divided by the circumference of a circle whose diameter is the mean of the diameters at the top and bottom of the thread. Then, for a square thread,

$$\text{Efficiency} = \frac{\tan\alpha(1 - f\tan\alpha)}{\tan\alpha + f} \tag{11}$$

in which f is the coefficient of friction of the threads. The coefficient of friction is not uniform and varies with the materials used, the relative smoothness of the contact surfaces, the quality of lubrication, and the unit pressure on the threads. With rubbing velocities less than 50 fpm, unit pressures less than 14,000 psi, and average lubrication, the coefficient of friction of threads for machine oil and graphite is 0.07; for lard oil it is 0.11; and for heavy machine oil it is 0.14. Figure 23 shows the relationship between the helix angle and the efficiency of square threads. Equation 11 and Fig. 23 are for the square threads only and do not consider the friction of the collar or step by which end thrust is resisted, and which further reduces the efficiency. Considering collar friction and for square threads,

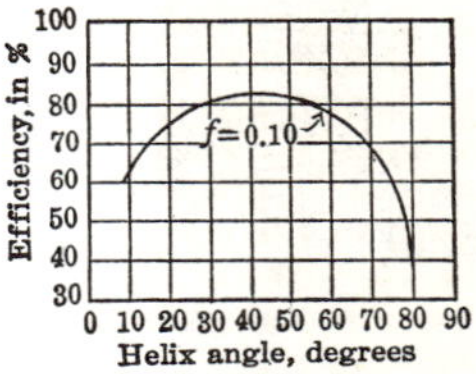

FIG. 23. Efficiency of screw threads (square) excluding collar friction.

$$\text{Efficiency} = \frac{\tan\alpha(1 - f\tan\alpha)}{\tan\alpha + f + \dfrac{f_cD_c}{D_m}(1 - f\tan\alpha)} \tag{12}$$

where f_c is the coefficient of friction of the collar, D_c is the mean diameter of the thrust collar, inches, and D_m is the mean diameter at the top and bottom of the thread, inches. C. W. Ham and D. G. Ryan (Ref. 3) give the following coefficients of friction for the thrust collars on power screws:

Material	Coefficient of Running Friction	Coefficient of Starting Friction
Soft steel on cast iron	.121	.170
Hardened steel on cast iron	.092	.147
Soft steel on bronze	.084	.101
Hardened steel on bronze	.063	.081

The efficiency of a V-thread is lower than that of a square thread. Neglecting collar friction, the efficiency of a V-thread is given by the equation

$$\text{Efficiency} = \frac{\tan\alpha(\cos\theta - f\sin\alpha)}{\tan\alpha\cos\theta + f\cos\alpha} \tag{13}$$

where $\cos\theta = \dfrac{1}{\sqrt{1 + \tan^2\phi + \tan^2\alpha}}$ and ϕ = one-half the included thread angle measured on a plane through axis, degrees. If the collar friction is considered,

$$\text{Efficiency} = \frac{\tan\alpha(\cos\theta - f\sin\alpha)}{\tan\alpha\cos\theta + f\cos\alpha + \dfrac{f_cD_c}{D_m}(\cos\theta - f\sin\alpha)} \tag{14}$$

THE WORM GEAR (Fig. 24), essentially an endless screw, is used to convert motion at high speed into motion at slow speed. When the handle P describes a complete circumference, the pitch line of the worm wheel moves through a distance equal to the pitch of the screw, and the weight W is lifted a distance equal to the pitch of the screw multiplied by the ratio of the diameter of the axle to the diameter of the pitch circle of the wheel. The ratio of the applied force to the weight lifted, or resistance, is inversely as the velocities, friction not being considered; but the friction in the worm gear is usually very great, amounting sometimes to three or four times the useful work done.

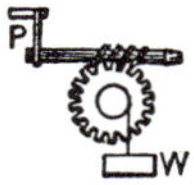

FIG. 24. Worm gear.

If v = the distance through which the force P acts in a given time, say one second, and V = distance the weight W is lifted in the same time, r = radius of the crank or wheel through which P acts, t = pitch of the screw, and also pitch of the teeth of the worm wheel, d = diameter of the axle, and D = diameter of the pitch line of the wheel,

$$v = \frac{6.283r}{t}\frac{D}{d}V \tag{15}$$

$$V = \frac{vtd}{6.283rD} \tag{16}$$

$$Pv = WV + \text{Friction} \tag{17}$$

COMPOUND AND DIFFERENTIAL SCREWS. **A compound screw** consists of a right-hand and a left-hand thread wound on the same or on different cylinders. When wound on the same cylinder (Fig. 25*a*) two nuts are employed. When wound on different cylinders (Fig. 25*b*) one nut is employed. Usually both threads have the same pitch. Each revolution of the screw causes the two nuts to move toward one another a distance equal to the sum of the pitches. Thus two fine threads will give the same relative advance as one coarse thread of double the pitch.

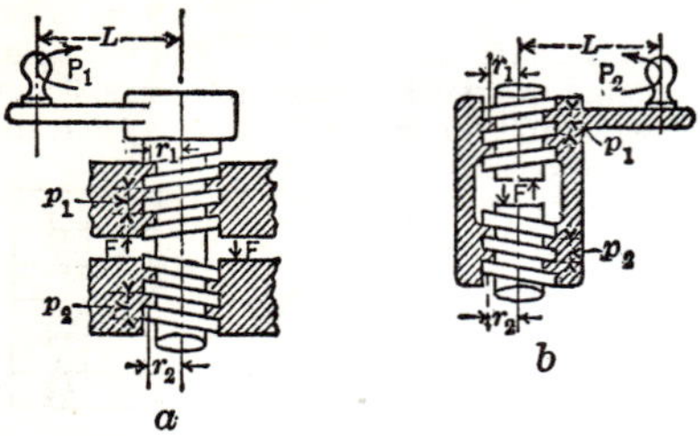

FIG. 25. Compound screw.

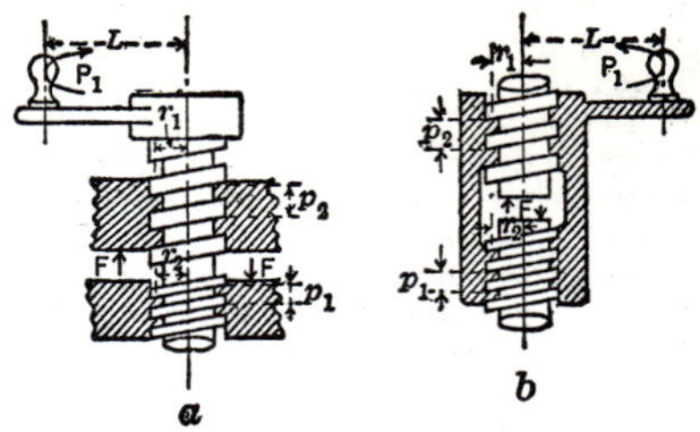

FIG. 26. Differential screw.

A differential screw comprises two threads of the same hand, but of different pitches, wound on the same or different cylinders. See Figs. 26*a* and 26*b*. One revolution of the screw causes the nuts (Fig. 26*a*) to move toward each other a distance equal to the difference of the pitches. One revolution of the nut (Fig. 26*b*) causes the screws to approach by the same distance. Thus the use of two coarse threads gives the same relative advance as one fine thread.

Efficiency. The following formulas give the efficiency of screws with square threads. In these formulas, r_1, r_2 = mean radius (to half depth of thread), inches; c_1, c_2 = mean circumference of screw, inches; p_1, p_2 = pitch of thread, inches; p = equivalent pitch, inches; f = coefficient of friction of threads; f' = coefficient of friction of collar; L = length of wrench, inches; P_1 = pull on wrench, pounds; F = force transmitted by screw, pounds; E = efficiency. Then

For simple screws:

To raise load (Fig. 27*a*),

$$P_1 = \frac{F}{L}\left(f'x + \frac{p_1 + fc_1}{c_1 - fp_1} r_1\right) \tag{18}$$

To lower load (Fig. 27*b*),

$$P_2 = \frac{F}{L}\left(f'x + \frac{fc_1 - p_1}{c_1 + fp_1} r_1\right) \tag{19}$$

in which x = mean radius of collar = $(k_2 + k_1)/2$. If $f = 0$,

$$P = \frac{F}{L}\frac{p_1}{6.2832} \quad \text{and} \quad E = \frac{P}{P_1} \tag{20}$$

For compound screws (see Fig. 25):

$$P_1 = \frac{F}{L}\left(\frac{p_1 + fc_1}{c_1 - fp_1} r_1 + \frac{p_2 + fc_2}{c_2 + fp_2} r_2\right) \tag{21}$$

If $f = 0$,

$$P = \frac{F}{L}\frac{p_1 + p_2}{6.2832} \quad \text{and} \quad E = \frac{P}{P_1} \tag{22}$$

For differential screws (see Fig. 26):

$$P_1 = \frac{F}{L}\left(\frac{p_1 + fc_1}{c_1 - fp_1} r_1 + \frac{fc_2 - p_2}{c_2 + fp_2} r_2\right) \tag{23}$$

If $f = 0$,

$$P = \frac{F}{L}\frac{p_2 - p_1}{6.2832} \quad \text{and} \quad E = \frac{P}{P_1} \tag{24}$$

For V-threads use $1.15f$ instead of f.

The load that may be lifted or the force that may be exerted by any of these screws is so much affected by friction that efficiency rarely is as high as 70%. It often is less than 30% for simple screws, and may be less than 5% for differential screws.

The chief use of the differential screw is to obtain very small movements in precision equipment. Its inefficiency as a means of transmitting power is illustrated by the differential screw punch of Fig. 28.

Differential Screw Punch. Figure 28 is a diagram of a differential screw punch in which the pitch of the outer thread is $1/3$ in. and that of the inner thread $2/7$ in., giving an equivalent pitch of $1/21$ in. By jamming the bottom of the outer screw against the top of the inner screw, the device operates as a simple screw punch, with an advance of $1/3$ in. per revolution. The punch was used both as a differential screw punch and a simple screw punch to punch $11/16$-in. holes in $1/4$-in. plate, requiring a force of 27,000 lb. The results with the device used as a differential screw punch were disappointing. The efficiency as computed by eqs. 23 and 24 and as actually measured were

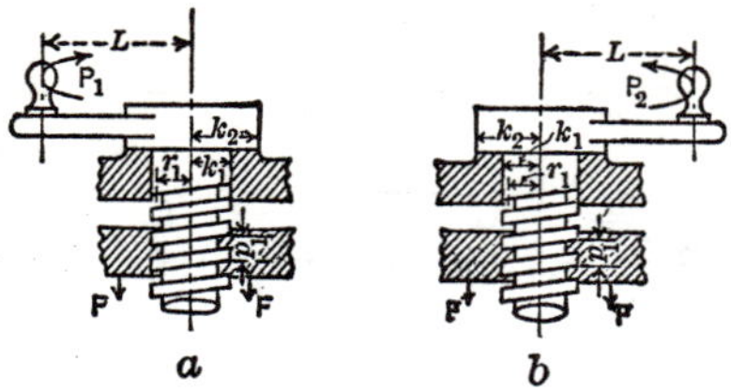

FIG. 27. Simple screw.

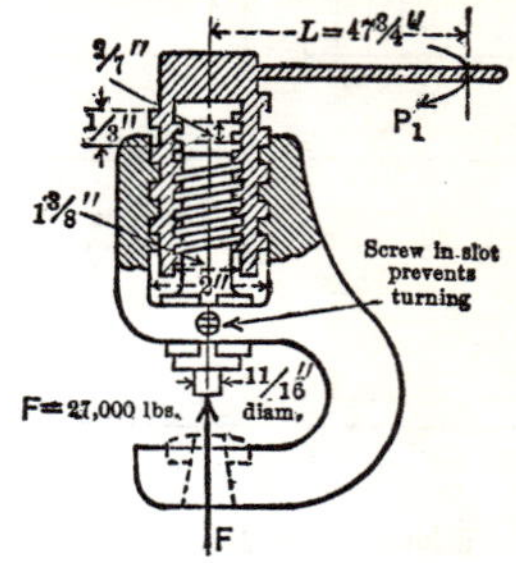

FIG. 28. Differential screw punch.

	Computed		Actual	
	Pull on Lever	Efficiency	Pull on Lever	Efficiency
As a differential punch	92 lb	4.6%	95 lb	4.5%
As a simple punch	83 lb	36.1%	82 lb	36.7%

4. CAMS

THE CAM is a revolving inclined plane. It may be either an inclined plane wrapped around a cylinder in such a manner that the height of the plane is radial to the cylinder, such as the ordinary lifting cam used in stamp mills (Fig. 29) or it may be an inclined plane curved edgewise and rotating in a plane parallel to its base (Fig. 30). The relation of the weight lifted to the applied force is calculated in the same manner as for the screw. (See p. 7-08.)

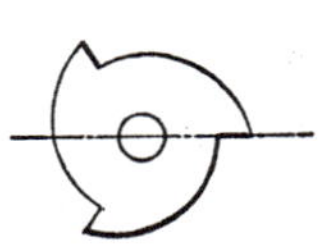

FIG. 29. Lifting cam.

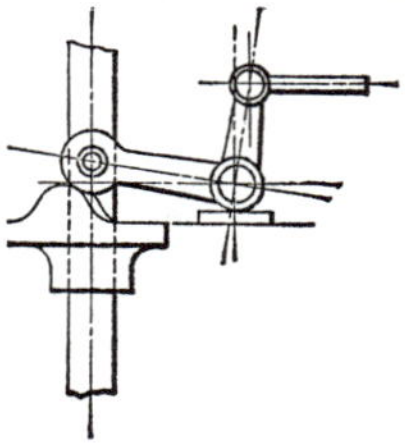

FIG. 30. Edgewise cam.

Cam Motions. The motion imparted by a cam to its follower may follow some law of motion or be such as to cause the follower to be at predetermined positions without regard to fulfilling some definite law of motion. A *displacement diagram*, with time or cam positions as abscissas and follower displacements as ordinates, is useful in studying the motions and in laying out the cam contour. If the motion of the follower does not follow a definite law of motion a displacement diagram is essential to obtain a curve of the motion between the points which represent the predetermined positions of the cam. When the motion imparted to a follower fulfills some law of motion no displacement curve is necessary for the laying out of the cam contour. Velocity and acceleration diagrams for the follower are useful in analyzing the motion and forces in the cam mechanism. With high speeds and accelerations, high forces will result, causing deflections in the linkage and poor operation of the parts controlled by cams. Cam motions are often modified so as to reduce the accelerations and the resulting shocks. Figure 31 shows the displacement, velocity, and acceleration diagrams for uniform motion, modified uniform motion, harmonic motion,

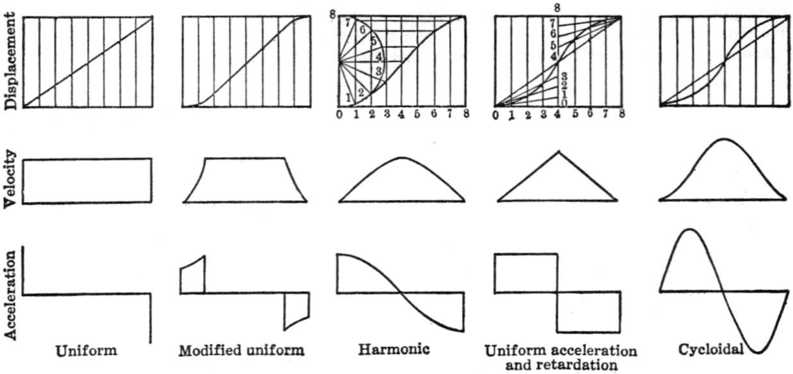

FIG. 31. Displacement, velocity, and acceleration diagrams for cam followers.

uniformly accelerated and retarded motion (often called gravitational motion or parabolic motion), and cycloidal motion.

Motion Layout. A cam in its 360-degree rotation may raise and/or lower its follower by imparting one or several types of motion during various increments of the 360-degree rotation of the cam. For example, the follower may be raised a distance by uniform motion for one-fourth turn of cam, then raised another distance one-fourth turn by harmonic motion, and then lowered to its original position during the last half turn of the cam by parabolic motion. In laying out the motions of the follower, a displacement diagram may be used or the motion may be laid off directly on the path of the follower by choosing the same number of increments of the motion of follower and angular increments of the cam during which the desired motion is imparted. Figure 32 shows the most common types of motions for followers. In each case 0–8 is the distance the follower is raised or lowered, the numerals 1, 2, 3, etc., are the positions of the follower at the corresponding angular position of the cam, and 1′, 2′, 3′, etc., are the divisions of motion laid off on the construction lines. Figure 32*a* shows the method of laying out uniform motion where equal distances are traveled in equal units of time. Figure 32*b* shows the method of laying

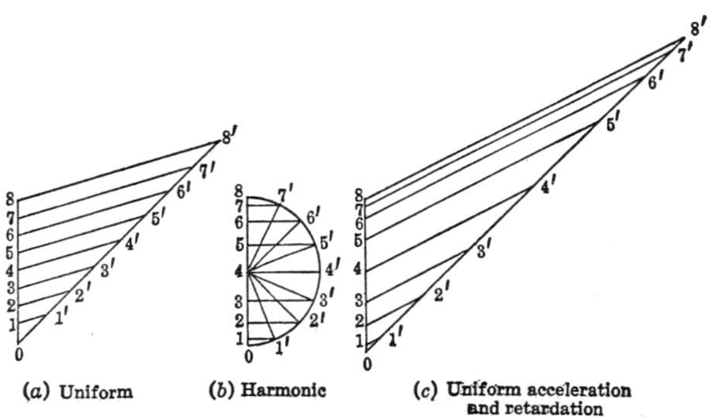

FIG. 32. Layout of uniform, harmonic, and uniform acceleration and retardation motions.

out simple harmonic motion; the follower is raised from 0 to 1 in one unit of time, from 1 to 2 in the second unit of time, etc., and the distance 0–8 is traversed with simple harmonic motion. Figure 32*c* shows the method of laying out uniformly accelerated and retarded motion where the follower is raised from its starting point as the square of the time ($s = at^2/2$, where s = displacement, a = constant acceleration, and t = time). The total spaces traveled will be proportional to 1, 4, 9, 16, etc., and successive spaces traveled in equal time units will be in the proportion 1, 3, 5, 7, etc. For retardation, the increments will be 7, 5, 3, 1.

Laying Out a Uniform-motion Cam. In Fig. 33 a circle of radius OC, equal to the smallest radius of the cam, is drawn. The vertical radius is prolonged beyond the circle to B. The distance 0–6, equal to the maximum movement of the follower, is laid off on OB from its intersection with the circle, and divided into any convenient number of equal parts. The semicircle to the right of OB is divided into the same number of equal arcs, and radii are drawn through the points of division. Then with O as a center, and with radii equal to O–1, O–2, O–3, etc., arcs are drawn intersecting the corresponding radii at a, b, c, etc. The smooth curve drawn through these points is the profile of the cam. The profile of the cam to the left of OC is identical, and is drawn in the same manner.

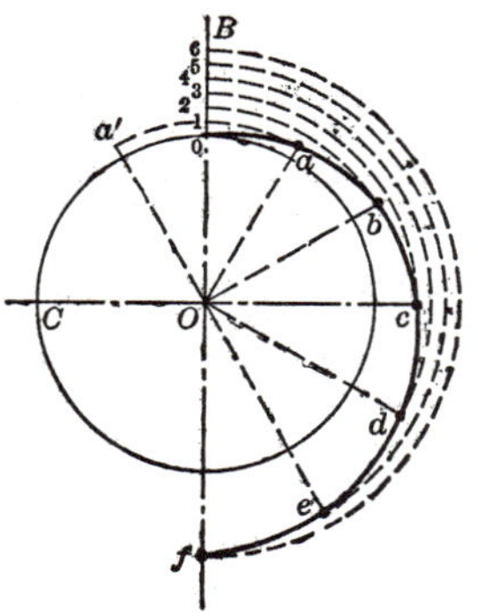

FIG. 33. Uniform-motion cam with point follower.

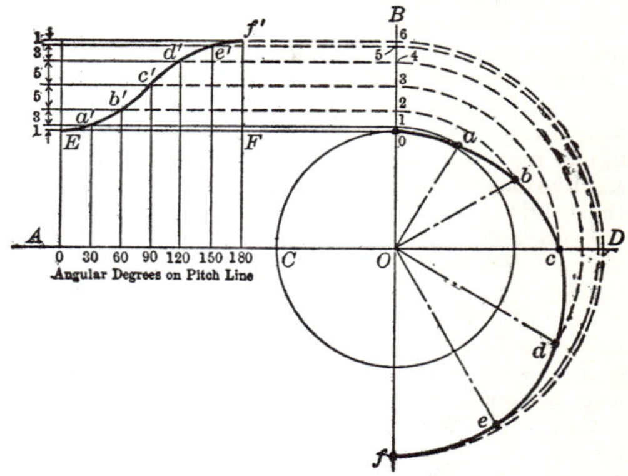

FIG. 34. Cam with uniform acceleration and retardation, employing point follower.

Laying Out a Cam with Uniform Acceleration and Retardation. (See Fig. 34.) Draw AD, and from O as a center, with radius OC equal to the smallest radius of the cam, draw a circle. Draw fOB perpendicular to AD. Draw the tangent $0FE$ parallel to AD. Divide the semicircle to the right of fOB into any convenient number of equal arcs, and draw radii through the points of division. Lay off on AC equal spaces corresponding to the arcs, and erect ordinates at the points of division. Assume a value for the increment or decrement of motion. In the present case it is 2. Hence, with the semicircle divided into 30-degree arcs, the relative motion of the follower over the successive arcs will be 1, 3, 5, 5, 3, 1. Note that the motion over the first and last arcs is one-half the value of the assumed increment. Make $f'F$ equal to the total movement of the follower. It also is equal to the sum of the increments and decrements of motion, viz., $1 + 3 + 5 + 5 + 3 + 1 = 18$. The length above EF of the several ordinates will be

Angular degrees	0	30	60	90	120	150	180
Length, fraction of fF	0	$^1/_{18}$	$^4/_{18}$	$^9/_{18}$	$^{14}/_{18}$	$^{17}/_{18}$	$^{18}/_{18}$

Denote these lengths by points a', b', c', etc., and draw $a'1$, $b'2$, etc., parallel to EF, intersecting OB at 1, 2, 3, etc. The procedure from this point is the same as for the uniform-motion cam.

Laying Out Cycloidal-motion Cam. Figure 35 shows the method of laying out a cam with cycloidal motion

$$s = \frac{L}{\pi}\left(\frac{\pi\theta}{\beta} - \frac{1}{2}\sin\frac{2\pi\theta}{\beta}\right)$$

where s = displacement of follower per interval of time, L = total displacement of follower, θ = angular displacement of cam from starting point, and β = total angle through which cam turns to cause the follower to move L distance.

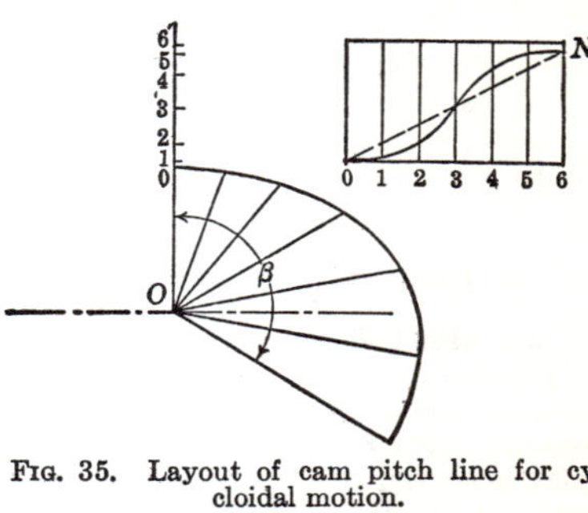

FIG. 35. Layout of cam pitch line for cycloidal motion.

A plate cam on a shaft whose axis is at O is to turn counterclockwise and cause motion to a follower. The follower is to move along the straight line from 0 to 6 while the cam turns 120 degrees. Let O–0 = 2 in. and 0–6 = L = 1 $^1/_2$ in. Since the motion of the follower takes place while the cam turns 120 degrees, $\beta = 2\pi/3$. When these values are substituted the equation becomes

$$s = \frac{1.5}{\pi}\left(\frac{\pi\theta}{2\pi/3} - \frac{1}{2}\sin\frac{2\pi\theta}{2\pi/3}\right) \tag{25}$$

$$= 0.478\left(\frac{3\theta}{2} - \frac{1}{2}\sin 3\theta\right)$$

Points on the cam pitch line will be found for each of six equal subdivisions of β, hence θ will have successive values of $(2\pi/3) \div 6$, $(4\pi/3) \div 6$, $(6\pi/3) \div 6$, ... The points O, 0, and 6 having been located on the drawing, the next step is to construct the displacement diagram with its x-axis perpendicular to 0–6 and passing through 0 in order that the ordinates may be projected directly on to the line 0–6.

To construct the displacement diagram it will be easier to use the two terms of eq. 25 separately. The first term, $0.478(3\theta/2)$, will give a straight line. Choose some distance 0–1 to represent the equal subdivisions of β and lay it off six times along the x-axis since six divisions of β are used. θ will have values of $\beta/6$, $2\beta/6$, $3\beta/6$, . . . , and since $\beta = 2\pi/3$ the values of θ will be $\pi/9$, $2\pi/9$, $\pi/3$, $4\pi/9$, $5\pi/9$, $2\pi/3$.

Erect ordinates at the points 1, 2, . . . and on the last ordinate, 6, measure up the distance $6\text{–}N = 0\text{–}6 = L$. Draw the straight line $0\text{–}N$. This is the curve for the term $0.478(3\theta/2)$. For division 1 the second term $0.478(-\,^1/_2\sin 3\theta)$ becomes $0.478(-\,^1/_2 \sin \pi/3) = -\,0.21$. That is, the actual curve cuts the ordinate at point 1 at a distance 0.21 in. below the line $0\text{–}N$. The other points on the curve are determined in like manner. The ordinates are now transferred to the line 0–6, and the pitch line of the cam is constructed as in previous examples.

Cam with Offset Follower (Fig. 36). Draw the path of the follower *AF* in its position relative to the center *O* of the cam, and divide it according to the character of motion desired. (See Figs. 31 and 32.) With *O* as a center and *AO* as a radius describe a circle. Starting at *A*, divide it at *a′*, *b′*, *c′*, etc., to correspond to the positions 1, 2, 3, etc., of the follower. With radius *OC* equal to the offset (perpendicular distance from *OB* to *AF*), draw the circle *ECG*. With *O* as a center, with radius equal to *O*1, draw from point 1 on the follower path the arc 1*a*. From point *a′* draw a tangent *a′a″* to circle *ECG*, intersecting the arc 1*a* at *a*. Similarly draw arcs 2*b*, 3*c*, etc., and tangents *b′b″*, *c′c″*, etc. The intersection of the arcs and tangents at *a*, *b*, *c*, etc., are points on the pitch line of the cam.

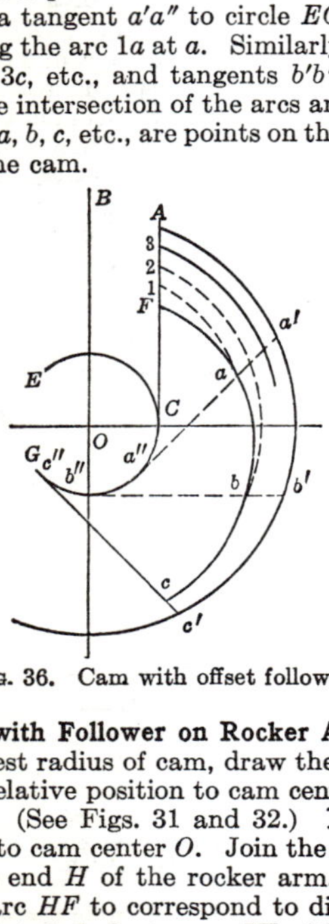

Fig. 36. Cam with offset follower.

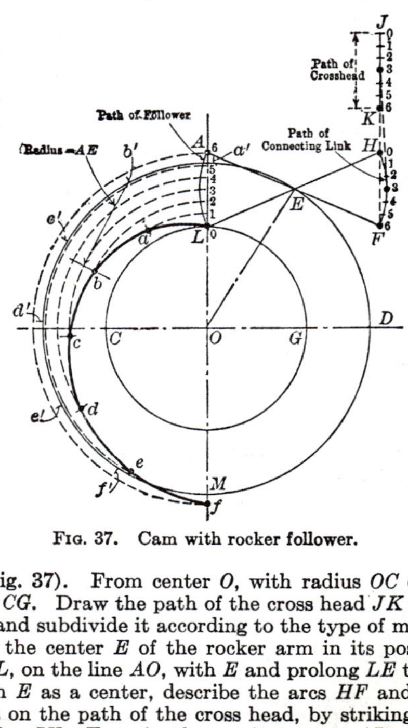

Fig. 37. Cam with rocker follower.

Cam with Follower on Rocker Arm (Fig. 37). From center *O*, with radius *OC* equal to smallest radius of cam, draw the circle *CG*. Draw the path of the cross head *JK* in its proper relative position to cam center *O*, and subdivide it according to the type of motion desired. (See Figs. 31 and 32.) Locate the center *E* of the rocker arm in its position relative to cam center *O*. Join the point *L*, on the line *AO*, with *E* and prolong *LE* to the opposite end *H* of the rocker arm. With *E* as a center, describe the arcs *HF* and *AL*. Divide arc *HF* to correspond to divisions on the path of the cross head, by striking arcs from points 0, 1, 2, etc., on *JK* with a radius *JH*. Transfer these points to arc *AL*. With

O as a center, and *OE* as a radius, draw circle *DEM*, and divide each semicircle at *a'*, *b'*, *c'*, etc., to correspond with the divisions on *JK*, starting the divisions at *E*, which corresponds to point 0 on *AL*. From *O* as a center, with radii of *O*1, *O*2, *O*3, etc., draw arcs 1*a*, 2*b*, 3*c*, etc. From points *a'*, *b'*, *c'*, etc., with radius equal to *LE*, strike arcs intersecting arcs 1*a*, 2*b*, 3*c*, etc., at *a*, *b*, *c*, etc. The intersections *a*, *b*, *c*, etc., are points on the pitch line of the cam.

NOTE. In the examples of cam layout, Figs. 36 and 37 a uniform-motion cam has been considered. The procedure for any other motion would be identical, except that the points of division of the path of the follower would be determined as illustrated in Figs. 31 and 32.

Correction for Roller Follower. A roller follower must be used with cams, unless the force to be transmitted is small. The pitch line of the cam as determined by the above methods then must be corrected. Figure 38 shows the method of making such corrections. The pitch line as determined above is *ab*. From the center *O* draw radii *Oc*, *Od*, etc. With the points of intersection of the radii with *ab* as centers, draw arcs of radius equal to the radius of the roller follower. The cam profile is a smooth curve drawn tangent to these arcs. In Fig. 38 the line *HG* is the profile of a plain cam, and the lines *HG* and *JK* are the boundaries of the slot of a slotted cam.

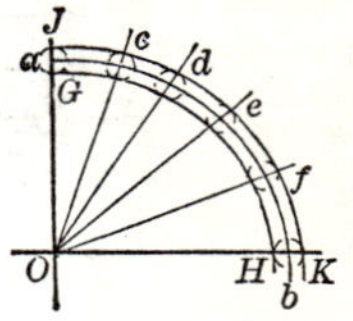

FIG. 38. Profile for roller follower.

Correction for Flat-faced Follower. Flat-faced followers are widely used. The pitch line of the cam as determined by the previous methods must be corrected. Figure 39 shows the method of making such corrections. Assume the pitch line *ab* to be determined as previously discussed. From the center *O* draw radii *Oc*, *Od*, etc. At the points of intersection of the radii with *ab*, draw the straight lines 1–2, 3–4, etc., making the same angles with the radii as the follower face *mn* makes with *Oa* (in this case 90 degrees). The surface of the cam will be the curve *ap* drawn tangent to the lines 1–2, 3–4, etc. If the flat face of the follower makes some other angle with the reference line *Oa*, the lines 1–2, 3–4, etc., should be drawn through the intersection of the radii *Oc*, *Od*, etc., and the pitch line at this same angle with *Oc*, *Od*, etc.

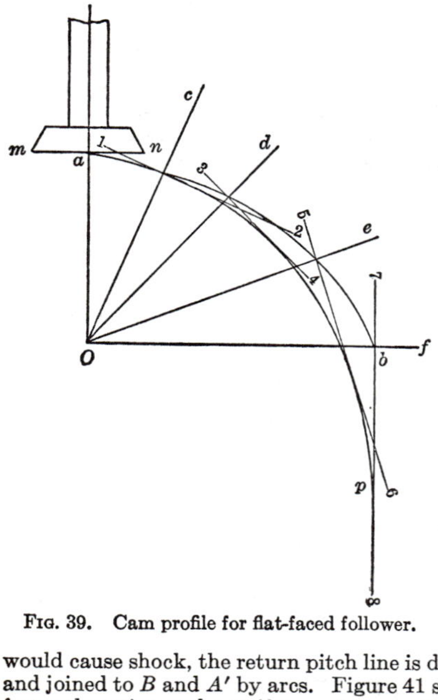

FIG. 39. Cam profile for flat-faced follower.

Cylindrical cams comprise a cylinder with a slot cut in the cylindrical surface. The variation in the slope of the slot causes a roller projecting into it to be moved as the cylinder rotates about its axis. Any desired variation in motion can be obtained by appropriate variation of the slot. Figure 40 shows the method of laying out a cam to give a uniformly accelerated and retarded forward motion and a uniform return motion in two revolutions of the cam. The line *AB* represents to scale the total movement in one direction of the follower. The line *AA'*, to the same scale, represents the circumference of the cam. The rectangle *AA'B'B* therefore represents the developed surface of the cam. Divide *AB* into the number of equal spaces necessary to represent the desired sequence of acceleration and deceleration increments. Erect ordinates at the division points and locate on them the points of the pitch line. (See Cams with Uniform Acceleration and Retardation, p. 7-13.) Draw the pitch line *AB'* for the forward motion of the follower. The pitch line of the return path is a straight line from *B* to *A'*. Inasmuch as strictly uniform motion would cause shock, the return pitch line is drawn between points slightly short of *B* and *A'*, and joined to *B* and *A'* by arcs. Figure 41 shows the layout of a cam that gives a harmonic forward motion and a uniform return motion to the follower in a single revolution.

Cylinder cams also can be designed to give a variety of motions in sequence, during several revolutions. The rectangle representing the developed surface of the cam is divided into half as many parts as the number of revolutions per cycle. The pitch line for that portion of the forward motion of the follower that is to take place during each revolution then is laid out, by the methods described above, in the portion of the rectangle corresponding to the revolution under consideration. Figure 42 shows the layout of a cam to give the following cycle of motions in four revolutions: (1) $^3/_4$ revolution, uniform acceleration and deceleration; (2) $^1/_4$ revolution, dwell; (3) $^3/_4$ revolution, uniform acceleration and deceleration; (4) $^1/_4$ revolution, dwell; (5) 2 revolutions, return.

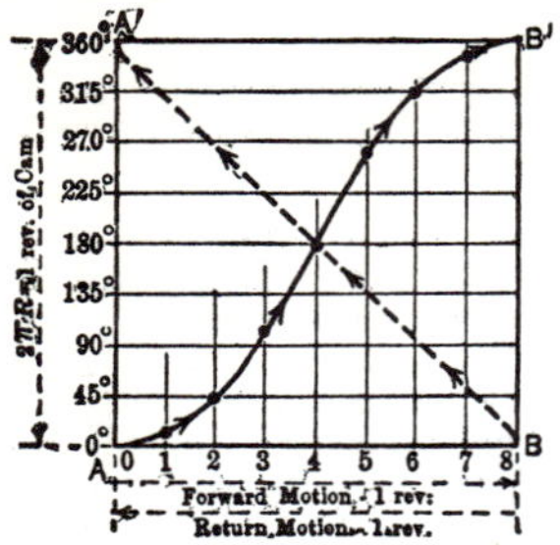

FIG. 40. Cylindrical cam layout, for two revolutions per cycle.

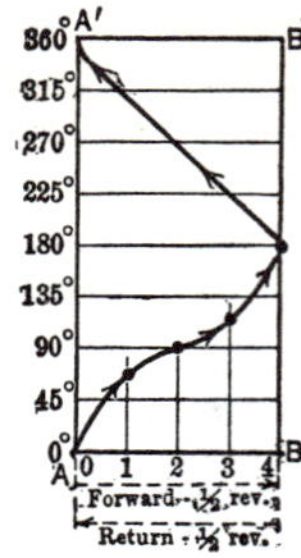

FIG. 41. Cylindrical cam layout, for one revolution per cycle.

Minimum radius of curvature of the pitch line at any point should not be less than $d + {}^1/_8$ in., where d = diameter of cam roller in inches. Otherwise there will be a loss of rise of the follower which will vary in accordance with the sharpness of the angle of intersection of two adjoining curves on the cam.

Rollers for Cylinder Cams. The preferred form of rollers that work in the slot of a cylinder cam is the frustum of a cone, with a corresponding taper of the cross section of the slot. Straight rollers do not have a true rolling action because of the difference in linear velocity of the top and the bottom of the slot. This sliding of a portion of the roller tends to wear both roller and slot, resulting in eventual backlash and faulty action.

The method of determining the taper of roller and slot is shown in Fig. 43. Draw AB equal to the total throw of the cam. Draw AC perpendicular to AB and of a length equal to the length of pitch line on the surface of the cam through which correct cam action is necessary; for example, lengths AD or BE in Fig. 42. Lay out CD equal to the length of the pitch line at the bottom of the slot. This may be determined by redrawing, for example, Fig. 42, making $AA' = 2\pi r$, where r = diameter of cam cylinder at bottom of slot. At D, draw DE parallel to AB and equal in length to AB. Then CB equals the

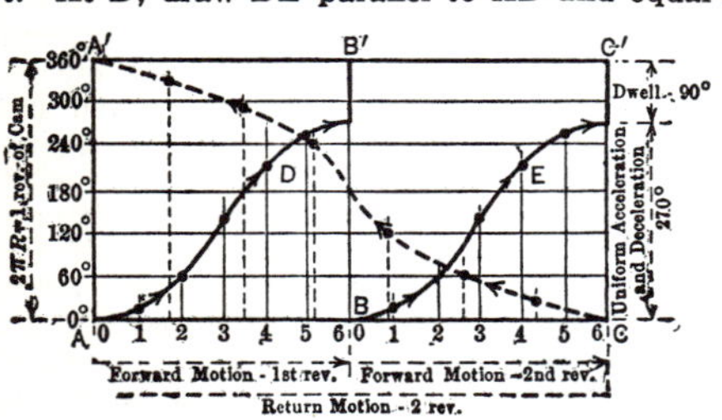

FIG. 42. Cylindrical cam layout, for four revolutions per cycle.

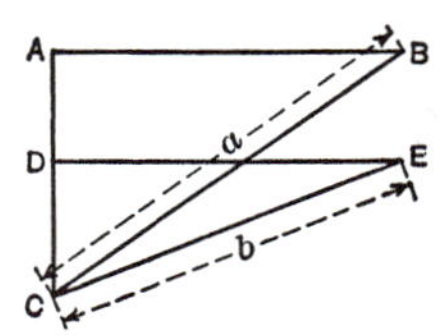

FIG. 43. Construction for determining cylindrical cam slot and roller dimensions.

developed length of the helix traversed by the top of the roller and CE equals the developed length of the helix traversed by the bottom of the roller. If $CB = a$, $CE = b$, D = diameter of roller at top, and d = diameter of roller at bottom, $d = bD/a$.

The head of the pin on which the roller is mounted should be recessed into the roller at least $^1/_{64}$ in. to avoid sliding contact with the bottom of the slot.

5. HOISTING TACKLE

Pulley Blocks. Let P = force applied, or pull; W = load lifted, or resistance. In the simple pulley A (Fig. 44A), the point P on the pulling rope descends the same amount that the load is lifted, therefore $P = W$. In Figs. 44B and 44C the point P moves twice as far as the load is lifted, therefore $W = 2P$. In Figs. 44B and 44C there is one movable block, and two plies of the rope engage with it. In Fig. 44D there are three sheaves in the movable block, each with two plies engaged, or six in all. Six plies of the rope, therefore, are shortened by the same amount that the load is lifted and the point P moves six times as far as the load; consequently $W = 6P$. In general, the ratio of W to P is equal to the number of plies of the rope that are shortened, and also is equal to the number of plies that engage the lower block. If the lower block has two sheaves and the upper three, the end of the rope is fastened to a hook in the top of the lower block; then there are five plies shortened instead of six, and $W = 5P$. If V = velocity of W and v = velocity of P, then in all cases $VW = vP$, whatever the number of sheaves or their arrangement. If the hauling rope, at the pulling end, passes first around a sheave in the upper or stationary block, the direction in which the rope is led from this block to the point at which the pull on the rope is applied is immaterial. If, however, it first passes around the movable block, it is necessary that the pull be exerted in a direction parallel to the line of action of the resistance, or a line joining the centers of the two blocks, in order to obtain the maximum effect. If the rope pulls on the lower block at an angle, the block will be pulled out of the line drawn between the load and the upper block, and the effective pull will be less than the actual pull on the rope in the ratio of the cosine of the angle the pulling rope makes with the vertical, or line of action of the resistance, to unity.

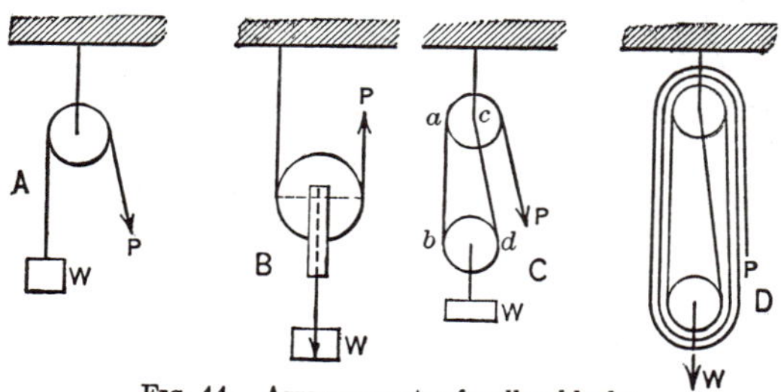

FIG. 44. Arrangements of pulley blocks.

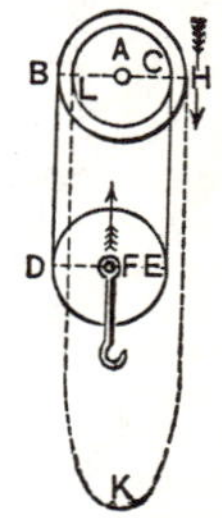

FIG. 45. Differential pulley.

Differential Pulley (Fig. 45). Two pulleys, B and C, of different radii, rotate as one piece about a fixed axis, A. An endless chain, $BDECLKH$, passes over both pulleys. The rims of the pulleys are shaped so as to hold the chain and prevent it from slipping. One of the bights or loops DE, in which the chain hangs, passes under and supports the running block F. The other loop or bight, HKL, hangs freely, and is called the hauling part. It is evident that the velocity of the hauling part is equal to that of the pitch circle of the pulley B.

In order that the velocity ratio may be exactly uniform, the radius of the sheave F should be an exact mean between the radii of B and C.

Consider that the point B of the cord BD moves through an arc whose length is AB; during the same time the point C of the cord CE will move downward a distance AC. The length of the bight or loop $BDEC$ will be shortened by $AB - AC$, which will cause the pulley F to be raised half this amount. If P is the pulling force on the cord HK, and W the weight lifted at F, then

$$P \times AB = W \times {}^1/_2\,(AB - AC) \tag{26}$$

The length of chain required for a differential pulley is half the circumference of A + half the circumference of B + half the circumference of F + twice the greatest distance of F from A + the least length of loop HKL. The last quantity is fixed according to convenience.

A wheel and axle, or windlass, resembles two pulleys on one axis having different diameters. If a weight is lifted by means of a rope wound over the axle, the force being applied at the rim of the wheel, the action is like that of a lever of which the shorter arm is equal to the radius of the axle plus half the thickness of the rope, and the longer arm is equal to the radius of the wheel. A wheel and axle is, therefore, sometimes classed as a perpetual lever. If P = the applied force, D = diameter of the wheel, W = the weight lifted, and d = diameter of the axle + the diameter of the rope, $PD = Wd$.

6. THE GYROSCOPE

The gyroscope is in effect a heavy disk spinning rapidly on its axis, and so mounted that its center is stationary while the axis may take any position whatever. See Fig. 46. A characteristic of the gyroscope, which is utilized in stabilizers to prevent the rolling of ships, in the gyro-compass, in monorail cars, etc., is that when the disk is spinning in a given plane, it resists any attempt to cause it to spin in a different plane. In Fig. 47, the disk of radius r spins about the axis ZZ. The disk lies in the plane $QRST$, and the axis ZZ in plane $ABCD$, perpendicular to $QRST$. Now let a force F be applied at one end of the axis ZZ, tending to move the disk about its stationary center O into a plane at an angle to $QRST$. The movement of the axis in the direction of F will set up a force G, perpendicular to plane $ABCD$ and parallel to plane $QRST$. If this *gyroscopic reaction* G is resisted by some means, as, for instance, having the ends of the axis bear against a guide parallel to $ABCD$, the resistance G', equal and opposite to G, will tend to impede movement in the direction of F. The direction of G is determined by considering F as a pull applied to the end of the axis. Then G will act in the same direction as a point on the disk, in the plane of F and on the same side of the axis, is moving.

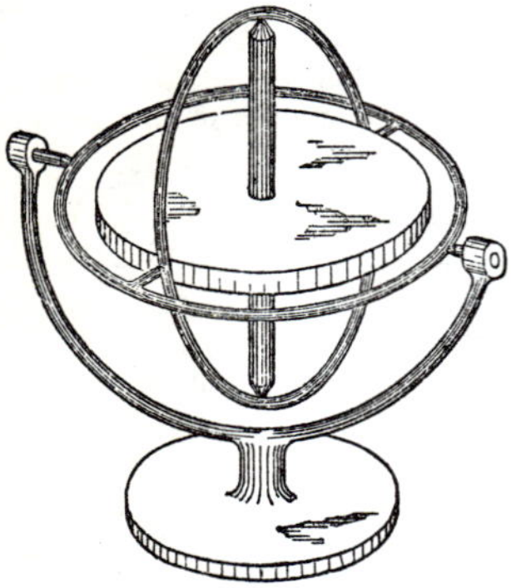

FIG. 46. Gyroscope.

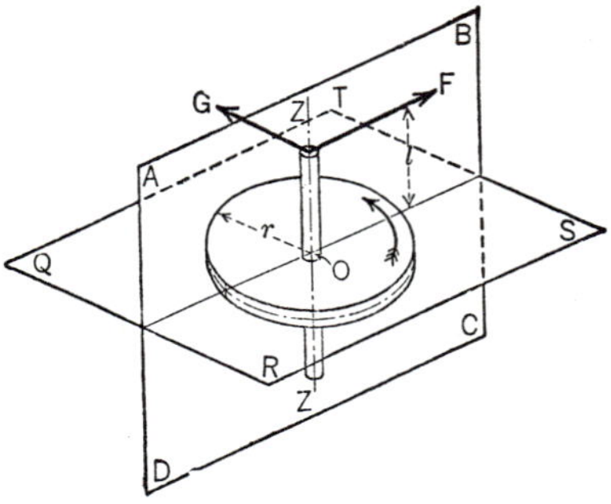

FIG. 47. Forces in the gyroscope.

The magnitude of the forces is as follows. Let F = force applied to move the axis, m = mass of the disk = w/g, R = its radius of gyration, a = acceleration at any instant due to application of force F, l = length of lever arm = OZ, N = rpm of disk, v = velocity corresponding to acceleration a. Then $F = {}^1/_2\, mR^2a/l^2$ and

$$Gl = mR^2 \left(\frac{\pi N}{30}\right) v/l$$

For a mathematical analysis of the forces in the gyroscope and a discussion of its applications, see Ref. 1, p. 378, and Ref. 4.

REFERENCES

1. F. B. Seely and N. E. Ensign, *Analytical Mechanics for Engineers*, 3rd ed., John Wiley and Sons, New York, 1941.
2. P. Schwamb, A. L. Merrill, W. H. James, and V. L. Doughtie, *Elements of Mechanism*, 6th ed., John Wiley and Sons, New York, 1947.
3. C. W. Ham and D. G. Ryan, *University of Illinois Bulletin*, Vol. 29, No. 81, June 7, 1932.
4. Deimel, *Mechanics of the Gyroscope*, The Macmillan Co., New York.

MECHANICS

In the design of structures and machines, certain equations and relations developed by the science of mechanics frequently must be used. The more commonly used formulas are given below. For a mathematical discussion of these relations, see Eshbach, *Handbook of Engineering Fundamentals*, Section 4.

7. CENTER OF GRAVITY

The center of gravity of a body or system of bodies rigidly connected is the point about which, if suspended, all parts will be in equilibrium, i.e., without tendency to rotate. In bodies of uniform density, the center of gravity is the center of magnitude.

The center of gravity of *two bodies* is at a point between them, whose distance from their respective centers of gravity is inversely proportional to their weights. The center of gravity of *more than two bodies* is located by determining the center of gravity of two of them; then considering these two as a single body whose weight is concentrated at their common center of gravity, the center of gravity with reference to the third body is located. The process is repeated until all the bodies in the system have been considered.

The center of gravity of regular figures, whether plane or solid, is the same as their geometrical center. The center of gravity of symmetrical figures lies on the axis of symmetry. The center of gravity may be determined by suspending a body or figure from two different points and dropping a plumb line from the point of suspension. The intersection of the plumb lines will be the center of gravity.

The location of the center of gravity of the more common lines, areas, and solids is:

LINES. Straight Line. At middle point of the line.

Semicircumference. On middle radius R, at distance $0.6366R$ above diameter.

Quadrant. On middle radius R, at distance $0.90028R$ from center.

Circular arc of subtended angle α, on middle radius R, at distance from center of $(R \sin {}^1/_2\alpha)/({}^1/_2\alpha)$, where α is in radians.

Cycloid, generated by circle of diameter D, on perpendicular to middle of base, and ${}^2/_3\, D$ above base.

PLANE FIGURES. Triangle. At the intersection of two medians L_1 and L_2, or at a distance of ${}^1/_3\, L$ above the base.

Parallelogram. At intersection of diagonals.

Trapezium or Trapezoid. At the intersection of the two lines joining the centers of gravity of the two triangles formed by drawing first one diagonal and then the other.

Sector of circle of radius R, chord C, arc L; on radius bisecting arc, at $(2CR/3L)$ from center.

Semicircle. On middle radius, $0.4244R$ from center.

Quadrant. On middle radius, $0.600R$ from center.

Segment of circle of chord C and area A, $C^3/12A$ from center.

Parabolic Surface. In the axis, three-fifths of its length from vertex.

Semiparabola (surface). Three-fifths of axis from vertex, and three-eighths of semibase from axis.

Quadrant of ellipse of semi-axes a and b. At intersection of parallels to the axes a and b, and distant from them, respectively, $[b - {}^4/_3\,(b/\pi)]$ and $[a - {}^4/_3\,(a/\pi)]$.

SOLID BODIES. Prism or Cylinder (parallel bases). At middle point of line joining centers of gravity of the bases.

Oblique frustum of right circular cylinder of radius R mean height H. In the plane of symmetry at $[{}^1/_2\, H + (R^2 \tan \alpha/8H)]$ above base, where α is angle of oblique section to the base, and at $({}^1/_4\, R^2 \tan \alpha/H)$ from the axis.

Cone or Pyramid. In the axis, one-fourth of its length above the base.

Paraboloid. In the axis, two-thirds of its length from the vertex.

Frustum of Cone or Pyramid. In axis, at distance above base of

$$\frac{a}{4} - \frac{3a'^3}{4(a^2 + aa' + a'^2)}$$

where a = length of line from vertex of complete cone to center of gravity of base and a' = that portion of it between vertex and top of frustum.

Hemisphere. On radius R, perpendicular to base, at ${}^3/_8\, R$ above base.

Hollow Hemisphere. Radii R and r. On radius perpendicular to base at

$$\frac{{}^3/_8\,(R^4 - r^4)}{R^3 - r^3} \text{ above base}$$

Segment of Sphere. Radius R; height of segment, H. On radius perpendicular to base, at $[{}^3/_4\,(2R - H)^2/(3R - H)]$.

Sector of Sphere. Radius R; height of segment subtended by sector, H. On the middle radius, at ${}^3/_8\,(2R - H)$ from center.

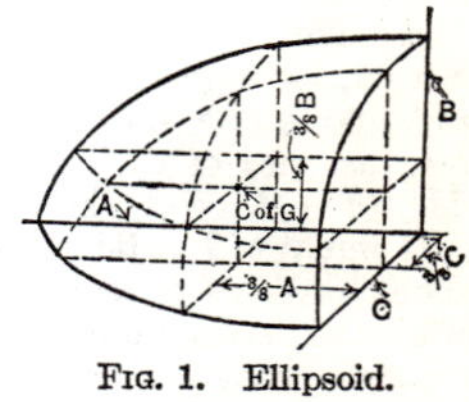

FIG. 1. Ellipsoid.

Ellipsoid of semi-axes, A, B, and C. If an ellipsoid is divided into eight parts by planes passing through each axis, the center of gravity of each part will lie in the point of intersection of three planes parallel to A, B, and C, and at a distance from them respectively, of ${}^3/_8\, B$, ${}^3/_8\, C$, and ${}^3/_8\, A$. See Fig. 1.

8. MOMENT OF INERTIA

THE MOMENT OF INERTIA OF A BODY with respect to an axis is the algebraic sum of the products of the weight of each elementary particle multiplied by the square of its distance from the axis. If $w_1, w_2 \cdots w_n$ represent the weights of the several particles, and $r_1, r_2 \cdots r_n$, their respective distances from axis XX, then the moment of inertia with respect to the axis XX is $I = w_1r_1^2 + w_2r_2^2 + \cdots w_nr_n^2$. The unit of the mass moment of inertia is pound-feet^2 if mass is expressed in pounds and dimensions of body in feet. I is least when the axis passes through the center of gravity of the body.

Moment of inertia as applied to areas, as the cross section of structural beams, equals ar^2, where a is any elementary area. The unit of the area moment of inertia is inches^4 or feet^4 if the dimensions of the body are expressed in inches or feet, respectively. Moments of inertia of the more commonly used bodies, about different axes, are given in Table 1.

Table 1. Moment of Inertia of Commonly Used Bodies

W = weight of body; w = weight of elementary unit of body; d = distance of center of gravity of body from axis.

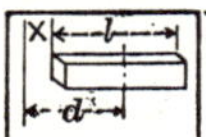	Solid rod, axis perpendicular to axis of rod, $I = W[(l^2/12) + d^2]$	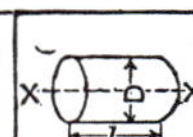	Solid cylinder, axis coincident with axis of cylinder, $I = WD^2/8$
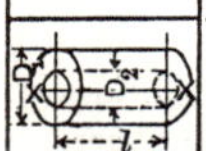	Hollow cylinder about its axis, $I = 1/8\,W(D_1^2 + D_2^2)$ Thin hollow cylinder, $I = 1/4\,WD_1^2$	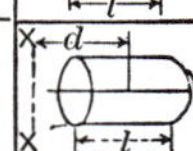	Solid cylinder, axis perpendicular to axis of cylinder, $I = W\left(\frac{D^2}{16} + \frac{l^2}{12} + d^2\right)$
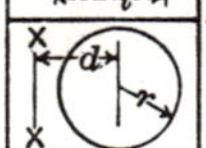	Thin circular plate, axis in its own plane, $I = W[(r^2/4) + d^2]$	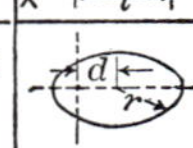	Thin circular plate, axis perpendicular to plate, $I = W[(r^2/2) + d^2]$
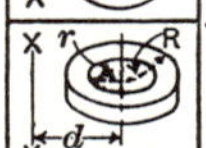	Circular ring, axis perpendicular to plane, $I = W[1/2(R^2 + r^2) + d^2]$	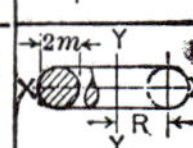	Circular ring about axis XX, $I = w\pi^2Rm^2(R^2 + 5/4m^2)$ About axis YY, $I = 1/2\,w\pi^2Rm^2(4R^2 + 3m^2)$
	Solid sphere about diameter, $I = 2/5\,Wr^2$		Hollow sphere about diameter, $I = 8/15[w\pi(R^5 - r^5)]$ Thin sphere, $I = 2/3\,WR2$
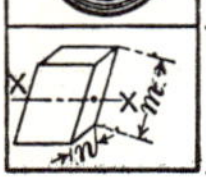	Rectangular prism, axis through c. of g. perpendicular to face mn, $I = 1/12\,W(m^2 + n^2)$	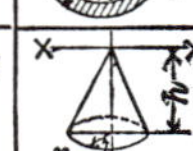	Right circular cone, axis XX, $I = 3/5\,W(1/4\,r^2 + h^2)$ About its axis of revolution, $I = 3/10\,Wr^2$

Moment of Inertia about Parallel Axes. If the moment of inertia I of an area or solid, about an axis through the center of gravity, is known, its moment of inertia I' about any parallel axis may be found by the formula $I' = I + (d^2 \times A)$, where d = distance between the two axes and A = the area or weight. Moments of inertia of sections are given on pp. 8-09 and 8-10.

Polar Moment of Inertia. If the moments of inertia of an area with respect to two rectangular axes in the same plane are known, its moment of inertia about an axis perpendicular to this plane, and passing through the intersection of the axes, is equal to the sum of the moments of inertia about the two rectangular axes. This latter, known as the polar moment of inertia, is designated by J. If I and I' are the moments of inertia about the rectangular axes XX and YY, $J = I + I'$.

Moment of Inertia about Any Axis. If b = breadth and h = depth of a rectangular section, its moment of inertia about its central axis (parallel to the breadth) is $1/12bh^3$; and about one side is $1/3bh^3$. If a parallel axis exterior to the section is taken, and d = distance of this axis from the farthest side and d_1 = its distance from the nearest side, $(d - d_1) = h$, the moment of inertia about this axis is $1/3b(d^3 - d_1^3)$.

Moment of Inertia of Compound Shapes. The moment of inertia of a compound shape about any axis equals the sum of the moments of inertia with reference to the same axis of all the rectangular portions composing it. It also equals the moment of inertia about a parallel axis through its center of gravity + (area of section × square of distance between axes). From this, the moments of inertia or radii of gyration of any single section being known, values may be found for any combination of these sections.

Table 2. Principal Radii of Gyration and Squares of Radii of Gyration

Surface or Solid	Radius of Gyration	Square of Radius of Gyration
Parallelogram: height h, axis at its base	$0.5773h$	$^1/_3\, h^2$
Parallelogram: height h, axis mid-height	$0.2886h$	$^1/_{12}\, h^2$
Straight rod: length l, or thin rectangular plate, axis at end	$0.5773l$	$^1/_3\, l^2$
Straight rod: length l, or thin rectangular plate, axis mid-length	$0.2886l$	$^1/_{12}\, l^2$
Rectangular prism: axes $2a$, $2b$, $2c$, referred to axis $2a$	$0.577\sqrt{b^2 + c^2}$	$(b^2 + c^2) \div 3$
Parallelopiped: length l, base b, axis at one end, at mid-breadth	$0.289\sqrt{4l^2 + b^2}$	$\dfrac{4l^2 + b^2}{12}$
Hollow square tube:		
Outer side h, inner h', axis mid-length	$0.289\sqrt{h^2 + h'^2}$	$(h^2 + h'^2) \div 12$
Very thin, side $= h$, axis mid-length	$0.408h$	$h^2 \div 6$
Thin rectangular tube: sides b, h, axis mid-length	$0.289\sqrt{\dfrac{h + 3b}{h + b}}$	$\dfrac{h^2}{12} \cdot \dfrac{h + 3b}{h + b}$
Thin circular plate: radius r, diameter h, axis on diameter	$^1/_2\, r$	$^1/_4\, r^2 = h^2 \div 16$
Flat circular ring: diameters h, h', axis on diameter	$^1/_4\sqrt{h^2 + h'^2}$	$(h^2 + h'^2) \div 16$
Solid circular cylinder: length l, axis on diameter at mid-length	$0.289\sqrt{l^2 + 3r^2}$	$\dfrac{l^2}{12} + \dfrac{r^2}{4}$
Circular plate: solid wheel of uniform thickness, or cylinder of any length, referred to axis of cylinder	$0.7071r$	$^1/_2\, r^2$
Hollow circular cylinder, or flat ring: l, length; R, r, outer and inner radii. Axis: 1, longitudinal axis;	$0.7071\sqrt{R^2 + r^2}$	$(R^2 + r^2) \div 2$
2, diameter at mid-length	$0.289\sqrt{l^2 + 3(R^2 + r^2)}$	$\dfrac{l^2}{12} + \dfrac{R^2 + r^2}{4}$
Same: very thin, axis its diameter	$0.289\sqrt{l^2 + 6R^2}$	$\dfrac{l^2}{12} + \dfrac{R^2}{2}$
Same: radius r; axis, longitudinal axis	r	r^2
Circumference of circle, axis its center	r	r^2
Circumference of circle, axis its diameter	$0.7071r$	$^1/_2\, r^2$
Sphere: radius r, axis its diameter	$0.6325r$	$^2/_5\, r^2$
Spheroid: equatorial radius r, revolving polar axis a	$0.6325r$	$^2/_5\, r^2$
Paraboloid: r = radius of base, revolving on axis	$0.5773r$	$^1/_3\, r^2$
Ellipsoid: semi-axes a, b, c; revolving on axis $2a$	$0.4472\sqrt{b^2 + c^2}$	$\dfrac{b^2 + c^2}{5}$
Spherical shell: radii R, r, revolving on its diameter	$0.6325\sqrt{\dfrac{R^5 - r^5}{R^3 - r^3}}$	$\dfrac{2}{5}\left(\dfrac{R^5 - r^5}{R^3 - r^3}\right)$
Same: very thin, radius r	$0.8165r$	$^2/_3\, r^2$
Solid cone: r = radius of base, revolving on axis	$0.5477r$	$0.3r^2$

9. CENTER AND RADIUS OF GYRATION

THE CENTER OF GYRATION with reference to an axis is that point at which the entire weight of a body may be concentrated without changing its moment of inertia. In a revolving body, it is the point at which the whole weight of the body may be conceived to be concentrated.

RADIUS OF GYRATION is the distance from the axis to the center of gyration. Let W = weight of the body; I = its moment of inertia; k = its radius of gyration. Then $I = Wk^2 = \Sigma wr^2$; $k = \sqrt{I/W} = \sqrt{\Sigma wr^2/W}$. The radius of gyration is least when the axis passes through the center of gravity, and is then known as the principal radius of gyration. Let k = principal radius of gyration; k' = any other radius of gyration; d = distance between the axes of k and k'. Then $k' = \sqrt{k^2 + d^2}$. Table 2 gives principal radii of gyration of the more common areas and solids, except structural shapes for which see Eshbach, *Handbook of Engineering Fundamentals*, Section 1. For radius of gyration of sections see pp. 8-09 and 8-10, this volume.

10. VELOCITY AND ACCELERATION

VELOCITY is the rate of motion, or speed of a body at any instant. If s = space, feet, passed over in t sec, and v = velocity, feet per second, when v is uniform, $s = vt$; $v = s/t$. If v varies uniformly, the mean velocity $v_m = {}^1/_2\,(v_1 + v_2)$, where v_1 and v_2 are the velocities at beginning and end of time t, respectively. $s = {}^1/_2\,(v_1 + v_2)t$.

If the velocity varies, but not uniformly, v for a very short interval of time $= s/t$, or in the calculus, $v = ds/dt$.

ACCELERATION, a, is the change in velocity occurring in unit time. The unit of a is 1 ft per sec per sec. For uniformly accelerated motion, a is constant. $a = (v_2 - v_1)/t$; $v_2 = (v_1 + at)$; $v_1 = (v_2 - at)$; $t = (v_2 - v_1)/a$.

If the body starts from rest, $v_1 = 0$, and the mean velocity $v_m = v_2/2$; $v_2 = 2v_m$; $a = v_2/t$; $v_2 = at$; $v_2 - at = 0$; $t = v/a$.

For acceleration due to gravity, and of falling bodies, see p. 7-25.

Deceleration or Retarded Motion. If a body starts with a velocity v_1 and comes to rest, $v_2 = 0$, and $s = {}^1/_2 v_1 t$. In any case, if the change in velocity is v, $s = {}^1/_2\, vt = v^2/2a = {}^1/_2\, at^2$.

For a body starting from or ending at rest, $v = at$; $s = {}^1/_2\, vt = {}^1/_2\, at^2$; $v^2 = 2as$.

CURVES OF VELOCITY AND ACCELERATION. If the velocity or acceleration are not uniform, or are represented by complex equations, they may be studied by plotting space, velocity, or acceleration against time. In Fig. 2, the space passed over is plotted as ordinates with seconds of time as abscissas. The angle made by the tangent to the curve at any point determines the velocity at that instant. Thus, the tangent BC at point A intersects the axis of abscissas at C and the velocity at A is $(AD/CD) = {}^3/_2$ ft per sec. Since the slope of the tangent is upward, the velocity is increasing, i.e., the point under consideration is being accelerated. The slope of tangent at F is downward, and hence the motion of the point is being retarded. If the curve at any point is parallel to the axis of abscissas, the motion is uniform.

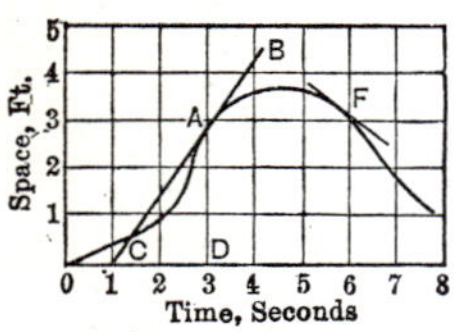

FIG. 2. Displacement-time diagram.

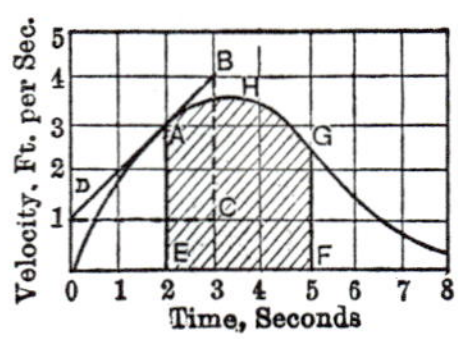

FIG. 3. Velocity-time diagram.

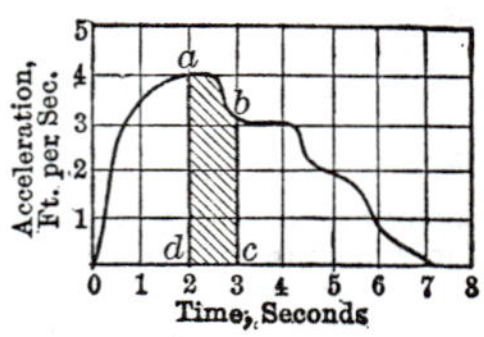

FIG. 4. Acceleration-time diagram.

In Fig. 3, velocity is plotted as ordinates and time as abscissas. The angle of the tangent to the curve at any point represents the acceleration at that instant. The acceleration at A is $(BC/CD) = {}^3/_3 = 1$ ft per sec per sec. As before, an upward slope represents acceleration, while a downward slope represents retardation or deceleration. The area under the curve between any two ordinates represents the space passed over in that interval of time. Thus area $AEFGH = 10$ ft.

In Fig. 4, acceleration is plotted as ordinates and time as abscissas. The angle of the tangent to the curve at any point represents the velocity at that instant, and the area under the curve between two ordinates ad and bc represents the change in velocity during the interval.

For *uniform motion* the space-time curve is a straight line, inclined to the axis of abscissas, and the velocity-time curve is a straight line parallel to the axis of abscissas.

For *constant acceleration* or *constant retardation* the space-time curve is a parabola, and the velocity-time curve is a straight line, inclined away from or toward the axis of abscissas, depending upon whether the velocity increases or decreases. The acceleration-time curve is a straight line, parallel to the axis of abscissas.

ANGULAR VELOCITY is the time rate of turning of a body about an axis, or the angle through which any radius of a body turns in one unit of time. The unit of angular velocity is any unit of angular displacement per any unit of time, as revolutions per minute, revolutions per second, radians per minute, and radians per second. One *radian* $= 180/\pi$ degrees $= 57.3$ degrees. One rpm $= {}^1/_{60}$ rps $= 2\pi$ radians per min $= \pi/30$ radians per sec.

LINEAR VELOCITY OF A POINT ON A TURNING BODY is the product of the angular velocity of the body in radians per unit time and the distance from the axis of rotation to the point. If ω = angular velocity, radians per second; R = distance from axis of rotation to point, feet; N = angular velocity, revolutions per minute; and V = linear velocity of the point, feet per second, then $V = \omega R = 2\pi NR/60$. If V is in feet per minute, then $V = 60\omega R = 2\pi NR$. The linear velocity of a point in a revolving body is proportional to its distance from the axis of rotation.

ANGULAR ACCELERATION is the time rate of change of the angular velocity. The most commonly used unit of angular acceleration is radians per second per second.

LINEAR ACCELERATION OF A POINT ON A TURNING BODY is the resultant of two components, the normal acceleration and the tangential acceleration of the point. The normal acceleration is in feet per second per second, $A^n = \omega^2 R$, and the tangential acceleration is in feet per second per second, $A^t = \alpha R$, where ω and R are as defined above and α is angular acceleration in radians per second per second. The resultant linear acceleration of the point in feet per second per second is $A = \sqrt{(A^n)^2 + (A^t)^2}$.

11. FORCE, WORK, AND ENERGY

UNIFORMLY ACCELERATED MOTION. Let f be a constant force, acting for t sec on a body, at rest at the beginning of the time, but free to move without friction, the weight of the body being w lb. The body will move at a gradually increasing velocity v, which varies as ft/w.

If v is in feet per second, f in standard pounds of force,* t in seconds, and w in pounds, the equations that govern the motion of the body are as follows.

Let K = a constant derived by experiment. Then $v = Kft/w$. The value of K at sea level at latitude 45 degrees is 32.1740, and is twice the distance that a body *in vacuo* would fall in 1 sec. It usually is designated as g, commonly called acceleration due to gravity. The fundamental equation then is

$$v = \frac{ftg}{w} \tag{1}$$

Let S = distance through which the body moves. If v varies directly as time (uniformly accelerated motion), then $S = v_m t$, where v_m = mean velocity during time t. If v_0 = initial velocity and v = final velocity at end of time t, $v_m = (v_0 + v)/2$. For a body starting from rest $v_0 = 0$, and $v_m = v/2$. Hence, $S = {}^1/_2\, vt$; $v = 2S/t$; $t = 2S/v$. Substituting $v = 2S/t$ in eq. 27,

$$S = \frac{ft^2g}{2w} \tag{2}$$

If $f = w$, which is true of a falling body at latitude 45 degrees at sea level, eq. 1 becomes $v = gt$. Substitute in this equation $t = S/v$; then

$$v = \frac{2gS}{v} \quad \text{and} \quad v = \sqrt{2gS} \tag{3}$$

In eq. 2, put $f = w$. Then $S = {}^1/_2\, gt^2$.

From eq. 1, by transposition,

$$ft = w\frac{v}{g} = v\frac{w}{g} \tag{4}$$

w/g, the mass of the body frequently is designated as m, whence eq. 4 becomes

$$ft = mv \tag{5}$$

Substitute in eq. 4, $t = 2S/v$. Then $f2S/v = v \times w/g$, whence

$$fS = {}^1/_2\, mv^2 \tag{6}$$

By definition acceleration $a = v/t$, whence, from eq. 5,

$$f = ma\,\dagger \tag{7}$$

If two forces f, f_1, act on a body at different times, producing accelerations a, a_1, then $f = ma$ and $f_1 = ma_1$, whence

$$\frac{f}{f_1} = \frac{a}{a_1} \tag{8}$$

The foregoing eight equations and their transformations will solve all problems of uniformly accelerated motion.

---

* The pound avoirdupois is the unit of force used by American and British engineers. It is the force with which gravity attracts 1 lb of matter at sea level at 45 degrees latitude, and which would give to 1 lb of matter an acceleration of 32.1740 ft per sec per sec.

In the centimeter-gram-second system, the unit of force is the *dyne*. It is that force which acting on a mass of 1 gram will produce in 1 sec a velocity of 1 cm per sec.

† Eq. 7 sometimes is read "force equals mass times acceleration," which is strictly true in the dyne-centimeter-gram-second, or "absolute" system of measurements, in which force is measured in dynes, but it is not true in the pound-foot-second system, unless it is understood that the word "mass" means the quotient of W divided by g, the unit of which is known as the "slug." A slug is a mass which requires a force of 1 lb to give it an acceleration of 1 ft per sec per sec. At sea level, at latitude 45 degrees, a slug weighs 32.1740 lb.

FORCE OF ACCELERATION. To produce an acceleration of g (i.e., 32.174 ft per sec per sec), force required, $f = mg = (w/g)g = w$, or the weight of the body. Since $f = ma$, $f = m(v_2 - v_1)/t$, where v_2 and v_1 = respectively, initial and final velocities at beginning and end of time t. Also $f = mg = (w/g)/[(v_2 - v_1)/t] = (w/g)a$; $f/w = a/g$. In problems where the local attraction of gravity is a factor, the local value of g must be used if accuracy is necessary.

EXAMPLE. Required the tension in a cord lifting 100 lb vertically through 80 ft, at velocity uniformly increasing from 0 to 40 at the end of 4 sec. Local value of g = 0.998 × standard value = 32.108; $v_m = 20$; final velocity $v_2 = 2v_m = 40$. Acceleration $a = v_2/4 = 10$. Force required to produce acceleration = $f_a = ma = wa/g$ = (100 × 10)/32.174 = 31.08 lb. The standard value of g(32.174) is used here, since force required for acceleration is independent of gravity. Force required to lift the weight without acceleration = f_w = 100 × 0.998 = 99.8 lb. The factor 0.998 is used in this equation because the force of gravity in the given locality is less than standard. Total tension in cord = $f_a + f_w$ = 130.88 lb.

Resistance to acceleration equals the force necessary to produce acceleration = $(w/g)[(v_2 - v_1)/t]$.

FORMULAS FOR ACCELERATED MOTION. For cases of uniformly accelerated motion, other than those of falling bodies, we have the formulas already given, $f = (w/g)a$; $f = (w/g)[(v_2 - v_1)/t]$. If the body starts from rest, $v_1 = 0$, $v_2 = v$, and $f = (w/g)(v/t)$; $fgt = wv$. We also have $S = {}^1/_2\, vt$. Transforming and substituting 32.174 for g,

$$f = \frac{wv^2}{64.35S} = \frac{wv}{32.17t} = \frac{wS}{16.09t^2} \tag{9}$$

$$w = \frac{32.17ft}{v} = \frac{64.35fS}{v^2} \tag{10}$$

$$S = \frac{wv^2}{64.35f} = \frac{16.09ft^2}{w} = \frac{vt}{2} \tag{11}$$

$$v = 8.02\sqrt{\frac{fS}{w}} = \frac{32.17ft}{w} \tag{12}$$

$$t = \frac{wv}{32.17f} = \frac{1}{4.01}\sqrt{\frac{wS}{f}} \tag{13}$$

For varying velocity,

$$f = w\,\frac{v_2^2 - v_1^2}{64.35S} \tag{14}$$

WORK OF ACCELERATION. The work P done in accelerating a body is $f \times S$, where f = force producing acceleration and S = distance covered in time t. The force, $f = ma = (w/g)[(v_2 - v_1)/t]$, and

$$P = \frac{w}{g}\,\frac{v_2 - v_1}{t}\,S \tag{15}$$

EXAMPLE. Work required to move horizontally 80 ft in 4 sec a frictionless body weighing 100 lb, at uniformly increasing velocity. Mean velocity, $v_m = 20$; initial velocity $v_1 = 0$; final velocity $v_2 = 2v_m = 40$; $a = (v_2 - v_1)t = 40/4 = 10$; $f = (w/g)a$ = (100/32.16) × 10 = 31.1 lb; $S = 80$; $P = fS$ = 31.1 × 80 = 2488 ft-lb. Energy stored in the body at the final velocity $v = 40$, equals the work of acceleration.

$$E = \frac{1}{2}mv^2 = \frac{1}{2}\frac{w}{g}v^2 = \frac{100 \times 40^2}{2 \times 32.16} = 2488 \text{ ft-lb}$$

WORK OF ACCELERATED ROTATION. Let ω = angular velocity, radians per second, of a particle of weight w, rotating at unit radius r around an axis. The velocity $v_1 = \omega_1 r$. If angular velocity is accelerated to ω_2, velocity of particle becomes v_2. Increase in velocity is $(v_2 - v_1) = r(\omega_2 - \omega_1)$, and the work of acceleration is

$$P_a = \frac{w}{g}\,\frac{v_2^2 - v_1^2}{2} = \frac{wr^2}{g}\,\frac{\omega_2^2 - \omega_1^2}{2} \tag{16}$$

The work of a rotating body around an axis is

$$\Sigma\,\frac{\omega}{g} \times \frac{1}{2}(v_2^2 - v_1^2) = \frac{\omega_2^2 - \omega_1^2}{2g \times \Sigma wr^2} \tag{17}$$

where Σwr^2 = moment of inertia of the body.

MEASURE OF ENERGY in a body is the product of its weight and the height from which it must fall to acquire actual velocity. If v = velocity, feet per second, h = height, feet, corresponding to velocity = $v^2/2g$ (see Falling Bodies below), and if w = weight, pounds, energy, foot-pounds, $E = {}^1/_2\, mv^2 = wv^2/2g = wh$. Since energy is the capacity for doing work, the units of work and energy are equivalent, or $FS = {}^1/_2\, mv^2 = wh$.

In a rotating body, $E = \omega^2 I/2g$, where ω = angular velocity, radians per second and I = moment of inertia, in pound-feet.^{2} Also, $E = wv^2/2g$, where v = velocity of center of gyration.

THE LAW OF GRAVITATION states that two bodies attract each other with a force directly proportional to their masses and inversely proportional to the square of the distance between them. Let F = force with which two bodies of masses m_1 and m_2 attract each other and d = distance between them. Then $F = Cm_1m_2/d^2 = C(w_1/g)(w_2/g)/d^2 = Cw_1w_2/d^2g^2$, where C = gravitation constant of two spheres of unit mass, i.e., the force with which they attract each other through unit distance. $C = 3.31 \times 10^{-11}$ if F, w, and d are in pounds and feet.

MOMENTUM. IMPULSE. Momentum is the product of the mass and velocity of a body, i.e., momentum $= mv = (w/g)v$. In eq. 5, $ft = mv$; the expression ft is sometimes termed the impulse, whence Impulse = Momentum.

FALLING BODIES. The formulas for falling bodies are derived from eqs. 9 to 14 by putting $f = w$. If g be taken as 32.16 (value at 40 degrees latitude), v = velocity, feet per second, at end of time t; h = height of fall, feet, then

$$v = gt = 32.16t = \sqrt{2gh} = 8.02\sqrt{h} = \frac{2h}{t} \tag{18}$$

$$h = \frac{gt^2}{2} = 16.08t^2 = \frac{v^2}{2g} = \frac{v^2}{64.32} = \frac{vt}{2} \tag{19}$$

$$t = \frac{v}{g} = \frac{v}{32.16} = \sqrt{\frac{2h}{g}} = \frac{\sqrt{h}}{4.01} = \frac{2h}{v} \tag{20}$$

u = space fallen through during Tth sec

$$= h_t - h_{t-1} = \frac{gt^2}{2} - \frac{g(t-1)^2}{2} = \frac{g}{2}(2t-1) = g(t - {}^1/_2) \tag{21}$$

Solving the above equations, the values of v, h, t, u, are:

$t =$	1	2	3	4	5	6	7	8	9	10
$g = 32.16 \times$	1	1	1	1	1	1	1	1	1	1
$v = 32.16 \times$	1	2	3	4	5	6	7	8	9	10
$u = (32.16/2) \times$	1	3	5	7	9	11	13	15	17	19
$h = (32.16/2) \times$	1	4	9	16	25	36	49	64	81	100

Value of g increases with latitude and decreases with elevation above sea level. At Philadelphia, latitude 40 degrees, $g = 32.16$. Everett's formula for g is

$$g = 32.173 - 0.082 \cos (2 \times \text{latitude}) - 0.000003 \text{ height in ft} \tag{22}$$

The standard value of g at latitude 45 degrees given by International Conference on Weights and Measures, Paris, 1901, is 32.1740. The value of $\sqrt{2g}$ decreases about 0.0004 per 1000-ft elevation above sea level. For all ordinary calculations in the United States, $g = 32.16$ and $\sqrt{2g} = 8.02$. In England, $g = 32.2$; $\sqrt{2g} = 8.025$. C. S. Pierce gives limiting values in the United States as: at latitude 49 degrees, at sea level, $g = 32.186$; at latitude 25 degrees, 10,000 ft above sea level, $g = 32.089$.

Local values of g are used in calculations involving local gravitational force, as falling bodies, lifting weights, and water powers. In all cases involving the expression w/g, the standard value, $g = 32.1740$ should be used.

CENTRIFUGAL FORCE. A body revolving in a curved path of radius = R ft exerts a force, called centrifugal force, on the arm or cord that restrains it from moving in a straight line. If W = weight of body, pounds; N = revolutions per minute; v = linear velocity of center of gravity of body, feet per second; $g = 32.174$,

$$v = \frac{2\pi RN}{60} \tag{23}$$

$$F = \frac{Wv^2}{gR} = \frac{Wv^2}{32.174R} = \frac{W4\pi^2RN^2}{3600g} = \frac{WRN^2}{2933.9} = 0.00034084WRN^2 \text{ lb} \tag{24}$$

If n = revolutions per second, $F = 1.2270WRn^2$.

The force that restrains the body from moving in a straight line, i.e., the tension in the cord or arm, is called *centripetal force*, and is equal to and opposite to the centrifugal force.

12. OSCILLATION AND PERCUSSION

CENTER OF OSCILLATION. A body oscillates about a fixed axis, outside its center of gravity. The center of oscillation is a point on the line drawn from the center of gravity of the body, perpendicular to the axis, whose motion is the same as it would be were the entire mass of the body there concentrated and allowed to vibrate as a pendulum.

RADIUS OF OSCILLATION is distance from center of oscillation to point of suspension $= r = R^2/l$; R = radius of gyration, l = distance from point of suspension to center of gravity. Centers of suspension and oscillation are mutually convertible. The center of oscillation of a straight line, or uniform thin bar or cylinder of length L suspended from and oscillating about one end, is $^2/_3\,L$ from the axis. If the point of suspension is $^1/_3\,L$, the center of oscillation is again $^2/_3\,L$ from the axis, i.e., at end of bar. The time of oscillation t is the same in both cases. If point of suspension is at center of gravity, in the simple equivalent pendulum $L = \infty$ and $t = \infty$.

A sphere of radius R is suspended by a cord. Let h = distance of axis of motion to center of sphere, h' = distance of center of oscillation to center of sphere, r = radius of oscillation $= h + h' = h + {}^2/_5\,(R^2/h)$. If the sphere oscillates about an axis tangent to the surface $h = R$ and $r = R + {}^2/_5\,R$.

For various plane figures the length of radius of oscillation is given in Table 3.

Table 3. Length of Radius of Oscillation of Plane Figures

Figure	Direction of Vibrations: Perpendicular to Plane of Figure	Direction of Vibrations: In Plane of Figure
Isosceles triangle	3/4 altitude	
Circle	5/8 diameter	3/4 height
Parabola	5/7 height	5/7 height + 1/3 parameter * 4/7 height + 1/2 parameter †
Rectangle		2/3 diagonal ‡

* Suspended by vertex. † Suspended by middle of base. ‡ Suspended by one angle.

THE PENDULUM. The *simple pendulum* is an ideal body concentrated at the center of oscillation, suspended from a fixed point by a weightless string. A *compound pendulum* is a body of any form suspended from a fixed axis about which it oscillates by force of gravity. The equivalent simple pendulum has the same weight and the same moment of inertia referred to an axis through point of suspension, and it vibrates in the same time. The ordinary pendulum of given length vibrates in equal times when the angle of vibration does not exceed 2 degrees or 2 1/2 degrees on each side of the vertical. This property is *isochronism.*

Let T = time of vibration, seconds, l = length of simple pendulum, g = acceleration due to gravity, which is constant at a given locality. $T = 2\pi\sqrt{l/g}$; since π = constant, $T \propto \sqrt{l/g}$, and in a given locality $T \propto \sqrt{l}$. If l = constant, then for any locality $T \propto \sqrt{g}$.

If T = constant, $gT^2 = \pi^2 l$; $l \propto g$, $g = \pi^2 l/T^2$. This equation gives the force of gravity at any locality if l of a simple pendulum vibrating seconds at that locality is known. At New York, l = 39.1017 in. = 3.2585 ft, and g = 32.16 ft. The time of vibration of any pendulum at New York $= T\sqrt{l/39.1017} = 0.16\sqrt{l}$. The length of a pendulum at New York, having a given time of vibration is $l = T^2 \times 39.1017$ in.

The time of vibration may be varied by adding a weight or *bob* above the point of suspension, which lengthens time T. By varying the height of the upper bob, T is varied.

Let W, w = weight of lower and upper bobs, pounds, respectively; D, d = distance, inches, of lower and upper bobs, respectively, from point of suspension. Then $w = W[(39.1D - D^2)/(39.1d - d^2)]$. Thus by adding a second bob, short pendulums may be made to vibrate as slowly as long ones.

Conical Pendulum. A weight suspended by a cord and revolving at a uniform speed in the circumference of a circular horizontal plane whose radius is r, the distance of the plane below the point of suspension being h, is held in equilibrium by three forces—the tension in the cord, the centrifugal force, which tends to increase the radius r, and the force of gravity acting downward. If v = velocity, feet per second of the center of gravity of the weight, as it describes the circumference, g = 32.16, and r and h are taken in feet, the time in seconds of performing one revolution is (at New York or other place where g = 32.16)

$$t = \frac{2\pi r}{v} = 2\pi\sqrt{\frac{h}{g}} \qquad h = \frac{gt^2}{4\pi^2} = 0.8146t^2$$

If t = 1 sec, h = 0.8146 ft = 9.775 in.

CENTER OF PERCUSSION is identical with center of oscillation and is the point at which the percussive action of a blow is the same as if the whole mass of the body were concentrated at the point.

IMPACT. Let m_1, m_2 = mass of two inelastic bodies; v_1, v_2 = their respective velocities. Then if these two bodies collide, they will move on as one mass with a common velocity v. $(m_1 + m_2)v = m_1v_1 + m_2v_2$, and $v = (m_1v_1 + m_2v_2)/(m_1 + m_2)$. If the bodies move in opposite directions, $v = (m_1v_1 - m_2v_2)/(m_1 + m_2)$. If $m_1v_1 = m_2v_2$ and the bodies impinge directly from opposite directions, $v = 0$; that is, the bodies will be brought to rest.

The loss of energy resulting from the impact of inelastic bodies is the sum of the energies lost and gained by the bodies respectively.

$$^1/_2\, m_1v_1^2 + {}^1/_2\, m_2v_2^2 - {}^1/_2\,(m_1 + m_2)v_2 = {}^1/_2\, m_1(v_1 - v)^2 + {}^1/_2\, m_2(v_2 - v)^2$$

EXAMPLE. Let $m_1 = 10$, $m_2 = 8$, $v_1 = 12$, $v_2 = 15$. $v = [(10 \times 12) - (8 \times 15)]/(10 + 8) = 0$.

$$\text{Energy loss} = (^1/_2 \times 10 \times 12^2) + (^1/_2 \times 8 \times 15^2)$$
$$= [^1/_2 \times 10 \times (12 - 0)^2] + [^1/_2 \times 8 \times (15 - 0)^2] = 1620 \text{ ft-lb}$$

The lost energy does internal work on the bodies by changing their shape and heating them.

For imperfectly elastic bodies of elasticity $e = f_r/f_c$,

$$v_1' = \frac{m_1v_1 + m_2v_2}{m_1 + m_2} - \frac{m_2e(v_1 - v_2)}{m_1 + m_2} \qquad v_2' = \frac{m_1v_1 + m_2v_2}{m_1 + m_2} + \frac{m_2e(v_1 - v_2)}{m_1 + m_2}$$

where v_1' and v_2' = respective velocities of the bodies after impact, f_r = internal force tending to restore shape of body after compression, f_c = force of compression, other notation being as before.

For perfectly elastic bodies, $v_1' - v_2' = v_2 - v_1$; that is, relative velocities before and after impact are the same, and $m_1v_1' + m_2v_2' = m_1v_1 + m_2v_2$.

13. FRICTION

FRICTION may be defined as the resisting force tangent to the surfaces of two bodies when one body moves or tends to move upon the other. The magnitude of the frictional force depends upon the force with which the bodies are pressed together. Friction may be either useful or harmful. In brakes, clutches, belts, bolted or riveted joints, and self-locking mechanisms, friction is useful. Friction is harmful in bearings, gears, and flow of fluids. If the resistance between two bodies prevents relative motion, the resistance is called friction of rest or static friction. When there is relative motion between the bodies, the resistance is called friction of motion or kinetic friction. Static friction increases until motion impends. Kinetic friction, except at very low velocities, is less than static friction when motion just impends. For a discussion of the principles and laws of friction see *Handbook of Engineering Fundamentals* by Eshbach (John Wiley and Sons).

COEFFICIENT OF FRICTION. The ratio of the force required to slide a body along a horizontal plane surface to the weight of the body is called the coefficient of friction. It is equivalent to the tangent of the *angle of repose*, which is the angle of inclination, to the horizontal, of an inclined plane on which the body will just overcome its tendency to slide. The angle usually is denoted by θ, and the coefficient by f or λ. $f = \tan\theta$. If the inclined surface (Fig. 5*a*) upon which a loaded block rests without motion is made gradually steeper until the block starts to slide, the angle at which the sliding starts is the angle of repose for static friction. After the motion has begun, if the angle is not reduced, the block usually will accelerate down the incline. If the angle is then reduced until the block slides with uniform velocity, the angle at which this occurs may be used to determine the friction opposing that particular velocity. The angle Δ shown in Fig. 5*b* is used as a measure of the manner in which granular materials may be heaped. This angle varies widely with the nature of the material.

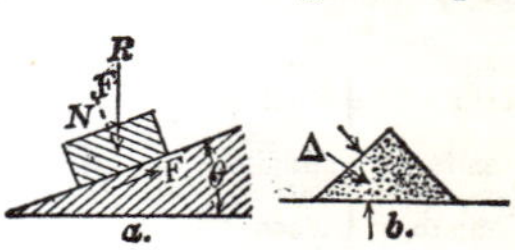

FIG. 5. Angle of repose.

Values of f for a wide variety of materials, both at rest and in motion, are given in Table 4. In Fig. 5*a*, F = frictional resistance = fN, where N is force normal to plane or force pressing two bodies together, = $R\cos\theta$, where R is weight of body.

Starting Friction of Newly Finished Bearing Surfaces. The next three paragraphs * summarize tests at Kingsbury Machine Works of newly finished surfaces to determine influence of method of finishing on

* By H. A. S. Howarth.

Table 4. Coefficients of Friction of Various Materials *

Material	In Contact with	At Rest or in Motion †	Lubrication	Coefficient of Friction *f*	Authority ‡
Metal and Metal					
Brass	Brass	M	Dry	.173	1
Brass	Cast iron	M	Dry	.188	2
Brass	Cast iron	M	Unct.	.111	2
Brass	Wrt. iron[1]	R	Dry	.136	1
Bronze	Bronze	M	Dry	.199	2
Bronze	Bronze	M	Unct.[42]	.133	2
Bronze	Cast iron	M	Dry	.213	2
Bronze	Cast iron	M	Unct.[42]	.098	2
Bronze	Wrt. iron[1]	M	Dry	.167	2
Cast iron	Bronze	M	Dry	.147	2
Cast iron	Bronze	M[2]	[3]	.054	2
Cast iron	Bronze	M	Unct.[42]	.132	2
Cast iron	Cast iron	R	Dry	.162	2
Cast iron	Cast iron	A	Dry	.48	1
Cast iron	Cast iron	M	Dry	.152	2
Cast iron	Steel[4]	R	Dry	.155	1
Cast iron	Wrt. iron[1]	R	Dry	.166	1
Cast iron[5]	Bronze	M	Unct.[42]	.140–.169	2
Cast iron[5]	Cast iron	.....	Ord.[6]	.073–.082	2
Cast iron[5]	Cast iron	M	Unct.[42]	.130–.140	2
Steel	Bronze	M	Dry	.152	2
Steel	Bronze[7]	M	Bath	.0004–.0008[8]	3
Steel	Bronze[7]	M[9]	Bath[10]	.0011	3
Steel	Bronze[7]	M[11]	Bath[10]	.0023	3
Steel	Bronze[7]	M[12]	Bath[10]	.0044	3
Steel	Bronze[7]	M[13]	Bath[10]	.0062	3
Steel	Cast iron	M	A †	.4	1
Steel	Cast iron[14]	M[15]	....	.0035–.0047	4
Steel	Cast iron[16]	M[17]	....	.036–.064	5
Steel	Cast iron	M	Dry	.204	2
Steel	Cast iron	M	Unct.[42]	.108	2
Steel	Steel	R	Dry	.149	2
Steel[4]	Wrt. iron[1]	R	Dry	.194	1
Steel	W. M.[18]	M[19]	....	.002–.030	6
Steel	Gun metal	M	S[20]	.0054–.0177	5
Steel	Gun metal	M	Pad	.0082–.0154	5
Steel	W M[18]	M[21]	....	.0016–.020	4
Steel	Gun metal	M[22]	Bath	.0011–.0070	5
Tin	Cast iron	M	Dry	.176	1
Wrt. iron[1]	Bronze	M	Unct.[42]	.160	2
Wrt. iron	Cast iron	M	Dry	.192	2
Wrt. iron	Cast iron	R	Dry	.190	2
Wrt. iron	Wrt. iron[1]	M	Dry	.138	2
Wrt. iron	Wrt. iron	A	Dry	.4	1
Wrt. iron	Wrt. iron	R	Dry	.152	1
Wrt. iron[5]	Wrt. iron	.....	Ord.[6]	.07–.08	2
Metal and Nonmetal					
Brass	Oak[23]	M	Dry	.616	2
Brass	Oak[23]	R	Dry	.619	2
Cast iron	Elm[23]	M	Dry	.193	2
Cast iron	Leather[24]	M	Wet	.237	2
Cast iron	Leather[24]	M	Dry	.221	1
Cast iron	Leather[24]	M	Wet	.284	1
Cast iron	Leather[24]	R	Wet	.613	2
Cast iron	Leather[25]	M	Dry	.561	2
Cast iron	Leather[25]	M	Gr.[26]	.229	2
Cast iron	Leather[25]	M	Wet	.359	2
Cast iron	Lig. vit.[27]	M	[28]	.093	2
Cast iron	Lig. vit.[27]	M	Unct.[42]	.100	2
Cast iron	Oak[23]	M	Dry	.37–.491	2
Cast iron	Oak[23]	M	Unct.[42]	.106	2
Cast iron	Oak[23]	M	Wet	.22	2
Cast iron	Oak[23]	R	Wet	.642	2
Metal	Hemp	.....	....	.279	9
Steel	Asb[29]	R	....	.27–.38	7
Steel	Cork	R	Dry	.455	8
Steel	Cork	R	Oily	.421	8
Steel	Cork	R	Wet[31]	.564	8
Steel	Cork	R	Wet	.694	8
Steel	Cork comp.[30]	R	Dry	.514	8
Steel	Cork comp.[30]	R	Gr.[26]	.55	8
Steel	Cork comp.[30]	R	Oily	.526	8
Steel	Cork comp.[30]	R	Wet	.894	8
Steel	Cork comp.[30]	R	Wet	.793	8
Steel	Ice	R	Dry	.266	1
Steel	Stone[31]	R	....	.420–.491	10
Wrt. iron[1]	Elm[23]	M	Dry	.252	2
Wrt. iron	Oak[23]	M	Dry	.605	2
Wrt. iron	Oak[23]	R	Dry	.619	2
Wrt. iron	Oak[23]	M	Soapy	.213	2
Wrt. iron	Oak[23]	M	Wet	.255	2
Nonmetal and Nonmetal					
Brick	Limestone[32]	R	Dry	.674	2
Brick	Limestone[4]	R	Dry	.665	2
Cork	Cork	R	Dry	.492	8
Cork	Glass	R	Dry	.516	8
Cork[30]	Pine[23]	R	....	.49	8
Cork comp[30]	Cork	R	....	.839	8
Cork comp[30]	Glass	R	....	.98	8
Earth	Earth	R	....	.25–1.0	11
Earth	Earth[32]	R	....	.81–1.11	11
Granite	Granite	M	Dry	.304	1
Hemp rope	Oak	R	Dry	.462	2
Hemp rope	Wood[33]	R	Dry	.331	10
Hemp rope	Wood[34]	R	Dry	.50	10
Leather	Oak	M	Dry	.35	8
Leather[24]	Oak	M	Dry	.335	2
Leather[24]	Oak	M	Wet	.291	2
Leather[24]	Oak	R	Dry	.43	2
Leather[24]	Oak	R	Wet	.79	2
Leather[25]	Oak	M	Dry	.516	2
Leather[25]	Oak	R	Dry	.595	2
Limestone[35]	Limestone[35]	M	Dry	.376	2
Limestone[35]	Limestone[35]	R	Dry	.702	2
Limestone[4]	Limestone[4]	R	Dry	.636–.744	2
Masonry	Brick	R	[36]	.741	11
Masonry	Brick	R	Dry	.65	11
Masonry	Clay[37]	R	Dry	.508	11
Masonry	Concrete	R	Dry	.785	10
Masonry	Earth[35]	R	....	.65	10
Masonry	Earth[35]	R	Wet	.30	10
Oak[38]	Limestone[35]	M	Dry	.381	2
Oak	Oak[39]	M	Dry	.48	2
Oak	Oak[39]	M	Soapy	.164	2
Oak	Oak[39]	M	[28]	.075	2
Oak	Oak[39]	R	Dry	.62	2
Oak	Oak[39]	R	Unct.[42]	.441	2
Oak	Oak[40]	M	Dry	.315	2
Oak	Oak[40]	M	Unct.[42]	.143	2
Oak	Oak[40]	R	Dry	.543	2
Oak	Oak[40]	R	Wet	.71	2
Oak	Oak[41]	M	Dry	.19	2
Oak	Oak[41]	R	Dry	.43	2
Sandstone	Sandstone	M	Dry	.367	1
Stone	Stone	R	Dry	.63	1
Stone[31]	Wood	R	Dry	.46–.60	10
Wood	Stone	R	....	.40	11
Wood	Wood	R	Soapy	.358	12
Woolen Cloth	Woolen Cloth	M	Dry	.439	1

Notes to Table 4

* Compiled from data assembled by J. F. Spiegel, Kingsbury Machine Works.

† A = abrasion; M = motion; R = rest.

‡ Authorities: (1) Rennie, Experiments on friction, *Phil. Trans. Roy. Soc. (London)*, Vol. 119, 1829. (2) Morin, Nouvelles expériences sur le frottement; faites à Metz, 1831 *Ann. mines*, 3d ser., Vol. 4; faites à Metz, 1832 *Ann. mines*, 3d ser., Vol. 6; faites à Metz, 1833 *Ann. mines* 3d ser., Vol. 10; or Paris, Bachelier 1832, 1834, 1835. Also *Nouvelles expériences sur l'adherence des pierres et des briques, le frottement des axes de rotation, la variation du tension des courroies ou cordes sans fin*, Paris, Carillan Goeury, 1838; Summary of the results of Coulomb, Rennie, and Morin may be found in the paper of Brix, Über die Reibung, *Verhandl. d. Vereins z. Forderung des Gewerbefleisses in Preussen*, Vol. 16, 1837. (3) Kingsbury, Experiments on the friction of screws, *Trans. ASME*, Vol. xvii, 1896. (4) Stribeck, Eigenschaften der Gelit- und Rollenlager, *Zeit. V.D.I.*, 1902. (5) Tower, First reports on friction experiments, *Proc. IME*, March 1891. (6) Lasche, Die Reibungsverhältnisse von Lagern mit hoher Umfangsgeschwindigkeit, *Zeit. V.D.I.*, 1902. (7) Chase, Practice and theory in clutch design, *Trans. SAE*, 1921, Pt. II. (8) Wenrich, University of Pittsburgh. (9) Unwin, *Elements of Machine Design.* (10) Hütte, *Ingenieurs Taschenbuch*, 26th ed. (11) Rankine, *Applied Mechanics.* (12). Lesley, *Trans. Soc. Naval Architects Marine Engrs.*, 1904.

[1] Wrought iron. [2] Slow. [3] Copious. [4] Soft. [5] Journal. [6] Ordinary. [7] Thrust bearing. [8] Low practical value. [9] 50 rpm. [10] Oil viscosity, 150 Saybolt universal sec. [11] 500 rpm. [12] 2000 rpm. [13] 4000 rpm. [14] Shell. [15] Velocity, 13.2 ft per sec; pressure, 14.2–356 psi. [16] Collar bearing. [17] 50–145 rpm; pressure, 75–90 psi. [18] W.M. = white metal. [19] Velocity, 3–66 ft per sec; pressure, 14.2–214 psi. [20] Siphon lubrication; rape oil. [21] Velocity, 13.2 ft per sec; pressure, 14.2–712 psi. [22] Velocity, 1.75–7.85 ft per sec; pressure, 100–520 psi. [23] On lengthwise grain. [24] On edge. [25] On flat. [26] Greasy. [27] Lignum vitae. [28] Tallow or lard. [29] Asbestos fabric. [30] Cork composition. [31] Stone, gravel. [32] Including shingle and gravel. [33] Polished. [34] Rough. [35] Hard. [36] Damp mortar. [37] Dry. [38] End grain. [39] Grains parallel. [40] Grains crossed. [41] End grain or parallel grain. [42] Unctuous.

---

the coefficient of starting friction in the absence of vibration. The tests were made with a babbitt-faced shoe, of 60.7 sq in. area, scraped to a surface plate, opposed to a close-grained cast-iron surface, finish turned and lapped with fairly coarse silicon carbide. The surfaces were immersed in light oil, and subjected to a load of over 300 psi for several hours. The coefficient of starting friction was as high as 0.66. It was reduced to as low as 0.21 by more refined finishes. The lowest coefficients were found when the cast-iron surfaces had been scrubbed by rotation, pressed by a leather-covered block, using lubricating oil impregnated with a very fine abrasive. The direction of motion during test (so long as it is concentric) seems to have little influence on the coefficient of friction.

Some of the values of starting coefficient of friction determined were: lard oil, load applied for 2 hr, $f = 0.18$; castor oil, load applied for 113 hr, $f = 0.25$; light mineral oil, viscosity 150 Saybolt universal seconds, load applied 14 to 18 hr, $f = 0.30$. Kerosene gave higher values than the mineral oil. The conclusions drawn from the test were: In the absence of vibration, the starting friction coefficient is a function of the newness of the surfaces, of the methods of finishing employed, of the time which the surfaces have been pressed together, of the pressure acting on them, of the nature and viscosity of the lubricating oil in which the surfaces are immersed, and of the area of the smaller of the two faces. When the surfaces are loaded to around 300 psi and have an area of 60 sq in., the coefficient of static friction in the absence of vibration may be expected to be below 0.30 for new faces, and below 0.20 for those in service. The influence of electrical vibration is believed to reduce these values 20% or more.

Influence of Surface Finish on Starting Friction and Friction at Low Speed. Howarth's experiments on a low speed and other tests of Kingsbury thrust bearings (*Trans. ASME*, xli, p. 685, 1919) show the variation, with speed and load, of friction of motion in a small thrust bearing of a mean diameter of 9.375 in., having four babbitt-faced shoes and a revolving cast-iron runner. The total shoe area was varied from 71.5 sq in. to 10 sq in. in order to test pressures ranging from 140 to 1000 psi. In a typical test, the load was 600 psi. Machine oil, of 5.345 Saybolt universal seconds viscosity at the test temperature, was used as lubricant, the surfaces being always submerged. The test was started at 16 rpm, speed being reduced gradually to 1/3 rpm. The friction coefficient increased gradually, but irregularly, from about 0.0015 at 16 rpm to about 0.132 at 1/3 rpm. The speed then was gradually increased without changing the load. The friction coefficients determined for the increasing speeds were much lower than that for decreasing speeds. This was explained by the improvement of bearing surfaces due to the long running at low speed. During these tests friction coefficients for starting from rest were found to vary from 0.15 to 0.20. These values, therefore, may be considered satisfactory for bearing surfaces that have been in operation for some time under favorable conditions in which a number of starts and stops have gradually brought the surfaces to a condition favorable to such starting when immersed in an oil bath.

Maximum Friction. When two metals are sweated or brazed together, the force required to separate them transversely at the joint is less a function of the pressure than of the transverse shearing strength of the metal bond. This might be a combination of the breaking of the bond between the soft and hard material and the actual shearing of the soft material on itself. The force required for such separation cannot be expressed consistently by means of a coefficient of friction.

INTERLOCKING FRICTION. If two surfaces interlock, as in a square thread screw (Fig. 6*a*), the force required to slide one on the other depends more on strength of the materials and proportion of interlocking than on the load acting between them. If the surfaces are free to move vertically against a load (Fig. 6*b*), the force required to slide them laterally would be a combination of sliding friction and lifting of the load. When an irregular hard surface presses against a soft one (Fig. 6*c*), the resistance offered to sliding one over the other will depend on the normal load, on the sliding friction, on resistance to deformation of the soft surface, and on the strength of the softer surface as it might be broken down because of interlocking. Figure 6*d* illustrates abrading frictional resistance to sliding of one surface over another, as a saw or file. The greater the normal pressure on the saw or file, the more will it cut into the lower block and the greater will be the resistance offered to transverse motion.

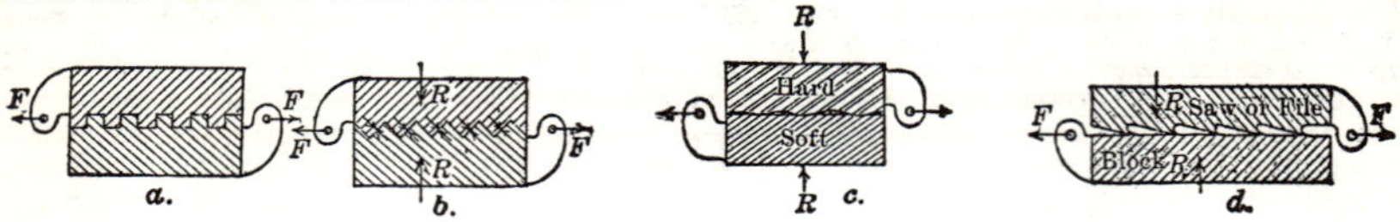

FIG. 6. Interlocking friction. (Courtesy of H. A. S. Howarth)

INFLUENCE OF VIBRATION. Tests to determine starting friction show coefficients of friction to be much higher in the absence of vibration than in its presence, and that friction decreases as the disturbance by vibration increases. Vibration similarly reduces the friction of motion so long as metallic contact is present in any considerable degree. With perfect lubrication, bearing surfaces are completely separated by lubricating films, and vibration does not appear to change the coefficient of friction of motion.

ROLLING FRICTION is the force required to roll a cylindrical or spherical body on a plane or on a curved surface. It depends on the nature of the surfaces and on the force with which they are pressed together, but is different from ordinary, or sliding, friction.

Rolling friction is a consequence of the irregularities of form and the roughness of surface of bodies rolling one over the other. Its laws are not yet definitely established in consequence of the uncertainty which exists in experiment as to how much of the resistance is due to roughness of surface, how much to original and permanent irregularity of form, and how much to distortion under the load.

Coefficients of Rolling Friction. Let R = resistance applied at the circumference of the wheel, W = total weight, r = radius of the wheel, and f = a coefficient which is a linear quantity expressed in the same units as r. Then $R = fW \div r$. The value of f is very variable. Probable values of f (in inches) based upon values given by Coulomb, Goodman, Pambour, and Weisbach are as follows:

Lignum vitae on oak	0.02
Elm on oak	0.032
Soft wood rollers on wood	0.06
Cast iron on cast iron	0.018
Cast iron or steel on steel	0.007 to 0.02
Cast iron or steel on wood	0.06 to 0.1
Cast iron or steel on macadam	0.05 to 0.2
Cast iron or steel on granite	0.085

The friction of pneumatic tires on roads is discussed in Kent's *Mechanical Engineers' Handbook—Power*, Section 14.

GRANULAR FRICTION. The friction required to move one surface over another when granular material is placed between them probably is a combination of several kinds of interaction. The simplest condition is that in which the granular particles turn over and over without interference (Fig. 7*a*). Figure 7*b* indicates how the granular particles

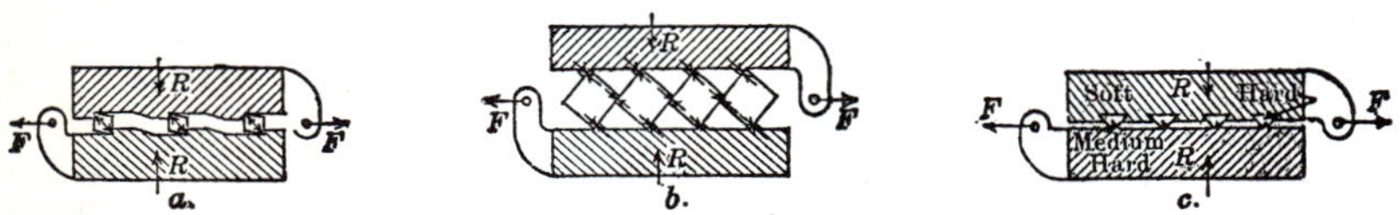

FIG. 7. Granular friction. (Courtesy of H. A. S. Howarth)

may become interlocked so that they will abrade each other and may cut both surfaces when motion occurs. Figure 7*c* represents granular particles embedded in the softer surface and acting as cutters when moved on the harder surface.

SKIDDING FRICTION results from slipping of rubber tires on a pavement. The friction depends on the type of tire surface and conditions of the pavement. With rough pavement and smooth tires, high spots on the pavement will penetrate any film that is formed and project into the rubber, thus diminishing tendency to skid. With a smooth pavement and tires, the film of water, clay, or pulpy material found on roads will completely separate the surfaces. The tire then will slide freely in any direction. The coefficient of friction of rubber tires on roads depends upon the condition of the road and the tire tread, speed, and inflation of tire. Values commonly used are $f = 0.7$ for dry asphalt, $f = 0.8$ for wet asphalt, and $f = 0.1$ for soft, slippery roads.

FLUID FRICTION includes the frictional resistance to motion of fluids in containers, as the flow of water or gas in pipes; of solid bodies completely separated by a film of fluid, as a perfectly lubricated bearing; of solids partially or wholly immersed in a fluid, as a boat moving through water. Thurston states that for all fluids, whether liquid or gaseous, the resistance is (1) independent of the pressure between the masses in contact; (2) directly proportional to the area of rubbing surface; (3) proportional to the square of the relative velocity at moderate and high speeds, and to the velocity nearly at low speeds; (4) independent of the nature of the surfaces of the solid against which the stream may flow, but dependent to some extent upon their degree of roughness; (5) proportional to the density of the fluid, and related in some way to its viscosity. Fluid friction, as related to the flow of liquids in pipes, is discussed in Kent's *Mechanical Engineers' Handbook—Power*, p. 6-35.

14. FRICTION IN MACHINE ELEMENTS

FRICTION OF BEARINGS. For a detailed discussion of the friction of bearings, see pp. 12-03 to 12-40. The following data, from various sources, are for general guidance only. For design, in cases where friction is an important factor, reference should be made to the pages cited.

FRICTION OF JOURNALS. The friction of journals usually is expressed in terms of units of projected area, that is, $d \times l$, where d = diameter and l = length of journal. Unwin states that the friction of journals must lie between $\pi fP/2$ and $4fP/\pi$, where f = coefficient of friction and P = load on journal. For journals in ordinary condition the value of fP usually is sufficiently accurate. Values of f vary according to different authorities. Unwin gives the following:

Authority	f	Conditions
Morin	0.05 –0.07	Good journals, well lubricated
Rühlmann and Waltjen	0.02 –0.03	Wrought-iron on gun metal
Kirchweger	0.009–0.01	Wrought-iron on white metal *
Kirchweger	0.014	Wrought-iron on gun metal *
Bokelberg and Welkner	0.0028	Small loads, low velocities †
Bokelberg and Welkner	0.013	Heavy loads, high velocities †

* Railway journals well lubricated.

† In these experiments f was less for gun metal than for white metal.

The work lost in friction in foot-pounds per minute, is $W = fP\pi n(d/12)$, where n = revolutions per minute, d = diameter of journal, inches, other notation as above.

The friction of a journal is influenced by the nature and quantity of lubrication.

Value of Antifriction Metals. According to Professor J. E. Denton the various white metals available for lining brasses do not afford coefficients of friction lower than can be obtained with bare brass, but they are less liable to "overheating," because of the superiority of such material over bronze in ability to permit of abrasion or crushing, without excessive increase of friction. Thurston says that gun bronze, babbitt, and other soft white alloys have substantially the same friction; in other words, the friction is determined by the nature of the unguent and not by that of the rubbing surfaces, when the latter are in good order. The soft metals run at higher temperatures than the bronze. This, however, does not necessarily indicate a serious defect, but simply defficient conductivity. The value of the white alloys for bearings lies mainly in their ready reduction to a smooth surface after any local or general injury by alteration of either surface of form. See Bearing Metals, pp. 4-73 to 4-78.

Ball Bearings. The friction of ball bearings varies with the design and construction of the bearing. In general, it decreases with increase of load and is independent of speed and temperature. Stribeck found the coefficient of friction in one type of bearing at 65 rpm to vary from $f = 0.0033$ with a load of 838 lb to $f = 0.0015$ with a load of 6614 lb. At 780 rpm the same bearings showed $f = 0.0037$ and $f = 0.0013$, respectively, at 838 lb and 6614 lb loads.

Goodman gives the following ranges for f in various types of ball bearings: cylindrical, 2-point bearing, $f = 0.0012$ to 0.0018; thrust, 2-point bearing, $f = 0.0008$ to 0.0012; thrust, 3-point bearing, $f = 0.0018$ (average); thrust, 4-point bearing, $f = 0.0055$.

Roller Bearings. The coefficient of friction of roller bearings decreases with increase of load and with increase of temperature. It seems to be independent of the speed. Goodman gives values at loads of 2000 and 10,000 lb respectively, of $f = 0.0131$ to 0.0072 for 40 rpm, and of $f = 0.0053$ to 0.0024 for 400 rpm.

GEARING. The friction of gearing depends on the type, the material of the gears, the quality of workmanship, the quality of lubrication, and the form of the teeth. Values given below should be used with caution, as they are applicable only under conditions identical with those under which the determination was made.

Spur Gears. Wilfred Lewis gives the following as the friction losses F of various types of spur gears, in which p = circular pitch; 20-degree stub gear, addendum of $0.24p$, $F < 1\%$; 22 1/2-degree gear, addendum of $0.28p$, $F = 1\%$ approx.; 14 1/2-degree gear, addendum of $0.32p$, $F = 1.3\%$ approx. Other investigators report friction losses of cut gears in the neighborhood of 1%. The length of addendum has greater effect on the friction than has the angle of obliquity.

Bevel Gears. Various investigators have reported friction losses ranging from 1% under exceptionally good conditions of workmanship and lubrication to 8.6% under less favorable conditions.

Helical Gears. Friction losses in helical gears are composed of the tooth friction losses, which correspond to those of spur gears, and the losses resulting from the transverse sliding of the teeth on each other. The ratio F_s of lost work, due to this latter cause, to the useful work may be expressed as

$$F_s = \frac{\sin \alpha(\beta + \theta)}{\sin \beta \sin (\alpha + \theta)} - 1$$

in which α and β are, respectively, the helical angles of the gears and θ is the angle of repose, determined from the coefficient of friction, i.e., $f = \tan \theta$.

Worm Gearing. Let β = lead angle of the worm, ϕ = pressure angle of the gear tooth, and f = coefficient of friction. Then

$$\text{Efficiency} = \frac{\tan \beta(\cos \phi - f \tan \beta)}{\cos \phi \tan \beta + f}$$

Inasmuch as the pressure angle ϕ is small, the cosine is almost unity ($\cos 14\,1/2° = 0.96815$ and $\cos 20° = 0.93969$) and the efficiency is approximately

$$\text{Efficiency} = \frac{\tan \beta - f \tan^2 \beta}{\tan \beta + f}$$

With f and $\tan \beta$ small, the term $f \tan^2 \beta$ is practically zero, hence

$$\text{Efficiency} \doteq \frac{\tan \beta}{\tan \beta + f}$$

where $\tan \beta = np/2\pi r$; n = number of threads on worm; p = linear pitch of worm; and r = pitch radius of worm.

The coefficient of friction f depends upon the material, the surface finish, the lubrication, and the rubbing velocity. Its value usually ranges from 0.02 to 0.05 but may be as high as 0.15 with poor lubrication and rubbing velocities below 40 ft per min. For a case-hardened, ground, and polished steel worm, mating with a bronze gear and well lubricated, the coefficient of friction is approximately $f = 1.28/V^{0.6}$, where V = rubbing velocity in feet per minute.

FRICTION OF HYDRAULIC PLUNGER PACKING. Hütte (Ref. 1, p. 202) says: "For stuffing-boxes with hemp, cotton or leather packing, with water pressures between 1 and 50 atmospheres, the frictional loss depends on the water pressure, the circumference of the packed surface, and a coefficient μ, which is constant for this range of pressure. The loss is independent of the depth of stuffing-box or leather ring, and is given by the formula $F = Kpd$, where F = total frictional loss, pounds, p = pressure, pounds per square inch, d = diameter of plunger, inches." K, a coefficient which depends on the kind and condition of the packing, is given as follows for various cases:

For cotton or hemp, loose or braided, dipped in hot tallow, plungers smooth, glands not pulled down too tight, packing therefore retaining its elasticity, dimensions such as usually occur, $K = 0.072$.

Same conditions, after packing is some months old, $K = 0.132$.

Materials the same, but with hard packing, unfavorable conditions, etc., K = as much as 0.299.

Leather packing, soft leather, well made, etc., $K = 0.036$ to 0.084.

Hard, stiffly tanned leather, $K = 0.12$ to 0.156.

Unfavorable conditions, rough plungers, gritty water, etc., K = up to 0.239.

Weisbach-Hermann (Ref. 2) gives a formula which, when translated into the same notation as the one in Hütte, is

$$F = 0.0312pd \text{ to } 0.0767pd$$

Since the total pressure on a plunger is $^1/_4\ \pi d^2 p$, the ratio of the loss of pressure to the total pressure is $Kpd/^1/_4\ \pi d^2 p$, or using the extreme values of K, 0.0312 and 0.299, the ratio ranges from $0.04/d$ to $0.38/d$, or from 4 to 38% divided by the diameter in inches.

FRICTION OF KEYS AND COTTERS. In a taper key, the angle between the tapered sides of which is α, the force required to drive the key home is

$$F = P[\tan(\theta + \alpha) + \tan\theta]$$

where P = pressure on the key normal to one of the tapered sides and θ = angle of repose of the surfaces in contact.

The force required to drive the key out is

$$F_1 = P[\tan(\theta - \alpha) + \tan\theta]$$

The limiting value of α is 2θ, for if $\alpha > 2\theta$ the key will not remain in position. If $f = 0.08$, $\theta = 4.5$ degrees. The greatest angle of taper, therefore, is 9 degrees or a taper of approximately 1 in 7. Similarly for cotters, where α_1, and α_2, respectively, are the angles made by the two edges of the cotter with a center line normal to the line of pressure,

$$F = P[\tan(\theta + \alpha_1) + \tan(\theta + \alpha_2)]$$

and

$$F_1 = P[\tan(\theta - \alpha_1) + \tan(\theta - \alpha_2)]$$

As before, the cotter will slip out if $(\alpha_1 + \alpha_2) > 2\theta$, which gives a limiting value of $(\alpha_1 + \alpha_2) = 9$ degrees when $f = 0.08$.

FRICTION OF ROPE SHEAVES. Unwin gives $f = 0.28$ for rope on a metal pulley. With ropes in grooves whose sides make an angle of 45 degrees, the pressure between the rope and the sides of the groove is greater than the force pressing the rope into the groove, the ratio being $(\operatorname{cosec}\theta/2) : 1$, where θ = angle of inclination of groove. The actual coefficient of friction of a rope in a sheave groove then is

$$f = 0.28 \times \operatorname{cosec} {}^1/_2\,\theta$$

giving for a 45-degree groove $f = 0.7$.

FRICTION OF V-GUIDES. Let P be the force pressing a V-shaped block into V-guides, the angle between the faces of the guides being β. The frictional resistance R to the movement of the block along the guides will be

$$R = \frac{fP}{\sin {}^1/_2\,\beta}$$

where f = coefficient of friction between the metals of the guides and blocks.

FRICTION IN LINKAGES may be analyzed graphically by use of the *friction circle.* If friction is neglected, the line of action of the forces is considered to be normal to the contacting surfaces. If friction is considered, the line of action of the forces is tangent to the friction circle whose diameter is the product of the coefficient of friction and the journal diameter. Figure 8 is a line drawing to scale of a slider-crank mechanism with

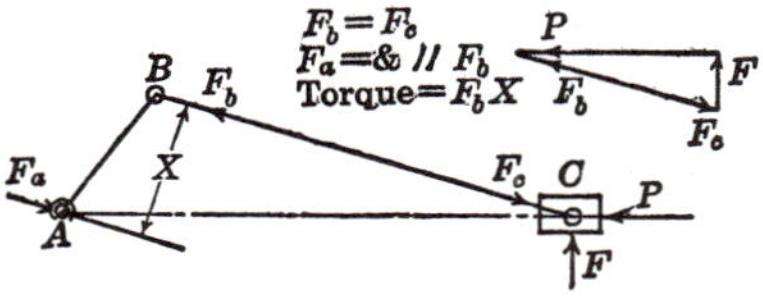

FIG. 8. Forces in a slider-crank mechanism, friction neglected.

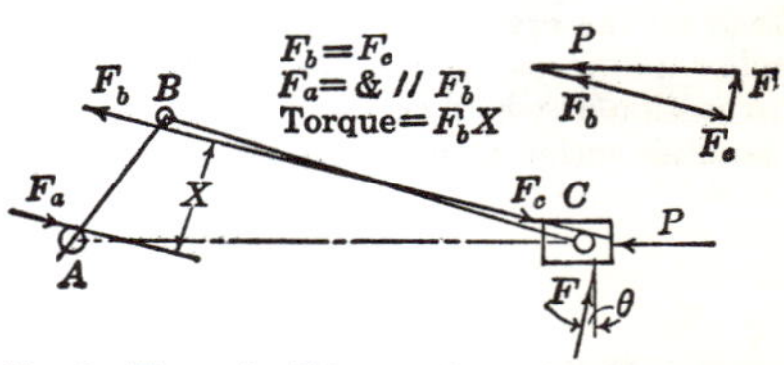

FIG. 9. Forces in slider-crank mechanism, friction considered.

friction neglected. The piston C is the driver with an instantaneous force P. All forces act normal to the contacting forces. Figure 9 is the same mechanism with friction considered. The circles at the connections are the friction circles drawn to the same scale as the drawing. θ is the friction angle for sliding friction. The lines of action of the forces are drawn tangent to the friction circles and in such a direction that friction opposes the motion.

REFERENCES

1. Hütte, *Ingenieurs Taschenbuch,* Vol. 1, 15th ed., W. Ernst und Sohn, Berlin.
2. Julius Weisbach and Gustav Hermann, *Mechanics in Hoisting Machinery,* The Macmillan Co., New York, 1893.

MACHINE ELEMENTS

15. CRANKS AND CRANKPINS

There are two types of cranks: the overhung or side crank (Figs. 1 and 2) and the center crank (Fig. 3). The proportions of each are different. In the side crank, the crankpin is a cantilever; in the center crank, the crankpin is a simple beam. The crankarm or web of a side crank must carry the same load as both arms of the center-crank type. The dimensions of the crankpin are determined by the bearing and lubrication requirements and then checked for strength and deflection. The proportions of the various parts are often empirical. Forged crankshafts are usually SAE 1020, SAE 1030, or SAE 2340 steel. Chrome-nickel steel, SAE 3140 and SAE 3240, are used in automobile and aircraft-engine crankshafts. Some of the special iron-copper alloys are also used in automobile crankshafts. Cast-iron cranks are used only in small inexpensive machinery; however, high-test cast iron is used for large Diesel engine crankshafts. Nickel alloy irons with tensile strengths of 60,000 to 75,000 psi have been used in casting large Diesel engine multicrankshafts weighing over a ton.

OVERHUNG CRANKS. Figures 1 and 2 show the proportions for cast-iron and forged-steel cranks. Although the sketches show different methods of securing the pin to the

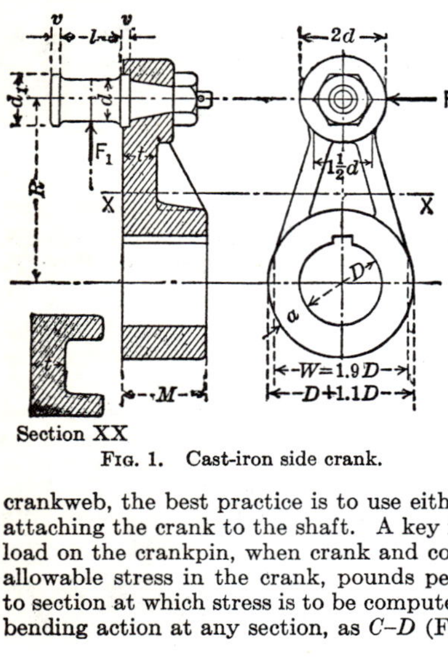

FIG. 1. Cast-iron side crank.

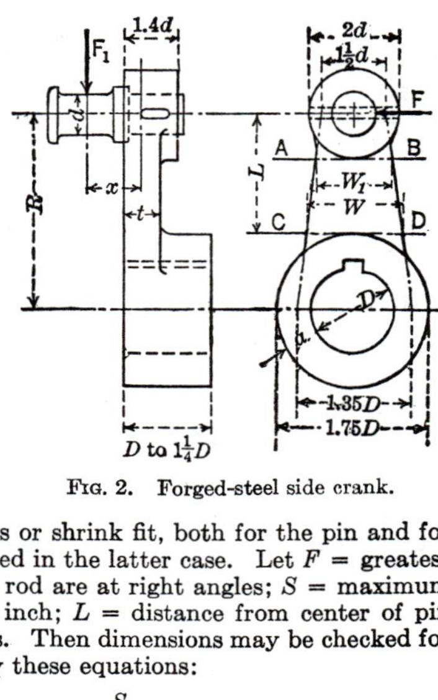

FIG. 2. Forged-steel side crank.

crankweb, the best practice is to use either a press or shrink fit, both for the pin and for attaching the crank to the shaft. A key is also used in the latter case. Let F = greatest load on the crankpin, when crank and connecting rod are at right angles; S = maximum allowable stress in the crank, pounds per square inch; L = distance from center of pin to section at which stress is to be computed, inches. Then dimensions may be checked for bending action at any section, as C–D (Fig. 2), by these equations:

$$\text{Bending moment} = FL = tW^2\frac{S}{6} \tag{1}$$

$$t = \frac{6FL}{W^2S} \tag{2}$$

$$W = \sqrt{\frac{6FL}{tS}} \tag{3}$$

Let F_1 = greatest force acting on the crankpin when the crank is on dead centers. The bending moment at the section A–B (Fig. 2) then is

$$F_1x = \frac{W_1t^2S}{6} \tag{4}$$

$$W_1 = \frac{6F_1x}{t^2S} \tag{5}$$

$$t = \sqrt{\frac{6F_1x}{W_1S}} \tag{6}$$

S varies from 5000 psi for cast steel to 10,000 psi for the better steels. For cast-iron cranks t may be taken as $0.6D$. The diameter of the crankpin bosses d_1 may be taken as $d_1 = 1.16d + 0.02$ in. to $1.24d + 0.02$ in. The thickness of the bosses, v, may be taken as $0.75\,(d_1 - d)$.

The thickness of metal in the hub or boss of the crank (dimension a, Fig. 1), varies with the length of the boss. Values of a given by Seaton (D being the diameter of the crankshaft) are

M	$= 1.0D$	$0.9D$	$0.8D$	$0.7D$
a, iron	$= 0.35D$	$0.38D$	$0.40D$	$0.41D$
a, steel	$= 0.30D$	$0.32D$	$0.33D$	$0.34D$

FORGED CRANKSHAFTS. D. A. Low and A. W. Bevis (Ref. 1, p. 196) give the dimensions shown in Fig. 3, based on a large number of examples taken chiefly from marine practice. Sometimes a hole one-third the diameter of the pin is drilled through the crankpin as shown.

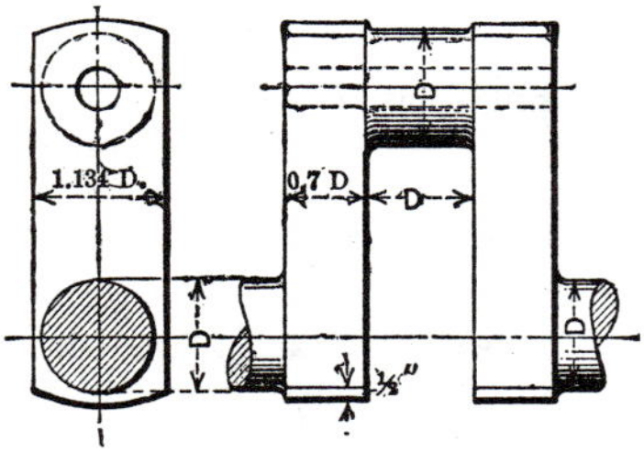

FIG. 3. Forged center crank.

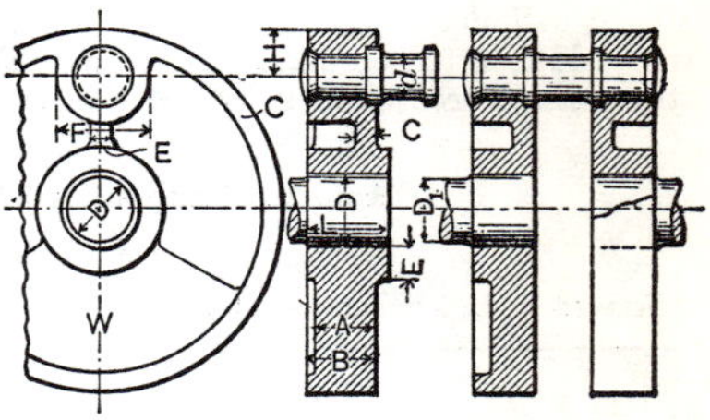

FIG. 4. Crankdisks.

CRANKDISKS (Fig. 4) are often used in small engines which run at high speed. The weight W, opposite the crankpin, is for the purpose of balancing the crankpin, and the part of the connecting rod which revolves with it. The following proportions may be used in designing crankdisks: $D_1 = D$ to $1.2D$; $B = 0.7D$ to D; $A = 0.9B$; $C = 0.25D$; $E = 0.4D$; $F = 2d$; $H = d$ to $1.5d$; $L = B$ to $1.25B$.

BALANCED CRANKS. Figure 5 shows a method of balancing cranks. When the size of the weight is determined by experiment the counterweight must be detachable as shown in Fig. 6. Figure 7 shows crankdisks where the center of mass of the balance weight does not fall opposite the crankpin.

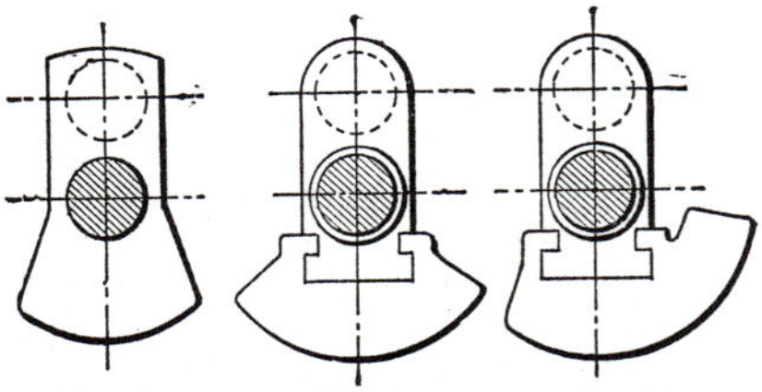

FIG. 5 FIG. 6 FIG. 7

Methods of balancing cranks.

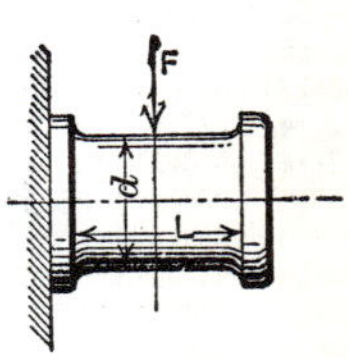

FIG. 8. Crankpin.

STRENGTH AND PROPORTIONS OF OVERHUNG CRANKPINS (Fig. 8). The dimensions of crankpins are based on the strength required to resist all the forces to which they will be subjected, and on the intensity of pressure permissible between pin and journal. This pressure must be kept low enough to avoid forcing lubricant out of the bearing, in which event heating and excessive wear would occur. Bending moment of the pin is $\frac{1}{2}FL$, where F = load on the pin, pounds, and L = length of pin, inches. If S = allowable maximum stress, pounds per square inch, and d = diameter of pin, inches, the resisting moment M_r, pound-inches, is $M_r = \pi d^3 S/32$. Hence

$$FL = \frac{d^3 S}{5.1} \tag{7}$$

and

$$d = 1.721\sqrt[3]{\frac{FL}{S}} \tag{8}$$

The pin is subjected to reversed stresses and possible shock. The value of S will depend upon the material used. Values of 6000 to 10,000 psi may be used. For a given ratio of $L/d = B$, eq. 8 becomes

$$d = \sqrt{\frac{5.1BF}{S}} \tag{9}$$

To avoid heating, pins are usually made of larger dimensions than are given by eq. 9. Heating depends on the projected area of the bearing, dL. If P = bearing pressure, pounds per square inch, $F = PLd$ and $L = F/Pd$. Then eq. 8 becomes

$$d = \sqrt[4]{\frac{5.1F^2}{PS}} \tag{10}$$

The deflection of pins should not exceed 0.001 in.

STRENGTH OF CENTER-CRANK CRANKPINS. These may be treated as simple beams with the load concentrated at the center. In multiple-cylinder engines the torque from each cylinder is transmitted along the shaft and subjects the crankpins to torsional shear in addition to bending. The crank webs are assumed to have no effect upon the pins and the shaft is treated as if it were straight. See Shaft Design, Section 15, Article 2.

Table 1. Tangential Effort and Velocity Factors

Factor = $\sec \phi \sin (\theta + \phi)$

Forward stroke is toward crankshaft. Velocity of cross head = velocity of crankpin × tabular value.

Crank Angle, degrees, Forward Stroke	Ratio (Length of Connecting Rod/Length of Crank)											Crank Angle, degrees, Return Stroke
	2	2 1/2	3	3 1/2	4	4 1/2	5	5 1/2	6	7	8	
5	0.13061	0.12191	0.11611	0.11197	0.10887	0.10645	0.10452	0.10295	0.10162	0.09956	0.09801	355
10	0.25948	0.24222	0.23075	0.22257	0.21644	0.21168	0.20787	0.20476	0.20216	0.19808	0.19503	350
15	0.38488	0.35936	0.34246	0.33044	0.32145	0.31447	0.30889	0.30432	0.30052	0.29456	0.29008	345
20	0.50512	0.47180	0.44986	0.43429	0.42266	0.41366	0.40644	0.40055	0.39567	0.38799	0.38223	340
25	0.61855	0.57807	0.55158	0.53286	0.51891	0.50811	0.49949	0.49247	0.48661	0.47743	0.47056	335
30	0.72361	0.67678	0.64639	0.62500	0.60911	0.59683	0.58704	0.57905	0.57242	0.56202	0.55423	330
35	0.81880	0.76659	0.73313	0.70966	0.69226	0.67884	0.66817	0.65947	0.65224	0.64092	0.63232	325
40	0.90278	0.84660	0.81082	0.78591	0.76751	0.75334	0.74209	0.73293	0.72533	0.71343	0.70454	320
45	0.97437	0.91562	0.87861	0.85291	0.83411	0.81961	0.80812	0.79878	0.79102	0.77890	0.76985	315
50	1.03257	0.97296	0.93581	0.91023	0.89147	0.87709	0.86570	0.85645	0.84879	0.83681	0.82788	310
55	1.07666	1.01807	0.98196	0.95723	0.93915	0.92534	0.91441	0.90554	0.89820	0.88674	0.87819	305
60	1.10622	1.05066	1.01678	0.99371	0.97691	0.96408	0.95396	0.94575	0.93896	0.92836	0.82047	300
65	1.12114	1.07070	1.04024	1.01961	1.00462	0.99320	0.98420	0.97691	0.97089	0.96147	0.95450	295
70	1.12174	1.07842	1.05250	1.03502	1.02235	1.01272	1.00514	0.99900	0.99393	0.98603	0.98015	290
75	1.10868	1.07434	1.05395	1.04024	1.03033	1.02281	1.01689	1.01210	1.00814	1.00199	0.99741	285
80	1.08305	1.05923	1.04516	1.03573	1.02890	1.02375	1.01969	1.01641	1.01370	1.00948	1.00635	280
85	1.04626	1.03406	1.02688	1.02207	1.01861	1.01598	1.01391	1.01225	1.01087	1.00873	1.00713	275
90	1.00000	1.00000	1.00000	1.00000	1.00000	1.00000	1.00000	1.00000	1.00000	1.00000	1.00000	270
95	0.94613	0.95833	0.96551	0.97032	0.97378	0.97641	0.97848	0.98014	0.98152	0.98366	0.98526	265
100	0.88657	0.91039	0.92446	0.93389	0.94070	0.94586	0.94994	0.95320	0.95591	0.96013	0.96327	260
105	0.82317	0.85751	0.87790	0.89161	0.90152	0.90904	0.91497	0.91975	0.92371	0.92987	0.93444	255
110	0.75765	0.80096	0.82688	0.84437	0.85703	0.86666	0.87425	0.88038	0.88546	0.89336	0.89924	250
115	0.69147	0.74192	0.77238	0.79301	0.80800	0.81941	0.82841	0.83570	0.84173	0.85113	0.85812	245
120	0.62583	0.68139	0.71527	0.73834	0.75514	0.76797	0.77809	0.78630	0.79309	0.80369	0.81158	240
125	0.56164	0.62023	0.65635	0.68107	0.69915	0.71297	0.72391	0.73276	0.74011	0.75157	0.76011	235
130	0.49952	0.55913	0.59628	0.62186	0.64062	0.65500	0.66639	0.67563	0.68330	0.69528	0.70421	230
135	0.43985	0.49859	0.53561	0.56124	0.58010	0.59460	0.60609	0.61544	0.62319	0.63531	0.64436	225
140	0.38279	0.43897	0.47475	0.49966	0.51804	0.53223	0.54348	0.55264	0.56025	0.57214	0.58104	220
145	0.32835	0.38048	0.41402	0.43740	0.45489	0.46831	0.47898	0.48751	0.49490	0.50623	0.51470	215
150	0.27639	0.32322	0.35361	0.37500	0.39089	0.40317	0.41296	0.42096	0.42758	0.43798	0.44577	210
155	0.22668	0.26717	0.29366	0.31238	0.32632	0.33712	0.34574	0.35277	0.35863	0.36780	0.37467	205
160	0.17892	0.21224	0.23418	0.24975	0.26138	0.27038	0.27759	0.28347	0.28837	0.29606	0.30181	200
165	0.13276	0.15828	0.17517	0.18719	0.19619	0.20317	0.20875	0.21331	0.21712	0.22308	0.22756	195
170	0.08782	0.10508	0.11655	0.12472	0.13086	0.13562	0.13943	0.14254	0.14514	0.14921	0.15223	190
175	0.04370	0.05241	0.05820	0.06234	0.06544	0.06786	0.06979	0.07137	0.07268	0.07475	0.07630	185
180	0.00000	0.00000	0.00000	0.00000	0.00000	0.00000	0.00000	0.00000	0.00000	0.00000	0.00000	180

16. FORCES ACTING ON CRANKS

TANGENTIAL FORCES. In engines, if the center line of the crankshaft lies in the same plane as the center of the piston and the cross head, the effective force of the crank acts in a line tangent to the crank circle at any given position of the crank. This effective force is

$$F_t = P \sec \phi \sin (\theta + \phi) \tag{11}$$

where F_t = effective tangential force on the crankpin, pounds; P = total force acting on the piston, pounds = area of piston, square inches, × pressure on piston, pounds per square inch; θ = angle between crank and the line between center of crank and center of piston, angle CAB in Fig. 8, p. 7-03; ϕ = angle between connecting rod and the line between center of crank and center of piston, angle BCA in Fig. 8, p. 7-03. Table 1 gives values of sec ϕ sin $(\theta + \phi)$ for various ratios of L/R, where L = length of connecting rod, inches, and R = radius of crank, inches. If the crank is used to impart motion to a cross head, as in a punch press, the resistance encountered by the cross head, i.e., the resistance of the metal to shearing or deformation, is taken as the value of P. The tangential force to be applied to the crank can then be computed, from which the horsepower required is determined.

Force acting along the center line of the connecting rod is

$$F_c = \frac{P}{\cos \phi} = \frac{P}{\sqrt{1 - \left(\frac{R \sin \theta}{L}\right)^2}} \tag{12}$$

The force acting on the guide is

$$F_g = P \tan \phi = \frac{P}{\sqrt{\left(\frac{L \csc \theta}{R}\right)^2 - 1}} \tag{13}$$

Note that P, pounds = (steam or gas pressure, pounds per square inch, + F_i from eq. 14) × A, square inches.

17. INERTIA FORCE OF PISTON

INERTIA FORCE OF PISTON in pounds per square inch of piston area is, approximately,

$$F_i = \frac{12WV_c^2}{gAR}\left(\cos \theta + \frac{R}{L} \cos 2\theta\right) \tag{14}$$

where all notation is as before and W = total weight of reciprocating parts, pounds; A = area of piston, square inches; g = acceleration due to gravity = 32.2 ft per sec per sec; and V_c = velocity of crankpin, feet per second. Table 2 gives values of $\cos \theta + (R/L) \cos 2\theta$ for various values of crank angles θ and ratios of L/R.

In eq. 14, F_i is actually $(W/gA) \times a$, where a is the acceleration of the piston, which may be obtained by double differentiation of eq. 2, p. 7-04. Equation 14 represents the first two terms of the Fourier series which may be used to express the acceleration. In balancing problems with multicylinder engines it is sometimes desired to know the values of additional harmonics, and eq. 14 may be written

$$F_i = \frac{12WV_c^2}{gAR}(\cos \theta + A_2 \cos 2\theta - A_4 \cos 4\theta + A_6 \cos 6\theta) \tag{14a}$$

where

$$A_2 = \frac{R}{L} + \frac{R^3}{4L^3} + \frac{15R^5}{128L^5}$$

$$A_4 = \frac{R^3}{16L^3} + \frac{R^5}{64L^5}$$

$$A_6 = \frac{R^5}{128L^5}$$

Table 2. Inertia and Acceleration Factors

Values of $\cos\theta + (R/L)\cos 2\theta$

Acceleration of crosshead = normal acceleration of crankpin × tabular value

Crank Angle, Forward Stroke, degrees	Ratio (Length of Connecting Rod/Length of Crank)											Crank Angle, Return Stroke, degrees
	2	2 1/2	3	3 1/2	4	4 1/2	5	5 1/2	6	7	8	
0	1.5000	1.4000	1.3333	1.2857	1.2500	1.2222	1.2000	1.1818	1.1667	1.1429	1.1250	360
5	1.4886	1.3901	1.3244	1.2776	1.2424	1.2148	1.1932	1.1752	1.1604	1.1369	1.1193	355
10	1.4547	1.3607	1.2980	1.2533	1.2197	1.1934	1.1728	1.1556	1.1415	1.1191	1.1022	350
15	1.3990	1.3123	1.2546	1.2134	1.1824	1.1582	1.1391	1.1234	1.1103	1.0897	1.0742	345
20	1.3227	1.2461	1.1950	1.1586	1.1312	1.1098	1.0929	1.0790	1.0674	1.0492	1.0355	340
25	1.2277	1.1634	1.1206	1.0900	1.0670	1.0490	1.0349	1.0232	1.0135	.9982	.9867	335
30	1.1160	1.0660	1.0327	1.0089	.9910	.9770	.9660	.9569	.9494	.9375	.9285	330
35	.9902	.9560	.9332	.9169	.9047	.8951	.8876	.8813	.8762	.8680	.8619	325
40	.8529	.8355	.8239	.8157	.8095	.8046	.8008	.7976	.7950	.7909	.7878	320
45	.7071	.7071	.7071	.7071	.7071	.7071	.7071	.7071	.7071	.7071	.7071	315
50	.5560	.5733	.5849	.5932	.5994	.6042	.6081	.6112	.6138	.6180	.6211	310
55	.4026	.4368	.4596	.4759	.4881	.4977	.5052	.5113	.5166	.5247	.5308	305
60	.2500	.3000	.3334	.3572	.3750	.3890	.4000	.4091	.4167	.4286	.4375	300
65	.1012	.1655	.2084	.2390	.2619	.2799	.2941	.3058	.3155	.3308	.3423	295
70	−.0410	.0356	.0867	.1232	.1505	.1720	.1888	.2028	.2143	.2326	.2463	290
75	−.1742	−.0876	−.0298	.0114	.0423	.0666	.0856	.1014	.1115	.1351	.1506	285
80	−.2962	−.2022	−.1396	−.0948	−.0613	−.0350	−.0143	.0028	.1700	.0394	.0562	280
85	−.4071	−.3068	−.2411	−.1942	−.1590	−.1315	−.1098	−.0919	−.0770	−.0536	−.0359	275
90	−.5000	−.4000	−.3333	−.2857	−.2500	−.2222	−.2000	−.1818	−.1667	−.1429	−.1250	270
95	−.5814	−.4811	−.4154	−.3685	−.3333	−.3058	−.2841	−.2661	−.2513	−.2279	−.2103	265
100	−.6435	−.5495	−.4869	−.4421	−.4090	−.3823	−.3620	−.3445	−.3303	−.3079	−.2911	260
105	−.6918	−.6052	−.5475	−.5062	−.4753	−.4511	−.4320	−.4163	−.4032	−.3826	−.3671	255
110	−.7350	−.6484	−.5973	−.5609	−.5335	−.5121	−.4952	−.4813	−.4697	−.4515	−.4378	250
115	−.7440	−.6797	−.6369	−.6063	−.5833	−.5653	−.5512	−.5395	−.5297	−.5145	−.5030	245
120	−.7500	−.7000	−.6667	−.6429	−.6250	−.6110	−.6000	−.5909	−.5834	−.5715	−.5625	240
125	−.7446	−.7104	−.6876	−.6713	−.6591	−.6495	−.6420	−.6358	−.6306	−.6225	−.6163	235
130	−.7396	−.7123	−.7007	−.6924	−.6862	−.6813	−.6775	−.6741	−.6717	−.6676	−.6645	230
135	−.7071	−.7071	−.7071	−.7071	−.7071	−.7071	−.7071	−.7071	−.7071	−.7071	−.7071	225
140	−.6792	−.6966	−.7082	−.7164	−.7226	−.7275	−.7313	−.7345	−.7371	−.7412	−.7443	220
145	−.6481	−.6823	−.7052	−.7214	−.7336	−.7432	−.7508	−.7570	−.7621	−.7703	−.7764	215
150	−.6160	−.6660	−.6994	−.7232	−.7410	−.7550	−.7660	−.7751	−.7827	−.7946	−.8035	210
155	−.5849	−.6492	−.6921	−.7227	−.7456	−.7636	−.7778	−.7895	−.7992	−.8145	−.8260	205
160	−.5567	−.6333	−.6844	−.7208	−.7481	−.7696	−.7865	−.8004	−.8120	−.8302	−.8439	200
165	−.5329	−.6159	−.6773	−.7185	−.7494	−.7737	−.7927	−.8085	−.8216	−.8422	−.8577	195
170	−.5150	−.6089	−.6716	−.7163	−.7499	−.7762	−.7969	−.8140	−.8282	−.8505	−.8674	190
175	−.5038	−.6023	−.6680	−.7148	−.7500	−.7778	−.7992	−.8172	−.8320	−.8555	−.8731	185
180	−.5000	−.6000	−.6667	−.7143	−.7500	−.7778	−.8000	−.8182	−.8333	−.8571	−.8750	180

18. CONNECTING RODS

Types of connecting rods for large engines are shown in Figs. 9 to 13. The solid rod shown in Fig. 9 is the strongest and most common form used with side-crank engines. Connecting rods for center-crank engines are made with the crankpin end in two or more pieces to facilitate assembly on the crankpin. Wedge blocks placed as shown in Fig. 11

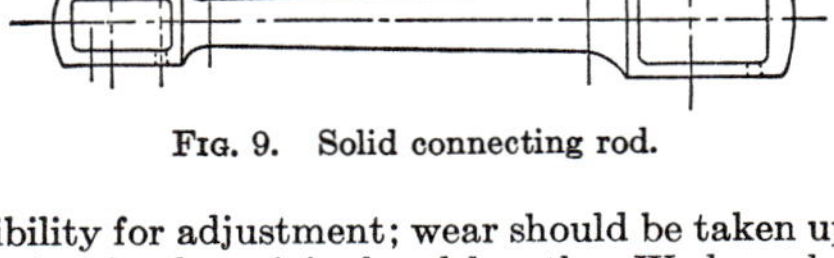

FIG. 9. Solid connecting rod.

have maximum accessibility for adjustment; wear should be taken up in the same direction on both bearings to maintain the original rod length. Wedges should always be placed so that the bolts pull the wedges down when tightened, reducing the danger of excessive end play and of piston striking cylinder head if the bolt should break. The crankpin end is generally provided with a babbitt-lined bearing box. Bronze bearings give better service at the wrist pin, where the pressures are high and the space limited. Rods for single-acting engines using trunk pistons are usually not made adjustable because of the inaccessibility.

DESIGN. The body of the rod is designed as a column with pin ends subjected to an additional bending load due to the inertia of the oscillating rod. The stress due to the column action is

$$s = \frac{F_c}{A\left[1 - \frac{s_y}{4n\pi^2 E}\left(\frac{L}{k}\right)^2\right]} \tag{15}$$

where F_c is obtained from eq. 12; A = cross-sectional area, square inches; s_y = yield stress of the material, pounds per square inch; n = column-end factor (unity for pin

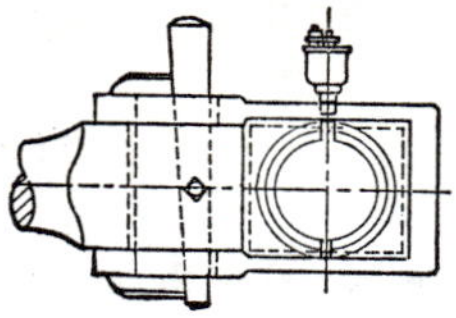

FIG. 10. Open-strap and key design of connecting-rod end for low-speed engines.

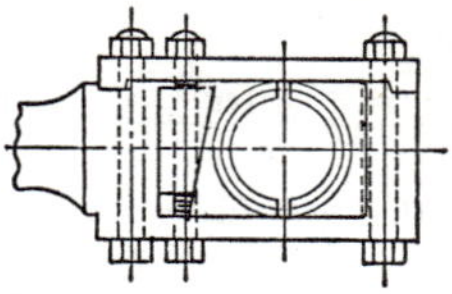

FIG. 11. Open-side and strap design of connecting-rod end.

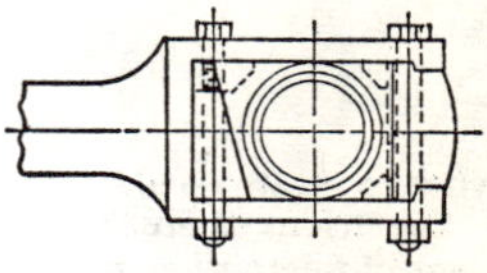

FIG. 12. Open-end and block design of connecting-rod end.

ends); E = modulus of elasticity; L = length center to center of pins, inches; k = radius of gyration of the cross section, inches. The rod may be assumed to be uniform in cross section, without any appreciable error, in which case the resultant inertia force due to the acceleration of the oscillating rod will act at a distance $2L/3$ from the wrist pin and will have a magnitude

$$F_1 = \frac{12Wv^2}{2gR}\sin\theta \tag{16}$$

where W = the weight of the body of the rod, pounds; v = the velocity of the crankpin, feet per second; R = the radius of the crank, inches; and θ = crank angle. The maximum bending moment, M, pound-inches, caused by this force will be at a distance $0.577L$ from the wrist pin, and its value will be

$$M = 0.774\frac{Wv^2}{g}\frac{L}{R}\sin\theta \tag{17}$$

The permissible stress should not exceed one-third the yield stress for single-acting engines, or one-sixth the yield stress for double-acting engines, since the stresses are repeated or reversed. The rod may be round, rectangular, or I-sectioned. The ratio of width to depth of a rectangular section should be 0.5 to 0.7.

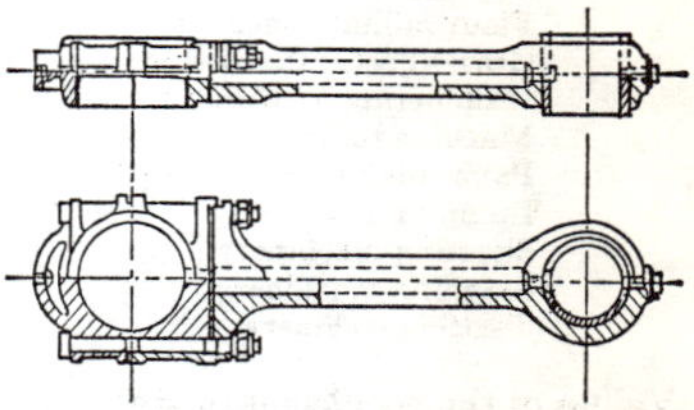

FIG. 13. Marine-type connecting rod.

Rods with I-sections must be proportioned for stiffness to prevent local buckling even when the column and bending equations indicate sufficient strength. I-section rods have a width 0.6 times the depth and a flange and web thickness 0.20 to 0.25 times the depth. Most rods are tapered so that the crankpin end is 1.10 to 1.15 times the depth computed for the center section. The outer end of solid rods and the cap of marine-type and split-end rods may be assumed to be simple beams supported at the center. The deflection computed for this beam should not exceed 0.001 in.

19. FLYWHEELS

A flywheel is primarily a rotating energy reservoir performing two distinct classes of service. First, the flywheel absorbs energy from a power source during the greater portion of the operating cycle and delivers a large amount of energy as useful work in a very short portion of the cycle. This class includes punch presses and shears. In the second class, including steam engines, internal-combustion engines, compressors, and reciprocating pumps, the flywheel smooths out the speed fluctuations caused by nonuniform flow of power from the pistons during each energy cycle.

FLYWHEEL ACTION. The kinetic energy E, foot-pounds, in the rim of a rotating flywheel is

$$E = \frac{Wv^2}{2g} \tag{18}$$

where W = weight of rim, pounds; v = mean velocity of rim, feet per second; and g = acceleration due to gravity = 32.2 ft per sec per sec. If the velocity changes from v_1 to v_2 the energy fluctuation is

$$\Delta E = \frac{W(v_1^2 - v_2^2)}{2g} \tag{19}$$

From eq. 19,

$$W = \frac{2g\Delta E}{v_1^2 - v_2^2} \tag{20}$$

$$\Delta E = \frac{W(v_1 + v_2)(v_1 - v_2)}{2g} = \frac{aWv^2}{g} \tag{21}$$

where $a = (v_1 - v_2)/v$ and is called the *coefficient of fluctuation.*

Coefficient of Fluctuation. The *coefficient of steadiness* b = the reciprocal of the coefficient of fluctuation $= v/(v_1 - v_2)$. From eq. 21,

$$W = \frac{g\Delta E}{av^2} = \frac{g\Delta E b}{v^2} \tag{22}$$

Since the linear velocity is $v = 2\pi kn/60$, where k = radius of gyration of rim, feet, and n = angular velocity of wheel, revolutions per minute:

$$\Delta E = \frac{Wk^2n^2}{91.2bg} = \frac{In^2}{91.2b} \tag{23}$$

where the mass moment of inertia about the axis of rotation is $I = Wk^2/g$, sometimes called the *flywheel effect.* In this case the coefficient of steadiness is $b = n/(n_1 - n_2)$.

From eq. 23,

$$W = \frac{91.2bg\Delta E}{n^2k^2} \tag{24}$$

Table 3. Values of Coefficient of Steadiness, *b*, for Flywheels

Engines Operating	b	Authority
Crushing machinery	5	Mayer
Electrical machinery	300	Unwin
Electrical machinery direct driven	150	Unwin
Engines with belt transmission	35	Mayer
Flour milling machinery	50	Mayer
Gear-wheel transmission	50	Mayer
Hammering machinery	5	Mayer
Machine tools	35	Unwin
Paper-making machinery	40	Mayer
Pumping machinery	20– 30	Mayer
Shearing machinery	20– 30	Mayer
Spinning machinery	50–100	Mayer; Unwin
Textile machinery	40	Unwin

Values of the coefficient of steadiness arc given in Table 3. Since the moment of inertia of the hub and arms is seldom over 5 to 10%, it is customary to neglect the weight of the hub and arms and assume the total weight of flywheel to be in the rim. The energy to be absorbed by a flywheel may be found either analytically or graphically. The latter method is the more convenient when the work input is variable, and data are available to draw an energy diagram.

STRESSES IN FLYWHEELS are rather complicated because of the effect of centrifugal force in the rim and arms and unknown shrinkage stresses caused by unequal cooling rates of the metal in the rim, arms, and hub. The rim is subjected to centrifugal force whose magnitude at the mean circumference is

$$F_c = \frac{12wbv^2}{gr} \tag{25}$$

where w = weight of material, pounds per cubic inch; b = rim thickness, inches; v = mean velocity, feet per second; r = mean radius, inches; and g = 32.2. This force F_c acting on a rotating rim without restraining arms produces hoop stress in pounds per square inch:

$$S_c = \frac{12wv^2}{g} \tag{26}$$

The section of the rim between the arms may be considered as a fixed-end beam with the centrifugal force uniformly distributed. The bending stress is

$$S_b = \frac{12\pi^2 Dwv^2}{gbn^2} \tag{27}$$

where D = mean diameter, inches, and n = number of arms.

Since the arms are not rigid, the actual stress in the rim at the junction with an arm is approximately

$$S = \frac{3S_c}{4} + \frac{S_b}{4} = \frac{wv^2}{g}\left(9 + \frac{3\pi^2 D}{bn^2}\right) \tag{28}$$

For cast iron S = 6000 psi and w = 0.26 lb per cu in. For cast steel S = 24,000 psi and w = 0.28 lb per cu in. For cast iron, rim velocities should be limited to 6000 ft per min when the rim, arms, and hubs are cast in one piece. For cast steel the velocity is limited to 8000 ft per min. In order to eliminate shrinkage stresses, the flywheel may be cast in sections and bolted together, or the rim may be cast one day with a number of wrought-steel spokes inserted in the mold and the hub cast on these the following day. Velocities up to 12,000 ft per min may be attained with the type. Flywheels with rims and hubs cast separately and bolted together by steel disks or webs may run at velocities up to 15,000 ft per min. The ratio of the width to height of rim section varies from 0.6 to 2. If a belt is run on flywheel the face must be at least 1 in. wider than the belt. The arms are designed to carry the full torque of the machine. Each arm is assumed to act as a cantilever beam fixed at the hub and carrying its proportionate share of the load concentrated at the free end at the rim. The moment $M = T(D - d)/D$, inch-pounds, and the stress is

$$S = \frac{T(D - d)}{N_a ZD} \tag{29}$$

where N_a = number of arms and Z = section modulus of the arm cross section at the hub. S for cast iron should not exceed 2000 psi and, if there is severe shock, as in pumps, not over 1000 psi. For an elliptical section $Z = \pi b^2 h/32$, where h = thickness and b = width of section. The arms should taper both in breadth and depth from the hub to the rim an amount of 1/4 in. per ft. The usual number of arms is six, but these may be eight, ten, or twelve, depending upon rim width and diameter of flywheel. For an elliptical section b/h = 2. The hub diameter = twice shaft diameter and hub length = 2 to 2.5 times shaft diameter.

FLYWHEEL CONSTRUCTION. Various types of flywheel construction are shown in Fig. 14. The solid rim wheel, usually made up to 8 ft in diameter, is the strongest. The

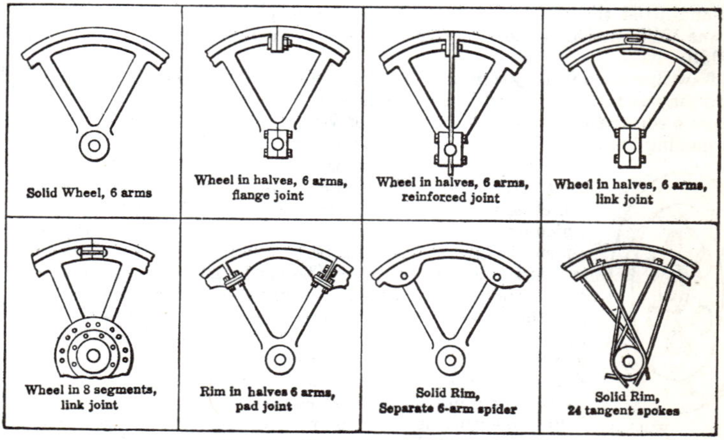

Fig. 14. Types of flywheel joints.

wheel made in halves with a flange joint between arms is the weakest. Flywheels made in two sections should be joined on an arm. Results of tests by Professor C. H. Benjamin

(Ref. 2) on the destruction of various types of flywheels shown in Fig. 14 are given in Table 4.

Table 4. Results of Tests to Destruction of Various Types of Flywheels

Type of wheel (see Fig. 14)	1	2	3	4	5	6	7	8
Rim speed at failure, ft per sec	395	194	225	305	256	223	393	424
Rim speed at failure, ft per min	23,700	11,640	13,500	18,300	15,360	13,380	23,580	25,440
Apparent rim stress at failure, psi	15,625	3,764	5,062	9,302	6,502	4,973	15,445	17,978
Comparative rim speeds at failure	100	49	57	77	65	56.5	100	107
Comparative rim stress at failure	100	24	32.5	60	42	32	99	115

The Haight construction is shown in Fig. 15. The rim section carrying the section necessary for the links joining the parts of the wheel is cut away around the entire circumference of the rim. The rim proper, therefore, is continuous and is not weakened by the drilling of holes or otherwise cutting it away for the reception of fastening members. The net result is to give a rim of I-section which is better fitted to resist stresses than the ordinary shape of rim. Furthermore, by having the joints come at the ends of the arms, which are made double at the joints, as shown in Fig. 15, the rim is continuous between arms and the tendency to failure at the joint is minimized.

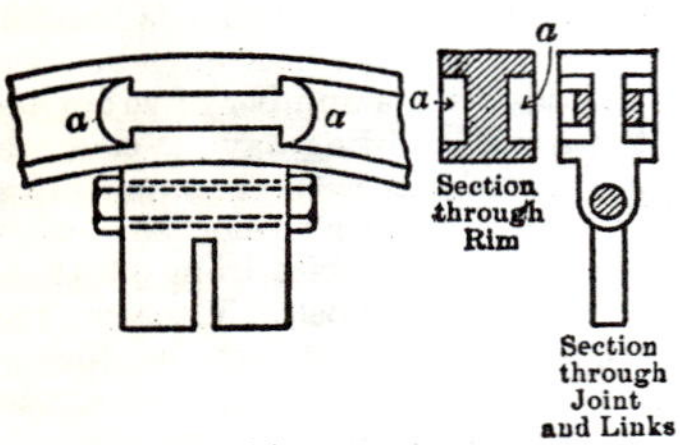

FIG. 15. Haight flywheel joint.

Various designs of flywheel have been made for different purposes. In one type of engine flywheel, the rim section is cast integral with the arms. The rim section is divided into two parts on a plane perpendicular to the axis of the wheel, each part having its own set of arms. The hub is cast separately, and the two halves of the rim section and the hub are fastened together by through bolts through the rim and the hub. Another type, for a rim speed of 15,000 ft per min, has been built with a rim formed of laminated plates. A cast-steel spider with twelve double arms had dovetail notches cut in its periphery. The laminated plates, cut in arcs of 30 degrees, were machined on the inside circumference to fit the notches on the spider, and assembled with broken joints. Side cover plates, machined with a groove into which the edge of the spider rim fits, were then placed outside the laminated plates, and the entire assembly bolted together with ten turned-and-fitted through bolts in each pair of side plates. The wheel was 13 ft outside diameter, 2 ft 2 in. face, and weighed about 100,000 lb.

FIG. 16. Cast-iron flywheel for punch press.

Bandsaw flywheels usually run at rim velocities of 10,000 ft per min and may at times run even higher. The Allis-Chalmers Company uses a construction in which the rim is of V-section and the arms are arranged diagonally, passing from one side of the hub to the opposite side of the rim, alternate arms being staggered with respect to each other. Another construction, largely used in bandsaw wheels, uses spokes composed of steel bars, either round, or square and set diagonal to the plane of rotation of the wheel, which are cast into the wheel and hub. Two rows of spokes are used, alternate spokes extending from the central plane of the rim to the opposite sides of the hub. The rim is poured first, and after it has cooled the hub is poured. Shrinkage strains in the rim that might pull the hub out of line thus are avoided. Construction of flywheels by welding steel plate not only eliminates blow holes often appearing in cast construction but also reduces cost of construction. Figure 16 shows a flywheel of cast construction. Figure 17 shows a flywheel for the same purpose, of welded steel.

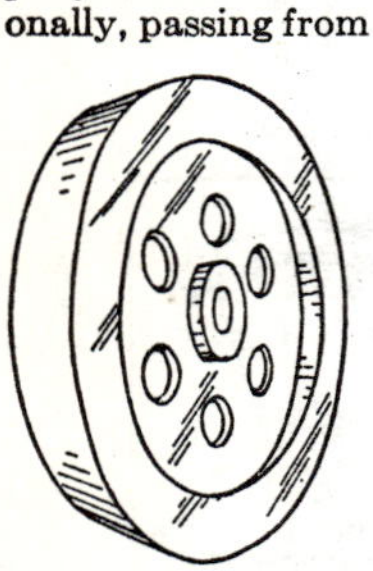

FIG. 17. Welded-steel flywheel to replace that shown in Fig. 16. (Courtesy of Lincoln Electric Co., Cleveland, Ohio)

For descriptions and illustrations of the construction details of a great variety of flywheels, including the above, for many different services, see Ref. 3.

ROTATING DISKS. Flywheels in the form of a disk, or those in which the depth of the rim is so great as compared to the rim that they approximate a disk form, cannot be designed by the methods used for the conventional type of flywheel. The complete determination of the stresses requires an involved mathematical analysis, which is complicated by the lack of experimental data. Kimball and Barr (Ref. 4) give the following equations for the radial and tangential stresses in a rapidly rotating plain disk: Let R, r = respectively, outer and inner radii of disk, inches; r_1 = radius at any point, inches; S_t, S_r = respectively, tangential and radial stresses at any radius r_1; w = weight of material, pounds per cubic inch; n = revolutions per minute; λ = Poisson's ratio = 0.3 for steel = 0.25 for cast iron. C = a constant = 0.00000355. Then for a solid disk,

$$S_t = Cwn^2[(3 + \lambda)R^2 - (1 + 3\lambda)r_1^2] \tag{30}$$

$$S_r = Cwn^2[(3 + \lambda)(R^2 - r_1^2)] \tag{31}$$

For a disk with a hole at the center,

$$S_t = Cwn^2\left[(3 + \lambda)\left(R^2 + r^2 + \frac{R^2r^2}{r_1^2}\right) - (1 + 3\lambda)r_1^2\right] \tag{32}$$

$$S_r = Cwn^2\left[(3 + \lambda)\left(R^2 + r^2 - \frac{R^2r^2}{r_1^2} - r_1^2\right)\right] \tag{33}$$

By taking r so small that r^2 is a negligible value, it can be shown that the tangential stress at the surface of the bore is twice that obtained by putting $r = 0$ in eq. 30. This shows that even a small hole at the center of a rotating disk will greatly increase the tangential stress in the disk.

Rotating disks in which the stresses are to be uniform at all points, as steam turbine disks, will be of varying cross section and thicker at the hub than at the periphery. The design of such disks involves complicated mathematical analysis. For one method of analysis, as used for steam turbine disks, see Kent's *Mechanical Engineers' Handbook —Power*, Section 8.

REFERENCES

1. D. A. Low and A. W. Bevis, *Manual of Machine Drawing and Design*, Longmans, Green and Co., London, 1922.
2. C. H. Benjamin, *Trans. Am. Soc. Mech. Engrs.*, Vol. 20, p. 209, 1899; Vol. 23, p. 168, 1902.
3. F. A. Halsey, *Handbook for Machine Designers*, McGraw-Hill Book Co., New York.
4. D. S. Kimball and J. H. Barr, *Elements of Machine Design*, 3rd ed., John Wiley and Sons, New York, 1935.

GENERAL REFERENCE

A. Vallance and V. L. Doughtie, *Design of Machine Members*, 2nd ed., McGraw-Hill Book Co., New York, 1943.

SECTION 8

STRENGTH OF MATERIALS

By

JOHN M. LESSELLS, *President, Lessells and Associates, Inc., Consulting Engineers, and Associate Professor of Mechanical Engineering, Massachusetts Institute of Technology*

G. S. CHERNIAK, *Vice President, Lessells and Associates, Inc., Consulting Engineers*

STATIC STRESSES

1. FUNDAMENTAL DEFINITIONS

TOTAL STRESS on a section *mn* through a loaded body is the resultant force S exerted by one part of the body on the other part in order to maintain in equilibrium the external loads acting on the part. Thus, in Figs. 1, 2, and 3, the total stress on section *mn* due to the external load P is S. The units in which it is expressed are those of load, i.e., pounds, tons, etc.

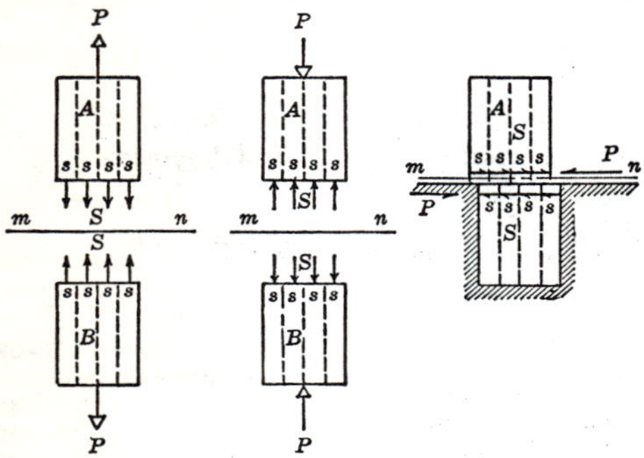

FIG. 1 FIG. 2 FIG. 3

FIG. 1. Tensile stress. FIG. 2. Compressive stress. FIG. 3. Shear stress.

UNIT STRESS more commonly called stress σ, is the total stress per unit of area at section *mn*. In general it varies from point to point over the section. Its value at any point of a section is the total stress on an elementary part of the area, including the point divided by the elementary area. If in Figs. 1, 2, and 3 the loaded bodies are one unit thick and four units wide, then when the total stress S is uniformly distributed over the area, $\sigma = P/A = P/4$. Unit stresses are expressed in pounds per square inch, tons per square foot, etc.

TENSILE STRESS OR TENSION is the internal total stress S exerted by the material fibers to resist the action of an external force P (Fig. 1), tending to separate the material into two parts along the line *mn*. For equilibrium conditions to exist, the tensile stress at any cross section will be equal and opposite in direction to the external force P. If the internal total stress S is distributed uniformly over the area, the stress can be considered as unit tensile stress $\sigma = S/A$.

COMPRESSIVE STRESS OR COMPRESSION is the internal total stress S exerted by the fibers to resist the action of an external force P (Fig. 2) tending to decrease the length of the material. For equilibrium conditions to exist, the compressive stress at any cross section will be equal and opposite in direction to the external force P. If the internal total stress S is distributed uniformly over the area, the unit compressive stress $\sigma = S/A$.

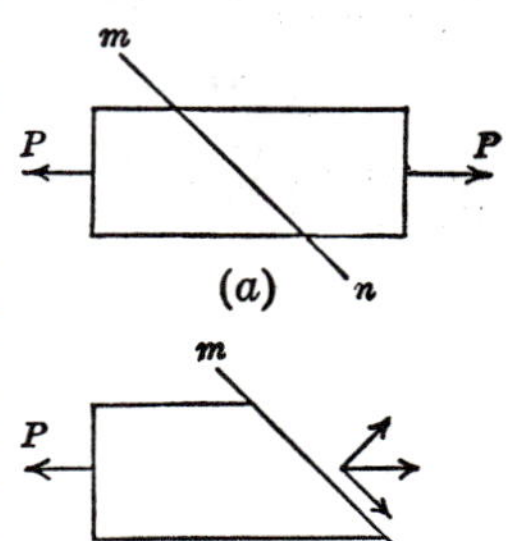

FIG. 4. Normal and shear stress components of resultant stress on section *mn*.

SHEAR STRESS is the internal total stress S exerted by the material fibers along the plane *mn* (Fig. 3) to resist the action of the external forces, tending to slide the adjacent parts in opposite directions. For equilibrium conditions to exist, the shear stress at any cross section will be equal and opposite in direction to the external force P. If the internal total stress S is uniformly distributed over the area, the unit shear stress $\tau = S/A$.

NORMAL STRESS is the component of the resultant stress which acts normal to the area considered (Fig. 4).

AXIAL STRESS is a special case of normal stress and may be either tensile or compressive. It is the stress existing in a straight homogeneous bar when the resultant of the applied loads coincides with the axis of the bar.

SIMPLE STRESS exists when either tension, compression, or shear is considered to operate singly on a body.

TOTAL STRAIN on a loaded body is the total elongation produced by the influence of an external load. Thus, in Fig. 5, the total strain is equal to δ. It is expressed in units of length, i.e., inches, feet, etc.

UNIT STRAIN or deformation per unit length is the total amount of deformation divided by the original length of the body before the load causing the strain was applied. Thus, if the total elongation is δ in an original gage length l, the unit strain $e = \delta/l$. Unit strains are expressed in inches per inch and feet per foot.

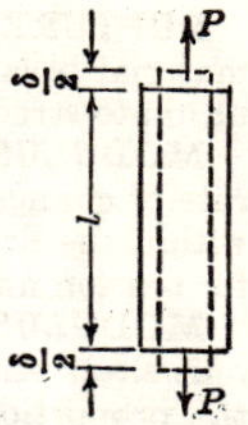

FIG. 5. Strain due to tension.

TENSILE STRAIN is the strain produced in a specimen by tensile stresses, which in turn are caused by external forces.

COMPRESSIVE STRAIN is the strain produced in a bar by compressive stresses, which in turn are caused by external forces.

SHEAR STRAIN is a strain produced in a bar by the external shearing forces.

POISSON'S RATIO is the ratio of lateral unit strain to longitudinal unit strain under the conditions of uniform and uniaxial longitudinal stress within the proportional limit. It serves as a measure of lateral stiffness. Average values of Poisson's ratio for the usual materials of construction are:

Material	Steel	Wrought Iron	Cast Iron	Brass	Concrete
Poisson's ratio	0.300	0.280	0.270	0.340	0.100

ELASTICITY is that property of a material which enables it to deform or undergo strain and return to its original shape upon the removal of the load.

HOOKE'S LAW states that within certain limits (not to exceed the proportional limit) the elongation of a bar produced by an external force is proportional to the tensile stress developed. Hooke's law gives the simplest relation between stress and strain.

PLASTICITY is that state of matter where permanent deformations or strains may occur without fracture. A material is plastic if the smallest load increment produces a permanent deformation. A perfectly plastic material is nonelastic and has no ultimate strength in the ordinary meaning of that term. Lead is a plastic material. A prism tested in compression will deform permanently under a small load and will continue to deform as the load is increased, until it flattens to a thin sheet. Wrought iron and steel are plastic when stressed beyond the elastic limit in compression. When stressed beyond the elastic limit in tension, they are partly elastic and partly plastic, the degree of plasticity increasing as the ultimate strength is approached.

STRESS-STRAIN RELATIONSHIP gives the relation between unit stress and unit strain when plotted on a stress-strain diagram in which the ordinate represents unit stress and the abscissa unit strain. Figure 6 shows a typical tension stress-strain curve for medium steel. The form of the curve obtained will vary according to the material, and the curve for compression will be different from the one for tension. For some materials like cast iron, concrete, and timber, no part of the curve is a straight line.

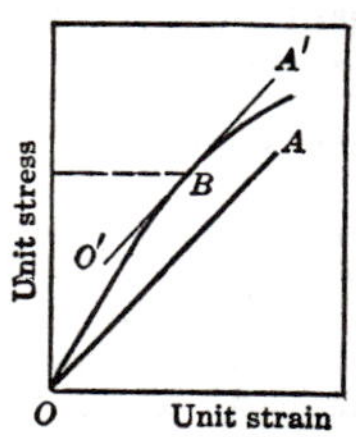

FIG. 6. Stress-strain relationship showing determination of apparent elastic limit.

PROPORTIONAL LIMIT is that unit stress at which unit strain begins to increase at a faster rate than unit stress. It can also be thought of as the greatest stress which a material can stand without deviating from Hooke's law. It is determined by noting on a stress-strain diagram the unit stress at which the curve departs from a straight line.

ELASTIC LIMIT is the least stress that will cause permanent strain, i.e., the maximum unit stress to which a material may be subjected and still be able to return to its original form upon removal of the stress.

Johnson's Apparent Elastic Limit. In view of the difficulty of determining precisely for some materials the proportional limit, J. B. Johnson proposed as the "apparent elastic limit" the point on the stress-strain diagram at which the rate of strain is 50% greater than at the original. It is determined by drawing OA (Fig. 6) with a slope with respect to the vertical axis 50% greater than the straight-line part of the curve; the unit stress at which the line $O'A'$ which is parallel to OA is tangent to the curve (point B, Fig. 6) is the apparent elastic limit.

YIELD POINT is the lowest stress at which strain increases without increase in stress. Only a few materials exhibit a true yield point. For other materials the term is sometimes used as synonymous with yield strength.

YIELD STRENGTH is the unit stress at which a material exhibits a specified permanent deformation or state. It is a measure of the useful limit of materials, particularly of those whose stress-strain curve in the region of yield is smooth and of gradual curvature.

ULTIMATE STRENGTH is the highest unit stress a material can sustain in tension, compression, or shear before rupturing.

RUPTURE STRENGTH OR BREAKING STRENGTH is the unit stress at which a material breaks or ruptures. It is observed in tests on steel to be slightly less than the ultimate strength because of a large reduction in area before rupture.

MODULUS OF ELASTICITY (Young's modulus) in tension and compression is the rate of change of unit stress with respect to unit strain for the condition of uniaxial stress within the proportional limit. For most materials the modulus of elasticity is the same for tension and compression.

MODULUS OF RIGIDITY (modulus of elasticity in shear) is the rate of change of unit shear stress with respect to unit shear strain for the condition of pure shear within the proportional limit. For metals it is equal to approximately 0.4 of the modulus of elasticity.

TRUE STRESS is defined as a ratio of applied axial load to the corresponding cross-sectional area. The units of true stress may be expressed in pounds per square inch, pounds per square foot, etc.

$$\sigma = \frac{P}{A}$$

where σ is the true stress, pounds per square inch, P is the axial load, pounds, and A is the smallest value of cross-sectional area existing under the applied load P, square inches.

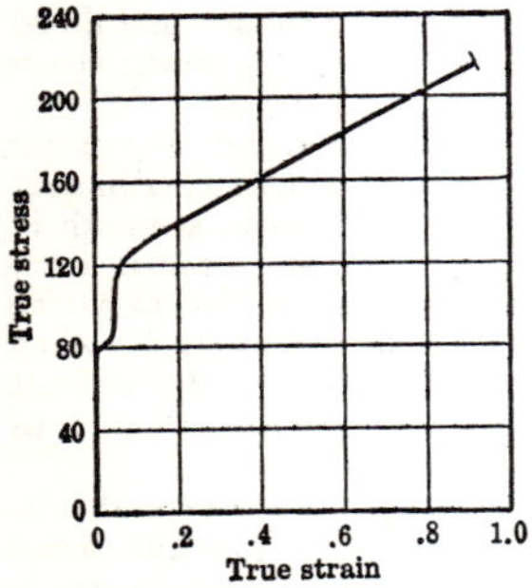

FIG. 7. True stress-strain relationship.

TRUE STRAIN is defined as a function of the original diameter to the instantaneous diameter of the test specimen:

$$q = 2 \log_e \frac{d_0}{d} \text{ in./in.}$$

where q = true strain, inches per inch, d_0 = original diameter of test specimen, inches, and d = instantaneous diameter of test specimen, inches.

TRUE STRESS-STRAIN RELATIONSHIP is obtained when the values of true stress and the corresponding true strain are plotted against each other in the resulting curve (Fig. 7). The slope of the nearly straight line leading up to fracture is known as the coefficient of strain hardening. It as well as the true tensile strength appear to be related to the other mechanical properties.

DUCTILITY is the ability of a material to sustain large permanent deformations in tension, as drawing into a wire.

MALLEABILITY is the ability of a material to sustain large permanent deformations in compression, as beating or rolling into thin sheets.

BRITTLENESS is that property of a material that permits it to be only slightly deformed without rupture. Brittleness is relative, no material being perfectly brittle, that is, capable of no deformation before rupture. Many materials are brittle to a greater or less degree, glass being one of the most brittle of materials. Brittle materials have relatively short stress-strain curves. Of the common structural materials, cast iron, brick, and stone are brittle in comparison with steel.

TOUGHNESS is the ability of the material to withstand high unit stress together with great unit strain, without complete fracture. The area *OAGH*, or *OJK*, under the curve of the stress-strain diagram (Fig. 8), is a measure of the toughness of the material. The distinction between ductility and toughness is that ductility deals only with the ability to deform, whereas toughness considers both the ability to deform and the stress developed during deformation.

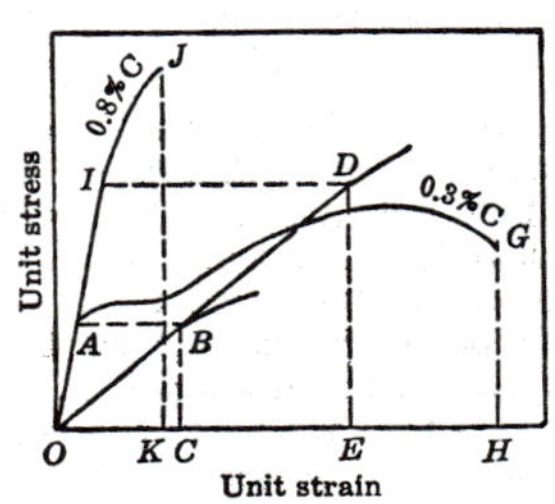

FIG. 8. Toughness comparison.

STIFFNESS is the ability to resist deformation under stress. The modulus of elasticity is the criterion of the stiffness of a material.

HARDNESS is the ability to resist very small indentations, abrasion, and plastic deformation. There is no single measure of hardness, as it is not a single property but a combination of several properties.

CREEP or flow of metals is a phase of plastic or inelastic action. Some solids, as asphalt or paraffin, flow appreciably at room temperatures under extremely small stresses; zinc, lead, and tin show signs of creep at room temperature under moderate stresses. At sufficiently high temperatures, practically all metals creep under stresses which vary with temperature, the higher the temperature the lower being the stress at which creep takes place. The deformation due to creep continues to increase indefinitely and becomes of extreme importance in members subjected to high temperatures, as parts in turbines, boilers, super-heaters, etc.

Creep limit is the maximum unit stress under which unit distortion will not exceed a specified value during a given period of time at a specified temperature. A value much used in tests, and suggested as a standard for comparing materials, is the maximum unit stress at which creep does not exceed 1% in 100,000 hours.

TYPES OF FRACTURE. A bar of brittle material, as cast iron, will rupture in a tension test in a clean sharp fracture with very little reduction of cross-sectional area and very little elongation (Fig. 9*a*). In a ductile material, as structural steel, the reduction of area and elongation are greater (Fig. 9*b*). In compression, a prism of brittle material will break by shearing along oblique planes; the greater the brittleness of the material, the more nearly will these planes parallel the direction of the applied force. Figures 10*a*, 10*b*, and 10*c*, arranged in order of brittleness, illustrate the type of fracture in prisms of brick, concrete, and timber. Figure 10*d* represents the deformation of a prism of plastic material, as lead, which flattens out under load without failure.

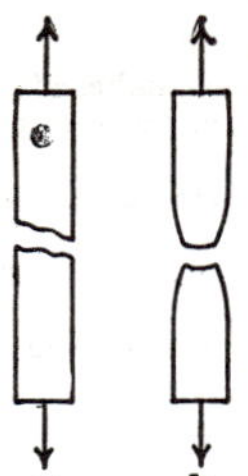

FIG. 9. Brittle (*a*) and ductile (*b*) fractures in tension.

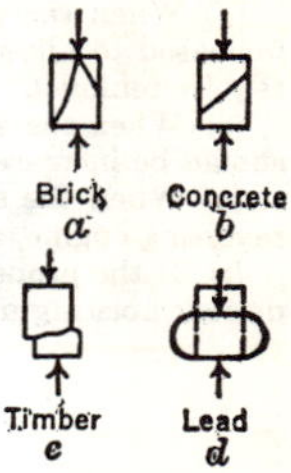

FIG. 10. Fractures in compression.

RELATIONS OF ELASTIC CONSTANTS

Modulus of elasticity, E

$$E = \frac{Pl}{Ae}$$

where P = load, pounds, l = length of bar, inches, A = cross-sectional area acted on by the axial load P, and e = total strain produced by axial load P.

Modulus of rigidity, G

$$G = \frac{E}{2(1 + \nu)}$$

where E = modulus of elasticity and ν = Poisson's ratio.

Bulk modulus, K, is the ratio of normal stress to the change in volume.

Relationships. The following relationships exist between the modulus of elasticity, E, the modulus of rigidity G, the bulk modulus of elasticity K, and Poisson's ratio, ν:

$$E = 2G(1 + \nu); \qquad G = \frac{E}{2(1 + \nu)}; \qquad \nu = \frac{E - 2G}{2G}$$

$$K = \frac{E}{3(1 - 2\nu)}; \qquad \nu = \frac{3K - E}{6K}$$

ALLOWABLE UNIT STRESS, also called allowable working unit stress, allowable stress, or working stress, is the maximum unit stress to which it is considered safe to subject a member in service. The term allowable stress is preferable to working stress, since the latter often is used to indicate the actual stress in a material when in service. Allowable unit stresses for different materials for various conditions of service are specified by different authorities on the basis of test or experience. In general, for ductile materials, allowable stress is considerably less than the yield point.

FACTOR OF SAFETY is the ratio of ultimate strength of the material to allowable stress. The term was originated for determining allowable stress. The ultimate strength of a given material divided by an arbitrary factor of safety, dependent on material and the use to which it is to be put, gives the allowable stress. In present design practice, it is customary to use allowable stress as specified by recognized authorities or building codes rather than an arbitrary factor of safety. One reason for this is that the factor of safety is misleading, in that it implies a greater degree of safety than actually exists. For example, a factor of safety of 4 does not mean that a member can carry a load four times as great as that for which it was designed. It also should be clearly understood that, even though each part of a machine is designed with the same factor of safety, the machine as a whole does not have that factor of safety. When one part is stressed beyond the proportional limit, or particularly the yield point, the load or stress distribution may be completely changed throughout the entire machine or structure, and its ability to function thus may be changed, even though no part has ruptured.

Although no definite rules can be given, if a factor of safety is to be used, the following circumstances should be taken into account in its selection:

(*a*) When the ultimate strength of the material is known within narrow limits, as for structural steel for which tests of samples have been made, when the load is entirely a steady one of a known amount and there is no reason to fear the deterioration of the metal by corrosion, the lowest factor that should be adopted is 3.

(*b*) When the circumstances of (*a*) are modified by a portion of the load being variable, as in floors of warehouses, the factor should not be less than 4.

(*c*) When the whole load, or nearly the whole, is likely to be alternately put on and taken off, as in suspension rods of floors of bridges, the factor should be 5 or 6.

(*d*) When the stresses are reversed in direction from tension to compression, as in some bridge diagonals and parts of machines, the factor should be not less than 6.

(*e*) When the piece is subjected to repeated shocks the factor should be not less than 10.

(*f*) When the piece is subjected to deterioration from corrosion the section should be sufficiently increased to allow for a definite amount of corrosion before the piece is so far weakened by it as to require removal.

(*g*) When the strength of the material or the amount of the load or both are uncertain the factor should be increased by an allowance sufficient to cover the amount of the uncertainty.

(*h*) When the strains are complex and of uncertain amount, such as those in the crankshaft of a reversing engine, a very high factor is necessary, possibly even as high as 40.

(*i*) If the property loss caused by failure of the part may be large or if loss of life may result, as in a derrick hoisting materials over a crowded street, the factor should be large.

2. BEAMS—THEORY OF FLEXURE

TYPES OF BEAMS. A beam is a bar or structural member subjected to transverse loads that tend to bend it. Any structural member acts as a beam if bending is induced by external transverse forces.

A simple beam (Fig. 11*a*) is a horizontal member that rests on two supports at the ends of the beam. All parts between the supports have free movement in a vertical plane under the influence of vertical loads.

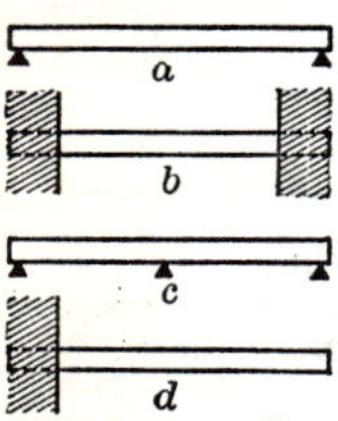

FIG. 11. Simple (*a*), constrained (*b*), continuous (*c*), and cantilever (*d*), beams.

A fixed beam, constrained beam, or restrained beam (Fig. 11*b*) is rigidly fixed at both ends or rigidly fixed at one end and simply supported at the other.

A continuous beam (Fig. 11*c*) is a member resting on more than two supports.

A cantilever beam (Fig. 11*d*) is a member with one end projecting beyond the point of support, free to move in a vertical plane under the influence of vertical loads placed between the free end and the support.

PHENOMENA OF FLEXURE. When a simple beam bends under its own weight the fibers on the upper or concave side are shortened, and the stress acting on them is compression; the fibers on the under or convex side are lengthened, and the stress acting on them is tension. In addition, shear exists along each cross section, the intensity of which is greatest along the sections at the two supports and zero at the middle section.

When a cantilever beam bends under its own weight the fibers on the upper or convex side are lengthened under tensile stresses; the fibers on the under or concave side are shortened under compressive stresses; the shear is greatest along the section at the support, and zero at the free end.

THE NEUTRAL SURFACE is that horizontal section between the concave and convex surfaces of a loaded beam, where there is no change in the length of the fibers and no tensile or compressive stresses acting upon them.

THE NEUTRAL AXIS is the trace of the neutral surface on any cross section of a beam. (See Fig. 12.)

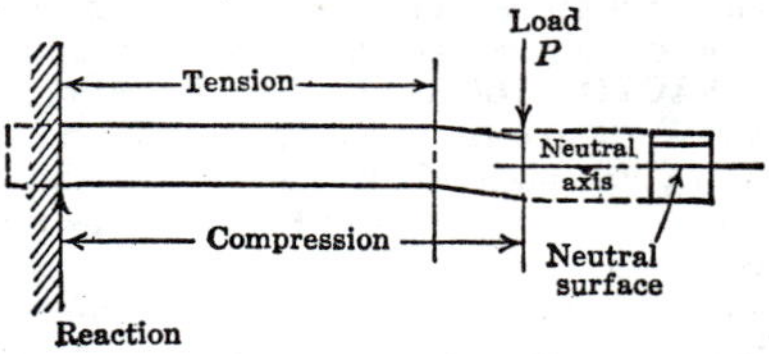

FIG. 12. Loads and stress conditions in a cantilever beam.

THE ELASTIC CURVE of a beam is the curve formed by the intersection of the neutral surface with the side of the beam, it being assumed that the longitudinal stresses on the fibers are within the elastic limit.

REACTIONS AT SUPPORTS. The reactions, or upward pressures at the points of support, are computed by applying the following conditions necessary for equilibrium of a system of vertical forces in the same plane: (1) The algebraic sum of all vertical forces must equal zero; i.e., the sum of the reactions equals the sum of the downward loads.

(2) The algebraic sum of the moments of all the vertical forces must equal zero. Condition (1) applies to cantilever beams and to simple beams uniformly loaded, or with equal concentrated loads placed at equal distances from the center of the beam. In the cantilever beam, the reaction is the sum of all the vertical forces acting downward, comprising the weight of the beam and the superposed loads. In the simple beam each reaction is equal to one-half the total load, consisting of the weight of the beam and the superposed loads. Condition (2) applies to a simple beam not uniformly loaded. The reactions are computed separately, by determining the moment of the several loads about each support. The sum of the moments of the load around one support is equal to the moment of the reaction of the other support around the first support.

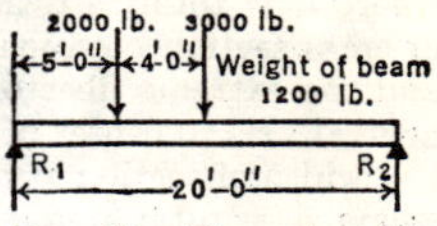

FIG. 13. Example of a simple loaded beam.

EXAMPLE. Determine reactions R_1, R_2 of a simple beam 20 ft long, loaded as in Fig. 13. The weight of the beam may be taken as a concentrated load applied at the center of gravity of the beam. Then, moments around left-hand and right-hand supports are, respectively,

Left-hand: $(2000 \times 5) + (3000 \times 9) + (1200 \times 10) - (R_2 \times 20) = 0;\quad R_2 = 2450$ lb
Right-hand: $(R_1 \times 20) - (2000 \times 15) - (3000 \times 11) - (1200 \times 10) = 0;\quad R_1 = 3750$ lb
$R_1 + R_2 = 6200$ lb = sum of all loads acting downward

CONDITIONS OF EQUILIBRIUM. The fundamental laws for the stresses at any cross section of a beam in equilibrium are: (1) Sum of horizontal tensile stresses = sum of horizontal compressive stresses. (2) Resisting shear = vertical shear. (3) Resisting moment = bending moment.

Vertical Shear. At any cross section of a beam the resultant of the external vertical forces acting on one side of the section is equal and opposite to the resultant of the external vertical forces on the other side of the section. These forces tend to cause the beam to shear vertically along the section. The value of either resultant is known as the vertical shear at the section considered. It is computed by finding the algebraic sum of the vertical forces to the left of the section; that is, it is equal to the left reaction minus the sum of the vertical downward forces acting between the left support and the section.

A shear diagram is a graphic representation of the vertical shear at all cross sections of the beam. Thus in the simple beam (Fig. 14) the ordinates to the line AOB represent to

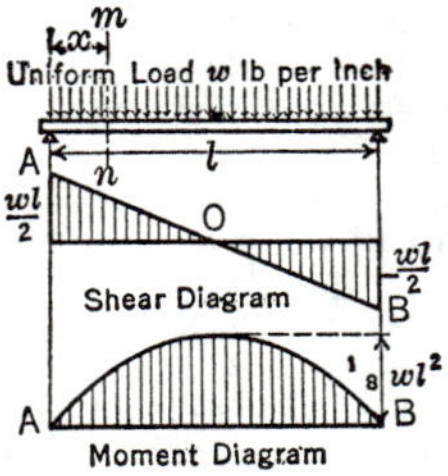

FIG. 14. Shear and moment diagrams, simple beam.

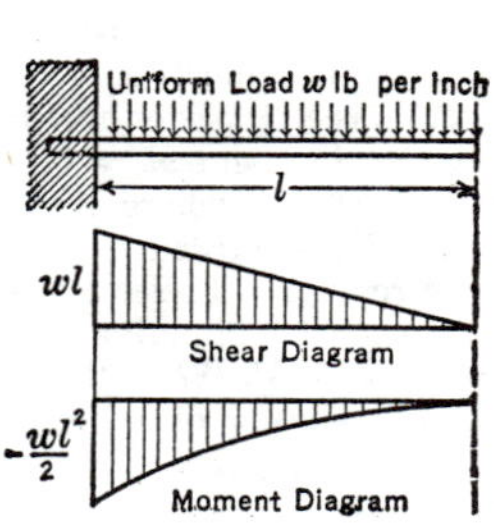

FIG. 15. Shear and moment diagrams, cantilever beam.

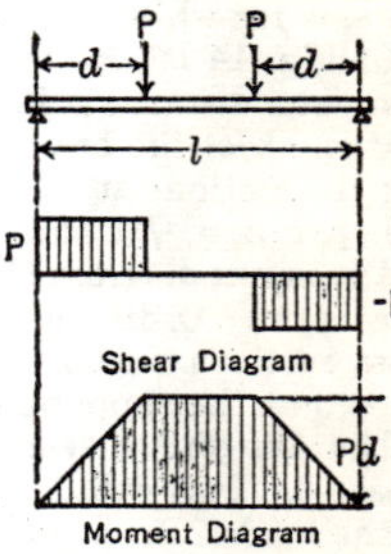

FIG. 16. Shear and moment diagrams, simple beam with two concentrated loads.

scale the intensity of the vertical shear at the corresponding sections of the beam. The vertical shear is greatest at the supports, where it is equal to the reactions, and it is zero at the center of the span. In the cantilever beam (Fig. 15) the vertical shear is greatest at the point of support, where it is equal to the reaction, and it is zero at the free end. Figure 16 shows graphically the vertical shear on all sections of a simple beam carrying two concentrated loads at equal distances from the supports, the weight of the beam being neglected.

Resisting Shear. The tendency of a beam to shear vertically along any cross section, due to the vertical shear, is opposed by an internal shearing stress at that cross section known as the resisting shear; it is equal to the algebraic sum of the vertical components of all the internal stresses acting on the cross section.

If V = vertical shear, pounds; V_r = resisting shear, pounds; τ = average unit shearing stress, pounds per square inch; and A = area of the section, square inches, then at any cross section

$$V_r = V = \tau A: \qquad \tau = \frac{V}{A} \tag{1}$$

The resisting shear is not uniformly distributed over the cross section, but the intensity varies from zero at the extreme fiber to its maximum value at the neutral axis.

At any point in any cross section the vertical unit shearing stress is

$$\tau = \frac{VA'c'}{It} \tag{2}$$

where V = total vertical shear in pounds for section considered; A' = area in square inches of cross section between a horizontal plane through the point where shear is being found and the extreme fiber on the same side of the neutral axis; c' = distance in inches from neutral axis to center of gravity of area A'; I = moment of inertia of the section, inches^4; t = width of section at plane of shear, inches. Maximum value of the unit shearing stress, where A = total area, square inches, of cross section of the beam, is

$$\text{For a solid rectangular beam:} \quad \tau = \frac{3V}{2A} \tag{3}$$

$$\text{For a solid circular beam:} \quad \tau = \frac{4V}{3A} \tag{4}$$

Horizontal Shear. In a beam, at any cross section where there is a vertical shearing force, there must be resultant unit shearing stresses acting on the vertical faces of particles that lie at that section. On a horizontal surface of such a particle, there is a unit shearing stress equal to the unit shearing stress on a vertical surface of the particle. Equation 2, therefore, also gives the horizontal unit shearing stress at any point on the cross-section of a beam.

Bending moment, at any cross section of a beam, is the algebraic sum of the moments of the external forces acting on either side of the section. It is positive when it causes the beam to bend convex downward, hence causing compression in upper fibers and tension in lower fibers of the beam. When the bending moment is determined from the forces that lie to the left of the section, it is positive if it acts in a clockwise direction; if determined from forces on the right side, it is positive if it acts in a counterclockwise direction. If the moments of upward forces are given positive signs, and the moments of downward forces negative signs, the bending moment will always have the correct sign, whether determined from the right or left side. The bending moment should be determined for the side for which the calculation will be simplest.

In Fig. 14 let M be the bending moment, pound-inches, at a section mn of a simple beam at a distance x, inches, from the left support; w = weight of beam per 1 in. of length; l = length of the beam, inches. Then the reactions are $\frac{1}{2}wl$, and $M = \frac{1}{2}wlx - \frac{1}{2}xwx$. For the sections at the supports, $x = 0$ or l and $M = 0$. For the section at the center of the span $x = \frac{1}{2}l$ and $M = \frac{1}{8}wl^2 = \frac{1}{8}Wl$, where W = total weight.

A moment diagram (Figs. 14, 15, and 16) shows the bending moment at all cross sections of a beam. Ordinates to the curve represent to scale the moments at the corresponding cross sections. The curve for a simple beam uniformly loaded is a parabola, showing $M = 0$ at the supports and $M = \frac{1}{8}wl^2 = \frac{1}{8}Wl$ at the center, M being in pound-inches.

The dangerous section is the cross section of a beam where the bending moment is greatest. In a cantilever beam it is at the point of support, regardless of the disposition of the loads. In a simple beam it is that section where the vertical shear changes from positive to negative, and it may be located graphically by constructing a shear diagram, or numerically by taking the left reaction and subtracting the loads in order from the left until a point is reached where the sum of the loads subtracted equals the reaction. For a simple beam, uniformly loaded, the dangerous section is at the center of the span.

Resisting Moment. The tendency to rotate about a point in any cross section of a beam is due to the bending moment at that section. This tendency is resisted by the resisting moment, which is the algebraic sum of the moments of all the horizontal stresses with reference to the same point.

FORMULA FOR FLEXURE. Let M = bending moment; M_r = resisting moment of the horizontal fiber stresses; σ = unit stress (tensile or compressive) on any fiber, usually that one most remote from the neutral surface; c = distance of that fiber from the neutral surface. Then

$$M = M_r = \frac{\sigma I}{c} \tag{5}$$

$$\sigma = \frac{Mc}{I} \tag{6}$$

where I = moment of inertia of the cross section with respect to its neutral axis. If σ is in pounds per square inch, M must be in pound-inches, I in inches^4, and c in inches.

Table 1. Elements of Sections

A = area of section
I = moment of inertia about axis I–I
c = distance from axis I–I to remotest point of section
I/c = section modulus
r = radius of gyration

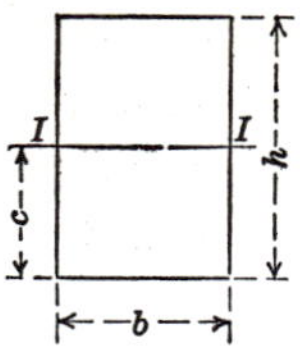

Rectangle

Axis through center

$A = bh$

$c = h/2$

$I = bh^3/12$

$I/c = bh^2/6$

$r = h/\sqrt{12} = 0.289h$

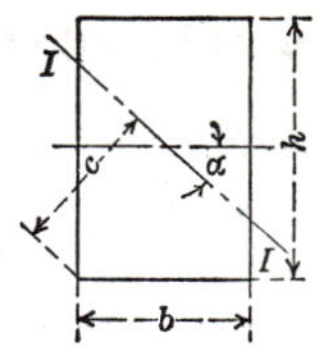

Rectangle

Axis any line through center of gravity

$A = bh$

$c = (b \sin \alpha + h \cos \alpha)/2$

$I = bh(b^2 \sin^2 \alpha + h^2 \cos^2 \alpha)/12$

$I/c = \dfrac{bh(b^2 \sin^2 \alpha + h^2 \cos^2 \alpha)}{6(b \sin \alpha + h \cos \alpha)}$

$r = \sqrt{(b^2 \sin^2 \alpha + h^2 \cos^2 \alpha)/12}$

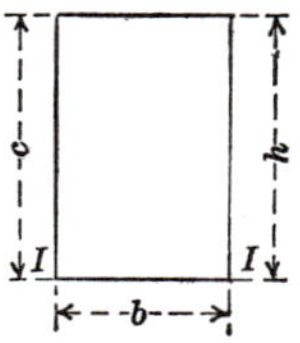

Rectangle

Axis on base

$A = bh$

$c = h$

$I = bh^3/3$

$I/c = bh^2/3$

$r = h/\sqrt{3} = 0.577h$

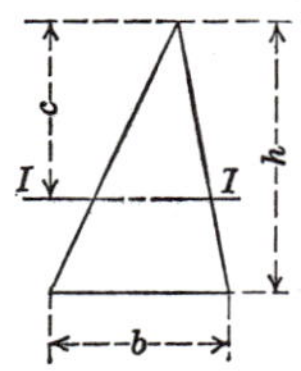

Triangle

Axis through center of gravity

$A = bh/2$

$c = 2/3\ h$

$I = bh^3/36$

$I/c = bh^2/24$

$r = h/\sqrt{18} = 0.236h$

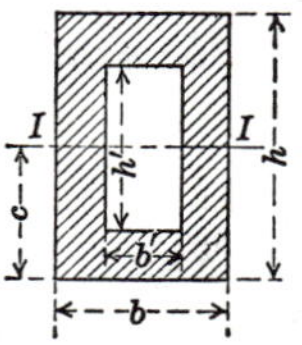

Hollow Rectangle

Axis through center

$A = bh - b_1h_1$

$c = h/2$

$I = (bh^3 - b_1h_1^3)/12$

$I/c = (bh^3 - b_1h_1^3)/6h$

$r = \sqrt{\dfrac{bh^3 - b_1h_1^3}{12(bh - b_1h_1)}}$

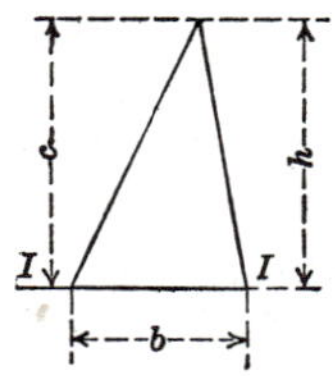

Triangle

Axis through base

$A = bh/2$

$c = h$

$I = bh^3/12$

$I/c = bh^2/12$

$r = h/\sqrt{6} = 0.408h$

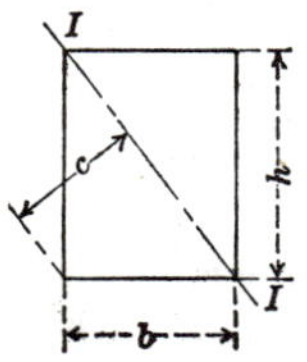

Rectangle

Axis on diagonal

$A = bh$

$c = bh/\sqrt{b^2 + h^2}$

$I = b^3h^3/6(b^2 + h^2)$

$I/c = b^2h^2/6\sqrt{(b^2 + h^2)}$

$r = bh/\sqrt{6(b^2 + h^2)}$

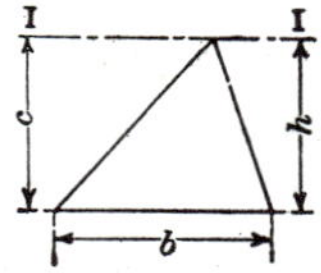

Triangle

Axis through apex

$A = bh/2$

$c = h$

$I = bh^3/4$

$I/c = bh^2/4$

$r = h/\sqrt{2} = 0.707h$

(*Table continued on p. 8-10*)

Table 1. Elements of Sections—*Continued*

Equilateral Polygon

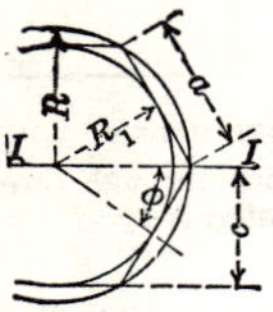

Axis through center, parallel to one side. n = number of sides

$A = nR_1^2 \tan \phi$

$c = a/2 \tan \phi = R_1$

$I = \{A(12R_1^2 + a^2)\}/48$

$I/c = \{A(12R_1^2 + a^2)\}/48R_1$

$r = \sqrt{(12R_1^2 + a^2)/48}$

Equilateral Polygon

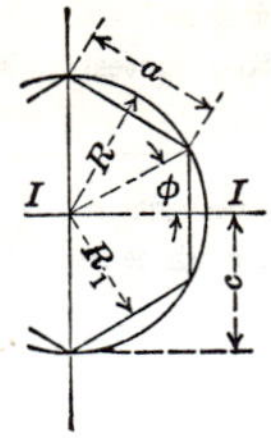

Axis through center, normal to side. n = number of sides

$A = nR_1^2 \tan \phi$

$c = a/(2 \sin \phi) = R$

$I = \{A(6R^2 - a^2)\}/24$

$I/c = \{A(6R^2 - a^2)\}/24R$

$r = \sqrt{(6R^2 - a^2)/24}$

Circle

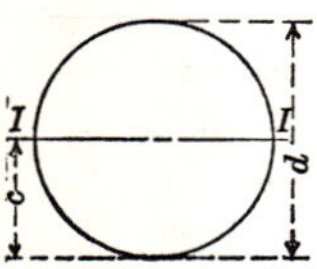

Axis through center

$A = \pi d^2/4 = 0.7854d^2$

$c = d/2$

$I = \pi d^4/64 = 0.0491d^4$

$I/c = \pi d^3/32 = 0.0982d^3$

$r = d/4$

Half Circle

Axis through center of gravity

$A = \pi d^2/8 = 0.3927d^2$

$c = \{d(3\pi - 4)\}/6\pi = 0.2878d.$

$I = \{d^4(9\pi^2 - 64)\}/1152\pi = 0.0068d^4$

$I/c = \dfrac{\{d^3(9\pi^2 - 64)\}}{\{192(3\pi - 4)\}} = 0.0238d^3$

$r = \{d\sqrt{(9\pi^2 - 64)}\}/12\pi = 0.1322d$

Hollow Circle

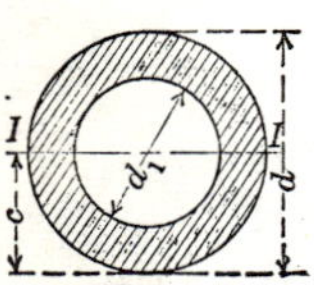

Axis through center

$A = \pi(d^2 - d_1^2)/4 = 0.7854(d^2 - d_1^2)$

$c = d/2$

$I = \pi(d^4 - d_1^4)/64 = 0.0491(d^4 - d_1^4)$

$I/c = \pi(d^4 - d_1^4)/32d = 0.0982(d^4 - d_1^4)/d$

$r = \sqrt{(d^2 + d_1^2)}/4$

Ellipse

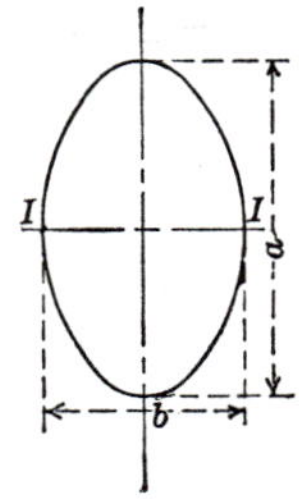

Axis through center

$A = \pi ab/4 = 0.7854ab$

$c = a/2$

$I = \pi a^3 b/64 = 0.0491a^3 b$

$I/c = \pi a^2 b/32 = 0.0982a^2 b$

$r = a/4$

Crossed Rectangles

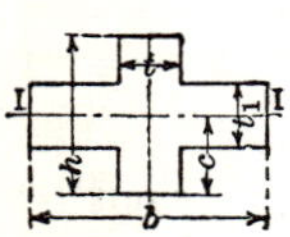

Axis through center

$A = th + t_1(b - t)$

$c = h/2$

$I = \{th^3 + t_1^3(b - t)\}/12$

$I/c = \{th^3 + t_1^3(b - t)\}/6h$

$r = \sqrt{\dfrac{th^3 + t_1^3(b - t)}{12\{th + t_1(b - t)\}}}$

Trapezoid

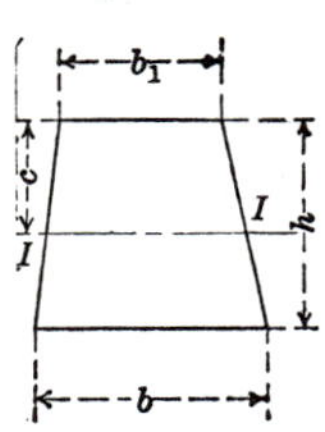

Axis through center of gravity

$A = \{(b + b_1)h\}/2$

$c = \dfrac{\{(b_1 + 2b)h\}}{\{3(b + b_1)\}}$

$I = \dfrac{h^3(b^2 + 4bb_1 + b_1^2)}{36(b + b_1)}$

$I/c = \dfrac{h^2(b^2 + 4bb_1 + b_1^2)}{12(b_1 + 2b)}$

$r = \dfrac{h}{6(b + b_1)}\sqrt{2(b^2 + 4bb_1 + b_1^2)}$

Equation 6 is the basis of the design and investigation of beams. It is true only when the maximum horizontal fiber stress σ does not exceed the proportional limit of the material.

MOMENT OF INERTIA is the sum of the products of each elementary area of the cross section multiplied by the square of the distance of that area from the assumed axis of rotation, or

$$I = \Sigma r^2 \,\Delta A = \int r^2 \, dA \tag{7}$$

where Σ is the sign of summation, ΔA is an elementary area of the section, and r is the distance of ΔA from the axis. The moment of inertia is greatest in those sections (such as I-beams) having much of the area concentrated at a distance from the axis. Unless otherwise stated, the neutral axis is the axis of rotation considered. I usually is expressed in in.^{4} See Table 1 for values of moments of inertia of various sections.

MODULUS OF RUPTURE is the term applied to the value of σ as found by eq. 6, when a beam is loaded to the point of rupture. Since eq. 6 is true only for stresses within the proportional limit, the value σ of the rupture strength so found is incorrect. However, the equation is used, as a measure of the ultimate load-carrying capacity of a beam. The modulus of rupture does not show the actual stress in the extreme fiber of a beam; it is useful only as a basis of comparison. If the strength of a beam in tension differs from its strength in compression, the modulus of rupture is intermediate between the two.

SECTION MODULUS, the factor I/c in flexure (eq. 5), is expressed in inches^3. It is the measure of a capacity of a section to resist a bending moment. For values of I/c for simple shapes, see Table 1. See Eshbach, *Handbook of Engineering Fundamentals* (John Wiley and Sons) for properties of standard steel and aluminum structural sections.

ELASTIC DEFLECTION OF BEAMS. When a beam bends under load, all points of the elastic curve except those over the supports are deflected from their original positions. The radius of curvature ρ of the elastic curve at any section is expressed as

$$\rho = \frac{EI}{M} \tag{8}$$

where E = modulus of elasticity of the material, pounds per square inch; I = moment of inertia, inches^4, of the cross section with reference to its neutral axis; M = bending moment, pound-inches, at the section considered. Where there is no bending moment, ρ is infinity and the curve is a straight line; where M is greatest, ρ is smallest and the curvature, therefore, is greatest.

If the elastic curve is referred to a system of coordinate axes in which x represents horizontal distances, y vertical distances, and l distances along the curve, the value of ρ is found, by the aid of the calculus, to be $d^3l/dx \cdot d^2y$. Differential eq. 9 of the elastic curve which applies to all beams when the elastic limit of the material is not exceeded is obtained by substituting this value in the expression $\rho = EI/M$ and assuming that dx and dl are practically equal:

$$EI\frac{d^2y}{dx^2} = M \tag{9}$$

Equation 9 is used to determine the deflection of any point of the elastic curve, by regarding the point of support as the origin of the coordinate axis, taking y as the vertical deflection at any point on the curve and x as the horizontal distance from the support to the point considered. The values of E, I, and M are substituted and the expression is integrated twice, giving proper values to the constants of integration, and the deflection y is determined for any point. See Table 2.

For example, the cantilever beam shown in Fig. 12 has a length = l, inches, and carries a load P, pounds, at the free end. It is required to find the deflection of the elastic curve at a point distant x, inches, from the support, the weight of the beam being neglected.

The moment $M = -P(l - x)$. By substitution in eq. 9, the equation for the elastic curve becomes $EI(d^2y/dx^2) = -Pl + Px$. By integrating and determining the constant of integration by the condition that $dy/dx = 0$ when $x = 0$, $EI(dy/dx) = -Plx + {}^1/_2 Px^2$ results. By integrating a second time and determining the constant by the condition that $x = 0$ when $y = 0$, $EIy = -{}^1/_2 Plx^2 + {}^1/_6 Px^3$, which is the equation of the elastic curve, results. When $x = l$, the value of y, or the deflection in inches at the free end, is found to be $-Pl^3/3EI$.

DEFLECTION DUE TO SHEAR. The deflection of a beam as computed by the ordinary formulas is that due to flexural stresses only. In short beams the deflection due to vertical shear sometimes is appreciable and may need to be considered. Because of the nonuniform distribution of the shear over the cross section of the beam, computing the

(Continued on p. 8-14)

Table 2. Bending Moment, Vertical Shear, and Deflection of Beams of Uniform Cross Section under Various Conditions of Loading

P = concentrated loads, lb
R_1, R_2 = reactions, lb
w = uniform load per unit of length, lb per in.
W = total uniform load on beam, lb
l = length of beam, in
x = distance from support to any section, in
E = modulus of elasticity, psi
I = moment of inertia, in.[4]
V_x = vertical shear at any section, lb
V = maximum vertical shear, lb
M_x = bending moment at any section, lb-in.
M = maximum bending moment, lb-in.
y = maximum deflection, in.

Simple Beam—Uniform Load

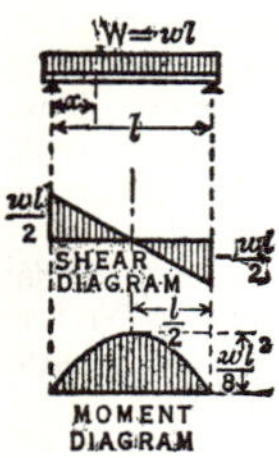

$$R_1 = R_2 = \frac{wl}{2}$$

$$V_x = \frac{wl}{2} - wx$$

$$V = \pm\frac{wl}{2}\left(\text{when}\begin{cases} x = 0 \\ x = l\end{cases}\right)$$

$$M_x = \frac{wlx}{2} - \frac{wx^2}{2}$$

$$M = \frac{wl^2}{8}\left(\text{when } x = \frac{l}{2}\right)$$

$$y = \frac{5Wl^3}{384EI} \text{ (at center of span)}$$

Simple Beam—Concentrated Load at Any Point

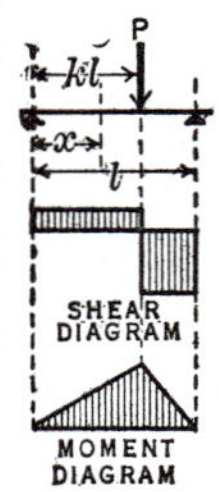

$R_1 = P(1 - k)$
$R_2 = Pk$
$V_x = R_1$ (when $x < kl$)
$= R_2$ (when $x > kl$)
$V = P(1 - k)$ (when $k < 0.5$)
$= -Pk$ (when $k > 0.5$)
$M_x = Px(1 - k)$ (when $x < kl$)
$= Pk(l - x)$ (when $x > kl$)
$M = Pkl(1 - k)$ (at point of load)

$$y = \frac{Pl^3}{3EI}(1 - k) \times (2/3k - 1/3k^2)^{3/2}$$

(at $x = l\sqrt{2/3k - 1/3k^2}$)

Simple Beam—Concentrated Load at Center

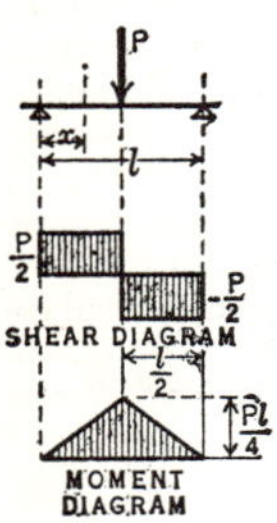

$$R_1 = R_2 = \frac{P}{2}$$

$$V_x = V = \pm\frac{P}{2}$$

$$M_x = \frac{Px}{2}$$

$$M = \frac{Pl}{4}\left(\text{when } x = \frac{l}{2}\right)$$

$$y = \frac{Pl^3}{48EI} \text{ (at center of span)}$$

Simple Beam—Two Equal Concentrated Loads at Equal Distances from Supports

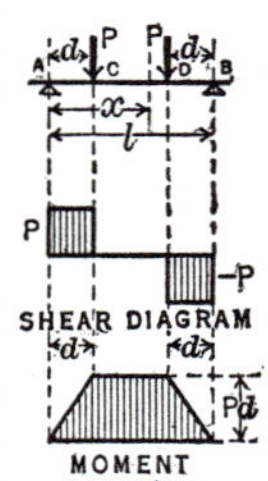

$R_1 = R_2 = P$		
$V_x = P$		for AC
$= 0$		for CD
$= -P$		for DB
$V = \pm P$		
$M_x = Px$		for AC
$= Pd$		for CD
$= P(l - x)$		for DB
$M = Pd$		

$$y = \frac{Pd}{24EI}(3l^2 - 4d^2)$$

(at center of span)

Simple Beam—Load Increasing Uniformly from Supports to Center of Span

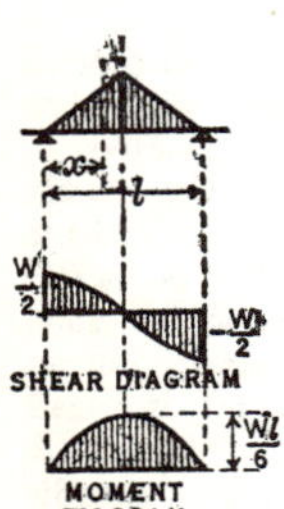

$$R_1 = R_2 = \frac{W}{2}$$

$$V_x = W\left(\frac{1}{2} - \frac{2x^2}{l^2}\right)$$

$$\left(\text{when } x < \frac{l}{2}\right)$$

$$V = \pm\frac{W}{2} \text{ (at supports)}$$

$$M_x = Wx\left(\frac{1}{2} - \frac{2x^2}{3l^2}\right)$$

$$M = \frac{Wl}{6} \text{ (at center of span)}$$

$$y = \frac{Wl^3}{60EI} \text{ (at center of span)}$$

Cantilever Beam—Load Concentrated at Free End

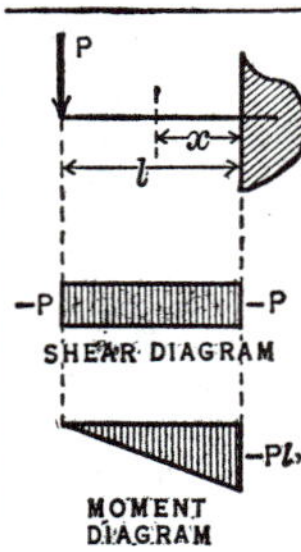

$$R = P$$

$$V_x = V = -P$$

$$M_x = -P(l - x)$$

$$M = -Pl(\text{when } x = 0)$$

$$y = \frac{Pl^3}{3EI}$$

Table 2. Bending Moment, Vertical Shear, and Deflection of Beams of Uniform Cross Section under Various Conditions of Loading—*Continued*

Simple Beam — Load Increasing Uniformly from Center to Supports

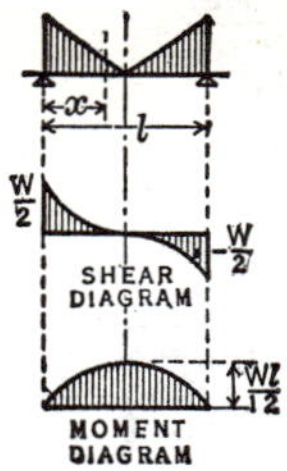

$$R_1 = R_2 = \frac{W}{2}$$

$$V_x = -W\left(\frac{2x}{l} - \frac{2x^2}{l^2} - \frac{1}{2}\right) \quad \left(\text{when } x < \frac{l}{2}\right)$$

$$V = \pm\frac{W}{2}$$

$$M_x = Wx\left(\frac{1}{2} - \frac{x}{l} + \frac{2}{3}\frac{x^2}{l^2}\right) \quad \left(\text{when } x < \frac{l}{2}\right)$$

$$M = \frac{Wl}{12} \text{ (at center of span)}$$

$$y = \frac{3}{320}\frac{Wl^3}{EI} \text{ (at center of span)}$$

Cantilever Beam—Uniform Load

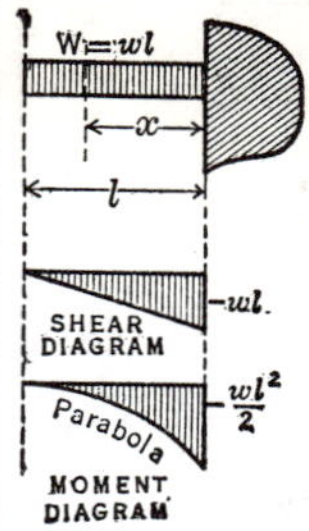

$$R = W = wl$$

$$V_x = -w(l - x)$$

$$V = -wl \text{ (when } x = 0)$$

$$M_x = -w(l - x)\left(\frac{l - x}{2}\right)$$

$$M = -\frac{wl^2}{2} \text{ (when } x = 0)$$

$$y = \frac{Wl^3}{8EI}$$

Simple Beam — Load Increasing Uniformly from One Support to the Other

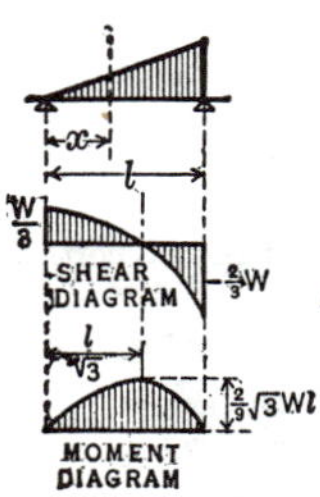

$$R_1 = \frac{W}{3}; \quad R_2 = \frac{2}{3}W$$

$$V_x = W\left(\frac{1}{3} - \frac{x^2}{l^2}\right)$$

$$V = -\frac{2}{3}W \text{ (when } x = l)$$

$$M_x = \frac{Wx}{3}\left(1 - \frac{x^2}{l^2}\right)$$

$$M = \frac{2}{9\sqrt{3}}Wl \quad \left(\text{when } x = \frac{l}{\sqrt{3}}\right)$$

$$y = \frac{0.01304}{EI}Wl^3$$

Cantilever Beam—Load Increasing Uniformly from Free End to Support

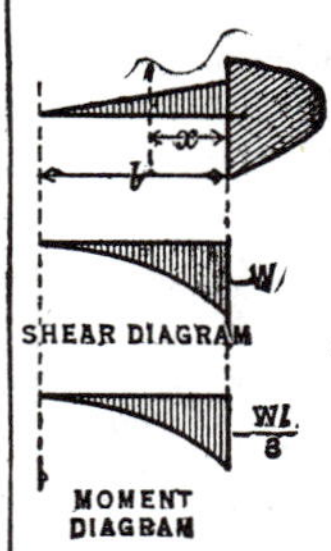

$$R = W$$

$$V_x = -W\frac{(l - x)^2}{l^2}$$

$$V = -W \text{ (when } x = 0)$$

$$M_x = -\frac{W}{3}\frac{(l - x)^3}{l^2}$$

$$M = -\frac{Wl}{3} \text{ (when } x = 0)$$

$$y = \frac{Wl^3}{15EI}$$

Fixed Beam — Concentrated Load at Center of Span

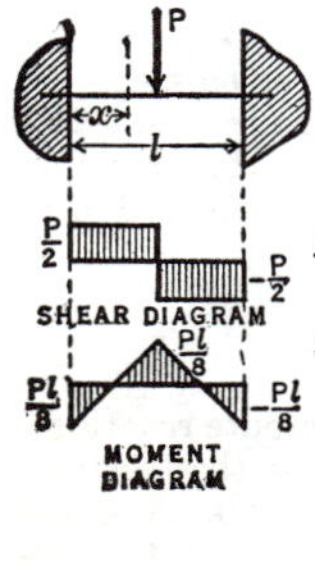

$$R_1 = R_2 = \frac{P}{2}$$

$$V_x = V = \pm\frac{P}{2}$$

$$M_x = P\left(\frac{x}{2} - \frac{l}{8}\right)$$

$$M_x = -\frac{Pl}{8}\left(\text{when }\begin{cases} x = 0 \\ x = l\end{cases}\right)$$

$$M = +\frac{Pl}{8} \text{ (at center of span)}$$

$$y = \frac{Wl^3}{192EI}$$

Fixed Beam—Uniform Load

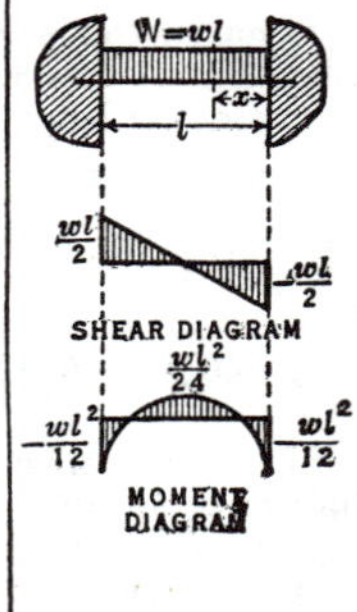

$$R_1 = R_2 = \frac{wl}{2} = \frac{W}{2}$$

$$V_x = \frac{wl}{2} - wx$$

$$V = \pm\frac{wl}{2} \text{ (at ends)}$$

$$M_x = -\frac{wl^2}{2}\left(\frac{1}{6} - \frac{x}{l} + \frac{x^2}{l^2}\right)$$

$$M = -{}^1/_{12}\,wl^2 \quad \left(\text{when }\begin{cases} x = 0 \\ x = l\end{cases}\right)$$

$$M = \frac{wl^2}{24}\left(\text{when } x = \frac{l}{2}\right)$$

$$y = \frac{Wl^3}{384EI}$$

(Table continued on p. 8-14)

Table 2. Bending Moment, Vertical Shear, and Deflection of Beams of Uniform Cross Section under Various Conditions of Loading—*Continued*

SIMPLE BEAM—DISTRIBUTED LOAD OVER PART OF BEAM

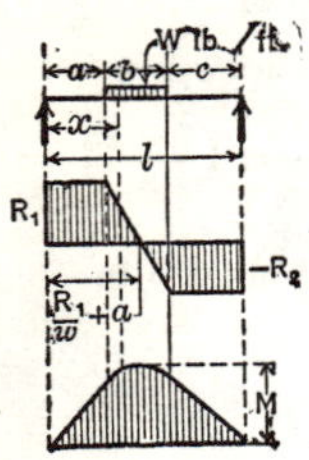

$$R_1 = \frac{wb(2c + b)}{2l}$$

$$R_2 = \frac{wb(2a + b)}{2l}$$

$$V_x = \frac{wb(2c+b)}{2l} - w(x - a)$$

$$V = R_1 \text{(when } a < c\text{)}$$

$$= R_2 \text{(when } a > c\text{)}$$

$$M_x = \frac{wbx(2c + b)}{2l} \text{ (when } x < a\text{)}$$

$$= R_1x - \frac{w(x - a)^2}{2} \text{ (when } a < x < a+b\text{)}$$

$$= R_2(l - x) \text{ (when } l - x < c\text{)}$$

$$M = \frac{wb(2c+b)[4al+b(2c+b)]}{8l^2}$$

BEAM SUPPORTED AT ONE END, FIXED AT OTHER—CONCENTRATED LOAD AT ANY POINT

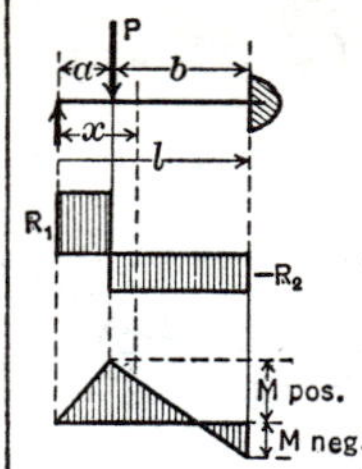

$$R_1 = \frac{Pb^2(2l + a)}{2l^3}$$

$$R_2 = P - R_1$$

$$V_x = R_1 \text{(when } x < a\text{)}$$

$$= R_2 \text{(when } x > a\text{)}$$

$$M_x = \frac{Pb^2x(2l + a)}{2l^3} \text{ (when } x < a\text{)}$$

$$= R_1x - P(x - a) \text{ (when } x > a\text{)}$$

$$M_{\text{positive}} = \frac{Pab^2(2l + a)}{2l^3} \text{ (when } x = a\text{)}$$

$$M_{\text{negative}} = -\frac{Pab(l + a)}{2l^2} \text{ (when } x = l\text{)}$$

FIXED BEAM—CONCENTRATED LOAD AT ANY POINT

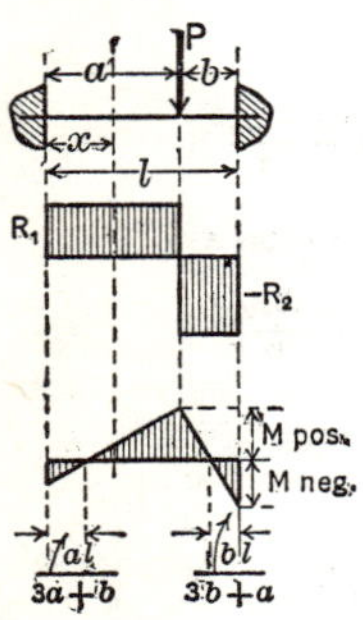

$$a > b$$

$$R_1 = Pb^2(l + 2a)/l^3$$

$$R_2 = Pa^2(l + 2b)/l^3$$

$$V_x = R_1 \text{(when } x < a\text{)}$$

$$= R_2 \text{(when } x > a\text{)}$$

$$V = R_2$$

$$M_x = R_1x - \frac{Pab^2}{l^2} \text{ (when } x < a\text{)}$$

$$= R_2(l - x) - \frac{Pa^2b}{l^2} \text{ (when } x > a\text{)}$$

$$M_{\text{positive}} = \frac{2Pa^2b^2}{l^3}$$

$$M_{\text{negative}} = -\frac{Pa^2b}{l^2}$$

$$y = -\frac{2Pa^3b^2}{3EI(3a + b)^2}$$

BEAM SUPPORTED AT ONE END, FIXED AT OTHER—DISTRIBUTED LOAD

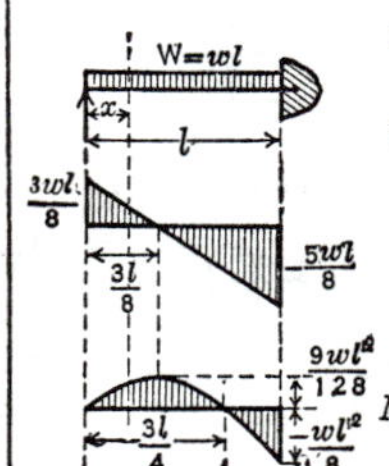

$$R_1 = \frac{3wl}{8}$$

$$R_2 = \frac{5wl}{8}$$

$$V_x = \frac{3wl}{8} - wx$$

$$V = \frac{3wl}{8} \text{ (at left support)}$$

$$= \frac{5wl}{8} \text{ (at right support)}$$

$$M_x = wx\left(\frac{3l}{8} - \frac{x}{2}\right)$$

$$M_{\text{positive}} = \frac{9wl^2}{128}$$

$$M_{\text{negative}} = -\frac{wl^2}{8}$$

$$y = -\frac{0.0054wl^4}{EI} \text{ (at 0.4215}l \text{ from } R_1\text{)}$$

deflection due to shear by exact methods is difficult. It may be approximated by $y_s = M/AE_s$, where y_s = deflection, inches, due to shear; M = bending moment, pound-inches, at the section where the deflection is calculated; E_s = modulus of elasticity in shear, pounds per square inch; A = area of cross section of beam, square inches. (See Ref. 1, p. 164.) For a rectangular section, the ratio of deflection due to shear, to deflection due to bending, will be less than 5% if the depth of the beam is less than one-eighth of the length.

3. DESIGN OF BEAMS

DESIGN PROCEDURE. In designing a beam the procedure is: (1) Compute reactions. (2) Determine position of the dangerous section and the bending moment at that section. (3) Divide the maximum bending moment (expressed in pound-inches) by the allowable unit stress (expressed in pounds per square inch) to obtain the minimum value of the section modulus. (4) Select a beam section with a section modulus equal to or slightly greater than the section modulus required.

WEB SHEAR. A beam designed in the foregoing manner is safe against rupture of the extreme fibers due to bending in a vertical plane, and usually the cross section will have sufficient area to sustain the shearing stresses with safety. For short beams carrying heavy loads, however, the vertical shear at the supports is large, and it may be necessary to increase the area of the section to keep the unit shearing stress within the limit allowed. For steel beams, the average unit shearing stress is computed by $\tau = V/A$, where V = total vertical shear, pounds; A = area of web, square inches.

MISCELLANEOUS CONSIDERATIONS. Other considerations which will influence the choice of section under certain conditions of loading are: (1) Maximum vertical deflection that may be permitted in beams coming in contact with plaster. (2) Danger of failure by sidewise bending in long beams, unbraced against lateral deflection. (3) Danger of failure by the buckling of the web of steel beams of short span carrying heavy loads. (4) Danger of failure by horizontal shear, particularly in wooden beams.

VERTICAL DEFLECTION. If a beam is to support or come in contact with materials like plaster, which may be broken by excessive deflection, it is usual to select such a beam that the maximum deflection will not exceed ($^1/_{360}$ × span). It may be shown that for a simple beam, supported at the ends, with a total uniformly distributed load W, pounds, the deflection, inches, is

$$y = \frac{30\sigma L^2}{Ed} \tag{10}$$

where σ = allowable unit fiber stress, pounds per square inch; L = span of beam, feet; E = modulus of elasticity, pounds per square inch; d = depth of beam, inches.

If the deflection of a steel beam is to be less than $^1/_{360}$ of the span, it may be shown from eq. 10 that, for a maximum allowable fiber stress of 18,000 psi, the limit of span in feet is approximately $1.8d$, where d = depth of the beam, inches.

For the deflection due to the impact of a moving load falling on a beam, see p. 8-40.

HORIZONTAL SHEAR IN TIMBER BEAMS. In beams of a homogeneous material which can withstand equally well shearing stresses in any direction, vertical and horizontal shearing stresses are equally important. In timber, however, shearing strength along the grain is much less than that perpendicular to the grain. Hence, the beams may fail owing to horizontal shear. Short wooden beams should be checked for horizontal shear in order that allowable unit shearing stress along the grain shall not be exceeded. (See Example below.)

RESTRAINED BEAMS. A beam is considered to be restrained if one or both ends are not free to rotate. This condition exists if a beam is built into a masonry wall at one or both ends, if it is riveted or otherwise fastened to a column, or if the ends projecting beyond the supports carry loads which tend to prevent tilting of the ends which would naturally occur as the beam deflects. The shears and moments given in Table 2 for fixed end conditions are seldom, if ever, attained, since the restraining elements themselves deform and and reduce the magnitude of the restraint. This reduction of restraint decreases the negative moment at the support and increases the positive moment in the central portion of the span. The amount of restraint which exists is a matter which must be judged for each case in the light of the construction used, the rigidity of the connections, and the relative sizes of the connecting members.

SAFE LOADS ON SIMPLE BEAMS. Equation 11 gives the safe loads on simple beams. This formula is obtained by substituting in the flexure eq. 5, the value of M for a simple beam, uniformly loaded, as given in Table 2. Let W = total load, pounds; σ = extreme fiber unit stress, pounds per square inch; S = section modulus, inches^3; L = length of span, feet. Then

$$W = \frac{2}{3}\sigma\frac{S}{L} \tag{11}$$

If σ is taken as a maximum allowable unit fiber stress, this equation gives the maximum allowable load on the beam. Most building codes permit a value of σ = 18,000 psi for quiescent loads on steel. For this value of σ, eq. 11 becomes

$$W = \frac{12{,}000\ S}{L} \tag{12}$$

If the load is concentrated at the center of the span, the safe load is one-half the value given by eq. 12. If the load is neither uniformly distributed nor concentrated at the center of the span, the maximum bending moment must be used. The foregoing equations are for beams laterally supported and are for flexure only. The other factors which influence the strength of the beam, as shearing, buckling, etc., must also be considered.

Table 3. Rectangular Beams of Uniform Strength *

σ = maximum fiber unit stress, pounds per square inch; E = modulus of elasticity; w = uniform load, pounds per inch; d = depth of beam, inches; b = width of beam, inches; y = maximum deflection, inches. All other dimensions in inches.

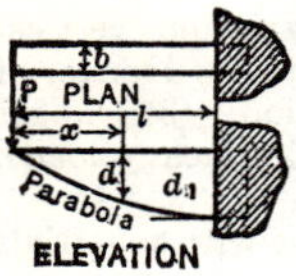

I. Cantilever Beam Loaded at Free End

Width is constant. Depth varies.

$$d = d_1\sqrt{x/l}$$
$$\sigma = 6Pl/bd_1^2$$
$$y = 8Pl^3/Ebd_1^3$$

Elevation is formed by a straight line and a parabola with its vertex at the loaded end.

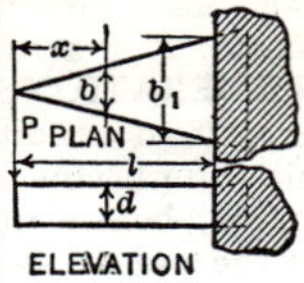

II. Cantilever Beam Loaded at Free End

Depth constant. Width varies.

$$b = b_1x/l$$
$$\sigma = 6Pl/b_1d^2$$
$$y = 6Pl^3/Eb_1d^3$$

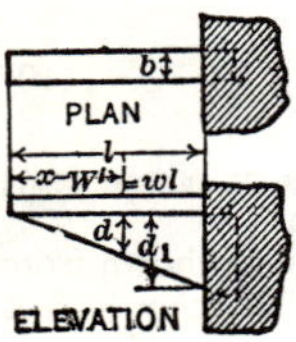

III. Cantilever Beam Uniformly Loaded

Width is constant. Depth varies.

$$d = (x/l)d_1$$
$$\sigma = 3wl^2/bd_1^2$$
$$y = 6wl^4/bEd_1^3$$

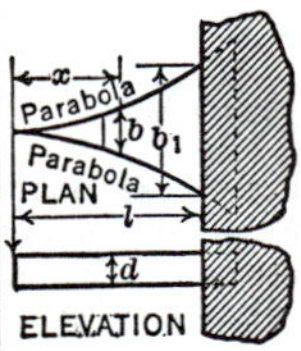

IV. Cantilever Beam Uniformly Loaded

Depth is constant. Width varies.

$$b = b_1x^2/l^2$$
$$\sigma = 3wl^2/b_1d^2$$
$$y = 3wl^4/b_1Ed^3$$

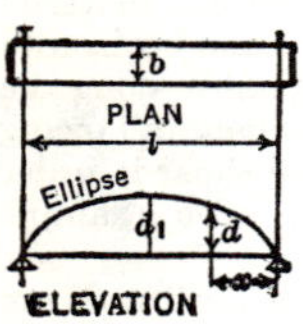

V. Simple Beam Uniformly Loaded

Width is constant. Depth varies.

$$d = \sqrt{\frac{4d_1^2(lx - x^2)}{l^2}}$$
$$\sigma = \frac{3wl^2}{4bd_1^2}$$

Elevation is formed by a straight line and an ellipse.

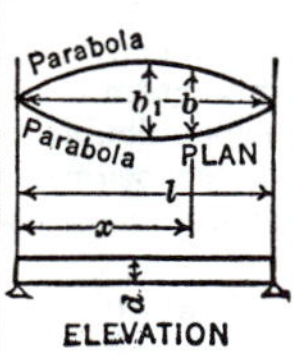

VI. Simple Beam Uniformly Loaded

Depth is constant. Width varies.

$$b = \frac{4b_1}{l^2}(lx - x^2)$$
$$\sigma = \frac{3}{4}\frac{wl^2}{b_1d^2}$$

Plan is two parabolas, with vertices at center of span.

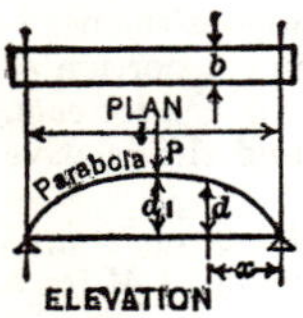

VII. Simple Beam Loaded at Center of Span

Width is constant. Depth varies.

$$d = d_1\sqrt{2x/l}$$
$$\sigma = \frac{3}{2}\frac{Pl}{bd_1^2}$$
$$y = \frac{1}{2}\frac{Pl^3}{Ebd_1^3}$$

Elevation is a parabola with vertices at points of support.

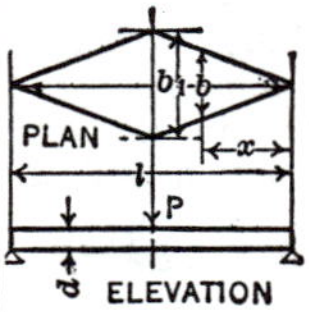

VIII. Simple Beam Loaded at Center of Span

Depth is constant. Width varies.

$$b = 2b_1x/l$$
$$\sigma = \frac{3}{2}\frac{Pl}{b_1d^2}$$
$$y = \frac{3}{8}\frac{Pl^3}{Eb_1d^3}$$

Plan is two triangles with vertices at points of support.

* The sections of the beams near the ends must be increased over the amounts shown to resist the vertical shear expressed by the formula $\sigma = 3/2\ V/A$.

USE OF TABLES IN DESIGN. The following is an example in the use of tables for the design of a wooden beam.

Example. Design a southern pine girder, of common structural grade, to carry a load of 9600 lb distributed uniformly over a 16-ft span in the interior of a building, the beam being a simple beam, freely supported at each end.

Solution. From Table 2, the bending moment of a simple beam uniformly loaded is $M = wl^2/8$ Since $W = wl$ and $l = 12L$,

$$M = 9600 \times 16 \times 12/8 = 230{,}400 \text{ lb-in.}$$

If the allowable unit stress on yellow pine is 1200 psi,

$$\frac{I}{c} = \frac{230{,}400}{1200} = 192 \text{ in.}^3$$

From Table 1, the section modulus of a rectangular section is $bd^2/6$. Assume $b = 8$ in. Then $8d^2/6 = 192$, and $d = \sqrt{144} = 12.0$ in. A beam 8 by 12 in. is selected tentatively, and checked for shear.

Maximum shearing stress (horizontal and vertical) is at the neutral surface over the supports. Equation 3 for horizontal shear in a solid rectangular beam is $\tau = 3V/2A$; $V = 9600/2 = 4800$, and $A = 8 \times 12 = 96$, whence $\tau = (3 \times 4800)/(2 \times 96) = 75$ psi.

If the safe horizontal unit shearing stress for common-grade southern yellow pine is 88 psi, and since the actual horizontal unit shearing stress is less than 88 lb, the 8 by 12 in. beam will be satisfactory.

A BEAM OF UNIFORM STRENGTH is one in which the dimensions are such that the maximum fiber stress σ is the same throughout the length of the beam. The form of the beam is determined by finding the areas of various cross sections from the flexure formula $M = \sigma I/c$, keeping σ constant and making I/c vary with M. For a rectangular section of width b and depth d, the section modulus $I/c = {}^1/_6\, bd^2$, and, therefore, $M = {}^1/_6\, \sigma bd^2$. By making bd^2 vary with M, the dimensions of the various sections are obtained. Table 3 gives the dimensions b and d, at any section, the maximum unit fiber stress σ and the maximum deflection y, of some rectangular beams of uniform strength. In this table, the bending moment has been assumed to be the controlling factor. On account of the vertical shear near the ends of the beams, the area of the sections must be increased over that given by an amount necessary to keep the unit shearing stress within the allowable unit shearing stress. The discussion of beams of uniform strength, though of considerable theoretical interest, is of little practical value since the cost of fabrication will offset any economy in the use of the material. A plate girder in a bridge or a building is an approximation in practice to a steel beam of uniform strength.

4. CONTINUOUS BEAMS

As in simple beams, the expressions $M = \sigma I/c$ and $\tau = V/A$ govern the design and investigation of beams resting on more than two supports. In the case of continuous beams, however, the reactions cannot be obtained in the manner described for simple beams. Instead, the bending moments at the various sections must be determined, and from these values the vertical shears at the sections and the reactions at the supports may be derived.

Consider the second span of length l_2, inches, of the continuous beam (Fig. 17). Vertical shear V_x at any section distant x, inches, from the left support of the span is equal to the

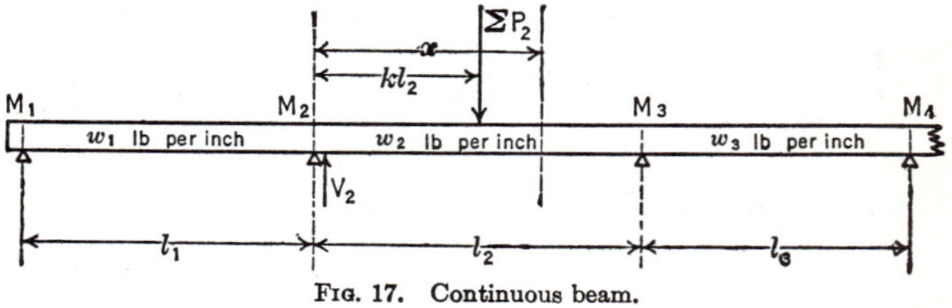

FIG. 17. Continuous beam.

algebraic sum of all the vertical forces on one side of the section. Thus, if V_2 = vertical shear at a section to the right of, but infinitely close to, the left support, w_2x = uniform load, and ΣP_2 = sum of the concentrated loads along the distance x, applied at a distance kl_2 from the left support, k being a fraction less than unity, then

$$V_x = V_2 - w_2x - \Sigma P_2 \tag{13}$$

At any section, distant x from the left support, the bending moment is equal to the algebraic sum of the moments of all forces on one side of the section. If M_2 is the moment, pound-inches, at the support to the left,

$$M_x = M_2 + V_2x - \frac{w_2x^2}{2} - \Sigma P_2(x - kl_2) \tag{14}$$

Assume that $x = l_2$. Then M_x becomes the moment M_3 at the next support to the right, and the expression may be written

$$V_2l_2 = M_3 - M_2 + \frac{w_2l_2^2}{2} + \Sigma P_2(l_2 - kl_2) \tag{15}$$

From eqs. 13, 14, and 15 it is evident that the bending moment M_x and the shear V_x at any section between two consecutive supports may be determined if the bending moments M_2 and M_3 at those supports are known.

To determine bending moments at the supports an expression known as the *theorem of three moments* is used. This gives the relation between the moments at any three consecutive supports of a beam. For beams with the supports on the same level, and uniformly loaded over each span, the formula is

$$M_1 l_1 + 2M_2(l_1 + l_2) + M_3 l_2 = -\tfrac{1}{4} w_1 l_1^3 - \tfrac{1}{4} w_2 l_2^3 \tag{16}$$

where M_1, M_2, and M_3 = moments of three consecutive supports; l_1 = length between first and second support; l_2 = length between second and third support; w_1 = uniform load per lineal unit over the first span; w_2 = uniform load per lineal unit over the second span. When both spans are of equal length and when the load on each span is the same, $l_1 = l_2$, $w_1 = w_2$, and eq. 16 reduces to

$$M_1 + 4M_2 + M_3 = -\tfrac{1}{2} wl^2 \tag{17}$$

which applies to most cases in practice.

Equations 16 and 17 are used as follows: For any continuous beam of n spans there are $(n + 1)$ supports. Assuming the ends of the beam to be simply supported without

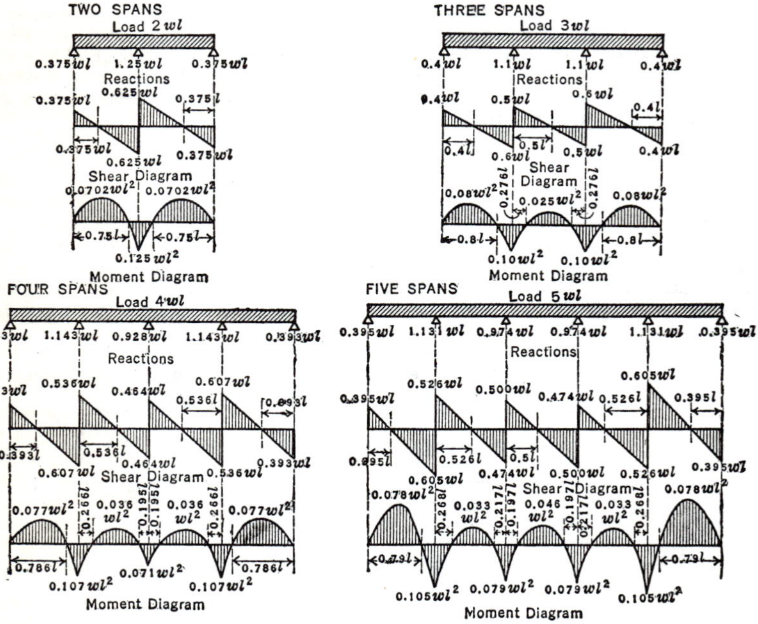

FIG. 18. Shear and moment diagrams of continuous beams.

any overhang, the moments at the end supports are zero, and there are, therefore, to be determined $(n - 1)$ moments at the other supports. This may be done by writing $(n - 1)$ equations of the form of eqs. 16 and 17 for each support. These equations will contain $(n - 1)$ unknown moments, and their solution will give values of M_1, M_2, M_3, etc., expressed as coefficients of wl^2. The shear V_1 at any support may be determined by substituting values of M_1 and M_2 in eq. 15, and the bending moment at any point in any span may be obtained by eq. 14. The shear at any point in any span may be determined from eq. 13.

Figure 18 gives values and diagrams for the reactions, shears, and moments at all sections of continuous beams uniformly loaded up to five spans. Note that the reaction at any support is equal to the sum of the shears to the right and to the left of that support.

5. AXIALLY END-LOADED BEAMS *

DEFLECTION OF BEAM NEGLIGIBLE. When a beam is subjected to an axial end load as well as to a transverse bending load, the resultant unit stress developed at any point in the beam is the algebraic sum of the unit stresses produced by each of the loads acting independently of each other. If the beam of cross-sectional area A, square inches, is considered so short that its deflection under load may be neglected, the unit stress produced at any cross section of the beam by an axial load P, pounds, is P/A. The maximum unit stress (tensile or compressive), pounds per square inch, due to the transverse or bending load is given by $\sigma = M/(I/c)$, where M, pound-inches, is the maximum bending moment due to this load; I, inches^4, is the moment of inertia of the cross section with respect to the neutral axis; c, inches, is the distance from the neutral axis to the extreme fiber. If the axial load is compressive, the maximum unit stress is compressive, occurring at the top of the beam, and is

$$\sigma = \frac{P}{A} + \frac{Mc}{I} \tag{18}$$

The unit stress on the bottom fiber may be either tensile or compressive, according as Mc/I is larger or smaller than P/A. If the longitudinal load P were a tensile load, the maximum unit stress would occur on the bottom fiber and would be a tensile stress.

DEFLECTION OF BEAM NOT NEGLIGIBLE. If the deflection of the beam is not negligible, the load P cannot be considered an axial load with respect to any cross sections except the end sections. The longitudinal load then has a moment arm equal to the deflection, and the stress due to this moment should be added algebraically to that caused by the other moments or loads. The stress due to deflection, y, inches (all other symbols as above), is Pyc/I. Thus, for a compressive longitudinal load, the maximum unit stress, pounds per square inch, is at the top of the beam and is

$$\sigma = \frac{P}{A} + \left(M + Py\right)\frac{c}{I} \tag{19}$$

Unit tensile stress, pounds per square inch, at the bottom of the beam for the same load is

$$\sigma = -\frac{P}{A} + \left(M + Py\right)\frac{c}{I} \tag{20}$$

Since the value of y depends on the total bending moment $(M + Py)$, and the bending moment depends, in turn, on the value of y, eqs. 19 and 20 usually are solved by a method of approximation. The value of y that would be caused by the cross-bending moment M, considered acting alone, is found first, and then is used in the expression, $M + Py$. This new value of the total bending moment is used to find a closer approximation to y. This operation may be repeated as many times as desired. When the deflection is small, as in comparatively short beams, the value of y as found from the cross-bending moment above often is used without any further approximation.

EXAMPLE. Find the stress at the bottom and at the top of a timber beam 6 in. wide, 8 in. deep, and 12 ft long, supported at each end, and carrying a distributed load of 400 lb per ft, and also an axial compressive load of 10,000 lb. The modulus of elasticity of the timber may be taken as 1,600,000.

Solution. Maximum unit stress will be at the center of the beam.

$$\frac{P}{A} = \frac{10{,}000}{48} = 208 \text{ psi}$$

$$\frac{Mc}{I} = \frac{400 \times 12 \times 144}{8} \times \frac{6}{6 \times 8^2} = 1350 \text{ psi}$$

See Table 2, p. 8-12.

$$y = \frac{5 \times 4800 \times 144^3 \times 12}{384 \times 6 \times 8^3 \times 1{,}600{,}000} = 0.456 \text{ in.}$$

$$\frac{Pyc}{I} = \frac{10{,}000 \times 0.456 \times 6}{6 \times 8^2} = 71 \text{ psi}$$

Total unit compressive stress (top) = 208 + 1350 + 71 = 1629 psi.

Total unit tensile stress (bottom) = −208 + 1350 + 71 = 1213 psi.

Here the stress due to deflection is small, as it usually will be in short deep beams.

---

* Article 5 is condensed from Seely, *Resistance of Materials*, 3rd Ed., Chapter VII, John Wiley and Sons, 1947.

Table 4. Values of Constant K for Curved Beams

Section	$\frac{R}{c}$	Values of K		$\frac{Y_0}{R}$ *	Section	$\frac{R}{c}$	Values of K		$\frac{Y_0}{R}$ *
		Inside Fiber	Outside Fiber				Inside Fiber	Outside Fiber	
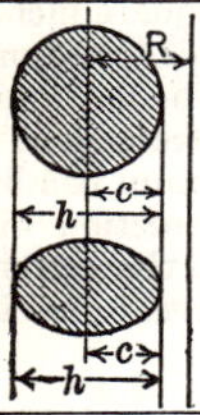	1.2	3.41	.54	.224	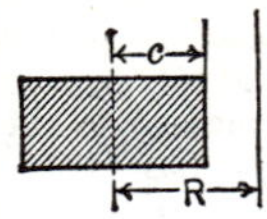	1.2	2.89	.57	.305
	1.4	2.40	.60	.151		1.4	2.13	.63	.204
	1.6	1.96	.65	.108		1.6	1.79	.67	.149
	1.8	1.75	.68	.084		1.8	1.63	.70	.112
	2.0	1.62	.71	.069		2.0	1.52	.73	.090
	3.0	1.33	.79	.030		3.0	1.30	.81	.041
	4.0	1.23	.84	.016		4.0	1.20	.85	.021
	6.0	1.14	.89	.0070		6.0	1.12	.90	.0093
	8.0	1.10	.91	.0039		8.0	1.09	.92	.0052
	10.0	1.08	.93	.0025		10.0	1.07	.94	.0033
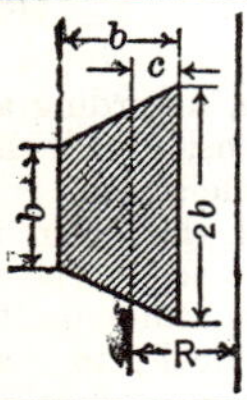	1.2	3.01	.54	.336	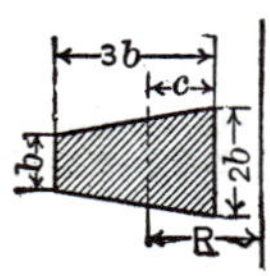	1.2	3.09	.56	.336
	1.4	2.18	.60	.229		1.4	2.25	.62	.229
	1.6	1.87	.65	.168		1.6	1.91	.66	.168
	1.8	1.69	.68	.128		1.8	1.73	.70	.128
	2.0	1.58	.71	.102		2.0	1.61	.73	.102
	3.0	1.33	.80	.046		3.0	1.37	.81	.046
	4.0	1.23	.84	.024		4.0	1.26	.86	.024
	6.0	1.13	.88	.011		6.0	1.17	.91	.011
	8.0	1.10	.91	.0060		8.0	1.13	.94	.0060
	10.0	1.08	.93	.0039		10.0	1.11	.95	.0039
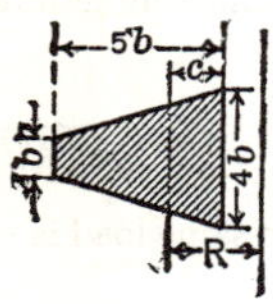	1.2	3.14	.52	.352	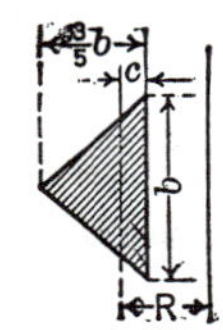	1.2	3.26	.44	.361
	1.4	2.29	.54	.243		1.4	2.39	.50	.251
	1.6	1.93	.62	.179		1.6	1.99	.54	.186
	1.8	1.74	.65	.138		1.8	1.78	.57	.144
	2.0	1.61	.68	.110		2.0	1.66	.60	.116
	3.0	1.34	.76	.050		3.0	1.37	.70	.052
	4.0	1.24	.82	.028		4.0	1.27	.75	.029
	6.0	1.15	.87	.012		6.0	1.16	.82	.013
	8.0	1.12	.91	.0060		8.0	1.12	.86	.0060
	10.0	1.11	.93	.0039		10.0	1.09	.88	.0039
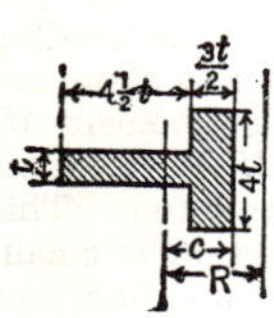	1.2	3.63	.58	.418	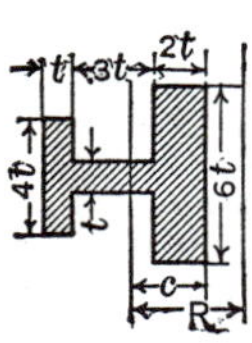	1.2	3.55	.67	.409
	1.4	2.54	.63	.299		1.4	2.48	.72	.292
	1.6	2.14	.67	.229		1.6	2.07	.76	.224
	1.8	1.89	.70	.183		1.8	1.83	.78	.178
	2.0	1.73	.72	.149		2.0	1.69	.80	.144
	3.0	1.41	.79	.069		3.0	1.38	.86	.067
	4.0	1.29	.83	.040		4.0	1.26	.89	.038
	6.0	1.18	.88	.018		6.0	1.15	.92	.018
	8.0	1.13	.91	.010		8.0	1.10	.94	.010
	10.0	1.10	.92	.0065		10.0	1.08	.95	.0065
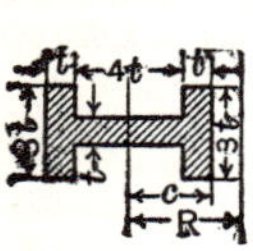	1.2	2.52	.67	.408	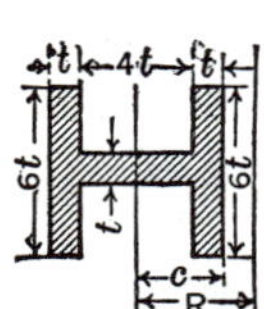	1.2	2.37	.73	.453
	1.4	1.90	.71	.285		1.4	1.79	.77	.319
	1.6	1.63	.75	.208		1.6	1.56	.79	.236
	1.8	1.50	.77	.160		1.8	1.44	.81	.183
	2.0	1.41	.79	.127		2.0	1.36	.83	.147
	3.0	1.23	.86	.058		3.0	1.19	.88	.067
	4.0	1.16	.89	.030		4.0	1.13	.91	.036
	6.0	1.10	.92	.013		6.0	1.08	.94	.016
	8.0	1.07	.94	.0076		8.0	1.06	.95	.0089
	10.0	1.05	.95	.0048		10.0	1.05	.96	.0057
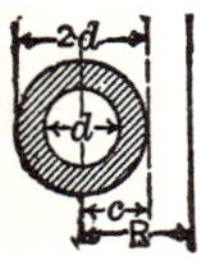	1.2	3.28	.58	.269	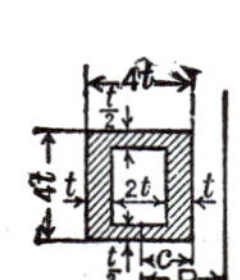	1.2	2.63	.68	.399
	1.4	2.31	.64	.182		1.4	1.97	.73	.280
	1.6	1.89	.68	.134		1.6	1.66	.76	.205
	1.8	1.70	.71	.104		1.8	1.51	.78	.159
	2.0	1.57	.73	.083		2.0	1.43	.80	.127
	3.0	1.31	.81	.038		3.0	1.23	.86	.058
	4.0	1.21	.85	.020		4.0	1.15	.89	.031
	6.0	1.13	.90	.0087		6.0	1.09	.92	.014
	8.0	1.10	.92	.0049		8.0	1.07	.94	.0076
	10.0	1.07	.93	.0031		10.0	1.06	.95	.0048

* Y_0 is distance from centroidal axis to neutral axis, where beam is subjected to pure bending.

6. CURVED BEAMS

The derivation of the flexure formula, $\sigma = Mc/I$, assumes that the beam is initially straight; therefore, any deviation from this condition introduces an error in the value of the stress. If the curvature is slight the error involved is not large, but in beams with a large amount of curvature, as hooks, chain links, and frames of punch presses, the error involved in the use of the ordinary flexure formula is considerable. The effect of the curvature is to increase the stress in the inside and to decrease it on the outside fibers of the beam and to shift the position of the neutral axis from the centroidal axis toward the concave or inner side.

The correct value for the unit fiber stress may be found by introducing a correction factor in the flexure formula, $\sigma = KMc/I$; the factor K depends on the shape of the beam and on the ratio R/c, where R = distance, inches, from the centroidal axis of the section to the center of curvature of the central axis of the unstressed beam; and c = distance, inches, of centroidal axis from the extreme fiber on the inner or concave side. Reference 2 has an analysis of curved beams as does Table 4, which gives values of K for a number of shapes and ratios of R/c. For slightly different shapes or proportions K may be found by interpolation.

7. COLUMNS—DEFINITIONS

A COLUMN OR STRUT is a bar or structural member under axial compression, which has an unbraced length greater than about eight or ten times the least dimension of its cross section. On account of its length, it is impossible to hold a column in a straight line under a load; a slight sidewise bending always occurs, causing flexural stresses in addition to the compressive stresses induced directly by the load. The lateral deflection will be in a direction perpendicular to that axis of the cross section about which the moment of inertia is the least. Thus in Fig. 19*A* the column will bend in a direction perpendicular to *aa*, in Fig. 19*B* it will bend perpendicular to *aa* or *bb*, and in Fig. 19*C* it is likely to bend in any direction.

RADIUS OF GYRATION of a section with respect to a given axis is equal to the square root of the quotient of the moment of inertia with respect to that axis, divided by the area of the section, that is

$$k = \sqrt{\frac{I}{A}}; \qquad \frac{I}{A} = k^2 \tag{21}$$

where I is the moment of inertia and A the sectional area. Unless otherwise mentioned, an axis through the center of gravity of the section is the axis considered. As in beams, the moment of inertia is an important factor in the ability of the column to resist bending, but for purposes of computation it is more convenient to use the radius of gyration.

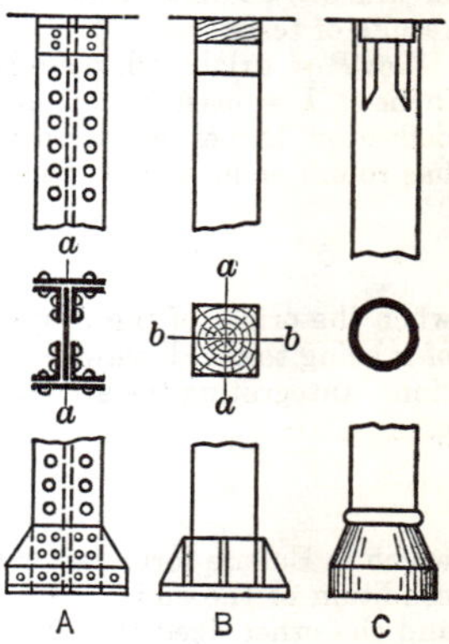

FIG. 19. Column end designs.

LENGTH OF A COLUMN is the distance between points unsupported against lateral deflection.

SLENDERNESS RATIO is the length l divided by the least radius of gyration k, both in inches. For steel, a *short column* is one in which $l/k < 20$ or 30, and its failure under load is due mainly to direct compression; in a *medium-length column*, l/k = about 30 to 175, and it will fail by a combination of direct compression and bending; in a *long column*, $l/k >$ about 175 to 200, and it will fail mainly by bending. For timber columns these ratios are about 0 to 30, 30 to 90, and above 90 respectively. The load which will cause a column to fail decreases as l/k increases. The above ratios apply to round-end columns. If the ends are fixed (see below), the effective slenderness ratio is one-half that for round-end columns, as the distance between the points of inflection is one-half the total length of the column. For flat ends it is intermediate between the two.

CONDITIONS OF ENDS. The various conditions which may exist at the ends of columns usually are divided into four classes: (1) Columns with round ends; the bearing at either end has perfect freedom of motion, as there would be with a ball-and-socket joint at each end. (2) Columns with hinged ends; they have perfect freedom of motion at the ends in one plane, as in compression members in bridge trusses where loads are transmitted through end pins. (3) Columns with flat ends; the bearing surface is normal to the axis of the column and of sufficient area to give at least partial fixity to the ends of the columns against lateral deflection. (4) Columns with fixed ends; the ends are rigidly secured, so that under any load the tangent to the elastic curve at the ends will be parallel to the axis in its original position.

Experiments prove that columns with fixed ends are stronger than columns with flat, hinged, or round ends, and that columns with round ends are weaker than any of the other types. Columns with hinged ends are equivalent to those with round ends in the plane in which they have free movement; columns with flat ends have a value intermediate between those with fixed ends and those with round ends. It often happens that columns have one end fixed and one end hinged, or some other combination. Their relative values may be taken as intermediate between those represented by the condition at either end. The extent to which strength is increased by fixing the ends depends on the length of column, fixed ends having a greater effect on long columns than on short ones.

8. THEORY OF COLUMNS

There is no exact theoretical formula which gives the strength of a column of any length under an axial load. Formulas involving the use of empirical coefficients have been deduced, however, and they give results which are consistent with the results of tests of full-sized members.

EULER'S FORMULA assumes that the failure of a column is due solely to the stresses induced by sidewise bending. This assumption is not true for very short columns, which fail mainly by direct compression, nor is it true for columns of medium length. The failure in such cases is by a combination of direct compression and bending. For columns in which $l/k > 200$, Euler's formula is approximately correct and agrees closely with the results of tests.

Let P = axial load, pounds; l = length of column, inches; I = least moment of inertia, inches^4; k = least radius of gyration, inches; E = modulus of elasticity; y = lateral deflection, inches, at any point along the column, that is caused by load P. If a column has round ends, so that the bending is not restrained, the equation of its elastic curve is

$$EI\frac{d^2y}{dx^2} = -Py \tag{22}$$

when the origin of the coordinate axes is at the top of the column, the positive direction of x being taken downwards and the positive direction of y in the direction of the deflection. Integrating the above expression twice and determining the constants of integration give

$$P = \Omega\pi^2\frac{EI}{l^2} \tag{23}$$

which is Euler's formula for long columns. The factor Ω is a constant depending on the condition of the ends. For round ends $\Omega = 1$; for fixed ends $\Omega = 4$; for one end round and the other fixed $\Omega = 2.05$. P is the load at which, if a slight deflection is produced, the column will not return to its original position. If P is decreased the column will approach its original position, but if P is increased the deflection will increase until the column fails by bending.

For columns with value of l/k less than about 150, Euler's formula gives results distinctly higher than those observed in tests. Euler's formula is now little used except for long members and as a basis for the analysis of the stresses in some types of structural and machine parts. It always gives an *ultimate* and never an allowable load.

SECANT FORMULA. The deflection of the column is used in the derivation of the Euler formula, but if the load were truly axial it would be impossible to compute the deflection. If the column is assumed to have an initial eccentricity of load of e in. (see Ref. 1, p. 241, for suggested values of e), the equation for the deflection y becomes

$$y_{max} = e\left(\sec\frac{l}{2}\sqrt{\frac{P}{EI}} - 1\right) \tag{24}$$

The maximum unit compressive stress becomes

$$\sigma = \frac{P}{A}\left(1 + \frac{ec}{k^2}\sec\frac{l}{2}\sqrt{\frac{P}{EI}}\right) \tag{25}$$

where l = length of column, inches; P = total load, pounds; A = area, square inches; I = moment of inertia, inches^4; k = radius of gyration, inches; c = distance from neutral axis to the most compressed fiber, inches; E = modulus of elasticity; both I and k are taken with respect to the axis about which bending takes place. Because the formula contains the secant of the angle $(l/2)\sqrt{P/EI}$ it is sometimes called the *secant formula.* It has been suggested by the Committee on Steel-Column Research (Ref. 3) that the best rational column formula can be constructed on the secant type, though of course it must contain experimental constants.

The secant formula can be used also for columns that are eccentrically loaded, if e is taken as the actual eccentricity plus the assumed initial eccentricity.

RANKINE'S FORMULA is a modification of an earlier formula by Gordon, and sometimes is called Gordon's formula. It applies to columns of slenderness ratios of 20 to 200, which are usually required in practice. Rankine's formula assumes that the maximum unit stress, σ, pounds per square inch, occurs in the fiber most remote from the axis on the concave side of the deflected column, and is equal to the sum of the direct unit compressive stress and the unit compressive stress due to bending alone; that is, $\sigma = (P/A) + \sigma_1$ where P = load, pounds; A = sectional area, square inches; σ_1 = unit compressive stress due to bending, pounds per square inch. The value of σ_1 in this equation may be expressed in terms of P by using the flexure formula, since the action is analogous to that of beams. Let l = length of the column, inches; I = least moment of inertia, inches^4; k = corresponding radius of gyration, inches; c = distance from axis to the remotest fiber on the concave side, inches; y = lateral deflection of column, inches. Substituting in the flexure formula, $\sigma_1 = Mc/I$, and by replacing M by its value Py and I by its value Ak^2, $\sigma_1 = Pyc/(Ak^2)$. Hence maximum unit compressive stress on the remotest fiber on the concave side is equal to $P/A + Pyc/(Ak^2)$. By analogy with the theory of beams the deflection y varies as l^2/c, so that by introducing an experimental coefficient ϕ, the expression for maximum unit stress may be written

$$\sigma = \frac{P}{A} + \phi\frac{Pl^2}{Ak^2} = \frac{P}{A}\left[1 + \phi\left(\frac{l}{k}\right)^2\right] \tag{26}$$

which is Rankine's formula for the investigation of columns. Variations in the value of ϕ have been determined by many experiments on columns of different materials and different conditions of ends. For further information see Ref. 1, p. 221.

The value to be taken for σ in Rankine's formula is the ultimate compressive stress of the material for rupture, and the allowable compressive unit stress for design. Rankine's formula may be used in the investigation of an existing column by comparing the computed value of σ with the proportional limit and with the ultimate strength of the material, thus determining the factor of safety. Similarly, the safe load P for an existing column may be computed by assuming an allowable unit stress σ.

STRAIGHT-LINE FORMULA. Plotted results of actual tests on columns show that the relation between ultimate load and l/k is fairly well represented by a straight line, for columns which fail by flexure of the whole column and not by local collapse. For a value of $l/k = 0$, average unit stress, P/A, on the section, is equal to the ultimate stress in compression for brittle materials and to the yield point in compression for ductile materials. From the point thus determined the straight line representing the average results of tests is drawn tangent to the curve of Euler's formula for the material. The equation of the straight line is

$$\frac{P}{A} = \sigma - \frac{Cl}{k} \tag{27}$$

where σ = ultimate unit stress in compression for a brittle material, or unit stress at the yield point in compression for a ductile material; C = an experimental constant. Table 5 gives average values of σ and C for structural steel, cast iron, and wood. The loads given

Table 5. Ultimate Strength Constants for the Straight-line Column Formula

Material	σ_1 psi	Character of Ends					
		Round	Fixed	1 round 1 fixed	Round	Fixed	1 round 1 fixed
		C			Limit of (l/k)		
Structural steel	35,000	150	75	100	160	320	240
Cast iron	34,000 *	175	88	116	90	160	115
Wood	5,000 *	40	20	30	75	150	112

* This is less than the ultimate in compression for small specimens of cast iron or wood, but from tests of full-size columns it seems to be the value to be used for full-size castings or timbers which may contain defects. See Timoshenko, *Theory of Elastic Stability,* Chapter III, McGraw-Hill Book Co., for further information.

by the straight-line formula using the constants in this table are *ultimate loads.* For design the straight-line formula usually is put in a form which gives allowable loads directly, constants σ and C being divided by a factor of safety. The straight-line formula is not suitable for investigating a column, i.e., determining values of σ due to given loads, as (Cl/k) is not a function of (P/A). It may be used to find the safe load for a given column under a given unit stress or to design a column for a given load and unit stress.

ECCENTRIC LOADS ON SHORT COMPRESSION MEMBERS. Where a direct push acting on a member does not pass through the centroid but at a distance e, inches, from it, both direct and bending stresses are produced. For short compression members in which column action may be neglected, the direct unit stress is P/A, where P = total load, pounds, and A = area of cross section, square inches. The bending unit stress is Mc/I, where $M = Pe$ = bending moment, pound-inches; c = distance, inches, from the centroid to the fiber in which the stress is desired; I = moment of inertia, inches^4. The total unit stress at any point in the section is $\sigma = P/A + Pec/I$, or $\sigma = (P/A)(1 + ec/k^2)$, since $I = Ak^2$, where k = radius of gyration, inches.

ECCENTRIC LOADS ON COLUMNS. Various column formulas must be modified when the loads are not balanced, that is, when the resultant of the loads is not in line with the axis of the column. If P = load, pounds, applied at a distance e in. from the axis, bending moment $M = Pe$. Maximum unit stress σ, pounds per square inch, due to this bending moment alone, is $\sigma = Mc/I = Pec/Ak^2$, where c = distance, inches, from the axis to the most remote fiber on the concave side; A = sectional area in square inches; k = radius of gyration in the direction of the bending, inches. This unit stress must be added to the unit stress that would be induced if the resultant load were applied in line with the axis of the column. The modified formulas, expressed in allowable load per unit of area, are:

Modified Rankine's formula for eccentric loads (see p. 8-23)

$$\frac{P}{A} = \frac{\sigma}{1 + \phi\left(\frac{l}{k}\right)^2 + \frac{ec}{k^2}} \tag{28}$$

Modified straight-line formula for eccentric loads (see p. 8-23):

$$\frac{P}{A} = \frac{\sigma - \frac{Cl}{k}}{1 + \frac{ec}{k^2}} \tag{29}$$

The secant formula, p. 8-22, also can be used for columns that are eccentrically loaded if e is taken as the actual eccentricity plus the assumed initial eccentricity.

COLUMN SUBJECTED TO TRANSVERSE OR CROSS-BENDING LOADS. A compression member that is subjected to cross-bending loads may be considered to be (1) a beam subjected to end thrust as discussed on p. 8-19 or (2) a column subjected to cross-bending loads, depending on the relative magnitude of the end thrust and cross-bending loads, and on the dimensions of the member. The various column formulas may be modified so as to include the effect of cross-bending loads. In this form they are:

Modified Rankine's formula for transverse loads:

$$\sigma = \frac{P}{A}\left(1 + \phi\frac{l^2}{k^2}\right) + \frac{Pyc}{Ak^2} + \frac{Mc}{Ak^2} \tag{30}$$

Modified straight-line formula for transverse loads:

$$\frac{P}{A} = \sigma - \frac{Cl}{k} - \frac{Pyc}{Ak^2} - \frac{Mc}{Ak^2} \tag{31}$$

Modified secant formula for transverse loads:

$$\sigma = \frac{P}{A}\left[1 + (e + y)\frac{c}{k^2}\sec\frac{l}{2k}\sqrt{\frac{P}{AE}}\right] + \frac{Mc}{Ak^2} \tag{32}$$

In these formulas, σ = maximum unit stress on concave side, pounds per square inch; P = axial end load, pounds; A = cross-sectional area, square inches; M = moment due to cross-bending load, pound-inches; y = deflection due to cross-bending load, inches; k = radius of gyration, inches; l = length of column, inches; e = assumed initial eccentricity, inches; c = distance, inches, from axis to the most remote fiber on the concave side; C = experimental constant (see Table 5). The modified straight-line formula, like the straight-line formula, cannot be used to solve for σ.

9. WOODEN COLUMNS

WOODEN COLUMN FORMULAS. One of the principal formulas is that formerly used by the AREA, $P/A = \sigma_1(1 - l/60d)$, where P/A = allowable unit load, pounds per square inch; σ_1 = allowable unit stress in direct compression on short blocks, pounds per square inch; l = length, inches; d = least dimension, inches. This formula is being replaced rapidly by formulas recommended by the ASTM and AREA. Committees of these societies, working with the U. S. Forest Products Laboratory, classified timber columns in three groups (ASTM Standards, 1937, D245-37):

1. Short Columns. The ratio of unsupported length to least dimension does not exceed 11. For these columns, the allowable unit stress should not be greater than the values given in Table 2, p. 5-84, under compression parallel to the grain.

2. Intermediate-length Columns. Where the ratio of unsupported length to least dimension is greater than 10, eq. 33, of the fourth power parabolic type, shall be used to determine allowable unit stress, until this allowable unit stress is equal to two-thirds of the allowable unit stress for short columns.

$$\frac{P}{A} = \sigma_1\left[1 - \frac{1}{3}\left(\frac{l}{Kd}\right)^4\right] \qquad (33)$$

where P = total load, pounds; A = area, square inches; σ_1 = allowable unit compressive stress parallel to grain, lb per sq inch (see Table 2, p. 5-84); l = unsupported length, inches; d = least dimension, inches; $K = l/d$ at the point of tangency of the parabolic and Euler curves, at which $P/A = {}^2/_3\,\sigma_1$. The value of K for any species and grade is $(\pi/2\sqrt{E/6\sigma_1})$, where E = modulus of elasticity.

3. Long Columns. Where P/A as computed by eq. 33 is less than ${}^2/_3\,\sigma_1$, eq. 34 of the Euler type, which includes a factor of safety of 3, shall be used:

$$\frac{P}{A} = \frac{1}{36}\left[\frac{\pi^2 E}{(l/d)^2}\right] \qquad (34)$$

Timber columns should be limited to a ratio of l/d equal to 50. No higher loads are allowed for square-ended columns. The strength of round columns may be considered the same as that of square columns of the same cross-sectional area.

USE OF TIMBER COLUMN FORMULAS. The values of E (modulus of elasticity) and σ_1 (compression parallel to grain) in the above formulas are given in Table 2, p. 5-84. Table 6 gives the computed values of K for some common types of timbers. These may

Table 6. Values of *K* for Columns of Intermediate Length

ASTM Standards, 1937, D245–37

Species	Continuously Dry		Occasionally Wet		Usually Wet	
	Select	Common	Select	Common	Select	Common
Cedar, western red	24.2	27.1	24.2	27.1	25.1	28.1
Cedar, Port Orford	23.4	26.2	24.6	27.4	25.6	28.7
Douglas fir, coast region	23.7	27.3	24.9	28.6	27.0	31.1
Douglas fir, dense	22.6	25.3	23.8	26.5	25.8	28.8
Douglas fir, Rocky Mountain region	24.8	27.8	24.8	27.8	26.5	29.7
Hemlock, west coast	25.3	28.3	25.3	28.3	26.8	30.0
Larch, western	22.0	24.6	23.1	25.8	25.8	28.8
Oak, red and white	24.8	27.8	26.1	29.3	27.7	31.1
Pine, southern	....	27.3	....	28.6	....	31.1
Pine, dense	22.6	25.3	23.8	26.5	25.8	28.8
Redwood	22.2	24.8	23.4	26.1	25.6	28.6
Spruce, red, white, Sitka	24.8	27.8	25.6	28.7	27.5	30.8

be substituted directly in eq. 33 for intermediate-length columns, or may be used in conjunction with Table 7, which gives the strength of columns of intermediate length, expressed as a percentage of strength (σ_1) of short columns. In the tables, the term "continuously dry" refers to interior construction where there is no excessive dampness or humidity; "occasionally wet but quickly dry" refers to bridges, trestles, bleachers, and grandstands; "usually wet" refers to timber in contact with the earth or exposed to waves or tidewater.

Table 7. Strength of Columns of Intermediate Length, Expressed as a Percentage of Strength of Short Columns

ASTM Standards, 1937, D245–37

Values for expression $\{1 - {}^1/_3(l/Kd)^4\}$ in eq. 33

K	Ratio of Length to Least Dimension in Rectangular Timbers, l/d																			
	12	13	14	15	16	17	18	19	20	21	22	23	24	25	26	27	28	29	30	31
22	97	96	95	93	91	88	85	81	77	72	67	...	...	...	...	...	...	...	...	...
23	98	97	95	94	92	90	87	84	81	77	72	67	...	...	...	...	...	...	...	...
24	98	97	96	95	93	92	89	87	84	80	76	72	67	...	...	...	...	...	...	...
25	98	98	97	96	94	93	91	89	86	83	80	76	72	67	...	...	...	...	...	...
26	99	98	97	96	95	93	92	91	89	86	83	80	76	72	67	...	...	...	...	...
27	99	98	98	97	96	95	93	92	90	88	85	82	79	74	71	67	...	...	...	...
28	99	98	98	97	96	95	94	93	91	89	87	85	82	79	75	71	67	...	...	...
29	99	99	98	98	97	96	95	94	92	91	89	87	84	82	79	75	71	67	...	...
30	99	99	98	98	97	97	96	95	94	92	90	88	86	84	81	78	75	71	67	...
31	99	99	99	98	98	97	96	95	94	93	92	90	88	86	84	81	78	75	71	67

Note. This table can also be used for columns not rectangular, the l/d being equivalent to $0.289l/k$, where k is the least radius of gyration of the section.

10. STEEL COLUMNS

TYPES. Two general types of steel columns are in use: (1) rolled shapes; (2) built-up sections. The rolled shapes are easily fabricated, accessible for painting, neat in appearance where they are not covered, and convenient in making connections. A disadvantage is the probability that thick sections are of lower-strength material than thin sections because of the difficulty of adequately rolling the thick material. For the effect of thickness of material on yield point, see Ref. 3, p. 1377.

GENERAL PRINCIPLES IN DESIGN. The design of steel columns is always a cut-and-try method, as no law governs the relation between area and radius of gyration of the section. A column of given area is selected, and the amount of load that it will carry is computed by the proper formula. If the allowable load so computed is less than that to be carried, a larger column is selected and the load for it computed, the process being repeated until a proper section is found.

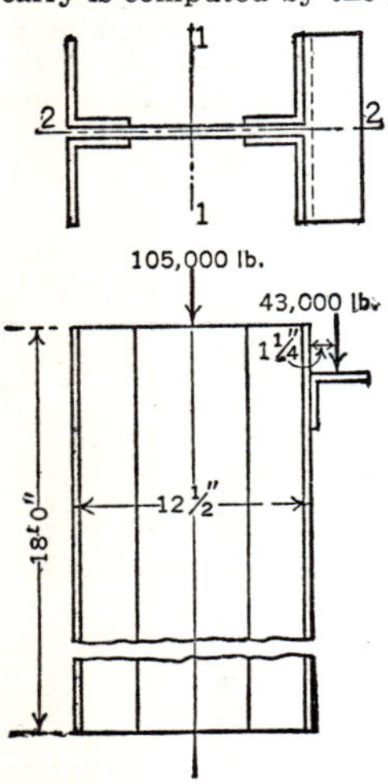

Fig. 20. Built-up column.

A few general principles should guide in proportioning columns. The radius of gyration should be approximately the same in the two directions at right angles to each other; the slenderness ratio of the separate parts of the column should not be greater than that of the column as a whole; the different parts should be adequately connected in order that the column may function as a single unit; the material should be distributed as far as possible from the center line in order to increase the radius of gyration.

STEEL COLUMN FORMULAS. A variety of steel column formulas are in use, differing mostly in the value of unit stress allowed with various values of l/k. See Ref. 4, p. 5-42, for a summary of the several formulas.

TESTS ON STEEL COLUMNS. After the collapse of the Quebec Bridge in 1907 as a result of a column failure, the ASCE, the AREA, and the U. S. Bureau of Standards cooperated in tests of full-size steel columns. The results of these tests are reported in Ref. 5, pp. 1583–1688, and are summarized in Ref. 4, pp. 5-43, 5-44. The tests showed that, for columns of the proportions commonly used, the effect of variation in the steel, kinks, initial stresses, and similar defects in the column was more important than the effect of length. They also showed that the thin metal gave definitely higher strength, per unit area, than the thicker metal of the same type of section.

Example. Design a steel column 18 ft long to carry an axial load of 105,000 lb and an additional load of 43,000 lb which comes from a crane runway, and which is 1 1/4 in. beyond the outside edge of the column.

The formula used will be the AISC formula (see Ref. 4),

$$\frac{P}{A} = \frac{18{,}000}{1 + \left(\dfrac{l^2}{18{,}000k^2}\right)}$$

The shape selected will be a plate and four angles, Fig. 20.

Using 10,000 psi as a preliminary estimate for the first selection, one plate 12 × 5/16 in. and four angles 5 × 3 1/2 × 3/8 in. will have an area of 15.95 sq in. The safe load for this column will now be found. The moment of inertia I_{1-1} is 412 in.^{4}, the radius of gyration k_{1-1} is 5.08 in., and the radius of gyration k_{2-2} is 2.08 in.

Considering the safe load with respect to axis 2–2 about which neither load is eccentric,

$$P = \frac{15.95 \times 18{,}000}{1 + \dfrac{(18 \times 12)^2}{18{,}000 \times (2.08)^2}} = 179{,}400 \text{ lb}$$

Consider now the safe load about axis 1–1, with respect to which the load of 105,000 lb is axial, and the load of 43,000 lb has an eccentricity of 7 1/2 in. The load of 43,000 lb will produce a stress due to its eccentricity, and this stress will reduce by that amount the unit load that the column can carry. The safe load will therefore be

$$P = 15.95\left(\frac{18{,}000}{1 + \dfrac{(18 \times 12)^2}{18{,}000 \times (5.08)^2}} - \frac{43{,}000 \times 7.5 \times 6.25}{412}\right) = 183{,}000 \text{ lb}$$

The area exceeds that needed by about 3 sq in.; the next smaller size is one plate 12 × 5/16 in. and four angles 5 × 3 1/2 × 5/16 in. The area is 13.99 sq in., moment of inertia, $I_{1-1} = 356$ in.^{4}, $k_{1-1} = 5.04$ in., and $k_{2-2} = 2.03$ in. With respect to the axis 2–2,

$$P = \frac{13.99 \times 18{,}000}{1 + \dfrac{(18 \times 12)^2}{18{,}000 \times (2.03)^2}} = 154{,}600 \text{ lb}$$

With respect to axis 1–1,

$$P = 13.99\left(\frac{18{,}000}{1 + \dfrac{(18 \times 12)^2}{18{,}000 \times (5.04)^2}} - \frac{43{,}000 \times 7.5 \times 6.25}{356}\right) = 149{,}300 \text{ lb}$$

As the load to be carried is 148,000 lb this column is satisfactory.

11. SHAFTS—DEFINITIONS

TORSIONAL STRESS. A bar is under torsional stress when it is held fast at one end, and a force acts at the other end to twist the bar. In a round bar (Fig. 21) with a constant force acting, the straight line *ab* becomes the helix *ad*, and a radial line in the cross section, *ob*, moves to the position *od*. The angle *bad* remains constant while the angle *bod* increases with the length of the bar. Each cross section of the bar tends to shear off the one adjacent to it, and in any cross section the shearing stress at any point is normal to a radial line drawn through the point. Within the shearing proportional limit, a radial line of the cross section remains straight after the twisting force has been applied, and the unit shearing stress at any point is proportional to its distance from the axis.

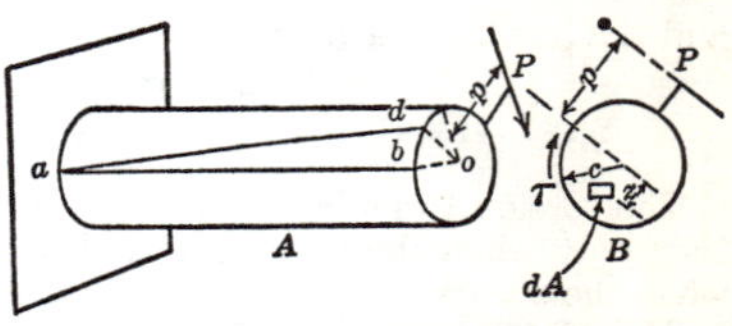

FIG. 21. Round bar subject to torsional stress.

TWISTING MOMENT, T, is equal to the product of the resultant, P, of the twisting forces, multiplied by its distance from the axis, p.

RESISTING MOMENT, T_r, in torsion, is equal to the sum of the moments of the unit shearing stresses acting along a cross section with respect to the axis of the bar. If dA is an elementary area of the section at a distance of z units from the axis of a circular shaft (Fig. 21B), and c is the distance from the axis to the outside of the cross section where the unit shearing stress is τ, then the unit shearing stress acting on dA is $(\tau z/c)\, dA$, its moment with respect to the axis is $(\tau z^2/c)\, dA$, and the sum of all the moments of the unit shearing stresses on the cross section is $\int (\tau z^2/c)\, dA$. In this expression the factor $\int z^2\, dA$ is the polar moment of inertia of the section with respect to the axis. Denoting this by J, the resisting moment may be written $\tau J/c$.

THE POLAR MOMENT OF INERTIA of a surface about an axis through its center of gravity and perpendicular to the surface is the sum of the products obtained by multiplying each elementary area by the square of its distance from the center of gravity of its surface; it is equal to the sum of the moments of inertia taken with respect to two axes in the plane of the surface at right angles to each other passing through the center of gravity. It is represented by J, inches^4. For the cross section of a round shaft,

$$J = {}^1/_{32}\, \pi d^4 \quad \text{or} \quad {}^1/_2\, \pi r^4 \tag{35}$$

for a hollow shaft,

$$J = {}^1/_{32}\, \pi(d^4 - d_1^4) \tag{36}$$

where d is the outside and d_1 the inside diameter, inches, or

$$J = {}^1/_2\, \pi(r^4 - r_1^4) \tag{37}$$

where r is the outside and r_1 the inside radius, inches.

THE POLAR RADIUS OF GYRATION, k_p, sometimes is used in formulas; it is defined as the radius of a circumference along which the entire area of a surface might be concentrated and have the same polar moment of inertia as the distributed area. For a solid circular section,

$$k_p^2 = {}^1/_8\, d^2 \tag{38}$$

for a hollow circular section,

$$k_p^2 = {}^1/_8\, (d^2 - d_1^2) \tag{39}$$

12. DETERMINATION OF TORSIONAL STRESSES IN SHAFTS

TORSION FORMULA FOR ROUND SHAFTS. The conditions of equilibrium require that the twisting moment, T, be opposed by an equal resisting moment, T_r, so that for the values of the maximum unit shearing stress, τ, within the proportional limit, the torsion formula for round shafts becomes

$$T_r = T = \tau \frac{J}{c} \tag{40}$$

If τ is in pounds per square inch, then T_r and T must be in pound-inches, J is in inches^4, and c is in inches. For solid round shafts having a diameter, d, inches,

$$J = {}^1/_{32}\, \pi d^4 \quad \text{and} \quad c = {}^1/_2\, d \tag{41}$$

and

$$T = {}^1/_{16}\, \pi d^3 \tau \quad \text{or} \quad \tau = \frac{16T}{\pi d^3} \tag{42}$$

For hollow round shafts,

$$J = \frac{\pi(d^4 - d_1^4)}{32} \quad \text{and} \quad c = {}^1/_2\, d \tag{43}$$

and the formula becomes

$$T = \frac{\tau\pi(d^4 - d_1^4)}{16d} \quad \text{or} \quad \tau = \frac{16Td}{\pi(d^4 - d_1^4)} \tag{44}$$

The torsion formula applies only to solid circular shafts or hollow circular shafts, and then only when the load is applied in a plane perpendicular to the axis of the shaft and when the shearing proportional limit of the material is not exceeded.

SHEARING STRESS IN TERMS OF HORSEPOWER. If the shaft is to be used for the transmission of power, the value of T, pound-inches, in the above formulas becomes $63{,}030H/N$, where H = horsepower to be transmitted; N = revolutions per minute. The maximum unit shearing stress, pounds per square inch, then is:

$$\text{For solid round shafts:} \quad \tau = \frac{321{,}000H}{Nd^3} \tag{45}$$

$$\text{For hollow round shafts:} \quad \tau = \frac{321{,}000Hd}{N(d^4 - d_1^4)} \tag{46}$$

If τ is taken as the allowable unit shearing stress, the diameter, d, inches, necessary to transmit a given horsepower at a given shaft speed can then be determined. These formulas give the stress due to torsion only, and allowance must be made for any other loads, as the weight of shaft and pulleys, and tension in belts.

Table 8. Formulas for Torsional Deformation and Stress

General formulas: $\theta = \dfrac{TL}{KG}$, $\tau = \dfrac{T}{Q}$, where θ = angle of twist, radians; T = twisting moment, in.-lb; L = length, in.; τ = unit shear stress, psi; G = modulus of rigidity, psi; K, in.^{4}; and Q, in.^{3} are functions of the cross section.

Shape	Formula for K in $\theta = \dfrac{TL}{KG}$	Formula for Shear Stress
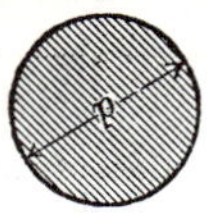	$K = \dfrac{\pi d^4}{32}$	$\tau = \dfrac{16T}{\pi d^3}$
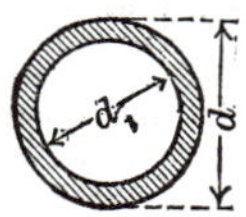	$K = 1/32\pi(d^4 - d_1^4)$	$\tau = \dfrac{16Td}{\pi(d^4 - d_1^4)}$
	$K = 2/3\ \pi r t^3$	$\tau = \dfrac{3T}{2\pi r t^2}$
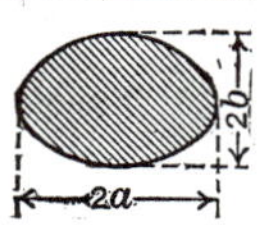	$K = \dfrac{\pi a^3 b^3}{a^2 + b^2}$	$\tau = \dfrac{2T}{\pi a b^2}$
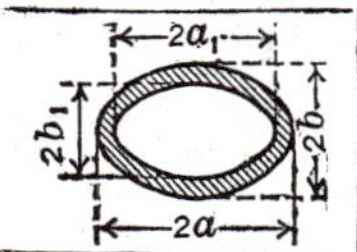	$K = \dfrac{\pi a_1^3 b_1^3}{a_1^2 + b_1^2}[(1+q)^4 - 1]$ $\quad q = \dfrac{a - a_1}{a_1}$, $q = \dfrac{b - b_1}{b_1}$	$\tau = \dfrac{2T}{\pi a_1 b_1^2[(1+q)^4 - 1]}$
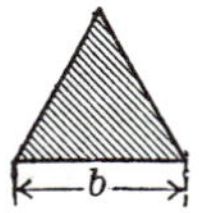	$K = \dfrac{b^4\sqrt{3}}{80}$	$\tau = \dfrac{20T}{b^3}$
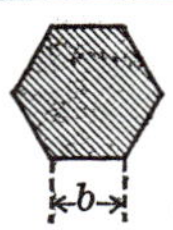	$K = 2.69b^4$	$\tau = \dfrac{1.09T}{b^3}$
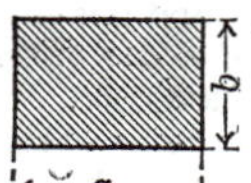	$K = \dfrac{ab^3}{16}\left[\dfrac{16}{3} - 3.36\dfrac{b}{a}\left(1 - \dfrac{b^4}{12a^4}\right)\right]$	$\tau = \dfrac{(3a + 1.8b)T}{a^2b^2}$
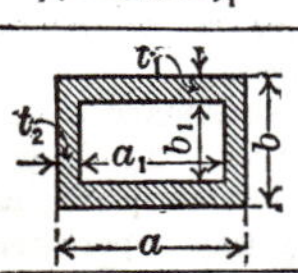	$K = \dfrac{2t_1t_2(a - t_2)^2(b - t_1)^2}{at_2 + bt_1 - t_2^2 - t_1^2}$	$\tau = \dfrac{T}{2t_2(a - t_2)(b - t_1)}$
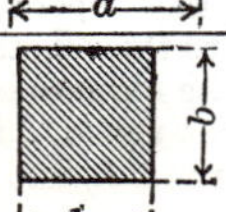	$K = 0.1406b^4$	$\tau = \dfrac{4.8T}{b^3}$

(Table continued on p. 8-30.)

Table 8. Formulas for Torsional Deformation and Stress—*Continued*

General formulas: $\theta = \frac{TL}{KG}$, $\tau = \frac{T}{Q}$, where θ = angle of twist, radians; T = twisting moment, in.-lb; L = length, in.; τ = unit shear stress, psi; G = modulus of rigidity, psi; K, in.^{4}; and Q, in.^{3} are functions of the cross section.

Shape	Formula for K in $\theta = \frac{TL}{KG}$	Formula for Shear Stress
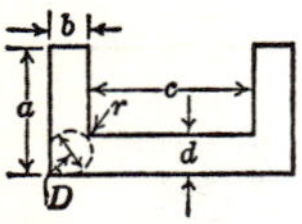	r = fillet radius D = diameter largest inscribed circle $K = 2K_1 + K_2 + 2\alpha D^4$ $K_1 = ab^3\left[\frac{1}{3} - 0.21\frac{b}{a}\left(1 - \frac{b^4}{12a^4}\right)\right]$ $K_2 = cd^3\left[\frac{1}{3} - 0.105\frac{d}{c}\left(1 - \frac{d^4}{192c^4}\right)\right]$ $\alpha = \frac{b}{d}\left(0.07 + 0.076\frac{r}{b}\right)$	For all solid sections of irregular form the maximum shear stress occurs at or very near one of the points where the largest inscribed circle touches the boundary, and of these, at the one where the curvature of the boundary is algebraically least. (Convexity represents positive, concavity negative, curvature of the boundary.) At a point where the curvature is positive (boundary of section straight or convex) this maximum stress is given approximately by: $\tau = G\frac{\theta}{L}c$ or $\tau = \frac{T}{K}c$ where $c = \frac{D}{1 + \frac{\pi^2 D^4}{16A^2}} \times \left[1 + 0.15\left(\frac{\pi^2 D^4}{16A^2} - \frac{D}{2r}\right)\right]$ where D = diameter of largest inscribed circle, r = radius of curvature of boundary at the point (positive for this case), A = area of the section.
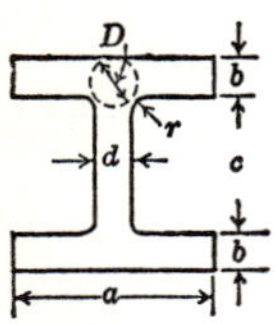	$K = 2K_1 + K_2 + 2\alpha D^4$ $K_1 = ab^3\left[\frac{1}{3} - 0.21\frac{b}{a}\left(1 - \frac{b^4}{12a^4}\right)\right]$ $K_2 = 1/3\, cd^3$ $\alpha = \frac{t}{t_1}\left(0.15 + 0.1\frac{r}{b}\right)$ $t = b$ if $b < d$ $t = d$ if $d < b$ $t_1 = b$ if $b > d$ $t_1 = d$ if $d > b$	
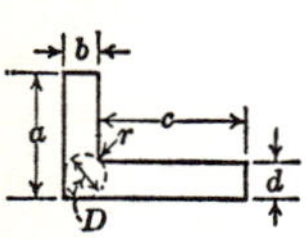	$K = K_1 + K_2 + \alpha D^4$ $K_1 = ab^3\left[\frac{1}{3} - 0.21\frac{b}{a}\left(1 - \frac{b^4}{12a^4}\right)\right]$ $K_2 = cd^3\left[\frac{1}{3} - 0.105\frac{d}{c}\left(1 - \frac{d^4}{192c^4}\right)\right]$ $\alpha = \frac{b}{d}\left(0.07 + 0.076\frac{r}{b}\right)$	

ANGLE OF TWIST. When the unit shearing stress τ does not exceed the proportional limit, the angle *bod* (Fig. 21) for a solid round shaft may be computed from the formula

$$\theta = \frac{Tl}{GJ} \tag{47}$$

where θ = angle in radians; l = length of shaft in inches; G = shearing modulus of elasticity of the material; T = twisting moment, pound-inches. Values of G for different materials are steel, 12,000,000; wrought iron, 10,000,000; and cast iron, 6,000,000.

When the angle of twist on a section begins to increase in a greater ratio than the twisting moment, it may be assumed that the shearing stress on the outside of the section has reached the proportional limit. The shearing stress at this point may be determined by substituting the twisting moment at this instant in the torsion formula.

TORSION OF NONCIRCULAR CROSS SECTIONS. The analysis of shearing stress distribution along noncircular cross sections of bars under torsion is complex. By drawing two lines at right angles through the center of gravity of a section before twisting, and observing the angular distortion after twisting, it has been found from many experiments that in noncircular sections the shearing unit stresses are not proportional to their distances from the axis. Thus in a rectangular bar there is no shearing stress at the corners of the sections, and the stress at the middle of the wide side is greater than at the middle of the narrow side. In an elliptical bar the shearing stress is greater along the flat side than at the round side.

It has been found by tests (Refs. 6 and 7) as well as by mathematical analysis that the torsional resistance of a section, made up of a number of rectangular parts, is approximately equal to the sum of the resistances of the separate parts. It is on this basis that nearly all the formulas for noncircular sections have been developed. For example, the torsional resistance of an I-beam is approximately equal to the sum of the torsional resistances of the web and the outstanding flanges. In an I-beam in torsion the maximum shearing stress will occur at the middle of the side of the web, except where the flanges are thicker than the web, and then the maximum stress will be at the midpoint of the width of the flange. Re-entrant angles, as those in I-beams and channels, are always a source of weakness in members subjected to torsion. Table 8 gives values of the maximum unit shearing stress τ and the angle of twist θ induced by twisting bars of various cross sections, it being assumed that τ is not greater than the proportional limit.

ULTIMATE STRENGTH IN TORSION. In a torsion failure, the outer fibers of a section are the first to shear, and the rupture extends toward the axis as the twisting is continued. The torsion formula for round shafts has no theoretical basis after the shearing stresses on the outer fibers exceed the proportional limit, as the stresses along the section then are no longer proportional to their distances from the axis. It is convenient, however, to compare the torsional strength of various materials by using the formula to compute values of τ at which rupture takes place. These computed values of the maximum stress sustained before rupture are somewhat higher for iron and steel than the ultimate strength of the materials in direct shear. Computed values of the ultimate strength in torsion are found by experiment to be: cast iron, 30,000 psi; wrought iron, 55,000 psi; medium steel, 65,000 psi; timber, 2000 psi. These computed values of twisting strength may be used in the torsion formula to determine the probable twisting moment that will cause rupture of a given round bar or to determine the size of a bar that will be ruptured by a given twisting moment. In design, large factors of safety should be taken, especially when the stress is reversed as in reversing engines and when the torsional stress is combined with other stresses as in shafting.

13. CYLINDERS AND PLATES

THIN CYLINDERS UNDER INTERNAL PRESSURE. A cylinder is regarded as thin when the thickness of the wall is small compared with the diameter. It is assumed that in such cases the tensile stress across the longitudinal section is uniformly distributed over the thickness of the wall. If P = internal pressure, l = length of cylinder, t = thickness of wall, d = diameter of cylinder, and σ = circumferential tensile stress,

$$Pdl = 2\sigma tl \quad \text{or} \quad \sigma = \frac{Pd}{2t} \tag{48}$$

For resistance to rupture in a circumferential seam, due to pressure on the ends of the cylinder, i.e., the tensile stress across a transverse section,

$$P\frac{\pi d^2}{4} = \sigma t\pi d \quad \text{or} \quad \sigma = \frac{Pd}{4t} \tag{49}$$

If the two equations are compared, it is seen that the resistance to rupture by circumferential stress (eq. 48) is one-half the resistance to rupture by longitudinal stress (eq. 49). For this reason cylindrical boilers are single riveted in the circumferential seams and double or triple riveted in the longitudinal seams.

In using the foregoing formulas to design cylindrical shells or piping, thickness t must be increased to compensate for rivet holes in the joints. Water pipes, particularly those of cast iron, require a high factor of safety, which results in increased thickness to provide security against shocks caused by water hammer or rough handling before they are laid.

Equation 49 applies also to the stresses in the walls of a thin hollow sphere, hemisphere, or dome. When holes are cut, the tensile stresses must be found by the method used in riveted joints. See Section 10, Article 7.

THIN CYLINDERS UNDER EXTERNAL PRESSURE. Equations 48 and 49 apply equally well to cases of external pressure if P is given a negative sign, but the stresses so found are significant only if the pressure and dimensions are such that no buckling can occur.

THICK CYLINDERS. When the thickness of the shell or wall is relatively large, as in guns, hydraulic machinery piping, and similar installations, the variation in stress from the inner surface to the outer surface is relatively large, and the ordinary formulas for

(*Continued on p. 8-35*)

Table 9. Formulas for Flat Plates

(By permission from *Formulas for Stress and Strain,* 2nd Ed., by R. J. Roark, copyrighted, 1943. McGraw-Hill Book Co., Inc.)

Notation: W = total applied load, lb; w = unit applied load, psi; t = thickness of plate, in.; σ = stress at surface of plate, psi; y = vertical deflection of plate from original position, in.; E = modulus of elasticity; m = reciprocal of ν, Poisson's ratio. q denotes any given point on the surface of plate; r denotes the distance of q from the center of a circular plate. Other dimensions and corresponding symbols are indicated on figures. Positive sign for σ indicates tension at upper surface and equal compression at lower surface; negative sign indicates reverse condition. Positive sign for y indicates upward deflection, negative sign downward deflection. Subscripts r, t, a, and b used with σ denote, respectively, radial direction, tangential direction, direction of dimension a, and direction of dimension b. All dimensions are in inches. All logarithms are to the base e ($\log_e x = 2.3026 \log_{10} x$).

Type of Load and Support	Formulas for Stress and Deflection
	Circular Flat Plates 
Outer edges supported. Uniform load over entire surface. 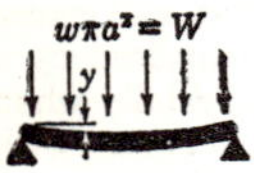	At center: $\max \sigma_r = \sigma_t = \frac{-3W}{8\pi m t^2}(3m+1)$ $\quad \max y = -\frac{3W(m-1)(5m+1)a^2}{16\pi E m^2 t^3}$ At q: $\sigma_r = -\frac{3W}{8\pi m t^2}\left[(3m+1)\left(1-\frac{r^2}{a^2}\right)\right]$ $\quad \sigma_t = -\frac{3W}{8\pi m t^2}\left[(3m+1)-(m+3)\frac{r^2}{a^2}\right]$ $y = -\frac{3W(m^2-1)}{8\pi E m^2 t^3}\left[\frac{(5m+1)a^2}{2(m+1)} + \frac{r^4}{2a^2} - \frac{(3m+1)r^2}{m+1}\right]$
Outer edges fixed. Uniform load over entire surface. 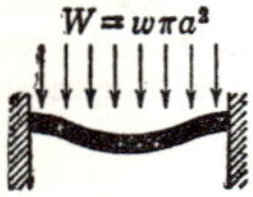	At center: $\sigma_r = \sigma_t = -\frac{3W(m+1)}{8\pi m t^2}$ $\quad \max y = -\frac{3W(m^2-1)a^2}{16\pi E m^2 t^3}$ At q: $\sigma_r = \frac{3W}{8\pi m t^2}\left[(3m+1)\frac{r^2}{a^2}-(m+1)\right]$ $\quad \sigma_t = \frac{3W}{8\pi m t^2}\left[(m+3)\frac{r^2}{a^2}-(m+1)\right]$ $y = \frac{-3W(m^2-1)}{16\pi E m^2 t^3}\left[\frac{(a^2-r^2)^2}{a^2}\right]$
Outer edges supported. Uniform load over concentric circular area of radius r_0. 	At q, $r < r_0$: $\sigma_r = -\frac{3W}{2\pi m t^2}\left[m + (m+1)\log\frac{a}{r_0} - (m-1)\frac{r_0^2}{4a^2} - (3m+1)\frac{r^2}{4r_0^2}\right]$ $\sigma_t = -\frac{3W}{2\pi m t^2}\left[m + (m+1)\log\frac{a}{r_0} - (m-1)\frac{r_0^2}{4a^2} - (m+3)\frac{r^2}{4r_0^2}\right]$ $y = -\frac{3W(m^2-1)}{16\pi E m^2 t^3}\left[4a^2 - 5r_0^2 + \frac{r^4}{r_0^2} - (8r^2+4r_0^2)\log\frac{a}{r_0} - \frac{2(m-1)r_0^2(a^2-r^2)}{(m+1)a^2} + \frac{8m(a^2-r^2)}{m+1}\right]$ At q, $r > r_0$: $\sigma_r = -\frac{3W}{2\pi m t^2}\left[(m+1)\log\frac{a}{r} - (m-1)\frac{r_0^2}{4a^2} + (m-1)\frac{r_0^2}{4r^2}\right]$ $\sigma_t = -\frac{3W}{2\pi m t^2}\left[(m-1) + (m+1)\log\frac{a}{r} - (m-1)\frac{r_0^2}{4a^2} - (m-1)\frac{r_0^2}{4r^2}\right]$ $y = -\frac{3W(m^2-1)}{16\pi E m^2 t^3}\left[\frac{(12m+4)(a^2-r^2)}{m+1} - \frac{2(m-1)r_0^2(a^2-r^2)}{(m+1)a^2} - (8r^2+4r_0^2)\log\frac{a}{r}\right]$ At center: $\max \sigma_r = \sigma_t = -\frac{3W}{2\pi m t^2}\left[m + (m+1)\log\frac{a}{r_0} - (m-1)\frac{r_0^2}{4a^2}\right]$ $\max y = -\frac{3W(m^2-1)}{16\pi E m^2 t^3}\left[\frac{(12m+4)a^2}{m+1} - 4r_0^2\log\frac{a}{r_0} - \frac{(7m+3)r_0^2}{m+1}\right]$

Table 9. Formulas for Flat Plates—*Continued*

Type of Load and Support	Formulas for Stress and Deflection
	Circular Flat Plates
Outer edges supported. Uniform load on concentric circular ring of radius r_0. 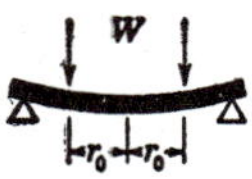 	At q, $r < r_0$: $\max \sigma_r = \sigma_t = -\frac{3W}{2\pi m t^2}\left[\frac{1}{2}(m-1) + (m+1)\log\frac{a}{r_0} - (m-1)\frac{r_0^2}{2a^2}\right]$ $y = -\frac{3W(m^2-1)}{2\pi E m^2 t^3}\left[\frac{(3m+1)(a^2-r^2)}{2(m+1)} - (r^2+r_0^2)\log\frac{a}{r_0} + (r^2-r_0^2) - \frac{(m-1)r_0^2(a^2-r^2)}{2(m+1)a^2}\right]$ At q, $r > r_0$: $\sigma_r = -\frac{3W}{2\pi m t^2}\left[(m+1)\log\frac{a}{r} + (m-1)\frac{r_0^2}{2r^2} - (m-1)\frac{r_0^2}{2a^2}\right]$ $\sigma_t = -\frac{3W}{2\pi m t^2}\left[(m-1) + (m+1)\log\frac{a}{r} - (m-1)\frac{r_0^2}{2r^2} - (m-1)\frac{r_0^2}{2a^2}\right]$ $y = -\frac{3W(m^2-1)}{2\pi E m^2 t^3}\left[\frac{(3m+1)(a^2-r^2)}{2(m+1)} - (r^2+r_0^2)\log\frac{a}{r} - \frac{(m-1)r_0^2(a^2-r^2)}{2(m+1)a^2}\right]$
Outer edges fixed. Uniform load over concentric circular area of radius r_0. 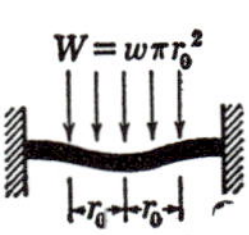 	At q, $r < r_0$: $\sigma_r = -\frac{3W}{2\pi m t^2}\left[(m+1)\log\frac{a}{r_0} + (m+1)\frac{r_0^2}{4a^2} - (3m+1)\frac{r^2}{4r_0^2}\right]$ $\sigma_t = -\frac{3W}{2\pi m t^2}\left[(m+1)\log\frac{a}{r_0} + (m+1)\frac{r_0^2}{4a^2} - (m+3)\frac{r^2}{4r_0^2}\right]$ $y = -\frac{3W(m^2-1)}{16\pi E m^2 t^3}\left[4a^2 - (8r^2+4r_0^2)\log\frac{a}{r_0} - \frac{2r^2r_0^2}{a^2} + \frac{r^4}{r_0^2} - 3r_0^2\right]$ At q, $r > r_0$: $\sigma_r = -\frac{3W}{2\pi m t^2}\left[(m+1)\log\frac{a}{r} + (m+1)\frac{r_0^2}{4a^2} + (m-1)\frac{r_0^2}{4r^2} - m\right]$ $\sigma_t = -\frac{3W}{2\pi m t^2}\left[(m+1)\log\frac{a}{r} + (m+1)\frac{r_0^2}{4a^2} - (m-1)\frac{r_0^2}{4r^2} - 1\right]$ $y = -\frac{3W(m^2-1)}{16\pi E m^2 t^3}\left[4a^2 - (8r^2+4r_0^2)\log\frac{a}{r} - \frac{2r^2r_0^2}{a^2} - 4r^2 + 2r_0^2\right]$ At center: $\sigma_r = \sigma_t = -\frac{3W}{2\pi m t^2}\left[(m+1)\log\frac{a}{r_0} + (m+1)\frac{r_0^2}{4a^2}\right] = \max \sigma_r$ when $r_0 < 0.588a$ $\max y = -\frac{3W(m^2-1)}{16\pi E m^2 t^3}\left[4a^2 - 4r_0^2\log\frac{a}{r_0} - 3r_0^2\right]$
Outer edges fixed. Uniform load on concentric circular ring of radius r_0. 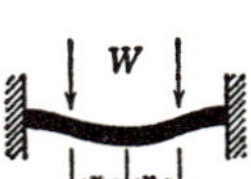 	At q, $r < r_0$: $\sigma_r = \sigma_t = -\frac{3W}{4\pi m t^2}\left[(m+1)\left(2\log\frac{a}{r_0} + \frac{r_0^2}{a^2} - 1\right)\right] = \max \sigma$ when $r < 0.31a$ $y = -\frac{3W(m^2-1)}{2\pi E m^2 t^3}\left[\frac{1}{2}\left(1 + \frac{r_0^2}{a^2}\right)(a^2-r^2) - (r^2+r_0^2)\log\frac{a}{r_0} + (r^2-r_0^2)\right]$ At q, $r > r_0$: $\sigma_r = -\frac{3W}{4\pi m t^2}\left[(m+1)\left(2\log\frac{a}{r} + \frac{r_0^2}{a^2}\right) + (m-1)\frac{r_0^2}{r^2} - 2m\right]$ $\sigma_t = -\frac{3W}{4\pi m t^2}\left[(m+1)\left(2\log\frac{a}{r} + \frac{r_0^2}{a^2}\right) - (m-1)\frac{r_0^2}{r^2} - 2\right]$ $y = -\frac{3W(m^2-1)}{2\pi E m^2 t^3}\left[\frac{1}{2}\left(1 + \frac{r_0^2}{a^2}\right)(a^2-r^2) - (r^2+r_0^2)\log\frac{a}{r}\right]$ At center: $\max y = -\frac{3W(m^2-1)}{2\pi E m^2 t^3}\left[\frac{1}{2}(a^2-r_0^2) - r_0^2\log\frac{a}{r_0}\right]$

(*Table continued on p. 8-34*)

Table 9. Formulas for Flat Plates—*Continued*

Type of Load and Support	Formulas for Stress and Deflection
	Circular Flat Plates with Concentric Circular Hole 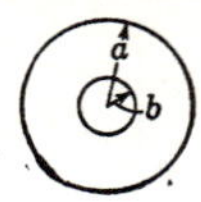
Outer edge supported. Uniform load over entire surface. 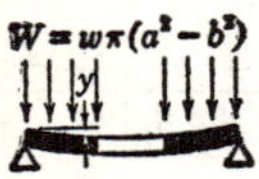	At inner edge: $\max \sigma = \sigma_t = -\frac{3w}{4mt^2(a^2-b^2)}\left[a^4(3m+1)+b^4(m-1)-4ma^2b^2-4(m+1)a^2b^2\log\frac{a}{b}\right]$ $\max y = -\frac{3w(m^2-1)}{2m^2Et^3}\left[\frac{a^4(5m+1)}{8(m+1)}+\frac{b^4(7m+3)}{8(m+1)}-\frac{a^2b^2(3m+1)}{2(m+1)}+\frac{a^2b^2(3m+1)}{2(m-1)}\log\frac{a}{b}-\frac{2a^2b^4(m+1)}{(a^2-b^2)(m-1)}\left(\log\frac{a}{b}\right)^2\right]$
Outer edge supported. Uniform load along inner edge. 	At inner edge: $\max \sigma = \sigma_t = -\frac{3W}{2\pi mt^2}\left[\frac{2a^2(m+1)}{a^2-b^2}\log\frac{a}{b}+(m-1)\right]$ $\max y = -\frac{3W(m^2-1)}{4\pi Em^2t^3}\left[\frac{(a^2-b^2)(3m+1)}{(m+1)}+\frac{4a^2b^2(m+1)}{(m-1)(a^2-b^2)}\left(\log\frac{a}{b}\right)^2\right]$
Supported along concentric circle near outer edge. Uniform load along concentric circle near inner edge. 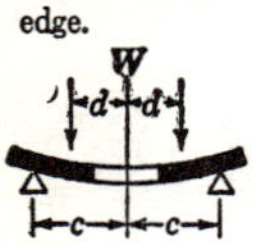	At inner edge: $\max \sigma = \sigma_t = -\frac{3W}{2\pi mt^2}\left[\frac{2a^2(m+1)}{a^2-b^2}\log\frac{c}{d}+(m-1)\frac{c^2-d^2}{a^2-b^2}\right]$
Inner edge supported. Uniform load over entire surface. 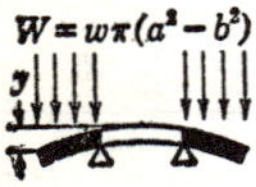	At inner edge: $\max \sigma = \sigma_t = \frac{3w}{4mt^2(a^2-b^2)}\left[4a^4(m+1)\log\frac{a}{b}+4a^2b^2+b^4(m-1)-a^4(m+3)\right]$ At outer edge: $\max y = \frac{3w(m-1)}{16Em^2t^3}\left[a^4(7m+3)+b^4(5m+1)-a^2b^2(12m+4)-\frac{4a^2b^2(3m+1)(m+1)}{(m-1)}\log\frac{a}{b}+\frac{16a^4b^2(m+1)^2}{(a^2-b^2)(m-1)}\left(\log\frac{a}{b}\right)^2\right]$
Outer edge fixed and supported. Uniform load over entire surface. 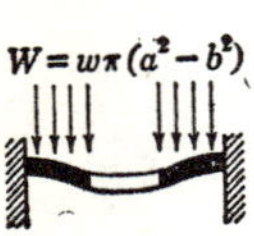	At outer edge: $\max \sigma_r = \frac{3w}{4t^2}\left[a^2-2b^2+\frac{b^4(m-1)-4b^4(m+1)\log\frac{a}{b}+a^2b^2(m+1)}{a^2(m-1)+b^2(m+1)}\right] = \nu \max \sigma$ At inner edge: $\max \sigma_t = -\frac{3w(m^2-1)}{4mt^2}\left[\frac{a^4-b^4-4a^2b^2\log\frac{a}{b}}{a^2(m-1)+b^2(m+1)}\right]$ $\max y = -\frac{3w(m^2-1)}{16m^2Et^3}\left[a^4+5b^4-6a^2b^2+8b^4\log\frac{a}{b}+\frac{\left\{[-8b^6(m+1)+4a^2b^4(3m+1)+4a^4b^2(m+1)]\log\frac{a}{b}-16a^2b^4(m+1)\left(\log\frac{a}{b}\right)^2+4a^2b^4-2a^4b^2(m+1)+2b^6(m-1)\right\}}{a^2(m-1)+b^2(m+1)}\right]$

Table 9. Formulas for Flat Plates—*Continued*

Type of Load and Support	Formulas for Stress and Deflection
	Circular Flat Plates with Concentric Circular Hole
Outer edge fixed and supported. Uniform load along inner edge. 	At outer edge: $\max \sigma_r = \frac{3W}{2\pi t^2}\left[1 - \frac{2mb^2 - 2b^2(m+1)\log\frac{a}{b}}{a^2(m-1)+b^2(m+1)}\right] = \max\sigma$ when $\frac{a}{b} < 2.4$ At inner edge: $\max \sigma_t = \frac{3W}{2\pi m t^2}\left[1 + \frac{ma^2(m-1) - mb^2(m+1) - 2(m^2-1)a^2\log\frac{a}{b}}{a^2(m-1)+b^2(m+1)}\right]$ $= \max\sigma$ when $\frac{a}{b} > 2.4$ $\max y = -\frac{3W(m^2-1)}{4\pi m^2 E t^3} \times \left[a^2 - b^2 + \frac{2mb^2(a^2-b^2) - 8ma^2b^2\log\frac{a}{b} + 4a^2b^2(m+1)\left(\log\frac{a}{b}\right)^2}{a^2(m-1)+b^2(m+1)}\right]$
Outer edge fixed. Uniform moment along inner edge. 	At inner edge: $\max \sigma_r = \frac{6M}{t^2}$ $\max y = \frac{6M(m^2-1)}{mEt^3}\left[\frac{a^2b^2 - b^4 - 2a^2b^2\log\frac{a}{b}}{a^2(m-1)+b^2(m+1)}\right]$ At outer edge: $\sigma_r = -\frac{6M}{t^2}\left[\frac{2mb^2}{(m+1)b^2+(m-1)a^2}\right]$
Outer edge supported. Unequal uniform moments along edges. 	At q: $\sigma_r = \frac{6}{t^2(a^2-b^2)}\left[a^2M_a - b^2M_b - \frac{a^2b^2}{r^2}(M_a - M_b)\right]$ $\sigma_t = \frac{6}{t^2(a^2-b^2)}\left[a^2M_a - b^2M_b + \frac{a^2b^2}{r^2}(M_a - M_b)\right]$ From outer edge level: $y = \frac{12(m^2-1)}{mEt^3(a^2-b^2)}\left[\frac{a^2-r^2}{2}\left(\frac{a^2M_a - b^2M_b}{m+1}\right) + \log\frac{a}{r}\left(\frac{a^2b^2(M_a-M_b)}{m-1}\right)\right]$

thin wall cylinders are no longer applicable. The maximum stresses can be computed from eqs. 50, 51, and 52.

$$\sigma_t = \frac{P_1r_1^2 - P_2r_2^2 + \frac{r_1^2r_2^2}{x^2}(P_1 - P_2)}{r_2^2 - r_1^2} \tag{50}$$

$$\sigma_r = \frac{P_2r_2^2 - P_1r_1^2 + \frac{r_1^2r_2^2}{x^2}(P_1 - P_2)}{r_2^2 - r_1^2} \tag{51}$$

With the maximum shear theory the criterion of failure, the maximum shear stress for the case of internal pressure only ($P_2 = 0$, $x = r_1$) is

$$\tau = P_1 \frac{r_2^2}{r_2^2 - r_1^2} \tag{52}$$

where σ_t = tangential stress, σ_r = radial stress, τ = shear stress, P_1 = internal pressure, P_2 = external pressure, r_1 = internal radius, r_2 = external radius, x = radial coordinate of any point in the wall; the minimum and maximum values for x are r_1 and r_2, respectively.

Table 10. Areas of Contact and Pressures with Two Surfaces in Contact

Poisson's ratio = 0.3; P = load, lb; P_1 = load per in. of length, lb; E = modulus of elasticity.

Character of Surfaces	Maximum Pressure, s, at Center of Contact, psi	Radius, r, or Width, b, of Contact Area, in.
Two spheres 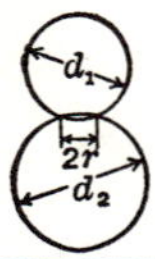	$s = 0.616 \sqrt[3]{PE^2 \left(\frac{d_1 + d_2}{d_1 d_2}\right)^2}$	$r = 0.881 \sqrt[3]{\frac{P}{E} \left(\frac{d_1 d_2}{d_1 + d_2}\right)}$
Sphere and plane 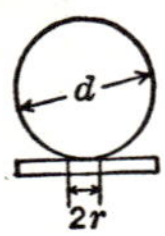	$s = 0.616 \sqrt[3]{\frac{PE^2}{d^2}}$	$r = 0.881 \sqrt[3]{\frac{Pd}{E}}$
Sphere and hollow sphere 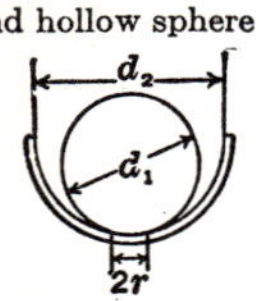	$s = 0.616 \sqrt[3]{PE^2 \left(\frac{d_2 - d_1}{d_1 d_2}\right)^2}$	$r = 0.881 \sqrt[3]{\frac{P}{E} \left(\frac{d_1 d_2}{d_2 - d_1}\right)}$
Cylinder and plane 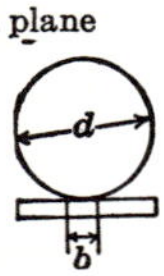	$s = 0.591 \sqrt{\frac{P_1 E}{d}}$	$b = 2.15 \sqrt{\frac{P_1 d}{E}}$
Two cylinders 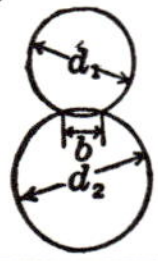	$s = 0.591 \sqrt{P_1 E \left(\frac{d_1 + d_2}{d_1 d_2}\right)}$	$b = 2.15 \sqrt{\frac{P_1}{E} \left(\frac{d_1 d_2}{d_1 + d_2}\right)}$
General case of two bodies in contact 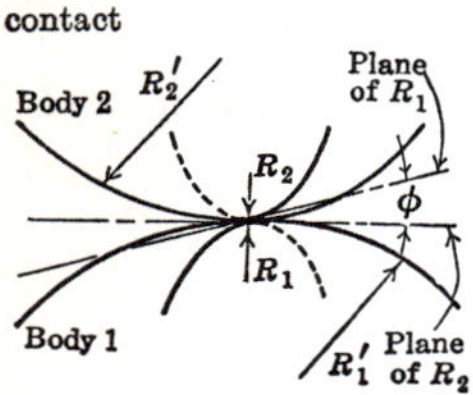	$s = \frac{1.5P}{\pi cd}$	$c = \alpha \sqrt[3]{\frac{P\delta}{K}}$ $d = \beta \sqrt[3]{\frac{P\delta}{K}}$ $\delta = \frac{4}{\frac{1}{R_1} + \frac{1}{R_2} + \frac{1}{R_1'} + \frac{1}{R_2'}}$ $K = \frac{8}{3} \frac{E_1 E_2}{E_2(1-\nu_1^2) + E_1(1-\nu_2^2)}$
	$\theta = \text{arc cos } \frac{1}{4} \delta \sqrt{\left\{ \left(\frac{1}{R_1} - \frac{1}{R_1'}\right)^2 + \left(\frac{1}{R_2} - \frac{1}{R_2'}\right)^2 + 2\left(\frac{1}{R_1} - \frac{1}{R_1'}\right)\left(\frac{1}{R_2} - \frac{1}{R_2'}\right) \cos 2\phi \right\}}$	

θ	0°	10°	20°	30°	40°	50°	60°	70°	80°	90°
α	∞	6.612	3.778	2.731	2.136	1.754	1.486	1.284	1.128	1.00
β	0	0.319	0.408	0.493	0.567	0.641	0.717	0.802	0.893	1.00

In *shrinkage fits* a hollow cylinder or collar is forced over a cylinder having an outer diameter slightly greater than the inside diameter of the outer cylinder. This will cause compressive stresses in the inner cylinder and tensile stresses in the outer cylinder. Timoshenko and Lessells (Ref. 8) give the equation for the pressure p, pounds per square inch, between the cylinders as

$$p = \frac{E\delta}{b}\frac{(b^2 - a^2)(c^2 - b^2)}{2b^2(c^2 - a^2)} \tag{53}$$

where E = modulus of elasticity, a = inner radius of inner cylinder, b = outer radius of inner cylinder, c = outer radius of outer cylinder, δ = amount by which the outer radius of the inner cylinder exceeds the inner radius of the outer cylinder. With the pressures known, external pressure on the inner cylinder and internal pressure on the outer cylinder, the stresses can be computed by means of eqs. 50 and 51.

PLATES. The formulas which apply for plates are based on the assumptions that the plate is flat, of uniform thickness, and of homogeneous isotropic material, thickness is not greater than one-fourth the least transverse dimension, maximum deflection is not more than one-half the thickness, all forces are normal to the plane of the plate, and the plate is nowhere stressed beyond the elastic limit. In Table 9 are formulas for deflection and stress for various shapes, forms of load and edge conditions. For further information see Ref. 9.

14. CONTACT STRESSES

The stresses caused by the pressure between elastic bodies are of importance in connection with the design or investigation of ball and roller bearings, trunnions, expansion rollers, track stresses, gear teeth, etc.

CONTACT STRESS THEORY. H. Hertz (Ref. 10) developed the mathematical theory for the surface stresses and the deformations produced by pressure between curved bodies, and the results of his analysis are supported by research. Formulas based on this theory give the maximum compressive stresses which occur at the center of the surfaces of contact, but do not consider the maximum subsurface shear stresses nor the maximum tensile stresses which occur at the boundary of the contact area. In Table 10 formulas are given for the elastic stress and deformation produced by bodies in contact. Numerous tests have been made to determine the bearing strength of balls and rollers, but there is difficulty in interpreting the results for lack of a satisfactory criterion of failure. One arbitrary criterion of failure is the amount of allowable plastic yielding. For further information on contact stresses see Refs. 9, 11, and 12.

REFERENCES

1. F. B. Seely, *Resistance of Materials*, 3rd ed., John Wiley and Sons, New York, 1947.
2. F. B. Seely, *Advanced Mechanics of Materials*, John Wiley and Sons, New York, 1932.
3. *Trans. Am. Soc. Civil Engrs.*, Vol. xcviii, 1933.
4. O. W. Eshbach (Editor), *Handbook of Engineering Fundamentals*, John Wiley and Sons, New York, 1936.
5. *Trans. Am. Soc. Civil Engrs.*, Vol. lxxxiii, 1919–1920.
6. Bach, *Elastizität u. Festigkeit.*
7. Young, *Bulletin* 4, School of Engineering Research, University of Toronto.
8. Timoshenko and Lessells, *Applied Elasticity*, Westinghouse Night School Press, 1925.
9. R. J. Roark, *Formulas for Stress and Strain*, 2d Ed., McGraw-Hill Book Co., New York, 1943.
10. H. Hertz, *Gesammelte Werke*, Vol. 1, Leipzig, 1895.
11. R. K. Allen, *Rolling Bearings*, Sir Isaac Pitman and Sons, London, 1945.
12. A. Palmgren, *Ball and Roller Bearing Engineering*, SKF Industries, Philadelphia, Pa., 1945.

DYNAMIC STRESSES

DYNAMIC STRESSES occur where the dimension of time is necessary in defining the loads. They include creep, fatigue, and impact stresses.

CREEP STRESSES occur when either the load or deformation progressively vary with time. They are usually associated with noncyclic phenomena.

FATIGUE STRESSES occur when the cyclic variation of either load or strain is coincident with respect to time.

IMPACT STRESSES occur from loads which are transient with time. The duration of the load application is of the same order of magnitude as the natural period of vibration of the specimen.

15. WORK AND RESILIENCE

EXTERNAL WORK: Let P = axial load, pounds, on a bar, producing an internal stress not exceeding the elastic limit; σ = unit stress produced by P, pounds per square inch; A = cross-sectional area, square inches; l = length of bar, inches; e = deformation, inches; E = modulus of elasticity; W = external work performed on bar, inch-pounds = $^1/_2\, Pe$. Then

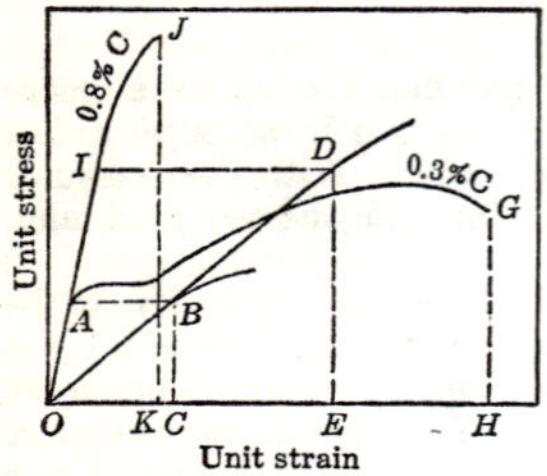

FIG. 1. Work areas on stress-strain diagram.

$$W = \frac{1}{2} A\sigma \left(\frac{\sigma l}{E}\right) = \frac{1}{2}\left(\frac{\sigma^2}{E}\right) Al \qquad (1)$$

The factor $(^1/_2)\ (\sigma^2/E)$ is the work required per unit volume, the volume being Al. It is represented on the stress-strain diagram by the area ODE or area OBC (Fig. 1), in which DE and BC are ordinates representing the unit stresses considered.

RESILIENCE is the strain energy which may be recovered from a deformed body when the load causing the stress is removed. Within the proportional limit, the resilience is equal to the external work performed in deforming the bar, and may be determined by eq. 1. When σ is equal to the proportional limit, the factor $(^1/_2)(\sigma^2/E)$ is the *modulus of resilience*, i.e., the measure of capacity of a unit volume of material to store strain energy up to the proportional limit. Average values of the modulus of resilience under tensile stress are given in Table 1.

Table 1. Average Values of Tensile Modulus of Resilience and Toughness

Material	Modulus of Resilience, in.-lb per cu in.	Relative Toughness (Area under Curve of Stress-deformation Diagram)
Gray cast iron	1.2	70
Malleable cast iron	17.4	3,800
Wrought iron	11.6	11,000
Low-carbon steel	15.0	15,700
Medium-carbon steel	34.0	16,300
High-carbon steel	94.0	5,000
Ni-Cr steel, hot-rolled	94.0	44,000
Vanadium steel, 0.98% C, 0.2% V, heat-treated	260.0	22,000
Duralumin, 17 ST	45.0	10,000
Rolled bronze	57.0	15,500
Rolled brass	40.0	10,000
Oak	2.3 *	13 *

* Bending.

The total resilience of a bar is the product of its volume and the modulus of resilience. These formulas for work performed on a bar, and its resilience, do not apply if the unit stress is greater than the proportional limit.

WORK REQUIRED FOR RUPTURE. Since beyond the proportional limit the strains are not proportional to the stresses, $^1/_2\, P$ does not express the mean value of the force acting. Equation 1, therefore, does not express the work required for strain after the proportional limit of the material has been passed, and cannot express the work required for rupture. The work required per unit volume to produce strains beyond the proportional limit or to cause rupture may be determined from the stress-strain diagram as it is measured by the area included between the axis of abscissas, and the stress-strain curve up to the strain in question, as $OAGH$ or OJK (Fig. 1). This area, however, does not represent the resilience, since part of the work done on the bar is present in the form of hysteresis losses and cannot be recovered.

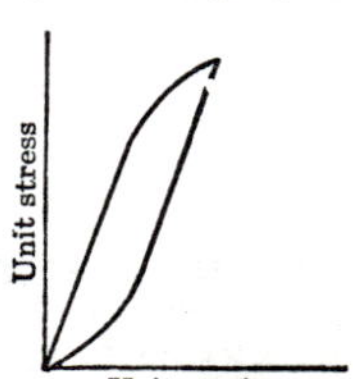

FIG. 2. Hysteresis loop for loading and unloading.

DAMPING CAPACITY (HYSTERESIS). Observations show that when a tensile load is applied to a bar it does not produce the complete elongation immediately, but there is a definite time lapse which depends on the nature of the material and the magnitude of the stresses involved. In parallel with this it is also noted that, upon unloading, complete recovery of energy does not occur. This phenomenon is variously termed *elastic hysteresis* or, for vibratory stresses, damping. Figure 2 shows a typical hysteresis loop obtained for one cycle of loading. The area of this hysteresis loop, representing the energy dissipated per cycle, is a measure of the damping properties of the material. While the exact mechanism of damping has not been fully investigated, it has been found that under vibratory conditions the energy dissipated in this manner varies approximately as the cube of the stress.

16. IMPACT STRESSES IN BARS AND BEAMS

EFFECT OF SUDDEN LOADS. If a sudden load P is applied to a bar, it will cause a deformation el, and the work done by the load will be Pel. Since the external work equals the internal work, $Pel = \sigma^2 Al/2E$, and since $e = \sigma/E$, $P = \sigma A/2$, or $\sigma = 2P/A$. The unit stress and also the unit strain are double those obtained by an equal load applied gradually. However, the bar does not maintain equilibrium at the point of maximum stress and strain. After a series of oscillations, however, in which the surplus energy is dissipated in damping, the bar finally comes to rest with the same strain and stress as that due to the equal static load.

STRESS DUE TO LIVE LOADS. In structural design two loads are considered, the dead load or weight of the structure and the live load or superimposed loads to be carried. The stresses due to the dead load and to the live load are computed separately, each being regarded as a static load. It is obvious that the stress due to the live load may be greatly increased, depending on the suddenness with which the load is applied. It has been shown above that the stress due to a suddenly applied load is double the stress caused by a static load. The term *coefficient of impact* is used extensively in structural engineering to denote the number by which the computed static stress is multiplied to obtain the value of the increased stress assumed to be caused by the suddenness of the application of the live load. If σ = static unit stress computed from the live load, and i = coefficient of impact, then the increase of unit stress due to sudden loading is $i\sigma$, and the total unit stress due to live load is $\sigma + i\sigma$. The value of i has been determined by empirical methods and varies according to different conditions.

In the building codes of most cities, specified floor loadings for buildings include the impact allowance, and no increase is needed for live loads except for special cases of vibration or other unusual conditions. For railroad bridges, the value of i depends upon the proportion of the length of the bridge which is loaded. No increase in the static stress is needed when the mass of the structure, as in monolithic concrete, is great. For machinery and for unusual conditions, such as elevator machinery and its supports, each structure should be considered by itself and the coefficient assumed accordingly. It should be noted that the meaning of the word impact used above differs somewhat from its strict theoretical meaning and as it is used in the next paragraph. The use of the terms impact and coefficient of impact in connection with live load stresses is, however, very general.

AXIAL IMPACT ON BARS. A load P dropped from a height h onto the end of a vertical bar of cross-sectional area A, rigidly secured at the bottom end, produces in the bar a unit stress which increases from 0 up to σ', with a corresponding total strain increasing from) up to e_1. The work done on the bar is $P(h + e_1)$, which, provided no energy is expended in hysteresis losses or in giving velocity to the bar, is equal to the energy $^1/_2\, \sigma' A e_1$ stored in the bar; that is,

$$P(h + e_1) = {}^1/_2\, \sigma' A e_1 \tag{2}$$

If e = strain produced by a static load P, within the proportional limit

$$\frac{e}{e_1} = \frac{P/A}{\sigma'} \tag{3}$$

Combining this with eq. 2 gives

$$\sigma' = \sigma + \sigma \sqrt{1 + 2\frac{h}{e}} \tag{4}$$

$$e_1 = e + e\sqrt{1 + 2\frac{h}{e}} \tag{5}$$

A wrought-iron bar 1 in. square and 5 ft long under a static load of 5000 lb will be shortened about 0.012 in., assuming no lateral flexure to occur; but, if a weight of 5000 lb drops on its end from a height of 0.048 in., a stress of 20,000 lb will be produced.

Equations 4 and 5 give values of stress and strain that are somewhat high because part of the energy of the applied force is not effective in producing stress, but is expended in overcoming the inertia of the bar and in producing local stresses. For light bars they give approximately correct results.

If the bar is horizontal and is struck at one end by a weight P, moving with a velocity V. the strain produced is e_1. Then, as before, $^1/_2\, \sigma' A e_1 = Ph$. In this case $h = V^2/2g$ =

height from which P would have to fall to acquire velocity V (g = acceleration due to gravity = 32.16 ft per sec²). Combining with eq. 3,

$$\sigma' = \sigma \sqrt{2\frac{h}{e}} \tag{6}$$

$$e_1 = e \sqrt{2\frac{h}{e}} \tag{7}$$

IMPACT ON BEAMS. If a weight P falls on a horizontal beam from a height h, producing a maximum deflection y and a maximum unit stress σ' in the extreme fiber, the values of σ' and y are given by

$$\sigma' = \sigma + \sigma \sqrt{1 + 2\frac{h}{y}} \tag{8}$$

$$y_1 = y + y \sqrt{1 + 2\frac{h}{y}} \tag{9}$$

where σ = extreme fiber unit stress and y = deflection due to P, considered as a static load. The value of σ may be obtained from the flexure formula (p. 8-08); that of y from the proper formula for deflection under static load.

If a weight P moving horizontally with a velocity V strikes a beam (the ends of which are secured against horizontal movement), the maximum fiber unit stress and the maximum lateral deflection are given by

$$\sigma' = \sigma \sqrt{2\frac{h}{y}} \tag{10}$$

$$y_1 = y \sqrt{2\frac{h}{y}} \tag{11}$$

where σ and y are as before and h is height through which P would have to fall to acquire the velocity V. These formulas, like those for axial impact on bars, give results higher than those observed in tests, particularly if the weight of the beam is great. For further discussion, see Ref. 1, Chapter XI.

RUPTURE FROM IMPACT. Rupture may be caused by impact provided the load has the requisite velocity. The above formulas, however, do not apply since they are valid only for stresses within the proportional limit. It has been found that the dynamic properties of a material are dependent on volume, velocity of the applied load, and material condition. If the velocity of the applied load is kept within certain limiting values, the total energy values for static and dynamic conditions are identical. If the velocity is increased, the impact values are considerably reduced. For further information see Ref. 2.

17. STEADY VIBRATORY STRESSES IN BARS AND BEAMS

For steady vibratory stresses the deflection of the bar, or beam, will be increased by the dynamic magnification factor (see Section 9 of this volume). The relation is given by

$$\delta_{\text{dynamic}} = \delta_{\text{static}} \times \text{dynamic magnification factor}$$

An example of the calculating procedure for the case of no damping losses is

$$\delta_{\text{dynamic}} = \delta_{\text{static}} \times \frac{1}{1 - \left(\dfrac{\omega}{\omega_n}\right)^2} \tag{12}$$

where ω is the frequency of oscillation of the load and ω_n the natural frequency of oscillation of the bar. ω_n is determined by

$$\omega_n = \sqrt{\frac{3EIg}{L^3W}} \tag{13}$$

where E = modulus of elasticity, I = moment of inertia, g = acceleration of gravity, L = length of bar, and W = weight of oscillating load.

18. FATIGUE

PARTS SUBJECTED TO REPEATED STRESS. Where parts of a structure or machine are subjected to varying or repeated loads, the ordinary methods of computing stresses and of determining the strength of materials under static load conditions are not satisfactory. In general, more than one-half to ten million repetitions of loading are necessary before the parts come under this classification. Such parts are crankshafts, shafts carrying rotating members (as in motors and generators), turbine blades, valve parts, piston rods, floor beams in elevated railroads, etc.

NATURE OF FAILURE. The failure of a ductile member caused by repeated loads is not accompanied by noticeable yielding; the failure is a gradual or progressive fracture. The fracture seems to start at some point in the member at which the stress is much larger than the calculated stress. This high localized stress seems to cause, or at least is accompanied by, a small crack which gradually spreads as the stress is repeated until the whole member ruptures without measurable yielding. The fatigue failure of a ductile material is similar to that of a static failure of a brittle material.

DETERMINATION OF ENDURANCE LIMIT. The highest unit stress σ at which a material can be subjected to a very large number of repetitions of load N without evidence of failure is the *endurance limit* of the material. It is usually found for any given material by constructing a σ-N curve for that particular metal from data obtained by tests. A piece of metal is loaded so as to produce a given stress, the load being repeated until failure occurs. Other specimens are then tested at different values of the unit stress until enough points have been obtained to plot a curve. This curve will incline steeply at the higher stresses and gradually flatten out at the lower stresses until, for steel, it approaches asymptotically a limiting stress called the endurance limit. For alloys and materials subjected to corrosive conditions, the σ-N curve continues to slope gradually down; the endurance limit is usually set at that stress which corresponds to approximately 10×10^6 cycles. The most common test is reversed bending by the rotating beam method since this specimen is most easily made and the testing machines are the least costly. Tests are also made in direct tension, direct shear, direct compression, reversed torsion, and in various combinations of them. Typical σ-N curves are shown in Fig. 3.

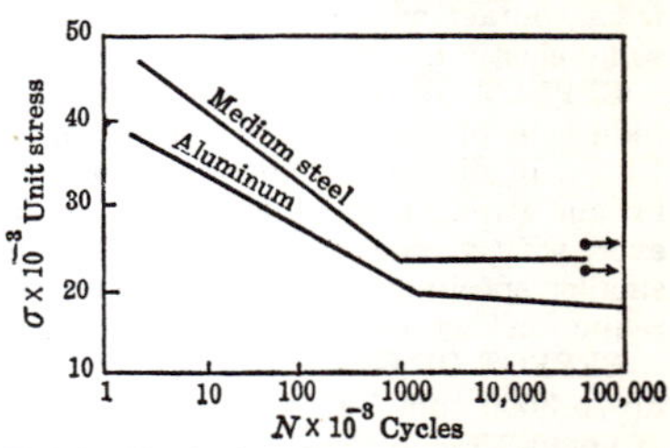

FIG. 3. Typical σ-N curves of fatigue resistance.

EFFECT OF SCRATCHES AND DEFECTS ON ENDURANCE LIMIT OF ROLLED OR FORGED FERROUS METALS. The beginning of rupture under repeated stress usually occurs at a point of discontinuity in the material. The same principle applies to a discontinuity on the outer surface where, in bending or torsion, the stress is greatest. The presence of scratches, sharp corners, or grooves reduces the endurance limit an appreciable amount. Moore and Kommers (Ref. 3) obtained the data given in Table 2. The tests were made on bars 0.40 in. in diameter. Horger and Niefert (Ref. 4) have made similar tests which corroborate the earlier work. It has also been found that the effect of scratches, grooves, etc., is not so pronounced in cast iron as in steel.

Table 2. Fatigue Endurance Limits

Data from *Bulletin* 124, University of Illinois Engineering Experiment Station.

0.49% C Quenched and Drawn 197 Brinell Steel

Finish	Endurance Limit, psi
High polish, longitudinal	51,000
00 Emery	48,000
Ground	45,000
Smooth turned	43,000
Rough turned	42,000

For keyways on shafts, Thum (Ref. 5) has determined that, for a soft steel of tensile strength 60,000 psi, an end-milled keyway reduced the strength to 20,500 psi, whereas a face-milled keyway reduced the strength to 28,000 psi.

Sharp corners (Ref. 6), oil holes (Ref. 7), press fits and surface rolling (Ref. 8), all produce marked reductions in the strength of the material. See also Article 7, Section 1 of this volume.

EFFECT OF RANGE OF STRESS. The endurance limit as determined by a rotating beam is the endurance limit for a complete reversal of stress from tension to compression and vice-versa. The endurance limit is less for a complete reversal than it is for those stress combinations where the stress goes only from a maximum to zero, or from a maximum to a minimum which is above zero. This is discussed further in Article 25 of this Section.

EFFECT OF CORROSION. The endurance limit of steels which have been treated (pickling) prior to use in an application of repeated stress with no further corrosion during stressing is affected considerably by roughening of the surface (i.e., production of pits and notches on the surface of the material). It is even more serious to have corrosion occur simultaneously with repeated stress. One explanation advanced is that a pit or notch formed by corrosion is opened up when tension is applied to the piece and it becomes filled with rust or other products of corrosion. When the tension is released and the pit closes down on the corrosion products it contains, these products exert a wedge action to produce cracking. The crack increases the local stress, and matters get progressively worse. Also, repeated stress tends to crack any protective coating of a film of corrosion products and thus allows continued corrosion at the cracks. This phenomenon is clearly outlined in the work of McAdam and Geil (Ref. 9).

Under corrosive conditions, steels have no endurance limit; the σ-N curve never becomes horizontal but continues to drop. Thus no corrodible steel can last indefinitely under corrosive conditions and repeated stress, no matter how small that stress is. (See also Ref. 10.)

The most effective method of protection is to keep the parts subjected to repeated loads from contact with corrosive agents. Where this is not possible, a protective coating of some elastic material such as varnish, cement, or enamel will aid. (See also Section 3.)

EFFECT OF SIZE. Tests made on specimens 0.3 in. in diameter and larger indicate a reduction of fatigue strength of approximately 10% for an increase in size from 0.3 to 1.0 in. in diameter. In specimens above 1 in. in diameter there was little variation in fatigue strength in geometrically similar specimens without stress concentration. However, with geometrically similar specimens having holes, fillets, and artificial cracks, the smaller specimens gave higher endurance limits than the larger ones (i.e., the strength-reduction factor is lower for the small specimens).

EFFECT OF SPEED. There seems to be little change in the fatigue strength for speeds up to 3000 rpm. However, from 3000 to 120,000 rpm, tests have indicated an increase of about 15% in the endurance limit and some increase in life.

EFFECT OF COLD WORKING. Cold drawing and cold working, as with static strength, increase the fatigue strength. The usefulness of this process is limited because of its tendency to introduce internal strains and minute cracks. Tests have indicated also that by understressing a specimen, that is, subjecting it to a certain safe range of stress before subjecting it to a higher range of stress, a permanent increase of as much as 30% has been obtained in the fatigue strength.

EFFECT OF TEMPERATURE. As the material is subjected to temperature above room temperature under repeated stress conditions, the fatigue strength drops. In one set of data reported the endurance limit dropped 10, 37, and 70% for temperature of 300, 400, and 500 F, respectively.

19. CREEP

INTRODUCTION. Materials subjected to a constant stress at elevated temperatures deform continuously with time, and the behavior under these conditions is different from the behavior at normal temperatures. This continuous deformation with time is called creep. In some applications the permissible creep deformations are critical; in others of no significance. But the existence of creep necessitates information on the creep deformations that may occur during the expected life of the machine.

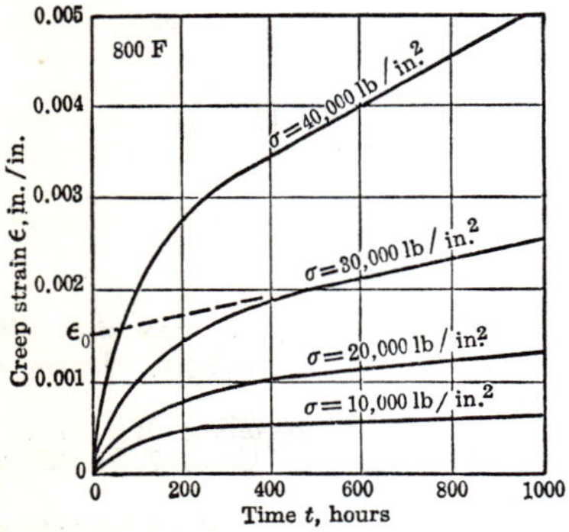

FIG. 4. Curves of creep strain for various stress levels.

MECHANISM OF CREEP FAILURE. There are generally four distinct phases distinguishable during the course of creep failure. The elapsed time per stage depends on the material, temperature, and stress condition. They are: (*a*) Initial phase—where the total deformation is partially elastic and partially plastic. (*b*) Second phase—where the creep rate decreases with time, indicating the effect of strain hardening. (*c*) Third phase—where the effect of strain hardening is counteracted by the annealing influence of the high temperature which produces a constant or minimum creep rate. (*d*) Final phase—where the creep rate increases until fracture occurs due to the decrease in cross-sectional area of the specimen.

CREEP EQUATIONS. In conducting a conventional creep test, curves of strain as a function of time are obtained for groups of specimens; each specimen in one group being subjected to a different constant stress, while all of the specimens in the group are tested at one temperature.

In this manner families of curves like those shown in Fig. 4 are obtained. Several methods have been proposed for the interpretation of such data. (See Ref. 11.) Two frequently used expressions of the creep properties of a material can be derived from the data in the following form:

$$C = B\sigma^m$$
$$\epsilon = \epsilon_0 + Ct \qquad (14)$$

where C = creep rate, B, m = experimental constants, σ = stress, ϵ = creep strain at any time t, ϵ_0 = zero time strain intercept, and t = time. See Fig. 5.

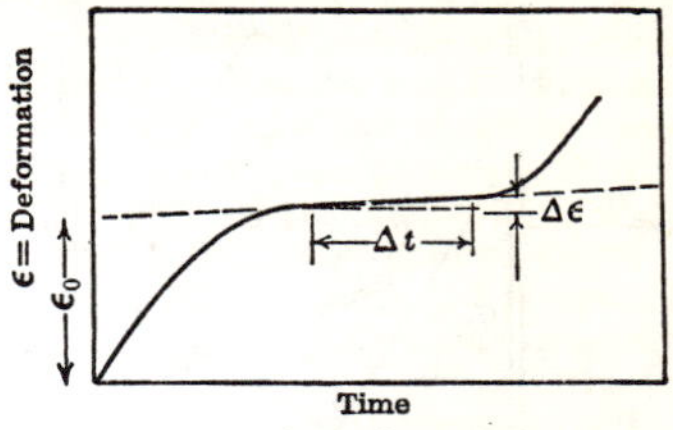

FIG. 5. Method of determining creep rate.

STRESS RELAXATION. Various types of bolted joints and shrink or press fit assemblies are applications of creep taking place with diminishing stress. This deformation tends to loosen the joint and produce a stress reduction or stress relaxation. The performance of a material to be used under diminishing creep-stress condition is determined by a tensile stress-relaxation test.

REFERENCES

1. F. B. Seely, *Resistance of Materials*, 3rd ed., John Wiley and Sons, New York, 1947.
2. H. C. Mann, articles, *Proc. Am. Soc. Testing Materials*, 1935, 1936, and 1937.
3. Moore and Kommers, *Bull.* 124, Engineering Experiment Station, University of Illinois.
4. Horger and Niefert, *Proc. Am. Soc. Testing Materials*, 1940.
5. Thum, *Trans. North East Coast Inst. Engrs. & Shipbuilders*, 1935.
6. Hendrickson, *SAE J.*, 1939.
7. Peterson and Wahl, *Trans. Am. Soc. Mech. Engrs.*, 1936.
8. Buckwalter and Horger, *Trans. Am. Soc. Metals*, 1937.
9. D. J. McAdam, Jr., and G. W. Geil, *J. Research Natl. Bur. Standards*, 1940.
10. Metallurgical Staff of Battelle Memorial Institute, *Prevention of the Failure of Metals under Repeated Stress*, John Wiley and Sons, New York, 1941.
11. J. Marin, *Mechanical Properties of Materials and Design*, McGraw-Hill Book Co., New York, 1942.

WORKING STRESSES

20. COMBINED STRESS

Under certain circumstances of loading a body is subjected to a combination of tensile. compressive, and/or shear stresses. For example, a shaft which is simultaneously bent and twisted is subjected to combined stresses, namely, longitudinal tension and compression and torsional shear. For the purposes of analysis it is convenient to reduce such systems of combined stresses to a basic system of stress coordinates known as principal stresses. These stresses act on axes which differ in general from the axes along which the applied stresses are acting and represent the maximum and minimum values of the normal stresses for the particular point considered.

DETERMINATION OF PRINCIPAL STRESSES. The expressions for the principal stresses in terms of the stresses along the x and y axes are

$$\sigma_1 = \frac{\sigma_x + \sigma_y}{2} + \sqrt{\left(\frac{\sigma_x - \sigma_y}{2}\right)^2 + \tau_{xy}^2} \qquad (1)$$

$$\sigma_2 = \frac{\sigma_x + \sigma_y}{2} - \sqrt{\left(\frac{\sigma_x - \sigma_y}{2}\right)^2 + \tau_{xy}^2} \qquad (2)$$

$$\tau_1 = \pm\sqrt{\left(\frac{\sigma_x - \sigma_y}{2}\right)^2 + \tau_{xy}^2} \qquad (3)$$

where σ_1, σ_2, and τ_1 are the principal stress components and σ_x, σ_y, and τ_{xy} the calculated stress components, all of which are determined at any particular point (Fig. 1).

GRAPHICAL METHOD OF PRINCIPAL STRESS DETERMINATION—MOHR'S CIRCLE. Let the axes x and y be chosen to represent the directions of the applied normal and shearing stresses, respectively (Fig. 2). Lay off to suitable scale distances $OA = \sigma_x$, $OB = \sigma_y$, and $BC = AD = \tau_{xy}$. With point E as a center construct the circle DFC. Then OF and OG are the principal stresses σ_1 and σ_2 respectively, and EC the maximum shear stress τ_1. The inverse also holds—that is, given the principal stresses, σ_x and σ_y can be determined on any plane passing through the point.

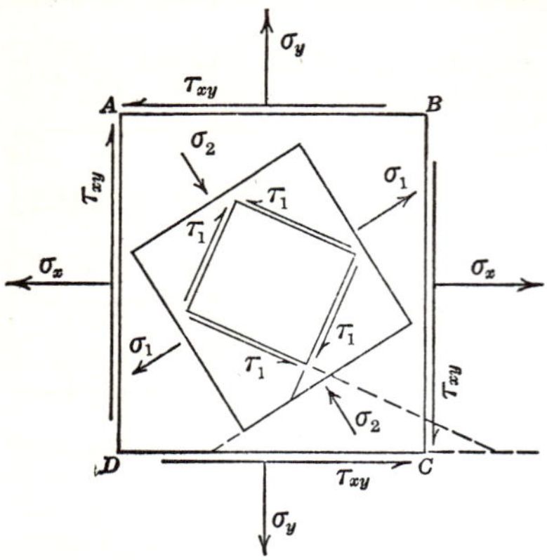

FIG. 1. Diagram showing relative orientation of stresses. (Reproduced by permission from *Mechanical Properties of Materials and Design*, by Joseph Marin, McGraw-Hill Book Co.)

STRESS–STRAIN RELATIONS. The linear relation between components of stress and strain is known as *Hooke's law*. This relation for the two-dimensional case can be expressed as

$$e_x = \frac{1}{E}(\sigma_x - \nu\sigma_y) \tag{4}$$

$$e_y = \frac{1}{E}(\sigma_y - \nu\sigma_x) \tag{5}$$

$$\gamma_{xy} = \frac{1}{G}\tau_{xy} \tag{6}$$

where σ_x, σ_y, and τ_{xy} are the stress components of a particular point, ν = Poisson's ratio, E = modulus of elasticity, G = modulus of rigidity, and e_x, e_y, and γ_{xy} = strain components.

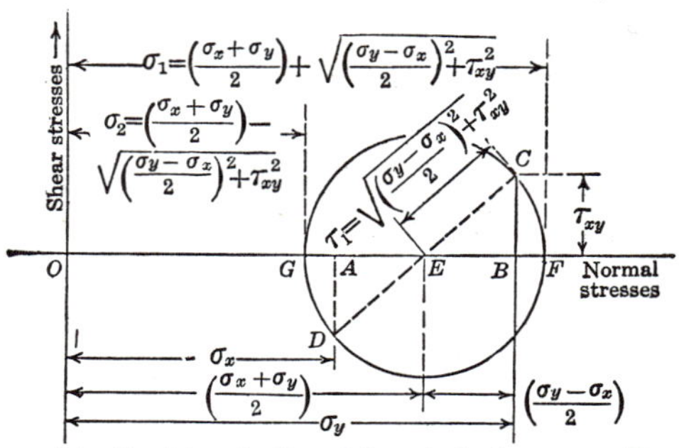

FIG. 2. Mohr's circle used for the determination of the principal stresses. (Reproduced by permission from Joseph Marin, *op. cit.*)

21. STRENGTH THEORIES

The determination of the magnitudes and directions of the principal stresses and strains and of the maximum shearing stresses is carried out for the purpose of establishing criteria of failure within the material under the anticipated loading conditions. To this end several theories have been advanced to elucidate these criteria. The more noteworthy ones are listed below. The theories are based on the assumption that the principal stresses do not change with time, an assumption that is justified since the applied loads in most cases are synchronous.

MAXIMUM-STRESS THEORY (RANKINE'S THEORY). This theory is based on the assumption that failure will occur when the maximum value of the greatest principal stress reaches the value of the maximum stress $\sigma_{\max}$ at failure in the case of simple axial loading. Failure is then defined as

$$\sigma_1 \text{ or } \sigma_2 = \sigma_{\max} \tag{7}$$

MAXIMUM-STRAIN THEORY (SAINT VENANT). This theory is based on the assumption that failure will occur when the maximum value of the greatest principal strain reaches the value of the maximum strain $e_{\max}$ at failure in the case of simple axial loading. Failure is then defined as

$$e_1 \text{ or } e_2 = e_{\max} \tag{8}$$

If e_{max} does not exceed the linear range of the material eq. 8 may be written as

$$\sigma_1 - \nu\sigma_2 = \sigma_{max} \tag{9}$$

MAXIMUM-SHEAR THEORY (GUEST). This theory is based on the assumption that failure will occur when the maximum shear stress reaches the value of the maximum shear stress at failure in simple tension. Failure is then defined as

$$\tau_1 = \tau_{max} \tag{10}$$

DISTORTION-ENERGY THEORY (HENCKY-VON MISES) (SHEAR ENERGY). This theory is based on the assumption that failure will occur when the distortion energy corresponding to the maximum values of the stress components equals the distortion energy at failure for the maximum axial stress. Failure is then defined as

$$\sigma_1^2 - \sigma_1\sigma_2 + \sigma_2^2 = \sigma_{max}^2 \tag{11}$$

STRAIN-ENERGY THEORY. This theory is based on the assumption that failure will occur when the total strain energy of deformation per unit volume in the case of combined stress is equal to the strain energy per unit volume at failure in simple tension. Failure is then defined as

$$\sigma_1^2 - 2\nu\sigma_1\sigma_2 + \sigma_2^2 = \sigma_{max}^2 \tag{12}$$

COMPARISON OF THEORIES. Figure 3 compares the five foregoing theories. In general the distortion-energy theory is the most satisfactory for ductile materials and the maximum-stress theory for brittle materials. The maximum-shear theory gives conservative results for both ductile and brittle materials. The conditions for yielding, according to the various theories, are given in Table 1, taking $\nu = 0.300$ as for steel.

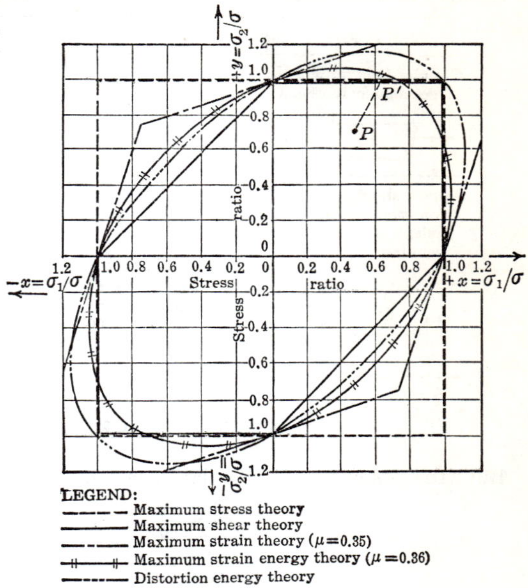

FIG. 3. Comparison of five theories of failure. (Reproduced by permission from Joseph Marin, *op. cit.*)

Table 1. Comparison of Stress Theories

$\tau = \sigma_{yp}$	(from the maximum-stress theory)
$\tau = 0.77\sigma_{yp}$	(from the maximum-strain theory)
$\tau = 0.50\sigma_{yp}$	(from the maximum-shear theory)
$\tau = 0.62\sigma_{yp}$	(from the maximum-strain-energy theory)

22. STRESS CONCENTRATIONS

The direct design procedure described above assumes no abrupt changes in cross section, discontinuities in the surface or holes, etc., through the member. In most structural parts this is not the case. The stresses produced at these discontinuities are different in magnitude from those calculated by the foregoing methods. The effect of the localized increase in stress, such as that caused by a notch, fillet, hole, or similar *stress raiser*, depends mainly on the type of loading, the geometry of the part, and the material. As a result, it is necessary to consider a stress-concentration factor K_t which is defined by the relationship

$$K_t = \frac{\sigma_{max}}{\sigma_{nominal}} \tag{13}$$

In general σ_{max} will have to be determined by the methods of experimental stress analysis or the theory of elasticity, and σ_{nominal} by a simple theory such as $\sigma = P/A$, $\sigma = Mc/I$, $\tau = Tc/J$ without taking into account the variations in stress conditions caused by geometrical discontinuities such as holes, grooves, and fillets. For ductile materials it is not customary to apply stress concentration factors to members under static loading. For brittle materials, however, stress concentration is serious and should be considered.

STRESS CONCENTRATION FACTORS FOR FILLETS, KEYWAYS, HOLES, AND SHAFTS. In Table 2 selected stress-concentration factors have been given from a complete table in Ref. 1.

Table 2. Stress-concentration Factors

(Adapted by permission from *Formulas for Stress and Strain* by R. J. Roark, McGraw-Hill Book Co.)

Type		K_t Factors					
Circular hole in plate or rectangular bar	$\frac{h}{a}$ =	0.67	0.77	0.91	1.07	1.29	1.56
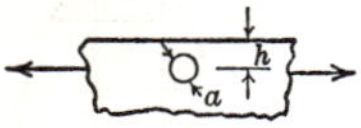	k =	4.37	3.92	3.61	3.40	3.25	3.16
Square shoulder with fillet for rectangular and circular cross sections in bending	$\frac{h}{r}$ \\ $\frac{r}{d}$	0.05	0.10	0.20	0.27	0.50	1.0
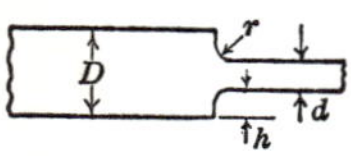	0.5	1.61	1.49	1.39	1.34	1.22	1.07
	1.0	1.91	1.70	1.48	1.38	1.22	1.08
	1.5	2.00	1.73	1.50	1.39	1.23	1.08
	2.0		1.74	1.52	1.39	1.23	1.09
	3.5		1.76	1.54	1.40	1.23	1.10

23. STATIC WORKING STRESSES

DUCTILE MATERIALS. For ductile materials the criteria for working stresses are

$$\sigma_w = \frac{\sigma_{yp}}{n} \qquad \text{(tension and compression)} \tag{14}$$

$$\tau_w = \frac{1}{2}\frac{\sigma_{yp}}{n} \tag{15}$$

BRITTLE MATERIALS. For brittle materials the criteria for working stresses are

$$\sigma_w = \frac{\sigma_{\text{ultimate}}}{K_t \times n} \qquad \text{(tension)} \tag{16}$$

$$\sigma_w = \frac{\sigma_{\text{compressive}}}{K_t \times n} \qquad \text{(compression)} \tag{17}$$

where K_t is the stress-concentration factor, n is the factor of safety, σ_w and τ_w are working stresses, and σ_{yp} is stress at the yield point.

WORKING-STRESS EQUATIONS FOR THE VARIOUS THEORIES:

Stress Theory

$$\sigma_w = \frac{\sigma_x + \sigma_y}{2} \pm \sqrt{\left(\frac{\sigma_x - \sigma_y}{2}\right)^2 + \tau_{xy}^2} \tag{18}$$

Shear Theory

$$\sigma_w = 2\sqrt{\left(\frac{\sigma_x - \sigma_y}{2}\right)^2 + \tau_{xy}^2} \tag{19}$$

Strain Theory

$$\sigma_w = (1 - \nu)\left(\frac{\sigma_x + \sigma_y}{2}\right) + (1 + \nu)\sqrt{\left(\frac{\sigma_x - \sigma_y}{2}\right)^2 + \tau_{xy}^2} \tag{20}$$

Distortion-energy Theory

$$\sigma_w = \sqrt{\sigma_x^2 - \sigma_x\sigma_y + \sigma_y^2 + 3\tau_{xy}^2} \tag{21}$$

Strain-energy Theory

$$\sigma_w = \sqrt{\sigma_x^2 - 2\nu\sigma_x\sigma_y + \sigma_y^2 + 2(1 + \nu)\tau_{xy}^2} \tag{22}$$

where σ_x, σ_y, τ_{xy} are the stress components of a particular point, ν is Poisson's ratio, and σ_w is working stress.

24. STRENGTH REDUCTION

If the loads acting on a member are cyclic with respect to time, the stress-concentration factor described in Article 22 can no longer be applied. It is now necessary to make fatigue tests on specimens to determine the strength-reduction factor K_f.

DEFINITION OF K_f:

$$K_f = \frac{\text{Endurance limit without stress concentration}}{\text{Endurance limit with stress concentration}} = \frac{\sigma_e}{\sigma_e'} \tag{23}$$

where σ_e = endurance limit stress. It has been found by test that generally for carbon steel $K_t > K_f$, but for alloy steel $K_t = K_f$. K_f has no meaning unless geometry, size, and material of the specimen are stated.

NOTCH SENSITIVITY INDEX. The notch sensitivity index q is a measure of the degree of agreement between K_f and K_t for a particular specimen, or member, of given size and material containing a stress concentrator of given size and shape:

$$q = \frac{K_f - 1}{K_t - 1} \tag{24}$$

Notch sensitivity varies between zero (where $K_f = 1$) and unity (where $K_f = K_t$). In general q is lower for cast alloys than for high-strength wrought materials.

25. FATIGUE WORKING STRESSES

The application of cyclic stress to a member changes the design calculations due to a reduction in the maximum allowable stress. (See Article 18.) In this case the strength theories discussed in Article 21 must be modified to take this into account.

DEFINITIONS:

Stress Cycle. A stress cycle is the smallest section of the stress-time function which is repeated identically and periodically as shown in Fig. 4.

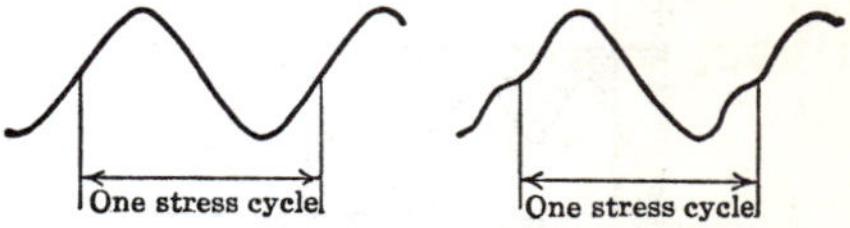

FIG. 4. Definition of one stress cycle.

Maximum Stress. $\sigma_{\max}$ is the greatest algebraic value of the stress in the stress cycle, being positive for a tensile stress and negative for a compressive stress.

Minimum Stress. $\sigma_{\min}$ is the smallest algebraic value of the stress in the stress cycle, being positive for a tensile stress and negative for a compressive stress.

Range of Stress. σ_r is the algebraic difference between the maximum and minimum stress in one cycle:

$$\sigma_r = \sigma_{\max} - \sigma_{\min} \tag{25}$$

For most cases of fatigue testing the stress varies about zero stress, but other types of variation may be experienced, as shown in Fig. 5.

Alternating-stress Amplitude (Variable Stress Component). σ_a is half the range of stress, $\sigma_a = \sigma_r/2$.

Mean-stress (Steady-stress Component). σ_m is the algebraic mean of the maximum and minimum stress in one cycle:

$$\sigma_m = \frac{\sigma_{\max} + \sigma_{\min}}{2} \tag{26}$$

Stress Ratio. R is the algebraic ratio of the minimum stress and the maximum stress in one cycle.

MODIFICATION OF THE STRENGTH THEORIES:

Stress Theory

$$\sigma_{\max} = (1 - P)\sigma_m \pm \sigma_e \tag{27}$$

where $P = \sigma_e/\sigma_{yp}$. This equation is for normal stresses only where σ_x and σ_y are principal stresses and $\tau_{xy} = 0$.

$$\pm\sigma_e = {}^1/_2 \{(\sigma_x' + \sigma_y') - (1 - P)(\sigma_x'' + \sigma_y'') \pm \sqrt{[(\sigma_y' - \sigma_x') - (1 - P)(\sigma_y'' - \sigma_x'')]^2 + 4[\tau_{xy}' - (1 - P)\tau_{xy}'']^2}\} \tag{28}$$

Shear Theory

$$\pm\sigma_e = \sqrt{[(1 - P)(\sigma_x'' - \sigma_y'') - (\sigma_x' - \sigma_y')]^2 + 4[(1 - P)\tau_{xy}'' - \tau_{xy}']^2} \tag{29}$$

Distortion-energy Theory

$$\sigma_e = \sqrt{(\sigma_x')^2 - \sigma_x'\sigma_y' + (\sigma_y')^2 + 3(\tau_{xy}')^2} - (1 - P)\sqrt{(\sigma_x'')^2 - \sigma_x''\sigma_y'' + (\sigma_y'')^2 + 3(\tau_{xy}'')^2} \quad (30)$$

Strain-energy Theory

$$\sigma_e = \sqrt{(\sigma_x')^2 - 2\nu\sigma_x'\sigma_y' + (\sigma_y')^2 + 2(1 + \nu)(\tau_{xy}')^2} - (1 - P)\sqrt{(\sigma_x'')^2 - 2\nu\sigma_x''\sigma_y'' + (\sigma_y'')^2 + 2(1 + \nu)(\tau_{xy}'')^2} \quad (31)$$

In the foregoing equations σ_e = endurance limit for complete reversal of stress (see Fig. 5),

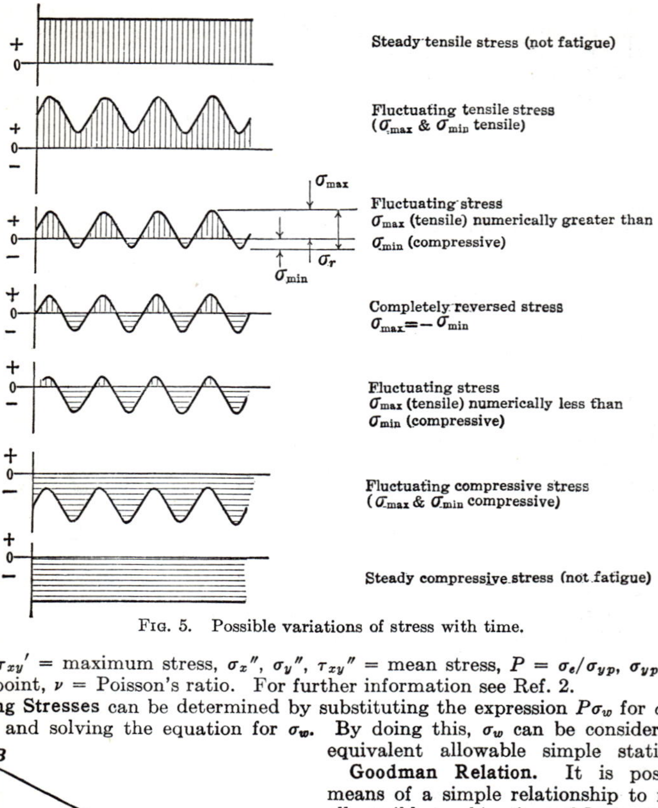

FIG. 5. Possible variations of stress with time.

σ_x', σ_y', τ_{xy}' = maximum stress, σ_x'', σ_y'', τ_{xy}'' = mean stress, $P = \sigma_e/\sigma_{yp}$, σ_{yp} = stress at yield point, ν = Poisson's ratio. For further information see Ref. 2.

Working Stresses can be determined by substituting the expression $P\sigma_w$ for σ_e in eqs. 27 to 31 and solving the equation for σ_w. By doing this, σ_w can be considered as an equivalent allowable simple static stress.

Goodman Relation. It is possible by means of a simple relationship to represent all possible combinations of fluctuating stress. It has been shown that the maximum tensile or compressive stress to which a member may be repeatedly subjected without causing fracture depends on the range of stress applied. The modified Goodman relation as shown in Fig. 6 can be used to predict either the cyclic stress σ_a or the mean stress σ_m which can be applied to the specified material. σ_m or σ_a for the respective cases must be known.

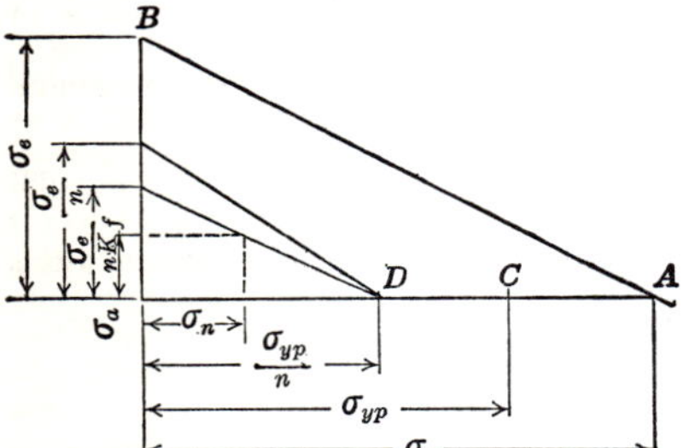

FIG. 6. Modified Goodman diagram.

26. EXPERIMENTAL STRESS ANALYSIS

The determination of static and dynamic strains in specimens and machine parts by means of physical measurements is defined as *experimental stress analysis*. Direct strain measurements are made possible by mechanical and electrical pickup devices. Mechanical and optical levers, brittle lacquers, photogrids, and cathetometers are mechanical systems. Electromagnetic, resistance, and capacitance methods are forms of electrical systems. Sometimes the nature of the installation and the operating conditions preclude direct strain measurements. Indirect methods such as high-speed photography of body displacements and velocity or acceleration measurements by means of moving coil pickups have proved to be acceptable. (See Section 18.)

DIRECT STRAIN MEASUREMENTS

Bonded wire strain gage (Fig. 7) consists of a grid of strain-sensitive wire protected by a paper sheath and cemented to the specimen. A change in the length of the grid due to strains in the specimen changes the resistance of the wire. This change in resistance is readily measured by means of suitable electrical circuits. The gage is suitable for measurement of dynamic as well as static strains.

Mechanical strain gage is used to measure static strains in a specimen under load. The instrument employs either the optical or the mechanical lever system to multiply the strain, which then may be read from a suitable scale.

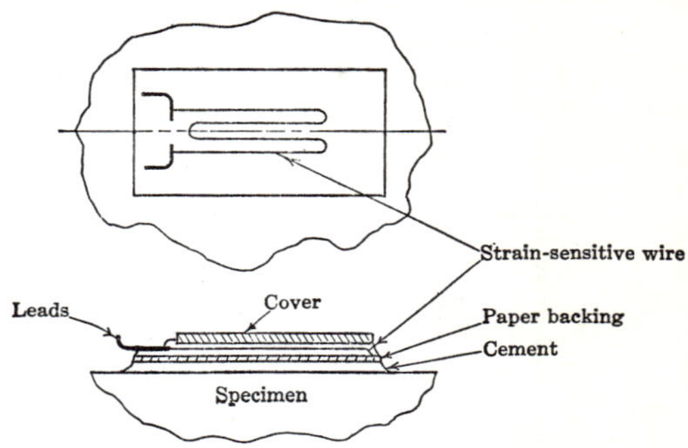

FIG. 7. Bonded wire strain gage.

Magnetic strain gage is based on the principle of a change in either self or mutual inductance of the gaging element. The simplest type is the variable inductance gage where the change in an air gap in a magnetic circuit varies the permeance of the circuit, thus giving an indication of the strains produced. See Fig. 8. This gage is used to measure both static and dynamic strains.

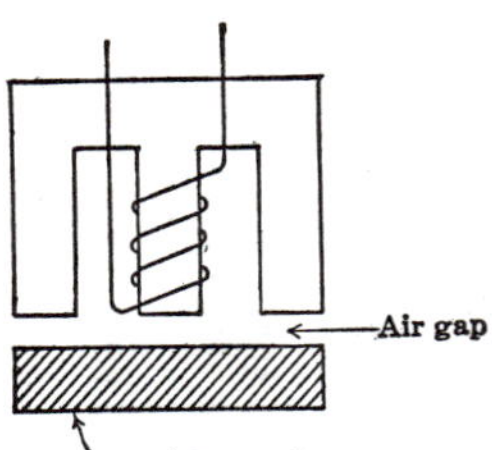

FIG. 8. Magnetic strain gage.

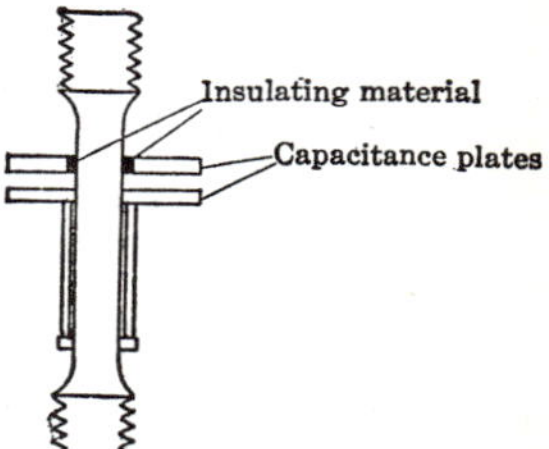

FIG. 9. Capacitance strain gage.

Capacitance strain gage is based on the principle of variation of capacitance by separation of elements, change in area of elements, or change in specific inductive capacity of a dielectric. The first of these methods (Fig. 9) is the most widely used, and the gage is so attached to the specimen that the strain produced causes a separation of the two plates. This type of gage is applicable for measuring both static and dynamic strains.

Acoustic strain gage employs a taut wire between two points mounted on the specimen, as shown in Fig. 10. The strain produced in the specimen alters the tension in the wire, thus changing its natural frequency. The natural frequency of the wire is determined by a variable frequency oscillator. Resonance is indicated by the output meter of the unit. By suitable methods the strain can be computed. This gage is suitable for measuring static and dynamic strains.

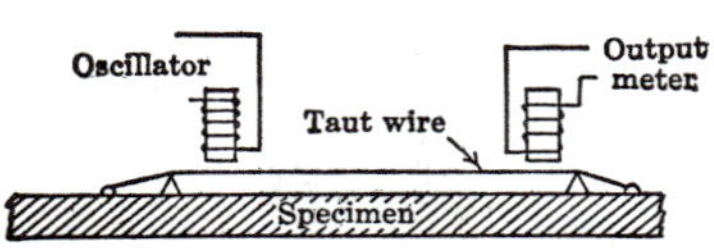

FIG. 10. Acoustic strain gage.

Brittle lacquer coatings are very convenient as they graphically present an overall strain picture, including the principal strain directions. They are revealed by cracks produced in the coating due to the strain in the specimen. The method consists of coating the specimen with lacquer, which becomes brittle on hardening. When the strains in the piece reach a prescribed magnitude, the lacquer cracks. This method, which gives quan-

(*Continued on p. 8-52*)

Table 3. Characteristics of Strain Measuring Devices

(Reprinted by permission from an article by Karl F. Smith in *Product Engineering*, Jan. 1947)

Name of Device	Type	Principle	Type Strains Best Adapted (in Steels)	Specimen Types	Usual Gage Length, In.	Approx. Smallest Measurable Strain, microinches per inch	Approx. Range of Strains, %	Adaptability to Recording	Reference
"SR-4" Bonded Electric Strain Gage	A-1	Electrical resistance	Elastic	Flats 0.5 × 1 in. or greater. Curved surfaces greater than 1 in. radius on length or 0.5 in. on width	1 (6 to 1/16 in. on other types)	0.5	1 (static loads)	Oscillograph, pen and ink recorders	Baldwin Locomotive Works, *Bulletin* 179
Tuckerman Strain Gage	0.2 in. lozenge	Optical lever	Elastic	Almost any round or flat surface, with suitable attachment	1 or 2	2 on 0.2 in. lozenge	0.25 on 2 in. g. l. (without reset)	Not possible, except by special adaptation	American Instrument Company, *Bulletin* 2084
General Motors Gage		Grids and photo-electric cell	Elastic	Any round or flat, with suitable attachment	1/16 to 1/4	2		Oscillograph	Gadd and Van de Grift *Trans. A.S.M.E.*, Vol. 64, p. A-15, 1942.
Interferometer Types		Light wave interference	Elastic	Depending on adaptation		2	About 1	Special adaptation required	
Statham Unbonded Wire Resistance Gages		Electrical resistance	Elastic	Any surface, with suitable attachment	1/2 to 6	4 on 2 in. g. l.	Depending on attachment	Oscillograph	The Statham Laboratories, 9328 Santa Monica Blvd., Beverly Hills, Calif., *Catalog*.
Magnetic Gage	Westinghouse, small size	Magnetic	Elastic	Any round or flat, with suitable attachment	3	5	About 1	Graphic meter or oscillograph	Langer (article), *Experimental Stress Analysis*, Vol. I, No. II, pp. 82–89
Whittemore	10 in. gage length	Mechanical lever	Elastic	Flats of over 10 in. in length	10	10	1	None	Baldwin Locomotive Works, *Bulletin* 167
Huggenberger Tensometer	A	Mechanical lever	Elastic	Any round or flat, with suitable attachment	1	10	1	None	Baldwin Locomotive Works, *Bulletin* 139
Templin	T-1	Mechanical lever (non-averaging)	Elastic	Rounds 0.05–0.505 in. dia., flats 0.01–0.50 in. thick × 0.50 in. wide	2	20 with 1000 magnification	2	Selsyn motors and rotating drum	Baldwin Locomotive Works, *Bulletin* 162

Peters	P-1	Mechanical lever (averaging)	Elastic	Rounds 0.505 in. dia.	2	20 with 1000 magnification	2	Selsyn motors and rotating drum	Baldwin Locomotive Works, *Bulletin* 162
Olsen Electronic	YUAAR	Magnetic	Elastic	Rounds or flats, 1/16 to 0.505 in.	2	20 with 1000 magnification	2	Special amplifier and revolving drum	Tinius Olsen Testing Machine Company, *Bulletin* 24
Porter-Lipp		Mechanical lever	Elastic	Any surface, with suitable attachment	1	20	0.5	Special adaptation required	P. L. Porter, 226 Oceano Drive, Los Angeles 24, California
Riehle	Averaging dial type	Dial gage	Elastic and early plastic (averaging)	Rounds or flats, 1/8 to 5/8 in.	2	100	2.5	Special adaptation required	American Machine and Metals, Inc., *Bulletins*
Berry Strain Gage		Mechanical lever	Elastic or plastic	Flats of over 10 in. in length	10	50 on 10 in. g. l.	About 1	None	Davis, Troxell, and Wiskocil
DeForest Scratch Gage		Mechanical trace	Elastic and early plastic	Any flat surface	2	100	2.5	Scratch pattern on special plate	Waugh Laboratories, 420 Lexington Ave., New York, N.Y., *Bulletin on Strain Measurements.*
Amsler		Dial gage	Elastic and early plastic	Rounds 0.02 to 1.5 in. dia. Flats to 1.5 in. wide and to 1.5 in. thick	2 to 10	100 on 2 in. g. l.	8	Special adaptation required	Alfred J. Amsler and Co., Schaffhouse, Switzerland, *Descriptive Bulletin.*
Kenyon-Burns-Young	10 : 1 wedge	Wedge	Plastic	Flats up to 1/8 in. thick and up to 1/2 in. wide	2	500 with 20 magnification	45	Line from wedge around drum	Baldwin Locomotive Works, *Bulletin* 128
Photogrids and Comparator		Microscope comparator	Plastic	Any surface	0.01 to 0.25	Depending on objective and test conditions	About 40	Photography of pattern	O'Haven and Harding (article) *Experimental Stress Analysis*, Vol. II, No. II, pp. 59–70
Micrometer Caliper		Graduated screw	Plastic	Rounds and flats for transverse measurements	0.5	500 on 0.505 in. round	No limit	Special adaptation required	....
Stresscoat	ST-103	Brittle coating	Elastic	Any surface	Very small	500	0.05 to 0.15	Photography of pattern	Magnaflux Corporation, ST-103 data sheet
Riehle Percentage Gage	2 in. gage length	Vernier	Plastic	Any surface prepared by punch marks	2	1000	70		American Machine and Metals, Inc., *bulletins*
Photoelasticity		Polarized light	Elastic	Transparent plastic model	Very small	Depends on model and testing conditions	Elastic region	Photography of pattern	Frocht, *Photoelasticity*, Vol. I, 1941, Vol. II, 1948, Wiley

titative results accurate to within ±10%, can be used to detect static and dynamic strains in tension or compression.

Photogrid method consists of coating the specimen with a photosensitive film on which a grid system is printed. The dimensional change in the grid system under load is a measure of the strain produced. This method is suitable for measuring permanent static strains only.

Cathetometer method consists of measuring the relative movement of two points on a test specimen under load by means of a microscope with a vernier scale attachment. Both static and dynamic strains can be measured by this method.

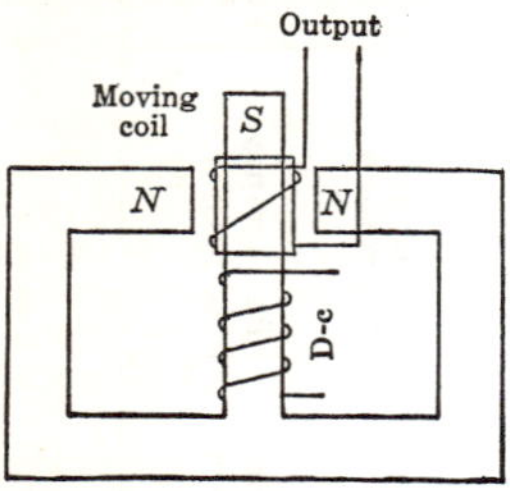

FIG. 11. Velocity pick-up.

INDIRECT STRAIN AND DISPLACEMENT MEASUREMENTS

Displacement pickups are used to detect the motion or displacement of the specimen. In general they consist of a seismic mass-spring system having suitable dynamic characteristics. The motion of the seismic element is used to produce an electrical signal by any of the electrical methods just described.

Velocity pickups usually employ the principle of a moving permanent magnet within a coil whose field strength is constant. The voltage induced in the coil is proportional to the relative velocity of the magnet and coil. By mechanical modification, the arrangement can serve as a generator where the voltage output is proportional to the velocity of the moving coil, as shown in Fig. 11. This pickup can be used to measure dynamic strains. By electrically integrating the output, a signal proportional to displacement can be obtained.

Acceleration can be detected by pickups similar to those used for displacement and velocity. The output can be integrated for convenience. In some designs the system consists of a piezoelectric crystal which, by virtue of its high natural frequency, permits the detection of sharp transients and high frequency accelerations.

Photoelastic methods involve the construction of a model of the prototype from an isotropic transparent material such as bakelite, celluloid, and glass. When these substances are subjected to stress they become doubly refracting, and plane polarized light on passing through the material is changed into two component mutually perpendicular rays which are parallel and perpendicular to the directions of stress. The emergent rays will be out of phase, depending on the stress and the thickness of the material. A photograph of the emergent rays gives light and dark fringes indicative of the distribution of the principal stresses. A schematic diagram of one method is shown in Fig. 12. This procedure is used to measure static stress.

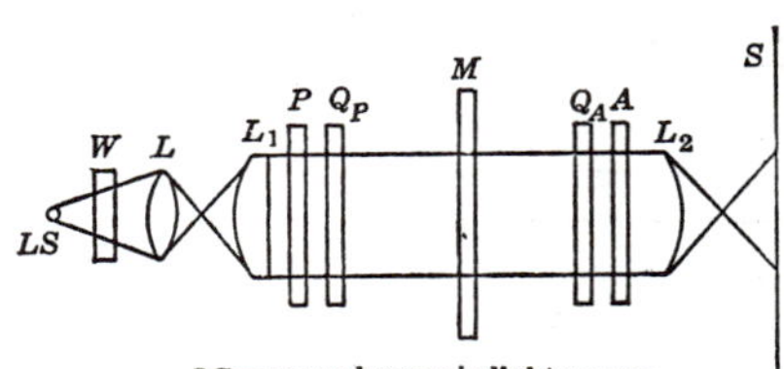

LS= monochromatic light source
L=collimating lense
L_1, L_2= lenses
P=polarizer
A=analyzer
Q_P, Q_A= quarter wave plates
M= stressed model
S=screen or camera

FIG. 12. Polariscope.

SUMMARY OF STRAIN MEASURING DEVICES. Table 3 presents a convenient summary of the more common commercial instruments. Detailed discussion of the various methods may be found in the references cited in the last column of Table 3, and also in Refs. 3 and 4.

REFERENCES

1. R. J. Roark, *Formulas for Stress and Strain*, 2d Ed., McGraw-Hill Book Co., New York, 1943.
2. J. Marin, *Mechanical Properties of Materials and Design*, McGraw-Hill Book Co., New York, 1942.
3. *Proc. Society for Experimental Stress Analysis.*
4. H. C. Roberts, *Mechanical Measurements by Electrical Methods*, Instruments Publishing Co.

TESTING OF MATERIALS

27. TESTING MACHINES

Machines for testing materials are of varying construction and capacity for testing specimens in tension, compression, torsion, impact, fatigue, and other specific purposes. Usually the mechanism for applying the load to the specimen is independent of the weighing apparatus. By this arrangement, the accuracy of the weighing is not affected by the deformation of the specimen which takes place as the load is applied.

TENSILE TESTING MACHINE, SCREW TYPE. This type of machine is in common use in the United States. It may be used for either tension or compression testing. The stress on the test piece is produced through the movement of platforms by screws, which, depending on the direction of turning, will either separate the platforms or draw them toward each other. A compound lever system transmits the platform force to a weighing beam kept in balance by moving the poise. Machines of this type are built by the Tinius Olsen Company in Philadelphia, Riehle Brothers in East Moline, Illinois, and Baldwin Southwark in Philadelphia.

TENSILE TESTING MACHINE, HYDRAULIC TYPE. A direct-acting hydraulic tensile testing machine has a head moved by hydraulic pressure, intensity of which is read on a pressure gage. The load on the test piece is determined by multiplying the pressure reading by a constant which depends upon the area of the cylinder. These machines are manufactured by Riehle in East Moline, Illinois, and Baldwin Southwark in Philadelphia.

TENSION CREEP-TESTING MACHINES. Since creep testing involves considerable time, most machines are built for multiple testing, and the tests can be carried out at any desired temperature. The test specimens are supported on knife edges at the top and loaded by lever systems. A common way to obtain the creep deformation is by a micrometer microscope.

TORSION-TESTING MACHINES. A torsion-testing machine consists essentially of a turning head which applies a torque to a test specimen. This torque is transmitted through a lever system to a hydraulic gage, the reading of which indicates the applied twisting force or torque. The angle of twist is measured directly on the turning head of the machine. The angle through which the specimen twists is recorded for each increment of torque until rupture occurs.

IMPACT-TESTING MACHINES:

Drop-weight Machines. The impact is obtained by a hammer of known weight falling from a predetermined height on the specimen. In one type of machine the absorption of kinetic energy by the specimen is determined from the deflection of a spring.

Charpy Pendulum Machine. This machine is used for testing small notched specimens for energy absorption (Fig. 1). The specimen is placed horizontally with the ends sup-

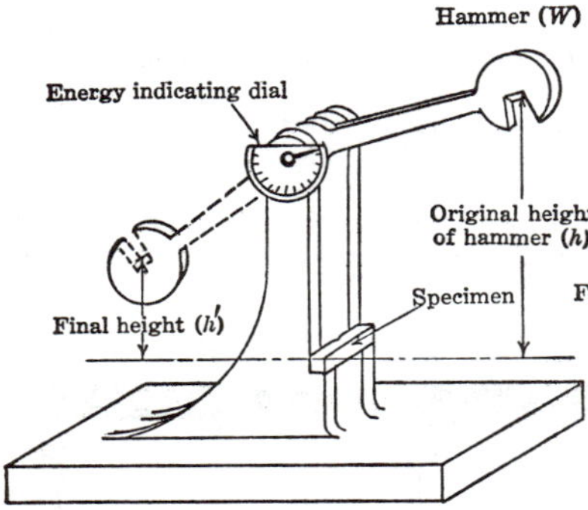

FIG. 1. Charpy impact testing machine.

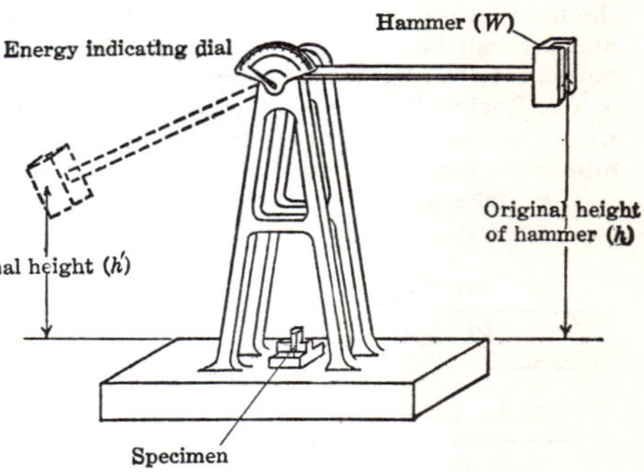

FIG. 2. Izod impact testing machine.

ported by an anvil. The hammer strikes the specimen midway between the supports opposite the notch. The energy expended in fracturing the specimen is the product of the weight of the pendulum and the difference between its height of fall and height of rise. This energy is expressed either in foot-pounds or in kilogram-meters.

The Izod machine is used for testing a notched specimen for energy absorption. The specimen is supported at its lower end in a vertical position as a cantilever (Fig. 2). The pendulum strikes the specimen at a definite distance above the notch on the same side as the notch. The energy used in breaking the specimen is the product of the weight of the pendulum and the difference between its height of fall and height of rise. This energy, expressed in foot-pounds, is known as the Izod value.

The rotary hammer impact machine is used to determine energy absorption of a test specimen. This is accomplished by means of a flywheel to which is attached a knife that can assume two positions, either flush with the rim of the wheel or projecting from the rim of the wheel. The latter position is required for breaking the test piece. The energy expended by the flywheel in breaking the specimen is measured by a tachometer tube.

The high-impact shock-testing machine consists of a large hammer hinged to swing in an arc (Fig. 3). It strikes the lower surface of a large anvil table on which equipment to

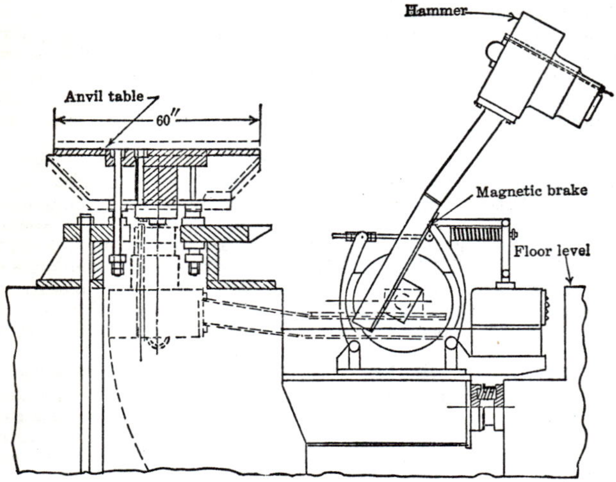

FIG. 3. High impact shock-testing machine for medium-weight equipment.

be shock tested is mounted. This machine is used for determining the vibratory motion and stresses set up in a piece of equipment under impact conditions.

HARDNESS-TESTING MACHINES (See also Section 1, Article 10):

The Brinell hardness-testing machine is used to measure the resistance of the material to the penetration of a hardened steel ball subjected to a standard load. The diameter of the indentation produced is measured with the aid of the microscope. A Brinell hardness number can be calculated or read from standard tables, and is used as a comparative measure of the hardness of materials.

The Rockwell hardness-testing machine is a direct-reading indentation type of machine where the ball or diamond point is impressed into the material. The Rockwell hardness number is read from the dial.

The Vickers hardness-testing machine follows the indentation principle, where, in this case, a pyramidal diamond indenter and standard load is used. A microscope measures the

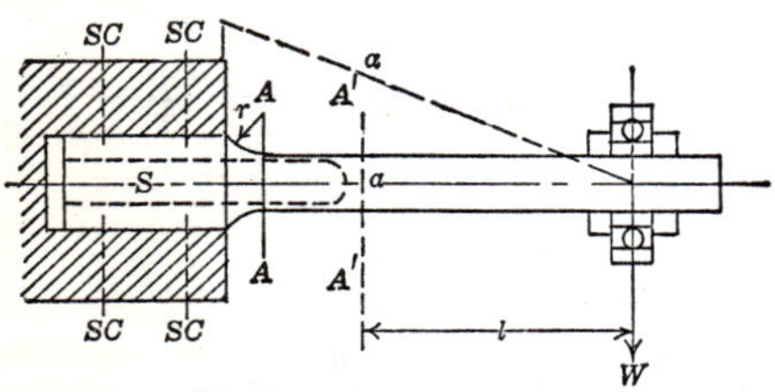

FIG. 4. Rotating cantilever bar fatigue test specimen.

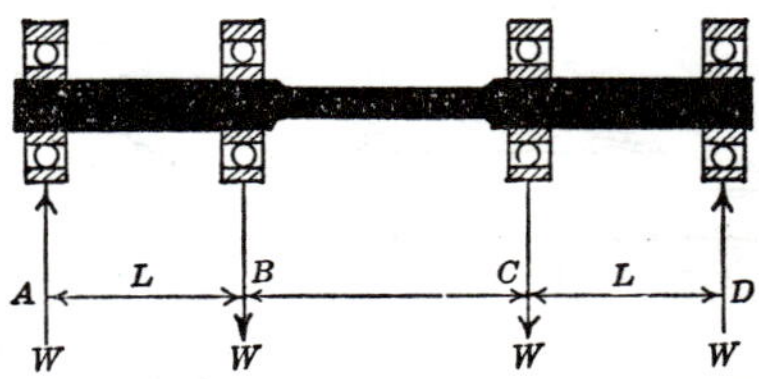

FIG. 5. Rotating beam bar fatigue test specimen.

diagonal of the indentation, and the Vickers hardness number is either computed or read from prepared charts.

The Shore scleroscope is a machine which measures the rebound of a falling weight dropped on the specimen. It is used primarily to give comparative values for checking uniformity of a product in production.

The Tukon hardness machine is a sensitive indenter type of machine of particular value in measuring hardness on individual grains and of certain small areas of specimens where the hardness of a zone having a certain microstructure is desired.

FATIGUE-TESTING MACHINES are classified as (*a*) rotary bending, (*b*) plane bending, (*c*) direct stress, and (*d*) torsional stress.

The rotary bending machine is the most widely used and the simplest type of fatigue-testing machine in which the specimen is in the form of a rotating bar. The rotating cantilever or the rotating beam form of loading may be applied. Figure 4 shows one of the modern forms of a rotating cantilever bar. The cantilever bar is end-loaded through a ball-bearing cage. In the rotating beam method the specimen is four-point loaded through ball bearings, as shown in Fig. 5. Machines are built to operate at speeds up to 30,000 rpm.

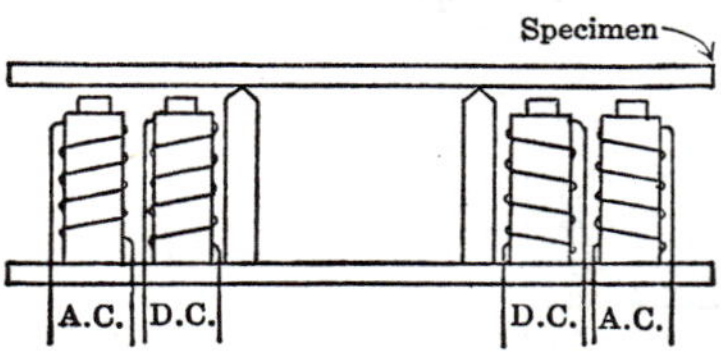

FIG. 6. Plane bending fatigue machine.

The plane bending machine subjects the specimen to reversed bending without rotation. The specimen is forced to vibrate by induced magnetism from electromagnets (Fig. 6). Frequencies up to 17,000 cpm have been employed.

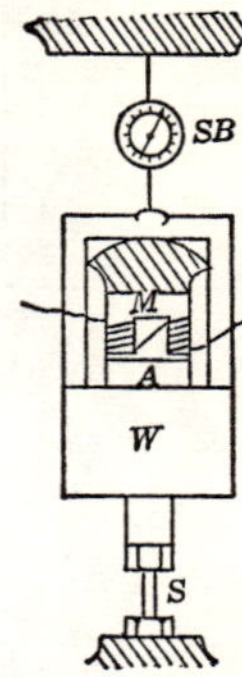

FIG. 7. Direct-stress fatigue machine.

The direct-stress method subjects the specimen to repeated tensile stress by means of the electromagnetic principle. A typical machine is shown in Fig. 7.

The torsional-stress machine applies repeated torsional stress to a specimen by means of inertia or resonant methods. Figure 8 shows a machine of the former type.

DAMPING-TESTING MACHINES. One type of machine used to determine the damping capacity of metals (Fig. 9) consists of a cylindrical specimen held rigidly at one end while a torque is applied to the opposite end. When the specimen is released, a record of its vibrations with time is made on sensitized photographic paper.

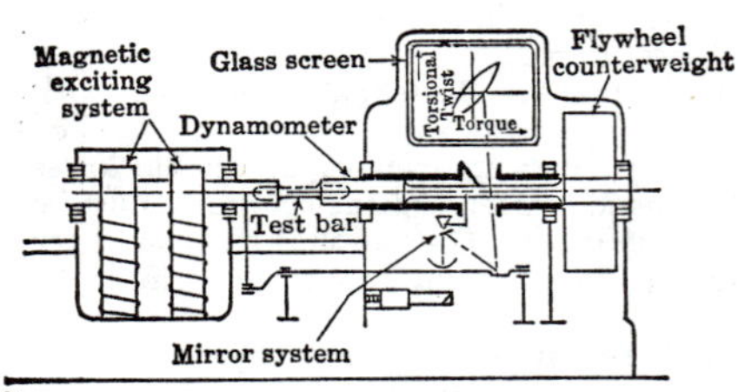

FIG. 8. Inertia torsional fatigue machine.

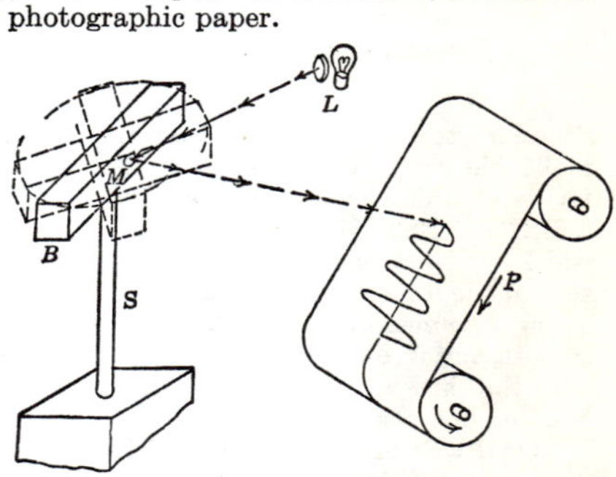

FIG. 9. Damping capacity machine.

28. STANDARD TEST PIECES

The fundamental data obtained in a test on material are affected by the method of testing and the size and shape of the specimen. To eliminate variations in results due to these causes, standards have been adopted by the ASTM, the ASME, and various associations of manufacturers. (See ASTM, ASME, etc., Standards for details.)

SPECIMENS FOR TENSION TESTS. Figure 10 shows a group of machined test bars in common use for testing steels and some nonferrous metals. The most common one, specimen *A*, is often referred to as the Standard 505 specimen. Its diameter is chosen to give a 0.2-square-inch area of section which is convenient in that the load in pounds on the specimen at any time during test can easily be converted to stress (pounds per square inch) by multiplying by 5. The subsize specimens (*B* to *E*) are used when the size of material available will not permit the use of specimen *A*. In all specimens, however.

it can be noted that the gage length (G.L.) over which the elongation measurements are taken is about four times the diameter of the reduced section. By keeping this similarity of specimens, the data obtained are reasonably comparable to those obtained on specimen *A*.

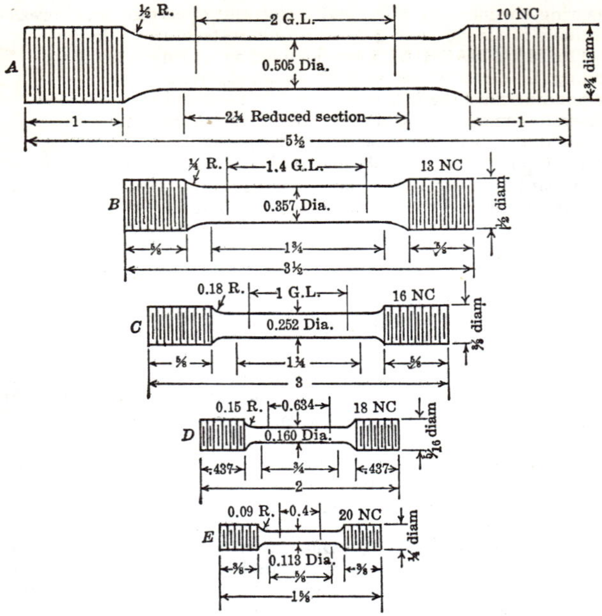

FIG. 10. Standard and subsize machined test specimens for cast, rolled, and forged steels, and nonferrous alloys.

Standard test bars for cast iron are shown in Fig. 3, p. 19-08, where specimen *A* is for malleable iron. Specimen *B* is the most common of the three standard test bars used for determining the properties of gray and alloy cast irons.

BEND-TEST SPECIMENS (ASTM Standards E16-39). Rectangular specimens with widths at least one and one-half times the thickness are used, the edges being smoothed so that the fracture will not take place at the edge. A radius up to one-eighth the thickness of the specimen may be used on the edges. The outside surfaces of the specimen must be smooth and free from local defects and transverse scratches. The length, which may vary with the thickness of the piece, is unimportant provided it is long enough to perform the bending operation.

COMPRESSION-TEST SPECIMENS (ASTM Tentative Standards E9-46T). It is recommended that standard compression-test specimens be in the form of circular cylinders and that the ends of the specimens shall be plain and normal to the longitudinal axis of the specimen. There are three types of compression-test specimens for metallic materials (short specimens, medium-length specimens, and long specimens). Short specimens are used for compression tests of such metals as bearing metals, which in service are used in the form of a thin plate or shell to carry load perpendicular to the surface. Medium-length specimens are used for determining the general compressive strength properties of metallic materials. Long specimens are best adapted for determining the modulus of elasticity in compression of metallic materials. Suggested dimensions for compression test specimens for general use are given in Table 1.

Table 1

	Diameter, *d*, in.	Length, *l*, in.
Short specimen	1 1/8 ± 0.01	1 ± 0.05
Medium-length specimens	0.798 ± 0.01	2 3/8 ± 1/8
	1 ± 0.01	3 ± 1/8
	1 1/8 ± 0.01	3 3/8 ± 1/8
Long specimens	0.798 ± 0.01	6 3/8 ± 1/8
	1 1/4 ± 0.01	12 1/2 min

IMPACT-TEST SPECIMENS (ASTM Tentative Standard E23-41T). The simple-beam (Charpy type) specimen and the cantilever-beam (Izod type) specimen are shown in Figs. 11 and 12, for use in the Charpy type and Izod type impact machines, respectively.

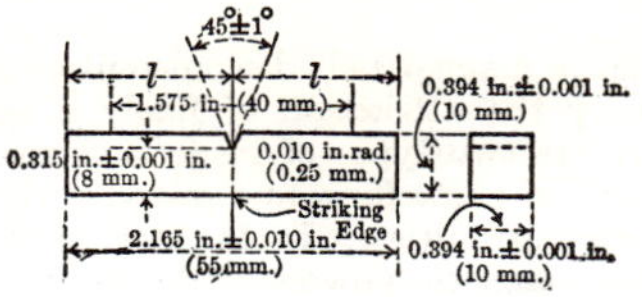

FIG. 11. Simple beam (Charpy) impact-test specimen.

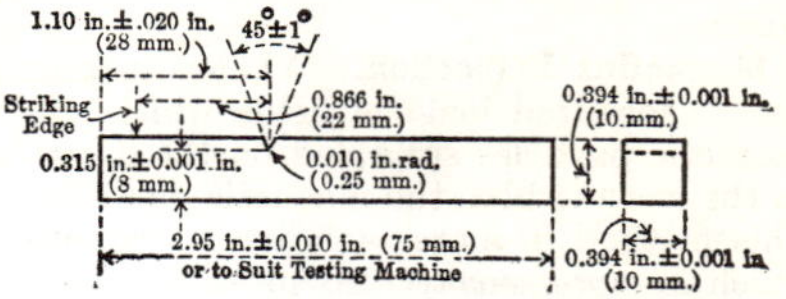

FIG. 12. Cantilever beam (Izod) impact-test specimen.

FATIGUE-TEST SPECIMENS. No specifications have been drawn up for standard fatigue test bars.

29. NONDESTRUCTIVE TESTING METHODS

Nondestructive testing is descriptive of test and inspection methods which, when applied to materials for purposes of detecting mechanical defects, variations in metallurgical condition, or composition do not impair the usefulness of the material or part.

SPARK TEST. The spark test is used to check material for chemical analysis. Since the sparks emitted from material when held to an abrasive wheel have a characteristic color, depending on the elements present in the steel, it is possible to classify unknown material rapidly by recognizing the color pattern produced.

RADIOGRAPHIC INSPECTION. The x-ray, because of the short wave length, can penetrate opaque materials; the depth of penetration depends upon the power of the x-ray beam. The sensitivity of the process is the ability of the method to reveal defects whose thickness or depth is expressed as a percentage of the metal thickness encountered. In other words, a sensitivity of 2% means that the defect revealed is $0.02t$ in depth, where t = metal thickness.

X-ray Radiography. When the x-ray passes through the material to the opposite side on which a special type sensitive film emulsion has been placed, a shadow picture is obtained wherein changes in density or thickness are revealed by light and dark areas. Voids, cracks, and holes appear as dark areas because they permit more rays to penetrate through them in a given time interval. This method of analysis is useful in checking welds, castings, forgings, plastic parts, and assemblies. Radiography can be used on light alloys, steel, other metals, and plastics. Under careful control a sensitivity of 2% is readily obtained which meets the requirements of the ASME boiler code. The standard equipment can be used for materials up to one foot thick.

Gamma-ray Radiography. Radium salts emit gamma-rays which are similar to x-rays but have a shorter wavelength, hence a greater penetrating power. Their advantage is that tests can be made without much expense since a capsule of radium salt is all that is required aside from the photographic equipment. The sensitivity is not so high as that of the x-ray for thicknesses less than 2 in. In general the sensitivity is approximately 3%.

Fluoroscopic Inspection. If the x-rays on emerging from the object being tested are allowed to fall on a fluorescent screen, an observer can view the shadow pattern directly. This procedure saves the time and expense of photographing the result. These units are particularly adaptable to conveyor feed systems. A sensitivity range of 5 to 15% is possible.

X-ray Diffraction. The nature and behavior of most substances depend on the arrangement of atoms and molecules in the structure. One of the properties of a material is its diffractive effect produced by x-radiation. Thus, if a beam of x-rays is passed through a crystal the beam will be bent, or redirected, into a series of emergent rays whose separation and intensities are characteristic of the material. With radiographic film a record is obtained of the location and intensity of the diffracted rays. These form a *fingerprint* of the substance, so called, because no two substances have been found to produce identical diffraction patterns. This fingerprint makes possible prediction of the behavior of substances when subjected to various treatments and also identification of what has happened to the substance up to the time of making the diffraction test.

This method is useful in determining the preferred crystal orientation, in deriving specifications for new materials, in discovering whether a steel has been adequately annealed for machining, in determining quality of case hardening, and in giving a quick and exact identification of a material and its prior processing.

MAGNETIC-PARTICLE INSPECTION. Although radiographic inspection will find deep-seated irregularities, other methods are required to find nonvisible, close-lipped discontinuities at or near the surface. The following types provide indications of cracks, semi-cracks, inclusions, and pores which may give rise to fatigue failures by acting as stress raisers.

Magnaflux Inspection. When a magnetic material is magnetized, discontinuities give rise to localized leakage fields which are revealed by the attraction of finely divided magnetic particles sprinkled on the material. There are two types of inspection media. In the wet method, the magnetic particles are prepared in paste form and mixed with oil. This method is more sensitive to minute surface discontinuities. In the dry method, which is more sensitive to subsurface discontinuities, magnetic powder is dusted on the magnetized material. Both direct and alternating currents are used for magnetization. The direct current is useful in finding significant subsurface discontinuities and is used for inspection of welds and castings. Because of *skin effect* the alternating current is limited to surface or very near surface discontinuities and is used for inspection of highly finished machine parts.

Magnaglo Inspection. This method employs fluorescent magnetic particles suspended in oil which give better contrast and extreme sensitivity. It is useful for material of complex contours such as splined shafts and milling cutters.

Zyglo Inspection. This fluorescent penetrant suspended in oil is used to determine discontinuities in nonmagnetic materials. Items like ball bearings are more easily inspected by this method. The material is usually dipped into the penetrant, which penetrates the very fine cracks. When viewed under "black" light the discontinuities are revealed.

SUPERSONIC INSPECTION. Longitudinal sound waves of extremely short wavelength travel through a given material at a velocity corresponding to the density and elastic properties of the substance. Thus a change in velocity or a loss of energy will occur when the sound beam strikes an irregularity such as a crack, separation, inclusion, or blowhole. By electrical pickup it is possible to detect an irregularity as small as 0.001 in.

Reflectoscope. Internal defects are detected by sending supersonic impulses from a crystal into the material and measuring the time required for these impulses to penetrate the material and be reflected from the opposite side of the defect. The operator views an oscilloscope on which the electrical impulses from the reflected vibrations are impressed. The scanning line indicates the initial impulse, any irregularities, and the back face reflection respectively. The main disadvantage is that the specimens should be finished to about 250 micro-inches rms in order to obtain satisfactory transmission of crystal vibration to the material. It is suitable for metals up to 28 ft thick, but plastics can be tested only to a thickness of 3 in. If plastics have low resin content, the depth is reduced to about $^1/_2$ in. because the material damps out the sound waves used in the test.

Brush Hypersonic Analyzer. This equipment is best suited to detect defects in plates, strips, and sheets. The ray is not reflected in this type, but pickup equipment is necessary on the opposite face. A pen and ink recorder is used to indicate changes in the sound energy picked up by the receiver. In general this test shows only the existence of the separation but not its nature or depth.

Sperry Supersonic Thruray. This equipment is used to compare production units with a standard unit for soundness of spot welds, the occurrence of laminations in sheet products, the quality of bond in bimetals, and the homogeneity of sintered materials. The principle of operation is similar to that described in the foregoing paragraph except that the units must be placed in a sound-transmitting fluid.

SPECTROGRAPHIC EXAMINATION. An accurate analysis of a material can be made rapidly either by spectrography or x-ray spectrometry. It is suitable for checking quality and structure of the material or as an aid in the development of materials and their control in manufacture. The metallic samples can be used as electrodes and an arc, or spark, passed between them. The arc spectrum is then analyzed for the component elements.

MAGNETIC ANALYSIS INSPECTION. The magnetic properties of irons and steels vary with their physical, chemical, and metallurgical properties. Therefore with suitable electronic devices it is possible to obtain comparative indications of structure, analysis, internal defects, case depth, and plating thickness. On bars and tubes defects 0.012 in. in depth can be detected.

Cyclograph. A change in structure which results in a significant change in physical properties or any distortion in the crystal lattice will produce some change in the magnetic and electrical properties of the metal. Internal stresses set up during quenching or cold working, residual stresses in shot-peened parts, and near surface variations in structure can be analyzed. The cyclograph does not reveal cracks and surface discontinuities.

For further information see the paper by Rupert Le Grand, Non-destructive Testing Methods, *American Machinist*, May 23, 1946.

SECTION 9

VIBRATION AND NOISE CONTROL

BY

R. O. FEHR, *General Engineering and Consulting Laboratory, General Electric Company*

VIBRATION

VIBRATION PREVENTION

NOISE

NOISE PREVENTION

Revision of material originally prepared by A. L. Kimball.

VIBRATION

FREE AND FORCED VIBRATIONS. RESONANCE. It is important to differentiate between free and forced vibrations. Free vibration occurs when the driving force is removed, dying out more or less gradually. Forced vibration is maintained by a driving force. The former has a frequency of its own; the latter can have any frequency, depending on that of the driving force. If the frequency of the driving force is in step with the natural or free period of the vibration, the forced vibration will build up to a large amplitude. The vibrating body is then in resonance with the exciting force.

Any body which vibrates freely when the exciting force is removed illustrates free vibration, i.e., in general, a weight vibrating on a spring. Free vibrations, as such, are not important in vibration prevention. Forced vibrations give trouble if an exciting stimulus is present. Resonant types, where the exciting force is in step with the free vibration period, are the most difficult. Knowledge of free vibration period thus is important. An example of a forced vibration is an unbalanced rotor which quivers at all speeds because of unbalanced centrifugal force; the frequency always equals the speed frequency. If speed is in step with the free vibration frequency of the rotor between bearings, vibration will reach a peak value and illustrate resonance. All critical speed vibrations are resonant. Calculation of critical speeds is the determination of free vibration periods to which the body resonates.

1. VIBRATION-FREQUENCY CALCULATION

The following simple cases of calculation of natural vibration frequency are given for reference. Two basic formulas are:

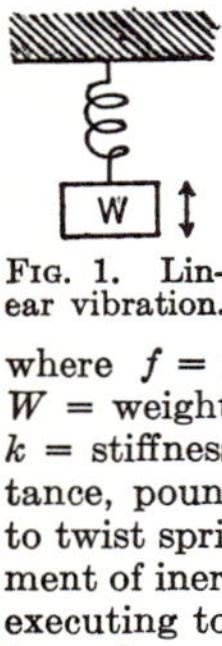

FIG. 1. Linear vibration.

Linear vibration for spring-supported weight (Fig. 1):

$$f = \frac{1}{2\pi}\sqrt{k/M} \qquad (1)$$

Torsional vibration for disk on rod (Fig. 2):

$$f = \frac{1}{2\pi}\sqrt{L_1/I} \qquad (2)$$

where f = frequency, cycles per second; M = mass, pound inch^{-1} second^2 = W/g; W = weight, pounds; g = acceleration of gravity = 386 inches per second^2; k = stiffness constant of spring = force required to stretch it to unit distance, pounds per inch; L_1 = same constant in torsion or torque required to twist spring or rod through 1 rad of arc, inch-pounds per radian; I = moment of inertia, pound-inch second^2 = WR^2/g; W = weight of body, pounds, executing torsional vibrations; R = radius of gyration, inches. Equation 1 is used more often than eq. 2, as most problems involve linear, rather than torsional, vibrations.

By substituting the value of g for the inch-pound-second system, eqs. 1 and 2 reduce to:

Linear vibrations:

$$f = 3.13\sqrt{k/W}\ \text{cps} \qquad (3)$$

Torsional vibrations:

$$f = 3.13\sqrt{L_1/WR^2}\ \text{cps} \qquad (4)$$

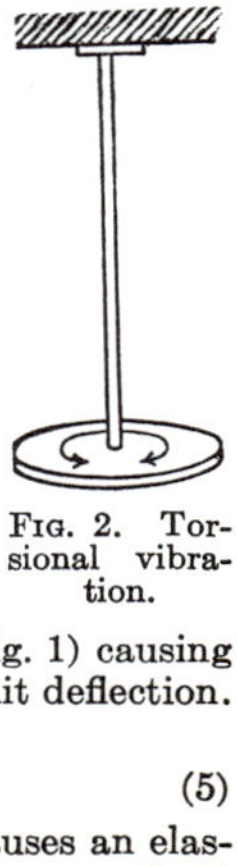
FIG. 2. Torsional vibration.

GRAVITY DEFLECTION FORMULA. For a spring carrying a weight (Fig. 1) causing extension d of the spring, elastic constant k must equal W/d = weight per unit deflection. Substituting this in eq. 3,

$$f = 3.13\sqrt{1/d} \qquad (5)$$

where f = cycles per second and d = inches deflection. Thus, if a weight causes an elastic support of relatively small weight to deflect downward through d inches, eq. 5, at once gives the up-and-down frequency. This formula is useful for quick, approximate estimates of certain vibration frequencies. Thus, a shaft carrying a relatively heavy flywheel has a deflection of d. The frequency and the critical speed (see p. 9-13) can be estimated without knowing weight of the flywheel. d is the deflection under one times gravity acceleration. When the system is subjected to accelerations of x-times gravity, the deflection of the system equals x times d.

TRANSVERSE VIBRATION OF UNIFORM BARS. For any vibrating bar of uniform cross section, the frequency is given by

$$f = (a/l^2)\sqrt{EI/W} \tag{6}$$

where a depends on shape of bar and how it is supported, and l = length, inches. For values of f, see Table 1. In this table, I = moment of inertia of cross section inches^4

Table 1. Frequencies for Circular and Rectangular Bars

Support and Mode		Uniform Bar, cps	Circular Section, cps	Rectangular Section, cps
Cantilever	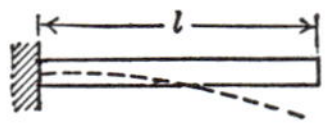	$(11/l^2)\sqrt{EI/W}$	$(2.7D/l^2)\sqrt{E/W_1}$	$(3.2h/l^2)\sqrt{E/W_1}$
		$(70/l^2)\sqrt{EI/W}$	$(17.6D/l^2)\sqrt{E/W_1}$	$(20.3h/l^2)\sqrt{E/W_1}$
		$(193/l^2)\sqrt{EI/W}$	$(48D/l^2)\sqrt{E/W_1}$	$(55.7h/l^2)\sqrt{E/W_1}$
Beam supported at ends	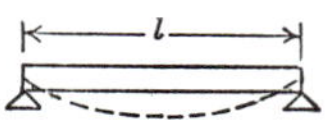	$(31/l^2)\sqrt{EI/W}$	$(7.7D/l^2)\sqrt{E/W_1}$	$(8.9h/l^2)\sqrt{E/W_1}$
	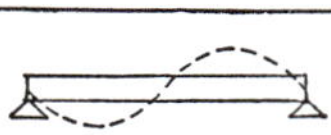	$(124/l^2)\sqrt{EI/W}$	$(31D/l^2)\sqrt{E/W_1}$	$(35.8h/l^2)\sqrt{E/W_1}$
		$(278/l^2)\sqrt{EI/W}$	$(69.5D/l^2)\sqrt{E/W_1}$	$(80.4h/l^2)\sqrt{E/W_1}$
Clamped-clamped beam*		$(70/l^2)\sqrt{EI/W}$	$(17.6D/l^2)\sqrt{E/W_1}$	$(20.3h/l^2)\sqrt{E/W_1}$
		$(193/l^2)\sqrt{EI/W}$	$(48D/l^2)\sqrt{E/W_1}$	$(55.7h/l^2)\sqrt{E/W_1}$
		$(379/l^2)\sqrt{EI/W}$	$(94.8D/l^2)\sqrt{E/W_1}$	$(109.5h/l^2)\sqrt{E/W_1}$

* Note that the free-free beam has the same frequencies as the clamped-clamped beam.

(not mass moment of inertia); W = weight, pounds per inch; W_1 = weight, pounds per cubic inch; l = length, inches; D = diameter, inches; h = depth, inches; E = modulus of elasticity, pounds per inch^2.

TRANSVERSE VIBRATION OF LOADED UNIFORM BARS. Table 2 gives frequencies of loaded bars, where weight of bar is neglected compared with weight it carries. Symbols have same meaning as in Table 1. These formulas are derived from eq. 5 by substituting proper value of d produced by a load W for each case. When weight of bar cannot be neglected, add 23% of weight of bar to W in case of cantilever and 50% of weight of bar in case of beam loaded in center of span.

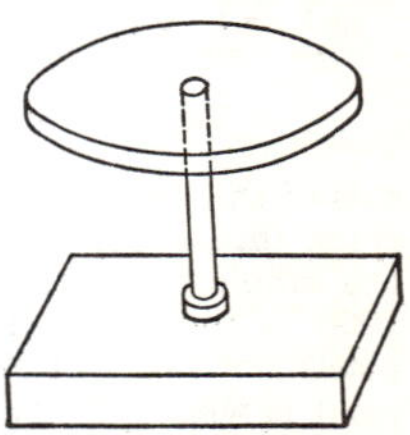

FIG. 3. Vibration of circular disk supported at center.

VIBRATION OF CIRCULAR DISK OF UNIFORM THICKNESS SUPPORTED AT THE CENTER (Fig. 3). Such a plate has exactly the same vibration frequencies as one perfectly free;

Table 2. Vibration of Loaded Uniform Bars

Type of Loading		Frequencies, cps	
		Circular Cross Section	Rectangular Cross Section
Loaded cantilever load on end	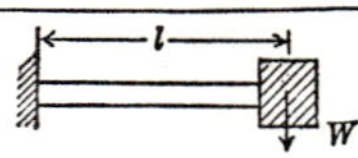	$1.2D^2\sqrt{E/Wl^3}$	$1.56\sqrt{Ebh^3/Wl^3}$
Loaded beam loaded in center of span	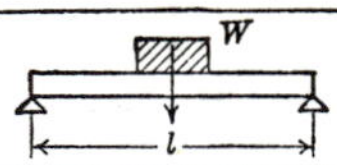	$4.8D^2\sqrt{E/Wl^3}$	$6.24\sqrt{Ebh^3/Wl^3}$
Loaded beam load off center of span		$1.2D^2\sqrt{\dfrac{El}{Wx^2y^2}}$	$1.56\sqrt{\dfrac{Ebh^3l}{Wx^2y^2}}$

the formula is:

$$f = K(h/r^2)\sqrt{(Eg/W_1)} \tag{7}$$

where h = thickness of disk, inches; r = radius of disk, inches; E = modulus of elasticity, pounds per inch^2; g = 386 inches per second^2; W_1 = weight of disk material, pounds per cubic inch; K = constant depending on the number of nodes n along disk edge; when $n = 4$, $K = 0.25$; $n = 6$, $K = 0.54$. This assumes the absence of ring nodes, which case will be omitted.

LONGITUDINAL VIBRATION OF UNIFORM BARS. Such a bar when clamped on both ends has the vibration frequency:

$$f = (n/2l)\sqrt{(Eg/W_1)}$$

where n = number of half waves along the length and where otherwise the same notations as above are used.

2. VIBRATION MEASUREMENT

A pure sinusoidal vibration is given by any two of the following four quantities: vibration displacement (double amplitude), d; vibration velocity, v; vibration acceleration, a; and frequency, f. They are interrelated by: $v_0 = \pi f d_0$ and $a_0 = 2\pi^2 f^2 d_0$, where d_0, v_0, and a_0 are maximum values. Hence, to measure vibration, any two of these quantities must first be determined; without knowledge of them, quantitative study is impossible. Also, form of path described, i.e., circular, elliptical, etc., and the phase relative to another vibration, sometimes may be necessary. (See Ref. 1.)

VIBRATION DISPLACEMENT MEASURING INSTRUMENTS. To measure displacement, a stationary reference point must be established. This point is either a structure mounted rigidly to the ground or a seismic mass. This is a mass suspended on a spring with a system natural frequency considerably below the vibration frequency. All measurements are being made relative to these points.

When the vibration displacement is above $^1/_8$ in., it can best be measured by placing a ruler parallel to the direction of vibration. In the majority of cases, the vibration is smaller than $^1/_8$ in.; then the motion must be amplified. This can be done either mechanically, optically, or electrically.

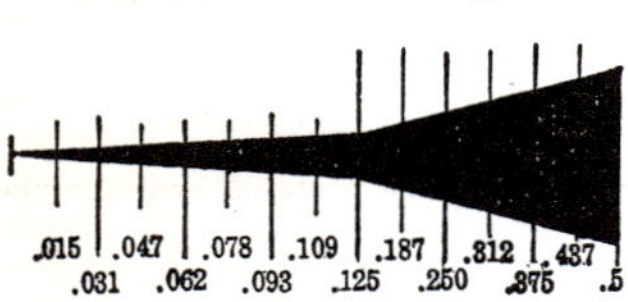

FIG. 4. V-scope.

V-scope. A wedge is printed on a piece of gummed paper approximately 2 in. long and $^1/_2$ in. wide (see Fig. 4). This paper is stuck to the vibrating surface. When the paper vibrates, a blurred triangle-shaped image can be seen. Lines perpendicular to the axis of the wedge are drawn on the paper and marked by figures. The figure on the line going through the apex of the triangle is the vibration displacement. This is probably the simplest and most inexpensive vibration indicator. The stickers are sold in rolls of 1000 by the Avery Adhesive Company in Los Angeles, Calif. They can measure a displacement range of 30 to 500 mils. Their accuracy for their present form and quality of print is within 10 mils below 100 mils, and within 10% above 100 mils. They are very useful for rough measurements on shake tables or other devices vibrating with larger displacements; their application is easy.

Dial-type Displacement Indicator. A dial gage, reading in mils, is set in a hollow block of metal weighing about 2 lb and supported by a spring (see Fig. 5). The axis of the indicator stem passes approximately through the center of gravity of the block. The base of the spring is placed on the vibrating surface with which the indicator stem then makes contact. The stem and the pointer are vibrating with the surface, while inertia keeps the mass and the face of the dial gage stationary. The width of the (blurred) image of the oscillating pointer, read on the mils scale, gives the vibration displacement. Higher accuracy of reading can be obtained when the indicator is observed under stroboscopic light of the vibration frequency.

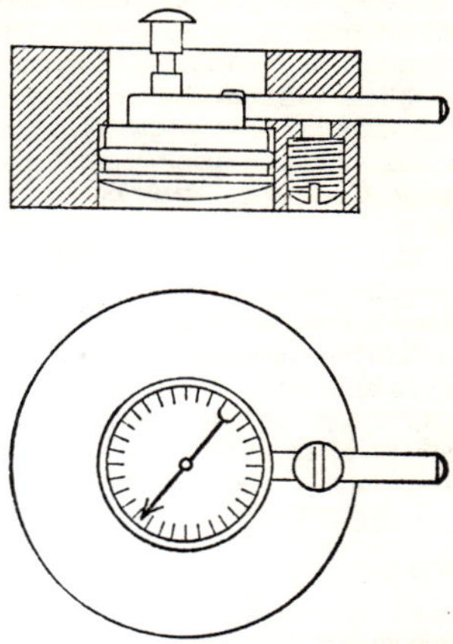

Fig. 5. Dial-type displacement indicator.

This instrument is portable and suitable for factory use where high accuracy is not required; accuracy is approximately 1 mil. Personnel can easily be trained to use this instrument. Its limitations are resonant frequency of the spring-mass system (low frequency limit) and the maximum acceleration which the indicator can follow (giving the maximum displacement for the various frequencies—3 mils at 60 cps, 10 mils at 30 cps, and 100 mils at 10 cps); chatter of the indicator button occurs above this acceleration. In special cases, the amplitude indicator without the inertia part can be rigidly attached to a supporting post if a convenient point of attachment, free from vibrations, is available.

THE VIBROGRAPH will make a complete graphical record of a vibration, giving a permanent record of both frequency and amplitude. The records are traced by means of a fine stylus on moving celluloid film or wax-paper band. Their range is from 0.5 to 125 mils displacement, and up to frequencies of 120 cps. As to limitations, the same considerations as for the dial-type displacement indicator apply, although their range is considerably wider. The mechanical amplification is of the order of 15 to 25, so that a 10 mils displacement vibration traces a sine wave approximately $^3/_{16}$ in. between peaks. A microscope or magnifying glass with a graduated scale is used to read records made of small displacement vibrations (below 10 mils). Vibrographs are very useful where wave shape of vibration or frequency components should be measured, and also to check natural frequencies of structures. They are easier to apply and more portable than electrical instruments, and often give the same information. They are manufactured by Cambridge Instrument Company, London and Cambridge, England; General Electric Company, Schenectady, N. Y., and Westinghouse Electric Corporation, Pittsburgh, Pa.

TORSIOGRAPH. To detect torsional oscillations in a revolving shaft, as a ship propeller shaft, or a Diesel engine shaft governed by a heavy flywheel, a Geiger (marketed by Commercial Engineering Laboratories, Detroit, Mich.), or other torsiograph may be used. This is an inertia instrument comprising essentially one part rigidly attached to the shaft, taking part in its torsional oscillations, and a flywheel, mounted in its own bearings and driven through a flexible spring at the shaft speed. The flywheel maintains substantially constant angular velocity. A multiplying lever mechanism responds to shaft oscillations with respect to the non-oscillating flywheel, and records oscillations in a graph (see Ref. 2). The vibrograph of the Cambridge Instrument Company can be adapted for torsional vibration measurements by addition of a flywheel. Another torsiograph is made by General Motors Research Laboratories, Detroit, Mich. The torsiographs are reported to give satisfactory readings up to approximately 100 cps.

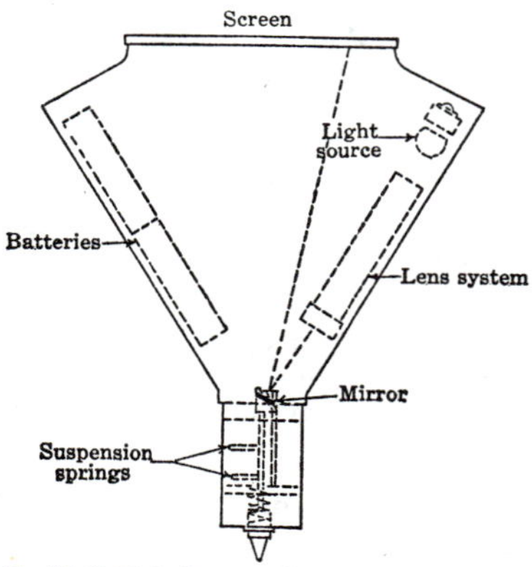

Fig. 6. Light-beam vibration-displacement indicator.

LIGHT-BEAM VIBRATION-DISPLACEMENT INDICATORS. These instruments use optical amplification, are easy to use, have a wide frequency range, and are very accurate. Vibration displacements as small as 0.1 mil can be measured. The indicators are relatively compact and of light weight, and they have been found to be very practical for vibration measurements on rotating machines in the factory as well as in the field. The vibration is transmitted from a stylus to a rotating mirror which reflects the light of a flashlight bulb on a scale engraved into an opaque glass plate. A light band is visible on the scale, and its width is proportional to vibration displacement (see Fig. 6). Light-beam vibration-displacement indicators are being manufactured by the Davey Vibroscope Company, New York, N. Y., and the General Electric Company, Schenectady, N. Y.

ELECTRICAL VIBRATION-MEASURING INSTRUMENTS. These instruments measure vibration displacement, velocity, and acceleration. They consist of a detector unit and an amplifier unit with indicating meter. The amplifier units are provided with an output plug for connecting oscillographs, chart recorders, earphones, or harmonic wave analyzers, so allowing recording or convenient analysis of the vibration. The most commonly used detectors are vibration velocity units and acceleration detectors. Both consist of spring-mass systems, the vibration velocity units working above their natural frequency, the accelerometers below their natural frequency. (See Ref. 3, p. 75.)

Velocity Units. In the vibration velocity unit a voltage proportional to vibration velocity is generated. A coil is attached to the housing of the unit which in turn is mounted to the vibrating surface. This coil moves within the air gap of a magnetic structure energized by a permanent magnet. Magnetic structure and permanent magnet are spring-suspended within the housing, and hence form a stationary reference point when the unit is vibrated. The output of the vibration velocity unit is of the order of 1 to 100 mv for most practical applications. In detectors which measure torsional vibration, the coil is attached to the housing of the detector which rotates with the oscillating shaft, and the permanent magnet and the magnetic circuit are part of a flywheel driven through springs. The flywheel is running at constant speed without any oscillations. The voltage generated in the coil is then proportional to the torsional vibration velocity.

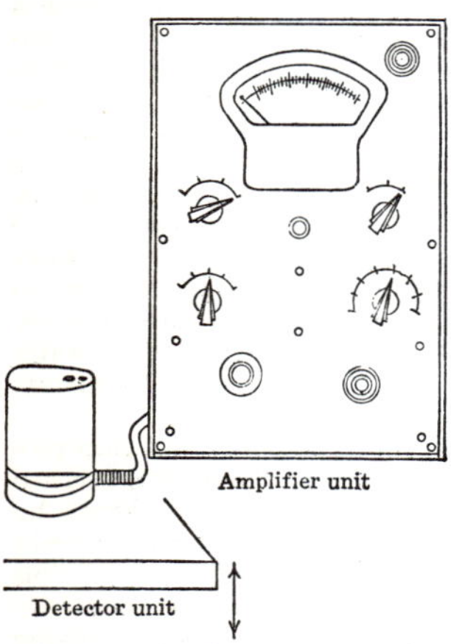

Fig. 7. Vibration velocity detector with amplifier unit for testing gas turbines.

Acceleration Detectors. In an acceleration detector, the mass is accelerated through a relatively stiff spring, whose deflection is proportional to the acceleration. This spring may be a piezoelectric crystal such as quartz, Rochelle salt, or ammonium dihydrogen phosphate (ADP), or a prestressed small-diameter wire (d = 0.001 in.) whose resistance is a function of its elongation. The output voltage of both types of detectors can be electrically (single or double) integrated or differentiated; hence measurements of displacement, velocity, or acceleration can be obtained with any of these detectors.

Range. The electrical instruments (see Fig. 7) cover a much wider range of displacement (0.0001 to 0.125 in.), acceleration (0.01g to 5000g), and frequency (0 to 10,000 cps) than the mechanical vibration instruments described above. They put less load on the vibrating surface, which is important when the mass of the vibrating object is small and when loading might change the vibration characteristic. They are being used where the operator cannot be close to the vibrating surface during its operation, as in airplane or gas-turbine vibration testing, or where permanent records over a longer period of time are required, as for turbine supervisory instruments. Depending on the recorders used, the charts will show either the wave shape of the vibration, or only its average or root mean square value. The output of these detectors is mostly of the order of millivolts, and care should be taken to avoid pickup of extraneous electrical signals. Operators of this equipment should be well trained in their use. Manufacturers are Consolidated Engineering Company, Pasadena, Calif.; General Electric Company, Schenectady, N. Y.; General Radio Company, Cambridge, Mass.; MB Manufacturing Company, New Haven, Conn.; Statham Laboratories, Pasadena, Calif.

Torsional Oscillations. To obtain records of torsional oscillations, it is in many cases possible to use the output of d-c machines which happen to be coupled rigidly to the oscillating shaft and participate in its motion, thus avoiding attachment of additional equipment. Circuit diagrams to be used are shown in Fig. 8. For the use of d-c machines, it is essential that the oscillation frequency be considerably smaller than the commutator ripple frequency. It is evident that direct-current generators can be attached to the shaft for torsional vibration measurements if no other equipment is available. Often this arrangement is more satisfactory than special torsiographs since there is no possibility of exciting and recording the natural frequency of the seismically driven flywheel. The output voltage due to torsional oscillations is proportional to the rated voltage of the machine and the ratio of the vibration velocity to the rotational velocity. This system was found especially useful for determining torsional oscillations of aircraft engines and their accessories.

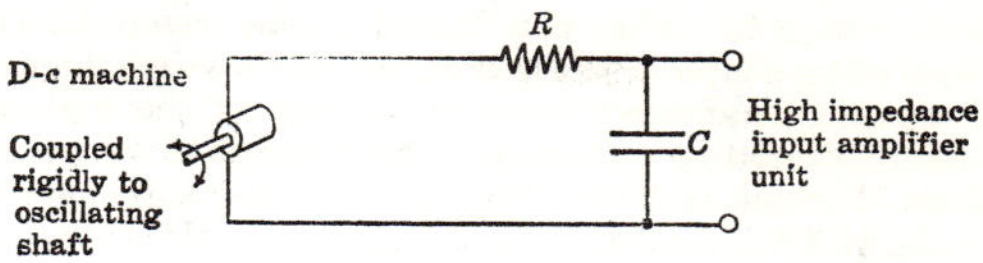

FIG. 8. Circuit for torsional vibration measurements with d-c machine.

REED FREQUENCY METER. The frequency components of a vibration can be determined with the Frahm tachometer (see Fig. 9). This meter has a large number of reeds carrying masses at their tips, and tuned to respond one after another over a range of frequencies. A convenient model about 6 in. long, made by James G. Biddle of Philadelphia, Pa., covers frequencies of 900 to 30,000 per min by means of three instruments. It is easily portable, readily applied, and shows rapidly varying frequencies as well as more than one frequency at a time if present. Above 30,000 cycles per min (500 per sec), sensitivity is lost because of stiffness of the reeds. For best response the meter should rest solidly against the vibrating body, direction of reed vibration being roughly parallel to direction of motion of exciting vibration. This instrument is an excellent substitute for speed counters and speed tachometers.

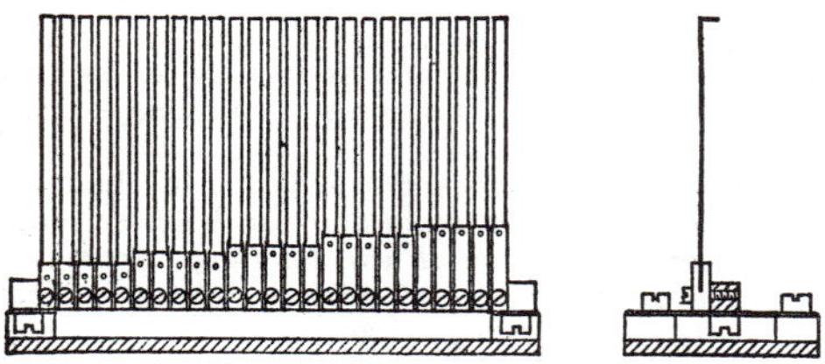

FIG. 9. Frahm tachometer.

To cover a wide frequency range, a number of Frahm tachometers has to be carried. (One tachometer covers approximately one octave.) Frequency meters of more convenient type have only one reed, but that is of adjustable length. The length is adjusted by a frequency calibrated micrometer screw until maximum vibration displacement is obtained. This type of meter is not as easy to use as the multiple-reed meter, but it has the advantage of greater compactness. The adjustable-type reed frequency meters are manufactured by the Davey Vibroscope Company, New York, N. Y., and by Westinghouse Electric Corporation, East Pittsburgh, Pa.

STROBOSCOPE. This device produces intermittent flashes of light. The duration of the flash is very short; the frequency of the flashes can be controlled by a frequency calibrated dial. When the frequency of the flashes is equal to the vibration frequency, the vibrating object appears to be standing still. *Caution:* Objects are also stopped when the flash frequency is a fraction of the vibration frequency. It is good practice to stop the object at several flash frequencies to be sure that the correct frequency is read. Stroboscopes are available from the General Radio Company, Cambridge, Mass., under the trade name Strobotac.

3. VIBRATION EXCITATION

Vibration motors are being used for fatigue testing, determination of natural frequencies and modes of vibration, for acceptance and performance testing where vibration specifications are given, and where vibration is to perform certain processes. Either vibration forces are applied to the object or the object is subjected to a vibration displacement. Vibration forces are applied by electromagnets or unbalanced rotating masses; the object is subjected to vibration displacements when it is attached to a vibrating platform. (See Ref. 1.)

MECHANICAL VIBRATION MOTOR. Exciters which apply vibration forces consist of one or two rotating unbalanced masses. The bearings supporting these masses are bolted to the object to be vibrated. The vibrating forces are uniform in one plane, but not in one direction, if only one unbalanced mass is being used. To obtain vibratory forces in only one direction, two unbalanced masses rotating in opposite directions must be used.

Vibrating platforms on which objects under test can be placed are driven by electric motors through crankshafts, scotch yokes, or eccentric drives. The mechanical vibration motors can be built to apply relatively large forces, and are therefore suitable to vibrate heavy objects. They are limited at the present in their upper frequency range to 60 cps. The wave shape of the vibrations of these machines often becomes poor after some use, because of damaged bearings caused by the high vibratory forces. These motors are manufactured by the Baldwin Southwark Company, Philadelphia, Pa., and by the American Machine and Tool Company, Chicago, Ill.

ELECTROMAGNETIC VIBRATION MOTOR. Some of these motors are built like electrodynamic loudspeakers, their a-c coils connected to power amplifiers which in turn are controlled by variable frequency oscillators. The vibratory force which can be obtained from these motors is often limited by the strength of the driving coils (approx. 400 lb). Objects under test can be placed directly on the motor or connected to the moving coils by means of a connecting link. This practice is being used for vibration testing of aircraft wings.

Another type of electromagnetic vibration motor employs an electromagnet whose poles are applied to the surface of the object under test. This method is especially useful for finding resonant frequencies in structures. When the surface of the structure is nonmagnetic, this method can still be applied when a strip of magnetic material is cemented to the surface. This method has been successfully used to determine resonant frequencies as high as 30,000 cps. A beam supported at its nodal point and excited by an electromagnet is being used to calibrate high g accelerometers. Vibration accelerations as high as 800g have been obtained at the ends of the beam.

Frequency and amplitude control is continuous and simple and is achieved by changing the oscillator controls. Frequencies can be held very accurately and can be closely adjusted, which is of importance for natural frequency testing of low damping mechanical structures. Fine adjustment of frequency is often difficult to achieve on mechanical vibration motors because energy is stored in the moving parts. Fatigue-testing machines to vibrate test specimens and special structures such as turbine blades or airplane propellers have been built using electromagnetic drives. Electromagnetic vibration motors can be obtained from the General Electric Company, Schenectady, N. Y., the Westinghouse Electric Corporation, East Pittsburgh, Pa., and MB Manufacturing Co., New Haven, Conn.

PNEUMATIC VIBRATION EXCITATION. An air stream is blown through a nozzle against the vibrating object which, in itself, functions as a valve. The air stream hits the specimen and bends it away from the nozzle; the specimen snaps back because of its elasticity, and the cycle is repeated. The frequency is equal to the resonant frequency of the specimen. Good stability and large amplitudes are obtained when the air column connected to the nozzle is tuned, like an organ pipe, to the frequency of the specimen. A piston-shaped armature attached to the specimen will improve performance. A fatigue-testing machine manufactured by the General Electric Company uses an arrangement with two nozzles located on each side of the specimen. These two nozzles are interconnected by a resonating air column. Force is applied to the specimen during each half cycle.

VIBRATION PREVENTION

Of the several methods of attack in vibration reduction the particular method or methods adopted depend on the individual problem. These methods are described briefly here.

REMOVAL OF EXCITING CAUSE. This method, when possible, is most satisfactory. Unbalance in rotating machinery is a common source of trouble. In rigid rotors it can be almost completely eliminated by balance weights. Gear-tooth frequency noise is best avoided by precision in manufacture.

TUNING is applicable only when vibration is distinctly of resonant type. Resonant vibrations usually are excited by a relatively small stimulus, which happens to be in step with a natural vibration frequency. The tuning of steam-turbine wheels is an example, as is also the avoidance of shaft critical speeds through selection of proper diameter at the operating speed.

ELASTIC SUSPENSION can be applied to two different problems. The first problem is the reduction of the forces transmitted by rotating or reciprocating machines, etc., into their bases. Industrial presses, for example, may be elastically mounted to prevent shaking of the building. The second problem is the vibration isolation of apparatus, such as isolation of delicate measuring equipment installed on a vibrating floor, or the isolation of car bodies from the movements of the wheels.

It should be noted that this type of isolation system does not necessarily have to absorb energy, as in case of friction damping. Springs (helical springs, rubber, etc.) put a high reactance (the same as an inductance in an electrical circuit) in the way of the energy flow, and hence prevent it from passing through the mount. The spring suspension of a railroad car isolates from the car body forced vibration imparted to the wheels; the leaf spring contains some friction damping which eliminates resonant bouncing, but has no part in isolation. It can be shown that such friction in springs and shock absorbers somewhat increases forced vibration transmission. The amount of increase is small, even when friction is considerable. Friction is necessary to avoid building up resonant vibrations.

An incorrectly designed elastic suspension may do more harm than good. The design of a good suspension is not always an easy matter and often requires knowledge of vibration principles as well as experience.

DAMPING, or reduction of vibration amplitude through action of friction, is a cure for resonant vibrations only. Forced vibrations are likely to result from powerful exciting forces which are not much reduced by application of friction devices. For example, gear-tooth frequency hum or whistle is a forced vibration, which is reduced but little by stuffing the wheel with a nonresonant damping substance with internal friction. On the other hand, the resonant ring of the wheel, because of its natural periods, may be greatly reduced by such treatment. Shock absorbers used in vehicle suspension are typical friction dampers which limit the amplitude of resonant vibrations excited by road irregularities.

4. REMOVAL OF EXCITING FORCE

MECHANICAL BALANCE OF MACHINERY. It is possible mechanically to balance completely any rigid rotor, whatever may be its unbalance or its shape, by only two weights. The problem is to determine the position and the size of these two weights. Balancing machines make this determination; trial-and-error methods will generally waste money. Balancing of rotors that cannot be considered rigid is a fine art and requires experienced operators.

STATIC AND DYNAMIC BALANCE. Mechanical unbalance in a rigid rotor always can be completely represented schematically as in Fig. 10. The weight A takes account of displacement of the center of gravity of the rotor from its true axis; weights B and C represent a couple, depending on how much the rotor is cocked or twisted sideways from its true axis. The three weights can be replaced by two by splitting A into two parts, placing one in the plane of B and the other in the plane of C. The two weights in each of these planes can then be replaced by a single one in each plane by applying the parallelogram of forces (centrifugal forces in this case). (See Ref. 4, pp. 85–94, and Ref. 2, pp. 62 ff.) That part of unbalance represented by A, Fig. 10, is static unbalance; it can be detected by a static or a gravity test by placing the rotor on ways and noting its heavy spot. The part represented by weights B and C is dynamic unbalance; it can be detected only by a rotational or dynamic test which develops a centrifugal couple. A rotational test, however, does not separate dynamic from static unbalance, but combines the effect of both; otherwise, mechanical balance would be comparatively simple. For this reason, specially designed machines and special methods described below must be used to determine the amount and position of balance weights.

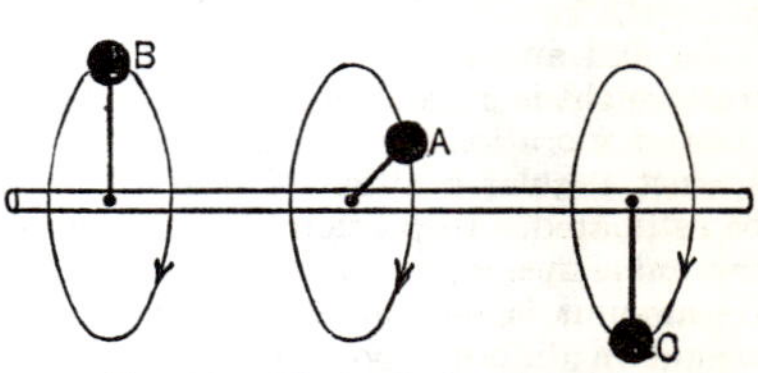

FIG. 10. Mechanical unbalance.

BALANCING MACHINES. Two types of balancing machines are available, stationary machines and portable machines. Stationary machines are used mostly for production balancing. Good general-purpose machines are marketed by the Gisholt Machine Company, Madison, Wis., and the Tinius Olsen Company, Philadelphia, Pa. A balancing machine designed especially for automobile wheels is made by the Bear Manufacturing Company. In many cases special machines are built to handle special jobs. They range from relatively simple machines indicating angular position and magnitude of balancing weights to rather complex automatic machines which balance rotors by drilling holes of the correct depths at the correct angular position.

Portable machines are being used to balance rotors in their final installations and under their actual operating conditions. They are also used for production purposes where the rotor is of such a size that it is not economical to build special bearings for the balancing operation. Portable machines have the disadvantage of requiring more time than the stationary machines to obtain the amounts and the angular positions of the balancing weights. It is the advantage of the portable machines that they balance the rotor under actual operating conditions which often change the balance obtained in the factory. Such changes might be due to temperature, speed, loading, or attaching of heavy couplings to the shaft. It is good practice to balance rotors in the factory production line and refine the balance after installation where only very small vibrations are permitted. It should be noted that vibrations below 0.0001 in. displacement are frequently caused by ball or roller bearing vibration and that further refinement of balancing does not result in any improvements. Portable dynamic balancing equipment is manufactured by the Davey Vibroscope Company, by the General Electric Company, and by the Westinghouse Electric Corporation. A portable machine for balancing of automobile wheels is made by the Stewart Warner Company.

BALANCING OF SPECIAL JOBS. Small Rigid Rotors. The following method of balancing small rigid rotors gives good results, with a minimum of apparatus, and is well applied where frequent balancing operations are not required. The rotor *R*, Fig. 11, is placed in a light wood frame *F*, resting in half bearings in the frame, and driven by motor *M* through belt *L*, whose tension is adjusted by screw *S*. The rotor is first pivoted in plane *C*, the pivots in plane *D* being withdrawn. A spring, not shown, is placed under the frame at bearing *B*, of sufficient stiffness to cause resonance at 700 to 800 rpm. The rotor being driven through the resonant frequency, the deflection indicator *E* is adjusted with its button resting on top of the frame to register vibration. The arc through which the needle oscillates approximately measures unbalanced moment about fulcrum in plane *C*.

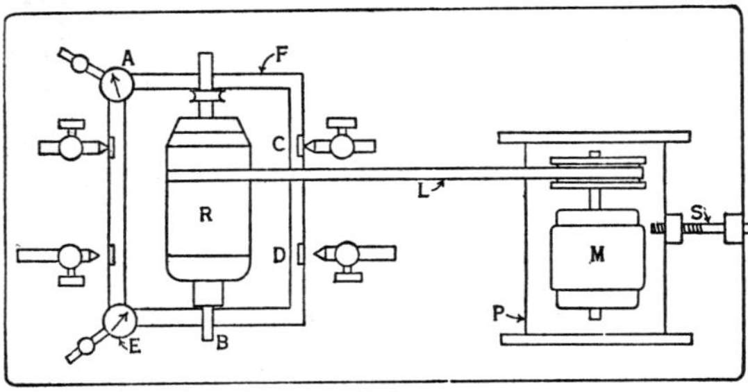

FIG. 11. Balancing set.

To find amount and position of the weight, vibration amplitude is read and a small trial weight is placed successively in four positions 90 degrees apart in plane *D*; the amplitude of vibration is read in each case. A curve of rise and fall of amplitude is plotted against angular position of trial weight, from which amount and position of weight can be estimated. Inspection of the curve shows position where weight subtracts most from the unbalance and adds most to it; the former is the correct position of balance weight. Its amount is estimated by comparing change of vibration amplitude produced by the weight in different positions with the original vibration amplitude, assuming vibration amplitude to be a direct measure of weight required. The trial weight for small rotors may be a piece of wax.

Besides having a light frame to insure large amplitudes, the outfit must not vibrate as a whole because its foundation is insecure. It must be placed where no floor vibrations are transmitted to it; if not, indicator readings will be inconsistent. Furthermore, the frame must be sufficiently stiff to resist all tendency to flexure. A good way to isolate the apparatus from such vibrations and prevent detrimental shaking is to place the outfit on a massive iron plate resting on commercially available vibration mounts or on an inflated automobile inner tube. The driving motor must be well balanced, and a light belt should be used to give a constant tension perpendicular to the oscillatory motion of the frame; the pull should not be too great. To balance the other end of the rotor, the process is repeated, with the other fulcrum. Weights so determined should produce complete balance so that the indicators show no oscillatory motion. This method can be refined by replacing the dial indicator with one of the more sensitive vibration-measuring instruments.

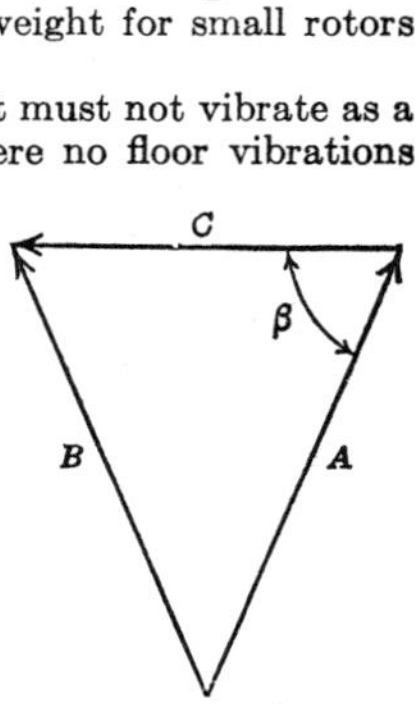

FIG. 12. Vectors representing unbalance.

Large rigid rotors are balanced mostly by portable equipment. They work on the principle that the bearing vibration is picked up by electrical vibration detectors whose generated voltage is proportional to the amount of the unbalance, and whose electrical phase angle corresponds to the angular position of the unbalance. The change in the generated voltage and in the electrical phase angle is measured when a trial weight is placed in the balancing plane of the rotor. The voltages, amplitudes, and phase can be represented by vectors as shown in Fig. 12 where vector A represents the output due to the unbalance and vector B the output after the trial weight has been attached. The vector C is then the effect of the trial weight. Balance is achieved when the vector C is equal and opposite in direction to vector A. This can be achieved by rotating the trial weight through the angle β and by increasing its weight in the ratio of A. (See Ref. 5.)

Airplane propeller unbalance may be made up of unbalanced air reaction and mass unbalance. If one blade exerts slightly greater traction on the air than the other, an unbalanced couple is present which revolves with the propeller, always tending to deflect the forward end of the propeller shaft toward the blade exerting the smaller traction, thus causing the engine to wobble. Since traction on one propeller blade is several hundred pounds, any dissymmetry between blades, either of shape or pitch, produces a relatively large unbalanced couple, and aerodynamic unbalance may far exceed pure mechanical unbalance. Hence extremely precise mass balance of airplane propellers may be disappointing. The slightest change in pitch of a blade noticeably affects plane vibration. The problem is complicated by blade deflections during operation which must react on the traction. Propellers of high-power planes are subject to severe engine torque pulsations with resulting vibration, but the exciting force comes from the engine. (See Ref. 6 for a detailed study of this problem.)

Reciprocating Engine Balance. Vibration of reciprocating engines is due not only to unbalanced inertia forces but also to gas pressure forces. The analysis of the inertia unbalance is more complex than for rotors, since reciprocating as well as rotating masses are present, and since vibrations of twice running speed frequency are caused by the second harmonic component of the piston acceleration. The rotating masses can be completely balanced, but not so the reciprocating masses; hence the horizontal forces can be made zero, but the vertical unbalanced forces exist. These unbalanced forces and resulting moments can be made to cancel each other in multicylinder engines. Care should be taken that cancellation is obtained for the forces of the running speed frequency as well as of the second harmonic frequency. A single-cylinder engine is inherently unbalanced, a two-cylinder engine with 180 degree crank angle has unbalanced forces of second harmonic frequency, and a longitudinal rotating moment of running frequency. Four-cylinder engines of 0, 90, 270, 180 degree crankshaft have unbalanced moments of running frequency; for 0, 180, 180, 0 degree crankshaft the forces and moments of second harmonic frequency are unbalanced. Six (0, 120, 240, 240, 120, 0 degree) and eight (0, 180, 90, 270, 270, 90, 180, 0 degree) cylinder engines have all forces and moments balanced. (See Ref. 3, pp. 213 ff.)

RULES FOR DESIGN. A good rule for design of rotating apparatus with moving parts, where unbalanced forces are present, is to keep the center of gravity of the com-

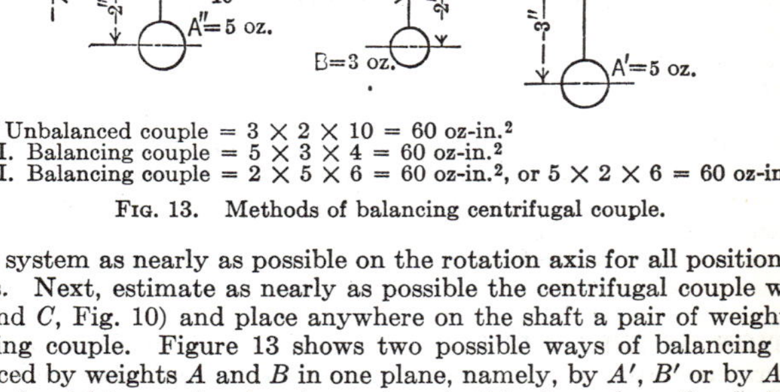

Unbalanced couple = 3 × 2 × 10 = 60 oz-in.^{2}
I. Balancing couple = 5 × 3 × 4 = 60 oz-in.^{2}
II. Balancing couple = 2 × 5 × 6 = 60 oz-in.^{2}, or 5 × 2 × 6 = 60 oz-in.^{2}

FIG. 13. Methods of balancing centrifugal couple.

plete moving system as nearly as possible on the rotation axis for all positions of the shaft as it revolves. Next, estimate as nearly as possible the centrifugal couple which remains (weights B and C, Fig. 10) and place anywhere on the shaft a pair of weights to produce a counteracting couple. Figure 13 shows two possible ways of balancing a centrifugal couple produced by weights A and B in one plane, namely, by A', B' or by A'', B''. Note that the pair of weights may be placed anywhere in this plane along the shaft if the couple produced is equal and opposite in magnitude to that of the original weights. The couple in each case is measured by the product of either one of the two weights times its distance from the shaft times its distance along the shaft to the second weight of the pair. The two weights do not have to be equal, but the products (magnitude of a weight times distance to shaft axis) must be equal. The use of unequal weights and axial distances is shown where the pair A'' and B'' are counterweights.

WORKMANSHIP. In symmetrical rotors with perfect workmanship and materials, there should be no unbalance. For a 1/2-hp, 1800-rpm motor, for example, an unbalanced weight of 0.05 oz will produce appreciable vibration of, say, 2 mils amplitude. So small an unbalance cannot be avoided by ordinary shop processes, even with the best supervision. It is cheaper to manufacture rapidly and to balance later.

MEASUREMENT OF AMOUNT OF UNBALANCE. It is sometimes necessary to form a quick estimate of the amount of unbalance in a machine. To measure this in an electric motor, place the motor on rubber pads which should compress at least 1/8 in. under the motor at usual speeds. Maximum vibration amplitude in a direction transverse to shaft is an approximate measure of the amount of unbalance. This is best determined by one of the vibration-measuring instruments.

Another form of suspension used by the author to estimate unbalance is to hang the motor by a relatively long spring through straps or a light cradle, preferably of wood, to reduce inertia. Weight of the motor should stretch the spring perceptibly. Vibration amplitude is measured as before by vibration indicator or meter. This method offers less restraint to the vibrating body than mounting on rubber pads.

The object of this test is to find how unbalanced apparatus vibrates when unrestrained by outside forces. The exciting stimulus acts almost entirely against the inertia of the body itself, and oscillations produced are proportional in amplitude to the exciting force. By these methods it is possible to select the best balanced among a number of similar motors quickly, if means for balancing are not available.

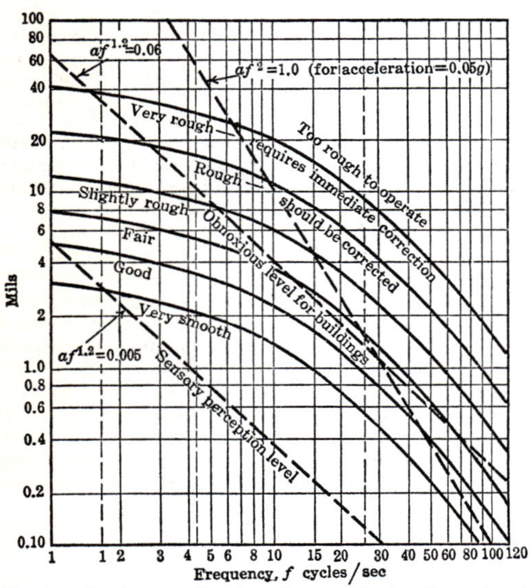

FIG. 14. Machinery vibration tolerance chart. (Reproduced by permission from Vibration Tolerance, by T. C. Rathbone, *Power Plant Engineering*, Nov. 1939)

Figure 14 is a chart showing vibration tolerance during operation (not special suspension). (See Ref. 7.)

In single-phase motors, there may be a vibration of considerable intensity due to a torque pulsation of twice the a-c frequency, i.e., 120 cps for a 60-cycle circuit (see Ref. 8). This frequency is much higher than ordinary rotation speeds, and thus much higher than vibration from unbalance which is always in step with rpm frequency. It shows up well with the foregoing suspensions; it is easily separated because of its higher frequency and its causing the motor to oscillate about the shaft as a stationary axis. Mechanical balance will not cure this electrical vibration.

Most motor hums from electrical causes are eliminated by elastic suspension. See Article 6.

Gas turbines vibrate in their test stand with a large number of frequencies. Some of them equal running speed, some of them equal the natural frequency of the mounting, and some of them equal wheel frequencies. To determine the unbalance, the vibration of running frequency only should be measured. The General Electric Company manufactures a vibration meter whose frequency range is restricted to the running-speed range. Other electrical vibration measuring instruments can be used in conjunction with electric band pass filters.

ELIMINATION OF SELF-INDUCED VIBRATIONS. A troublesome type of vibration that occurs in rotating machinery operating above the critical speed is self-induced vibration. Its frequency is always the same as the critical-speed frequency, which may be lower than the rpm frequency of the shaft in the same way that the revolution frequency of the earth around the sun is much lower than its rotation frequency about its axis.

Whirling Due to Oil Action in Journal Bearings. This type of self-induced vibration arises only when the rotation of the machine is more than approximately 1.7 the critical-speed frequency. It is produced by wedges of oil driving the journal around in the bearing and exciting a whirl. If a well-balanced machine tends to unstable operation, and vibration appears when the rotation frequency is more than 1.7 the critical speed, this kind of whirl is the probable cause. The best method of elimination is to insure that the critical-speed frequency is more than 60% the rpm frequency of the shaft. Another remedy, in smaller rotors, is to replace journal bearings by ball bearings.

Whirling Due to Frictional Action within the Rotor. Any rotor with self-contained friction produced by flexure while revolving will build up a whirl at any rotational speed higher than the critical. Friction of this type may be produced by rings so shrunk on a shaft that working may occur between the bore of the rings and the surface of the shaft. Frictional working also can arise in laminated structures, carried by the shaft; the structures flex and rotate at the same time.

These phenomena usually occur in motors with flexible shafts that operate far above the critical speed, and in such devices as extractor-type washing machines, which contain rotating parts operating well above the lower critical speed.

Wing Flutter. This phenomenon occurs when vibration of the wing is started by exterior forces such as shock loading, and when energy is added to the system during the vibration (positive damping). In subsonic range flutter can appear when the wing vibrates simultaneously in a twisting as well as an up and down mode. Flutter can be overcome by increasing the natural frequency of one of the modes, or by shifting the elastic axis fore or aft. (See *NACA Report* 496, 1935, by Theodore Theodorsen.)

5. VIBRATION ELIMINATION THROUGH TUNING

SHAFT WHIRLING. Critical Speeds. When a shaft or rotor, revolving in bearings, speeds up to where centrifugal force tending to whip it sideways just balances the elastic stiffness tending to keep it straight, the slightest mechanical unbalance builds up a whirling motion, which may be of destructive amplitude. The rotor then is revolving at a *critical speed.* It is inadvisable to run any rotor or shaft at or near critical speed for any length of time, regardless of perfection of balance. Critical speeds usually are avoided by designing the rotor shaft sufficiently stiff to hold its operating speed well below its critical speed. This sometimes is impossible for very heavy or very high speed rotors of long span between bearings. The shaft then is made sufficiently flexible to insure operation well above critical speed. This practice is to be avoided if possible, as vibration may develop from other causes in operation above critical speeds, although, so far as unbalance is concerned, operation is stable except close to critical speed.

Critical-speed Calculation. Critical speeds of a rotor are almost identical with natural vibration frequencies which the stationary rotor may assume.

Uniform Shaft. The first critical speed is given by the equation

$$f = 7.7(d/l^2)\sqrt{E/w_1} \tag{8}$$

where f = critical frequency, cycles per second; d = shaft diameter, inches; l = span between bearings, inches; E = modulus of elasticity shaft material, psi; w_1 = weight of 1 cu in. of shaft material, pounds. This formula is given in Table 1 (p. 9-03) for a circular bar vibrating between two end supports. For steel, the formula becomes

$$f = 79{,}000\frac{d}{l^2} \tag{9}$$

At the first critical speed, the shaft whirls in one loop with nodes at the bearings; at the second critical speed, in two loops with a node in the middle, at four times the first speed; at the third critical speed, in three loops at nine times the first critical speed, etc. Upper critical speeds, which enter only in high-speed installations, are special problems in themselves.

Shafts with Concentrated Loads. A useful approximate formula derived by Dunkerley from analysis, checked by careful tests on models, for a uniform shaft carrying concentrated loads (Fig. 15) is

$$\frac{1}{f^2} = \frac{1}{f_s^2} + \frac{1}{f_1^2} + \frac{1}{f_2^2}, \cdots \tag{10}$$

where f is critical frequency of the whole system; f_s that of shaft *by itself* as calculated from eq. 8; f_1 that of shaft considered as weightless, with load W_1 by itself, given by the last formula in column 3 of Table 2 (p. 9-04), etc., for the other concentrated loads. It is evident that values for x and y in the latter formula depend on location of W_1, W_2, etc., on the shaft. Quantities are inches, pounds, and seconds. This formula should not be used for three-bearing sets.

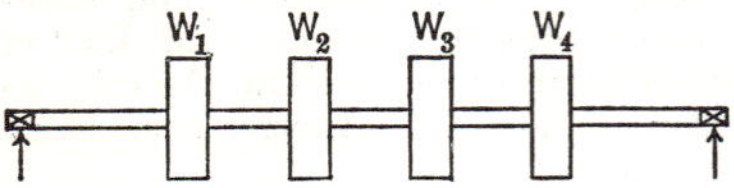

FIG. 15. Shaft with concentrated loads.

Shaft of Varying Cross Section and Load. When cross section of shaft is not uniform, and the load is distributed along it, as in most rotors, the problem is not so simple. It can be solved approximately from knowledge of the static load-deflection curve, as vibration frequency calculated on the basis of this deflection curve is only slightly higher than that based on the true deflection curve for actual vibration.

When static deflection curve is known, vibration frequency or critical-speed frequency is calculated easily from

$$f = 3.13\sqrt{\frac{W_1y_1 + W_2y_2 + W_3y_3 + \cdots}{W_1y_1^2 + W_2y_2^2 + W_3y_3^2 + \cdots}} \tag{11}$$

where shaft and rotor are divided into a series of slices of convenient length of weight W_1, W_2, etc. (including loads), and corresponding deflections y_1, y_2, etc. (See Fig. 16.) All quantities are inches, pounds, and seconds. The problem is primarily one of static deflection-curve determination, which, in cases similar to Fig. 16, requires using graphical statics. For the solution of such a problem, see Ref. 2.

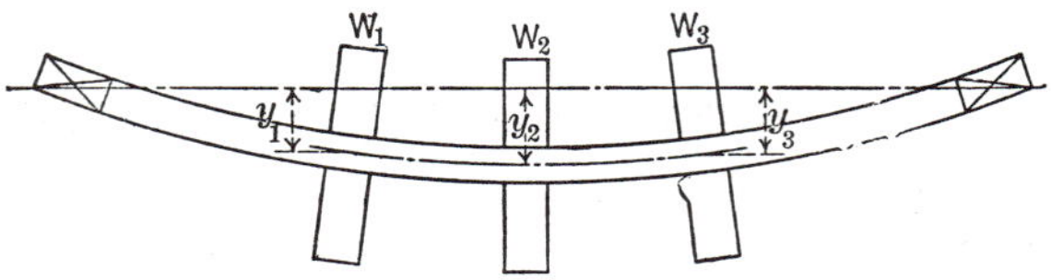

FIG. 16. Transverse vibration of multimass shaft system.

Three-bearing Sets. A three-bearing problem can be solved by eq. 11. Both first and second critical speeds should be found, as they may lie not far apart, whereas the second critical speed of a rotor with only one span will be at least twice the first. At the first critical speed of a three-bearing set, one span deflects upward when the other deflects downward; in finding static deflection curve, loads on one span must be taken as acting upward. Deflection of one span influences that of the other through a bending moment across the middle bearing. The deflection-curve determination must take account of this. The second critical speed occurs at a somewhat higher frequency where both spans deflect in the same direction, with a point of inflection in the shaft near the middle bearing. The static deflection curve then is found from loads acting in one direction, but, again, bending moment is transmitted across the middle bearing. The best procedure for this special problem is to refer to more advanced treatments noted in preceding paragraph.

Bearing Elasticity. All critical speeds are actually somewhat lower than calculation indicates, as no bearing is perfectly rigid. Also, up-and-down stiffness of a bearing is likely to be much greater than its sidewise stiffness. Because of this likelihood, two critical speeds may be present. One involves an elliptical whirl where the motion is mostly up and down; the other, at a speed a few per cent lower, a sidewise, elliptical motion. The lower frequency results from greater sideways elastic yielding of the bearings. For shafts of high natural frequency, critical speed may be 50% lower than calculated for rigid bearings since static deflection of bearing may be comparable to static deflection of shaft.

Elliptical Shafts. If a shaft is not circular, but elliptical or flattened on opposite sides, being stiffer in one axial plane than in one 90 degrees away, bad performance must be expected in operation near critical speed or half critical speed. The rotor evidently will vibrate in one plane at a somewhat lower frequency than in the plane for which it is stiffer. Ideally, with no damping, operation is unstable between these two frequencies. A model shaft and flywheel, with the shaft flattened on two sides, vibrates violently near the critical speed, and must be restricted by a guard ring. Rotors that perform badly near the critical speed are open to suspicion of being built stiffer in one plane than in another. An analysis is given in Ref. 3, p. 408.

TORSIONAL VIBRATION OF LOADED SHAFTS. Critical torsional vibration frequencies as well as transverse criticals may occur in shafts carrying flywheels and subject to periodic torque impulses, such as occur in Diesel engines. Torsional critical speeds are more insidious than transverse, as they are not always evidenced in operation by excessive vibration, as are the transverse criticals. Since dissipation is small, operation on a torsional critical speed causes large torsional stresses which often result in fatigue failure.

In ship's drives torsional vibrations lead often to tooth separation of the gears when the oscillating torque is larger than the driving torque. This separation will cause impacts on the teeth of the gears, making replacements necessary after short periods of operation.

Loaded shafts only are treated below. Torsional inertia of shaft itself is neglected, as it usually is small. It cannot be neglected in all cases, but in many problems nearly all torsional inertia is concentrated in certain planes, as in shafts carrying flywheels and crankshafts. See Refs. 9 and 10.

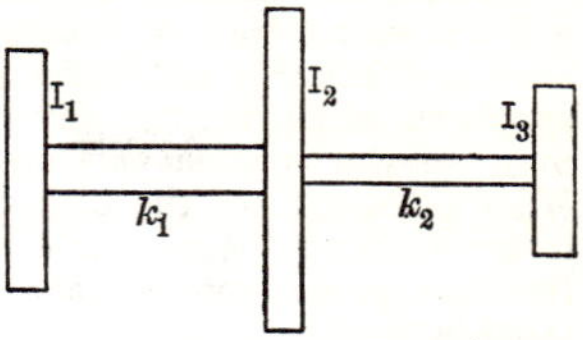

FIG. 17. Shaft carrying three flywheels.

Shaft Carrying Three Flywheels. Figure 17 shows a shaft carrying three flywheels, one of which may be the equivalent of a crankshaft. This system has two critical frequencies, given by the expression.

$$f = 3.13\sqrt{\frac{-B \pm \sqrt{(B^2 - 4AC)}}{2A}} \qquad (12)$$

where $A = (I_1I_2I_3)/(k_1k_2)$; $B = -[\{(I_1I_2 + I_1I_3)/k_1\} + \{(I_2I_3 + I_1I_3)/k_2\}]$; $C = I_1 + I_2 + I_3$; I_1, I_2, I_3 = moments of inertia of the three flywheels; and k_1, k_2 = torsional elastic constants of the two shaft lengths between the flywheels, inch-pounds per radian of twist.

For a disk-shaped flywheel of weight W pounds and radius r inches,

$$I = \frac{Wr^2}{2} \text{ pound-inches}^2 \qquad (13)$$

If the shafting that spans distance between two flywheels is of uniform diameter for a given span,

$$K = \frac{\pi N a^4}{2L} \qquad (14)$$

where N = rigidity or shear modulus of shaft material; a = shaft radius, inches and L = length of span, inches. The lower of the two frequencies given by eq. 12 is that for which the inside radical is negative. The first and last flywheels vibrate in opposite phase with a node somewhere between. If the system is made up of three identical flywheels connected by two identical spans the node is exactly at the middle flywheel, which remains stationary while the two end ones vibrate against each other. For the upper of the two frequencies, the end flywheels vibrate in phase; the middle one is in phase opposition, with a node somewhere along each connecting shaft length, making two nodes in all. No more resonant frequencies are possible with this system.

Shaft with one flywheel at each end is solved by eqs. 12, 13, and 14, where $I_2 = 0$, in which case eq. 12 reduces to the simple form

$$f = 3.13\sqrt{\frac{(I_1 + I_3)k}{I_1I_3}} \qquad (15)$$

where $k = k_1k_2/(k_1 + k_2)$. This system contains one resonant frequency. Torsional elastic constants k_1 and k_2 are calculated by eq. 14 for shaft sections of uniform diameter as before, where L_1 and L_2 are lengths of the two shaft sections, and all quantities are in inch-pound-second systems. If the shaft is of uniform diameter from I_1 to I_3, the constant k can be found directly by eq. 14.

Design Considerations. The torsional stiffness of shaft and inertia loads must be such that none of the calculated natural frequencies is in step with any periodic impulse from the driving engine. The torque impulses from an engine may be of the order of $1/2$, 1, $1\ 1/2$, 2, $2\ 1/2$, 3 etc., i.e., the lowest frequency impulse is one-half the running speed frequency. For airplane engines and for ship drives the propeller-blade frequency is one of the stimulating frequencies, and often the one which causes the largest stresses in the shaft. It is desirable, but often not practical, to make the system so stiff that all its critical frequencies are above the highest torsional impulse frequency. Friction devices sometimes are used for smaller engines, whereby one or more inertia parts are permitted to turn on the shaft against friction clamps. When torsional oscillations develop, the clamps damp them out. Friction devices of this kind often give trouble.

Most actual cases cannot be treated by the simple analysis given above because of varying shaft sections or the presence of more than three inertia parts, etc. For more complicated problems in this field, as those on automobile crankshaft design, see Ref. 2?

Ref. 3, p. 232; and Ref. 11. An understanding of general principles, however, should enable a designer to approach problems of this kind intelligently and to guard against torsional vibration.

TURBINE-WHEEL VIBRATION. Cause. The edge or periphery of an elastic disk will transmit a train of flexural waves around it at a characteristic speed determined by mass per unit area and stiffness of the disk. As the boundary of the disk has a finite circumferential length, the wavelengths must have certain fixed values; e.g., the disk edge may carry a continuous train of 2, 4, 6, or 8, etc., waves, whose wavelength is $^1/_2$, $^1/_4$, $^1/_6$, $^1/_8$, etc., of the disk circumference. Each wave train has a different velocity, longer wavelengths moving at considerably higher speeds than shorter. In a disk of uniform thickness, supported at the center, wave speed, revolutions per second, can be calculated by

$$\text{Revolutions per second} = K(h/r^2)\,\frac{\sqrt{Eg/W_1}}{2n} \tag{16}$$

where K depends on n, the number of nodal points along the disk edge. For values of K and meaning of symbols, see p. 9-04.

Turbine wheels are not of uniform thickness and must be calculated by special methods. For detailed discussion of turbine-wheel vibration and design, see Ref. 12.

Critical speeds exist for turbine wheels at which operation is hazardous. These critical speeds are equal to the several wave-train speeds given by eq. 16 except for a correction necessary to allow for stiffening effect of centrifugal force, which makes wave speeds for a revolving wheel somewhat higher than those given by eq. 16 for a stationary wheel. See Ref. 12.

When a turbine wheel runs at one of its criticals, the wave train at whose speed the wheel revolves will be stationary in space, because the wave train moves backward in the wheel at the same speed that the wheel turns forward. A small axial pressure on the wave zone of the wheel near its outer edge, i.e., on the bucket zone, may sustain and build up such a standing wave train. Thus only the wave train moving backward in the wheel is built up, but the corresponding forward-moving wave train is not maintained.

Long waves, corresponding to 4 or 6 radial nodes, build up more easily than wave trains of shorter wavelength. Large, relatively thin wheels are more susceptible to wave-building action because of the greater area on which steam acts and their greater flexibility, permitting small initial disturbances from which wave trains build up to be more easily excited. For the same reason large-amplitude waves build up easily, with resulting fatigue failures of disk material.

Higher radial modes can be dangerous although their amplitudes are relatively small. The distance along the periphery between the nodal diameters is rather small, and then small amplitudes cause considerable stresses in the wheel. Failures near the bucket roots are known to have occurred in high-speed wheels for high-pressure turbines for vibrations with as many as 80 nodal diameters, at frequencies in the thousands of cycles per second.

Prevention. The remedy is twofold. (1) Design the wheels to avoid operation at all critical speeds or wave-train speeds. For small turbines, this usually can be done; for large wheels, it is not practical for the highest-speed wave train, unless the wheels are excessively heavy. The next best remedy is to place operating speed halfway between those of two consecutive-node wave trains. (2) Do not depend too much on avoiding critical speeds, but keep the buckets and wheels sufficiently stiff, based on past successful operation.

Precautions. It is not safe to depend on calculation alone. Measured frequencies, from which critical speeds can be calculated, may diverge 10% from the best calculations, because of initial stresses in the disk itself or to cramping of buckets. A hoop tension or compression along the outer zone of the disk will alter frequency materially. For details see Ref. 12.

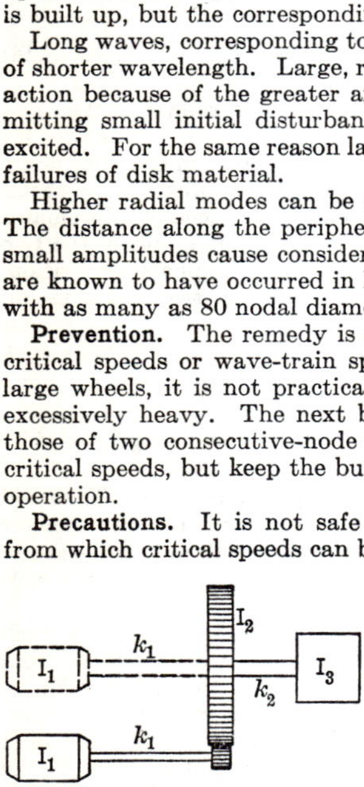

FIG. 18. Torsional oscillation in gear train.

GEAR-WHEEL VIBRATIONS. Torsional Oscillations in Gear Trains. Figure 18 illustrates a system of gears and flywheels subject to torsional oscillation. The two frequencies of this system are best calculated by reducing to a system equivalent to Fig. 17, by replacing I_1 and K_1 (Fig. 17) by I_1' and K_1', where $I_1' = n^2I_1$ and $K_1' = n^2K_1$, n being gear ratio. The inertia of the pinion has been neglected. If moment of inertia, I_2, of the gear wheel is small compared with I_1 and I_3 of the flywheels, I_2 may be taken as zero. There then is but one resonant frequency, found by

$$f = \sqrt{\left(\frac{I_1' + I_3}{I_1'I_3}\right)\left(\frac{K_1'K_2}{K_1' + K_2}\right)} \tag{17}$$

It is remarkable that in this case the pair of gears gives the same frequency that would obtain if the shafts were fixed end to end with no gears between, regardless of value of gear ratio n. This statement can be checked by substituting $I = I'/n^2$ and $K = K'/n^2$ in eq. 17, and comparing with result obtained when I' and K' are used. The relative vibration *amplitudes*, however, depend directly on n; that is, in Fig. 18, when gear train is present for a given oscillation amplitude of I_3, that of I_1 is $1/n$ times its value for the shafts directly connected without the intermediate gears. For more than one pair of gears the problem is more complex. See Ref. 2, p. 256, for a mathematical analysis of this problem.

Vibration of Gears Due to Impact. Practically all gears produce a hum of gear-tooth frequency, which in high-speed gearing under load may be very penetrating. This is primarily a forced vibration, and will be little helped by tuning. The only sure cure is precision of manufacture, together with a design which permits perfectly constant angular velocity for all tooth positions. Sometimes a resonant ring in gear wheels also is present. This type of noise can be largely eliminated by methods given under the heading of vibration damping (see Article 7).

VIBRATION OF STRUCTURES. All structures and buildings are subject to resonant vibrations of several frequencies; the lower frequencies arise from a swinging of the whole structure; the higher from vibrations of component parts. The best way to eliminate such vibrations, when set up by a periodic force of known frequency, is by some method of tuning.

FOUNDATION VIBRATIONS OF MACHINERY. Foundations of heavy machinery may be structures of considerable size. Sidewise and up-and-down frequencies in a steam turbine should be roughly estimated in design and compared with the revolution frequency or, for other types of machinery, with the predominant exciting frequency. Accurate calculations of frequencies of such structures are too complex to be worth while unless a thorough study is contemplated. Rough estimates based on common-sense application of vibration principles give almost as reliable results as more refined calculations.

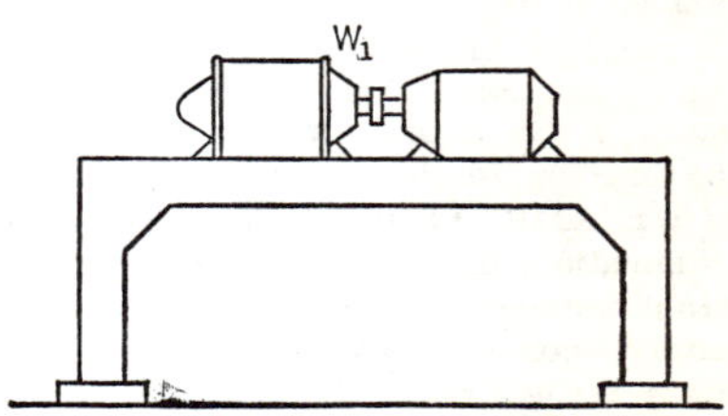

FIG. 19. Structure supporting rotating machinery.

Procedure. Reduce the system to a single lumped mass vibrating against single stiffness constants, one for up-and-down motion, one for sidewise motion, etc., and use eq. 3. For example, a structure (Fig. 19) carries rotating machinery of weight W_1. The total weight used to calculate up-and-down vibration frequency includes, say, half the structure floor, as only the middle portions vibrate up and down to any extent. For sidewise vibration, the total floor weight and part of the columns is included, as the whole floor and column tops move through approximately full amplitude. Constant k is found in each case by estimating the downward force on the platform and the sidewise force, respectively, in pounds required to produce 1 in. of deflection. This estimate requires some ingenuity, but with a little practice it is not difficult to obtain an approximate result. Frequency depends on the square root of mass and stiffness, and error is minimized in the final result.

Even if the principal modes of vibration have frequencies a safe distance from the operating frequencies, individual parts of the structure may resonate, especially when built of steel members. To cure resonance of individual members, stiffen or load them in different ways until vibration disappears. This cure usually is not difficult. It evidently is impractical to attempt to predetermine frequencies of *all* the individual members of such a structure.

BUILDING VIBRATIONS. Ordinary buildings have sidewise natural frequencies of 2 or 3 per sec, as is clearly shown during slight earthquakes, or when delicate apparatus is being mounted.

GROUND VIBRATIONS. Heavy machinery causes local vibration of its foundation, and also at times earth waves radiating in all directions to a considerable distance. If the foundation rests on bedrock, such transmission is negligible; with a foundation on clay or sand, and a periodic force of high intensity and low frequency, surface waves can be detected $^1/_2$ mile or more away in clay or sand beds. Ground waves may be avoided by placing heavy machinery foundations on bedrock and by avoiding clay or sand flats, as well as unbalanced exciting forces. Bedrock, however, transmits objectionable sound vibrations of higher frequency better than clay.

RESONANCE IN HELICAL SPRINGS. Helical springs subject to periodic compression and extension, as automobile valve springs, may resonate badly, with resulting fatigue failures. Longitudinal resonant frequencies in helical springs fixed at each end are calculated with considerable precision by

$$f = (n/2l)\sqrt{kg/w} = (9.8n/l)\sqrt{k/w} \tag{18}$$

where n = number of half waves between fixed ends; l = length of spring, inches; w = weight per inch of axial length, pounds; k = elastic constant of spring *per unit length*, i.e., force, pounds, necessary to double the length of spring; and f = cycles per second. Thus, the lowest frequency, where $n = 1$, corresponds to one-half wave between ends; the next, where $n = 2$, to two half waves, etc. (See Fig. 20.) To avoid these resonances, the spring must be designed so that none of these frequencies is in the operating range. If possible, its stiffness should be such that the lowest resonant frequency for which $n = 1$ is above the operating range of frequencies.

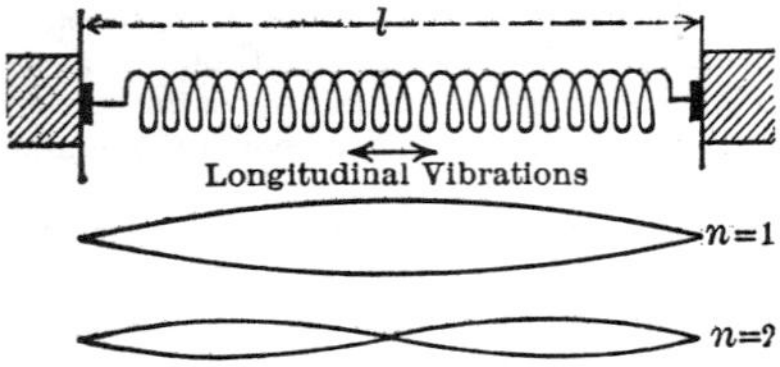

Fig. 20. Vibration of helical spring.

SHIP VIBRATION. Hulls of ships, particularly steel ships, are subject to resonant vibrations. A long ship has several resonant frequencies in which flexural vibrations of the ship as a whole occur, with two or more transverse nodal lines. These frequencies can be calculated approximately and should not be in step with exciting forces. The principal periodic exciting force to be avoided is propeller blade frequency, i.e., revolutions per second times the number of blades. This force comes from a periodic reaction of water from the propeller on parts of the stern, which is difficult to eliminate completely. Such a condition of resonance may be tuned out to a large extent by altering cargo distribution, thus altering natural frequency

DYNAMIC VIBRATION ABSORBER. Force elimination, or change of mass and spring constant to avoid resonant frequency vibration, is not always possible, especially when the vibrating force varies over a wide frequency range. The dynamic vibration absorber might then find application. It consists of a small spring-mass system attached to the machine (large spring-mass system) and tuned to the disturbing frequency (see Fig. 21). The amplitude of vibration of the machine under a vibrating force of its original resonant frequency is then exceedingly small; however, two new resonant frequencies occur whose frequencies are a function of the mass ratios and frequency. When a damping member is paralleled to the spring of the small mass-spring system, the maximum amplitude of the big mass which is obtained for excitation over the complete frequency range is $\epsilon = \sqrt{1 + (2/\mu)}$ for the optimum condition where the tuning f_m of the absorber is $f_m = 1/(1 + \mu)$. μ is the ratio of the absorber mass to the mass of the machine, ϵ is the ratio of the main mass deflection to the base deflection, and f_m is the ratio of the natural frequency of the absorber system to the main system.

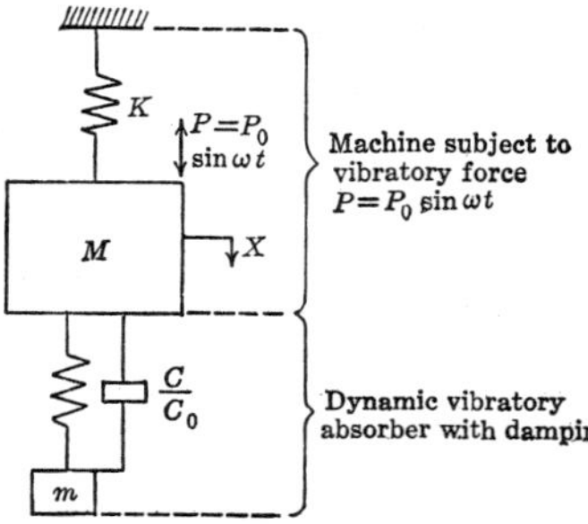

Fig. 21. Idealized dynamic vibration absorber.

Internal-combustion Engines. Dynamic vibration absorbers are being used to reduce torsional oscillations. Instead of small spring-mass system, a pendulum is used. The natural frequency of the pendulum is directly proportional to the rotational speed of the shaft when centrifugal forces are acting on the pendulum mass. By proper design the pendulum has a natural frequency which is always equal to rotational frequency or its multiples. For details of these centrifugal pendulums, see Ref. 3, p. 274.

Ship Stabilizers. The Frahm tank installed in ships to reduce rolling works on the principle of the dynamic vibration absorber. The idealized Frahm tank consists of a U-tube filled with water. The water level oscillates when the U-tube is vibrated. Damping is realized by installation of throttles at the openings of the tubes. See Ref. 3, p. 136. Similar effects are obtained with a gyro stabilizer, which is a large gyroscope coupled to the ship by suitable damping.

6. VIBRATION REDUCTION BY ISOLATION

Forced vibrations may be cured by resorting to isolation; they do not yield readily to damping and cannot be tuned out. The standard method of isolation is to use an elastic suspension.

ELASTIC SUSPENSIONS function through ability to reduce the periodic force reaction of a vibrating body on its support. Figure 22 shows the simplest possible spring suspension, where a weight W, with an up-and-down forced vibration of frequency f imposed on it, is supported by a vertical spring. Transmissibility of such a spring support is given by

$$\text{Transmissibility} = \epsilon = \frac{f_0^2}{f^2 - f_0^2} \qquad (19)$$

where f_0 is natural or resonant up-and-down frequency of spring support as given by eqs. 3 and 5. Transmissibility may be defined as the ratio of vibration forces actually transmitted to those which would be transmitted without spring suspension. Thus, to be of use, ϵ must be less than unity; experience shows that in most cases to be really effective ϵ should be as small as one-thirtieth to one-fortieth. This means that f_0 should be $^1/_5 f$ to $^1/_6 f$. Figure 23 shows a graph of eq. 19 as a function of $f/f_0 = r$.

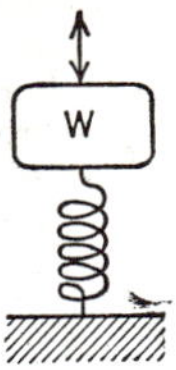

FIG. 22. Simple elastic suspension.

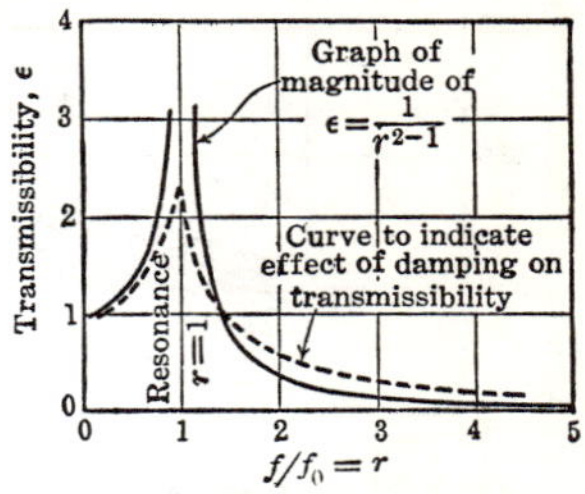

FIG. 23. Transmissibility with (dotted curve) and without (solid curve) damping.

PRECAUTIONS IN USE OF ELASTIC SUSPENSIONS. All elastic suspensions necessarily have their own resonances. Thus if a suspension is designed to isolate a frequency of $f = 120$ cps with a resonant frequency $f_0 = 30$ cps, no 30-cycle exciting frequency must be present in addition to the 120-cycle one. This is especially true for rotating machinery, whose rotation speed is usually in that field. Figure 23 shows that transmissibility is greater than unity when f approaches f_0, and may be very large near resonance. Furthermore, ϵ is always greater than unity when f is less than f_0. Nearly all elastic suspensions have resonance frequencies other than f_0 corresponding to the particular mode of vibration for which they are designed. This fact must be considered in design.

Elastic suspensions do not damp a vibration by friction, but prevent transmission by power to isolate. Contrary to general belief, damping increases transmissibility, although the increase is small with a moderate amount of damping. (See Ref. 4, pp. 96–107.) A moderate amount of damping is of advantage as it safeguards against excessive vibration near resonance, but not because it improves isolation. Resonant frequency of the suspension is excited when the machine is started or stopped, or in case of shock excitation, as for punch presses. Elastic suspensions for punch presses should have a considerable amount of damping. See eq. 5 for frequency calculation from gravity deflection d.

ELASTIC SUSPENSION MATERIALS. Steel, rubber, and cork are commonly used for spring suspensions.

Steel has the advantage of being very durable. It contains practically no damping, and therefore is subject to possible resonances. It has also the disadvantage that very high frequencies are transmitted through the steel mounting in the longitudinal direction just as a listening rod held against a motor bearing transmits the high frequency bearing noises even though its transverse natural frequency is very low. It is good practice to place rubber pads in series with the steel suspension when frequencies above 200 cps are present. These rubber pads can be rather stiff since they have to isolate only the higher frequency components.

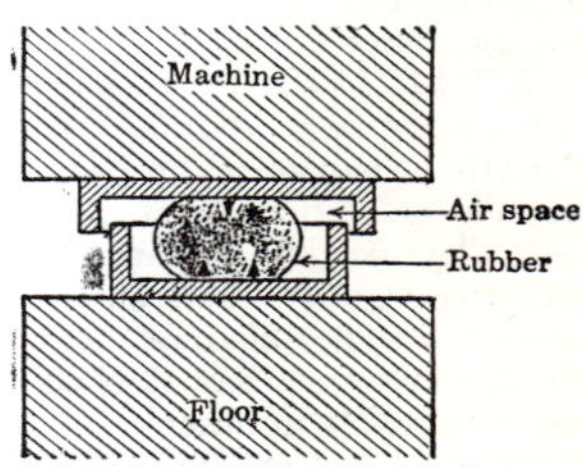

FIG. 24. Rubber support.

Rubber is more flexible than steel from the design standpoint. It has some damping. To avoid age hardening, only the highest grade, produced for this special purpose, should be used. Long life is then possible, even under severe conditions. Rubber should not be loaded more heavily than 80 psi. It is highly incompressible and yields almost entirely through shear. In designing rubber supports, space for elastic flow must be maintained, or they may be too rigid (see Fig. 24). Rubber pads usually should be of small area so that they can freely squeeze out sideways under load. For a given thickness the hardness of a rubber pad increases disproportionately under compression with increase of area (see Fig. 25).

Rubber must not be used at temperatures over 120 F and below −20 F. When applications beyond this temperature range are required it is necessary to take into account the considerable change in modulus of elasticity. Natural rubber must not come in contact with oil. Coating with oil-resistant varnish affords some protection. Some synthetic rubbers are oil resistant, however, but are frequently inferior because of lower tensile strength, higher damping, greater change of elasticity modulus as function of temperature, and creep. For long-time applications there is some creep in natural rubber mounts. Commercially, rubber mounts are available for the majority of applications, using, almost exclusively, rubber loaded in shear and bonded to metal adapters which can be readily attached to the vibrating structure (see Fig. 26 for typical mounts). Manufacturers of mounts are the Lord Manufacturing Company, Erie, Pa.; MB-Manufacturing Company, New Haven, Conn.; L. N. Barry Company, Cambridge, Mass.; and Vibration Eliminator Company, Long Island City, N. Y.

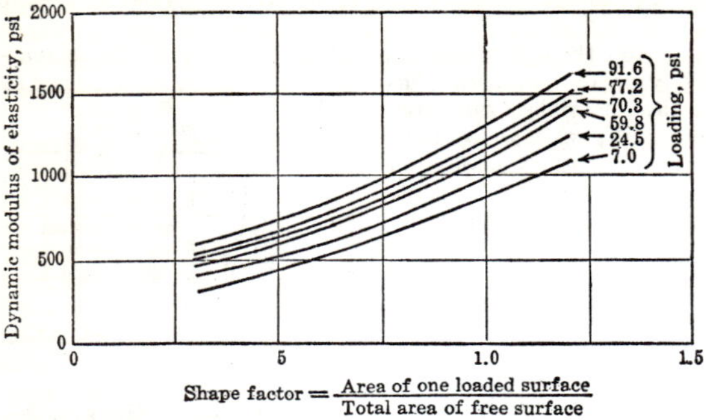

FIG. 25. Modulus of elasticity as a function of shape for various loadings, rubber compression mountings.

Cork differs from rubber in that it is directly compressible, and space for elastic flow is not needed. The stiffness of cork increases directly as its area when used in large pads. Cork pads of comparatively large area thus may be effective in isolation, particularly for higher frequencies in the sound range. Cork has the disadvantage of gradually compressing under load after initial loading, resulting in gradual loss of isolating power for lower frequencies. Despite this fact, cork has performed successfully for many years in some installations. Cork mounts are being superseded today by rubber mounts except for a few special applications. Mounts using cork are available from the Korfund Company and Armstrong Cork Company.

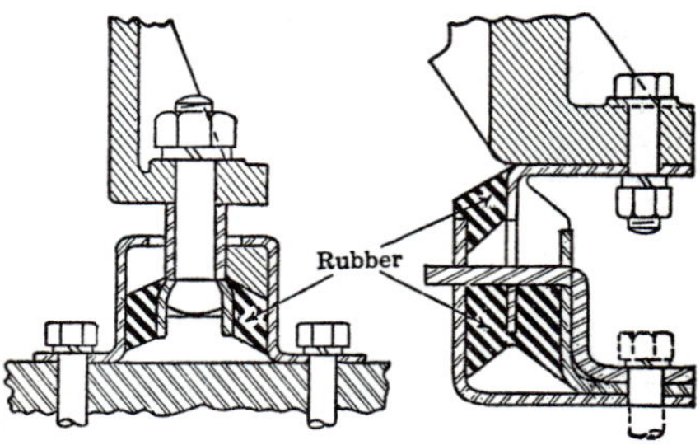

FIG. 26. Shear mountings employing rubber bonded to channel or angle bars.

SUSPENSION FREQUENCY CALCULATIONS for up-and-down frequencies are in all cases made by eq. 5, but for rubber and cork actual frequency is higher than calculated, because of imperfect elasticity. For some materials, gravity compression d must be greater than required by eq. 5. Cork, to obtain required suspension frequency, may be two or three times as great as for perfectly elastic material like steel.

APPLICATIONS. General application of elastic suspensions requires considerable knowledge of vibration. Elastic suspensions usually have several modes of vibration. Exciting frequencies, other than the one being isolated, probably will be present, e.g., revolution frequency of a rotating machine operating over a wide range. A suspension must be designed so that all its vibration frequencies avoid all exciting frequencies. An otherwise good elastic suspension may have an objectionable low-frequency quiver. A suspension, flexible in some directions, often must be rigid against belt pull, in which case rubber is better than steel springs. Steel springs sufficiently flexible in one direction may be too flexible in others. Flexible mountings operating on a floor that is both light and flexible, e.g., spring-suspended machinery on spring-supported car bodies or airplane engines on the relatively light structure have to be determined by analysis of the compound elastic systems with more than one mass and elastic connection.

There is a general tendency to assume that a spring-mounted machine will have only one vertical natural frequency. Since a mounted object will usually have at least six natural frequencies, such a procedure may be embarrassing and dangerous. The following modes of vibration can be expected for a machine supported by elastic mounts at its four corners: (1) Vertical vibration in the vertical plane. (2) Torsional vibration around the vertical axis. (3) A low rocking mode in the vertical plane around a front-rear axis. (4) A high rocking mode in the same plane as 3. (5) A low rocking mode in the vertical plane around a left-side–right-side axis. (6) A high rocking mode in the same plane as 5.

If the ratio of the horizontal to vertical spring constant is less than 0.5, the mounts can generally be arranged so that all frequencies are nearly equal or less than the vertical frequency. If possible, the line of action of the mounting should pass through the center of gravity of the mounted equipment, so avoiding coupled vibrations, i.e., a translatory force will excite translatory motion only, a torque or a moment will excite only rotation about one of the axes. (See Ref. 4, pp. 96-107, and Ref. 13.)

FLOOR SUSPENSIONS. Some machinery can be isolated by mounting on a comparatively flexible floor frame, elastically supported on the solid floor below. This prevents the transmission of rumbles, grinding noises, and hums through the building, to radiate sound from panels, etc. Springs or rubber pad suspensions used at intervals of about 1 ft, depending on the floor load, will perform well, steel being more permanent. The flooring should yield perceptibly, say 0.1 in., under machinery load. A difficulty is the presence of possible side swings and low-frequency quivers, which are best cured by rearranging supports, rather than by calculation. Printing-press vibration, for example, has been so isolated. Machinery requiring exact alignment, as an accurate lathe, cannot be placed directly on a flexible floor, but must be set on a rigid, spring-supported platform. Such work is done best by concerns specializing in construction of elastic floorings.

If a cork layer is used instead of a spring-supported flexible floor, low-frequency rumbling noises may be transmitted, if not the higher frequency notes, unless a layer is used that is thick enough to permit appreciable compression ($^1/_{16}$ to $^1/_8$ in.) under load.

MOTOR SUSPENSIONS. **Single-phase electric motors** are subject to torque pulsation of twice the a-c supply frequency, i.e., 120 cps for a 60-cycle source, causing a 120-cycle torsional forced vibration of both stator and rotor about the motor axis. This 120-cycle vibration causes the hum noticed in domestic use of such motors, which a well-designed elastic suspension will completely eliminate. To isolate such 120-cycle vibrations, a suspension of torsion flexibility is required, whose natural frequency f_0 about the motor axis is, say, 20 cps; since $f = 120$, $f/f_0 = 6$, and, from eq. 19, $\epsilon = ^1/_{35}$, giving fair isolation. This support must have considerable transverse rigidity to resist belt pull, and also to prevent resonating to the speed frequency due to slight mechanical unbalance. Figure 27 shows a satisfactory elastic suspension using rubber.

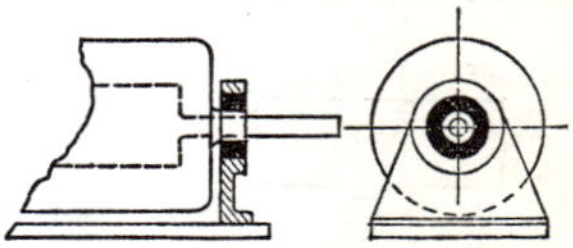

FIG. 27. Elastic motor suspension using rubber.

Polyphase electric motors with good elastic suspension can be used instead of d-c motors where noiseless operation is required. Although subject to much less vibration than single-phase, no polyphase motor is free from electrical hums which can be transmitted to surrounding structures. Especially designed elastic bases are furnished by the manufacturer. The design of such a base is not easy, but if correctly designed with rubber pad suspensions, its performance is excellent.

AUTOMOBILE-ENGINE SUSPENSION. Flexible supports of comparatively low resonant frequency are being used on many types of automobile engines with apparent success. Avoidance of suspension resonances is difficult with a wide operating speed range and consequent wide range of exciting frequency. The principal resonant frequency is low and is easily passed through in accelerating the engine. At normal operating speeds, isolation is excellent, and engine vibration is hardly perceptible in the car. The chief disturbance is the variation in torque reaction, and the mountings are applied so as to afford torsional resilience about a longitudinal axis. The axis about which the moment of inertia is least is chosen, and virtual center of gravity suspension is attained.

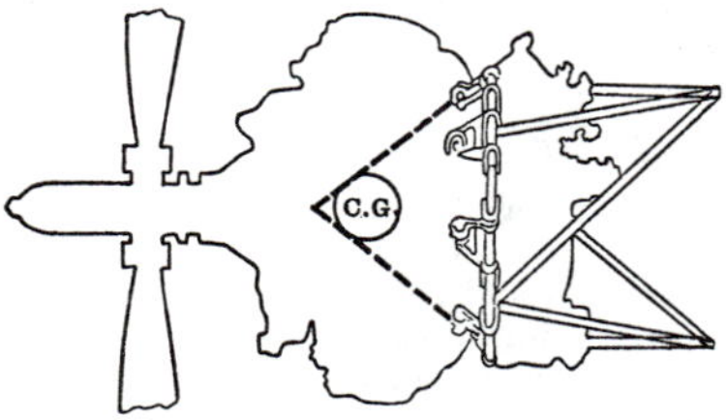

FIG. 28. Dynafocal airplane engine mounting.

AIRPLANE-ENGINE VIBRATION. For in-line engines the same principles as for automobile-engine suspension apply. For radial engines supported from one side only, the Dynafocal system (see Fig. 28) is being used. This system provides center-of-gravity suspension, the individual mountings being inclined so that they point toward a focal point just beyond the center of gravity.

ISOLATION FROM VIBRATION. The isolation of certain bodies from forced vibrations and waves of outside origin sometimes is necessary. This problem is the reciprocal of that of vibration elimination through tuning, and the same principles and formulas apply.

Isolation of delicate apparatus from building tremors requires an idea of the approximate frequencies present, obtained by observing quiver of the surface of a dish of mercury, a Frahm tachometer, or any other vibration instrument permitting recording of the vibration wave shape. Building and ground tremors (from vehicles, etc.) contain low-frequency harmonics which can be isolated only by a suspension having frequency less than 1 cps. This follows from transmissibility eq. 19, where f is frequency of the building tremor and f_0 that of the suspension. Evidently both the up-and-down oscillating frequency and the sidewise swing of the suspension must be very low. The instrument to be isolated should be hung by long springs (helical) which extend several inches under the load. (See eq. 5.) Since a seconds pendulum is about 3 ft long, suspension springs and wires must be at least several feet long to insure a low enough frequency of sidewise swing.

Several pieces of apparatus may be isolated together on a relatively massive platform hung either from the ceiling or a davit, using springs extending several inches under the load.

Good isolation sometimes can be obtained by placing the apparatus on a heavy metal plate resting on an inflated automobile inner tube. This is not sufficient for very low-frequency disturbances, because to screen from the latter the suspension must yield several inches under load.

All helical springs used in suspensions should be loosely stuffed or wrapped with damping material to prevent local spring resonant oscillations. It is important not to use too much damping, as in a spring suspension it increases the transmissibility.

When frequencies above 5 cps are to be isolated, a suspension using rubber pads or commercially available rubber mounts is often cost and space saving.

VEHICLE SUSPENSION. A successful vehicle suspension should be as flexible or of as low frequency as is consistent with stability. It must have some damping. The lower the frequency the better the isolation from road impulses. Damping is essential in vehicle suspensions, to limit unavoidable resonant bounding oscillations due to the great range of exciting frequencies produced by various road conditions and speeds.

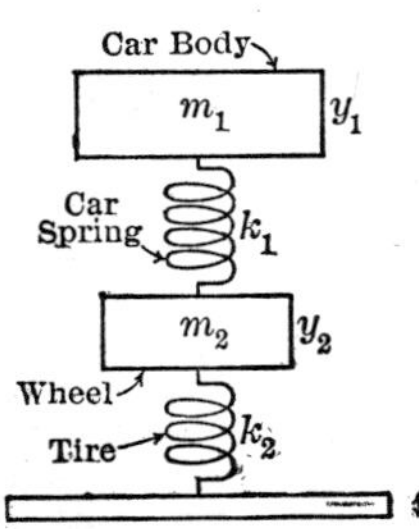

FIG. 29. Schematic representation of vehicle suspension.

In railroad cars, the use of built-up leaf springs provides damping by friction between individual leaves. In automobiles, leaf spring friction is not sufficient. So-called shock absorbers are necessary, whose function is to reduce transmissibility at or near suspension resonance points. For other exciting frequencies, transmissibility is slightly increased by such devices.

Automobile suspensions must isolate against more severe vibration excitation than railroad car suspensions. They are compound, with one elastic support between wheel and road and another between wheel and car body. Figure 29 represents schematically such a system with damping omitted; m_1 represents car body mass; k_1, car spring elastic constant; m_2, wheel mass; k_2, tire elastic constant. Actually the system is more complex, but the scheme of Fig. 29 yields the two principal modes of vibration of ordinary automobile suspensions.

Because of its compound character, transmissibility is given by an expression which contains both its resonant frequencies, f_1 and f_2, as well as exciting frequency, f, from the road. Where y_1 and y_3 are amplitudes, this expression is

$$\text{Transmissibility} = \frac{y_1}{y_3} = \frac{1}{[(f^2 - f_1^2)/f_1^2][(f^2 - f_2^2)/f_2^2]} \tag{20}$$

The higher resonant frequency f_1 is excited by "washboard" roads which cause the wheels to resonate up and down between the road and stops on the vehicle body. This condition arises when $f = f_1$, and without shock absorbers would be excessive. The lower resonance occurs when $f = f_2$, where the whole car body swings up and down in step with long waves in the road. At this frequency shock absorbers act again. This compound system reduces to a simple one, with transmissibility given by eq. 19 when a rigid connection replaces k_1. In this case, f_1 becomes so large compared with f that the first factor $(f^2 - f_1^2)/f_1^2$ can be replaced by unity.

This discussion brings out the principal factors in vehicle suspensions. Frequencies f_1 and f_2 should be as low as possible so that ϵ is as small as consistent with stability between resonant points. At the same time, ϵ is limited at these points by shock absorbers which are not so energy absorbent that ϵ is appreciably increased between resonant points. For derivation of eq. 20, see Ref. 4, pp. 107–109. See also Ref. 14.

SHOCK MOUNTS. Punch presses should be isolated against transmission of impacts into the floor. The exciting force has a multitude of frequency components. It is not practical to mount for the lowest frequency component applying the principles mentioned above, since this would lead to a rather unstable support that would oscillate at its resonant frequency after each impulse with rather large amplitude. A natural frequency of approximately 30 cps has been found suitable for this type of mount. Shock mounts should have a considerable amount of damping. Rubber is used mostly as resilient material.

7. VIBRATION DAMPING

Vibration damping is reduction of vibration by energy absorption such as friction. Neither avoidance of resonance by tuning nor prevention of transmission by spring suspensions are damping processes. Vibration damping is of maximum effect in curing resonant vibrations where the rate of energy supply exciting vibration is relatively small.

SLIDING FRICTION DAMPING exists in built-up leaf springs whose leaves slip against each other, creating friction when the spring oscillates. This friction is important in limiting resonant vibrations of railroad cars on their springs. Sliding friction damping is used in some forms of automobile snubbers.

Torsional oscillations in shafts may be limited by a heavy flywheel, attached to the shaft by sliding friction, which slips against oscillations of the shaft. Critical speed vibrations of shafts can be eliminated by spring-supporting bearings in a collar that permits slipping against a friction clamp. Sliding friction dampers often give trouble from variable friction due to variation of pressure of the clamps, to scoring, or to the presence of foreign materials like oil.

Sliding friction in the joints is important in limiting vibration of steel structures because of the extremely low internal friction of steel. Turbine wheels with separately made buckets have far more resistance to vibration than those with buckets integral with the disk.

IMPACT DAMPING arises from the impact of loose parts of a structure in an effort to vibrate independently. Such impact has a strong dissipating effect and rapidly damps vibration. A steel gear which otherwise would ring under a hammer blow will sound quite dead if a steel ring is snapped into a slot under the rim on each side. Impact damping is effective in dissipating high-frequency vibrations of small amplitude.

FLUID DAMPING differs from sliding friction damping in that the resisting force rapidly increases with the velocity of motion; with sliding friction resisting force is independent of velocity. Dashpots utilizing fluid flow have powerful damping action for large amplitudes; the damping action falls off rapidly with slow motion at small amplitudes.

SOLID DAMPING. All solid materials have more or less internal friction damping resistance to periodic deflection. The most perfect tuning fork loses its vibration amplitude from this cause when all external friction effects, as air resistance, have been removed. Softer elastic materials, as rubber, have considerable internal friction. Steel has so little solid damping that structures built up of steel members owe nearly all their vibration resistance, as a whole, to sliding and impact friction joints. Individual members and panels, however, may be damped largely through what internal friction exists. Individual members in a steel structure may be highly resonant.

Concrete has ten times as much damping as steel, and concrete structures are not particularly resonant. The overall damping in a steel structure does not differ greatly from that of a similar concrete structure, because of dissipation in friction and vibration interference between individual members.

8. METHODS FOR SOLVING VIBRATION PROBLEMS

Calculations of the natural frequencies of coupled systems become rather complex. These problems are often solved by punch cards that are fed through tabulating machines. Such a system is widely used for the calculation of the natural frequencies of shafts supported by several bearings. (See Ref. 15.)

Electromechanical analogies are frequently used where mechanical quantities such as force and velocity are set equivalent to electrical quantities such as voltage and current. (See Ref. 16.) An electrical circuit representing the mechanical constants is often found to be useful. Difficulties in the application of this analogy, however, arise because the damping of the electrical circuit often cannot be made low enough for the analogy. The electromechanical analogy is especially useful when coupled mechanical and electrical systems are to be analyzed, such as servo systems and electromechanical transducers.

Differential analyzers have been found to be very useful for solving problems of torsional vibration. (See Ref. 17.)

REFERENCES

1. R. O. Fehr, Vibration testing, *Gen. Elec. Rev.*, Dec. 1942.
2. S. Timoshenko, *Vibration Problems in Engineering*, D. Van Nostrand Co., New York, 1937.
3. J. P. Den Hartog, *Mechanical Vibrations*, McGraw-Hill Book Co., New York, 1947.
4. A. L. Kimball, *Vibration Prevention in Engineering*, John Wiley and Sons, New York, 1932.
5. E. L. Thearle, A new type of dynamic balancing machine, Paper APM-54-12, *Trans. Am. Soc. Mech. Engrs.*, 1932.
6. S. G. Best, Propeller balancing problems, *SAE J.*, Nov. 1945, p. 648.
7. T. C. Rathbone, Vibration tolerance, *Power Plant Eng.*, Nov. 1939.
8. A. L. Kimball and P. L. Alger, Single-phase motor torque pulsations, *Trans. Am. Inst. Elec. Engrs.*, 1924, p. 730.
9. W. A. Tuplin, *Torsional Vibration*, John Wiley and Sons, New York, 1934. Out of print.
10. W. Ker Wilson, *Practical Solution of Torsional Vibration Problems*, 2nd ed., John Wiley and Sons, New York, 1940.
11. R. E. Root, *Dynamics of Engine and Shaft*, John Wiley and Sons, New York, 1932.
12. W. Campbell, Protection of steam turbine disc wheels from axial vibration, *Trans. Am. Soc. Mech. Engrs.*, Vol. 46, 1924, p. 31.
13. J. N. MacDuff, Isolation of vibration in spring mounted apparatus, *Product Eng.*, July and Aug. 1946.
14. B. S. Cain, *Vibration of Rail and Road Vehicles*, Pitman Publishing Corp., New York, 1940.
15. A. W. Rankin, Calculation of the multiple span critical speed of flexible shafts by means of punched card machines, *Trans. Am. Soc. Mech. Engrs.*, Vol. 68, June 1946, p. A117.
16. H. F. Olson, *Dynamical Analogies*, D. Van Nostrand Co., New York, 1943.
17. C. Concordia, Torsional oscillations of ship drives, *Soc. Naval Architects and Marine Engrs.*, 1943, p. 51.

NOISE

9. NOISE IN GENERAL

DEFINITION. Noise is defined as an undesirable sound. When vibrations of bodies are transmitted into the air as sound waves of different frequencies, the heterogeneous sound resulting is noise. Noise reduction is important in engineering.

VELOCITY OF SOUND is a wave phenomenon in the atmosphere. It has constant velocity V of transmission, regardless of its frequency or wavelength, of about 1100 ft per sec.

$$V = \sqrt{1.4pg/w} \tag{1}$$

where V = velocity, inches, per second; p = atmospheric pressure, pounds per square inch; g = acceleration of gravity, inch-seconds^{-2} = 386; and w = weight of 1 cu in. of air. Since w varies as p, V is unaltered by pressure change. When temperature increases, however, w alone decreases. Thus while not altered by pressure change, transmission velocity is proportional to the square root of absolute temperature, and varies about 5% in ordinary temperature ranges.

PERCEPTION OF SOUND. The human ear can perceive as tones, noise frequencies from 25 to over 16,000 cps. In its most sensitive range, 500 to 6000 cps, the ear is extraordinarily sensitive, the wave amplitude of easily detected sounds being of molecular dimensions. In this range, amplitudes of a millimeter are heard as terrific noises. Sounds in this range are easily excited and difficult to eliminate.

FREQUENCY AND WAVELENGTH. The frequency of a sound is the number of complete waves transmitted per second. High frequencies correspond to short wavelengths and vice versa.

$$\text{Wavelength} = \lambda = \frac{V}{f} \tag{2}$$

where, if V = inches per second, and f = cycles per second, λ is in inches. Table 1 compares frequencies and wavelengths in the audible range.

Table 1. Sound Frequencies and Wavelengths

Frequency f, cycles per sec	100	200	500	1000	2000	4000	6000	10,000
Wavelength, in.	132	66	26.4	13.2	6.6	3.3	2.2	1.3

A heterogeneous sound may be made up of wavelengths from this entire range, superimposed upon each other, the smaller traveling over the larger. Individual frequencies are called components. In noise work, sound should be considered in terms of its frequency.

SOUND VELOCITY IN SOLIDS is much higher than in air. In steel, it is about 17,000 ft per sec; in other building materials it is less, but much greater than in air.

FREQUENCY AND CHARACTER OF SOUND. Rumbles and boomings are of long wavelength. Ordinary whistling sounds are of intermediate wavelength. Tinkling noises and rattling of paper are of higher frequency. The whirring of a fan is a very complex noise in which, for small fans, higher frequencies predominate; for large fans and blowers a considerable proportion of low-frequency harmonics is present.

10. NOISE MEASUREMENT

SOUND-LEVEL METERS afford the best means of measuring noise. Specifications for sound-level meters have been established by the American Standards Association in *American Standard for Sound Level Meters for Measurement of Noise and Other Sounds*, Z24.3, 1944. Sound-level meters consist essentially of a microphone of uniform directional characteristic, a vacuum-tube amplifier, and a voltmeter. The meter reads sound level directly in decibels (see below). The amplifier has three networks corresponding to the ear response at 40 db, 70 db, and 100 db. At 100 db, the response is essentially flat. Commercial sound-level meters cover a range of 20 to 130 db. For the analysis of the sound the output of the sound-level meter can be connected to earphones, oscillographs, and recorders, or to harmonic analyzers (see below). Sound-level meters are manufactured by the General Electric Company, Schenectady, New York, the General Radio Company, Cambridge, Mass., and the Western Electric Company, New York, N. Y.

HARMONIC ANALYZER. A separate analyzer to determine frequency components can be used with the sound-level meter. Frequency components and their relative amplitudes are indicated. These analyzers are also useful for vibration analysis, when used with the vibration detectors mentioned before. They are made by the General Radio Company, the Hewlett Packard Company, Palo Alto, Calif., and the Western Electric Company.

TUNING FORKS. Much time can be saved in a frequency determination by using a set of tuning forks and noting beats. To determine, for example, whether or not a whistle from a gear is tooth-frequency noise, the note frequency is compared with the known frequency of gear-tooth impacts.

STETHOSCOPES. A physician's stethoscope, with a sharp point about 1 in. long fixed to the center of the diaphragm, placed against a vibrating structure is a very sensitive detector of noise transmission which otherwise cannot be heard. By tilting the point in different directions, the approximate direction of maximum amplitude can be detected. This instrument can locate node and loop areas on a sounding body of small vibration amplitude.

A steel diaphragm and point near a-c electrical apparatus will give a false indication. Disk and point and holder should be of phosphor bronze or other nonferrous metals.

VIBRATION DETECTORS. In connection with a sound analysis it is often necessary to study vibration patterns. Any of the vibration-measuring instruments mentioned on page 9-06 may be used. Many of the electrical pickups can be used with the sound-level meter, and very small vibration amplitudes, associated with sound production, can be directly read.

SOUND-LEVEL UNITS—DECIBELS. The sound-level unit almost universally adopted in commercial work is the decibel (db). Table 2 shows the meaning of the decibel scale in terms of: (*a*) everyday sounds; (*b*) vibrational energy in sound waves; (*c*) amplitude of motion of air particles in a 100- and a 1000-cycle wave; and (*d*) rise and fall of pressure in sound waves. The type of noise scale in Table 2 is constructed so that every increase of 10 db corresponds to ten times the noise intensity as shown in the third column. Thus a very loud noise of 115 db is ten billion times as energetic as a very low sound in perfectly quiet surroundings. This scale covers the whole audible sound range in 100 units, and distinguishes noises more nearly as the ear actually does than would a scale based upon energy or pressure units, as given in columns 3 and 6 of Table 2.

For practical purposes it is useful to remember that every 3-db increase in noise means that the intensity almost exactly doubles. If one noise is 12 db greater than another, the first noise intensity must be doubled four times, i.e., multiplied by 16, to equal that of the second in intensity, since $^{12}/_{3} = 4$. Two noises of equal intensity placed side by side give twice the intensity, or a total noise 3 db greater than either of them by itself regardless of their separate noise levels. Examples are shown in Table 3.

Table 2. Decibel Scales in Various Units

Decibels	Everyday Sounds	Sound Intensity (Threshold = 15 db)	Half Amplitude of Wave Motion, cm — Cycles 100	Half Amplitude of Wave Motion, cm — Cycles 1000	Rise and Fall of Pressure in 1000-cycle Wave, dynes per sq cm
15	Very low	1	....	....	10^{-3}
25	Rustle of leaves	10	10^{-7}	10^{-8}	....
35	Whisper	10^2	....	....	10^{-2}
45	Quiet residence	10^3	10^{-6}	10^{-7}	....
55	Average residence	10^4	....	....	10^{-1}
65	Average business office	10^5	10^{-5}	10^{-6}	....
75	Passenger automobile, 15 to 50 ft	10^6	....	....	1
85		10^7	10^{-4}	10^{-5}	....
95	Motor trucks, 15 to 20 ft	10^8	....	....	10
105	Heavy street traffic in city	10^9	10^{-3}	10^{-4}	....
115	Subway express passing through	10^{10}	....	....	10^2

Table 3. Comparison of Noise Values from Several Sources

Source 1	Source 2	Source 3	All Sources Together. Intensity of Source 1×	Source 1	Source 2	Source 3	All Sources Together. Intensity of Source 1×
35 db	35 db	0 db	2 = 38 db	40 db	40 db	0 db	2 = about 43 db
35 db	35 db	35 db	3 = about 40 db	40 db	40 db	43 db	4 = about 46 db
35 db	35 db	38 db	4 = about 41 db				

MATHEMATICAL DEFINITION. The difference in decibels between two noises of intensities I_1 and I_2 is defined by the mathematical relation:

$$\text{Difference in decibels} = 10 \log_{10} \frac{I_1}{I_2} \tag{3}$$

$$= 10 \log_{10} \frac{p_1^2}{p_2^2} = 20 \log_{10} \frac{p_1}{p_2} \tag{4}$$

where p_1 and p_2 are sound pressure amplitudes corresponding to I_1 and I_2. Equation 4 is multiplied by twice as large a constant as that of eq. 3 because sound intensity is proportional to the square of pressure amplitude of sound waves.

Equations like 3 and 4 involve only pressure and intensity ratios, and give decibel differences. It is thus necessary to define a zero point to fix scale values. Zero decibel's loudness level corresponds to sound produced by a 1000-cycle tone with a pressure amplitude of 0.0002 dyne per sq cm, or an intensity of 10^{-16} watt per sq cm.

At the time of selection of the reference level it was believed that this sound was just audible in a perfectly quiet room. On such a scale, sounds below audibility give negative decibel values. Table 3 is derived from these equations, as are the various relations brought out in the discussion of sound-level units above. (See Ref. 1.)

NOISE PREVENTION

A noise that cannot be prevented or greatly reduced at the source, by design or by damping, will radiate into the air either directly from the surface of the vibrating body or from

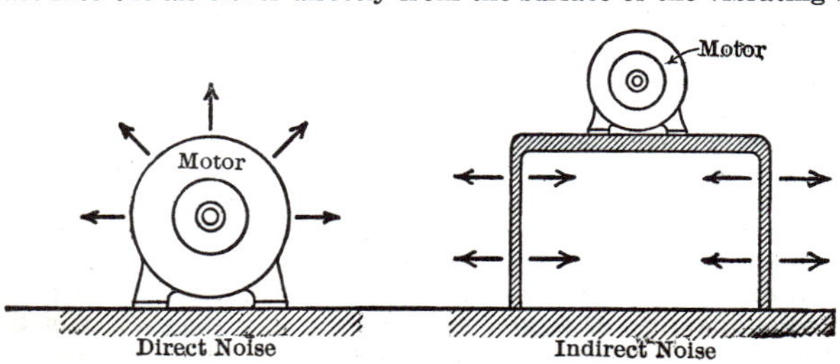

FIG. 1. Direct and indirect noise.

parts of the structure to which it is attached. Such noise radiation is indirect noise. See Fig. 1, where part of the motor noises radiate directly from the motor and part from panels of the room below, which may resonate to the frequency transmitted.

The following classification is useful when applying a preventive for noise:

Classification	Method		
Prevention at source	Design changes Damping		
Not readily preventable at source	Direct noise	Screening Absorption Filtration	Higher frequency range, 300 cycles and up
	Indirect noise	Isolation by suspension	Lower frequency range, under 300 cycles

A surface vibrating at small amplitude will radiate high frequencies directly into the air much more strongly than low. The latter require much larger vibrating surface area for direct radiation. Hence, in general, direct noises are of higher frequency than indirect; the 300-cycle limit in the classification is an average value.

11. NOISE PREVENTION AT SOURCE

DESIGN CHANGES involve the field of noisy apparatus. It is too large to be considered here, other than to cite a few instances. Remedies must come from studies by manufacturers.

Car-wheel Noise. Special car wheels with steel rims isolated from the rest of the wheel by rubber are being used on PCC trolley cars. Such wheels have greatly contributed to reducing the noise of cars.

Electric Motors. Special studies have produced designs with much lower a-c electrical hums.

Blower. Whistles from blowers in ventilating ducts usually are blade-frequency noises that can be eliminated at the source by design changes.

Gear Whistles. High-speed gearing can be greatly improved by accurate design to avoid impacts and by using helical instead of spur gears.

DAMPING. A sounding body, like a gear casing or small blower casing, can be quieted by covering it with a suitable damping material that has the ability to reduce vibrations.

Putty Lagging. A layer of nondrying putty 1 in. or more thick, covered with varnished cambric or similar material to prevent gradual drying of the putty, will sometimes greatly reduce noise. The putty exerts a powerful damping action, especially on high-frequency whistles and ringing noises, by reducing amplitude of case vibration. Lower-frequency rumbling noises are less strongly damped. The higher the frequency, the thinner may be the putty. Suitable nondrying putty can be made by any good chemist. It should have fairly high density and a binder to prevent its falling apart. Before application, the surface should be rubbed with a small amount of putty to improve adhesion.

Other Types of Lagging. Feltlike materials, when applied as lagging, have far less ability than putty to damp vibrations at the source, although they are sometimes effective as noise reducers. This effectiveness is due to their high sound-absorption coefficients, reducing the ability of a surface to radiate noise, and not to the fact that they reduce the amplitude of vibration. Although these materials are easier to apply, putty is more effective in such applications than any other material used by the author because of its power actually to reduce the surface vibration amplitude.

Resinous and tarlike materials, when applied as lagging, offer great friction to slow vibratory deflections of large amplitude. They are of relatively little use in damping high-frequency vibrations of small amplitude, aside from the increase in surface mass that they produce. These vibrations are so rapid that flow, with its attendant friction, does not occur. The relative vibration-absorption ability of different laggings is tested by coating the material on a steel diaphragm welded to a short pipe to support its edges. If the diaphragm is resonated by an electromagnet of variable frequency, it emits a loud sound unless the lagging is an effective damper. Impacting with a hammer is a good test if an electromagnet is not available. Commercially available materials are made by the Minnesota Mining & Manufacturing Co., and the Pittsburgh Plate Glass Co.

The correct use of acoustic felts is for the absorption of sound after sound waves are on the air. In this application, putty is of small value, compared to felt and similar materials.

12. SCREENING AND TRANSMISSION

SCREENING OF SOUND. Noise is screened either to confine it within a given space or to maintain a given space noise free. In either case the principle used is the same, i.e., to construct screens or walls highly resistant to noise transmission. The problem thus is one of noise transmission.

PROCESS OF TRANSMISSION. Sound is transmitted through a wall by air-pressure oscillations produced on its farther side by leakage or, more commonly, by oscillations produced by vibrations of the wall itself. This latter phenomenon usually occurs regardless of the presence of the former. Porous, feltlike materials transmit largely through leakage of wave oscillations.

Ordinary wood and plaster walls transmit by motion of the wall itself, or by vibration of portions of the wall. This is evident from Table 1, showing wavelengths of various sound frequencies. These are much in excess of the thickness of most walls, except those of high frequency, which are not readily transmitted. Sound waves of a given frequency in wall materials will be several times longer than in air, in proportion to the higher sound-transmission velocity of these materials. Only the highest frequencies in the audible range can exist as complete individual waves during transmission through ordinary wall.

Leakage. Small holes or cracks in a wall, otherwise highly resistant to noise transmission, will pass a surprising amount of sound, unless the crack is deep. Considerable damping action then is introduced by air friction on the crack sides. A soundproof room or box must have tight-fitting doors.

MEASUREMENT OF SOUND TRANSMISSION involves resonant effects depending upon mass, elastic characteristics and structure of the partition as determined by its dimensions, and reflection effects influencing the observations of the transmitted sound.

In engineering, a sound meter of the type described on p. 9-25 will indicate the intimate effect of the acoustics of the immediate surroundings. For example, the noise level in an ordinary room may fluctuate widely from place to place because of systems of standing sound waves which may be present. The ear will detect nodes and loops when the head is swinging back and forth if one or two fairly strong tones are present in a room.

FACTORS INFLUENCING TRANSMISSION. Noise is transmitted primarily by vibratory motion of a wall as a whole. Important factors in transmission prevention are mass per unit area and stiffness.

Stiffness is a trouble rather than an aid in the transmission resistance of walls, unless the wall is thick, of relatively small area, and unable to resonate to any objectionable noise harmonics. A small rigid metal case may resist noise transmission largely because of rigidity rather than mass per unit area.

Ordinary walls have various flexural resonant frequencies well below some objectionable noise harmonics. Stiffness can never aid such cases. It slightly increases wall-vibration amplitude, but this effect is negligible for relatively high frequencies.

A third important factor is wall structure. A double wall, with air space unbridged by rigid members, may resist noise transmission far more than a single wall of the same total weight per square foot.

DESIGN OF WALLS OF LOW TRANSMISSIBILITY. A double wall is more desirable than a single wall, being cheaper and less massive for the same transmission loss. A comparatively rigid skeleton of ordinary 2 by 4 in. studding should be used, with inner and outer walls hung from it by spring connections. These walls, of Celotex or similar material, should weigh 1 to 5 lb per sq ft. A thin layer of plasterlike reinforcement is helpful. Besides adding mass, it decreases porosity, which is an advantage. The Celotex may be reinforced by thin sheet iron. Sheet iron must not be used without Celotex, as it is subject to resonances.

Spring connectors eliminate the bridging effect of studding that results when the walls are directly nailed to it. Steel-spring connectors should be set on felt or rubber pads to reduce possible high-frequency vibration transmission through the springs. Such a wall has an air space of about 5 in., which is fair spacing. Air spacings of the order of 1 in. are much less effective, while spacing of more than 6 in. is usually unnecessary. (See Fig. 2.) If sound resistance required is not high, Celotex may be nailed directly to studding on both sides, but the advantage of the air space, which is large, is then much reduced.

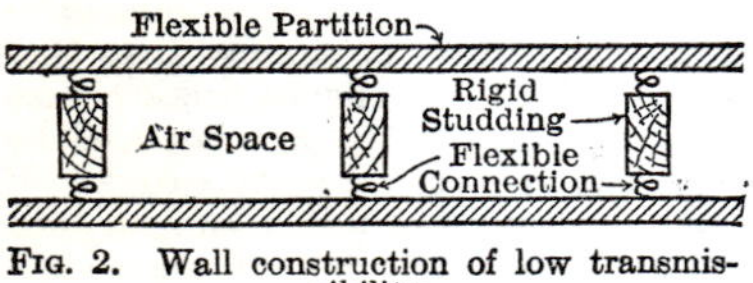

FIG. 2. Wall construction of low transmissibility.

A single wall should be of brick or masonry to have enough mass to be effective. A heavy masonry wall has high transmission resistance, but is expensive and cumbersome.

This type of wall is not particularly susceptible to resonances, its transmission resistance being due almost entirely to its mass.

Double walls of comparatively small mass, as recommended above, should have some damping, such as is contained in a material like Celotex, to avoid resonances. (See Ref. 2.)

TRANSMISSION LOSS. The aim of a soundproof wall or partition is to have sufficient transmission loss to be effective. This loss increases as the frequency rises. Table 4 gives an average transmission loss over the audible frequency range. It is of about the same order as that for a pure 500-cycle sound. Transmission loss is measured by the observed drop in sound intensity in decibels produced by the wall and is independent of the intensity of sound being screened. A wall causing a drop in noise level from 64 to 34 db will also cause a drop from 54 to 24 db, i.e., transmission loss is 30 db. For a double wall of the type first recommended the transmission loss is high, 40 to 60 db.

Table 4. Transmission Loss Produced by Various Partitions *

No.	Description	Transmission Loss, db	Lb per sq ft
1	Steel, 1/4 in. thick	30.0	....
2	Solid oak, 1 3/4 in. thick	20.9	....
3	Plate glass, single pane, 1/4 in. thick	22.8	....
4	Plate glass, 2 panes, 3/16 in. thick	19.7	....
5	Hair felt, 3 in. thick	11.8	....
6	3 layers 1-in. felt alternated with 4 layers building paper	27.9	....
7	2-in. gypsum tile, unplastered	20.2	10.4
8	3-in. gypsum tile, unplastered	23.7	14.2
9	4-in. clay tile, unplastered	28.8	17.0
10	4-in. clay tile, 1/2-in. plaster	30.8	22.0
11	8-in. brick wall, 1-in. plaster	46.1	88.0
12	Double 2-in. solid gypsum tile, unplastered unbridged, 4-in. separation	51.8	20.4
13	Same as 12, but bridged at top and bottom	46.9	20.4
14	Same as 12, inner faces lined with 1-in. felt	57.3	22.3
15	Staggered 2 by 4 in. wood studs, metal lath, 1/2-in. gypsum plaster	39.1	19.8
16	Same as 15, with 1/2-in. Celotex instead of metal lath	36.6	13.0

* Taken from work done under direction of Dr. Paul Sabine at Riverbank Acoustical Laboratory.

PRECAUTIONS. Several precautions are necessary in the design of transmission-resisting walls and screens after the design of the wall itself has been decided.

Transmission from Footing. Studding should be set on a layer of yielding material to prevent transmission to wall from ground or other support. A layer of rubber under the sill is good, but by itself becomes rather hard. By the principles of vibration isolation (see p. 9-19), a gravity elastic deflection of 1/8 in. or more is desirable.

Transmission through Pipes and Ducts. Metallic pipes are effective transmitters of sound. Two or three pipes in a well-designed sound screen may greatly reduce its effectiveness. Transmission along a metallic pipe can be greatly reduced by rubber joints to prevent metallic contact of pipe sections. For air ducts, heavy canvas connections are effective. Direct transmission of sound through ventilating ducts may be considerable unless the inside of the duct is lined with sound-absorbing materials. (See below.)

Edges of Double Walls. Edges of the sound-insulating properties of elastically suspended wall partitions used with double walls must be free from rigid contacts with the frame, at the same time avoiding cracks through which sound can pass. Cracks can be stuffed with felt or similar material.

13. ABSORPTION OF SOUND

If directly radiated noise, as the noise in a business office or in a factory, cannot be screened, a decided reduction in the noise level can be made by means of sound-absorbing materials. Sound absorption takes place during wave reflection. It is measured by the fraction of the incident energy which disappears in a single reflection. An open window space reflects practically no sound and represents a maximum obtainable reduction.

Table 5 gives sound-absorption coefficients of various materials based on unity, as absorption of an open window. (See Ref. 3.)

Feltlike materials have high sound-absorbing power, as have certain porous materials. A good absorbent covering has a smooth surface layer of oilcloth or thin metal perforated with many holes, under which is a porous material. Such covering gathers less dust than a feltlike surface and is just as effective. An inexpensive method to increase the absorp-

Table 5. Sound-absorption Coefficients of Different Materials at Three Frequencies

Pitch

Description	Coefficient		
	128 cps	512 cps	2048 cps
Open window	1.0	1.0	1.0
Brick wall, painted	0.012	0.017	0.023
unpainted	0.024	0.03	0.049
Carpet, unlined	0.09	0.20	0.27
felt lined	0.11	0.37	0.27
Fabrics, hung straight	0.04 to 0.1	0.11 to 0.5	0.30 to 0.82
Floors, concrete	0.01	0.015	0.02
wood	0.05	0.03	0.03
linoleum	0.03 to 0.08	0.03 to 0.08	0.03 to 0.08
Glass	0.035	0.027	0.02
Marble	0.01	0.01	0.015
Plaster, smooth finish on tile or brick	0.013	0.025	0.04
Plaster, smooth on lath	0.02	0.03	0.04
Plaster, rough on lath	0.039	0.06	0.054
Wood paneling	0.08	0.06	0.06
Fiberglas acoustical tile 1 in. thick mounted 10 1/2 in. from wall	0.67	0.82	0.91
Acousti-Celotex 1 in. thick cemented to wall	0.08	0.92	0.71
Mounted to 2 by 2 in. furring strips	0.41	0.68	0.75

Reprinted by permission from Sound Absorption Coefficients of Architectural Acoustical Materials, *Bulletin IX*, published by the Acoustical Materials Association, 1947.

tion in a room is hanging of double cones made of sound-absorbing materials (obtainable from Johns-Manville Corporation). The density to be used is approximately one absorber per 4 sq ft. These double cones are, per area of sound-absorbing material, more efficient than materials placed on walls (see Ref. 4). Manufacturers of sound-absorbing materials are the Armstrong Cork Company, the Celotex Corporation, the Johns-Manville Corporation, the National Gypsum Company, the Owens-Corning Fiberglas Corporation, and the United States Gypsum Company.

The sound-absorbing materials are either cemented directly to the plaster or the concrete, or they are nailed to wood furring strips about 2 by 2 in. and 12 to 24 in. off center. The space between the sound-absorbing material and the wall may be filled with mineral wool. To obtain high absorption coefficients at low frequencies the spacing between sound-absorbing materials and wall should be increased. To obtain rooms with better than 90% sound absorption at frequencies below 120 cps, wedges of sound-absorbing materials approximately 3 ft by 4 ft long must cover the walls. (See Ref. 5.) The use of the correct amount of absorption on the interior walls of lecture and music halls comprises an important field of acoustic engineering, known as architectural acoustics. (See Refs. 6 and 7.)

Sound absorption and transmission resistance are entirely different processes. A good sound absorber, like felt, is a poor transmission preventive. Such material may be effectively used, however, as a lining of an enclosed space to reduce further the final noise level obtained by screening. If no sound-resisting walls were used with the absorbing layer, however, the noise level would be very much raised.

CALCULATION OF NOISE REDUCTION THROUGH ABSORPTION. When the noise level in a room and the sound absorption coefficients of its wall (Table 5) are known, the possible reduction in noise level through the use of more absorbent materials on the walls can be approximately calculated. From eq. 4,

$$\text{Decibels} = 10 \log_{10} \frac{a_2}{a_1} \tag{5}$$

To find a_1, multiply wall area of each part of the room by its absorption coefficient. If S = total area and if the coefficient of $^1/_6S = 0.03$, of $^1/_3S = 0.01$, and of $^1/_2S = 0.02$, then $a_1 = [(S/6) \times 0.03] + [(S/3) \times 0.01] + [(S/2) \times 0.02] = (S/6) \times 0.11$. Assume that the part of the surface whose coefficient is 0.02 is to be covered with an absorbent with a coefficient of 0.5. Then $a_2 = [(S/6) \times 0.03] + [(S/3) \times 0.01] + [(S/2) \times 0.50] = (S/6) \times 1.55$, and $a_2/a_1 = {}^{155}/_{11} = 14.1$. From eq. 5, db $= 10 \log 14.1 = 10 \times 1.492 = 11.49$. If all the walls and the floor were covered with such a material, $a_2 = (S/6) \times 3.00$; $a_2/a_1 = {}^{300}/_{11} = 29.2$, and db $= 10 \log 29.2 = 10 \times 1.4654 = 14.65$. Thus covering half the whole area would reduce the noise level by 11.49 db, but covering the whole area would reduce it only a little over 3 db more.

The conclusion is that, in reducing noise level by sound-absorbing materials on the walls, the greatest reduction is effected by the first part of the walls covered. After this, less and less improvement is obtained from a given area of absorption covering. It usually is sufficient to treat only part of the wall area of a noisy room. Sound-absorbing materials should be used in interiors so that they can reduce the noise by many back and forth reflections.

NOISE ABSORPTION IN DUCTS. Noise in ventilating ducts can be greatly reduced by linings of acoustic absorbent materials. This is particularly true of the higher-frequency rushing noises and hisses. As Table 5 shows, most available materials are effective only in the medium and high frequency ranges. For low-frequency absorption, Celocrete blocks have been found effective down to 30 cps. (See Acoustic Filters, below.)

14. FILTRATION OF SOUND

ACOUSTIC FILTERS IN GENERAL. When direct noise, containing definite notes or harmonics, is transmitted along ducts or through openings, it often is possible to elim-

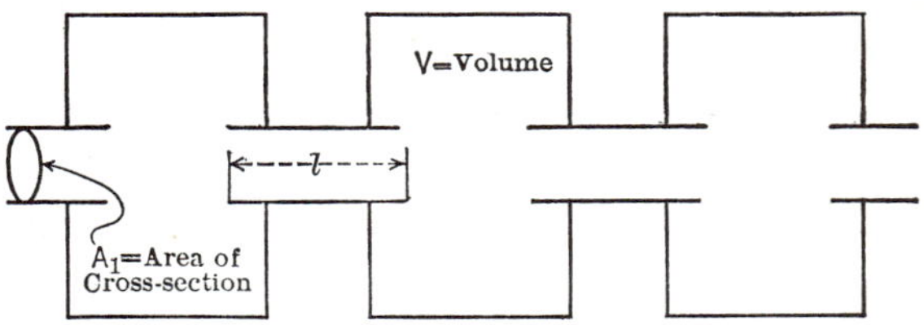

FIG. 3. Type 1, Low-pass acoustic filter.

inate objectionable harmonics completely by an acoustic filter that obstructs but little air flow or ventilation. Sound filters are of two classes: (1) low-pass filters which transmit low frequencies, but filter out the higher ones; (2) high-pass filters which do the opposite. A low-pass filter prevents transmission of the relatively high whistle often present in a ventilating duct. Low frequencies, predominant in explosive discharges, are filtered out by a high-pass filter, as an automobile muffler or Maxim silencer.

Space is the usual limitation in filter applications, restricting for low-pass filters less than high-pass. Low-pass filters return sound energy to the source, where it dissipates

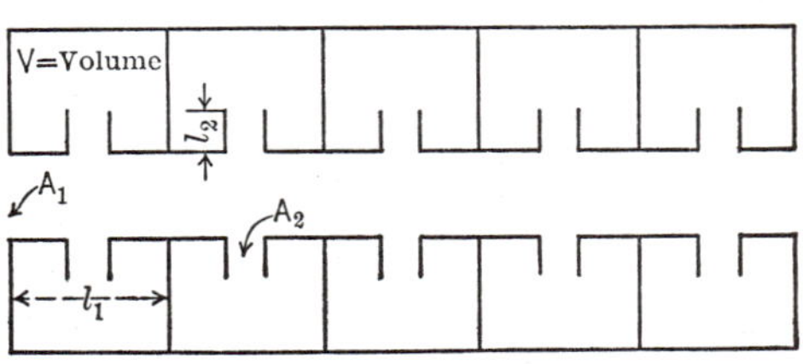

FIG. 4. Type 2, Low-pass acoustic filter.

itself through reflections, etc. High-pass filters must themselves dissipate the sound, this being an important reason for their comparative bulk. The individual stages of a filter must be considerably shorter than the wavelength of sound being filtered, each stage producing a definite amount of attenuation, as illustrated by the specific types of design which follow.

LOW-PASS FILTERS may be divided into two classes of design. Figure 3 shows the first, and Fig. 4 the second. In designing a filter, the lower limit of frequency, f_0, is determined, below which it will transmit, and above which the filtering action begins.

For Type 1, Fig. 3,

$$f_0 = 4200\sqrt{\frac{A_1}{lV}} \tag{6}$$

where A_1 = area of cross section, square inches; l = length, inches; and V = volume, cubic inches, the symbols referring to Fig. 3.

For Type 2, Fig. 4,

$$f_0 = 4200 \sqrt{\frac{1}{V\{l_1/A_1 + 4[(l_2 + \frac{\pi}{4}D)/nA_2]\}}} \tag{7}$$

where D = width of orifice. Equation 7 can in most cases be replaced by the more approximate formula

$$f_0 = 2100 \sqrt{nA_2/V(l_2 + \frac{\pi}{4}D)} \tag{8}$$

where the symbols refer to Fig. 4, and n = number of passages per stage of sectional area A_2. If l_2 is very small, such that A_2 is a simple orifice, the ratio $A_2/[l_2 + (\pi D/4)]$ is replaced by D, the diameter of the orifice. If there are several passages or orifices in each stage, $A_2/[l_2 + (\pi D/4)]$ or D must be multiplied by that number, n. Equation 8 becomes

$$f_0 = 2100\sqrt{nD/V}$$

For further details see Ref. 8.

Example of Design of Type 1 Filter. A whistle of about 400 cycles frequency is emitted from a tube of 1/16 sq in. cross-sectional area. A Type 1 filter (Fig. 3) should eliminate this sound. A_1 = 1/16 sq in.; l = 3 in.; V = 3 cu in. Figure 5 is a section of the filter in three stages, the two end tubes being shortened. From eq. 6, $f_0 = 4200\sqrt{1/16 \times 1/3 \times 1/3} = 350$ cycles. The lower frequency

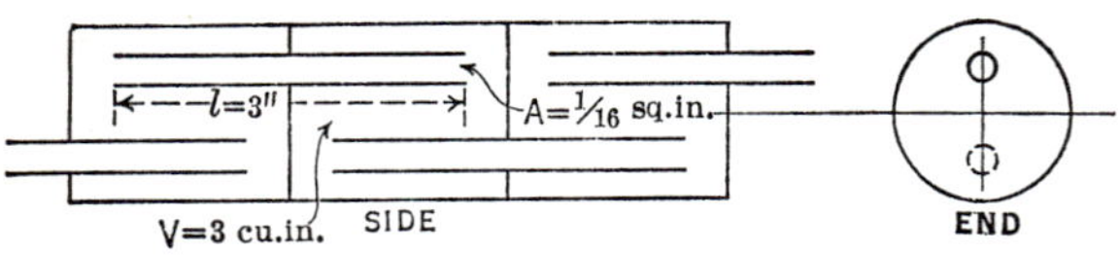

Fig. 5. Special filter design.

limit of this filter being 350 cycles, the 400-cycle frequency should be eliminated. The wavelength of this sound is (1100/400) × 12 = 33 in., which is greater than the filter stage length. For higher frequencies, where the wavelength is below 6 in., the filtering action will fall off.

Example of Design of Type 2 Filter. Apply this type of design to the previous problem. Type 2 filters are more selective than type 1, *because filtering action is efficient only for frequencies near and just above the lower limiting frequency* f_0. Figure 6 is a design which should be satisfactory. Applying approximate eq. 8 and design data from Fig. 6, $f_0 = 2100\sqrt{1/4 \times 2 \times 1/16} = 372$ cycles. The

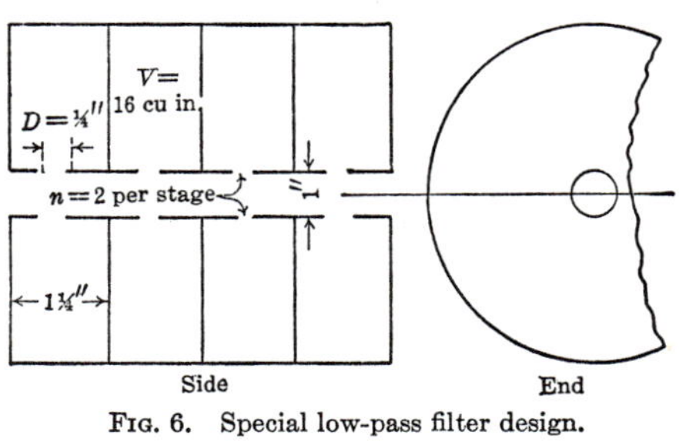

Fig. 6. Special low-pass filter design.

more exact eq. 7 will give a slightly lower frequency, but one sufficiently near 400 cycles to have a strong attenuating action on it. This design also consists of stages that are short compared with the wavelength of the sound to be filtered. The filtering action of type 1 low-pass filter covers a broader range of frequency than type 2 and is very efficient. Type 2, however, can sometimes be designed to occupy considerably less space than type 1. The preferred type depends on conditions to be met.

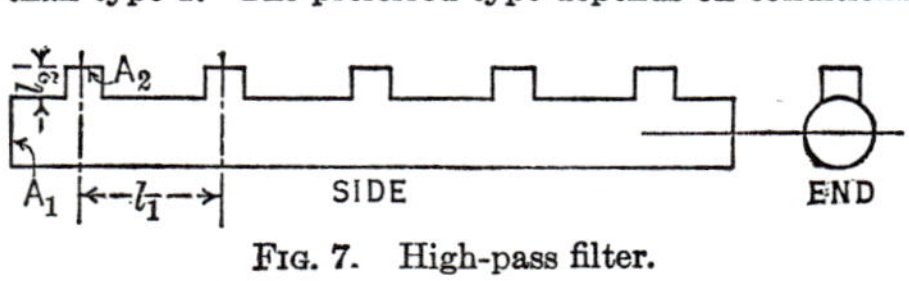

Fig. 7. High-pass filter.

HIGH-PASS FILTERS are not so efficient in performance as low-pass filters and are, as a rule, more bulky. Figure 7 is a typical high-pass filter of simple design. The limiting frequency, above which such a filter ceases to perform, is

$$f_0 = 2100 \frac{\sqrt{nA_2}}{A_1 l_1 l_2} \tag{9}$$

The symbols are shown in Fig. 7, and, as before, $A_2/l_2 = D$ = hole diameter when l_2 is small, and n = number of holes of area A_2 per stage.

This type of filter functions through successive side leakage of pressure waves as they pass through the tube. When the wave escapes from the tube end, its pressure amplitude and its intensity are greatly reduced. An appreciable amount of sound, however, escapes through the side holes. Good high-pass filters require an outer shell of considerably

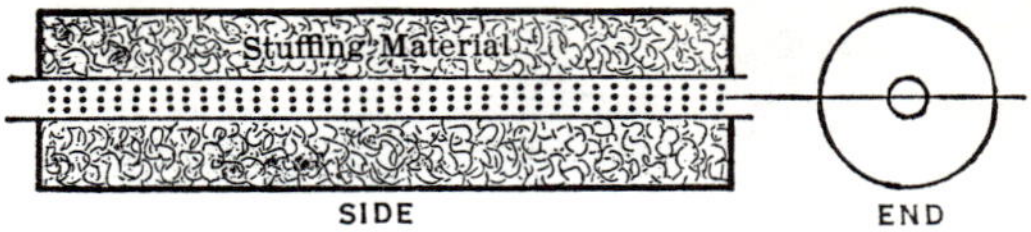

FIG. 8. Explosion muffler.

larger diameter than the inner tube, the interspace being stuffed with a damping material such as metal wool or shavings. Steel wool will rust in gas-engine exhaust mufflers.

Figure 8 is a cross section of an efficient muffler that has considerably less back pressure than the baffle-type automobile muffler. For high attenuation, many small holes are better than the few large ones of Fig. 6. The use of eq. 9 is not necessary for explosion mufflers, because these filters always attenuate low-frequency impulse waves. This equation will show that the limit f_0, above which attenuation ceases, is well above these low frequencies, being seldom below 100 cps.

PERFORMANCE. An ideal sound filter produces a certain percentage drop in sound intensity per stage for every frequency within its range of attenuation. Thus, a six-stage low-pass filter may attenuate a 300-cycle tone by 4 db per stage, or 24 db in all. A sharply tuned type 2 low-pass filter will attenuate much more. Attenuation in decibels per stage is independent of the initial sound intensity.

Performance of a low-pass filter sometimes is disappointing. Tube-resonance effects must be avoided. A tube of length equal to a multiple of one-half the wavelength of sound of a given frequency will resonate to that frequency. An otherwise well-designed filter may not perform well if such effects enter; placing it in different positions in the duct usually will eliminate them. High-pass filters, as mufflers, are less subject to this trouble. A fairly good knowledge of the elementary principles of sound is desirable in making sound-filter applications. For a discussion of sound filtration, see Refs. 9 and 10.

15. INDIRECT NOISE

Indirect noise is the result of transmission, through a structure, of sound vibration from an exciting source which may be quite remote. An ordinary structure has ability to transmit vibration waves a long distance almost perfectly noiselessly to some point. For example, a panel with the same natural frequency as the wave may resonate with surprising strength. The source of vibration sometimes is hard to find. The treatment of indirect noise is to isolate the source by elastic suspension. Vibration isolation is discussed on p. 9-19. The same methods apply to cases of noise isolation.

REFERENCES

1. W. Mikelson, Terminology and measurement of sound, *Gen. Elec. Rev.*, Dec. 1942.
2. A. L. Kimball, Theory of transmission of plane sound waves through multiple partitions, *J. ASA*, Jan. 1936, p. 222.
3. Sound absorption coefficients of architectural acoustical materials, *Acoustical Materials Assoc. Bulletin* IX, Chicago, Ill., 1947.
4. H. F. Olson, Functional sound absorbers, *RCA Rev.*, Dec. 1946, p. 503.
5. P. J. Mills, Construction and design of Parmly laboratory and anechoic chamber, *J. ASA*, Nov. 1941, p. 988.
6. P. E. Sabine, *Acoustics and Architecture*, Chap. VIII, McGraw-Hill Book Co., New York, 1932.
7. Olson, *Elements of Acoustical Engineering*, D. Van Nostrand Co., New York, 1947.
8. H. L. Mason, *Electromechanical Transducers and Wave Filters*, D. Van Nostrand Co., New York, 1943, pp. 128 and 140.
9. H. F. Olson, *Dynamical Analogies*, D. Van Nostrand Co., New York, 1943.
10. Stewart and Lindsay, *Acoustics*, Chap. VII, D. Van Nostrand Co., New York, 1930.

SECTION 10

FASTENERS

Acknowledgments: The editor acknowledges with thanks the assistance of the Industrial Fasteners Institute, William C. Stewart, technical adviser, and of Mr. William G. Waltermire, chief products engineer, Lamson and Sessions Co., in furnishing material and advice for the revision of Articles 1, 2, 3, and 4. Mr. L. J. Markwardt, assistant director, U. S. Forest Products Laboratory, helped with material for Articles 8 and 9.

SCREWS AND SCREW FASTENERS

CLASSIFICATION. Screws and bolts may be classified, according to use, in two general groups: A. Screws for transmitting power. B. Screw fastenings. Class B may be further subdivided into (1) bolts, (2) studs, (3) machine screws, (4) cap screws or tap bolts, (5) set screws, (6) T-head bolts, (7) coupling bolts, (8) hook bolts, (9) automobile bolts, (10) carriage bolts, (11) stove bolts, (12) expansion bolts, (13) foundation bolts, (14) eye bolts, (15) stay bolts, (16) patch bolts, (17) wood screws, and (18) lag screws and other special screws and bolts.

FORM OF THREAD depends on the service required and on the engineering standards of the country in which the thread is used. In general, the full V thread and the modified V thread are best adapted for fastenings. These combine strength and cheapness; the inefficiency of the V thread in transmitting power lessens the liability of unscrewing. Power transmission or translating threads in general are provided with smaller thread angles to increase the efficiency.

1. TRANSLATING THREADS

TRANSLATING THREAD STANDARDS. Five basic translating screw thread series are covered by ASA standards: Acme, Acme stub, 60-degree stub, modified square, and buttress.

Acme Screw Threads. Originally developed for the purpose of producing transverse motions on machine tools, Acme screw threads are now used extensively for a variety of purposes. Good current practice provides for two types of applications: general purpose and centralizing (see Fig. 1). Basic dimensions of the Acme form of thread are (see Fig. 1): angle of thread, $2\alpha = 29$ degrees; basic height of thread, $h = 0.5p$; thickness of thread at pitch line, $t = 0.5p$; basic width of flat, $F = 0.3707p$; where p = pitch, inches. The most generally used pitches are listed in Table 1. Basic proportions of recommended diameter and pitch combinations using Acme threads are given in Table 2.

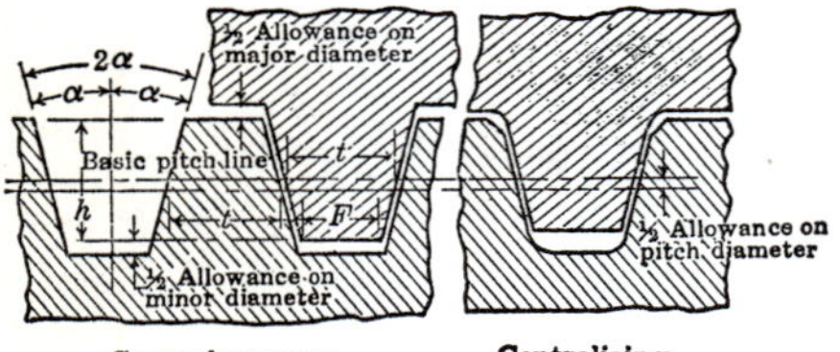

General purpose — Centralizing

Fig. 1. Acme thread forms.

Table 1. General-purpose Acme Threads, Standard and Special Pitches *

Threads per Inch	Pitch, p, in.	Threads per Inch	Pitch, p, in.	Threads per Inch	Pitch, p, in.
16	0.06250	6	0.16667	2	0.50000
14	0.07143	5	0.20000	1 1/2	0.66667
12	0.08333	4	0.25000	1 1/3	0.75000
10	0.10000	3	0.33333	1	1.00000
8	0.12500	2 1/2	0.40000		

* Recommended diameter and pitch combinations are given in Table 2.

General-purpose Acme threads are of three classes, 2G, 3G, and 4G, which differ from each other in pitch diameter allowances and tolerances (see American Standard). Class 2G is preferred for general-purpose assemblies, classes 3G and 4G being intended for applications requiring less backlash and endplay.

Centralizing Acme threads are of five classes, 2C, 3C, 4C, 5C, and 6C. Class 2C provides the maximum endplay or backlash. For centralizing threads the allowance at the major diameter is smaller than that at the pitch or minor diameters, the concentricity of screw and nut thus being controlled at the major diameter. For allowances, tolerances, and other details, see American Standard, *Acme Screw Threads.*

Acme Stub Thread. Intended for unusual applications where a coarse-pitch thread of shallow depth is required because of mechanical or metallurgical considerations, this

Table 2. Basic Dimensions of Acme and Stub Acme Thread Series
(All dimensions in inches)

Nominal Size	Threads per Inch, $1/p$	Acme Threads: Basic Height of Thread, h	General-purpose (All Classes) and Centralizing Classes 2C, 3C, and 4C: Basic Major Diameter, D	General-purpose (All Classes) and Centralizing Classes 2C, 3C, and 4C: Helix Angle at Basic Pitch Diameter, λ	Centralizing Classes 5C and 6C: Basic Major Diameter, B	Centralizing Classes 5C and 6C: Helix Angle at Basic Pitch Diameter, λ	Stub Acme Threads: Basic Height of Thread, h'	Stub Acme Threads: Helix Angle at Basic Pitch Diameter
1/4	16	.03125	0.2500	5° 12′	....	....	.01875	4° 54′
5/16	14	.03571	0.3125	4° 42′	....	....	.02143	4° 28′
3/8	12	.04167	0.3750	4° 33′	....	....	.02500	4° 20′
7/16	12	.04167	0.4375	3° 50′	....	....	.02500	3° 41′
1/2	10	.05000	0.5000	4° 3′	0.4823	4° 13′	.03000	3° 52′
5/8	8	.06250	0.6250	4° 3′	0.6052	4° 12′	.03750	3° 52′
3/4	6	.08333	0.7500	4° 33′	0.7284	4° 42′	.05000	4° 20′
7/8	6	.08333	0.8750	3° 50′	0.8516	3° 57′	.05000	3° 41′
1	5	.10000	1.0000	4° 3′	0.9750	4° 10′	.06000	3° 52′
1 1/8	5	.10000	1.1250	3° 33′	1.0985	3° 39′	.06000	3° 25′
1 1/4	5	.10000	1.2500	3° 10′	1.2220	3° 15′	.06000	3° 4′
1 3/8	4	.12500	1.3750	3° 39′	1.3457	3° 44′	.07500	3° 30′
1 1/2	4	.12500	1.5000	3° 19′	1.4694	3° 23′	.07500	3° 12′
1 3/4	4	.12500	1.7500	2° 48′	1.7169	2° 52′	.07500	2° 43′
2	4	.12500	2.0000	2° 26′	1.9646	2° 29′	.07500	2° 22′
2 1/4	3	.16667	2.2500	2° 55′	2.2125	2° 58′	.10000	2° 50′
2 1/2	3	.16667	2.5000	2° 36′	2.4605	2° 39′	.10000	2° 32′
2 3/4	3	.16667	2.7500	2° 21′	2.7085	2° 23′	.10000	2° 18′
3	2	.25000	3.0000	3° 19′	2.9567	3° 22′	.15000	3° 12′
3 1/2	2	.25000	3.5000	2° 48′	3.4532	2° 51′	.15000	2° 43′
4	2	.25000	4.0000	2° 26′	3.9500	2° 28′	.15000	2° 22′
4 1/2	2	.25000	4.5000	2° 8′	4.4470	2° 10′	.15000	2° 6′
5	2	.25000	5.0000	1° 55′	4.9441	1° 56′	.15000	1° 53′

For general-purpose and centralizing classes 2C, 3C, and 4C, basic pitch diameter $E = D - h$; basic minor diameter $K = D - 2h$.

For centralizing classes 5C and 6C, basic pitch diameter $E = B - h$; basic minor diameter $K = B - 2h$.

For stub Acme, basic pitch diameter $E = D - h'$; basic minor diameter $K = D - 2h'$.

series has the same angle of thread (29 degrees) as the Acme, but the basic depth is reduced to $h = 0.3p$. Basic width of flat is $F = 0.4224p$.

Recommended combinations of nominal diameter and pitch for stub Acme screws and nuts are the same as for the Acme, Table 2. Only one class of thread is established for general usage, that corresponding to class 2G for general-purpose Acme threads (see the standard). If a fit with less backlash is required, tolerances and allowances used in other classes of Acme thread may be used.

Two modified stub Acme thread forms are recognized in the Standard. Form 1 employs a basic thread height $h = 0.375p$ and Form 2 a basic thread height $h = 0.25p$. Basic widths of flat at crest of internal thread are, respectively, $F = 0.4030p$ and $F = 0.4353p$ for Forms 1 and 2.

60-Degree Stub Thread. Angle between sides of thread is $2\alpha = 60$ degrees, and basic depth is $h = 0.433p$. Basic width of flat at crest of internal thread is $F = 0.250p$. (See ASA B1.3-1941.)

Modified Square Thread. Included angle is $2\alpha = 10$ degrees, and basic depth of thread is $h = 0.5p$ (same as the Acme). Basic width of flat is $F = 0.4563p$. The 10-degree angle results in a thread which, for practical purposes, is equivalent to a "square" thread but is capable of economical production. (See ASA B1.3-1941.) The square thread ($2\alpha = 0$, $h = 0.5p$, $t = 0.5p$) is expensive to make and is not easily compensated for wear.

Buttress thread (Fig. 2) is a nonsymmetrical thread intended for applications with exceptionally high stresses in one direction only, along the thread axis. Because the thrust side of the thread is very nearly perpendicular to the thread axis, the radial component of thrust is a minimum, making this form of thread particularly suitable for tubular members which have to be screwed together. Applications include breech mechanisms of large guns and airplane propeller hubs.

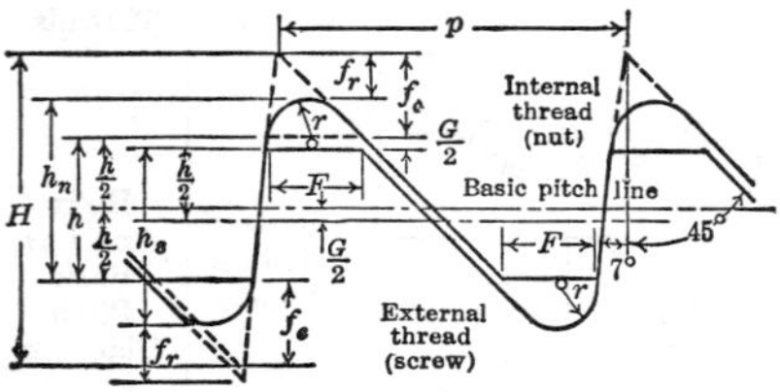

Fig. 2. Basic form of buttress thread.

Thread form is shown in Fig. 2, the dimensions being as follows: height of sharp V thread $H = 0.89064p$; basic height of thread $h = 0.4p$; root radius $r = 0.12055p$; root truncation $f_r = 0.13946p$; crest truncation $f_c = 0.24532p$; crest width $F = 0.27544p$; G = allowance; height of thread in nut h_n = height of thread in screw $h_s = 0.50586p$. Pressure flank angle, measured in an axial plane, is 7 degrees from the normal to the axis, and the trailing flank angle is 45 degrees.

Inasmuch as buttress threads are applied mainly to specially designed components, there is no standard series of diameter-pitch combinations. However, it is recommended that nominal major diameters be selected from the geometric series:

1	1 3/4	3	5 1/2	10	18
1 1/8	2	3 1/2	6	11	20
1 1/4	2 1/4	4	7	12	22
1 3/8	2 1/2	4 1/2	8	14	24
1 1/2	2 3/4	5	9	16	

Pitches, likewise, should be selected from the geometric series having the following threads per inch:

20	10	5	2 1/2	1 1/4
16	8	4	2	1
12	6	3	1 1/2	

It is suggested that diameter and pitch combinations be within the following ranges, using the foregoing preferred numbers:

Diameter Range, inches	Associated Pitches, threads per inch	Diameter Range, inches	Associated Pitches, threads per inch
1 to 2 1/2	20 to 6	6 to 10	12 to 2 1/2
2 1/2 to 4	20 to 5	10 to 16	8 to 1 1/2
4 to 6	16 to 4	16 to 24	6 to 1

A modified form of the foregoing buttress thread may be employed where manufacture with a single-point cutting is general practice or where the 7-degree pressure flank may produce high bursting stresses. It conforms as closely as practicable to the standard form swung through an angle of 7 degrees to give a pressure flank angle of zero and an included angle of 50 degrees.

EFFICIENCY OF SCREW THREADS. If a torque T applied to a screw produces an axial force F, the efficiency is

$$e = \frac{Fr \tan \lambda}{T} \tag{1}$$

where r is the pitch radius of the screw and λ is the lead or helix angle, calculated from the relation

$$\tan \lambda = \frac{L}{2\pi r} = \frac{np}{2\pi r} \tag{2}$$

where L is the lead $= np$, p being the pitch and n the number of thread starts, if a multiple thread. For a single thread, $n = 1$. The relation between torque and force is

$$\frac{T}{Fr} = \frac{\dfrac{f_1}{\cos \phi} + \tan \lambda}{1 - \dfrac{f_1}{\cos \phi} \tan \lambda} + f_2 \frac{r_c}{r} \tag{3}$$

where r_c is the radius of the thrust collar, f_1 is the coefficient of friction between screw and nut, f_2 is the coefficient of friction at the thrust collar, and ϕ is the pressure flank angle. For symmetrical threads, $\phi = \alpha = 14^1/_2$ degrees in the Acme series, and 30 degrees in the Unified and American screw thread series. Values of λ are given in Tables 2 to 8 for various thread series. Pitch radius $r = E/2$.

2. SCREW THREAD STANDARDS FOR FASTENERS

Screw thread standards in the United States, Canada, and Great Britain were unified through prolonged joint efforts which culminated in the signing of a declaration of accord in 1948. The *Unified and American Screw Threads* standard (ASA B1.1-1949) is closely similar to the American National standard formerly used in the United States. Thread angle and depth of V thread are identical, as is the truncation ($h/8$) at the major diameter (Fig. 3). Basic pitch diameters therefore are identical for the same diameter and pitch. Threads made to old and new standards are interchangeable.

The new standard emphasizes the desirability of a rounded root contour for the external or bolt thread (as in the British Whitworth standard) but permits a flat root when using unworn tools (as in the American National Standard). Crest of the external thread may be flat or rounded whereas the internal or nut threads have flat crests, and flat roots which may become rounded as tools wear.

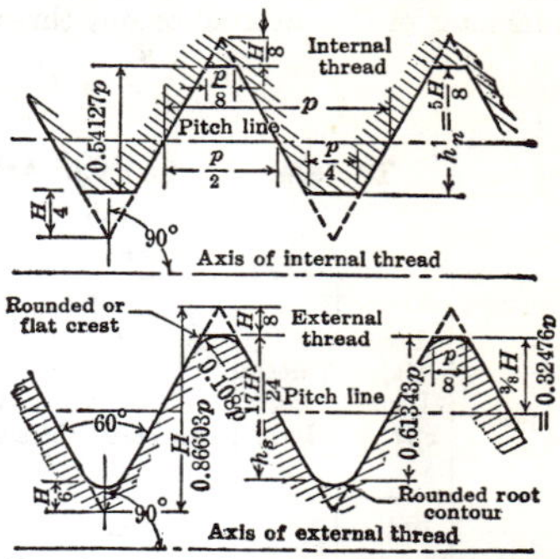

FIG. 3. Unified and American internal and external screw thread design forms, maximum metal condition.

For details of the old American National thread see p. 10-28, or the eleventh edition of Kent's *Mechanical Engineers' Handbook* (*Design, Shop Practice* Volume), p. 9-04, or ASA B1.1-1935.

UNIFIED FORM OF THREAD. The unified form of thread is illustrated in Fig. 3. The included angle between the flanks of the thread, measured in an axial plane, is 60 degrees, and the line bisecting this 60-degree angle is perpendicular to the axis of the screw thread. Formulas for the basic thread dimensions are given in Fig. 3. Basic depth of thread engagement is $h_e = 0.54127p$

THREADED PARTS have either an external thread (screw) or an internal thread (nut). Basic sizes of the principal dimensions are given by the following formulas which, in conjunction with Tables 3, 4, 5, 6, 7, 8, and 9, may be used to calculate basic pitch diameters and minor diameters. For limiting dimensions, see Tables 10, 11, 12, 13, and 14.

Table 3. Unified and American Coarse-thread Series—UNC and NC

(ASA B1.1-1949. All dimensions in inches.)

For limiting dimensions, see Tables 10, 11, 12, 13, 14.

Sizes	Basic Major Diameter, D	Threads per Inch, n	Lead Angle at Basic Pitch Diameter, λ		Minor Diameter Internal Threads, K_n	Stress Area *	Sizes	Basic Major Diameter, D	Threads per Inch, n	Lead Angle at Basic Pitch Diameter, λ		Minor Diameter Internal Threads, K_n	Stress Area *
	Inches		Deg	Min	Inches	Sq In.		Inches		Deg	Min	Inches	Sq In.
No. 1 †	0.0730	64	4	31	0.0561	0.0026	3/4	0.7500	10	2	40	0.6417	0.3340
No. 2 †	0.0860	56	4	22	0.0667	0.0036	7/8	0.8750	9	2	31	0.7547	0.4612
No. 3 †	0.0990	48	4	26	0.0764	0.0048	1	1.0000	8	2	29	0.8647	0.6051
No. 4 †	0.1120	40	4	45	0.0849	0.0060	1 1/8	1.1250	7	2	31	0.9704	0.7627
No. 5 †	0.1250	40	4	11	0.0979	0.0079	1 1/4	1.2500	7	2	15	1.0954	0.9684
No. 6 †	0.1380	32	4	50	0.1042	0.0090	1 3/8	1.3750	6	2	24	1.1946	1.1538
No. 8 †	0.1640	32	3	58	0.1302	0.0139	1 1/2	1.5000	6	2	11	1.3196	1.4041
No. 10 †	0.1900	24	4	39	0.1449	0.0174	1 3/4	1.7500	5	2	15	1.5335	1.8983
No. 12 †	0.2160	24	4	1	0.1709	0.0240	2	2.0000	4 1/2	2	11	1.7594	2.4971
1/4	0.2500	20	4	11	0.1959	0.0317	2 1/4	2.2500	4 1/2	1	55	2.0094	3.2464
5/16	0.3125	18	3	40	0.2524	0.0522	2 1/2	2.5000	4	1	57	2.2294	3.9976
3/8	0.3750	16	3	24	0.3073	0.0773	2 3/4	2.7500	4	1	46	2.4794	4.9326
7/16	0.4375	14	3	20	0.3602	0.1060	3	3.0000	4	1	36	2.7294	5.9659
1/2 †	0.5000	13	3	7	0.4167	0.1416	3 1/4	3.2500	4	1	29	2.9794	7.0992
1/2	0.5000	12	3	24	0.4098	0.1374	3 1/2	3.5000	4	1	22	3.2294	8.3268
9/16	0.5625	12	2	59	0.4723	0.1816	3 3/4	3.7500	4	1	16	3.4794	9.6546
5/8	0.6250	11	2	56	0.5266	0.2256	4	4.0000	4	1	11	3.7294	11.0805

For pitch diameter, see eq. 4; for basic minor diameter of external threads, see eq. 5.

* See eq. 8.

† Not included in the Unified standard.

Major diameter (D) of a threaded screw or nut is the largest diameter of the screw thread. The basic major diameter, which is the minimum major diameter of the internal or nut thread, is equal to the nominal size.

Pitch diameter (E) is the diameter of an imaginary cylinder whose surface would pass through the thread profiles at such points as to make the width of the groove equal to one-half the basic pitch. On a perfect thread this occurs at the point where the widths of the thread and groove are equal. The basic pitch diameter, which is the minimum pitch diameter of the internal or nut thread, is given by the formula

$$E = D - 0.64952p = D - \frac{0.64952}{n} \tag{4}$$

Table 4. Unified and American Fine Thread Series—UNF and NF

(ASA B1.1-1949. All dimensions in inches.)

For limiting dimensions, see Tables 10, 11, 12, 13, 14.

Sizes	Basic Major Diameter, D	Threads per Inch, n	Lead Angle at Basic Pitch Diameter, λ		Minor Diameter Internal Threads, K_n	Stress Area *	Sizes	Basic Major Diameter, D	Threads per Inch, n	Lead Angle at Basic Pitch Diameter, λ		Minor Diameter Internal Threads, K_n	Stress Area *
	Inches		Deg	Min	Inches	Sq In.		Inches		Deg	Min	Inches	Sq In.
No. 0 †	0.0600	80	4	23	0.0465	0.0018	3/8	0.3750	24	2	11	0.3299	0.0876
No. 1 †	0.0730	72	3	57	0.0580	0.0027	7/16	0.4375	20	2	15	0.3834	0.1185
No. 2 †	0.0860	64	3	45	0.0691	0.0039	1/2	0.5000	20	1	57	0.4459	0.1597
No. 3 †	0.0990	56	3	43	0.0797	0.0052	9/16	0.5625	18	1	55	0.5024	0.2026
No. 4 †	0.1120	48	3	51	0.0894	0.0065	5/8	0.6250	18	1	43	0.5649	0.2555
No. 5 †	0.1250	44	3	45	0.1004	0.0082	3/4	0.7500	16	1	36	0.6823	0.3724
No. 6 †	0.1380	40	3	44	0.1109	0.0101	7/8	0.8750	14	1	34	0.7977	0.5088
No. 8 †	0.1640	36	3	28	0.1339	0.0146	1	1.0000	12	1	36	0.9098	0.6624
No. 10	0.1900	32	3	21	0.1562	0.0199	1 1/8	1.1250	12	1	25	1.0348	0.8549
No. 12 †	0.2160	28	3	22	0.1773	0.0257	1 1/4	1.2500	12	1	16	1.1598	1.0721
1/4	0.2500	28	2	52	0.2113	0.0362	1 3/8	1.3750	12	1	9	1.2848	1.3137
5/16	0.3125	24	2	40	0.2674	0.0579	1 1/2	1.5000	12	1	3	1.4098	1.5799

For pitch diameter, see eq. 4; for basic minor diameter of external threads, see eq. 5.

* See eq. 8.

† Not included in the Unified standard.

Table 5. American Extra-fine Thread Series—NEF

(ASA B1.1-1949. All dimensions in inches.)

For allowances and tolerances, see Table 15.

Sizes	Basic Major Diameter, D	Threads per Inch, n	Lead Angle at Basic Pitch Diameter, λ		Minor Diameter Internal Threads, K_n	Stress Area *	Sizes	Basic Major Diameter, D	Threads per Inch, n	Lead Angle at Basic Pitch Diameter, λ		Minor Diameter Internal Threads, K_n	Stress Area *
	Inches		Deg	Min	Inches	Sq In.		Inches		Deg	Min	Inches	Sq In.
No. 12	0.2160	32	2	55	0.1822	0.0269	1 1/16	1.0625	18	0	59	1.0024	0.7973
1/4	0.2500	32	2	29	0.2162	0.0377	1 1/8	1.1250	18	0	56	1.0649	0.8993
5/16	0.3125	32	1	57	0.2787	0.0622	1 3/16	1.1875	18	0	53	1.1274	1.0074
3/8	0.3750	32	1	36	0.3412	0.0929	1 1/4	1.2500	18	0	50	1.1899	1.1216
7/16	0.4375	28	1	34	0.3988	0.1270	1 5/16	1.3125	18	0	48	1.2524	1.2420
1/2	0.5000	28	1	22	0.4613	0.1695	1 3/8	1.3750	18	0	45	1.3149	1.3684
9/16	0.5625	24	1	25	0.5174	0.2134	1 7/16	1.4375	18	0	43	1.3774	1.5010
5/8	0.6250	24	1	16	0.5799	0.2676	1 1/2	1.5000	18	0	42	1.4399	1.6397
11/16	0.6875	24	1	9	0.6424	0.3280	1 9/16	1.5625	18	0	40	1.5024	1.7846
3/4	0.7500	20	1	16	0.6959	0.3855	1 5/8	1.6250	18	0	38	1.5649	1.9357
13/16	0.8125	20	1	10	0.7584	0.4578	1 11/16	1.6875	18	0	37	1.6274	2.0929
7/8	0.8750	20	1	5	0.8209	0.5352	1 3/4	1.7500	16	0	40	1.6822	2.2382
15/16	0.9375	20	1	0	0.8834	0.6194	2	2.0000	16	0	35	1.9323	2.9501
1	1.0000	20	0	57	0.9459	0.7095							

For pitch diameter, see eq. 4; for basic minor diameter of external threads, see eq. 5.

* See eq. 8.

Table 6. American 8-pitch Thread Series—8N

$n = 8$

(ASA B1.1-1949. All dimensions in inches.)

For limiting dimensions, see Tables 11, 13, and 14.

Sizes	Basic Major Diameter, D	Lead Angle at Basic Pitch Diameter, λ		Minor Diameter, Internal Threads, K_n	Stress Area *	Sizes	Basic Major Diameter, D	Lead Angle at Basic Pitch Diameter, λ		Minor Diameter, Internal Threads, K_n	Stress Area *
Inches	Inches	Deg	Min	Inches	Sq In.	Inches	Inches	Deg	Min	Inches	Sq In.
1	1.0000	2	29	0.8647	0.6051	3	3.0000	0	47	2.8647	6.4957
1 1/8	1.1250	2	11	0.9897	0.7896	3 1/4	3.2500	0	43	3.1147	7.6738
1 1/4	1.2500	1	57	1.1147	0.9985	3 1/2	3.5000	0	40	3.3647	8.9504
1 3/8	1.3750	1	46	1.2397	1.2319	3 3/4	3.7500	0	37	3.6147	10.3249
1 1/2	1.5000	1	36	1.3647	1.4899	4	4.0000	0	35	3.8647	11.7975
1 5/8	1.6250	1	29	1.4897	1.7723	4 1/4	4.2500	0	33	4.1147	13.3683
1 3/4	1.7500	1	22	1.6147	2.0792	4 1/2	4.5000	0	31	4.3647	15.0372
1 7/8	1.8750	1	16	1.7397	2.4107	4 3/4	4.7500	0	29	4.6147	16.8042
2	2.0000	1	11	1.8647	2.7665	5	5.0000	0	28	4.8647	18.6694
2 1/8	2.1250	1	7	1.9897	3.1469	5 1/4	5.2500	0	26	5.1147	20.6330
2 1/4	2.2500	1	3	2.1147	3.5519	5 1/2	5.5000	0	25	5.3647	22.6945
2 1/2	2.5000	0	57	2.3647	4.4352	5 3/4	5.7500	0	24	5.6147	24.8541
2 3/4	2.7500	0	51	2.6147	5.4164	6	6.0000	0	23	5.8647	27.1118

For pitch diameter, see eq. 9; for basic minor diameter of external threads, see eq. 10.
* See eq. 8.

Table 7. American 12-pitch Thread Series—12N

$n = 12$

(ASA B1.1-1949. All dimensions in inches.)

For allowances and tolerances, see Table 16.

Sizes	Basic Major Diameter, D	Lead Angle at Basic Pitch Diameter, λ		Minor Diameter Internal Threads, K_n	Stress Area *	Sizes	Basic Major Diameter, D	Lead Angle at Basic Pitch Diameter, λ		Minor Diameter Internal Threads, K_n	Stress Area *
Inches	Inches	Deg	Min	Inches	Sq In.	Inches	Inches	Deg	Min	Inches	Sq In.
1/2	0.5000	3	24	0.4098	0.1374	2 3/8	2.3750	0	39	2.2848	4.1291
9/16	0.5625	2	59	0.4723	0.1816	2 1/2	2.5000	0	37	2.4098	4.5916
5/8	0.6250	2	40	0.5348	0.2319	2 5/8	2.6250	0	35	2.5348	5.0784
11/16	0.6875	2	24	0.5973	0.2883	2 3/4	2.7500	0	34	2.6598	5.5900
3/4	0.7500	2	11	0.6598	0.3508	2 7/8	2.8750	0	32	2.7848	6.1259
13/16	0.8125	2	0	0.7223	0.4195	3	3.0000	0	31	2.9098	6.6865
7/8	0.8750	1	51	0.7848	0.4943	3 1/8	3.1250	0	30	3.0348	7.2714
15/16	0.9375	1	43	0.8473	0.5753	3 1/4	3.2500	0	29	3.1598	7.8812
1	1.0000	1	36	0.9098	0.6624	3 3/8	3.3750	0	27	3.2848	8.5152
1 1/16	1.0625	1	30	0.9723	0.7556	3 1/2	3.5000	0	26	3.4098	9.1740
1 1/8	1.1250	1	25	1.0348	0.8550	3 5/8	3.6250	0	26	3.5348	9.8570
1 3/16	1.1875	1	20	1.0973	0.9604	3 3/4	3.7500	0	25	3.6598	10.5649
1 1/4	1.2500	1	16	1.1598	1.0721	3 7/8	3.8750	0	24	3.7848	11.2970
1 5/16	1.3125	1	12	1.2223	1.1898	4	4.0000	0	23	3.9098	12.0540
1 3/8	1.3750	1	9	1.2848	1.3138	4 1/4	4.2500	0	22	4.1598	13.6411
1 7/16	1.4375	1	6	1.3473	1.4438	4 1/2	4.5000	0	21	4.4098	15.3265
1 1/2	1.5000	1	3	1.4098	1.5800	4 3/4	4.7500	0	19	4.6598	17.1099
1 5/8	1.6250	0	58	1.5348	1.8701	5	5.0000	0	18	4.9098	18.9916
1 3/4	1.7500	0	54	1.6598	2.1853	5 1/4	5.2500	0	18	5.1598	20.9717
1 7/8	1.8750	0	50	1.7848	2.5250	5 1/2	5.5000	0	17	5.4098	23.0496
2	2.0000	0	47	1.9098	2.8892	5 3/4	5.7500	0	16	5.6598	25.2257
2 1/8	2.1250	0	44	2.0348	3.2779	6	6.0000	0	15	5.9098	27.4998
2 1/4	2.2500	0	42	2.1598	3.6914						

For pitch diameter, see eq. 12; for basic minor diameter of external threads, see eq. 13.
* See eq. 8.

where p is the pitch, inches; and n is the number of threads per inch. See Tables 3 to 8 for values of D and n for standard series. Table 9 lists recommended combinations of D and n for special, nonstandard applications.

Minor diameter (K) is the smallest diameter of the screw thread. Basic minor diameter of the *external* or screw thread (K_s) is given by the formula

$$K_s = D - 1.22687p = D - \frac{1.22687}{n} \tag{5}$$

Basic minor diameter of the *internal* or nut thread (K_n) is given by the formula

$$K_n = D - 1.08253p = D - \frac{1.08253}{n} \tag{6}$$

See Tables 3 to 8 for values of K_n for standard series.

Lead angle (λ) is the angle made by the helix of the thread at the pitch line with a plane perpendicular to the axis. Lead angle at the basic pitch diameter is given by the formula

$$\lambda = \tan^{-1}\frac{p}{\pi E} = \tan^{-1}\frac{1}{n\pi E} \tag{7}$$

See Tables 3 to 8 for values of λ for standard thread series.

Stress area (A_s) is the assumed area of an externally threaded part, used for the purpose of computing tensile strength. It is given by the formula

$$A_s = 3.1416\left(\frac{E_m + K_m}{4}\right)^2 \tag{8}$$

Table 8. American 16-pitch Thread Series—16N

$n = 16$

(ASA B1.1-1949. All dimensions in inches.)

For allowances and tolerances, see Table 17.

Sizes	Basic Major Diameter, D	Lead Angle at Basic Pitch Diameter, λ		Minor Diameter Internal Threads, K_n	Stress Area *	Sizes	Basic Major Diameter, D	Lead Angle at Basic Pitch Diameter, λ		Minor Diameter Internal Threads, K_n	Stress Area *
Inches	Inches	Deg	Min	Inches	Sq In.	Inches	Inches	Deg	Min	Inches	Sq In.
3/4	0.7500	1	36	0.6823	0.3724	2 5/16	2.3125	0	30	2.2448	3.9780
13/16	0.8125	1	29	0.7448	0.4429	2 3/8	2.3750	0	29	2.3073	4.2018
7/8	0.8750	1	22	0.8073	0.5197	2 7/16	2.4375	0	29	2.3698	4.4319
15/16	0.9375	1	16	0.8698	0.6025	2 1/2	2.5000	0	28	2.4323	4.6682
1	1.0000	1	11	0.9323	0.6916	2 5/8	2.6250	0	26	2.5573	5.1790
1 1/16	1.0625	1	7	0.9948	0.7867	2 3/4	2.7500	0	25	2.6823	5.6745
1 1/8	1.1250	1	3	1.0573	0.8880	2 7/8	2.8750	0	24	2.8073	6.2143
1 3/16	1.1875	1	0	1.1198	0.9955	3	3.0000	0	23	2.9323	6.7789
1 1/4	1.2500	0	57	1.1823	1.1090	3 1/8	3.1250	0	22	3.0573	7.3678
1 5/16	1.3125	0	54	1.2448	1.2287	3 1/4	3.2500	0	21	3.1823	7.9814
1 3/8	1.3750	0	51	1.3073	1.3545	3 3/8	3.3750	0	21	3.3073	8.6194
1 7/16	1.4375	0	49	1.3698	1.4865	3 1/2	3.5000	0	20	3.4323	9.2821
1 1/2	1.5000	0	47	1.4323	1.6246	3 5/8	3.6250	0	19	3.5573	9.9691
1 9/16	1.5625	0	45	1.4948	1.7687	3 3/4	3.7500	0	18	3.6823	10.6809
1 5/8	1.6250	0	43	1.5573	1.9191	3 7/8	3.8750	0	18	3.8073	11.4170
1 11/16	1.6875	0	42	1.6198	2.0757	4	4.0000	0	17	3.9323	12.1779
1 3/4	1.7500	0	40	1.6823	2.2382	4 1/4	4.2500	0	16	4.1823	13.7730
1 13/16	1.8125	0	39	1.7448	2.4070	4 1/2	4.5000	0	15	4.4323	15.4662
1 7/8	1.8750	0	37	1.8073	2.5819	4 3/4	4.7500	0	15	4.6823	17.2575
1 15/16	1.9375	0	36	1.8698	2.7629	5	5.0000	0	14	4.9323	19.1470
2	2.0000	0	35	1.9323	2.9501	5 1/4	5.2500	0	13	5.1823	21.1350
2 1/16	2.0625	0	34	1.9948	3.1434	5 1/2	5.5000	0	13	5.4323	23.2208
2 1/8	2.1250	0	33	2.0573	3.3427	5 3/4	5.7500	0	12	5.6823	25.4047
2 3/16	2.1875	0	32	2.1198	3.5483	6	6.0000	0	11	5.9323	27.6868
2 1/4	2.2500	0	31	2.1823	3.7601						

For pitch diameter, see eq. 15; for basic minor diameter of external threads, see eq. 16.

* See eq. 8.

Table 9. Recommended Special Diameter and Pitch Combinations *

Diameter	Threads per Inch					
No. 10 (0.190)	56	48	40	36	28	
No. 12 (0.216)	56	48	40	36		
1/4	56	48	40	36	24	
5/16	48	40	36	28	20	
3/8	40	36	28	20	18	
7/16	32	24	18	16		
1/2	32	24	18	16	14	
9/16	32	28	20	16	14	
5/8	32	28	20	16	14	
3/4	32	28	24	18	14	
7/8	32	28	24	18	10	
1	32	28	24	18	14	10
1 1/8	28	24	20	14	10	
1 1/4	24	20	14	10		
1 3/8	24	20	14	10		
1 1/2	24	20	14	10		
1 5/8	24	20	14	10	7	6
1 3/4	20	18	14	10	7	6
1 7/8	20	18	14	10	7	6
2	20	18	14	10	7	6
2 1/4	20	18	14	10	7	6
2 1/2	20	18	14	10	7	6
2 3/4	18	14	10	7	6	
3	18	14	10	7	6	
3 1/4	18	14	10	7	6	
3 1/2	18	14	10	7	6	
3 3/4	18	14	10	7	6	
4	14	10	7	6		
4 1/4	14	10	7	6	4	
4 1/2	14	10	7	6	4	
4 3/4	14	10	7	6	4	
5	14	10	7	6	4	
5 1/4	14	10	7	6	4	
5 1/2	14	10	7	6	4	
5 3/4	14	10	7	6	4	
6	14	10	7	6	4	

* UNS, UN, and NS designations not included in Tables 3, 4, 5, 6, 7, or 8.

where E_m = mean pitch diameter and K_m = mean minor diameter, both based on class 3 tolerances. See Tables 3 to 8 for values of A_s for standard thread series.

SCREW THREAD SERIES. Six standard series of threads are included in the *Unified and American Screw Threads* standard, ASA B1.1-1949. Of these only two, the coarse and fine thread series, are included in the Unified standard used also in Canada and Great Britain. Following are the six series.

Coarse-thread Series (UNC and NC). Diameter-pitch combinations of the Unified and American coarse-thread series are given in Table 3. The series is recommended for general use in engineering work where conditions are favorable to quick and easy assembly of parts with bolts, screws, and other threaded components, and where a fine thread is not required. It is the former United States Standard supplemented in the sizes below 1/4 in. by a part of the Standard Proportions for Machine Screws established by the ASME in 1907. Limiting dimensions are given in Tables 10, 11, 12, 13, and 14.

Fine-thread Series (UNF and NF). Table 4 lists the diameter-pitch combinations for this series, which is recommended for general use in automotive and aircraft work and where special conditions require a fine thread. The fine-thread series is the former Regular Screw Thread Series established by the SAE in 1911, supplemented below 1/4 in. by sizes of the ASME Standard Proportions for Machine Screws. Limiting dimensions are given in Tables 10, 11, 12, 13, and 14.

Extra-fine-thread Series (NEF). This series (see Table 5) is intended for special uses where thin-walled material is to be threaded, where thread depth of nuts clearing ferrules,

(*Continued on p. 10-23*)

Table 10. Limiting Dimensions, Coarse- and Fine-thread Series, Classes 1A and 1B

(ASA B1.1-1949. All dimensions in inches.)

Designation		External Thread Limits of Size, Class 1A						Internal Thread Limits of Size, Class 1B					
Size	Threads per Inch	Thread Symbol	Major Diameter Limits		Pitch Diameter Limits		Minor Diameter, max	Thread Symbol	Minor Diameter Limits		Pitch Diameter Limits		Major Diameter, min
			Max	Min	Max	Min			Min	Max	Min	Max	
1/4	20	UNC-1A	0.2489	0.2367	0.2164	0.2108	0.1876	UNC-1B	0.1959	0.2067	0.2175	0.2248	0.2500
1/4	28	UNF-1A	0.2490	0.2392	0.2258	0.2208	0.2052	UNF-1B	0.2113	0.2190	0.2268	0.2333	0.2500
5/16	18	UNC-1A	0.3113	0.2982	0.2752	0.2691	0.2431	UNC-1B	0.2524	0.2630	0.2764	0.2843	0.3125
5/16	24	UNF-1A	0.3114	0.3006	0.2843	0.2788	0.2603	UNF-1B	0.2674	0.2754	0.2854	0.2925	0.3125
3/8	16	UNC-1A	0.3737	0.3595	0.3331	0.3266	0.2970	UNC-1B	0.3073	0.3182	0.3344	0.3429	0.3750
3/8	24	UNF-1A	0.3739	0.3631	0.3468	0.3411	0.3228	UNF-1B	0.3299	0.3472	0.3479	0.3553	0.3750
7/16	14	UNC-1A	0.4361	0.4206	0.3897	0.3826	0.3485	UNC-1B	0.3602	0.3717	0.3911	0.4003	0.4375
7/16	20	UNF-1A	0.4362	0.4240	0.4037	0.3975	0.3749	UNF-1B	0.3834	0.3916	0.4050	0.4131	0.4375
1/2	13	NC-1A	0.4985	0.4822	0.4485	0.4411	0.4041	NC-1B	0.4167	0.4284	0.4500	0.4597	0.5000
1/2	12	UNC-1A	0.4985	0.4813	0.4444	0.4367	0.3963	UNC-1B	0.4098	0.4223	0.4459	0.4559	0.5000
1/2	20	UNF-1A	0.4987	0.4865	0.4662	0.4598	0.4374	UNF-1B	0.4459	0.4537	0.4675	0.4759	0.5000
9/16	12	UNC-1A	0.5609	0.5437	0.5068	0.4990	0.4587	UNC-1B	0.4723	0.4843	0.5084	0.5186	0.5625
9/16	18	UNF-1A	0.5611	0.5480	0.5250	0.5182	0.4929	UNF-1B	0.5024	0.5106	0.5264	0.5353	0.5625
5/8	11	UNC-1A	0.6234	0.6052	0.5644	0.5561	0.5119	UNC-1B	0.5266	0.5391	0.5660	0.5767	0.6250
5/8	18	UNF-1A	0.6236	0.6105	0.5875	0.5805	0.5554	UNF-1B	0.5649	0.5730	0.5889	0.5980	0.6250

3/4	10	UNC-1A	0.7482	0.7288	0.6832	0.6744	0.6255	UNC-1B	0.6417	0.6545	0.6850	0.6965	0.7500
3/4	16	UNF-1A	0.7485	0.7343	0.7079	0.7004	0.6718	UNF-1B	0.6823	0.6908	0.7094	0.7192	0.7500
7/8	9	UNC-1A	0.8731	0.8523	0.8009	0.7914	0.7368	UNC-1B	0.7547	0.7681	0.8028	0.8151	0.8750
7/8	14	UNF-1A	0.8734	0.8579	0.8270	0.8189	0.7858	UNF-1B	0.7977	0.8068	0.8286	0.8392	0.8750
1	8	UNC-1A	0.9980	0.9755	0.9168	0.9067	0.8446	UNC-1B	0.8647	0.8797	0.9188	0.9320	1.0000
1	12	UNF-1A	0.9982	0.9810	0.9441	0.9353	0.8960	UNF-1B	0.9098	0.9198	0.9459	0.9573	1.0000
1 1/8	7	UNC-1A	1.1228	1.0982	1.0300	1.0191	0.9475	UNC-1B	0.9704	0.9875	1.0322	1.0463	1.1250
1 1/8	12	UNF-1A	1.1232	1.1060	1.0691	1.0601	1.0210	UNF-1B	1.0348	1.0448	1.0709	1.0826	1.1250
1 1/4	7	UNC-1A	1.2478	1.2232	1.1550	1.1439	1.0725	UNC-1B	1.0954	1.1125	1.1572	1.1716	1.2500
1 1/4	12	UNF-1A	1.2482	1.2310	1.1941	1.1849	1.1460	UNF-1B	1.1598	1.1698	1.1959	1.2079	1.2500
1 3/8	6	UNC-1A	1.3726	1.3453	1.2643	1.2523	1.1681	UNC-1B	1.1946	1.2146	1.2667	1.2822	1.3750
1 3/8	12	UNF-1A	1.3731	1.3559	1.3190	1.3096	1.2709	UNF-1B	1.2848	1.2948	1.3209	1.3332	1.3750
1 1/2	6	UNC-1A	1.4976	1.4703	1.3893	1.3772	1.2931	UNC-1B	1.3196	1.3396	1.3917	1.4075	1.5000
1 1/2	12	UNF-1A	1.4981	1.4809	1.4440	1.4344	1.3959	UNF-1B	1.4098	1.4198	1.4459	1.4584	1.5000
1 3/4	5	UNC-1A	1.7473	1.7165	1.6174	1.6040	1.5019	UNC-1B	1.5335	1.5575	1.6201	1.6375	1.7500
2	4 1/2	UNC-1A	1.9971	1.9641	1.8528	1.8385	1.7245	UNC-1B	1.7594	1.7861	1.8557	1.8743	2.0000
2 1/4	4 1/2	UNC-1A	2.2471	2.2141	2.1028	2.0882	1.9745	UNC-1B	2.0094	2.0361	2.1057	2.1247	2.2500
2 1/2	4	UNC-1A	2.4969	2.4612	2.3345	2.3190	2.1902	UNC-1B	2.2294	2.2594	2.3376	2.3578	2.5000
2 3/4	4	UNC-1A	2.7468	2.7111	2.5844	2.5686	2.4401	UNC-1B	2.4794	2.5094	2.5876	2.6082	2.7500
3	4	UNC-1A	2.9968	2.9611	2.8344	2.8183	2.6901	UNC-1B	2.7294	2.7594	2.8376	2.8585	3.0000
3 1/4	4	UNC-1A	3.2467	3.2110	3.0843	3.0680	2.9400	UNC-1B	2.9794	3.0094	3.0876	3.1088	3.2500
3 1/2	4	UNC-1A	3.4967	3.4610	3.3343	3.3177	3.1900	UNC-1B	3.2294	3.2594	3.3376	3.3591	3.5000
3 3/4	4	UNC-1A	3.7466	3.7109	3.5842	3.5674	3.4399	UNC-1B	3.4794	3.5094	3.5876	3.6094	3.7500
4	4	UNC-1A	3.9966	3.9609	3.8342	3.8172	3.6899	UNC-1B	3.7294	3.7594	3.8376	3.8597	4.0000

Table 11. Limiting Dimensions, Coarse-, Fine-, and 8-thread Series, Classes 2A and 2B

(ASA B1.1-1949. All dimensions in inches.)

Designation		External Thread Limits of Size, Class 2A							Internal Thread Limits of Size, Class 2B					
Size	Threads per Inch	Thread Symbol	Major Diameter Limits Max *	Major Diameter Limits Min 1	Major Diameter Limits Min 2	Pitch Diameter Limits Max *	Pitch Diameter Limits Min	Minor Diameter, max *	Thread Symbol	Minor Diameter Limits Min	Minor Diameter Limits Max	Pitch Diameter Limits Min	Pitch Diameter Limits Max	Major Diameter, min
No. 0	80	NF-2A	0.0595	0.0563	....	0.0514	0.0496	0.0442	NF-2B	0.0465	0.0514	0.0519	0.0542	0.0600
No. 1	64	NC-2A	0.0724	0.0686	....	0.0623	0.0603	0.0532	NC-2B	0.0561	0.0623	0.0629	0.0655	0.0730
No. 1	72	NF-2A	0.0724	0.0689	....	0.0634	0.0615	0.0554	NF-2B	0.0580	0.0635	0.0640	0.0665	0.0730
No. 2	56	NC-2A	0.0854	0.0813	....	0.0738	0.0717	0.0635	NC-2B	0.0667	0.0737	0.0744	0.0772	0.0860
No. 2	64	NF-2A	0.0854	0.0816	....	0.0753	0.0733	0.0662	NF-2B	0.0691	0.0753	0.0759	0.0786	0.0860
No. 3	48	NC-2A	0.0983	0.0938	....	0.0848	0.0825	0.0727	NC-2B	0.0764	0.0845	0.0855	0.0885	0.0990
No. 3	56	NF-2A	0.0983	0.0942	....	0.0867	0.0845	0.0764	NF-2B	0.0797	0.0865	0.0874	0.0902	0.0990
No. 4	40	NC-2A	0.1112	0.1061	....	0.0950	0.0925	0.0805	NC-2B	0.0849	0.0939	0.0958	0.0991	0.1120
No. 4	48	NF-2A	0.1113	0.1068	....	0.0978	0.0954	0.0857	NF-2B	0.0894	0.0968	0.0985	0.1016	0.1120
No. 5	40	NC-2A	0.1242	0.1191	....	0.1080	0.1054	0.0935	NC-2B	0.0979	0.1062	0.1088	0.1121	0.1250
No. 5	44	NF-2A	0.1243	0.1195	....	0.1095	0.1070	0.0964	NF-2B	0.1004	0.1079	0.1102	0.1134	0.1250
No. 6	32	NC-2A	0.1372	0.1312	....	0.1169	0.1141	0.0989	NC-2B	0.1042	0.1140	0.1177	0.1214	0.1380
No. 6	40	NF-2A	0.1372	0.1321	....	0.1210	0.1184	0.1065	NF-2B	0.1109	0.1186	0.1218	0.1252	0.1380
No. 8	32	NC-2A	0.1631	0.1571	....	0.1428	0.1399	0.1248	NC-2B	0.1302	0.1389	0.1437	0.1475	0.1640
No. 8	36	NF-2A	0.1632	0.1577	....	0.1452	0.1424	0.1291	NF-2B	0.1339	0.1416	0.1460	0.1496	0.1640
No. 10	24	NC-2A	0.1890	0.1818	....	0.1619	0.1586	0.1379	NC-2B	0.1449	0.1555	0.1629	0.1672	0.1900
No. 10	32	NF-2A	0.1891	0.1831	....	0.1688	0.1658	0.1508	NF-2B	0.1562	0.1641	0.1697	0.1736	0.1900
No. 12	24	NC-2A	0.2150	0.2078	....	0.1879	0.1845	0.1639	NC-2B	0.1709	0.1807	0.1889	0.1933	0.2160
No. 12	28	NF-2A	0.2150	0.2085	....	0.1918	0.1886	0.1712	NF-2B	0.1773	0.1857	0.1928	0.1970	0.2160
1/4	20	UNC-2A	0.2489	0.2408	0.2367	0.2164	0.2127	0.1876	UNC-2B	0.1959	0.2067	0.2175	0.2223	0.2500
1/4	28	UNF-2A	0.2490	0.2425	....	0.2258	0.2225	0.2052	UNF-2B	0.2113	0.2190	0.2268	0.2311	0.2500
5/16	18	UNC-2A	0.3113	0.3026	0.2982	0.2752	0.2712	0.2431	UNC-2B	0.2524	0.2630	0.2764	0.2817	0.3125
5/16	24	UNF-2A	0.3114	0.3042	....	0.2843	0.2806	0.2603	UNF-2B	0.2674	0.2754	0.2854	0.2902	0.3125
3/8	16	UNC-2A	0.3737	0.3643	0.3595	0.3331	0.3287	0.2970	UNC-2B	0.3073	0.3182	0.3344	0.3401	0.3750
3/8	24	UNF-2A	0.3739	0.3667	....	0.3468	0.3430	0.3228	UNF-2B	0.3299	0.3372	0.3479	0.3528	0.3750
7/16	14	UNC-2A	0.4361	0.4258	0.4206	0.3897	0.3850	0.3485	UNC-2B	0.3602	0.3717	0.3911	0.3972	0.4375
7/16	20	UNF-2A	0.4362	0.4281	....	0.4037	0.3995	0.3749	UNF-2B	0.3834	0.3916	0.4050	0.4104	0.4375

1/2	13	NC-2A	0.4985	0.4876	0.4822	0.4485	0.4435	0.4041	NC-2B	0.4167	0.4284	0.4500	0.4565	0.5000
1/2	12	UNC-2A	0.4985	0.4871	0.4813	0.4444	0.4393	0.3963	UNC-2B	0.4098	0.4223	0.4459	0.4525	0.5000
1/2	20	UNF-2A	0.4987	0.4906	....	0.4662	0.4619	0.4374	UNF-2B	0.4459	0.4537	0.4675	0.4731	0.5000
9/16	12	UNC-2A	0.5609	0.5495	0.5437	0.5068	0.5016	0.4587	UNC-2B	0.4723	0.4843	0.5084	0.5152	0.5625
9/16	18	UNF-2A	0.5611	0.5524	....	0.5250	0.5205	0.4929	UNF-2B	0.5024	0.5106	0.5264	0.5323	0.5625
5/8	11	UNC-2A	0.6234	0.6113	0.6052	0.5644	0.5589	0.5119	UNC-2B	0.5266	0.5391	0.5660	0.5732	0.6250
5/8	18	UNF-2A	0.6236	0.6149	....	0.5875	0.5828	0.5554	UNF-2B	0.5649	0.5730	0.5889	0.5949	0.6250
3/4	10	UNC-2A	0.7482	0.7353	0.7288	0.6832	0.6773	0.6255	UNC-2B	0.6417	0.6545	0.6850	0.6927	0.7500
3/4	16	UNF-2A	0.7485	0.7391	....	0.7079	0.7029	0.6718	UNF-2B	0.6823	0.6908	0.7094	0.7159	0.7500
7/8	9	UNC-2A	0.8731	0.8592	0.8523	0.8009	0.7946	0.7368	UNC-2B	0.7547	0.7681	0.8028	0.8110	0.8750
7/8	14	UNF-2A	0.8734	0.8631	....	0.8270	0.8216	0.7858	UNF-2B	0.7977	0.8068	0.8286	0.8356	0.8750
1	8	UNC-2A	0.9980	0.9830	0.9755	0.9168	0.9100	0.8446	UNC-2B	0.8647	0.8797	0.9188	0.9276	1.0000
1	12	UNF-2A	0.9982	0.9868	....	0.9441	0.9382	0.8960	UNF-2B	0.9098	0.9198	0.9459	0.9535	1.0000
1 1/8	7	UNC-2A	1.1228	1.1064	1.0982	1.0300	1.0228	0.9475	UNC-2B	0.9704	0.9875	1.0322	1.0416	1.1250
1 1/8	8	N-2A	1.1229	1.1079	1.1004	1.0417	1.0348	0.9695	N-2B	0.9897	1.0047	1.0438	1.0528	1.1250
1 1/8	12	UNF-2A	1.1232	1.1118	....	1.0691	1.0631	1.0210	UNF-2B	1.0348	1.0448	1.0709	1.0787	1.1250
1 1/4	7	UNC-2A	1.2478	1.2314	1.2232	1.1550	1.1476	1.0725	UNC-2B	1.0954	1.1125	1.1572	1.1668	1.2500
1 1/4	8	N-2A	1.2479	1.2329	1.2254	1.1667	1.1597	1.0945	N-2B	1.1147	1.1297	1.1688	1.1780	1.2500
1 1/4	12	UNF-2A	1.2482	1.2368	....	1.1941	1.1879	1.1460	UNF-2B	1.1598	1.1698	1.1959	1.2039	1.2500
1 3/8	6	UNC-2A	1.3726	1.3544	1.3453	1.2643	1.2563	1.1681	UNC-2B	1.1946	1.2146	1.2667	1.2771	1.3750
1 3/8	8	N-2A	1.3728	1.3578	1.3503	1.2916	1.2844	1.2194	N-2B	1.2397	1.2547	1.2938	1.3031	1.3750
1 3/8	12	UNF-2A	1.3731	1.3617	....	1.3190	1.3127	1.2709	UNF-2B	1.2848	1.2948	1.3209	1.3291	1.3750
1 1/2	6	UNC-2A	1.4976	1.4794	1.4703	1.3893	1.3812	1.2931	UNC-2B	1.3196	1.3396	1.3917	1.4022	1.5000
1 1/2	8	N-2A	1.4978	1.4828	1.4753	1.4166	1.4093	1.3444	N-2B	1.3647	1.3797	1.4188	1.4283	1.5000
1 1/2	12	UNF-2A	1.4981	1.4867	....	1.4440	1.4376	1.3959	UNF-2B	1.4098	1.4198	1.4459	1.4542	1.5000
1 5/8	8	N-2A	1.6228	1.6078	1.6003	1.5416	1.5342	1.4694	N-2B	1.4897	1.5047	1.5438	1.5535	1.6250
1 3/4	5	UNC-2A	1.7473	1.7268	1.7165	1.6174	1.6085	1.5019	UNC-2B	1.5335	1.5575	1.6201	1.6317	1.7500
1 3/4	8	N-2A	1.7477	1.7327	1.7252	1.6665	1.6590	1.5943	N-2B	1.6147	1.6297	1.6688	1.6786	1.7500
1 7/8	8	N-2A	1.8727	1.8577	1.8502	1.7915	1.7838	1.7193	N-2B	1.7397	1.7547	1.7938	1.8038	1.8750
2	4 1/2	UNC-2A	1.9971	1.9751	1.9641	1.8528	1.8433	1.7245	UNC-2B	1.7594	1.7861	1.8557	1.8681	2.0000
2	8	N-2A	1.9977	1.9827	1.9752	1.9165	1.9037	1.8443	N-2B	1.8647	1.8797	1.9188	1.9289	2.0000
2 1/8	8	N-2A	2.1226	2.1076	2.1001	2.0414	2.0335	1.9692	N-2B	1.9897	2.0047	2.0438	2.0540	2.1250
2 1/4	4 1/2	UNC-2A	2.2471	2.2251	2.2141	2.1028	2.0931	1.9745	UNC-2B	2.0094	2.0361	2.1057	2.1183	2.2500
2 1/4	8	N-2A	2.2476	2.2326	2.2251	2.1664	2.1584	2.0942	N-2B	2.1147	2.1297	2.1688	2.1792	2.2500

(Table continued on p. 10-14)

Table 11. Limiting Dimensions, Coarse-, Fine-, and 8-thread Series, Classes 2A and 2B—*Continued*

(ASA B1.1-1949. All dimensions in inches.)

Designation		External Thread Limits of Size, Class 2A							Internal Thread Limits of Size, Class 2B					
Size	Threads per Inch	Thread Symbol	Major Diameter Limits			Pitch Diameter Limits		Minor Diameter, max *	Thread Symbol	Minor Diameter Limits		Pitch Diameter Limits		Major Diameter, min
			Max *	Min 1	Min 2	Max *	Min			Min	Max	Min	Max	
2 1/2	4	UNC-2A	2.4969	2.4731	2.4612	2.3345	2.3241	2.1902	UNC-2B	2.2294	2.2594	2.3376	2.3511	2.5000
2 1/2	8	N-2A	2.4976	2.4826	2.4751	2.4164	2.4082	2.3442	N-2B	2.3647	2.3797	2.4188	2.4294	2.5000
2 3/4	4	UNC-2A	2.7468	2.7230	2.7111	2.5844	2.5739	2.4401	UNC-2B	2.4794	2.5094	2.5876	2.6013	2.7500
2 3/4	8	N-2A	2.7475	2.7325	2.7250	2.6663	2.6580	2.5941	N-2B	2.6147	2.6297	2.6688	2.6796	2.7500
3	4	UNC-2A	2.9968	2.9730	2.9611	2.8344	2.8237	2.6901	UNC-2B	2.7294	2.7594	2.8376	2.8515	3.0000
3	8	N-2A	2.9974	2.9824	2.9749	2.9162	2.9077	2.8440	N-2B	2.8647	2.8797	2.9188	2.9299	3.0000
3 1/4	4	UNC-2A	3.2467	3.2229	3.2110	3.0843	3.0734	2.9400	UNC-2B	2.9794	3.0094	3.0876	3.1017	3.2500
3 1/4	8	N-2A	3.2474	3.2324	3.2249	3.1662	3.1575	3.0940	N-2B	3.1147	3.1297	3.1688	3.1801	3.2500
3 1/2	4	UNC-2A	3.4967	3.4729	3.4610	3.3343	3.3233	3.1900	UNC-2B	3.2294	3.2594	3.3376	3.3519	3.5000
3 1/2	8	N-2A	3.4974	3.4824	3.4749	3.4162	3.4074	3.3440	N-2B	3.3647	3.3797	3.4188	3.4303	3.5000
3 3/4	4	UNC-2A	3.7466	3.7228	3.7109	3.5842	3.5730	3.4399	UNC-2B	3.4794	3.5094	3.5876	3.6021	3.7500
3 3/4	8	N-2A	3.7473	3.7323	3.7248	3.6661	3.6571	3.5939	N-2B	3.6147	3.6297	3.6688	3.6805	3.7500
4	4	UNC-2A	3.9966	3.9728	3.9609	3.8342	3.8229	3.6899	UNC-2B	3.7294	3.7594	3.8376	3.8523	4.0000
4	8	N-2A	3.9973	3.9823	3.9748	3.9161	3.9070	3.8439	N-2B	3.8647	3.8797	3.9188	3.9307	4.0000
4 1/4	8	N-2A	4.2472	4.2322	4.2247	4.1660	4.1567	4.0938	N-2B	4.1147	4.1297	4.1688	4.1809	4.2500
4 1/2	8	N-2A	4.4972	4.4822	4.4747	4.4160	4.4066	4.3438	N-2B	4.3647	4.3797	4.4188	4.4310	4.5000
4 3/4	8	N-2A	4.7471	4.7321	4.7246	4.6659	4.6564	4.5937	N-2B	4.6147	4.6297	4.6688	4.6812	4.7500
5	8	N-2A	4.9971	4.9821	4.9746	4.9159	4.9062	4.8437	N-2B	4.8647	4.8797	4.9188	4.9314	5.0000
5 1/4	8	N-2A	5.2471	5.2321	5.2246	5.0659	5.0561	5.0937	N-2B	5.1147	5.1297	5.1688	5.1815	5.2500
5 1/2	8	N-2A	5.4970	5.4820	5.4745	5.4158	5.4059	5.3436	N-2B	5.3647	5.3797	5.4188	5.4317	5.5000
5 3/4	8	N-2A	5.7470	5.7320	5.7245	5.6658	5.6558	5.5936	N-2B	5.6147	5.6297	5.6688	5.6818	5.7500
6	8	N-2A	5.9970	5.9820	5.9745	5.9158	5.9056	5.8436	N-2B	5.8647	5.8797	5.9188	5.9320	6.0000

* The maximum dimensions of threads which are electroplated or have coatings of similar thickness will be increased by the amount of the allowance.

1 For semifinished and finished screws and bolts, threaded portion only.

2 For unfinished hot rolled material threaded portion only.

Table 12. Limiting Dimensions, Coarse- and Fine-thread Series, Classes 3A and 3B

(ASA B1.1-1949. All dimensions in inches.)

Designation		External Thread Limits of Size, Class 3A						Internal Thread Limits of Size, Class 3B					
Size	Threads per Inch	Thread Symbol	Major Diameter Limits		Pitch Diameter Limits		Minor Diameter, max	Thread Symbol	Minor Diameter Limits		Pitch Diameter Limits		Major Diameter, min
			Max	Min	Max	Min			Min	Max	Min	Max	
1/4	20	UNC-3A	0.2500	0.2419	0.2175	0.2147	0.1887	UNC-3B	0.1959	0.2067	0.2175	0.2211	0.2500
1/4	28	UNF-3A	0.2500	0.2435	0.2268	0.2243	0.2062	UNF-3B	0.2113	0.2190	0.2268	0.2300	0.2500
5/16	18	UNC-3A	0.3125	0.3038	0.2764	0.2734	0.2443	UNC-3B	0.2524	0.2630	0.2764	0.2803	0.3125
5/16	24	UNF-3A	0.3125	0.3053	0.2854	0.2827	0.2614	UNF-3B	0.2674	0.2754	0.2854	0.2890	0.3125
3/8	16	UNC-3A	0.3750	0.3656	0.3344	0.3311	0.2983	UNC-3B	0.3073	0.3182	0.3344	0.3387	0.3750
3/8	24	UNF-3A	0.3750	0.3678	0.3479	0.3450	0.3239	UNF-3B	0.3299	0.3372	0.3479	0.3516	0.3750
7/16	14	UNC-3A	0.4375	0.4272	0.3911	0.3876	0.3499	UNC-3B	0.3602	0.3717	0.3911	0.3957	0.4375
7/16	20	UNF-3A	0.4375	0.4294	0.4050	0.4019	0.3762	UNF-3B	0.3834	0.3916	0.4050	0.4091	0.4375
1/2	13	NC-3A	0.5000	0.4891	0.4500	0.4463	0.4056	NC-3B	0.4167	0.4284	0.4500	0.4548	0.5000
1/2	12	UNC-3A	0.5000	0.4886	0.4459	0.4421	0.3978	UNC-3B	0.4098	0.4223	0.4459	0.4509	0.5000
1/2	20	UNF-3A	0.5000	0.4919	0.4675	0.4643	0.4387	UNF-3B	0.4459	0.4537	0.4675	0.4717	0.5000
9/16	12	UNC-3A	0.5625	0.5511	0.5084	0.5045	0.4603	UNC-3B	0.4723	0.4843	0.5084	0.5135	0.5625
9/16	18	UNF-3A	0.5625	0.5538	0.5264	0.5230	0.4943	UNF-3B	0.5024	0.5106	0.5264	0.5308	0.5625
5/8	11	UNC-3A	0.6250	0.6129	0.5660	0.5619	0.5135	UNC-3B	0.5266	0.5391	0.5660	0.5714	0.6250
5/8	18	UNF-3A	0.6250	0.6163	0.5889	0.5854	0.5568	UNF-3B	0.5649	0.5730	0.5889	0.5934	0.6250
3/4	10	UNC-3A	0.7500	0.7371	0.6850	0.6806	0.6273	UNC-3B	0.6417	0.6545	0.6850	0.6907	0.7500
3/4	16	UNF-3A	0.7500	0.7406	0.7094	0.7056	0.6733	UNF-3B	0.6823	0.6908	0.7094	0.7143	0.7500
7/8	9	UNC-3A	0.8750	0.8611	0.8028	0.7981	0.7387	UNC-3B	0.7547	0.7681	0.8028	0.8089	0.8750
7/8	14	UNF-3A	0.8750	0.8647	0.8286	0.8245	0.7874	UNF-3B	0.7977	0.8068	0.8286	0.8339	0.8750
1	8	UNC-3A	1.0000	0.9850	0.9188	0.9137	0.8466	UNC-3B	0.8647	0.8797	0.9188	0.9254	1.0000
1	12	UNF-3A	1.0000	0.9886	0.9459	0.9415	0.8978	UNF-3B	0.9098	0.9198	0.9459	0.9516	1.0000

(Table continued on p. 10-16)

Table 12. Limiting Dimensions, Coarse- and Fine-thread Series, Classes 3A and 3B—*Continued*

(ASA B1.1-1949. All dimensions in inches.)

Designation		External Thread Limits of Size, Class 3A						Internal Thread Limits of Size, Class 3B					
Size	Threads per Inch	Thread Symbol	Major Diameter Limits Max	Major Diameter Limits Min	Pitch Diameter Limits Max	Pitch Diameter Limits Min	Minor Diameter, max	Thread Symbol	Minor Diameter Limits Min	Minor Diameter Limits Max	Pitch Diameter Limits Min	Pitch Diameter Limits Max	Major Diameter, min
1 1/8	7	UNC-3A	1.1250	1.1086	1.0322	1.0268	0.9497	UNC-3B	0.9704	0.9875	1.0322	1.0393	1.1250
1 1/8	12	UNF-3A	1.1250	1.1136	1.0709	1.0664	1.0228	UNF-3B	1.0348	1.0448	1.0709	1.0768	1.1250
1 1/4	7	UNC-3A	1.2500	1.2336	1.1572	1.1517	1.0747	UNC-3B	1.0954	1.1125	1.1572	1.1644	1.2500
1 1/4	12	UNF-3A	1.2500	1.2386	1.1959	1.1913	1.1478	UNF-3B	1.1598	1.1698	1.1959	1.2019	1.2500
1 3/8	6	UNC-3A	1.3750	1.3568	1.2667	1.2607	1.1705	UNC-3B	1.1946	1.2146	1.2667	1.2745	1.3750
1 3/8	12	UNF-3A	1.3750	1.3636	1.3209	1.3162	1.2728	UNF-3B	1.2848	1.2948	1.3209	1.3270	1.3750
1 1/2	6	UNC-3A	1.5000	1.4818	1.3917	1.3856	1.2955	UNC-3B	1.3196	1.3396	1.3917	1.3996	1.5000
1 1/2	12	UNF-3A	1.5000	1.4886	1.4459	1.4411	1.3978	UNF-3B	1.4098	1.4198	1.4459	1.4522	1.5000
1 3/4	5	UNC-3A	1.7500	1.7295	1.6201	1.6134	1.5046	UNC-3B	1.5335	1.5575	1.6201	1.6288	1.7500
2	4 1/2	UNC-3A	2.0000	1.9780	1.8557	1.8486	1.7274	UNC-3B	1.7594	1.7861	1.8557	1.8650	2.0000
2 1/4	4 1/2	UNC-3A	2.2500	2.2280	2.1057	2.0984	1.9774	UNC-3B	2.0094	2.0361	2.1057	2.1152	2.2500
2 1/2	4	UNC-3A	2.5000	2.4762	2.3376	2.3298	2.1933	UNC-3B	2.2294	2.2594	2.3376	2.3477	2.5000
2 3/4	4	UNC-3A	2.7500	2.7262	2.5876	2.5797	2.4433	UNC-3B	2.4794	2.5094	2.5876	2.5979	2.7500
3	4	UNC-3A	3.0000	2.9762	2.8376	2.8296	2.6933	UNC-3B	2.7294	2.7594	2.8376	2.8480	3.0000
3 1/4	4	UNC-3A	3.2500	3.2262	3.0876	3.0794	2.9433	UNC-3B	2.9794	3.0094	3.0876	3.0982	3.2500
3 1/2	4	UNC-3A	3.5000	3.4762	3.3376	3.3293	3.1933	UNC-3B	3.2294	3.2594	3.3376	3.3484	3.5000
3 3/4	4	UNC-3A	3.7500	3.7262	3.5876	3.5792	3.4433	UNC-3B	3.4794	3.5094	3.5876	3.5985	3.7500
4	4	UNC-3A	4.0000	3.9762	3.8376	3.8291	3.6933	UNC-3B	3.7294	3.7594	3.8376	3.8487	4.0000

Table 13. Limiting Dimensions, Coarse-, Fine-, and 8-thread Series, Class 2
(ASA B1.1-1949. All dimensions in inches.)

Designation			Limits of Size										
			External Threads						Internal Threads				
Size	Threads per Inch	Thread Symbol	Major Diameter			Pitch Diameter		Minor Diameter	Minor Diameter		Pitch Diameter		Major Diameter
			Max	Min 1	Min 2	Max	Min	Max	Min	Max	Min	Max	Min
No. 0	80	NF-2	0.0600	0.0566	....	0.0519	0.0502	0.0447	0.0465	0.0514	0.0519	0.0536	0.0600
No. 1	64	NC-2	0.0730	0.0692	0.0678	0.0629	0.0610	0.0538	0.0561	0.0623	0.0629	0.0648	0.0730
No. 1	72	NF-2	0.0730	0.0694	....	0.0640	0.0622	0.0560	0.0580	0.0634	0.0640	0.0658	0.0730
No. 2	56	NC-2	0.0860	0.0820	0.0804	0.0744	0.0724	0.0641	0.0667	0.0737	0.0744	0.0764	0.0860
No. 2	64	NF-2	0.0860	0.0822	....	0.0759	0.0740	0.0668	0.0691	0.0746	0.0759	0.0778	0.0860
No. 3	48	NC-2	0.0990	0.0946	0.0928	0.0855	0.0833	0.0734	0.0764	0.0841	0.0855	0.0877	0.0990
No. 3	56	NF-2	0.0990	0.0950	....	0.0874	0.0854	0.0771	0.0797	0.0856	0.0874	0.0894	0.0990
No. 4	40	NC-2	0.1120	0.1072	0.1052	0.0958	0.0934	0.0813	0.0849	0.0938	0.0958	0.0982	0.1120
No. 4	48	NF-2	0.1120	0.1076	....	0.0985	0.0963	0.0864	0.0894	0.0960	0.0985	0.1007	0.1120
No. 5	40	NC-2	0.1250	0.1202	0.1182	0.1088	0.1064	0.0943	0.0979	0.1062	0.1088	0.1112	0.1250
No. 5	44	NF-2	0.1250	0.1204	....	0.1102	0.1079	0.0971	0.1004	0.1068	0.1102	0.1125	0.1250
No. 6	32	NC-2	0.1380	0.1326	0.1304	0.1177	0.1150	0.0997	0.1042	0.1145	0.1177	0.1204	0.1380
No. 6	40	NF-2	0.1380	0.1332	....	0.1218	0.1194	0.1073	0.1109	0.1179	0.1218	0.1242	0.1380
No. 8	32	NC-2	0.1640	0.1586	0.1564	0.1437	0.1410	0.1257	0.1302	0.1384	0.1437	0.1464	0.1640
No. 8	36	NF-2	0.1640	0.1590	....	0.1460	0.1435	0.1299	0.1339	0.1402	0.1460	0.1485	0.1640
No. 10	24	NC-2	0.1900	0.1834	0.1808	0.1629	0.1596	0.1389	0.1449	0.1559	0.1629	0.1662	0.1900
No. 10	32	NF-2	0.1900	0.1846	....	0.1697	0.1670	0.1517	0.1562	0.1624	0.1697	0.1724	0.1900
No. 12	24	NC-2	0.2160	0.2094	0.2068	0.1889	0.1856	0.1649	0.1709	0.1801	0.1889	0.1922	0.2160
No. 12	28	NF-2	0.2160	0.2098	....	0.1928	0.1897	0.1722	0.1773	0.1835	0.1928	0.1959	0.2160
1/4	20	NC-2	0.2500	0.2428	0.2398	0.2175	0.2139	0.1887	0.1959	0.2060	0.2175	0.2211	0.2500
1/4	28	NF-2	0.2500	0.2438	....	0.2268	0.2237	0.2062	0.2113	0.2173	0.2268	0.2299	0.2500
5/16	18	NC-2	0.3125	0.3043	0.3011	0.2764	0.2723	0.2443	0.2524	0.2630	0.2764	0.2805	0.3125
5/16	24	NF-2	0.3125	0.3059	....	0.2854	0.2821	0.2614	0.2674	0.2739	0.2854	0.2887	0.3125
3/8	16	N-2	0.3750	0.3660	0.3624	0.3344	0.3299	0.2983	0.3073	0.3184	0.3344	0.3389	0.3750
3/8	24	NF-2	0.3750	0.3684	....	0.3479	0.3446	0.3239	0.3299	0.3364	0.3479	0.3512	0.3750
7/16	14	NC-2	0.4375	0.4277	0.4235	0.3911	0.3862	0.3499	0.3602	0.3721	0.3911	0.3960	0.4375
7/16	20	NF-2	0.4375	0.4303	....	0.4050	0.4014	0.3762	0.3834	0.3906	0.4050	0.4086	0.4375

(Table continued on p. 10-18)

Table 13. Limiting Dimensions, Coarse-, Fine-, and 8-thread Series, Class 2—*Continued*

(ASA B1.1-1949. All dimensions in inches.)

Designation			Limits of Size										
			External Threads						Internal Threads				
Size	Threads per Inch	Thread Symbol	Major Diameter			Pitch Diameter		Minor Diameter	Minor Diameter		Pitch Diameter		Major Diameter
			Max	Min [1]	Min [2]	Max	Min	Max	Min	Max	Min	Max	Min
1/2	12	NC-2	0.5000	0.4888	0.4842	0.4459	0.4403	0.3978	0.4098	0.4225	0.4459	0.4515	0.5000
1/2	13	NC-2	0.5000	0.4896	0.4852	0.4500	0.4448	0.4056	0.4167	0.4290	0.4500	0.4552	0.5000
1/2	20	NF-2	0.5000	0.4928	....	0.4675	0.4639	0.4387	0.4459	0.4531	0.4675	0.4711	0.5000
9/16	12	NC-2	0.5625	0.5513	0.5467	0.5084	0.5028	0.4603	0.4723	0.4850	0.5084	0.5140	0.5625
9/16	18	NF-2	0.5625	0.5543	....	0.5264	0.5223	0.4943	0.5024	0.5100	0.5264	0.5305	0.5625
5/8	11	NC-2	0.6250	0.6132	0.6080	0.5660	0.5601	0.5135	0.5266	0.5397	0.5660	0.5719	0.6250
5/8	18	NF-2	0.6250	0.6168	....	0.5889	0.5848	0.5568	0.5649	0.5725	0.5889	0.5930	0.6250
3/4	10	NC-2	0.7500	0.7372	0.7316	0.6850	0.6786	0.6273	0.6417	0.6553	0.6850	0.6914	0.7500
3/4	16	NF-2	0.7500	0.7410	....	0.7094	0.7049	0.6733	0.6823	0.6903	0.7094	0.7139	0.7500
7/8	9	NC-2	0.8750	0.8610	0.8550	0.8028	0.7958	0.7387	0.7547	0.7689	0.8028	0.8098	0.8750
7/8	14	NF-2	0.8750	0.8652	....	0.8286	0.8237	0.7874	0.7977	0.8062	0.8286	0.8335	0.8750
1	8	NC-2	1.0000	0.9848	0.9778	0.9188	0.9112	0.8466	0.8647	0.8795	0.9188	0.9264	1.0000
1	12	NF-2	1.0000	0.9888	....	0.9459	0.9403	0.8978	0.9098	0.9188	0.9459	0.9515	1.0000
1 1/8	7	NC-2	1.1250	1.1080	1.1002	1.0322	1.0237	0.9497	0.9704	0.9858	1.0322	1.0407	1.1250
1 1/8	8	N-2	1.1250	1.1098	....	1.0438	1.0359	0.9716	0.9897	1.0045	1.0438	1.0517	1.1250
1 1/8	12	NF-2	1.1250	1.1138	....	1.0709	1.0653	1.0228	1.0348	1.0438	1.0709	1.0765	1.1250
1 1/4	7	NC-2	1.2500	1.2330	1.2252	1.1572	1.1487	1.0747	1.0954	1.1108	1.1572	1.1657	1.2500
1 1/4	8	N-2	1.2500	1.2348	....	1.1688	1.1605	1.0966	1.1147	1.1295	1.1688	1.1771	1.2500
1 1/4	12	NF-2	1.2500	1.2388	....	1.1959	1.1903	1.1478	1.1598	1.1688	1.1959	1.2015	1.2500
1 3/8	6	NC-2	1.3750	1.3548	1.3460	1.2667	1.2566	1.1705	1.1946	1.2126	1.2667	1.2768	1.3750
1 3/8	8	N-2	1.3750	1.3598	....	1.2938	1.2852	1.2216	1.2397	1.2545	1.2938	1.3024	1.3750
1 3/8	12	NF-2	1.3750	1.3638	....	1.3209	1.3153	1.2728	1.2848	1.2938	1.3209	1.3265	1.3750

1 1/2	6	NC-2	1.5000	1.4798	1.4710	1.3917	1.3816	1.2955	1.3196	1.3376	1.3917	1.4018	1.5000
1 1/2	8	N-2	1.5000	1.4848	....	1.4188	1.4098	1.3466	1.3647	1.3795	1.4188	1.4278	1.5000
1 1/2	12	NF-2	1.5000	1.4888	....	1.4459	1.4403	1.3978	1.4098	1.4188	1.4459	1.4515	1.5000
1 5/8	8	N-2	1.6250	1.6098	....	1.5438	1.5345	1.4716	1.4897	1.5045	1.5438	1.5531	1.6250
1 3/4	5	NC-2	1.7500	1.7268	1.7162	1.6201	1.6085	1.5046	1.5335	1.5551	1.6201	1.6317	1.7500
1 3/4	8	N-2	1.7500	1.7348	....	1.6688	1.6591	1.5966	1.6147	1.6295	1.6688	1.6785	1.7500
1 7/8	8	N-2	1.8750	1.8598	....	1.7938	1.7838	1.7216	1.7397	1.7545	1.7938	1.8038	1.8750
2	4 1/2	NC-2	2.0000	1.9746	1.9632	1.8557	1.8430	1.7274	1.7594	1.7835	1.8557	1.8684	2.0000
2	8	N-2	2.0000	1.9848	....	1.9188	1.9084	1.8466	1.8647	1.8795	1.9188	1.9292	2.0000
2 1/8	8	N-2	2.1250	2.1098	....	2.0438	2.0331	1.9716	1.9897	2.0045	2.0438	2.0545	2.1250
2 1/4	4 1/2	NC-2	2.2500	2.2246	2.2132	2.1057	2.0930	1.9774	2.0094	2.0335	2.1057	2.1184	2.2500
2 1/4	8	N-2	2.2500	2.2348	....	2.1688	2.1578	2.0966	2.1147	2.1295	2.1688	2.1798	2.2500
2 1/2	4	NC-2	2.5000	2.4720	2.4592	2.3376	2.3236	2.1933	2.2294	2.2564	2.3376	2.3516	2.5000
2 1/2	8	N-2	2.5000	2.4848	....	2.4188	2.4071	2.3466	2.3647	2.3795	2.4188	2.4305	2.5000
2 3/4	4	NC-2	2.7500	2.7220	2.7092	2.5876	2.5736	2.4433	2.4794	2.5064	2.5876	2.6016	2.7500
2 3/4	8	N-2	2.7500	2.7348	....	2.6688	2.6564	2.5966	2.6147	2.6295	2.6688	2.6812	2.7500
3	4	NC-2	3.0000	2.9720	2.9592	2.8376	2.8236	2.6933	2.7294	2.7564	2.8376	2.8516	3.0000
3	8	N-2	3.0000	2.9848	....	2.9188	2.9058	2.8466	2.8647	2.8795	2.9188	2.9318	3.0000
3 1/4	4	NC-2	3.2500	3.2220	3.2092	3.0876	3.0736	2.9433	2.9794	3.0064	3.0876	3.1016	3.2500
3 1/4	8	N-2	3.2500	3.2348	....	3.1688	3.1556	3.0966	3.1147	3.1295	3.1688	3.1820	3.2500
3 1/2	4	NC-2	3.5000	3.4720	3.4592	3.3376	3.3236	3.1933	3.2294	3.2564	3.3376	3.3516	3.5000
3 1/2	8	N-2	3.5000	3.4848	....	3.4188	3.4055	3.3466	3.3647	3.3795	3.4188	3.4321	3.5000
3 3/4	4	NC-2	3.7500	3.7220	3.7092	3.5876	3.5736	3.4433	3.4794	3.5064	3.5876	3.6016	3.7500
3 3/4	8	N-2	3.7500	3.7348	....	3.6688	3.6554	3.5966	3.6147	3.6295	3.6688	3.6822	3.7500
4	4	NC-2	4.0000	3.9720	3.9592	3.8376	3.8236	3.6933	3.7294	3.7564	3.8376	3.8516	4.0000
4	8	N-2	4.0000	3.9848	....	3.9188	3.9053	3.8466	3.8647	3.8795	3.9188	3.9323	4.0000
4 1/4	8	N-2	4.2500	4.2348	....	4.1688	4.1551	4.0966	4.1147	4.1295	4.1688	4.1825	4.2500
4 1/2	8	N-2	4.5000	4.4848	....	4.4188	4.4050	4.3466	4.3647	4.3795	4.4188	4.4326	4.5000
4 3/4	8	N-2	4.7500	4.7348	....	4.6688	4.6549	4.5966	4.6147	4.6295	4.6688	4.6827	4.7500
5	8	N-2	5.0000	4.9848	....	4.9188	4.9048	4.8466	4.8647	4.8795	4.9188	4.9328	5.0000
5 1/4	8	N-2	5.2500	5.2348	....	5.1688	5.1547	5.0966	5.1147	5.1295	5.1688	5.1829	5.2500
5 1/2	8	N-2	5.5000	5.4848	....	5.4188	5.4046	5.3466	5.3647	5.3795	5.4188	5.4330	5.5000
5 3/4	8	N-2	5.7500	5.7348	....	5.6688	5.6545	5.5966	5.6147	5.6295	5.6688	5.6831	5.7500
6	8	N-2	6.0000	5.9848	....	5.9188	5.9044	5.8466	5.8647	5.8795	5.9188	5.9332	6.0000

[1] For semifinished and finished screws and bolts, threaded portion only.
[2] For unfinished hot rolled material threaded portion only.

Table 14. Limiting Dimensions, Coarse-, Fine-, and 8-thread Series Class 3

(ASA B1.1-1949. All dimensions in inches.)

Designation			Limits of Size										
			External Threads					Internal Threads					
Size	Threads per Inch	Thread Symbol	Major Diameter		Pitch Diameter		Minor Diameter Max	Minor Diameter		Pitch Diameter		Major Diameter Min
			Max	Min	Max	Min		Min	Max	Min	Max	
No. 0	80	NF-3	0.0600	0.0566	0.0519	0.0506	0.0447	0.0465	0.0514	0.0519	0.0532	0.0600
No. 1	64	NC-3	0.0730	0.0692	0.0629	0.0615	0.0538	0.0561	0.0623	0.0629	0.0643	0.0730
No. 1	72	NF-3	0.0730	0.0694	0.0640	0.0627	0.0560	0.0580	0.0634	0.0640	0.0653	0.0730
No. 2	56	NC-3	0.0860	0.0820	0.0744	0.0729	0.0641	0.0667	0.0737	0.0744	0.0759	0.0860
No. 2	64	NF-3	0.0860	0.0822	0.0759	0.0745	0.0668	0.0691	0.0746	0.0759	0.0773	0.0860
No. 3	48	NC-3	0.0990	0.0946	0.0855	0.0839	0.0734	0.0764	0.0841	0.0855	0.0871	0.0990
No. 3	56	NF-3	0.0990	0.0950	0.0874	0.0859	0.0771	0.0797	0.0856	0.0874	0.0889	0.0990
No. 4	40	NC-3	0.1120	0.1072	0.0958	0.0941	0.0813	0.0849	0.0938	0.0958	0.0975	0.1120
No. 4	48	NF-3	0.1120	0.1076	0.0985	0.0969	0.0864	0.0894	0.0960	0.0985	0.1001	0.1120
No. 5	40	NC-3	0.1250	0.1202	0.1088	0.1071	0.0943	0.0979	0.1062	0.1088	0.1105	0.1250
No. 5	44	NF-3	0.1250	0.1204	0.1102	0.1086	0.0971	0.1004	0.1068	0.1102	0.1118	0.1250
No. 6	32	NC-3	0.1380	0.1326	0.1177	0.1158	0.0997	0.1042	0.1145	0.1177	0.1196	0.1380
No. 6	40	NF-3	0.1380	0.1332	0.1218	0.1201	0.1073	0.1109	0.1179	0.1218	0.1235	0.1380
No. 8	32	NC-3	0.1640	0.1586	0.1437	0.1418	0.1257	0.1302	0.1384	0.1437	0.1456	0.1640
No. 8	36	NF-3	0.1640	0.1590	0.1460	0.1442	0.1299	0.1339	0.1402	0.1460	0.1478	0.1640
No. 10	24	NC-3	0.1900	0.1834	0.1629	0.1605	0.1389	0.1449	0.1559	0.1629	0.1653	0.1900
No. 10	32	NF-3	0.1900	0.1846	0.1697	0.1678	0.1517	0.1562	0.1624	0.1697	0.1716	0.1900
No. 12	24	NC-3	0.2160	0.2094	0.1889	0.1865	0.1649	0.1709	0.1801	0.1889	0.1913	0.2160
No. 12	28	NF-3	0.2160	0.2098	0.1928	0.1906	0.1722	0.1773	0.1835	0.1928	0.1950	0.2160
1/4	20	NC-3	0.2500	0.2428	0.2175	0.2149	0.1887	0.1959	0.2060	0.2175	0.2201	0.2500
1/4	28	NF-3	0.2500	0.2438	0.2268	0.2246	0.2062	0.2113	0.2173	0.2268	0.2290	0.2500
5/16	18	NC-3	0.3125	0.3043	0.2764	0.2734	0.2443	0.2524	0.2630	0.2764	0.2794	0.3125
5/16	24	NF-3	0.3125	0.3059	0.2854	0.2830	0.2614	0.2674	0.2739	0.2854	0.2878	0.3125

3/8	16	NC-3	0.3750	0.3660	0.3344	0.3312	0.2983	0.3073	0.3184	0.3344	0.3376	0.3750
3/8	24	NF-3	0.3750	0.3684	0.3479	0.3455	0.3239	0.3299	0.3364	0.3479	0.3503	0.3750
7/16	14	NC-3	0.4375	0.4277	0.3911	0.3875	0.3499	0.3602	0.3721	0.3911	0.3947	0.4375
7/16	20	NF-3	0.4375	0.4303	0.4050	0.4024	0.3762	0.3834	0.3906	0.4050	0.4076	0.4375
1/2	12	N-3	0.5000	0.4888	0.4459	0.4419	0.3978	0.4098	0.4225	0.4459	0.4499	0.5000
1/2	13	NC-3	0.5000	0.4896	0.4500	0.4463	0.4056	0.4167	0.4290	0.4500	0.4537	0.5000
1/2	20	NF-3	0.5000	0.4928	0.4675	0.4649	0.4387	0.4459	0.4531	0.4675	0.4701	0.5000
9/16	12	NC-3	0.5625	0.5513	0.5084	0.5044	0.4603	0.4723	0.4850	0.5084	0.5124	0.5625
9/16	18	NF-3	0.5625	0.5543	0.5264	0.5234	0.4943	0.5024	0.5100	0.5264	0.5294	0.5625
5/8	11	NC-3	0.6250	0.6132	0.5660	0.5618	0.5135	0.5266	0.5397	0.5660	0.5702	0.6250
5/8	18	NF-3	0.6250	0.6168	0.5889	0.5859	0.5568	0.5649	0.5725	0.5889	0.5919	0.6250
3/4	10	NC-3	0.7500	0.7372	0.6850	0.6805	0.6273	0.6417	0.6553	0.6850	0.6895	0.7500
3/4	16	NF-3	0.7500	0.7410	0.7094	0.7062	0.6733	0.6823	0.6903	0.7094	0.7126	0.7500
7/8	9	NC-3	0.8750	0.8610	0.8028	0.7979	0.7387	0.7547	0.7689	0.8028	0.8077	0.8750
7/8	14	NF-3	0.8750	0.8652	0.8286	0.8250	0.7874	0.7977	0.8062	0.8286	0.8322	0.8750
1	8	NC-3	1.0000	0.9848	0.9188	0.9134	0.8466	0.8647	0.8795	0.9188	0.9242	1.0000
1	12	NF-3	1.0000	0.9888	0.9459	0.9419	0.8978	0.9098	0.9188	0.9459	0.9499	1.0000
1 1/8	7	NC-3	1.1250	1.1080	1.0322	1.0263	0.9497	0.9704	0.9858	1.0322	1.0381	1.1250
1 1/8	8	N-3	1.1250	1.1098	1.0438	1.0383	0.9716	0.9897	1.0045	1.0438	1.0493	1.1250
1 1/8	12	NF-3	1.1250	1.1138	1.0709	1.0669	1.0228	1.0348	1.0438	1.0709	1.0749	1.1250
1 1/4	7	NC-3	1.2500	1.2330	1.1572	1.1513	1.0747	1.0954	1.1108	1.1572	1.1631	1.2500
1 1/4	8	N-3	1.2500	1.2348	1.1688	1.1630	1.0966	1.1147	1.1295	1.1688	1.1746	1.2500
1 1/4	12	NF-3	1.2500	1.2388	1.1959	1.1919	1.1478	1.1598	1.1688	1.1959	1.1999	1.2500
1 3/8	6	NC-3	1.3750	1.3548	1.2667	1.2596	1.1705	1.1946	1.2126	1.2667	1.2738	1.3750
1 3/8	8	N-3	1.3750	1.3598	1.2938	1.2877	1.2216	1.2397	1.2545	1.2938	1.2999	1.3750
1 3/8	12	NF-3	1.3750	1.3638	1.3209	1.3169	1.2728	1.2848	1.2938	1.3209	1.3249	1.3750
1 1/2	6	NC-3	1.5000	1.4798	1.3917	1.3846	1.2955	1.3196	1.3376	1.3917	1.3988	1.5000
1 1/2	8	N-3	1.5000	1.4848	1.4188	1.4125	1.3466	1.3647	1.3795	1.4188	1.4251	1.5000
1 1/2	12	NF-3	1.5000	1.4888	1.4459	1.4419	1.3978	1.4098	1.4188	1.4459	1.4499	1.5000
1 5/8	8	N-3	1.6250	1.6098	1.5438	1.5373	1.4716	1.4897	1.5045	1.5438	1.5503	1.6250
1 3/4	5	NC-3	1.7500	1.7268	1.6201	1.6119	1.5046	1.5335	1.5551	1.6201	1.6283	1.7500
1 3/4	8	N-3	1.7500	1.7348	1.6688	1.6620	1.5966	1.6147	1.6295	1.6688	1.6756	1.7500
1 7/8	8	N-3	1.8750	1.8598	1.7938	1.7868	1.7216	1.7397	1.7545	1.7938	1.8008	1.8750

(Table continued on p. 10-22)

Table 14. Limiting Dimensions, Coarse-, Fine-, and 8-thread Series, Class 3—*Continued*

(ASA B1.1-1949. All dimensions in inches.)

Designation			Limits of Size									
			External Threads					Internal Threads				
Size	Threads per Inch	Thread Symbol	Major Diameter Max	Major Diameter Min	Pitch Diameter Max	Pitch Diameter Min	Minor Diameter Max	Minor Diameter Min	Minor Diameter Max	Pitch Diameter Min	Pitch Diameter Max	Major Diameter Min
2	4 1/2	NC-3	2.0000	1.9746	1.8557	1.8468	1.7274	1.7594	1.7835	1.8557	1.8646	2.0000
2	8	N-3	2.0000	1.9848	1.9188	1.9115	1.8466	1.8647	1.8795	1.9188	1.9261	2.0000
2 1/8	8	N-3	2.1250	2.1098	2.0438	2.0363	1.9716	1.9897	2.0045	2.0438	2.0513	2.1250
2 1/4	4 1/2	NC-3	2.2500	2.2246	2.1057	2.0968	1.9774	2.0094	2.0335	2.1057	2.1146	2.2500
2 1/4	8	N-3	2.2500	2.2348	2.1688	2.1611	2.0966	2.1147	2.1295	2.1688	2.1765	2.2500
2 1/2	4	NC-3	2.5000	2.4720	2.3376	2.3279	2.1933	2.2294	2.2564	2.3376	2.3473	2.5000
2 1/2	8	N-3	2.5000	2.4848	2.4188	2.4106	2.3466	2.3647	2.3795	2.4188	2.4270	2.5000
2 3/4	4	NC-3	2.7500	2.7220	2.5876	2.5779	2.4433	2.4794	2.5064	2.5876	2.5973	2.7500
2 3/4	8	N-3	2.7500	2.7348	2.6688	2.6601	2.5966	2.6147	2.6295	2.6688	2.6775	2.7500
3	4	NC-3	3.0000	2.9720	2.8376	2.8279	2.6933	2.7294	2.7564	2.8376	2.8473	3.0000
3	8	N-3	3.0000	2.9848	2.9188	2.9096	2.8466	2.8647	2.8795	2.9188	2.9280	3.0000
3 1/4	4	NC-3	3.2500	3.2220	3.0876	3.0779	2.9433	2.9794	3.0064	3.0876	3.0973	3.2500
3 1/4	8	N-3	3.2500	3.2348	3.1688	3.1595	3.0966	3.1147	3.1295	3.1688	3.1781	3.2500
3 1/2	4	NC-3	3.5000	3.4720	3.3376	3.3279	3.1933	3.2294	3.2564	3.3376	3.3473	3.5000
3 1/2	8	N-3	3.5000	3.4848	3.4188	3.4095	3.3466	3.3647	3.3795	3.4188	3.4281	3.5000
3 3/4	4	NC-3	3.7500	3.7220	3.5876	3.5779	3.4433	3.4794	3.5064	3.5876	3.5973	3.7500
3 3/4	8	N-3	3.7500	3.7348	3.6688	3.6594	3.5966	3.6147	3.6295	3.6688	3.6782	3.7500
4	4	NC-3	4.0000	3.9720	3.8376	3.8279	3.6933	3.7294	3.7564	3.8376	3.8473	4.0000
4	8	N-3	4.0000	3.9848	3.9188	3.9093	3.8466	3.8647	3.8795	3.9188	3.9283	4.0000
4 1/4	8	N-3	4.2500	4.2348	4.1688	4.1592	4.0966	4.1147	4.1295	4.1688	4.1784	4.2500
4 1/2	8	N-3	4.5000	4.4848	4.4188	4.4091	4.3466	4.3647	4.3795	4.4188	4.4285	4.5000
4 3/4	8	N-3	4.7500	4.7348	4.6688	4.6590	4.5966	4.6147	4.6295	4.6688	4.6786	4.7500
5	8	N-3	5.0000	4.9848	4.9188	4.9089	4.8466	4.8647	4.8795	4.9188	4.9287	5.0000
5 1/4	8	N-3	5.2500	5.2348	5.1688	5.1589	5.0966	5.1147	5.1295	5.1688	5.1787	5.2500
5 1/2	8	N-3	5.5000	5.4848	5.4188	5.4088	5.3466	5.3647	5.3795	5.4188	5.4288	5.5000
5 3/4	8	N-3	5.7500	5.7348	5.6688	5.6587	5.5966	5.6147	5.6295	5.6688	5.6789	5.7500
6	8	N-3	6.0000	5.9848	5.9188	5.9086	5.8466	5.8647	5.8795	5.9188	5.9290	6.0000

Table 15. Extra-fine-thread Series (UN and NEF) Allowances and Tolerances
(ASA B1.1-1949. All dimensions in inches.)

Size	Allowance *	Tolerances							
		Major Diameter External Threads†		Pitch Diameter				Minor Diameter Internal Threads	
	Class 2A a	Class 2A t_{ds}	Classes 2 and 3 t_{ds}	Class 2A t_{es}	Class 2B t_{en}	Class 2 t_{es}, t_{en}	Class 3 t_{es}, t_{en}	Class 2B t_{kn}	Classes 2 and 3 t_{kn}
No. 12 (*0.216*)	.0009	.0060	.0054	.0031	.0041	.0031	.0022	.0073	.0046
1/4	.0010	.0060	.0054	.0032	.0042	.0032	.0022	.0067	.0046
5/16	.0010	.0060	.0054	.0032	.0042	.0033	.0023	.0060	.0046
3/8	.0010	.0060	.0054	.0034	.0044	.0034	.0024	.0057	.0046
7/16 ‡	.0011	.0065	.0062	.0036	.0046	.0036	.0025	.0063	.0053
1/2 ‡	.0011	.0065	.0062	.0037	.0048	.0037	.0026	.0063	.0053
9/16	.0012	.0072	.0066	.0039	.0051	.0040	.0028	.0070	.0061
5/8	.0012	.0072	.0066	.0040	.0052	.0041	.0029	.0070	.0061
11/16	.0012	.0072	.0066	.0040	.0052	.0041	.0029	.0070	.0061
3/4 ‡	.0013	.0081	.0072	.0044	.0057	.0046	.0032	.0078	.0068
13/16 ‡	.0013	.0081	.0072	.0044	.0057	.0046	.0032	.0078	.0068
7/8 ‡	.0013	.0081	.0072	.0044	.0057	.0047	.0033	.0078	.0068
15/16 ‡	.0014	.0081	.0072	.0045	.0059	.0047	.0033	.0078	.0068
1 ‡	.0014	.0081	.0072	.0045	.0059	.0048	.0034	.0078	.0068
1 1/16	.0014	.0087	.0082	.0047	.0062	.0051	.0036	.0078	.0075
1 1/8	.0014	.0087	.0082	.0047	.0062	.0052	.0036	.0078	.0075
1 3/16	.0015	.0087	.0082	.0049	.0063	.0052	.0036	.0078	.0075
1 1/4	.0015	.0087	.0082	.0049	.0063	.0053	.0037	.0078	.0075
1 5/16	.0015	.0087	.0082	.0049	.0063	.0053	.0037	.0078	.0075
1 3/8	.0015	.0087	.0082	.0049	.0063	.0054	.0038	.0078	.0075
1 7/16	.0015	.0087	.0082	.0050	.0065	.0054	.0038	.0078	.0075
1 1/2	.0015	.0087	.0082	.0050	.0065	.0055	.0038	.0078	.0075
1 9/16	.0015	.0087	.0082	.0050	.0065	.0055	.0039	.0078	.0075
1 5/8	.0015	.0087	.0082	.0050	.0065	.0056	.0039	.0078	.0075
1 11/16	.0015	.0087	.0082	.0051	.0066	.0056	.0039	.0078	.0075
1 3/4 ‡	.0016	.0094	.0090	.0053	.0069	.0059	.0041	.0085	.0080
2 ‡	.0016	.0094	.0090	.0054	.0070	.0061	.0043	.0085	.0080

Values are based on a length of engagement of nine threads.
* Allowances are zero for classes 2 and 3.
† Major diameter of internal threads may extend to a $p/24$ flat.
‡ Unified threads, in classes 2A and 2B.

couplings, flanges, etc., must be held to a minimum, and where maximum practicable number of threads is required within a given thread length. The series is the same as the former SAE fine series and the present SAE extra-fine series. Allowances and tolerances for various fits are given in Table 15. Limiting dimensions may be calculated from eqs. 4 to 6 and 18 to 27, or may be obtained from the standard (ASA B1.1-1949).

Eight-thread Series (8N). This series (see Table 6) has application for fastenings which require that an initial tension be set up by elastic deformation of the fastening and of the components which it holds together. Examples are bolts for high-pressure pipe flanges and cylinder-head studs which hold against pressure. To secure proper initial tension the pitch should not increase with diameter; otherwise the torque required to assemble the fastening would become excessive in the larger sizes. To meet this need, the 8-thread series has come into general use in many classes of engineering work as a substitute for the coarse-thread series.

For the 8-thread series, eqs. 4, 5, and 6 become, respectively,

$$E = D - 0.0812 \tag{9}$$

$$K_s = D - 0.1534 \tag{10}$$

$$K_n = D - 0.1353 \tag{11}$$

Limiting dimensions are given in Tables 11, 13, and 14.

Twelve-thread Series (12N). This series (see Table 7) is used in machine construction for thin nuts on shafts and sleeves. It is the coarsest thread in general use that will permit a threaded collar to slip over a shaft and screw onto a threaded shoulder, the difference in diameter between shoulder and shaft being $^1/_8$ in. Sizes of 12-pitch threads from $^1/_2$ in. up to and including 1 $^3/_4$ in. are used in boiler practice, which requires that worn stud holes be retapped with a tap of the next larger size, the increment being $^1/_{16}$ in. throughout most

Table 16. 12-pitch Thread Series (12UN and 12N) Allowances and Tolerances
(ASA B1.1-1949. All dimensions in inches.)

Size	Allowance *	Tolerances							
		Major Diameter External Threads†		Pitch Diameter				Minor Diameter Internal Threads	
	Class 2A a	Class 2A t_{ds}	Classes 2 and 3 t_{ds}	Class 2A t_{es}	Class 2B t_{en}	Class 2 t_{es}, t_{en}	Class 3 t_{es}, t_{en}	Class 2B t_{kn}	Classes 2 and 3 t_{kn}
1/2 ‡,§	.0015	.0114	.0112	.0051	.0066	.0056	.0040	.0125	.0127
9/16 ‡,§	.0016	.0114	.0112	.0052	.0068	.0056	.0040	.0120	.0127
5/8	.0016	.0114	.0112	.0054	.0071	.0056	.0040	.0115	.0090
11/16	.0016	.0114	.0112	.0054	.0071	.0056	.0040	.0112	.0090
3/4	.0017	.0114	.0112	.0055	.0072	.0056	.0040	.0109	.0090
13/16	.0017	.0114	.0112	.0055	.0072	.0056	.0040	.0106	.0090
7/8	.0017	.0114	.0112	.0055	.0072	.0056	.0040	.0104	.0090
15/16 ‡	.0017	.0114	.0112	.0057	.0074	.0056	.0040	.0102	.0090
1 ‡,§	.0018	.0114	.0112	.0059	.0076	.0056	.0040	.0100	.0090
1 1/16 ‡	.0017	.0114	.0112	.0057	.0074	.0056	.0040	.0100	.0090
1 1/8 ‡,§	.0018	.0114	.0112	.0060	.0078	.0056	.0040	.0100	.0090
1 3/16 ‡	.0017	.0114	.0112	.0058	.0075	.0056	.0040	.0100	.0090
1 1/4 ‡,§	.0018	.0114	.0112	.0062	.0080	.0056	.0040	.0100	.0090
1 5/16 ‡	.0017	.0114	.0112	.0058	.0075	.0056	.0040	.0100	.0090
1 3/8 ‡,§	.0019	.0114	.0112	.0063	.0082	.0056	.0040	.0100	.0090
1 7/16 ‡	.0018	.0114	.0112	.0059	.0076	.0056	.0040	.0100	.0090
1 1/2 ‡,§	.0019	.0114	.0112	.0064	.0083	.0056	.0040	.0100	.0090
1 5/8	.0018	.0114	.0112	.0059	.0076	.0064	.0045	.0100	.0090
1 3/4 ‡	.0018	.0114	.0112	.0060	.0078	.0065	.0046	.0100	.0090
1 7/8	.0018	.0114	.0112	.0060	.0078	.0066	.0046	.0100	.0090
2 ‡	.0018	.0114	.0112	.0061	.0079	.0067	.0047	.0100	.0090
2 1/8	.0018	.0114	.0112	.0061	.0079	.0068	.0048	.0100	.0090
2 1/4 ‡	.0018	.0114	.0112	.0061	.0079	.0069	.0048	.0100	.0090
2 3/8	.0019	.0114	.0112	.0062	.0081	.0070	.0049	.0100	.0090
2 1/2 ‡	.0019	.0114	.0112	.0062	.0081	.0071	.0049	.0100	.0090
2 5/8	.0019	.0114	.0112	.0062	.0081	.0071	.0050	.0100	.0090
2 3/4 ‡	.0019	.0114	.0112	.0062	.0081	.0072	.0050	.0100	.0090
2 7/8	.0019	.0114	.0112	.0063	.0082	.0073	.0051	.0100	.0090
3 ‡	.0019	.0114	.0112	.0063	.0082	.0074	.0051	.0100	.0090
3 1/8	.0019	.0114	.0112	.0063	.0082	.0074	.0052	.0100	.0090
3 1/4 ‡	.0019	.0114	.0112	.0063	.0082	.0075	.0052	.0100	.0090
3 3/8	.0019	.0114	.0112	.0064	.0084	.0076	.0053	.0100	.0090
3 1/2 ‡	.0019	.0114	.0112	.0064	.0084	.0076	.0053	.0100	.0090
3 5/8	.0019	.0114	.0112	.0064	.0084	.0077	.0054	.0100	.0090
3 3/4 ‡	.0019	.0114	.0112	.0064	.0084	.0078	.0054	.0100	.0090
3 7/8	.0020	.0114	.0112	.0065	.0085	.0078	.0055	.0100	.0090
4 ‡	.0020	.0114	.0112	.0065	.0085	.0079	.0055	.0100	.0090
4 1/4 ‡	.0020	.0114	.0112	.0065	.0085	.0080	.0056	.0100	.0090
4 1/2 ‡	.0020	.0114	.0112	.0065	.0085	.0081	.0057	.0100	.0090
4 3/4 ‡	.0020	.0114	.0112	.0067	.0087	.0083	.0058	.0100	.0090
5 ‡	.0020	.0114	.0112	.0067	.0087	.0084	.0059	.0100	.0090
5 1/4 ‡	.0020	.0114	.0112	.0067	.0087	.0085	.0059	.0100	.0090
5 1/2 ‡	.0020	.0114	.0112	.0067	.0087	.0086	.0060	.0100	.0090
5 3/4 ‡	.0021	.0114	.0112	.0069	.0090	.0087	.0061	.0100	.0090
6 ‡	.0021	.0114	.0112	.0069	.0090	.0088	.0062	.0100	.0090

* Allowances are zero for classes 2 and 3.
† Major diameter of internal threads may extend to a $p/24$ flat.
‡ Unified threads, in classes 2A and 2B.
§ These are standard sizes of the UNC or UNF series and tolerances are the same as for those series, being based on a length of engagement equal to the nominal diameter. All other classes 2A and 2B tolerances and all classes 2 and 3 tolerances for sizes larger than 1 $^1/_2$ in. are based on a length of engagement of 9 threads.

of the range. Die-head chasers for sizes up to 3 in. are readily available. The 12-thread series also provides continuation of the fine-thread series (UNF) in diameters larger than 1 1/2 in. Allowances and tolerances for various fits are given in Table 16.

Table 17. 16-pitch Thread Series (16UN and 16N) Allowances and Tolerances

(ASA B1.1-1949. All dimensions in inches.)

Size	Allowance *	Tolerances							
		Major Diameter External Threads†		Pitch Diameter				Minor Diameter Internal Threads	
	Class 2A a	Class 2A t_{ds}	Classes 2 and 3 t_{ds}	Class 2A t_{es}	Class 2B t_{en}	Class 2 t_{es}, t_{en}	Class 3 t_{es}, t_{en}	Class 2B t_{kn}	Classes 2 and 3 t_{kn}
3/4 ‡	.0015	.0094	.0090	.0050	.0065	.0045	.0032	.0085	.0080
13/16 ‡	.0015	.0094	.0090	.0049	.0063	.0051	.0035	.0085	.0080
7/8 ‡	.0015	.0094	.0090	.0049	.0063	.0051	.0036	.0085	.0080
15/16 ‡	.0015	.0094	.0090	.0050	.0065	.0052	.0036	.0085	.0080
1 ‡	.0015	.0094	.0090	.0050	.0065	.0052	.0037	.0085	.0080
1 1/16 ‡	.0015	.0094	.0090	.0050	.0065	.0053	.0037	.0085	.0080
1 1/8 ‡	.0015	.0094	.0090	.0050	.0066	.0054	.0038	.0085	.0080
1 3/16 ‡	.0015	.0094	.0090	.0051	.0066	.0054	.0038	.0085	.0080
1 1/4 ‡	.0015	.0094	.0090	.0051	.0066	.0055	.0038	.0085	.0080
1 5/16 ‡	.0015	.0094	.0090	.0051	.0066	.0055	.0039	.0085	.0080
1 3/8 ‡	.0015	.0094	.0090	.0051	.0066	.0056	.0039	.0085	.0080
1 7/16 ‡	.0016	.0094	.0090	.0052	.0068	.0056	.0040	.0085	.0080
1 1/2 ‡	.0016	.0094	.0090	.0052	.0068	.0057	.0040	.0085	.0080
1 9/16	.0016	.0094	.0090	.0052	.0068	.0058	.0040	.0085	.0080
1 5/8	.0016	.0094	.0090	.0052	.0068	.0058	.0041	.0085	.0080
1 11/16	.0016	.0094	.0090	.0053	.0069	.0058	.0041	.0085	.0080
1 3/4 ‡	.0016	.0094	.0090	.0053	.0069	.0059	.0041	.0085	.0080
1 13/16	.0016	.0094	.0090	.0053	.0069	.0059	.0042	.0085	.0080
1 7/8	.0016	.0094	.0090	.0053	.0069	.0060	.0042	.0085	.0080
1 15/16	.0016	.0094	.0090	.0054	.0070	.0060	.0042	.0085	.0080
2 ‡	.0016	.0094	.0090	.0054	.0070	.0061	.0043	.0085	.0080
2 1/16	.0016	.0094	.0090	.0054	.0070	.0061	.0043	.0085	.0080
2 1/8	.0016	.0094	.0090	.0054	.0070	.0062	.0043	.0085	.0080
2 3/16	.0016	.0094	.0090	.0054	.0070	.0062	.0043	.0085	.0080
2 1/4 ‡	.0016	.0094	.0090	.0054	.0070	.0062	.0044	.0085	.0080
2 5/16	.0017	.0094	.0090	.0055	.0072	.0063	.0044	.0085	.0080
2 3/8	.0017	.0094	.0090	.0055	.0072	.0063	.0044	.0085	.0080
2 7/16	.0017	.0094	.0090	.0055	.0072	.0064	.0045	.0085	.0080
2 1/2 ‡	.0017	.0094	.0090	.0055	.0072	.0064	.0045	.0085	.0080
2 5/8	.0017	.0094	.0090	.0055	.0072	.0065	.0045	.0085	.0080
2 3/4 ‡	.0017	.0094	.0090	.0055	.0072	.0066	.0046	.0085	.0080
2 7/8	.0017	.0094	.0090	.0056	.0073	.0066	.0046	.0085	.0080
3 ‡	.0017	.0094	.0090	.0056	.0073	.0067	.0047	.0085	.0080
3 1/8	.0017	.0094	.0090	.0056	.0073	.0068	.0047	.0085	.0080
3 1/4 ‡	.0017	.0094	.0090	.0056	.0073	.0069	.0048	.0085	.0080
3 3/8	.0017	.0094	.0090	.0058	.0075	.0069	.0048	.0085	.0080
3 1/2 ‡	.0017	.0094	.0090	.0058	.0075	.0070	.0049	.0085	.0080
3 5/8	.0017	.0094	.0090	.0058	.0075	.0071	.0049	.0085	.0080
3 3/4 ‡	.0017	.0094	.0090	.0058	.0075	.0071	.0050	.0085	.0080
3 7/8	.0018	.0094	.0090	.0059	.0076	.0072	.0050	.0085	.0080
4 ‡	.0018	.0094	.0090	.0059	.0076	.0072	.0051	.0085	.0080
4 1/4 ‡	.0018	.0094	.0090	.0059	.0076	.0074	.0052	.0085	.0080
4 1/2 ‡	.0018	.0094	.0090	.0059	.0076	.0075	.0053	.0085	.0080
4 3/4 ‡	.0018	.0094	.0090	.0061	.0079	.0076	.0053	.0085	.0080
5 ‡	.0018	.0094	.0090	.0061	.0079	.0077	.0054	.0085	.0080
5 1/4 ‡	.0018	.0094	.0090	.0061	.0079	.0078	.0055	.0085	.0080
5 1/2 ‡	.0018	.0094	.0090	.0061	.0079	.0079	.0056	.0085	.0080
5 3/4 ‡	.0019	.0094	.0090	.0062	.0081	.0080	.0056	.0085	.0080
6 ‡	.0019	.0094	.0090	.0062	.0081	.0081	.0057	.0085	.0080

Values for classes 2A and 2B are based on a length of engagement equal to the nominal diameter for the 3/4 in. size and a length of engagement of 9 threads for larger sizes, and for classes 2 and 3 on a length of engagement of 9 threads.

* Allowances apply to external threads class 2A only.

† Major diameter of internal threads may extend to a $p/24$ flat.

‡ Unified threads, in classes 2A and 2B.

For the 12-thread series eqs. 4, 5, and 6 become, respectively,

$$E = D - 0.0541 \tag{12}$$

$$K_s = D - 0.1022 \tag{13}$$

$$K_n = D - 0.0902 \tag{14}$$

Limiting dimensions may be calculated from the foregoing equations in conjunction with eqs. 18 to 27 and Table 16, or may be obtained from the standard (ASA B1.1-1949).

Sixteen-Thread Series (16N). This series (see Table 8) is intended for applications which require a relatively fine thread such as threaded adjusting collars and bearing retaining nuts. It also provides continuation of the extra-fine series (NEF) in diameters larger than 2 in. Allowances and tolerances for various fits are given in Table 17.

For the 16-thread series, eqs. 4, 5, and 6 become, respectively,

$$E = D - 0.0406 \tag{15}$$

$$K_s = D - 0.0767 \tag{16}$$

$$K_n = D - 0.0677 \tag{17}$$

Limiting dimensions may be calculated from the foregoing equations in conjunction with eqs. 18 to 27 and Table 17. Tabulated values are given in ASA B1.1-1949.

Special Threads (UNS, UN, or NS). Diameter and pitch combinations not included in the coarse-thread (UNC and NC), fine-thread (UNF and NF), and 8-thread (8N) series are designated as special threads. Special threads, therefore, include the national extra-fine (NEF), the 12-thread (12N), and 16-thread (16N) series as well as any combinations of diameter and pitch not covered by the six series listed in Tables 3, 4, 5, 6, 7, and 8.

Where requirements cannot be met by any of the combinations in these six series, it is recommended that one of the combinations listed in Table 9 be selected. Basic dimensions for special combinations may be calculated from eqs. 4, 5, 6, 7, and 8. The standard (ASA B1.1-1949) lists limiting dimensions of threads conforming to Tables 6, 7, 8, and 9 for classes 2A, 2B, 2, and 3 with thread engagement equal to $9p$.

Table 18. Limiting Dimensions of Screw Threads for High-strength Bolting, Class 7 Fits

(ASA B1.4-1945. All dimensions in inches.)

Size	Threads per Inch	Screw					Nut				
		Major Diameter		Pitch Diameter *		Minor Diameter, Max	Major Diameter, Min	Pitch Diameter *		Minor Diameter	
		Max	Min	Max	Min			Min	Max	Min	Max
1/4	20	0.2490	0.2418	0.2165	0.2139	0.1877	0.2500	0.2175	0.2211	0.1959	0.2060
5/16	18	0.3114	0.3032	0.2753	0.2723	0.2432	0.3125	0.2764	0.2805	0.2524	0.2630
3/8	16	0.3737	0.3647	0.3331	0.3299	0.2970	0.3750	0.3344	0.3389	0.3073	0.3184
7/16	14	0.4362	0.4264	0.3898	0.3862	0.3486	0.4375	0.3911	0.3960	0.3602	0.3721
1/2	13	0.4985	0.4881	0.4485	0.4448	0.4041	0.5000	0.4500	0.4552	0.4167	0.4290
9/16	12	0.5609	0.5497	0.5068	0.5028	0.4587	0.5625	0.5084	0.5140	0.4723	0.4850
5/8	11	0.6233	0.6115	0.5643	0.5601	0.5118	0.6250	0.5660	0.5719	0.5266	0.5397
3/4	10	0.7481	0.7353	0.6831	0.6786	0.6254	0.7500	0.6850	0.6914	0.6417	0.6553
7/8	9	0.8729	0.8589	0.8007	0.7958	0.7366	0.8750	0.8028	0.8098	0.7547	0.7689
1	8	0.9978	0.9826	0.9166	0.9112	0.8444	1.0000	0.9188	0.9264	0.8647	0.8795
1 1/8	8	1.1226	1.1074	1.0414	1.0359	0.9692	1.1250	1.0438	1.0517	0.9897	1.0045
1 1/4	8	1.2475	1.2323	1.1663	1.1605	1.0941	1.2500	1.1688	1.1771	1.1147	1.1295
1 3/8	8	1.3725	1.3573	1.2913	1.2852	1.2191	1.3750	1.2938	1.3024	1.2397	1.2545
1 1/2	8	1.4973	1.4821	1.4161	1.4098	1.3439	1.5000	1.4188	1.4278	1.3647	1.3795
1 5/8	8	1.6222	1.6070	1.5410	1.5345	1.4688	1.6250	1.5438	1.5531	1.4897	1.5045
1 3/4	8	1.7471	1.7319	1.6659	1.6591	1.5937	1.7500	1.6688	1.6785	1.6147	1.6295
1 7/8	8	1.8720	1.8568	1.7908	1.7838	1.7186	1.8750	1.7938	1.8038	1.7397	1.7545
2	8	1.9969	1.9817	1.9157	1.9084	1.8435	2.0000	1.9188	1.9292	1.8647	1.8795
2 1/8	8	2.1218	2.1066	2.0406	2.0331	1.9684	2.1250	2.0438	2.0545	1.9897	2.0045
2 1/4	8	2.2467	2.2315	2.1655	2.1578	2.0933	2.2500	2.1688	2.1798	2.1147	2.1295
2 1/2	8	2.4965	2.4813	2.4153	2.4071	2.3431	2.5000	2.4188	2.4305	2.3647	2.3795
2 3/4	8	2.7463	2.7311	2.6651	2.6564	2.5929	2.7500	2.6688	2.6812	2.6147	2.6295
3	8	2.9962	2.9810	2.9150	2.9058	2.8428	3.0000	2.9188	2.9318	2.8647	2.8795
3 1/4	8	3.2461	3.2309	3.1649	3.1556	3.0927	3.2500	3.1688	3.1820	3.1147	3.1295
3 1/2	8	3.4960	3.4808	3.4148	3.4055	3.3426	3.5000	3.4188	3.4321	3.3647	3.3795

* Pitch diameter tolerances include errors of lead and angle.

High-strength bolting threads (ASA B1.4-1945) employ the coarse-thread series for sizes 1 in. and smaller, and the 8-thread series for larger sizes (see Table 18). Developed to provide a thread that could readily be machined from alloy steels and with an allowance permitting the application of lubricant during assembly, this thread has been applied to many services, both hot and cold, where high-tensile stresses are produced when the joints are made up.

SCREW THREAD DESIGNATION. For brevity on drawings, etc., screw threads are designated by the nominal size (diameter in inches, or screw number) followed by the number of threads per inch, the initial letters of the thread series, and the screw-thread class (see next paragraph) with or without the pitch diameter tolerances or limits of size. For example,

$$^1/_4 - 20 \text{ UNC} - 2\text{A}$$

denotes a $^1/_4$-in. nominal-size screw having 20 threads per inch, which is in the coarse-thread series (UNC). The term 2A indicates class 2A tolerance and allowance, this being an external thread. Unless specifically designated by the letters LH for left-hand, the screw is understood to be right-handed. If a standard thread is modified, the word "modified" should be added with an asterisk, and the nonstandard feature or dimension of the thread enclosed in brackets and likewise marked with an asterisk.

SCREW THREAD CLASSES are distinguished from each other by the amount of tolerance and allowance specified. The unified standard includes classes 1A, 2A, and 3A for external threads and classes 1B, 2B, and 3B for internal threads. In addition, the American standard includes classes 2 and 3 for both external and internal threads, a holdover from the old National standard.

Classes 2A and 2B are the recognized standards for normal production of screws, bolts, and nuts and are suitable for a wide range of applications. See Fig. 4. The allowances and tolerances provide a clearance under all conditions; the minimum clearance is such as to minimize galling and seizure in assembly and use. Plating applied to external threads may project into or fill this space.

Classes 2A and 2B have been established for all thread series. Limiting dimensions of coarse, fine, and 8-thread series of these classes are given in Table 11. Allowances and tolerances for other series are given in Tables 15, 16, and 17.

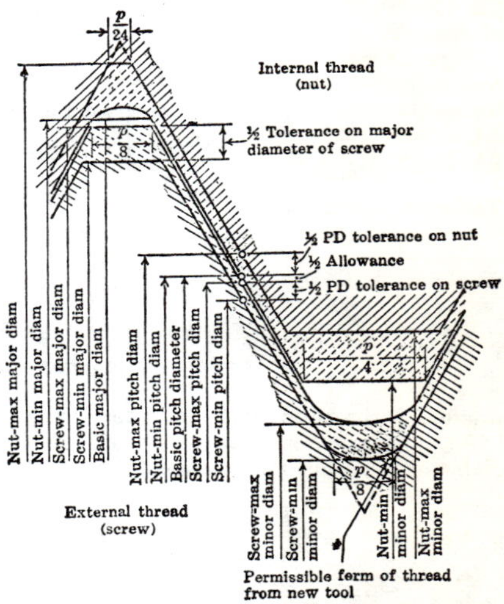

FIG. 4. Tolerances, allowance, and crest clearances for Unified and American classes 1A, 2A, 1B, and 2B. For classes 3A, 3B, 2, and 3 the allowance is zero.

Classes 1A and 1B in combination provide a fit suitable for threaded parts where quick and easy assembly is necessary, even when the threads are bruised or dirty, as in ordnance and certain special applications. See Fig. 4. Tolerances are 50% greater than on classes 2A and 2B, but the allowance on 1A is the same as on 2A. Classes 1A and 1B have been established for the coarse- and fine-thread series only, Table 10.

Classes 3A and 3B are suitable for applications requiring close fit and accuracy of lead and angle of thread. Such fits are obtainable only by the use of high-quality production equipment supported by efficient gaging and inspection. No allowance is provided on external threads (3A), and the tolerances are 75% of those for classes 2A and 2B. Classes 3A and 3B have been established for the coarse- and fine-thread series only, Table 12.

Classes 2 and 3, held over from the earlier National standard during the transition to the unified standard, correspond in general application to classes 2A and 2B, and 3A and 3B, respectively. However, no allowance is provided on either class 2 or class 3. Tolerances for class 3 are approximately 70% of class 2 tolerances. Limiting dimensions of coarse, fine, and 8-pitch threads of class 2 are given in Table 13, and of class 3 in Table 14. Allowances and tolerances for other series are given in Tables 15, 16, and 17.

Class 4 Fit. Included in the American National standard, ASA B1.1-1935, but not in the Unified and American Standard, ASA B1.1-1949, class 4 results in selective assembly type of fit. It is not adaptable to quantity production and was intended only for unusual and exacting requirements. With a maximum screw and a minimum nut there is actual interference of metal.

Class 5 Fit. Based on experimental data developed by the National Bureau of Standards for permanent assembly of steel studs in hard material, class 5 results in interference. It is included in the Bureau's *Handbook H-28*, 1944 edition.

Class 7 Fit. This designation is applied to the screw-thread fit used on High Strength Bolting (ASA B1.4-1945). The nuts are identical with class 2 nuts of equal diameter and pitch. Allowance on the bolts is obtained by making the maximum pitch diameter of the screw smaller than the minimum pitch diameter of the nut by an amount equal to the difference between the class 2 and class 3 tolerances. The screw has a class 3 tolerance so that the minimum pitch diameter of the screw is equal to the minimum pitch diameter of a class 2 screw. Limiting dimensions of High Strength Bolting screws and nuts employing the class 7 fit are given in Table 18.

Limiting dimensions for all thread series other than the coarse, fine, and 8-pitch (which are given in Tables 10, 11, 12, 13, and 14) may be calculated from the basic dimensions (eqs. 4, 5, and 6) in conjunction with Tables 15, 16, and 17, which give the allowances and tolerances. The following equations give the relationships between basic dimensions, allowances and tolerances, and limiting dimensions (see also Fig. 4):

For external threads:

$$\text{Major diameter, max} = D - a \tag{18}$$

$$\text{Major diameter, min} = D - a - t_{ds} \tag{19}$$

$$\text{Pitch diameter, max} = E - a \tag{20}$$

$$\text{Pitch diameter, min} = E - a - t_{es} \tag{21}$$

$$\text{Minor diameter, max} = K_s - a \tag{22}$$

Minor diameter, min Not specified

For internal threads:

$$\text{Minor diameter, min} = K_n \tag{23}$$

$$\text{Minor diameter, max} = K_n + t_{kn} \tag{24}$$

$$\text{Pitch diameter, min} = E \tag{25}$$

$$\text{Pitch diameter, max} = E + t_{en} \tag{26}$$

$$\text{Major diameter, min} = D \tag{27}$$

Major diameter, max Not specified

where D is the basic major diameter (Tables 3, 4, 5, 6, 7, 8, and 9); E, K_s, and K_n are the basic dimensions calculated from eqs. 4, 5, and 6; a is the allowance; t_{ds} is the major diameter tolerance for external (screw) thread; t_{es} is the pitch diameter tolerance for external (screw) thread; t_{en} is the pitch diameter tolerance for internal (nut) thread; and t_{kn} is the minor diameter tolerance for internal (nut) thread.

For complete data on tolerances, allowances, and limiting dimensions of threads other than those covered by Tables 3, 4, 5, 6, 7, and 8, the standard (ASA B1.1-1949) should be consulted.

In applying the foregoing formulas it should be noted that the allowance a applies only to classes 1A and 2A. The allowance is zero for class 3A, and also for classes 2 and 3.

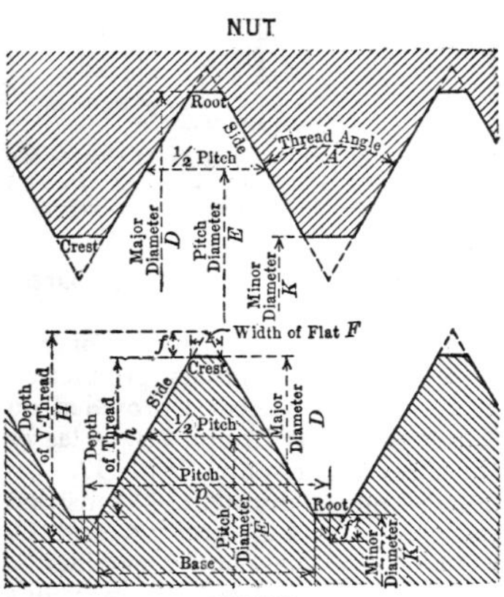

Fig. 5. American National thread form.

AMERICAN NATIONAL form of thread, which the Unified and American thread form replaces, is that previously known as the United States Standard or Sellers profile (Fig. 5). Basic angle of thread, measured in an axial plane, is 60 degrees; basic width of flat at the root and crest of the thread is $F = 0.125p = 0.125/n$; basic depth of the thread form is $h = 0.649519p = 0.649519/n$, where p = pitch, inches, and n = number of threads per inch. Complete information is given in ASA B1.1-1935.

Clearance at Minor Diameter. Clearance is provided at the minor diameter of the screw thread in the nut by truncation such that the basic depth of thread is reduced by one-sixth or more, depending on the size and pitch.

Clearance at Major Diameter. Clearance at the major diameter of the screw thread in the nut is provided by decreasing the depth of the truncation triangle any desired amount down to one-third its theoretical value.

Interchangeability. American National threads are interchangeable with the Unified and American threads. Where American National threads are specified it is recommended that users accept the new classes as optional, and vice versa during the transition.

DARDELET THREAD (Lock Thread Corporation). Similar to the Acme, with an included thread angle of 29 degrees, the Dardelet thread has smaller depth and thickness in relation to the pitch, as well as inclined flats at the roots of the bolt thread and inclined crests on the nut threads (Fig. 6). Approximate proportions are as follows, where n = number of threads per inch: pitch $p = 1/n$; $B_1 = 0.290p$; $B_2 = 0.272p$; $B_3 = 0.282p$; $H = 0.349p$; $H_1 = 0.238p$; and $F = 0.282p$. The 6-degree angle of inclination provides a wedging action when the nut is tightened, locking the two elements against vibration. Dardelet threads are made in a coarse-thread and a fine-thread series (Table 19).

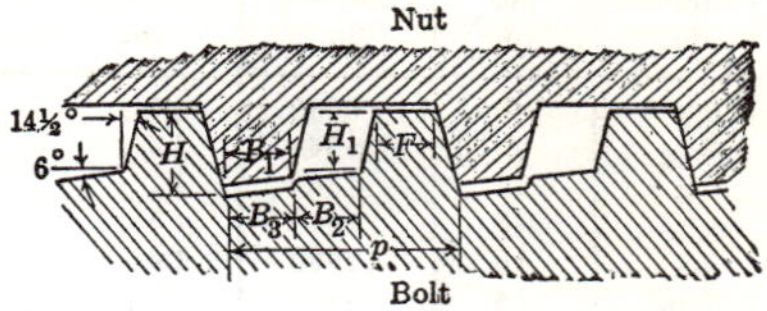

FIG. 6. Dardalet relieved thread (unlocked).

Table 19. Diameter and Pitch Combinations Using Dardelet Threads

Size or Diameter	Threads per Inch Coarse	Threads per Inch Fine	Size or Diameter	Threads per Inch Coarse	Threads per Inch Fine
No. 8	..	24	5/8	8	13
No. 10	..	24	11/16	8	12
No. 12	..	24	3/4	8	12
No. 14	..	20	13/16	7	12
1/4	16	20	7/8	7	11
5/16	14	18	1	6	10
3/8	12	16	1 1/8	5	8
7/16	11	16	1 1/4	5	8
1/2	10	16	1 3/8	5	7
9/16	9	14	1 1/2	4	6

LOK-THRED (Lock Thread Corp.). Similar to the Unified and American Standard, the included angle of the thread being 60 degrees, this thread differs in having smaller depth and slightly less thickness in relation to the pitch, as well as 6-degree inclined spiral flats at the root of the screw thread (Fig. 7). Thread depth, Y, of the external (screw) thread is either 60 or 70% of that of the American National standard, depending on the application. The 70% ($Y = 0.7h$) standard is recommended for cap screws and bolts and the 60% ($Y = 0.6h$) for studs. It will be noted from Fig. 7 that the width of root, T, for the two series is $0.5p$ minimum and $0.425p$ minimum, respectively, as against $0.125p$ for the Unified thread.

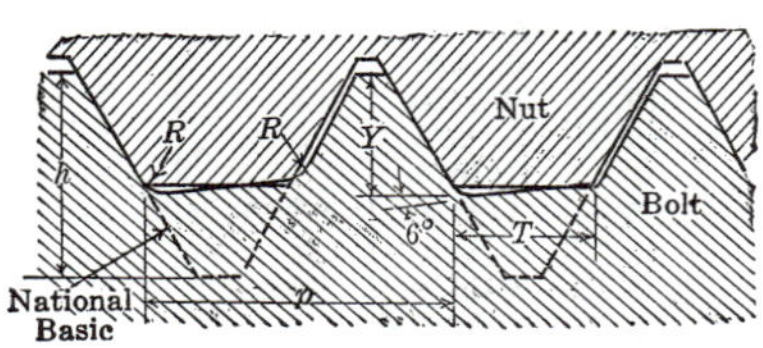

FIG. 7. Lok-Thred for 70% thread: $Y = 0.7h$, $T = 0.5p$, and $R = 0.137p$. For 60% thread, $Y = 0.6h$, $T = 0.425p$, and $R = 0.137p$.

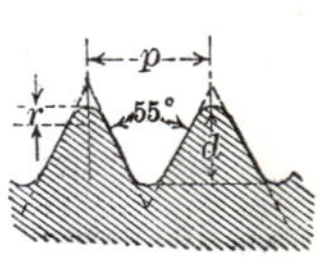

FIG. 8. Whitworth thread.

Either design provides greater cross-sectional area at the root of the thread than the Unified form of thread. Interference between the crest of the internal thread and the tapered flats of the external thread provides a locking and sealing action when assembled. Nuts are Unified and American standard with tap-drilled minor diameters held to limits to insure uniform assembly torque. Lok-Threads are reusable and can replace Unified and American threads. Lead threads are provided for ease of starting.

BRITISH STANDARD WHITWORTH (BSW) threads employ a 55-degree included angle (Fig. 8). Actual depth of the thread is two-thirds the height of the sharp V-thread and is $d = 0.640327p$; radius at crest and root is $r = 0.137329p$, where p = pitch. Diameter and pitch combinations and basic sizes of bolts are listed in Table 20. Effective diameter (equivalent to pitch diameter in American nomenclature) is the mean of the full (major) and the core (minor) diameters. Corresponding dimensions of nuts are greater

Table 20. Proportions of British Standard Whitworth Screw Threads

Full Diameter, in.	Threads per in.	Core Diameter, in.	Full Diameter, in.	Threads per in.	Core Diameter, in.	Full Diameter, in.	Threads per in.	Core Diameter, in.
1/8	40	.0930	1 1/4	7	1.0670	3 1/4	3.25	2.8560
3/16	24	.1341	1 3/8 *	6	1.1616	3 3/8 *	3.25	2.9810
1/4	20	.1860	1 1/2	6	1.2866	3 1/2	3.25	3.1060
5/16	18	.2413	1 5/8 *	5	1.3689	3 5/8 *	3.25	3.2310
3/8	16	.2950	1 3/4	5	1.4938	3 3/4	3	3.3232
7/16	14	.3461	1 7/8 *	4.5	1.5904	3 7/8 *	3	3.4481
1/2	12	.3932	2	4.5	1.7154	4	3	3.5732
9/16	12	.4557	2 1/8 *	4.5	1.8404	4 1/8 *	3	3.6981
5/8	11	.5086	2 1/4	4	1.9298	4 1/4 *	2.875	3.8046
11/16 *	11	.5711	2 3/8 *	4	2.0548	4 3/8 *	2.875	3.9296
3/4	10	.6220	2 1/2	4	2.1798	4 1/2	2.875	4.0546
13/16 *	10	.6844	2 5/8 *	4	2.3048	4 3/4 *	2.75	4.2843
7/8	9	.7328	2 3/4	3.5	2.3840	5	2.75	4.5344
15/16 *	9	.7952	2 7/8 *	3.5	2.5091	5 1/4 *	2.625	4.7621
1	8	.8400	3	3.5	2.6340	5 1/2	2.625	5.0122
1 1/8	7	.9420	3 1/8 *	3.5	2.7591	6	2.5	5.4878

* For general use these sizes should not be used.

by 0.002. Tables of close, medium, and free fits are published in the British Standards Institution *Report* 84-1940

BRITISH STANDARD FINE (BSF) threads are of the same form as the Whitworth (Fig. 8). Diameter and pitch combinations and basic sizes of bolts are listed in Table 21. Corresponding dimensions of nuts are greater by 0.002 in. For sizes above 4 1/4 in., 4 threads per inch are recommended.

Table 21. Proportions of British Standard Fine Threads

Full Diameter, in.	Threads per in.	Core Diameter, in.	Full Diameter, in.	Threads per in.	Core Diameter, in.	Full Diameter, in.	Threads per in.	Core Diameter, in.
3/16	32	.1475	3/4	12	0.6432	2 1/4	6	2.0366
7/32	28	.1730	13/16	12	0.7057	2 1/2	6	2.2866
1/4	26	.2008	7/8	11	0.7586	2 3/4	6	2.5366
9/32	26	.2320	1	10	0.8720	3	5	2.7438
5/16	22	.2543	1 1/8	9	0.9828	3 1/4	5	2.9938
3/8	20	.3110	1 1/4	9	1.1078	3 1/2	4.5	3.2154
7/16	18	.3663	1 3/8	8	1.2150	3 3/4	4.5	3.4654
1/2	16	.4200	1 1/2	8	1.3400	4	4.5	3.7154
9/16	16	.4825	1 5/8	8	1.4650	4 1/4	4	3.9298
5/8	14	.5336	1 3/4	7	1.5670			
11/16	14	.5961	2	7	1.8170			

BRITISH ASSOCIATION (BA) threads have an included angle of $47^1/_2$ degrees (Fig. 9). Actual depth of the thread is $0.6p$, approximately, and the radius at crest and root is $r = {}^2/_{11}\, p$, approximately. Diameter and pitch combinations and basic sizes are listed in Table 22. The BA thread is recommended by the British Standards Institution for screws below 1/4 in., except the 7/32-in. BSF size, which is recommended in preference to the No. 0 BA size.

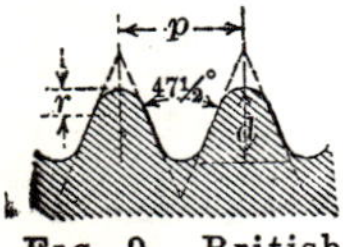

FIG. 9. British Association thread.

INTERNATIONAL STANDARD (SI) metric threads employ a 60-degree included angle and proportions similar to the American National thread. Crests are flat, and roots may be flat or rounded, the exact shape being left to the manufacturer. Depth of engagement is $0.6495p$, which is the same as the basic depth of thread form of the American National thread. Crest clearance adds $0.045p$ (max $0.054p$, min $0.036p$) to the height, making the actual thread depth $0.7p$ (max $0.7035p$, min $0.6855p$). Diameter and pitch combinations and basic sizes of metric threads are listed in Table 23.

TAPS. Standard dimensions and tolerances for cut and ground thread taps are covered by American Standard ASA B5.4-1948. Cut thread taps made to these specifications should, when used under normal conditions, produce holes within class 2 tolerances. Ground thread taps similarly used may be expected to produce holes within class 3 tol-

Table 22. Proportions of British Association Threads

No.	Full Diameter, in.	Threads per in.	Core Diameter, in.	No.	Full Diameter, in.	Threads per in.	Core Diameter, in.
0	.2362	25.4	.1890	12	.0511	90.9	.0378
1	.2087	28.2	.1661	13	.0472	102.0	.0352
2	.1850	31.4	.1468	14	.0394	109.9	.0280
3	.1614	34.8	.1269	15	.0354	120.5	.0250
4	.1417	38.5	.1106	16	.0311	133.3	.0220
5	.1260	43.0	.0981	17	.0276	149.0	.0197
6	.1102	47.9	.0852	18	.0244	169	.0173
7	.0984	52.9	.0758	19	.0213	181	.0146
8	.0866	59.1	.0663	20	.0189	212	.0134
9	.0748	65.1	.0564	21	.0165	231	.0114
10	.0669	72.6	.0504	22	.0146	259	.0098
11	.0591	81.9	.0443	23	.0110	282	.0087

Table 23. International Standard (Metric) Threads

(Dimensions in millimeters)

Outside Diameter	Pitch	Nut Minor Diameter	Outside Diameter	Pitch	Nut Minor Diameter	Outside Diameter	Pitch	Nut Minor Diameter
6	1.0	4.70	22	2.5	18.75	48	5.0	41.50
7	1.0	5.70	24	3.0	20.10	52	5.0	45.50
8	1.25	6.38	27	3.0	23.10	56	5.5	48.86
9	1.25	7.38	30	3.5	25.45	60	5.5	52.86
10	1.5	8.05	33	3.5	28.45	64	6.0	56.21
12	1.75	9.73	36	4.0	30.80	68	6.0	60.21
14	2.0	11.40	39	4.0	33.80	72	6.0	64.21
16	2.0	13.40	42	4.5	36.15	76	6.0	68.21
18	2.5	14.75	45	4.5	39.15	80	6.0	72.21
20	2.5	16.75						

erances. However, on account of varying conditions, the size of a tapped hole cannot be guaranteed.

The standard includes a list of definitions and tap terminology, from which the following paragraphs are excerpted.

Regular (Standard) Hand Taps. A short tap with thread and shank approximately the same length, the latter having a square to accommodate a tap wrench or other driving mechanism. Made in fractional sizes only. Developed for use by hand but used largely in machines at the present time. Furnished in taper, plug, and bottoming types, the difference being in the number of threads chamfered, as follows: bottoming, 1 to 1 1/2 threads; plug, 2 1/2 to 5 threads; taper, 7 to 9 threads.

Three Fluted Hand Taps. Identical with regular (standard) hand taps except they have a smaller number of flutes. Suitable for machine use in tapping tough stringy metal.

Spiral Pointed Hand Taps. Regular (standard) hand taps having a fewer number of flutes and wider lands; cutting face of the first few threads ground at an angle to force the chips ahead to prevent clogging in the flutes.

Serial Hand Taps. Regular (standard) hand taps in sets of three, No. 1 being smaller than No. 2, and No. 2 smaller than No. 3 in pitch diameter, so that the work is distributed among three taps. For very tough metal where a full thread cannot be cut at one pass. Distinguished by rings on the shank near the square.

Regular (Standard) Machine Screw Taps. Similar to the regular (standard) hand taps except that they are made in so-called numbered or machine screw sizes.

Spiral Pointed Machine Screw Taps. Regular (standard) machine screw taps having fewer flutes, wider lands, and cutting face of the first few threads ground at an angle to force the chips ahead to prevent clogging in the flutes.

Nut Taps. Designed for use of machine shops which tap their own nuts. Customarily the entering threads for approximately 75% of the thread length have a slight taper in both pitch diameter and root diameter to insure tapping a nut with a full thread form. Length overall, length of thread, and length of shank are appreciably longer than on a regular (standard) hand tap.

Straight Shank Tapper Taps. Used by nut manufacturers in tapping nuts in vertical spindle machines. Made 12 or 15 in. long, the nuts being allowed to accumulate on the shank during tapping. Shanks are machined to fit the various spindles in ordinary use. Made in both fractional and machine screw sizes.

Bent Shank Tapper Taps. Designed for use in an automatic tapping machine. Nuts are fed to the tap by a hopper, there being continuous production of the tapped product without stopping or reversing. Made in both fractional and machine screw sizes.

Pulley Taps. Originally designed for tapping set screw holes in the hubs of steel pulleys. Thread length is approximately equal to that of a regular (standard) hand tap; shank is full diameter of thread, in various lengths. Meet demand for hand taps of longer length.

Boiler Taps. Hand taps having 12 threads per inch for use by locomotive builders and railroad shops for tapping holes in boiler plate. Available straight, and tapered ($^3/_4$ in. per ft).

Mud or Washout Taps. A hand tap having $1^1/_4$ in. taper per foot and 12 threads per inch. For tapping and retapping holes in mud drums of boilers.

Staybolt Taps. Have 12 threads per inch and a long shank and reamer section. Used, generally in an air drill, to tap boiler sheets preparatory to inserting staybolts.

Spindle Staybolt Taps. A short tap with a hole drilled through its entire length through which a guide spindle fits. Used for tapping holes from the inside of locomotive fireboxes.

Taper Pipe Taps. A hand tap used for tapping American Standard taper pipe fittings, either by hand or by machine.

Straight Pipe Taps. A hand tap identical with the taper pipe tap but with threaded portion straight and of proper size for tapping American Standard straight pipe thread.

Materials. Cut thread taps may be carbon steel or high-speed steel. Ground thread taps, both commercial and precision ground, are high-speed steel.

TAP DRILLS. Selection of tap drill size depends on the percentage of full thread desired or considered practical from a production standpoint. The percentage is figured on the basic height of the original American National thread form (Fig. 5), which is $h = 0.64952p$ or $0.64952/n$, where n is the number of threads per inch. Basic height of the Unified and American thread form (Fig. 3) for an internal thread is $5H/8$ or $0.54127/n$; hence the percentage of full thread is 0.54127/0.64952 or 83.3%, for a thread pair in the maximum metal condition.

For any thread pair having major diameter of the external thread equal to D and minor diameter of the internal thread K, the actual thread height is $(D - K)/2$. Dividing the actual thread height by the basic thread height, $0.64952/n$, and multiplying by 100 gives the percentage of full thread:

$$T = 77n(D - K) \tag{28}$$

Transposing eq. 28 gives a convenient formula for calculating the size of tapped hole for any desired percentage of full thread:

$$K = D - \frac{T}{77n} \tag{29}$$

Determination of tap drill size may be made by first finding the value of K_n in Tables 3, 4, 5, 6, 7, or 8 if a close approximation to a standard thread is desired, or by calculating K from eq. 29 if some other percentage T is desired. The nearest standard size drill may then be selected, referring to Table 1 in Article 23, Section 23, for numbered, lettered and fractional size drills. If the nearest standard size differs markedly from the desired size, eq. 28 may be used to check the actual percentage of full thread, allowing for accuracy of drilling.

3. BOLTS, NUTS, AND SCREWS

Screw fasteners in commonest use may be broadly classified according to style and application as (1) bolts, (2) studs, (3) cap screws, (4) machine screws, (5) set screws, (6) carriage bolts, and (7) wood screws. Dimensions of the various styles of heads, points, and nuts have been standardized by the ASA and are summarized in Tables 24 to 49 inclusive. For complete details of manufacturing tolerances, the appropriate standard which is identified at the top of each table should be consulted.

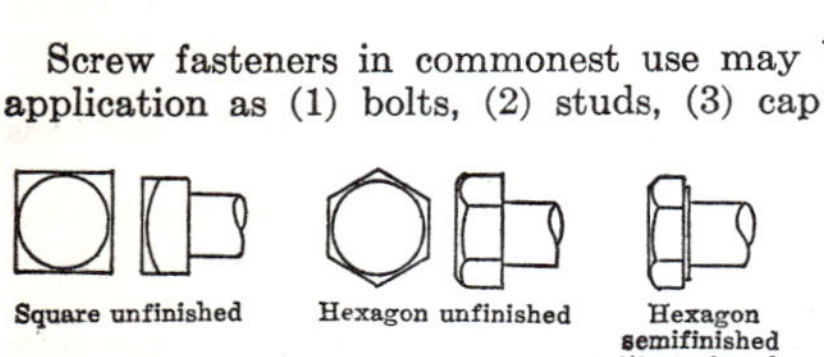

FIG. 10. Bolt heads. See Table 24.

BOLTS, used preferably in through holes with the appropriate nut, are furnished with square or hexagon heads, *unfinished*, *semifinished*, or *finished* (see Fig. 10, and Table 24).

Unfinished Bolt Heads (see Fig. 10 and Table 24) have flat tops and are chamfered to an angle of 25 degrees for square and 30 degrees for hexagon heads, the angle being meas-

Table 24. Dimensions of American Standard Regular and Heavy Bolt Heads
(ASA B18.2-1941. All dimensions in inches.)

Nominal Size or Basic Major Diameter of Thread	Regular Bolt Heads, Fig. 10			Heavy Bolt Heads, Fig. 10		
	Unfinished, Square, and Hexagon		Semi-finished, Hexagon	Unfinished, Square and Hexagon		Semi-finished, Hexagon
	Width across flats, (Basic), max	Height, max	Height, max	Width across Flats (Basic), max	Height, max	Height, max
1/4	0.3750	0.188	0.172	....	....	....
5/16	0.5000	0.220	0.205	....	....	....
3/8	0.5625	0.268	0.252	....	....	....
7/16	0.6250	0.316	0.300	....	....	....
1/2	0.7500	0.348	0.317	0.8750	0.458	0.426
9/16	0.8750	0.396	0.365	0.9375	0.490	0.459
5/8	0.9375	0.444	0.413	1.0625	0.553	0.522
3/4	1.1250	0.524	0.493	1.2500	0.649	0.618
7/8	1.3125	0.620	0.589	1.4375	0.745	0.714
1	1.5000	0.684	0.622	1.6250	0.840	0.778
1 1/8	1.6875	0.780	0.718	1.8125	0.936	0.874
1 1/4	1.8750	0.876	0.813	2.0000	1.032	0.970
1 3/8	2.0625	0.940	0.878	2.1875	1.128	1.065
1 1/2	2.2500	1.036	0.974	2.3750	1.224	1.161
1 5/8	2.4375	1.132	1.069	2.5625	1.319	1.257
1 3/4	2.6250	1.196	1.134	2.7500	1.415	1.352
1 7/8	2.8125	1.292	1.230	2.9375	1.511	1.448
2	3.0000	1.388	1.263	3.1250	1.606	1.482
2 1/4	3.3750	1.548	1.423	3.5000	1.798	1.673
2 1/2	3.7500	1.708	1.583	3.8750	1.990	1.864
2 3/4	4.1250	1.889	1.744	4.2500	2.181	2.056
3	4.5000	2.060	1.935	4.6250	2.373	2.248

Tolerances on bolt head dimensions are fairly liberal, and the standard should be consulted if actual values of limiting dimensions are needed. Bolt head thickness total tolerances range from approximately 1/32 in. for 1/4-in. nominal size to approximately 9/64 in. for 4-in. size. Minimum width across flats is 2 1/2 to 3 1/2% less than the maximum given. Allowing for slight rounding of corners, width across corners is approximately 1.14 times width across flats for hexagon, and 1.375 times width across flats for square bolt heads and nuts.

ured from the top surface. Diameter of the circle is the maximum width across the flats within a tolerance of minus 15%. Maximum radius under heads is 1/32 in. for sizes 1/4 to 1/2 in.; 1/16 in. for sizes 9/16 to 1 in.; 1/8 in. for sizes 1 1/8 to 2 in.; and 3/16 in. for sizes 2 1/4 to 3 in.

Semifinished bolt heads are machined or otherwise formed or treated on the bearing surface so as to provide a washer face (Fig. 10). Washer face thickness is approximately 1/64 in., included in the head height; diameter is the maximum width across flats within a tolerance of minus 5%. Maximum radius under head and chamfer of top are the same as for unfinished bolt heads.

Finished bolt heads are the same as semifinished except that the surfaces other than the bearing surface are treated to provide a special appearance. The finish desired on all nonbearing surfaces should be specified by the purchaser.

Bolt heads also are classified as *regular* and *heavy series*. Regular bolt heads are for general use. Heavy bolt heads are for use where greater bearing surface is necessary, that is, where a large clearance between the bolt and hole is considered essential. They are also used where greater wrench bearing surface is desired.

Automotive hexagon-head bolts (American Standard) have greater head depth than regular bolts and, in sizes above 1/2 in., less width across flats. Maximum head dimensions are listed in Table 25. For minimum limits the standard should be consulted. Top of the head is flat and chamfered, the angle with the top surface being 30 degrees, and the bearing surface is washer faced (Fig. 10). Threads are either coarse or fine series, class 2A. Minimum thread length (extreme point of bolt to last complete thread) of bolts is $2D + 1/4$ in. for bolts up to and including 6 in. length; and $2D + 1/2$ in. for longer bolts.

Table 25. Automotive Hexagon Head Bolts

(All dimensions in inches)

Nominal Size Major Diameter	Width across Flats, max	Height, max	Nominal Size Major Diameter	Width across Flats, max	Height, max	Nominal Size Major Diameter	Width across Flats, max	Height, max
1/4	.4375	.194	9/16	0.8125	.433	1 1/8	1.5000	0.864
5/16	.5000	.242	5/8	0.8750	.481	1 1/4	1.6875	0.959
3/8	.5625	.289	3/4	1.0000	.577	1 3/8	1.8750	1.055
7/16	.6250	.338	7/8	1.1250	.672	1 1/2	2.0625	1.151
1/2	.7500	.386	1	1.3125	.768			

Tap bolts are bolts used without a nut in the same manner as a cap screw, a threaded hole being provided in one of the parts to be joined. Their use should be confined to locations where neither a through bolt nor a stud can be used. With soft or brittle materials an insert should be employed, especially if occasional removal is expected.

NUTS, like bolts, are classified as unfinished, semifinished, and finished (see Fig. 11, and Table 26).

Unfinished nuts are threaded only. Semifinished nuts are threaded and finished on the bearing surface either by turning a washer face or by chamfering (Fig. 11). Washer face thickness is 1/64 in., approximately, included in the nut thickness, and diameter is maximum width across flats within a tolerance of minus 5%. Jam nuts (Table 26) are finished on both faces. Tops of nuts are flat and chamfered 30 degrees with the top surface; diameter of top circle is maximum width across flats within a tolerance of minus 15%.

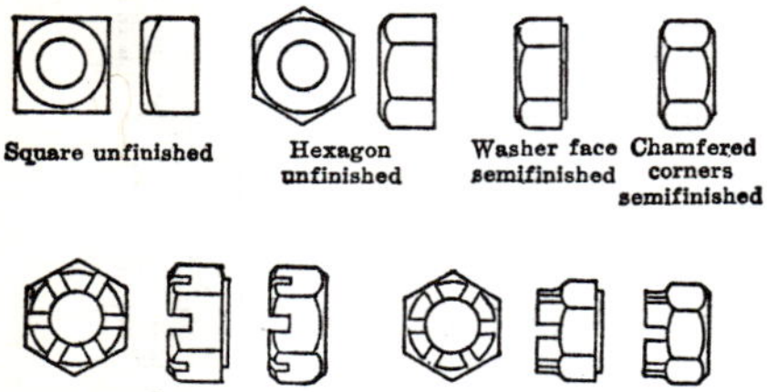

Fig. 11. Nut styles. See Table 26.

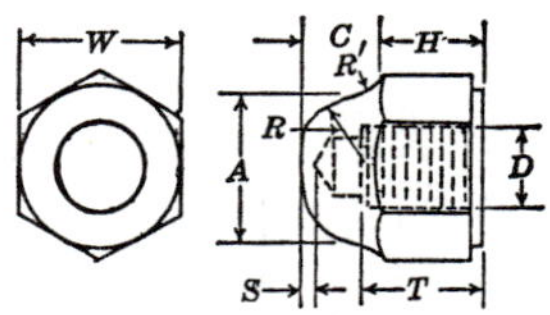

Fig. 12. Cap or acorn nut. See Table 27.

Nuts are furnished in *regular*, *heavy*, and *light series* (Table 26). Regular- and heavy-series nuts are used for the same purposes as the corresponding bolt series. Light-series nuts have smaller dimensions across flats than regular series.

Slotted and castle nuts are illustrated in Fig. 11, standard dimensions being given in Table 26.

Cap, or acorn, nuts are illustrated in Fig. 12. SAE recommended practice for this type of nut is given in Table 27. In addition to the regular and light series, a low-type nut is included. Cap nut construction may be either solid or two piece.

Machine Screw Nuts. Standard dimensions are listed in Table 28. Hexagon machine screw nuts have tops similar to regular series nuts. Bottoms are flat, but for special purposes may be chamfered or washer faced. Square machine screw and stove bolt nuts have tops and bottoms flat without chamfer.

STUD BOLTS. Inasmuch as total lengths and thread lengths are generally determined by the requirements of a particular job, there is no American standard for stud bolts. The following includes recommendations by the Industrial Fasteners Institute.

Tap-end stud bolts are chamfered on the tap end and rounded on the nut end. Length of thread on tap end = $D + 1/8$ in., measured from the end of the stud to the end of the imperfect thread, which should not exceed 2 1/2 threads. Minimum length of thread on nut end = $2D + 1/4$ in. Length of unthreaded portion or shoulder = D except for stud bolts shorter than $4D + 3/8$ in., in which case shoulder length may be decreased or eliminated. For very short studs the bolt end thread length may be reduced, but not the tap end.

Thread fit in the tapped hole is wrench-tight, the preferred method of assembly being to depend on thread tightness to hold the stud, rather than to allow jamming at the imperfect threads or at the bottom of the tapped hole. The patented Lok-Thred provides interference at the minor diameter and is used for studs in certain applications. Inserts ranging

Table 26. Dimensions of American Standard Nuts

(ASA B18.2-1941. All dimensions in inches.)

Diameter, *D*	Regular Nuts (Fig. 11): Unfinished, Square, and Hexagon: Width across Flats (Basic), max	Regular Nuts: Unfinished, Square, and Hexagon: Thickness, max	Regular Nuts: Semifinished, Hexagon: Thickness, max	Heavy Nuts (Fig. 11): Unfinished, Square, and Hexagon: Width across Flats (Basic), max	Heavy Nuts: Unfinished, Square, and Hexagon: Thickness, max	Heavy Nuts: Semifinished, Hexagon: Thickness, max	Heavy Jam and Heavy Slotted Nuts (Fig. 11): Unfinished: Width across Flats (Basic), max	Heavy Jam/Slotted: Unfinished: Jam Nuts, Thickness, max	Heavy Jam/Slotted: Semifinished: Jam Nuts, Thickness, max	Heavy Jam/Slotted: Semifinished: Slotted Nuts, Thickness, max	Regular Jam and Regular Slotted Nuts (Fig. 11): Unfinished: Width across Flats (Basic), max	Regular Jam/Slotted: Unfinished: Jam Nuts, Thickness, max	Regular Jam/Slotted: Semifinished: Jam Nuts, Thickness, max	Regular Jam/Slotted: Semifinished: Slotted Nuts, Thickness, max	Light Jam, Light and Light Thick Slotted, and Light Castle Nuts, Semifinished Only (Fig. 11): Width across Flats (Basic), max	Light: Jam Nuts, Thickness, max	Light: Slotted Nuts, Thickness, max	Light: Thick Slotted and Castle Nuts, Thickness, max	Castle Nuts Only: Nominal Height of Flats	Slot, All Slotted Nuts: Width	Slot, All Slotted Nuts: Depth
1/4	0.4375	0.235	0.219	0.5000	0.266	0.250	0.5000	0.204	0.188	0.250	0.4375	0.172	0.157	0.219	0.4375	.163	0.226	0.288	3/16	5/64	3/32
5/16	0.5625	0.283	0.267	0.5938	0.330	0.314	0.5938	0.236	0.220	0.314	0.5625	0.204	0.189	0.267	0.5000	.195	0.273	0.336	15/64	3/32	3/32
3/8	0.6250	0.346	0.330	0.6875	0.393	0.377	0.6875	0.268	0.252	0.377	0.6250	0.237	0.221	0.330	0.5625	.227	0.337	0.415	9/32	1/8	1/8
7/16	0.7500	0.394	0.378	0.7812	0.456	0.441	0.7812	0.300	0.285	0.441	0.7500	0.269	0.253	0.378	0.6250	.260	0.385	0.463	19/64	1/8	5/32
1/2	0.8125	0.458	0.442	0.8750	0.520	0.504	0.8750	0.332	0.317	0.504	0.8125	0.332	0.317	0.442	0.7500	.323	0.448	0.573	13/32	5/32	5/32
9/16	0.8750	0.521	0.505	0.9375	0.584	0.568	0.9375	0.365	0.349	0.568	0.8750	0.365	0.349	0.505	0.8750	.324	0.496	0.621	27/64	5/32	3/16
5/8	1.0000	0.569	0.553	1.0625	0.647	0.631	1.0625	0.397	0.381	0.631	1.0000	0.397	0.381	0.553	0.9375	.387	0.559	0.731	1/2	3/16	7/32
3/4	1.1250	0.680	0.665	1.2500	0.774	0.758	1.2500	0.462	0.446	0.758	1.1250	0.462	0.446	0.665	1.0625	.389	0.670	0.827	9/16	3/16	1/4
7/8	1.3125	0.792	0.776	1.4375	0.901	0.885	1.4375	0.526	0.510	0.885	1.3125	0.526	0.510	0.776	1.2500	.454	0.782	0.922	21/32	3/16	1/4
1	1.5000	0.903	0.887	1.6250	1.028	1.012	1.6250	0.590	0.575	1.012	1.5000	0.590	0.575	0.887	1.4375	.518	0.893	1.018	23/32	1/4	9/32
1 1/8	1.6875	1.030	0.999	1.8125	1.155	1.139	1.8125	0.655	0.639	1.139	1.6875	0.655	0.639	0.999	1.6250	.582	1.004	1.176	13/16	1/4	11/32
1 1/4	1.8750	1.126	1.094	2.0000	1.282	1.251	2.0000	0.782	0.751	1.251	1.8750	0.782	0.751	1.094	1.8125	.647	1.116	1.272	7/8	5/16	3/8
1 3/8	2.0625	1.237	1.206	2.1875	1.409	1.378	2.1875	0.846	0.815	1.378	2.0625	0.846	0.815	1.206	2.0000	.774	1.227	1.399	1	5/16	3/8
1 1/2	2.2500	1.348	1.317	2.3750	1.536	1.505	2.3750	0.911	0.880	1.505	2.2500	0.911	0.880	1.317	2.1875	.838	1.338	1.526	1 1/16	3/8	7/16
1 5/8	2.4375	1.460	1.429	2.5625	1.663	1.632	2.5625	0.976	0.944	1.632	2.4375	0.976	0.944	1.429	......	......	......	......	......	3/8	7/16
1 3/4	2.6250	1.571	1.540	2.7500	1.790	1.759	2.7500	1.040	1.009	1.759	2.6250	1.040	1.009	1.540	......	......	......	......	......	7/16	1/2
1 7/8	2.8125	1.683	1.651	2.9375	1.917	1.886	2.9375	1.104	1.073	1.886	2.8125	1.104	1.073	1.651	......	......	......	......	......	7/16	9/16
2	3.0000	1.794	1.763	3.1250	2.044	2.013	3.1250	1.169	1.138	2.013	3.0000	1.169	1.138	1.763	......	......	......	......	......	7/16	9/16
2 1/4	3.3750	2.017	1.970	3.5000	2.298	2.251	3.5000	1.298	1.251	2.251	3.3750	1.298	1.251	1.970	......	......	......	......	......	7/16	9/16
2 1/2	3.7500	2.240	2.193	3.8750	2.552	2.505	3.8750	1.552	1.505	2.505	3.7500	1.552	1.505	2.193	......	......	......	......	......	9/16	11/16
2 3/4	4.1250	2.462	2.415	4.2500	2.806	2.759	4.2500	1.681	1.634	2.759	4.1250	1.681	1.634	2.415	......	......	......	......	......	9/16	11/16
3	4.5000	2.685	2.638	4.6250	3.060	3.013	4.6250	1.810	1.763	3.013	4.5000	1.810	1.763	2.638	......	......	......	......	......	5/8	3/4
3 1/4	......	......	......	5.0000	3.314	3.252	5.0000	1.939	1.876	3.252	......	......	......	......	......	......	......	......	......	5/8	3/4
3 1/2	......	......	......	5.3750	3.568	3.506	5.3750	2.068	2.006	3.506	......	......	......	......	......	......	......	......	......	5/8	3/4
3 3/4	......	......	......	5.7500	3.822	3.760	5.7500	2.197	2.134	3.760	......	......	......	......	......	......	......	......	......	5/8	3/4
4	......	......	......	6.1250	4.076	4.014	6.1250	2.326	2.264	4.014	......	......	......	......	......	......	......	......	......	5/8	3/4

Tolerances on nut dimensions are fairly liberal, and the standard should be consulted if actual values of limiting dimensions are needed. Nut thickness total tolerances range from approximately 1/32 in. for 1/4-in. nominal size to approximately 9/64 in. for 4-in. size. Minimum width across flats is 2 1/2 to 3 1/2% less than the maximum given. Allowing for slight rounding of corners, width across corners is approximately 1.14 times width across flats for hexagon, and 1.375 times width across flats for square nuts.

Table 27. Dimensions of Cap and Acorn Nuts

(SAE Recommended Practice. All dimensions in inches. See Fig. 12.)

Size, D	C	W Light	W Regular	R	S Min	T Regular	Size, D	C	W Light	W Regular	R	S Min	T Regular
No. 6	3/16	5/16	......	9/64	1/32	9/64	1/2	3/8	3/4	13/16	5/16	3/64	1/2
No. 8	3/16	5/16	......	9/64	1/32	5/32	9/16	7/16	7/8	7/8	3/8	1/16	9/16
No. 10	15/64	3/8	......	11/64	1/32	3/16	5/8	7/16	15/16	1	3/8	1/16	5/8
No. 12	15/64	3/8	......	11/64	1/32	7/32	3/4	17/32	1 1/16	1 1/8	7/16	1/16	3/4
1/4	1/4	7/16	7/16	3/16	3/64	1/4	7/8	39/64	1 1/4	1 5/16	33/64	3/32	7/8
5/16	17/64	1/2	9/16	13/64	3/64	5/16	1	11/16	1 7/16	1 1/2	19/32	3/32	1
3/8	19/64	9/16	5/8	15/64	3/64	3/8	1 1/8	25/32	1 5/8	1 11/16	43/64	3/32	1 1/8
7/16	11/32	5/8	3/4	9/32	3/64	7/16	1 1/4	27/32	1 13/16	1 7/8	47/64	3/32	1 1/4

Other dimensions (Fig. 12) are as follows:
$A = (W - 1/32)$ for sizes 6 to 7/16, incl.; $A = (W - 1/16)$ for sizes 1/2 to 1 1/4, incl.
$H = D$ for sizes 6 to 10, incl.; $H = (7/8)\,D$ for sizes 12 to 1 1/4, incl.
$H' = (3/4)D$ for low nuts, light series only.
$R' = 1/64$ in. for sizes 6 to 7/16, incl.; $R' = 1/32$ in. for sizes 1/2 to 1 1/4, incl.
For low-type nuts $T = T_{reg} - (H - H')$.

Table 28. American Standard Square and Hexagon Machine-screw and Stove-bolt Nuts

(ASA B18.2-1941. All dimensions in inches. See Fig. 11.)

Diameter of Screw, D		Width across Flats, max (Basic)	Width across Corners, min		Thickness, max	Diameter of Screw, D		Width across Flats, max (Basic)	Width across Corners, min		Thickness, max
			Square	Hex					Square	Hex	
No. 0	.0600	.1562	.206	.171	.050	No. 8	.1640	.3438	.456	.378	.130
No. 1	.0730	.1562	.206	.171	.050	No. 10	.1900	.3750	.497	.413	.130
No. 2	.0860	.1875	.247	.205	.066	No. 12	.2160	.4375	.581	.482	.161
No. 3	.0990	.1875	.247	.205	.066	1/4	.2500	.4375	.581	.482	.193
No. 4	.1120	.2500	.331	.275	.098	5/16	.3125	.5625	.748	.621	.225
No. 5	.1250	.3125	.415	.344	.114	3/8	.3750	.6250	.833	.692	.257
No. 6	.1380	.3125	.415	.344	.114						

For limiting dimensions, the standard should be consulted. Nut thickness total tolerances range from 0.007 in. for No. 0 size to 0.018 in. for 3/8 in. size. Minimum width across flats is approximately 3 to 4% less than the maximum.

from simple threaded bushings through a variety of patented designs are often used with softer materials, especially where occasional removal of the stud is anticipated. With soft materials such as aluminum the thread on the tap end is sometimes larger and coarser than that on the bolt end.

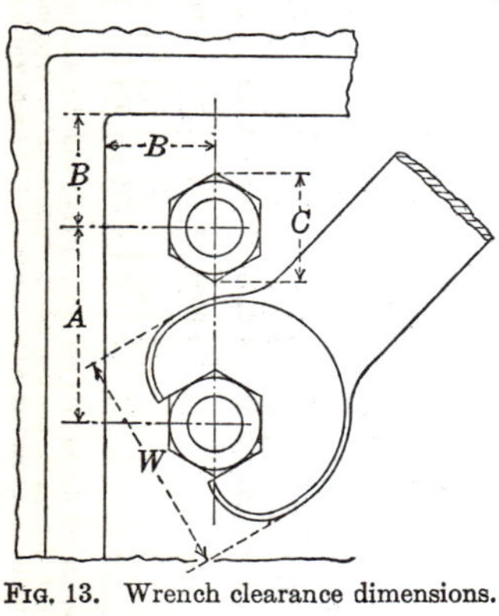

FIG. 13. Wrench clearance dimensions.

Double-end stud bolts (rod) are symmetrical, both ends having rounded points and thread length $2D +$ 1/4 in., measured from the end of the stud to the end of the imperfect thread, which should not exceed 2 1/2 threads. Threads may be rolled or cut.

Continuous thread stud bolts (flange studs) are chamfered at both ends and have a continuous thread from end to end.

WRENCH OPENINGS for standard bolts and nuts are listed in Table 29. The minimum wrench openings provide a positive clearance between maximum nut and minimum wrench opening. The plus tolerance is as great as possible without danger of deforming the hexagon corners under extreme conditions.

Bolt Spacing for Wrench Clearance. Space between bolts, and between bolts and adjacent projections, may be estimated from these formulas, given in *Machinery's Handbook* (see Fig. 13 for notation):

$$A = 1/2\,(W + C + d) \qquad B = 0.48W$$

where d is the nominal bolt diameter. Value of C, the distance across the corners of the nut or bolt head, is approximately 1.14 times the distance across the flats, which may be obtained from Tables 24, 25, 26, 27, or 28. Maximum width of wrench head, W, may be obtained from Table 30.

WASHERS. Dimensions of standard plain washers for general industrial purposes are listed in Table 31. They include light, medium, heavy, and extra heavy series washers in a composite listing.

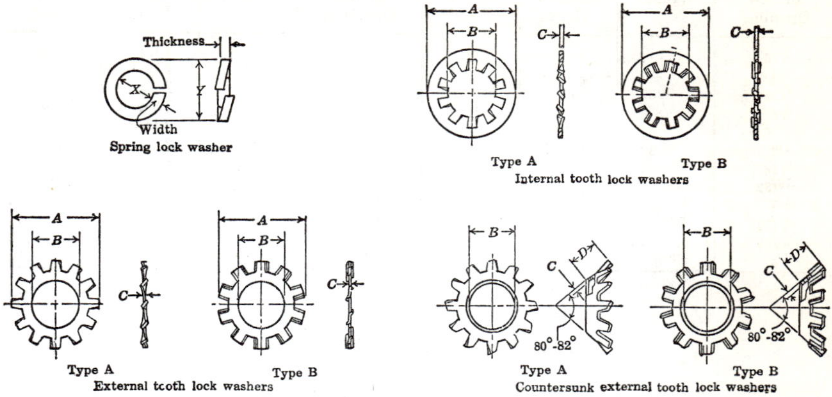

FIG. 14. Lock washers. See Tables 32 and 33.

Spring lock washers, illustrated in Fig. 14, have standard dimensions listed in Table 32. Spring lock washers are made from carbon steel fabricated and treated to a hardness of 45 to 53 Rockwell C.

Tooth lock washers are illustrated in Fig. 14, and their leading dimensions are listed in Table 33. The number of teeth, length of teeth, width of rim, and thickness of washer

Table 29. Dimensions of American Standard Wrench Openings

(ASA B18.2-1941. All dimensions in inches.)

Nominal Size of Wrench, also Basic Width across Flats of Bolt Heads and Nuts	Allowance between Bolt Head or Nut and Jaws of Wrench	Wrench Openings		Nominal Size of Wrench, also Basic Width across Flats of Bolt Heads and Nuts	Allowance between Bolt Head or Nut and Jaws of Wrench	Wrench Openings		Nominal Size of Wrench, also Basic Width across Flats of Bolt Heads and Nuts	Allowance between Bolt Head or Nut and Jaws of Wrench	Wrench Openings	
		Min	Max			Min	Max			Min	Max
5/32	.002	0.158	0.163	1 1/16	.006	1.068	1.077	2 5/8	.014	2.639	2.656
3/16	.002	0.190	0.195	1 1/8	.007	1.132	1.142	2 3/4	.014	2.766	2.783
1/4	.002	0.252	0.257	1 1/4	.007	1.257	1.267	2 13/16	.015	2.827	2.845
5/16	.003	0.316	0.322	1 5/16	.008	1.320	1.331	2 15/16	.016	2.954	2.973
11/32	.003	0.347	0.353	1 3/8	.008	1.383	1.394	3	.016	3.016	3.035
3/8	.003	0.378	0.384	1 7/16	.008	1.446	1.457	3 1/8	.017	3.142	3.162
7/16	.003	0.440	0.446	1 1/2	.008	1.508	1.520	3 3/8	.018	3.393	3.414
1/2	.004	0.504	0.510	1 5/8	.009	1.634	1.646	3 1/2	.019	3.518	3.540
9/16	.004	0.566	0.573	1 11/16	.009	1.696	1.708	3 3/4	.020	3.770	3.793
19/32	.004	0.598	0.605	1 13/16	.010	1.822	1.835	3 7/8	.020	3.895	3.918
5/8	.004	0.629	0.636	1 7/8	.010	1.885	1.898	4 1/8	.022	4.147	4.172
11/16	.004	0.692	0.699	2	.011	2.011	2.025	4 1/4	.022	4.272	4.297
3/4	.005	0.755	0.763	2 1/16	.011	2.074	2.088	4 1/2	.024	4.524	4.550
25/32	.005	0.786	0.794	2 3/16	.012	2.200	2.215	4 5/8	.024	4.649	4.676
13/16	.005	0.818	0.826	2 1/4	.012	2.262	2.277	5	.026	5.026	5.055
7/8	.005	0.880	0.888	2 3/8	.013	2.388	2.404	5 3/8	.028	5.403	5.434
15/16	.006	0.944	0.953	2 7/16	.013	2.450	2.466	5 3/4	.030	5.780	5.813
1	.006	1.006	1.015	2 9/16	.014	2.576	2.593	6 1/8	.032	6.157	6.192

Table 30. Maximum Width of Wrench Heads

(Federal Standard Stock Catalog. All dimensions in inches)

Nominal Size of Wrench (Width of Jaw Opening)	Dimension *W* (Fig. 13)			Nominal Size of Wrench (Width of Jaw Opening)	Dimension *W* (Fig. 13)		
	Box Wrench	Engineer's Wrenches			Box Wrench	Engineer's Wrenches	
		Regular Head	Spear Head			Regular Head	Spear Head
3/16	....	....	15/32	13/16	1 5/16	2 1/4	1 5/8
1/4	....	....	1/2	7/8	1 3/8	2 3/16	1 3/4
5/16	17/32	....	11/16	15/16	1 7/16	2 3/8	1 15/16
3/8	5/8	....	13/16	1	1 1/2	2 3/8	2 1/8
7/16	47/64	1 5/16	15/16	1 1/16	1 19/32	2 13/16	2 1/8
1/2	27/32	1 7/16	1 1/16	1 1/8	1 13/16	2 3/4	2 3/8
9/16	61/64	1 1/2	1 3/16	1 1/4	1 7/8	3 3/16	2 3/8
19/32	....	1 1/2	....	1 5/16	2 1/16	3 3/16	2 3/8
5/8	1 1/16	1 3/4	1 5/16	1 3/8	2 1/16	....	....
11/16	1 1/8	1 7/8	1 7/16	1 7/16	2 5/32	3 7/8	....
25/32	....	2 3/16	1 9/16	1 1/2	2 5/32	3 7/8	....
3/4	1 3/8	2 3/16	1 5/8	1 5/8	2 3/8	4	....

Table 31. Dimensions of Plain Washers

(All dimensions in inches.)

Inside Diameter	Outside Diameter	Thickness		Inside Diameter	Outside Diameter	Thickness		Inside Diameter	Outside Diameter	Thickness	
		Max	Min			Max	Min			Max	Min
5/64	3/16	.025	.016	7/16	7/8	.104	.064	15/16	3 3/8	.213	.153
3/32	7/32	.025	.016	7/16	1	.104	.064	1 1/16	2	.160	.108
3/32	1/4	.025	.016	7/16	1 3/8	.104	.064	1 1/16	2 1/4	.192	.136
1/8	1/4	.028	.017	15/32	59/64	.080	.051	1 1/16	2 1/2	.192	.136
1/8	5/16	.040	.025	1/2	1 1/8	.104	.064	1 1/16	3 7/8	.280	.210
5/32	5/16	.048	.027	1/2	1 1/4	.104	.064	1 3/16	2 1/2	.192	.136
5/32	3/8	.065	.036	1/2	1 5/8	.104	.064	1 1/4	2 3/4	.192	.136
11/64	13/32	.065	.036	17/32	1 1/16	.121	.074	1 5/16	2 3/4	.192	.136
3/16	3/8	.065	.036	9/16	1 1/4	.132	.086	1 3/8	3	.192	.136
3/16	7/16	.065	.036	9/16	1 3/8	.132	.086	1 7/16	3	.213	.153
13/64	15/32	.065	.036	9/16	1 7/8	.132	.086	1 1/2	3 1/4	.213	.153
7/32	7/16	.065	.036	19/32	1 3/16	.121	.074	1 9/16	3 1/4	.213	.153
7/32	1/2	.065	.036	5/8	1 3/8	.132	.086	1 5/8	3 1/2	.213	.153
15/64	17/32	.065	.036	5/8	1 1/2	.132	.086	1 11/16	3 1/2	.213	.153
1/4	1/2	.065	.036	5/8	2 1/8	.160	.108	1 3/4	3 3/4	.213	.153
1/4	9/16	.065	.036	21/32	1 5/16	.121	.074	1 13/16	3 3/4	.213	.153
1/4	9/16	.080	.051	11/16	1 1/2	.160	.108	1 7/8	4	.213	.153
17/64	5/8	.065	.036	11/16	1 3/4	.160	.108	1 15/16	4	.213	.153
9/32	5/8	.080	.051	11/16	2 3/8	.192	.136	2	4 1/4	.213	.153
5/16	3/4	.080	.051	13/16	1 1/2	.160	.108	2 1/16	4 1/4	.213	.153
5/16	7/8	.080	.051	13/16	1 3/4	.177	.122	2 1/8	4 1/2	.213	.153
11/32	11/16	.080	.051	13/16	2	.177	.122	2 3/8	4 3/4	.248	.193
3/8	3/4	.080	.051	12/16	2 7/8	.192	.136	2 5/8	5	.280	.210
3/8	7/8	.104	.064	15/16	1 3/4	.160	.108	2 7/8	5 1/4	.310	.228
3/8	1 1/8	.080	.051	15/16	2	.192	.136	3 1/8	5 1/2	.327	.249
13/32	13/16	.080	.051	15/16	2 1/4	.192	.136				

Diameter tolerances are ±0.010-in., except the ID tolerances on 5/64 to 7/32 in., and 1/4 in. where OD is 9/16 in., which are ±0.005 in.

Thicknesses are Birmingham gage. Limits given represent tolerance of plus or minus one gage, or the spread from the minimum of one gage minus to the maximum of one gage plus.

Table 32. Dimensions of Spring Lock Washers

(ASA B27.1-1944. All dimensions in inches. See Fig. 14.)

Nominal Size	Inside Diameter, min, X	Light			Medium			Heavy			Extra Heavy		
		Width, min, W	Thickness,* min	Outside Diameter, max, Y	Width, min, W	Thickness,* min	Outside Diameter, max, Y	Width, min, W	Thickness,* min	Outside Diameter, max, Y	Width, min, W	Thickness,* min	Outside Diameter, max, Y
No. 2	0.088	.030	.015	0.165	.035	.020	0.175	.040	.025	0.185	.053	.027	0.211
No. 3	0.102	.035	.020	0.188	.040	.025	0.198	.047	.031	0.212	.062	.034	0.242
No. 4	0.115	.035	.020	0.202	.040	.025	0.212	.047	.031	0.226	.062	.034	0.256
No. 5	0.128	.040	.025	0.225	.047	.031	0.239	.055	.040	0.255	.079	.045	0.303
No. 6	0.141	.040	.025	0.237	.047	.031	0.251	.055	.040	0.267	.079	.045	0.315
No. 8	0.168	.047	.031	0.280	.055	.040	0.296	.062	.047	0.310	.096	.057	0.378
No. 10	0.194	.055	.040	0.323	.062	.047	0.337	.070	.056	0.353	.112	.068	0.437
No. 12	0.221	.062	.047	0.364	.070	.056	0.380	.077	.063	0.394	.130	.080	0.500
1/4	0.255	.107	.047	0.489	.109	.062	0.493	.110	.077	0.495	.132	.084	0.539
5/16	0.319	.117	.056	0.575	.125	.078	0.591	.130	.097	0.601	.143	.108	0.627
3/8	0.382	.136	.070	0.678	.141	.094	0.688	.145	.115	0.696	.170	.123	0.746
7/16	0.446	.154	.085	0.780	.156	.109	0.784	.160	.133	0.792	.186	.143	0.844
1/2	0.509	.170	.099	0.877	.171	.125	0.879	.176	.151	0.889	.204	.162	0.945
9/16	0.573	.186	.113	0.975	.188	.141	0.979	.193	.170	0.989	.223	.182	1.049
5/8	0.636	.201	.126	1.082	.203	.156	1.086	.210	.189	1.100	.242	.202	1.164
11/16	0.700	.216	.138	1.178	.219	.172	1.184	.227	.207	1.200	.260	.221	1.266
3/4	0.763	.233	.153	1.277	.234	.188	1.279	.244	.226	1.299	.279	.241	1.369
13/16	0.827	.249	.168	1.375	.250	.203	1.377	.262	.246	1.401	.298	.261	1.473
7/8	0.890	.264	.179	1.470	.266	.219	1.474	.281	.266	1.504	.322	.285	1.586
15/16	0.954	.277	.191	1.562	.281	.234	1.570	.298	.284	1.604	.345	.308	1.698
1	1.017	.289	.202	1.656	.297	.250	1.672	.319	.306	1.716	.366	.330	1.810
1 1/16	1.081	.301	.213	1.746	.312	.266	1.768	.338	.326	1.820	.389	.352	1.922
1 1/8	1.144	.314	.224	1.837	.328	.281	1.865	.356	.345	1.921	.411	.375	2.031
1 3/16	1.208	.324	.234	1.923	.344	.297	1.963	.373	.364	2.021	.431	.396	2.137
1 1/4	1.271	.336	.244	2.012	.359	.312	2.058	.393	.384	2.126	.452	.417	2.244
1 5/16	1.335	.346	.254	2.098	.375	.328	2.156	.410	.403	2.226	.472	.438	2.350
1 3/8	1.398	.356	.264	2.183	.391	.344	2.253	.427	.422	2.325	.491	.458	2.453
1 7/16	1.462	.366	.273	2.269	.406	.359	2.349	.442	.440	2.421	.509	.478	2.555
1 1/2	1.525	.375	.282	2.352	.422	.375	2.446	.458	.458	2.518	.526	.496	2.654

* Thickness tabulated is the nominal mean thickness of the trapezoidal section, that is, $(T + t)/2$, where T is the thickness at the inner periphery and t the thickness at the outer periphery (Fig. 14). T is larger than t by an amount ranging between $0.032W$ and $0.064W$.

over the teeth (free height) are optional with the manufacturer. Washers are carbon steel fabricated and heat treated to a hardness of 40 to 50 Rockwell C.

INTERNAL WRENCHING BOLTS. High-strength, high-temperature internal wrenching bolts have cylindrical heads with hexagon sockets (Fig. 15). Leading dimensions are listed in Table 34. Length, L, increases by 1/8-in. increments up to 1 in., by 1/4-in. increments over 1 in. to 4 in., and by 1/2-in. increments over 4 in. in length. Thread tolerances conform to ASA class 7 (see Table 18), and the internal mating threads should conform to class 2 tolerances (see Table 13). The heads are larger than those of socket head cap screws (Fig. 17) to provide greater bearing surface under the head and to insure head and wrench strength.

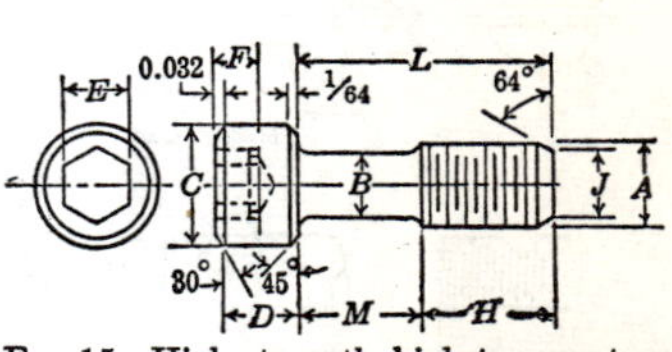

Fig. 15. High-strength, high-temperature internal wrenching bolt. See Table 34.

MACHINE SCREWS. Figure 16 illustrates the various types of machine screws standardized by the ASA. Tables 35, 36 and 37, list their leading dimensions. For details of specification and manufacturing tolerances the standard (ASA B18.6-1947) should be consulted. Referring to the designation of machine screws by number, it is convenient

(Continued on p. 10-43)

Table 33. Dimensions of Tooth Lock Washers
(All dimensions in inches. See Fig. 14.)

Nominal Size	Internal and External Tooth				Heavy Internal Tooth		Countersunk External Tooth		
	B min	*C* max	Internal Tooth *A* max	External Tooth *A* max	*C* max	*A* max	*B* min	*C* max	*D* max
No. 2	0.089	.015	0.200	....	....	....	....	....	....
No. 3	0.102	.019	0.232	....	....	....	....	....	....
No. 4	0.115	.019	0.270	0.290	....	....	.113	.019	.065
No. 5	0.129	.021	0.280	....	....	....	....	....	....
No. 6	0.141	.021	0.295	0.320	....	....	.140	.021	.092
No. 8	0.168	.023	0.340	0.381	....	....	.167	.021	.105
No. 10	0.195	.025	0.381	0.410	....	....	.195	.025	.099
No. 12	0.221	.025	0.410	0.475	....	....	.220	.025	.128
1/4	0.256	.028	0.478	0.510	.045	0.536	.255	.025	.128
No. 16	....	....	....	....	....	....	.273	.028	.147
5/16	0.320	.034	0.610	0.610	.050	0.607	.318	.028	.192
3/8	0.384	.040	0.692	0.694	.050	0.748	.383	.034	.255
7/16	0.448	.040	0.789	0.760	.067	0.858	.448	.045	.270
1/2	0.512	.045	0.900	0.900	.067	0.924	.512	.045	.304
9/16	0.576	.045	0.985	0.985	.067	1.034	....	....	....
5/8	0.640	.050	1.071	1.070	.067	1.135	....	....	....
11/16	0.704	.050	1.166	1.155	....	....	....	....	....
3/4	0.769	.055	1.245	1.260	.084	1.265	....	....	....
13/16	0.832	.055	1.315	1.315	....	....	....	....	....
7/8	0.894	.060	1.410	1.410	.084	1.447	....	....	....
1	1.019	.067	1.637	1.620	....	....	....	....	....
1 1/8	1.144	.067	1.830	....	....	....	....	....	....
1 1/4	1.275	.067	1.975	....	....	....	....	....	....

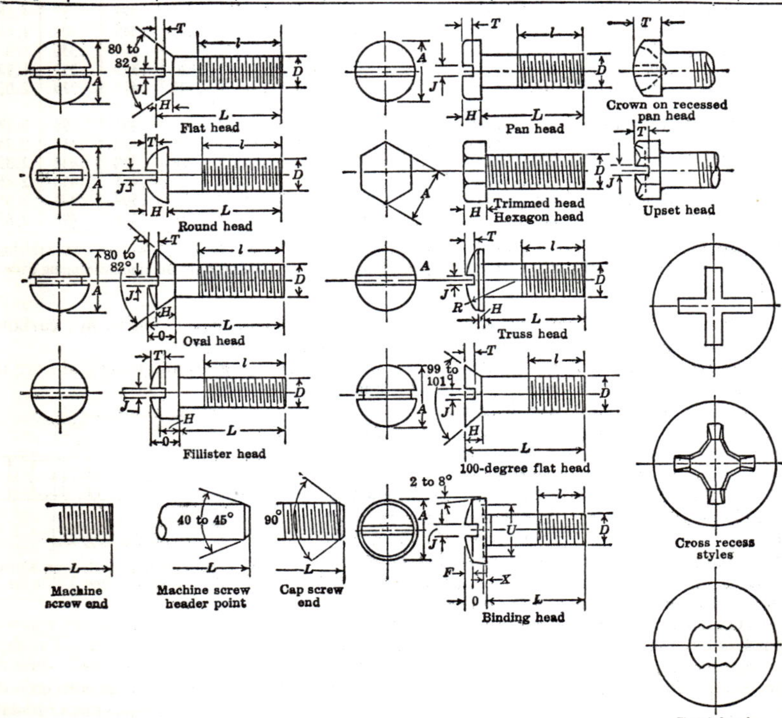

Fig. 16. Machine screws and slotted-head cap screws. See Tables 35, 36, 37, and 38

Table 34. High-strength, High-temperature Internal Wrenching Bolts

(All dimensions in inches. See Fig. 15.)

Screw Diameter, *A*	Threads per Inch	Body Diameter Unthreaded Portion, max, *B*	Head Diameter, max, *C*	Head Height, max, *D*	Socket Dimensions: Across Flats, min, *E*	Socket Dimensions: Depth, min, *F*	Thread Length, max, *H*	Screw End Diameter, max, *J*	Shank Length, min, *M*
3/8	16	0.325	0.750	0.375	0.375	0.227	0.750	0.278	0.188
1/2	13	0.440	0.875	0.500	0.500	0.290	0.875	0.376	0.250
5/8	11	0.496	1.000	0.625	0.562	0.352	1.000	0.483	0.312
3/4	10	0.608	1.125	0.750	0.625	0.414	1.125	0.562	0.375
7/8	9	0.718	1.312	0.875	0.750	0.477	1.250	0.668	0.438
1	8	0.825	1.500	1.000	0.875	0.539	1.375	0.756	0.500
1 1/4	8	1.074	1.875	1.250	1.125	0.665	1.625	1.006	0.625
1 1/2	8	1.323	2.250	1.500	1.375	0.790	1.875	1.256	0.750
1 3/4	8	1.572	2.625	1.750	1.625	0.915	2.125	1.506	0.875
2	8	1.822	3.000	2.000	1.875	1.040	2.375	1.756	1.000

Table 35. Dimensions of American Standard Machine Screw Heads

(ASA B18.6-1947. All dimensions in inches. See Fig. 16.)

Nominal Size	Max Diameter, *D*	Head Diameter, max, *A*	Height of Head, max, *H*	Width of Slot, min, *J*	Depth of Slot, min, *T*	Total Height of Head, max, *O*	Head Diameter, max, *A*	Height of Head, max, *H*	Width of Slot, min, *J*	Depth of Slot, min, *T*	Total Height of Head, max, *O*
		Flat Head					Round Head				
No. 0	.060	0.119	.035	.016	.010	....	0.113	.053	.016	.029	....
No. 1	073	0.146	.043	.019	.012	....	0.138	.061	.019	.033	....
No. 2	.086	0.172	.051	.023	.015	....	0.162	.069	.023	.037	....
No. 3	.099	0.199	.059	.027	.017	....	0.187	.078	.027	.040	....
No. 4	.112	0.225	.067	.031	.020	....	0.211	.086	.031	.044	....
No. 5	.125	0.252	.075	.035	.022	....	0.236	.095	.035	.047	....
No. 6	.138	0.279	.083	.039	.024	....	0.260	.103	.039	.051	....
No. 8	.164	0.332	.100	.045	.029	....	0.309	.120	.045	.058	....
No. 10	.190	0.385	.116	.050	.034	....	0.359	.137	.050	.065	....
No. 12	.216	0.438	.132	.056	.039	....	0.408	.153	.056	.072	....
1/4	.250	0.507	.153	.064	.046	....	0.472	.175	.064	.082	....
5/16	.3125	0.635	.191	.072	.058	....	0.590	.216	.072	.099	....
3/8	.375	0.762	.230	.081	.070	....	0.708	.256	.081	.117	....
7/16	.4375	0.812	.223	.081	.066	....	0.750	.328	.081	.148	....
1/2	.500	0.875	.223	.091	.065	....	0.813	.355	.091	.159	....
9/16	.5625	1.000	.260	.102	.077	....	0.938	.410	.102	.183	....
5/8	.625	1.125	.298	.116	.088	....	1.000	.438	.116	.195	....
3/4	.750	1.375	.372	.131	.111	....	1.250	.547	.131	.242	....
		Oval Head					Fillister Head				
No. 0	.060	0.119	.035	.016	.025	.056	0.096	.045	.016	.015	.059
No. 1	.073	0.146	.043	.019	.031	.068	0.118	.053	.019	.020	.071
No. 2	.086	0.172	.051	.023	.037	.080	0.140	.062	.023	.025	.083
No. 3	.099	0.199	.059	.027	.043	.092	0.161	.070	.027	.030	.095
No. 4	.112	0.225	.067	.031	.049	.104	0.183	.079	.031	.035	.107
No. 5	.125	0.252	.075	.035	.055	.116	0.205	.088	.035	.040	.120
No. 6	.138	0.279	.083	.039	.060	.128	0.226	.096	.039	.045	.132
No. 8	.164	0.332	.100	.045	.072	.152	0.270	.113	.045	.054	.156
No. 10	.190	0.385	.116	.050	.084	.176	0.313	.130	.050	.064	.180
No. 12	.216	0.438	.132	.056	.096	.200	0.357	.148	.056	.074	.205
1/4	.250	0.507	.153	.064	.112	.232	0.414	.170	.064	.087	.237
5/16	.3125	0.635	.191	.072	.141	.290	0.518	.211	.072	.110	.295
3/8	.375	0.762	.230	.081	.170	.347	0.622	.253	.081	.133	.355
7/16	.4375	0.812	.223	.081	.174	.345	0.625	.265	.081	.135	.368
1/2	.500	0.875	.223	.091	.176	.354	0.750	.297	.091	.151	.412
9/16	.5625	1.000	.260	.102	.207	.410	0.812	.336	.102	.172	.466
5/8	.625	1.125	.298	.116	.235	.467	0.875	.375	.116	.193	.521
3/4	.750	1.375	.372	.131	.293	.578	1.000	.441	.131	.226	.612

Edges of head on flat- and oval-head machine screws may be rounded.

Radius of fillet at base of flat- and oval-head machine screws shall not exceed twice the pitch of the screw thread.

Radius of fillet at base of round- and fillister-head machine screws shall not exceed one-half the pitch of the screw thread.

All four types of screws in this table may be furnished with cross-recessed heads of either style shown in Fig. 15.

Fillister-head machine screws in sizes No. 2 to 3/8-in., inclusive, may be furnished with a drilled hole through the head along a diameter at right angles to the slot but not breaking through the slot.

Table 36. Dimensions of American Standard Machine Screw Heads—Pan, Hexagon, Truss, and 100-degree Flat Heads

(ASA B18.6-1947. All dimensions in inches. See Fig. 16.)

Nominal Size	Max Diam, *D*	Head Diam, max, *A*	Height of Slotted Head, max, *H*	Width of Slot, min, *J*	Depth of Slot, min, *T*	Radius, *R*	Height of Recessed Head, max, *O*	Head Diam, max, *A*	Height of Head, max, *H*	Width of Slot, min, *J*	Depth of Slot, min, *T*
		PAN HEAD						HEXAGON HEAD			
No. 2	.086	0.167	.053	.023	.023	0.035	.062	.125	.050	....	....
No. 3	.099	0.193	.060	.027	.027	0.037	.071	.187	.055	....	....
No. 4	.112	0.219	.068	.031	.030	0.042	.080	.187	.060	.031	.025
No. 5	.125	0.245	.075	.035	.032	0.044	.089	.187	.070	.035	.030
No. 6	.138	0.270	.082	.039	.038	0.046	.097	.250	.080	.039	.033
No. 8	.164	0.322	.096	.045	.043	0.052	.115	.250	.110	.045	.052
No. 10	.190	0.373	.110	.050	.050	0.061	.133	.312	.120	.050	.057
No. 12	.216	0.425	.125	.056	.060	0.078	.151	.312	.155	.056	.077
1/4	.250	0.492	.144	.064	.070	0.087	.175	.375	.190	.064	.083
5/16	.3125	0.615	.178	.072	.092	0.099	.218	.500	.230	.072	.100
3/8	.375	0.740	.212	.081	.113	0.143	.261	.562	.295	.081	.131
		TRUSS HEAD						100-DEG FLAT HEAD			
No. 2	.086	0.194	.053	.023	.022	0.129	....	....	....	....	....
No. 3	.099	0.226	.061	.027	.026	0.151	....	....	....	....	....
No. 4	.112	0.257	.069	.031	.030	0.169	....	.225	.048	.031	.017
No. 5	.125	0.289	.078	.035	.034	0.191	....	....	....	....	....
No. 6	.138	0.321	.086	.039	.037	0.211	....	.279	.060	.039	.022
No. 8	.164	0.384	.102	.045	.045	0.254	....	.332	.072	.045	.027
No. 10	.190	0.448	.118	.050	.053	0.283	....	.385	.083	.050	.031
No. 12	.216	0.511	.134	.056	.061	0.336	....	....	....	....	....
1/4	.250	0.573	.150	.064	.070	0.375	....	.507	.110	.064	.042
5/16	.3125	0.698	.183	.072	.085	0.457	....	.635	.138	.072	.053
3/8	.375	0.823	.215	.081	.100	0.538	....	.762	.165	.081	.064
7/16	.4375	0.948	.248	.081	.116	0.619	....	....	....	....	....
1/2	.500	1.073	.280	.091	.131	0.701	....	....	....	....	....
9/16	.5625	1.198	.312	.102	.146	0.783	....	....	....	....	....
5/8	.625	1.323	.345	.116	.162	0.863	....	....	....	....	....
3/4	.750	1.573	.410	.131	.182	1.024	....	....	....	....	....

Radius of fillet at base of truss- and pan-head machine screws shall not exceed one-half the pitch of the screw thread.

Truss-, pan-, and 100-degree flat-head machine screws may be furnished with cross-recessed heads of either style shown in Fig. 15.

Hexagon-head machine screws are usually not slotted; the slot is optional. Also optional is an upset head type for hexagon-head machine screws of sizes 4, 5, 8, 12, and 1/4 in.

Table 37. Dimensions of American Standard Machine Screw Heads—Binding Head

(ASA B18.6-1947. All dimensions in inches. See Fig. 16.)

Nominal Size	Max Diameter, *D*	Head Diameter, max, *A*	Total Height of Head, max, *O*	Width of Slot, min, *J*	Depth of Slot, min, *T*	Height of Oval, max, *F*	Diameter of Undercut,* min, *U*	Depth of Undercut, min, *X*
No. 2	.086	.181	.046	.023	.024	.018	.124	.005
No. 3	.099	.208	.054	.027	.029	.022	.143	.006
No. 4	.112	.235	.063	.031	.034	.025	.161	.007
No. 5	.125	.263	.071	.035	.039	.029	.180	.009
No. 6	.138	.290	.080	.039	.044	.032	.199	.010
No. 8	.164	.344	.097	.045	.054	.039	.236	.012
No. 10	.190	.399	.114	.050	.064	.045	.274	.015
No. 12	.216	.454	.130	.056	.074	.052	.311	.018
1/4	.250	.513	.153	.064	.088	.061	.360	.021
5/16	.3125	.641	.193	.072	.112	.077	.450	.027
3/8	.375	.769	.234	.081	.136	.094	.540	.034

Binding-head machine screws may be furnished with cross-recessed heads of either style shown in Fig. 16.

* Use of undercut is optional.

to note that a No. 5 screw has a major diameter 1/8 in. (0.125) and that the difference in diameter between successive numbers is 0.013 in.

Machine screws are regularly supplied with plain sheared ends, not pointed (Fig. 16). Header points with included angle of chamfer 40 to 45 degrees are obtainable up to certain lengths. On machine screws up to 2 in. long, complete threads extend to within two threads of the bearing surface of the head; longer screws have minimum complete thread length l = 1 3/4 in. Diameter of the unthreaded portion of machine screws is not less than the minimum pitch diameter or more than the maximum major diameter of the thread.

Screw threads on machine screws are Unified or American National Coarse or Fine thread series, class 2 or 2A. They are regularly supplied with a naturally bright finish, not heat treated.

Machine Screw Heads. With the exception of the hexagon, all heads illustrated in Fig. 16 may be obtained with a screwdriver slot, as shown, or with one of the two styles of cross recess illustrated. The "clutch head" style shown is available commercially. Slot dimensions are included in Tables 35, 36 and 37; for cross-recess dimensions the standard should be consulted. Complete details of cross recesses and drivers may be obtained from manufacturers licensed to produce them.

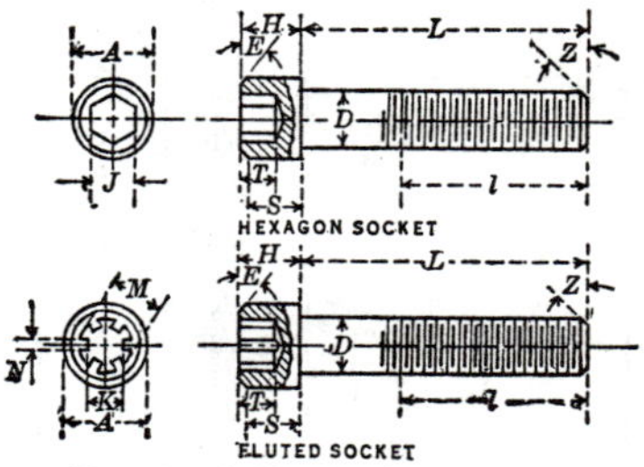

FIG. 17. Socket-head cap screws.

Binding-head screws (Table 37) are usually undercut as shown in Fig. 16, but the undercut is optional.

CAP SCREWS (Figs. 16 and 17) have chamfered points. Leading dimensions are listed in Tables 38, 39 and 40. For details of specification and manufacturing tolerances, the appropriate standards should be consulted.

Table 38. Dimensions of American Standard Slotted-head Cap Screws

(ASA B18.6-1947. All dimensions in inches. See Fig. 16.)

Nominal Size (Body Diameter, max), D	Width of Slot, min, J	Fillister Head				Flat Head			Round Head		
		Head Diameter, max, A	Height of Head, max, H	Total Height of Head, max, O	Depth of Slot, min, T	Head Diameter, max, A	Height of Head, average, H	Depth of Slot, min, T	Head Diameter, max, A	Height of Head, max, H	Depth of Slot, min, T
1/4	.064	0.375	.172	.216	.077	0.500	.140	.046	0.437	.191	.097
5/16	.072	0.437	.203	.253	.090	0.625	.176	.057	0.562	.246	.126
3/8	.081	0.562	.250	.314	.113	0.750	.210	.069	0.625	.273	.138
7/16	.081	0.625	.297	.368	.133	0.8125	.210	.069	0.750	.328	.167
1/2	.091	0.750	.328	.412	.148	0.875	.210	.069	0.812	.355	.179
9/16	.102	0.812	.375	.466	.169	1.000	.245	.080	0.937	.410	.208
5/8	.116	0.875	.422	.521	.190	1.125	.281	.092	1.000	.438	.220
3/4	.131	1.000	.500	.612	.233	1.375	.352	.115	1.250	.547	.277
7/8	.147	1.125	.594	.720	.264	1.625	.423	.139	....	....	....
1	.166	1.312	.656	.802	.292	1.875	.494	.162	....	....	....

For fillister- and round-head cap screws, radius of fillet at base of head is 0.016 to 0.031 in. for sizes 1/4 to 3/8 in., incl.; 0.016 to 0.047 in. for sizes 7/16 to 9/16 in., incl.; 0.031 to 0.062 in. for sizes 5/8 to 1 in., incl.

For flat-head cap screws the radius of the fillet at the base of the head shall not exceed twice the pitch of the screw thread.

Table 39. Dimensions of American Standard Socket-head Cap Screws
(ASA B18.3-1947. All dimensions in inches. See Fig. 17.)

Nominal Size	Body Diameter, max, *D*	Head Diameter, max, *A*	Head Height, max, *H*	Head Side Height, max, *S*	Hexagon * Socket Width across Flats, min, *J*	Fluted Socket Number of Flutes	Fluted Socket Socket Diameter Minor, min, *K*	Fluted Socket Socket Diameter Major, min, *M*	Fluted Socket Width of Socket Land, max, *N*
No. 2	0.0860	.140	.086	0.0803	1/16	6	0.063	0.073	.016
No. 3	0.0990	.161	.099	0.0923	5/64	6	0.080	0.097	.022
No. 4	0.1120	.183	.112	0.1043	5/64	6	0.080	0.097	.022
No. 5	0.1250	.205	.125	0.1163	3/32	6	0.096	0.113	.025
No. 6	0.1380	.226	.138	0.1284	3/32	6	0.096	0.113	.025
No. 8	0.1640	.270	.164	0.1522	1/8	6	0.126	0.147	.032
No. 10	0.1900	5/16	.190	0.1765	5/32	6	0.161	0.186	.039
No. 12	0.2160	11/32	.216	0.2005	5/32	6	0.161	0.186	.039
1/4	0.2500	3/8	1/4	0.2317	3/16	6	0.188	0.219	.050
5/16	0.3125	7/16	5/16	0.2894	7/32	6	0.219	0.254	.060
3/8	0.3750	9/16	3/8	0.3469	5/16	6	0.316	0.377	.092
7/16	0.4375	5/8	7/16	0.4046	5/16	6	0.316	0.377	.092
1/2	0.5000	3/4	1/2	0.4620	3/8	6	0.383	0.460	.112
9/16	0.5625	13/16	9/16	0.5196	3/8	6	0.383	0.460	.112
5/8	0.6250	7/8	5/8	0.5771	1/2	6	0.506	0.601	.138
3/4	0.7500	1	3/4	0.6920	9/16	6	0.531	0.627	.149
7/8	0.8750	1 1/8	7/8	0.8069	9/16	6	0.600	0.705	.168
1	1.0000	1 5/16	1	0.9220	5/8	6	0.681	0.797	.189
1 1/8	1.1250	1 1/2	1 1/8	1.0372	3/4	6	0.824	0.966	.231
1 1/4	1.2500	1 3/4	1 1/4	1.1516	3/4	6	0.824	0.966	.231
1 3/8	1.3750	1 7/8	1 3/8	1.2675	3/4	6	0.824	0.966	.231
1 1/2	1.5000	2	1 1/2	1.3821	1	6	1.003	1.271	.298

* Maximum socket depth (*T*, Fig. 17) should not exceed three-fourths of minimum head height (*H*).

Head chamfer angle *Z* (Fig. 17) is 28 to 32 degrees, the edge between flat and chamfer being slightly rounded.

Screw point chamfer angle *Z* (Fig. 17) is 35 to 40 degrees, the chamfer extending to the bottom of the thread. Edge between flat and chamfer is slightly rounded.

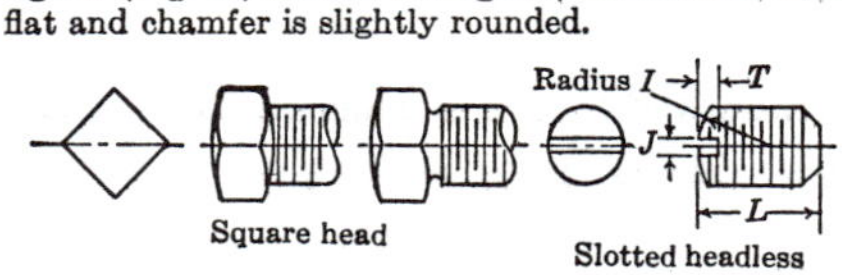

FIG. 18. Square-head set screws (Table 40) and slotted, headless set screws (Table 41).

Table 40. Dimensions of American Standard Hexagon-head Cap Screws and Square-head Set Screws
(ASA B18.2-1941. All dimensions in inches.)

Diameter of Screw, *D*	Hexagon-head Cap Screws (Fig. 10) Width across Flats, max (Basic)	Hexagon-head Cap Screws (Fig. 10) Height, max	Square-head Set Screws (Fig. 18) Width across Flats, max (Basic)	Square-head Set Screws (Fig. 18) Height, max	Square-head Set Screws (Fig. 18) Diameter of Neck, max	Diameter of Screw, *D*	Hexagon-head Cap Screws (Fig. 10) Width across Flats, max (Basic)	Hexagon-head Cap Screws (Fig. 10) Height, max	Square-head Set Screws (Fig. 18) Width across Flats, max (Basic)	Square-head Set Screws (Fig. 18) Height, max	Square-head Set Screws (Fig. 18) Diameter of Neck, max
1/4	0.4375	.194	0.2500	0.196	0.185	3/4	1.0000	.577	0.7500	0.582	0.620
5/16	0.5000	.242	0.3125	0.245	0.240	7/8	1.1250	.672	0.8750	0.678	0.731
3/8	0.5625	.289	0.3750	0.293	0.294	1	1.3125	.768	1.0000	0.774	0.838
7/16	0.6250	.338	0.4375	0.341	0.345	1 1/8	1.5000	.864	1.1250	0.870	0.939
1/2	0.7500	.386	0.5000	0.389	0.400	1 1/4	1.6875	.959	1.2500	0.966	1.064
9/16	0.8125	.433	0.5625	0.437	0.454	1 3/8	....	....	1.3750	1.063	1.159
5/8	0.8750	.481	0.6250	0.485	0.507	1 1/2	....	....	1.5000	1.159	1.284

Threads on cap screws are Unified or American National Coarse or Fine series.

Slotted-head cap screws are available in flat, fillister, and round head styles (Fig. 16 and Table 38). Minimum thread length is $2D + 1/4$ in. with a plus tolerance of $2\ 1/2$ times the pitch. When too short for the specified minimum thread length, cap screws have complete threads extending within $2\ 1/2$ threads of the head.

Chamfered points have 90-degree included angle; maximum diameter of the point is 91.5% of the basic minor diameter plus 0.001 in.; minimum is 83.5% of basic minor diameter minus 0.001 in.

Socket-head cap screws are available with hexagon or fluted sockets (Table 39). For dimensions of keys fitting these screws, the standard (ASA B18.3-1947) should be consulted. Minimum thread length for coarse threads is $l = 2D + 1/2$ in. or $1/2L$, whichever is greater. Thread length for fine threads is $l = 1\ 1/2D + 1/2$ in. or $3/8L$, whichever is greater. Screws too short to allow application of the formulas are threaded as close to the head as practicable.

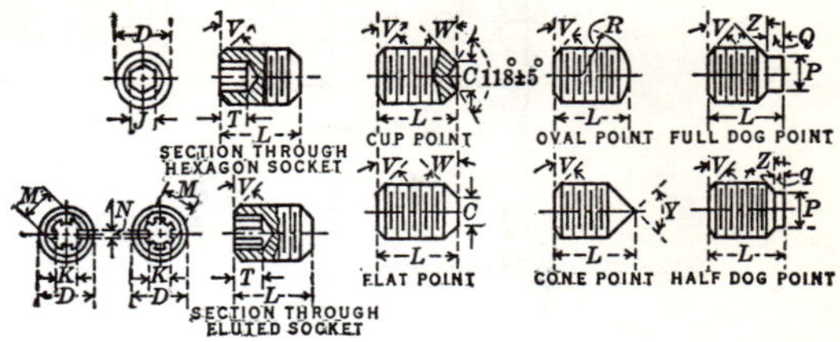

FIG. 19. Headless, socket set screws.

Screw points are flat and chamfered to an angle of 35 to 40 degrees with the flat. Chamfer extends to the bottom of the thread, and edge between flat and chamfer is slightly rounded.

Hexagon-head cap screws (Table 40) have heads with a washer face (see Fig. 10) and

Table 41. Dimensions of American Standard Headless Set Screws

(ASA B18.6-1947 and B18.3-1947. All dimensions in inches. See Fig. 19.)

Nominal Size, D	Cup and Flat Point Diameter, C	Oval Point Radius, R	Dog Point Diameter, max, P	Height of Dog		Hexagon Socket Width across Flats, min, J	Fluted Socket (Fig. 19)			Slotted (Fig. 18)		
				Full Q	Half q		Minor Diameter, min, K	Major Diameter, min, M	Land Width, max, N	Radius of Slotted End, I	Width of Slot, J	Depth of Slot, T
No. 5	0.057	3/32	.083	.06	.03	1/16	0.052	0.070	.022	.125	.023	.031
No. 6	0.064	7/64	.092	.07	.03	1/16	0.055	0.078	.023	.138	.025	.035
No. 8	0.076	1/8	.109	.08	.04	5/64	0.080	0.097	.022	.164	.029	.041
No. 10	0.088	9/64	.127	.09	.04	3/32	0.096	0.113	.025	.190	.032	.048
No. 12	0.101	5/32	.144	.11	.06	3/32	0.096	0.113	.025	.216	.036	.054
1/4	0.118	3/16	5/32	1/8	1/16	1/8	0.126	0.147	.032	.250	.045	.063
5/16	0.156	15/64	13/64	5/32	5/64	5/32	0.161	0.186	.039	.313	.051	.078
3/8	0.194	9/32	1/4	3/16	3/32	3/16	0.188	0.219	.050	.375	.064	.094
7/16	0.232	21/64	19/64	7/32	7/64	7/32	0.219	0.254	.060	.438	.072	.109
1/2	0.270	3/8	11/32	1/4	1/8	1/4	0.252	0.296	.068	.500	.081	.125
9/16	0.309	27/64	25/64	9/32	9/64	1/4	0.252	0.296	.068	.563	.091	.141
5/8	0.347	15/32	15/32	5/16	5/32	5/16	0.316	0.377	.092	.625	.102	.156
3/4	0.425	9/16	9/16	3/8	3/16	3/8	0.383	0.460	.112	.750	.129	.188
7/8	0.502	21/32	21/32	7/16	7/32	1/2	0.506	0.601	.138	....	....	....
1	0.579	3/4	3/4	1/2	1/4	9/16	0.531	0.627	.149	....	....	....
1 1/8	0.655	27/32	27/32	9/16	9/32	9/16	0.600	0.705	.168	....	....	....
1 1/4	0.733	15/16	15/16	5/8	5/16	5/8	0.681	0.797	.189	....	....	....
1 3/8	0.808	1 1/32	1 1/32	11/16	11/32	5/8	0.740	0.865	.207	....	....	....
1 1/2	0.886	1 1/8	1 1/8	3/4	3/8	3/4	0.824	0.966	.231	....	....	....
1 3/4	1.039	1 5/16	1 5/16	7/8	7/16	1	1.003	1.271	.298	....	....	....
2	1.193	1 1/2	1 1/2	1	1/2	1	1.003	1.271	.298	....	....	....

Chamfers and point angles are (see Fig. 19):

W = 45 to 50 degrees

Y = 116 to 120 degrees when L equals or is less than the nominal diameter

Y = 88 to 92 degrees when L is greater than the nominal diameter

Z = 35 to 40 degrees

V = 35 to 40 degrees

Number of flutes is 4 for sizes 5 and 6. All other sizes have 6 flutes.

are usually finished all over. Length of thread is $2D + 1/4$ in., minimum thread length being measured from the end of the screw to the last complete thread.

SET SCREWS are made with square heads, or headless with a slot for a screwdriver (Fig. 18), or with a socket into which a key can be inserted (Fig. 19).

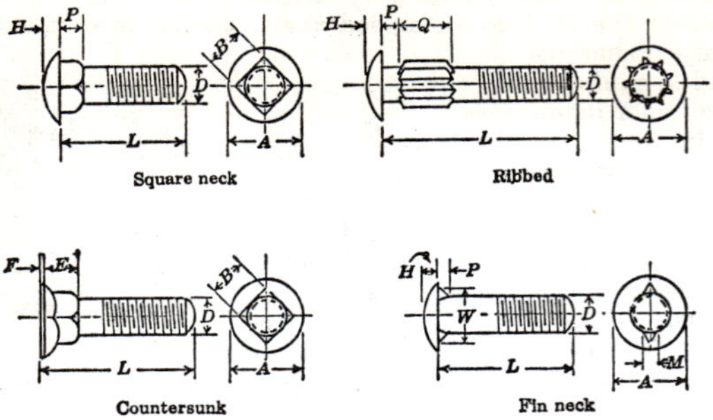

FIG. 20. Carriage bolts. See Tables 42 and 43.

Square-head Set Screws. Standard dimensions are listed in Table 40. Width of the neck under the head (Fig. 18) is not more than two times the pitch of the thread. Under

Table 42. Dimensions of American Standard Carriage Bolts—Square Neck and Ribbed Neck

(ASA B18.5–1939. All dimensions in inches.)

Nominal Diameter of Bolt, D	Diameter of Head, max, A	Height of Head, max, H	Square Neck (Fig. 20) Depth of Square, max P	P	Width of Square, max, B	Ribbed Neck (Fig. 20) Ribs below Head, P $L \leq 7/8$	$L \geq 1$	Length of Ribs, Q $L \leq 7/8$	$L = 1$ $L = 1\,1/8$	$L \geq 1\,1/4$	Number of Ribs
No. 10	0.469	.114	.125 ($L \leq 1\,1/8$)	.219 ($L \geq 1\,1/4$)	0.199	.031	.063	.188	.313	.500	9
1/4	0.594	.145	.156 ($L \leq 1\,1/4$)	.250 ($L \geq 1\,3/8$)	0.260	.031	.063	.188	.313	.500	10
5/16	0.719	.176	.187 ($L \leq 1\,1/4$)	.281 ($L \geq 1\,3/8$)	0.324	.031	.063	.188	.313	.500	12
3/8	0.844	.208	.219 ($L \leq 1\,1/2$)	.312 ($L \geq 1\,5/8$)	0.388	.031	.063	.188	.313	.500	12
7/16	0.969	.239	.250 ($L \leq 1\,1/2$)	.344 ($L \geq 1\,5/8$)	0.452	.031	.063	.188	.313	.500	14
1/2	1.094	.270	.281 ($L \leq 1\,7/8$)	.375 ($L \geq 2$)	0.515	.031	.063	.188	.313	.500	16
9/16	1.219	.312	.312 ($L \leq 1\,7/8$)	.406 ($L \geq 2$)	0.579	.094	.094	.188	.313	.500	18
5/8	1.344	.344	.344 ($L \leq 1\,7/8$)	.437 ($L \geq 2$)	0.642	.094	.094	.188	.313	.500	19
3/4	1.594	.406	.406 ($L \leq 1\,7/8$)	.500 ($L \geq 2$)	0.768	.094	.094	.188	.313	.500	22
7/8	1.844	.469	.469 ($L \leq 1\,7/8$)	.562 ($L \geq 2$)	0.895	......	......	......	......	......	......
1	2.094	.531	.531 ($L \leq 1\,7/8$)	.625 ($L \geq 2$)	1.022	......	......	......	......	......	......

Table 43. Dimensions of American Standard Carriage Bolts—Countersunk and Fin Neck

(ASA B18.5-1939. All dimensions in inches.)

Nominal Diameter of Bolt, D	Countersunk (Fig. 20) Diameter of Head, max, A	Feed Thickness, F	Depth of Square and Countersink, max, E	Width of Square, max, B	Fin Neck (Fig. 20) Diameter of Head, max, A	Height of Head, max, H	Depth of Fins, max, P	Distance across Fins, max, W	Thickness of Fins, max, M
No. 10	0.520	.016	.250	.199	0.489	.098	.088	.395	.098
1/4	0.645	.016	.312	.260	0.614	.129	.104	.458	.114
5/16	0.770	.031	.375	.324	0.739	.161	.135	.551	.145
3/8	0.895	.031	.437	.388	0.864	.192	.151	.645	.161
7/16	1.020	.031	.500	.452	0.989	.223	.182	.739	.192
1/2	1.145	.031	.562	.515	1.114	.254	.198	.833	.208
9/16	1.275	.031	.625	.579	....	....	....	....	....
5/8	1.400	.031	.687	.642	....	....	....	....	....
3/4	1.650	.047	.812	.768	....	....	....	....	....

surface of the head is beveled not more than 40 degrees, and the top or crown of the head is rounded to a radius 2 1/2 times the major diameter of the thread. Point styles are illustrated in Fig. 19, point dimensions being given in Table 41.

Headless set screws are threaded the entire length. Screw threads are Unified or American National Coarse or Fine series, class 2 or 2A, unless otherwise specified. Standard dimensions are listed in Table 41, and the various styles are illustrated in Figs. 18 and 19. When the usable length of thread is less than the nominal diameter, the half-dog point is used. Slotted-type set screws are regularly furnished case hardened.

Dimensions of keys for hexagon and fluted sockets may be found in the standard, ASA B18.3-1947.

CARRIAGE BOLTS, used where the head bears against wood, are illustrated in Fig. 20. Leading dimensions are given in Tables 42 and 43. Screw threads conform to the Unified or American National Coarse thread series and may be either cut or rolled. Leading dimensions of countersunk, button head, and step bolts are given in Table 44 and the styles are illustrated in Fig. 21.

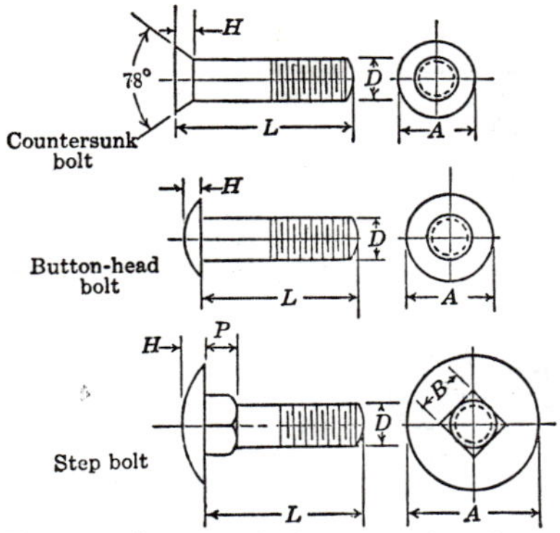

Fig. 21. Countersunk, button-head, and step bolts. See Table 44.

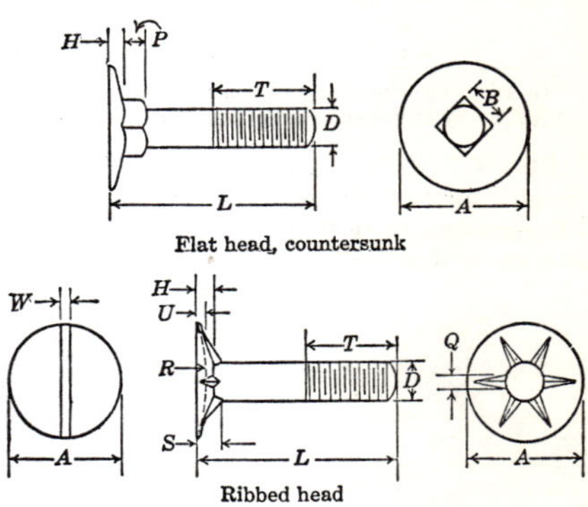

Fig. 22. Elevator bolts. See Table 45.

ELEVATOR BOLTS are illustrated in Fig. 22, and leading dimensions are given in Table 45. The oval-head elevator bolt, or step bolt, is illustrated in Fig. 21, dimensions being given in Table 44.

Table 44. Dimensions of American Standard Button-head, Step, and Countersunk Bolts

(ASA B18.5–1939. All dimensions in inches.)

Countersunk Bolt (Fig. 21)			Nominal Diameter of Bolt, *D*	Button Head (Fig. 21)		Step Bolt (or Oval-head Elevator Bolt), Fig. 21				
Nominal Diameter of Bolt, *D*	Head Diameter, max, *A*	Head Depth, *H*		Head Diameter, max, *A*	Head Height, max, *H*	Head Diameter, max, *A*	Head Height, max, *H*	Depth of Square, max, *P*	Depth of Square, max, *P*	Width of Square, max, *B*
1/2	0.936	0.250	No. 10	0.469	.114	0.656	.114	.125 ($L \leq 1\ 1/8$)	.219 ($L \geq 1\ 1/4$)	.199
9/16	1.049	0.281	1/4	0.594	.145	0.844	.145	.156 ($L \leq 1\ 1/4$)	.250 ($L \geq 1\ 3/8$)	.260
5/8	1.194	0.313	5/16	0.719	.176	1.031	.176	.187 ($L \leq 1\ 1/4$)	.281 ($L \geq 1\ 3/8$)	.324
3/4	1.421	0.375	3/8	0.844	.208	1.219	.208	.219 ($L \leq 1\ 1/2$)	.312 ($L \geq 1\ 5/8$)	.388
7/8	1.647	0.438	7/16	0.969	.239	1.406	.239	.250 ($L \leq 1\ 1/2$)	.344 ($L \geq 1\ 5/8$)	.452
1	1.873	0.500	1/2	1.094	.270	1.594	.270	.281 ($L \leq 1\ 7/8$)	.375 ($L \geq 2$)	.515
1 1/8	2.114	0.563	9/16	1.219	.312	......	......	......	......	......
1 1/4	2.341	0.625	5/8	1.344	.344	......	......	......	......	......
1 3/8	2.567	0.688	3/4	1.594	.406	......	......	......	......	......
1 1/2	2.793	0.750	7/8	1.844	.469	......	......	......	......	......
1 5/8	3.019	0.813	1	2.094	.531	......	......	......	......	......
1 3/4	3.262	0.875	......	......	......	......	......	......	......	......
1 7/8	3.488	0.938	......	......	......	......	......	......	......	......
2	3.714	1.000	......	......	......	......	......	......	......	......

Table 45. Dimensions of Elevator Bolts

(All dimensions in inches)

Nominal Diameter of Bolt, *D*	Flat Head (Fig. 22)				Ribbed Head (Fig. 22)					
	Diameter of Head * (Basic), *A*	Height of Head, max, *H*	Depth of Square, max, *P*	Width of Square, max, *B*	Diameter of Head,* (Basic), *A*	Height of Head, max, *H*	Width of Ribs at Body, *Q*	Depth of Ribs from Top Surface, *S*	Width of Slot,† *W*	Center Depth of Slot,† *U*
No. 10	0.750	.083	.125	.199	....	....	....	....	....	....
No. 12	0.781	.083	.219	.226	0.594	.098	0.078	0.156	0.078	0.031
1/4	0.969	.098	.250	.260	0.719	.114	0.094	0.172	0.094	0.047
5/16	1.188	.114	.281	.324	0.875	.145	0.109	0.203	0.094	0.063
3/8	1.313	.145	.281	.388	1.000	.176	0.109	0.234	0.094	0.078
7/16	1.438	.176	.281	.452	....	....	....	....	....	....
1/2	1.563	.176	.281	.515	....	....	....	....	....	....

* In order to make the edge as thin as possible, the circumference of the head may be somewhat irregular.
† Heads may be slotted or unslotted, as specified. Radius of slot, *R*, is 1 1/2 in. (See Fig. 22).

PLOW BOLTS. Two types, the No. 3 and the No. 7, illustrated in Fig. 23, are recommended for new designs. Other types are included in the new American standard. The No. 3 has an 80-degree round countersunk head and a square neck. It is economical to produce and use but is not well suited to application in thin materials; the square weakens

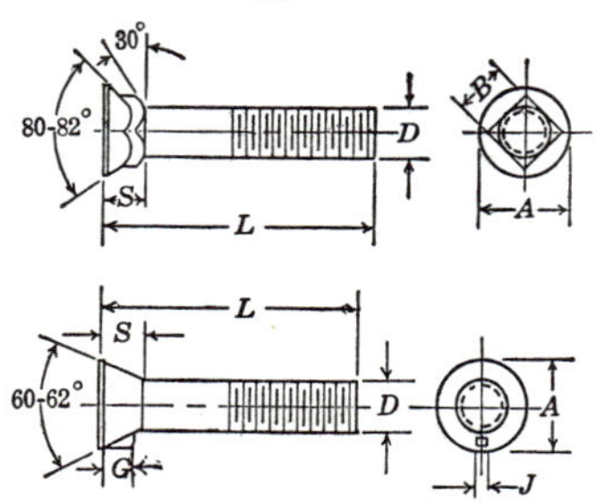

Fig. 23. Plow bolts No. 3 (upper) and No. 7 (lower). See Table 46.

the bolted part more than a round hole, making it less desirable for use close to an edge. The No. 7 has a 60-degree round countersunk head and a reverse key. It can be used in thinner sections than the No. 3. Threads are UNC or NC class 2A. Dimensions of regular-head plow bolts No. 3 and No. 7 are given in Table 46; repair head bolts have head dimensions, in general, slightly less than the regular head.

Table 46. Dimensions of Plow Bolts, Regular Head *

(All dimensions in inches. See Fig. 23.)

Nominal Diameter of Bolt, *D*	No. 3 Plow Bolt				No. 7 Plow Bolt				
	Head Diameter, max, *A*	Feed Thickness, max, *F*	Depth of Square and head, max, *S*	Width of Square, max, *B*	Head Diameter, max, *A*	Feed Thickness, max, *F*	Head Height, max, *S*	Key Width, max, *J*	Key Length, max, *G*
5/16	0.605	.025	.269	0.325	0.592	.025	.233	.156	.200
3/8	0.708	.031	.312	0.387	0.661	.031	.239	.156	.202
7/16	0.826	.036	.364	0.450	0.776	.036	.282	.156	.242
1/2	0.945	.042	.417	0.515	0.892	.042	.328	.156	.282
9/16 †	1.045	.045	.461	0.578	1.021	.045	.383	.156	.336
5/8	1.147	.050	.506	0.640	1.121	.050	.414	.156	.363
3/4	1.303	.050	.541	0.765	1.277	.050	.440	.156	.390
7/8	1.512	.063	.626	0.906	....	....	....	....	....
1	1.700	.063	.690	1.031	....	....	....	....	....

* Dimensions in table are for regular heads. Repair head dimensions are given in the standard.
† This size not recommended.

SHOULDER SCREWS, used as pivots for oscillating or rotating parts, are illustrated in Fig. 24. Leading dimensions of socket-head shoulder screws are listed in Table 47.

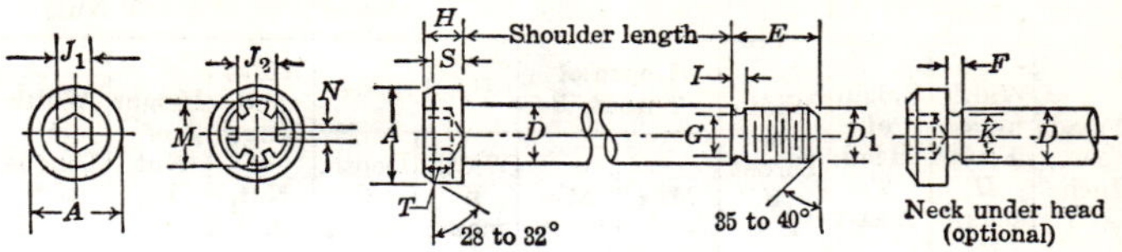

FIG. 24. Shoulder screws (stripper bolts). See Table 47.

T BOLTS. Standard dimensions of T bolts, T slots, and nuts are listed in Table 48 and illustrated in Fig. 25. Screw threads on T bolts and studs usually are Unified or

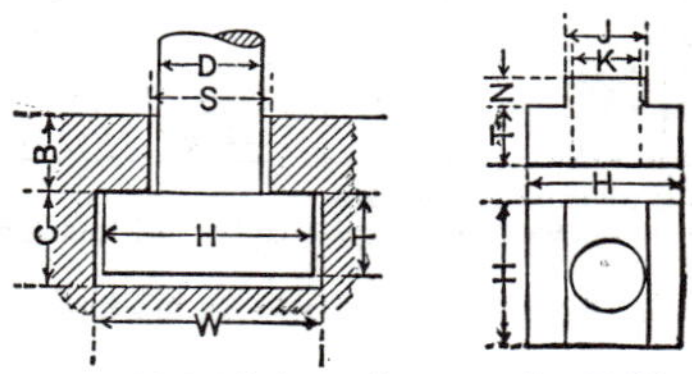

FIG. 25. T-slot bolts and nuts. See Table 48.

American coarse series, class 3 or 2A. The standard for T slots (ASA B5.1-1941) should be consulted for the dimensions of T-slot cutters and of inserted tongues and tongue seats.

TAPPING SCREWS are of three types (Fig. 26 and Table 49). Length of tapping

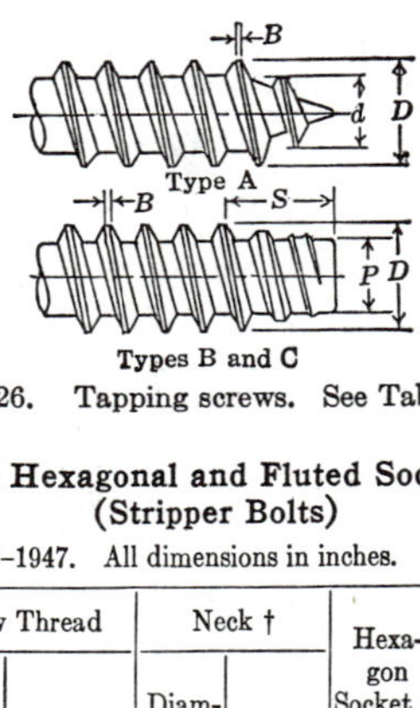

FIG. 26. Tapping screws. See Table 49.

Table 47. Dimensions of Hexagonal and Fluted Socket-head Shoulder Screws (Stripper Bolts)

(ASA B18.3–1947. All dimensions in inches. See Fig. 24.)

Nominal Size or Shoulder Diameter,* D	Head			Screw Thread		Neck †		Hexagon Socket,‡ Width, min, J_1	Fluted Socket ‡			Neck §	
	Diameter, max, A	Height, max, H	Side Height, max, S	Diameter,‖ D_1	Length,¶ E	Diameter, max, G	Width, I		Minor Diameter, min, J_2	Major Diameter, max, M	Width of Land, max, N	Diameter, max, K	Width, F
1/4	3/8	3/16	.1641	10–24	3/8	.145	1/16	1/8	.126	.149	.032	0.243	3/32
5/16	7/16	7/32	.1899	1/4–20	7/16	.196	5/64	5/32	.161	.188	.039	0.305	3/32
3/8	9/16	1/4	.2155	5/16–18	1/2	.252	5/64	3/16	.188	.221	.050	0.368	3/32
1/2	3/4	5/16	.2668	3/8–16	5/8	.307	3/32	1/4	.252	.298	.068	0.493	3/32
5/8	7/8	3/8	.3181	1/2–13	3/4	.417	1/8	5/16	.316	.380	.092	0.618	3/32
3/4	1	1/2	.4341	5/8–11	7/8	.526	9/64	3/8	.383	.463	.112	0.743	3/32
1	1 5/16	5/8	.5366	3/4–10	1	.642	5/32	1/2	.531	.631	.149	0.993	1/8
1 1/4	1 3/4	3/4	.6386	7/8–9	1 1/8	.755	11/64	5/8	.681	.801	.189	1.243	1/8

* Maximum diameter ($D - 0.002$-in.), minimum diameter ($D - 0.004$-in.).

† Minimum radius, 0.03 in. at corners of neck bottom.

‡ Maximum socket depth, $T = 0.75H$ (see Fig. 22).

§ Neck under head optional with manufacturer.

‖ Screw threads are National Coarse, class 3 (NC-3).

¶ Length tolerance, ±0.010 in. for E = 3/8 to 1/2 in., incl.; ±0.015 in. for E = 5/8 to 1 1/8 in., incl.

Table 48. Dimensions of Standard T Bolts, T Slots, and T Nuts

(ASA B5.1-1941. All dimensions in inches. See Fig. 25.)

Diameter of Bolt, D	T Bolt			T Slot					T Nuts			
	Threads per Inch	Width across Flats, H, max *	Height of Head, T, max †	Width of Throat, S	Depth of Throat, B		Head Space		Width and Length of Nut, H *	Height of Nut, T ‖	Width of Tongue, J ¶	Thickness of Tongue, N
					Max	Min	Width, W, max ‡	Depth, C max §				
1/4	20	15/32	5/32	9/32	3/8	1/8	9/16	15/64	9/16	3/16	0.330	3/32
5/16	18	9/16	3/16	11/32	7/16	5/32	21/32	17/64	11/16	1/4	0.418	1/8
3/8	16	11/16	1/4	7/16	9/16	7/32	25/32	21/64	7/8	5/16	0.543	7/32
1/2	13	7/8	5/16	9/16	11/16	5/16	31/32	25/64	1 1/8	13/32	0.668	7/32
5/8	11	1 1/8	13/32	11/16	7/8	7/16	1 1/4	31/64	1 5/16	17/32	0.783	1/4
3/4	10	1 5/16	17/32	13/16	1 1/16	9/16	1 15/32	5/8	1 11/16	11/16	1.033	5/16
1	8	1 11/16	11/16	1 1/16	1 1/4	3/4	1 27/32	53/64	2 1/16	15/16	1.273	3/8
1 1/4	7	2 1/16	15/16	1 5/16	1 9/16	1	2 7/32	1 3/32	2 1/2	1 3/16	1.523	7/16
1 1/2	6	2 1/2	1 3/16	1 9/16	1 15/16	1 1/4	2 21/32	1 11/32	......	......	......	......

* A tolerance of −0.031 in. is allowed on all sizes.

† A tolerance of −0.016 in. is allowed on sizes 1/4 to 5/8 in., inclusive, and of −0.031 in. on sizes 3/4 to 1 1/2 in., inclusive.

‡ A tolerance of −0.063 in. is allowed on sizes 1/4 to 5/8 in., inclusive, and of −0.094 in. on sizes 3/4 to 1 1/2 in., inclusive.

§ A tolerance of −0.031 in. is allowed on sizes 1/4 to 3/4 in., inclusive, of −0.047 in. on 1 in. size, and of −0.063 in. on 1 1/4 and 1 1/2 in. size.

‖ A tolerance of −0.016 in. is allowed on sizes 1/4 to 1/2 in., inclusive, and of −0.031 in. on sizes 5/8 to 1 1/4 in.

¶ A tolerance of −0.010 in. is allowed on sizes 1/4 to 5/8 in., inclusive, and of −0.015 in. on sizes 3/4 to 1 1/4 in.

Table 49. Dimensions of American Standard Tapping Screws

(ASA B18.6-1947. All dimensions in inches. See Fig. 26.)

Nominal Size	Type A			Type B				Type C		
	Threads per Inch	Major Diameter, max, D	Minor Diameter, min, d	Threads per Inch	Major Diameter, max, D	Minor Diameter, min, d	Point Diameter, max, P	Threads per Inch	Major Diameter, max, D	Point Diameter, max, P
No. 0	40	.060	.039	48	.060	.033	.033	......	......	......
No. 1	32	.075	.048	42	.075	.046	.046	......	......	......
No. 2	32	.088	.056	32	.088	.060	.055	......	......	......
No. 3	28	.101	.071	28	.101	.071	.069	......	......	......
No. 4	24	.114	.078	24	.114	.082	.078	40	.1120	.086
No. 4	......	......	......	......	......	......	......	48	.1120	.090
No. 5	20	.130	.090	20	.130	.090	.084	40	.1250	.099
No. 5	......	......	......	......	......	......	......	44	.1250	.101
No. 6	18	.141	.096	20	.139	.099	.093	32	.1380	.106
No. 6	......	......	......	......	......	......	......	40	.1380	.112
No. 7	16	.158	.108	19	.154	.109	.101	......	......	......
No. 8	15	.168	.116	18	.166	.116	.107	32	.1640	.132
No. 8	......	......	......	......	......	......	......	36	.1640	.135
No. 10	12	.194	.126	16	.189	.135	.130	24	.1900	.147
No. 10	......	......	......	......	......	......	......	32	.1900	.158
No. 12	11	.221	.155	14	.215	.157	.146	24	.2160	.173
No. 12	......	......	......	......	......	......	......	28	.2160	.179
No. 14	10	.254	.178	......	......	......	......	......	......	......
No. 16	10	.280	.189	......	......	......	......	......	......	......
No. 18	9	.306	.209	......	......	......	......	......	......	......
No. 20	9	.333	.226	......	......	......	......	......	......	......
No. 24	9	.390	.282	......	......	......	......	......	......	......
1/4	......	......	......	14	.246	.185	.174	20	.2500	.198
1/4	......	......	......	......	......	......	......	28	.2500	.213
5/16	......	......	......	12	.315	.236	.235	18	.3125	.255
5/16	......	......	......	......	......	......	......	24	.3125	.269
3/8	......	......	......	12	.380	.299	.298	16	.3750	.310
3/8	......	......	......	......	......	......	......	24	.3750	.332
7/16	......	......	......	10	.440	.349	.348	......	......	......
1/2	......	......	......	10	.504	.413	.411	......	......	......

screws is measured from the largest diameter of the bearing surface of the head to the extreme point. All types are threaded to the head and are regularly furnished case hardened. Tapping screws are available with heads of the following types: truss, pan, round, flat, oval, and hexagon (see Fig. 16).

4. LOCKING FASTENERS

A nut and bolt properly tightened will not loosen provided sufficient pressure is maintained between the threads and at the nut and bolt-head seats. Under vibrating loads these pressures may be periodically relieved, permitting the nut or screw to loosen. Because so many machinery applications involve vibration a wide demand for locking fasteners exists. Many commercial fasteners are now available, with new ones continually being developed, and no attempt will be made to cover all types. A few general locking methods will be discussed, the principles employed in classifying locking fasteners will be outlined, and brief mention will be made of some of the more widely used commercial fasteners available in 1949. For latest information on locking fasteners the reader is referred to current technical periodicals and to manufacturers' catalogs.

COTTER PINS secured in holes drilled in the bolts, with either plain or castle nuts (see Table 26, p. 10-35), positively prevent the fastener assembly from falling apart, and are widely used in the interior of machines such as engines, where complete separation of fastener elements could have disastrous results. The method is not well adapted to quick assembly and does not necessarily prevent relaxation (loss of bolt tension).

WIRES threaded through holes drilled in the heads of two or more cap screws serve the same purpose as cotter pins with nuts, preventing the screw from falling out or turning more than a few degrees but not necessarily preventing relaxation.

LOCKING PLATES attached to the bolted part and bearing against the nut or bolt head also are employed to prevent turning.

JAM NUTS used in pairs or in conjunction with plain nuts (see Table 26, p. 10-35) develop internal "locked-up" stresses when tightened against each other, furnishing the necessary pressure against the threads regardless of bolt load. Load is carried by the outer nut.

LOCK WASHERS (Fig. 14, p. 10-37) provide spring take up to compensate for developed looseness between component parts of an assembly. *Spring lock washers* (Table 32) also serve as hardened thrust bearings to facilitate assembly and disassembly by decreasing the frictional resistance between the bolted surface and the bearing face of the bolt head or nut. *Tooth lock washers* (Table 33) provide increased friction between fasteners and assembly.

Lock Washer and Machine Screw Assemblies. For mass production and other operations where speed and convenience are paramount factors, lock washers are assembled permanently to machine screws. The inside diameter of the washers are of a size to permit free rotation but to prevent their coming off the screw after the thread has been rolled. Spring lock washers are standardized in association with pan, fillister, truss, round, and hexagon head machine screws (Fig. 16, p. 10-40). Tooth lock washers are standardized in association with flat, oval, pan, truss, round, and hexagon head machine screws.

LOCKNUTS are of many and varied designs, which have been classified by the Industrial Fasteners Institute. The *tension* class develops at least normal tension load unaided, whereas the *shear* class does not.

Each of these classifications can be further subdivided into *load-locking* and *self-locking* types. The load-locking type requires a base load to develop locking whereas the self-locking type locks whether or not it engages a bearing surface.

Load-locking types in turn may be termed *compression lock* when the locking is due to elastic action which unlocks when the base load is removed, or *bearing set* when the locking is due to permanent deformation which maintains locking even after the base load is removed.

Self-locking types in turn can be *inherent*, requiring a single assembly operation consisting of wrenching to position, or *manual*, requiring multiple assembly operation.

Of the locknuts commercially available in 1949, the following are representative.

Anco-Nut (Automatic Nut Co., Inc., Lebanon, Pa.). This nut has a locking pin that impinges at an angle against the bolt at the base of the threads. This impingement acts in a similar manner to a ratchet, permitting tightening by a wrench but resisting loosening of the nut. Reverse wrenching changes the angle of impingement and permits removal of the nut.

An-cor-lox (An-cor-lox Division of Laminated Shim Co., Glenbrook, Conn.). This nut uses a relatively soft, nonspringing metal ring at the base of the nut to provide locking action. Normally made of bronze, the ring is permanently inserted and is an integral part of the nut. When torque is applied, the ring is flattened and forced into the thread of the bolt.

Bellows (Boots Aircraft Nut Corp., Stamford, Conn.). This is a spring-type locknut in which the locking and load sections are integral. Material is aluminum or Duronze. The bolt threads establish spring tension in the locking section, resulting in a differential thrust in the fastener against the bolt.

Drake (The National Screw and Manufacturing Co., Cleveland, Ohio). This wedge-type fastener is composed of two pieces, a prime nut or lower member which tightens the work to the exact adjustment required, and a locking member which is applied by a firm wrenching action. Contact between the two members takes place on mating tapered sections, the locking member having a male taper extending into the lower member. Each member turns freely on the bolt until seated or locked in position.

Dynamic (The National Screw and Manufacturing Co., Cleveland, Ohio). A groove in the top of this nut and a recess in its base provide a resilient central section which acts as a diaphragm. When the nut is tightened to the adjustment desired, an outward axial spring action develops. This tension in the upper diaphragm causes a constant pressure on the upper flanks of the nut threads and the mating lower flanks of the bolt threads, locking them together.

Esna (Elastic Stop Nut Corporation of America, Union, N. J.). Locking element in this reusable, interference-type nut is a nonmetallic compression collar locked into the head of the nut. The nut is made so that the center opening of the compression collar is somewhat smaller than the bolt, thus giving a friction grip on the threads of the bolt.

Flexloc (Standard Pressed Steel Co., Jenkintown, Pa.). Top of this one-piece, all-metal, reusable-type stop nut has six radial slots which provide six resilient sections to do the actual locking. Pressure due to deflecting these sections creates a firm grip on the bolt threads.

Gripco (Grip Nut Co., Chicago, Ill.). Three triangular depressions on one face of the one-piece nut deform the adjacent threads so as to cause pressure on the bolt threads. The resulting friction locks the nut on the bolt. Nut is reusable.

Huglock (National Machine Products Co., Detroit, Mich.). This one-piece, reusable, all-metal locknut is formed with a slotted tapered top and is threaded the full length of the nut. The slotted portions are curved in so that the six tapered segments press inwardly against the bolt, creating a friction lock.

Lamson (The Lamson and Sessions Co., Cleveland, Ohio). This one-piece, reusable, spring stopnut has a collar on top and is threaded in the usual way. The locking collar is distorted from the round on opposite sides, causing it to grip the bolt threads when assembled. Permanent spring action is given the collar by heat treatment.

Loktite (Lewis Bolt and Nut Co., Minneapolis, Minn.). A curved arch on the bearing face of this nut flattens as the nut starts to bear. This causes the thread to become deformed, creating frictional tension and spring action which locks the engaging threads of the bolt and nut.

Marsden (American Marsden Co.). Upper portion of this one-piece, reusable nut is slotted, forming six threaded segments, and the bottom face is recessed so that a continuous seating surface is provided outside the thread pitch line. When the nut is tightened to a seat, the threaded upper segments move radially inward, causing the nut threads to align themselves securely with their mating bolt threads, and producing a friction lock.

M-F Lock Nut No. 1 (MacLean-Fogg Lock Nut Co., Chicago, Ill.). Applied over a common nut to hold it in place, this is a free-spinning, seating-lock, reusable-type fastener. The nut takes a permanent set and must be wrenched for further movement. Both faces of the nut are arched (concave). When the arched surface contacts the top of the common nut and flattens due to additional tightening, the nut threads are deflected and bind against the bolt.

M-F Speed Holding Nut No. 2 (MacLean-Fogg Lock Nut Co., Chicago, Ill.). Lower surface of this nut is arched (concave), and the upper half has a wide recess extending across the nut and approximately half way through the thickness. When the lower arched face is flattened by tightening against the face of the work, the top portion deflects inward, binding the nut threads against the bolt threads. The nut is used alone and can be applied to bolts which cannot be held against rotation.

M-F Unitary Nut No. 3 (MacLean-Fogg Lock Nut Co., Chicago, Ill.). The top two or three threads of this reusable-type lock nut are deflected slightly out of the true helix. The nut starts freely on the bolt but requires wrenching to final position, the induced friction at the distorted threads clamping the nut in place. The nut is used alone.

Nylok (Nylok Corp., New York, N. Y.). Two types of self-locking torque nuts are available: the E type has a nylon plug through one of the hexagon sides acting as a wedge; the N series has on top an annular nylon insert which is slightly undersize, causing lateral pressure and friction against the bolt.

Palnut (the Palnut Co., Irvington, N. J.). This is a single-thread spring steel locknut usually formed in a hexagon shape to standard nut dimensions. The formed sides are turned up to provide wrench-engaging surfaces, and the hole is spirally formed to engage the bolt thread. When wrench tightened, the inner section yields with spring forces reacting downward on the work surface, upward on the bolt thread, and inward on the root of the bolt thread, resulting in secure locking action.

Roltop (Boots Aircraft Nut Corp., Stamford, Conn.). This one-piece, reusable, spring-type locknut has a resilient locking section at the top which eliminates axial thread play. Spring tension established in the locking section results in downward thrust by the nut threads of the locking section against the bolt threads, and upward pressure of the nut threads in the load section against the bolt threads.

Speed Nut (Tinnerman Products, Inc., Cleveland, Ohio). This reusable, seating-lock type fastener is made of flat spring steel with an arched base and two arched prongs which engage the bolt threads. As the nut is tightened, the two arched prongs are forced down and inward to engage and lock against the root of the screw thread. In addition, the compression of the arch in both prongs and base gives a strong upward thrust against the screw threads.

Stover (Stover Lock Nut and Machinery Corp., Easton, Pa.). The locking characteristics of this reusable, one-piece, spring-type nut are obtained by compressing two opposite flats of the nut so that the top becomes elliptical in shape while the bottom remains circular. The elliptical section provides a heavy spring grip on two opposite sides of the bolt.

Tri-Lok (Boots Aircraft Nut Corp., Stamford, Conn.). Locking action of this spring-type nut is achieved by radial compression at three points at the top of the nut, by shearing the nut walls. A slight downward out-of-phase set is also established in the shearing operation.

Wing Nut (Boots Aircraft Nut Corp., Stamford, Conn.). Made of sheet metal (steel and aluminum), these nuts are designed primarily for weight saving. Their action is similar to that of the Bellows and Roltop locknuts.

SELF-LOCKING THREADS. Dardelet and Lok-Thred threads (see p. 10-29) lock by wedging action of tapered surfaces at the crest of the nut threads and root of the bolt threads. Dardelet is a load-locking type, and Lok-Thred a self-locking. Both are reusable.

5. STRENGTH OF SCREW FASTENERS

By William G. Waltermire

STRENGTH OF THREADED ASSEMBLY. Where both elements of the assembly are of the same material, the screw or bolt will fail by breaking or stripping when loaded to failure. The ultimate load of the screw or bolt, therefore, becomes the proof load of the nut. The proof load of the bolt or screw is based on the yield strength of its material. Table 50 shows recommended classes of bolt and nut material. All values are based on the stress area listed in Tables 3, 4, 5, 6, 7, and 8. This is the area at the mean of the pitch and minor diameters.

Table 50. Grades of Bolts

(From SAE Recommended Practice)

	Minimum Tensile, psi	Elastic Proof Load, psi	Brinell	Rockwell
Commercial bolts	55,000	....	207 max	95 B max
Low-carbon, Cold-headed, Stress-relieved Bolts				
1/2 and under	68,000	....	241 max	100 B max
Over 1/2 to 3/4 in., incl.	64,000	....	241 max	100 B max
Medium-carbon, Quenched and Tempered				
1/2 in. and under	125,000	90,000	255–321	25–34C
Over 1/2 to 3/4 in., incl.	120,000	85,000	255–321	25–34C
Over 3/4 to 1 in., incl.	110,000	78,000	223–285	19–30C
Over 1 to 1 1/2 in., incl.	105,000	74,000	223–285	19–30C
Medium-carbon Alloy, Quenched and Tempered				
1/2 in. and under	150,000	120,000	302–352	32–38C
Over 1/2 to 1 in., incl.	140,000	110,000	285–352	30–38C
Over 1 to 1 1/2 in., incl.	125,000	95,000	255–321	25–34C

Note: Nuts must meet minimum tensile requirements of bolts.

The action of the threads in engagement accounts for the ability of a nut to fail a bolt of the same material. As a result, nut material with a lower modulus of elasticity than the bolt may be used and is recommended. This permits the threads in the nut to conform more readily to the bolt threads and maintains more uniform contact pressure, thus raising the fatigue strength of the bolt.

SELECTION OF THREAD CLASS AND FIT. As illustrated in Fig. 27, when the externally threaded part of a joint is the weakest, as is generally the case with a bolt and nut, a fine thread will develop the greatest static tensile strength. However, where, as is often the case, the material around the tapped hole is the weakest, a coarse thread will develop the greatest static strength of the joint.

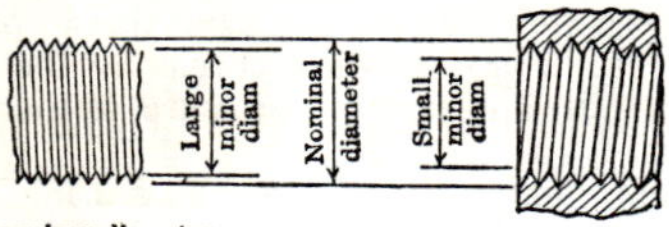

FIG. 27. Fine-thread bolt (*left*) and coarse-thread tapped hole (*right*).

It is generally agreed that coarse threads develop at least as high a fatigue strength under highly stressed vibratory loading as fine threads. A loaded nut dilates at the bottom and compresses axially, thus reducing the pitch. At the same time the bolt contracts radially and stretches, thus increasing its pitch. This action is accompanied by flexion of the triangular section of the threads. The result produces a progressive decrease of the unit load on the threads until the top thread carries proportionally less of the average load per thread.

As a result of this action, some authorities caution about the use of the relatively shallow fine-thread series on large-size bolts and screws, as the threads are more readily disengaged by the "breathing" action of the joint.

It is now generally agreed that a close thread fit does not improve the strength of a joint or its resistance to loosening under loading. On the other hand, it increases the danger of metal interference due to chips or burrs, plating, and heat of driving.

EFFECT OF FASTENER DESIGN AND LOCATION ON STRENGTH. The strength of a screw or bolt in tension is determined by its smallest cross-sectional area, which, on conventional bolts, is in the threads. Hence the unit stress is higher and the consequent deformation is higher than in other portions. If the unit stress in the body of the bolt or screw is made equal to that in the threaded portion, the deformation will be uniform throughout the whole length and its capacity to meet shock loads within the limits of elasticity of the fastener will be correspondingly increased. This may be done by using a bolt of reduced body diameter, as indicated in Fig. 28 or a Place bolt as shown in Fig. 29. The latter has a hollow head and a counterbored bearing face so proportioned that they provide a highly resilient head which is still stronger than the body of the bolt.

FIG. 28. Bolt of reduced body diameter for increased shock resistance.

Increasing the thread length increases fatigue life and may be used in lieu of reducing the body diameter as long as the fastener does not function in shear. Any of these means increase the total stretch in the bolt and hence make it more resistant to loosening.

While the nut could be redesigned to accomplish the same end, using tapered threads, annular notches around the bearing face, or other means, cost considerations for one make it generally inadvisable.

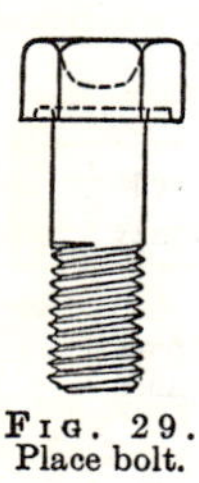

FIG. 29. Place bolt.

For best results the angle of thread runout should be 15 degrees or less (30-degree included angle), and the radius at the root of the threads ought to be rounded. Commercial roll-threaded bolts and screws normally have rounded roots.

For a given material, it has been found that screws with threads rolled on after heat treatment have a much higher fatigue strength than screws with cut threads or with threads rolled on before heat treatment. In other words, heat treatment will not increase the fatigue strength of threaded parts.

In Figs. 30, 31, and 32, J. O. Almen (Ref. 1) illustrates the effect of elastic elements in a design on strength. Good design requires the maximum stiffness of the bolted material and the maximum elasticity of the bolt.

Bruno and Schmittner (Ref. 2) studied the effect on loading of two steam chests, whose walls were located relative to the bolt or stud centerlines, as shown in Fig. 33. They found that, as the wall was brought out with respect to the bolt or stud centerline, the moment caused by the load, tending to open the joint at the flange, diminished and then reversed in sign; and the tests bore this out. No leakage was found with the wall positioned as shown at *C*.

SHEAR LOAD. If the load applied to the screw or bolt in service produces a shearing action, the shank should be full size so as to fit the hole as accurately as practicable and, where P is the load per fastener, d, the fastener diameter, and S_s, the shearing stress, $p = (\pi/4)d^2S_s$ and $S_s = 4p/\pi d^2$.

The clamping force induced by adequate prestress, according to F. B. Frankland

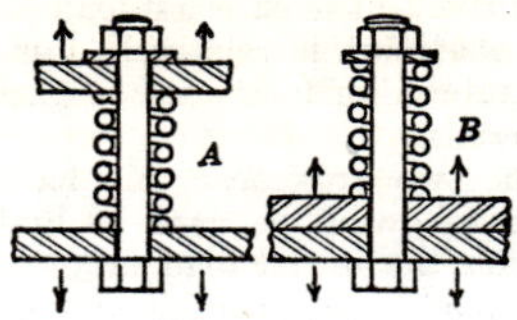

Fig. 30. Effect of design on bolt strength. In *A*, spring is part of abutments and bolt is weak. In *B*, spring is part of bolt, and bolt is strong. (Reproduced by permission from Tightening Is Vital Factor in Bolt Endurance, by J. O. Almen, in *Machine Design*, Feb. 1944)

Fig. 31. Elastically deformed plates and gasket increase maximum and cyclic bolt loads. (Reproduced by permission from *Machine Design, loc. cit.*)

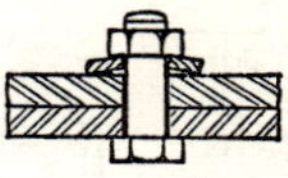

Fig. 32. Spring washer can augment bolt elasticity, resulting in longer bolt life. (Reproduced by permission from *Machine Design, loc. cit.*)

(Ref. 3) will account for as much as 15% increase in resistance to shear, although this is not taken into account by conventional design methods.

BENDING STRESSES. The effect of bending in increasing the combined stresses in a fastener, thus reducing the useful tensile stress it can develop, has long been recognized.

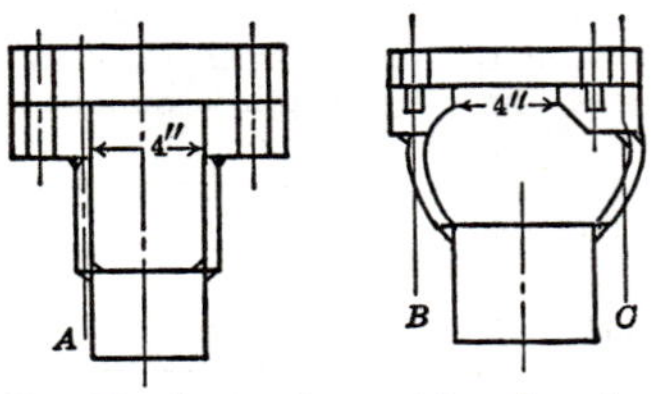

Fig. 33. Steam chests with wall center line (*A*) inside bolt center line, (*B*) in line with bolt center line, and (*C*) outside bolt center line. (Reproduced by permission from Studies on the Bolting of a Fabricated Steam Chest, by Bruno and Schmittner, in *Fasteners*, Vol. 4, No. 2)

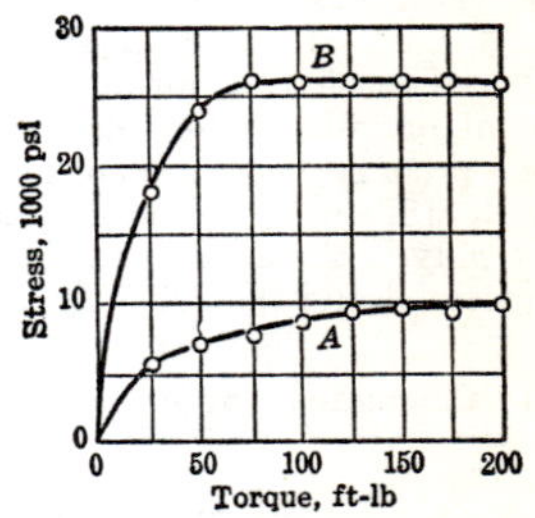

Fig. 34. Stress in a $^7/_8$-in. bolt having a nut (*A*) true with its mating surface and (*B*) with a 0.015-in. clearance at one side. (Reproduced by permission from *Fasteners, loc. cit.*)

Bending can be caused by actual bend in the bolt, runout (eccentricity) of a screw used in a tapped hole, or by angularity between the face of a bolt head or nut and their mating surface. The latter condition was demonstrated by Bruno and Schmittner (Ref. 2) in the case of a $^7/_8$-in. nut which had 0.015-in. clearance at one side when the nut was finger-tight. Curve *A* (Fig. 34) is for a nut which was apparently true with its mating surface. Curve *B* was for a nut with 0.015-in. clearance at one side.

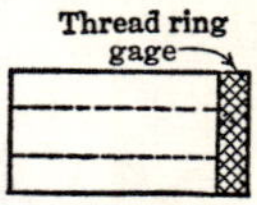

Fig. 35. Runout gage.

The term *runout* includes eccentricity of the thread and shank, bend and out of roundness. It is particularly important on cap screws and fasteners used in tapped holes. The automobile industry has generally adopted a sleeve gage, as shown in Fig. 35, to control runout. The internal diameter of the sleeve gage is opened to equal the maximum body diameter plus the runout allowance. At the end of the gage a thread ring gage, set to basic pitch diameter and centered by a setting plug, is located. Standard runout allowance for cap screws is 0.010 in. per inch of screw length.

PRESTRESSING. In the left view of Fig. 36 the fastener is shown as having a spring body for purposes of the illustration, and the bolted material *B* is assumed to be rigid. It is obvious, therefore, that when the nut is tightened, a load is induced under the nut at *X* and at the face of the joint at *Y* that is due to stretching the spring (bolt). Furthermore, if an external load is applied at *A* as shown, it will not move the assembly, that is, further stretch the spring (bolt) until it is at least as great as the induced load. For example, if *Y* is 2000 lb any external load less than 2000 lb would not stretch the spring.

The latter would be stretched as the load exceeded 2000 lb. Thus, prestressing provides freedom from vibration within certain limits.

In the right side of Fig. 36 the conditions of relative stiffness are reversed. The bolt is shown as rigid, and the bolted material, *B*, is represented by a spring. In this case it is obvious that screwing up the nut will not provide the same immunity to vibration as before, but that instead much of the prestress will be additive to the external load. This condition is approached when the joint is relatively less stiff than the fastener owing to its material or to the use of gaskets, springs, or resilient sealing materials.

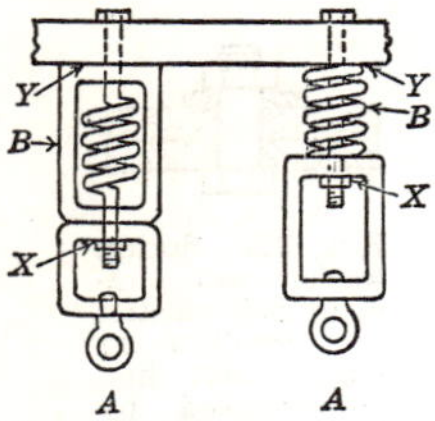

Fig. 36. Representation of resilient bolt with stiff connected parts (*left*), and a stiff bolt with resilient parts (*right*).

These situations represent the two extremes. Kimball and Barr (Ref. 4) suggest that the following be used as limiting guides in deciding the proportion the initial load or prestress ought to bear to the external load. The combined load, of course, cannot exceed the rated proof stress of the bolt or screw.

1. If a bolt is manifestly very much more yielding than the connected members, design the bolt simply for the initial load or for the external load, whichever is greater.
2. In case of extreme yielding of the connected members, assume the resultant load to be nearly the sum of the initial and external loads.

AMOUNT OF PRESTRESS. Most of the ordinary usages fall in condition 1 above. The author recommends in this case that the prestress be limited to 80% of the value of the proof strength. This is to allow for some addition of external load, for errors and variations in driver torques, for uncontrolled variations in frictional conditions of driving, and other conditions affecting stress and torque. It has been found that when bolts fail during driving (Ref. 5) they fail at induced tensions as low as 68% of the rated tensile strength. If driven just short of this point of failure, they still develop 100% of the tensile strength on direct pull. Assuming a proof strength 85% of tensile, the 80% of proof strength recommended checks exactly with Hill's 68% of tensile, since 80% of 85% is 68%.

Theoretically, a threaded joint should be prestressed to or above the yield point of the weakest material and then slightly loosened. This sets up residual stresses at the points

Table 51. Calculated Torque to Produce 60,000 psi Tensile Stress with No Lubrication

Size and Pitch	Torque, inch-pounds				
	Unplated Regular Hexagon Semifinished Nut on Unplated Bolt	Unplated Cap Screws in Spot-faced Tapped Holes in Steel			
		Hexagon Head	Flat Head	Round Head	Fillister Head
NC thread					
1/4–20	100	102	136	102	97
5/16–18	206	200	277	208	191
3/8–16	356	346	490	358	346
7/16–14	572	544	756	577	544
1/2–13	856	797	1120	861	797
9/16–12	1214	1202	1633	1260	1200
5/8–11	1611	1629	2252	1699	1629
3/4–10	2920	2834	4012	3042	2834
7/8–9	4700	4499	6486		4499
1–8	7014	6779	7647		6779
NF thread					
1/4–28	112	113	152	112	107
5/16–24	223	217	302	226	207
3/8–24	391	380	543	394	380
7/16–20	623	592	829	630	592
1/2–20	936	872	1236	944	872
9/16–18	1320	1296	1776	1359	1300
5/8–18	1850	1787	2493	1867	1787
3/4–16	3172	3070	4388	3304	3070
7/8–14	5056	4839	7031		4839
1–12	7669	7201	8147		7201

of greatest stress concentration, such as the roots of threads, and these are beneficial when the joint is subjected to fatigue.

AMOUNT OF TORQUE. Torque, rather than bolt strain or angular movement of a nut or screw head, is commonly used as a means of evaluating the stress induced in the fastener. This is because it lends itself to practical use. In practice the variables of frictional conditions and commercial torque-measuring equipment must be accepted.

For bolts tightened with nuts the formula $PL = RDT$ has been found to be reasonably accurate, where PL = torque, inch-pounds; R = a torque coefficient depending on frictional conditions; D = bolt diameter, inches; and T = tensile load, pounds.

The frictional conditions depend on class of thread, thread fit, size of bearing area of the nut or bolt head, whichever is driven, smoothness of the bearing areas in engagement, frictional characteristics of the materials engaged, the presence and kind of platings or lubricants, temperature, and other factors. For ordinary steel nuts and bolts, driven dry in steel, R approximates 0.2. The column for nuts in Table 51 is calculated on this basis. The foregoing formula may be used to derive torques for sizes outside the scope of the table.

By "driven dry" is meant the use of commercial bolt and nut finishes without actually lubricating the assembly. Some materials, like stainless steel, and some platings, like zinc, tend to increase the coefficient of friction; other materials, like aluminum, and other platings, like cadmium, tend to decrease it.

Even when known, the resultant torque is nominal only, and as much as plus and minus 30% variation has been observed in commercial nut and bolt torques driven under such conditions of uniformity as would be obtained in production use.

When screws are torqued by the head, the additional variables of screw head shape and of the material of the tapped hole complicate the problem of estimating torque.

For cap screw torques the following formulas, used in calculating the last four columns of Table 51, apply:

Hexagon, round, and fillister heads:

$$\frac{PL}{W} = \mu_b R_b + R_t \left(\mu_t \sec \alpha + \frac{1}{2\pi N R_p} \right)$$

For 80-degree countersunk heads:

$$\frac{PL}{W} = 1.524 \mu_b R_b + R_t \left(\mu_t \sec \alpha + \frac{1}{2\pi N R_p} \right)$$

where P = force applied to wrench, pounds
L = effective length of wrench, inches
W = total induced tension force in bolt or screw shank, pounds
μ_b = effective coefficient of friction on bearing face
= 0.15 for a spot-faced steel application without lubrication
R_b = effective radius of action of frictional forces on bearing face of bolt, nut, or screw, inches
$= \dfrac{2}{3} \dfrac{R_o^3 - R_i^3}{R_o^2 - R_i^2}$, where R_o = outer and R_i = inner radius
R_t = effective radius of action of frictional forces on thread contact faces, inches (= pitch radius, R_p)
μ_t = effective coefficient of friction on contacting surfaces of thread flanks (assumed equal to μ_b)
α = angle between mating faces of threads and a normal to the thread axis
= 30 degrees
N = number of threads per inch
R_p = pitch radius of threads, inches

For both bolts and cap screws, as an alternative to using Table 51, it is recommended that torque tests be conducted under assembly conditions and drivers set to about 80% of the *lowest* breaking torque found on bolts of minimum tensile strength. This insures adequate prestress but almost entirely eliminates the danger of a partial or complete failure during assembly. Partial failures are particularly dangerous as they cannot be detected.

Torques for some of the most widely used fasteners are shown in Table 51. They are not directly applicable if conditions obtain other than those specified.

Machine Screws. In the numbered sizes of the machine-screw range, variation in the proportion of thread depth to diameter for different sizes greatly affects torque conditions. One authority (Ref. 6) lists torques for machine screws tightened with a nut and no lubrication. In Table 52 these have been converted to the torque necessary to induce a combined tensile stress of 60,000 psi in the screw, and can be applied directly. For example, a 10–24 machine screw of common steel (see Table 50) is rated at 55,000 psi tensile strength. The torque to prestress is therefore $^{55}/_{60}$ of the 23.3 in.-lb value listed in Table 52, which is 21.3 in.-lb.

Table 52. Torque to Produce 60,000 psi Maximum Combined Tensile Stress in Machine Screws Tightened with a Nut and No Lubrication

NC Threads				NF Threads			
Size and Pitch	Torque, in.-lb	Size and Pitch	Torque, in.-lb.	Size and Pitch	Torque, in.-lb	Size and Pitch	Torque, in.-lb
....	...	5–40	7.9	0–80	0.8	5–44	8.7
1–64	1.4	6–32	9.1	1–72	1.6	6–40	11.1
2–56	2.3	8–32	17.4	2–64	2.6	8–36	19.0
3–48	3.3	10–24	23.3	3–56	3.7	10–32	29.6
4–40	4.8	12–24	38.9	4–48	5.6	12–28	44.2

As an alternative to Table 52 for machine screws, it is recommended that tests be conducted under actual conditions of assembly and that drivers be set to about 70% of the lowest breaking torque found with screws of minimum strength.

REFERENCES

1. J. O. Almen, Tightening is vital factor in bolt endurance, *Machine Design*, Feb. 1944.
2. Bruno and Schmittner, Studies on the bolting of a fabricated steam chest, *Fasteners*, Vol. 4, No. 2, published by the American Institute of Bolt, Nut, and Rivet Manufacturers (now the Industrial Fasteners Institute).
3. F. B. Frankland, *Eng. News-Record*, March 7, 1946.
4. D. S. Kimball and J. H. Barr, *Elements of Machine Design*, 3rd ed., John Wiley and Sons, New York, 1935.
5. Hill, Does torque weaken bolts?, *Fasteners*, Vol. 4, No. 1.
6. Korn, Torque-stress load relationship in bolts and screws, *Product Engineering*, Nov. 1943.

RIVETS AND RIVETED JOINTS

GENERAL CONSIDERATIONS. In making a riveted joint, the head is formed with a set or forming tool, by hand or by hydraulic or pneumatic machines. Machine riveting usually is better than hand work, as the pressure of the machine causes the rivet shank to fill completely the rivet hole. A rivet driven hot shrinks in cooling, pinching the riveted parts and making a firm joint. The cooling also leaves a tensile stress in the rivet, the amount of which cannot be measured. The pinching due to cooling of the rivet causes a frictional resistance between the riveted parts which Stoney found to be about 10,000 psi of rivet area. American and English practice takes no account of this frictional resistance but bases design on the tensile and shearing strength of the plates and rivets. Some French and German authorities base design entirely on the frictional resistance.

6. RIVETS

RIVET HEADS. Rivet heads are of many different forms, the particular form being determined by the use of the joint or fastening. Heads of greatest stiffness and strength are used when the rivet is to resist a tensile stress. Countersunk rivets weaken the plate and should be used only where projecting heads are objectionable.

Large rivets ($^1/_2$ in. nominal diameter and over) are manufactured to proportions given in ASA B18.4-1937. The styles and basic proportions are shown in Fig. 1. Fillet radius under head may be up to $^1/_{16}$ in. For manufacturing tolerances, material specifications, weight, and other details the standard should be consulted. These rivets are designed for use in structural work of bridges, buildings, etc., and in ship and boiler construction. Sizes range from $D = {}^1/_2$ to $1\,^3/_4$ in. by $^1/_8$-in. increments.

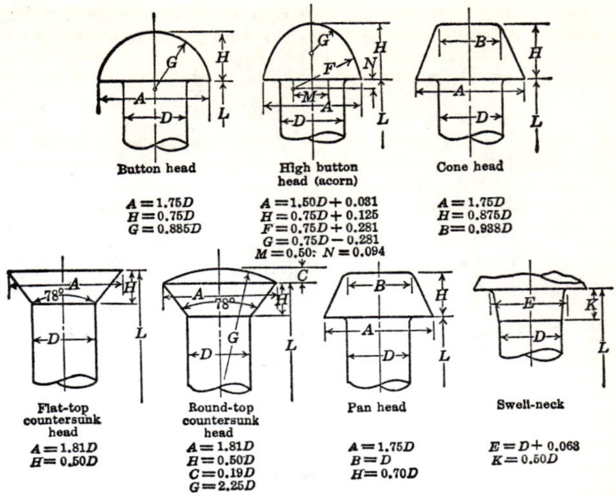

Fig. 1. Proportions of large rivets (D = $^1/_2$ in. and over) after driving. ASA B18.4-1937.

Driven heads, and manufactured heads after driving, have proportions as follows, the notation being the same as in Fig. 1:

Button head: $A = 1.827D$; $H = 0.688D$; $G = 0.954D$.
High button head (acorn): $A = 1.5D + 0.125$; $H = 0.425A$; $F = 1.5H$; $G = 0.425A$.
Cone head: $A = 1.827D$; $B = 0.938D$; $H = 0.828D$.
Pan head: $A = 1.827D$; $B = D$; $H = 0.656D$.

Rivet set and hold-on (dolly bar) impressions for the foregoing rivet heads are as follows, the prime (′) being added to the identifying letter of Fig. 1 to indicate the corresponding impression dimensions:

Button head: $A' = 1.800D$; $H' = 0.641D$; $G' = 0.954D$.
High button head (acorn): $A' = 1.485D + 0.124$; $H' = 0.4A$; $F' = 1.5H$; $G' = 0.425A$.
Cone head: $A' = 1.781D$; $B' = 0.938D$; $H' = 0.781D$.
Pan head: $A' = 1.781D$; $B' = D$; $H' = 0.609D$.

Sharp corners on the exterior of the set or hold-on are rounded to $^1/_{16}$-in. radius.

Boiler Rivets. The rivet proportions after driving, shown in Fig. 2, developed by the Hartford Steam Boiler Inspection and Insurance Co., have been adopted as standard in the ASME Boiler Code, in addition to the ASA standards shown in Fig. 1.

Small rivets ($^7/_{16}$ in. nominal diameter and under) have the proportions indicated in Fig. 3. Fillets under the head may be up to $^1/_{32}$ in. radius, except for countersunk heads. For manufacturing tolerances, materials specifications, weight, and other details, the standard (ASA B18a-1927) should be consulted. Standard sizes range from $D = {}^3/_{32}$ to $^7/_{16}$ in., by increments of $^1/_{32}$ in., omitting $^{13}/_{32}$ in.

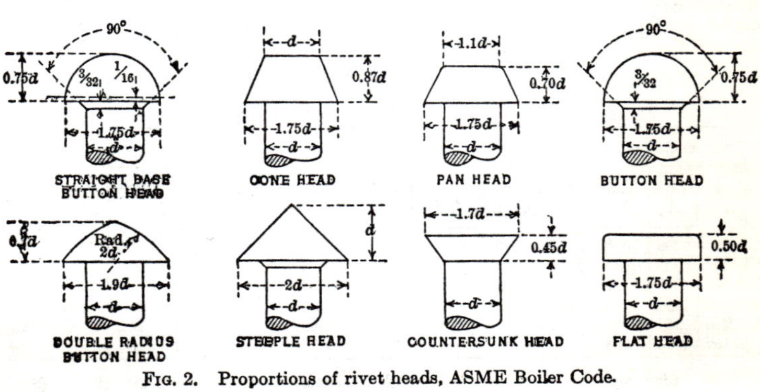

Fig. 2. Proportions of rivet heads, ASME Boiler Code.

Tinners' rivets are similar to the flat-head type small rivet shown in **Fig. 3**, but the basic proportions are different. Table 1 lists the dimensions of standard tinners' rivets.

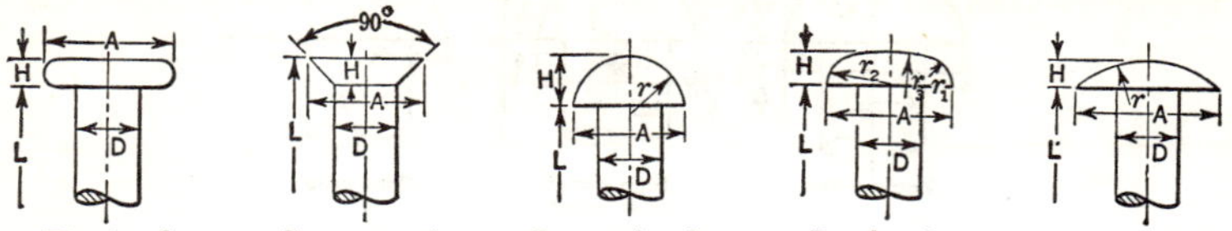

Flat head	Countersunk head	Button head	Pan head	Truss or wagon box head
$A = 2.00D$	$A = 1.850D$	$A = 1.750D$	$A = 1.720D$	$A = 2.500D$
$H = 0.33D$	$H = 0.425D$	$H = 0.750D$	$H = 0.570D$	$H = 0.330D$
		$r = 0.885D$	$r_1 = 0.314D$	$r = 2.512D$
			$r_2 = 0.850D$	
			$r_3 = 3.430D$	

FIG. 3. Proportions of small rivets, D = 7/16-in. and under. ASA B18a-1927.

Table 1. Tinners' Rivets

(ASA B18g-1929 [revised 1942]. Dimensions in inches.)

Size No.*	Head Diam.	Head Thick.	Body Diam.†	Body Length	Size No.*	Head Diam.	Head Thick.	Body Diam.†	Body Length	Size No.*	Head Diam.	Head Thick.	Body Diam.†	Body Length
8 oz.	.207	.027	.089	.16	2 1/2 lb	.333	.044	.148	.28	10 lb	.535	.071	.238	.47
12 "	.236	.031	.105	.19	3 "	.360	.048	.160	.31	12 "	.582	.077	.259	.50
1 lb	.249	.033	.111	.20	4 "	.396	.052	.176	.34	14 "	.639	.085	.284	.52
1 1/2 "	.292	.039	.130	.23	6 "	.456	.060	.203	.39	16 "	.675	.090	.300	.53
2 "	.324	.043	.144	.27	8 "	.504	.067	.224	.44					

* Size No. is the trade name or weight of 1000 rivets. † Body diameter tolerances are: 8 oz–2 1/2 lb, +0.002, −0.004; 3–10 lb, +0.003, −0.006; 12–16 lb, +0.004, −0.008.

Coopers' rivets are illustrated in Fig. 4, the basic dimensions being listed in Table 2.

Belt rivets are similar to the coopers' rivets (Fig. 4), except the proportions are such that circumference A is sharp instead of cylindrical. Dimensions are listed in Table 3.

RIVET HOLES. The rivet holes in a plate may be either punched or drilled. Drilling is preferred as the action of the punch causes an injurious lateral flow of metal. It is current practice to make holes 1/16 in. larger than the cold rivet diameter, except that for rivets not over 7/8 in. in diameter a hole of 1/32 in. greater diameter than the rivet is preferred. Where work is to be reamed, it is usual to punch the holes 1/8 or 3/16 in. less than the finished diameter, and to drill or ream to the proper size after the parts are assembled.

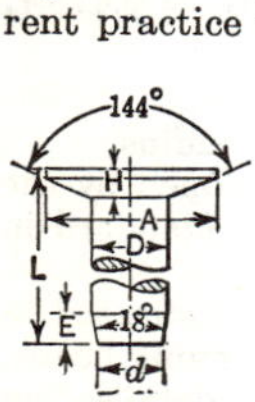

FIG. 4. Proportions of coopers' and belt rivets. See Tables 2 and 3.

It is good practice to remove the sharp corner at both ends of the reamed holes, so that a fillet will form where the body joins the head in finished rivets. Also, annealing will equalize the internal stresses caused by punching, and thus restore the strength of the plate.

The ASME Boiler Code Committee recommends for boiler work that: Rivet holes, except for attaching stays or angle bars to heads, shall be drilled full size with plates, butt-straps, and heads bolted in position; or they may be punched not to exceed 1/4 in. less than full diameter for plates over 5/16 in. thick, and 1/8 in. less than full diameter for plates not

Table 2. Coopers' Rivets

(ASA B18g-1929 [revised 1942]. See Fig. 4.)

Size No.*	Diameter, in. Body, D†	Diameter, in. Head, A	Diameter, in. Tip, d	Length, in. Body, l	Length, in. Over-all, L	Length, in. Head, H	Length, in. Cham-fer, E	Size No.*	Diameter, in. Body, D†	Diameter, in. Head, A	Diameter, in. Tip, d	Length, in. Body, l	Length, in. Over-all, L	Length, in. Head, H	Length, in. Cham-fer, E
1 lb	.109	.245	.098	.187	.219	.032	.043	6 lb	.203	.456	.182	.406	.466	.060	.081
1 1/2 "	.127	.285	.114	.218	.256	.038	.050	8 "	.238	.535	.214	.500	.571	.071	.095
2 "	.141	.317	.126	.250	.292	.042	.056	10 "	.250	.562	.225	.531	.606	.075	.100
2 1/2 "	.148	.333	.133	.281	.325	.044	.059	12 "	.259	.582	.233	.531	.608	.077	.103
3 "	.156	.351	.140	.312	.358	.046	.062	14 "	.271	.609	.243	.562	.643	.081	.108
4 "	.165	.371	.148	.343	.392	.049	.066	16 "	.281	.632	.252	.593	.677	.084	.112

* Size No. is trade name or weight of 1000 rivets. † Tolerances are the same as for tinners' rivets. (See Table 1.)

exceeding $^5/_{16}$ in. thick, and then drilled or reamed to full diameter with plates, butt-straps, and heads bolted in position. After drilling rivet holes, the plates and butt-straps shall be separated and the burrs removed.

Table 3. Belt Rivets

(ASA B18g-1929 [revised 1942]. See Fig. 4.)

Size No.*	Diameter, in.			Length, in.			Size No.*	Diameter, in.			Length, in.		
	Body,† D	Head, A	Tip, d	Overall, L	Head, H	Cham-fer, E		Body,† D	Head, A	Tip, d	Overall, L	Head, H	Cham-fer, E
7	.180	.504	.162	From 3/8 to	.054	.072	11	.120	.336	.108	From 3/8 to	.036	.048
8	.165	.462	.148	3/4 in. by	.049	.066	12	.109	.305	.098	3/4 in. by	.032	.043
9	.148	.414	.133	1/8-in. in-	.044	.059	13	.095	.266	.085	1/8-in. in-	.028	.038
10	.134	.375	.120	crements	.040	.053					crements		

* Stubs Iron Wire Gage No. of stock used in body. † Body diameter tolerances are: +0.002, −0.004.

PRESSURE REQUIRED TO DRIVE RIVETS. According to Wilfred Lewis, the pressure required to drive a hot rivet is 150,000 psi or 75 tons per square inch of rivet cross-sectional area. For cold riveting, the values are twice as great.

7. RIVETED JOINTS

RIVET SPACING. Rivet spacing has been the subject of much controversy and experiment. The rivets must be spaced far enough apart to allow proper clearance for use of the set during driving; $^3/_{16}$ in. is the minimum clearance. However, rivets must not be spaced at such a distance that the fastened parts open between rivets, allowing leakage in a fluid pressure joint, or the accumulation of dirt or water to aggravate rusting in a structural joint. With the usual proportions of riveted joints, a distance from the center of the rivet to the edge of the plate equal to (rivet diameter × 1 1/2) is sufficient to prevent tearing or shearing.

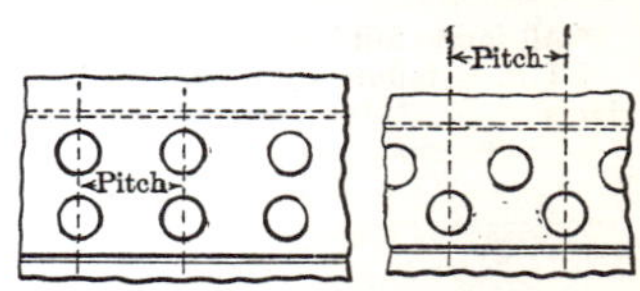

FIG. 5. Chain and staggered riveting.

ARRANGEMENT OF RIVETS. When several rows of rivets are used in lap joints, the rivets may be arranged as (1) chain riveting; (2) staggered riveting. (Fig. 5.) See Figs. 6 to 12 for examples of the use of the two types in boiler work.

Boilers and Pressure Vessels. Proper spacing of rivets for joints in boilers and pressure vessels is given in Figs. 6 to 12 and in Tables 5 to 10 of this article.

Structural Joints. According to the specifications of the American Institute of Steel Construction, minimum rivet distances should be as shown in Table 4.

Table 4. Minimum Rivet Distances for Structures

Rivet diameter, in.	1/2	5/8	3/4	7/8	1	1 1/8	1 1/4
Between centers of rivet holes, minimum	1 1/2	1 7/8	2 1/4	2 5/8	3	3 3/8	3 3/4
Preferred minimum	1 3/4	2	2 1/2	3	3 1/2	4	4 1/2
Edge to center of punched holes							
In sheared edge	1	1 1/8	1 1/4	1 1/2	1 3/4	2	2 1/4
In rolled edge, parallel flanges	7/8	1	1 1/8	1 1/4	1 1/2	1 3/4	2
In rolled edge, sloping flanges *	3/4	7/8	1	1 1/8	1 1/4	1 1/2	1 3/4

* May be decreased 1/8 in. when holes are near end of beam.

Maximum pitch in the line of stress of compression members composed of plates and shapes should not exceed sixteen times the thickness of the thinnest outside plate or shape, or twenty times the thinnest enclosed plate or shape, with a maximum of 12 in. Distance between lines of rivets at right angles to the direction of stress should not exceed thirty times the thickness of the thinnest shape or plate. For angles in built sections with two gage lines, with staggered rivets, the maximum pitch in the line of stress in each gage line should not exceed twenty-four times the thickness of the thinnest plate, with a maximum of 18 in.

Pitch of rivets at the ends of built compression members should not exceed four times the rivet diameter for a length equal to one and one-half times the maximum width of the member. In tension members composed of two angles, a pitch of 42 in. is permissible and

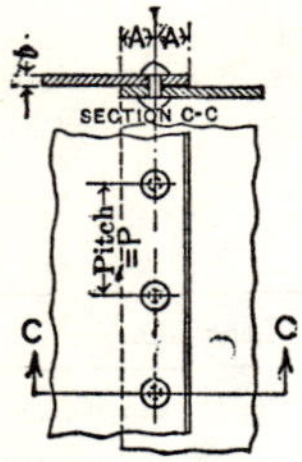

FIG. 6. Single-riveted lap joint. See Table 5.

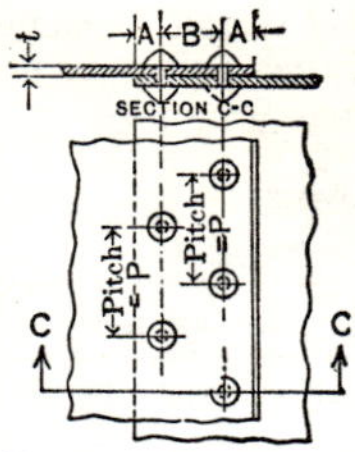

FIG. 7. Double-riveted lap joint. See Table 5.

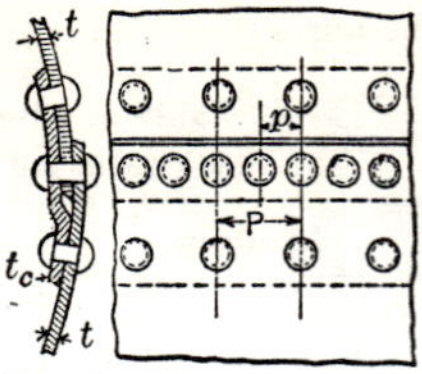

FIG. 8. Single-riveted lap joint with inside cover plate.

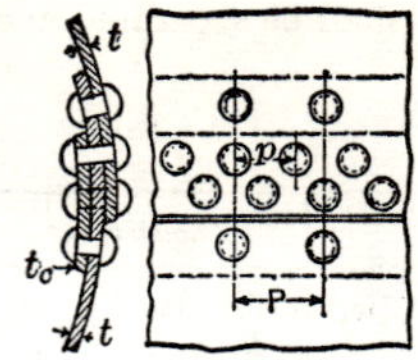

FIG. 9. Double-riveted lap joint with inside cover plate.

Table 5. Dimensions and Efficiency of Double- and Single-riveted Lap Joints for Longitudinal Seams

(See Figs. 6 and 7)

Thickness of Plate, in.	Diam. of Rivet Holes, in.	Single-riveted Joints				Double-riveted Joints *			
		Efficiency, %	Pitch, in.	A, in.	Method of Failure †	Efficiency, %	Pitch, in.	A, in.	B, in.
1/4	11/16	60.7	1 3/4	1 1/16	T.P.	69.5	2 1/4	1 1/16	1 3/4
5/16	13/16	59.4	2	1 1/4	T.P.	69.1	2 5/8	1 1/4	1 7/8
3/8	15/16	58.3	2 1/4	1 7/16	T.P.	68.9	3	1 7/16	2
7/16	1 1/16	57.5	2 1/2	1 5/8	T.P.	68.5	3 3/8	1 5/8	2 1/8
1/2	1 1/16	56.7	2 1/2	1 5/8	S.R.	68.5	3 3/8	1 5/8	2 1/8

* All joints fail by tearing net section of plate between rivet holes.

† T.P. = failure by tearing plate in net section between rivet holes; S.R. = failure by shearing of rivets.

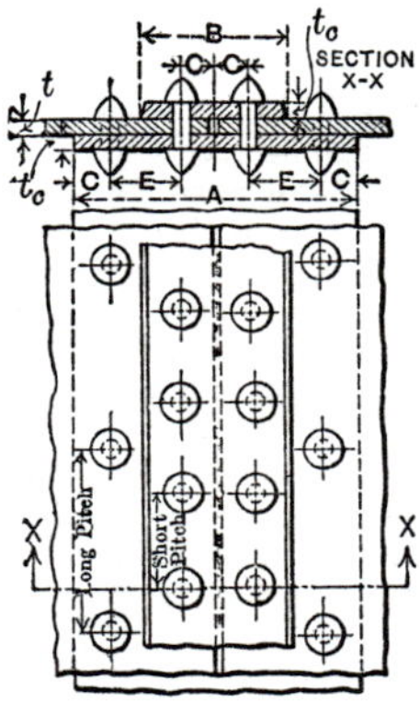

FIG. 10. Double-riveted butt joint. See Table 6.

Table 6. Dimensions and Efficiency of Double-riveted Butt Joints

(See Fig. 10)

All joints fail by tearing the plate between rivet holes in the outer row.

Thickness of Plate, in.	Thickness of Straps, in.	Diam. of Holes, in.	Efficiency, %	Long Pitch, in.	Short Pitch, in.	A, in.	B, in.	C, in.	D, in.
1/4	1/4	11/16	82.8	4	2	8 1/2	4 1/4	1 1/16	2 1/8
5/16	9/32	13/16	81.9	4 1/2	2 1/4	9 7/8	5	1 1/4	2 7/16
3/8	5/16	13/16	81.9	4 1/2	2 1/4	9 7/8	5	1 1/4	2 7/16
7/16	3/8	15/16	81.3	5	2 1/2	11 1/4	5 3/4	1 7/16	2 3/4
1/2	7/16	15/16	81.3	5	2 1/2	11 1/4	5 3/4	1 7/16	2 3/4

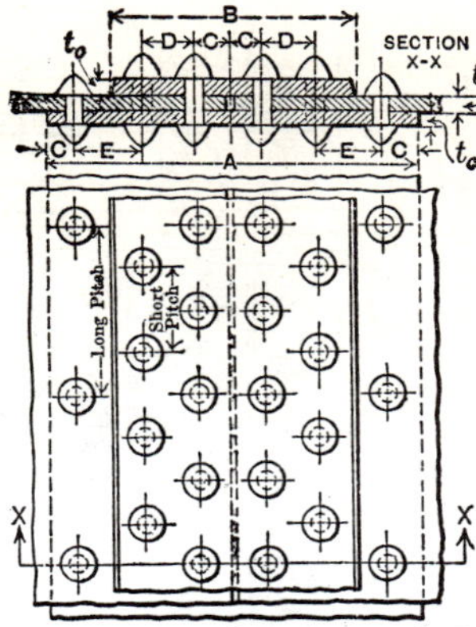

FIG. 11. Triple-riveted butt joint. See Table 7.

Table 7. Dimensions and Efficiency of Triple-riveted Butt Joints

(See Fig. 11)

Joints for plate thicknesses of 3/4 in., and 7/8 to 1 in. (both inclusive) fail by tearing the plate between rivet holes in the second row and shearing a rivet in the outer row. All other joints fail by tearing the plate between rivet holes in the outer row.

Thickness of Plate, in.	Thickness of Straps, in.	Diam. of Holes, in.	Efficiency, %	Long Pitch, in.	Short Pitch, in.	*A*, in.	*B*, in.	*C*, in.	*D*, in.	*E*, in.
1/4	1/4	11/16	87.5	5 1/2	2 3/4	12	7 3/4	1 1/16	1 3/4	2 1/8
5/16	9/32	13/16	87.5	6 1/2	3 1/4	13 5/8	8 3/4	1 1/4	1 7/8	2 7/16
3/8	5/16	13/16	88.4	7	3 1/2	13 5/8	8 3/4	1 1/4	1 7/8	2 7/16
7/16	3/8	15/16	87.9	7 3/4	3 7/8	15 1/4	9 3/4	1 7/16	2	2 3/4
1/2	7/16	15/16	88.3	8	4	15 1/4	9 3/4	1 7/16	2	2 3/4
9/16	7/16	1 1/16	86.7	8	4	17	11	1 5/8	2 1/4	3
5/8	1/2	1 1/16	86.7	8	4	17	11	1 5/8	2 1/4	3
11/16	1/2	1 3/16	85.6	8 1/4	4 1/8	18 1/2	12	1 13/16	2 3/8	3 1/4
3/4	1/2	1 3/16	85.5	8 1/4	4 1/8	18 1/2	12	1 13/16	2 3/8	3 1/4
13/16	9/16	1 5/16	84.6	8 1/2	4 1/4	20 1/4	13 1/4	2	2 5/8	3 1/2
7/8	5/8	1 5/16	84.1	8 3/4	4 3/8	20 1/4	13 1/4	2	2 5/8	3 1/2
15/16	11/16	1 5/16	83.7	9	4 1/2	20 1/4	13 1/4	2	2 5/8	3 1/2
1	3/4	1 7/16	83.4	9 1/2	4 3/4	22	14 1/2	2 3/16	2 7/8	3 3/4

in compression members 24 in.; but the ratio l/r for each angle between the rivets should be not more than three-fourths of that for the whole member. Maximum distance from the center of any rivet to the near edge should be twelve times the thickness of the plate but should not exceed 6 in.

STRENGTH AND EFFICIENCY OF RIVETED JOINTS. In calculating the strength of riveted joints the stress conditions which would exist at the yield point are assumed, that is, plastic deformation permits each rivet shearing area to take an equal share of the total load.

Shearing Strength. Total resistance to shearing of the rivets is equal to the number of shearing areas times the cross-sectional area of each rivet, $\pi d^2/4$, times the shearing stress, S_s. ASME boiler code values of S_s are 44,000 psi for steel rivets and 38,000 psi for iron rivets. These are ultimate values; with a factor of safety of 5 the working stresses are 8800 psi and 7600 psi, respectively. For structural rivets the working stresses in shear are 10,000 psi for hand-driven rivets and 13,500 psi for power-driven rivets.

For aluminum rivets, ultimate shearing strengths are: 17S-T, 33,000 psi; 24S-T, 42,000 psi (cold driven); 53S-T, 23,000 psi (cold driven); 53S-W, 18,000 to 24,000 psi (hot driven), depending on temperature. Cold-driven rivets are assumed to have had proper quenching treatment immediately prior to driving.

Crushing Strength. Joint failure may result from crushing between the rivet and the plate. Resistance to crushing at one rivet is equal to the projected bearing area, dt, times the crushing stress, S_c, where t is the least thickness of plate or plates through which the load is transferred to the rivet. ASME boiler code prescribes S_c = 95,000 psi for steel

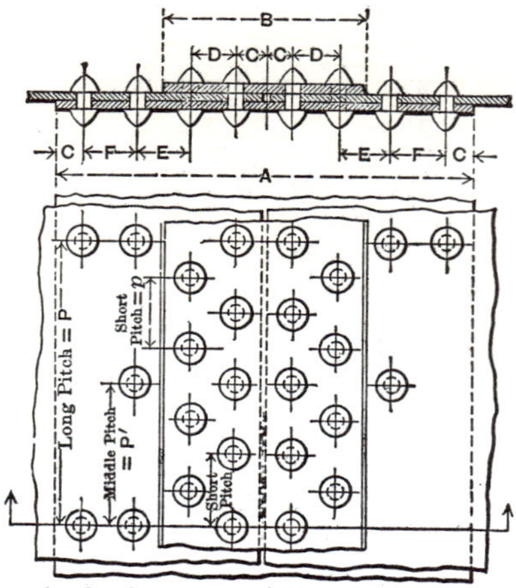

FIG. 12. Quadruple-riveted butt joint. See Table 8.

Table 8. Dimensions and Efficiency of Quadruple-riveted Butt Joints

(See Fig. 12)

Joints for 3/4-in. and 7/8-in. to 1 1/4-in. plates (both inclusive) fail by tearing the plate between rivet holes in the third row and shearing the rivets in the two outer rows. All other joints fail by tearing the plate between rivet holes in the outer row.

Thickness of Plate, in.	Thickness of Straps, in.	Diam. of Holes, in.	Efficiency, %	Long Pitch, in.	Middle Pitch, in.	Short Pitch, in.	*A*, in.	*B*, in.	*C*, in.	*D*, in.	*E*, in.	*F*, in.
1/4	1/4	11/16	93.8	11	5 1/2	2 3/4	16 1/2	7 3/4	1 1/16	1 3/4	2 1/8	2 1/4
5/16	9/32	13/16	93.8	13	6 1/2	3 1/4	18 7/8	8 3/4	1 1/4	1 7/8	2 7/16	2 5/8
3/8	5/16	13/16	94.2	14	7	3 1/2	19 1/8	8 3/4	1 1/4	1 7/8	2 7/16	2 3/4
7/16	3/8	15/16	94.0	15 1/2	7 3/4	3 7/8	21 3/8	9 3/4	1 7/16	2	2 3/4	3 1/16
1/2	7/16	15/16	94.1	16	8	4	21 1/2	9 3/4	1 7/16	2	2 3/4	3 1/8
9/16	7/16	1 1/16	93.4	16	8	4	23 5/8	11	1 5/8	2 1/4	3	3 5/16
5/8	1/2	1 1/16	93.4	16	8	4	23 5/8	11	1 5/8	2 1/4	3	3 5/16
11/16	1/2	1 3/16	92.8	16 1/2	8 1/4	4 1/8	25 5/8	12	1 13/16	2 3/8	3 1/4	3 9/16
3/4	1/2	1 3/16	92.7	16 1/2	8 1/4	4 1/8	25 5/8	12	1 13/16	2 3/8	3 1/4	3 9/16
13/16	9/16	1 5/16	92.3	17	8 1/2	4 1/4	27 7/8	13 1/4	2	2 5/8	3 1/2	3 13/16
7/8	5/8	1 5/16	91.2	17 1/2	8 3/4	4 3/8	28	13 1/4	2	2 5/8	3 1/2	3 7/8
15/16	11/16	1 5/16	90.1	18	9	4 1/2	28 1/8	13 1/4	2	2 5/8	3 1/2	3 15/16
1	3/4	1 7/16	90.2	19	9 1/2	4 3/4	30 1/2	14 1/2	2 3/16	2 7/8	3 3/4	4 1/4
1 1/16	3/4	1 7/16	89.0	19	9 1/2	4 3/4	30 1/2	14 1/2	2 3/16	2 7/8	3 3/4	4 1/4
1 1/8	3/4	1 7/16	88.0	19	9 1/2	4 3/4	30 1/2	14 1/2	2 3/16	2 7/8	3 3/4	4 1/4
1 3/16	13/16	1 7/16	87.7	20	10	5	30 5/8	14 1/2	2 3/16	2 7/8	3 3/4	4 5/16
1 1/4	7/8	1 7/16	86.8	20	10	5	30 5/8	14 1/2	2 3/16	2 7/8	3 3/4	4 5/16

Table 9. Dimensions and Efficiency of Lap-riveted Girth Joints

(For cylindrical vessels with solid unstayed heads)

Thickness of Plate, in.	Diam. of Rivet Holes, in.	Single-riveted Joint		Double-riveted Joint		Thickness of Plate, in.	Diam. of Rivet Holes, in.	Double-riveted Joint	
		Pitch of Rivets, in.	Efficiency, %	Pitch of Rivets, in.	Efficiency, %			Pitch of Rivets, in.	Efficiency, %
1/4	11/16	1 3/4	60.7	....	....	13/16	1 5/16	4	66.6
5/16	13/16	1 7/8	56.7	....	....	7/8	1 5/16	4	61.9
3/8	13/16	1 7/8	56.7	....	....	15/16	1 5/16	4	57.7
7/16	15/16	2 1/8	55.9	....	....	1	1 7/16	4 1/2	57.7
1/2	15/16	2 1/8	52.0	....	....	1 1/16	1 7/16	4 1/2	54.3
9/16	1 1/16	2 3/8	53.3	3 1/4	67.3	1 1/8	1 7/16	4 1/2	51.3
5/8	1 1/16	....	....	3 1/4	67.3	1 3/16	1 7/16	4 1/2	48.6
11/16	1 3/16	....	....	3 3/4	68.3	1 1/4	1 7/16	4 1/2	46.2
3/4	1 3/16	....	....	3 3/4	63.0	....	....	....	....

Table 10. Lap-riveted Joints for Girth Seams of Horizontal Tubular Boilers

	Thickness of Plate, in.					
	1/4	5/16	3/8	7/16	1/2	9/16
Diameter of holes, in.	11/16	13/16	13/16	15/16	15/16	1 1/16
Pitch of rivets, in.	1 3/4	1 7/8	1 7/8	2 1/8	2 1/8	2 3/8
Efficiency of joint, %	60.7	56.7	56.7	55.9	52.0	53.3
Shearing strength of rivets, % of solid plate	67.9	70.8	59.1	59.4	52.0	53.3

plates, corresponding to 19,000 psi working stress with safety factor 5. Allowable crushing stresses in structural joints are 40,000 psi for rivets in double shear, 32,000 psi for rivets in single shear. With aluminum alloys the crushing or bearing strengths are dependent on the plate or shape material, and its heat treatment. Values range from 34,000 psi ultimate in soft material to 129,000 psi ultimate in hard material (see *Alcoa Structural Handbook*).

Tearing Strength. A joint may fail by tearing of the plate between the rivet holes. Resistance to tearing at any section is equal to the net area remaining after the holes have been made times the tensile strength of the plate, S_t. ASME code value for S_t is 55,000 psi ultimate, or 11,000 psi allowable with factor of safety 5. For structural steel the allowable is 20,000 psi.

Joint Analysis. Actual strength of a riveted joint is found by applying the foregoing principles. Total load required to cause failure in each mode is calculated. Lowest of these values, expressed as a percentage of the load to cause failure of the solid plate, is called the efficiency.

Double and triple riveted joints may fail by a combination of shearing, crushing, and tearing. Tables 5 to 10 show the dimensions and efficiencies of standard boiler joints, as published in the *Boiler Book* of the Hartford Steam Boiler Inspection and Insurance Company. The dimensions D, E, and F in these tables may be increased for convenience in riveting, without affecting the efficiency of the joint. They should not be decreased.

Notation. a = diameter of rivet hole, usually $1/16$ in. greater than rivet, t = thickness of plate, t_c = thickness of cover plates, p = pitch of inner row of rivets, P = pitch of outer row of rivets, all in inches; and let s_s = shearing strength of rivets; s_t = tensile strength of plate, s_c = crushing strength of rivet or plate, all in pounds per square inch. Then the resistance for the various joints are as follows.

Single-riveted Lap Joint (Fig. 6). Resistance to shearing one rivet = $(1/4\pi d^2)s_s$.

Resistance to tearing plate between rivets = $(P - d)ts_t$.
Resistance to crushing of rivet or plate = dts_c.

Double-riveted Lap Joint (Fig. 7). Resistance to shearing two rivets = $2(1/4\pi d^2)s_s$.

Resistance to tearing between two rivets = $(P - d)ts_t$.
Resistance to crushing in front of two rivets = $2dts_c$.
Strength of solid plate = Pts_t.

Single-riveted Lap Joint, Inside Cover Plate (Fig. 8). Resistance to tearing between outer row of rivets = $(P - d)ts_t$.

Resistance to tearing between inner row of rivets, and shearing outer row of rivets = $(P - 2d)ts_t + (1/4\pi d^2)s_s$.
Resistance to shearing three rivets = $3(1/4\pi d^2)s_s$.
Resistance to crushing in front of three rivets = $3tds_c$.
Resistance to tearing at inner row of rivets and crushing in front of one rivet in outer row = $(P - 2d)ts_t + tds_c$.
Strength of solid plate = Pts_t.

Double-riveted Lap Joint with Inside Cover Plate (Fig. 9). Resistance to tearing at outer row of rivets = $(P - d)ts_t$.

Resistance to shearing four rivets = $4(1/4\pi d^2)s_s$.
Resistance to tearing at inner row and shearing outer row of rivets = $(P - 1\,1/2d)ts_t + (1/4\pi d^2)s_s$.
Resistance to crushing in front of four rivets = $4tds_c$.
Resistance to tearing at inner row of rivets, and crushing in front of one rivet = $(P - 1\,1/2d)ts_t + tds_c$.
Strength of solid plate = Pts_t.

Double-riveted Butt Joint (Fig. 10). Resistance to tearing at outer row of rivets = $(P - d)ts_t$.

Resistance to shearing two rivets in double shear and one in single shear = $5(^1/_4\pi d^2)s_s$.

Resistance to tearing at inner row of rivets and shearing one of the outer row of rivets = $(P - 2d)ts_t + (^1/_4\pi d^2)s_s$.

Resistance to crushing in front of three rivets = $3tds_c$.

Crushing in front of two rivets and shearing one rivet = $2tds_c + (^1/_4\pi d^2)s_s$.

Strength of solid plate = Pts_t.

Triple-riveted Butt Joint (Fig. 11). Resistance to tearing at outer row of rivets = $(P - d)ts_t$.

Resistance to shearing four rivets in double shear and one in single shear = $9(^1/_4\pi d^2)s_s$.

Resistance to tearing at middle row of rivets and shearing one rivet = $(P - 2d)ts_t + (^1/_4\pi d^2)s_s$.

Resistance to crushing in front of four rivets and shearing one rivet = $4dts_c + (^1/_4\pi d^2)s_s$.

Resistance to crushing in front of five rivets = $5dts_c$.

Strength of solid plate = Pts_t.

Quadruple-riveted Butt Joint with Two Cover Plates (Fig. 12). Resistance to tearing at outer row of rivets = $(P - d)ts_t$.

Resistance to tearing at the third row and shearing one rivet in the outer row = $(P - 2d)ts_t + (^1/_4\pi d^2)s_s$.

Resistance to tearing at the second row of rivets and shearing three rivets = $(P - 4d)ts_t + 3(^1/_4\pi d^2)s_s$.

Resistance to double-shearing eight rivets and single-shearing three rivets = $19(^1/_4\pi d^2)s_s$.

Resistance to crushing in front of eight rivets and single-shearing three rivets = $8dts_c + 3(^1/_4\pi d^2)s_s$.

Resistance to crushing in front of eleven rivets = $11dts_c$.

Resistance to tearing at the third row of rivets and crushing in front of one rivet in the outer row = $(P - 2d)ts_t + dts_c$.

Resistance to tearing at the second row of rivets and crushing in front of three rivets = $(P - 4d)ts_t + 3dts_c$.

Strength of the solid plate = Pts_t.

Lap-riveted Girth Joints. (*Boiler Book* of Hartford Steam Boiler Inspection & Insurance Co.). In horizontal tubular boilers, tanks, and similar vessels, it is customary to use rivets of the same size in both girth and longitudinal seams. If the heads are not stayed by tubes or through braces, the strength of the girth joints should be at least 50% of that required for the longitudinal joints. The joints in Table 9 are designed to meet these requirements; a higher efficiency could be obtained in some cases by using a rivet of different size. It is assumed in each case that the efficiency of the corresponding longitudinal joint is not greater than that of the standard design of quadruple-riveted butt joints given in Table 8.

Where 50% or more of the load which would act on an unstayed head of the same diameter as the shell is relieved by the effect of tubes or through stays, as in tubular boilers, the strength of the girth joints should be at least 35% of that required in longitudinal joints. When the girth joints are exposed to products of combustion, the shearing strength of the rivets should be not less than 50% of the full strength of the plate corresponding to the thickness at the joint. Table 10 is designed for such conditions, rivets of the same size as standard butt joints being used. Plates thicker than $^9/_{16}$ in. are not advised if the girth seams are exposed to fire or products of combustion. In those portions of the seams exposed to fire, the riveting should be started at the bottom of the boiler and carried up continuously on both sides, reaming the holes anew as the riveting is carried up to the full diameter of the boiler.

Precautions to Be Observed in Making Riveted Joints. Plates must be brought into close contact before the rivet is driven, either by bolts close to the rivet hole, firmly drawn up, or by a power-actuated holding device. The latter sometimes is incorporated in power-driven riveting machines.

Rivet holes must be perfectly aligned in the members that are to be joined. This is best insured by drilling of the members. Punched holes, usually used in structural work, are difficult of alignment, and have a stress set up around the hole which seriously impairs the strength of the plate unless it is subsequently relieved by reaming, drilling, or annealing. Punched plates or structural members in which the holes do not align never should be brought into alignment by drifting with taper pins. This practice usually causes serious injury to the plates or structural members. The best practice with punched members is to punch the holes smaller than required, and, after assembly, ream or drill them to the correct size.

The rivet must completely fill the hole, and consequently must be upset throughout its entire length. Otherwise, an undue concentrated stress, either in bearing or in shearing, may be brought upon adjoining rivets. To insure complete upsetting, all the larger sizes of rivets should be power driven. Care must be taken in machine riveting to avoid applying a riveting pressure that will set up stresses in the plate tending to crush it. It is essential in machine riveting that the pressure be so applied that the heads are concentric. Pressure should be maintained on the rivet until it has cooled sufficiently to hold the plate firmly.

Aluminum Rivets. Squeeze-type riveters should be used on aluminum rivets where possible. Pneumatic hammers and back-up tools should be heavier than for steel rivets of the same size. The flat-cone-type driven head shown in Fig. 13 facilitates driving and has a good appearance.

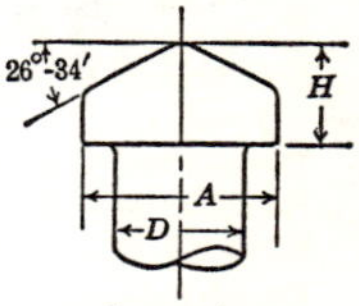

FIG. 13. Preferred form of driven head for aluminum rivets.

Cold driving is used for 53S-T61 and A17S-T alloy rivets as received from the manufacturer. The flat-cone head (Fig. 13) is readily driven with a pneumatic hammer in sizes up to $^5/_8$ in. in diameter. Larger rivets should be squeeze driven. Cold driving is also used for 17S and 24S alloy rivets where proper heat-treating equipment is available. Such rivets should be heat treated, quenched, and driven immediately to take advantage of the workability of the metal before age hardening occurs. This "as quenched" workability can be retained for 8 to 10 hr by refrigerating at 32 F or less and storing at this temperature until use. Such cold driving normally is restricted to sizes less than $^1/_2$ in. diameter.

Hot driving is often desirable for rivets over $^1/_2$ in. in diameter because the driving pressures are only one-third to one-fifth as great as for corresponding cold rivets. For hot driving, 53S rivets should be brought to 1030 to 1050 F in a furnace having accurate temperature control. Rivets must be transferred to the work as quickly as possible and driven at once, quenching being obtained by contact with cold metal and tools. Rivets of 17S alloy can be hot driven, procedure being similar, but with a heating range of 930 to 950 F.

FORM OF RIVETED JOINTS FOR MORE THAN TWO PLATES. In tanks, boilers, and all pressure vessels requiring tight seams, the problem often is to join three or four plates. Figure 14 shows method of making such joints.

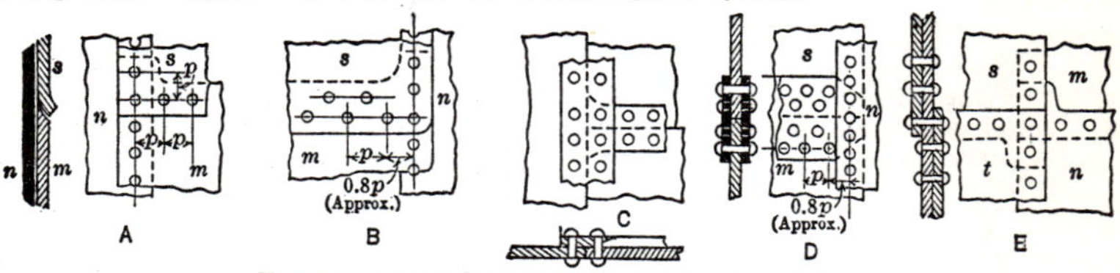

FIG. 14. Riveted joints for more than two plates.

Figure 14*A* is a junction of three plates, *m*, *n*, and *s*, where both seams are single-riveted lap joints. The corner of *m* is drawn to an edge and tucked under *s*.

Figure 14*B* is a junction of three plates; one seam is single riveted and the other is a double-riveted lap joint. As before, the corner of *m* is drawn down and tucked under *s*.

Figure 14*C* is a junction of three plates with butt straps; the longitudinal butt strap is planed down and tucked under the cross butt strap.

Figure 14*D* is the junction of three plates; the cross seam is a single-riveted lap joint, and the longitudinal joint is a double-riveted butt joint with double cover plates. The upper cover plate is planed on the end so that it can be tightly calked where it abuts against plate *n*.

Figure 14*E* is a method of joining four plates; both seams are single-riveted lap joints, *n* and *s*, both being drawn down as shown.

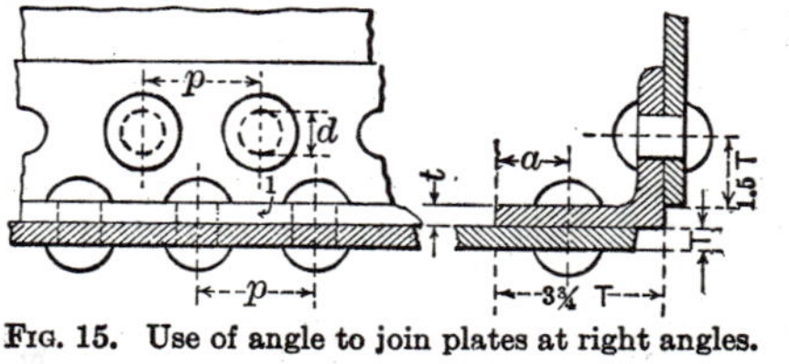

FIG. 15. Use of angle to join plates at right angles.

Junction of plates not in the same plane usually is effected by means of a rolled structural form. Figure 15 shows an angle used with plates at right angles to each other. If

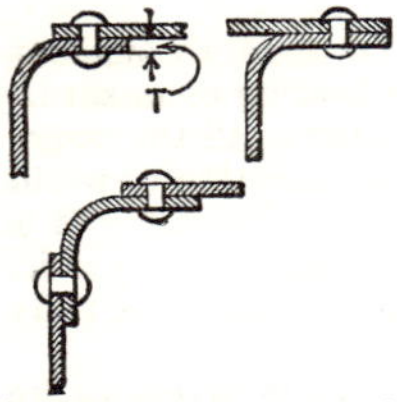

Fig. 16. Use of flanged plates for joining plates in different planes.

T = thickness of plate and t = mean thickness of angle leg, then $t = T + 1/16$ in. The width w of each leg of the angle = $3.75T$; the diameter d of the rivet = $2T$ and $a = 1/2(w - t)$, where w = width of angle.

Figure 16 shows methods of connecting plates by flanging the plates themselves instead of using angles. Care should be taken not to use too sharp a radius of curvature, and the flange should be annealed after it is bent. The inside radius of the flange must be greater than the thickness of the plate even with the mildest steel. Unwin recommends an inside radius of at least $4T$.

FASTENERS FOR WOOD JOINTS

8. NAILS AND SCREWS

NAILS. Table 1 lists the sizes, lengths, and numbers to the pound of the commoner standard steel wire nails. Decimal sizes are listed in Table 2.

Holding Power of Nails. Withdrawal resistance tests at the U. S. Forest Products Laboratory indicate that holding power of wire nails and spikes depends on density of the wood, diameter of the nail, and depth of penetration. By applying a factor of safety of 6, the safe load may be expressed as

$$P = 1150G^{5/2}D \tag{1}$$

where P is the safe load per lineal inch of penetration into the side grain of seasoned wood, pounds; G is the specific gravity of the wood; and D is the diameter of the nail, inches. Table 3 lists values of specific gravity, and also $G^{5/2}$, for various wood species. Tables 1 and 2 list values of D for various standard wire nail sizes.

Table 1. Sizes, Dimensions, and Number to the Pound of Steel Wire Nails
(Data courtesy of American Steel and Wire Co., Cleveland, Ohio)

Size	Length, in.	Common Nails (Flat head)			Finishing Nails (Brad head *)			Casing Nails (Deep countersunk 32° head *)			Smooth Box Nails (Large flat head)		
		Shank Diameter, gage †	Head Diameter, in.	Approx. No. per lb	Shank Diameter, gage †	Head Diameter, gage †	Approx. No. per lb	Shank Diameter, gage †	Head Diameter, gage †	Approx. No. per lb	Shank Diameter, gage †	Head Diameter, in.	Approx. No. per lb
2d	1	15	11/64	847									
3d	1 1/4	14	13/64	543	15 1/2	12 1/2	880				14 1/2	7/32	588
4d	1 1/2	12 1/2	1/4	294	15	12	630	14	11	489	14	7/32	453
5d	1 3/4	12 1/2	1/4	254							14	7/32	389
6d	2	11 1/2	17/64	167	13	10	288	12 1/2	9 1/2	244	12 1/2	17/64	225
7d	2 1/4	11 1/2	17/64	150							12 1/2	17/64	200
8d	2 1/2	10 1/4	9/32	101	12 1/2	9 1/2	196	11 1/2	8 1/2	147	11 1/2	19/64	136
9d	2 3/4	10 1/4	9/32	92									
10d	3	9	5/16	66	11 1/2	8 1/2	124	10 1/2	7 1/2	96	10 1/2	5/16	90
12d	3 1/4	9	5/16	61									
16d	3 1/2	8	11/32	47				10		73	10	11/32	69
20d	4	6	13/32	29							9	3/8	50
30d	4 1/2	5	7/16	22									
40d	5	4	15/32	17									
50d	5 1/2	3	1/2	13									
60d	6	2	17/32	10									

All nails listed in this table have diamond points.

* Finishing nails and casing nails can be furnished with cup head style to accommodate nail set.

† Wire sizes are expressed in terms of Steel Wire Gage (formerly known as Washburn and Moen gage and American Steel and Wire Co. gage). See Table 2.

Table 2. Decimal Equivalents of Gage Sizes for Steel Wire Nails

Gage	In.	Gage	In.	Gage	In.	Gage	In.	Gage	In.
2	.2625	6	.1920	10	.1350	11 1/2	.1130	14	.0800
3	.2437	7	.1770	10 1/4	.1314	12	.0985	14 1/2	.0760
4	.2253	8	.1620	10 1/2	.1278	12 1/2	.1035	15	.0720
5	.2070	9	.1483	11	.1205	13	.0915	15 1/2	.0673

Table 3. Factors for Calculating Strength of Nails, Wood Screws, and Lag Screws

Species	Specific Gravity, G,* and Powers				Safe Lateral Load Constant			Basic Stresses for Bolts, psi	
	G	$G^{3/2}$	G^2	$G^{5/2}$	Wire Nails, K_1	Wood Screws, K_2	Lag Bolts, K_3	Parallel with Grain	Perpendicular to Grain
Hardwoods									
Alder, red	.43	.28	.18	.12	1250	2900		1200	250
Ash, black	.53	.39	.28	.20	1250	2900	1900	925	175
Ash, white	.64	.51	.41	.33	1700	4000	2200	1500	400
Aspen	.40	.25	.16	.10	900	2100	1700	925	175
Basswood	.40	.25	.16	.10	900	2100	1700	925	175
Birch, yellow	.66	.54	.44	.35	1700	4000	2200	1500	400
Butternut	.40	.25	.16	.10	900	2100		925	175
Cherry, black	.53	.39	.28	.20	1700	4000		1500	400
Chestnut	.45	.30	.20	.14	900	2100	1700	925	175
Cottonwood, eastern	.43	.28	.18	.12	900	2100	1700	925	175
Elm, American	.55	.41	.30	.22	1250	2900	1900	1200	250
Elm, rock	.66	.54	.44	.35	1700	4000	2200	1500	400
Gum, black	.55	.41	.30	.22	1250	2900	1900	1200	250
Gum, red	.53	.39	.28	.20	1250	2900	1900	1200	250
Gum, tupelo	.52	.37	.27	.19	1250	2900	1900	1200	250
Hackberry	.56	.42	.31	.23	1250	2900		1200	250
Hickory	.72	.61	.52	.44	1700	4000	2200	1500	400
Locust, black	.71	.60	.50	.42	1700	4000		1500	400
Magnolia, cucumber	.52	.37	.27	.19	1250	2900		1200	250
Magnolia, evergreen	.53	.39	.28	.20	1250	2900		1200	250
Maple, hard	.68	.56	.46	.38	1700	4000	2200	1500	400
Maple, soft	.51	.36	.26	.19	1250	2900	1900	1200	250
Oak	.68	.56	.46	.38	1700	4000	2200	1500	400
Poplar, yellow	.43	.28	.18	.12	900	2100	1700	925	175
Sycamore	.54	.40	.29	.21	1250	2900	1900	1200	250
Walnut, black	.56	.42	.31	.23	1700	4000		1500	400
Conifers									
Cedar, northern white	.32	.18	.10	.06	900	2100	1500	800	150
Cedar, Port Orford	.44	.29	.19	.13	1125	2700	1700	1000	200
Cedar, western red	.34	.20	.12	.07	1125	2700	1700	1000	200
Cypress, southern	.48	.33	.23	.16	1125	2700	1700	1300	275
Douglas fir (coast)	.51	.36	.26	.19	1375	3300	1900	1300	275
Fir, commercial white	.42	.27	.18	.11	900	2100	1500	800	150
Hemlock, eastern	.43	.28	.18	.12	900	2100	1500	800	150
Hemlock, western	.44	.29	.19	.13	1125	2700	1700	1000	200
Larch, western	.59	.45	.35	.27	1375	3300	1900	1300	275
Pine, northern white	.37	.22	.14	.08	900	2100	1500	800	150
Pine, ponderosa	.42	.27	.18	.11	900	2100	1500	800	150
Pine, southern yellow	.58	.44	.34	.26	1375	3300	1900	1300	275
Redwood	.39	.25	.15	.09	1125	2700	1700	1300	275
Spruce	.39	.25	.15	.09	900	2100	1500	800	150
Tamarack	.56	.42	.31	.23	1125	2700	1700	1300	275

* Based on oven-dry weight and volume.

Nails driven parallel to the grain show a withdrawal resistance only 75% or even 50% of that for nails driven into the side grain. Nails driven into green wood have higher resistance while the wood is green, but will lose a large part of their holding power as the wood seasons. Some wood species indicate lower or higher resistance than shown by eq. 1, and coatings may affect holding power either favorably or adversely. See Ref. 1 for further data. In general, a long sharp point gives higher holding power than a common diamond point, particularly with softer woods. However, if the wood tends to split, this tendency will be increased by the sharp point.

Lateral Resistance of Nails and Wire Spikes. For nails driven at right angles to the grain, safe lateral resistance may be expressed by the equation

$$P = K_1 D^{3/2} \tag{2}$$

where P is the safe lateral load per nail, pounds; D is the nail diameter, inches; and K_1 is a constant which depends on the wood species (see Table 3). Equation 2 is based on a depth of penetration, into the block receiving the point, of not less than ten times (for dense woods) to fourteen times (for soft woods) the nail diameter. If the nails are holding metal to wood, the safe lateral load can be increased about 25%. When nails are driven parallel to the grain, the loads should be reduced to 60% of the values given by eq. 2.

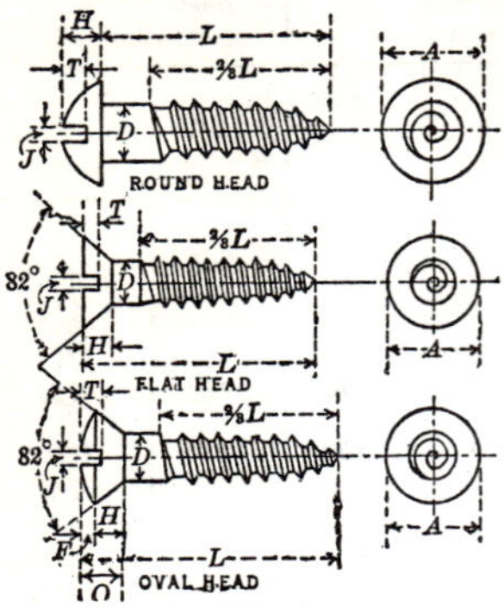

FIG. 1. Wood screws.

Resistance of Drift Bolts. Resistance to withdrawal of round drift bolts driven perpendicular to the grain into holes having 1/8 in. less diameter than the bolt is

$$P = 6000G^2D \qquad (3)$$

where P is the maximum withdrawal load per lineal inch of penetration, pounds; G is the specific gravity of the wood species (see Table 3 for G and G^2); and D is the drift bolt diameter, inches.

WOOD SCREWS are standardized in three styles shown in Fig. 1, with either style of cross recess (Fig. 16, p. 10-40) available as alternatives to the screwdriver slot. Dimensions are given in Table 4.

Table 4. Dimensions of American Standard Wood Screws

(ASA B18.6-1947. All dimensions in inches. See Fig. 1.)

Nominal Size	Diameter, D	Threads per Inch	Round Head: Head Diameter, max, A	Round Head: Height of Head, max, H	Round Head: Width of Slot, min, J	Round Head: Depth of Slot, min, T	Flat and Oval Heads: Head Diameter, max, A	Flat and Oval Heads: Height of Head, max, H	Flat and Oval Heads: Width of Slot, min, J	Depth of Slot T: Flat Head, min	Depth of Slot T: Oval Head, min	Total Height of Oval Head, max, O
0	.060	32	.113	.053	.016	.029	.119	.035	.016	.010	.025	.056
1	.073	28	.138	.061	.019	.033	.146	.043	.019	.012	.031	.068
2	.086	26	.162	.069	.023	.037	.172	.051	.023	.015	.037	.080
3	.099	24	.187	.078	.027	.040	.199	.059	.027	.017	.043	.092
4	.112	22	.211	.086	.031	.044	.225	.067	.031	.020	.049	.104
5	.125	20	.236	.095	.036	.047	.252	.075	.035	.022	.055	.116
6	.138	18	.260	.103	.039	.051	.279	.083	.039	.024	.060	.128
7	.151	16	.285	.111	.039	.055	.305	.091	.039	.027	.066	.140
8	.164	15	.309	.120	.045	.058	.332	.100	.045	.029	.072	.152
9	.177	14	.334	.128	.045	.062	.358	.108	.045	.032	.078	.164
10	.190	13	.359	.137	.050	.065	.385	.116	.050	.034	.084	176
12	.216	11	.408	.153	.056	.073	.438	.132	.056	.039	.096	.200
14	.242	10	.457	.170	.064	.080	.491	.148	.064	.044	.108	.224
16	.268	9	.506	.187	.064	.087	.544	.164	.064	.049	.120	.248
18	.294	8	.555	.204	.072	.094	.597	.180	.072	.054	.132	.272
20	.320	8	.604	.220	.072	.101	.650	.196	.072	.059	.144	.296
24	.372	7	.702	.254	.081	.116	.756	.228	.081	.069	.168	.344

Withdrawal Resistance of Wood Screws. Safe load for a common wood screw inserted into the side grain of seasoned wood is

$$P = 1700G^2D \qquad (4)$$

where P is the safe load (assuming safety factor 6), pounds; G is the specific gravity of the wood (see Table 3 for G and G^2); and D is the diameter of the screw (see Table 4). Equation 4 may be considered accurate to within plus or minus 10 to 15%, and is based on depth of penetration, into the block receiving the point, not less than two-thirds the length of the screw, the screws being inserted into holes having diameter about 70% (for softwoods) to 90% (for hardwoods) of the root diameter of the screw. If the screws are driven parallel to the grain and no splitting occurs, the safe loads are about 75% of those given by eq. 4. Equation values are applicable to the following screw lengths and gages:

Screw length, in.	1/2	3/4	1	1 1/2	2	2 1/2	3
Gage limits	1–6	2–11	3–12	5–14	7–16	9–18	12–20

For lengths and gages outside these limits, safe loads are likely to be less, especially in denser woods where screw failures rather than wood failures may occur.

Lateral Resistance of Wood Screws. Safe lateral load for a wood screw may be expressed by an equation of the form

$$P = K_2D^2 \quad (5)$$

where P is the safe lateral load per screw, pounds; D is the diameter of the shank (see Table 4), pounds; and K_2 is a constant which depends on the wood species (see Table 3). Equation 5 is based on a depth of penetration, into the block receiving the point, not less than seven times the screw diameter. If the screws are holding metal to wood, the safe lateral resistance may be increased 25% above the values given by eq. 5.

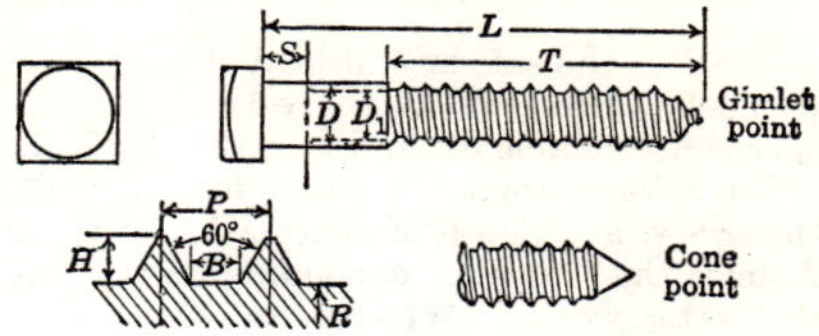

FIG. 2. Lag bolt. See Table 5.

LAG BOLTS are made with a cone point or a gimlet point, and a square head (Fig. 2). Leading dimensions are listed in Table 5.

Table 5. Dimensions of Lag Bolts

(Industrial Fasteners Institute. See Fig. 2. All dimensions in inches.)

Diameter of Bolt, D	Threads per Inch	Thread Dimensions				Length of Shoulder for Rolled Thread		
		Pitch, P	Flat at Root, B	Depth of Thread, H	Root Diameter, R	To 1 1/2 Long	1 5/8 1 3/4 Long	2 and Longer
						S		
No. 10	11	.091	.039	.035	0.120	3/32	3/16	3/16
1/4	10	.100	.043	.039	0.173	3/32	3/16	3/16
5/16	9	.111	.048	.043	0.227	1/8	1/4	1/4
3/8	7	.143	.062	.055	0.265	1/8	1/4	1/4
7/16	7	.143	.062	.055	0.328	5/32	1/4	5/16
1/2	6	.167	.072	.064	0.371	5/32	1/4	5/16
9/16	6	.167	.072	.064	0.435		5/16	3/8
5/8	5	.200	.086	.077	0.471		5/16	3/8
3/4	4 1/2	.222	.096	.085	0.579		3/8	1/2
7/8	4	.250	.108	.096	0.683		3/8	1/2
1	3 1/2	.286	.123	.110	0.780			5/8
1 1/8	3 1/4	.308	.133	.119	0.887			5/8
1 1/4	3 1/4	.308	.133	.119	1.012			5/8

Threads shall be rolled or cut. Diameter of rolled thread shank is not dimensioned, but is indicated by D_1 in Fig. 2. All rolled-thread lag bolts shall be provided with shoulder, S.

For minimum thread lengths, see below.

Dimensions of heads are the same as those for regular, unfinished, square-head machine bolts (see p. 10-33, Table 24).

Bolt length, L	1	1 1/2	2	2 1/2	3	4	5	6	7	8	9	10
Thread length, T	3/4	1 1/8	1 1/2	1 3/4 *	2	2 1/2	3	3 1/2	4	4 1/2	5	5 1/4

* For No. 10 and 1/4-in. sizes, T = 1 1/2; for 5/16 and 3/8-in. sizes, T = 1 5/8.

For bolt lengths greater than 10 in., thread is about one-half the length of the bolt.

Lag Bolt Joint Assembly. Prebored holes should be used except for small screws in soft wood. Lead hole for the shank should be the same diameter as the shank; lead hole for the threaded portion should be 40 to 70% of shank diameter in such species as northern white pine, 60 to 75% in Douglas fir and southern yellow pine, and 65 to 85% in white oak. The larger figures in each range apply to larger screws. Soap or other lubricant should be used on screws, and lead holes may be slightly larger than indicated if the lag bolts are of excessive length.

Withdrawal Resistance of Lag Bolts. Safe load for a lag bolt inserted into the side grain of the wood is given by

$$P = 1500D^{3/4}G^{3/2} \quad (6)$$

where P is safe load (based on safety factor 5) per inch of penetration, pounds; D is diameter of screw, inches (see Table 5 for usual sizes); and G is specific gravity of wood species (see Table 3 for values of G and $G^{3/2}$). For screws inserted into the end grain the safe load is about 75% of the value in eq. 6.

Lateral resistance of lag bolts may be expressed by an equation of the form

$$P = K_3D^2 \tag{7}$$

where P is the safe lateral load of a lag-screw joint of dry wood, with the bearing surfaces parallel to the grain of the wood, pounds; K_3 is a constant which depends on the wood species (see Table 3); and D is the shank diameter of the screw, inches (see Table 5).

Equation 7 applies to joints having cleat thicknesses 3 1/2 times the shank diameter of the screw, and depth of penetration of the screw in the main members varying from 11 to 7 times the diameter, depending on the wood species. The higher penetration applies to the softer woods. When metal cleats are used, the loads may be increased 25%. For further data on lag-screw joints, see Ref. 2.

9. TIMBER-CONNECTOR JOINTS

METHODS OF FAILURE. Wood-framed structures, especially roof trusses, towers, bridges, and trestles, tend to fail at bolted joints. Figure 3 shows the method of failure when the wood is stronger than the bolt. In Fig. 4, where the bolt is stronger than the wood, the joint fails by deforming and splitting the wood. Drawing the bolt up tight creates friction between the surfaces and retards slipping. Shrinkage of the wood decreases the friction.

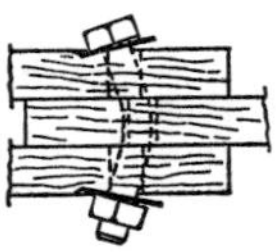

FIG. 3. Failure when wood is stronger than bolt.

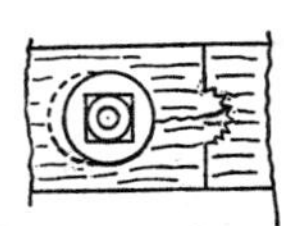

FIG. 4. Failure when bolt is stronger than wood.

BOLTS. Bearing stresses in the wood under a bolt are not uniform but are concentrated at the edges of the timbers, the effect becoming more pronounced as the ratio of length of bolt in the timber to its diameter increases, and as the ratio of wood strength to bolt strength increases. Basic stresses given in the last two columns of Table 3 may be used for L/D ratios up to 2 with any combination of bolt and wood species. With the softer wood species and higher strength bolts, these basic stresses can be used for higher L/D ratios. Beyond the limiting L/D ratios modifying factors must be used. For complete data, see Ref. 1.

TIMBER CONNECTORS. Modern connectors of many types and variations have been patented. Their purpose is to furnish a positive resistance to slip between the faces of the joint, like a dowel or key. The more important of these are illustrated in Figs. 5, 6, and 7. The split-ring connector (Fig. 5) may be straight sided or beveled and is fitted into precut grooves in the adjoining faces of the joint. The toothed or Alligator joint (Fig. 6) is forced into the timbers as the members are pressed or clamped together. Claw plates (Fig. 7) fit into prebored recesses and have short teeth which are forced further into the wood. They are used singly in making timber-to-metal connections, or in matched pairs (male and female) for timber-to-timber connections. Data on the strength of these connectors are found in Ref. 3.

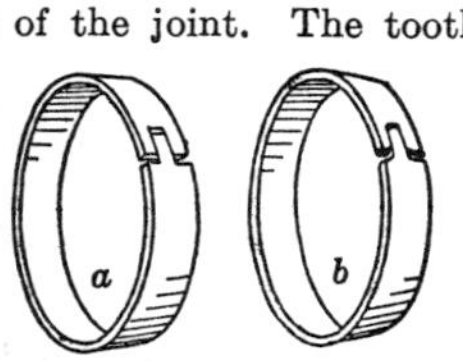

FIG. 5. Split-ring connector.

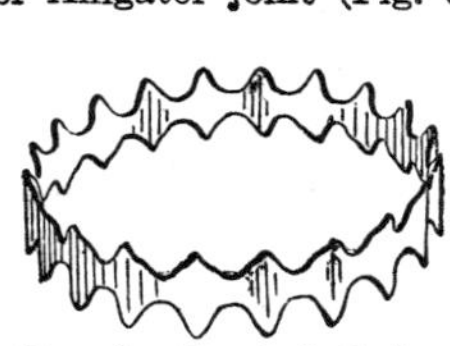

FIG. 6. Toothed timber connector.

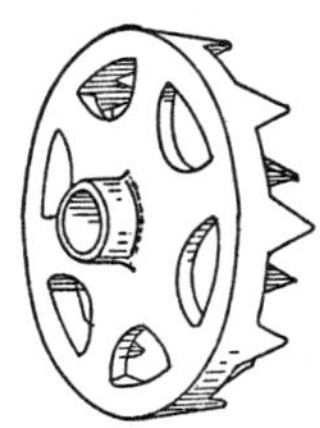

FIG. 7. Claw-plate timber connector.

REFERENCES

1. *Wood Handbook*, Forest Products Laboratory, U. S. Department of Agriculture, Washington, D. C.
2. J. A. Newlin and J. M. Gahagan, *Lag-screw Joints: Their Behavior and Design*, U. S. Department of Agriculture, Washington, D. C., 1938.
3. John A. Scholten, *Timber-connector Joints; Their Strength and Design*, U. S. Department of Agriculture, Washington, D. C., 1944.

GENERAL REFERENCES

Howard J. Hansen, *Timber Engineers' Handbook*, John Wiley and Sons, New York, 1948.
Howard J. Hansen, *Modern Timber Design*, 2nd ed., John Wiley and Sons, New York, 1948.
National Design Specifications, National Lumber Manufacturers Association, Washington, D. C.

SECTION 11

MECHANICAL SPRINGS

By

A. M. WAHL *Research Laboratories, Westinghouse Electric Corporation*

MECHANICAL SPRINGS

DEFINITIONS AND TYPES OF SPRINGS. A mechanical spring may be defined as an elastic body or resilient member whose primary function is to deflect or distort under load; it recovers its original shape when released after being distorted. The more important types of springs are:

Helical compression or tension springs are made of bar stock or wire coiled into a helical form, the load being applied along the helix axis. In a compression spring, the helix is compressed; in a tension spring, it is extended.

Helical torsion springs are similar in form to the helical compression spring but loaded by a torque about the helix axis.

Leaf springs are composed of flat bars of varying lengths clamped together so as to obtain greater efficiency and resilience (automobile leaf springs). Such springs may be *full-elliptic*, *semi-elliptic*, or *cantilever*.

Spiral springs consist of flat strip wound in the form of a spiral (clock springs) and are loaded in torsion.

Belleville springs are composed essentially of coned disks which may be stacked up to give a variety of spring load-deflection characteristics.

1. HELICAL COMPRESSION SPRINGS

BASIC THEORY AND FORMULAS. Probably the most widely used type of spring in the mechanical engineering field is the helical round-wire compression spring (Fig. 1). Among the reasons for the wide use of this type of spring are low cost, compactness, and efficient use of material.

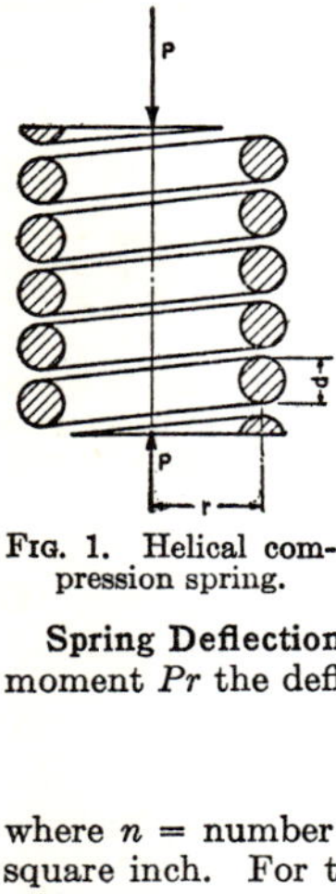

FIG. 1. Helical compression spring.

As shown in Fig. 1, a spring loaded by an axial load P may be considered essentially as a bar under a torsion moment Pr. On this basis, the torsional shearing stress S_s will be equal to the moment divided by the section modulus. This gives

$$S_s = \frac{16Pr}{\pi d^3} \tag{1}$$

where r = mean coil radius and d = bar or wire diameter. This stress, which is widely used in spring design, is frequently termed the *uncorrected* stress, since it does not correct for effects of bar or wire curvature or for the effects of direct shear loading (which will be discussed further below). In spite of its simplicity, this formula is surprisingly good as a criterion of static load-carrying ability without excessive set.

Spring Deflection. If the spring is considered essentially as a bar subjected to a torsion moment Pr the deflection δ becomes

$$\delta = \frac{64Pr^3n}{Gd^4} \tag{2}$$

where n = number of active coils and G = modulus of rigidity of material, pounds per square inch. For the usual compression spring with ends squared and ground, approximately one inactive coil may be taken at each end, giving a value for n equal to the total number of coils minus 2. However, slightly better accuracy will be obtained in most cases by using a figure of 1.75 instead of 2 (Ref. 1, p. 158).

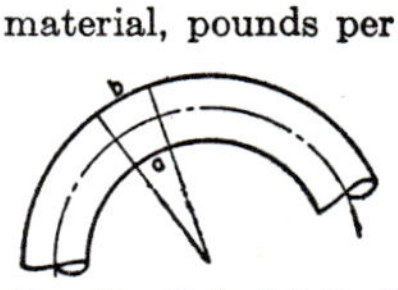
FIG. 2. Coil of helical spring.

For most carbon steel springs (music wire, oil-tempered wire) the modulus G may be taken as about 11.5×10^6 psi whereas for hot-wound springs, heat-treated after coiling, $G = 10.5 \times 10^6$ psi is usually preferred. (See Table 8, p. 11-18, for further values of G for different spring materials.)

The *spring rate* (or scale) in pounds per inch deflection is

$$\frac{P}{\delta} = \frac{Gd^4}{64r^3n} \tag{3}$$

Parts of the data in this section have been abstracted from the author's *Mechanical Springs* (Ref. 1).

Correction for Curvature and Direct Shear. In deriving eq. 1, neither the effect of the direct shear load P on the spring nor the effect of curvature of the bar or wire is considered. Referring to Fig. 2, which represents a plan view of the coil of a helical spring, the fiber length at a on the inside of the coil is considerably shorter than that at b on the outside.

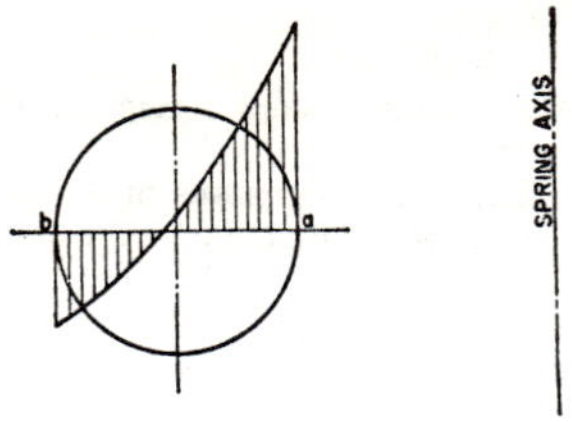

FIG. 3. Elastic shearing stress distribution across transverse diameter of helical spring of small index.

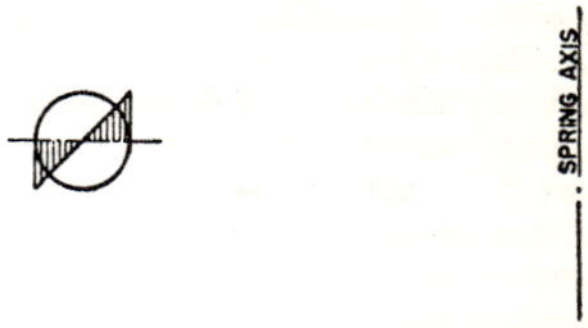

FIG. 4. Elastic shearing stress distribution across transverse diameter of helical spring of large index.

Consequently when two neighboring radial cross sections are rotated with respect to each other, the unit shearing displacements are much greater at a than at b. Thus *below the elastic limit*, the stress distribution across the transverse diameter of the spring will be something like that in Fig. 3, with a peak at the inside of the coil at a for small index springs, i.e., springs with a small ratio of coil-to-wire diameter. For a spring of large index the torsional stress distribution will approximate the linear type for a straight bar in torsion (Fig. 4).

It should be noted that the elastic stress distributions shown in Figs. 3 and 4 apply strictly only to cases where the springs are initially free of stress and where the loading is such that the elastic limit of the material is not exceeded. Where springs are cold-set (as is often done in manufacture) yielding of the material occurs so that at high loads the stress distribution is somewhat as shown in Fig. 5, the result being a more uniform distribution of stress. At the same time residual stresses are induced which subtract from the load stress at lower loads.

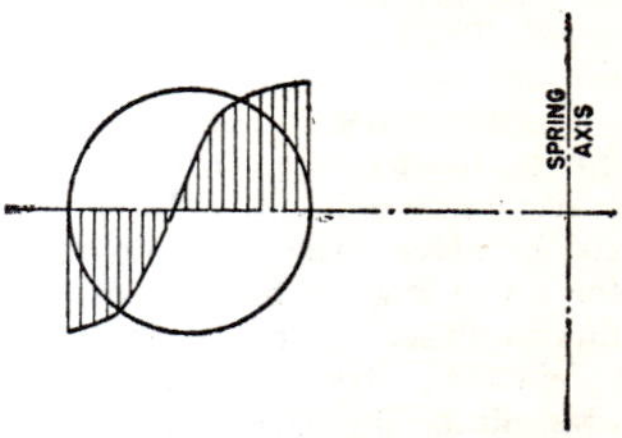

FIG. 5. Stress distribution across transverse diameter in small index spring at high loads after yielding has occurred.

To calculate the elastic stress S_s' at point a (Fig. 3) at the inside of the coil, the stress given by the usual formula (eq. 1) may be multiplied by a factor K (usually known as a curvature correction factor) which depends on the spring index $c = 2r/d$ (Ref. 2). This gives

$$S_s' = K\,\frac{16Pr}{\pi d^3} \tag{4}$$

where

$$K = \frac{4c - 1}{4c - 4} + \frac{0.615}{c} \tag{5}$$

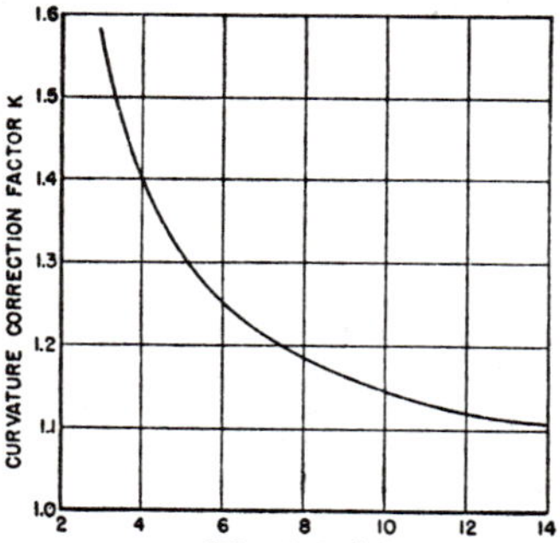

FIG. 6. Correction factor K for helical springs.

A plot of the correction factor K versus spring index $2r/d$ is given in Fig. 6. The stress calculated in this way is frequently called the *corrected* stress to distinguish it from the *uncorrected* stress figured from eq. 1.

More accurate equations have been developed to take into account the elastic effects due to curvature, direct shear, and pitch angle (Ref. 3). For most practical springs these differ but slightly from the results given by eq. 4.

In manufacture, compression springs are frequently *cold-set*, i.e., coiled to a greater free height than is specified and then compressed sufficiently to cause a permanent set and a reduction in free height to the specified value. This operation, also known as *scragging* or *presetting*, results in yielding and plastic flow at points of maximum stress. Hence the corrected stress calculated by using eq. 4 will not usually exist in cold-set springs, i.e., the actual peak stress is lower than the calculated value. However, where fatigue loading is involved, if P is taken as load *range*, i.e., the difference between minimum and maximum load during a cycle, the corrected stress S_s' of eq. 4 will be equal to the stress *range*. This is of importance where fatigue loading is present.

If the average stress due to the direct shear load P is added to the uncorrected stress (eq. 1), the resultant stress S_s'' is

$$S_s'' = K_s \frac{16Pr}{\pi d^3} = K_s S_s \tag{6}$$

where $K_s = 1 + (0.5/c)$ and c = spring index. This formula, sometimes used for static loading, yields somewhat more conservative results than eq. 1.

DESIGN METHODS AND WORKING STRESSES. For purposes of design, helical-spring applications may conveniently be divided into three groups, each of which requires a somewhat different method of approach for best overall results. They are:

(1) Static loading (load on spring steady or applied less than, say, 1000 times).
(2) Intermediate loading (load applied, say, 1000 to 100,000 times during the life of the spring).
(3) Fatigue or endurance loading (load applied a million times or more).

(1) Static Loading (Load Steady or Infrequently Repeated). This category includes cases where the load is steady, or applied less than, say, 1000 times during the life of the spring. Such applications include circuit-breaker springs operating infrequently, safety-valve springs, and springs used to apply gasket pressure. In such cases failure by fatigue fracture is not a problem, the chief concern being to avoid excessive set or creep. The choice of working stress thus depends to some extent on whether a small amount of set may be tolerated; if so, a higher working stress may be used than would be the case otherwise. The problem of creep or set is also aggravated by temperatures above about 250 F for ordinary steel springs (p. 11-05).

At the present time, it is the consensus of most spring authorities that the stress augments due to curvature and direct shear may be neglected when calculating springs for static loading, i.e., that the *uncorrected* stress (eq. 1) may be used. Where somewhat more conservative design is required and the spring index is small, it is suggested that stresses for static loading be figured by using eq. 6, which takes into account the stress augment due to direct shear loading but not that due to curvature.

Working Stresses Used in Practice—Static Loading and Normal Temperature. As an example of the extremely high working stresses for helical springs which can be used in practice where space is at a premium, the curves of Fig. 7 have been taken from data published by the Society of Automotive Engineers (Ref. 4). They represent *uncorrected* allowable stress values (eq. 1) for various spring materials and various wire or bar sizes for helical compression springs used in ordnance applications, where a relatively short spring life is ordinarily required. These curves apply only under the following conditions: springs cold-set and shot-peened for sizes over $^1/_{16}$ in.; normal temperature operation; spring index between 4 and 9. Where these conditions are not fulfilled stresses should be lower.

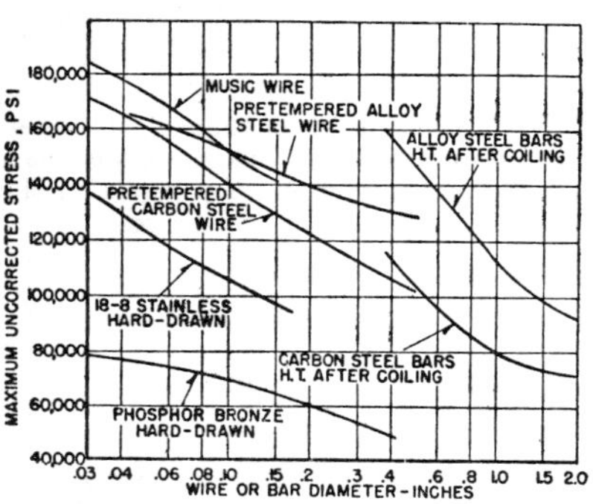

Fig. 7. Stresses in compression springs for ordnance applications where space is limited and only short life is required. (Data from *SAE Manual*, Ref. 4)

It should be emphasized that the stress values of Fig. 7 are to be considered as *maximum* values which may be used where necessary when space is at a premium and where only limited life is expected. Where fatigue loading is present (as in automotive valve springs) the use of such stresses would generally result in early fatigue failure. For industrial applications where more space is usually available, it is suggested that about 25% lower stresses than those given in Fig. 7 be used even for infrequent or static loading.

As a further example of working stress values used in practice, suggested stress values by Associated Spring Corp. (Ref. 5) are given in Table 1 for various spring materials. These values apply to springs under occasional loading at normal temperature. In this case the working stress values are taken 20% higher when using the corrected stress formula (eq. 4).

It may be noted that the uncorrected stress values given in Table 1 are, in general, about 20 to 25% lower than those listed in Fig. 7.

Table 1. Safe Working Stress for Helical Springs

(Under occasional loading only)

Material	Wire Size, in.	Corrected Stress, psi (eq. 4)	Uncorrected Stress, psi (eq. 1)
Music wire	.015	180,000	150,000
	.060	155,000	129,000
	.156	128,000	107,000
Chrome-vanadium wire	.040	157,000	131,000
	.100	140,000	117,000
	.500	115,000	96,000
Oil-tempered or carbon valve spring wire	.020	164,000	137,000
	.100	127,000	106,000
	.500	92,000	77,000
18-8 Stainless wire	.025	122,000	102,000
	.150	93,000	78,000
	.500	70,000	58,000
Phosphor-bronze wire	.025	77,000	64,000
	.150	65,000	54,000
	.500	46,000	38,000

Static Loading—Elevated Temperature. It has been found that at relatively low operating temperatures (as low as 250 F) considerable relaxation or set, depending on stress and temperature, will take place in helical springs loaded continuously. Thus a spring clamped between parallel plates at elevated temperature will gradually lose part of its load over a period of time. In such cases very much lower stresses than those given in Table 1 must be used.

Tests by Zimmerli (Ref. 6) indicate that at 80,000 psi corrected stress (72,000 uncorrected) music-wire springs will lose around 2 to 3% of their initial loads at 250 F and 7% at 350 F within three days. At 100,000 psi (90,000 uncorrected), load losses in three days of about 5% at 250 F and 10% at 350 F may be expected. Extrapolated relaxation curves (Ref. 7) for music-wire springs indicate that in a year or so two or more times as much relaxation will occur as occurred in three days. Hence it appears that for operation in the range of 250 F to 350 F and uncorrected stresses around 90,000 psi music-wire springs may lose around 20% of initial load in about a year's time because of relaxation. For this reason it is in general desirable to use springs of 18-8 stainless steel wire above 250 F since tests show that they will have much lower load losses than the music-wire springs.

At temperatures of 350 to 500 F the superiority of stainless steel over music-wire or carbon-steel springs is even more pronounced. Extrapolated relaxation data available (Ref. 7) indicate that at 300 F and 90,000 psi uncorrected stress 18-8 stainless-steel springs will lose around 8% load in one year as compared with about 4% in three days. At 450 to 500 F this loss may be as high as 12 to 14% in one year. This relaxation should be considered in design, or steps should be taken to reduce it.

Thus, for applications at temperature above 500 F, the use of an alloy such as Z-nickel or Inconel may be advisable. The spring manufacturer should be consulted in such cases. The relaxation or set occurring during operation may be reduced by loading the springs under temperature at a load somewhat above the working load for a few days. In this manner a considerable part of the initial set is removed; hence less will occur under subsequent loading.

(2) Intermediate Loading (Load Applied 1000 to 100,000 Times). Perhaps the majority of spring applications fall in this category, which includes springs subject to the peak load, say, 1000 to 100,000 cycles during their lives. Examples are automobile suspension springs and springs used in circuit-breaker mechanisms. Such springs would be considered to be under a more severe application than where only static loading is involved; at the same time the condition is not so severe as where fatigue loading is involved.

An estimate of the amount of increase in allowable stress where springs are subject to 100,000 or fewer cycles of load application may be obtained by using the fatigue-test results reported by Edgerton (Ref. 8). These test results (Fig. 8) were obtained on carbon-steel springs ($^3/_4$ bar diameter, index 5) tested in a zero to maximum stress range. In this curve the corrected stress range is plotted against number of cycles to failure (log scale), the trend of the average being indicated by the full line. From this it is seen that for a life of 100,000 cycles the allowable stress is about 1.5 times the stress for a life of one million cycles. Higher values would be expected for centerless-ground bar springs, shot-peened, of index 5, and lower values for springs of higher indexes.

For applications involving a limited number of cycles, the following conservative method of design may be used. The stress at maximum load should be figured as uncorrected stress (i.e., without curvature correction) and should not exceed the allowable value for static loading (see p. 11-04). The stress range, obtained from the upper and lower limits of the load range, should be figured as a corrected stress (eq. 4), and this should be compared with the endurance range for the required number of cycles using curves of the type shown in Fig. 9. For small numbers of cycles and for small indexes (i.e., larger curvature correction factors) this method will probably err on the safe side, since as the number of

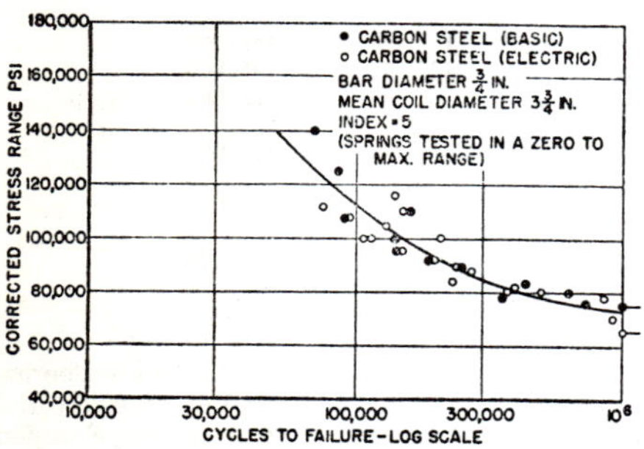

FIG. 8. Results of fatigue tests of hot-wound heavy helical springs. (Edgerton, *Trans. ASME*, Oct. 1937, and *Progress Report* 3 of ASME Spring Committee, Ref. 8)

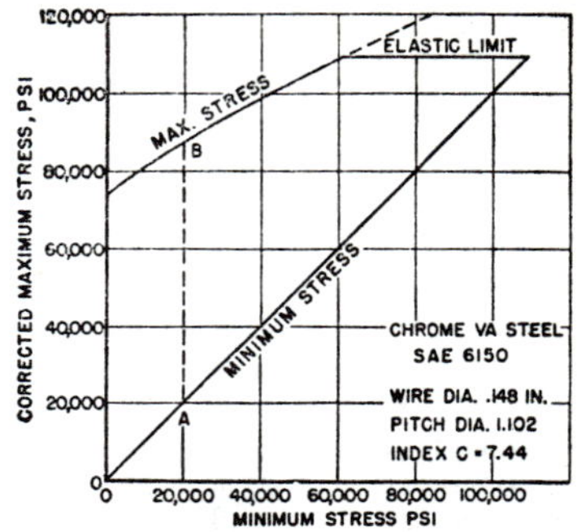

FIG. 9. Typical endurance diagram for helical compression springs. (Zimmerli, *University of Michigan Engineering Research Bull.* 26, Ref. 10)

cycles is reduced the condition tends to approach a static load condition where curvature correction may be neglected. Further test data are needed to throw light on this question.

An example of working-stress values suggested for average service conditions is given in Table 2, based on data published by American Steel and Wire Company (Ref. 9).

Table 2. Recommended Maximum Corrected Stresses for Helical Compression and Extension Springs Under Average Service Conditions *

	Kind of Wire (Plain Carbon Steels)			
Wire Diameter, in.	Music	Tempered	Hard-drawn	
0.020 to 0.030	100,000	100,000	90,000	
0.031 to 0.092	90,000	100,000	80,000	
0.093 to 0.176	90,000	90,000	80,000	
0.177 to 0.282	....	90,000	70,000	
0.283 to 0.436	....	85,000	......	
0.437 to 0.624	....	80,000		If OD greater than 3 in.—(cold-wound)
	....	90,000		If OD less than 3 in.—(hot-wound)
0.625 to 0.874	....	90,000		Hot-wound
0.875 to 1.249	....	80,000		Hot-wound
1.250 to 1.5	....	80,000		Hot-wound

* Defined as noncorrosive atmosphere, temperatures not over 150 F, slowly varying or static loads. All stresses in pounds per square inch.

Average service conditions in this case (Table 2) are defined as noncorrosive atmosphere, temperature not over 150 F, and under relatively slowly varying loads. Since curvature correction is included in the stress figures of Table 2, it will be noted that for average spring indexes where curvature correction is about 20%, these values are equivalent to lower uncorrected stresses than those of Fig. 7 for ordnance applications.

EXAMPLE 1. COMPRESSION SPRING FOR CIRCUIT BREAKER. A spring used in a circuit breaker is designed for a life of approximately 100,000 cycles. Dimensions are: OD = 5/8; d = 0.071 (music wire); r = 0.277 in.; index = 7.8; 9 active turns; free length 2 1/4 in.; closed length 1 3/32 in. In operation the load varies from a minimum of 11.7 lb to a maximum of 27.3 lb. From eq. 1, the *uncorrected* stress at maximum load is

$$S_s = \frac{16 \times 27.3 \times 0.277}{\pi(0.071)^3} = 108,000 \text{ psi}$$

The load *range* between maximum and minimum load is 27.3 − 11.7 = 15.6 lb, and the corrected stress *range* from eq. 3, taking $K = 1.19$ from Fig. 6 for $c = 7.8$, is

$$S_s' = \frac{16 \times 15.6 \times 0.277 \times 1.19}{\pi(0.071)^3} = 73{,}000 \text{ psi}$$

In this case, it is seen that the peak uncorrected stress is somewhat below that listed in Table 1 for this size wire. Also the stress range is below that to be expected from endurance tests of this size wire for 100,000 cycles life.

EXAMPLE 2. CUP SPRING FOR TURBINE LOCOMOTIVE DRIVE. An example of the design of heavy springs for intermediate loading conditions is given by the springs used in a turbine locomotive as the flexible element between the arms of a spider pressed on the axle and the gear driven by the turbine. These springs are hot wound from chrome-vanadium steel bars, centerless-ground, shot-peened, and magnaflux-tested. Dimensions are: OD 3.56 in.; bar diameter 0.870 in.; index 3.1; free height 10.51 in.; 8 3/4 active turns; solid load 7650 lb; solid deflection 1.59 in. Maximum uncorrected stress on the spring with coils solid is 80,000 psi, corresponding to 57% adhesion between wheel and rail, which occurs only at rare intervals. For normal starting conditions, where 30% maximum adhesion between wheel and rail is assumed, the uncorrected stress will be 42,000 psi and the corrected stress 65,000 psi. This stress range of 63,000 psi is around half the expected endurance range for 100,000 cycles for shot-peened springs of index 3 (Fig. 8). Under normal operating conditions with 18% adhesion the corrected stress in this spring will be 39,000 psi.

(3) Fatigue or Endurance Loading (Load Applied a Million Times or More). Examination of springs which have failed due to fatigue or repeated loading shows that in most instances the fracture starts at the inside of the coil or at the point where the range of stress is a maximum. The stress *range* at this point may be calculated with sufficient accuracy using the corrected stress formula (eq. 4), taking P equal to load range, i.e., difference between maximum and minimum load. On this basis it appears reasonable, where fatigue loading is involved, to calculate the stress range as a corrected stress.

Because of lack of sensitivity of some materials to stress concentration (as represented by curvature effects) the endurance ranges calculated by this method will probably be higher for the lower-index springs (small $2r/d$) than for the high-index springs, the difference depending on the material. Available data indicate that this difference will be considerably greater for heavy hot-wound springs (Ref. 8) than for small springs of valve-spring wire. Thus if the endurance range calculated as a corrected stress is determined from tests of large index springs it will, in general, be safe to apply the results to lower index springs of the same material and wire size. However, caution should be exercised in applying the results of fatigue tests on low- or medium-index springs to those of high index of the same wire size, since the latter will frequently show lower endurance range values. Also in general the endurance range of the material decreases somewhat as the maximum stress of the range increases, and for best results this should be considered in design. In other words, for a load range from zero to maximum, the endurance range will be appreciably greater than for a load range from, say, half the maximum to the maximum.

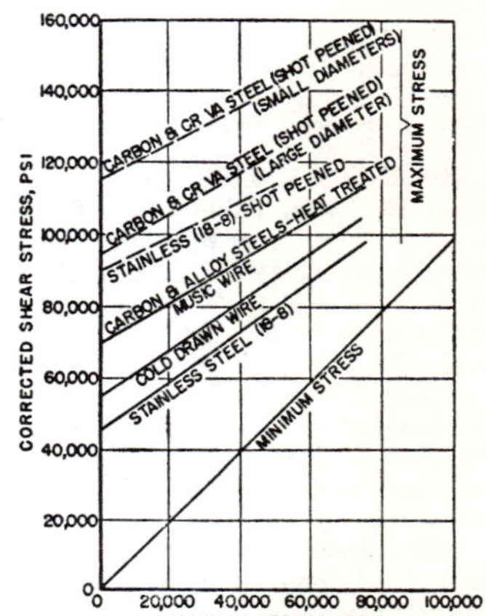

FIG. 10. Approximate endurance diagrams for good-quality helical compression springs.

A typical endurance diagram for a helical spring made of chrome vanadium steel wire of 0.148 in. diameter, index 7.44, is shown in Fig. 9. This curve, based on tests by Zimmerli (Ref. 10), represents the maximum and minimum corrected stress limits for a spring under fatigue loading. The limiting stress range is read vertically between the line representing maximum stress and one representing minimum stress. Thus a spring operating in a fatigue range between 20,000 and 87,000 psi corrected stress would be represented by the vertical dashed line AB. A range within these limits (say from 20,000 to 70,000 psi) would not be expected to cause failure. It should be noted that this applies to one case (wire size 0.148 in., index $c = 7.44$). For other wire sizes or indexes the curve could be appreciably different.

Approximate endurance diagrams which may be expected from good-quality helical compression springs of different materials are shown in Fig. 10. In all cases, the limiting endurance range is read vertically between the lower line representing minimum stress and the upper line corresponding to the given spring material. These curves hold roughly for springs having indexes between 5 and 10. Since these curves represent rough average values, considerable deviation in individual instances may be noted. Sometimes higher values than those of Fig. 10 may be obtained for small-index springs. The effect of shot peening in raising the endurance limit is shown by the three upper dashed lines. It should be emphasized that in practical design lower stress ranges than those indicated by these endurance values should be used in order to provide a factor of safety. This factor of safety may range from 1.25 to 2, depending on the seriousness of failure and on the accuracy of the fatigue-test data available.

Several examples of working stress ranges where fatigue loading is involved (automotive valve springs) are given in Ref. 10. The corrected stress ranges in such applications are around 30,000 to 50,000 psi while the peak stress may vary from 60,000 to 83,000 psi, corrected, for springs not shot-peened. (Actually the above figures are nominal stresses based on valve lift; at high speeds of operation the actual peak values may be greater due to vibration and surging.) By using shot-peened springs a higher factor of safety is obtained.

As a further example of working stresses used in practice where fatigue is involved, the stresses given in Table 3 are used as a basis for design for severe service involving fatigue loading by Westinghouse Electric Corporation. These stresses, which are corrected values, apply to music or oil-tempered wire in the smaller sizes and to hot-wound springs, heat-treated after forming in the larger sizes. This table is intended to be used primarily as a guide in spring selection; in many cases higher values may be used if a careful analysis of the application is carried out.

Table 3. Working Stresses for Severe Service—Helical Compression Springs *

Diameter of Wire, in.	Working Stresses
Up to .085	60,000
Above .085 to 0.185	55,000
Above .185 to 0.320	48,000
Above .320 to 0.530	42,000
Above .530 to 0.970	36,000
Above .970 to 1.5	32,000

* For springs made of music or oil-tempered wire, in smaller sizes and hot-wound springs, heat-treated after forming, in larger sizes. For phosphor-bronze 50% of these values, and for stainless steel 75% of these values, are used. Table does not apply where corrosion effects are present. Values represent *corrected* stresses.

SPRING TABLES. Based on Corrected Stress. To facilitate spring selection in practical design Table 4 may be used (Ref. 11). For each OD and wire size two figures are given: an upper one representing load in pounds at 100,000 psi *corrected* stress and a lower one representing spring scale in pounds per inch per single coil based on a shear modulus G of 11.5×10^6 psi. For any actual spring, the spring scale per coil given in the tables must be divided by the number of active coils. This scale is independent of the load. To obtain the spring scale per coil at any modulus G, the values in the table should be multiplied by $11.5 \times 10^6/G$.

It should be noted that the figure of 100,000 psi corrected stress is used mainly for convenience and is not necessarily the working stress. For any other corrected stress S_s', the loads given may be multiplied by the ratio $S_s'/100{,}000$. Also the loads given in the table may be considered to represent load *ranges* at corrected stress *ranges* of 100,000 psi.

If it is desired to determine the load at an *uncorrected* stress of 100,000 psi, the loads given in Table 4 should be multiplied by the correction factor K (Fig. 6) corresponding to the spring index.

EXAMPLE 3. Assume that a compression spring is required for a mechanism to give 200 lb at a deflection of 1.0 in. operating under fatigue conditions. The required spring scale is thus 200/1.0 = 200 lb per in. The space available is such that an OD of 2 in. may be used. From Table 4, for 0.283-in. wire and 2 in. OD, load at 100,000 psi corrected stress is 415 lb, and spring scale per coil is 1820 lb per in. Since this spring is to operate under severe conditions, the allowable stress will be taken from Table 3, and is 48,000 psi for 0.283 wire. Load at 48,000 psi corrected stress is 0.48 (415) = 200 lb. Since the scale wanted is 200 lb per in., number of active coils = 1820/200 or approximately 9. Total number of coils would be around 10 1/2 to 11. This would require a spring of about 1.1 (10.5)(0.283) + 1.0 = 4.27 in. free length, allowing 10% extra length for space between the turns when the spring is compressed at a load of 200 lb. The length at a load of 200 lb would be 4.27 − 1.0 or 3.27 in. and the solid length about 10% less or about 2.94 in.

Tables for Static Loading. Where springs are *statically loaded* Table 5, based on 100,000 psi stress calculated using the factor K_s, may be used. (This factor takes into account direct shear but not curvature effects.) Loads P and deflections per turn y at a stress of 100,000 psi are given for various wire sizes and outside coil diameters. The deflection y is based on a shear modulus of 11.4×10^6 psi. Although stresses of 100,000 psi may be used in many cases, this figure is used in Table 5 merely for convenience. For any other stress

Table 4. Load and Spring Scale per Coil—Helical Compression or Extension Springs

H. F. Ross, *Trans. ASME*, Oct. 1947.

(Based on 100,000 psi—corrected stress)

Wire Diam., *d*	Outside Diameter of Coil, in.																
	1/8	5/32	3/16	7/32	1/4	9/32	5/16	11/32	3/8	13/32	7/16	15/32	1/2	17/32	9/16	5/8	11/16
.010	.305 9.47	← Load in lb at 100,000 psi—corrected stress ← Spring scale per single coil															
.011	.401 14.1	.325 6.90															
.012	.522 20.7	.421 10.0	.350 5.43														
.014	.823 40.3	.669 19.4	.560 10.5	.480 6.45	.422 4.18												
.016	1.21 72.9	.983 34.4	.824 18.4	.714 11.3	.626 7.37	.561 5.05											
.018	1.71 124	1.40 57.4	1.17 30.7	1.01 18.6	.889 12.1	.793 8.30											
.020	2.32 200	1.89 91.3	1.60 48.6	1.39 29.3	1.22 18.9	1.08 12.8	.972 9.06										
.022	3.03 306	2.50 140	2.12 73.9	1.83 44.0	1.60 28.2	1.44 19.3	1.30 13.7	1.18 10.1									
.024	3.90 464	3.22 208	2.72 108	2.36 64.3	2.09 41.4	1.87 28.2	1.68 19.9	1.53 14.6	1.41 11.1	1.31 8.62	1.21 6.67	1.14 5.42					
.026	4.88 678	4.06 299	3.44 155	2.98 91.1	2.63 58.4	2.35 39.6	2.14 28.0	1.94 20.5	1.79 15.5	1.65 12.0	1.54 9.43	1.44 7.55	1.35 6.18	1.28 5.11			
.029	6.61 1160	5.55 496	4.71 253	4.10 149	3.64 94.1	3.25 63.7	2.94 44.6	2.71 32.9	2.50 24.6	2.29 18.9	2.13 14.9	2.00 12.0	1.87 9.75	1.77 8.07	1.66 6.66		
.032		7.30 793	6.25 398	5.47 231	4.84 146	4.35 97.8	3.95 68.0	3.60 49.9	3.33 37.5	3.07 28.7	2.85 22.4	2.67 17.9	2.51 14.8	2.37 12.1	2.24 10.0	2.03 7.27	
.035		9.34 1213	8.10 604	7.10 346	6.32 217	5.66 145	5.14 101	4.72 74.0	4.32 55.0	4.02 42.2	3.73 33.0	3.50 26.3	3.29 21.4	3.08 17.7	2.93 14.7	2.64 10.4	2.41 7.72
.039		12.6 2066	10.9 1009	9.66 572	8.63 354	7.76 234	7.03 162	6.45 118	5.93 87.9	5.52 67.3	5.14 52.8	4.80 41.1	4.52 34.1	4.27 28.0	4.06 23.3	3.66 16.6	3.33 12.2
.043			14.4 1618	12.7 901	11.4 559	10.3 368	9.33 249	8.60 183	7.91 134	7.36 103	6.82 80.0	6.43 63.7	6.07 51.6	5.69 42.6	5.36 34.9	4.89 25.0	4.44 18.3
.047			18.3 2507	16.4 1378	14.7 840	13.3 547	12.1 373	11.1 270	10.3 199	9.54 151	8.90 117	8.33 93.3	7.83 75.6	7.39 61.9	6.99 51.0	6.35 36.2	5.81 26.6
.051				20.4 2061	18.5 1233	16.8 800	15.3 543	14.2 389	13.0 288	12.1 217	11.3 168	10.6 134	10.0 107	9.45 88.3	8.97 73.0	8.01 51.1	7.37 37.5
.055					22.8 1767	20.8 1143	18.9 768	17.6 553	16.2 401	15.1 304	14.1 234	13.3 187	12.4 149	11.7 122	11.2 101	10.0 70.4	9.23 51.9
.063						30.3 2196	27.9 1461	25.8 1032	23.8 721	22.4 562	20.9 429	19.7 338	18.6 272	17.5 221	16.6 181	15.1 128	13.8 92.6
.071							38.7 2563	36.2 1783	33.4 1295	31.5 972	29.4 741	27.8 580	26.1 463	24.8 375	23.6 307	21.3 215	19.5 156
.080								50.3 3245	47.1 2300	44.1 1705	41.3 1285	39.2 1000	36.9 795	34.5 640	33.4 525	30.2 364	27.8 262
.090									65.2 4075	61.3 2990	57.9 2236	54.8 1734	51.7 1368	49.2 1101	46.9 890	42.9 617	39.0 442
.100										81.8 5018	77.5 3726	73.5 2860	69.7 2248	66.6 1800	63.3 1449	58.0 993	53.0 707
.112												100 4950	95.6 3886	91.5 3081	87.4 2462	79.7 1678	73.8 1186
.125													130 6701	124 5232	119 4205	109 2817	101 1977
.135															147 6080	135 4060	125 2820
.148																175 6350	163 4380
.162																	207 6800

(Table continued on p. 11-10)

Table 4. Load and Spring Scale per Coil—Helical Compression or Extension Springs— *Continued*

H. F. Ross, *Trans. ASME*, Oct. 1947.
(Based on 100,000 psi—corrected stress)

Wire Diam., *d*	Outside Diameter of Coil, in.													
	3/4	13/16	7/8	15/16	1	1 1/16	1 1/8	1 3/16	1 1/4	1 3/8	1 1/2	1 5/8	1 3/4	1 7/8
.039	3.04 9.30	2.82 7.20	← Load in lb at 100,000 psi—corrected stress ← Spring scale per single coil											
.043	4.10 14.0	3.76 10.8	3.52 8.60											
.047	5.31 20.1	4.93 15.6	4.57 12.4	4.27 9.89	4.00 8.09									
.051	6.78 28.6	6.29 22.1	5.79 17.3	5.44 14.0	5.14 11.4	4.82 9.43								
.055	8.50 39.5	7.83 30.1	7.32 23.9	6.77 19.0	6.40 15.6	6.02 12.9	5.70 10.7							
.063	12.7 70.0	11.7 53.7	10.9 42.3	10.2 33.7	9.60 27.5	9.00 22.6	8.55 18.8	8.07 15.8	7.75 13.6					
.071	18.0 117	16.6 88.9	15.5 70.6	14.5 56.2	13.6 45.3	12.9 37.4	12.2 31.5	11.6 26.4	11.1 22.4	9.97 16.4	9.21 12.6			
.080	25.7 196	23.7 150	22.2 117	20.8 93.7	19.5 75.6	18.3 62.1	17.3 51.4	16.4 42.9	15.7 37.0	14.2 27.0	13.1 20.5	12.2 16.0		
.090	36.2 329	33.6 250	31.2 195	29.4 155	27.6 125	26.1 103	24.7 85.1	23.3 71.3	22.2 60.3	20.2 44.3	18.6 33.5	17.2 25.9		
.100	49.2 525	45.6 397	42.6 309	40.1 245	37.6 197	35.5 161	33.5 132	31.9 111	30.5 94.5	27.7 69.0	25.5 52.4	23.6 40.4	22.1 32.1	20.4 25.5
.112	68.1 871	63.5 657	59.3 509	55.7 401	52.2 323	49.6 263	47.0 218	44.6 182	42.4 152	38.6 112	35.8 84.5	33.2 65.4	30.7 51.7	28.8 41.5
.125	93.8 1436	87.2 1083	81.9 832	76.8 651	72.5 525	68.2 426	65.0 350	61.7 292	58.8 246	53.9 180	49.4 135	45.7 104	42.6 82.0	39.9 65.6
.135	116 2035	109 1535	102 1176	96.3 919	90.8 739	85.5 600	81.4 493	77.2 410	73.5 344	67.2 251	62.2 188	57.6 145	53.3 113	50.0 90.7
.148	151 3166	142 2343	133 1802	125 1403	119 1119	112 896	106 741	101 611	96.3 514	88.0 375	81.2 279	75.7 214	70.5 168	65.7 133
.162	195 4887	183 3581	172 2734	162 2112	153 1685	146 1353	139 1114	131 918	126 769	115 554	106 414	98.5 317	91.5 247	86.1 196
.177	250 7485	235 5504	222 4150	209 3215	198 2526	188 2035	178 1664	170 1367	163 1145	149 819	137 609	128 468	119 361	112 289
.192		293 8184	277 6128	262 4712	249 3694	236 2950	226 2407	215 1982	205 1640	190 1184	174 873	162 666	151 515	142 408
.207			341 8857	322 6765	307 5266	293 4210	279 3419	267 2796	255 2320	235 1658	217 1218	202 930	188 716	177 569
.225				408 10175	388 7918	371 6246	353 5072	339 4109	323 3407	299 2415	276 1770	258 1344	241 1043	226 822
.244					484 11805	461 9257	444 7462	426 6051	406 4988	376 3498	349 2562	326 1935	306 1488	287 1179
.250						494 10466	474 8389	455 6842	436 5640	401 3953	375 2874	349 2169	328 1667	307 1306
.283								639 12432	614 10199	574 7086	534 5154	499 3836	467 2919	440 2281
.3125										753 11444	706 8143	662 6051	623 4611	586 3586
.343										972 18101	915 12833	858 9429	806 7139	761 5535
.362											1058 16794	997 12293	940 9207	886 7162
.375											1166 19966	1099 14556	1038 10903	980 8426
.394												1262 18641	1190 13967	1127 10602
.4375													1587 23202	1504 17715

Table 4. Load and Spring Scale per Coil—Helical Compression or Extension Springs— *Continued*

H. F. Ross, *Trans. ASME*, Oct. 1947.

(Based on 100,000 psi—corrected stress)

Wire Diam., *d*	Outside Diameter of Coil, in.																
	2	2 1/8	2 1/4	2 3/8	2 1/2	2 5/8	2 3/4	2 7/8	3	3 1/8	3 1/4	3 3/8	3 1/2	3 3/4	4	4 1/4	4 1/2
.100	19.3 20.9	18.1 17.2	← Load in lb at 100,000 psi—corrected stress ← Spring scale per single coil														
.112	27.0 33.5	25.4 27.7	24.1 23.1														
.125	37.6 53.3	37.7 51.3	33.4 36.6	31.6 30.8	30.2 26.2	28.7 22.5											
.135	47.2 73.9	44.3 60.9	41.9 50.4	40.0 42.5	37.9 36.2	36.0 31.0	34.5 26.6										
.148	62.0 109	57.9 88.8	55.1 74.3	52.6 62.8	49.7 53.1	47.7 45.6	45.3 39.1	43.3 34.1	41.7 29.7	40.0 26.2							
.162	80.8 161	76.1 131	72.0 109	68.4 90.9	65.1 77.9	62.3 66.4	59.1 56.9	57.2 50.0	54.6 43.5	52.2 38.0	50.5 33.5	48.7 30.0	46.8 26.7				
.177	105 234	99.1 191	94.0 159	89.5 133	84.7 113	81.1 96.7	77.2 83.3	74.3 72.3	71.0 63.2	68.3 54.9	65.5 48.4	63.0 43.0	61.2 38.4	57.0 30.9			
.192	134 331	126 269	120 226	113 186	108 160	103 135	98.1 117	94.1 101	90.2 88.6	86.9 77.2	83.6 68.6	80.8 60.4	78.0 54.2	72.5 43.5	68.4 35.5		
.207	166 457	158 375	150 310	141 258	134 218	129 187	122 160	118 139	113 121	109 107	104 93.6	101 83.1	97.2 74.1	91.2 59.5	85.2 48.5	80.7 40.1	76.0 33.5
.225	212 659	201 536	191 445	181 371	173 314	165 266	157 229	151 198	144 173	139 151	134 134	129 118	124 104	116 83.6	110 68.6	103 56.5	97.9 47.2
.244	269 941	255 764	241 631	229 526	218 443	208 376	199 322	192 280	183 243	177 213	170 188	165 167	158 148	148 118	139 95.9	132 79.4	125 66.7
.250	290 1051	273 853	260 700	247 586	235 492	225 419	215 359	205 310	198 270	191 237	183 208	177 184	170 163	159 131	150 106	141 87.9	134 73.3
.283	415 1820	393 1474	374 1210	355 1007	338 849	322 718	309 616	296 530	285 460	276 405	263 352	256 314	247 276	228 219	216 180	204 148	194 123
.3125	555 2842	525 2313	499 1892	473 1568	453 1311	432 1085	413 947	397 816	381 705	368 617	355 541	341 478	330 424	309 338	290 273	274 224	258 187
.343	725 4370	684 3515	655 2869	620 2375	593 1982	570 1675	544 1425	522 1226	501 1059	483 925	466 810	451 714	433 633	408 503	382 404	362 333	341 278
.362	844 5653	801 4518	765 3655	729 3026	692 2535	665 2131	634 1814	613 1555	589 1346	567 1171	547 1025	525 905	508 800	478 634	452 518	426 423	403 350
.375	932 6652	883 5281	844 4302	804 3551	767 2960	737 2497	704 2124	679 1820	652 1575	628 1368	605 1195	585 1053	567 933	529 739	497 597	471 492	445 406
.394	1068 8350	1022 6684	974 5420	926 4476	884 3714	849 3117	816 2656	781 2268	750 1955	722 1704	701 1487	677 1307	651 1159	612 917	575 740	542 604	515 503
.4375	1433 13859	1363 11010	1307 8861	1249 7257	1201 6005	1148 5013	1102 4243	1063 3644	1019 3132	980 2718	951 2367	919 2082	888 1835	835 1450	783 1166	738 952	704 785
.500	2073 26645	1976 20999	1897 16817	1819 13718	,742 11268	1675 9358	1617 7880	1555 6694	1500 5736	1451 4971	1396 4325	1356 3783	1310 3330	1229 2619	1160 2097	1092 1705	1041 1405
.5625			2623 29977	2508 24232	2426 19853	2337 16388	2251 13751	2177 11654	2096 9957	2020 8541	1956 7412	1898 6471	1832 5686	1727 4442	1628 3546	1541 2868	1468 2359
.625					3239 33289	3115 27445	3010 22872	2921 19268	2825 16386	2722 14045	2649 12135	2565 10551	2491 9240	2344 7192	2221 5710	2101 4608	1997 3773

S_s, loads P and deflections y should be multiplied by the ratio $S_s/100{,}000$, and for any other modulus G, deflections y should be multiplied by $11.4 \times 10^6/G$. Interpolation for intermediate wire sizes or OD values may be used with sufficient accuracy for most purposes.

In general, for most springs, loads and deflections per turn as given in Table 5 are about 5 to 10% lower than they would be if calculated by the uncorrected stress formula (eq. 1). This table thus represents slightly more conservative ratings for the smaller indexes than if the uncorrected stress values were used. To find loads and deflections at 100,000 psi uncorrected stress, values of P and y given in the table should be multiplied by the factor $K_s = 1 + (0.5/c)$, where c = spring index.

Table 5. Loads * and Deflections per Turn for Statically Loaded Helical Springs

Adapted by permission from *Machine Design*, July 1942.

Wire Diam., in.	Outside Diameter of Springs, in.														
	1/8	5/32	3/16	1/4	5/16	3/8	7/16	1/2	5/8	3/4	7/8	1	1 1/8	1 1/4	1 3/8
.014	.912 .0228	.720 .0378	.596 .0570	.441 .106	.352 .172	.293 .253	← Load, P lb, at 100,000 psi ← Deflection per turn, y in., at 100,000 psi								
.016	1.372 .0189	1.09 .0319	.900 .0478	.664 .0910	.529 .1465	.438 .218	.376 .302								
.018	1.98 .0162	1.55 .0273	1.275 .0413	.952 .0794	.753 .1285	.625 .191	.538 .266	.470 .352							
.020	2.73 .0139	2.15 .0238	1.76 .0365	1.31 .0702	1.04 .114	.860 .169	.735 .234	.644 .312							
.022	3.66 .0119	2.87 .0207	2.37 .0319	1.75 .0620	1.395 .102	1.145 .151	.981 .208	.855 .279	.686 .449						
.024	4.77 .0104	3.76 .0183	3.07 .0284	2.28 .0557	1.81 .0918	1.50 .137	1.27 .191	1.115 .254	.888 .410						
.026	6.15 .00915	4.80 .0163	3.97 .0257	2.92 .0499	2.31 .0830	1.91 .125	1.62 .173	1.42 .231	1.13 .373						
.028	7.76 .00811	6.08 .0146	4.93 .0228	3.66 .0457	2.90 .0756	2.39 .114	2.03 .160	1.77 .214	1.41 .344	1.18 .506					
.030	9.60 .00715	7.5 .0130	6.12 .0206	4.52 .0416	3.59 .0685	2.93 .1045	2.52 .1475	2.19 .198	1.74 .317	1.445 .468					
.034	11.4 .00574	11.1 .0106	9.07 .0172	6.64 .0351	5.22 .0591	4.32 .0899	3.68 .1265	3.23 .170	2.54 .275	2.12 .407	1.80 .563	1.56 .762			
.038		15.6 .00869	12.8 .01435	9.32 .0299	7.30 .0508	6.05 .0782	5.15 .1105	4.52 .150	3.54 .241	2.97 .361	2.54 .500	2.20 .662	1.96 .842		
.042		21.6 .0073	17.5 .0121	12.8 .0259	9.80 .0439	8.24 .0685	7.00 .0978	6.06 .132	4.84 .217	3.99 .320	3.40 .445	2.98 .591	2.65 .760	2.37 .941	
.046			23.2 .0104	16.8 .0224	13.2 .0389	10.85 .0605	9.21 .0858	8.01 .118	6.35 .193	5.27 .289	4.48 .401	3.92 .533	3.48 .685	3.14 .856	
.051			32.5 .00856	23.1 .0189	18.1 .0338	14.9 .0525	12.6 .0752	10.9 .103	8.68 .170	7.20 .256	6.13 .356	5.35 .476	4.74 .611	4.27 .763	3.89 .937
.059				36.5 .0147	28.4 .0268	23.3 .0426	19.8 .0620	17.1 .0853	13.5 .143	11.1 .212	9.56 .300	8.35 .403	7.37 .518	6.63 .649	6.01 .782
.067				54.8 .0117	42.1 .0218	34.7 .0354	29.3 .0519	25.2 .0715	19.9 .121	16.4 .183	14.1 .258	12.2 .347	10.8 .447	9.70 .557	8.82 .686
.074					57.4 .0183	46.8 .0302	39.7 .0446	34.2 .0621	27.1 .106	22.3 .161	18.9 .229	16.5 .307	14.7 .398	13.2 .503	11.9 .610
.082					80.2 .0152	64.7 .0252	54.8 .0382	47.0 .0535	37.1 .0925	30.4 .141	26.0 .202	22.5 .272	20.0 .353	18.0 .444	16.3 .548
.090						83.5 .0212	70.3 .0326	60.8 .0462	47.5 .0808	39.2 .124	33.2 .178	28.8 .241	25.7 .314	23.2 .397	21.9 .490
.106							122. .0246	104. .0356	81.2 .0630	66.8 .0987	57.0 .143	49.2 .196	43.4 .255	38.9 .324	35.4 .402
.121								159. .0283	121. .0516	101. .082	85.5 .120	73.5 .164	65.3 .217	58.0 .273	52.9 .342
.135								224. .0230	174. .0431	141. .0695	120. .102	103. .141	91.0 .187	81.7 .239	73.6 .297
.148									231. .0367	188. .060	159. .0891	137. .124	121. .165	108. .212	98. .264
.162									309. .0311	249. .0517	211. .0778	183. .110	160. .145	142. .186	128. .235
.177										330. .0442	266. .0668	240. .0955	210. .127	187. .165	168. .208
.207													343. .101	303. .132	275. .167
.225														394. .116	355. .147
.263															572. .115

Table 5. Loads * and Deflections per Turn for Statically Loaded Helical Springs— *Continued*

Adapted by permission from *Machine Design*, July 1942.

Wire Diam., in.	Outside Diameter of Springs, in.														
	1 1/2	1 5/8	1 3/4	1 7/8	2	2 1/4	2 1/2	2 3/4	3	3 1/2	4	4 1/2	5	5 1/2	6
.051	3.55 1.12	← Load, P lb, at 100,000 psi ← Deflection per turn, y in,, at 1000.000 psi													
.059	5.50 .955	5.08 1.13													
.067	8.06 .828	7.45 .985	6.90 1.15	6.45 1.32											
.074	10.9 .739	10.0 .876	9.33 1.03	8.68 1.19	8.12 1.36										
082	14.8 .656	13.6 .780	12.7 .915	11.8 1.06	11.1 1.22										
.090	19.1 .592	17.6 .702	16.2 .825	15.0 .946	14.7 1.09	12.6 1.40									
.106	32.5 .487	29.9 .582	27.6 .681	25.7 .790	24.1 .905	21.2 1.15	19.1 1.46	17.3 1.78							
.121	48.4 .415	44.4 .494	41.1 .582	38.4 .677	36.1 .780	31.7 1.00	28.5 1.25	25.9 1.53	23.8 1.85						
.135	66.7 .367	62.0 .434	57.5 .510	53.6 .595	50.0 .685	44.4 .890	39.5 1.11	35.9 1.36	32.9 1.63						
.148	88.7 .320	81.5 .385	75.5 .455	70.8 .534	66.0 .612	59.1 .797	52.5 1.00	47.5 1.22	43.5 1.47						
.162	117. .285	107. .342	100. .405	92.5 .473	87.3 .550	77.2 .715	69.0 .903	62.5 1.11	57.1 1.33						
.177	154. .255	141. .307	131. .364	121. .425	114. .493	101. .642	90.5 .811	81.8 .995	74.8 1.20						
.207	250. .206	228. .247	211. .296	196. .347	183. .403	161. .525	145. .670	130. .829	121. 1.01	103. 1.41					
.225	321. .183	297. .223	271. .264	252. .311	236. .361	208. .475	187. .600	170. .746	154. .905	132. 1.27	114. 1.69				
.263	520. .145	477. .177	437. .212	410. .252	381. .295	335. .386	300. .492	270. .610	249. .748	211. 1.05	184. 1.41	164. 1.83			
.283	653. .129	600. .159	550. .190	513. .226	478. .264	422. .351	375. .447	339. .556	311. .680	266. .962	232. 1.30	204. 1.67	183. 2.10		
.307		780. .140	709. .169	660. .201	615. .235	543. .313	483. .401	434. .501	396. .611	339. .870	296. 1.17	262. 1.52	236. 1.93	212. 2.36	193. 2.82
.331			897. .150	833. .180	775. .210	682. .282	608. .364	547. .454	498. .557	426. .793	370. 1.07	329. 1.39	294. 1.75	268. 2.16	244. 2.60
.363				1110. .154	1030. .182	908. .246	811. .320	730. .402	668. .494	563. .702	487. .950	436. 1.24	390. 1.57	356. 1.94	323. 2.34
.394					1330. .161	1170. .218	1040. .282	938. .358	858. .442	724. .631	630. .859	558. 1.12	499. 1.42	453. 1.76	416. 2.13
.430						1530. .189	1370. .249	1230. .315	1115. .391	950. .565	824. .768	725. 1.00	652. 1.28	591. 1.58	540. 1.92
.460							1690. .224	1510. .286	1380. .354	1170. .515	1010. .700	892. .922	797. 1.17	725. 1.45	663. 1.77
.500							2170. .196	1960. .251	1780. .311	1510. .456	1310. .630	1150. .827	1030. 1.05	930. 1.31	854. 1.59

* Stress of 100,000 psi is used for convenience. For, any other stress S_s, values of P (load in pounds) and y (deflection per turn) should be multiplied by $S_s/100{,}000$. Deflections are based on a torsion modulus $G = 11.4 \times 10^6$ psi. For any other value of G, deflections given should be multiplied by the ratio $11{,}400{,}000/G$.

To facilitate approximate determination of loads and deflections the charts of Figs. 11 and 12 may be used. In Fig. 11 the ordinate represents load at 100,000 psi torsion stress (calculated from eq. 6, which includes effects of direct shear but not of curvature) while the abscissa represents wire diameter. Each curve represents a given outside diameter of the spring. Thus for a wire size of 0.090 in. and an outside diameter of 1/2 in. the load at 100,000 psi stress is 61 lb. In Fig. 12 the deflections per turn, represented by the ordinates, are plotted against wire diameter for various outside coil diameters. Thus in this figure for a wire size of 0.106 in. and a coil outside diameter of $^1/_2$ in., the deflection per turn is 0.036 in. Results from the charts of Figs. 11 and 12 usually show a small error, but they are sufficiently accurate for most practical purposes.

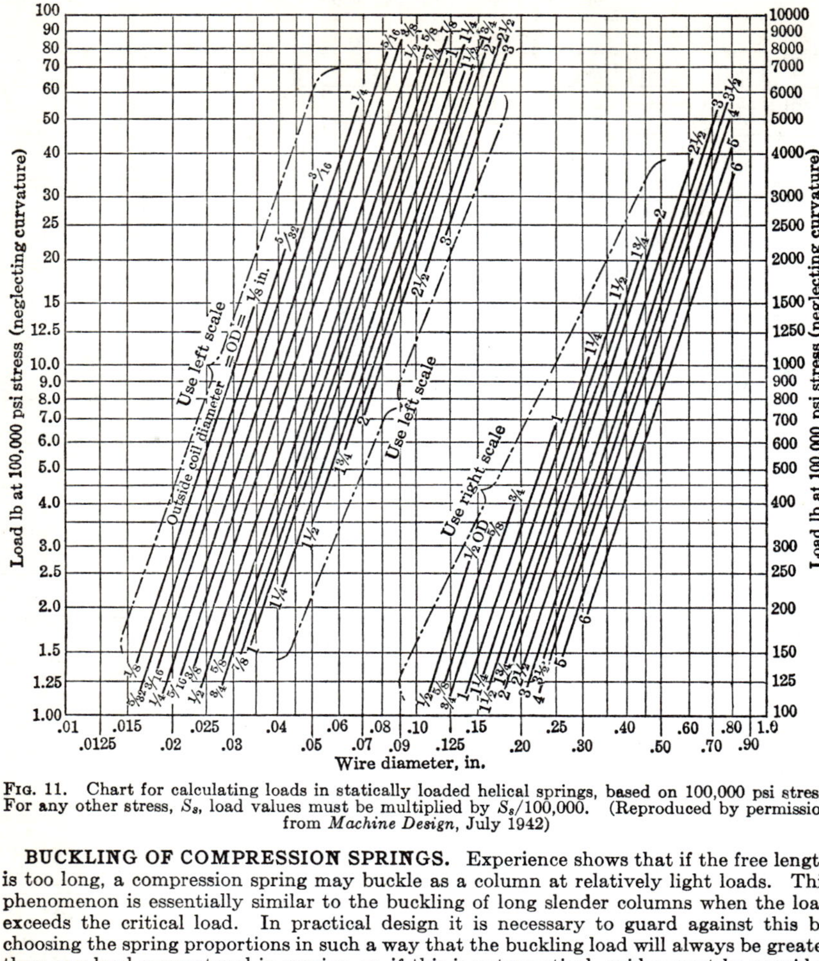

FIG. 11. Chart for calculating loads in statically loaded helical springs, based on 100,000 psi stress. For any other stress, S_s, load values must be multiplied by $S_s/100{,}000$. (Reproduced by permission from *Machine Design*, July 1942)

BUCKLING OF COMPRESSION SPRINGS. Experience shows that if the free length is too long, a compression spring may buckle as a column at relatively light loads. This phenomenon is essentially similar to the buckling of long slender columns when the load exceeds the critical load. In practical design it is necessary to guard against this by choosing the spring proportions in such a way that the buckling load will always be greater than any load encountered in service, or, if this is not practical, guides must be provided to prevent sidewise motion.

The calculation of critical buckling loads for helical springs involves essentially the same methods as those used in column theory. However, the analysis in the case of the spring differs from that used in ordinary column theory in that it is necessary to consider the decrease in length under load. In the case of the usual steel column, on the other hand, this decrease is small and may be neglected. Besides the change in length under load, it is also necessary to consider the deformations of the spring due to lateral shearing forces. An analysis taking these factors into account has been carried out (Ref. 12). This analysis shows that the critical deflection at which buckling occurs depends on the ratio between free length l_0 and mean coil diameter $2r$ and on the method of fastening the spring ends, i.e., whether fixed or hinged. In Fig. 13a a spring with fixed ends is indicated; in Fig. 13b, one with hinged ends. In Fig. 14, curve a shows the ratio of the *critical buckling deflection* to *free length* plotted against ratio of free length l_0 to mean coil diameter, for a spring with hinged ends (Ref. 12). Curve b represents the built-in end condition. Actual helical springs compressed between plates which are approximately parallel will usually show results intermediate between these two cases. For this reason a conservative method of design is to use the hinged-end curve although this may yield somewhat low values for the buckling load. The critical buckling load for any spring is the critical deflection multiplied by the spring rate (eq. 3). Tests (Ref. 12) show good agreement with the curve

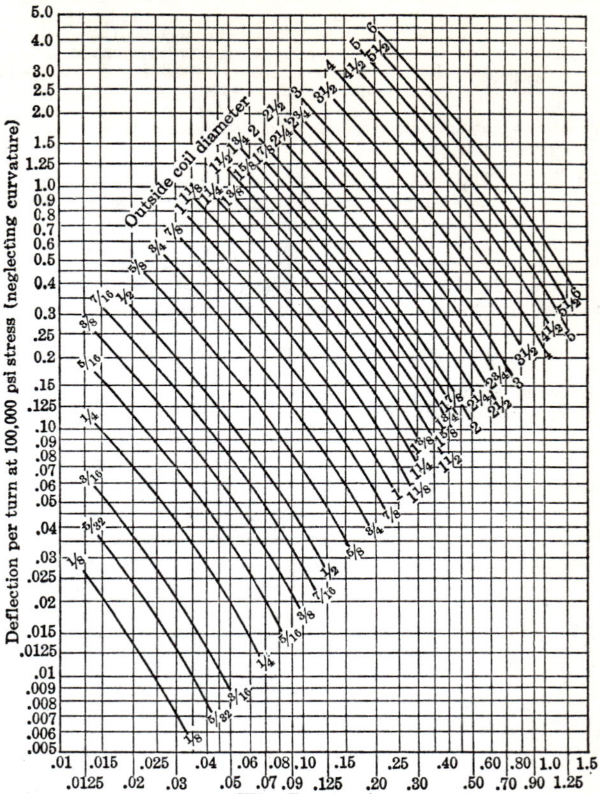

FIG. 12. Chart for calculating deflections in statically loaded helical springs of round wire, based on 100,000 psi stress and 11.4×10^6 psi torsional modulus. For any other stress, S_s, deflection values must be multiplied by $S_s/100{,}000$; for any other modulus, G, values must be multiplied by $11{,}400{,}000/G$. (Reproduced by permission from *Machine Design*, July 1942)

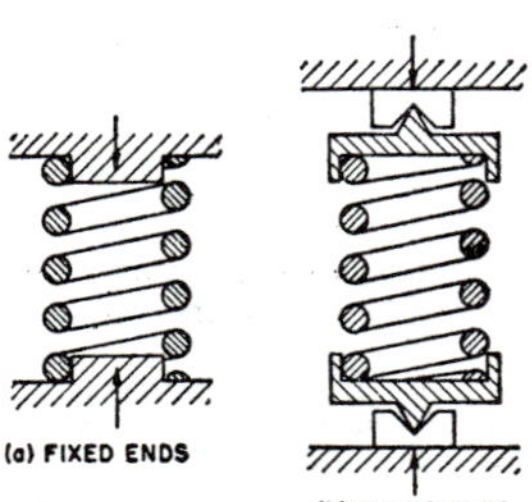

FIG. 13. Springs with fixed and hinged ends.

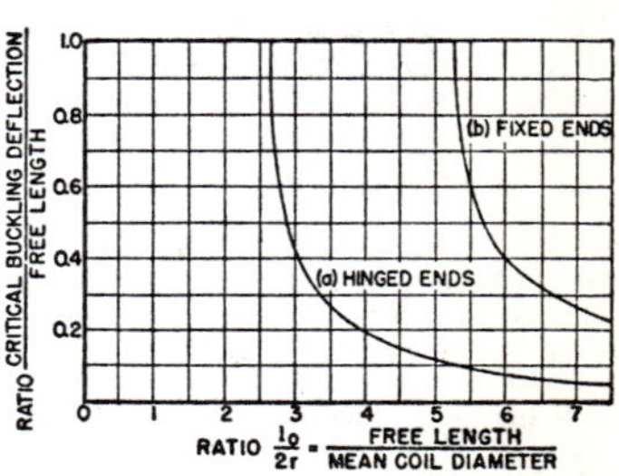

FIG. 14. Curves for finding critical buckling deflection. Buckling load = buckling deflection × spring rate. (Data from Haringx, Ref. 12)

for springs with hinged ends. In practical springs some deviation from these theoretical curves must be expected because of uncertainty as to the end condition and the effects of eccentricity of loading. From Fig. 14 it is seen that if the ratio $l_0/2r$ between free length and mean coil diameter is less than about 2.7 for the hinged-end condition, no buckling will occur before the spring is compressed solid. For the built-in end condition, buckling will not occur for a ratio $l_0/2r$ less than about 5.3.

Example 4. As an example of the use of the curve of Fig. 14 for determining danger of buckling, assume a steel helical compression spring of the following dimensions: free length, $l_0 = 6$ in., mean coil radius $r = 0.75$ in., $l_0/2r = 4$, outside diameter = 1.75 in., wire diameter 0.25 in., active turns $n = 12$. From Table 4, p. 11-10, for these dimensions the spring rate is 1667/12 = 139 lb per in. Assuming that the spring has hinged ends, from Fig. 14 the ratio of critical deflection to free length for $l_0/2r = 4$ is 0.2. Thus buckling for such a spring would occur at a deflection $0.2l_0 = 0.2 \times 6 = 1.2$ in. or at a load of 1.2 × 139 lb per in. = 167 lb. However, if this spring were rigidly clamped to two parallel plates and then loaded so that the built-in end condition were realized, no buckling would be expected at any practical deflection.

LATERALLY LOADED HELICAL SPRINGS. In practice helical springs are sometimes loaded laterally while being compressed, as indicated in Fig. 15. Examples of such applications are certain types of railway trucks and spring supports for refrigerator compressors. For such springs it will be found that in general the lateral deflection δ_1 due to the load Q, increases as the axial load P increases. To calculate this lateral deflection δ_1, the following formula may be used. This applies to steel springs where the modulus of elasticity E is 30×10^6, and modulus of rigidity $G = 11.5 \times 10^6$ psi.

$$\delta_1 = \frac{CQnr}{10^6 d^4}[0.408l^2 + 2.12r^2] \qquad (7)$$

where n = number of active coils; r = mean coil radius; d = wire diameter; l = compressed length; Q = lateral load (Fig. 15); C = a factor which depends on the ratio of free length to mean coil diameter and on the ratio δ/l_0 between compression due to axial load P and free length. A chart for determining the factor C is shown in Fig. 16 (Ref. 12). It should be emphasized that because of imperfect clamping of the ends and other deviations from the idealized case assumed in deriving eq. 7 and Fig. 16, considerable difference may occur between test and theoretical values for lateral deflections. (See Ref. 13.) Also because of these lateral deflections, additional stresses are set up (Ref. 1, p. 180) which should be allowed for in design.

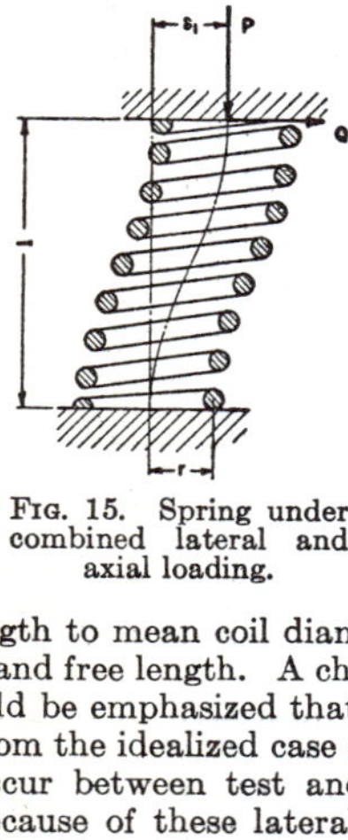

Fig. 15. Spring under combined lateral and axial loading.

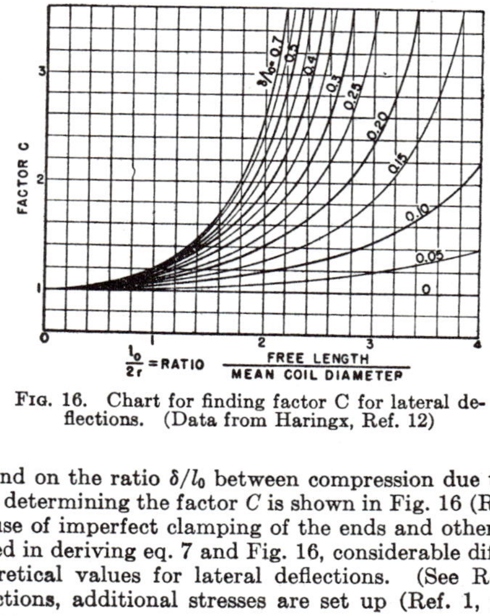

Fig. 16. Chart for finding factor C for lateral deflections. (Data from Haringx, Ref. 12)

Example 5. A laterally loaded steel spring has the following dimensions: OD = 5 in.; mean coil diameter $2r = 4.25$ in.; bar diameter $d = 3/4$ in.; free length $l_0 = 9.5$; $l_0/2r = 2.24$; and active coils $n = 8$. Using these dimensions in eq. 3, the spring rate P/δ is 732 lb per in.

Assume that the axial load P on this spring is 2400 lb and the lateral load $Q = 200$ lb. Then the axial deflection δ at this load is 2400/732 = 3.28 in. Compressed length $l = 9.5 - 3.28 = 6.22$ in. The ratio δ/l_0 between deflection and free length is 3.28/9.5 = 0.345. From the chart of Fig. 16 for $\delta/l_0 = 0.345$ and $l_0/2r = 2.24$ the factor $C = 2.15$. Using this value and taking $Q = 200$ lb, we find from eq. 7 the lateral deflection $\delta_1 = 0.59$ in. In this case the lateral deflection is about twice the value obtained for a very small axial load where C will be approximately 1.

OTHER DESIGN CONSIDERATIONS. Allowance for End Coils. In forming the usual compression spring, about three quarters of a turn is tapered near each end. The end may then be squared and ground, or if the taper has been made by forging it is merely squared. For such ends, tests have shown that around 0.8 to 1.0 turn near each end is inactive, or a total of 1.6 to 2 turns. Thus if a spring has 12 turns tip to tip of bar, the active coils may vary from 10 to 10.4 turns. At higher loads the number of inactive turns is reduced because of seating of the end coils. For practical design, an average figure of 1 3/4 inactive coils will usually give satisfactory results (Ref. 1, p. 158). The solid height of the spring will be approximately equal to the number of solid coils (see below) times the bar or wire diameter.

Eccentricity of Loading. If a compression spring of usual design is compressed between two parallel plates as in a testing machine, it will be found that in general the resultant load is not directed along the spring axis but instead is slightly eccentric. The effect of this eccentric loading is to increase the torsional stress on one side of the spring and decrease it on the other. An analysis of the effect of this load eccentricity has been carried out (Ref. 14). The results indicate that the load eccentricity e depends primarily on the number of coils and may be expressed as

$$e = 1.12r\left[\frac{0.504}{N} + \frac{0.121}{N^2} + \frac{2.06}{N^3}\right] \tag{8}$$

In this r = mean coil radius and N = number of solid coils or total number of coils minus 1/2 for the usual type of squared and ground ends. The stress in the spring will be increased approximately in the ratio $1 + e/r$ as compared with the stress in a spring under a purely axial load. This correction is of importance for a spring with a small number of active coils. For example, in the case of a spring with five total coils the correction amounts to about 15%. Because of inaccuracies in forming the end turns, eccentricity values in different springs may vary over a wide range, and eq. 8 should be considered to be very rough.

Effect of Variations in Spring Dimensions. Another factor of importance is the effect of slight variations in spring-wire or coil diameters on load-deflection characteristic of springs. Analysis shows that a 1% change in the mean coil diameter means a 3% change in the load-deflection characteristic, whereas a 1% change in the wire diameter results in a 4% change in the deflection characteristic.

In general it is not practical to hold load-deflection characteristics closer than about 5 to 10% in practical springs except at extra cost.

Allowable variations in commercial spring-wire diameters as listed in ASTM standards for various materials are given in Table 6.

Table 6. Allowable Variations in Commercial Spring-wire Sizes

Material	ASTM Specification Number	Wire Diameter, in.	Permissible Variation, in.
Hard-drawn spring wire	A-227-39T	.028 to .072	±.001
Chrome-vanadium spring wire	A-229-39T	.073 to .375	±.002
Oil-tempered wire	A-231-39T	.376 and over	±.003
Music wire	A-228-39T	.026 and under	±.0003
		.027 to .063	±.0005
		.064 and over	±.001
Carbon-steel valve spring wire	A-230-39T	.093 to .148	±.001
		.149 to .177	±.0015
Chrome-vanadium valve spring wire	A-232-39T	.178 to .250	±.002

In winding springs cold, there is always some "spring-back." In other words, the inside diameter of the spring after winding will be slightly greater than the diameter of the mandrel as a consequence of the elastic and plastic properties of the material; this results in unavoidable variations in coil diameters.

Table 7. Tolerances on Spring-coil Diameters (Close Cold-wound Helical Springs)

Mean Coil Diameter, D, in.	Variations in Diameter, D, in. $D/d = 4$	$D/d = 8$	$D/d = 12$
1/8	±.003	±.0035	±.005
1/8 to 3/16	.0035	.005	.0065
3/16 to 1/4	.005	.0065	.0085
1/4 to 3/8	.0065	.0085	.0105
3/8 to 1/2	.0085	.0105	.0130
1/2 to 3/4	.0105	.0130	.0155
3/4 to 1	.0130	.0155	.0230
1 to 1 1/2	.0155	.0205	.0318
1 1/2 to 2	.0185	.0313	.0408
2 to 3	.030	.0430	.0528
3 to 4	.....	.0550	.0730
4 to 5	.....	.0725	.095
5 to 6	.....	.....	.125
6 to 7	.....	.....	.165
7 to 8	.....	.....	.210

As an example of actual variations in coil diameters to be expected in practice, the tolerances given by one spring manufacturer are listed in Table 7 (Ref. 9). It may be seen that these tolerances depend both on the spring index D/d and on the mean diameter D.

Effect of Modulus of Rigidity. An accurate calculation of the deflection of actual springs requires that the modulus of rigidity of the spring material be known with good accuracy. However, modulus values reported in the literature vary considerably for different materials. In particular, the effect of a decarburized layer only a few mils thick may reduce the effective modulus by several per cent. On the basis of available data a good average figure for modulus of rigidity for carbon and alloy steels is 11.5×10^6 psi. However, for hot-wound carbon-steel springs made from hot-rolled material in the larger sizes, some manufacturers recommend a modulus figure of 10.5×10^6 psi. Average values of the modulus for various spring materials are given in Table 8. However, it should be noted that in individual cases deviations of several per cent may occur from these values.

Table 8. Average Values for Modulus of Rigidity—Spring Materials

Material	Modulus of Rigidity, psi
Music wire under 0.015 diameter	12×10^6
Music wire over 0.015 diameter Carbon and chrome-vanadium steels	11.5×10^6
Hard-drawn stainless steel	10.5×10^6
Monel metal	9×10^6
Phosphor bronze	6×10^6

For springs operating at elevated temperature, the modulus is reduced. Test data indicate that for carbon, alloy, or stainless steels this reduction will be about 5% at 300 F and about 10% at 600 F (Ref. 1, p. 81).

Stress at Solid Compression. In the design of compression springs it is usually desirable to choose the coil pitch so that when the spring is compressed solid, no appreciable permanent set will occur. The reason for this is that in operation the spring may at times be compressed solid, and, if under these conditions it takes a set, the load at working deflections will be changed; hence the spring will no longer have its initial characteristics. Thus the stress at solid compression should not in general exceed certain limits.

Suggested corrected stresses at solid compression for various spring materials as given by Associated Spring Corp. (Ref. 15) are listed in Table 9. In the second column is given the stress at solid compression up to which it is not necessary to remove set, while in the last column is given the stress at solid compression which may be used after all set is removed by means of repeated compressions of the spring in the cold-setting operation.

Table 9. Suggested Corrected Stresses at Solid Compression—Helical Springs

Material	Stress at Solid Compression up to Which It Is Not Necessary to Remove Set, psi	Maximum Stress at Solid Compression with All Set Removed, psi
Music Wire		
up to 0.032 diam.	130,000	180,000
0.032 to 0.062	110,000	170,000
0.062 to 0.125	100,000	160,000
0.125 and over	90,000	150,000
Hard-drawn Spring Wire		
up to 0.032 diam.	120,000	170,000
0.032 to 0.062	100,000	160,000
0.062 to 0.125	90,000	150,000
0.125 and over	80,000	140,000
Oil-tempered Wire		
0.125 and over	80,000	140,000
18-8 Stainless (Hard-drawn)		
up to 0.125	85,000	140,000
over 0.125	75,000	120,000
Phosphor Bronze		
(general sizes)	40,000	70,000

VIBRATION AND SURGING OF HELICAL SPRINGS. When helical springs are subject to rapid alternation of load, vibration (or surging, as it is also called) of the coils may occur due to dynamical effects. Examples of such springs are valve springs for internal-combustion engines and recoil springs for guns. Particularly severe dynamical effects may occur in such cases when a harmonic in the motion of the spring end (or in the valve lift curve of valve springs) has a frequency which coincides with one of the natural frequencies of the spring (corresponding to one of the modes of vibration). For example, harmonics of very high order (up to the fifteenth or twentieth) may have to be considered in valve spring design. Usually the lowest natural frequency is of the most importance. For a spring compressed between parallel plates the first mode of vibration (corresponding to the lowest natural frequency) will consist of a motion of the middle part of the spring with the ends remaining stationary. The second mode of vibration (corresponding to a higher frequency) will have a node (or point of zero motion of the coils) in the middle of the spring, while maximum motion of the coils occurs at points one-fourth and three-fourths of the length distant from a given end of the spring.

Natural Frequencies. The lowest natural frequency f in cycles per second for a helical spring clamped between two parallel plates (Ref. 1, p. 232) is given by

$$f = \frac{d}{2\pi r^2 n}\sqrt{\frac{Gg}{32\gamma}} \text{ cycles per second} \tag{9}$$

where d = wire size, inches; r = mean coil radius, inches; G = modulus of rigidity, pounds per square inch; g = acceleration of gravity, inches per second^2 (g = 386 inches per second^2); γ = specific weight of material, pounds per inch^3; and n = number of active coils.

For usual steel springs having both ends clamped, where $G = 11.5 \times 10^6$ psi and $\gamma = 0.285$ lb per inch^3, the lowest natural frequency f becomes

$$f = \frac{3510d}{r^2 n} \text{ cycles per second} \tag{10}$$

Frequencies in the higher modes of vibration will be 2, 3, 4, etc., times this frequency.

For a spring with one end free and the other clamped, the lowest natural frequency would be equal to that of a similar spring twice as long but with both ends clamped. For such a spring eq. 10 may be used if the number of turns is taken as twice the actual number in the spring.

Example 6. As an example of the use of these equations in calculating natural frequency, assume a steel spring with d = 0.3 in.; r = 1 in.; n = 6. Using eq. 10, the lowest natural frequency becomes

$$f = \frac{3510 \times 0.3}{(1)^2 \times 6} = 175 \text{ cycles per second}$$

In the second mode of vibration the spring frequency will be double this (or 350 cycles per second).

Methods of Reducing Surge or Vibration. To reduce stresses due to resonant vibrations in valve springs several methods are useful. The natural frequency of the spring should be made as high as possible so that resonance will occur only for the higher order harmonics (which are usually of lower amplitude). The cam contour may also be shaped so as to reduce the amplitudes of the harmonics which are of importance in the speed range within which the engine is to be used. In many cases it is possible by a change in the cam contour to reduce the magnitudes of the critical harmonics to low values within the speed ranges of importance. Reducing or varying the pitch of the coils near the ends of the spring will often result in an improvement. The reason for this is that if resonance occurs with one harmonic, the end coils will close up, thus changing the natural frequency of the spring and tending to throw it out of resonance. Friction dampers, consisting of a three-pronged device with the prongs pressing against the center coils of the spring, have also been used to damp out resonant oscillations.

2. HELICAL TENSION SPRINGS

In designing tension springs the weakening effect of the end coils should be considered. Unless special care is taken in the manufacture of such springs a fairly sharp curvature of the wire or bar at the point where the hook joins the body of the spring may occur.

Thus if a half-end turn of a tension spring is bent up sharply during manufacture, a sharp curvature at point A (Fig. 17) tends to result in a high concentration of stress at this point with the result that most failures of tension springs occur at such points. For this reason, and also because tension springs cannot usually be cold set without leaving excessive space between turns, a somewhat lower working stress (about 20 to 25% lower) is usually recommended as compared with those used in compression springs. The stress-concentration effect may be reduced (and the strength increased) by shaping the end turn so that the radius of curvature is as large as possible.

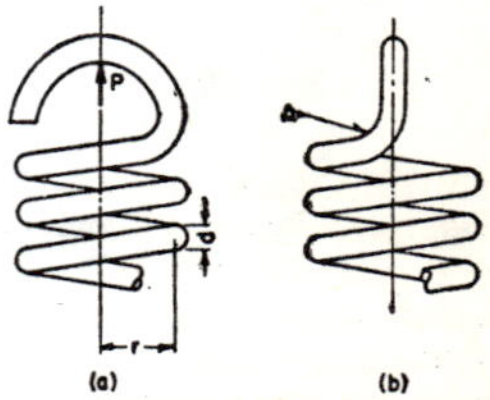

Fig. 17. Tension spring with sharp curvature at end loop

INITIAL TENSION. In cold-wound tension springs it is possible by partly overlapping the wire as it is fed to the coiling machine or mandrel to produce *initial tension*, i.e., a certain initial load must be applied to the spring before the coils start to separate. After separation of the coils begins, the spring rate or slope of the load-deflection diagram is the same as that which would be obtained for a spring with no initial tension. This effect is important in the operation of many mechanisms.

Values of initial tension load P_1 which can be obtained without difficulty in practice are given by the formula (Ref. 16):

$$P_1 = \frac{\pi S_s d^3}{16r} \tag{11}$$

where S_s = torsion stress due to initial tension, d = wire diameter, and r = mean coil radius.

Values of S_s depend on the spring index $2r/d$ and may be taken from Table 10.

Table 10. Torsional Stress Corresponding to Initial Tension

Spring Index, $2r/d$	Initial Tension Stress, psi
3	24,000
4	22,500
5	20,000
6	18,000
7	16,200
8	14,500
9	13,000
10	11,600
11	10,600
12	9,700
13	8,800
14	7,900
15	7,000

EXAMPLE 7. Assume a tension spring of 2-in. OD, 1/4-in. wire diameter, and index $2r/d = 7$. When wound with initial tension, from Table 10 the stress due to this tension will be S_s = 16,200 psi. From eq. 11 the initial tension load becomes

$$P_1 = \frac{\pi(16{,}200)(0.25)^3}{16 \times (0.875)} = 57 \text{ lb}$$

By changes in the method of winding, values of initial tension greater than this may also be obtained but usually at extra expense.

TYPES OF SPRING ENDS. Usually the end turns of tension springs are made in the simple forms indicated in Fig. 18, which also shows some of the variations possible. Some of these designs may, however, result in a considerably greater stress in the spring than that calculated on the basis of an axial load. For example, if the loop is at the side, the moment arm of the load (on which the maximum torsion stress in the springs depends) is practically twice that which would exist if a purely axial load were applied to the same spring. This means a doubling of the stress.

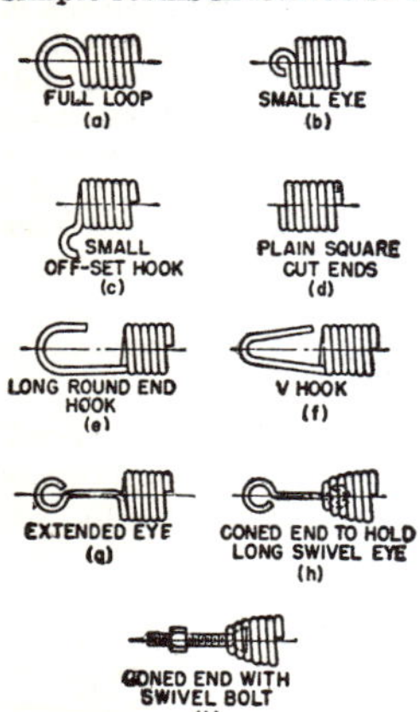

FIG. 18. Types of tension spring ends.

Sometimes in the actual loading of tension springs, even if the usual type of end loop is used, the line of action of the load may still be displaced by an appreciable amount from the axis of the spring. In this case a considerable increase in stress may also occur and should usually be considered.

Often tension springs are made with plain ends. Special fixtures called *spring ends* are attached to these as indicated in Fig. 19*a*. When using these, the spring is close wound and the ends of the spring are spread apart by screwing the spring into the holes. In this case an initial stress corresponding to the spreading apart of the turns near the ends is set up. This initial stress will correspond to a load equal to the initial tension in the spring plus the load corresponding to the distance the end coils are spread apart. A second type of spring end is shown in Fig. 19*b*. This is screwed into the ends of the spring coiled without end loops.

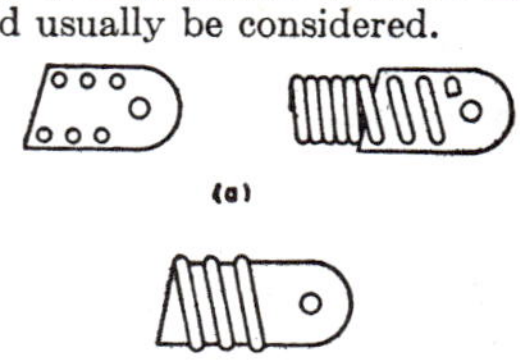

FIG. 19. Spring ends.

Designs shown in Fig. 18*h* and *i* will have relatively low stress in the end loops.

WORKING STRESSES FOR TENSION SPRINGS. An example of possible working stresses in tension springs for ordnance applications where space is restricted and only a limited life is required is shown in Fig. 20 for various materials (data from SAE Spring Committee, Ref. 4). In general, where more space is available and longer life is desired, stresses around 20 or 25% lower than these values should be used.

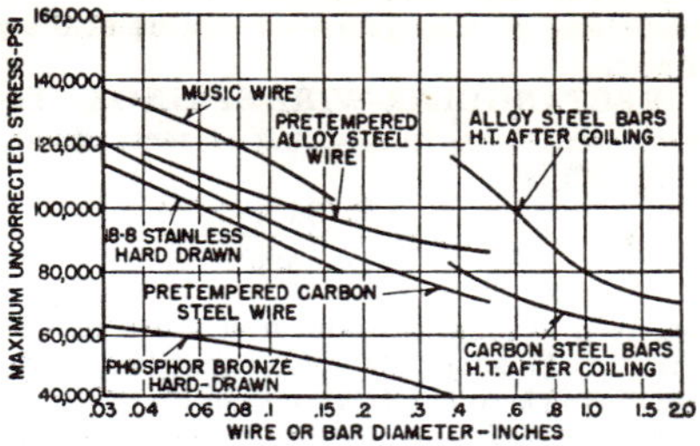

FIG. 20. Stresses for tension springs in ordnance applications where space is limited, and where only short life is required. (Data from *SAE Manual*, Ref. 4)

If fatigue is present, it is suggested that the design be carried out in the same manner as for compression springs (p. 11-07) using stresses about 20% lower.

3. SQUARE- OR RECTANGULAR-BAR SPRINGS

Helical tension or compression springs of rectangular bar section are sometimes used to provide a maximum amount of energy storage within a given space. Theoretically, a square-bar spring having the same material, bar size, and OD as a round-bar spring will be able to carry about 27% more load before complete yielding occurs over the cross section. This advantage is partially offset by the fact that it is difficult to obtain material in the square or rectangular section of equal quality to that in the round section. Also the stress distribution under elastic conditions is not as uniform in the square or rectangular section as it is in the round. This nonuniformity may be of importance where repeated loading is present.

SQUARE WIRE. Neglecting correction for curvature and direct shear, the uncorrected stress in a helical spring of square section is

$$S_s = \frac{4.8Pr}{a^3} \tag{12}$$

where P = load, a = side of square cross section, and r = mean coil radius.

Deflection δ of a square-bar helical spring is given by

$$\delta = \frac{44.6Pr^3n}{Ga^4} \tag{13}$$

where n = number of active coils and G = modulus of rigidity. This formula may be around 2 to 4% in error for indexes between three and four, but for most practical cases it is sufficiently accurate.

For square wire the *corrected* stress is found by multiplying the stress of eq. 12 by a factor K' (Ref. 3), where

$$K' = 1 + \frac{1.2}{c} + \frac{0.56}{c^2} + \frac{0.5}{c^3} \tag{14}$$

In this equation $c = 2r/a$ = spring index. This factor is slightly below the factor K for round wire (eq. 5).

In general when a square-wire spring is wound, as a consequence of plastic deformation during coiling, the section becomes trapezoidal as indicated in Fig. 21. In such cases an approximation may be had by taking an average value of a equal to

$$a = \frac{b_1 + b_2 + 2a_1}{4}$$

and taking the spring index equal to $2r/a_1$ for finding K'.

Table 5 and Figs. 11 and 12 which apply to round-wire helical springs may also be used for an approximate calculation of loads and deflections in square-wire springs at given uncorrected stresses. It is merely necessary to calculate the load and deflection at the given working stress in the corresponding round-wire spring, i.e., one having the same outside coil diameter and number of turns and a wire diameter equal to the average side of the square cross section. The loads thus found are multiplied by the factor 1.06 and the deflections by 0.738 to find those for the square-wire spring at the given stress.

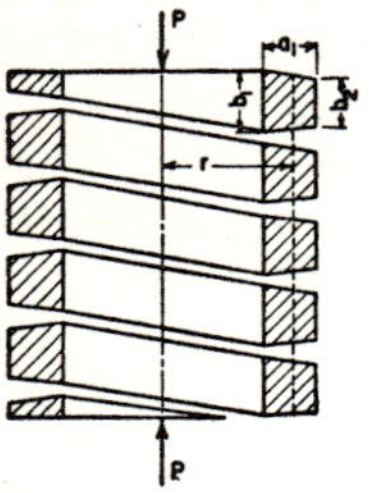

FIG. 21. Helical spring of square wire axially loaded. The wire becomes somewhat trapezoidal due to the coiling operation.

RECTANGULAR-WIRE SPRINGS. The uncorrected stress in a spring of rectangular wire (Fig. 22) is

$$S_s = \frac{Pr(3a + 1.8b)}{a^2b^2} \tag{15}$$

where a = long side and b = short side of rectangle. This formula holds regardless of whether the spring is coiled flatwise as in Fig. 22 or lengthwise with the long side of the section parallel to the spring axis. However, the correction for curvature is different in the two cases.

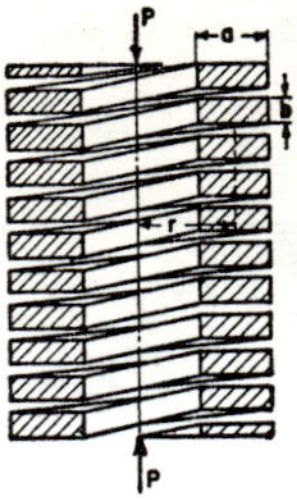

FIG. 22. Helical spring of rectangular wire coiled flatwise.

To calculate corrected stress S_s' in rectangular-bar springs. the chart of Fig. 23 may be used (Ref. 17). This gives

$$S_s' = \beta \frac{Pr}{ab\sqrt{ab}} \tag{16}$$

Note that in this case a and b are sides perpendicular and parallel respectively to the spring axis (Fig. 23) and β is a factor depending on the ratio a/b or b/a and on the spring index $2r/a$. (Note that in this case b always represents the side parallel to the spring axis.)

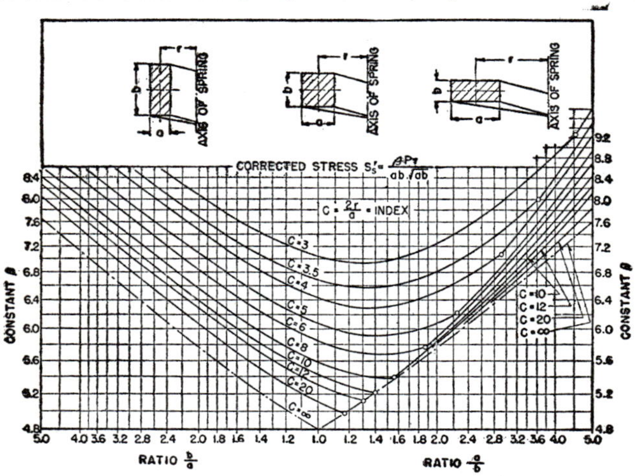

FIG. 23. Curves for finding stress factor β for rectangular bar springs. (Data from charts by Liesecke Ref. 17, p. 425)

EXAMPLE 8. Spring coiled flatwise as in Fig. 22. Dimensions are: $a = 1/2$ in., $b = 1/4$ in., $a/b = 2$ $r = 1.5$, $c = 2r/a = 6$, $P = 300$ lb. From Fig. 23 for $a/b = 2$, $c = 6$, $\beta = 5.88$. Corrected stress S_s is (from eq. 16)

$$S_s' = \beta \frac{Pr}{ab\sqrt{ab}} = \frac{5.88(300)1.5}{(0.25)(0.5)\sqrt{0.125}} = 60{,}000 \text{ psi}$$

To calculate deflections in rectangular-bar springs having large indexes (say greater than 8), the following formula, based on torsion of a straight bar of rectangular section, will yield results accurate to within a few percent where the pitch angle is not large:

$$\delta = \frac{19.6Pr^3n}{Gb^3(a - 0.56b)} \tag{17}$$

In this formula a represents the long side and b the short side of the cross section (Fig. 22).

For both small and large indexes, the deflection of a rectangular-bar spring may be calculated by using the chart of Fig. 24 (Ref. 17). Deflection is

$$\delta = \frac{8\gamma P r^3 n}{a^2 b^2 G} \tag{18}$$

where the constant γ depends on the ratio a/b or b/a and may be taken from Fig. 24.

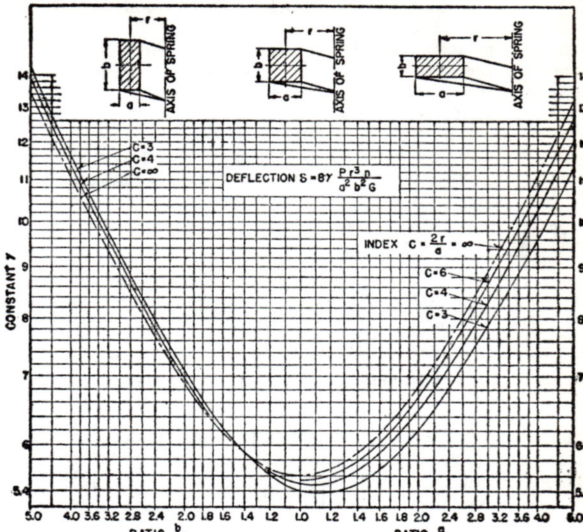

Fig. 24. Curves for finding deflection factor γ for rectangular bar springs. (Data from charts by Liesecke, Ref. 17, p. 892)

Here b is the side parallel to the spring axis, and a the side perpendicular thereto. If the spring is coiled flatwise, the ratio a/b is taken; if it is coiled edgewise, a ratio b/a is taken. (See sketches on Fig. 24.)

Example 9. A spring is coiled flatwise with $a = 1/2$ in., $b = 1/4$ in., $r = 1.5$ in., $P = 300$ lb, number of active turns $n = 5$, and $G = 11.5 \times 10^6$ lb per sq in. (steel). From Fig. 24 the constant $\gamma = 6.7$ for $a/b = 2$, $2r/a = 6$. From eq. 18

$$\delta = \frac{8\gamma P r^3 n}{a^2 b^2 G} = \frac{6.7 \times 8 \times 300 \times 3.37 \times 5}{0.25 \times 0.0625 \times 11.5 \times 10^6} = 1.51 \text{ in.}$$

4. HELICAL TORSION SPRINGS

Helical torsion springs have essentially the same shape as helical compression or tension springs, except that the ends are formed in such a way that the spring may be subject to torque about the coil axis instead of an axial load. Because of the mode of stressing of such springs the primary stress is *flexural*, in contrast to the helical compression or tension spring where the primary stress is *torsional*. Typical shapes of ends for torsion springs are shown in Fig. 25. They are made primarily for the purpose of transmitting external torque to the spring. Examples of torsion spring applications include door-hinge springs, springs for starters in automobiles, and springs for brush holders in electric motors.

A typical method of loading a torsion spring is illustrated in Fig. 26, where the spring is wound around a rod, one end of it being fastened to the rod while the other end has a straight portion projecting radially. If the spring is loaded by a force P at a radius R from the axis so as to wind the spring, the moment tending to twist the spring will be PR as indicated. Because of friction between the spring and guide the actual moment may decrease along the spring so that an exact calculation becomes rather involved.

In most cases it is advisable to load torsion springs in such a way that the spring tends to wind up as the load is applied. The reason for this is that the residual stresses set up as a consequence of cold winding are in such a direction as to subtract from the peak stress due to the loading, provided the load is in a direction tending to wind up the spring. If the direction of loading is such as to unwind the spring, a low-temperature heat treatment to remove such stresses is advisable.

Another reason for loading the spring in this manner is (see Fig. 26 for a load tending to wind up the spring) that the reaction will be against the arbor, and the peak bending moment in the spring will be PR. However, if the load is in the opposite direction to that shown, the peak moment on the wire will be $P(R + r)$, which may be considerably greater, r being the mean radius of the coil.

Often, if the ends of the spring are clamped or if special ends are used, some stress concentration may be expected near the ends. These stress concentrations should be carefully considered, particularly if the spring is subject to repeated loading.

Because a torsion spring (for usual applications) tends to wind up with load, its diameter decreases. In design it is important that sufficient clearance be allowed between the arbor or rod about which the spring is wound and the inner diameter of the spring. If this is not done, the spring may bind or wrap around the arbor and high stresses will be set up.

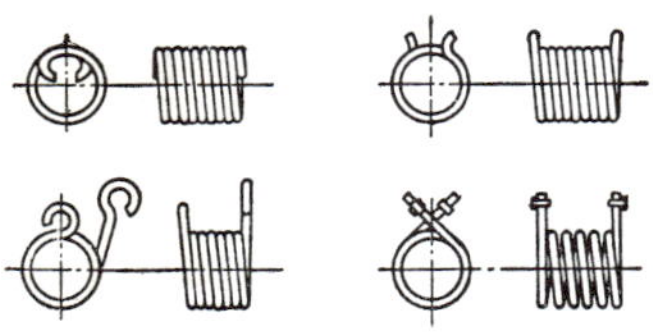

FIG. 25. Torsion spring ends.

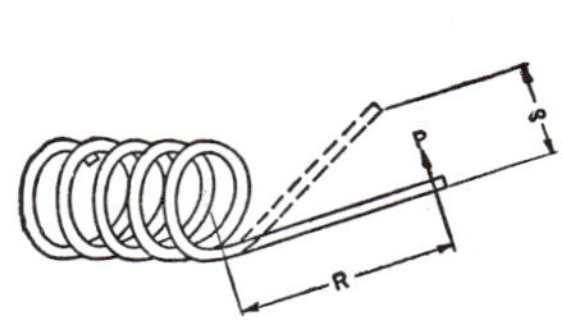

FIG. 26. Typical method of loading a torsion spring.

The clearance necessary may be estimated from the calculated deflection of the ends of the spring as given by eq. 21. Thus if the spring end deflects 90 degrees or one-fourth turn and the spring has 8 turns, the diameter will change in the ratio of $^1/_4$ to 8 or by about 3%, which should be allowed for in design. Likewise if the spring fits inside a tube and is loaded so as to unwind, sufficient clearance must be allowed between the outside diameter of the spring and the inside diameter of the tube.

Where torsion springs are relatively long, there is always the possibility of torsional buckling if the torque exceeds a certain value. This may be avoided by providing such springs with guides (such as rods or tubes). By properly clamping the ends or by applying an initial tension load, buckling may sometimes be avoided, and guides may not be required.

Usually, for manufacturing reasons, torsion springs are made of round wire. However, where maximum energy storage is required in a given space the use of square or rectangular wire may be advisable since the material is stressed more efficiently in such sections. Also for a given spring index, more material may be provided within a given outside diameter in the case of rectangular wire; this further increases the amount of energy which may be stored in a given space.

DESIGN FORMULAS FOR TORSION SPRINGS. Stresses and Deflections. Uncorrected stress for *circular* wire is

$$S = \frac{32M}{\pi d^3} \tag{19}$$

where M = bending moment and d = wire diameter.

Since this spring is essentially a curved bar, there will be a stress augment due to bar curvature which may be calculated from curved bar theory (Ref. 1). The corrected stress S' then becomes

$$S' = K_1 \frac{32M}{\pi d^3} \tag{20}$$

where K_1 depends on the spring index $2r/d$ and may be taken from the curve of Fig. 27.

Angular deflection ϕ in radians due to a given moment M is

$$\phi = \frac{128Mrn}{Ed^4} \tag{21}$$

where r = mean coil radius, n = number of active coils, and E = modulus of elasticity

For a load P at the end of the spring as in Fig. 26, neglecting friction, the deflection δ along the arc will be

$$\delta = \frac{128PR^2rn}{Ed^4} \tag{22}$$

where R = lever arm of load P.

For *rectangular-bar* torsion springs, the corresponding formulas are:

$$\text{Uncorrected stress: } S = \frac{6M}{bh^2} \tag{23}$$

where b = width, and h = depth of section, and

$$\text{Corrected stress: } S' = K_2\left(\frac{6M}{bh^2}\right) \tag{24}$$

where K_2 depends on the ratio $2r/h$ (Fig. 27).

Angular deflection ϕ in radians is

$$\phi = \frac{24\pi Mrn}{Ebh^3} \tag{25}$$

Deflection δ along arc due to load P (Fig. 26) is

$$\delta = \frac{24\pi PR^2rn}{Ebh^3} \tag{26}$$

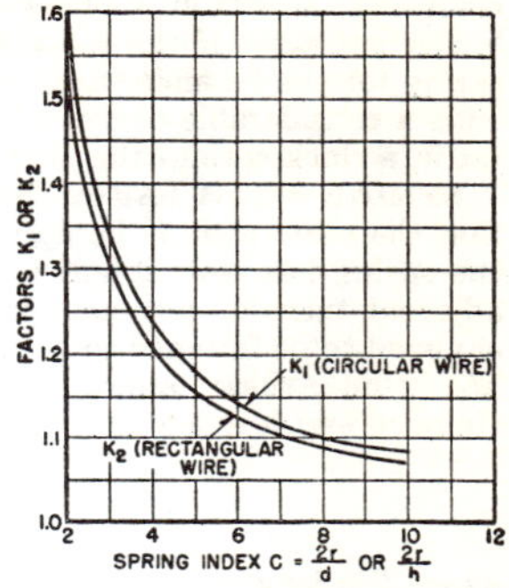

FIG. 27. Stress concentration factors K_1 and K_2 for torsion springs of circular and rectangular wire.

EXAMPLE 10. A brush-holder spring for small motors is loaded as indicated in Fig. 26. Dimensions are: load arm $R = 7/8$ in.; mean coil radius $r = 3/16$ in.; wire size $d = 0.04$ in.; spring index $c = 2r/d = 9.4$. The load P at the end of arm is 1 1/4 lb. From Fig. 27 the factor K_1 for $c = 9.4$ is 1.08. Using eq. 20, the corrected stress is (taking $M = PR$)

$$S' = K_1\frac{32M}{\pi d^3} = \frac{1.08 \times 32 \times 1.25 \times 0.875}{\pi \times (0.04)^3} = 188{,}000 \text{ psi}$$

If there are ten active turns, from eq. 21 the deflection in radians is (for $E = 30 \times 10^6$ psi)

$$\phi = \frac{128Mrn}{Ed^4} = \frac{128 \times 1.25 \times 0.875 \times 0.187 \times 10}{30 \times 10^6 \times (0.04)^4} = 3.4 \text{ radians}$$

This corresponds to a deflection of slightly more than 180 degrees.

Working Stresses. Maximum design stresses for torsion springs of different wire sizes and materials for average service conditions suggested by American Steel and Wire Company are given in Table 11 (Ref. 9). These conditions are defined as noncorrosive atmosphere, temperatures not exceeding 150 F, and relatively slowly varying or static loads.

Where the spring is subject to repeated or fatigue stresses through a considerable range, in general it will be necessary to use much lower working stresses than those suggested in Table 11. This is also true if the spring is subject to high temperatures or corrosion conditions. In such cases, consultation with the spring manufacturer is advisable.

Table 11. Recommended Maximum Design Stresses *—Helical Torsion Springs

(For average service conditions)

	Kind of Wire (Plain Carbon Steels)		
Diameter d	Hard-drawn	Tempered	Music
.004–.009	....	....	280,000
.010–.020	....	....	270,000
.021–.040	160,000	....	240,000
.041–.060	160,000	180,000	220,000
.061–.080	160,000	180,000	210,000
.081–.100	140,000	180,000	200,000
.101–.150	130,000	165,000	185,000
.151–.225	110,000	145,000	165,000
.226–.400	....	135,000	....
.401–.625	....	125,000	....
	Stainless steel 18-8	140,000	
	Monel metal	60,000	
	Brass	30,000	

* Computed by taking curvature effects into account.

5. SPIRAL SPRINGS

Spiral springs consist essentially of flat strip wound to form a spiral shape (Fig. 28). Usually the inner end is clamped to an arbor while the outer end may be either pinned or clamped. Common applications are springs for clocks and watches and motor brush-holder springs. If the individual turns of the spring do not come in contact, as in the spring for the balance wheel of a watch, the analysis of such springs may be carried out with a considerable degree of accuracy. However, when the spring is wound up tightly (as in a clock spring) the analysis is different. These cases will be treated separately.

SPRING WITH LARGE NUMBER OF TURNS. Clamped Outer End. It is assumed that the outer end of the spring is clamped or built in as indicated in Fig. 28, and that the spring has several turns or more which are, however, separated sufficiently so that adjacent turns do not come in contact during operation. The inner end of the spring is assumed to be fastened to an arbor which pivots about point O and is acted on by moment M_0. Under these assumptions the angular deflection ϕ of the arbor with respect to the clamped end is given by

$$\phi = \frac{M_0 l}{EI} \text{ radians} \tag{27}$$

where l = length of spiral, E = modulus of elasticity, $I = bh^3/12$ moment of inertia of section. To obtain the degrees of rotation, eq. 27 must be multiplied by 57.3.

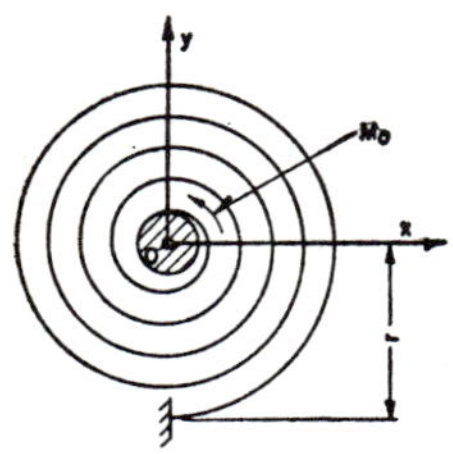

FIG. 28. Spiral spring with large number of turns (clamped outer end).

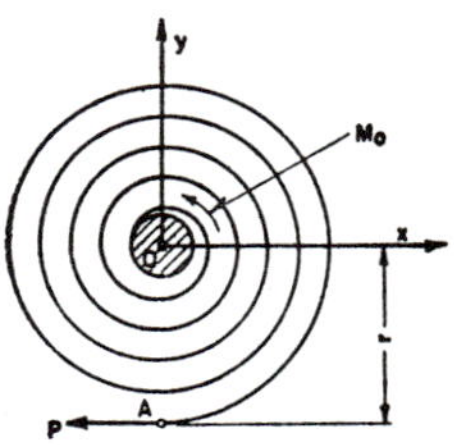

FIG. 29. Spiral spring with large number of turns (pinned outer end).

Since the moment is constant along the length, the stress for a spiral spring with the end conditions of Fig. 28 will be given by

$$S = \frac{6M_0}{bh^2} \tag{28}$$

where b = width of spring and t = thickness. Usually, however, there is some stress concentration at the clamped ends of the spring. Where fatigue loading is present this should be taken into account (see p. 11-38).

For a spring with a small number of turns, say less than two, the angular deflection ϕ will in general be around 3 to 15% less than that given by eq. 27. The peak stress will also be considerably greater than the value given by eq. 28 if the number of turns is less than about 3. For example, for 1 turn the ratio of the peak stress to the value given by eq. 28 varies from 1.35 to about 1.6, depending on the ratio of inside to outside diameter of the wound portion; for three turns this ratio varies from 1.12 to 1.19 (Ref. 18). These values assume a ratio of inside to outside diameter of the wound portion varying between 0.4 and 0.6.

Pinned Outer End. Frequently in practice the outer end A (Fig. 29) of the spiral spring is hinged instead of being clamped. In this case the external moment M_0 may be taken as Pr, and an analysis shows a value for angular deflection about 25% greater than for the clamped-end condition (Fig. 28) for the same applied moment and assuming a large number of turns.

The maximum stress S in this case is

$$S = \frac{12M_0}{bh^2} \tag{29}$$

Although this stress is twice that for a spring with a clamped outer edge (eq. 28) it occurs at a point where there is no stress concentration.

Example 11. A steel spiral spring having a pinned end (Fig. 29) is subject to a torque M_0 equal to 25 in.-lb. The outer diameter is 2 in., the strip section is 1/2 by 0.060 in., and the total length of the spiral is 15 in. Required: the stress and the deflection.

$$I = \frac{bh^3}{12} = 9.0 \times 10^{-6}$$

$$E = 30 \times 10^6 \text{ psi}$$

From eq. 27, for a clamped end, $\phi = \frac{M_0 l}{EI} = \frac{25 \times 15}{30 \times 10^6 \times 9 \times 10^{-6}} = 1.39$ radians. For a pinned end, this will be 25% greater, or $1.25 \times 1.39 = 1.74$ radians. Since 1 radian = 57.3 degrees, this means a deflection at this torque equal to $57.3 \times 1.74 = 100$ degrees.

From eq. 29 the maximum stress is

$$S = \frac{12M_0}{bh^2} = \frac{12 \times 25}{0.5 \times (0.06)^2} = 167{,}000 \text{ psi}$$

The nominal stress at the clamped end O where stress concentration occurs will be but half of this, or 84,000 psi.

Working Stresses. Working stresses in spiral springs may run as high as 175,000 psi for 1/8-in. thick steel strip to 240,000 psi for 1/32-in. strip, in applications where fatigue conditions are not a factor. For example, an ordinary clock spring during its life may be subject to less than 5000 cycles, and hence may be stressed much higher than if millions of cycles were involved. When fatigue conditions are present (as, for example, in the spiral spring for the balance wheel of a watch) the stress range should be kept well below the endurance range of the material, stress concentration conditions at the clamped edges being considered (see p. 11-38).

SPIRAL SPRINGS WITH LARGE DEFLECTIONS. Coils in Contact. The previous discussion of spiral springs has been based on the assumption that individual coils do not touch each other. This condition, however, does not apply in many cases, as for example in the power spring of a phonograph. In such cases the spring is usually placed inside a hollow case as indicated in Fig. 30. Here the spring is shown wound up on the arbor.

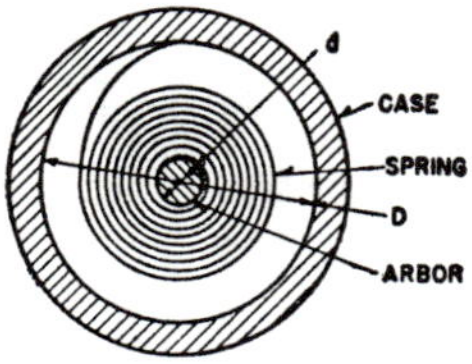

Fig. 30. Spiral spring wound on arbor.

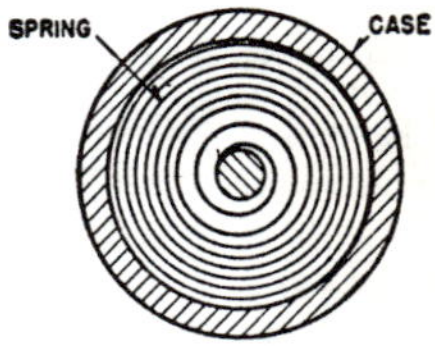

Fig. 31. Spiral spring unwound and resting against case.

When the spring is unwound it rests against the inside of the case as indicated in Fig. 31.

To find the total number of turns N delivered by the spring in unwinding from the wound-up position of Fig. 30 to the unwound position of Fig. 31, the following expression may be used (Ref. 19):

$$N = \frac{K}{2h}\left(\sqrt{\frac{4}{\pi}lh + d^2} + \sqrt{D^2 - \frac{4}{\pi}lh} - D - d\right) \tag{30}$$

where l = total length of spring, h = strip thickness, and d, D are dimensions given in Fig. 30. The factor K depends on the ratio m given by

$$m = \frac{\pi}{4lh}(D^2 - d^2) \tag{31}$$

Values of K for various values of m are given in Table 12.

Table 12

m	5	4	3	2	1.5
K	0.672	0.702	0.739	0.796	0.85

In practice, to avoid excessive stress, the arbor diameter is usually made about 15 to 25 times the strip thickness.

Example 12. A spiral spring is made of 0.015 × 3/4 in. strip steel, 100 in. long. Arbor diameter is 3/8 in., and inner diameter of the case is 2 1/4 in. The problem is to find the number of turns delivered from the solid to the free condition. Thus $h = 0.015$, $l = 100$, $D = 2.25$, and $d = 0.375$ in.

From eq. 31,

$$m = \frac{\pi}{4(0.015)(100)}(\overline{2.25^2} - \overline{0.375^2}) = 2.58$$

From Table 12, by interpolation, for $m = 2.58$, $K = 0.76$.

If the given values of K, l, D, and d are used in eq. 30, $N = 14.9$ turns which will be delivered by the spring in unwinding.

6. FLAT SPRINGS

SIMPLE CANTILEVER SPRING. Springs made from flat strip or bar stock are usually called flat springs even though they may be formed into more or less complicated shapes. Frequently, however, such springs may be considered as simple cantilevers (Fig. 32).

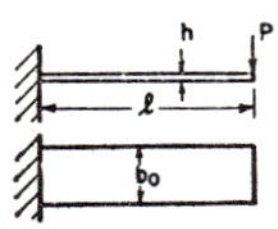

Fig. 32. Simple cantilever spring.

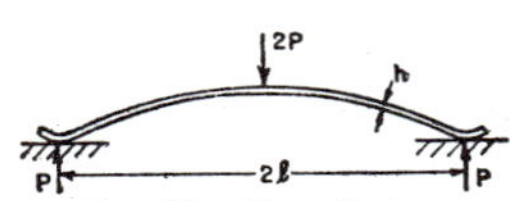

Fig. 33. Bowed strip.

Thus the spring shown in Fig. 33, which consists of a single bowed strip, may be considered essentially as a cantilever spring of length l loaded by a force P.

The formula for deflection of a simple cantilever spring is

$$\delta = \frac{4Pl^3}{Eb_0h^3} \tag{32}$$

where l = length (Fig. 32), b_0 = width, h = thickness, and E = modulus of elasticity of the material, pounds per square inch.

Where the thickness h is small compared to the width b_0, as in most practical cases, somewhat more accuracy, in general, will be obtained by multiplying the deflection figured by the simple cantilever formula by a factor 0.91. This factor takes into account the stiffening effect which occurs because lateral expansion or contraction of elements near the surface of the strip is prevented.

The maximum nominal stress is given by

$$S = \frac{6Pl}{b_0h^2} \tag{33}$$

Note that, if repeated loading is present, this stress may be augmented by stress concentration effects (p. 11-29).

TRAPEZOIDAL PROFILE SPRING. Somewhat more efficient use of material is obtained from the cantilever spring of trapezoidal profile (Fig. 34). Also leaf springs of

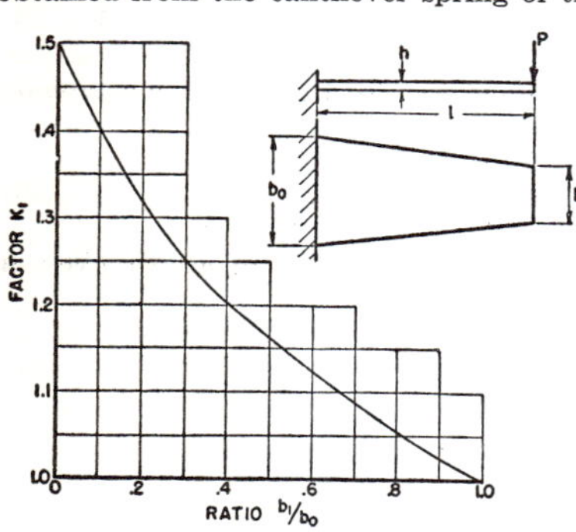

Fig. 34. Curve for finding factor K_1 (trapezoidal profile spring).

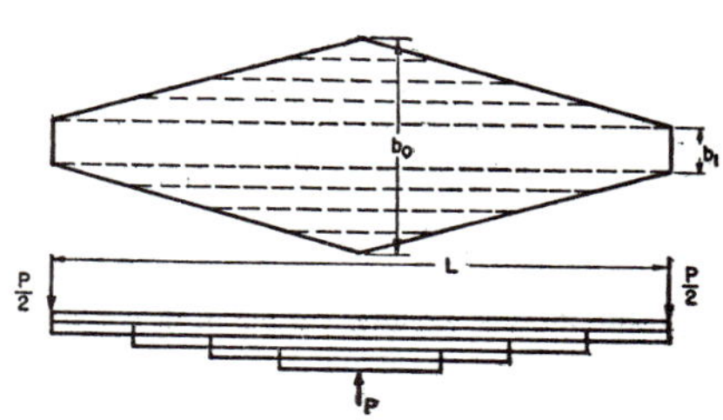

Fig. 35. Semi-elliptic leaf spring reduced to two trapezoidal cantilever springs.

the usual shape may, for practical purposes of analysis, be frequently considered as cantilever springs of trapezoidal profile as shown in Fig. 35. For a given load P the maximum stress is again given by eq. 33; in this case b_0 is the width at the built-in end

However, analysis based on ordinary beam theory shows that for a trapezoidal profile spring, the deflection is:

$$\delta = K_1 \left(\frac{4Pl^3}{Eb_0h^3}\right) \tag{34}$$

where

$$K_1 = \frac{3}{\left(1 - \frac{b_1}{b_0}\right)^3}\left[\frac{1}{2} - 2\frac{b_1}{b_0} + \left(\frac{b_1}{b_0}\right)^2\left(\frac{3}{2} - \log_e \frac{b_1}{b_0}\right)\right]$$

and b_1 = width at oaded end (Fig. 34). The factor K_1 may be taken from the curve of Fig. 34 for various values of b_1/b_0. It is thus seen that the deflection of a trapezoidal profile spring is equal to that of a rectangular profile spring multiplied by the factor K_1. Theoretically the most efficient spring is obtained where $b_1/b_0 = 0$ since, other things being equal, this value gives the maximum deflection for a given value of load. Practical considerations, however, usually dictate a value of b_1/b_0 greater than zero. When b_0 is very large compared to the thickness, the deflection given by eq. 34 should be multiplied by a factor 0.91, as explained previously.

STRESS-CONCENTRATION EFFECTS IN FLAT SPRINGS. (a) Sharp Bends. Frequently flat springs are formed with sharp bends (Fig. 36) which produce stress-concentration effects. To obtain the corrected stress in such cases the nominal stress calculated from eq. 33 must be multiplied by a stress-concentration factor K_2, which depends on the ratio $2r/h$ and may be taken from the curve marked K_2 in Fig. 27. Where fatigue loading is involved, it is advisable to calculate the stress *range* using the factor K_2; for static or infrequently repeated loading, K_2 may be assumed as unity, i.e., the uncorrected stress may be used. In general, sharp bends should be avoided as far as possible not only to reduce stress concentration but also for manufacturing reasons.

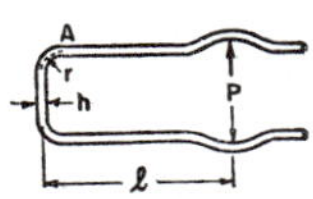

FIG. 36. Spring clip. Stress concentration is present due to sharp curvature of bend at *A*.

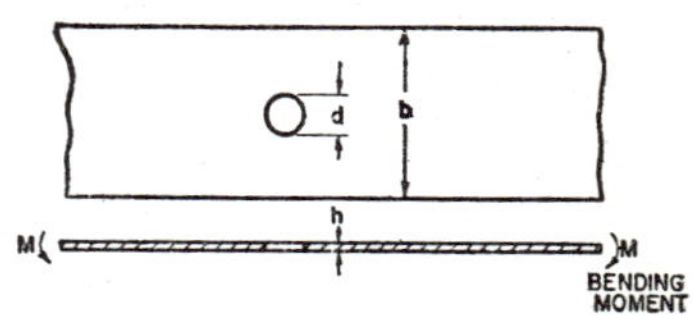

FIG. 37. Flat spring with hole large in diameter compared to thickness.

(b) Holes or Notches. A hole in a flat strip (Fig. 37) will result in a theoretical stress-concentration factor K_t equal to 1.85 if the hole diameter is large compared to the thickness h but small compared to the width b, i.e., if h/d and d/b are small compared to unity. Thus in this case the corrected stress would be $1.85[6M/(b - d)h^2]$, where M = applied bending moment. Where $h/d = 1/2$, the factor approximates 2; and where $h/d = 1$, $K_t = 2.25$, d/b being assumed small in both cases (Ref. 20). For very small d/h and d/b a theoretical value of K_t equal to 3 is obtained. These values are reduced somewhat if the hole diameter becomes an appreciable fraction of the width (Ref. 1, p. 300).

For semicircular notches of diameter d which are small compared to the width of the strip, values of K_t may be taken approximately the same as those for holes of the same diameter, and assuming the same ratios d/h.

Where fatigue loading is present it is advisable to use K_t in figuring the stress range.

WORKING STRESSES. For static loads or loads infrequently repeated, working stresses for flat springs made of 0.70 to 0.80 carbon steel (45 to 48 Rockwell C) may run as high as 180,000 psi for 0.125-in. thick strip to 240,000 psi for 0.025-in. thick strip (Ref. 5, p. 76). Where fatigue or repeated loading is present, much lower stresses than these should generally be used, and the corrected stress range should be kept well below the endurance range of the material.

7. LEAF SPRINGS

Sketches of a typical semi-elliptic leaf spring and of some typical ends and center clamps are shown in Figs. 38, 39, and 40. Rebound and alignment clips are illustrated in Fig. 41.

SYMMETRIC SEMI-ELLIPTIC LEAF SPRING. A semi-elliptic leaf spring of length L, loaded centrally by a force P (Fig. 42*a*), may as an approximation be considered as a trapezoidal cantilever spring (Fig. 35). Thus if there are n leaves (total) of width b, and

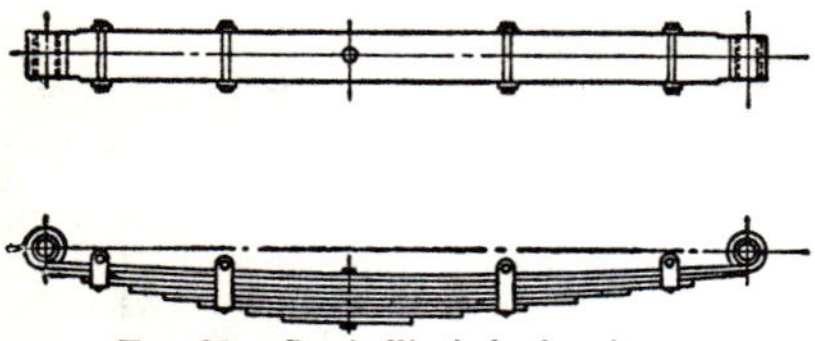

FIG. 38. Semi-elliptic leaf spring.

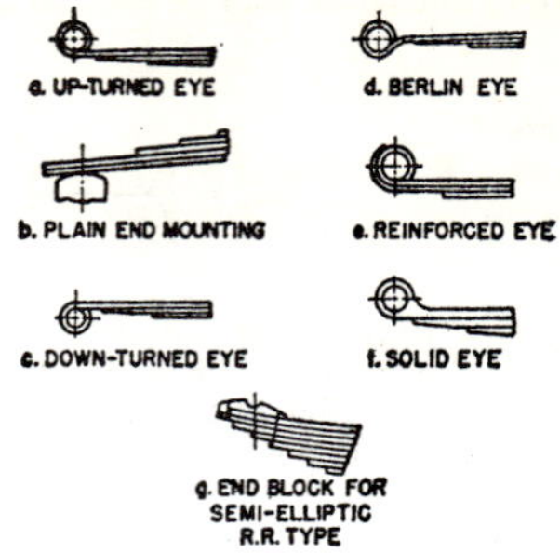

FIG. 39. Spring ends.

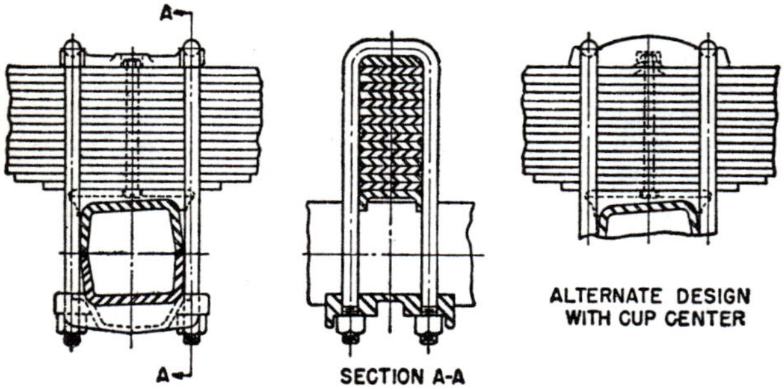

FIG. 40. Center clamps.

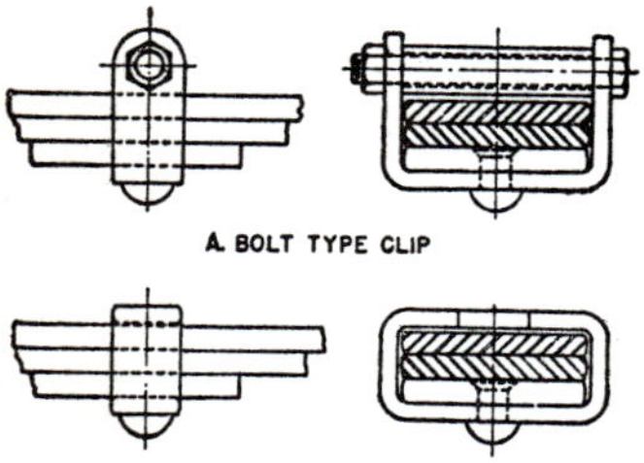

FIG. 41. Rebound clips.

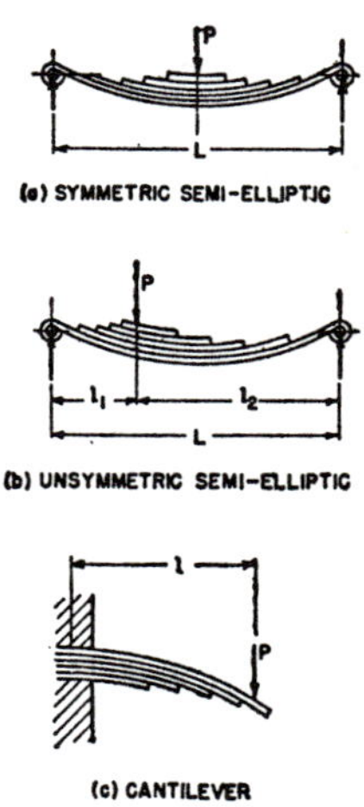

FIG. 42. Leaf springs.

n_1 leaves of length L, then $b_0 = nb$ and $b_1 = n_1b$. The ratio b_1/b_0 will be n_1/n, and the curve of Fig. 34 may be used to find K_1. In this case the deflection δ at load P is

$$\delta = K_1 \frac{PL^3}{4Enbh^3} \tag{35}$$

where E = modulus of elasticity and h = thickness of leaves (assumed uniform).

Spring rate in pounds per inch is

$$\frac{P}{\delta} = \frac{4Enbh^3}{K_1L^3} \tag{36}$$

If there is only one main leaf (Fig. 42*a*), the spring may frequently be considered approximately as a beam of uniform strength, $K_1 = 1.5$. In this case the spring rate in pounds per inch is

$$\frac{P}{\delta} = \frac{8Enbh^3}{3L^3} \tag{37}$$

Maximum stress in both cases is

$$S = \frac{3PL}{2nbh^2} \tag{38}$$

In practical springs where a center clamp is used, the length L in eqs. 35 to 38 is to be taken as the total length minus that of the center clamp.

UNSYMMETRIC SEMI-ELLIPTIC LEAF SPRING. For a leaf spring loaded unsymmetrically (Fig. 42*b*), all leaves being of equal thickness, approximate formulas for spring rate and stress, based on the assumption of a beam of uniform strength, are (Ref. 21)

$$\text{Rate } \frac{P}{\delta} = \frac{Ebnh^3L}{6l_1^2l_2^2} \tag{39}$$

$$\text{Stress } S = \frac{6Pl_1l_2}{Lnbh^2} \tag{40}$$

CANTILEVER SPRING (Fig. 42*c*), having n leaves, l being the length of the longest leaf, assuming a beam of uniform strength.

$$\text{Rate } \frac{P}{\delta} = \frac{Ebnh^3}{6l^3} \tag{41}$$

$$\text{Stress } S = \frac{6Pl}{nbh^2} \tag{42}$$

Equations 37 to 42, although based on the assumption that the spring may be considered as a beam of uniform strength (triangular profile), are frequently used in practical design even where these conditions are not strictly fulfilled. Effects which may cause deviations from the ideal conditions assumed are use of leaves of different thicknesses, use of more than one main leaf, interleaf friction, effects of spring shackles which result in angular loading at the ends. These effects may result in rather large deviations from the values calculated from the formulas (eqs. 37 to 42). For a further discussion, see Ref. 21.

WORKING STRESSES. Design stresses suggested by the SAE Spring Committee for automotive and railroad springs are given in Fig. 43 (Ref. 21). In calculating these stresses, eqs. 37 to 42 may be used, assuming that all leaves have equal thickness and free curvature. If the design is based on the static loading condition, as is frequently done, the design stress increases as the static deflection increases (lower curves of Fig. 43). Where possible, however, the design should be based on the maximum stress which occurs under conditions of maximum deflection, a value of 130,000 psi being suggested for carbon steel and 140,000 psi for alloy steel.

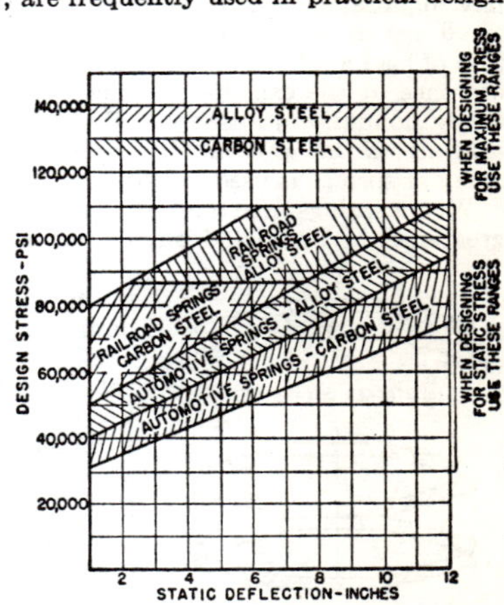

Fig. 43. Design stresses for leaf springs (SAE Spring Committee, Ref. 21)

Use of these stresses in general will result in a life of about 100,000 full-load cycles of stress (assuming that the springs have been shot-peened and cold-set). Also, a set of

3 to 5% of full-load deflection may be expected in service at these stresses. Where the spring operates under conditions of appreciable negative loading in rebound (i.e., when the stress may be reversed in direction) lower stresses than those of Fig. 43 must be used. Also, where other stresses than those due to vertical loading (such as wind-up torque stresses) are present, they must be considered in calculating maximum stress, or lower stresses must be used if the design is based on the static loading condition.

MATERIALS commonly used for leaf springs include SAE 1095, 4063, 5150, 6150, and 9260. In general, alloy steels are used for automotive springs and carbon steels are used for railway springs. Heat treatments giving Brinell hardnesses 415 to 460 will usually yield satisfactory results for spring steels. Also, shot-peening and cold-setting operations are desirable for satisfactory life in service.

8. INITIALLY CONED (BELLEVILLE) SPRINGS

The Belleville spring consists essentially of a coned or dished disk having a diametral cross section as shown in Fig. 44, the load being applied in the direction indicated by the arrow. On loading, the disk tends to flatten out, spring action being thus obtained. Changes in load-deflection characteristics obtained by varying the ratio h/t between cone height (or dish) and thickness are shown in Fig. 45. For ratios h/t around 0.5 the curve approximates a straight line, whereas for h/t equal to 1.5, the load is nearly constant (within a few percent) over a considerable range of deflection. For this reason springs having ratios approximating this value are frequently called *constant-load* or *zero-rate* springs.

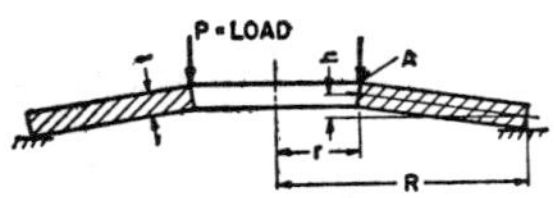

FIG. 44. Belleville spring.

Belleville springs may be used singly, stacked in parallel as shown in Fig. 46*a*, or stacked in series as shown in Fig. 46*b*. By stacking in parallel, load capacity is increased in proportion to number of disks; in addition, damping is obtained by friction between disks. By stacking in series the deflection for a given load is increased in proportion. Obviously with this latter arrangement deflections greater than the cone height cannot be obtained unless spacers are used. Belleville springs have high capacity and a relatively small space requirement in direction of load application, but this is obtained at the expense of a relatively nonuniform stress distribution across the section. Where a stack of springs is used, guides are usually necessary to avoid lateral buckling.

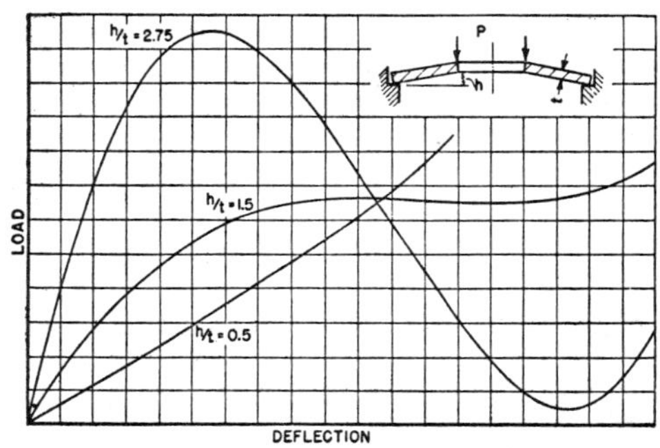

FIG. 45. Shapes of load-deflection diagrams obtained by varying ratio h/t.

Where springs are stacked in series (Fig. 46*b*) ratios between cone height and thickness greater than 1.3 should not in general be used, since instability and an irregular load-deflection characteristic may result. A ratio of outside to inside diameter between 1.5 and 2.5 will usually prove most desirable in practice.

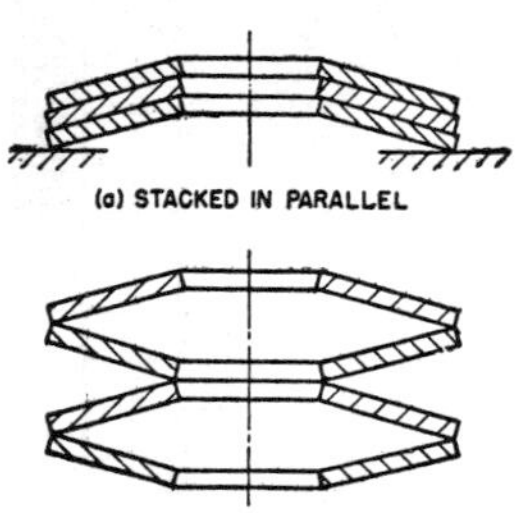

FIG. 46. Methods of stacking Belleville springs.

FORMULAS. The following approximate formulas for calculating stress and deflections in Belleville springs are based on the assumption that radial cross sections of the spring do not distort during deflection (Ref. 22). The results are in approximate agreement with available test data. However, high accuracy should not be expected since frictional and other effects not taken into account by these formulas can result in appreciable deviation between test and calculated results.

Load and Deflections. Load P at deflection δ is

$$P = \frac{1.1E\delta C}{R^2}\left[(h - \delta)\left(h - \frac{\delta}{2}\right)t + t^3\right] \tag{43}$$

where C depends on the ratio R/r (Fig. 47); R = outside radius; r = inside radius; h = cone height; t = thickness (Fig. 44); and E = modulus of elasticity of material.

This formula may be simplified (Ref. 1, p. 249) to the following:

$$P = \frac{CC_1Et^4}{R^2} \tag{44}$$

where C_1 is a load factor which depends on the ratios δ/t and h/t, given in the chart of Fig. 48.

It should be noted that the shapes of the curves for C_1 also represent the shapes of the load-deflection diagrams for the various ratios h/t. Where h/t is in the neighborhood of 1.4 to 1.6, approximately constant load results over a rather wide range of deflection. Theoretical load P_1 required to flatten spring is obtained from eq. 43, taking $\delta = h$. This gives

$$P_1 = \frac{1.1ECht^3}{R^2} \tag{45}$$

If the spring is tested against a flat plate, actual loads will be higher than P_1 because of the tendency of the spring to stiffen up near the flattened position.

Stress formulas are (Ref. 22) outlined below.

Maximum compression stress S_c at upper inner edge A (Fig. 44) at deflection δ is

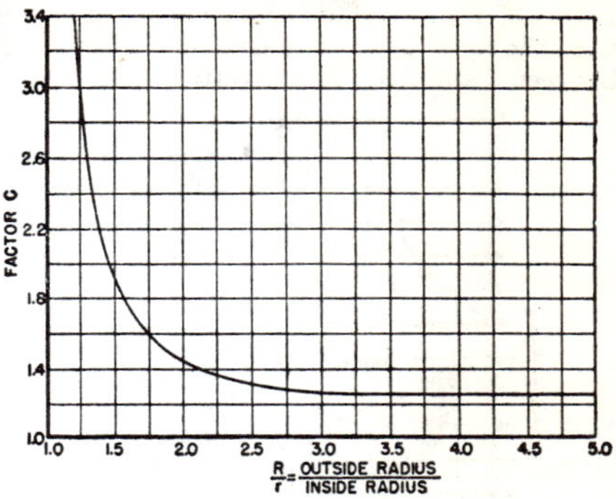

FIG. 47. Factor C, for Belleville springs.

$$S_c = \frac{-1.1E\delta C}{R^2}\left[C_2\left(h - \frac{\delta}{2}\right) + C_3t\right] \tag{46}$$

(Note that in general this will be the maximum stress in the spring provided δ is less than $2h$.) The factors C_2 and C_3 depend on the ratio R/r and may be taken from the curve of Fig. 49.

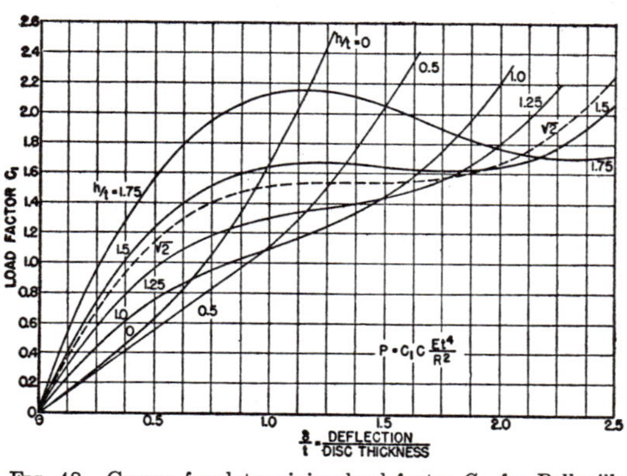

FIG. 48. Curves for determining load factor C_1, for Belleville springs.

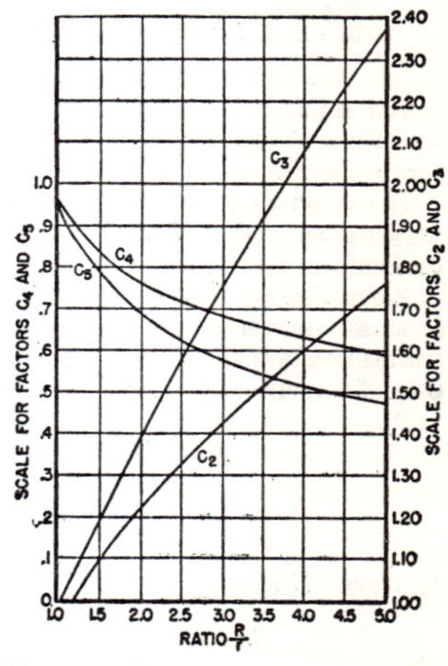

FIG. 49. Curves for finding C_2, C_3, C_4, and C_5, for Belleville springs.

Stress S_{t1} (normally tension) at lower inner edge of the spring is

$$S_{t1} = \frac{1.1E\delta C}{R^2}\left[-C_2\left(h - \frac{\delta}{2}\right) + C_3t\right] \tag{47}$$

Tension stress S_{t2} at lower outer edge of the spring is

$$S_{t2} = \frac{1.1E\delta C}{R^2}\left[C_4\left(h - \frac{\delta}{2}\right) + C_5t\right] \tag{48}$$

where C_4 and C_5 may be taken from the curve of Fig. 49. Since fatigue fractures frequently start at one of the lower edges, depending on the ratio h/t, either of the stresses S_{t1} or S_{t2} may be critical even though numerically smaller than the compression stress S_c.

In all cases in eqs. 46 to 48 a negative sign indicates compression and a positive sign tension.

It should be noted that if $\delta > 2h$, i.e., for deflections over twice the cone height, the stress at the lower inner edge S_{t1} (eq. 47) will become numerically larger than S_c.

Equation 46 for calculating S_c may be simplified to the following:

$$S_c = \frac{KEt^2}{R^2} \tag{49}$$

In this K is a stress factor depending on the ratios R/r, h/t, and δ/t, which may be taken from the lower group of curves of Fig. 50 for $R/r = 1.5$ and from those of Fig. 51 for R/r between 2 and 2.5. The negative value of K in these curves signifies a compression stress.

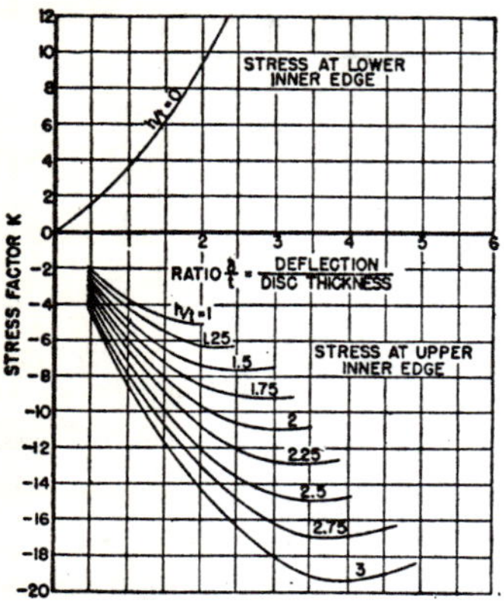

FIG. 50. Curves for determining stress factor K for Belleville springs (R/r = 1.5).

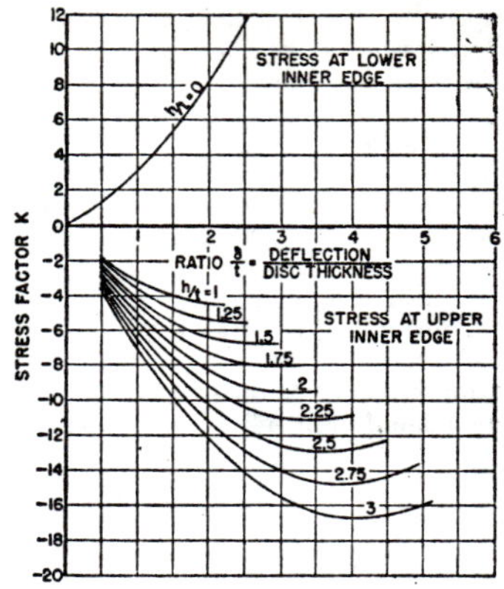

FIG. 51. Curves for determining stress factor K for Belleville springs (R/r = 2 to 2.5).

For a flat plate where $h/t = 0$, the upper curve of Fig. 50 may be used for $R/r = 1.5$ and that of Fig. 51 for $R/r = 2.0$ to 2.5. In this case, using K in eq. 49, the value of the tension stress at the lower inner edge is obtained.

It should be noted that there will usually be a small error involved in using the charts of Figs. 50 and 51, particularly for small values of h/t and δ/t. If greater accuracy is desired eq. 46 should be used.

Formulas for Small h/t Ratios. Where the ratio h/t between cone height and thickness is small (say less than about 0.5), the formulas for stress and deflection may be taken from flat plate theory to simplify the results. It is also assumed that δ/t is less than about 0.5. In such cases approximate deflection and stress are given by

$$\delta = K_1 \frac{PR^2}{Et^3} \tag{50}$$

$$S = K_2 \frac{P}{t^2} \tag{51}$$

In these K_1 and K_2 are factors depending on R/r and may be taken from Table 13.

Table 13. Factors K_1 and K_2 for Calculating Belleville Springs

(h/t and δ/t less than 0.5)

R/r	1.25	1.5	2	3	4
K_1	0.341	0.519	0.672	0.734	0.724
K_2	1.10	1.26	1.48	1.88	2.17

Nominal Stress across Diametral Section. Assuming that the spring is approximately flat when loaded, as is true in most practical cases, the nominal stress S_n across a diametral section at any load P as calculated by dividing the external bending moment by the section modulus is

$$S_n = \frac{0.955P}{t^2} \tag{52}$$

Derived in this way, this formula yields a kind of average stress, the stress concentration effect due to the presence of the hole being neglected. It is believed to offer a somewhat better indication of the load-carrying capacity of the spring under static loading conditions than the more commonly used eqs. 46 or 49.

EXAMPLE 13. A Belleville spring is to be designed for a gasket application where the load is to be held approximately constant at 5000 to 6000 lb and where the deflection range due to temperature expansions will be around 0.09 in. Type of load deflection curve desired is that for $h/t = 1.5$ (Fig. 45). Space available will permit using an 8 1/2-in. OD and 4 1/4-in. ID spring. Maximum compression stress S_c (eq. 49) is limited to 220,000 psi.

In order to avoid the steep rise in the load near the flattened position, maximum deflection will be taken as 1.3 times the thickness, or $\delta/t = 1.3$. Taking $R = 4\,1/4$ in., $r = 2\,1/8$ in., $R/r = 2$, from Fig. 51, for $h/t = 1.5$, $\delta/t = 1.3$, $K = -5.1$. Taking $S_c = -220{,}000$ psi (compression), $E = 30 \times 10^6$ psi for steel, by solving eq. 49 for t and substituting these values, the required thickness t becomes

$$t = R\sqrt{\frac{S_c}{KE}} = 4.25\sqrt{\frac{-220{,}000}{-5.1 \times 30 \times 10^6}} = 0.161 \text{ in.}$$

From Fig. 48 for $h/t = 1.5$, $\delta/t = 1.3$, the load factor $C_1 = 1.67$, and from Fig. 47 for $R/r = 2$, the factor $C = 1.45$. Load per disk at $\delta = 1.3t$ or 0.21 in. is

$$P = \frac{C_1 C E t^4}{R^2} = 2700 \text{ lb}$$

Using two springs in parallel, a load of twice this or 5400 lb is obtained which will be approximately constant within the deflection range desired.

WORKING STRESSES. Static Loading. Where Belleville springs are subject only to static or infrequently repeated loading, experience indicates that calculated stresses S_c ranging from 200,000 psi to as high as 375,000 psi for alloy steels as calculated from eq. 46 or eq. 49 may be used. Although these values seem extremely high, it should be remembered that such calculated stresses in most cases are localized near the upper inner edge of the spring (near A, Fig. 44). Consequently local yielding which will occur at such points when the load is applied will allow a redistribution of stresses so that the remaining portions of the spring can take a greater share of the load.

If the stress is figured using the nominal stress formula (eq. 52), the value should be kept well below the yield point of the material.

Fatigue Loading. For springs subject to fatigue or repeated loading, lower stresses should be used than for those subject to static loading. Although not many data are available, the indications are that Belleville springs made of good-quality spring steel will withstand repeated loading up to perhaps 500,000 cycles even for stresses as high as 200,000 psi calculated from eq. 49 (Ref. 5). For longer spring life or higher operating temperatures, it is necessary to use lower stresses.

It should be noted that the fatigue strength of Belleville springs may in general be increased by shot peening and presetting, and also by grinding the surfaces after heat treatment. Care must be taken to avoid surface cracking during this grinding operation.

MATERIALS used in the smaller sizes are SAE 1074, 1085, and 1095. For larger thicknesses and for highest quality, silico-manganese steel SAE 9260 or chrome-vanadium steel SAE 6150 are preferred. Phosphor bronze (spring temper) may also be used, in which case stresses about half those allowable for steel springs should be used and the modulus E should be taken as 15×10^6 psi.

9. SPRING MATERIALS AND PROPERTIES

The approximate chemical composition of various commonly used spring steels is given in Table 14.

Table 14. Composition of Various Spring Materials

(Reprinted by permission from *Mechanical Springs*, by A. M. Wahl, Penton Publishing Co., Cleveland, 1944.)

(ASTM Standard Specifications)

Material	ASTM Specification	Carbon, %	Manganese, %	Phosphorus, max %	Sulfur, max %	Silicon, %	Chromium, %	Vanadium, %
Music wire	A228-41	.70 to 1	.20 to .60	.03	.03	.12 to .30		
Oil-tempered wire over 7/32 diam. (comp. A)	A229-41	.55 to .75	.80 to 1.20	.045	.050	.10 to .30		
Oil-tempered wire under 7/32 diam. (comp. B)	A229-41	.55 to .75	.60 to .90	.045	.050	.10 to .30		
Hard-drawn spring wire	A227-41	.45 to .75	.60 to 1.20	.045	.050	.10 to .30		
Hot-wound carbon-steel helical springs *	A68-39	.90 to 1.05	.25 to .50	.05	.05	.15		
Chrome-vanadium valve spring wire †	A232-41	.45 to .55	.60 to .90	.03	.03	.12 to .30	.80 to 1.10	.15 to .25

* Heat-treated after forming.

† For ordinary chrome-vanadium wire, somewhat higher amounts of phosphorus (.04%) and sulfur (.05%) are allowed.

Values of typical physical properties of commonly used spring steels and of nonferrous materials are given in Tables 15 and 16.

Table 15. Tensile Properties of Typical Spring Steels

(Reprinted by permission from *Mechanical Springs*, by A. M. Wahl, Penton Publishing Co., Cleveland, 1944.)

Material	Ultimate Tensile Strength, psi	Elastic Limit in Tension, psi	Modulus of Elasticity, psi	Elongation in 2 in., %	Modulus of Rigidity, psi
Hard-drawn spring wire	160,000 to 310,000 *	60% of T.S. †	30×10^6	5	11.4×10^6
Oil-tempered spring wire	170,000 to 310,000 *	70 to 85% of T.S.	30×10^6	8	11.4×10^6
Music wire	255,000 to 440,000 *	60 to 75% of T.S.	30×10^6	8	11.5×10^6
Annealed, high-carbon wire	250,000 to 300,000	200,000 to 275,000	30×10^6	7	11.5×10^6
Hot-rolled, high-carbon steel	175,000 to 195,000	75 to 85% of T.S.	29×10^6	6	10.5×10^6 ‡
Chrome-vanadium steel	210,000 to 300,000	80 to 90% of T.S.	30×10^6	10	11.5×10^6

* Depending on size.
† T.S. = Tensile strength.
‡ Low value probably due to surface decarburization.

Table 16. Typical Physical Properties of Stainless Steel and Nonferrous Metals

(Reprinted by permission from *Mechanical Springs*, by A. M. Wahl, Penton Publishing Co., Cleveland, 1944.)

(As used in Springs)

Kind of Material	Ultimate Tensile Strength, psi	Elastic Limit in Tension, psi	Modulus of Elasticity, psi	Elongation in 2 in., %	Endurance Limit in Bending, psi	Modulus of Rigidity, psi
Stainless steel (18-8)	160,000 to 320,000 *	70,000 to 230,000	$26\text{–}28 \times 10^6$	....	....	$10 \text{ to } 11.2 \times 10^6$
Phosphor bronze (4–6% tin)	100,000 to 150,000	50,000 to 110,000	15×10^6	2–3	....	6×10^6
K-Monel (spring temper, heat-treated)	160,000 to 200,000	110,000 to 140,000	26×10^6	3–8	50,000	9.5×10^6
Z-nickel (spring temper, heat-treated)	180,000 to 230,000	130,000 to 170,000	30×10^6	5–10	....	11×10^6
Beryllium copper	160,000 to 200,000	100,000 to 150,000	$16\text{–}19 \times 10^6$ †	1–2	40,000	$6 \text{ to } 7.5 \times 10^6$ †
Spring brass	90,000 to 130,000	30,000 to 60,000	15×10^6	5	20,000 to 25,000	5.5×10^6

* Depending on size.
† Depending on heat treatment.

Data on endurance ranges and physical properties for spring materials in torsion and bending obtained by various investigators are given in Tables 17 and 18. On each table, in addition to the limiting endurance range, pertinent information is given including kind of material, heat treatment, surface condition (i.e. whether ground and polished, or untouched), ultimate and yield strengths in tension, modulus of rupture, and yield strength in torsion. An endurance range of 0 to 110,000 psi as listed means that the bar will withstand indefinitely a stress range between these limits while completely reversed stress is indicated by the sign $\pm$. For further information, refer to the footnotes below the tables.

Data given in Tables 17 and 18 indicate that, where the test specimen is ground and polished, a considerably higher endurance limit is obtained than where the surface is in the "as received" condition. This difference in endurance limits may probably be attributed largely to surface decarburization due to the heat-treatment process, although other factors (such as surface flaws or defects) may also be present. The various commonly used spring materials are described in the remaining pages of this chapter.

MUSIC WIRE. A high-quality carbon steel, this wire is widely used for small-sized helical springs, particularly those subject to severe stress conditions. The high strength of this material is obtained by using a steel of about 0.70 to 1.00% carbon, patenting and cold drawing to size. See Tables 14 and 15 for composition and properties. The tensile strength of music wire may vary from 255,000 psi for the larger wire sizes (0.16-in. diameter) to 440,000 psi for the smaller (0.005-in. diameter). In forming helical springs of music wire, the winding is done cold over a mandrel. After winding, it is advisable to give the springs a low-temperature heat treatment to relieve coiling stresses. This is usually done by heating the springs to a temperature of around 500 F for 1 hr for the larger sizes and for 15 to 30 min for the smaller sizes.

Table 17. Endurance Limits and Physical Properties of Spring Materials in Torsion

(Reprinted by permission from *Mechanical Springs*, by A. M. Wahl, Penton Publishing Co., Cleveland, 1944.)

Material	Specimen or Wire Diam., in.	Heat Treatment	Condition of Surface	Hardness	Ult. Strength (Tension), psi	Yield Point (Tension), psi	Mod. of Rupture (Torsion), psi	Yield Point (Torsion),* psi	Limiting Endurance Range in Torsion,† psi	Investigator
Tempered Swedish steel wire	.125		As received	45 Rockwell C	226,000	203,000	166,000	139,000	0 to 110,000 ‡	Weibel [1]
.60% C	.187		As received	43 Rockwell C	211,000	190,000	162,000	125,000	0 to 107,000 ‡	Weibel [1]
.35% Mn (approx.)	.225		As received	42 Rockwell C	204,000	181,000	166,000	120,000	±56,000 0 to 100,000 ‡	Weibel [1]
.6% C spring steel	.13 §	O.Q. 950 C T. 500 C	Ground and polished	365 Brinell	177,000	161,000	150,000		±52,600 4,500 to 103,000 55,000 to 125,000	Hankins [2]
Cr-Va spring steel	.13 §	O.Q. 850 C T. 600 C	Ground and polished	385 Brinell	177,000	168,000	148,000		±56,000 10,000 to 111,000 57,000 to 131,000	Hankins [2]
Cold-drawn steel wire	.160		As received	350 Brinell	202,000		155,000		±36,400 ‖	Lea and Dick[3]
High-carbon O.H. steel, .91% C, .38% Mn	.25	O.Q. 1575 F D. 940 F	Ground and polished	438–450 Brinell	225,000	179,000	173,000	118,000	±52,000 0 to 102,000	Johnson [4]
Cr-Va electric, .52% C, .88% Cr, .21% Va	.25	O.Q. 1600 F D. 810 F	Ground and polished	477–488 Brinell	237,000	229,000	183,000	141,000	±75,000 0 to 128,000	Johnson [4]
High-carbon electric steel	.25	O.Q. 1550 F D. 800 F	Ground and polished	430–470 Brinell	237,000	194,000	194,000	126,000	0 to 123,000	Johnson [4]
Beryllium bronze	.25			303 Brinell	166,000	132,000	110,000	95,000	±16,000 0 to 30,000	Johnson [4]

* Based on .2% plastic strain.

† Where more than one figure is listed for a given material, this means that tests were made at different stress ranges, i.e., ± or completely reversed stress, 0 to maximum, or intermediate to maximum stress as indicated.

‡ These figures probably represent ideal conditions when no flaws are present.

§ Specimen turned down to this diameter.

‖ Low value attributed by investigators to defects set up by drawing operation.

[1] *Trans. A.S.M.E.*, Nov. 1935, p. 501.

[2] *Eng. Research Special Report* 9, Dept. of Scientific and Industrial Research (British).

[3] *Proc. Inst. Mech. Engrs.* 1931, p. 661.

[4] *Iron Age*, March 15, 1934, p. 12.

Table 18. Endurance Limits and Physical Properties of Spring Materials in Bending

(Reprinted by permission from *Mechanical Springs*, by A. M. Wahl, Penton Publishing Co., Cleveland, 1944.)

(Round Specimens)

Material	Diam. of Specimen or Wire, in.	Heat Treatment *	Condition of Surface	Hardness	Ult. Strength (Tension), psi	Yield Point (Tension), psi	Elongation, %	Endurance Limit in Reversed Bending, psi	Investigator
.6% C steel		O.Q. 950 C T. 500 C	Ground and polished	365 Brinell	177,000	161,000	7 †	±85,000	Hankins [1]
Cr-Va spring steel		O.Q. 850 C T. 600 C	Ground and polished	385 Brinell	177,000	168,000	7 †	±95,000	Hankins [1]
.65% C steel	.162		As recd.		221,000			± 76,000	Shelton, Swanger [2]
.65% C steel	.130		Ground and polished		221,000			±126,000	Shelton, Swanger [2]
.65% C steel	.148		As recd.		217,000			±65,000	Shelton, Swanger [2]
Tempered Swedish valve spring wire	.225		As recd.	42 Rockwell C	204,000	181,000	10.5 ‡	±67,000	Weibel [3]
Tempered Swedish valve spring wire	.225		Rough ground 60 grit wheel 1 pass	42 Rockwell C	204,000	181,000	10.5 ‡	±81,000	Weibel [3]
Tempered Swedish valve spring wire	.225		Shot-blasted	42 Rockwell C	204,000	181,000	10.5 ‡	±85,000	
High-carbon O.H. steel, .91% C, .38% Mn	.273	O.Q. 1575 F D. 940 F	Ground and polished	438–450 Brinell	225,000	179,000	7 ‡	±80,000	Johnson [4]
Cr-Va steel, .52% C, .88% Cr, .21% Va		O.Q. 1600 F D. 810 F	Ground and polished	477–488 Brinell	237,000	229,000	11 ‡	±104,000	Johnson [4]
High-carbon electric steel, 1.04% C, .36% Mn		O.Q. 1550 F D. 800 F	Ground and polished	430–470 Brinell	237,000	194,000	5 ‡	±98,000	Johnson [4]

* O.Q. = oil quenched. T. = tempered. D. = drawn.
† In 8 in.
‡ In 2 in.
[1] Dept. of Sci. and Ind. Research, *Eng. Research*, *Special Report* 9.
[2] *J. Research Natl. Bur. Standards*, Vol. 14, 1935, RP 754, pp. 17–32 .
[3] *Trans. ASME*, Nov. 1935, p. 501.
[4] *Iron Age*, March 15, 1934, p. 12.

OIL-TEMPERED SPRING WIRE. This is a good-quality high-carbon steel wire, made by the open-hearth or electric-furnace process. In manufacturing, the wire is cold drawn to size and then heat treated. Tensile ultimate strengths vary from around 170,000 psi for the larger sizes (0.5-in. diameter) to around 310,000 for smallest sizes (0.025-in. diameter). Springs made of this wire are wound cold and then usually given a thermal treatment.

HARD-DRAWN SPRING WIRE. A carbon steel of lower quality than music or oil-tempered wire, this wire is used where the stresses are low or where a high degree of uniformity is not so essential. This material has tensile-strength properties slightly below that of oil-tempered wire. It should not, in general, be used where fatigue loading is involved.

ANNEALED HIGH-CARBON WIRE. This material, which has high ductility in the annealed state, is used where severe forming operations are necessary in the manufacture of the spring, for example in torsion springs with certain shapes of end turns. The springs are first formed from the annealed wire and then heat treated after winding to obtain the required physical properties. A typical composition calls for carbon 0.80 to 0.95%, manganese 0.50% max. See Table 15 for properties.

HOT-WOUND HELICAL SPRINGS HEAT TREATED AFTER FORMING. For the larger sizes of helical springs (say over $^3/_8$ to $^1/_2$ in. wire diameter) it is not practical to wind the springs cold. In such cases, the spring may be made from carbon or alloy steel bars wound hot and then heat treated. For winding these springs, ASTM specification A125-39 calls for heating to a temperature of 1700 F and coiling on a preheated mandrel. The springs are then allowed to cool uniformly to a black heat, after which they are heated uniformly to a temperature sufficient to refine the grain, and then oil quenched. After quenching they are tempered by heating to a temperature below the critical for a sufficient time to yield the required hardness values.

CHROME-VANADIUM STEEL WIRE. In the past this alloy steel wire has been frequently specified where a high-quality material is needed and where temperatures are somewhat higher than normal, as for automotive valve springs. It may be obtained either in the annealed or in the heat-treated condition. When wound from annealed wire, the springs must be heat treated after coiling. After winding from oil-tempered chrome-vanadium wire, a low-temperature heat treatment at around 500 to 700 F should be given, the higher bluing temperatures being preferred for applications involving elevated temperatures.

STAINLESS-STEEL (18-8) SPRING WIRE. Stainless steels having a composition of about 18% chromium and 8% nickel are widely used for springs subject to corrosion conditions, and are also of value for elevated temperature conditions. A typical specification for this material calls for the following composition:

Carbon, max	0.15%
Chromium	16.00 to 20.00%
Nickel	8.00 to 12.00%
Nickel plus chromium	26% min

The tensile strength of this wire is developed by cold drawing and may vary from 160,000 psi in 0.307-in. diameter size to 320,000 psi in the 0.01-in. diameter size. Springs are wound cold and stress relieved at 750 F for 15 min. to an hour, depending on wire size.

PHOSPHOR BRONZE. This material finds its greatest use where a spring with good electrical conductivity is desired; it is also used for applications where corrosion resistance is important. Typical physical properties are given in Table 16.

BERYLLIUM COPPER is an alloy consisting essentially of about 2% beryllium and the rest copper, together with small amounts of other alloys. It has the advantage of having a high electrical conductivity while not requiring any tin in its manufacture. In general, wire made from this material is quenched from 1475 F and then cold drawn to increase the hardness. After coiling, it is then heat treated to increase the physical properties. This heat treatment may also be varied to change the modulus of elasticity or the amount of drift or creep. Further data on the properties of this material are given in Table 16.

K-MONEL is a copper-nickel alloy to which 2 to 4% of aluminum has been added. A typical composition is copper 29%, nickel 66%, aluminum 2.75%. Wire made from this material is given a solution heat treatment and then cold drawn. After winding, the springs are given a final heat treatment to increase the hardness and strength. By this means ultimate strengths around 170,000 to 200,000 psi can be obtained. Springs made of this alloy are used for corrosion conditions and for resistance to elevated temperatures.

Z-NICKEL. A corrosion-resistant alloy containing about 98% nickel, Z-nickel is used for springs subject to elevated temperatures. Because it has a fair ductility after heat treatment, springs made of this material may often be wound from heat-treated wire. Further data are given in Table 16.

REFERENCES

1. A. M. Wahl, *Mechanical Springs*, Penton Publishing Co., Cleveland, 1944.
2. A. M. Wahl, Paper APM 51-17, Stresses in heavy closely coiled helical springs, *Trans. Am. Soc Mech. Engrs.*, Vol. 51, p. 191, 1929. Also Ref. 1, p. 33.
3. O. Göhner, articles in *Ingenieur-Archiv*, Vol. 1, p. 619, 1930; Vol. 2, p. 1, 1931. See also Ref. 1, p. 38, and *Z. Ver. deut. Ing.*, Vol. 76, p. 269.
4. *Manual on Design and Application of Helical and Spiral Springs for Ordnance*, published by Society of Automotive Engineers, New York, 2nd ed., 1945.
5. *Mechanical Springs, Their Engineering and Design*, published by Wallace Barnes Co., Div. of Assoc. Spring Corp., 1944.
6. F. P. Zimmerli, Effects of temperature on coiled steel springs at various loadings, *Trans. Am. Soc. Mech. Engrs.*, p. 363, May 1941.
7. C. T. Eakin, unpublished test data, Westinghouse Electric Corp.
8. Heavy helical springs, *Progress Report* 3, ASME Special Research Committee on Mechanical Springs; abstract by C. T. Edgerton, *Trans. Am. Soc. Mech. Engrs.*, p. 609, Oct. 1937.
9. *Manual of Spring Engineering*, published by American Steel and Wire Co., Cleveland, 1941.
10. F. P. Zimmerli, Permissible stress range for small helical springs, *University of Michigan Engineering Research Bulletin* 26, 1934.
11. H. F. Ross, Application of tables for helical compression and extension spring design, *Trans. Am. Soc. Mech. Engrs.*, p. 725, Oct. 1947.
12. J. A. Haringx, On the buckling and lateral rigidity of helical compression springs, *Proc. Nederl. Akademie van Wetenschappen*, Vol. XLV, No. 6 and No. 7, 1942.
13. W. E. Burdick, F. S. Chaplin, and W. L. Sheppard, Deflection of helical springs under transverse loadings, *Trans. Am. Soc. Mech. Engrs.*, p. 623, Oct. 1939.
14. H. C. Keysor, Calculation of the elastic curve of a helical compression spring, *Trans. Am. Soc. Mech. Engrs.*, p. 319, May 1940.
15. *The Mainspring*, published by Wallace Barnes Co., Div. of Assoc. Spring Corp., Feb. 1941.
16. *The Mainspring*, published by Wallace Barnes Co., Div. of Assoc. Spring Corp., April 1941.
17. Liesecke, *Z. Ver. deut. Ing.*, Vol. 77, 1933, pp. 892, 425.
18. R. P. Kroon and Davenport, Spiral springs with small number of turns, *J. Franklin Inst.*, p. 171. 1938.
19. *The Mainspring*, published by Wallace Barnes Co., Div. of Assoc. Spring Corp., Aug. 1937.
20. Discussion by D. C. Drucker, *J. Applied Mechanics*, p. A251, Sept. 1946. Also article by E. Reissner, ibid., June 1945, p. A-69.
21. *Manual on Design and Application of Leaf Springs*, published by the Society of Automotive Engineers, 1944.
22. J. O. Almen and A. Laszlo, The uniform section disk spring, *Trans. Am. Soc. Mech. Engrs.*, p. 305, 1936.

SECTION 12

BEARINGS AND LUBRICATION

By

H. A. STEVENS HOWARTH, *Consulting Engineer on Oil Film Bearings, Philadelphia, Pa., and Research Engineer, Franklin Institute Laboratories for Research and Development*

PAUL G. EXLINE, *Exline Engineering Co., Tulsa, Okla.*

GUNNAR PALMGREN, *Chief Engineer, SKF Industries, Inc.*

BRYCE RULEY, *Senior Engineer, SKF Industries, Inc.*

E. STANLEY AULT, *Professor of Machine Design, Purdue University*

CLASSIFICATION OF BEARINGS

THE LOADING AND FRICTION OF BEARINGS ON FILMS OF LUBRICANT

By H. A. S. Howarth

LUBRICANTS

By Paul G. Exline

ROLLING BEARINGS

By Gunnar Palmgren and Bryce Ruley

STRAIGHT, NEEDLE, AND TAPERED ROLLER BEARINGS

By E. Stanley Ault

CLASSIFICATION OF BEARINGS

Bearings may be classified according to function, form, design, and lubrication.

CLASSIFICATION BY FUNCTION. Bearings may be defined according to function as follows:

Radial Bearing. A bearing applied to a rotating shaft to hold its axis in line, and prevent movement in a radial direction.

Thrust Bearing. A bearing applied to a shaft to prevent free endwise movement.

Angular Bearing. A bearing to constrain a shaft against both radial and axial movement; the load is received in an angular direction and has radial and axial components.

Guides or Ways. Bearings to permit and control the rectilinear movement of a sliding machine element, as a ram or crosshead, or slide. Guides usually are designed to prevent rotational movement of the sliding element.

CLASSIFICATION BY FORM. Bearings may be divided into the following forms, depending on the way in which friction is produced in them:

Sliding surface bearings include all that depend on lubricating films, perfect or imperfect, of oil, grease, graphite or other material.

Rolling contact bearings include all forms of ball and roller bearings.

Sliding rolling face bearings include those found in gear teeth, cams, and some forms of roller bearings.

CLASSIFICATION BY DESIGN may be made as follows:

Self-aligning bearings are those in which the parts will automatically align themselves when assembled and loaded.

Rigid Bearings. Bearings in which bearing surfaces are practically invariable in form and position after installation. Also known as nonaligning bearings.

Pivoted Segment Bearings. Bearings in which one of the active surfaces is made up of segments so freely mounted as to permit the formation of natural wedge-shaped lubricating films.

Elastic Bearings. Those in which one of the active surfaces is elastic or elastically supported so as to flex with respect to the other for the purpose of alignment and film formation.

CLASSIFICATION BY LUBRICATION. Bath Lubricated Bearings. Bearings in which active surfaces always are submerged in a well of lubricant.

Automatically Lubricated or Self-oiling Bearings. Bearings in which the rotation of the shaft or movement of parts causes lubricant to be applied continuously to the bearing surfaces.

Independent Pressure-lubricated Bearings. Bearings lubricated from a pressure source, independent of the bearing mountings, which may be operated before the machine is started, or may be geared into the machine, operating only when the machine runs. The system is then known as *automatic pressure lubrication.*

Periodically-lubricated bearings include all those using any form of intermittent application, by whatever means the lubricant is applied.

FACTORS IN BEARING DESIGN AND OPERATION. The several factors listed below are important to the designer and user of bearings, from the standpoint of permissible loads, lubrication, etc.

Surface speed rarely is a limitation with perfectly lubricated bearings. See Cooling of Bearings, Article 5 of this section.

Straight-line motion of sliding bearing surfaces may be continuous in one direction, as wire in a drawing die. It may be intermittent in one direction as in the ram of a ratchet jack. It may be reciprocating motion, as in crossheads or machine guides.

Circular sliding motion, as in journal, thrust, and angular bearings may be continuous in one direction, or it may be reversible, i.e., rotation in opposite directions for equal or unequal periods of time. It may be oscillating, as in a shaft carrying a rocking lever, or intermittent as in bearings for a ratchet wheel shaft. The angle of rotation is important in the last two classes.

Approach and recession of bearing faces frequently is superimposed on sliding motion, as in rocking levers, wrist pins, and reversing shafts. It is caused by change in magnitude of bearing load, by reversal, shock, or other cycle of load variation.

Rotating-radial load is found more often in vertical than in horizontal shafts. It is an unusual condition, generally due to lack of balance, accidental or otherwise, but it may result as in rock crushers from the process of manufacture carried on in the machine.

THE LOADING AND FRICTION OF BEARINGS ON FILMS OF LUBRICANT

By H. A. S. Howarth

1. PERFECT LUBRICATION OF BEARINGS

VISCOUS FLOW. Of the two opposite kinds of fluid flow *viscous* flow is simpler and easier to express mathematically, whereas *turbulent* flow is the more complex. Viscous flow of a fluid through a straight tube of uniform bore is straight-line flow, parallel with the axis of the bore.

Coefficient of viscosity is the measure of the resistance the fluid offers to the distortion its body undergoes under the action of the forces that cause the flow through the tube. Usually the term viscosity is used by itself. For mathematical calculations the viscosity is represented by the symbol μ, meaning the coefficient of viscosity. However, when the viscosity is said to be 200 SUS it is understood that 200 seconds were required for a standard quantity to pass by gravity through a short tube or jet known as the Saybolt universal viscometer. Such a measurement can be transformed into the absolute coefficient of viscosity by means of tables or charts and knowledge of the density of the fluid. (See p. 12-63.) Herein, however, the character μ will be assumed to represent the coefficient of viscosity.

For a discussion of *Reynolds' number* and its use in determining whether a fluid will have viscous or turbulent flow under the problem conditions, see p. 13-02.

Viscosity Units. Figure 1 illustrates an element of the fluid within a film whose flow is viscous. The film, of which the element is but a small part, can be thought of as flowing between two parallel plates whose distance apart is h. One plate may be thought of as stationary and the other as moving over it with some known velocity U.

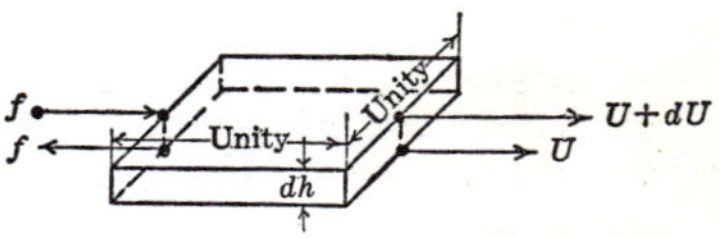

FIG. 1. Forces and velocities in viscous flow.

The element to be examined is assumed to have a square unit of area, parallel with the plates, and a differential thickness dh. The forces acting on the element give one of its faces a velocity that is higher than the other by a differential amount dU in the direction indicated. The force that causes this velocity increase of the unit area is represented by f. The ratio dU/dh is called the rate of distortion of the fluid. The coefficient of viscosity is defined as the ratio of the force f, upon a unit area, to the rate of distortion it produces in the fluid. Thus

$$\mu = \frac{f}{dU/dh} \tag{1}$$

If the rate of distortion is uniform throughout the film thickness h, the value of μ could be determined by eq. 1. Conversely, when μ is known, f can be found from $f = \mu U/h$.

For the convenience of those who wish to convert the coefficient of viscosity μ from one system of units to another, the following relationships may be used. The three units commonly employed vary in magnitude, from greatest to least, from the Reyn, to the poise, to the centipoise, in the ratio 6,900,000 to 100 to 1. The Reyn expresses viscosity in pounds, seconds, per square inch.

One Reyn equals 69,000 poises or 6,900,000 centipoises. Therefore if the viscosity of an oil is 3.4×10^{-6} Reyns, it is 0.2345 poise, or 23.45 centipoises.

Commercial Viscosities. Viscosities are commonly measured commercially by instruments that use the weight of the oil as the means of forcing it through a jet. Such readings involving volume and time are convertible directly into what is termed kinematic viscosity, whose units, *stoke* and *centistoke*, parallel the poise and the centipoise. The density of the liquid multiplied by the kinematic viscosity gives the absolute coefficient of viscosity in units that are required for engineering calculations. The Reyn, the poise, and the centipoise are units of absolute viscosity (viscosity coefficients) and may be used in engineering calculations without correction for density. For commercial viscosities, see p. 12-63.

Viscous Flow through Tubes (Fig. 2). Assuming no molecular slippage at the tube surface and not including entrance and discharge losses, the laws of viscous flow yield the following formulas: V_r = volume of flow through tube; t = time required, seconds; P_1 and P_2 = respectively, the pressures acting upon the liquid at entering and discharging ends of tube; h = diameter of the tube bore; L = length of tube; u = linear velocity of flow at a distance y from the tube bore radially.

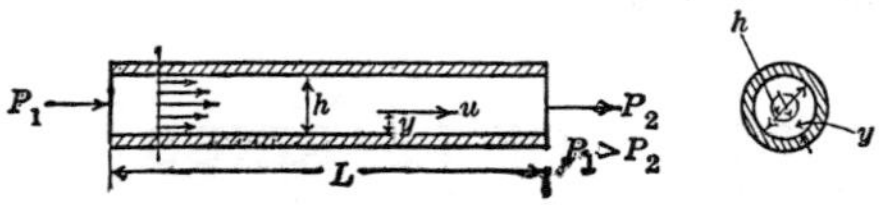

FIG. 2. Viscous flow through tube.

$$\frac{V_r}{t} = \frac{\pi h^4 (P_1 - P_2)}{128 \mu L} \tag{2}$$

$$u = \frac{(P_1 - P_2)(hy - y^2)}{4 \mu L} \tag{3}$$

Viscous Flow between Stationary Parallel Plates (Fig. 3). Assuming no molecular slippage at the plate surfaces and not including entrance or discharge losses, the laws of viscous flow yield the following formulas: V_p = volume of flow through space between plates, neglecting influence of the sides; t = time required, seconds; P_1 and P_2 = respectively, the pressures at entrance and discharge; h = uniform distance between inner faces of plates; L = length of plates in direction of flow; b = their width; y = a distance from one plate face at which the velocity of flow u is to be determined.

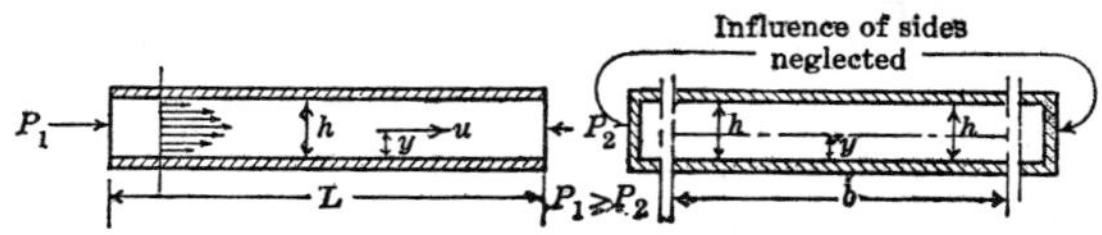

FIG. 3. Viscous flow between parallel plates.

$$\frac{V_p}{t} = \frac{bh^3 (P_1 - P_2)}{12 \mu L} \tag{4}$$

$$u = \frac{(P_1 - P_2)(hy - y^2)}{2 \mu L} \tag{5}$$

Viscous Flow from Approaching Parallel Plates (Fig. 4). The time rate at which two parallel flat surfaces will approach each other when the space between them is filled with a viscous liquid, when the plates are elliptical with no central cavity, and when no oil is added, and when the upper plate moves downward under a constant normal load R, was found by Osborne Reynolds to be:

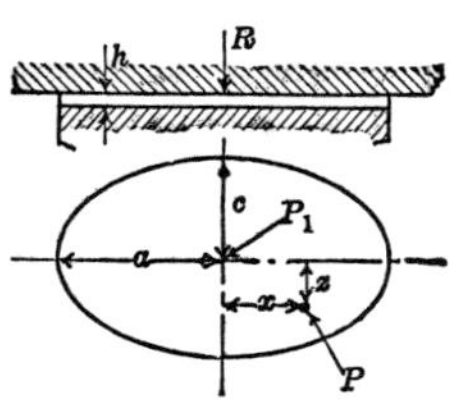

FIG. 4. Approaching parallel faces with oil film between them.

$$t = \frac{3\mu\pi a^3 c^3}{2R(a^2 + c^2)} \left\{ \frac{1}{h_1^2} - \frac{1}{h_2^2} \right\} \tag{6}$$

$$P_4 = \frac{2R}{\pi ac} \tag{7}$$

$$P_5 = -\frac{2R}{\pi ac} \left\{ \frac{x^2}{a^2} + \frac{z^2}{c^2} - 1 \right\} \tag{8}$$

The symbols a and c are, respectively, the semimajor and semiminor axes of the elliptical faces. x and z are the coordinates of any point measured respectively from the c and a axes. R is the total load on the plate. h_1 and h_2 are the distances between the plates at the start and the finish of the time interval t. P_4 is the pressure at the center of the plates. P_5 is the pressure at any point whose coordinates are x and z.

Separation of Circular Plates by High-pressure Oil (Fig. 5). One plate may be considered to be flat and to extend beyond the other, but to press upon it with a load R. The lower plate is assumed to be circular and flat, with a smaller circular recess at its center, and into which oil is assumed to be forced from below with a pressure sufficient to cause a radially outward flow between the parallel faces of the plates that will hold them apart a distance h. The pressure of the oil supply to the recess is P_1. The recess radius is r_1. The outer radius of the circular face is r_2. At any other radius r the pressure will be P. The average pressure over the surface πr_2^2 will be $P_0 = R/\pi r_2^2$. The gallons of oil

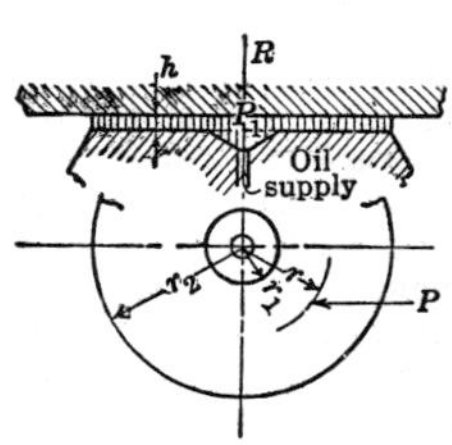

FIG. 5. Separation of parallel faces by high-pressure oil.

per minute supplied to the recess to hold the faces apart a distance h will be G. The coefficient of viscosity of the lubricant, μ, will be expressed in pounds per square inch (Reyns). Linear dimensions are in inches and pressures in pounds per square inch in the formulas that follow.

CASE 1. The load is carried by a continuous flow of oil:

$$G = \frac{0.0867h^3R}{\mu(r_2^2 - r_1^2)} \tag{9}$$

$$P_1 = \frac{2R\log_e\left(\frac{r_2}{r_1}\right)}{\pi(r_2^2 - r_1^2)} = \frac{7.35\mu G\log_e\left(\frac{r_2}{r_1}\right)}{h^3} \tag{10}$$

$$P = \frac{2R\log_e\left(\frac{r_2}{r}\right)}{\pi(r_2^2 - r_1^2)} \tag{11}$$

$$\frac{P_1}{P_0} = \frac{2r_2^2\log_e\left(\frac{r_2}{r_1}\right)}{r_2^2 - r_1^2} \tag{12}$$

CASE 2. The oil supply is stopped and the plates move toward each other. In time t seconds the distance between the plates reduces from h_2 to h_1 under load R.

$$t = \frac{3\mu\pi(r_2^4 - r_1^4)}{4R}\left\{\frac{1}{h_1^2} - \frac{1}{h_2^2}\right\} \tag{13}$$

$$P_1 = \frac{2R}{\pi(r_2^2 - r_1^2)} \tag{14}$$

$$P = \frac{2R(r_2^2 - r^2)}{\pi(r_2^4 - r_1^4)} \tag{15}$$

Friction and Viscosity in Films of Uniform Thickness. The relation between friction and the viscosity of the lubricant in bearings seems to have been first pointed out in 1883 by Petroff (Ref. 1), who assumed that a journal runs concentrically in a sleeve bearing. He suggested that there might be slippage between the lubricant and the bearing surfaces. Hence the formula he proposed was not so simple as the one that follows, in which slippage is taken as zero.

At high speed a journal, if lightly loaded, will run very nearly concentric in a 360° bearing sleeve uninterrupted by grooves, if copiously supplied with oil (Fig. 6). If the journal radius is a, the radial clearance h, the sleeve width b, and the rpm is N, the viscous tangential frictional resistance to rotation will be F as follows:

$$F = 2\pi abf \tag{16}$$

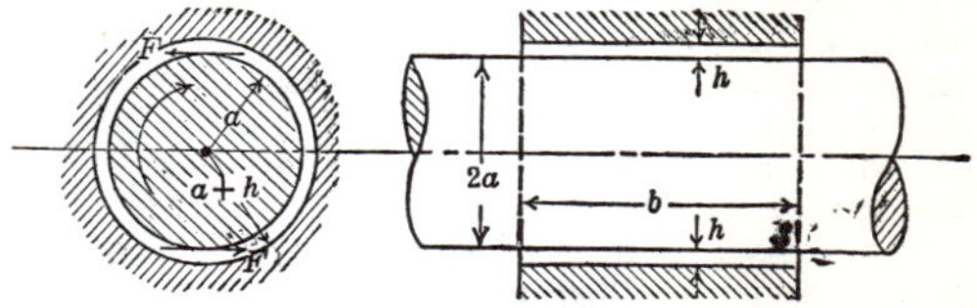

FIG. 6. A journal, central within its bearing.

It was pointed out that, when the viscosity coefficient μ is known, the relation $f = (\mu U/h)$ evolves from eq. 1. In this equation $U = (2\pi aN/60)$. If these are combined with eq. 16 there emerges the simple 360° bearing friction formula:

$$F = \frac{\mu\pi^2a^2bN}{15h} \tag{17}$$

This equation leaves out of account the frictional resistance offered by the oil-flow due to the pressure gradient. With concentric operation, constant viscosity, and no load, there would be no pressure gradient.

If the viscosity coefficient μ in eq. 17 is expressed in Reyns (pounds, seconds, per square inch) and the linear dimensions are in inches, F will be in pounds. If the viscosity is expressed in poises (dynes, seconds, per square centimeter) and linear dimensions in centimeters, the force F will be in dynes. If the viscosity is expressed in centipoises and linear dimensions are in centimeters, the force F will be hundredths of dynes.

ANALYSES OF THE WEDGE-SHAPED OIL FILM. Oil films in bearings are not so simple in shape as the one of uniform thickness described above. It has long been known that the film in a bearing should be wedge shaped in order best to develop pressures necessary to support a substantial load without metallic contact during rotation. (See Refs. 2, 3, 4, 5.)

Continuity of Flow with Various Film Shapes, without Side Loss. The theory underlying film form in bearings is really not difficult to understand when examined from the viewpoint of continuity of forward viscous flow, with absence of side leakage loss of lubricant. There are of course such side losses, which will be discussed later. The present approach, however, is simpler without them.

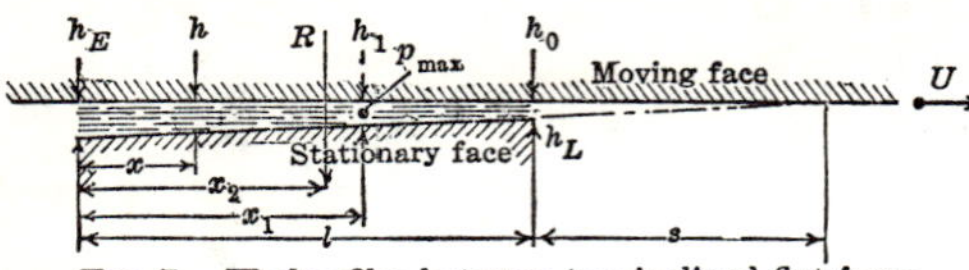

Fig. 7. Wedge film between two inclined flat faces.

In the following résumé, covering several film forms whose cross sections are seen in Figs. 7 to 16, the film is assumed to be held stationary for observation of its contour, the study of its pressure distribution, and the positioning of the line of action of the load it carries. The velocities of the bearing faces are represented by vectors for direction and relative magnitude. The positions indicated for load line R, least film thickness h_0, and the film thickness h_1 at the point of maximum pressure are approximate. All films are shown for the same length l, throughout which the pressure rises from P_E at h_E to its maximum at h_1 and then falls off to P_L at h_L. Initial and final film pressures depend upon the environment in which the film functions and is required to carry its load.

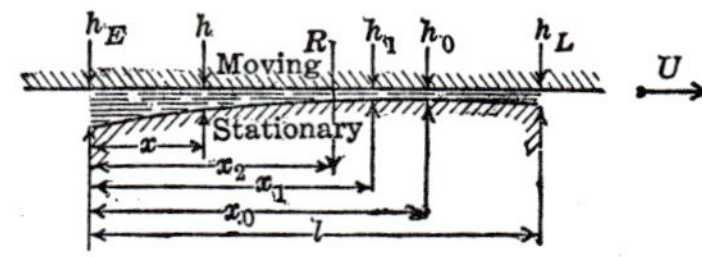

Fig. 8. Wedge film between a flat and a curved face.

Film faces may be flat or curved or a combination of the two. For the purpose of analysis any film having two curved faces can be shown as having one flat and one curved. However this résumé shows the films in their actual forms, with thicknesses highly exaggerated.

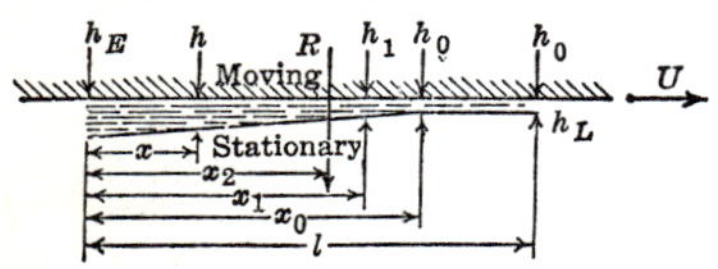

Fig. 9. Combined flat-face wedge and parallel-face film.

In Figs. 7, 8, and 9 only one face is shown to be moving. Its velocity is U with respect to the load line R. If the other face has any velocity it moves with the load line and relative to it has a velocity of zero. However, when both surfaces of the film have velocities with respect to the load line, with which the pressure film must remain, both those velocities must be considered. They appear as U_1 and U_2 in Figs. 10 to 16.

Reynolds (Ref. 3) investigated the cases of Figs. 7 and 8 in his study of Tower's experiments, but his theory can be applied equally well to the others. It is conceivable for purpose of comparison that the films of Fig. 8 and of Figs. 10 to 16 are alike in the lettered dimensions; they may be so assumed for the present. A simple analysis will then indicate their relative abilities to support a load.

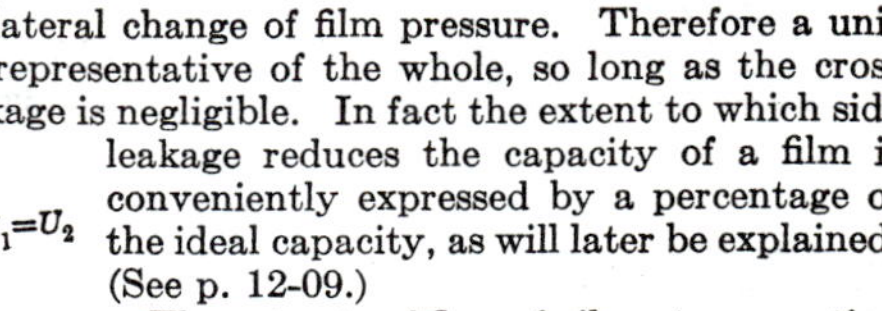

Fig. 10. Film separating rolling faces.

The volume of oil that enters a film in a given time will pass the region of greatest fluid pressure in the same time if there is no change of density and no lateral change of film pressure. Therefore a unit width of the film can be considered as representative of the whole, so long as the cross section is uniform and so long as side leakage is negligible. In fact the extent to which side leakage reduces the capacity of a film is conveniently expressed by a percentage of the ideal capacity, as will later be explained. (See p. 12-09.)

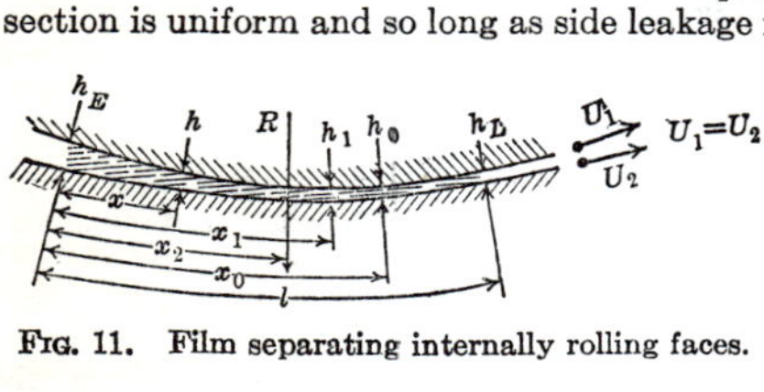

Fig. 11. Film separating internally rolling faces.

The net rate of flow of oil past any section of the film will be equal to the algebraic sum of the flows due to the several causes of flow. Two such causes are, first, the velocities of the film faces with respect to the load line and, second, the pressure gradient. At the region of maximum pressure the gradient will be zero and the total flow will be due to relative motion of the film faces. Thus the following equation can be written:

$$\frac{Q_1}{t} = \frac{Q_m}{t} + \frac{Q_p}{t} \tag{18}$$

where Q_1 is the rate of flow where the pressure is greatest; at this point the film thickness is h_1, which lies at distance x_1 from the beginning of the film where the thickness is marked h_E. At any other distance x, where the film thickness is h, the forward flow will be $Q_m/t + Q_p/t$. Thus eq. 18 is a statement of the law of continuity of flow in simple terms.

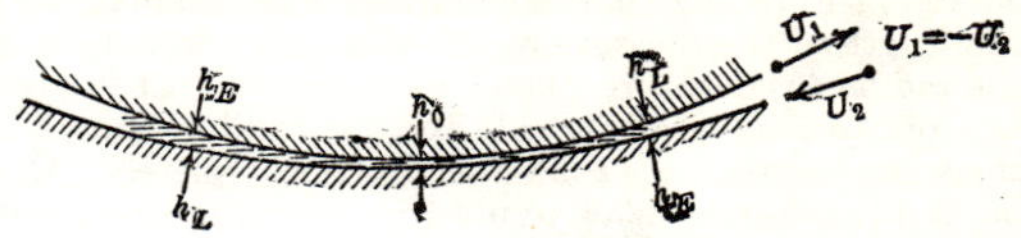

FIG. 12. Film between two faces having equal but opposite speeds past line of action of the load R.

The rate of flow Q_m/t, due to relative motion, will obviously be $Uh/2$ per unit of film width for films of Figs. 7, 8, and 9. The rate of flow Q_p/t, due to the pressure gradient, can be taken from the previous study of parallel plate flow (p. 12-04) placing width b at unity. Thus

$$\frac{Q_p}{t} = -\frac{h^3}{12\mu}\left\{\frac{dp}{dx}\right\} \tag{19}$$

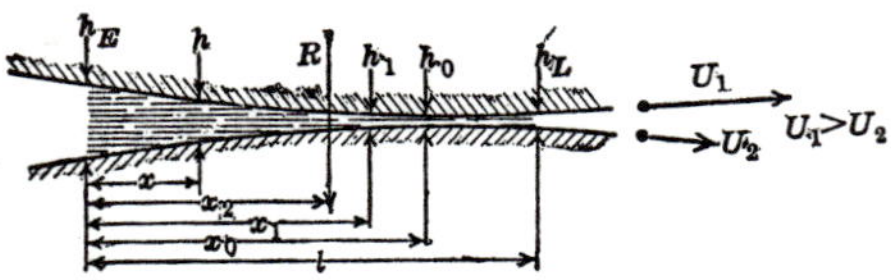

FIG. 13. Film separating faces that roll and slide.

Inserting these values in eq. 18 yields the following equation when solved for dp/dx.

$$\frac{dp}{dx} = \frac{6\mu U(h - h_1)}{h^3} \tag{20}$$

The law of variation of h with x is clear for the film of Fig. 7. If C is the slope factor, $h = C(l + s - x)$ and from eq. 20 there emerges

$$\frac{dp}{dx} = \frac{6\mu U(x_1 - x)}{C^2(l + s - x)^3} \tag{21}$$

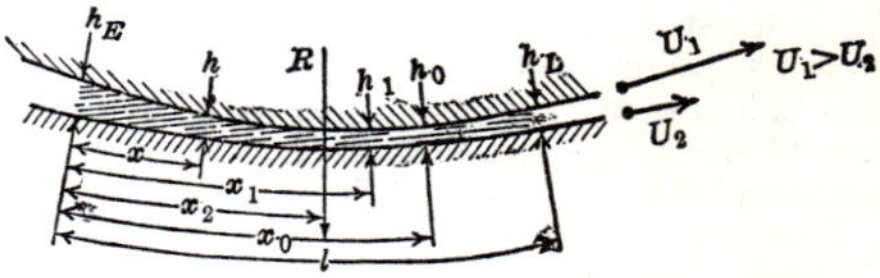

FIG. 14. Film between faces that roll and slide internally.

In order to obtain similar definite equations for other films than Fig. 7 it is necessary likewise to define the relation of film thickness to its position in the film. Equation 20 is therefore general for films like Figs. 7, 8, and 9, where U is the only velocity. Extensive analyses of films like Fig. 8 have been made as will appear later in data on journal bearings with running clearance and on those whose surfaces are fitted. (See p. 12-16.) The data that appear later (p. 12-33) on pivoted shoes were obtained from an extension of the study of Fig. 7 and its eq. 21.

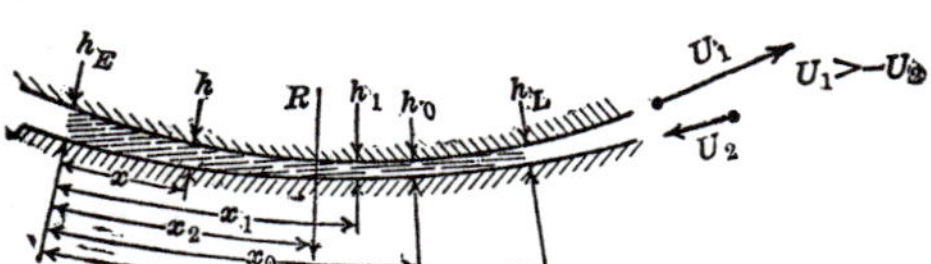

FIG. 15. Film between two faces having unequal but opposite speeds past line of action of the load R.

Before proceeding with examination of more complicated films it will be profitable to study Fig. 9. In that figure a film is illustrated that has a beveled or wedge-shaped beginning and a pair of flat parallel faces for an ending, with no interruption where they join. Between the flat faces of the wedge portion the relation (eq. 19) will hold. Beyond h_{01}, i.e., in the parallel portion, the flow will be of uniform rate, which is the sum of the two forward flows, the total equaling $Uh_1/2$. They are $Uh_0/2$ and $\dfrac{h_0^3}{12\mu}\dfrac{(P_{01} - P_{02})}{(l - x_0)}$. Therefore

$$\frac{Uh_1}{2} = \frac{Uh}{2} - \frac{h^3}{12\mu}\left(\frac{dp}{dx}\right)$$

$$= \frac{Uh_0}{2} + \left(\frac{P_{01} - P_{02}}{l - x_0}\right)\frac{h_0^3}{12\mu} \tag{22}$$

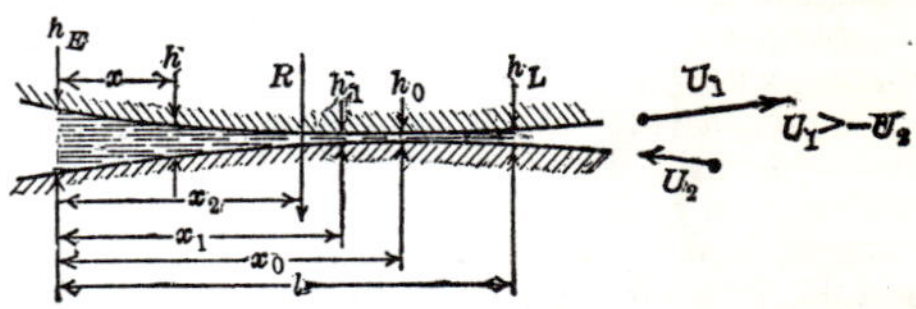

FIG. 16. Film between two convex cylindrical faces having unequal opposite speeds past line of action of the load R.

Ordinarily P_{02} would be zero and P_{01} only a little less than P_{max}. In this particular combined film the maximum pressure will be much closer to the thin end of the flat wedge than in the plain flat-wedge film of Fig. 7. Obviously this film approaches the full flat wedge in form as the flat parallel-faced portion becomes shorter.

Thus far we have been dealing with films having one fixed face and one face that moves over it with velocity U. However, it is possible to have pressure developed in films between two faces that move in the same general direction with equal speed past the load line R and the zone of closest approach. This is most easily visualized by two slightly curved faces rolling upon each other but separated by a film of oil as in Fig. 10. Obviously the length of the pressure region of the film would be short relative to the radii. The less the radius of curvature, the shorter the film. If the centers of curvature are on the same side of the film as in Fig. 11, and the radii are only slightly different, the film might fill the space between the rolling members completely. One member would be rolling around inside the other, or both would turn about separate but parallel noncoincident axes.

For the general case of Figs. 10 and 11 the first fundamental eq. 18 will serve as a starting point, but the rate of flow at the region of maximum pressure will be twice as great as for Figs. 7, 8, and 9. Q_1/t will be equal to $\frac{(U_1 + U_2)}{2} h_1$, instead of $Uh_1/2$, when $U = U_1 = U_2$. Therefore the equations that apply to Figs. 10 and 11 will be:

$$\frac{(U_1 + U_2)h_1}{2} = \frac{(U_1 + U_2)h}{2} - \frac{h^3}{12\mu}\left(\frac{dp}{dx}\right) \tag{23}$$

$$\frac{dp}{dx} = \frac{6\mu(U_1 + U_2)(h - h_1)}{h^3} \tag{24}$$

Comparison of eqs. 20 and 24 shows the pressure gradients to have the ratio of $U/(U_1 + U_2)$. This means that when two surfaces roll together without slip, and without contact, their ability to carry a load, upon a film of given viscosity and dimensions, will be in the ratio $(U_1 + U_2)/U$ to the ability of the same film when one of its faces moves with velocity U and the other is stationary, with respect to the load line R.

Figures 13 and 14 show that their velocity vectors indicate combined rolling and sliding. U_1 is drawn equal to $2U_2 = 2U$. Hence the pressure gradient would be three times that of eq. 20. However, when the velocity vectors are not pointed in the same direction their sum is less in the positive direction, or it is zero, or it is net negative. Thus in Figs. 15 and 16 the sum is net positive, whereas in Fig. 12 it is zero. The film of Fig. 12 would carry no load, i.e., it would develop no pressure.

Dynamic Loads upon Bearings. The film forms shown by Figs. 7 to 16, while not exhaustive, cover most of the best films available to the machine designer for preventing the wearing away of the bearing surfaces if they can be adequately lubricated. Such films form between the surfaces that move across the load line with suitable velocities, and they obviously vary in length inversely with the rate of change of film thickness, if not interrupted by termini.

When films are studied in dynamically loaded bearings, in which the load line changes location rapidly and in cycles, with respect to other motions of the bearing faces, all motions must be determined with respect to the load direction and position. Such a study of the bearings of the crank-engine linkage is very important and is being made the subject of careful research. Important papers have been presented in recent years, as, for example, paper by J. M. Stone and A. F. Underwood (Ref. 6) of General Motors Research Division, and paper by Professor J. T. Burwell (Ref. 7) of Massachusetts Institute of Technology. To supplement these analytical studies extensive tests have been made and are continuing to be run in an effort to bring experience and analysis into agreement.

Pivoted Shoe Bearings. Before the integration of the fundamental equations are discussed, the independent inventions of pivoted shoe bearings should be noted, by Kingsbury first and by Michell second, a very few years apart, the one at Durham, New Hampshire, U.S.A., and the other at Melbourne, Australia. Both inventors acknowledge indebtedness to Osborne Reynolds' mathematical analysis of the flat-wedge film in connection with his study of Tower's experiments.

There is a vast difference between a fixed bevel thrust bearing and the variable bevel obtained with pivoted thrust shoes. The latter can adjust its wedge taper and film thickness to match the requirements of varying load, speed, and oil viscosity, whereas the fixed bevel thrust bearing cannot vary its wedge taper, and therefore is not so widely useful. When it slows down to a stop its film loses pressure rapidly and comes to a stop suddenly. Conversely, when started up it has to reach its designed speed before its film will have its best form.

Capacity and Friction of Flat-wedge Films. The integration of the pressure gradient equation for the flat-wedge film, made by Reynolds himself, is summarized and set down accurately by Kingsbury (see Ref. 8). Kingsbury pointed out, as others had done, that Reynolds ignored the pressure gradient effect upon friction, and that the friction equation should be:

$$f = \frac{\mu U}{h} + \frac{h}{2}\left(\frac{dp}{dx}\right) \tag{25}$$

The friction is seen to be due to the sum of two rates of distortion of the film at the bearing or journal surface. The first and larger (usually) is the distortion due to relative motion of the bearing faces and the second is the distortion due to the forward or backward flow of the lubricant due to the pressures within the film.

For the flat-wedge film of Fig. 7 the friction F for a unit width is readily found to be

$$F = \frac{\mu U s}{h_0}\left[4 \log_e\left(\frac{l+s}{s}\right) - \frac{6l}{l+2s}\right] \tag{26}$$

The preliminary analysis of the flat-faced wedge film made by Osborne Reynolds has become the basis for the mathematical calculations required for the application of such films to straight-line motion as in cross heads and planers, and to rotary motion as in thrust bearings, especially those of the Kingsbury and Michell types.

It must be borne in mind that the terms "flat-wedge film" and "pivoted shoe film" are not necessarily synonymous, although it is usual to assume that the pivoted shoe films have the flat-faced wedge form, and to apply to them the formulas developed for it. Flat-wedged films are extensively examined in subsequent discussions.

Capacity and Friction of Journal-bearing Films. Reynolds next applied his analysis to Tower's experimental partial journal bearing which was assumed to have running clearance (see Ref. 3). Reynolds showed that his hydrodynamical theory explained Tower's results, as to friction and film pressures, provided the film of lubricant had a curved wedge form.

The analytical studies of journal-bearing films of various angular extents was carried forward by later investigators (see Refs. 9, 10, 11).

THE EFFECT OF SIDE LEAKAGE UPON CAPACITY AND FRICTION. Up to this point of our discussion the effect of side leakage has been neglected and our formulas have been given for a unit of width of a film so wide (infinite) that only forward or backward flow need be considered. The actual flow in a film of finite width is best illustrated by means of a flat-wedge film, three views of which are shown in Fig. 17. In the plan view the oil flow is indicated by curved arrow-headed lines. Obviously the narrower the width b relative to a given length l the greater will be the proportionate oil loss at the sides of the film, and the less will be the pressures developed in the film.

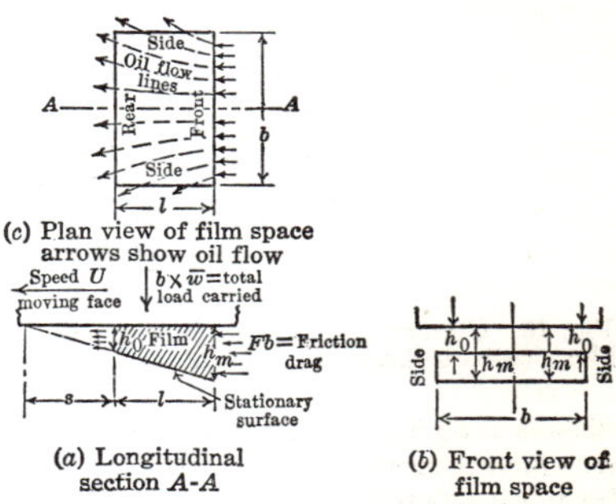

FIG. 17. The flat-wedge film. (H. A. S. Howarth, The Loading and Friction of Thrust and Journal Bearings with Perfect Lubrication, *ASME Trans.*, MSP-57-2, 1935)

Kingsbury proposed the character L_1 for representing the ratio of actual to ideal carrying capacity of a film. He also proposed that F_1 represent the ratio of the actual to the ideal friction loss of the film. Both these "correction factors," as they are called, vary with the ratio of film length to film width represented by the characters l/b.

In Fig. 18 Kingsbury's factors for L_1 and F_1 are shown for easy reference in the solving of problems that will be given later with the study of charts of bearing capacities. The side leakage curves by Needs (Ref. 12) in Fig. 19 were prepared specifically for 120° centrally loaded journal bearings and may be used for correction factors when solving problems involving such bearings, as will also appear later.

OPTIMUM CHARACTERISTICS OF OIL FILMS IN BEARINGS. Effects of Deviations. Reynolds pointed out in his early studies that a certain shape of wedge film would carry the greatest load. This is true also for the curved wedge films of journal bearings. Special studies of optima have shown that there are others than that for capacity. It has also been learned how far these optima may be deviated from with impunity. A study of optima for the flat-wedge film follows.

Optima for Flat-wedge Films. The flat-wedge film itself is illustrated by three views in Fig. 17. Its characteristics of load and friction with relation to film shape are shown in Fig. 20.

Variation in film form can best be visualized by keeping the length l and the minimum thickness h_0 unchanged and then changing the distance s. This will of course change h_m. Such changes in the film form will change its load-carrying capacity and the frictional resistance offered by the film to motion of the moving face, for cases where there is no other change of condition such as viscosity or speed.

If the load-carrying capacity, for example, should vary, under the above changing con-

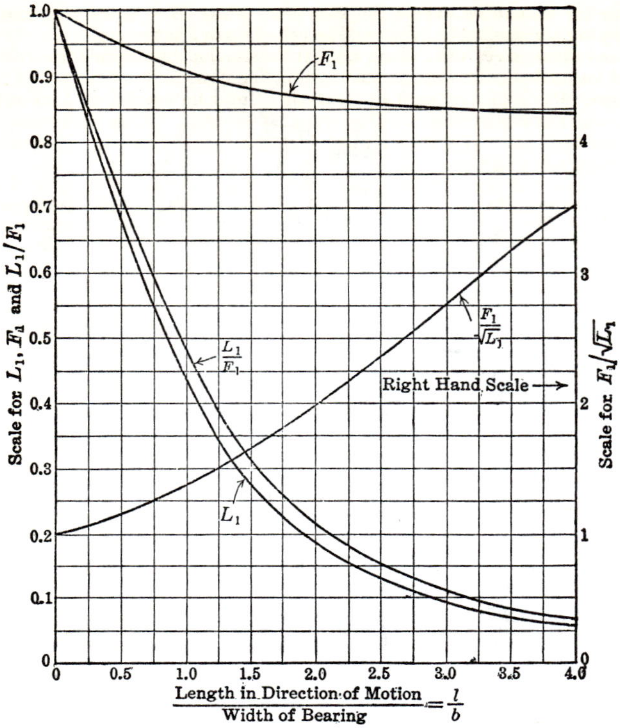

Fig. 18. Correction factors for influence of side leakage on load-carrying capacity and friction of fluid film. (Albert Kingsbury, Optimum Conditions in Journal Bearings, *ASME Trans.*, RP-54-7, 1932)

ditions as to film form, in such a way that it passes through a maximum value, the length s at which it occurs, when divided by the length of the shoe l, will give a ratio M that may be called an optimum. Curve 2 in Fig. 20 has such a maximum ordinate when $M = 0.841$. It belongs to the group of variables given in the equation for curve 2, whose ordinate Y is seen to be proportional to the square root of the load (Fig. 20). When $M = 0.841$ the film will carry the greatest load for a minimum thickness h_0. This will be called Case 2 optimum.

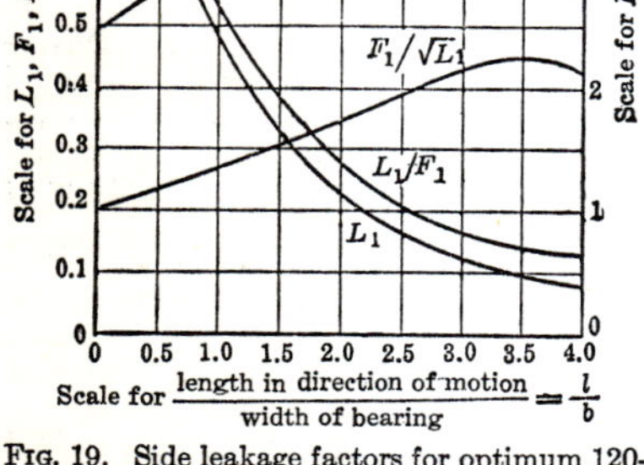

Fig. 19. Side leakage factors for optimum 120-degree central bearings for minimum coefficient of friction. (S. J. Needs, Effects of Side Leakage in 120-degree Centrally Supported Journal Bearing, *ASME Trans.*, APM-56-16, 1934)

If the total frictional resistance offered to the carrying of a given load should vary in such a way that it passes through a minimum value, as the ratio M varies, the particular M ratio at which this minimum occurs may be considered desirable and be named an optimum one. In curve 1 there is seen to be such a minimum ordinate when $M = 0.484$. It belongs to the group of variables given in the equation for curve 1 whose ordinate Z is seen to be proportional to the friction. When $M = 0.484$ the film will carry a given load with the least frictional resistance. Nothing is said about the film thickness, but it is obvious from the preceding paragraph that for this value of $M = 0.484$ the thickness h_0 must be less than it was when $M = 0.841$. Thus the film form for least frictional resistance will not be the same as the film form for maximum capacity. However, when $M = 0.484$ the friction will be least for carrying a given load. This will be called Case 1 optimum.

There is one other curve (No. 3) in Fig. 20 that has a minimum point. It represents the group of variables involving the coefficient of friction and a function of the angle of the film wedge. For a film of length l the ratio of friction coefficient λ to minimum film thickness h_0 will have a minimum value when $M = 0.652$. This minimum is called by Kingsbury the optimum friction coefficient condition and Case 3. It should be noted that it is not the same as Case 1.

Figure 20 shows that the minimum film thickness h_0 will vary directly as Y for a series of films of same length l carrying the same load at the same speed, the films having the

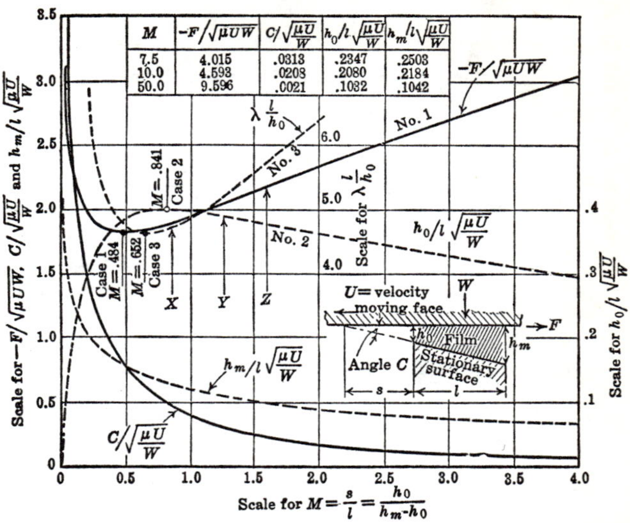

Fig. 20. **Effect of deviations from optimum conditions for a bearing with flat-wedge films.** (H. A. S. Howarth, *ASME Trans.*, MSP-57-2, 1935)

same average viscosity. Hence for Cases 1, 2, and 3, already discussed, the thicknesses h_0 are not very different, but where $M = 50$ (see table at top of Fig. 20), the thickness h_0 is about one-fourth of the average of the first three values.

Case 4 is interesting because it has a convenient value of M which is unity. Hence its mean thickness would be 50% greater than its h_0, and its maximum thickness would be twice its minimum, i.e., $h_m = 2h_0$. This case and the examples tabulated at the top of Fig. 20 are of interest in any study of the effects of deviations from "optimum" conditions.

In the solving of certain bearing problems the mean film thickness is often needed. Values are listed below for the four cases most frequently referred to:

Case 1	Case 2	Case 3	Case 4
$h_1 = 2.033h_{01}$	$h_2 = 1.594h_{02}$	$h_3 = 1.765h_{03}$	$h_4 = 1.5h_{04}$

The meaning of the subscripts in this tabulation and in Table 1 is, for example, that for Case 2 the mean film thickness, represented by h_2, is 1.594 times the least film thickness represented by h_{02}.

Application of Side Leakage Correction Factors. In formulas given for X, Y, and Z (Fig. 20) no correction has been made for side leakage. Hence it is necessary to introduce into them in the proper manner the correction factors L_1 and F_1. In Case 2 it is found convenient to square the terms. The corrected formulas are:

$$\frac{\lambda l}{h_0} = \frac{XF_1}{L_1} \tag{27}$$

$$\frac{h_0^2 W}{l^2 \mu U} = Y^2 L_1 \tag{28}$$

$$\frac{F}{\sqrt{\mu UW}} = \frac{ZF_1}{\sqrt{L_1}} \tag{29}$$

There have been grouped together in Table 1 certain formulas useful for solving problems involving flat-wedge films, especially those of pivoted shoes when arranged in a ring whose mean diameter is d. The only new term to appear besides d is P_0, which means the average pressure acting on the face of a pivoted shoe. The mean circumferential length of one such shoe would be l. The load carried by a circumferential band of unit radial width on diameter d would be

$$lP_0 = W \tag{30}$$

The velocity U at the mean diameter d can be related to rpm by the relation in eq. 31.

$$U = \frac{\pi dN}{60} \tag{31}$$

The optimum formulas are given in Table 1 for Cases 1, 2, 3, and 4.

Table 1. Formulas for Flat-wedge Films

Case 1

$$\lambda_1 = 1.815 \frac{F_1}{\sqrt{L_1}} \sqrt{\frac{\mu U}{p_0 l}} \quad (32) \qquad \lambda_1 = 0.415 \frac{F_1}{\sqrt{L_1}} \sqrt{\frac{\mu dN}{p_0 l}} \quad (33)$$

$$h_{01}^2 = \frac{0.00763 L_1 \mu d l N}{p_0} \quad (34) \qquad h_1^2 = \frac{0.0315 L_1 \mu d l N}{p_0} \quad (35)$$

Case 2

$$\lambda_2 = 1.890 \frac{F_1}{\sqrt{L_1}} \sqrt{\frac{\mu U}{p_0 l}} \quad (36) \qquad \lambda_2 = 0.432 \frac{F_1}{\sqrt{L_1}} \sqrt{\frac{\mu dN}{p_0 l}} \quad (37)$$

$$h_{02}^2 = \frac{0.00837 L_1 \mu d l N}{p_0} \quad (38) \qquad h_2^2 = \frac{0.0213 L_1 \mu d l N}{p_0} \quad (39)$$

Case 3

$$\lambda_3 = 1.845 \frac{F_1}{\sqrt{L_1}} \sqrt{\frac{\mu U}{p_0 l}} \quad (40) \qquad \lambda_3 = 0.422 \frac{F_1}{\sqrt{L_1}} \sqrt{\frac{\mu dN}{p_0 l}} \quad (41)$$

$$h_{03}^2 = \frac{0.00815 L_1 \mu d l N}{p_0} \quad (42) \qquad h_3^2 = \frac{0.0254 L_1 \mu d l N}{p_0} \quad (43)$$

Case 4

$$\lambda_4 = 1.950 \frac{F_1}{\sqrt{L_1}} \sqrt{\frac{\mu U}{p_0 l}} \quad (44) \qquad \lambda_4 = 0.446 \frac{F_1}{\sqrt{L_1}} \sqrt{\frac{\mu dN}{p_0 l}} \quad (45)$$

$$h_{04}^2 = \frac{0.00830 L_1 \mu d l N}{p_0} \quad (46) \qquad h_4^2 = \frac{0.0187 L_1 \mu d l N}{p_0} \quad (47)$$

The foregoing study of flat-wedge films has been fairly comprehensive because of their simplicity and fundamental importance. It is, however, only in straight-line motion that the film can have a true flat-wedge shape. In rotary motion the film form, in its circular section, will follow the sine curve wedge form in thrust bearings and in journal bearings with clearance, and the fitted surface wedge form in fitted cylindrical journal bearings. Two such film forms are shown in Figs. 21 and 22 as they are found in common types of journal bearings not having pivoted shoes.

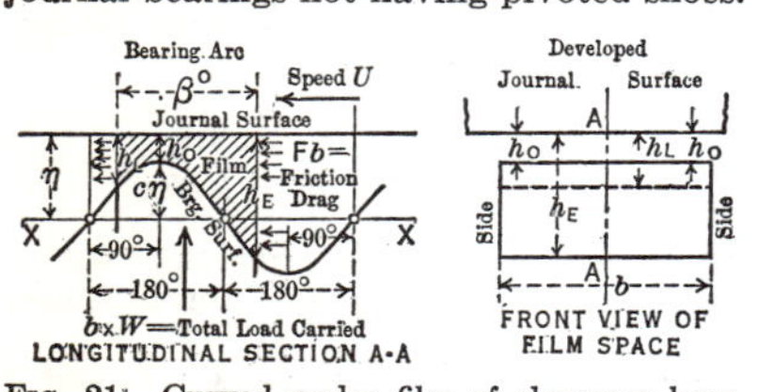

FIG. 21. Curved wedge film of clearance bearing. (H. A. S. Howarth, *ASME Trans.*, MSP-57-2, 1935)

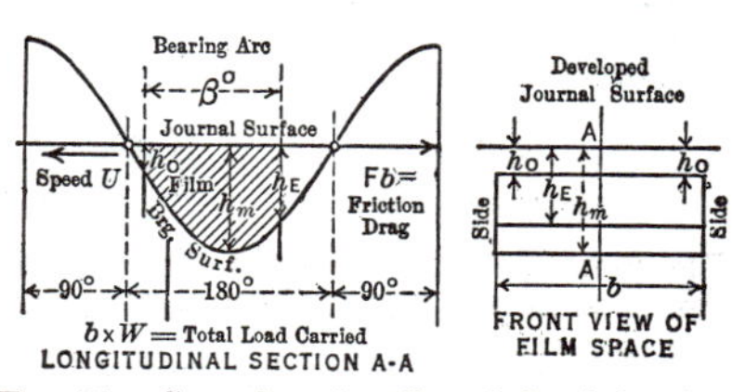

FIG. 22. Curved wedge film of fitted bearing. (H. A. S. Howarth, *ASME Trans.*, MSP-57-2, 1935)

In thrust bearings of the pivoted shoe type the sine curve shape of the film wedge section varies in thickness and node range, not only with the radius to the circular section examined, but also with changes in speed of rotation, or of viscosity of lubricant. For example, as a pivoted shoe thrust bearing slows down, the surfaces come closer together until they eventually meet with the fit originally provided, usually flat to flat. Obviously one important virtue of the pivoted shoe thrust bearing is that ability to adjust its film thickness and form (wedge angle) to suit all operating conditions, from zero velocity to high speed, from low to high film viscosity, and also from light to heavy load. The study of pivoted shoes is continued in more detail later, and loading tables will be given together with safe minimum film thicknesses.

An ordinary journal bearing with running clearance, however, must come to rest on the line of contact between two surfaces of different curvature. Hence its carrying capacity at zero speed would be expected to be less than that of fitted surfaces. Journal bearings are therefore frequently made with fitted surfaces, or with combined fitted and clearance surfaces, as well as with true cylindrical surfaces with running clearance.

CAPACITY AND FRICTION DETERMINATION FOR JOURNAL BEARINGS. As already stated, Osborne Reynolds made the first analysis of such a bearing in his hydrodynamical study of Tower's test bearing. The latter was assumed by Reynolds to be what is now generally called a *partial bearing with running clearance.* In practice such bearings have film arcs varying from 60 to 360°, the latter being called a *full bearing.* When the load upon such a bearing bisects the film the bearing is said to be *centrally loaded.* When the leading angle of the film is longer than the trailing angle the bearing is said to be *eccentrically loaded.* There are widely differing series of eccentrically loaded bearings. They are defined by the α/β ratio, α being the leading angle of the film and β its total angle. When $\alpha/\beta = 0.5$ the bearing is centrally loaded.

Sommerfeld (Ref. 9), using polar coordinates, analyzed a 360°, or full, clearance bearing and also a 180° centrally loaded bearing.

Harrison (Ref. 10) later made an analysis of the 360° clearance bearing, in rectangular coordinates, for the purpose of studying Kingsbury's air bearing test results, as already mentioned. Harrison's equations led Howarth to make his graphical analysis (Ref. 11) based upon them, covering bearing film arcs less than 360°.

Graphical Integrations for General Characteristics. This graphical analysis deserves careful note because it points the way for integration of any differential ratio, especially of one whose exact mathematical integration is either formidable or impossible. The pressure gradient eq. 21 can be applied to a full or 360° clearance bearing as was done by Harrison, who took as the origin for the angular positions of any film thickness h, the location of maximum value of h. Then when $x = a\theta$ and $h = \eta(1 + c\cos\theta)$, eq. 21 becomes

$$\frac{dP}{d\theta} = \frac{6\mu Uac(\cos\theta - \cos\theta_1)}{\eta^2(1 + c\cos\theta)^3} \tag{48}$$

In eq. 48 a is the journal radius and $(a + \eta)$ the radius of the bore of the bearing, while c is the eccentricity factor for the journal running position.

For graphical integration of eq. 48 see Fig. 23. When $dP/d\theta$ is made the ordinate and θ the abscissa, the area under the curve, for a given set of constants, when integrated from

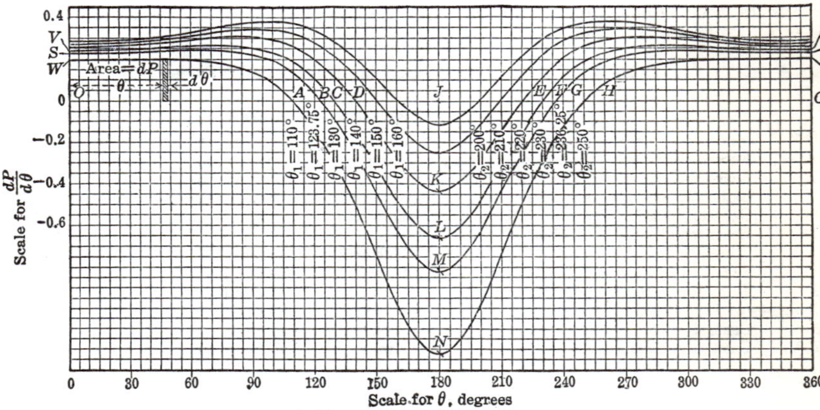

FIG. 23. Graph of eq. 48, assuming $\frac{6\mu Ua}{\eta^2} = 1$ and $c = 0.4$. (H. A. S. Howarth, *ASME Trans.*, Ref. 11)

$\theta = 0°$ to $\theta = 30°$, will be a measure of the pressure in the film at 30°, and so on. It will be noted that the curves and scales are based on the assumption that $c = 0.4$ and $(6\mu Ua)/\eta^2 = 1$. Several values of θ_1 are assumed, and a curve is drawn for each. For only one such curve, $\theta_1 = 123.75$, do the areas under the curve and above the axis of abscissas OO' equal the area under it. Hence in 360° or full bearings, when $\theta_1 = 123.75°$ and $c = 0.4$, the pressure within the film begins at 0 (zero) when $\theta = 0°$, reaches a positive maximum when $\theta = 123.75°$, reaches zero again when $\theta = 180°$, passes to a negative minimum which it reaches at 236.25°, and then rises to become zero again when $\theta = 360°$.

It can readily be understood how integrations between other limits can be made to suit the film conditions existing at the limits, whether they are alike or different. For example, the bearing film might be much less than 360° in extent, and the initial pressure might be well above atmospheric, due to impact, while the final pressure might be atmospheric. For a full description of this graphical process see Ref. 11.

Over-center Running of Journals in Bearings. "Over-center" or "above-center" running is possible, under Reynolds' theory, and is characteristic of bearings whose leading film angle considerably exceeds the trailing film angle. (See Ref. 13.)

A way to prevent the undue rise of the journal, suggested by Howarth, is by offsetting a cap under which a similar counter-acting film will automatically develop. This cap may be mounted with freedom to develop automatically the restraining film, or it may be fixed in a suitable offset position. This method may be used whether the bearing has running clearance or is of the fitted type.

OIL-FILM INSTABILITY. Under certain conditions a phenomenon known as oil-film whirl may occur in journal bearings. According to B. L. Newkirk (Ref. 14):

> The oil-film whirl arises from a potential instability of the oil film which supports a rotating journal. The instability develops into whirling or vibration in co-operation with elastic action of the rotor, elasticity of the supporting structure being also a factor. The development of the whirl from a state of steady running seems to occur as follows: A slight shock causes the rotor to vibrate at its natural frequency, which is numerically equal to the critical or resonant speed. This vibration of the rotor causes periodic displacement of the journal from its equilibrium position on the oil film. When the journal is thus displaced the restoring force that develops in the oil film is not directed exactly toward the equilibrium position. There is a component of the restoring force which sets up a forward whirl of the journal about its equilibrium position. If the speed of the rotor is equal to or greater than some value which is approximately twice the whirl frequency the whirl builds up, otherwise it dies out. The whirl occurs at a resonant frequency of the rotor (usually the lowest resonant frequency) for all speeds of rotation above the lowest speed at which whirling develops.

Whirling is preventable by proper design of oil-film bearings that operate at high speed, as by a combination plain and pivoted-shoe bearing for each journal. See also the preceding discussion of over-center running, which is cured in a similar manner, and Ref. 15.

2. LOADING AND FRICTION CHARTS AND TABLES FOR JOURNAL BEARINGS

The charts, tables, and examples that follow are placed in two groups: (1) general characteristics, and (2) optimum characteristics. The classes of bearings covered are lettered A, B, C, D, E, and F. (See Table 2.)

Table 2. Key to Journal Bearing Charts and Tables

Class	Description	Examined by	Characteristics Determined	Data Given in Tables	Data Given in Figures
A	Eccentrically loaded clearance bearing	Kingsbury	Two optima	3, 4	31–34
B	Eccentrically loaded fitted bearing	Kingsbury	Two optima	5, 6	35–38
C	Centrally loaded clearance bearing	Kingsbury	Two optima	7, 8	39–42
C	Centrally loaded clearance bearing	Howarth	General		24
D	Eccentrically loaded (offset) clearance bearing	Howarth	General		25, 26
E	Eccentrically loaded fitted bearing	Howarth	General		27
F	Full (360°) clearance bearing	Howarth	General		28
All classes of 120° clearance bearing		Howarth	General		29
120°	Centrally loaded clearance bearing	Kingsbury	General		30
120°	Centrally loaded clearance bearing	Kingsbury	Three optima		30

NOTATION

In the discussion which follows, the significance of the symbols used is given below. Unless otherwise specified in the text, all linear dimensions are in inches, and forces (loads) are in pounds.

a = radius of journal.

A = least angle from line of centers to leading edge of partial bearing, measured in direction of rotation. See exception in study of fitted bearings (Fig. 35).

b = width of bearing surface measured along the shaft axis in a journal bearing or, in general, at right angles to direction of motion.

c = eccentricity factor. When multiplied by the radial clearance η, the product $c\eta$ equals distance from center of bearing to center of journal.

F = total tangential friction force at surface of a journal, or at mean diameter of a thrust bearing, per unit of width of bearing surface.

F_1 = correction factor for friction, *with side leakage*.

h = film thickness at any angular position θ in journal bearings, or any distance x for flat tapered film. For fitted journal bearings $h = \epsilon \sin \theta$, and for clearance journal bearings $h = \eta(1 + c \cos \theta)$.

h_0 = minimum thickness of oil film in a bearing, usually. When the line of centers for a clearance bearing does not intersect the partial bearing surface, h_0 is measured where the bearing surface, if extended, would cross that line.

h_1 = thickness of oil film at region of maximum pressure.

h_A = thickness of oil film at entering edge of bearing film, if less than h_m.

h_E = film thickness at trailing edge of partial clearance bearing, when greater or less than h_0.

h_m = maximum thickness of oil film.

H = horsepower friction developed by a journal bearing, or thrust bearing, per unit of bearing width.

Hp = horsepower.

l = length of a bearing face in the direction of motion. In a partial journal bearing, whose angular extent β is expressed in degrees, $l = \pi\beta a/180$. This is useful for determining length-width ratio, l/b, used in selecting side leakage correction factors.

L_1 = correction factor for load carrying capacity, *with side leakage*.

N = revolutions per minute.

n = number of pivoted segments in a thrust bearing.

O = center of curvature of journal when running on oil film.

O' = center of curvature of bearing surface supporting journal.

P = pressure within the oil film where the thickness is h.

p_0 = nominal mean pressure acting on bearing. In journal bearings, p_0 is found by dividing total applied load by (journal diameter $\times$ axial bearing width), except where otherwise stated. It also is expressed by $W/2a$. W is taken per unit of width. In thrust bearings, $p_0 = R/nlb$.

P_m or $P_{\max}$ = maximum pressure within oil film of a bearing.

R = total load applied by a journal to a bearing, or by a collar to a thrust bearing, except where otherwise stated.

R' = total reaction of bearing against the journal.

s = a film form dimension (see Figs. 7 and 17).

U = relative velocity of the bearing surfaces (usually inches per second). In thrust bearings it is measured on the mean diameter.

V = relative velocity of bearing surfaces, feet per minute.

w = nominal mean unit load a bearing would carry if side leakage could be neglected. See p_0. The relation is $p_0 = L_1 w$.

W = actual load applied by a journal to its bearing, per unit of axial width = R/b.

Z = character expressing viscosity coefficient in centipoises. See μ.

α = leading angle for partial bearing, measured in direction of motion of journal from leading edge of bearing to line of action of load.

β = angular extent of bearing surface over which perfect film lubrication is maintained. Usually expressed in degrees.

ϵ = eccentricity of journal axis when running in a fitted bearing.

η = radial clearance provided in a clearance bearing. When a = journal radius, $(a + \eta)$ = radius of curvature of a bearing surface, and $c\eta$ = eccentricity of running position of journal axis in a clearance bearing.

θ = angle from reference diameter to point within oil film where thickness is h.

θ_1 = angle θ corresponding with location of maximum unit pressure within the oil film.

λ = coefficient of friction for journal surface. In thrust bearings, λ is the coefficient for mean diameter of thrust collar face. $\lambda = (F_1/L_1)\lambda_H$.

λ' = coefficient of friction for bearing surface.

λ_H = uncorrected coefficient of journal friction in charts.

μ = coefficient of viscosity. When not otherwise specified, μ is expressed in pound-seconds per square inch (Reyns). The term *Reyn* is applied to the coefficient when expressed in English units. The cgs unit for μ is the *Poise* measured in dyne-seconds per square centimeter. One poise = 0.0000145 Reyn; 1 Reyn = 69,000 poises.

ϕ = angle relating load line to line of centers of journal and bearing in Kingsbury's optimum condition. See also bearing cross sections (Figs. 31, 35, and 39).

ϕ_H = angle relating load line to line of centers of journal and bearing in bearing cross sections in Figs. 24 to 29.

GENERAL CHARACTERISTICS OF JOURNAL BEARINGS. General characteristics mean the relations between load-carrying capacity, and also friction, and the operating position of the journal upon the bearing film, shown by the eccentricity factor c in clearance bearings, for a wide range of eccentricities or running positions of the journal.

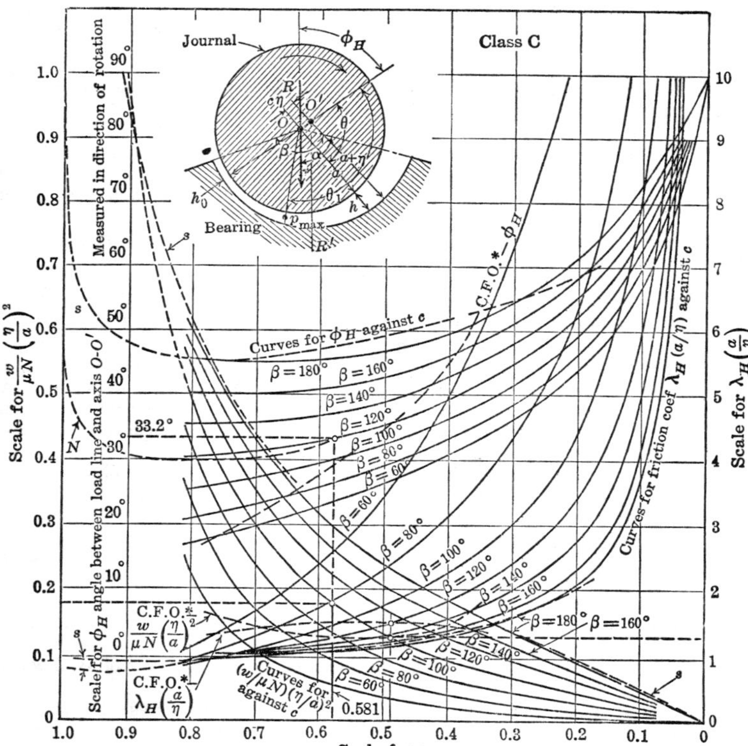

FIG. 24. General characteristics of centrally loaded clearance bearings, Class C, assuming constant velocity and no side leakage. C.F.O.* = Class C friction optima. Where the dotted curves cut the corresponding full line curves the values are identical with the optimum conditions for minimum friction coefficient in Figs. 40 and 41 and Table 8. (H. A. S. Howarth, *ASME Trans.*, MSP-57-2, 1935)

Class C. Centrally Loaded Partial Bearings with Clearance. The general characteristics of these bearings are given in Fig. 24, in which there is also a cross section of the journal and the bearing. Data are given for bearing film arcs ranging from 60° to 180°, covering the load group $(w/\mu N)(\eta/a)^2$, the friction group $\lambda_H(a/\eta)$, and the angular position ϕ_H of the line through the centers of the journal and the bearing, all with relation to the eccentricity factor c, which varies from near zero to about 0.8 (80%). Some data extend to $c = 1.0$. The full lines are those obtained by Howarth, using graphical integration. The dotted curves S are based on Sommerfeld's calculations. Those marked N are based upon Needs' (Ref. 12) calculations. The dotted curves, marked C.F.O., are plotted from data in Kingsbury's table of optimum friction coefficient condition of operation (Table 8)

In this case there is one such optimum eccentricity factor c for each film angle. This chart (Fig. 24) is one of the most complete and the most useful available.

The side leakage correction factors can be applied to the ordinates of the curves of this chart in the manner already explained. The symbol w that appears in the capacity group is intended to convey the fact that the ordinates of the capacity curves are uncorrected for side leakage. As the character p_0 is ordinarily thought of as being the actual pressure due to applied load, the relationship $p_0 = L_1 w$ would hold and w can be determined from it. By this means the eccentricity of operation can be determined. Once that value of c is known, it must be held and the corresponding values of the other symbol groups found. When they are so found the correction factors may be applied to their ordinates. Thus, applying L_1 to p_0, we find $w = (p_0/L_1)$. Entering the chart with the symbol group $(w/\mu N)(\eta/a)^2$, we find the ordinate of the proper curve leads us to the base scale, where we find and read the value of c. Holding this value, we extend the ordinate to the friction

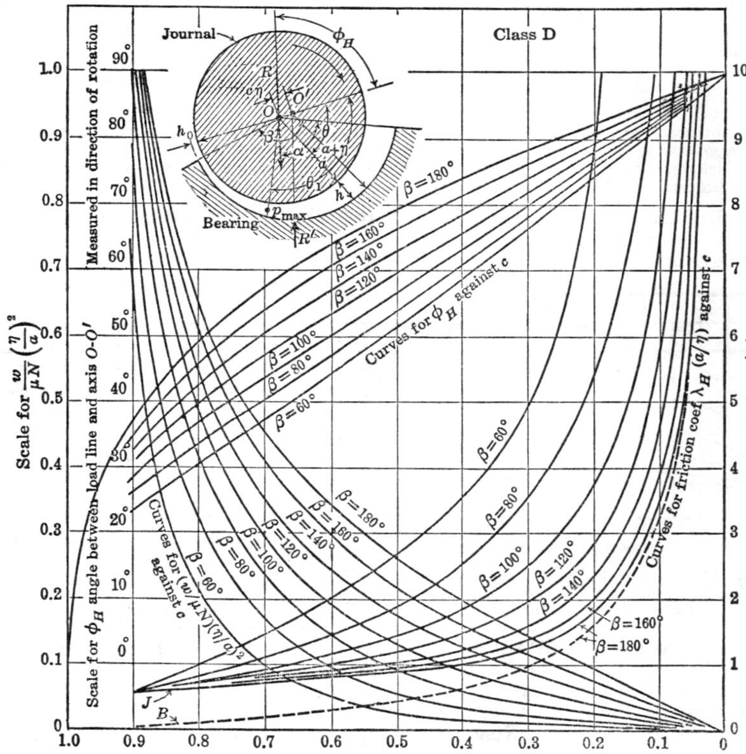

FIG. 25. General characteristics of offset clearance bearings, Class D, assuming constant viscosity and no side leakage. See Fig. 26 for corresponding ratios of α/β. This class of bearings differs somewhat from the eccentrically loaded bearings in Class A. (H. A. S. Howarth, *ASME Trans.*, MSP-57-2, 1935)

coefficient symbol group curve. Then we read the value of that new ordinate on the $\lambda_H(a/\eta)$ scale. Then we place the correction factor F_1/L_1 before the ordinate value and find the true value of $\lambda_H(a/\eta)$.

EXAMPLE 1. Given a centrally loaded 120° bearing with a total load $R = 10{,}000$ lb and actual nominal pressure $p_0 = 125$ psi. Journal diameter is 8 in., and bearing width is 10 in. Clearance ratio $\eta/a = 0.001$, and speed $N = 400$ rpm. Determine the running position of the journal when using two oils of viscosity coefficient $\mu = 3.4 \times 10^{-6}$ Reyn, and 10.2×10^{-6} Reyn, respectively.

Solution. In Fig. 24 the scale at the left represents a symbol group in which all values are known except w, which is the ideal nominal unit pressure that the bearing would sustain if side leakage could be neglected. This ideal pressure is reduced to actual pressure p_0 by side leakage, and $p_0 = L_1 w$. Correction factor L_1 is found from Fig. 18. The ratio $l/b = [(120/360)\pi 8]/10 = 0.839$. When Fig. 18

is entered at $l/b = 0.839$, the correction factors are found to be: $L_1 = 0.51$; $L_1/F_1 = 0.555$; $F_1 = 0.92$. Then $W = 125/0.51 = 245$, and

$$\left(\frac{w}{\mu N}\right)\left(\frac{\eta}{a}\right)^2 = \frac{245 \times 1 \times 10^{-6}}{\mu \times 400} = \frac{0.613 \times 10^{-6}}{\mu}$$

The next step is to substitute in the above expression the two values of μ given in the statement of the problem, and with the resulting values enter Fig. 24 at the left. First substituting $\mu = 3.4 \times 10^{-6}$, $(w/\mu N)(\eta/a)^2 = 0.180$. Locating this point on the load-capacity curve, $\beta = 120°$ (the group of curves starting at the lower right), find $c = 0.581$. The ordinate of $c = 0.581$ intersects the friction coefficient curve for $\beta = 120°$ (group of curves starting at the lower left) at $\lambda_H a/\eta = 1.29$. This ordinate also intersects the ϕ_H curve (group of curves starting at the upper right) for $\beta = 120°$ at $\phi_H = 33.2°$.

The actual friction coefficient will be $\lambda = (F_1/L_1)\lambda_H = 1.29 \times 10^{-3}/0.555 = 0.00233$.

The journal running position will be on the line $\phi_H = 33.2°$, and its distance from the center of the bearing will be $c\eta = 0.581 \times (8/2) \times 0.001 = 0.002324$ in. The minimum film thickness will be $h_0 = (1 - c)\eta = 0.001676$ in. Optimum eccentricity is shown by the dotted C.F.O. loading curve $(w/\mu N)(\eta/a)^2$ to be 0.49 for a 120° bearing, with a corresponding value of $h_0 = 0.00204$. Hence, with $\mu = 3.4 \times 10^{-6}$, this bearing is not quite so safe as the optimum would be.

Solving the problem with $\mu = 10.2 \times 10^{-6}$, the value of $(w/\mu N)(\eta/a)^2$ for entering the chart is 0.060. The following values then are established:

$$c = 0.28; \quad \phi_H = 46°; \quad \lambda_H \frac{a}{\eta} = 2.54; \quad \lambda = 0.00458; \quad h_0 = 0.00288$$

Friction and minimum film thickness are greater with the heavier oil. An intermediate oil would cause the journal to run with optimum eccentricity. The decision then rests on which of the three oil film thicknesses $h_0 = 0.00168$, 0.00204, or 0.00288 in. best suits the required shaft alignment and other conditions. Heavier oil is best where shaft alignment conditions are not ideal.

Class D. Eccentrically Loaded (Offset) Partial Bearings with Clearance. This series is somewhat different from Kingsbury's optimum eccentrically loaded bearings. However, it must be remembered that in Kingsbury's tables only one eccentricity is given as optimum whereas in Fig. 25, by Howarth, the characteristics of each bearing angle are given for eccentricities ranging from zero to 0.9 (90%). One thing to note about this series is that the leading angle varies with the eccentricity for a given total film angle. This is seen in Fig. 26.

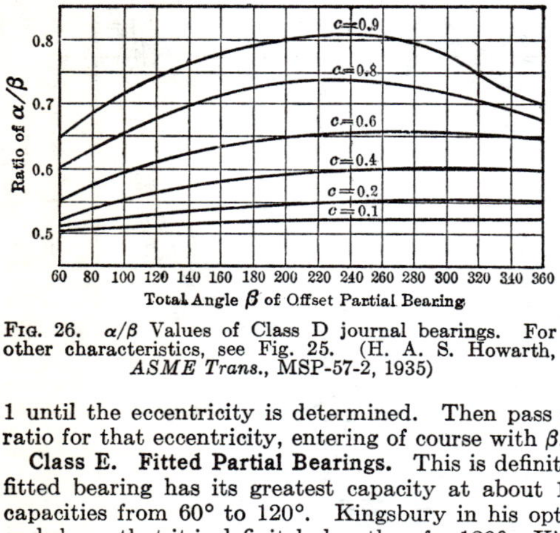

FIG. 26. α/β Values of Class D journal bearings. For other characteristics, see Fig. 25. (H. A. S. Howarth, *ASME Trans.*, MSP-57-2, 1935)

Howarth thought that this Class D series was close to the optimum, although he did not seek an exactly optimum one. How close it came to Kingsbury's later determined optima may be seen in Fig. 29, which will be discussed later.

When solving problems in this series proceed as for Class C Example 1 until the eccentricity is determined. Then pass from Fig. 25 to 26 and find the α/β ratio for that eccentricity, entering of course with β.

Class E. Fitted Partial Bearings. This is definitely an eccentrically loaded series. A fitted bearing has its greatest capacity at about 120° film angle. Howarth gives the capacities from 60° to 120°. Kingsbury in his optima gives the capacity also for 150° and shows that it is definitely less than for 120°. Kingsbury's and Howarth's values differ only slightly. The Kingsbury values for optimum friction coefficient are shown by dotted lines on Howarth's chart (Fig. 27).

EXAMPLE 2. Given a 100° fitted bearing of journal diameter 5 in., and width 6 in., carrying a total load R of 3000 lb, running at $N = 300$ rpm, and using oil of viscosity $\mu = 3 \times 10^{-6}$ Reyn. Find the minimum film thickness.

Solution. $l/b = (5\pi/6) \times (100/360) = 0.727$, from which (see Fig. 18) $L_1 = 0.565$, and $w = 3000/(6 \times 5 \times 0.565) = 177$. Then $(w/\mu N) \times 10^{-6} = (177 \times 10^{-6})/(3 \times 10^{-6} \times 300) = 0.197$, which value is used to enter Fig. 27. The intersection of this value with the lower diagonal $\beta = 100$ gives a value of ϵ/a of 0.0008. The minimum thickness of oil film h_0 will be at the trailing edge of the bearing and is, numerically, $h_0 = \epsilon \sin \theta$. From inset in Fig. 27, $\theta = 270° + \beta - (\phi_H + \alpha) = 157.6°$. Hence $h_0 = 0.0008 \times 2.5 \times 0.381 = 0.00076$.

Other data that may be obtained from Fig. 27 are: uncorrected coefficient of friction $\lambda_H = 0.00082$ Applying correction factor $L_1/F_1 = 0.61$ (Fig. 18), the actual coefficient of friction

$$\lambda = \frac{\lambda_H}{L_1/F_1} = \frac{0.00082}{0.61} = 0.00134$$

Class F. Full (360°) Bearings with Running Clearance. This bearing, which is shown in the inset in Fig. 28, is treated as a half bearing so far as carrying capacity and running position are concerned. As regards friction, it is treated as a full bearing in which the running position of the journal is the same as for a half bearing. The line of application of the load when the film is in running position would not pass through the center of the supporting film arc. The useful film of the bearing, therefore, may be treated as that of an offset bearing, Class D. In Fig. 28, the load capacity curve is taken from Fig. 25 for $\beta = 180°$, Class D bearings. The pressure film in full bearings is not likely to exceed 180° in angular extent. The experiments of Stanton (Ref. 16), Bradford (Ref. 17), and McKee (Ref. 18) indicate shorter films for certain conditions. For other arcs than 180° the appropriate curve for load capacity in Fig. 25 can be used to find the eccentricity. Figure 28 then can be entered with the eccentricity so found, the approximate total friction determined, and the horsepower loss estimated. The friction coefficient curve in Fig. 28 can be used only with the load capacity curve in the same illustration. Example 3 shows the use of side leakage correction factors in a full bearing.

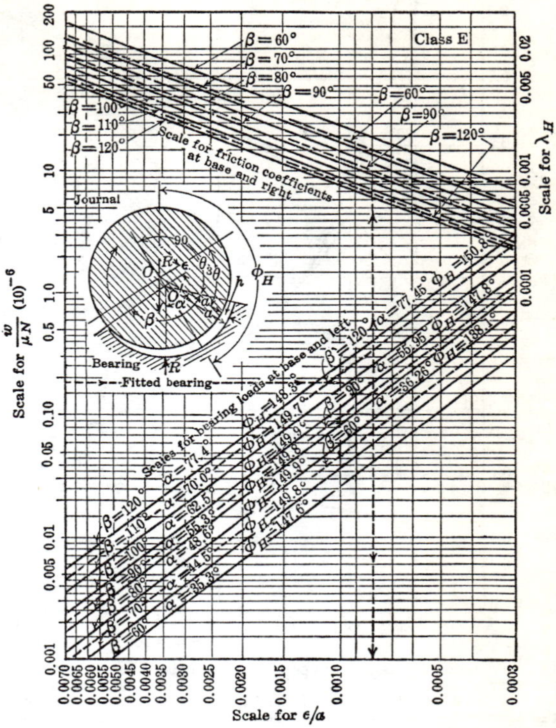

FIG. 27. General characteristics of fitted bearings, Class E. This series differs slightly from Class B as indicated by the broken lines for $\beta = 60°$, $\beta = 90°$, and $\beta = 120°$, which represent the optima for minimum friction coefficient. (H. A. S. Howarth, *ASME Trans.*, MSP-57-2, 1935)

EXAMPLE 3. Determine the friction power loss and friction coefficient of a full bearing of diameter $2a = 4$ in., width $b = 8$ in., running at $N = 3600$ rpm, and carrying a total journal load $R = 3600$ lb. The clearance ratio $\eta/a = 0.003$, and the oil viscosity $\mu = 2 \times 10^{-6}$ Reyn.

Solution. In order to find L_1, assume $\beta = 180°$. Then $l/b = (4\pi/8) \times (180/360) = 0.79$, and from Fig. 18 $L_1 = 0.53$. $F_1 = 0.925$, and $L_1/F_1 = 0.575$.

$$w = \frac{R}{2a \times b \times L_1} = \frac{3600}{4 \times 8 \times 0.53} = 212 \text{ psi}$$

Figure 28 is entered with the value of the symbol group

$$\left(\frac{w}{\mu N}\right)\left(\frac{\eta}{a}\right)^2 = \frac{212 \times 9 \times 10^{-6}}{2 \times 10^{-6} \times 3600} = 0.265$$

This value on the load capacity curve corresponds to a value of eccentricity factor $c = 0.5$. For this value of c the value of $\lambda_H a/\eta$ is read from the friction coefficient curve as 1.88. The total friction factor is read from the total friction curve as 3.2.

$$\text{Total power loss} = \text{Total friction factor} \times \left(\frac{\mu b}{2}\right)\left(\frac{a}{\eta}\right) a^2 F_1 N^2 \times 10^{-5} = 4.1 \text{ hp}$$

$$\text{Actual coefficient of friction } \lambda = \frac{\lambda_H}{L_1/F_1} = \frac{\lambda_H}{0.583} = 0.0098$$

If λ_H is checked back to the total power loss, the latter is found to be the same if the correction factor is properly introduced in $\lambda = (F_1/L_1)\lambda_H$.

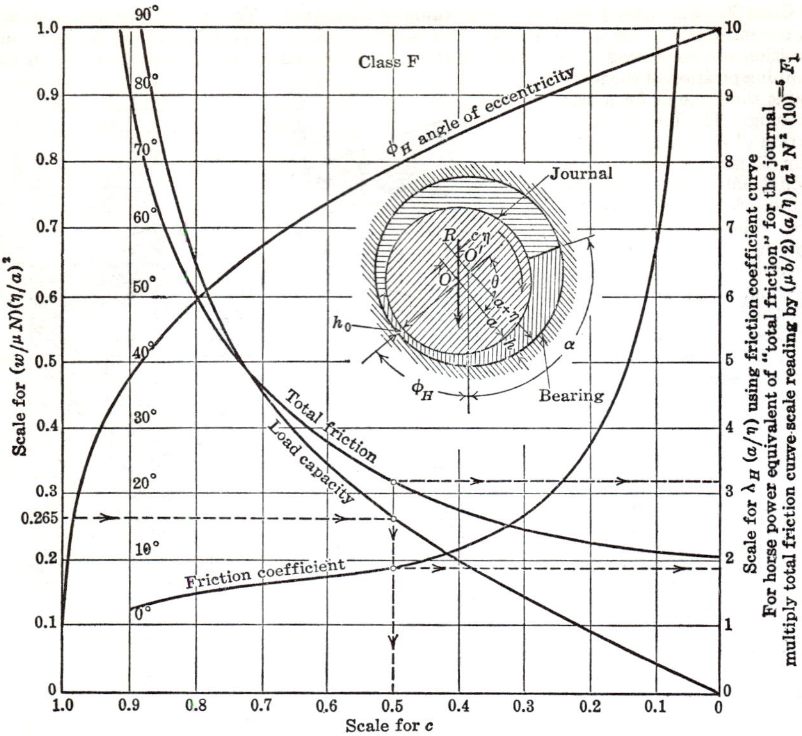

FIG. 28. General characteristics of full bearings, Class F, with running clearance, assuming constant viscosity and no side leakage. The load capacity curve corresponds to that for an offset (Class D) bearing for which $\beta = 180°$. The friction curves assume that for the purpose of calculating this loss the bearing film may be considered as extending all around the journal. This assumption is arbitrary but appears to be well on the safe side. (H. A. S. Howarth, *ASME Trans.*, MSP-57-2, 1935)

All Classes of 120° Partial Bearings with Clearance. In Fig. 29 the general carrying capacity characteristics of this common bearing film angle are given more fully than for any other angle. As heretofore stated this investigation was made to find out whether a journal center would rise above the bearing center under any condition, when the load acts directly downward. Such over-center running was found to be possible and also preventable. (See Ref. 13.)

Aside from its original purpose, Fig. 29 shows many other things about oil film bearings. For example, it shows the conditions under which negative pressure may be expected at the trailing end of a film and also when it may be looked for at the leading end.

If a line is drawn through the abscissa peaks of the eccentricity factor c curves, it will pass through the point marked ALO. That mark means "Class A load optimum" as determined by Kingsbury for a 120° Class A clearance bearing.

The dot and dash line marked "offset series, variable α/β" describes the 120° bearing of Howarth's offset series of clearance bearings, Class D. (See Figs. 25 and 26.) It was said above in connection therewith that Kingsbury's optimum "points" were not far from the "offset series." Note where those points ALO and AFO lie in Fig. 29 with respect to the "offset series" line.

The CLO and CFO points are, as expected, on the curve for which $\alpha/\beta = 0.5$ (by definition). Figure 29 also gives the angular location, θ_1, of the maximum film pressure. These curves may be used to determine whether there is a possibility of negative pressures existing in the oil film. All pressures in the film are positive between the lines M and N. An unpublished investigation by S. J. Needs shows that negative pressure is to be expected in a finite bearing where the bearing of infinite width would show all pressures to be positive. Hence, when the data of a problem show that the running position approaches the upper limit line of Fig. 29, some negative pressure is probable. If such conditions exist, it is advisable to place the problem in some other class of bearing, using $\alpha = 60°$ as suggested below.

If conditions of operation are such that negative pressures may occur at the trailing end, 120° bearings are best studied as if they were Class A, eccentrically loaded bearings, or Class D, offset bearings with $\beta < 120°$. In both cases, leading angle α should be taken as 60°, allowing the trailing angle $(\beta - \alpha)$ to be determined by the data in the class used.

EXAMPLE 4. Determine whether negative pressure would exist in a journal of diameter $2a = 6$ in., and width $b = 10$ in., carrying a total load $R = 600$ lb at $N = 1000$ rpm, taking α/β ratios as follows: (1) $\alpha/\beta = 0.44$; (2) $\alpha/\beta = 0.52$; (3) $\alpha/\beta = 0.62$. Clearance ratio $\eta/a = 0.002$, and the oil viscosity $\mu = 4.0 \times 10^{-6}$ Reyn.

Solution. Determine correction factor L_1 corresponding to $l/b = (6\pi/10) \times (120/360) = 0.628$. From Fig. 18, $L_1 = 0.613$, and $w = 600/(6 \times 10 \times 0.613) = 16.30$. Since $\mu = 4.0 \times 10^{-6}$, the group

$$\left(\frac{w}{\mu N}\right)\left(\frac{\eta}{a}\right)^2 = \frac{16.3}{4 \times 10^{-6} \times 1000} \times 0.002^2 = 0.0163$$

If an ordinate is erected at 0.0163 on the scale $(w/\mu N)(\eta/a)^2$ in Fig. 29 it will intersect the curves for $\alpha/\beta = 0.44$, 0.52, and 0.62 at the points 1, 2, and 3 respectively. Hence, with ratios of α/β of 0.44 and 0.62, negative pressure is quite likely to exist at the trailing and leading ends, respectively, of the bearing. With a ratio of $\alpha/\beta = 0.52$, the bearing probably will not have negative pressure.

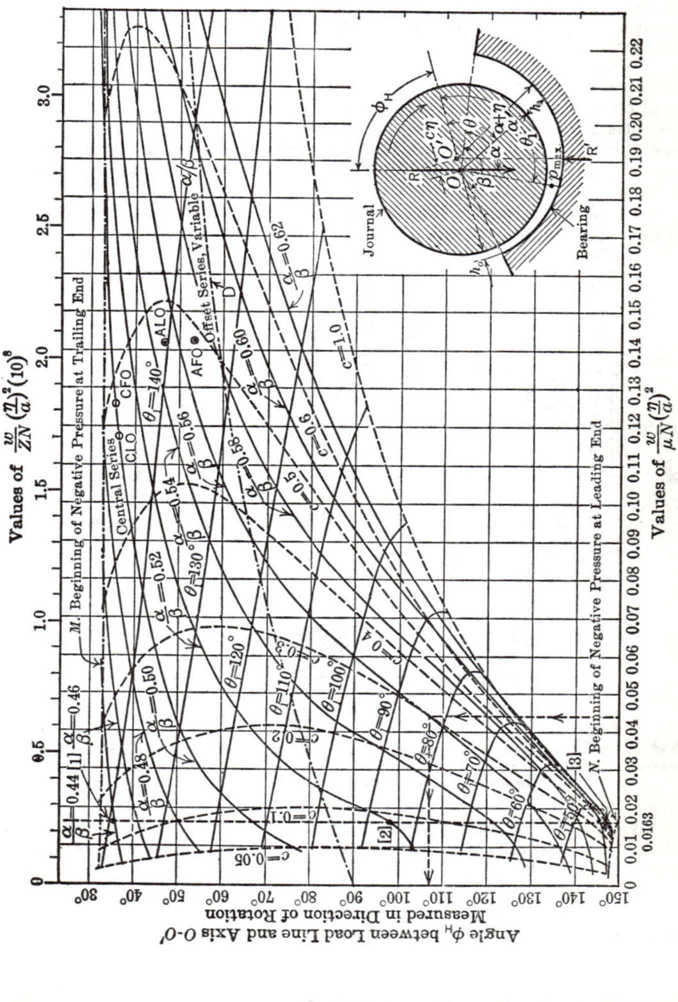

FIG. 29. General capacity characteristics of 120° clearance bearings, including Classes A, C, and D, assuming constant viscosity and no side leakage. Angular length of bearing $\beta = 120°$. (H. A. S. Howarth, *ASME Trans.*, MSP-57-2, 1935)

Other items that may be determined from Fig. 29 are the total lift of the journal and the point of maximum pressure in the oil film.

EXAMPLE 5. Determine the lift of the journal when in running position, and the point of maximum pressure in the oil film of a centrally loaded, type Class C, 120° clearance turbine bearing without a cap, with ratio of $\alpha/\beta = 0.56$; journal diameter $2a = 15$ in.; length $b = 30$ in.; vertical load $R = 67{,}500$ lb; speed $N = 1500$ rpm. Clearance ratio $\eta/a = 0.001$ in. The oil at operating temperature has a viscosity $\mu = 3.4 \times 10^{-6}$ Reyn.

Solution. Ratio $l/b = (15\pi/30) \times (120/360) = 0.524$, whence from Fig. 18, $L_1 = 0.673$.

$$w = \frac{R}{2a \times b \times L_1} = \frac{67{,}500}{15 \times 30 \times 0.673} = 223$$

$$\left(\frac{w}{\mu N}\right)\left(\frac{\eta}{a}\right)^2 = \frac{223}{3.4 \times 10^{-6} \times 1500} \times 0.001^2 = 0.0437$$

with which value Fig. 29 is entered. The ordinate erected at $(w/\mu N)(\eta/a)^2 = 0.0437$ intersects the curve $\alpha/\beta = 0.56$ at $\phi_H = 107°$, which corresponds to a value of $c = 0.3$.

$$\text{Lift} = \eta(1 - c \cos \phi_H) = 0.075[1 - \{0.3 \times (-0.29237)\}] = 0.0684 \text{ in.}$$

The point of maximum pressure P_m is the value of θ_1 at the point where the ordinate $(w/\mu N)(\eta/a)^2 = 0.0437$ intersects the curve $\alpha/\beta = 0.60$, which is $\theta_1 = 83°$. P_m, therefore, will be on the radius of the journal in running position, 83° in the direction of rotation from the line $O-O'$ joining the center of the journal and the center of curvature of the bearing.

Optima and Deviations for a 120° Class C Journal Bearing. Just as Fig. 20 was applied to the study of optimum conditions for flat-wedge films, so may Fig. 30 be applied to the finding of the optima for the curved wedge of a 120° journal bearing of the centrally loaded clearance type (Class C). The scale for M needs to have a somewhat different but parallel meaning because of the curvatures and the fact that the least thickness h_0 does not necessarily lie at the terminus of the film.

It is seen in Fig. 30, as it was in Fig. 20, that the value of M is different for the three separate optima. See points of maxima and minima of curves 1, 2, and 3.

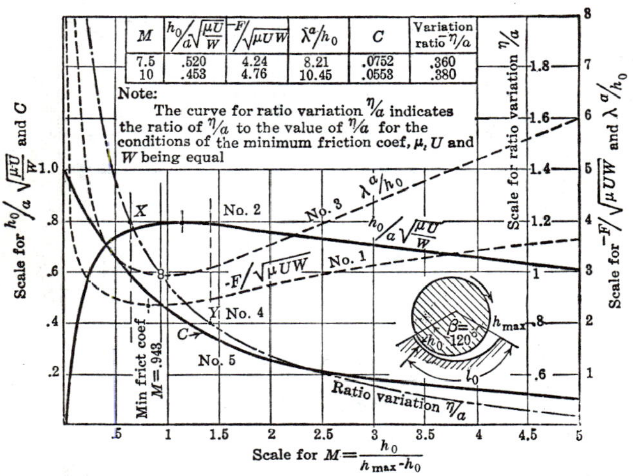

FIG. 30. Effect of deviation from optimum friction-coefficient conditions in a 120° centrally loaded (Class C) clearance bearing. This chart is comparable with Fig. 20 for flat bearing faces. (H. A. S. Howarth, *ASME Trans.*, MSP-57-2, 1935)

The data contained in Kingsbury's tables of optimum conditions that are to follow were determined from points of inflection like the above for curves 2 (maximum capacity for a given minimum thickness h_0) and 3 (least friction coefficient for a given minimum thickness h_0).

Many series of clearance journal bearing films are possible. In the present study the film of Fig. 30 is a curved wedge, extending from h_{max} at the entering edge to and beyond

h_0, where the film thickness is least, for which the wedge length is l_0. Length l_0 usually will be a little less than 120°, depending on operating conditions. The length l_0 would equal the film length l when h_0 lies at the end of the film.

The series of curves, showing the relations of dimensionless groups, will be useful in arranging tolerances for journal and bearing diameters. The curves that are plotted against a scale of $M = h_0/(h_{\max} - h_0)$ are: (1) Friction drag $= -F/\sqrt{\mu UW}$. (2) Bearing capacity, $h_0/a\sqrt{(\mu U/W)}$. (3) Friction coefficient, $\lambda(a/h_0)$. (4) Clearance ratio η/a. (5) Eccentricity factor, c. (See pp. 12-14 to 12-16 for notation.) The application of the curves can be illustrated by an example.

EXAMPLE 6. Determine the effect of an increase of 20% and a decrease of 20% in clearance ratio from optimum for optimum friction condition of a 120° clearance bearing.

Solution. Minimum friction coefficient ratio is shown by curve 3 to correspond to a value of $M = 0.943$. Optimum clearance ratio η/a corresponding to this has been taken as unity in plotting curve 4. An increase of 20% corresponds to point X on curve 4 with a corresponding value of $M = 0.65$. The ordinate through X intersects the other curves to give the following values: friction drag, curve 1, 2.4; bearing capacity, curve 2, 0.773; friction coefficient, curve 3, 3.1; eccentricity factor, curve 5, 0.587.

A decrease of 20% in clearance ratio (η/a) corresponds to the point Y on curve 4, with a corresponding value of $M = 1.42$. From the intersection of this ordinate with the several curves, the following values are found: Friction drag, 2.53; bearing capacity, 0.788; friction coefficient ratio, 3.1; eccentricity factor, 0.348.

The tabulation at the top of Fig. 30 permits the extension of the curves beyond the limits of the chart. With a value of $M = 10$, the other values are such that an accurately made, properly designed bearing would function with pure film lubrication.

OPTIMUM CHARTS AND TABLES FOR JOURNAL BEARINGS. The charts, tables, and figures that follow are given for two of the three sets of optimum operating conditions that were discussed in connection with Fig. 30 for curved-wedge films and with Fig. 20 for flat-wedge films. In each of those figures curve 2 showed the maximum load conditions and curve 3 those for minimum friction coefficient, both with relation to a chosen minimum film thickness h_0 and a film length. For the flat wedge the length appears directly as l, but for the curved wedge it usually appears indirectly as the film arc β and the journal radius a.

As side leakage factors enter into the examples that follow and also into Figs. 34, 38, and 42, they should be applied as explained below. If this influence be neglected in the solution of bearing problems, the data of the charts will not agree well with actual conditions, and side leakage factors, therefore, must be included in the bearing problem data. Figure 18 gives factors applying to optimum film proportions, but they are fairly dependable when applied to conditions not too far different from optimum. These are the best generally applicable data available for side leakage corrections. When applied to abnormal conditions, they tend to bring results toward correct values, and usually are on the safe side.

Figure 18 is used as follows: In the dynamical tables and curves following which apply to the various classes of bearings, as enumerated above, numerical coefficients are for infinitely wide conditions, under which side leakage is zero. For bearings of finite width and a given ratio of (length/width) the numerical coefficients are to be multiplied by a correction factor from Fig. 18. The proper correction curve in Fig. 18 to use for the various symbols or symbol groups of the dynamical tables and curves is:

Symbol group	W	F	λ or $\dfrac{F}{W}$	H	$\dfrac{H}{W}$	$\dfrac{W}{H}$	$\dfrac{\lambda}{\sqrt{\mu N/p_0}}$
Correction factor	L_1	F_1	$\dfrac{F_1}{L_1}$	F_1	$\dfrac{F_1}{L_1}$	$\dfrac{L_1}{F_1}$	$\dfrac{F_1}{\sqrt{L_1}}$

Data given below are useful in the design of optimum bearings of classes A, B, and C, respectively. They show the relation of proportion, dimensions, power loss, film thickness, speed, and load for a variety of conditions with each type. Total load R on the bearing, and speed, usually are the known factors, and the minimum thickness of the oil film h_0 may be assumed. (See examples below.) The permissible loading of the bearing p_0 also may be assumed. The projected area of the bearing then is R/p_0; it also is $2a^2\,(b/a)$. The ratio of width to radius of journal, i.e., b/a, when decided, may be substituted in $2a^2\,(b/a)$. By equating the two expressions for projected area, the value of a can be determined. Then from R and b the value of W is obtained.

Class A. Optima for Eccentrically Loaded Clearance Bearings (Fig. 31). Table 3 gives optimum values for maximum load.

Table 4 and Figs. 32, 33, and 34 give optimum values for minimum friction coefficient.

(*Continued on p. 12-30*)

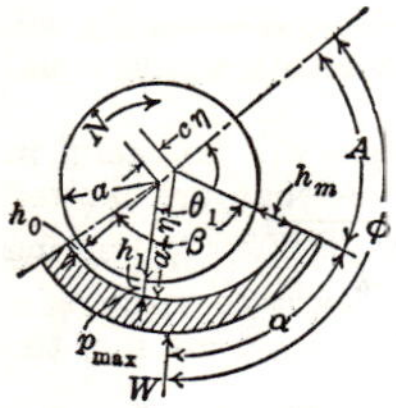

FIG. 31. Eccentrically loaded clearance bearing, $\alpha > \beta/2$, Class A. (Albert Kingsbury, *ASME Trans.*, RP-54-7, 1932)

Table 3. Optimum Values for Maximum Load W of Class A Journal Bearings (Fig. 31)

Geometrical Optima							Dynamical Optima						
Symbol Group or Curve No.		Values of β					Symbol Group or Curve No.		Values of β				
		60°	120°	180°	270°	360°			60°	120°	180°	270°	360°
1	A	122.3°	66.33°	12.17°	−62.48°	−118.6°	10	$W\frac{h_0^2}{\mu N a^3}$	0.0201	0.0759	0.1543	0.2635	0.2939
2	α	32.94°	66.49°	101.3°	156.6°	208.6°	11	$F\frac{h_0}{\mu N a^2}$	0.0936	0.1863	0.2768	0.4080	0.5379
3	α/β	0.5490	0.5541	0.5627	0.5799	0.5796	12	$\lambda\frac{a}{h_0}$	4.660	2.453	1.794	1.549	1.830
4	c	0.7481	0.4821	0.3655	0.3310	0.3379	13	$\lambda\frac{l}{h_0}$	4.880	5.138	5.637	7.297	11.50
5	h_m/h_0	2.382	2.305	2.139	1.990	2.021	14	$H\frac{h_0(10)^6}{\mu N^2 a^3}$	1.485	2.956	4.392	6.473	8.534
6	h_1/h_0	1.200	1.209	1.226	1.257	1.266	15	$H\frac{(10)^5}{WNh_0}$	7.394	3.893	2.847	2.457	2.904
7	η/h_0	3.970	1.931	1.576	1.495	1.510	16	$\frac{WNh_0}{H(10)^5}$	0.1352	0.2569	0.3513	0.4070	0.3444
8	θ_1	158.8°	140.8°	127.4°	118.8°	118.6°	17	$\lambda/\sqrt{\frac{\mu N}{p_0}}$	0.4671	0.4780	0.4983	0.5620	0.7016
9	ϕ	155.3°	132.8°	113.5°	94.09°	90.00°							

Table 4. Optimum Values for Minimum Friction Coefficients of Class A Journal Bearings (Fig. 31)

Geometrical Optima. (See Fig. 32)							Dynamical Optima. (See Fig. 33)						
Symbol Group or Curve No.		Values of β					Symbol Group or Curve No.		Values of β				
		60°	120°	180°	270°	360°			60°	120°	180°	270°	360°
1	A	117.9°	58.04°	1.119°	−72.15°	−121.0°	10	$W\frac{h_0^2}{\mu N a^3}$	0.0197	0.0742	0.1507	0.2602	0.2925
2	α	33.89°	68.54°	104.7°	161.4°	211.0°	11	$F\frac{h_0}{\mu N a^2}$	0.0902	0.1790	0.2658	0.3985	0.5327
3	α/β	0.5649	0.5711	0.5815	0.5977	0.5860	12	$\lambda\frac{a}{h_0}$	4.584	2.413	1.763	1.531	1.821
4	c	0.7496	0.4911	0.3844	0.3520	0.3660	13	$\lambda\frac{l}{h_0}$	4.801	5.054	5.540	7.217	11.44
5	h_m/h_0	2.586	2.476	2.249	2.086	2.155	14	$H\frac{h_0(10)^6}{\mu N^2 a^3}$	1.432	2.840	4.217	6.323	8.452
6	h_1/h_0	1.248	1.263	1.283	1.292	1.280	15	$H\frac{(10)^5}{WNh_0}$	7.274	3.829	2.798	2.430	2.890
7	η/h_0	3.986	1.965	1.624	1.543	1.577	16	$\frac{WNh_0}{H(10)^5}$	0.1375	0.2612	0.3574	0.4115	0.3460
8	θ_1	156.4°	136.7°	123.1°	117.5°	121.0°	17	$\lambda/\sqrt{\frac{\mu N}{p_0}}$	0.4548	0.4647	0.4841	0.5524	0.6965
9	ϕ	151.8°	126.6°	105.8°	89.24°	90.0°							

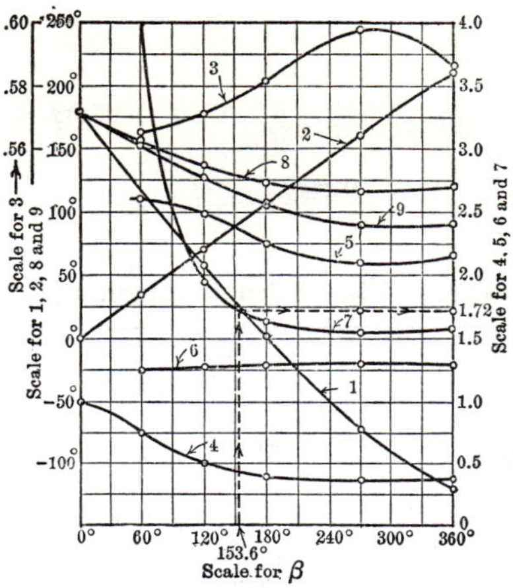

FIG. 32. Geometrical optima for minimum friction coefficient. See Table 4. (Albert Kingsbury, *ASME Trans.*, RP-54-7, 1932)

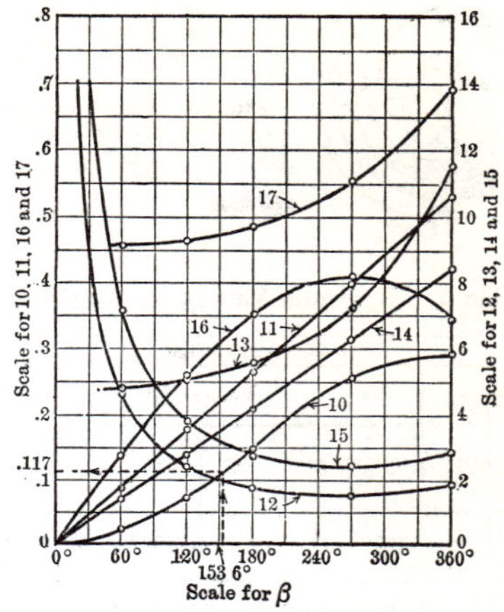

FIG. 33. Dynamical optima for minimum friction coefficient. See Table 4. (Albert Kingsbury, *ASME Trans.*, RP-54-7, 1932)

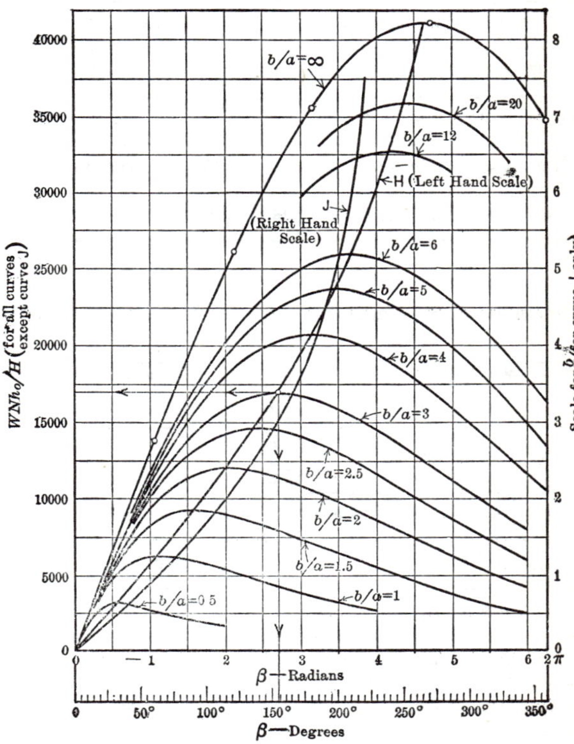

FIG. 34. Side leakage chart for Class A journal bearings for minimum friction coefficient. See Fig. 31.

This chart shows values of nondimensional group (WNh_0/H) as affected by side leakage, with film forms for minimum coefficient of friction. Curve J gives values of b/a at which group (WNh_0/H) is greatest for each value of β. See Figs. 32 and 33 and Tables 3 and 4 for optimum conditions, assuming constant viscosity and no side leakage. When a curve for a given ratio of b/a is chosen, each ordinate of it will be correct for the optimum relation of the symbols in this group. Hence the correction factors for side leakage are already allowed for in the heights of these ordinates. (Albert Kingsbury, *ASME Trans.*, RP-54-7, 1932)

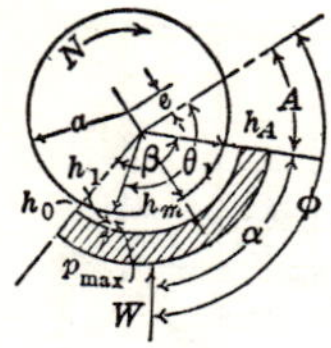

FIG. 35. Fitted partial bearing, $\alpha > \beta/2$, Class B. (Albert Kingsbury, *ASME Trans.*, RP-54-7, 1932)

Table 5. Optimum Values for Maximum Load W of Class B Journal Bearings (Fig. 35)

Geometrical Optima								Dynamical Optima							
Symbol Group or Curve No.		Values of β						Symbol Group or Curve No.		Values of β					
		0°	30°	60°	90°	120°	150°			0°	30°	60°	90°	120°	150°
1	A	180°	120.1°	92.21°	64.30°	40.67°	19.39°	9	$W\frac{10h_0^2}{\mu Na^3}$	0	0.0451	0.1685	0.3309	0.4534	0.3924
2	α	0°	17.47°	35.79°	55.72°	77.76°	101.1°	10	$F\frac{h_0}{\mu Na^2}$	0	0.0407	0.0774	0.1073	0.1284	0.1339
3	α/β	0.5774	0.5822	0.5964	0.6191	0.6480	0.6739	11	$\lambda\frac{a}{h_0}$	∞	9.027	4.597	3.242	2.831	3.413
4	h_A/h_0	2.189	2.176	2.143	2.078	1.969	1.804	12	$\lambda\frac{l}{h_0}$	4.706	4.727	4.814	5.093	5.930	8.935
5	h_m/h_0	2.189	2.176	2.143	2.306	3.021	5.433	13	$H\frac{h_0(10)^6}{\mu N^2a^3}$	0	0.6455	1.229	1.702	2.037	2.125
6	h_1/h_0	1.373	1.398	1.480	1.624	1.843	2.131	14	$H\frac{(10)^5}{WNh_0}$	∞	14.32	7.293	5.144	4.492	5.415
7	θ_1	180°	150.1°	136.4°	135.2°	142.4°	156.9°	15	$\frac{WNh_0}{H(10)^5}$	0	0.0698	0.1371	0.1944	0.2226	0.1847
8	ϕ	180°	146.6°	128.0°	120.0°	118.4°	120.5°	16	$\lambda\sqrt{\frac{\mu N}{p_0}}$	0.4311	0.4285	0.4219	0.4171	0.4263	0.4781

Table 6. Optimum Values of Minimum Friction Coefficient for Class B Journal Bearings (Fig. 35)

Geometrical Optima (See Fig. 36)								Dynamical Optima (See Fig. 37)							
Symbol Group or Curve No.		Values of β						Symbol Group or Curve No.		Values of β					
		0°	30°	60°	90°	120°	150°			0°	30°	60°	90°	120°	150°
1	A	180°	133.0°	95.59°	66.29°	41.73°	20.08°	9	$W\frac{10h_0^2}{\mu Na^3}$	0	0.0443	0.1663	0.3281	0.4504	0.3876
2	α	0°	17.84°	36.26°	55.95°	77.45°	99.41°	10	$F\frac{h_0}{\mu Na^2}$	0	0.0393	0.0755	0.1054	0.1267	0.1307
3	α/β	0.5914	0.5946	0.6043	0.6217	0.6455	0.6627	11	$\lambda\frac{a}{h_0}$	∞	8.877	4.538	3.214	2.813	3.372
4	h_A/h_0	2.533	2.501	2.408	2.276	2.122	1.993	12	$\lambda\frac{l}{h_0}$	4.622	4.648	4.752	5.049	5.892	8.829
5	h_m/h_0	2.533	2.501	2.408	2.486	3.189	5.804	13	$H\frac{h_0(10)^6}{\mu N^2a^3}$	0	0.6238	1.197	1.673	2.010	2.074
6	h_1/h_0	1.434	1.457	1.528	1.662	1.873	2.165	14	$H\frac{(10)^5}{WNh_0}$	∞	14.08	7.201	5.100	4.463	5.351
7	θ_1	180°	154.8°	140.8°	138.1°	144.0°	158.1°	15	$\frac{WNh_0}{H(10)^5}$	0	0.0710	0.1389	0.1961	0.2240	0.1869
8	ϕ	180°	150.8°	131.9°	122.2°	119.2°	119.5°	16	$\lambda\sqrt{\frac{\mu N}{p_0}}$	0.4194	0.4177	0.4138	0.4117	0.4221	0.4695

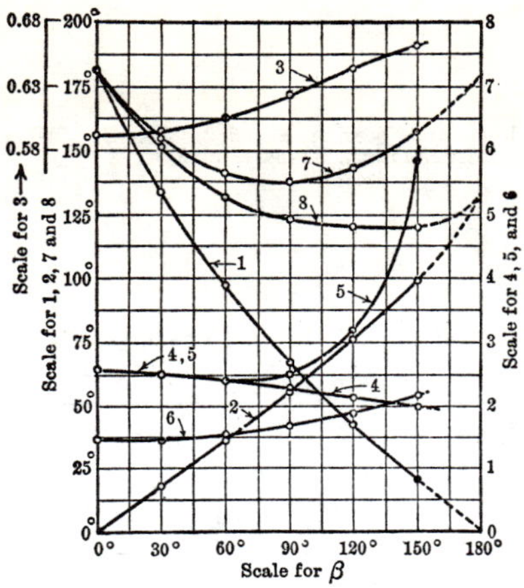

FIG. 36. Geometrical optima for minimum friction coefficients for Class B journal bearings. See Table 6. (Albert Kingsbury, *ASME Trans.*, RP-54-7, 1932)

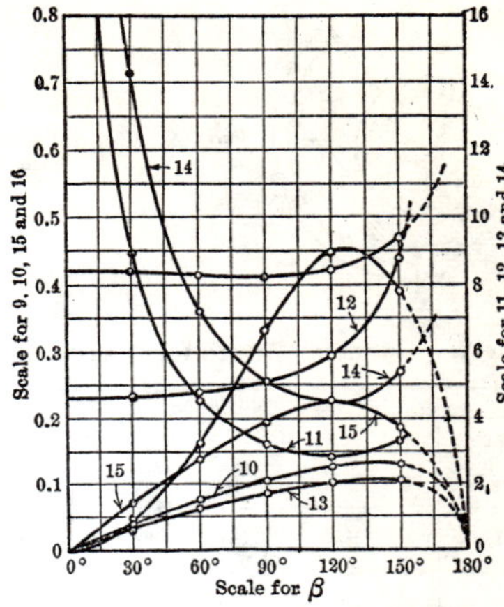

FIG. 37. Dynamical optima for minimum friction coefficients for Class B journal bearings. See Table 6. (Albert Kingsbury, *ASME Trans.*, RP-54-7, 1932)

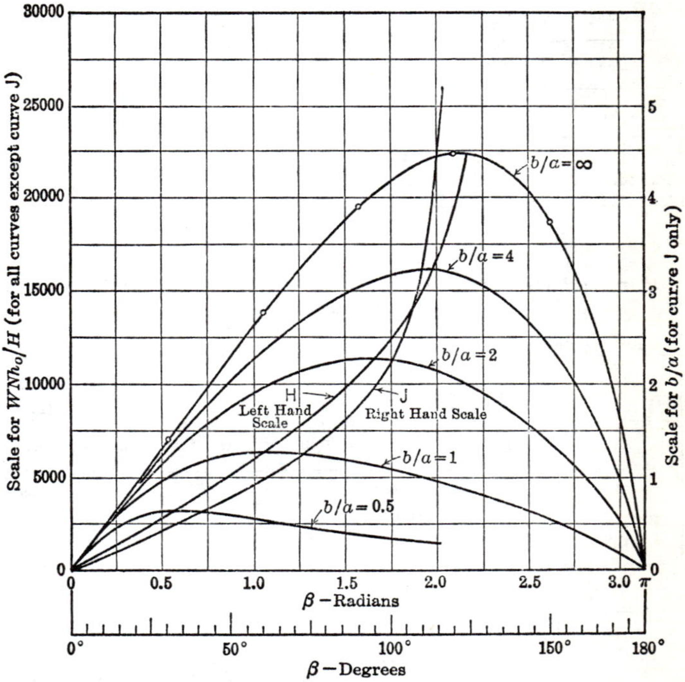

FIG. 38. Side leakage chart for Class B journal bearings for minimum friction coefficients. See Fig. 35. (Albert Kingsbury, *ASME Trans.*, RP-54-7, 1932)

This chart shows values of the nondimensional group (WNh_0/H) as affected by side leakage with film forms for the minimum coefficient of friction. Curve J gives the value of b/a at which group (WNh_0/H) is greatest for each value of β. See Figs. 36 and 37 and Tables 5 and 6 for optimum conditions, assuming constant viscosity and no side leakage. When the curve for a given ratio of b/a is chosen, each ordinate of it will be correct for the optimum relation of the symbols in this group. Hence the correction factors for side leakage are already allowed for in the heights of those ordinates.

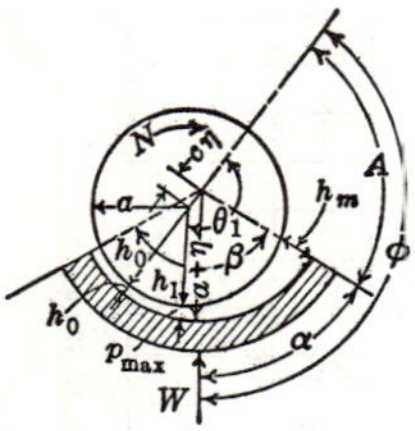

FIG. 39. Centrally loaded clearance bearing, $\alpha = \beta/2$, Class C. (Albert Kingsbury, *ASME Trans.*, RP-54-7, 1932)

Table 7. Optimum Values for Maximum Load W of Class C Journal Bearings (Fig. 39)

Geometrical Optima					Dynamical Optima				
Symbol Group or Curve No.		Values of β			Symbol Group or Curve No.		Values of β		
		60°	120°	180°			60°	120°	180°
1	A	131.2°	83.66°	37.83°	10	$W \frac{h_0^2}{\mu N a^3}$	0.0180	0.0673	0.1352
2	α	30°	60°	90°	11	$F \frac{h_0}{\mu N a^2}$	0.0982	0.1954	0.2902
3	α/β	0.50	0.50	0.50	12	$\lambda \frac{a}{h_0}$	5.464	2.906	2.147
4	c	0.7466	0.4688	0.3437	13	$\lambda \frac{l}{h_0}$	5.722	6.087	6.744
5	h_m/h_0	2.004	1.980	1.937	14	$H \frac{h_0(10)^6}{\mu N^2 a^3}$	1.559	3.101	4.605
6	h_1/h_0	1.137	1.144	1.158	15	$H \frac{(10)^5}{W N h_0}$	8.670	4.611	3.406
7	η/h_0	3.946	1.883	1.524	16	$\frac{W N h_0}{H(10)^5}$	0.1153	0.2169	0.2936
8	θ_1	162.5°	146.8°	134.3°	17	$\lambda / \sqrt{\frac{\mu N}{p_0}}$	0.5181	0.5329	0.5581
9	ϕ	161.2°	143.7°	127.8°					

Table 8. Optimum Values for Minimum Friction Coefficient of Class C Journal Bearings (Fig. 39)

Geometrical Optima (See Fig. 40)					Dynamical Optima (See Fig. 41)				
Symbol Group or Curve No.		Values of β			Symbol Group or Curve No.		Values of β		
		60°	120°	180°			60°	120°	180°
1	A	131.6°	84.39°	38.71°	10	$W \frac{h_0^2}{\mu N a^3}$	0.0180	0.0671	0.1349
2	α	30°	60°	90°	11	$F \frac{h_0}{\mu N a^2}$	0.0980	0.1948	0.2890
3	α/β	0.50	0.50	0.50	12	$\lambda \frac{a}{h_0}$	5.458	2.902	2.142
4	c	0.7615	0.4904	0.3645	13	$\lambda \frac{l}{h_0}$	5.716	6.078	6.731
5	h_m/h_0	2.072	2.056	2.021	14	$H \frac{h_0(10)^6}{\mu N^2 a^3}$	1.555	3.091	4.586
6	h_1/h_0	1.140	1.148	1.163	15	$H \frac{(10)^5}{W N h_0}$	8.660	4.604	3.399
7	η/h_0	4.193	1.962	1.574	16	$\frac{W N h_0}{H(10)^5}$	0.1155	0.2172	0.2942
8	θ_1	163.0°	147.8°	135.7°	17	$\lambda / \sqrt{\frac{\mu N}{p_0}}$	0.5171	0.5317	0.5564
9	ϕ	161.6°	144.4°	128.7°					

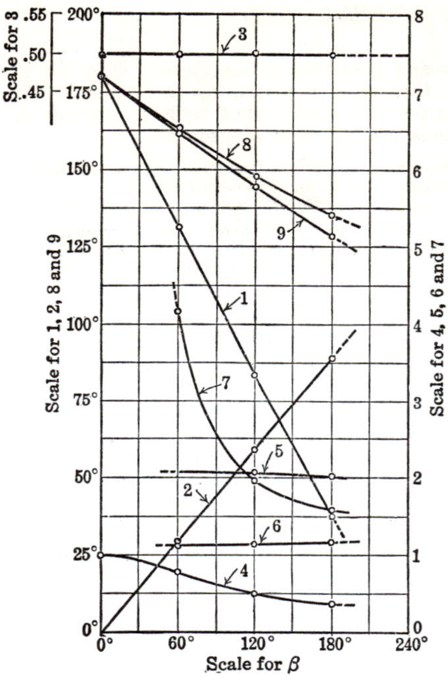

FIG. 40. Geometrical optima for minimum friction coefficient for Class C journal bearings. See Table 8. (Albert Kingsbury, *ASME Trans.*, RP-54-7, 1932)

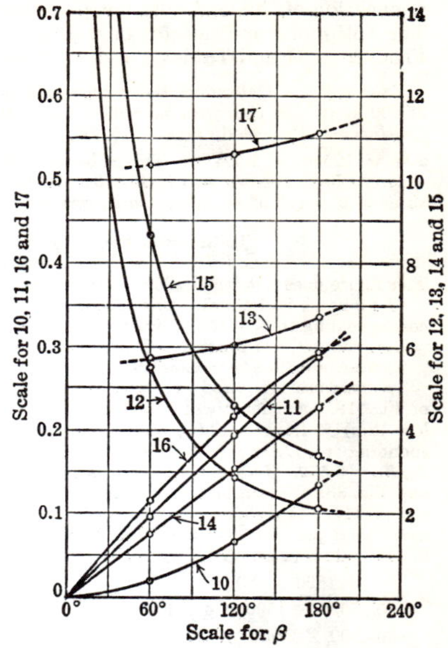

FIG. 41. Dynamical optima for minimum friction coefficient for Class C journal bearings. See Table 8. (Albert Kingsbury, *ASME Trans.*, RP-54-7, 1932)

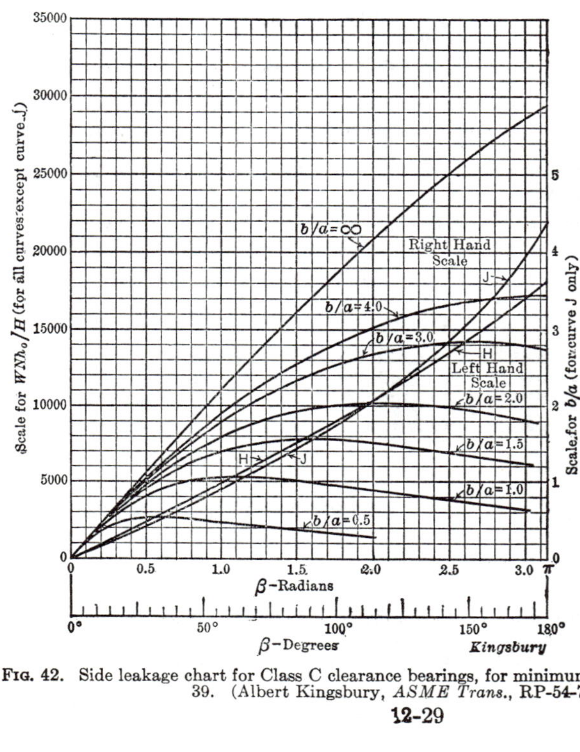

This chart shows the values of the nondimensional group (WNh_0/H) as affected by side leakage with film forms for the minimum coefficient of friction. Curve J gives the value of b/a at which group (WNh_0/H) is greatest for each value of β. See Figs. 40 and 41 and Tables 7 and 8 for optimum conditions, assuming constant viscosity and no side leakage. When the curve for a given ratio of b/a is chosen, each ordinate of it will be correct for the optimum relation of the symbols in this group. Hence the correction factors for side leakage are already allowed for in the heights of those ordinates.

FIG. 42. Side leakage chart for Class C clearance bearings, for minimum friction coefficients. See Fig. 39. (Albert Kingsbury, *ASME Trans.*, RP-54-7, 1932)

The radius of curvature of the bearing is greater than that of the journal. The line of application of the load does not pass through the center of the bearing arc, and $\alpha > \beta/2$. The optimum condition for best ratio of α/β for values of β from 0 to 360° are given in Figs. 32 to 34 and Tables 3 and 4. Their use is best shown by an example.

EXAMPLE 7. Determine the proportions of a bearing of minimum friction to support a load of 25,000 lb at 1500 rpm with $h_0 = 0.003$ in., with a pressure of 200 psi of projected area.

Solution. Projected area = 25,000/200 = 125 sq in. = $2a^2(b/a)$. Assume $b/a = 3$, whence $a = \sqrt{(125/6)} = 4.564$ in. This value should be checked for strength in bending and in torsion.

From the curve $b/a = 3$ in Fig. 34, the minimum power loss H per unit of width b occurs with a value of $\beta = 153.6° = 2.68$ radians, corresponding to a value of $WNh_0/H = 16{,}900$, whence

$$\text{Total power loss } (H \times b) = 25{,}000 \times 1500 \times \frac{0.003}{16{,}900} = 6.66 \text{ hp}$$

This figure does not include losses in any clearance spaces outside of the arc of contact β.

Clearance η is determined as follows: The ratio of η/h_0 is found from curve 7, Fig. 32, to be 1.72, corresponding to $\beta = 153.6$, when $\eta = 1.72h_0 = 1.72 \times 0.003 = 0.00516$ in.

Bore diameter of bearing $= 2(a + \eta) = 2(4.564 + 0.00516) = 9.138$.

Clearance ratio $\eta/a = 0.00516/4.564 = 0.00113$.

To ascertain required oil viscosity coefficient, correction factors must be determined by means of Fig. 18. Arc length of bearing surface $l = \beta$ (radians) $\times a = 2.68 \times 4.564 = 12.23$ in. Width $b = (b/a) \times 4.564 = 13.69$, and $l/b = 0.893$. At the points on the several curves in Fig. 18 corresponding to $l/b = 0.893$, the following correction factors are found: $L_1 = 0.487$; $F_1 = 0.916$; $L_1/F_1 = 0.532$. To ascertain the mean viscosity in the oil film, refer to group 10 and curve 10, Table 4 and Fig. 33. For an infinitely wide bearing and $\beta = 153.6$, curve 10 shows the value of $W(h_0^2/\mu Na^3)$ to be 0.117. Applying the correction factor, $W(h_0^2/\mu Na^3)$, $L_1 = 0.117 \times 0.487 = 0.05698$. The actual mean load per unit width of bearing is $W = 25{,}000/13.69 = 1826$ lb. Substituting this and other known values in group 10, we find

$$\mu = \frac{1826 \times 0.003^2}{0.05698 \times 1500 \times 4.564^3} = 2.022 \times 10^{-6} \text{ Reyn}$$

$$= 69{,}000 \times 2.002 \times 10^{-6} = 0.1395 \text{ poise, corresponding to a light machinery oil at 130 to 140 F}$$

For corresponding commercial viscosities, see p. 12-03.

The coefficient of friction λ is found by means of curve 12 and group 12, Fig. 33, and Table 4. Let m = the value on curve 12 corresponding to value of β. In the present case, $\beta = 153.6$, whence $m = 2.00$. Then

$$\lambda = \frac{F_1}{L_1} \times m \times \frac{h_0}{a} = \frac{2.00 \times 0.003}{0.532 \times 4.564} = 0.00247$$

Inspection of Fig. 34 shows that, other things being equal, an increase of the ratio b/a will reduce the relative power loss. (See Example 8.) The possible increase in b/a is limited by considerations of space, cost, and strength or elasticity of the journal. In Example 1, β might be increased, following the curve $b/a = 3$, Fig. 34, to obtain a greater bearing area, with, however, a relatively greater friction loss. Other data for the bearing in question can be found from Figs. 32 and 33, and Tables 3 and 4; for example, leading angle α from group 3 and curve 3.

EXAMPLE 8. Assume the same data as for Example 7, excepting that $b/a = 4$. Then using the same curves and tables, we have,

$$a = \sqrt{\frac{125}{8}} = 3.953 \text{ in.}; \beta = 180.8° = 3.155 \text{ radians for least power loss}$$

$$\text{Total power loss} = \frac{25{,}000 \times 1500 \times 0.003}{20{,}700} = 5.435 \text{ hp}$$

Clearance $\eta = 1.622h_0 = 0.004866$ in. $= 0.00123$ in. per inch of diameter
Length $l = 3.155 \times 3.953 = 12.47$ in. Width $b = 4 \times 3.953 = 15.81$ in.
Ratio $l/b = 0.789$. This determines the correction factors:

$$L_1 = 0.533;\ F_1 = 0.924;\ \frac{L_1}{F_1} = 0.577$$

$$W = \frac{25{,}000}{15.8} = 1581 \text{ lb}$$

$$\mu = \frac{1581 \times 0.003^2}{0.1517 \times 0.533 \times 1500 \times 3.953^3} = 1.90 \times 10^{-6} \text{ Reyn, corresponding to 0.131 poise}$$

$$\lambda = \frac{F_1}{L_1} \times m \times \frac{h_0}{a} = \frac{1.762 \times 0.003}{0.577 \times 3.953} = 0.002318.$$

The charts and tables also may be used to determine proper radial clearance or oil viscosity for a known set of other conditions. (See Example 9.)

Class B. Optima for Fitted Partial Bearings (Fig. 35). Table 5 gives optimum values for maximum load.

Table 6 and Figs. 36, 37, and 38 give optimum values for minimum friction coefficient.

The journal and bearing have the same radius of curvature. The line of application of the load does not pass through the center of the bearing arc, and $\alpha > \beta/2$. Figures 36 and 37 and Tables 5 and 6 show optimum conditions for values of β from 0 to 150°. The procedure followed in design is the same as that for Class A bearings, except that reference is made to the tables and curves for Class B bearings corresponding to those used for Class A bearings in Examples 7 and 8. (See also Example 2.)

Class C. Optima for Centrally Loaded Clearance Bearings (Fig. 39). Table 7 gives optimum values for maximum load.

Table 8 and Figs. 40, 41, and 42 give optimum values for minimum friction coefficient.

The radius of curvature of the bearing is greater than that of the journal. The line of application of the load passes through the center of the bearing arc, and $\alpha = \beta/2$. Figures 40 to 42 and Tables 7 and 8 show optimum conditions for values of β from 0 to 180°. Bearing problems may be solved by the same procedure as used for Examples 7 and 8 except that reference should be made to the curves and tables of Class C bearings. Example 9 shows the application of the curves to the selection of the correct radial clearance η.

Example 9. Determine the best radial clearance η for a bearing in which the total load $R = 25{,}000$ lb, $N = 1500$ rpm, $\beta = 120°$, $P_0 = 125$ psi, ratio of width to radius, $b/a = 4$, oil viscosity, $\mu = 3.4 \times 10^{-6}$ Reyn. Bearing is centrally loaded.

Solution. By procedure outlined in Example 7 and from the foregoing data, $2a = 10$ in., $b = 20$ in., $W = 1250$, $l = 10.47$ in., $l/b = 0.524$.

Correction factors from Fig. 18 will be $L_1 = 0.672$, $F_1 = 0.947$, $L_1/F_1 = 0.710$. Then from group 10 in Table 8, by introducing correction factor L_1, we have $Wh_0^2/\mu Na^3 = 0.0672L_1 = 0.0451$. Substituting the known values in this equation, and solving for h_0, we find $h_0 = 0.0048$. In group 7, Table 8, the value of η/h_0 for $\beta = 120°$ is 1.962, whence $\eta = 1.962 \times h_0 = 1.962 \times 0.0048 = 0.0094$. Clearance ratio $= \eta/a = 0.0094/5 = 0.00188$. Friction coefficient factor $\lambda a/h_0$ is found from group 12, Table 8. to be 2.902, to which the correction factor L_1/F_1 must be applied. Hence,

$$\lambda = \frac{h_0}{a} \times 2.902 \times \frac{1}{L_1/F_1} = \frac{0.0048}{5} \times 2.902 \times \frac{1}{0.710} = 0.00392$$

3. LOADING AND FRICTION OF PIVOTED SHOE BEARINGS

Pivoted shoes are applicable to straight-line motion as in cross heads, and to rotary motion as in both journal and thrust bearings. In rotary motion they may be fitted to any surface of revolution, whether flat (Fig. 43), cylindrical (Fig. 44), conical, or spherical

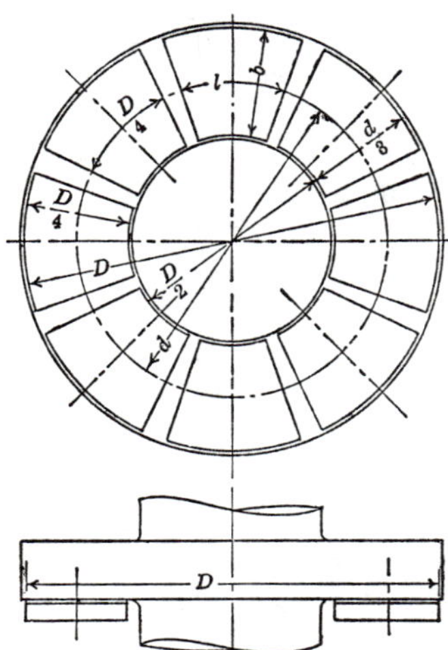

Fig. 43. Thrust bearing with flat pivoted shoes.

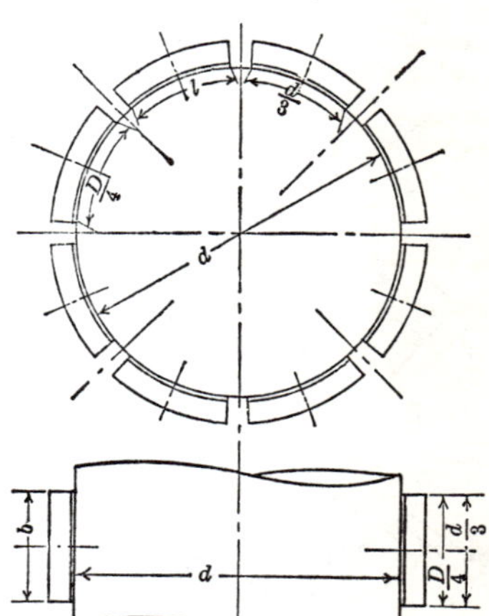

Fig. 44. Journal bearing with pivoted shoes fitted to cylindrical surface.

(Figs. 45 and 46). The mean diameter d is purposely made common to all these figures, being the journal diameter for some and the mean thrust face diameter for the others.

Although the shoe faces in the foregoing figures are shown to be "square" (with mean length equal to width), they can and do have other proportions than $l/b = 1$. This variation is seen in thrust bearings in Fig. 47, which illustrates the use of six and eight shoes, with the same inside, mean, and outside diameters. The usual standards of those diameters are two to one, the mean being three-fourths the outside diameter. The standard sum of the face angles is as shown, equal to 306°, the remaining 54° being equally distributed for oil channels.

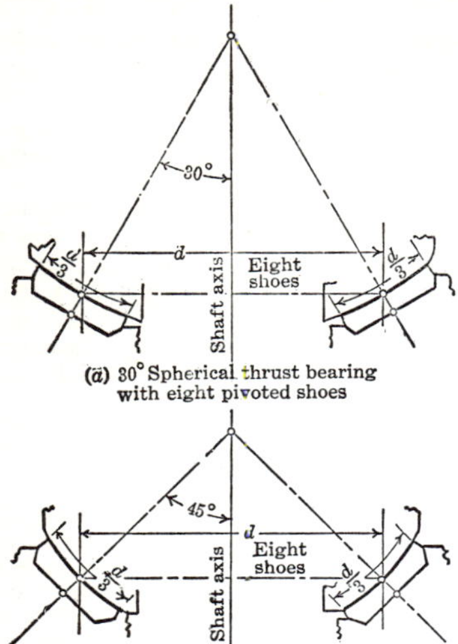

(a) 30° Spherical thrust bearing with eight pivoted shoes

(b) 45° Spherical thrust bearing with eight pivoted shoes

FIG. 45. Spherical thrust bearings for suspended loads. Lateral capacity depends on the angularity and the applied thrust load.

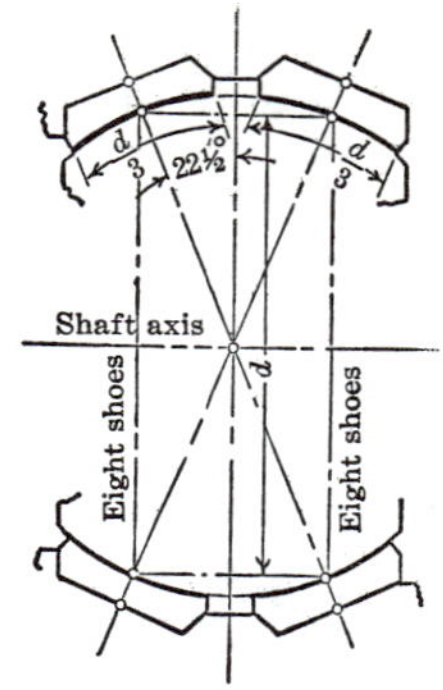

FIG. 46. Spherical journal bearing with two balanced sets of eight pivoted shoes. These bearings have an axial capacity proportional to angularity.

The mathematical analysis of the relative carrying capacities and friction of pivoted shoes, of varying length-width ratios, has been made for straight-line motion, such as that of a cross head, but with continuous uniform speed. See Fig. 48 for the midsection of a pivoted shoe and its film, supporting a collar. The position of the pivot must be right, both radially and circumferentially, in order to balance the pressures of a correctly shaped film.

The fundamental relationship of the variables involved in the study of straight-line motion of pivoted shoes carrying a load upon flat-wedge films of lubricant, without metallic contact, is given in the following equation for conditions known as Case 2 optimum load. (See discussion of optima on pp. 12-09 ff.)

$$h_0^2 = \frac{1.92\mu v l L_1}{p_0} \tag{49}$$

In this formula v is velocity, feet per second; μ is the viscosity, Reyns; l is the mean length of the shoe face, inches; and p_0 is the average film pressure, pounds per square inch.

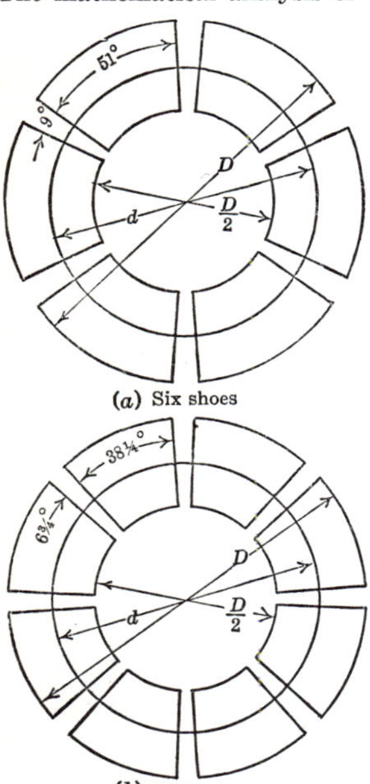

FIG. 47. Comparison of thrust bearings with six and eight shoes.

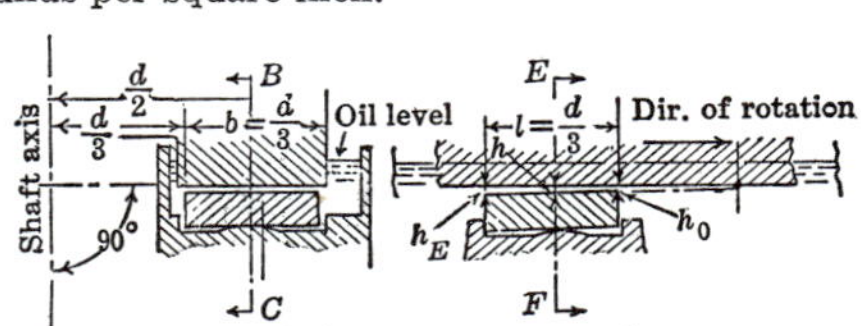

FIG. 48. One of a set of pivoted bearing shoes, showing pivot and constraints.

The factor L_1, called the side-leakage correction factor for pressure, has been found to vary with shoe face proportions according to laws determined by Kingsbury, Martin

Needs, and others. The Kingsbury values of Fig. 18 will be used in this discussion.

COMPARATIVE CAPACITIES FOR DIFFERENT SHOE FACE PROPORTIONS. Several relationships of the variables in eq. 49 have been compared for length-width ratios ranging from 0.5 to 2.0, in order to find the best ones for general applications.

The purpose of the comparison was: (1) to evolve a rational system of loading of pivoted shoes, (2) to determine the relative merits of the different proportions as to capacity and friction, and (3) to obtain the basis for the loading tables (see Tables 10 and 11).

The proportions $l/b = 4/3$ have been common to six-shoe Kingsbury thrust bearings for many years. In Table 9 other shoe faces are compared with this face in various ways, the most interesting of which is p_0. Because both experience and engineering practice call for thinner films for shoes of smaller area than for those of larger area, other conditions being the same, a study was made of loading practice common over many years. The result is indicated by eq. 50, which will apply to a uniform series of similar standard thrust bearing shoes whose $l/b = 4/3$.

$$h_0 = (0.675 \times 10^{-10} vl)^{3/8} \tag{50}$$

By comparing similar total areas of eight different shapes, as they would apply to thrust bearings with full complements of shoes, the relative capacities $nR/6$ in Table 9 indicate that the greatest capacity is had by twelve shoes having $l/b = 0.667$ and the least by four shoes having $l/b = 2.000$. The relationships found in Table 9 form the basis for the loading tables for pivoted shoes.

Table 9. Comparison of Pivoted Shoe Bearing Performance

l/b	0.500	0.667	0.800	1.000	1.143	1.333	1.600	2.000
A	0.375	0.500	0.600	0.750	0.857	1.000	1.200	1.500
p_0	1.171	1.208	1.198	1.152	1.115	1.000	0.903	0.772
R	0.438	0.604	0.719	0.865	0.955	1.000	1.083	1.158
n	16	12	10	8	7	6	5	4
$nR/6$	1.171	1.208	1.198	1.152	1.115	1.000	0.903	0.772

Notation: l/b = length/width ratio; A = relative area per shoe; p_0 = relative mean pressure on shoe; R = relative capacity per shoe; n = number of shoes in bearing.

Assumed conditions: Viscosity, speed, and widths constant. Film thicknesses vary according to eq. 50.

LOADING, FRICTION, AND FILM THICKNESS TABLES FOR PIVOTED SHOE BEARINGS. Tables for Thrust Bearings. Because of the wide use of six-shoe thrust bearings like those shown in Fig. 47*a*, Table 10 was prepared. In Table 11 similar data are given for eight-shoe bearings (Fig. 47*b*), which, for a given occupied space, have greater mechanical strength as well as greater film capacity. For simplicity the data are given for one shoe whose area in square inches is tabulated under A_s. The mean speeds v_d are those for the mean diameter d. The rpm tabulated opposite N_d is that which will produce the mean surface speed v_d feet per second for the mean diameter d inches. The average pressure, in pounds per square inch over the shoe face, is tabulated opposite p_0. Hence the load carried by one shoe will be the product of A_s and p_0, pounds. The horsepower friction loss per shoe will be as marked opposite H_s when the mean viscosity of the film is 3.4×10^{-6} Reyns (pound-seconds per square inch), equivalent to 0.2345 poise or to 23.45 centipoises. For SUS Saybolt universal seconds equivalents, see Table 1, Article 13, of this section.

For the six-shoe bearings (Table 10) the following equations may be used to calculate sizes not included in the table:

$$p_0 = \frac{0.2705 \mu v d}{h_0^2} \tag{51}$$

$$h_0 = (0.3 \times 10^{-10} vd)^{3/8} \tag{52}$$

$$H_s = 1.465 \times 10^{-2} \mu \frac{A_s v^2}{h_0} \tag{53}$$

For the eight-shoe bearings (Table 11) the equations are

$$p_0 = \frac{0.28 \mu v d}{h_0^2} \tag{54}$$

$$h_0 = (0.2616 \times 10^{-10} vd)^{3/8} \tag{55}$$

$$H_s = 1.494 \times 10^{-2} \mu \frac{A_s v^2}{h_0} \tag{56}$$

The equations for p_0 and H_s include the side leakage factors L_1 and F_1 found from Fig. 18.

(*Continued on p. 12-36*)

Table 10. Load Capacity and Friction Horsepower of 51° Thrust Shoes (Fig. 47a)

For six-shoe bearing, total thrust = $6A_s p_0$ and total friction hp = $6H_s$

(Reprinted by permission from *Machine Design*, July 1948.)

D, in.	d, in.	A_s, in.²	v_d, ft/sec	5	10	20	40	60	80	100	120	140
6	4 1/2	3	p_0 (psi)	156	186	221	263	291	313	331	346	359
			h_0 (in.)	.000364	.000472	.000612	.000794	.000924	.00103	.00112	.00120	.00127
			N_d (rpm)	254	509	1018	2036	3053	4071	5089	6107	7124
			H_s (hp)	.01025	.0317	.0977	.3013	.5825	.929	1.335	1.794	2.305
12	9	12	p_0 (psi)	186	221	263	313	346	372	393	412	428
			h_0 (in.)	.000472	.000612	.000794	.00103	.00120	.00133	.00145	.00155	.00165
			N_d (rpm)	127	254	509	1018	1526	2036	2544	3053	3562
			H_s (hp)	.0317	.0977	.3010	.9290	1.794	2.880	4.125	5.555	7.100
18	13 1/2	27	p_0 (psi)	205	244	291	346	383	412	434	456	473
			h_0 (in.)	.000549	.000713	.000924	.00120	.00140	.00155	.00169	.00181	.00192
			N_d (rpm)	85	170	339	679	1018	1357	1696	2036	2375
			H_s (hp)	.06125	.1887	.5830	1.795	3.460	5.555	7.966	10.70	13.74
24	18	48	p_0 (psi)	221	263	313	371	411	441	467	490	513
			h_0 (in.)	.000612	.000794	.00103	.00133	.00155	.00173	.00188	.00202	.00214
			N_d (rpm)	64	127	254	509	763	1018	1272	1527	1781
			H_s (hp)	.0977	.3010	.9300	2.877	5.555	8.850	12.72	17.05	21.9
30	22 1/2	75	p_0 (psi)	233	278	331	393	434	468	495	518	537
			h_0 (in.)	.000665	.000863	.00112	.00145	.00169	.00188	.00205	.00219	.00232
			N_d (rpm)	51	102	204	407	611	814	1018	1221	1425
			H_s (hp)	.1405	.4330	1.335	4.125	7.966	12.73	18.24	24.60	31.60
36	27	108	p_0 (psi)	244	291	346	411	455	490	518	542	563
			h_0 (in.)	.000713	.000924	.00120	.00155	.00181	.00202	.00219	.00235	.00249
			N_d (rpm)	42.3	85	170	339	509	679	848	1018	1187
			H_s (hp)	.1887	.5825	1.795	5.555	10.70	17.05	24.60	33.0	42.35
42	31 1/2	147	p_0 (psi)	254	302	359	428	473	511	537	564	584
			h_0 (in.)	.000755	.000979	.00127	.00165	.00192	.00214	.00232	.00249	.00263
			N_d (rpm)	36.4	73	145	291	436	582	727	872	1018
			H_s (hp)	.2427	.7475	2.308	7.105	13.74	21.90	31.60	42.37	54.6
48	36	192	p_0 (psi)	263	313	372	441	490	526	555	583	604
			h_0 (in.)	.000794	.00103	.00133	.00173	.00202	.00224	.00244	.00261	.00277
			N_d (rpm)	31.8	64	127	254	382	509	636	763	890
			H_s (hp)	.3012	.9300	2.880	8.850	17.05	27.35	39.22	52.83	67.74
54	40 1/2	243	p_0 (psi)	270	322	383	455	504	542	572	600	622
			h_0 (in.)	.000830	.00108	.00140	.00181	.00211	.00235	.00255	.00273	.00289
			N_d (rpm)	28.3	57	113	226	339	452	565	679	792
			H_s (hp)	.3650	1.120	3.460	10.70	20.68	33.0	47.5	63.8	82.1
60	45	300	p_0 (psi)	278	331	393	467	518	555	587	616	639
			h_0 (in.)	.000863	.00112	.00145	.00188	.00219	.00244	.00265	.00284	.00301
			N_d (rpm)	25.5	51	102	204	305	407	509	611	712
			H_s (hp)	.4330	1.334	4.125	12.73	24.60	39.22	56.4	75.8	97.3
66	49 1/2	363	p_0 (psi)	285	338	403	478	531	569	602	631	654
			h_0 (in.)	.000894	.00116	.00150	.00195	.00227	.00253	.00275	.00295	.00312
			N_d (rpm)	23.1	46.3	92	185	278	370	463	555	647
			H_s (hp)	.5060	1.557	4.825	14.83	28.68	45.76	65.8	88.35	113.7
72	54	432	p_0 (psi)	291	346	412	490	542	583	615	645	670
			h_0 (in.)	.000924	.00120	.00155	.00202	.00235	.00261	.00284	.00304	.00322
			N_d (rpm)	21.2	42.3	85	170	254	339	424	509	594
			H_s (hp)	.5830	1.795	5.555	17.05	33.00	52.83	75.8	102.0	131.0
78	58 1/2	507	p_0 (psi)	297	353	420	500	553	593	626	658	682
			h_0 (in.)	.000952	.00123	.00160	.00208	.00242	.00269	.00293	.00314	.00332
			N_d (rpm)	19.6	39.2	78	156	235	313	392	470	548
			H_s (hp)	.6630	2.053	6.320	19.44	37.60	60.15	86.2	115.8	149.2
84	63	588	p_0 (psi)	302	359	428	511	563	604	639	670	695
			h_0 (in.)	.000979	.00127	.00165	.00214	.00249	.00277	.00301	.00322	.00342
			N_d (rpm)	18.2	36.3	73	145	218	291	364	436	509
			H_s (hp)	.7490	2.307	7.105	21.90	42.35	67.74	97.4	131.1	168.0

Tabular values are based on $\mu = 3.4 \times 10^{-6}$ lb-sec per sq in. (Reyns), equivalent to 0.2345 poise or 23.45 centipoises. When loading shoe higher than tabular value of p_0, increase viscosity in same ratio to maintain film thickness. Horsepower loss will increase in same ratio.

Table 11. Load Capacity and Friction Horsepower of $38\frac{1}{4}°$ Thrust Shoes (Fig. 47b)

For eight-shoe bearing, total thrust = $8A_sP_0$ and total friction hp = $8H_s$

(Reprinted by permission from *Machine Design*, July 1948.)

D, in.	d, in.	A_s, in.²	v_d, ft/sec	5	10	20	40	60	80	100	120	140
6	4 1/2	2 1/4	p_0 (psi)	180	214	255	303	336	361	382	399	414
			h_0 (in.)	.000346	.000448	.000582	.000754	.000878	.000980	.00106	.00114	.00121
			N_d (rpm)	254	509	1018	2036	3053	4071	5089	6107	7124
			H_s (hp)	.00825	.0256	.0787	.243	.470	.748	1.075	1.446	1.855
12	9	9	p_0 (psi)	214	255	303	361	399	429	453	475	493
			h_0 (in.)	.000448	.000582	.000754	.000980	.00114	.00126	.00138	.00147	.00157
			N_d (rpm)	127	254	509	1018	1526	2036	2544	3053	3562
			H_s (hp)	.0256	.0787	.243	.748	1.446	2.320	3.325	4.476	5.720
18	13 1/2	20 1/4	p_0 (psi)	236	281	336	399	442	475	500	525	545
			h_0 (in.)	.000522	.000677	.000878	.00114	.00133	.00147	.00161	.00172	.00182
			N_d (rpm)	85	170	339	679	1018	1357	1696	2036	2375
			H_s (hp)	.0493	.152	.470	1.446	2.788	4.475	6.420	8.62	11.08
24	18	36	p_0 (psi)	255	303	361	428	474	508	538	565	592
			h_0 (in.)	.000582	.000754	.000980	.00126	.00147	.00164	.00179	.00192	.00203
			N_d (rpm)	64	127	254	509	763	1018	1272	1527	1781
			H_s (hp)	.0787	.243	.750	2.319	4.476	7.130	10.25	13.74	17.65
30	22 1/2	56 1/4	p_0 (psi)	269	321	382	453	500	539	571	597	619
			h_0 (in.)	.000632	.000820	.00106	.00138	.00161	.00179	.00195	.00208	.00220
			N_d (rpm)	51	102	204	407	611	814	1018	1221	1425
			H_s (hp)	.1132	.349	1.075	3.323	6.420	10.26	14.70	19.83	25.46
36	27	81	p_0 (psi)	281	336	399	474	525	565	597	625	649
			h_0 (in.)	.000678	.000878	.00114	.00147	.00172	.00192	.00208	.00223	.00237
			N_d (rpm)	42.3	85	170	339	509	679	848	1018	1187
			H_s (hp)	.152	.4695	1.446	4.476	8.62	13.74	19.82	26.60	34.12
42	31 1/2	110 1/4	p_0 (psi)	293	348	414	493	545	589	619	650	673
			h_0 (in.)	.000718	.000930	.00121	.00157	.00182	.00204	.00220	.00237	.00250
			N_d (rpm)	36.4	73	145	291	436	582	727	872	1018
			H_s (hp)	.1952	.603	1.860	5.720	11.08	17.65	25.47	34.15	44.00
48	36	144	p_0 (psi)	303	361	429	508	565	606	640	672	696
			h_0 (in.)	.000754	.000980	.00126	.00164	.00192	.00213	.00232	.00248	.00263
			N_d (rpm)	31.8	64	127	254	382	509	636	763	890
			H_s (hp)	.2426	.749	2.320	7.130	13.74	21.07	31.60	42.60	54.60
54	40 1/2	182 1/4	p_0 (psi)	311	371	442	525	581	625	659	692	717
			h_0 (in.)	.000789	.00103	.00133	.00172	.00200	.00223	.00242	.00259	.00275
			N_d (rpm)	28.3	57	113	226	339	452	565	679	792
			H_s (hp)	.294	.902	2.788	8.62	16.67	26.60	38.25	51.45	66.20
60	45	225	p_0 (psi)	321	382	453	538	597	640	677	710	736
			h_0 (in.)	.000820	.00106	.00138	.00179	.00208	.00232	.00252	.00270	.00286
			N_d (rpm)	25.5	51	102	204	305	407	509	611	712
			H_s (hp)	.349	1.074	3.325	10.26	19.83	31.60	45.50	61.20	78.50
66	49 1/2	272 1/4	p_0 (psi)	329	390	465	551	612	656	693	727	753
			h_0 (in.)	.000850	.00110	.00143	.00185	.00216	.00240	.00261	.00280	.00296
			N_d (rpm)	23.1	46.3	92	185	278	370	463	555	647
			H_s (hp)	.4075	1.254	3.890	11.95	23.10	36.90	53.10	71.20	91.7
72	54	324	p_0 (psi)	336	399	475	565	625	672	708	743	772
			h_0 (in.)	.000878	.00114	.00147	.00192	.00223	.00248	.00270	.00289	.00306
			N_d (rpm)	21.2	42.3	85	170	254	339	424	509	594
			H_s (hp)	.470	1.446	4.476	13.74	26.60	42.60	61.20	82.20	105.5
78	58 1/2	380 1/4	p_0 (psi)	342	407	484	576	638	684	722	758	786
			h_0 (in.)	.000905	.00117	.00152	.00198	.00230	.00256	.00278	.00298	.00316
			N_d (rpm)	19.6	39.2	78	156	235	313	392	470	548
			H_s (hp)	.534	1.654	5.090	15.66	30.30	48.50	69.50	93.20	120.2
84	63	441	p_0 (psi)	348	414	493	589	649	696	737	772	801
			h_0 (in.)	.000930	.00121	.00157	.00204	.00237	.00263	.00286	.00306	.00325
			N_d (rpm)	18.2	36.3	73	145	218	291	364	436	509
			H_s (hp)	.603	1.860	5.720	17.64	34.15	54.60	78.50	105.6	135.4

Tabular values are based on $\mu = 3.4 \times 10^{-6}$ lb-sec per sq in. (Reyns), equivalent to 0.2345 poise or 23.45 centipoises. When loading shoe higher than tabular value of p_0, increase viscosity in same ratio to maintain film thickness. Horsepower loss will increase in same ratio.

It must be remembered that the pressures tabulated for the various conditions in Tables 10 and 11 are based on a uniform viscosity. When loading the shoe more heavily than the tabular p_0 the viscosity should be increased in the same proportion, in order to maintain the tabular minimum film thickness. This obviously increases the power loss H_s in the same direct proportion. Hence the designer can load a bearing up to any desired pressure within the mechanical strength of the bearing elements, if he is careful to lubricate it with oil of the required viscosity range and to cool that oil so that it will have the desired mean viscosity at the mean operating temperature of the film.

Tables for Journal Bearings. When the journal is surrounded by a set of equally spaced similar pivoted shoes, as suggested by Fig. 44 for eight square-faced shoes, the shoe loads, because of film pressure generated by rotation, are equal only when all the films

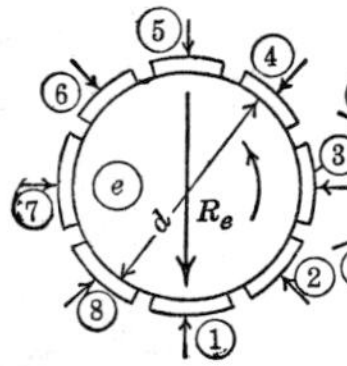

FIG. 49

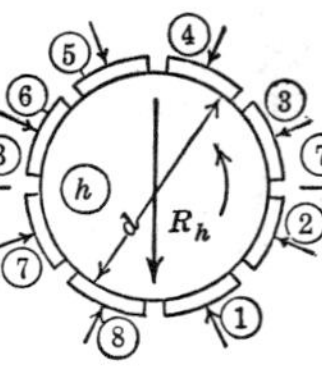

FIG. 50

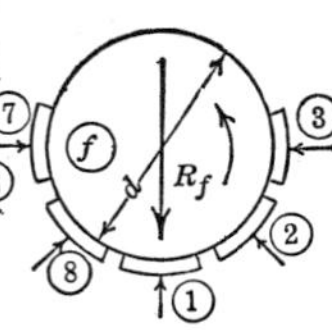

FIG. 51

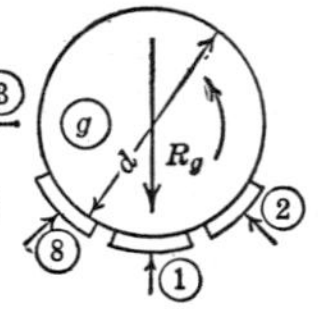

FIG. 52

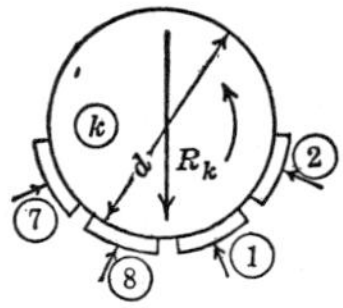

FIG. 53

have the same thickness, as they do when the journal surface is equidistant from all their faces. This can occur only when there is no externally applied radial load, because the result of such a load is to thin the films on the loaded side of the journal and to thicken those on the other. If the bearing is to carry practically no radial load, the radial running clearance, for Case 2 optimum operation (see p. 12-11), could be as low as 1.594 times the h_0 value given by Table 11 for a mean diameter d equal to the journal diameter. With no radial load the film viscosity is useful only as a damping means and could be anything within reason. The lower the viscosity, the less would be the power loss. However, it is the present problem to learn how such bearings react to definite radial loads and to find out what the running clearance should be when thus loaded.

In vertical shaft service the journal is normally surrounded by pivoted shoes as in Fig. 44. There are, therefore, three simple ways such a journal can present its load to the pivoted shoes. One way is seen in Fig. 49 with the load always acting toward one shoe. A second way is seen in Fig. 50, with the load always acting toward the space between two adjacent shoes. The third way is for the load direction to follow the rotation of the shaft.

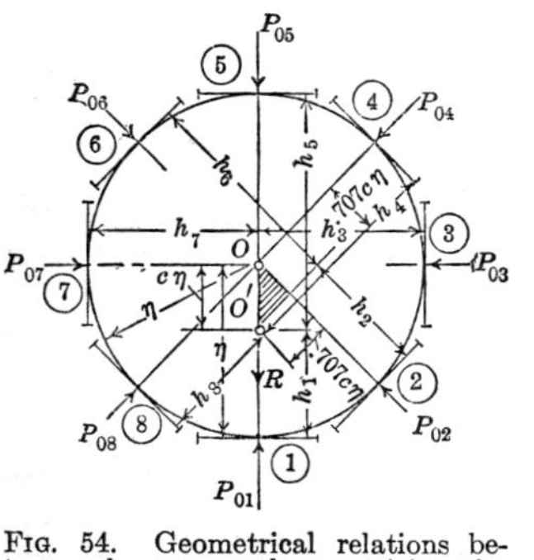

FIG. 54. Geometrical relations between clearance and eccentricity of a pivoted-shoe journal bearing.

In horizontal shaft service full sets of shoes are also used, but other arrangements are feasible and common as seen in Figs. 51, 52, and 53. The load direction is indicated in each figure.

As film thicknesses are minute compared with their journal diameters it is customary in studying them to represent the bearing by a circle whose radius is the radial clearance η and to indicate the journal as a point O' within that circle, with more or less eccentricity c. This method may be applied to pivoted journal shoes by representing the latter as a series of tangents to the clearance circle, as seen in Fig. 54.

The journal position is shown at O', a distance $c\eta$ below the clearance circle center O, Fig. 54; c is the eccentricity factor. The clearance circle has eight tangents representing eight pivoted shoes which are numbered 1 to 8 counterclockwise. The distance from the journal to each shoe, which is the mean film thickness, is shown by h with the shoe number as a subscript.

Radial Load Acting toward One Shoe. In Fig. 54 the mean film thicknesses h will be

$$h_1 = (1 - c)\eta \qquad h_2 = h_8 = (1 - c\cos 45°)\eta \qquad h_3 = h_7 = \eta$$

$$h_4 = h_6 = (1 + c\cos 45°)\eta \qquad h_5 = (1 + c)\eta$$

It is known from eq. 49 that

$$p_0 h_0^2 = 1.92\mu v l L_1$$

As the shoes of this journal bearing have the same square proportions as those of the eight-shoe bearings of Fig. 47b, the data given in Table 11 can be used for reference. The general formula given above can be put into the more useful arrangement, $p_0h^2 = K$, by replacing h_0 by the mean thickness h, which equals $1.594h_0$ for Case 2 optimum load (see p. 12-11). The value of K will therefore be as given in the following equation for square-faced shoes ($l/b = 1$) whose mean diameter is $d = 3l$ and for which $L_1 = 0.437$:

$$p_0h^2 = K = 0.7105\mu vd \tag{57}$$

Analysis of a bearing like Figs. 49 and 54 shows that there is a best eccentricity, so far as carrying capacity is concerned, when the least film thickness has the value tabulated in the thrust bearing (Table 11).

The relations between eccentricity, radial clearance, and mean film thickness have already been given. It therefore remains to determine the average radial pressure p_0 for each of the eight-shoe faces. Because all faces have equal areas, it is not necessary to introduce them here.

When the radial pressures are known, their components parallel with the journal load may be found and lettered P'_0 with a subscript after the subscript zero to show the shoe number. Hence, for Fig. 54,

$$p'_{01} = p_{01} \qquad p'_{02} = p'_{08} = p_{02}\cos 45° \qquad p'_{03} = p'_{07} = 0$$

$$p'_{04} = p'_{06} = -p_{04}\cos 45° \qquad p'_{05} = -p_{05}$$

Radial Load Acting toward Space between Two Shoes. Similar analysis shows that if the radial load acts toward the space between two shoes, as in Fig. 50, the radial capacity is substantially the same as for Fig. 49, for the same radial clearance and the same eccentricity. Hence, if a shaft is out of balance, and if that is the cause of the radial load on the bearing, the film resistances to radial displacement will be substantially constant throughout the revolution of the shaft, whose axis will describe a circle with radius of $c\eta$.

The shoe arrangements in Figs. 44, 49, and 50 are usual in vertical shafts. Those of Figs. 51, 52, and 53 are suitable for horizontal shafts. The two side shoes three and seven of Fig. 51 do not add to the vertical capacity of the bearing but do add to the journal stability where there is some change of load direction, not exceeding 30° approximately. At the same time, however, they increase the friction loss.

Friction Losses in Pivoted Shoe Journal Bearings. The friction losses in journal bearings having eight square pivoted shoes may now be investigated. The loss per shoe is determinable from eq. 56 for H_s. It is seen to vary inversely as the minimum film thickness and directly as the viscosity. The mean thickness $h = 1.594h_0$ can be introduced giving the following equations, in which the viscosity is assumed to be $\mu = 3.4 \times 10^{-6}$ Reyns:

$$H_s = \frac{8.1 \times 10^{-8} A_s v^2}{h} \tag{58}$$

$$H_s = \frac{2.382 \times 10^{-2} \mu A_s v^2}{h} \tag{59}$$

Loading and Friction Tables for Journal Shoes. In order that a wide variety of problems involving pivoted shoe journal bearings like those shown in Figs. 49, 51, and 52 may be readily solved, Table 12 has been prepared. For sizes and conditions not included in the table, eqs. 57 and 59 may be employed. Comparing this table with Table 11 shows that the first three tabular values of each block are alike, as would be expected. The fourth entry in the blocks of eight values is the radial clearance for greatest radial capacity for eight square-faced shoes. As $c = 0.5$, the least mean film thickness, which is over shoe 1, will be $h_1 = (1 - c)\eta = 0.5\eta$ (see Fig. 54). Also, from the Case 2 optimum load requirement that $h_1 = 1.594h_{01}$, $\eta = 3.188h_{01}$. Because the film thicknesses, pressures, etc., vary from shoe to shoe in a loaded radial bearing, the subscript 1 is added to p_0 and h_0, making them p_{01} and h_{01} to show that the values tabulated refer only to the most heavily loaded shoe 1.

The next pair of tabular values, R_8 and H_8, refer to a bearing like that of Fig. 49. They are the total load, and the friction for all eight shoes, when the minimum film thickness of the nearest shoe 1 has the value tabulated opposite h_{01} and when the average pressure over the face of that nearest shoe 1 has the value tabulated opposite p_{01}.

The last pair of tabular values R_5 and H_5, in the blocks of eight in Table 12, refer to a bearing like Fig. 51. They are totals for all five shoes. The values tabulated for P_0 and h_{01} refer to shoe 1 of the group of five. When the side shoes 3 and 7 are omitted, as in Fig. 52, the total load carried by the remaining three is equal to R_5. The total friction for the three will be equal to $0.717H_5$.

Load capacity and friction for bearings like those shown in Figs. 50 and 53 have not been tabulated but may be calculated by following the procedure described. Performance of a bearing like that of Fig. 50 is approximately the same as performance of one like that of Fig. 49.

The Loading and Friction of 38¼° Journal Shoes Wider and Narrower than Square. Table 12 for square journal bearing shoes, i.e., the ones for which $l/b = 1$, can be used to obtain the capacity and other data for *wider* and *narrower* shoes of the same angular

Table 12. Load Capacity and Friction Horsepower of 38¼-degree Pivoted Shoe Journal Bearings

(Reprinted by permission from *Machine Design*, August 1948.)

d, in.	A_s, sq in.	v_d, ft per sec	5	10	20	40	60	80	100	120	140
4 1/2	2 1/4	p_{01} (psi)	180	214	255	303	336	361	382	399	414
		h_{01} (in.)	.000346	.000448	.000582	.000754	.000878	.000980	.00106	.00114	.00121
		N_d (rpm)	254	509	1,018	2,036	3,053	4,071	5,089	6,107	7,124
		η (in.)	.001102	.001428	.001855	.002404	.002800	.003125	.003380	.003635	.003860
		R_8 (lb)	626	744	887	1,053	1,167	1,255	1,328	1,386	1,440
		H_8 (hp)	.0380	.1180	.3630	1.120	2.166	3.450	4.960	6.670	8.530
		R_5 (lb)	749	891	1,061	1,261	1,397	1,500	1,590	1,660	1,720
		H_5 (hp)	.0292	.0907	.2790	.860	1.665	2.650	3.810	5.120	6.550
9	9	p_{01} (psi)	214	255	303	361	399	429	453	475	493
		h_{01} (in.)	.000448	.000582	.000754	.000980	.00114	.00126	.00138	.00147	.00157
		N_d (rpm)	127	254	509	1,018	1,526	2,036	2,544	3,053	3,562
		η (in.)	.001428	.001855	.002404	.003125	.003635	.004020	.004400	.004685	.005010
		R_8 (lb)	2,980	3,550	4,220	5,020	5,550	5,970	6,300	6,610	6,860
		H_8 (hp)	0.1180	0.3630	1.120	3.450	6.670	10.68	15.33	20.64	26.40
		R_5 (lb)	3,560	4,250	5,050	6,020	6,650	7,150	7,550	7,910	8,220
		H_5 (hp)	0.0907	0.2790	0.860	2.650	5.120	8.20	11.80	15.90	20.30
13 1/2	20 1/4	p_{01} (psi)	236	281	336	399	442	475	500	525	545
		h_{01} (in.)	.000522	.000677	.000878	.00114	.00133	.00147	.00161	.00172	.00182
		N_d (rpm)	85	170	339	679	1,018	1,357	1,696	2,036	2,375
		η (in.)	.001664	.002160	.002800	.003635	.004240	.004685	.005135	.005485	.005810
		R_8 (lb)	7,390	8,800	10,520	12,500	13,840	14,870	15,650	16,440	17,060
		H_8 (hp)	0.227	0.701	2.170	6.670	12.85	20.64	29.60	39.70	51.10
		R_5 (lb)	8,840	10,510	12,580	14,940	16,540	17,770	18,730	19,650	20,400
		H_5 (hp)	0.174	0.539	1.670	5.120	9.87	15.90	22.75	30.50	39.30
18	36	p_{01} (psi)	255	303	361	428	474	508	538	565	592
		h_{01} (in.)	.000582	.000754	.000980	.00126	.00147	.00164	.00179	.00192	.00203
		N_d (rpm)	64	127	254	509	763	1,018	1,272	1,527	1,781
		η (in.)	.001855	.002404	.003125	.004020	.004685	.005230	.005715	.006125	.006475
		R_8 (lb)	14,200	16,850	20,100	23,800	26,400	28,300	29,950	31,450	32,950
		H_8 (hp)	0.363	1.120	3.460	10.68	20.64	32.90	47.25	63.30	81.40
		R_5 (lb)	16,980	20,200	24,060	28,500	31,600	33,860	35,850	37,650	39,450
		H_5 (hp)	0.279	0.860	2.660	8.20	15.90	25.30	36.30	48.70	62.50
22 1/2	56 1/4	p_{01} (psi)	269	321	382	453	500	539	571	597	619
		h_{01} (in.)	.000632	.000820	.00106	.00138	.00161	.00179	.00195	.00208	.00220
		N_d (rpm)	51	102	204	407	611	814	1,018	1,221	1,425
		η (in.)	.002015	.002615	.003380	.004400	.005135	.005715	.006220	.006630	.007020
		R_8 (lb)	23,400	27,900	33,200	39,400	43,500	46,850	49,650	51,900	53,800
		H_8 (hp)	0.522	1.610	4.960	15.33	29.60	47.25	67.8	91.5	117.4
		R_5 (lb)	28,000	33,400	39,750	47,200	52,000	56,100	59,400	62,150	64,400
		H_5 (hp)	0.401	1.240	3.810	11.80	22.75	36.30	52.1	70.3	90.2
27	81	p_{01} (psi)	281	336	399	474	525	565	597	625	649
		h_{01} (in.)	.000678	.000878	.00114	.00147	.00172	.00192	.00208	.00223	.00237
		N_d (rpm)	42.3	85	170	339	509	679	848	1,018	1,187
		η (in.)	.002161	.002800	.003635	.004685	.005485	.006125	.006630	.007115	.007560
		R_8 (lb)	35,200	42,100	50,000	59,400	65,750	70,750	74,750	78,250	81,300
		H_8 (hp)	0.701	2.164	6.670	20.64	39.70	63.30	91.40	122.5	157.4
		R_5 (lb)	42,100	50,300	59,800	71,000	78,700	84,700	89,500	93,700	97,200
		H_5 (hp)	0.539	1.664	5.120	15.90	30.50	48.70	70.20	94.2	120.6
31 1/2	110 1/4	p_{01} (psi)	293	348	414	493	545	589	619	650	673
		h_{01} (in.)	.000718	.000930	.00121	.00157	.00182	.00204	.00220	.00237	.00250
		N_d (rpm)	36.4	73	145	291	436	582	727	872	1,018
		η (in.)	.002290	.002966	.003860	.005010	.005810	.006510	.007020	.007560	.007975
		R_8 (lb)	49,900	59,300	70,600	84,000	92,900	100,300	105,400	110,600	114,600
		H_8 (hp)	0.900	2.780	8.570	26.40	51.10	81.40	117.40	157.4	203.0
		R_5 (lb)	59,700	71,000	84,400	100,400	111,000	120,000	126,000	132,400	137,000
		H_5 (hp)	0.692	2.14	6.580	20.30	39.30	62.50	90.20	120.6	156.0

R_8 and H_8 denote total radial load capacity and total friction horsepower, respectively, for an eight-shoe bearing (Fig. 49). R_5 and H_5 denote corresponding values for five-shoe bearing (Fig. 51).

Table is based on $\mu = 3.4 \times 10^{-6}$ lb-sec per sq in. (Reyns) equivalent to 0.2345 poise or 23.45 centipoise. When loading higher than the tabular value of p_{01} or H is required, increase the viscosity in the ratio of pressure in order to maintain film thickness. Horsepower loss will increase in the same ratio.

length of 38 1/4°. This will be explained below, taking account of the differences in the influence of side leakage upon capacity and friction, and also of the requirements of the film thickness philosophy already explained.

The problem is illustrated in Figs. 55 and 56 with a journal having diameter of $d = 3.00$, surrounded by eight shoes of length $l = 1.0$. They are assumed to be pivoted members, suitably mounted. In Fig. 56 several different vertical sections of shoes for such a bearing are shown, the shaded portions indicating a width $b = 1.00$, which also equals the shoe face length l of Fig. 55. However, Fig. 55 might represent shoes of any width b. Eight different values of b are indicated in Fig. 56, ranging from 3.0 down to 0.75. They will now be compared.

The philosophy of loading underlying pivoted thrust shoe practice, explained heretofore and applied to the derivation of the thrust shoe loading tables and to the choice of a best shoe form, becomes of new interest now. Equation 50 is re-expressed below as applied to the comparison, in minimum film thickness, of two shoes of uniform proportions, operating at the same surface speed. The areas have been introduced because they will vary as the square of the similar linear dimensions:

$$\frac{h_0}{h_{01}} = \left(\frac{l}{l_1}\right)^{3/8} = \left(\frac{A_s}{A_{s1}}\right)^{3/16} \tag{60}$$

By the above equation we can find the relative minimum thickness for square shoe faces. Our loading philosophy proposes the same thickness of film for two shoes of the same area but with different face proportions.

With the relative minimum thicknesses for the films for the eight shoe faces the capacities of those shoes can be found by first getting the average pressure p_0 from eq. 49, p. 12-32. It applies to Case 2 optimum films.

After p_0 is found, the capacity of the face can be determined by multiplying p_0 by A_s. Thus by means of eqs. 60 and 61 we can determine the relative capacities and suitable film thicknesses for the lower (most heavily loaded) shoes of eight shoe bearings (Fig. 55)

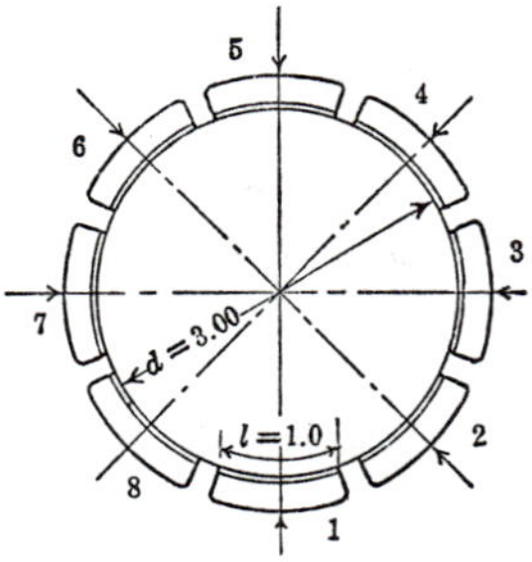

FIG. 55. Eight-shoe journal bearing having shoes of constant length l.

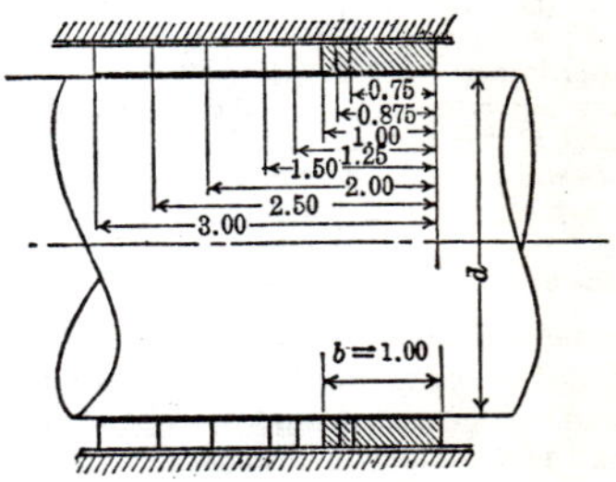

FIG. 56. Shoes of varying width b, for bearing in Fig. 55.

whose shoe face widths are narrower or wider than square, as indicated in Fig. 56. The comparisons are made for common operating conditions as to viscosity and surface speed.

$$\frac{p_0}{p_{01}} = \left(\frac{h_{01}}{h_0}\right)^2 \frac{(L_1)}{(L_1)_1} \tag{61}$$

The general friction equation giving horsepower loss per shoe, with side leakage correction factor F_1, is as follows for Case 2 optimum:

$$H_s = 1.645 \times 10^{-2} \mu F_1 \left(\frac{A_s v^2}{h_0}\right) \tag{62}$$

This may be applied to obtain the power loss of the most heavily loaded shoe of the set whose least film thickness was found by eq. 60 as described just above.

Equation 62 can be used as follows for comparison of different shoes, just as was eq. 60, for constant μ and v:

$$\frac{H_s}{H_{s1}} = \frac{(F_1)}{(F_1)_1} \frac{A_s}{A_{s1}} \frac{h_{01}}{h_0} \tag{63}$$

In Table 13 the different shoe faces are compared with a square face

Table 13. Comparison of Capacities and Friction for Journal Shoes Not Square

l/b	0.333	0.400	0.500	0.667	0.800	1.000	1.143	1.333
A	3.000	2.500	2.000	1.500	1.250	1.000	0.875	0.750
h_0	1.229	1.187	1.139	1.079	1.043	1.000	0.975	0.948
p_0	1.183	1.200	1.207	1.162	1.097	1.000	0.937	0.806
Ap_0	3.549	3.000	2.414	1.743	1.371	1.000	0.820	0.604
H	2.595	2.226	1.836	1.429	1.216	1.000	0.889	0.774

Notation: l/b = length/width ratio; A = relative area per shoe; h_0 = relative minimum film thickness; p_0 = relative mean pressure on shoe; H = relative horsepower.

Assumed conditions: Minimum film thicknesses from eq. 60. Mean pressure from eq. 61. Horsepower from eq. 63.

Now, if the minimum film thickness of one square shoe, of any area, is known its area may be treated as unity, and the corresponding thicknesses for larger and smaller square shoes found. The new thicknesses may then be applied to the wider or narrower shoes.

EXAMPLE. An eight-shoe journal bearing like the one shown in Fig. 55, when d = 18 in., is required to have shoe faces that are 12 in. wide, i.e., twice as wide as the 6-inch square shoe for which data are given in Table 12. The speed is to be 10 ft per sec and $\mu = 3.4 \times 10^{-6}$ Reyns. For this problem the relative area = 2 and relative minimum film thickness = 1.139 (Table 13), and they will apply to the lower or most heavily loaded shoe of the set of eight.

This means that the proper minimum film thickness for the 6- by 12-inch shoe will be 1.139 times that tabulated in Table 12 for the 6-inch square shoe or 1.139 × 0.000754 = 0.000858. The relative average pressures on the two shoes will then be as seen in Table 13, i.e., 1.207 × 303 = 365.7 psi. Hence the radial load upon the lower 12-inch wide shoe would be 72 × 365.7 = 26,330 lb.

It is apparent that the total radial capacity of the eight 12-inch wide shoes of Fig. 55 for d = 18 in. will be determined by the running clearance. The eccentricity factor should still be c = 0.5. The radial clearance η = 0.002404, given in Table 12, must therefore be increased to 0.002738, i.e., in the same proportion (1.139) as was the minimum film thickness h_0 over the lower shoe, which was raised from 0.000754 to 0.000858.

The net capacity R_8 of the eight 6- by 6-in. shoes in Table 12 will have to be increased, first by the pressure factor 1.207 and second by the area factor 2.0 in order to obtain the actual capacity, 2.414 × 16,850 = 40,676 lb.

To obtain the power loss, the tabular loss H_8 must be increased by the factor 1.836 due to the combination of greater area of shoe faces, better face proportions, and greater film thickness. Therefore actual power loss = 1.836 × 1.12 = 2.06.

The foregoing values cover a full complement of eight shoes. When the upper three are removed the net capacity of the bearing will be increased from 40,676 to 2.414 × 20,200 = 48,763 lb, but the friction loss will be reduced from 2.06 to 1.58 hp. (See Table 12.)

Thus a typical block of factors has been found with the resulting tabular values that would apply to a journal bearing supported by 38 1/4° pivoted shoes whose axial width is twice as great as the width of the square-faced shoes covered by Table 12.

The capacities thus far tabulated for pivoted shoes are based on the use of a light oil. If the film viscosity should be increased tenfold, the film capacity would be increased tenfold. Whether or not a pivoted bearing is mechanically strong enough to sustain such a heavy load is a matter of design.

4. CONSTRUCTION AND STANDARDIZATION OF JOURNAL BEARINGS

JOURNAL BEARING ELEMENTS. Classification. Figures 57 to 75 serve to illustrate and classify the usual ways that journal bearings are applied to heavy machinery. For ordinary service where axial loads are light the plain cylindrical journal bearing serves well.

Running friction is less than for so-called antifriction bearings for moderate speeds, and *durability* greater for high speeds.

Starting Friction. When the loads in a machine are created by its operation the starting friction is negligible. When the loads are heavy before starting, the starting friction is high and it is sometimes necessary to make provision for lessening it. Several effective ways are available, and manufacturers should be consulted about them. One means frequently resorted to is high-pressure oil that forces into the bearing a preliminary film upon which the starting is very easy indeed, easier than for any other type of bearing. See Fig. 5 and text for the laws of fluid flow involved.

Standardization of plain journal bearings has not taken place effectively because they are so simple and easy to make. Hence suppliers of bushings carry enormous stock lists of sizes. There is a strong tendency toward babbitt-lined shells of steel or bronze of light accurate construction which are entirely practicable when the housing in which they are mounted is stiff. Many solid bronze bearing shells are now also made of pressed powdered metals.

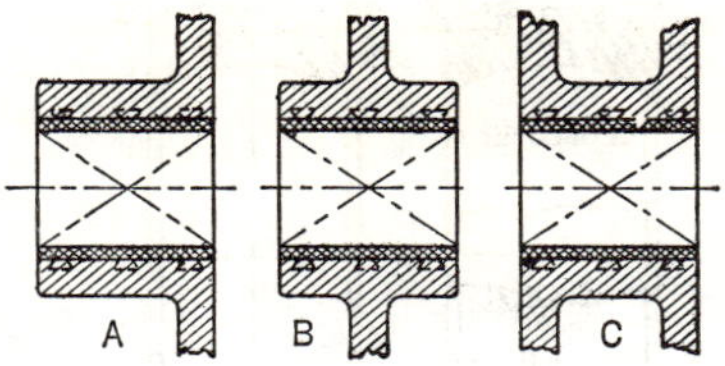

Fig. 57. Integral bearings.

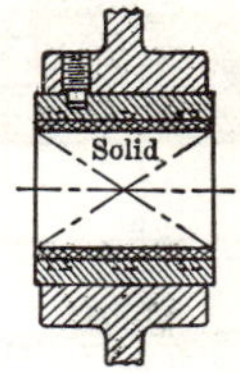

Fig. 60. Plain renewable shell.

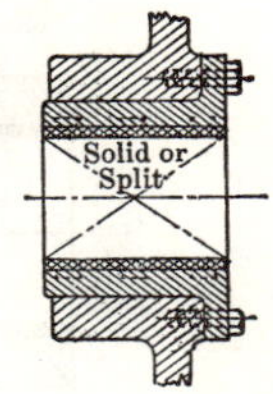

Fig. 61. Single-flanged shell.

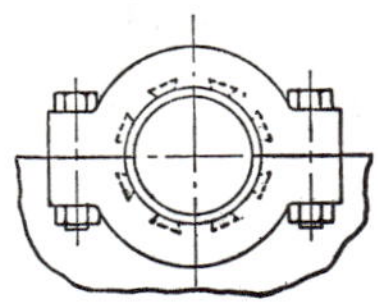

Fig. 58. Simple split bearing.

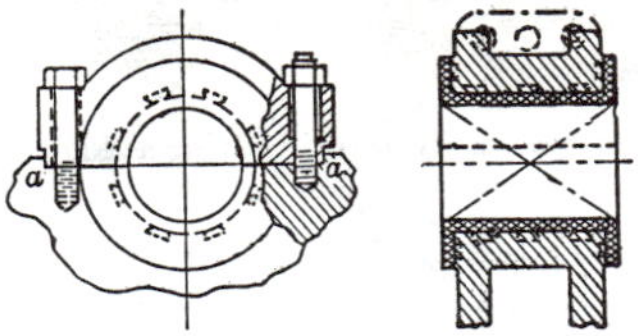

Fig. 59. Integral bearing for heavy machinery.

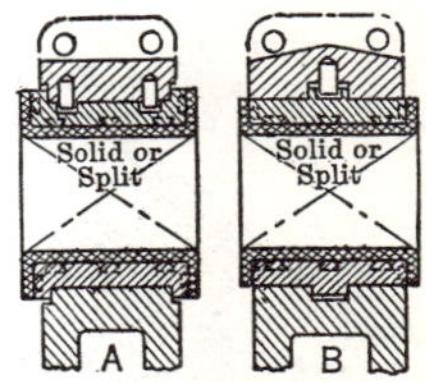

Fig. 62. Shells with babbitt-faced ends.

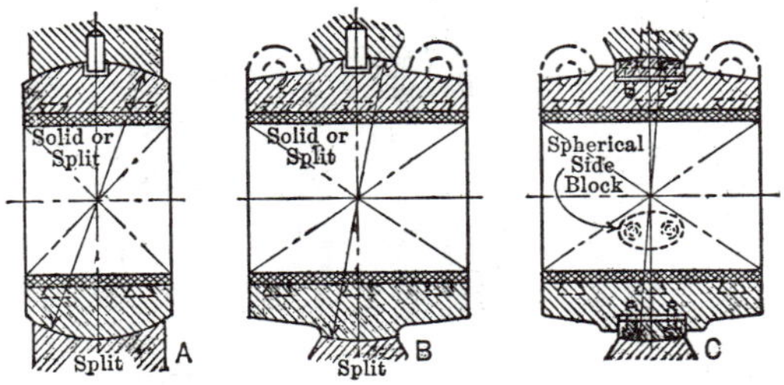

Fig. 63. Spherical-seated self-aligning shells.

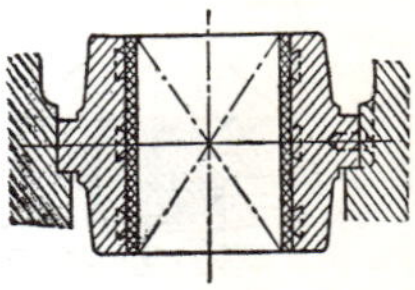

Fig. 64. Vertical journal bearing.

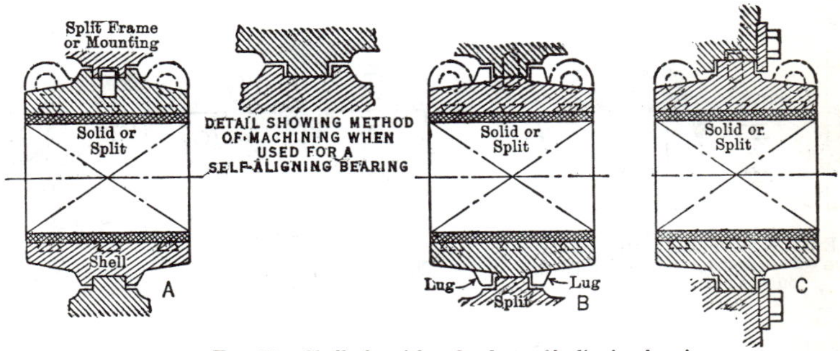

Fig. 65. Shells for either fixed or self-aligning bearings.

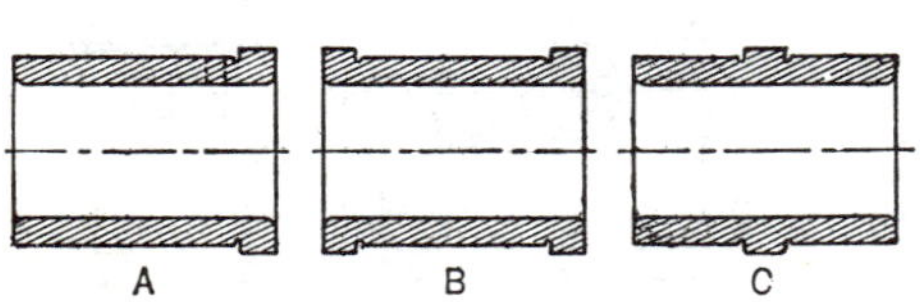

Fig. 66. Types of solid renewable bearing shells.

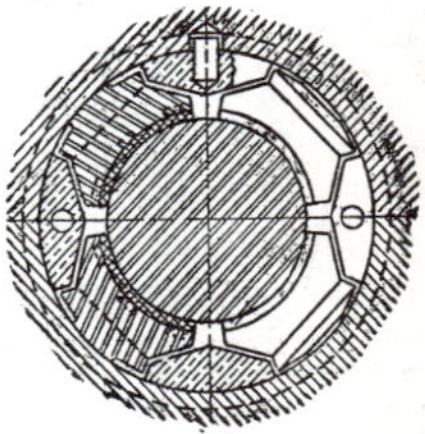

Fig. 67. Pivoted-shoe shell, self-aligning.

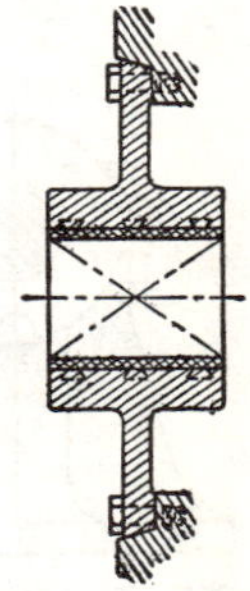

Fig. 68. Self-aligning web-mounted shell.

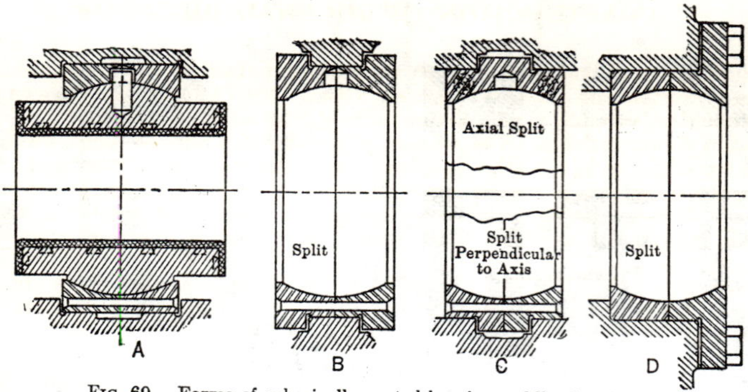

FIG. 69. Forms of spherically seated bearing saddles for shells.

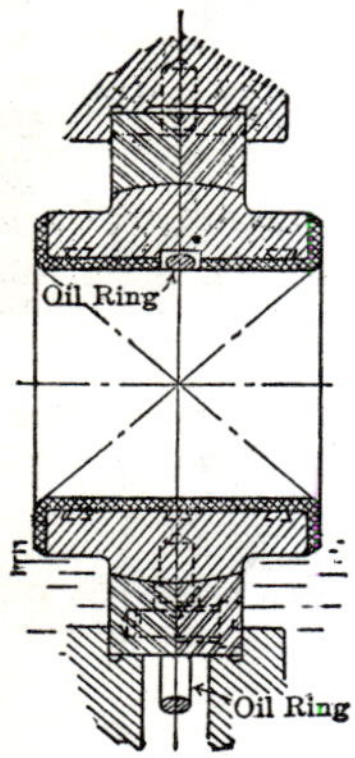

FIG. 70. Spherically seated shell and saddle with ring oiling.

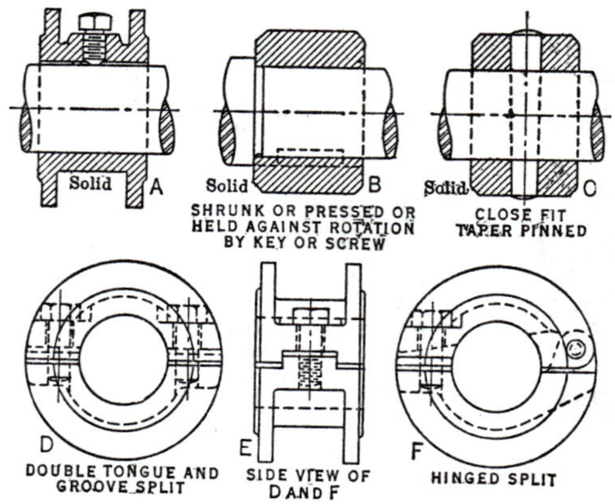

FIG. 71. Types of shaft collars.

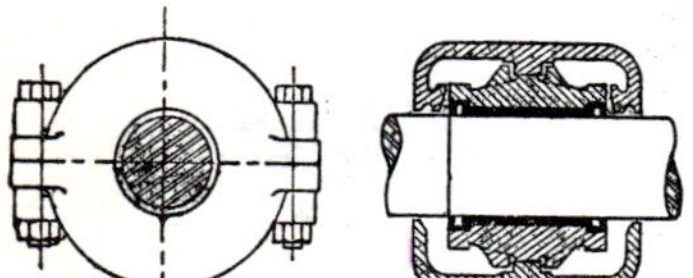

FIG. 72. End view and axial section of barrel portion of mounting for self-aligning journal bearings.

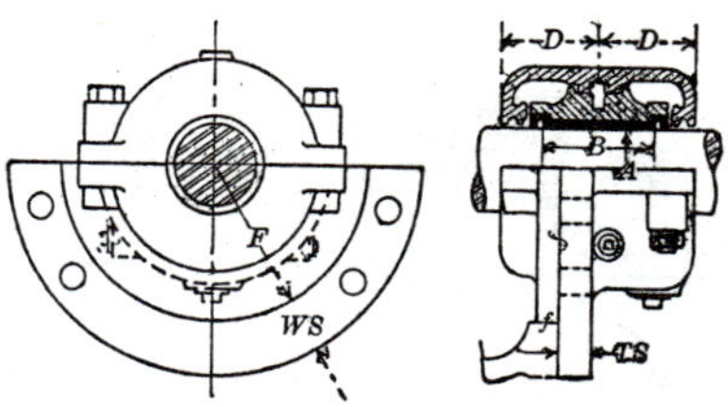

FIG. 73. End view and part axial section of semicircular flanged mounting.

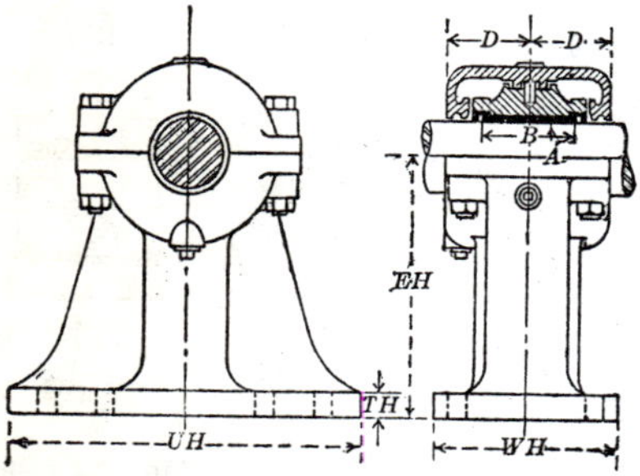

FIG. 74. End view and part axial section of high pedestal mounting.

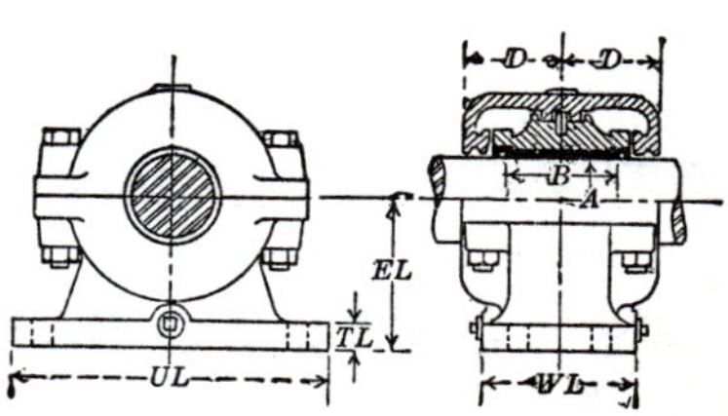

FIG. 75. End view and part axial section of low pedestal mounting.

Standardization of mountings is even more difficult than for the bearing shells. The direction which this might take is indicated in Figs. 72 to 75, in which the barrel and shell portion would be common to all. See methods of lubrication, Article 5.

Operating Characteristics. The relation of running clearance to speed, viscosity, and load has been discussed at length heretofore. In Article 9, Optimum Loading and Current Practice, this question is answered: How thin should the film be allowed to become during operation? Therein the wide variation of clearance met with in practice is also shown. With this information available there should be a gradual approach toward greater uniformity of design.

OIL GROOVES IN JOURNAL BEARINGS. Oil should be applied to a journal bearing at the point where there is no load, or where the load is lightest if it changes its radial direction. Grooves, therefore, may be used for the following purposes: (1) the application of oil, (2) the collection of side leakage, and (3) the pumping of oil.

Application grooves should be parallel with the shaft. The length may vary from an oil hole diameter to the full length of the bearing surface. Such grooves usually do not extend to the sides. (See Fig. 76.) While two such grooves frequently are cut diametrically opposite, one usually is enough. The grooves should be rounded on the edges parallel to the shaft. If the bearing revolves about a fixed shaft, the groove should be in the shaft on the unloaded side.

If a load exerted by a shaft turning in a fixed bearing acts successively in all radial directions with a known portion of the shaft journal subjected to little or no load, the oil grooves should be located in the shaft at that place if practicable. Oil may be supplied to the groove through the shaft or from a ring groove at one side of the bearing, registering with one end of the groove in the shaft. Figure 77 shows grooving for use where the direction of the radial load is unknown. This arrangement presents a fair bearing surface to the journal whenever the region of closest approach of the bearing surfaces passes over an oil groove. With pressure oiling, the circumferential groove may be omitted, thereby increasing the carrying capacity of the bearing.

High-speed journals are cooled by continuously changing the oil. It then is necessary to provide oil outlets at the ends of the application grooves, as in Fig. 78. The grooves must be large enough to pass the desired quantity of oil with the pressure available. In Fig. 77 outlets would be provided at the ends of the laterals. The cross section of the application groove may be, in general, similar to that shown in Fig. 76. If the grooves are used to limit the areas of the bearing surface, they sometimes are enlarged as shown in Fig. 79.

Collector grooves are circumferential grooves at the sides of automatically or pressure-oiled bearings. They collect side leakage and return it to the oil well to prevent its being wasted by flow along the shaft. Such grooves usually are single, but frequently double. They should have an adequate discharge hole at the bottom. They should be continuously circumferential without a break. Oil application grooves should not enter the collector grooves. (See Fig. 80.)

Pumping grooves are used both to pump and to apply oil. The most common type is the helical groove used with a vertical shaft to lift oil from the well to the upper end of a bearing that may be higher than the surface of oil in the well. The most effective pumping groove is circumferential, if it stops short of extending all the way around the shaft, and is cut in the stationary bearing surface. The inlet end must open to the oil in the well in a nanner that will readily receive oil. The other end must be dammed in the circumferential direction. From this point, suitable passages lead oil to any location within the limits of the available pressure. It simply may be led along the shaft through application grooves. Pumping grooves should be completely submerged in oil to avoid aeration of the oil bath. Aeration is, at best, hard to control in high-speed bearings.

The cross section of a pumping groove should be shallow radially. The shallower the depth the greater will be the pressure developed, but the less the discharge. The wider the axial width the greater will be the rate of discharge. For calculations involved in design, see Viscous Flow, p. 12-03. Ordinarily a simple groove of cross section shown in Fig. 76, but shallower, will be satisfactory.

Many self-contained systems have been devised to lubricate the journal and thrust bearings of a machine continuously, which cause oil to flow to all surfaces needing it, and then return it to the well. Bearings so lubricated are durable and require a minimum of care. (See Figs. 81 and 89.)

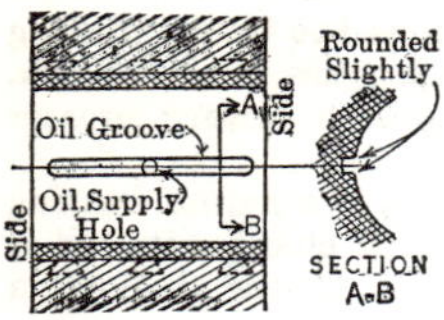

FIG. 76. Application oil groove.

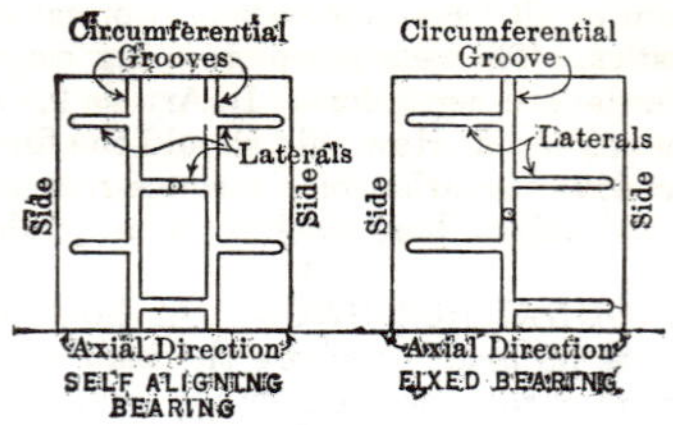

FIG. 77. Grooving where direction of radial load is unknown.

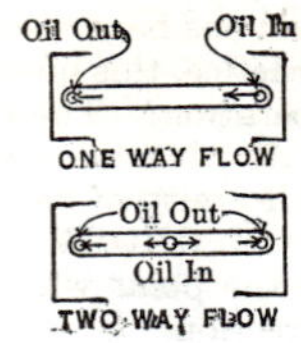

FIG. 78. Oil grooves with outlets.

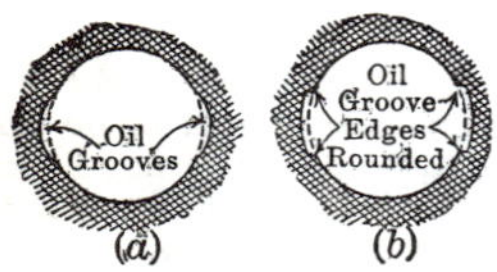

FIG. 79. Methods of enlarging oil grooves.

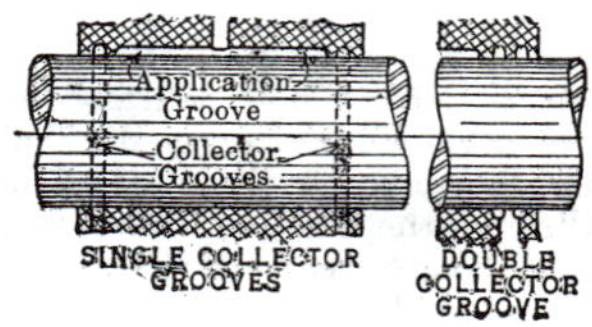

FIG. 80. Collector grooves.

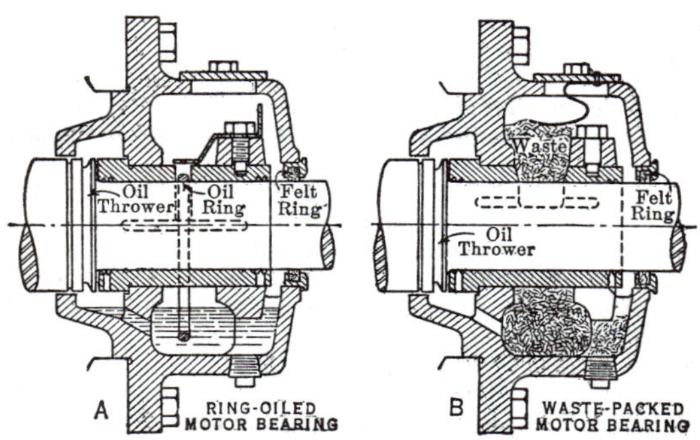

FIG. 81. Motor bearings.

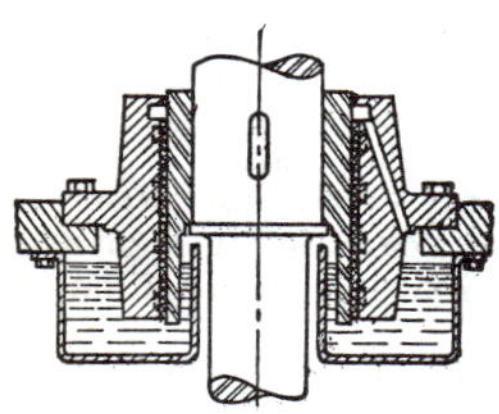

FIG. 82. Lubricating system for vertical journal bearing, spiral groove not shown.

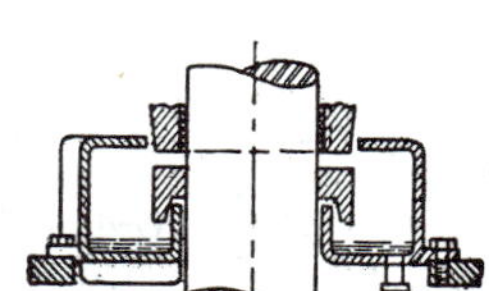

FIG. 83. Collar and pan for collecting oil below a vertical journal bearing.

5. LUBRICATION AND COOLING OF BEARINGS

LUBRICATION. Bearings are lubricated by these systems: (1) automatic, (2) intermittent, (3) bath, (4) splash, (5) pressure, and (6) mechanical.

Automatic Lubrication. The rotation of the shaft causes oil to be fed to the bearing surfaces. Lubrication ceases when the shaft stops turning. The systems in common use are ring oiling, chain oiling, disk oiling, capillary oiling, waste pack oiling, vacuum oiling, and oiling by means of a shaft-driven pump of any type. All these systems lift oil to the bearing from a well below it and are capable of continuous operation for long periods, because oil leaking from the ends of the bearing or elsewhere returns to the well and is used over again. (See Fig. 81.)

Intermittent Lubrication. Oil is applied to the bearing at intervals from a source independent of the rotation of the shaft. The simplest, but least dependable system, is hand oiling from an oil can. Cup oiling is better, as it has the possibility of being practically continuous and uniform, with visible and adjustable rate. Wick oiling, with textile or metal wicks, is, next to hand oiling, the simplest and cheapest system. The number of wick strands determines the rate of supply. The wicks filter the oil. The general faults of intermittent systems include usually complete loss of oil after it leaves the bearing, frequent refilling of cups and wells, and inadequate or wasteful oiling by reason of ignorance or carelessness.

Bath Lubrication. The bearing faces are completely immersed in oil at all times. This system is used for vertical shafts, and is simplest for the lower end, where thrust and radial bearings can be combined. At other places on the shaft, the well may be formed by a revolving bowl, or the well may be like the one shown in Fig. 82. This last system is commonly used for thrust bearings placed anywhere along a vertical shaft. Packing may be used in stuffing boxes to hold oil in any kind of a recess about a bearing, horizontal or vertical, that will make it a bath-oiled bearing. Oil will leak away, even though slowly, from such a bearing. There also is the danger of setting up the packing so tightly that it will run hot and the shaft will be worn within it.

Splash lubrication was common in the crank cases of internal-combustion engines and still is considered a simple way to oil all the bearings of a trunk-piston engine. The splash is produced by the lower ends of the connecting rods dipping into the oil in the crank case. There are various refinements to control the splash. This might be considered a continuous system, because most of the oil is returned to the crank case to be used again, although partly contaminated or broken down by exposure to cylinder walls. The system is used in compressors and other crank machines.

Pressure Lubrication. Pipes lead oil to the bearings from a source of oil under pressure. The pressure is available at the point of application of oil to the bearing and will therefore tend to keep oil passages open. Bearings with this system usually are simple when the whole machine is so enclosed that oil will drain to the sump from which an oil pump redistributes it to the bearings. The rate of oil supply to each bearing may be regulated by making the size of the supply hole to suit the pressure in the system. The source of oil may be independent of the operation of the machine, as when a separate motor-driven oil pump is used. It may otherwise be driven by some shaft in the machine, and, therefore, operates only when the machine is running. Each method has its drawbacks, and judgment must be used in selection.

Mechanical Oilers. A variety of mechanical devices are on the market for producing what may be styled centrally controlled, intermittent lubrication of bearings. They eliminate oil cups, but introduce a multitude of small oil pipes, through which oil is fed or forced intermittently, in measured amounts. These devices usually operate by ratchets driven from the shaft of the machine whose bearings they oil.

Lubrication of Vertical Journal Bearings. With pressure lubrication, the simplest bearing form may be used. The journal surface may be on the shaft itself. To collect the oil flowing down the shaft below the bearing, a collar on the shaft throws the oil into a stationary pan, whence it flows to the sump of the oiling system. (See Fig. 83.)

Vertical journals may be on the surface of sleeves shrunk or otherwise fastened on the shaft. (See Fig. 82.) The lower end of the sleeve, which is bored out larger than the shaft, projects into the stationary oil pan. Oil may be taken into the bearing film by capillarity, by viscosity pumping, centrifugally, or by combinations of these methods.

If the shaft terminates at the bottom in a vertical journal bearing, lubrication is a relatively simple problem. With a through shaft, simple constructions also may be used, if lubrication is by cup, and if some waste of oil or grease is unimportant. Leakage along the shaft may be minimized by a stuffing box if the speed of rotation is not so high as to cause heating of the shaft by the packing.

COOLING OF BEARING OIL. To keep oil in a bearing at a safe temperature, the heat of friction due to operation must be dissipated. Safe temperature depends on the kind of oil used and the conditions against which the bearing must be protected. Thickness of oil film and heat of friction both increase with viscosity of oil.

Direct Air Cooling. The oil housing surrounding the bearing should be sealed tight to prevent leakage. The oil then is allowed to warm up sufficiently to dissipate friction heat to the surrounding air by radiation and convection. As the oil heats it becomes less viscous and the friction loss in the bearing decreases. The rate of radiation and convection increases as the oil gets warmer. When heat generation and heat dissipation become equal the oil temperature will remain quite uniform, being affected only by changes in the surrounding air temperature. In air-cooled bearings the oil level should be watched and new oil added, as required, to replace that lost by leakage and evaporation.

Direct water cooling is applied in three ways: (1) water-cooling coils in the oil within the housing, (2) water-cooled shells or shoes, (3) water-cooled shafts or collars. With all these applications water must be circulated fast enough to remove the heat of friction from the oil and keep it at a safe temperature. When water-cooling coils are used in the housing, the oil should have a pour test point considerably lower than the temperature of the coldest cooling water, to prevent congealing on the coils.

Self-contained Unit Circulating System. Each machine in which bearings are installed has an individual oiling system that circulates oil through all bearings. Within the system are means for air-cooling or water-cooling the oil, depending on the amount of friction heat to be dissipated. Siphoning of oil back through the inlet pipe should be prevented by placing its opening above the oil level or by providing an air vent hole.

Central Station System. With this system oil should be circulated through the housing of each bearing at the rate recommended for each case, to lubricate the bearing and carry away the friction heat developed. The piping in such a system often is so extensive that the oil will be sufficiently air cooled in making the circuit. When necessary, however, the system is provided with water-cooling coils in the central station.

6. BEARING MATERIALS

METALS (see also Section 4, p. 4-73). Hardened steel will work well on hardened steel at low speed under very heavy loads. It is fairly durable even if imperfectly lubricated. The same metals work well at high speeds when carrying moderate loads if well designed, finished, and lubricated. Hard-surface bearings are least subject to abrasion by operation under adverse conditions. They are much used for small parts, but are too costly for large ones.

Gray cast iron is useful for bearing faces, especially if chill-cast. For moderate loads, such iron is excellent for both bearing surfaces, even for high-speed bearings, if properly lubricated. Cast iron works well with softer bearing materials, as babbitt metals and bronzes when properly lubricated, and can be made to carry high pressures.

Soft steel will run well with babbitt metals and other white metal bearings, and will carry considerable pressure if properly lubricated. It also will run directly on cast iron. Soft steel on soft steel is not recommended, because of the tendency to seizure. Hard steel on soft steel sometimes is useful. A carbon content of 0.30 to 0.40 is recommended for shafts running on soft metal bearings. Soft copper-base alloys containing lead are useful for moderate speed and heavy pressure. There is a tendency, however, to line brass and bronze bearings with a thin coating of white bearing metal. This avoids the tendency of high-copper alloys, when overheated in bearings, to destroy the finish of the opposing cast-iron or steel surface by roughening, plating it, or producing surface cracks.

Bearing metal alloys are discussed fully on pp. 4-73 to 4-78.

NONMETALS. Fiber (pressed paper) washers often are used for thrust bearings for low speed and considerable load, or high speed with light load when oil lubricated. **Jewels** of ruby or other crystals are used with hardened polished steel pivots in timepieces and instruments, oil lubricated. **Ivory** bushings are found in the best old clocks, with hardened steel pivots. **Hardwood** bearings have been used, lubricated by tallow and soaps.

Lignum Vitae. In the best applications lignum vitae is used with grain perpendicular to the bearing face. It is used to line stern tubes of ships, with fresh or salt water lubricant. The bearing in a vertical-shaft hydraulic turbine often has a lining of lignum vitae blocks in longitudinal grooves in the bearing shell. Adjustment is required to compensate wear. In this service they are lubricated with fresh water.

Rubber is used to line bearings where only water is available as lubricant and where bearing pressure is light. It is useful in guide bearing bushings for vertical revolving shafts, and in stern tube bearings in ships.

Resin-bonded Bearings. L. M. Tichvinsky gives the following résumé of literature on this bearing material:

The rapid development of the phenolytic or resin-bonded bearings was directly connected with the steel mill industry. According to their structure and manufacture, the synthetic resin-bonded bearings can be classified into three groups: (1) manufactured from plain or graphitic molding powders, (2) manufactured from various felts of impregnated nonoriented fibers, (3) manufactured from laminated products.

The bearings of group 1 are manufactured in great quantities for low load application, and for guide bearings of high-speed machines. High brittleness limits their application.

Because of different bases, bearings of group 2 are tougher and more rigid than bearings of group 1. Their application is also for low or medium loads. Group 3 covers the most important type of bearings bonded with synthetic resin. They also are called laminated bearings because of their base consisting of woven textile fabrics. Occasionally, for less important performances, they are made from laminated paper products. In manufacturing these bearings high pressures and high temperatures are applied for a predetermined period of time during which the powdered synthetic resin liquefies. On cooling, a well-bonded product is obtained. These bearings are marketed under various trade names.

For more severe operations, as in steel mills, these bearings perform most efficiently with water alone for both lubricating and cooling. The average physical and mechanical properties of a good laminated bearing material are: specific gravity, 1.25 to 1.40; weight (approx.), 0.05 lb per cu in.; Brinell hardness, 30 to 42; tensile strength, pounds per square inch (along grain), 8000–15,000; (across grain), 6000–11,000; modulus of elasticity, 1,000,000; compressive strength, pounds per square inch (perpendicular to laminae) 40,000–45,000; (parallel to laminae) 20,000–24,000; water absorption, 7-day test, 1–3%; oil absorption, negligible.

OIL-LESS BEARINGS. Although bearings are offered that are lubricated by other media than oil or grease, it is assumed here that the term oil-less bearing applies only to bearings to which no lubricant is expected to be applied. These comprise a wide variety of bearing shell materials with which lubricating substances are combined by one means or another in the process of manufacture. Among them are: (1) fiber, plain or impregnated with lubricant; (2) wood, plain or impregnated with lubricant, or with cavities filled with hard lubricant; (3) special highly graphitic alloys; (4) metal bushings with insets of hard lubricant; (5) press-molded mechanical mixtures of bearing materials and lubricants. Some of these materials will carry light loads at moderate speeds without serious wear. They are best suited to intermittent service or for very low speeds. Manufacturers should be consulted for detailed information concerning them.

7. MISCELLANEOUS JOURNAL BEARINGS

"INTERCHANGEABLE WITH BALL AND ROLLER BEARINGS." Study of the oil film type of journal bearing has indicated the possibility of their successful application within dimensions identical with those now used for ball and roller bearings. In the Nomy bearing, separate pivoted shoes are mounted in a ring that revolves with the shaft. The shoes bear on the inside of a stationary outer race. Provision is made for automatic lubrication. The Fast bearing has shoes integrally formed on the inner ring and connected thereto by stems. The bearing is quite rugged.

In neither of these bearings is the revolving element integral with the shaft. It, therefore, must be carefully fitted and clamped against the shaft, usually against a shoulder, and keyed or otherwise held to prevent slippage at starting. Otherwise these bearings are mounted in the same way as ball or roller bearings. A comparison of a bearing of this type and the more usual construction in which a bearing shell about one diameter long supports the shaft journal directly shows the space occupied to be about the same in both cases for similar service. Manufacturers should be consulted for detailed information.

ROLL-NECK BEARINGS. Roll necks in mills have long presented a serious problem, especially in steel mills. Early practice, still widely followed, was crude and very wasteful of power. The bearings were undependable. Wear on necks and bearings was considerable. The bearing equipment acted as a bar to the construction of larger mills. To remove this handicap, antifriction bearings have been used with considerable success, but have exhibited certain limitations.

Since 1930, the oil film type of bearing has been successfully applied to roll necks. One of these, the Morgoil bearing (Ref. 19), comprises a ground and polished steel sleeve or journal fitted over the roll neck, the latter being tapered at an angle of about 12°. The journal thus can be removed easily from the neck, which also can be made of larger diameter than is ordinarily possible. The sleeve is keyed to the neck. The bearing comprises a casing, into which is inserted a steel or bronze liner, faced with centrifugally cast babbitt. Oil for lubrication is pumped into the bearing through holes in the casing, and flows through appropriate grooves to a drain pipe. Tests described in the reference show this bearing to be capable of withstanding loads in excess of the roll-neck strength, and to effect considerable savings in power.

In Kingsbury roll-neck bearings installed on a wide strip mill of the Inland Steel Company, load was carried directly on the neck of the roll. The bearing shell was water cooled. Oil was fed to the bearing by a gravity system to make up leakage at the seals. High-pressure oil was introduced to reduce starting friction, permitting operation at any speed, irrespective of the load on the bearings. The friction was about the same as for roller bearings for all conditions.

8. MAKING AND FINISHING OF METAL BEARING SURFACES

SOFT METAL LININGS. Anchoring. For holding a soft metal lining to harder bearing surfaces the ordinary dovetail circular anchor is much used, but the T-head anchor is better because shrinkage of its neck holds the metal faces together. Anchors should be small and sparingly used, because they remove the backing of the soft bearing face by shrinking away from it. Sometimes they are omitted entirely. Small T-slot grooves are useful, especially if machined. Simple anchors sometimes are formed by a drill point or by plain grooves.

Tinning. In order that heat may pass freely from the bearing lining to its backing, the latter often is tinned. The surface to be tinned should be machined or ground smooth and must be clean. The shell should be heated, flux applied, and the tinning metal then spread thoroughly by scrubbing. It is best to tin the surface with the lining material. Usually with the residual preheat the shell is then ready to receive its soft metal lining. Some metals tin more easily than others. Steel and ordinary bronze are more easily tinned than cast iron and the aluminum- and other special bronzes. Copper plating before tinning improves the bond.

Peening. The value of peening or burnishing in the preparation of soft metal bearing faces is an open question. The resulting texture may be less uniform, even though sometimes harder. Peening holds the lining tightly to dovetail anchors, but peening a face that has been applied to a tinned surface is apt to loosen the tinned bond. Some alloys are hardened more by peening or burnishing than are others.

FITTING UNDER LOAD is useful for large vertical thrust bearings and often is used to compensate deflections produced by loads. In this process, a few turns at very low speed are made to spot the softer surfaces by the harder. A fairly heavy coating of lubricant will prevent overheating during the process. The hard face then is lifted and the rubbed spots on the soft face scraped down lightly. High spots on the harder face may be reduced by fine stoning. Usually all fitting is done on the softer face. The process is repeated until a fit of the desired extent and accuracy is obtained. The oil used with this fitting should be only heavy enough to get the desired marking on the bearing surface without injury or waste of time.

RUNNING-IN of bearings often is used as a system of inspection or of preliminary service. It sometimes is rather brutal, the journal being used to ream or burnish the bearing. It should not be taken as a substitute for good design and workmanship. See Fitting under Load, above. Running-in, if carefully done, will improve the bearing surface smoothness. It may be used in lieu of fitting under load with small machines. It is best, however, to premachine the bearing elements as nearly correct as possible, and to use designs and clearances suitable for immediate service.

FORMING. The hardest surfaces can be finished by grinding, and softer materials by machining, grinding, or scraping. The smoother and truer the surface the better will the bearing operate with proper lubrication. Lapping, after grinding, or over a smooth tool finish, is one of the best ways to produce accurately cylindrical, flat, or spherical surfaces on steel and cast iron. (See Section 23, Article 30.) Final finishing may be by rotation under a lightly pressing block of wood faced with leather, using fine abrasive and oil. Scraping, stoning, and filing are all useful for finishing and for fitting one surface to another. Smooth file finishing on a revolving shaft is common, but it will not true a surface initially out of round. Careful final finishing or fitting of the bearing surface is essential. A very smooth and accurate tool or grinder finish may not need further treatment. In important work, fitting of the softer bearing surface to the other or to a master surface by hand scraping is common. Thus, a bearing may be fitted to a mandrel of a diameter equal to that of the journal plus the desired running clearance.

EFFECT OF FINISH ON STARTING FRICTION. See Section 7, Article 13.

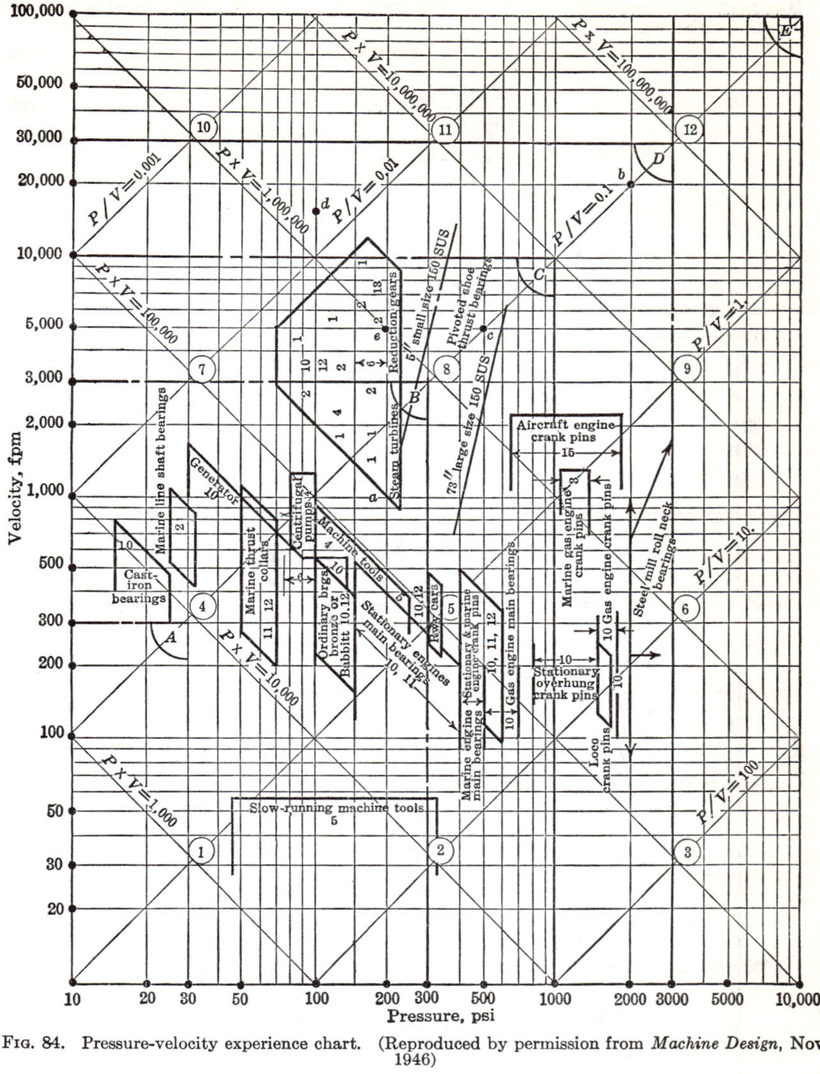

FIG. 84. Pressure-velocity experience chart. (Reproduced by permission from *Machine Design*, Nov. 1946)

9. OPTIMUM LOADING AND CURRENT PRACTICE

The current practice charted in Fig. 84 as to pressure and velocity is supplemented by running clearance data of Figs. 85 and 86. As little is said about the viscosity of the lubricant at operating temperature the designer is naturally in a quandary. Since it is the purpose of this review to help the designer, a table of least film thicknesses is proposed for properly designed journal bearings (Table 14).

It is assumed that the bearing film will extend over an angle of 120° and be of the Class C type (see p. 12-16) with running clearance. If operation is to be that for greatest capacity the characteristics will be those of Table 7 and Fig. 39 (p. 12-28) for $\beta = 120°$. If these relationships are maintained they will probably not agree with each of Figs. 84, 85, and 86.

Maximum load operation requires an eccentricity of $c = 0.469$. The following formulas are based on this condition:

$$\frac{Pa}{V\mu}\left(\frac{\eta}{a}\right)^2 = 0.228L_1 \tag{64}$$

$$\frac{H}{PV}\frac{a}{\eta}\frac{10^5}{a} = 9.36\frac{F_1}{L_1} \tag{65}$$

$$\frac{\lambda}{\frac{\eta}{a}} = 1.542\frac{F_1}{L_1} \tag{66}$$

$$\frac{\lambda}{\sqrt{\frac{\mu N}{P}}} = 0.533\frac{F_1}{\sqrt{L_1}} \tag{67}$$

For meanings of symbols see Article 2, p. 12-14, except that P is used instead of p_0.

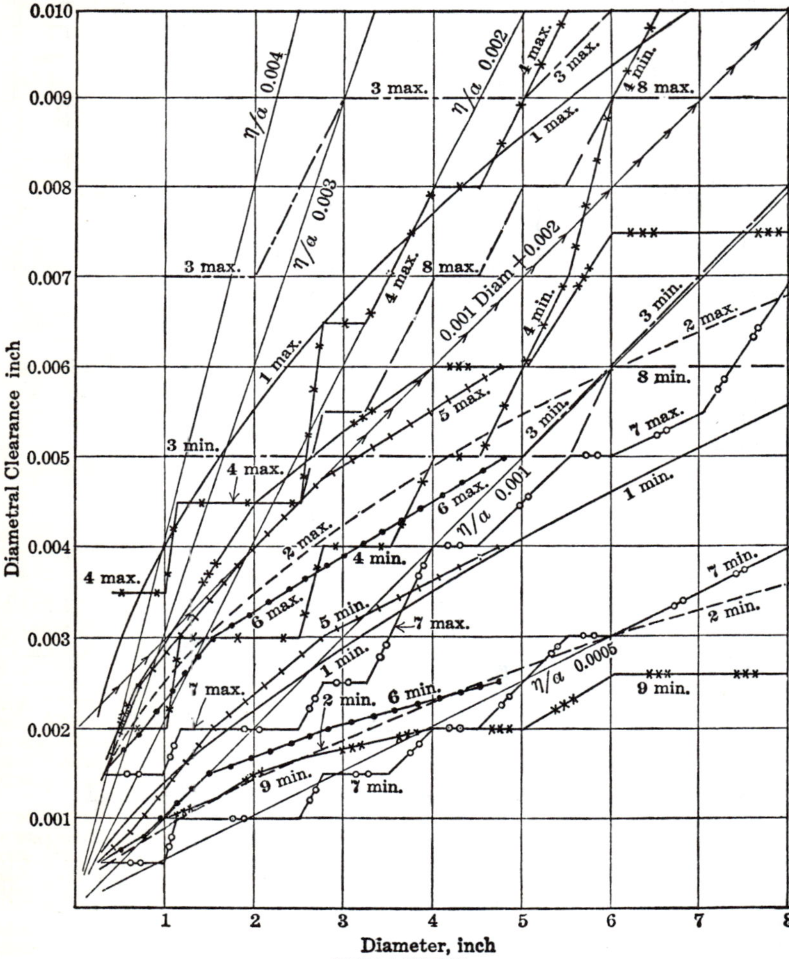

KEY TO CURVES

1. A.S.A., Free Fit, over 600 r.p.m. ———
2. A.S.A., Medium Fit, under 600 r.p.m. - - - -
3. Marine Bearings —·—·—
4. Horizontal Bearings —✱—✱—
5. Brown & Sharpe, over 600 r.p.m. —+++—
6. Brown & Sharpe, under 600 r.p.m. —•—•—
7. Step Bearings —○○—
8. Vertical Bearings — —
9. British Standards —✱✱✱—

FIG. 85. Accepted practice in bearing clearance, shaft diameter, 0 to 8 in. See Fig. 86 for shafts 6 to 24 in. diameter.

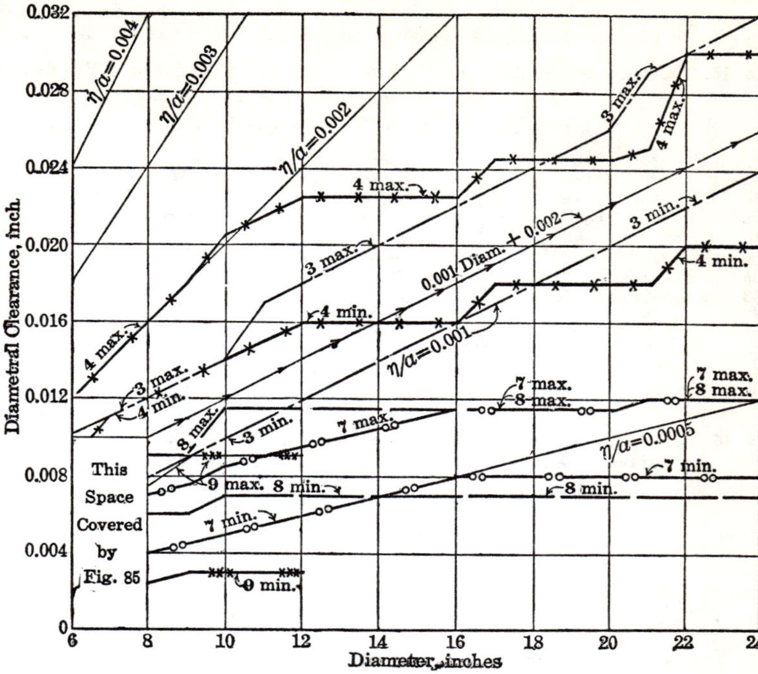

FIG. 86. Accepted practice in bearing clearance, shaft diameter, 6 to 24 in. See Fig. 85 for shaft diameters 0 to 8 in., and key to curves.

Table 14. Proposed Minimum Film Thicknesses for 120° Centrally Loaded Clearance Bearings under Optimum Capacity Conditions

(Reprinted by permission from *Machine Design*, Oct. 1946.)

Journal Diam., d, in.	Proj. Area, A, sq in.	When Surface Speed Is								
		5 fps 300 fpm	10 fps 600 fpm	20 fps 1200 fpm	40 fps 2400 fpm	60 fps 3600 fpm	80 fps 4800 fpm	100 fps 6000 fpm	120 fps 7200 fpm	140 fps 8400 fpm
		Minimum Film Thickness (h_0 inches) Should Be								
1 1/2	2 1/4	.000461	.000597	.000776	.001005	.001171	.001307	.001413	.001520	.001613
3	9	.000597	.000776	.001005	.001307	.001520	.001680	.001840	.001960	.002093
4 1/2	20 1/4	.000696	.000903	.001171	.001520	.001773	.001960	.002147	.002293	.002427
6	36	.000776	.001005	.001307	.001680	.001960	.002187	.002387	.002560	.002707
7 1/2	56 1/4	.000843	.001093	.001413	.001840	.002147	.002387	.002600	.002773	.002933
9	81	.000904	.001171	.001520	.001960	.002293	.002560	.002773	.002973	.003160
10 1/2	110 1/4	.000957	.001240	.001613	.002093	.002427	.002720	.002967	.003160	.003333
12	144	.001005	.001307	.001680	.002187	.002560	.002840	.003093	.003307	.003507
13 1/2	182 1/4	.001052	.001373	.001773	.002293	.002667	.002973	.003227	.003453	.003667
15	225	.001093	.001413	.001840	.002387	.002773	.003093	.003360	.003600	.003813
16 1/2	272 1/4	.001133	.001467	.001907	.002467	.002880	.003200	.003480	.003733	.003947
18	324	.001171	.001520	.001960	.002560	.002973	.003307	.003600	.003853	.004080
19 1/2	380 1/4	.001207	.001560	.002027	.002640	.003067	.003413	.003707	.003973	.004213
21	441	.001240	.001613	.002093	.002720	.003160	.003507	.003813	.004080	.004333

The side leakage factors L_1 and F_1 in Table 15 were taken from Fig. 19, p. 12-10, which was prepared for Class C, 120° clearance bearings (Ref. 12). The values given for length-width ratio of 1.00 were used by the author in preparing Table 16 and Fig. 87.

Table 15. Performance Factors for Various Length-width Ratios of 120° Centrally Loaded Clearance Bearings

(Reprinted by permission from *Machine Design*, Nov. 1946.)

Length-width Ratio, l/b	Leakage Correction (Fig. 19) Friction Factor, F_1	Load Factor, L_1	Eq. 64 $\frac{Pa}{V\mu}\left(\frac{\eta}{a}\right)^2$	Eq. 65 $\frac{H}{PV}\frac{a}{\eta}\frac{10^5}{a}$	Eq. 66 $\frac{\lambda}{\eta/a}$	Eq. 67 $\frac{\lambda}{\sqrt{\mu N/P}}$	Bearing Width, b	Bearing Area, $A = bd$
2.000	0.820	0.223	0.0508	34.45	5.67	0.926	$1/2\ d$	$1/2\ d^2$
1.600	0.860	0.303	0.0692	26.6	4.38	0.833	$5/8\ d$	$5/8\ d^2$
1.333	0.886	0.372	0.0848	22.3	3.68	0.774	$3/4\ d$	$3/4\ d^2$
1.143	0.905	0.432	0.0985	19.6	3.235	0.735	$7/8\ d$	$7/8\ d^2$
1.000	0.920	0.481	0.1096	17.9	2.95	0.7075	d	d^2
0.800	0.938	0.558	0.1272	15.7	2.59	0.669	$1\ 1/4\ d$	$1\ 1/4\ d^2$
0.667	0.948	0.613	0.1398	14.5	2.38	0.645	$1\ 1/2\ d$	$1\ 1/2\ d^2$
0.500	0.963	0.689	0.1571	13.1	2.16	0.618	$2d$	$2d^2$

Table 16. Clearance, Friction, and Load-carrying Factors for 120° Centrally Loaded Clearance Bearings under Optimum Capacity Conditions ($b = 2a$)

(Reprinted by permission from *Machine Design*, Oct. 1946.)

η/a	λ	$\sqrt{\mu N/P}$	$\frac{\mu N}{P} \times 10^6$	ZN/P
.0001	.000297	.000421	0.177	1.224
.0002	.000594	.000842	0.709	4.890
.0003	.000891	.001263	1.595	11.00
.0004	.001188	.001684	2.830	19.54
.0005	.001485	.002105	4.430	30.60
.0006	.001782	.002526	6.375	44.00
.0007	.002079	.002947	8.685	59.90
.0008	.002376	.003368	11.35	78.30
.0009	.002673	.003789	14.38	99.20
.0010	.002970	.004210	17.72	122.40
.0012	.003564	.005052	25.50	176.0
.0014	.004158	.005894	34.75	240.0
.0016	.004752	.006736	45.40	313.0
.0018	.005346	.007578	57.50	397.0
.0020	.005940	.008420	70.90	489.0

When Table 15 is used for length-width ratios other than 1.00 the least film thickness should be chosen not according to journal diameter but from the projected area column of Table 14. Hence when $l/b = 0.5$, the axial width of the bearing will be about twice the journal diameter and its projected area twice that in Table 14 for that diameter. For example, if the journal diameter were 9 in. its width would be 18 in. with projected area of 162 sq in. The least film thickness for 1200 fpm would therefore be interpolated between 0.00168 and 0.001773, which would yield $h_0 = 0.00173$. Dividing this by 0.531 would give the radial clearance of $\eta = 0.00326$ in. Following this through, the operating viscosity of the lubricant would depend upon the load to be carried.

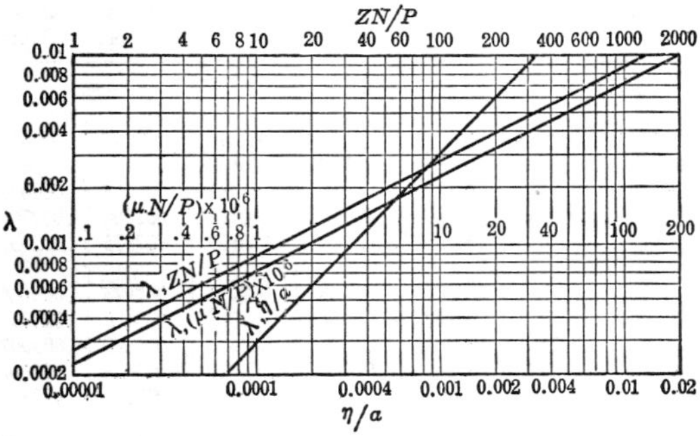

FIG. 87. Clearance, friction, and load-capacity factors for 120° centrally loaded clearance bearing with a width-diameter ratio of $b/d = 1$. (Reproduced by permission from *Machine Design*, Oct. 1946)

The current practice chart for loading (Fig. 84) is divided into zones representing degrees of conservatism, A to E, the letters appearing in a quadrant at the upper right corner of the zone. Zone A is obviously ultraconservative whereas zone D has few data points in it in the high-speed area but has many in the high-pressure area. Zone E may be said to be beyond the pale at present, judging by the dearth of plotted data.

For operating conditions other than optimum, reference must be made to the charts of general conditions. For centrally loaded clearance bearings, see Article 2, p. 12-14.

For a more comprehensive description of the tables and figures of this chapter, see Ref. 20.

BEARING CLEARANCES. A study of bearing clearances indicates wide divergences in practice. Figure 85 gives clearance data for marine bearings, horizontal bearings, vertical bearings, step bearings, Brown & Sharpe practice for over and under 600 rpm, respectively, the American Standards Association standards and British standards, up to a total clearance of 0.010 in. and diameters up to 8 in. Figure 86 extends this range to larger clearances and diameters. The line marked 0.001 diam. + 0.002 shows a rule used by Kingsbury for standard duty journal bearings. In order that the other data may be interpreted with reference to the clearance ratio η/a, reference values of it are given from 0.0005 to 0.004. If clearances are chosen from these curves, they must, for best results, be supplemented by a suitable oil.

10. SHAFT DEFLECTIONS AND BEARING PERFORMANCE

Shaft deflections within journal bearings, which may be considerable when compared with the thickness of the oil film, should be investigated, particularly with long spans between bearings and with relatively slender shafts. For the purpose of determining shaft deflections and the general direction of journal center lines with respect to the bearings, loads may be assumed concentrated at the midpoint of the bearing. To study oil film shape in a self-aligning bearing, or where shaft deflection is so little out of line as not to require self-aligning bearings, bearing reaction may be assumed uniformly distributed.

As a first approximation it may be assumed that all the deflection occurs in the shaft, the supporting framework of the machine being relatively stiff. The bearings, if rigid, therefore remain aligned while the shaft is subject to deflection. Misalignment of journal and bearing is the product of the slope of the deflection curve and the length of the bearing. The general method of calculating deflections and slopes of shafts subject to bending is explained in Section 8, Article 2, p. 8-06. If the deflection is excessive relative to the oil film thickness, it is necessary to provide a self-aligning bearing. Even here the shaft may become slightly curved within the bearing due to the distributed load, and should be checked.

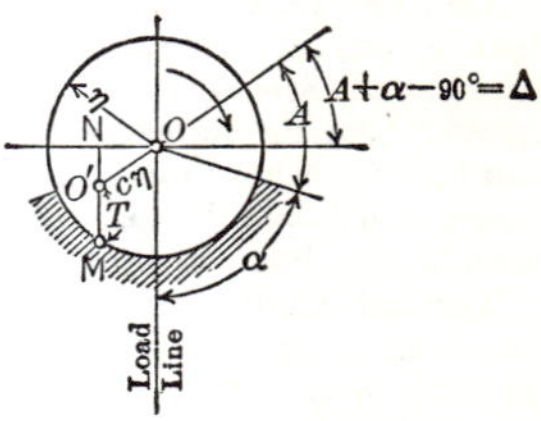

FIG. 88. Shaft deflection and shift of shaft center.

In calculating shaft deflection in a bearing, the position of the center of the shaft, when running, is important. It shifts to the point O' (Fig. 88) within the clearance circle of radius η. The plane of shaft deflection cuts this circle at M, which is the point at which the deflected shaft will make contact with the bearing, assuming that the latter does not deflect. The following relations (Fig. 88) hold: Let $(A + \alpha - 90°) = \Delta$. Then $\overline{ON} = c\eta \cos \Delta$; $MN = \sqrt{\eta^2 - \overline{ON}^2}$; $TO' = MN - c\eta \sin \Delta$. The deflection should be compared with distance $O'M$, and should not be an unreasonable percentage, say 30%, of $O'M$. The shaft usually should be examined for both strength and stiffness.

11. CONSTRUCTION AND STANDARDIZATION OF THRUST BEARINGS

THE THRUST PROBLEM. More or less axial load is encountered with every vertical shaft. With horizontal shafts, when there is no thrust load, only shaft shoulders are needed to hold the machine parts in proper axial relation.

Heavy horizontal loads are encountered in unbalanced centrifugal pumps and steam turbines, in marine propeller shafts, in industrial machines, and in some machine tools. When thrust loads are encountered at very low speed, the wear and the power lost in friction often are ignored and very inefficient bearings used, although it is better practice to use pivoted shoe thrust bearings and heavy oil, or to use rolling contact bearings. With relatively high speeds and large shafts, as for marine propellers, the power loss is important, and efficient durable thrust bearings are used. For very high speeds with heavy thrust loads, as in steam turbines, the thrust bearing must also be efficient and durable.

Light Horizontal Loads. Horizontal shafts, with little or no thrust load, often require close control of endwise position. This may be done by a collar on the shaft end as in Fig. 89. One collar in one bearing mounting then controls end play. Using the shaft shoulder, as in Fig. 81, requires two bearing mountings. See Thrust Bearing Design and Construction below.

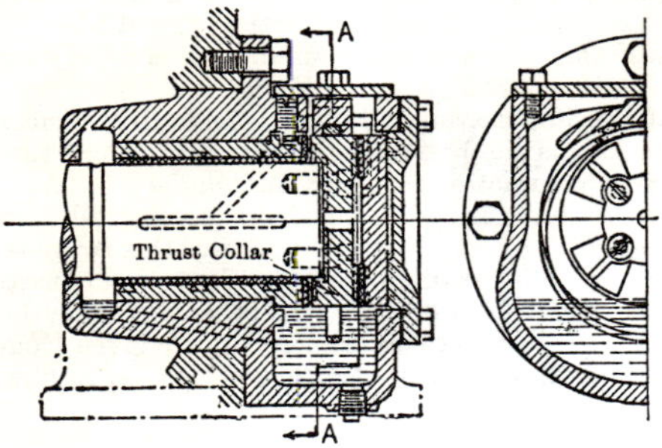

Fig. 89. Ring-oiled journal and thrust bearing.

Vertical Loads. A vertical shaft in a machine always has a thrust load, because of the weight of the machine elements that in a horizontal shaft would be carried by journal bearings. However, a vertical shaft has no side loads on its journals except those produced by the operation of the machine itself. The thrust load on a vertical shaft may be either increased or reduced by loads produced by operation of the machine. Thus, in a hydroelectric generating unit or a vertical centrifugal pump, hydraulic thrust in the runner may add to the load resulting from the weight of the revolving parts. The importance of the thrust problem depends on the combination of load and speed.

THRUST BEARING DESIGN. Thrust bearings may be divided into two main groups, viz., imperfectly lubricated and perfectly lubricated. The degrees of imperfection or perfection will vary. Among bearings that are perfectly lubricated, a bearing of one design may run with several times more friction than another, under otherwise identical operating conditions, and with no metallic contact of bearing faces during operation. Similarly, with imperfect lubrication, some bearings will run with higher friction than others, with the same oil, load, and speed conditions, but with more or less metallic contact of the bearing faces always present.

IMPERFECTLY LUBRICATED THRUST BEARINGS. Imperfect lubrication implies lack of provision for the functioning of adequate wedge-shaped films rather than lack of lubricant. See Perfect Lubrication, p. 12-03. For very low speed, heavy oils and hardness of bearing materials have been relied on to keep the rate of wear within reasonable limits. Such bearings should be placed in oil baths, if possible, so that lubricant may be continuously available. Grooving may be designed to keep oil from wasting or to produce circulation that will continually change the oil on the bearing faces.

Vertical Shafts. For this service the plain pivot, the multiple-disk-supported pivot, the plain ring thrust, the washer-supported ring thrust (Figs. 90 to 93) are used. The grooves in these bearings, if submerged in oil, should be carefully formed, usually in the upward faces, and with rounded edges, as in Fig. 94, in the harder surface. If oil is present

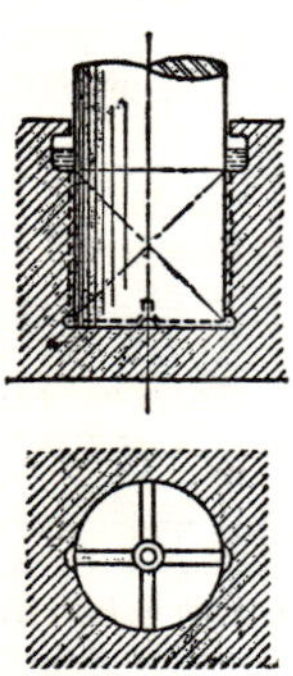

Fig. 90. Plain pivot thrust bearing.

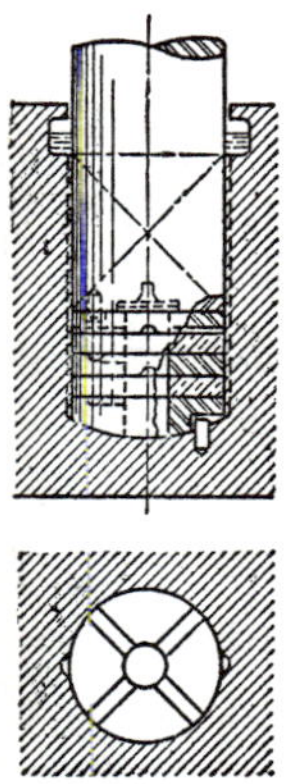

Fig. 91. Multiple-disk-supported pivot thrust bearing.

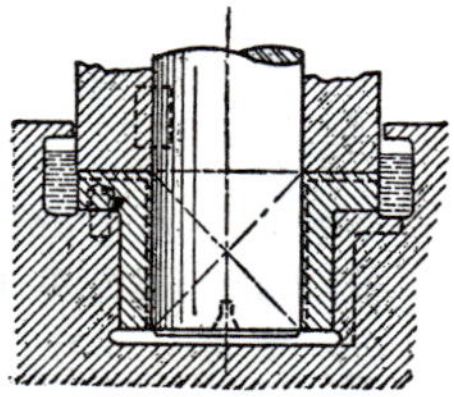

Fig. 92. Plain ring thrust bearing.

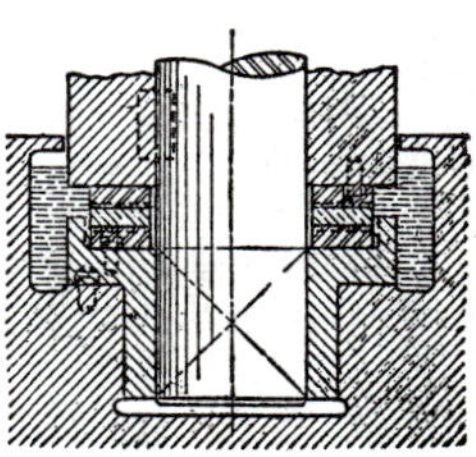

Fig. 93. Washer-supported ring thrust bearing.

Fig. 94. Form of oil groove.

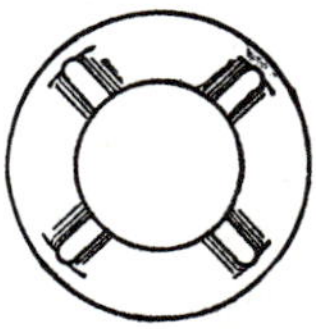

Fig. 95. Arrangement of groove in thrust washer.

in only minute quantities, grooves should be in the downward face. Further improvement can be made by slightly tapering the grooved bearing face for a distance at least equal to the width of the oil groove, on each side of the groove, but using not more than half the total surface for grooves and tapers. Two to four oil grooves usually are enough. For large diameters, or areas where radial width is relatively small compared with the circumference, more grooves may be used. The size of the groove will vary, increasing with size of the bearing, sluggishness of the lubricant, and speed of rotation. If an oil bath cannot be used, it is best to stop the grooves short of the circumference of the disk or washer and feed oil to the grooves from the center outward. (See Fig. 95.)

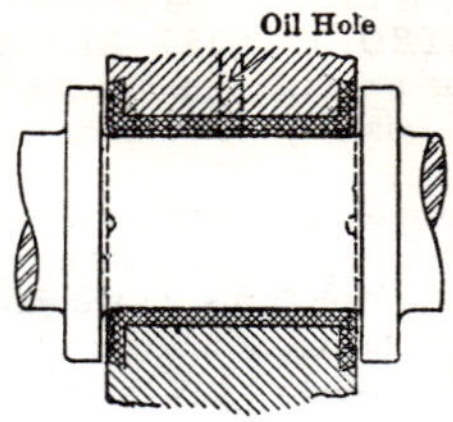

FIG. 96. Bearing to take light thrust in two directions.

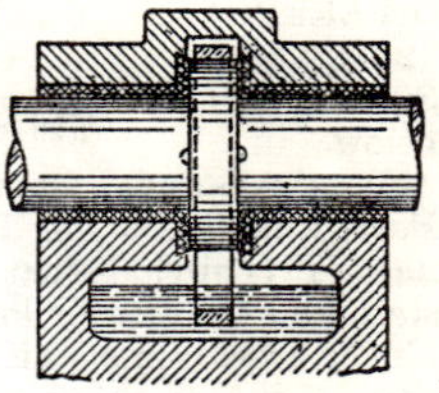

FIG. 97. Ring-oiled two-direction bearing for light thrust load.

Horizontal Shafts. Conditions usually are quite different from those of vertical shafts. An oil bath is less easily provided. Also, because of end play, it is difficult to prevent oil leaking away from the thrust surface, especially when thrust action is intermittent. For a simple construction like Fig. 96, thrust face grooves may be most easily cut in the stationary bearing face, although the harder face is preferable. Grooves should stop short of the outer diameter of the thrust collar. Sometimes a thrust collar, integral with or fastened to the shaft, is placed between two short journal bearings whose adjacent sides are enlarged to present reasonable thrust surfaces to the collar. If of sufficient diameter, this collar may be used to lubricate all bearing surfaces from a well below the shaft. Otherwise a ring may be used as in Fig. 97.

For data on friction coefficients of imperfectly lubricated thrust bearings, see p. 7-28. In such bearings the coefficient will, if bearing is adequately oiled, vary from 0.20 downward to about 0.02. This lower value would be considered very high for a perfectly lubricated bearing surface.

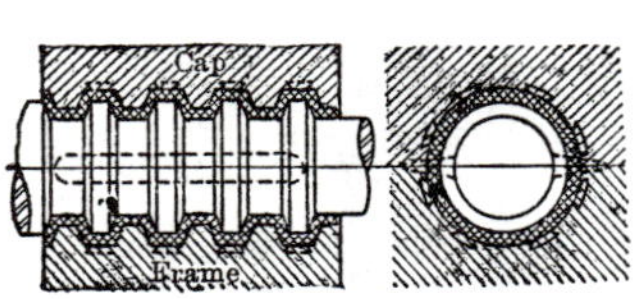

FIG. 98. Multiple-collar bearing sometimes used in woodworking machinery to position spindle.

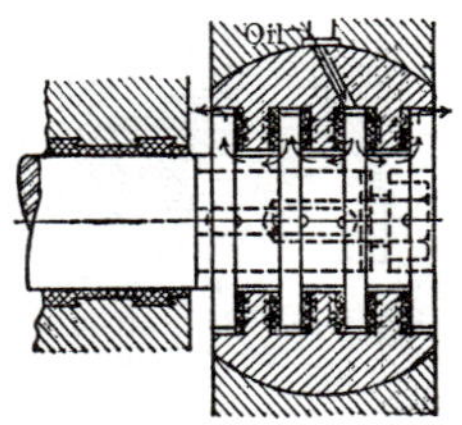

FIG. 99. Multi-collar thrust bearing for light loads.

Multi-collar thrust bearings are used in several forms, more or less well made. In woodworking machinery, a bearing-metal lining is poured into the frame and cap, around a grooved shaft (Fig. 98). Oil is best applied at the sides. Such a bearing if lightly loaded, and not too tightly clamped on the shaft, may serve as a combined journal and thrust bearing, but it is not considered good practice. Sometimes the cap of such a bearing is made axially adjustable for control of end play. Figure 99 shows a multi-collar bearing used only to carry light thrust loads. It is adjacent to an adequate journal bearing. Oil under pressure may be applied as marked, and caused to flow through suitable grooves back and forth across the faces of the collars. If carefully made, and when oil grooves are properly designed with adjacent surfaces properly beveled, these bearings can carry moderate loads even at high speed.

One serious difficulty with all multi-collar thrust bearings is the inequality of distribution of load among the collars. To help overcome this in large bearings for marine service, inverted U-shaped elements, called horseshoes, with soft metal faces, are placed between the shaft collars. Sometimes grease is used for lubrication, but usually oil is supplied from a bath below the shaft into which the collars dip. The shoes sometimes are water-cooled by passages, but for high-speed bearings a water coil usually is placed in the bath below the collars. Horseshoes in marine practice are loaded at 30 to 70 psi. They now are seldom used, because more efficient and more durable types of bearings are available that operate with about one-tenth the power loss. (See pp. 12-56 to 12-60.)

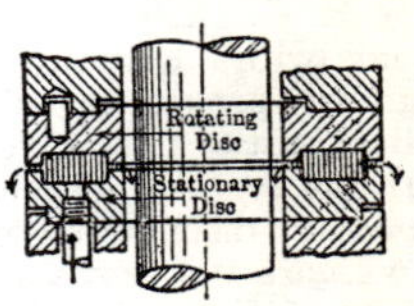

FIG. 100. Oil pressure thrust bearing.

THE OIL PRESSURE THRUST BEARING employs two disks constructed as in Fig. 100. Oil is pumped under high pressure into the stationary disk at a definite rate, lifting the upper or rotating disk until oil discharge rate equals supply rate. Accumulators are required to protect against stoppage of the pumps. Rollers sometimes are provided to prevent contact of narrow faces of disks when oil supply fails. If the disks are permitted to rub they will "flash" and be damaged. Friction in the bearing itself is very low, but considerable power is required to drive the pumps.

PERFECTLY LUBRICATED THRUST BEARINGS. See Perfect Lubrication, p. 12-03, for a discussion of the principles underlying the formation of the tapered oil film. Such films, in various forms, exist between the bearing faces of the thrust bearings described below.

The fixed-wedge film type of thrust bearing was suggested in 1888 by Professor Archibald Barr, when discussing Beauchamp Tower's then recent tests on plain collar thrust bearings. It is obvious that part of the thrust bearing face must be flat to permit the collar to start and stop under load without undue concentration of pressure. The wedge-shaped film in a bearing of this kind is unable to lift its load until a fair speed is attained. Conversely, it is likely to stop suddenly when slowing down. The fixed wedge may be combined with a flat surface, by using alternate bevels and flats with a supply groove *G* at the thick end of the bevel, in the direction of motion (Fig. 101). This construction is

FIG. 101. Fixed wedge-film Gibbs thrust bearing.

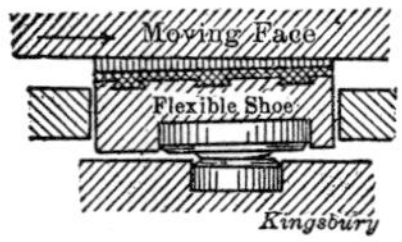

FIG. 102. Free wedge-film thrust bearing.

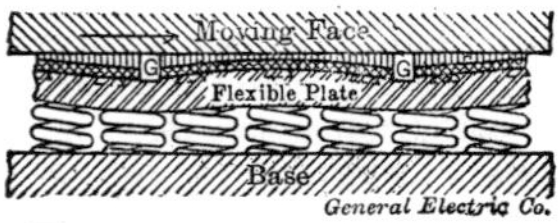

FIG. 103. Reist thrust bearing.

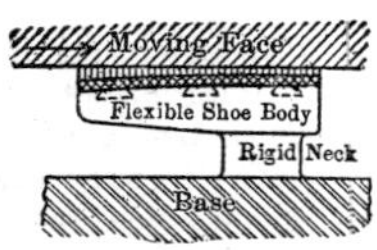

FIG. 104. Thrust bearing with flexible shoe on rigid neck.

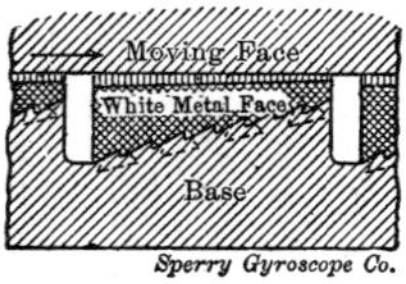

FIG. 105. Restrained wedge-film bearing.

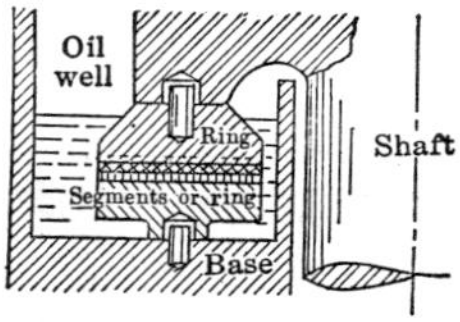

FIG. 106. Radial section of thrust bearing indicating radial tilting.

used in the Gibbs thrust bearing, built by S. Morgan Smith Company, York, Pa. Bearing pressures range from 200 to 300 psi.

A characteristic of these bearings is that the slope of the wedge remains constant for all conditions of load, speed, oil viscosity, and temperature for a given bearing. The film thickness, however, will vary automatically as these conditions change, provided the speed is great enough to separate the flat faces.

The free-wedge film (Fig. 102) type of thrust bearing (see Pivoted Shoe Bearings, p. 12-57) was conceived, constructed, and tested in 1896 by Professor Albert Kingsbury. A parallel, but independent, development of the similar Michell thrust bearing has taken place in England. The bearing members (shoes) in this type are tiltably mounted, and are free to form automatically the most efficient natural oil film required by each different combination of load, speed, viscosity, and temperature to which the bearing is subjected.

The restrained-wedge film type of thrust bearing appears in several forms. In this type the elements forming the films are not fully free, but do respond, in a limited degree, to change of load, speed, viscosity, and temperature. One of these (Fig. 103) was invented by H. G. Reist about 1915. Its stationary element is a babbitted, relatively thin, steel ring, with one radial gap, mounted on many small stiff precompressed springs. Radial grooves are cut in both the opposing bearing faces to supply cool oil to the bearing faces. Both stationary and rotating grooves are chamfered to direct the oil between the surfaces. In the event of misalignment, the flexible support automatically deflects to equalize the pressure distribution over the entire bearing.

Figure 104 shows a second form of restrained-wedge film bearing, inventor unknown. Restraint of wedge film formation is imposed by the rigid neck and the stiff portions of the adjacent shoe body. This is an integral structure. The bearing faces of the shoe bodies are finished flat. The flexing of the overhung portion takes place under the pressures incident to the formation of the film.

A third and rather unusual restrained wedge film, invented by Elmer D. Sperry, is shown in Fig. 105. The supporting base is faced with bearing metal of thickness that varies considerably. Babbitt metal has a coefficient of elasticity much greater than iron or steel. Hence if the thickness is greater where the pressures are lower, a wedge-shaped film may be formed.

Radial tiltability may be given to relatively thick bearing rings by applying the load to their backs through annular supporting ribs. (See Fig. 106.) Also the lower ring may be made in separate segments that have freedom only for radial tilting, in which case the ribs supporting the segments may be tangential. Such elements may have fixed bevel bearing faces or flexible leading portions, as already described.

Film Forms. When the relative motion is circular, instead of straight line, the flat-faced wedge film is not geometrically a flat wedge in the developed section of the path of relative motion but is a sine curve wedge, as in journal bearings. There also are changes in film shape due to elasticity of materials, and to temperature variations. Thrust bearing films, therefore, usually are of a composite nature. Occasionally a composite form is purposely given the film, as when beveling is added to the leading or trailing portions of the bearing face of the freely tilting shoe. A variable composite film appears in the Reist bearing (Fig. 103), in which both bearing faces are provided with oil grooves. The double grooving tends alternately to build up and to break down the pressure in the films, but it helps the radial flow of oil across the bearing faces.

CONSTRUCTION DETAILS. Reversible Bearings. In general a thrust bearing may be made suitable for both directions of rotation, i.e., reversible, by making its supporting feature symmetrical with respect to the bearing face segments or grooves. For example, the free-wedge type may have its shoes supported in the middle, and it loses little efficiency thereby. The fixed-wedge type requires the provision of an additional beveled portion, tapered in the opposite direction. When the limit of dead load supportable by the flat parallel portion of the bearing face is reached, the reversible fixed-wedge bearing area would have to be larger than if made for one direction only.

ALIGNING AND EQUALIZING MEANS FOR THRUST BEARINGS. When the machine is so accurately made that alignment of shafting is perfect within the limits of the film thicknesses in the bearings and the elasticity of the component parts, no aligning means need be provided. This also is true when the bearings are so placed with respect to the machine frame that the slight elasticity of the frame itself enables the bearings to preserve their proper alignment. However, it often is desirable to provide some aligning means within the thrust bearing, and the methods cited below are employed.

The design of aligning means should be governed by the misalignment to be corrected, as initial, subsequent, or continuous or rapidly recurring. The durability of the aligning means, therefore, is an important consideration in the selection of a thrust bearing. Special forms have been devised for such service as gyro-stabilizers and for thrust collars that are out of square with the shaft.

The spherical washer (Fig. 109) is the commonest form and has several advantages, one of which is ruggedness. When used with three freely tilting shoes it tends to equalize the load distribution among them. For best efficiency, the mean angle of the spherical faces should be about 45°. Even so, when once well loaded, an eccentricity of the load equal to more than half the projected radius to the mean zone of the spherical fit is required before slipping will take place. Hence, this method of equalizing or aligning is not sensitively responsive to variations in alignment under load.

Adjustable Bearings. When freely tilting shoes are mounted on jack screws, the latter may be used for initial setting of height, alignment, and equalization of load, but are not useful for preserving alignment when the supporting structure shifts.

Elastic support is an approximate aligning means, and also an approximate equalizing means for various portions of a bearing when little correction need be made. With the stiff springs necessarily employed, careful preliminary setting of the thrust deck is essential. Means to prevent or control vibration must be employed.

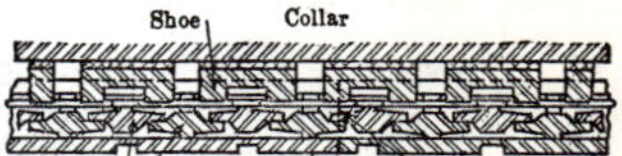

FIG. 107. Equalizing supports of Kingsbury thrust bearing.

Equalizing Supports. Systems of rocking levers were introduced by Kingsbury for initial alignment and equalization. They continue effective, even in the presence of considerable subsequent displacement of the thrust deck. (See Fig. 107.) This form has been adapted, with roller contacts, to very large hydroelectric bearings.

PIVOTED SHOE THRUST BEARINGS. Kingsbury began to standardize pivoted shoe thrust bearings when he applied the adjustable type to vertical shafted hydroelectric units 1912 to 1915. This type ran to large sizes (e.g., Fig. 108) and did not lend itself readily to small diameters. The application of his self-aligning and equalizing thrust bearings to the geared turbine drives of ships led to the Kingsbury standards listed in Table 17

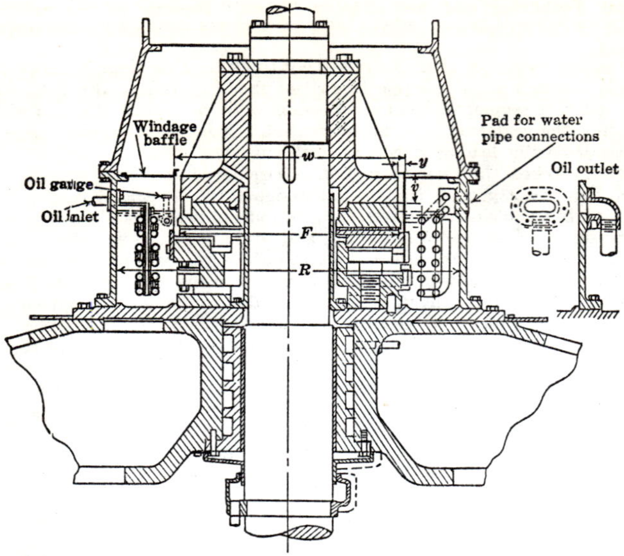

FIG. 108. Section of mounting for Kingsbury adjustable thrust bearing showing water coil for moderate speed, and inlet and overflow pipes for oil circulation. (Courtesy of Kingsbury Machine Works)

Table 17. Dimensions of Standard Three-shoe and Six-shoe Thrust Bearings

See Figs. 109–112

(Kingsbury Machine Works, Inc., Philadelphia)

Nominal Size, *A*	Net Area, sq in.		*B* (Bore)	Dimensions, in.								
	6 shoes	3 shoes		*C*	*D*	*G*	*H*	*Q*	*U*	*V*	*W*	*X*
5	12.5	6.3	1.750	5.375	2 5/8	5 1/8	1 3/4	2 3/4	6.000	7/8	2 3/8	7/8
6	18.0	9.0	2.125	6.375	3 1/16	6 1/8	2 1/16	3 1/4	7.000	1	2 7/8	1
7	24.5	12.3	2.500	7.375	3 5/8	7 1/8	2 3/8	3 3/4	8.000	1 1/4	3 3/8	1 1/4
8	32.0	16.0	3.000	8.375	4 1/16	8 1/8	2 11/16	4 5/16	9.125	1 3/8	3 7/8	1 3/8
9	40.5	20.3	3.500	9.375	4 1/2	9 1/8	3	4 7/8	10.125	1 1/2	4 3/8	1 1/2
10 1/2	55.1	27.6	4.125	11.000	5 1/8	10 11/16	3 3/8	5 11/16	11.750	1 3/4	5 1/16	1 3/4
12	72.0	36.0	4.750	12.500	5 3/4	12 3/16	3 3/4	6 1/2	13.375	2	5 13/16	2
13 1/2	91.1	45.6	5.375	14.000	6 1/2	13 11/16	4 1/4	7 5/16	14.875	2 1/4	6 9/16	2 1/4
15	112.5	56.3	6.000	15.500	7 1/8	15 3/16	4 5/8	8 1/8	16.500	2 1/2	7 5/16	2 1/2
17	144.5	72.3	6.625	17.625	8	17 1/4	5 1/4	9 3/16	18.625	2 3/4	8 1/4	2 7/8
19	180.0	......	7.500	20.250	7 3/4	19 1/4	4 3/4	10 5/8	....	3	9 1/4	3 1/4
21	220.0	......	8.500	22.250	8 1/4	21 1/4	5 1/4	11 3/4	....	3	10 1/4	3 5/8
23	264.0	......	9.375	24.500	8 11/16	23 1/4	5 11/16	12 3/4	....	3	11 1/4	3 7/8
25	312.0	......	10.000	26.500	9 9/16	25 1/4	6 3/16	14	....	3 3/8	12 1/4	4 1/4
27	364.0	......	11.000	28.750	10 1/16	27 1/4	6 11/16	15	....	3 3/8	13 1/4	4 5/8
29	420.0	......	11.750	30.750	10 1/2	29 3/8	7 1/8	16 1/4	....	3 3/8	14 1/8	5
31	480.0	....	12.500	33.000	11	31 3/8	7 5/8	17 1/4	....	3 3/8	15 1/8	5 1/4
33	545.0	......	13.375	35.000	11 7/8	33 3/8	8 1/8	18 1/2	....	3 3/4	16 1/8	5 5/8
37	684.0	......	15.000	40.000	12 3/4	37 1/2	9	20 3/4	....	3 3/4	18	6 3/8
41	840.0	......	16.750	43.750	14	41 1/2	10	23	....	4	20	7
45	1012.0	....	18.500	48.000	15 1/8	45 1/2	11	25	....	4 1/8	22	7 5/8

For capacity see Table 10, p. 12–34.

Horizontal Equalizing and Self-aligning Types. Because of the relatively recent development of the Kingsbury pivoted shoe oil film bearings their standardization is not widely known.

Internal Parts. The self-aligning and equalizing types of Kingsbury thrust bearings consist mainly of the shoes and their supporting structures shown in Figs. 109 and 110.

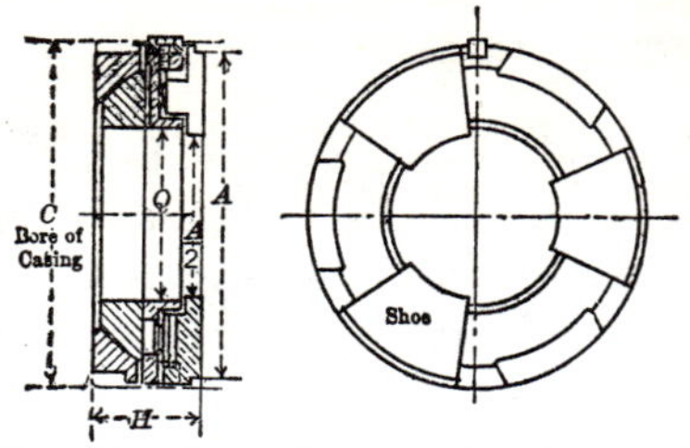

FIG. 109. Three-shoe base for horizontal or vertical thrust bearing.

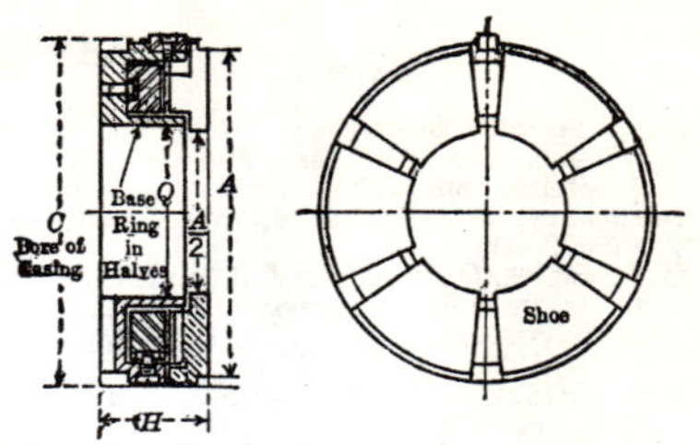

FIG. 110. Six-shoe base for horizontal or vertical thrust bearing.

The overall dimensions of these 3-shoe and 6-shoe elements are made the same for interchangeability. (See Table 17.) These elements may be used either horizontally or vertically, but usually they rely on a surrounding wall of the mounting to hold the shoes radially in place.

Commonly used combinations of these elements are (1) 3-shoe base and 6-shoe base with thrust collar between, (2) double 3-shoe base with collar between, (3) double 6-shoe bearing for two-way thrust with separate collar (see Fig. 111), and (4) double 6-shoe bearing with collar integral with shaft (see Fig. 112). In this construction the 6-shoe bases are made in halves.

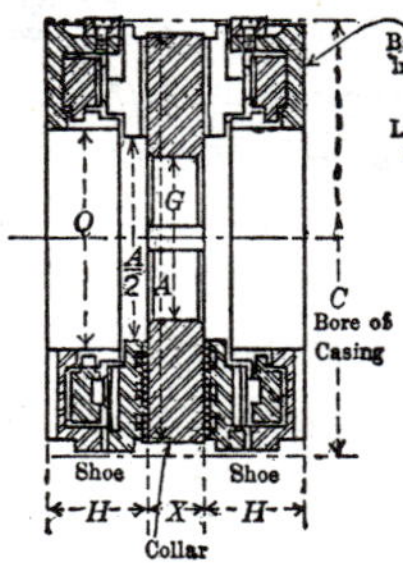

FIG. 111. Double thrust bearing with 6-shoe bases and separate collar.

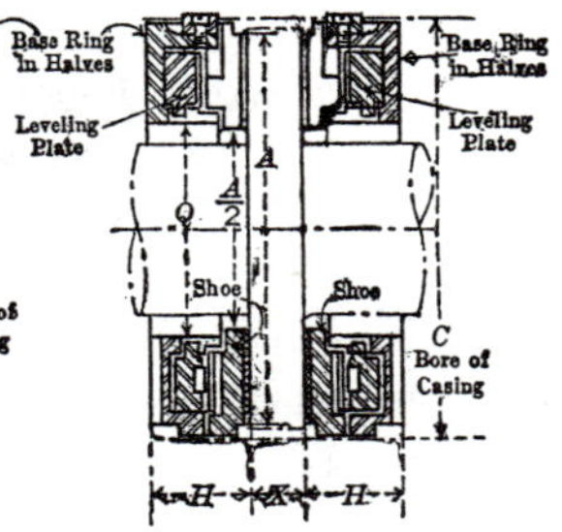

FIG. 112. Double horizontal thrust bearing with 6-shoe bases and integral collar.

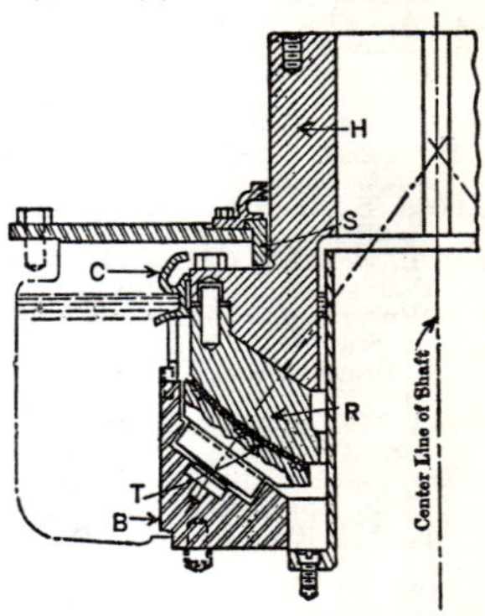

FIG. 113. Spherical thrust bearing.

SPHERICAL THRUST BEARING (Fig. 113) performs in one unit the functions of both flat thrust and guide bearing. Six pivoted shoes T are seated in the solid base ring B, the shoe faces being concave against the spherical-faced runner R. The bearing will sustain a radial load of about 15% of the downward load, and, unlike a cylindrical guide bearing, it requires no radial running clearance. The shaft, therefore, will run accurately centered. For high speeds, an air seal ring C around the runner minimizes foaming. A narrow bronze bushing S surrounding, and loose on, the thrust block hub H holds the parts together in handling. For temporary radial overloads, or lifting of the shaft, an auxiliary guide bearing may be applied above the spherical thrust bearing. The guide bearing shoes then are bored with a larger running clearance, and act only if the shaft lifts. Four 30-in. bearings of this type, operating at 500 rpm, are used in condensers at the Leaside Station, Toronto, Ont.

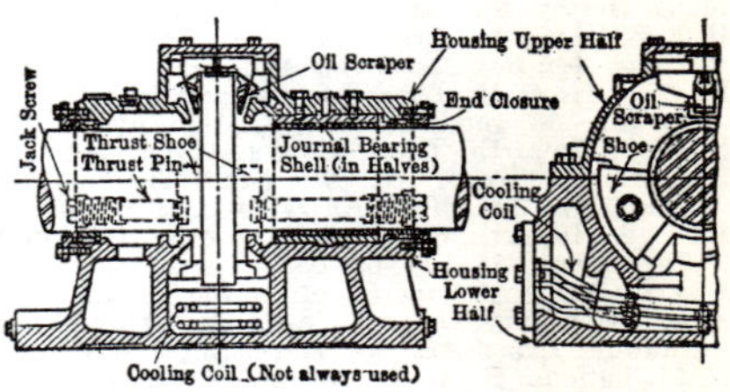

FIG. 114. Marine type automatically lubricated thrust bearing.

MARINE TYPE AUTOMATICALLY LUBRICATED THRUST MOUNTING (Fig. 114) is commonly used with Kingsbury thrust bearings abaft ship propulsion machinery.

where the shaft diameter is large and the full area of the thrust collar is not required to carry the load. Two thrust shoes in each end of the mounting are adjusted by jack screws and hardened thrust pins. The adjacent journal bearing has a removable split shell. End closures commonly are of nonwearing types. Oil lifted by the thrust collar is directed to the bearing surfaces.

REFERENCES

1. N. P. Petroff, New theory of friction, *Eng. J.* (Russian), Petrograd, 1883, pp. 71–140, 228–279, 377–436, 534–564, Abs., *Fortschr. Physik*, pp. 250–252. German translation by L. Voss, Hamburg, 1887. French translation, *Rev. mécanique*, Vol. 7, 1900, pp. 571–602.
2. Beauchamp Tower, First report of friction experiments, *Proc. Inst. Mech. Engrs.* (London), 1883, pp. 632–659.
3. O. Reynolds, On the theory of lubrication and its application to Mr. Beauchamp Tower's experiments, including an experimental determination of the viscosity of olive oil, *Phil. Trans. Roy. Soc.* (London), Vol. 177, Pt. 1, 1886, pp. 157–234. *Papers on Mechanical and Physical Subjects*, Vol. 2, 1901, pp. 228–310.
4. Albert Kingsbury, Experiments with an air lubricated journal, *J. Am. Soc. Naval Engrs.*, Vol. 9, 1897, pp. 267–292.
5. W. J. Harrison, Hydrodynamical theory of lubrication, *Trans. Cambridge Phil. Soc.*, Vol. 22, No. 3, 1913, pp. 39–54.
6. J. M. Stone and A. F. Underwood, Load-carrying capacity of journal bearings, *Quarterly Trans. SAE*, Vol. 1, Jan. 1947, pp. 56–67.
7. J. T. Burwell, The calculated performance of dynamically loaded sleeve bearings, *J. Applied Mechanics*, ASME, Vol. 14, No. 3, Sept. 1947, pp. A.231–A.245. Presented at Schenectady, N. Y., June 1947.
8. Albert Kingsbury, Optimum conditions in journal bearings, *Trans. Am. Soc. Mech. Engrs.*, RP-54-7, 1932, pp. 123– .
9. A. Sommerfeld, Zur Hydrodynamischen Theorie der Schmiermittelreibung, *Z. Math. u. Physik*, Vol. 50, 1904, pp. 97–155.
10. W. J. Harrison, Hydrodynamical theory of lubrication of a cylindrical bearing under variable load, and of a pivot bearing, *Trans. Cambridge Phil. Soc.*, Vol. 22, 1920, pp. 373–388.
11. H. A. S. Howarth, *A Graphical Analysis of Journal Bearing Lubrication*, pamphlet ASME, 1927. Thirty-five page reprint containing three papers, with discussions, from *Trans. Am. Soc. Mech. Engrs.* as follows: Vol. 45, No. 1906 (1923), pp. 421–448; Vol. 46, No. 1936 (1924), pp. 809–832; Vol. 47, No. 1988 (1925), pp. 1073–1099. Also certain relevant material from *J. Ind. Eng. Chem.*, Vol. 18, No. 5, May 1926, pp. 453–460.
12. S. J. Needs, Effects of side leakage in 120-deg. centrally supported journal bearing, *Trans. Am. Soc. Mech. Engrs.*, APM-56-16, 1934, pp. 721– .
13. H. A. S. Howarth, Journal running positions, *Trans. Am. Soc. Mech. Engrs.*, APM-51-3, 1929, pp. 21– .
14. B. L. Newkirk, General discussion on lubrication, and lubricants, Group I, *Journal and Thrust Bearings*, p. 205, Institution of Mechanical Engineers, London, 1938.
15. D. Robertson, Whirling of a journal in a sleeve bearing, *Phil. Mag.*, Vol. 15, 1933, pp. 113–130.
16. T. E. Stanton, *Friction*, Longmans, Green and Co., London, 1923.
17. L. J. Bradford and L. J. Grunder, Oil film pressures in complete bearings, *Pa. State College Bull.* 39, 1930.
18. S. A. McKee and T. R. McKee, Pressure distribution in oil films of journal bearings, *Trans. Am. Soc. Mech. Engrs.*, RP-54-8, 1932, pp. 149– .
19. F. P. Dahlstrom, Morgoil roll-neck bearing, *Trans. Am. Soc. Mech. Engrs.*, IS-55-2, 1933.
20. H. A. S. Howarth, Design of fluid-film bearings for optimum loading, *Machine Design*, Oct. 1946, pp. 101–103, and Nov. 1946, pp. 151–154.

GENERAL REFERENCES

R. O. Boswall, *Theory of Film Lubrication*, Longmans, Green and Co., New York, 1928.

G. Duffing, Die Schmiermittelreibung bei Gleitflächen von endlicher Breite, Vol. 5, pp. 839–850, Auerbach-Hort, *Handbuch d. physik, u. tech. Mechanik*, J. A. Barth, Leipzig, 1930.

A. Fogg, Film lubrication of parallel thrust surfaces, *Engineering* (London), Vol. 159, No. 4127, Feb. 16, 1945, pp. 138–140; *Proc. Inst. Mech. Engrs.* (London), Vol. 155, 1946 (War Emergency Issue, No. 14), pp. 49–53 (d.), 53–67.

M. D. Hersey, Fundamental action of lubricants, *Am. Machinist*, June 13 and 20, 1929.

H. A. S. Howarth, Current practice in pressures, speeds, clearances, and lubrication of oil-film bearings, *Trans. Am. Soc. Mech. Engrs.*, MSP-56-2, 1934, pp. 891– .

H. A. S. Howarth, Load capacity and friction of pivoted-shoe bearings, *Machine Design*, July 1948, pp. 151–154, and Aug. 1948, pp. 149–151.

H. A. S. Howarth, The loading and friction of thrust and journal bearings with perfect lubrication. Reprint of *Trans. Am. Soc. Mech. Engrs.*, MSP-57-2, May 1935, pp. 169–187, with discussion, *Trans. Am. Soc. Mech. Engrs.*, Feb. 1936, pp. 122–126.

H. A. S. Howarth, Slow-speed and other tests of Kingsbury thrust bearings, *Trans. Am. Soc. Mech. Engrs.*, Vol. 41, 1919, pp. 685– .

Albert Kingsbury, Experiments on the friction of screws, *Trans. Am. Soc. Mech. Engrs.*, Vol. xvii, 1896.

Albert Kingsbury, On problems in the theory of fluid-film lubrication, with an experimental method of solution, *Trans. Am. Soc. Mech. Engrs.*, PME-53-5, 1931, pp. 59– .

C. M. Laffoon, R. A. Baudry, and P. R. Heller, Performance of vertical water wheel thrust bearings during the starting period, *Trans. Am. Soc. Mech. Engrs.*, May 1947, pp. 371–379. Includes discussion.

H. M. Martin, The theory of the Michell thrust bearing, *Engineering* (London), Feb. 20, 1920.

A. G. M. Michell, The lubrication of plane surfaces, *Z. Math. u. Physik*, Vol. 52, 1905, pp. 123–137.

M. Muskat and F. Morgan, *Studies in Lubrication*, a series of five papers published in *J. Applied Phys.*, Vol. 9, 1938, and Vol. 10, 1939.

M. C. Shaw, An analysis of the parallel-surface thrust bearing, *Trans. Am. Soc. Mech. Engrs.*, May 1947, pp. 381–387. Includes discussion.
M. C. Shaw and C. D. Strang, Jr., The hydrosphere—a new hydrodynamic bearing, *J. Applied Mechanics*, Vol. 15, No. 2, June 1948, pp. 137–145.
H. W. Swift, Fluctuating loads in sleeve bearings, *J. Inst. Civil Engrs.* (London), Vol. 5, 1937, pp. 161–195.

Note. The *Engineering Index* contains abstracts of most of the foregoing references and of many others, covering all phases of film-type bearing design, materials, manufacture, operation, lubrication, and lubricants.

LUBRICANTS

By Paul G. Exline

A lubricant is a material interposed between two relatively moving machine elements to reduce friction and wear. Liquid, plastic, and solid materials are used as lubricants although sound design requires the application of liquid lubricants wherever possible.

12. TYPES OF LUBRICANTS

MINERAL OILS. Mineral oils are by far the most commonly used liquid lubricants, primarily because of their cheapness and stability. Their current consumption is at the rate of approximately 2.25 billion gallons per year. They are complex mixtures of paraffinic, naphthenic, aromatic, and unsaturated hydrocarbons distinguished from other crude oil fractions by their viscosity.

In general, three classes of distillates are made in separating the lubricating cuts from the crude oil: (1) Wax distillates, dewaxable by filter-pressing, from which neutrals and paraffin oils are derived. (2) Unpressable distillates, which are also noncentrifugal and must be dewaxed by the use of solvents. (3) Overhead cylinder stocks, representing the last high-boiling constituents of a crude oil except a usually asphaltic residuum.

Residual materials made are usually of two kinds: (1) Long residua, including everything in the crude oil after the wax distillates have been removed. (2) Cylinder stocks, including everything after the unpressable distillates have been taken overhead. Both long residua and cylinder stocks are dewaxed by centrifuging.

By redistillation, refining, and blending of these five type stocks, lubricating oils are produced in wide range, from spindle oils hardly more viscous than kerosene to highly viscous and heat-resistant airplane engine and steam cylinder oils.

Modern methods provide the refiner with greater independence of the crude oils than was formerly true, since certain constituents can be removed from a naphthenic stock leaving a material which approximates that derived from an intermediate base crude. Likewise a product can be obtained from an intermediate-base oil having the properties of a high-grade paraffinic material.

Compounding of mineral oils is commonly practiced to secure properties of the finished product not obtainable by blending and refining the different stocks. Vegetable and animal oils are added to obtain greater lubricating action where conditions of boundary lubrication may be met. Instead of 10 to 20% of the fixed oils, smaller percentages of the fatty acids may be used. In many cases small quantities of synthetic addition agents may be used to produce specific effects such as to retard oxidation, to provide detergency, or to control emulsification.

ANIMAL AND VEGETABLE OILS. Animal, vegetable, and fish oils were among the earliest known lubricants and are still widely used both alone and as compounding agents. They are commonly referred to as fixed oils because of their tendency to decompose before distillation. Their lubricating properties are due to varying percentages of glycerides of the fatty acids: linoleic, oleic, palmitic, stearic, and others.

Typical oils of this class in common usage are castor, peanut, palm, and rapeseed of the vegetable oils, neatsfoot, lard, and tallow oils of animal origin, and whale and sperm oils. Sperm oil differs from the others in that the fatty acids are in combination with alcohol rather than glycerol.

SYNTHETIC LIQUIDS. During the early 1940's impetus was given to the development of certain synthetic lubricants.

One group of the synthetic liquids are the polyorganosiloxanes or silicone fluids. They are similar in molecular structure to the mineral oil hydrocarbons, but certain of the carbon atoms are replaced by silicon atoms. They can be manufactured in a wide range of viscosities and with an unusually flat viscosity-temperature curve. (See Fig. 1.) The value of the silicones as lubricants is more than usually dependent upon the bearing materials, and their use should be referred to the manufacturer. Their stability at high temperatures has enabled them to be used as journal bearing lubricants at 300 to 375 F for long periods of time. Their low pour point, less than −100 F in some of the low viscosity grades, permits their use at lower temperatures than the temperatures at which mineral oils can be used.

Other synthetic lubricants of interest are oils obtained by the Fischer Tropsch process, the polyglycal ethers, and the aliphatic diester oils. Some of the synthetics have unusually small temperature coefficients of viscosity and others extremely low freezing points. In general, their high cost limits their application to places where mineral oils cannot be used.

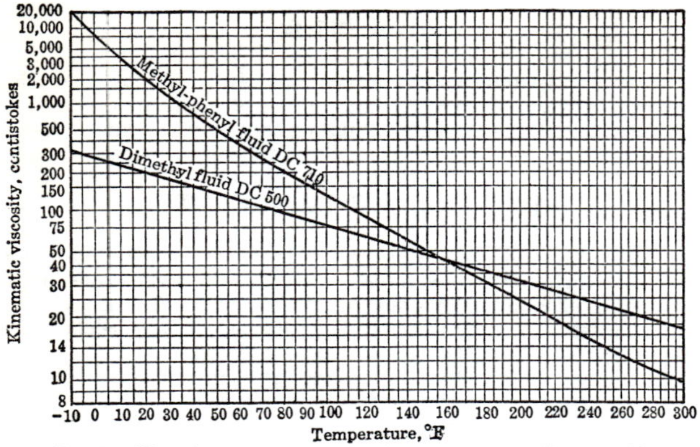

Fig. 1. Viscosity-temperature relationships of two silicone fluids.

MISCELLANEOUS FLUIDS. Various fluids have been used as lubricants in special cases; such as water for stern tube bearings and air for high-speed centrifuge bearings. In certain chemical processes it is desirable to use the process liquid as the lubricant. In such cases the compatibility of the lubricant with the bearing materials should be very carefully considered.

GREASES. Greases are usually suspensions or gels of metallic soaps dispersed in a lubricating liquid. The calcium and sodium soaps of the common fatty acids are largely used, although aluminum, lead, and lithium soaps are employed to a lesser degree. The consistency of greases varies from nearly liquid to the hard blocks used for locomotive crosshead lubrication, depending upon kind and percentage of the soap content and viscosity of the base oil. In general, a grease is used for relatively slow speed systems which at the same time offer difficulty in the retention of liquid lubricants, or in systems where high temperatures are encountered.

Synthetic resins or even rubber may be added to greases to increase their stickiness or retentivity.

SOLID LUBRICANTS. The most commonly used solid lubricant is graphite, both in powdered and colloidal form. Its stability at elevated temperatures and its low coefficient of friction make it desirable where organic materials cannot be applied. Graphite is also used as an additive to both oils and greases. Soapstone, talc, and waxes are used as lubricants for certain applications.

Under conditions of extremely high pressures, tungsten disulfide, molybdenum disulfide, and silver sulfate have been found to be effective solid lubricants.

13. PROPERTIES OF LUBRICANTS

Under conditions of thick film lubrication, the viscosity of a lubricant is its only important property. Since many applications of lubricants are where thick film conditions do not exist, and since even bearings designed for thick film lubrication will occasionally operate in the thin film or boundary condition, the oiliness of a lubricant becomes of importance. As a lubricant is quite often required to operate in a system for long periods of time and occasionally at elevated temperatures, its oxidation and its thermal stability are of interest. Other properties of lubricants such as acidity, saponification number, color, cloud and pour points, sedimentation, flash and fire points, emulsification, iodine number, and specific gravity are commonly determined by the refinery laboratories as specification tests to control the uniformity of the products. Their value in determining the relative worth of lubricants is minor or nonexistent.

VISCOSITY. The viscosity of a liquid is defined as the ratio of the shearing stress to the rate of shear. In the cgs system the unit of viscosity is the poise or one dyne sec/cm^2. The centipoise or 0.01 poise is more frequently used because of its size. One centipoise or 0.01 dyne sec/cm^2 = 2.09 × 10^{-5} lb sec/ft^2 = 1.45 × 10^{-7} lb sec/in.^{2} (Reyn) = 2.42 × 10^{-9} lb min/in.^{2} Because certain viscometers measure the ratio of viscosity to specific gravity, this term has been called the stoke, and one-hundredth its value the centistoke. It is expressed in cgs units.

It is common practice in the American petroleum industry to express viscosity in Saybolt universal seconds (SUS). This is the time of efflux, in seconds, necessary for 60 ml of the liquid to drain through a tubular orifice in the bottom of a vessel whose dimensions and construction are rigidly specified. As it is impossible to use Saybolt seconds in numerical computations of bearing performance, they must first be converted to absolute units (see Table 1).

Table 1. Values for Converting Saybolt Viscosity to Kinematic Viscosity

Kinematic Viscosity, centistokes	Equivalent SUS at 100 F	Equivalent SUS at 210 F
2	32.6	32.8
2.5	34.4	34.6
3	36.0	36.3
3.5	37.6	37.9
4	39.1	39.4
4.5	40.7	41.0
5	42.3	42.6
6	45.5	45.8
7	48.7	49.0
8	52.0	52.4
9	55.4	55.8
10	58.8	59.2
15	77.2	77.7
20	97.5	98.2
25	118.9	119.7
30	140.9	141.9
40	185.7	187.0
50	231.4	233.0
60	277.4	279.3
70	323.4	325.7
Over 70	SUS = CS × 4.620 CS = SUS × 0.2165	SUS = CS × 4.652 CS = SUS × 0.2150

Note. To obtain SUS equivalent to a kinematic viscosity at t F, multiply the equivalent SUS at 100 F by $1 + (t - 100) \times 6.4 \times 10^{-5}$.

Since the Saybolt viscometer is insensitive to differences in viscosity in the low viscosity range, readings below 32 sec are not taken. The practice is becoming widespread of using the modified Ostwald viscometer, which gives results directly in kinematic viscosity. Unfortunately, oil companies still convert these values to SUS for their published data, so that it is necessary to convert the data back to absolute viscosity before they can be used.

Mineral oils decrease rapidly in viscosity with increase in temperature. This decrease may be from 0.6 to 6.0% per °F. The most convenient method of expressing the variation of viscosity of lubricating oil with temperature is the ASTM chart, which is based on an assumed equation that agrees very well with results of tests and gives a straight line graph when a function of viscosity is plotted against a function of temperature. The ASTM chart is available with temperature in °F as the abscissa and either kinematic viscosity in centistokes or SUS as the ordinate. With this chart, a straight line can be drawn through two points, indicating the viscosity at two temperatures, so that the viscosity at any temperature within a reasonable range can be determined.

The equation on which the ASTM chart is based is

$$\log_{10} \log_{10} (\nu + 0.8) = n \log_{10} T + C$$

where ν is the kinematic viscosity, centistokes; T the absolute temperature, Fahrenheit degrees; and n and C are constants for any given oil. This expression is very unwieldy where it is necessary to express the viscosity as a function of temperature in mathematical analysis. For this purpose, a much more convenient equation is based on the formula of Slotte

$$\mu = at^{-b}$$

where μ is the absolute viscosity, centipoise; t the temperature, °F; and a and b are constants for any given oil. If the constants are evaluated for measurements made at 100 and 210 F, the equation is quite satisfactory for the range 60 to 300 F.

Dean and Davis have devised a system of rating lubricating oils by means of an arbitrarily determined viscosity index (VI) based on a change of SUS for a given temperature

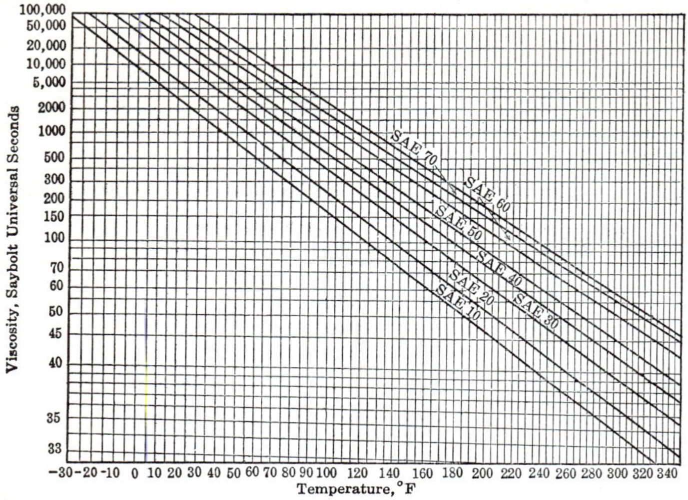

FIG. 2. Viscosity-temperature relationship of SAE grades of petroleum oils having 100 viscosity index

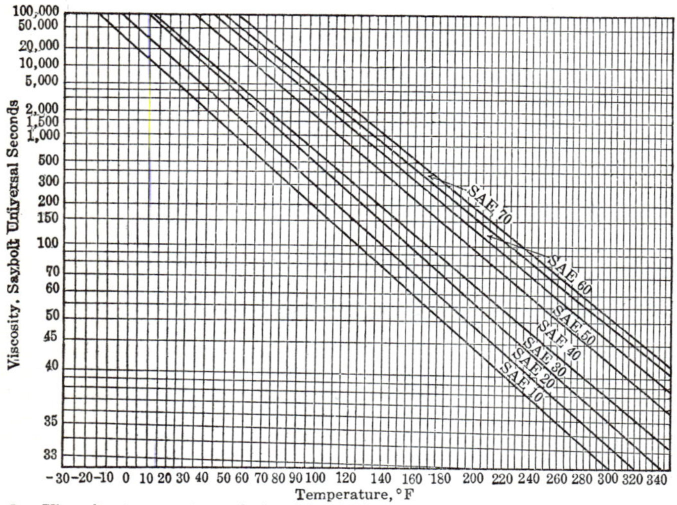

FIG. 3. Viscosity-temperature relationship of SAE viscosity grades of zero viscosity index.

change in °F. Two groups of mineral oils were used for establishing this rating; one group was from the Pennsylvania fields, and the other from the Gulf Coast. The Pennsylvania oils were given a rating of 100 VI, and the Gulf Coast oils a rating of zero VI, and a method was set up for determining the VI of any other oil based on its Saybolt viscosity at 100 F and at 210 F. This irrational system gives a somewhat deceptively large numerical change for what are actually relatively small viscosity changes. Application of VI numbers to some of the newer synthetics having quite low changes of viscosity with temperature are particularly confusing and misleading.

The Society of Automotive Engineers has developed a numerical system for expressing the different viscosity grades of lubricants. Although some overlapping is encountered, it provides a useful method of selecting viscosity within commercial tolerances. Figures 2 and 3 show the viscosity ranges for the different SAE numbers for a viscosity index of 100 and 0, respectively. Figure 4 shows the ratios of kinematic viscosities at 100 and 210 F as a function of VI for oils of various viscosities at 210 F.

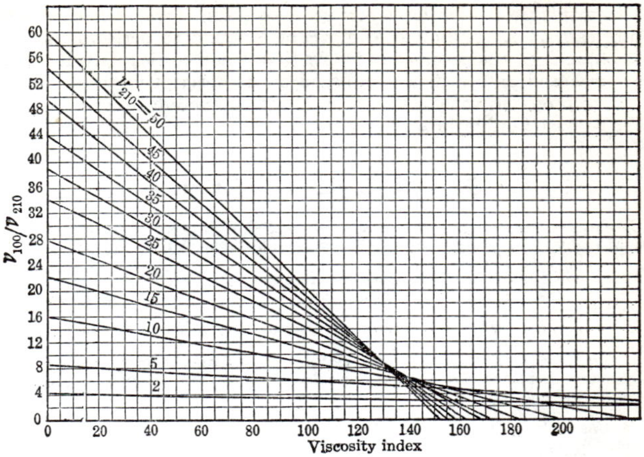

FIG. 4. Viscosity ratios versus viscosity index, at various viscosities expressed in centistokes. ν_{100} = viscosity at 100 F, ν_{210} = viscosity at 210 F.

SPECIFIC GRAVITY. The density or specific gravity of a petroleum oil has no correlation with its lubricating value. It is generally included in the description of a lubricant and is useful in converting from kinematic viscosity to absolute viscosity. Specific gravity is generally given as the ratio of the density of the lubricant at 60 F to the density of water at 60 F. In the petroleum industry, the term "API gravity" is preferred to specific gravity and is expressed in terms of API degrees. The relation between the two is

$$°\text{API} = \frac{141.5}{\text{Specific gravity 60/60 F}} - 131.5$$

or

$$\text{Specific gravity 60/60 F} = \frac{141.5}{°\text{API} + 131.5}$$

Experimental data on the expansion of petroleum oils indicate that their coefficient of expansion is a function of the density (at a fixed temperature) only. This permits a fortunate simplification of all density corrections, as specification of the specific gravity at a single temperature is sufficient for its calculation at any other temperature. Figure 5 shows the coefficient of expansion K plotted as a function of the specific gravity 60/60 F. The specific gravity, S_t, at any temperature t F can be computed by this equation:

$$S_t = S_{60}(1 - K[t - 60])$$

SPECIFIC HEAT. A knowledge of the specific heat of a lubricant is necessary in computing heat balances in force-fed bearings. The data on petroleum hydrocarbons have been summarized in the following equation, which is accurate to within 5%.

$$C = \frac{1}{S}(0.388 + 0.00045t)$$

where C = specific heat Btu/lb/ F; S = specific gravity 60/60 F; and t = temperature F.

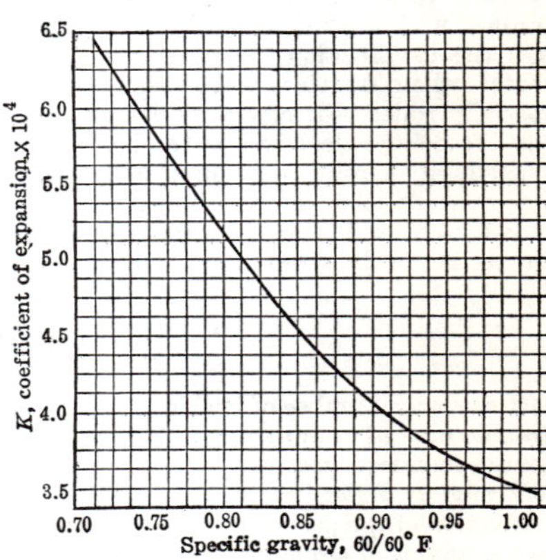

FIG. 5. Coefficient of expansion of petroleum oils versus specific gravity at 60°/60° F.

SELECTION OF MINERAL LUBRICANTS. Mineral lubricants are selected principally on the basis of their viscosity and their temperature coefficient of viscosity. Many cases arise where other properties become of importance, such as oxidation stability, oiliness, and film strength. Not only are these properties poorly susceptible of definition but they are also not a function of the lubricant alone, being dependent upon the conditions of use and the materials in contact. There is not nearly enough reliable information now available about test methods to permit writing workable oil-purchase specifications that would within themselves, without performance tests, serve as a reliable guide in the selection of lubricants. Even though a lubricant is found to perform satisfactorily in a given application, it does not necessarily follow that other lubricants will be equally satisfactory when purchased on specifications based on its descriptive tests. Purchase specifications should be prepared with the assistance of a competent oil chemist, and wherever possible should be based on performance tests.

GENERAL REFERENCES

Archbutt and Deeley, *Lubrication and Lubricants*, 5th ed., C. Griffin and Co. Ltd., London, 1927.

Atkins, Baker, Murphy, and Zisman, Synthetic lubricant fluids from branched-chain diesters, *Ind. Eng. Chem.*, Vol. 39, April 1947, p. 491.

Battle, *The Handbook of Industrial Oil Engineering*, Philadelphia, Pa., 1926.

Boyd and Robertson, *The Friction Properties of Various (Solid) Lubricants at High Pressures*, ASME Paper 44-SA 8, 1944.

Brophy, Larson, and Militz, *High Temperature Performance of Silicone Fluids in Journal Bearings*, ASME Paper 47-A-114, New York, 1947.

Brophy, Militz, and Zisman, Dimethyl silicone polymer fluids and their performance characteristics in unilaterally loaded bearings, *Trans. Am. Soc. Mech. Engrs.*, Vol. 68, p. 355, May 1946.

Bureau of Standards Circular 410.

Cragoe, *Natl. Bur. Standards Misc. Publ.* 97, 1929.

Dean and Davis, *Chem. Met. Eng.*, Vol. 36, 1929, p. 618.

Ferris, Birkhimer, and Henderson, *Ind. Eng. Chem.*, Vol. 23, 1931, p. 753.

Gruse and Stevens, *The Chemical Technology of Petroleum*, McGraw-Hill Book Co., New York, 1942.

Hain, Jones, Merker, and Zisman, Synthetic low temperature greases from aliphatic diesters, *Ind. Eng. Chem.*, Vol. 39, April 1947, p. 500.

Kauppi and Pedersen, Silicones as lubricants, *Natl. Petroleum News* (Technical Section), Vol. 37, p. r 944, Dec. 5, 1945.

Klemgard, *Lubricating Greases*, Reinhold Publishing Corp., New York, 1937.

National Standard Petroleum Oil Tables, National Bureau of Standards, 1936.

ROLLING BEARINGS

By Gunnar Palmgren and Bryce Ruley

Ball or roller bearings support a loaded moving part on a set of hardened steel rolling elements. These rolling elements usually roll between two smooth hardened steel rings, although in some cases the rolling elements may roll directly on the shaft or in the housing bore. Normally, a cage or a separator is used to keep the rolling elements properly spaced about the bearing pitch circumference. All parts are manufactured to very close tolerances and to exacting material specifications.

Standard rolling bearings are manufactured as interchangeable units and can be quickly obtained and inserted as a unit if their replacement is necessary. Ball and roller bearings have the low friction characteristic of rolling motion, this friction being only slightly greater when starting than under uniform speed of rotation. Anti-friction bearings, especially roller bearings, are relatively insensitive to moderate overloads or shock loads, their life being calculated on the basis of mean effective load rather than maximum load.

Although in rolling bearings sliding motion is essentially replaced by rolling motion, some miscellaneous sliding does occur (e.g., between balls and cage, between rollers and guide flange, etc.). Some lubrication is therefore necessary, but in most cases the lubrication needs of rolling bearings are moderate compared to the needs of other bearing types.

14. BEARING TYPES

Of the large number of bearing types which have been invented, only a few types, which combine ease of manufacture with some inherent advantage in practice, have survived as modern standard bearings. Some modern types of ball bearings are shown in Fig. 1 and some modern roller bearing types in Fig. 2.

Selection of the proper bearing for a given application may be a difficult problem, requiring the advice of a qualified bearing engineer, but some general principles can be set down.

Ball bearings are usually the most economical choice for the smaller sizes and lighter loads; roller bearings are usually more satisfactory for the larger sizes and heavier loads. Roller bearings have considerably less sensitivity to shock loading than ball bearings. Most bearings lose capacity if misaligned any measurable amount, but self-aligning bearings

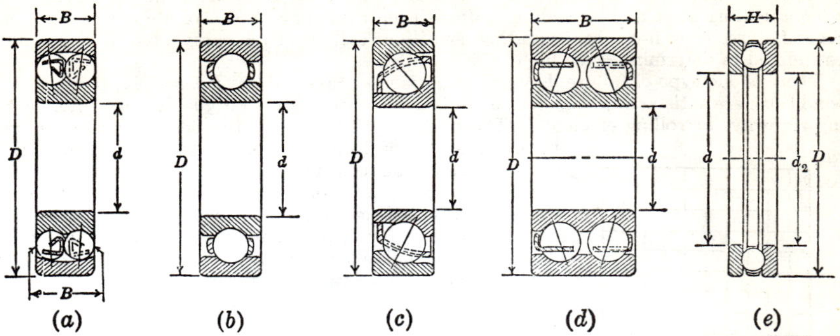

Fig. 1. Types of ball bearings: (*a*) self-aligning; (*b*) deep-groove; (*c*) angular contact; (*d*) double-row; (*e*) thrust.

(spherical roller bearings, Fig. 2*d*, spherical roller thrust bearings, Fig. 2*e*, and self-aligning ball bearings, Fig. 1*a*) can take two or three degrees of misalignment without loss in capacity.

A major influence on bearing choice is the direction and magnitude of the load. **Cylindrical roller bearings** (Fig. 2*a*,*b*,*c*), for example, are designed to carry radial loads, some of them being made without flanges on one ring so that they can carry no thrust load, whereas others are able to take the light thrust loads necessary to stabilize the shaft.

Self-aligning ball bearings (Fig. 1*a*) are designed primarily to carry light and moderate radial loads but can carry some thrust load in either direction. **Deep-groove ball bearings** (Fig. 1*b*) and **double-row ball bearings** (Fig. 1*d*) are designed to carry light, medium, or heavy radial or combined loads. These bearings can carry substantial thrust loads but are not usually the best choice for pure thrust loads.

Angular contact ball bearings (Fig. 1*c*) are designed to carry thrust loads or combined loads which are predominantly thrust. They can carry radial loads if two of them are mounted opposed, but they are not recommended for the purpose. **Ball thrust bearings** (Fig. 1*e*) can carry pure thrust load only. Any radial load present must be carried by a

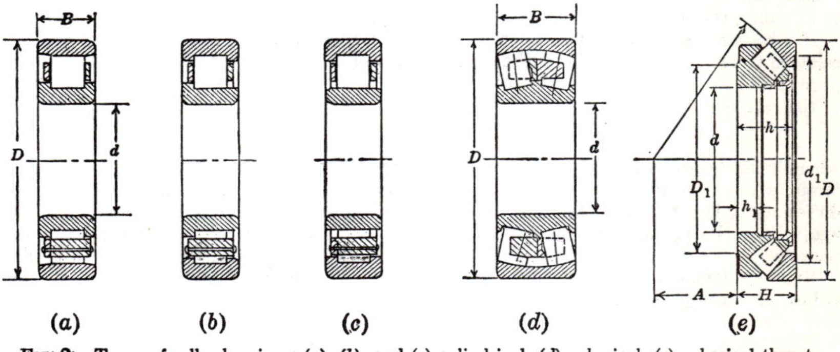

Fig. 2. Types of roller bearings: (*a*), (*b*), and (*c*) cylindrical; (*d*) spherical; (*e*) spherical thrust.

separate radial bearing. Ball thrust bearings can carry fairly heavy thrust loads but their use is normally limited to low and medium speeds. For high-speed applications angular contact ball bearings are usually more satisfactory, even for pure thrust loads.

Spherical roller bearings (Fig. 2*d*) are a self-aligning type designed to carry medium or heavy radial or combined loads. Although spherical roller bearings can carry heavy thrust loads, a related type, the **spherical roller thrust bearing** (Fig. 2*e*), is usually a better choice if the load is pure thrust or if the radial component of load is always less than two-thirds the thrust component.

15. BEARING LIFE

Selection of the bearing on an engineering basis has usually eliminated any chance of excessive or improper loading, and satisfactory operating conditions can be insured by adequate lubrication and by sealing against the entrance of corrosive or abrasive materials. Since failures from improper operating conditions can usually be prevented, they are not considered in determining the life of the bearing.

There is one type of failure, however, that is not preventable. Ultimately, if operating conditions are satisfactory, the bearing fails from a progressive fatigue flaking of the bearing raceways or rolling elements. The number of revolutions the bearing makes before the first evidence of fatigue can be detected is defined as the life of the individual bearing.

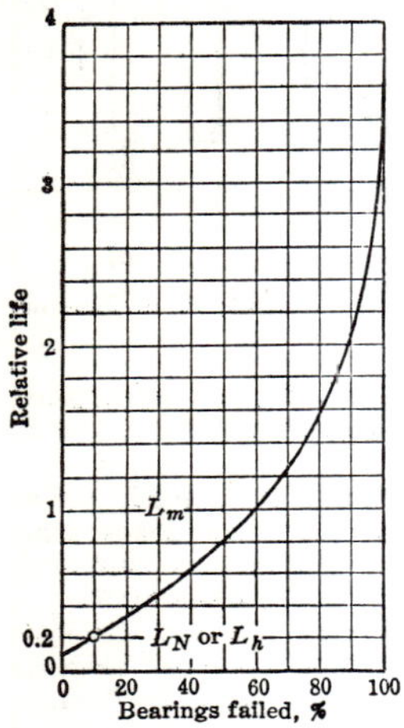

FIG. 3. Life dispersion of rolling bearings.

When a large number of bearings, of the same type and size, are run to fatigue failure under equal loads, a considerable dispersion is found in the individual lives as shown in Fig. 3. Since reliability is a major requirement for bearings in most applications, bearings are rated on the basis of the life which 90% of the bearings in a group will equal or exceed before fatigue failure. This is called the rated life or simply the life of the bearing. The average life of a bearing is approximately five times its rated life.

If two groups of bearings, of the same type and size, are run to fatigue failure, the first group under a load F_1 giving a rated life L_1 and the second group under a load F_2 giving a rated life L_2,

$$\frac{L_1}{L_2} = \left(\frac{F_2}{F_1}\right)^3 \tag{1}$$

This is sometimes referred to as the fundamental law and may be stated: *The rated life of a bearing varies inversely as the cube of the load.*

If the rated life for some given load is known, the rated life for any other load can be calculated by means of the fundamental law. A convenient reference load is the *dynamic reference capacity*, C, defined for radial bearings as that pure radial load which will give the bearing a rated life of one million revolutions (e.g., 500 hours at 33.3 rpm) when the bearing inner ring rotates relative to the direction of load. Similarly for thrust bearings, the dynamic reference capacity is defined as the pure thrust load which would give the bearing a rated life of one million revolutions. Numerical values of dynamic reference capacities are given in the bearing tables (Tables 5 to 13).

If a radial bearing is loaded by a thrust or combined load, or if the outer ring moves relative to the direction of load, the equivalent radial load under normal conditions (e.g., pure radial load and inner ring rotating relative to the load direction) is found by means of the relation

$$P = XF_r + YF_a \tag{2}$$

where P is the equivalent radial load under standard conditions, pounds; X is a rotation factor which depends on the bearing type and on which ring rotates relative to the load direction; Y is a thrust factor which depends on the bearing type and sometimes on the load; F_r is the radial component of load, pounds; F_a is the thrust component of load, pounds.

Numerical values of X are given in Table 1 for various bearing types and numerical values of Y are given in Tables 5 to 13.

Table 1. Rotation Factors

Bearing Type	X Factor for Rotating Inner Ring	X Factor for Rotating Outer Ring
Self-aligning ball bearing	1	1
Single row angular contact ball bearing	0.5*	0.7†
All other radial bearings	1	1.4

* If $P < F_r$, set $P = F_r$.
† If $P < 1.4F_r$, set $P = 1.4F_r$.

Most thrust bearings cannot carry any radial load. For those which can (e.g., spherical roller thrust bearings) the equivalent thrust load can be calculated from the equation

$$P_a = F_a + Y_a F_r \tag{3}$$

where P_a is the equivalent thrust load, pounds; and Y_a is a rotation factor, which, for spherical roller thrust bearings, is 1.8 for inner ring rotation and 2.1 for outer ring rotation.

After the equivalent load has been calculated by eq. 2 or eq. 3, the rated life may be calculated by the fundamental law, eq. 1. In practice, however, this law is modified. Since the life in number of rotations can be expressed as a function of the life in hours and the speed in rpm, eq. 1 can be expressed as

$$C = \frac{f_h P}{f_n} \tag{1a}$$

or

$$f_h = \frac{f_n C}{P} \tag{1b}$$

where C = the dynamic reference capacity, pounds (see Tables 5 to 13); P = the equivalent radial (or equivalent thrust) load; f_n = a speed factor (found from scale 1, Fig. 4) = $\sqrt[3]{33.3/n}$; and f_h = a life factor (found from scale 2, Fig. 4) = $\sqrt[3]{L_h/500}$.

The following example illustrates the use of this method.

Example. What is the rated life in hours for a medium series deep-groove ball bearing (Fig. 1*b*) with a 50-mm bore under a radial load of 1000 lb and a thrust load of 500 lb at 600 rpm, if the inner ring rotates relative to the direction of load?

Since the bearing has a moderate load a Y factor of 1.2 can be assumed. Then the equivalent radial load at 750 rpm is found from eq. 2:

$$P = 1 \times 1000 + 1.2 \times 500 = 1600 \text{ lb}$$

Since the speed is known, the value of f_n can be read from scale 1. The value which corresponds to 750 rpm is $f_n = 0.354$. The value of C for this bearing is found from Table 6 to be 10,400 lb. Then, by eq. 1*b*,

$$f_h = \frac{0.354 \times 10{,}400}{1600} = 2.30$$

By scale 2, the rated life is found to be 6100 hours.

As an indication of the rated life desired in various types of machines, Table 2 may be used.

At times the load acting on the bearings is not constant. If the cyclic changes in the load are known and if the load cycle is not a multiple of the time necessary for the bearing to make one rotation, then the *mean effective load* can be found from the equation

$$P_m = \sqrt[3]{\frac{P_1^3 N_1 + P_2^3 N_2 + \cdots + P_n^3 N_n}{N_1 + N_2 + \cdots + N_n}} \tag{4}$$

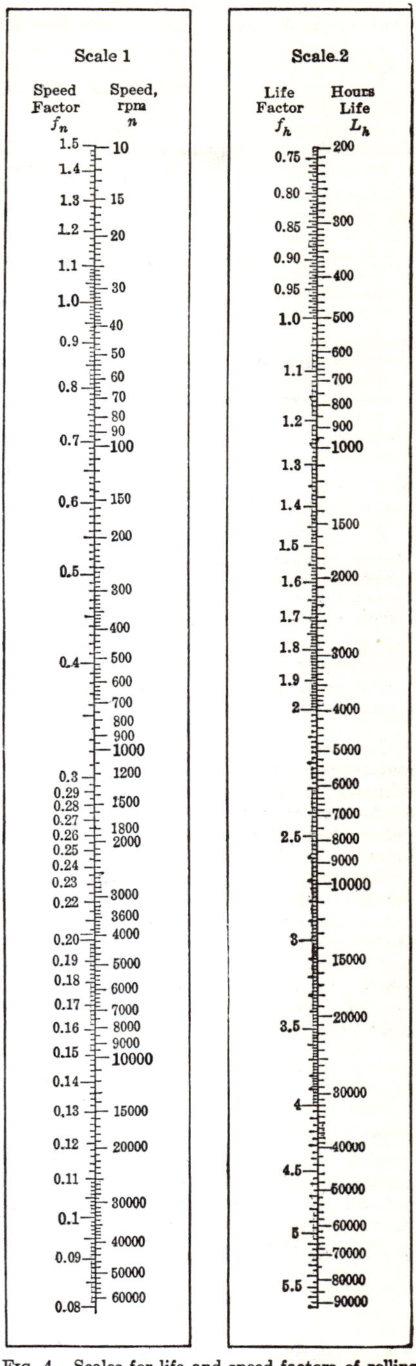

Fig. 4. Scales for life and speed factors of rolling bearings.

where P_1 is the equivalent radial load during N_1 revolutions, P_2 the equivalent radial load during N_2 revolutions $\cdots$, and P_n the equivalent radial load during N_n revolutions. The periods 1, 2, $\cdots$, n are periods within the cycle when the load stays essentially constant.

Table 2. Selection of Bearing Life

Type of Machine	f_h	L_h
Instruments and apparatus which are only infrequently used: demonstration apparatus, devices for operation of sliding doors	1	500
Aircraft engines	1 –1.5	500– 2,000
Machines for service of short duration or intermittent operation, where service interruptions are of minor importance: hand tools, lifting tackle in machine shops, hand-driven machines in general, farm machinery, assembly cranes, charging machines, foundry cranes, household machines	2 –2.5	4,000– 8,000
Machines for intermittent service where dependable operation is of great importance: auxiliary machines in power stations, work moving devices in production lines, elevators, general cargo cranes, machine tools less frequently used	2.5–3	8,000– 14,000
Machines for 8-hour service which are not always fully utilized: stationary electric motors, gear drives for general purposes	3 –3.5	14,000– 20,000
Machines for 8-hour service which are fully utilized: machines in general in the mechanical industries, cranes for continuous service, blowers, jackshafts	3.5–4	20,000– 30,000
Machines for continuous operation (24-hour service): separators, compressors, pumps, main line shafting, roller beds and conveyor rollers, mine hoists, stationary electric motors, continuously working machines aboard warships	4.5–5	50,000– 60,000
Machines for 24-hour service where dependability is of great importance: cellulose and paper machines, public power stations, mine pumps, public pumping stations, machines for continuous service aboard merchant ships	6 –7	100,000–200,000

Occasionally the question is asked: How much load can be placed on the bearing while stationary if the bearing is subsequently to rotate without excessive noise or vibration? Although the answer to this question depends on the type of bearing application as well as the type of bearing, a convenient reference point is that load which causes a *permanent* deformation, at the most heavily loaded ball or roller, of 0.0001 of the ball or roller diameter. The load which corresponds to this is about half the dynamic reference capacity for self-aligning ball bearings and approximately equal to the dynamic reference capacity for other bearing types.

For bearings where vibration must be held to a minimum the foregoing limits apply. On other applications which are not so sensitive, either because of low speed of rotation or less accurate operating requirements, several times this load can be applied. For example, aircraft control pulley bearings can support about four times the load given above. Danger of bearing fracture due to overload will occur only after the bearing is loaded to eight times the above limit.

16. FRICTION AND LIMITING SPEEDS

The friction in bearings is compounded from a number of factors, among them the fluid friction of the lubricant, the sliding friction between rolling elements and cage, the sliding friction within the rolling contact areas, and, in roller bearings, the sliding friction between rollers and guide flanges. Slight changes in any of these factors can change the bearing friction considerably, but under ordinary conditions the friction coefficients given in Table 3 can be used and the friction torque found from the equation

$$T = \frac{\mu P d}{2} \tag{5}$$

where μ is the coefficient of friction (Table 3); P is the equivalent radial load, pounds d is the bearing bore, inches; T is the friction torque, inch-pounds.

Table 3. Coefficients of Bearing Friction

Self-aligning ball bearings	$\mu = 0.0010$
Cylindrical roller bearings with flange-guided short rollers	$\mu = 0.0011$
Thrust ball bearings	$\mu = 0.0013$
Single-row ball bearings	$\mu = 0.0015$
Spherical roller bearings	$\mu = 0.0018$

The speed at which a bearing can safely be run is limited by the amount of heat generated as friction and by the ability of the lubricant to dissipate this heat or to stand the resulting temperature rise without breaking down. In general, grease is a satisfactory lubricant at low speeds. Oil bath lubrication is satisfactory at medium speeds, and circulating oil or special lubrication is necessary at high speeds. While no definite dividing line exists between these different cases, Table 4 gives conservative speed limits for grease and oil bath lubrication. This limit is given in the form of a *dn* factor which is the product of the bearing bore in *millimeters* by the speed in rpm. Since these limits are approximate only, there are conditions which permit grease or oil bath lubrication at somewhat higher speeds and, if oil mist or circulating oil is used, considerably higher speeds can be reached. For example, deep-groove ball bearings and cylindrical roller bearings have been run with circulating oil lubrication at dn = 1,000,000. If the limits given in the table are to be exceeded, however, the lubrication and heat dissipation characteristics should be checked by a competent bearing engineer.

Table 4. Tentative Limiting Speeds for Grease and Oil Bath Lubrication

Type of Bearing	Type of Cage	Normal Limit for the *dn* Value When Using: Grease	Oil
Self-aligning ball bearings	Pressed steel	200,000	300,000
Single-row deep-groove ball bearings	Pressed steel	200,000	300,000
	Pressed bronze ball centered	200,000	400,000
Double-row ball bearings	Pressed steel	150,000	225,000
Angular contact ball bearings Large angle	Bronze; Phenolic composition	200,000	300,000
Angular contact ball bearings Small angle	Machined phenolic composition, ring-centered	200,000	500,000
		300,000	700,000
Cylindrical roller bearings	Machined bronze cage, roller-centered	150,000	300,000
Spherical roller bearings	Pressed steel, ring-centered	100,000	150,000
	Bronze pressed or machined, ring-centered	100,000	200,000
Thrust ball bearings	Pressed or machined	150,000	225,000

In most cases the friction which must be considered is not the bearing friction alone but the friction of the total application. If friction type seals are used (see Article 17) the friction of the application will be at least three to five times the friction of the bearing alone. If labyrinth type seals are used, the seal friction is negligible. In general, when friction is a problem the friction of the seals must be considered in the light of seal efficiency and cost and a satisfactory compromise reached.

17. LUBRICATION AND SEALS

Lubrication of bearings may be discussed by dividing the lubrication according to type of lubricant and lubricating device. Although the two common types of seals, *friction* and *labyrinth*, find use with both oil and grease lubrication, there are significant changes in their design for different types of lubricant.

Grease lubrication is used in many small or low-speed jobs where the dissipation of the heat due to bearing and seal friction involves no great difficulty. The grease chosen should be free of acid and corrosive factors and it should be of the proper consistency to give good service under operating conditions (e.g., not so hard that it channels too much and not so soft that it escapes through the seals). The housing and bearing cavity should not be filled more than two-thirds full of grease in order that the bearing friction may not

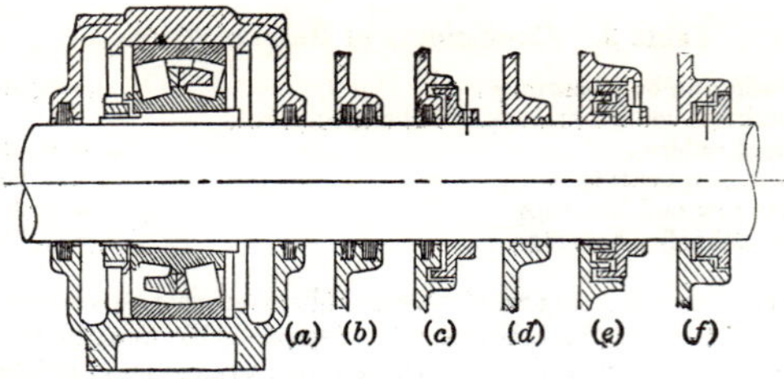

FIG. 5. Seals for grease lubrication: (*a*), (*b*) friction seals; (*c*) combination friction and labyrinth seals; (*d*), (*e*), and (*f*) labyrinth seals.

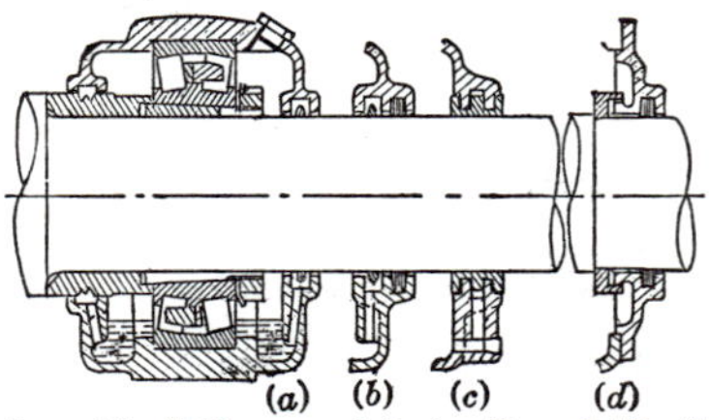

FIG. 6. Seals for oil lubrication: (*a*) oil flinger and drain; (*b*) and (*d*) oil flingers and friction seal; (*c*) labyrinth ring and drain.

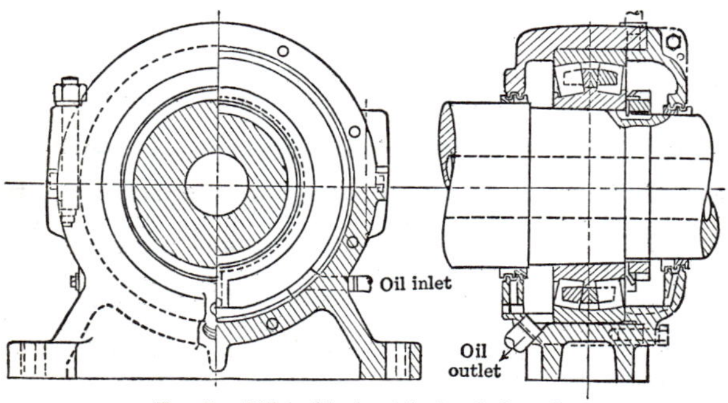

FIG. 7. Pillow block with circulating oil.

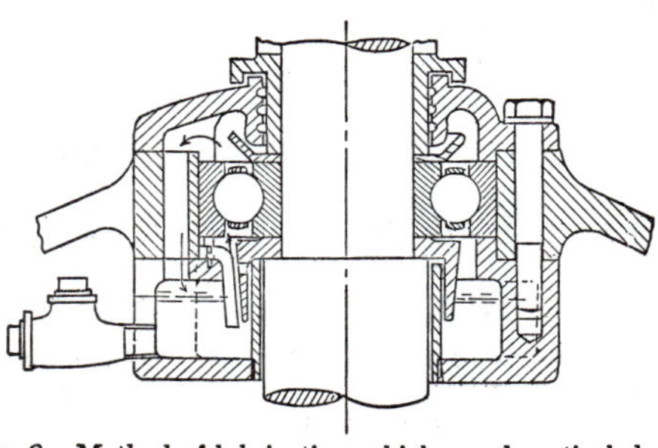

FIG. 8. Method of lubricating a high-speed vertical shaft.

become too high because of excessive churning. If grease is to be injected after a period of operation, care should be taken to see that the new grease reaches a point close to the raceways, and provision should be made for the removal of old grease.

Since grease approaches a solid in consistency, it acts to improve the efficiency of the seals. For that reason, and because of the moderate speed at which most grease-lubricated bearings operate, grease seals are usually fairly simple.

Both friction and labyrinth type seals are in common use with grease lubricated bearings, as is shown in Fig. 5.

Oil lubrication is used when the bearing speed exceeds that for which grease lubrication is satisfactory. It may also be used at lower speeds if more convenient. Oil bath lubrication is used if the speed is low enough so that the oil bath can dissipate the heat of the bearing friction without sufficient temperature rise to cause the oil to carbonize or deteriorate. The level of the oil when stationary should be even with the center of the lowermost ball or roller. A higher oil level causes excessive friction. The viscosity of the oil is determined by the nature of the bearing application. Oil in the bath should have a viscosity of at least 70 Saybolt universal seconds at operating temperature. A somewhat heavier oil is, as a rule, permissible and preferable.

Seals for oil bath lubrication are usually more complicated than for grease lubrication since the oil, being a liquid, flows easily through any simple nonfriction seal. In most cases friction seals tend to lose their efficiency after a short period and allow oil to leak out. Figure 6 shows some labyrinth seals and some combined labyrinth and friction type seals which have been found satisfactory for some oil bath lubricated bearing applications.

For speeds of rotation above that for which oil bath lubrication is permissible, two general methods are in use to insure adequate lubrication. For lightly loaded bearings it may be practical to reduce the bearing friction, and therefore the heat to be dissipated, by blowing an air current which carries a mist of oil to the bearing. This oil is just enough to give the necessary lubrication. By this means the heat of friction is held to a minimum and may be dissipated by ordinary means.

When this method of high-speed lubrication is not satisfactory, a circulating oil system is necessary. Here the heat of friction is much greater since more churning of the lubricant takes place, but this heat is rapidly removed from the bearing by the oil and carried to a cooling tank where the heat is dissipated. Another place where circulating oil may be necessary is in vertical mountings. Figure 7 shows a pillow block and bearing with adequate seals for circulating oil. Figure 8 shows a vertical mounting where an oil supply is brought to the bearing by running up the taper of the rotating shaft and returned to the oil supply by a gravity return.

18. STANDARDS AND TOLERANCES

Standard types of ball and roller bearings are interchangeable parts made to the same dimensions by all manufacturers. The internal design of the bearing may vary slightly with different manufacturers. This interchangeability of standard bearings facilitates their rapid replacement when necessary and is one of the strongest reasons why they should be used in preference to special bearings whenever possible. External dimensions of some standard bearings are given in Tables 5 to 13 and tolerances for these dimensions in Tables 14 and 15.

(*Continued on p. 12-86*)

Table 5. External Dimensions and Radial Capacities of Self-aligning Ball Bearings

(SKF Industries, Inc., Philadelphia, Pa.)

Nominal Bearing Dimensions (Fig. 1a)						Maximum Fillet Radius on Shaft	Thrust Factor	Dynamic Reference Capacity
d (bore)		*D* (OD)		*B* (width)				
mm	in.	mm	in.	mm	in.	in.	*Y*	*C*, lb
				Light Series				
10	0.3937	30	1.1811	9	0.3543	.024	2.3	865
12	0.4724	32	1.2598	10	0.3937	.024	2.2	915
15	0.5906	35	1.3780	11	0.4331	.024	2.3	1,250
17	0.6693	40	1.5748	12	0.4724	.039	2.4	1,400
20	0.7874	47	1.8504	14	0.5512	.039	2.7	1,830
25	0.9843	52	2.0472	15	0.5906	.039	2.8	2,240
30	1.1811	62	2.4409	16	0.6299	.039	3.1	3,100
35	1.3780	72	2.8346	17	0.6693	.039	3.3	3,350
40	1.5748	80	3.1496	18	0.7087	.039	3.4	4,250
45	1.7717	85	3.3465	19	0.7480	.039	3.5	4,750
50	1.9685	90	3.5433	20	0.7874	.039	3.7	5,100
55	2.1654	100	3.9370	21	0.8268	.059	3.9	6,200
60	2.3622	110	4.3307	22	0.8661	.059	4.1	7,100
65	2.5591	120	4.7244	23	0.9055	.059	4.4	7,650
70	2.7559	125	4.9213	24	0.9449	.059	4.2	8,500
75	2.9528	130	5.1181	25	0.9843	.059	4.3	9,300
80	3.1496	140	5.5118	26	1.0236	.079	4.7	9,800
85	3.3465	150	5.9055	28	1.1024	.079	4.4	11,800
90	3.5433	160	6.2992	30	1.1811	.079	4.5	12,900
100	3.9370	180	7.0866	34	1.3386	.079	4.3	16,300
110	4.3307	200	7.8740	38	1.4961	.079	4.3	20,400
				Medium Series *				
12	0.4724	37	1.4567	12	0.4724	.039	2.2	1,500
15	0.5906	42	1.6535	13	0.5118	.039	2.2	1,630
17	0.6693	47	1.8504	14	0.5512	.039	2.3	2,120
20	0.7874	52	2.0472	15	0.5906	.039	2.6	2,240
25	0.9843	62	2.4409	17	0.6693	.039	2.7	3,250
30	1.1811	72	2.8346	19	0.7480	.039	2.9	4,050
35	1.3780	80	3.1496	21	0.8268	.059	3.1	4,800
40	1.5748	90	3.5433	23	0.9055	.059	3.1	6,100
45	1.7717	100	3.9370	25	0.9843	.059	3.0	7,650
50	1.9685	110	4.3307	27	1.0630	.079	3.2	8,650
55	2.1654	120	4.7244	29	1.1417	.079	3.2	10,400
60	2.3622	130	5.1181	31	1.2205	.079	3.3	12,000
65	2.5591	140	5.5118	33	1.2992	.079	3.3	12,700
70	2.7559	150	5.9055	35	1.3780	.079	3.4	15,300
75	2.9528	160	6.2992	37	1.4567	.079	3.4	16,300
80	3.1496	170	6.6929	39	1.5354	.079	3.5	18,000
85	3.3465	180	7.0866	41	1.6142	.098	3.5	20,000
90	3.5433	190	7.4803	43	1.6929 †	.098	3.4	22,800
95	3.7402	200	7.8740	45	1.7717 ‡	.098	3.3	25,500
100	3.9370	215	8.4646	47	1.8504 §	.098	3.2	27,500
110	4.3307	240	9.4488	50	1.9685 ‖	.098	3.4	33,500
				Wide, Light Series				
20	0.7874	47	1.8504	18	0.7087	.039	1.6	2,200
25	0.9843	52	2.0472	18	0.7087	.039	1.7	2,320
30	1.1811	62	2.4409	20	0.7874	.039	1.9	3,050
35	1.3780	72	2.8346	23	0.9055	.039	2.0	4,250
40	1.5748	80	3.1496	23	0.9055	.039	2.3	4,650
45	1.7717	85	3.3465	23	0.9055	.039	2.5	5,100

* Also available in tapered bore.

† Also B_1 = 1 49/64. ‡ Also B_1 = 1 57/64. § Also B_1 = 2 3/64. ‖ Also B_1 = 2 5/32.

Table 5. External Dimensions and Radial Capacities of Self-aligning Ball Bearings—*Continued*

(SKF Industries, Inc., Philadelphia, Pa.)

Nominal Bearing Dimensions (Fig. 1*a*)						Maximum Fillet Radius on Shaft	Thrust Factor	Dynamic Reference Capacity
d (bore)		*D* (OD)		*B* (width)				
mm	in.	mm	in.	mm	in.	in.	*Y*	*C*, lb
				WIDE, LIGHT SERIES—*Continued*				
50	1.9685	90	3.5433	23	0.9055	.039	2.6	5,300
55	2.1654	100	3.9370	25	0.9843	.059	2.7	6,100
60	2.3622	110	4.3307	28	1.1024	.059	2.7	7,650
65	2.5591	120	4.7244	31	1.2205	.059	2.7	9,500
70	2.7559	125	4.9213	31	1.2205	.059	2.8	9,800
75	2.9528	130	5.1181	31	1.2205	.059	3.0	10,400
				WIDE, MEDIUM SERIES				
17	0.6693	47	1.8504	19	.7480	.039	1.5	2,320
20	0.7874	52	2.0472	21	.8268	.039	1.5	3,050
25	0.9843	62	2.4409	24	.9449	.039	1.6	4,050
30	1.1811	72	2.8346	27	1.0630	.039	1.7	5,400
35	1.3780	80	3.1496	31	1.2205	.059	1.6	6,700
40	1.5748	90	3.5433	33	1.2992	.059	1.7	8,000
45	1.7717	100	3.9370	36	1.4173	.059	1.8	9,500
50	1.9685	110	4.3307	40	1.5748	.079	1.8	11,200
55	2.1654	120	4.7244	43	1.6929	.079	1.8	12,900
60	2.3622	130	5.1181	46	1.8110	.079	1.9	15,000
65	2.5591	140	5.5118	48	1.8898	.079	2.0	16,300
70	2.7559	150	5.9055	51	2.0079	.079	2.0	18,300
75	2.9528	160	6.2992	55	2.1654	.079	2.0	20,400
80	3.1496	170	6.6929	58	2.2835	.079	2.0	22,400
85	3.3465	180	7.0866	60	2.3622	.098	2.0	23,600

Table 6. External Dimensions and Radial Capacities of Single-row Deep-groove Ball Bearings

(SKF Industries, Inc., Philadelphia, Pa.)

Nominal Bearing Dimensions (Fig. 1b)						Maximum Fillet Radius on Shaft	Thrust Factor	Dynamic Reference Capacity
d (bore)		D (OD)		B (width)				
mm	in.	mm	in.	mm	in.	in.	Y *	C, lb
Light Series								
10	0.3937	30	1.1811	9	0.3543	.024	1–1.5	750
12	0.4724	32	1.2598	10	0.3937	.024	1–1.5	1,140
15	0.5906	35	1.3780	11	0.4331	.024	1–1.5	1,250
17	0.6693	40	1.5748	12	0.4724	.039	1–1.5	1,600
20	0.7874	47	1.8504	14	0.5512	.039	1–1.5	2,160
25	0.9843	52	2.0472	15	0.5906	.039	1–1.5	2,320
30	1.1811	62	2.4409	16	0.6299	.039	1–1.5	3,250
35	1.3780	72	2.8346	17	0.6693	.039	1–1.5	4,300
40	1.5748	80	3.1496	18	0.7087	.039	1–1.5	4,900
45	1.7717	85	3.3465	19	0.7480	.039	1–1.5	5,500
50	1.9685	90	3.5433	20	0.7874	.039	1–1.5	5,850
55	2.1654	100	3.9370	21	0.8268	.059	1–1.5	7,350
60	2.3622	110	4.3307	22	0.8661	.059	1–1.5	8,800
65	2.5591	120	4.7244	23	0.9055	.059	1–1.5	9,650
70	2.7559	125	4.9213	24	0.9449	.059	1–1.5	10,400
75	2.9528	130	5.1181	25	0.9843	.059	1–1.5	11,000
80	3.1496	140	5.5118	26	1.0236	.079	1–1.5	12,200
85	3.3465	150	5.9055	28	1.1024	.079	1–1.5	13,700
90	3.5433	160	6.2992	30	1.1811	.079	1–1.5	15,600
95	3.7402	170	6.6929	32	1.2598	.079	1–1.5	17,600
100	3.9370	180	7.0866	34	1.3386	.079	1–1.5	19,600
105	4.1339	190	7.4803	36	1.4173	.079	1–1.5	21,600
110	4.3307	200	7.8740	38	1.4961	.079	1–1.5	24,000
Medium Series								
10	0.3937	35	1.3780	11	0.4331	.024	1–1.5	1,460
12	0.4724	37	1.4567	12	0.4724	.039	1–1.5	1,800
15	0.5906	42	1.6535	13	0.5118	.039	1–1.5	1,960
17	0.6693	47	1.8504	14	0.5512	.039	1–1.5	2,360
20	0.7874	52	2.0472	15	0.5906	.039	1–1.5	2,750
25	0.9843	62	2.4409	17	0.6693	.039	1–1.5	3,650
30	1.1811	72	2.8346	19	0.7480	.039	1–1.5	4,800
35	1.3780	80	3.1496	21	0.8268	.059	1–1.5	5,700
40	1.5748	90	3.5433	23	0.9055	.059	1–1.5	6,950
45	1.7717	100	3.9370	25	0.9843	.059	1–1.5	9,000
50	1.9685	110	4.3307	27	1.0630	.079	1–1.5	10,400
55	2.1654	120	4.7244	29	1.1417	.079	1–1.5	12,000
60	2.3622	130	5.1181	31	1.2205	.079	1–1.5	13,700
65	2.5591	140	5.5118	33	1.2992	.079	1–1.5	15,300
70	2.7559	150	5.9055	35	1.3780	.079	1–1.5	17,000
75	2.9528	160	6.2992	37	1.4567	.079	1–1.5	18,600
80	3.1496	170	6.6929	39	1.5354	.079	1–1.5	20,800
85	3.3465	180	7.0866	41	1.6142	.098	1–1.5	22,400
90	3.5433	190	7.4803	43	1.6929	.098	1–1.5	24,500
95	3.7402	200	7.8740	45	1.7717	.098	1–1.5	26,500
100	3.9370	215	8.4646	47	1.8504	.098	1–1.5	30,500
105	4.1339	225	8.8583	49	1.9291	.098	1–1.5	32,500
110	4.3307	240	9.4488	50	1.9685	.098	1–1.5	37,500
Heavy Series								
25	0.9843	80	3.1496	21	0.8268	.059	1–1.5	6,300
30	1.1811	90	3.5433	23	0.9055	.059	1–1.5	7,500
35	1.3780	100	3.9370	25	0.9843	.059	1–1.5	9,500
40	1.5748	110	4.3307	27	1.0630	.079	1–1.5	11,000
45	1.7717	120	4.7244	29	1.1417	.079	1–1.5	13,200
50	1.9685	130	5.1181	31	1.2205	.079	1–1.5	14,600
55	2.1654	140	5.5118	33	1.2992	.079	1–1.5	17,300
60	2.3622	150	5.9055	35	1.3780	.079	1–1.5	19,000
65	2.5591	160	6.2992	37	1.4567	.079	1–1.5	20,800
70	2.7559	180	7.0866	42	1.6535	.098	1–1.5	26,000
75	2.9528	190	7.4803	45	1.7717	.098	1–1.5	28,000

* As $\frac{C}{F_r + 1.5F_a}$ varies from 3 to 15, Y varies from 1 to 1.5.

Table 7. External Dimensions and Radial Capacities of Angular Contact Ball Bearings
(SKF Industries, Inc., Philadelphia, Pa.)

Nominal Bearing Dimensions (Fig. 1c)						Maximum Fillet Radius on Shaft	Thrust Factor	Dynamic Reference Capacity
d (bore)		D (OD)		B (width)				
mm	in.	mm	in.	mm	in.	in.	Y *	C, lb
EXTRA LIGHT, SMALL ANGLE								
35	1.3780	62	2.4409	14	0.5512	.039	.9–1.3	3,150
40	1.5748	68	2.6772	15	0.5906	.039	.9–1.3	3,450
45	1.7717	75	2.9528	16	0.6299	.039	.9–1.3	4,150
50	1.9685	80	3.1496	16	0.6299	.039	.9–1.3	4,400
55	2.1654	90	3.5433	18	0.7087	.039	.9–1.3	5,850
60	2.3622	95	3.7402	18	0.7087	.039	.9–1.3	6,000
65	2.5591	100	3.9370	18	0.7087	.039	.9–1.3	6,400
70	2.7559	110	4.3307	20	0.7874	.039	.9–1.3	7,800
75	2.9528	115	4.5276	20	0.7874	.039	.9–1.3	8,300
80	3.1496	125	4.9213	22	0.8661	.039	.9–1.3	10,200
90	3.5433	140	5.5118	24	0.9449	.059	.9–1.3	12,200
100	3.9370	150	5.9055	24	0.9449	.059	.9–1.3	13,200
110	4.3307	170	6.6929	28	1.1024	.079	.9–1.3	17,300
120	4.7244	180	7.0866	28	1.1024	.079	.9–1.3	18,300
130	5.1181	200	7.8740	33	1.2992	.079	.9–1.3	22,000
140	5.5118	210	8.2677	33	1.2992	.079	.9–1.3	22,800
150	5.9055	225	8.8583	35	1.3780	.079	.9–1.3	25,500
160	6.2992	240	9.4488	38	1.4961	.079	.9–1.3	29,000
LIGHT SERIES, SMALL ANGLE								
12	0.4724	32	1.2598	10	0.3937	.024	.9–1.3	1,220
15	0.5906	35	1.3780	11	0.4331	.024	.9–1.3	1,400
17	0.6693	40	1.5748	12	0.4724	.039	.9–1.3	1,800
20	0.7874	47	1.8504	14	0.5512	.039	.9–1.3	2,400
25	0.9843	52	2.0472	15	0.5906	.039	.9–1.3	2,750
30	1.1811	62	2.4409	16	0.6299	.039	.9–1.3	3,800
35	1.3780	72	2.8346	17	0.6693	.039	.9–1.3	5,100
40	1.5748	80	3.1496	18	0.7087	.039	.9–1.3	6,100
45	1.7717	85	3.3465	19	0.7480	.039	.9–1.3	6,700
50	1.9685	90	3.5433	20	0.7874	.039	.9–1.3	6,950
55	2.1654	100	3.9370	21	0.8268	.059	.9–1.3	8,650
60	2.3622	110	4.3307	22	0.8661	.059	.9–1.3	10,400
65	2.5591	120	4.7244	23	0.9055	.059	.9–1.3	11,200
70	2.7559	125	4.9213	24	0.9449	.059	.9–1.3	12,200
75	2.9528	130	5.1181	25	0.9843	.059	.9–1.3	13,400
80	3.1496	140	5.5118	26	1.0236	.079	.9–1.3	15,000
85	3.3465	150	5.9055	28	1.1024	.079	.9–1.3	16,600
90	3.5433	160	6.2992	30	1.1811	.079	.9–1.3	19,300
95	3.7402	170	6.6929	32	1.2598	.079	.9–1.3	20,800
100	3.9370	180	7.0866	34	1.3386	.079	.9–1.3	23,200
LIGHT SERIES								
25	0.9843	52	2.0472	15	0.5906	.039	.48	2,280
30	1.1811	62	2.4409	16	0.6299	.039	.48	3,200
35	1.3780	72	2.8346	17	0.6693	.039	.48	4,250
40	1.5748	80	3.1496	18	0.7087	.039	.48	5,100
45	1.7717	85	3.3465	19	0.7480	.039	.48	5,700
50	1.9685	90	3.5433	20	0.7874	.039	.48	6.000
55	2.1654	100	3.9370	21	0.8268	.059	.48	7,350
60	2.3622	110	4.3307	22	0.8661	.059	.48	8,800
65	2.5591	120	4.7244	23	0.9055	.059	.48	10,000
70	2.7559	125	4.9213	24	0.9449	.059	.48	11,000
75	2.9528	130	5.1181	25	0.9843	.059	.48	11,400
80	3.1496	140	5.5118	26	1.0236	.079	.48	12,700

* For small-angle series, Y varies from 0.9 to 1.3 as $\frac{C}{0.5F_r + 1.3F_a}$ varies from 5 to 20.

(*Table continued on p. 12-78*)

Table 7. External Dimensions and Radial Capacities of Angular Contact Ball Bearings—*Continued*

(SKF Industries, Inc., Philadelphia, Pa.)

Nominal Bearing Dimensions (Fig. 1c)						Maximum Fillet Radius on Shaft	Thrust Factor	Dynamic Reference Capacity
d (bore)		*D* (OD)		*B* (width)				
mm	in.	mm	in.	mm	in.	in.	*Y* *	*C*, lb
Light Series								
85	3.3465	150	5.9055	28	1.1024	.079	.48	14,300
90	3.5433	160	6.2992	30	1.1811	.079	.48	16,600
95	3.7402	170	6.6929	32	1.2598	.079	.48	17,600
100	3.9370	180	7.0866	34	1.3386	.079	.48	19,600
110	4.3307	200	7.8740	38	1.4961	.079	.48	24,000
120	4.7244	215	8.4646	40	1.5748	.079	.48	27,500
130	5.1181	230	9.0551	40	1.5748	.098	.48	31,000
140	5.5118	250	9.8425	42	1.6535	.098	.48	33,500
150	5.9055	270	10.6299	45	1.7717	.098	.48	39,000
160	6.2992	290	11.4173	48	1.8898	.098	.48	41,500
170	6.6929	310	12.2047	52	2.0472	.118	.48	43,000
Medium Series								
20	0.7874	52	2.0472	15	0.5906	.039	.48	2,700
25	0.9843	62	2.4409	17	0.6693	.039	.48	3,800
30	1.1811	72	2.8346	19	0.7480	.039	.48	4,800
35	1.3780	80	3.1496	21	0.8268	.059	.48	5,700
40	1.5748	90	3.5433	23	0.9055	.059	.48	6,950
45	1.7717	100	3.9370	25	0.9843	.059	.48	9,000
50	1.9685	110	4.3307	27	1.0630	.079	.48	10,400
55	2.1654	120	4.7244	29	1.1417	.079	.48	12,000
60	2.3622	130	5.1181	31	1.2205	.079	.48	13,700
65	2.5591	140	5.5118	33	1.2992	.079	.48	15,300
70	2.7559	150	5.9055	35	1.3780	.079	.48	17,000
75	2.9528	160	6.2992	37	1.4567	.079	.48	19,000
80	3.1496	170	6.6929	39	1.5354	.079	.48	20,800
85	3.3465	180	7.0866	41	1.6142	.098	.48	22,800
90	3.5433	190	7.4803	43	1.6929	.098	.48	24,500
95	3.7402	200	7.8740	45	1.7717	.098	.48	26,500
100	3.9370	215	8.4646	47	1.8504	.098	.48	32,500
105	4.1339	225	8.8583	49	1.9291	.098	.48	34,500
110	4.3307	240	9.4488	50	1.9685	.098	.48	39,000
130	5.1181	280	11.0236	58	2.2835	.118	.48	48,000
150	5.9055	320	12.5984	65	2.5591	.118	.48	58,500
Heavy Series								
25	0.9843	80	3.1496	21	0.8268	.059	.48	6,200
30	1.1811	90	3.5433	23	0.9055	.059	.48	7,350
35	1.3780	100	3.9370	25	0.9843	.059	.48	9,300
40	1.5748	110	4.3307	27	1.0630	.079	.48	10,600
45	1.7717	120	4.7244	29	1.1417	.079	.48	12,900
50	1.9685	130	5.1181	31	1.2205	.079	.48	14,300
55	2.1654	140	5.5118	33	1.2992	.079	.48	16,600
60	2.3622	150	5.9055	35	1.3780	.079	.48	18,300
65	2.5591	160	6.2992	37	1.4567	.079	.48	20,000
70	2.7559	180	7.0866	42	1.6535	.098	.48	25,500
75	2.9528	190	7.4803	45	1.7717	.098	.48	27,000

* For small-angle series, Y varies from 0.9 to 1.3 as $\dfrac{C}{0.5F_r + 1.3F_a}$ varies from 5 to 20.

Table 8. External Dimensions and Radial Capacities of Double-row Ball Bearings

(SKF Industries, Inc., Philadelphia, Pa.)

Nominal Bearing Dimensions (Fig. 1*d*)						Maximum Fillet Radius on Shaft	Thrust Factor	Dynamic Reference Capacity
d (bore)		*D* (OD)		*B* (width)				
mm	in.	mm	in.	mm	in.	in.	*Y*	*C*, lb
LIGHT SERIES								
10	0.3937	30	1.1811	14.3	0.5625	.024	1.6	1,220
12	0.4724	32	1.2598	15.9	0.6250	.024	1.6	1,860
15	0.5906	35	1.3780	15.9	0.6250	.024	1.6	2,040
17	0.6693	40	1.5748	17.5	0.6875	.039	1.6	2,600
20	0.7874	47	1.8504	20.6	0.8125	.039	1.6	3,550
25	0.9843	52	2.0472	20.6	0.8125	.039	1.6	3,800
30	1.1811	62	2.4409	23.8	0.9375	.039	1.6	5,300
35	1.3780	72	2.8346	27.0	1.0625	.039	1.6	7,100
40	1.5748	80	3.1496	30.2	1.1875	.039	1.6	8,000
45	1.7717	85	3.3465	30.2	1.1875	.039	1.6	9,000
50	1.9685	90	3.5433	30.2	1.1875	.039	1.6	9,650
55	2.1654	100	3.9370	33.3	1.3125	.059	1.6	11,800
60	2.3622	110	4.3307	36.5	1.4375	.059	1.6	14,300
65	2.5591	120	4.7244	38.1	1.5000	.059	1.6	15,600
70	2.7559	125	4.9213	39.7	1.5625	.059	1.6	17,000
75	2.9528	130	5.1181	41.3	1.6250	.059	1.6	18,000
80	3.1496	140	5.5118	44.4	1.7500	.079	1.6	19,600
85	3.3465	150	5.9055	49.2	1.9375	.079	1.6	20,400
90	3.5433	160	6.2992	52.4	2.0625	.079	1.6	23,200
95	3.7402	170	6.6929	55.6	2.1875	.079	1.6	28,500
100	3.9370	180	7.0866	60.3	2.3750	.079	1.6	32,000

mm	in.	mm	in.	mm	in.	in.	*Y*	*C*, lb
MEDIUM SERIES								
17	0.6693	47	1.8504	22.2	0.8750	.039	1.6	3,800
20	0.7874	52	2.0472	22.2	0.8750	.039	1.6	4,500
25	0.9843	62	2.4409	25.4	1.0000	.039	1.6	6,000
30	1.1811	72	2.8346	30.2	1.1875	.039	1.6	7,800
35	1.3780	80	3,1496	34.9	1.3750	.059	1.6	9,300
40	1.5748	90	3.5433	36.5	1.4375	.059	1.6	11,200
45	1.7717	100	3.9370	39.7	1.5625	.059	1.6	14,600
50	1.9685	110	4.3307	44.4	1.7500	.079	1.6	17,000
55	2.1654	120	4.7244	49.2	1.9375	.079	1.6	19,300
60	2.3622	130	5.1181	54.0	2.1250	.079	1.6	22,000
65	2.5591	140	5.5118	58.7	2.3125	.079	1.6	25,000
70	2.7559	150	5.9055	63.5	2.5000	.079	1.6	27,500
75	2.9528	160	6.2992	68.3	2.6875	.079	1.6	30,500
80	3.1496	170	6.6929	68.3	2.6875	.079	1.6	33,500
85	3.3465	180	7.0866	73.0	2.8750	.098	1.6	33,500
90	3.5433	190	7.4803	73.0	2.8750	.098	1.6	35,500

Table 9. External Dimensions and Thrust Capacities of Ball Thrust Bearings

(SKF Industries, Inc., Philadelphia, Pa.)

Nominal Bearing Dimensions (Fig. 1e)								Maximum Fillet Radius on Shaft	Dynamic Reference Capacity *
d (bore)		d_2		D (OD)		H (height)			
mm	in.	mm	in.	mm	in.	mm	in.	in.	C, lb
Light Series									
35	1.3780	36	1.417	53	2.0866	12	0.472	.024	3,200
40	1.5748	42	1.654	60	2.3622	13	0.512	.024	4,300
45	1.7717	47	1.850	65	2.5591	14	0.551	.024	4,550
50	1.9685	52	2.047	70	2.7559	14	0.551	.024	4,900
55	2.1654	57	2.244	78	3.0709	16	0.630	.024	6,000
60	2.3622	62	2.441	85	3.3465	17	0.669	.039	7,100
65	2.5591	67	2.638	90	3.5433	18	0.709	.039	7,350
70	2.7559	72	2.835	95	3.7402	18	0.709	.039	7,650
75	2.9528	77	3.031	100	3.9370	19	0.748	.039	8,150
80	3.1496	82	3.228	105	4.1339	19	0.748	.039	8,300
85	3.3465	88	3.465	110	4.3307	19	0.748	.039	8,650
90	3.5433	93	3.661	120	4.7244	22	0.866	.039	11,000
100	3.9370	103	4.055	135	5.3150	25	0.984	.039	15,300
110	4.3307	113	4.449	145	5.7087	25	0.984	.039	16,300
120	4.7244	123	4.843	155	6.1024	25	0.984	.039	17,000
130	5.1181	133	5.236	170	6.6929	30	1.181	.039	19,300
140	5.5118	143	5.630	180	7.0866	31	1.220	.039	20,000
Medium Series									
15	0.5906	16	0.630	35	1.3780	14	0.551	.024	2,750
20	0.7874	21	0.827	40	1.5748	14	0.551	.024	3,100
25	0.9843	26	1.024	48	1.8898	15.5	0.610	.024	4,000
30	1.1811	31	1.220	53	2.0866	15.5	0.610	.024	4,400
35	1.3780	36	1.417	62	2.4409	18	0.709	.024	6,100
40	1.5748	42	1.654	64	2.5197	18	0.709	.024	5,700
45	1.7717	47	1.850	73	2.8740	22	0.866	.024	7,650
50	1.9685	52	2.047	78	3.0709	22	0.866	.024	8,800
55	2.1654	57	2.244	88	3.4646	24.5	0.965	.024	10,800
60	2.3622	62	2.441	90	3.5433	24.5	0.965	.024	9,800
65	2.5591	67	2.638	100	3.9370	27	1.063	.039	12,200
70	2.7559	72	2.835	103	4.0551	27	1.063	.039	12,500
75	2.9528	77	3.031	110	4.3307	27	1.063	.039	12,900
80	3.1496	82	3.228	115	4.5276	29	1.142	.039	15,000
85	3.3465	88	3.465	125	4.9213	30.5	1.201	.039	18,000
90	3.5433	93	3.661	135	5.3150	30.5	1.201	.039	21,200
95	3.7402	98	3.858	140	5.5118	32.5	1.280	.039	22,800
100	3.9370	103	4.055	150	5.9055	32.5	1.280	.039	25,500
105	4.1339	108	4.252	155	6.1024	40	1.575	.039	28,500
115	4.5276	118	4.646	165	6.4961	43	1.693	.039	29,000
125	4.9213	128	5.039	175	6.8898	46.5	1.831	.039	31,500
130	5.1181	133	5.236	180	7.0866	46.5	1.831	.039	30,500
140	5.5118	143	5.630	200	7.8740	52	2.047	.039	39,000

* Dynamic reference capacity is in thrust load.

Table 10. External Dimensions and Radial Capacities of Single-row Cylindrical Roller Bearings

(SKF Industries, Inc., Philadelphia, Pa.)

Nominal Bearing Dimensions (Fig. 2*a*)						Maximum Fillet Radius on Shaft	Dynamic Reference Capacity
d (bore)		*D* (OD)		*B* (width)			
mm	in.	mm	in.	mm	in.	in.	*C*, lb
			Light Series				
25	0.9843	52	2.0472	15	0.5906	.039	2,450
30	1.1811	62	2.4409	16	0.6299	.039	4,050
35	1.3780	72	2.8346	17	0.6693	.039	5,850
40	1.5748	80	3.1496	18	0.7087	.039	7,650
45	1.7717	85	3.3465	19	0.7480	.039	8,000
50	1.9685	90	3.5433	20	0.7874	.039	8,300
55	2.1654	100	3.9370	21	0.8268	.059	10,200
60	2.3622	110	4.3307	22	0.8661	.059	12,500
65	2.5591	120	4.7244	23	0.9055	.059	14,600
70	2.7559	125	4.9213	24	0.9449	.059	14,600
75	2.9528	130	5.1181	25	0.9843	.059	18,000
80	3.1496	140	5.5118	26	1.0236	.079	19,600
85	3.3465	150	5.9055	28	1.1024	.079	22,400
90	3.5433	160	6.2992	30	1.1811	.079	28,500
95	3.7402	170	6.6929	32	1.2598	.079	31,500
100	3.9370	180	7.0866	34	1.3386	.079	34,500
110	4.3307	200	7.8740	38	1.4961	.079	46,500
120	4.7244	215	8.4646	40	1.5748	.079	50,000
			Medium Series				
25	0.9843	62	2.4409	17	0.6693	.039	5,200
30	1.1811	72	2.8346	19	0.7480	.039	6,800
35	1.3780	80	3.1496	21	0.8268	.059	8,800
40	1.5748	90	3.5433	23	0.9055	.059	10,400
45	1.7717	100	3.9370	25	0.9843	.059	14,300
50	1.9685	110	4.3307	27	1.0630	.079	17,000
55	2.1654	120	4.7244	29	1.1417	.079	20,800
60	2.3622	130	5,1181	31	1.2205	.079	23,200
65	2.5591	140	5.5118	33	1.2992	.079	26,000
70	2.7559	150	5.9055	35	1.3780	.079	30,500
75	2.9528	160	6.2992	37	1.4567	.079	36,500
80	3.1496	170	6.6929	39	1.5354	.079	36,500
85	3.3465	180	7.0866	41	1.6142	.098	44,000
90	3.5433	190	7.4803	43	1.6929	.098	47,500
95	3.7402	200	7.8740	45	1.7717	.098	54,000
100	3.9370	215	8.4646	47	1.8504	.098	63,000
110	4.3307	240	9.4488	50	1.9685	.098	81,500
120	4.7244	260	10.2362	55	2.1654	.098	98,000
130	5.1181	280	11.0236	58	2.2835	.118	116,000
140	5.5118	300	11.8110	62	2.4409	.118	129,000
150	5.9055	320	12.5984	65	2.5591	.118	140,000

Note. Type listed has no flanges on the outer ring (Fig. 2*a*). The bearing is also available with no flanges on the inner ring (Fig. 2*c*), and one flange on inner ring, two on outer (Fig. 2*b*).

Table 11. External Dimensions and Radial Capacities of Double-row Cylindrical Roller Bearings

(SKF Industries, Inc., Philadelphia, Pa.)

Nominal Bearing Dimensions (Fig. 2*d*)						Maximum Fillet Radius on Shaft	Dynamic Reference Capacity
d (bore)		*D* (OD)		*B* (width)			
mm	in.	mm	in.	mm	in.	in.	*C*, lb
25	0.9843	47	1.8504	16	0.6299	.024	3,800
30	1.1811	55	2.1654	19	0.7480	.039	4,550
35	1.3780	62	2.4409	20	0.7874	.039	5,850
40	1.5748	68	2.6772	21	0.8268	.039	6,800
45	1.7717	75	2.9528	23	0.9055	.039	8,000
50	1.9685	80	3.1496	23	0.9055	.039	8,500
55	2.1654	90	3.5433	26	1.0236	.039	11,000
60	2.3622	95	3.7402	26	1.0236	.039	11,800
65	2.5591	100	3.9370	26	1.0236	.039	12,500
70	2.7559	110	4.3307	30	1.1811	.039	15,600
75	2.9528	115	4.5276	30	1.1811	.039	15,600
80	3.1496	125	4.9213	34	1.3386	.039	19,300
85	3.3465	130	5.1181	34	1.3386	.039	20,400
90	3.5433	140	5.5118	37	1.4567	.059	23,600
95	3.7402	145	5.7087	37	1.4567	.059	24,500
100	3.9370	150	5.9055	37	1.4567	.059	26,000
105	4.1339	160	6.2992	41	1.6142	.079	32,500
110	4.3307	170	6.6929	45	1.7717	.079	38,000
120	4.7244	180	7.0866	46	1.8110	.079	40,000
130	5.1181	200	7.8740	52	2.0472	.079	50,000
140	5.5118	210	8.2677	53	2.0866	.079	52,000
150	5.9055	225	8.8583	56	2.2047	.079	58,500
160	6.2992	240	9.4488	60	2.3622	.079	65,500
170	6.6929	260	10.2362	67	2.6378	.079	81,500

Note. Available in tapered bore only; *d* dimension is minimum bore.

Table 12. External Dimensions and Radial Capacities of Spherical Roller Bearings
(SKF Industries, Inc., Philadelphia, Pa.)

Nominal Bearing Dimensions (Fig. 2d)						Maximum Fillet Radius on Shaft	Thrust Factor	Dynamic Reference Capacity
d (bore)		D (OD)		B (width)				
mm	in.	mm	in.	mm	in.	in.	Y	C, lb
Dimension Series 30								
120	4.7244	180	7.0866	46	1.8110	.079	4.0	48,000
130	5.1181	200	7.8740	52	2.0472	.079	3.8	61,000
140	5.5118	210	8.2677	53	2.0866	.079	3.9	64,000
150	5.9055	225	8.8583	56	2.2047	.079	4.1	73,500
160	6.2992	240	9.4488	60	2.3622	.079	4.1	85,000
170	6.6929	260	10.2362	67	2.6378	.079	3.9	102,000
180	7.0866	280	11.0236	74	2.9134	.079	3.7	122,000
190	7.4803	290	11.4173	75	2.9528	.079	3.9	129,000
200	7.8740	310	12.2047	82	3.2283	.079	3.8	150,000
220	8.6614	340	13.3858	90	3.5433	.098	3.7	180,000
240	9.4488	360	14.1732	92	3.6220	.098	4.0	193,000
260	10.2362	400	15.7480	104	4.0945	.118	3.8	240,000
280	11.0236	420	16.5354	106	4.1732	.118	4.0	260,000
300	11.8110	460	18.1102	118	4.6457	.118	3.9	325,000
320	12.5984	480	18.8976	121	4.7638	.118	4.0	340,000
340	13.3858	520	20.4724	133	5.2362	.157	4.0	415,000
360	14.1732	540	21.2598	134	5.2756	.157	4.1	430,000
380	14.9606	560	22.0472	135	5.3150	.157	4.2	450,000
400	15.7480	600	23.6220	148	5.8268	.157	4.1	520,000
420	16.5354	620	24.4094	150	5.9055	.157	4.2	550,000
440	17.3228	650	25.5906	157	6.1811	.197	4.2	600,000
460	18.1102	680	26.7717	163	6.4173	.197	4.3	655,000
480	18.8976	700	27.5591	165	6.4961	.197	4.4	670,000
500	19.6850	720	28.3465	167	6.5748	.197	4.5	710,000
Dimension Series 31								
110	4.3307	180	7.0866	56	2.2047	.079	3.1	58,500
120	4.7244	200	7.8740	62	2.4409	.079	3.1	73,500
130	5.1181	210	8.2677	64	2.5197	.079	3.2	80,000
140	5.5118	225	8.8583	68	2.6772	.079	3.2	91,500
150	5.9055	250	9.8425	80	3.1496	.079	3.0	118,000
160	6.2992	270	10.6299	86	3.3858	.079	3.0	137,000
170	6.6929	280	11.0236	88	3.4646	.079	3.1	146,000
180	7.0866	300	11.8110	96	3.7795	.098	3.0	176,000
190	7.4803	320	12.5984	104	4.0945	.098	3.0	196,000
200	7.8740	340	13.3858	112	4.4094	.098	2.9	224,000
220	8.6614	370	14.5669	120	4.7244	.118	3.0	265,000
240	9.4488	400	15.7480	128	5.0394	.118	3.1	300,000
260	10.2362	440	17.3228	144	5.6693	.118	3.0	360,000
280	11.0236	460	18.1102	146	5.7480	.157	3.1	390,000
300	11.8110	500	19.6850	160	6.2992	.157	3.1	465,000
320	12.5984	540	21.2598	176	6.9291	.157	2.9	550,000
340	13.3858	580	22.8346	190	7.4803	.157	2.9	640,000
360	14.1732	600	23.6220	192	7.5591	.157	3.0	670,000
380	14.9606	620	24.4094	194	7.6378	.157	3.1	710,000
400	15.7480	650	25.5906	200	7.8740	.197	3.2	765,000
420	16.5354	700	27.5591	224	8.8189	.197	3.1	915,000
Dimension Series 22								
80	3.1496	140	5.5118	33	1.2992	.079	4.0	20,800
85	3.3465	150	5.9055	36	1.4173	.079	3.9	27,000
90	3.5433	160	6.2992	40	1.5748	.079	3.7	34,000
95	3.7402	170	6.6929	43	1.6929	.079	3.7	40,000
100	3.9370	180	7.0866	46	1.8110	.079	3.7	46,500

Note. Also available in tapered bore. (*Table continued on p. 12-84*)

Table 12. External Dimensions and Radial Capacities of Spherical Roller Bearings— *Continued*

(SKF Industries, Inc., Philadelphia, Pa.)

Nominal Bearing Dimensions (Fig. 2d)						Maximum Fillet Radius on Shaft	Thrust Factor	Dynamic Reference Capacity
d (bore)		D (OD)		B (width)				
mm	in.	mm	in.	mm	in.	in.	Y	C, lb
				DIMENSION SERIES 22				
110	4.3307	200	7.8740	53	2.0866	.079	3.5	61,000
120	4.7244	215	8.4646	58	2.2835	.079	3.5	75,000
130	5.1181	230	9.0551	64	2.5197	.098	3.6	93,000
140	5.5118	250	9.8425	68	2.6772	.098	3.6	106,000
150	5.9055	270	10.6299	73	2.8740	.098	3.6	118,000
160	6.2992	290	11.4173	80	3.1496	.098	3.6	143,000
170	6.6929	310	12.2047	86	3.3858	.118	3.5	163,000
180	7.0866	320	12.5984	86	3.3858	.118	3.5	166,000
190	7.4803	340	13.3858	92	3.6220	.118	3.5	183,000
200	7.8740	360	14.1732	98	3.8583	.118	3.4	204,000
220	8.6614	400	15 7480	108	4.2520	.118	3.5	260,000
				DIMENSION SERIES 32				
100	3.9370	180	7.0866	60.3	2.3750	.079	2.9	67,000
110	4.3307	200	7.8740	69.8	2.7500	.079	2.9	83,000
120	4.7244	215	8.4646	76	2.9921	.079	2.7	102,000
130	5.1181	230	9.0551	80	3.1496	.098	2.7	114,000
140	5.5118	250	9.8425	88	3.4646	.098	2.7	134,000
150	5.9055	270	10.6299	96	3.7795	.098	2.7	160,000
160	6.2992	290	11.4173	104	4.0945	.098	2.6	183,000
170	6.6929	310	12.2047	110	4.3307	.118	2.7	208,000
180	7.0866	320	12.5984	112	4.4094	.118	2.7	220,000
190	7.4803	340	13.3858	120	4.7244	.118	2.7	255,000
200	7.8740	360	14.1732	128	5.0394	.118	2.7	285,000
220	8.6614	400	15.7480	144	5.6693	.118	2.6	335,000
240	9.4488	440	17.3228	160	6.2992	.118	2.5	405,000
260	10.2362	480	18.8976	174	6.8504	.157	2.5	480,000
280	11.0236	500	19.6850	176	6.9291	.157	2.7	510,000
300	11.8110	540	21.2598	192	7.5591	.157	2.7	620,000
320	12.5984	580	22.8346	208	8.1890	.157	2.6	695,000
				DIMENSION SERIES 23				
40	1.5748	90	3.5433	33	1.2992	.059	2.4	13,700
45	1.7717	100	3.9370	36	1.4173	.059	2.4	17,600
50	1.9685	110	4.3307	40	1.5748	.079	2.4	24,000
55	2.1654	120	4.7244	43	1.6929	.079	2.4	28,500
60	2.3622	130	5.1181	46	1.8110	.079	2.4	34,000
65	2.5591	140	5.5118	48	1.8898	.079	2.7	36,500
70	2.7559	150	5.9055	51	2.0079	.079	2.7	49,000
75	2.9528	160	6.2992	55	2.1654	.079	2.7	51,000
80	3.1496	170	6.6929	58	2.2835	.079	2.7	61,000
85	3.3465	180	7.0866	60	2.3622	.098	2.8	65,500
90	3.5433	190	7.4803	64	2.5197	.098	2.7	78,000
95	3.7402	200	7.8740	67	2.6378	.098	2.7	85,000
100	3.9370	215	8.4646	73	2.8740	.098	2.7	100,000
110	4.3307	240	9.4488	80	3.1496	.098	2.7	122,000
120	4.7244	260	10.2362	86	3.3858	.098	2.8	150,000
130	5.1181	280	11.0236	93	3.6614	.118	2.7	173,000
140	5.5118	300	11.8110	102	4.0157	.118	2.7	190,000
150	5.9055	320	12.5984	108	4.2520	.118	2.7	212,000
160	6.2992	340	13.3858	114	4.4882	.118	2.7	232,000
170	6.6929	360	14.1732	120	4.7244	.118	2.7	270,000
180	7.0866	380	14.9606	126	4.9606	.118	2.7	290,000
190	7.4803	400	15.7480	132	5.1968	.157	2.7	320,000
200	7.8740	420	16.5354	138	5.4331	.157	2.7	340,000
220	8.6614	460	18.1102	145	5.7087	.157	2.9	400,000

Note. Also available in tapered bore.

Table 13. External Dimensions and Thrust Capacities of Spherical Roller Thrust Bearings

Nominal Bearing Dimensions (Fig. 2e)						Maximum Fillet Radius	Dynamic Reference Capacity *
d (bore)		D (OD)		H (height)			
mm	in.	mm	in.	mm	in.	in.	C, lb
Medium							
110	4.3307	190	7.4803	48	1.890	.079	86,500
130	5.1181	225	8.8583	58	2.283	.079	122,000
150	5.9055	250	9.8425	60	2.362	.079	140,000
170	6.6929	280	11.0236	67	2.638	.098	170,000
190	7.4803	320	12.5984	78	3.071	.118	232,000
220	8.6614	360	14.1732	85	3.346	.118	275,000
240	9.4488	380	14.9606	85	3.346	.118	280,000
260	10.2362	420	16.5354	95	3.740	.157	360,000
300	11.8110	480	18.8976	109	4.291	.157	450,000
320	12.5984	500	19.6850	109	4.291	.157	455,000
Heavy							
90	3.5433	190	7.4803	60	2.362	.079	120,000
100	3.9370	210	8.2677	67	2.638	.098	143,000
110	4.3307	230	9.0551	73	2.874	.098	173,000
120	4.7244	250	9.8425	78	3.071	.118	196,000
130	5.1181	270	10.6299	85	3.346	.118	228,000
150	5.9055	300	11.8110	90	3.543	.118	270,000
170	6.6929	340	13.3858	103	4.055	.157	340,000
190	7.4803	380	14.9606	115	4.528	.157	415,000
200	7.8740	400	15.7480	122	4.803	.157	465,000
220	8.6614	420	16.5354	122	4.803	.197	475,000

* Dynamic reference capacity is in thrust load.

Table 14. Tolerances for Standard Radial Bearings

Tolerance Values in 0.0001 in.

Inner Ring								
Bore Diam. d, mm		Tolerances for Bore				Tolerances for Width		Maximum Radial Runout
		d_{mean}		d_{min}	d_{max}			
Over	Incl.	Lower Limit	Upper Limit			Upper Limit	Lower Limit	
0	9	−3	+0	−4	+1	+0	−50	3
9	18	−3	+0	−4	+1	+0	−50	4
18	30	−4	+0	−5	+1	+0	−50	5
30	50	−5	+0	−7	+2	+0	−50	6
50	80	−6	+0	−8	+2	+0	−50	8
80	120	−8	+0	−11	+3	+0	−50	10
120	180	−10	+0	−13	+3	+0	−50	12
180	250	−12	+0	−16	+4	+0	−100	16
250	315	−14	+0	−18	+4	+0	−100	20
315	400	−16	+0	−20	+4	+0	−160	24
400	500	−18	+0	−23	+5	+0	−180	26
500	630	−20	+0	....	....	+0	−220	28
630	800	−30	+0	....	....	+0	−300	32
800	1000	−40	+0	....	....	+0	−400	35

(*Table continued on p. 12-86*)

Table 14. Tolerances for Standard Radial Bearings—*Continued*

Tolerance Values in 0.0001 in.

OUTER RING

Outside Diam. D, mm		Tolerances for Outside Diameter				Tolerances for Width		Maximum Radial Runout
		Dmean		Dmax	Dmin			
Over	Incl.	Upper Limit	Lower Limit			Upper Limit	Lower Limit	
0	18	+0	−4	+1	−5	+0	Identical to that of inner ring of same bearing.	6
18	30	+0	−4	+1	−5	+0		6
30	50	+0	−5	+2	−7	+0		8
50	80	+0	−5	+2	−7	+0		10
80	120	+0	−6	+3	−9	+0		14
120	150	+0	−8	+3	−11	+0		16
150	180	+0	−10	+3	−13	+0		18
180	250	+0	−12	+4	−16	+0		20
250	315	+0	−14	+4	−18	+0		24
315	400	+0	−16	+4	−20	+0		28
400	500	+0	−18	+5	−23	+0		32
500	630	+0	−20	+6	−26	+0		40
630	800	+0	−30	....	....	+0		47
800	1000	+0	−40	....	....	+0		55

Table 15. Tolerances for Standard Thrust Bearings

Tolerance Values in 0.0001 in.

Bore or Outside Diam., mm		Tolerances for Bore				Tolerances for Outside Diam. Dmean		Maximum Axial Runout of Either Ring
		dmean		dmin	dmax			
Over	Incl.	Lower Limit	Upper Limit			Upper Limit	Lower Limit	
0	30	−4	+0	−5	+1	+0	−4	4
30	50	−5	+0	−7	+2	+0	−5	5
50	80	−6	+0	−8	+2	+0	−6	6
80	120	−8	+0	−10	+2	+0	−8	7
120	180	−10	+0	−13	+3	+0	−10	7
180	250	−12	+0	−16	+4	+0	−12	8
250	315	−14	+0	−18	+4	+0	−14	10
315	400	−16	+0	−20	+4	+0	−16	12
400	500	−18	+0	−23	+5	+0	−18	14
500	630	−20	+0	....	....	+0	−24	16
630	800	−30	+0	....	....	+0	−30	18
800	1000	−40	+0	....	....	+0	−40	20

As can be seen from the tables, standard bearings are manufactured to very precise tolerances. In some cases, however, even greater accuracy is necessary (for example, in machine tool spindles). To meet these needs, bearings of extra- or super-precision tolerances are manufactured in some types and sizes. These bearings may be regarded as semi-standard, in that they are manufactured in fairly large quantities, but they are more costly than standard bearings so that their use is justified only if the rest of the machine is designed for extremely accurate performance. Bearings of extra-precision tolerances are manufactured in deep-groove ball, angular contact ball, ball thrust, and cylindrical roller types. The tolerances are given in Table 16.

Table 16. Tolerances for Extra Precision Radial Bearings

ABEC 5 Tolerances in 0.0001 in.

Inner Ring

Bore Over	Bore Incl.	Bore Tolerances d_{min}	Bore Tolerances d_{max}	Radial Runout	Parallelism of Sides	Side Runout with Bore	Groove Parallelism with Side
0	9	−2	+0	2	2	3	3
9	18	−2	+0	2	2	3	3
18	30	−2	+0	2	2	3	3
30	50	−2	+0	2	2	3	3
50	80	−3	+0	2	2	3	3
80	120	−3	+0	3	3	3	4

Outer Ring

OD Over	OD Incl.	OD Tolerances D_{max}	OD Tolerances D_{min}	Radial Runout	Parallelism of Sides	OD Square with Side	Groove Parallelism with Side
0	18	+0	−2	2	2	3	3
18	30	+0	−2	2	2	3	3
30	50	+0	−2	2	2	3	3
50	80	+0	−3	3	2	3	4
80	120	+0	−3	4	3	3	5
120	150	+0	−4	4	3	4	5
150	180	+0	−5	5	3	4	6
180	250	+0	−5	5	4	4	6

ABEC 7 Tolerances in 0.0001 in.

Inner Ring

Bore Over	Bore Incl.	Bore Tolerances d_{min}	Bore Tolerances d_{max}	Radial Runout	Parallelism of Sides	Side Runout with Bore	Groove Parallelism with Side
0	9	−1 1/2	+0	1	1	1	1
9	18	−1 1/2	+0	1	1	1	1
18	30	−1 1/2	+0	1 1/2	1	1 1/2	1 1/2
30	50	−2	+0	1 1/2	1	1 1/2	1 1/2
50	80	−2	+0	1 1/2	1 1/2	2	1 1/2
80	120	−2 1/2	+0	2	1 1/2	2	2

Outer Ring

OD Over	OD Incl.	OD Tolerances D_{max}	OD Tolerances D_{min}	Radial Runout	Parallelism of Sides	OD Square with Side	Groove Parallelism with side
0	18	+0	−2	2	1	1 1/2	2
18	30	+0	−2	2	1	1 1/2	2
30	50	+0	−2	2	1	1 1/2	2
50	80	+0	−2	2	1	1 1/2	2
80	120	+0	−3	2	2	2	2
120	150	+0	−4	3	2	2	3
150	180	+0	−4	3	2	2	3
180	250	+0	−4	4	3	3	4

Width tolerances are the same as for standard bearings.

In order to prevent creep and subsequent work erosion between the bearing ring and the shaft or housing, the bearing ring which rotates relative to the direction of load should be mounted with a tight fit. Recommendations for shaft and housing fits are given in Tables 17 and 18. The numerical amounts and limits of these fits are given in Tables 19 and 20.

Table 17. Selection of Shaft Tolerances

SHAFTS FOR RADIAL BEARINGS

Conditions		Examples	Ball Bearings	Shaft Diameter, mm: Cylindrical Roller Bearings	Shaft Diameter, mm: Spherical Roller Bearings	Tolerance Symbol According to ISA	Remarks
BEARINGS WITH CYLINDRICAL BORES							
Fixed Inner Ring Load	The inner ring to be easily displaced on the shaft	Wheel on stationary shaft	All diameters			g6	
	The inner ring does not need to be easily displaced on the shaft	Tension pulleys and rope sheaves				h6	
Rotating Inner Ring Load or Indeterminate Load Direction	Light and variable loads	Electrical apparatus, machine tools, pumps, ventilators, industrial trucks	$\leqq 18$	...	...	h5	By light and variable loads are meant those which, as a rule, do not exceed 6 to 7% of the specific capacity *C*. In very accurate applications j5, k5, and m5 are used instead of j6, k6, and m6 respectively
			18 ... 100	$\leqq 40$	$\leqq 40$	j6	
			>100	40 ... 160	40 ... 100	k6	
			...	160 ... 225	100 ... 225	m6	
	Normal and heavy loads	Applications in general, electrical motors, turbines, pumps, combustion engines, gear transmissions, woodworking machines	$\leqq 18$	...	...	j5	
			18 ... 100	$\leqq 40$	$\leqq 40$	k5	
			>100	40 ... 160	40 ... 100	m5	
			...	160 ... 225	100 ... 200	n6	
			...	225 ... 400	200 ... 355	p6	
			...	...	355 ... 500	r6	
			...	...	>500	r7	
	Very heavy loads and shock loads	Journal boxes for locomotives and other heavy rail vehicles, traction motors	...	60 ... 160	60 ... 100	n6	Bearings with greater clearance than normal must be used
			...	160 ... 225	100 ... 200	p6	
			...	...	200 ... 355	r6	
			...	...	355 ... 500	r7	
Pure thrust load		Applications of all kinds	All diameters			j6	

SHAFTS FOR THRUST BEARINGS

Conditions		Shaft Diameter, mm	Tolerance Symbol According to ISA
Pure Thrust Load		All diameters	j6
Combined load on spherical roller thrust bearings	Fixed load on inner ring	All diameters	j6
	Rotating load on inner ring or indeterminate load direction	$\leqq 200$	k6
		200 ... 400	m6
		>400	n6

Table 18. Selection of Bearing Housing Tolerances

Housings of Cast Iron or Steel for Radial Bearings *

Conditions			Examples	Tolerance Symbol According to ISA	Remarks †
One-piece Bearing Housing	Rotating Outer Ring Load	Heavy loads on bearings in thin-walled housings	Supporting wheels in cranes, wheel hubs with roller bearings, crank bearings	P7	The outer ring is not displaceable axially
		Normal and heavy loads	Wheel hubs with roller bearings, crank bearings	N7	
		Light and variable loads	Conveyor rollers, rope sheaves, tension pulleys	M7	
	Indeterminate Load Direction	Heavy shock loads	Electrical traction motors		
		Heavy and normal loads, axial displacement of outer ring not required	Electrical motors, pumps, crankshaft main bearings	K7	The outer ring is, as a rule, not displaceable axially
Split or One-piece Bearing Housing		Normal and light loads, axial displacement of the outer ring desirable	Electrical motors, pumps, crankshaft main bearings	J7	The outer ring is, as a rule, displaceable axially
	Fixed Outer Ring Load	Shock loads, temporary complete unloading	Journal boxes for rail vehicles		
		All loads	Bearing applications in general, journal boxes for rail vehicles	H7	The outer ring is easily displaced axially
		Normal and light loads, loads under simple operating conditions	Line shaftings	H8	
		Heat supplied through the shaft	Dryer cylinders	G7	
One-piece Bearing Housing	Applications Requiring Particular Accuracy	Very accurate running and small deflections under variable loads	Roller bearings for main spindles in machine tools: $D>250$ mm; $250 \geq D > 125$ mm; $D \leq 125$ mm	P6 N6 M6	The outer ring is not displaceable axially
		Very accurate running under light loads and indeterminate load direction	Ball bearings on the working side of grinding spindles, held bearings in high-speed centrifugal compressors	K6	The outer ring, as a rule, is not displaceable axially
		Very accurate running, axial displacement of outer ring desirable	Ball bearings at the driven end in grinding spindles, floating bearings in high-speed centrifugal compressors	J6	The outer ring is displaceable axially

* For housings of light metal, tolerances are generally selected which give a slightly tighter fit than those given in the table.

† The notes regarding axial displaceability of the outer ring serve as a guide for judgment of the suitability of the fit, when self-contained bearings are used as floating bearings.

Housings for Thrust Bearings

Conditions		Tolerance Symbol According to ISA	Remarks
Pure thrust load	Thrust ball bearings	H8	In less accurate applications, the housing ring or spherical washer, respectively, is mounted with radial play
	Spherical roller thrust bearing, companion bearing to carry radial load	....	The outer ring is mounted with radial play
Combined load on spherical roller thrust bearings	Fixed load on outer ring	J7	
	Rotating load on outer ring or indeterminate load direction	K7	In general
		M7	In case of comparatively heavy radial loads

Table 19. Deviation of Shaft Diameters from Nominal Dimensions in Inches

Nominal Diam., mm		g6	h6	h5	j5	j6	k5	k6	m5	m6	n6	p6	r6	r7
Over	Incl.													
3	6	−.0002 −.0005	0 −.0003	0 −.0002	+.0002 −.0000									
6	10	−.0002 −.0006	0 −.0004	0 −.0002	+.0002 −.0001	+.0003 −.0001								
10	18	−.0002 −.0007	0 −.0004	0 −.0003	+.0002 −.0001	+.0003 −.0001	+.0004 +.0000	+.0005 +.0000						
18	30	−.0003 −.0008	0 −.0005	0 −.0004	+.0002 −.0002	+.0004 −.0002	+.0004 +.0001	+.0006 +.0001	+.0007 +.0003	+.0008 +.0003	+.0011 +.0006			
30	50	−.0004 −.0010	0 −.0006	0 −.0004	+.0002 −.0002	+.0004 −.0002	+.0005 +.0001	+.0007 +.0001	+.0008 +.0004	+.0010 +.0004	+.0013 +.0007	+.0017 +.0010		
50	80	−.0004 −.0011	0 −.0007	0 −.0005	+.0002 −.0003	+.0005 −.0003	+.0006 +.0001	+.0008 +.0001	+.0009 +.0004	+.0012 +.0004	+.0015 +.0008	+.0020 +.0013		
80	120	−.0005 −.0013	0 −.0009	0 −.0006	+.0002 −.0004	+.0005 −.0004	+.0007 +.0001	+.0010 +.0001	+.0011 +.0005	+.0014 +.0005	+.0018 +.0009	+.0023 +.0015		
120	180	−.0006 −.0015	0 −.0010	0 −.0007	+.0003 −.0004	+.0006 −.0004	+.0008 +.0001	+.0011 +.0001	+.0013 +.0006	+.0016 +.0006	+.0020 +.0011	+.0027 +.0017		
180	200												+.0042 +.0030	
200	225	−.0006 −.0017	0 −.0011	0 −.0008	+.0003 −.0005	+.0006 −.0005	+.0009 +.0002	+.0013 +.0002	+.0015 +.0007	+.0018 +.0007	+.0024 +.0012	+.0031 +.0020	+.0043 +.0031	
225	250												+.0044 +.0033	

Fit, Inner Ring to Shaft		Push Fit	Push Fit to Wringing Fit	Wringing Fit	Drive Fit		Light Force Fit		Force Fit		Heavy Force Fit			
250	280	−.0007 −.0019	0 −.0013	0 −.0009	+.0003 −.0006	+.0006 −.0006	+.0011 +.0002	+.0014 +.0002	+.0017 +.0008	+.0020 +.0008	+.0026 +.0013	+.0035 +.0022	+.0050 +.0037	
280	315												+.0051 +.0039	+.0059 +.0039
315	355	−.0007 −.0021	0 −.0014	0 −.0010	+.0003 −.0007	+.0007 −.0007	+.0011 +.0002	+.0016 +.0002	+.0018 +.0008	+.0022 +.0008	+.0029 +.0015	+.0039 +.0024	+.0057 +.0043	+.0065 +.0043
355	400												+.0059 +.0045	+.0067 +.0045
400	450	−.0008 −.0024	0 −.0016	0 −.0011	+.0003 −.0008	+.0008 −.0008	+.0013 +.0002	+.0018 +.0002	+.0020 +.0009	+.0025 +.0009	+.0031 +.0016	+.0043 +.0027	+.0065 +.0050	+.0074 +.0050
450	500												+.0068 +.0052	+.0077 +.0052
500	560		0 −.0017	0 −.0012	+.0003 −.0009	+.0009 −.0008	+.0014 +.0002	+.0019 +.0002	+.0022 +.0010	+.0027 +.0010	+.0034 +.0017	+.0046 +.0030	+.0074 +.0057	+.0083 +.0057
560	630												+.0077 +.0060	+.0087 +.0060
630	710		0 −.0019	0 −.0013	+.0004 −.0009	+.0009 −.0009	+.0015 +.0002	+.0020 +.0002	+.0024 +.0011	+.0030 +.0011	+.0038 +.0019	+.0051 +.0033	+.0085 +.0067	+.0097 +.0067
710	800												+.0088 +.0069	+.0099 +.0069
800	900		0 −.0020	0 −.0015	+.0004 −.0011	+.0010 −.0010	+.0017 +.0002	+.0023 +.0002	+.0027 +.0013	+.0033 +.0013	+.0042 +.0021	+.0056 +.0036	+.0098 +.0077	+.0110 +.0077
900	1000												+.0102 +.0081	+.0114 +.0081
1000	1120		0 −.0023	0 −.0016	+.0004 −.0012	+.0011 −.0011	+.0018 +.0002	+.0025 +.0002	+.0030 +.0014	+.0037 +.0014	+.0046 +.0024	+.0062 +.0039	+.0112 +.0089	+.0126 +.0089

Table 20. Deviation of Housing Bores from Nominal Dimensions in Inches

Nominal Diam., mm		G7	H8	H7	J7	J6	K6	K7	M6	M7	N6	N7	P6	P7
Over	Incl.													
10	18	+.0002 +.0009	0 +.0011	0 +.0007	−.0003 +.0004	−.0002 +.0002	−.0004 +.0001	−.0005 +.0002	−.0006 −.0002	−.0007 0	−.0008 −.0004	−.0009 −.0002	−.0010 −.0006	−.0011 −.0004
18	30	+.0003 +.0011	0 +.0013	0 +.0008	−.0004 +.0005	−.0002 +.0003	−.0004 +.0001	−.0006 +.0002	−.0007 −.0002	−.0008 0	−.0009 −.0004	−.0011 −.0003	−.0012 −.0007	−.0014 −.0006
30	50	+.0004 +.0013	0 +.0015	0 +.0010	−.0004 +.0006	−.0002 +.0004	−.0005 +.0001	−.0007 +.0003	−.0008 −.0002	−.0010 0	−.0011 −.0005	−.0013 −.0003	−.0015 −.0008	−.0017 −.0007
50	80	+.0004 +.0016	0 +.0018	0 +.0012	−.0005 +.0007	−.0002 +.0005	−.0006 +.0002	−.0008 +.0004	−.0009 −.0002	−.0012 0	−.0013 −.0006	−.0015 −.0004	−.0018 −.0010	−.0020 −.0008
80	120	+.0005 +.0019	0 +.0021	0 +.0014	−.0005 +.0009	−.0002 +.0006	−.0007 +.0002	−.0010 +.0004	−.0011 −.0002	−.0014 0	−.0015 −.0006	−.0018 −.0004	−.0020 −.0012	−.0023 −.0009
120	180	+.0006 +.0021	0 +.0025	0 +.0016	−.0006 +.0010	−.0003 +.0007	−.0008 +.0002	−.0011 +.0005	−.0013 −.0003	−.0016 0	−.0018 −.0008	−.0020 −.0005	−.0024 −.0014	−.0027 −.0011
180	250	+.0006 +.0024	0 +.0028	0 +.0018	−.0006 +.0012	−.0003 +.0009	−.0009 +.0002	−.0013 +.0005	−.0015 −.0003	−.0018 0	−.0020 −.0009	−.0024 −.0006	−.0028 −.0016	−.0031 −.0013
250	315	+.0007 +.0027	0 +.0032	0 +.0020	−.0006 +.0014	−.0003 +.0010	−.0011 +.0002	−.0014 +.0006	−.0016 −.0004	−.0020 0	−.0022 −.0010	−.0026 −.0006	−.0031 −.0019	−.0035 −.0014
315	400	+.0007 +.0030	0 +.0035	0 +.0022	−.0007 +.0015	−.0003 +.0011	−.0011 +.0003	−.0016 +.0007	−.0018 −.0004	−.0022 0	−.0024 −.0010	−.0029 −.0006	−.0034 −.0020	−.0039 −.0016
400	500	+.0008 +.0033	0 +.0038	0 +.0025	−.0008 +.0017	−.0003 +.0013	−.0013 +.0003	−.0018 +.0007	−.0020 −.0004	−.0025 0	−.0026 −.0011	−.0031 −.0007	−.0037 −.0022	−.0043 −.0018
500	630	+.0009 +.0035	0 +.0041	0 +.0027	−.0609 +.0018	−.0003 +.0014	−.0014 +.0003	−.0019 +.0008	−.0022 −.0005	−.0027 0	−.0029 −.0012	−.0034 −.0007	−.0041 −.0024	−.0046 −.0020
630	800	+.0009 +.0039	0 +.0045	0 +.0030	−.0009 +.0020	−.0004 +.0015	−.0015 +.0004	−.0020 +.0009	−.0024 −.0006	−.0030 0	−.0032 −.0014	−.0038 −.0008	−.0046 −.0027	−.0051 −.0021
800	1000	+.0010 +.0043	0 +.0051	0 +.0033	−.0010 +.0023	−.0004 +.0017	−.0017 +.0004	−.0023 +.0010	−.0027 −.0007	−.0033 0	−.0036 −.0015	−.0042 −.0009	−.0050 −.0030	−.0056 −.0023
1000	1250	+.0011 +.0047	0 +.0057	0 +.0037	−.0011 +.0025	−.0004 +.0019	−.0018 +.0005	−.0025 +.0011	−.0030 −.0007	−.0037 0	−.0039 −.0017	−.0046 −.0010	−.0055 −.0032	−.0062 −.0026
1250	1600	+.0011 +.0053	0 +.0063	0 +.0041	−.0013 +.0028	−.0004 +.0021	−.0020 +.0005	−.0028 +.0013	−.0033 −.0008	−.0041 0	−.0044 −.0019	−.0052 −.0011	−.0061 −.0035	−.0069 −.0028
Fit Housing—Outer Ring		Close Running Fit	Slide Fit		Push Fit		Wringing Fit		Drive Fit		Heavy Drive Fit		Light Force Fit	

19. MOUNTING AND MAINTENANCE

The bearing should be mounted so that the ring which rotates relative to the direction of load has a tight fit on the shaft or in the housing, as the case may be. In the majority of applications a tight fit on the shaft is required. The bearing may be pressed on the shaft by the use of an arbor press, tapped on the shaft by a mounting tube and hammer, shrunk on the shaft after heating, or forced on with mounting nuts. When the bearing is tight on the shaft, it is usually a slide or push fit in the housing and may be pushed in without any trouble.

If the bearing is tight in the housing, the bearing may be mounted by pressing the bearing in the housing. In this case the mounting on the shaft is usually a push fit which can be made without difficulty.

Bearings can usually be removed with the help of bearing pullers and presses. For heavy fits it may be necessary to design into the bearing mounting special methods for removing

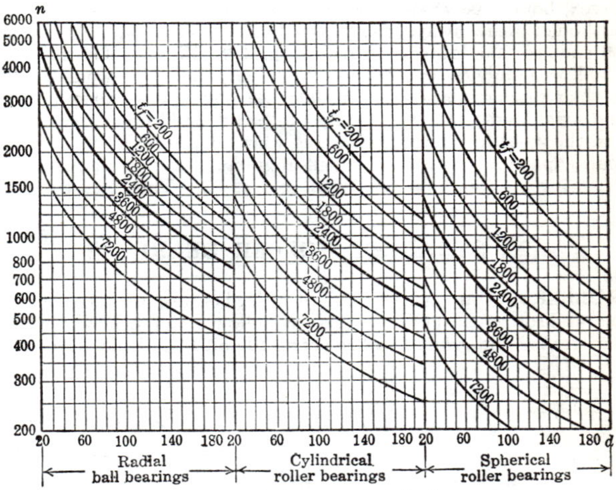

FIG. 9. Recommended grease relubrication periods.

the bearing. They may consist of removal nuts, hydraulic pressure chambers, etc. A new method which allows an oil film to be pressed between the bearing ring and the shaft or housing breaks the fit and allows the bearing to be pulled off by a comparatively small force.

New bearings do not in general need cleaning before using. The best policy is to keep the bearing wrapped until ready to mount. Then remove the bearing from the wrapping, check dimensions and mount, taking care to see that the surroundings are as clean as possible so that no dirt gets in the bearing. Then lubricant should be added and the bearing covers closed so that the bearing is exposed to contamination for as short a time as possible.

When bearings must be removed for cleaning because of caked grease, the bearing should if possible be removed from shaft and housing and soaked in clean cold petroleum solvent or kerosene. If this soaking fails to remove the caked material, soak again in a hot light oil, agitating the bearing container from time to time, and then clean out any remaining material with a hand brush. After cleaning, spin in light oil and relubricate immediately to protect the metal surface. Do not use air to clean bearings. Do not spin bearings until clean and relubricated.

If the bearing cannot be removed easily from the machine, it may be cleaned by flushing hot oil through the bearing. After cleaning, this oil should be completely removed from the bearing and new lubricant inserted.

For a more extensive discussion of bearing mounting and maintenance, consult *Anti-Friction Bearing Maintenance*, compiled by the Anti-Friction Bearing Manufacturers' Association, Inc., 60 East 42nd St., New York 17, N. Y.

If the bearing has oil lubrication, the oil needs replacing only if leakage occurs or if the oil is carbonized by heat. Grease lubrication should be replaced at intervals, depending on the speed and bearing size and type. Suggested grease relubrication periods are given in Fig. 9.

GENERAL REFERENCES

R. K. Allan, *Rolling Bearings*, Sir Isaac Pitman & Sons, Ltd., London, 1945.
Anti-Friction Bearing Manufacturers' Association, *Anti-Friction Bearing Maintenance.*
W. Jurgensineyer, *Die Walzlager*, Julius Springer, Berlin, 1937.
A. Palmgren, *Ball and Roller Bearing Engineering*, SKF Industries, Philadelphia, 1945.
SKF Industries, *A Guide to Better Bearing Lubrication.*
SKF Industries, *SKF General Catalogue and Engineering Data.*

STRAIGHT, NEEDLE, AND TAPERED ROLLER BEARINGS

By E. Stanley Ault

Roller bearings use rollers which are straight or cylindrical, tapered or conical, barrel shaped or convex, hour-glass shaped or concave. The straight roller is suitable for only radial loads, although the raceways may be constructed so that the rollers bear on shoulders at one or both ends to take locating thrusts (Fig. 2, p. 12-67). The rollers may be solid cylinders or hollow cylinders, constructed from strip helically wound. Rollers long in proportion to their diameter are called needle bearings. The tapered roller can sustain both radial and axial loads, and the bearing can be adjusted upon assembly. The convex roller is used in the so-called spherical bearing where it provides some axial load capacity with radial load capacity. So used it is self-aligning. Further discussion and tables on the spherical bearing are found on pp. 12-67 and 12-83. The concave roller is also self-aligning.

20. STRAIGHT ROLLER BEARINGS

WOUND ROLLERS. One form of straight roller bearing made by the Hyatt Bearings Division, General Motors Corporation, uses a hollow roller helically wound from alloy steel strip, heat treated and ground to size (Figs. 10 and 11). Rollers wound right and left

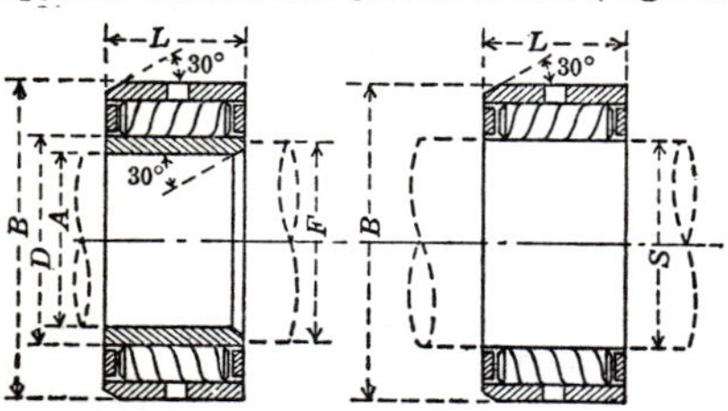

FIG. 10. Wound roller type roller bearing with solid outer race.

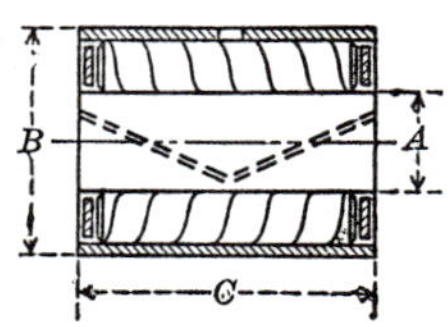

FIG. 11. Wound roller type roller bearing with split outer race.

hand are assembled alternately in the bearing and held in proper alignment by a cage retainer. Bearings are made with both solid inner and outer races, with solid outer race and no inner race, and with a split outer race and no inner race for use directly on the surface of commercial shafting. Tables 21 and 22 give dimensions and load rating based upon an operating temperature of not more than 350 F; operation at temperatures above 500 F is not recommended. For load ratings required for an average life expectancy other than 5000 hr multiply by factor from Table 23.

Temperature Factor (*c*). For temperatures above 350 F the bearing load should be multiplied by one of the following factors to obtain the required rated load:

Maximum temperature, °F	350	400	450	500
Temperature factor, *c*	1.00	1.15	1.37	1.67

Service Factor (*d*). For certain classes of service, depending upon load fluctuations, shock, overload, etc., the bearing loads should be multiplied by one of the following service factors (*d*):

(*Table 21*) Agricultural equipment, trailer trucks, 0.67–1.0; hand-operated hoists, 0.80–1.0; power-house cranes, 0.80; centrifugal extractors, electric motors, industrial locomotive journals and transmissions, 1.0–1.33; mining machinery, except mine cars, 1.33; air compressors, hot saws, sprayers, and dusters, 1.33–2.0; dairy machinery, 1.0–2.0; house-lighting units, 2.0; tractor crawler tracks, 5.0.

(*Continued on p. 12-98*)

Table 21. Dimensions and Load Capacities of Roller Bearings with Wound Rollers with Solid Outer and Inner Races and without Inner Race (See Fig. 10)

(Hyatt Bearings Division, General Motors Corporation, Harrison, N. J.)

Inner Race Inside Diam., *A*		Roller Assembly Inside Diam., *D*, or Shaft Diam., *S*	Outer Race Outside Diam., *B*	Min. Diam. of Shaft Shoulder, *F*	Bearing Width, *L*	Roller Diam.	Outer Race Hole Diam.	Load Rating in Pounds, 5000 hr, Ave. Life Expectancy. The load rating of the bearing selected must be at least equal to $L \times a \times b \times c \times d$, where L = imposed bearing load, a = shaft hardness factor, b = stationary shaft factor, c = temperature factor, and d = service factor. Values of these factors are given. Shaft hardness Rockwell C - 54. Revolutions per Minute						
mm	in.	in.	in.	in.	in.	in.	in.	100	200	300	500	1000	1500	2000
25	0.9843	1 1/4	2.4409	1 3/16	1 1/8	7/16	1/4	1700	1300	1100	840	560	440	355
30	1.1811	1 1/2	2.4409	1 7/16	13/16	5/16	1/4	1240	900	735	560	370	280	230
30	1.1811	1 1/2	2.4409	1 7/16	1 1/8	5/16	1/4	1940	1430	1160	880	580	445	365
30	1.1811	1 1/2	2.8346	1 7/16	1 3/16	1/2	1/4	2230	1710	1425	1100	720	560	455
30	1.1811	1 1/2	2.8346	1 7/16	15/16	1/2	1/4	1550	1190	980	760	510	385	310
35	1.3780	1 3/4	2.8346	1 5/8	15/16	3/8	1/4	1620	1215	990	730	480	380	295
35	1.3780	1 3/4	2.8346	1 5/8	1 3/16	3/8	1/4	2300	1710	1400	1050	680	530	420
35	1.3780	1 3/4	3.1496	1 5/8	1 3/8	1/2	1/4	2900	2195	1800	1380	900	695	570
35	1.3780	1 3/4	3.1496	1 5/8	1	1/2	1/4	1840	1400	1150	880	575	440	355
40	1.5748	2	3.1496	1 7/8	1	3/8	1/4	1960	1440	1175	880	560	435	340
40	1.5748	2	3.1496	1 7/8	1 3/8	3/8	1/4	3080	2230	1800	1360	890	670	525
40	1.5748	2	3.5433	1 7/8	1 7/16	9/16	5/16	3750	2810	2300	1750	1140	855	665
40	1.5748	2	3.5433	1 7/8	1 1/4	9/16	5/16	3070	2290	1890	1445	920	705	580
45	1.7717	2 3/16	3.3465	2 1/16	1 1/8	3/8	1/4	2380	1740	1400	1050	660	500	395
45	1.7717	2 3/16	3.3465	2 1/16	1 9/16	3/8	1/4	3710	2690	2160	1630	1040	790	615
45	1.7717	2 1/4	3.9370	2 1/8	1 9/16	5/8	5/16	4550	3390	2800	2120	1360	1030	815
45	1.7717	2 1/4	3.9370	2 1/8	1 5/16	5/8	5/16	3590	2690	2195	1660	1065	800	630
50	1.9685	2 3/8	3.5433	2 1/4	1 1/4	3/8	5/16	3000	2150	1750	1300	820	610	475
50	1.9685	2 3/8	3.5433	2 1/4	1 3/4	3/8	5/16	4700	3325	2700	2000	1260	945	740
50	1.9685	2 3/8	3.5433	2 1/4	3 1/2	3/8	5/16	9400	6650	5400	4000	2520	1890	1480
50	1.9685	2 1/2	4.3307	2 3/8	1 3/4	11/16	5/16	5410	4050	3330	2530	1600	1160	910
50	1.9685	2 1/2	4.3307	2 3/8	1 7/16	11/16	5/16	4125	3070	2510	1910	1225	930	725
55	2.1654	2 5/8	3.9370	2 1/2	1 5/16	7/16	5/16	3500	2520	2040	1525	900	700	.....
55	2.1654	2 5/8	3.9370	2 1/2	1 13/16	7/16	5/16	5400	3900	3100	2300	1460	1070	.....
55	2.1654	2 5/8	3.9370	2 1/2	3 5/8	7/16	5/16	10800	7800	6200	4600	2920	2140	.....
55	2.1654	2 3/4	4.7244	2 5/8	1 15/16	11/16	3/8	6540	4950	4000	3000	1880	1375	.....
60	2.3622	2 7/8	4.3307	2 3/4	1 7/16	1/2	5/16	4340	3150	2550	1875	1170	855	.....
60	2.3622	2 7/8	4.3307	2 3/4	1 15/16	1/2	5/16	6300	4625	3750	2800	1720	1260	.....
60	2.3622	2 7/8	4.3307	2 3/4	3 7/8	1/2	5/16	12600	9250	7500	5600	3440	2525	.....
60	2.3622	3	5.1181	2 13/16	2 1/8	3/4	3/8	8000	6000	4850	3620	2250	1650	.....
65	2.5591	3 1/8	4.7244	3	1 1/2	1/2	3/8	4980	3580	2850	2120	1300	950	.....
65	2.5591	3 1/8	4.7244	3	2 1/16	1/2	3/8	7500	5400	4300	3200	1950	1380	.....
65	2.5591	3 1/8	4.7244	3	4 1/8	1/2	3/8	15000	10800	8600	6400	3900	2760	.....
65	2.5591	3 1/4	5.5118	3 1/16	2 5/16	13/16	7/16	10100	7500	6050	4500	2780	2040	.....
70	2.7559	3 5/16	4.9213	3 3/16	1 5/8	1/2	3/8	5650	4000	3210	2380	1450	.....	.....
70	2.7559	3 5/16	4.9213	3 3/16	2 3/8	1/2	3/8	8700	6220	5000	3650	2250	.....	.....
70	2.7559	3 5/16	4.9213	3 3/16	4 3/4	1/2	3/8	17440	12440	10000	7300	4500	.....	.....
70	2.7559	3 1/2	5.9055	3 5/16	2 1/2	7/8	7/16	11400	8300	6630	4900	2940	.....	.....
75	2.9528	3 1/2	5.1181	3 3/8	1 3/4	1/2	3/8	6050	4275	3410	2530	1550	.....	.....
75	2.9528	3 1/2	5.1181	3 3/8	2 5/8	1/2	3/8	10000	7150	5700	4200	2600	.....	.....
75	2.9528	3 1/2	5.1181	3 3/8	5 1/4	1/2	3/8	20000	14300	11400	8400	5200	.....	.....
75	2.9528	3 3/4	6.2992	3 9/16	2 11/16	15/16	7/16	13100	9850	7950	5860	3550	.....	.....
80	3.1496	3 3/4	5.5118	3 5/8	1 13/16	9/16	7/16	7200	5110	4100	3050	1870	.....	.....
80	3.1496	3 3/4	5.5118	3 5/8	2 5/8	9/16	7/16	11250	8150	6500	4800	2800	.....	.....
80	3.1496	3 3/4	5.5118	3 5/8	5 1/4	9/16	7/16	22500	16300	13000	9600	5600	.....	.....
80	3.1496	4	6.6929	3 3/4	2 11/16	1	7/16	14000	10450	8450	6150	3650	.....	.....
85	3.3465	4	5.9055	3 13/16	1 15/16	5/8	7/16	8300	5875	4700	3500	2080	.....	.....
85	3.3465	4	5.9055	3 13/16	2 3/4	5/8	7/16	13000	9200	7300	5300	3200	.....	.....
85	3.3465	4	5.9055	3 13/16	5 1/2	5/8	7/16	26000	18400	14600	10600	6400	.....	.....
85	3.3465	4 1/4	7.0866	4	2 7/8	1	9/16	16300	11950	9600	7050	4250	.....	.....

(Table continued on p. 12-96)

Table 21. Dimensions and Load Capacities of Roller Bearings with Wound Rollers with Solid Outer and Inner Races and without Inner Race (See Fig. 10)—*Continued*

(Hyatt Bearings Division, General Motors Corporation, Harrison, N. J.)

Inner Race Inside Diam., *A*		Roller Assembly Inside Diam., *D*, or Shaft Diam., *S*	Outer Race Outside Diam., *B*	Min. Diam. of Shaft Shoulder, *F*	Bearing Width, *L*	Roller Diam.	Outer Race Hole Diam.	Load Rating in Pounds, 5000 hr Ave. Life Expectancy. The load rating of the bearing selected must be at least equal to $L \times a \times b \times c \times d$, where L = imposed bearing load, a = shaft hardness factor, b = stationary shaft factor, c = temperature factor, and d = service factor. Values of these factors are given. Shaft hardness Rockwell C-54. Revolutions per Minute						
mm	in.	in.	in.	in.	in.	in.	in.	100	200	300	500	1000	1500	2000
90	3.5433	4 1/4	6.2992	4 1/16	2 1/16	11/16	7/16	8800	6210	5000	3670	2200	.....	.....
90	3.5433	4 1/4	6.2992	4 1/16	2 13/16	11/16	7/16	13500	9500	7500	5600	3400	.....	.....
90	3.5433	4 1/4	6.2992	4 1/16	5 5/8	11/16	7/16	27000	19000	15000	11200	6800	.....	.....
90	3.5433	4 1/2	7.4803	4 1/4	3	1 1/16	9/16	17000	12500	10000	7450	4400	.....	.....
95	3.7402	4 1/2	6.6929	4 5/16	2 3/16	3/4	7/16	10700	7600	6050	4420	2600	.....	.....
95	3.7402	4 1/2	6.6929	4 5/16	3	3/4	7/16	15700	11300	9000	6500	3900	.....	.....
95	3.7402	4 1/2	6.6929	4 5/16	6	3/4	7/16	31400	22600	18000	13000	7800	.....	.....
95	3.7402	4 3/4	7.8740	4 1/2	3 1/16	1 1/8	9/16	18100	13600	11000	8000	4600	.....	.....
100	3.9370	4 3/4	7.0866	4 9/16	2 5/16	3/4	9/16	11000	7850	6250	4530	2660	.....	.....
100	3.9370	4 3/4	7.0866	4 9/16	3 1/4	3/4	9/16	17300	12400	9800	7000	4100	.....	.....
100	3.9370	4 3/4	7.0866	4 9/16	6 1/2	3/4	9/16	34600	24800	19600	14000	8200	.....	.....
100	3.9370	5	8.4646	4 11/16	3 1/4	1 1/4	9/16	19400	15300	12400	9000	5200	.....	.....
110	4.3307	5 1/4	7.8740	5	2 9/16	7/8	9/16	14000	10100	7900	5700	.....	.....	.....
110	4.3307	5 1/4	7.8740	5	3 1/2	7/8	9/16	21000	14800	11750	8500	.....	.....	.....
110	4.3307	5 1/4	7.8740	5	7	7/8	9/16	42000	29600	23500	17000	.....	.....	.....
110	4.3307	5 1/2	9.4488	5 3/16	3 5/8	1 3/8	9/16	20300	16500	13800	10400	.....	.....	.....
120	4.7244	5 5/8	8.4646	5 3/8	2 13/16	15/16	9/16	16000	11300	8900	6400	.....	.....	.....
120	4.7244	5 5/8	8.4646	5 3/8	3 7/8	15/16	9/16	24000	17100	13500	9700	.....	.....	.....
120	4.7244	5 5/8	8.4646	5 3/8	7 3/4	15/16	9/16	48000	34200	27000	19400	.....	.....	.....
120	4.7244	6 1/16	10.2362	5 11/16	4 1/8	1 3/8	9/16	25500	20 000	16600	12300	.....	.....	.....
130	5.1181	6 1/16	9.0551	5 13/16	3 1/8	1	9/16	20900	14750	11200	7980	.....	.....	.....
130	5.1181	6 1/16	9.0551	5 13/16	4 1/4	1	9/16	30000	21200	16100	11450	.....	.....	.....
130	5.1181	6 1/16	9.0551	5 13/16	8 1/2	1	9/16	60000	42400	32200	22900	.....	.....	.....
130	5.1181	6 9/16	11.0236	6 3/16	4 3/8	1 1/2	9/16	29300	23100	19100	14000	.....	.....	.....
140	5.5118	6 5/8	9.8425	6 5/16	3 1/4	1 1/16	9/16	22300	15900	12500	8800	.....	.....	.....
140	5.5118	6 5/8	9.8425	6 5/16	4 3/4	1 1/16	9/16	36000	25500	19050	14400	.....	.....	.....
140	5.5118	6 5/8	9.8425	6 5/16	9 1/2	1 1/16	9/16	72000	51000	38100	28800	.....	.....	.....
150	5.9055	7 1/16	10.6299	6 3/4	3 1/2	1 3/16	5/8	26400	18500	14300	10500	.....	.....	.....
150	5.9055	7 1/16	10.6299	6 3/4	4 3/4	1 3/16	5/8	37500	26250	20300	14950	.....	.....	.....
160	6.2992	7 5/8	11.4173	7 1/4	3 7/8	1 1/4	5/8	28400	20800	16300	11300	.....	.....	.....
160	6.2992	7 5/8	11.4173	7 1/4	4 7/8	1 1/4	5/8	39000	27900	22000	15300	.....	.....	.....
160	6.2992	7 5/8	11.4173	7 1/4	9 3/4	1 1/4	5/8	78000	55800	44000	30600	.....	.....	.....
170	6.6929	8 1/16	12.2047	7 11/16	5 3/8	1 3/8	11/16	46000	35300	26000	18800	.....	.....	.....
170	6.6929	8 1/16	12.2047	7 11/16	10 3/4	1 3/8	11/16	92000	70600	52000	37600	.....	.....	.....
180	7.0866	8 15/32	12.5984	8 1/16	5 7/8	1 3/8	11/16	53000	40000	31500	21800	.....	.....	.....

Hardness and Stationary Shaft Factors *a* and *b*

Shaft Hardness	Rockwell Brinell	Bearing with Inner race	C-54 534	C-50 484	C-46 437	C-42 393
Rotating Shaft	Hardness Factor, *a*	1.00	1.00	1.00	1.23	1.54
Stationary Shaft	Stationary Factor, *b*	1.33	1.33	1.54	1.82	2.00

Table 22. Dimensions and Load Capacities of Roller Bearings for Wound Rollers with Split Outer Race and No Inner Race (See Fig. 11)

(Hyatt Bearings Division, General Motors Corporation, Harrison, N. J.)

Nominal Shaft Diam., in.	Bearing Length, in.	Nominal Outside Diam. of Bearing, in.	Shaft Limits		Limits for Housing Bore		Rollers		Load Ratings in Pounds, 5000 hr, Ave. Life Expectancy. Based on rotating shaft of Rockwell hardness B-86. For stationary shafts and other hardness, use factor at foot of table. Load rating from $L \times a \times b \times c \times d$. (See Table 21) Revolutions per Minute				
			High Limit, in.	Low Limit, in.	High Limit, in.	Low Limit, in.	No.	Diam., in.					
A	B	C							25	100	300	500	1000
1	1	1 15/16	1.000	0.997	1.940	1.938	7	3/8	130	120	105	90	65
1	4	1 15/16	1.000	0.997	1.940	1.938	7	3/8	780	730	610	530	400
1 1/8	2	2 3/16	1.125	1.122	2.190	2.188	8	7/16	390	365	305	265	200
1 1/8	4	2 3/16	1.125	1.122	2.190	2.188	8	7/16	865	810	680	590	440
1 3/16	2	2 3/8	1.188	1.185	2.377	2.375	7	1/2	405	380	315	275	200
1 3/16	4	2 3/8	1.188	1.185	2.377	2.375	7	1/2	910	855	710	610	455
1 1/4	2	2 7/16	1.250	1.247	2.440	2.438	8	1/2	425	400	330	285	210
1 1/4	4	2 7/16	1.250	1.247	2.440	2.438	8	1/2	950	890	740	635	470
1 3/8	2	2 9/16	1.375	1.372	2.565	2.563	8	1/2	465	435	355	305	225
1 3/8	4	2 9/16	1.375	1.372	2.565	2.563	8	1/2	1040	970	800	690	500
1 7/16	2	2 3/4	1.438	1.435	2.752	2.750	8	9/16	485	450	370	315	235
1 7/16	5	2 3/4	1.438	1.435	2.752	2.750	8	9/16	1390	1295	1060	905	670
1 1/2	2	2 13/16	1.500	1.497	2.815	2.813	8	9/16	505	470	385	330	240
1 1/2	5	2 13/16	1.500	1.497	2.815	2.813	8	9/16	1440	1340	1090	935	690
1 5/8	2	2 15/16	1.625	1.622	2.940	2.938	8	9/16	540	500	405	350	255
1 5/8	4	2 15/16	1.625	1.622	2.940	2.938	8	9/16	1210	1125	915	785	575
1 11/16	2	3	1.688	1.685	3.003	3.000	9	9/16	560	520	420	360	260
1 11/16	4	3	1.688	1.685	3.003	3.000	9	9/16	1260	1170	945	810	585
1 3/4	2	3 1/16	1.750	1.747	3.066	3.063	9	9/16	580	535	435	370	265
1 3/4	5	3 1/16	1.750	1.747	3.066	3.063	9	9/16	1650	1515	1240	1060	765
1 7/8	4	3 3/16	1.875	1.872	3.191	3.188	10	9/16	1380	1280	1030	870	625
1 7/8	5	3 3/16	1.875	1.872	3.191	3.188	10	9/16	1770	1640	1320	1120	800
1 15/16	2	3 7/16	1.938	1.935	3.441	3.438	10	5/8	635	585	470	400	285
1 15/16	6	3 7/16	1.938	1.935	3.441	3.438	10	5/8	2210	2045	1640	1390	1000
2	3	3 1/2	2.000	1.997	3.503	3.500	9	5/8	1060	980	785	665	475
2	5	3 1/2	2.000	1.997	3.503	3.500	9	5/8	1870	1725	1380	1170	840
2 3/16	3	4 3/16	2.188	2.185	4.191	4.188	8	7/8	1150	1060	845	715	510
2 3/16	5	4 3/16	2.188	2.185	4.191	4.188	8	7/8	2020	1860	1480	1250	890
2 1/4	3	4 1/4	2.250	2.247	4.253	4.250	8	7/8	1180	1090	870	730	515
2 1/4	6	4 1/4	2.250	2.247	4.253	4.250	8	7/8	2500	2315	1850	1550	1100
2 7/16	5	4 9/16	2.438	2.435	4.566	4.563	8	15/16	2230	2055	1625	1350	950
2 1/2	3	4 5/8	2.500	2.497	4.628	4.625	8	15/16	1300	1200	940	780	550
2 1/2	7	4 5/8	2.500	2.497	4.628	4.625	8	15/16	3280	3025	2380	1980	1385
2 3/4	4	4 7/8	2.750	2.747	4.878	4.875	9	15/16	1940	1775	1385	1160	790
2 3/4	6	4 7/8	2.750	2.747	4.878	4.875	9	15/16	3000	2750	2150	1800	1225
3	4	5 1/4	3.000	2.996	5.254	5.250	9	1	2100	1915	1480	1290	845
3	7	5 1/4	3.000	2.996	5.254	5.250	9	1	3840	3505	2710	2260	1525

(*Table continued on p. 12-98*)

Table 22. Dimensions and Load Capacities of Roller Bearings for Wound Rollers with Split Outer Race and No Inner Race (See Fig. 11)—*Continued*

(Hyatt Bearings Division, General Motors Corporation, Harrison, N. J.)

Nominal Shaft Diam., in.	Bearing Length, in.	Nominal Outside Diam. of Bearing, in.	Shaft Limits		Limits for Housing Bore		Rollers		Load Ratings in Pounds, 5000 hr, Ave. Life Expectancy. Based on rotating shaft of Rockwell hardness B-86. For stationary shafts and other hardness, use factor at foot of table. Load rating from $L \times a \times b \times c \times d$. (See Table 21) — Revolutions per Minute				
			High Limit, in.	Low Limit, in.	High Limit, in.	Low Limit, in.	No.	Diam., in.					
A	*B*	*C*							25	100	300	500	1000
3 1/2	4	6 1/8	3.500	3.496	6.129	6.125	12	1 1/8	2860	2570	1980	1650	.....
3 1/2	7	6 1/8	3.500	3.496	6.129	6.125	12	1 1/8	5240	4740	3630	3020	.....
4	4	6 5/8	4.000	3.996	6.629	6.625	14	1 1/8	3370	3040	2310	1885	.....
4	7	6 5/8	4.000	3.996	6.629	6.625	14	1 1/8	6170	5570	4225	3460	.....
4 1/2	4	7 3/8	4.500	4.496	7.380	7.375	14	1 1/4	3730	3300	2510	2050	.....
4 1/2	7	7 3/8	4.500	4.496	7.380	7.375	14	1 1/4	6800	6100	4580	3740	.....
5	4	7 7/8	5.000	4.995	7.880	7.875	15	1 1/4	3940	3520	2610	2110	.....
5	7	7 7/8	5.000	4.995	7.880	7.875	15	1 1/4	7200	6400	4780	3870	.....
6	5	8 7/8	6.000	5.995	8.880	8.875	18	1 1/4	6370	5600	4050	3200	.....
6	7	8 7/8	6.000	5.995	8.880	8.875	18	1 1/4	9170	7900	5830	4450	.....
7	7	10 1/8	7.000	6.995	10.130	10.125	18	1 3/8	9520	8000	5900	4500	.....

Hardness and Stationary Shaft Factors *a* and *b*

Shaft Hardness	Rockwell	B-81	B-83	B-85	B-87	B-93	C-24	C-32	C-43	C-51
	Brinell	150	156	162	168	200	250	300	400	500
Rotating Shaft	Hardness Factor, *a*	1.15	1.09	1.04	1.00	0.83	0.67	0.56	0.56	0.56
Stationary Shaft	Stationary Factor, *b*	....	....	....	1.67	1.19	0.71	0.56	0.40	0.30

Table 23. Factors for Various Desired Life Expectancies

(Hyatt Bearings Division, General Motors Corporation)

Desired Ave. Life, hr	Life Factor	Desired Ave. Life, hr	Life Factor	Desired Ave. Life, hr	Life Factor
1500	0.548	5000	1.000	15000	1.732
2000	0.633	6000	1.096	20000	2.000
2500	0.707	7000	1.183	25000	2.236
3000	0.775	8000	1.265	30000	2.450
4000	0.894	9000	1.342	40000	2.828
4500	0.949	10000	1.414	50000	3.162

(*Table 22*) Pintle bearings, 0.25; hand trucks, 0.4–0.67; industrial car caster wheels, 0.5–0.67; hand-operated cranes and hoists, 0.80–1.0; steel mill cars, charging cars, ingot cars, 0.67; agricultural equipment, gravity conveyors, industrial cars, contractors' cars, 0.67–1.0; hand-operated hoists, 0.80–1.0; mining machinery, except cars, 1.33; belt conveyor idlers, rotating shafts, 2.0–2.8; soaking pit covers, 3.0; belt conveyor idlers, stationary shafts, 3.0–4.0.

For unusual conditions the manufacturer should be consulted for the proper load rating.

Stationary Shaft Factor (*b*). Required rated load for stationary shaft in bearings with raceways is obtained by multiplying bearing load by 1.33. For bearings without inner raceway, use factors from table.

Lubrication of wound roller bearings depends upon the speed and temperature. The Hyatt Bearings Division gives a chart showing the range of speed for various shaft diameters by which it may be classified as operating at low, medium, or high speed. From this chart the following figures are taken:

Shaft diameter, in.	1	$1\,^1/_2$	2	$2\,^1/_2$	3	4	5
Low speed, max rpm	550	400	300	230	200	150	120
Medium speed, max rpm	1400	1140	940	800	670	510	400

Table 24 gives the viscosities recommended for these bearings by the manufacturer. Ordinarily the lightest oil that will stay in the housing should be used. When leakage

Table 24. Viscosity of Oils for Wound Roller Bearings

(Hyatt Bearings Division, General Motors Corporation)

Atmospheric Temperature, °F	Viscosity, SUS		
	High Speed	Medium Speed	Low Speed
−20 to 32	100–160 @ 100 F	100–250 @ 100 F	160–325 @ 100 F
32 to 150	130–225 @ 130 F	60– 70 @ 210 F	70– 80 @ 210 F
150 to 250	70– 80 @ 210 F	70– 80 @ 210 F	70– 80 @ 210 F
250 to 500	80–105 @ 210 F	80–125 @ 210 F	105–150 @ 210 F

prevents the use of oil, grease may be used. The grease should be a well-blended mixture of highly refined oils and carefully prepared soap, without filler. Calcium or lime soap greases may be used at temperatures below 150 F and speeds below 300 rpm. Soda-lime or mixed base greases are satisfactory up to 250 F and at higher speeds.

SOLID ROLLER BEARINGS. Solid roller bearings comprise a series of solid cylindrical rollers confined between inner and outer raceways, and held in position relative to each

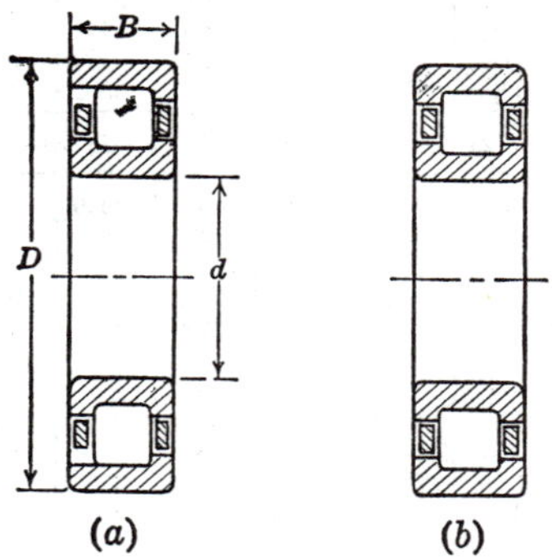

FIG. 12. Solid roller bearings: (*a*) one-lip, (*b*) two-lip type.

other by cages or spacers. The inner raceway frequently is grooved to prevent endwise movement of the rollers (Fig. 2*a*). In some cases the inner raceway may be the shaft or spindle itself. The outer raceway may be a plain ring. The rollers then are free to move endwise across the face of the raceway, thus permitting endwise movement of the shaft, or the outer raceway may have a single lip or two lips (Fig. 2*b* and *c*), forming a groove in the raceway, thus locating the shaft endwise. The outer raceway may be turned to the surface of a sphere and enclosed in a ring with a spherical interior surface, making a self-aligning bearing.

Typical forms of solid cylindrical roller bearings are shown in Fig. 2 (*a*, *b*, and *c*) and in Fig. 12. Dimensions and ratings of typical metric bearings are listed in Table 10. Inch-series bearings are listed in Table 25. While the ratings in Table 10 are for one million revolutions with the inner ring rotating relative to the load direction, those in Table 25 are based on an average life expectancy of 10,000 hr. To get the required rating for other desired life expectancies multiply the bearing load by the proper life factor listed at the bottom of the table.

Table 25. Dimensions and Load Capacities of Solid Cylindrical Roller Bearings

(Norma-Hoffman Bearings Corporation, Stamford, Conn.)

Bore	Outside	Width	Rollers		Load Rating, lb. Based upon rotating shaft for 10,000 hr average life expectancy. For outer race rotation multiply speed by following factor before selecting rating: heavy series, 1.7; medium, 1.6; light 1.5; all others, 1.2.								
A, in.	Diam. *B*, in.	*C*, in.	No.	Diam. *D*, in.	25	50	100	200	450	900	1800	3600	7200
Standard Type (Fig. 2*a*)—Extra Light Series													
1.5000	2.6870	0.563	18	1/4	2730	2170	1710	1370	1040	825	655	520	412
1.6250	2.8745	0.563	18	1/4	2730	2170	1710	1370	1040	825	655	520	412
1.7500	2.9993	0.563	20	1/4	2900	2300	1820	1440	1100	880	700	550	435
1.8750	3.1868	0.625	18	5/16	4200	3330	2650	2100	1600	1280	1010	800	635
2.0000	3.3118	0.625	18	5/16	4200	3330	2650	2100	1600	1280	1010	800	635
2.2500	3.5618	0.625	20	5/16	4500	3600	2850	2260	1710	1380	1090	860	682
2.5000	3.8743	0.688	20	11/32	5500	4330	3440	2740	2090	1670	1320	1050	.....
2.7500	4.1243	0.688	22	11/32	5700	4500	3600	2850	2170	1730	1370	1090	.....
3.0000	4.4993	0.750	22	3/8	6800	5350	4260	3390	2580	2050	1630	1295	.....
3.2500	4.7493	0.750	22	3/8	6800	5350	4260	3390	2580	2050	1630	1295	.....
3.5000	4.9990	0.750	24	3/8	7100	5600	4490	3570	2700	2160	1700	1350	.....
3.7500	5.2490	0.750	26	3/8	7500	5950	4720	3750	2860	2300	1800	1430	.....
4.0000	5.6240	0.875	24	7/16	9800	7700	6150	4900	3700	2950	2340	1860	.....
4.2500	5.9990	0.875	26	7/16	10400	8200	6490	5150	3930	3150	2500	1985	.....
4.5000	6.2490	0.875	28	7/16	10900	8600	6810	5400	4100	3300	2600	2065	.....
4.7500	6.4990	0.875	30	7/16	11200	8800	7000	5560	4210	3380	2680	2130	.....
5.0000	6.9990	1.000	26	1/2	13300	10500	8300	6600	5050	4000	3190	2530	.....
Standard (Fig. 2*a*) and One-lipped (Fig. 12*a*) Types—Light Series													
0.5000	1.3122	0.375	10	3/16	1110	875	696	550	420	335	265	210	170
0.6250	1.5622	0.438	9	1/4	1930	1530	1220	960	735	585	465	365	290
0.7500	1.8747	0.563	10	9/32	2550	2010	1600	1280	960	770	610	485	380
0.8750	1.9995	0.563	11	9/32	2700	2130	1690	1350	1020	813	640	510	405
1.0000	2.2495	0.625	11	5/16	3300	2610	2090	1650	1260	1000	790	630	500
1.2500	2.7495	0.688	11	3/8	4730	3730	2990	2370	1800	1430	1140	900	715
1.5000	3.2493	0.750	12	13/32	5800	4570	3620	2900	2200	1750	1390	1100	870
1.7500	3.7493	0.813	14	7/16	7400	5860	4660	3700	2820	2250	1790	1410	1120
2.0000	3.9993	0.813	15	7/16	7500	5930	4720	3730	2850	2280	1800	1430	1135
2.2500	4.4993	0.875	15	1/2	10000	7860	6260	5000	3780	3020	2400	1900	.....
2.5000	4.9990	0.938	16	9/16	13100	10300	8190	6500	4950	3950	3120	2475	.....
2.7500	5.2490	0.938	17	9/16	13500	10700	8440	6700	5100	4100	3230	2565	.....
3.0000	5.7490	1.063	18	9/16	13900	11000	8700	6900	5260	4200	3330	2645	.....
3.2500	5.9990	1.063	16	5/8	15800	12500	9950	7850	6000	4780	3770	2995	.....
3.5000	6.4990	1.125	16	11/16	19200	15200	12100	9550	7260	5800	4600	3650	.....
3.7500	6.7490	1.125	16	11/16	19200	15200	12100	9550	7260	5800	4600	3650	.....
4.0000	7.2490	1.250	18	3/4	24800	19600	15600	12300	9350	7460	5930	4720	.....
4.2500	7.4990	1.250	18	3/4	24800	19600	15600	12300	9350	7460	5930	4720	.....
4.5000	7.9990	1.313	18	13/16	29100	23000	18300	14700	11100	8800	7000	5560	.....
4.7500	8.2490	1.313	19	13/16	30000	23600	18800	15000	11400	9000	7130	......	.....
5.0000	8.9990	1.375	19	7/8	34600	27300	21700	17200	13200	10500	8230	......	.....
Standard (Fig. 2*a*) and One-lipped (Fig. 12*a*) Types—Medium Series													
0.5000	1.6247	0.625	8	1/4	1810	1430	1140	900	685	545	435	345	270
0.6250	1.8122	0.625	10	9/32	2620	2060	1640	1300	990	790	625	495	395
0.7500	1.9995	0.688	10	5/16	3200	2520	2000	1600	1210	965	765	605	480
0.8750	2.2495	0.688	10	11/32	3900	3100	2460	1940	1490	1190	935	740	590
1.0000	2.4995	0.750	10	3/8	4650	3680	2920	2320	1770	1410	1120	880	700
1.2500	3.1243	0.875	10	1/2	8300	6550	5210	4150	3150	2510	1990	1580	1255
1.5000	3.7493	0.938	12	1/2	9100	7200	5700	4510	3430	2750	2190	1720	1365
1.7500	4.2493	1.063	11	5/8	13700	10900	8550	6800	5150	4110	3260	2600	.....
2.0000	4.4993	1.063	11	5/8	13700	10900	8550	6800	5150	4110	3260	2600	.....
2.2500	4.9990	1.250	12	11/16	16900	13400	10700	8400	6400	5100	4040	3200	.....

Table 25. Dimensions and Load Capacities of Solid Cylindrical Roller Bearings—*Continued*

(Norma-Hoffman Bearings Corporation, Stamford, Conn.)

Bore	Outside	Width	Rollers		Load Rating, lb								
					Based upon rotating shaft for 10,000 hr average life expectancy. For outer race rotation multiply speed by following factor before selecting rating: heavy series, 1.7; medium, 1.6; light 1.5; all others, 1.2.								
A, in.	Diam. *B*, in.	*C*, in.	No.	Diam. *D*, in.	25	50	100	200	450	900	1800	3600	7200
STANDARD (FIG. 2*a*) AND ONE-LIPPED (FIG. 12*a*) TYPES—MEDIUM SERIES													
2.5000	5.4990	1.250	12	25/32	22000	17400	13800	11000	8300	6700	5260	4175	.....
2.7500	6.2490	1.375	12	7/8	28000	22000	17500	13900	10600	8460	6700	5320	.....
3.0000	6.9990	1.563	12	1	36500	29000	23000	18200	13900	11100	8750	6940	.....
3.2500	7.4990	1.563	13	1	37500	29700	23600	18700	14200	11400	9000	7140	.....
3.5000	8.1240	1.750	12	1 1/8	45700	36100	28900	23000	17300	13900	11000	8730	.....
3.7500	8.2490	1.750	13	1 1/8	48500	38300	30400	24200	18400	14700	11700	......	.....
4.0000	8.4990	1.750	13	1 1/8	48500	38300	30400	24200	18400	14700	11700	......	.....
4.2500	8.7490	1.750	14	1 1/8	50000	39600	31400	25000	19000	15200	12000	......	.....
4.5000	9.3740	2.000	14	1 1/4	62000	49000	38900	30900	23500	18800	14900	......	.....
4.7500	9.9990	2.000	15	1 1/4	63500	50500	40000	31800	24100	19200	15230	......	.....
5.0000	9.9990	2.000	15	1 1/4	63500	50500	40000	31800	24100	19200	15230	......	.....
TWO-LIPPED TYPE (FIG. 12*b*)—LIGHT SERIES													
0.7500	1.8747	0.563	8	9/32	2190	1730	1380	1100	830	665	525	420	330
0.8750	1.9995	0.563	9	9/32	2360	1860	1480	1180	900	720	560	450	355
1.0000	2.2495	0.625	9	5/32	2900	2300	1830	1450	1110	880	700	550	440
1.2500	2.7495	0.688	9	3/8	4170	3300	2620	2080	1590	1270	1000	790	630
1.5000	3.2493	0.750	10	13/32	5060	4000	3200	2530	1930	1550	1220	960	760
1.7500	3.7493	0.813	11	7/16	6300	5000	3970	3130	2400	1900	1510	1200	950
2.0000	3.9993	0.813	12	7/16	6450	5100	4070	3200	2450	1950	1550	1230	975
2.2500	4.4993	0.875	11	1/2	8100	6400	5110	4060	3090	2470	1940	1540	.....
2.5000	4.9990	0.938	11	9/16	10400	8200	6510	5150	3930	3140	2500	1985	.....
2.7500	5.2490	0.938	12	9/16	10800	8400	6670	5300	4030	3230	2560	2030	.....
3.0000	5.7490	1.063	13	9/16	11200	8800	7000	5550	4230	3390	2680	2130	.....
3.2500	5.9990	1.063	12	5/8	13100	10400	8200	6500	5000	4000	3130	2490	.....
3.5000	6.4990	1.125	12	11/16	15900	12600	10000	7900	6000	4800	3800	3020	.....
3.7500	6.7490	1.125	12	11/16	15900	12600	10000	7900	6000	4800	3800	3020	.....
4.0000	7.2490	1.250	13	3/4	19900	15700	12500	9900	7500	6000	4750	3770	.....
4.2500	7.4990	1.250	13	3/4	19900	15700	12500	9900	7500	6000	4750	3770	.....
4.5000	7.9990	1.313	13	13/16	23300	18500	14700	11700	8800	7060	5600	4440	.....
4.7500	8.2490	1.313	13	13/16	23300	18500	14700	11700	8800	7060	5600	4440	.....
5.0000	8.9990	1.375	13	7/8	26900	21100	16900	13400	10200	8100	6400	......	.....
TWO-LIPPED TYPE (FIG. 12*b*)—MEDIUM SERIES													
0.5000	1.6247	0.625	7	1/4	1650	1310	1040	820	620	500	396	313	250
0.6250	1.8122	0.625	8	9/32	2250	1790	1410	1130	850	685	540	430	340
0.7500	1.9995	0.688	8	5/16	2730	2170	1720	1370	1040	830	655	520	413
0.8750	2.2495	0.688	8	11/32	3390	2670	2120	1690	1290	1020	810	640	510
1.0000	2.4995	0.750	8	3/8	4000	3170	2520	2000	1520	1220	960	760	605
1.2500	3.1243	0.875	8	1/2	7200	5700	4500	3600	2730	2190	1720	1370	1090
1.5000	3.7493	0.938	9	1/2	7500	5930	4710	3750	2850	2300	1800	1430	1135
1.7500	4.2493	1.063	8	5/8	11000	8700	6910	5460	4160	3330	2630	2090	.....
2.0000	4.4993	1.063	9	5/8	11800	9260	7350	5860	4450	3570	2830	2230	.....
2.2500	4.9990	1.250	9	11/16	14200	11200	8850	7000	5300	4300	3400	2680	.....
2.5000	5.4990	1.250	9	25/32	18100	14300	11400	9000	6860	5500	4330	3440	.....
2.7500	6.2490	1.375	9	7/8	23000	18200	14500	11600	8700	7000	5500	4370	.....
3.0000	6.9990	1.563	8	1	28000	22200	17600	14000	10700	8500	6700	5320	.....
3.2500	7.4990	1.563	9	1	29800	23500	18700	14900	11300	9000	7150	5675	.....
3.5000	8.1240	1.750	9	1 1/8	37500	29800	23700	18800	14300	11500	9000	7150	.....
3.7500	8.2490	1.750	9	1 1/8	37500	29800	23700	18800	14300	11500	9000	7150	.....
4.0000	8.4990	1.750	9	1 1/8	37500	29800	23700	18800	14300	11500	9000	7150	.....
4.2500	8.7490	1.750	10	1 1/8	40000	31400	25000	19800	15000	12000	9500	......	.....
4.5000	9.3740	2.000	9	1 1/4	46000	36500	29000	23000	17500	14000	11100	......	.....
4.7500	9.9990	2.000	10	1 1/4	48500	38300	30500	24300	18400	14800	11750	......	.....
5.0000	9.9990	2.000	10	1 1/4	48500	38300	30500	24300	18400	14800	11750	......	.....

(*Table continued on p. 12-102*)

Table 25. Dimensions and Load Capacities of Solid Cylindrical Roller Bearings—*Continued*

Life Factors for Various Desired Average Life Expectancies

Desired Ave. Life, hr.	1000	1500	2000	2500	3000	3500	4500	5500	6500
Life Factor	0.47	0.54	0.59	0.63	0.67	0.71	0.77	0.82	0.87
Desired Ave. Life, hr.	8000	10000	12000	15000	20000	25000	30000	40000	50000
Life Factor	0.93	1.00	1.06	1.15	1.26	1.36	1.44	1.59	1.71

21. NEEDLE BEARINGS

Needle bearings are bearings using solid cylindrical rollers whose length to diameter ratio is relatively large, recommended to be between 6 to 1 and 10 to 1. The rollers are generally assembled without a cage, thus allowing the use of a large number of rollers. These bearings are suitable for only radial load and are particularly useful where diametral space is limited. The long rollers are kept in alignment by the outer race. Axial restraint is provided by flanges on the race. The needle bearing is well adapted to low-speed or oscillating operation but can be designed for successful operation at high speeds. Rollers may operate directly on shaft and bearing seat or be contained in a manufactured bearing assembly. Load capacity is greatly affected by rigidity of raceway and supports.

LOOSE-ROLLER BEARINGS. The capacity of the rollers is greater than that of the raceways. The Torrington Company recommends a raceway hardness of Rockwell C 60 minimum. Hardness less than this results in marked reduction of carrying capacity.

A roller diameter equal to 10% of pitch diameter gives a well-proportioned bearing; higher-speed rotating shafts use a higher ratio and lower-speed oscillating shafts a lower ratio (to 2 1/2%). Load for shaft rotation and dimensional relationships recommended are:

$$R = \frac{32{,}000 D_p L}{\sqrt[3]{S}} \quad (1) \qquad \text{or} \qquad R = 2200\, NLd \text{ at 100 rpm} \quad (2)$$

$$D_p = Kd + \frac{N(cc)}{\pi} \quad (3) \qquad D = D_p - d - (dc) \quad (4)$$

$$B = D_p + d \quad (5) \qquad N = \frac{\pi D_p}{d + (cc)} \quad (6)$$

where R = rated load, pounds; S = speed of bearing, rpm; L = effective roller length, inches; D_p = pitch diameter, inches; D = shaft diameter, inches; B = housing bore diameter, inches; N = number of rollers; (cc) = circumferential clearance per roller, inches; (dc) = diametral clearance, inches; $K = 1 \div \sin 180°/N$. A convenient table of values of K, $N(cc)/\pi$, and their combination is given in the Torrington Company Catalog. Selected values of K are reproduced in Table 26.

The capacity for oscillating shafts is taken the same as for rotation at 100 rpm.

Table 26. Values of $K = 1 \div \sin 180°/N$

(The Torrington Company, South Bend, Ind.)

N	K	N	K	N	K	N	K	N	K	N	K
10	3.23607	27	8.61379	44	14.01754	61	19.42549	78	24.83488	95	30.24495
11	3.54947	28	8.93141	45	14.33559	62	19.74366	79	25.15311	96	30.56320
12	3.86370	29	9.24907	46	14.65365	63	20.06183	80	25.47134	97	30.88146
13	4.17858	30	9.56677	47	14.97171	64	20.38001	81	25.78956	98	31.19971
14	4.49395	31	9.88452	48	15.28979	65	20.69819	82	26.10779	99	31.51797
15	4.80973	32	10.20230	49	15.60788	66	21.01638	83	26.42603	100	31.83623
16	5.12583	33	10.52011	50	15.92597	67	21.33457	84	26.74426	101	32.15449
17	5.44219	34	10.83795	51	16.24408	68	21.65277	85	27.06250	102	32.47275
18	5.75877	35	11.15582	52	16.56219	69	21.97096	86	27.38074	103	32.79101
19	6.07553	36	11.47372	53	16.88031	70	22.28917	87	27.69898	104	33.10927
20	6.39245	37	11.79163	54	17.19844	71	22.60737	88	28.01722	105	33.42753
21	6.70951	38	12.10957	55	17.51657	72	22.92558	89	28.33546	106	33.74579
22	7.02667	39	12.42752	56	17.83471	73	23.24379	90	28.65370	107	34.06406
23	7.34394	40	12.74550	57	18.15286	74	23.56201	91	28.97195	108	34.38232
24	7.66130	41	13.06349	58	18.47101	75	23.88022	92	29.29020	109	34.70059
25	7.97873	42	13.38149	59	18.78916	76	24.19844	93	29.60845	110	35.01885
26	8.29623	43	13.69951	60	19.10732	77	24.51666	94	29.92670		

Table 27. Dimensions of Rounded-end Rollers for Needle Bearings

All dimensions in inches. Diametral tolerance = −0.0002 in., length tolerance = 0.020 in.

(The Torrington Company, South Bend, Ind.)

Diam.	Length	Diam.	Length	Diam.	Length	Diam.	Length	Diam.	Length	Diam.	Length
.0310	.298	.0984	.370	.0901	.975	.1250	0.707	.1582	0.562	.1875	1.375
.0627	.250	.0937	.416	.0939	.975	.1250	0.750	.1582	0.740	.1875	1.740
.0627	.370	.0940	.494	.1040	.396	.1250	0.875	.1582	1.014	.2180	1.750
.0627	.490	.0940	.540	.1040	.509	.1250	0.938	.1875	0.527	.2500	1.000
.0627	.620	.0957	.625	.1012	.572	.1250	1.000	.1875	0.745	.2500	1.250
.0627	.666	.0938	.750	.1250	.385	.1250	1.125	.1881	0.923	.2500	1.500
.0627	.750	.0939	.796	.1262	.520	.1250	1.250	.1875	1.000	.2500	1.750
.0772	.955			.1250	.615			.1875	1.187		

Table 28. Dimensions and Load Capacities of Needle Bearings with Small Rolls without Inner Race

(The Torrington Company, Torrington, Connecticut)

Nom'l Shaft Diam., in.	Nom'l OD, in.	L'gth 0.000 −0.001, in.	Inside Depth, in.	Rolls		Load Rating, lb — Based upon rotating shaft for 3000 hr average life expectancy, with shaft hardened to Rockwell C 52. For outer race rotation use 85% of values listed. For oscillating shaft use values under 500 rpm.									
				No.	Size, in.	Static	50	100	300	500	1000	1500	2000	3600	5000
5/32	9/32	1/4	0.20	14	.0446	125	50	40	28	23	18	16	15	12	11
3/16	11/32	3/8	0.31	14	.0535	407	163	133	92	76	60	52	48	39	35
1/4	7/16	7/16	0.36	15	.0655	670	268	212	147	124	99	87	79	65	58
5/16	1/2	7/16	0.36	18	.0655	803	321	255	177	149	118	103	94	77	69
5/16	1/2	9/16	0.49	18	.0655	1170	469	372	258	218	173	151	138	113	101
3/8	9/16	5/8	0.55	21	.0655	1593	637	505	350	295	234	205	187	154	137
7/16	5/8	1/2	0.42	24	.0655	1285	514	408	283	238	189	165	150	123	111
1/2	11/16	1/2	0.42	27	.0655	1480	592	470	326	274	218	191	174	143	128
1/2	11/16	3/4	0.67	27	.0655	2590	1037	823	571	481	382	334	304	250	224
9/16	3/4	1/2	0.42	30	.0655	1675	670	531	369	310	247	216	197	162	145
9/16	3/4	3/4	0.67	30	.0655	2880	1150	914	633	534	424	371	338	278	248
5/8	13/16	1/2	0.42	33	.0655	1808	722	573	398	334	266	232	212	174	156
5/8	13/16	3/4	0.67	33	.0655	3170	1268	1005	697	587	467	408	372	306	274
11/16	7/8	3/4	0.67	36	.0655	3440	1375	1090	768	638	506	442	401	330	296
3/4	1	5/8	0.53	28	.0944	2700	1080	857	594	500	398	348	317	260	233
3/4	1	3/4	0.66	28	.0944	3540	1412	1120	777	654	520	455	414	340	304
13/16	1 1/16	1	0.91	30	.0947	4000	1600	1270	880	743	590	515	470	386	346
7/8	1 1/8	1/2	0.41	33	.0918	2140	856	679	471	396	315	275	253	208	185
15/16	1 3/16	1	0.91	34	.0952	6130	2450	1989	1380	1136	902	788	715	588	527
1	1 1/4	1/2	0.41	37	.0925	2420	967	768	533	448	356	311	284	233	209
1	1 1/4	3/4	0.66	37	.0925	4560	1827	1450	1005	847	673	588	537	441	395
1	1 1/4	1	0.91	37	.0925	6740	2690	2135	1480	1245	990	865	789	649	581
1 1/8	1 3/8	3/4	0.66	41	.0931	5080	2015	1600	1110	935	743	646	593	486	435
1 1/8	1 3/8	1	0.91	41	.0931	7500	3000	2380	1650	1390	1105	965	880	723	648
1 1/4	1 1/2	3/4	0.66	45	.0936	5560	2240	1780	1230	1040	825	721	654	537	482
1 1/4	1 1/2	1 1/4	1.16	45	.0936	10900	4360	3460	2430	2020	1605	1400	1270	1045	940
1 3/8	1 5/8	3/4	0.66	51	.0900	6100	2440	1940	1340	1130	900	785	713	585	525
1 3/8	1 5/8	1 1/4	1.16	51	.0900	11170	4460	3540	2455	2065	1640	1440	1310	1077	962
1 1/2	1 7/8	7/8	0.77	35	.1476	7375	2950	2340	1623	1367	1085	949	862	708	636
1 1/2	1 7/8	1	0.87	35	.1476	9000	3600	2875	1980	1630	1326	1160	1050	865	774
1 3/4	2 1/8	1 1/2	1.37	41	.1449	17580	7030	5660	3930	3250	2580	2250	2055	1690	1513
1 7/8	2 1/4	1/2	0.38	45	.1475	3125	1250	990	685	580	460	400	365	300	269
2 1/8	2 1/2	1 1/2	1.37	49	.1454	21450	8570	6810	4730	3980	3160	2760	2523	2075	.....
2 5/8	3	1	0.88	59	.1475	14920	5970	4730	3280	2770	2200	1920	1745	......	.....
2 3/4	3 1/8	1 1/4	1.12	61	.1491	21300	8530	6770	4715	3960	3150	2740	2510	......	.....
3 1/2	4	3/4	......	55	.2118	12550	5025	4000	3775	2335	1850	......	......	......	.....

Circumferential clearance (*cc*), except for special circumstances, is 0.0001 in. per roller plus tolerances. For severe operating conditions this clearance would be reduced.

Diametral clearance (*dc*) for rotating shaft should be 0.0005 in., minimum, seldom greater than 0.002 in. with maximum tolerances.

Raceway tolerances are plus zero, minus 0.0005 in. for inner or shaft and minus zero, plus 0.0005 in. for outer raceway bore.

Roller tolerances are minus 0.0002 to 0.0004 in. for diameter and minus 0.020 in. for length. Minimum roller end play against end retainers is 0.010 in. For moderate service, retainers may be flat, split rings, although heavy retainers increase capacity.

Table 27 gives stock sizes of needle rollers with rounded ends. Where load is greater on one end of bearing, or greater length is required, two or more rows of rollers may be used with hardened, ground spacers between the roll ends.

UNIT-ASSEMBLY BEARINGS. The Torrington Company manufactures unit assemblies with hardened, pressed-steel housing for press fit in a bored housing, assemblies with outer race and heavy end guards, and inner races for use with soft shafts.

Table 28 gives sizes and rated capacity at 100 rpm of shaft based upon 3000 hours average life expectancy when used on a shaft hardened to Rockwell C 52. Capacity for oscillating shafts is taken same as that at 500 rpm. For outer race rotation use 85% of listed load values.

22. TAPERED ROLLER BEARINGS

The tapered roller bearing is constructed upon the principle of rolling cones, the elements of rollers and raceways converging to a common intersection on the axis of the bearing (Fig. 13). True rolling motion is obtained along the cone elements, and the

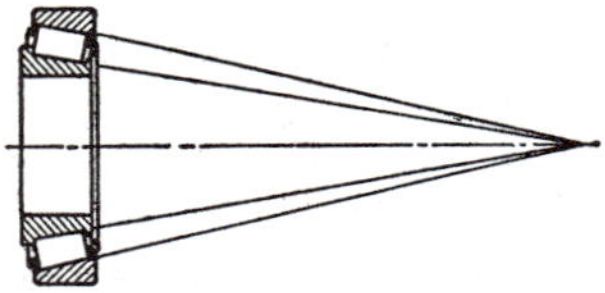

FIG. 13. Basic principle of tapered roller bearing.

bearing will sustain radial or thrust loads, or any combination of them. The essential parts are an inner raceway or cone, an outer raceway or cup, tapered rollers, and a cage or roller retainer (Fig. 14). The end thrust on the rollers is balanced by a shoulder on the cone, and the retainer acts merely as a spacer, and sustains no load.

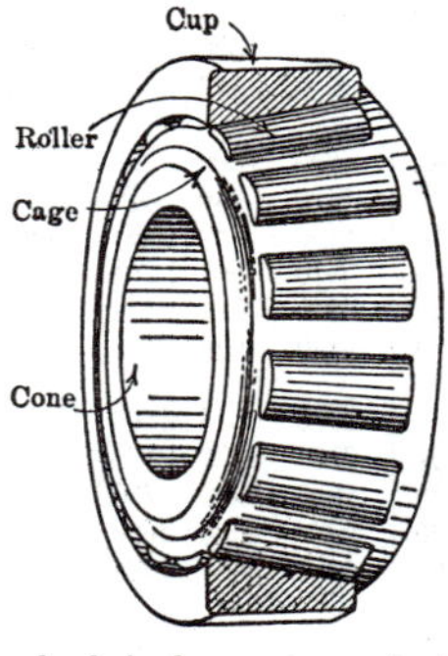

FIG. 14. Standard single-row tapered roller bearing.

BEARING SELECTION. Dimensions and rated capacity of selected bearings of the various types are listed in Tables 29 to 33. Data for other sizes and types may be obtained from the manufacturers.

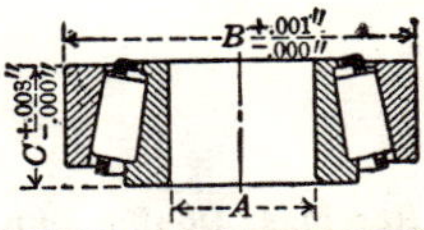

Table 29. Dimensions and Load Capacities of Standard Single-row Tapered Roller Bearings—Type TS

(Timken Roller Bearing Company, Canton, Ohio)

Bore	Heavy Series				Medium Series				Light Series			
	Dimensions, in.		Load Rating, lb, at 500 rpm		Dimensions, in.		Load Rating, lb, at 500 rpm		Dimensions, in.		Load Rating, lb, at 500 rpm	
A	*B*	*C*	Radial	Thrust	*B*	*C*	Radial	Thrust	*B*	*C*	Radial	Thrust
0.375	....	....	....	....	....	....	....	....	1.2595	0.3940	315	250
0.500	....	....	....	....	....	....	....	....	1.3775	0.4330	360	320
0.625	....	....	....	....	....	....	....	....	1.6250	0.5625	615	375
0.750	....	....	....	....	1.9380	0.7813	1175	615	1.8504	0.5662	725	510
0.875	....	....	....	....	2.2400	0.7625	1245	750	2.0470	0.5910	760	600
1.000	2.6150	0.9375	2030	1015	2.3750	0.7813	1240	805	2.3125	0.7500	1240	805
1.125	2.8593	1.1875	2685	1755	2.6150	0.9375	2030	1015	2.5000	0.8125	1395	960
1.250	3.1250	1.1563	2830	2030	2.7500	0.9375	2130	1145	2.7170	0.7813	1520	1140
1.375	3.3125	1.1563	3085	1655	3.0000	0.9375	2230	1330	2.8440	0.8125	1460	1150
1.500	3.4375	1.1875	3080	1855	3.1250	1.0000	2230	1330	3.0000	0.8125	1660	1310
1.625	3.7500	1.2500	3235	1810	3.3750	1.1875	3100	2450	3.1496	0.8268	2155	1155
1.750	4.0000	1.3750	4520	2530	3.4375	1.1875	3080	1855	3.3464	0.8125	2230	1345
1.875	4.2500	1.4375	4745	2755	3.6718	1.1875	3365	2240	3.5000	0.8125	2250	1415
2.000	4.3750	1.5000	4745	2755	3.8750	1.1875	3365	2240	3.6718	0.8125	2345	1560
2.250	4.8750	1.5000	5180	3525	4.1250	1.1875	3540	2335	3.9370	0.8268	2430	1690
2.500	5.3750	1.6250	6605	4695	4.4375	1.1875	3820	3020	4.3307	0.8661	2810	2220
2.750	5.9090	1.7500	8770	5605	4.8750	1.1875	4000	3020	....	....	....	....
3.000	6.3750	1.8750	9150	6130	5.3750	1.1875	4425	3860	5.0000	1.0625	2790	2470
3.250	6.6250	1.8750	9150	6130	5.5115	1.4375	5580	4425	5.3750	1.1875	4425	3860
3.500	7.1250	1.8750	9505	7200	6.0000	1.5625	6160	5350	....	....	....	....
3.750	7.5000	2.2500	13005	8555	6.3750	1.5625	6330	5905	5.8750	1.2500	4675	4520
4.000	7.8750	2.4375	13005	8555	6.6250	1.6250	7590	7015	....	....	....	....
4.250	8.3750	2.6250	17125	10955	....	....	....	....	....	....	....	....

Table 30. Dimensions and Capacities of Steep Angle Tapered Roller Bearings—Single Row, Type TSS

(Timken Roller Bearing Company, Canton, Ohio)

Dimensions, in. (Fig. 15)			Load Rating, lb, at 500 rpm		Dimensions, in. (Fig. 15)			Load Rating, lb, at 500 rpm	
A	*B*	*C*	Radial	Thrust	*A*	*B*	*C*	Radial	Thrust
0.750	2.1250	0.8750	1280	1485	1.750	3.750	1.2188	2,835	4,120
1.000	2.5625	0.8750	1410	2020	2.000	4.375	1.1875	3,010	5,225
1.125	2.8593	0.9688	1925	2280	2.500	5.513	1.4375	4,990	8,490
1.250	3.1250	1.0000	2145	2815	3.000	6.375	1.9375	7,820	10,890
1.500	3.4834	1.0000	2210	3390	3.500	7.874	2.0772	11,745	14,585

Table 31. Dimensions and Load Capacities of Flanged-cup Roller Bearings, Type TSF

(Timken Roller Bearing Company, Canton, Ohio)

Dimensions, in. (Fig. 16)						Load Rating, lb, at 500 rpm	
A	*B*	*E*	*G*	*H*	*J*	Radial	Thrust
3.6250	6.0000	1.1875	0.2500	6.2500	0.6250	6,160	5,350
3.7500	6.6250	1.1875	0.2813	6.9063	0.7188	7,590	7,015
3.8750	6.3750	1.0313	0.2500	6.6250	0.6563	6,330	5,905
4.0000	7.1250	1.5000	0.3125	7.4375	0.6875	9,505	7,200
4.0000	7.5000	1.7500	0.3750	7.8750	0.8750	13,005	8,555
4.2500	8.3750	2.1250	0.4375	8.8125	0.9375	17,125	10,955
4.5000	7.5000	1.3750	0.3125	7.8125	0.8125	11,255	9,195
5.0000	8.5000	1.3750	0.3125	8.8125	0.8125	11,495	11,005
5.0000	9.2500	1.9375	0.4375	9.6875	1.0000	18,490	13,415
5.5000	9.0000	1.7500	0.3750	9.3750	0.8750	15,000	12,395
5.7500	10.0000	1.8750	0.4375	10.4360	1.1875	18,440	14,690
6.5000	11.3750	1.8750	0.4375	11.8125	1.0625	20,935	19,250
7.5000	12.5000	1.8125	0.4375	12.9380	1.1250	22,630	23,265

Table 32. Dimensions and Load Capacities of Two-row Roller Bearings, Double Cup, Single Cones—Type TDO

(Timken Roller Bearing Company, Canton, Ohio)

Dimensions, in. (Fig. 17)				Load Rating, lb, at 500 rpm		Dimensions, in. (Fig. 17)				Load Rating, lb, at 500 rpm	
A	B	C	E	Radial	Thrust	A	B	C	E	Radial	Thrust
1.2500	2.7500	2.6250	2.2500	4,260	1,145	6.0000	12.1250	7.8750	5.7500	56,160	17,990
1.5000	3.1510	1.8125	1.3750	3,320	1,310	6.5000	11.3750	5.6250	4.3750	41,870	19,250
1.7500	3.5480	2.0000	1.7500	4,460	1,345	7.0000	11.2500	5.3750	3.6250	30,920	13,155
2.0000	3.6718	2.5625	2.0625	6,730	2,240	8.0000	12.5000	5.0000	3.5000	30,070	9,255
2.2500	3.9370	1.9375	1.5625	4,860	1,690	9.0000	14.0000	6.0000	4.3750	50,620	16,420
2.5000	4.8750	3.1250	2.5000	10,360	3,525	10.0000	16.6250	7.0312	5.5000	70,480	23,020
2.7500	5.0000	3.1875	2.5625	10,780	3,850	11.0000	18.5000	7.8750	5.8750	91,300	33,890
3.0000	6.0000	3.7500	3.0000	14,240	5,720	12.0000	17.4980	6.0000	5.0000	53,340	19,785
4.0000	7.1250	4.1250	3.3750	19,010	7,200	13.0000	19.0000	5.2500	3.5000	40,270	19,810
4.2500	6.2500	2.1250	1.5625	6,270	3,735	14.0000	19.7500	6.1250	4.2500	57,950	25,020
4.5000	7.5000	4.1875	3.1875	22,510	9,195	16.0000	22.6250	5.9375	3.9375	63,430	30,980
4.7500	9.8750	4.8750	3.5000	31,090	12,950	18.0000	23.5000	6.5000	4.7500	87,350	34,720
5.0000	9.2500	5.6250	4.5000	36,980	13,415	20.0000	29.0000	7,3426	4.5000	123,610	57,535
5.5000	9.5000	5.1875	4.1875	29,930	12,990						

Table 33. Dimensions and Load Capacities of Two-row Roller Bearings, Double Cone, Single Cups—Type TDI

(Timken Roller Bearing Company, Canton, Ohio)

Dimensions, in. (Fig. 18)				Load Rating, lb, at 500 rpm		Dimensions, in. (Fig. 18)				Load Rating, lb, at 500 rpm	
A	B	C	D	Radial	Thrust	A	B	C	D	Radial	Thrust
1.4375	2.6875	1.4492	1.5000	2,510	1,095	8.0000	12.5000	5.2500	5.2500	45,260	23,265
1.6875	3.1875	1.3750	1.2500	2,580	1,340	9.0000	14.0000	4.7500	4.7500	50,620	16,420
2.0000	3.6718	1.9690	2.0940	4,650	1,545	10.0000	14.3720	3.6500	3.6560	33,530	11,950
2.5000	5.5130	2.6020	2.5980	9,980	8,490	11.0000	18.5000	6.6875	6.5625	91,300	33,890
3.0000	6.3750	4.0000	4.2500	18,300	6,130	12.0000	17.2500	4.7500	3.8750	36,620	15,765
3.5000	7.5000	4.6250	5.0000	26,010	8,555	13.0000	19.0000	4.1250	3.7500	40,270	19,810
4.0000	8.3750	5.6250	6.0000	31,375	10,040	14.0000	20.2500	5.0000	4.3750	57,960	25,080
4.5000	9.2500	5.5000	6.0000	36,980	13,415	16.0000	22.6250	5.1563	4.5000	63,430	30,980
5.0000	11.6250	5.8750	6.2500	54,710	21,385	18.0000	33.9960	14.5000	14.5000	387,240	136,960
5.5000	8.7500	2.3850	2.1250	13.040	5,600	20.0000	33.0000	10.5000	10.5000	259,060	122,045
6.0000	11.7500	4.2500	4.3750	39,910	13,035	26.0000	42.0000	12.6563	12.2812	486,400	147,960
7.0000	11.3750	6.2500	6.2500	41,870	19,250	30.0000	48.0000	15.6250	15.2500	667,970	233,230

The bearings are rated for both pure radial load, BRR, and for pure thrust load, BTR, for 3000 hours B-10 (or minimum) life expectancy at 500 rpm of shaft rotation. Average life is about five times B-10 life. The required radial rating at 500 rpm of the bearing is $R \times LF \times AF \times SF$, where R is the calculated or equivalent radial load, LF is the life factor from Table 34, AF is the application factor, and SF is the speed factor, obtainable from Table 35.

Table 34. B-10 Life Factors for Tapered Roller Bearings

(Timken Roller Bearing Company, Canton, Ohio)

Desired Life (B-10), hr	LF	Desired Life (B-10), hr	LF	Desired Life (B-10), hr	LF	Desired Life (B-10), hr	LF	Desired Life (B-10), hr	LF	Desired Life (B-10), hr	LF	Desired Life (B-10), hr	LF
500	.584	1000	.719	2000	0.886	3250	1.024	5000	1.166	8,000	1.342	20,000	1.767
600	.617	1200	.760	2250	0.917	3500	1.047	5500	1.199	10,000	1.437	25,000	1.889
700	.646	1400	.796	2500	0.947	3750	1.070	6000	1.231	12,500	1.537	30,000	1.995
800	.672	1600	.828	2750	0.974	4000	1.090	6500	1.261	15,000	1.623	40,000	2.175
900	.697	1800	.858	3000	1.000	4500	1.130	7000	1.289	17,500	1.689	50,000	2.326

Table 35. Speed Factors for Tapered Roller Bearings

(Timken Roller Bearing Company, Canton, Ohio)

rpm	*SF*	rpm	*SF*	rpm	*SF*	rpm	*SF*	rpm	*SF*	rpm	*SF*	rpm	*SF*
25	.407	200	.759	500	1.000	1000	1.232	1500	1.391	2000	1.515	3000	1.712
50	.501	250	.807	600	1.056	1100	1.267	1600	1.418	2200	1.560	3500	1.792
100	.617	300	.857	700	1.106	1200	1.300	1700	1.443	2400	1.600	4000	1.866
125	.660	350	.898	800	1.151	1300	1.332	1800	1.468	2600	1.639	4500	1.934
150	.697	400	.935	900	1.193	1400	1.363	1900	1.493	2800	1.678	5000	1.995

For stationary shaft, radial load on bearing must be multiplied by 1.7, thrust load by 1.00.

Application factor is frequently unity, values greater than unity being 1.33 for farm tractor wheels, feed grinders, hammer mills, paper mill machinery, commercial stokers; 1.5 for printing machinery, rubber machinery, dredge pumps; 1.75 for automotive wheels and differentials; 2.00 for farm machinery and highway trailers.

Where hours of life are not specified but are dependent upon type of equipment, the following values may be used for life factor: 1.00 for automotive, meat slicers, farm machinery, and tractor transmissions, household appliances, lawn mowers, portable tools, spray and household pumps, road machinery, household stokers, highway trailers; 1.25 for steel mill machinery; 1.5 for amusement devices, belt drives, clay-working machinery, farm and industrial engines, tractor wheels, gear reduction units, feed grinder and hammer mills, I-beam trolleys, pulverizing and screening equipment, railway section cars, and commercial stokers; 2.0 for dough mixers, elevators, industrial and oil-field engines, glass-making and grain-milling machinery, cranes and hoists, ice-making, laundry, and leather-working machinery, locomotives, machine tools, cement plant, and printing equipment, deep-well, fire, and dredge pumps, railway car drives, rubber, wood-working, textile machinery; 3.0 for large stationary compressors, paper mill pumps, and machinery.

For thrust load the bearing is selected from the thrust rating in the same manner.

A radial load on a tapered roller bearing induces a thrust equal to $0.47/K$, where

$$K = \frac{\text{Radial load rating of bearing}}{\text{Thrust load rating of bearing}} = \frac{\text{BRR}}{\text{BTR}}$$

Values of K for a bearing may be computed from the given ratings; usual values of K are for standard bearings about 1.5; for steep-angle bearings about 0.75.

For a bearing simultaneously subjected to a radial load R and a thrust load T, the equivalent radial load,

$$ER = 0.53R + KT \text{ for rotating shaft}$$

and

$$ER = 1.23R + KT \text{ for stationary shaft}$$

When the applied radial load is greater than the equivalent radial load, the radial load is used.

When a shaft is supported on two opposed bearings A and B so that A carries practically the whole radial load, the equivalent radial load on B is then

$$ER_B = 0.53R_B + 0.47R_A\left(\frac{K_B}{K_A}\right) \text{ for rotating shaft}$$

$$= 1.23R_B + 0.47R_A\left(\frac{K_B}{K_A}\right) \text{ for stationary shaft}$$

If bearing B carries a thrust load T_B also,

$$ER_B = 0.53R_B + K_B\left(T_B + \frac{0.47R_A}{K_A}\right) \text{ for rotating shaft}$$

$$= 1.23R_B + K_B\left(T_B + \frac{0.47R_A}{K_A}\right) \text{ for stationary shaft}$$

THE STANDARD BEARING is made in three basic series, heavy, medium, and light. They are established by maintaining fundamental relations between the various components of the bearing. The standard or heavy series has wide races and long rollers; the light series has narrow races and short rollers; the medium series is an approximate mean between the other two. (See Table 29.) The standard type of bearing is a general service bearing. It is widely used in automotive vehicles, machinery, farm implements, mining machinery, etc.

THE STEEP ANGLE BEARING (Fig. 15) is used in power take-offs, worms and bevel gear shafts, etc., where the thrust load predominates, although there is an appreciable radial load. It is similar to the standard bearing except that the angles of both the raceway and rollers are quite steep. (See Table 30.)

FLANGED CUP BEARINGS (Fig. 16) are used where extreme precision is required as on machine tool spindles. The bearing housing is bored straight through, and a seat faced

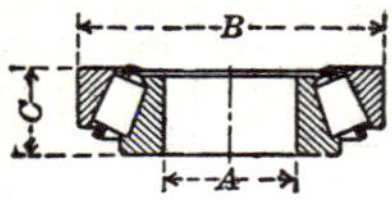

Fig. 15. Single-row steep angle tapered roller bearings.

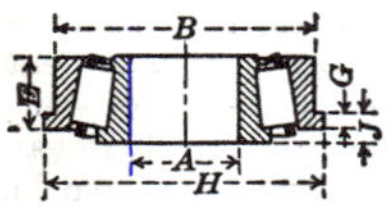

Fig. 16. Single-row flanged-cup tapered roller bearings.

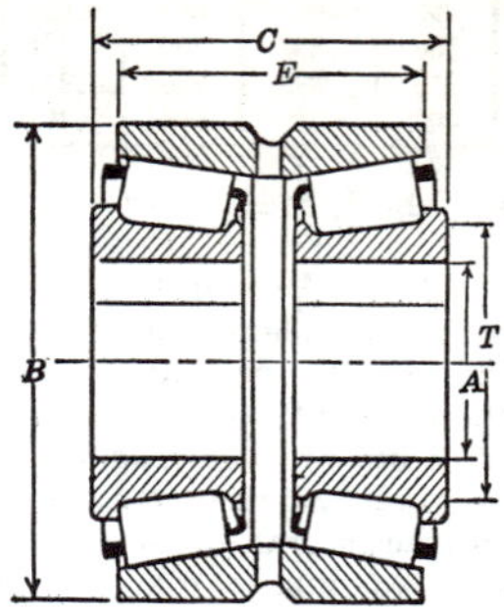

Fig. 17. Two-row double-cup single-cone tapered roller bearing.

at right angles to the bore. This face provides a backing for the flange of the cup. This bearing also is used where there is insufficient space for conventional bearing-mounting methods, and where the housing is of manganese steel or other material difficult to machine. Table 31 gives dimensions and capacities.

MULTIPLE-ROW BEARINGS provide maximum lo -carrying capacity in small space, resist thrust loads from both directions, and maintain a precise and rigid mounting or a floating assembly as desired. They are made in several types as follows:

Two-row Double-cup Single-cone Bearings (Fig. 17). Three assemblies are possible. (1) A spacer is introduced between the cones to establish the set-up or running clearances at the time the bearing is assembled. (2) The cones are butted together, thus establishing the proper running clearance. (3) No cone spacer is used, and the proper clearance is obtained by some form of adjusting device. Types 1 and 2 usually are preferred if quick removal and replacement of bearings are necessary, or if operating conditions do not necessitate take-up after a long period of service. With this type of bearing, shaft expansion is provided for by locking the cone in position and permitting the cup to float in the housing. The opposite end of the shaft usually is fixed in position, and movement is confined to one end. Both conventional and steep-angle assemblies are possible with this bearing. (See Table 32.)

Two-row double-cone single-cup bearings (Fig. 18) are used when the capacity of a two-row bearing is required, with a simpler mounting than two standard single-row bear-

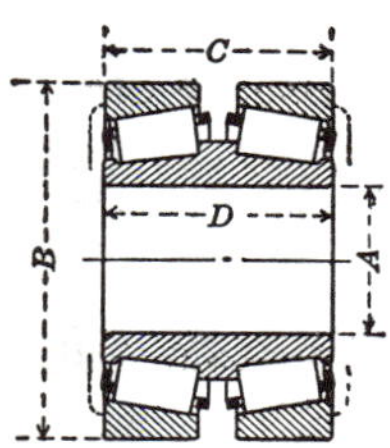

Fig. 18. Two-row double-cone single-cup tapered roller bearings.

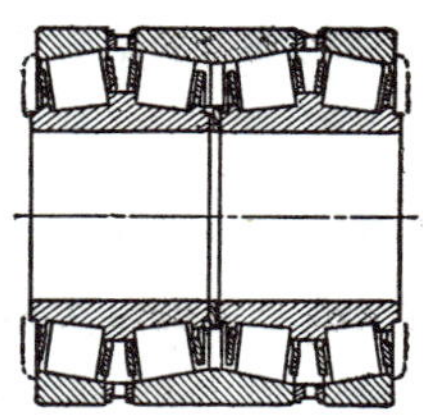

Fig. 19. Four-row tapered roller bearing.

ings. They also are used on the fixed end of a shaft when allowance must be made for shaft expansion. The bearing used on the opposite end then usually is a floating double-cup unit. (See Table 33.)

Four-row bearings (Fig. 19) are used where maximum capacity is required in a limited space, as in rolling mills, cement plants, rubber mills, and other units which develop extreme pressures. It comprises four sets of rollers. The arrangement is similar to two double-cone bearings set side by side with a double cup carrying the load from the two inner sets of rollers and a single cup carrying the load imposed on the two outer sets of rollers. Spacer rings are recommended in all applications. Manufacturers should be consulted regarding application of this type of bearing.

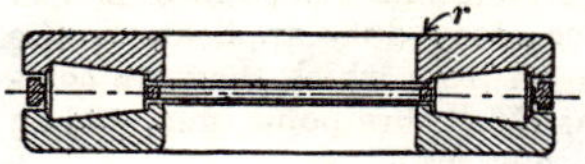

FIG. 20. Tapered roller thrust bearing.

Thrust bearings to withstand thrust loads only without radial loads are shown in Fig. 20.

METHODS OF MOUNTING TAPERED ROLLER BEARINGS. Figure 21 shows various methods of mounting tapered roller bearings. Design 1 is for those cases where the maximum possible shaft extension diameter is desired. In design 2 it is possible to use a smaller size nut than in design 1, and yet have the necessary cone backing. Design 3 uses a single lock nut and tongued washer to clamp another member through the cones. A small spacer is necessary before the front face of the cone to prevent interference by the washer with the bearing cage. The diameter of the spacer must be such as to provide cage clearance. In design 4 the nut should provide sufficient backing to clamp the assembly solidly. Design 5 is a typical automotive application. A hardened tongued washer is used between the inner nut and cone face. The length of thread and keyway must be

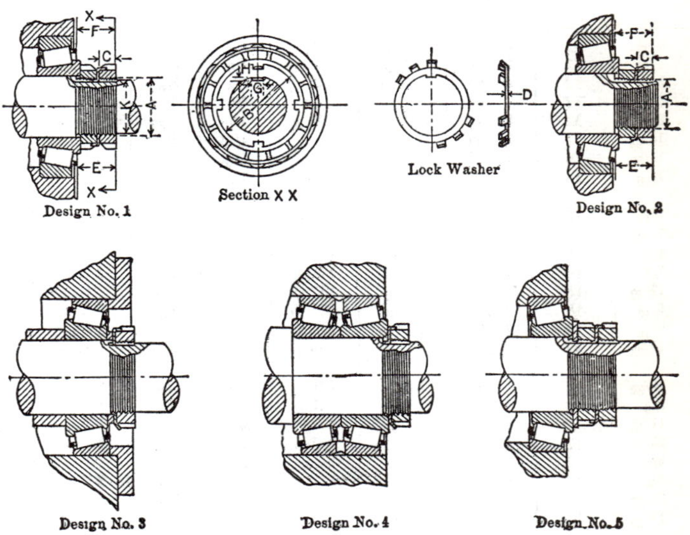

FIG. 21. Timken lock nuts.

increased by the thickness of the hardened washer. Table 36 gives dimensions of lock nuts and tongued washers. For data concerning the theory of tapered roller bearings and their application to various services consult the Timken *Engineering Journal* published by the Timken Roller Bearing Company, Canton, Ohio.

LUBRICATION. Table 37 gives the specifications of lubricants recommended for tapered roller bearings by the Timken Roller Bearing Company. In general, bearings less than 6 in. outside diameter, operating at speeds of less than 1000 rpm, without unusual conditions may be lubricated with a lime or soda base grease of medium consistency. With moisture present, only a lime base should be used. For speeds above 1000 rpm, oil is more satisfactory, but grease may be used on the smaller bearings. Grease is recommended for bearings over 6 in. outside diameter only up to 500 rpm. High-speed bearings are best lubricated with light oil and a drop feed or wick system with provision to drain excess oil to avoid heat due to churning. The drain should be located so that the lowest point of the bearing is not immersed more than $^1/_8$ in. A circulating oil system is recommended for conditions of extreme heat. For unusual conditions, advice as to lubrication should be obtained from the makers of the bearings.

For general machinery applications, grease should be compounded from a high-grade soap and a refined mineral oil, free from acid, alkali, and fillers that would tend to score the bearing. A bright copper or steel plate, submerged in the grease for 24 hr, should show no discoloration. Ash and moisture should not exceed 2%. Oil from which the grease is compounded should have a minimum flash point of 340 F, a minimum fire point of 380 F, and a Saybolt universal viscosity of 200 sec at 100 F. For heavy-duty applications the greases should follow the same general specifications. The oil from which they are compounded should conform to the following: flash point, min, 375 F; fire point, min, 425 F; min Saybolt universal viscosity, 75 sec at 210 F; pour test, max 40 F.

Table 36. Shaft Lock Nuts and Washers for Tapered Roller Bearings

See Fig. 21. Thread tolerances and fits in accordance with ASA coarse-thread series, medium fit, Class 3

(Timken Roller Bearing Co., Canton, Ohio)

Lock Nut Dimensions, in.			Shaft Dimensions, in.						
Outside Diam., *B*	Width, *C*	Threads per Inch	Outside Diam. of Threads, *A*	Thread Length, *E*	Length of Full Depth of Keyway, *F*	Width of Key-way, *G*	Depth of Key-way, *H*	Max. Diam. of Shaft Extension, *I*	Lock Washer Thickness, in.
3/4	7/32	32	0.391	19/32	19/32	1/8	3/32	5/16	.042
7/8	5/16	32	0.469	25/32	25/32	1/8	3/32	13/32	.042
1	5/16	32	0.586	13/16	13/16	1/8	3/32	1/2	.058
1 1/8	11/32	32	0.664	7/8	7/8	1/8	3/32	9/16	.058
1 3/8	3/8	32	0.781	29/32	29/32	3/16	3/32	23/32	.042
1 9/16	13/32	32	0.969	1	1	3/16	1/8	7/8	.062
1 3/4	13/32	18	1.173	1	1	3/16	1/8	1 1/16	.062
2 1/16	7/16	18	1.376	1 1/16	1 1/16	3/16	1/8	1 1/4	.062
2 1/4	7/16	18	1.563	1 1/16	1 1/16	5/16	1/8	1 15/32	.072
2 17/32	7/16	18	1.767	1 1/16	1 1/16	5/16	1/8	1 11/16	.072
2 11/16	1/2	18	1.968	1 3/16	1 3/16	5/16	1/8	1 7/8	.072
2 21/32	1/2	18	2.157	1 3/16	1 3/16	5/16	1/8	2 1/16	.072
3 5/32	17/32	18	2.360	1 9/32	1 9/32	5/16	5/32	2 1/4	.082
3 3/8	9/16	18	2.548	1 11/32	1 11/32	5/16	5/32	2 7/16	.082
3 5/8	9/16	18	2.751	1 11/32	1 11/32	5/16	5/32	2 5/8	.082
3 7/8	19/32	12	2.933	1 13/32	1 13/32	5/16	3/16	2 25/32	.095
4 5/32	19/32	12	3.137	1 13/32	1 13/32	3/8	3/16	3	.095
4 13/32	5/8	12	3.340	1 15/32	1 15/32	3/8	3/16	3 3/16	.095
4 21/32	11/16	12	3.527	1 5/8	1 5/8	3/8	7/32	3 3/8	.125
4 15/16	23/32	12	3.730	1 11/16	1 11/16	3/8	7/32	3 9/16	.125
5 3/16	3/4	12	3.918	1 3/4	1 3/4	3/8	7/32	3 25/32	.125
5 7/16	3/4	12	4.122	1 3/4	1 3/4	3/8	7/32	3 15/16	.125
5 23/32	25/32	12	4.325	1 13/16	1 13/16	3/8	7/32	4 3/16	.140
6 1/8	13/16	12	4.716	1 29/32	1 29/32	3/8	1/4	4 9/16	.165
6 3/4	7/8	12	5.106	2 1/32	2 1/32	1/2	1/4	4 15/16	.165
7 3/32	1 3/16	12	5.497	2 21/32	2 21/32	5/8	1/4	5 5/16	.165
7 11/16	1 1/4	12	5.888	2 13/16	2 13/16	5/8	9/32	5 23/32	.203
8 1/16	1 9/32	8	6.284	2 7/8	2 7/8	5/8	5/16	6 1/16	.203
8 21/32	1 11/32	8	6.659	3	3	3/4	5/16	6 7/16	.203
9 1/16	1 13/32	8	7.066	3 1/8	3 1/8	3/4	5/16	6 7/8	.203
9 15/32	1 13/32	8	7.472	3 1/8	3 1/8	3/4	5/16	7 1/4	.203
9 27/32	1 1/2	8	7.847	3 5/16	3 5/16	7/8	5/16	7 5/8	.203

Table 37. Lubricants for Tapered Roller Bearings

Oil			Grease			
Specification No.	Viscosity, Saybolt Universal		Specification No.	Base	Consistency, Karns-Maag Method	
	sec	°F			mm	°F
1	75–500	100	10	Lime	20–40	75
2	100–700	100	11	Lime	15–30	75
3	60–100	210	12	Lime or soda	20–50	75
4	100 & Up	210	13	Lime or soda	20–40	75
5	75–300	100	14	Soda	15–30	75
6 *	75–100	210	15	Soda; short fiber	30–50	75
7 *	100–175	210	16	Soda	20–40	75
8 *	100–250	210	17 *	Water resistant	20–30	75
9 *	175–250	210	18 *	Lime or soda	20–30	75
....	....	....	19 *	Water resistant	15–25	75
....	....	....	20 *	Lime or soda	15–25	75
....	....	....	21 *	Lime or soda	30–45	75

Oil Lubrication—Recommended Specification

Speed, rpm	Bearings up to 6 in. Outside Diam.				Bearings over 6 in. Outside Diam.			
	Maximum Temperature, °F				Maximum Temperature, °F			
	32	125	180	Over 180	32	125	180	Over 180
	Recommended Specification No.				Recommended Specification No.			
0–500 †	....	....	....	....	6 ‡	7	8	9
0–500	1	2	3	4	1 ‡	2	3	4
500–1000	1	2	3	4	1	2	3	4
Over 1000	5	2	3	4	5	2	3	4

Grease Lubrication—Recommended Specification

Speed, rpm	Moisture Present								No Moisture Present							
	Bearings up to 6 in. Outside Diam.				Bearings over 6 in. Outside Diam.				Bearings up to 6 in. Outside Diam.				Bearings over 6 in. Outside Diam.			
	Maximum Temperature, °F								Maximum Temperature, °F							
	32	125	180	Over 180	32	125	180	Over 180	32	125	180	Over 180	32	125	180	Over 180
	Recommended Specification No.								Recommended Specification No.							
0–500 †	....	....	....	....	....	17	19	§	....	....	....	....	21	18	20	§
0–500	....	10	11	§	....	10	11	§	12	13	14	§	12	13	14	§
500–1000	....	10	11	§	§	§	§	§	12	13	14	§	§	§	§	§
Over 1000	....	10	11	§	§	§	§	§	15	16	14	§	§	§	§	§

* Extreme pressure lubricant. † Heavy duty applications. ‡ No moisture present.
§ Not recommended.

SECTION 13

MACHINE HYDRAULICS

By

L. S. LINDEROTH, JR., *Professor of Mechanical Engineering, Iowa State College of Agriculture and Mechanic Arts*

MACHINE HYDRAULICS

It is the purpose of this chapter to describe the various hydraulic principles and elements used in the hydraulic actuation of machinery. The detailed design of the various elements does not come within the scope of this discussion, but references are given to books and articles that will serve to amplify this text.

The basic principles on which our modern hydraulic machines are founded were discovered by early physicists, Archimedes, Galileo, Torricelli, Pascal, and others. From early practical use of hydraulics the art developed with the invention of better pumps, accumulators, control valves, and suitable hydraulic fluids. Until the end of the nineteenth and the early years of the twentieth century hydraulics was used chiefly for pressing, hoisting, and elevating. The use of hydraulics in machine tools for feeds, clamps, drives, and controls is a relatively recent development.

1. FLUID FLOW AS APPLIED TO MACHINE HYDRAULICS

The flow of a hydraulic fluid in a pipe line or through a valve is a function of pressure differential, temperature, viscosity, surface condition, length of the flow path, and Reynolds' number. Viscosity is a function of temperature and inherent characteristics of the fluid. It will be discussed separately.

REYNOLDS' NUMBER is the nondimensional product of pipe or orifice diameter, fluid velocity, and the reciprocal of kinematic viscosity:

$$N_R = DV\frac{1}{\nu} \tag{1}$$

where N_R is Reynolds' number; D is diameter of pipe, feet or inches; V is average fluid velocity, feet or inches per second; and ν is kinematic viscosity, feet^2 per second or inches^2 per second. In viscosity tables ν is usually given in stokes or centistokes in the cgs system. The conversion is

$$\nu \text{ in ft}^2/\text{sec} = \nu \text{ in stokes} \times 0.001077 \tag{2}$$

$$1 \text{ stoke} = 100 \text{ centistokes}$$

$$= 1 \text{ cm}^2/\text{sec} = 0.001077 \text{ ft}^2/\text{sec}$$

When the Reynolds' number of a given flow condition is less than 2000 the flow is viscous or streamline and the "friction factor" in a smooth pipe is

$$f = \frac{64}{N_R} \tag{3}$$

This is known as the Hagen-Poiseuille law. Above 4000 the friction factor follows no simple equation. Values of f for three types of pipe are given in the Stanton diagram, Fig. 1. Those values are for flow in a substantially straight pipe of uniform cross section and surface condition.

In the range between 2000 and 4000 the type of flow is indeterminate. If the flow starts at a Reynolds' number of less than 2000 it will remain viscous until it reaches 4000. Likewise if the flow starts at a Reynolds' number in excess of 4000 the flow will remain turbulent as the Reynolds' number decreases until approximately 2000 has been reached. The point of change from turbulent to viscous or vice versa is not exact and may not be the same for two successive runs in the same systems.

PRESSURE LOSS. In calculating the pressure loss in a pipe line the friction factor f from Fig. 1 is used in Fanning's equation:

$$\frac{P_1 - P_2}{L} = \frac{f\rho V^2}{2D} \tag{4}$$

$$= \frac{f\omega V^2}{2gD} \tag{5}$$

where L = length of pipe, feet; ρ = absolute density, slugs per foot^3; V = average velocity, feet per second; D = internal diameter of pipe, feet; P_1 and P_2 = pressure or

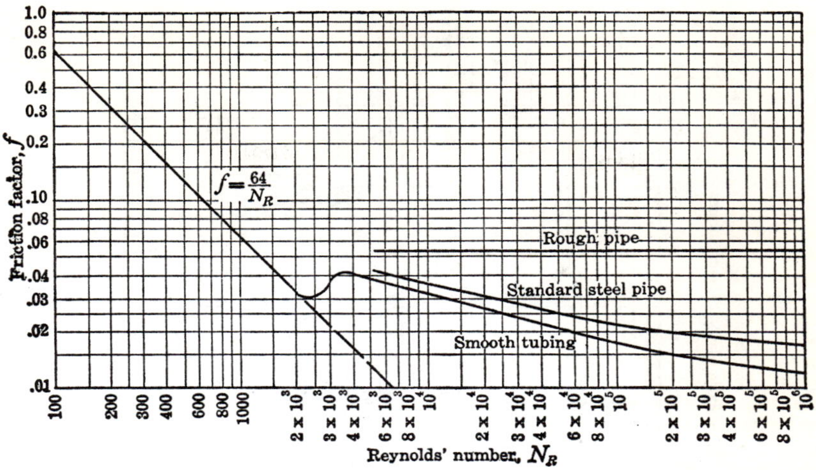

FIG. 1. Friction factor as a function of Reynolds' number for straight pipe of uniform cross section and surface condition. (Reproduced by permission from *Heat Exchanger Tube Manual*, Scovill Manufacturing Co.)

head, pounds per square foot; ω = weight of fluid, pounds per foot^3; g = gravity constant, feet per second^2; and f = friction factor.

Equation 4 can be simplified for practical use by substituting for f its equivalent in eq. 3. This gives for viscous flow:

$$P_1 - P_2 = L \frac{64}{N_R} \frac{\rho V^2}{2D} \tag{6}$$

which, when DV/ν is substituted for N_R, gives

$$P_1 - P_2 = \frac{L32\rho V\nu}{D^2} \tag{7}$$

$$= \frac{L32\omega V\nu}{gD^2} \tag{8}$$

For turbulent flow the Blasius approximation for smooth pipes can be used, where

$$f = \frac{0.316}{\sqrt[4]{N_R}} \tag{9}$$

Giving:

$$P_1 - P_2 = \frac{L0.316\rho V^2}{2D\sqrt[4]{N_R}} \tag{10}$$

$$= \frac{0.158L\rho V^{7/4}\nu^{1/4}}{D^{5/4}} \tag{11}$$

VISCOSITY is the measure of internal friction in the fluid. A high viscosity number indicates a fluid that flows with more difficulty than one with a low viscosity number. *Absolute viscosity* is measured in poises in the cgs system of units. A poise has the dimension of M/LT or grams mass per centimeter × seconds. In the engineering system of units this would be slugs per foot-second.

$$1 \text{ poise} = 0.00209 \text{ slug/ft-sec} \tag{12}$$

A more common measure of viscosity is *kinematic viscosity*. This is an arbitrary unit which is the quotient of absolute viscosity divided by absolute density:

$$\frac{\mu \text{ in slugs/ft-sec}}{\rho \text{ in slugs/ft}^3} = \nu \text{ in ft}^2\text{/sec} \tag{13}$$

Its dimension is therefore L^2/T.

Since in the cgs system absolute density and specific gravity are numerically equal it is convenient to change absolute viscosity in poises to kinematic viscosity in stokes before

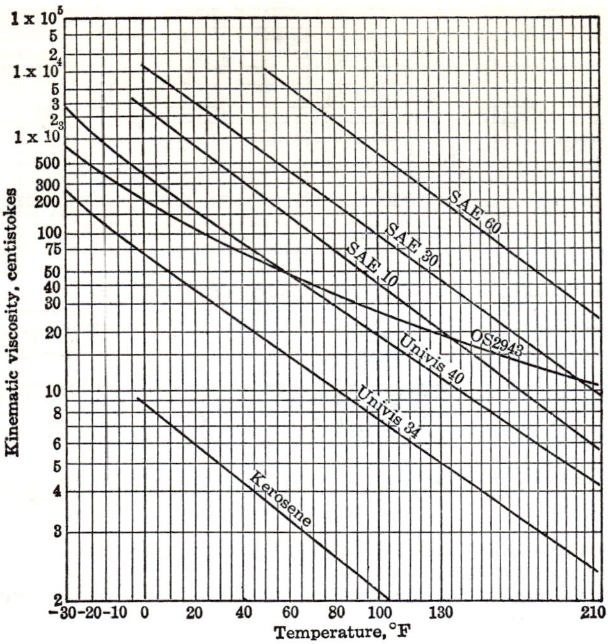

FIG. 2. Kinematic viscosity of typical hydraulic fluids.

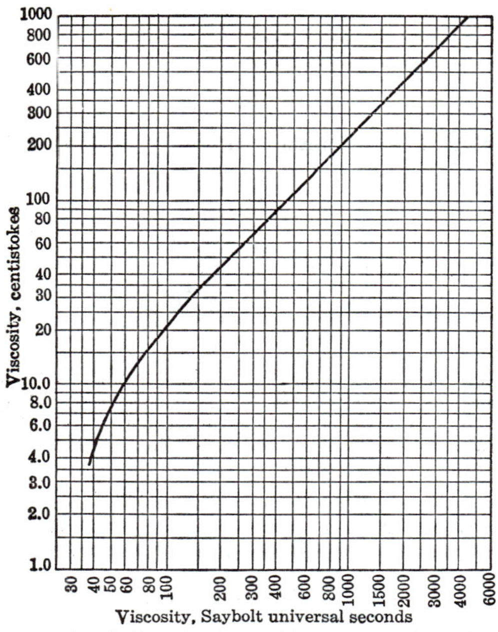

FIG. 3. Chart for converting Saybolt viscosity to kinematic viscosity. (Courtesy of Fischer and Porter Co.)

conversion to the engineering units of slugs, feet, and seconds. Note the conversion factor in eq. 2. Graphs of viscosity versus temperature for various hydraulic oils are shown in Fig. 2. The viscosity of hydraulic fluid is frequently given in units of Saybolt universal seconds at a specific temperature. Viscosity in these units can be converted to kinematic viscosity in centistokes by using the conversion chart in Fig. 3.

ORIFICE CALCULATIONS. The flow of fluid through an orifice is expressed by the relation

$$Q = kA\sqrt{\Delta p} \tag{14}$$

where Q = quantity flowing in unit time, foot³ per second; k = orifice constant; A = area of orifice, feet²; and Δp = pressure drop across orifice, pounds per inch².

Equation 14 is based on the basic formula for velocity of a fluid:

$$V = \sqrt{2gh} \tag{15}$$

and since

$$Q = AV_1 \tag{16}$$

where V = actual velocity at given point in the fluid and V_1 = average velocity over orifice. Then

$$Q = k_1AV \tag{17}$$

$$= k_1A\sqrt{2gh} \tag{18}$$

where k_1 is a velocity correction constant, g is the gravity constant, and h is head in feet.

In order to make eq. 18 more convenient to use, it is often written in the form of eq. 14 in which the constant k contains the constant k_1 of eq. 18, the factor $2g$, and a factor converting h to Δp in pounds per square inch. Constant k is not nondimensional and Δp is not in fundamental units, so care must be exercised not to use this equation in developing others unless proper regard is paid to its factors. It is, nevertheless, most useful in orifice flow calculations.

Orifices are used for measuring fluid flow by measuring the value Δp on a manometer or a differential pressure gage. A typical installation is diagrammed in Fig. 4, h in the diagram being the measure of h in eq. 18. It must be converted by multiplying by the ratio of the specific gravity of the manometer fluid to the fluid being measured.

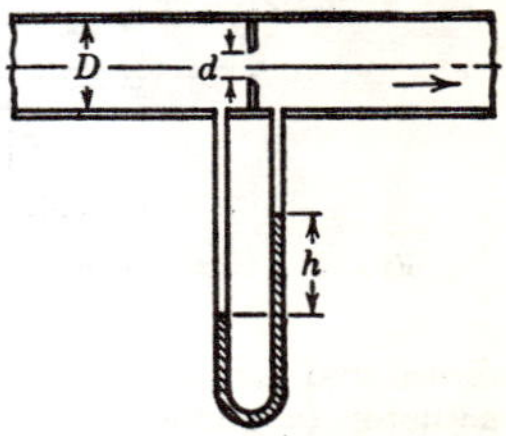

FIG. 4. Flowmeter employing orifice and differential pressure measurement.

Figure 5 illustrates the effect of viscosity on the orifice constant by showing its variation with Reynolds' number. Viscosity influences the orifice constant because of the proximity of the tube or pipe wall.

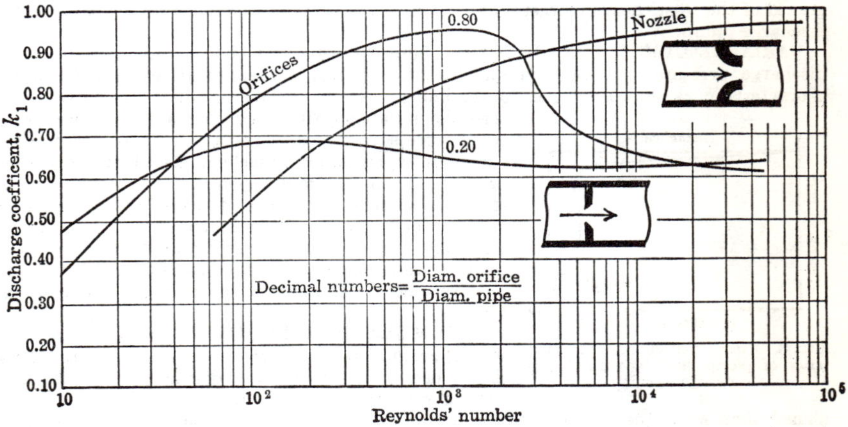

FIG. 5. Discharge coefficients of nozzles and orifices as a function of Reynolds' number. (Courtesy of Fischer and Porter Co.)

COMPRESSIBILITY OF HYDRAULIC FLUIDS. Hydraulic fluids are commonly thought of as incompressible but they actually have relatively low moduli of elasticity.

Mineral oil commonly used in hydraulic circuits has a modulus of elasticity on the order of 200,000 psi; water has a modulus of 300,000 psi. These are bulk moduli.

Because of this compressibility of hydraulic fluids they cannot be depended upon to locate accurately a piston under varying loads or give uniform velocity to a motor under varying torque. Pressure surges and vibration in the fluid are possible because of this compressibility.

2. HYDRAULIC CIRCUITS

CONSTANT-PRESSURE CIRCUIT. The simplest type of hydraulic circuit is the constant-pressure circuit. This is illustrated in Fig. 6. A constant-discharge hydraulic pump delivers oil to a main line to the directional control valve. In this main line is a pressure-regulator valve and a pressure-indicating gage. In the circuit illustrated, the

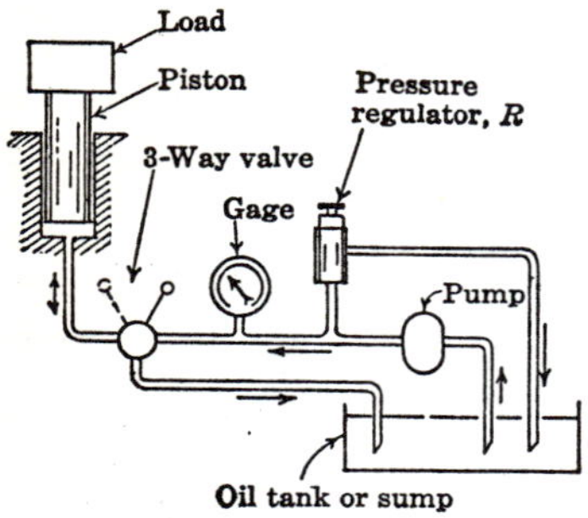

FIG. 6. Constant-pressure hydraulic circuit.

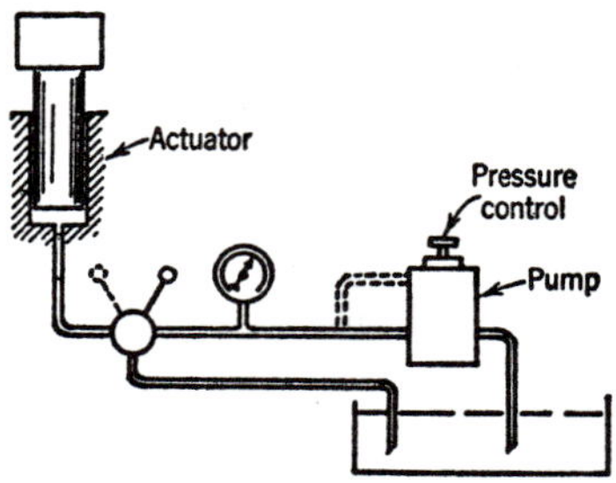

FIG. 7. Variable-discharge hydraulic circuit.

directional control valve is a three-way type and directs the flow of oil to and from the actuator (a piston and cylinder or a hydraulic motor). This simple type of system operating at constant pressure is inefficient. The pump continues to operate at full flow, pressure, and power even though the actuator has reached the end of its stroke or has stalled. The power consumed by the pump is dissipated in the form of heat in the oil; this heat must be taken out of the oil by radiation from the various components in the circuit.

This general type of simple hydraulic system is used in small installations where the power requirements do not exceed a few horsepower. It becomes exceedingly difficult to get rid of the waste heat in larger units and it becomes more economical to use a variable-discharge pump.

VARIABLE-DISCHARGE CIRCUIT. Figure 7 illustrates a circuit utilizing a variable-discharge pressure-controlled pump. This pump has a theoretical pressure-discharge characteristic as shown by the dotted line in Fig. 8. When the discharge pressure is zero the pump adjusts itself to maximum discharge. As the pressure increases there is a small loss in discharge until pressure P is reached. This is the operating pressure for the hydraulic system. As the pressure is increased slightly to P_2, when the load on the actuator increases or stalls, the pump discharge is adjusted automatically to zero. The power required to operate the pump at zero discharge is very low and the oil remains relatively cool. The rest of the circuit is the same as Fig. 6.

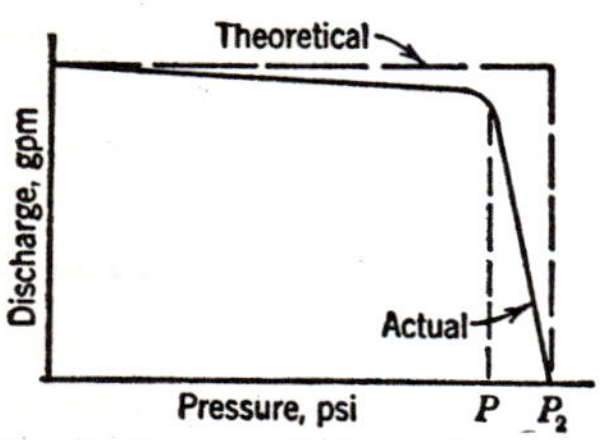

FIG. 8. Pressure-discharge characteristics of pump used in variable-discharge circuit.

ACCUMULATOR CIRCUIT. Figure 9 illustrates a simple accumulator-operated hydraulic circuit. This circuit uses a simple constant discharge pump in connection with an unloading valve. This valve is so designed that when the pressure in the accumulator reaches the desired value the valve shifts and allows the oil to by-pass back to the tank at substantially zero pressure. When the directional control valve is operated to direct oil into the actuator, the oil in the accumulator is consumed; and when the accumulator reaches the end of its stroke, the resultant pressure drop causes the unloading valve to shift, closing its by-pass port and again directing oil from the pump to the accumulator and the main circuit. Since the pump is operating at substantially zero pressure for a large part of the time very little power is consumed and consequently little heat must be dissipated.

This is a more expensive circuit to use since the accumulator and the unloading valve are relatively costly parts. It is, however, a good circuit to use where the hydraulic power is applied over a short period of the entire cycle.

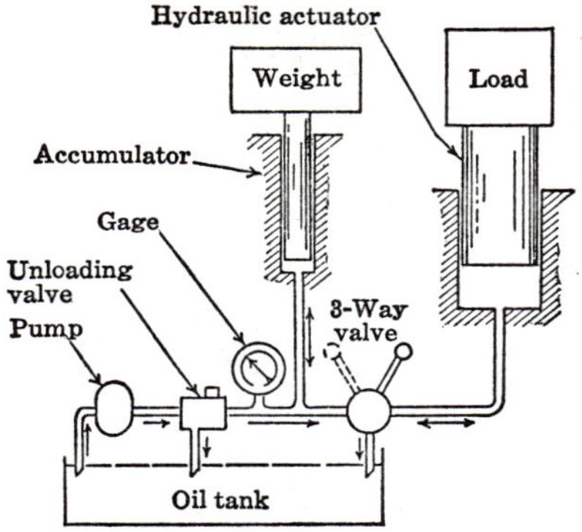

FIG. 9. Accumulator circuit, employing constant discharge pump. (Reproduced by permission from *Machine Design*)

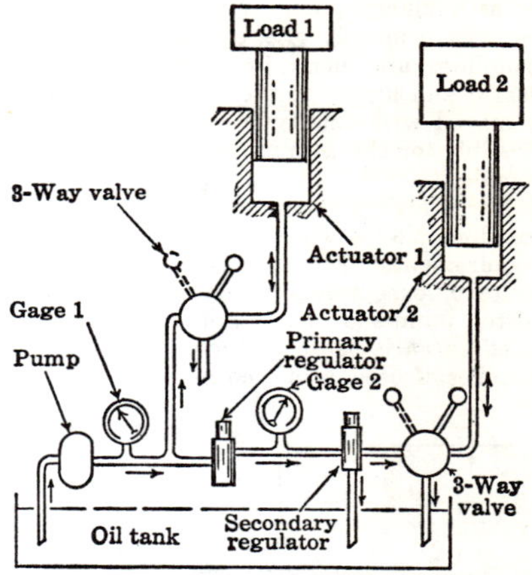

FIG. 10. Multiple-pressure hydraulic circuit employing pressure-regulating valves. (Reproduced by permission from *Machine Design*)

MULTIPLE-PRESSURE CIRCUIT. Figure 10 illustrates a multiple-pressure system in which pressure-regulating valves are placed in series, the discharge from one feeding the next pressure circuit. In this type of circuit each succeeding pressure cannot be made higher than the preceding one. It is obvious that as the last valve pressure is increased it will increase the pressure by a like amount in the preceding circuit unless specially designed valves are used. If the primary regulator is properly designed, the variations in the downstream pressure regulated by the secondary regulator will not affect the upstream pressure until the two are equal. Any increase in the downstream pressure above this point regulates the pressure in the entire circuit.

One or more actuators can be operated from each pressure circuit. The number of circuits that can be operated in this cascade fashion is unlimited.

MANIFOLD CIRCUIT. Figure 11 illustrates a multiple-pressure system in which a regulating valve R_s regulates a maximum manifold pressure and pressure-reducing valves R_1, R_2, etc., determine the pressure in branch lines from the manifold. This type of circuit must be used when the various pressures must be entirely independent of each other.

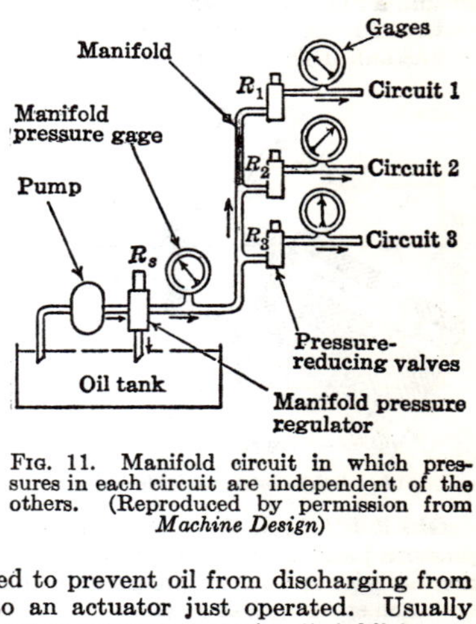

FIG. 11. Manifold circuit in which pressures in each circuit are independent of the others. (Reproduced by permission from *Machine Design*)

Since the fundamental circuit involving the pump, manifold, and regulating valve is identical to that in Fig. 6, the same faults apply to this one. It is inefficient and tends to cause excessive heating of the oil. It is therefore limited to relatively low powers. Here again, a variable-discharge pump can be used to advantage.

The branch lines may consist of one or a number of actuators. It should be noted that where two or more actuators receive their energy from a single source, means must be provided to prevent oil from discharging from one or a number of actuators under pressure to an actuator just operated. Usually a check valve in the branch supply line is sufficient to prevent excessive "wink" in one actuator when another is connected on the line.

3. HYDRAULIC FLUID PUMPS

PISTON PUMPS. The simplest and earliest type of fluid-pressure pump is the single-cylinder, single-acting, hand-operated pump, such as is used on common automotive jacks or as auxiliary pumping equipment on aircraft. This unit is illustrated in Fig. 12. It consists primarily of a cylinder body, containing the inlet and discharge valves, oil-line couplings, and means for mounting the actuating handle. The piston is usually internally sealed and linked to the actuating handle. Pressures up to several thousand psi can be achieved with this type of pump but the volumetric displacement is low since it is not possible for the average man to exert a continuous force of over 50 lb on the actuating handle.

Early high-pressure pumps were limited to bulky "triplex" units used extensively in molding plants where an accumulator was used to stabilize the momentary high-flow requirements. Present-day piston pumps include small multicylinder radial- or axial-type units of exceptionally high volumetric and mechanical efficiency. Fixed-displacement piston pumps are made but since the piston pump lends itself to variable-displacement design, by far the largest number of designs offered for sale are of this type.

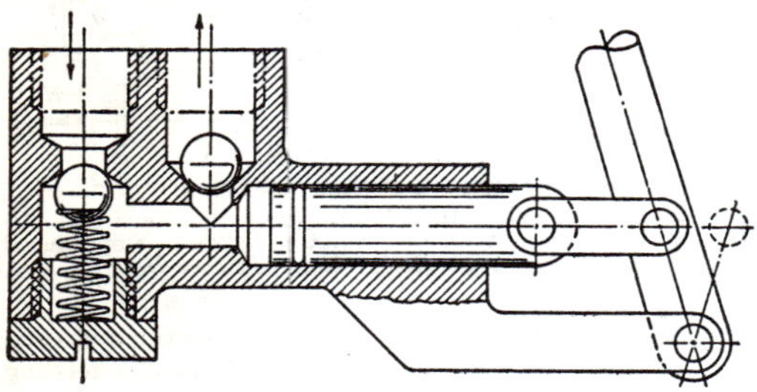

Fig. 12. Cross section through single-cylinder, single-acting, hand-operated pump.

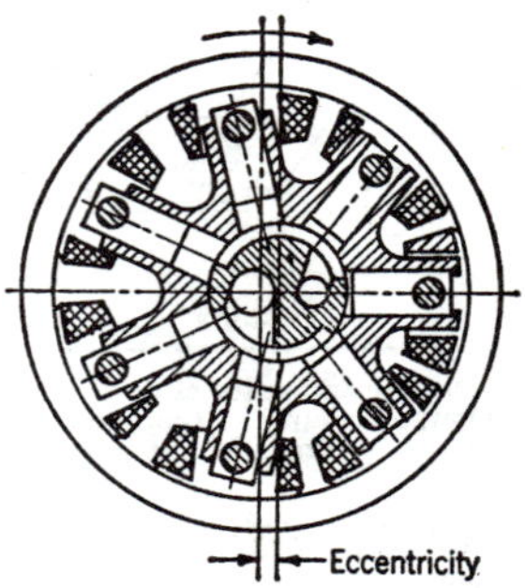

Fig. 13. Radial-piston seven-cylinder variable-displacement pump.

Radial Piston Pumps. Figure 13 illustrates a typical radial-piston seven-cylinder variable-displacement unit. The pistons are moved toward the center of rotation by cam action and returned outwardly by supercharging pressure provided by a gear pump. Centrifugal force and cams have been used for this outward motion but the supercharging design gives the highest overall efficiency. The pistons and cylinders rotate as a unit around a fixed pintle which acts as a valve as well as a bearing for the cylinder block.

Axial Piston Pumps. In axial pumps as illustrated in Fig. 14 the pistons are actuated by a crank disk. The angle between the axis of this disk and the axis of rotation of the cylinders determines the piston displacement, which may be fixed or variable.

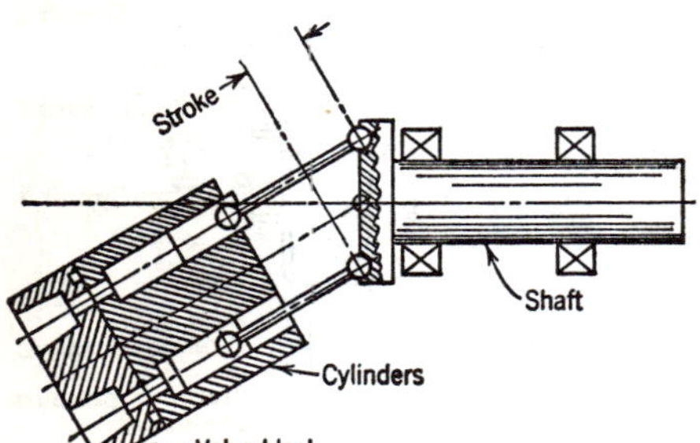

Fig. 14. Axial-piston variable-displacement pump.

The number of pistons in a pump is generally seven or nine or multiples of these numbers. Less than seven cylinders results in unsteady discharge and pressure ripples, and more than nine does not seem practical. An odd number of cylinders makes for smoother valve action, discharge, and pressure characteristics.

Piston pumps can be designed for pressures of 10,000 psi or more and are always found in high-pressure equipment.

GEAR PUMPS. The gear pump in its simplest form is the best known of all fluid-pressure pumps. It consists essentially of two meshing gears in a suitable housing containing inlet and outlet ports and connections. Each tooth in the gear acts as a piston, and each space acts as a cylinder. As the tooth pulls out of its space as a result of rotation the space is filled with fluid which is carried around the periphery of the gear confined by the housing until on the other side a tooth again enters the space and displaces the oil.

This is done in the line of the discharge port, and oil is thereby carried from inlet to discharge.

Efficiency of a gear pump is determined by the accuracy with which the various components fit together. Oil allowed to leak back through the mesh of the two gears or between the gears and the housing causes a loss and a lowering of efficiency. This flow or "slip" is usually expressed as a volume per pounds per square inch. The pump speed has little or no effect on slip, but outlet pressure has a very great effect. If s is the slip factor, the discharge of the pump can be written:

$$Q = (\text{Pump factor} \times \text{rpm}) - (s \times \text{psi}) \tag{19}$$

External Gear Pump. The designs of gear pumps may be divided into three groups: external or internal gear designs, and conjugate curve designs. Figure 15 illustrates a simple external gear pump. The pump can be run in either direction and the ports become inlet or outlet ports depending on the direction of rotation. External gear pumps are frequently made with helical or herringbone teeth in order to prevent trapping of oil in the tooth space and consequent pounding of the gears. By proper selection of materials and design proportions, gear pumps can be operated at pressures up to 2000 psi and possibly higher. The difficulty in maintaining a high volumetric efficiency precludes their use at higher pressures.

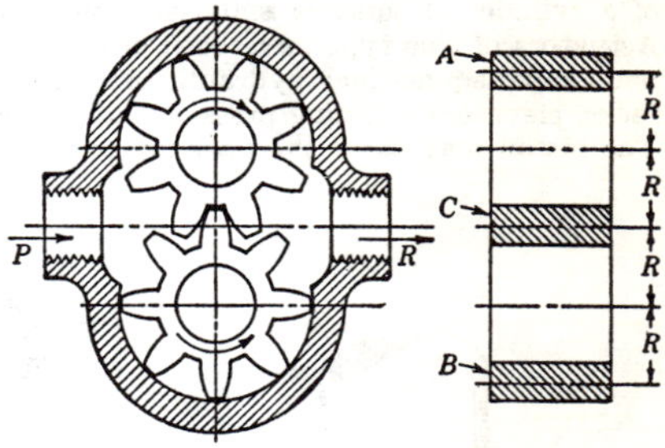

FIG. 15. Simple external gear pump.

Internal Gear Pump. Figure 16 illustrates a typical internal gear pump A pinion acts as a driver of an internal ring gear rotating in a housing. The ring gear is usually made only slightly larger in pitch diameter than the pinion gear. A crescent-shaped block fills the space between the two gears and the ports are located at the points of the crescent. These pumps find general use where compactness is required and the maximum required pressure is well under 500 psi. A modification of this design is found in the *conjugate curve pumps* illustrated in Fig. 17. In these pumps the internal member drives the outer member. The spaces between the high points of the inner member as they contact the outer conjugate curve change in volume and thereby give rise to a pumping action. Porting can be axial, as shown, or radial through holes drilled in the outer member. These pumps have been made with great accuracy, high efficiency, and for pressures up to 2000 psi. They are relatively quiet and compact.

In the *screw pump* (not illustrated) two parallel screws of meshing-tooth form are rotated in a housing closely surrounding both screws. Crescents of fluid travel axially

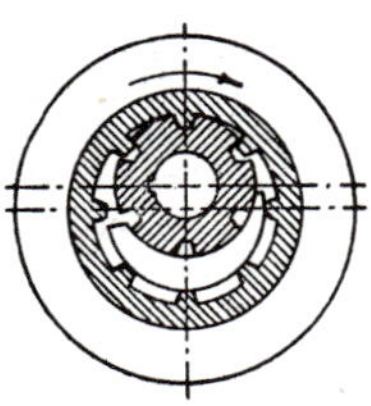
FIG. 16. Internal gear pump. (Courtesy of Sundstrand Machine Tool Co.)

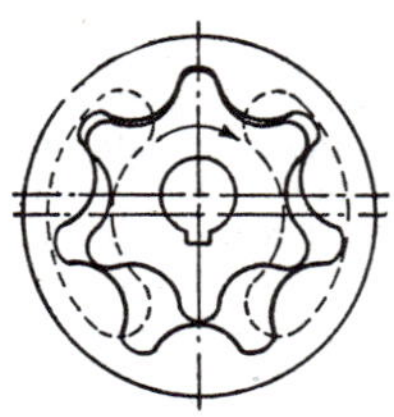
FIG. 17. Conjugate-curve internal gear pump. (Courtesy of Gerotor May Corp.)

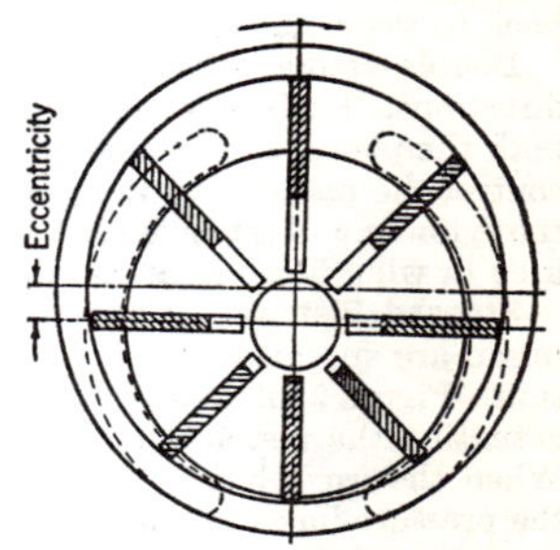

FIG. 18. Vane pump.

along each screw as they are rotated together. These pumps can also be made with three screws and are made with flow from one end to the other, which might be termed single-acting, or with the flow from both ends toward the center. This latter form gives a balanced rotor with a practically zero end thrust on the bearings. Screw pumps have been made for pressures up to 1000 psi.

VANE PUMPS. Figure 18 illustrates a simple vane pump. In this design the pumping action is brought about by the change in volume of the spaces as the blades or vanes move in or out in their radial slots. Usually the fluid pressure is used to hold the vanes against the cam surface which causes their radial motion. For greatest efficiency the vanes have no radial movement while under a differential pressure. The radial motion takes place while the vane is in line with the inlet or discharge port. Vane pumps are frequently built with pressure-balanced rotors. They have two inlet and two outlet ports diametrically opposite each other. Such rotors operate with less friction and simpler bearings. Vane pumps must be carefully made of proper materials to insure minimum leakage across the barrier created by the vane. These pumps are made for pressures up to 2000 psi.

4. HYDRAULIC ACTUATORS

The simplest form of hydraulic actuator is illustrated in Fig. 19. It consists essentially of a cylinder of suitable material with an open bore into which a long piston is fitted. Actuators of this type have been designed for pressures up to several hundred thousand pounds per square inch hydraulic pressure. Actuators of this type are found in hydraulic jacks, elevators, molding presses, and wherever pressure is exerted in one direction only. The piston may carry the packing, or the end of the cylinder bore may be fitted with a gland to hold the packing. It is also practical in low-pressure (under 500 psi) designs to use this type of actuator without a packing, if means are provided to drain oil leakage back to the reservoir.

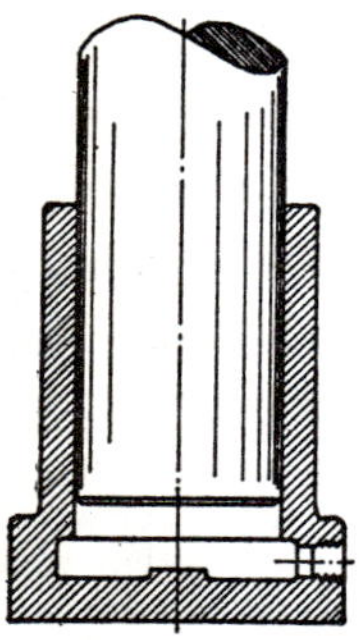

Fig. 19. Simple single-acting hydraulic actuator.

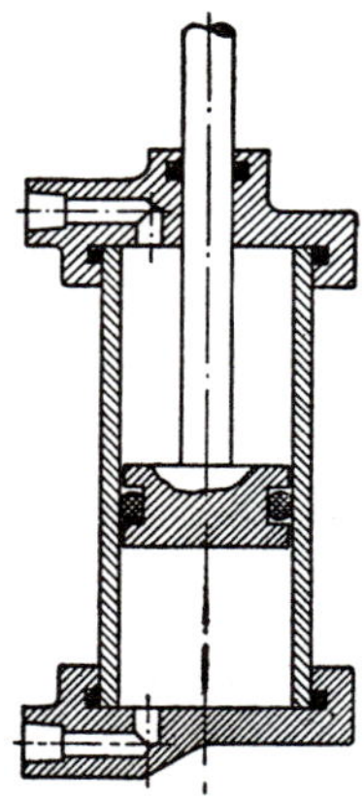

Fig. 20. Double-acting hydraulic actuator.

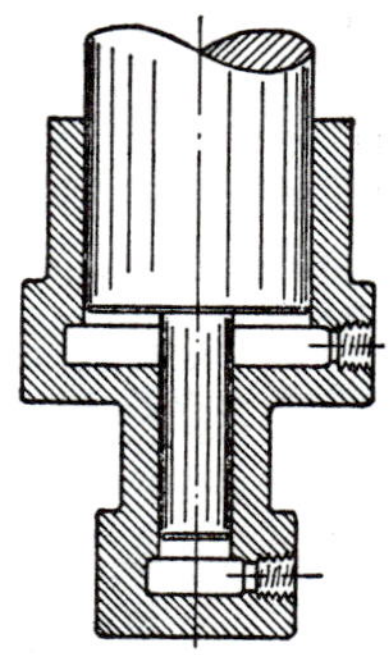

Fig. 21. Stepped-piston hydraulic actuator.

Double-acting actuators are used where force is required in two diametrically opposite directions. Such a design is illustrated in Fig. 20. For medium- and high-pressure work both the piston and the piston rod are packed. When for reasons of hydraulic balance or control the piston area in either direction must be equal, the piston rod extends through the cylinder and is packed in both cylinder caps. Double-acting cylinders have been widely used in wing flap actuators, airplane landing gear, planer and milling table feeds, etc.

Stepped Pistons. Where high actuating speed, large final force, and a relatively small pump are specified, the use of stepped or multiple-diameter pistons yields a simple solution. Figure 21 illustrates a design in which the oil is first introduced to the small cylinder actuating the piston which causes the large cylinder to act as a pump and fill with oil. When the work has been contacted, valves are shifted to connect the large cylinder to the pressure line and cut off the suction line to the large cylinder. Thus the output force is considerably increased.

FLUID MOTORS are used wherever continuous rotary motion is required. These motors may be of the **gear type** as illustrated in Fig. 15. With pressure at P and return at R, the pump will rotate in the direction of the arrows. The theoretical torque of such a motor is

$$P \times (\text{Area } A + \text{Area } B - \text{Area } C) \times R = T \tag{20}$$

and since A, B, and C are equal, this simplifies to

$$PRC = T \tag{21}$$

in which P = pressure, pounds per square inch; R = one-half center distance of gears; C = area of interference of gear OD's (tooth engagement); and T = theoretical torque.

The actual torque is considerably less than this and in well-designed and accurately fitted motors may reach 70% of theoretical. The average well-built gear pump can be used as a motor. Gear motor efficiency can be increased by proper porting and hydraulic balancing of the rotating elements. Gear motors are fixed displacement units.

PISTON MOTORS may be either axial- or radial-type units. In either case the motors may be of fixed or variable displacement. Figure 14 illustrates a design of axial motor. In this design cylinders, pistons, connecting rods, crank disk, and shaft all rotate about their respective center lines at the same speed. As the pistons in the cylinder block move about the block center line, the pistons are caused to move inwardly or outwardly by the angular displacement of the shaft center line, giving a stroke which is proportional to the sine of the angle α (Fig. 22). Angle α can be made fixed or variable as desired. The number of pistons is made either seven or nine. The torque of this type of motor is not uniform, being made up of the sum of the torques from each separate cylinder. Each cylinder through 180 degrees of shaft rotation produces torque proportional to the cosine of the angle of rotation. This can be written:

$$\text{Torque} = PAr \sin \alpha \cos \beta \qquad (22)$$

in which P = pressure of hydraulic fluid; A = area of piston; α = angular displacement of center lines of shaft and cylinder block; β = angular rotation of shaft (Fig. 22); and r = radius of crank.

The remaining 180 degrees of shaft rotation discharges the oil or fluid in the cylinder and results in a small negative torque of the same characteristics as the torque in the first 180 degrees of rotation.

The total torque of the axial piston motor can be determined by calculation of the torque produced by one cylinder at points every 10 or 15 degrees for 360 degrees. Plot these values on graph paper using torque as ordinates and degrees of rotation as abscissas. Then plot a second curve the same as the first but displaced a number of degrees from the first equal to the number of cylinders divided into 360 degrees. Do this for each cylinder and then add the curves graphically. The net result should show a torque for a seven- or nine-cylinder motor with practically no ripple.

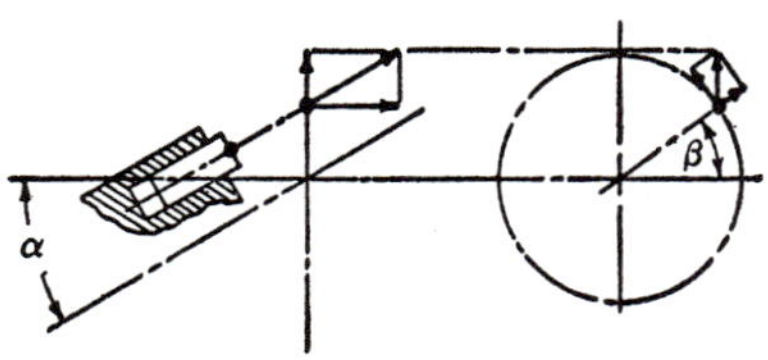

FIG. 22. Forces acting in axial piston hydraulic motor.

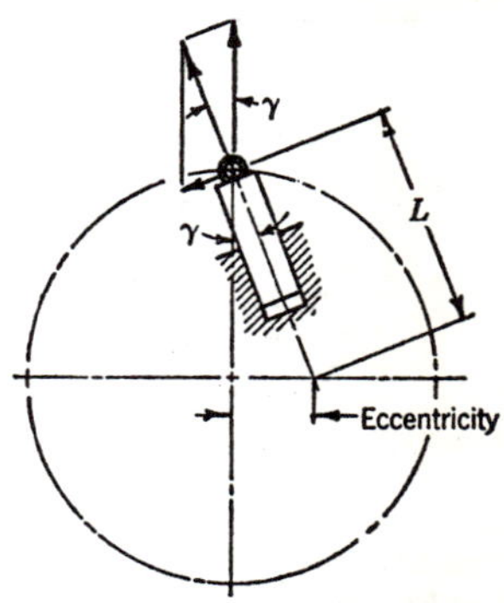

FIG. 23. Forces acting in a radial piston hydraulic motor.

Figure 13 illustrates a radial motor in which the pistons act in cylinders forming a central hub rotating on a fixed pintle. The end of the piston away from the cylinder contacts a circular ring and slides or rolls along its surface. This ring is eccentrically located with respect to the pintle center. This eccentricity may be fixed or variable. The torque of a radial piston motor is, as in the axial piston motor, equal to the sum of the torques of each piston. Figure 23 illustrates the geometry of the torque calculation. It is more complicated to calculate the torques in radial motors since the effective radius, the normal force, and the angle between the piston center line and the eccentric radius are constantly changing as the cylinder rotates around the pintle.

The torque of a radial piston motor may be written as follows for one cylinder rotating about the shaft for 180 degrees ' power stroke:

$$T = APL \tan \gamma \qquad (23)$$

in which A = area of piston; P = pressure; γ = angle between radius of ring at point of piston contact and center line of piston and cylinder; and L = moment arm of piston center line.

VANE MOTORS are similar in construction to vane pumps. The best designs have balanced rotors and double ports. The torque of a vane motor is the area of the vane unbalance times its effective radius. This is most easily visualized by referring to Fig. 24 in which a simple eccentric-type motor is illustrated diagrammatically. Oil enters under pressure at P and causes rotation in the direction of the arrow. The fluid then exhausts through port E. The maximum vane area is shown at A and the minimum at B. Their effective radii to the center line of rotation are R and r respectively. It follows, therefore, that the torque can be determined from the following equation:

$$\text{Torque} = P(AR - Br) \tag{24}$$

In some vane-motor designs, area B is made zero so that eq. 24 then becomes, for single-lift unbalanced rotors,

$$\text{Torque} = PAR \tag{25}$$

When a double-lift balanced rotor is used the torque can be written:

$$T = 2PARE \tag{26}$$

in which T = torque, inch-pounds; P = pressure, pounds per square inch; A = area, square inches; R = radius, inches; and E = mechanical efficiency (usually around 70%).

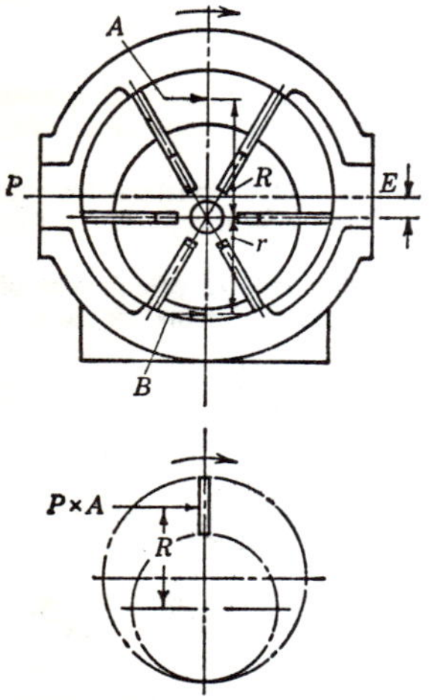

FIG. 24. Torque-producing forces in a vane motor.

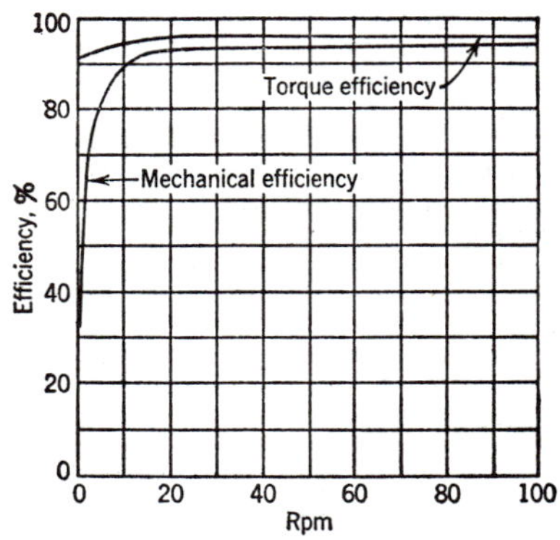

FIG. 25. Characteristics of a typical vane-type motor. (Courtesy of Sundstrand Machine Tool Co.)

The speed of any fluid motor is a function of its displacement and the amount of fluid available to operate it. Commercial motors are usually rated in theoretical displacement per revolution at zero pressure, and a curve of volumetric efficiency plotted against operating pressure is also given to enable calculations of exact requirements in terms of gallons per minute or any other convenient units to be made. Figure 25 illustrates the characteristics of a typical motor.

5. SPEED OF OPERATION

CYLINDER BORE AND STROKE. The work to be done by a hydraulic actuating cylinder may be specified exactly. This in no way limits the size and shape of the cylinder. Usually the stroke is fixed by the specifications and, unless stroke amplification or reduction through levers or gears is called for or is practical, the cylinder will be designed around this stroke. Likewise, the required force to be applied by the cylinder is usually specified. It becomes then a choice relating the force specified to the piston area and the hydraulic pressure, i.e.,

$$F = PA \tag{27}$$

There is an infinite number of combinations of P and A whose product will equal F. In general it is customary to fix an arbitrary value for P and calculate A. If the size of A then gives a reasonable and practical size to the cylinder the choice of P was a valid one. If A is too big or too small, a new value of P must be assumed and a new size A calculated. In practical machine designs pressure P can be several thousand pounds per square inch, but when P is made larger than 1000 psi the cost of the various hydraulic elements begins to rise appreciably. It is best to use as small a value of P as is consistent with the physical dimension limitations of the actuating cylinder. An allowance of 10% is usually added to the force to be exerted by a hydraulic cylinder to allow for friction of piston, rod, and packings. This is an arbitrary amount for pressures up to 1000 psi. For higher pressures, packing friction may absorb more of the available force than 10%.

CALCULATING OIL FLOW DESIRED. The flow of oil Q required to actuate a cylinder in a given time t through a stroke s when the piston area is a can be determined by the following equation:

$$Q = \frac{as}{t} \times 0.26 \tag{28}$$

where Q = gallons per minute; a = piston area, square inches; s = stroke, inches; and t = time, seconds.

ACCELERATING AND DECELERATING FLOW IN PIPES. Accelerated flow occurs when a control valve is opened and a pressure in excess of that required to balance the force of the actuator piston is available. This causes a considerable pressure drop and results in an acceleration of the fluid. Acceleration will become zero (velocity has reached a maximum) when the flow reaches the maximum delivered by the pump.

Deceleration occurs when a piston meets increasing resistance, causing a lowering of the pressure drop. Deceleration will continue until the piston load is equal to the pressure on the piston times its effective area. Another way of describing this is to say that deceleration continues as long as the pressure differential continues to decrease.

6. HYDRAULIC VALVES

DIRECTIONAL CONTROL VALVES. In designing directional control valves it is a good policy to limit the fluid velocity through the valve ports to not more than 10 ft per sec. It would be even better to reduce this to 5 if space considerations permit. Valves have been designed and used successfully with fluid velocities in excess of 10 ft per sec but at a sacrifice in power loss due to high pressure drops through the valve.

Commercial valves are available in the various types listed in the following paragraph. Wherever it seems more practical to do so, valves can be designed as integral parts of other hydraulic elements such as pumps, manifolds, and actuating cylinders.

The basic valve types are piston, plug, poppet, gate, and globe valves. *Piston* valves are used primarily when the actuating means is reciprocating. *Plug* valves are used when the actuating means is rotary, either through part or all of a revolution. *Poppet* valves are used when extreme tightness against high pressures is required. The poppets are usually cam-actuated. *Gate* valves and *globe* valves are usually used as manually operated shut-off valves.

Figure 26 illustrates the difference between two-, three-, and four-way valves. A *two-way* valve is essentially a shut-off valve and is used as such. A *three-way* valve is used

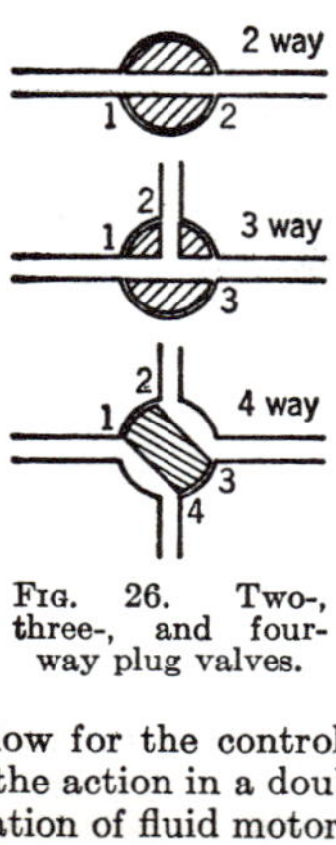

FIG. 26. Two-, three-, and four-way plug valves.

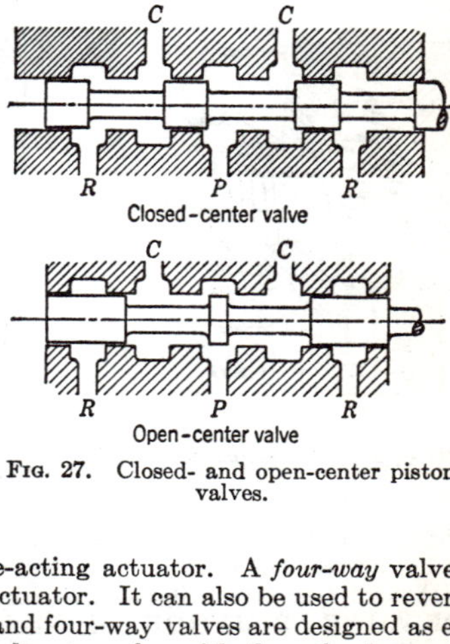

FIG. 27. Closed- and open-center piston valves.

to direct flow for the control of a single-acting actuator. A *four-way* valve is one used to control the action in a double-acting actuator. It can also be used to reverse the direction of rotation of fluid motors. Three- and four-way valves are designed as either closed- or open-center units (Fig. 27). In a *closed-center* valve with the valve in neutral position, the actuator and pressure lines are blocked off; the discharge lines remain open. In an *open-center* valve, the actuator and pressure lines are interconnected; the drain lines still take care of leakage. Another design of open-center valve interconnects all lines so that the pressure is unloaded and the pump runs under substantially zero load.

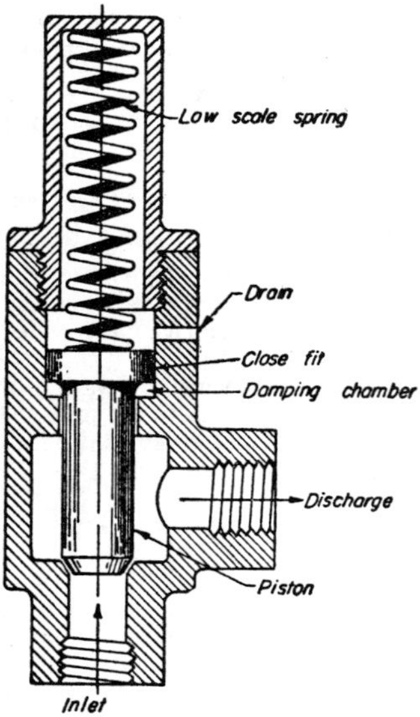

FIG. 28. Hydraulic sequence valve. (Reproduced by permission from *Machine Design*)

MISCELLANEOUS VALVES. Sequence valves are used to guarantee the completion of one actuator cycle before another begins. Such a valve is illustrated in Fig. 28. It depends for its operation on the pressure in the hydraulic fluid. The direction of flow through the valve is shown by arrows. When inlet pressure is great enough to overcome the spring force on the valve piston, the piston will lift and allow fluid to pass through the valve to an actuator or other hydraulic device on the downstream side of the valve. As discharge pressure increases it has no effect on the piston since the piston diameter is the same as the orifice. As inlet pressure approaches discharge there is little, if any, change in the position of the piston. The valves are usually designed with a dashpot to avoid possible unstable operation of the piston. The pressure characteristics of the sequence valve must be compatible with those of the pressure regulator in the system to insure that all the oil delivered by the pump will be available in the secondary circuit. The pressure in the primary circuit must, therefore, be higher than the final secondary pressure by an amount sufficient to lift the piston to full flow position at full pump pressure against the valve spring force.

Check valves are used to insure flow in a given circuit in one direction only. They should be designed to give a minimum of resistance to flow and should be simple to service. They may be of ball, poppet, or swing type. Figure 29 illustrates some common designs.

Timing or flow-control valves are designed to pass a specified amount of fluid in a unit of time. They are essentially refined needle valves, with or without viscosity and pressure compensation. In some designs the needle valve is replaced by a tapered orifice uncovered by a sliding piston. In calculating the basic dimensions of a timing valve the orifice coefficients of a simple orifice may be used as a first approximation. Leeway should be allowed in both directions from the specified time constant in order to insure that the range of flow through the valve will cover the desired range.

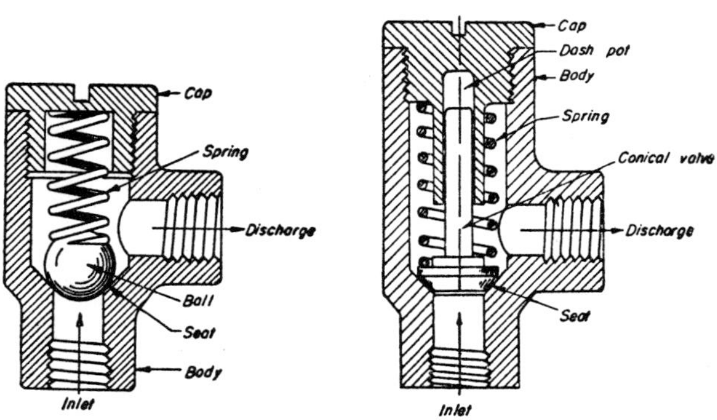

FIG. 29. Ball-type (*left*) and cone-type (*right*) check valves. (Reproduced by permission from *Machine Design*)

MEANS OF VALVE ACTUATION. Hydraulic control valves can be actuated manually, mechanically, electrically, or hydraulically.

Manually operated control valves are generally equipped with levers, push-pull handles, or handwheels to actuate either the rotary or sliding action of the piston or valve stem. The amount of force required to operate a valve depends on the refinements in the hydraulic balance and the minimizing of packing friction. It rarely needs to exceed 10 lb pull or 50 lb-in. of torque. If pressures are so high or valves so large that larger operating forces are required it becomes necessary to use power assist.

Mechanically operated valves are usually cam- or crank-operated from a suitable rotating shaft which is a part of the complete machine or unit. A series of cam-operated valves can be used to control a sequence of operations in a machine. Usually in such a case a complete revolution of the valve cam shaft completes one machine cycle. The cam shaft may be driven from the hydraulic power source by means of a suitable hydraulic motor, or it may be geared to some other rotating part of the machine of which it is a part.

Where only small forces are available to actuate a valve it is possible to use **hydraulically** operated valves in which small hydraulic cylinders actuate the main valve elements, and the flow of the hydraulic fluid to these valve-actuating (pilot) cylinders is controlled by a small direct-actuated four-way valve in which the valve-actuating force is relatively small.

Electrically operated valves are solenoid actuated where high speed and simplicity are the criteria. These valves must be so designed that the solenoid can complete its stroke and close its magnetic circuit. Otherwise, excessive noise in the solenoid and overheating will occur. Solenoid-operated valves are very bulky, particularly in the high-voltage designs, because many electrical codes require boxes to enclose connecting leads and solenoid. Solenoid valves are frequently used with micro switches to form safety interlocks to prevent hydraulic actuation before certain mechanical positioning has taken place. The micro switches are connected in series and all must be closed before the valve will function.

PRESSURE-CONTROL VALVES are designed to regulate the pressure in a hydraulic circuit or a part of such a circuit. They are so made that they prevent the hydraulic fluid from passing through them until a desired pressure has been reached. At this pressure the valve begins to open and continues to do so until the required flow is passed by the valve.

In the simple hydraulic circuit illustrated in Fig. 6 the pressure regulator, R, is designed to pass the entire flow from the fixed displacement pump. Such a valve is illustrated in Fig. 30. As the pressure reaches the required amount the valve piston starts to uncover the ports. This pressure is indicated by P_1 in Fig. 31. When the entire flow, Q, is passing through the valve the piston has moved up to increase the area of the ports. In order to do this the pressure had to increase to value P_2 to compress the spring a sufficient amount. This difference in pressure with flow variations is known as "regulation" or "control band." It is represented in Fig. 31 by the solid line when the valve is opening and the dotted line when closing.

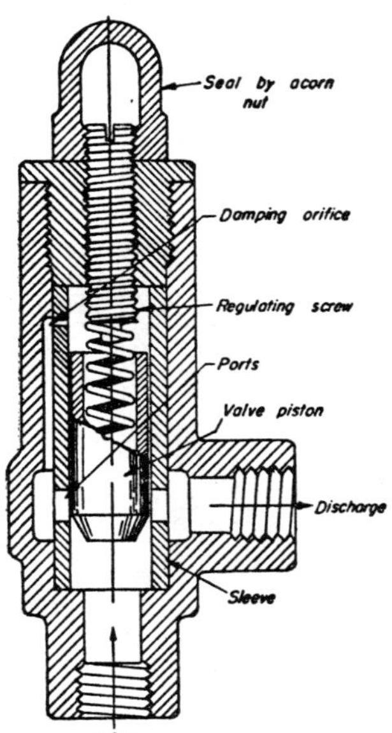

FIG. 30. Pressure-regulating valve for continuous operation. (Reproduced by permission from *Machine Design*)

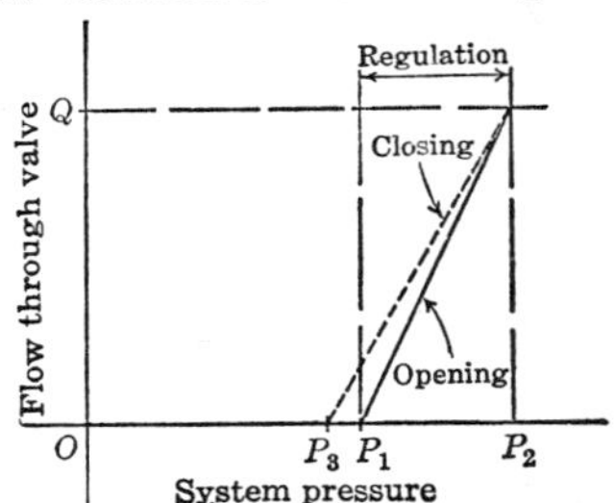

FIG. 31. Characteristics of pressure-regulating valve shown in Fig. 30. (Reproduced by permission from *Machine Design*)

Figure 32 diagrammatically illustrates a means of reducing regulation substantially to zero. Flow through the valve is regulated by the position of the main piston. The location of the main piston in relation to the orifice is achieved by controlling the pressure on top of this piston to balance the pressure in the system. A small flow of oil is by-passed through an orifice and controlled by a simple spring-loaded pilot valve. A change in flow causes a change in pressure P_1 causing the main piston to move up or down to achieve

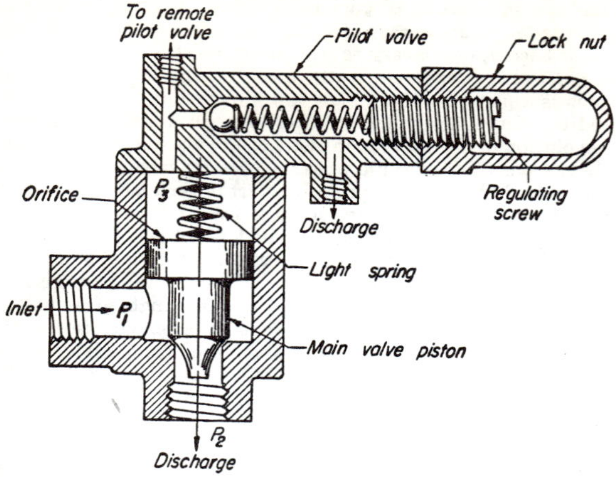

FIG. 32. Pressure-control valve with practically zero regulation. (Reproduced by permission from *Machine Design*)

equilibrium with pressure P_3 and produce pressure P_1. Figure 33 illustrates the pressure characteristics of such a valve. As long as the flow is greater than that required by the pilot valve the output pressure will remain constant.

Figure 34 illustrates the theory of operation of a quick-acting unloading valve. Oil from the pump enters the valve at the lower left-hand corner in the diagram. It can flow either up through the check valve to the accumulator, or through the main piston valve back to the tank or reservoir. If the pressure in the accumulator is high enough, it will hold the auxiliary piston valve in the position shown, and the main valve piston will allow oil from the pump to return to the tank. As soon as the pressure in the accumulator falls to a pre-set figure the detent can no longer hold the auxiliary piston valve against the

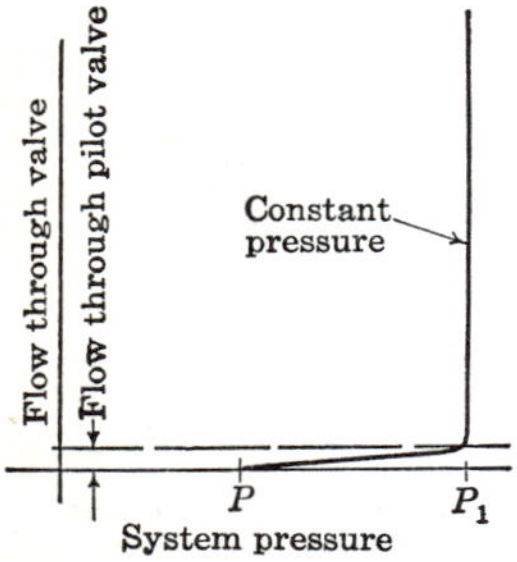

FIG. 33. Characteristics of the valve shown in Fig. 32. (Reproduced by permission from *Machine Design*)

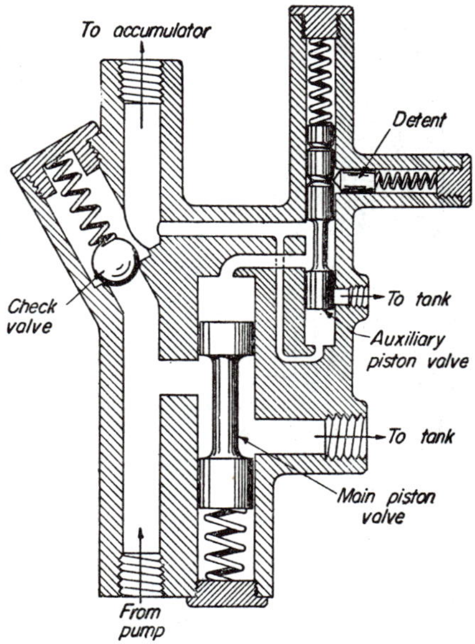

FIG. 34. Quick-acting unloading valve. (Reproduced by permission from *Machine Design*)

spring. It shifts with a snap action, in turn causing the main piston to shift and block the exit of oil from the pump, thereby causing the oil to flow to the accumulator and increase pressure in the hydraulic circuit.

Figure 35 illustrates a form of pressure-reducing valve. These valves are designed to reduce the hydraulic pressure from a high pressure to a lower pressure, and to maintain the lower pressure regardless of flow. The regulating spring holds the piston in the open position until pressure builds up against the diaphragm to overcome the spring force and close the valve piston. As flow varies, the position of the piston will adjust itself to maintain equilibrium. This valve is used as illustrated in Fig. 11.

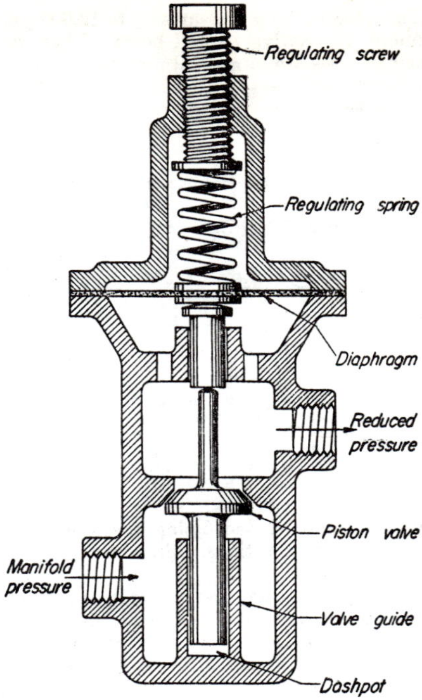

FIG. 35. Pressure-reducing valve. (Reproduced by permission from *Machine Design*)

Vibration or "whistling" is a common fault of simple pressure regulators. Means must be provided to cause damping of the valve vibrations, which is usually accomplished by the use of suitable dashpots.

7. HYDRAULIC PIPING, FITTINGS, AND FLUIDS

HYDRAULIC PIPING may be divided into two major groups, rigid and flexible. **Rigid piping** may be steel or iron pipe, either seamless or welded, using standard screw thread fittings. It may also be this same steel or iron pipe using brazed or welded fittings. Most modern machines use steel, brass, or aluminum tubing with a minimum of fittings, bending the tubing with hand or machine-bending tools. Table 1 gives a list of common tube sizes, standard bending radii, and other pertinent data.

Table 1.

Tube, OD	Wall	Maximum Flow, gal per min	Bending Radii to ℄ of Tube: Hand Bending	Bending Radii to ℄ of Tube: Machine Bending
1/8	.020	0.25	3/8	.....
3/16	.020	0.75	7/16	.....
1/4	.028	1.2	9/16	.....
5/16	.028	2.3	11/16	.....
3/8	.032	3.5	15/16	15/16
1/2	.035	6.0	1 1/2	1 1/4
5/8	.042	10.5	2	1 1/2
3/4	.065	16.0	2 1/2	1 3/4
1	.072	29.0	3 1/2	3
1 1/4	.083	50.0	.....	3 3/4
1 1/2	.095	80.0	.....	5
1 3/4	.095	100.0	.....	7

FITTINGS. A wide variety of hydraulic tubing fittings have been developed and are commercially available. Figure 36 illustrates various combinations of fitting ends. They are available in most of the commercial designs. Figure 37 illustrates two designs of hydraulic pressure fittings: one, a flare type in which the tube end must be flared when the fitting is assembled; the other, a ferrule type in which the tube needs only to be cut to length.

Flexible tubes are available for working pressures up to 12,000 psi. These high-pressure tubes are limited in diameter to relatively small bores. They gain their strength from built-in steel braid. Flexible tubes are in general not readily flexed when under pressure and as a result their life is short. It is best to use a minimum of flexible tubing in hydraulic machine design. Fittings on the flexible tubing are generally swaged or clamped on and the tubing will burst before a well-mounted fitting blows off. Fittings are available to match most commercial rigid tubing fittings. Standard pipe thread fittings are also available.

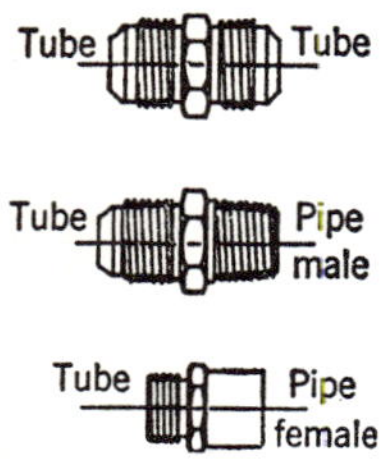

FIG. 36. Typical combinations of fitting ends.

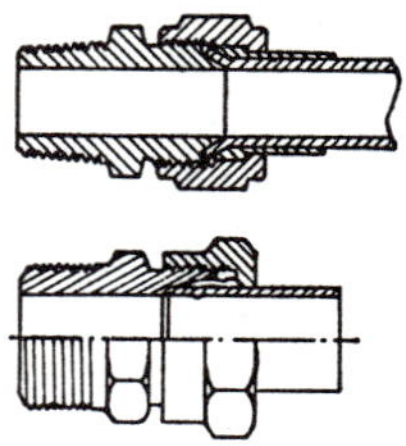

FIG. 37. Flare-type fitting (Parker Appliance Co.), upper view; ferrule-type fitting (Weatherhead Co.), lower view.

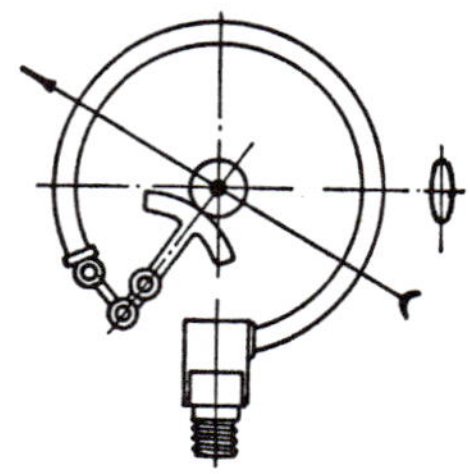

FIG. 38. Bourdon tube pressure gage.

PRESSURE INDICATORS. Hydraulic pressure is generally measured by a pressure gage consisting essentially of a bourdon tube and a suitable means for converting the deflection of the bourdon tube into motion of a pointer along a scale. For moderate pressure, the bourdon tube is made of steel. Figure 38 illustrates the principal parts of a simple pressure gage. Pressure gages should be calibrated on a dead-weight testing device.

The deflection under pressure of bellows or diaphragms can also be used as a means for measuring pressure. Here again, the deflection is converted by mechanical means to a suitable pointer and scale.

Pressure gages of the bourdon tube or bellows design are suitable only for static pressure measurements and for the measurement of relatively slow pressure changes. The lag and damping effect of the masses in the pointer actuating mechanism prevent rapid pressure changes from being indicated, and the rapid pointer motion cannot be followed by the human eye.

For measuring dynamic pressure some means of recording the pressure fluctuations is necessary. The best way of doing this is by a gage measuring the deflection in some part of the hydraulic circuit and recording this change on an oscillograph. (See Section 18, Article 7.)

HYDRAULIC FLUIDS. Water was one of the most widely used hydraulic fluids in early hydraulic machinery. It has the advantage of being inexpensive, readily available, and it has a very flat viscosity curve. Its disadvantages are that it is a poor lubricant, is corrosive to steel and iron, and cannot be used below 32°F unless additives are used. In spite of its disadvantages it is still used in large central hydraulic pressure systems, such as are found in rubber plants and plastic molding plants.

Mineral oil has displaced water as a hydraulic fluid in hydraulic machine design. Mineral oil is a good lubricant, gives good packing life, and (if proper oils are selected) can be used successfully at subfreezing temperatures. Mineral oil has the disadvantage of having poor temperature-viscosity characteristics. Various additives have been included in mineral oils to improve their viscosity characteristics. They are reflected in the flatter curves illustrated in Fig. 2.

Organic oils have received wide use as hydraulic fluids, notably in automobile brakes. This has been brought about by the use of natural rubber for lines and packings. Natural rubber is not affected by organic oils. These oils oxidize rapidly and must have additives to minimize these tendencies.

In recent years **synthetic oils,** notably the silicones, have been offered for hydraulic service. They are excellent for this use but cost more than mineral oils. Their advantage lies in their freedom from sludge-forming components and their very flat viscosity characteristics.

HYDRAULIC FLUID FILTERS. In order to insure minimum wear and long life of the various elements in a hydraulic circuit it is necessary to maintain absolutely clean fluid. Clean fluid also helps to maintain reliable action of controls and other closely fitted parts. Small particles of dirt can cause malfunctions and sticking of valve stems and poppets.

For best results the so-called micronic filters will be required. These filters will pass no particles over a few microns in major dimension. They must be large enough in area to pass the entire flow of oil with only a few pounds per square inch pressure drop across the filter element.

So-called edge-type filters are next in cleaning ability. They will pass particles substantially the same size as the plate or flat wire spacing, depending upon which type is used. Commercial filters are usually constructed with a by-pass set at a pressure below which the pump should operate. If the filter plugs up, the by-pass will allow the oil to flow past the filter without building up a dangerous pressure. Filters are made to be used in the pressure side of a circuit as well as in the suction line to the pump. If the tank or reservoir is not tightly sealed, dirt can enter it, and it is best then to have the filter on the suction pipe. If, on the other hand, the entire hydraulic system is tightly sealed the filter may be installed in any part of the circuit through which substantially all the oil is flowing.

Filters should be cleaned regularly and new elements installed. Self-clearing types of filters are not as good as the micronic designs. Great care should be taken in the assembly of hydraulic circuits to prevent foreign material from getting into them. This will minimize the work the filter will have to do.

HYDRAULIC FLUID RESERVOIR. The tank or reservoir is an important detail of a hydraulic circuit that is frequently neglected by the designer. It should contain enough oil so that heat generated in the circuit can be dissipated adequately by radiation and conduction. In general, if the capacity of the tank in gallons is equal to the number of gallons pumped in one minute and the tank is not too cubical in design enough tank surface will be available for good radiation. This relation will be valid only for pressures of 1000 psi or less.

The tank should if possible be compartmented. Oil returned to the tank should enter one compartment and then should flow over a dam and under a baffle to a second compartment containing the suction line near the bottom. The purpose of the dam and baffle is to surface entrained bubbles of air and to present only solid oil to the pump suction line. Means should be provided to clean out the tank and to drain off oil from its lowest point.

GENERAL REFERENCES

R. A. Dodge and M. J. Thompson, *Fluid Mechanics*, McGraw-Hill Book Co., New York, 1937.

L. Prandtl and O. G. Tietjens, *Hydro- and Aero-Mechanics*, 2 vols., McGraw-Hill Book Co., New York, 1934.

Hydraulics, Ford Trade School publication.

Harold Adams, *Aircraft Hydraulics*, McGraw-Hill Book Co., New York.

Anthony Zino, What to look for in hydraulic oils, *Am. Machinist*, 1947–1948.

W. S. Owen, Viscosity, *Product Eng.*, September 1937.

Richard Schmoyer, Compressibility of liquids, *Machine Design*, August 1947.

C. E. Grosser, articles in *Machine Design*, May, July, September 1941, December 1942.

H. Ziebolz, Designing hydraulic servos, *Machine Design*, July 1947.

J. C. Van Grundy, Hydraulic oils for industrial equipment, *Machine Design*, August 1946.

L. S. Linderoth, Jr., articles in *Machine Design*, Sept. 1944, April 1948.

Theory of the Rotameter, Fischer and Porter Co., Hatboro, Pa.

SECTION 14

GEARING

BY

S. LEROY CRAWSHAW, *Manager, Engineering and Sales, Western Gear Works*

GEARING

1. THE INVOLUTE SYSTEM OF TOOTHED WHEEL GEARING

The data in this section are limited to toothed wheel gearing of the involute form. Since the earliest appearance of geared mechanisms, when wooden pegs were set in the periphery of wooden wheels, many curves have been used as the basis for the forms of the gear teeth. In all cases curves were selected to provide constant relative velocity of the gear tooth faces (Fig. 1). The essential property common to these curves is such that

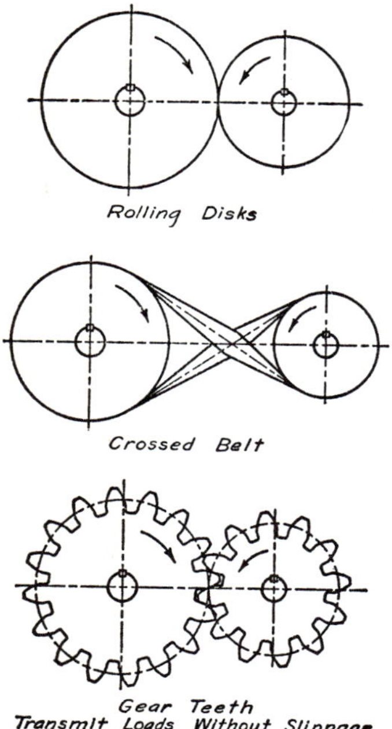

FIG. 1. Comparison of disks, belts, and gearing.

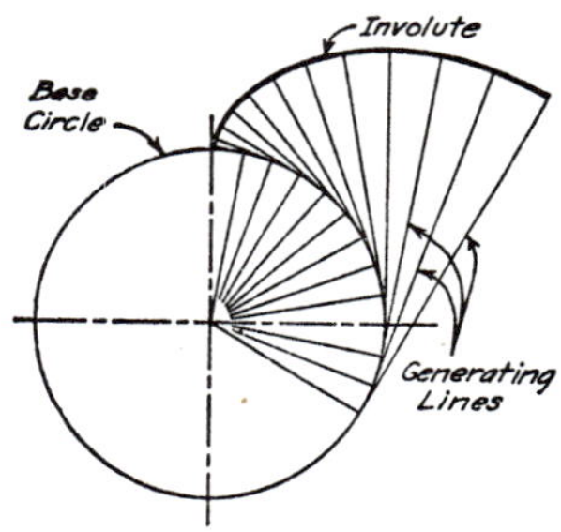

FIG. 2. Typical involute curve generation.

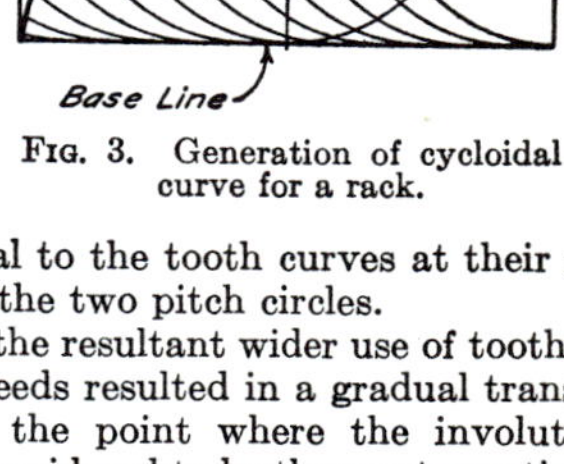

FIG. 3. Generation of cycloidal curve for a rack.

when two teeth are in contact, the common normal to the tooth curves at their point of contact must pass through the point of contact of the two pitch circles.

The introduction of gear-cutting machinery and the resultant wider use of toothed gearing to transmit heavier loads at ever-increasing speeds resulted in a gradual transition of tooth forms to the point where the involute curve (Fig. 2) is now considered to be the most practical curve for gear teeth. A few applications still employ curves of the cycloidal type (Fig. 3), and the pin-tooth type may be found in clock mechanisms, but general industry today relies on the involute curve system.

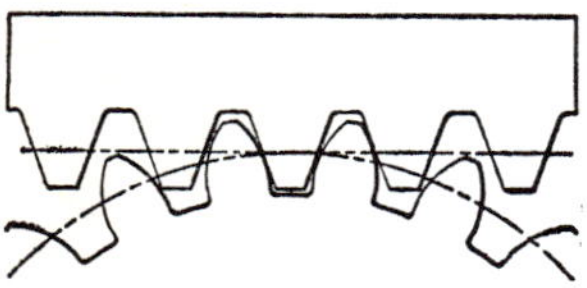

FIG. 4. Straight-sided rack used with involute gearing.

The basic rack form of the involute gear tooth has straight lines and is consequently relatively simple to produce. Present-day power transmission demands precision gearing, and the involute system based on a straight-sided rack (Fig. 4) provides a means of controlling quality to meet exacting requirements. In addition, gears of the involute type may be operated on variable center distances without affecting the theoretically correct action of the teeth in mesh (Fig. 5).

The above factors, and others, have resulted in universal acceptance of the involute curve system as the basis for gearing. The American Standards Association (ASA), as well as the American Gear Manufacturers Association (AGMA), uses the involute system in its standardization program.

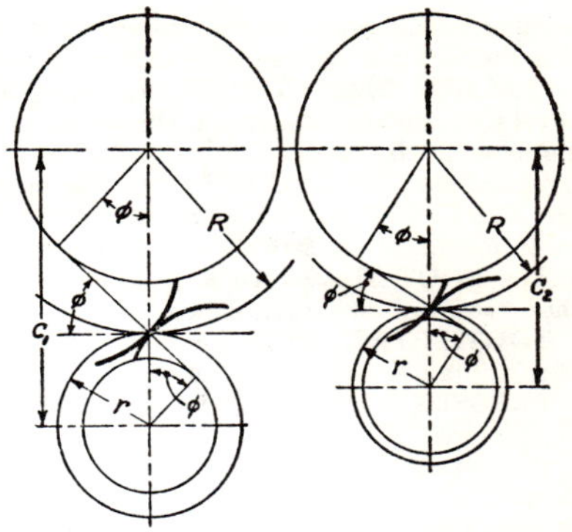

FIG. 5. Effect of varying center distance on involute action. Pressure angle changes but action is still correct.

2. NOMENCLATURE AND GEOMETRY

The technological advancement in the art of gearing has created its own unique division of mathematics. The AGMA has established a standard nomenclature for gearing and in collaboration with the ASA has arrived at a series of engineering symbols with the object of providing a common ground for all those interested in this specialized field of power transmission.

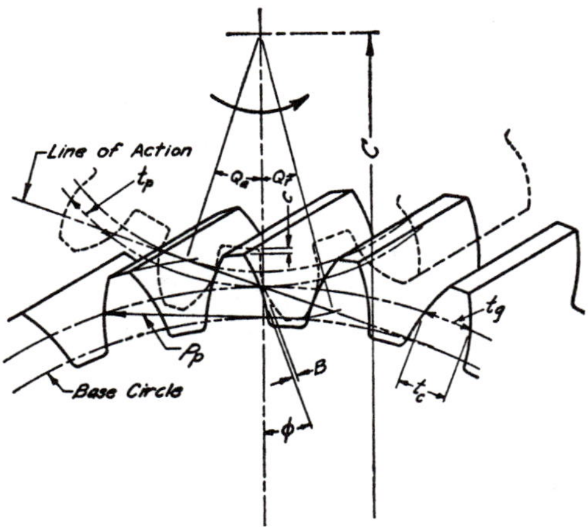

FIG. 6. Nomenclature and geometry terms for gearing.

The following more commonly used definitions are from the AGMA data (Figs. 6 and 7).

PITCH DIAMETER (D, d) is the diameter of the pitch circle. Unless otherwise specified, it is the *standard pitch diameter* obtained by dividing the number of teeth by the diametral pitch or by multiplying the circular pitch by the number of teeth divided by π (3.1416). In parallel-shaft gears, the pitch diameters can be determined directly from the number of teeth and the center distance. *Operating pitch diameter* is the pitch diameter at which the gear operates. *Generating pitch diameter* is the pitch diameter at which a gear is generated. In bevel gears the pitch diameter is understood to be at the outer ends of the teeth unless otherwise specified.

ADDENDUM (a) is the height by which a tooth projects beyond the pitch circle; it is also the radial distance between the pitch circle and the addendum circle.

DEDENDUM (b) is the depth of a tooth space below the pitch circle; it is also the radial distance between the pitch circle and the root circle.

CLEARANCE (d) is the amount by which the dedendum in a given gear exceeds the addendum of its mating gear.

ROOT DIAMETER (D_R, d_R) is the diameter of the root circle.

THE BASE CIRCLE is the circle from which involute tooth profiles are derived.

WORKING DEPTH (h_k) is the depth of engagement of two gears, that is, the sum of their addendums.

CENTER DISTANCE (C) is the distance between the parallel axes of spur gears and

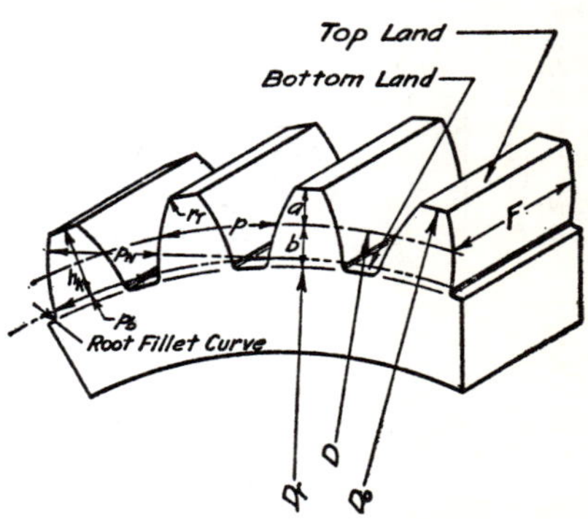

FIG. 7. Gear tooth terminology.

parallel helical gears or the crossed axes of crossed helical gears and worm gears. Also it is the distance between the centers of the pitch circles.

OUTSIDE DIAMETER (D_o, d_o) is the diameter of the addendum (outside) circle. In a bevel gear it is the diameter at the crown circle. In a throated worm gear it is the maximum diameter of the blank.

CHORDAL THICKNESS (t_c) is the length of the chord subtending a circular-thickness arc. **CIRCULAR THICKNESS** (t_g, t_p) is the length of arc between the two sides of a gear tooth on the pitch circle unless otherwise specified.

The **ROOT FILLET CURVE** is the concave portion of the tooth profile where it joins the bottom of the tooth space.

TOOTH TIP RADIUS (r_T) is the radius of the circular arc used to join a side-cutting edge and an end-cutting edge in gear-cutting tools.

PRESSURE ANGLE (ϕ) is the angle between a tooth profile and the radial line at its pitch point. Unless otherwise specified, it is the standard pressure angle at the standard pitch diameter. The *operating pressure angle* is the pressure angle at the pitch diameter at which the pair of gears operates. The *generating pressure angle* is the pressure angle at the pitch diameter at which the gear is generated. Still other pressure angles are used in gear calculations. In various gear-cutting tools the pressure angle is the angle between the cutting edge and the principal direction. For a spur gear or straight bevel gear which has tooth profiles only in the transverse section the general term "pressure angle" may be used without qualifications.

PROFILE RADIUS OF CURVATURE (ρ_p) is the radius of curvature of a tooth profile, usually at the pitch point or a point of contact.

CIRCULAR PITCH (p) is the distance along the pitch circle or pitch line between corresponding profiles of adjacent teeth.

ANGLE OF APPROACH (Q_a) is the arc of the pitch circle through which a tooth profile moves from its beginning of contact until the point of contact arrives at the pitch point.

ANGLE OF RECESSION (Q_r) is the arc of the pitch circle through which a tooth profile moves from contact at the pitch point until contact ends.

A **POINT OF CONTACT** is any point at which two tooth profiles touch each other.

The **TOOTH SURFACE** forms the side of a gear tooth.

TOP LAND is the surface of the top of a tooth.

BOTTOM LAND is the surface at the bottom of a tooth space adjoining the fillet.

BACKLASH (B) is the amount by which the width of a tooth space exceeds the thickness of the engaging tooth on the pitch circles. As actually indicated by measuring devices, backlash may be determined variously in the transverse, normal, or axial planes, and either in the direction of the pitch circles, or on the line of action. Such measurements should be corrected to corresponding values on transverse pitch circles for general comparisons.

Table 1. Relation of Dimensions of Parts of Gears, Composite and Full-depth Systems

Notations: P = diametral pitch; p = circular pitch; N = number of teeth; D = pitch diameter; D_o = outside diameter; a = addendum; t = thickness of tooth at pitch line; r = root; c = clearance; W = working depth; H = whole depth.

Diametral Pitch			Circular Pitch		
Given	To Find	Formula	Given	To Find	Formula
p	P	$P = 3.1416 \div p$	P	p	$p = 3.1416 \div P$
D, N	P	$P = N \div D$	D, N	p	$p = D \div 0.3183N$
D_o, N	P	$P = (N + 2) \div d$	D_o, n	p	$p = D_o \div (0.3183N + 2)$
N, P	D	$D = N \div P$	N, p	D	$D = Np \times 0.3183$
N, D_o	D	$D = D_oN \div (N + 2)$	N, D_o	D	$D = D_oN \div (N + 2)$
D_o, P	D	$D = D_o - (2 \div P)$	D_o, p	D	$D = D_o - 0.6366p$
a, N	D	$D = aN$	a, N	D	$D = aN$
N, P	D_o	$D_o = (N + 2) \div P$	N, p	D_o	$D_o = 0.3183p(N + 2)$
D, P	D_o	$D_o = D + (2 \div P)$	D, p	D_o	$D_o = D + 0.6366p$
D, N	D_o	$D_o = (N + 2) \div (N \div D)$	N, a	D_o	$D_o = a(N + 2)$
N, a	D_o	$D_o = (N + 2)a$	D, p	N	$N = 3.1416D \div p$
D, P	N	$N = DP$	p	t	$t = p \div 2$
D_o, P	N	$N = D_oP - 2$	p	a	$a = D \div N = 0.3183p$
P	t	$t = 1.5708 \div P$	p	r	$r = a + c = 0.3683p$
P	a	$a = 1 \div P = D \div N$	p	W	$W = 0.6366p$
P	r	$r = a + c = 1.157 \div P$	p	H	$H = W + c = 0.6866p$
P	W	$W = 2 \div P$	p	c	$c = 0.05p$
P	H	$H = W + c = 2.157 \div P$	t	c	$c = 0.1t$
P	c	$c = 0.157 \div P$			
t	c	$c = 0.1t$			

The **LINE OF ACTION** is the path of contact in involute gears. It is the straight line passing through the pitch point and tangent to the base circles.

FACE WIDTH (F) is the length of the teeth in an axial plane.

BASE PITCH (p_b) in an involute gear is the pitch on the base circle or along the line of action. Corresponding sides of involute gear teeth are parallel curves, and the base pitch is the constant and fundamental distance between them along a common normal in a plane of rotation.

CONTACT RATIO (m_c), which is the ratio of the arc of action to the circular pitch, is sometimes thought of as an average number of teeth in contact. For involute gears, the contact ratio is obtained most directly as the ratio of the length of action to the base pitch.

DIAMETRAL PITCH (P or P_d) is the ratio of the number of teeth to the number of inches in the pitch diameter.

BRITISH MODULE PITCH (m) is the ratio of the pitch diameter in inches to the number of teeth. It is the reciprocal of the diametral pitch. **METRIC MODULE PITCH** (m) is the ratio of the pitch diameter in millimeters to the number of teeth.

GEAR RATIO (m_G) is the ratio of the larger to the smaller number of teeth in a pair of gears.

3. GEAR TOOTH FORM SYSTEMS

Within the broad terminology of the involute system, there are many variations possible in the tooth form produced. Many applications in industry employ special combinations of pressure angles or variable proportions of addendum and dedendum to achieve the ultimate in performance. The factors governing them are so varied as to make it impossible to delineate them clearly here. The general run of gear requirements can usually be satisfied by one of the following four systems which have been standardized by the ASA under the joint sponsorship of the AGMA and the ASME:

A. The 14 1/2-degree composite system.
B. 14 1/2-degree full-depth tooth system.
C. The 20-degree full-depth tooth system.
D. The 20-degree stub-tooth system.

In the following paragraphs on the tooth form systems, the data given apply to the finished form. Manufacturing procedures such as grinding and shaving often require special tooth forms to be used in the primary production stages to allow finish stock on certain portions of the profile.

14 1/2-DEGREE COMPOSITE SYSTEM, developed originally by Brown & Sharpe Manufacturing Company for use with its formed gear-milling cutters, has been commonly known as the Brown & Sharpe 14 1/2-degree involute system. It has proved to be a very satisfactory gear-tooth system for general purposes. Figure 8 shows the form of the basic rack and the standard tooth proportions. Gears of this system are generally produced by formed milling cutters, but hobs made to this basic rack form are sometimes used.

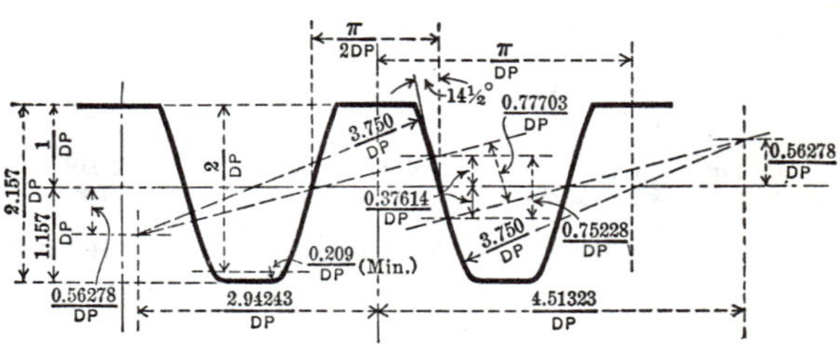

FIG. 8. Basic rack for 14 1/2-degree composite system.

Formed Milling Cutters for 14 1/2-Degree Composite System. Theoretically a different cutter is required for every different diameter of gear, even though the pitch remains the same. In practice, however, it has been found that the same cutter can be used for several sizes of gears having the same pitch, without serious error (see Table 2). By this plan 8 cutters will cut all gears of 12 teeth and over, of any one pitch. Thus, it takes 8 cutters to cut all 4-pitch gears of 12 teeth or more. A No. 8, 5-pitch cutter cuts only 5-pitch gears having 12 or 13 teeth. On each cutter the numbers of teeth at the limits of its range and the number of the cutter are stamped. When required, the error introduced by the use

Table 2. List of Gear Cutters and Numbers of Teeth Cut

(Brown & Sharpe Manufacturing Co., Providence, R. I.)

Cutter No.	No. of Teeth	Cutter No.	No. of Teeth	Cutter No.	No. of Teeth	Cutter No.	No. of Teeth
1	135–rack	3	35–54	5	21–25	7	14–16
2	55–134	4	26–34	6	17–20	8	12–13

of range cutters can be minimized by cutting the teeth in mating gears to slightly different depths. See Ref. 1, Chapter IV.

14 1/2-Degree full-depth tooth involute system was developed for use with gear hobs, for both spur and helical gears. As the useful tooth form cannot exist below the base circle of the involute form, this system should not be used for small numbers of teeth unless a suitable change is made in the tooth proportions of mating gears. See Ref. 1, Chapter V. This system has proved to be a satisfactory form for general purposes for both spur and helical gears. The form of the basic rack and the standard tooth proportions are shown in Fig. 9. The standard tooth proportions of this system are the same as those for the 14 1/2-degree composite system. (See Table 3.) When the number of teeth in any gear of this system is less than 35, undercut will be present. When the number of teeth is less than 22, the amount of undercut may be enough to reduce the angle of action appreciably.

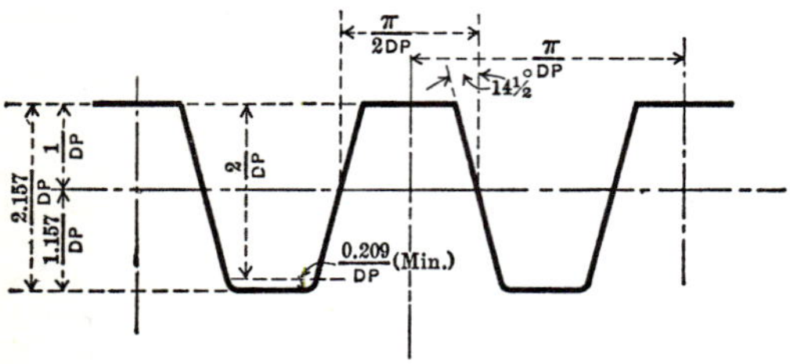

FIG. 9. Basic rack for 14 1/2-degree full depth involute system.

20-DEGREE FULL-DEPTH TOOTH INVOLUTE SYSTEM was developed for use with hobs and also for use with Fellows gear shapers. The increased pressure angle results in stronger tooth forms, and its use is rapidly increasing for general industrial purposes.

Table 3. Tooth Proportions for Spur Gears *

N = number of teeth; DP = diametral pitch; CP = circular pitch

Part	Basic Rack for 14 1/2-Degree Full-depth Involute, 14 1/2-Degree Composite		Basic Rack for 20-Degree Full-depth Involute		Basic Rack for 20-Degree Stub-tooth Involute	
	In Terms of Diametral Pitch, in.	In Terms of Circular Pitch, in.	In Terms of Diametral Pitch, in.	In Terms of Circular Pitch, in.	In Terms of Diametral Pitch, in.	In Terms of Circular Pitch, in.
1. Addendum	1/DP	.3183 × CP	1/DP	.3183 × CP	0.8/DP	.2546 × CP
2 Minimum dedendum †	1.157/DP	.3683 × CP	1.157/DP	.3683 × CP	1/DP	.3183 × CP
3. Working depth	2/DP	.6366 × CP	2/DP	.6366 × CP	1.6/DP	.5092 × CP
4. Minimum total depth †	2.157/DP	.6866 × CP	2.157/DP	.6866 × CP	1.8/DP	.5729 × CP
5. Pitch diameter	N/DP	.3183 × N × CP	N/DP	.3183 × N × CP	N/DP	.3183 × N × CP
6. Outside diameter	(N + 2)/DP	.3183 × (N + 2) × CP	(N + 2)DP	.3183 × (N + 2) × CP	(N + 1.6)/DP	(.3183 × N + .5092) × CP
7. Basic tooth thickness on pitch line	1.5708/DP	.5 × CP	1.5708/DP	.5 × CP	1.5708/DP	.5 × CP
8. Minimum clearance †,‡	0.157/DP	.05 × CP	0.157/DP	.05 × CP	0.2/DP	
9. Radius of fillet	1 1/3 × clearance		1 1/2 × clearance		1 1/2 × clearance	

* The term Diametral Pitch is used up to 1 DP inclusive and the term Circular Pitch is used for 3 in. CP and over.

† A suitable working tolerance should be considered in connection with all minimum recommendations.

‡ Minimum clearance refers to the clearance between the top of the gear tooth and the bottom of the mating gear space, and is specified as minimum to allow for necessary cutter clearance for all methods of producing gears.

It is also widely used in automotive construction. The form of the basic rack and the standard tooth proportions are shown in Fig. 10. The standard tooth proportions of this system are the same as those for the 14 1/2-degree composite system. (See Tables 1 and 3.) When the number of teeth in any gear of this system is less than 17, undercut will be present. When the number of teeth is less than 14, this undercut may be enough to reduce the angle of action appreciably.

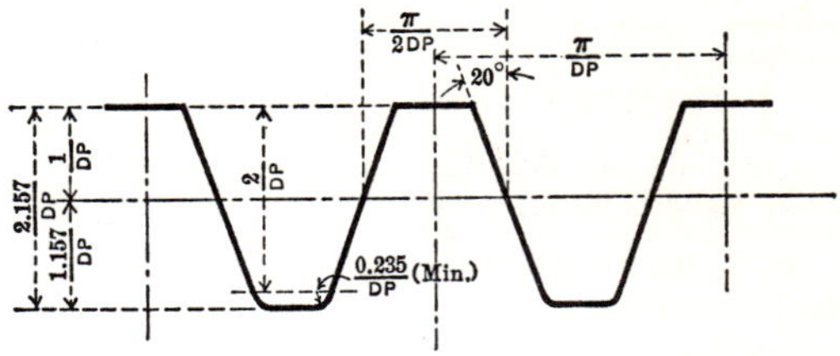

Fig. 10. Basic rack for 20-degree full depth involute system.

20-DEGREE STUB-TOOTH SYSTEM originally was developed for heavy-duty, generally slow-speed, machinery applications in order to obtain the strongest practical tooth forms. It was later adopted for automotive uses where the numbers of teeth were small. Several different stub-tooth systems, differing but slightly from each other, have been developed. The basic rack form and tooth proportions shown in Fig. 11 will interchange and operate with all of the 20-degree stub-tooth systems in general use, the only difference being in the amount of clearance. The dimensions of the tooth parts of this system are given in Table 4.

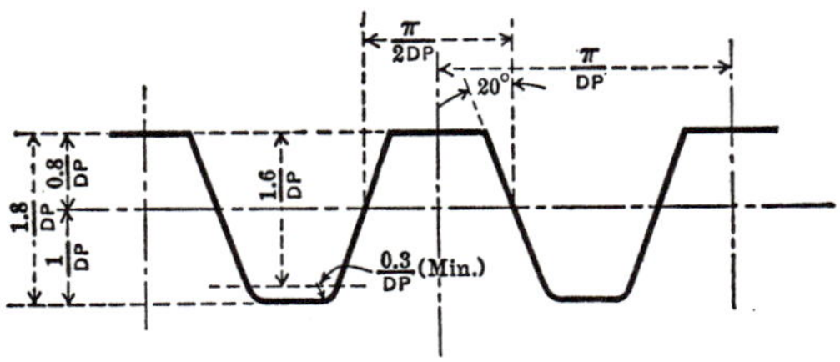

Fig. 11. Basic rack for 20-degree stub involute system.

Table 4. Dimensions of Tooth Parts. 20-Degree Stub-tooth System

Diametral Pitch	Circular Pitch	Thickness of Tooth on Pitch Line	Addendum	Working Depth of Tooth	Depth of Space below Pitch Line	Whole Depth of Tooth
1	3.1416	1.5708	.8000	1.6000	1.0000	1.8000
1 1/4	2.5133	1.2566	.6400	1.2800	0.8000	1.4400
1 1/2	2.0944	1.0472	.5333	1.0667	.6667	1.2000
1 3/4	1.7952	0.8976	.4571	0.9143	.5714	1.0285
2	1.5708	.7854	.4000	.8000	.5000	0.9000
2 1/4	1.3963	.6981	.3556	.7111	.4444	.8000
2 1/2	1.2566	.6283	.3200	.6400	.4000	.7200
3	1.0472	.5236	.2667	.5333	.3333	.6000
3 1/2	0.8976	.4488	.2286	.4571	.2857	.5143
4	.7854	.3927	.2000	.4000	.2500	.4500
4 1/2	.6981	.3491	.1778	.3556	.2222	.4000
5	.6283	.3142	.1600	.3200	.2000	.3600
6	.5236	.2618	.1333	.2667	.1667	.3000
7	.4488	.2244	.1143	.2286	.1429	.2572
8	.3927	.1963	.1000	.2000	.1250	.2250
9	.3491	.1745	.0889	.1778	.1111	.2000
10	.3142	.1571	.0800	.1600	.1000	.1800
12	.2618	.1309	.0667	.1333	.0833	.1500
14	.2244	.1122	.0571	.1143	.0714	.1286
16	.1963	.0982	.0500	.1000	.0625	.1125
18	.1745	.0873	.0444	.0889	.0556	.1000
20	.1571	.0785	.0400	.0800	.0500	.0900
24	.1309	.0654	.0333	.0667	.0417	.0750
28	.1122	.0561	.0286	.0571	.0357	.0643
32	.0982	.0491	.0250	.0500	.0313	.0563
36	.0873	.0436	.0222	.0444	.0278	.0500
40	.0785	.0393	.0200	.0400	.0250	.0450
48	.0654	.0327	.0167	.0333	.0208	.0375
56	.0561	.0280	.0143	.0286	.0179	.0322
64	.0491	.0245	.0125	.0250	.0156	.0281
72	.0436	.0218	.0111	.0222	.0139	.0250
80	.0393	.0196	.0100	.0200	.0125	.0225

FELLOWS STUB-TOOTH GEAR SYSTEM. The 20-degree stub-tooth gear system developed by the Fellows Gear Shaper Company has been widely used for automobile transmissions. The proportions of the teeth are based on two diametral pitches (Fig. 12), one of which is used to obtain the dimensions of the addendum and dedendum, whereas the other is used for the dimensions of the pitch circle and tooth thickness. These gears are designated by a fraction, as $^4/_5$, $^{10}/_{12}$ pitch, etc. The numerator designates the pitch determining the number of teeth and tooth thickness. The denominator designates the pitch determining the height of the teeth. The clearance is 0.25/diametral pitch. Table 5 gives dimensions of Fellows stub-tooth gears.

Table 5. Fellows Stub-tooth System

Diametral Pitch	Thickness of Tooth, in.	Addendum, in.	Working Depth, in.	Depth of Space below Pitch Line, in.	Clearance, in.	Whole Depth of Tooth, in.
4/5	.3927	.2000	.4000	.2500	.0500	.4500
5/7	.3142	.1429	.2858	.1786	.0357	.3214
6/8	.2618	.1250	.2500	.1562	.0312	.2812
7/9	.2244	.1111	.2222	.1389	.0278	.2500
8/10	.1963	.1000	.2000	.1250	.0250	.2250
9/11	.1745	.0909	.1818	.1136	.0227	.2045
10/12	.1571	.0833	.1667	.1041	.0208	.1875
12/14	.1309	.0714	.1429	.0893	.0179	.1607

LONG AND SHORT ADDENDUM GEARING. In the discussions of the $14^1/_2$-degree and 20-degree full-depth systems, mention was made of the possibilities of undercut existing in low numbers of teeth. Varying proportions of addendum (Fig. 13) are commonly resorted to in order to obtain the maximum efficiency of the tooth form selected.

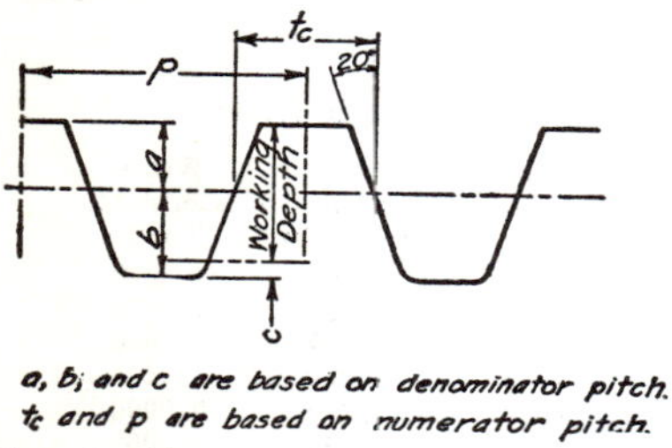

FIG. 12. Basic rack tooth form proportions for Fellows stub system.

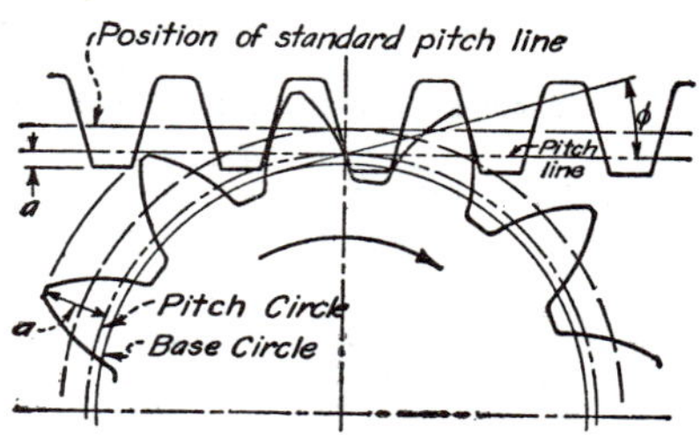

FIG. 13. Long and short addenda teeth.

Where the tooth form selected indicates undercutting, decreasing the outside diameter of the gear until it falls inside the point of tangency of the line of action with the pinion base circle and increasing the outside diameter of the pinion a like amount will eliminate the condition. The teeth in the modified blanks can be readily produced with standard hobs and assembled on standard centers.

Additional advantages in modifying the addendum proportions include increased strength of the pinion teeth and smoother tooth action.

METRIC OR MODULE SYSTEM. The word "module" is sometimes used in connection with the British module term. which is the ratio of the pitch diameter in inches to the number of teeth. (It is the reciprocal of the diametral pitch.) Most generally, it is used to designate the metric system.

Metric or module system of gear teeth is a system in which the diametral pitch is not used, but instead the dimensions of gear teeth are expressed by reference to the module of the gear. The module in this case is equal to the pitch diameter in millimeters divided by the number of teeth in the gear. For example, if the pitch diameter of a gear is 50 mm and the number of teeth is 25, the module equals $50 \div 25 = 2$. Table 6 gives a comparison between the module and the corresponding diametral pitch of gears.

The module is also equal to the circular pitch in millimeters divided by 3.1416. Either rule gives the same result. For changing module pitch to circular pitch: $p =$ module $\times$ 0.123684792. For changing module pitch to diametral pitch: $P = 1 \div$ (module $\times$ 0.03937).

Table 7 shows equivalents of diametral, circular, and module pitches. Standard pitches are shown in bold type and the equivalents in light-face type.

Table 6. Tooth Parts for Metric or Module System Gear Teeth

Module	Circular Pitch, in.	Thickness of Tooth at Pitch Line, in.	Addendum, in.	Whole Depth of Tooth, in.
1 1/4	.1546	.0773	.0492	.1061
1 1/2	.1854	.0927	.0591	.1273
1 3/4	.2164	.1082	.0686	.1486
2	.2472	.1236	.0787	.1698
2 1/4	.2784	.1392	.0886	.1911
2 1/2	.3092	.1546	.0984	.2123
2 3/4	.3402	.1701	.1083	.2336
3	.3710	.1855	.1181	.2548
3 1/2	.4330	.2165	.1378	.2973
4	.4948	.2474	.1575	.3397
4 1/2	.5568	.2784	.1772	.3822
5	.6186	.3093	.1969	.4247
5 1/2	.6802	.3401	.2165	.4670
6	.7420	.3710	.2362	.5095

Table 7. Equivalent Diametral, Circular, and Module Pitches

(Standard pitches in bold type; equivalents in light face)

Diametral	Circular, in.	Module, mm	Diam-etral	Circular, in.	Module, mm	Diam-etral	Circular, in.	Module, mm
1/2	6.2832	50.7991	2.2848	**1 3/8**	11.1168	10.0531	**5/16**	2.5265
0.5080	6.1844	**50**	2.3090	1.3606	**11**	10.1598	.3092	**2 1/2**
.5236	**6**	48.5095	**2 1/2**	1.2560	10.1598	**11**	.2856	2.3090
.5644	5.5659	**45**	2.5133	**1 1/4**	10.1062	**12**	.2618	2.1166
.5712	**5 1/2**	44.4671	2.5400	1.2369	**10**	12.5664	**1/4**	2.0212
.6283	**5**	40.4246	**2 3/4**	1.1424	9.2362	12.6998	.2474	**2**
.6350	4.9475	**40**	2.7925	**1 1/8**	9.0955	**13**	.2417	1.9538
.6981	**4 1/2**	36.3822	2.8222	1.1132	**9**	**14**	.2244	1.8143
.7257	4.3290	**35**	**3**	1.0472	8.4665	**15**	.2094	1.6933
3/4	4.1888	33.8661	3.1416	**1**	8.0849	**16**	.1963	1.5875
.7854	**4**	32.3397	3.1749	0.9895	**8**	16.7562	**3/16**	1.5159
.8378	**3 3/4**	30.3185	3.3510	**15/16**	7.5796	16.9330	.1855	**1 1/2**
.8467	3.7106	**30**	**3 1/2**	.8976	7.2570	**17**	.1848	1.4941
.8976	**3 1/2**	28.2972	3.5904	**7/8**	7.0743	**18**	.1745	1.4111
.9666	**3 1/4**	26.2760	3.6285	.8658	**7**	**19**	.1653	1.3368
1	3.1416	25.3995	3.8666	**13/16**	6.5690	**20**	.1571	1.2700
1.0160	3.0922	**25**	3.9076	.8040	**6 1/2**	**22**	.1428	1.1545
1.0472	**3**	24.2548	**4**	.7854	6.3499	**24**	.1309	1.0583
1.1424	**2 3/4**	22.2335	4.1888	**3/4**	6.0637	**25**	.1257	1.0160
1 1/4	2.5133	20.3196	4.2333	.7421	**6**	25.1327	**1/8**	1.0106
1.2566	**2 1/2**	20.2123	4.5696	**11/16**	5.5584	25.3995	.1237	**1**
1.2700	2.4737	**20**	4.6181	.6803	**5 1/2**	**26**	.1208	0.9769
1.3963	**2 1/4**	18.1911	**5**	.6283	5.0799	**28**	.1122	.9071
1.4111	2.2264	**18**	5.0265	**5/8**	5.0531	**30**	.1047	.8467
1 1/2	2.0944	16.9330	5.0799	.6184	**5**	**32**	.0932	.7847
1.5708	**2**	16.1698	5.5851	**9/16**	4.5478	**34**	.0924	.7470
1.5875	1.9790	**16**	5.6443	.5566	**4 1/2**	**36**	.0873	.7055
1.6755	**1 7/8**	15.1592	**6**	.5236	4.2333	**38**	.0827	.6684
1.6933	1.8553	**15**	6.2832	**1/2**	4.0425	**40**	.0785	.6350
1 3/4	1.7952	14.5140	6.3499	.4947	**4**	**42**	.0748	.6048
1.7952	**1 3/4**	14.1486	**7**	.4488	3.6285	**44**	.0714	.5773
1.8143	1.7316	**14**	7.1808	**7/16**	3.5372	**46**	.0683	.5522
1.9333	**1 5/8**	13.1380	7.2530	.4329	**3 1/2**	**48**	.0654	.5292
1.9538	1.6079	**13**	**8**	.3927	3.1749	**50**	.0628	.5080
2	1.5708	12.6998	8.3776	**3/8**	3.0318	50.2655	**1/16**	.5053
2.0944	**1 1/2**	12.1274	8.4665	.3711	**3**	50.7991	.0618	**1/2**
2.1166	1.4842	**12**	**9**	.3491	2.8222	**56**	.0561	.4536
2 1/4	1.3963	11.2887	**10**	.3142	2.5400	**60**	.0524	.4233

4. SPUR, HELICAL, AND HERRINGBONE GEARING

The usual types of gearing connecting parallel shafts may be described as having spur, single helical, double helical, or herringbone teeth (Fig. 14), and the gearing may be either external or internal. The process used to produce the gear teeth may require slight variations in construction, but the involute curve and a common basic rack provide interchangeability.

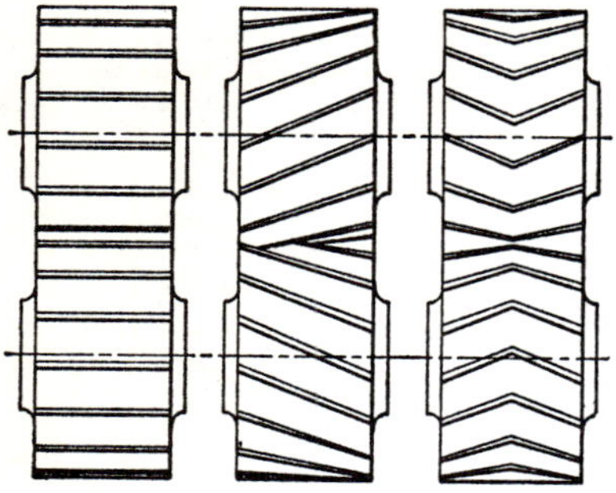

FIG. 14. Types of gears for parallel shafting.

SPUR GEARING is the term for designating designs using gear teeth whose faces are parallel to the axis of the blank. One of the principal advantages in spur gearing is its ability to transmit motion without end thrust, and without axial displacement limits as long as the face width in contact meets the load-carrying requirement.

The contact conditions in straight spur gearing are such that the load is carried on an area (Fig. 15) equal to the face times the width of the compressively flattened area parallel to the axis. Generally speaking, the number of teeth in contact in the plane of rotation may be assumed to divide the total transmitted load among several teeth at the point of transfer of the contact from one tooth (Fig. 16) in contact to two teeth in contact. The entire load is applied to one tooth.

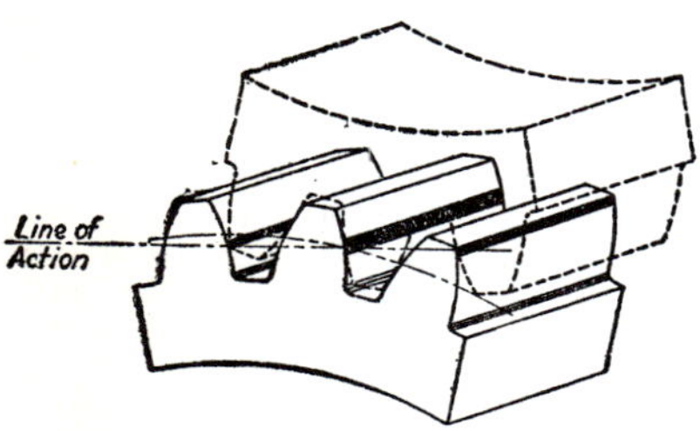

FIG. 15. Load zones in spur gearing.

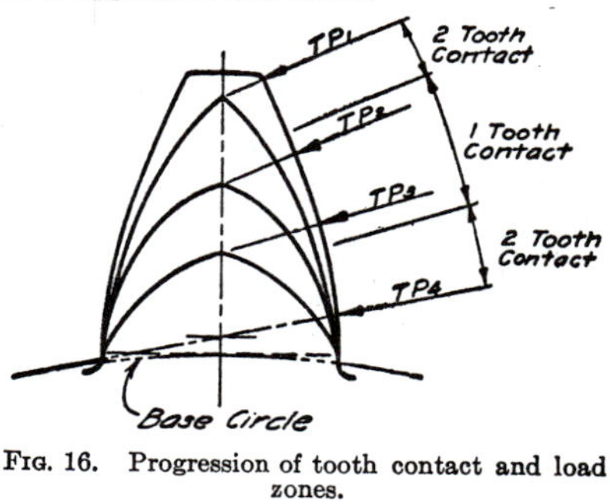

FIG. 16. Progression of tooth contact and load zones.

In large slow-moving spur gearing, a staggered tooth arrangement (Fig. 17) is sometimes resorted to in an effort to divide spacing errors and obtain smoother tooth-to-tooth action. With modern methods and machinery, this load transfer condition is minimized through control of tooth profiles, spacing, and backlash.

It is apparent that the problem of maintaining accuracy of tooth forms is simpler in straight spur gearing than in others; consequently many precision gear trains employ this type of gearing in preference to other types.

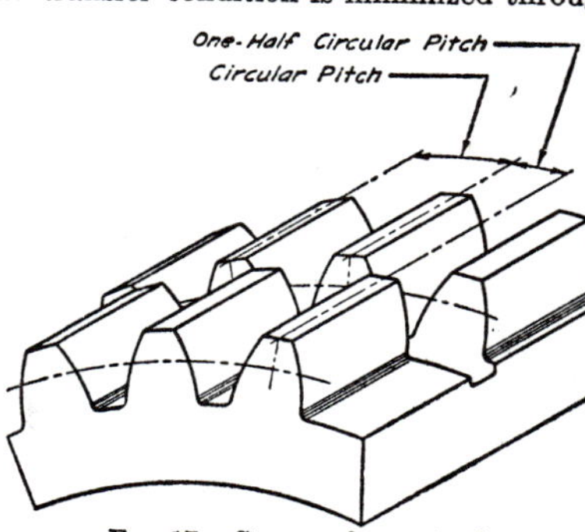

FIG. 17. Staggered spur teeth.

HELICAL GEARING. If we were to take an infinite number of spur gears of infinitesimal thickness and place them together with each lamination uniformly displaced ahead of the preceding one, we would have a "stepped gear" approaching helical gear construction. If the position of the first lamination is such that it is just entering contact, the successive laminations will be found to be in contact in uniformly advancing positions along the line of action.

Modern helical gears, made popular through improved gear-cutting machinery and economical thrust bearings, utilize this improved contact condition by twisting the elements of a spur-gear tooth uniformly across the face into a helix. Helical gears using the same basic spur gear normal section will have varying pressure angles, depending on the helix angle. The angle of the tooth (at its pitch line) with the axis of the gear is called the helix angle (Fig. 18).

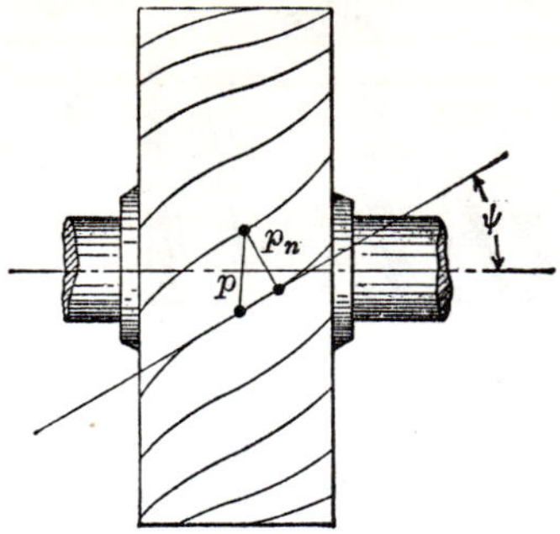

FIG. 18. Helix angle ψ, circular pitch p, normal circular pitch p_n.

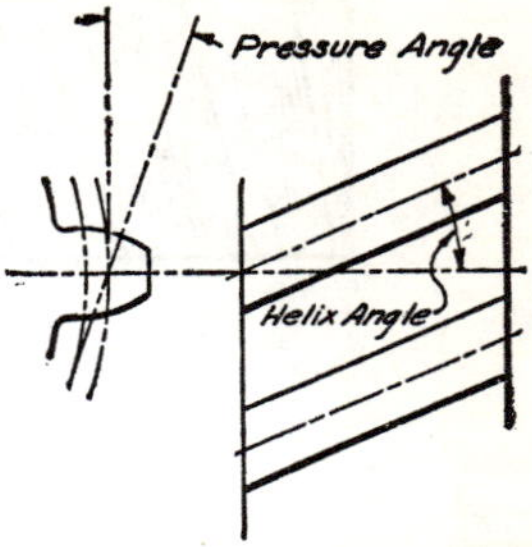

FIG. 19. Helical gear spur proportions in plane of rotation.

Helical gears may be cut with spur-gear dimensions in the plane of rotation (Fig. 19), and thus be interchangeable on centers with spur gears of the same pitch and numbers of teeth, or spur-gear cutters may be used to produce the spur-gear tooth parts in the normal plane (Fig. 20), thereby increasing the diameters, tooth thicknesses of the gears, and the center distance by relation to the helix angle.

The same basic rack forms are used for spur gears and for helicals, although special forms and proportions are often used for specific application conditions. Although the conjugate action between the tooth profiles in the plane of rotation is identical to that on spur gears, there is, in addition, a tendency for continuous action along the helices of mating teeth. While it has been shown that any helix angle develops this tendency, it is obvious that as the helix angle increases, more teeth are brought into the zone of action, and the action becomes smoother.

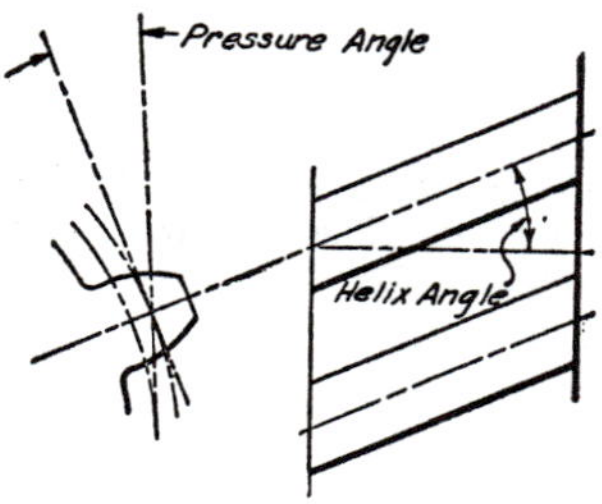

FIG. 20. Helical gear spur proportions in normal plane.

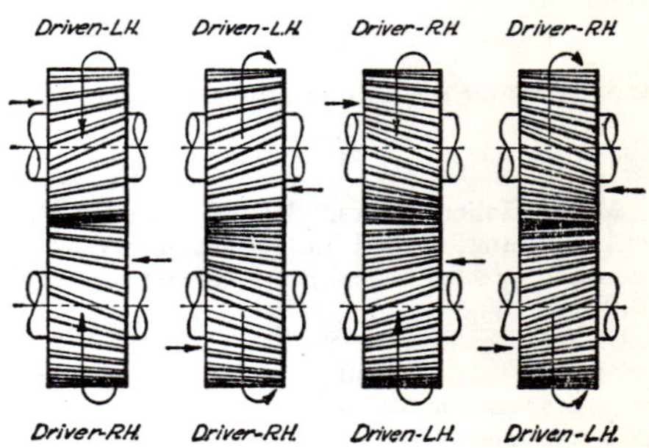

FIG. 21. Thrust conditions in helical gears. Arrows show direction of thrust.

Since single helical gears develop an end thrust in proportion to the tangent of the helix angle (neglecting friction), it is customary to use reasonably low helix angle ranges (under 25 degrees) unless the developed thrust is being used to offset external reactions. The direction of the thrust depends on the hand of the helix and the direction of rotation (Fig. 21).

Formulas for helical gearing are the same as for spur gearing where the cutters used produce the spur-gear tooth proportions in the plane of rotation (Fig. 22). Where tooth proportions in the plane normal to the helix are identical to spur teeth, the following formulas may be used to determine the plane of rotation proportions:

$$D = \frac{N}{P_{nd} \cos \psi}$$

where D = pitch diameter, N = number of teeth, P_{nd} = normal diametral pitch, and

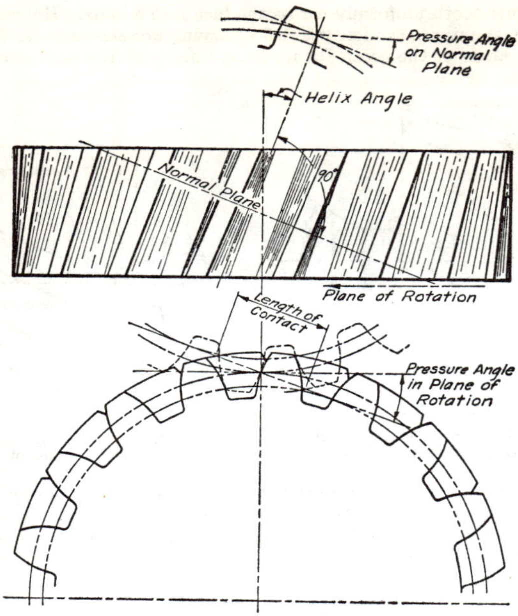

FIG. 22. Helical gear nomenclature.

ψ = helix angle. Also the diametral pitch in the plane of rotation is

$$P_d = \frac{P_{nd}}{\cos\psi}$$

and the center distance of a pair of helical gears is

$$C = \frac{N_G + N_p}{2P_{nd}\cos\psi}$$

Milled Helical Gears. Helical gears may be produced by milling on a milling machine with the same formed milling cutters as are used for spur gears. The general formulas required for such helical gears are shown in Table 8.

Table 8. General Formulas for Helical Gearing

To Find	Symbol	Formulas
Number of teeth in gears *	N_G, N_p	$N = DP_n \cos\psi$
Center distance	C	$C = (N_G + N_p)/2P_{nd}\cos\psi$
Circular pitch (circumferential)	p	$p = 3.1416D/N = p_n/\cos\psi$
Normal circular pitch	p_n	$p_n = p\cos\psi$
Diametral pitch	P	
Normal diametral pitch	P_{nd}	$P_{nd} = 3.1416/p_n$ (pitch of cutter)
Pitch diameter	D	$D = pN \div 3.1416 = p_nN/3.1416\cos\psi = ND_o \div (N + 2\cos\psi)$
Outside diameter	D_o	$D_o = D + 2a = D + (2/P_{nd})$
Angle of teeth with axis	ψ	$\cos\psi = N \div (P_{nd}D)$
Thickness of tooth	t	$t = p_n/2$
Addendum	a	$a = p_n/3.1416 = 1/P_{nd} = 1/2(D_o - D)$
Whole depth of tooth	H	$H = 2a + 0.1t$
Number of teeth marked on cutter when teeth are to be cut in milling machines †	T	$T = (N/\cos^3\psi) + (p_n \times D_c \times \tan^2\psi)$ ‡
Lead (exact) of spiral on pitch surface	L_1	$L_1 = Np/\tan\psi = N \times 3.1416/P\tan\psi$
Lead (approx.) of spiral on pitch surface	L_2	$L_2 = 10WG_2 \div SG_1$ §

* The subscripts G and p appearing in notations refer to gear G and pinion p, respectively.

† Cutters of composite system. See Table 2.

‡ D_c = diameter of milling cutter to middle of tooth profile.

§ W = number of teeth on worm; G_1, G_2 = teeth in first and second gears on stud, respectively; S = teeth in gear on screw, on spiral heads of Brown & Sharpe milling machines. The formula may not apply to spiral heads of different construction.

HERRINGBONE GEARING is sometimes called double helical gearing since the construction of a herringbone gear is essentially the same as two single helical gears having opposite hand helices and placed back to back. Herringbone gearing has various constructions. The gear tooth faces may be continuous, separated only sufficiently for cutter clearance, widely separated to span secondary gear trains, or staggered (Fig. 23).

The action of herringbone gearing is identical to that of single helical gearing, except that the thrust developed from the helix angle is self-contained within the shaft assembly, since the helix angle of the two tooth surfaces are opposite hand.

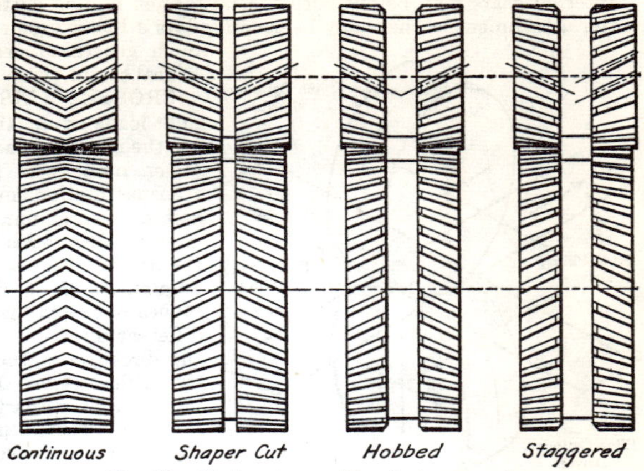

FIG. 23. Various types of herringbone gearing.

INTERNAL GEARING. An internal gear is one with the teeth cut into the bore instead of outside the rim, as is the case with a conventional spur or helical gear (Fig. 24). The internal gear possesses several advantages when properly applied and is frequently used for reduction gear and planetary gear combinations. In many respects, it can be treated as a conventional spur or helical gear. The same basic rack forms can be used, and the mating pinion is a conventional spur or helical gear. The center distance on an internal-gear drive is equal to the difference between the pitch radius of the internal gear and the pitch radius of the mating pinion. With an internal-gear drive, both members rotate in the same direction. With its shorter center distance the internal gear is more compact than the corresponding external-gear combination. Also, the tooth action is inherently smoother than that of a conventional spur or helical-gear drive. The tooth forms of the internal gear are stronger than those of the corresponding external toothed gears; they are stronger than the basic rack tooth form. For convenience in computing their beam strength, however, it is satisfactory to use the rack tooth form factor.

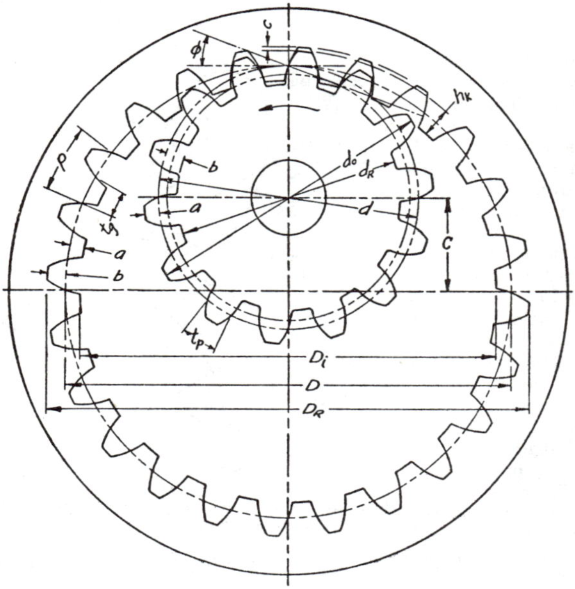

FIG. 24. Internal gearing.

When the internal-gear drive is used as a reduction drive, employing a small pinion and a large internal gear, the question of design as regards the tooth shape and action does not require special consideration. It is only when the number of teeth in the pinion closely approaches the number of teeth in the gear that it is necessary to lay out the teeth to see that proper tooth action is present without interference as the teeth come out of engagement. As a general rule, the smallest permissible difference between the number of teeth in the pinion and internal gear, to avoid this interference, is seven teeth for the stub-tooth form and twelve teeth for the full-depth tooth forms. Differences may sometimes be smaller, but this requires special design of tooth forms and proportions.

Internal gears may be cut with special formed milling cutters and an internal milling cutter attachment or they may be generated on a Fellows gear shaper or Sykes generator. Such generated gears may have straight or helical teeth. (See Ref. 12.)

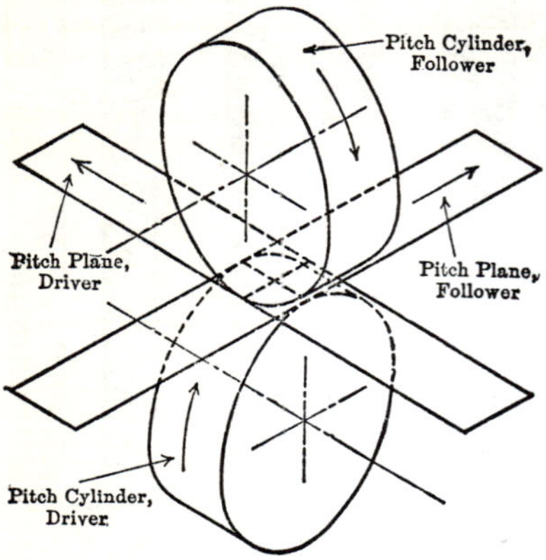

FIG. 25. Pitch surfaces of crossed-axis helical gears, pitch planes coinciding.

CROSSED-AXIS HELICAL GEARS, for light loads within their load-carrying capacity, are the most satisfactory type of screw gearing. When made with reasonable accuracy, they transmit motion smoothly and quietly, the result of several unique characteristics of such drives. Theoretically, it is impossible to spoil the nature of the action by mislocation or misalignment. As ordinarily designed, the pitch planes of mating gears coincide but travel in different directions (see Fig. 25), depending on the direction of the planes of rotation of the two helical gears. If the axes of the gears are moved farther apart, the sizes of the pitch cylinders will remain unchanged, the corresponding pitch planes will remain tangent to their respective pitch cylinders, and they will not coincide. (See Fig. 26.) The conjugate actior between the tooth profiles, however, would remain theoretically correct. The gears could be designed so that the axes would be shifted to bring the entire tooth profiles between the two pitch planes, the only result being a relatively small reduction in the duration of contact.

A similar shifting of the pitch surfaces will occur if the angular setting of the axes of the gears is changed. The circular pitches on the pitch planes then would change, as these pitches depend on the direction of the plane of rotation of the mating gear in relation to their axial position, and this change would cause the size of the mating pitch cylinder to change accordingly. The results of these characteristics are that errors in the position or alignment of involute helical gears used on nonparallel shafts have no effect on the correctness of tooth action.

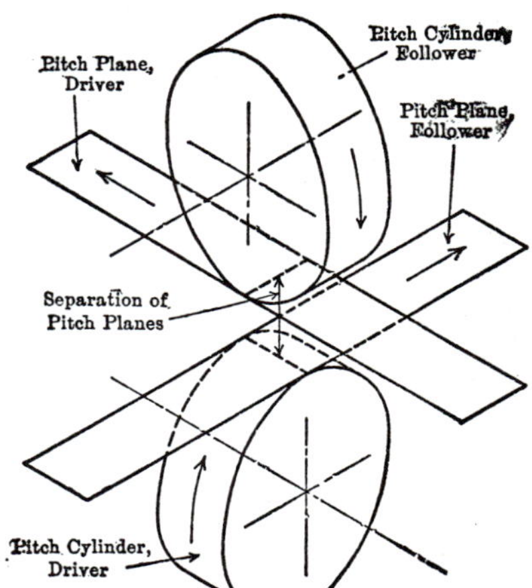

FIG. 26. Pitch surfaces of crossed-axis helical gears, pitch planes separated.

Generally, the helix angles on a pair of such gears are of the same hand. If the angle between shafts is small, however, the helix angles may be of opposite hand. Except when both helix angles are the same, there is no direct solution for the helix angles and pitch diameters of such gears, unless the center distance may be varied to suit the particular combination of helix angles which are selected arbitrarily. The final result in all other cases must be found by trial. When the two gears have helix angles of the same hand, the one with the greater helix, or lesser lead, angle should always be the driver. Except with gears of the same helix angle, the diameters of the gears are not directly proportional to the numbers of teeth. The speed ratio is established by the numbers of teeth and not by the diameters. Also the terms driver and follower are commonly used in place of the terms pinion and gear. Crossed-axis helical gears are sometimes known as spiral gears.

Let C = center distance, inches; P_n = normal diametral pitch (or diametral pitch of basic rack); n = number of teeth in driver; N = number of teeth in follower; α = helix angle of driver; β = helix angle of follower; γ = angle of axes; d = pitch diameter of driver, inches; D = pitch diameter of follower, inches. Then $d = n/P_n \cos \alpha$ and $D = N/P_n \cos \beta$.

When the two pitch planes coincide, $2C = d + D$.

When both helix angles are of the same hand, $\gamma = \alpha + \beta$.

When the helix angles are of opposite hand, γ = difference between α and β.

When L_1 = lead of driver, inches, and L_2 = lead of follower, inches, $L_1 = 3.1416d/\tan\alpha$, and $L_2 = 3.1416D/\tan\beta$.

The foregoing equations give all the basic relationships required to solve any spiral-gear problem. As noted before, in many cases the mathematical solution is indeterminate and must be solved by trial. The following simple graphical solution has been devised by J. K. Olsen, chief draftsman of the Stewart-Warner Speedometer Corporation.

Referring to Fig. 27, the number of teeth and the normal diametral pitch are selected arbitrarily. Circles are drawn representing the pitch diameters of the corresponding spur gears at the required center distance. Two intersecting straight lines are drawn on tracing paper, the angle between them corresponding to the angle between the shafts. This tracing is then placed over the two drawn circles and adjusted until both lines are tangent to their respective circles, their intersection being on the common center line of the two circles. The respective distances from the centers of the driver and follower to the intersection of the two lines on the common center line are the pitch radii of driver and follower. The angle between the vertical center line of each circle and the line tangent to that circle is the helix angle of that member. The values of the helix angle and the pitch radii would be measured from the layout, and should be checked by the foregoing equations. If the sum of the two pitch diameters so calculated is less than twice the specified center distance by not more than the depth of the tooth, no further refinement is necessary. The leads should be calculated from the formulas, using the measured values of the helix angles and the calculated values of the pitch diameters. The outside diameters of the gear blanks would then be chosen so that their sum equals twice the center distance plus twice the working depth of the tooth of the gear system used. The teeth would be cut to the standard depth in these blanks. The exact position of the tooth form in relation to the pitch cylinders is of little importance.

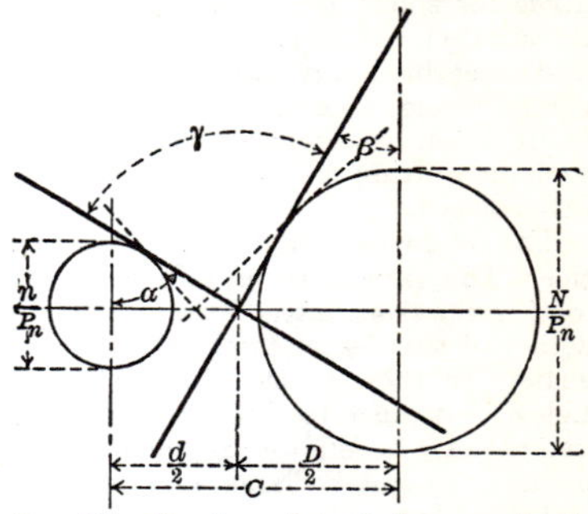

FIG. 27. Number of teeth in crossed-axis helical gears.

Figure 27 applies to the conditions when the two helix angles are of the same hand. In most cases, two points will be found on the common center line which meet the foregoing requirements. The point which gives the larger pitch radius and helix angle to the driver should be used. Sometimes the initial selection of tooth numbers and pitch results either in a condition where the intersection of the two lines on the tracing paper cannot touch the common center line, or the diameters so obtained are not satisfactory. In the first case, a new selection should be made which will reduce the size of the original circles. In the second case, a new selection should be made in which the size of the circles will be increased.

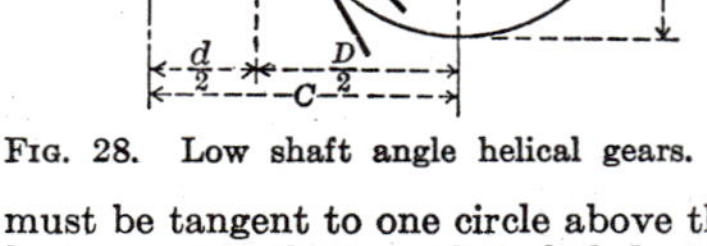

FIG. 28. Low shaft angle helical gears.

When the angle between the shafts is small, and the helix angles are to be of opposite hand, the procedure is as indicated in Fig. 28. The procedure is the same as before, except that one line must be tangent to one circle above the common center line, whereas the other line must be tangent to the second circle below the common center line.

Tooth Forms for Spiral Gears. Either the 14 1/2-degree composite system for milled gears or the 14 1/2-degree full-depth involute system for generated gears gives the most satisfactory tooth forms for spiral-gear drives. The selection of the proper milling cutter, etc., for milled spiral gears is identical to that for helical gears as shown in Table 8.

Load-carrying Ability of Spiral Gears. The nature of the contact between spiral gears is point contact. Because of this point contact, relatively light tooth loads set up very high compressive stresses in the material at the point of contact. With spiral gears, excessive wear due to pitting or abrasion is usually the most critical condition to be avoided, as the teeth will wear excessively under much smaller loads than those required to break the teeth.

Practically no reliable information is available on the safe loads for wear on spiral gears. The limit load for wear on such gears is very uncertain. If the gears are operated in their actual working position under a light load for a period of time until a thin polished line or band appears on the teeth where contact occurs, they can carry appreciably greater loads without excessive wear than they could if assembled and operated under full load without the preliminary polishing run. Again, if the gears are carefully run in under increasing loads until a polished band of appreciable width appears on the surfaces of the gear teeth where contact takes place, they can carry very much higher loads without excessive wear than they can carry after only a short polishing run. In fact, the longer a pair of spiral gears runs without abrasive wear on the tooth surfaces, the greater will be the loads that they can carry without excessive wear. It should be apparent, therefore, that load factors given for spiral gears depend on the care with which they have been run in after their installation. With proper care, a load that would cause excessive wear on the gears when first assembled may often be increased to several times the original limit load, because of the influence of careful running in, without resulting in appreciable wear. On the other hand, when excessive abrasive wear has once started, it is almost impossible to stop it without shifting the position of the gears and thus bringing a new part of the tooth surface into contact.

The following formulas and load factors have proved to be reasonably reliable in practice. They are based on initial running in as indicated. The relationship of the dynamic load to the transmitted load is still an open question. In general, because of the screwing action of spiral-gear teeth, their operation is much smoother than that of spur or bevel gears. In the absence of definite information, the following relationship may be used. Let W = transmitted tooth load, pounds; W_d = dynamic tooth load, pounds; V = pitch line velocity of follower in plane of rotation, feet per minute; d = pitch diameter of driver, inches; D = pitch diameter of follower, inches; K = load factor, Table 9, and Q = ratio factor. Then

$$W_d = \left(\frac{1200 + V}{1200}\right) W; \qquad Q = \left(\frac{2 \times D}{d + D}\right)^2$$

$$W_w = d^2KQ$$

where W_w = limit load for wear in pounds.

Table 9. Values of *K* for Spiral Gears

Materials in Gears	After Short Polishing Run	After Careful Polishing Run
Steel and steel (250 Brinell)	2	5
Steel (250 Brinell) and bronze	4	12
Steel (500 Brinell) and bronze	5	20
Steel (500 Brinell) and cast iron	6	20
Steel (500 Brinell) and steel (500 Brinell)	7	15
Cast iron and cast iron	8	20
Nonmetallic and steel or cast iron	10	25

In general, to avoid sparking, bronze or nonmetallic materials or hardened steel should be used when the pitch-line velocity of either member in their plane of rotation is much in excess of 1000 ft per min. Bronze never should be used in combination with the nonmetallic materials for spiral gears because of the excessive abrasion or lapping that takes place on the bronze member.

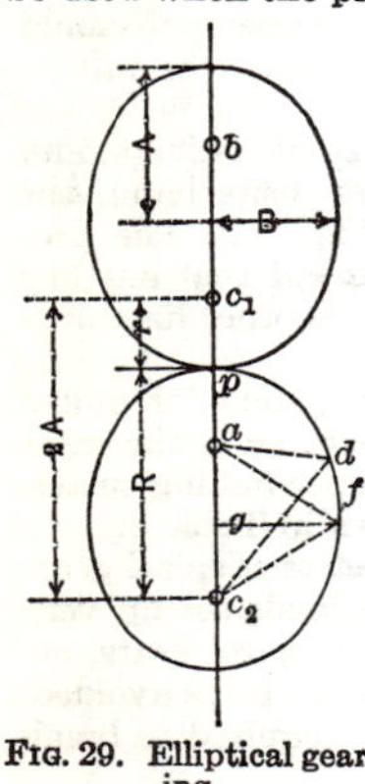

FIG. 29. Elliptical gearing.

ELLIPTICAL GEARS give a variable velocity ratio throughout the revolution. They represent a cheap way of obtaining a quick-return motion, and are so used on planers, shapers, slotters, and similar cutting tools, and also on pumps, shears, punches, and other machines which do their work mostly on the forward stroke of a reciprocating member. (See Refs. 2 and 3.) To determine the pitch ellipses for any given set of conditions, let c_1c_2 (Fig. 29) be the given distance between shaft centers. Divide c_1c_2 at p so that c_1p/pc_2 equals the extreme velocity ratio. Lay off $pa = pc_1$ for the other focus of the lower ellipse. As the ellipses must be equal, lay off $c_1b = c_2a$, and b is the other focus of the upper ellipse. The extremity f of the minor axis is found by making $af = fc_2 = {}^1/_2c_1c_2$. Any other point on the ellipse, such as d, is found by making $ad + dc_2 = c_1c_2$. The two free foci a and b remain at a constant distance apart during rotation, and are sometimes connected by a link. If $V = R^2/r^2$, A = one-half the major axis of the ellipse and B = one-half the minor axis, an equation which gives the ratio of the semimajor and semiminor axes for a given range of velocity ratio is

$$\frac{B}{A} = 2\sqrt[4]{V} \div (1 + \sqrt{V})$$

VARIABLE STANDARD CENTER DISTANCE GEARING. In some cases it is necessary to design gear trains to suit specific locations of driving and/or driven members. Often such requirements result in the center distance of a gear set varying from normal standard centers to such a degree that standard tooth proportions are not suitable.

The approach to this type of gear problem should preferably be made using full-depth generating types of cutters, although rotary cutters can be specially designed for each situation.

The best recommended practice is to increase (or decrease) the diameters of both mating members to suit the nonstandard center distance, the amount of increase or decrease for each being proportioned to the numbers of teeth. There are design limitation guides for this analysis, most important of which are given below.

For Increased Centers:

a. Check the number of teeth in contact at the maximum operating center distance and proportion addenda circles so as to obtain at least 1.1 teeth in contact.

b. Where extreme enlargement of the pinion teeth is required, a check should be made of the top land width so as to avoid a pointed condition or, in some cases, an insufficient cross section for heat treatment.

c. Sufficient running clearances, between outside diameter of each member and the root diameter of its mate, should be provided for minimum operating centers.

For Decreased Centers:

a. Undercutting should be avoided on the pinion teeth in order to maintain maximum strength characteristics.

b. Where undercutting of either gear or pinion cannot be avoided, contact conditions should be analyzed to avoid loss of continuous tooth action.

c. Where diameters must be reduced a substantial amount, the amount of material between the root diameter and the bore (or top of keyway or spline) should be checked.

5. BEVEL GEARING

TERMS AND DEFINITIONS. The terms *gear* and *pinion* when applied to bevel gears refer to the members with the larger and smaller number of teeth in the pair respectively. A *crown gear* has a pitch angle of 90 degrees. *Miter gears* have equal numbers of teeth and operate with axes at right angles. The angle between the axis of a bevel gear and an element of its pitch cone is called the *pitch angle*. Other terms applying to bevel gears are shown in Fig. 30. Note that the face-cone apex does not coincide with the pitch-cone apex. Instead the face cone of the blank is made parallel to the root cone of its mate, thereby giving constant clearance. This is preferred practice.

TYPES OF BEVEL GEARS. There are three types of bevel gears in general use.

Straight bevel gears have teeth which are straight and tapered and, if extended inward, would intersect at the axis.

Spiral bevel gears have curved oblique teeth on which contact begins gradually and continues smoothly from end to end.

Zerol bevel gears have curved teeth which are in the same general direction as straight teeth. They are spiral bevel gears with zero spiral angle.

There are also skew bevel gears having straight oblique teeth, but they are seldom used in the United States.

USES. The selection of the type of gear for a specific application is made according to operating requirements and mounting conditions.

Straight and **Zerol** bevel gears are satisfactory for peripheral speeds up to approximately 1000 ft per min where maximum smoothness and quietness are not of prime importance. In these applications plain bearings may be used, although antifriction bearings are always preferred.

Straight bevel gears, being the simplest to calculate and produce, are suitable for small lots where fixed charges must be kept to a minimum.

Zerol bevel gears are recommended in place of straight bevel gears (1) where hardened gears of extreme accuracy are required, since Zerol gears may have their teeth ground; (2) when the producer has only spiral-type equipment.

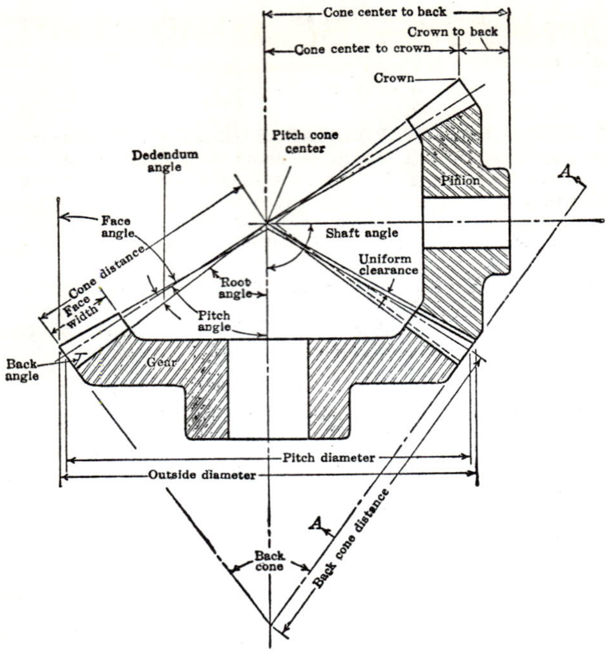

FIG. 30. Axial section of bevel gears showing terms.

Spiral bevel gears should always be used for speeds in excess of 1000 ft per min or 1000 rpm. In many instances they may be used to advantage at lower speeds, especially in applications where smoothness and quietness are important. Spiral bevel gears should, in general, be mounted on antifriction bearings because of the increased axial thrust of oblique teeth. Reference should be made to the manuals published by the bearing manufacturers to determine the loads and bearing sizes.

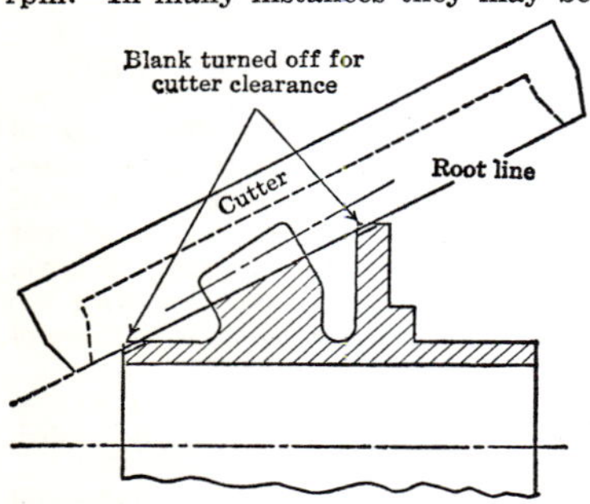

FIG. 31. Root-line interference of bevel gear cutter.

Caution. In designing spiral bevel and Zerol bevel gears, all front and rear hub projections which extend above the root line must be eliminated. (See Fig. 31.)

STRAIGHT BEVEL GEARS (Fig. 32). Gears produced on the older bevel gear-cutting machines may have contact extending the full length of the tooth. Gears cut on the most recent machines have a localized tooth bearing and are called Coniflex gears. This localization permits slight misalignment in assembly and some displacement due to deflections under operating loads without causing the concentration of load on the ends of the teeth.

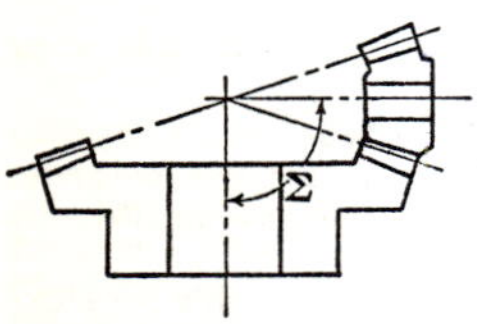

FIG. 32. Right-angle drive.

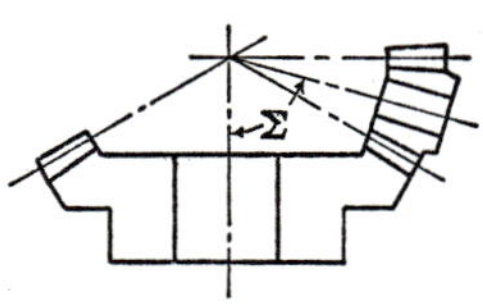

FIG. 33. Acute-angle drive.

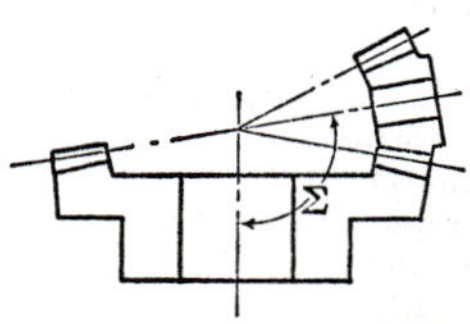

FIG. 34. Obtuse-angle drive.

Tables 10, 11, 12, 13, and 14 of recommended proportions for the tooth design of generated straight bevel gears present data from the Gleason Bevel Gear System as revised in 1949.

Table 10. Proportions of Generated Straight Bevel Gears *

(All linear dimensions in inches)

Working depth	$\frac{2.000}{DP}$
Whole depth	$\frac{2.188}{DP} + 0.002$

* With axes at right angles and thirteen or more teeth in pinion. For general industrial drives, both peed-increasing and speed-decreasing.

Table 11. Gear Addendum for One Diametral Pitch

(To obtain addendum, select from table value corresponding to ratio given by the formula *)

$$\text{Ratio} = \frac{\text{Number of teeth in gear}}{\text{Number of teeth in pinion}}$$

Ratios From	Ratios To	Addendum	Ratios From	Ratios To	Addendum
1.00	1.00	1.000	1.42	1.45	.760
1.00	1.02	0.990	1.45	1.48	.750
1.02	1.03	.980	1.48	1.52	.740
1.03	1.04	.970	1.52	1.56	.730
1.04	1.05	.960	1.56	1.60	.720
1.05	1.06	.950	1.60	1.65	.710
1.06	1.08	.940	1.65	1.70	.700
1.08	1.09	.930	1.70	1.76	.690
1.09	1.11	.920	1.76	1.82	.680
1.11	1.12	.910	1.82	1.89	.670
1.12	1.14	.900	1.89	1.97	.660
1.14	1.15	.890	1.97	2.06	.650
1.15	1.17	.880	2.06	2.16	.640
1.17	1.19	.870	2.16	2.27	.630
1.19	1.21	.860	2.27	2.41	.620
1.21	1.23	.850	2.41	2.58	.610
1.23	1.25	.840	2.58	2.78	.600
1.25	1.27	.830	2.78	3.05	.590
1.27	1.29	.820	3.05	3.41	.580
1.29	1.31	.810	3.41	3.94	.570
1.31	1.33	.800	3.94	4.82	.560
1.33	1.36	.790	4.82	6.81	.550
1.36	1.39	.780	6.81	∞	.540
1.39	1.42	.770			

* If there is a choice, use the larger addendum.

Table 12. Values of *K* for Circular Thickness Formula in Table 14

(Select value according to ratio and number of teeth in pinion.*)

Number of Teeth in Pinion	Ratios 1.000 to 1.020	1.020 to 1.075	1.075 to 1.140	1.140 to 1.260	1.260 to 1.855	1.855 to 2.250	2.250 to 2.645	2.645 to 3.105	3.105 to 3.650	3.65 to 4.35	4.35 to 5.21	5.21 to 6.25	6.25 to 7.58	7.58 to 9.35	9.35 to 11.50
	Values of *K* (in Inches) for Different Ratios														
13 to 14	......	......	......	......	+.075	+.060	+.040	+.020	+.005	−.015	−.035	−.050	−.070	−.090	−.110
15 to 16	+.000	+.020	+.035	+.055	+.070	+.060	+.045	+.025	+.010	−.010	−.030	−.045	−.065	−.080	−.100
17 to 21	+.000	+.030	+.035	+.050	+.070	+.060	+.045	+.025	+.010	−.005	−.025	−.040	−.060	−.075	−.095
22 to 26	+.000	+.015	+.030	+.045	+.060	+.050	+.035	+.020	+.005	−.005	−.025	−.040	−.060	−.075	−.095
27 to 35	+.000	+.015	+.025	+.040	+ 050	+.040	+.030	+.015	+.000	−.010	−.030	−.045	−.060	−.075	−.095
36 to 45	+.000	+.010	+.020	+.030	+.040	+.030	......	......	......	......	......	......	......	......	......

* In case of choice, use smaller value.

Table 13. Recommended Backlash

D.P.	Backlash
1.00 to 1.25	0.020–0.030
1.25 to 1.50	0.018–0.026
1.50 to 1.75	0.016–0.022
1.75 to 2.00	0.014–0.018
2.00 to 2.50	0.012–0.016
2.50 to 3.00	0.010–0.013
3.00 to 3.50	0.008–0.011
3.50 to 4.00	0.007–0.009
4 to 5	0.006–0.008
5 to 6	0.005–0.007
6 to 8	0.004–0.006
8 to 10	0.003–0.005
10 to 20	0.002–0.004
20 and finer	0.001–0.003

The table at left gives the recommended backlash for gears assembled ready to run. Because of manufacturing tolerances and changes resulting from heat treatment, it is frequently necessary to subtract more than half the recommended value from the theoretical tooth thickness of each member in order to obtain the correct backlash in assembly. In case of choice, use the smaller backlash tolerances.

Table 14. Typical Straight Bevel-gear Dimensions

(All linear dimensions in inches)

1. Number of pinion teeth	n (= 16)	5. Working depth h_k (Table 10)	(= 0.400)
2. Number of gear teeth	N (= 49)	6. Whole depth h_t (Table 10)	(= 0.440)
3. Diametral pitch	P (= 5)	7. Pressure angle ϕ	(= 20°)
4. Face width	F (= 1.5)		

	Pinion		Gear	
8. Pitch diameter	$d = \frac{n}{P}$	(= 3.2000)	$D = \frac{N}{P}$	(= 9.8000)
9. Cone distance	$A = \frac{D}{2 \sin \Gamma}$		(= 5.1546)	
10. Circular pitch	$p = \frac{3.1416}{P}$		(= 0.6283)	
11. Addendum	$a_P = h_k - a_G$	(= 0.284)	$a_G = \frac{\text{(Table 11)}}{P}$	(= 0.116)
12. Dedendum	$b_P = h_t - a_P$	(= 0.154)	$b_G = h_t - a_G$	(= 0.322)
13. Clearance	$c = h_t - h_k$		(= 0.038)	
14. Pitch angle	$\gamma = \tan^{-1} \frac{n}{N}$	(= 18° 5′)	$\Gamma = 90° - \gamma$	(= 71° 55′)
15. Dedendum angle	$\delta_P = \tan^{-1} \frac{b_P}{A}$	(= 1° 42′)	$\delta_G = \tan^{-1} \frac{b_G}{A}$	(= 3° 34′)
16. Face angle of blank	$\gamma_o = \gamma + \delta_G$	(= 21° 39′)	$\Gamma_o = \Gamma + \delta_P$	(= 73° 37′)
17. Root angle	$\gamma_R = \gamma - \delta_P$	(= 16° 23′)	$\Gamma_R = \Gamma - \delta_G$	(= 68° 21′)
18. Outside diameter	$d_o = d + 2a_p \cos \gamma$	(= 3.740)	$D_o = D + 2a_G \cos \Gamma$	(= 9.872)
19. Cone center to crown	$\frac{D}{2} - a_p \sin \gamma$	(= 4.812)	$\frac{d}{2} - a_G \sin \Gamma$	(= 1.490)
20. Circular thickness	$t = p - T$	(= 0.3703)	$T = \frac{p}{2} - (a_P - a_G) \tan \phi + \frac{K \text{ (Table 12)}}{P}$	(= 0.2580)
21. Backlash	(Table 13)		(= 0.005 to 0.007)	

NOTE. These remarks apply to the formulas in Table 14. *Item 4:* It is strongly recommended that the face width of bevel gears shall not exceed one-third the cone distance or $10/P$, whichever is smaller. In best design the ratio of face width to cone distance will be 0.25 to 0.30. Increasing the face width over recommended proportions results in manufacturing difficulties. *Item 7:* The pressure angle has been standardized at 20 degrees. *Item 12:* The dedendum value computed here is used in subsequent calculations. However, the actual dedendum will be greater by 0.002 in. *Item 20:* The circular thickness formula given here is for the theoretical value before deducting the allowance for backlash.

Angular Gears (Figs. 33 and 34). The formulas in Table 14 are not directly applicable for angular gears. Required changes are listed below by item number.

Item 14. Pitch Angles

Case 1. Shaft angle Σ is less than 90 degrees.

$$\tan \Gamma = \frac{\sin \Sigma}{\frac{n}{N} + \cos \Sigma}; \qquad \tan \gamma = \frac{\sin \Sigma}{\frac{N}{n} + \cos \Sigma}$$

$$\text{Check: } \Gamma + \gamma = \Sigma$$

Case 2. Shaft angle Σ is greater than 90 degrees, and Γ is less than 90 degrees.

$$\tan \Gamma = \frac{\sin (180° - \Sigma)}{\frac{n}{N} - \cos (180° - \Sigma)}; \qquad \tan \gamma = \frac{\sin (180° - \Sigma)}{\frac{N}{n} - \cos (180° - \Sigma)}$$

$$\text{Check: } \Gamma + \gamma = \Sigma$$

Case 3. Shaft angle Σ is greater than 90 degrees, and Γ is greater than 90 degrees (internal gear). In most cases this type of gear cannot be produced.

Item 11. Tooth Proportions

First compute equivalent 90-degree bevel-gear ratio:

$$\frac{N_{90}}{n_{90}} = \sqrt{\frac{N \cos \gamma}{n \cos \Gamma}} = \tan \Gamma_{90}$$

This value should be used as the ratio when determining the gear addendum for 1 DP in Table 11.

Item 19. Cone Center to Crown

$$\text{Pinion} = A \cos \gamma - a_P \sin \gamma$$

$$\text{Gear} = A \cos \Gamma - a_G \sin \Gamma$$

Item 20. Circular Thickness

To determine the value of K from Table 12, it is necessary to use equivalent 90-degree bevel-gear ratio given above (Item 11) and number of teeth in equivalent 90-degree bevel pinion:

$$\text{Equivalent } 90° \text{ ratio} = \tan \Gamma_{90}$$

$$n_{90} = \frac{n \sin \Gamma_{90}}{\cos \gamma}$$

Tooth Design for Bevel-gear Planers. Gears which are to be produced on a gear planer, where a former or template is used, should be designed with equal addendum teeth. Calculations for gear blanks which are to be cut by this method will be the same as given in Table 14, with the following exceptions:

Item 6: $\text{Whole depth} = \frac{2.188}{P}$ in.

Item 7: Pressure angle = $14\ ^1/_2°$ or $20°$

Item 11: Pinion addendum = gear addendum = $\frac{1.000}{P}$ in.

Item 16: On planer-cut gears it has been customary to make the face-cone apex coincide with the pitch-cone apex. It is here recommended that the face cone of the blank be made parallel to the root cone of its mate. Accordingly, the formula given in Table 14 should be used.

Item 20: Pinion circular thickness = gear circular thickness = $\frac{p}{2}$

SPIRAL BEVEL GEARS. One of the principal features associated with spiral bevel gears is the complete control of the localized tooth contact. (See Fig. 35.) This localization permits some mounting deflection without concentrating the load dangerously near either end of the tooth.

Selection of Spiral Angle. Figure 36 shows how the spiral angle is measured. For best design practice, the spiral angle should be sufficient to give a face contact ratio of at least 1.25. The design chart (Fig. 37) may be used. The series of curves are for face contact ratios of 1.0 to 2.5. The spiral angle is plotted in terms of the product of face width and diametral pitch.

Tooth Design for Spiral Bevel Gears. Because the tooth proportions for spiral bevel gears are dependent upon the method of cutting, no formulas are included here. Where such information is desired, it is recommended that the application be discussed with a manufacturer of this type of gearing.

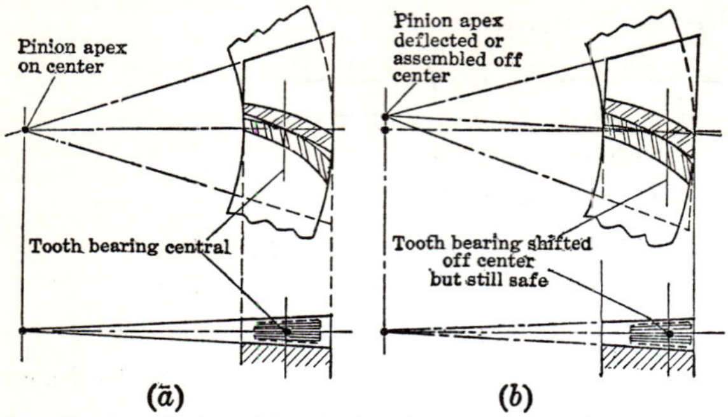

FIG. 35. Localized tooth bearing in spiral bevel gears, showing effect of displacement.

ZEROL BEVEL GEARS can be manufactured on the same machines as are used for producing spiral bevel gears. All Zerol bevel gears are manufactured with a localized tooth contact. They have the same end thrust and tooth action as straight bevel gears, and may be used in the same mountings.

Tooth Design for Zerol Bevel Gears. Like spiral bevel gears, the tooth proportions depend on the cutting method. Problems involving this type of gear should be reviewed with a gear manufacturer for recommendations.

Hypoid Gears closely resemble spiral bevel gears but have the pinion axis offset from the gear axis. If there is sufficient offset, the shafts may pass one another, and, in general, a compact straddle mounting can be used on both the gear and pinion. Hypoid gears are usually designed so that the pinion has a larger spiral angle than the gear, making the hypoid pinion larger in diameter and, therefore, stronger than the corresponding spiral bevel pinion.

In general, the shaft offset should not exceed 40% of the equivalent bevel-gear cone distance, and on very heavily loaded gears, such as for truck and tractor axles, the offset should be limited to 20%.

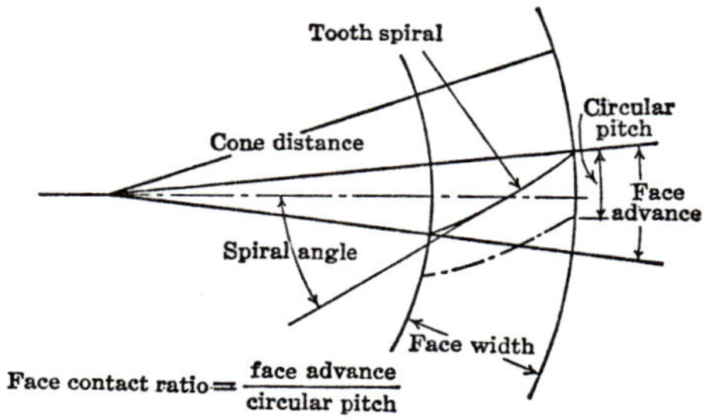

FIG. 36. Diagram defining spiral angle and face contact ratio.

The tooth surfaces of hypoid gears have some lengthwise sliding, the amount being a function of the offset. Hypoid gears are even smoother and quieter than spiral bevel gears, and for this reason are almost universally used in passenger automobile rear axles.

These gears may be produced on spiral bevel gear machines having vertical adjustment of the work spindle.

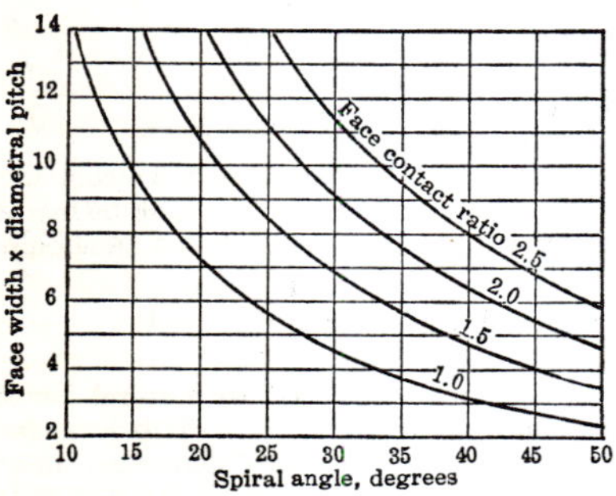

FIG. 37. Face contact ratio graph.

Tooth Design for Hypoid Gears. The tooth proportions for hypoid gears are related to the spiral angles and shaft offset. For this reason standardization has been impractical. Problems involving this type of gear should be reviewed with a gear manufacturer for recommendations.

GRINDING. Machinery is available for grinding the teeth of spiral bevel, Zerol bevel, and hypoid gears. The advantages of grinding are (1) increase in accuracy of hardened gears and (2) smooth blending of tooth profiles, fillets, and bottoms.

STRENGTH. For **general industrial** purposes, the beam strength of straight and spiral bevel gearing can be computed, using the following equation for peak strength rating from the AGMA standard rating practice:

$$P = \frac{snD_pFYC_v}{126{,}000C_BP_d}$$

where s = allowable bending stress = 250 times Brinell hardness number for untreated and heat-treated materials, or 300 times Bhn for case-hardened steels; n = pinion revolutions per minute; D_p = pinion pitch diameter, inches; F = face width, inches; Y = tooth form factor based on 1 DP with load at top of tooth (for spiral bevel gearing, use the tooth section in the transverse plane); C_v = velocity factor = $78/(78 + \sqrt{v})$; C_B = bevel gear factor = $A/(A - 0.5F)$, where A is the outside cone distance in inches; P_d = diametral pitch.

If the gearing is **critically loaded** or if the designer is interested in making a more detailed investigation of the stresses in bevel-gear teeth, the General Motors method of determining fatigue life should be referred to. Since this method involves considerable detail, those interested should review and analyze the complete method. Rather than an attempted briefing of the procedure, the following remarks are included to indicate only the theory.

It is a modification of the well-known Lewis formula arranged to determine the exact location of the point of application of the greatest load.

The general procedure is to determine the stress (following the detailed method developed by the Gleason Works) in the mean transverse section of the tooth when the total load is carried on one tooth in such a position that the tooth is under greatest stress. By means of layout and calculation, the stress in the weakest section is determined. (See Ref. 19.)

Comparing the calculated maximum stress with the General Motors stress-cycle graph establishes an estimated fatigue life. The General Motors data were established from records of life tests on hundreds of automotive spiral bevel gear sets; hence the extension of the life curve to industrial applications is accompanied by certain modifications of calculated values.

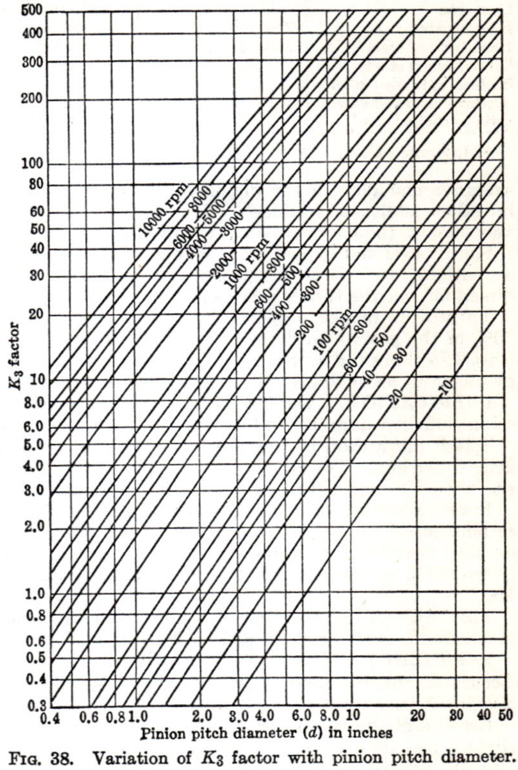

FIG. 38. Variation of K_3 factor with pinion pitch diameter.

Surface Durability. Except for cast-iron gears, surface-hardened gears, or gears of brittle material, surface durability is usually the determining factor for load capacity. The rating for surface durability is obtained from the following equation:

$$P = FC_MC_SK_3$$

where P = maximum continuous horsepower transmitted (instantaneous peak horsepower not to exceed twice this value); F = face width; C_M = material factor from Table 15; C_S = factor for contact ratio ($C_S = \sqrt{0.4m_F + m_p}$), where m_F = face contact ratio from the chart (Fig. 37), its value being zero for straight bevels and Zerol bevels; m_p = profile contact ratio, which is 1.2 for spiral bevels and 1.5 for straight bevels and Zerol bevels; and K_3 = combined factor for pinion diameter, velocity factor, revolutions per minute, and allowable unit load (Fig. 38).

MOUNTINGS. The proper design of shafts, bearings, and supporting structures is essential to successful gear performance. Rigid mountings should be provided; otherwise the gears will frequently fail prematurely.

The hand of spiral of a pair of spiral bevel gears should be selected carefully. In general, a left-hand spiral pinion driving clockwise (viewed from the back) or a right-hand pinion driving counterclockwise tends to move axially away from the cone center. If there is

Table 15. Material Factors for Bevel Gearing

Gear			Pinion			C_M for Dura-bility
	Brinell	Rock-well "C"		Brinell	Rock-well "C"	
Cast iron	....	....	Cast iron	....	....	0.30
Cast iron	....	....	Annealed steel	160–200	....	0.30
Cast iron	....	....	Surface-hardened steel	....	50 *	0.40
Cast iron	....	....	Case-hardened steel	....	55 *	0.40
Heat-treated steel	210–245	....	Heat-treated steel	245–280	....	0.35
Heat-treated steel	250–300	....	Case-hardened steel	....	55 *	0.50
Oil-hardened steel	....	....	Oil-hardened steel	....	....	0.65
Surface-hardened steel	....	50 *	Surface-hardened steel	....	50 *	1.00
Surface-hardened steel	....	50 *	Case-hardened steel	....	55 *	1.00
Case-hardened steel	....	55 *	Case-hardened steel	....	55 *	1.00

* Minimum values.

excessive end play in the pinion shaft because of faulty assembly, the movement will introduce additional backlash. With the opposite hand of spiral, the pinion will move in and take up the backlash, and may even wedge.

Spiral bevel and hypoid gears should, in general, be mounted on antifriction bearings in an oiltight case. Both the gears and bearings should be locked against thrust in either direction, and the thrust bearings should be preloaded.

To provide proper support, the distance between bearings should never be less than 70% of the gear pitch diameter, and, on an overhung pinion, the distance should be at least 2 1/2 times the overhang. The shaft diameter should be equal to or greater than the overhang in order to provide sufficient shaft stiffness.

LUBRICATION. The lubricating system should furnish a continuous supply of clean oil in sufficient quantity to lubricate and adequately cool the gears and bearings.

Splash lubrication is generally satisfactory for applications up to peripheral speeds of 2000 ft per min. The oil level should cover the full face of the lowest gear, and the quantity should be sufficient to maintain the oil temperature within recommended limits.

Pressure lubrication is used for velocities above 2000 ft per min. Spray lubrication is recommended with the jets located to direct the stream over the full length of the teeth on both members, preferably close to the mesh point on the leaving side. Jets should also be provided to lubricate the bearings. In general, oil temperatures should be kept near 150 F.

The "AGMA Specification on Lubrication of Enclosed and Open Gearing" is recommended as a guide to the type and grade of oil for various operating conditions.

EFFICIENCY. Various tests on bevel gears have shown efficiencies of over 99%. Accurately manufactured gears properly assembled in their mountings should easily show a gear efficiency of 98%.

Because of the lengthwise sliding encountered in hypoid gears, their efficiency will be slightly less than that of bevel gears. With the exception of high-ratio hypoids (over 10-to-1), efficiencies of 96 to 98% should be obtained.

6. LOAD-CARRYING CAPACITY OF GEARING

The ultimate in load-carrying abilities of gear teeth has so far been an elusive problem. The technological advancements in the art of producing tooth forms are improving the possibility of evaluating such factors as surface finish and load distribution on capacity determinations.

As of this writing, various experiments are being carried on to isolate and formulate many factors that are currently classed as "indeterminate" and, as a consequence, given an arbitrary value.

The load-carrying ability of gearing has been recognized for some time as contingent on contact conditions under load. As a consequence, in fields of high productions such as automotive and aircraft transmissions, complete assemblies are life-tested to provide "design constants" related to specific related conditions. Such values do not lend themselves to general use, and, as a consequence, there may appear to the uninitiated to be several "schools of thought" in the gearing field, whereas it will be found on closer inspection that the final design data in use stem from essentially the same basic considerations.

There are five principal elements to be dealt with in designing gear trains: first, resistance to breakage; second, resistance to surface wear (due to localized high unit stresses); third, resistance to scoring (due to high sliding velocities or a combination of high unit pressures and relatively high velocities); fourth, dynamic load conditions as they affect tooth-contacting conditions and/or the intensity of the imposed load; and, last, the indicated load-distribution characteristics of the mounting.

Several years ago someone said that "involute gears wear out—until they wear in—and then never wear out." Thus one observer recognized the effect of load-distribution qualities of the final setup of a gear train on its operating qualities, provided, however, that the gears were properly manufactured and given reasonable maintenance.

From a design viewpoint, then, it would appear necessary only to balance the strength, surface, contact, and deflection characteristics of a pair of mating gears to obtain a perfect solution to each power transmission problem.

The degree to which we analyze the various details connected with this phase of the design must be determined from the requirements of the mechanism. It is a sound conclusion that the best gear for a given application is the one that will satisfactorily perform all the functions for which it is intended, for the length of time specified and for the lowest possible cost of manufacture. So it is with capacity formulas. Where simple checks make it possible to proportion gear trains, they should be used. On the other hand, where the application demands that the most be made of available materials, the tooth proportions must be analyzed in great detail in order to achieve the maximum in load-carrying ability.

TOOTH-STRENGTH FORMULAS. The formula most generally referred to in strength calculations is a combination of the work of Wilfred Lewis and Carl G. Barth. The Lewis formula, presented in a paper read before the Engineers' Club of Philadelphia on October 15, 1892, was the first (see Ref. 1) to recognize the relationship of tooth form to gear-tooth strength. This formula is:

$$W = spfy$$

where W = transmitted tooth load, pounds; s = safe working stress of the material, pounds per square inch; p = circular pitch, inches; f = width of the gear-tooth face, inches; y = tooth-form factor, obtained by considering the gear tooth as a beam, fixed at one end and loaded at the other.

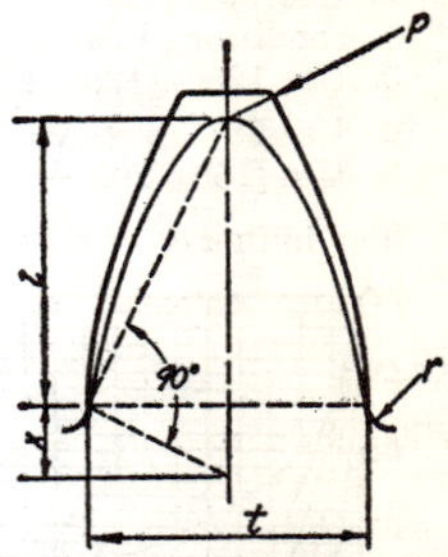

FIG. 39. Determination of y factor in Lewis equation.

The y factor is generally established from a layout as shown in Fig. 39. In considering the gear tooth as a beam for the establishing of a tooth-form factor, the analogy and procedure may be summarized in the following way.

The formula for the strength of a beam under comparable loading conditions (see Fig. 39) is

$$W = \frac{sft^2}{6l}$$

where W = load, pounds; s = maximum permissible fiber stress, pounds per square inch; f = width of beam, inches; t = thickness of beam (at critical section), inches; l = length of beam (from critical section to point of load application).

The layout shows that $x = t^2/4l$, and if this value is substituted in the beam equation, we have

$$W = sf\frac{2x}{3}$$

and for a gear tooth

$$W = Spf\frac{2x}{3p}$$

with the proportion $2x/3p$ becoming the y factor. The value of y for several gear-tooth systems is assembled for ready reference in Table 16.

The Barth contribution consisted of an approximation of safe working stresses in relation to pitch-line velocities in the form of the following equation:

$$S = \frac{600s}{600 + V}$$

in which S becomes the safe working stress, pounds per square inch; s is safe static stress, pounds per square inch (usually taken as one-third the elastic limit of the material); and V is pitch-line velocity, feet per minute.

It has since been recognized that the combination of the Lewis and Barth equations makes no allowance for the nature of the load, the influence of the amount of error in

Table 16. Values of y in Lewis Formula

No. of Teeth	14 1/2°	20° Full Depth	20° Stub	No. of Teeth	14 1/2°	20° Full Depth	20° Stub
12	.067	.078	.099	27	.100	.111	.136
13	.070	.083	.103	30	.102	.114	.139
14	.072	.088	.108	34	.104	.118	.142
15	.075	.092	.111	38	.107	.122	.145
16	.077	.094	.115	43	.110	.126	.147
17	.080	.096	.117	50	.112	.130	.151
18	.083	.098	.120	60	.114	.134	.154
19	.087	.100	.123	75	.116	.138	.158
20	.090	.102	.125	100	.118	.142	.161
21	.092	.104	.127	150	.120	.146	.165
23	.094	.106	.130	300	.122	.150	.170
25	.097	.108	.133	Rack	.124	.154	.175

gear-tooth action, and the influence of the rotating masses. Present-day usage of the Lewis formula includes variations of the basic formula, some of which are basically sound; others contain an arbitrary factor founded on experience. For instance, it is not uncommon to use varying velocity factors to represent varying degrees of gear-tooth action. Following are typical examples that have been used.

1. Limit the $600/(600 + V)$ factor to spur gears of commercial quality.
2. Use $900/(900 + V)$ for low-angle helicals or spur gearing where improved contact conditions justify.
3. Use $1200/(1200 + V)$ for high-angle helicals and herringbone gearing.
4. Use $78/(78 + \sqrt{v})$ for precision gearing.
5. Use $[150/(200 + V) + 0.25]$ for nonmetallic materials.

The influence of these modifications can be observed in the charted data of Fig. 40.

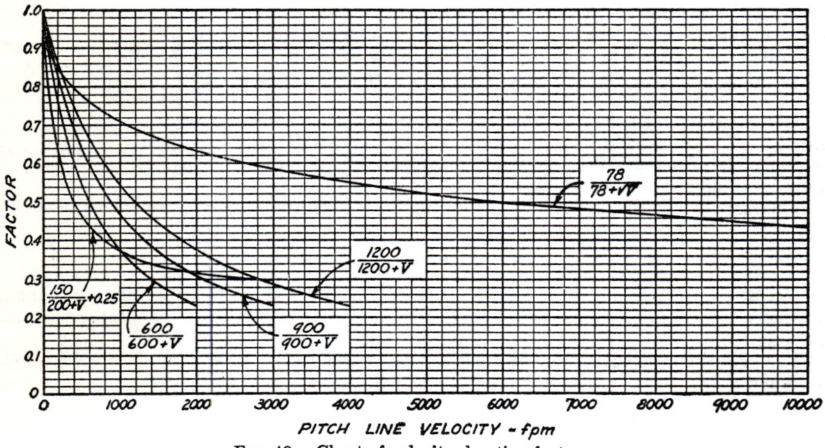

FIG. 40. Chart of velocity derating factors.

In addition to varying the velocity factor to suit the conditions, McMullen and Durkan (Ref. 22), in an analysis of the contact conditions in gear teeth, proposed a system of bevel gearing in which the point of application of the maximum load on the gear-tooth profile was used in arriving at strength ratings.

DYNAMIC-LOAD METHOD. In May 1926, Professor Earle Buckingham presented a spur-gear theory to the AGMA in which the load-carrying abilities of gear teeth were related to maximum specific compressive stresses from the Hertz equation

$$S^2 = \frac{0.35W\left(\dfrac{1}{R_1} + \dfrac{1}{R_2}\right)}{L\left(\dfrac{1}{E_1} + \dfrac{1}{E_2}\right)}$$

where W = load; R_1 and R_2 = radius of cylinder; E_1 and E_2 = modulus of elasticity of material; and L = length of cylinder. Also included was a method of evaluating the dynamic-loading condition based on errors of profile and pitch-line *velocity*.

The dynamic load is the maximum momentary load set up by the operating or dynamic conditions:

$$W_d = \frac{0.05V(FC + W)}{(0.05V + \sqrt{FC + W})} + W$$

where W_d = total dynamic tooth load, pounds; F = width of face, inches; and C = deformation constant from Table 17.

Table 17. Values of C *

Materials	Tooth Form	Error in Action of Gears, in. 0.0005	0.001	0.002	0.003	0.004	0.005
Cast iron and cast iron	14 1/2°	400	800	1600	2400	3200	4000
Cast iron and steel	14 1/2°	550	1100	2200	3300	4400	5500
Steel and steel	14 1/2°	800	1600	3200	4800	6400	8000
Cast iron and cast iron	20° full depth *	415	830	1660	2490	3320	4150
Cast iron and steel	20° full depth	570	1140	2280	3420	4560	5700
Steel and steel	20° full depth	830	1660	3320	4980	6640	8300
Cast iron and cast iron	20° stub	430	860	1720	2580	3440	4300
Cast iron and steel	20° stub	590	1180	2360	3540	4720	5900
Steel and steel	20° stub	860	1720	3440	5160	6880	8600

* The values given for cast iron also are used for bronze, since the modulus of elasticity is practically the same.

In order to use the foregoing formula, it is necessary to have some measure of the error in action of the gears. From tests of gears made in many plants, it appears that this error on well-cut industrial gears ranges from about 0.002 in. on gears of 6-diametral pitch and finer, to about 0.004 in. on gears of 2-diametral pitch. On gears cut with particular care, these errors are reduced by about one-half the above. On ground gears made with extreme care, and on cut gears where every effort is made to secure accuracy, these errors are reduced to even smaller amounts. The foregoing conditions of error are tabulated in Table 18 and classified as Class 1, Class 2, and Class 3, respectively.

Table 18. Maximum Error in Action between Gears

Diametral Pitch	Class 1	Class 2	Class 3	Diametral Pitch	Class 1	Class 2	Class 3
1	.0048 in.	.0024 in.	.0012 in.	4	.0026 in.	.0013 in.	.0007 in.
2	.0040 in.	.0020 in.	.0010 in.	5	.0022 in.	.0011 in.	.0006 in.
3	.0032 in.	.0016 in.	.0008 in.	6 and finer	.0020 in.	.0010 in.	.0005 in.

As a matter of fact, noise of operation is usually a very good test of the accuracy of gears. For example, gears to run at pitch-line velocities of 5000 ft per min and over cannot have errors much greater than about 0.0005 in., else the noise of operation will render them unsatisfactory. Table 19 gives some measure of the accuracy required at the different velocities. This table is a guide for the selection of the proper class of gear for specific speed conditions. For extreme quietness of operation, a higher order of accuracy than shown is necessary, even at the lower speeds. The values given should keep the noise of operation and the dynamic load within reasonable limits.

Table 19. Maximum Permissible Error in Action between Gears

Velocity, ft per min	Error, in.	Velocity, ft per min	Error, in.	Velocity, ft per min	Error, in.
250	.0037	1750	.0017	3250	.0009
500	.0032	2000	.0015	3500	.0008
750	.0028	2250	.0014	4000	.0007
1000	.0024	2500	.0012	4500	.0006
1250	.0021	2750	.0011	5000	.0005
1500	.0019	3000	.0010	Over 5000	.0005

Helical Gearing. As with all other types of gears, the load-carrying ability of helical gears may be limited by the heat of operation, the beam strength of the teeth, or the surface endurance limits of the materials. The dynamic load is the maximum momentary load set up by the operating or dynamic conditions. Where W = transmitted load at pitch line, pounds; V = pitch-line velocity, feet per minute; W_d = dynamic tooth load, pounds; F = width of active face, inches; C = deformation constant from Table 17; and α = helix angle of gears,

$$W_d = \frac{0.05V(W + FC\cos^2\alpha)\cos\alpha}{(0.05V + \sqrt{W + FC\cos^2\alpha})} + W$$

Helical and herringbone gears often are made with much wider faces than those commonly used on spur gears. With such wide-faced gears, the load across the face is not uniform because of the torsional deformation. The following approximation gives some measure of the intensity of this load concentration at one end of the face. When W_m = maximum unit dynamic load, pounds per inch of face, and D = pitch diameter of pinion, inches,

$$W_m = \frac{W_d}{F}\left(1 + \frac{F}{4D}\right)$$

For herringbone gears, the value of F in the above equation should be taken as the width of face of either half of the herringbone pinion because of the centering influence of the herringbone design. If the torsional deflection were uniform across the face, the load would vary uniformly across the face, the load being greatest where the torsional deflection is least. In such a case, the average unit load would exist at the middle of the face, with a maximum and minimum unit load at opposite ends, each varying the same amount from the average. This probably is a condition that never exists because the varying load will introduce a varying torsional deflection angle. For the sake of simplicity, the foregoing approximation is based on uniform torsional deflection across the face.

If W_w = limiting tooth load for wear, pounds (not to be less than the dynamic load for continuous service without appreciable wear); D = pitch diameter of pinion, inches; n = number of teeth in pinion; N = number of teeth in gear; α = pressure angle of gears; F = face width of gears, inches; Q = ratio factor; K = load-stress factor; E_1, E_2 = moduli of elasticity of materials; S_w = surface endurance limit of material, pounds per square inch,

$$Q = \frac{2N}{N + n} \quad \text{(for spur gears)}$$

$$Q = \frac{2N}{N - n} \quad \text{(for internal gears)}$$

$$K = \frac{S_w^2 \sin\alpha}{1.40}\left(\frac{1}{E_1} + \frac{1}{E_2}\right)$$

the Hertz equation becomes

$$W_w = DFKQ$$

The value of S_w for steel appears to vary quite consistently with its Brinell hardness numbers. Tests give, for steel, the values in Table 20.

Table 20. Values of S_w for Steel

Brinell Hardness No.	S_w	Brinell Hardness No.	S_w	Brinell Hardness No.	S_w	Brinell Hardness No.	S_w
150	50,000	300	110,000	450	170,000	550	210,000
200	70,000	350	130,000	500	190,000	600	230,000
250	90,000	400	150,000				

Table 21 gives values of K for various combinations of materials for 14 1/2-degree and 20-degree gears, taking into consideration the cold working received in operation. Some of the high-test semisteels and other special alloys of cast iron have a greater modulus of elasticity than cast iron, and the value of K for such materials must be calculated directly from their own specific properties.

Limit Load for Wear of Helical-gear Teeth. Let D = pitch diameter of pinion, inches; F = width of active face, inches; K = load-stress factor (same as for spur gears, Table 21); Q = ratio factor (same as for spur gears); α = helix angle; W_w = limiting load for wear, pounds. Then

$$W_w = \frac{DFKQ}{\cos^2\alpha}$$

The limiting load for wear should not be less than the dynamic load for continuous service without appreciable wear. The types of physical wear on helical gears are identical to those on spur gears.

Table 21. Values of *K*

Pinion Material	Pinion Brinell Hardness No.	Gear Material	Gear Brinell Hardness No.	S_w	K, 14 1/2°	K, 20°
Steel	150	Steel	150	50,000	30	41
Steel	200	Steel	150	60,000	43	58
Steel	250	Steel	150	70,000	58	79
Steel	200	Steel	200	70,000	58	79
Steel	250	Steel	200	80,000	76	103
Steel	300	Steel	200	90,000	96	13[illegible]
Steel	250	Steel	250	90,000	96	13[illegible]
Steel	300	Steel	250	100,000	119	162
Steel	350	Steel	250	110,000	144	196
Steel	300	Steel	300	110,000	144	196
Steel	350	Steel	300	120,000	171	233
Steel	400	Steel	300	125,000	186	254
Steel	350	Steel	350	130,000	201	275
Steel	400	Steel	350	140,000	233	318
Steel	450	Steel	350	145,000	250	342
Steel	400	Steel	400	150,000	268	366
Steel	500	Steel	400	155,000	286	391
Steel	600	Steel	400	160,000	305	417
Steel	450	Steel	450	170,000	344	470
Steel	600	Steel	450	175,000	364	497
Steel	500	Steel	500	190,000	430	588
Steel	600	Steel	600	230,000	630	861
Steel	150	Cast iron	...	50,000	44	60
Steel	200	Cast iron	...	70,000	87	119
Steel	250 and higher	Cast iron	...	90,000	144	196
Steel	150	Ph. bronze	...	50,000	46	62
Steel	200	Ph. bronze	...	70,000	87	119
Steel	250 and higher	Ph. bronze	...	85,000	135	204
Cast iron	...	Cast iron	...	90,000	193	284

In papers presented by Heymans and Kimball (Refs. 23 and 24) and by Timoshenko and Baud (Ref. 25), data were given from studies of gear-tooth stresses, using the photoelastic method of analysis. The results of studies clearly indicated the influence of the size of the fillet at the root of a gear tooth on the magnitude of the maximum stress developed in the tooth by the applied load. In evaluating stress concentration factors K, the ratio of the fillet radius r to the thickness of the tooth t_b at the base above the fillet is used. Figure 41 represents this relationship as currently used, based on photoelastic values arrived at in studies by Frocht.

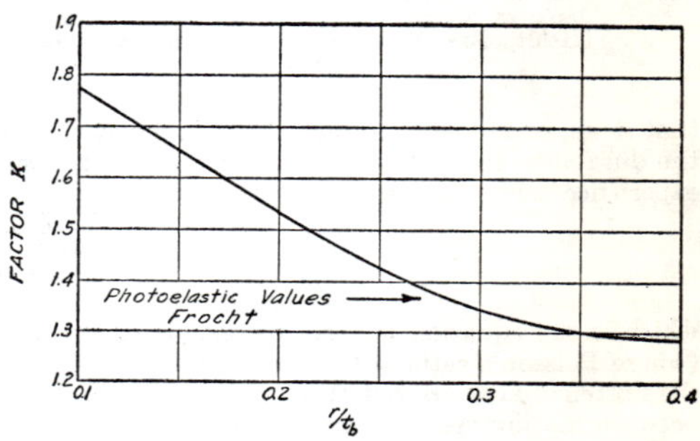

Fig. 41. Stress concentration factors for root fillet. (AGMA)

A paper by Frehse (Ref. 26) on automobile transmissions explores the method of computing stresses in helical gear teeth in which the effect of helix overlap on the contact conditions is recognized. J. O. Almen and Straub (Refs. 27 and 28) have given the gear industry relationships between computed stresses in automotive gearing and fatigue life (Fig. 42), data that have advanced the art materially.

Heldt (Ref. 4, p. 108) summarizes beam strength calculations in a similar manner, determining first the tooth carrying the maximum load (see Fig. 43) and then applying the beam strength formula as follows:

$$S = \frac{Fq}{wpz}$$

where S = stress in the weakest section; q = stress concentration factor for root fillet; w = face width, inches; p = circular pitch, inches; z = a tooth-form factor = $x/1.5p$ (Fig. 44); F = tangential tooth load at pitch line.

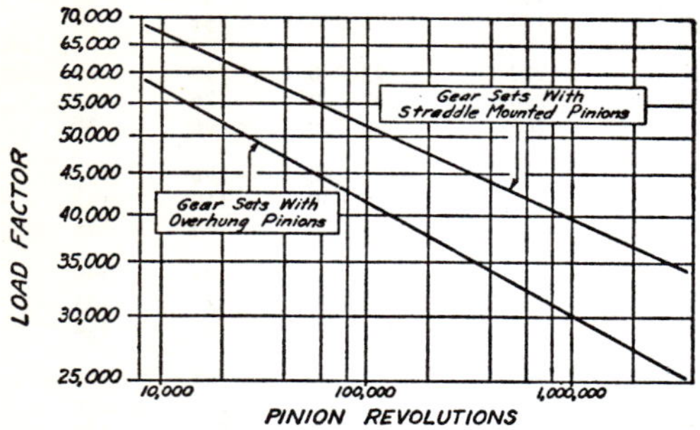

FIG. 42. Computed stress versus fatigue life for carburized spiral bevel gears. (Almen and Straub, *Automotive Industries*, Refs. 27, 28)

WEAR RATING METHODS. As speeds of gearing have increased, and methods of producing gear tooth forms have improved, it has become more and more apparent that more gears fail because of wear or surface destruction than by breakage. A review of established practice reveals many industries in which satisfactory gear performance was correlated to a load per inch of face per inch of diameter without reference to tooth proportion. Obviously, the gear teeth were checked for strength, but the loading factor widely known as the K factor was imposed as an overall limit.

Among the experimenters who have worked to establish optimum values for wear is S. Way of the Westinghouse Research Laboratories, who has made studies of the so-called pitting limit of gear teeth. In Ref. 29 and in an article appearing in the April 1936 issue of *The Electric Journal*, it was shown that the subsurface shear theory, using the Hertz method of computing stresses, resulted in relating maximum allowable compressive stresses to Brinell hardness. The exact relationship depends greatly on surface finish so

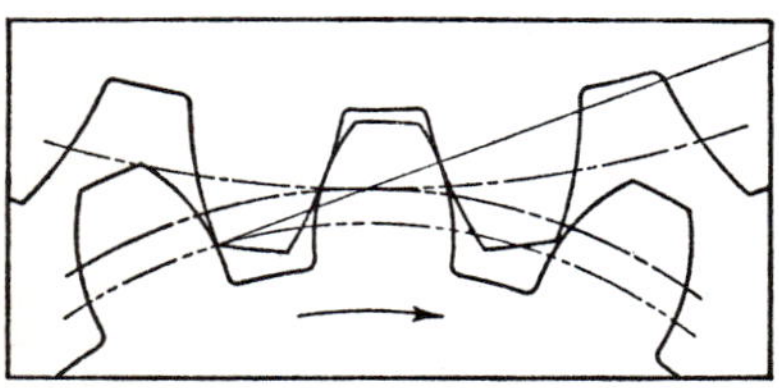

FIG. 43. Diagram showing highest point on flank at which one tooth carries the load. (Heldt, *Torque Converters and Transmissions*, Ref. 4)

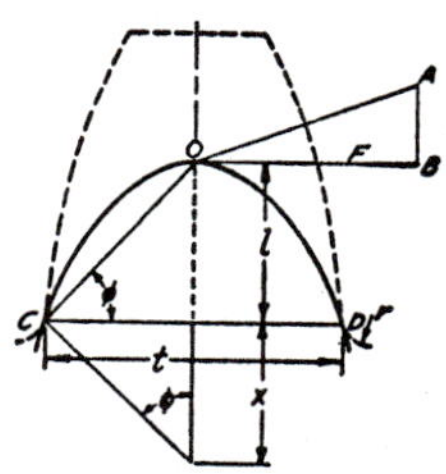

FIG. 44. Tooth-form factor z equals $x/1.5p$. (Heldt, *op. cit.*)

that it appears unwise to include an index at this time. In the AGMA rating practices, the durability ratings for wear are based on an assumed endurance limit curve related to experience using the formula

$$S_s = 958\sqrt{\frac{P(R_1 + R_2)}{2R_1R_2}}$$

which is the equation for the magnitude of the maximum shear stress in steel cylinders (where Poisson's ratio = 0.25) occurring at a distance equal to 0.393 times the width of the flattened area arrived at in the Hertz formula. Arbitrary combinations of hardnesses between mating members are used to provide for the difference in number of stress cycles in the two members.

F. R. McFarland (Ref. 30), has outlined analyses made on aircraft engine gears in which surface stresses computed by the Hertz method and sliding velocities were correlated (Fig. 45) on Almen's PV basis to explain operating performance. In these studies

$$P = \sqrt{\frac{0.35 \times \text{tooth load} \times \left(\frac{1}{R_1} + \frac{1}{R_2}\right)}{\text{Face width} \times \left(\frac{1}{E_1} + \frac{1}{E_2}\right)}}$$

V = instantaneous tangential velocities (see Fig. 46)

$$PV = P \times (V_2 - V_1) = \frac{\text{lb}}{\text{in.}^2} \times \frac{\text{ft}}{\text{sec}}$$

Heldt (Ref. 4), dealing principally with automotive transmission gearing, applies the compressive stress method of evaluating load-carrying abilities of surfaces of gear teeth

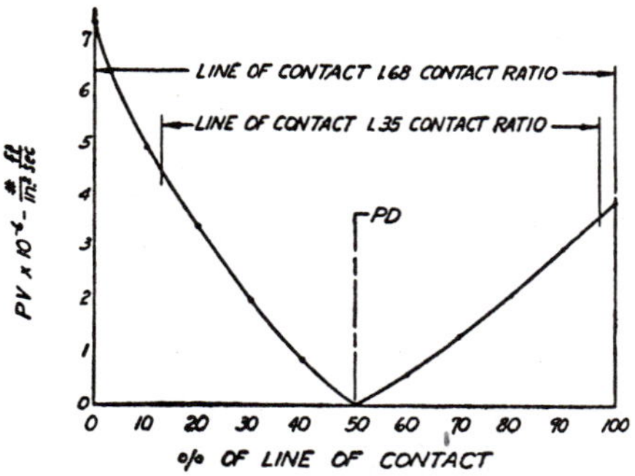

FIG. 45. Typical PV chart for gear teeth. (McFarland, *SAE Journal*, Ref. 30)

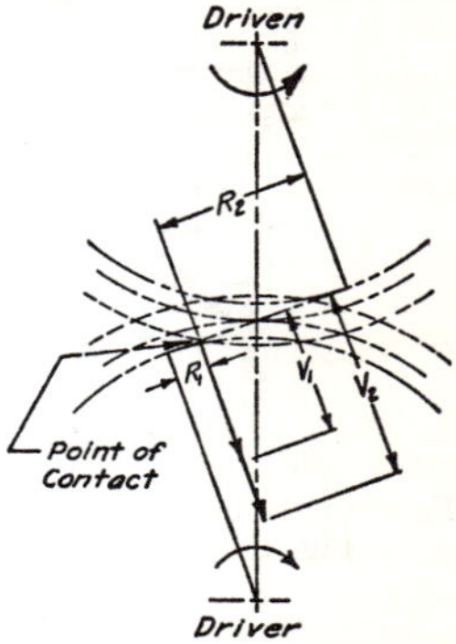

FIG. 46. Instantaneous sliding velocities at point of contact. (McFarland, *op. cit.*)

and comments on the relationship between surface loading and rubbing speed. His formula for gear teeth is:

$$S_c = \sqrt{\frac{PE(p+1)}{3.5fR \sin 2\phi}}$$

where P = tangential load at pitch line, E = modulus, p = gear ratio, R = pitch radius of larger gear, and ϕ = pressure angle.

Values of S_c used in automotive practice in which allowance is made for dynamic loading are:

Case-hardened 3 1/2 % nickel steel	200,000 psi
Oil-hardened SAE 2350 and 3250	160,000 psi
Case-hardened SAE 4615 and 2315	200,000 psi
Case-hardened SAE 4815 and 2512	220,000 psi

Where the service conditions require operating under continuous loads, it is recommended that the limiting compressive stress values be reduced approximately 25%.

AGMA PROCEDURES—Strength Rating. The AGMA has adopted standard rating practices for various types of gears. Methods are given for beam-strength calculations and surface durability that have been related to the accumulated experience of its members. These data are:

1. "Strength of Spur Gear Teeth"—AGMA Standard 220.01.

$$P = \frac{sF_iY_kD_G}{P_d}$$

2. "Strength of Helical and Herringbone Gear Teeth"—AGMA Standard 221.01.

$$P = SY_{hk}FZM_nD_m$$

3. "Beam Strength of Straight Bevel and Spiral Bevel Gear Teeth"—AGMA Standard 222.01.

$$P = \frac{snD_pFYC_v}{126{,}000C_BP_d}$$

In the foregoing strength-rating practices, the symbols used refer to the following:

P = peak horsepower rating for strength
s = allowable bending stress (see Table 22)
F_i = face-width factor (see chart, Fig. 47)

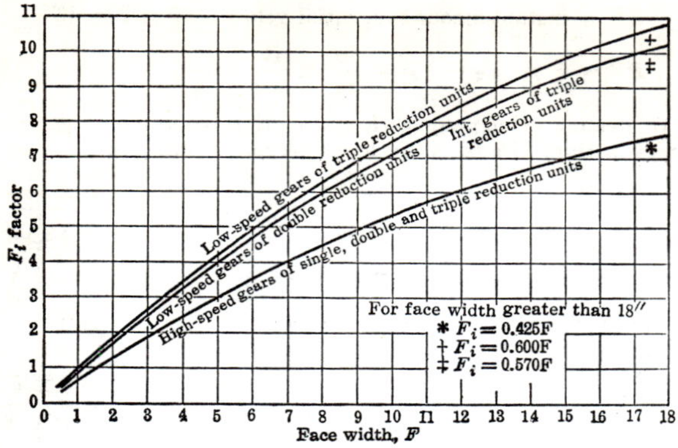

FIG. 47. Chart for face-width factor, F_i. (AGMA)

Y_k = tooth-form factor for spur gearing calculated from the following formula and Fig. 48:

$$Y_k = \frac{1}{K_c \dfrac{\cos \phi_L}{\cos \phi}\left(\dfrac{1.5}{X} - \dfrac{\tan \phi_L}{t}\right)}$$

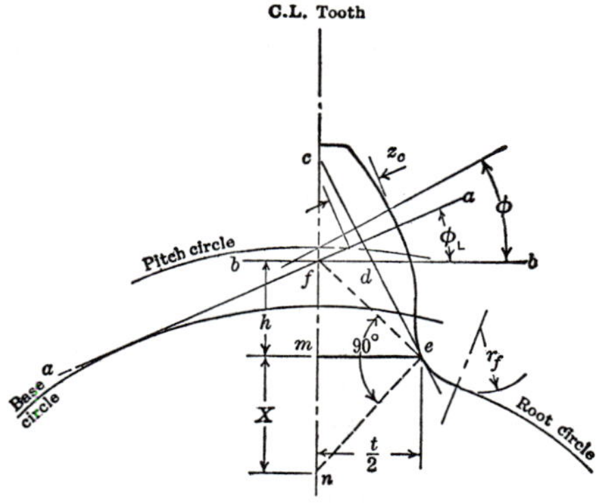

FIG. 48. Tooth-form factor Y_k determination. (AGMA)

K_c = fillet stress concentration factor, Fig. 41
D_g = diameter, velocity, and speed factor from this formula:

$$D_g = \frac{D_p C_{vs} n}{126{,}000}, \text{ where}$$

$$C_{vs} = 1 - \frac{\sqrt{v}}{84}$$

n = number of rpm—pinion

Y_{hk} = tooth-form factor for helical and herringbone gearing taken in the normal plane with $Y = \dfrac{2X}{3}$ (see Fig. 49) and k, the stress-concentration factor from curve (Fig. 41)

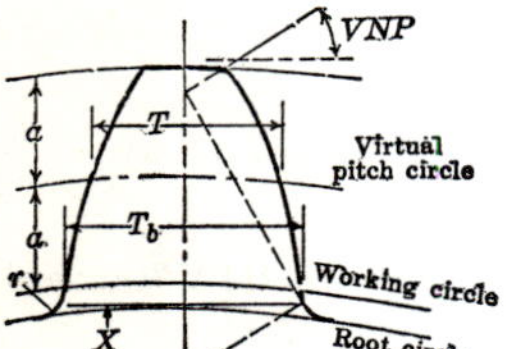

FIG. 49. Tooth-form factor Y_{hk} determination. (AGMA)

Z = length of the line of action

M_h = angle factor = $\dfrac{\cos \psi}{\cos \phi}$

D_m = diameter, velocity, and rpm factor

$$D_m = \frac{D_p C_v n}{395{,}000}, \text{ where } C_v = \frac{78}{78 + \sqrt{v}}$$

C_B = bevel factor = $\dfrac{A}{A - 0.5F}$, where A = outside cone distance in inches

Table 22. Allowable Bending Stresses AGMA Strength Horsepower Formulas

Cast iron	14,000 psi
160 Bhn steel	40,000 psi
210 Bhn steel	50,000 psi
245 Bhn steel	60,000 psi
270 Bhn steel	65,000 psi
315 Bhn steel	70,000 psi
335 Bhn steel	77,000 psi
360 Bhn steel	83,000 psi
440 Bhn steel and higher	90,000 psi

Wear Rating. For surface durability (wear rating), the AGMA data available are:

1. "Surface Durability of Spur Gears"—AGMA Standard 210.01.

$$P = F_i K_r D_o C_f$$

for 20° stub or 20° full-depth teeth. Where 14 1/2° full-depth teeth, a 0.75 multiplier is used.

2. "Surface Durability of Helical and Herringbone Gears"—AGMA Standard 211.01.

$$P = F_i K_r D_s$$

3. "Surface Durability of Straight Bevel and Spiral Bevel Gears"—AGMA Standard 212.01.

$$P = C_{mb} D_b F$$

In the foregoing formulas, the symbols used refer to the following:

P = allowable horsepower rating for service duty involving uniform loads and not over eight hours a day

F_i = face-width factor for spur, helical, and herringbone gears (see Fig. 47)

K_c = combined factor for spur, helical, and herringbone gearing covering material tooth-contact relationship and ratio factors (see charts, Figs. 50 and 51)

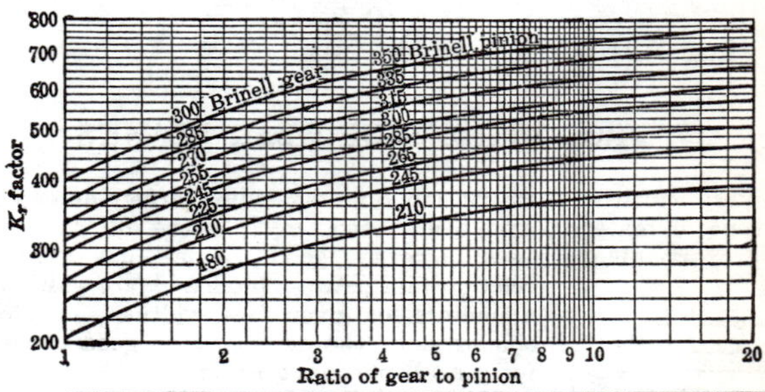

	Case-hardened and through hardened steel						Surface hardened	Cast iron	Bronze
Brinell gear	575	500	350	335	315	225	440	200	40,000 psi
Brinell pinion	575	500	450	380	360	450	440	210	180
K_r	$1530\,C_r^{0.9}$	$1350\,C_r^{0.9}$	$1060\,C_r^{0.9}$	$973\,C_r^{0.95}$	$870\,C_r^{0.95}$	$608\,C_r^{0.95}$	$890\,C_r^{0.9}$	$344\,C_r^{1.0}$	$274\,C_r^{1.0}$

$$\text{Gear ratio} = \frac{\text{Teeth in gear}}{\text{Teeth in pinion}} \qquad C_r = \frac{\text{Gear ratio}}{\text{Gear ratio} + 1}$$

FIG. 50. Factor K_r for external gearing. (AGMA)

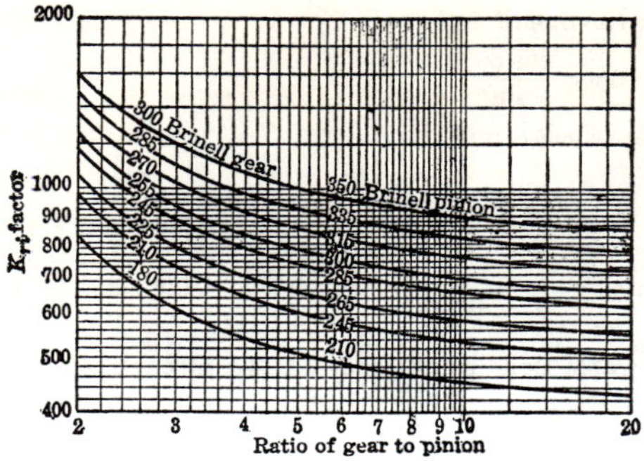

	Case-hardened and through hardened steel						Surface hardened	Cast iron	Bronze
Brinell gear	575	500	350	335	315	225	440	200	40,000 psi
Brinell pinion	575	500	450	380	360	450	440	210	180
K_{ri}	1530 C_{ri} 0.9	1350 C_{ri} 0.9	1060 C_{ri} 0.9	973 C_{ri} 0.95	870 C_{ri} 0.95	608 C_{ri} 0.95	890 C_{ri} 0.9	344 C_{ri} 1.0	274 C_{ri} 1.0

$$\text{Gear ratio} = \frac{\text{Teeth in gear}}{\text{Teeth in pinion}} \qquad C_{ri} = \frac{\text{Gear ratio}}{\text{Gear ratio} - 1}$$

FIG. 51. Factor K_{ri} for internal gearing. (AGMA)

D_o = diameter and speed factor for spur gearing from the formula $D_o = \dfrac{D_p^2 C_{vs} n}{158{,}000}$

C_f = modification factor for increased stress in spur gearing at the point where single tooth contact occurs. Tables for this value are supplied in AGMA data; however,

$C_f = \dfrac{\rho_G \rho_p}{r_G r_p}$, where

ρ = radius of curvature at start of single tooth action and r = radius of curvature at the pitch point

D_s = diameter and speed factor for helical and herringbone gearing from

$$D_s = \frac{D_p^2 C_v n}{126{,}000}, \text{ where } C_v = \frac{78}{78 + \sqrt{v}}$$

C_{mp} = materials factor for bevel gearing (see Table 15).

D_b = diameter and speed factor for bevel gearing, from

$$D_b = \sqrt{\frac{D_p D_{pm} C_v}{233}}$$

7. PROPORTIONS OF GEAR BLANKS

Gear blanks vary widely in their construction, depending largely on end use, mounting arrangement, type of material, and overall economy. There are no "standard rules" governing the proportioning of gears; however, "good practice" has established desirable relationships between pitch and face width, also maximum diameters for given face widths. The following notes should guide the designer into good overall proportions for general-purpose gearing.

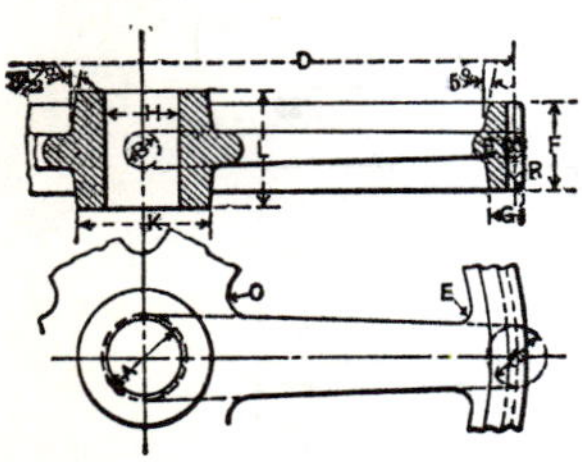

FIG. 52. Machine tool gear blank proportions.

MACHINE TOOL GEARING (Brown & Sharpe Manufacturing Co., Providence). A regular system of proportioning gears must be followed, if the gears are to have good proportions, sufficient strength without excessive weight, and uniform appearance when the gears are used in sets. The following formulas and Table 23 refer to Fig. 52, and are for gears of 4 in. diameter, 20 diametral pitch, to 72 in. diameter, 1 1/2 diametral pitch. D = pitch diameter, P = diametral pitch, d = dedendum, f = addendum, and t = thickness of tooth at pitch line. The significance of the other factors in the formulas is shown in Fig. 52.

Table 23. Formulas for Proportions of Spur Gears

Proportions of Arm	Proportions of Rim
$A = 0.138D\,[(0.95/P) + 0.28] + 0.67$	$*F = (8/P) + 0.25$
$a = 2/3\ A$	$E = 1/4\ A$
$B = 0.4A$	$G = (d + f) + t + 0.06$
$b = 0.4a$ or $2/3\ B$	R = 1/32 in. (to 20 pitch incl.)
$C = 1/4\ A$	3/64 in. (to 14 pitch incl.)
	1/16 in. (to 8 pitch incl.)
Proportions of Hub *	3/32 in. (to 4 pitch incl.)
$K = 1.5H + 1/4$ in.† $L = 1.5H$	1/8 in. (above 4 pitch)

Values of $(0.95/P) + 0.28$ for Finding A

Diametral pitch	1	1 1/2	2	2 1/2	3	4	5	6	7	8	9
$(0.95/P) + 0.28$	1.23	0.915	0.755	0.660	0.597	0.520	0.470	0.44	0.42	0.40	0.39
Diametral pitch	10	11	12	13	14	15	16	17	18	19	20
$(0.95/P) + 0.28$	0.38	0.37	0.36	0.354	0.352	0.345	0.340	0.337	0.334	0.332	0.33

* Suitable for general practice. For light change gears, F should be less than given by formula. The hole H should vary to suit conditions, but can usually be made to agree with one of the following formulas:

$$H = \frac{4}{P} + 0.5; \qquad = \frac{5}{P} + 0.5; \qquad = \frac{6}{P} + 0.5$$

† Where heavy service is required, or large set screws or keyways are to be used, make $K = 1.625H + 1/4$ or reinforce hub over keyway or around set screw.

TO DETERMINE WHETHER GEARS SHALL BE SOLID, WEB, OR ARM. Gears to be used together should be consistent as to construction. Table 24 shows the maximum diameter that should be used in each style of gear. Diameters greater than those shown in the "web" column should be made with six arms, of the proportions given by the formula above.

Table 24. Maximum Permissible Diameter of Solid and Web Gears

Diametral Pitch	Solid Gears		Web Gears		Diametral Pitch	Solid Gears		Web Gears	
	Change, in.	Combination, in.	Change, in.	Combination, in.		Change, in.	Combination, in.	Change, in.	Combination, in.
4	5.50	6.00	9.00	9.50	10	3.80	3.80	5.80	6.00
5	4.80	5.00	7.50	8.00	12	3.67	3.67	5.67	5.75
6	4.50	4.75	7.30	7.75	14	3.29	3.29	5.30	5.50
7	4.30	4.57	6.86	7.00	16	3.13	....	4.15	....
8	4.25	4.25	6.75	6.88	18	3.00	....	4.55	....
9	4.00	4.00	6.67	6.75	20	2.80	....	4.40	....

For ordinary service, "combination" gears should be used, the "change" gears being adapted only for light service.

ARM STYLES AND PROPORTIONS FOR GENERAL-PURPOSE GEARING. Suitable proportions of sections for gear wheel arms are shown in Fig. 53 (see also Ref. 5). It is also suggested that the section modulus of the arm at the center of the hub be computed for the formula $I/c = WR/nS$, where W = load on the gear; R = length of arm, or radius of the gear, inches; S = allowable fiber stress of the material; n = number of arms in the gear; I = moment of inertia of section; and c = distance of extreme fiber from axis. This formula for the elliptical section gives

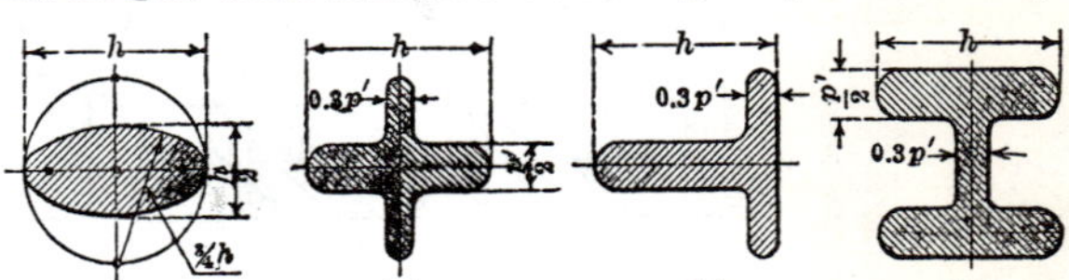

Fig. 53. General-purpose gear blank arms.

$$\frac{I}{c} = h\sqrt[3]{\frac{20WR}{nS}}$$

The dimensions of the arm at the pitch line can be approximately 0.7 of those at the center

The number of arms suggested are: for gears up to 16 to 20 in. diameter, 4 or 5 arms; for gears 20 to 60 in. diameter, 6 arms; for gears 60 to 96 in. diameter, 8 arms; for gears over 96 in. diameter, 10 or 12 arms. For small gears, web centers of thickness approximately one-half the circular pitch are recommended.

H-ARMS FOR HEAVY-DUTY GEARING. The axial thrust loads in helical gearing and the heavier loads imposed by heavy-duty service have increased the popularity of the H-arm to the point where additional guidance on its proportions is outlined herewith.

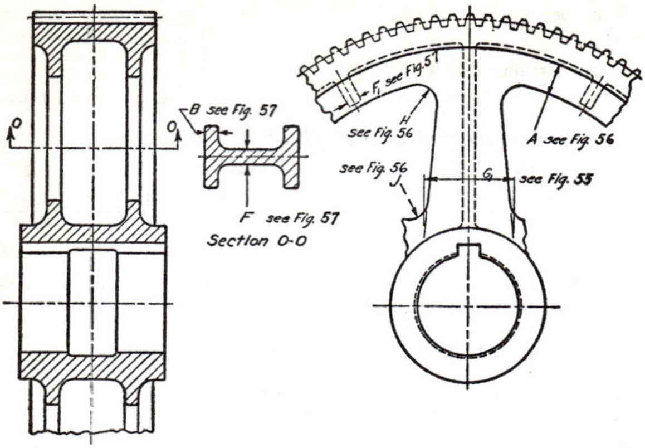

Fig. 54. Typical H-arm blank for heavy-duty gearing.

Figure 54 shows a typical H-arm gear blank design in which the proportions are based on the following assumptions:

Maximum Face Widths

6, 5 and 4 DP	10 in.
3 1/2 and 3 DP	12 in.
2 1/2 and 2 DP	15 in.
1 3/4 and 1 1/2 DP	18 in.
1 1/4 and 1 DP	21 in.

Number of Arms

Up to 30 in. diameter	4 arms
31 to 45 in. diameter	5 arms
46 to 120 in. diameter	6 arms

Hub diameter in relation to the bore varying in relation to bore size and pitch of teeth (index of capacity) from 1.7 to 1 for heavy pitches, and 1.5 to 1 for lighter pitches.

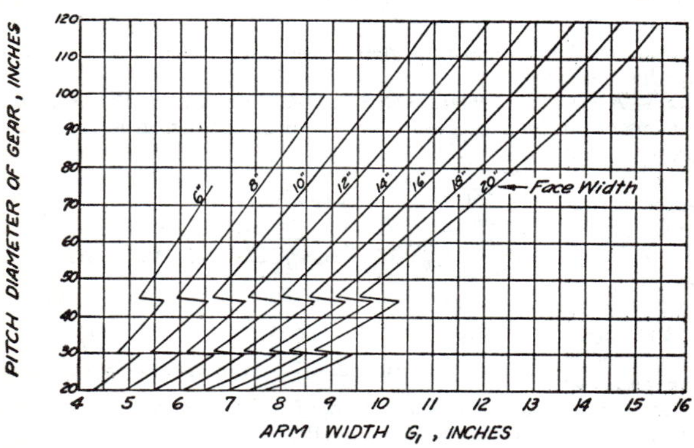

Fig. 55. H-arm width proportions for heavy-duty gearing. See dimension G_1, Fig. 54.

Hubs of blanks counterbored wherever possible for approximately one-third the hub length and a depth of 1/32 in. for bores up to 12 in., and 1/16 in. above. As a guide, counter-

bores should be provided where the hub length is greater than two times bore diameter up to 5 in. bore size; or 1 $^{1}/_{2}$ times bore diameter in bore sizes 5 in. to 14 in.; and in all bore sizes above 14 in. diameter.

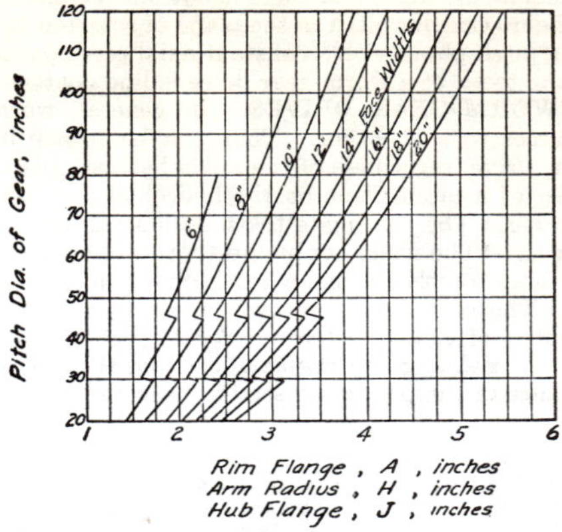

Fig. 56. Rim flange depth, and radii, for heavy-duty H-arm gearing. See Fig. 54.

Arm width proportions—Fig. 55
Flange depth—Fig. 56
Web thickness—Fig. 57
Rim thickness under the tooth equal to the whole depth of full-depth teeth of the pitch involved.

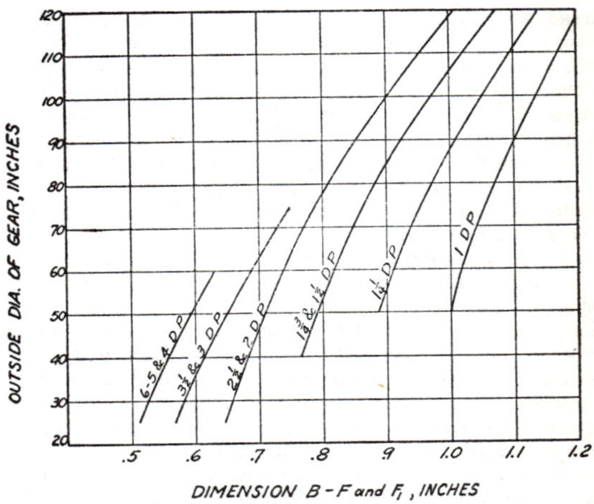

Fig. 57. Web thicknesses for heavy-duty H-arm gearing. See Fig. 54.

8. WORM GEARING

A common conception of a worm-gear drive is a screw thread driving a circular nut. A better mental image is obtained by picturing such a drive as the development of a spiral-gear drive where one member has been made to envelop the other in order to obtain line contact instead of point contact. Either member may be made to envelop the other. For expediency, the more common practice has been to make the larger member envelop the smaller. Until recently, the use of a smaller member enveloping the larger one has been exceptional, and such a construction has been erroneously called an hourglass or globoid worm drive. A worm-gear drive has one member with a uniform helical lead of tooth along its axis, while the mating member is made to envelop it. The enveloping member retains its cylindrical pitch surface, but loses its pitch plane with the loss of its uniform

helical lead of tooth along its axis. The helicoidal member retains its pitch plane but loses its pitch cylinder because the mating pitch plane has been eliminated. Thus the pitch surfaces of a worm drive consist of a pitch cylinder on the enveloping member and a pitch plane on the helicoidal member, and there is but one common circular pitch for both members. There can be no circular pitch in the plane of rotation of the helicoidal member unless there is a mating pitch plane with a constant axial pitch on the contacting member. The contact between the teeth of a worm-gear drive is line contact.

CONVENTIONAL WORM-GEAR DRIVES. In general, worm drives are used to drive shafts at 90 degrees to each other. They can be substituted, however, for any crossed-axis helical gear drive, regardless of the angle between the shafts.

Pitch surfaces consist of a pitch plane on the helicoidal member and a pitch cylinder on the enveloping member. The conjugate tooth action can best be studied as a special case of a rack and a gear. The same conjugate tooth action exists when the helicoidal member is moved, without revolving in an axial direction, as exists when the helicoidal member revolves. The rack form, however, is not constant across the face of the enveloping member, but is of changing form. (See Fig. 58.) The amount of change in form depends upon the thread angle in the axial section, the lead of the helicoid and its diameter, and the distance of the particular section from the axial plane.

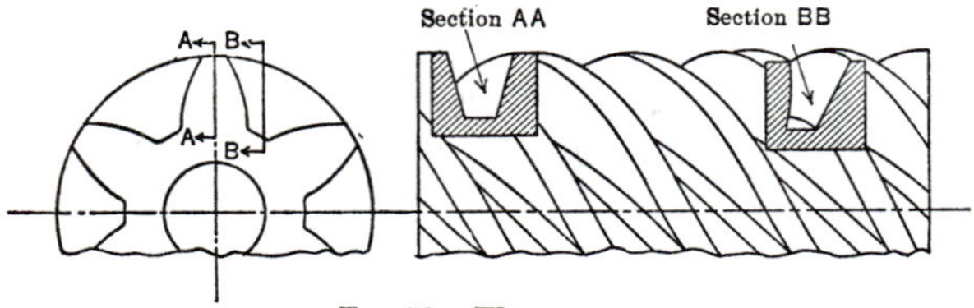

FIG. 58. Worm gear.

Lead and Thread Angles. Because of these conditions, higher thread angles are required for effective tooth action as the lead angle is increased. Narrower faces for the enveloping member are sometimes desirable for the higher lead angles, particularly if the thread angle in the axial section is relatively small. Thus a 14 1/2-degree thread angle should not be used for lead angles greater than about 15 degrees; a 20-degree thread angle should not be used for lead angles greater than about 25 degrees; a 25-degree thread angle should not be used for lead angles greater than about 35 degrees; and a 30-degree thread angle is desirable for lead angles up to 45 degrees.

Profile Variations. Many types of helicoids are used for worms. Where the lead angle is small, the actual forms produced by different methods of cutting the thread are similar, although they are different types of helicoids. As the lead angles increase, however, marked differences become apparent. For example, the axial form of the thread profile produced by a straight-sided lathe tool, tipped to the lead angle, is concave. That produced by a straight-sided milling cutter set to the lead angle generally is convex; although in extreme cases where the diameter of the cutter is small in relation to the depth of the thread and the lead angle is large, the axial form may be of double curvature, convex near the root and concave near the outside or tip of the thread. The larger the diameter of the cutter, the greater the amount of curvature in the axial profile of the thread, because of the side-cutting action of the cutter. If the milling cutter axis is parallel to the axis of the helicoid or worm, a condition that exists when hobbing a thread with an annular cutter, the form produced is that of an involute helicoid which is that of an involute helical gear. This same involute helicoid also is produced when the flat side of a cutter or grinding wheel, set to both the thread angle and lead angle, is used. Figure 59 shows the differences in axial profile produced on a worm with a lead angle of nearly 45 degrees by different tools of the same form of cutting edge. These tools include a straight-sided threading tool set to the lead angle, a 4-in. diameter milling cutter, a 12-in. diameter grinding wheel, and the involute helicoid which would be produced by the flat side of a milling cutter or grinding wheel.

All these forms of helicoids have been commonly used as thread forms for worms. As far as theoretical conditions are concerned, one type does not appear to have any great inherent advantage over another. The major consideration is that the form of helicoid used for the worm should be reproduced on the cutting tool which is to be used to generate the enveloping member. In general, the best form to use is largely a matter of expediency, controlled by the type of equipment available to produce the helicoidal member. Hence the toolmaker must know the method to be used to produce the helicoidal member in order to make correctly the cutting tool which is to be used to generate the enveloping member.

Involute Helicoid. For many practical reasons, the involute helicoid would appear to be the best form to use where the design is not restricted by existing equipment. With this helicoid, the same hob or generating cutter which is used to generate the enveloping member also may be used to generate the worm or helicoidal member. Again, with this helicoid, the helicoidal members would be involute helical gears, and the same hob or other generating cutter could be used to generate all worms of the same normal diametral pitch of all tooth numbers. Under such conditions, the practice and technique already developed in the making of tools for helical and spur gears could be applied directly to

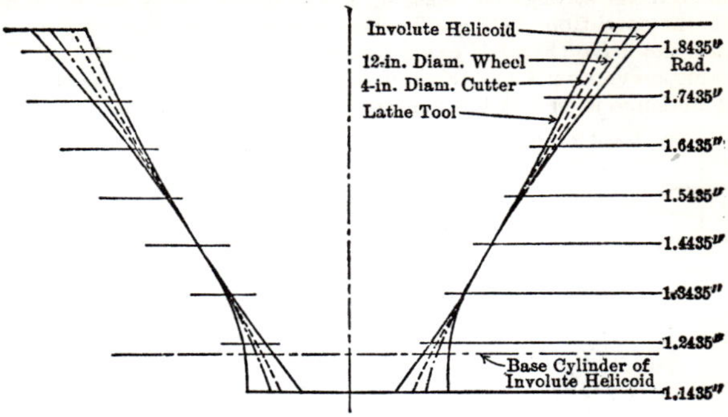

FIG. 59. Axial sections of milled and chased worms.

the making of tools for worms and worm gears. Furthermore, this involute helicoidal form, where ground worms are required, is the simplest one to grind and control; the size of the wheel has no influence on the form produced; and several different grinding methods are available.

In further discussion, the angle of the thread form will refer to the tangent to the thread profile at or near the middle of the working depth, regardless of the type of helicoid involved.

CONTACT ON WORM-GEAR DRIVES. The analysis of worm drives is a very complex operation. In the majority of cases where such analyses have been made, only the conjugate action between parallel sections of the worm, treating them as rack sections meshing with the enveloping member or gear, has been considered. In addition, it is necessary to have some measure of the nature and position of the actual contact lines between mating teeth, duration and field of contact between them, in order to determine

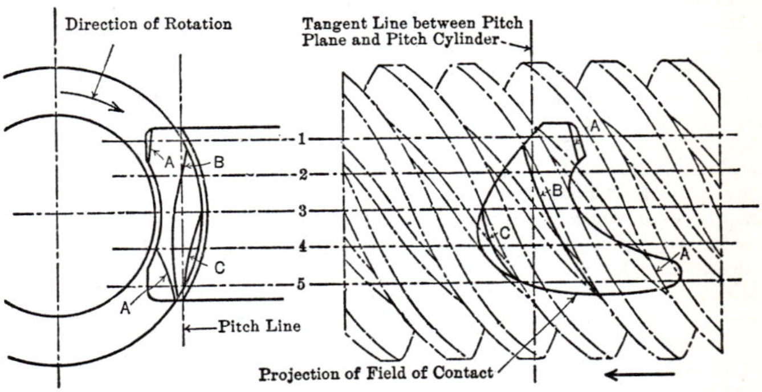

FIG. 60. Contact lines in worm gearing.

even approximately the intensity of the stresses and probable load-carrying capacities of such drives. Fortunately all types of helicoids follow certain general laws so that workable approximations are possible in all except the extremely critical conditions.

Considerable uncertainty exists as to the most effective position of the pitch plane in relation to the tooth on the worm. The conventional design places it at the middle of the working depth of the tooth or thread on the worm. The practice in some plants is to place it nearer the root of the thread. On all types of helicoids, the tendency of the contact lines is to converge toward the pitch line on the leaving side of the worm. (See Fig. 60.)

This shows a six-thread worm with straight sides of thread in an axial section, meshing with a forty-tooth gear. The heavy outline on the plan view of the worm shows the projection of the field of contact. In the relative position of the two members as shown, contact exists on three threads of the worm as shown by heavy lines A, B, and C.

With this same worm, when the pitch line is dropped to the outside diameter of the gear, by simply closing in the center distance the amount that the pitch plane is moved, the contact conditions are as shown in Fig. 61. Here the contact lines tend more to sweep from the bottom to the top of the worm thread. This change in position of the contact lines causes much lower stresses to be set up in the materials by the same applied load, as compared with the conditions in Fig. 60. Because of the relative curvature conditions at the contact lines the drive (Fig. 61) can safely carry about double the loads that the drive (Fig. 60) can carry, even with a portion of the gear face undercut as shown. With lower lead angles, the change is less with a change in the position of the pitch plane.

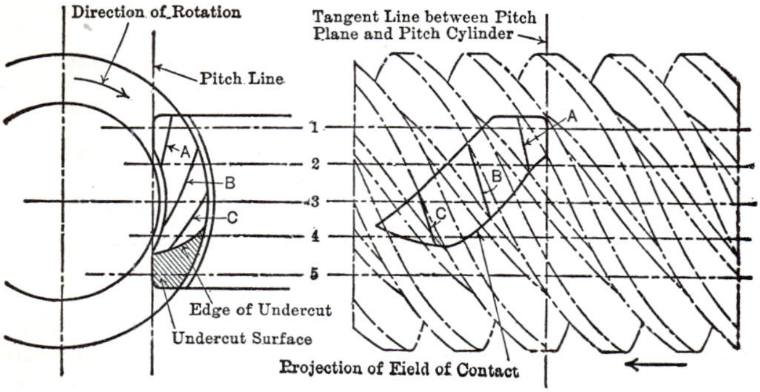

FIG. 61. Contact lines and field of contact with pitch line moved to outside diameter of gear.

In the design of a worm drive, the center distance can be varied a limited amount, and the pitch plane may be placed anywhere between the middle of the working depth of the thread on the worm and the outside diameter of the gear to suit other conditions, such as a prefixed center distance. On the other hand, the lower the pitch plane is placed on the worm thread, the greater will be its load-carrying capacity and the smoother the action.

DESIGN OF WORMS. There is considerable difference of opinion as to whether the form of the thread in the axial section or that in the so-called "normal section" should be specified. As a matter of fact, there is no "normal section" as generally implied on any helicoid. As noted before, the essential information is best specified by stating the shape, type, and size of the tool to be used to finish the worm. The AGMA recommends that the thread form to be regarded as standard will be the form produced by a straight-sided milling cutter having a diameter not less than the outside diameter, nor greater than $1\,^1/_4$ times the outside diameter, of the worm; also, that the following linear pitches be regarded as standard: $^1/_4$, $^5/_{16}$, $^3/_8$, $^1/_2$, $^5/_8$, $^3/_4$, 1, $1\,^1/_4$, $1\,^1/_2$, $1\,^3/_4$, and 2 in.

Since the pitch surface of the worm or helicoidal member is a plane, and since the pitch on such a plane is constant at all distances from the axis, the term "pitch diameter of a worm" is misleading. Considering it as on a screw thread, where the same term is used to define that diameter where the thread thickness is equal to the width of space, or where both distances are equal to one-half the pitch, it might best be called the nominal pitch diameter or, better, the effective diameter.

The AGMA also recommends that the top edge of the worm thread should always be rounded to avoid cutting, and that the amount of this rounding be made equivalent to 0.05 linear pitch. When the pitch plane of the worm is dropped to about its root diameter or lower, and the worm is the driving member, the conjugate tooth action at this edge always is recess action, and under such conditions this rounding is not necessary.

The AGMA recommends the use of a $14\,^1/_2$-degree and a 20-degree thread angle for worms from single to quadruple threads. As noted before, the lead angle and not the number of threads is the major factor that should determine the selection of the thread angle. To these two angles could be added a 25-degree and a 30-degree thread angle to cover all lead angles up to 45 degrees or higher. If a single thread angle is to be used, the 30-degree thread angle can be effectively used on the lower lead angles, and several manufacturers of worm gears have adopted this angle as standard for all worms, with very satisfactory results.

MILLED AND CHASED WORMS. For milled and chased worms, where definite axial pitches are maintained, a choice between two procedures is necessary. (1) If fixed tooth heights are used for the same pitch, regardless of the lead angle, a different cutter will be required for each different lead angle. (2) If the same milling cutter or chasing tool is to be used for all worms of the same axial pitch, regardless of the lead angle, a different tooth height must be calculated for each lead angle. In general, as regards the standardization of gears, greatest benefits and economies will be secured by standardizing the tools used to produce them, together with the reduction of variety as far as possible. To use a single cutter for all worms of a given pitch evidently would be the most economical procedure. In such a case, let p = axial pitch of worm, inches; α = half included angle of cutter; β = lead angle of worm; L = lead of worm, inches; f = width of flat at tip of cutter, inches; c = clearance, inches; D = nominal pitch diameter of worm, inches; a = nominal addendum, inches; W = working depth of thread, inches; H = whole depth of thread, inches; d = outside diameter of worm, inches; and r = root diameter of worm, inches. Then, when $c = a/6$,

$$W = \left(\frac{3}{7 \tan \alpha}\right)(p \cos \beta - 2f)$$

$$a = \frac{W}{2} = \left(\frac{3}{14 \tan \alpha}\right)(p \cos \beta - 2f)$$

$$H = W + c = \frac{13W}{12}$$

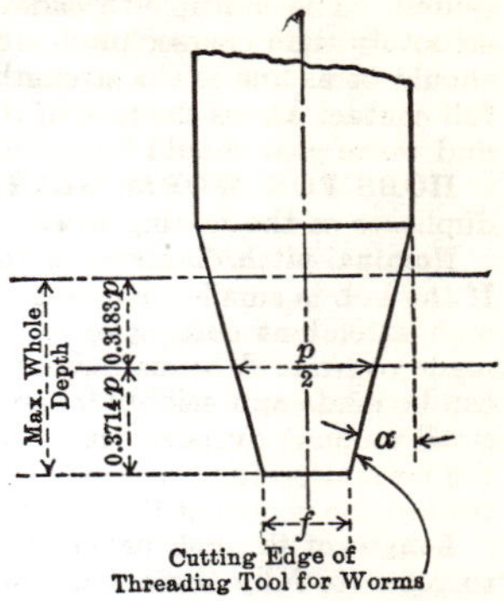

FIG. 62. Forms of threading cutters.

Figure 62 and Table 25 show satisfactory forms for threading cutters of the several thread angles.

Table 25. Dimensions of Threading Cutters (See Fig. 62)

Thread Angle, α	14 1/2°	20°	25°	30°
Tan α	.25862	.36397	.46631	.57735
Maximum whole depth	.6897p	.6897p	.6897p	.6897p
Maximum lead angle, β	15°	25°	35°	45°
Width of flat, f	.3079p	.2297p	.1537p	.0712p
$2f$	.6158p	.4594p	.3074p	.1424p
$3/(14 \tan \alpha)$	.82857	.58875	.45953	.37115

For worms which are to be made integral with the shaft, for lead angles less than 45 degrees, and where a shank hob is used to generate the gear, the formula $D = 2p + 0.5$ may be used. The AGMA recommends for worms to be mounted on the shaft, $D = 2.4p + 1.1$. In both cases, $d = D + W$; $\tan \beta = L/3.1416D$; $r = d - 2H$.

HOBBED WORMS. With the increasing use of multiple-threaded worms, where the number of threads on the worm may be 24 or more, the hobbing of worms is growing in favor. In this case, there is no reason why the standard spur and helical gear hobs should not be used to produce such worms, where lead angles are low. For higher lead angles, involute gear hobs with a pressure angle of 30 degrees would prove effective. The form of the basic rack could be the same as shown for the form of the 30-degree threading tool in Fig. 62. The sizes of such worms would be computed exactly the same as for helical gears, and the generating tool for the worm gear should be made to suit the worm thread form.

DESIGN OF WORM GEARS. The pitch diameter of the worm gear depends on the number of teeth in the gear and the axial pitch of the worm. Let D_g = pitch diameter of worm gear, inches; N = number of teeth in gear; and p = circular pitch (axial pitch of worm), inches. Then $D_g = pN/3.1416$. All other dimensions of the worm gear depend primarily upon the size of the worm and the center distance. When D = nominal pitch diameter of worm, inches; r = root diameter of worm, inches; c = clearance, inches; W = working depth of worm thread, inches; and C = center distance, inches, the dimensions shown in Fig. 63 become

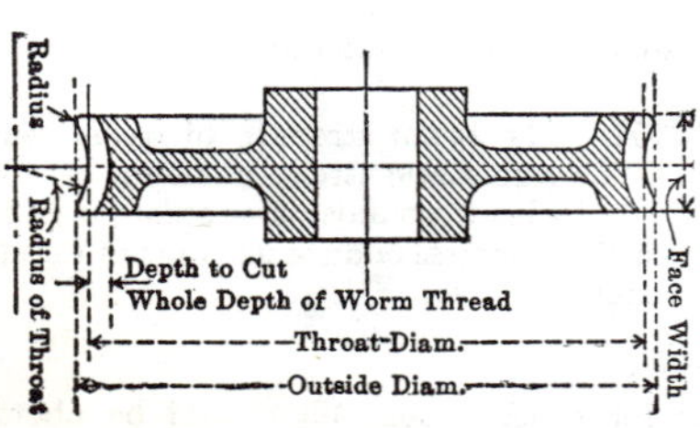

FIG. 63. Worm gear blank design.

$$\text{Throat diameter} = 2C - (r + 2c)$$

$$\text{Throat radius} = \frac{r}{2} + c$$

$$\text{Outside diameter} = \text{throat diameter} + \frac{W}{2}$$

Face width varies from 0.70D to 0.50D, depending on the rubbing speed of the worm. With increasing rubbing speeds, the face diameter should be reduced. The pitch diameter of the worm gear should lie between the diameter equal to $2C - D$ and the outside diameter of the worm gear. As noted before, the lower the pitch plane is placed on the worm thread, the greater is the load-carrying capacity and the smoothness of the action.

In general, the pitches commonly used on worm drives are much coarser than are required. The multiple-threaded and finer pitch worms are more efficient and run more smoothly than coarser pitch worms. To obtain the most effective worm drive, the pitch should be as fine as the strength of the gear teeth permits. For effective tooth action and full contact across the face of the worm gear, the sum of the number of teeth in the worm and worm gear should be not less than forty, approximately.

HOBS FOR WORM GEARS. The hob for a worm gear should be substantially a duplicate of the mating worm, but two precautions must be observed.

Nominal pitch diameter of the hob must never be less than that of the mating worm. If the hob is smaller, contact will be concentrated on the edges of the face of the worm, with attendant cutting or scoring and rapid wear. The amount which the hob can be made oversize depends largely on the lead angle of the worm. With low lead angles, it can be made appreciably larger, but with the high lead angles, it can be made only a very small amount oversize, unless the axis of the hob is to be twisted slightly in relation to the normal position of the axis of the worm. Such twisting will throw the initial contact toward the center of the worm face.

Length of the hob never should be less than the length of the mating worm in order to generate fully the contacting surfaces. This applies particularly when the worms are relatively short. In general, a hob whose length is six or seven times the axial pitch of the worm will be sufficiently long for all purposes. If the hobs are shorter, the worm should be made somewhat shorter than the hob. If a differential feed of the hob across the periphery of the worm gear is used, shorter hobs are satisfactory. Such hobs often are made as tapered hobs, thus combining the roughing and finishing operation. In such cases, the length of the full diameter of the hob plus the distance of differential feed should be slightly greater than the length of the mating worm.

LOAD-CARRYING CAPACITY OF WORM DRIVES. As with all other types of gearing, there are three limiting factors, any one or more of which may limit the capacity of the drive. These factors are heat of operation, beam strength of teeth, and surface endurance limits of materials. With worm drives, heat of operation and the surface endurance limits of the materials are the principal limiting factors.

On worm drives, rubbing velocity and pitch-line velocity must be considered. For simplicity, consider rubbing velocity to be equal to peripheral velocity at the nominal pitch diameter of the worm, although it varies at different points and may be greater or less than that used, depending upon the nature of the contact. Whence,

$$\text{Rubbing velocity in feet per minute} = 0.2618D \times \text{rpm}$$

where D = nominal pitch diameter of worm, inches, and rpm = revolutions per minute of worm.

$$V = \frac{L \times \text{rpm}}{12}$$

where V = pitch line velocity of worm gear, feet per minute, and L = lead of worm, inches.

The actual intensity of the dynamic load on worm-gear teeth is an open question. In the absence of more definite information, the following relation may be used: W = transmitted tooth load, pounds, and W_d = dynamic tooth load, pounds. Then

$$W_d = W \frac{1200 + V}{1200} \quad \text{for metal gears}$$

$$= W \frac{200 + V}{200 + 0.25V} \quad \text{for phenolic laminated material}$$

BEAM STRENGTH OF WORM-GEAR TEETH. The beam strength of worm-gear teeth may be determined by the Lewis equation. The safe static strength should be not less than the dynamic load. Let W_s = safe static tooth load, pounds; p = axial pitch of worm, inches; F = face width of worm gear, inches; S = flexural endurance limit of material, pounds per square inch; and y = tooth-form factor. Then

$$W_s = SpFy$$

For steady loads W_s should be about $1.25W_d$; for shock loads, W_s should be about $1.50W_d$.

With the changing forms and heights of the worm-gear teeth across the face, an exact tooth-form factor is a difficult value to obtain. For all practical purposes, where the sum of the number of teeth in the worm and worm gear is not less than 40, a constant value of y for each thread angle can be used:

Thread angle =	$14\,^1/_2$ degrees	20 degrees	25 degrees	30 degrees
y =	0.100	0.125	0.150	0.175

A hobbed worm gear never is generated completely. Its final finishing is obtained by its actual operation in position with its mating worm. For this reason, the materials used for worm gears should be such that they will conform to the worm without galling or excessive abrasion. Furthermore, the material of the worm should be smooth and hard enough to retain its form through such running-in or burnishing. A steel worm, therefore, to transmit power should have a Brinell hardness of not less than 250, and it should be as much harder as practicable.

Cast iron, manganese bronze, phosphor bronze, and some nickel bronzes, and several of the aluminum alloys have proved to be satisfactory materials for worm gears; the phenolic laminated materials prove very effective for the smaller gears and high-speed applications. Cast iron should not be used for rubbing velocities greater than about 1000 ft per min because of the danger of sparking. High-grade bronzes and aluminum alloys are essential for satisfactory results. A poorly alloyed material is practically worthless for a worm-gear material. Values for S are given in Table 26.

Table 26. Values of *S* for Worm Gears

Materials	S	Materials	S
Gray iron	12,000	Manganese bronze	17,000
Phenolic laminated material	6,000	Phosphor bronze	24,000

Values for nickel bronze and for the various alloys of aluminum suitable for worm-gear materials should be obtained from the manufacturers of the materials. It should be appreciated that increased physical strength obtained at the expense of lesser ductility of the material is of questionable advantage in materials to be used for worm gears. Also, the physical properties and structure of these materials depend more on the care and skill with which they have been alloyed, poured, and heat treated than upon the chemical analysis alone. Alloys of the same chemical analysis may have strength and wear values varying as much as two to one because of the difference in making and the difference in the resulting structure.

LIMIT LOADS FOR WEAR. Let W_w = limiting load for wear, pounds; D_g = pitch diameter of worm gear, inches; F = face width of worm gear, inches; K = load-stress value (Table 27). Then

$$W_w = D_g F K$$

The values of W_w should not be less than the dynamic load.

Table 27. Values of *K* for Worm Gears

Running with steel worms of 250 Brinell or higher

Material in Worm Gear	K	Material in Worm Gear	K
Cast iron	50	Phosphor bronze	100
Manganese bronze	80	Phenolic laminated material	125

EFFICIENCY OF WORM-GEAR DRIVES depends on several factors, such as the lead angle of the worm, the nature of the contact, intensity of the load, and speed of rubbing. When E = efficiency, β = lead angle of worm, f = coefficient of friction,

$$E = \frac{\tan \beta(1 - f \tan \beta)}{\tan \beta + f}$$

This applies to the efficiency at the tooth mesh only. The overall efficiency of a complete drive also must include the power losses of the thrust and radial bearings of both members of the drive.

The coefficient of friction will vary with the rubbing speed and with the load, and it follows closely the conditions existing on plain cylindrical bearings, except that the unit loads on worm-gear drives are much greater than those on plain bearings. The values in Table 28 represent average conditions. Table 29 shows the pronounced influence of the lead angle of the worm on the efficiency of the drive. These values are calculated from the foregoing equation, using the values of f given in Table 28.

Table 28. Values of the Coefficient of Friction f in Worm Drives

Rubbing Speed, ft/min	f	Rubbing Speed, ft/min	f	Rubbing Speed ft/min	f
5	.146	150	.045	1000	.041
10	.116	200	.042	1250	.046
20	.090	300	.040	1500	.050
40	.070	400	.036	1750	.054
80	.054	500	.033	2000	.058
100	.050	750	.036	2500	.065

A worm with a lead angle of 30 degrees is practically as efficient as one with 45 degrees. Lead angles less than 10 degrees are very inefficient.

The heat of operation of worm-gear drives would be estimated in the same manner as shown for spur gears, by determining values for the power loss in foot-pounds per minute and the exposed area of the case in square feet.

Table 29. Relation of Lead Angle, Speed, and Efficiency of Worm-gear Drives

Rubbing Velocity, ft/min	Lead Angle of Worm 5°	10°	15°	25°	35°	45°	Rubbing Velocity, ft/min	Lead Angle of Worm 5°	10°	15°	25°	35°	45°
	Efficiency							Efficiency					
5	.37	.53	.62	.71	.74	.75	400	.71	.82	.87	.91	.93	.93
10	.42	.59	.68	.76	.79	.79	500	.72	.84	.88	.92	.93	.94
20	.49	.65	.73	.80	.83	.83	750	.71	.82	.87	.91	.93	.93
40	.55	.71	.78	.84	.86	.87	1000	.68	.80	.86	.90	.92	.92
80	.61	.76	.82	.87	.89	.90	1250	.65	.79	.84	.89	.91	.91
100	.63	.77	.83	.88	.90	.90	1500	.63	.77	.83	.88	.90	.90
150	.66	.79	.85	.89	.91	.91	1750	.61	.76	.82	.87	.89	.90
200	.67	.80	.85	.90	.91	.92	2000	.60	.74	.81	.86	.88	.89
300	.68	.81	.86	.90	.92	.92	2500	.57	.72	.79	.85	.87	.88

BRITISH STANDARD. In worm gearing, the mating parts are so closely related that it is practically impossible to specify the tooth form of the gear without referring to the mating worm. The generating tool for the worm gear must be in all respects a counterpart of the worm. This problem has led to many controversies about the possibility or desirability of establishing design standards. The British Standards Institution established in 1937 a standard specification (B.S. 721–1937) for worm gearing based on a normal pressure angle of 20 degrees of the worm-thread form. Considerable leeway is given the designer in selection of proportions, such as number of teeth, number of threads, and the ratio of worm diameter to pitch selected, but formulas are given for strength and wear for each member.

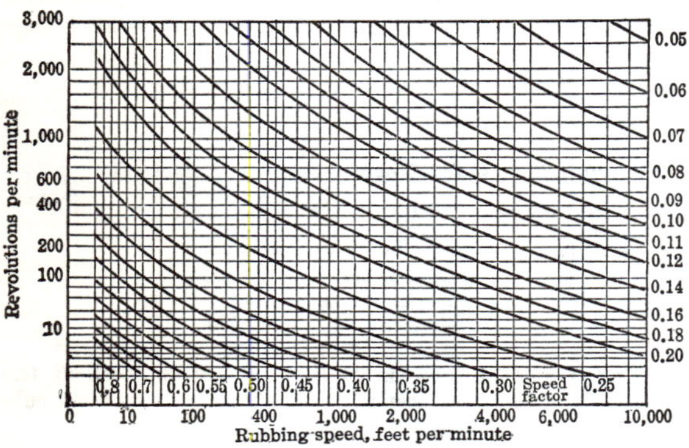

FIG. 64. Speed factors X_{cp} and X_{cw} for worm gearing. (From British Standards Institution Specification B.S. 721–1937, by permission)

The following formulas are taken by permission from the July 1941 revision of British Standard 721–1937, p. 22, and are applicable to 20-degree pressure angle worm gearing for a period of 12 hours a day under uniform load conditions:

Permissible Load for Wear

$$\text{For worm} = 0.18 S_{cp} X_{cp} f_{ew} D^{1.8} \text{ lb-in.}$$

$$\text{gear} = 0.18 S_{cw} X_{cw} f_{ew} D^{1.8} \text{ lb-in.}$$

where S_{cp} and S_{cw} are surface stress factors from Table 30.

X_{cp} and X_{cw} are speed factors reflecting revolutions per minute and rubbing speed. See Fig. 64.

$$f_{ew} = \text{maximum effective face width (see Fig. 65)}$$
$$= 2m\sqrt{q+1}$$

where m = module = 0.3183 × axial pitch; q = worm pitch diameter per module.

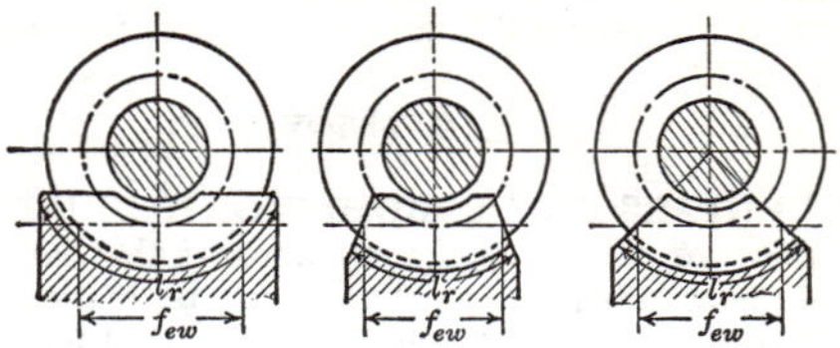

Fig. 65. Face factor, f_{ew}, for worm gearing. (From British Standards Institution Specification B.S. 721–1937, by permission)

Permissible Load for Strength

$$\text{For worm} = 0.625 S_{bp} X_{bp} l_r m D \cos\lambda \text{ lb-in.}$$
$$\text{gear} = 0.625 S_{bw} X_{bw} l_r m D \cos\lambda \text{ lb-in.}$$

where S_{bp} and S_{bw} are bending stress factors for the material as listed in Table 30.

Table 30. Stress Factors for Worm Gears

	Material	Bending Stress Factor per sq in. S_b lb	Surface Stress Factor (S_c) When Running with A	B	C	D	E
A	Phosphor bronze, sand cast	7,000	.....	600 *	600	700	1,400
A	Phosphor bronze, chill cast	8,500	.....	800 *	800	900	1,600
A	Phosphor bronze, centrifugally cast	10,000	.....	1,000 *	1,000	1,100	2,000
B	Cast iron (gray)	6,000	900 *	600 *	600 †	600 †	750 †
C	0.4% Carbon steel normalized	20,000	1,400	900 †	.....	.....	.....
D	0.55% Carbon steel normalized	22,000	2,000	1,100 †	.....	.....	.....
E	Low-carbon case-hardening steel	27,000	6,000	4,000 †	.....	.....	.....
E	3 1/2% Nickel case-hardening steel	40,000	6,500	4,000 †	.....	.....	2,000 †
E	5% Nickel case-hardening steel	47,000	7,000	4,000 †	.....	.....	2,000 †
E	3 1/2% Nickel chromium case-hardening steel	47,000	7,000	4,000 †	.....	.....	2,000 †
E	High nickel chromium case-hardening steel	47,000	8,000	4,000 †	.....	.....	2,000 †

* Maximum permissible rubbing speed = 500 ft per min.
† Should not be used except for hand motions.

X_{bp} and X_{bw} are speed factors for the worm and wheel to be selected from the charted data shown as Fig. 66.

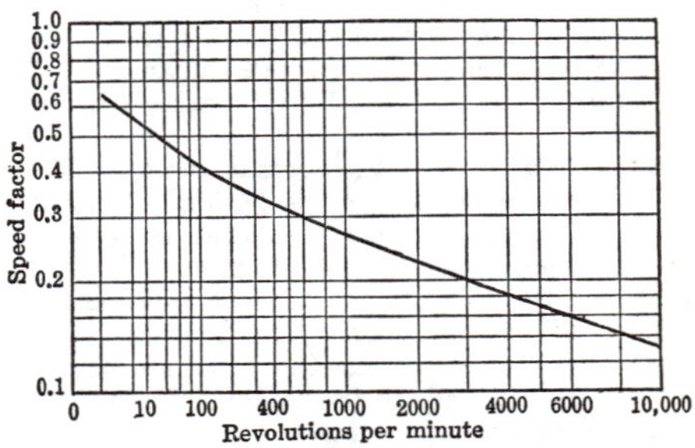

Fig. 66. Speed factors X_{bp} and X_{bw} for worm gearing. (From British Standards Institution Specification B.S. 721–1937, by permission)

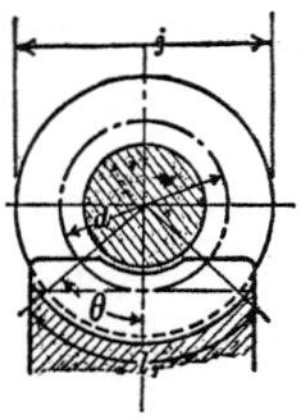

FIG. 67. Length of root of worm-wheel teeth, l_r. (From British Standards Institution Specification B.S. 721–1937, by permission)

l_r is the length of the root of worm-wheel teeth on cross section of rim (see Fig. 67) and may be calculated from this formula:

$$l_r = (j + 2c) \sin^{-1} [f_w/(j + 2c)]$$

where j = outside diameter of worm; $c = 0.2m \cos \lambda$; f_w = face width of wheel; and λ = lead angle of worm threads.

Horsepower rating for the gear set is then established by using the smallest of the permissible torque ratings designated as M_w in the formula

$$\text{Horsepower} = \frac{M_w N}{63{,}000}$$

HOURGLASS OR GLOBOID WORM DRIVES. In an hourglass or globoid worm drive, neither member of the pair has a uniform helical lead of tooth along its axis. One or both members may be made to envelop the other. The contact is line contact, and the pitch surfaces are indeterminate. No fixed pitch point exists. Each different design must be analyzed individually to establish the nature of the action and position of the contact. The motion is primarily that of a barrel cam operating against a series of cam followers free to rotate about an axis.

The earliest type of such a drive is the Hindley worm. Many other later drives of this type have been designed. Each different drive of this type has so many individual features that they require individual analyses to determine the exact nature of the contact, etc. Also their production requires special manufacturing facilities. In all such drives, however, the maximum theoretical contact that can be obtained is line contact.

Double-enveloping Gearing

CONE-DRIVE GEARS. Cone-Drive (Ref. 31) is the trade name for a right-angle drive double-enveloping gearing. The basic principle of the Cone-Drive embodies a rational method of producing a double-enveloping gear drive having full mating capacity without modifications or limitations (Fig. 68).

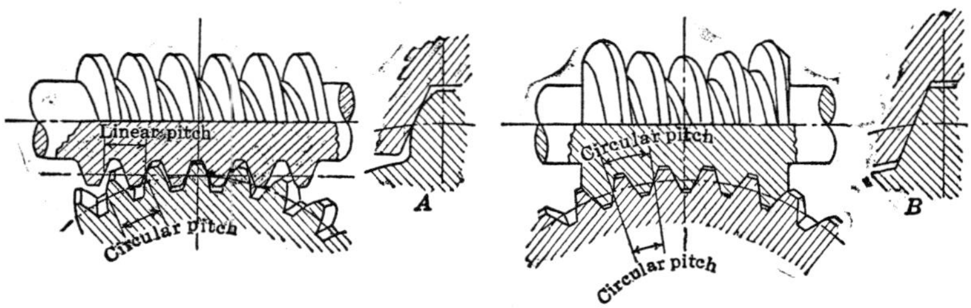

FIG. 68. Cone-Drive gearing compared with conventional worm gear. (Courtesy of Michigan Tool Co.)

The manufacturing process is such that the hob or cutter is in the same relative position to the part being generated as the mating part will occupy when the set is mounted in operating position. Thus pinion and gear teeth are generated on operating centers in geared time relation. The hob and cutter have thinner teeth than the space to be generated, and use side feeding to develop proper tooth thickness.

The general design characteristics of Cone-Drive gearing (Fig. 69) may be summarized as follows.

Pressure angle is always 20 degrees in the normal section. This has been established after consideration of the effect of pressure angle on load distribution in the pinion root, gear teeth, and bearings. Thus no consideration need be given to changes in pressure angle when working on manufacturing processes or tools.

Number of teeth selected for the gear for a given ratio and center distance has no effect except on economy of manufacture. The load-carrying capacity, mechanical strength, and total area of contact of a Cone-Drive remain unchanged, regardless of the number of teeth in the gear and threads in the pinion. Thus the number of teeth can be selected on the most economical basis.

Circular pitch is not considered in designing a Cone-Drive gear set. The "correct" pitch is automatically obtained by designing the gearing from the standpoint of correct proportioning, ratio, stresses, and number of teeth desired. It is a result, not a specification.

Tooth thickness in both the gear and the pinion is controlled by the manufacturing operation and is independent of hob or cutter design. It can be varied at will. Tooth thickness, therefore, can be and is varied to balance stresses resulting from the use of different materials in the gear or pinion.

A so-called soft pinion is used in Cone-Drives. This pinion, which is the result of exhaustive tests and experiments, is in line with the principle that when wear occurs in this drive, each member tends to regenerate the correct form in the other. The pinion

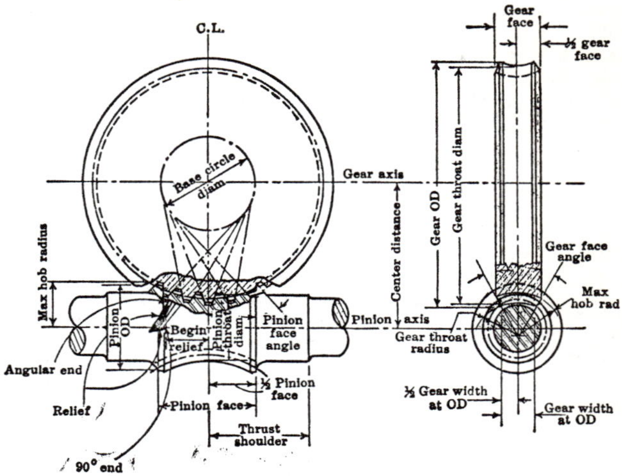

FIG. 69. Nomenclature of Cone-Drive gearing. (Courtesy of Michigan Tool Co.)

is made of deep-hardening alloy steel which is heat treated to 35-38 Rockwell C scale before the threads are generated. This eliminates heat-treating errors and gives a satisfactory coefficient of friction.

Load-carrying ability of Cone-Drive gearing is based on a combination of factors including pinion revolutions per minute, gear ratio, velocity factor, and pressure constants. Stress analysis of gears at rated load shows a fiber stress in the gear teeth of approximately 8000 psi at zero rpm, and a corresponding stress in the pinion root under 30,000 psi. When designing for static load conditions or extremely low speeds, the capacity may, therefore, be materially increased to the point where the fiber stresses in the gear teeth approach 75% of the yield point of the material.

Horsepower is calculated from the formula

$$\text{Horsepower} = \frac{\text{Speed}}{\text{Gear ratio}} \times KNf$$

where K is a pressure constant ranging from 0.12 for a 3-in. center distance to 500 for a 40-in. center distance gear set; N is a ratio correction factor which ranges from 0.70 to 0.76 for ratios of 10 : 1 to 100 : 1; and f is a velocity correction factor which varies with center distance and speed.

Efficiency of Cone-Drive gearing is dependent primarily on the helix angle of the pinion. Between 10 and 55 degrees helix angle, efficiency ranges from 91 to 97.5%. From 5 to 10 degrees helix angle, the efficiency ranges from 83 to 91%. These figures are for pinion speeds up to 2400 rpm. Below 1500 rpm the efficiency drops approximately 1% for each 200 rpm, and below 1000 rpm the efficiency drops about 1% for each 100 rpm.

The double-enveloping characteristic of Cone-Drive gearing usually results in a resemblance to hour-glass worm gearing. In some cases of extreme ratios the drive may resemble right-angle helical gearing, although even in these constructions the double-enveloping characteristics are retained.

Analysis of contact lines shows that every point on the pinion (within the unrelieved load-carrying section) contacts some portion of the gear tooth in addition to the midplane through which all points on the pinion pass. In its passage across the gear tooth some point of this section of the pinion also touches every point on the gear-tooth face.

The materials ordinarily used for Cone-Drive gearing up to pinion speeds of approximately 2000 fpm consist of a high-quality gear bronze which develops minimum physica properties on the order of 50,000 psi tensile strength. The pinion is usually made from a chrome-nickel-molybdenum steel, heat treated to a minimum hardness of 32 Rockwell C and 140,000 psi tensile strength.

At slow speeds (under 250 ft per min rubbing velocity), the material combinations may use a pinion of aluminum bronze, running with a 0.40 to 0.50% carbon steel gear, or an electric furnace iron casting.

Where pinion speeds in excess of 2000 fpm are encountered, a special gear bronze, containing 10% lead, is recommended.

For general service conditions up to 2000 ft per min rubbing speed of the pinion a pure Pennsylvania oil having a viscosity of 150 SSV at 210 F is recommended. The addition of 5 to 7% acidless animal fat is said to improve the operating characteristics of this lubricant. The design of Cone-Drive gearing is such that the entering thread provides a tapering oil film wedge ahead of the load-carrying contact area, thus providing improved lubrication.

9. GEAR TRAINS AND APPLICATION DATA

SIMPLE GEAR TRAINS are usually referred to as being one pair of mating gears having essentially one-tooth mesh area. In parallel externally toothed gearing (Fig. 14) the direction of rotation of the mating elements is opposite and the speed ratio is in inverse proportion to the numbers of teeth in each member. Where an internal gear (Fig. 24) is used, the direction of rotation is the same for each member, and the speed ratio is again in inverse proportion to the numbers of teeth in the mating gears.

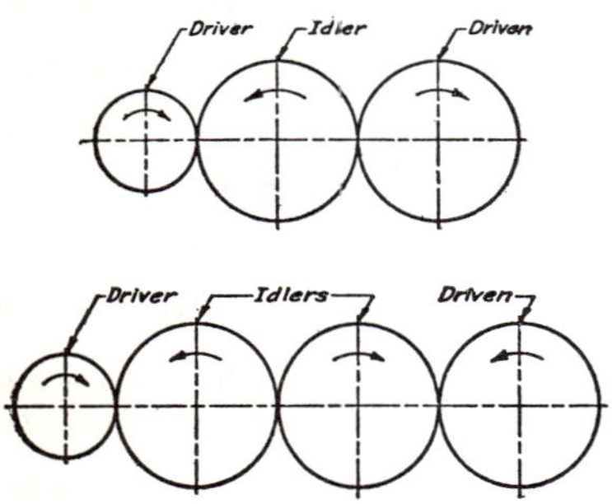

FIG. 70. Idler gear trains.

IDLER GEAR TRAINS (Fig. 70) are often used where it is necessary to maintain direction of rotation between the driving and driven members or, in other cases, where the center distances between the shafts suggest the introduction of an idler to maintain size and velocity of the gear members within reasonable limits.

The speed ratio of gear trains using idlers is in proportion to the driving and driven members only. Where multiple idlers are used in even numbers, the direction of rotation of the driven gear is opposite the driver; and when the number of idlers is odd, the direction of rotations is the same.

RIGHT-ANGLE DRIVES employing bevel gearing follow the same pattern as simple gear trains, as described above, but, in addition (Figs. 32, 33, 34), each mesh point changes the angle between the supporting shafts.

MULTIPLE REDUCTIONS are obtained by compounding series of simple trains as shown in Fig. 71. To obtain overall speed ratios and directions of rotation, it is necessary only to consider each set as a simple train and to follow the motion from driving member through to the driven member.

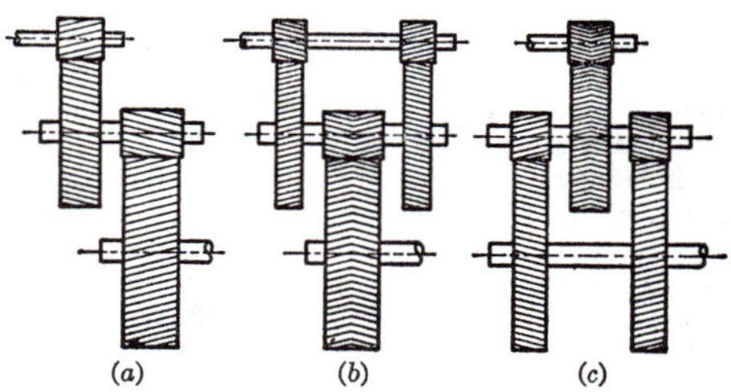

FIG. 71. Multiple-reduction gear trains: (a) single helical gears; (b) herringbone with divided high-speed gear set; (c) herringbone with divided low-speed gear set.

HELICAL OR BEVEL GEAR types require consideration of the amount and direction of the end thrust. Where double-reduction gear sets of the single-helical type are involved (see Fig. 21), the hand of the helix of the driven member of the high-speed train and the driving member of the low-speed set should be the same so as to offset the thrust developed in each.

HERRINGBONE gear trains should be arranged so that the apex of the Vee formed by the two helices is in the direction of rotation (see Fig. 23), thus avoiding excessive corner contact loading that might be caused by deflections.

MULTIPLE-TOOTH CONTACT gear trains are finding many practical uses, particularly where space and weight requirements justify the added cost related to precision work. Where multiple-contact systems are used without an internal gear (Fig. 72), they are often called multiple jack-shaft gear trains. They should be analyzed, using the formulas applying to simple trains, except in load-carrying ability where the division of the load between the various contacts must be studied in arriving at tooth contact stresses. It is generally thought practical to apply a factor of 0.90 to 0.95 (depending on accuracy) to the loading factor, based on uniform load distribution.

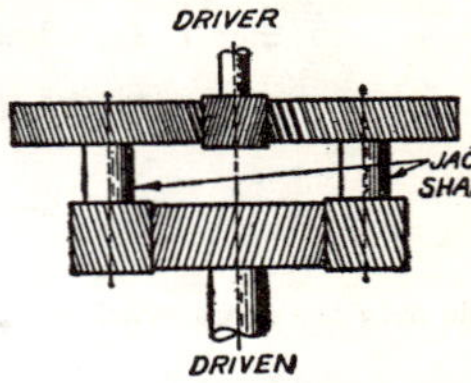

FIG. 72. Multiple-contact gear trains.

PLANETARY OR EPICYCLIC GEAR TRAINS may be defined as trains of wheels in which some or all of the wheels have a motion compounded of a rotation about an axis and a revolution of that axis. They are employed to produce large speed ratios with few parts. The link, or planet carrier arm, usually rotates about the axis of one of the gears, which may be fixed. The output gear may or may not be placed on this axis, depending upon whether pure rotation or a cycloidal output motion is desired. The name epicyclic used in connection with planetary gear trains has no connection with the form of the gear teeth, which may be cycloidal, involute, or any other system.

Speed calculations for planetary trains are complicated by the fact that the line of centers of the gears usually rotates. A variety of methods has been developed, of which the principal ones are outlined in the following pages.

Relative Angular Velocity Method. Either a tabular or a formula method may be used to solve directly for angular velocity ratios.

In the tabular method the planetary train is subjected to two imaginary operations: (1) the entire unit, with the gears locked against relative rotation, is turned through one complete revolution; and (2) with the carrier arm held stationary and the gears unlocked, the gear which is intended to be fixed in actual operation is turned back one revolution, thus restoring it to its original position. At the same time the corresponding rotations of the other gears in the train are calculated from their arrangements and number of teeth (see p. 14-48). Both steps are recorded in tabular form and the net effect determined by algebraic summation.

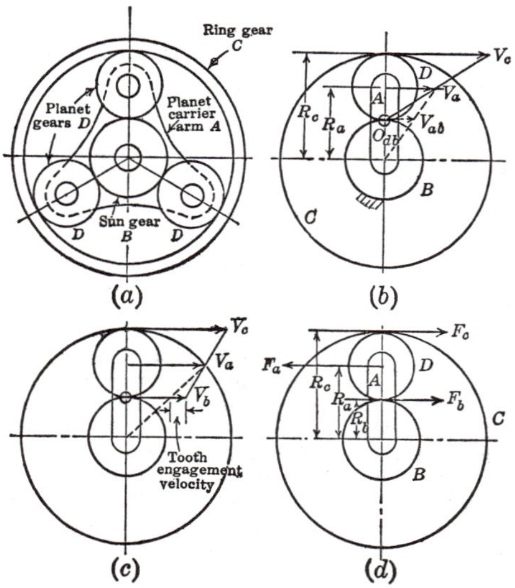

FIG. 73. Simple internal-external planetary system.

EXAMPLE 1. In a simple planetary train (Fig. 73*a*) it is assumed that sun gear B is fixed to the frame of the machine, ring gear C is the input, and planet carrier arm A is the output. It is required to find the speed ratio of input to output. The results of the two-step procedure just described are shown in the accompanying table, in which the number of teeth in each gear or pinion is denoted by the identifying letter on the figure; thus, B denotes the number of teeth in gear B. In completing the second line it is important to place the proper sign in front of each term. This is found by inspection of the figure and observation of the relative direction of rotation of each gear. Thus both ring gear C and planet pinion D rotate opposite to sun gear B (with respect to the arm A) and carry a plus sign since the rotation of B carries a minus sign. From the last line of the table, while arm A rotates $+1$ revolution the ring gear C rotates $1 + (B/C)$ revolution, which is the required ratio.

	A	B	C	D
Step 1. Train locked	$+1$	$+1$	$+1$	$+1$
Step 2. Train unlocked, carrier arm fixed	0	-1	$+\frac{B}{C}$	$+\frac{B}{D}$
Resultant	$+1$	0	$1+\frac{B}{C}$	$1+\frac{B}{D}$

EXAMPLE 2. In the compound planetary train shown in Fig. 74*a* the sun gear B is the input, the outer ring gear C is fixed, and the inner ring gear G is the output. The carrier arm A rotates freely. It is desired to calculate the ratio of input speed (of B) to output speed (of G). In the accompanying

	A	B	C	G
Step 1. Train locked	$+1$	$+1$	$+1$	$+1$
Step 2. Train unlocked, arm fixed	0	$+\frac{C}{B}$	-1	$-\frac{CE}{DG}$
Resultant	$+1$	$1+\frac{C}{B}$	0	$1-\frac{CE}{DG}$

table only the gears which are involved in the problem are listed. Gear C being fixed is given one minus turn in the second line while the arm A is held stationary. From the last line of the table the required speed ratio is

$$\frac{1+\dfrac{C}{B}}{1-\dfrac{CE}{DG}} = \frac{1+\dfrac{120}{12}}{1-\dfrac{120 \times 43}{54 \times 109}} = 89.2$$

A general formula may be written relating the speed or angular velocity of the planet carrier arm A to any two of the gears on the same axis. With the two gears denoted by B and C, the formula is

$$\omega_{af} = \frac{\omega_{bf} - \left(\dfrac{\omega_{ba}}{\omega_{ca}}\right)\omega_{cf}}{1 - \left(\dfrac{\omega_{ba}}{\omega_{ca}}\right)} = \frac{\omega_{bf} - k\omega_{cf}}{1-k} \qquad (1)$$

where ω_{af} = speed of carrier arm A relative to frame of the machine, F; ω_{bf} = speed of gear B relative to frame; ω_{cf} = speed of gear C relative to frame; ω_{ba} = speed of gear B relative to arm A; ω_{ca} = speed of gear C relative to arm A; $k = \omega_{ba}/\omega_{ca}$ = speed ratio of the two gears calculated with the carrier arm stationary. In applying the formula it is not necessary to calculate ω_{ba} and ω_{ca} individually; only their ratio k is needed. It will be observed that the calculation of k corresponds to Step 2 in the tabular solution already described. If there are only two gears of the sun or ring type it is convenient to designate them B and C so that the formula can be applied direct. If there are more than two such gears so that another letter is needed it is necessary only to substitute the new letter for either b or c in each of the subscripts. In such event, of course, k has a different value and should be denoted by k' or k''. The formula takes care of only two gears at a time, hence must be rewritten for each additional gear. This is illustrated in Example 4, p. 14-51.

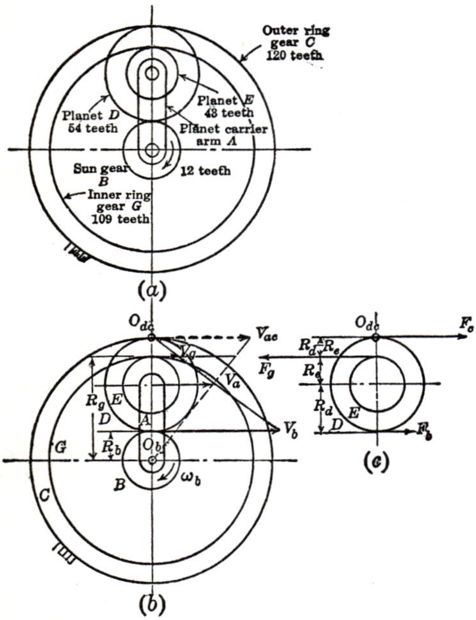

FIG. 74. Compound internal-external planetary system.

EXAMPLE 3. If the formula is applied to the problem stated in Example 1, it is noted that $\omega_{bf} = 0$, ω_{cf} is unknown input, $\omega_{af} = 1$, and $k = -C/B$. Substituting,

$$1 = \frac{0 - (-C/B)\omega_{cf}}{1 - (-C/B)} \quad \text{from which } \omega_{cf} = 1 + \frac{B}{C} \text{ as before}$$

The formula method is particularly convenient if both gears are driving, as in differential drives. Thus, if in this example gear B turns at 1000 rpm and gear C at 600 rpm, the resultant speed of arm A can be determined by direct substitution in the formula, letting $\omega_{bf} = 1000$ and $\omega_{cf} = 600$, k being already known.

EXAMPLE 4. If eq. 1 is applied to the problem stated in Example 2, it is noted that $\omega_{cf} = 0$, hence

$$\omega_{af} = \frac{\omega_{bf}}{1 - \dfrac{\omega_{ba}}{\omega_{ca}}}$$

This gives the speed of the arm but not that of the ring gear G, which is the output. It is therefore necessary to rewrite the equation to bring in the gear G. This can be done by simply substituting g for b in the subscripts of the equation, giving

$$\omega_{af} = \frac{\omega_{gf}}{1 - \dfrac{\omega_{ga}}{\omega_{ca}}}$$

Since the arm A rotates freely it is not necessary to know ω_{af} for present purposes; hence it may be eliminated by equating the two foregoing expressions, as shown below. By inspection of Fig. 74a it is evident that $\omega_{ba}/\omega_{ca} = -C/B$, and $\omega_{ga}/\omega_{ca} = +CE/DG$. Substitution in the two preceding formulas gives the result

$$\frac{\omega_{bf}}{\omega_{gf}} = \frac{1 - \dfrac{\omega_{ba}}{\omega_{ca}}}{1 - \dfrac{\omega_{ga}}{\omega_{ca}}} = \frac{1 + \dfrac{C}{B}}{1 - \dfrac{CE}{DG}} = \frac{1 + \dfrac{120}{12}}{1 - \dfrac{120 \times 43}{54 \times 109}} = 89.2$$

Instant Center and Linear Velocity Method. This method (Ref. 32) permits either a graphical or an analytical solution and in addition readily furnishes information on pitch line velocities of tooth engagement. It has the further advantage of affording a visual check of the directions of motion of each element.

Figure 73b, which represents the mechanism of Fig. 73a in skeleton form, shows that the pitch point O_{db} of the planet pinion D and the fixed sun gear B is the instant center of rotation of the planet pinion. Therefore the linear velocity of each point on the planet pinion is proportional to its distance from O_{db}. If the pitch line velocity of the ring gear C ($V_c = \omega_c R_c$) is known, the simple graphical construction shown gives, by similar triangles, the velocity of any other point, such as the center of the planet pinion on the carrier arm, V_a. Angular velocity of the planet carrier arm is V_a/R_a, and the speed ratio of ring gear to carrier arm therefore is $(V_c/R_c) \div (V_a/R_a) = (V_c/V_a)(R_a/R_c)$. The ratios of velocity and radii can be readily determined from the geometry of the figure if an analytical solution is desired.

Pitch line velocity of engagement of the teeth is found by constructing a radial line (shown dotted in Fig. 73b) from the center of rotation of the carrier arm through the tip of the V_a vector. A vector V_{ab} drawn from the pitch point O_{db} represents the velocity of the carrier arm at the pitch radius of the fixed sun gear B and therefore represents the pitch line velocity of engagement.

When both sun and ring gears are driving, as in a differential drive, it is necessary only to calculate the linear velocities of each, plot the vectors V_b and V_c at the two pitch points, and determine the velocity of the planet pinion center, V_a, as shown in Fig. 73c. The pitch line velocity of tooth engagement is found as before by drawing a radial line from the center of rotation of the carrier arm (shown dotted). A vector from the pitch point to this dotted line represents the velocity of the carrier arm at this radius; therefore the pitch line velocity of tooth engagement is the vector difference between this velocity and V_b.

EXAMPLE 5. Solution of the problem stated in Example 2, by the method of instant centers and linear velocities, is indicated in Fig. 74b. Vector $V_b = \omega_b R_b$ is drawn to represent the pitch line velocity of the sun gear B. The instant center of the planet cluster being at O_{dc}, the velocity at any point on the cluster is found by drawing a straight line from O_{dc} to the tip of the V_b vector and constructing a vector line parallel to V_b. The angular velocity of the output gear G is found from the figure, $\omega_g = V_g/R_g$, and the required angular velocity ratio is $\omega_b/\omega_g = (V_b/V_g)(R_g/R_b)$. Ratios of velocities and radii are readily determined from the geometry of the figure.

To find pitch line velocities of engagement a radial line (shown dotted) is drawn through the tip of the V_a vector from the center of rotation of the carrier arm. Since ring gear C is fixed, the velocity of the carrier arm at this radius, V_{ac}, represents the engagement velocity of gear C. The engagement velocity of gear B is the vector difference between V_b and the arm velocity at the radius of gear B. Inspection of the figure will show that this is the same as the engagement velocity for C, which is as it should be since both B and C engage the same planet pinion D. In like manner the engagement velocity of gear G is the vector difference between the velocity of the arm at radius R_g and the velocity V_g.

Forces and Moments. Static force and torque relationships furnish another viewpoint on planetary gearing, and can be employed to determine speed ratios. In the simple system of Fig. 73d it is evident that for the planet to be in equilibrium the forces at the two pitch points must be equal to each other and together equal and opposite to the force at the pinion center, that is, $F_b = F_c$ and $F_b + F_c = F_a$. Torque on each element is equal to the force times the radius to the center of rotation; hence the input torque is $T_c = F_cR_c$, the output torque is $T_a = -F_aR_a$, and the reaction torque on the fixed sun gear is $T_b = F_bR_b$. Torque ratio between output and input is $T_a/T_b = F_aR_a/F_bR_b$, which may be written $(F_a/F_b)(R_a/R_b)$. The force ratios are readily determined from the equilibrium of the planet, and the radius ratios from the geometry of the planetary set. Torque ratio of output to input is also equal to speed ratio of input to output, assuming no internal losses; hence the torque calculations offer a method of calculating speeds.

Example 6. Solving by force and torque relationships the problem stated in Example 2, the free-body diagram for the pinion cluster is shown in Fig. 74c. Taking moments about pitch pint O_{dc}, $F_b \times 2R_d = F_g(R_d - R_e)$ from which $F_g = F_b \times 2R_d/(R_d - R_e)$. The output torque is F_gR_g and the ratio of output to input torques is $T_g/T_b = F_gR_g/F_bR_b = (F_g/F_b)(R_g/R_b)$. From the analysis of pinion equilibrium, $F_g/F_b = 2R_d/(R_d - R_e)$; hence $T_g/T_b = 2R_dR_g/R_b(R_d - R_e)$, which may be calculated from the geometry of the figure. Pitches being equal for all gears, the radii may be replaced by the numbers of teeth, or $T_g/T_b = 2 \times 54 \times 109/12(54 - 43) = 89.2$.

LIMITATIONS ON PLANETARY GEAR RATIOS.* In a simple system of planetary gears, such as is shown in Fig. 73*a*, three alternatives are possible. Sun gear *B* can be stationary with ring gear *C* driving the carrier arm *A* in the same direction; ring gear *C* can be stationary with sun gear *B* driving the carrier arm in the same direction; or carrier arm *A* can be stationary with ring gear *C* and sun gear *B* rotating in opposite directions. With the notation *B* = number of teeth in gear *B*, etc., the following formulas give the speed ratios for each alternative arrangement:

Ring gear stationary $$\frac{\omega_b}{\omega_a} = k_1 = 1 + \frac{C}{B}$$

Sun gear stationary $$\frac{\omega_c}{\omega_a} = k_2 = 1 + \frac{B}{C}$$

Arm stationary $$\frac{\omega_b}{\omega_c} = k_3 = -\frac{C}{B}$$

A reversing drive (arm stationary) always has a speed ratio numerically smaller by one than a unidirectional drive with gears of similar proportions, and ring gear stationary.

Maximum Number of Planet Pinions. To avoid interference between the teeth of adjacent planet pinions, the speed ratio with standard addendum teeth (addendum = one/diametral pitch) must be less than

$$k_{1\max} = \frac{2B - 4}{B\left[1 - \sin\left(\frac{180}{n}\right)\right]} \tag{2}$$

where n is the number of equally spaced planet pinions. Table 31 lists maximum ratios k_1 computed for two to six planet pinions and with 12, 15 or 20 teeth in the sun gear *B*.

Table 31. Maximum Ratio k_1 † for Simple Planetary

No. of Planet Gears	Ratio k_1: $B = 12$	$B = 15$	$B = 20$
2	∞	∞	∞
3	12.37	12.93	13.43
4	5.68	5.92	6.14
5	4.04	4.21	4.37
6	3.33	3.47	3.60

† k_1 is the ratio of sun gear speed to carrier arm speed when the ring gear is stationary.

Equal Spacing of Planet Pinions. For equal spacing the following relation between the numbers of teeth in the sun gear and the planet pinion must be observed:

$$\frac{2(B + D)}{n} = \text{an integer} \tag{3}$$

where n is the number of equally spaced planet pinions, *B* is the number of teeth in the sun gear, and *D* is the number of teeth in the planet pinion. Number of teeth, *C*, in the ring gear is given by the relation $C = 2D + B$.

* Contributed by George S. Hoell.

Table 32 gives the speed ratio k_1 for various combinations of sun and planet gears, together with the number of equally spaced planet pinions that can be used with each. The ratio k_1 is the speed of the sun gear if the carrier arm speed is unity and the ring gear is fixed. For values beyond the range of the table the equation $k_1 = 2[1 + (D/B)]$ may be used. If more than two planets must be used the limiting ratio due to possible interference should be checked by Eq. 2.

Example 7. It is desired to obtain a speed reduction k_1 equal to approximately 4.2, using as many planet pinions as possible to carry the load. From Table 31 the sun gear must have at least 15 teeth if it is desired to use five equally spaced planet gears. If four planets are sufficient, the sun gear may have fewer teeth. Table 32 allows the use of four planets with 20 teeth in the sun gear and 22 in each of the planet pinions, at a ratio of 4.2. If the ratio requirement is not exact and five equally spaced planets are desired, Table 32 shows that 19 teeth in the sun gear and 21 in each of the planet gears give a ratio of 4.211 and permit use of five planets.

Table 32. Ratio k_1 and Allowable Number of Planets *

(Courtesy of *Machine Design*)

No. of Teeth in One Planet Gear D	Number of Teeth in Sun Gear B								
	12	13	14	15	16	17	18	19	20
12	4.000 2–3–4	3.846 2	3.714 2–4	3.600 2–3	3.500 2–4	3.411 2	3.333 2–3–4–5	3.263 2	3.200 2–4
13	4.167 2	4.000 2–4	3.857 2–3	3.733 2–4	3.625 2	3.529 2–3–4–5	3.444 2	3.368 2–4	3.300 2–3
14	4.333 2–4	4.154 2–3	4.000 2–4	3.867 2	3.750 2–3–4–5	3.647 2	3.556 2–4	3.474 2–3	3.400 2–4
15	4.500 2–3	4.308 2–4	4.143 2	4.000 2–3–4–5	3.875 2	3.765 2–4	3.667 2–3	3.579 2–4	3.500 2–5
16	4.667 2–4	4.462 2	4.286 2–3–4	4.133 2	4.000 2–4	3.882 2–3	3.778 2–4	3.684 2–5	3.600 2–3–4
17	4.833 2	4.615 2–3–4	4.429 2	4.267 2–4	4.125 2–3	4.000 2–4	3.889 2–5	3.789 2–3–4	3.700 2
18	5.000 2–3–4	4.769 2	4.571 2–4	4.400 2–3	4.250 2–4	4.118 2–5	4.000 2–3–4	3.895 2	3.800 2–4
19	5.167 2	4.923 2–4	4.714 2–3	4.533 2–4	4.375 2	4.235 2–3–4	4.111 2	4.000 2–4	3.900 2–3
20	5.333 2–4	5.077 2–3	4.857 2–4	4.667 2	4.500 2–3–4	4.353 2	4.222 2–4	4.105 2–3	4.000 2–4–5
21	5.500 2–3	5.231 2–4	5.000 2	4.800 2–3–4	4.625 2	4.471 2–4	4.333 2–3	4.211 2–4–5	4.100 2
22	5.667 2–4	5.385 2	5.143 2–3–4	4.933 2	4.750 2–4	4.588 2–3	4.444 2–4	4.316 2	4.200 2–3–4
23	5.833 2	5.538 2–3–4	5.286 2	5.067 2–4	4.875 2–3	4.706 2–4	4.556 2	4.421 2–3–4	4.300 2
24	6.000 2–3–4	5.692 2	5.429 2–4	5.200 2–3	5.000 2–4	4.823 2	4.667 2–3–4	4.526 2	4.400 2–4
25	6.167 2	5.846 2	5.571 2–3	5.333 2–4	5.125 2	4.941 2–3–4	4.778 2	4.632 2–4	4.500 2–3

* The upper figure of each pair is the ratio k_1; the lower figures indicate the number of planets which may be used equally spaced.

WORM-GEAR TRAINS provide large reductions in a single gear set inasmuch as the speed ratio of this type of gearing is the ratio between the number of teeth in the wheel and the number of threads in the worm.

GEARED POWER TRANSMISSIONS have become very popular, and the AGMA has established recommended practices for the proportions, rating, and application of the more popular types, including helical and herringbone spur gear, planetary, combination helical and bevel, and worm gear constructions. The combination of a geared reduction unit and an electric motor, generally known as a gearmotor or a motorized speed reducer, has given industry a compact unit, representative of the application possibilities of geared mechanisms.

STANDARD SPEED-REDUCING mechanisms of many types are available, and it will be the purpose of the following brief outlines of typical constructions to familiarize the reader with some of the more popular design arrangements.

Single-reduction parallel shaft types are available with single helical or herringbone gear trains and with shafts in either the horizontal or vertical planes. Ratios generally are under 10 to 1, and standard sizes run upwards of 500 hp.

Double-reduction parallel shaft types (Fig. 71) employ various construction arrangements. Some designs use single helical gears in both sets, Fig. 71*a*. Others use herringbone gear sets in tandem. Generally speaking, the most popular type because of its balanced loading is that shown in Fig. 71*c*, wherein the low-speed double helical gear set (essentially herringbone in action) spans the high-speed herringbone train. Ratios for standard reducers of the double-reduction type are usually limited to a range between 10 to 1 minimum and 60 or 90 to 1 maximum. Horsepower capacities approach 500 for standard makes.

Triple-reduction designs follow the double-reduction gearing arrangements with the additional gear set often nested between the high-speed and intermediate shafts of the double trains, thus conserving space. Speed reduction ratios for these units run between 80 to 1 and 300 to 1, with horsepower capacities of over 100 being available in standard units.

Gearmotors or motorized speed reducers are available from many sources in horsepower ranges up to 75 and in output speeds down to 13.5 rpm. Generally speaking, the standard models consist of single-, double-, and triple-reduction constructions.

Other Types. It is obviously impossible to cover in a handbook all the constructions used in standard industry practice. We might call attention to the availability of vertical geared units, worm-gear units, combination helical and right-angle drives, and many other ingenious arrangements that have become widely used. The reader interested in data on these equipments is referred to the general catalogs of industry.

10. FINE-PITCH GEARING

Fine-pitch gears have been defined as gears of 20 diametral pitch and finer, by the AGMA. Although there is no mathematical difference between fine- and coarse-pitch gears of involute form, fine-pitch gears require special considerations for the following reasons.

A. Greater clearances are required than for coarse-pitch.

B. Consideration must be given to point widths of gear cutting tools and to top lands of gears, especially to top lands of enlarged heat-treated pinions.

C. Fine-pitch gears require different production and inspection techniques from coarse-pitch gears, except in the coarser range of fine-pitch gears. Their use is generally for the transmission of motion rather than power. Surface durability is of far greater importance than beam strength.

The AGMA has developed a number of standards to meet the requirements of fine-pitch gears. Standard 470.01 establishes clearances for all 20-degree pressure angle fine-pitch gears of 20 DP and finer, except bevel gears. Greater clearances than for coarse-pitch gears are required to allow for

A. Greater wear on point of generating tool.

B. Fillet radius produced by dull generating tool is greater than derived from formula 0.157/DP.

C. Accumulation of foreign matter at bottom of teeth.

AGMA Standard 207.01 establishes the tooth form for fine-pitch involute spur and helical gears of 20-degree normal pressure angle. Table 33 gives the standard tooth pro-

Table 33. Standard Tooth Proportions for Fine-pitch Gears

N = number of teeth in gear n = number of teeth in pinion
P = diametral pitch Dimensions are in inches.

Addendum	$\frac{1.000}{P}$	Whole depth	$\frac{2.200}{P} + 0.002$	Pitch diameter	$\frac{N}{P} : \frac{n}{P}$
Dedendum	$\frac{1.200}{P} + 0.002$	Clearance	$\frac{0.200}{P} + 0.002$	Outside diameter	$\frac{N+2}{P} : \frac{n+2}{P}$
Working depth	$\frac{2.000}{P}$	Tooth thickness on pitch diameter	$\frac{1.5708}{P}$	Center distance	$\frac{N+n}{2P}$

portions for this system. Pinions having seventeen or fewer teeth are enlarged in accordance with Table 34. Two options are given in this standard:

A. Standard center distance system (long and short addendum).

B. Enlarged center distance system.

Table 34. 20-Degree Pressure Angle Involute Fine-pitch System for Spur Gears

Dimensions Required When Using Enlarged Pinions

The outside diameters of small pinions are enlarged to avoid undercut. To maintain standard center distances when using enlarged pinions, the mating-gear diameters must be decreased by the amount of the pinion enlargement. If mating gears are made with standard tooth proportions, the center distances must be increased as shown below. Tabular values are in inches for 1 DP. For other pitches, divide tabular values by the diametral pitch.

Number of Teeth	Pinion Dimensions		Standard Center Distance System (Long and Short Addendum) Gear Dimensions				Enlarged Center Distance System	
							Standard Mating Gear Diameter	
	Outside Diameter	Cir. Tooth Thickness at Standard Pitch Diameter	Decrease in Standard Outside Diameter	Cir. Tooth Thickness at Standard Pitch Diameter	Recommended Minimum No. of Teeth (N_g)	Contact Ratio N_p Mating with N_g	Increase Over Standard Center Distance	Contact Ratio N_p Mating Itself
10	12.8302	1.8730	.8302	1.2686	33	1.419	.4151	1.135
11	13.7132	1.8304	.7132	1.3112	30	1.450	.3566	1.186
12	14.5963	1.7878	.5963	1.3538	27	1.473	.2982	1.238
13	15.4793	1.7452	.4793	1.3964	25	1.493	.2397	1.290
14	16.3623	1.7027	.3623	1.4389	23	1.508	.1812	1.344
15	17.2453	1.6601	.2453	1.4815	21	1.516	.1227	1.398
16	18.1284	1.6175	.1284	1.5241	19	1.519	.0642	1.436
17	19.0114	1.5749	.0114	1.5667	18	1.522	.0057	1.511

Each system has some advantages and disadvantages over the other. The standard does not recommend the use of pinions having fewer than ten teeth. Almost all the advantages of the involute curve are lost when pinions having fewer than ten teeth are enlarged to avoid objectionable undercut.

ENLARGEMENT OF HELICAL PINIONS OF 20-DEGREE NORMAL PRESSURE ANGLE. The enlargement of helical pinions to avoid undercut cannot be tabulated as

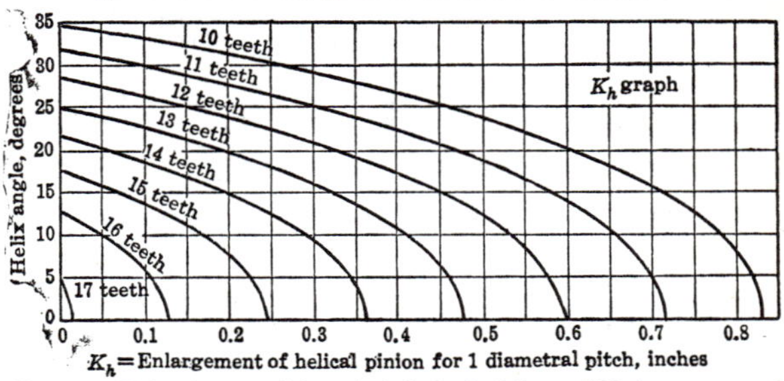

FIG. 75. Enlargement of fine-pitch helical pinions of 20-degree normal pressure angle. (AGMA Standard 207.01)

simply as for spur pinions because of the wide range of helix angles used. It is recommended that helical pinions be enlarged in accordance with Fig. 75.

Notation

n = number of teeth
P_n = normal diametral pitch
P_t = transverse diametral pitch
ψ = helix angle
d = pitch diameter
d_o = outside diameter
K_h = enlargement of helical pinion for 1 normal DP from graph (Fig. 75)

Formulas

$$P_t = P_n \cos \psi \quad (1) \qquad d = \frac{n}{P_t} \quad (2) \qquad d_o = d + \frac{2 + K_h}{P_n} \quad (3)$$

Example. *Given:* 12 teeth, 32 normal DP 20-degree normal pressure angle, and 18-degree helix angle

$$P_t = 32 \times 0.95106 = 30.4339 \quad (1) \qquad K_h = 0.388 \text{ (from graph, Fig. 75)}$$

$$d = \frac{12}{30.4339} = 0.3943 \quad (2) \qquad d_o = 0.3943 + \frac{2 + 0.388}{32} = 0.468 \quad (3)$$

When an enlarged helical pinion is used, either the mating gear must be reduced in diameter or the center distance must be increased to correspond with the enlargement, as in spur gears.

INSPECTION AND TOLERANCES OF FINE-PITCH GEARS. AGMA Standard 236.01 covers the inspection and tolerances of fine-pitch gears. This standard was developed to meet the diverse applications of fine-pitch gears and safeguard the interests of quality control. Fine-pitch gears lend themselves readily and economically to inspection methods in which measurements are made under conditions closely approaching those of their operation. The gear is rotated in intimate contact with a master, both mounted on a variable center-distance fixture. The displacements or variations in center distance are measured or charted by suitable means, giving a composite check. (See Fig. 76.)

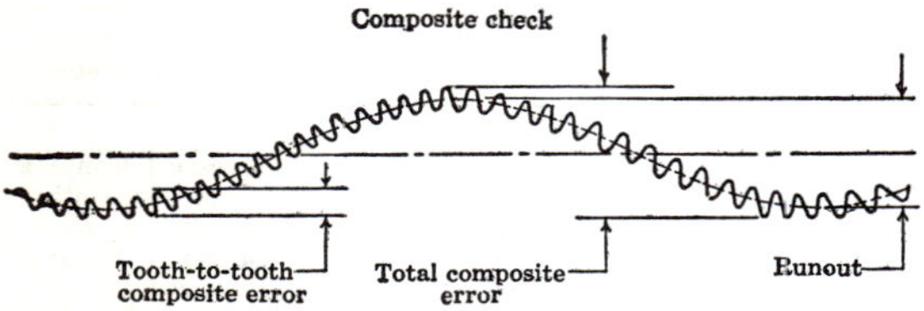

Fig. 76. Displacement variation with fine pitch gears. (AGMA Standard 236.02)

Quantity specifications are made on the basis of the results of the composite check, which do not involve backlash since backlash does not directly affect the accuracy of the gear operation and may be produced or eliminated in the mountings. A choice of backlash to suit the particular application is provided in this standard by giving each class group several ranges of backlash.

Tolerances for Commercial Fine-pitch Gears

Class	Total Composite Error, in.	Tooth-to-tooth Composite Error, in.
Commercial 1	.006	.002
Commercial 2	.004	.0015
Commercial 3	.002	.001

Tolerances for Precision Fine-pitch Gears

Class	Total Composite Error, in.	Tooth-to-tooth Composite Error, in.
Precision 1	.001	.0004
Precision 2	.0005	.0003
Precision 3	.00025 *	.0002

* This possibly would be the result of selection and segregation.

Backlash is specified by a letter suffix to the class number, such as Commercial 1B or Precision 3C.

Backlash for Different Classes of Gears

Diametral Pitch	Backlash, in.
Class A	
20 to 45	.004 to .006
46 to 70	.003 to .005
71 to 90	.002 to .0035
Class B	
20 to 60	.002 to .004
61 to 120	.0015 to .003
121 and finer	.001 to .002
Class C	
20 to 60	.001 to .002
61 to 120	.0007 to .0015
121 and finer	.0005 to .001
Class D	
No measurable backlash at any pitch.	

FINE PITCH WORM GEARING (AGMA Standard 374.01) covers eight axial worm pitches (p_x):

0.030 in.	0.080 in.
0.040 in.	0.100 in.
0.050 in.	0.130 in.
0.065 in.	0.160 in.

and fifteen lead angles (λ):

0.5 degree	4.0 degrees	14.0 degrees
1.0 degree	5.0 degrees	17.0 degrees
1.5 degrees	7.0 degrees	21.0 degrees
2.0 degrees	9.0 degrees	25.0 degrees
3.0 degrees	11.0 degrees	30.0 degrees

Worm pitch diameter (d) is calculated from the equation

$$d = \frac{l}{\pi \tan \lambda} = \frac{np_x}{\pi \tan \lambda}$$

where l is the lead and n is the number of threads in the worm. Minimum recommended worm pitch diameter is 0.250 in., and the maximum is 2.000 in. Values of d are given in Table 35.

Worm gear pitch diameter (D) is calculated from the equation

$$D = \frac{Np}{\pi}$$

where N is the number of teeth in the worm gear and p is the circular pitch of the worm gear. Pitch relations are expressed by the equation:

$$p_n = p_x \cos \lambda = p \cos \chi$$

where p_n = normal circular pitch of worm and worm gear and χ = helix angle of worm gear. If the worm and worm gear shafts are at right angles, $\lambda = \chi$, and $p_x = p$.

Tooth Proportions. Pressure angle in the normal plane (ϕ_n) is approximately 20 degrees and the tooth proportions are as follows (see also Table 36):

$$\begin{aligned} &\text{Addendum } a &&= 0.3183p_n \\ &\text{Whole depth } h_t &&= 0.7003p_n + 0.002 \text{ in.} \\ &\text{Working depth } h_k &&= 0.6366p_n \\ &\text{Clearance } c &&= h_t - h_k \\ &\text{Tooth thickness } t &&= 0.5p_n \end{aligned}$$

Backlash in fine-pitch worm gearing is produced by thinning the worm threads by a suitable amount. The worm gear is cut so that its tooth thickness and space width are equal at the standard pitch diameter.

Worm Dimensions. Outside diameter of the worm (d_o) is

$$d_o = d + 2a$$

and safe minimum length of threaded portion (F_W) is

$$F_W = \sqrt{D_o^2 - D^2}$$

Worm Gear Dimensions. Two forms of worm gear blanks are used: a throated blank (for power drives), and a nonthroated blank (for transmission of motion). These are illustrated in Fig. 77.

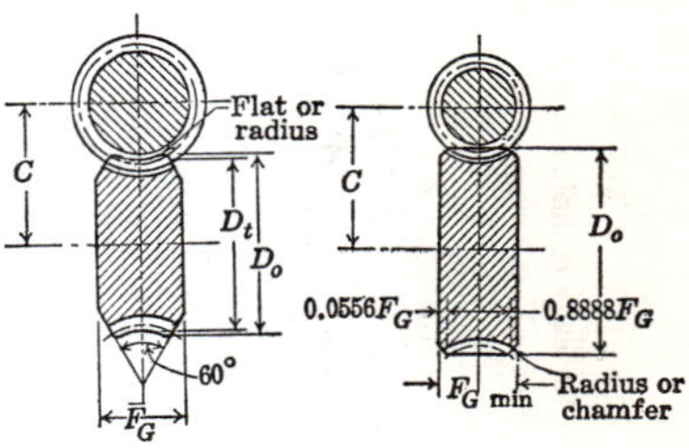

FIG. 77. Fine pitch worm gears: (a) throated gear; (b) nonthroated gear. Type a is suitable for power drives, type b for transmission of motion.

For the throated blank (Fig. 77a) the recommended dimensions are:

$$\begin{aligned} &\text{Outside diameter } D_o &&= 2C - 0.891d + 1.782a \\ &\text{Face width } F_G &&= 0.57735d_o \\ &\text{Throat diameter } D_t &&= D + 2a \end{aligned}$$

where C = center distance = $\frac{1}{2}(d + D)$.

For the nonthroated blank (Fig. 77b) the recommended dimensions are:

$$\begin{aligned} &\text{Outside diameter } D_o &&= 2C - d + 2a \\ &\text{Minimum face width } F_{G\,\min} &&= 1.125\sqrt{(d_o + 2c)^2 - (d - 4a)^2} \end{aligned}$$

Other proportions are shown in Fig. 77.

(Continued on p. 14-61)

Table 35. Pitch Diameters of Fine-pitch Worms

(This table gives the pitch diameter for each combination of lead and lead angle, together with the number of threads for a particular lead and diameter.)

n	Lead, in.	Lead Angles in Degrees														
		0.5	1.0	1.5	2.0	3.0	4.0	5.0	7.0	9.0	11.0	14.0	17.0	21.0	25.0	30.0
1	0.030	1.0937	0.5472	0.3647	0.2735											
1	0.040	1.4583	0.7297	0.4863	0.3646	0.2429										
1	0.050	1.8228	0.9121	0.6079	0.4558	0.3037	0.2276									
2	0.060	2.1874	1.0945	0.7295	0.5469	0.3644	0.2731									
1	0.065		1.1857	0.7903	0.5925	0.3948	0.2959	0.2365								
1,2	0.080		1.4593	0.9726	0.7293	0.4859	0.3641	0.2911								
3	0.090		1.6417	1.0942	0.8204	0.5466	0.4097	0.3274	0.2333							
1,2	0.100		1.8242	1.2158	0.9116	0.6073	0.4552	0.3638	0.2592							
3,4	0.120		2.1890	1.4590	1.0939	0.7288	0.5462	0.4366	0.3111	0.2412						
1,2	0.130			1.5805	1.1851	0.7896	0.5917	0.4730	0.3370	0.2613						
3,5	0.150			1.8237	1.3674	0.9110	0.6828	0.5457	0.3889	0.3015	0.2456					
1,2,4	0.160			1.9453	1.4585	0.9718	0.7283	0.5821	0.4148	0.3216	0.2620					
6	0.180			2.1884	1.6408	1.0932	0.8193	0.6549	0.4667	0.3618	0.2948					
3	0.195				1.7776	1.1843	0.8876	0.7095	0.5055	0.3919	0.3193	0.2490				
2,4,5	0.200				1.8232	1.2147	0.9104	0.7276	0.5185	0.4020	0.3275	0.2553				
7	0.210				1.9143	1.2754	0.9559	0.7640	0.5444	0.4221	0.3439	0.2681				
3,6,8	0.240				2.1878	1.4576	1.0924	0.8732	0.6222	0.4823	0.3930	0.3064	0.2499			
5	0.250					1.5184	1.1380	0.9096	0.6481	0.5024	0.4094	0.3192	0.2603			
2,4	0.260					1.5791	1.1835	0.9459	0.6741	0.5225	0.4258	0.3319	0.2707			
9	0.270					1.6398	1.2290	0.9823	0.7000	0.5426	0.4421	0.3447	0.2811			
7	0.280					1.7006	1.2745	1.0187	0.7259	0.5627	0.4585	0.3575	0.2915			
3,6,10	0.300					1.8220	1.3656	1.0915	0.7778	0.6029	0.4913	0.3830	0.3123	0.2488		
2,4,8	0.320					1.9435	1.4566	1.1642	0.8296	0.6431	0.5240	0.4085	0.3332	0.2654		
5	0.325					1.9739	1.4794	1.1824	0.8426	0.6532	0.5322	0.4149	0.3384	0.2695		
7	0.350					2.1257	1.5932	1.2734	0.9074	0.7034	0.5731	0.4468	0.3644	0.2902		

9	0.360						1.6387	1.3098	0.9333	0.7235	0.5895	0.4596	0.3748	0.2985	0.2457	
3,6	0.390						1.7752	1.4189	1.0111	0.7838	0.6387	0.4979	0.4060	0.3234	0.2662	
4,5,8,10	0.400						1.8207	1.4553	1.0370	0.8039	0.6550	0.5107	0.4165	0.3317	0.2730	
9	0.450						2.0483	1.6372	1.1666	0.9044	0.7369	0.5745	0.4685	0.3732	0.3072	.2481
7	0.455							1.6554	1.1796	0.9144	0.7451	0.5809	0.4737	0.3773	0.3106	.2509
3,6	0.480							1.7463	1.2444	0.9647	0.7860	0.6128	0.4998	0.3980	0.3277	.2646
5,10	0.500							1.8191	1.2963	1.0049	0.8188	0.6383	0.5206	0.4146	0.3413	.2757
4,8	0.520							1.8919	1.3481	1.0451	0.8515	0.6639	0.5414	0.4312	0.3550	.2867
7	0.560							2.0374	1.4518	1.1255	0.9170	0.7149	0.5830	0.4644	0.3823	.3087
9	0.585								1.5166	1.1757	0.9580	0.7469	0.6091	0.4851	0.3993	.3225
6	0.600								1.5555	1.2059	0.9825	0.7660	0.6247	0.4975	0.4096	.3308
4,8	0.640								1.6592	1.2863	1.0480	0.8171	0.6663	0.5307	0.4369	.3529
5,10	0.650								1.6852	1.3064	1.0644	0.8298	0.6767	0.5390	0.4437	.3584
7	0.700								1.8148	1.4068	1.1463	0.8937	0.7288	0.5805	0.4778	.3859
9	0.720								1.8666	1.4470	1.1791	0.9192	0.7496	0.5971	0.4915	.3970
6	0.780								2.0222	1.5676	1.2773	0.9958	0.8121	0.6468	0.5324	.4300
5,8,10	0.800									1.6078	1.3101	1.0213	0.8329	0.6634	0.5461	.4411
9	0.900									1.8088	1.4738	1.1490	0.9370	0.7463	0.6144	.4962
7	0.910									1.8289	1.4902	1.1618	0.9474	0.7546	0.6212	.5017
6	0.960									1.9294	1.5721	1.2256	0.9995	0.7961	0.6553	.5293
10	1.000									2.0098	1.6376	1.2767	1.0412	0.8292	0.6826	.5513
8	1.040										1.7031	1.3277	1.0828	0.8624	0.7099	5734
7	1.120										1.8341	1.4299	1.1661	0.9287	0.7645	.6175
9	1.170										1.9160	1.4937	1.2181	0.9720	0.7987	.6451
8	1.280										2.0961	1.6341	1.3327	1.0614	0.8738	.7057
10	1.300											1.6597	1.3535	1.0780	0.8874	.7167
9	1.440											1.8384	1.4993	1.1941	0.9830	.7939
10	1.600											2.0427	1.6658	1.3268	1.0922	.8821

Table 36. Fine-pitch Worm Gear Tooth Proportions Based on Normal Pitch for All Combinations of Standard Axial Pitches and Lead Angles

Standard Axial Pitch, in.	Tooth Proportions	Lead Angle in Degrees														
		0.5	1	1.5	2	3	4	5	7	9	11	14	17	21	25	30
.030	a	.0095	.0095	.0095	.0095	.0095	.0095	.0095	.0095	.0094	.0094	.0093	.0091	.0089		
	h_t	.0229	.0229	.0229	.0229	.0229	.0229	.0229	.0229	.0227	.0227	.0225	.0220	.0216		
	p_n	.0300	.0300	.0300	.0300	.0300	.0299	.0299	.0298	.0296	.0294	.0291	.0287	.0280		
.040	a	.0127	.0127	.0127	.0127	.0129	.0127	.0127	.0126	.0126	.0125	.0124	.0122	.0119	.0115	
	h_t	.0299	.0299	.0299	.0299	.0299	.0299	.0299	.0297	.0297	.0295	.0293	.0288	.0282	.0273	
	p_n	.0400	.0400	.0400	.0400	.0399	.0399	.0398	.0397	.0395	.0393	.0388	.0383	.0373	.0363	
.050	a	.0159	.0159	.0159	.0159	.0159	.0159	.0159	.0158	.0157	.0156	.0154	.0152	.0149	.0144	.0138
	h_t	.0370	.0370	.0370	.0370	.0370	.0370	.0370	.0368	.0365	.0363	.0359	.0354	.0348	.0337	.0324
	p_n	.0400	.0500	.0500	.0500	.0499	.0499	.0498	.0496	.0494	.0491	.0485	.0478	.0467	.0453	.0433
.065	a		.0207	.0207	.0207	.0207	.0206	.0206	.0205	.0204	.0203	.0201	.0198	.0193	.0188	.0179
	h_t		.0475	.0475	.0475	.0475	.0473	.0473	.0471	.0469	.0467	.0462	.0456	.0445	.0434	.0414
	p_n		.0650	.0650	.0650	.0649	.0648	.0648	.0645	.0642	.0638	.0631	.0622	.0607	.0589	.0563
.080	a		.0255	.0255	.0254	.0254	.0254	.0254	.0253	.0252	.0250	.0247	.0244	.0238	.0231	.0221
	h_t		.0581	.0581	.0579	.0579	.0579	.0579	.0577	.0574	.0570	.0563	.0557	.0544	.0528	.0506
	p_n		.0800	.0800	.0800	.0799	.0798	.0797	.0794	.0790	.0785	.0776	.0765	.0747	.0725	.0693
.100	a		.0318	.0318	.0318	.0318	.0318	.0317	.0316	.0314	.0312	.0309	.0304	.0297	.0288	.0276
	h_t		.0720	.0720	.0720	.0720	.0720	.0717	.0716	.0711	.0706	.0700	.0689	.0673	.0654	.0627
	p_n		.1000	.1000	.0999	.0999	.0998	.0996	.0993	.0988	.0982	.0970	.0956	.0934	.0906	.0866
.130	a			.0414	.0414	.0413	.0413	.0412	.0411	.0409	.0406	.0402	.0396	.0386	.0375	.0358
	h_t			.0931	.0931	.0929	.0929	.0926	.0924	.0920	.0913	.0904	.0891	.0869	.0845	.0808
	p_n			.1300	.1299	.1298	.1297	.1295	.1290	.1284	.1276	.1261	.1243	.1214	.1178	.1126
.160	a			.0509	.0509	.0509	.0508	.0507	.0506	.0503	.0500	.0494	.0487	.0475	.0462	.0441
	h_t			.1140	.1140	.1140	.1138	.1135	.1133	.1127	.1120	.1107	.1091	.1065	.1036	.0990
	p_n			.1599	.1599	.1598	.1596	.1594	.1588	.1580	.1571	.1552	.1530	.1494	.1450	.1386

Manufacture. The standard recognizes that worm profiles depend on method of manufacture, and depart from the calculated shape when production methods differ, even for the same worm. It is emphasized that manufacturing information concerning the work must be furnished when ordering gear cutting tools in order to obtain gears which are conjugate to worms. The standard includes a table of profile deviations and pressure angle changes covering the entire range of the standard.

11. MATERIALS USED IN GEARING

Materials, as such, are covered extensively in Sections 2, 3, 4, and 5 of this handbook. The problem of gearing with highly stressed surfaces and tooth structures should impress the reader with the necessity of using good-quality materials in gear design.

In the AGMA recommended practices, steels softer than 180 Brinell are not listed since good practice has been to run them with harder mates so that work hardening in service brings their "gear qualities" up close to those listed for 180 Brinell.

Referring to Fig. 50, we note that differences in hardness between the pinion (smaller member) and the gear are arbitrarily given. As has been mentioned throughout this chapter, some values are selected from experience, and this particular hardness relationship is one of them. Where exceptionally high ratios are encountered, such as in ball mill, rod mill, and kiln drives, the suggestion has been made that differing degrees in hardness might be used. In such cases the K_r factor is based (Fig. 78) on the hardness of the gear material, and a modifying factor, W_H (Fig. 79), is applied in arriving at wear-capacity ratings.

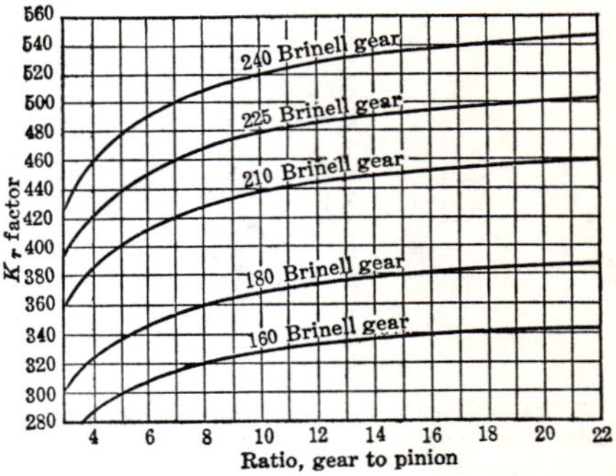

FIG. 78. Factor K_r for open large reduction gearing.

The importance of high-strength materials coupled with the awareness of load distribution has caused the gear designer to explore many treatments that promise the best compromise to these often troublesome factors. As a result, flame hardening, nitriding, shot peening of stressed areas, burnishing or cold working, and many other processes are being employed in the gear-manufacturing industry.

AGMA recommended practices for materials are covered in their Standard 241.01 for steels and 242.02 for cast iron.

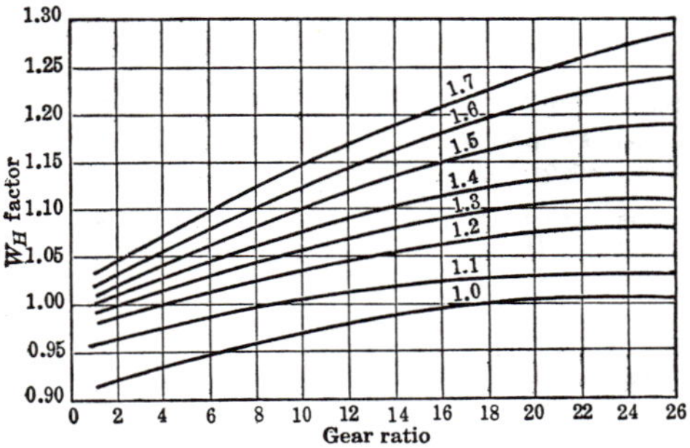

FIG. 79. Variation of W_H factor according to ratio of pinion Brinell hardness to gear hardness.

12. LUBRICATION OF GEARING

The lubrication of enclosed and open gearing requires special attention, since the areas of contact on gear teeth (regardless of type of construction) are relatively small, and the unit pressures developed in transmitted rated loads are relatively high. It is, therefore, essential to use a lubricant of sufficient film strength to withstand the localized pressure during the short period of contact.

Unfortunately, lubricating-oil specifications have not been developed to the point where they can be referred to as such. As a consequence, certain general characteristics to be looked for in selecting a gear lubricant are summarized in the following paragraphs.

Lubricating oils for geared mechanisms should be high grade, high quality, and well refined. They must not be corrosive to gear materials, and should be neutral in reaction and free from grit and abrasives. In addition, gear lubricants should have good defoaming properties, and, where high operating temperatures are encountered, their resistance to oxidation should be checked.

Straight mineral types of lubricants are usually recommended for all types of gears except those of the worm and hypoid construction. For worm gearing, additions of 3 to 10% acidless tallow or animal fats to straight mineral lubricant is usually found to be desirable. Hypoid gears require special attention, for which acceptable grades of extreme pressure lubricants can be selected.

Tables 37, 38, 39, 40, and 41 are taken from the AGMA Standard Specification 250.01, covering lubrication of enclosed and open gearing.

Table 37. Viscosity Range for Various AGMA Lubricants

AGMA Lubricant Number	Viscosity Range, SUV seconds At 100 F	At 210 F
1	180 to 240	....
2	280 to 360	....
3	490 to 700	....
4	700 to 1000	....
5	....	80 to 105
6	....	105 to 125
7	....	125 to 150
7 Comp.*	....	125 to 150
8	....	150 to 190
8 Comp.*	....	150 to 190
9	....	350 to 550
10	....	900 to 1200
11	....	1800 to 2500

* The oils marked "Comp." are compounded with 3 to 10% of acidless tallow or other suitable animal fat. Oils 9, 10, and 11 should be of a tacky adhesive nature so that they will adhere well to the tooth surfaces.

Table 38. Recommended Lubricants for Enclosed Units of All Types except Worm Gears

Type of Units	Size of Units	Ambient Temperature, °F 0–40 Use AGMA No.	41–100 Use AGMA No.	101–150 Use AGMA No.
Parallel shaft units	Low-speed, centers up to 20 in.	2	4	5
	Low-speed, centers over 20 in.	3	5	6
Planetary gear units	OD of housing up to 16 in.	2	3	4
	OD of housing over 16 in.	3	4	5
Gear motors	All sizes	2	4	5
Spiral or straight bevel gear units	Cone distance up to 12 in.	2	4	5
	Cone distance over 12 in.	3	5	6
High-speed units	All sizes	1	2	3

Table 39. Recommended Lubricants for Enclosed Worm-gear Units Only

Service and Worm Speed	Ambient Temperature, °F 0–40 Use AGMA No.	41–90 Use AGMA No.	91–120 Use AGMA No.
Intermittent operation—all worm speeds *	5	5	7 Comp.
Continuous operation—worm speeds below 600 rpm	7 Comp. diluted †	8 Comp.	8 Comp.
Continuous operation—worm speeds 600 rpm and over	7 Comp. diluted †	7 Comp.	7 Comp.

* Where period of operation is insufficient to produce any appreciable rise in oil bath temperature.

† Diluted AGMA 7 comp. oil should be diluted with a lighter oil, preferably not exceeding 500 sec viscosity at 100 F, until the desired fluidity is obtained. The lubricant used for dilution should be of the same basic crude as that of the recommended oil. The lubricant supplier should be consulted if there is any doubt.

Table 40. Recommended Lubricants for Open Gearing of All Types except Worm Gears

Method	Ambient Temperature, °F 0–40 Use AGMA No.	41–100 Use AGMA No.	101–150 Use AGMA No.
Slush pans	3	5	6
Hot—by brush or paddle	10	10	11
Cold—by brush or paddle	6	8	10
Hand-oiled	4	6	8

Table 41. Recommended Lubricants for Open Worm Gears Only

Method	Ambient Temperature, °F 0–40 Use AGMA No.	41–100 Use AGMA No.	101–150 Use AGMA No.
Cold or hot—by brush or paddle	6	8	..

Lubrication Maintenance. (*a*) Ambient temperature, which is here defined as the temperature of the air in the vicinity of the gears or gear unit, has considerable bearing on the viscosity of oil to be used. In general, the viscosity of mineral oil varies inversely with the temperature. This makes it desirable to use an oil of heavier grade at high ambient temperatures than would be used for the same application operating in a low ambient temperature. Also, for low temperatures, an oil must be selected which has a pour point lower than the minimum temperature expected for the application.

(*b*) Sludge in gear units may be caused by metallic particles ground off the gears, by dust and dirt, and by the presence of moisture or chemical fumes. Therefore, every precaution should be taken to prevent water and foreign particles from entering the gear case.

(*c*) Open gears operating in dusty atmospheres should be periodically cleaned with kerosene or other suitable solvent. The dirt and grit adhering to them can seriously damage the tooth surfaces if it continuously passes through the mesh.

ACKNOWLEDGMENTS. Much of the material on conventional worm gearing, spiral gearing, and elliptical gearing has been reproduced from Professor Earle Buckingham's material used in the previous edition.

The data on bevel gearing are taken from material supplied by Mr. F. L. Knowles of the Gleason Works, who is chairman of the Bevel Gear Committee of the AGMA.

Louis D. Martin, gear engineer of Eastman Kodak and chairman of the Fine-Pitch Committee of the AGMA, assisted in culling out of AGMA data, pertinent information for this section.

The AGMA was most cooperative in locating and supplying information on industry standards.

Dean McCurdy, a senior at the University of Washington, prepared the majority of the new illustrations used in this section.

REFERENCES

Note. This list of references, although numbered, is more general than specific. Only a few of the items are referred to in the text.

Books

1. Earle Buckingham, *Spur Gears*, McGraw-Hill Book Co., New York.
2. W. C. Unwin, *The Elements of Machine Design*, Longmans, Green & Co., London.
3. C. H. Logue, *American Machinist Gear Book*, McGraw-Hill Book Co., New York.
4. P. M. Heldt, *Torque Converters and Transmissions*, P. M. Heldt, Nyack, N. Y.
5. O. A. Leutwiler, *Elements of Machine Design*, McGraw-Hill Book Co., New York.
6. *Maintenance of Gear Drives*, Mill and Factory Maintenance Manual 18.
7. Earle Buckingham, *Manual of Gear Design*, Sect. 1, Mathematical Tables; Sect. 2, Spur and Internal Gears, *Machinery*, New York.
8. H. E. Merritt, *Gears*, Sir Isaac Pitman & Sons, London.
9. Reginald Trautschold, *Standard Gear Book*, McGraw-Hill Book Co., New York.
10. F. H. Colvin and F. A. Stanley, *Gear Cutting Practices*, McGraw-Hill Book Co., New York.

Manufacturers' Literature

11, 12, 13. Fellows Gear Shaper Co., *The Involute Gear, The Internal Gear, The Helical Gear.*
14. Michigan Tool Co., *Involutrometry and Trigonometry.*
15, 16. Brown and Sharpe Manufacturing Co., *Practical Treatise on Gearing, Formulas in Gearing.*
17, 18, 19. Gleason Works, *The Straight Bevel Gear System, The Spiral Bevel Gear System, Stress Determination and Fatigue Life of Generated Bevel Gears.*
20. Socony Vacuum Oil Co., *Gears and Their Lubrication.*
21. The Texas Co., *Industrial Gears—Designing for Lubrication*, Sept. 1945.

Articles

22. McMullen and Durkan, The Gleason Works system of bevel gears, *Machinery, N. Y.*, June 1922, p. 788.
23. Paul Heymans and A. L. Kimball, Jr., Photoelastic method of studying gear tooth stresses, *Trans. Am. Soc. Mech. Engrs.*, Dec. 1922.

24. Paul Heymans, Mathematical theory of dynamic stresses in rotating gear pinions, *Mech. Eng.*, Vol. 46, pp. 583–587, 1924.
25. S. Timoshenko and R. V. Baud, Strength of gear teeth, *AGMA*, May 1926, and *Trans. Am. Soc. Mech. Engrs.*, Nov. 1926.
26. A. W. Frehse, Some thoughts on present-day automobile transmissions, *S.A.E. Journal*, Jan. 1936.
27. J. O. Almen and J. C. Straub, Durability of spiral bevel gears for automobiles, *Automotive Ind.*, Nov. 16 and Nov. 23, 1935.
28. J. O. Almen and J. C. Straub, Durability of automobile transmission gears, *Automotive Ind.*, Sept. 25 and Oct. 9, 1937.
29. Stewart Way, Roller tests to determine pitting fatigue strength, *AGMA*, Sept. 1937.
30. F. R. McFarland, Aircraft engine gears, *S.A.E. Journal*, Sept. 1945.
31. F. E. Birtch, Cone-Drive, *AGMA*, Oct. 1947.
32. Guy J. Talbourdet, Graphical analysis of planetary gear systems, *Machine Design*, Aug., Sept., and Oct. 1942.
33. L. J. Collins, Gearing for steel mill auxiliary cranes, *Iron Steel Engr.*, May 1947.
34. A. H. Candee, Involute gear calculations simplified, *Am. Machinist*, Sept. 13, 1945.
35. Ernest Wildhaber, Gear tooth curvature treated simply, *Am. Machinist*, Aug. 30, 1945.
36. The elements of gearing series, *The Tool Engineer*, Dec. 1945 through April 1946.
37. W. P. Schmitter, Determining the capacity of helical and herringbone gearing, *Machine Design*, June and July 1934.

American Gear Manufacturers Association Standards

111.01—Letter Symbols for Gear Engineering.
206.02—Fine-Pitch Straight Bevel Gears.
207.01—20° Involute Fine-Pitch System for Spur Gears.
210.01—Surface Durability of Spur Gears.
211.01—Surface Durability of Helical and Herringbone Gears.
212.01—Surface Durability of Straight Bevel Gears and Spiral Bevel Gears.
220.01—Strength of Spur Gear Teeth.
221.01—Strength of Helical and Herringbone Gear Teeth.
222.01—Beam Strength of Straight Bevel and Spiral Bevel Gears.
231.01—Gear Tolerances and Inspection—Spur and Helical Gears.
232.01—Gear Tolerances and Inspection—Bevel and Hypoid Gears.
233.01—Gear Tolerances and Inspection—Backlash.
235.01—Gear Tolerances and Inspection—Master Gears.
236.01—Gear Tolerances and Inspection—Fine-Pitch Gears.
250.01—Lubrication of Enclosed and Open Gearing.

American Standards Association

Gear Tolerances and Inspection (B.6-1946).
Spur Gear Tooth Form (B6.1-1932).
Gear Materials and Blanks (B6.2-1933).
Letter Symbols for Gear Engineering (B6.5-1943).
Milling Cutters (B5.c-1930).

British Standards Institution Specifications*

436—Helical and Straight Spur.
545—Bevel (with Helical, Curve, and Straight Teeth).
721—Worm Gearing.

* Copies of the British Standards Specifications may be obtained from the British Standards Institution, 28 Victoria Street, London, S.W. 1, England.

SECTION 15

MECHANICAL POWER TRANSMISSION

By

A. F. GAGNE, JR., *Development Engineering Division, E. I. du Pont de Nemours and Co.*

R. M. NARDONE, *Chief Engineer, Victory Engineering Corp.*

W. A. WILLIAMS, *Chief Engineer, The American Pulley Co.*

M. L. PRICE, *Professor of Machine Design, Worcester Polytechnic Institute*

PHILIP G. RHOADS, *J. E. Rhoads and Sons*

E. L. SCHWARZ-KAST, *University of Chicago*

H. A. WILSON, *Allis-Chalmers Manufacturing Co.*

SHAFTS, KEYS, AND SPLINES

1. SHAFTING

STANDARD SIZES for finished shafting are given in Table 1. Diameters tabulated under Transmission Shafting agree with the commercial sizes of bearings, couplings, etc., and are usually supplied in the full stock lengths of 16, 20, or 24 ft, as desired. The series designated Machinery Shafting, which includes a larger number of sizes in the same range of diameter, is intended for machine construction, etc., and is generally desired in shorter lengths. Three types of shafting are contemplated within the standard: (1) cold drawn, (2) turned and polished, and (3) turned, ground, and polished.

Table 1. American Standard Finished Steel Shafting

(ASA B17.1-1943)

Stock Diameter		Tolerance on Diameter, in.	Stock Diameter		Tolerance on Diameter, in.	Stock Diameter		Tolerance on Diameter, in.
Transmission Shafting Sizes, in.	Machinery Shafting Sizes, in.		Transmission Shafting Sizes, in.	Machinery Shafting Sizes, in.		Transmission Shafting Sizes, in.	Machinery Shafting Sizes, in.	
....	1/2	−.002	....	1 13/16	−.003	....	3 3/4	−.004
....	9/16	−.002	....	1 7/8	−.003	....	3 7/8	−.004
....	5/8	−.002	1 15/16	1 15/16	−.003	3 15/16	4	−.004
....	11/16	−.002	....	2	−.003	....	4 1/4	−.005
....	3/4	−.002	....	2 1/16	−.004	4 7/16	4 1/2	−.005
....	13/16	−.002	....	2 1/8	−.004	....	4 3/4	−.005
....	7/8	−.002	2 3/16	2 3/16	−.004	4 15/16	5	−.005
15/16	15/16	−.002	....	2 1/4	−.004	....	5 1/4	−.005
....	1	−.002	....	2 5/16	−.004	5 7/16	5 1/2	−.005
....	1 1/16	−.003	....	2 3/8	−.004	....	5 3/4	−.005
....	1 1/8	−.003	2 7/16	2 7/16	−.004	5 15/16	6	−.005
1 3/16	1 3/16	−.003	....	2 1/2	−.004	....	6 1/4	.006
....	1 1/4	−.003	....	2 5/8	−.004	6 1/2	6 1/2	.006
....	1 5/16	−.003	....	2 3/4	−.004	....	6 3/4	.006
....	1 3/8	−.003	....	2 7/8	−.004	7	7	.006
1 7/16	1 7/16	−.003	2 15/16	3	−.004	....	7 1/4	.006
....	1 1/2	−.003	....	3 1/8	−.004	7 1/2	7 1/2	.006
....	1 9/16	−.003	....	3 1/4	−.004	....	7 3/4	.006
....	1 5/8	−.003	....	3 3/8	−.004	8	8	.006
1 11/16	1 11/16	−.003	3 7/16	3 1/2	−.004	....	....	....
....	1 3/4	−.003	....	3 5/8	−.004	....	....	....

SHAFT HANGERS are made of cast iron or pressed steel in a variety of forms by different manufacturers. Typical designs are illustrated in Figs. 1, 2, and 3. The frames will accommodate various types of bearings such as ball-and-socket bearings, ball or roller bearings, ring or capillary oiling bearings, and plain babbitted bearings. Hangers usually have adjusting screws to adjust the bearing in a vertical plane, to conform to the desired shaft height.

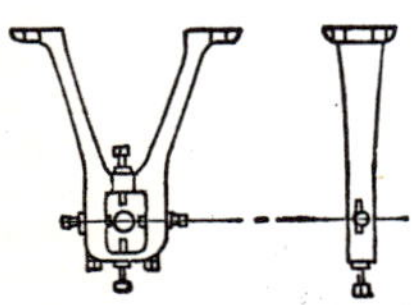

Fig. 1. Shaft hanger.

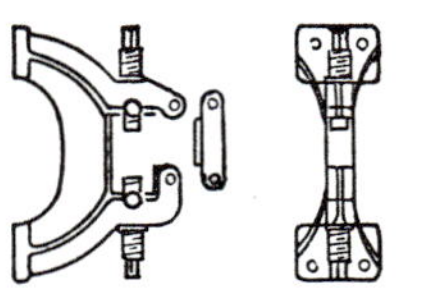

Fig. 2. Adjustable ball-and-socket post hanger.

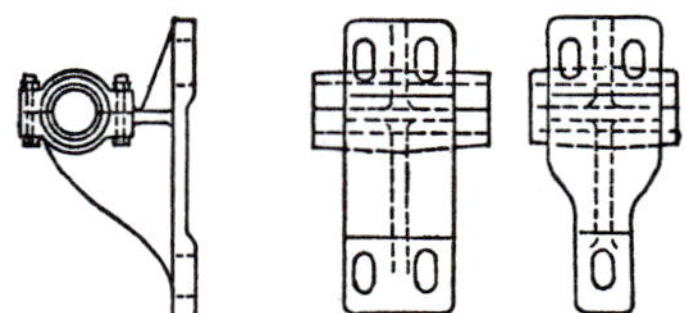

Ring oiling bearings $1\frac{7}{16}$-$4\frac{15}{16}$ in. Babbitted bearings over 3 in.

Babbitted bearings up to 3 in. shaft

Fig. 3. Rigid post bearing.

DISTANCE BETWEEN BEARINGS. The spacing of hangers for line shafting depends on the size of shaft and the points at which load is applied. The following figures represent average practice with pulleys placed near the bearings.

Shaft size, in.	15/16–1 7/16	1 11/16–2 3/16	2 7/16–2 15/16	3 7/16–3 15/16	4 7/16–4 15/16
Distance between bearings, ft	6	8	10	12	14

Linear Deflection. For continuous line shafting, it is considered good practice to limit deflection to a maximum of 0.01 in. per foot of length. The weight of bare shafting, in pounds, is $W = 2.6d^2L$. When the shaft is fully loaded with pulleys as is usual in practice, with an allowance of 40 lb per inch of width for the vertical pull of belts, the load in pounds will be about $W = 13d^2L$. With a modulus of transverse elasticity of 26,000,000, the following equations determine length between bearings to avoid excessive deflection.

For bare shafting, $L = \sqrt[3]{873d^2}$ $\quad d = \sqrt{L^3/873}$ (1)

For shafting carrying pulleys, etc., $L = \sqrt[3]{175d^2}$ $\quad d = \sqrt{L^3/175}$ (2)

In the foregoing equations L = maximum distance in feet between bearings for continuous shafting subjected to bending stress only; d = diameter of shafting in inches.

FLEXIBLE SHAFTING consists of an inner core of built-up wire coils, brazed or soldered into end fittings, surrounded by a protecting casing.

2. SHAFT DESIGN

HORSEPOWER TRANSMITTED BY SHAFTING. Formulas relating shaft diameter, shearing stress, torque, and horsepower are given in Section 8, Article 12. In practice, the following factors of safety are generally used for transmission shafting: for head shafts, 15; for line shafts carrying pulleys, 10; for small, short shafts, countershafts, etc., 7. For steel shafting of 60,000 psi tensile strength, the allowable stress with the foregoing factors would be 4,000, 6,000, and 8,500 psi, respectively, whence the diameters of shafts would be

Head shafts $d = \sqrt[3]{80\ \text{hp}/N}$ (3)

Line shafts $d = \sqrt[3]{53.5\ \text{hp}/N}$ (4)

Short shafts $d = \sqrt[3]{38\ \text{hp}/N}$ (5)

TORSIONAL DEFLECTION. For the same horsepower, torsional stress is inversely proportional to the speed of rotation, but bending stress will not be reduced in the same ratio. Therefore, it is not practicable to write an equation covering the whole problem which is sufficiently simple for practical use. The following rules, however, are correct within the range of velocities that usually occur under shop conditions. For continuous shafting, so proportioned as to deflect not more than 0.01 in. per ft of length, allowance being made for the weakening effect of keyseats,

For bare shafts, $d = \sqrt[3]{50\ \text{hp}/N}$ $\quad L = \sqrt[3]{720d^2}$ (6)

For shafts carrying pulleys, etc., $d = \sqrt[3]{70\ \text{hp}/N}$ $\quad L = \sqrt[3]{140d^2}$ (7)

CODE FOR DESIGN OF TRANSMISSION SHAFTING. The American Standards Association has issued a code for the design of transmission shafting, in which all commonly used theories are discussed and a rational procedure for the design of shafts under all conditions of loading is developed. The following notes are adapted from the code.

Notation. A = area of cross section, square inches; α = ratio of maximum to average intensity of stress, resulting from axial loading only; D = outside diameter of shaft, inches; d = inside diameter of hollow shaft, inches; E = modulus of elasticity, approximately 30,000,000; F = axial tensile or compressive load, pounds; $K = d/D$ = ratio of inside diameter to outside diameter of hollow shafts; K_m = shock and fatigue factor to be applied to the computed bending moment; K_t = shock and fatigue factor to be applied to the computed torsional moment; k = radius of gyration of transverse cross section of the shaft, inches; l = length of shaft between centers of supports, inches; M = maximum bending moment, inch-pounds; N = revolutions per minute; P = maximum power, transmitted by the shaft, horsepower; S = maximum flexural stress, either tension or compression, pounds per square inch, 16,000 for commercial shafting without allowance for keyways, 12,000 with allowance for keyways;* S_s = maximum shearing stress in shaft, pounds per square inch, 8000 for commercial shafting without allowance for keyways, 6000 with allowance for keyways;* T = maximum torsional moment, inch-pounds; V = transverse shear, pounds; W = transverse load supported by shaft, pounds.

* S may be taken as 60% of the elastic limit in tension, but not more than 36% of the ultimate tensile strength of steel purchased under definite physical specifications. Similar values for S_s are 30% of elastic limit and 18% of ultimate strength.

SOLID SHAFTS. The formulas recommended differ from the formulas heretofore in general use in that they introduce factors K_m and K_t for shock and fatigue. These factors vary with the nature of the load. Values are given in Table 2.

Table 2. Shock and Fatigue Factors for Shafts

	K_m	K_t
Stationary shafts		
Gradually applied loads	1.0	1.0
Suddenly applied loads	1.5–2.0	1.5–2.0
Rotating shafts		
Gradually applied or steady loads	1.5	1.0
Suddenly applied loads, minor shocks only	1.5–2.0	1.0–1.5
Suddenly applied loads, heavy shocks	2.0–3.0	1.5–3.0

Shafts in Pure Torsion.

$$K_tT = 0.1963S_sD^3 \tag{8}$$

$$D = \sqrt[3]{321{,}000K_tP/S_sN} \tag{9}$$

Shafts Subjected to Bending.

$$D = \sqrt[3]{32K_mM/\pi S} \tag{10}$$

Shafts Subjected to Combined Torsion and Bending.

$$D = \sqrt[3]{(16/\pi S_s)\sqrt{(K_mM)^2 + (K_tT)^2}} \tag{11}$$

If the diameter to transmit a given horsepower is desired, it may be found by substituting for the term $(K_tT)^2$ in eq. 11 its value in terms of horsepower, as follows:

$$(K_tT)^2 = \left(\frac{396{,}000K_tP}{2\pi N}\right)^2 \tag{12}$$

Shafts Subjected to Torsion, Flexure, and Axial Loading. This is the general case, and all other shaft formulas may be derived from it by the omission of the factors that do not apply to the case under consideration.

$$D = \sqrt[3]{\frac{16}{\pi S_s}\sqrt{\left(K_mM + \frac{\alpha FD(1 + K^2)}{8}\right)^2 + (K_tT)^2} \times \frac{1}{\sqrt[3]{1 - K^4}}} \tag{13}$$

To determine the diameter of a shaft in terms of horsepower transmitted, substitute in equation 13 the value of $(K_tT)^2$ as found by equation 12.

Shafts in Bending over Short Spans with Heavy Transverse Shear.

$$D = \sqrt{16V/3\pi S_s} \tag{14}$$

Shafts Subjected to Combined Torsion and Heavy Transverse Shear.

$$S_s = (16T/\pi D^3) + (16V/3\pi D^2) \tag{15}$$

Diagrams for Solution of Shafting Problems. All equations given above, except equations 14 and 15, can best be solved by the diagrams in Figs. 4 and 5. In these diagrams, the heavy arcs represent standard transmission shafting sizes, and the lighter arcs the standard machinery shafting sizes. The scale of abscissas represents values of K_mM; the scale of ordinates at the right represents values of K_tT; the scale of ordinates at the left represents the values of K_tP, where P = horsepower at 100 rpm. The diagrams are applicable to steady, fluctuating, reversing, and gradually or suddenly applied loads, with or without minor or heavy shocks, provided that correct values of K_m and K_t are used. They are not applicable to shafts in bending over short spans with heavy transverse shear, or to shafts subjected to combined torsion and heavy transverse shear.

The diagrams are based on a maximum permissible shearing stress of 6000 psi but can be used for other stresses by multiplying the shaft diameter found from the diagram by an appropriate factor as follows:

Shearing stress, psi	4000	5000	6000	7000	8000	9000
Factor	1.14	1.06	1.00	0.95	0.91	0.87

EXAMPLE 1. Required: the size of shaft to withstand a maximum bending moment M of 9350 in-lb, and a maximum torsional moment T of 9600 in-lb, the allowable unit shearing stress being 8000 psi, and the load suddenly applied with minor shock. From Table 2, values of K_m and K_t may be taken as 1.5 and 1.25, respectively. Then $K_mM = 1.5 \times 9350 = 13{,}975$, and $K_tT = 9600 \times 1.25 = 12{,}000$. In Fig. 4, the intersection of $K_mM = 13{,}975$, and of $K_tT = 12{,}000$ lies on the arc representing a shaft 2 1/2 in. in diameter, with a unit stress of S_s of 6000 psi. Since the allowable unit torsional stress in the present case is 8000 psi, the correct diameter for the conditions given is found by applying the factor for the higher stress. $2.5 \times 0.91 = 2.275$. The next higher standard size of shaft is 2 5/16 in.

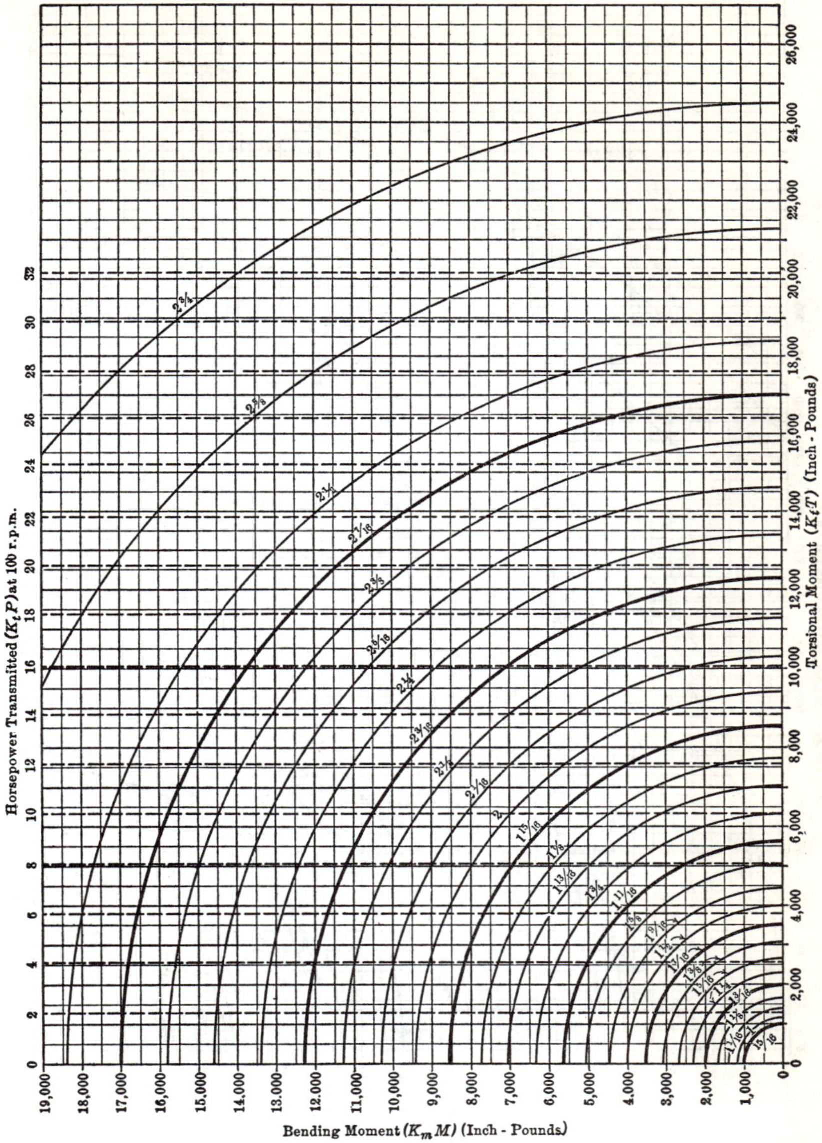

FIG. 4. Shaft selection chart.

EXAMPLE 2. Required: the size of an alloy steel shaft to transmit 210 hp at 150 rpm and at the same time sustain a bending moment of 90,000 in.-lb. The unit allowable shearing stress may be taken as 9000 psi. At 150 rpm, 210 hp is equivalent to 210 × (100/150) = 140 hp at 100 rpm. The intersection of 140 hp and K_mM = 90,000 in Fig. 5 shows that a shaft 4 3/4 in. in diameter would be required at a unit shearing stress of 6000 psi. Applying the factor 0.87 for the higher unit stress, 4.75 × 0.87 = 4.1325 in. The next larger standard shaft is 4 1/4 in. in diameter.

EXAMPLE 3. Required: the size of a nonrotating shaft to support a steady load of 4000 lb, the supports being spaced 3 ft apart. From the equation for simple beams (see p. 8-12), $M = Pl/4$ = 4000 × (3 × 12)/4 = 48,000 in.-lb. From Table 2, $K_m = 1$, and K_mM = 48,000. Since there is no rotation, $T = 0$, and $K_tT = 0$. In Fig. 5, on the abscissa of $K_tT = 0$, at a value of K_mM = 48,000, the nearest size of shaft that will be suitable at a unit shearing stress of 6000 psi is 3 1/2 in.

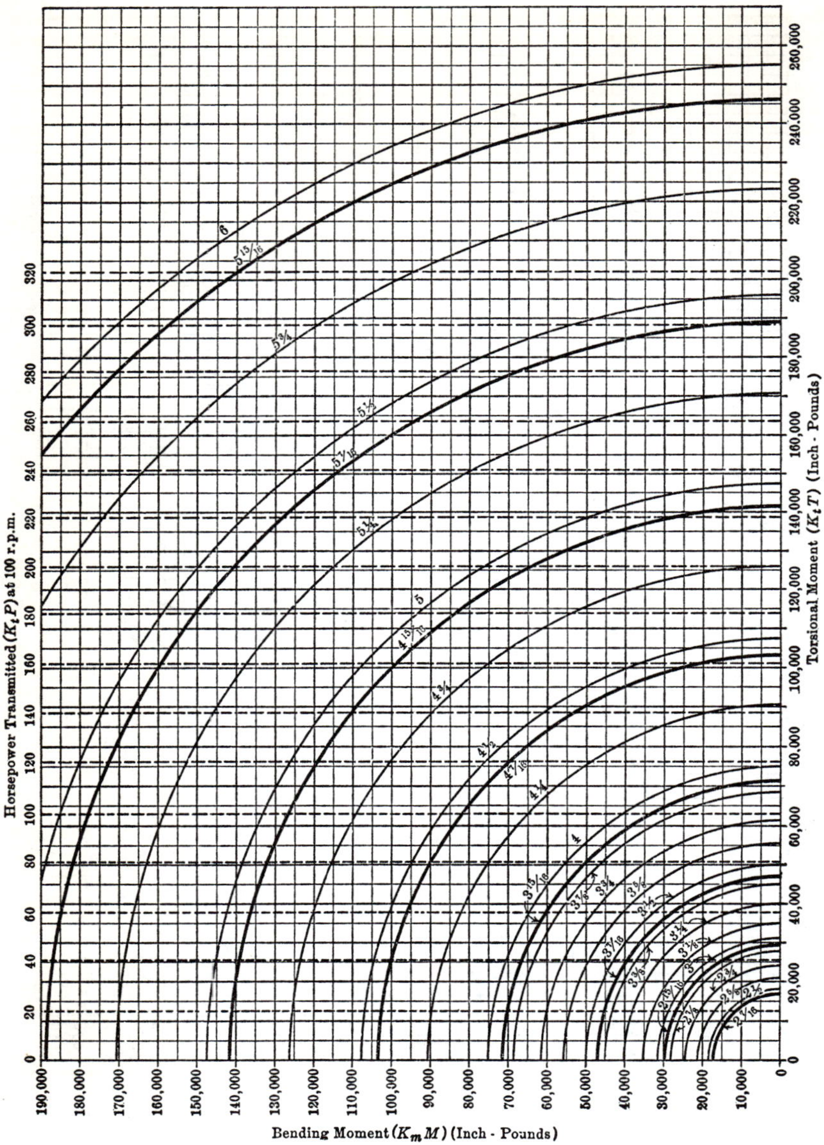

FIG. 5. Shaft selection chart.

HOLLOW SHAFTS. All the foregoing equations are applicable to hollow shafts, when multiplied by a factor equivalent to $(1/\sqrt[3]{1 - K^4})$, in which $K = d/D$, that is, the ratio of the inside and outside diameters. Values of $(1/\sqrt[3]{1 - K^4})$ are as follows:

K	0.95	0.90	0.85	0.80	0.75	0.70	0.65	0.60	0.55	0.50
$\sqrt[3]{1 - K^4}$	0.57	0.70	0.78	0.84	0.88	0.91	0.93	0.95	0.97	0.98
$1/\sqrt[3]{1 - K^4}$	1.75	1.43	1.28	1.19	1.14	1.10	1.07	1.05	1.03	1.02

The procedure is to determine the diameter of a solid shaft for the given conditions and then to multiply the diameter so found by the factor corresponding to the desired value of d/D.

EXAMPLE 4. Determine the size of a hollow shaft to be used under the conditions given in Example 1, assuming the value of K as 0.70. As determined in Example 1, the diameter of a solid shaft required would be 2.275 in. The factor for $K = 0.70$ is 1.10. Therefore, the outside diameter of an equivalent hollow shaft would be $2.275 \times 1.10 = 2.502$ in. The inside diameter would be $2.502 \times 0.70 = 1.75$ in.

SHAFTS UNDER AXIAL LOADING. The maximum intensity of stress resulting from a force F applied axially is $S = \alpha F/A$, where α = ratio of maximum to average intensity of stress resulting from the axial load only, and A = cross-sectional area of the shaft, square inches. If the shaft is rigidly supported at short intervals and virtually acts as a column under compressive loads, $\alpha = 1$. Under all tensile axial loads, also, α is 1. If the shaft is not rigidly supported, it acts as a long column, and α should be computed by one of the long-column formulas. Values of α, for ratios of l/k up to 115, computed according to the formula

$$\alpha = 1/[1 - 0.0044(l/k)] \tag{16}$$

are given below. For ratios of l/k greater than 115, use Euler's equation (see p. 8-22).

Ratio, l/k	0	25	50	75	100	115
α	1.00	1.12	1.28	1.49	1.78	2.02

The value of α so determined then should be used in equation 13 to determine the size of shaft required.

SHAFTS UNDER ALTERNATING STRESS. Shafts subjected to torsion and bending encounter reversal of stress at each revolution and should be calculated by means of equation 11. The values to be ascribed to the shock and fatigue factors, K_t and K_m, which take into account the stress reversals, depend on the conditions under which the shaft operates. No general statement can be made concerning the values to be used, since the shock factor, in particular, depends on the resiliency of the shaft system as a whole. For complete stress reversal, K_t and K_m should be 1.5.

CRITICAL SPEEDS. Shafts carrying masses such as pulleys, flywheels, gears, etc., may be subject to transverse or torsional vibrations. These vibrations may become critical when the speed of rotation is equal to a natural frequency of free vibration. Torsional vibrations may be critical at several speeds, owing to exciting harmonics in the transmitted torque. For methods of calculation, see Section 9, Article 5.

3. KEYS

Keys are used to prevent relative movement of rotating members and the shafts or spindles on which they are mounted. They are of various forms, depending on the class of service in which they are to be used.

SUNK KEYS are made in various forms, as square, flat, tapered, etc. Square and flat keys preferably should be fitted on all four sides. See Figs. 6, 7, 8, and 9.

A key fitting tightly on top and bottom of the keyway drives partly by friction. If fitted only on the sides of the keyway, it exerts a prying action on the hub and shaft and is subjected to severe bending and shearing stresses.

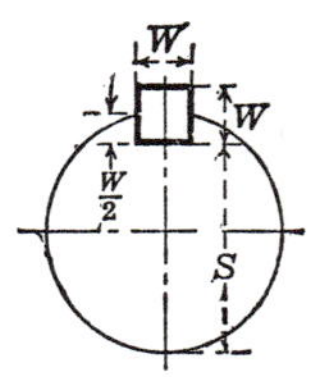

FIG. 6. Square plain parallel stock key. See Table 3.

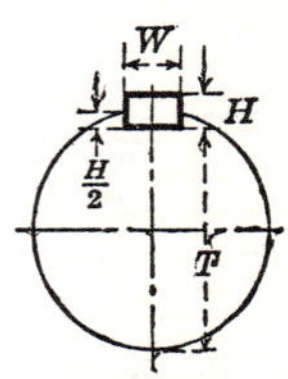

FIG. 7. Flat parallel stock key. See Table 4.

Taper keys (Fig. 9) are tapered on the upper surface only, i.e., on the surface projecting into the hub. A corresponding taper is cut in the keyway in the hub. When the key is driven, hub and shaft are drawn together on the side opposite the key. A frictional resistance thus is established which tends to prevent axial movement of the hub on the shaft.

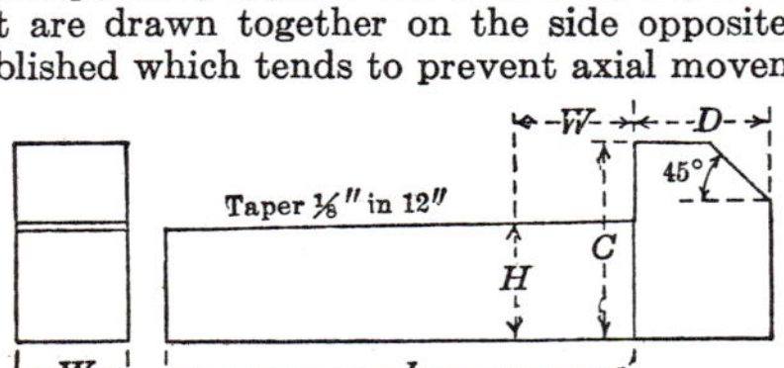

FIG. 8. Gib-head taper stock key. See Table 5.

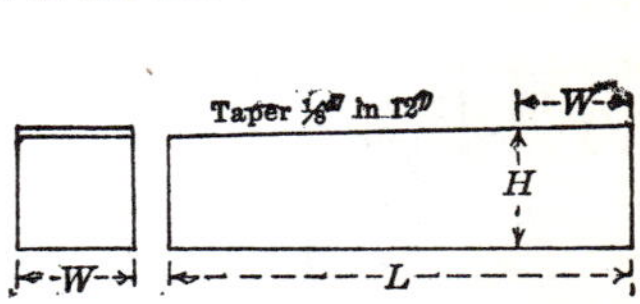

FIG. 9. Plain taper stock key. See Table 6.

If the key is more tightly fitted at one end than at the other, or if the bore of the hub is tapering, the hub may be so drawn up that its axis is at an angle to the axis of the shaft. The Woodruff key (Fig. 10) adjusts itself to taper in the hub and thus avoids misalignment of shaft and hub.

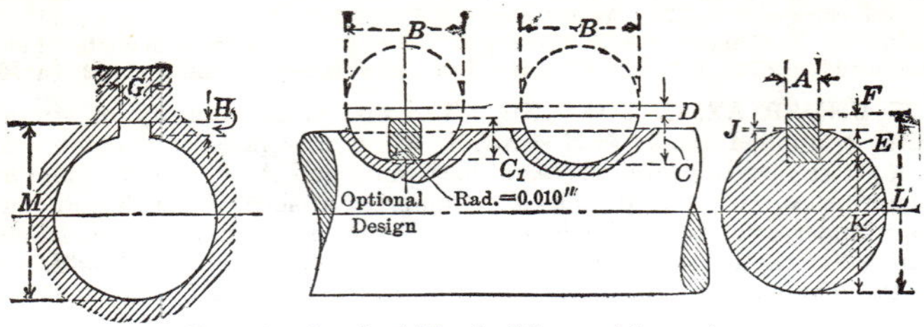

FIG. 10. Standard Woodruff key and keyseat.

FORMS OF KEYS. Standard Stock Keys. The American Standards Association has established standard sizes for plain parallel and taper stock keys, square and flat, and square and flat taper gib-head stock keys. Dimensions of these keys are given in Tables 3, 4, 5, and 6. These keys are intended to be cut from cold-finished stock and used without further machining, the dimensions and tolerances having been fixed with this consideration in mind.

Depth of Keyseats. In addition to giving the key dimensions, Tables 3 and 4 give the distance from the bottom of the keyseat to the opposite side of the shaft, in order to facilitate gaging for the correct depth of keyseat when machining. This dimension is calculated on the basis of the keyseat being cut to a depth equal to one-half of the height of the key, this depth being measured at the side of the keyseat. For shaft and key dimensions other than those in the tables, S and T may be calculated from the following equations:

$$S = d - A - \frac{W}{2} \tag{17}$$

and

$$T = d - A - \frac{H}{2} \tag{18}$$

where $A = (d - \sqrt{d^2 - W^2})/2$, which is equal to the height of the arc projecting above

Table 3. Dimensions of Square Plain Parallel Stock Keys

(ASA B17.1-1943. All dimensions in inches. See Fig. 6.)

Shaft Diameter	Width and Thickness of Key, W *	Bottom of Keyseat to Opposite Side of Shaft, S	Shaft Diameter	Width and Thickness of Key, W *	Bottom of Keyseat to Opposite Side of Shaft, S	Shaft Diameter	Width and Thickness of Key, W *	Bottom of Keyseat to Opposite Side of Shaft, S	Shaft Diameter	Width and Thickness of Key, W *	Bottom of Keyseat to Opposite Side of Shaft, S
1/2	1/8	0.430	1 7/16	3/8	1.225	2 3/8	5/8	2.021	3 7/8	1	3.309
9/16	1/8	0.493	1 1/2	3/8	1.289	2 7/16	5/8	2.084	3 15/16	1	3.373
5/8	3/16	0.517	1 9/16	3/8	1.352	2 1/2	5/8	2.148	4	1	3.437
11/16	3/16	0.581	1 5/8	3/8	1.416	2 5/8	5/8	2.275	4 1/4	1	3.690
3/4	3/16	0.644	1 11/16	3/8	1.479	2 3/4	5/8	2.402	4 7/16	1	3.881
13/16	3/16	0.708	1 3/4	3/8	1.542	2 7/8	3/4	2.450	4 1/2	1	3.944
7/8	3/16	0.771	1 13/16	1/2	1.527	2 15/16	3/4	2.514	4 3/4	1 1/4	4.042
15/16	1/4	0.796	1 7/8	1/2	1.591	3	3/4	2.577	4 15/16	1 1/4	4.232
1	1/4	0.859	1 15/16	1/2	1.655	3 1/8	3/4	2.704	5	1 1/4	4.296
1 1/16	1/4	0.923	2	1/2	1.718	3 1/4	3/4	2.831	5 1/4	1 1/4	4.550
1 1/8	1/4	0.986	2 1/16	1/2	1.782	3 3/8	7/8	2.880	5 7/16	1 1/4	4.740
1 3/16	1/4	1.049	2 1/8	1/2	1.845	3 7/16	7/8	2.944	5 1/2	1 1/4	4.803
1 1/4	1/4	1.112	2 3/16	1/2	1.909	3 1/2	7/8	3.007	5 3/4	1 1/2	4.900
1 5/16	5/16	1.137	2 1/4	1/2	1.972	3 5/8	7/8	3.140	5 15/16	1 1/2	5.091
1 3/8	5/16	1.201	2 5/16	5/8	1.957	3 3/4	7/8	3.261	6	1 1/2	5.155

* Tolerance on W is −0.0020 in. for shafts 1/2 to 1 3/4 in., inclusive; −0.0025 in. for shafts 1 13/16 to 3 1/4 in., inclusive; −0.0030 in. for shafts 3 3/8 to 6 in., inclusive.

the intersection of the side of the keyway with the circumference of the shaft and d = shaft diameter.

The Woodruff key is extensively used in machine tools, and in the automotive industry. The keyslot is cut in the shaft with a milling cutter. The hub is forced on the shaft over the key. Dimensions of standard Woodruff keys are given in Table 7. The dimensions of J, K, L, and M, Fig. 10, which are not given in Table 7, can be computed by means of the following formulas, in which d = diameter of the shaft, other factors being as shown in Fig. 10.

Versed sine,

$$J = {}^1/_2 d - \sqrt{(d^2 - A^2)/4} \tag{19}$$

Bottom of keyseat to opposite side of shaft,

$$K = (d - J - E) \tag{20}$$

Top of key to opposite side of shaft,

$$L = (F + E + K) \tag{21}$$

Bottom of keyway to opposite side of bore,

$$M = (d - J + H) \tag{22}$$

Keys and Keyways for Arbors and Milling Cutters. Table 8 gives dimensions and tolerances of keys and keyways for milling cutters and arbors, as developed by the American Standards Association.

The Kennedy Key. The Kennedy key, largely used in rolling mill work, is shown in Fig. 12. In these keys, $w = t = {}^1/_4 D$. They are tapered $^1/_8$ in. per ft on top, and the sides are a neat fit. The keys are so set in the shaft that diagonals through them intersect at the axis of the shaft. If, in assembling, the hub is to be slid along the shaft for a considerable distance, the hub is bored for a press fit and then is rebored eccentrically about $^1/_{64}$ D off center. The keyways are cut in the eccentric side. General practice is to use single keys for diameters up to and including 6 in. where the torque is constant and the power transmitted always in one direction. For shafts above 6 in. in diameter, double keys should be used, and, if the torque is intermittent and in alternate directions, double keys should be used down to shaft diameters of 4 in.

Table 4. Dimensions of Flat Plain Parallel Stock Keys

(ASA B17.1-1943. All dimensions in inches. See Fig. 7.)

Shaft Diameter	Width, W *	Thickness, H *	Bottom of Keyseat to Opposite Side of Shaft, T	Shaft Diameter	Width, W *	Thickness, H *	Bottom of Keyseat to Opposite Side of Shaft, T	Shaft Diameter	Width, W *	Thickness, H *	Bottom of Keyseat to Opposite Side of Shaft, T
1/2	1/8	3/32	0.445	1 3/4	3/8	1/4	1.605	3 3/8	7/8	5/8	3.005
9/16	1/8	3/32	0.509	1 13/16	1/2	3/8	1.590	3 7/16	7/8	5/8	3.069
5/8	3/16	1/8	0.548	1 7/8	1/2	3/8	1.654	3 1/2	7/8	5/8	3.132
11/16	3/16	1/8	0.612	1 15/16	1/2	3/8	1.717	3 5/8	7/8	5/8	3.259
3/4	3/16	1/8	0.676	2	1/2	3/8	1.781	3 3/4	7/8	5/8	3.386
13/16	3/16	1/8	0.739	2 1/16	1/2	3/8	1.843	3 7/8	1	3/4	3.434
7/8	3/16	1/8	0.802	2 1/8	1/2	3/8	1.908	3 15/16	1	3/4	3.498
15/16	1/4	3/16	0.827	2 3/16	1/2	3/8	1.971	4	1	3/4	3.562
1	1/4	3/16	0.890	2 1/4	1/2	3/8	2.034	4 1/4	1	3/4	3.815
1 1/16	1/4	3/16	0.954	2 5/16	5/8	7/16	2.051	4 7/16	1	3/4	4.006
1 1/8	1/4	3/16	1.017	2 3/8	5/8	7/16	2.114	4 1/2	1	3/4	4.069
1 3/16	1/4	3/16	1.081	2 7/16	5/8	7/16	2.178	4 3/4	1 1/4	7/8	4.229
1 1/4	1/4	3/16	1.144	2 1/2	5/8	7/16	2.242	4 15/16	1 1/4	7/8	4.420
1 5/16	5/16	1/4	1.169	2 5/8	5/8	7/16	2.368	5	1 1/4	7/8	4.483
1 3/8	5/16	1/4	1.232	2 3/4	5/8	7/16	2.495	5 1/4	1 1/4	7/8	4.733
1 7/16	3/8	1/4	1.288	2 7/8	3/4	1/2	2.575	5 7/16	1 1/4	7/8	4.927
1 1/2	3/8	1/4	1.351	2 15/16	3/4	1/2	2.639	5 1/2	1 1/4	7/8	4.991
1 9/16	3/8	1/4	1.415	3	3/4	1/2	2.702	5 3/4	1 1/2	1	5.150
1 5/8	3/8	1/4	1.478	3 1/8	3/4	1/2	2.829	5 15/16	1 1/2	1	5.341
1 11/16	3/8	1/4	1.542	3 1/4	3/4	1/2	2.956	6	1 1/2	1	5.405

* Tolerance on W and H is −0.0020 in. for shafts 1/2 to 1 3/4 in., inclusive; −0.0025 in. for shafts 1 13/16 to 3 1/4 in. inclusive; −0.0030 in. for shafts 3 3/8 to 6 in., inclusive.

Table 5. Dimensions of Square and Flat Taper Gib-head Stock Keys

(ASA B17.1-1943. All dimensions in inches. See Fig. 8.)

Shaft Diameter (inclusive)	Square Type					Flat Type					Stock Length, L		
	Key		Gib Head			Key		Gib Head					
	Maximum Width, W *	Height at Large End,† H *	Height, C	Length D	Height, Edge of Chamfer, E	Maximum Width, W *	Height at Large End,† H *	Height, C	Length, D	Height, Edge of Chamfer, E	Minimum	Maximum	Advancing by Increments of
1/2 to 9/16	1/8	1/8	1/4	7/32	5/32	1/8	3/32	3/16	1/8	1/8	1/2	2	1/4
5/8 to 7/8	3/16	3/16	5/16	9/32	7/32	3/16	1/8	1/4	3/16	5/32	3/4	3	3/8
15/16 to 1 1/4	1/4	1/4	7/16	11/32	11/32	1/4	3/16	5/16	1/4	3/16	1	4	1/2
1 5/16 to 1 3/8	5/16	5/16	9/16	13/32	13/32	5/16	1/4	3/8	5/16	1/4	1 1/4	5 1/2	5/8 ‡
1 7/16 to 1 3/4	3/8	3/8	11/16	15/32	15/32	3/8	1/4	7/16	3/8	5/16	1 1/2	6	3/4
1 13/16 to 2 1/4	1/2	1/2	7/8	19/32	5/8	1/2	3/8	5/8	1/2	7/16	2	8	1
2 5/16 to 2 3/4	5/8	5/8	1 1/16	23/32	3/4	5/8	7/16	3/4	5/8	1/2	2 1/2	10	1 1/4
2 7/8 to 3 1/4	3/4	3/4	1 1/4	7/8	7/8	3/4	1/2	7/8	3/4	5/8	3	12	1 1/2
3 3/8 to 3 3/4	7/8	7/8	1 1/2	1	1	7/8	5/8	1 1/16	7/8	3/4	3 1/2	14	1 3/4
3 7/8 to 4 1/2	1	1	1 3/4	1 3/16	1 3/16	1	3/4	1 1/4	1	13/16	4	16	2
4 3/4 to 5 1/2	1 1/4	1 1/4	2	1 7/16	1 7/16	1 1/4	7/8	1 1/2	1 1/4	1	5	20	2 1/2
5 3/4 to 6	1 1/2	1 1/2	2 1/2	1 3/4	1 3/4	1 1/2	1	1 3/4	1 1/2	1 1/4	6	24	3

* Tolerance: −0.0020 on width, and +0.0020 on height for shafts 1/2 to 1 3/4 in. diameter; −0.0025 on width and +0.0025 on height for shafts 1 13/16 to 3 1/4 in. diameter; −0.0030 on width and +0.0030 on height for shafts 3 3/8 to 6 in. diameter.

† This height of the key is measured at the distance W, equal to the width of key, from the gib head.

‡ Final increment, 1 1/8 in.

Table 6. Dimensions of Square and Flat Taper Stock Keys

(ASA B17.1-1943. All dimensions in inches. See Fig. 9.)

Shaft Diameter (inclusive)	Square Type		Flat Type		Tolerance		Stock Length, *L*		
	Maximum Width, *W*	Height at Large End, *H* *	Maximum Width, *W*	Height at Large End, *H* *	On Width	On Height	Minimum	Maximum	Advancing by Increments of
1/2 to 9/16	1/8	1/8	1/8	3/32	−.0020	+.0020	1/2	2	1/4
5/8 to 7/8	3/16	3/16	3/16	1/8	−.0020	+.0020	3/4	3	3/8
15/16 to 1 1/4	1/4	1/4	1/4	3/16	−.0020	+.0020	1	4	1/2
1 5/16 to 1 3/8	5/16	5/16	5/16	1/4	−.0020	+.0020	1 1/4	5 1/4	5/8 †
1 7/16 to 1 3/4	3/8	3/8	3/8	1/4	−.0020	+.0020	1 1/2	6	3/4
1 13/16 to 2 1/4	1/2	1/2	1/2	3/8	−.0025	+.0025	2	8	1
2 5/16 to 2 3/4	5/8	5/8	5/8	7/16	−.0025	+.0025	2 1/2	10	1 1/4
2 7/8 to 3 1/4	3/4	3/4	3/4	1/2	−.0025	+.0025	3	12	1 1/2
3 3/8 to 3 3/4	7/8	7/8	7/8	5/8	−.0030	+.0030	3 1/2	14	1 3/4
3 7/8 to 4 1/2	1	1	1	3/4	−.0030	+.0030	4	16	2
4 3/4 to 5 1/2	1 1/4	1 1/4	1 1/4	7/8	−.0030	+.0030	5	20	2 1/2
5 3/4 to 6	1 1/2	1 1/2	1 1/2	1	−.0030	+.0030	6	24	3

* This height of the key is measured at the distance *W*, equal to the width of the key, from the large end.

† 4 1/2 in. length instead of 4 3/8 in.

Figure 13 shows a key devised by Wilfred Lewis, which is similar in principle to the Kennedy key and which is used for lighter service, as machine tool work. It is subject to compression only, but it has the disadvantage of being expensive and difficult to fit.

The Barth Key. The key shown in Fig. 14 was devised by Carl G. Barth to combine the advantages of the Lewis key with those of the ordinary rectangular key. The Barth key is rectangular, with one-half of both sides beveled at 45 degrees. The key does not need to fit tightly, as pressure tends to drive it into its seat. There is no tendency to turn it, and the only stress to which it is subject is compression. This key has been used in many cases as a feather to replace rectangular feather keys which have given trouble. It has found wide application as a feather key in drill sockets and drill shanks, reamers, etc., which are commonly driven with a tang.

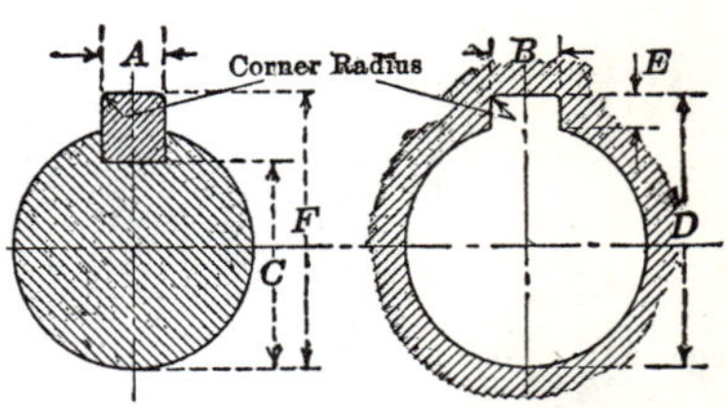

FIG. 11. Keys and keyways for arbors and milling cutters.

The Nordberg Key. The Nordberg Manufacturing Co. has adopted for the ends of shafts the round keys shown in Fig. 15. The advantages of this key are: there is no tendency toward deformation; they are a driven fit in the direction of the shear; they are always in true shear and are cheaper than the square key. In manufacturing, a hole *A* is drilled in the joint, and next a hole *B* as large as the size of the keyway will admit is drilled in the shaft in order to avoid the tendency of the drill used for drilling the keyway to size to crowd into the soft cast iron. The taper is 1/16 in. per ft., measured on the diameter.

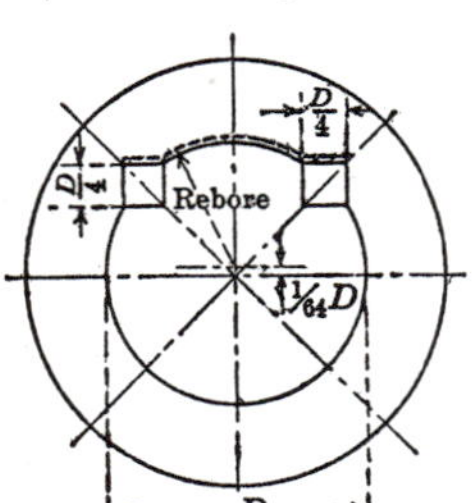

FIG. 12. Kennedy key.

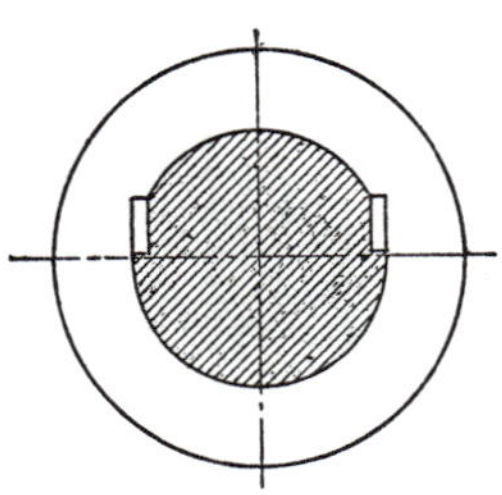

FIG. 13. Lewis key.

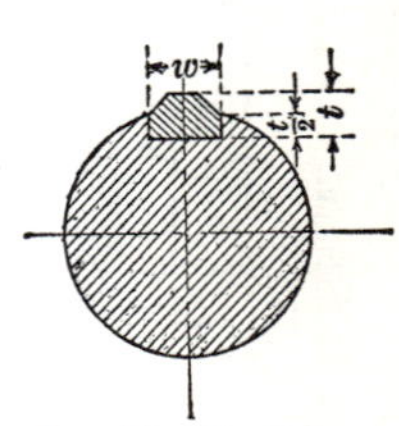

FIG. 14. Barth key.

Table 7. Dimensions of Woodruff Keys, Keyslots, and Keyways

(SAE Standard. All dimensions in inches. See Fig. 10.)

SAE Nominal Size	Key						Key above Shaft	Keyslot			Keyway	
	A +.001 −.000	*B* +.000 −.010	*C* +.000 −.005	C_1 +.000 −.006	*D* Nominal	*L* * +.000 −.010	*F* ±.005	*A* † Min	*E* +.005 −.000	*B* ‡ Min	*G* +.002 −.000	*H* +.005 −.000
1/16 x 1/4	.0625	0.250	.109	....	1/64	....	.0312	.0615	.0728	0.255	.0635	.0372
1/16 x 5/16	.0625	0.312	.140	....	1/64	....	.0312	.0615	.1038	0.317	.0635	.0372
3/32 x 5/16	.0938	0.312	.140	....	1/64	....	.0469	.0928	.0882	0.317	.0948	.0529
1/16 x 3/8	.0625	0.375	.172	....	1/64	....	.0312	.0615	.1358	0.380	.0635	.0372
3/32 x 3/8	.0938	0.375	.172	....	1/64	....	.0469	.0928	.1202	0.380	.0948	.0529
1/8 x 3/8	.1250	0.375	.172	....	1/64	....	.0625	.1240	.1045	0.380	.1260	.0685
1/16 x 1/2	.0625	0.500	.203	.194	3/64	....	.0312	.0615	.1668	0.510	.0635	.0372
3/32 x 1/2	.0938	0.500	.203	.194	3/64	....	.0469	.0928	.1511	0.510	.0948	.0529
1/8 x 1/2	.1250	0.500	.203	.194	3/64	....	.0625	.1240	.1355	0.510	.1260	.0685
3/32 x 5/8	.0938	0.625	.250	.240	1/16	....	.0469	.0928	.1981	0.635	.0948	.0529
1/8 x 5/8	.1250	0.625	.250	.240	1/16	....	.0625	.1240	.1825	0.635	.1260	.0685
5/32 x 5/8	.1563	0.625	.250	.240	1/16	....	.0781	.1553	.1669	0.635	.1573	.0841
3/16 x 5/8	.1875	0.625	.250	.240	1/16	....	.0937	.1863	.1513	0.635	.1885	.0997
1/8 x 3/4	.1250	0.750	.313	.303	1/16	....	.0625	.1240	.2455	0.760	.1260	.0685
5/32 x 3/4	.1563	0.750	.313	.303	1/16	....	.0781	.1553	.2299	0.760	.1573	.0841
3/16 x 3/4	.1875	0.750	.313	.303	1/16	....	.0937	.1863	.2143	0.760	.1885	.0997
1/4 x 3/4	.2500	0.750	.313	.303	1/16	....	.1250	.2487	.1830	0.760	.2510	.1310
5/32 x 7/8	.1563	0.875	.375	.365	1/16	....	.0781	.1553	.2919	0.887	.1573	.0841
3/16 x 7/8	.1875	0.875	.375	.365	1/16	....	.0937	.1863	.2763	0.887	.1885	.0997
7/32 x 7/8	.2188	0.875	.375	.365	1/16	....	.1093	.2175	.2607	0.887	.2198	.1153
1/4 x 7/8	.2500	0.875	.375	.365	1/16	....	.1250	.2487	.2450	0.887	.2510	.1310
3/16 x 1	.1875	1.000	.438	.428	1/16	....	.0937	.1863	.3393	1.012	.1885	.0997
7/32 x 1	.2188	1.000	.438	.428	1/16	....	.1093	.2175	.3237	1.012	.2198	.1153
1/4 x 1	.2500	1.000	.438	.428	1/16	....	.1250	.2487	.3080	1.012	.2510	.1310
5/16 x 1	.3125	1.000	.438	.428	1/16	....	.1562	.3111	.2768	1.012	.3135	.1622
3/16 x 1 1/8	.1875	1.125	.484	.475	5/64	....	.0937	.1863	.3853	1.137	.1885	.0997
7/32 x 1 1/8	.2188	1.125	.484	.475	5/64	....	.1093	.2175	.3697	1.137	.2198	.1153
1/4 x 1 1/8	.2500	1.125	.484	.475	5/64	....	.1250	.2487	.3540	1.137	.2510	.1310
5/16 x 1 1/8	.3125	1.125	.484	.475	5/64	....	.1562	.3111	.3228	1.137	.3135	.1622
3/16 x 1 1/4	.1875	1.250	.547	.537	5/64	....	.0937	.1863	.4483	1.265	.1885	.0997
7/32 x 1 1/4	.2188	1.250	.547	.537	5/64	....	.1093	.2175	.4327	1.265	.2198	.1153
1/4 x 1 1/4	.2500	1.250	.547	.537	5/64	....	.1250	.2487	.4170	1.265	.2510	.1310
5/16 x 1 1/4	.3125	1.250	.547	.537	5/64	....	.1562	.3111	.3858	1.265	.3135	.1622
3/8 x 1 1/4	.3750	1.250	.547	.537	5/64	....	.1875	.3735	.3545	1.265	.3760	.1935
1/4 x 1 3/8	.2500	1.375	.594	.584	3/32	....	.1250	.2487	.4640	1.390	.2510	.1310
5/16 x 1 3/8	.3125	1.375	.594	.584	3/32	....	.1562	.3111	.4328	1.390	.3135	.1622
3/8 x 1 3/8	.3750	1.375	.594	.584	3/32	....	.1875	.3735	.4015	1.390	.3760	.1935
1/4 x 1 1/2	.2500	1.500	.641	.631	7/64	....	.1250	.2487	.5110	1.515	.2510	.1310
5/16 x 1 1/2	.3125	1.500	.641	.631	7/64	....	.1562	.3111	.4798	1.515	.3135	.1622
3/8 x 1 1/2	.3750	1.500	.641	.631	7/64	....	.1875	.3735	.4485	1.515	.3760	.1935
3/16 x 2 1/8	.1875	2.125	.406	.396	21/32	1.380	.0937	.1863	.3073	2.125	.1885	.0997
1/4 x 2 1/8	.2500	2.125	.406	.396	21/32	1.380	.1250	.2487	.2760	2.125	.2510	.1310
5/16 x 2 1/8	.3125	2.125	.406	.396	21/32	1.380	.1562	.3111	.2448	2.125	.3135	.1622
3/8 x 2 1/8	.3750	2.125	.406	.396	21/32	1.380	.1875	.3735	.2135	2.125	.3760	.1935
3/16 x 2 1/8	.1875	2.125	.531	.521	17/32	1.723	.0937	.1863	.4323	2.125	.1885	.0997
1/4 x 2 1/8	.2500	2.125	.531	.521	17/32	1.723	.1250	.2487	.4010	2.125	.2510	.1310
5/16 x 2 1/8	.3125	2.125	.531	.521	17/32	1.723	.1562	.3111	.3698	2.125	.3135	.1622
3/8 x 2 1/8	.3750	2.125	.531	.521	17/32	1.723	.1875	.3735	.3385	2.125	.3760	.1935
1/4 x 2 3/4	.2500	2.750	.594	.584	25/32	2.000	.1250	.2487	.4640	2.750	.2510	.1310
5/16 x 2 3/4	.3125	2.750	.594	.584	25/32	2.000	.1562	.3111	.4328	2.750	.3135	.1622
3/8 x 2 3/4	.3750	2.750	.594	.584	25/32	2.000	.1875	.3735	.4015	2.750	.3760	.1935
7/16 x 2 3/4	.4375	2.750	.594	.584	25/32	2.000	.2187	.4360	.3703	2.750	.4385	.2247
1/2 x 2 3/4	.5000	2.750	.594	.584	25/32	2.000	.2500	.4985	.3390	2.750	.5010	.2560
1/4 x 2 3/4	.2500	2.750	.750	.740	5/8	2.317	.1250	.2487	.6200	2.750	.2510	.1310
5/16 x 2 3/4	.3125	2.750	.750	.740	5/8	2.317	.1562	.3111	.5888	2.750	.3135	.1622
3/8 x 2 3/4	.3750	2.750	.750	.740	5/8	2.317	.1875	.3735	.5575	2.750	.3760	.1935
7/16 x 2 3/4	.4375	2.750	.750	.740	5/8	2.317	.2187	.4360	.5263	2.750	.4385	.2247
1/2 x 2 3/4	.5000	2.750	.750	.740	5/8	2.317	.2500	.4985	.4950	2.750	.5010	.2560
3/8 x 3 1/2	.3750	3.500	.938	.927	13/16	2.880	.1875	.3735	.7455	3.500	.3760	.1935
7/16 x 3 1/2	.4375	3.500	.938	.927	13/16	2.880	.2187	.4360	.7143	3.500	.4385	.2247
1/2 x 3 1/2	.5000	3.500	.938	.927	13/16	2.880	.2500	.4985	.6830	3.500	.5010	.2560
9/16 x 3 1/2	.5625	3.500	.938	.927	13/16	2.880	.2812	.5610	.6518	3.500	.5635	.2872
5/8 x 3 1/2	.6250	3.500	.938	.927	13/16	2.880	.3125	.6235	.6205	3.500	.6260	.3185
11/16 x 3 1/2	.6875	3.500	.938	.927	13/16	2.880	.3437	.6860	.5893	3.500	.6885	.3497
3/4 x 3 1/2	.7500	3.500	.938	.927	13/16	2.880	.3750	.7485	.5580	3.500	.7510	.3810

* Length. Smaller keys have length equal to chord (Fig. 10); larger keys have corners trimmed to length *L*.

† Tolerances are +.0015 for widths up to .1562 in.; +.0017 for .1863 in. width; +.0018 for .2175 and .2487-in. width; +.0019 for .3111-in. width; and +.0020 for .3735-in. width.

‡ Tolerances are +.005 for sizes up to 1.515 in.; 0.010 for larger sizes.

Table 8. Keys and Keyways for Arbors and Milling Cutters

(ASA B5c-1930. All dimensions in inches. See Fig. 11.)

Diameter of Arbor	Nominal Size, Square Key	*A* Key-seat,* max	*A* Key,† max	*B* Keyway Max	*B* Keyway Min	*C* Max †	*D* Bottom of Keyway to Opposite Side of Bore, min	*E* Nominal	*F* Top of Key to Opposite Side of Shaft ‡	Corner Radius
1/2	3/32	0.0947	0.0932	0.106	0.099	0.4531	0.5578	3/64	0.5468	0.020
5/8	1/8	0.126	0.1245	0.137	0.130	0.5625	0.6985	1/16	0.6875	1/32
3/4	1/8	0.126	0.1245	0.137	0.130	0.6875	0.8225	1/16	0.8125	1/32
7/8	1/8	0.126	0.1245	0.137	0.130	0.8125	0.9475	1/16	0.9375	1/32
1	1/4	0.251	0.2495	0.262	0.255	0.8438	1.104	3/32	1.094	3/64
1 1/4	5/16	0.3135	0.3120	0.325	0.318	1.063	1.385	1/8	1.375	1/16
1 1/2	3/8	0.376	0.3745	0.410	0.385	1.281	1.666	5/32	1.656	1/16
1 3/4	7/16	0.4385	0.4370	0.473	0.448	1.500	1.948	3/16	1.938	1/16
2	1/2	0.501	0.4995	0.535	0.510	1.687	2.198	3/16	2.188	1/16
2 1/2	5/8	0.626	0.6245	0.660	0.635	2.094	2.733	7/32	2.718	1/16
3	3/4	0.751	0.7495	0.785	0.760	2.500	3.265	1/4	3.250	3/32
3 1/2	7/8	0.876	0.8745	0.910	0.885	3.000	3.890	3/8	3.875	3/32
4	1	1.001	0.9995	1.035	1.010	3.375	4.390	3/8	4.375	3/32
4 1/2	1 1/8	1.126	1.1245	1.160	1.135	3.813	4.953	7/16	4.938	1/8
5	1 1/4	1.251	1.2495	1.285	1.260	4.250	5.515	1/2	5.500	1/8

* Tolerance allowed of −0.001 in. from maximum.
† Tolerance allowed of −0.005 in. from maximum.
‡ Tolerance allowed of −0.006 in. from maximum.

LIGHT KEYS. Where only small amounts of power are to be transmitted and it is impracticable or inadvisable to use a sunk key, a key driving on a flat or a saddle key may be used. These two types of key drive mainly by friction.

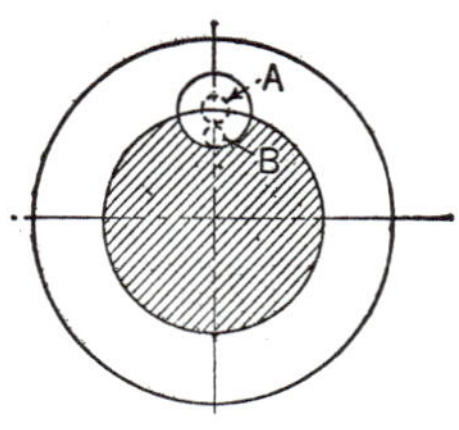

FIG. 15. Nordberg key.

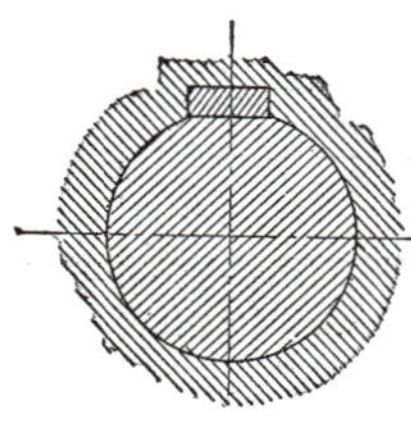

FIG. 16. Key on flat.

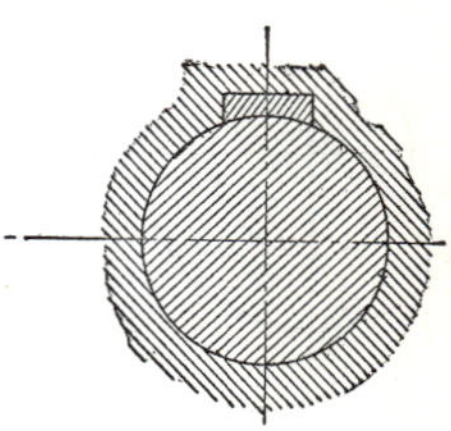

FIG. 17. Saddle key.

Key on Flat. If a small flat be milled or filed on the shaft, a flat key with a slight taper may be driven into a keyway in the hub opposite the flat on the shaft. See Fig. 16. This type of key is suitable for transmission of light powers and has the advantage of ease of assembling. Such a key should not be subjected to sudden changes in torque, as it then may slip on the shaft and seriously damage it.

Saddle Key. The saddle key, Fig. 17, depends entirely on friction to prevent motion of the hub relative to the shaft. The key is driven into a keyway in the hub, and the face of the key bearing on the shaft is made with a radius somewhat smaller than the radius of the shaft. The key should have a slight taper.

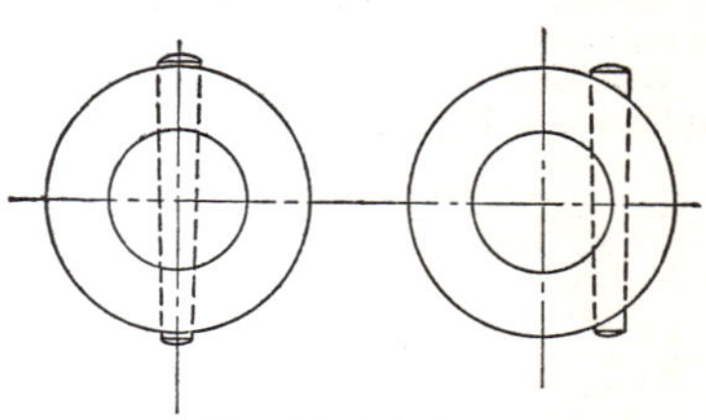

FIG. 18. Pin keys.

Pin keys are adapted to transmit power with light loads. Figure 18 shows two methods of using taper pins as keys. Table 9 gives the dimensions of standard taper pins.

Groov-Pins are parallel pins having a shallow axial split extending along part or all of the length. This split is widened so as to enlarge the effective diameter of the pin.

Table 9. Dimensions of Taper Pins

Taper 1/4 in. in 12 in. All dimensions in inches. See Fig. 19.

(Morse Twist Drill and Machine Co., New Bedford, Mass.)

No.	Diameter, Large End *	Approximate Fractional Size	Radius, R	Length, L		Advancing by Increments of	No.	Diameter, Large End *	Approximate Fractional Size	Radius, R	Length, L		Advancing by Increments of
				Min	Max						Min	Max	
0 000 000	0.0625	1/16	1/16	1/4	3/4	1/8	5	0.289	19/64	7/32	3/4	4	1/4
000 000	0.078	5/64	1/16	1/4	1	1/8	6	0.341	11/32	7/32	3/4	5	1/4
00 000	0.094	3/32	1/16	1/2	1	1/8	7	0.409	13/32	1/4	1	5	1/4
0 000	0.109	7/64	1/16	1/2	1	1/8	8	0.492	1/2	3/8	1 1/4	5	1/4
000	0.125	1/8	1/16	1/2	1	1/8	9	0.591	19/32	1/2	1 1/2	6	1/4
00	0.141	9/64	1/16	1/2	2 1/2	1/4	10	0.706	23/32	5/8	1 1/2	6	1/4
0	0.156	5/32	5/64	1/2	3	1/4	11	0.860	55/64	7/8	2	9	1/2
1	0.172	11/64	3/32	1/2	3	1/4	12	1.032	1 1/32	1 1/16	2 1/2	9	1/2
2	0.193	3/16	7/64	3/4	3 1/2	1/4	13	1.241	1 1/4	1 1/4	3	11	1/2
3	0.219	7/32	5/32	3/4	3 1/2	1/4	14	1.523	1 33/64	1 1/2	4	13	1/2
4	0.250	1/4	3/16	3/4	4	1/4							

* Tolerance (to No. 5) + 0.001 in., −0.000; (No. 6 to No. 14) + 0.002 in., −0.000.

When driven into a drilled hole the pin holds tight by virtue of the pressure required to close the split. (Groov-Pin Corp., Union City, N. J.)

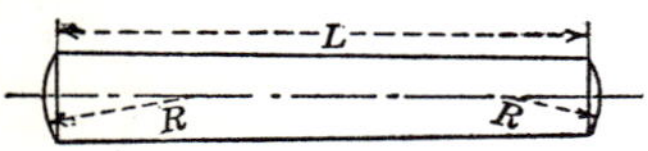

FIG. 19. Taper pin.

FEATHER KEYS are used when the hub must slide axially along the shaft but at the same time rotate with it. The feather key is a sunk key. It may be either dovetailed into the shaft or fastened in place by screws. Figure 20 shows various forms. If the key is required to slide with the hub, the construction shown in Fig. 20*c* may be used. Instead of feather keys, spline shafts now are largely used.

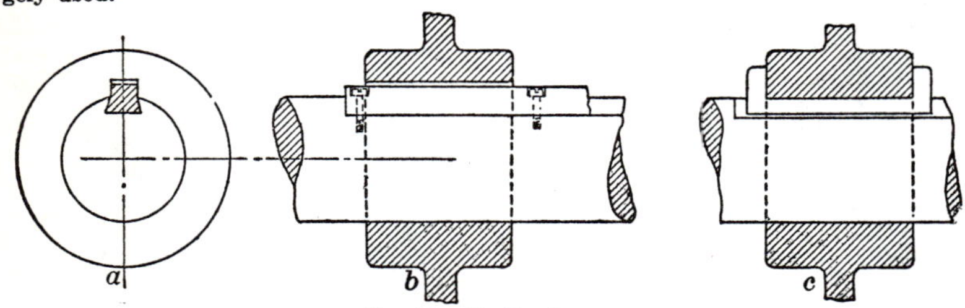

FIG. 20. Feather keys.

4. SPLINES AND SERRATIONS

Instead of keys, spline shafts may be used for the transmission of power from a shaft to a hub or vice versa. Two types of splines have been standardized—square and involute.

SQUARE SPLINES are employed in multiple-spline fittings having 4, 6, 10, or 16 splines (see Fig. 21). They have become established as basic standards for a great many

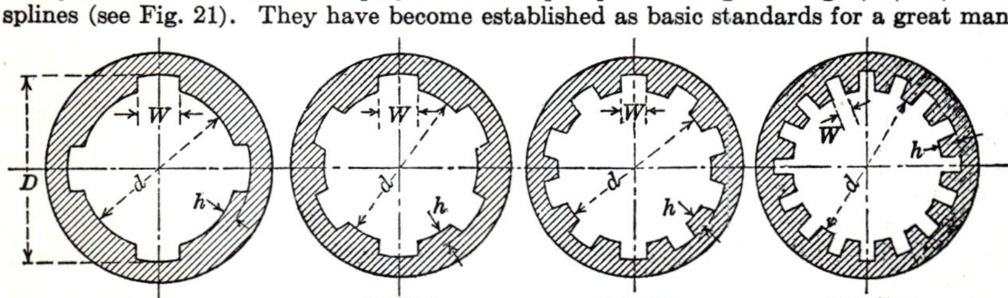

FIG. 21. Spline fittings.

applications in the automotive, machine tool, and other industries. Standardized by the SAE, their proportions are intended to apply only to soft broached holes, the shaft dimensions depending on the shape and material of the parts, their heat treatment, methods of machining, etc., to give the required fit. Dimensions of these standardized fittings are given in Table 10. Nominal values of D range from $^3/_4$ in. to 1 $^3/_4$ in. by increments of $^1/_8$ in., from 1 $^3/_4$ in. to 2 $^1/_2$ in. by increments of $^1/_4$ in., and from 2 $^1/_2$ to 6 in. by increments of $^1/_2$ in. Three types of fits provide for (A) permanent fit, (B) sliding when not under load, and (C) sliding under load.

Torque capacity of spline fittings may be determined by the formula

$$T = 1000NrhL$$

where N = number of splines; r = mean radial distance from center of hole to center of spline = $(D + d)/4$; h = depth of spline; and L = length of spline bearing surface; all in inches. This gives torque in inch-pounds and is based on a pressure of 1000 psi on the sides of the spline.

Table 10. Formulas for Spline Proportions

(SAE Standard. See Fig. 21.)

No. of Splines	W For All Fits	Permanent Fit		To Slide When Not under Load		To Slide under Load	
		h	d	h	d	h	d
Four	.241D	.075D	.850D	.125D	.750D	....	....
Six	.250D	.050D	.900D	.075D	.850D	.100D	.800D
Ten	.156D	.045D	.910D	.070D	.860D	.095D	.810D
Sixteen	.098D	.045D	.910D	.070D	.860D	.095D	.810D

For D and d, tolerance allowed of -0.001 in. for shafts $^3/_4$ to 1 $^3/_4$ in., inclusive; -0.002 for shafts 2 to 3 in., inclusive; -0.003 in. for shafts 3 $^1/_2$ to 6 in., inclusive, for 4-, 6-, and 10-spline fittings; tolerance of -0.003 in. allowed for all sizes of 16-spline fittings.

For W, tolerance allowed of -0.002 in. for shafts $^3/_4$ to 1 $^3/_4$ in., inclusive; -0.003 in. for shafts 2 to 6 in., inclusive, for 4-, 6-, and 10-spline fittings; tolerance of -0.003 allowed for all sizes of 16-spline fittings.

INVOLUTE SPLINES. American Standard. Involute splines are multiple keys in the general form of internal and external involute gear teeth, used to prevent relative rotation of cylindrically fitted machine parts. Involute splines can be machined by the same methods and equipment as those customarily used in the production of gears. The nomenclature agrees as nearly as possible with the standard gear nomenclature adopted by the American Gear Manufacturers Association and the American Standards Association, except that the words "root diameter," "outside diameter," and "inside diameter" have been changed to "major diameter" and "minor diameter" to avoid confusion. The pressure angle is 30 degrees, splines are based on diametral pitch, involute form, having addendum and dedendum each equal to $0.5/DP$, and circular tooth thickness of $1.5708/DP$. (For nomenclature, see Fig. 22.)

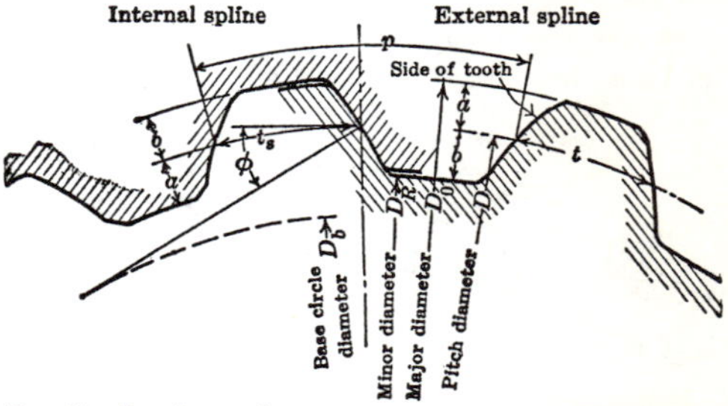

FIG. 22. Involute spline nomenclature. (ASA B5.15-1946)

Because of the almost universal practice of broaching the internal spline, it has been held to basic dimensions and the external spline dimensions varied to control the fit. This corresponds to the general practice of fitting to a reamed or drilled hole, because of the fixed condition of the finishing tools and gages.

Tooth size is designated by two numbers such as $^8/_{16}$, where the first number (8 in this case) is the diametral pitch (DP) used in all calculations. The following pitches are included in the standard: 1/2, 2.5/5, 3/6, 4/8, 5/10, 6/12, 8/16, 10/20, 12/24, 16/32, 20/40, 24/48, 32/64, 40/80, and 48/96. Table 11 gives formulas for the leading dimensions of splines, which may have numbers of teeth ranging from 6 to 50. Clearances and tolerances being determined by manufacturing experience and not by direct proportions, space does not permit inclusion here of all the manufacturing dimensions and inspection data for all sizes. These may be found in the complete standard, ASA B5.15-1946, *Involute Splines.*

Table 11. Basic Formulas for Involute Spline Proportions
(See Fig. 22)

Dimension	Symbol	Formula
Circular pitch	p	$\frac{3.1416}{DP}$
Circular tooth thickness	t	$\frac{1.5708}{DP}$
Pitch diameter	D	$\frac{N}{DP}$
Base circle diameter	D_s	$0.8660D$
Addendum, dedendum	a, b	$\frac{0.500}{DP}$
Major diameter	D_O	$\frac{N+1}{DP}$
Minor diameter	D_R	$\frac{N-1}{DP}$

Types of Fits. There are three methods of fitting spline members together: *major diameter*, where fit is controlled by varying the major diameter of the external splines; *sides of the teeth*, where fit is controlled by varying tooth thickness (it is customary to use this fit on full fillet-type splines); and *minor diameter*, where fit is controlled by varying the minor diameter of the internal splines (requires special tooling).

The three methods are each divided into three classes: sliding, close, and press fits. Sliding fits must have clearance at all points; close fits must be close at one point; and press fits must involve an interference at only one point.

Length of Splines. For carrying any load, or to obtain stability, it is seldom necessary to have a spline longer than its pitch diameter. If the strength of the shaft in shear is equated to that of the spline in shear, allowing for only 25% effective tooth contact, the following equations result for spline length:

With a solid shaft of diameter D,

$$L = \frac{D_R^3}{D^2}$$

With a hollow shaft of external diameter D and internal diameter D_I,

$$L = \frac{D_R^4 - D_I^4}{D^2 D_R}$$

INVOLUTE SERRATIONS. SAE Standard. Involute serrations are multiple keys in the general form of internal and external involute gear teeth, used for permanent fits between shafts and parts mounted on them. The standard provides a uniform, easily fabricated set of serrations that can be made by several manufacturing processes. The nomenclature agrees with that of the ASA standard for involute splines (see preceding paragraphs and Fig. 23).

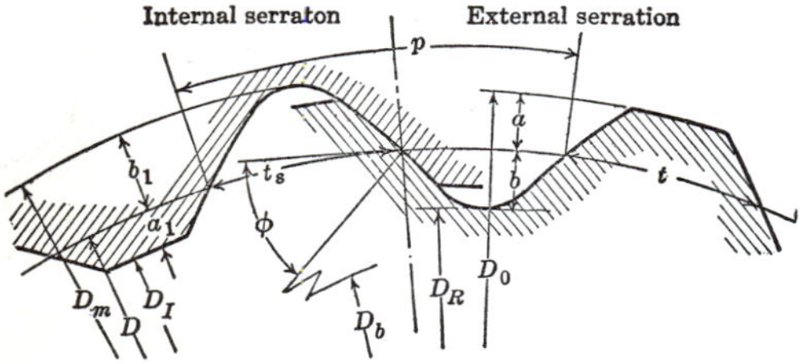

FIG. 23. Involute serration nomenclature. (Reproduced by permission from the *SAE Handbook*, 1949 edition)

The pressure angle is 45 degrees, and the serrations are based on involute form, as generated with a straight-sided hob of the form included in the standard. Serrations may be parallel or tapered (recommended taper, 0.750 in. per ft included angle, measured on major diameter).

The scope of the standard includes the pitches given in Table 12; the first number in the pitch designation (e.g., 24 in 24/48) is the DP used in all calculations. For each pitch, the range of numbers of teeth is also given.

Table 12

Pitch	Range of N	Pitch	Range of N	Pitch	Range of N
10/20	6–100	32/64	21–65	64/128	21–34
16/32	6–100	40/80	32–55	80/160	14–27
24/48	6–100	48/96	24–46	128/256	11–24

Formulas for the leading dimensions are given in Table 13 (see Fig. 23 for nomenclature).

Table 13. Formulas for Involute Serration Proportions

(See Fig. 23)

Dimension	Symbol	Formula
Circular pitch	p	$\frac{3.1416}{DP}$
Circular tooth thickness	t	$\frac{1.3708}{DP}$
Circular space width	t_s	$\frac{1.7708}{DP}$
Pitch diameter	D	$\frac{N}{DP}$
Base circle diameter	D_b	$0.7071D$
External serration		
Addendum	a	$\frac{0.500}{DP}$
Dedendum	b	$\frac{0.500}{DP}$
Major diameter	D_O	$\frac{N+1}{DP}$
Minor diameter	D_R	$\frac{N-1}{DP}$
Internal serration		
Addendum	a_1	$\frac{0.300}{DP}$
Dedendum	b_1	$\frac{0.700}{DP}$
Major diameter	D_m	$\frac{N+1.40}{DP}$
Minor diameter	D_I	$\frac{N-0.60}{DP}$

Fits and Tolerances. Three classes of fits are designated in the standard: loose (A); close (B); and press (C). The dimensions for the internal serrations are held the same on all fits, the external serration being varied to obtain the desired fit. Modifications are made to obtain equal-strength teeth in both members and at the same time retain even steps of major diameter in well-known fractions. The standard also specifies tolerances for space width and tooth circular thickness and should be consulted for all manufacturing and inspection data.

COUPLINGS

By A. F. Gagne, Jr.

Couplings are used to join lengths of shafting, which must often be sectionalized for practicability and economy in manufacture and shipping or for purposes of ready installation. This applies to long transmission shafting, to shafts of separately built driving and driven machine units, and, on occasion, to short shafts within machines. Couplings of special types are also used to join shafts at angles, to prevent the transmission of overloads, to secure controlled slippage, and to engage or disengage power at will. Couplings of the last type are commonly known as clutches and are discussed in the succeeding chapter.

5. RIGID COUPLINGS

Rigid couplings are used to make a solid connection between shafts. They are of various types.

THE SLEEVE COUPLING forms a simple, low-cost, rigid joint between two shafts. Also known as a "box" or "muff" coupling, it consists of a substantial sleeve fitted snugly over the shafts and secured with keys, taper pins, or setscrews. Figure 1 gives approximate proportions of commercial setscrew couplings, suitable for light duty. Sleeve couplings are commonly applied from the very smallest sizes up to 2 in., although a heavy-duty "OK" coupling has been developed by SKF for large marine drives which dispenses with keys and uses a patented high-pressure oil injection technique to expand the sleeve for movement on or off. The threaded solid coupling, another heavy-duty type, is often used on the vertical drive shaft of deep well pumps in capacities up to 1000 hp at 1200 rpm.

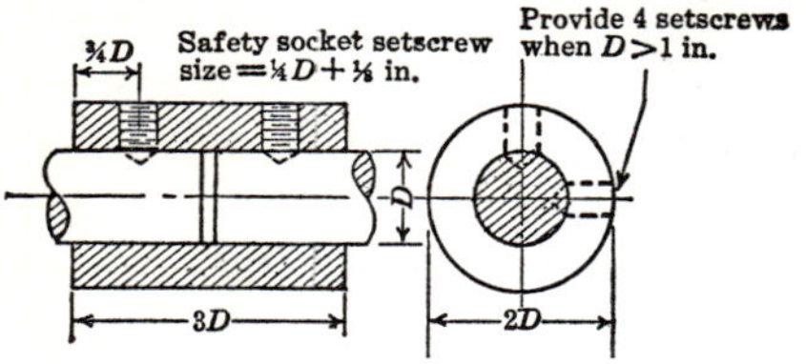

FIG. 1. Set-screw sleeve coupling.

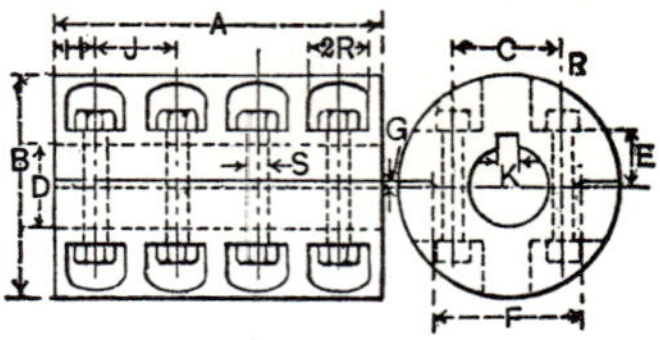

FIG. 2. Clamp shaft coupling. See Table 1.

THE CLAMP SHAFT COUPLING is essentially a split and bolted sleeve coupling, proportioned to clamp firmly on the shafts. This construction permits removal without interfering with adjoining bearings or requiring separation of the shafts. These couplings are made for the heaviest duty, and because of convenience and low relative cost are often used at moderate speeds in sizes up to 5 in. When the coupling must carry a tensile load across the shafts, split ring keys can be inserted in matching grooves machined on the shafts and in the coupling halves. *Halsey's Handbook* gives the typical dimensions shown in Fig. 2 and Table 1 for the standard cast-iron design.

Table 1. Dimensions of Clamp Shaft Couplings

(Representative listing, abbreviated)

Dimensions, in. (see Fig. 2)												Weight, lb.
D	*A*	*B*	*C*	*E*	*F*	*G*	*H*	*J*	*K*	*R*	*S*	
1 7/16	6	4 1/4	2 1/2	1	3 3/8	1/8	1 1/2	....	5/16	15/16	5/8	18.7
1 15/16	8	5 1/2	3 1/4	1 3/8	4 1/4	1/4	2	....	7/16	1 1/16	3/4	42.7
2 15/16	12	8	4 1/4	2	5 3/8	1/4	1 1/2	3	11/16	1 1/16	3/4	125.2
3 15/16	16	10 1/4	5 1/2	2 3/4	7	1/4	2	4	13/16	1 5/16	1	277.9
4 15/16	20	12 5/8	6 3/4	3 3/8	8 1/2	1/4	2 1/2	5	13/16	1 9/16	1 1/4	496.7

FLANGE COUPLINGS are commonly used in permanent installations for heavy loads and large sizes and particularly for vertical drives, as agitators. Although somewhat larger in diameter, they are shorter and, in typical designs, 20% lighter than other heavy-duty types. Their dimensions have not been standardized by the various manufacturers, but Fig. 3 and Table 2 are representative. Alignment is obtained by piloting the end of one shaft one-eighth diameter into the opposing flange as shown or, at extra cost, by making the flange halves male and female, a shallow rabbet on one flange fitting into a recess in the other flange near its periphery. For best results, the flanges should be shrunk on the shafts and finish machined. Uniform bolt load distribution is obtained by reaming both flanges together for a body fit on the bolts.

Flange couplings may be disconnected when required, provided the pilot shaft end is lubricated and the flanges are separated to prevent rubbing. A removable split plate is useful in securing separation without disturbing the fit of flange to shaft.

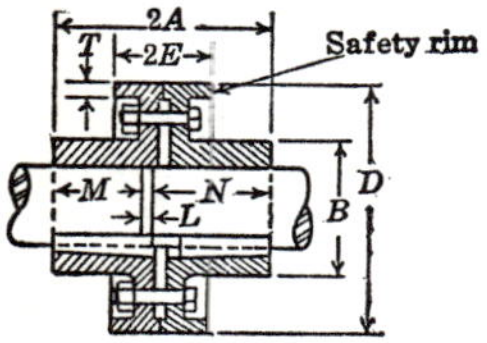

FIG. 3. Flange coupling. See Table 2.

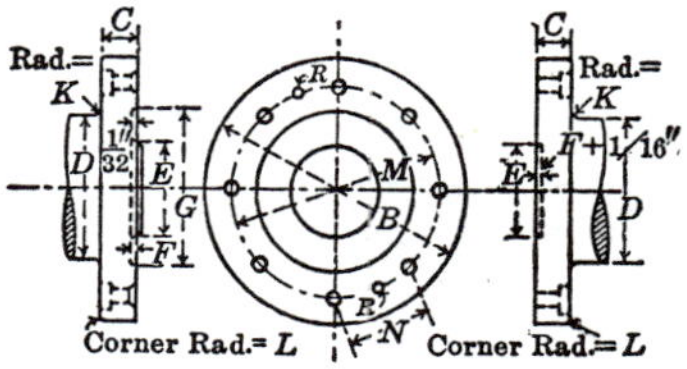

FIG. 4. Integrally forged hydroelectric unit coupling. See Table 3.

HYDROELECTRIC UNIT COUPLINGS. A complete series of integrally forged, small-diameter, heavy-duty flange couplings has been standardized by the ASA, as indicated in Fig. 4 and Table 3. Minimum weight has been achieved by reducing flange torsional and bending strength to match that of the shaft.

Table 2. Dimensions of Flange Couplings

(Representative listing, abbreviated)

Shaft Size	Dimensions, in. (see Fig. 3)								Bolts			Keyway		Average Weight, lb
	A	*B*	*D*	*E*	*L*	*M*	*N*	*T*	Diameter Bolt Circle, in.	No.	Size, in.	Width, in.	Depth, in.	
1 3/16	2 1/4	2 1/4	6	1	1/8	1 13/16	2 9/16	3/16	4	3	1/2	5/16	5/32	10
1 15/16	3 1/4	3 7/8	7 3/4	1 1/2	1/8	2 11/16	3 11/16	1/4	5 5/8	4	5/8	1/2	1/4	27
3 15/16	5 1/4	7 1/8	12 5/16	2 1/4	5/16	4 7/16	5 3/4	3/8	9 3/4	4	7/8	1	1/2	115
5 15/16	8 1/8	10 1/2	17 3/4	2 15/16	3/8	7 3/8	8 1/2	1/2	13 3/4	6	1 1/4	1 1/2	3/4	375
8	10 1/8	14 5/8	24	4 1/4	1/2	8 7/8	10 7/8	11/16	18 1/2	8	1 1/2	1 3/4	7/8	997
12	16	20	33 1/4	5 9/16	3/4	14 1/4	17	1	26	8	1 3/4	2 1/4 *	1	2350

* Provide two keys 90 degrees apart for shafts 10 in. in diameter and over.

Table 3. Dimensions of Integrally Forged Shaft Couplings for Hydroelectric Units

(ASA B49.1-1947, abbreviated. Dimensions in inches. See Fig. 4.)

Shaft Diameter, *D*	Flange								Bolts					
	Outside Diameter, *B*	Thickness, *C*		Rabbet Diameter, *E*	Height, *F*	Relief Diameter, *G*	Fillet Radius, *K*	Corner Radius, *L*	No.	Diameter	Bolt Circle Diameter, *M*	Center Distance of Bolts, *N*	Size of Nuts	Threads per Inch
		Shaft Subject to Bending	Shaft Not Subject to Bending											
3 1/2	7 1/2	7/8	7/8	2 1/8	5/16	2 1/2	3/16	1/16	6	7/8	5 5/8	2.81	7/8	8
5	10 1/4	1 1/4	1 1/4	3	5/16	4	3/16	1/16	6	1 1/4	7 3/4	3.87	1 1/4	8
7	13 1/2	1 3/4	1 3/4	4 1/4	5/16	6	3/8	3/32	8	1 1/2	10 1/2	4.02	1 1/2	8
9	16 1/4	2 1/4	2	5 3/8	5/16	8	5/8	3/32	10	1 3/4	13 1/4	4.09	1 1/2	8
12	22	3	2 1/2	7 1/4	3/8	11	1	1/8	10	2 1/4	18	5.56	2	8
16	29	4	3 1/4	9 1/2	3/8	15	1 1/2	3/16	12	2 3/4	24	6.21	2 1/2	8
20	35 1/4	5	4	12	3/8	19	1 3/4	1/4	14	3	29 3/4	6.62	2 3/4	8
24	42	6	4 3/4	14 1/2	1/2	23	2	1/4	14	3 3/4	35	7.79	3 1/2	8
32	55	8	6 1/2	19 1/4	5/8	31	3	3/8	16	4 1/2	46 1/4	9.02	4 1/4	8
40	68	10	8	24	5/8	39	3 1/2	3/8	18	5 1/4	57 3/4	10.03	5	8
48	80	12	9 1/2	28	5/8	47	4	3/8	18	6 1/4	67 1/2	11.72	6	8

Note: Rabbets are turned and coupling bolt holes reamed for standard wring fits. Safety flanges can be provided over the bolts. Three to four jack screws are provided, ranging in size from 3/4″-10NC, for the smallest coupling to 1 1/2″-8N, for the largest.

The values in the table are based on the following approximate proportions:

Flange diameter, $B = 1.75D$

Flange thickness, $C = 0.25D$, if subject to bending

Flange thickness, $C = 0.20D$, if not subject to bending

Rabbet diameter, $E = 0.60D$

Permissible bolt stress = 1.20 × shaft stress

KEYLESS COMPRESSION COUPLINGS. The common flange design consists of a slotted, double-taper sleeve, clamped on the shafts by bolted taper flanges as in Fig. 5 and Table 4. Jackscrews are used to disassemble. This design offers the advantages of adjustability and low installed cost, since no machining of the shafts is required. It is not suitable for heavy or shock loads.

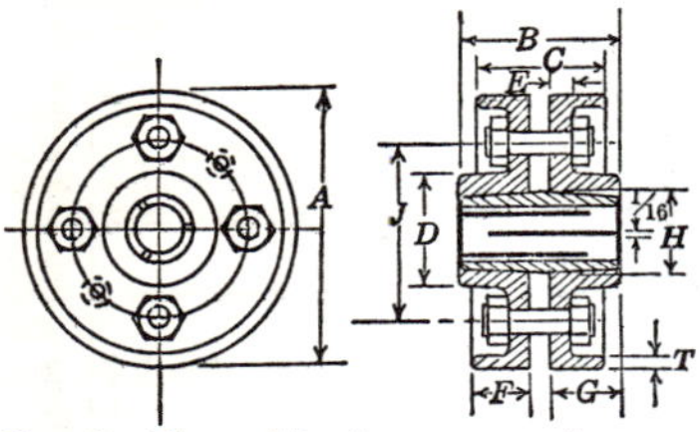

FIG. 5. Flanged keyless compression coupling. See Table 4.

CURVIC COUPLINGS maintain firm alignment in built-up crankshafts such as are used on large aircraft engines. The construction involves the machining of accurately matching crown gear teeth into the two mating faces of the hubs to be joined. The assembly is ordinarily held together by a central bolt but can be separated for clutching action if mounted on a pilot shaft. An adaptation is seen in Fig. 22.

Table 4. Dimensions of Flanged Keyless Compression Couplings

(Dodge Manufacturing Corp., Mishawaka, Ind.)

Shaft Size, in.	Dimensions, in. (see Fig. 5)										Bolts		Average Weight, lb
	A	*B*	*C*	*D*	*E*	*F*	*G*	*H* *	*J*	*T*	No.	Size, in.	
15/16	5	3	2 3/8	2	3/8	1	1 5/16	1.424	3 1/4	3/16	3	7/16	6
1 3/16	5 1/2	3 3/4	2 1/2	2 3/8	3/8	1 1/16	1 11/16	1.736	3 3/4	3/16	3	7/16	9
1 15/16	7 1/4	6	2 7/8	3 3/4	9/16	1 1/4	2 13/16	2.802	5 3/8	1/4	5	7/16	25
2 15/16	10 1/4	9	3 3/4	5 1/2	3/4	1 5/8	4 1/4	4.171	7 5/8	3/8	6	5/8	72

Note: Slotted reducing thimbles furnished for either shaft as required. Intermediate sizes available.

* Maximum flange diameter, plus 0.004 in., minus 0.000 in. This is 0.012 in. smaller than maximum sleeve diameter. Taper is 3/4 in. per ft, diametral.

6. FLEXIBLE COUPLINGS

When three or more bearings are located closely in line on one shaft or on shafts of two machine units connected by a rigid coupling, there is, except for special constructions, a considerable likelihood of bearing trouble and loss of power. These result from misalignment, which also produces fatigue-inducing alternations of tensile and compressive stress in the shaft. The misalignment is typically caused by inadvertent installation errors or by subsequent wear, thermal expansion, machine deflection under load, foundation settling, or loosening of bolts.

The ill effects of such misalignment are commonly avoided by inserting a flexible coupling between shafts, which also lessens alignment labor. A well-chosen flexible coupling will operate with negligible maintenance cost and power absorption and will introduce no harmful vibrations or resonances into the system.

In general, flexibility is attained in one or more of the following three ways:

1. Loosely fitted hard parts. Such couplings are suitable for low speeds and light loads only.
2. Hard parts arranged for lubricated sliding action and made with close fits.
3. Flexible or resilient parts, which do not require lubrication but may place some load on the bearings if misaligned.

In applying flexible couplings, it must be borne in mind that most types are of very limited flexibility and accordingly will suffer damage or rapid wear if due care is not taken at installation to assure an alignment well within the limits recommended by the manufacturer. In general, good installation practice on ordinary couplings has been considered a lateral alignment within 0.002 to 0.004 in. for high-speed drives such as turbines, and within 0.01 in. for low-speed drives.

There are various methods of securing alignment accuracy. One of the best aids is a straight edge laid across the shafts or coupling halves at several points around the periphery and checked with feeler gages. At the same time, sufficient clearance must be allowed between flanges to accommodate all motor and machine end-play, as cramping of the flexible coupling action and destructive hammering will otherwise occur. During the service which follows, alignment and clearance should be checked periodically and corrected as required to stay within the manufacturer's recommendation.

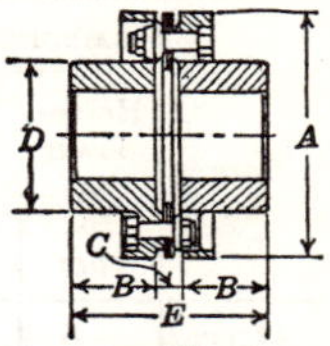

FIG. 6. Single-engagement Thomas flexible coupling. See Table 5.

In view of the large variety of flexible couplings commercially available, they are classified as follows according to the amount and degree of freedom offered.

LONGITUDINAL FLEXIBILITY is usually necessary to accommodate thermal expansion and motor end play. Together with other essential characteristics, it is found in most flexible coupling designs.

ANGULAR FLEXIBILITY. Figure 6 and Table 5 illustrate a single-engagement Thomas coupling, used effectively between two bearings of a combination of three bearings in line (Fig. 7); for example, a fan motor and a self-aligning outboard bearing. The laminated metal ring permits a maximum angular deflection of 1/2 degree and a longitudinal displacement of 1% of the shaft diameter yet

Table 5. Dimensions and Capacities of Laminated Steel Ring Flexible Coupling
(Type ST—The Thomas Flexible Coupling Co., Warren, Pa.)

Maximum				Dimensions, in. (see Fig. 6)				
Bore, in.	Rpm	Horsepower at 100 rpm *	*P*,† lb	*A*	*B*	*C*	*D*	*E*
1 5/8	5800	0.8	150	4 19/32	1 3/4	13/32	2 3/4	3 29/32
2	4650	1.5	225	5 3/4	2 1/8	7/16	3 1/2	4 11/16
3 1/2	2900	8.5	900	9 1/8	3 3/4	11/16	6	8 3/16
5	2000	33.5	2400	13 7/16	5	1 1/8	8 3/8	11 1/8
8	1200	177	7700	22 3/8	7 3/4	1 7/8	13 5/8	17 3/8

Note: Other types available include flange adapter for flywheels and double engagement for greater flexibility, in sizes from 1/16 to 13 in. Intermediate sizes available.
* With full lateral load. Double capacity without lateral load.
† *P* is the lateral load or pull on coupling in pounds.

provides full support for rotor weight. The coupling is quiet at high speeds and does not require lubrication. Its torsional rigidity adapts it well to cam and other closely timed drives.

The double-engagement style, consisting of two flexible rings bolted compactly to a central spider, is available when lateral flexibility is indicated, as discussed below.

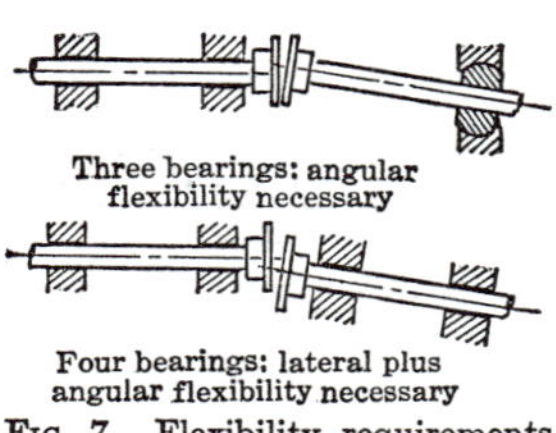

FIG. 7. Flexibility requirements of different bearing combinations.

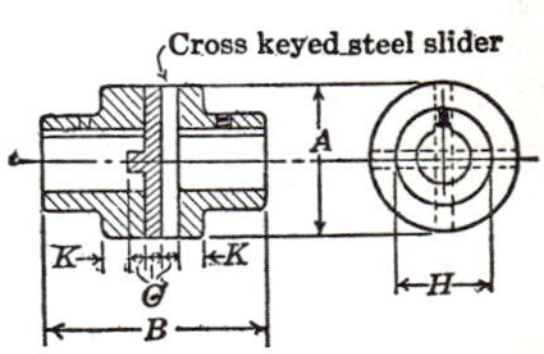

FIG. 8. Oldham coupling. See Table 6.

LATERAL, OR PARALLEL, FLEXIBILITY is required in addition to angular flexibility when two pairs of bearings are connected. (See Fig. 7.) On lightly loaded shafts under 1 in. in diameter and operating at speeds under 100 rpm, adequate lateral and angular flexibility can be secured at low cost by a loose-fitting *tongue and groove joint* in the shaft ends or by a standard *jaw clutch* made with considerable clearance.

The well-known *Oldham coupling*, Fig. 8 and Table 6, eliminates the need for large clearances and the resultant noisy backlash by providing a double-tongued central slider fitting between two flanges slotted at right angles to each other. It allows considerable end play and lateral misalignment (up to 5% of shaft diameter under favorable condi-

Table 6. Dimensions and Capacities of Oldham Coupling

(Palmer-Bee Co., Detroit, Mich.)

Maximum			Dimensions, in. (see Fig. 8)						Weight, Each, lb
Bore, in.	Horsepower at 100 rpm	Torque, inch-lb	*A*	*B*	*C*	*H*	*K*	Standard Key Seat	
1 7/16	5	2700	4 1/2	6 3/8	5/8	2 5/8	15/16	3/8 × 3/16	19
1 15/16	10	6200	5 3/4	8 1/4	3/4	3 1/2	1 1/8	1/2 × 1/4	40
2 15/16	34	21000	8 1/2	12	1	5 1/4	1 1/2	3/4 × 3/8	110
3 15/16	80	50000	10 3/4	15 3/4	1 1/4	7	1 7/8	1 × 1/2	220
5 15/16	270	170000	16	23 1/4	1 3/4	10 1/2	2 5/8	1 1/2 × 3/4	650

tions) plus angular misalignment of 1 degree. It is often used for speeds under 100 rpm and for heavy loads up to the full shaft capacity, as in connecting a gear reducer to the driven machine. Lubrication is required, and the coupling must be protected from grit and dirt. Each shaft should be supported by a bearing near the coupling.

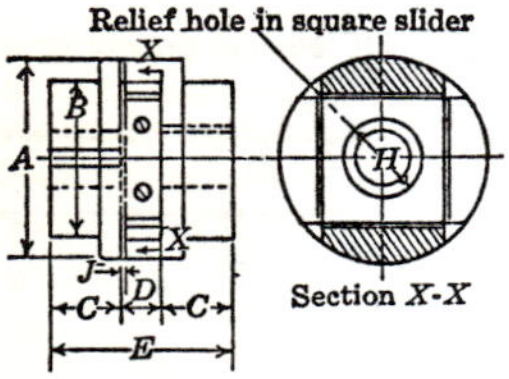

FIG. 9. C-flange coupling. See Table 7.

Figure 9 and Table 7 illustrate a modern development of the Oldham coupling adapted by its large and well-lubricated bearing area to general-purpose work at speeds up to those of standard motors. Torque is transmitted by C-shaped jaw flanges through a square, sliding center member which is surfaced with replaceable bearing strips lubricated from a reservoir within. This coupling will compensate for misalignments as high as 10% lateral and 3-degree angular without inducing reactionary stresses in the shafts and bearings, but for maximum life it should be accurately aligned at installation, like most other couplings. When the coupling is abused by excessive misalignment such as might appear after some time, all wear is taken by the easily replaceable bearing strips.

Table 7. Dimensions and Capacities of C-Flange Coupling

(American Flexible Coupling Co., Erie, Pa.)

Maximum				Dimensions in inches (see Fig. 9)							Approx. Shipping Weight, lb
Bore, in.	Speed, rpm	Horsepower at 100 rpm *	Torque, in.-lb *	*A*	*B*	*C*	*D*	*E*	*H*	*J* †	
1 1/4	6300	1.1	1250	3	2 3/8	1 3/8	13/16	3 9/16	15/16	1/16	6
2	4160	5.4	6120	5 1/4	3 5/8	2 1/4	1 3/16	5 11/16	1 11/16	1/16	23
4	1870	42	47600	10 1/2	7 1/4	4	2 1/16	10 1/16	4	1/16	138
8	880	292	331000	19	13 3/4	6 1/2	3 7/8	16 7/8	7 1/4	1/8	775
13	500	1240	1410000	30	21	12	6 11/16	30 11/16	11	3/16	3200

Note: Intermediate and fractional horsepower sizes available.

* For Series B, nonmetallic bearing strips on hard chrome-plated jaws. One-half for Series A, with nonmetallic bearing strips on plain jaws. Double capacity for Series C, using metallic, nonferrous bearing strips on hard chrome-plated jaws.

† Increase for high angular misalignments or greater end-play allowance.

TORSIONAL RESILIENCE. In addition to misalignment compensation, a small degree of torsional flexibility is often desirable in a high-speed flexible coupling in order to lessen noise and to prevent transmission of starting and running shocks to the motor or gear reducer. This is provided in various amounts by a wide variety of couplings based on the fundamental construction of two flanges connected by resilient spring or nonmetallic elements. The Ajax style, Fig. 10 and Table 8, cushions by means of steel pins bolted alternately to one flange and sliding in self-lubricated bronze bushings, rubber-cushioned in the opposite flange. The manufacturer recommends limiting misalignment to 1/2 degree angular and 0.5% lateral. Torsional deflection under full load is 1/4 degree. The construction permits free axial movement to accommodate motor end play and is especially

Table 8. Dimensions and Capacities of Coupling with Rubber-Bushed Studs

(Type A—Ajax Flexible Coupling Co. Inc., Westfield, N. Y.)

Maximum			Dimensions, in. (see Fig. 10)								Studs		Approx. Weight, lb
Bore, in.	Rpm	Horse-power at 100 rpm	*A*	*B*	*C*	*D*	*E*	*J*	*K*	*L*	No.	Size, in.	
7/8	7000	1.5	3 15/16	1 11/16	1 5/16	1/16	2 11/16	5/8	11/16	2 21/32	6	7/16	4
2	3800	7.8	6 15/16	3 7/16	2 1/2	1/8	5 1/8	1 1/4	1 1/4	4 7/8	6	3/4	25
4	2200	57.0	12 1/4	7 1/8	4 1/2	1/8	9 1/8	2 1/2	2	9 5/16	10	1	135
8 1/4	1400	266.0	19 1/2	13 1/4	7 1/2	3/16	15 3/16	4 1/2	3	15 7/8	14	1 1/4	730
15	840	1700.0	32	22	12	1/4	24 1/4	7	5	28	24	1 1/2	3000

Note: Other styles available include brake drum, shear pin, mill motor, and flywheel adapter. Intermediate sizes available.

suited to damping shock and momentary overload. It also affords electrical insulation, thus preventing such dangers as electrolysis in direct motor-driven pumps.

A frequently seen variation made by a number of manufacturers provides alternate pins or lugs on the face of each flange, fitting loosely in a leather or composition disk so as to place this material in compression (Fig. 11, Table 9). The driving and driven members should be aligned carefully for best results. The nonmetallic material is sometimes used in the form of replaceable cushions or plugs in pockets, a construction permitting brake

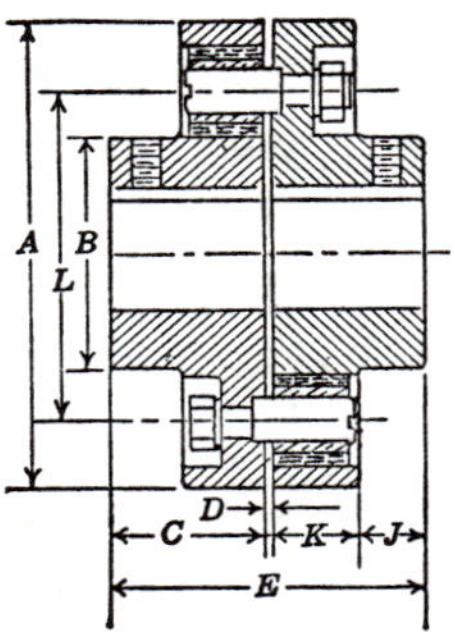

Fig. 10. Coupling with rubber-bushed studs. See Table 8.

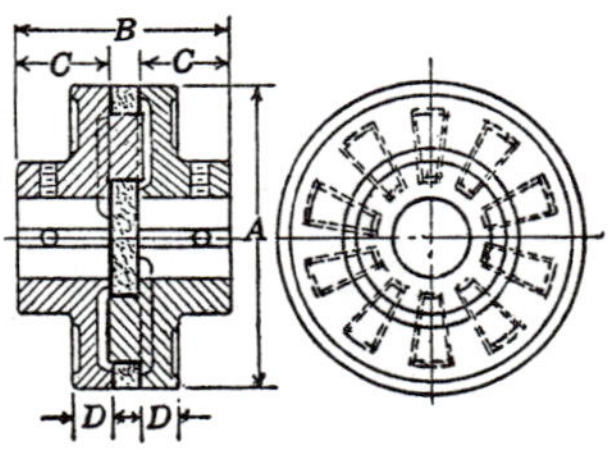

Fig. 11. Leather disk coupling. See Table 9.

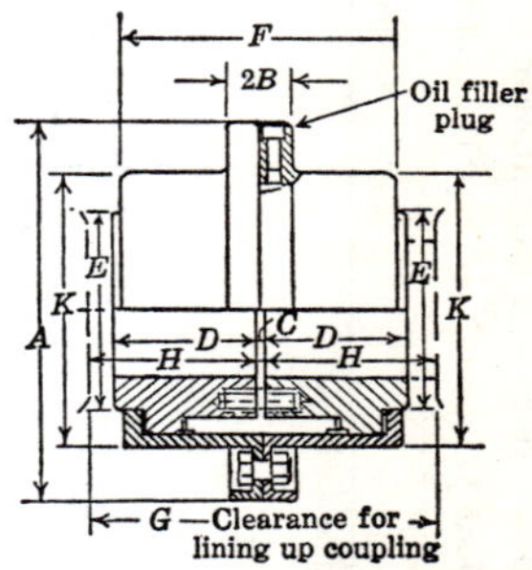

Fig. 12. Double-engagement gear-type coupling. See Table 10.

lining for shock loads at low speeds and high temperatures; leather for sustained loads; and armored rubber for high flexibility, high speeds, and chemicals or wet conditions. In case of cushion failure, such couplings continue to drive but give audible warning.

In still other types, an *endless leather belt* is interwoven between flange pins, or tangential *leather links* carry the load across flange bolts. A more modern construction uses a composition fabric disk bolted in alternation to flanges on each side, similar to Fig. 6.

Table 9. Dimensions and Capacities of Leather Disk Coupling

(T. B. Wood Sons Co., Chambersburg, Pa.)

Maximum				Dimensions, in. (see Fig. 11)						Set Screw Size, in.	No. of Driv-ing Lugs	Weight, lb
Bore, in.		Rpm	Horse-power at 100 rpm	*A*	*B*	*C*	*D*	Hub Diameter, *H*				
Standard Hub	Large Hub							Standard	Large			
3/4	1	6300	3/4	3	3 3/4	1 9/16	9/16	1 1/2	1 3/4	3/8	2	3 1/2
2	2 1/2	2400	14	8	6 1/2	2 7/8	7/8	4	4 1/2	1/2	4	30
3	4 1/4	1600	50	12	9 1/8	4 1/16	1 1/16	5 1/2	6 3/4	3/4	6	80
5 1/2	7	1000	150	18	15	6 15/16	1 1/2	9 1/2	11 1/4	7/8	8	325

Note: Intermediate sizes available.

Table 10. Dimensions and Capacities of Gear-Type Coupling

(John Waldron Corp., New Brunswick, N. J.)

Size and Nominal Bore, in.	Maximum				Dimensions, in. (see Fig. 12)									Weight, lb
	Bore, in.	Rpm	Horsepower at 100 rpm	Parallel Misalignment, in.	*A*	*B*	*C*	*D*	*E*	*F*	*G*	*H*	*K*	
1 1/4 *A*	1 3/8	12000	3 1/2	0.030	4 9/16	5/8	1/8	1 1/2	2	2 7/8	4	1 15/16	2 11/16	10
2 *A*	2 1/8	10000	12	0.040	6	3/4	1/8	2 1/16	2 7/8	4	5 1/8	2 1/2	3 13/16	20
4 *A*	4 1/4	6000	130	0.080	11	1 1/8	1/4	4 3/16	5 3/4	8 1/16	11	5 3/8	7 11/16	100
6 *A*	6 5/8	4000	550	0.120	16 9/16	1 1/2	5/16	6 5/8	8 3/4	12 5/8	16 13/16	8 1/4	12 9/16	400

Note: Other styles available include heavy-duty mill, quick disengaging, marine thrust, continuous lubrication, and vertical. Intermediate sizes available.

Table 11. Dimensions and Capacities of Falk "Steelflex" Couplings

(The Falk Corp., Milwaukee, Wis.)

Maximum						Dimensions, in. (see Fig. 15)											Grid Member			
Bore, in.	Rpm	Horsepower at 100 rpm	Angular Misalignment *	Lateral Misalignment, in.*	Total Endwise Movement, in.	*A*	*B*	*C*	*D*	*F*	*G*	*H*	*J*	*M* †	*S*	Normal Gap	Total Rungs	Outside Diameter, in.	WR^2 lb-in.^{2} ‡	Weight, lb
1	6000	0.6	2° 0′	.015	1/4	3 3/4	3 3/8	1 5/8	1 9/16	2 19/32	51/64	9/16	7/8	1 11/16	1 1/8	1/8	20	2.375	....	3.6
1 7/8	6000	6	2° 0′	.020	5/16	5 5/8	4 3/8	2 1/8	2 15/16	4 15/32	1 3/64	9/16	1 1/8	2 3/16	1 3/8	1/8	36	4.250	....	14
4	2700	70	1° 30′	.040	39/64	10 15/16	7 15/16	3 7/8	6 1/8	9 1/16	2 1/64	11/16	1 15/16	3 3/4	......	3/16	52	8.750	....	97
8	2000	490	1° 0′	.045	5/8	21 1/2	15 1/4	7 1/2	12	16 1/4	3 31/32	1 3/16	3 5/8	7 1/8	......	1/4	72	15.750	14406	644
12	1100	2200	1° 0′	.060	15/16	35 1/2	25 1/2	12 1/2	18 1/2	30	5 21/32	1 3/4	7 1/16	12 3/16	......	1/2	80	28.687	206844	3208

Note: Larger sizes available to 70,000 hp at 100 rpm, also intermediate sizes.

* Misalignment values shown can be accommodated by sliding of the teeth on the grid member without strain on the coupling shafts. However, operation at maximum values will accelerate wear on teeth and grid member; hence, couplings should be closely aligned for long life and minimum maintenance.

† Clearance *M* required to allow installation or removal of grid member.

‡ Includes one hub and half the total grid, grease, and cover weight.

A moderate-cost, heavy-duty, *gear-type coupling*, made by a number of manufacturers, consists of a spur gear on one hub free to slide axially in a mating internal gear on the other hub. The gear teeth on the hub are usually crowned and made with sufficient clearance to provide a $^{3}/_{4}$-degree free rocking action for axial shaft misalignment. Ordinarily, lateral misalignment must also be accommodated, and the double-engagement style is specified as in Fig. 12 and Table 10. In extreme cases, two single-engagement types may be separated by a short floating shaft. Lubrication is provided by a dust- and grit-proof casing filled with heavy oil distributed to all teeth by centrifugal force. The large area of lubricated surface also cushions shock and permits quiet high-speed operation.

These couplings are commonly used for paper mills, jordans, and rolling mills, but they are being specified for an increasing number of general-purpose jobs. Being all metal, they are suitable for service at elevated temperatures. They are built for horizontal operation ordinarily, but suitable supports and lubricant retainers can be provided for vertical operation. The largest standard size is 30 in. (120,000 hp at 200 rpm maximum speed, Fast's Coupling Dept. of Koppers Co., Inc.).

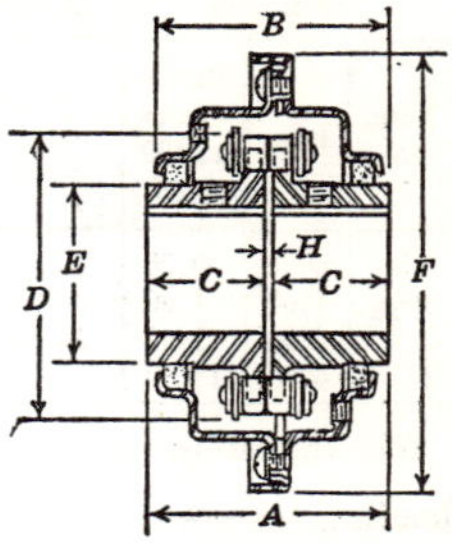

FIG. 13. Roller chain flexible coupling. See Table 12.

The chain coupling, Fig. 13 and Table 12, provides limited cushioning through slight spring action. It is assembled from two sprockets connected by roller or silent chain and

Table 12. Dimensions and Capacities of Roller Chain Flexible Coupling

(Link-Belt Co., Philadelphia, Pa.)

Maximum			Dimensions, in. (see Fig. 13)							Weight, lb
Bore, in.	Rpm	Horse-power at 100 rpm	*A*	*B*	*C*	*D*	*E*	*F*	*H*	
3/4	5000	1.5	2 5/16	2 1/4	1 1/8	2 19/32	1 5/16	4 1/8	1/16	1 1/2
1 3/4	5000	7.2	2 15/16	2 1/2	1 7/16	3 3/4	2 17/32	5 3/4	1/16	3 1/2
3 15/16	2000	49	8 3/8	5 5/8	4 1/8	12 1/16	6 1/4	13 17/32	1/8	88
7 7/8	1500	90.6	9 3/8	5 5/8	4 5/8	14 5/8	10 3/4	16 1/8	1/8	153
12	900	186	12 5/8	6 1/2	6 1/8	19 15/16	15 1/16	21 3/8	1/8	316

Note: One-degree maximum permissible angular misalignment. Intermediate sizes available.

is preferably covered with a protective and grease-retaining shield. Because lateral flexibility derives from small clearances between sprockets and chain, the coupling requires careful alignment. A relatively new type, the Baldwin-Duckworth, minimizes backlash and provides up to 6 degrees of angular flexibility through the use of spherical rollers on one or both sprockets. Silent chain couplings (Morse) are made for capacities up to 1800 hp at 300 rpm, maximum. Principal advantages are low cost and ready disconnection.

HIGH TORSIONAL FLEXIBILITY is sometimes necessary to isolate extreme shock loads or vibration. Together with good lateral and angular flexibility, this is provided in fractional horse-power sizes ($^{3}/_{16}$- to 1-in. shafts) by the inexpensive bonded-rubber flexible couplings made by the Lord Manufacturing Co. Full-load torsional deflection is 15 degrees.

The Airflex coupling, made in sizes ranging from 1 $^{1}/_{4}$ to 17 in. (4700 hp at 100 rpm) consists of an inflated rubber tube fitting inside the driven member. With a strong damping action and a full-load twist angle of 1 degree, it is well suited to high-speed Diesel engine or reciprocating pump and compressor drives. Permissible lateral misalignment is $^{1}/_{16}$ in.; angular misalignment is 1 degree. The coupling may be disconnected by deflating.

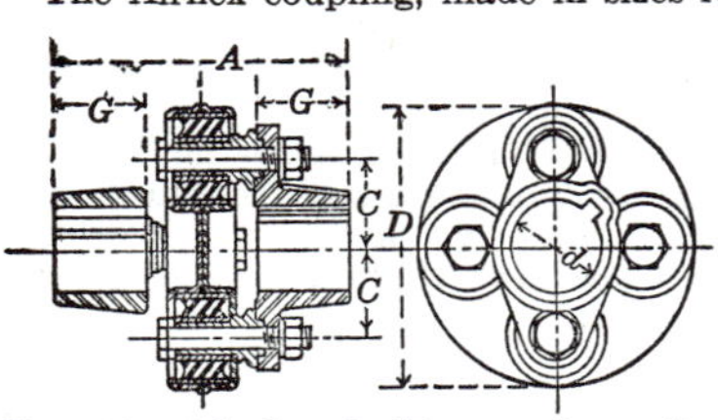

FIG. 14. Morflex flexible coupling. See Table 13.

The *Morse Morflex coupling*, Fig. 14, secures high flexibility by means of a floating center spider carrying four heavy neoprene cushions bolted to alternate flanges. Full-load twist averages 3 degrees, and maximum permissible angular deflection is 5 degrees. Lateral misalignments less than 1% of the shaft diameter are carried without excessive bearing

Table 13. Dimensions and Capacities of Morflex Flexible Couplings

(Morse Chain Co., Ithaca, N. Y.)

Maximum			Dimensions in inches (see Fig. 14)				Stiffness Factors					
Bore, in.	Rpm	Horsepower at 100 rpm	*A*	*C*	*D*	*G*	Torsional, ft-lb/deg	Lateral, lb/in.	Axial, lb/in.	Angular, ft-lb/deg	Wr^2, lb-in.^{2}	Weight lbs.
3/4	6000	0.25	2 1/4	15/16	2 5/8	3/4	6.00	6250	807	1.55	....	3/4
1 3/8	4500	1.60	4 7/8	1 7/8	5 1/16	1 7/8	23.65	6390	1270	9.42	8.8	5
2	4200	5.40	7 5/16	2 3/4	7	2 7/16	122.50	18120	1800	24.50	52.1	15
3	4000	13.80	11	3 7/8	10	3 1/8	200.00	14420	2222	62.00	301.0	43

Note: Intermediate and fractional horsepower sizes available.

loading, as seen in Table 13, which gives stiffness factors in addition to dimensions of selected sizes. The Morflex coupling requires no lubrication and may be operated in oily or grimy conditions at temperatures up to 180 F; hence it is well suited for portable construction and oil-field equipment.

Because of high torsional deflection, destructive resonance can occur on pulsing drives, and the ratio of *critical speeds* must be checked as with any coupling having a full load deflection over 1 degree. The manufacturer recommends that the natural frequency of the coupling and rotating part system be made less than 40% or more than 200% of the exciting frequency. This can be done by selecting a larger coupling to raise the natural system frequency or by using two couplings in tandem to lower it. Auxiliary damping can be provided where the range of natural frequencies is too great. Morse also manufactures a high-damping *Radial coupling* having a full-load torsional deflection of only 1 degree, which may be a better choice for installations requiring less misalignment compensation. Sizes range from 1 3/8 to 8 in. (262 hp at 100 rpm).

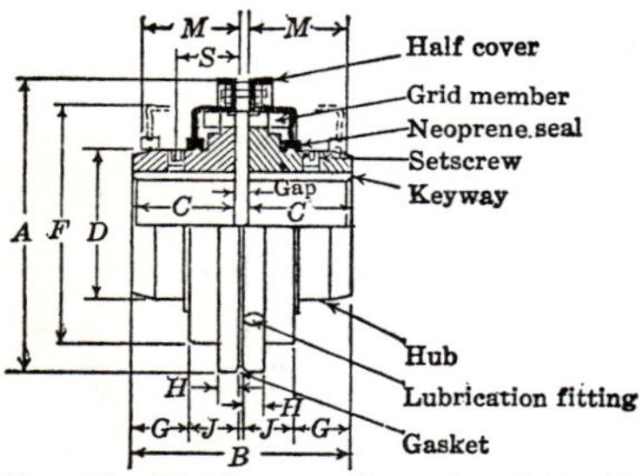

FIG. 15. Steelflex coupling. See Table 11.

There are several all-metal high-flex couplings particularly useful for heavy duty or higher temperatures. *Flexible shafting* is excellent for small sizes, although in larger sizes it is necessary to provide flange-type couplings with some form of spring connection. The *Falk Steelflex coupling*, Fig. 15, comprises a flat steel spring wound in the form of a grid over specially shaped grooves cut into each hub. A housing retains grease and protects the coupling from grit. Table 11 gives dimensions and permissible misalignment values for standard couplings. Because of its nonlinear spring rate and grease filling, this coupling has good damping characteristics and is useful in avoiding certain types of tooth chatter. It should be checked for critical speeds, however.

DOUBLE ENGAGEMENT (FLOATING-SHAFT) COUPLINGS. As previously indicated, when an unusual amount of flexibility is required in all directions, for example by the very flexible frame found in most vehicles, two high-flex couplings can be connected in tandem. These must be types that give full lateral support to the intermediate shaft or element. The Oldham as such would be quite impractical. Again, provided that floating shaft length and speed are not so great as to produce whipping, the double coupling construction is often used to eliminate intermediate bearings when the driving and driven machines are separated by several feet, or by a wall.

SERVICE FACTORS. In order to assure adequate coupling life on pulsating loads, it is necessary to increase the size of the coupling over that determined on a basis of maximum shaft size or rated horsepower capacity. The service factors given in Table 14 are representative of values given in most catalogs and are to be used in the following formula. which allows for fatigue and overheating at higher speeds.

$$\text{Nominal horsepower capacity at 100 rpm} = (\text{Service factor})\ (\text{required horsepower}) \left(\frac{100}{\text{required rpm}}\right)^{3/4}$$

Note: The manufacturer should be consulted regarding critical or unusual conditions.

Table 14. Service Factors

Type of Load	Electric Motor or Steam Turbine	Steam Engine or 4-, 6-, or 8-Cylinder Gasoline Engine	Gas or Oil Engine
Uniform (fans, centrifugal pumps, agitators, generators, line shafts)	1	1 1/2	2
Moderate shock (beaters, grinders, hoists, ball mills, reciprocating pumps or compressors with 3 or more cylinders)	1 1/2	2	2 1/2
Heavy shock (boat propellers, 1- to 2-cylinder compressors and pumps, crushers, hammer and rolling mills)	2	2 1/2	3

Note: Multiply by 1.5 for operation 24 hr per day.

SELECTION FACTORS. The following factors should be checked in addition to the usual considerations of cost, size, weight, and limiting speed for a coupling of adequate capacity:

Special requirements, as frequent disengagement, high end play, thrust or lateral support, and electrical insulation.

Torsional Stiffness or Flexibility. Check on advantages of vibration isolation versus possibilities of critical speeds.

Accessibility. Some couplings must be accessible for inspection, lubrication, or replacement of perishable parts.

Service Conditions. Will the coupling operate satisfactorily when wet, oily, dirty, hot, or cold, as the case may be? Is it suited to shock or reversing loads, and is there a sufficient margin of flexibility to provide for probable alignment drifts with time? The greater the speed, the more margin desirable.

Means of Installation. Light machinery—types with spring resistance to lateral deflection aid line-up at assembly. Heavy machinery—it should be possible to lower the machinery directly into place and then assemble the coupling halves after aligning, without further movement of the machinery.

Failure. Will the coupling give warning before failing completely?

7. UNIVERSAL JOINTS

Universal joints allow positive transmission of power at larger angles than are permissible with flexible couplings. Such joints are sometimes competitive with bevel gearing, although they differ in that the angle may be varied at will but the speed ratio must remain unity.

THE HOOKE OR CARDAN JOINT is the commonest universal joint. It is formed of two U-shaped yoke hubs joined by a cross-shaped piece with four trunnion bearings, one fitting in each yoke arm. Table 15 gives proportions of selected sizes, and Fig. 16 illustrates the *Bocorselski* construction, equivalent to the *Curtis*, in which the smaller trunnion pin fits through the larger to lock it in place. Made of alloy steel, this sturdy, compact style has been extensively used in aircraft. Various other constructions are available for particular applications, as the *Apex* rubber-booted joint or the *Mechanics* and *Blood-Brothers* sealed, heavy-duty, roller-bearing joints, made in sizes up to 50,000 ft-lb torque.

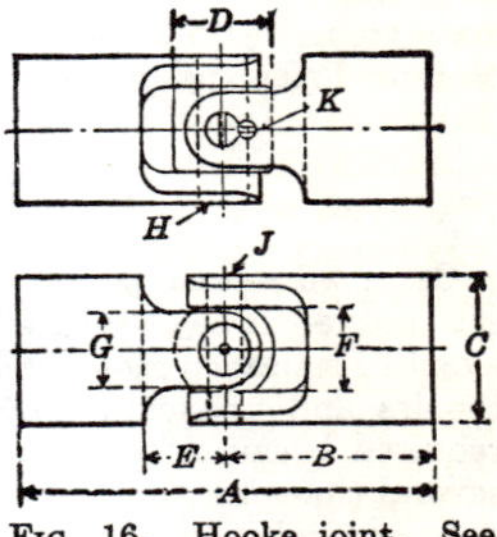

FIG. 16. Hooke joint. See Table 15.

The Hooke joint has one basic defect, a cyclic pulsation of the angular velocity of the driven shaft. The total variation is 3% at 10 degrees, 13% at 20 degrees, 29% at 30 degrees, 55% at 40 degrees, and 71% at 45 degrees. Accordingly, the working angle is kept below 15 degrees if possible, and angles over 30 degrees up to a maximum of 45 degrees are permissible only for manual or very slow speed motions.

When required, uniform angular velocity can be obtained by compensating the motion of one joint with a second; the double joint also considerably increases maximum allowable working angles. The two Hooke joints are connected by a plate or short intermediate shaft, often made telescoping as when powering table feeds on machine tools. In order

Table 15. Proportions and Capacities of Hooke Universal Joints

(Baush Machine Tool Co., Springfield, Mass.)

Maximum		Dimensions, in. (see Fig. 16)							Diameters, in.			Approx. Weight, lb
Bore, in.	Horsepower at 100 rpm *	*A*	*B*	*C*	*D*	*E*	*F*	*G*	*H*	*J*	*K*	
3/16	0.07	1 3/4	7/8	3/8	17/64	7/32	7/32	3/16	0.1065	0.0595	...	0.05
1/2	2.0	3 3/8	1 11/16	1	11/16	9/16	9/16	1/2	5/16	7/32	1/8	0.67
1	16	5 7/16	2 23/32	2	1 7/16	1 3/32	1 3/16	1 1/16	11/16	7/16	1/8	4.0
2 1/4	103	10 5/8	5 5/16	4	2 3/4	2 3/16	2 1/16	1 7/8	1 1/16	3/4	1/4	31

* Safety factor of 2; 0 to 5 degrees operating angle.

to secure perfect uniformity of motion, it is essential that the trunnion pins of the two inner yokes be assembled parallel and that the intermediate shaft make the same angle with both driving and driven shafts.

CONSTANT VELOCITY irrespective of shaft angle is also characteristic of three heavy-duty universal joints originally developed to secure a compact means of powering automotive front-drive steering axles. These joints are typically capable of a 37-degree maximum deflection and are lubricated and sealed for high-speed operation under adverse conditions. Standard sizes range from 15/16 to 2 1/2 in.

The *Bendix-Weiss joint* (Fig. 17) consists of four steel balls placed in intersecting curved races cut in the joint members, the races guiding the balls to precisely half the shaft angle. A central ball absorbs thrust. The *Rzeppa joint*, made by the Gear Grinding Co., is somewhat similar but requires a cage to control ball position. It consists of a spherical, internally grooved driving member fitting about a star-shaped driven member in the manner of a ball-and-socket joint. Six steel balls fit in the mating half-grooves in each member and transmit torque by a free rolling action. The *Tracta joint*, made by the New Process Gear Co., is essentially a double Hooke joint compacted by ingenious construction.

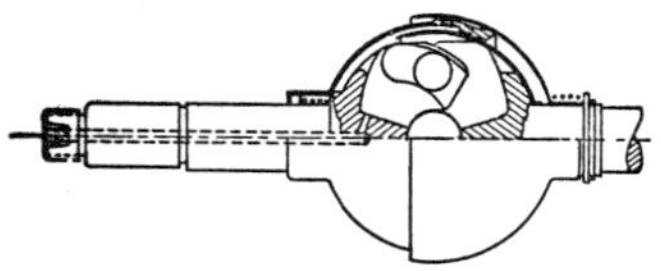

FIG. 17. Bendix-Weiss constant-velocity universal joint.

Another type of constant velocity universal joint built for low cost is typified by the *Brooks hinged joint*, in which power is transmitted between two bevel gears of alternate ball-and-socket form supported in a casing hinged on a line through the ball centers in the zone of contact. The angle between shafts may be varied from 0 to 92 degrees. The joint was originally developed for long-distance manual operation of valves, etc., but is now recommended for light power transmission at speeds up to 150 rpm. It is made in sizes from 1/2 to 2 1/2 in. An equivalent joint using rounded crown teeth is made by *Gray & Prior* and operates through an angle of 0 to 136 degrees.

8. OVERLOAD RELEASE COUPLINGS

Overload release couplings are devices which will transmit no more than a certain preset torque. They are used for protection of driving or driven machine elements in the event of a sudden stop or jam and are a valuable and inexpensive insurance against costly repairs and production outages. Basically, mechanical overload release protection is required because the momentary overload capacity of the standard motor and drive is several times that of critical machine members or of fragile articles being processed.

In general, overload protection is recommended for the following:

(1) Automatic machinery in which the motor armature inertia is a major portion of the total machine inertia, particularly if a serious breakdown could result from poor timing or a failure of any individual part or motion.

(2) Light accessory drives taking power from a heavy-duty main drive.

(3) Machinery processing or handling damageable articles, as hoppers and packaging machines.

(4) Machinery subject to damage from tramp iron, oversize material, or other obstruction, as conveyors, crushers, winches, and dredge pumps.

(5) Limit releases on traverse drives for doors, machine tools, etc.

(6) Feeds and drives for valuable machine-tool cutters subject to breakage from jams, as tapping heads, milling cutters, and broaches.

Overload release type devices are also used for special purposes, including nut runners and wire or sheet windups. Various multipurpose forms are available, incorporating special features as may be required, such as cushioned or flexible coupling action, controlled slippage for easy starts and speed or tension control, and overspeed or underspeed release. Such protection is installed as near the load as possible to minimize parasitic inertia.

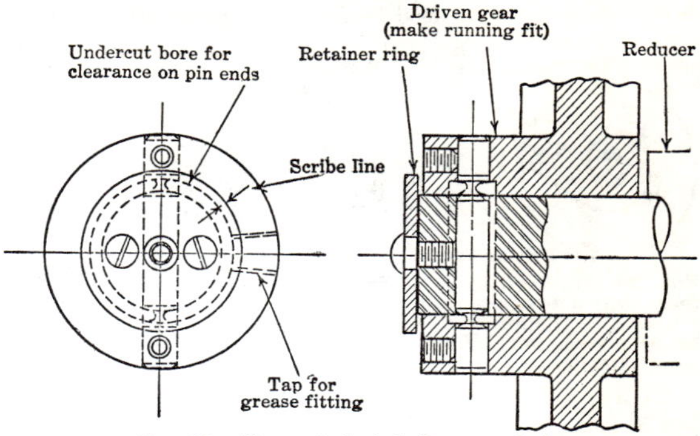

FIG. 18. Shear pin installed on gear hub.

THE SHEAR PIN is readily applied to most drive arrangements at a minimum of cost and is reliable when properly designed. Because of fatigue, it cannot be made very sensitive, however, and replacement is a nuisance. Therefore, use should be restricted to drives subject to overloading at very infrequent intervals only, where the design breaking load can be at least double the normal running torque.

Several important design points are illustrated in Fig. 18, which shows a noncommercial application made on a gear hub in place of a taper pin.

Proper Control of Fracture. The shear pin body is made double or triple the required diameter and the shear point or points are U-notched to suit the load. This also permits torque adjustment.

Pin Material. Should be predictable and low in notch sensitivity, although wood has been used successfully in addition to brass, bronze, mild and stainless steels, and cast iron.

Free Rotation after Fracture. Assured by providing lubrication plus grooves in the hub and shaft to clear the ends of the pin after shearing. Individual setscrews on each pin section prevent possible shifting and jamming before or after shearing.

Easy Pin Replacement. All edges and shoulders are well chamfered, and the rotating parts are scribed to aid positioning.

This design contains one possible fault: Unauthorized personnel may replace the necked-down pin with one of full diameter or even of hardened steel in order to stop breakages, even though the shear pin may be preventing serious damage. For this reason, some engineers prefer to anticipate by designing for a hardened pin of full diameter and accepting the possibility that undirected pin breakage will cause seizing.

The following formula can be used for the double-shear design of Fig. 18·

$$\text{Torque} = D \times S_s \times 0.7854 \times d^2 \text{ in.-lb}$$

where D = shaft diameter, inches; S_s = shear strength of pin material, pounds per square inch, and d = diameter of necked-down section of pin, in.

Shear-pin protection is supplied on request by most manufacturers of power transmission equipment, typically in the form of "breaking-pin hubs" on gears, sprockets, or flexible couplings. Two or more pins are provided, sharing the load to improve consistency: a maximum variation of 10% can be expected with six or more pins under nonfatigue conditions, as contrasted to the 50% variation typical of single pins. The multiple pins are placed parallel to the shaft in a ring around the flange and are fitted in hardened steel bushings to prevent flange mutilation.

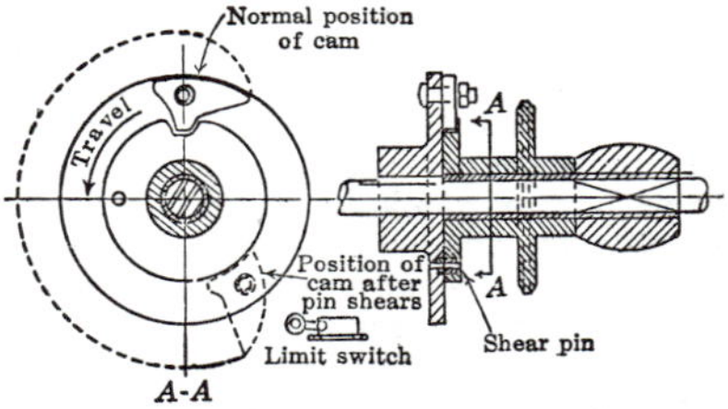

FIG. 19. Jeffrey telltale shear-pin hub.

A unique *telltale shear-pin hub* patented by *Jeffrey* is shown in Fig. 19. The limit switch stops the motor and gives warning through a bell or light if desired.

THE SAFETY WASHER, a variation of the shear pin, is sometimes applied for protection of worm gear drives. It consists of a cast-iron support cup which will break and permit the worm to screw itself away from a jaw coupling. A preloaded spring can also be used to support the worm.

A DETENT RELEASE COUPLING (noncommercial) is shown in Fig. 20. Relatively simple, this device has proved much more useful than the shear-pin coupling because it can be adjusted to within 5% of the load and resets itself without attention or time loss. A turnbuckle can be used to facilitate spring adjustment. If desired, the motor can be

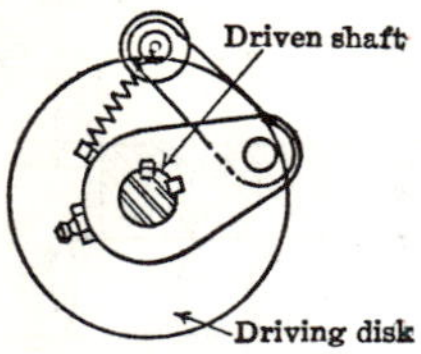

FIG. 20. Spring-loaded roller type detent release coupling. (Reproduced by permission from *Machine Design*)

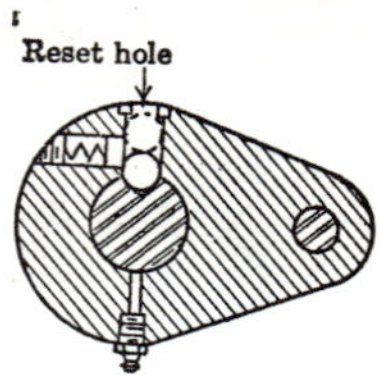

FIG. 21. Spring-loaded ball detent release coupling. (Reproduced by permission from *Machine Design*)

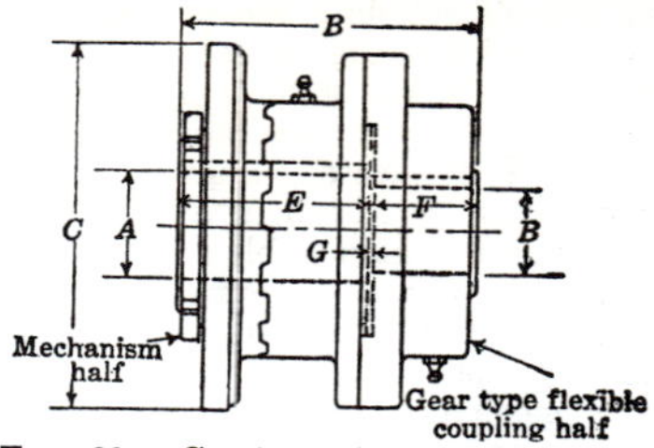

FIG. 22. Curvic tooth overload release coupling. See Table 16.

shut off automatically by driving the roller arm instead of the disk and placing a limit switch where it will be struck by the roller when out of the notch. Figure 21 illustrates another style suitable for light loads.

Figure 22 and Table 16 illustrate the Hilliard Curvic tooth overload release coupling. As a result of the snap action of the Belleville spring, this coupling releases completely and

Table 16. Proportions and Capacities of Curvic Tooth Overload Release Couplings
(The Hilliard Corp., Elmira, N. Y.)

Maximum		Dimensions, in. (see Fig. 22)						
Horsepower at 100 rpm	Torque, lb-ft	*A*	*B*	*C*	*D*	*E*	*F*	*G*
1/2	26.3	3/4	3/4	3 3/4	3 1/8	2	1	1/8
6	315	2 3/16	1 3/4	8	6 1/2	3 15/16	2 7/16	1/8
40	2100	3 5/16	3 3/16	12	9 7/8	6	3 5/8	1/4
400	21000	8 7/16	7 1/2	30	23 3/4	15	8 3/8	3/8

Note: Intermediate sizes also available.

does not hammer. It is reset by force from the end, preferably by a built-in fork lever or by an air cylinder in the large sizes. In order to insure a return to the original relationship between shafts at reset, essential in timed mechanisms, one tooth can be removed and the opposite space filled in. This and other commercial overload release couplings can be obtained with sleeves for pulleys, gears, or sprockets as well as in the shaft cut-off design shown.

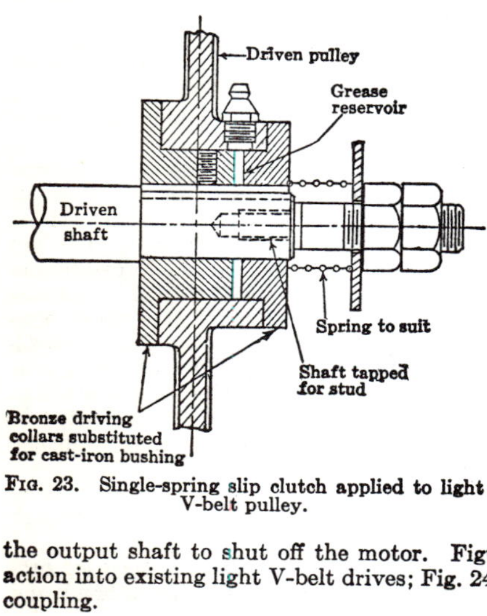

FIG. 23. Single-spring slip clutch applied to light V-belt pulley.

SLIP COUPLINGS are effective protection, particularly on a shock or pulsating load, because they can be set to the normal peak running load, yet will slip momentarily on abnormal peaks or fast starts—a cushioning action—without releasing completely. Slip couplings are also used to simplify construction and assure full travel in traverse drives and to provide controlled slippage in windup equipment.

Where the continued application of torque after slipping is objectionable or hazardous, a spring flapper can be built into the clutch to sound a warning, or a standard plugging switch can be belted to the output shaft to shut off the motor. Figure 23 illustrates a method for building slip action into existing light V-belt drives; Fig. 24 and Table 17 give dimensions of a standard coupling.

Overload slip action of varying sensitivity can be produced in most clutch types which do not operate on positive wedge or toggle locking principles, particularly in centrifugal, pneumatic, hydraulic, and magnetic clutches and couplings. (See discussion under

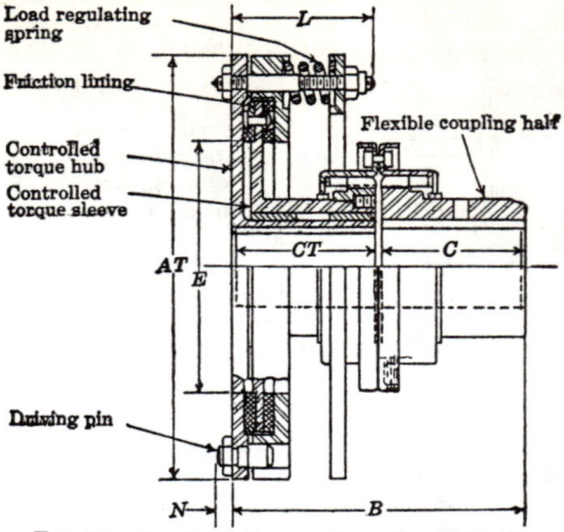

FIG. 24. Overload slip coupling. See Table 17.

Clutches.) Other power transmission elements with slip action include belts, particularly flat, round, and spring types, and certain friction variable-speed transmissions.

Table 17. Proportions and Capacities of Falk Overload Slip Coupling

(The Falk Corp., Milwaukee, Wis.)

Maximum				Dimensions, in. (see Fig. 24)							Approx. Weight, lb
Bore, Torque Hub, in.*	Bore, Coupling Hub, in.†	Rpm	Torque in.-lb	*AT*	*B*	*C*	*CT*	*E*	*L*	*N*	
5/8	1	3700	250	6 7/8	4 3/16	1 5/8	2 7/16	3 7/8	3 1/16	3/16	15
1 7/16	1 7/8	2600	2500	9 1/2	5 1/8	2 1/8	2 7/8	5 3/4	3 11/16	5/16	34
3 3/16	4	1300	30000	18 1/4	9 1/16	3 7/8	5	11 1/8	5 1/16	1/2	183
6 1/2	8	700	....	33 1/4	16 3/4	7 1/2	9	20 1/2	8 3/4	3/4	1125
9 1/4	12	450	....	48	29	12 1/2	16	24	10	1 1/8	3950

* Controlled torque hub may be mounted on either driving or driven shaft. Ordinarily it is mounted on smaller shaft.
† Same as standard flexible coupling hub. See Table 11 for complete coupling dimensions in these bore sizes.

CLUTCHES

By R. M. Nardone

Clutches are used to connect shafts to a driving mechanism such as a motor, engine, or line shaft, and to disconnect them at will. They may be classed as (1) friction clutches in which the driving force is transmitted by the friction of surfaces in contact, (2) jaw clutches in which jaws, dogs, or teeth in the two elements interlock, (3) hydraulic clutches (fluid couplings) in which torque is transmitted hydrodynamically through change of momentum of a moving fluid, and (4) electromagnetic clutches in which torque is transmitted across an air gap through the creation of an electromagnetic field. In addition, there are overrunning or freewheeling clutches which permit torque transmission in one direction only.

9. FRICTION CLUTCHES

MULTIPLE-DISK CLUTCHES are usually of the multiple-plate type in which one set of disks is splined to one shaft and the other set, alternating with the first, is splined to the other shaft (Fig. 1). Pressure on the clutch pack is usually applied by one or more springs which are either adjusted to impose a steady pressure on the disks or which may be regulated for varying pressure. Let P = total spring force, pounds; r = mean radius of disks, inches; n = number of slipping surfaces; μ = coefficient of friction; and T = torque capacity of clutch, lb-in. Then

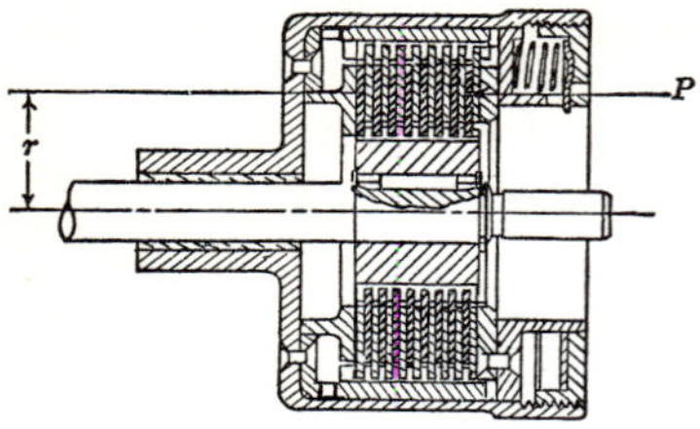

FIG. 1. Multiple-disk clutch.

$$T = P \cdot \mu \cdot n \cdot r \tag{1}$$

Slipping surfaces are usually steel against bronze but may be asbestos against steel, cork on steel, etc. The coefficient of friction of bronze on steel with grease lubrication, is approx. 0.15. Without lubrication, μ varies between 0.2 and 0.25. For an asbestos surface, the friction coefficient is usually 0.3 to 0.4.

Heating of the clutch depends on the amount of slipping or the energy absorbed during operation. The temperature rise is

$$(t_1 - t_2) = \frac{H}{WC} = \frac{E}{778.3WC} \tag{2}$$

where t_1 = maximum temperature of the clutch disks, degrees F; t_2 = initial disk temperature; H = heat absorbed by clutch, (Btu); E = energy, foot-pounds; W = weight of clutch plates, pounds; and C = specific heat of clutch plate material.

The amount of energy absorbed by the clutch is equal to

$$E = \frac{T \times 2\pi}{12} \times N = \frac{TN\pi}{6} \tag{3}$$

where N = number of revolutions of slip. A temperature t_1 of 500 F should not be exceeded.

Capacity. All types of friction clutches should have torque capacities of double the torque delivered to them in order to start the load from rest without excessive slip. That is, the torque to be transmitted should be multiplied by a static load factor of 2. If the torque delivered to the clutch is irregular, as in certain types of internal combustion engines, the static load factor should be increased from 2 to possibly 3 or 4.

Additional factors should be applied depending on the character of the load. Table 1 gives factors recommended by the Twin Disc Clutch Co. as a guide in the selection of a clutch based on the characteristics of the driven mechanism.

Table 1. Application Factors for Dry-plate Friction Clutches

Driven Mechanism	Factor	Driven Mechanism	Factor
Air compressors (depending on cycles)	2.00–2.50	Generator, steady load	1.00
Ammonia compressors	2.50	Hammer mills	2.00
Blowers	1.00	Hoists, elevators, cranes, shovels	2.00
Brick-making and similar machinery	3.00	Line shaft (driven by belt or gears)	1.50
Centrifugal pumps	1.00	Printing presses, textile looms, etc.	1.50
Crushers, veneers, hogs, ball mills, etc.	2.00	Pulp grinders, screens, etc.	1.75
Double-acting pump	1.50	Rock and stone crushers	3.00
Driers and calenders	2.00	Single-acting pump	2.00
Generator, fluctuating load	1.50	Woodworking machinery, general	1.75

CENTRIFUGAL CLUTCHES. A centrifugal clutch produces its torque by virtue of the centrifugal force of weights pressing against the driving or frictionally driven member. The simplest form is shown in Fig. 2 where two weights A are shown connected to and driven by motor B and separated from each other by springs C so as to press against the inside diameter of the driving drum E.

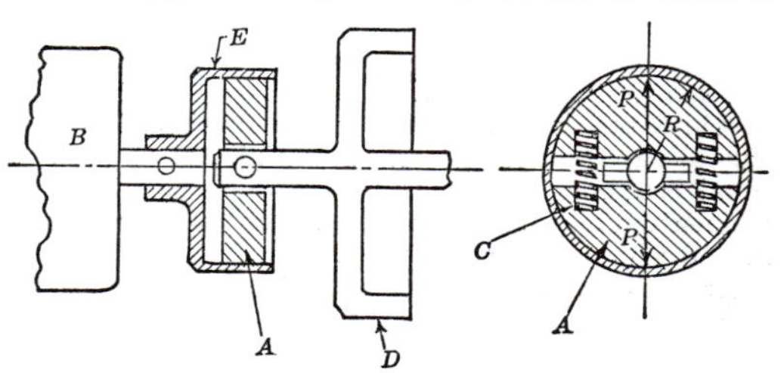

FIG. 2. Centrifugal clutch.

Referring to Fig. 2, let R = inside radius of drum E, inches; r = radius of gyration of weights A, inches; w = weight of the centrifugal member A, pounds; μ = coefficient of friction; F_c = total centrifugal force of weights A, pounds; P = spring force on each weight, pounds; T_1 = torque due to spring force, pound-feet; T_2 = torque due to centrifugal action of weights A, pound-feet; T = total torque at any speed, pound-feet; t = time required to accelerate flywheel D, seconds; ω = speed of flywheel, radians per second; n = number of centrifugal weights; I = total moment of inertia of driven members, pound-feet second squared; and α = acceleration, radians per second per second. Then

$$T_1 = \frac{P\mu Rn}{12} \tag{4}$$

$$F_c = \frac{wr\omega^2 n}{12g} \tag{5}$$

$$T_2 = F_c \times \frac{\mu R}{12} = \frac{wr\omega^2 Rn\mu}{g \times 12 \times 12} \tag{6}$$

$$T = T_1 + T_2 = \frac{P\mu Rn}{12} + \frac{Wr\omega^2\mu Rn}{144g} \tag{7}$$

But $T = I\alpha$ = torque resistance of flywheel D. Therefore

$$I\alpha = \frac{P\mu Rn}{12} + \frac{wr\omega^2\mu Rn}{144g}$$

and

$$\alpha = \frac{P\mu Rn}{12I} + \frac{wr\omega^2\mu Rn}{144gI} \tag{8}$$

Let

$$C = \frac{P\mu Rn}{12I} \quad \text{and} \quad K = \frac{wr\mu Rn}{144gI}$$

Then

$$\alpha = C + K\omega^2 = \frac{d\omega}{dt}$$

$$dt = \frac{d\omega}{C + K\omega^2}$$

$$t = \int \frac{d\omega}{C + K\omega^2} = \frac{1}{\sqrt{KC}} \times \tan^{-1} \omega \sqrt{\frac{K}{C}} \text{ seconds} \tag{9}$$

EXAMPLE: Required: to accelerate flywheel D to 6000 rpm. Motor speed and torque are at all times greater than flywheel speed and transmitted torque, respectively. How long will it take? Proportions are as follows: I = 0.007 lb-ft sec^2; n = 2; P = 1 lb; μ = 0.2; R = 2 in; w = 1 lb, each; and r = 1.5 in. Then

$$C = \frac{P\mu Rn}{12I} = \frac{1 \times 0.2 \times 2 \times 2}{12 \times 0.007} = 9.52; \qquad K = 0.037; \qquad \sqrt{KC} = 0.593$$

$$\frac{1}{\sqrt{KC}} = 1.685; \qquad \sqrt{\frac{K}{C}} = 0.0624$$

$$t = 1.685 \tan^{-1} \omega \times 0.0624$$

$$\omega = 630 \text{ radians per sec (approximately)}$$

$$t = 1.685 \tan^{-1} 630 \times 0.0624$$

$$= 1.685 \tan^{-1} 39.3$$

$$\tan^{-1} 39.3 = 88.54 \text{ degrees} = 1.55 \text{ radians}$$

Therefore

$$t = 1.685 \times 1.55 = 2.61 \text{ sec}$$

CONE CLUTCHES. The conical friction clutch consists of the frustum of a cone, so fitted to a shaft by means of a feather key that it can be pushed into an opposite engaging surface rigidly attached to the other shaft. Such clutches require some force to put them into gear. This force, acting parallel to the shaft, produces an undesirable end thrust. Cone clutches may be faced with leather or asbestos fabric or equipped with cork inserts to increase their driving capacity. The inclination α of the slant side is highly important in order to avoid "sticking" on the one hand, and too sudden seizure on the other. α usually ranges from 7 1/2 to 13 degrees. Proportions may be taken as follows (see Fig. 3): $B = 2d + 1$; $C = 1.5d$; $F = 1.8d$; $H = 0.5d$; $E = 0.4d + 0.4$; $G = 0.3d + 0.3$; $t = 0.2d + 0.1$; $L = 2d$.

Formulas for Cone Clutches (Fig. 3). Let M = moment clutch is capable of transmitting, inch-pounds; hp = horsepower transmitted; N = revolutions per minute; r_2 = small radius of friction cone, inches; r_1 = large radius of friction cone, inches; r = mean radius of friction cone, inches; V = velocity of a point at distance r from center, feet per minute; T = total tangential force acting at radius r, pounds; P_n = total normal pressure between

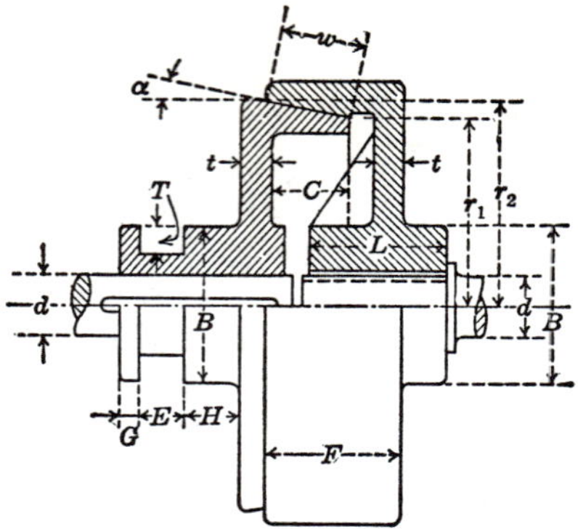

FIG. 3. Conical friction clutch.

cone surfaces, pounds; p = allowable pressure, pounds per square inch of contact surface; P_s = total axial or spring pressure, pounds; α = angle of clutch surface with axis of shaft; w = width of cone surface, inches; and μ = coefficient of friction. Then

$$r = {}^1/_2(r_1 + r_2); \qquad M = Tr; \qquad T = \mu P_n; \qquad V = {}^1/_{12}(2\pi r N); \qquad w = \frac{P_n}{2\pi r p}$$

For engagement without slip,

$$\text{Hp} = \frac{P_n \mu r N}{63{,}030}$$

$$P_s = \frac{T}{\mu}(\sin\alpha + \mu\cos\alpha) = \text{hp} \times 63{,}030\left(\frac{\sin\alpha + \mu\cos\alpha}{\mu r N}\right)$$

and after engagement

$$P_n = \frac{P_s}{\sin\alpha}$$

$$P_s = \text{hp} \times \frac{63{,}030}{\mu r N}\sin\alpha$$

The allowable normal pressure p may be taken as 7 or 8 psi for leather bands, according to *Machinery's Handbook;* Smith and Marx give values of 30 to 40 psi for leather or cork, and 35 to 45 psi for maple wood. For values of friction coefficient, see Table 4, p. 7-28.

MAGNETIC CLUTCHES are a form of friction clutch in which the friction surfaces are brought into contact by an electromagnet instead of by mechanical means. They may be operated at higher rotative speeds than mechanical clutches and are recommended for heavy duty where mechanical clutches are impractical, for high-speed drives, and where the motor cannot start a heavy load. The principle of the clutch is shown in Fig. 4. A magnet coil A, carried by one member of the clutch, attracts to itself the armature B, carried by a spring plate C, which flexes when the magnet is energized. A friction ring D, carried by the armature, is faced with friction lining F. When the coil A is energized, the friction lining is drawn into contact with field friction ring E. The friction between E and F transmits the torque from one clutch member to the other. For heavy duty, two friction linings often are used. See Fig. 5.

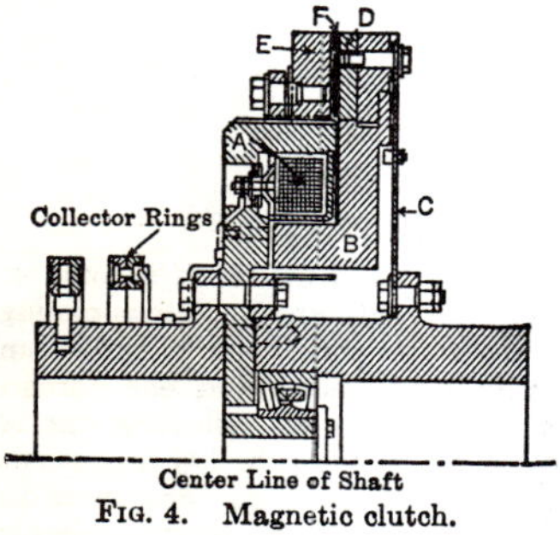

FIG. 4. Magnetic clutch.

The clutch is engaged by energizing the magnet coil; it is disengaged by de-energizing the coil. The spring plate then pulls the friction lining away from the friction ring until there is an idling clearance of about 1/16 in. between them. The current in the magnet coil is varied by a manually operated controller, which permits the torque to be built up gradually. Heavy loads thus can be started without shock.

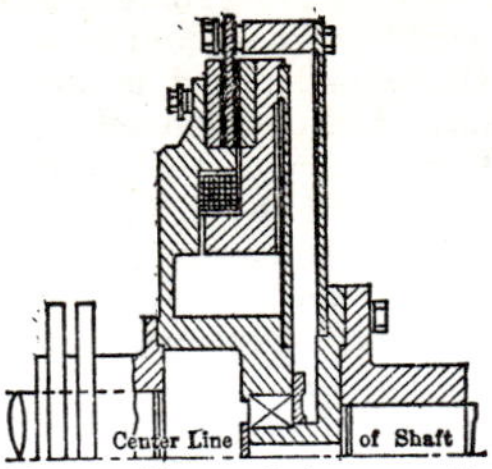

FIG. 5. Heavy duty magnetic clutch.

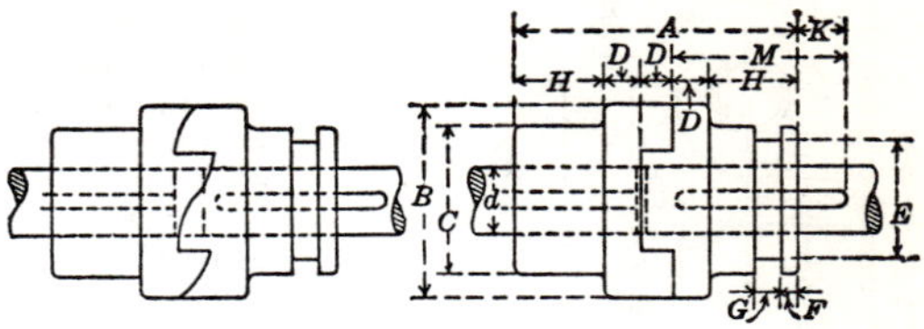

FIG. 6. Jaw clutches.

10. POSITIVE-ENGAGEMENT CLUTCHES

JAW CLUTCHES (Fig. 6) are used for slow-moving shafts, where sudden starting action is not objectionable and when the inertia of the driven parts is relatively small. The various forms of such clutches differ only in the shapes or angles of the engaging surfaces. One-half of the coupling is fitted with a feather key, permitting it to be engaged with or disengaged from the other half at will.

If N = number of teeth, R = mean radius of clutch tooth, A = area of engaging face of one tooth, A' = area of the tooth subjected to shear, s_c = allowable unit crushing stress of the clutch jaw, and s_s = allowable shearing stress of the shaft, then the relation between strength of the shaft and strength of clutch is

$$\tfrac{1}{16}\, s_s \pi d^3 = RNAs_c = RNA's_s \tag{10}$$

The spiral form (Fig. 6) is more easily put into gear, but the motion of the shaft always must be in the same direction. Table 2 gives the proportions used by one maker for both square and spiral jaw clutches.

Table 2. Dimensions of Square and Spiral Jaw Clutches

(Data from Dodge Mfg. Co., Mishawaka, Ind.)

Shaft Diam., in.	Dimensions, in. (see Fig. 6)										Weight, lb
d	A	B	C	D	E	F	G	H	K	M	
15/16	4 3/4	3	2	1/2	1 5/8	1/2	1/4	1 5/8	3/4	2 7/8	18
1 3/16	5 3/8	3 1/2	2 1/2	5/8	2 1/8	1/2	1/4	1 3/4	1	3 3/8	22
1 7/16	6 1/4	4 1/8	3	3/4	2 1/2	9/16	3/8	2	1 1/8	3 7/8	28
1 11/16	6 7/8	4 3/4	3 1/2	7/8	2 7/8	9/16	3/8	2 1/8	1 3/8	4 3/8	35
1 15/16	7 1/2	5 3/8	4	1	3 1/4	5/8	3/8	2 1/4	1 1/2	4 3/4	44
2 3/16	8 1/8	6	4 1/2	1 1/8	3 5/8	5/8	3/8	2 3/8	1 3/4	5 1/4	58
2 7/16	8 3/4	6 3/4	5	1 1/4	4 1/8	5/8	7/16	2 1/2	1 7/8	5 5/8	75
2 11/16	9 3/8	7 1/8	5 1/2	1 3/8	4 5/8	11/16	7/16	2 5/8	2	6	81
2 15/16	10	7 3/4	6	1 1/2	5 1/8	11/16	1/2	2 3/4	2 1/4	6 1/2	104
3 7/16	11 1/2	8 1/4	7	1 3/4	5 7/8	3/4	1/2	3 1/8	2 1/2	7 3/8	136
3 15/16	12 1/2	9 1/2	8	2	6 3/4	13/16	5/8	3 1/4	2 7/8	8 1/8	190

Synchronized Engagement. Another form of jaw or dog clutch is used in automotive transmissions. The clutch elements comprise an external and internal gear or spline, each element having the same number of teeth. One member slides on a spline connection into and out of mesh with the other, at the will of the operator. Clashing due to speed differences is prevented by synchronizing clutches, usually consisting of small friction clutches which engage through a detent mechanism prior to the tooth engagement. To make the speed synchronization possible, one of the elements must be free to accelerate or decelerate to the speed of the other during engagement. In automobiles this is done by releasing the main friction clutch.

11. HYDRAULIC COUPLINGS OR FLUID CLUTCHES

In a fluid clutch, an impeller and a runner are enclosed in a housing filled with a light oil. The two members are the shape of two hollow doughnuts, one within the other and split in half along the circular axis. The impeller is connected to the driving shaft and the runner to the driven shaft. (See Fig. 7.) The liquid is set in motion by the radial blades of the impeller and is dashed against similar blades in the runner. A drag is thereby produced on the runner, causing it to rotate with the impeller.

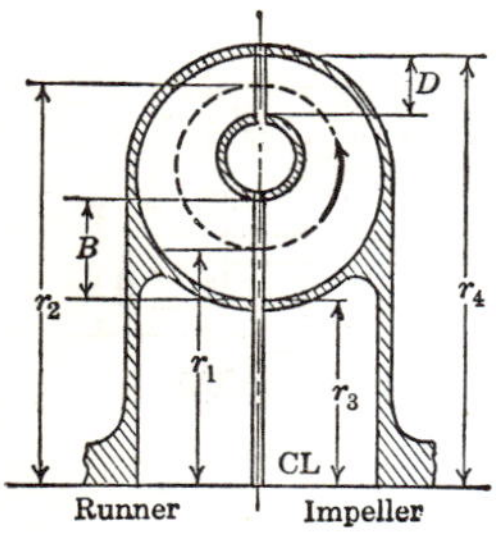

FIG. 7. Fluid clutch.

The liquid carried by the coupling around the shaft axis has centrifugal force imparted to it. This force pushes the liquid out of the impeller into the slower rotating runner, and the difference in centrifugal force causes rapid circulation of the fluid in the torus parallel to the radial vanes.

Power is transmitted almost wholly through the loss of kinetic energy caused by the difference in the linear tangential velocity of the liquid at the outlet D of the impeller and the lower tangential velocity at outlet B of the runner. The impeller horsepower is usually greater than that of the runner due to the differences in their speeds. However, the torque on both members is always equal, that is, impeller torque equals runner torque.

If the physical dimensions of the coupling, and the impeller and runner speeds are known, the loss in kinetic energy, transmitted horsepower, and torque may be calculated. Most couplings are designed to have the cross-sectional area of the liquid path constant. Therefore, radial width B equals $D \times r_2/r_1$. Design formulas are given in Table 3.

Table 3. Fluid Clutch Design Formulas

(See Fig. 7 and nomenclature below)

$$B = \frac{r_2}{r_1} D \tag{11}$$

$$Q = \frac{a \cdot 2\pi(r_2 + r_1)}{2} \times 0.85 \quad \text{(0.85 is factor to compensate for volume of radial blades)} \tag{12}$$

$$W = Qd \tag{13}$$

$$F_i = \frac{W}{4g} \cdot \frac{(r_2 + r_1)}{2 \times 12} \cdot \left(\frac{2\pi N}{60}\right)^2 = \frac{W(r_2 + r_1)N^2}{282{,}000} \tag{14}$$

$$F_r = \frac{W}{4g} \cdot \frac{(r_2 + r_1)}{2 \times 12} \cdot \left(\frac{2\pi n}{60}\right)^2 = \frac{W(r_2 + r_1)n^2}{282{,}000} \tag{15}$$

$$F = F_i - F_r \tag{16}$$

$$L = \frac{\pi(r_2 - r_1)}{12} \tag{17}$$

$$H = \frac{F}{A \cdot d \cdot 12} \tag{18a}$$

$$H = f\left(\frac{12L}{D}\right)\frac{V^2}{2g} \tag{18b}$$

$$A = 2\pi r_2 \cdot D \tag{19}$$

$$P = 12VZAd \tag{20}$$

$$KE_i = \frac{Pv_i^2}{2g} \tag{21}$$

$$KE_r = \frac{Pv_r^2}{2g} \tag{22}$$

$$KE = KE_i - KE_r \tag{23}$$

$$v_i = \frac{2\pi r_2}{12} \times \frac{N}{60} \tag{24}$$

$$v_r = \frac{2\pi r_1}{12} \times \frac{n}{60} \tag{25}$$

$$\text{HP}_i = \frac{KE}{550} \tag{26}$$

$$T = \frac{\text{HP} \times 5250}{N} \tag{27}$$

In Table 3,

N = impeller speed, revolutions per minute.
n = runner speed, revolutions per minute.
F_i = centrifugal force on liquid in impeller, pounds.
F_r = centrifugal force on liquid in runner, pounds.
F = net centrifugal force causing vortex circulation, pounds.
d = density of liquid = .032 lb per cu in. for usual oils of 0.875 specific gravity.
s = specific gravity of liquid = 0.875, average.
μ = absolute viscosity of liquid, poises.
L = circuit length, feet; dotted line in Fig. 7.
H = hydraulic head, feet.

V = velocity of liquid along path L, feet per second.
v_i = linear velocity tangential to shaft rotation of liquid at outlet of impeller, feet per second.
v_r = linear velocity tangential to shaft rotation of liquid at outlet of runner or inlet to impeller, feet per second.
W = total weight of liquid circulating, pounds.
A = mean average area, transverse to vortex circuit at outlet of impeller, square inches.
f = friction factor in formula $H = f\left(\frac{12L}{D}\right)\frac{V^2}{2g}$ (see Fig. 8).
D = radial width of vortex path at impeller outlet, inches.
B = radial width of vortex path at impeller inlet, inches.
K = criterion used to find $f = DVs/100\mu$ (see Fig. 8).
Z = turbulence friction factor = 0.80 for high slip to 0.95 for viscous flow.
a = area of liquid circulating around circuit L, square inches.
Q = total volume of liquid circulating, cubic inches.
r_2 = equivalent mean radius of outflow from impeller, inches.
r_1 = equivalent mean radius of inflow to impeller, inches.
g = acceleration of gravity = 32.2 ft per sec^2.
P = fluid circulating rate, pounds per second.
KE_i = kinetic energy delivered by impeller, foot-pounds per second.
KE_r = kinetic energy received by impeller, foot-pounds per second.
KE = $KE_i - KE_r$ = difference in kinetic energy, foot-pounds per second.
HP_i = horsepower delivered by impeller.
HP_r = horsepower delivered by runner to output shaft.
T = torque, pound-feet.

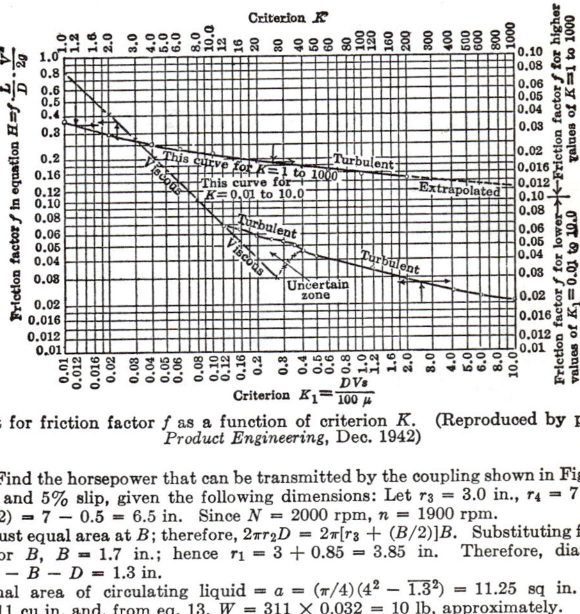

Fig. 8. Chart for friction factor f as a function of criterion K. (Reproduced by permission from *Product Engineering*, Dec. 1942)

Example. Find the horsepower that can be transmitted by the coupling shown in Fig. 7 at 2000-rpm impeller speed and 5% slip, given the following dimensions: Let $r_3 = 3.0$ in., $r_4 = 7$ in., $D = 1$ in., $r_2 = r_4 - (D/2) = 7 - 0.5 = 6.5$ in. Since $N = 2000$ rpm, $n = 1900$ rpm.

Area at D must equal area at B; therefore, $2\pi r_2 D = 2\pi[r_3 + (B/2)]B$. Substituting for r_2, r_3, and D, and solving for B, $B = 1.7$ in.; hence $r_1 = 3 + 0.85 = 3.85$ in. Therefore, diameter of inner ring $= r_4 - r_3 - B - D = 1.3$ in.

Cross-sectional area of circulating liquid $= a = (\pi/4)(4^2 - \overline{1.3}^2) = 11.25$ sq in. From eq. 12, volume $Q = 311$ cu in, and, from eq. 13, $W = 311 \times 0.032 = 10$ lb, approximately.

The centrifugal force on $W/2$ lb of the liquid in the impeller is, from eq. 14,

$$F_i = \frac{10 \times (6.5 + 3.85) \times 2000^2}{282{,}000} = 1470 \text{ lb}$$

and the centrifugal force on the liquid in the runner is, from eq. 15,

$$F_r = \frac{10 \times (6.5 + 3.85) \times 1900^2}{282{,}000} = 1330 \text{ lb}$$

The difference in centrifugal force is, from eq. 16, $F = 1470 - 1330 = 140$ lb. The mean average area $A = 2\pi \times 6.5 \times 1 = 40.9$ sq in. from eq. 19; therefore, the hydraulic head $H = 140/(40.9 \times 0.032 \times 12) = 8.9$ ft, from eq. 18*a*.

For trial purposes, assume a reasonable value for the velocity of liquid circulation in the vortex, V. Using $V = 30$, $s = 0.875$ specific gravity, and $\mu = 1$ absolute viscosity, criterion K (Fig. 8) is $K = (DVs)/100\mu = (1 \times 30 \times 0.875)/(100 \times 1) = 0.2625$.

From Fig. 8, $f = 0.056$. Using eq. 18*b*, calculate V. If the value so found comes within 5% of the assumed value, use it in succeeding calculations. Otherwise, recalculate.

From eq. 17, $L = \pi(6.5 - 3.85)/12 = 0.695$ ft.

Substituting in eq. 18*b*, $8.9 = 0.056 \times 12 \times 0.695 \times V^2/64.4$, from which $V^2 = 1210$ and $V = 35$. With $V = 35$, $K = 0.36$, $f = 0.054$, and $V = 35.5$ ft per sec.

Pounds of fluid circulating, taking turbulence friction factor $Z = 0.85$, is, from eq. 20, $P = 12 \times 35.5 \times 0.85 \times 40.9 \times 0.032 = 475$ lb per sec.

The kinetic energy of the liquid in the impeller is calculated from eq. 21, where v_i, from eq. 24, is $(2\pi \times 6.5 \times 2000)/(12 \times 60) = 113.7$ ft per sec. Equation 21 therefore gives $KE_i = (475 \times \overline{113.7^2})/64.4 = 95{,}500$ ft-lb per sec. In like manner, eq. 25 gives $v_r = (2\pi \times 3.85/12) \times (1900/60) = 63.9$ ft per sec, and eq. 22 gives $KE_r = (475 \times \overline{63.9^2})/64.4 = 30{,}100$ ft-lb per sec.

Difference in $KE = 95{,}500 - 30{,}100 = 65{,}400$ ft-lb per sec. Therefore, impeller horsepower $Hp_i = 65{,}400/550 = 118.5$; transmitted torque $T = (118.5 \times 5250)/2000 = 310$ lb-ft; and output shaft horsepower $Hp_r = (310 \times 1900)/5250 = 112.5$ hp.

Applications. Hydraulic couplings are commercially available (see Article 29 in this section) and are used for a variety of purposes as clutches (by emptying and filling as required), as variable-speed drives (by partial emptying), and to minimize shock and vibration.

ELECTROMAGNETIC CLUTCHES. If the basic elements (field and armature) of an electric motor are so mounted that one is attached to a driving shaft and the other to the driven, the result is a unit which may function as a clutch. When energized by an external source of d-c excitation the clutch transmits torque, but when de-energized it disconnects the driving and driven shafts by removing the electromagnetic field across the air gap.

Speed-torque characteristics of such a clutch are somewhat similar to those of a fluid coupling. They are used in similar applications as clutches, as variable-speed drives, and as shock-absorbing couplings. When they are operated at less than full output speed, power is dissipated. The coupling must be well ventilated under such conditions.

12. OVERRUNNING CLUTCHES

Freewheeling clutches are employed when one shaft is intended to drive another but the second shaft cannot drive the first in the same direction of rotation.

ROLLER CLUTCH. Figure 9 shows a roller assembly of one type of roller clutch in which A is the driving member, B the driven number, and C one of a number of similar rollers equally spaced about the cam member. The roller contacts the inner diameter of A and a flat or cam surface on B. A spring is normally used to keep the roller in light contact with both driving and driven members.

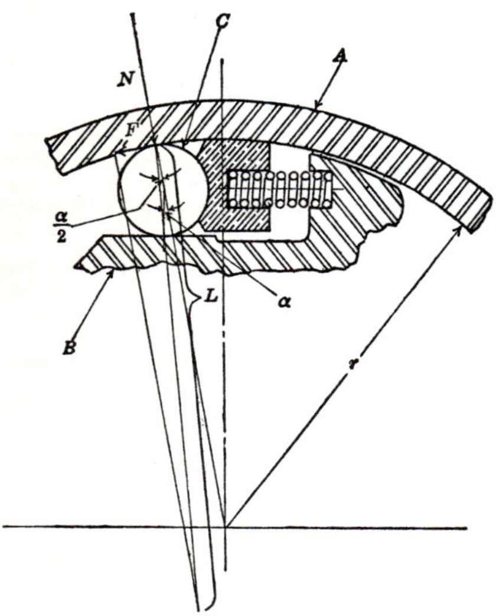

FIG. 9. Roller clutch.

Referring to Fig. 9, let T = total driving torque, pound-inches; F = driving force due to T, pounds; N = normal force on roller, pounds; r = internal radius of A, inches; α = angle formed by a radius passing through the center of the roller and a perpendicular from roller center to its point of contact with B, degrees; L = resultant of F and N, pounds; l = length of roller, inches; and n = number of rollers. Then $L = F/[\sin(\alpha/2)]$ pounds; $F = T/nr$ pounds from which $L = T/[nr \sin(\alpha/2)]$ pounds (force at contact points of roller).

Then L/l, which is the pounds per inch of roller length, should not exceed 25,000 under worst condition of operation.

The half angle $\alpha/2$ should be between 2 and 6 degrees.

SPRING CLUTCHES. A number of different types and forms of spring clutches are in use. The external type shown in Fig. 10 is in effect a band brake connecting two cylinders placed end to end and rotating on a common axis. The spring is assumed fastened to the left-hand arbor and slips on the right-hand arbor when that arbor is rotated clockwise. The spring is initially wound somewhat smaller as to its inner diameter than the outside diameter of the arbor. The inner radial force per unit length of the spring on the arbor is assumed to be constant when no torque is applied. This force f_0 depends on the amount the spring has to be opened up in order to slide it over the arbor.

Referring to Figs. 10 and 11, let f_0 = radial force of spring on arbor when no torque is applied, pounds per inch of contact line; r_2 = radius of arbor, inches; T = torque transmitted, inch-pounds; e = base of Napierian system of logarithms = 2.71828; N = number of turns on right-hand arbor or driving drum; μ = coefficient of friction (approximately 0.15); R_2 = radius to neutral axis of wire when spring is on arbor, inches; R_1 = radius to neutral axis, spring free, inches; E = Young's modulus for the spring material, pounds per square inch; and I = the area moment of the wire section, inches^4. Then $f_0 = EI[(R_2 - R_1)/R_1R_2^2r_2)]$ pounds per inch, and $T = r_2^2f_0(e^{2\pi N\mu} - 1)$ inch-pounds. The torque increases rapidly with an increase in N. Limiting torque is the product of r_2 and the breaking strength of the spring wire.

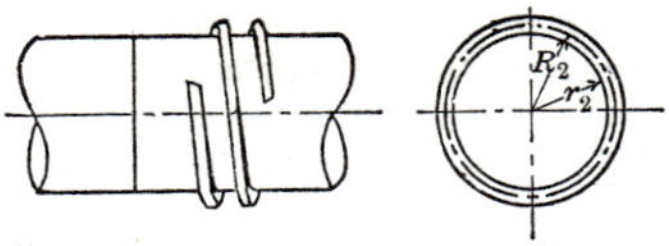

FIG. 10. Overrunning spring clutch, external type.

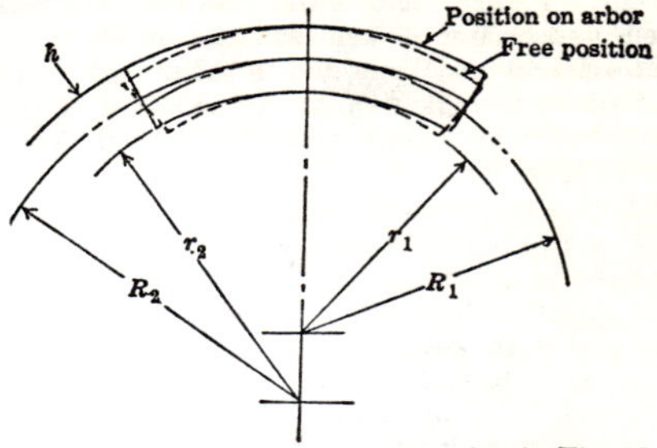

FIG. 11. Enlarged segment of spring in Fig. 10.

PULLEYS

By W. A. Williams

13. CAST-IRON PULLEYS

Cast-iron pulleys, as contrasted with wood or steel pulleys, are particularly adapted for service in plants where excessive dampness or acid fumes may be prevalent, such as in chemical plants, tanneries, dye houses, paper mills, textile plants, etc. Because of their relatively greater weight, they lend themselves to installation wherever kinetic energy or flywheel effect is desired in the transmission system.

Pulleys for the transmission of power or for use on conveyors have been standardized by manufacturers, and stock pulleys are available for practically all installations. The design of pulleys, except for special conditions, is not necessary. It usually will prove more economical to consult manufacturers of pulleys for the possible adaptation of a stock pulley to meet the special condition.

SIZES AVAILABLE. Cast-iron pulleys are manufactured in sizes ranging from 3-in. diameter by 2-in. face to 72-in. diameter by 40-in. face for transmission purposes. For conveyor applications, widths up to 75 in. are available.

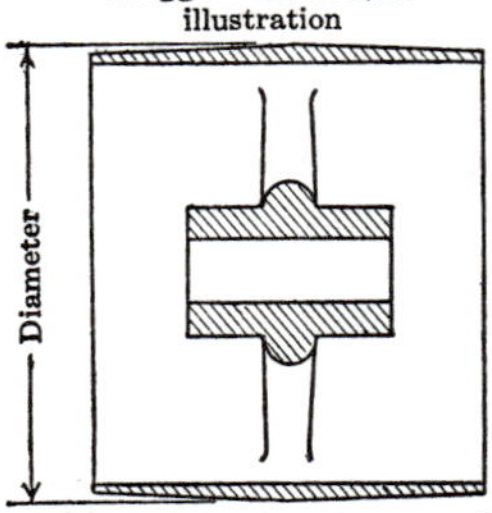

FIG. 1. Standard crown and diameter measurement of pulleys.

TYPES OF PULLEYS. A wide variety of types are manufactured to meet practically every conceivable condition. Cast-iron pulleys are made with solid rim and hub, solid rim and split hub, and two types of split rim and split hub. Specific design depends on the manufacturer and size of the pulley. For light-duty service, six-arm split pulleys are made by fracturing the rim so that when installed and tightened on the shaft they have a smooth, uninterrupted face, run true, and can be balanced as accurately as solid pulleys.

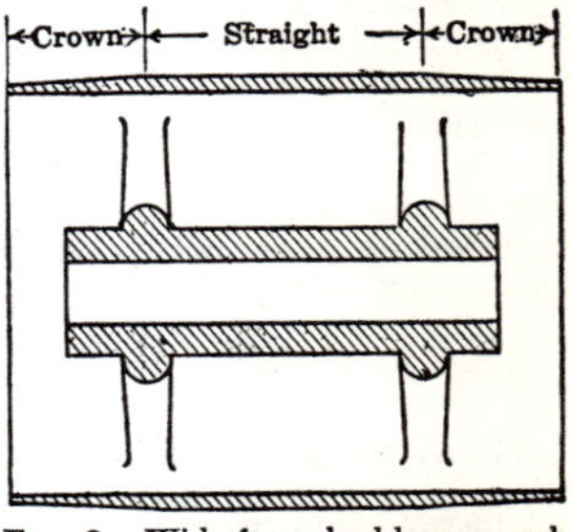

FIG. 2. Wide-face double-arm pulley with flat center portion.

Pulleys can be furnished with either flat, crown, or flanged rims. Pulleys are crowned to keep the belt on the pulley. There has been a tendency to reduce the crown on pulleys, since severe crown is detrimental to belts, and no amount of crown will correct poor alignment. The amount of crown used on most transmission pulleys does not exceed $^1/_8$ in. to the foot. Wide-face pulleys usually require less taper, or the center half of the pulley is left flat. Figure 1 shows the crown on single-arm pulleys, and Fig. 2 shows the flat crown often used on double-arm wide-face pulleys.

FLANGES may be added to the outside of the pulley at extremities of the rim to retain the belt; or a single flange may be added to the center of the pulley to separate two belts. Flanges may be placed on the inside of the rim to stiffen the rim and to permit bolting two pulleys together.

HUBS are usually made in one of three styles. Straight solid hubs are provided with a keyway and two setscrews, one over the key and one 90 degrees away from it. Split hubs are either split on one side only and provided with two bolts to tighten the pulley to the shaft, or are split on both sides and provided with four bolts. They may or may not be equipped with keyways, depending on the magnitude of the load. On light loads the compression or clamping action provided by the bolts is sufficient to keep the pulley tight on the shaft. Pulleys may also be equipped with one of various designs of tapered bushings which permit the use of pulleys on any shaft size within the limits of the pulley bore. These bushings permit the omission of a key except on applications where severe loads or shock loads are encountered.

PROPORTIONS OF CAST-IRON PULLEYS. The specific proportions of cast-iron pulleys vary somewhat among various manufacturers, depending on the type of industry the pulleys were originally designed for and on the particular foundry and manufacturing methods in use. Commercial stock pulleys are made with three or four arms up to 10 in. in diameter, six arms from 11 to 74 in. in diameter, and eight arms for larger diameters. Pulleys with two or three sets of arms may be considered as two or three separate pulleys combined in one. Releaux states in such cases that the proportions of the arms should be 0.8 or 0.7 those of single-arm pulleys.

The arms on most cast-iron pulleys, flywheels, and sheaves are elliptical in section. This shape combines neatness in appearance, low windage loss, and high strength in the direction of rotation. It is also a section that is easily molded and cast. For minimum weight with equivalent strength, an H section is preferable, but some sacrifice in appearance and increase in windage must be expected.

ALLOWABLE RIM SPEED. The following rim speeds are the maximum speeds advisable for cast-iron pulleys:

Regular iron pulleys	Solid or split	3750 ft per min
Regular iron pulleys	Specially balanced solid or split	4500 ft per min
Split-arm iron pulleys	Six-arm pulleys through two opposing arms, rather than between arms specially balanced	6000 ft per min
Specially designed solid iron pulleys	Specially balanced	7000 ft per min

Should rim speeds above 7000 ft per min be necessary, the manufacturer should be consulted and material other than cast iron should be selected for the pulley construction.

CAPACITY OF STANDARD PULLEYS. As regularly listed, cast-iron pulleys are designed and built to withstand the stresses which may be placed upon them by medium double leather belts, or their equivalent, as figured on a basis of 65 lb of effective tension per inch of width. Pulleys may also be obtained for heavy double-belt service based on an effective tension of 90 lb per inch of width, or for triple-belt service based on an effective tension of 120 lb per inch of width. For extreme service conditions, pulleys can be furnished for effective tensions up to 210 lb per inch of width.

14. STEEL PULLEYS

STEEL SPLIT PULLEYS. The vast majority of pulleys used in flat-belt power transmission are steel split pulleys. The reasons for this are their low cost, light weight, greater availability, and the fact that they are made with interchangeable bushings which permits their use on any shaft up to the maximum bore of the pulley.

Steel split pulleys are manufactured in standard sizes from 3-in. diameter by 2-in. face up to 144-in. diam. by 36-in. face. Standard pulleys are manufactured in diameters 1 in. apart up to 16 in. and 2 in. apart up to 72 in. Pulleys above 72 in. are made on order only.

TYPES OF CONSTRUCTION. Details of construction by different manufacturers vary with diameter, face width, and duty. In general, a two-arm construction is used on small pulleys, six-arm on intermediate sizes up to 44 in., and eight-arm on pulleys over 44 in.

Heavy-duty pulleys are made with the disk center construction, or with heavier arms with additional support gussets between the arms and the rims, and with heavier hub construction.

Pulleys can be obtained with single, double, or triple flanges.

ALLOWABLE RIM SPEEDS. Recommended rim speeds vary between 5000 and 6000 ft per min, depending on pulley sizes, construction, and manufacturer. In general, the manufacturer should be consulted when rim speeds of over 5000 ft per min are contemplated.

SOLID-STEEL CONVEYOR PULLEYS. A relatively new type of steel pulley now being manufactured is the solid-steel conveyor pulley illustrated by Fig. 3. These pulleys, with interchangeable cast-iron hubs, are manufactured in the following standard diameters: 6, 8, 10, 12, 14, 16, 18, 20, 24, 30, 36, 42, 48, 54, and 60 in. Standard face widths are: 12, 14, 16, 18, 20, 22, 24, 26, 30, 32, 36, 38, 42, 44, 48, 51, 63, 70, and 76. On these pulleys, the actual face should exceed the belt width by 2 in. for belts up to 42 in.; by 3 in. for belts of 48, 54, and 60 in.; and by 4 in. for belts of 66 and 72 in.

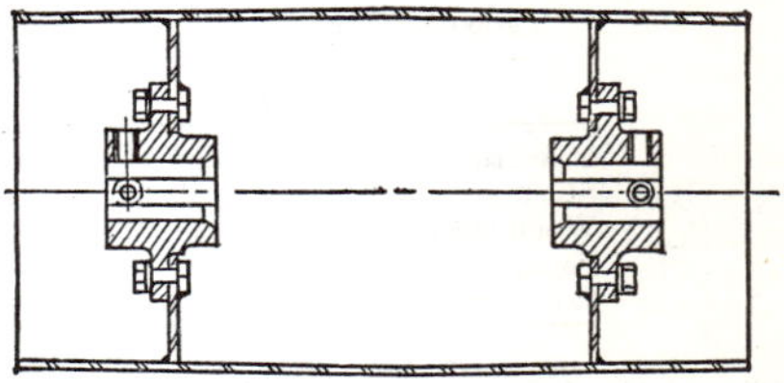

Fig. 3. Steel conveyor pulley.

The crown on these pulleys is held to 1/16 to 1/8 in. of increment of pulley diameter per foot of total pulley width on pulleys up to 26-in. face width, inclusive; 3/64 to 5/64 in. of increment of pulley diameter per foot of total pulley width on pulleys over 26-in. face width.

The above standards were adopted by the Power Transmission Association as applicable to all manufacturers making steel conveyor pulleys who are members of that Association. These pulleys have found wide acceptance on all classes of conveyor application and can also be used for power transmission service where heavy-duty conditions, at relatively slow speeds, are encountered.

IRON SPIDER STEEL RIM PULLEYS. A design that eliminates the shrinkage strains inherent in cast-iron pulleys is made with cast-iron spider to which a rolled steel rim is riveted. These pulleys are made for heavy-duty service and can be made for exceptionally high-speed service.

15. NONMETALLIC PULLEYS

WOOD PULLEYS. Pulleys made entirely of wood are available for service where lightness is important and where moisture and fire hazard are not prevalent. Claims are made that wood pulleys provide a higher coefficient of friction against a leather belt than cast-iron or steel, but available test data do not seem conclusive.

In addition to all-wood pulleys, wood-rim pulleys with cast-iron spiders are also available.

A special form of wood pulley used mostly for motor pulleys is made of laminated layers of soft wood cemented together and compressed under heavy pressure until the thickness has been substantially reduced, thereby increasing the density of the wood. The laminated layers make it possible to insure that end-grain wood is uniformly distributed over the face of the pulley, to provide uniform wear resistance. These pulleys have ribbed cast-iron hubs which are forced into undersized holes in blocks of the laminated wood. They are available in either bushed or on-bore construction, in sizes from 1 1/2 by 1 1/2 in. up to 24 by 24 in. Faces may be either crown, straight, stepped, or flanged in any combination. With standard pulleys, belt speeds up to 6000 ft per min are common, and with specially balanced pulleys, 10,000 ft per min.

PAPER PULLEYS. Paper pulleys made of impregnated paper board are used particularly for motor pulleys. They are also popular in any service requiring high coefficient of friction. In small sizes the hub is cast within the fiber; in medium sizes a tapered hub is pressed into the material; and in larger sizes rings of paper board are assembled onto a cast-iron spider. Other pulleys are made with a web center. Recommended rim speed is 5000 ft per min for standard paper pulleys, and 10,000 ft per min for specially balanced pulleys.

RELATIVE CAPACITY OF DIFFERENT TYPES OF PULLEYS. The capacity of a pulley depends on so many factors other than the actual make-up of the pulley itself that it is unwise to accept the results of any test as conclusive unless all variables have been considered. Factors which can influence performance are: type of belt (hard, soft), age of belt, type and amount of belt dressing used, crown of pulleys, age of pulleys (if impregnated type), belt speed, size of pulleys, and load per inch on belt in comparison with recommended rating. After consideration of all the above factors, it will still be necessary to arrive at an acceptable criterion of performance.

In an investigation to ascertain the value of compressed spruce pulleys, Professor G. S. Wilson (*Bulletin* 10, Univ. of Wash. Eng. Expt. Sta., 1920) found the horsepower capacity of several different types of pulleys to be as given in Table 1. In this investigation the slip of the belt was measured at belt tensions ranging from 37 1/2 to 187 1/2 psi of belt section, and the horsepower determined by a Prony brake. The pulleys used in the investigation were 18-in. diameter and 6-in. face.

Table 1. Horsepower Capacities per Inch Width of Belt of Various Pulleys

Type of Pulley	Weight, lb	Slip of Belt			Relative Capacity at 2% Slip
		1%	1 1/2%	2%	
		Horsepower			
Cast iron	65.0	.58	.69	.74	100.0
Steel	40.0	.71	.78	.81	109.4
Wood split	21.5	.63	.77	.82	110.8
Paper	41.5	.62	.76	.83	112.1
Compressed spruce	22.0	.76	.81	.84	113.5

A paper by Lawrence Whitcomb (National Association of Cotton Manufacturers, 1909) gives the results of tests made at the Lowell Textile School on the relative transmitting capacity of wood-rim, wood-rim with cork inserts, wood-rim lagged with leather, and steel pulleys, all 18 in. diameter by 6-in. face. The results of the tests are summarized in Table 2.

Table 2. Tests of Lagged and Cork Insert Pulleys

Type of Pulley	Horsepower at 2% Slip	Maximum Horsepower Transmitted	Slip, % at Maximum Horsepower	Relative Value at 2% Slip
Wood rim, cork inserts	11.8	13.2	4.6	100%
Wood rim, leather-lagged	8.7	12.4	4.7	73%
Steel	8.3	12.7	6.0	70%
Wood rim	7.8	9.6	4.3	65%

Recent tests conducted by Professor M. L. Price of Worcester Polytechnic Institute on various pulleys and belts under identical horsepower conditions show the wide variation that can be expected in results. Tests were conducted on a drive having a 4.0-in. diameter motor pulley, a 16-in. diameter driven pulley, 24-in. center distance, with an 1140-rpm motor giving a belt speed of 1200 ft per min. The horsepower was maintained constant, and total belt tension varied over a wide range to obtain different values of

Table 3. Horsepower Factors for Various Drives at 2% Slip

Type of Pulley	Belt	Horsepower Factor
Laminated wood, treated	Soft tan leather	95.5
Cast iron	Leather with dressing	95.0
Laminated wood, treated	Rubber fabric	92.0
Paper	Tan leather	91.0
Laminated wood, untreated	Soft tan leather	79.0
Cast iron	Soft tan leather	75.0
Paper	Rubber fabric	73.0
Laminated wood, untreated	Rubber fabric	71.5
Cast iron	Rubber fabric	68.0
Cast iron, glazed	Rubber fabric	55.0
Cast iron	Leather, oily	32.0

slip. The final performance is rated at 2% slip with a horsepower factor obtained from the formula:

$$\frac{T_1 - T_2}{T_1} \times 100 \tag{1}$$

where T_1 is tight side tension and T_2 is slack side tension. A perfect belt would be one in which the slack side tension reached zero, indicating a positive drive. Results of the tests are shown in Table 3.

CONE PULLEYS

Cone pulleys for driving machine tools should be so designed that the speeds will be in geometrical progression, that is, each speed is the product of the next lower speed and a constant multiplier. According to Carl G. Barth (*Am. Machinist*, Jan. 11, 1912), the ideal ratio of speeds is $\sqrt[4]{2} = 1.189$.

There are three general designs for cone pulley drives: (1) crossed belts; (2) open belts where the pulleys are so far apart that it is unnecessary to compensate for the tendency of the changing angle to alter the length of the belt; and (3) open belts with the pulleys so close together that the change in the angle tends to alter the length of the belt, which tendency must be compensated. Inasmuch as the length of the belt is constant in crossed belts and nearly so in open belts with long center distances, it is only necessary that the sum of the diameters of the corresponding steps on the mating cones be constant. In case 3, the length of the belt is a function of both the sum and difference of the diameters of the pulleys, and a simple solution is not possible. A large number of graphical methods for determining the diameters of the several steps of the pulleys are given in Halsey's *Handbook for Machine Designers*. However, a slight error in a graphical solution on a small scale may result in a serious error in the size of the pulleys, and a mathematical solution is preferable.

The equation for belt length (see page 15-51, eq. 12) may be written in the following form:

$$L_b = 2C + 1.571(r + 1)d + \frac{(r - 1)^2 d^2}{4C} \tag{2}$$

where L_b = length of belt, C = center distance between pulley shafts, d = diameter of smaller pulley, and r = ratio of diameter of larger pulley to that of smaller pulley = D/d. When the two pulleys are equal ($D = d = d_0$), the belt-length equation reduces to the simple form

$$L_b = 2C + 3.14d_0 \tag{3}$$

from which suitable preliminary values of L_b, C, and d_0 may be determined, in conjunction with the belt rating tables and formulas given on page 15-46. Having established belt length, center distance, and pulley ratio r at each step, the numbers may be substituted into eq. 2, which then reduces to a quadratic with unknown d, which may be solved by the standard formula

$$d = \frac{-b + \sqrt{b^2 - 4ac}}{2a} \tag{4}$$

where, in this case, $a = (r - 1)^2/4C$, $b = 1.571(r + 1)$, and $c = 2C - L_b$.

SPEEDS IN GEOMETRICAL PROGRESSION. If it is desired that the speed ratios shall increase by a constant percentage, or in geometrical progression, then $r_2/r_1 = r_3/r_2 = r_n/r_{n-1} = c$, a constant.

$$\frac{r_n}{r_1} = c^{n-1}; \qquad c = \sqrt[n-1]{\frac{r_n}{r_1}}$$

BELT DRIVES

By M. L. Price

16. THEORY OF POWER TRANSMISSION BY BELTING

GENERAL BELT FORMULA. Although many types of belts and pulleys are used for transmission of mechanical power, the same general principles are applicable to any particular combination.

Since a belt drive is fundamentally a friction drive, the most important factor involved is the relationship between belt tensions, which produce pressure between belt and pulley as well as the means of developing pulley torque, and friction coefficient.

For static conditions, friction coefficient f is used to define the limiting value of friction force F_f, available to prevent sliding between two flat surfaces, and normal force N which holds the two surfaces together.

$$F_f = fN \tag{1}$$

Applied to the dynamic conditions and curved surfaces involved in belt drives, this very fundamental law of mechanics becomes

$$\frac{T_1 - T_c}{T_2 - T_c} = e^{f\theta} \tag{2}$$

in which T_1 is the tension, on the tight side of the belt, pounds; T_2 is the tension, on the slack side of the belt, pounds; T_c is the tension developed in the belt by centrifugal action, pounds; e is the base of the natural or Napierian logarithms, 2.71828; f is the friction coefficient between belt and pulley; and θ is the arc of contact between belt and pulley, radians.

The centrifugal tension T_c may be neglected for belt speeds below 2000 ft per min, in which the general equation becomes

$$\frac{T_1}{T_2} = e^{f\theta} \tag{3}$$

For belt speeds in the region from 3000 to 5000 ft per min, the centrifugal stress developed within the belt becomes an appreciable portion of the total working stress. Belt drives should not be designed to operate at belt speeds greater than 5000 ft per min without a very careful consideration of the centrifugal effect. The significance of the centrifugal tension term, T_c, is that a part of the gross tensions T_1 and T_2 is used up in forcing the belt to follow a curved path around the pulley and is thus not available to produce pressure between the belt and pulley.

The magnitude of the centrifugal tension is given by the formula

$$T_c = \frac{\rho b t V^2}{300g} \text{ lb} \tag{4}$$

where ρ is the density of the belt, pounds per cubic inch, usually taken as 0.035 to 0.045 for leather or fabric belts; b is the width of the belt, inches; t is the thickness of the belt, inches; V is the belt speed, feet per minute; and g is the gravitational constant, 32.2 ft per sec per sec.

The term, $e^{f\theta}$, in which θ is expressed in radians, may be transformed to the common logarithmic system and the use of arc of contact in degrees as follows,

$$e^{f\theta} = 10^{0.00758f\alpha} \tag{5}$$

The general belt formula may thus be interpreted as follows: The ratio of the available or effective belt tensions is the number whose common logarithm is $0.00758f\alpha$.

BELT TENSIONS. Equation 2 gives the limiting condition of effective or available tension ratio in terms of friction coefficient and arc of contact between belt and pulley. Design values for this ratio vary from 2 to 6, depending on the nature of the belt and pulley and on the arc of contact. Atmospheric conditions and degree of cleanliness of the drive should be taken into account when design belt tension ratios are selected.

The tension in the tight side strand of the belt should not be so great as to give a unit tensile stress greater than a safe design value. For leather belts, the design stress may be taken as 200 to 400 psi of belt section, based on an ultimate tensile strength of 2000 to 5000 psi and a joint strength of 50% for wire lacing to 100% for cemented joints.

The tension in the slack side strand of the belt should not be too low or the stability of the drive will be lost.

Sufficient tension may be provided by center distances of at least 20 to 25 ft and a reasonable belt sag, by weighted idler pulleys, pivoted motor bases, and spring loaded sliding bases, or, in the case of fixed center distances, by belt clamps and spring scales to insure sufficient initial tension when belts are made endless. One empirical formula used by many belt manufacturers states that the initial belt tension, to insure good drive conditions, should be about 36 lb per inch of width, per ply, per strand (tight and slack side). Thus, for a 4-in., double-ply leather belt, the correct total tension, $T_1 + T_2$, should be about 576 lb. Since the drive is so dependent on having sufficient belt tension to develop traction between belt and pulley, the use of some sort of automatic tensioning device is to be recommended wherever possible.

ARC OF CONTACT. The exponential function $e^{f\theta}$ gives the ratio of effective tensions under limiting conditions (eq. 2) and must be as high as possible in order to develop maximum difference in tension, $T_1 - T_2$, for maximum pulley torque.

It is desirable to have the arc of contact as near 180 degrees as possible and, except in very special cases, not less than 120 to 130 degrees. The arc of contact, expressed in degrees, is given by the expression

$$\theta = 180 \pm 2 \sin^{-1}\left(\frac{D - d}{2C}\right) \tag{6}$$

where D and d represent the diameters of the large and small pulleys, respectively, and C represents the center distance between pulley shafts. The minus sign is to be used in getting the arc of contact for the small pulley, and the plus sign is used for the large pulley.

FRICTION COEFFICIENT. A number of empirical relationships have been developed for the purpose of obtaining a design value for the friction coefficient under different operating conditions. However, none of the available formulas are adequate, since they do not take into account a sufficient number of variables. It is far safer to use values obtained by actual laboratory tests which have been carried on under conditions similar to those in the drive being designed.

Some of the factors influencing the friction coefficient are type and condition of pulley surface, type and condition of belt surface, cleanliness of surrounding air, temperature, humidity, belt speed, area of surface of contact between belt and pulley, thickness of belt compared to pulley diameter, and percentage of slip between belt and pulley as expressed in terms of belt speed.

For design purposes, the following friction coefficients may be used unless more specific values can be obtained by direct tests.

Table 1. Values of Friction Coefficient for Belt Drives

Pulley Material	Belt Material	Friction Coefficient
Cast iron or steel	Leather	0.40
Cast iron or steel	Rubberized canvas	0.25
Paper or wood (treated)	Leather	0.50
Paper or wood (treated)	Rubberized canvas	0.40
For very oily or wet conditions, any type of pulley or belt		0.15–0.20

17. HORSEPOWER OF LEATHER BELTS

THEORETICAL DESIGN. The general belt formula given in eq. 2 may be expressed in the following form:

$$(T_1 - T_2) = bt\left(s - \frac{\rho V^2}{300g}\right)\left(\frac{e^{f\theta} - 1}{e^{f\theta}}\right) \tag{7}$$

where the symbols are the same as those used in eqs. 2 and 4. Equation 7 gives the limiting friction force or tension difference available for producing pulley torque without an excessive amount of belt slip; it may be used either for determining the horsepower capacity of a given belt and pulley combination, or for the selection of a belt width for a given horsepower and drive pulley speed, and for an assumed belt thickness t and design stress s. The value of thickness of leather belts is given at the top of Table 3.

HORSEPOWER FORMULA. The horsepower capacity of a belt drive may be expressed in terms of the allowable value of the tension difference, as given by eq. 7, and the belt speed in feet per minute, as follows:

$$\text{Horsepower} = \frac{(T_1 - T_2)V}{33{,}000} \tag{8}$$

where V is found from eq. 9.

BELT SPEED. The speed of the belt, V ft per min, may be obtained by the formula

$$V = \frac{D}{12}\pi N \tag{9}$$

in which D is the pulley diameter in inches, π is 3.1416, and N is the pulley speed in revolutions per minute.

MANUFACTURER'S RATINGS OF LEATHER BELTS. In an attempt to simplify and standardize the selection of leather belts, the Engineering Committee of the American Leather Belting Association (ALBA) has issued copyrighted horsepower rating tables for oak-tanned flat leather belting used with electric motor drives. These tables (Tables 2, 3, 4, 5, and 6) are used as follows:

Table 2 gives the belt speed in feet per minute corresponding to the pulley diameter at its specified speed in revolutions per minute.

Table 2. Belt Speeds in Feet per Minute for Different Pulley Diameters and Revolutions per Minute

(American Leather Belting Assoc.)

Pulley Diameter, in.	Full-load Motor Speed							
	3450 rpm	1750 rpm	1150 rpm	850 rpm	690 rpm	575 rpm	490 rpm	435 rpm
	Belt Speed, ft/min							
2	1805	916	602					
2 1/2	2257	1145	753					
3	2708	1374	903	667				
3 1/2	3160	1603	1054	779	632			
4	3611	1832	1204	890	721	602		
4 1/2	4062	2062	1355	1001	810	677		
5	4514	2291	1505	1113	902	752	642	
5 1/2	4965	2520	1656	1224	992	828	706	
6	5417	2749	1806	1335	1082	903	770	684
6 1/2	5868	2978	1957	1447	1171	978	834	741
7	6319	3207	2107	1558	1261	1054	899	798
8	7222	3665	2408	1780	1441	1204	1027	912
9	8125	4124	2709	2003	1622	1355	1155	1026
10		4582	3010	2225	1800	1505	1284	1140
11		5040	3311	2448	1980	1656	1412	1254
12		5498	3612	2670	2161	1806	1541	1368
13		5956	3913	2893	2342	1957	1669	1482
14		6414	4214	3115	2522	2107	1797	1596
15		6873	4515	3338	2705	2258	1926	1710
16		7331	4816	3560	2882	2409	2054	1824
17		7789	5117	3783	3062	2559	2182	1937
18			5418	4006	3240	2710	2311	2051
19			5719	4228	3422	2860	2439	2165
20			6020	4451	3605	3011	2568	2279
21			6321	4673	3793	3161	2696	2393
22			6622	4896	3965	3312	2824	2507
23			6923	5118	4150	3462	2953	2621
24			7224	5341	4330	3613	3081	2735
25				5568	4520	3766	3210	2849
26				5790	4700	3917	3338	2963
27				6013	4881	4068	3466	3077
28					5062	4218	3595	3191
29					5243	4369	3723	3305
30					5423	4520	3851	3419

Table 3 gives the value of K, the theoretical horsepower capacity of the belt per inch of width for different thicknesses at the speeds indicated. Minimum recommended pulley diameters are shown at the bottom of the table.

Table 3. Horsepower per Inch of Width (Factor K)

(American Leather Belting Assoc.)

	Single Ply		Double Ply			Triple Ply	
	11/64 in.	13/64 in.	18/64 in.	20/64 in.	23/64 in.	30/64 in.	34/64 in.
Belt Speed, ft/min	Med.	Heavy	Light	Med.	Heavy	Med.	Heavy
600	1.1	1.2	1.5	1.8	2.2	2.5	2.8
800	1.4	1.7	2.0	2.4	2.9	3.3	3.6
1000	1.8	2.1	2.6	3.1	3.6	4.1	4.5
1200	2.1	2.5	3.1	3.7	4.3	4.9	5.4
1400	2.5	2.9	3.5	4.3	4.9	5.7	6.3
1600	2.8	3.3	4.0	4.9	5.6	6.5	7.1
1800	3.2	3.7	4.5	5.4	6.2	7.3	8.0
2000	3.5	4.1	4.9	6.0	6.9	8.1	8.9
2200	3.9	4.5	5.4	6.6	7.6	8.8	9.7
2400	4.2	4.9	5.9	7.1	8.2	9.5	10.5
2600	4.5	5.3	6.3	7.7	8.9	10.3	11.4
2800	4.9	5.6	6.8	8.2	9.5	11.0	12.1
3000	5.2	5.9	7.2	8.7	10.0	11.6	12.8
3200	5.4	6.3	7.6	9.2	10.6	12.3	13.5
3400	5.7	6.6	7.9	9.7	11.2	12.9	14.2
3600	5.9	6.9	8.3	10.1	11.7	13.4	14.8
3800	6.2	7.1	8.7	10.5	12.2	14.0	15.4
4000	6.4	7.4	9.0	10.9	12.6	14.5	16.0
4200	6.7	7.7	9.3	11.3	13.0	15.0	16.5
4400	6.9	7.9	9.6	11.7	13.4	15.4	16.9
4600	7.1	8.1	9.8	12.0	13.8	15.8	17.4
4800	7.2	8.3	0.1	12.3	14.1	16.2	17.8
5000	7.4	8.4	0.3	12.5	14.3	16.5	18.2
5200	7.5	8.6	10.5	12.8	14.6	16.8	18.5
5400	7.6	8.7	10.6	12.9	14.8	17.1	18.8
5600	7.7	8.8	10.8	13.1	15.0	17.3	19.0
5800	7.7	8.9	10.9	13.2	15.1	17.5	19.2
6000	7.8	8.9	10.9	13.2	15.2	17.6	19.3
Minimum Pulley Diameter, in. — Belts Under 8 in. Wide	3	5	6	8	12	20	24
Minimum Pulley Diameter, in. — Belts 8 in. Wide and Over			8	10	14	24	30

These are the minimum allowable pulleys for the belts of the above thicknesses.

Note: For belt speeds of over 6000 ft per min, consult the leather belting manufacturer.

Table 4 shows motor-load correction factor M, which is dependent upon the starting characteristics of the electric motor used.

Table 4. Correction Factors M for Type of Motor and Starting Methods Used

(American Leather Belting Assoc.)

Motor Type and Starting Method	Correction Factor, M
Squirrel cage, compensator start	1.5
Squirrel cage, line start	2.0
Slip ring and high starting torque	2.5

Table 5 gives pulley correction factor P for the diameter of the smaller pulley.

Table 5. Correction Factors P for Diameters of Smaller Pulley

(American Leather Belting Assoc.)

Diameter of Small Pulley, in.	Correction Factor, P
4 and under	0.50
4 1/2 to 8	0.60
9 to 12	0.70
13 to 16	0.80
17 to 30	0.90
Over 30	1.00

Width or horsepower rating of belt are found from the formulas

$$b = \frac{R \times M}{K \times P} \tag{10}$$

$$\text{HP} = \frac{b \times K \times P}{M} \tag{11}$$

where b = width of belt, inches; R = name-plate horsepower rating of electric motor; HP = horsepower rating of belt; M = motor-load factor, Table 4; K = theoretical belt capacity factor, Table 3; and P = pulley correction factor, Table 5.

For special operating conditions, multiply the belt width given by eq. 10 by a factor F selected from Table 6.

Table 6. Correction Factors F for Special Operating Conditions

(American Leather Belting Assoc.)

Operating Conditions	Factor F
Oily, wet, or dusty atmosphere	1.35
Vertical drives	1.2
Jerky loads	1.2
Shock and reversing loads	1.4

Example 1. Squirrel cage induction motor rated 15 hp at 1750 rpm, across the line starting. Motor pulley, 8 in. in diameter; driven pulley, 16 in. in diameter. Required: thickness and width of belt to be used for this drive.

Solution. Belt speed is 3665 ft per min, as determined from Table 2. Belt thickness is medium double, which is suitable for 8-in. diameter pulley, as determined from bottom of Table 3. Belt capacity factor K is 10.2, as determined from Table 3 by interpolation. Motor correction factor M is 2.0, as determined from Table 4. Pulley correction factor P for 8-in. diameter pulley is 0.60 from Table 5. From eq. 10

$$b = \frac{R \times M}{K \times P} = \frac{15 \times 2.0}{10.2 \times 0.6} = 4.9 \text{ in.}$$

Use a medium double belt 5 in. wide unless special operating conditions exist, in which case the width should be increased by the factor indicated in Table 6.

Example 2. Determine the horsepower rating of an 8-in. heavy double leather belt operating between a 48-in. diameter cast-iron pulley mounted on a line shaft turning at 200 rpm and a 16-in. diameter cast-iron fan pulley turning at 600 rpm. The drive is horizontal and the atmosphere is dusty.

Solution. Belt speed may be determined from Table 2 by simple proportion. The belt speed for a 16-in. pulley and 435 rpm is 1824 ft per min. At 600 rpm, the belt speed would be

$$\frac{600}{435} \times 1824 = 2515 \text{ ft per min}$$

Belt capacity factor K from Table 3, corresponding to a belt speed of 2515 ft per min and heavy double leather belt, is 8.6 hp per inch of width. Pulley correction factor P for a 16-in. pulley is 0.80, as obtained from Table 5. Special operating condition factor F for dusty atmosphere is 1.35, as shown in Table 6. Horsepower rating, as given by eq. 11, is

$$Hp = \frac{b \times K \times P}{M \times F} = \frac{8 \times 8.6 \times 0.80}{1.0 \times 1.35} = 40.8$$

18. FLAT FABRIC AND RUBBERIZED CANVAS BELTS

FABRIC BELTS. Although leather has for many years occupied a position of prominence in the field of flat belting, woven fabric belts of one type or another also are used in many applications and many special claims are made for them. Most of these fabric belts are fabricated from woven cotton duck or canvas, although several types of hair belts, known generally as camel's hair, are used. Most fabric belts are made up of layers of woven material and are held together and protected as well as "frictionized" by means of gums, rubber, or synthetic resins.

STITCHED CANVAS. Belts made up of layers of canvas duck to provide sufficient strength and thickness and held together by stitching are inexpensive for applications where little protection against moisture and oil is required. The friction coefficient is low, usually about 0.20, but the resistance to higher operating temperatures is good.

Because of low friction, care must be exercised so that machine bearing pressures are not excessive.

Stitched canvas belts are made of 24- to 48-oz. duck, with from 3 to 12 plies, and in widths from 1 to 60 in.

Strength of stitched canvas belts compares favorably with a leather belt of equivalent thickness and width, but the horsepower capacity must either be reduced considerably because of lower friction coefficient, or higher total belt tensions must be employed. (See eq. 7.)

Stitched canvas belts are sometimes made moisture resistant by painting them with moisture-proofing compounds.

RUBBERIZED CANVAS BELTS. Several methods are used in the fabrication of rubberized canvas belts. In nearly all, the individual layers of duck are first impregnated under pressure with a slow-aging rubber or synthetic rubber compound. Desired belt thickness is built up either by building one layer on another, followed by vulcanization and slitting to proper width, or by folding and vulcanizing under pressure. The latter method provides smooth edges. Tension or load-carrying cords are often employed to reduce stretch and to increase strength. Such cords usually are of prestretched cotton, rayon, glass, or flexible steel-wire rope.

Rubberized canvas belts have the advantage of high resistance to moisture, oil, and higher operating temperatures. The main disadvantage of this type of belt lies in the low coefficient of friction after the surface coating of rubber has been worn down to present a smooth glazed surface on the canvas. Higher traction can be provided with paper or wood pulleys which have been impregnated and surface-coated with compounds of highly viscous nature.

BALATA BELTING. Canvas belts are sometimes impregnated with balata gum, similar to gutta-percha, obtained from trees in South America. These belts have high resistance to moisture and animal oils but do not stand up under action of mineral oils or temperatures above 120 F. The strength of balata belting depends on the canvas and its method of fabrication so that the horsepower capacity is about the same as for rubberized belts.

HORSEPOWER RATING OF CANVAS BELTS. Published tables vary considerably regarding horsepower capacity of canvas and rubberized canvas belting. It must be kept in mind that the coefficient of friction between worn rubberized canvas and cast-iron or steel pulleys is considerably less than that for leather and cast-iron or steel. Greater total belt pull is therefore required for the transmission of a given amount of power and pulley diameters, and care must be taken to prevent overloading of bearings.

Table 7. Horsepower Rating of Rubberized Canvas Belting, per Inch of Width

(180-degree arc of contact)

(Reprinted from *Power*, Feb. 1937, by permission)

No. of Plies	Belt Speed, ft/min 1000	1500	2000	2500	3000	3500	4000	4500	5000	5500
3	1.2	1.8	2.4	2.9	3.5	4.0	4.3	4.6	4.9	5.2
4	1.7	2.5	3.2	4.0	4.7	5.4	5.8	6.2	6.6	7.0
5	2.1	3.1	4.1	5.0	5.9	6.8	7.5	7.9	8.4	8.8
6	2.7	3.9	5.2	6.4	7.5	8.5	9.4	10.2	10.8	11.1
7	3.1	4.5	6.0	7.4	8.7	10.0	10.7	11.6	12.7	13.0
8	3.5	5.2	6.9	8.5	10.0	11.4	12.4	13.3	14.1	14.8
9	4.0	5.8	7.7	9.6	11.2	12.9	14.0	15.0	15.9	16.7
10	4.4	6.5	8.6	10.6	12.4	14.3	15.5	16.7	17.7	18.6
12	5.3	7.8	10.3	12.8	15.0	17.2	18.6	20.0	21.2	22.3

The largest possible pulley diameters allowed by space limitations should be used so as to reduce required effective belt pull or friction force. Table 7 gives representative horsepower ratings for regular rubberized canvas belts not having special cord or wire rope reinforcement.

SELECTION OF BELT. After pulleys have been selected to give correct speed ratio, using the largest pulleys consistent with space limitations and a belt speed that does not exceed 5000 to 6000 ft per min, consult Table 8 for a tentative selection of number of

Table 8. Minimum Recommended Pulley Diameters for Rubberized Canvas Belts Operating at Different Speeds

(Reprinted from *Power*, Feb. 1937, by permission)

No. of Plies	Belt Speed, ft/min: Under 2000	2000–4000	Over 4000
	Minimum pulley diameter, in.		
3	3	4	5
4	5	6	8
5	7	8	10
6	10	12	14
7	14	16	18
8	18	20	22
9	22	24	28
10	28	32	36
12	36	42	48

plies. From Table 7, obtain horsepower rating corresponding to number of plies and belt speed. Compute arc of contact by eq. 6, and obtain correction factor from Table 9. Multiply horsepower rating by correction factor. Divide peak or starting horsepower to

Table 9. Correction for Arc of Contact, Rubberized Canvas Belts

When arc of contact on small pulley is less than 180 degrees, multiply ratings in Table 7 by the following:

Arc of Contact, degrees	180	170	160	150	140	130	120
Factor	1.0	0.95	0.90	0.85	0.80	0.75	0.70

be transmitted by corrected horsepower rating to get required width of belt. If required width does not agree favorably with the values listed in Table 10, repeat the selection, using either more or less plies as needed.

Table 10. Recommended Number of Plies for Rubberized Canvas Belts for Different Widths

Belt Width, in.	Minimum No. of Plies	Maximum No. of Plies	Belt Width, in.	Minimum No. of Plies	Maximum No. of Plies
2	3	5	16	5	8
3	3	6	18	5	8
4	3	6	20	5	9
5	4	6	22	5	9
6	4	6	24	6	10
8	4	6	26	6	10
10	4	7	30	6	11
12	4	7	36	7	12
14	5	8	42	7	12

19. BELT INSTALLATION AND MAINTENANCE

PULLEY ALIGNMENT. Great care should be exercised to see that pulleys and shafts are in correct alignment. Shafts should be parallel, and edges of pulleys should be lined up with a straightedge or strings.

BELT LENGTH. The best way to determine the correct length of belt is by a steel tape wrapped around the pulleys.

If belts are to be spliced and cemented, an additional amount of belt length must be provided for making the lap.

If the center distance between pulley shafts cannot be adjusted, the correct initial belt tension can be provided by subtracting from 1/16 to 1/8 in. per foot of belt length, depending on the stretchiness of the belt material. The exact amount to be subtracted must be determined by experience or can be predicted by the belt manufacturer.

Several formulas have been developed for those cases where it is impossible to determine belt length by direct measurement, but the formula now in almost universal use is

$$L_b = 2C + 1.571(D + d) + \frac{(D - d)^2}{4C} \tag{12}$$

where L_b = length of belt, C = center distance between pulley shafts, D = diameter of larger pulley, and d = diameter of smaller pulley.

It is recommended that all dimensions be in inches and that the length of belt be given in inches when ordering from a manufacturer or distributor. Also state whether the belt is to be made endless, whether the allowance for the splice has been provided, and whether the manufacturer is to deduct enough belt to provide initial tension.

BELT JOINTS. Belt manufacturers strongly recommend that endless belts be used wherever possible. Leather belts which have been spliced and cemented by use of a high quality waterproof cement will be just as strong at the joint as in the rest of the belt. Some applications require some sort of lacing so that the ends of the belt may be quickly joined together whenever the belt must be shortened to take care of stretch. The oldest form of lacing consists of rawhide strips which are passed through holes punched in the ends of the belt. Although this form of lacing produces a safe and flexible joint, strength is sacrificed.

Rubberized canvas belts may be made endless by splicing and vulcanizing for maximum strength and smoothness of joint.

Either leather or rubberized canvas belts may be joined by a wide variety of wire or metal hooks, clips, clamps, and plates. All the various methods for lacing belts reduce the net strength of the belt joint, and certain types introduce a real safety hazard. Belts jointed by metal lacing should be surrounded by guards.

BELT DRESSING. Most manufacturers of leather belting introduce various oils and greases into new belts to lubricate the leather fibers and to control flexibility. In some cases, other compounds are incorporated to increase friction coefficient. If sufficient tension is provided and if the belt and pulley surfaces are kept clean, little need exists for the dressing of a belt to make it pull properly. When a belt slips excessively, the chances are that the tension is not sufficient or that the belt has become caked with oil, grease, or dirt. Belts should be regularly cleaned with a mixture of carbon tetrachloride and unleaded gasoline, in the proportions of from 25% to 50% carbon tetrachloride. After such a cleaning, at intervals ranging from 3 months to a year, depending upon operating conditions, type of service, and atmospheric conditions the belt should be dressed with a dressing obtained from a reputable belt manufacturer. Good dressings are those designed primarily for the lubrication of the belt fibers. Resinous dressings of a tacky nature should be avoided; they are unnecessary if proper tension is provided and they do not provide the proper lubricating action.

Dressing should be applied uniformly but sparingly so that no excess dressing remains on the belt and pulley surfaces.

20. ARRANGEMENT OF BELT DRIVES

At one stage in the development of our machine age, most power was transmitted by long line shafts and many belts and pulleys. Quarter-turn drives were very common, and special arrangements to take care of nonintersecting shafts were often required. This type of drive is not so common today, and the following discussion is considerably condensed. More complete coverage of guide pulleys, quarter-twist belts, and mule pulley stands will be found in the 11th edition of Kent's *Mechanical Engineers' Handbook, Design and Shop Practice* volume.

ARRANGEMENT OF PULLEYS. If possible, machinery driven from overhead shafting should be so arranged that the belt will make an angle from the vertical of not less than 45 degrees. Never, unless it is absolutely impossible to avoid it, should driving and driven pulleys be in the same vertical line. Care should be taken to arrange the pulleys in line with each other, the tops of the crowns being aligned. Otherwise the belt will tend to stretch unevenly and to run crooked. Machinery should be so placed that the tight strand of the belt will be the bottom strand. However, if machines are so placed that the pull of the belts at alternate line-shaft pulleys is in opposite directions, the wear and strain on bearings will be reduced. This arrangement of machinery will make one-half of the belts drive with the top strand tight.

The distance between shafts has considerable influence on the life of belts. Distances should, as a rule, be made as great as possible up to reasonable limits. For narrow belts on small pulleys, a good average figure is 15 ft; for larger belts, the center-line distance should be from 20 to 25 ft; and for main belt drives using wide and heavy belts, pulleys should be from 25 to 30 ft apart. If the distance is made too great the belt will have a tendency to flap and injure itself.

IDLER PULLEYS. Belts which are run on pulleys located close together are liable to stretch rapidly and thus lose the tension necessary to the transmission of the desired amount of power. For such cases, and also for belts which are required to run for long periods without stopping, an idler pulley, which can be moved against the belt to take up the slack and thus increase the tension, will be found desirable. The idler should be located about one-quarter of the distance from the driving pulley and should bear against the slack side of the belt. It increases not only the tension but also the arc of contact. The idler may be so constructed that it can be moved against the belt at will, or it may act automatically by means of a spring or gravity. Such an arrangement is shown in Fig. 1.

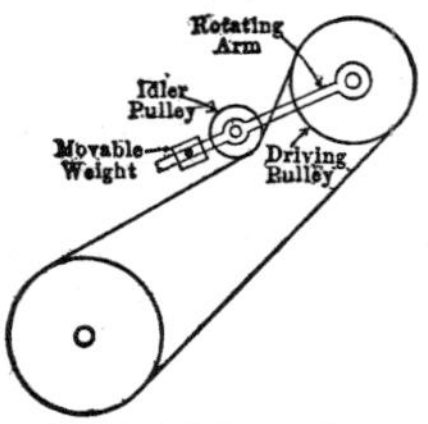

FIG. 1. Idler pulley.

GUIDE PULLEYS. Where two shafts that are not parallel and that do not intersect are to be connected by belting, and the conditions are such as to make a quarter-twist belt inadvisable, guide pulleys are used (Fig. 2). The function of the guide pulley is to alter the direction of the belt and cause it to run squarely on to the receiving pulley. The location of the guide pulley with reference to the main pulleys is determined by the intersection *CD* of the planes passing through the center of the face of the main pulleys and perpendicular to their axes. The guide pulleys should be located at the intersection of the tangents *HE* and *EJ* and *FG* and *FK*, respectively, and the plane through the center of the face of each pulley should coincide with the planes formed by these tangents. Other arrangements will suggest themselves according to the needs of each particular case. It is only necessary to observe the precaution that the guide pulleys be so placed that they will receive the belt squarely from the main pulley and deliver it squarely to the next one. If this is done, the pulleys may be placed in practically any location and at practically any angle.

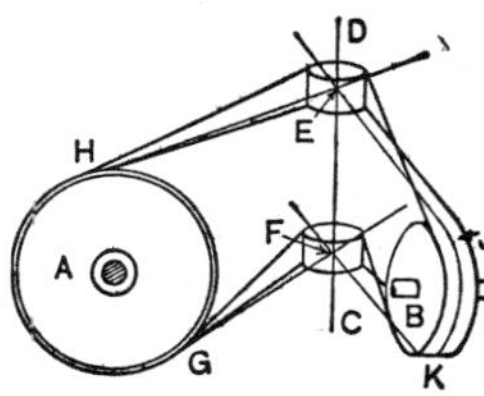

FIG. 2. Guide pulleys.

QUARTER-TWIST BELTS. One of two shafts at right angles to each other and in different planes may be driven from the other by means of a belt with a quarter twist in it, providing that the distance between the shafts is not too short. The belt must run squarely on to each pulley, and the point at which the belt leaves each pulley must lie in the straight line formed by the intersection of the planes passing through the center line of the face of each pulley and perpendicular to their axes, as shown in Fig. 3. The driving pulley is *AD*, and the belt runs in the direction shown by the arrows. The point *C* at the end of the horizontal diameter of the driven pulley at which the belt leaves it is set so that it is vertically above the point *A* at the end of the horizontal diameter of the driving pulley where the belt leaves that pulley. The quarter-twist has been used with belts as wide as 30 in. There should be a considerable distance between the shafts, however. For heavy belts and short center-line distances, guide pulleys or mule pulley stands should be used. A quarter-twist belt will drive in one direction only.

Quarter-twist belts may be used to drive shafts which make angles with one another of other than 90 degrees, if the rule is observed that the points at which the belt leaves the pulleys are in the straight line determined as explained previously.

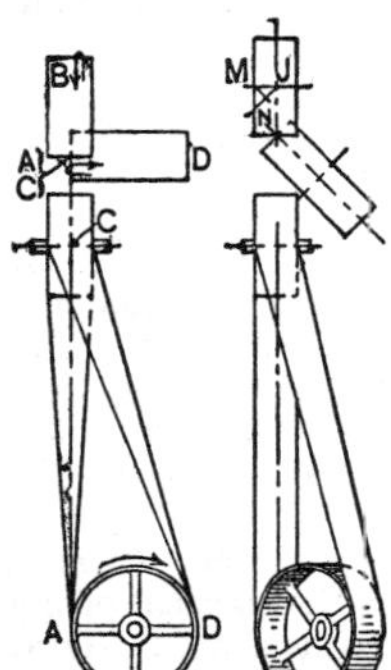

FIG. 3. Quarter-twist belt.

MULE PULLEY STANDS. A common arrangement of guide pulleys, known as a mule pulley stand, is shown in Fig. 4. If it be borne in mind that the governing principle of mule pulley drives, and in fact of all drives where the belt joins two pulleys that are not in the same plane, is that the belt must be delivered squarely on the pulleys, but that it can leave them at any angle within reasonable limits, it is evident that the mule pulleys can be placed in any location desired. If local conditions make it advisable, the mule pulleys can be so placed as to require the belt to leave either or both of the main driving pulleys in a horizontal position. The method of laying out the drive is the same in any event.

The fact that it is possible to vary the position of the mule pulleys without affecting the working of the drive makes it possible to place both mule pulleys on the same shaft.

With reference to Fig. 4, if the belt travels in the direction of the arrows, it is evident that, if the lower mule pulley be moved in the direction that the belt travels, the belt will still be delivered squarely to pulley *B*. It will, of course, leave pulley *A* at an angle, but this will have no effect if the mule pulley is maintained at the same angle as before, since the belt will still be received squarely by the mule pulley. If the distance between the axes of

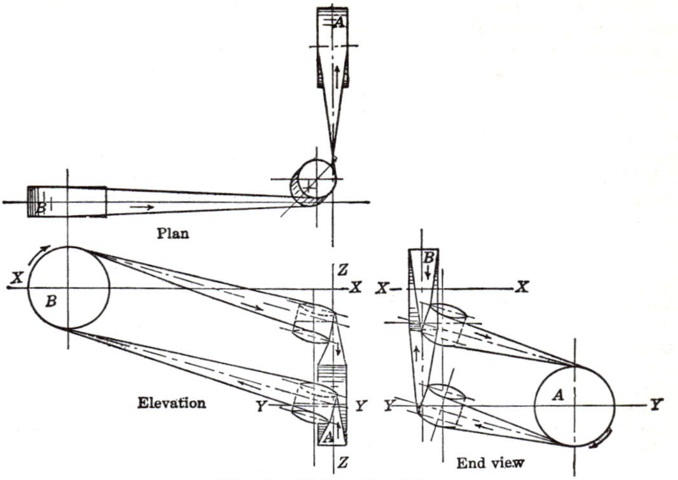

FIG. 4. Mule pulley drive.

the mule pulley shafts is not excessive, the lower mule pulley may be moved until the two shafts are in line, as regards the elevation. They will, however, still be out of line as regards the end view. By the same process of reasoning as before, however, it is evident that the upper mule pulley can be similarly moved in the direction of the belt travel until its axis is in line with that of the lower pulley in the end view. The axes of the two mule pulleys then will coincide, and in their new locations they can be mounted on the same shaft.

The foregoing considerations apply to drives where the main pulleys are of the same diameter. If the diameters are unequal, the shafts of the mule pulleys will lie at different angles and the pulleys must be mounted on separate shafts.

21. V-BELT DRIVES

By H. A. Wilson

The V-belt drive consists of one or more endless belts constructed of cords (cotton, rayon, synthetic, or steel) and fabric impregnated with and embedded in rubber, and of trapezoidal cross section. The belts run on sheaves which have V-shaped grooves on the periphery. The angle of the groove depends on the size of the sheave and on the service in which the belt is to be used. Manufacturers of V-belts should be consulted concerning details of a particular drive.

The drives are so designed that the belt is approximately flush with the top of the groove. A clearance is provided at the bottom of the grooves in the sheaves to insure contact with the belt only on the sides of the grooves. These drives will run in either direction with slack either on top or bottom strands. They are especially adapted to reversing service.

The V-belt was first developed by the Allis-Chalmers Manufacturing Company whose product is sold under the name of Texrope Drives. Other manufacturers are Dodge Manufacturing Corporation, Mishawaka, Ind.; Worthington Pump and Machinery Corporation, Buffalo, N. Y.; The American Pulley Company, Philadelphia, Pa.; and T. B. Woods's Sons Company, Chambersburg, Pa. Belts are manufactured by B. F. Goodrich Company, Akron, Ohio; Gates Rubber Company, Denver, Col.; Dayton Rubber Company, Dayton, Ohio; and Goodyear Rubber Company, Akron, Ohio.

OPERATION. In the operation of V-belt drives, it is important to maintain proper alignment. The driving and driven shafts should be parallel, and the V-belts should be at right angles to the shafts. Every V-belt drive installation should include some method of adjusting tension on the belts. This can be accomplished by an adjustable base where electric motors are used, or by an idler arrangement when it is impossible to provide adjustment otherwise. Idlers should preferably operate on the inside of the V-belt.

V-belt drives should not be operated at temperatures in excess of 160 F. *Standard* V-belts should be protected from oil. *Special* oil-resisting V-belts are available for drives which must operate under conditions where oil may come in contact with the belts. Special belts are also available for installation where static charges built up on the belts must be dissipated. Belt dressing should never be used on any V-belt drive. V-belts should be stored in a cool, dry place, protected from light. When a V-belt drive requires replacement of belts for the first time, a complete new set should be installed; the belts which are still in good condition can then be used as future replacements.

BELT SIZES. Standard V-belts are manufactured in five different cross sections and various lengths. Table 11 shows the cross section dimensions, serial number, and pitch length of the belt.

Table 11. Standard V-belts

Section A 1/2 by 11/32 in.		Section B 21/32 by 7/16 in.		Section C 7/8 by 17/32 in.		Section D 1 1/4 by 3/4 in.		Section E 1 1/2 by 1 in.	
Serial No.	Pitch Length	Serial No.	Pitch Length	Serial No.	Pitch Length	Serial No.	Pitch Length	Serial No.	Pitch Length
A26	26.9	B35	36.1	C51	52.6	D120	122.4	E180	183.3
A31	31.9	B38	39.1	C60	61.4	D128	130.4	E195	198.5
A35	35.9	B42	43.1	C68	69.4	D144	146.4	E210	213.6
A38	38.9	B46	47.1	C75	76.6	D158	160.4	E240	240.0
A42	42.9	B51	52.1	C81	82.6	D162	164.4	E270	270.0
A46	46.9	B55	56.4	C85	86.2	D173	175.4	E300	300.0
A51	51.9	B60	61.0	C90	91.9	D180	182.6	E330	330.0
A60	60.9	B68	69.3	C96	97.9	D195	197.6	E360	360.0
A68	68.9	B75	76.3	C105	106.9	D210	213.1		
A75	75.9	B81	82.3	C112	113.9	D240	240.0		
A80	81.1	B85	86.5	C120	122.0	D270	270.0		
A85	86.1	B90	91.5	C128	130.0	D300	300.0		
A90	91.1	B97	99.0	C144	146.0	D330	330.0		
A96	97.1	B105	106.5	C158	160.0	D360	360.0		
A105	106.1	B112	114.0	C162	164.0				
A112	113.1	B120	121.5	C173	175.0				
A120	121.1	B128	129.5	C180	182.0				
A128	129.1	B144	145.5	C195	197.0				
		B158	159.5	C210	212.0				
		B173	174.5	C240	240.0				
		B180	181.6	C270	270.0				
		B195	196.3	C300	300.0				
		B210	211.3	C330	330.0				
		B240	240.0	C360	360.0				
		B270	270.0						
		B300	300.0						

HORSEPOWER RATING. Selection of a V-belt drive to transmit power economically and satisfactorily depends on the proper consideration of the maximum load to be transmitted and the characteristics of the component machines. Service factors have been developed (Table 17) to aid in the selection of the proper drive.

The horsepower tables (Tables 12 to 16) list the power transmitted by one belt with velocity and sheave size as the determining factor in power transmitted.

(*Continued on p. 15-60*)

Table 12. Horsepower Ratings for Section A V-belts

Velocity, ft/min	Pitch Diameter, in.						
	2.6	3.0	3.4	3.8	4.2	4.6	5.0 and Larger
1000	0.5	0.7	0.8	0.9	0.9	1.0	1.0
1100	0.6	0.7	0.9	1.0	1.0	1.1	1.1
1200	0.6	0.8	1.0	1.0	1.1	1.2	1.2
1300	0.7	0.9	1.0	1.1	1.2	1.3	1.3
1400	0.7	0.9	1.1	1.2	1.3	1.4	1.4
1500	0.8	1.0	1.1	1.3	1.4	1.4	1.5
1600	0.8	1.0	1.2	1.3	1.4	1.5	1.6
1700	0.9	1.1	1.3	1.4	1.5	1.6	1.7
1800	0.9	1.2	1.4	1.5	1.6	1.7	1.8
1900	0.9	1.2	1.4	1.6	1.7	1.8	1.9
2000	1.0	1.3	1.5	1.6	1.8	1.9	2.0
2100	1.0	1.3	1.5	1.7	1.9	2.0	2.1
2200	1.1	1.4	1.6	1.8	1.9	2.0	2.2
2300	1.1	1.4	1.6	1.8	2.0	2.1	2.3
2400	1.1	1.4	1.7	1.9	2.0	2.2	2.3
2500	1.1	1.5	1.7	2.0	2.1	2.3	2.4
2600		1.6	1.8	2.0	2.2	2.3	2.5
2700		1.6	1.8	2.1	2.3	2.4	2.6
2800		1.6	1.9	2.1	2.3	2.5	2.6
2900		1.6	1.9	2.2	2.4	2.6	2.7
3000			2.0	2.2	2.4	2.6	2.7
3100			2.0	2.3	2.5	2.6	2.8
3200			2.0	2.3	2.5	2.7	2.9
3300			2.1	2.4	2.6	2.7	2.9
3400				2.4	2.7	2.8	3.0
3500				2.5	2.7	2.8	3.0
3600				2.5	2.7	2.9	3.1
3700				2.5	2.8	2.9	3.1
3800					2.8	3.0	3.2
3900					2.8	3.0	3.2
4000					2.8	3.0	3.3
4100					2.9	3.1	3.3
4200						3.1	3.3
4300						3.1	3.3
4400						3.2	3.4
4500						3.2	3.4
4600							3.4
4700							3.4
4800							3.4
4900							3.4
5000							3.4

Table 13. Horsepower Ratings for Section B V-belts

Velocity, ft/min	Pitch Diameter, in.					
	5.0	5.4	5.8	6.2	6.6	7.0 and Larger
1000	1.3	1.4	1.5	1.6	1.7	1.8
1100	1.4	1.5	1.6	1.7	1.8	1.9
1200	1.5	1.6	1.7	1.8	1.9	2.0
1300	1.6	1.8	1.9	2.0	2.1	2.2
1400	1.7	1.9	2.0	2.1	2.2	2.3
1500	1.8	2.0	2.1	2.3	2.4	2.5
1600	1.9	2.1	2.3	2.4	2.5	2.6
1700	2.0	2.2	2.4	2.5	2.7	2.8
1800	2.2	2.4	2.5	2.7	2.8	2.9
1900	2.3	2.5	2.6	2.8	3.0	3.1
2000	2.4	2.6	2.8	2.9	3.1	3.2
2100	2.4	2.7	2.9	3.1	3.2	3.4
2200	2.5	2.8	3.0	3.2	3.4	3.5
2300	2.6	2.9	3.1	3.3	3.5	3.6
2400	2.7	3.0	3.2	3.4	3.6	3.7
2500	2.8	3.1	3.3	3.5	3.7	3.9
2600	2.9	3.2	3.4	3.6	3.8	4.0
2700	3.0	3.3	3.5	3.7	3.9	4.1
2800	3.0	3.3	3.6	3.8	4.1	4.3
2900	3.1	3.4	3.7	3.9	4.2	4.4
3000	3.2	3.5	3.8	4.0	4.3	4.5
3100	3.2	3.6	3.9	4.1	4.4	4.6
3200	3.3	3.6	4.0	4.2	4.5	4.7
3300	3.3	3.7	4.0	4.3	4.6	4.8
3400	3.4	3.7	4.1	4.4	4.7	4.9
3500	3.4	3.8	4.1	4.4	4.7	4.9
3600	3.4	3.8	4.2	4.5	4.8	5.0
3700	3.5	3.9	4.3	4.5	4.8	5.1
3800	3.5	3.9	4.3	4.6	4.9	5.2
3900	3.5	3.9	4.3	4.6	4.9	5.2
4000	3.5	4.0	4.4	4.7	5.0	5.3
4100	3.5	4.0	4.4	4.7	5.0	5.3
4200	3.5	4.0	4.4	4.7	5.1	5.4
4300	3.5	4.0	4.4	4.7	5.1	5.4
4400	3.5	4.0	4.4	4.8	5.1	5.4
4500	3.5	4.0	4.4	4.8	5.1	5.4
4600	3.4	4.0	4.4	4.8	5.1	5.5
4700	3.4	3.9	4.4	4.8	5.2	5.5
4800	3.4	3.9	4.4	4.8	5.2	5.5
4900	3.3	3.9	4.3	4.8	5.1	5.5
5000	3.3	3.8	4.3	4.7	5.1	5.5

Table 14. Horsepower Ratings for Section C V-belts

Velocity, ft/min	Pitch Diameter, in.					
	7.0	8.0	9.0	10.0	11.0	12.0 and Larger
1000	2.0	2.5	2.8	3.1	3.4	3.6
1100	2.2	2.7	3.1	3.5	3.7	3.9
1200	2.4	2.9	3.4	3.8	4.0	4.3
1300	2.6	3.2	3.7	4.1	4.4	4.6
1400	2.7	3.4	3.9	4.4	4.7	5.0
1500	2.9	3.6	4.2	4.6	5.0	5.3
1600	3.1	3.9	4.5	4.9	5.3	5.7
1700	3.3	4.1	4.7	5.3	5.6	6.0
1800	3.4	4.3	5.0	5.5	5.9	6.3
1900	3.6	4.5	5.2	5.8	6.2	6.6
2000	3.7	4.7	5.5	6.1	6.5	7.0
2100	3.9	4.9	5.7	6.3	6.8	7.3
2200	4.0	5.1	5.9	6.6	7.1	7.6
2300	4.2	5.3	6.2	6.8	7.4	7.9
2400	4.3	5.5	6.4	7.1	7.7	8.2
2500	4.5	5.7	6.6	7.3	8.0	8.5
2600	4.6	5.8	6.8	7.6	8.2	8.8
2700	4.7	6.0	7.0	7.8	8.4	9.0
2800	4.8	6.2	7.2	8.0	8.7	9.3
2900	4.9	6.3	7.4	8.3	9.0	9.6
3000	5.0	6.5	7.6	8.5	9.2	9.8
3100	5.1	6.6	7.8	8.7	9.5	10.1
3200	5.2	6.7	7.9	8.9	9.7	10.3
3300	5.3	6.9	8.1	9.1	9.9	10.6
3400		7.0	8.3	9.3	10.1	10.8
3500		7.1	8.4	9.4	10.3	11.0
3600		7.2	8.5	9.6	10.5	11.2
3700		7.3	8.7	9.8	10.7	11.4
3800		7.4	8.8	9.9	10.9	11.6
3900			8.9	10.1	11.0	11.8
4000			9.0	10.2	11.2	12.0
4100			9.1	10.3	11.3	12.2
4200			9.2	10.4	11.5	12.3
4300			9.3	10.5	11.6	12.5
4400				10.6	11.7	12.6
4500				10.7	11.8	12.7
4600				10.8	11.9	12.9
4700				10.8	12.0	13.0
4800				10.9	12.1	13.3
4900					12.1	13.2
5000					12.2	13.2

Table 15. Horsepower Ratings for Section D V-belts

Velocity, ft/min	Pitch Diameter, in.					
	12.0	13.0	14.0	15.0	16.0	17.0 and Larger
1000	4.5	5.1	5.6	6.1	6.5	6.8
1100	4.9	5.8	6.2	6.7	7.2	7.5
1200	5.3	6.1	6.7	7.3	7.7	8.2
1300	5.7	6.5	7.2	7.8	8.4	8.8
1400	6.1	7.0	7.8	8.4	9.0	9.4
1500	6.5	7.5	8.2	9.0	9.6	10.1
1600	6.9	7.9	8.8	9.5	10.2	10.7
1700	7.3	8.4	9.3	10.1	10.8	11.4
1800	7.7	8.8	9.8	10.6	11.3	12.0
1900	8.0	9.2	10.2	11.1	11.9	12.6
2000	8.4	9.7	10.7	11.7	12.5	13.2
2100	8.8	10.1	11.2	12.2	13.0	13.8
2200	9.1	10.5	11.7	12.7	13.6	14.4
2300	9.5	10.9	12.1	13.2	14.1	14.9
2400	9.8	11.3	12.6	13.7	14.6	15.5
2500	10.1	11.6	12.9	14.1	15.1	16.0
2600	10.4	12.0	13.4	14.6	15.6	16.6
2700	10.6	12.3	13.8	15.0	16.1	17.1
2800	10.9	12.7	14.1	15.5	16.6	17.6
2900	11.2	13.0	14.5	15.9	17.0	18.1
3000	11.4	13.3	14.9	16.3	17.5	18.6
3100	11.7	13.6	15.2	16.7	17.9	19.0
3200	11.9	13.9	15.6	17.1	18.4	19.5
3300	12.1	14.1	15.9	17.4	18.9	19.9
3400	12.2	14.4	16.2	17.7	19.1	20.3
3500	12.4	14.6	16.5	18.1	19.5	20.7
3600	12.6	14.8	16.7	18.4	19.8	21.1
3700	12.7	15.0	17.0	18.7	20.2	21.5
3800	12.8	15.2	17.2	18.9	20.4	22.2
3900		15.3	17.4	19.2	20.7	22.3
4000		15.5	17.6	19.4	21.0	22.5
4100		15.6	17.8	19.6	21.3	22.8
4200			17.9	19.8	21.5	23.0
4300			18.0	20.0	21.8	23.0
4400			18.1	20.1	21.9	23.3
4500				20.3	22.1	23.5
4600				20.4	22.3	23.9
4700				20.5	22.4	24.1
4800					22.5	24.2
4900					22.6	24.3
5000					22.6	24.4

Table 16. Horsepower Ratings for Section E V-belts

Velocity, ft/min	Pitch Diameter, in.								
	20.0	21.0	22.0	23.0	24.0	25.0	26.0	27.0	28.0 and Larger
1000	8.0	8.5	9.0	9.4	9.8	10.2	10.5	10.9	11.1
1100	8.7	9.3	9.8	10.3	10.8	11.2	11.6	11.9	12.2
1200	9.5	10.1	10.7	11.2	11.7	12.2	12.6	13.0	13.3
1300	10.2	10.9	11.6	12.1	12.7	13.1	13.6	14.0	14.4
1400	11.0	11.7	12.4	13.0	13.6	14.1	14.6	15.0	15.4
1500	11.7	12.5	13.2	13.9	14.5	15.1	15.6	16.1	16.5
1600	12.4	13.3	14.1	14.8	15.4	16.0	16.6	17.1	17.5
1700	13.1	14.0	14.9	15.6	16.3	16.9	17.5	18.1	18.6
1800	13.8	14.8	15.7	16.5	17.2	17.9	18.5	19.0	19.6
1900	14.5	15.5	16.4	17.3	18.0	18.8	19.4	20.0	20.6
2000	15.2	16.2	17.2	18.1	18.9	19.6	20.3	21.0	21.6
2100	15.8	17.0	18.0	18.9	19.7	20.5	21.2	21.9	22.5
2200	16.5	17.6	18.7	19.7	20.6	21.4	22.1	22.8	23.5
2300	17.1	18.3	19.4	20.4	21.4	22.2	23.0	23.7	24.4
2400	17.7	18.9	20.1	21.2	22.2	23.1	23.9	24.6	25.3
2500	18.3	19.6	20.8	21.9	23.0	23.9	24.7	25.5	26.3
2600	18.9	20.2	21.5	22.6	23.7	24.7	25.5	26.4	27.1
2700	19.4	20.8	22.1	23.3	24.4	25.4	26.4	27.2	28.0
2800	19.9	21.4	22.8	24.0	25.1	26.2	27.1	28.0	28.9
2900	20.4	22.0	23.4	24.7	25.8	26.9	27.9	28.8	29.7
3000	20.9	22.7	24.0	25.3	26.5	27.6	28.6	29.6	30.5
3100	21.4	23.0	24.5	26.0	27.2	28.3	29.4	30.4	31.3
3200	21.8	23.5	25.1	26.5	27.8	29.0	30.1	31.1	32.0
3300	22.3	24.0	25.6	27.1	28.4	29.6	30.7	31.8	32.8
3400	22.6	24.5	26.1	27.6	29.0	30.2	31.4	32.5	33.5
3500	23.0	24.9	26.6	28.1	29.5	30.8	32.0	33.1	34.2
3600	23.4	25.3	27.0	28.6	30.0	31.4	32.6	33.8	34.8
3700	23.7	25.6	27.4	29.1	30.6	31.9	33.2	34.4	35.5
3800	24.0	26.0	27.8	29.5	31.0	32.4	33.7	34.9	36.1
3900	24.2	26.3	28.2	29.9	31.5	32.9	34.3	35.5	36.6
4000	24.5	26.6	28.5	30.3	31.9	33.4	34.7	36.0	37.2
4100		28.8	28.8	30.6	32.3	33.8	35.2	36.5	37.7
4200		27.1	29.1	30.9	32.6	34.2	35.6	37.0	38.2
4300			29.3	31.2	33.0	34.6	36.0	37.4	38.7
4400			29.5	31.5	33.3	34.9	36.4	37.8	39.1
4500				31.7	33.5	35.2	36.7	38.2	39.5
4600				31.9	33.7	35.5	37.0	38.5	39.9
4700					33.9	35.7	37.3	38.8	40.2
4800					34.1	35.9	37.5	39.1	40.5
4900						36.0	37.7	39.3	40.7
5000						36.2	37.9	39.5	40.9

SERVICE FACTORS FOR V-BELT DRIVES. The horsepower ratings (Tables 12 to 16) are maximum ratings. Since V-belts are used on every type of machinery, conditions may exist which cannot be covered by horsepower ratings. For that reason the service factors (extra drive capacity) in Table 17 should be used. This extra capacity should be above the name-plate rating of the motor.

Table 17. Service Factors

Agitators, paddle-propeller	1.0–1.2
Bakery machinery	1.0–1.2
Brick and clay machinery	1.2–1.6
Compressors	1.2–1.4
Conveyors	1.0–1.6
Crushing machinery	1.4–1.6
Fans and blowers	1.2–2.0
Flour-feed and cereal mill machinery	1.0–1.4
Generators and exciters	1.2
Laundry machinery	1.2
Line shafts	1.4–2.0
Machine tools	1.0–1.4
Mills	1.4–1.6
Oil-field machinery	1.2–1.6
Paper machinery	1.2–1.6
Printing machinery	1.2
Pumps	1.2–2.0
Rubber plant machinery	1.4–2.0
Screens	1.2–1.4
Textile machinery	1.2–1.8

SELECTION of a V-belt drive should bé based on nature of the load, type of driving unit, horsepower rating, and speeds of driving and driven units. To obtain the horsepower capacity for the drive, multiply the rated horsepower (name-plate rating) of the driving unit by the recommended service factor (Table 17). From Table 18 select the belt section

Table 18. General Belt Selection Table

Motor Horsepower	Motor Speed, rpm				
	1800	1200	900	720	600 and below
1/2–3/4	A	A	A		
1–1 1/2	A	A	A		
2–3	A or B	A or B	A or B	A or B	
5	A or B	A or B	A or B	B	
7 1/2	A or B	A or B	B	B or C	B or C
10	A or B	B	B	B or C	C
15	B	B or C	B or C	C	C
20	B or C	C	C	C	C
25	C	C	C	C	C
30	C	C	C	C	C
40	C	C or D	C or D	C or D	D
50	C	C or D	C or D	D	D
60	C	C or D	C or D	D	D
75	C	C or D	C or D	D or E	D or E
100	C	D	D	D or E	D or E
125		D	D	D or E	D or E
150		D	D	D or E	E
200		D	D	E	E
250 and over				E	E

Note: When two cross sections are shown: If sheave pitch diameters are limited, use the smaller cross section. If sheave face widths are limited, use the larger cross section. If there are no limitations, use cross section providing the most economical drive.

to be used. Sheave diameter should not be less than the recommended minimum diameter (Table 19) for maximum life of belts. Correction for arc of contact on small sheave is determined from Table 20, the arc being given by the approximate formula:

$$\text{Arc of contact} = 180 - \frac{(D - d)\,60}{C}$$

Table 19. Minimum Sheave Sizes

Belt Section	Minimum Recommended Pitch Diameter, in.
A	3.0
B	5.4
C	9.0
D	13.0
E	21.0

where D = diameter of large sheave, inches; d = diameter of small sheave, inches; and C = center distance of drive inches. See Table 20 for correction factors. Belt velocity should be held under 5000 ft per min.

Table 20. Correction Factors Corresponding to Various Degrees of Contact on the Small Sheave

Arc of Contact, degrees	Correction Factor	Arc of Contact, degrees	Correction Factor
180	1.00	130	0.86
170	0.98	120	0.83
160	0.95	110	0.79
150	0.92	100	0.74
140	0.89	90	0.69

CENTER DISTANCE. If the sheave sizes required for a drive and the approximate center distance are known, the belt length can be estimated by the approximate formula

$$L = 2C + \frac{\pi}{2}(D + d)$$

Select the nearest belt length from Table 11.

The center distance can be calculated from the formula

$$C = \frac{L - \left[\frac{\pi}{2}(D + d) + \frac{(D - d)^2}{L}\right]}{2}$$

For small ratios and long centers, reasonably accurate centers can be found by the formula

$$C = \frac{L - 1.57(D + d)}{2}$$

The pitch length of the belt given in Table 11 is used in all calculations.

Example. A fan operating continuously at 525 rpm is to be driven by a 15-hp 1750-rpm motor. Center distance is approximately 36 in. From Table 18, a B belt is recommended for this drive. Table 17 shows service factor of 1.2 to 2.0. For continuous operation with squirrel cage motor, the service factor will be 1.6. Horsepower of drive will be $1.6 \times 15 = 24.0$ hp. Speed ratio $1750/525 = 3.33$. Sheave sizes of 6 and 20 in. will give the desired ratio, and 6 in. is above the minimum for B belts. Sheaves are stock sizes (consult manufacturer's catalog for stock sizes). Belt length $L = 2C + (\pi/2)(D + d) = 112.8$. Belt required is B112, length of which is 114 in. (Table 11). The center distance will be found to be 36 in. Belt velocity will be 2745 ft per min. From Table 13, the horsepower per belt = 3.6. Correcting for arc of contact on small sheave, correction factor = 0.94. Corrected horsepower = $0.94 \times 3.6 = 3.38$. Number of belts = $24/3.38 = 7.1$. Therefore, the drive would have 8 belts.

V-FLAT DRIVES, in which a small grooved sheave is used with a large flat pulley and a set of V-belts, are most successful when the following conditions are satisfied: (1) The speed ratio is 3 to 1 or more. (2) The center distance is less than the diameter of the large pulley. Centers three-fourths the diameter of the large pulley are desirable and should not exceed 1 1/2 times the diameter of the large pulley. (3) The large pulley has a flat face (no crown). (4) The face width of the flat pulley is at least 1 in. wider than the face width of the grooved sheave. For large drives a 10% increase in face width should be used. Face widths of grooved sheaves are given in Table 21.

Table 21. Face Width of Stock and Standard Sheaves

(Inches)

Belt Section	Number of Grooves																								
	1	2	3	4	5	6	7	8	9	10	11	12	13	14	15	16	17	18	19	20	21	22	23	24	25
A	3/4	1 3/8	2	2 5/8	3 1/4	3 7/8	4 1/2	5 1/8	5 3/4	6 3/8	7	7 5/8	8 1/4	8 7/8	9 1/2	10 1/8	10 3/4	11 3/8	12	12 5/8	13 1/4	13 7/8	14 1/2	15 1/8	15 3/4
B	1	1 3/4	2 1/2	3 1/4	4	4 3/4	5 1/2	6 1/4	7	7 3/4	8 1/2	9 1/4	10	10 3/4	11 1/2	12 1/4	13	13 3/4	14 1/2	15 1/4	16	16 3/4	17 1/2	18 1/4	19
C	1 3/8	2 3/8	3 3/8	4 3/8	5 3/8	6 3/8	7 3/8	8 3/8	9 3/8	10 3/8	11 3/8	12 3/8	13 3/8	14 3/8	15 3/8	16 3/8	17 3/8	18 3/8	19 3/8	20 3/8	21 3/8	22 3/8	23 3/8	24 3/8	25 3/8
D	1 3/4	3 3/16	4 5/8	6 1/16	7 1/2	8 15/16	10 3/8	11 13/16	13 1/4	14 11/16	16 1/8	17 9/16	19	20 7/16	21 7/8	23 5/16	24 3/4	26 3/16	27 5/8	29 1/16	30 1/2	31 15/16	33 3/8	34 13/16	36 1/4
E	2 1/4	4	5 3/4	7 1/2	9 1/4	11	12 3/4	14 1/2	16 1/4	18	19 3/4	21 1/2	23 1/4	25	26 3/4	28 1/2	30 1/4	32	33 3/4	35 1/2	37 1/4	39	40 3/4	42 1/2	44 1/4

Table 22. Correction Factors for V-flat Drives

Arc of Contact, degrees	Correction Factor	Arc of Contact, degrees	Correction Factor
180	.74	130	.86
170	.77	120	.83
160	.79	110	.79
150	.82	100	.74
140	.84	90	.69

Table 23. V-flat Drives

Belt Section	Add to Diameter of Flat Pulley
A	0.35 in.
B	0.45 in.
C	0.50 in.
D	0.75 in.
E	1.00 in.

Table 24. Deep-groove Sheave Face Width

Number of Grooves	A Belt	B Belt	C Belt	D Belt	E Belt
	Sheave Face Width, in.				
1	3/4	1	1 3/8	1 7/8	2 1/8
2	1 3/8	1 7/8	2 9/16	3 1/2	4 1/8
3	2	2 3/4	3 3/4	5 1/8	6 1/8
4	2 5/8	3 5/8	4 15/16	6 3/4	8 1/8
5	3 1/4	4 1/2	6 1/8	8 3/8	10 1/8
6	3 7/8	5 3/8	7 5/16	10	12 1/8
7	4 1/2	6 1/4	8 1/2	11 5/8	14 1/8
8	5 1/8	7 1/8	9 11/16	13 1/4	16 1/8
9	5 3/4	8	10 7/8	14 7/8	18 1/8
10	6 3/8	8 7/8	12 1/16	16 1/2	20 1/8

Flat pulleys of cast iron or pressed steel are generally used. If crown face pulleys are used, there will be unequal tension between the center and outer belts, which may cause the belts to "crowd" toward the center of the crown.

Selecting a V-flat Drive. Use the correction factors for arc of contact in Table 22. The pitch diameter of the flat pulley (for calculation purposes) is found by adding to the diameter of the flat pulley the amount shown in Table 23. All other data required are the same as for a V-belt drive.

QUARTER-TURN DRIVES are used to transmit power from a horizontal shaft to a vertical shaft, or vice versa. Speed ratios should not exceed 2 1/2 to 1. Horsepower ratings of belts are 75% of those listed in Tables 12 to 16. The arc of contact correction factor is disregarded. The center distance should not be less than $5(P + W)$, where P = pitch diameter of larger sheave, inches, and W = face width of large sheave, inches. The center line of the horizontal shaft should be half the diameter of the large sheave above the center line of the vertical shaft, and the vertical shaft should be offset half the diameter of the small sheave. Special deep-groove sheaves are required. Face widths of deep-groove sheaves are given in Table 24.

SHORT-CENTER BELT DRIVES

By Philip G. Rhoads

Belt drives with flat belts, between shafts whose center lines are but a short distance apart, usually prove unsatisfactory unless provision is made to increase the tension in the belt, give a greater arc of contact, or increase the grip of the belt on the driving pulley. Increased tension and greater arc of contact may be obtained by a weighted floating idler (Fig. 1, p. 15-52). Increased tension which automatically adjusts itself to the load is obtained by means of a pivoted motor base (Figs. 6 and 10). The design shown in Fig. 14

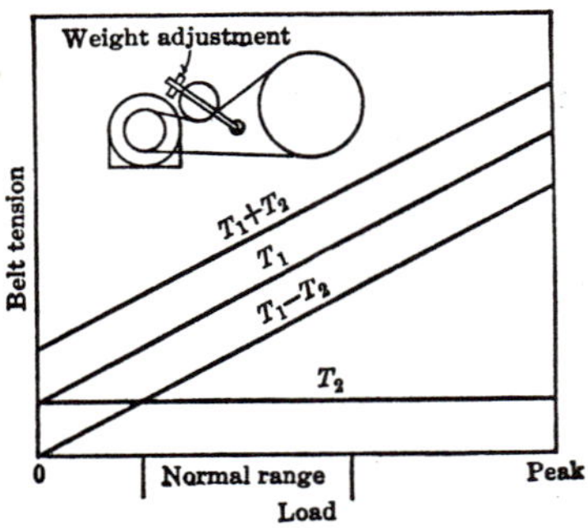

Fig. 1. Gravity idler drive. *Advantages:* Easy initial tension setting. Good automatic adjustment with load. Good automatic belt take-up. Good arc of contact. *Disadvantages:* Reverse bend in belt shortens belt life. Additional running bearing adds maintenance difficulties. Alignment difficult to maintain. Limited to rotation of motor in one direction.

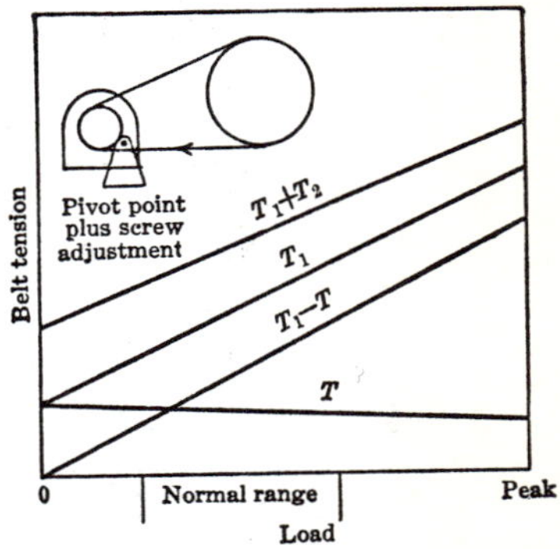

Fig. 2. Econ-O-Matic base. *Advantages:* Good automatic tension adjustment with load. Limited automatic belt take-up. Moderate maintenance. Easy adjustment for alignment. *Disadvantages:* Change in belt length affects tensions substantially. Limited to rotation of motor in one direction.

adds the elasticity of springs in the base to the elasticity of the belt, thereby absorbing minor changes in belt length without introducing as much belt creep as would occur if the belt were made very elastic. V-rubber belt drives offset the low coefficient of friction of rubberized cotton on cast iron or steel by using the belt tension to wedge the belts into grooves and by using the expansion of the portion of the belt in compression, where bent, to bind the belts in the grooves. Good grip of flat leather belt on pulleys is obtained by use of fiber pulleys, special tanned belts, and proper belt dressing.

The accompanying charts, Figs. 1–5, show characteristic changes in belt tensions with changes in load for different types of short center drives. All charts give the same tension characteristics at peak load, which is the load for which the drive must be designed. Lowered tensions at normal and fractional loads are favorable to long belt and bearing

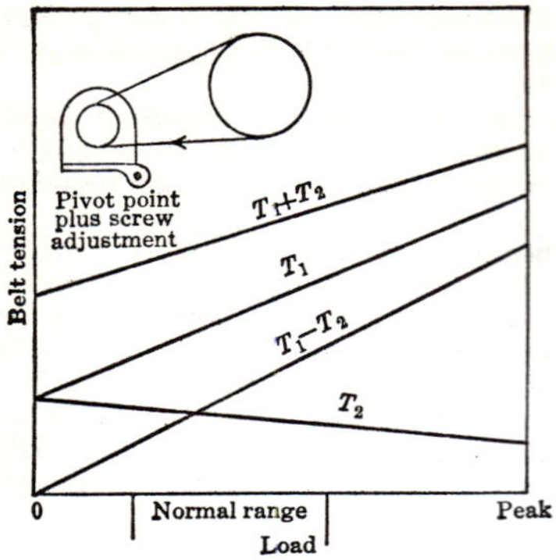

FIG. 3. Rockwood base. *Advantages:* Fair automatic tension adjustment with load. Good automatic belt take-up. Easy maintenance. Easy adjustment for alignment. *Disadvantages:* Inadequate tension available for some steep angle drives. Limited to rotation of motor in one direction.

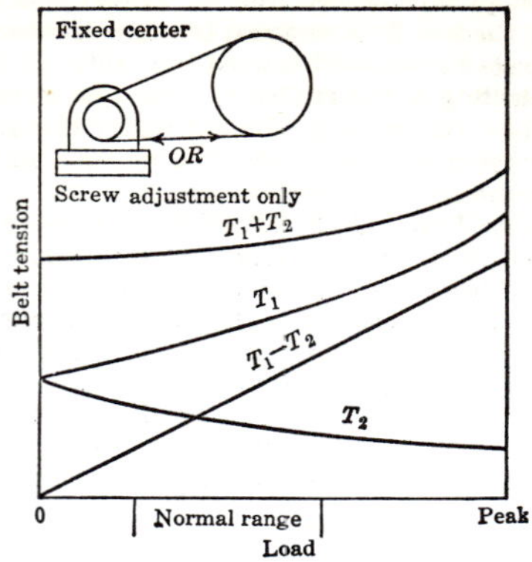

FIG. 4. Slide-rail base. *Advantages:* Easy installation. Motor can rotate in either direction. *Disadvantages:* No automatic tension adjustment except from elasticity of belt. No automatic belt take-up.

life and greater drive efficiency. Each method has characteristic advantages and disadvantages. The following notation is used: T_1 = tight side belt tension, pounds; T_2 = slack side belt tension, pounds; $T_1 - T_2$ = effective belt tension; and $T_1 + T_2$ = total belt tension.

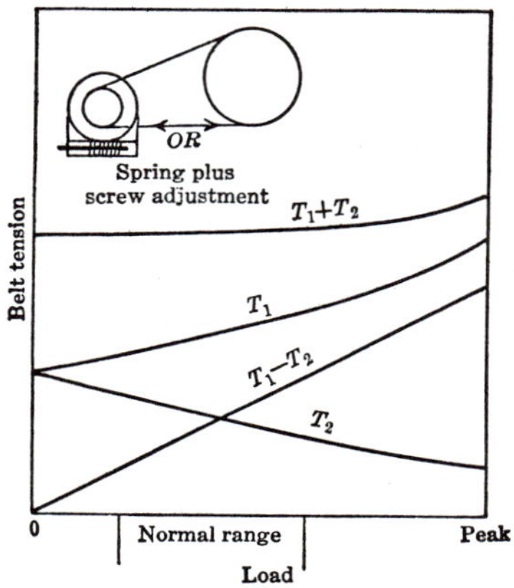

FIG. 5. Automatic base. *Advantages:* Easy application of initial tension. Easy adjustment of tension after installation. Moderate automatic belt take-up. Motor can rotate in either direction. *Disadvantages:* Sacrifices some of automatic tension adjustment from belt elasticity. Larger than necessary tension at light and normal loads.

22. PIVOTED MOTOR DRIVES

Pivoted motor-base drives are applicable for transmitting power through belts on short centers from motor to machine or counter shaft. The motor may be mounted on machine, floor, wall, or ceiling.

THE ROCKWOOD MOTOR BASE, made by the Rockwood Manufacturing Company, Inc., Indianapolis, Ind., replaces the usual slide rails and consists of a cast-iron or welded structural-steel base, a pivot shaft, and overhanging arms. (See Fig. 6.) The motor is mounted on the arms and is free to rotate to a limited extent on the pivot shaft. The overhang is adjustable and is set so that the weight of the motor maintains the correct belt tension irrespective of small changes in belt length, as caused by changes in load, centrifugal tension, changes in humidity, or permanent belt stretch.

Figure 7 shows the arrangement of base for various mountings. Note that the pivot shaft is near the tight strand of the belt. When so placed, the belt tension automatically increases with increased load on the motor and continuously adjusts itself to the load, relieving belt and bearings from peak tensions that are needed only for momentary peak loads or when starting.

Standard base sizes are available for common NEMA motor sizes. Semistandard bases are built for two and three bearing motors up to 400 hp. Bases of special design are available for ceiling mounting, Fig. 7*a* and mounting the motor above the driven pulley, Fig. 7*b*. There are also bases of special design for loom drives, spinning frame drives, agricultural drives, fractional hp motors, and other special services.

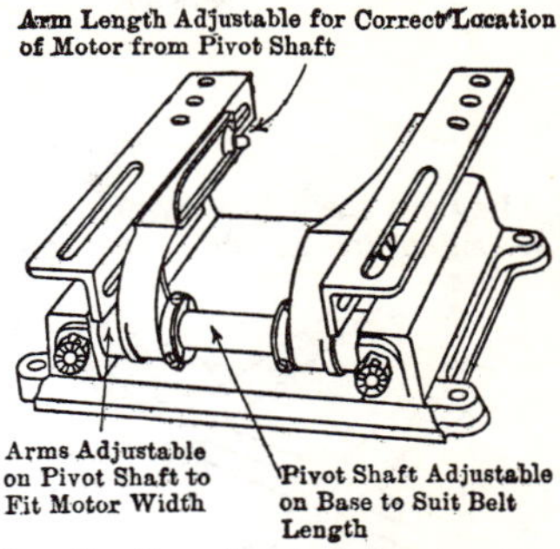

FIG. 6. Base for pivoted motor drive (Rockwood).

Rockwood bases can be advantageously used with V-rubber and flat rubber belts but are used more extensively with flat leather belts. Belts and pulleys should be selected as for a fixed center drive with proper consideration for maximum torque of motor, minimum pulley diameters, atmospheric conditions, and load characteristics. See Article 17 in this section on Horsepower of Leather Belts.

Correct belt tension is the minimum tension at which the belt will transmit the peak load without excessive slip. On pivoted motor drives the tension in the belt is regulated by the overhang of the motor. Instructions for adjusting the overhang by trial may be obtained from the manufacturers.

The required overhang for any drive may also be obtained by the graphical method shown in Fig. 8. The dimensions of the motor, motor base, and pulleys, the weight and

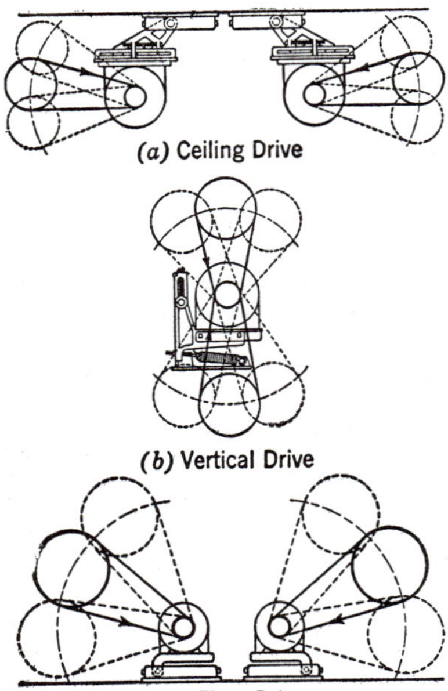

FIG. 7. Alternative mounting arrangements of Rockwood motor base.

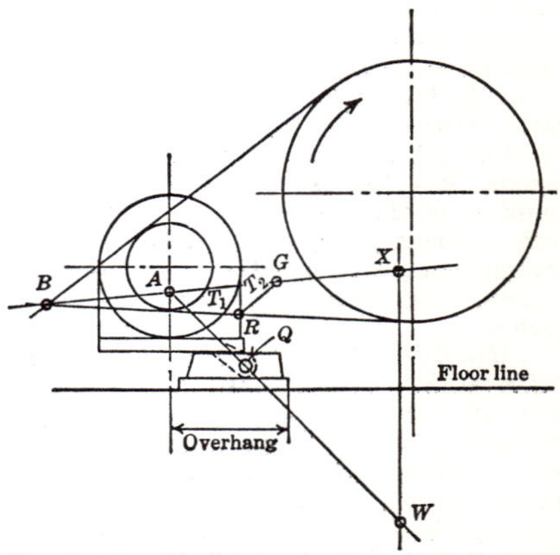

FIG. 8. Graphical determination of overhang on Rockwood motor base.

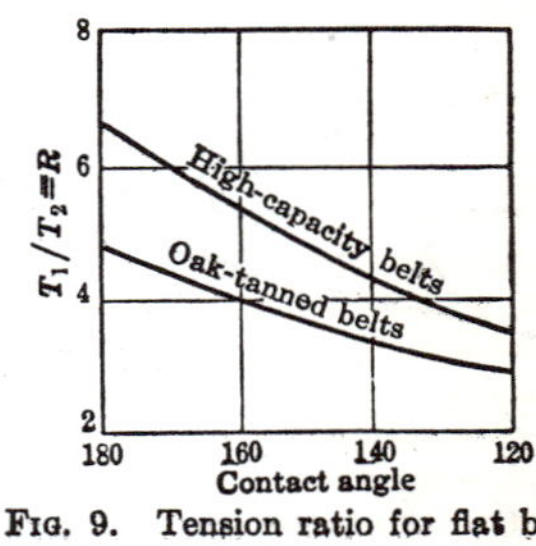

FIG. 9. Tension ratio for flat belts.

speed of the motor, the peak load to be transmitted, and the limiting safe belt tension ratio are the data required. For safe practice, peak load $P = 1.5 \times$ motor name-plate rating for reduced-voltage starting, and $2.5 \times$ motor name-plate rating for line-start motors. Calculate the belt speed V (ft per min). Calculate T_1 and T_2 from the equations.

$$T_1 - T_2 = \frac{P \times 33{,}000}{V}$$

$$\frac{T_1}{T_2} = R = \text{Limiting tension ratio (see Fig. 9)}$$

$$T_2 = \frac{P \times 33{,}000}{V(R-1)}$$

$$T_1 = RT_2$$

Lay out the side view of the drive to any convenient scale. First draw the floor, wall or ceiling line on which the base is to be mounted. Show any posts, pipes, etc., which limit the location of the drive. Extend the lines of the belt strands to their point of intersection B.

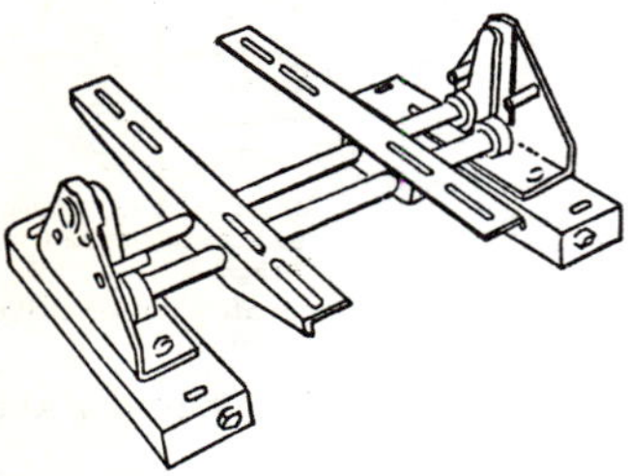

FIG. 10. Econ-O-Matic motor base. (Courtesy of American Pulley Co.)

Note. When the pulleys are of nearly equal diameter, the belt strand lines may intersect beyond the paper. In this case, the resultant tension may be assumed to be $T_1 + T_2$. A practical solution for the location of the resultant is to divide the chord between the points of tangency of the belt on each pulley equal to the ratio of T_1 and T_2 with the point of division nearest T_1. Draw a line through these points and lay off $T_1 + T_2$ to scale on this line.

From this point of intersection B, using an engineer's scale, lay off $BR = T_1$ to any convenient decimal along the tight strand. At R, lay off $RG = T_2$ parallel to the slack strand, using the same scale as before. Draw and scale the resultant belt tension BG.

Drop a vertical line from the center of the motor pulley and find its intersection at A with the resultant belt tension line BG; prolong BG to X, making $AX = BG$.

At X, lay off the motor weight equal to XW vertically downward, using the same scale. Draw the diagonal from W to A.

Draw a horizontal line at the level of the pivot shaft. Its intersection Q with the diagonal WA locates the desired position of the pivot shaft, and the overhang is the scaled horizontal distance from Q to the vertical line through the motor pulley.

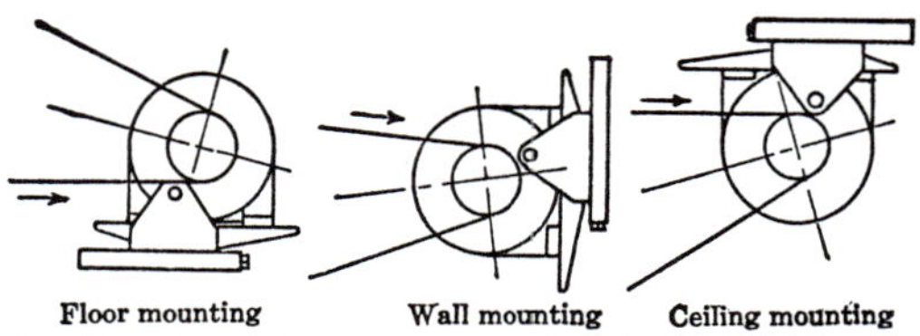

FIG. 11. Alternative mountings of Econ-O-Matic motor base.

THE ECON-O-MATIC MOTOR BASE* is illustrated in Fig. 10. Its only moving element is a "cradle" made up of cast- or forged-steel cranks, solid-steel cross rods, and pressed-steel motor base arms, on which the motor is mounted. This "cradle" is pivoted in bronze bearings of the ball-and-socket self-aligning type except in the two smallest sizes in which plain bearings are used.

By utilizing the reverse torque action of the motor, it is possible, by selection of the proper motor pulley, pivoted base, and motor, not to require any initial tension on the belt and to have exactly the required tension at full or peak load. This is an ideal condition which is not actually realized in practice. In normal installations an initial belt tension of about 15% of the peak load tension is required.

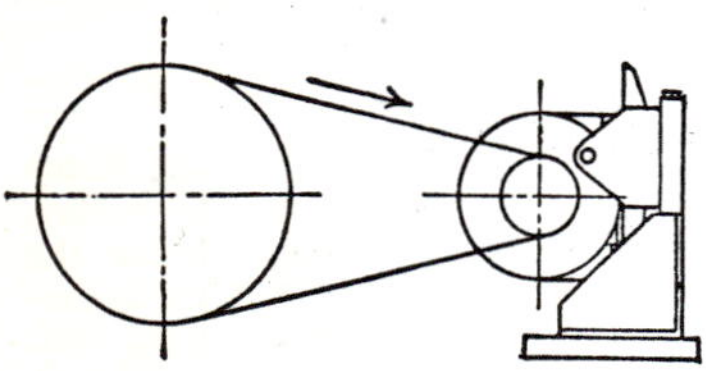

FIG. 12. Tight-side-top mounting of Econ-O-Matic motor base.

The standard Econ-O-Matic motor base shown in Fig. 10 can be utilized for floor, wall, and ceiling mounting as shown in Fig. 11. Note that in each mounting the direction of rotation is such that the tight side of the belt is closest to the pivot point of the motor base. Because reaction torque provides the greater portion of the belt tension, it is necessary that reaction torque work to increase the tension rather than to decrease it. When the tight side of the belt is away from the pivot point, the effect

* Contributed by W. A. Williams.

is for the reaction torque to decrease the belt tension with increase in load rather than to increase it. In applications requiring the tight side of the belt on top, it is necessary to employ the base shown in Fig. 12 which is called a tight-side-top base in which the pivot is carried above the motor pulley rather than below it and which fulfills the requirement that the pivot be closest to the tight side of the belt.

On vertical downdrives, a special base is required to provide initial tension against the weight of the motor. This is accomplished by a base equipped with longer arms and a counterweight which overcomes the weight of the motor and provides initial tension adequate for the drive. This type of base is illustrated in Fig. 13. Note that the tight side is closest to the pivot point. Vertical updrives are taken care of with standard floor mounting by moving the motor as far out on the arms as possible.

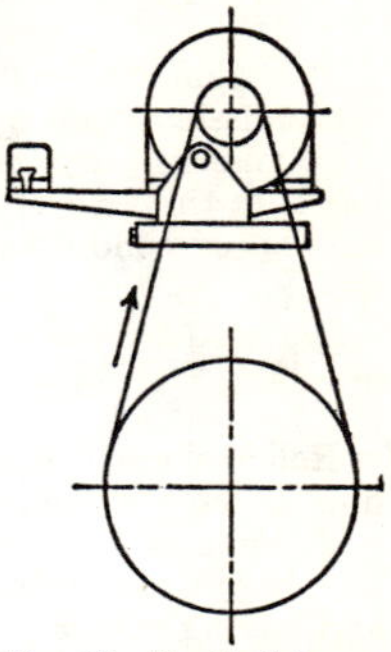

FIG. 13. Vertical downdrive mounting for Econ-O-Matic motor base.

The majority of motor-base installations utilize the standard floor mounting. For these, drive tables have been developed which show the proper horizontal distance between the pivot point and the center of the motor shaft, for various angles, motor pulley sizes, and motor frame sizes. This dimension is approximately the same for a given frame size, regardless of horsepower and speed.

23. SPRING-TENSION MOTOR-BASE DRIVES

The Automatic motor base, manufactured by the Automatic Motor Base Company, Paterson, N. J., is particularly suited for use with smaller motors on belt drives. It consists of a pressed steel base, in which two parallel rails are rigidly mounted, and a plate which can slide or, in the larger sizes, roll along the rails (Fig. 14). The motor is bolted to the plate, and belt tension is applied through springs that are readily adjusted to give the desired tension.

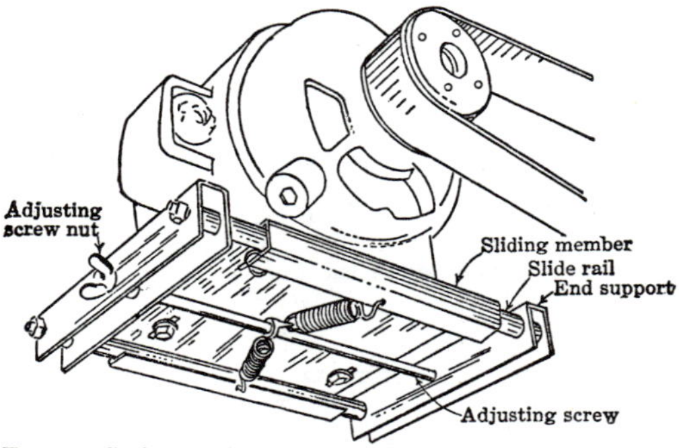

FIG. 14. Spring-tension motor base. (Courtesy of Automatic Motor Base Co.)

The base can be mounted in any position, and the operation of the base is independent of the direction of rotation of the motor. It can be used with flat leather, flat rubber, or V-rubber belts, which should be selected on the same basis as for a fixed center drive with proper consideration for peak starting load of the motor and other conditions.

The Automatic base is made in twelve sizes to accommodate motors from 1/6 to 100 hp. The NEMA frame number determines the size of the base to be used. Extra-heavy springs can be supplied when the motor is mounted vertically above the driven pulley.

The actual installation is a simple operation as illustrated by the following brief instructions:

Turn the adjusting screw until the sliding member is located at the end of the frame nearest the driven pulley. This will permit maximum take-up as belt stretch occurs. Bolt the motor to the sliding member. Position the motor and base assembly so that the belt or belts, in the case of a multiple V drive, go over the pulleys readily. Bolt the assembly in place. Turn the spring adjusting screw until a slight tension is applied to the belt. Start the motor and make the final adjustment, providing only as much spring tension as is necessary for the belt to handle the maximum load without slippage. More tension than is necessary serves no useful purpose and merely puts an undue stress on belts and bearings.

CHAIN DRIVES *

Precision steel roller chains and inverted tooth or silent chains are classified as power transmission chains. Block chains and malleable chains are sometimes used for power transmission purposes at very slow speeds or at very light loads. However, their major use is in the field of conveyor applications. For the transmission of power at average to high speed conditions, it is essential that roller or inverted tooth chains be used.

24. ROLLER CHAINS

Roller chain is a series of alternately assembled roller links and pin links in which the pins articulate inside the bushings and the rollers are free to turn on the bushings. Pins and bushings are press fit in their respective side plates. The chain may be single strand having one row of roller links, or multiple strand having more than one row of roller links and having center plates located between the strands of roller links.

Precision steel transmission roller chains have been standardized by the Association of Roller and Silent Chain Manufacturers and conform to the American Standard B29a-1930 (now being revised and restandardized). The following material has been supplied by the Association, and most of it is as proposed for revision of these ASA standards.

GENERAL CHAIN PROPORTIONS AND DESIGNATIONS.

Roller diameter is approximately five-eighths of the pitch.

Chain width is defined as the minimum distance between the roller link plates. This width is the nearest binary fraction to five-eighths of the pitch.

Pin diameter is approximately five-sixteenths of the pitch or one-half of the roller diameter.

Thickness of link plates for the standard series is approximately one-eighth of the pitch. Thickness of link plates for the heavy series chain of any pitch is approximately that of the next larger pitch standard series chain.

Maximum width of roller link plates = 0.95 of the pitch.

Maximum width of pin link plates = 0.82 of the pitch.

Minimum ultimate tensile strength of standard series chains = 12,500 × $(\text{pitch})^2$ per strand of chain.

Measuring load is the load under which a chain should be measured for length. It is equal to 125 × $(\text{pitch})^2$, with a minimum of 18 lb.

Standard Chain Numbers. The right-hand digit in the chain designation is zero for roller chains of the usual proportions, 1 for a lightweight chain, and 5 for a rollerless bushing chain. (See Table 1.) The numbers to the left of the right-hand figure denote

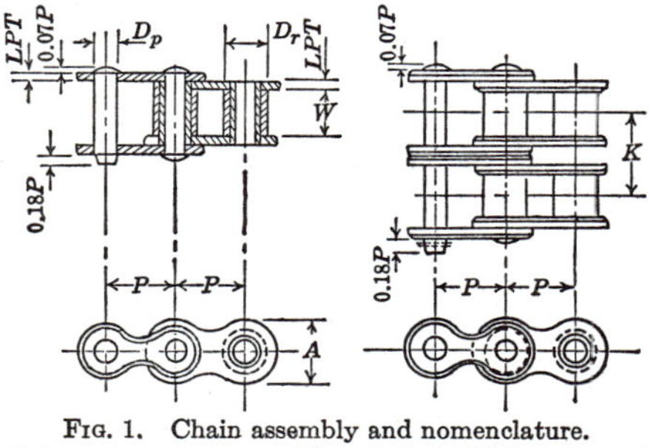

FIG. 1. Chain assembly and nomenclature.

P = chain pitch.
D_r = roller diameter.
LPT = link plate thickness.
A = link plate width.
W = chain width between roller link plates.
D_p = pin diameter between pin link plates.
K = transverse pitch for multiple chains.

the number of 1/8 inches in the pitch. The letter H following the chain number denotes the heavy series. The hyphenated number 2 suffixed to the chain number denotes a double strand, 3 a triple strand, 4 a quadruple strand chain, and so forth.

Heavy series chains made in 3/4-in. and larger pitches differ from the standard series in thickness of link plates. Their value is only in the acceptance of higher tensile or jerk loads during operation at lower speeds.

* Contributed by the Association of Roller and Silent Chain Manufacturers.

Table 1. General Chain Dimensions in Inches

(See Fig. 1)

Standard Series

Standard Chain No.	Pitch, P	Roller Diameter, D_r	Width, W	Pin Diameter, D_p	Link Plate Thickness, LPT	Link Plate Width, A	Transverse Pitch for Multiple Strands, K	Average Ultimate Strength, lb *
25	1/4	0.130 †	1/8	.0905	.030	0.237	0.252	800
35	3/8	0.200 †	3/16	.141	.050	0.356	0.399	2,100
40	1/2	5/16	5/16	.156	.060	0.475	0.566	3,700
50	5/8	0.400	3/8	.200	.080	0.594	0.713	6,100
60	3/4	15/32	1/2	.234	.094	0.713	0.897	8,500
80	1	5/8	5/8	.312	.125	0.950	1.153	14,500
100	1 1/4	3/4	3/4	.375	.156	1.187	1.408	24,000
120	1 1/2	7/8	1	.437	.187	1.425	1.789	34,000
140	1 3/4	1	1	.500	.219	1.663	1.924	46,000
160	2	1 1/8	1 1/4	.562	.250	1.900	2.305	58,000
180	2 1/4	1 13/32	1 13/32	.687	.281	2.138	2.592	73,000
200	2 1/2	1 9/16	1 1/2	.781	.312	2.375	2.817	90,000
240	3	1 7/8	1 7/8	.937	.375	2.850	3.493	130,000

* Average ultimate strength for multiple-strand chains is the average ultimate strength for one strand multiplied by the number of strands.

† Made only without roller.

Table 2. Recommended Maximum Revolutions per Minute of Sprockets for American Standard Roller Chain

Chain No.	25	35	41	40	50	60	80	100	120	140	160	180	200	240
Pitch, in.	1/4	3/8	1/2	1/2	5/8	3/4	1	1 1/4	1 1/2	1 3/4	2	2 1/4	2 1/2	3
Teeth														
11	4310	2260	1020	1690	1220	920	580	415	325	235	200	165	145	110
12	4960	2590	1170	1940	1400	1050	670	475	375	270	230	190	165	125
13	5540	2900	1310	2180	1570	1180	750	535	415	305	260	215	185	140
14	6070	3170	1430	2380	1720	1290	820	585	455	335	280	235	205	155
15	6530	3420	1540	2560	1850	1390	880	630	490	360	305	255	220	165
16	6940	3630	1630	2720	1960	1480	935	670	520	380	325	270	235	175
17	7290	3810	1720	2860	2060	1550	985	700	550	400	340	285	245	185
18	7590	3970	1790	2980	2150	1610	1020	730	570	415	355	295	255	195
19	7840	4100	1850	3080	2220	1670	1060	755	590	430	365	305	265	200
20	8050	4210	1890	3160	2280	1720	1090	775	605	440	375	315	270	205
21	8230	4300	1940	3230	2330	1750	1110	790	620	450	385	320	280	210
22	8370	4380	1970	3290	2370	1780	1130	805	630	460	390	325	280	215
23	8480	4430	2000	3330	2400	1800	1150	815	640	465	395	330	285	215
24	8560	4480	2020	3360	2420	1820	1160	825	645	470	400	330	290	220
25	8610	4510	2030	3380	2440	1830	1160	830	650	475	400	335	290	220
30	8580	4490	2020	3370	2430	1830	1160	825	645	470	400	335	290	220
35	8200	4290	1930	3220	2320	1740	1110	790	615	450	380	320	275	210
40	7580	3970	1780	2970	2140	1610	1020	730	570	415	355	295	255	195
45	6820	3570	1600	2670	1930	1450	920	655	515	375	320	265	230	175
50	5950	3110	1400	2330	1680	1270	805	575	450	325	275	230	200	150
55	5010	2620	1180	1970	1420	1070	675	480	375	275	235	195	170	125
60	4020	2100	950	1580	1140	860	545	390	305	220	185	155	135	100

Lightweight machinery chain, designated as No. 41, is 1/2 in. in pitch and 1/4 in. wide. It has 0.306-in. diameter rollers, 0.141-in. pin diameter, and 0.050-in.-thick link plates. The measuring load is 18 lb.

Tolerance for Chain Length. New chains, under standard measuring load, may be over-length 1/64 in. per ft but must not be under-length.

Standard offset links are made to accommodate chains having roller link plates with a maximum width, after being beveled, equal to 95% of the pitch, and pin link plates with a maximum width, after being beveled, equal to 82% of the pitch. Therefore the standard minimum values of x and y (Fig. 2) are:

$$x \text{ (min)} = 0.41P + 0.008 \text{ in.}$$
$$y \text{ (min)} = 0.475P + 0.008 \text{ in.}$$

where P = chain pitch.

SELECTION OF PRECISION STEEL ROLLER CHAINS FOR POWER TRANSMISSION PURPOSES. For each pitch of chain there is a maximum sprocket revolutions per minute which should not be exceeded without consulting the chain manufacturer. Because of this, it is well to begin the selection of the drive with consideration of the fastest rotating shaft of the drive. Next, refer to Table 2, and select the maximum pitch chain which will operate at the speed of the fastest turning shaft. In the selection it is always advisable to have at least 17 teeth in the smaller sprocket and preferably 24. A smaller pitch chain than selected may be found necessary if the number of teeth is too few or if extreme quietness is desired.

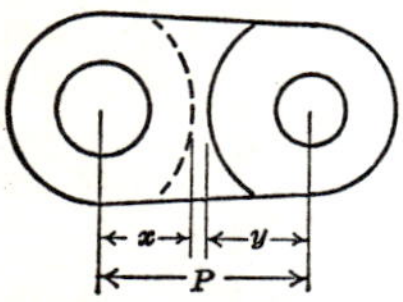

Fig. 2. Offset link plate.

After selection of the chain pitch and number of teeth in the smallest sprocket, check the sprocket for accommodation of the shaft and keyway. The bore should not exceed approximately seven-tenths of the maximum hub diameter (see Fig. 3).

The life of a roller chain drive can be affected adversely by any one of the following causes: (1) small sprockets operating at high speeds; (2) insufficient and improper lubrication; (3) inaccurate sprockets; (4) misalignment of sprockets; (5) overtensioning of chain during adjustment of centers or adjustment of idler sprocket.

The number of strands of chain required are determined from service factors and horsepower ratings, as explained in the following.

HORSEPOWER RATINGS OF AMERICAN STANDARD SINGLE-STRAND ROLLER CHAIN, condensed in Tables 3 to 13, have been established on a life expectancy of approximately 15,000 hr under optimum drive conditions and a service factor of 1. Principles and service factors subsequently outlined must be adhered to in making selections from the rating tables. Failure to do so may result in unsatisfactory service from abnormal joint wear, accelerated chain elongation, and serious damage to sprockets.

The horsepower ratings are for drives with two sprockets. For design of installation having more than two sprockets, idlers, and unusual conditions or requirements, engineers of chain manufacturers should be consulted.

Table 3. Horsepower Ratings, Standard Single-strand Roller Chain #25, 1/4-in. Pitch

No. of Teeth, Small Sprocket	Revolutions per Minute, Small Sprocket										
	100	200	600	1000	1400	1800	2200	2600	4000	6000	8000
11	.051	.08	0.23	0.35	0.44	0.50	0.55	0.59	0.62		
13	.061	.10	0.30	0.44	0.56	0.66	0.73	0.79	0.90		
16	.076	.13	0.38	0.57	0.73	0.86	0.97	1.06	1.26	1.29	
20	.095	.17	0.47	0.72	0.93	1.10	1.25	1.37	1.68	1.78	1.57
25	.118	.22	0.58	0.90	1.15	1.37	1.55	1.71	2.09	2.22	1.95
40	.187	.33	0.89	1.35	1.73	2.03	2.26	2.46	2.85	2.65	
60	.274	.51	1.28	1.87	2.30	2.65	2.93	3.11	3.07		
Lubrication	Type I		Type II					Type III			

Table 4. Horsepower Ratings, Standard Single-strand Roller Chain #35, 3/8-in. Pitch

No. of Teeth, Small Sprocket	Revolutions per Minute, Small Sprocket										
	50	100	200	400	800	1200	1600	2000	2400	3200	4000
11	.087	.165	0.303	0.54	0.89	1.15	1.31	1.40			
13	.105	.198	0.369	0.66	1.13	1.48	1.73	1.91	2.02		
16	.130	.248	0.463	0.84	1.46	1.93	2.30	2.58	2.79	2.98	
20	.164	.312	0.585	1.07	1.86	2.49	2.97	3.37	3.65	4.01	4.06
25	.204	.389	0.730	1.33	2.31	3.09	3.71	4.19	4.54	4.98	5.04
40	.322	.613	1.14	2.05	3.50	4.59	5.39	5.96	6.32	6.58	
60	.474	.890	1.64	2.89	4.76	6.00	6.78	7.14			
Lubrication		Type I		Type II				Type III			

Table 5. Horsepower Ratings, Standard Single-strand Roller Chain #40, $^{1}/_{2}$-in. Pitch

No. of Teeth, Small Sprocket	Revolutions per Minute, Small Sprocket										
	50	100	200	400	600	800	1000	1400	1800	2200	3000
11	0.202	0.379	0.687	1.19	1.59	1.89	2.14	2.44			
13	0.243	0.458	0.842	1.48	2.00	2.44	2.79	3.31	3.59		
16	0.303	0.572	1.06	1.89	2.59	3.17	3.67	4.44	4.96	5.24	
20	0.380	0.720	1.34	2.40	3.29	4.07	4.72	5.76	6.50	6.98	7.22
25	0.475	0.902	1.67	3.00	4.11	5.07	5.90	7.19	8.10	8.67	8.95
40	0.748	1.413	2.60	4.59	6.22	7.57	8.67	10.31	11.23	11.61	
60	1.095	2.05	3.71	6.39	8.45	10.04	11.22	12.57			
Lubrication		Type I			Type II			Type III			

Table 6. Horsepower Ratings, Standard Single-strand Roller Chain #50, $^{5}/_{8}$-in. Pitch

No. of Teeth, Small Sprocket	Revolutions per Minute, Small Sprocket										
	50	100	200	400	600	800	1000	1200	1400	1800	2200
11	0.385	0.72	1.29	2.19	2.85	3.33	3.68	3.88			
13	0.467	0.87	1.59	2.74	3.65	4.36	4.91	5.30	5.57		
16	0.582	1.09	2.00	3.52	4.74	5.73	6.55	7.19	7.67	8.24	
20	0.732	1.38	2.54	4.48	6.07	7.38	8.46	9.35	10.05	10.96	11.30
25	0.914	1.72	3.17	5.59	7.58	9.20	10.55	11.64	12.50	13.62	14.02
40	1.44	2.70	4.91	8.51	11.33	13.57	15.28	16.57	17.49	18.13	
60	1.96	3.90	6.96	11.71	15.15	17.54	19.10				
Lubrication	Type I		Type II					Type III			

Table 7. Horsepower Ratings, Standard Single-strand Roller Chain #60, $^{3}/_{4}$-in. Pitch

No. of Teeth, Small Sprocket	Revolutions per Minute, Small Sprocket										
	50	100	200	400	600	800	1000	1200	1400	1600	1800
11	0.66	1.21	2.15	3.58	4.56	5.21					
13	0.79	1.48	2.65	4.52	5.91	6.92	7.62				
16	0.99	1.85	3.36	5.82	7.73	9.21	10.34	11.16	11.70		
20	1.25	2.34	4.26	7.41	9.92	11.91	13.46	14.64	15.48	16.00	
25	1.56	2.92	5.32	9.27	12.37	14.84	16.76	18.21	19.24	19.89	20.18
40	2.45	4.55	8.20	13.99	18.35	21.57	23.86	25.42	26.08	26.03	
60	3.58	6.55	11.57	19.03	24.04	27.13					
Lubrication	Type I		Type II					Type III			

Table 8. Horsepower Ratings, Standard Single-strand Roller Chain #80, 1-in. Pitch

No. of Teeth, Small Sprocket	Revolutions per Minute, Small Sprocket										
	25	50	100	200	300	400	500	600	700	800	1000
11	0.810	1.517	2.76	4.78	6.36	7.60	8.56				
13	0.972	1.833	3.36	5.93	8.03	9.75	11.17	12.30	13.21		
16	1.212	2.293	4.24	7.56	10.34	12.70	14.69	16.37	17.78	18.92	
20	1.522	2.886	5.36	9.60	13.19	16.26	18.90	21.16	23.08	24.70	27.08
25	1.902	3.606	6.69	11.98	16.45	20.28	23.57	26.38	28.76	30.75	33.68
40	2.998	5.654	10.39	18.36	24.88	30.28	34.73	38.35	41.23	43.46	46.28
60	4.393	8.198	15.03	25.57	33.83	40.14	44.89				
Lubrication		Type I			Type II				Type III		

Table 9. Horsepower Ratings, Standard Single-strand Roller Chain #100, 1 1/4-in. Pitch

No. of Teeth, Small Sprocket	Revolutions per Minute, Small Sprocket										
	10	25	50	100	150	200	300	400	500	600	700
11	0.663	1.56	2.88	5.19	7.10	8.78	11.4	13.6			
13	0.792	1.87	3.52	6.35	8.82	10.9	14.6	17.6	19.6		
16	0.987	2.34	4.37	8.07	11.2	14.1	18.9	23.2	26.1	28.7	
20	1.24	2.94	5.52	10.1	14.3	17.9	24.3	29.8	33.9	37.3	40.2
25	1.54	3.66	6.90	12.7	17.8	22.4	30.3	36.8	42.2	46.4	50.0
40	2.44	5.77	10.8	19.6	27.3	34.0	44.3	54.2	61.0	66.3	70.8
60	3.61	8.44	15.6	27.8	38.1	46.8	60.5				
Lubrication		Type I			Type II				Type III		

Table 10. Horsepower Ratings, Standard Single-strand Roller Chain #120, 1 1/2-in. Pitch

No. of Teeth, Small Sprocket	Revolutions per Minute, Small Sprocket										
	10	25	50	100	150	200	250	300	400	500	600
11	1.14	2.63	4.85	8.61	11.6	14.3	16.5	18.2			
13	1.35	3.17	5.90	10.6	14.6	18.1	21.0	23.6	27.7		
16	1.68	3.97	7.42	13.4	18.6	23.3	27.4	31.0	36.9	41.3	
20	2.12	4.99	9.37	17.0	23.8	29.7	34.9	39.7	47.6	53.9	58.5
25	2.64	6.26	11.7	21.3	29.7	37.1	43.6	49.5	59.5	67.1	73.0
40	4.18	9.80	18.2	32.8	45.2	56.0	65.3	73.4	86.1	95.3	
60	6.16	14.3	26.3	46.3	62.7	76.0	87.2	96.2			
Lubrication		Type I			Type II				Type III		

Table 11. Horsepower Ratings, Standard Single-strand Roller Chain #140, 1 3/4-in. Pitch

No. of Teeth, Small Sprocket	Revolutions per Minute, Small Sprocket										
	10	25	50	100	150	200	250	300	350	400	450
11	1.79	4.12	7.55	13.3	17.8	21.5					
13	2.13	5.07	9.19	16.3	22.3	27.3	31.6	35.2			
16	2.66	6.38	11.6	20.8	28.6	35.4	41.3	46.4	50.7		
20	3.34	8.01	14.6	26.4	36.4	45.3	53.0	59.8	65.6	70.7	
25	4.17	10.40	18.2	32.9	45.5	56.5	66.1	74.5	81.7	88.0	93.4
40	6.58	15.68	28.3	50.6	69.1	84.9	98.2	109.1	118.2	126.7	
60	9.67	22.68	40.7	70.9	94.8	113.9					
Lubrication		Type I		Type II					Type III		

Table 12. Horsepower Ratings, Standard Single-strand Roller Chain #160, 2-in. Pitch

No. of Teeth, Small Sprocket	Revolutions per Minute, Small Sprocket										
	10	20	40	80	120	160	200	240	280	320	360
11	2.64	4.97	9.14	16.1	21.9	26.5	30.4				
13	3.16	6.00	11.1	19.9	27.2	33.6	39.0	43.6			
16	3.93	7.49	14.0	25.3	34.9	43.4	50.9	57.3	62.9	67.9	
20	4.94	9.41	17.6	32.0	44.3	55.3	65.0	73.6	81.4	87.8	93.4
25	6.17	11.8	22.0	40.0	55.4	69.1	81.2	91.9	101.2	109.5	116.5
40	9.74	18.5	34.3	61.5	84.5	104.2	121.1	135.4	148.3	157.2	
60	14.3	26.9	49.3	86.5	116.5	140.7					
Lubrication		Type I		Type II				Type III			

Table 13. Horsepower Ratings, Standard Single-strand Roller Chain #200, 2 1/2-in. Pitch

No. of Teeth, Small Sprocket	Revolutions per Minute, Small Sprocket										
	10	20	40	60	80	100	120	160	200	240	280
11	5.06	9.46	17.2	23.9	29.8	35.1	39.7				
13	6.08	11.4	21.0	29.5	37.1	44.0	50.3	60.9			
16	7.56	14.3	26.5	37.4	47.2	56.3	64.6	79.3	91.7		
20	9.51	18.0	33.5	47.3	60.0	71.7	82.3	101.5	118.2	132.3	
25	11.8	22.5	41.8	59.1	74.9	89.6	102.9	126.8	147.2	164.8	179.7
40	18.7	35.3	64.9	91.0	114.8	135.8	155.5	189.2	214.0	231.4	
60	27.4	51.2	92.8	128.6	159.9	187.2	211.4				
Lubrication		Type I				Type II				Type III	

SERVICE FACTORS as given in the following tabulation should be used in connection with the horsepower ratings, the load being multiplied by the factor to obtain the required chain capacity:

Type of Load	Conditions of Service 10-hr Day	24-hr Day
Uniform load, average conditions	1.0	1.2
Moderate shock, abnormal conditions	1.2	1.4
Heavy shock, abnormal conditions	1.4	1.7

LUBRICATION of roller chains is essential for effectively minimizing metal-to-metal contact of pin-bushing joints of the chain. Oil should be applied to link plate edges as explained under Type I method of lubrication subsequently, since access to pin-bushing clearances is possible only through clearances between pin link plates and roller link plates of single chains, and, in addition, between center link plates and roller link plates of multiple chains. Oil applied on center line of rollers cannot reach pin-bushing joints and therefore cannot retard chain wear elongation.

A good grade of mineral oil, without additives, of medium or light consistency, free flowing at the prevailing temperature, should be used.

Drives should be protected against dirt and moisture. The method of lubrication, which is influenced by the speed of the chain and the amount of power transmitted, should be in accordance with one of the following:

Type I. Drip (4 to 10 drops per minute), shallow bath, or manual with oil applied frequently with a brush or spout can to upper edges of all link plates in the lower span of chain.

Type II. Rapid drip (20 drops per minute minimum) or continuous with shallow bath, disk, or slinger. A circulating pump may be required when transmitted horsepower is substantial and center distance is short.

Type III. Continuous, with disk, slinger, or circulating pump.

The choice of method or type of lubrication should in general be guided by the following:

Chain Speed	Method of Lubrication
Up to 600 ft per min	Type I
600 to 1500 ft per min	Type II
Above 1500 ft per min	Type III

Heavy oils and greases are not recommended for lubrication of roller chains except under unusual conditions of service because they are generally too stiff to enter and fill the small clearances between the chain parts. Engineers of chain manufacturers should be consulted in regard to lubrication of roller chain drives for which lubrication other than as outlined might seem to be desirable.

SPROCKETS should have tooth shape, thickness, profile, and diameters conforming to the standards. Smaller sprockets, 25 teeth and less, are preferably of steel, 180 Brinell minimum for speeds to about 600 ft per min. For speeds greater than 600 ft per min, either .20 carbon steel, carburized, hardened and drawn, or .40 or higher carbon steel, heat treated and drawn, are generally recommended, the hardness usually being between 300 Brinell minimum and 550 Brinell maximum depending upon the type of service.

Larger sprockets may be made from unhardened steel plates, bars, castings, or forgings, or cast iron, depending on the duty imposed.

Sprocket dimensions are illustrated in Fig. 3. Using the notation P = pitch, inches; PD = pitch diameter, inches; D_r = roller diameter, inches; and N = number of teeth, the following equations relate the various diameters:

$$\text{Bottom diameter of sprocket} = PD - D_r$$

$$\text{Pitch diameter of sprocket} = \frac{P}{\sin \dfrac{180}{N}}$$

$$\text{Caliper diameter for } \textit{even} \text{ number of teeth} = \text{Bottom diameter}$$

$$\text{Caliper diameter for } \textit{odd} \text{ number of teeth} = PD\left(\cos \frac{90°}{N}\right) - D_r$$

Tolerances on caliper diameter of sprockets:

$$\text{Plus tolerance} = 0.000 \text{ in.}$$

$$\text{Minus tolerance} = 0.001P\sqrt{N} + 0.003 \text{ in.}$$

$$\text{Approximate outside diameter of sprocket} = P\left(0.6 + \cot \frac{180}{N}\right)$$

Maximum hub and groove diameter (MHD) of sprockets

$$= P\left(\cot \frac{180}{N} - 1\right) - 0.030 \text{ in.}$$

Types of Sprockets. The designations for the four principal types of sprockets are as follows (see Fig. 4):

Type A—plain plate
Type B—hub on one side only
Type C—hub on both sides
Type D—hub detachable

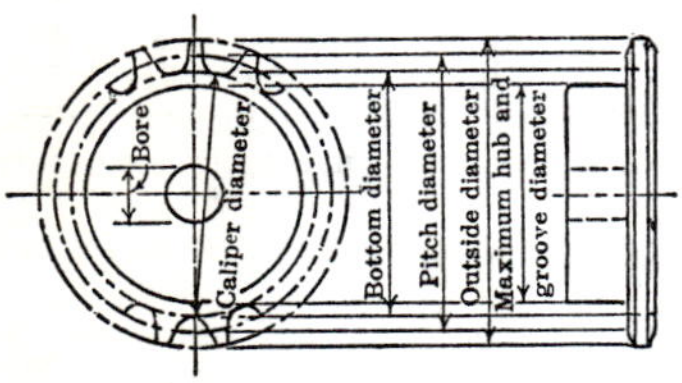

FIG. 3. Chain sprocket dimensions.

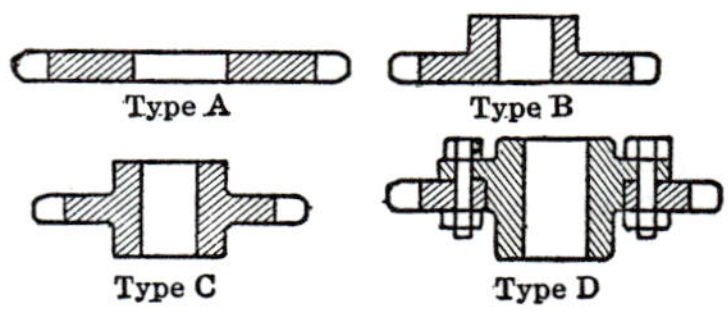

FIG. 4. Four principal types of chain sprockets.

The section profile (Fig. 5) shows the recommended chamfering of sprocket teeth for roller chains and tooth flange location for multiple strand chains. Dimensions are given in Tables 14 and 15. Other dimensions are given by the following relationships: P = pitch of chain; w = chain width; LPT = nominal thickness of link plates; $k = W + (4.22$

Table 14. Sprocket Flange Thicknesses in Inches

(See Fig. 5)

Width of Chain, W	Maximum Sprocket Flange Thickness, t			Minus Tolerance on t and M_2, M_3, M_4, etc.	Maximum Variation of t on each Flange
	Single Strand	Double and Triple Strand	Quadruple Strand and Over		
1/8	0.110	0.107	0.096	.007	.004
3/16	0.168	0.162	0.149	.008	.004
1/4	0.227			.009	.004
5/16	0.284	0.275	0.256	.009	.005
3/8	0.343	0.332	0.311	.010	.005
1/2	0.459	0.444	0.418	.011	.006
5/8	0.575	0.557	0.526	.012	.006
3/4	0.692	0.669	0.633	.014	.007
1	0.924	0.894	0.848	.016	.008
1 1/4	1.156	1.119	1.063	.019	.009
1 13/32	1.301	1.259	1.197	.020	.010
1 1/2	1.389	1.344	1.278	.021	.011
1 7/8	1.738	1.682	1.601	.025	.012

Table 15. Sprocket Tooth Section Profile Dimension in Inches

(See Fig. 5)

Pitch, P	Depth of Chamfer, h	Width of Chamfer, g	Minimum Radius, R_c	Transverse Pitch, Standard Series, k
1/4	1/8	1/32	0.265	0.252
3/8	3/16	3/64	0.398	0.399
1/2	1/4	1/16	0.531	0.566
5/8	5/16	5/64	0.664	0.713
3/4	3/8	3/32	0.796	0.897
1	1/2	1/8	1.062	1.153
1 1/4	5/8	5/32	1.327	1.408
1 1/2	3/4	3/16	1.593	1.789
1 3/4	7/8	7/32	1.858	1.924
2	1	1/4	2.124	2.305
2 1/4	1 1/8	9/32	2.392	2.592
2 1/2	1 1/4	5/16	2.654	2.817
3	1 1/2	3/8	3.187	3.458

$\times$ LPT) = transverse pitch for multiple strand chains; M_2, M_3, M_4, etc. = k (strand multiple − 1) + t; R_c min = $1.063P$ (approximately tangent to side) = chamfer radius; and r_f max = $0.04P$ for maximum hub diameter = fillet radius.

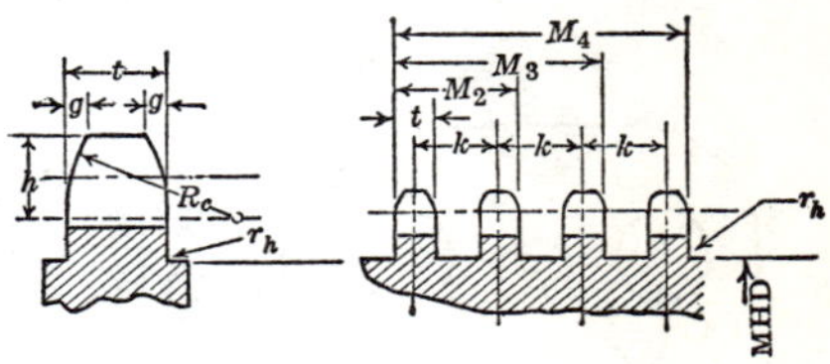

FIG. 5. Profile of sprocket teeth.

Cutters for Sprocket Teeth. Three types of cutters are used for cutting sprocket teeth, viz., space cutters, straddle cutters, and hobs. Space cutters will cut the following ranges of teeth: 7 to 8, 9 to 11, 12 to 17, 18 to 34, 35 and over. These cutters are of the common type, designed to cut the spaces between the teeth. Straddle cutters are designed to straddle the teeth and cut a part of two spaces at the same time. Two straddle cutters will fill the requirements for all numbers of teeth. Hobs are made to cut any required number of teeth of a given pitch and roller diameter. Proportions of cutters are given in American Standard B29a-1930 (now in process of revision) or may be secured from roller chain manufacturers.

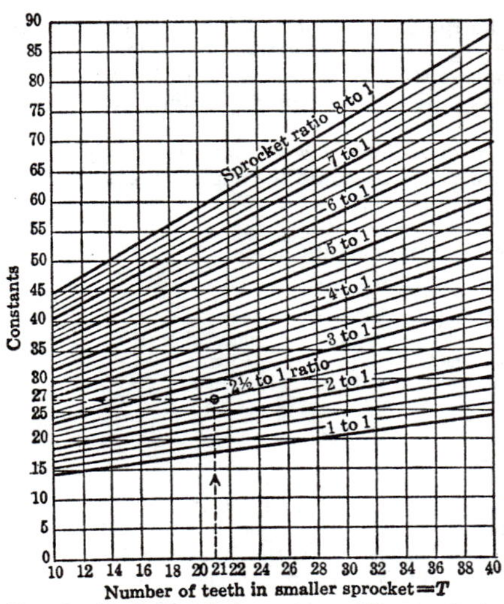

FIG. 6. Desirable chain centers. The constants multiplied by pitch give approximate center distance in inches. (Courtesy of Link-Belt Co.)

ALIGNMENT of shafting and sprocket tooth faces must provide distribution of the load across the entire chain width. Shaftings, bearings, and foundations should be suitable to maintain the initial static alignment. The horsepower tables are directly applicable only if the sprockets are on parallel, horizontal shafts.

CENTER DISTANCE between sprockets should be sufficient to provide approximately 135 degrees minimum arc of meshing of the chain on the smaller sprocket. Approximate centers may be obtained from Fig. 6. For low powers and speeds the distances given by Fig. 6 can be reduced to a point where the sprockets just clear on low ratios, and where arc of contact is no less than about 100 degrees on small sprockets with large ratios. Centers and sprockets selected should allow adequate clearance for slack strand of drive. Means for adjustment of center distance is recommended especially if there is fluctuation of load or if the sprockets are on vertical centers. Centers should be selected for chain length of an even number of pitches.

LENGTH OF CHAIN for two-sprocket drives is given exactly by the following formula

$$L = \frac{N + n}{2} + \phi \frac{N - n}{180} + 2C \cos \phi \tag{1}$$

where N and n are, respectively, the numbers of teeth in the large and small sprockets, C = center distance, and $\phi = \sin^{-1}[(N - n)/2\pi C]$ measured in degrees. Lengths L and C are measured in terms of pitches. For most purposes, the following approximation will serve:

$$L = \frac{N + n}{2} + \frac{(N - n)^2}{4\pi^2 C} + 2C \tag{2}$$

Although chain lengths may be calculated by the foregoing formulas, it is recommended that a careful layout be made to scale, full size, or double size, from which length may be measured by stepping off with dividers. Such layouts may be studied so that proper ratios, adjustment, etc., can be provided and should be submitted to the chain manufacturer for checking and approval. For drives employing more than two sprockets, the layout method is the fastest for determining chain length.

25. SILENT CHAINS

A silent or inverted-tooth chain consists of a series of links of the shape shown in Fig. 7, which are assembled over pins to form various widths of chain, as shown in Fig. 8. The chain engages with the sprocket as shown in Fig. 9. As the pitch lengthens, owing to wear, the links adjust themselves to a larger pitch circle as shown in Fig. 10, thus automatically

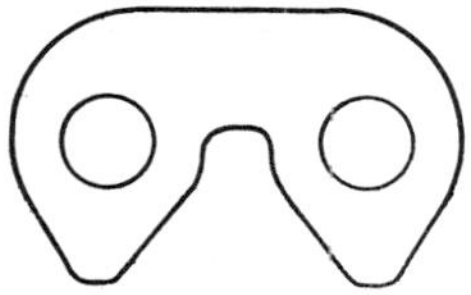

FIG. 7. Elementary link of silent chain.

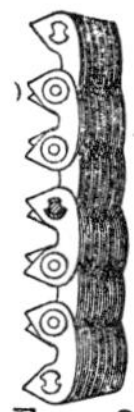

FIG. 8. Silent chain.

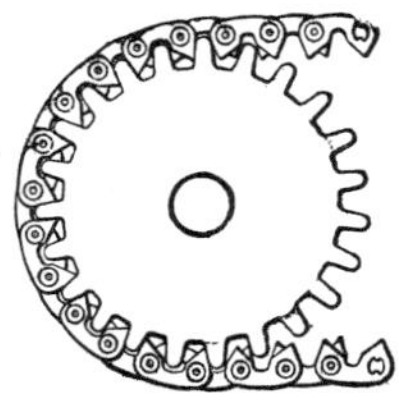

FIG. 9. New chain on sprocket.

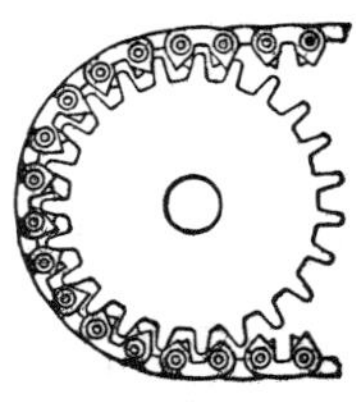

FIG. 10. Worn chain on sprocket.

compensating for wear. These chains can be run in either direction at velocities up to 4,500 ft per min, quietly and smoothly, on sprockets of suitable numbers of teeth.

DESIGN OF CHAIN DRIVES. Design must always be based on nature of load, horsepower (hp) to be transmitted, and revolutions per minute (rpm) of driving and driven units.

Before the design is selected, the nominal or given horsepower should be multiplied by a service factor in accordance with Table 16, which is based on the necessary extra capacity required for severity of conditions.

Table 16. Service Factors *K* for Silent Chains

	10-hour Day or Less		24-hour Day	
Nature of Load	Motor or Turbine Drive	Internal-Combustion Engine Drive	Motor or Turbine Drive	Internal-Combustion Engine Drive
Uniform	1	1.3	1.3	1.7
Moderate shock	1.3	1.7	1.7	2
Heavy shock	1.7	2	2	2.5

Pitch of chain should be selected first on the basis of values given in Table 17. The higher shaft speed is the limiting factor and indicates the largest pitch chain that can be used. Also, the diameter of this shaft must be checked with the largest recommended bore that can be used in the sprocket selected. In general, the largest pitch gives the best drive, but limitations of sprocket diameters and necessity for quiet operation often dictate smaller pitches. For the quietest drive, the smallest pitch that will carry the load should be chosen. Large pitch chain should be used for slow speeds and larger power.

Table 17. Silent Chain Drive Data

Pitch, in.	Pitches per Foot	Chain Widths, in.	Approx. Weight per Foot, 1-in. Wide, lb	Small Sprocket Driver: Normal Revolutions per Minute	Maximum Revolutions per Minute *	Recommended Minimum No. of Teeth on Driver †
3/8	32	3/8–4	0.75	4000	6000	19
1/2	24	1/2–7	1.00	2600	4000	19
5/8	19.2	1–8	1.25	2300	3500	19
3/4	16	1–10	1.50	1650	2500	21
1	12	1 1/2–14	2.00	1300	2000	21
1 1/4	9.6	2 1/2–20	2.50	1000	1500	23
1 1/2	8	3–24	3.00	800	1200	23
2	6	4–30	4.00	600	900	25

* Under favorable conditions these values may be exceeded, and under unfavorable conditions they must be decreased.

† Smaller numbers of teeth on the driver may be used under certain conditions. For these drives, consult manufacturer.

Speed ratio is the basis of selection of driving and driven sprockets. The smaller sprocket is usually selected first on the following basis:

Speed Ratio	1 : 1	2 : 1	3 : 1	4 : 1	5 : 1
Number of Teeth of Small Sprocket	25–35	19–29	19–25	19–23	19–21

Ratios as high as 8 to 1 are possible, but the use of sprockets of less than 17 teeth or more than 120 teeth are to be avoided in the interests of long life and efficient operation. See Table 18, listing desirable sprocket combinations and giving various ratios.

Table 18. Ratio Table

Ratio	Pinion	Sprocket	Ratio	Pinion	Sprocket
1.00	25	25	2.71	21	57
1.09	23	25	3.00	19	57
1.10	21	23	3.04	25	76
1.11	19	21	3.3	23	76
1.19	21	25	3.62	21	76
1.21	19	23	3.80	25	95
1.32	19	25	4.00	19	76
1.52	25	38	4.13	23	95
1.65	23	38	4.52	21	95
1.81	21	38	4.56	25	114
2.00	19	38	4.95	23	114
2.28	25	57	5.00	19	95
2.48	23	57	5.43	21	114

Chain speed is determined by the equation

$$\text{Chain speed (fpm)} = \frac{\text{Number of teeth} \times \text{rpm} \times \text{pitch (in.)}}{12}$$

Chain width is determined on the basis of values of horsepower per inch of width of chain, given in Table 19. Then

$$\text{Chain width} = \frac{\text{Horsepower} \times K}{\text{Horsepower per inch of width}}$$

where K is the service factor, Table 16. Available chain widths may be obtained from the manufacturers' catalogs.

It is usually possible to arrive at two or more different sound design selections for the same set of data. Try to select a pitch and sprocket size that will allow a chain width of less than eight times the pitch. Manufacturers of silent chain should be consulted concerning drives involving unusual conditions.

Chain length is determined by either of the equations 1 or 2, page 15-76. Arrange centers for an even number of pitches where possible, thus avoiding the use of offset links in the chain.

(Continued on p. 15-80)

Table 19. Horsepower Rating of Silent Chain

(Per inch of Width)

3/8-in. Pitch
Maximum Width of Chain, 4 in.

No. of Teeth of Small Sprocket	Revolutions per Minute, Small Sprocket							
	500	1000	1500	2000	2500	3000	3500	4000
19	2.0	3.8	5.2	6.3	7.3	7.9	8.3	8.4
21	2.2	4.1	5.8	7.2	8.3	9.1	9.6	9.9
23	2.4	4.5	6.4	8.0	9.2	10	11	11
25	2.6	4.9	7.0	8.8	10	11	12	12
27	2.8	5.4	7.6	9.5	11	12	13	14
29	3.0	5.8	8.2	10.3	12	13	14	15
33	3.5	6.6	9.4	12	14	15	17	17
35	3.7	7.0	10	13	15	16	18	18
40	4.2	8.1	12	14	17	19	20	21
45	4.8	9.1	13	16	19	21	22	23

1-in. Pitch
Maximum Width of Chain, 14 in.

No. of Teeth of Small Sprocket	Revolutions per Minute, Small Sprocket							
	200	300	500	700	1000	1200	1500	1800
19	6	8	12	16	20	21	22	....
21	6	9	14	18	23	25	26	26
23	7	10	15	20	25	28	30	30
25	7	11	17	22	28	31	33	33
27	8	12	19	24	31	34	37	37
29	9	13	20	26	33	37	40	41
33	10	14	23	30	39	43	47	47
35	10	15	24	32	41	45	49	50
40	12	18	28	36	47	52	56	....
45	13	20	31	41	52	57	61	....

1/2-in. Pitch
Maximum Width of Chain, 7 in.

No. of Teeth of Small Sprocket	Revolutions per Minute, Small Sprocket							
	500	700	1000	1800	2000	2500	3000	3500
19	3	4	6	9	10	11	11	11
21	4	5	7	11	11	12	13	13
23	4	6	8	12	13	14	15	15
25	4	6	8	13	14	16	17	17
27	5	7	9	14	15	17	19	19
29	5	7	10	15	17	19	20	20
33	6	8	11	18	19	22	23	24
35	6	9	12	19	20	23	25	25
40	7	10	14	22	23	26	28	28
45	8	11	15	24	24	29	31	....

1 1/4-in. Pitch
Maximum Width of Chain, 20 in.

No. of Teeth of Small Sprocket	Revolutions per Minute, Small Sprocket							
	200	300	500	600	700	800	1000	1200
19	8	12	19	21	23	25	27	28
21	9	14	21	24	26	29	32	33
23	10	15	23	27	29	32	36	37
25	11	16	25	29	32	35	40	42
27	12	18	28	32	35	39	43	46
29	13	19	30	34	38	42	47	50
33	15	22	34	39	44	48	55	58
35	16	23	36	42	47	51	58	62
40	19	27	42	48	54	59	66	70
45	21	30	47	54	60	65	73	....

5/8-in. Pitch
Maximum Width of Chain, 8 in.

No. of Teeth of Small Sprocket	Revolutions per Minute, Small Sprocket							
	500	700	1000	1200	1800	2000	2500	3000
19	5	7	10	11	13	14	14	....
21	6	8	10	12	15	16	16	16
23	6	9	12	13	17	18	19	18
25	7	9	13	15	19	20	21	21
27	8	10	14	16	21	22	23	23
29	8	11	15	17	22	24	25	25
33	9	13	17	20	26	27	29	29
35	10	13	18	21	27	29	31	31
40	11	15	21	24	31	33	35	....
45	13	17	23	27	35	37	....	....

1 1/2-in. Pitch
Maximum Width of Chain, 24 in.

No. of Teeth of Small Sprocket	Revolutions per Minute, Small Sprocket							
	200	300	500	600	700	800	900	1000
19	12	17	25	28	31	32	33	34
21	13	19	29	32	35	37	39	39
23	15	21	32	36	39	42	44	45
25	16	23	35	40	44	47	49	52
27	18	25	38	43	48	51	54	56
29	19	27	41	47	52	56	59	60
33	22	31	47	54	60	64	68	70
35	23	33	50	57	63	68	72	74
40	26	38	58	66	72	78	84	....
45	30	43	65	74	81	86	90	....

3/4-in. Pitch
Maximum Width of Chain, 10 in.

No. of Teeth of Small Sprocket	Revolutions per Minute, Small Sprocket							
	500	700	1000	1200	1500	1800	2000	2500
19	7.4	10	12	14	16	17	17	....
21	8	11	14	16	18	19	20	19
23	9	12	16	18	20	22	22	22
25	10	13	17	20	23	25	25	24
27	11	14	19	22	25	27	28	28
29	12	16	21	24	27	29	30	30
33	13	18	24	27	31	34	35	35
35	14	19	25	29	33	36	37	37
40	15	22	29	33	38	41	42	42
45	18	24	32	37	42	45	46	....

2-in. Pitch
Maximum Width of Chain, 30 in.

No. of Teeth of Small Sprocket	Revolutions per Minute, Small Sprocket							
	100	200	300	400	500	600	700	800
19	11	21	29	35	40	43	45	....
21	12.5	23	32	40	42	50	52	....
23	13.5	26	36	44	51	56	59	60
25	14	28	39	49	56	62	66	68
27	16	30	43	53	62	68	73	75
29	17	33	46	58	67	74	79	82
33	20	37	53	66	77	85	91	94
35	21	40	57	70	82	91	97	100
40	24	46	65	81	94	103	110	113
45	27	51	72	90	105	115	121	....

Center Distance. Minimum center-to-center distance between sprockets is the sum of the sprocket diameters for drives, under normal conditions. Long centers are practical on drives where provision is made for adjustment for chain wear or "stretch."

SILENT CHAIN SPROCKETS.

$$\text{Pitch diameter} = \frac{\text{Pitch}}{\sin \dfrac{180°}{N}}$$

Outside diameter is approximately that of the pitch diameter. To determine the approximate diameter of the chain when wrapped around the sprocket, add 1.25 times chain pitch to pitch diameter.

$$\text{Approximate weight in pounds of steel pinions} = T^2 \times (F + 1) \times c$$

where T = number of teeth, F = face in inches, and c = pitch constant given below.

Pitch	3/8	1/2	5/8	3/4	1	1 1/4	1 1/2	2
	.003	.0045	.0063	.009	.016	.025	.035	.028

$$\text{Approximate weight in pounds of spoke-type sprockets} = T \times F \times C$$

where T = number of teeth, F = face in inches, and C = pitch constant given below.

Pitch	3/8	1/2	5/8	3/4	1	1 1/4	1 1/2	2
C	0.10	0.16	0.25	0.35	0.60	0.75	1.00	2.00

For approximate weight of *split* sprockets, add 25%.

PROTECTION AND LUBRICATION. Silent chain drives, wherever possible, should be operated in an oiltight casing with provision for lubrication.

When the chain speed is less than 1500 ft per min, a drip or sight-feed oiler may be used. The oil should be applied to the inner surface of the chain at the rate of about one drop per minute per inch of width.

When the chain speed is from 1500 to 3000 ft per min, bath lubrication is recommended. The oil is maintained at a level in the lower portion of the casing so that the chain, while running, will be immersed to a depth of approximately one-half the pitch.

The viscosity of the oil depends somewhat upon temperature conditions and the chain speed but in general should be a straight mineral oil of a viscosity of about SAE 30. Heavy-bodied oil such as SAE 50 or 60 may be used on very slow-moving chains. Greases of any kind should be avoided as they cannot at ordinary temperatures penetrate to the "joint" of the chain where lubrication is most needed.

ROPE DRIVING

Since electrical transmission of energy has largely displaced rope driving as a power transmission medium, only a brief discussion of rope drives is included here. More detailed information may be obtained from the eleventh edition of Kent's *Mechanical Engineer's Handbook, Design and Shop Practice* volume.

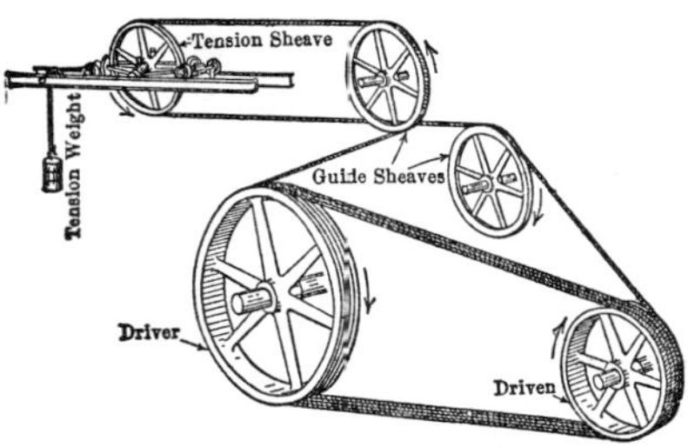

FIG. 1. Continuous rope drive.

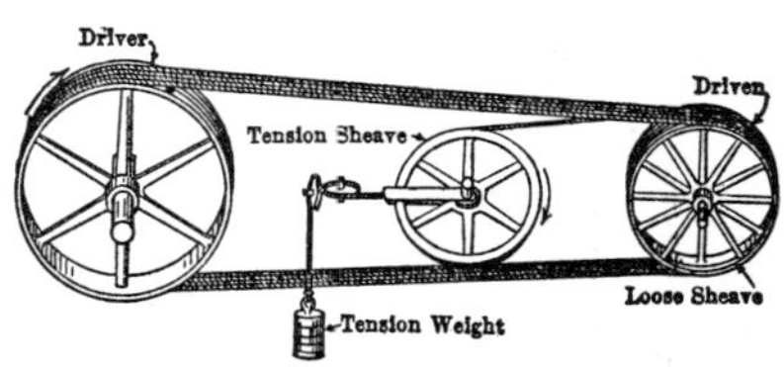

FIG. 2. Continuous rope drive with loose sheave.

26. MANILA AND COTTON ROPE DRIVES

SYSTEMS OF ROPE DRIVING. Two systems of rope driving are in use: (1) the American or continuous system (Figs. 1 and 2) in which a single rope passes over the sheaves several times, the slack being taken up by a tension carriage, which also transfers the rope from the outside groove of the driven sheave to the opposite outside groove of the driving sheave; (2) the English multiple system, comprising one or more separate ropes alongside of each other in the pulley sheaves.

The continuous system is adapted to drives where the shafts are not parallel, to vertical drives, and to drives where the rope is exposed to the weather. A disadvantage, however, is that the breakage of a single strand puts the entire drive out of service.

The arrangement of the tension carriage is important. The slack preferably should be taken up on the slack side of the drive at the driving sheave. (See Fig. 1.) If this is not possible, it may be taken up at the driven sheave. (See Fig. 2.) The outside groove of the driven sheave then should be a loose sheave, and the rope led to it from the tension sheave, whence it is returned to the opposite outside groove on the driving sheave. The weight applied to the tension carriage should be just sufficient to prevent any tendency for the rope to slip. It can be approximated from the value of the tension in the slack side, as calculated by formula 4.

The English multiple system, using separate parallel ropes, has no tension carriage but depends on the weight of the ropes to provide the necessary tension. The power transmitted by a given drive may be increased at any time by the installation of additional ropes. An advantage of this system is that the breakage of a single rope does not put the drive out of service. Disadvantages are the difficulty of maintaining uniform tension on the several ropes, with resultant excessive wear on the more heavily loaded ones; the necessity of protecting the drive from the weather; the inability to use it for vertical drives; and the requirement that the distance between sheaves must be sufficiently long to give the necessary tension on the ropes.

HORSEPOWER OF ROPE DRIVES. Let T = working tension in rope, pounds; T_1 and T_2 = the total tension in tight and slack sides of rope, respectively; t_1 and t_2 = effective tension in tight and slack sides of rope, respectively, pounds; t_c = tension due to centrifugal force; V and v = velocity of rope, feet per minute and feet per second, respectively; w = weight of rope per foot, pounds; d = diameter of rope, inches; S = breaking strength of rope, pounds. Then

$$T = (t_1 - t_2) \tag{1}$$

$$t_c = \frac{wv^2}{g} = \frac{wv^2}{32.2} \tag{2}$$

$$\text{Horsepower} = \frac{TV}{33{,}000} \tag{3}$$

Tension t_1 may be considered as made up of (1) tension doing useful work, (2) tension to balance the tension necessary for adhesion, and (3) tension due to centrifugal force. Tension t_2 may be considered as made up of (1) tension necessary for adhesion T_a and (2) tension due to centrifugal force t_c.

Tension on the tight side is assumed to be double that on the slack side. Then $T_a/T = {}^1/_2$, and $T_a = {}^1/_2\,T$. Since $T_a = t_2 - t_c$, $t_2 = t_c + {}^1/_2\,T$, and

$$T = t_1 - t_c - {}^1/_2\,T \tag{4}$$

from which

$$T = {}^2/_3\,(t_1 - t_c) \tag{5}$$

Substituting the value of T from 5 in eq. 3,

$$\text{Horsepower} = \frac{{}^2/_3\,(t_1 - t_c)V}{33{,}000} \tag{6}$$

According to the American Manufacturing Co., t_1 should be limited to ${}^1/_{35}\,S$, and S may be taken as $7000d^2$, whence $t_1 = 200d^2$. Substituting the values $t_1 = 200d^2$ and $t_c = wv^2/32.2$ in eq. 6, we have

$$\text{Horsepower} = \frac{{}^2/_3\,V\left(200d^2 - \dfrac{wv^2}{32.2}\right)}{33{,}000}$$

$$= 0.0000222\,V\left(200\,d^2 - \frac{wv^2}{32.2}\right) \tag{7}$$

For theoretical discussion of the effect of arc of contact and coefficient of friction on the transmission of power by belts, bands, ropes, etc., see Article 16 in this section.

Since the value of T_c increases with the velocity of the rope, it is obvious that there is a certain speed at which the maximum horsepower will be transmitted. For a tension in the tight side of 200 psi, this is in the neighborhood of 80 ft per sec as shown by Fig. 3.

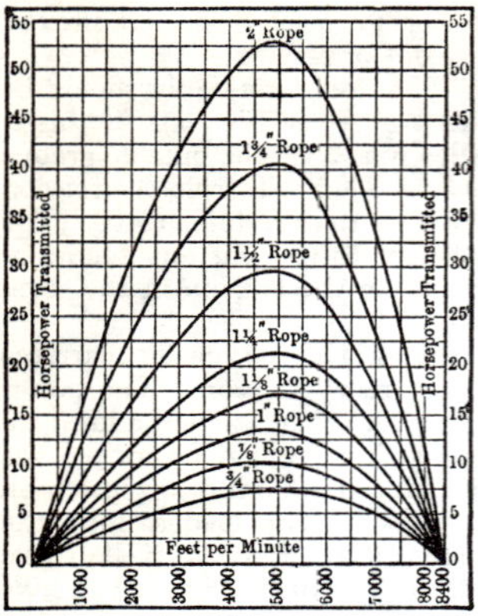

FIG. 3. Horsepower transmitted by manila rope.

Sag of Rope. The catenary of the driving side is not affected by the speed or by the diameter of the rope. The deflection of the rope between the pulleys on the slack side varies with each change of load or speed. The deflections D and D_1, or sag in the tight and slack sides, respectively, for horizontal drives may be approximated from eqs. 8 and 9, t_1 being taken as $200d^2$, and L being distance between sheaves, feet.

$$D = \frac{wL^2}{8t_1} \qquad (8)$$

$$D_1 = \frac{wL^2}{8t_2} \qquad (9)$$

The value of t_2 is found from eqs. 4 and 5. Equations 8 and 9 will closely approximate the sag on the driving side for both multiple and continuous systems, and also the sag on the slack side in the continuous system. In the multiple system the equation for the slack side is true only when the ropes are first spliced on the pulleys, as the stretch in new ropes, when they are put in motion, rapidly increases the sag.

SHEAVES. The larger the sheaves, the less the internal wear of the rope. Sheaves should not be less than forty times the diameter of the rope for economical wear, and as much larger as it is possible to make them. This rule applies also to idler and tension sheaves.

Figure 4 shows the form of the grooves in driving and driven sheaves, for the continuous system. The grooves for the multiple-rope system are of similar form, but deeper, and the angle of the grooves is 30 degrees, instead of 45 degrees. Figure 5 shows the form of grooves recommended for carrier and idler sheaves in the continuous-rope system. Grooves should be well polished to avoid wear on the ropes. They also should be free from sand- or blowholes, which tend to cut the rope rapidly.

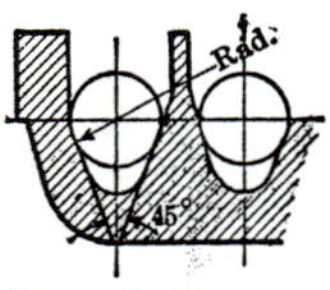

FIG. 4. Sheave grooves for continuous system.

ROPE MATERIALS. Cotton ropes are largely used in Great Britain but only to a limited extent in the United States where the preference is for manila rope. The American Manufacturing Co. states that cotton ropes are advantageously used as bands or cords on the smaller machine appliances. The fiber, being softer and more flexible than manila hemp, gives good results for small sheaves; but for large drives, where power transmitted is in considerable amounts, cotton rope, as compared with manila, is hardly to be considered, on account of the following disadvantages: It is less durable; it is injuriously affected by the weather, so that for exposed drives, paper-mill work, or use in water-wheel pits, it is unsatisfactory; it is difficult, if not impossible, to splice uniformly; even the best quality cotton rope is much inferior to manila in strength—the breaking strain of the highest grade being but 4000 × diameter² as against 7000 × diameter² for manila, whereas, for the transmission of equal powers, the cost of a cotton rope varies from one-third to one-half more than manila. See also Article 5, Section 6.

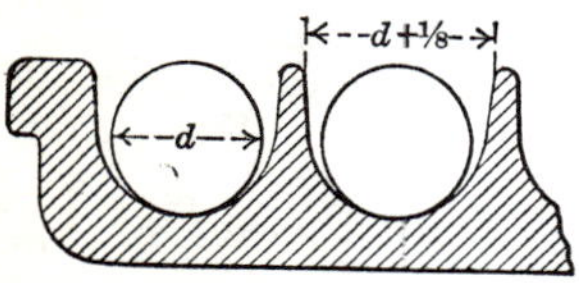

FIG. 5. Idler and carrier sheave grooves.

SHEAVE CENTER DISTANCES. The distance between the axes of the sheaves of rope drives, where the tight strands are on the bottom, depends on the sheave diameters. The slack strands should clear the tight lower strands by at least $0.1L$, where L = center-to-center distance of the sheaves. Where the tight strand is on top, L may be as great as 100 ft, provided that shafts and bearings are sufficiently rugged to withstand the strain of the extra weight of rope. It is not advisable to make the distance between the faces of the sheaves less than 25 ft.

27. WIRE ROPE TRANSMISSION

For data on the various types of wire rope available for power transmission, including construction and breaking strength, see pp. 6-09 to 6-20.

WIRE ROPE SHEAVES. Sheaves used with wire rope for the transmission of power are of the form shown in Fig. 6. The bottom of the sheave is filled with blocks of rubber or leather. The minimum diameter of sheave that should be used with 6 × 7 wire rope is $100d$, where d = diameter of rope. With 6 × 19 rope, the minimum sheave diameter should be $66d$. Sheaves must be set true on the shafts and at right angles to the line of transmission to avoid wear on the rope and loss of power. Guide sheaves and supporting sheaves should not be used unless absolutely necessary, for the same reasons. In designing wire rope transmissions to transmit a given horsepower, a choice frequently can be made between small sheaves running at high speed and large sheaves running at a lower speed. Since the wear on the rope always is greater in bending over small sheaves, economy in life of rope will be served by selecting the larger sheaves and lower speeds.

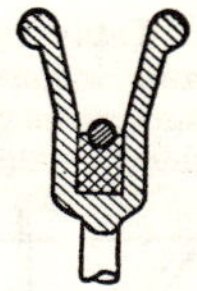

FIG. 6. Wire rope sheave groove.

FRICTION GEARING

FRICTION GEARING. If two cylinders with parallel axes are in contact and one of them rotates upon its axis, the friction between the surfaces will drive the other cylinder. The driving capacity depends upon the coefficient of friction of the material of the wheels,

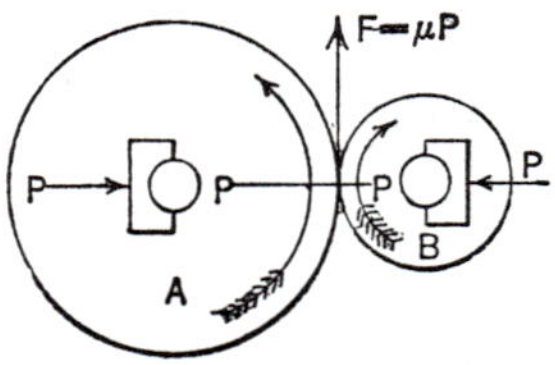

FIG. 1. Forces in plain friction gearing.

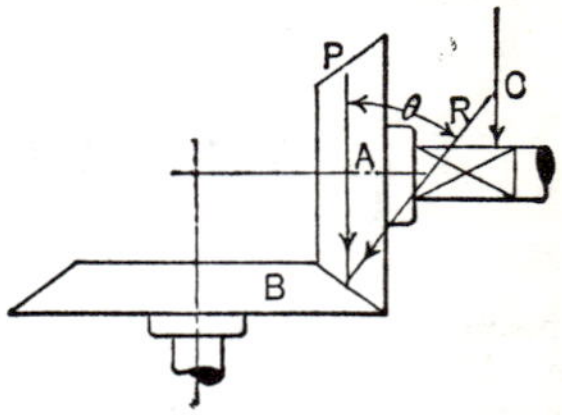

FIG. 2. Forces in bevel friction gearing.

the pressure with which they are held in contact, and the length of the cylinders. In order to secure good frictional conditions, one member only of a friction gearing is of metal; the other is of a composition giving a high coefficient of friction with metal. The composition wheel is always the driving member, and its slip on the driven tends to distribute the wear uniformly over its surface. Friction gearing cannot be used if positive driving is essential, as it is always liable to slip. It may be used where the power to be transmitted is not very great; where the speed is so high that toothed wheels would be noisy; where the shafts are frequently to be put in and out of gear or to have their relative directions of motion reversed; or where the velocity ratio is to be changed while the machinery is in motion.

POWER TRANSMITTED BY FRICTION WHEELS. The tangential force that can be transmitted by two friction gears is equal to the product of the total normal pressure and the coefficient of friction. Thus, in Fig. 1 let P = total normal pressure, pounds, and μ = the coefficient of friction; then the resultant tangential force $F = \mu P$. If V = velocity of the surface of the friction wheels, feet per minute, the horsepower transmitted = $\mu PV/33000$. If d = diameter of the driver, inches; l = length of the face, inches; w = allowable load per inch of face, pounds; N = revolutions per minute,

$$\text{Horsepower} = \frac{\mu wl \times 3.1416dN}{12 \times 33000} = 0.000008\mu wldN$$

In *bevel wheels*, Fig. 2, the resultant R of the applied force P presses the wheels together, and $R = P \div \cos\theta$. The velocity of the mean circumference of the driver may be taken as the velocity of transmission.

Table 1. Pressure and Friction Coefficient of Friction Gear Materials

Material	w	μ with Cast Iron	Material	w	μ with Cast Iron	μ with Aluminum
Cast iron	3000	.15	Straw fiber	150	.15	.16
Cork composition	50	.21	Leather fiber	240	.18	.18
Paper	150	.15	Tarred fiber	240	.09	.11
Rubber	100	.20	Sulfite fiber	140	.20	.19
Wood	150	.15	Leather	150	.08	.13

Table 1 gives allowable load per inch of face, w, for various materials, the values being approximately one-fifth of the load which would cause failure. The friction coefficients shown in the same table are approximately 60% of the maximum and represent safe design values with allowance for deterioration in service, decrease in normal pressure, etc.

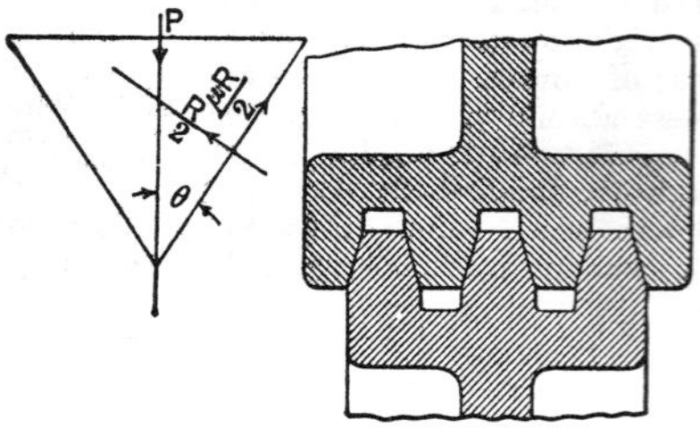

FIG. 3. Forces in grooved friction wheels.

WEDGE-FACED OR GROOVED FRICTION WHEELS give greater resistance to slipping with a given radial pressure. In Fig. 3, P = the radial force applied to the grooved surfaces, pounds; F = the tangential force transmitted, pounds; $R/2$ = the reaction on each face; and 2θ = the angle of the groove. The wedge is held in equilibrium by the force P, the reactions $R/2$, and the frictional resistance $\mu R/2$ due to the wedging action. Then $R = F/\mu$; $P = R \sin \theta + \mu R \cos \theta$; $F = \mu P/(\sin \theta + \mu \cos \theta)$. As before, the tangential force that can be transmitted by the gears is the product of the total normal pressure and the coefficient of friction. The number of wedges has no influence on the power transmitted but decreases the wear by distributing it over several surfaces.

The angle 2θ of the grooves varies from 30 to 40 degrees, but to avoid sticking it should never be less than 30 degrees.

VARIABLE-SPEED DRIVES

By E. L. Schwarz-Kast

Every machine has an optimum speed at which operation is most efficient or production maximum. With the same machine this optimum speed may differ for various circumstances or products over a wide range and can seldom be accurately predetermined. Frequently, speed changes are necessary during operation. Often a run with creeping speed prior to the actual operation is required for adjustments, or for positioning of tools and workpieces. These requirements were recognized in the early days of primitive machines, and much effort has been directed toward finding convenient methods of speed control. Step pulleys, cone pulleys, friction disks, variable pitch pulleys, and shifting gears were steps in this development.

The following check list is recommended as an aid in selection of the most appropriate system for a particular drive:

CHECK LIST

1. Necessary speed range.
2. Are several definite speed steps or continuous speed variations needed?
3. Speed regulation at any adjusted speed step:

$$\text{Speed regulation } (\%) = \frac{(\text{No load speed} - \text{full load speed})}{\text{Full load speed}} \times 100$$

Sometimes a constant speed or a very small regulation is required. At other times a falling speed characteristic is desirable so that overload will cause a drop in speed, thus protecting motor and gears against peak loads.

4. Torque-speed characteristics of the load should be known. The following three groups of loads are quite common: (*a*) torque constant over the entire speed range, the horsepower varying proportionally with the speed (for example, pure friction load or reciprocating compressors); (*b*) horsepower required constant, the torque varying inversely as the speed (e.g., some machine tools); (*c*) variable torque (e.g., fans, propellers, centrifugal pumps, where the torque varies as the square and the horsepower as the cube of the speed).

Torque and horsepower requirements of the load should be known over the entire speed range. Selection must be based on the maximum torque and the speed at which this torque is required. Furthermore, the variable-speed device must also start the load and has to carry at start the actual load torque and, in addition, the necessary torque required to accelerate the load. The following formula is convenient for calculation of the accelerating torque:

$$T = \frac{WR^2 \times \text{revolutions per minute}}{308 \times t}$$

where T = accelerating torque, pound-feet; WR^2 = total moment of inertia of the load, pound-feet^2, reduced to the power input shaft; and t = total time, seconds, for acceleration of the load from zero to full speed.

This check on starting torque is especially important for start of heavy inertia loads.

5. Overall efficiency, power losses, eventual heat dissipation.
6. Continuous or intermittent duty cycle.
7. Simplicity and reliability of operation and speed adjustment.
8. Life. As a rule the lifetime should not be less than that of the machine to be driven.
9. Space requirement.
10. Weight.
11. Speed of response and inertia of the device.
12. Cost.

For good planning a check against a list of this kind is recommended, simply to be sure that nothing of importance is overlooked. In some instances only a few of the items apply, and the others may be unimportant.

Variable-speed devices are divided into three groups: mechanical, hydraulic, and electrical. A brief description of some typical devices in each group is given in the following articles.

28. MECHANICAL VARIABLE-SPEED DRIVES

THE P.I.V. GEAR is a variable-speed device using a chain drive to transmit power from the input to the output shaft. It uses a unique self-tooth-forming chain that runs on the grooved faces of opposing conical surfaced wheels, as shown in Fig. 1. The hand-

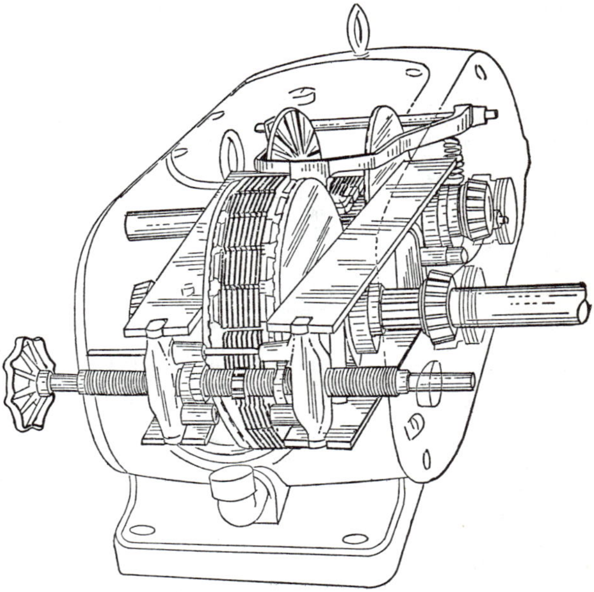

FIG. 1. P.I.V. speed changer. See Table 1.

wheel moves the control levers in and out of pivots. This movement alters the effective diameters of the opposing wheels and establishes a positive no-slip transmission that gives continuously variable speed. The entire gear is fully inclosed, and all moving parts operate in oil. The only external parts are the speed selection indicator and the control handwheel.

Table 1 shows the speed ratios, horsepower, and torque ratings of the various sizes, which are available in both horizontal and vertical units. Other standard assemblies include motorized and geared units for output reduction or step-up gearing, and also a bare unit to be built into a machine housing.

The P.I.V. gear is built by the Link Belt Co., Chicago, Ill.

Table 1. Speed and Horsepower Ratings of P.I.V. Gears

(Link-Belt Co., Chicago, Ill.)

Size	Ratio of Speed Variation	Maximum Speeds and Horsepower			Speeds and Horsepower Output per 100 Rpm		Output Torque, lb-in.		Maximum Horsepower of Motor
		Input, rpm	Output						
			Maximum Horsepower at Revolutions per Minute	Minimum Horsepower at Revolutions per Minute	Maximum Horsepower at Revolutions per Minute	Minimum Horsepower at Revolutions per Minute	At Maximum Speed	At Minimum Speed	
0	2 : 1	620	0.45 at 876	0.32 at 438	0.073 at 141	0.052 at 71	30	45	1/2
	3 : 1	620	0.45 at 1074	0.26 at 358	0.073 at 173	0.042 at 58	25	45	1/2
1/2	2 : 1	690	1.35 at 976	0.95 at 488	0.195 at 141	0.138 at 71	90	125	1 1/2
	3 : 1	690	1.35 at 1194	0.78 at 398	0.195 at 173	0.113 at 58	70	125	1 1/2
	4 : 1	690	1.35 at 1380	0.68 at 345	0.195 at 200	0.099 at 50	60	125	1 1/2
1	2 : 1	860	2.7 at 1216	1.9 at 608	0.314 at 141	0.221 at 71	140	200	3
	3 : 1	860	2.7 at 1491	1.6 at 497	0.314 at 173	0.186 at 58	115	200	3
	4 : 1	860	2.7 at 1720	1.4 at 430	0.314 at 200	0.163 at 50	100	200	3
	5 : 1	690	1.8 at 1545	0.80 at 309	0.261 at 224	0.116 at 45	75	165	2
	6 : 1	690	1.8 at 1692	0.73 at 282	0.261 at 245	0.106 at 41	65	165	2
2	2 : 1	860	4.5 at 1216	3.2 at 608	0.523 at 141	0.372 at 71	235	330	5
	3 : 1	860	4.5 at 1491	2.6 at 497	0.523 at 173	0.302 at 58	190	330	5
	4 : 1	860	4.5 at 1720	2.3 at 430	0.523 at 200	0.267 at 50	165	330	5
	5 : 1	690	2.7 at 1545	1.2 at 309	0.391 at 224	0.174 at 45	110	245	3
	6 : 1	690	2.7 at 1692	1.1 at 282	0.391 at 245	0.159 at 41	100	245	3
3	2 : 1	860	6.75 at 1216	4.8 at 608	0.785 at 141	0.558 at 71	350	495	7 1/2
	3 : 1	860	6.75 at 1491	3.9 at 497	0.785 at 173	0.453 at 58	285	495	7 1/2
	4 : 1	860	6.75 at 1720	3.4 at 430	0.785 at 200	0.395 at 50	250	495	7 1/2
	5 : 1	690	4.5 at 1545	2.0 at 309	0.652 at 224	0.29 at 45	180	405	5
	6 : 1	690	4.5 at 1692	1.8 at 282	0.652 at 245	0.26 at 41	165	405	5
4	2 : 1	690	9.0 at 976	6.4 at 488	1.304 at 141	0.928 at 71	580	820	10
	3 : 1	690	9.0 at 1194	5.2 at 398	1.304 at 173	0.754 at 58	475	820	10
	4 : 1	690	9.0 at 1380	4.5 at 345	1.304 at 200	0.652 at 50	410	820	10
	5 : 1	580	6.75 at 1295	3.0 at 259	1.163 at 224	0.517 at 45	325	730	7 1/2
	6 : 1	580	6.75 at 1422	2.8 at 237	1.163 at 245	0.482 at 41	300	730	7 1/2
5	2 : 1	580	13.5 at 820	9.5 at 410	2.328 at 141	1.638 at 71	1035	1470	15
	3 : 1	580	13.5 at 1005	7.8 at 335	2.328 at 173	1.345 at 58	845	1470	15
	4 : 1	580	13.5 at 1160	6.8 at 290	2.328 at 200	1.172 at 50	735	1470	15
	5 : 1	500	9.0 at 1120	4.0 at 224	1.8 at 224	0.80 at 45	505	1130	10
	6 : 1	500	9.0 at 1224	3.7 at 204	1.8 at 245	0.74 at 41	460	1130	10
6	2 : 1	500	22.5 at 708	15 at 354	4.5 at 141	3.2 at 71	2010	2850	25
	3 : 1	500	22.5 at 867	13 at 289	4.5 at 173	2.6 at 58	1645	2850	25
	4 : 1	500	22.5 at 1000	11.3 at 250	4.5 at 200	2.3 at 50	1425	2850	25
	5 : 1	450	18 at 1005	8 at 201	4 at 224	1.8 at 45	1115	2500	20
	6 : 1	450	18 at 1104	7.3 at 184	4 at 245	1.6 at 41	1020	2500	20

VARIABLE-PITCH PULLEYS. The Reeves variable-speed transmission (Fig. 2) consists of two pairs of cone-shaped disks mounted on spline shafts. The disks are carried in bearings on the levers *V*, pivoted at *P*. The handwheel actuates the right- and left-hand screws, causing the levers to rotate about *P*, and thus causing the disks of one pair to approach and those of the other pair to recede from each other as desired. One shaft is connected to the input shaft; the other is connected to the driven shaft. A rubber belt with blocks of hard wood bolted to each face of the belt connects the two pairs of disks. The ends of these blocks are finished at the same angle as that of the disks. The driving force is transmitted by friction between the blocks and the disk faces. When the disks of one pair are widely separated, the diameter of the circle of contact with the blocks is small; it is correspondingly larger on the other pair. By varying the distance between the disks, any desired speed ratio between the driving and the driven shaft can be obtained, within the range of the transmission.

These transmissions (Reeves Pulley Co., Columbus, Ind.) are available in both horizontal and vertical types and both open and enclosed. Capacities range up to 100 hp. Table 2 lists data for selection of the proper size for a given set of conditions. As may be seen from the table, the horsepower rating varies with the speed of the driven shaft. Consequently, for constant-horsepower drives requiring full horsepower at the lowest speed, selection of proper size has to be based on the horsepower requirement at the lowest speed.

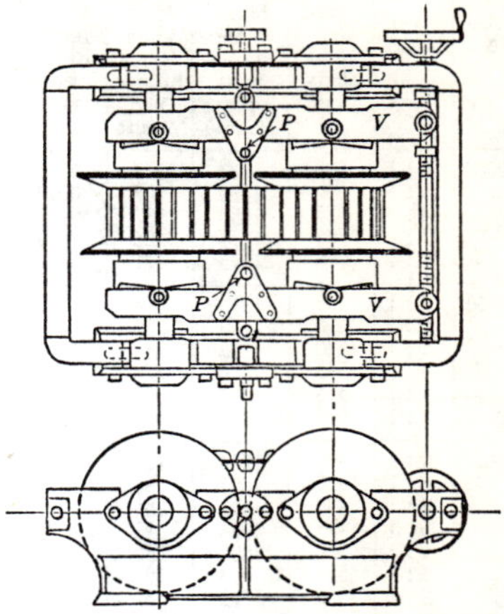

FIG. 2. Reeves variable-speed transmission. See Table 2.

The Reeves Vari-Speed pulley (Fig. 3) consists of an assembly of two, facing cone-shaped disks, one stationary, the other sliding laterally, and a compression spring. The whole assembly is mounted on any standard motor shaft. A motor base has an adjusting handwheel by which the motor and disk assembly can be moved forward or backward, and a Reeves V-belt transmits the power to the driven machine. When the motor is nearest to the driven pulley, the V-belt is at largest contact arc over the disks and maximum speed is developed. Turning the handwheel moves the motor away from the driven pulley, the V-belt runs over a smaller arc of contact, and the output speed is reduced. Capacities range from 1/8 to 15 hp at 1800 rpm motor speed, with speed ratios up to 4 : 1.

The Vari-Speed Jr., with capacities up to 1 1/2 hp, operates on a similar principle but employs standard A- or B-section V-belts instead of special wide belts.

The Vari-Speed motor drive, Fig. 4, combines a standard constant-speed motor, a speed-control mechanism, and, when required, speed-reduction gears. Of the two disks on the motor shaft one is fixed to the shaft, the other being axially adjustable by a handwheel operating through a screw. One of the disks on the variable-speed shaft is fixed to its shaft, the other being axially movable against a spring as shown in Fig. 4. The spring keeps the disk in proper contact with the V-belt,

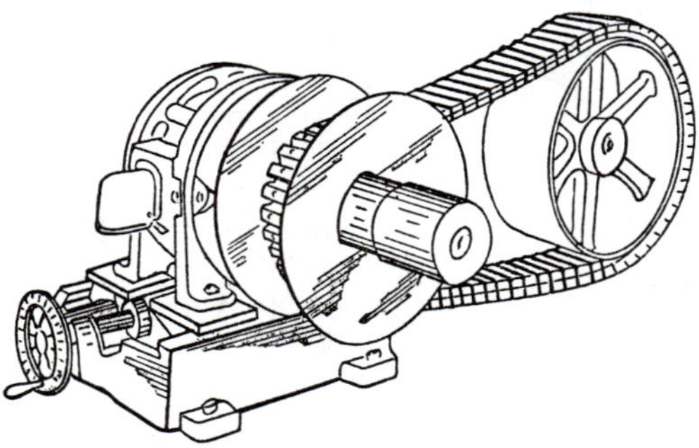

FIG. 3. Variable-speed pulley. (Courtesy of Reeves Pulley Co.)

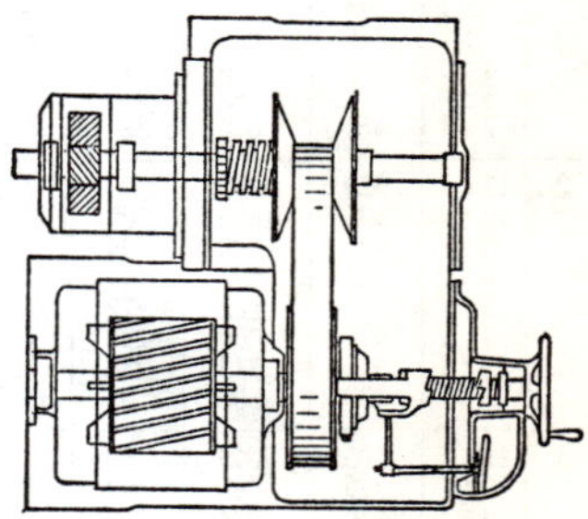

FIG. 4. Combined variable-speed and motor drive. (Courtesy of Reeves Pulley Co.)

providing correct belt tension. This unit can be furnished in vertical or horizontal types, in capacities from 1/4 to 20 hp, and with speed variation up to 6 : 1 when a single-speed motor is used. A two-speed motor extends the speed range to 12 : 1.

Vari-Pitch drive (Allis-Chalmers Mfg. Co., Milwaukee) consists of a Texrope V-belt drive with one or both sheaves arranged so that the grooves can be adjusted to provide a larger or smaller pitch diameter, producing an increase or decrease in speed of the driven shaft. The range of speed variation is from 15 to 20% per sheave at capacities from 1 to 125 hp and larger. For drives requiring greater speed range, wide-range Vari-Pitch sheaves have been developed which provide up to 100% variation at capacities from fractional up to 30 hp.

Vari-Drive (U. S. Electrical Motors, Inc., Chicago, Ill.) is a simple squirrel-cage motor built integral with a variable speed Vari-belt transmission. This transmission consists of a pair of dual disks, and the speed is changed by varying the pitch diameter of the disks.

Table 2. Sizes, Speed Ratios and Capacities of Reeves Variable-speed Transmissions

(Reeves Pulley Co., Columbus, Ind.)

Size No.	Speed Range	Recommended Speed Constant Shaft, rpm	Variable-Speed Shaft: Maximum Revolutions per Minute	Variable-Speed Shaft: Minimum Revolutions per Minute	Horsepower Output: Maximum Revolutions per Minute	Horsepower Output: Minimum Revolutions per Minute
0000	6 : 1	605	1482	247	0.25	0.10
	5 : 1	640	1433	286	0.27	0.12
	4 : 1	675	1350	337	0.30	0.15
	3 : 1	705	1219	406	0.35	0.20
	2 : 1	740	1043	521	0.39	0.27
000	8 : 1	465	1316	164	1.12	0.40
	6 : 1	490	1200	200	1.20	0.49
	5 : 1	520	1165	233	1.35	0.60
	4 : 1	545	1090	272	1.50	0.75
	3 : 1	575	994	331	1.73	1.00
	2 : 1	600	845	422	1.93	1.37
00	10 : 1	400	1264	126	1.4	0.44
	8 : 1	425	1202	150	1.5	0.53
	6 : 1	450	1102	184	1.6	0.65
	5 : 1	475	1064	213	1.8	0.80
	4 : 1	500	1000	250	2	1
	3 : 1	525	908	303	2.3	1.33
	2 : 1	550	775	387	2.6	1.85
0	10 : 1	390	1232	123	2.2	0.69
	8 : 1	410	1160	145	2.4	0.85
	6 : 1	430	1050	175	2.6	1.06
	5 : 1	445	996	199	2.9	1.3
	4 : 1	465	930	232	3.24	1.62
	3 : 1	485	840	280	3.68	2.12
	2 : 1	500	705	352	4.15	2.94
1	12 : 1	350	1211	101	3.2	0.925
	10 : 1	365	1153	115	3.45	1.09
	8 : 1	380	1070	134	3.7	1.3
	6 : 1	395	966	161	4	1.63
	5 : 1	415	925	185	4.5	2
	4 : 1	430	860	215	5	2.5
	3 : 1	445	768	256	5.7	3.3
	2 : 1	460	648	324	6.4	4.5
2	12 : 1	320	1107	92	4.75	1.37
	10 : 1	335	1058	105	5.1	1.6
	8 : 1	350	990	124	5.5	1.95
	6 : 1	365	894	149	6	2.45
	5 : 1	385	862	172	6.7	3
	4 : 1	400	800	200	7.5	3.75
	3 : 1	415	715	238	8.5	4.9
	2 : 1	430	606	303	9.6	6.8
3	12 : 1	305	1055	89	7	2
	10 : 1	315	995	99	7.5	2.3
	8 : 1	330	934	116	8.1	2.8
	6 : 1	345	845	141	8.8	3.5
	5 : 1	360	800	160	10	4.5
	4 : 1	375	750	188	10.8	5.4
	3 : 1	390	670	223	12.5	7.25
	2 : 1	405	571	285	14	10
4	12 : 1	285	986	82	9.4	2.7
	10 : 1	295	932	93	10	3.16
	8 : 1	310	877	109	11.25	3.9
	6 : 1	320	784	130	11.5	4.7
	5 : 1	335	750	150	13	5.8
	4 : 1	350	700	175	14.5	7.25
	3 : 1	360	621	207	16.5	9.5
	2 : 1	375	528	264	18.75	13.3

Size No.	Speed Range	Recommended Speed Constant Shaft, rpm	Variable-Speed Shaft: Maximum Revolutions per Minute	Variable-Speed Shaft: Minimum Revolutions per Minute	Horsepower Output: Maximum Revolutions per Minute	Horsepower Output: Minimum Revolutions per Minute
5	16 : 1	235	940	58	10.3	2.58
	14 : 1	245	916	65	10.8	2.88
	12 : 1	255	880	73	11.5	3.3
	10 : 1	270	850	85	12.5	3.95
	8 : 1	285	800	100	13.6	4.8
	6 : 1	295	722	120	14.5	5.9
	5 : 1	310	695	139	16.25	7.25
	4 : 1	320	640	160	18.25	9.12
	3 : 1	335	579	193	21	12
	2 : 1	345	486	243	23	16.3
6	16 : 1	210	840	52	12	3
	14 : 1	220	822	58	13	3.5
	12 : 1	235	813	67	14	4
	10 : 1	245	774	77	15	4.7
	8 : 1	255	720	90	16.5	5.8
	6 : 1	265	650	108	17.5	7.1
	5 : 1	280	627	125	19.5	8.7
	4 : 1	290	580	145	21.5	10.7
	3 : 1	300	519	173	25	14.5
	2 : 1	310	437	218	28	20
6 1/2	8 : 1	240	680	85	22	7.8
	6 : 1	250	612	102	23.4	9.5
	5 : 1	260	582	116	26	11.6
	4 : 1	270	540	135	29.5	14.7
	3 : 1	280	485	162	33.5	19.3
	2 : 1	290	410	205	37	26
6 3/4	8 : 1	240	680	85	29.5	10.4
	6 : 1	250	612	102	31	12.65
	5 : 1	260	582	116	35	14.5
	4 : 1	270	540	135	39	19.5
	3 : 1	280	485	162	44	25.4
	2 : 1	290	410	205	50	35
7	8 : 1	225	632	79	38	13.4
	6 : 1	235	576	96	41.5	17
	5 : 1	245	545	109	45	20
	4 : 1	255	508	127	51	25.5
	3 : 1	265	459	153	58	33.5
	2 : 1	275	388	194	66	46.5
7 1/2	8 : 1	225	632	79	42	14.5
	6 : 1	235	576	96	46.5	19
	5 : 1	245	545	109	50	22
	4 : 1	255	508	127	57	28.5
	3 : 1	265	459	153	65	37.5
	2 : 1	275	388	194	75	53
7 3/4	8 : 1	205	580	72.5	49	17.3
	6 : 1	215	525	88	53	21.5
	5 : 1	225	505	101	59.5	24.4
	4 : 1	235	470	117	68.5	34.2
	3 : 1	245	424	141	78	45
	2 : 1	255	360	180	87	62

By turning a handwheel the required disk diameter, the proper distance of the disks and the correct belt tension are simultaneously adjusted. This drive is rated in sizes up to 10 hp and a speed variation up to 7 : 1.

Variable V planetary speed selector (Speed Selector, Inc., Cleveland, Ohio) uses a planetary system, with V-belts and four variable pitch sheaves instead of gears. By the planetary motion combined with the variable-pitch V-belt sheaves, any speed from zero to 800 rpm can be provided at constant torque. Three different sizes are available: 0.5-hp, 1-hp, and 2-hp output at maximum speed of 800 rpm. The speed selector can be installed at any convenient location where a belt from the speed selector to the motor, and another to the driven shaft, can be provided.

Lewellen. The Lewellen Manufacturing Co., Columbus, Ind., has developed variable-pitch pulleys and variable-speed transmissions consisting of one set of pitch-adjustable V-belt disks. The V-belt is a special double block belt. The device is rated in sizes from fractional to 60 hp, vertical or horizontal arrangement, and open or enclosed.

The All-Speed selector (Worthington Pump and Machinery Corp., Harrison, N. J.) uses a combination of two sets of variable-pitch V-belt transmissions. Two sizes, A and C, are available. Model A is rated from 1/4- to 3/4-hp output and has a ratio up to 16 : 1, with a speed range from one-eighth to twice the motor speed. Model C is rated from 1- to 3-hp output and has a ratio up to 9 : 1, with a speed range from two-ninths to twice the motor speed. Provision is made for maintaining the proper belt tension. The belts are Goodyear stretchless steel cable V belts.

Lovejoy. The Lovejoy Flexible Coupling Company, Chicago, Ill., developed variable-pitch V-belt pulleys from fractional up to 8 hp and a maximum speed ratio of 3 : 1. The device consists of a spring-operated variable-pitch pulley, a V-belt, and an adjustable motor base.

Another type is the Select-O-Speed transmission of the same company. Two variable-pitch pulleys are mounted on a common shaft, which runs in ball bearings. The pitch of both pulleys can be changed simultaneously by a common control lever. This device can be inserted in any ordinary V-belt drive and gives a speed ratio up to 10 : 1. It is available in sizes from fractional up to 7 1/2 hp.

VARIABLE-RATIO ROLLING CONTACT DRIVES. The Graham variable-speed transmission has a range up to 1 hp. Its operation is based on the principle of a ring of fixed diameter in contact with conical rollers, the speed ratio depending on the ratio of the diameter of the ring to the diameter of the rollers at the point of contact. Figure 5 shows a diagram of the drive. Three conical rollers *A* spaced 120 degrees apart are carried in the carrier *B*. The rollers *A* bear with fixed pressure against the contact ring *C* which surrounds the carrier. The position of *C* can be varied axially by means of the control cam *D*; the slot in the cam is spiral in shape to give a linear relation between speed and cam-shaft movement, thus allowing the dial to be calibrated in revolutions per minute, with a uniform scale. The carrier *B* is attached to the drive or input shaft. The rollers *A* carry, at their small ends, pinions *F* which mesh with an internal gear *G*, attached to the driven or output shaft. In operation, as the carrier *B* is rotated, it causes the rollers *A* to revolve around the center line of the carrier. The friction between the rollers and the contact ring *C* causes the rollers to rotate about their own axes and in turn to drive the output shaft through the pinions and gear, *F* and *G*. The rate of rotation of the rollers, and consequently the speed of the output shaft, depends on the position of the contact ring *C*. The speed of the driven shaft is lowest when the ring is in contact with the smallest diameter of the rollers, and highest when it is in contact with the largest diameter. An indicator on the front of the case shows the position of the contact ring and the speed ratio at which the drive is operating.

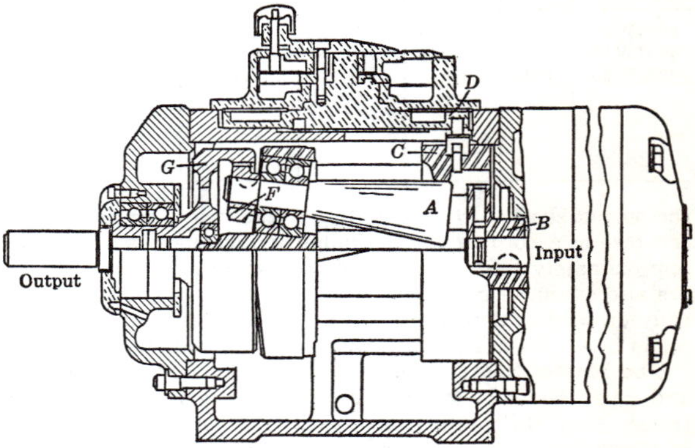

FIG. 5. Variable-ratio rolling contact drive. (Courtesy of Graham Transmissions, Inc.)

Table 3 lists the speed range and the capacity of drives up to 1 hp. In the standard unit, the speed variation is from maximum through zero and slightly into reverse. In the B style (which differs from the standard only in the number of teeth in the pinions and ring gear), equal speeds are had, forward and reverse, without reversing the motor. The drive may be operated by a built-in motor or by a motor coupled to the input shaft. It is also made with built-in-spur-geared heads on the output end which reduce the speed of

Table 3. Speed, Torque Range, and Overall Dimensions of Graham Variable-speed Transmissions

(Graham Transmissions, Inc., Milwaukee, Wis.)

Model	Speed, rpm		Torque, in.-lb		Motor Horsepower	
	Input	Output Range	At Maximum Speed	At Minimum Speed		
15 40 40L	3600 3600 1800	1100/0 1100/0 450/0	3.5 30 35	7.5 60 60	1/15 to 1/8 1/4 to 3/4 1/4 to 1/2	For coupled motor or offset drive.
15M 40T	3600 3600	1100/0 1100/0	3.5 45	7.5 90	1/15 to 1/8 1/2 to 1	With built-in motor especially designed to form an integral part of the drive.
15MR5 40MR5 40LMR5 40TMR7 15MS2.5 40MS2.8	3600 3600 1800 3600 3600 3600	220/0 220/0 90/0 160/0 2750/0 3100/0	15 140 160 300 1.3 10	35 280 300 600 2.8 20	1/15 to 1/8 1/4 to 3/4 1/4 to 1/2 1/2 to 1 1/15 to 1/8 1/4 to 3/4	With built-in motor and built-in spur reduction or step-up.
15MW20 15MW40	3600 3600	55/0 28/0	35 60	75 120	1/15 to 1/8 1/15 to 1/8	With built-in motor and built-in worm reduction, Model 15. Output shaft may extend horizontally to either side, or vertically up or down. Additional ratios available—6 : 1, 10 : 1, 15 : 1, 30 : 1, and 60 : 1.
40MW6 40MW18 40MW60	3600 3600 3600	190/0 60/0 19/0	120 360 800	250 720 1700	1/4 to 3/4 1/4 to 3/4 1/4 to 3/4	With built-in motor and built-in worm reduction, Model 40. Output shaft extends both sides horizontally or vertically up or down. Additional ratios available—9 : 1, 12 : 1, 36 : 1 and 47 : 1.

the output shaft to $1/_7$, $1/_5$, or $1/_{2.8}$ of the values given by the standard drives and increase the torque accordingly. A similar arrangement may be had with step-up gearing with approximately 2.5 : 1 ratio and corresponding decrease in torque. Worm-geared heads are also available in various ratios, with the output shaft extending horizontally or vertically upward or downward. Table 3 also lists the main overall dimensions. The drive is made by Graham Transmissions, Inc., Milwaukee, Wis.

The Master Speed Ranger (Master Electric Co., Dayton, Ohio) is an all-metal design and consists mainly of a pair of driver cones and a pair of driven cones with a metal ring transmitting the power from the driving cones to the driven cones. Each pair of cones consists of one stationary and one sliding cone. The speed is varied by moving the sliding cones, thus changing the radii of contact with the transmitting ring. The whole unit operates in oil. The units are available from $1/_6$ to 5 hp and a speed ratio up to 8 : 1, with some units even 12 : 1 and 15 : 1.

29. HYDRAULIC VARIABLE-SPEED DRIVES

There are two main systems of hydraulic variable-speed devices, the hydrostatic and the hydrodynamic types.

HYDROSTATIC TYPES use a variable displacement pump as a source of fluid power and a constant displacement or variable displacement hydraulic motor. Both pump and motor are of the piston type.

Power is transmitted by the fluid as a result of the pump delivery and the hydraulic pressure. The kinetic energy of the fluid is not used since it is very small because of the low velocity and small quantity involved; also, there is no change in velocity, the fluid leaving the motor with the same velocity as that with which it entered.

The main characteristics of the positive displacement piston-type variable-speed devices are as follows:

1. Output revolutions per minute varies directly as the volume rate of flow of fluid.
2. Assuming constant fluid pressure, motor torque is constant and horsepower varies proportionally with speed.
3. Torque is directly proportional to fluid pressure.
4. With a given volume of fluid, motor horsepower varies directly with fluid pressure. The drive can be laid out for constant horsepower, constant torque or any combination of the two.

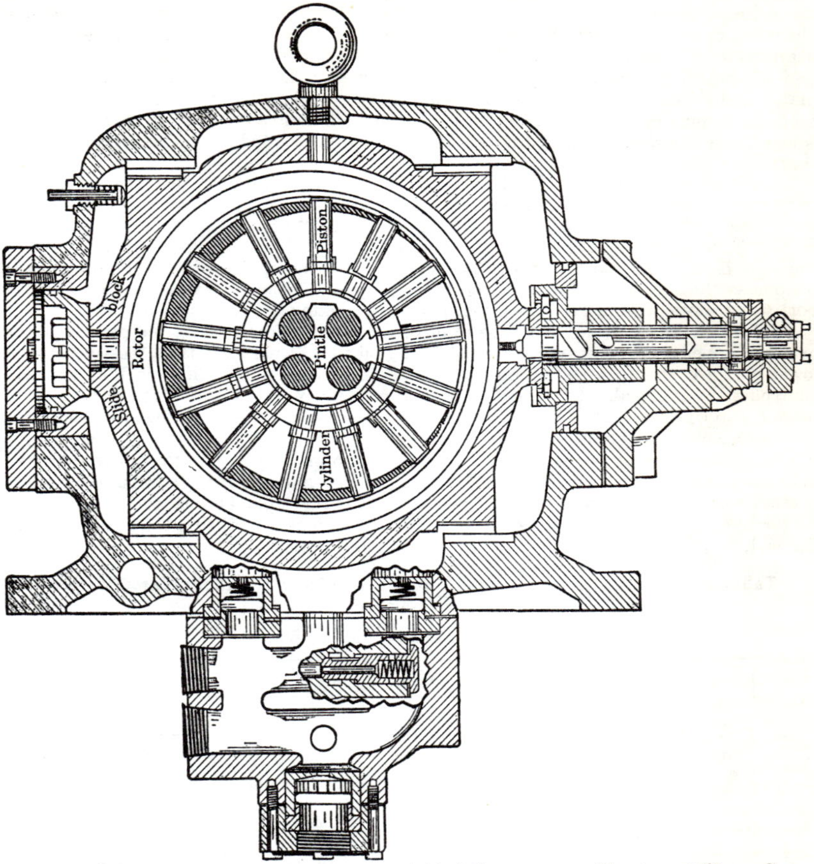

FIG. 6. Cross section through radial-piston variable-delivery pump. (Courtesy of Oilgear Co.)

5. Minimum speed depends on many factors, such as nature of source of fluid power, leakage, number of pistons, and type of load. In general, the usable minimum speed is somewhere between 1 and 30 rpm.
6. Torque efficiency is 80% or more and can easily reach 90% or more at rated pressure. It is substantially constant from stalled to full rated speed.
7. Although the delivery of a pump may be varied from zero to maximum displacement, it is not possible, owing to mechanical inefficiencies, to vary the displacement of a piston motor more than about 4 : 1. Therefore, the combination variable-displacement motor and constant-displacement pump is used in exceptional cases only.
8. The most outstanding feature of piston-type devices is their low inertia. The ratio of rated torque over inertia can hardly be exceeded by other methods. This system is one of the best for applications where rapid response or greatest reversal rate are essential.
9. This hydraulic drive has an inherent overload protection; it can be safely stalled under load.

A radial piston variable-delivery pump (Oilgear Co., Milwaukee, Wis.) is shown in Fig. 6. It consists essentially of a heavy-duty, multiple, radial rolling piston pumping unit, which delivers any volume from zero to maximum in either direction; an integral device for actuation of the slide block manually, hydraulically, or electrically to select the volume and direction of oil flow; a small low-pressure gear pump to supply oil for control operation, to lubricate the radial piston pump when operating in neutral and partially to supercharge the radial piston pump; and a drive shaft and housing assembly and associated relief and check valves, all compactly arranged in a sturdy case. The pintle is pressed into the case. The cylinder rotates on the closely fitted tapered pintle. The shaft drives the cylinder through a floating coupling. The pistons are closely fitted in the cylinder. Reaction rings in the rotor confine the pistons. The rotor is mounted on antifriction bearings in an adjustable-stroke slide block and is rotated with the cylinder through contact of the rolling piston against the reaction ring. The slide block is guided on four plates and carries the rotor from a point concentric with the cylinder to any eccentric point up to maximum stroke on either side of the cylinder center line. Two opposing pistons actuate the slide block hydraulically to a position selected by the helix on the control shaft.

In operation, the shaft drives the cylinder clockwise, and the spot contacts of the convex piston heads on the concave reaction rings cause the rotor to rotate at the same speed as the shaft. When the center lines of the cylinder and the rotor coincide, no reciprocating motion is imparted to the pistons as the unit rotates. Therefore no oil is delivered. As the slide block is moved to the left (see Fig. 6), reciprocating motion is imparted to the pistons so that those passing over the lower port in the pintle are delivering oil to that port while those passing over the upper ports are sucking or filling up with oil.

As the slide block is moved to the right of the cylinder center line, the pistons passing over the upper port in the pintle are delivering oil to that port and those passing over the lower port are sucking or filling up with oil. The position of the slide block is accurately controlled hydraulically through the helix-type control shaft and vertical hole in the large control piston. The oil delivery can be varied over a stepless range in either direction from zero to maximum.

The construction of the constant-displacement motor is similar to that of the variable-delivery pump except that the slide block is omitted and the rotor is mounted on antifriction bearings in the motor case and set at maximum stroke on one side of the cylinder center line. The available sizes of Oilgear transmissions and their ratings are shown in Table 4.

Table 4. Horsepower and Torque Ratings of Oilgear Hydraulic Variable-speed Transmissions

(The Oilgear Co., Milwaukee, Wis.)

Size Pump	Size Motor	Output							Input	
		Horsepower		Torque, in.-lb			Speed, rpm		Speed, rpm	Horsepower at Normal Torque Full Speed
		Normal,* Continuous	Peak,* 10% of Time	Starting	Normal *	Peak *	Maximum, ±2%	Minimum		
211	211	2.3	3.4	188	135	202	1090	5–30	1140	2.8
411	411	4.2	6.2	341	246	367	1090	5–30	1140	5.1
811	811	8.5	12.3	680	490	732	1090	5–30	1140	10.3
1211	1211	14.5	21.2	1170	840	1260	1090	5–30	1140	17.4
2017	2017	21.7	31.8	1815	1295	1960	1060	5–30	1140	25.9
3517	3517	34.5	50.1	3800	2715	4110	800	5–30	860	41.1
6017	6017	63.5	92.4	7010	5010	7570	800	5–30	860	75.5
10017	10017	103.0	150.0	11400	8125	12300	800	5–30	860	122.6

* Assumes 4 ft of pipe between units. Values reduce by amount of additional pipe line losses if longer pipes are used.

An axial piston variable-speed transmission (Waterbury Tool Division of Vickers, Inc., Waterbury, Conn.) is shown in Fig. 7. It consists of a variable-displacement pump, called the A-End, and a constant displacement motor, called the B-End. These ends are of the positive displacement type, using oil as the hydraulic fluid, and may be closely coupled (Type C), as shown, or connected by external piping (Type K) at any angle, at any level, and over considerable distances.

The A-End is driven at constant speed by an electric motor or other prime mover. The output of B-End speed can be varied continuously from zero to the speed of the A-End in either direction. The rated torque can be transmitted at all speeds down to 1 rpm and can even be exerted at zero speed. Full torque loads can be rotated from a standstill position. These units are particularly adapted to carry accelerating and decelerating loads, and momentary shock or overloads.

The output shaft may be brought up to speed, or reversed, in a few seconds or less while carrying the load, according to the type of control used, which may be manual, rotary (as shown), sliding, lever, selsyn, servo, follow-up, and other hydraulic, electrical, or mechanical types.

In the A-End, an odd number of axially displaced pistons and ball-ended connecting rods are driven in and out of cylinders in a rotating cylinder barrel by a socket ring revolving in a plane inclined to the driving shaft. The socket ring is coupled to the shaft through a universal joint. The piston thrust is taken up by roller thrust and radial bearings mounted in a trunnioned box, which can be tilted by means of the controls. When the plane of the socket ring is vertical, the pump is in the neutral or off stroke position, and no pumping occurs. Movement of the control causes the tilting box and socket ring

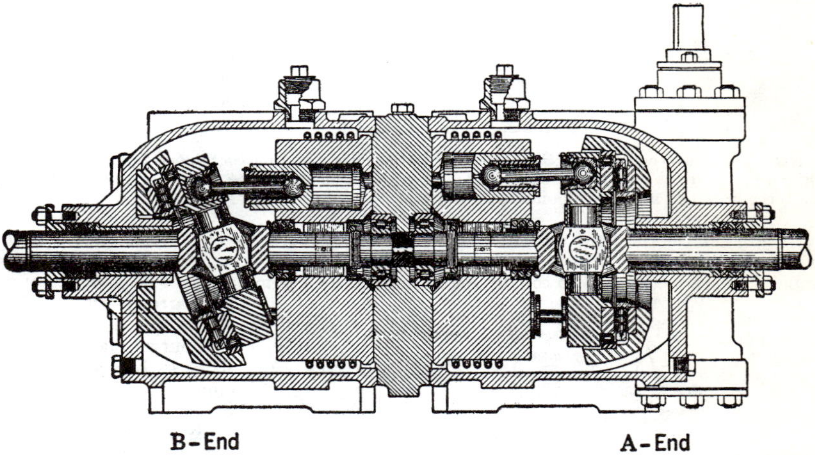

Fig. 7. Section through axial-piston variable-speed transmission. (Courtesy of Waterbury Tool Division of Vickers, Inc.)

to tilt at an angle to the shaft. The pump is then on stroke, and pumping occurs in proportion to the angle of tilt. The tilting angle is limited to 20 degrees. The B-End is similar to the A-End, but it has the box fixed at 20 degrees.

Oil is pumped from elongated ports in the A-End cylinders into a common pressure port in a member, called the valve plate, against which the face of the cylinder barrel rotates. This pressure port is connected directly through the valve plate to the B-End cylinder barrel for closely coupled units, as shown, or by external piping to the B-End valve plate. The valving of the ports is done automatically by the valve plate design. The pressure oil pumped into the B-End acts on the pistons to impart a thrust through the connecting rods to the inclined socket ring, and, because of the inclination, causes the socket ring, and hence the output shaft, to rotate. The oil in the B-End cylinders is carried around to the low-pressure port of the valve plate and is pumped back into the A-End by the return stroke of the B-End pistons. The output speed depends on the volume of oil being pumped. Table 5 shows the available sizes, speeds, torque, and horsepower ratings of the Waterbury variable-speed transmission.

The transmissions described are representative of the field. Among other prominent manufacturers of this type of equipment are the following:

Vickers, Inc., Detroit, Mich. Variable-delivery piston-type pumps and constant- and variable-displacement motors from fractional to several hundred horsepower.

Northern Pump Company, Minneapolis, Minn. Variable-delivery piston-type pumps and constant-displacement motors from 1 to several hundred horsepower.

Sundstrand Pump Division, Rockford, Ill. Small sizes, variable-delivery piston-type pumps and constant-displacement motors up to 2 hp.

HYDRODYNAMIC TYPE. One development of this type is the variable-speed hydrodynamic fluid drive of the American Blower Corporation, Detroit, Mich.

Variable-speed fluid drives are intended for use with constant-speed a-c motors to provide stepless adjustment of speed for industrial machinery. These transmissions depend on

Table 5. Horsepower and Torque Ratings of Waterbury Hydraulic Variable-speed Transmissions

(Waterbury Tool Division of Vickers, Inc., Waterbury, Conn.)

Size No.	A-ends or Pump Units			B-ends or Motor Units		
	Horsepower Input at Normal Torque and Rate of Speed	Working Speed, rpm	Working Horsepower, Continuous Duty	Speed Range, rpm	Working Horsepower, Continuous Duty at Maximum Speeds	Working Torque, in.-lb
2	10	900	8	0–900	8	580
5	15	600	12	0–600	12	1,260
10	30	550	24	0–550	24	2,520
20	50	500	40	0–500	40	5,040
50	100	400	80	0–400	80	12,600
150	200	250	160	0–250	160	37,800
172	350	350	300	0–350	300	54,000

Note: The above normal working horsepower ratings are based on continuous 24-hr daily service. Intermittent ratings may be increased 25% and, for occasional services, more; for momentary overloads the working torque may be increased 150%.

turbine action of the fluid (oil) to transmit power. There are two rotors with straight radial vanes which form a torus ring, as illustrated in the cross section (Fig. 8). The impeller which is attached to the driving motor acts as a pump and sets up a vortex of oil (see arrow) which drives the runner smoothly and without mechanical means like a turbine wheel. The runner is connected to the driven machine.

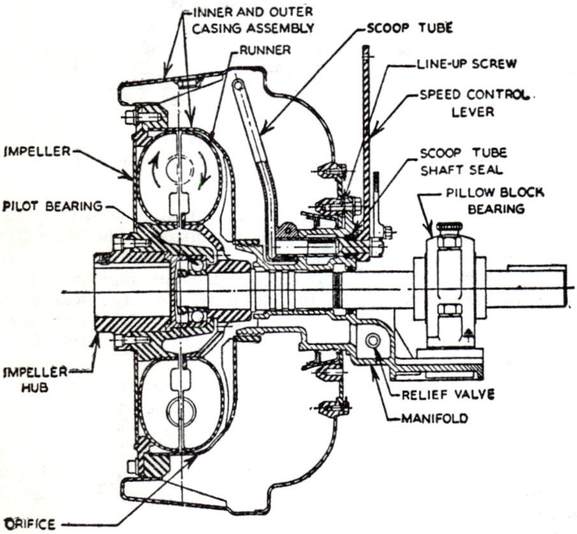

FIG. 8. Variable-speed fluid drive. (Courtesy of American Blower Corp.)

By changing the quantity of oil in the vortex, it is possible to adjust the output speed. The various means of adjusting oil quantity (and speed) are described below. The efficiency of power transmission when the drive is operating at full output speed is approximately 96 to 98%.

Lever-controlled Fluid Drive. This design (see Fig. 8) has a large rotating casing attached to the impeller, the purpose of the casing being to store (centrifugally) the idle oil which is not in the vortex between the runner and the impeller. A continuous flow of oil drains from the vortex into this casing through calibrated orifices (nozzles) and forms an annular ring. The rotating ring of oil impinges on a stationary scoop tube and is thus caused to flow through a cooler circuit and back into the vortex. The position of the scoop tube is adjusted by the external speed-control lever so that any desired amount of oil is carried in the vortex. By this means, speed is varied. Remote manual or automatic control devices are readily adaptable.

Field of Application. This unit is designed for machinery drives ranging from 5 to 350 hp. It is intended to be hub-mounted on the driving motor.

Pump-controlled Fluid Drive. This design has an external reservoir to store the oil which is not in the working circuit. To increase speed, a small oil pump transfers oil from the reservoir to the working circuit through a stationary manifold. Oil is removed by a stationary scoop tube inserted between the rotating casings and attached to the manifold. When no speed change is required, the scoop tube continually circulates oil through a cooler and back to the rotors, there being no change in the quantity of oil in the working circuit. But when a decrease in speed is required, the oil from the scoop tube can be diverted to the external reservoir (by a valve or by reversing the oil pump). The amount of oil in the rotors is thus decreased. It is a simple matter to adapt automatic control to operate the control pump or the valves.

Field of Application. This unit is designed for fan and centrifugal pump drives ranging from 100 to 2000 hp. It is also used for various machinery drives including rotary kilns, centrifugal compressors, and large test stands. It is a self-contained unit with its own supporting bearings. The input and output shafts are connected to the motor and driven machine by flexible couplings.

Enclosed Fluid Drive. This design has a stationary housing surrounding the rotors and the bearings. The bearings and the housing are split for inspection (or removal) of the rotors and shafts.

The essential differences between the control of this unit and the control of the other two designs are (1) the absence of nozzles to regulate the oil flow from the vortex and (2) the use of the oil pump to supply oil continuously to the rotors. The oil distributes itself centrifugally throughout the vortex *and* the space between the casings, forming an annular ring of fluid. The speed controller tube is inserted between the casings and acts as a weir as it skims off the rotating ring of oil according to its position. The speed controller thus sets the level of oil *in the vortex as well as in the casings* and controls output speed and torque. It discharges directly to the reservoir in the lower part of the housing. By fully inserting the speed controller, all the oil can be removed. Any degree of engagement is possible by adjusting its position. It is readily adaptable to automatic control.

Field of Application. The enclosed fluid drive is designed for drives of 100 hp and above, for power plant service (mechanical draft fans and boiler feed pumps), and for general industrial drives (centrifugal pumps and compressors, ball mills, test stands, etc.).

Ratings. In general, horsepower varies as $\overline{\text{rpm}^3}$ and $\overline{\text{diameter}^5}$. The horsepower capacity is taken with the normal slip of 3% or less between runner and impeller speeds.

The speed-control range of these fluid drives depends on the application. The speed can be varied as follows: (1) *Variable torque loads* (fans, centrifugal pumps, centrifugal compressors). The horsepower of the load falls off very rapidly (approximately $\overline{\text{rpm}^3}$) at reduced speeds. With this type of load, the output speed can be reduced to one-fifth the driving motor speed, or lower in special cases. (2) *Constant torque loads with high inertia and with negligible load fluctuation* (for example, printing presses). The output speed can be reduced to one-third the driving motor speed. By hand control, it is possible to run at lower speeds for short periods of time. (3) *Constant torque loads with some load fluctuations* (for example, rotary kilns). The output speed can be reduced to one-half the driving motor speed for normal operation, lower for starting duty only.

30. ELECTRICAL METHODS OF SPEED CONTROL

A great variety of electrical methods of speed control are available. In general, the electrical methods are convenient, simple, and reliable, provided that current, motor type, motor size, and control method are properly selected for the particular speed-torque characteristics and the speed range required.

Table 6 shows a brief survey of the conventional electrical systems of variable speed drives. For the sake of simplicity, methods not commonly used are omitted, as, for example, primary resistors and speed governors for fractional horsepower motors or intricate methods as the Scherbius, Kraemer, or Rosman system, which have been developed for special purposes and high power only. Furthermore, the list gives the characteristics for positive loads only; overhauling loads are not considered.

COMBINATION OF ELECTRICAL AND HYDRAULIC OR MECHANICAL METHODS. Frequently a convenient solution of a specific speed-control problem can be effected by combining one of the electrical methods with one of the other methods, for example, a combination of an electrical with a mechanical or an electrical with a hydraulic system. In this way, almost any required speed range or speed-torque characteristic can be achieved.

Table 6. Conventional Electrical Method of Speed Control

Motor Type	Method Used	Practical Speed Range, Basic Speed = 100%	Regulation Adjusted Speed	Rated Torque at Changed Speed	Efficiency	Notes
D-c series	Series resistance	20–100	Varies	Constant	Low	
D-c series	Series resistance Arm shunt	10–100	Varies	Constant	Low	
D-c series	Field shunt	100–130	Varies	Decreases with speed increase	High	Limited by commutation
D-c shunt or compound	Series resistance	20–100	Varies	Constant	Low	
D-c series	Two-motor series, parallel	50–100	Varies	Constant	High	
D-c shunt or compound	Field control	100–400	Constant	Decreases with speed increase	High	Conventional design
D-c shunt or compound	Field control	100–800	Constant	Decreases with speed increase	High	Special field structure Westinghouse *
A-c polyphase squirrel cage	By frequency and voltage change	10–400 or more	Constant	Constant	High	Maximum speed limited by mechanical conditions
A-c polyphase squirrel cage	Consequent poles	50, 100	Constant	Constant torque or constant horsepower or variable torque	High	Two speeds
A-c polyphase squirrel cage	Two windings and consequent poles	33, 66, 100, 200	Constant	Constant torque or constant horsepower or variable torque	High	Four speeds– other ranges possible
A-c polyphase wound rotor	Secondary resistance	25–100	Varies	Constant	Low	
A-c polyphase wound rotor	Two motors in concatenation	25–100	Constant	Constant	High	Just four economic speeds
A-c supply conversion to variable voltage and d-c shunt motors	By variable voltage	10–100	Fairly constant	Constant	High	Conventional
	By field control	100–400	Constant	Decrease with speed increase		
	By variable voltage	5–100	Constant	Constant	High	Using regulating exciters and closed cycle speed control
	By field control	100–400	Constant	Decrease with speed increase		
A-c polyphase shunt (Schraege)	Brush shifting	25–100	Fairly constant	Constant	High	

* See *Westinghouse Engineer*, November 1947.

HYDRAULIC TORQUE CONVERTERS

The hydraulic torque converter is a power transmission element which provides an output speed which varies with the torque requirements. The speed ratio is controlled not by the operator but automatically by the torque. To obtain accurate control over output speed, therefore, the torque converter must be used with a variable-speed driving element such as an internal combustion engine.

In its simplest form the hydraulic torque converter consists of three elements: a pump, a turbine, and a reaction member (stator) (Fig. 1). The pump element, which corresponds to the impeller of a centrifugal pump, delivers hydraulic fluid to the turbine runner, in which the fluid flows radially inward. Fluid leaving the turbine flows through a series of guide vanes in the stationary reaction member and again enters the pump. This stationary reaction member is an essential feature of a torque converter. By redirecting the flow of fluid, it absorbs the difference between the input and output torques, and without it no change in torque could be effected.

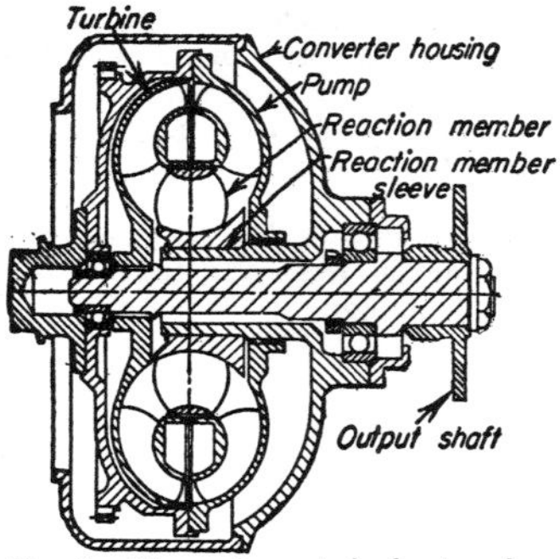

FIG. 1. Three-element single-stage hydraulic torque converter. (Reproduced by permission from *Machine Design*, May 1946)

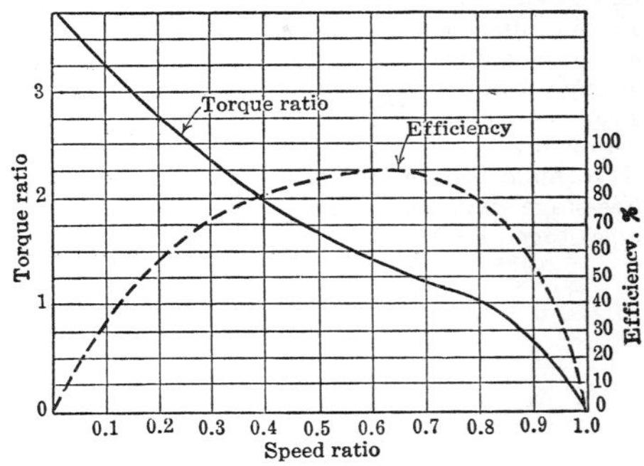

FIG. 2. Torque ratio and efficiency of three-element single-stage hydraulic torque converter. (Reproduced by permission from *Machine Design*, May 1946)

Performance of a torque converter such as shown in Fig. 1 may be represented by a pair of curves showing efficiency and torque ratio plotted against speed ratio (Fig. 2). These three quantities are related by the equation

$$\frac{T_2}{T_1} = \frac{N_1}{N_2} \times \frac{e}{100}$$

where T_2/T_1 is the ratio of output to input torques; N_2/N_1 is the ratio of output to input speeds; and e is the efficiency. The actual torque capacity is roughly proportional to the square of the input speed and to the fifth power of the converter diameter. It is also affected by the speed ratio, as shown in Fig. 3, in which primary torque is defined as the input torque required to maintain constant input speed.

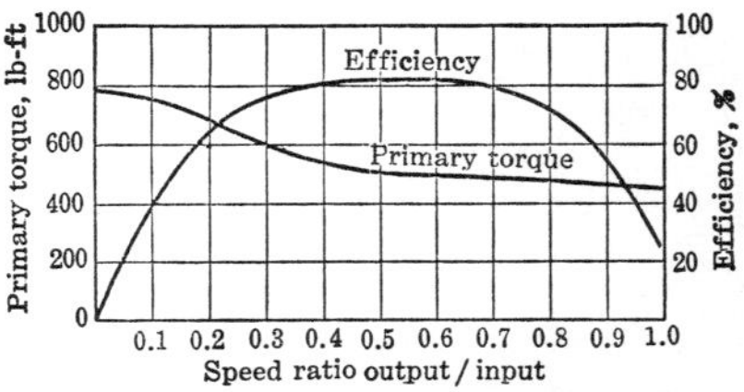

FIG. 3. Primary torque and efficiency of a three-stage hydraulic torque converter. (Reproduced by permission from *Machine Design*, April 1948)

As Fig. 2 shows, torque multiplication is maximum at zero output speed (the runner being stalled). As output speed increases, torque ratio falls off until at a certain speed ratio which usually lies between 0.65 and 0.85, depending on the design, it becomes unity. This is called the "clutch point" because under this condition the converter acts merely as a coupling or fluid clutch; there is no torque conversion and no torque reaction on the stationary member.

If the speed ratio were allowed to exceed that at the clutch point, the torque ratio would drop below unity and torque reaction on the stationary member would be negative. If it is desired to operate above this point, the resulting loss of power may be avoided by doing either of two things: The reaction member may be "cut loose" so that it is free to revolve with the fluid and exerts no drag; or the input and output shafts may be directly coupled mechanically, thus by-passing the converter. In the first method the stator is mounted on a free-wheeling clutch which automatically allows it to revolve in the direction of motion as soon as the torque reaction reverses direction. Above the clutch point, therefore, the unit functions as a fluid coupling, and its efficiency is practically equal to the speed ratio.

Many combinations of torque converter elements with clutches and gears, including planetary gears, are possible, permitting a wide variety of torque and speed characteristics. Chief applications are in conjunction with internal combustion engines. The result is a drive which is highly flexible, both as to output speed and output torque, and which is well adapted to such service as automotive, hoisting, oil-well drilling, etc.

Torque multiplication can be increased by using multistage elements in place of the single stage shown in Fig. 1. To increase the speed range over which high efficiency can be maintained, the stator and turbine elements may be split into two or more sections, some of which are free to freewheel whenever torque reaction becomes zero. In this way the angle through which the element diverts the fluid adjusts itself according to the requirements imposed by the torque-speed relationships, a larger angle being required at high ratios.

Hydraulic torque converters available for industrial uses include the Lysholm-Smith three-stage unit (made by the Twin Disc Clutch Co., Racine, Wis.) and the Schneider single-stage unit (made by the Torcon Corp., Chagrin Falls, Ohio).

SECTION 16

ELECTRIC MOTORS AND THEIR CONTROL

By

M. N. HALBERG, *Industrial Power Division, General Electric Co., Schenectady, N. Y.*

ELECTRIC MOTORS AND THEIR CONTROL

1. DEFINITIONS

ALTERNATING CURRENT. A current that alternately increases and decreases in magnitude with respect to time, whose values recur at equal intervals of time and whose average value is zero. Abbreviation, a-c.

ARMATURE. The part of a dynamoelectric machine that includes the main current-carrying winding. In d-c machines and in a-c commutator machines, the armature winding is connected to the commutator, and the armature is the rotating member. In a-c machines without commutators, the armature may be either the rotating member or the stationary member.

ASYNCHRONOUS MACHINE. A dynamo-electric machine in which the speed of operation is not exactly proportional to the frequency of the system to which it is connected.

AUTOMATIC TRIPPING. The opening of a circuit breaker under predetermined or other conditions without the intervention of an operator.

AUTOTRANSFORMER. A form of transformer that has one continuous winding, part of which is common to both primary and secondary circuits.

BRANCH CIRCUIT. That portion of a wiring system extending beyond the final automatic overload protective device of the circuit.

CAPACITOR (Static Condenser). A device whose primary purpose is to store electric charges. It commonly is used in a-c circuits to change the relative time phase between voltage applied and current flowing in the circuit. The usual form consists of two conducting plates, usually of aluminum foil, separated by a thin layer of insulation (dielectric), the combination being hermetically sealed in a steel container.

CIRCUIT BREAKER. A device for interrupting a circuit between separable contacts under normal or abnormal conditions. Ordinarily circuit breakers are required to operate only infrequently, although some classes of breakers are suitable for frequent operation.

CIRCULAR MIL. The area of a circle of 1-mil diameter, or $(\pi/4) \times (0.001)^2$ sq in. = 0.000 000 7854 sq in.

CLASS A INSULATION consists of: (1) cotton, silk, paper, and similar organic materials when either impregnated or immersed in a liquid dielectric; (2) molded and laminated materials with cellulose filler, phenolic resins, and other resins of similar properties; (3) films and sheets of cellulose acetate and other cellulose derivatives of similar properties; and (4) varnishes (enamel) as applied to conductors.

CLASS B INSULATION consists of inorganic materials, such as mica and asbestos in built-up form, combined with binding substances. If Class A material is used in small quantities in conjunction with Class B, for structural purposes only, the combined material may be considered Class B, provided the electrical and mechanical properties of the insulated winding are not impaired by the application of the temperature permitted for Class B material. (The word "impair" is here used in the sense of causing any change that could disqualify the insulating material for continuous service.)

CLASS C INSULATION consists of inorganic materials such as pure mica, porcelain, quartz, etc.

CLASS H INSULATION consists of (1) mica, asbestos, Fiberglas, and similar inorganic materials in built-up form with binding substances composed of silicone compounds, or materials with equivalent properties; (2) silicone compounds in rubbery or resinous forms, or materials with equivalent properties. A minute proportion of Class A materials may be had only where essential for structural purposes during manufacture, providing the electrical and mechanical properties of the insulated winding are not impaired by the application of the temperature permitted for Class H material.

CLASS O INSULATION consists of cotton, silk, paper, and similar organic materials when neither impregnated nor oil-immersed.

COMMUTATING POLE (Interpole). An auxiliary pole placed between the main poles of a commutating dynamoelectric machine. Its exciting winding carries a current proportional to the load current and produces a flux in such a direction and phase as to assist the reversal of current in the short-circuited coil.

COMMUTATOR. An annular ring or disk assembly of conducting members, individually insulated in a supporting structure, with an exposed surface for contact with current-collecting brushes, and ready for mounting on an armature spider, shaft, or quill.

CONTROLLER. A device, or group of devices, that governs, in some predetermined manner, the electric power delivered to the apparatus to which it is connected.

DIRECT CURRENT. A unidirectional current in which the changes in value are either zero or so small that they may be neglected. Abbreviation, d-c.

EFFICIENCY of a device with respect to a physical quantity that may be stored, transferred, or transformed by the device is the ratio of the useful output of the quantity to its total input. Unless specifically stated otherwise, the term *efficiency* means efficiency with respect to power.

ELECTRICAL DEGREE. The 360th part of the angle subtended, at the axis of the machine, by two consecutive field poles of like polarity. One *mechanical degree* is thus equal to as many electrical degrees as there are pairs of poles in the machine.

EXCITER. An auxiliary generator that supplies electric current to excite the field of another electrical machine.

FREQUENCY, of an Alternating Current. The number of cycles through which the current passes per second; frequency is the reciprocal of the *period.*

GENERATOR, Double-current. A machine that supplies both direct and alternating currents from the same armature.

GENERATOR, Electric. A machine that transforms mechanical energy to electrical energy.

GENERATOR, Induction. An induction machine (see below) driven above synchronous speed by an external source of mechanical power, and returning energy to the circuit.

GENERATOR, Magnetoelectric. An electric generator, the field poles of which are permanent magnets.

GENERATOR, Synchronous. A synchronous machine that transforms mechanical power into electric power.

GROUND. A conducting connection, whether intentional or accidental, between an electric circuit or equipment and earth, or to some conducting body that serves in place of the earth.

INDUCTION MACHINE. An asynchronous a-c machine that comprises a magnetic circuit interlinked with two electric circuits, rotating with respect to each other, and in which power is transferred from one circuit to another by electromagnetic induction. Examples of induction machines are induction motors, induction generators, and certain types of frequency converters and phase converters.

KILOWATT. A unit of electrical power equal to 1000 watts or 1.34 hp.

KILOWATT-HOUR. The work performed by 1 kw acting for 1 hr.

LOAD FACTOR. The ratio of the average load over a designated period to the peak load occurring in that period.

MAGNETIC CONTROLLER. A controller all of whose basic functions are performed by electromagnets.

MANUAL CONTROLLER. A controller all of whose basic functions are performed by hand.

MOTOR, Adjustable-speed. See pp. 16-12, 16-51.

MOTOR, Capacitor. See pp. 16-21, 16-26.

MOTOR, Capacitor-start. See p. 16-21.

MOTOR, Compound-wound. A d-c motor with two separate field windings connected—*a*, in parallel with the armature circuit, and, *b*, in series with the armature circuit. Winding *a* usually is the predominant field.

MOTOR, Constant-speed. See p. 16-12.

MOTOR, General-purpose. Any motor of 200 hp or less and 450 rpm or more, having a continuous rating, and designed, listed, or offered in standard ratings for use without restriction to a particular application.

MOTOR, Induction. An induction machine that converts electrical energy delivered to the primary circuit into mechanical energy by the electromagnetic interaction between the primary and secondary circuits. The secondary circuit is short-circuited or closed through a suitable circuit.

MOTOR, Large-power (Integral Horsepower). A motor built in a frame having a continuous rating of 1 hp, open type, at 1700–1800 rpm or in a larger frame.

MOTOR, Nonexcited Synchronous. A synchronous motor without a d-c field winding.

MOTOR, Permanent-split Capacitor. See p. 16-21.

MOTOR, Reactor-start. A form of split-phase motor (see below) designed to start with a reactor in series with the main winding. The reactor is short-circuited or otherwise made ineffective, and the auxiliary circuit is opened when the motor has attained a predetermined speed.

MOTOR, Repulsion. A single-phase a-c motor with a stator winding arranged for connection to the source of supply, and a rotor winding connected to a commutator. Brushes on the commutator are short-circuited and so set that the line connecting the positive and negative brushes is at an angle to the neutral axis of the stator winding. This type of motor has a varying speed characteristic.

MOTOR, Repulsion-induction. A form of repulsion motor which has a squirrel-cage winding in the rotor, in addition to the repulsion motor winding. This type of motor may have constant speed or varying speed characteristics. See pp. 16–21, 16–26.

MOTOR, Repulsion-start Induction. A single-phase motor having the same windings as a repulsion motor, but in which, at a predetermined speed, the rotor winding is short-circuited or otherwise connected to give the equivalent of a squirrel-cage winding. This type of motor starts as a repulsion motor but operates as an induction motor with constant speed characteristics. See pp. 16–21, 16–26.

MOTOR, Resistance-start. A form of split-phase motor (see below) with a resistance connected in series with the auxiliary winding. The auxiliary circuit is opened when the motor has attained a predetermined speed.

MOTOR, Series-wound. A commutator motor in which the field circuit and armature circuit are connected in series.

MOTOR, Shaded-pole. See p. 16-21.

MOTOR, Shunt-wound. A d-c motor in which the field circuit and armature circuit are connected in parallel.

MOTOR, Slip-ring. See Motor, wound-rotor induction.

MOTOR, Small-power (Fractional Horsepower). A motor built in a frame smaller than that having a continuous rating of 1 hp, open type, at 1700–1800 rpm.

MOTOR, Special-purpose. An industrial power motor specifically designated and listed for a particular power application, where load requirements and duty cycle are definitely known.

MOTOR, Split-phase. A single-phase induction motor equipped with an auxiliary winding, which is displaced in magnetic position from, and connected in parallel with, the main winding. The auxiliary circuit is opened when the motor has attained a predetermined speed. See p. 16-21.

NOTE: The term split-phase motor, when unqualified, describes a motor to be used without impedance other than that offered by the motor windings themselves.

MOTOR, Squirrel-cage Induction. An induction motor in which the secondary circuit consists of conductors suitably disposed in slots in the secondary core. See pp. 16–22, 16–35.

MOTOR, Synchronous. A synchronous machine that transforms a-c electrical power to mechanical power, and that has a speed strictly proportional to the frequency of the operating current. See p. 16-22.

MOTOR, Synchronous Induction. A wound-rotor induction motor to which d-c excitation is supplied when it approaches rated speed, enabling it to start as an induction motor and operate as a synchronous motor. See pp. 16–22, 16–47.

MOTOR, Two-value Capacitor. See p. 16-21.

MOTOR, Universal. A series-wound or compensated series-wound motor that will operate either on direct current or single-phase alternating current (of approximately the same voltage and a frequency not greater than 60 cps) at approximately the same speed and output. See p. 16-22.

MOTOR, Wound-rotor Induction. An induction motor in which the secondary circuit consists of a polyphase winding, or coils whose terminals are short-circuited or closed through suitable circuits. (When provided with collector or slip rings, a wound-rotor induction motor also is known as a slip-ring induction motor.) See p. 16-37.

OPEN MACHINE (or Motor). A self-ventilated machine having no restriction to ventilation other than that necessitated by mechanical construction.

POWER. The rate of transfer of energy. In the case of an a-c circuit the word power generally denotes the average value of the power over a cycle. The power in an electric circuit at any instant is equal to the product of the values of the current and voltage at that instant, and generally is called instantaneous power.

POWER FACTOR. The ratio of the power to the apparent power, that is, kw/kva. The maximum power factor is 1 or unity. It may be lagging or leading, depending on whether the current wave lags or leads the voltage wave. See *Power* Volume of Kent's *Mechanical Engineers' Handbook*, Section 16, Articles 12 and 13.

-PROOF (as a suffix). A machine (or apparatus) is designated as splashproof, moistureproof, explosionproof, etc., when so constructed, protected, or treated that subjection to the specified material or condition will not interfere with its successful operation.

RATING of a Machine, Apparatus, or Device. A designated limit of operating characteristics based on definite conditions. Such operating characteristics as load, voltage, frequency, etc., may be given in the rating.

RATING, Continuous. The rating that defines the load which can be carried for an indefinitely long time.

RATING, Short-time. The rating that defines the load which can be carried for a short and definitely specified time, the machine, apparatus, or device being at approximately room temperature at the time load is applied.

-RESISTING (as a suffix). A machine (or apparatus) is designated as moisture-resisting, fume-resisting, etc., when so constructed, protected, or treated that it will not be readily injured when subjected to the specified material.

SEMIMAGNETIC CONTROLLER. A controller having part of its basic functions performed by electromagnets and part by other means.

SINGLE-PHASE CIRCUIT. A circuit energized by a single alternating electromotive force. (A single-phase circuit usually comprises two wires. The currents in these two wires, counted outwards from the source, differ in phase by 180 degrees or a half cycle.)

SLIP (of an Induction Machine). The ratio, expressed as a percentage, of the difference between operating speed and synchronous speed to the synchronous speed.

SWITCH, General Use. A switch for general distribution and branch circuits. It is rated in amperes and is capable of interrupting its rated current at its rated voltage.

SWITCH, Isolating. A switch for isolating a circuit from its source of power. It is to be operated only when the circuit has been opened by some other means.

SWITCH, Motor-circuit. A switch for a motor branch circuit. It is rated in horsepower and is capable of interrupting the maximum operating overload current of a motor of the same horsepower at the rated voltage. The maximum operating overload of a-c motors is six times the rated current, and of d-c motors is four times the rated current.

SYNCHRONOUS MACHINE. A dynamoelectric machine in which the speed of normal operation is exactly proportional to the frequency of the current.

THREE-PHASE CIRCUIT. A combination of circuits energized by alternating electromotive forces that differ in phase by one-third of a cycle, i.e., 120 degrees.

-TIGHT (as a suffix). A machine is designated as watertight, dust-tight, etc., when so constructed that the enclosing case will exclude the specified material.

TORQUE. See p. 16-14.

TWO-PHASE CIRCUIT. A combination of circuits energized by alternating electromotive forces that differ in phase by a quarter of a cycle; i.e., 90 degrees.

UNDERVOLTAGE PROTECTION. The effect of a device, operative on the reduction or failure of voltage, to cause and maintain the interruption of power to the main circuit.

UNDERVOLTAGE PROTECTION, Time-delay. The effect of a device, operative on the reduction or failure of voltage for a definite period of time, to cause and maintain the interruption of power to the main circuit.

UNDERVOLTAGE RELEASE. The effect of a device, operative on the reduction or failure of voltage, to cause the interruption of power to the main circuit, but not to prevent the re-establishment of the main circuit on return of voltage.

2. APPLICATION OF MOTORS AND CONTROL

FACTORS AFFECTING MOTOR SELECTION. The various factors to be considered in selecting a motor and controller for a given application are summarized and discussed on pp. 16-06 to 16-10. These are broadly classified in accordance with their relation to (1) power supply, (2) driven machine, and (3) location.

Power Supply	Driven Machine	Location
Characteristics	Mechanical arrangement	Ambient temperature
D-c—voltage	Horizontal or vertical shaft	Altitude
A-c—voltage phases and frequency	Tilting and rolling	Surrounding atmosphere
Power rates	Inverted operation	Moisture
Energy charge	Stationary or portable	Dust
Power-factor clause	Method of connecting motor: direct-connected, built-in, belt, chain, or gear drive	Corrosive or solvent
Demand charge	Space limitations	Hazardous dust or gas
Voltage regulation	Speed, speed range, regulation	Outdoor conditions
Starting current	Direction of rotation, reversing	Underwater operation
Current pulsation	Synchronized operation	Effect of noise
Continuity of power	Horsepower, rms, peak	Codes and ordinances
Short-circuit current	Torques	
Voltage surges	Breakaway, accelerating, pull-in	
	Torque pulsation	
	Wk^2 of rotating parts	
	Thrust	
	Balance	
	End play	
	Frequency of starting	
	Deceleration	
	Operating schedule	
	Initiation of operating sequence	

Power Supply

VOLTAGE of a d-c power supply and voltage, number of phases, and frequency of an a-c power supply must be known in order to select a correspondingly rated motor and controller. In most cases these conditions are fixed and readily determined. However, sometimes a choice is available or a conversion from one set of conditions to another is desirable. For example, for economical use of small motors it often is necessary to convert a high a-c voltage to a lower value by means of transformers. Again, in some applications, as those involving operation over a wide speed range or severe reversing duty, it is desirable to convert alternating current to direct current, by means of a motor-generator set or a rectifier, to allow the use of d-c motors.

POWER RATES may affect the choice of motor and controller, since the efficiencies of various types of motor differ appreciably, which means a corresponding difference in the power used. This is particularly true of adjustable speed motors with wide variations in efficiency, which must be balanced against wide differences in cost. Many a-c power-rate schedules incorporate a *power-factor clause*, by which the billing is adjusted in accordance with the power factor of the current. This may affect the choice of motor. For example, it may justify the selection of a synchronous motor instead of an induction motor. Finally, most rate schedules embody a *demand charge;* that is, the maximum rate at which energy is used for a given period of time, as well as the total amount consumed, determines the power bill. This fact makes it desirable to maintain as uniform a load as possible at all times. In some applications it will justify the use of an adjustable-speed or multi-speed motor instead of a constant-speed machine. See Purchased Power in Industry, Article 1, and Power Factor Improvement, Articles 12 and 13, in Section 16 of the *Power* Volume of Kent's *Mechanical Engineers' Handbook.*

VOLTAGE REGULATION of the power supply represents the drop in voltage that occurs with a certain increase in the current drawn. This must be considered, first, because the torques developed by motors vary with the impressed voltage, so that the actual voltage at the motor terminals under the various operating conditions must be known. With a-c motors, the starting torques vary as the square of the voltage. Since these motors draw a relatively high starting current, the effect of the voltage regulation of the power supply is particularly important. Second, if lights are supplied from the same power circuit, a drop in voltage of about 2% causes a noticeable flicker of these lights. Finally, a drop in voltage of 30% or more may cause other motors on the system to stall or be disconnected from the line by their undervoltage protective device. Power companies usually establish limits to the starting current that can be drawn from the line by motors, in order to prevent excessive drop in line voltage. These factors often necessitate reduced-voltage controllers where full-voltage starting otherwise would be acceptable. In some cases, a different type of motor must be used—for example, a wound-rotor induction motor instead of a squirrel-cage induction motor—in order to obtain sufficient starting torque while keeping within the allowable starting current.

Pulsating Current. When the load is such that the motor draws a pulsating current from the line, corresponding variations in line voltage occur. If these have a frequency lying in a certain range (see Fig. 1), they can be of very small magnitude and yet produce a noticeable flickering of lights supplied from the circuit. The usual limit established for current pulsation is 66% of motor full-load current. This usually does not cause light flicker, but sometimes the regulation of the system may require a much lower limit of pulsation, such as 30 or 40%.

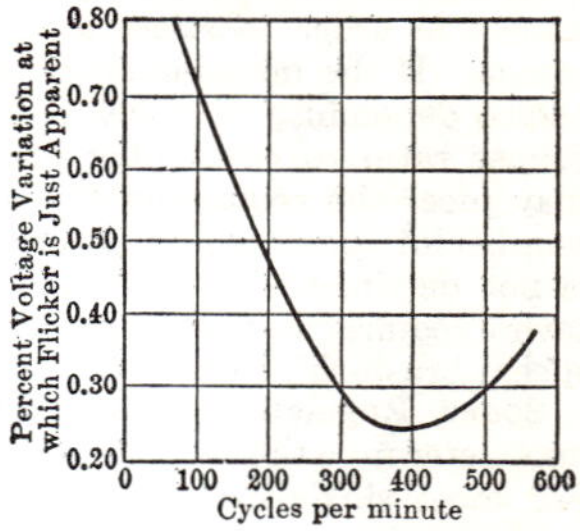

FIG. 1. Voltage variation causing light flicker at various frequencies.

CONTINUITY OF POWER SUPPLY may affect the choice of motors and control. Thus, if momentary interruptions or *dips* in line voltage are likely to occur, the use of time-delay rather than instantaneous undervoltage protection may be justified. See p. 16-59. In special cases, control equipment may be required that automatically will transfer the motor from one source of power to another, in the event of failure of the first.

SHORT-CIRCUIT CURRENT of the power system may affect the choice of control equipment. The line switch of a controller must be capable of interrupting the short-circuit current or must withstand this current without damage until it is interrupted by a circuit breaker or fuses connected between the power source and the controller.

VOLTAGE SURGES. When a motor is supplied from overhead power lines, either directly or through a transformer, lightning disturbances in the surrounding area may produce dangerous voltage surges at the motor terminals. Lightning arresters and surge protective capacitors may be required to protect the motor insulation. Direct-current motors on trolley circuits may be subjected to momentary overvoltages that require special design or external protective equipment.

Driven Machine

MECHANICAL ARRANGEMENT of the driven machine is the first consideration. Whether it is driven through a horizontal or a vertical shaft determines which of these two types of motor should be selected. Horizontal motors are more generally available and less expensive. Most grease-lubricated ball-bearing motors and small-power motors with waste-packed sleeve bearings will operate in either position.

Motors operated with a tilted shaft, if the tilt is momentary, require special construction of sleeve-bearing housings and, for long periods of tilted operation, may need bearings suitable for end thrust. If the motor operates with a roll (changing angle around horizontal shaft) of more than 10 or 12 degrees, a ball-bearing machine is usually necessary. Within this angle sleeve-bearing motors with modified oil gages are applicable.

Method of connecting motor to the driven machine must be considered. If the speed corresponds to that of an economical motor speed, a direct-connected motor normally is used. Otherwise, speed transformation by belt, chain, or gears must be employed. Maintenance, efficiency, and space requirements as well as initial cost enter into this choice. A direct-connected motor economically may be *built-in*. This type usually involves a special motor without shaft or bearings, arranged to be mounted on the shaft of the driven machine. Occasionally, the motor stator also is arranged to be supported from the driven machine. Compactness is one of the principal advantages of this construction. When the motor drives through a flat belt, multiple-V-belt, or chain, the dimensions of pulleys or sprockets and the center distance must be known in order to determine the side pull on the motor shaft. In some cases, the motor must have an extended shaft and outboard bearing to limit bearing pressure and shaft deflection. A sliding base or other means for adjusting the position of the motor to maintain proper tension in the driving member is necessary with belt or chain drive.

Space available may affect the choice of motor. If space is limited, a direct-connected or built-in motor usually is most suitable. Where the speed of the driven machine prohibits this choice, a gear motor often is the best solution. In some cases forced ventilation of a motor by an external blower may effectively reduce the motor size. Similar results may be obtained from a motor with Class B insulation, operating at a correspondingly higher temperature rise.

SPEEDS. The operating speed, speeds, or speed range of the driven machine must be known in order to select a corresponding single-speed, multispeed, or adjustable-speed motor. If the machine operates at a single speed, any type of motor may be used, the choice depending on other factors. When two, three, or four fixed speeds bearing the proper ratio to each other are acceptable, a multispeed squirrel-cage induction motor may meet the requirements. If the speed range does not exceed 2 : 1 and a reasonable number of speed points is satisfactory, and if considerable variation in speed with load is not detrimental, the wound-rotor induction motor should be considered. For more severe requirements a d-c motor with field or variable-voltage control, or an a-c adjustable-speed, brush-shifting motor may be necessary.

Speed Regulation. The allowable speed regulation (variation in speed with load) may determine the type of motor, particularly when adjustable speed operation is involved. For most single-speed applications, the squirrel-cage induction motor or shunt-wound d-c motor is satisfactory, but in a few applications zero speed regulation is required. This necessitates a synchronous motor. For adjustable-speed service, the wound-rotor induction motor and the d-c motor with armature resistance control often are eliminated because of their high-speed regulation.

DIRECTION OF ROTATION. Certain types of motors require adjustment or reconnection of windings to change the direction of rotation, which can be most conveniently done in the factory. Hence, the direction of rotation should be specified. If the motor is to run in both directions, a motor suitable for reversing and a reversing controller must be provided.

SYNCHRONIZED OPERATION. Sometimes two or more widely separated motor-driven machines must be kept in synchronism, i.e., must operate as if their shafts were directly coupled. This synchronization may require power selsyns (a wound-rotor induction motor on each drive with their rotor windings connected in parallel) in addition to the driving motors. In other cases, two or more motors must operate over a wide speed range but with a fixed ratio maintained between the speeds of the various motors. This operation usually requires adjustable-speed motors with suitable automatic control for maintaining the desired speed relationship.

HORSEPOWER required to drive the load determines the motor rating. Where the load is not constant but varies with time, the complete duty cycle (horsepower vs. time curve) should be obtained. From this curve, the peak horsepower, which determines the maximum torque of motor, and the root-mean-square (rms) horsepower can be calculated. The latter is the square root of the mean value (averaged against time) of the square of the horsepower needed. It is an approximate indication of the horsepower rating of the driving motor required from the standpoint of heating. However, where large variations in load occur or where the motor is shut down, accelerating or decelerating an appreciable portion of the time, the rms horsepower does not necessarily give a true indication of the equivalent continuous load, and it is advisable to consult the motor manufacturer. If the load continues at a certain value for an extended period of time (15 min to 2 hr, depending on the size of the motor), the horsepower rating of the motor should be not less than this value, regardless of any subsequent reduction in load. Where the motor operates at more than one speed, the horsepower required at each speed must, of course, be known.

TORQUES necessary to start and drive a load deserve careful consideration. Of particular importance is the breakaway torque requirement, since this often is considerably higher than the running torque. Furthermore, reduced-voltage starting may be necessary to meet starting current limitations, and this reduces the torque delivered by the motor. Where a synchronous motor is being considered, the torque required at full speed under starting conditions is equally important, since it determines the pull-in torque. Pull-in torque is not necessarily equal to full-load torque, as the machine may be unloaded for starting or may be required to start when the load is temporarily higher than normal. It is often desirable or essential to know the load torque at all speeds from standstill to full speed (speed-torque curve). This affects the accelerating time and hence the heating of the motor and control equipment. In particular, squirrel-cage induction motors, synchronous motors, and starting resistors, reactors, and autotransformers may have to be of special design if the accelerating time is unusually long. Where reduced-voltage starting is employed, the speed-torque curve of the load determines the speed to which the motor will accelerate on reduced voltage and hence determines the current inrush when the motor is transferred to full voltage.

Running Torque. The horsepower required and the speed of the driven machine determine the running torque. Certain driven machines, of which reciprocating compressors are typical, have a load torque which pulsates periodically. To prevent excessive pulsation in the current drawn by the motor proper, flywheel effect, Wk^2, must be provided in the motor or driven machine. With synchronous motor drives, the flywheel effect must be such that the natural frequency of the motor is not close to the frequency of any impulses in the load torque. The former can be calculated as follows: Let F = natural frequency of motor, cycles per minute; Wk^2 = (weight, pounds) $\times$ (radius of gyration, feet)^2 of the rotating parts of motor and load; N = revolutions per minute of motor; f = line frequency; and P_r = synchronizing power of motor, kilowatts per electrical radian. Then $F = (35{,}200/N)\sqrt{(P_r \times f)/Wk^2}$.

Wk^2 OF THE ROTATING PARTS of the driven machine affects the accelerating time and hence affects the heating during the starting period of motors and control equipment. Its value is particularly important when the motor is started frequently or when reversing duty is involved. The Wk^2 always should be known if a synchronous motor is to be employed, since the pull-in torque of the motor varies approximately as the square root of the total Wk^2 of motor and load. If a rotating member of the driven machine operates at a different speed from that of the motor, its Wk^2 can be converted to an equivalent value at the motor shaft by multiplying by

$$\left(\frac{\text{Rpm of rotating member}}{\text{Rpm of motor}}\right)^2$$

THRUST. Some driven machines (certain types of pumps and fans are typical) impose a thrust on the motor in a direction parallel to the shaft. Horizontal motors with thrust bearings may be required. Vertical motors may need an oversized thrust bearing if considerable external thrust is imposed.

BALANCE. Some driven machines such as precision grinders require motors specially balanced to give very small amplitudes of vibration.

END PLAY. Certain driven machines require motors with limited or zero end play and may need motors with preloaded ball bearings or sleeve bearings with end-play limiting devices.

FREQUENCY OF STARTING may affect the choice of motor and controller, since each start results in additional heating of the motor windings, starting resistors, etc. This must be considered in connection with the load torques and load Wk^2, which determine the accelerating time and hence the heating during each start.

In some applications, **rapid deceleration** of the motor and load is essential. With induction motors it may be necessary to provide *plugging control*, and with synchronous or d-c motors, dynamic braking control (see p. 16-64).

Plugging consists of reversing two leads of a motor while the motor is running, so as to reverse the phase rotation. The control thus is similar to that provided for reversing service. A relay, mounted on the motor shaft, usually is included to disconnect the motor from the line when it stops and to prevent it from starting in the reverse direction.

THE OPERATING SCHEDULE of a machine, i.e., the total number of hours of operation per year, should be known. If this figure is high, motors having maximum efficiency are more readily justified. If the load or speed varies, the operating time under each condition should be considered in order to permit proper evaluation of the motor characteristics.

Initiating Operating Sequence. If it is acceptable for the operator to start and stop the motor at the controller, the manual type of controller is permissible. Where remote control or automatic operation in response to temperature, pressure, liquid level, etc., is required, a magnetic controller must be used.

Location

THE AMBIENT TEMPERATURE, if above 40 C, and the **altitude,** if greater than 3300 ft above sea level, may affect the choice of motor. See p. 16-09. If a motor is to operate in a very moist atmosphere, special insulation or, under severe conditions, totally enclosed or enclosed and separately ventilated construction may be required. Extreme conditions of dirt or corrosive atmosphere also require totally enclosed or enclosed and separately ventilated motors. For locations where explosive dust (Class II, Group G) or gases (Class I, Group D) may be present, a corresponding type of explosionproof motor should be selected.

Outdoor application usually necessitates totally enclosed construction for protection against rain and snow. Splashproof or protected-type motors should be used only if they are low-voltage machines (below 600 volts) and are located in a mild climate. Special weather-resisting finishes are always necessary.

Underwater operation over long periods usually requires special enclosed motors, with counter air pressure to prevent entrance of water. For occasional flooding, extremely tight, totally enclosed, nonventilated motors equipped with a stuffing box at the shaft entrance may suffice. Where "hosing down" or splashing of water is involved, splashproof motors will meet all but extreme conditions, the latter requiring enclosed motors.

Noise. Installations in public buildings, such as schools, theaters, hospitals, stores, etc., and in residences, unless isolated from occupied sections, may require quiet operating motors to avoid objectionable noise. Suitable resilient mountings to prevent transmission of vibration through the building structure are often necessary. Finally, local regulations such as codes and ordinances must be considered. Some regulations demand special enclosures on motors and control equipment to reduce the hazard to operators.

ELECTRIC MOTORS

3. GENERAL CHARACTERISTICS

POWER SUPPLY. Practically all public utility power systems distribute a-c power, although d-c power is provided in the downtown areas of certain large cities. Alternating-current motors are, in general, less costly, require less maintenance, and meet the requirements of the large majority of applications. However, for certain loads d-c motors have more suitable characteristics and are used even though a-c to d-c conversion equipment is required. Direct-current motors must be designed for the proper line voltage, and a-c motors for the proper voltage, phase, and frequency.

Standard motor voltages are as follows: direct current—32, 115, 230, 550 volts; single-phase alternating current—115, 230 volts; polyphase alternating current—110, 208(60 cycles only), 220, 440, 550, 2300, 4000, 6600, 11,000, 13,200 volts. In some industrial plants, power is generated or purchased at a voltage higher than can be economically employed directly on the motors, so that a voltage transformation is necessary. If such a voltage change is economical, selection of the proper motor voltage (or voltages) depends on many factors which vary with each plant. However, it is seldom economical to select a voltage such that the horsepower ratings of the motors are outside the limits shown in Table 1. Certain types of motors are available with a *dual voltage* feature: i.e., sufficient winding terminals are provided so that the motor can be connected for either one of two voltages, such as 115 and 230 volts or 220 and 440 volts.

Table 1. Motor Horsepower Limits at Various Voltages

Single-phase Alternating Current			Polyphase Alternating Current			Direct Current		
Voltage	Horsepower		Voltage	Horsepower		Voltage	Horsepower	
	Min	Max		Min	Max		Min	Max
115	No limit	1	110	1/6	15	115	No limit	30
230	No limit	10	208–220	1/6	200	230	No limit	500
			440–550	1/6	500	550	1/2	No limit
			2300	75	No limit			
			4000	150	No limit			
			6600	400	No limit			
			11000	1000	No limit			
			13200	1000	No limit			

Frequency. In the United States a-c frequencies of 60 and 25 cycles are standard. The latter, primarily because of the limitations it imposes on the selection of motor speeds and its inferiority for lighting, has largely been superseded by the former. Frequencies of 50, 40, and 30 cycles also are in use in certain sections of the United States but are being superseded by 60 cycles. In Europe and European dependencies, 50 cycles is the general standard. In Canada, except for parts of Ontario, 60 cycles is standard.

Phase. Single-phase, three-phase, and, to a limited extent, two-phase power is distributed and used commercially in the United States. Three-phase energy is most satisfactory for power purposes and is the accepted standard. In general, single-phase motors cost more and have less satisfactory operating characteristics than three-phase machines. However, single-phase power is employed for lighting and in residential and rural districts often is the only power available. Hence, single-phase motors commonly are used on household appliances and other small equipment to be installed in such districts. Power companies place a limit on the size of motor which may be connected to single-phase lines. This limit usually falls between $^1/_4$ and $^3/_4$ hp for 110-volt circuits and between 3 and 10 hp for 220-volt circuits.

HORSEPOWER RATING. Motors usually are rated in horsepower, the rating representing their power output. Very small motors (less than $^1/_{20}$ hp) usually are rated in watts output (746 watts = 1 hp) or in millihorsepower (mhp) where 1000 mhp = 1 hp. Standard ratings have been established for the various types of motors.

Most motors are rated on a continuous basis; that is, they will carry rated load continuously. However, many types, such as crane and hoist motors, are given short time ratings such as 15-min, $^1/_2$-hr, or 1-hr ratings, indicating that they can carry their rated load for the designated period of time, after which they must be allowed to cool to room temperature. Some motors are given several different ratings, each for a different period of time. Short-time-rated motors are used on intermittent loads, the time rating being chosen to give the motor heating equivalent to that produced by the actual duty cycle of the load.

TEMPERATURE RISE. Motors are designated 40 C, 50 C, etc., to indicate the temperature rise in degrees centigrade that will not be exceeded when the motor carries rated load continuously (or for a specified time in the case of short-time-rated motors). The temperature rise refers to the windings, core, and other parts adjacent to the insulation. Some motors are given different temperature guarantees for different parts, such as armature, field, etc. Temperatures usually are determined by thermometer, although temperature detectors embedded in the winding are used on large motors. In some cases, the temperature rise of a winding is determined by measuring the change in its resistance.

Temperature rise guarantees usually are on the basis of normal conditions of ventilation, an altitude of not more than 1000 meters (3300 ft), and an ambient temperature of 40 C. Less effective dissipation of heat at higher altitudes (a result of reduced density of air) causes a higher temperature rise. It is assumed that the temperature rise as measured at altitudes below 3300 ft is 1% less than the temperature rise at a higher altitude for each 330 ft by which this altitude exceeds 3300 ft. Expressed as an equation, this becomes

$$t_h = \frac{t_s}{1.1 - (h/33{,}000)}$$

where t_h = temperature rise at altitude h, t_s = temperature rise at sea level, h = altitude, feet. Thus a motor having a temperature rise of 40 C at sea level will have at an altitude of 9900 ft a temperature rise of

$$t_h = \frac{40}{1.1 - (9900/33{,}000)} = 50 \text{ C}$$

The type of insulation has an important bearing on the allowable temperature rise. For Class A insulation (see p. 16-02) the upper temperature limit is placed at 105 C. If the insulation is allowed to reach a higher temperature, unduly rapid deterioration is expected. For Class B insulation the upper temperature limit is placed at 125 C. Class A insulation is standard for most types of motors, although Class B insulation usually can be obtained at increased cost. Since it is impractical to measure the temperature of a motor at the *hottest spot*, standard practice limits the temperature as measured to a lower value.

If the temperature is measured by thermometer, it is assumed that the thermometer reading will be 15 C below the hottest-spot temperatures for open motors and 10 C for totally enclosed or totally enclosed, fan-cooled machines. If temperature detectors are used, the differential is taken as 5 C, and with the resistance method the differential is taken as 5 C for open and totally enclosed fan-cooled machines (zero differential on totally enclosed motors.) On this basis the temperature limits and temperature rises (based on 40 C ambient) are as given in Table 2. If a motor is to operate in an ambient temperature in excess of 40 C, a motor should be chosen with a temperature rise correspondingly lower.

Many types of motors are rated on the basis of a temperature rise less than allowable for the type of insulation employed. For example, a 40 C rise by thermometer is a common rating for motors with Class A insulation, whereas in a 40 C ambient temperature a rise of 50 C is allowable. Such motors are frequently guaranteed to have a certain overload capacity, which is not available in motors rated at the maximum allowable temperature. Temperatures and temperature rises usually are given in degrees centigrade. For conversion of centigrade to Fahrenheit and of Fahrenheit to centigrade, see pp. 28–53 and 28–54.

Table 2. General Temperature Limits for Motors

	Class A Insulation, °C	Class B Insulation, °C
Measured by thermometer method		
Temperature limit, open motors	90	110
Temperature limit, totally enclosed and totally enclosed fan-cooled motors	95	115
Temperature rise (40 C ambient), open motors	50	70
Temperature rise (40 C ambient), totally enclosed and totally enclosed fan-cooled motors	55	75
Measured by temperature detectors		
Temperature limit	100	120
Temperature rise (40 C ambient)	60	80
Measured by resistance method		
Temperature limit, open and totally enclosed fan-cooled motors	100	120
Temperature limit, totally enclosed motors	105	125
Temperature rise (40 C ambient), open and totally enclosed fan-cooled motors	60	80
Temperature rise (40 C ambient), totally enclosed motors	65	85

SPEED CHARACTERISTICS. All motors may be classified according to their speed characteristics as follows: (1) constant-speed, (2) varying-speed, (3) adjustable-speed, (4) adjustable-varying-speed, and (5) multispeed.

A constant-speed motor is one in which the speed is practically constant regardless of load. A change of a few per cent in speed between no load and full load is characteristic, except with synchronous motors. Induction motors with small slip and d-c shunt motors are constant-speed machines.

A varying-speed motor is one in which speed varies with load, ordinarily decreasing as load increases. The series d-c motor and the high-slip, squirrel-cage induction motor are typical.

An adjustable-speed motor is one in which the speed can be varied gradually over a considerable range but, being unaffected by the load, remains practically constant when once adjusted. The d-c shunt-wound motor with field resistance control, designed for a considerable range of speed adjustment, is an example of this type of motor.

An adjustable-varying-speed motor is one in which the speed can be varied gradually over a considerable range, but the speed, when once adjusted to a given load, will vary in considerable degree with change in load; for example, a wound-rotor induction motor with secondary resistance control.

A multispeed motor is one which can be operated at any one of several definite speeds, each being practically independent of the load. An induction motor with a primary winding capable of various pole groupings is an example.

Operating Speeds. A characteristic of all constant-speed a-c motors is that their operating speed is equal to or slightly less than one of the *synchronous speeds* corresponding to the frequency of the power supply (see Table 3). Multispeed a-c motors operate at two or more of these speeds. Synchronous motors, as their name implies, operate at exact synchronous speed. Induction motors operate at synchronous speed under no load, but under load they operate at a slightly lower speed. The difference between synchronous speed and full-load speed is known as full-load slip; it varies between 1 and 5% of synchronous speed for constant-speed induction motors. Direct-current motors can be designed for any required speed, subject to mechanical limitations. However, for the purpose of standardization, constant-speed d-c motors usually are designed for full-load speeds approximately equal to the full-load speeds of 60-cycle induction motors.

In some cases, the speed classification of a motor depends entirely on the control used with it. For example, a wound-rotor induction motor may be constant speed, varying speed, or adjustable varying speed with no structural change in the motor whatsoever. A constant-speed d-c motor may become, with no change in the motor itself, a wide-range adjustable-speed motor by the provision of means for varying the voltage impressed on its armature.

DIRECTION OF ROTATION. The standard direction of rotation for all nonreversing motors, with the exception of polyphase induction motors, is counterclockwise as one faces the end of the machine opposite the drive. Unless otherwise specified, standard motors are designed or connected for the standard direction of rotation. In polyphase a-c motors

Table 3. Synchronous Speeds Possible at Various Frequencies

No. of Poles	60 Cycles	50 Cycles	40 Cycles	25 Cycles	No. of Poles	60 Cycles	50 Cycles	40 Cycles
2	3600	3000	2400	1500	52	138.5	115.4	92.3
4	1800	1500	1200	750	54	133.3	111.1	88.9
6	1200	1000	800	500	56	128.6	107.2	85.7
8	900	750	600	375	58	124.1	103.5	82.8
10	720	600	480	300	60	120	100	80
12	600	500	400	250	62	116.1	96.8	77.4
14	514.2	428.6	343	214.3	64	112.5	93.7	75
16	450	375	300	187.5	66	109	90.8	72.7
18	400	333.3	266.6	166.6	68	105.9	88.2	70.6
20	360	300	240	150	70	102.8	85.7	....
22	327.2	272.7	218.1	136.3	72	100	83.3	....
24	300	250	200	125	74	97.3	81	....
26	277	230.8	184.5	115.4	76	94.7	78.9	....
28	257.1	214.2	171.5	107.1	78	92.3	76.9	....
30	240	200	160	100	80	90	75	....
32	225	187.5	150	93.7	82	87.8	73.2	....
34	212	176.5	141.1	88.2	84	85.7	71.4	....
36	200	166.6	133.3	83.3	86	83.7	....	....
38	189.5	157.9	126.3	78.9	88	81.8	....	....
40	180	150	120	75	90	80	....	....
42	171.5	142.8	114.2	71.4	92	78.3	....	....
44	163.5	136.3	109	....	94	76.6	....	....
46	156.6	130.5	104.3	....	96	75	Lower speeds are limited by cost rather than by design.	
48	150	125	100	....	98	73.5		
50	144	120	96	....	100	72		

the terminals must be connected to the power lines in accordance with their phase rotation marking. This rule does not apply to polyphase induction motors whose terminals are not marked for any particular direction of rotation. They must be connected to the power lines, and, if the direction of rotation is wrong, two leads are reversed. Most motors can be reversed either by a change in connections or by mechanical adjustments.

EFFICIENCY. The efficiency of a motor is defined as the ratio of useful power output to the total power input. Since power input is usually given in kilowatts,

$$\text{Efficiency} = \frac{0.746 \times \text{Horsepower output}}{\text{Kilowatts input}}$$

$$\text{Kilowatts input} = \frac{0.746 \times \text{Horsepower output}}{\text{Efficiency}}$$

Motors of the same type and rating usually differ but little in efficiency. There may be considerable difference in the efficiency of motors of different types; this is particularly true of adjustable speed motors operating at reduced speeds. The efficiency of motors also varies considerably with the load and usually is highest at three-fourths or full load. At light loads the efficiency is quite low.

POWER FACTOR. The power factor of a motor is defined as the ratio of the kilowatts input to the kilovolt-amperes input where

$$\text{Kilovolt-amperes} = \frac{\text{Volts} \times \text{Amperes}}{1000} \quad \text{for single-phase motors}$$

$$= \frac{1.73 \times \text{Volts} \times \text{Amperes}}{1000} \quad \text{for three-phase motors}$$

$$= \frac{2 \times \text{Volts} \times \text{Amperes}}{1000} \quad \text{for two-phase motors}$$

Thus the kilovolt-amperes input to a motor equals kilowatts input/power factor.

Direct-current motors and certain types of a-c motors have a power factor of unity (100%); that is, kilowatts input equals kilovolt-amperes input. However, the more common types of a-c motors, such as the induction motor, have a *lagging* power factor of less than unity. For the effect of low power factor and for the means of improving it, see the *Power* Volume of Kent's *Mechanical Engineers' Handbook*, Section 16, Articles 12 and 13.

The product of the efficiency and the power factor of a motor is the *apparent efficiency.* Thus

$$\text{Kilovolt-ampere input} = \frac{\text{Kilowatts}}{\text{Power factor}} = \frac{0.746 \times \text{hp}}{E \times \text{Power factor}}$$

$$= \frac{0.746 \times \text{hp}}{E_a}$$

where E = efficiency and E_a = apparent efficiency.

FULL-LOAD CURRENT. The full-load current of a motor is the current drawn from the line when the motor is carrying full load, with rated voltage and frequency (a-c motors) applied. It may be calculated as follows, $I_{\max}$ being the full-load current.

Direct-current Motors

$$I_{\max} = \frac{746 \times \text{Horsepower}}{\text{Efficiency} \times \text{Voltage}}$$

Single-phase A-c Motors

$$I_{\max} = \frac{746 \times \text{Horsepower}}{\text{Efficiency} \times \text{Voltage} \times \text{Power factor}}$$

Three-phase A-c Motors

$$I_{\max} = \frac{746 \times \text{Horsepower}}{1.73 \times \text{Efficiency} \times \text{Voltage} \times \text{Power factor}}$$

Two-phase A-c Motors, 4-wire *

$$I_{\max} = \frac{746 \times \text{Horsepower}}{2 \times \text{Efficiency} \times \text{Voltage} \times \text{Power factor}}$$

Approximate values of full-load current for popular types and ratings of motors are given in Table 4.

TORQUES. The full-load torque of a motor is fixed by its horsepower and speed rating.

$$\text{Full-load torque in pound-feet} = 5250 \times \frac{\text{Horsepower}}{\text{Rpm}}$$

This is the basic torque of a motor. Other torques (see below) are often expressed as a percentage of full-load torque.

Other motor torques of importance are (1) starting torque, (2) pull-up torque (a-c motors), (3) breakdown torque (a-c motors), (4) pull-in torque (synchronous motors), and (5) pull-out torque (synchronous motors).

Starting torque of a motor is the torque developed at zero speed. In a-c motors it is the minimum torque for all angular positions of the rotor, with rated voltage and frequency applied to the motor.

Pull-up torque is the minimum external torque developed by the motor during the period of acceleration from rest to the speed at which breakdown torque occurs. For motors that do not have a definite breakdown torque, the pull-up torque is the minimum torque developed up to rated speed.

Rated breakdown torque of an a-c motor is the maximum torque the motor will carry, with rated voltage and frequency applied to the motor, without an abrupt drop in speed. It indicates the peak load the motor can carry without stalling or decelerating to a low speed.

Pull-in torque of a synchronous motor is the maximum constant torque under which the motor will pull its connected inertia load into synchronism, at rated voltage and frequency, when its field excitation is applied. The speed to which a motor will bring its load depends on the power required to drive it; and whether the motor can pull the load into step from this speed depends on the inertia of the revolving parts. Hence, pull-in torque cannot be determined unless the Wk^2 of the external load and the torque of the load are known.

Pull-out torque of a synchronous motor is the maximum sustained torque the motor will develop at synchronous speed for 1 min, with rated voltage applied at rated frequency and with normal excitation.

* Where two-phase motors are supplied from a three-wire circuit, the current in the common wire is 1.41 times the current in either other wire.

Table 4. Approximate Full-load Currents of Motors

Horse-power of Motor	Full-load Current—Amperes																			
	Direct-current Motors		Alternating-current Motors																	
			Single-phase Motors		Three-phase															
					Induction Motors								Synchronous Motors							
					Squirrel-cage				Wound-rotor				1.0 Power Factor				0.8 Power Factor			
	115-volt	230-volt	115-volt	230-volt	220-volt	440-volt	550-volt	2200-volt	220-volt	440-volt	550-volt	2200-volt	220-volt	440-volt	550-volt	2200-volt	220-volt	440-volt	550-volt	2200-volt
1/6	2.0	1.0	3.4	1.7	0.90	0.45	0.36													
1/4	2.6	1.3	4.0	2.0	1.16	.58	.48													
1/3	3.2	1.6	6.0	3.0	1.4	.70	.56													
1/2	4.6	2.3	8.0	4.0	1.9	.95	.76													
3/4	6.4	3.2	12	6.0	2.6	1.3	1.04													
1	8.2	4.1	14	7.0	3.4	1.7	1.36	......	5.4	2.7	2.2									
1 1/2	12.4	6.2	18	9.0	5.0	2.5	2.0	......	6.8	3.4	2.7									
2	16.2	8.1	22	11	6.2	3.1	2.5	......	8.0	4.0	3.2									
3	24	12	31	16.5	9.0	4.5	3.6	......	10.5	5.3	4.2									
5	40	20	......	24	14.5	7.2	5.7	......	16	8.0	6.4									
7 1/2	60	30	......	38	21	10.5	7.3	......	23	11.5	9.2									
10	78	39	......	45	26	13.5	11	......	29	14.5	10.5									
15	114	57	......	......	40	20	16	......	42	21	17									
20	150	75	......	......	52	26	21	......	54	27	23	......	46	23	18.5	4.6	62	31	25	6.2
25	186	93	......	......	65	32	26	7.0	68	34	27	7.7	58	29	23	5.8	74	37	30	7.4
30	225	112	......	......	78	39	31	8.1	80	40	32	8.8	68	34	27	6.8	88	44	35	8.8
40	295	147	......	......	102	51	41	10.5	104	52	42	11	88	44	35	8.8	115	57	46	11.5
50	365	182	......	......	126	63	51	12.5	128	64	51	13.5	110	55	44	11	140	70	56	14
60	......	218	......	......	152	76	60	15	154	77	62	15.5	130	65	52	13	170	84	67	17
75	......	270	......	......	188	94	75	18.5	188	94	75	19	162	81	65	16	210	104	83	21
100	......	355	......	......	250	125	100	24.5	250	125	100	25	216	108	86	22	280	140	112	28
125	......	445	......	......	310	155	125	30	310	155	125	31	268	134	107	27	340	170	136	34
150	......	530	......	......	370	185	145	36	370	185	145	37	320	160	128	32	410	205	164	41
200	......	700	......	......	490	245	195	49	490	245	195	49	420	210	168	42	540	270	216	54

STARTING CURRENT. It is characteristic of most motors that if full-line voltage is impressed on their terminals at start (rotor at standstill) a current considerably in excess of the normal running current will be drawn from the line. This current usually is designated *locked rotor current* and is defined as the current taken from the line with the rotor locked and with rated voltage applied (as well as frequency in the case of a-c motors).

For d-c motors, starting current must be held to certain limits (usually 150 to 300% of full load current) to prevent excessive burning of the motor commutator. This prevents starting these motors on full voltage except in the smaller sizes. Reduced-voltage starting is generally recommended for motors larger than about 1 hp, although 115- and 230-volt motors rated 1750 rpm and below may be started on full voltage in sizes up to 1 1/2 hp (shunt wound) or 5 hp (series or compound wound) with some reduction in brush and commutator life. Reduced-voltage starting is usually accomplished by means of a resistor temporarily inserted in series with the armature winding.

The full-voltage starting current of a-c motors varies over a wide range for the different types of motors and may be as high as ten times normal full-load current. Hence, reduced-voltage starting sometimes is necessary to prevent distortion of the windings and resultant failure of the insulation. However, with a few exceptions a-c motors have the windings adequately braced, and full-voltage starting is permissible so far as the motor is concerned. But, in any case, reduced-voltage starting frequently is necessary to prevent undue voltage disturbance on the power lines. The starting current, flowing through transformers, transmission lines, and feeders, may cause a drop in voltage sufficient to dim the lights supplied from the circuit or to cause other motors to stall or trip off the line.

Most power companies have established regulations governing the starting current that may be drawn by motors connected to their distribution systems.

The following regulations for single-phase motors have been adopted by some utilities and are becoming recognized as reasonable limitations:

Rule 1. Automatically controlled single-phase motorized equipment designed for general use shall be equipped with motors having locked-rotor currents at rated voltage designed not to exceed the following: (1) 20 amperes at 115 volts; (2) 25 amperes at 230 volts. A tolerance of +15% of these current values taken at 25 C is acceptable in testing any individual motor. Note: Exceptions may be made for automatically started equipment whose operating cycles will normally involve long runs with infrequent starting.

Rule 2. Manually controlled single-phase motorized equipment designed for general use shall be equipped with motors having locked-rotor currents at rated voltage designed not to exceed the following: (1) 40 amperes at 115 volts. (2) 50 amperes at 230 volts. A tolerance of +15% of these current values taken at 25 C is acceptable in testing any individual motor.

Rule 3. Single-phase motorized equipment, to be used where industrial, commercial, or other large loads or special conditions exist, may, upon approval of the electric utility, be equipped with motors having locked-rotor currents in excess of those allowed by Rule 1 and Rule 2.

For polyphase motors, the maximum starting currents as given below are allowed by some utilities. These limits generally permit full-voltage starting.

Horsepower	Amperes at 220 Volts *	Horsepower	Amperes at 220 Volts *	Horsepower	Amperes at 220 Volts *
1 or less	24 †	15	220	75	1085
1 1/2	35	20	290	100	1450
2	45	25	365	125	1815
3	60	30	435	150	2170
5	90	40	580	200	2900
7 1/2	120	50	725		
10	150	60	870		

* Current at other voltages is inversely proportional to the voltage.
† Amperes per horsepower (this rating only).

Where no power company regulation exists, the limits of motor starting current are determined by the voltage variation permissible and the drop in voltage that results from a given increase in current. To avoid any perceptible light flicker it is necessary to prevent a sudden change in voltage in excess of approximately 2% of normal voltage; but in many cases a considerably greater voltage change, particularly if it occurs at rare intervals, will not be objectionable. So far as motors are concerned, a 10 or 15% momentary drop in voltage usually is not serious, although in some special applications the resulting change in motor speed may cause difficulty. The drop in voltage caused by a given increase in current is determined by the rating and characteristics of the generating equipment and by the resistance and reactance of transformers and feeders supplying the motor.

All a-c motors rated 1/20 hp and larger, except polyphase wound-rotor motors, now have on their name plate a "Code" letter selected from Table 5. This letter indicates the locked-rotor kilovolt-ampere per horsepower drawn by the motor on full voltage (NEMA standard).

Table 5. Code Letters for Locked-rotor Kilovolt-ampere

(NEMA Standard)

Code Letter	Kilovolt-amperes/hp	Code Letter	Kilovolt-amperes/hp	Code Letter	Kilovolt-amperes/hp
A	0 –3.15	H	6.3– 7.1	P	12.5–14.0
B	3.15–3.55	J	7.1– 8.0	R	14.0–16.0
C	3.55–4.0	K	8.0– 9.0	S	16.0–18.0
D	4.0 –4.5	L	9.0–10.0	T	18.0–20.0
E	4.5 –5.0	M	10.0–11.2	U	20.0–22.4
F	5.0 –5.6	N	11.2–12.5	V	22.4–and up
G	5.6 –6.3				

It is characteristic of most motors that if the starting current is reduced, for example, by means of a reduced voltage starter, the starting torque will also be reduced. Hence, the starting torque of the motor and the torque required by the load must be considered, as well as the starting current. A high ratio of starting torque to starting current (called *torque efficiency*) obviously is desirable when severe starting-current restrictions exist. This ratio varies over a wide range for the different types of motors and may vary with the type of control equipment employed.

MECHANICAL FORMS. In its simplest form a motor may consist simply of the stator and rotor windings with their necessary magnetic structures. Certain types of small motors often are in this form. The rotor then is mounted on a shaft of the driven machine, which also is provided with a suitable support for the stator. However, a motor usually is furnished with shaft, two bearings, and mounting feet forming part of the stator frame. The bearings usually are carried by *end shields* that bolt to each end of the frame. Very large motors often have pedestal-type bearings. In this case, a base may be provided which supports the stator and the bearing pedestals, or individual sole plates may be provided for each bearing pedestal and for each side of the stator frame.

Motors sometimes are supplied without mounting feet, being supported from the end shield on the drive end. This shield has a machined flange or face for bolting to the driven machine. This same construction may be used to support the driven machine by the motor wholly or partially. In this case the motor feet are retained.

Motors with belt drives usually have a sliding base that permits adjustment of the motor position, in a direction perpendicular to the shaft, for tightening the belt. The smaller standard two-bearing motors are suitable for belt drive. In the larger sizes, an outboard bearing is provided.

Good practice under normal operating conditions for flat belt, V-belt, gear, and chain drive on motors without outboard bearings is shown in Table 6.

Table 6. Speed Limitations—Two-bearing Motors for Flat-belt, V-belt, Gear, and Chain Drives

(NEMA Standard)

Full-load Rpm of Motor		Maximum Horsepower Rating of Motor			
Above	Including	Flat-belt Drive	V-belt Drive	Gear Drive	Chain Drive
2400	3600	20	20	..	20
1800	2400	30	40	..	40
1500	1800	40	75	5	75
1200	1500	40	75	10	75
900	1200	75	125	25	125
750	900	125	200	50	200
720	750	150	250	75	250
560	720	200	300	75	300

All motors are available in the standard horizontal-shaft construction. Most types also are obtainable in the vertical-shaft construction. In this case the motor is supported by feet or a ring base which forms part of, or is bolted to, the lower end shield. Two types of vertical motors are available. The *normal-thrust type* is designed to carry a moderate thrust load (280 to 2100 lb, depending on rating) in addition to the weight of the motor rotor and coupling. The *high-thrust type* is designed to carry substantially higher thrust loads. Certain types of vertical motors can be obtained in *hollow-shaft construction.* These are used primarily for deep-well pumping, as the hollow shaft allows the pump shaft to be brought to the top of the motor for convenient coupling to the motor shaft and for adjustment of its length.

The combination of a motor with an enclosed speed-reducing gear, built as an integral unit and designated *gear motor*, is offered by several manufacturers. The sizes available range from $^1/_6$ to 75 hp, with output shaft speeds of approximately 13.5 to 780 rpm. An 1800 rpm motor usually is employed. Various types of gear reduction units are used, including the internal planetary with concentric output shaft, the worm and wheel type with right-angle shaft, and the straight helical gear reduction with parallel offset shaft. The gear motor has a considerably lower cost than that of a conventional motor of corresponding output-shaft speed.

The type of bearings—sleeve, ball, or roller—varies with the size and type of motor. Sleeve bearings are oil lubricated, with oil rings or waste packing to carry the oil from the reservoir to the bearing surface. Ball and roller bearings are generally grease lubricated, although oil lubrication often is employed for large, high-speed motors and for the thrust bearing of large vertical motors.

ENCLOSING AND PROTECTIVE FEATURES. A motor normally requires the circulation of air through its interior to remove the heat resulting from the motor losses. If there is no restriction to the flow of ventilating air other than that necessitated by mechanical construction, the motor is designated as *open type*. Although this type is most common and the lowest in cost, a number of other types are available, offering various degrees of enclosure.

A protected motor is one in which all ventilating openings in the frame are protected with a wire screen, expanded metal, or perforated covers, with openings not exceeding $^1/_2$ sq in. in area and are shaped so as not to permit the passage of a rod larger than $^1/_2$ in. in diameter. If the distance of exposed live parts from the guard is more than 4 in., openings may be $^3/_4$ in. in diameter. This construction is designed to prevent accidental contact with rotating or live parts. Its use is confined largely to d-c motors.

A semiprotected motor is one in which some of the ventilating openings in the frame, usually those in the top half, are protected as in a protected machine, the others being left open.

A drip-proof motor is one in which the ventilating openings are so constructed that drops of liquid or solid particles falling on the machine at any angle not greater than 15 degrees from the vertical cannot enter the machine either directly or by striking and running along a horizontal or inwardly inclined surface.

A splashproof motor is one in which the ventilating openings are so constructed that drops of liquid or solid particles falling on the machine or coming towards it in a straight line at any angle not greater than 100 degrees from the vertical cannot enter the machine either directly or by striking and running along a surface.

A totally enclosed motor is one enclosed to prevent exchange of air between the inside and the outside of the case but not sufficiently enclosed to be termed airtight. It protects the motor from dirt, moisture, chemical fumes, or other harmful ingredients of the surrounding atmosphere.

A totally enclosed fan-cooled motor is a totally enclosed machine equipped for exterior cooling by means of a fan or fans integral with the machine but external to the enclosing parts. It serves the same purpose as the totally enclosed motor. The fan-cooling feature is desirable on larger motors to dissipate the heat.

An enclosed self-ventilated motor is one having openings for the admission and discharge of ventilating air, which is circulated by means integral with the machine, which is otherwise totally enclosed. The openings are so arranged that inlet and outlet ducts or pipes may be connected to them. This construction serves the same purpose as the totally enclosed type in which air free from harmful ingredients can be obtained within a reasonable distance. It also prevents heat, resulting from the motor losses, from being discharged into the motor room.

An enclosed separately ventilated motor is one having openings for the admission and discharge of the ventilating air, which is circulated by means external to and not a part of the machine, which otherwise is totally enclosed. These openings are so arranged that inlet and outlet ducts or pipes may be connected to them. This construction, rather than the enclosed self-ventilated construction, is employed on low-speed machines whose rotor fans do not develop sufficient pressure to circulate the ventilating air through the required duct length.

Both the enclosed self-ventilated and the enclosed separately ventilated machines can be provided with a closed system of ventilation; that is, the hot air discharged from the machine can be cooled and then recirculated through the machine. Cooling usually is accomplished by water circulated through surface air coolers. The closed system of ventilation generally is restricted to rather large machines.

An explosionproof motor is one in an enclosing case which is designed and constructed to withstand an explosion of a specified gas or dust that may occur within it and to prevent the ignition of the specified gas or dust surrounding the motor by sparks, flashes, or explosions that may occur within the machine casing. Such motors are designed for certain hazardous locations.

The National Electrical Code (Article 500) classifies hazardous locations as follows.

Class I. Locations in which flammable gases or vapors may be present in the air in sufficient quantities to produce explosive or ignitable mixtures. Subdivisions of this class are: Group A, atmospheres containing acetylene; Group B, atmospheres containing hydrogen or gases or vapors of equivalent hazard such as manufactured gas; Group C, atmospheres containing ethyl ether vapor; Group D, atmospheres containing gasoline, petroleum, naphtha, alcohols, acetone, lacquer solvent vapors, propane, butane, and natural gas.

Class II. Locations which are hazardous because of the presence of combustible dust. Subdivisions of this class are: Group E, atmospheres containing metal dust; Group F, atmospheres containing carbon black, coal, or coke dust; Group G, atmospheres containing grain dust.

Class III. Locations which are hazardous because of the presence of easily ignitable fibers or flyings, but in which such fibers or flyings are not likely to be in suspension in air in sufficient quantities to produce ignitable mixtures.

The code also recognizes two divisions of each class of hazardous locations, designated Division 1 and Division 2:

Class I, Division 1 locations are those in which hazardous concentrations of gas or vapor (1) exist under normal operating conditions, (2) exist frequently because of repair or maintenance operations or leakage, or (3) exist as a result of breakdown or faulty operation of the equipment which might also cause simultaneous failure of the electrical equipment.

Class I, Division 2 locations are those in which flammable gases or vapors can exist only in case of (1) accidental rupture or breakdown of the container or system in which they are normally confined, (2) a failure or abnormal operation of ventilating equipment which normally prevents hazardous concentrations of the gas or vapor, or (3) occasional communication of the gas or vapor from an adjacent Class I, Division 1 location.

Class II, Division 1 locations are those in which (1) explosive or ignitable mixtures of dust may occur under normal operating conditions, (2) such mixtures are produced in the vicinity of electrical equipment under abnormal operation or failure of equipment, or (3) dusts of an electrically conducting nature may be present.

Class II, Division 2 locations are those in which explosive or ignitable mixtures of the dust normally will not be in suspension in the air during normal operation or as a result of abnormal operation or failure of the equipment, but (1) where deposits of dust may be sufficient to interfere with the safe dissipation of heat from electrical equipment, or (2) where such deposits of dust in the vicinity of the electrical equipment might be ignited by arcs, sparks, or burning material from such equipment.

Class III, Division 1 locations are those in which easily ignitable fibers or materials producing combustible flyings are handled, manufactured, or used.

Class III, Division 2 locations are those in which ignitable fibers are stored or handled (except in process of manufacture).

For **Class I, Division 1** locations, the code specifies explosionproof motors approved for Class I locations. For **Class I, Division 2** locations, it permits either explosionproof motors or open motors with no sliding contacts, switching mechanism, or integral resistance unit (or with such devices provided with an explosionproof enclosure approved for Class I locations).

For **Class II, Division 1** locations, totally enclosed (not ventilated, pipe-ventilated, or fan-cooled) motors approved for Class II locations are specified except where dust from the production of magnesium, aluminum, or aluminum bronze powders may be present. In the latter case totally enclosed (not ventilated or fan-cooled) motors specifically approved for such locations are required. For **Class II, Division 2** locations, totally enclosed (not ventilated, pipe-ventilated, or fan-cooled) motors are specified. However, in locations where, in the judgment of the code-enforcing authority, only moderate accumulations of nonconducting, nonabrasive dust are likely to occur and where the equipment is readily accessible for routine cleaning and maintenance, self-cleaning textile motors of the squirrel-cage type or standard open-type machines without sliding contacts, switching mechanisms, or integral resistances (or else with such devices enclosed in tight metal housings without openings) may be installed. Protected or splashproof motors are not to be used.

For **Class III, Division 1** locations, totally enclosed (not ventilated, pipe-ventilated, or fan-cooled) motors are specified. However, in locations where, in the judgment of the code-enforcing authority, only moderate accumulations of lint and flyings will be likely to collect on, in, or near a motor and where the motor is readily accessible for routine cleaning and maintenance, self-cleaning textile motors of the squirrel-cage type or standard open-type machines without sliding contacts, switching mechanisms, or integral resistances (or else with such devices enclosed in tight metal housings without openings) may be used. For **Class III, Division 2** locations, totally enclosed (not ventilated, pipe-ventilated, or fan-cooled) motors are specified. Protected and splashproof motors are not permitted for Class III locations.

At present no motors are available for Class I, Division 1, Groups A and B locations. Motors in sizes up to 20 hp (at 3600 rpm) are obtainable for Class I, Division 1, Group C locations and in sizes up to 500 hp (at 3600 and 1800 rpm) for Class I, Division 1, Group D locations. Motors are also available in sizes up to 200 hp (at 3600 and 1800 rpm) for Class II, Division 1, Group E locations, and in sizes up to 500 hp (at 3600 and 1800 rpm) for Class II, Division 1, Groups F and G conditions.

The Underwriters' Laboratories, Inc., established and maintained by the National Board of Fire Underwriters, offers a testing and inspection service available to manufacturers of equipment for hazardous locations. Motors that have been tested and inspected by them are provided with a label so indicating. Since most inspection authorities look to the Underwriters' Laboratories for guidance as to the suitability of equipment for hazardous locations, motors with this label generally are acceptable for the specified atmospheres anywhere in the United States.

Motors for installation in mines come under the jurisdiction of the U. S. Bureau of Mines. This body does not issue approvals covering a motor or controls but approves only a complete power-using assembly consisting of a machine, such as pump, coal cutter, conveyor, etc., combined with the motor that drives it and the complete control equipment and wiring. However, certain types of motors are available which have passed the inspection and tests of the U. S. Bureau of Mines, indicating that they are eligible for ultimate approval, without further tests, as part of a complete power-using assembly.

4. SMALL-POWER MOTORS

Small-power or fractional-horsepower motors are motors built on frames having a continuous rating of less than 1 hp (open type) at 1700 to 1800 rpm. Standard horsepower and speed ratings of small-power motors are given in Table 7. The various types in general use are as follows.

1. Single-phase motors: (*a*) shaded-pole, (*b*) split-phase, (*c*) repulsion-start induction (and repulsion-induction), (*d*) capacitor, (*e*) series-wound, and (*f*) synchronous (nonexcited).
2. Polyphase motors: (*a*) squirrel-cage induction and (*b*) synchronous (nonexcited).
3. Direct-current motors: (*a*) shunt-wound, (*b*) series-wound, and (*c*) compound-wound.
4. Universal motors (operate from either alternating current or direct current); (*a*) series-wound, and (*b*) compensated series-wound.

Table 7. Standard Horsepower and Speed Ratings of Small-power Motors

Speed in Rpm

60 Cycles		50 Cycles		25 Cycles		Direct Current	Horsepower * (max)
Synchronous	Approximate Full Load	Synchronous	Approximate Full Load	Synchronous	Approximate Full Load	Approximate Full Load	
3600	3450	3000	2850			3450	1
1800	1725	1500	1425	1500	1425	1725	3/4
1200	1140	1000	950			1140	1/2
900	850					850	1/3

* Standard horsepower ratings are 1/20, 1/12, 1/8, 1/6, 1/4, 1/3, 1/2, 3/4, and 1.

SINGLE-PHASE MOTORS are available in a variety of types, but practically all operate essentially as squirrel-cage induction motors. Since a normal single-phase induction motor does not develop any starting torque, it must be provided with an auxiliary starting-winding or phase-shifting device. The essential difference between the various types of single-phase motors lies in the type of starting-winding or phase-shifting device employed.

Shaded-pole motors have a squirrel-cage rotor winding and a concentrically wound single-phase stator winding. A portion of each stator pole is "shaded" by a single turn of heavy copper wire. Current that is out of phase with the current in the main winding is induced in the shading coil. A revolving magnetic field results, similar to that of a polyphase induction motor, and hence starting torque is developed. Motors of this type are inefficient and have rather low starting torque. However, in the small sizes they are the most economical design. They are used in ratings up to about $^1/_{25}$ hp (40 mhp) for small fans, timing devices, instruments, relays, toys, sign flashers, small motion-picture projectors, and miscellaneous control devices. They are designed for a particular direction of rotation that cannot be changed after the motor is built.

Split-phase motors have squirrel-cage rotors and two-phase stators with one of the phases wound for relatively high resistance. Since the ratios of resistance to reactance of the two phases are not equal, the currents in the phases are not of the same time phase and approach the relations of a true two-phase motor. Hence, a revolving magnetic field is produced and starting torque is developed. The high-resistance winding, because of its high losses, must be disconnected when the motor gets up to speed. The disconnecting switch usually is operated by a centrifugal mechanism on the rotor. These motors are used in ratings from $^1/_{60}$ to $^1/_3$ hp for a variety of applications whose power requirements fall within this range. Typical applications are advertising devices, fans, blowers, oil burners, kitchen appliances, machine tools, mechanical office equipment, pumps, textile and laboratory equipment, and washing and ironing machines.

Repulsion-start induction and repulsion-induction motors have a single-phase distributed winding on the stator, and a commutated winding, similar to that of a d-c machine, on the rotor. The brushes bearing on the commutator are spaced 180 electrical degrees apart and are short-circuited. Repulsion-start induction motors also have a centrifugal device which short-circuits all commutator segments when the motor has accelerated to a certain speed, converting the rotor winding to a squirrel-cage winding. Thus the motor starts as a repulsion motor but runs as an induction motor. The centrifugal device sometimes is arranged to lift the brushes of the motor as well as to short-circuit the commutator segments. Repulsion-induction motors, on the other hand, do not have any centrifugal device but instead have a squirrel-cage winding on the rotor in addition to the commutated winding. The squirrel-cage winding is placed below the commutated winding and is arranged to have a high reactance. At low motor speeds (high slip frequency) very little current flows in the squirrel-cage winding, and the motor starts essentially as a repulsion motor. When the motor reaches full speed, the frequency of the induced currents is low, so that considerable current flows in the squirrel-cage winding and the motor operates essentially as an induction motor. Repulsion-start induction or repulsion-induction motors in ratings of $^1/_6$ hp and larger have been used in such devices as pumps, compressors, machine tools, hoists, conveyors, garage and service station equipment, and woodworking machinery. However, capacitor-start motors are now more generally employed.

Capacitor motors are of three types. The *capacitor-start* motor has a squirrel-cage rotor and a main winding and an auxiliary winding on the stator. A capacitor (frequently mounted on top of the motor) is connected in series with the auxiliary winding to provide the necessary shift in time phase of the current flowing through it. Since the capacitor is intermittently rated, it must be disconnected for normal operation. Disconnection usually is accomplished by a switch operated by a centrifugal mechanism on the rotor. Capacitor-start motors are available in ratings from $^1/_6$ to 1 hp and are used on fans, pumps, compressors, coal stokers, machine tools, and in other industrial applications.

The *permanent-split* capacitor motor is similar to the capacitor-start motor except that in the permanent-split type the capacitor in series with the auxiliary winding is continuously rated and so is left in the circuit during running as well as starting. Furthermore, the capacitor is selected to give best operation at full speed (maximum efficiency and minimum noise and vibration) at a sacrifice in starting characteristics. As a result, these motors develop only 40 to 60% starting torque and can be applied only to easily started loads such as direct-connected fans and blowers. Two-speed or adjustable-speed operation of fans and blowers is often provided by controllers that in effect vary the voltage impressed on the motor. This variation raises or lowers the motor's speed-torque curve, which then intersects the blower- or fan-torque curve at different speeds.

The *two-value capacitor* motor uses different values of effective capacitance in series with the auxiliary winding during starting and running. It usually has a continuously rated capacitor that remains in the circuit during starting and running and an intermittently rated capacitor that is in the circuit during starting only. Thus this motor has a high starting torque as well as good running characteristics. However, it is seldom employed in small-power ratings.

The series-wound single-phase motor is similar in construction and principle of operation to the series-wound d-c motor. It employs a stationary field winding and a commutated rotor (armature winding), the two being connected in series. Like the series d-c motor, it is of the varying-speed type, that is, its speed increases as the load decreases. Usual full-load speeds range from 5000 to 10,000 rpm; no-load speeds range from 150 to 200% of full-load speeds. The principal advantages of series motors are their extreme compactness and light weight in relation to the power delivered. Hence, they are particularly suitable for portable devices, such as vacuum cleaners and portable tools of all types. Since they usually are built in the power-using device, they frequently are supplied as "motor parts," that is, the bare stator and rotor, without bearings or feet, the bearings and stator support being incorporated in the driven machine.

Synchronous motors in the small-power sizes usually are built as *reluctance motors.* The stators are similar to those of single-phase induction motors and may be of the shaded-pole, split-phase, or capacitor type. The rotor has a squirrel-cage winding, and the core is shaped to provide projections (salient poles) corresponding to the number of poles for which the stator is wound. The motor starts as an induction motor, but, upon reaching a speed near synchronism, it "pulls into step," owing to the effect of the salient poles, and operates at exactly synchronous speed. Unlike the large-power synchronous motor, which has a field winding on the rotor supplied with d-c excitation and which operates at unity or leading power factor with high efficiency, the reluctance motor operates at a lagging power factor and has a rather low efficiency. Hence, it is used only where exact synchronous speed is required. Typical applications include electric clocks, time switches, relays, and recording meters.

Another type of synchronous motor available in the small-power sizes is the *hysteresis motor.* Its construction is similar to that of the reluctance motor except that the rotor is perfectly cylindrical and does not have a squirrel-cage winding. Its operation depends on the permanent magnetism induced in the rotor by the magnetic field of the stator. It develops a constant torque from zero to synchronous speed. It is used principally for phonograph-turntable drive.

Also available are **synchronous inductor** motors, which are essentially inductor alternators operated as motors. They operate at low speeds such as 75 or 100 rpm and with a built-in gear reduction have output shaft speeds of 1 to 4 rpm. They are used on recording instruments, small regulating or controlling devices, and other applications that require low speeds in exact synchronism with the power-supply frequency.

POLYPHASE MOTORS. Squirrel-cage induction motors meet the requirements of practically all applications in the small-power field. They develop a high starting torque with a starting current well within the limits of polyphase systems. Their efficiencies are slightly higher than those of single-phase motors. Polyphase motors are generally available in sizes of $^1/_6$ hp and larger at 110, 220, 440, and 550 volts.

Synchronous motors of the reluctance type are employed when exact synchronous speed is required. They are similar to the single-phase motors of this type except that no starting-winding or phase-splitting device is necessary.

High-frequency polyphase motors often are used on portable tools, textile machinery, and other applications in which the number of high-speed motors required justifies the installation of special generating or conversion equipment to supply the high-frequency power. A frequency of 180 cycles usually is employed, giving a synchronous speed (2-pole motors) of 10,800 rpm. The advantages over 60-cycle induction motors are savings in cost, size, and weight. As compared to the series motor, the high-frequency polyphase motor requires less maintenance, since it has no brushes or commutator and has a higher efficiency and better speed regulation.

DIRECT-CURRENT MOTORS of the **shunt-wound** or **compound-wound** type are used for most applications. In the smaller sizes (approximately $^1/_{20}$ hp and below) shunt-wound motors are standard, whereas in the larger sizes the standard is the compound-wound. The **series-wound** motor is used where its varying-speed characteristics and high no-load speed are not objectionable. It develops a very high starting torque but has a limited output at the higher speeds.

UNIVERSAL MOTORS, which operate from either single-phase a-c or from d-c power, are all of the series-wound type; that is, they have a field winding on the stator which is connected in series with a commutated winding on the rotor. The **straight series-wound** motor has a salient pole field winding on the stator, like the ordinary d-c motor, whereas the **compensated series-wound** motor has a distributed winding arranged in slots around the periphery of the stator. The latter has the advantages of better speed regulation and commutation and higher starting torque; the former is simpler in construction and more easily ventilated. In general, the straight series-wound motor is used in sizes up to approximately $^1/_3$ hp, and the compensated series-wound motor is employed in the larger sizes. It is not feasible to build universal motors for low-speed operation, the minimum full-load speed ranging from 1800 rpm on the larger sizes to 5000 rpm on the smaller sizes. Typical full-load speed ratings are 5000, 7500, and 10,000 rpm, with no-load speeds ranging from 12,000 to 18,000 rpm. Universal motors have a variety of applications including portable tools, office appliances, vacuum cleaners, motion-picture projectors, kitchen appliances, sewing machines, and portable equipment in general.

Operating Characteristics

TEMPERATURE RISE. The standard temperature rise of small-power motors, measured by thermometer, is 40 C for general-purpose (open-type) motors, 55 C for all totally enclosed machines, and 40 or 50 C for special-purpose (open or drip-proof) motors, depending on the application. The temperature rise of a-c motors may, at the option of the manufacturer, be determined by the resistance method, in which case a 10 C higher temperature rise is allowed.

EFFICIENCY. Typical full-load efficiencies and power factors of general-purpose 60-cycle 1725-rpm motors are shown in Fig. 2. The efficiencies of d-c motors are comparable with those of single-phase a-c motors.

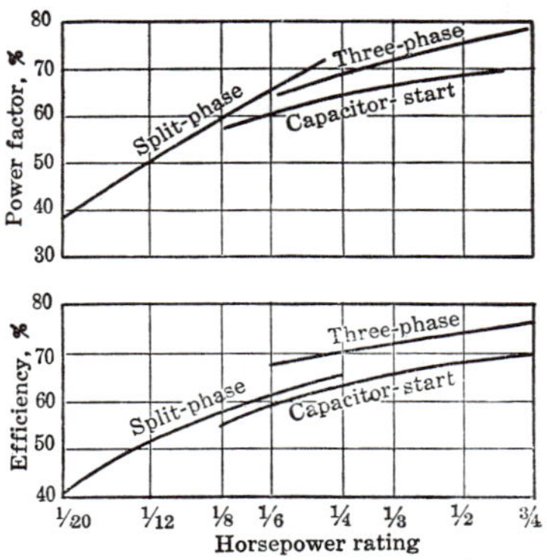

FIG. 2. Full-load efficiencies and power factors of 60-cycle 1725-rpm motors.

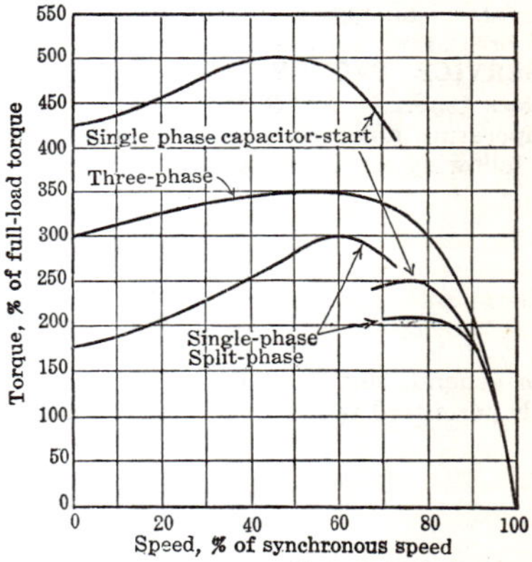

FIG. 3. Typical speed-torque curves for general-purpose small-power motors.

STARTING CURRENT. The maximum locked-rotor starting currents of single-phase motors as established by NEMA are given in Table 8. The locked-rotor currents of polyphase squirrel-cage induction motors vary between 400 and 800% of full-load current.

Table 8. Maximum Locked-rotor Starting Currents of Single-phase Small-power Motors, 60 Cycles, 900 to 3600 Rpm Inclusive *

Horsepower	1/6 and smaller	1/4	1/3	1/2	3/4
			Current in Amperes		
110 volts	20	23	31	45	61
220 volts	10	11 1/2	15 1/2	22 1/2	30 1/2

* Except the split-phase motors used on washing and ironing machines.

TORQUES. The breakdown torques for general-purpose single-phase motors standardized by NEMA are given in Table 9. The breakdown torques of special-purpose motors lie between those listed in Table 9 for the horsepower rating involved and those listed for the next smaller horsepower rating. If a higher breakdown torque is required, a motor of larger horsepower rating should be specified. The breakdown torques of squirrel-cage polyphase motors are generally equal to or greater than those standardized for single-phase motors of corresponding ratings. Typical speed-torque curves for general-purpose motors are shown in Fig. 3. Direct-current motors develop 500% or more starting torque on full voltage.

VOLTAGE VARIATION. Small-power motors should operate successfully at rated load (and rated frequency for a-c motors) at any voltage not more than 10% above or below normal, but not necessarily in accordance with the standards of performance established for operation at normal rating. An exception to this rule is the series-type

Table 9. Breakdown Torques of General-purpose Single-phase Motors

(NEMA Standard)

Synchronous Rpm	3600	3000	1800	1500	1200	1000	900
Approximate Full-load Rpm	3450	2850	1725	1425	1140	950	850
Horsepower Rating	Breakdown Torque, ounce-feet						
1/20	3.7	4.4	7.1	8.5	10.4	12.4	13.5
1/12	6.0	7.2	11.5	13.8	16.5	19.8	21.5
1/8	8.7	10.5	16.5	19.8	24.1	28.9	31.5
1/6	11.5	13.8	21.5	25.8	31.5	37.8	40.5
1/4	16.5	19.8	31.5	37.8	44.0	53.0	58.0
1/3	21.5	25.8	40.5	48.5	58.0	69.5	77.0
1/2	31.5	37.8	58.0	69.5	82.5	99.0	
3/4	44.0	53.0	82.5	99.0			
1	58.0	69.5					

universal motor, for which the maximum allowable voltage variation is 6%. The starting currents of all types are proportional to the applied voltage, and the starting, pull-up, and breakdown torques are approximately proportional to the square of the voltage.

SERVICE FACTOR. Manufacturers of general-purpose (open-type) small-power motors generally authorize operation (at rated voltage and frequency and in an ambient temperature not exceeding 40 C) at a continuous load equal to rated load multiplied by the following service factors.

Horsepower	Service Factor
1/20, 1/12, and 1/8	1.4
1/6, 1/4, and 1/3	1.35
1/2 and 3/4	1.25
1 (3600 rpm only)	1.25

In general, slight differences in efficiencies and power factors can be expected at the loads permitted by the service factor as compared with those at rated load.

Mechanical Features

Small-power motors can be obtained as bare rotors and stators (with or without feet) for mounting directly on the driven machine. However, the usual construction consists of a stator frame provided with mounting feet and end-shield-type bearings. Motors with the drive end shield machined for bolting to the driven machine are available. Vertical motors, as well as the standard horizontal motors, are available in all types.

Sleeve bearings packed with wool yarn are standard for most open or drip-proof horizontal motors, whereas grease-lubricated ball bearings are usually furnished on enclosed or vertical motors. Ball bearings may also be employed on horizontal motors that are required to take considerable axial thrust.

Single-phase induction motors inherently produce a pulsating torque, resulting in a vibration of the motor frame which, if transmitted to the motor support, may cause an objectionable noise. Hence, such motors frequently are furnished with a resilient rather than a rigid connection between the motor frame and the motor base. This resilient connection may consist of rubber cushions, or of metal springs of sufficient flexibility to prevent or greatly reduce the transmission of vibration to the supporting structure. Some manufacturers offer motors with a resilient mounting combined with an automatic-belt-tightener base, which maintains a belt tension just sufficient to prevent slippage at all loads throughout the operating range.

Gear motors of the various electrical types are available in ratings from 1/6 to 3/4 hp with output shaft speeds of 5.7 to 520 rpm. Both concentric and right-angle shaft types are obtainable. The latter employ worm-and-wheel reduction units and have lower costs but considerably lower efficiency. The horsepower output rating of a gear motor may be much less than the rating of the motor itself, the difference depending on the type of gear and the speed ratio.

Standard motors are of open or drip-proof construction, but practically all types can be obtained in the totally enclosed construction. Most types are also available in explosionproof construction for Class I, Group D, or Class II, Group E, F, and G atmospheres. (See p. 16-19.)

WEIGHTS AND DIMENSIONS. Figure 4 gives three sets of mounting dimensions standardized by NEMA for small-power motors rated 1/20 hp and larger. Each set is identified by a frame number equal to sixteen times the D dimension. The frame for any

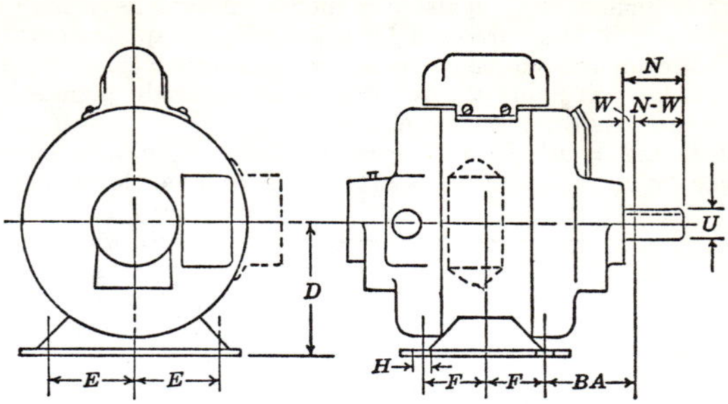

All dimensions in inches

Frame No.	*D*	*E*	*F*	*BA*	*H*		*U*	*N–W*	Flat	Keyway		
										Width	Depth	Length
42	2 5/8	1 3/4	27/32	2 1/16	9/32	slot	3/8	1 1/8	3/64	....	....	....
56	3 1/2	2 7/16	1 1/2	2 3/4	11/32	slot	5/8	1 7/8	.....	3/16	3/32	1 3/8
66	4 1/8	2 15/16	2 1/2	3 1/8	13/32	slot	3/4	2 1/4	.....	3/16	3/32	1 7/8

FIG. 4. Standardized dimensions of small-power motors, 1/20 hp and larger.

particular rating is determined by the motor manufacturer. Table 10, however, shows typical frame assignments for the popular ratings of general-purpose open-type motors. Approximate weights for the various frames are also given in Table 10.

Table 10. Typical Frame Sizes of General-purpose Open-type Small-power Motors

Approximate Full-load Rpm	Horsepower Ratings Frame 42	Frame 56	Frame 66
3450 (60 cycles, and d-c)	1/12	1/3, 1/2	3/4, 1
2875 (50 cycles)	1/12	1/3, 1/2	3/4, 1
1725 (60 cycles, and d-c)	1/20	1/6, 1/4, 1/3	1/2, 3/4
1425 (50 and 25 cycles)	1/20	1/6, 1/4, 1/3	1/2, 3/4
1140 (60 cycles, and d-c)	...	1/6, 1/4	1/3, 1/2
Approximate weight, pounds	10–15	22–38	46–77

5. SINGLE-PHASE LARGE-POWER MOTORS

Standard horsepower and speed ratings of single-phase large-power motors are given in Table 11. The following types are available: (1) repulsion-start induction (and repul-

Table 11. Standard Horsepower and Speed Ratings of Single-phase Large-power Motors

Horse-power Rating	60 Cycles, Synchronous Rpm				50 Cycles, Synchronous Rpm				25 Cycles, Synchronous Rpm	
1/2	....	....	....	900	....	....	....	750	....	......
3/4	....	....	1200	900	....	....	1000	750	....	......
1	....	1800	1200	900	....	1500	1000	750	1500	......
1 1/2	3600	1800	1200	900	3000	1500	1000	750	1500	......
2	3600	1800	1200	900	3000	1500	1000	750	1500	750
3	3600	1800	1200	900	3000	1500	1000	750	1500	750
5	3600	1800	1200	900	3000	1500	1000	750	1500	750
7 1/2	3600	1800	1200	900	3000	1500	1000	750	1500	750
10	3600	1800	1200	900	3000	1500	1000	750	1500	750
15	3600	1800	1200	900	3000	1500	1000	750	1500	750
20	3600	1800	1200	900	3000	1500	1000	750	1500	750
25	3600	1800	1200	900	3000	1500	1000	750	1500	750

sion-induction), (2) capacitor, (3) brush-shifting repulsion, and (4) polyphase induction with phase-shifting starter.

For details of construction and principle of operation of types 1 and 2, see p. 16-21.

Repulsion-start induction and repulsion-induction motors have high starting torque and low starting current. They are used for practically all applications requiring single-phase motors. However, the simple construction, quiet operation, and low maintenance of the capacitor motor has made the latter type increasingly popular, and it is being successfully used in the majority of applications.

Capacitor motors are available in normal-starting-torque and high-starting-torque types. The former are usually capacitor-start motors for fans, blowers, centrifugal pumps, and many general-purpose applications with moderate starting-torque requirements. The high-starting-torque motors are either capacitor-start or two-value capacitor motors used in applications requiring high starting torque, such as reciprocating pumps, reciprocating compressors, and loaded conveyors.

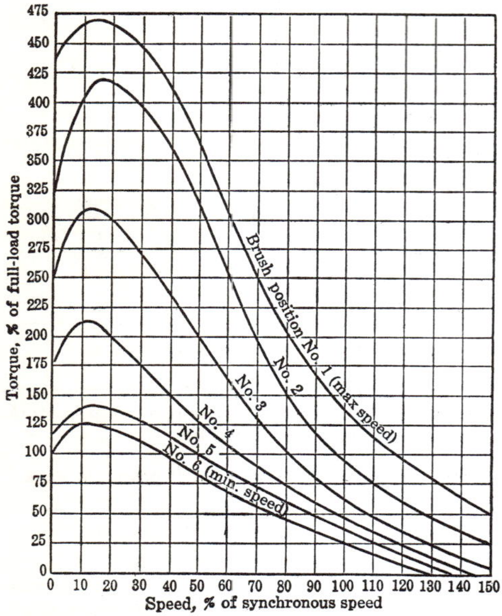

FIG. 5. Speed-torque curves of brush-shifting repulsion motor.

The brush-shifting repulsion motor is a repulsion motor provided with a mechanism for shifting the angular position of the brushes that bear on the commutator of the motor. Shifting modifies the speed-torque curves of the motor, as shown in Fig. 5. As the motor has a varying speed characteristic, the speed obtained with a given brush setting varies over a wide range, depending on the torque required by the load. Thus, to obtain the expected speed range of the motor, it is necessary to match closely the motor rating with power requirement of the load. Ratings available range from $^1/_4$ to 3 hp at 1800 to 600 rpm.

Single-phase motors are seldom used in sizes above 10 hp. Hence, none of the types described above is available in the higher ratings. When a larger single-phase motor is required, it is usual to use a polyphase induction motor with a split-phase starter of the resistance-reactance type. The horsepower output of a polyphase-induction motor operating single-phase is approximately one-half its normal rating.

Operating Characteristics

TEMPERATURE RISE. The standard temperature rise of large-power single-phase motor, measured by thermometer, is 40 C for general-purpose (open-type or drip-proof) motors, 55 C for totally enclosed and totally enclosed fan-cooled machines, and 40 or 50 C for special-purpose motors. The temperature rise may, at the option of the manufacturer, be determined by the resistance method, in which case a 10 C higher temperature rise is allowed for open and totally enclosed motors, and a 5 C higher temperature rise for totally enclosed fan-cooled machines.

For motors having a temperature rise of 40 C, manufacturers usually authorize operation (at rated voltage and frequency and in an ambient temperature not exceeding 40 C) at a continuous load equal to rated load multiplied by the following service factors.

Horsepower	Service Factor
$^1/_2$, $^3/_4$, 1	1.25
1 $^1/_2$, 2	1.20
3 and larger	1.15

EFFICIENCY. Approximate full-load efficiencies and power factors of popular ratings of normal-starting-torque capacitor motors are shown in Fig. 6.

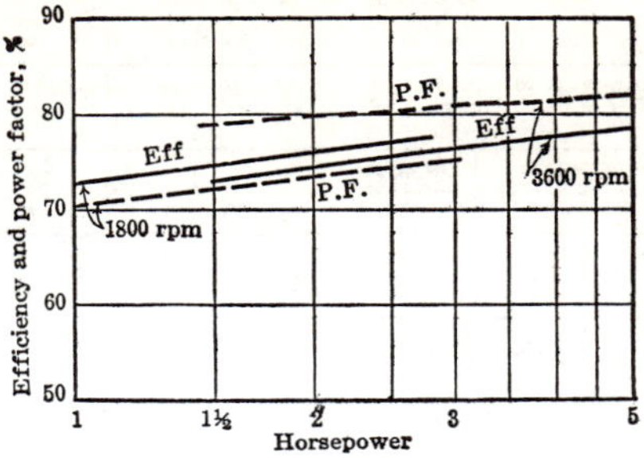

Fig. 6. Approximate full-load efficiency and power factor of normal-starting-torque capacitor motors.

TORQUES AND STARTING CURRENT. Breakdown torques of general-purpose single-phase (large-power) motors as standardized by NEMA are given in Table 12.

Table 12. Breakdown Torques of General-purpose Single-phase Motors

(NEMA Standard)

Synchronous Rpm	3600	3000	1800	1500	1200
Frequency	60	50	60	50	60
Horsepower			Torque in Pound-feet		
3/4	....	....	....	....	6.9
1	....	....	6.8	8.2	9.2
1 1/2	4.6	5.5	10.1	12.1	13.8
2	6.0	7.2	13.0	15.6	18.0
3	8.6	10.2	19.0	22.8	25.8
5	13.5	16.2	30.0	36.0	40.5
7 1/2	20.0	24.0	45.0	54.0	60
10	27.0	32.4	60.0	72.0	....

Starting torques range from approximately 160 to 200% of full-load torque for normal-torque capacitor motors and from approximately 250 to 300% for repulsion-induction, repulsion-start induction, and high-starting-torque capacitor motors. Brush-shifting repulsion motors have at least 300% starting torque and 300% breakdown torque with the brushes in the maximum-speed position. The maximum *starting currents* of large-power single-phase Design M motors as standardized by NEMA are as follows:

Horsepower Rating	Locked Rotor Current, Amperes 115 Volts	230 Volts
1	70	35
1 1/2	..	40
2	..	50
3	..	70
5	..	100
7 1/2	..	150
10	..	200
15	..	300
20	..	400

Design M designates motors that meet these starting currents and the breakdown torques of Table 12. Motors with higher starting currents and the same breakdown torques are designated Design L.

VOLTAGE VARIATION. Large-power, single-phase motors should operate successfully at rated load and frequency at any voltage not more than 10% above or below normal but not necessarily in accordance with the standards of performance established for operation at normal rating. The starting currents are proportional to the applied voltage, and the starting, pull-up, and breakdown torques are proportional to the square of the voltage.

Table 13. Standard Dimensions and Tolerances for NEMA Standardized Frames for D-c and A-c Motors

All dimensions in inches

Frame Number	Key		*AA*, Size of Conduit Minimum	*A* (max)	*B* (max)	*D* *	*E*	*F*	*BA*	*H*	*N-W*	*U*	*V* (min)	*AL*	*AM*	*AO*	*AR*	*AU*	*AW*	*AX*	*AY*, Bases	*AY*, Rails (max)	*BT*
	Width	Thickness																					
203	3/16	3/16	3/4	10	7 1/2	5	4	2 3/4	3 1/8	13/32	2 1/4	3/4	2	14	11	5	4 3/4	1/2	3/4	1 3/4	1/2	4	3
204	3/16	3/16	3/4	10	8 1/2	5	4	3 1/4	3 1/8	13/32	2 1/4	3/4	2	14	12	5	5 1/4	1/2	1 1/4	1 3/4	1/2	4	3
224	1/4	1/4	3/4	11	8 3/4	5 1/2	4 1/2	3 3/8	3 1/2	13/32	3	1	2 3/4	15 1/2	12 1/4	5 1/2	5 3/8	1/2	1 3/8	1 3/4	1/2	4 1/4	3 1/2
225	1/4	1/4	3/4	11	9 1/2	5 1/2	4 1/2	3 3/4	3 1/2	13/32	3	1	2 3/4	15 1/2	13	5 1/2	5 3/4	1/2	1 3/4	1 3/4	1/2	4 1/4	3 1/2
254	1/4	1/4	1	12 1/2	10 3/4	6 1/4	5	4 1/8	4 1/4	17/32	3 3/8	1 1/8	3 1/8	17 3/4	15 1/8	6 1/4	6 5/8	5/8	1 5/8	2	5/8	5	4
284	1/4	1/4	1	14	12 1/2	7	5 1/2	4 3/4	4 3/4	17/32	3 3/4	1 1/4	3 1/2	19 3/4	16 7/8	7	7 1/2	5/8	2	2	5/8	5 3/4	4 1/2
324	3/8	3/8	1 1/4	16	14	8	6 1/4	5 1/4	5 1/4	21/32	4 7/8	1 5/8	4 5/8	22 3/4	19 1/4	8	8 1/2	3/4	2	2 1/2	3/4	6 1/2	5 1/4
326	3/8	3/8	1 1/4	16	15 1/2	8	6 1/4	6	5 1/4	21/32	4 7/8	1 5/8	4 5/8	22 3/4	20 3/4	8	9 1/4	3/4	2 3/4	2 1/2	3/4	6 1/2	5 1/4
364	1/2	1/2	1 1/2	18	15 1/4	9	7	5 5/8	5 7/8	21/32	5 5/8	1 7/8	5 3/8	25 1/2	20 1/2	9	9 1/8	3/4	2 1/8	2 1/2	3/4	7 1/2	6
364S	3/8	3/8	1 1/2	18	15 1/4	9	7	5 5/8	5 7/8	21/32	3 1/4	1 5/8	3	......	......	......	......	......	......	......	......	......	......
365	1/2	1/2	1 1/2	18	16 1/4	9	7	6 1/8	5 7/8	21/32	5 5/8	1 7/8	5 3/8	25 1/2	21 1/2	9	9 5/8	3/4	2 5/8	2 1/2	3/4	7 1/2	6
365S	3/8	3/8	1 1/2	18	16 1/4	9	7	6 1/8	5 7/8	21/32	3 1/4	1 5/8	3	......	......	......	......	......	......	......	......	......	......
404	1/2	1/2	2	20	16 1/4	10	8	6 1/8	6 5/8	13/16	6 3/8	2 1/8	6 1/8	28 3/4	22 3/8	10	9 7/8	7/8	2 3/8	3	7/8	7 7/8	7
404S	1/2	1/2	2	20	16 1/4	10	8	6 1/8	6 5/8	13/16	3 3/4	1 7/8	3 1/2	......	......	......	......	......	......	......	......	......	......
405	1/2	1/2	2	20	17 3/4	10	8	6 7/8	6 5/8	13/16	6 3/8	2 1/8	6 1/8	28 3/4	23 7/8	10	10 5/8	7/8	3 1/8	3	7/8	7 7/8	7
405S	1/2	1/2	2	20	17 3/4	10	8	6 7/8	6 5/8	13/16	3 3/4	1 7/8	3 1/2	......	......	......	......	......	......	......	......	......	......
444	5/8	5/8	2	22	18 1/2	11	9	7 1/4	7 1/2	13/16	7 1/8	2 3/8	6 7/8	31 1/4	24 5/8	11	11	7/8	3 1/2	3	7/8	8 3/8	7 1/2
444S	1/2	1/2	2	22	18 1/2	11	9	7 1/4	7 1/2	13/16	4 1/4	2 1/8	4	......	......	......	......	......	......	......	......	......	......
445	5/8	5/8	2	22	20 1/2	11	9	8 1/4	7 1/2	13/16	7 1/8	2 3/8	6 7/8	31 1/4	26 5/8	11	12	7/8	4 1/2	3	7/8	8 3/8	7 1/2
445S	1/2	1/2	2	22	20 1/2	11	9	8 1/4	7 1/2	13/16	4 1/4	2 1/8	4	......	......	......	......	......	......	......	......	......	......
504U	3/4	3/4	2 1/2	25	21	12 1/2	10	8	8 1/2	15/16	8 5/8	2 7/8	8 3/8	35	28	12 1/2	12 1/2	1	3 1/2	3 1/2	1	9	8
504S	1/2	1/2	2 1/2	25	21	12 1/2	10	8	8 1/2	15/16	4 1/4	2 1/8	4	......	......	......	......	......	......	......	......	......	......
505	3/4	3/4	2 1/2	25	23	12 1/2	10	9	8 1/2	15/16	8 5/8	2 7/8	8 3/8	35	30	12 1/2	13 1/2	1	4 1/2	3 1/2	1	9	8
505S	1/2	1/2	2 1/2	25	23	12 1/2	10	9	8 1/2	15/16	4 1/4	2 1/8	4	......	......	......	......	......	......	......	......	......	......

Note 1. For significance of the above letters see Fig. 7.

Note 2. For tolerances on shaft extension diameters and keyways, see Table 17.

* Dimension *D* will never be greater than the above values, but it may be less so that shims are usually required for coupled or geared machines. When exact dimension is required, shims up to 1/32 in. may be necessary on frame sizes whose dimension *D* is 8 in. and less; on larger frames shims up to 1/16 in. may be necessary.

Mechanical Features

Large-power single-phase motors usually are furnished with feet on the stator frame and with end-shield bearings. They are also available with the drive end shield machined for bolting to the driven machine. Vertical motors, as well as standard horizontal motors, can be obtained.

Sleeve bearings, wool-yarn-packed or oil-ring-lubricated, or grease-lubricated ball bearings are standard for horizontal open or drip-proof motors. Ball bearings are standard

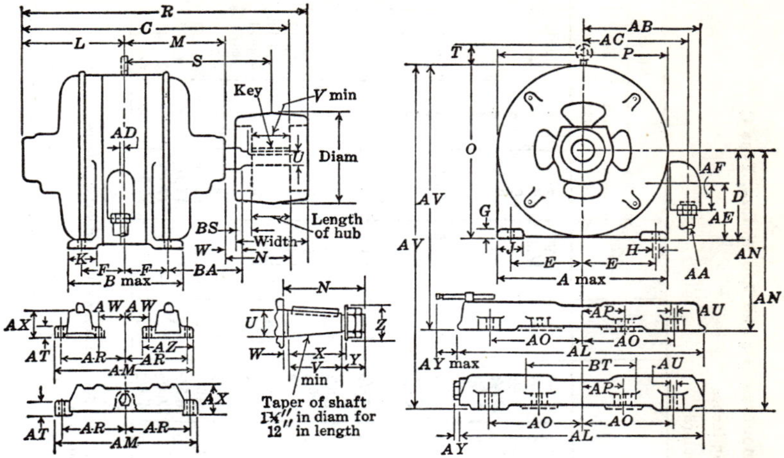

FIG. 7. Standard lettering used in designating dimensions of horizontal motors. See Table 13.

on vertical and totally enclosed motors. All motors can be obtained with pulleys and sliding bases for two-bearing belt drive. Gear motors are available in ratings from 1 to 3 hp with output-shaft speeds of 13.5 to 780 rpm.

Standard motors are of the open or drip-proof type, but totally enclosed and explosion-proof (Class I, Group D) or dust-explosion-proof (Class II, Groups E, F, and G) motors are available.

WEIGHT AND DIMENSIONS. Mounting dimensions with corresponding frame numbers standardized by NEMA for large-power motors are shown in Table 13 and Fig. 7 (horizontal motors), Table 14 and Fig. 8 (horizontal or vertical type-C face-mounted motors), Table 15 and Fig. 9 (horizontal or vertical type-D flange-mounted motors), and

(*Continued on p. 16-33*)

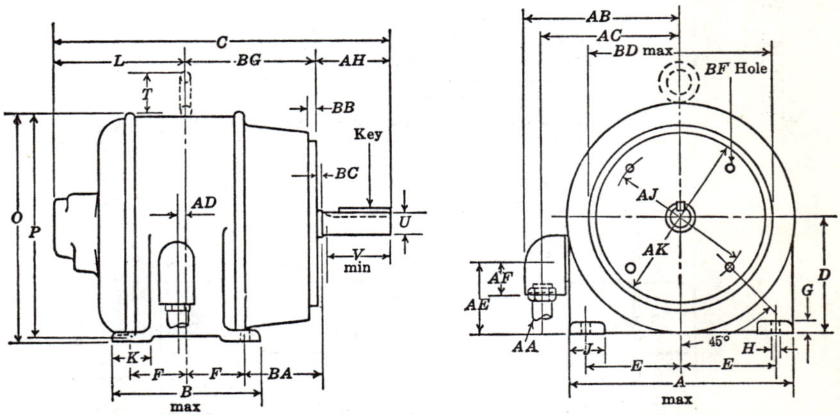

FIG. 8. Standard lettering for dimensions of Type C face-mounted motors with feet. See Table 14.

Table 14. Standard Dimensions for Face-mounted D-c and A-c Motors, Type C

All dimensions in inches

(NEMA Standard)

Frame Number	A (max)	B (max)	D *	E	F	H, Hole	U	AH	AJ	AK †	BA	BB	BC	BD (max)	BF Hole: Number	BF Hole: Tap Size	BF Hole: Min Depth	Key: Width	Key: Thickness
203C	10	7 1/2	5	4	2 3/4	13/32	3/4	2 5/16	7 1/4	8 1/2	3 1/8	1/4	1/16	9	4	1/2–13	3/4	3/16	3/16
204C	10	8 1/2	5	4	3 1/4	13/32	3/4	2 5/16	7 1/4	8 1/2	3 1/8	1/4	1/16	9	4	1/2–13	3/4	3/16	3/16
224C	11	8 3/4	5 1/2	4 1/2	3 3/8	13/32	1	3 1/8	7 1/4	8 1/2	3 1/2	1/4	1/8	9	4	1/2–13	3/4	1/4	1/4
225C	11	9 1/2	5 1/2	4 1/2	3 3/4	13/32	1	3 1/8	7 1/4	8 1/2	3 1/2	1/4	1/8	9	4	1/2–13	3/4	1/4	1/4
254C	12 1/2	10 3/4	6 1/4	5	4 1/8	17/32	1 1/8	3 3/4	7 1/4	8 1/2	4 1/4	1/4	3/8	10	4	1/2–13	3/4	1/4	1/4
284C	14	12 1/2	7	5 1/2	4 3/4	17/32	1 1/4	4 7/16	9	10 1/2	4 3/4	1/4	11/16	11 1/4	4	1/2–13	3/4	1/4	1/4
324C	16	14	8	6 1/4	5 1/4	21/32	1 5/8	5 5/16	9	10 1/2	5 1/4	1/4	7/16	12 1/2	4	1/2–13	3/4	3/8	3/8
326C	16	15 1/2	8	6 1/4	6	21/32	1 5/8	5 5/16	9	10 1/2	5 1/4	1/4	7/16	12 1/2	4	1/2–13	3/4	3/8	3/8
364C	18	15 1/4	9	7	5 5/8	21/32	1 7/8	5 3/8	11	12 1/2	5 7/8	1/4	−1/4	14	4	5/8–11	15/16	1/2	1/2
365C	18	16 1/4	9	7	6 1/8	21/32	1 7/8	5 3/8	11	12 1/2	5 7/8	1/4	−1/4	14	4	5/8–11	15/16	1/2	1/2
404C	20	16 1/4	10	8	6 1/8	13/16	2 1/8	6 1/8	11	12 1/2	6 5/8	1/4	−1/4	15 1/2	4	5/8–11	15/16	1/2	1/2
405C	20	17 3/4	10	8	6 7/8	13/16	2 1/8	6 1/8	11	12 1/2	6 5/8	1/4	−1/4	15 1/2	4	5/8–11	15/16	1/2	1/2
444C	22	18 1/2	11	9	7 1/4	13/16	2 3/8	6 7/8	14	16	7 1/2	1/4	−1/4	17 1/2	4	5/8–11	15/16	5/8	5/8
445C	22	20 1/2	11	9	8 1/4	13/16	2 3/8	6 7/8	14	16	7 1/2	1/4	−1/4	17 1/2	4	5/8–11	15/16	5/8	5/8
364SC	18	15 1/4	9	7	5 5/8	21/32	1 5/8	3	11	12 1/2	5 7/8	1/4	−1/4	14	4	5/8–11	15/16	3/8	3/8
365SC	18	16 1/4	9	7	6 1/8	21/32	1 5/8	3	11	12 1/2	5 7/8	1/4	−1/4	14	4	5/8–11	15/16	3/8	3/8
404SC	20	16 1/4	10	8	6 1/8	13/16	1 7/8	3 1/2	11	12 1/2	6 5/8	1/4	−1/4	15 1/2	4	5/8–11	15/16	1/2	1/2
405SC	20	17 3/4	10	8	6 7/8	13/16	1 7/8	3 1/2	11	12 1/2	6 5/8	1/4	−1/4	15 1/2	4	5/8–11	15/16	1/2	1/2
444SC	22	18 1/2	11	9	7 1/4	13/16	2 1/8	4	14	16	7 1/2	1/4	−1/4	17 1/2	4	5/8–11	15/16	1/2	1/2
445SC	22	20 1/2	11	9	8 1/4	13/16	2 1/8	4	14	16	7 1/2	1/4	−1/4	17 1/2	4	5/8–11	15/16	1/2	1/2

Note 1. See Fig. 8 for the meaning of the letter dimensions.

Note 2. These dimensions may also be applied to vertical motors, in which case the frame designation may have suffix letters CV or SCV.

Note 3. Dimensions *A*, *B*, *D*, *E*, *F*, and *H* do not apply to machines furnished without feet.

* Dimension *D* will never be greater than the above values. When exact dimension is required, shims up to 1/32 in. may be necessary.

† The tolerance of the *AK* dimension is +0.000, −0.003 in., for frames 203–326, inclusive, and +0.000, −0.005 in., for frames 364–445, inclusive.

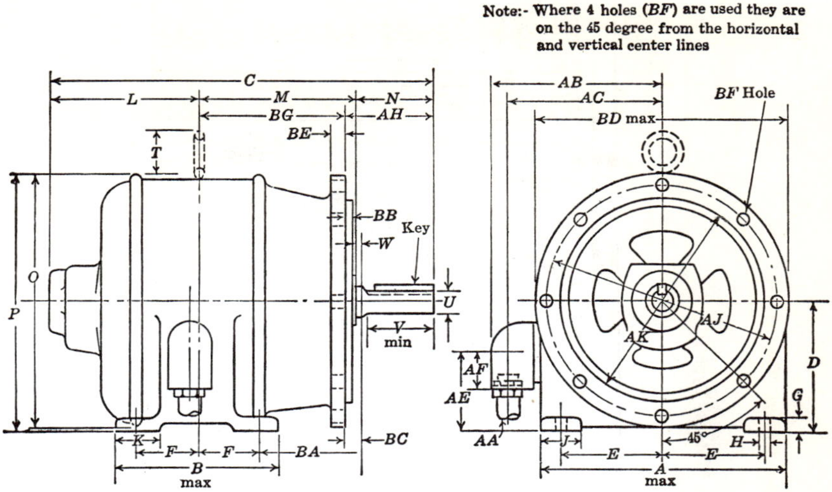

FIG. 9. Standard lettering for dimensions of Type D flange-mounted motors with feet. See Table 15.

Table 15. Standard Dimensions for Flange-mounted D-c and A-c Motors, Type D

All dimensions in inches

(NEMA Standard)

Frame Number	*AK* *	*AJ*	*BD* (max)	*AH*	*BC*	*U*	Key		Clearance Holes *BF*	
							Width	Thickness	Size	Number
203D	9	10	11	2 1/4	0	3/4	3/16	3/16	17/32	4
204D	9	10	11	2 1/4	0	3/4	3/16	3/16	17/32	4
224D	9	10	11	3	0	1	1/4	1/4	17/32	4
225D	9	10	11	3	0	1	1/4	1/4	17/32	4
254D	9	10	11	3 3/8	0	1 1/8	1/4	1/4	17/32	4
284D	11	12 1/2	14	3 3/4	0	1 1/4	1/4	1/4	13/16	4
324D	11	12 1/2	14	4 7/8	0	1 5/8	3/8	3/8	13/16	4
326D	11	12 1/2	14	4 7/8	0	1 5/8	3/8	3/8	13/16	4
364D	14	16	18	5 5/8	0	1 7/8	1/2	1/2	13/16	4
365D	14	16	18	5 5/8	0	1 7/8	1/2	1/2	13/16	4
404D	14	16	18	6 3/8	0	2 1/8	1/2	1/2	13/16	4
405D	14	16	18	6 3/8	0	2 1/8	1/2	1/2	13/16	4
444D	18	20	22	7 1/8	0	2 3/8	5/8	5/8	13/16	8
445D	18	20	22	7 1/8	0	2 3/8	5/8	5/8	13/16	8
504D	18	22	25	8 5/8	0	2 7/8	3/4	3/4	13/16	8
505D	18	22	25	8 5/8	0	2 7/8	3/4	3/4	13/16	8
364SD	14	16	18	3 1/4	0	1 5/8	3/8	3/8	13/16	4
365SD	14	16	18	3 1/4	0	1 5/8	3/8	3/8	13/16	4
404SD	14	16	18	3 3/4	0	1 7/8	1/2	1/2	13/16	4
405SD	14	16	18	3 3/4	0	1 7/8	1/2	1/2	13/16	4
444SD	18	20	22	4 1/4	0	2 1/8	1/2	1/2	13/16	8
445SD	18	20	22	4 1/4	0	2 1/8	1/2	1/2	13/16	8
504SD	18	22	25	4 1/4	0	2 1/8	1/2	1/2	13/16	8
505SD	18	22	25	4 1/4	0	2 1/8	1/2	1/2	13/16	8

Note 1. See Fig. 9 for the meaning of the letter dimensions.

Note 2. These dimensions may also be applied to vertical motors, in which case the frame designation may have the suffix letters DV or SDV.

Note 3. For dimensions *A*, *B*, *D*, *E*, *F*, *BA*, *H*, and *V* of Fig. 9, see Table 13.

* The tolerance for the *AK* dimension shall be +0.000, −0.003 in., for frames 203–326, inclusive, and +0.000, −0.005 in., for frames 364–505, inclusive.

Table 16. Standard Dimensions for Vertical Hollow-shaft Motors, Type P

All dimensions in inches

(NEMA Standard)

Frame Numbers	*BD*	*AK* *	*AJ*	*BB*	*BF* Hole			Standard Clutch						Maximum Bore Clutch					
								BX ‡	*BY* Hole		*BZ*	Key		*BX* ‡	*BY* Hole		*BZ*	Key	
					Number	Tap Size †	Clearance †		Number	Tap Size		Width	Thickness		Number	Tap Size		Width	Thickness
203P, 204P, 224P, 225P	10	8 1/4	9 1/8	3/16	4	3/8–16	7/16	0.751	4	.190–32	1 3/8	3/16	3/16	0.751	4	.190–32	1 3/8	3/16	3/16
254P, 284P	10	8 1/4	9 1/8	3/16	4	3/8–16	7/16	0.751	4	.190–32	1 3/8	3/16	3/16	1.001	4	.190–32	1 3/8	1/4	1/4
324PH, 326PH	12	8 1/4	9 1/8	3/16	4	3/8–16	7/16	1.001	4	.190–32	1 3/8	1/4	1/4	1.188	4	.190–32	1 3/4	1/4	1/4
324P, 326P	16 1/2	13 1/2	14 3/4	1/4	4	5/8–11	11/16	1.001	4	.190–32	1 3/8	1/4	1/4	1.188	4	.190–32	1 3/4	1/4	1/4
364P	16 1/2	13 1/2	14 3/4	1/4	4	5/8–11	11/16	1.188	4	.190–32	1 3/4	1/4	1/4	1.188	4	.190–32	1 3/4	1/4	1/4
365P	16 1/2	13 1/2	14 3/4	1/4	4	5/8–11	11/16	1.188	4	.190–32	1 3/4	1/4	1/4	1.501	4	1/4–20	2 1/8	3/8	3/8
404P, 405P	20	13 1/2	14 3/4	1/4	4	5/8–11	11/16	1.501	4	1/4–20	2 1/8	3/8	3/8	1.501	4	1/4–20	2 1/8	3/8	3/8
444P, 445P	20	13 1/2	14 3/4	1/4	4	5/8–11	11/16	1.501	4	1/4–20	2 1/8	3/8	3/8	1.751	4	1/4–20	2 1/2	3/8	3/8
504P, 505P	24 1/2	13 1/2	14 3/4	1/4	4	5/8–11	11/16	1.938	4	1/4–20	2 1/2	1/2	1/2	1.938	4	1/4–20	2 1/2	1/2	1/2
150–200 §	24 1/2	13 1/2	14 3/4	1/4	4	5/8–11	11/16	1.938	4	1/4–20	2 1/2	1/2	1/2	2.126	4	3/8–16	3 1/4	1/2	1/2
250–300 §	30 1/2	22	26	1/4	4	3/4–10	13/16	2.188	4	3/8–16	3 1/4	1/2	1/2	2.501	4	3/8–16	3 1/4	5/8	5/8
350–500 §	30 1/2	22	26	1/4	4	3/4–10	13/16	2.438	4	3/8–16	3 1/4	5/8	5/8	2.501	4	3/8–16	3 1/4	5/8	5/8

Note 1. See Fig. 10 for meaning of letter dimensions.
Note 2. Dimensions *BD*, *AK*, *AJ*, *BB*, and *BF* (clearance hole) may also be applied to vertical solid-shaft motors.
* The tolerance for the *AK* dimension is −0.000, +0.005 in.
† Either tap size hole or clearance hole should be specified.
‡ The tolerance for the BX dimension is −0.000, +0.001 in., up to and including 1 1/2-in. diameter; −0.000, +0.0015 in., 1 1/2-in. to 2 7/16-in. diameter.
§ Horsepower at 1800 rpm.

Table 16 and Fig. 10 (vertical hollow-shaft motors, type P). The tolerances on shaft-extension diameters and keyways are given in Table 17.

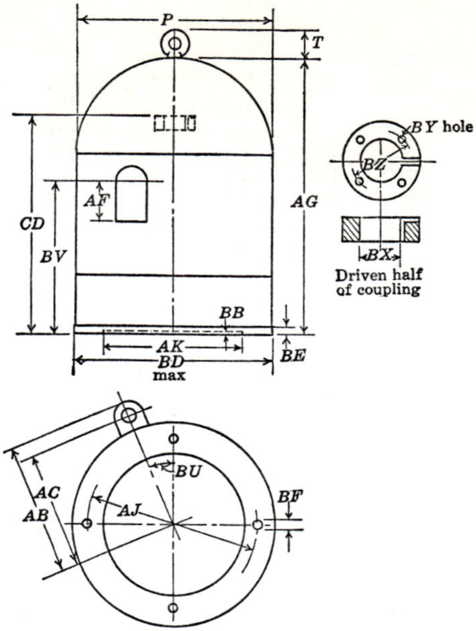

FIG. 10. Standard lettering for dimensions of vertical hollow-shaft motors, Type P. See Table 16.

Table 17. Tolerances on Shaft Extension Diameters and Keyways

(NEMA Standard)

1. Tolerance on shaft extension diameters
   Shaft diameters, 1/4 to 1 1/2 in., inclusive: +0.000, −0.0005 in.
   Shaft diameters, 1 5/8 to 3 3/8 inches inclusive: +0.000, −0.001 in.
2. Tolerance on width of keyways
   Width, 3/16 to 3/4 in., inclusive: +0.000, −0.002 in.
3. Tolerance for making keyways parallel to center line of shaft
   Shaft diameters, 3/4 to 1 1/2 in., inclusive: 0.002 in.
   Shaft diameters, 1 5/8 to 3 3/8 in., inclusive: 0.003 in.
4. Tolerance on depth of keyways:

Shaft Diameter, inches	Key Dimensions, inches	Bottom of Keyseat to Opposite Side of Shaft, inches
5/8	3/16 × 3/16	0.517–0.502
3/4	3/16 × 3/16	0.644–0.629
1	1/4 × 1/4	0.859–0.844
1 1/8	1/4 × 1/4	0.986–0.971
1 1/4	1/4 × 1/4	1.112–1.097
1 5/8	3/8 × 3/8	1.416–1.401
1 3/4	3/8 × 3/8	1.542–1.527
1 7/8	1/2 × 1/2	1.591–1.576
2 1/8	1/2 × 1/2	1.845–1.830
2 3/8	5/8 × 5/8	2.021–2.006
2 5/8	5/8 × 5/8	2.275–2.260
2 7/8	3/4 × 3/4	2.450–2.435

The frame sizes assigned by NEMA to open, totally enclosed, and totally enclosed fan-cooled 60-cycle single-phase motors are given in Table 18. Approximate net weights are given in Table 19.

Table 18. Frame Sizes of Horizontal and Vertical 115- and 230-volt Single-phase 60-cycle Motors

(NEMA Standard)

Synchronous Rpm	Horsepower 1/2	3/4	1	1 1/2	2	3	5	7 1/2
				Open Motors				
3600	...	...	...	203	204	224	225	254
1800	...	...	203	204	224	225	254	...
1200	...	204	224	225	254	...	...	...
900	224	225	254	254	...	...	...	...
				Totally Enclosed Motors				
1800	...	...	224	225	254	...	...	...
1200	...	224	225	254	...	...	...	...
900	224	225	254	...	...	...	...	...
				Totally Enclosed Fan-cooled Motors				
3600	...	...	...	...	...	224	225	254
1800	...	...	...	...	224	225	254	...
1200	...	...	224	225	254	...	...	...
900	224	225	254	254	...	...	...	...

Table 19. Approximate Weights of Single-phase Motors

Frame Size	Approximate Net Weight, pounds: Open Motor	Base	Pulley
203	72	19	1
204	80	20	1
224	105	25	3
225	130	25	3
254	205	30	4

Table 20. Standard Horsepower and Speed Ratings of Large-power Polyphase Induction Motors

(NEMA Standard)

Horsepower	Synchronous Speed in Rpm: 60 Cycles								25 Cycles		
1/2	......	....	....	900	720	600	514	450	......	750	500
3/4	......	....	1200	900	720	600	514	450	......	750	500
1	......	1800	1200	900	720	600	514	450	1500 *	750	500
1 1/2	3600 *	1800	1200	900	720	600	514	450	1500 *	750	500
2	3600 *	1800	1200	900	720	600	514	450	1500 *	750	500
3	3600 *	1800	1200	900	720	600	514	450	1500 *	750	500
5	3600 *	1800	1200	900	720	600	514	450	1500 *	750	500
7 1/2	3600 *	1800	1200	900	720	600	514	450	1500 *	750	500
10	3600 *	1800	1200	900	720	600	514	450	1500 *	750	500
15	3600 *	1800	1200	900	720	600	514	450	1500 *	750	500
20	3600 *	1800	1200	900	720	600	514	450	1500 *	750	500
25	3600	1800	1200	900	720	600	514	450	1500 *	750	500
30	3600	1800	1200	900	720	600	514	450	1500 *	750	500
40	3600	1800	1200	900	720	600	514	450	1500 *	750	500
50	3600	1800	1200	900	720	600	514	450	1500	750	500
60	3600	1800	1200	900	720	600	514	450	1500	750	500
75	3600	1800	1200	900	720	600	514	450	1500	750	500
100	3600	1800	1200	900	720	600	514	450	1500	750	500
125	3600	1800	1200	900	720	600	514	450	1500	750	500
150	......	1800	1200	900	720	600	514	450	1500	750	500
200	......	1800	1200	900	720	600	514	...	1500	750	...

Note 1. Ratings below the heavy lines are for direct connection only.

Note 2. Standard speed ratings at 50 cycles are 3000, 1500, 1000, and 750 rpm, and the standard horsepower ratings at each speed are the same as for the 60-cycle speeds of 3600, 1800, 1200, and 900 rpm, respectively.

* Applies to squirrel-cage motors only.

6. POLYPHASE INDUCTION LARGE-POWER MOTORS

Horsepower and speed ratings of large-power polyphase induction motors as standardized by NEMA are given in Table 20. Larger and lower-speed motors are also available. Motors larger than 200 hp are generally given horsepower ratings varying in steps of 50 to 500 hp, steps of 100 to 1000 hp, steps of 250 to 2500 hp, and steps of 500 to 6000 hp. The synchronous speeds available with various power-supply frequencies are given in Table 3.

TYPES of polyphase induction motors in general use are (1) squirrel-cage motors (single-speed and multispeed), (2) wound-rotor motors, and (3) adjustable-speed brush-shifting motors. Special designs of these basic types are available for specific applications including (1) crane and hoist motors, (2) elevator motors, (3) shell-type motors, (4) loom motors, and (5) screenless open textile motors.

Squirrel-cage Motors. Five classes of large-power, squirrel-cage induction motors are defined by NEMA and are designated designs A, B, C, D, and F. All designs withstand full-voltage starting but differ in their torque and/or starting-current characteristics.

Design A motors have normal starting torque (in accordance with Table 21), high breakdown torque (in excess of values shown in Table 22), high starting current (higher than values shown in Table 23), and low full-load slip (less than 5% except for motors having 10 or more poles, in which case the slip may be slightly greater than 5%). These motors are suitable for general application but often require reduced-voltage starting because of their relatively high starting current.

Table 21. Locked Rotor Torque of 60- and 50-cycle Polyphase Squirrel-cage Motors Designs A and B

(NEMA Standard)

Synchronous Rpm—60 Cycles	3600	1800	1200	900	720	600	514	450
Synchronous Rpm—50 Cycles	3000	1500	1000	750	600	500	428	375
Horsepower	Torque, % of Full-load Torque							
1/2				150	150	115	110	105
3/4			175	150	150	115	110	105
1		275	175	150	150	115	110	105
1 1/2	175	265	175	150	150	115	110	105
2	175	250	175	150	145	115	110	105
3	175	250	175	150	135	115	110	105
5	150	185	160	130	130	115	110	105
7 1/2	150	175	150	125	120	115	110	105
10	150	175	150	125	120	115	110	105
15	150	165	140	125	120	115	110	105
20	150	150	135	125	120	115	110	105
25	150	150	135	125	120	115	110	105
30	150	150	135	125	120	115	110	105
40	135	150	135	125	120	115	110	105
50	125	150	135	125	120	115	110	105
60	125	150	135	125	120	115	110	105
75	110	150	135	125	120	115	110	105
100	110	125	125	125	120	115	110	105
125	100	110	125	125	120	115	110	105
150	100	110	125	125	120	115	110	105
200	100	100	125	125	120	115	110	105

Table 22. Breakdown Torques of 60- and 50-cycle Polyphase Squirrel-cage Motors, Design B

(NEMA Standard)

Synchronous Rpm	3600–3000	1800–1500	1200–1000	900–750	Below 750
Horsepower	Torque, % of Full-load Torque				
1/2				250	200
3/4			275	250	200
1		300	275	250	200
1 1/2	275	300	275	250	200
2	250	275	250	225	200
3	250	275	250	225	200
5	225	225	225	225	200
7 1/2	215	215	215	215	200
10 and larger	200	200	200	200	200

Table 23. Locked-rotor Current of 60-cycle Three-phase Squirrel-cage Motors, Designs B, C, and D

(NEMA Standard)

(Current in Amperes at 220 Volts *)

Horsepower	Amperes	Horsepower	Amperes	Horsepower	Amperes
1 or less	24 †	15	220	75	1085
1 1/2	35	20	290	100	1450
2	45	25	365	125	1815
3	60	30	435	150	2170
5	90	40	580	200	2900
7 1/2	120	50	725		
10	150	60	870		

* Current at other voltages is inversely proportional to the voltage.
† Amperes per horsepower (this rating only).

Design B motors have normal starting torque (in accordance with Table 21), high breakdown torque (in accordance with Table 22), low starting current (in accordance with Table 23), and low full-load slip (less than 5% except for motors having 10 or more poles, in which case the slip may be slightly greater than 5%). Design B motors are the most popular of all types since their well-balanced characteristics make them suitable for the majority of applications.

Design C motors have high starting torque (in accordance with Table 24), normal breakdown torque (in accordance with Table 25), low starting current (in accordance with

Table 24. Locked-rotor Torque of 60- and 50-cycle Polyphase Squirrel-cage Motors, Design C

(NEMA Standard)

Synchronous Rpm—60 Cycles	1800	1200	900
Synchronous Rpm—50 Cycles	1500	1000	750
Horsepower	Torque, % of Full-load Torque		
3	...	250	225
5	250	250	225
7 1/2	250	225	200
10	250	225	200
15	225	200	200
20 and larger	200	200	200

Table 25. Breakdown Torque of 60- and 50-cycle Polyphase Squirrel-cage Motors, Design C

(NEMA Standard)

Synchronous Rpm—60 Cycles	1800	1200	900
Synchronous Rpm—50 Cycles	1500	1000	750
Horsepower	Torque, % of Full-load Torque		
3	...	225	200
5	200	200	200
7 1/2 and larger	190	190	190

Table 23), and low full-load slip (less than 5%). They are used on such devices as reciprocating compressors started under load, which require high starting torque with good running characteristics.

Design D motors have high starting torque (275% for 60- and 50-cycle, four-, six-, and eight-pole motors), low starting current (in accordance with Table 23), and high full-load slip (5% or more). No breakdown-torque values are specified, since these motors have no sharply defined breakdown point. They are used principally for drives imposing rapidly recurring peak loads (such as punch presses) and for drives involving a heavy starting duty but light or intermittent running duty. In the former application a flywheel frequently is provided to equalize the load peaks. This type of motor should not be used to drive a continuous steady load, as its efficiency is quite low.

Design F motors have low starting torque (125% for 60- and 50-cycle, four- and six-pole motors, rated 30 hp and larger), low breakdown torque (135% for motors rated 30 hp and larger), low starting current (9 amp per horsepower at 220 volts, and inversely proportional to the voltage at other voltages, for motors rated 30 hp and larger) and low full-load slip (less than 5%). These motors have a limited field of application because of their low breakdown torque. They are used occasionally when limitations of the power system make their low starting current desirable and when the driven machine has moderate starting-torque requirements and a uniform running load.

Figure 11 shows typical speed-torque curves, and Fig. 12 shows typical speed-current curves of each of the five types of squirrel-cage induction motors.

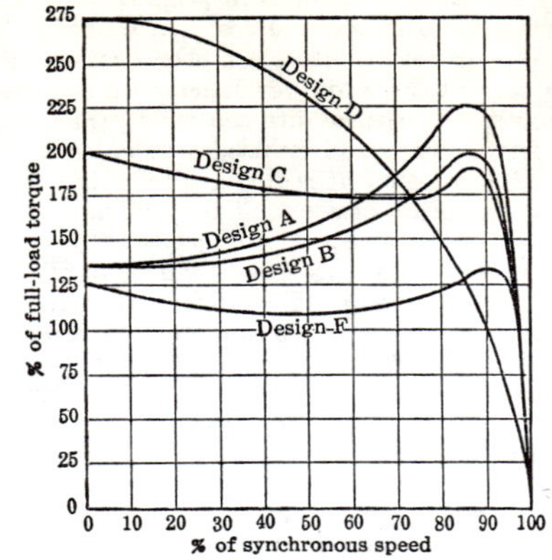

FIG. 11. Typical speed-torque curves of squirrel-cage induction motors.

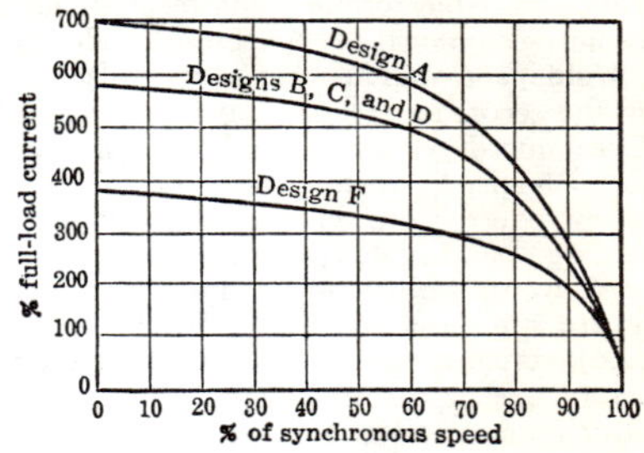

FIG. 12. Typical speed-current curves of squirrel-cage induction motors.

Multispeed squirrel-cage motors may be provided with (1) a single stator winding connected to give either one of two speeds, (2) two or more stator windings giving a corresponding number of speeds, or (3) a combination of (1) and (2). The single-winding motor is lowest in cost and has the best characteristics, but it gives only two speeds having a ratio of 2 : 1. Two or more windings must be employed where more than two speeds or other speed ratios are required.

Multispeed motors may be classified into three general types. (1) *Variable-torque motors* have horsepower ratings at each speed proportional to the square of the speed (*viz.*, 20/5 hp, 1800/900 rpm) and are used for driving centrifugal pumps, fans, and agitators. (2) *Constant-torque motors* have horsepower ratings at each speed directly proportional to the speed (*viz.*, 30/20/15/10 hp, 1800/1200/900/600 rpm) and are used for conveyors, stokers, reciprocating compressors, printing presses, and dough mixers. (3) *Constant-horsepower motors* have the same horsepower rating at all speeds (*viz.*, 5 hp, 1800/600 rpm). Typical applications are lathes, boring mills, planers, radial drills, grinders, milling machines, winches, and dragline scrapers. Multispeed squirrel-cage motors may have electrical characteristics corresponding to Classes A, B, C, or D, but in general the same characteristics are obtained at all speeds.

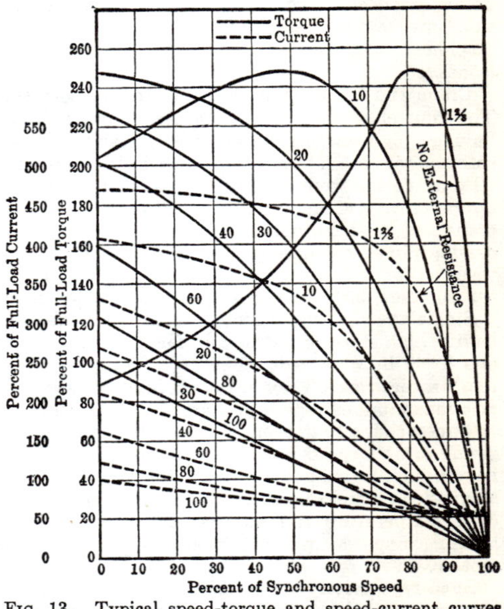

FIG. 13. Typical speed-torque and speed-current curves of wound-rotor induction motors with different external resistances in series with rotor winding.

The wound-rotor induction motor has an insulated three-phase winding on the rotor, whose terminals are connected to three slip rings. Stationary brushes bear on the latter, through which the rotor circuit is completed external to the motor. By inserting various amounts of resistance in series with the rotor winding, the torque and current characteristics of the motor can be varied as shown in Fig. 13. High starting torque thus can be obtained with very low starting current. With all resistance cut out, the characteristics of this motor are similar to those of the Class A squirrel-cage motors, but with somewhat lower efficiency. With suitable control, wound-rotor motors can operate continuously at reduced speeds down to 50% speed, but the efficiency is reduced in proportion to the speed. At reduced speeds the motor has varying speed characteristics; i.e., a change in load results in considerable change in speed. Wound-rotor motors are used occasionally where a high ratio of starting torque to starting current (high torque efficiency) is required or where severe accelerating duty is met. However, these motors are used principally on applications requiring an adjustable speed drive. They can be used economically for driving variable torque loads (centrifugal pumps, fans, etc.) at reduced speeds, but they are not economical for continuous reduced-speed operation of constant torque loads.

Multispeed wound-rotor motors also are available. When speed adjustment (by varying the secondary resistance) is required on all speeds, only two speeds can be obtained. These must have a ratio of 2 : 1. In many applications, speed adjustment is required on only one speed, in which case two-, three-, or four-speed motors, having any synchronous speeds required, can be obtained. Except on the speed which is to be adjusted, the rotor winding operates short-circuited as a squirrel-cage motor. Typical applications of multispeed wound-rotor motors include hoists for dragging and lifting service, conveyors, elevators, oil-well pumps, and rotary car dumpers.

Adjustable-speed brush-shifting motors are similar in construction to wound-rotor motors, except that the primary winding is on the rotor, the secondary winding is on the stator, and an additional adjusting winding connected to a commutator is provided on the rotor. Power is supplied to the primary winding through collector rings, and the secondary (stator) winding is connected to brushes that bear on the commutator. The brushes may be shifted around the commutator for speed adjustment. The speed range is usually 3 : 1, and popular ratings are 3 to 15 hp (at top speed), 2500/833 rpm, and 20 to 50 hp (at top speed), 1650/550 rpm. The motors are constant-torque machines, so the horsepower output is directly proportional to the speed. At a given brush setting the change in speed between no load and full-load is 5 to 10% at maximum speed and 15 to 30% at minimum speed. At the low-speed setting (normal starting position) the starting torque is about 150%, the starting current 200%, and the breakdown torque 175 to 200%. Although these motors are higher in cost than wound-rotor or multispeed squirrel-cage motors, they have the advantages of a wide speed range with an infinite number of speed points, good efficiency at all speed points, and good speed regulation. Typical applications are textile machinery, stokers, draft fans, conveyors, cement kilns, small paper machines, wire-drawing machines, laundry flat-plate ironers, traveling plate ovens, lathes, etc.

Crane and hoist motors are available in both the squirrel-cage and wound-rotor types. The squirrel-cage motors are similar to Design D or (in the smaller sizes) Design C motors, having a high starting torque and low starting current. Both the squirrel-cage and wound-rotor motors are intermittently rated (15 or 30 min for the squirrel-cage motors and 30 or 60 min for the wound-rotor motors) and are available in open construction with a 50 C temperature rise, and in totally enclosed construction with a 55 C temperature rise. All crane and hoist motors can be obtained with a shaft extension on both ends, one for the driving pinion or coupling, the other for a brake. In addition to crane and hoist duty, these motors often are used for such applications as bascule lift and swing bridges, rotary car dumpers, larry cars, coke pushers, and hydraulic gates.

Single-speed elevator motors of the squirrel-cage type are similar to Design D motors, having high starting torque and low starting current. They are available in ratings from one to 60 hp at speeds of 600 to 1800 rpm and are intermittently rated 30 min, 50 C temperature rise. They are used on low-speed freight and passenger elevators operating at a maximum car speed of about 150 ft per min.

Two-speed elevator motors also are available. They may employ a single winding giving a 2 : 1 speed ratio, or separate windings giving a speed ratio up to 6 : 1. They have a high resistance rotor winding giving Design D characteristics. These motors are available in ratings up to 100 hp with a top speed of 1200 or 900 rpm and are rated 1-hour 50 C temperature rise on the high speed. Two-speed motors are used on elevators having car speeds up to 550 ft per min. Higher-speed elevators usually require d-c driving motors.

Shell-type motors are squirrel-cage induction motors furnished with stators and rotors only, i.e., without shaft, end shields, or bearings. The stator "frame" consists of a cylindrical steel shell wrapped around the magnetic core. In some cases the steel shell is omitted and the laminations forming the magnetic core are welded together. The rotor is mounted on the shaft of the driven machine, which must also be provided with a suitable support for the stator. Shell-type motors are available in ratings up to 30 hp (at 3600 rpm) and are used for wood-working and machine tool applications such as boring machines, grinders, tapping machines, and multiple-spindle drills. They are frequently operated from a special high-frequency power supply (such as 120 or 180 cycles per sec), giving synchronous speeds up to 10,800 rpm for two-pole motors.

Loom motors are squirrel-cage motors of totally enclosed construction, exclusively for driving textile looms. The popular ratings range from 1 to 2 hp at 1800 and 1200 rpm. They have higher breakdown torque and slightly higher full-load slip than the standard Design B motors. Higher than normal Wk^2 is often provided either in the motor rotor itself or in the form of a flywheel mounted on a motor shaft extension opposite the drive end. The motor bearings are either the oil-lubricated waste-packed sleeve type or the grease-lubricated antifriction type.

Screenless open textile motors are squirrel-cage motors having electrical characteristics similar to the standard Design B machines but differing in their mechanical construction. They are open ball-bearing machines with large unobstructed passages for ventilating air which, by means of special fans on the rotor, prevents the rapid accumulation of dust or lint inside the motor. These motors have a variety of applications in textile plants, including spinning frames, cards, cleaners, draw frames, and roving frames.

Operating Characteristics

TEMPERATURE RISE. The standard temperature rise of large-power polyphase induction motors, measured by thermometer, is 40 C for general-purpose open-type motors, 50 C for splashproof motors, and 55 C for totally enclosed and totally enclosed fan-cooled machines. The temperature rise may, at the option of the manufacturer, be determined by the resistance method, in which case a 5 C higher temperature rise is allowed for totally enclosed fan-cooled motors and a 10 C higher temperature rise for other types.

For motors having a temperature rise of 40 C, manufacturers usually authorize operation (at rated voltage and frequency and in an ambient temperature not exceeding 40 C) at a continuous load equal to rated load multiplied by the following service factors.

Horsepower	Service Factor
1/2, 3/4, 1	1.25
1 1/2, 2	1.20
3 and larger	1.15

EFFICIENCY. Typical full-load efficiencies and power factors of standard Design A squirrel-cage induction motors are given in Figs. 14 and 15, respectively. Design B

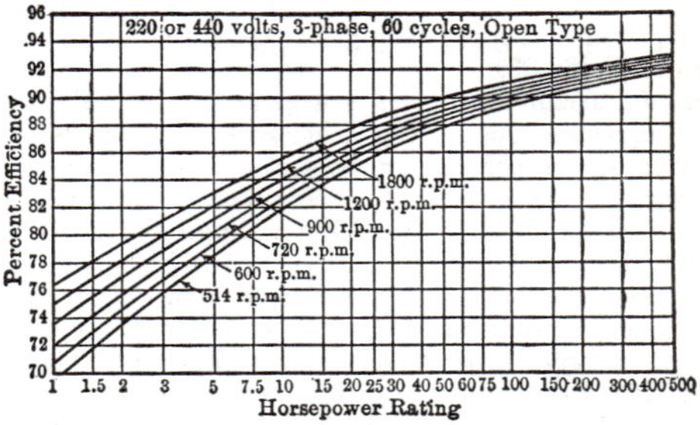

FIG. 14. Approximate full-load efficiencies of Design A squirrel-cage induction motors.

motors have comparable efficiencies, Design C motors slightly lower efficiencies, and Design D motors substantially lower efficiencies than Design A machines. Design B and Design C motors also have somewhat lower power factors than Design A machines. Wound-rotor motors operating at full speed have efficiencies slightly lower than those of Design A or B squirrel-cage motors, and at reduced speed their efficiency is reduced approximately in proportion to the reduction in speed.

TORQUE AND STARTING CURRENT. The starting and breakdown torques and starting currents of squirrel-cage induction motors are given under the description of the various types.

As shown in Fig. 13, the starting torque and starting current of wound-rotor motors vary over a wide range, depending on the amount of external resistance in the rotor circuit. It is usual to start these motors with sufficient external resistance to give 100% torque and current at standstill, the resistance being short-circuited in steps as the motor comes

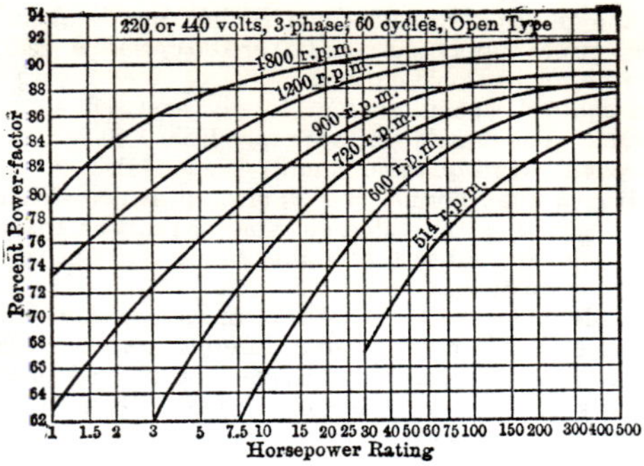

FIG. 15. Approximate full-load power factors of Design A squirrel-cage induction motors.

up to speed. The minimum breakdown torque for wound-rotor motors is 200% of full-load torque.

VOLTAGE VARIATION. All polyphase induction motors should operate successfully at rated load and frequency at any voltage not more than 10% above or below normal, but not necessarily in accordance with the standards of performance established for operation at normal rating. The starting currents are approximately proportional to the applied voltage, and the starting and breakdown torques are approximately proportional to the square of the voltage.

Mechanical Features

Large-power polyphase induction motors usually are furnished with feet on the stator frame and with end-shield-type bearings, although pedestal-type bearings, with or without base, may be used in sizes above 1 hp per rpm. The smaller motors also can be obtained with a machined face on the drive end shield or with a flange for bolting to the driven machine.

Vertical motors with either solid or hollow shafts are available. The solid-shaft motors are usually furnished with a type P base (Fig. 10) or a D flange base (Fig. 9), but the smaller sizes can be obtained with a face-type base (Fig. 8). Hollow-shaft motors generally have a type P base. Horsepower and speed ratings of vertical hollow-shaft motors standardized by NEMA are given in Table 33.

Open and drip-proof horizontal motors are generally furnished with ball bearings in sizes up to and including frame 326. In the larger sizes both sleeve- and ball-bearing motors are available, but the sleeve-bearing machines generally are slightly less expensive. Furthermore, ball bearings are not usually recommended in motor sizes above 150 hp at 3600 rpm, 250 hp at 1800 rpm, 350 hp at 1200 rpm, and 400 hp at lower speeds. Splashproof motors generally have ball bearings, as have totally enclosed and totally enclosed fan-cooled motors in sizes up to 60 hp at 3600 rpm, and 300 or 350 hp at lower speeds. Vertical motors are generally provided with ball or roller bearings except in rather large sizes, which employ plate-type thrust bearings and sleeve guide bearings.

Motor ratings suitable for flat-belt, multiple-V-belt, chain, and gear drives with two bearings are indicated in Table 6. Larger ratings require an extended base and shaft with an outboard bearing. Motors for belt or chain drive are usually furnished with a sliding base or slide rails for belt adjustment.

Gear motors are available in ratings up to 50 hp with output-shaft speeds of 13.5 to 780 rpm. High-speed (1800 rpm at 60 cycles) motors are usually employed.

Standard motors are of open or drip-proof construction, but splashproof, totally enclosed, totally enclosed fan-cooled, and explosionproof motors (Class I, Group D, or Class II, Groups E, F, and G) are also available. Totally enclosed motors are available in sizes up to 15 hp (at 1800 to 900 rpm), and totally enclosed fan-cooled motors are available in sizes ranging from 250 hp at 600 rpm to 500 hp at 1800 and 3600 rpm.

WEIGHTS AND DIMENSIONS. Frame numbers assigned to the various types of continuously rated squirrel-cage and wound-rotor motors are given in Tables 26 to 32 inclusive. Standardized mounting dimensions corresponding to the various frame numbers are given in Tables 13, 14, and 15. Approximate net weights are given in Table 34.

Table 26. Frame Sizes for Two- and Three-phase, Squirrel-cage General-purpose Open-type Motors, 40 C, 60 Cycles

(NEMA Standard)

Synchronous Rpm	3600	1800	1200	900	720	600	514	450	
Horsepower					Frame Number *				
1/2	...	...	...	204	224	225	...	...	110, 208, 220, 440, and 550 volts
3/4	...	...	203	224	225	254	284	284	
1	...	203	204	225	254	254	284	324	
1 1/2	203	204	224	254	254	284	324	326	
2	204	224	225	254	284	324	326	365	208, 220, 440, and 550 volts
3	224	225	254	284	324	326	365	404	
5	225	254	284	324	326	364	404	405	
7 1/2	254	284	324	326	364	365	405	444	
10	284	324	326	364	365	404	444	445	
15	324	326	364	365	404	405	445	504U	
20	326	364	365	404	405	444	504U	505	
25	364S	364	404	405	444	445	505	...	
30	364S	365	405	444	445	504U	...	...	
40	365S	404	444	445	504U	504	...	...	
50	404S	405S	445	504U	505	...	...	...	
60	405S	444S	504U	505	...	...	...	...	
75	444S	445S	505	...	...	...	...	...	
100	445S	504S	...	...	...	...	...	...	
125	504S	505S	...	...	...	...	...	...	
150	505S	...	...	...	...	...	...	...	
40		...	445	504U					2300 volts, three-phase
50		445S	504U	505					
60		445S	505	...					
75		504S	...	...					
100		505S	...	...					

Note: Frame numbers above the jagged line also apply to splashproof 50C motors.
* Dimensions for these frame numbers are given in Tables 13, 14, and 15.

Table 27. Frame Sizes for Two- and Three-phase Wound-rotor General-purpose Open-type Motors, 40 C, 60 Cycles

(NEMA Standard)

Synchronous Rpm	1800	1200	900	720	600	
Horsepower			Frame Number *			
1	...	...	225	...	...	110, 208, 220, 440 and 550 volts
1 1/2	...	...	254	...	...	
2	224	225	254	324	...	
3	225	254	284	326	365	
5	254	284	324	365	404	208, 220, 440 and 550 volts
7 1/2	284	324	326	404	405	
10	324	326	365	405	444	
15	326	365	404	444	445	
20	364	404	405	445	504U	
25	365	405	444	504U	505	
30	404	444	445	505	...	
40	405	445 †	504U †	...	...	
50	444S †	504U †	505 †	...	...	
60	445S †	505 †	...	...	...	
75	504S †	...	...	...	...	
100	505S †	...	...	...	...	

* Dimensions for these frame numbers are given in Tables 13, 14, and 15.
† These frame numbers also apply to 2300-volt three-phase motors

Table 28. Frame Sizes for Two- and Three-phase Squirrel-cage Motors, Totally Enclosed and Totally Enclosed Fan-cooled, 55 C, 60 Cycles, 550 Volts or Less

(NEMA Standard)

Synchronous Rpm	3600	1800	1200	900	720	600	514	450
Horsepower	Frame Number *—Totally Enclosed							
1/2	...	...	...	204				
3/4	...	...	203	224				
1	...	203	204	225				
1 1/2	204	204	224	254				
2	224	224	254	284				
3	284	284	324	324				
5	326	326	365	365				
7 1/2	...	365	404	404				
10	...	405	444	444				
	Frame Number *—Totally Enclosed Fan-cooled							
1/2	...	...	...	...	224	225	...	...
3/4	...	...	...	224	225	254	284	284
1	...	...	...	225	254	254	284	324
1 1/2	...	...	224	254	254	284	324	326
2	...	224	225	254	284	324	326	365
3	224	225	254	284	324	326	365	...
5	225	254	284	324	326	365	...	...
7 1/2	254	284	324	326	365	...	...	...
10	...	324	326	365	...	...	...	...
15	...	326	365	...	...	...	...	...
20	...	364	...	...	...	...	...	...
25	...	365	...	...	...	...	...	...

* Standard dimensions for these frame numbers are given in Tables 13, 14, and 15.

Table 29. Frame Sizes for Three-phase Two-speed Single-winding Squirrel-cage Induction Motors, Open Type, 60 Cycles, 550 Volts or Less

(NEMA Standard)

Variable-torque Motors						Constant-torque Motors						Constant-horsepower Motors				
Horsepower at		Rpm				Horsepower at		Rpm				Horse-power	Rpm			
High Speed	Low Speed	3600 and 1800	1800 and 900	1200 and 600	900 and 450	High Speed	Low Speed	3600 and 1800	1800 and 900	1200 and 600	900 and 450		3600 and 1800	1800 and 900	1200 and 600	900 and 450
Frame Numbers *																
...	...	...	...	...	...	1/6	1/12	...	...	...	204	1/6	...	...	204	225
...	...	...	...	...	...	1/4	1/8	...	...	...	224	1/4	...	...	224	225
...	...	...	...	...	...	1/3	1/6	...	...	204	224	1/3	...	...	224	254
1/2	0.12	...	...	...	204	1/2	1/4	...	...	224	225	1/2	...	204	225	254
3/4	0.19	...	...	204	224	3/4	3/8	...	...	224	254	3/4	...	224	254	284
1	0.25	204	204	204	225	1	1/2	204	204	225	254	1	204	225	254	324
1 1/2	0.37	204	224	224	254	1 1/2	3/4	204	224	254	284	1 1/2	224	254	284	326
2	0.50	224	225	225	254	2	1	224	225	254	324	2	225	254	324	365
3	0.75	224	225	254	284	3	1 1/2	224	254	284	326	3	225	284	326	404
5	1.2	225	254	284	324	5	2 1/2	225	284	326	365	5	254	324	365	405
7 1/2	1.9	254	284	324	326	7 1/2	3 3/4	254	284	364	404	7 1/2	284	326	404	444
10	2.5	284	324	326	365	10	5	284	324	365	405	10	324	365	405	445
15	3.7	324	326	365	404	15	7 1/2	324	364	404	444	15	326	404	444	504
20	5.0	326	364	404	405	20	10	326	365	405	445	20	364	405	445	505
25	6.2	364	365	405	444	25	12 1/2	364	404	444	504	25	365	444	504	...
30	7.5	365	405	444	445	30	15	365	405	445	504	30	405	445	505	...
40	10	404	444	445	504	40	20	404	444	504	505	40	444	504	...	...
50	12	405	445	504	505	50	25	405	445	505	...	50	445	505	...	...
60	15	444	504	505	...	60	30	444	504	...	...	60	504	...	...	...
75	19	445	505	...	...	75	37	445	505	...	...	75	505	...	...	...
100	25	504	...	...	...	100	50	504	...	...	...	...	...	...	...	...
125	31	505	...	...	...	125	62	505	...	...	...	...	...	...	...	...

* Standard mounting dimensions for these frames are given in Tables 13, 14, and 15.

Table 30. Frame Sizes for Variable Torque Two- and Three-phase Two-speed Two-winding Squirrel-cage Induction Motors, Open Type, 60 Cycles, 550 Volts or Less

(NEMA Standard)

1800–1200 Rpm		1800–900 Rpm		1800–600 Rpm		1200–900 Rpm	
Horsepower	Frame No.*	Horsepower	Frame No.*	Horsepower	Frame No.*	Horsepower	Frame No.*
....	.....	....	.....	....	.....	1/2 – 0.3	204
3/4 – 0.33	204	3/4 – 0.19	204	....	.....	3/4 – 0.4	224
1 – 0.5	204	1 – 0.25	204	....	.....	1 – 0.6	225
1 1/2 – 0.7	224	1 1/2 – 0.37	224	1 1/2 – 0.17	224	1 1/2 – 0.8	254
2 – 0.9	225	2 – 0.5	225	2 – 0.22	225	2 – 1.1	254
3 – 1.3	254	3 – 0.75	254	3 – 0.33	254	3 – 1.7	284
5 – 2.2	284	5 – 1.25	284	5 – 0.56	284	5 – 2.7	324
7 1/2 – 3.3	324	7 1/2 – 1.9	324	7 1/2 – 0.83	324	7 1/2 – 4.2	326
10 – 4.4	326	10 – 2.5	326	10 – 1.1	326	10 – 5.6	365
15 – 6.7	364	15 – 3.7	364	15 – 1.7	364	15 – 8.4	404
20 – 8.9	365	20 – 5	365	20 – 2.2	365	20 – 11.3	405
25 – 11	405	25 – 6.2	405	25 – 2.8	405	25 – 14	444
30 – 13	444	30 – 7.5	444	30 – 3.3	444	30 – 17	445
40 – 18	445	40 – 10	445	40 – 4.4	445	40 – 22	504
50 – 22	504	50 – 13	504	50 – 5.6	504	50 – 28	505
60 – 27	505	60 – 15	505	60 – 6.7	505	....	.....

1200–720 Rpm		1200–600 Rpm		900–720 Rpm		900–600 Rpm	
Horsepower	Frame No.*	Horsepower	Frame No.*	Horsepower	Frame No.*	Horsepower	Frame No.*
1/2 – 0.2	204	1/2 – 0.12	204	....	.....	....	.....
3/4 – 0.3	224	3/4 – 0.19	224	3/4 – 0.48	225	....	.....
1 – 0.4	225	1 – 0.25	225	1 – 0.64	254	....	.....
1 1/2 – 0.5	254	1 1/2 – 0.37	254	1 1/2 – 0.96	254	1 – 0.5	254
2 – 0.7	254	2 – 0.5	254	2 – 1.3	284	1 1/2 – 0.7	254
3 – 1.1	284	3 – 0.75	284	3 – 1.9	324	2 – 0.9	284
5 – 1.8	324	5 – 1.2	324	5 – 3.2	326	3 – 1.3	324
7 1/2 – 2.7	326	7 1/2 – 1.9	326	7 1/2 – 4.8	365	5 – 2.2	326
10 – 3.6	365	10 – 2.5	365	10 – 6.4	404	7 1/2 – 3.3	365
15 – 5.4	404	15 – 3.7	404	15 – 9.6	405	10 – 4.4	404
20 – 7.2	405	20 – 5	405	20 – 13	444	15 – 6.7	405
25 – 9	444	25 – 6.2	444	25 – 16	445	20 – 8.9	444
30 – 11	445	30 – 7.5	445	30 – 19	504	25 – 11	445
40 – 15	504	40 – 10	504	40 – 26	505	30 – 13	504
50 – 18	505	50 – 12	505	....	.....	40 – 18	505

900–450 Rpm		720–600 Rpm	
Horsepower	Frame No.*	Horsepower	Frame No.*
1 – 0.25	254	1 – 0.7	254
1 1/2 – 0.37	254	1 1/2 – 1	284
2 – 0.5	284	2 – 1.4	324
3 – 0.75	324	3 – 2.1	326
5 – 1.2	326	5 – 3.5	365
7 1/2 – 1.9	365	7 1/2 – 5.2	404
10 – 2.5	404	10 – 7	405
15 – 3.7	405	15 – 10	444
20 – 5	444	20 – 14	445
25 – 6.2	445	25 – 17	504
30 – 7.5	504	30 – 21	505
40 – 10	505	....	.....

See Tables 13, 14 and 15 for standard mounting dimensions.

Table 31. Frame Sizes of Constant-torque and Constant-horsepower Two- and Three-phase Two-speed Two-winding Squirrel-cage Induction Motors, Open Type, 60 Cycles, 550 Volts or Less

(NEMA Standard)

Constant-torque Motors

1800–1200 Rpm		1800–900 Rpm		1200–900 Rpm		1200–600 Rpm	
Horsepower	Frame No.*	Horsepower	Frame No.*	Horsepower	Frame No.*	Horsepower	Frame No.*
.....	.....	.....	.....	1/3 – 1/4	204	.....	.....
.....	.....	.....	.....	1/2 – 3/8	224	.....	.....
3/4 – 1/2	204	3/4 – 3/8	204	3/4 – 9/16	224	.....	.....
1 – 2/3	224	1 – 1/2	224	1 – 3/4	225	.....	.....
1 1/2 – 1	224	1 1/2 – 3/4	224	1 1/2 – 1 1/8	254	1 – 1/2	254
2 – 1 1/3	225	2 – 1	225	2 – 1 1/2	254	1 1/2 – 3/4	254
3 – 2	254	3 – 1 1/2	254	3 – 2 1/4	284	2 – 1	284
5 – 3 1/3	284	5 – 2 1/2	284	5 – 3 3/4	324	3 – 1 1/2	324
7 1/2 – 5	324	7 1/2 – 3 3/4	324	7 1/2 – 5 2/3	326	5 – 2 1/2	326
10 – 6 2/3	326	10 – 5	326	10 – 7 1/2	365	7 1/2 – 3 3/4	365
15 – 10	365	15 – 7 1/2	365	15 – 11	404	10 – 5	404
20 – 13	404	20 – 10	404	20 – 15	405	15 – 7 1/2	405
25 – 17	405	25 – 12	405	25 – 19	444	20 – 10	444
30 – 20	444	30 – 15	444	30 – 22	445	25 – 12	444
40 – 27	445	40 – 20	445	40 – 30	504	30 – 15	445
50 – 33	504	50 – 25	504	50 – 37	505	40 – 20	504
60 – 40	505	60 – 30	505	.....	.....	50 – 25	505

Constant-torque Motors

900–600 Rpm		900–450 Rpm	
Horsepower	Frame No.*	Horsepower	Frame No.*
1 – 2/3	284	1 – 1/2	284
1 1/2 – 1	284	1 1/2 – 3/4	324
2 – 1 1/3	324	2 – 1	326
3 – 2	326	3 – 1 1/2	365
5 – 3 1/3	365	5 – 2 1/2	404
7 1/2 – 5	404	7 1/2 – 3 3/4	405
10 – 6 2/3	405	10 – 5	444
15 – 10	444	15 – 7 1/2	445
20 – 13	445	20 – 10	504
25 – 17	504	25 – 12	505
30 – 20	505	.....	.....

Constant-horsepower Motors

Horse-power	1800–1200 Rpm	1800–900, 1200–900 Rpm	1800–600, 1200–600, 900–600 Rpm	1800–450, 900–450, 600–450 Rpm	Horse-power	1800–1200 Rpm	1800–900, 1200–900 Rpm	1800–600, 1200–600, 900–600 Rpm	1800–450, 900–450, 600–450 Rpm
	Frame Number *					Frame Number *			
1/6	.....	.....	224	224	5	324	326	404	444
1/4	.....	.....	224	225	7 1/2	326	365	405	445
1/3	.....	204	225	254	10	365	404	444	504
1/2	204	224	254	284	15	404	405	445	505
3/4	224	225	284	324	20	405	444	504	.....
1	225	254	284	326	25	444	445	505	.....
1 1/2	254	254	324	365	30	445	504	.....	.....
2	254	284	326	404	40	504	505	.....	.....
3	284	324	365	405	50	505	.....	.....	.....

* See Tables 13, 14 and 15 for standard mounting dimensions.

Table 32. Frame Sizes for Three-phase Four-speed Two-winding Squirrel-cage Induction Motors, Open Type, 60 Cycles, 550 Volts or Less

(NEMA Standard)

Variable-torque Motors				Constant-torque Motors				Constant-horsepower Motors			
1800–1200–900–600 Rpm		1200–900–600–450 Rpm		1800–1200–900–600 Rpm		1200–900–600–450 Rpm		1800–1200–900–600 Rpm		1200–900–600–450 Rpm	
Horsepower	Frame No.*	Horsepower	Frame No.*	Horsepower	Frame No.*	Horsepower	Frame No.*	Horsepower	Frame No.*	Horsepower	Frame No.*
....	......	....	......	....	......	....	......	1/4	225	1/4	254
....	......	....	......	....	......	....	......	1/3	254	1/3	254
....	......	....	......	....	......	....	......	1/2	254	1/2	284
3/4 – 0.33 – 0.19 – 0.08	224	3/4 – 0.42 – 0.19 – 0.11	224	3/4 – 1/2 – 3/8 – 1/4	225	3/4 – 1/2 – 3/8 – 1/4	254	3/4	284	3/4	324
1 – 0.44 – 0.25 – 0.11	224	1 – 0.56 – 0.25 – 0.14	225	1 – 2/3 – 1/2 – 1/3	254	1 – 3/4 – 1/2 – 3/8	254	1	284	1	326
1 1/2 – 0.67 – 0.37 – 0.17	225	1 1/2 – 0.84 – 0.37 – 0.21	254	1 1/2 – 1 – 3/4 – 1/2	254	1 1/2 – 1 – 3/4 – 1/2	284	1 1/2	324	1 1/2	365
2 – 0.89 – 0.5 – 0.22	254	2 – 1.1 – 0.50 – 0.27	254	2 – 1 1/3 – 1 – 2/3	284	2 – 1 1/2 – 1 – 3/4	324	2	326	2	404
3 – 1.3 – 0.75 – 0.33	254	3 – 1.7 – 0.75 – 0.42	284	3 – 2 – 1 1/2 – 1	284	3 – 2 1/4 – 1 1/2 – 1	326	3	365	3	405
5 – 2.2 – 1.2 – 0.55	284	5 – 2.8 – 1.2 – 0.70	324	5 – 3 1/3 – 2 1/2 – 1 2/3	324	5 – 3 3/4 – 2 1/2 – 1 3/4	365	5	404	5	444
7 1/2 – 3.3 – 1.9 – 0.83	324	7 1/2 – 4.2 – 1.9 – 1.1	326	7 1/2 – 5 – 3 3/4 – 2 1/2	326	7 1/2 – 5 1/2 – 3 3/4 – 2 3/4	404	7 1/2	405	7 1/2	445
10 – 4.4 – 2.5 – 1.1	326	10 – 5.6 – 2.5 – 1.4	365	10 – 6 2/3 – 5 – 3 1/3	365	10 – 7 1/2 – 5 – 3	405	10	444	10	504
15 – 6.7 – 3.7 – 1.7	365	15 – 8.4 – 3.7 – 2.1	404	15 – 10 – 7 1/2 – 5	404	15 – 11 – 7 1/2 – 5 1/2	444	15	445	15	505
20 – 8.9 – 5 – 2.2	404	20 – 11 – 5 – 2.7	405	20 – 13 – 10 – 6 2/3	405	20 – 15 – 10 – 7 1/2	445	20	504	......	......
25 – 11 – 6.2 – 2.7	405	25 – 14 – 6 – 3.5	444	25 – 17 – 12 – 8 1/3	444	25 – 19 – 12 – 9 1/2	504	25	504	......	......
30 – 13 – 7.6 – 3.2	444	30 – 17 – 7.5 – 4.2	445	30 – 20 – 15 – 10	445	30 – 22 – 15 – 11	505	......	......	......	......
40 – 18 – 10 – 4.5	445	40 – 22 – 10 – 5.5	504	40 – 27 – 20 – 13	504	....	......	......	......	......	......
50 – 22 – 12 – 5.5	504	50 – 28 – 12 – 7	505	50 – 33 – 25 – 17	505	....	......	......	......	......	......
60 – 27 – 15 – 6.7	505	....	......	....	......	....	......	......	......	......	......

* See Tables 13, 14 and 15 for standard mounting dimensions.

Table 33. Horsepower and Speed Ratings for Vertical Hollow-shaft Motors, Two- or Three-phase, 60/50 Cycles, Types P and PH, 110, 208, 220, 440, 550, 2300 Volts

(NEMA Standard)

Frame Number	*BD* Dimensions, inches	Rpm—60 Cycles / Rpm—50 Cycles: 3600 / 3000	1800 / 1500	1200 / 1000	900 / 750	
				Horsepower		
203P	10	1 1/2	1	3/4	....	110, 208,* 220, 440, and 550 volts
204P	10	2	1 1/2	1	1/2	
224P	10	3	2	1 1/2	3/4	
225P	10	5	3	2	1	
254P	10	7 1/2	5	3	1 1/2 and 2	
284P	10	10	7 1/2	5	3	
324P or 324PH	12 or 16 1/2	15	10	7 1/2	5	
326P or 326PH	12 or 16 1/2	20	15	10	7 1/2	
364P	16 1/2	25 and 30	20 and 25			208,* 220, 440, and 550 volts
365P	16 1/2	40	30	20	15	
404P	20		40	25	20	
405P	20		50	30	25	
444P	20		60	40	30	
445P	20		75	50	40	
504P	24 1/2		100	60	50	
505P	24 1/2		125	75	60	
....	24 1/2		150	100	75	
....	24 1/2		200	125	....	
....	24 1/2		....	150	100	
....	30 1/2		250	200	125	
445P	20		50 and 60	40	....	2300 volts, three phase
504P	24 1/2		75	50	40	
505P	24 1/2		100	60	50	
....	24 1/2		125	75	60	
....	24 1/2		150	100	75	
....	24 1/2		200	125	....	
....	24 1/2		....	150	100	
....	30 1/2		250	......	....	
....	30 1/2		300	200	150	
....	30 1/2		400	250	200	

Note. Refer to Table 16 and Fig. 10 for dimensions of these motors.
* Applies to 60-cycle units only.

Table 34. Approximate Weights of Polyphase Induction Motors Built in NEMA Standardized Frames

Frame Number	Approximate Weight, pounds: Squirrel Cage — Open	Squirrel Cage — Totally Enclosed	Squirrel Cage — Totally Enclosed Fan-cooled	Wound-rotor, Open	Base	Pulley
203	67	84	....	....	19	1
204	74	92	110	....	20	1
224	92	115	145	120	22	3
225	105	130	160	135	23	3
254	141	175	215	170	28	3 1/2
284	192	210	300	235	33	5
324	256	285	375	310	45	9
326	288	340	400	370	53	15
364	380	....	635	460	74	21
365	430	....	685	500	87	21
404	620	....	825	760	103	32
405	665	....	895	800	106	32
444	875	....	1105	1040	125	47
445	1010	....	1210	1170	129	47
504	1285	....	1440	1460	162	97
505	1425	....	1625	1600	166	97

POLYPHASE SYNCHRONOUS LARGE-POWER MOTORS

Standard horsepower and speed ratings of large-power polyphase synchronous motors are given in Table 35.

Table 35. Standard Horsepower and Speed Ratings of Large-power Polyphase Synchronous Motors

(NEMA Standard)

Horsepower Ratings, High- and Low-speed Motors											
20	40	75	150	225 *	350	500	800	1250	2000	3000	4500
25	50	100	175 *	250	400	600	900	1500	2250	3500	5000
30	60	125	200	300	450	700	1000	1750	2500	4000	....

Speed Ratings

High-speed Motors				Low-speed Motors											
No. of Poles	Cycles 60	Cycles 50	Cycles 25	No. of Poles	Cycles 60	Cycles 50	Cycles 25	No. of Poles	Cycles 60	Cycles 50	Cycles 25	No. of Poles	Cycles 60	Cycles 50	Cycles 25
	Rpm				Rpm				Rpm				Rpm		
2	3600	3000	1500	8	....	....	375	26	277	231	115	56	128	107	.....
4	1800	1500	750	10	....	....	300	28	257	214	107	60	120	100	.....
6	1200	1000	500	12	....	....	250	30	240	200	100	66	109	91	.....
8	900	750	.....	14	....	428	214	32	225	188	94	72	100	83	.....
10	720	600	.....	16	450	375	188	36	200	167	83	76	95	79	.....
12	600	500	.....	18	400	333	166	40	180	150	75	80	90	.....	.....
14	514	.....	.....	20	360	300	150	44	164	136	.....	84	86	.....	.....
......	.....	.....	.....	22	327	273	136	48	150	125	.....	90	80	.....	.....
......	.....	.....	.....	24	300	250	125	52	138	115	.....	......	.....	.....	.....

* Low-speed motor only.

TYPES in general use are as follows. (1) High-speed motors: (*a*) general-purpose and (*b*) large high-speed. (2) Low-speed motors. (3) Special high-torque motors.

High-speed motors operate at speeds of 500 rpm or higher. The *general-purpose* group includes those rated 200 hp or less, with speed ratings of 500 to 1800 rpm, inclusive. The *large high-speed* group includes motors rated 250 hp or more, and all two-pole ratings. (See Table 36.) The general-purpose group has somewhat higher torques, but otherwise both groups have similar characteristics. High-speed synchronous motors have somewhat the same applications as general-purpose squirrel-cage induction motors, but their use is confined largely to ratings of 50 hp or above. Except in the large sizes, the synchronous motors are more costly than the induction motors, but their high efficiency and unity or leading power factor frequently justify their use. Typical applications include pumps, motor-generator sets, fans, blowers, reciprocating and centrifugal compressors, rubber mills, and line shafts.

Low-speed motors operate at speeds of 450 rpm or below. In general, these motors have lower costs than corresponding ratings of induction motors, as well as higher efficiencies and unity or leading power factor. Typical applications include reciprocating compressors; reciprocating, centrifugal, and screw-type pumps; Jordan engines; ball mills; metal rolling mills; electroplating generators; and line shafts.

High-torque motors. Synchronous motors can be obtained with sufficient starting torque for any application and with moderate starting current. To meet the requirements of high starting torque with very low starting current, several special types of synchronous motors are available.

The *synchronous induction* motor is essentially a wound-rotor induction motor which starts in the normal manner, but after it is up to speed d-c excitation is applied to the rotor winding. This causes the motor to pull into step and operate as a synchronous motor. Thus the high starting torque with low starting current of the wound-rotor induction motor is obtained. Disadvantages of this type of motor are that its efficiency is somewhat lower than for other types of synchronous motors, and a high-current, low-voltage exciter is required for excitation.

Synchronous motors with *phase-wound* amortisseur windings are also available. These have the usual salient-pole field, but the amortisseur or starting winding, instead of being

Table 36. Approximate Full-load Efficiencies of General Purpose and Large High-speed Synchronous Motors, Unity Power Factor, 3-Phase, 60 Cycles

Horse-power	Voltage	Rpm 514	600	720	900	1200	1800
		Approximate Full-load Efficiency					
20	220, 440, 550	85.6	86.5	87.5	88.3	89.0	89.4
	2300	84	85	86	86.5	87.5	87.8
25	220, 440, 550	86.5	87.3	88.2	88.9	89.6	89.8
	2300	85.4	86.1	87	87.7	88.4	88.8
30	220, 440, 550	87.2	87.9	88.7	89.4	90	90.3
	2300	86.5	87	87.8	88.5	89.1	89.4
40	220, 440, 550	88.2	88.8	89.6	90.1	90.6	90.9
	2300	87.8	88.2	89	89.6	90.1	90.4
50	220, 440, 550	88.9	89.6	90.3	90.6	91.2	91.4
	2300	88.8	89.2	89.8	90.3	90.8	91
60	220, 440, 550	89.5	90.1	90.6	91.1	91.5	91.7
	2300	89.4	89.8	90.5	90.8	91.3	91.4
75	220, 440, 550	90.2	90.8	91.1	91.5	92	92.1
	2300	90.2	90.8	91.1	91.4	91.8	91.9
100	220, 440, 550	91	91.4	91.8	92.1	92.5	92.6
	2300	91	91.4	91.8	92.1	92.5	92.5
125	220, 440, 550, 2300	91.6	92	92.2	92.5	92.9	92.9
150		92	92.3	92.6	92.9	93.1	93.2
200		92.7	93	93.1	93.4	93.6	93.6
250		93.2	93.4	93.6	93.7	93.9	93.9
300		93.5	93.7	93.8	94	94.1	94.2
350		93.8	94	94.1	94.2	94.3	94.3
400		94.1	94.2	94.3	94.4	94.5	94.5
450		94.3	94.4	94.5	94.6	94.6	94.6
500		94.5	94.6	94.6	94.7	94.8	94.8
600		94.8	94.8	94.9	95	95	95
700		95	95	95.1	95.2	95.1	95.1
800		95.2	95.2	95.2	95.3	95.3	95.3
900		95.4	95.4	95.4	95.4	95.4	95.4
1000		95.5	95.5	95.5	95.6	95.5	95.5

of the squirrel-cage type, is insulated, phase-wound, and brought out to collector rings through which the circuit may be completed external to the motor. Some motors are built with a revolving armature and a stationary field, eliminating the necessity for collector rings for the field or amortisseur winding, but requiring collector rings for the a-c armature winding. In any case the motor is started with a high resistance in the amortisseur winding circuit, which is short-circuited in steps as the motor comes up to speed. Thus high starting torque with relatively low starting current is obtained.

These special types of synchronous motors are considerably more costly than the conventional type and are justified only in particular applications involving severe starting duty and low starting current.

Operating Characteristics

POWER FACTOR. Standard power-factor ratings of synchronous motors are unity and 0.8 leading, but motors rated at leading power factors down to 0.2 are available. The unity-power-factor machines are least costly and have the highest efficiency, but the leading-power-factor machines often are justified when additional leading kilovolt-amperes are required.

TEMPERATURE RISE. The temperature rise of standard open-type unity-power-factor high-speed synchronous motors is 40 C, measured by thermometer, for the armature winding, and 50 C, measured by resistance, for the field winding; for leading-power-factor high-speed motors it is 40 C, by thermometer, for the armature winding, and 60 C, by resistance, for the field winding. Splashproof motors are rated 50 C rise, by thermometer, for the armature winding, and 60 C rise, by resistance, for the field winding. The temperature rise of standard open-type unity- and 0.8-power-factor low-speed synchronous motors is 50 C, by thermometer, for the armature winding, and 60 C, by resistance, for the field winding.

High-speed synchronous motors rated 40 C usually are given a service factor of 1.15; i.e., they may be operated at 1.15 times the rated load without injurious heating.

EFFICIENCY. Typical full-load efficiencies of unity-power-factor high-speed motors are given in Table 36, and of low-speed synchronous motors in Fig. 16. Efficiencies for machines without bearings do not include windage or friction losses. Motors rated 0.8 power factor have efficiencies ranging from 0.5 to 4% lower than for unity power-factor machines.

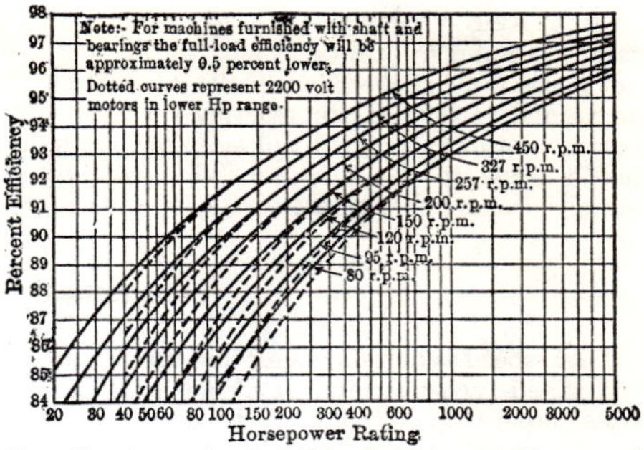

FIG. 16. Approximate full-load efficiency of low-speed synchronous motors.

A synchronous motor operating at full load with rated excitation delivers *leading reactive kilovolt-amperes* to the power system. The value is determined by

$$\text{Kilovolt-amperes} = \frac{0.746 \times \text{Horsepower rating}}{\text{Efficiency} \times \text{Power factor}} \sqrt{1 - (\text{Power factor})^2}$$

When the motor operates at partial loads with rated excitation maintained, the power factor becomes more leading, and more leading kilovolt-amperes are supplied to the system. (See Fig. 17.)

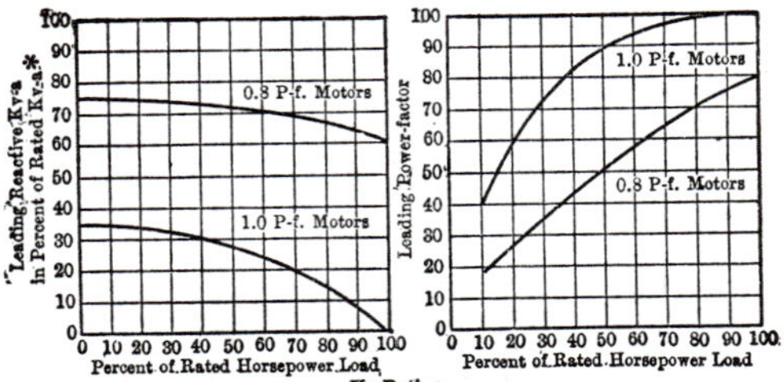

$*$Rated Kv-a $= 0.746 \times \dfrac{\text{Hp Rating}}{\text{Full-load Eff.} \times \text{P-f. Rating}}$

FIG. 17. Power-factor relations of synchronous motors.

TORQUE. Minimum values of starting, pull-in, and pull-out torques for general-purpose large high-speed and low-speed synchronous motors are given in Table 37. The pull-in torque of a synchronous motor varies with the flywheel effect (Wk^2) of its connected load; the higher the Wk^2, the lower the load which the motor is capable of pulling into synchronism. The values of pull-in torque given in Table 37 are based on loads having a normal Wk^2. However, motors with considerably higher starting pull-in and pull-out torques can be obtained, though usually at increased cost.

Table 37. Minimum Starting, Pull-in and Pull-out Torques of Standard Synchronous Motors

Speed Range, rpm	Horsepower Range	Unity Power Factor — Torque, % of Full-load Torque: Starting	Pull-in *	Pull-out	0.8 Power Factor — Torque, % of Full-load Torque: Starting	Pull-in *	Pull-out
			General-purpose Motors				
514–1200	....	110	110	175	125	125	250
1800	....	110	110	150	125	125	200
			Large High-speed Motors				
514–1800	200–500	....	....	....	125	125	200
514–1800	250–500	110	110	150	....	....	....
514–1800	600 and over	85	85	150	100	100	200
			Low-speed Motors				
....	....	40	30	150	40	30	200

* Based on normal load Wk^2.

STARTING CURRENT. Full-voltage starting currents range from 500 to 800% for high-speed motors, and from 250 to 500% for low-speed motors.

VOLTAGE VARIATION. All synchronous motors should operate successfully at rated load and frequency at any voltage not more than 10% above or below normal, but not necessarily in accordance with the standards of performance established for operation at normal rating. Their starting currents and pull-out torques are approximately proportional to the applied voltage; their starting and pull-in torques are approximately proportional to the square of the voltage. Their pull-out torques also are approximately proportional to the excitation current.

Mechanical Features

High-speed synchronous motors usually are furnished with feet on the stator frame and end-shield-type bearings. Pedestal-type bearings, with or without base, may be used in sizes above 1 hp per rpm. A direct-connected exciter often is supplied. This is overhung

Table 38. Approximate Weights and Dimensions of High-speed Synchronous Motors, Unity Power Factor, Open Type, 60 Cycles

(Dimensions in inches. See Fig. 18.)

Horsepower	Weight, pounds	*C*		*O*	*P*	Weight, pounds	*C*		*O*	*P*
		With Exciter	Without Exciter				With Exciter	Without Exciter		
	1800 Rpm					1200 Rpm				
50	1050	47	37	22	22	1305	51	41	25	25
60	1200	49	39	22	22	1465	53	43	25	25
75	1475	50	40	25	25	1465	53	43	25	25
100	1625	52	42	25	25	1705	53	43	29	29
125	1775	54	44	29	29	1905	55	45	29	29
150	1905	55	45	29	29	2275	58	48	29	29
200	2325	58	48	29	29	2855	60	50	34	34
250	3075	62	52	34	34	3075	62	52	34	34
300	3475	64	54	34	34	3500	65	54	34	34
350	3975	67	57	34	34	3500	65	54	34	34
400	6515	80	69	43	57	4000	68	57	34	34
450	6515	80	69	43	57	5380	74	62	40	39
500	6515	80	69	43	57	5380	74	62	40	39
	900 Rpm					720 Rpm				
50	1465	53	43	25	25	1805	62	50	29	29
60	1745	60	50	29	29	2005	64	52	29	29
75	1970	63	52	29	29	2375	67	55	29	29
100	1970	63	52	29	29	2935	68	55	34	34
125	2340	66	55	29	29	3030	70	55	34	34
150	2915	62	50	34	34	3250	72	57	34	34
200	3135	64	52	34	34	3650	73	59	34	34
250	3535	66	54	34	34	4990	74	59	40	39
300	4895	79	59	40	39	5590	79	62	40	39
350	5415	75	62	40	39	5590	79	62	40	39
400	5510	77	62	40	39	6360	82	65	40	39
450	6280	80	65	40	39	7420	80	63	50	50
500	6280	80	65	40	39	7420	80	63	50	50
	600 Rpm					514 Rpm				
50	2005	64	52	29	29	....	....	....	....	....
60	2005	64	52	29	29	....	....	....	....	....
75	2935	68	55	34	34	....	....	....	....	....
100	3030	70	55	34	34	3330	73	57	34	34
125	3330	73	57	34	34	4730	81	64	40	39
150	3650	73	59	34	34	5225	82	66	40	39
200	4650	79	64	40	39	5715	77	62	40	39
250	5070	76	59	40	39	5715	77	62	40	39
300	5590	79	62	40	39	7545	80	64	50	50
350	6360	82	65	40	39	7545	80	64	50	50
400	7420	80	63	56	50	7545	80	64	50	50
450	7420	80	63	50	50	8585	84	64	50	50
500	8350	83	66	50	50	8585	84	64	50	50

on an extension of the motor shaft on the collector-ring end (opposite the drive end). The exciter frame is bolted to the corresponding end shield or is supported from the base of pedestal-bearing machines.

Low-speed synchronous motors commonly are supplied without base, shaft, or bearings, the rotor being mounted directly on the shaft of the driven machine. In this case, a completely split-rotor spider, or split hub spider, may be furnished to facilitate assembly at the installation. Foundation caps usually are provided for support of the stator frame. A shaft, pedestal-type bearings, and base or sole plates are provided if a self-contained unit is required.

Direct-connected exciters seldom are used with low-speed synchronous motors, since higher-speed belt- or motor-driven exciters are more economical.

Sleeve bearings, oil-ring-lubricated, are standard for most horizontal synchronous motors, although grease-lubricated ball bearings can be obtained at increased cost. Vertical motors in the smaller sizes usually are furnished with ball bearings, and in the larger sizes with oil-lubricated plate-type bearings.

The horsepower limits for two-bearing belt drives are essentially the same as for induction motors (Table 6). Pulleys and sliding bases are available. Vertical synchronous motors, as well as the standard horizontal machines, can be obtained.

In addition to the open motors, dripproof, splashproof, totally enclosed fan-cooled, enclosed self-ventilated, and enclosed separately ventilated motors are available.

WEIGHTS AND DIMENSIONS. Approximate overall dimensions and net weights of high-speed synchronous motors are given in Table 38 and Fig. 18. Low-speed synchronous motors generally are specially designed for the particular application, with resultant wide variation in weights and dimensions for each rating.

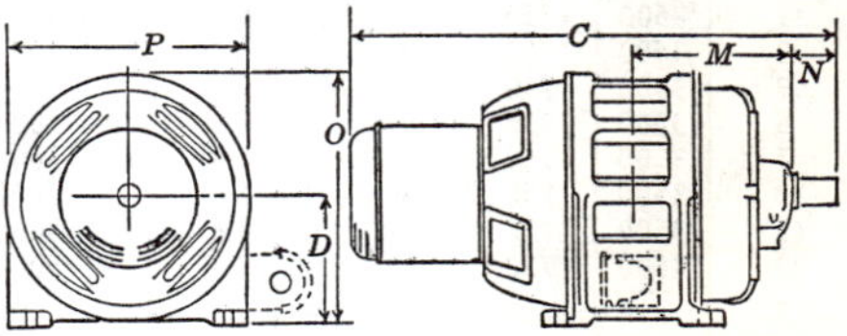

FIG. 18. Dimensions of high-speed synchronous motors. See Table 38.

8. DIRECT-CURRENT LARGE-POWER MOTORS

The following types of large-power d-c motors are in general use: (1) general-purpose constant-speed, (2) adjustable-speed, (3) large d-c, (4) crane and hoist, (5) mill-type, (6) metal rolling-mill, (7) elevator, and (8) low-speed engine-type.

General-purpose Motors. Standard horsepower and speed ratings for general-purpose constant-speed motors are given in Table 39. These motors are available with shunt-, series-, or compound-wound field windings. The shunt-wound motors have the lowest speed regulation (change in speed with change in load), are lowest in cost, and have the widest range of application. The compound-wound motors, which have somewhat higher speed regulation, are used on loads requiring high starting torque and on those subject to torque pulsation. Series-wound motors have very high speed regulation and should not be employed with belt drives or on any load in which the torque might drop below 15% of full-load torque, since a dangerously high speed would be developed. Series motors are suitable for loads requiring a very high starting torque or severe accelerating duty and where high-speed regulation is advantageous or of no consequence.

Shunt-wound constant-speed motors, having a full-load speed of 1750 rpm or below, usually can be operated at some increase in speed, without increase in horsepower, by inserting a resistance in series with the field winding. The maximum increase in speed permissible varies from about 10% for the higher speed ratings to 100% for the lower speed ratings. Speed variation by shunt field control generally is not advisable on compound-wound motors, since, with a weakened field, the motor characteristics approach those of a series motor and high speed regulation results.

Adjustable-speed Motors. These motors are shunt-wound machines designed particularly for adjustable-speed operation by varying the strength of the shunt field. Standard horsepower and minimum (basic) speed ratings and the maximum speed for each rating as standardized by NEMA are given in Table 40. As indicated, three different horsepower ratings may be given to adjustable-speed d-c motors: (1) a standard continuous horsepower rating over entire speed range, with a temperature rise of 50 C from basic speed to 150% of basic speed, and a temperature rise of 40 C from 150% of basic speed to the maximum speed; (2) a standard intermittent horsepower rating over the entire speed range, with a temperature rise of 50 C in 1 hr; (3) a *tapered rating* consisting of a certain standard horsepower from basic speed to 150% of basic speed, the next larger standard horsepower at three times basic speed and higher, and a horsepower varying along a straight line from the lower value to the upper value as the speed varies from 150% of basic speed to three times basic speed. The temperature rise is 50 C continuous from basic speed to 150% of basic speed, and 40 C continuous from 150% of basic speed to maximum speed.

In general, a motor rated in accordance with (1) will have the same cost as a motor of the next larger horsepower rated in accordance with (2), and the same cost as a tapered

Table 39. Standard Horsepower and Speed Ratings for General-purpose Constant-speed D-c Motors

(Ratings above the horizontal lines are suitable for two-bearing belt drive.)

Horsepower	Full-load Speed in Rpm														
1/2	......	......	......	850	690	575	500	450	400	350	300	250	200	150	100
3/4	......	......	1150	850	690	575	500	450	400	350	300	250	200	150	100
1	......	1750	1150	850	690	575	500	450	400	350	300	250	200	150	100
1 1/2	3500	1750	1150	850	690	575	500	450	400	350	300	250	200	150	100
2	3500	1750	1150	850	690	575	500	450	400	350	300	250	200	150	100
3	3500	1750	1150	850	690	575	500	450	400	350	300	250	200	150	100
5	3500	1750	1150	850	690	575	500	450	400	350	300	250	200	150	100
7 1/2	3500	1750	1150	850	690	575	500	450	400	350	300	250	200	150	100
10	3500	1750	1150	850	690	575	500	450	400	350	300	250	200	150	100
15	3500	1750	1150	850	690	575	500	450	400	350	300	250	200	150	100
20	3500	1750	1150	850	690	575	500	450	400	350	300	250	200	150	100
25	3500	1750	1150	850	690	575	500	450	400	350	300	250	200	150	100
30	3500	1750	1150	850	690	575	500	450	400	350	300	250	200	150	100
40	3500	1750	1150	850	690	575	500	450	400	350	300	250	200	150	100
50	3500	1750	1150	850	690	575	500	450	400	350	300	250	200	150	100
60	......	1750	1150	850	690	575	500	450	400	350	300	250	200	150	100
75	......	1750	1150	850	690	575	500	450	400	350	300	250	200	150	100
100	......	1750	1150	850	690	575	500	450	400	350	300	250	200	150	100
125	......	1750	1150	850	690	575	500	450	400	350	300	250	200	150	100
150	......	1750	1150	850	690	575	500	450	400	350	300	250	200	150	100
200	......	1750	1150	850	690	575	500	450	400	350	300	250	200	150	100

Table 40. Standard Horsepower and Speed Ratings of Adjustable-speed D-c Motors

Horsepower			Base Speed												
			1150	850	690	575	500	450	400	350	300	250	200	150	100
Continuous	1 Hour	Tapered	Maximum Speed												
1/2	3/4	1/2– 3/4	.....	3400	3450	3450	3000	2700	2400	2100	1800	1500	1200	900	600
3/4	1	3/4– 1	3450	3400	3450	3450	3000	2700	2400	2100	1800	1500	1200	900	600
1	1 1/2	1 – 1 1/2	3450	3400	3450	2875	2500	2700	2400	2100	1800	1500	1200	900	600
1 1/2	2	1 1/2– 2	3450	2550	2760	2875	2500	2250	2400	2100	1800	1500	1200	900	600
2	3	2 – 3	3450	2550	2760	2875	2500	2250	2400	2100	1800	1500	1200	900	600
3	5	3 – 5	.....	2550	2300	2300	2500	2250	2400	2100	1800	1500	1200	900	600
5	7 1/2	5 – 7 1/2	.....	.....	2300	2300	2000	2250	2000	2100	1800	1500	1200	900	600
7 1/2	10	7 1/2– 10	.....	.....	2300	2300	2000	1800	2000	2100	1800	1500	1200	900	600
10	15	10 – 15	.....	.....	.....	2300	2000	1800	2000	1750	1800	1500	1200	900	600
15	20	15 – 20	.....	.....	.....	1800	1800	1800	1600	1750	1500	1500	1200	900	600
20	25	20 – 25	.....	.....	.....	1800	1800	1800	1600	1750	1500	1500	1200	900	600
25	30	25 – 30	.....	.....	.....	1800	1500	1500	1600	1500	1500	1200	1200	900	600
30	40	30 – 40	.....	.....	.....	1800	1500	1500	1600	1500	1500	1200	1200	900	600
40	50	40 – 50	.....	.....	.....	.....	1500	1500	1600	1500	1200	1200	1200	900	600
50	60	50 – 60	.....	.....	.....	.....	1500	1500	1200	1200	1200	1200	1200	900	....
60	75	60 – 75	.....	.....	.....	.....	1500	1350	1200	1200	1200	1200	1200	900	....
75	100	75 –100	.....	.....	.....	.....	.....	.....	1200	1200	1200	1200	1200	....	....
100	125	100 –125	.....	.....	.....	.....	.....	.....	1200	1200	1200	1200	.....	....	....
125	150	125 –150	.....	.....	.....	.....	.....	.....	1200	1200	1200	.....	.....	....	....
150	200	150 –200	.....	.....	.....	.....	.....	.....	1200	1200	.....	.....	.....	....	....

rating motor whose lower horsepower is the same. Adjustable-speed d-c motors are used most extensively on machine tools.

Large d-c motors have the following standard horsepower ratings: 250, 300, 350, 400, 500, 600, 700, 800, 900, 1000, then in steps of 250 hp up to 2500 hp, and then in steps of 500 hp up to 6000 hp. Standard speed ratings in rpm are: 1150, 850, 690, 575, 500, 450, 400, 350, 300, 250, 225, 200, 175, 150, 125, 110, 100, 90, 80, 70, 65, 60, 55, and 50. The motors may be constant speed or adjustable speed (constant horsepower). The maximum speed range varies from about 5 : 1 for the smaller low-speed ratings to 1.5 : 1 for the larger high-speed ratings. Typical applications include pumps, fans, paper machines, rubber mills, and certain marine applications.

Crane and hoist motors are series- or compound-wound totally enclosed or semi-enclosed heavy-duty machines specially designed for crane and hoist service. The totally enclosed motors are usually given a 15- or 30-min rating with a 55 C temperature rise, and the semi-enclosed machines are given a 1-hr or a continuous rating with a 55 C temperature rise. The motors are available in ratings up to about 165 hp, with speed ranging from 425 to 1450 rpm. Shaft extensions at both ends, for brake and drive, respectively, and pinion, gear, and reduced-speed axle mounted integrally with the motor are available. Although higher in cost than standard motors, their compactness, rugged mechanical construction, and low-inertia armature justify their use on cranes, hoists, and similar loads involving intermittent service and severe reversing duty.

Mill motors are heavy-duty machines designed particularly for such services as steel-mill auxiliaries, oil-well drilling rigs, power shovels, cranes, bridges, hoists, and dredges. The basic motor is series-wound, totally enclosed, intermittently rated as shown in Table 41. Ratings from 7 1/2 to 200 hp are available with mounting dimensions standard-

Table 41. Standard Horsepower and Speed Ratings of Mill-type D-c Motors

(Series-wound, totally enclosed, 1 hr, 75 C temperature rise)

Horsepower	5	7 1/2	10	15	25	35	50	75	100	150	200	250	350
Full-load, rpm	900	800	725	650	575	525	500	475	460	450	410	375	375

ized by the AISE. In addition to the series-wound motors, compound-wound and shunt-wound (constant- and adjustable-speed) motors are available, as are enclosed motors rated 30 min, protected self-ventilated motors rated 1 hr, and protected self-ventilated and force-ventilated motors with continuous ratings. In general, the horsepower and speed ratings for such motors differ from those of the basic motors. All motors have Class B insulation, roller or double-row ball bearings, and shaft extensions on both ends. A pinion, gear, and reduced-speed axle with sleeve or roller bearings also can be obtained. Rugged mechanical construction, low-inertia armatures, compactness, and accessibility of all parts are features of these motors.

Metal rolling-mill motors are heavy-duty machines designed particularly for metal rolling-mill drive. Standard horsepower and speed ratings are given in Table 42. The

Table 42. Standard Horsepower and Speed Ratings of Adjustable-speed Metal Rolling-mill D-c Motors

Standard Basic Speed Ratings, rpm

1150, 850, 690, 575, 500, 450, 400, 350, 300, 250, 225, 200, 175, 150, 125, 110, 100, 90, 80, 70, 65, 60, 55, 50

Standard Horsepower Ratings

Horsepower	Basic Speeds, rpm	Horsepower	Basic Speeds, rpm	Horsepower	Basic Speeds, rpm
250	100 to 1150	800	60 to 850	2250	50 to 350
300	100 to 1150	900	55 to 850	2500	50 to 350
350	100 to 1150	1000	50 to 850	3000	50 to 300
400	100 to 1150	1250	50 to 575	3500	50 to 250
500	90 to 1150	1500	50 to 500	4000	50 to 225
600	80 to 850	1750	50 to 500	4500	50 to 200
700	70 to 850	2000	50 to 400	5000	50 to 175

motors are usually arranged for adjustable-speed operation, the maximum speed varying from basic speed for the larger higher-speed ratings, to five times basic speed for the smaller lower-speed ratings. The motors normally are furnished with a base and pedestal-type bearings.

Geared elevator hoisting machines employ high-speed motors similar to the general-purpose motors. They usually are adjustable-speed machines with shunt field control or variable armature voltage control. Gearless elevators require a special low-speed driving motor. The hoisting drum is mounted on the motor shaft between the two bearings. A brake wheel for a solenoid brake usually is cast integrally with the drum. Heavy shafts and large bearings are provided. The motors are of the adjustable-speed type with armature voltage control.

Low-speed engine-type d-c motors are available in ratings from 100 to 250 hp at speeds of 150 to 327 rpm. The standard speeds are the same as for low-speed 60-cycle synchronous motors (Table 35). The motors are compound-wound and are furnished without base, shaft, or bearings. They are used to drive low-speed machines, such as reciprocating compressors, the rotor being mounted on the shaft of the driven machine.

Operating Characteristics

TEMPERATURE RISE of standard general-purpose motors is 40 C, measured by thermometer for open motors, 50 C for semi-enclosed, drip-proof, and splashproof machines, and 55 C for totally enclosed and totally enclosed fan-cooled machines. Open motors rated 40 C usually are given a service factor of 1.15; that is, the manufacturers guarantee successful operation at 1.15 times rated load without injurious heating. Direct-current motors generally are capable of delivering 50% overload in torque momentarily, without temperature guarantee.

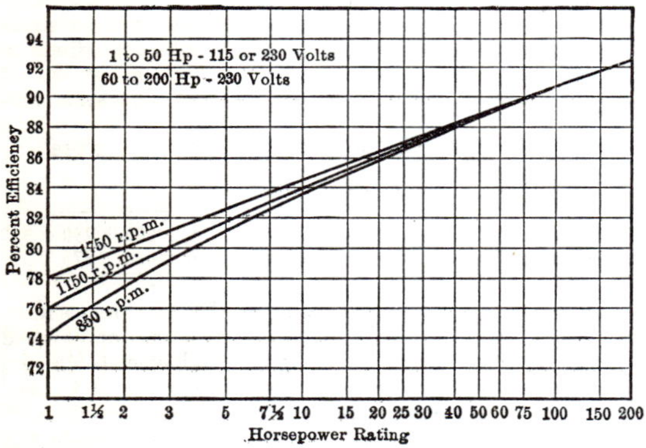

Fig. 19. Approximate full-load efficiency of general-purpose constant-speed, shunt-wound direct-current motors, open type.

EFFICIENCY. Typical approximate full-load efficiencies of general-purpose constant-speed shunt-wound open-type motors are given in Fig. 19.

TORQUE. With full voltage on the shunt field (if provided) and with the armature current limited to full-load current, the starting torque of any type of d-c motor is 100% of full-load torque. For shunt-wound motors the starting torque varies directly with the armature current; for series-wound motors it varies approximately as the square of the armature current; for compound-wound motors the variation lies between these two limits, depending on the ratio of shunt and series windings. The armature current, in turn, is determined by the amount of resistance inserted in series with this winding at start. Standard starters usually limit the current to from 125 to 200% of full-load current.

SPEED REGULATION from full load to no load, hot, based on full-load speed should not exceed the following:

Constant-speed motors, shunt-wound: 5 hp and smaller, 12%; 7 1/2 to 200 hp, inclusive, 10%; 250 hp and larger, 6%; compound-wound, constant-speed motors, 25%.

Adjustable-speed motors, shunt-wound: 2 to 5 hp, inclusive, 22%; 7 1/2 to 200 hp, inclusive, 15%; 250 hp and larger, at maximum speed, 4%; at basic speed, 6 to 15%.

At normal operating temperature, rated load, and voltage, the variation above or below rated speed should not exceed the following:

Constant-speed motors, shunt- or compound-wound, 7 1/2 hp at 1150 rpm or smaller, 7 1/2%; larger than 7 1/2 hp, at 1150 rpm, 5%.

Adjustable-speed motors, shunt-wound, 5 hp, 650 rpm (minimum speed) or smaller, 7 1/2%; larger than 5 hp, 650 rpm, 5%.

The variation in speed from full load, cold, to full load, hot, should not exceed 10%, based on full-load speed for open motors, and 15% for enclosed motors.

VOLTAGE VARIATION. All d-c motors should operate successfully at normal rated load at any voltage not more than 10% above or below normal, but not necessarily in accordance with the standards of performance established for operation at normal rated voltage. In general, with a given load, an increase in voltage results in an increase in speed and vice versa.

Any shunt-wound motor may be operated as an adjustable-speed machine by maintaining constant voltage on the shunt field while varying the voltage on the armature. The speed will be approximately proportional to the armature voltage. The basic speed of such a machine is the maximum speed (full voltage on the armature). At reduced speeds, the horsepower is reduced approximately in proportion to the speed. Thus *variable voltage control* gives constant-torque characteristics, as contrasted with speed variation by field control which gives essentially constant-horsepower characteristics. If the source of variable voltage direct current is such that the voltage is essentially independent of the current drawn, the motor is truly an adjustable-speed machine; i.e., its speed when once adjusted to a given value remains essentially constant, regardless of the load. A d-c generator, driven at constant speed with voltage variation by field control, meets this requirement and commonly is used. On the other hand, if the applied armature voltage varies appreciably with the current drawn, as it would if a variable resistance were used to reduce the voltage, the drive would be of the adjustable-varying-speed type; i.e., the speed would vary appreciably with the load.

Mechanical Features

Large-power d-c motors in sizes up to about 200 hp usually have feet on the stator frame and end-shield-type bearings. In the larger sizes, a base with pedestal-type bearings usually is supplied. Vertical motors, as well as the standard horizontal motors, can be obtained.

Sleeve bearings, oil-ring-lubricated, are standard for most horizontal machines. Ball bearings are standard on the totally enclosed, the splashproof, and the smaller vertical motors. They can be obtained on all motors at increased cost. Ball bearings usually are grease-lubricated.

Motors that can be belted with two bearings are indicated in Table 6. Sliding bases and pulleys are available. Larger ratings require an extended base and shaft with outboard bearing. Slide rails usually are supplied for belt adjustment.

Gear motors are generally available in sizes up to 7 1/2 hp with output-shaft speeds of 13.5 to 780 rpm. High-speed (1750 rpm) motors are usually employed.

In addition to the standard open-type motors, protected, semi-protected, drip-proof, splashproof, totally enclosed, totally enclosed fan-cooled, and explosionproof (Class I, Group D, and Class II, Groups E, F, and G) motors are available.

WEIGHTS AND DIMENSIONS. Frame numbers assigned to general-purpose constand-speed d-c motors are given in Table 43. Approximate net weights are given in Table 44.

Table 43. Frame Sizes for General-purpose Constant-speed D-c Motors, Open Type

(NEMA Standard)

Full-load Speed, rpm	3500	1750	1150	850	690	575	500
Horsepower	Frame Size *						
1/2	...	...	...	204	224	225	...
3/4	...	...	203	224	225	254	284
1	...	203	204	225	254	254	284
1 1/2	203	204	224	254	254	...	...
2	204	224	225	254	284	...	...
3	224	225	254	284	...	...	...
5	225	254	284	...	...	...	...
7 1/2	254	284	...	...	...	...	...
10	284	...	...	...	...	...	...

* For standard mounting dimensions, see Tables 13, 14, and 15.

Table 44. Approximate Weights of D-c motors Built in NEMA Standardized Frames

(Weight in pounds)

Frame Number	Open and Totally Enclosed Motors	Totally Enclosed Fan-cooled Motors	Base	Pulley
203	95	...	19	1
204	105	...	20	1
224	145	164	22	3
225	160	179	23	3
254	210	235	28	4
284	280	313	33	5

9. MOTOR APPLICATIONS

The types of motors commonly selected for various applications are given in Table 45. The items listed in the table are those of general application for which conditions are well standardized. It is not possible to list the unusual conditions under which individual motors operate. Since many factors affecting the choice of motor vary with each installation, the recommendations given should be considered typical and not necessarily applicable to every case. Manufacturers of motors should be consulted with reference to the selection of the proper motor for unusual or nonstandard conditions.

Table 45. Types of Motors Commonly Selected for Various Applications

Application	Power Supply		
	Single Phase Alternating Current	Polyphase Alternating Current	Direct Current
	Type of Motor Commonly Used *		
Air compressors, centrifugal	....	Sq-In, Design B	Gp, Cs, Sh
Air compressors, reciprocating			
3 hp and below	Cp	Sq-In, Design B	Gp, Cs, Cw
5 to 10 hp	Rp-In	Sq-In, Design C	Gp, Cs, Cw
15 to 75 hp	....	Sq-In, Design C	Gp, Cs, Cw
100 hp and larger	....	Syn, Ls	Et, Ls
Blowers, positive displacement, rotary			
1 to 3 hp	Cp	Sq-In, Design B	Gp, Cw, Sh
5 to 100 hp	....	Sq-In, Design B	Gp, Cw, Sh
125 hp and larger	....	Syn, Ls or Hs	Gp, Cw, Sh
Conveyors			
3 hp and below	Cp	Sq-In, Design B	Gp, Cs, Cw
5 hp and larger	Rp-In	Sq-In, Design C	Gp, Cs, Cw
Fans			
1/25 hp and below	Shp	....	....
1/20 to 1/3 hp	Sp	Sq-In, Design B	Gp, Cs, Sh
1/2 to 3 hp	Cp	Sq-In, Design B	Gp, Cs, Sh
5 to 10 hp	Rp-In	Sq-In, Design B	Gp, Cs, Sh
15 to 100 hp	....	Sq-In, Design B	Gp, Cs, Sh
125 hp and larger	....	Syn, Hs	Gp, Cs, Sh
Generators, d-c			
3/4 to 5 kw	Cp or Rp-In	Sq-In, Design B	Gp, Cs, Sh
7 1/2 to 50 kw	....	Sq-In, Design B	Gp, Cs, Sh
60 kw and larger	....	Syn, Hs	Gp, Cs, Sh
Hoists	Rp-In, V	Sq or Wr-In, Ch	Sr, Ch
Machine Tools			
Bolt cutters	....	Sq-In, Design B, Us or Ms	Sh, Cs or As
Bending brakes	....	Sq-In, Design D	Cw, Cs
Bolt heading, upsetting, and forging machines	....	Sq-In, Design D or Wr	Cw, Cs
Bolt pointers	....	Sq-In, Design B	Sh, Cs
Boring or turning mills	....	Sq-In, Design B, Us or Ms	Sh, Cs or As
Buffing machines	....	Sq-In, Design B	Sh, Cs
Bulldozers	....	Sq-In, Design B or Wr	Cw, Cs
Drilling machines	....	Sq-In, Design B, Us or Ms	Sh, Cs or As
Gear cutters	....	Sq-In, Design B, Us or Ms	Sh, Cs or As
Grinding wheels	....	Sq-In, Design B	Sh, Cs or As
Hammers	....	Sq-In, Design D	Cw, Cs
Lathes	....	Sq-In, Design B, Us or Ms	Sh, Cs or As
Milling machines	....	Sq-In, Design B, Us or Ms	Sh, Cs or As
Nut facers	....	Sq-In, Design B	Sh, Cs
Nut tappers	....	Sq-In, Design B, Us or Ms	Sh, Cs or As
Pipe threading or cutting-off machines	....	Sq-In, Design B, Us or Ms	Sh, Cs or As
Planers	....	D-c motor supplied by a-c to d-c motor generator set	Sh, As
Polishing machines	....	Sq-In, Design B	Sh, Cs
Punch presses	....	Sq-In, Design D, or Wr	Cw, Cs
Rolls, bending and straightening	....	Wr-In	Cw, Cs
Saws	....	Sq-In, Design B	Sh, Cs
Shears	....	Sq-In, Design D, or Wr	Cw, Cs
Slotters and key seaters	....	Sq-In, Design B, Us or Ms	Sh, Cs or As

Table 45. Types of Motors Commonly Selected for Various Applications—*Continued*

Application	Power Supply: Single Phase Alternating Current	Power Supply: Polyphase Alternating Current	Power Supply: Direct Current
	Type of Motor Commonly Used *		
Pumps, centrifugal			
1/4 hp and less	Sp	Sq-In, Design B	Gp, Cs, Sh
1/3 to 3 hp	Cp	Sq-In, Design B	Gp, Cs, Sh
5 to 10 hp	Rp-In	Sq-In, Design B	Gp, Cs, Sh
15 to 100 hp	....	Sq-In, Design B	Gp, Cs, Sh
125 hp and larger	....	Syn, Hs or Ls	Gp, Cs, Sh
Refrigerating compressors, reciprocating			
3 hp and less	Cp	Sq-In, Design B	Gp, Cs, Cw
5 to 10 hp	Rp-In	Sq-In, Design C	Gp, Cs, Cw
15 to 75 hp	....	Sq-In, Design C	Gp, Cs, Cw
100 hp and larger	....	Syn, Ls	Et, Ls
Refrigerating compressors, centrifugal	....	Wr	Sh, Cs or As
Textile machinery			
Cards	....	Sot	....
Cleaners	....	Sot	....
Drawframes	....	Sot	....
Finishing Processes	....	D-c motors with a-c to d-c conversion equipment	Sh, As
Intermediates	....	Sot	....
Looms	....	Lm	....
Pickers	....	Sot	....
Roving frames	....	Sot	....
Spinning frames			
Constant speed	....	Sot	....
Variable speed	....	As, Bs	....
Vacuum pumps			
1 to 3 hp	Cp	Sq-In, Design B	Gp, Cs, Cw
5 to 10 hp	Rp-In	Sq-In, Design B	Gp, Cs, Cw
15 to 75 hp	....	Sq-In, Design B	Gp, Cs, Cw
100 hp and larger	....	Syn, Ls	Et, Ls
Valves	Cp or Rp-In	Sq-In, Design C	Sr

* As = adjustable speed. Bs = brush shifting. Ch = crane and hoist. Cp = capacitor. Cs = constant speed. Cw = compound wound. Et = engine type. Gp = general purpose. Hs = high speed. In = induction. Lm = loom motor. Ls = low speed. Ms = multispeed. Rp = repulsion. Sh = shunt wound. Shp = shaded pole. Sot = screenless open textile. Sp = split phase. Sq = squirrel cage. Sr = series wound. Syn = synchronous. Us = single speed. V = reversible. Wr = wound rotor.

ELECTRIC MOTOR CONTROL

10. GENERAL CONSIDERATIONS

FUNCTIONS. The basic function of all motor control is to provide a means for making and breaking the connections between the motor windings and the power lines, thereby starting and stopping the motor. Practically all controllers also provide overload protection to the motor. As will be described, many other functions are provided by certain types of control.

CLASSIFICATION. Motor controllers may be classified as (1) full-voltage controllers or (2) reduced-voltage controllers. The former connect the motor directly to the power lines, that is, provide full voltage at the motor terminals during the starting period, as well as for running. The latter connect the motor to a voltage-reducing device during the starting period, and, after the motor has come up to speed, transfer the connections to the power lines. Autotransformers commonly are used to obtain the reduced voltage in a-c motor starters, although resistors and reactors frequently are employed. Resistors always are used for this purpose in d-c motor starters. Full-voltage controllers are the simplest and least expensive, but reduced-voltage controllers often are required to limit the current drawn from the power lines at start.

Motor controllers also may be classified according to the degree of automatic operation. On this basis, there are three standard types: (1) manual controllers, (2) magnetic controllers, and (3) semimagnetic controllers. In type 1 controllers, all connections to the motor are made by hand-operated switches. In type 2, the switches are electrically operated. To start the motor it is only necessary to close a pilot circuit by means of a push button, thermostat, or similar device that may be located at some distance from the controller. Reduced-voltage magnetic controllers have a relay that automatically effects the transfer from starting to running connections. In type 3 controllers, part of the starting operation is performed by hand-operated switches and the remainder by electrically operated devices.

LINE SWITCHES on low-voltage controllers (600 volts and below) usually are of the *air-break* type, i.e., the circuit is broken in air. Both air-break and oil-immersed switches are employed for the higher voltages, with the former becoming increasingly popular. Oil-immersed switches are also used in special low-voltage controllers in which the complete controller is oil-immersed. These are suitable outdoors or in corrosive atmospheres. Electromagnets, either the solenoid or clapper type, usually serve to close electrically operated switches, although motor-operated mechanisms sometimes are used on circuit breakers.

OVERLOAD PROTECTION. Thermal-type overload relays usually are employed. The most common form consists of a heater element that carries line current (or a current proportional to line current, obtained through a current transformer) mounted adjacent to a bimetallic strip. An overload on the motor causes an increase in line current, which increases the temperature of the heater element. The heat radiated to the bimetallic strip causes it to deflect and open a contact that disconnects the motor from the line. The time required for the relay to trip varies inversely with the magnitude of the overload, as does the length of time the motor can operate without injurious heating. With proper design, the relay will follow closely the heating characteristics of the motor. It thus adequately protects the motor and yet will not take it off the line unnecessarily (see Fig. 20). After the relay has tripped, it must be allowed to cool before it can be reset and the motor restarted. This time lapse also allows the motor to cool.

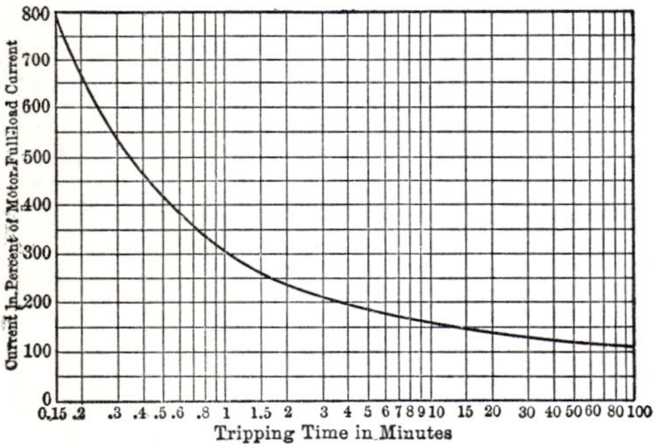

FIG. 20. Tripping characteristics of typical temperature-overload relay.

Another form of thermal overload relay, used on small motor starters, is the *solder-film* type. In this, the tripping mechanism normally is held to a stationary element by a thin film of solder. The heater that carries line current will, on an overload of sufficient duration, melt the solder film, thus releasing the tripping mechanism. After tripping the solder cools and solidifies, allowing the tripping mechanism to be reset.

Nonthermal overload relays also are used. They incorporate a time-delay feature, as an oil or air dashpot, to vary the tripping time with the magnitude of the overload. A disadvantage of this type is that when once tripped the relay can be reset immediately, and the motor started before it has had time to cool off.

Fuses. Ordinary fuses do not, in general, give satisfactory overload protection. Their time lag is so short that they usually must be oversize to prevent tripping on the starting current of the motor or on high overloads of short duration. As a result they do not give adequate protection against moderate overloads of long duration.

Relay Resets. Most overload relays are of the "hand-reset" type; that is, after tripping they must be reset manually before the motor can be restarted. In a few applications, particularly those in which the motor starts and stops automatically (for example, in response to a thermostat), it may be desirable to use an automatically reset overload relay that resets itself after a time delay of sufficient duration to permit the motor to cool. Relays of this type are available for small motors. They frequently are mounted on the motor itself and, in fractional-horsepower single-phase motors, may be designed to open directly the power circuit to the motor.

Two overload relay elements placed in different phases normally are used with three-phase and two-phase motors, whereas a single element is used with single-phase and d-c motors.

UNDERVOLTAGE PROTECTION AND RELEASE. Most controllers are so arranged that on failure of line voltage the circuit to the motor is opened. Manual starters for small motors are an exception. If the line switch is held in the closed position by a magnet energized from the line, this feature is obtained automatically. If the line switch is held closed by a mechanical latch, an *undervoltage device* must be provided, which is energized from the line and, when de-energized, operates to trip the latch. If the control is so arranged that on return of voltage the line switch recloses, *undervoltage release* is obtained. On the other hand if the line switch does not reclose on return of voltage, *undervoltage protection* is obtained. The latter is more common. The former should be used only when no hazard is involved in the motor's starting unexpectedly after a voltage failure.

In many cases, voltage failures are of very short duration, viz., 1 1/2 sec or less. Thus it is often desirable to provide control that will restart the motor if the voltage failure does not last over a certain predetermined time, but that will disconnect it permanently if the voltage failure is of longer duration. Control so arranged provides *time-delay undervoltage protection.* If the line switch is held closed by a magnet energized from the line, a time-delay push-button station generally is used. This is so arranged that its contacts are held closed for a definite time (1 to 4 sec) after a failure of voltage but can be opened instantaneously by the *stop* button. When the line switch is held closed by a mechanical latch, an undervoltage tripping mechanism incorporating a time delay in its action must be provided.

SHORT-CIRCUIT PROTECTION. The current that flows through a controller, due to a short circuit on the motor side, depends upon the characteristics of the power system to which the equipment is connected and may be a great many times the maximum current flow under normal conditions. The controller should be capable of interrupting the short-circuit current or safely withstanding this current until it is interrupted by some other device such as fuses or a circuit breaker in the circuit between the controller and the power source. With low-voltage controllers, fuses or air-circuit breakers in the branch circuit to the controller are usually provided to interrupt the short-circuit current. With high-voltage controllers, special "current-limiting" fuses are often used in a combination starter. These fuses limit the short-circuit current by interrupting it before it has built up to its peak value and thus prevent damage to the contactor and other parts of the controller. Both low- and high-voltage controllers employing a high-interrupting-capacity contactor or circuit breaker are also available.

ENCLOSURES. Open-type controllers (no enclosing case) are used where they are under competent supervision and are installed where accessible only to authorized persons. All switches and devices whose voltage rating is in excess of 600 volts should be mounted on the back of the panel or suitably enclosed to prevent accidental contact.

Controllers for small motors almost always are enclosed; enclosures can be obtained for all sizes. A number of different types of enclosure are available. The standard *general-purpose enclosure* is designed primarily to prevent accidental contact; it also protects against dust and light indirect splashing, but it is not dust-tight. It usually is of sheet-metal construction. The *driptight enclosure* is similar to the general-purpose type with the addition of drip shields or their equivalent. It is used where condensation may be severe, as in cooling rooms and laundries.

Weather-resisting enclosures are of cast construction or of suitably rigid sheet metal, with proper protection against weather hazards, as rain and sleet. They are installed on outdoor applications, as on docks, canal locks, and construction work, and also in subways and tunnels. A *watertight enclosure* is so designed that a stream of water from a hose (not less than 1 in. in diameter) under a head of 35 ft can be played on the case from a distance of 10 ft for several minutes without leakage. It is suitable for outdoor applications and for dairies, breweries, and similar locations. *Submersible enclosures* are so designed that they may be submerged in water, as in quarries, mines, and manholes. *Dust-tight enclosures* are provided with gaskets or their equivalent to exclude dust and are used in such situations as often occur in steel and cement mills.

Controllers for Class I, Group D, hazardous locations (see p. 16-19) may be of the air-break or oil-immersed type. Air-break controllers depend on the strength of the enclosing case to resist the force of an internal explosion, and on long, close fits at all joints to cool the flame and prevent ignition of the surrounding atmosphere. In oil-immersed controllers, all contacts and connections are immersed in oil to prevent ignition of the surrounding atmosphere. This construction is preferable where the control is placed in a corrosive atmosphere. Controllers for Class II, Group G, conditions, and others designed to meet the requirements of the U. S. Bureau of Mines, also are available.

Combination magnetic starters consist of a full-voltage magnetic starter combined with a manually operated motor circuit switch (fused or unfused), or a circuit breaker, mounted in a single enclosing case. Thus, means are provided for disconnecting the motor and starter, and (with fuses or circuit breaker) protection is supplied against short circuits in the starter, motor, or wiring on the motor side of the switch. General-purpose, dust-tight and watertight enclosing cases are available. Oil-immersed switches suitable for corrosive atmospheres, or for Class I, Group D hazardous locations, also can be obtained.

Frequently, a saving in space and in installation expense is obtained with a combination magnetic starter instead of the two separate devices. Combination starters are available in ratings up to 100 hp at 220 volts, and 200 hp at 440 or 550 volts.

LOCATION. In general, controllers should be placed within sight of the motor they control. If this is not feasible, the controller or disconnecting device should be provided with means for locking it in the open position, or a manually operated switch, which will prevent restarting, should be placed within sight of the motor. This switch may be located in the pilot circuit of the controller.

11. CONTROL FOR SINGLE-PHASE MOTORS

TYPES. The following types of control for single-phase motors are in common use: (1) manual full-voltage, (2) magnetic full-voltage, (3) manual reduced-voltage, (4) adjustable-speed (for capacitor motors), and (5) reversing.

Manual full-voltage controllers are available in sizes up to 1 1/2 hp at 110 volts, 3 hp at 220 volts, and 5 hp at 440 and 550 volts. They consist of a single- or double-pole line switch and an overload relay, usually of the solder-film type, mounted on a compound base. They may be obtained open, or enclosed in a sheet-metal case. Explosionproof enclosures with air-break switches for Class I, Group D conditions also are available. The smaller controllers (3/4 hp and below) usually are arranged so they can be flush mounted, using a sectional switch box and tumbler-switch flush plate. Small single-phase motors (3/4 hp and below) can be obtained with a tumbler switch mounted on the motor to serve as a starter. These do not provide overload protection and should not be used where the switch may be called upon to interrupt the stalled current of the motor.

Manual, full-voltage controllers are the lowest in cost of all types and are used where full-voltage starting is acceptable and where remote control or automatic operation is not required. They do not provide undervoltage protection.

Magnetic full-voltage controllers are available in sizes up to 7 1/2 hp at 110 volts, 15 hp at 220 volts, and 25 hp at 440 and 550 volts. They consist of a magnetic contactor mounted on a compound base. A thermal-type overload relay usually is provided in sizes above 1 hp at 110 volts and 1 1/2 hp at 220 volts. Various types of enclosures are available, including general-purpose, dust-tight, watertight, and explosionproof for Class I, Group D conditions (both air-break and oil-immersed). In the smaller sizes, enclosing cases with start-stop push buttons in the cover can be obtained. Otherwise, a separately mounted push-button station, either instantaneous or time-delay type, must be used. These are also available with general-purpose, dust-tight, watertight, and explosionproof enclosures. Frequently a thermostat, pressure switch, or similar device is used for pilot-circuit control, providing undervoltage release.

Since full-voltage starting of single-phase motors generally is acceptable and remote or automatic control often is necessary, full-voltage magnetic starters have a wide field of application covering practically all types of drive.

Manual reduced-voltage controllers consist of a rheostat, which at start is inserted in series with the motor winding, thus limiting the current drawn from the line. As the motor comes up to speed, the operating handle of the rheostat is turned, which short-circuits the resistance step by step until the motor is on full voltage. A magnet, whose coil is energized from the line voltage, holds the operating handle in the running position. If voltage fails, the handle is released and the motor is disconnected from the line. Overload protection ordinarily is not included. Manual reduced-voltage starters usually are provided with a general-purpose enclosing case. They are used only when the full-voltage starting current of the motor would cause objectionable voltage disturbance. Since the starting torque of the motor is reduced in proportion to the square of the current drawn, it is apparent that a reduction in the current drawn from the line will be obtained only if the load does not require the full-voltage torque of the motor to start.

Adjustable-speed controllers for capacitor fan motors usually consist of ring-wound autotransformers with a movable tap by which the voltage impressed on the motor is varied, thus changing the motor speed. They are available in sizes up to 3/4 hp.

Reversing controllers usually are of the full-voltage type. Manually operated drum or cam switches are available for this service, in sizes up to 1 1/2 hp at 110 volts, 3 hp at 220 volts, and 5 hp at 440 or 550 volts. These controllers do not provide overload or undervoltage protection. Magnetic reversing switches are obtainable in sizes up to 7 1/2 hp at 110 volts, 15 hp at 220 volts, and 25 hp at 440 and 550 volts. They consist of two magnetic contactors, mechanically interlocked to prevent closing of both at the same time. Both overload and undervoltage protection are provided. A three-button push-button station is required. Enclosing cases of the general-purpose type usually are supplied, although watertight and dust-tight enclosing cases also are available.

12. CONTROL FOR POLYPHASE INDUCTION MOTORS

TYPES. The following types of control for polyphase induction motors are in general use:

Squirrel-cage induction motors: (1) manual full-voltage, (2) magnetic full-voltage, (3) manual reduced-voltage (autotransformer type), (4) manual reduced-voltage (resistor type), (5) magnetic reduced-voltage (autotransformer type), (6) magnetic reduced-voltage (resistor type), (7) multispeed, and (8) reversing.

Wound-rotor induction motors: (1) manual primary and secondary, (2) magnetic primary, manual secondary, and (3) magnetic primary and secondary.

Squirrel-cage Motors

Manual full-voltage controllers of the air-break type for squirrel-cage induction motors are available in sizes up to 5 hp at 220 volts, and 7 1/2 hp at 440 and 550 volts. They consist of a manually operated switch and a two-element, thermal-type, overload relay, mounted in an enclosing case. Undervoltage protection is not provided.

Standard enclosing cases are of the general-purpose type, but dust-tight and watertight enclosure and enclosures suitable for Class I, Group D or Class II, Group G hazardous locations are available.

Magnetic full-voltage controllers are obtainable in all sizes. They are the most popular of all types. The common form consists of a magnetic contactor combined with a thermal-type overload relay mounted in an enclosing case. Air-break-type contactors are standard for voltages of 600 and below, but completely oil-immersed starters are available for corrosive atmospheres. For higher voltages both air-break and oil-immersed contactors are employed. A start-stop push-button station or similar device for closing and opening the pilot circuit also is required. Enclosing cases of the general-purpose, dust-tight, and watertight type are available. Explosionproof controllers for Class I, Group D conditions can be obtained in both the air-break and oil-immersed types. In the smaller sizes of low-voltage controllers, enclosing cases of the general-purpose, dust-tight, watertight, and explosionproof (air-break) type with start-stop push buttons mounted in the cover are available.

Combination magnetic starters are also available. In the higher voltages (2300 and 4160 volts) these starters consist of magnetic contactors and disconnecting-type current-limiting fuses, making them suitable for installations on power systems having a high short-circuit current. Controllers employing a circuit breaker for the line switch and having whatever current-interrupting capacity is required can also be obtained.

Manual reduced-voltage autotransformer controllers are available in sizes up to 300 hp at 220 volts, 700 hp at 440 or 550 volts, and 900 hp at 2000 to 5000 volts. They consist of an autotransformer with two or three sets of low-voltage taps and an oil-immersed double-throw manually operated switch. In starting, the operating handle is first moved in one direction connecting the motor to a set of reduced-voltage taps of the autotransformer and held in that position until the motor has come up to speed, then the handle is moved quickly in the opposite direction to connect the motor directly to the line. The operating handle usually is interlocked mechanically, so that it must be moved to the start position before it can be moved to the run position. Overload and undervoltage (instantaneous or time-delay) protection are provided. General-purpose enclosing cases are standard, but dust-tight cases also are available.

The manual autotransformer starters are, in general, the least costly of the various reduced-voltage types. They are used whenever manual control is acceptable, and a reduction is desirable in the current drawn, or torque developed, by the motor on full voltage.

The current drawn from the line is reduced in proportion to the square of the voltage tap. For example, if the motor is connected to the 80% voltage tap, the current drawn from the line will be approximately 64% of the full-voltage starting current of the motor. Since the torque developed by an induction motor also varies as the square of the voltage impressed on its terminals, the ratio of starting torque to starting current (*torque efficiency*) remains essentially the same as with full-voltage starting.

Manual reduced-voltage resistor controllers are available in ratings up to 15 hp at 220 volts, and 20 hp at 440 and 550 volts. They consist of a rotating drum switch and a resistor. Rotation of the drum switch connects the motor to the line through the entire resistor and then successively short-circuits sections of the resistor until the motor is connected directly to the line. Neither overload nor undervoltage protection are provided. These controllers are used largely for motors driving machine tools, hoists, small cranes, and similar devices requiring manual control of the torque delivered by the motor.

Magnetic reduced-voltage autotransformer controllers are available in all sizes. These controllers are similar to the manual autotransformer type, except that, in place of the manual switch, magnetic contactors and a relay, usually of the definite-time type, effect the transfer from reduced voltage to full voltage. Overload and undervoltage protection are provided. The latter may be obtained by a time-delay push-button station, or, in the case of contactors held in by a mechanical latch, a time-delay undervoltage tripping mechanism may be provided on the contactor itself. General-purpose and dust-tight enclosing cases are available. Completely oil-immersed controllers, both low-voltage and high-voltage, can be obtained for corrosive atmospheres.

Magnetic reduced-voltage controllers of the autotransformer or reactor type with current-limiting fuses are available in sizes up to 700 hp at 2300 volts, and 1250 hp at 4160 volts, for use where high current-interrupting capacity is required.

Magnetic reduced-voltage resistor controllers are obtainable in sizes up to 250 hp at 220 volts, and 500 hp at 440 and 550 volts. They are similar to the autotransformer type except that a resistor, inserted in series with the motor windings at start, serves, instead of an autotransformer, to obtain the reduction in voltage. Since the resistor is simply short-circuited as the motor comes up to speed, the circuit to the motor is not opened during the starting period. Hence this type of controller is well adapted for use where the starting current limitations do not limit the total current that can be drawn, provided the current is built up in steps of a certain magnitude without opening the circuit between steps. Frequently, a number of accelerating contactors and relays are provided. These short-circuit the starting resistor in steps in order to limit the current increment on each step to the required value.

Unlike autotransformer starting, resistor starting results in a reduction in the starting torque efficiency, since the current drawn from the line is reduced in proportion to the voltage impressed on the motor and the starting torque is reduced in proportion to the square of the voltage.

Multispeed Controllers. Controllers for multispeed induction motors are available in both the manual and magnetic type. The manual type consists of a rotating drum switch and can be obtained in sizes up to 10 hp at 220, 440, and 550 volts for two-, three-, or four-speed motors. It does not provide overload or undervoltage protection. The magnetic type is available in all sizes. It is made up of magnetic contactors, one for each speed, suitably interlocked so that only one can be closed at a time. An overload relay for each speed is required to give complete overload protection. With a momentary-contact push-button station (one start button for each speed) undervoltage protection is provided. Controllers for two-speed motors can be provided with a *compelling relay*, which makes it necessary to start the motor on low speed only, or with a *sequence relay* that starts the motor on low speed regardless of which start button is pressed but that automatically transfers to high speed after a definite time delay, when the high-speed button is pressed. General-purpose enclosing cases normally are supplied.

A multispeed motor also may be controlled by a manually operated switch, which makes the proper connections for each speed, plus a line switch, manual or magnetic, for making and breaking the circuit to the motor. Since the manual switch is operated only when the line switch is open, it does not have to open or close on current, and a knife switch or inexpensive type of drum switch can be employed. Undervoltage protection may be provided in the line switch. In place of the line switch, a reduced-voltage controller may be employed to reduce the starting current.

Reversing controllers may be of the manual or magnetic type. The manual type consists of drum switches and is available in sizes up to 10 hp at 220, 440, or 550 volts. It does not provide overload or undervoltage protection. The magnetic type is made up of two magnetic contactors and is available in all sizes. The contactors usually are interlocked mechanically, so that both cannot be closed at the same time. Overload and undervoltage protection normally are provided. A three-button push-button station is required. General-purpose, watertight, and dust-tight enclosing cases are available. Reversing control also may be provided by a manually operated drum switch (in sizes up to 75 hp at 220 volts, and 150 hp at 440 or 550 volts) or a knife switch, which serves to reverse the connections to the motor, plus a manual or magnetic full-voltage or reduced-voltage controller for making and breaking the circuit to the motor.

Wound-rotor Motors

Wound-rotor induction motors require control for both the primary (stator) winding and the secondary (rotor) winding.

The primary control connects the primary winding to the source of power. It is similar to squirrel-cage motor control of the full-voltage type and may be either manual or magnetic.

The secondary control varies the resistance of the secondary circuit of the motor, i.e., provides a high resistance at the start and decreases it as the motor comes up to speed. It consists of a resistor connected across the slip rings of the motor and manual or magnetic switches for short-circuiting the resistor in steps as the motor comes up to speed. The resistor usually is designed to limit, to the full-load values, the current drawn and the torque developed by the motor. As the resistor is short-circuited the starting torque increases, up to the maximum torque of the motor, with a corresponding increase in current. If a portion of the resistor is allowed to remain in the circuit, the motor will operate at reduced speed. In this case the resistor must be designed for continuous duty. Before closing the primary switch, the full value of secondary resistance should be inserted in the circuit.

Manual primary and secondary control may consist of a single drum switch with contacts for closing the primary circuit as well as for short-circuiting the secondary resistor. A separate resistor must be provided. These drum switches are available in sizes up to 100 hp at 220 volts, and 200 hp at 440 or 550 volts. They frequently are arranged to reverse the motor as well by changing the primary connections. Manual primary and secondary control, in sizes up to 500 hp, may be supplied by a face-plate-type rheostat or drum switch, with resistor for secondary control, combined with a manual full-voltage controller for primary control. If an oil-circuit breaker is used for the latter, it generally can be provided with a *magnetic lock*, which prevents the breaker from being closed unless the secondary controller is in the starting position, i.e., unless all the resistance is in the circuit.

Magnetic primary and manual secondary control is similar to the type just described except that a magnetic controller is used for primary control. The secondary controller usually is so arranged that moving the operating handle to the first position closes the control circuit to the primary switch, causing it to close. Thus complete control of the motor is obtained at the secondary controller.

Magnetic primary and secondary control employs magnetic contactors for the primary switch and either a motor-operated drum switch or magnetic contactors for short-circuiting the secondary resistor. Definite-time relays or current relays which operate in response to the drop in secondary current as the motor comes up to speed are used to close the several secondary contactors in proper sequence. Overload and undervoltage protection normally are provided. The controllers may be obtained in the open type or with general-purpose enclosures. Although more costly than other types, these controllers give complete control from a remotely mounted push button. They also can be obtained with an extra contactor for automatic reversing.

13. CONTROL FOR DIRECT-CURRENT MOTORS

TYPES. Direct-current motor controllers of the following types are in general use: (1) Manual controllers: (*a*) full voltage, constant-speed, (*b*) reduced-voltage, constant-speed, (*c*) reduced-voltage, adjustable-speed, by field control, (*d*) reduced-voltage, adjustable-speed, by armature control, (*e*) reduced-voltage, adjustable-speed, by armature and field control, (*f*) drum controllers. (2) Magnetic controllers: (*a*) full-voltage, constant-speed, (*b*) reduced-voltage, constant-speed, and (*c*) reduced-voltage, adjustable-speed, by field control.

Manual full-voltage constant-speed controllers are available in ratings up to 1 hp at 115 and 230 volts. They consist of a single-pole or double-pole switch and a single-element thermal-type overload relay mounted in a general-purpose or explosionproof (Class I, Group D, conditions) enclosing case. These simple and low-cost controllers are applicable to small motors in which full-voltage starting and manual control are acceptable.

Manual reduced-voltage constant-speed controllers are available in ratings up to 75 hp at 115 volts, and 150 hp at 230 and 550 volts. They consist essentially of a dial-type switch and a resistor. As the operating handle is turned, the switch first connects the motor armature to the line through the entire resistor and then short-circuits the resistor in steps until the motor is on full voltage. Full voltage is applied to the field winding throughout. A retaining magnet, which holds the operating arm in the running position, releases the arm and disconnects the motor on failure of voltage. Controllers rated 550 volts and the larger 115- and 230-volt controllers usually have a magnetic contactor connected in series with the circuit to the motor. This interrupts the current so that no current is broken on the dial-switch contacts. The controllers are generally furnished with general-purpose enclosing cases.

Manual reduced-voltage adjustable-speed controllers with field-control speed adjustment are available in ratings up to 30 hp at 115 volts and 50 hp at 230 volts. They are similar to the constant-speed controllers described above, with the addition of a resistor, inserted in series with the field winding, and means for short-circuiting the resistor in steps, thereby varying the field current and the speed of the motor. The dial switch has an extra set of segments or buttons by means of which the field current is varied. The control is so arranged that all the resistance in the field circuit must be short-circuited before the motor can be started.

Armature-control speed adjustment controllers are available in ratings up to 15 hp at 115 volts, 25 hp at 230 volts, and 5 hp at 550 volts. They are similar to the constant-speed controllers, except that they have a continuously rated armature resistor and the operating handle of the dial switch can be set in various positions to leave a certain portion of the resistor in the circuit. Thus the voltage impressed on the armature of the motor can be varied, with consequent variation in speed. Speed reduction down to 50% of full speed is standard. With this type of control the speed of the motor varies considerably with the load. It cannot be used where close speed regulation is required.

Armature and field-control speed adjustment controllers are available in ratings up to 15 hp at 115 volts, 25 hp at 230 volts, and 5 hp at 550 volts. They consist of a combination of the two types described above. A variable armature resistance, sufficient to give 50% speed reduction, and a variable field resistance to give 25% speed increase are standard.

Manual drum controllers of the full-voltage type are available in ratings up to 3 hp at 115 and 230 volts. The reduced-voltage type can be obtained in ratings up to 35 hp at 115 volts, 65 hp at 230 volts, and 100 hp at 550 volts. Both types can be obtained for reversing as well as for nonreversing duty. The reduced-voltage type also is suitable for adjustable-speed operation either by field or by armature control. If field control is provided, the field resistor usually is mounted on the side of the controller, and the drum switch has segments for short-circuiting the resistor in steps as the operating handle is turned. The armature resistor generally is mounted separately.

Drum controllers arranged to provide *dynamic braking* also can be obtained. The motor is connected as a generator and a resistor is connected across the armature so that, when the motor is driven by an overhauling load, power is delivered to the resistor, which holds back the load.

Since drum controllers do not furnish overload or undervoltage protection, a *protective panel*, consisting of a magnetic contactor and a temperature overload relay mounted in an enclosing case, often is used. The contactor will open on failure of voltage, disconnecting the motor, after which the drum controller must be returned to the off position in order to reclose the contactor. In the event of injurious overload, the relay will open the contactor. An open-close push-button station is recommended. This may be located at any convenient place and provides an additional means for stopping the motor. Drum controllers are employed with motors driving machine tools, cranes, hoists, and similar devices, when a convenient and compact manual controller is required.

Magnetic full-voltage constant-speed controllers, both reversing and nonreversing, are available in ratings up to 1 1/2 hp at 115 volts, and 2 hp at 230 volts. They consist of a magnetic contactor and a single-element overload relay. Enclosing cases can be obtained in general-purpose, watertight, dust-tight, and explosionproof (Class I, Group D) construction.

Magnetic reduced-voltage constant-speed controllers are available in all ratings. They consist of a starting resistor, two or more magnetic contactors, and relays, usually of the definite-time type, for closing the contactors in proper sequence to short-circuit the resistor as the motor comes up to speed. Some types have accelerating contactors, incorporating a time-delay feature. These eliminate the necessity for a separate relay. A single-element temperature-overload relay and a general-purpose enclosing case usually are provided. Controllers arranged for dynamic braking with or without reversing features also are available.

Magnetic reduced-voltage adjustable-speed controllers with speed adjustment by *field control* are available in all sizes. They are similar to the type described above, with the addition of a field rheostat for varying the field current, and hence the speed, of the motor. A *field-accelerating relay* often is added, which prevents high armature current when the motor accelerates rapidly from basic speed (full field) to a higher speed (weak field). A "field-protective" relay may be installed to protect the motor from excessive speed if the field circuit is opened accidentally.

14. CONTROL FOR SYNCHRONOUS MOTORS

The control for the armature (primary) winding of a synchronous motor is similar to that provided for a squirrel-cage induction motor. (See p. 16-61.) It may be of the full-voltage or reduced-voltage type, either manual or magnetic. In addition, control for the field winding must be provided. This control usually is of the magnetic type, so that after the primary control has connected the armature winding to the a-c power lines direct current automatically is applied to the field winding when the motor has accelerated to the proper speed, i.e., 95 to 98% of synchronous speed. When the magnetic field control is combined with magnetic primary control the complete controller is designated *magnetic*, and when combined with manual primary control it is designated *semimagnetic*.

A field contactor and *field-application relay* are required to apply field excitation automatically. The latter may be of the definite-time type or it may be of a speed-responsive type operating from changes in the electrical characteristics of the motor as it approaches synchronous speed. In addition, a *discharge resistor* must be connected across the field winding during the starting period in order to limit the voltage induced in that winding. The discharge-resistor circuit is opened when field excitation is applied, either by normally closed contacts on the field contactor or by a separate field discharge contactor actuated by the field-application relay. Synchronous motor controllers also include a *field-removal relay*, which operates automatically to open the field contactor and remove excitation whenever the motor pulls out of synchronism. The latter condition may occur as a result of a high overload or a drop in line voltage. By removing field excitation, after the load or voltage returns to normal, the motor may be able to re-accelerate, and the field application relay will re-apply field so that normal operation is resumed automatically. If field excitation is not removed on pull-out, the motor is apt to stall and, in addition, is subjected to violent torque and current surges. If the driven machine must be unloaded when starting, in order that the motor will accelerate and synchronize, it often is possible to provide an automatic, electrically operated unloader, which can be connected to the field contactor. Then, on pull-out, the automatic removal of field also will unload the machine, so that the motor can resynchronize. The load is then automatically re-applied. The field-removal relay may operate on the abrupt drop in the power factor of the current drawn by the motor as it pulls out of step, or on the increase in line current under this condition, or on the current induced in the field winding as the motor speed drops below synchronous speed.

Means for adjusting d-c excitation applied to a synchronous motor also should be provided. If an exciter is used to supply excitation to a single synchronous motor, adequate control of the excitation can be obtained by means of the exciter field rheostat. However, if a single d-c generator supplies excitation to two or more synchronous motors, a motor field rheostat or tapped resistor must be connected in series with the field winding of each motor for adjustment of the excitation current. Synchronous motor controllers are usually arranged for mounting an exciter field and, where required, a tapped motor field resistor or (on open-type controllers) a motor field rheostat. If a motor field rheostat is required and the control is of the enclosed type, a separately mounted, motor-operated rheostat with only its control switch mounted on the controller is generally recommended.

Usual practice provides synchronous-motor controllers with two ammeters, an a-c ammeter for reading line current, and a d-c ammeter for reading field current. Instruments, such as an a-c voltmeter, indicating wattmeter, or watthour meter, are often valuable additions.

Like induction-motor controllers, synchronous-motor controllers normally have a thermal-type *overload relay*. Time-delay *undervoltage protection* is standard for controllers rated 2200 volts and above; for controllers of lower voltage rating, instantaneous undervoltage protection is standard, but the time-delay type can be obtained.

Controllers of the open type are available, but those furnished with general-purpose enclosing cases are becoming increasingly popular.

TYPES. Synchronous-motor controllers of the following types are commonly used: (1) magnetic full-voltage, (2) magnetic reduced-voltage autotransformer reactor or resistor, (3) semimagnetic reduced-voltage autotransformer, and (4) magnetic part-winding.

Types 1 and 2 are similar to corresponding types of magnetic squirrel-cage induction-motor controllers with the addition of the automatic field control and ammeters described above. Type 3 is similar to the corresponding type of manual squirrel-cage induction-motor controllers with the same additional features.

Magnetic part-winding controllers are arranged to start a motor by energizing only a portion of the complete armature winding and then, after the motor has accelerated, energizing the remainder of the winding in one or more steps. The motor must be designed especially for this type of starting, with a *multiple-circuit* armature winding and sufficient terminal leads to allow energization of the various circuits independently. The effect of this starting method is to reduce the current drawn from the line and the torque developed by the motor at start, giving the effect of reduced-voltage starting without the use of an autotransformer, resistor, or other voltage reducing device. The amount of reduction in torque and current depends on the percentage of the winding energized and also on the characteristics of the individual motor. With some motors the torque is reduced in proportion to the reduction in current, whereas in others a considerably greater reduction in torque occurs. Like reduced-voltage resistor starting, part-winding starting has the advantage that the circuit to the motor is not opened during the starting period, making the method particularly applicable where the starting current limitations are of the increment type. Motors of relatively high voltage in relation to their horsepower cannot be economically designed for part-winding starting. The method is most applicable to low-speed machines of large horsepower and moderate voltage. Most motors and controllers for part-winding starting are arranged to give just one reduced kilovolt-ampere step (two-point starting), but in many cases they can be provided with two, three, or more reduced kilovolt-ampere steps.

SECTION 17

CONTROL MECHANISMS

By

ED S. SMITH, *Ballistician, Aberdeen Proving Ground*

HERBERT HARRIS, JR., *Sperry Gyroscope Co., Inc.*

L. W. HERCHENROEDER, *Westinghouse Electric Corp.*

R. M. NARDONE, *Victory Engineering Corp.*

AUTOMATIC REGULATORS

By Ed S. Smith

SERVOMECHANISMS

By Herbert Harris, Jr.

INDUSTRIAL ELECTRONIC CONTROL

By L. W. Herchenroeder

BRAKES

Revised by R. M. Nardone

AUTOMATIC REGULATORS

By ED S. SMITH

1. TERMINOLOGY

Control and Regulation. Loose use of the terms *controller* and *regulator* as full equivalents is fairly common (Refs. 1–7). However, to avoid the ambiguity resulting from such use, the following terminology is used in this section: *control* is the basic or generic term, thus a valve controls the flow of a fluid through it; *regulation* is control which tends to maintain a given value, whether this is set constant or varied in any manner as, according to a program or schedule, a ratio to some other value, or in any desired manner. For example, the common pressure regulator acts to maintain pressure at a given value, a time-temperature regulator acts to cause the regulated temperature to follow a reference value according to a predetermined schedule, and a flow-ratio regulator acts to control one flow in accordance with the rate of another to maintain their ratio constant.

A typical regulator, diagrammatically shown in Fig. 1, consists basically of a *meter* sensitive to the value of the variable to be regulated, a *control valve* for the rate of a *control agent* affecting the value of said variable, and means connecting the meter with the valve so that the meter governs the valve's position to tend to cause the valve to keep the measurement at a set reference value. Completing the control loop between the control valve and the meter is a *plant* in which there is usually some storage of the control agent, or other source of lag or retardation of the response of the measurement to a change of position of the control valve. Also the basic need for a regulator arises from "disturbances" or changes in "load" of the plant.

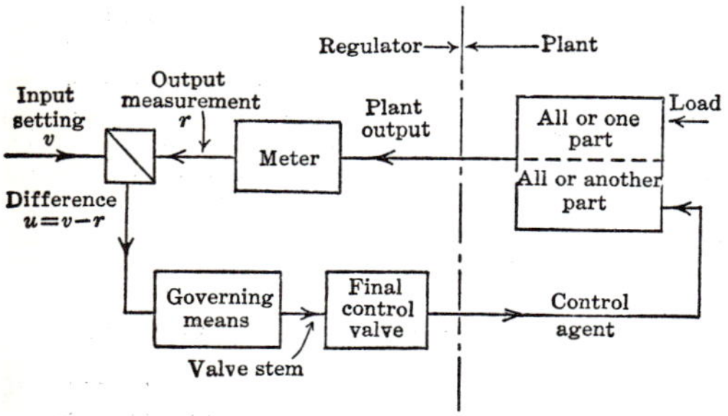

Fig. 1. Block diagram of typical regulated system.

Self-operated and Servo-operated Regulators. In a *self-operated regulator* the meter actuates the control valve whereas, in a *servo-operated regulator*, the meter actuates a pilot or relay which, in turn, controls a servomotor that actuates the final control valve.

A servomechanism (Refs. 8–10), as commonly used, is a regulator-type system in which the emphasis is on having the output position, θ_o, of the final controlled element accurately follow θ_i, that of a variable setting, instead of compensating for the effect of load changes, although such compensation is occasionally also an important factor. Servomechanism theory advanced rapidly during World War II, mainly on the basis of continuously minimizing the difference $\theta = \theta_i - \theta_o$ between the position θ_i of an input or setting element, e.g., a telescopic sight, and θ_o, that of an output or final controlled element such as a gun. This theory, which is more or less directly applicable also to self-operated, but otherwise similar, mechanisms and to regulated systems generally, is treated in the following chapter on servomechanisms.

Treatment of the theory of automatic control or regulation in general in the present chapter is properly preliminary, especially for nonelectrical engineers, to the chapter on servomechanisms. Furthermore, there are differences in practice since, in servomechanisms, the plant usually consists of a single element, such as a gun, having relatively high inertia, whereas in process control it usually contains several "capacities" for storing the control agent and occasionally includes several additional sources of lags or retardations.

2. STATIC RELATIONS

The static, ultimate or "potential," effect of a change of load on a regulated plant is independent of any lag or retardation in the system as long as the system is not thereby prevented from settling to rest. Figure 2 shows the static relations for a typical simple regulated process system, comprising, for example, a heating plant in which a temperature rise of T degrees Fahrenheit of a water throughput load of L lb per sec is produced by a heat input $Q = LT$ Btu per sec. System is controlled by a simple proportional, or corresponding, regulator according to the relation $Q = M(T_i - T)$, where T_i is the cut-off value for zero throughput load.

In Fig. 2, the *plant* characteristic appears as a warped surface which has the same shape as the familiar thermodynamic gas surface, i.e., hyperbolic, because $Q = LT$, the *regulator* characteristic appears as an inclined plane, and their intersection is the characteristic of the regulated system. Figure 2 shows that "droop" or "load error" is inherent in proportional regulation, indicated by the heavy dash projection of the intercept on the L : T plane.

Although this three-dimensional presentation shows the static or ultimate relations involved, there is a (fourth) dimensional difficulty involved in showing the transient

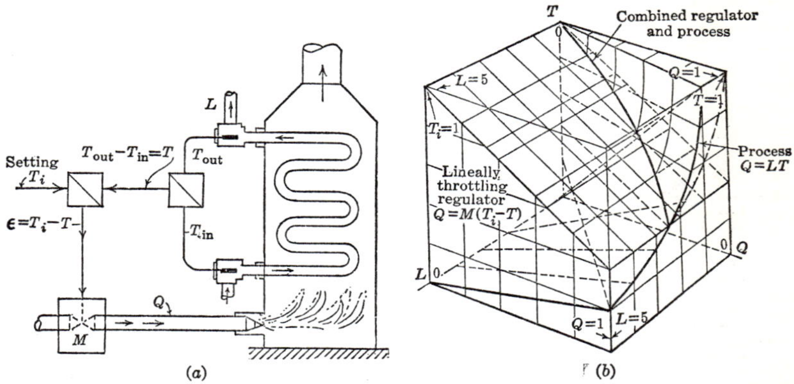

FIG. 2. Simple regulated system (*a*), and three-dimensional representation (*b*). (Fig. 2*b* reproduced by permission from *Temperature, Its Measurement and Control in Science and Industry*, Reinhold Publishing Corp., 1941)

relations following a change of load on this basis. A working picture of the relations involved may be obtained by plotting the transient response with the time-axis normal to the $Q : T$ plane. It is commoner practice, however, to plot the $L : t$ and $Q : t$ relations on a common grid, which directly relates the cause with its transient response or effect, with still further loss of the physical picture involved.

Fairchild has shown (Ref. 11) that constant static sensitivity over any desired operating range may be obtained for this and related common regulated process systems, both continuous and batch, by a proportional log-log regulator, for which $Q = M_1/(T_i - T)^s$, where s is a function of said sensitivity alone. A semilog proportional regulator, for which $Q = M_2 \log (T_i - T)$, provides an approximation to the theoretically required log-log proportional regulator for constant overall sensitivity over a wide range, which is generally accepted in practice as satisfactory for process control. The semilog regulator is preferable for flow control, especially where there is appreciable resistance in series with the control valve.

The lineal proportional regulator of Fig. 2 is well suited for the control of liquid level in a reservoir having vertical sides. It also gives closer control than the semilog regulator, where the control valve is by-passed by a fixed resistor (Ref. 12, pp. 48–49).

In throttling control, where the meter and governing means are lineal, the modification of the control valve which suits the plant for a proportional regulator is also suitable for a proportional-speed floating regulator, the stability otherwise generally falling off at low loads. In a floating regulator, the speed of moving the valve is normally slow enough to let the valve settle at its proper control position. In a proportional-speed floating regulator, the speed of the valve is proportional to the error, or difference, between the measurement and the set value to be maintained.

3. TRANSIENT PERFORMANCE

TYPICAL PLANTS. A few of these are included here to bring out their characteristics qualitatively, mostly as to lags, retardations, and attenuations which affect the stability and type of control attainable with conventional regulators. Before taking up these effects, a few of their causes are considered.

Storage of the control agent may exist in the plant. Taking for example level control of a single capacity or reservoir as in Fig. 3, it is inherent that this storage exerts both retardation and attenuation of the response to a movement of its control valve. If the movement of the control valve is sinusoidal, the response of the head or level lags 90 degrees in the absence of "self-regulation" (explained below) and is attenuated to be inversely proportional to the size, i.e., area of air-liquid surface in this example. In other words, a large capacity just before the meter exerts a smoothing effect.

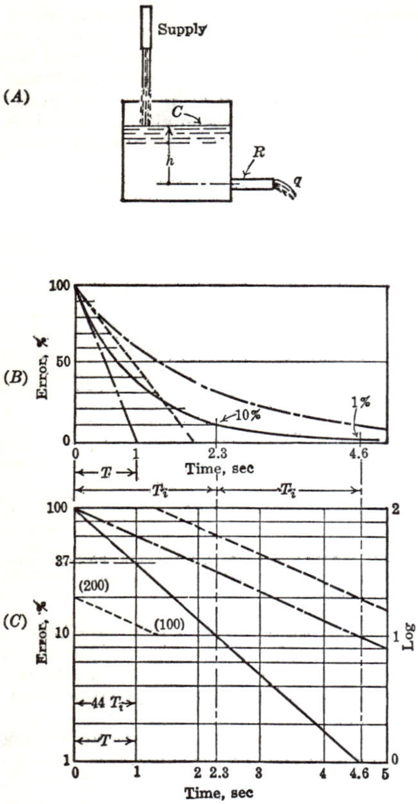

FIG. 3. (A) Capacity and resistor, a single-capacity plant. (B) Fall of head h after the supply is shut off, even-graduations graph. (C) Fall of head h after the supply is shut off, semilog graph. (From *Automatic Control Engineering*, by Ed S. Smith, McGraw-Hill Book Co., 1944, by permission)

Self-regulation is said to exist when an increase in the inflow to such a system causes the level to rise toward a definite head instead of increasing indefinitely. This can have several causes: flow over a weir or through a resistor with the flow increasing with head. Such a resistor R introduces an in-phase component for the foregoing oscillated-control-valve, single-capacity example, which reduces the phase lag below the 90 degrees due to the capacity C.

Referring to Fig. 3, if the outflow is directly proportional to head, then $q = h/R$. If supply is completely shut off, the head decreases at such a rate that $C(dh/dt)$ is equal to the outflow. Equating these two values,

$$\frac{h}{R} = -C\frac{dh}{dt} \quad \text{or} \quad \frac{dt}{RC} = -\frac{dh}{h}$$

Integrating between the limits $t = 0$, $h = h_o$, and $t = t$, $h = h$,

$$\frac{h}{h_o} = e^{-t/RC}$$

For a sudden, or step, change of the valve position, the retardation may be expressed by the *time constant* $RC = T$, which is the time it would take for the ultimate value to be reached if the initial rate were continued (Fig. 3*B* and *C*). For this value of t the corresponding actual value of the head would be, from the relation just derived, $h = h_oe^{-1} = h_o/2.7183 = 0.368h_o$. In other words, this time constant is the same time that is required for the difference from the ultimate value to be reduced to 36.8% of its initial value. (See Fig. 4.) This time constant is also equal to the lag ultimately approached for the same single-capacity system with a constant-velocity, instead of a step, change. In general, self-regulation alone cannot provide the needed accuracy of control but occasionally it is strong enough, with a sufficiently low time constant, to enable stable control to be obtained with a somewhat simpler or less expensive floating control than the proportional control that would otherwise be required. In floating control, the final control valve moves gradually to its proper position, instead of only nearly to it, as it would in proportional control.

Where the plant includes a number of capacities, or storages separated by resistances, their distribution affects the *controllability* of the plant. For example, a two-capacity plant with unequal capacities has more attenuation and hence smoother control when the larger capacity is next to the meter than when it is adjacent to the control valve. A special case, approximated where the components are nearly alike, is one in which the several time constants are equal. This case is generally difficult to control, especially where damped oscillations, rather than an asymptotic or monotonic approach to a final value, occur following an upset. For a two-capacity plant, the contribution of each of the two time constants to the plant's response to a step change is illustrated in Fig. 5.

With a plant having a large number of capacities in series, there is a tendency, as distributed capacity and resistance is approached, to have a virtually direct time lag of

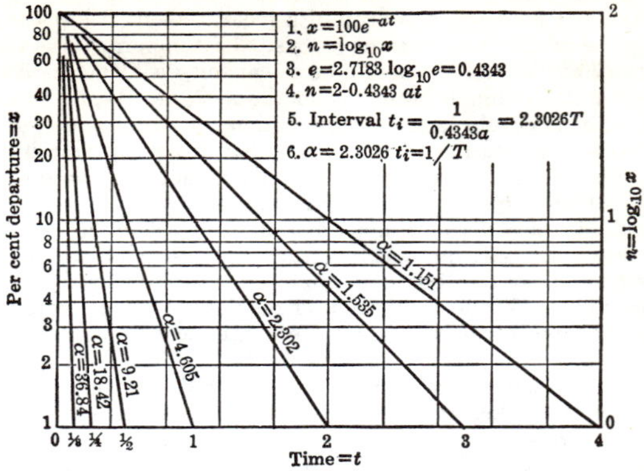

Fig. 4. Asymptotic return paths for various values of $\alpha = 1/T$ for single-capacity plants. (From Automatic Regulators, Their Theory and Application, by Ed S. Smith, in *Trans. ASME*, Vol. 9, No. 4, pp. 291–303, May 1936)

the response at the meter to a step movement of the control valve. This outright delay makes stable control difficult or even impossible to achieve with conventional continuous regulators. The initial overshoot, before a counteracting control effect is initiated, tends to be proportional to the dead time. Figure 6 shows that the dead time t_{L3} produced by a dead zone tends to increase directly with the number of capacities.

A *direct time lag* also occurs in plants in which a lag exists due to a physical transportation of the control agent from the control valve to the point at which it becomes effective. This is known as "transportation" or "distance/velocity" lag since the delay is equal to the distance divided by the velocity with which the agent is transported.

And, in some plants, the response may be erratic for some reason, making the plant hard to control. This may be the case where a thermocouple is exposed to a fluid at different temperatures, which depend upon the vagaries of an unguided thermal circulation.

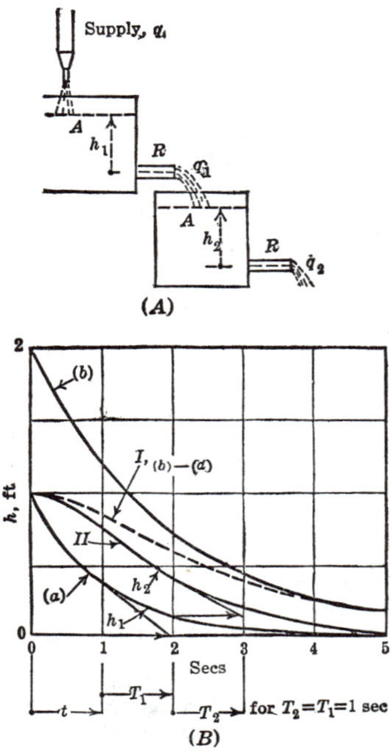

Fig. 5. (*A*) Cascaded tanks, a two-capacity plant, with equal capacities and equal resistors; (*B*) performance as it is not in curve I, and as it is in curve II. (From *Automatic Control Engineering*, by Ed S. Smith, by permission)

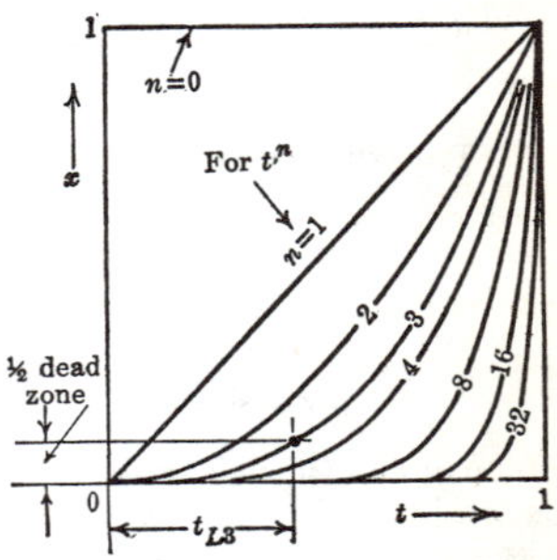

Fig. 6. Initial reaction t^n to a step change. (From *Automatic Control Engineering*, by Ed S. Smith, by permission)

4. TYPICAL REGULATORS

In practice, regulation is not always accurate to a set value, due to the characteristics of the regulator and/or plant, as is brought out in the following typical cases.

On-and-off regulators convert each drop of a measurement below the set value into an "on" heating impulse which continues as long as the temperature is too low. This sort of regulation is especially effective where a large capacity is located adjacent to the meter and where there is not too much other lag in the system. While this sort of regulator hunts continuously, a relatively large capacity adjacent to the meter is usually provided to keep the oscillations of the measured variable within acceptable limits. Also, in on-and-off control, the full-on initial action following each switch operation provides a *liveliness of control* which helps to offset lag in the plant.

The test for stability of a discontinuous or nonlinear regulated system is not whether oscillations completely die out, but whether they drop to unobjectionably low value, an "unstable" discontinuous system being one in which the oscillations build up to a steady objectionably high value. While an on-and-off regulator itself operates to a point, a load produces a droop or change of the average value of the measurement for a given set control value.

Since the amplitude of the oscillations of the measured variable tends to vary inversely with their frequency, a marked improvement in control accuracy can often be obtained by oscillating the control-setting reference itself at a higher frequency than that of the system when the latter is regulated by a simple on-and-off regulator. This improved mode has long been known as Gouy regulation (Ref. 13).

Two-position regulation (discontinuous) is also often used, with a space between the "low" and "high" contacts at the meter providing a dead zone in which oscillations can die out and normally do. When this type of regulation is used on an actual single-capacity heating plant, there is a saw-tooth action from a succession of the initial portions of heating and cooling curves between the contact limits. But upon the addition of a second thermal capacity and resistance, a hunting is set up which tends to pass far beyond these limits (Ref. 12, pp. 28–30). If a floating action is produced by using a constant-speed motor to actuate the control valve, the width of the dead zone can be narrowed by lowering the speed of moving the control valve. However, in practice a limit is soon reached at which such regulation is too sluggish to meet the changes in load on the plant.

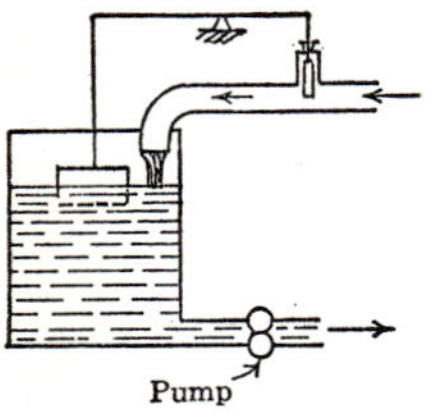

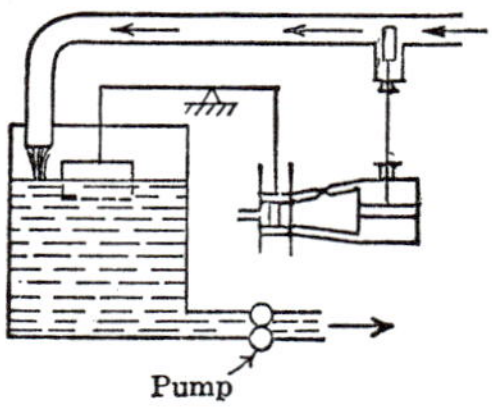

Fig. 7 (*left*). Proportional Control. Stable asymptotic regulation of a single-capacity plant without self-regulation. (From *Automatic Control Engineering*, by Ed S. Smith, by permission)

Fig. 8 (*right*). Floating Control. Unstable, steady-hunting regulation of a single-capacity plant without self-regulation. (From *Automatic Control Engineering*, by Ed S. Smith, by permission)

Proportional regulation (continuous), Fig. 7, powerfully tends to stabilize a regulated plant at the expense of load error or droop. In other words, an error is inherent since the movement of the control valve is proportional to such error and consequently cannot completely remove it. The regulator's sensitivity, or ratio of movement of the control valve to the amount of error, may be raised quite high, e.g., of the order of 20, when the plant lag is not unduly large or unfortunately distributed. From one point of view, this droop of the regulator is a useful substitute for inadequate self-regulation in the plant.

Proportional-speed floating regulation (continuous), Fig. 8, is commonly useful where the plant has adequate self-regulation and the lag in the plant is not too great. With a sinusoidal oscillation of the meter, the control valve of such a floating regulator lags 90 degrees behind the measurement (since it follows the *integrated* error), in comparison with the substantially zero lag of a proportional regulator. The importance of this lag may appear from the fact that instability may occur when there is a lag of as much as 180 degrees in the entire regulated system. In other words, such a floating regulator cannot be relied upon to produce stable control of a plant in which the lag can be 90 degrees or more.

Follow-up Link. By the use of a follow-up link, as in Fig. 9, the movement of the control valve is carried back to the pilot valve, with the stabilizing effect of dropping the sensitivity to a finite value as with proportional control. This is substantially attained, with a reduction of the servo lag from 90 degrees for floating control to nearly the zero lag for proportional control, under steady oscillations.

Another factor in stability, loosely expressed, is the sensitivity of the regulator. In the effort to reduce load error it ordinarily is increased nearly to the point (at 180-degree lag) which causes an oscillation circulating around the control loop to increase successively with each time around. The sensitivity of a proportional regulator is theoretically constant, regardless of the frequency of oscillation, while the effective sensitivity of the floating regulator varies inversely with the frequency, which is likewise the case for most plants.

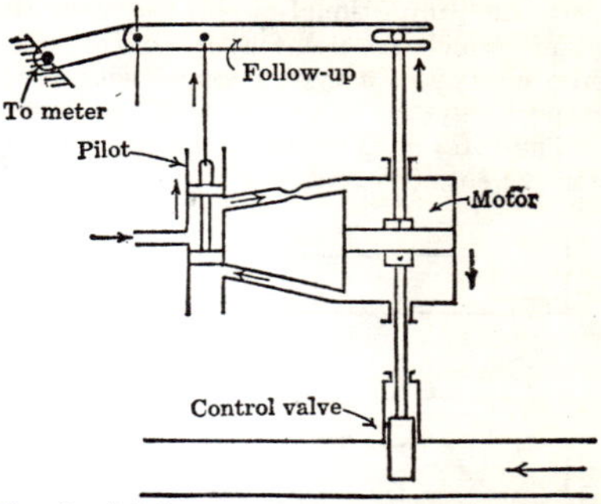

Fig. 9. A follow-up servo regulator; the control valve lags only a few degrees in the normal operation range under steady oscillation of the measurement. (From *Automatic Control Engineering*, by Ed S. Smith, by permission)

Stability. The philosophy underlying stability of regulation may be somewhat loosely expressed by briefly stating that instability tends to exist when, in passing around a broken control loop, the lag (1) exceeds 180 degrees and the overall sensitivity, or ratio of effect/cause at the ends of the broken control loop, (2) exceeds unity, and (3) decreases with an increase of frequency. This statement is seen to agree with common experience in regulation.

A conventional air-operated regulator is shown diagrammatically in Fig. 10. A metering bellows moves a vane relative to a nozzle which is supplied with air through a feed restriction to determine the pressure in the line governing the control valve. When the left-hand end of the vane is pivoted in a fixed journal (not shown), the sensitivity of control is so very high with this essentially proportional control that it breaks down into on-and-off steadily hunting control in the presence of any appreciable lag in the plant.

A relay (not shown) is usually included to provide a larger supply of air to the diaphragm of the control valve than can be controlled directly by the usually small nozzle required for high sensitivity and hence high accuracy of regulation. A "valve positioner" (not shown) is sometimes provided to insure smooth operation of the control valve in spite of mechanical friction; it includes a relay and a follow-up link to the valve stem.

Stable control can usually be obtained at the expense of an increase of load error, by converting this regulator to give low-sensitivity proportional control by pivoting the left-hand end of the vane on proportional bellows connected to the governing pressure. The stiffness or spring action of the bellows insures proportional movements of the pivot and control valve (the reset bellows being omitted).

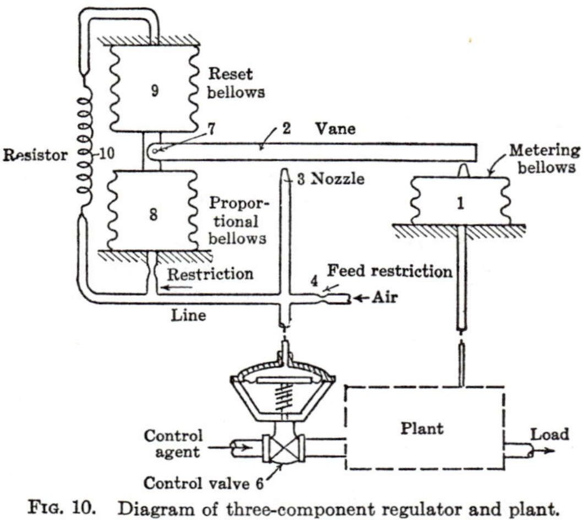

Fig. 10. Diagram of three-component regulator and plant.

A reset bellows may be connected through a resistor with the line to provide for the gradual elimination of substantially all the load error resulting from the follow-up movement of the pivot. Thus, following a step change of either the measurement or setting, there is an initial low-sensitivity proportional response which keeps the control valve from overshooting, followed by a further reset (or *integrating*) response during which the control valve asymptotically approaches its final position. On starting up after a shutdown, the reset may cause objectionable overshooting unless it is provided with special limiting or reversing means (not shown).

By "reciprocation," the addition of a resistor (10) connecting the proportional bellows with the line provides a **rate** (or *derivative*) action, the resistance of this restrictor being considerably smaller than that of the one connected to the reset bellows, with the result that the time constant of the proportional bellows is much smaller than that of the reset bellows.

The proportional-plus-reset regulator uses the proportional component to obtain stability and the reset component, which is slow relative to the plant, to wipe out gradually the load error due to the proportional component. If the reset speed is unduly increased, the stability is decreased as the relatively lower stability of floating control is approached. But the proportional-plus-reset regulator successfully handles many industrial process plants which are not readily controlled. Figures 11*A* and *B* compare the response to a step change of a typical plant with proportional and proportional-plus-reset regulators, respectively.

The rate, or derivative, component is occasionally added to the proportional and reset components to produce a control effect which occurs sooner than that due to proportional control, this earlier effect being sometimes said to provide "anticipatory" control. A rate component may be considered as, in effect, peeling off one capacity from a regulated multicapacity system (Fig. 6). On a steadily hunting system, or on one with an oscillating input, the rate component removes up to 90 degrees from the phase lag of the system. For one example, a finite rate of change of head occurs in a single-capacity plant as soon as its outflow changes, and the rate component can cause a substantial correcting movement of its inflow control valve before the head has changed enough to build up much of a proportional component. For another example, after an increase of load in a heating system in which considerable metering lag exists, a temperature deficiency accumulates during the lag before the meter responds; then the rate component provides a heat *impulse* to offset the accumulated deficiency, thus leaving the proportional and/or floating components to function as described earlier.

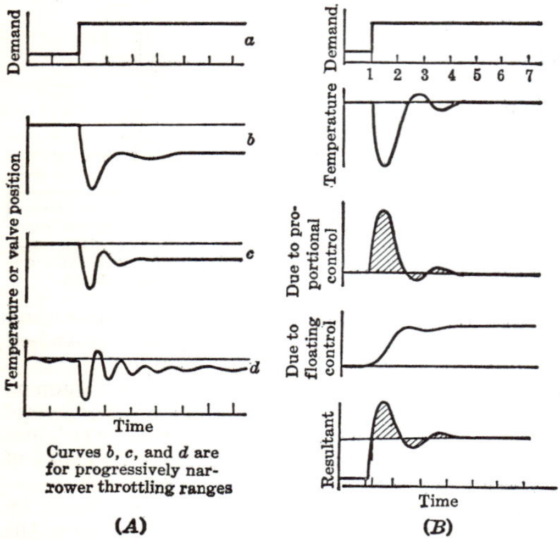

FIG. 11. Response to a step disturbance with proportional (*A*) and proportional-plus-floating (*B*) controls. (From Some Fundamental Considerations in the Application of Automatic Control to Continuous Processes, by E. S. Bristol and J. C. Peters, *Trans. ASME*, Vol. 60, pp. 641–650, 1938)

The function of the reset component remains of course unchanged in a three-component regulator. As is shown later, each of the three components is occasionally needed to stabilize a process plant that is unusually difficult to control. A rate component, or some element having a generally similar effect, is needed with most servomechanisms, the rate of movement of the final control element being often used instead of that of the meter.

TYPICAL REGULATED SYSTEMS. The ideal three-component regulator follows the relation

$$-y = R\frac{dx}{dt} + Px + F\int x\,dt \tag{1}$$

where R, P, and F are the rate, proportional, and floating speed component constants. For example, let y be a force exerted on a plant consisting of a damped sprung mass and x be the displacement of the mass M from its position of rest; then

$$y = M\frac{d^2x}{dt^2} + N\frac{dx}{dt} + Bx \tag{2}$$

where N is the damping factor and B the spring stiffness in the plant. Then, after a disturbance of the system,

$$M\frac{d^3x}{dt^3} + (N + R)\frac{d^2x}{dt^2} + (B + P)\frac{dx}{dt} + Fx = 0 \tag{3}$$

after differentiation to convert the integro-differential equation into a third-order differential equation in which, for stability, the product of the inner pair of coefficients must exceed that of the outer, or $(N + R)(B + P) > MF$.

Figure 12 shows the performance of different combinations of inertia M, damping N, and spring stiffness B in the plant with different combinations of rate R, proportional P, and floating speed F components of the regulator, the presence of the different plant components being shown as the abscissa and of those of the regulator as the ordinate. For example, an M plant, i.e., one having no damping or self-regulation, is seen to be unstably controlled by a regulator having only the F, or floating, component (Fig. 13). This instability is readily checked since the product of the inner pair of coefficients is zero, which is obviously not greater than the product MF as required for stability.

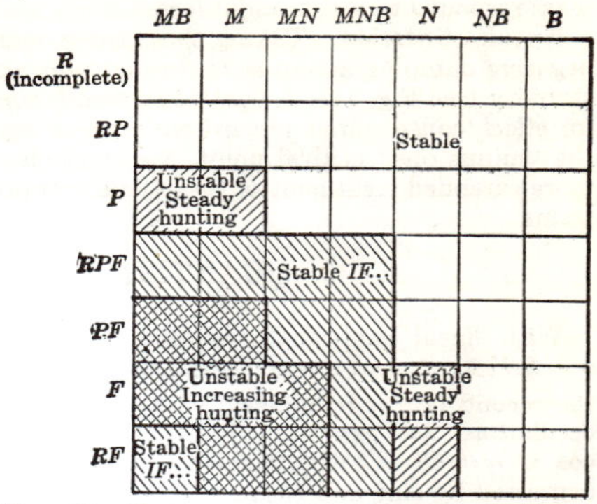

FIG. 12. Performance of a damped-sprung-mass plant with a three-component regulator. (From *Automatic Control Engineering*, by Ed S. Smith, by permission)

Transient Performance. Given the characteristics of such a system, its transient performance may be determined directly from Liu's graph (Ref. 12, pp. 101, 102, 235–244), and vice versa from Kochenberger's (Ref. 12, pp. 244–246), a cubic auxiliary equation being involved.

FREQUENCY RESPONSE ANALYSIS. For a more general approach, that of Nyquist, disconnect the link between the final control element and its actuator and manually oscillate the element at a given amplitude and frequency. On completing the loop through the plant, the corresponding oscillations of the actuator will be of the same frequency and of relative amplitude and phase, which may be used in determining stability.

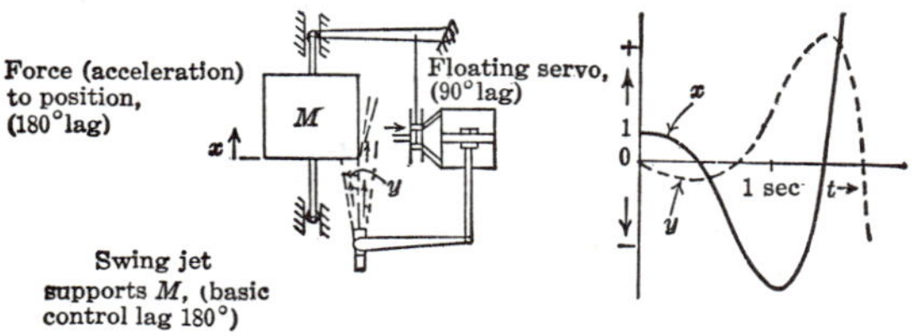

(*a*) Two-capacity plant with floating control, lags stated are for steady hunting

(*b*) Operation of the system of (*a*), no steady hunting but highly unstable with increasing hunting

FIG. 13. A two-capacity system and its performance. (From *Automatic Control Engineering*, by Ed S. Smith, by permission)

Consider the case of **steady hunting,** the critical case between instability and stability. Each successive cycle is of the same amplitude, and the length of the corresponding vector does not increase with time. For this case and following one impulse (at the vector's time position) around the loop, it is manifest that the result must occur just 360 degrees later in the cycle for the stated steady hunting, i.e., at the resonant frequency. Since the control action for regulation is necessarily opposite to its cause (a rise of temperature causing a control action which tends to lower the temperature), an inherent 180 degrees of direction or phase difference, the complemental time lag must be 180 degrees for steady hunting. Referring to the broken control loop having equal amplitudes and 180 degree lag, the broken ends can be connected so that the cyclical impulses are self-energizing and continue without change.

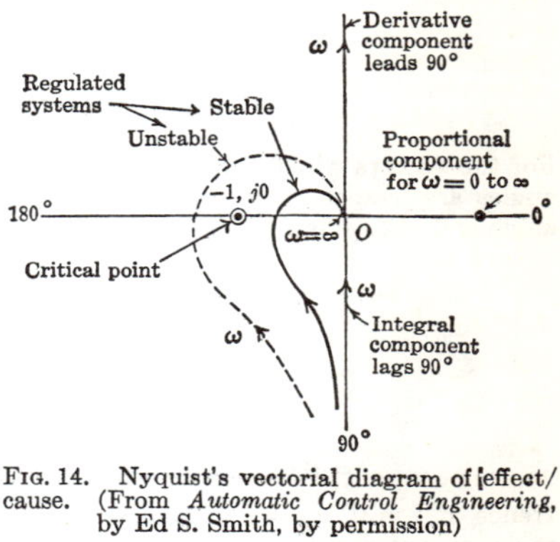

FIG. 14. Nyquist's vectorial diagram of effect/cause. (From *Automatic Control Engineering*, by Ed S. Smith, by permission)

What is true for one impulse part of a sinusoidal wave can be extended by superposition to the others represented by the vector. To have the vector length increase in each cycle is necessary for instability. And an increase of lag beyond 180 degrees is no exception to the general rule that the tendency toward instability increases with increase of lag. In any actual physical system, an increase in frequency tends, as the limit of infinite frequency is approached, to result in a reduction of amplitude toward zero, as can be seen where a mass responds to a sinusoidal force.

Nyquist Criterion. Taking these three considerations together for a system having no negative damping amounts to a practically useful statement of the Nyquist criterion for stability (see Fig. 14): A control system is unstable if the locus of the broken-loop vector for effect/cause spirals inward with increasing frequency in a clockwise direction beyond the "minus one" critical point, the clockwise direction being that for increasing lag. A more extended treatment of the Nyquist approach is given in the chapter on servomechanisms.

5. QUANTITATIVE CONSIDERATIONS

With lineal controlled systems, the response follows e^{-st}, where the natural base $e = 2.7183\ldots$ and the complex quantity $s = r + j\omega$, where the first term (r) is real and the second ($j\omega$) is imaginary, with $j = \sqrt{-1}$. Thus the damping is expressed by r, and oscillations are expressed by the angular frequency ω in radians per second, since $\cos \omega t = (e^{j\omega t} + e^{-j\omega t})/2$ and $\sin \omega t = (e^{j\omega t} - e^{-j\omega t})/2j$ and any phase (ϕ) relation is expressed by sine and cosine components.

In operational terms, s or p is commonly used for (d/dt) since $de^{pt}/dt = pe^{pt}$. And, since $de^{j\omega t}/dt = j\omega e^{j\omega t}$, we can write for the steady oscillation case that $d/dt \cong p \cong j\omega$. The differentiation corresponds to a 90-degree phase lead, while an integration accordingly has a 90-degree phase lag. It may be noted here that the Laplace transforms are $1/p$ times those of Heaviside, the step function being expressed by $1/p$ and **1**, respectively, in the two systems.

Consider a forcing function denoted by $f(t)$ impressed on a unit system whose "impedance" $Z(p)$ determines the response or output $i(t) = f(t)/Z(p)$, just as electric current equals voltage divided by resistance, or $I = E/R$. Instead of impedance, its reciprocal, the transfer function $Y(p)$, may be used, in which event $i(t) = Y(p)f(t)$.

For steady oscillations, the particular solution is as stated above, and for $f(t) = e^{at}$ this response becomes, for all but the special equal roots case (which can be approximated by nearly equal roots), $i(t) = e^{at}/Z(a)$. For a forcing function consisting of a number of terms of the e^{at} sort, the steady-state response of a linear system is, by superposition, simply the sum of the individual respective responses to each of said e^{at} terms.

For example (Ref. 12, pp. 276–279), consider a system for which the constants lead to the equation

$$\frac{d^2x}{dt^2} + 2\frac{dx}{dt} + 7x = 2\cos 10t = e^{j10t} + e^{-j10t}$$

Writing $p = d/dt$, the left-hand side of the equation becomes

$$p^2x + 2px + 7x = (p^2 + 2p + 7)x$$

and the impedance is $Z(p) = p^2 + 2p + 7$. The system is subjected to the sinusoidal forcing function for which $a_1 = j10t$ and $a_2 = -j10t$. Then the steady-state response is

$$\begin{aligned} x_p &= \frac{e^{j10t}}{(j10)^2 + 2(j10) + 7} + \frac{e^{-j10t}}{(-j10)^2 + 2(-j10) + 7} \\ &= -\frac{e^{j10t}}{93 - j20} - \frac{e^{-j10t}}{93 + j20} \end{aligned}$$

For the vectors which the denominators denote, the radius is $\sqrt{93^2 + 20^2} = 95.1$ and the angles are, respectively, $\tan^{-1}(-20/93)$ and $\tan^{-1}(+20/93)$. Numerical value of each angle is 0.212 radian = 12.1 degrees; hence the steady-state response may be written

$$\begin{aligned} x_p &= -\frac{e^{j10t}}{95.1e^{-j0.212}} - \frac{e^{-j10t}}{95.1e^{+j0.212}} \\ &= -\frac{2}{95.1}\,\frac{e^{j(10t+0.212)} + e^{-j(10t+0.212)}}{2} \\ &= -0.021\cos(10t + 0.212) = +0.021\cos(10t + 0.212 + \pi) \\ &= 0.021\cos(10t + 3.354) \end{aligned}$$

Hence at $t = 0$, $x_p = 0.021\angle 3.354$ radians $= 0.021\angle -167.9$ degrees. In other words, the resultant displacement x lags 167.9 degrees behind the imposed sinusoidal force.

In following a sinusoidal forcing function around a control loop, each transfer function $Y(p)$ is taken in turn and, for this series case, we multiply together the amplitudes or radii of the vectors and add their phase angles algebraically, each amplitude being the ratio of output to input for its respective element. Thus if a regulator G creates a torque $Q(t)$ which acts upon a plant H, the overall loop transfer characteristic function is simply $G(p)H(p)$, or GH as usually written, since $Q(t) = G(p)(t)$ and $\theta_o(t) = H(p)Q(t) = G(p)H(p)(t)$ or GH and $GH = \theta_o/\theta$, where the deviation (*not* error) is $\theta = \theta_i - \theta_o$. This θ convention takes care of the "minus one" control relation and leaves only time, or phase, *lags* in the results of the analysis.

The transfer function $Y(p)$ is seen to be the "admittance" or inverse of impedance $Z(p)$. For a loop in which part of the system is in shunt, we can simply add vectorially the transfer function $Y(p)$, or $Y(j\omega)$, vectors for the shunt portion, and proceed as before for any series portion, following conventional rules for shunt and series impedance networks with due regard for the stated inversion. Using $j\omega$ instead of p in the overall loop transfer function θ_o/θ permits us to use the available network laws, including Kirchhoff's (Ref. 14) and that of duality (Ref. 14, pp. 63–67), in handling the more complicated multiple-loop systems. Some algebraic simplification results from using θ/θ_o, the inverse of the above expression, in complicated networks.

SERVOMECHANISM EXAMPLE. Consider an example (Ref. 15) (Fig. 15) of a simple servo system in which a rotatable mass is positioned by an electric motor whose supply voltage v_a is governed by a proportional regulator having an amplifier that is sensitive to a voltage v proportional to the positional deviation θ. Going around the loop: the amplifier has a static sensitivity or "gain" of 41.5 and a time constant of 0.04 sec so that its transfer function is $v_a/(v - v_s) = 41.5/(1 + 0.04p)$, where the time constant has the usual sense of the time required to complete a response if the initial rate persisted.

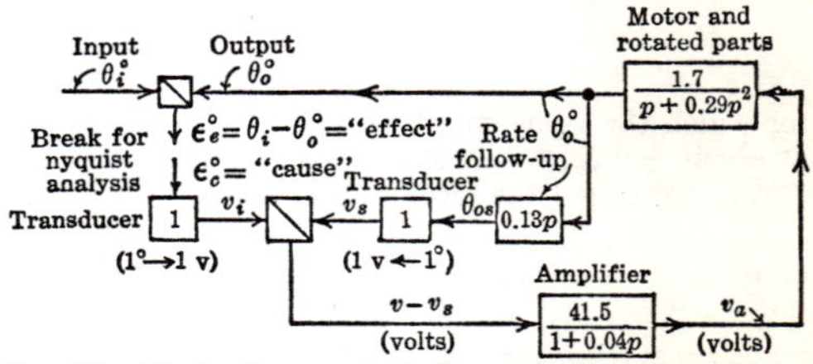

FIG. 15. Block diagram, including output-rate stabilizing means. (From *A Method for the Design of Automatic Controls*, by J. D. Peterson, ASME, Ref. 15)

The inertia of the motor, rotating mass, and their connecting gearing can be lumped under a p^2 term representing the acceleration produced by the torque q. The motor produces a static torque proportional to the voltage v_a, which is determinable from a test at stall. When running, the available torque of a typical servo motor falls off in proportion to the speed, as expressed by a subtracted p term, also determinable from knowledge of the full speed at no load. The transfer function for this idealized lumped motor, rotatable mass and gearing unit, is $1.7/(p + 0.29p^2) = \theta_o/v_a$ or $-\theta/v_a$ for a zero value of θ_i.

In the system of Fig. 15 less the "rate follow-up," breaking the control loop between the rotating mass and the amplifier, and feeding in as a "cause" an oscillating θ_c of one radian amplitude through a transducer to the amplifier at a frequency of 10 radians per sec, the voltage output v_a follows from

$$\frac{v_a}{\theta_c} = \frac{41.5}{1 + 0.04(j10)} = \frac{41.5}{1 + j0.4} = \frac{41.5}{1.076\angle 22.0^\circ} = 38.5\angle -22.0^\circ ,$$

having a lag of 22.0 degrees, and the "effect" on the plant (motor-gearing) output follows the relation

$$\frac{-\theta_e}{v_a} = \frac{-1.7}{j10 + 0.29(j10)^2} = \frac{-1.7}{j10 - 29} = \frac{-1.7}{30.7\angle -19.1^\circ} = -0.0554\angle 19.1^\circ$$

Hence the overall transfer function, except for the control reversal of sign, is $-\theta_e/\theta_c = (-0.0554\angle 19.1^\circ)(38.5\angle -22.0^\circ) = -2.133\angle -2.9^\circ$, or $2.133\angle -182.9^\circ$, or a *lag* of slightly more than 180 degrees. If the $\theta = \theta_i - \theta_o$ unit is included, the "effect" θ_e is seen to occur 362.9 degrees later than its "cause" θ_c.

And for $\omega = 10$ radians per sec, the locus point is at about 180 degrees lag, with the effect $-\theta_e$ equal to about twice its cause θ_c. And likewise for $\omega = 15$ radians per sec, $-\theta_e/\theta_c = 0.91\angle -198.1^\circ$ so that, for about the same value of effect and cause, the effect *lags* more than 180 degrees behind its cause. Since these values show the overall loop transfer locus to be spiraling in toward the center and outside the "minus one" point, which is said to be "enclosed" (Fig. 14), the Nyquist rule establishes that this system is unstable.

The overall loop transfer function for this unstable case may be written directly for convenience in plotting the locus, as $\dfrac{-\theta_e}{\theta_c} = \dfrac{70.5}{p(1 + 0.29p)(1 + 0.04p)}$. The system may be stabilized by either (1), at the expense of control speed, increasing the gear ratio to, say, thirteen times the original value when, neglecting any change in the gearing's inertia, the overall gain becomes 70.5/13 = 5.42, or (2), at the expense of an increase of load error, decreasing the amplifier gain to 41.5/13 = 3.19.

Alternatively, the system may be stabilized with better performance by adding a follow-up loop for the motor speed (see Fig. 15), in which a speedometer element moves 0.133 degrees for 1 degree per sec velocity of the rotating mass, for which $\theta_{os} = 0.133p\theta_o$ and the transfer function is $\theta_{os}/\theta_o = 0.133p$. Since this ratio is dimensionless and p has the dimensions of inverse seconds ($\sec^{-1}$) for (d/dt), that of 0.133 is seconds. A "transducer" means is provided for converting θ_{os} to a corresponding voltage v_s. Thus, however obtained, a voltage is generated proportional to the output shaft velocity, and imposed on the portion of the amplifier which receives the voltage proportional to the "error."

Since this rate component θ_{os} is subtracted from the error $-\theta_e$ and the difference fed to the amplifier,

$$\theta_o = \left(\frac{1.7}{p + 0.29p^2}\right)\left(\frac{41.5}{1 + 0.04p}\right)(-\theta_e - 0.133p\theta_o)$$

$$\frac{\theta_o}{-\theta_e} = \frac{\dfrac{70.5}{p(1 + 0.29p)(1 + 0.04p)}}{1 + \dfrac{9.38}{(1 + 0.29p)(1 + 0.04p)}}$$

for which the locus spirals in inside the "minus one" point, crossing the 180-degree axis at about −0.25, which shows that the system is now highly stable.

Another use of the locus diagram, proposed by Harris (Ref. 16), is as follows: On it are plotted circles for constant values of $M = \left|\frac{\theta_o}{\theta_i}\right| = \sqrt{\frac{R_o^2 + x_o^2}{R_i^2 + x_i^2}}$ for $\frac{\theta_e}{\theta_i} = \frac{R_o + jx_o}{R_i + jx_i} = \frac{\theta_o/\theta}{1 + \theta_o/\theta} = \frac{GH}{1 + GH}$ and circles for the θ_o/θ_i phase angle $\phi = \tan^{-1}(x_o/R_o) - \tan^{-1}(x_i/R_i)$. These circles give the value of M and ϕ wherever crossed by the locus. Where the locus is tangent to an M circle near the minus one point, there is a tendency for the system to oscillate at the frequency for the locus at the point of tangency. For example, for the stable single-loop system discussed earlier having a value of overall gain of 5.42, the locus is tangent to an M circle at $\omega = 4$ radians per sec, which is the angular frequency of the damped oscillatory transient of this system. For a given class of system, the amount of a transient's initial overshoot following a step change is proportional to the value of M of the thus-tangent M circle.

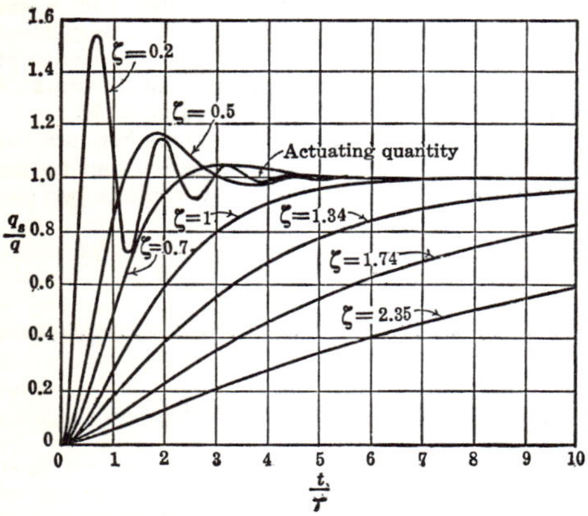

FIG. 16. Instrument response to a step function. (From Design Factors Controlling the Dynamic Performance of Instruments, by C. S. Draper and G. P. Bentley, *Trans. ASME*, Vol. 62, pp. 421–432, July 1940)

RESONANCE. Figure 16 shows the transient response of a damped (N) sprung (B) mass (M) to a step change of force, where the damping ratio $\zeta = \frac{N}{N_c} = N/2\sqrt{MB}$ and the characteristic time $\tau = 2M/N$. Figure 17 shows the transient performance of this same plant or instrument to a suddenly applied, lineally increasing force, a much less drastic change. Figure 18 shows the steady-state response of this same plant to a sinusoidal force, with the noteworthy change of phase from 0 to 180 degrees as resonance is passed with increasing frequency for an undamped mass.

Analogs in control are especially useful in working with systems containing nonlinearities. They are alternate physical means for obtaining the same mathematical relations as for the prototype. For example, the mechanical system having the force equation

$$M\frac{d^2x}{dt^2} + N\frac{dx}{dt} + Bx = f$$

has for an analog an electrical system having the voltage equation

$$L\frac{d^2q}{dt^2} + R\frac{dq}{dt} + \frac{1}{C}q = e$$

or

$$L\frac{d^2i}{dt^2} + R\frac{di}{dt} + \frac{1}{C}i = \frac{de}{dt}$$

where q and i are respectively expressed in coulombs and amperes (Refs. 14, pp. 67–70, and 12, pp. 120–125). The correspondence of the different physical elements is exact when the proper dual (Ref. 14, pp. 63–67) is chosen.

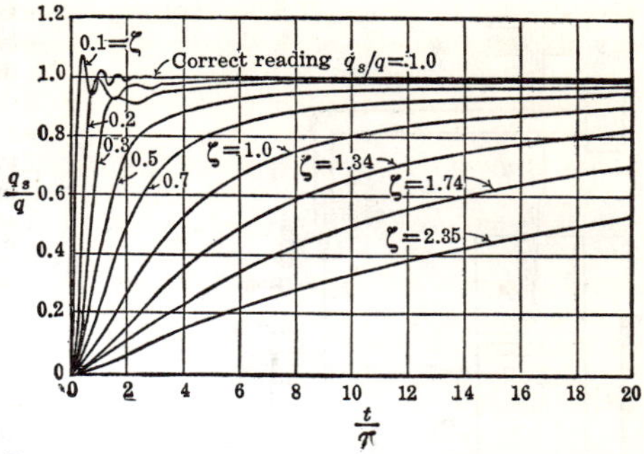

FIG. 17. Lineal forcing function; variation with time of actual reading as fraction of true value of actuating quantity. (From C. S. Draper and G. F. Bentley, *loc. cit.*)

In a servo-controlled system, especially with a proportional speed floating servo, the stabilizing effect of a dead zone is small since the interruption of hunting oscillations occurs in each cycle only near the point where the velocity of the final control valve is a minimum (Ref. 12, pp. 84–85). But such a system may be strongly stabilized by interrupting the motion of this control valve for a like interval in each swing when there is a reversal of trend of the measured variable, since this coincides with the maximum velocity of the control valve for an integrating servo of the type mentioned above, which lags 90 degrees behind the measurement.

Another powerful stabilizing means is by temporarily shifting the control point in the direction of a change so that the meter tends to stop the final control valve when the measurement is still appreciable but the measured variable has actually returned to the set value.

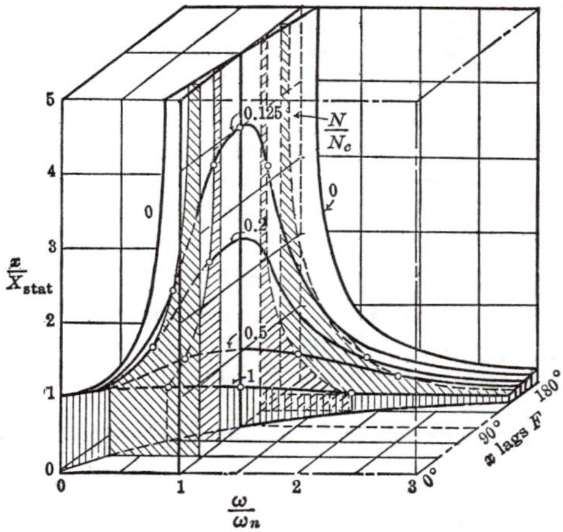

FIG. 18. Performance of a damped-sprung mass under an oscillating force input. x is the amplitude of the response. (From *Mechanical Vibrations*, by J. P. Den Hartog, McGraw-Hill Book Co., 1944, by permission)

6. PROCESS CONTROL

GENERAL. To illustrate the nature of the problem which a regulator has to solve, consider the process control system of Fig. 19 in which the load $L(t)$ affects a part H_a of the plant H, with a resultant effect at the meter of $x_L = H_aL$ if the regulator did *not* operate. This x_L can be considered as equivalent to a change of the input x_i so that the net input is $x_{in} = x_i - x_L$. The effect at the meter due to the difference θ (following the servomechanism convention) in passing around the loop of the system is simply $x_s = -GH\theta$. Hence for $\theta = x_{in} - x_s = x_i - x_L - x_s$ and, letting the setting x_i be steady at a zero reference value,

$$\theta(t) = -H_aL(t) - GH\,\theta(t) \quad \text{or}$$

$$-\theta(t) = \frac{1}{1 + GH}H_aL(t) = \frac{1}{1 + GH}x_L$$

the stability being determined by the loop term (GH) above as for other systems. For purposes of analysis, some nonlinear systems may be virtually linearized by throwing their nonlinearities into the load. Dr. Claude E. Shannon has noted that a switching system may

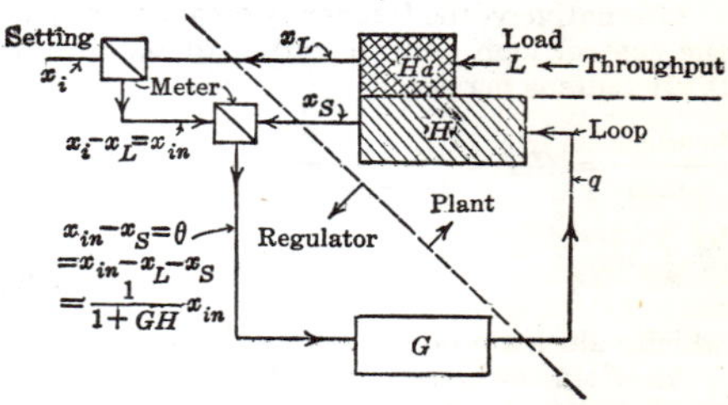

FIG. 19. Block diagram of regulated system under load.

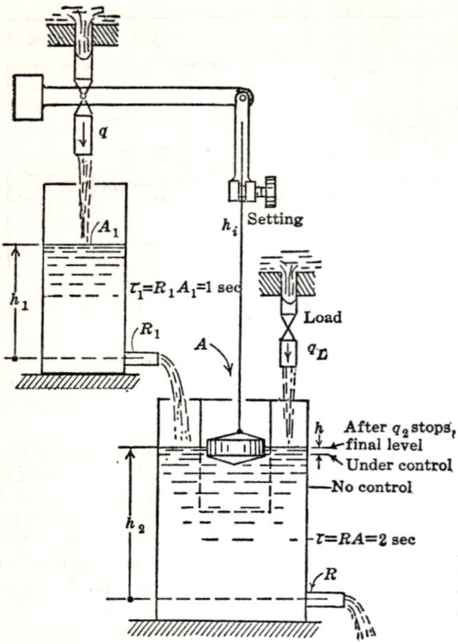

FIG. 20. Diagram of two-capacity plant with proportional regulator for h_2.

be replaced by an otherwise similar continuous one with a square-wave load of different frequencies and "on" to "off" duration ratios.

EXAMPLE. The plant H may consist of two tanks and associated resistors in series or cascade (Ref. 12, pp. 22 and 93–95), with the load an independently variable flow q_L into the second tank whose level is to be regulated. If, as in Fig. 20, there is no appreciable reaction of the second tank's head on either the flow from the first, or of the first tank's head on the controlled flow q, the relations are relatively simple. Note that this case also corresponds to that of an idealized governed engine which has been extensively treated elsewhere (Ref. 17).

The upstream tank has area $A_1 = 1$ ft^2 and resistance $R_1 = 1$ sec per ft^2, with a resultant time constant $\tau_1 = 1$ sec. Likewise, the second tank has $A = 2$ ft^2, $R = 1$ sec per ft^2, and $\tau = 2$ sec. In both cases, the outflow is substantially proportional to the head. A 1-ft change of level in the second tank operates the control valve to cause a change of the flow rate q of 8 ft^3 per sec. The load q_L is a 1 ft^3 per sec flow rate into the second tank.

TRANSIENT ANALYSIS. Starting at $t = 0$ with the system in equilibrium, q_L is suddenly stopped. When this occurs, the level response in the second tank would follow a $1e^{-t/2}$ curve if the control valve did not operate.

Since $G = 8$ and $H = \dfrac{1}{(1 + p)(1 + 2p)}$, by superposition, with the control valve operating and the head setting unchanged, the deviation from the final position is $h(t) = \dfrac{1}{1 + GH} e^{-t/2} = \dfrac{1}{1 + \dfrac{8}{(1 + p)(1 + 2p)}} e^{-t/2} = \dfrac{p^2 + p + 0.5}{p^2 + 1.5p + 4.5} e^{-0.5t}$. The numerator operates directly on the exponential term by simple differentiation since $p \cong d/dt$, without departing from real time constants.

But the denominator indicates that *damped oscillations are present*, since it corresponds with $p^2 + 2\zeta\omega p + \omega_n^2$, where the undamped natural frequency $\omega_n = \sqrt{4.5} = 2.121$ radians per sec and the damping ratio $\zeta = 1.5/(2 \times 2.121) = 0.3537$, so that the oscillation $e^{-\zeta\omega t} \cos(\omega t + \phi)$ becomes $e^{-0.3537 \times 2.121 t} \cos(2.12t + \phi)$ or $e^{-0.75t} \cos(2.12t + \phi)$, neglecting the usually slight increase of the actual frequency due to damping. The relatively high value of 8 for the gain or sensitivity is seen to be the cause of the oscillations, and the *static error* is $1/(8 + 1)$ times that for the system if the regulator did not operate.

While the foregoing "crowbar" analysis gives a generally adequate idea as to the nature of the transient response and stability, the complete solution may be obtained by any familiar method of solving differential equations.

Alternatively, the **frequency response** may be used to determine the stability. Breaking the control loop and oscillating the control valve at the system's natural frequency of 2.121 radians per sec,

$$\frac{-\text{effect}}{\text{cause}} = GH = \frac{8}{(1 + 2j\omega)(1 + j\omega)} = \frac{8}{(1 + j4.242)(1 + j2.121)} =$$

$$\frac{8}{(4.36 \angle 76.7^\circ)(2.345 \angle 64.8^\circ)} = 0.783 \angle -141.5^\circ$$

which values are on the stable side of the Nyquist "minus one" point.

An effort has been made in this treatment of control to cover the general theory common to all types of regulators and servomechanisms, as well as enough instances of practical use to facilitate its application in any particular field. In its short space, considerable ground is covered and a number of references included so that its user can follow the subject as far as is necessary. As it indicates, regulation or control is not a science but, instead, one of the more difficult engineering arts (Refs. 18–20).

REFERENCES

1. *Automatic-Control Terms*, American Society of Mechanical Engineers, New York, March 1946.
2. *Sources of Information on Instruments*, American Society of Mechanical Engineers, New York, June 1945.
3. Ed S. Smith, Control bibliography, *Automatic Control Engineering*, McGraw-Hill Book Co., New York, 1944.
4. Donald P. Eckman, *Principles of Industrial Process Control*, John Wiley and Sons, New York, 1945.
5. Thomas J. Rhodes, *Industrial Instruments for Measurement and Control*, McGraw-Hill Book Co., New York, 1941.
6. W. Trinks, early engine governor bibliography, *Governors and the Governing of Prime Movers*, D. Van Nostrand Co., New York, 1919. Out of print.
7. R. R. Batcher and W. Moulic, *The Electronic Control Handbook*, Caldwell-Clements, New York, 1946.
8. LeRoy A. MacColl, *Fundamental Theory of Servomechanisms*, D. Van Nostrand Co., New York, 1945.
9. H. Lauer, R. Lesnick, and L. E. Matson, *Servomechanism Fundamentals*, McGraw-Hill Book Co., New York, 1947.
10. W. R. Ahrendt and J. F. Taplin, servomechanism bibliography, *Automatic Regulation*, Ahrendt & Taplin, P.O. Box 4673, Washington 20, D. C., 1947.
11. C. O. Fairchild, Elementary theory of automatic temperature regulation (control), p. 601, *Temperature, Its Measurement and Control in Science and Industry*, Reinhold Publishing Corp., New York, 1941; and U.S. Patent 2,274,266.
12. Ed S. Smith, *Automatic Control Engineering*, McGraw-Hill Book Co., New York, 1944.
13. C. O. Fairchild, Note on the throttling of electric heat, *Instruments*, pp. 305, 317, Dec. 1937.
14. W. C. Johnson, *Mathematical and Physical Principles of Engineering Analysis*, pp. 23, 45 ff., McGraw-Hill Book Co., New York, 1944.
15. Joel D. Peterson, *A Method for the Design of Automatic Controls*, paper presented before the ASME, Boston, Nov. 1946.
16. Herbert Harris, Jr., The frequency response of automatic control systems, *Am. Inst. Elec. Engrs. Technical Paper* 46–114, presented May 1946.
17. W. R. Ahrendt and J. F. Taplin, *Automatic Regulation*, pp. 56 ff., 61 (Fig. 3:13), Ahrendt & Taplin, P.O. Box 4673, Washington 20, D. C., 1947.
18. N. Minorsky, *Introduction to Non-Linear Mechanics*, J. W. Edwards Bros., Ann Arbor, Mich., 1947.
19. G. S. Brown and D. P. Campbell, *Principles of Servomechanisms*, John Wiley and Sons, New York, 1948.
20. I. A. Greenwood, Jr., J. V. Holdam, Jr., and D. MacRae, Jr., *Electronic Instruments*, McGraw-Hill Book Co., New York, 1948.

SERVOMECHANISMS

By HERBERT HARRIS, JR.

A SERVOMECHANISM is an automatic system used for the control of mechanical position. Its functions are *torque amplification*, *remote data transmission*, and the *conversion of physical measurements into mechanical position*. It is distinguished from other possible systems which meet this same definition by its closed cycle character, illustrated by the block diagram of Fig. 1 showing the components and direction of signal flow.

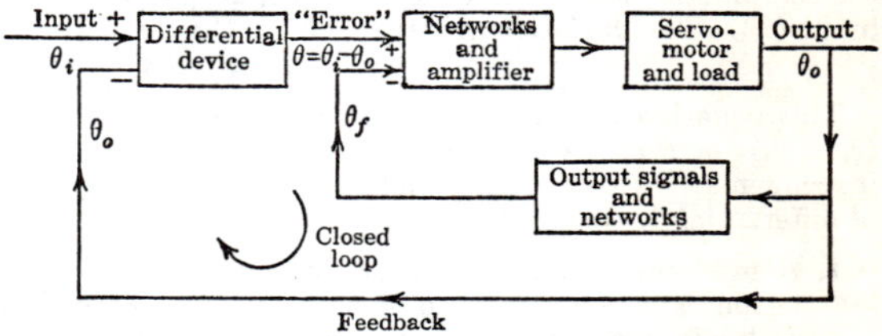

FIG. 1. Servomechanism block diagram.

The *output* or mechanical load is driven by the *servomotor*. The torque of the servomotor is controlled by the output of the amplifier which receives several signals as an input, the principal one being that from the differential device. The *differential device* or *error-measuring device* takes the difference between the actual output position and the desired or *input* position. This difference is called the *deviation* or "*error*." It is this return of the output position measurement through the differential device and then on to control the motor which forms the *closed loop* or *closed cycle* which is characteristic of servomechanisms. Various networks may be interposed in the amplifier or other signals added to its input to provide stability and other special characteristics as described later.

TYPICAL APPLICATIONS (Ref. 1).

1. In some industrial recorders the output or pen is positioned by a servomotor under the control of an amplified voltage. This voltage is the difference between an input

voltage, say from a thermocouple, and a potentiometer voltage which is proportional to pen position. This is an example of the conversion of a physical measurement into a mechanical position.

2. A profile milling machine has the work piece positioned by a servomotor in such a way that it follows a model or pattern over which a light feeler travels. This is an example of a *torque amplifier* or *follow-up*.
3. Guns weighing many tons are positioned very precisely by an operator turning a light handwheel. When, as sometimes happens, the operator or an automatic gun position computer is at a remote position the servo system also includes a data-transmission feature.

7. SERVOMOTORS

The servomotor which provides the mechanical force for positioning the output is usually especially designed so that large output power may be controlled by a low-power signal or force. Among the numerous types are the following:

1. Two-phase motors are controlled by placing one winding across the a-c line and driving the other by an electronic amplifier with a 90-degree out-of-phase voltage. These motors are usually used in the range up to about $^1/_{20}$ horsepower for applications requiring high acceleration capabilities.
2. Alternating- and direct-current commutator-type motors are used by controlling the field excitation through an electronic amplifier. These motors are used most often in the $^1/_{100}$ to $^1/_5$ horsepower range.
3. Motor-generator sets are used for the control of a commutator-type motor with fixed field excitation. The armature current is supplied by a generator which is field-excited through an electronic amplifier. These sets are used from about $^1/_4$ horsepower into the multihorsepower region.
4. Hydraulic or pneumatic pistons are used as servomotors by valving high-pressure oil or air to one side or the other. The valves can be moved by small solenoids or electric motors under electronic control since the pressures on the valve are balanced so that only small driving forces are required. Pistons are used for applications which do not require continuous motion in one direction.
5. Variable-displacement hydraulic pumps and motors are common in the multi-horsepower region. The displacement of such a pump is controlled by varying the stroke of the pistons or vanes by mechanical means. An electric motor or solenoid, sometimes followed by a small hydraulic servo, is used for this purpose.

8. DIFFERENTIAL DEVICES

The function of a differential device (Ref. 1) is measurement of "error" or deviation of the servomotor's actual position from the position it should have. To simplify terminology, the servomotor position is called the output (θ_o) while the desired position is called the input (θ_i). The deviation or "error" (θ) is the difference between the two,

$$\theta = \theta_i - \theta_o \tag{1}$$

This equation is conventional for servomechanism theory. The deviation of "error" ($\theta_i - \theta_o$) for servomechanisms is the negative of the error ($\theta_o - \theta_i$) generally used for instruments. "Error" is hereinafter used without this distinction as to sign. Examples of differential devices include the following:

1. A mechanical planetary gear differential is often used when both the input and output are easily coupled shaft rotations. Potentiometers or inductance variation pick-offs are used for converting the resulting mechanical motion into various electrical signals.
2. Variation of the airgap of inductive or capacitive elements is used when the two members may be placed on the input and output respectively and the differential movement is small (0.00001 in. to 0.1 in.). This is a particularly useful pick-off or take-off for a differential device such as a gyroscope, which must be lightly loaded. A variety of circuits may be used for the conversion of the parameter variations into voltage changes.
3. When the input and output are physically separated by such a great distance that mechanical coupling is not desirable, some means for remote data transmission must be included in the differential device. *Synchronous units* or *synchros* (Refs. 1 and 2), manufactured under trade names such as *Autosyns*, *Diehlsyns*, and *Selsyns*, provide this function and are commonly used.

Synchronous units are small motor-type units with a single-winding rotor and a three-winding stator wound similarly to that of a three-phase motor. Two such units are used in transmitting data. The two stator windings are connected together, one rotor being connected across the a-c line and the other providing the electrical error signal for the servo amplifier. One rotor is rotated by the input and the other by the output.

When connected as described, the rotor connected to the a-c line sets up a magnetic field which induces voltages in the three stator windings. These voltages cause currents to flow in the stator of the other unit and set up a parallel magnetic field. The second rotor then has an induced voltage proportional to the cosine of the angle between its axis and that of the magnetic field. The zero-error position is taken at the zero-voltage point when the two rotors are 90 degrees apart. Thus the two rotors must be turned in synchronism in order to maintain the zero-error voltage. The voltage output is proportional to the sine of the error angle, and the direction of error may be determined by the phase relative to that of the a-c line. Alternating-current motors reverse torque with a reversal of the control-voltage phase, but d-c motors require the use of an electronic circuit known as a phase-sensitive detector or "discriminator."

These units are manufactured with a large variety of characteristics. In determining which units to use, it is necessary to specify line voltage, line frequency, maximum speed, number of outputs to be connected to one input, and accuracy. Normally the error of these units is about $^1/_4$ degree, but at least one manufacturer has been able to improve this by a large factor.

When greater accuracy is required, it is common to use the *fine* and *coarse* system of data transmission. Two additional synchro units, wired in the same way as the first, are geared to the input and output to turn faster by some factor such as 25 : 1. Thus a small error provides a signal 25 times as great from these *fine* units, and an accuracy improvement by a factor of 25 is achieved. However, these *fine* units have 24 spurious zero error positions occurring at each revolution when the error is a multiple of 360/25 degrees. These spurious positions are eliminated by the original or *coarse* set of synchros which is connected directly to the input and output shaft. A mixing circuit is used in conjunction with these two signals so that the servomechanism ignores the *fine* signal until the *coarse* signal is near zero and then the *fine* signal is used for the precise setting at zero error.

Synchronous units of a special design may also be used for the direct transmission of mechanical position. If both rotors are connected to the a-c line and the two stators are connected together, the two rotors will tend to line up with a force proportional to the sine of any angular difference between the two. This system is used only when the precision and large output force of a servomechanism are not needed.

9. TRANSIENT RESPONSE

As shown in Fig. 1, the simplest servomechanism would have the servomotor torque controlled by the error signal alone, either by a continuous or an on-and-off action.

TWO-POSITION CONTROL (Ref. 3) refers to a system which makes use of switch contacts in such a way that the servomotor exerts its maximum corrective torque whenever the error becomes greater than some preset value. Between the two switch points controlling the opposing directions of torque is a dead zone which does not call for any torque. Servomechanisms of this type have the serious disadvantage of hunting or oscillating when the dead zone is made small, in order to provide more precise control. If one contact is closed long enough for the motor to build up an appreciable velocity, the load inertia will serve to carry it past the dead zone and to enter the region of the opposing contact at high speed. The torque then reverses, but again the motor will build up speed in the new direction and run across the dead zone. Unless friction is present, which produces large power loss, hunting may continue indefinitely. Many early servomechanisms were of this type but they have been gradually displaced by continuous control for better performance. By the addition of devices which move the contact points back and forth as a function of motor speed and vibrate them to remove errors due to the dead zone, it is possible to retain the advantages of contacts in providing simple control of large power and still obtain some of the properties of continuous control.

PROPORTIONAL CONTROL (Refs. 4 and 5), the simplest example of a continuous control, provides servomotor torque proportional to the error signal. In order to keep the error small, the proportionality constant must be made large so that large corrective and accelerating torques are available. Unless sufficient damping friction is present, this system will also oscillate just as a spring-mass combination would.

A servomechanism with proportional control alone is rarely used except for applications in which the motor is only lightly loaded and the input does not experience high accelerations. The proportionality constant cannot be increased beyond the point where the oscillations do not damp out sufficiently fast, and then the response is sluggish and external loads produce large errors. In order to make full use of the possibilities of continuous control, the torque must be made a function of signals other than just the error. The following analysis illustrates this.

Consider a servomotor and load having a moment of inertia J which requires a torque $J(d^2\theta_o/dt^2)$ for acceleration; also, either the load has friction which increases with speed, or the motor has a speed-torque characteristic which drops with speed such that the torque available for acceleration is reduced by $f(d\theta_o/dt)$. Then if the controlled torque $K_o\theta$ is proportional to the error, the following equation applies: *

$$K_o\theta = J\frac{d^2\theta_o}{dt^2} + f\frac{d\theta_o}{dt} \tag{2}$$

If this is combined with eq. 1 relating the input, output, and error,

$$J\frac{d^2\theta}{dt^2} + f\frac{d\theta}{dt} + K_o\theta = J\frac{d^2\theta_i}{dt^2} + f\frac{d\theta_i}{dt} \tag{3}$$

Equation 3 allows the error to be determined for any desired input motion. Many classical and operational methods (Refs. 5 and 6) are available for the solution, involving the solution of the *characteristic equation* obtained by setting $\theta_i = 0$ and replacing the operation (d/dt) by the operator p. This yields

$$p^2J + pf + K_o = 0 \tag{4}$$

The roots of this equation determine the modes of response for the system. In this case the roots are

$$p = -\frac{f}{J} \pm \sqrt{\left(\frac{f}{2J}\right)^2 - \frac{K_o}{J}} = -\zeta\omega_n \pm j\omega_n\sqrt{1-\zeta^2} \tag{5}$$

where

$$\omega_n = \sqrt{\frac{K_o}{J}} \quad \text{and} \quad \zeta = \frac{f}{2\sqrt{K_oJ}} \tag{6}$$

With these roots the response for θ will have a transient term $Ae^{-\zeta\omega_n t}\cos(\sqrt{1-\zeta^2}\omega_n t + \phi)$ in addition to a particular solution determined by the input motion. Both A and ϕ are determined by the initial conditions.

ω_n is called the *undamped natural frequency* since it is the frequency of oscillation that occurs when the damping is zero ($f = 0$).

ζ, called the *damping ratio*, expresses the ratio of damping present to that required for *critical damping* which exists when there are two equal real roots. When ζ is greater than unity, the system is overdamped, with real unequal roots, and no oscillatory term is present.

Figure 2 shows the response of this servomechanism in returning from an initial error, as might occur after forcing the output shaft out of position and then releasing it. For a motor and load with a given J and f, the response becomes faster as K_o is increased, since ζ decreases and ω_n increases, thereby shortening the time scale. Eventually, however, ζ will become too small, and the response will be oscillatory and have too much overshoot.

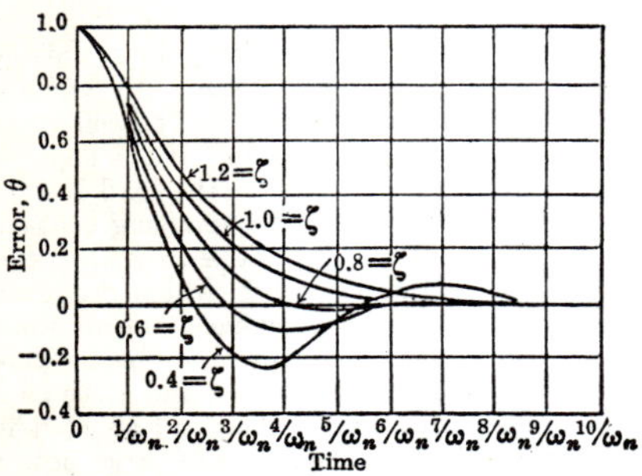

Fig. 2. Response to an initial error.

If the input is rotated at constant speed, the output will follow at the same speed after a transient similar to that just described has died out. However, a fixed error, known as the *constant velocity error* or *speed lag*, will remain. It is generally measured as a time constant defined as the lag per unit of velocity and in this case is equal to f/K_o.

Another error that may be present, called the *static error*, represents the size of error that must be present before the servomotor torque ($K_o\theta$) can overcome the dry friction F which is present. The error may have any value within the static-error zone F/K_o without producing any corrective action.

* This analysis and the remainder of the chapter assume that, for small motions, the system may be considered linear.

The design for the simple proportional-control servomechanism requires the choice of a value for K_o. Increasing the sensitivity K_o will reduce the static and constant velocity errors as well as the response time, but too large a value will cause an oscillatory response. A satisfactory compromise cannot generally be found for applications other than light instrument servomechanisms. An increase in the damping coefficient f is not often useful since it represents power loss as well as an increase in speed lag.

For most applications a more complex control system is necessary such as the use of a derivative control to provide the necessary damping without the disadvantages of friction. If the control torque is made equal to $K_o\theta + K_1(d\theta/dt)$, the second term is called a *derivative control.* The characteristic equation then becomes $p^2J + p(f + K_1) + K_o = 0$ so that K_1 acts similarly to the friction for damping. However, K_1 does not add to the speed lag. Thus the sensitivity may be increased and nonoscillatory response still achieved. The derivative control term may be obtained by measuring the speeds of the input and output shafts, but, more often, a network operating on the error signal is used to produce an approximate derivative.

The increase of sensitivity and provision of damping by a derivative control has its limits also. Eventually the response becomes so fast that various other factors must be included in setting up the equations of motion, since they then begin to have an appreciable effect. In particular, various response retardations or delays, such as those occasioned by the time required to build up current in a motor field due to its inductance, become important. When these delays and others due to the introduction of demodulators and networks in the amplifier are taken into account, the degree of the characteristic equation becomes higher than the second, and the addition of a derivative control does not always provide the necessary damping. This is also true of more complicated servomechanisms such as hydraulic servomechanisms which have an electric motor as well as a hydraulic torque amplifier in the loop ahead of the servomotor.

A slow-response type of control is often used to reduce the speed lag and static error without increasing the speed of response to the point that internal response delays become important. If the control torque is $K_o\theta + K_1(d\theta/dt) + K_{-1}\int \theta\, dt$, the last term is called an *integral control.* For any steady errors, the torque gradually builds up in magnitude and can provide the necessary steady driving torque even after θ becomes zero. In practical cases, the integral cannot be taken exactly and the control acts more like a slow increase in K_o and the steady error cannot become exactly zero. Figure 3 shows a circuit that gives this integral type of control.

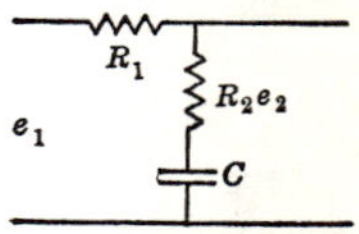

$$\frac{e_2}{e_1} = \frac{1 + R_2Cp}{1 + (R_1 + R_2)Cp}; \ R_1 \gg R_2$$

FIG. 3. Integrating type network.

With a characteristic equation of high degree, the concepts discussed in the foregoing still apply, but they are not easily worked into a design procedure. The characteristic equation may be factored into a series of quadratics, each of which has its own undamped natural frequency and damping ratio. Unfortunately, above the fourth degree this factorization can be carried out only numerically and even then only with considerable time expenditure. One set of relations between the coefficients, known as *Routh's criterion* (Ref. 6), is often used to determine if any of the damping ratios are negative, a negative damping ratio indicating an *unstable* system in which the oscillations build up rather than die out.

In order to facilitate the transient analysis of complicated systems, differential analyzers (Refs. 7 and 8), which automatically solve the equations of motion, are finding considerable use. However, for the synthesis and design of servomechanisms, another approach through the concept of frequency response is most important. (See Article 10 and Refs. 4, 9, 10, 11, and 12.)

10. FREQUENCY RESPONSE

If the input is oscillated sinusoidally at constant amplitude and frequency there will be a transient which dies out and then there will remain a steady-state response with the output, error, and all components also oscillating sinusoidally. The ratio of the amplitude of output oscillation to that of the input oscillation is a function of frequency and is called the *magnitude of the system transfer function,* denoted by the symbol M. Similarly the difference in phase of the two sinusoids is called the *phase of the system transfer function* and denoted by the symbol ϕ.

The magnitude and phase of the system transfer function are easily computed by the following steps illustrated by the simple example used for eq. 2.

1. Set up the differential equation for the system.

2. Replace d/dt by p, and $\int ()dt$ by $1/p$:

$$Jp^2\theta_o + fp\theta_o = K_o(\theta_i - \theta_o)$$

3. Solve the equation for θ_o/θ_i. This is the *system transfer function in operational form.*
4. Replace p by $(j\omega)$ where $j = \sqrt{-1}$ and $\omega = 2\pi \times$ frequency.

$$\frac{\theta_o}{\theta_i} = \frac{K_o}{(K_o - \omega^2 J) + j\omega f} = \frac{R_N + jX_N}{R_D + jX_D} \tag{7}$$

where R and X denote numbers found from the constants of the system and the frequency, the subscripts standing for numerator and denominator.

This is the *system transfer function in frequency form.* It is a complex number with the magnitude M and phase angle ϕ. Thus

$$M = \left|\frac{\theta_o}{\theta_i}\right| = \frac{\sqrt{R_N^2 + X_N^2}}{\sqrt{R_D^2 + X_D^2}} \tag{8}$$

and

$$\phi = \angle\frac{\theta_o}{\theta_i} = \tan^{-1}\frac{X_N}{R_N} - \tan^{-1}\frac{X_D}{R_D} \tag{9}$$

Figure 4 shows M and ϕ plotted as functions of frequency for this servomechanism. Since they correspond to the same system as the transient responses of Fig. 2, they illustrate certain general interpretations that may be applied to the frequency response:

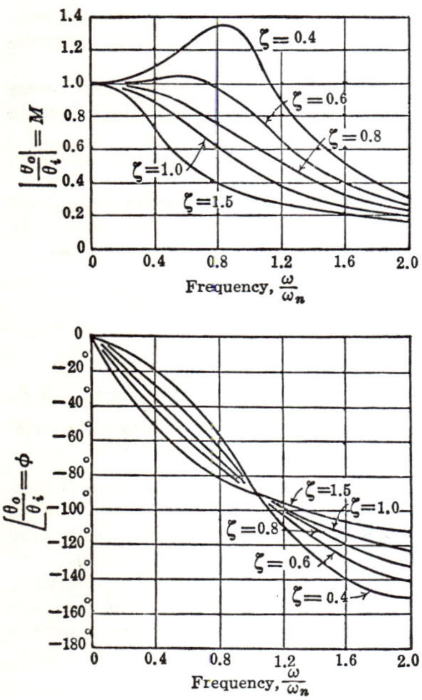

FIG. 4. Frequency response.

1. A large resonant peak for M corresponds to a poorly damped or oscillatory transient response. For good design, the peak value of M is usually limited to less than 1.5, although it is often better to keep M less than 1.25.
2. Large phase shift at low frequency corresponds to sluggish response.
3. The time and frequency scales vary inversely (see Figs. 2 and 4) so that a quick time response corresponds to a high-frequency response.
4. One way to specify the characteristics of a servomechanism without giving all details is to give the peak value of M and the frequency at which the phase ϕ is some specific value, say 45 degrees. Thus for $\zeta = 0.6$, the peak of M is 1.05 and ϕ is 45 degrees at $\omega = 0.55\ \omega_n$. With few exceptions, the fastest servomechanisms have the 45-degree phase-shift point below 20 radians per sec if M is under 1.5.

For purposes of synthesis, this means of calculation and interpretation is still not ideal since the effect on the system of each parameter is not easily visualized. The following methods, which have been derived from principles and theories originally applied to electrical feedback amplifiers, are superior.

These methods are based on the following algebraically true relation (Fig. 1):

$$\frac{\theta_i}{\theta_o} = \frac{\theta}{\theta_o} + 1 + \frac{\theta_f}{\theta_o} \tag{10}$$

Each output-to-input ratio is a transfer function and each input-to-output ratio is a reciprocal or *inverse transfer function* of some part of the system. Thus θ/θ_o is the inverse transfer function of the forward path of the closed loop and $(1 + \theta_f/\theta_o)$ is the transfer function of the return or feedback path.

Since these transfer functions, when in frequency form, have a magnitude and phase angle (or a real and imaginary component) they may be plotted at each frequency as a complex number vector from the coordinate origin. As frequency varies, the vector extremity plots out a locus. Thus a transfer function or its inverse may be plotted as a single curve, such as those in Fig. 5, with the numbers along the curve showing the frequency at each point.

In addition to being plotted as vectors, transfer functions have similar mathematical properties. If a number of networks and devices are placed in series, with the input of each fed by the output of the preceding unit, the magnitude of the overall transfer function is equal to the product of all the individual magnitudes. Also, the total phase shift is the sum of the individual phase shifts. These are the same as the properties of two-dimensional vectors used in plotting complex numbers. Similarly, two units in parallel call for vector addition of transfer functions. This property makes it easy to calculate transfer functions and to separate out the effect of each component, particularly in the case of θ/θ_o and θ_f/θ_o which usually consist of many elements in series.

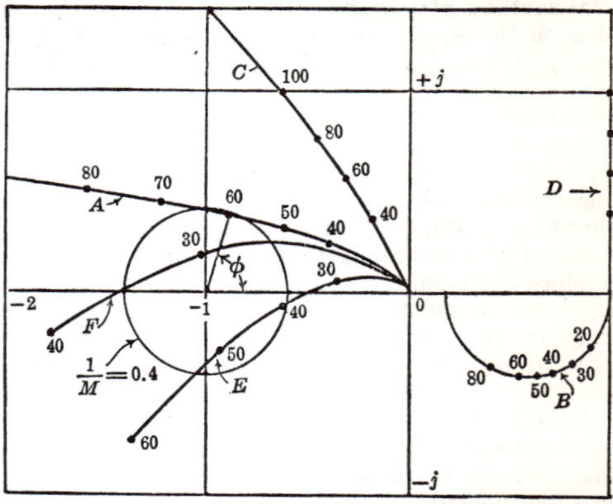

FIG. 5. Transfer functions plotted as the loci of vector extremities.

Curve A of Fig. 5 shows such a locus for θ/θ_o. If there is no feedback θ_f, it is apparent that θ_i/θ_o is represented by the same plot if a new origin is taken at -1. Thus the amplitude ratio M is measured as the reciprocal of the distance from the -1 point to the locus. The phase at any frequency is the angle of the locus point measured from the real axis about the -1 point. Thus, in order to keep the resonance peak low, the θ/θ_o locus must be kept away from the -1 point (Refs. 9 and 10).

The following example illustrates the use of such a locus. Consider a servomotor with an inertia $J = 0.005$ in.-oz per radian per sec^2 and a speed-torque characteristic with $f = 0.125$ in.-oz per radian per sec. A 10-in.-oz dry friction load and a static error specification of $^1/_2$ radian sets the sensitivity K_o at 20 in.-oz per radian (see statement preceding eq. 2). Since $\theta/\theta_o = (\theta_i - \theta_o)/\theta_o = (\theta_i/\theta_o) - 1$, substituting in eq. 7 gives $\theta/\theta_o = -0.00025\omega^2 + j0.00625\omega$. This equation is plotted as curve A in Fig. 5. This locus approaches within 0.4 unit of the -1 point, showing a resonance value of $M = 2.5$.

To remove this resonance, a *lead network* such as is shown in Fig. 6 may be placed in the amplifier. Curve B of Fig. 5 plots its inverse transfer function (e_1/e_2), and, when the corresponding vectors of curves A and B are multiplied, curve C results. This circuit is called a lead network because its transfer function produces leading phase angles. It approximates a true derivative for low frequencies. In this example, the introduction of the lead network has reduced M_{max} to about 1.14 at $\omega = 50$ radians per sec, the minimum distance from the -1 point to the curve C being 0.88 unit.

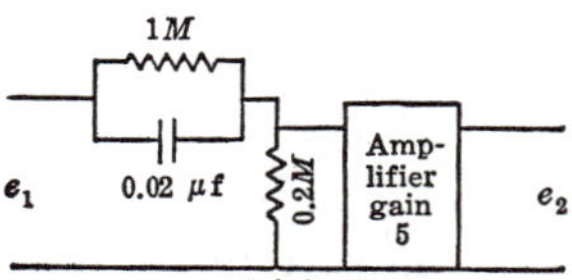

$$\frac{e_2}{e_1} = \frac{1 + j0.02\omega}{1 + j0.004\omega}$$

FIG. 6. Lead network.

Thus far only lead networks in the forward loop have been used to deform the locus away from the -1 point. As an alternative, feedback signals contributing to the θ_f term can do the same thing. For example, if the output speed were fed back so that $\theta_f/\theta_o = K_f j\omega$, the locus would be raised vertically and away from the -1 point. This is one form of an *output feedback or control*.

Instability is easily spotted on these diagrams; it occurs if the locus moves past the -1 point in the clockwise direction as illustrated by curve E, Fig. 5. This transfer function is the result of adding to curve A a motor delay due to field inductance with a 0.02-second time constant, having an inverse transfer function $1 + j0.02\omega$ as shown by curve D. This unstable system can be made stable by decreasing K_o and thereby expanding the locus to curve F. The motor delays often determine the ultimate limit to the possible increase of K_o.

Such stability criteria, attributed to Nyquist, may be proved and rigorously stated by use of the Cauchy theorem (Refs. 13, 14, 15, and 16) concerning the number of poles and zeros of an analytic function within a contour.

These same concepts are used with other types of plots which have advantages under some circumstances. With these other methods θ_f is considered only as a local feedback which modifies θ_o/θ, and the overall feedback as just the return of θ_o to the differential device. The basic equation uses transfer functions only:

$$\frac{\theta_o}{\theta_i} = \frac{\dfrac{\theta_o}{\theta}}{1 + \dfrac{\theta_o}{\theta}} \tag{11}$$

One method plots θ_o/θ in the vector locus form and makes use of loci of constant M and ϕ on the same coordinates to determine θ_o/θ_i. This method is generally used only when $\theta_f = 0$, and its only advantage is that it deals with transfer functions rather than their inverses.

A second method (Refs. 14, 15, 17) plots $20 \log_{10} | \theta_o/\theta |$, or "gain" in decibels, against $\log \omega$ on rectangular coordinates. Bode has shown that the phase and magnitude curves are related so that stability can be assured by giving the magnitude alone a proper shape. A logarithmic scale is used so that component transfer functions in series may be added, rather than multiplied, thereby simplifying calculations.

Instead of determining the peak value for M, it is sometimes easier to specify merely the phase and gain margins for θ_o/θ. The phase margin is the angle of θ_o/θ when $| \theta_o/\theta | = 1$, it being good practice to have this between 40 and 60 degrees. The gain margin is the magnitude of θ_o/θ when the angle is 180 degrees, it being good practice to make this over 2 or 3. These margins determine how much parameters may change before instability results, but they usually limit the peak value of M at the same time.

DESIGN PROCEDURE. In the actual design of a servomechanism, it is necessary first to set down specifications concerning static error, speed lag, errors due to accelerations, load errors, maximum velocity of output, maximum acceleration of output, load forces, and inertia.

On the basis of these specifications the servomotor can be chosen which satisfies the torque, speed, and acceleration requirements. In addition, it must have sufficiently small response delays to allow the design of a nonoscillatory system.

The final step is the design of amplifiers and stabilizing and feedback circuits to give the desired frequency response and still meet the error specifications.

REFERENCES

1. H. Lauer, R. Lesnick, and L. E. Matson, *Servomechanism Fundamentals*, McGraw-Hill Book Co., New York, 1947.
2. R. Goertz, Synchro controls for motors and servos, *Electronic Ind.*, Vol. 4, No. 9, pp. 78–83, 180, 184, 188, 192, 194, Sept. 1945.
3. H. L. Hazen, Theory of servomechanisms, *J. Franklin Inst.*, Vol. 218, pp. 279–330, Sept. 1934.
4. G. S. Brown and A. C. Hall, Dynamic behaviour and design of servomechanisms, *Trans. Am. Soc. Mech. Engrs.*, Vol. 68, pp. 503–504, July 1946.
5. H. Harris, The analysis and design of servomechanisms, *Trans. Am. Soc. Mech. Engrs.*, Vol. 69, pp. 267–280, April 1947.
6. M. F. Gardner and J. L. Barnes, *Transients in Linear Systems*, John Wiley and Sons, New York, 1942.
7. G. O. McCann, S. W. Herwald, and H. S. Kirschbaum, Electrical analogy methods applied to servomechanism problems, *Elec. Eng., AIEE Trans.* Section, Vol. 65, p. 91, Feb. 1946.
8. J. R. Ragazzini, R. H. Randall, and F. A. Russell, Analysis of problems in dynamics by electronic circuits, *Proc. Inst. Radio Engrs. and Waves and Electrons*, Vol. 35, p. 444, May 1947.
9. H. Harris, The frequency response of automatic control systems, *Elec. Eng., AIEE Trans.* Section, Vol. 65, Sept. 1946.
10. H. Harris, A comparison of two basic servomechanism types, *AIEE Paper* No. 47-12 (1947 Winter Convention).
11. A. C. Hall, Application of circuit theory to the design of servomechanisms, *J. Franklin Inst.*, Vol. 242, Oct. 1946.
12. H. T. Marcy, Parallel circuits in servomechanisms, *Elec. Eng., AIEE Trans.* Section, Vol. 65, pp. 521–529, Sept. 1946.
13. H. Nyquist, Regeneration theory, *Bell System Tech. J.*, Jan. 1932.
14. L. A. MacColl, *Fundamental Theory of Servomechanisms*, D. Van Nostrand Co., New York, 1945.
15. H. W. Bode, *Network Analysis and Feedback Amplifier Design*, D. Van Nostrand Co., New York, 1945.
16. W. R. Ahrendt and J. F. Taplin, *Automatic Regulation*, Ahrendt and Taplin, P.O. Box 4673, Washington, D. C., 1947.
17. R. E. Graham, Linear servo theory, *Bell System Tech. J.*, Vol. 25, pp. 616–651, Oct. 1946.

INDUSTRIAL ELECTRONIC CONTROL *

By L. W. HERCHENROEDER

Electronics is the science of conduction of electricity through a gas or vacuum. Industrial electronic equipment finds its place in those applications where it can do a better job than other equipment and in the many applications where no other equipment is available for doing the job. Electronic equipment, long the foundation of the communications industry, has become an accepted part of the large family of general industrial equipment.

11. ELECTRONIC TUBES

Tubes are the heart of electronic control. An electronic tube consists of a glass or metal envelope containing a gas or vacuum and having two or more elements. A two-element tube has a *cathode* and an *anode.* The cathode provides a source of electrons and, when the anode is at a positive potential with respect to the cathode, the electrons are attracted to the anode. Since the cathode is capable of releasing electrons into the space within the tube, a flow of electrons from the cathode to the anode results when the anode is positive. If the cathode is made positive and the anode negative, there is no continuous flow of electrons from the anode to the cathode because the anode does not supply electrons to the space in the tube. Thus the electronic tube is a rectifier because it permits flow of electrons in one direction only. The conventional direction of current flow is from positive to negative, which is opposite to the actual flow of electrons. When a tube is conducting, the current flows, in the conventional sense, from the anode to the cathode.

The cathode is caused to emit electrons by various means. The most important is *thermionic emission,* in which electrons are aided in their escape from the cathode by heating the cathode material. In some tubes the filament is used as the cathode; in others the cathode is indirectly heated by the filament. Some tubes have cold cathodes, and electrons are forced to leave the cathode by a high voltage gradient. Phototubes utilize photoelectric emission. Photosensitive materials on the cathode emit electrons upon exposure to light.

Tubes may be broadly classified as *vacuum* tubes and *gas* tubes. Vacuum tubes are characterized by low current-carrying capacity and high internal voltage drop while gas tubes are characterized by high current-carrying capacity and low internal voltage drop. The vacuum tube is capable of high amplification with a fine degree of control while the gas tube is relatively coarse in its action. Thus the vacuum tube finds use as a regulating or controlling element while the gas-filled tube does the heavy work. Vacuum and gas tubes take many forms. The tubes most widely used in industrial control are kenotrons, pliotrons, phanotrons, thyratrons, ignitrons, and phototubes.

KENOTRONS, often called **diodes,** are two-element vacuum tubes with hot cathodes, used mainly as rectifiers to supply relatively small currents. They are capable of supplying as high as 100,000 volts for electrostatic dust precipitation, x-ray, and radio.

Kenotrons have a high internal voltage drop determined to a large extent by the space charge, which is caused by the cloud of electrons in the space between the cathode and anode. The charge due to the free electrons alters the potential distribution between the anode and cathode so that the potential near the cathode may actually be more negative than the cathode. Electrons being emitted are then repelled, and those emitted at the lower velocities will be forced to return to the cathode. Under these conditions the current is said to be space-charge-limited.

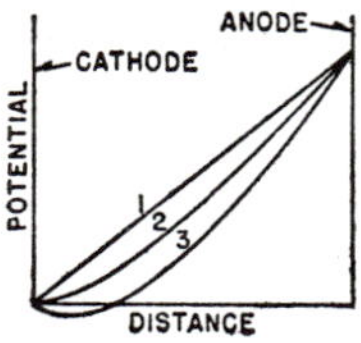

FIG. 1. Potential distribution in a kenotron.

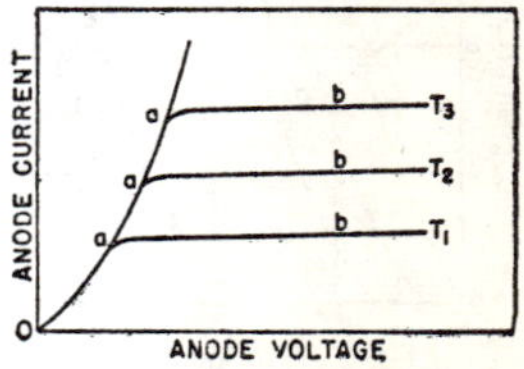

FIG. 2. Typical kenotron characteristic for several filament temperatures.

Figure 1 shows the change in potential distribution in the tube as the cathode emission is changed. On curve 1 the cathode is cold, there is no emission of electrons, and the potential distribution is a straight line. On curve 2 the cathode is heated, some emission is obtained, and the effect of space charge lowering the potential near the cathode is shown. Emission is increased further for curve 3 and results in a negative potential near the cathode. Electrons emitted from the cathode are at first repelled by the field and only those emitted with more than a

* In preparing this chapter the author has consulted the first two references at the end of the chapter (p. 17-37) for much of the basic information presented.

certain velocity can pass through the negative field and be accelerated toward the anode. Electrons emitted with less velocity are repelled and return to the cathode.

Figure 2 shows a typical kenotron voltage-current characteristic for several filament temperatures. From O to a on the various curves, the anode current is limited by the space-charge effect. From a to b the anode current is limited by the number of electrons the cathode can emit. The anode current varies approximately as the 1.5 power of the anode voltage over the range where the space-charge effect predominates.

GAS-FILLED TUBES have considerably different characteristics. Electrons passing from the cathode to the anode collide with gas particles and tear away electrons, thus making the gas particles positively charged ions. The positive ions are attracted toward the cathode; however, since they have much greater mass than electrons, their speed is much slower. While traveling toward the cathode, the positive ions neutralize the charge of many of the fast-moving electrons, thus neutralizing the negative space charge.

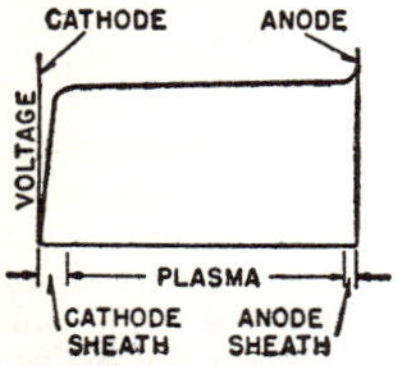

Fig. 3. Phanotron voltage distribution.

Figure 3 shows that very little voltage drop occurs in the plasma, which fills most of the space between cathode and anode, or in the anode sheath, a thin layer near the anode. Most of the voltage drop occurs in the cathode sheath, a thin layer near the cathode. The drop across a gas-filled tube is of the order of 15 volts and is practically constant over the normal range of load current.

PHANOTRONS are two-element gas-filled tubes used mainly as rectifiers for voltages up to approximately 20,000 and medium currents. They have hot cathodes and are usually filled with mercury vapor. In starting up phanotron rectifiers, the application of anode voltage must be delayed until the cathode comes up to normal temperature in order to prevent destruction of the cathode by positive-ion bombardment. Rectigons and tungars are low-voltage gas-filled tubes used for charging batteries. They have hot cathodes and are usually filled with argon gas.

PLIOTRONS are vacuum tubes with hot cathodes and having three or more elements. There are many varieties of pliotrons; those having three elements are called triodes; those with four elements, tetrodes; and those with five elements, pentodes. Amplification, the characteristic that has made the pliotron so useful, can best be described by considering the triode.

TRIODES are merely kenotrons with the addition of a third element called the grid. The grid is a conductor inserted between the cathode and anode. Usually it is a helix of wire concentric with the cathode. A given change in the input to the grid of a triode will result in a much greater change in the anode circuit, which results in amplification.

The grid controls the anode current by its effect on the space charge. The anode current of a triode is limited by the space-charge effect. Enough anode current will flow to produce a space charge that will result in a slight negative potential near the cathode similar to that shown for a kenotron in Fig. 1, curve 3. By varying the grid voltage, the potential distribution, neglecting the space charge, can be varied, thus varying the anode current required to produce the space charge necessary to give a slight negative potential near the cathode.

The grid is normally operated at a negative potential. Electrons, therefore, are not attracted to the grid so that grid current and input power to the tube are negligible.

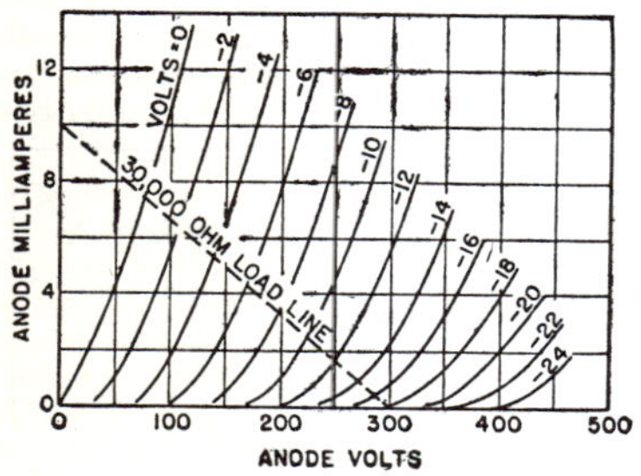

Fig. 4. Anode characteristic for a triode with 30,000-ohm load line.

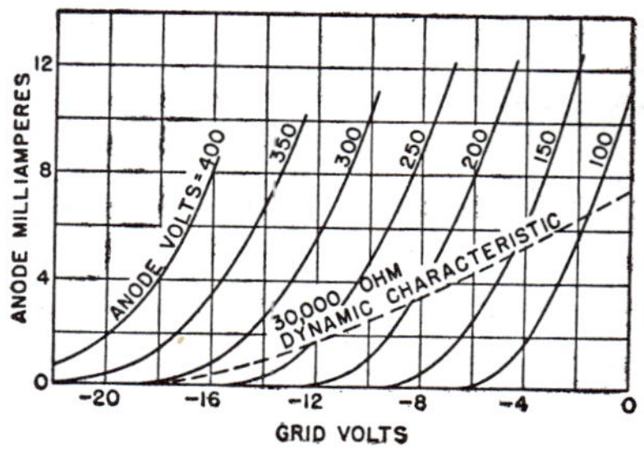

Fig. 5. Transfer characteristic for a triode with 30,000-ohm dynamic transfer characteristic.

Figure 4 shows the *anode characteristic* of a typical triode in which anode current is plotted against anode voltage for various values of grid voltage. The *transfer characteristic*, in which anode current is plotted against grid voltage for various values of anode voltage, is shown in Fig. 5. The data on the two sets of curves are identical; they are merely plotted in different ways.

Several coefficients or "constants" descriptive of the performance of a tube may be obtained from Figs. 4 and 5. The *amplification factor* μ is the rate of change of anode voltage for a change in grid voltage, with the anode current being constant. For example, in Fig. 4, at 4-ma anode current and −6-volt grid voltage, the anode voltage is 165. At 4 ma and −4-volt grid voltage, the anode voltage is 125. Therefore,

$$\text{Amplification factor} = \mu = \frac{\text{Change in anode voltage}}{\text{Change in grid voltage}} = \frac{165 - 125}{6 - 4} = 20$$

Anode resistance is the rate of change of anode voltage with respect to anode current, the grid voltage being constant. In Fig. 4 for a grid voltage of −6 volt and 6-ma anode current, the anode voltage is 185. At 4 ma the anode voltage is 165. Therefore,

$$\text{Anode resistance} = R_A = \frac{\text{Change in anode voltage}}{\text{Change in anode current}} = \frac{185 - 165}{0.006 - 0.004} = 10{,}000 \text{ ohms}$$

Mutual conductance is the rate of change of anode current with respect to grid voltage, the anode voltage being constant. In Fig. 4 for anode voltage of 150 and −4 grid volts, the anode current is 6.6 ma. For −6 volts the anode current is 2.6 ma.

$$\text{Mutual conductance} = G_M = \frac{\text{Change in anode current}}{\text{Change in grid voltage}} = \frac{0.0066 - 0.0026}{6 - 4} = 0.002 \text{ mho}$$

The above calculations are only approximate as small increments of voltage and current should be taken for accurate results. To be theoretically correct, the increments should be infinitesimal in magnitude. Amplification factor, anode resistance, and mutual conductance have the following relationship:

$$\mu = R_A G_M$$

Thus, if any two are known, the third may be found. Figure 6 shows a plot of the three functions against anode current. It will be noted that the amplification factor is practically constant and that the anode resistance and mutual conductance vary considerably.

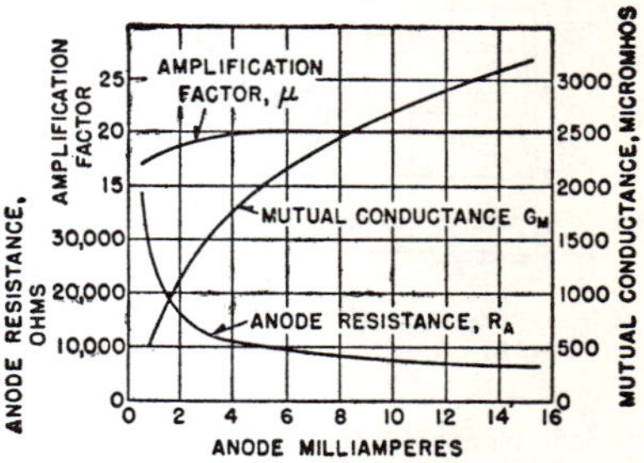

FIG. 6. Tube parameters for a triode.

The curves in Fig. 5 are plotted for constant values of anode voltage. In practice a load resistance will be connected in series with the anode and the constant voltage supply. When the grid voltage is changed, the anode current is changed, and the anode voltage must change due to the change in drop across the load resistor.

To take account of the change in anode voltage, the *dynamic transfer characteristic* should be constructed. Assume that the anode supply is 300 volts and the load resistance 30,000 ohms. If the anode current is zero, the entire 300 volts will appear at the anode. If all the voltage is across the resistor, the anode current will be 10 ma and the anode voltage zero. Plot these two points on the anode characteristic and join them with a straight line, as shown by the dotted line in Fig. 4. It is known as the *load line*. Pick off the intersections of the load line with the anode characteristic curves and plot them on the transfer characteristic as shown by the dotted line in Fig. 5. The dotted line of Fig. 5 is called the dynamic transfer characteristic. It applies only for a load resistance of 30,000 ohms and an anode supply of 300 volts.

TETRODES are four-element vacuum tubes having a cathode, control grid, screen grid, and anode. They are the equivalent of a triode with an extra grid, called a screen grid, located between the control grid and the anode. The screen grid improves the performance of the tube by reducing the effect of the change in anode voltage on the cathode space charge and by shielding the grid from the anode, thus reducing the capacitance between the grid and anode. A positive potential is applied to the screen grid, and, since it is nearer the cathode, the screen grid has much greater control over the potential distribution near the cathode than does the anode.

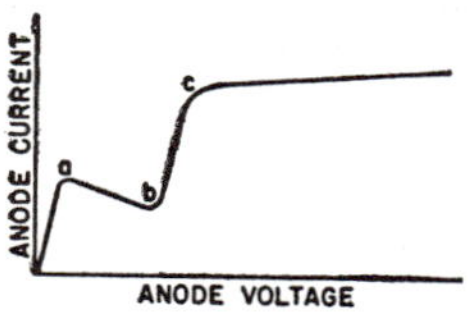

FIG. 7. Anode characteristic for a tetrode.

Figure 7 shows the anode characteristic of a typical tetrode and shows that the anode current is practically independent of anode voltage in the normal operating range to the right of point *c*. The peculiar characteristic to the left of *c* is due to the anode being less positive than the screen grid. When electrons arrive at the anode with greater than a certain velocity, electrons are emitted from the anode. This phenomenon, called *secondary emission*, occurs in all tubes. In a triode the anode is the only positive element, and the electrons due to secondary emission are attracted back to the anode. When the screen grid in a tetrode is more positive than the anode these electrons are attracted to the screen grid. The result is distortion or unstable operation when the anode voltage is less than the screen-grid voltage.

Tetrodes provide much higher amplification than triodes; however, they have become practically obsolete because of the superior characteristics of pentodes.

PENTODES are five-element vacuum tubes containing cathode, anode, and three grids. Essentially the pentode is a tetrode with an additional grid, called the suppressor grid, between the screen grid and the anode. The purpose of the suppressor grid is to prevent the undesirable operating characteristics of the tetrode for anode voltages less than the screen-grid voltage. The suppressor grid is usually connected to the cathode and, therefore, produces a low potential near the anode. Electrons released from the anode by secondary emission find themselves in a space in which the anode is the most positive element, and they return to the anode. The anode voltage can, therefore, swing below the screen-grid voltage without obtaining adverse tube characteristics. Figure 8, which is a typical pentode anode characteristic, shows that the anode current is practically independent of anode voltage.

VARIABLE-MU TUBES have high amplification for small negative-grid voltages and low amplification for high negative-grid voltages. Figure 9 is the anode characteristic of a

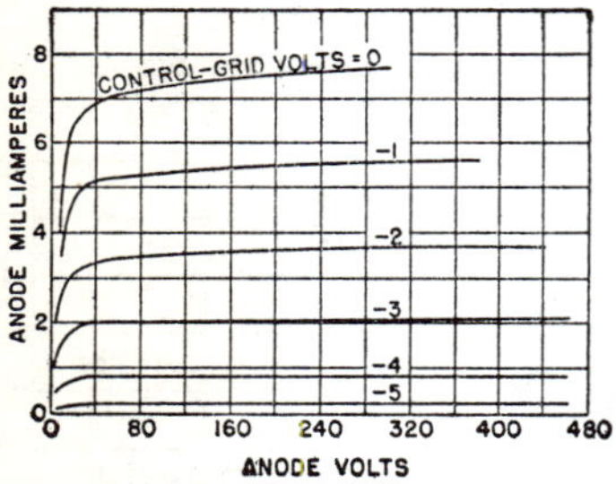

FIG. 8. Anode characteristic for a conventional pentode.

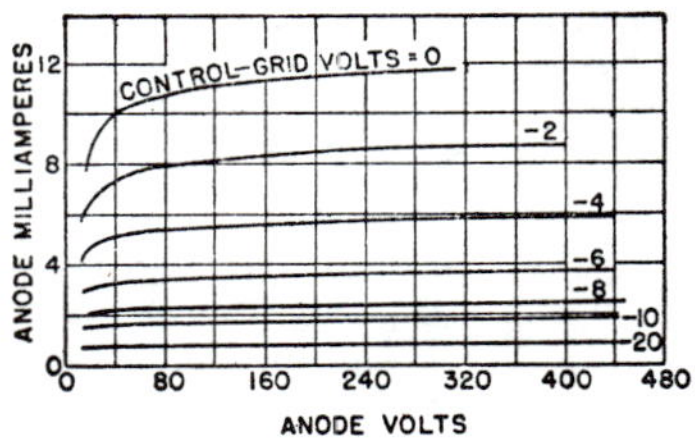

FIG. 9. Anode characteristic for a variable-mu or remote cut-off pentode.

typical variable-mu pentode. Variable-mu tubes are also called *supercontrol* or *remote-cut-off* tubes as contrasted to *sharp-cut-off* tubes illustrated by Fig. 8. Cut-off refers to the value of grid voltage that reduces the anode current to zero.

BEAM POWER TUBES are tetrodes having characteristics similar to pentodes. Electrodes at cathode potential focus the electron stream in a beam directed at the anode. Near the anode a zone is created having a concentration of electrons sufficient to create a potential less positive than the anode. The zone of concentration of electrons has properties similar to the suppressor grid in a pentode. Thus electrons due to secondary emission of the anode are attracted back to the anode. Beam tubes are ordinarily used as power-output tubes, and pentodes are used as voltage amplifiers and for low power output.

THYRATRONS are hot-cathode gas-filled tubes having one or more grids. Thus a three-element thyratron is a phanotron with the addition of a control grid. The action of the control grid in a thyratron is quite different from that of the grid in a pliotron. The control grid of a pliotron has control of the anode current at all times. The control grid of a thyratron has only a triggering action. It can prevent the tube from conducting but once it initiates an anode current, the grid loses control. The anode current can then be stopped only by reducing the anode voltage below the arc drop in the tube, either by opening the anode circuit with a switch or by otherwise reducing the anode voltage to a value below the arc drop. In practice, thyratrons are usually operated with an alternating anode voltage. The anode voltage becomes zero every half cycle, at which time the grid regains control of the tube. For most industrial applications the grid, therefore, gains control of the anode current often enough to give the equivalent of continuous control.

The action of the control grid can be explained as follows. As long as the tube is not conducting, the control grid can be maintained at a potential that will prevent flow of electrons from the cathode to the anode. If the grid voltage is made less negative or more positive, a point is reached (called the firing point) where sufficient electrons are emitted from the cathode to cause ionization of the gas in the tube. The positive ions thus formed permit the flow of large currents through the tube with a low voltage drop, as was described for the phanotron. Positive ions form a sheath around the grid, giving it a positive potential and rendering it ineffective in controlling the anode current. Figure 10 shows the firing characteristics of a mercury vapor thyratron. The shaded area shows the variation in firing point that may be expected because of temperature and other variations that occur over the life of the tube.

The addition of a shield grid to the three-element thyratron improves the operating characteristics in several ways. The shield grid almost surrounds and isolates the cathode, anode, and control grid. The capacity between the anode and control grid is minimized, thus reducing the chance of loss of control due to transients in the anode circuit. The control-grid current is reduced, and the firing characteristics can be modified by adjusting the potential on the shield grid. Two signals can be used to control the tube—one operating the control grid and the other the shield grid.

When thyratrons are used in a-c circuits, the average anode current can be controlled by controlling the angle at which the tube is fired. Figure 11 illustrates one method by which this can be accomplished. An alternating voltage is applied to the control grid, and some means of shifting the phase of the grid voltage with respect to the anode voltage is provided. The tube will fire at the point of intersection of the grid voltage with the firing voltage characteristic. Figure 11*A* shows the grid voltage in phase with the anode voltage, resulting in flow of anode current over most of the positive half cycle. Figure 11*B* shows the grid voltage shifted so that the firing of the tube is delayed almost 90 degrees, and anode current flows over slightly more than half the positive half cycle of the anode voltage. There are several other means of controlling the firing angle. One involves permanently shifting the grid voltage 90 degrees and adding a d-c component to shift the a-c wave up and down to vary the point of intersection with the firing voltage characteristic. Another method involves applying a sharp pulse of voltage to the control grid every half cycle and providing means of shifting the phase of the pulse to control the firing point.

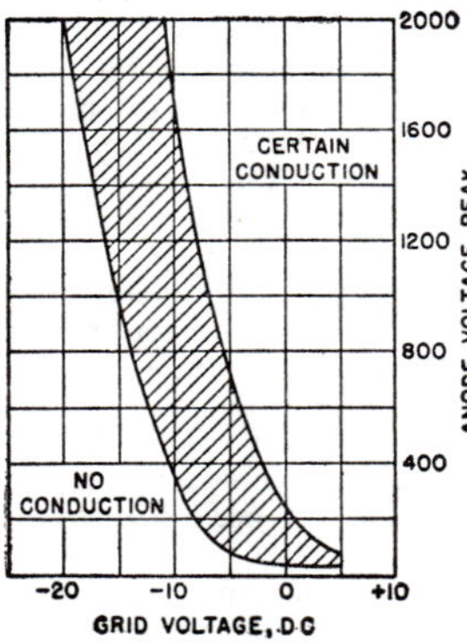

FIG. 10. Firing characteristic for a mercury vapor thyratron.

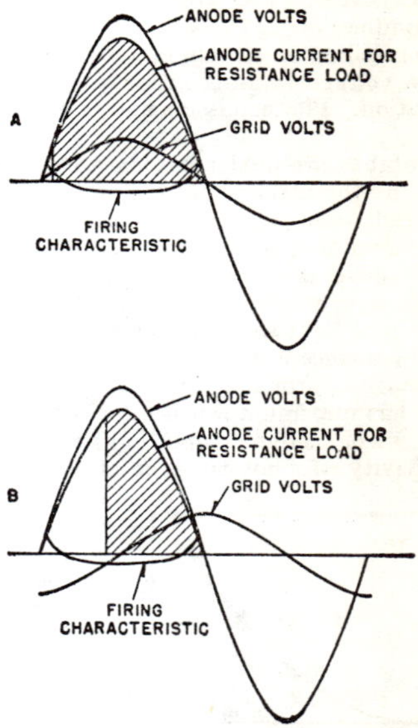

FIG. 11. Control of firing of a thyratron by shifting the phase angle of the grid voltage.

IGNITRONS are mercury pool cathode tubes containing an element called an ignitor. The function of the ignitor is similar to that of the grid in a thyratron, i.e., it controls the initiation of flow of current in the tube but has no control once the anode current is flowing. The cathode of an ignitron consists of a pool of mercury in the bottom of the tube. The tip of the ignitor electrode projects into the mercury pool. Passing current through the ignitor causes an arc to form between the ingitor and mercury pool. If the anode voltage is sufficiently positive, the arc will spread to the anode and the tube will continue to conduct as long as the anode voltage remains sufficiently positive to maintain the arc. Thus, by supplying a pulse of current to the ignitor at the proper time, the tube can be caused to conduct over the desired portion of a cycle in a manner similar to the control of conduction of a thyratron by means of the control grid.

Ignitrons are capable of carrying heavy currents because of the mercury pool cathode, which is capable of supplying an almost unlimited quantity of electrons once the arc has been established. Sealed-off tubes are available with average current-carrying capacities up to 400 amperes. Larger ratings are available in pumped-type tubes which are used for large-capacity rectifier substations.

Ignitron tubes have metal walls and are provided with water jackets or cooling coils to remove the heat generated by the heavy current flowing in the tube. Tubes for rectifier or other continuous service are provided with a baffle and splash ring to protect the anode from mercury dispersed by agitation of the mercury pool. The larger tubes have an anode shield to shorten the de-ionization time of the tube.

Anode firing is a simple method of initiating conduction of an ignitron. A phanotron is connected between the anode and ignitor. When the anode voltage becomes sufficiently positive, the phanotron conducts current to the ignitor, causing the ignitron to conduct. The phanotron can be replaced by a thyratron, and, by controlling the control grid of the thyratron, conduction of the ignitron can be controlled.

More positive firing can be obtained by using a firing circuit separate from the anode circuit. One such circuit uses a rectifier to charge a capacitor, which discharges through a thyratron and the ignitor. By controlling the firing point of the thyratrons, the point on the voltage wave at which the ignitron fires can be controlled.

PHOTOELECTRIC DEVICES may be classified as photoemissive, photovoltaic, or photoconductive. Photoemissive devices operate on the principle that light or other radiation falling on certain materials causes electrons to be emitted. Photovoltaic devices produce voltage while the resistance of photoconductive devices changes upon exposure to radiation. Photoemissive and photovoltaic devices are used on most industrial applications.

Phototubes are diodes having a cathode with a photosensitive surface. The electrons emitted by the cathode depend on the cathode material and the wavelength and intensity of the radiation falling on the sensitized surface. Cathode photosensitive surfaces are made of a number of materials that are sensitive over different wavelength bands of radiant energy. Several surface designations are used as unofficial standards. The S_1 surface has maximum sensitivity in the red and infrared bands and is quite suitable for use with incandescent light sources. The S_2 surface has almost the same sensitivity characteristics as the S_1 surface and will probably be abandoned. Because the S_3 surface has a sensitivity range approximately that of the human eye it is desirable for colorimetry. The S_4 surface has maximum sensitivity in the ultraviolet and is very sensitive to sunlight and fluorescent or mercury-vapor light sources.

Sensitivity of phototubes is often expressed in microamperes of anode current per lumen of light. Since phototubes are sensitive to radiation outside the visible light band, a more scientific measure of sensitivity is the ratio of microamperes of anode current to microwatts of radiant energy for a definite wavelength band. For convenience, the sensitivity is often expressed as microamperes per lumen from a given light source.

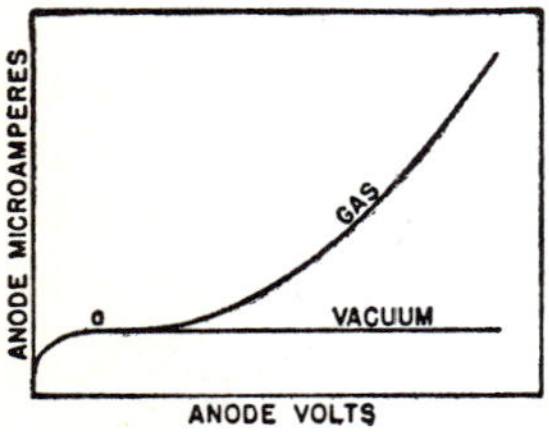

FIG. 12. Characteristics of vacuum and gas phototubes.

Phototubes may be of the vacuum or gas type. Figure 12 shows that, above *a*, the anode current remains constant with increasing anode voltage for the vacuum phototube. The anode current of the gas phototube increases with increasing anode voltage because of ionization of the gas. Because of its greater sensitivity the gas phototube is used on the majority of industrial applications.

In spite of its inferior sensitivity, the vacuum phototube has several advantages over the gas phototube that are important in some applications. The output of the vacuum phototube is much more stable than that of the gas phototube. The vacuum phototube output is a linear function of the illumination. The output of the gas phototube is not linear. The vacuum phototube has no measurable time lag in its operation. The gas phototube has a time lag in its response to illumination which results in a decrease of sensitivity for high-frequency illumination. The dynamic sensitivity of gas phototubes is substantially the same as the static sensitivity up to approximately 1000 cycles per sec; however, above 10,000 cycles per sec, the reduction in dynamic sensitivity may be serious. Gas phototubes are subject to destructive glow discharges if more than a certain anode voltage is used. Vacuum phototubes are not subject to this effect.

Conventional phototubes provide small output current even at high illumination. The multiplier phototube utilizes secondary emission to amplify greatly the weak photoelectric current. The tube consists of a photosensitive cathode, an anode, and a series of elements (usually nine) called dynodes. Electrons emitted by the cathode are focused on the first dynode, where an amplified number of electrons are emitted by secondary emission. These electrons are focused on the second dynode, where secondary emission also occurs, resulting in further amplification. The electrons are focused from dynode to dynode, secondary emission and amplification occurring at each step, and are finally collected by the anode greatly amplified in number over the electrons originally emitted from the cathode.

Photovoltaic cells differ from phototubes in that they generate a voltage when exposed to illumination. Their output is high enough to operate an indicating instrument. The principal application of photovoltaic cells is for photographic exposure meters and for illumination studies.

12. RECTIFIERS

Rectifiers are used to convert a-c power to d-c power. The amount of power converted may vary from less than a watt to thousands of kilowatts. Since the electronic tube allows flow of current in one direction only, it is much used in rectifying circuits.

Rectifier circuits vary considerably, depending on the amount of power required and the type of load. Table 1 shows several common rectifier circuits used for small- and medium-power output. The simplest is the single-phase half-wave rectifier consisting of a transformer and a single diode tube in series with the load. The tube permits current to flow during the half cycle when the anode is positive and prevents current flow during the half cycle when the anode is negative. The result is a pulsating unidirectional output voltage and current. The unidirectional current also flows in the transformer and tends to saturate the core. This must be taken into account in the design of the transformer and results in a more expensive transformer. The pulsating output voltage can be smoothed by use of a suitable filter; however, the single-phase half-wave rectifier is usually limited to small-power outputs.

The single-phase full-wave rectifier consists of two tubes and a transformer with a center-tapped secondary as shown in Table 1. When the anode of one tube is positive, the other anode is negative, and vice versa. One tube conducts current for one half of the cycle, and the other tube conducts the other half of the cycle. The current through the load is in the same direction regardless of which tube is conducting. The wave shape is

Table 1. Rectifier Circuits and Characteristics

	Single Phase			3 Phase Zigzag
	Half Wave	Full Wave	Full-wave Bridge	
Circuit	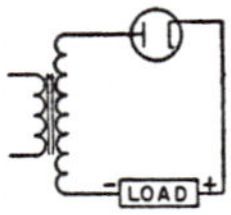	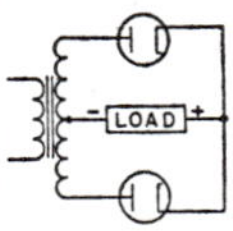	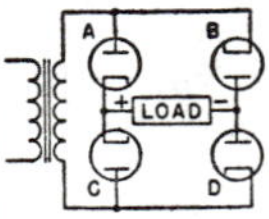	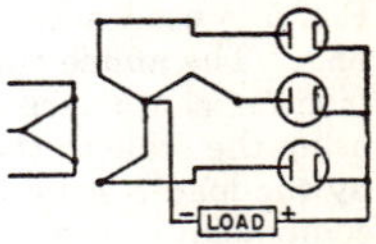
Output Voltage Wave Shape	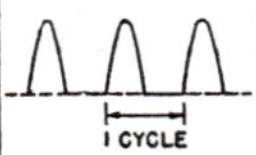	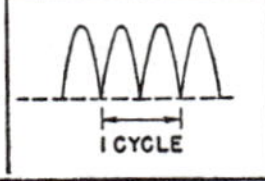	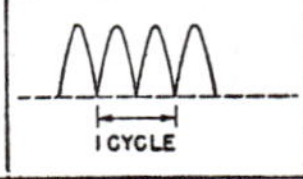	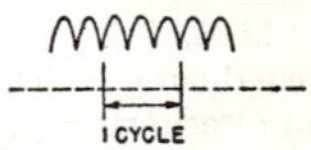
Average d-c volts	1.00	1.00	1.00	1.00
Peak d-c volts	3.14	1.57	1.57	1.21
Peak inverse volts	3.14	3.14	1.57	2.09
Anode current per tube				
Average	1.00	0.50	0.50	0.33
Peak (resistance load)	3.14	1.57	1.57	1.21
Peak (inductance load)		1.00	1.00	1.00
Transformer				
Rms sec volts/leg	2.22	1.11	1.11	0.854
Secondary volt-ampere	3.49	1.57	1.11	1.71
Primary volt-ampere	3.49	1.11	1.11	1.21
Ripple				
Major frequency	f	$2f$	$2f$	$3f$
Major voltage rms	1.11	0.472	0.472	0.177

much more continuous than that of the single-phase half-wave rectifier, and, since current is conducted for both halves of the cycle, saturation of the transformer does not occur. A variation of this circuit, replacing the two tubes by a single tube having two anodes and a common cathode, is widely used to supply d-c power in electronic equipment.

The single-phase full-wave bridge circuit consists of a transformer and four tubes as shown in Table 1. Tubes A and D conduct over half the cycle, and tubes B and C conduct over the other half of the cycle. The wave shape of the output is the same as the full-wave center-tap rectifier. The full-wave bridge has the advantage of using a conventional transformer; however, four tubes are required. The full-wave bridge circuit is used extensively with metallic rectifiers of the copper oxide or selenium type as there is no disadvantage in building them in four legs.

The three-phase half-wave zigzag circuit requires three tubes, each conducting over one-third the cycle. The output is more smooth than that of the single-phase rectifiers. The zigzag connection of the transformer secondary avoids saturation of the core. Many other rectifier circuits are used, depending on the application. Most rectifiers for large-power outputs use six- or twelve-phase circuits.

13. AMPLIFIERS

Amplification is probably the most useful property of the electronic tube. Input signals to electronic devices are often minute and must be amplified many times to perform the desired task. Figure 13 shows a typical single-stage amplifier circuit using a triode. An input voltage e_g in series with a grid bias voltage E_{cc} is applied between the grid and cathode. Anode supply voltage E_{bb} is connected between the anode and cathode in series with the load resistor R_L. The heater supply is not shown. Amplification of an alternating input voltage is illustrated in Fig. 14. The dynamic transfer characteristic of Fig. 5 is used with 30,000 ohms load resistance, 300 volts anode supply, and 8 volts grid bias. The anode current can be obtained by projecting the input voltage on the dynamic transfer characteristic. The output curve may also be used to give the load voltage by using the scale to the right, since the load voltage is equal to the anode current multiplied by the load resistance. The amplification may be obtained by comparing the alternating component of the load voltage with the input voltage. The total input change is 12 volts, compared to a total output change of 160 volts, resulting in amplification of 13.3.

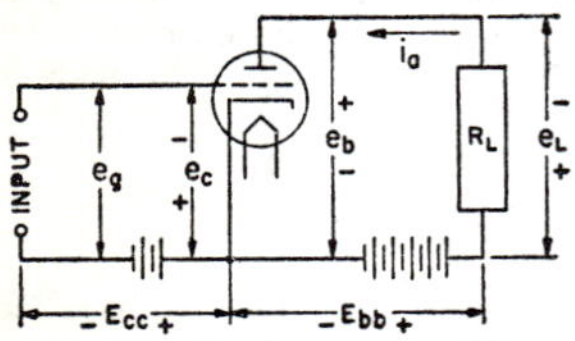

FIG. 13. Single stage amplifier.

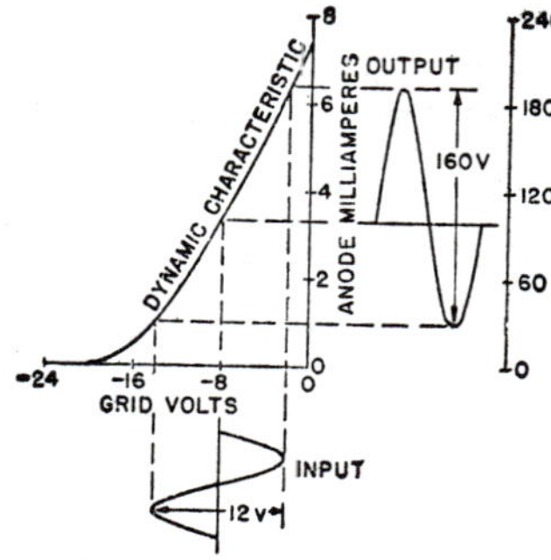

FIG. 14. Class A amplification.

AMPLIFIERS are divided into several classifications, depending on the portion of the cycle during which anode current is allowed to flow. Class A amplifiers have a grid bias and alternating grid voltage such that anode current flows over the entire cycle, as shown in Fig. 14. Class AB amplifiers have a grid bias and alternating grid voltage such that anode current flows for more than half but less than the entire cycle. Class B amplifiers have the grid biased to cut-off so that anode current flows over half of each cycle, as shown in Fig. 15. Class C amplifiers have the grid biased beyond cut-off so that anode current flows for less than half of each cycle. The subscript 1 may be used to indicate that no appreciable grid current flows during any part of the cycle, and the subscript 2 may be used to indicate that appreciable grid current flows during a portion of the cycle. Practically, the subscript 1 indicates that the grid is negative over the entire cycle, and 2 indicates that the grid is positive during a portion of the cycle.

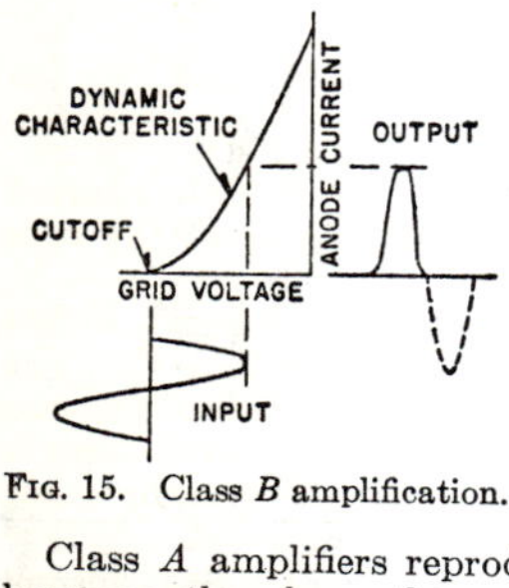

FIG. 15. Class B amplification.

Class A amplifiers reproduce the shape of the input signal with the least distortion; however, they have the poorest efficiency of the various classes. Class AB, B, and C amplifiers have, in the order named, higher efficiency and greater distortion of the input signal. For communication applications Class AB and Class B amplifiers are used in push-pull circuits to reduce distortion. Distortion is often of little importance in industrial applications, and Class B or Class C operation may be used. In many cases an alternating anode and grid bias voltage are used so that anode current flows only when the anode is positive and the grid voltage is above cut-off. Photoelectric circuits are often operated in this manner.

Amplifiers may also be classified as voltage or power amplifiers. The difference is a matter of design emphasis. Amplifiers usually have several stages in cascade, the last of which is a power amplifier and all the rest of which are voltage amplifiers. The circuit elements of a voltage amplifier are selected to amplify the voltage of the signal with little regard for the power output. Tubes with high amplification factor and low anode current capacity are used. The power amplifier is used to supply some device requiring a relatively large amount of power, such as a relay or the field of a generator or motor. Power-output tubes are designed to provide a certain power output with little regard being given to amplification of voltage.

Direct-coupled Amplifiers. Many circuits have been devised for coupling successive stages of cascaded amplifiers, a few of which will be described. The simplest is the direct-coupled amplifier, shown in Fig. 16. The input is amplified by tube T_1 and appears across resistor R_{L1}. The grid of T_2 is connected to the anode of T_1 through the battery E_{cc2}. The grid circuit of T_2 includes the anode supply E_{bb1} of T_1 and a high-voltage bias E_{cc2} is required to bring the grid voltage down to the normal operating range. The direct-coupled amplifier is extensively used for the amplification of d-c and low-frequency a-c signals; however, it has several disadvantages. The power supply is considerably more involved and expensive than it is for other types of coupling. A serious disadvantage is the tendency to drift with changing tube characteristics and changes in power-supply voltage. Many elaborate circuits have been devised to counteract the drift of the simple direct-coupled amplifier.*

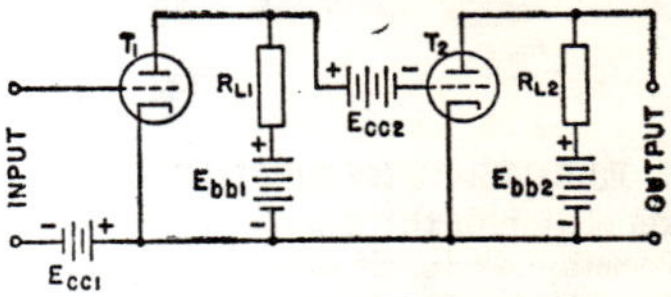

FIG. 16. Direct-coupled amplifier.

Resistance-capacitance-coupled amplifiers (often called resistance-coupled amplifiers) are widely used in electronic circuits for the amplification of alternating voltages. Figure 17 shows the connections between successive stages. The alternating input is amplified by tube T_1 and appears across resistor R_{L1} along with a d-c component as shown in Fig. 14. The alternating component of this voltage causes an alternating current to flow through the capacitor C and the grid resistor R_{g2}. An alternating voltage thus appears across R_{g2} and serves as the input to the tube T_2. The capacitor C blocks the d-c anode supply of T_1 from the grid of T_2, and a common anode supply can be used on all tubes. The amplification of the resistance-capacitance-coupled amplifier can be made fairly constant for intermediate frequencies (roughly 100 to 10,000 cycles per sec). At lower frequencies the

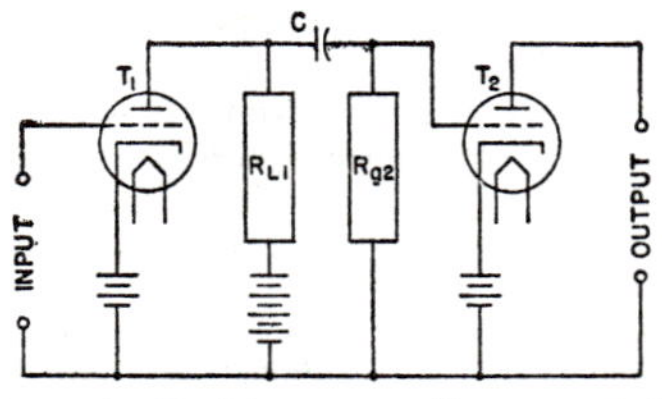

FIG. 17. Resistance-capacitance-coupled amplifier.

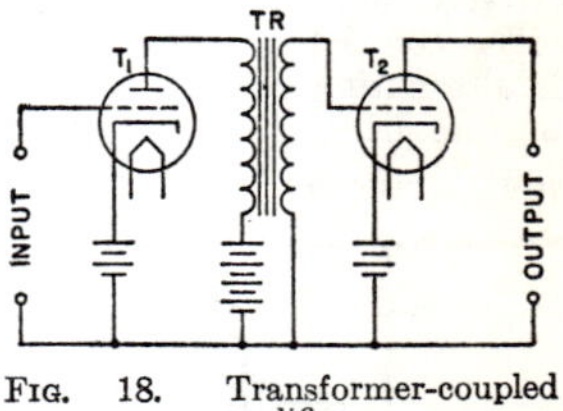

FIG. 18. Transformer-coupled amplifier.

amplification drops off because of the increased impedance of capacitor C, and at higher frequencies the amplification drops off because of the internal capacitance of the tube elements.

Transformer-coupled amplifiers may be used for amplifying alternating voltages. Figure 18 shows the primary of a transformer connected in the anode circuit of tube T_1. The secondary of the transformer is connected in the grid circuit of tube T_2. Alternating components of the output of tube T_1 are coupled to the grid of T_2, while the d-c component is not. Transformer coupling has the advantage of introducing a lower d-c drop in the anode circuit than resistance coupling. Also, the voltage amplification can be increased approximately in proportion to the turns ratio of the secondary to primary. The transformer becomes expensive, however, if good frequency response is required, and transformer coupling is little used for voltage amplifiers.

Push-pull amplifiers are often used in power output stages in order to obtain high output power with low distortion. A push-pull circuit with transformer input and output is

* Maurice Artzt, Survey of DC Amplifiers, *Electronics*, pp. 112–118, Aug. 1945.

shown in Fig. 19. The input transformer TR_1 has a center-tapped secondary and applies a grid voltage on T_1 that is 180 degrees out of phase with the grid voltage on T_2. Thus the grid of T_1 is at maximum voltage when the grid of T_2 is at minimum voltage. The primary of the output transformer TR_2 has a center tap to which the anode supply voltage is connected. The effect of the d-c component of the anode currents of the two tubes will cancel in the transformer since they flow in opposite directions in the primary. The fundamental frequency components of the anode currents will add since they are 180 degrees out of phase with respect to the center tap but flow in opposite directions in the winding. The second harmonic is most important from the standpoint of causing distortion and is canceled along with all other even harmonics generated in the tubes.

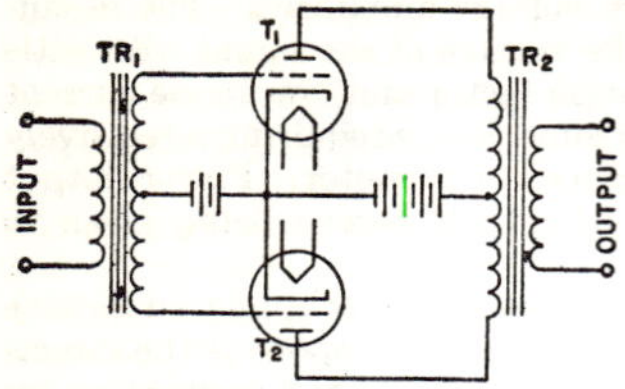

FIG. 19. Push-pull amplifier.

14. CONTROL APPLICATIONS

ELECTRONIC TIMING is widely used because of easy adjustment and accuracy of repetition of the timing interval. Most timing circuits are based on the time constant of a resistance-capacitance circuit, a typical example of which is shown in Fig. 20. The circuit is arranged to energize relay *CR* a certain time after switch *S* is closed. With *S* open, the grid of the thyratron functions as an anode in the circuit from line 1 through a portion of P_1, R_1, grid, cathode, and R_2 to line 2. The current flowing through this circuit causes a charge to build up on C_1 with the negative polarity on the grid side. When switch *S* is closed, the circuit to relay *CR* is completed through the thyratron, but the grid is negative because of the charge on C_1, and the tube does not fire. The grid voltage consists of an a-c component across a portion of P_1 and a d-c component across C_1. The capacitor C_1 discharges through R_1, and, when the voltage across C_1 becomes low enough, the a-c component of grid voltage will cause the tube to conduct and *CR* will pick up. The time interval may be adjusted by means of potentiometer P_1 which determines the voltage to which C_1 is charged when *S* is open.

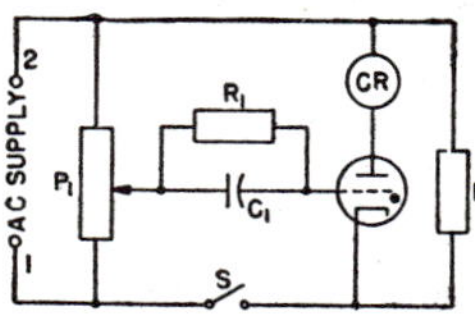

FIG. 20. Timing relay.

PHOTOELECTRIC RELAYS are available in several standard forms for general use in industry. Figure 21 shows the schematic diagram of a commercial photoelectric relay. It is relatively insensitive but is widely used when adequate illumination is available. The circuit is connected to energize the relay *CR* when the phototube is dark. The anode supply is alternating current for both the phototube and thyratron. The grid voltage controlling the firing of the thyratron is composed of the voltages across R_3 and R_2. An alternating voltage appears across R_3 that leads the thyratron anode voltage by approximately 90 degrees because of the phase shift circuit consisting of R_3 and capacitor C_2. The phototube conducts current only when its anode is positive. The capacitor C_1 is therefore charged by the phototube for half a cycle and discharges through resistor R_2 the other half cycle. The voltage across R_2 provides a negative bias on the thyratron grid, the value depending on the light falling on the phototube. When the light on the phototube is low, the voltage across R_2 is low, and the sum of the voltages across R_2 and R_3 is sufficient to cause the thyratron to conduct. When the light is bright, the voltage across R_2 is sufficiently high to reduce the grid voltage below that necessary to fire the thyratron. The coil of relay *CR* is in series with the anode circuit of the thyratron. Since current can flow only on the positive half cycle, the capacitor C_3 is used to prevent chatter of the relay. R_4 limits the charging current required of the thyratron. Potentiometer P_1 provides adjustment of the level of illumination at which the relay operates.

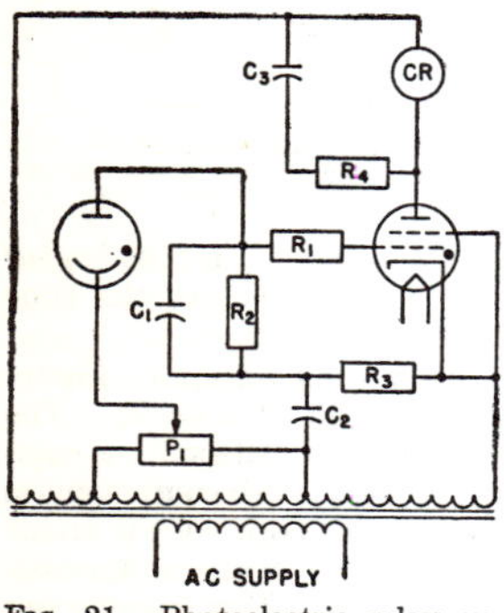

FIG. 21. Photoelectric relay using a-c anode supply.

The simple system described above, using an a-c anode supply, requires an illumination at the phototube of 40 to 50 foot-candles. Using a d-c anode supply for the phototube reduces the illumination requirement to 5 to 10 foot-candles. Adding a stage of d-c amplification between the phototube and thyratron reduces the requirements to 1 foot-candle or less. Making the system sensitive to rate of change of illumination (as described below) reduces the requirements to approximately 0.02 or 0.03 foot-candle.

Where background light is troublesome, a light source producing "chopped light" is desirable. A perforated disk is placed in front of the light source and driven by a synchronous motor, interrupting the light beam 540 times per sec. The amplifier circuit is made sensitive to the rate of change of light instead of the total change of light. Greater sensitivity is obtained and an a-c amplifier can be used. The "chopped light" system is of particular advantage where the light source is at a distance from the phototube or for outdoor applications where background illumination is difficult to eliminate.

The optical system is as much a part of the application as the photoelectric relay itself. The most common system is to direct the beam from the light source directly on the phototube. An example is automatic weight checking. The light source is placed on one side of the scale and the phototube on the other. When the weight is correct the scale pointer interrupts the light beam and operates the photoelectric relay.

A reflecting surface, either specular or diffuse, may be inserted in the light path. Diffuse reflection usually is preferred. Refraction of a liquid or solid is sometimes used. Radiation from a heated body is sometimes used to indicate temperatures above 1500 F.

There are several points to watch on photoelectric applications. The light source should be capable of giving sufficient illumination under the most unfavorable operating conditions, and the beam should cover as much of the phototube cathode as possible. The percentage and actual value of change in illumination on the phototube should be as high as possible with 50% change or more representing good practice. The optical system should be as short as possible, and extraneous phototube illumination should be kept to a minimum. Undue supply voltage fluctuation may require the use of a voltage-regulating transformer. Leads between the phototube and amplifier should be as short as possible to avoid picking up false signals. Preferably, the phototube should be mounted in the amplifier enclosure.

ELECTRONIC ADJUSTABLE-SPEED DRIVES are available in several standard forms. The electronic drives are comparable to the conventional adjustable voltage or Ward Leonard drives in which a motor-generator set provides adjustable voltage power to the armature of a d-c motor in order to obtain an adjustable-speed drive. The electronic motor control essentially substitutes a controlled thyratron rectifier for the motor-generator set; however, certain refinements such as speed regulation and current limit acceleration are included which are not as easily obtained with the motor-generator set drive.

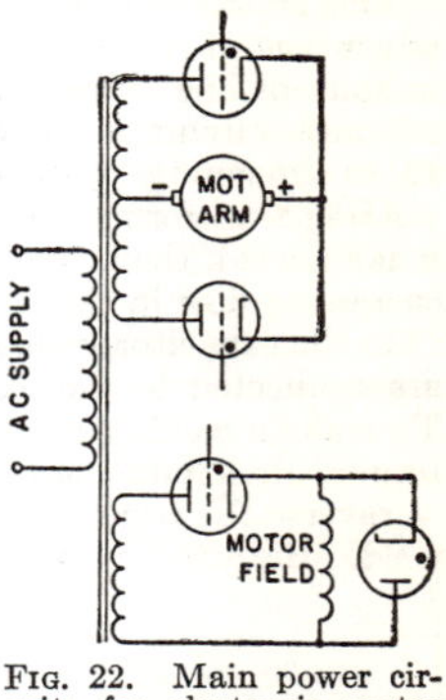

FIG. 22. Main power circuits for electronic motor control.

Figure 22 shows the main power circuits used in a typical electronic motor control. A single-phase full-wave rectifier supplies power for the armature of the d-c motor. By controlling the grids of the thyratrons, the voltage applied to the motor armature and, therefore, the speed of the motor, can be controlled. The motor field is supplied from a separate single-phase half-wave thyratron rectifier. A phanotron is used as a field-discharge tube which allows the current to flow through the highly inductive motor field during the half cycle that the thyratron is not conducting. Control of the thyratron grid results in control of the average motor-field current and provides adjustment of the motor speed by field weakening in addition to the speed adjustment by armature voltage control. When adjustment of speed by motor-field control is not desired, the thyratron may be replaced by a phanotron.

The speed of a d-c motor having constant field current is essentially proportional to the armature voltage minus the voltage drop due to the resistance of the armature circuit. To maintain constant speed with change in motor load, the output voltage of the armature rectifier is regulated to equal a constant value plus the resistance drop in the motor. For applications requiring higher accuracy, a tachometer generator is coupled to the d-c motor, and the voltage output of the armature rectifier is regulated to maintain constant voltage output of the tachometer generator which results in constant motor speed.

Current-limit control is provided to protect the motor and tubes from excessive current and to provide smooth acceleration. A measure of current supplied to the motor armature is obtained by a current transformer or by other means. A regulating circuit is used that is inactive as long as the motor current is less than the desired amount. However, when the motor current attempts to exceed this amount, the grid circuit of the thyratrons is phased back to reduce the voltage output of the rectifier sufficiently to reduce the motor current to a safe value.

The motor can be reversed by means of contactors in the motor armature circuit. Control features such as dynamic braking and inching are provided by conventional means. Electronic control permits the use of small speed-control potentiometers that can be mounted in an ordinary pushbutton station.

CONTACTORS. The ignitron contactor consists of two ignitrons connected "back to back" as shown in Fig. 23. Each ignitron conducts on alternate half cycles, and alternating current flows through the load. The ignitrons have the function of an ordinary contactor, conducting current to the load when the relay contacts *CR* in the ignitor circuits are closed and preventing current flow when the relay contacts are open. The ignitron contactor can be operated frequently while it is handling large currents without the mechanical wear and burning of contacts associated with frequent operation of conventional magnetic contactors. Ignitron contactors have been used for motor control and show great promise where motors must be started frequently.

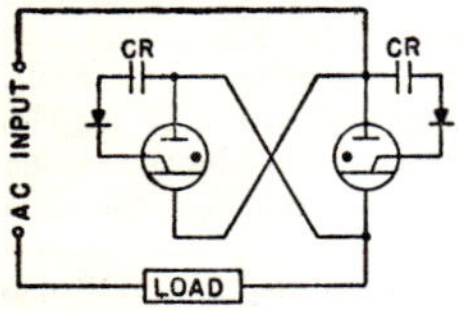

FIG. 23. Ignitron contactor.

Resistance welding has been the largest field for the ignitron contactor where it is used for controlling the primary current of a welding transformer. Resistance welding controls may be classified as nonsynchronous, synchronous precision, or energy storage.

Nonsynchronous control involves an ignitron contactor in the primary circuit of a welding transformer and an electronic timer to control the duration of the weld.

Synchronous precision control has several refinements which provide more uniform welds. Firing of the ignitron contactor is initiated at the natural power factor angle (the point on the voltage wave corresponding to zero current) of the welding transformer circuit. The polarity of the initial current wave is arranged to be opposite that of the last current wave of the preceding weld. Transient transformer currents are thus eliminated and each cycle of the current wave has a definite heating effect.

Heat control is obtained electronically by controlling the firing angle of the ignitrons. Firing the ignitrons at the natural power factor angle of the transformer circuit results in continuous flow of current and provides maximum heat in the weld. Increasing the delay of firing results in discontinuous current flow and reduces the heat in the weld.

Energy storage control is most frequently used on aluminum and its alloys and has the advantage of reducing current peaks taken from the line and providing consistent welds in spite of line voltage variations. Two systems are used, one storing the energy in the magnetic circuit of the welding transformer and the other storing the energy in capacitors. In the magnetic system an ignitron rectifier is connected to the primary of a special welding transformer and builds up flux in the transformer at a relatively slow rate. To make a weld, the primary circuit is opened and the decay of flux in the core causes a high current to flow in the transformer secondary.

In the capacitor system a rectifier is used to charge a bank of capacitors. The capacitors are connected to the welding transformer primary in series with an ignitron contactor. To make a weld, the capacitors are discharged through the primary of the welding transformer, thus inducing the welding current in the secondary. An auxiliary rectifier is used to reverse the transformer residual flux while the capacitors are being charged for the next weld. Both rectifiers are prevented from conducting during the weld cycle.

15. REGULATOR APPLICATIONS

REGULATION is a closed-cycle process as compared with control, which is an open-cycle process. The heating of a home with an ordinary furnace is an example of an open-cycle process. The amount of fuel and draft are set, and the furnace delivers enough heat to hold the house at the desired temperature under the existing conditions. If the outside temperature increases, the temperature in the house also increases because no information is sent back to the furnace to tell it that less heat is required to hold the desired temperature.

The addition of a thermostat and its associated equipment produces a closed cycle or regulating system. The thermostat measures the temperature in the room, compares it with the desired temperature, and causes the furnace to supply heat when the temperature is too low or shuts the furnace off when the temperature is too high. Variation in the outside temperature merely changes the percentage of time that the furnace is on.

Regulating systems comprise a reference quantity, a regulated quantity, an error detector that compares these two quantities and produces an error signal, and an amplifier that amplifies the error and causes an error corrector to operate on the regulated quantity so as to reduce the error. A block diagram of the elements of a regulating system is shown in Fig. 24. In the furnace regulator, the reference quantity is the desired temperature set on the thermostat, the regulated quantity is the temperature of the room, the error detector is the temperature-measuring bimetal in the thermostat, the amplifier consists of the thermostat contacts and relay and fuel-control device, and the error corrector is the furnace.

Regulating systems usually pose questions of stability and speed of response. Many regulating systems can be stabilized by reducing the amplification sufficiently; however, this remedy often defeats the purpose for which the regulator is desired. Antihunting and anticipating means are often used in regulating systems requiring high accuracy and speed of response. For a discussion of the general characteristics of regulating systems, see the chapter, "Automatic Regulators," in this Section.

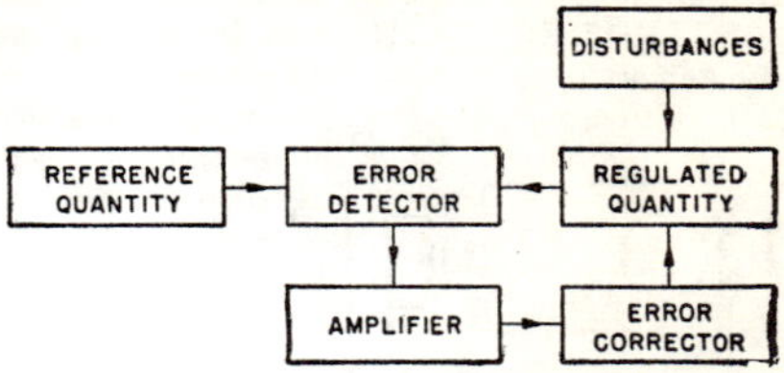

FIG. 24. Regulator block diagram.

Electronic equipment has several characteristics that make it highly desirable for use in regulating systems. High amplification can be obtained with negligible time delay. Because almost all other amplifiers have appreciable time delay, an electronic amplifier often simplifies the stability problem. Another advantage is the ease with which antihunting and anticipating circuits can be introduced and adjusted. The power input required by an electronic amplifier is very low. In many systems the power available for operating the regulator is so low that only electronic equipment can be used.

A TEMPERATURE REGULATOR for an electric furnace is shown in block diagram form in Fig. 25. The furnace is heated by resistor-type heaters connected to the power supply in series with a saturable reactor. The saturable reactor has a-c and d-c windings. When the current in the d-c winding is high, the reactor core is saturated, the reactance of the a-c winding is low, and maximum voltage appears across the furnace-heating elements, thus providing maximum heat input to the furnace. By decreasing the current in the d-c winding, the reactor core becomes less saturated, the reactance of the a-c winding increases, and less heat is delivered to the furnace.

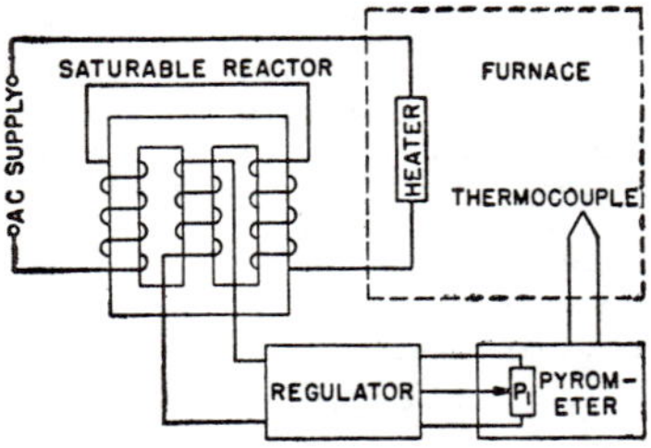

FIG. 25. Block diagram of electric furnace temperature regulator.

A thermocouple is located in the furnace, and the temperature is measured by a pyrometer having a motor-driven potentiometer P_1. The potentiometer arm, which is in the center when the temperature is correct, moves toward one end when the temperature is high and toward the other end when the temperature is low. The potentiometer controls the firing of thyratrons in the regulator, which supply direct current for the d-c winding of the saturable reactor.

ELECTRONIC VOLTAGE REGULATORS are often used for accurate regulation of generator voltage. Figure 26 illustrates the main elements of a voltage regulator for a d-c generator. A portion of the generator voltage appears across potentiometer P_1 and is balanced against a reference voltage provided by a battery. The difference between the two voltages is the error voltage which is amplified and applied across resistor R_1. The thyratron grid circuit includes the voltage across R_1 and an a-c voltage that lags the anode voltage by 90 degrees. Firing of the thyratron is controlled by the magnitude of the voltage across R_1. The thyratron supplies direct current to the generator field. If the generator voltage becomes too high, the voltage across R_1 increases, thus decreasing the portion of the cycle that the thyratron conducts. This decrease results in lower generator field current and, therefore, lower generator voltage. If the generator voltage becomes too low, the opposite sequence of events occurs and the regulator tends to increase the generator voltage to the desired value. The output voltage of the generator can be adjusted by means of potentiometer P_1. For simplicity the antihunting circuit is omitted in Fig. 26, and a single-phase half-wave thyratron rectifier is shown supplying the generator field.

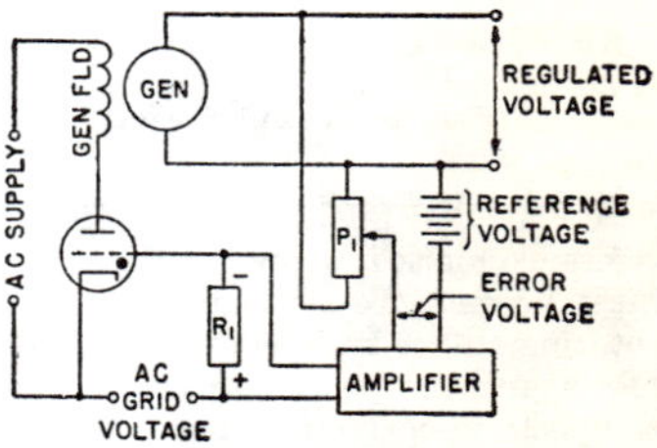

FIG. 26. Electronic voltage regulator for a d-c generator.

One commercial regulator of this type uses three thyratrons in a three-phase half-wave zigzag circuit to supply the generator field. In many cases the output of the rectifier is not great enough to supply the field of the main generator. An exciter is then used to supply the generator field, and the regulator supplies the exciter field. An a-c generator can also be used with this type of regulator by employing a voltage-indicating unit which rectifies the alternating voltage, smoothes it, and supplies a direct voltage input to the regulator.

SPEED REGULATION can be accomplished in a similar manner, as shown in Fig. 27. A d-c generator and motor are connected as an adjustable voltage or Ward Leonard drive in which the speed of the d-c motor is controlled by the voltage output of the generator. A d-c tachometer generator is coupled to the d-c motor. The fields of the motor and tachometer generator are supplied from a constant-voltage source. The speed of the d-c motor is then roughly proportional to the generator voltage, and the tachometer generator voltage is proportional to the speed of the motor. The output of the tachometer is fed into the regulator which supplies the field of the d-c generator. The regulator adjusts the generator voltage so that the speed of the motor is such that the tachometer voltage is held constant. Constant tachometer voltage results in constant speed of the d-c motor. The voltage regulator in Fig. 26 and the speed regulator in Fig. 27 are similar in that they both regulate for constant voltage output of a generator. However, the speed regulator presents a more difficult stability problem because of the added time delay introduced by the mechanical inertia of the d-c motor and load.

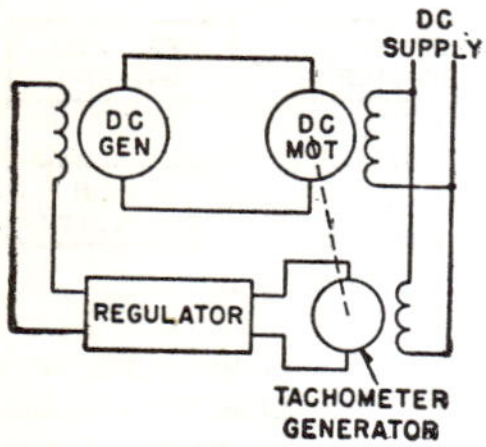

Fig. 27. Electronic speed regulator.

POSITION REGULATION is illustrated in Fig. 28. The d-c motor is required to follow the motion of the shaft represented by the handcrank. The output synchro is connected to the d-c motor and gives an indication of its position which, when compared with the input synchro, results in an error voltage that is proportional to the error for small errors. The error voltage is fed into the amplifier, which has a push-pull type output stage that supplies two bucking fields on the d-c generator. For zero error, the currents in the two fields are equal, the generator voltage is zero, and the motor does not run. If there is an error, the current through one of the fields will be greater than the other, resulting in a net excitation on the generator which produces a voltage, making the motor run in a direction to reduce the error.

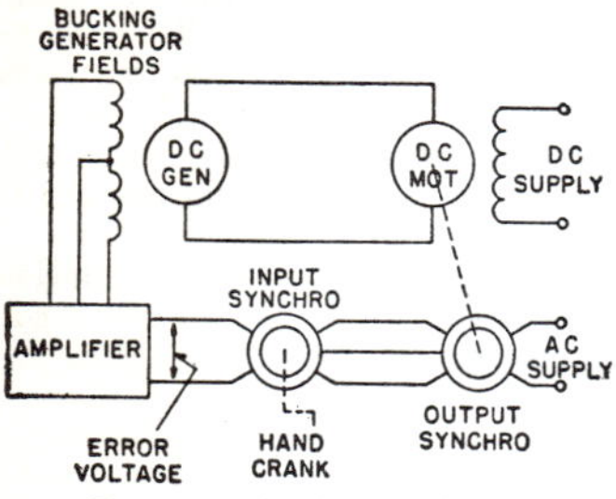

Fig. 28. Position regulator.

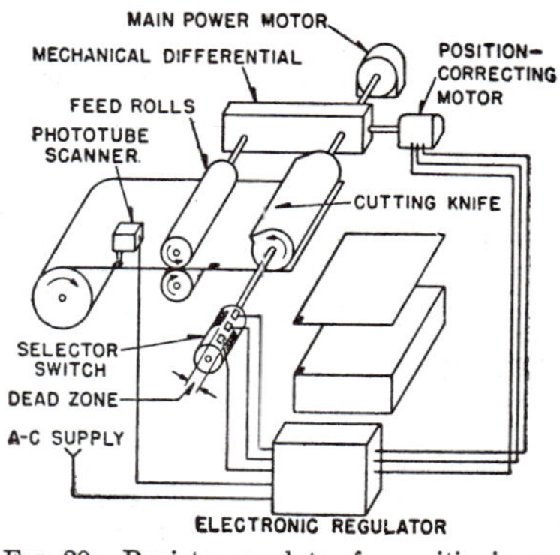

Fig. 29. Register regulator for positioning a printed sheet with respect to a cutting knife.

A PHOTOELECTRIC REGISTER REGULATOR for positioning a sheet relative to a reference mark is shown in Fig. 29. The roll of paper has been printed with a design and must be cut into separate sheets at the proper place with respect to the printing. The rotating cutter knife is driven by the main motor. The feed rolls are driven by the main motor and the position-correcting motor through a mechanical differential. Rotation of the position-correcting motor in one direction advances the reference mark, and rotation in the other direction retards the reference mark with respect to the cutter knife.

The photoelectric scanner consists of a light source, projecting and receiving lenses, phototube, and amplifier. Light is projected onto the sheet and is reflected back to the phototube. The phototube circuit is arranged to work on rate of change of illumination. When a reference mark passes under the light beam, the phototube current changes suddenly and a pulse is generated in the amplifier output. The output of the amplifier passes through the rotating selector switch and is introduced in the grid circuits of two thyratrons which control a set of reversing contactors. They, in turn, control the position-correcting motor. The selector switch is driven by the cutting knife shaft and has a dead zone in which the phototube amplifier output is connected to neither of the thyratron grid circuits. Rotation of the selector switch to one side of the dead zone connects the phototube amplifier output into the grid circuit of one thyratron, and rotation to the other side of the dead zone introduces the output to the grid circuit of the other thyratron. The regulating system is set up so that the phototube pulse occurs with the selector switch in the dead-zone position when the printed design is in the desired position with respect to the cutting knife. Under these conditions, neither thyratron conducts, and the position-correcting motor does not rotate.

Assume that the paper slips behind so that it is cut too soon. The phototube pulse will be connected by the selector switch to the grid of the thyratron controlling the contactor that connects the position-correcting motor to run in a direction that speeds up the feed rolls. The position-correcting motor will continue to run until the phototube pulse again occurs in the dead zone and the paper is in register. If the printed design gets ahead of the cutting knife, the selector switch sets up an opposite sequence of events which slows down the feed rolls, thus bringing the web into register.

The simple register regulator described above will regulate the position on the printed material within $\pm 1/32$ in. More elaborate regulators such as multicolor press register regulators maintain an accuracy of 0.003 in. so that successive printing of the basic colors in multicolor printing can be accomplished without blurring.

GENERAL REFERENCES

Articles describing applications of industrial electronic control appear frequently in many trade magazines and engineering publications. Books covering industrial electronic control in addition to several general books on electronic control are listed below.

Industrial Electronic Control

Westinghouse Electric Corp., *Industrial Electronics Reference Book*, John Wiley and Sons, New York, 1948.
W. I. Bendz and C. A. Scarlott, *Electronics for Industry*, John Wiley and Sons, New York, 1947.
W. Richter, *Fundamentals of Industrial Electronic Circuits*, McGraw-Hill Book Co., New York, 1947.
G. M. Chute, *Electronics in Industry*, McGraw-Hill Book Co., New York, 1946.
W. D. Cockrell, *Industrial Electronic Control*, McGraw-Hill Book Co., New York, 1944.
F. H. Gulliksen and E. H. Vedder, *Industrial Electronics*, John Wiley and Sons, New York, 1935.

General Electronic Theory

H. J. Reich, *Theory and Application of Electron Tubes*, McGraw-Hill Book Co., New York, 1944.
A. V. Eastman, *Fundamentals of Vacuum Tubes*, McGraw-Hill Book Co., New York, 1941.
M. I. T. Electrical Engineering Staff, *Applied Electronics*, John Wiley and Sons, New York, 1943.
W. G. Dow, *Fundamentals of Engineering Electronics*, John Wiley and Sons, New York, 1937.

BRAKES

Revised by R. M. Nardone

FRICTION BRAKES are used (1) to slow or stop a machine by converting its energy of motion into heat and (2) to control the speed of a descending load. Brakes may be classified as block brakes, band brakes, axial brakes, and load brakes.

The following points must be considered in the design of brakes: (1) Each part must have sufficient mechanical strength to withstand safely the maximum load that will come upon it during operation. (2) The pressure per unit area of surface must not be taken so high that excessive wear will result. (3) The coefficient of friction should be carefully considered; its value depends on the materials in contact, on the presence or absence of lubricating matter, and on the rate of slippage. With metal on metal, with increased velocity of slip, the coefficient of friction decreases from its value as the coefficient of friction of rest; the reverse, however, is true of leather. (4) The heat generated by the rubbing surfaces must be dissipated so as not to injure the brake.

16. BLOCK BRAKES

RESISTANCE OF BLOCK BRAKES. The simplest form of brake is the block brake, commonly used for railway car wheels, and in some forms for hoisting engines. In the car-wheel brake the resistance is not only that of sliding friction but also the resistance of cutting or grinding away the surface of the metals in contact. If P = total pressure acting normal to the sliding surface, pounds, μ = coefficient of friction, V = velocity of sliding, feet per minute, then the energy absorbed by the brake is $PV\mu$ ft-lb per min. Values of μ may be taken from Table 4, pp. 7-28 and 7-29. It should be noted that the value of μ is different for lubricated and unlubricated surfaces. If pressure P is great enough to squeeze out the lubricant, the value of μ increases, and the surfaces will cut and wear with an increase of temperature.

Block Brake Formulas. Various arrangements of block brakes are possible. Equations for different arrangements of single-block brakes are given in Table 1. If the faces of

Table 1. Block Brake Formulas

P = force at end of brake lever, pounds; F = tangential force at rim of brake wheel, pounds; R = reaction between brake wheel and block, pounds; μ = coefficient of friction; $P(a + b) = Rb$.

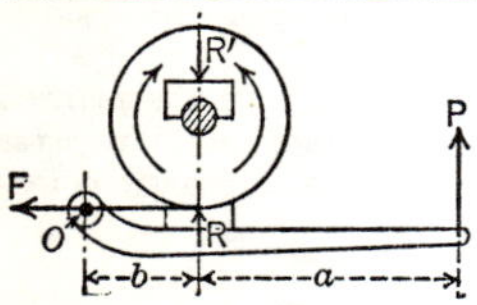	Line of action of F passing through fulcrum O For rotation in either direction $F = \mu R = \dfrac{\mu P(a+b)}{b}$ (1) $P = \dfrac{Fb}{(a+b)}\dfrac{1}{\mu}$ (2)
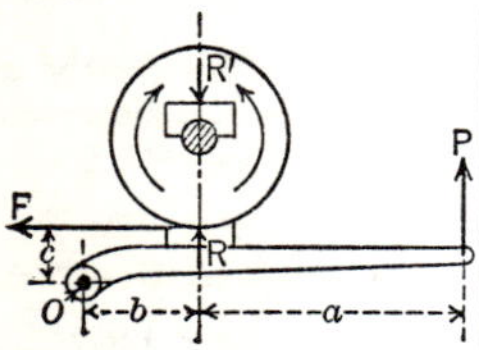	Line of action of F passing above fulcrum O $F = \mu R = \dfrac{P(a+b)}{b\left(\dfrac{1}{\mu} \mp \dfrac{c}{b}\right)}$ (3) $P = \dfrac{Fb}{(a+b)}\left(\dfrac{1}{\mu} \mp \dfrac{c}{b}\right)$ (4) The minus sign is used for clockwise, and the plus sign for counterclockwise rotation.
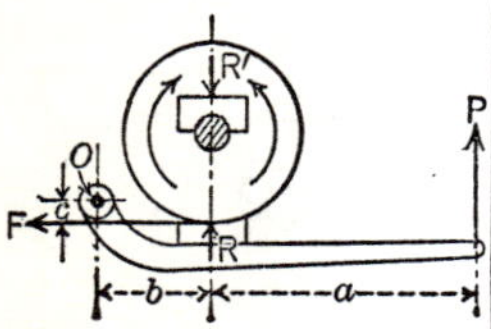	Line of action of F passing below fulcrum O $F = \mu R = \dfrac{P(a+b)}{b\left(\dfrac{1}{\mu} \pm \dfrac{c}{b}\right)}$ (5) $P = \dfrac{Fb}{(a+b)}\left(\dfrac{1}{\mu} \pm \dfrac{c}{b}\right)$ (6) The plus sign is used for clockwise, and the minus sign for counterclockwise rotation.
	For grooved brake wheel and block, Substitute $\dfrac{\mu}{\sin\alpha + \mu\cos\alpha}$ for μ in the above equations α = 1/2 included angle of groove faces.

the sliding surfaces are grooved, the values given by eqs. 1 to 6 should be corrected as shown in the table. In the single-block brake, the reaction R of the brake shoe is opposed by an equal and opposite reaction R' at the bearing. The braking effect resulting from the friction of the journal can be neglected in designing the brake but should be considered in the design of the bearing. The single-block brake tends to bend the shaft. In the double-block brake (Fig. 1) reactions R of the two blocks are equal and opposite, and tendency to bend the shaft is eliminated. The braking force is equal to twice the value given by eq. 1. If the line of action of F passes either inside or outside of fulcrums O, the braking load is not the same at the two blocks, and a tendency to bend the shaft exists. The total braking load is the sum of the loads given by eqs. 3 or 5 as the case may be.

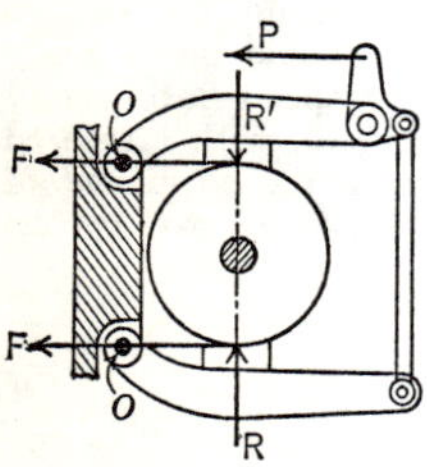

FIG. 1. Double-block brake.

17. BAND BRAKES

In wrapping band brakes, the torque exerted on the drum by friction is equal to the algebraic sum of the moments of the external loads applied to the band. The resultant load, represented by the difference between the tension on the tight side and the tension on the slack side, multiplied by the radius of the band, gives the torque or braking capacity.

Referring to Figs. 2 and 3,

$$F = T_1 - T_2$$
$$T_1 = T_2 e^{\mu\alpha}$$
$$Q = FR = (T_1 - T_2)R = T_2 R(e^{\mu\alpha} - 1) = T_1 R\left(\frac{e^{\mu\alpha} - 1}{e^{\mu\alpha}}\right)$$

where F = total friction of the band on the brake wheel; T_1 = tension or pull on tight side of band, pounds; T_2 = tension or pull on slack side of band, pounds; R = radius of band,

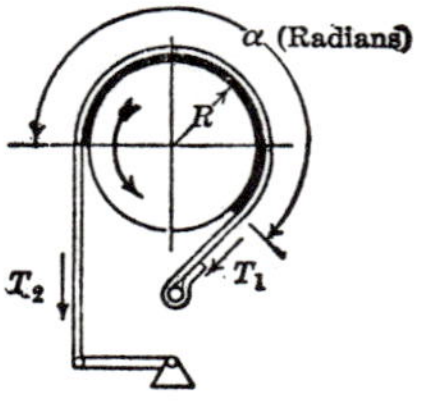

FIG. 2. Band brake, actuated on slack side.

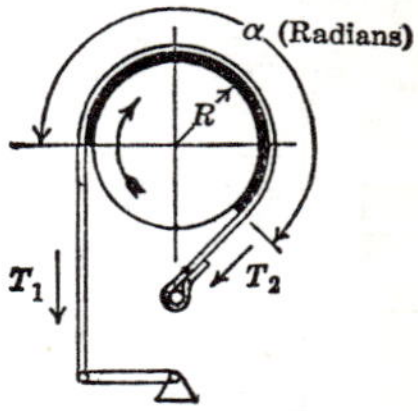

FIG. 3. Band brake, actuated on tight side.

feet; Q = torque developed by brake, pound-feet; e = 2.718; μ = coefficient of friction for lining material on drum; α = arc of contact in radians of band between points where T_1 and T_2 are taken.

Values of $e^{\mu\alpha}$ are given in Table 2. Table 3 shows various arrangements of band brakes and the formulas applicable thereto.

Table 2. Values of $e^{\mu\alpha}$

Proportion of Circumference	Radians, α	$°\theta$	Values of μ										
			0.1	0.2	0.25	0.3	0.35	0.4	0.45	0.5	0.6	0.7	0.8
			Values of $e^{\mu\alpha}$										
.1	0.6283	36	1.065	1.134	1.170	1.207	1.246	1.286	1.327	1.369	1.458	1.552	1.653
.2	1.2566	72	1.134	1.286	1.369	1.458	1.552	1.653	1.760	1.874	2.126	2.410	2.733
.25	1.5708	90	1.170	1.369	1.481	1.602	1.733	1.874	2.028	2.193	2.566	3.003	3.514
.3	1.8850	108	1.207	1.458	1.602	1.760	1.934	2.126	2.336	2.566	3.099	3.742	4.518
.4	2.5133	144	1.286	1.653	1.874	2.126	2.410	2.733	3.099	3.514	4.518	5.809	7.468
.425	2.6704	153	1.306	1.706	1.950	2.228	2.546	2.910	3.326	3.801	4.964	6.484	8.468
.45	2.8274	162	1.327	1.760	2.028	2.336	2.690	3.099	3.569	4.111	5.455	7.237	9.602
.475	2.9845	171	1.348	1.816	2.109	2.448	2.842	3.300	3.831	4.447	5.994	8.078	10.888
.5	3.1416	180	1.369	1.874	2.193	2.566	3.003	3.514	4.111	4.811	6.586	9.017	12.346
.525	3.2987	189	1.391	1.934	2.281	2.690	3.173	3.742	4.412	5.204	7.237	10.067	13.999
.55	3.4558	198	1.413	1.996	2.373	2.820	3.352	3.984	4.736	5.629	7.593	11.235	15.873
.575	3.6128	207	1.435	2.060	2.467	2.956	3.541	4.243	5.082	6.089	8.738	12.541	17.999
.6	3.7699	216	1.458	2.126	2.566	3.099	3.742	4.518	5.455	6.586	9.602	13.999	20.409
.65	4.0841	234	1.504	2.263	2.776	3.405	4.176	5.123	6.283	7.707	11.594	17.442	26.241
.7	4.3982	252	1.552	2.410	3.003	3.742	4.662	5.809	7.237	9.017	13.999	21.733	33.739
.75	4.7124	270	1.602	2.566	3.248	4.111	5.204	6.586	8.336	10.551	16.903	27.078	43.379
.8	5.0266	288	1.653	2.733	3.514	4.518	5.809	7.468	9.602	12.346	20.409	33.739	55.774

In general, for satisfactory results, the width of the band should be made approximately one-seventh the brake drum diameter. This diameter should range from $\sqrt[3]{2.4Q}$ to $\sqrt[3]{3Q}$ inches.

HEATING OF BAND BRAKES. The heat generated in a brake is a function of the braking torque and the number of revolutions of slip. The energy absorbed is

$$E = 2\pi Qn \text{ ft-lb}$$

Table 3. Formulas for Band Brakes

P = force at end of brake lever, pounds; T_1 = tension in tight side of band, pounds; T_2 = tension in slack side of band, pounds; $F = T_1 - T_2$ = tangential force at rim of brake wheel, pounds; μ = coefficient of friction between band and wheel; e = base of Naperian system of logarithms = 2.71828; α = angle of contact of band with wheel, radians (1 radian = 57.296 degrees).

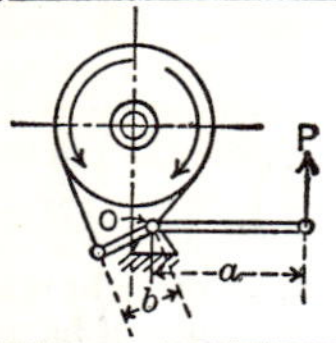

$$T_1 = F\frac{e^{\mu\alpha}}{e^{\mu\alpha}-1}; \qquad T_2 = F\frac{1}{e^{\mu\alpha}-1} \tag{7}$$

For clockwise rotation, $P = \dfrac{Fb}{a}\dfrac{e^{\mu\alpha}}{e^{\mu\alpha}-1}$ (8)

For counterclockwise rotation, $P = \dfrac{Fb}{a}\dfrac{1}{e^{\mu\alpha}-1}$ (9)

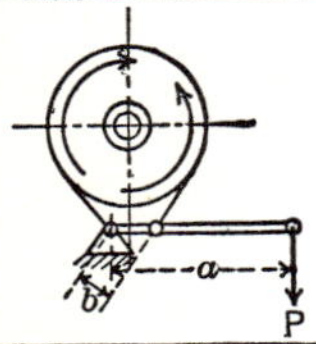

For clockwise rotation, $P = \dfrac{Fb}{a}\dfrac{1}{e^{\mu\alpha}-1}$ (10)

For counterclockwise rotation, $P = \dfrac{Fb}{a}\dfrac{e^{\mu\alpha}}{e^{\mu\alpha}-1}$ (11)

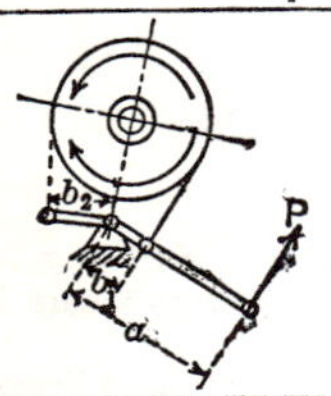

For clockwise rotation, $P = \dfrac{F}{a}\dfrac{b_2e^{\mu\alpha}-b_1}{e^{\mu\alpha}-1}$ (12)

For counterclockwise rotation, $P = \dfrac{F}{a}\dfrac{b_2-b_1e^{\mu\alpha}}{e^{\mu\alpha}-1}$ (13)

Unless $b_2 > b_1e^{\mu\alpha}$ the band will automatically grip the wheel and cause rupture of the band or some other part.

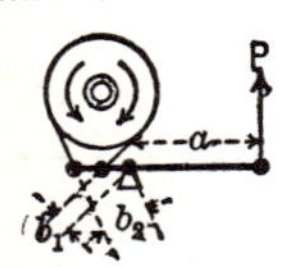

For clockwise rotation, $P = \dfrac{F}{a}\dfrac{b_2e^{\mu\alpha}+b_1}{e^{\mu\alpha}-1}$ (14)

For counterclockwise rotation, $P = \dfrac{F}{a}\dfrac{b_1e^{\mu\alpha}+b_2}{e^{\mu\alpha}-1}$ (15)

If $b_1 = b_2 = b$, for both cases, $P = \dfrac{Fb}{a}\dfrac{e^{\mu\alpha}+1}{e^{\mu\alpha}-1}$ (16)

If a weight is lowered through a given distance, by slipping of the brake, the energy absorbed by the brake is

$$E = Lh \text{ ft-lb}$$

The heat equivalent of energy is

$$H = \frac{E}{778.3}\text{ Btu}$$

and the temperature rise is

$$t_i = \frac{H}{WC} \tag{17}$$

where L = load, pounds; h = distance load is lowered, feet; n = revolutions of brake slip; t_i = increase in rim temperature, degrees Fahrenheit; W = weight of rim, pounds; c = specific heat of metal in drum; (c = 0.13 for cast iron and 0.116 for steel).

For best grades of asbestos friction lining, the permissible maximum temperature is 500 F. The temperature rise t_i will, therefore, depend on the initial or average low temperature of the drum.

For repeatedly applied brakes, the rate of heat dissipation must be equal to or greater than the rate of heat generation. The rate of dissipation per square inch of radiating area, denoted by $(t_1 - t_a)\,Xa$, varies between 0.15 and 0.36 Btu per min for t_i between 200 F and 400 F, respectively. Expressed in symbols,

$$H_d \geqq H_g$$

$$H_d = (t_1 - t_a)XaA_r$$

$$H_g = 42.4qMA_b$$

Therefore

$$(t_1 - t_a)XaA_r \geqq 42.4qMA_b$$

and

$$A_r \geqq \frac{42.4qMA_b}{(t_1 - t_a)Xa} \tag{18}$$

where t_a = temperature of surrounding air, °F; t_1 = maximum rim temperature, °F; X = cooling coefficient; a = temperature factor; M = horsepower capacity per square inch of brake; q = load factor, or portion of time that brake is in continuous use; A_b = total brake area, square inches; A_r = total radiating area, square inches; H_g = Btu generated per minute in brake; H_d = Btu dissipated per minute by the brake. Values for M depend on the service. A safe figure is 0.25.

18. CONE AND DISK BRAKES

CONE BRAKES. Figure 4 illustrates one form of cone brake. Force P applied to the brake lever L causes axial movement of the brake cone into its mating shoe, thus creating resistance to the rotation of the member to which the shoe is attached. Let M = twisting moment on brake shaft, inch-pounds; r = mean radius of brake wheel, inches; P = pull on lever, pounds; $F = M/r$ = force at the circumference of the brake wheel, pounds; α = half angle of the cones; μ = coefficient of friction between brake wheel and shoe; A = axial pressure, pounds, along the cone shaft forcing the cones together; N = normal pressure on the cone, in order that the braking action shall overcome the turning moment; L = length of lever, inches; l = distance from fulcrum of brake lever to axis of cone. Then

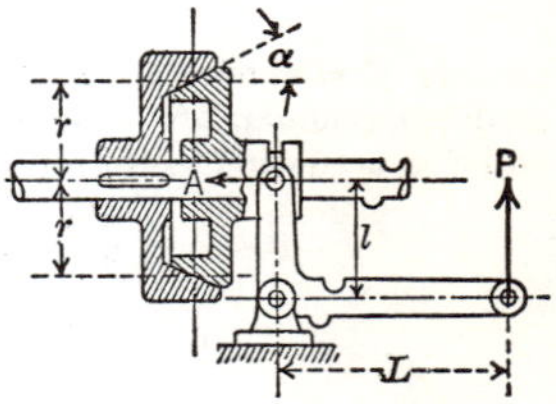

FIG. 4. Cone brake.

$$F \gtreqless 2\mu N \tag{19}$$

$$A = \frac{PL}{l} \tag{20}$$

$$N = \frac{A}{2(\sin\alpha + \mu\cos\alpha)} \quad (21) \quad \text{or} \quad \frac{A}{2\sin\alpha} \tag{22}$$

$$F \gtreqless \frac{\mu A}{\sin\alpha + \mu\cos\alpha} \quad (23) \quad \text{or} \quad \frac{\mu A}{\sin\alpha} \tag{24}$$

$$P \lesseqgtr \frac{Fl(\sin\alpha + \mu\cos\alpha)}{\mu L} \quad (25) \quad \text{or} \quad \frac{Fl\sin\alpha}{\mu L} \tag{26}$$

Angle α usually ranges from 10 to 18 degrees. For cast iron on cast iron, μ can be taken as 0.12; for one surface of wood and the other of cast iron, $\mu = 0.25$ may be used.

From an experimental investigation of a cone brake by Professor H. Bonte (*Z. Ver. deut. Ing.*, Dec. 15, 1915), Professor Leutwiler (*Elements of Machine Design*, McGraw-Hill Book Co., p. 419) derives the following conclusions concerning eqs. 23 and 24: (1) When α is 15 degrees, the error introduced by using eq. 23 is large, while the agreement between the experimental results and those obtained by using eq. 24 is very close. (2) When α is 30 degrees, the experimental results lie between those obtained by eqs. 23 and 24. (3) When plotted, the curve representing the results obtained by eq. 24 lies above, but much closer to the experimental curve than that obtained by eq. 23. (4) For the angles 45 degrees and 60 degrees, the experimental points lie above those obtained by eq. 24, which in turn lie above the points obtained by eq. 23. (5) Apparently, the coefficient of friction is not constant, as generally assumed, but varies slightly with the pressure. As a result of this experimental investigation, Professor Bonte recommends that eq. 23, which apparently is incorrect, should not be used in the design of cone brakes or clutches.

DISK BRAKES. Like the cone brake, the disk brake depends on relative axial movement between fixed and moving elements. Where high torque is applied over a limited period of time, as in airplane landing wheel brakes, multiple-disk brakes serve well. Design principles are similar to those governing disk clutches (see Section 15, Article 9). In brakes the cooling problems will, in general, be more acute than in clutches.

19. INTERNAL BRAKES

Internal shoe brakes are used principally for automotive vehicles. The inertia of the vehicle is utilized to obtain the braking torque with a minimum of operating effort.

SINGLE-SHOE BRAKE. Referring to Fig. 5, let T = braking torque, pound-inches; f = coefficient of friction; r = radius of drum, inches; P = cam force on brake shoe,

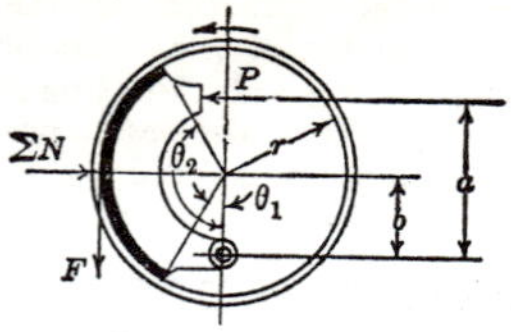

Fig. 5. Single-shoe internal brake.

pounds; F = sum of all frictional forces on drum, pounds; ΣN = sum of all normal forces on drum, pounds; $\Sigma N'$ = sum of all normal forces on drum if there were no friction, pounds; and S = actuation factor represented by $\Sigma N/\Sigma N'$:

$$A = \int_{\theta_1}^{\theta_2} \cos\theta \sin\theta \, d\theta = -{}^1/_4 (\cos 2\theta_2 - \cos 2\theta_1)$$

$$B = \int_{\theta_1}^{\theta_2} \sin^2\theta \, d\theta = -{}^1/_4 (\sin 2\theta_2 - \sin 2\theta_1) + \frac{\theta_2}{2} - \frac{\theta_1}{2}$$

$$C = \int_{\theta_1}^{\theta_2} \sin\theta \, d\theta = -(\cos\theta_2 - \cos\theta_1)$$

a, b θ_1 and θ_2 as shown in Fig. 5. Then

$$\Sigma N = P \times \frac{aC}{bB - f(bA + rC)} \text{ pounds}$$

$$\Sigma N' = P \times \frac{aC}{bB} \text{ pounds}$$

$$S = \frac{\Sigma N}{\Sigma N'} = \frac{bB}{bB - f(bA + rC)}$$

$$F = f\Sigma N \text{ pounds}$$

$$T = Fr, \text{ pound-inches}$$

Example 1. Find ΣN, F, and T for the brake shown in Fig. 5 when $a = 10$, $\theta_1 = 25$ degrees, $b = 5$, $\theta_2 = 159$ degrees, $r = 8$, and $f = 0.3$.

Substituting in the foregoing formulas,

$$C = -(\cos 159° - \cos 25°) = 1.84$$

$$B = -{}^1/_4 (\sin 318° - \sin 50°) + \frac{159\pi}{360} - \frac{25\pi}{360} = 1.529$$

$$A = -{}^1/_4 (\cos 318° - \cos 50°) = -0.0251$$

Then

$$\Sigma N = P \times \frac{10 \times 1.84}{5 \times 1.529 - 0.3(5 \times -0.0251 + 8 \times 1.840)} = 5.64P \text{ lb}$$

and

$$\Sigma N' = P \times \frac{18.4}{7.645} = 2.405P \text{ lb}$$

$$S = \frac{\Sigma N}{\Sigma N'} = 2.34$$

$$F = 5.64 \times 0.3P = 1.692P \text{ lb}$$

$$T = 1.692P \times 8 = 13.536P \text{ lb-in.}$$

If $P = 500$ lb, $T = 6750$ lb-in.

With a symmetrical shoe in which a perpendicular to a radius passing through the anchor pin center divides the shoe in half for equal arcs of contact, the formula ΣN is somewhat simplified. In that case, since $A = 0$,

$$\Sigma N = P \times \frac{aC}{bB - frC}$$

The actuation factor S increases rapidly with an increase in coefficient of friction f. If too high a value of f is used, the brake is likely to grab.

TWO-SHOE BRAKE. Both shoes are shown pivoted at the same point in Fig. 6.

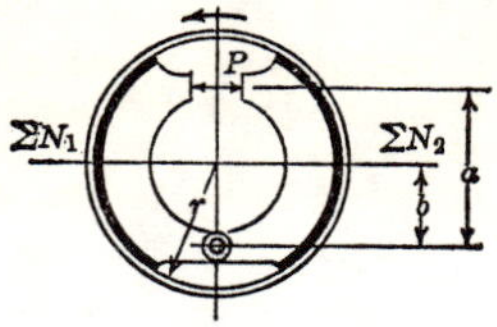

FIG. 6. Double-shoe internal brake.

For the drum rotation shown, the left-hand brake shoe is self-actuated while the right-hand shoe is not.

$$\Sigma N_1 = P \times \frac{aC}{bB - f(bA + rC)} \text{ lb}$$

$$\Sigma N_2 = P \times \frac{aC}{bB + f(bA + rC)} \text{ lb}$$

The sum of all normal forces is $\Sigma N = \Sigma N_1 + \Sigma N_2$. Also,

$$\Sigma N_1' = \Sigma N_2' = P\frac{aC}{bB}$$

For the left shoe,

$$S_1 = \frac{\Sigma N_1}{\Sigma N_1'}$$

$$F_1 = f\Sigma N_1$$

$$T_1 = F_1 r$$

For the right shoe,

$$S_2 = \frac{\Sigma N_2}{\Sigma N_1}$$

$$F_2 = f\Sigma N_2$$

$$T_2 = F_2 r$$

Total torque

$$T = T_1 + T_2$$

Example 2. Shoes symmetrical (Fig. 6): $\theta_1 = 30$ degrees, $\theta_2 = 150$ degrees, $a = 10$, $b = 5$, $r = 8$, $f = 0.3$.

For a symmetrical shoe:

$$\Sigma N_1 = P \times \frac{aC}{bB - frC}$$

$$C = -(\cos 150° - \cos 30°) = 1.732$$

$$B = -1/4(\sin 300° - \sin 60°) + \frac{150\pi}{360} - \frac{30\pi}{360} = 1.4802$$

Therefore,

$$\Sigma N_1 = 5.33P$$

$$\Sigma N_2 = 1.5P$$

$$\Sigma N_1' = 2.34P$$

$$S_1 = 2.28$$

$$F_1 = 1.599P$$

$$T_1 = 12.792P$$

$$S_2 = 0.281$$

$$F_2 = 0.45P$$

$$T_2 = 3.60P$$

$$T = 16.392P$$

From the foregoing, it will be seen that the braking torque of the assisted or self-actuated shoe is several times greater than the braking torque of the opposed shoe. Therefore, the assisted shoe will wear faster than the opposed shoe. If the cam which applies the actuating force on the shoes is fixed, P_1 on the assisted shoe will be less than P_2 on the opposed shoe and ΣN_1 will equal ΣN_2. In this case both shoes will wear equally.

SECTION 18

MACHINE COMPONENTS AND TESTING

By

L. S. LINDEROTH, JR., *Professor of Mechanical Engineering, Iowa State College of Agriculture and Mechanic Arts*

P. E. NOKES, *United Shoe Machinery Corp.*

ROBERT B. SOSMAN, *Professor of Ceramics, Rutgers University*

W. J. ECKERT, *Director of Pure Science, International Business Machines Corp.*

PRESSURE SEALS

By L. S. Linderoth, Jr.

Loss of fluid and loss of pressure in hydraulic and pneumatic devices are prevented by the proper design and use of seals, packings, and gaskets.

CLASSIFICATION. Seals fall into two major classes: static and dynamic. Static seals or gaskets are used to prevent the flow of fluid across a static joint between two parts of a pressure vessel. A typical example is the gasket between the cylinder head and cylinder block of an automobile engine. Dynamic seals are used to prevent the flow of fluid across a sliding or rotating joint between two members of a hydraulic or pneumatic device. Typical examples are the piston packing in a cylinder and the shaft packing of a rotary pump.

A further classification divides seals into metallic and nonmetallic seals. Metallic packings or seals are illustrated by sheet lead gaskets and soft copper or aluminum washers. Nonmetallic packings are illustrated by sheet cork, sheet asbestos, leather, rubber, etc.

Subclassifications of nonmetallic seals are organic and inorganic seals, typical examples of which are leather, flax, and cotton for organic seals and asbestos and synthetic rubber for inorganic seals.

Seals may also be classified as to whether they are molded, sheet, or braided packings. Typical molded packings are cup packings, V rings, U cups, and O rings. Sheet packings are usually rolled from a mass of granulated or shredded packing material in a suitable binder.

The definitions of the words seals, packings, and gaskets are not exact. Generally speaking, any means to prevent the flow of fluid across a joint between adjacent members is referred to as a seal. If the seal is static it is frequently called a gasket. Means to prevent leakage in dynamic seals are most frequently referred to as packings. The synthetic metal-cased low-pressure shaft seals are referred to as "seals."

1. STATIC SEALS (GASKETS)

SIMPLE GASKET. Figure 1 illustrates the simplest type of gasket. It is cut by hand or blanked in a die from sheet stock to fit the particular joint it is to be used on. Gaskets are made from all kinds of material, such as paper, sheet cork, rubber, and asbestos. They must have some yield or compressibility in order to conform to the surface irregularities

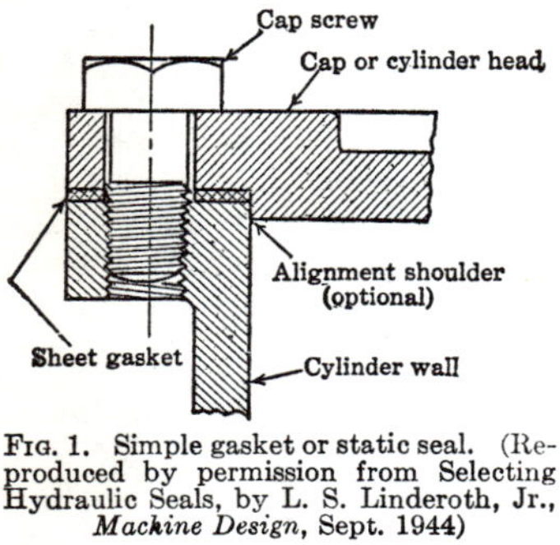

FIG. 1. Simple gasket or static seal. (Reproduced by permission from Selecting Hydraulic Seals, by L. S. Linderoth, Jr., *Machine Design*, Sept. 1944)

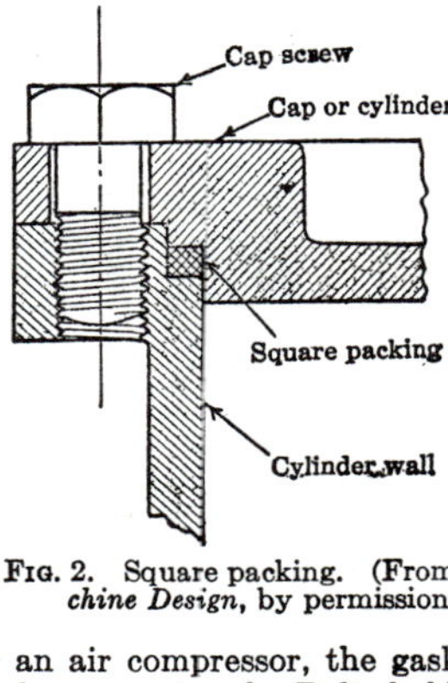

FIG. 2. Square packing. (From *Machine Design*, by permission)

If heat is present, as in an automobile engine or an air compressor, the gasket must be able to stand up under the highest temperature to be encountered. Bolts holding a cylinder head in place with a yielding gasket will have a load equal to gasket pressure plus cylinder pressure.

SQUARE PACKING. Figure 2 illustrates a form of gasket that gives a minimum load on the cylinder head bolts or manhole cover, as the case may be. A recess is designed into the joint between the two members, and a ring of cork is compressed in this space. Cork is the best material for this type of joint since it can be compressed to 75% of its original volume. Recommended dimensions are given in Table 1.

Table 1. Recommended Sections for Square Packing

Packing Ring Diameter, in.	Packing Thickness, in.
Up to 1	3/32
1 to 2	1/8
2 to 3	5/32
3 to 5	3/16
5 and over	1/4

The joints in Figs. 1 and 2 are used when disassembly is not a frequent occurrence. Assembly of a joint like Fig. 1 requires a new gasket for reliable results. Disassembly of a joint like Fig. 2 is difficult to accomplish without spoiling the gasket, particularly if the assembly is an old one.

Since the joint in Fig. 2 involves metal-to-metal contact the relation between the head and cylinder is accurate and can be predetermined. The joint in Fig. 1 depends on the

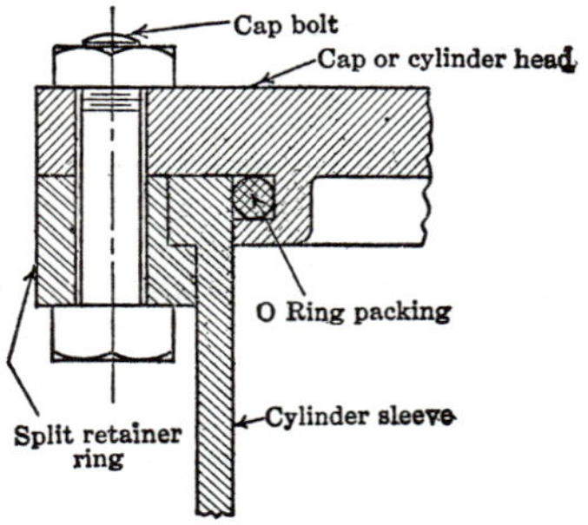

FIG. 3. O ring gasket. (From *Machine Design*, by permission)

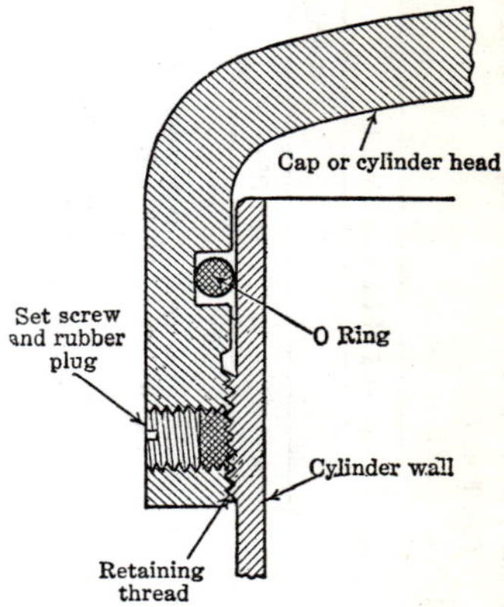

FIG. 4. O ring gasket. (From *Machine Design*, by permission)

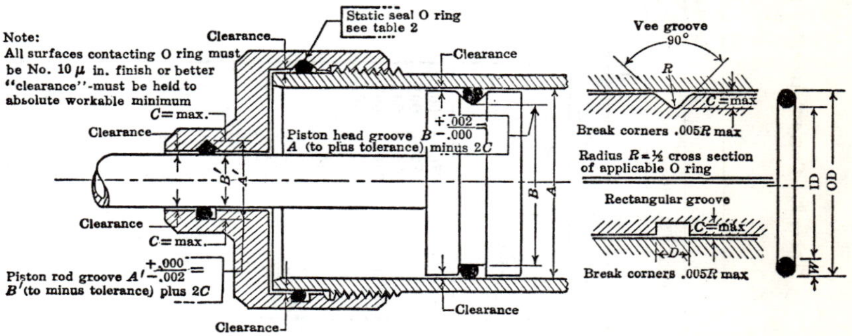

FIG. 5. AN standards for O ring packing application. Rectangular grooves are preferred for all applications. Sides of these grooves are perpendicular to the direction of motion with a tolerance of +10°, −0°. See Table 2.

bolt tension and the yield of the gasket material for the location of the head relative to the cylinder. This is not an accurate or a permanent relation.

O RING GASKET. Figure 3 illustrates the design of a seal for frequent disassembly. A standard O ring is used in an appropriate groove so laid out that assembly of the parts will not pinch the ring. Both surfaces against which the O ring seals must be as smooth as possible (mirror finish). The cylinder sleeve should have a slight bevel and rounded edge to prevent the O ring from being cut in assembly. Figure 4 is a variation of this design. The size of the O ring slot can be calculated from data in Fig. 5 and Table 2.

(*Continued on p. 18-07*).

Table 2. Current Accepted Army-Navy Standards for Operating O Ring Application

(See Fig. 5)

(All dimensions in inches)

AN 6227 Dash No.	Operating O Ring Packings and Machining Details										
	O Ring Specifications					Machining Details					
	Nominal Size			Actual Size		A and A' ± .001	B and B' ± .001	C		D	R*
	W	ID	OD	W	ID						
	Cross Section	Inside Diameter	Outside Diameter	Cross Section	Inside Diameter	Inside Diameter of Cylinder or Outside Diameter of Gland Groove	Diameter of Piston Head Groove or Diameter of Piston Rod	See Note Cross Section of Recess	Minimum Squeeze	Rectangular Groove Length	Vee Groove Radius
1	1/16	1/8	1/4	.070 ± .003	.114 ± .005	.250	.123	.057	.010	3/32	.0570
2	1/16	5/32	9/32	.070 ± .003	.145 ± .005	.281	.154	"	"	"	"
3	1/16	3/16	5/16	.070 ± .003	.176 ± .005	.312	.185	"	"	"	"
4	1/16	7/32	11/32	.070 ± .003	.208 ± .005	.344	.217	"	"	"	"
5	1/16	1/4	3/8	.070 ± .003	.239 ± .005	.375	.248	"	"	"	"
6	1/16	5/16	7/16	.070 ± .003	.301 ± .005	.4375	.310	"	"	"	"
7	1/16	3/8	1/2	.070 ± .003	.364 ± .005	.500	.373	"	"	"	"
8	3/32	3/8	9/16	.103 ± .003	.362 ± .005	.5625	.373	.090	.010	9/64	.0900
9	3/32	7/16	5/8	.103 ± .003	.424 ± .005	.625	.435	"	"	"	"
10	3/32	1/2	11/16	.103 ± .003	.487 ± .005	.6875	.498	"	"	"	"
11	3/32	9/16	3/4	.103 ± .003	.549 ± .005	.750	.560	"	"	"	"
12	3/32	5/8	13/16	.103 ± .003	.612 ± .005	.8125	.623	"	"	"	"
13	3/32	11/16	7/8	.103 ± .003	.674 ± .005	.875	.685	"	"	"	"
14	3/32	3/4	15/16	.103 ± .003	.737 ± .005	.9375	.748	"	"	"	"
15	1/8	3/4	1	.139 ± .004	.734 ± .006	1.001	.747	.123	.012	3/16	.1230
16	1/8	13/16	1 1/16	.139 ± .004	.796 ± .006	1.063	.809	"	"	"	"
17	1/8	7/8	1 1/8	.139 ± .004	.859 ± .006	1.126	.872	"	"	"	"
18	1/8	15/16	1 3/16	.139 ± .004	.921 ± .006	1.186	.934	"	"	"	"
19	1/8	1	1 1/4	.139 ± .004	.984 ± .006	1.251	.997	"	"	"	"
20	1/8	1 1/16	1 5/16	.139 ± .004	1.046 ± .006	1.313	1.059	"	"	"	"

21	1/8	1 1/8	1 3/8	.139 ± .004	1.109 ± .006	1.376	1.122	"	"	"	"
22	1/8	1 3/16	1 7/16	.139 ± .004	1.171 ± .006	1.438	1.184	"	"	"	"
23	1/8	1 1/4	1 1/2	.139 ± .004	1.234 ± .006	1.501	1.247	"	"	"	"
24	1/8	1 5/16	1 9/16	.139 ± .004	1.296 ± .006	1.563	1.309	"	"	"	"
25	1/8	1 3/8	1 5/8	.139 ± .004	1.359 ± .006	1.626	1.372	"	"	"	"
26	1/8	1 7/16	1 11/16	.139 ± .004	1.421 ± .006	1.688	1.434	"	"	"	"
27	1/8	1 1/2	1 3/4	.139 ± .004	1.484 ± .006	1.751	1.457	"	"	"	"
28	3/16	1 1/2	1 7/8	.210 ± .005	1.475 ± .010	1.876	1.497	.1875	.017	9/32	.1875
29	3/16	1 5/8	2	.210 ± .005	1.600 ± .010	2.001	1.622	"	"	"	"
30	3/16	1 3/4	2 1/8	.210 ± .005	1.725 ± .010	2.126	1.747	"	"	"	"
31	3/16	1 7/8	2 1/4	.210 ± .005	1.850 ± .010	2.251	1.872	"	"	"	"
32	3/16	2	2 3/8	.210 ± .005	1.975 ± .010	2.376	1.997	"	"	"	"
33	3/16	2 1/8	2 1/2	.210 ± .005	2.100 ± .010	2.501	2.122	"	"	"	"
34	3/16	2 1/4	2 5/8	.210 ± .005	2.225 ± .010	2.626	2.247	"	"	"	"
35	3/16	2 3/8	2 3/4	.210 ± .005	2.350 ± .010	2.751	2.372	"	"	"	"
36	3/16	2 1/2	2 7/8	.210 ± .005	2.475 ± .010	2.876	2.497	"	"	"	"
37	3/16	2 5/8	3	.210 ± .005	2.600 ± .010	3.001	2.622	"	"	"	"
38	3/16	2 3/4	3 1/8	.210 ± .005	2.725 ± .015	3.126	2.747	"	"	"	"
39	3/16	2 7/8	3 1/4	.210 ± .005	2.850 ± .015	3.251	2.872	"	"	"	"
40	3/16	3	3 3/8	.210 ± .005	2.975 ± .015	3.377	2.996	"	"	"	"
41	3/16	3 1/8	3 1/2	.210 ± .005	3.100 ± .015	3.502	3.121	"	"	"	"
42	3/16	3 1/4	3 5/8	.210 ± .005	3.225 ± .015	3.627	3.246	"	"	"	"
43	3/16	3 3/8	3 3/4	.210 ± .005	3.350 ± .015	3.752	3.371	"	"	"	"
44	3/16	3 1/2	3 7/8	.210 ± .005	3.475 ± .015	3.877	3.496	"	"	"	"
45	3/16	3 5/8	4	.210 ± .005	3.600 ± .015	4.002	3.621	"	"	"	"
46	3/16	3 3/4	4 1/8	.210 ± .005	3.725 ± .015	4.127	3.746	"	"	"	"
47	3/16	3 7/8	4 1/4	.210 ± .005	3.850 ± .015	4.252	3.871	"	"	"	"
48	3/16	4	4 3/8	.210 ± .005	3.975 ± .015	4.377	3.996	"	"	"	"
49	3/16	4 1/8	4 1/2	.210 ± .005	4.100 ± .015	4.502	4.121	"	"	"	"
50	3/16	4 1/4	4 5/8	.210 ± .005	4.225 ± .015	4.627	4.246	"	"	"	"
51	3/16	4 3/8	4 3/4	.210 ± .005	4.350 ± .015	4.752	4.371	"	"	"	"
52	3/16	4 1/2	4 7/8	.210 ± .005	4.475 ± .015	4.877	4.496	"	"	"	"
53	1/4	4 5/8	5 1/8	.275 ± .006	4.600 ± .015	5.128	4.621	.250	.019	3/8	.2400
54	1/4	4 3/4	5 1/4	.275 ± .006	4.725 ± .015	5.253	4.746	"	"	"	"
55	1/4	4 7/8	5 3/8	.275 ± .006	4.850 ± .015	5.378	4.871	"	"	"	"

(Table continued on p. 18-06)

Table 2. Current Accepted Army-Navy Standards for Operating O Ring Application—*Continued*

(See Fig. 5)

(All dimensions in inches)

AN 6227 Dash No.	Operating O Ring Packings and Machining Details										
	O Ring Specifications					Machining Details					
	Nominal Size			Actual Size		A and A' ± .001	B and B' ± .001	C		D	R*
	W	ID	OD	W	ID			See Note			
	Cross Section	Inside Diameter	Outside Diameter	Cross Section	Inside Diameter	Inside Diameter of Cylinder or Outside Diameter of Gland Groove	Diameter of Piston Head Groove or Diameter of Piston Rod	Cross Section of Recess	Minimum Squeeze	Rectangular Groove Length	Vee Groove Radius
56	1/4	5	5 1/2	.275 ± .006	4.975 ± .015	5.503	4.996	.250	.019	3/8	.2400
57	1/4	5 1/8	5 5/8	.275 ± .006	5.100 ± .023	5.628	5.121	"	"	"	"
58	1/4	5 1/4	5 3/4	.275 ± .006	5.225 ± .023	5.753	5.246	"	"	"	"
59	1/4	5 3/8	5 7/8	.275 ± .006	5.350 ± .023	5.878	5.371	"	"	"	"
60	1/4	5 1/2	6	.275 ± .006	5.475 ± .023	6.003	5.496	"	"	"	"
61	1/4	5 5/8	6 1/8	.275 ± .006	5.600 ± .023	6.128	5.621	"	"	"	"
62	1/4	5 3/4	6 1/4	.275 ± .006	5.725 ± .023	6.253	5.746	"	"	"	"
63	1/4	5 7/8	6 3/8	.275 ± .006	5.850 ± .023	6.378	5.871	"	"	"	"
64	1/4	6	6 1/2	.275 ± .006	5.975 ± .023	6.503	5.996	"	"	"	"
65	1/4	6 1/4	6 3/4	.275 ± .006	6.225 ± .023	6.753	6.246	"	"	"	"
66	1/4	6 1/2	7	.275 ± .006	6.475 ± .023	7.003	6.496	"	"	"	"
67	1/4	6 3/4	7 1/4	.275 ± .006	6.725 ± .023	7.253	6.746	"	"	"	"
68	1/4	7	7 1/2	.275 ± .006	6.975 ± .023	7.503	6.996	"	"	"	"
69	1/4	7 1/4	7 3/4	.275 ± .006	7.225 ± .030	7.753	7.246	"	"	"	"
70	1/4	7 1/2	8	.275 ± .006	7.475 ± .030	8.003	7.496	"	"	"	"
71	1/4	7 3/4	8 1/4	.275 ± .006	7.725 ± .030	8.253	7.746	"	"	"	"
72	1/4	8	8 1/2	.275 ± .006	7.975 ± .030	8.503	7.996	"	"	"	"
73	1/4	8 1/2	9	.275 ± .006	8.475 ± .030	9.003	8.496	"	"	"	"
74	1/4	9	9 1/2	.275 ± .006	8.975 ± .030	9.503	8.996	"	"	"	"
75	1/4	9 1/2	10	.275 ± .006	9.475 ± .030	10.003	9.496	"	"	"	"
76	1/4	10	10 1/2	.275 ± .006	9.975 ± .030	10.503	9.996	"	"	"	"
77	1/4	10 1/2	11	.275 ± .006	10.475 ± .030	11.003	10.496	"	"	"	"
78	1/4	11	11 1/2	.275 ± .006	10.975 ± .030	11.503	10.996	"	"	"	"

Note. C is computed to include clearance.

*** V grooves are almost obsolete. See Air-Force-Navy Specification AN-P-74b for complete details.**

SOFT METAL GASKET. Figure 6 illustrates the use of a soft metal (copper or annealed aluminum) gasket. The two mating surfaces of the joint are serrated with sharp 60-degree vees directly opposite each other. These serrations sink into the soft metal gasket and effect a seal.

EXTRA-HIGH-PRESSURE SEALS. Figure 7 illustrates an extra-high-pressure seal. A wedge-shaped groove in the mating surfaces holds a soft metal ring. The bolts seat the ring into the grooves. Such a seal can be used for pressures up to 20,000 psi. Figure 8 illustrates a form of extreme pressure packing. As can be seen in the cross section, this type of packing is complex. It was developed by Dr. P. W. Bridgman at Harvard University for experiments at pressures on the order of a million psi.

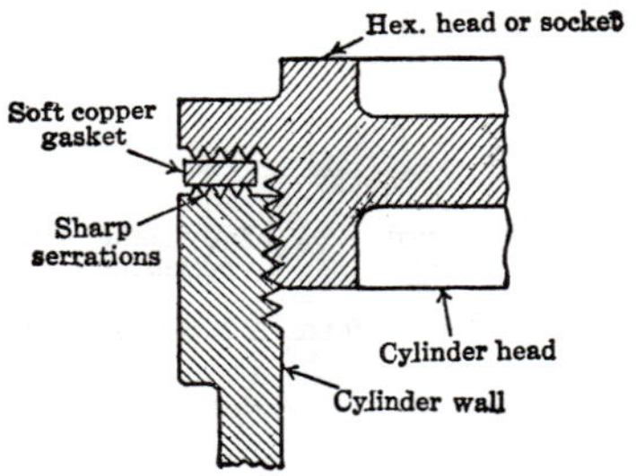

FIG. 6. Soft metal gasket. (From *Machine Design*, by permission)

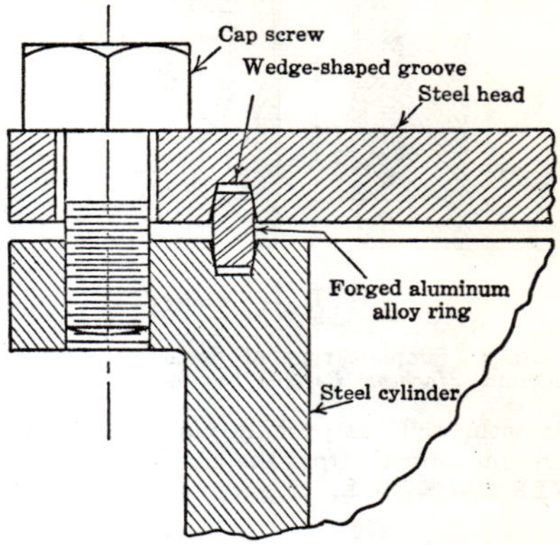

FIG. 7. Extra-high-pressure seal. (From *Machine Design*, by permission)

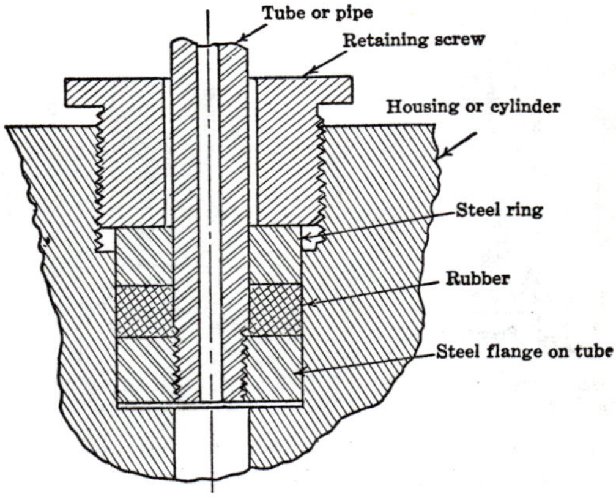

FIG. 8. Extreme pressure packing. (From *Machine Design*, by permission)

Mention should be made of various plastic compounds that are used to make gaskets. They are marketed in collapsible lead tubes for ease of handling. The paste or plastic is extruded from the tube and spread over the surface of the joint with a spatula or other suitable tool. Materials are available that will stand high temperatures, will act as a cement, and will remain plastic, depending on the particular requirements. These pastes can be had both waterproof and/or oilproof.

2. DYNAMIC SEALS (PACKINGS)

SIMPLE PACKINGS. Figure 9 illustrates the simplest type of dynamic seal. It may be either reciprocating or rotating. The packing material can be cotton, flax, or other fiber, either twisted or braided. The proportions of the gland depend on the diameter of the rod, a 1/2-in. rod having a gland of 1/4-in. radial dimension and a 2-in. rod one of 1/2-in. radial dimension.

An improvement over the internally threaded gland is shown in Fig. 10. Here the gland

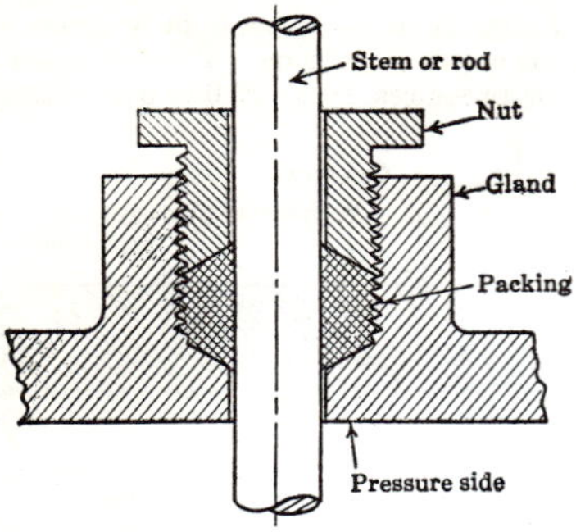

FIG. 9. Simplest type of dynamic seal. (From *Machine Design*, by permission)

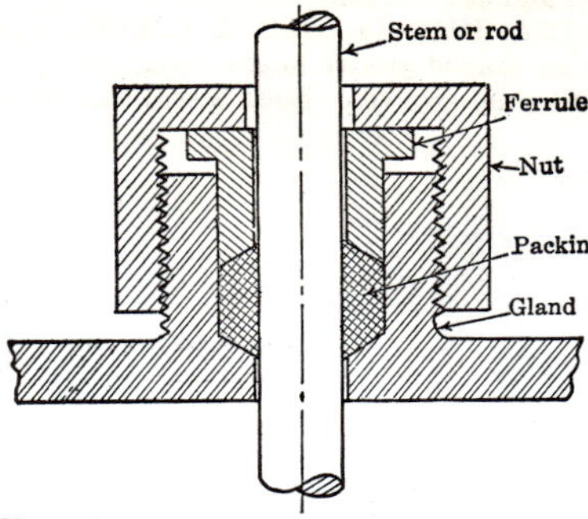

FIG. 10. Improved type of dynamic seal. (From *Machine Design*, by permission)

is smooth, and the packing material can flow more readily to form a tight seal with a minimum pressure from the gland nut.

VEE PACKINGS. Figure 11 illustrates a dynamic seal using a preformed or molded

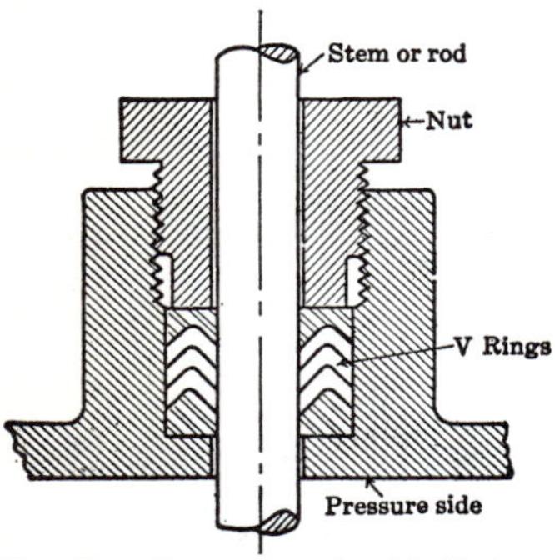

FIG. 11. Dynamic seal with V rings. (From *Machine Design*, by permission)

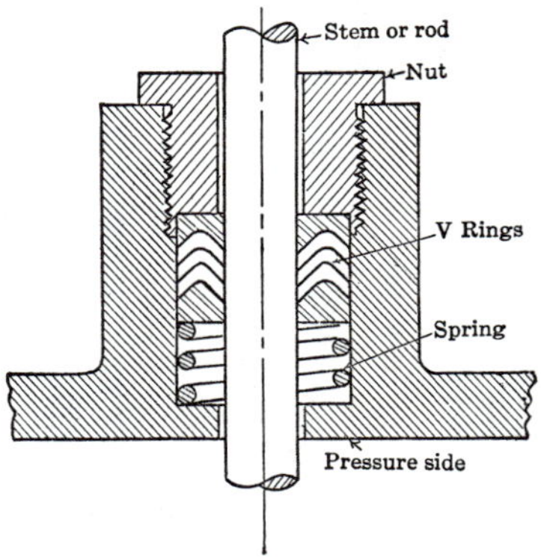

FIG. 12. Self-adjusting V ring seal design. (From *Machine Design*, by permission)

packing ring with a V-shaped cross section. This type of packing reduces the friction between packing and moving rod. By using three or more V rings the pressure drop against which each ring must operate is greatly diminished, resulting in longer life and lower friction. Likewise, by using multiple layers of V rings they can be cut and assembled over the rod without the need of a free end on the rod. The joints are staggered. Figure 12 is a spring-loaded self-adjusting design of V ring assembly. Details of size and design of V packings are shown in Fig. 13.

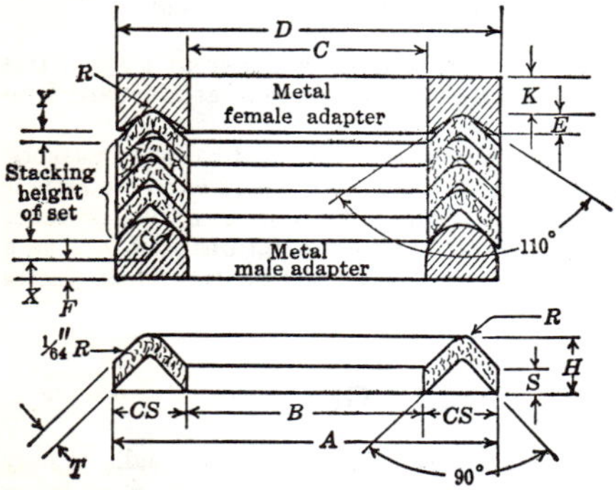

FIG. 13. Dimensions of Vee packing. (From *Mechanical Packings Manual*, by permission)

Cross-sectional width is $CS = (A - B)/2$, where A is the outside diameter of the stuffing box or packing recess or inside diameter of the cylinder and B is the outside diameter of the piston rod or inside diameter of the machined recess on the piston head (see Fig. 13). Ten cross-sectional widths have been agreed upon as standard by the Mechanical Leather Packings Manufacturers division of American Leather Belting Association: $^3/_{16}$, $^1/_4$, $^5/_{16}$, $^3/_8$, $^7/_{16}$, $^1/_2$, $^5/_8$, $^3/_4$, $^7/_8$, and 1 in. For new designs the six sizes $^1/_4$ to $^5/_8$ in. should be adhered to as they are the sizes generally used.

Thickness. Correct thickness T (Fig. 13) is $T = CS/4$, the absolute maximum being $CS/3$. For almost any job $^1/_8$-in. thickness is sufficiently large, though $^5/_{32}$- or $^3/_{16}$-in. can be used if cross section CS allows.

Dimensions. Formulas for the dimensions in Fig. 13 are:

S = stack height of one packing = $1.414T$
H = height of one packing = $CS/2 + T - {}^1/_{64}$ in.
R = outside radius of packing = radius in female adapter = $T + {}^1/_{64}$ in.
G = radius for male adapter = $CS/2 = (A - B)/4$
X = sealing clearance = $0.207CS$
Y = flexing allowance = $0.15CS$
E = female adapter groove depth = $0.7G - 0.22R$
F = optional (preferably equal to T)
K = optional ($CS/2$ is desirable)

Dimensions C and D, the inside and outside diameters of the female adapter, should have as close a fit as possible to the mating parts.

When the female adapter is made as shown in Fig. 13 the space created by the differential in angles of the Vee and adapter allows the legs to flex toward the surfaces to be sealed.

On initial installation it is sometimes necessary to tighten the follower onto the set to get a slight initial spreading of Vees to seal properly.

The differential of angles between the Vees and female adapter and the shape of the male adapter shown on the left promote the spreading action, leave room for the legs to flex toward the surfaces to be sealed, and generally improve the performance of the installation.

When the Vee packings have served their normal service life, but a few hours more run is required; by tightening the follower and compressing the Vees, the set will respond to compression.*

Molded synthetic V rings have been standardized by the Army and Navy under AN6225. Design details of these rings and adapters are given in Fig. 14. The AN specification material from which these rings are molded is used for operating temperatures ranging from −65 to +160 F. Other compounds can, of course, be used to give operating characteristics suitable for higher tempera-

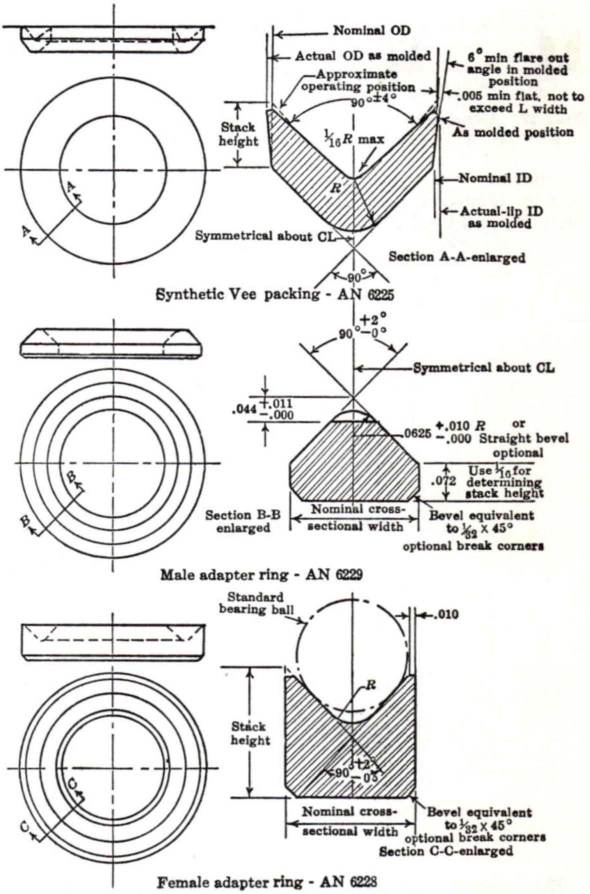

Fig. 14. Design of molded synthetic V rings standardized by the Army and Navy.

* From *Mechanical Packings Manual*, published by Graton and Knight Co., Worcester, Mass.

tures. For complete data see the Army-Navy standard AN6225, or manufacturers' catalogs.

FLANGE PACKING. Figure 15 illustrates a flange packing. This molded or preformed packing is available in standard dimensions. It is frequently installed with a means for insuring pressure of the packing lip against the piston, rod, or shaft. The use of garter springs and flat leaf springs for this purpose is quite common.

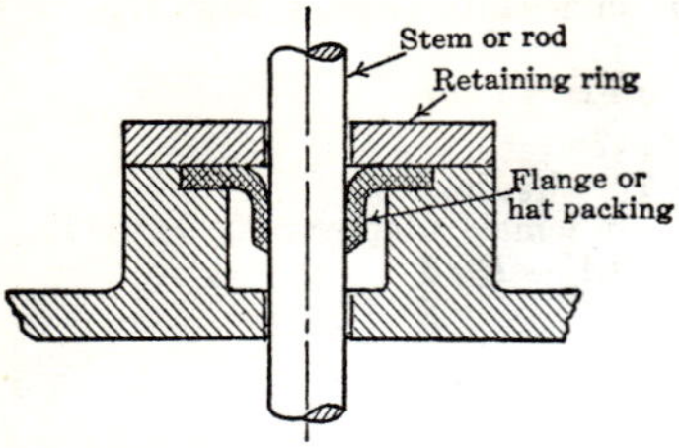

FIG. 15. Flange packing. (From *Machine Design*, by permission)

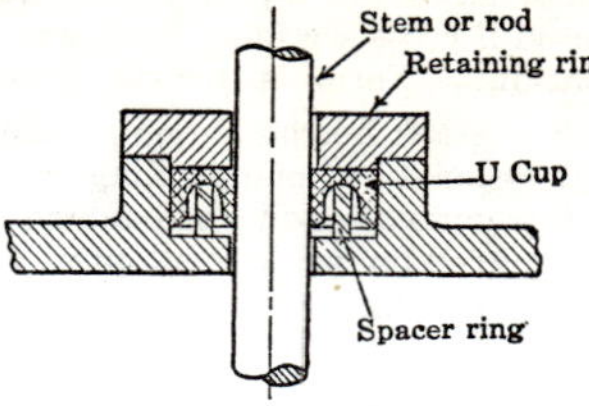

FIG. 16. Dynamic seal using molded U cup. (From *Machine Design*, by permission)

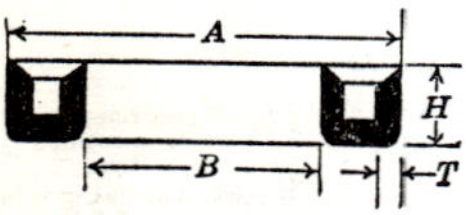

FIG. 17. Dimensions of U packing. See Table 3.

U PACKINGS. Figure 16 shows a shaft or piston rod seal using a molded U cup. It is fitted in a gland of suitable proportions and held in place by a permanently located cap or retaining ring. The U cup is supported by a spacing ring as illustrated. Leather U cups are molded in sizes shown in Fig. 17 and Table 3. The gland dimensions and spacer are made to suit these sizes.

Molded U cups are available in synthetic rubber for use in oil hydraulic systems. They conform to U. S. Army and Navy specifications, but can also be made in special sizes. See catalogs by manufacturers such as Graton and Knight, E. F. Houghton and Co., Linear Inc., and Victor Manufacturing and Gasket Co.

Table 3. Recommended Proportions for U Packings

See Fig. 17

(Reprinted by permission from *Mechanical Packings Manual*, Graton and Knight)

A Outside Diameter (Packing Carried in Gland or Recess), in.	*B* Inside Diameter, in.	*H* Height of Packing, in.	*T* Thickness of Leather, in.
1 3/4	1	3/8	1/16
3	2	7/16	3/32
4	3	1/2	1/8
5	4	5/8	1/8
6 1/4	5	5/8	5/32
7 1/4	6	3/4	5/32
8 1/2	7	3/4	3/16
9 1/2	8	7/8	3/16
10 1/2	9	7/8	3/16
11 1/2	10	1	3/16
13 1/2	12	1	3/16
15 1/2	14	1	3/16
17 1/2	16	1	3/16
19 1/2	18	1 1/4	3/16
21 1/2	20	1 1/4	3/16
23 1/2	22	1 1/4	3/16
25 1/2	24	1 1/4	3/16
29 3/4	28	1 1/4	3/16
33 3/4	32	1 1/4	3/16
37 3/4	36	1 1/4	3/16
42	40	1 1/2	3/16
46	44	1 1/2	3/16
50	48	1 1/2	3/16

The above table is sound practice, but greater heights can be furnished as required for existing equipment. It should be remembered that friction is directly proportional to the area of the U contacting the surface to be sealed and that unnecessary height increases friction but adds nothing to the performance or life of the packings. On new designs, adhere to values given in this table.

HIGH-PRESSURE PACKINGS. The packing in Fig. 18 seals relatively high pressure and occupies a minimum of space. It is the simple O ring, a patented device that is available in a large range of sizes that were standardized under Army and Navy specifications, AN-6227. It is usually molded from a synthetic rubber composition suitable for use at temperatures ranging from −65 to + 160 F. Figure 5 and Table 2 give the dimensions of the groove for all standard sizes of the packing. The important point to watch in using O-ring packings is the production of extra smooth, 10 microinch or better, surfaces. These smooth surfaces can be produced inexpensively with modern honing and superfinishing methods. The assembly of units using O rings must be done with extra care to prevent the ring's being damaged by sharp corners. Rings should be wet with oil before assembly is attempted.

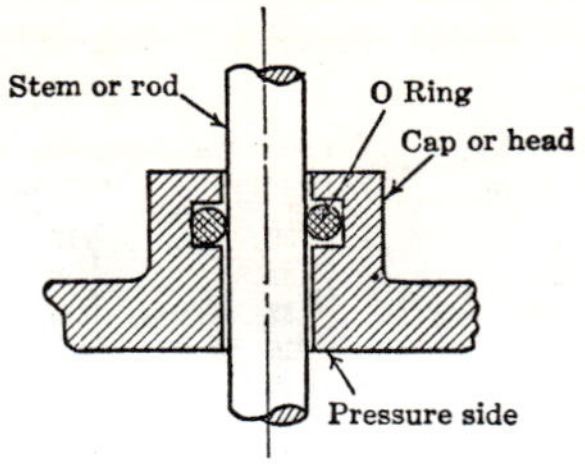

FIG. 18. O ring dynamic seal for high pressure. (From *Machine Design*, by permission)

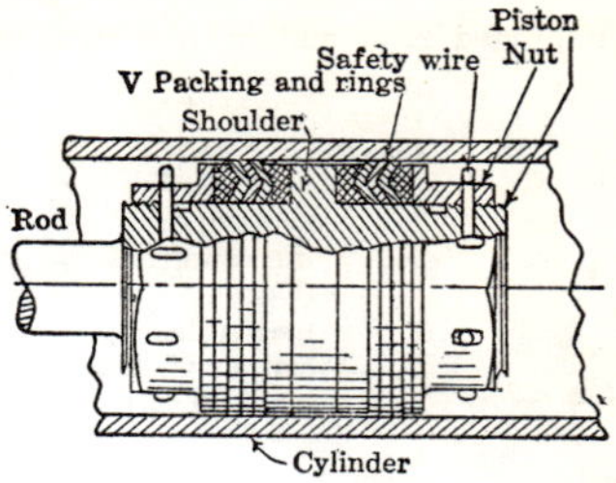

FIG. 19. Double-acting piston with V ring packing.

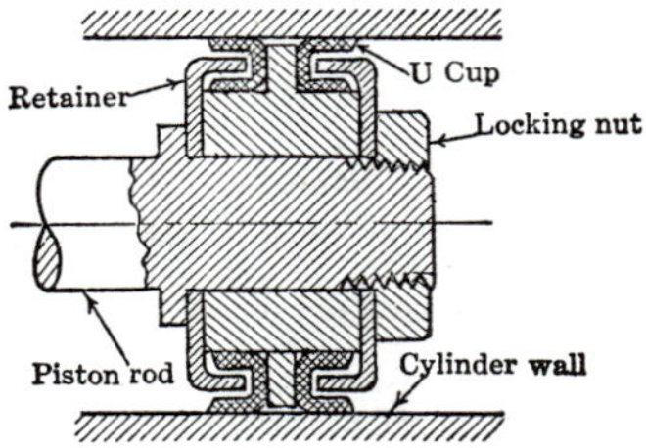

FIG. 20. Double-acting piston with U cup ring packing. (From *Machine Design*, by permission)

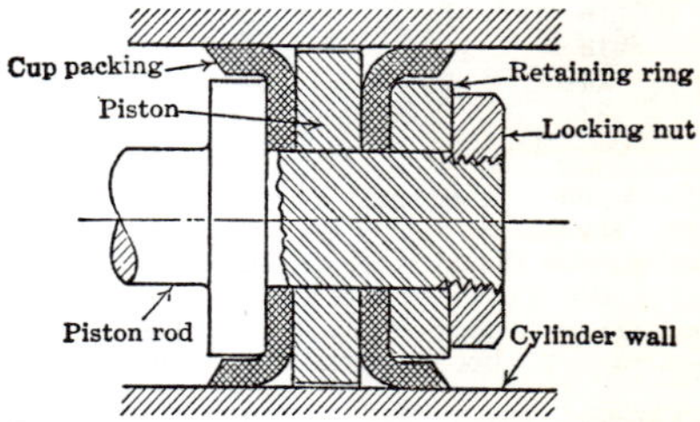

FIG. 21. Double-acting piston with cup packing. (From *Machine Design*, by permission)

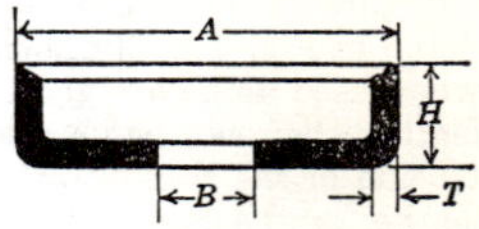

FIG. 22. Dimensions of cup packings. See Table 4.

Figures 19, 20, and 21 show, respectively, the assembly of V rings, U cups, and cup packings on double-acting pistons. The V ring assembly shows two sets of packings facing in opposite directions. The chevrons formed by the V section face away from each other so that no pressure can build up between the two sets. Means are provided to lock the adjustment since once the assembly is complete it is difficult to reach this adjusting nut.

Figure 20 is shown assembled in a fixed relation. Fluid pressure is depended on to exert pressure against the packing lip to effect a seal. This is possible with a U cup or cup packing whose lip is flexible but is not practical with a chevron-type packing. Figure 22 and Table 4 give dimensions of the cup packing illustrated in Fig. 21. This

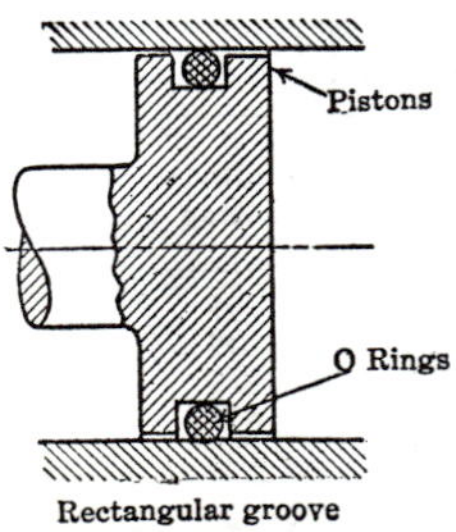

FIG. 23. Application of O ring in piston packing. (From *Machine Design*, by permission)

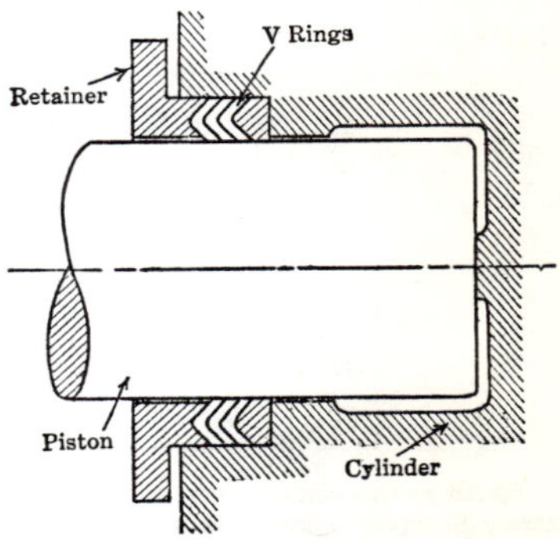

FIG. 24. Outside-packed single-acting piston. (From *Machine Design*, by permission)

Table 4. Recommended Cup-packings Sizes

See Fig. 22

(Reprinted by permission from *Mechanical Packings Manual*, Graton and Knight)

A Outside Diameter, in.	*B* Hole	*H* Height of Packing, in.	*T* Thickness of Leather, in. Minimum	Maximum
1/2	To suit operat-	1/4	1/32	1/16
3/4	ing conditions.	5/16	1/32	1/16
1	(Consult	3/8	1/32	3/32
2	mechanical	1/2	1/16	1/8
3	leather packings	5/8	3/32	5/32
4	manufacturer.)	5/8	3/32	5/32
5		3/4	3/32	5/32
6		3/4	1/8	3/16
7		3/4	1/8	3/16
8		1	1/8	3/16
9		1	1/8	3/16
0		1	1/8	3/16
12		1 1/4	1/8	3/16
14		1 1/4	1/8	3/16
16		1 1/2	1/8	3/16
18		1 1/2	1/8	3/16
20		1 3/4	1/8	3/16
22		1 3/4	1/8	3/16
Over 24		2	1/8	3/16

Table 4 gives the generally accepted proportions for cup packings as accepted by the Mechanical Leather Packings Division of the American Leather Belting Association.

The height *H* should never exceed values shown. Increased height does not improve seal, but in creases friction on the cylinder wall.

The diameter of the hole or holes and their location are optional, but care must be exercised to leave ample area on the bottom of the cup to anchor the cup securely to the piston head or plunger.

cup packing molded from properly treated leather makes the best piston packing for water pumps and is widely used on pumps for domestic water supplies on farms and suburban homes. It is also frequently used on the pistons of small air pumps such as bicycle pumps and atomizing spray equipment. It must be properly supported by the piston and a retaining ring or washer.

Figure 23 shows the use of O rings in piston packings. This type of packing is shown in detail in Fig. 5. All dimensions conform to Army and Navy standards AN-6227. O rings when used alone are good for pressures up to 1500 psi. When used with back-up rings of leather they can be used up to 3000 psi. The problem at high pressures is to prevent extrusion of the ring into the piston-cylinder clearance. Various materials such as bakelite, leather, and soft metals have been tried for back-up rings. When hard materials are used the ring must be split or the piston assembled from parts having the groove in them. Rectangular groove dimensions for use with O rings and back-up washers are listed in Table 5.

Only one O ring is recommended in any one installation, it should be noted. Oil will be trapped between rings used in multiple, thereby increasing the packing friction and de-

Table 5. Proportions of Rectangular Grooves for O Rings and Back-up Washers

Dash No.	Section Diameter Nominal	Gland Width *C* max (Fig. 5)		Minimum Squeeze	Groove Length +.005 −.000	
		Dynamic	Static		One Back-up Ring	Two Back-up Ring
1–7	1/16	.057	.052	.010	.139	.206
8–14	3/32	.090	.083	.010	.171	.238
15–27	1/8	.123	.113	.012	.208	.275
28–52	3/16	.188	.173	.017	.311	.410
53–88	1/4	.240	.220	.027	.408	.538

Piston groove diameter equals bore diameter (at plus tolerance) minus 2*C*. Rod gland groove diameter equals piston groove diameter (at minus tolerance) plus 2*C*.

Groove length *D* = actual *W* of O ring + nominal thickness of leather or leathers + 1/2 (plus tolerance of all components).

creasing its life. The rectangular groove will give slightly lower starting friction and is easier to produce to the required dimensions. It is the preferred form, and the V-groove for O rings is becoming obsolete.

In the design of single-acting pistons a choice of packing designs is indicated in Figs. 24 and 25. Pistons may be "outside" packed as in Fig. 24, having the advantage of ease of

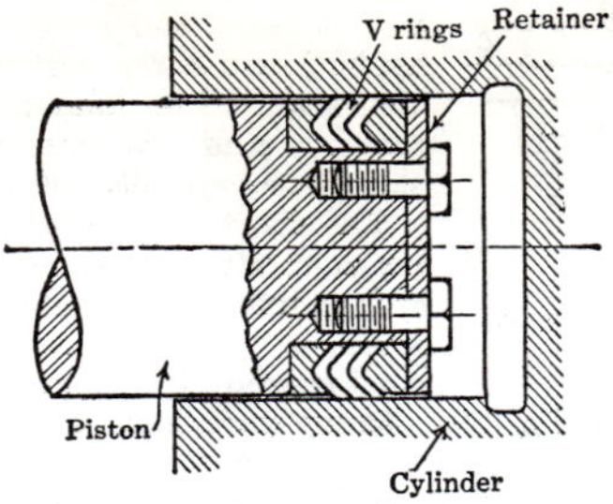

FIG. 25. Inside-packed single-acting piston. (From *Machine Design*, by permission)

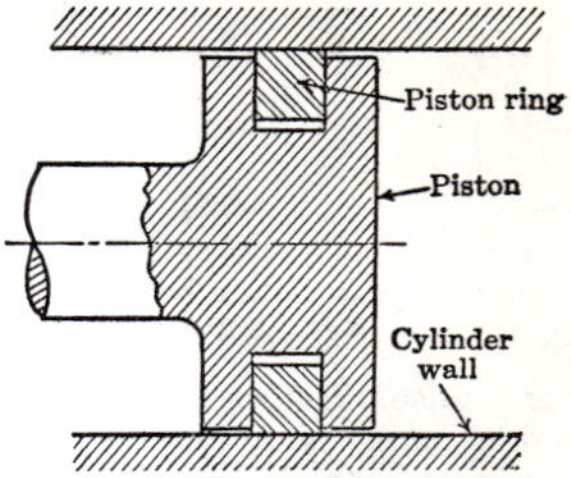

FIG. 26. Piston ring used as dynamic seal. (From *Machine Design*, by permission)

maintenance and adjustment; or "inside" packed as in Fig. 25, having the advantage of a smaller overall cylinder size, but difficult maintenance features. The choice of which design to use will depend on the circumstances and factors affecting the particular situation under consideration: space, weight, leeway from a perfect seal, etc.

Figure 26 illustrates the use of a piston ring as a dynamic seal. Single or multiple rings may be used, depending on the pressure to be sealed. They are used primarily under high-temperature conditions where nonmetallic packings would quickly fail. It is also used because of its low packing friction.

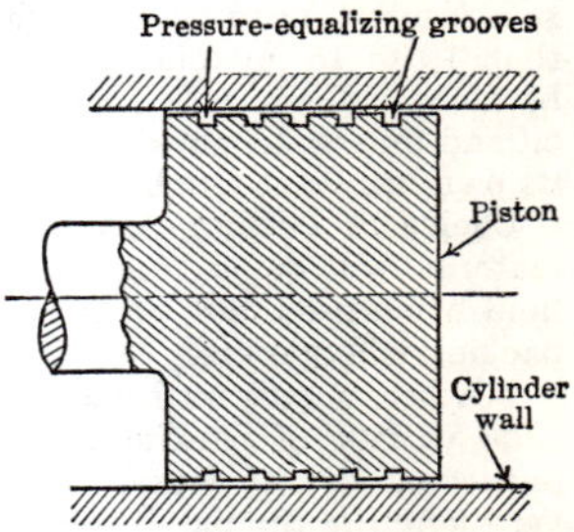

FIG. 27. Low-friction piston design where slight leakage is permissible. (From *Machine Design*, by permission)

Figure 27 shows a piston design that is used when minimum friction is required and some leakage can be tolerated. Small annular grooves are turned or ground in the cylindrical surface of the piston to insure equal distribution of pressure around the piston, thereby causing it to stay in the center of the cylinder. These grooves are made about $^1/_{16}$ in. wide by $^1/_{32}$ in. deep for a 1-in. piston, proportionately larger for larger piston diameters.

The leakage past such a piston can be accurately calculated if the dimensions of the cylinder and piston are known accurately; and if the pressure drops across the piston, the total effective length of the piston and the absolute viscosity of the oil are known. The basic formula * used for flow between two parallel planes is

$$Q = \frac{1}{12} \frac{C^3 \times \Delta p \times W}{\mu \times l}$$

in which Q is flow, cubic feet per second; C is space between planes, feet; W is width of flow path, feet; Δp is pressure drop, pounds per square feet between points; l is distance in feet along the path of flow; and μ is the absolute viscosity in slugs/foot-second. For a piston in a cylinder l is the length of the piston, C the radial clearance, and W the circumference ($2\pi R$). The formula would then become

$$Q = \frac{1}{12} \frac{C^3 \times \Delta p \times 2\pi R}{\mu \times l} = \frac{\pi}{6} \times \frac{C^3 \times \Delta p \times R}{\mu \times l}$$

Q is then multiplied by a factor 2.5 to give the flow for maximum eccentricity between piston and cylinder bore. Any consistent units of measurement can be used in place of those indicated.

SHAFT SEALS have gained considerable popularity for use in connection with ball-bearing installations. They are of three basic designs, as shown in Fig. 28, namely, leaf spring (*a*), garter spring (*b*), and self-sealing (*c*). There is little choice between reliable makes of shaft seals.

* C. E. Grosser, Hydraulic leakage can be controlled, *Machine Design*, Aug. 1942.

The sealing lip is compounded to suit the average operating conditions. The primary criterion is surface speed under the sealing lip. It is important that the seal is designed for the shaft or surface speed at which it will be used. If a slow-speed design is used for high shaft speeds, excessive friction will develop and result in overheating and early destruction of the seal. The manufacturer's ratings and recommendations should be carefully followed.

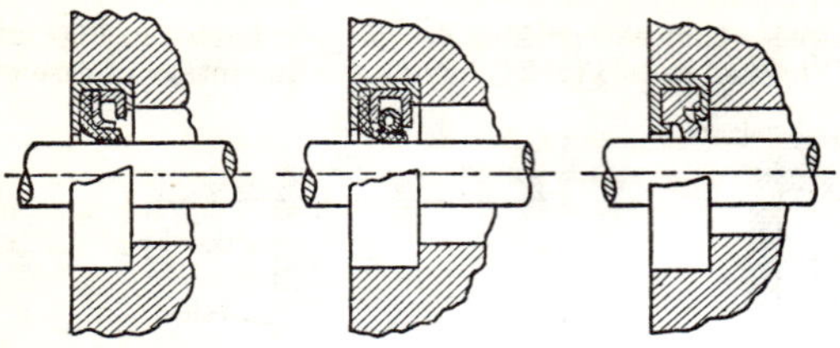

(a) (b) (c)

FIG. 28. Seals for antifriction bearings. (From *Machine Design*, by permission)

For shaft installations requiring a relatively permanent seal and minimum maintenance to preserve perfect sealing, the steel-to-graphite lapped seal is the proper choice. Such a seal is illustrated in Fig. 29. Many variations of this design are available, some using bellows and some rubber as the means of sealing the graphite or bronze ring to the shaft. The main operating feature is the use of lapped surfaces of mutually low coefficient friction materials for the sliding surfaces.

FACTORS TO BE CONSIDERED IN PACKING DESIGN. Surface Finish. Too much cannot be said regarding the importance of having smooth surfaces for packings to operate against. In order that dynamic seals may have a maximum life every effort must be made to have and to maintain smooth surfaces over which the packing slides. The ideal surface is one machined to a finish of 10 to 15 microinches rms. This can be done by fine grinding or honing. It is interesting to note that if the surface is smoother than 10 to 15 microinches, lubrication of the packing becomes difficult as the surface is too smooth to hold oil and unless the packing material is porous and holds its own lubricant trouble may result.

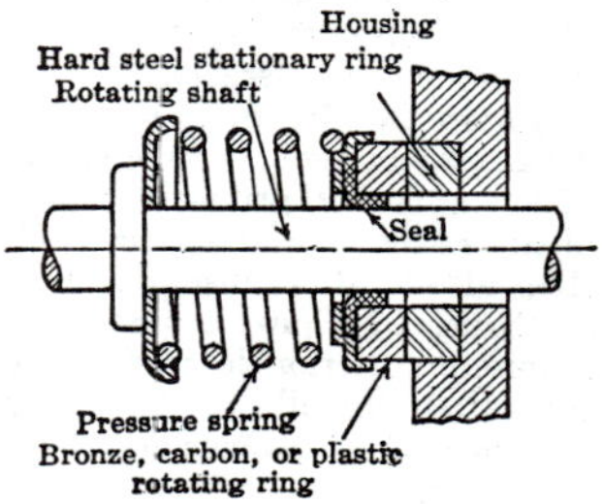

FIG. 29. Lapped seal. (From *Machine Design*, by permission)

Choice of Packing Material. Choice of a packing material will depend on the type of installation and fluid being used, also on the temperature at which the packing will operate.

Leather, a universal packing material, is used for a wide variety of packings. It is usually molded and impregnated with waxes or oils best suited for a particular service condition. It has a very long life compared to rubber, but does not give as good an initial seal.

Synthetic rubber is not as porous as leather, and for this reason it has many applications where leather cannot be used. Rubber tends to cold bond to metal, and if dynamic rubber seals stand idle too long they may resist motion to the point of breaking machine parts. This tendency has been overcome in recent synthetic compounds. Reference is made to packing manufacturers' manuals which cover the choice of materials more completely.

Pneumatic devices in general are best designed with a porous packing, such as leather which can be self-lubricating for dynamic seals.

Where temperatures are high, asbestos compounds are needed. Recent developments in silicone rubber should yield good high-temperature packings. However, silicone rubber so far is good only for static seals because of its low tensile strength and poor wearing quality.

Valve stems are usually packed with some form of wick or braided packing impregnated with a suitable grease or graphite to act as a lubricant.

Rotating pump shafts are sometimes packed in a manner similar to valve stems. More recent designs employ molded V rings or rotating seals similar to those shown in Fig. 30. The rotating metallic and graphite seals, while more expensive in the initial installation, give less trouble and have lower maintenance costs.

Proper Support for the Packing or Seal. The finest packing material will fail if not properly used in a given installation. The backing rings and adapters for V rings, V cups, and other molded packings must be of proper dimensions to support the packing materials and allow proper distribution of the pressure load on the packing. Clearance between cylinder and piston and between rod and bearing must be at a minimum to avoid excessive packing extrusion. Groove dimensions for O rings must be held to proper tolerances. Packing manuals and manufacturers' specifications should be carefully followed.

Assembly of Packings. Great care should be taken during the assembly of packings into hydraulic or pneumatic units to insure that these packings are clean and free from defects. Extreme care must be exercised in passing packings over threads on a piston rod or across sharp valve ports in a cylinder to avoid cutting the packing. Shim steel ferrules used over the threads safeguard the packings. This advice applies equally to rubber or leather packings. On assembling mating surfaces with a gasket, the gasket should be free from tears or breaks and the metal surfaces must be clean. The thinnest possible gasket that will do the work should be used.

Care and Maintenance. Thought should be given in the design of dynamic seals to the amount of servicing they will require. This amount will be greatly influenced by operating conditions: temperature, cleanliness, ratio of operating to shut-down time, lubrication of packing, etc. If adjustments are necessary in a given packing design, they should be accessible and easy to make. The ability to replace worn packings without considerable disassembly of mechanisms is of great value.

New Data. Because of rapid advances in the art of packing design, the development of new materials, and the standardization of data throughout the hydraulic industry, it is recommended that engineers check with packings manufacturers to be sure of using the latest available information. Publications of the Joint Industry Conference (JIC) in particular should be consulted.

GENERAL REFERENCES

Army and Navy Drawings: AN-6225; AN-6226; AN-6227; AN-6228; AN-6229; AN-6230; AN-902.
Army and Navy Specifications: AN-HH-P-114; AN-H-2.
P. W. Bridgman, *Physics of High Pressure*, The Macmillan Co., New York.
Catalog F, 1940, The Garlock Packing Co., Palmyra, N. Y.
Design and materials for hydraulic packings, *Aero Digest*, April 1942, p. 119.
Gaskets, Packings, and Seals, a bulletin published by the Armstrong Cork Co., Lancaster, Pa.
Hydraulic and Pneumatic Packings, published by E. F. Houghton & Co., Philadelphia, Pa.
L. S. Linderoth, Jr., Selecting hydraulic seals, *Machine Design*, Sept. 1944, pp. 119–128.
Stanley L. Lopata, Practical approach to packing of chemical pumps, *Chem. Metallurgical Eng.*, Dec. 1944.
H. C. Matheis, *Design and Application of Leather U Packings*, The Modern Industrial Press.
Mechanical Packings Manual, published by Graton and Knight Co., Worcester, Mass.
W. J. Montgomery, Chromatic gasketing, *Sheet Metal Worker*, Feb. 1945.
O-ring packings simplify lightweight hydraulic equipment, *Product Eng.*, Sept. 1944.
F. C. Thorn, Gaskets, *Ind. Eng. Chem.*, Feb. 1936, pp. 164–170.
Handbook on Leather Packings, Handbook of Synthetic Packings, E. F. Houghton and Co., Philadelphia.

COTTERED JOINTS, PIN JOINTS, AND TAPERS

3. COTTERED JOINTS

A cotter is a form of key used to join parts subjected to a tension or compression, tending to separate them. The taper of cotters should not exceed $^1/_2$ in. per ft of length unless a locking device is used. Various forms of cotters are given in Figs. 1 to 7 and in Table 1,

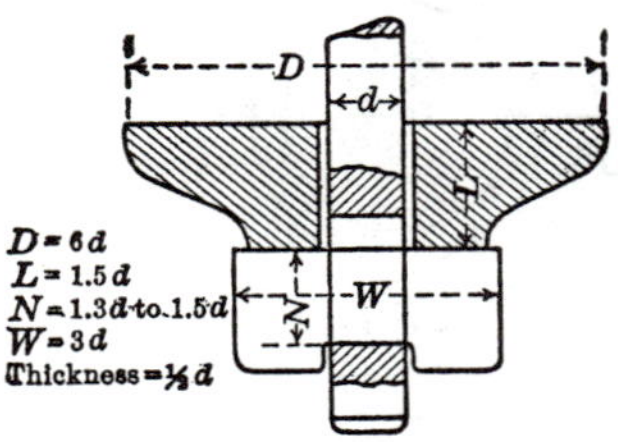

FIG. 1

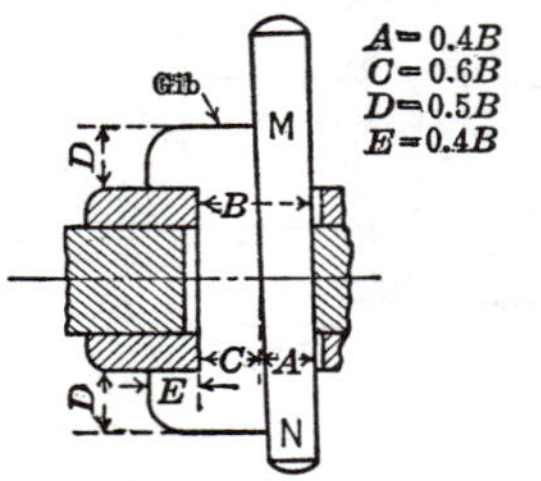

FIG. 2

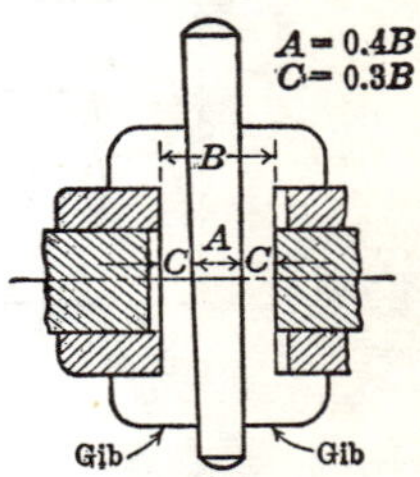

FIG. 3

which shows the proportions of the joint. The function of the gib, in the gib and cotter joint (Figs. 2 and 3) is to prevent spreading of the strap. When a gib is used, the sides of the hole are made parallel, the taper being given to the gib. Where two cotters are used, as in Fig. 3, the taper is divided equally between them. The tapered joint (Fig. 6) usually is of permanent character. The cotter is removed only when the members which it holds are to be repaired or replaced. It is largely used for joining pistons and piston rods.

One form of grip, taken from *Machine Design Construction*, by Spooner, is shown in Fig. 7. The joint may fail in any one of ten ways.

Table 1. Dimensions of Cottered Joints

(See Figs. 4–6)

	Rod	Cotter	Dimensions								
			A	*B*	*C*	*D*	*E*	*F*	*G*	*H*	*J*
Fig. 4	Wrought iron	Steel	0.3*d*	1.21*d*	1.75*d*	1.5*d*	2.4*d*	0.42*d*	0.8*d*	1.6*d*	0.8*d*
Fig. 5	{ Steel { Wrought iron	Steel } Wrought iron }	0.3*d*	1.21*d*	....	1.5*d*	......	0.5*d*	......	1.6*d*	1.0*d*
Fig. 6	Steel	Steel	0.25*d*	....	....	......	......	......	0.9*d*	0.9*d*	0.9*d*

Notation. s_t = ultimate strength of rod ends in tension; s_s = ultimate strength of cotter in shear; s_b = ultimate strength of bearing surfaces = say, $2s_t$; s'_s = ultimate strength of steel or iron to resist shear in direction parallel with fibers; all in pounds per square inch; F = axial force causing rupture of the joint. The significance of the other quantities in the equations is shown in Fig. 7, all dimensions being taken in inches.

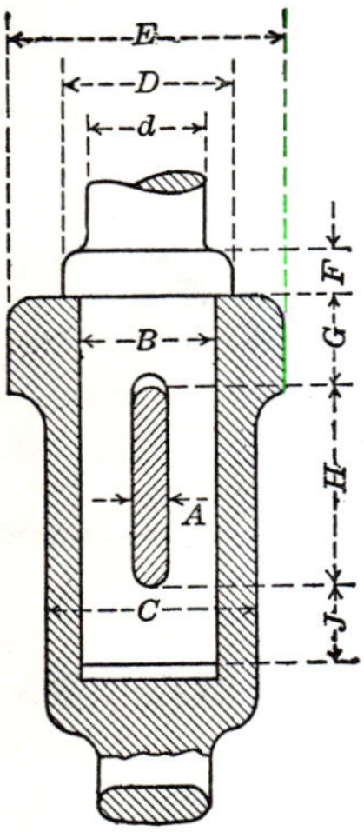

FIG. 4

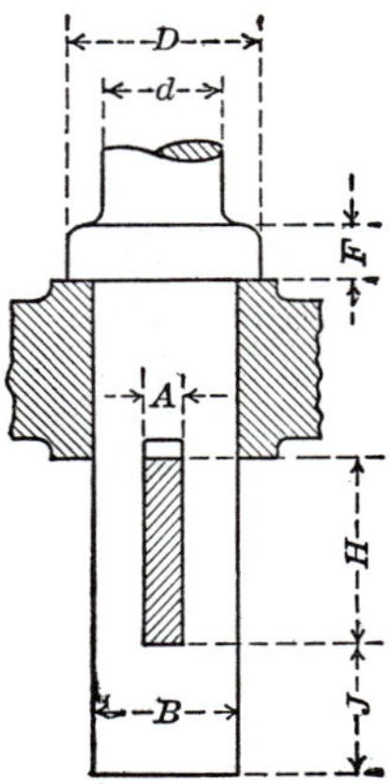

FIG. 5

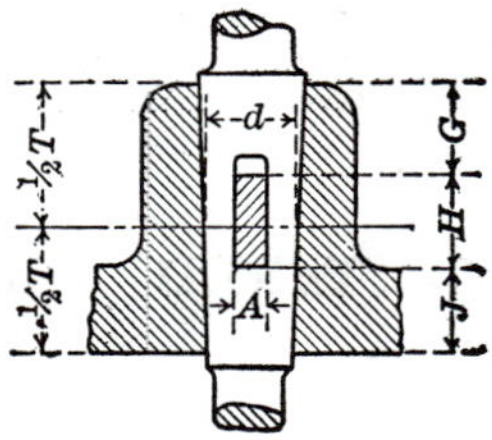

FIG. 6

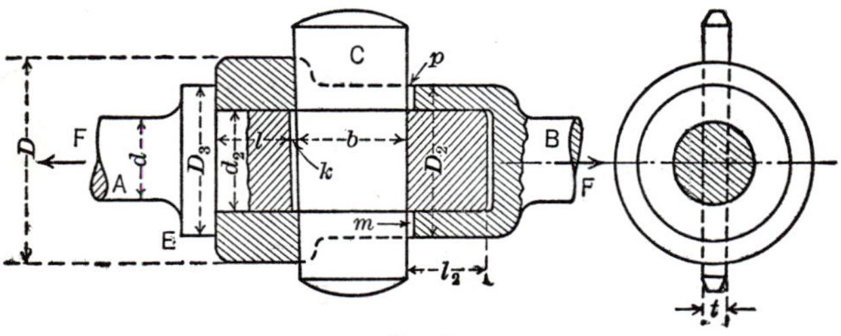

FIG. 7

1. The rods may fail in tension at A or B, in which case the external load

$$F = {}^1/_4\ \pi d^2 s_t$$

2. By the rod end tearing across the cotter hole; $F = ({}^1/_4\ \pi d_2^2 - d_2 t)s_t$. Equate cases 1 and 2 and let $t = {}^1/_4\ d_2$; then

$$d_2 = 1.21d, \quad d = 0.82d_2, \quad \text{and} \quad t = 0.3d$$

For joints with all parts either wrought iron or steel, the breadth of the cotter usually is about 1.6*d*.

3. By the socket tearing across the cotter hole; then

$$F = [{}^1/_4\,\pi(D_2^2 - d_2^2) - (D_2 - d_2)t]s_t$$

whence by equating case 1, $D_2 = 1.63d$, or allowing for clearance, say, $D_2 = 1.75d$.

4. By double shearing of the cotter,

$$F = 2bts_s \quad \text{and} \quad b = 1.31ds_t/s_s$$

or $1.06d_2s_t/s_s$, where b = mean breadth of the cotter.

5. By crushing the cotter bearing surface in the rod end; $F = d_2ts_b$.

6. By crushing the cotter bearing surface in socket; $F = t(D - d_2)s_b$ and $D = 2.42d$.

7. By shearing socket end; $F = 2\,[l(D - d_2)s'_s]$.

8. By shearing rod end; $F = 2l_2d_2s'_s$. $l = l_2 = 0.648d$; make l from $0.75d$ to $1.0d$.

9. By crushing the collar E; $F = {}^1/_4\,\pi(D_3^2 - d_2^2)s_b$ and $D_3 = 1.3915d$, say, $1.4d$.

10. By shearing the collar E off the rod end; $F = d_2\pi t_2s'_s$ and $t_2 = 0.413d$, say, $t_2 = 0.42d$, where t_2 is thickness of collar.

4. PIN JOINTS

STRENGTH OF KNUCKLE JOINT PIN. (Spooner, *Machine Design.*) The proportions of a forked pin joint or knuckle joint are shown in Fig. 8. If there is no bending

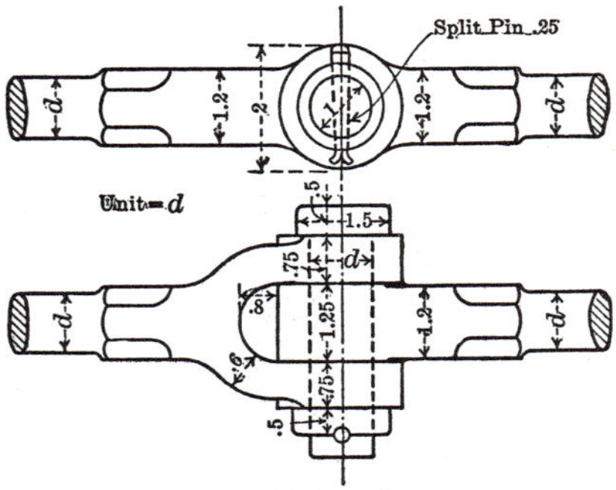

FIG. 8. Knuckle joint.

action, the pin will fail in double shear. If d = diameter of rod, d_1 = diameter of the pin, s_t = tensile strength of rod, s_s = shearing strength of pin, pounds per square inch, = say, $0.8s_t$, then

$$ {}^1/_4\,\pi d^2s_t = 2({}^1/_4\,\pi d_1^2s_s) \quad \text{or} \quad d_1 = 0.79d$$

Unwin shows that where bending occurs and the joint is subjected to stresses reversing in sign, a skin stress of 5400 psi, due to bending, will not be exceeded if the diameter of the pin $= 0.0224\sqrt{F}$, where F = axial force in the rod.

SAE STANDARD ROD ENDS AND PINS. Proportions of adjustable and plain yokes, and eyes and pins, are given in Fig. 9 and Table 2.

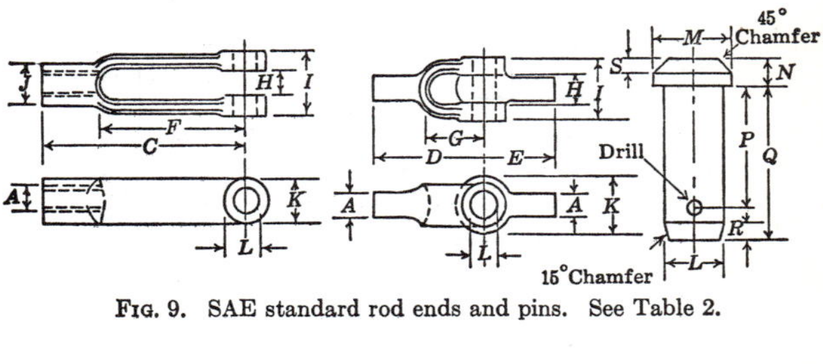

FIG. 9. SAE standard rod ends and pins. See Table 2.

5. TAPERS

AMERICAN STANDARD TAPER. The ASA standards for self-holding tapers comprise twenty-two sizes. The three smallest sizes (Table 3) are taken from the Brown and

(Continued on p. 18-20)

Table 2. Rod Ends and Pins

(SAE Standard. See Fig. 9.)

Series	A (Fine Thread Screw)	C	D	E	F	G	H *	I †	J	K	L: Hole Diameter, Nominal	L: Hole Diameter, Tolerance +0.001 Minus	L: Pin Diameter, Maximum	L: Pin Diameter, Minimum	M	N	P	Q	R	S	Drill
Light	No. 10	1 9/16	1 1/4	1 1/4	1	7/16	3/16	7/16	5/16	3/8	3/16	.001	.186	.181	5/16	1/16	31/64	19/32	3/64	1/64	5/64
	1/4	2	1 3/4	1 1/4	1 1/4	5/8	9/32	5/8	7/16	1/2	1/4	.001	.248	.243	3/8	3/32	43/64	51/64	1/16	1/32	5/64
	5/16	2 1/4	2	1 3/8	1 7/16	3/4	11/32	3/4	1/2	19/32	5/16	.001	.311	.306	7/16	3/32	13/16	31/32	5/64	1/32	7/64
	3/8	2 1/2	2 1/8	1 1/2	1 5/8	27/32	7/16	7/8	5/8	11/16	3/8	.001	.373	.368	1/2	1/8	15/16	1 3/32	5/64	1/32	7/64
	7/16	2 7/8	2 1/4	1 1/2	1 7/8	1	1/2	1	23/32	13/16	7/16	.001	.436	.431	9/16	5/32	1 1/16	1 15/64	3/32	3/64	7/64
	1/2	3	2 1/2	1 3/4	1 7/8	1 1/8	9/16	1 1/8	13/16	15/16	1/2	.002	.496	.491	5/8	5/32	1 13/64	1 27/64	7/64	3/64	9/64
Heavy	1/2	4 3/16	2 1/2	1 3/4	3 1/16	1 1/8	9/16	1 1/8	13/16	15/16	1/2	.002	.496	.491	5/8	5/32	1 13/64	1 27/64	7/64	3/64	9/64
	5/8	4 15/16	2 7/8	2	3 11/16	1 7/16	11/16	1 3/8	1 1/16	1 3/16	5/8	.002	.621	.616	13/16	13/64	1 15/32	1 23/32	9/64	1/16	9/64
	3/4	6 1/16	3 5/8	2 3/8	4 9/16	1 11/16	13/16	1 5/8	1 1/4	1 7/16	3/4	.002	.746	.741	15/16	1/4	1 3/4	2 3/64	5/32	5/64	11/64
	7/8	7 1/8	4	2 3/4	5 1/4	2	15/16	1 7/8	1 7/16	1 11/16	7/8	.002	.871	.866	1 1/16	5/16	2 1/64	2 11/32	3/16	3/32	11/64
	1	8	4 1/2	3	6	2 1/2	1 1/16	2 1/8	1 5/8	1 15/16	1	.002	.996	.991	1 3/16	11/32	2 17/64	2 5/8	7/32	7/64	11/64

* Fork tolerance ±0.010; eye tolerance +0.000, −0.010.

† Tolerance ±0.010.

Table 3. Machine Tapers, Self-holding, Tang Drive

(ASA B5.10-1943. All dimensions in inches.)

No. of Taper	Taper per Foot	Diameter at Gage Line	Shank Retained by Friction (See Fig. 10)												Shank Retained by Key (See Fig. 11)					
			Shank		Tang					Socket		Knockout Keyway *			Shank			Socket		
			Overall Length	Exposed Length	Thickness	Length	Radius of Mill	Diameter	Radius	Depth	Gage Line to Keyway	Width	Length	Shank End to Back of Keyway	Gage Line to Bottom of Keyway	Length	Width	Gage Line to Front of Keyway	Length	Width
		A	B	C	E	F	G	H	J	K	M	N	O	P	Y′	X	N′	Y	Z	N′
.239	.50200	0.23922	1 9/32	3/32	0.125	3/16	3/16	11/64	1/32	1 1/16	15/16	0.141	3/8	1/8						
.299	.50200	0.29968	1 19/32	3/32	0.1562	1/4	3/16	7/32	1/32	1 5/16	1 11/64	0.172	1/2	11/64						
.375	.50200	0.37525	1 31/32	3/32	0.1875	5/16	3/16	9/32	3/64	1 5/8	1 15/32	0.203	5/8	7/32						
1	.59858	0.47500	2 9/16	1/8	0.2031	3/8	3/16	11/32	3/64	2 3/16	2 1/16	0.213	3/4	3/8						
2	.59941	0.70000	3 1/8	3/16	0.250	7/16	1/4	17/32	1/16	2 21/32	2 1/2	0.260	7/8	7/16						
3	.60235	0.93800	3 7/8	3/16	0.3125	9/16	9/32	23/32	5/64	3 5/16	3 1/16	0.322	1 3/16	9/16	1 1/32	1 1/8	.260	1 1/8	1 3/16	.260
4	.62326	1.23100	4 7/8	1/4	0.4687	5/8	5/16	31/32	3/32	4 3/16	3 7/8	0.478	1 1/4	1/2	1 13/32	1 3/16	.385	1 1/2	1 1/4	.385
4 1/2	.62400	1.50000	5 3/8	1/4	0.5625	11/16	3/8	1 13/64	1/8	4 5/8	4 5/16	0.573	1 3/8	9/16	1 23/32	1 1/4	.447	1 13/16	1 3/8	.447
5	.63151	1.74800	6 1/8	1/4	0.625	3/4	3/8	1 13/32	1/8	5 5/16	4 15/16	0.635	1 1/2	9/16	2	1 3/8	.510	2 1/8	1 1/2	.510
6 †	.62565	2.49400	8 9/16	5/16	0.750	1 1/8	1/2	2	5/32	7 13/32	7	0.760	1 3/4	1/2	2 1/8	1 5/8	.635	2 1/4	1 3/4	.635
7 †	.62400	3.27000	11 5/8	3/8	1.125	1 3/8	3/4	2 5/8	3/16	10 5/32	9 1/2	1.135	2 5/8	7/8	2 1/2	1 11/16	.760	2 5/8	1 13/16	.760

* Corners at entrance side of keyways shall be chamfered at 45 deg as follows: No. 3, 3/64 in.; all other sizes 1/16 in. deep.

† These sizes are continued in the tang drive series for the present to meet special needs.

Tolerances are:

For diameter of shank at gage line (A): All sizes, $+0.002 - 0.000$.

For diameter of hole at gage line (A): All sizes, $+0.000 - 0.002$.

For thickness of tang (E): Up to and including No. 5, $+0.000 - 0.006$. Larger than No. 5, $+0.000 - 0.008$.

For width of knockout keyway (N): Up to and including No. 5, $+0.006 - 0.000$. Larger than No. 5, $+0.008 - 0.000$.

For concentricity of tang (E) with center line of taper: Up to and including No. 5, 0.0035 (indicator reading). Larger than No. 5, 0.005 (indicator reading).

For concentricity of knockout keyway (N) with center line of taper: Up to and including No. 5, 0.0035 (indicator reading). Larger than No. 5, 0.005 (indicator reading)

Tolerances on fractional dimensions, ± 0.010 unless otherwise specified.

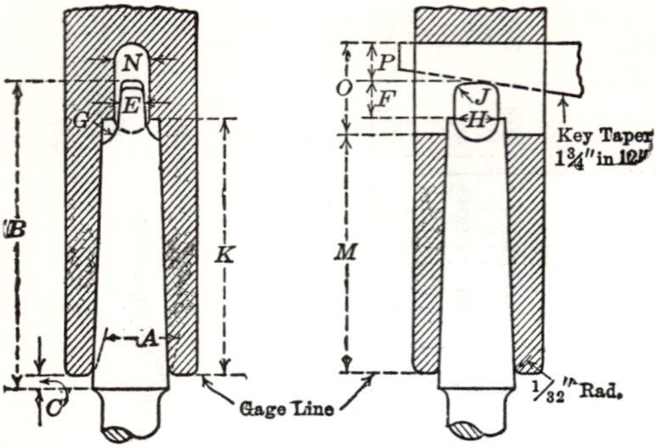

FIG. 10. Taper with tang drive, shank retained by friction. See Table 3.

Sharpe series and have a nominal taper of 1/2 in. per ft. The next eight sizes (Table 3) are taken from the Morse series and have a nominal taper of 5/8 in. per ft. The eleven largest sizes (Table 4) are based on a taper of 3/4 in. per ft.

The eleven smaller sizes (Table 3) employ a tang drive and may be retained entirely by friction (Fig. 10). Of these the six largest may be retained by a holdback key (Fig. 11). Removal of the shank from the socket is effected by starting with a drift key.

The eleven largest sizes (Table 4) employ a key drive. The shank may be retained in the socket either by a holdback key (Fig. 12) or a drawbolt (Fig. 13).

Steep Taper. Tapers listed in Table 5 have an angle sufficiently large to permit self-release. Retention of the shank in the socket is dependent on a positive locking device. The taper fit serves only to maintain alignment.

OTHER TAPERS. Prior to the adoption of the American Standard tapers several tapers were in use for spindles and drill shanks. The more important are listed in the following paragraphs.

Reed Lathe Center Taper. This comprises a series of ten tapered lathe centers of standard dimensions. Size is based on the size of the lathe.

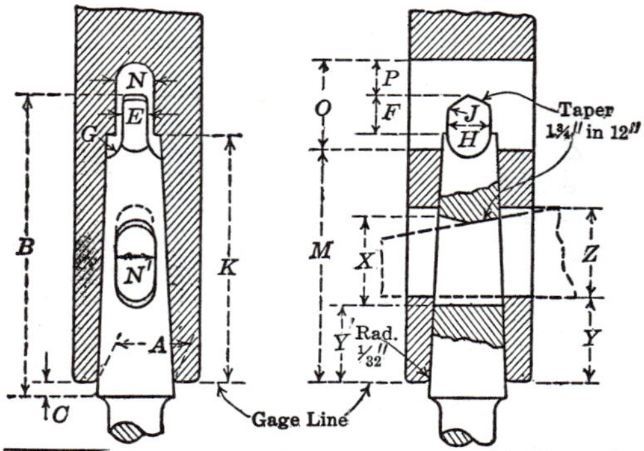

FIG. 11. Taper with tang drive, shank retained by key. See Table 3

Standard Tool Co. Taper. Used for twist drill shanks, this series comprises seven tapers numbered from 1 to 7. Taper ranges from 0.6 to 0.625 in. per ft. Dimensions are approximately similar to those of the corresponding Morse tapers.

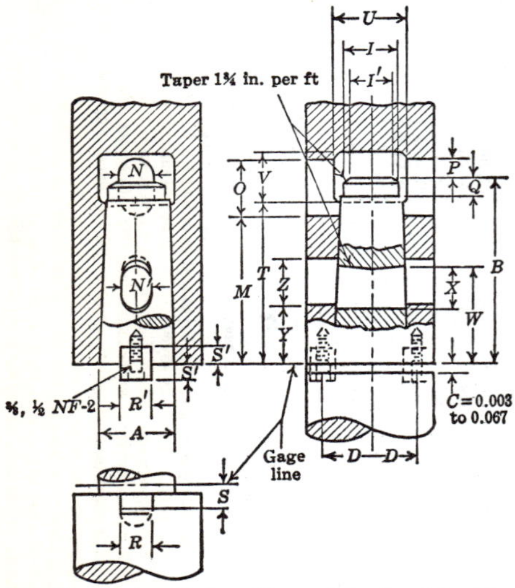

FIG. 12. Taper with key drive and shank retained by key. See Table 4.

Morse Tapers. Original series comprises eight tapers numbered from 0 to 7. The American Standard includes those from 1 to 7 with the addition of No. 4 1/2 (see Table 3).

Cleveland Twist Drill Tapers. This series comprises seven tapers numbered from 0 to 6, with taper ranging from 0.6 to 0.63 in. per ft. Dimensions are approximately the same as the corresponding Morse taper numbers.

Sellers Taper. This series consists of a set of seven sizes of standard shank-and-socket parts with a taper of 3/4 in. per ft. Intermediate sizes can be fitted to each other, and keys are provided for driving and for knockout.

Brown and Sharpe Taper. Complete series consists of eighteen tapers numbered from 1 to 18. The American Standard includes the first three of these (see Table 3).

Table 4. Machine Tapers, Self-holding, Key Drives

(ASA B5.10-1943. All dimensions in inches.)

No. of Taper †	Diameter at Gage Line	General Dimensions of Key Drive Tapers (See Figs. 12 and 13)													Shank Retained by Holdback Key (See Fig. 12)								Shank Retained by Drawbolt (See Fig. 13)		
		Shank						Drive Key		Socket					Holdback Keyway in Shank			Knockout * Keyway		Holdback Keyway * in Socket			Bolt	Shank	Socket
		Length from Gage Line	Length of Relief	Diameter of Flat	Diameter of Relief	Drive Keyway Width	Drive Keyway Depth	Center Line to Center of Screw	Size of Screw (*NF*)	Gage Line to Front of Relief	Diameter Relief	Depth Relief	Drive Keyway Width	Drive Keyway Depth	Gage Line to Back of Keyway	Length	Width	Length ‡	Shank End to Back of Keyway	Gage Line to Keyway	Gage Line to Front of Keyway	Length §	Diameter (*NC*)	Depth of Thread	Diameter of Draw Bolt Hole
	A	*B′*	*Q*	*I′*	*I*	*R*	*S*	*D*		*T*	*U*	*V*	*R′*	*S′*	*W*	*X*	*N′*	*O*	*P*	*M*	*Y*	*Z*			*d*
200	2.000	5 1/8	1/4	1 1/8	1 5/8	1.005	9/16	1 13/32	3/8	4 3/4	1 13/16	1	1.000	1/2	3 7/16	1 9/16	0.656	1 9/16	15/16	4 1/2	2	1 11/16	7/8	1 3/4	1
250	2.500	5 7/8	1/4	1 1/8	2 1/16	1.005	9/16	1 21/32	3/8	5 1/2	2 1/4	1	1.000	1/2	3 11/16	1 9/16	0.781	1 15/16	1 1/4	5 3/16	2 1/4	1 11/16	7/8	1 3/4	1
300	3.000	6 5/8	1/4	1 3/8	2 1/2	2.005	9/16	2 1/4	3/8	6 1/4	2 3/4	1	2.000	1/2	4 1/16	1 9/16	1.031	2 3/16	1 1/2	5 15/16	2 5/8	1 11/16	1	2	1 1/8
350	3.500	7 7/16	5/16	2	2 15/16	2.005	9/16	2 1/2	3/8	6 15/16	3 3/16	1 1/4	2.000	1/2	4 7/8	2	1.031	2 3/16	1 1/2	6 3/4	3	2 1/8	1	2	1 1/8
400	4.000	8 3/16	5/16	2 3/8	3 5/16	2.005	9/16	2 3/4	3/8	7 11/16	3 5/8	1 1/4	2.000	1/2	5 5/16	2 1/4	1.031	2 3/16	1 1/2	7 1/2	3 1/4	2 3/8	1 1/4	2 1/2	1 3/8
450	4.500	9	3/8	2 3/8	3 13/16	3.005	13/16	3	1/2	8 3/8	4 3/16	1 1/2	3.000	3/4	5 7/8	2 7/16	1.031	2 3/4	1 3/4	8	3 5/8	2 9/16	1 1/4	2 1/2	1 3/8
500	5.000	9 3/4	3/8	2 1/2	4 1/4	3.005	13/16	3 1/4	1/2	9 1/8	4 5/8	1 1/2	3.000	3/4	6 7/16	2 5/8	1.031	2 3/4	1 3/4	8 3/4	4	2 3/4	1 1/4	2 1/2	1 3/8
600	6.000	11 5/16	7/16	2 5/8	5 3/16	3.005	13/16	3 3/4	1/2	10 9/16	5 1/2	1 3/4	3.000	3/4	7 7/16	3	1.281	3 1/4	2 1/16	10 1/8	4 5/8	3 1/4	1 1/2	3	1 5/8
800	8.000	14 3/8	1/2	3 1/2	7	4.010	1 1/16	4 3/4	1/2	13 1/2	7 3/8	2	4.000	1	9 9/16	4	1.781	4 1/4	2 3/4	12 7/8	5 3/4	4 1/4	1 1/2	3	1 5/8
1000	10.000	17 7/16	5/8	4 1/2	8 3/4	4.010	1 1/16			16 5/16	9 3/16	2 1/2	4.000	1	11 1/2	4 3/4	2.031	5	3 5/16	15 3/4	7	5	2	4	2 1/4
1200	12.000	20 1/2	3/4	5 3/8	10 1/2	4.010	1 1/16			19	11	3	4.000	1	13 3/4	5 3/4	2.531	6	4	18 1/2	8 1/4	6	2	4	2 1/4

* Corners at entrance side of keyways shall be chamfered at 45 degrees as follows: No. 200 to 350, inclusive, 1/16 in. deep; No. 400 to 600, inclusive, 3/32 in. deep; No. 800 to 1200, inclusive, 1/8 in. deep.

† Taper is 0.75 in. per ft for all sizes in this table.

‡ Width is $N = N'$.

§ Width is equal to N'.

Tolerances are:

For diameter of shank at gage line (A): All sizes, +0.002 − 0.000.
For diameter of hole at gage line (A): All sizes, +0.000 − 0.002.
For width of drive keyway (R'): In socket, +0.000 − 0.001.
For width of drive keyway (R): In shank, +0.010 − 0.000.
For width of knockout keyway (N): +0.008 − 0.000.
For concentricity of drive keyway (R): With center line of spindle, 0.003.
For concentricity of drive keyway (R'): With center line of spindle, 0.001.
For concentricity of knockout keyway (N): With center line of spindle, 0.007.
Tolerances on fractional dimensions: ±0.010 unless otherwise specified.

Jarno Taper. This series comprises nineteen tapers numbered from 2 to 20, with a taper of 0.6 in. per ft. The number of a taper is its diameter in tenths of an inch at the small end, in eighths of an inch at the large end, and the length in halves of an inch.

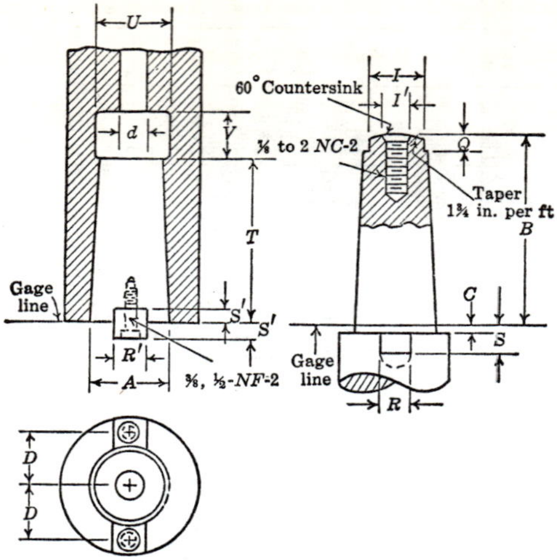

FIG. 13. Taper with key drive and shank retained by drawbolt. See Table 4.

Table 5. Dimensions of Steep Machine Tapers

(ASA B5.10–1943. All dimensions given in inches.)

No. of Taper	Taper per Foot *	Diameter at Gage Line		Length Along Axis	
5	3.500	1/2	0.500	11/16	0.6875
10	**3.500**	**5/8**	**0.625**	**7/8**	**0.8750**
15	3.500	3/4	0.750	1 1/16	1.0625
20	**3.500**	**7/8**	**0.875**	**1 5/16**	**1.3125**
25	3.500	1	1.000	1 9/16	1.5625
30	**3.500**	**1 1/4**	**1.250**	**1 7/8**	**1.8750**
35	3.500	1 1/2	1.500	2 1/4	2.2500
40	**3.500**	**1 3/4**	**1.750**	**2 11/16**	**2.6875**
45	3.500	2 1/4	2.250	3 5/16	3.3125
50	**3.500**	**2 3/4**	**2.750**	**4**	**4.0000**
55	3.500	3 1/2	3.500	5 3/16	5.1875
60	**3.500**	**4 1/4**	**4.500**	**6 3/8**	**6.3750**

* This taper corresponds to an included angle of 16° 35′ 33.4″.

The tapers numbered 10, 20, 30, 40, 50, and 60 that are printed in bold-face type are designated as the "Preferred Series." The tapers numbered 5, 15, 25, 35, 45, and 55 that are printed in light-face type are designated as the "Intermediate Series."

MACHINE-PERFORMANCE TESTING

By P. E. Nokes

DEFINITION. Machines on which performance tests are required may be divided into two broad classifications: (1) Machines whose output is power, as such, and whose efficiency is therefore of primary importance. (2) Machines whose output is one of a diversity of articles of manufacture. In these, quality and quantity of product are usually the prime considerations. Power efficiency may be a secondary or even negligible factor.

This chapter will deal entirely with class 2. For performance testing of machines of class 1, see *Power Volume* of Kent's *Mechanical Engineers' Handbook*.

SCOPE OF THE WORK. It is the principal duty of the engineer in charge of performance testing to obtain specific, detailed, and accurate answers to the following questions:

1. Does the machine accurately and consistently perform the operations that are expected of it?
2. Will the machine produce the expected quantity of the product in the specified time?
3. Will the machine and its component parts have a reasonable life expectancy? Are expendable parts of low enough cost and easy to replace?
4. Is the machine properly safe for a person to operate?

Should the answer to any of these questions be "no," the performance test engineer must supply reasons and sufficiently detailed data to aid in correcting the difficulties.

6. METHOD OF PROCEDURE

To evaluate properly the performance of a machine, the test engineer must:

1. Know precisely what the machine was expected to accomplish and in what manner it was expected to operate. These things are by no means always obvious. If specifications are not available or not clear, the engineer may have to write his own, after suitable study.
2. Know in detail the construction of the machine. This should include a thorough knowledge of the expected behavior of all parts, in particular the instrumentalities which actually do work on the article produced.
3. Be prepared to measure and describe the behavior of any part whose performance might be open to suspicion.

In checking over work to be done, a regular procedure may be followed. Below is a check list which will be found to fit the average machine. It may be expanded or contracted to fit special cases. Starting at the point where driving power enters the machine, we have the following:

A. Input (motor—lineshaft belt, or other)
   (1) Power—average—any excessive peaks?
   (2) Speed—is it correct—is it maintained during machine cycle?
   (3) Bearings—excessive temperatures—excessive bearing pressures—lubrication?

B. Input end drive (direct—belts—gears, etc.)
   (1) Has it sufficient power-transmitting capacity?
   (2) Slippage (particularly of belt drives)—is it excessive?
   (3) Flywheel (if present)—is it necessary—is its speed drop during delivery of power reasonable? If not present at all, should a flywheel be used?
   (4) Clutch and brake mechanism—do these bring machine to speed and stop it in specified time—is excessive slip present with machine idling, is any excessive drag present (as evidenced by temperatures reached)—will clutch or brake be adversely affected by oil on surfaces?—are associated mechanisms, such as operating and nonrepeat devices, properly designed and do they perform as intended?
   (5) Bearings—are these in sufficient number and properly disposed—are bearing pressures reasonable—is there adequate provision for lubrication?

C. Machine proper
   (1) Principle of operation—do parts actually perform the service intended—are linkage paths as anticipated—do cams and their associated linkwork give the specified end result?
   (2) Interferences—are there any actual or potential interferences between parts?
   (3) Stresses—any overstressed parts?
   (4) Deformations—are any parts elastically or permanently deformed in service so as to render them unserviceable, or so as to produce undesirable secondary effects?
   (5) Bearing loads and pressures—too great at any point?
   (6) Gear and cam loads (should be given particular attention).
   (7) Lubrication—design of system—adequate supply—proper lubricants—reliability—ease of maintenance and servicing?
   (8) Vibration and noise—present—can they be eliminated?

D. Working instrumentalities (i.e., the parts which work directly on "the product")
   (1) Specific action of each on the product—is the action of each as intended—does each produce the desired result? (Note. Both these questions should be answered; they are not, as might appear, mutually exclusive.)
   (2) Paths of motion of these parts—are they correct?
   (3) Loads, forces, and stresses, in and on these parts—too great in any case—if so, can they be corrected?
   (4) Wear—is this a potential source of trouble?
   (5) Expendable parts—present—have they been made as simple and inexpensive as possible—short life?

To obtain authoritative answers to the questions contained in the foregoing list (revised as necessary), numerous tests must be made, the nature of each of which will be apparent from the question. Since, however, it is almost axiomatic that machines "at speed" do not perform in the same manner as when "turned over by hand," it will be found desirable in most cases to test at actual working speeds, and under as nearly actual service conditions as possible. Static testing will be found to be the exception rather than the rule.

7. TEST EQUIPMENT

The most useful and versatile pieces of apparatus for dynamic testing of machinery in general are the modern electric strain gage set-up and the high-speed motion picture camera. The camera, with its accessories, will cover all cases where the phenomenon, except for speed, could be viewed properly by the eye. The strain gage apparatus will handle almost all cases where instrument observation is necessary.

MOTION PICTURES. In normal motion-picture work, the individual pictures comprising the complete film are taken and projected at the same speed (i.e., same number of pictures or "frames" per second). This number is usually either 16 or 24 frames per second. If a larger number of frames per second is used in taking than in projecting, the motions photographed will appear to be "slowed down" by the ratio of taking speed to projecting speed. For example, if taken at 64 frames per second and projected at 16, motions will be apparently slowed by a factor of 4.

Cameras are available capable of taking motion pictures at as high as 8000 frames per second on 8-mm film, 4000 frames per second on 16-mm film, and about 1200 frames per second on 35-mm film. All these may be projected at 16 frames per second, giving "slow-down" ratios of 500, 250, and 75, respectively. The camera speed is in all cases adjustable.

Cameras are of two general types. In one type (Edgerton, produced by General Radio Corp.) a bank of stroboscopic lamps, synchronized with the camera, provides both illumination of sufficient power and the necessary separation between frames. Its advantage is short exposure time (1 to 10 microseconds) and extreme motion-stopping power. Its disadvantages are first cost and power consumption. A second type of camera (Fastax, by Western Electric, and the Eastman Kodak Company's high-speed camera) depends on a small revolving prism in the optical system of the camera. The prism causes the image to follow the film during exposure. The exposure time of this camera is longer than that of the Edgerton type. It is lower in first cost, however, and uses more conventional illumination. Most mechanical cases can be adequately handled by either type. In both, development and printing of the resultant film follows conventional photographic practice.

Projection equipment is also standard. However, it is well to provide a projector capable of showing single frames and also of being reversed. Reversal (i.e., running the film backward) will, while unconventional, often provide additional information, while projection of single frames allows measurement of the screen image. Such measurements are of great value in obtaining motion paths of machine parts, if care is taken to account for the magnification ratio of the system. A scale included in the original picture will, of course, provide for this point.

If time is an important consideration, it is not possible to depend on film speed in the camera. A clock should be included in the picture. A suitable clock for this purpose may be constructed by placing a properly marked dial directly on the shaft of a small 1800 or 3600 rpm synchronous motor. If great precision is necessary, the motor may be driven, through amplifiers, by a precise tuning fork.

The speed at which pictures should be taken in any case will be governed by the time duration of the phenomenon under observation. During this time, a sufficient number of frames should be exposed to detail the cycle properly without missing essential parts.

Some trial may be necessary to determine both speed and optimum lighting in a particular case.

ELECTRIC STRAIN-GAGE APPARATUS. A complete set-up of electric strain-gage apparatus consists of (for one channel or one measured value) a strain-gage bridge, a power supply for this bridge, an amplifier, and some form of indicating or recording device. This apparatus is shown diagrammatically in Fig. 1.

All the diagram to the right of line *B–B* in Fig. 1 constitutes the permanent section of the strain-gage apparatus. It consists of one-half a Wheatstone bridge (adjustable resistances r_1 and r_2), with addition of condenser *C* and switch *S* for balancing cable capacities,

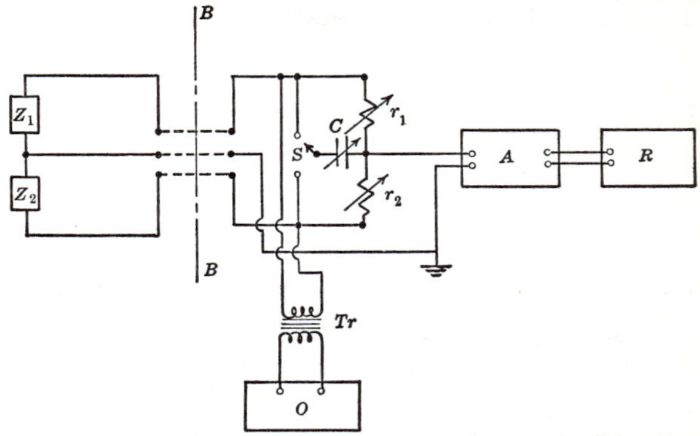

FIG. 1. Strain gages with associated bridge circuit, amplifier, and recorder.

oscillator (*O*) for supplying power to the bridge, amplifier (*A*), and, finally, recorder (*R*). This section of the apparatus may be obtained from several sources (General Electric, Hathaway Instrument Co., etc.). Its construction is not within the scope of this chapter. For mechanical testing, however, and as a guide to purchasing, experience dictates that this part of the apparatus should have these characteristics: (1) A frequency range extending from zero to about 1000 cycles per sec. (2) Wide adjustability in amplification. (3) Adaptability to use in simultaneous multichannel form. (4) Produce good-sized legible records. (5) Be provided with an indicating meter for use with slow phenomena and where a permanent record is not required. (6) Be usable with all current forms of electric strain gage.

The diagram to the left of line *B–B* in Fig. 1 constitutes the sensory element (or elements) of the apparatus and the other half of the strain-gage bridge. It consists of two electrical impedances (resistance, inductance, or capacity) made up, often quite crudely, to suit the job on hand. Either or both may be active gages. If only one is active, the other is present as a "dummy" to complete the circuit and/or to maintain thermal balance. The simplest way to maintain proper balance is, in any case, to make Z_1 and Z_2 physically identical. Since Z_1 and/or Z_2 are the sensory elements, they are, of course, located on or adjacent to the element to be tested. Shielded cable is used to connect them to the permanent part of the apparatus. The two most useful forms of Z_1 and Z_2, namely, resistance and inductance, will be discussed in more detail subsequently.

For the recorder (*R*) of Fig. 1, a multi-element bifilar oscillograph will be found most useful and versatile for mechanical testing. For a detailed description, the reader is referred to any electrical engineering handbook. The device produces a record by moving a light beam over photographic paper or film, which is subsequently developed in the usual manner. The oscillograph may be obtained in a form giving as many as forty-eight simultaneous records, although six or twelve is more usual and will be found most useful for this class of work.

This type of recorder may be used for static, as well as dynamic, work where necessary and is recommended unless no dynamic work at all is contemplated.

The two most useful forms of Z_1 and Z_2, of Fig. 1, are the resistance wire strain gage and the magnetic (or inductance) gage. A more detailed discussion of these forms follows.

RESISTANCE WIRE GAGES. The resistance-wire strain gage is based primarily on the principle that when a wire is stretched, its length increases and its diameter decreases. Both these actions cause an increase in its electrical resistance, and this increase, though small, is detected, amplified, and recorded by the apparatus previously described. A convenient form of the gage (see Fig. 2) may be obtained in various sizes from Baldwin Locomotive Works. It consists of a small piece of thin paper (*a*), to which is affixed, with cement, a grid (*b*) of very small diameter wire (usually 0.001 in.). Larger terminal leads (*c*) are provided to which connecting wires may be soldered. In use, the entire assembly is cemented firmly to a machine part at the desired point, and since its structure is insufficiently stiff to resist, partakes of the surface strain at that point. Stress may be computed from this strain measurement by the usual methods.

Characteristically, the gages are small in size and extremely light in weight. Consequently, they may be affixed in small and relatively inaccessible quarters, and do not add appreciably to the mass of even small parts. The gage is known to follow stress fluctuation to at least 30,000 cycles per sec.

The use of resistance wire gages for the measurement of stress as such is a subject of considerable magnitude and will not be discussed in detail here. References are given at

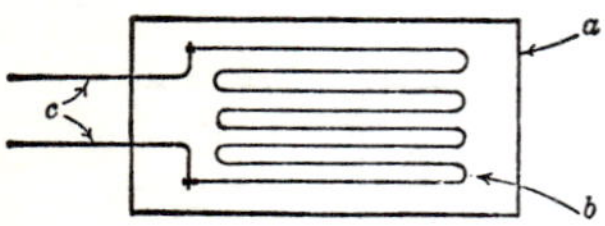

Fig. 2. Strain gage element.

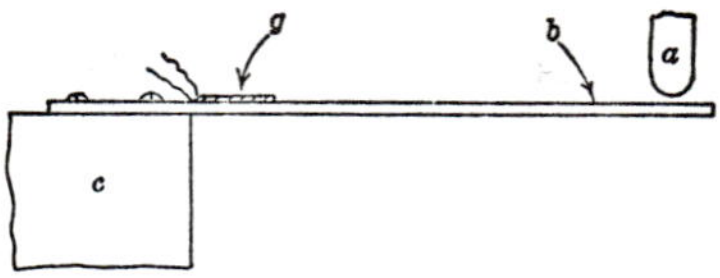

Fig. 3. Application of strain gage to space-time measurement.

the end of the chapter which discuss the detailed use of the gages for this purpose. This information should be followed in obtaining information on strength of machine frames and parts under working conditions. Other quantities may be measured as follows.

Torque and power measurements may be made using resistance wire gages. For this purpose, one or two gages (refer again to Z_1 and Z_2 in Fig. 1) are placed at an angle of 45 degrees on a shaft either existing in the machine or supplied for the purpose. Connections, in this case, are made through slip rings and brushes. A plain brass slip ring, with phosphor-bronze or beryllium-copper brushes will be found excellent. Do not use carbon brushes. Calibration may be obtained most readily by twisting the shaft in a torsion testing machine, although in some cases a spring scale and Stillson wrench will serve. A separate and simultaneous record of speed (see below) will be required to obtain instantaneous power. If desired, compact torque-measuring units of the type described above may be purchased from Baldwin Locomotive Works.

Space-Time Measurements. An excellent device for measuring the path of parts at moderate speeds is shown in Fig. 3. A strip of thin, flat spring stock (*b*) is clamped by some convenient means to machine frame *c*, and allowed to rest against part *a* (the part whose path is to be described). A wire strain gage *g* forms either Z_1 or Z_2 of Fig. 1. Calibration is obtained by moving part *a* slowly and making measurements with a rule or micrometer.

Pressure Measurement (hydraulic or pneumatic). Devices for this purpose, using wire-strain gages, may be purchased or may be constructed as shown in Fig. 4. A flattened piece of thin-walled tubing, closed at the outer end, is soldered to a standard pipe fitting of suitable size. A wire-strain gage is cemented around the tube at *g*. This gage is again

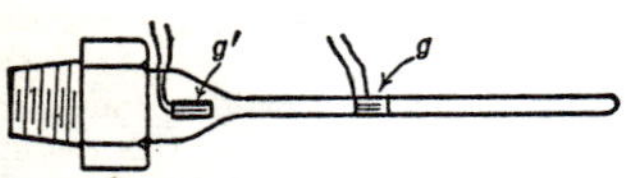

Fig. 4. Strain gages applied to pressure measurement.

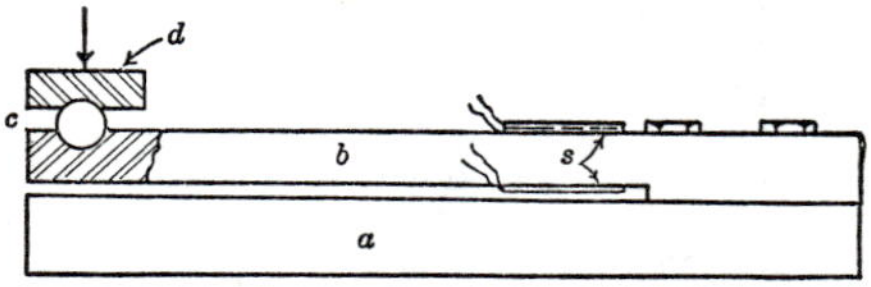

Fig. 5. Device for measuring force transmitted by a machine to the floor.

Z_1 of Fig. 1. If temperature compensation is necessary, a gage at *g′* or some other unstressed part of the system forms Z_2 of Fig. 1. This device may be varied in range by varying wall thickness of the tubing; and for pressures above about 500 psi, the tube need not be flattened. A suitable calibrating system consists of a hand pump and a conventional pressure gage of good quality.

Force. The force exerted by or on a machine part may often be measured by determining the stresses existing in the part or in adjacent parts. The case exists, however, in which it is necessary to know the force exerted externally by the machine on some object. A case in point is the dynamic force transmitted by the machine to the floor on which it stands. This can be done advantageously by supporting the machine on a suitable number of devices as shown in Fig. 5.

To a steel base plate (a) is bolted a stiff cantilever beam (b). A large ball-bearing ball (c) transmits load from pad d (fixed to the machine). A pair of wire-strain gages (s), affixed to the beam, form Z_1 and Z_2 of Fig. 1. The device is calibrated in a conventional testing machine. The device must, of course, be proportioned to support the weight of the machine plus any expected dynamic loads, and should be as stiff as possible in order to avoid changes in transmission characteristics due to spring loading.

Acceleration. Rather simple devices are available for measuring acceleration if the curve representing it is periodic and of essentially sinusoidal form. Such conditions exist in vibration measurements. The more usual case, in machine testing, is the one in which the acceleration is nonrepetitive and is complex in form. In this case, such approaches are useless, and, if possible, acceleration should be deduced from force and mass considerations, force being measured as previously described. If direct measurement is desired, a small cantilever beam erected on the part whose acceleration is to be measured may be equipped with wire-strain gages, and the stress therein used as a measure of acceleration. Considerable care must be used in design, as the basic vibrational frequency of the cantilever must be several times (say five) the highest frequency to be recorded. An enclosure to contain damping oil may be necessary. For large accelerations, a tension rod may advantageously be substituted for the cantilever. Several commercial embodiments of the above apparatus are currently appearing.

MAGNETIC GAGES. Impedances Z_1 and Z_2 of Fig. 1 may be inductances instead of resistance as in the wire gage. An inductance for this purpose consists of a coil of wire wound on an iron or steel core. The prototype of these gages, for use with a strain gage setup, is shown in Fig. 6. It consists of an iron core (a), with a thin air gap at b, and a coil (c). The phenomenon under test is in some fashion caused to vary the width of air gap (b), thus varying the inductive impedance of the coil. This coil, and an identical one for balancing, form the impedances Z_1 and Z_2 of Fig. 1.

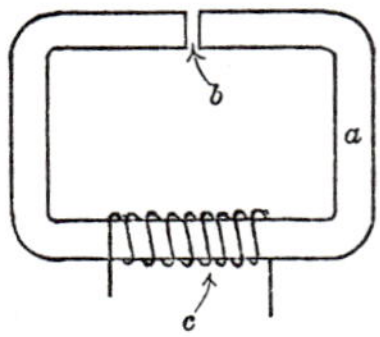

FIG. 6. Elementary magnetic gage.

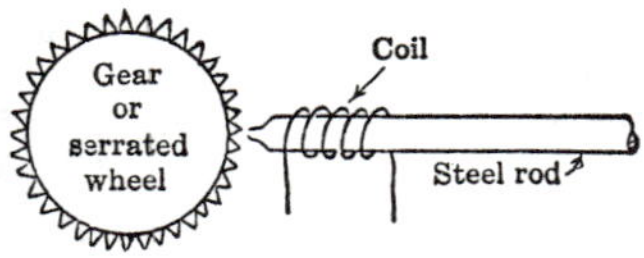

FIG. 7. Device for instantaneous speed measurement.

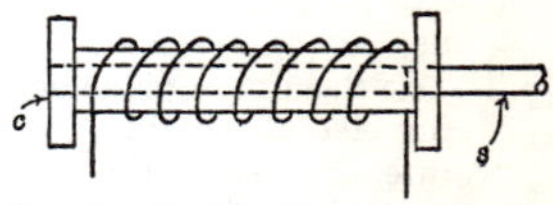

FIG. 8. Application of magnetic gage for measurement of linear motion.

Gages of this type, specifically adapted for strain measurement, may be obtained commercially (General Electric Co., Hathaway Instrument Co.) and may in some cases be extended to measure other quantities. A discussion of other specific types for various purposes appears below.

Speed Measurements. An excellent device of the inductive impedance type for the measurement of rpm may be constructed as in Fig. 7. A ferrous metal gear or serrated wheel (it may be part of the machine under test) has a steel rod with chisel-pointed end arranged to just clear its teeth. On this rod is wound a coil which forms the impedance Z_1 of Fig. 1. The coil may have about 500 turns. Wire size is not important. The device produces an oscillograph record containing one peak for each tooth of the wheel. In conjunction with a timing wave on the record, it will yield information on rapidly varying angular velocities. The number of teeth in the wheel may vary, depending on the detail desired. The device is simple and inexpensive for measurement of such factors as clutch slippage and the like.

Space-time Measurements. For the measurement of small, rapidly varying dimensions such as shaft deflection and "whip" under working conditions, the steel rod and coil combination of Fig. 7 may again be used. The rod end is arranged to just clear the shaft or other piece as it rotates. Shaft deflection appears as a steady signal, while "whip" produces a repeated wave, often a sine function.

Large Motions. Figure 8 shows an apparatus for the study of larger motions under working conditions. A steel rod (s) is allowed to move through a coil wound onto a preferably nonmetallic core (c), which may also serve as a guide for the rod. Either coil or rod may move in response to the motion under examination, and may be of any length desired. The signal is quite linear, only provided the rod end is not allowed to approach too close (say $^1/_8$ in.) to either end of the coil at the limits of its motion.

Calibration and Balancing. The usual method of obtaining a balance between Z_1 and Z_2 of Fig. 1 for inductance gages consists of providing two almost identical parts and coils, one used for Z_1 and one for Z_2. Either may be the active gage. Since, in the usual strain-gage amplifier, no provision is made for inductive balance, it is necessary first to make a resistance and capacity balance, then remove or add turns to one of the coils until a balance is obtained. Resistance and capacity balance are finally corrected to obtain a balance of the whole system.

Since all these devices read statically or dynamically, calibration may be obtained by means of feeler gages, micrometer, or rule, as suits the case.

OTHER DEVICES FOR STRAIN-GAGE APPLICATION. The foregoing paragraphs on the use of the strain-gage oscillograph combination gives only the general picture, plus a few of the useful measuring devices with which it may be used. With a little experience, many more (particularly special ones) will suggest themselves to the test engineer. The principles shown above may be extended indefinitely. A careful perusal of the references at the end of this chapter will also show many applications.

STROBOSCOPES. The stroboscope, like the high-speed motion picture camera, is used for the direct observation of fast-moving mechanisms. Its use, however, is confined to mechanisms whose motions repeat themselves at least eight or ten times per second. In principle, the stroboscope allows the eye periodically a very short-time view of the object under observation. If, for example, the eye is allowed exactly one look per revolution of a rotating part, the part will appear stationary. Obviously, the period of observation must be synchronized in some manner with the periodicity of the mechanism, while the observation period must be sufficiently brief so that the mechanism does not move appreciably during this time.

Commercial stroboscopes may be obtained in two types:

1. A rotating shutter type, in which a shutter, often clockwork driven, is held before the observer's eyes. Adjustment of speed is provided for synchronization. This type affords excellent portability (one type will fit in a coat pocket), but the length of exposure gives some blur to the field. Some of these devices are speed calibrated for use as observational tachometers.
2. The flashing light type, of which excellent examples may be obtained from the leading electrical manufacturers. These are based on the periodic production of a brief, intense flash of light. Since the duration of flash may be as short as a few microseconds, definition is excellent. Some are driven by adjustable calibrated oscillators and may be used as tachometers. Usually, a contactor mechanism is provided to allow synchronization from the machine under observation.

Some care is required to obtain accurate observations and speed readings with a stroboscope. If the periodicity of the phenomenon under observation is n per second, stationary images will also be obtained with $2n$, $3n$, etc., observations per second, although these will be double, triple, etc., images. Likewise, synchronization will be obtained with $^1/_2n$, $^1/_3n$, etc., observations per second, and the images will be natural (i.e., not double, triple, etc.). Practice will show that the last synchronization before a double image appears is the fundamental and correct one. An excellent practice object is a revolving white cardboard disk with one black spot on it.

GENERAL REFERENCES

Proceedings of the Society for Experimental Stress Analysis, the Addison-Wesley Press, Cambridge, Mass.

Mechanical Measurements by Electrical Methods, Howard C. Roberts, The Instruments Publishing Co., Inc., Pittsburgh, Pa.

Testing Topics, published by the Baldwin Locomotive Works, Philadelphia, Pa.

DYNAMOMETERS

Dynamometers are instruments used for measuring power. They are of several classes: (1) Traction dynamometers, used for determining the power required to pull a car or other vehicle, or a plow or harrow. (2) Brake or absorption dynamometers, in which the power of a rotating shaft or wheel is absorbed or converted into heat by the friction of a brake or into electrical energy. (3) Transmission dynamometers, in which the power in a rotating shaft is measured during its transmission through a belt or other connection to another shaft, without being absorbed.

TRACTION DYNAMOMETERS generally contain two principal parts: (1) A spring or series of springs, through which the pull is exerted, the extension of the spring measuring the pulling force. (2) A paper-covered drum, rotated either at a uniform speed by clockwork, or at a speed proportional to the speed of the traction, through gearing, on which the extension of the spring is registered by a pencil. The average height of the diagram above the zero line gives the average pulling force in pounds. This multiplied by the distance traversed, in feet, gives the work done, in foot-pounds. The product divided by the time in minutes and by 33,000 gives the horsepower.

ABSORPTION DYNAMOMETERS may be mechanical, hydraulic, or electrical. Power developed is wasted in the process of measuring, usually by conversion into heat which must be dissipated.

The Prony brake is the typical form of absorption dynamometer. Primarily this consists of a lever so connected to a revolving shaft or pulley that the friction induced between the surfaces in contact will tend to rotate the arm in the direction in which the shaft revolves. This rotation is counterbalanced by a weight-measuring apparatus, as a platform scale, one or more spring balances, or simply by weights hung in a pan at the end of the lever arm attached to the friction surfaces. (See Fig. 1.) The friction blocks *B* are of wood, and are confined by the strap which can be tightened on them, to increase the friction on the wheel, by means of the bolt *C*. The lever *D* is bolted to the band *A* and bears on the scale as shown. The heat generated by the friction is carried off by water introduced inside the rim, and removed by the outlet pipe which is shaped so as to scoop the water from the rotating wheel. Alternatively, the water may be allowed to evaporate, the outlet scoop being eliminated. The downward pressure of the lever arm is balanced by the weights *W* on the platform scale.

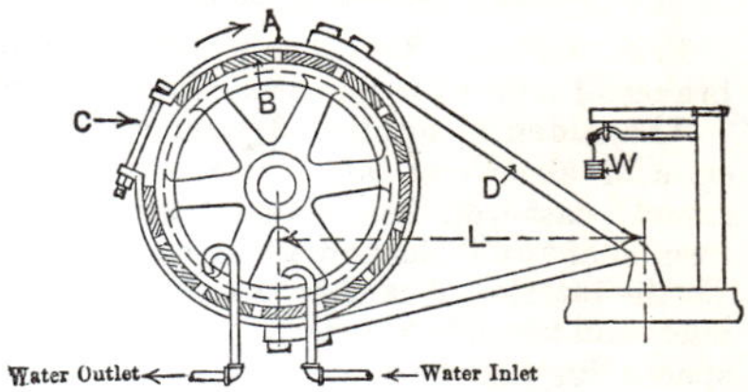

FIG. 1. Prony brake.

If L = horizontal length of lever arm, feet; N = rpm of wheel; W = weight required to balance the downward pressure of the lever arm, pounds; w = tare of brake, pounds, = weight on scales with apparatus at rest; and Hp = the horsepower of the shaft or mechanism to which the brake is attached,

$$\text{Hp} = \frac{2\pi LN(W - w)}{33{,}000} = \frac{(W - w)N}{K} \qquad (1)$$

where $K = 33{,}000/2\pi L$. If $L = 33/2\pi = 5\,^1/_4$ ft (approx.), $\text{Hp} = (W - w)N/1000$.

If the rubbing surface is too small, the resulting friction will show great irregularity; probably because of insufficient lubrication, the blocks will seize the pulley and thus produce shocks and sudden vibrations of the lever arm.

Soft woods, as bass, plane tree, beech, poplar, or maple, are to be preferred to the harder woods for brake blocks. The blocks should be spaced at least $^1/_4$ in. apart. The rubbing surface should be well lubricated with a heavy grease.

Another form of Prony brake is a pair of blocks or jaws, gripping the shaft. The blocks should extend around the greater part of the circumference. A pair of bolts through the blocks on either side of the shaft serves to tighten the blocks on the shaft. A ever arm of any desired length is attached to the upper block and bears on a platform scale or is supported by a spring balance. The horsepower is computed by eq. 1.

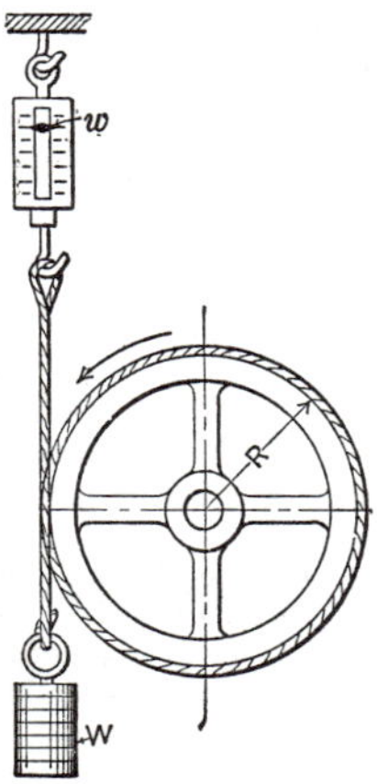

FIG. 2. Rope brake.

Rope Brake. For small amounts of power, the Prony brake may consist simply of a rope or strap wrapped around the brake wheel, one end being fastened to the floor or some other immovable part of the structure, through a spring balance, and the other end attached to a second spring balance or weighted. (See Fig. 2.) If W and w are, respectively, the net applied weight and the net spring balance reading and R is the radius of the brake wheel in feet,

$$\text{Hp} = \frac{2\pi RN(w - W)}{33{,}000} \qquad (2)$$

Prony brakes are suitable only for the measurement of relatively small powers, say up to about 100 hp and for speeds of not over 1000 rpm. For greater horsepower and higher speeds, a water brake or electric dynamometer should be used.

Capacity of Friction Brakes. W. W. Beaumont (Ref. 1) deduced a formula by means of which the relative capacity of brakes can be compared, based on the amount of horsepower ascertained by their use.

Let W = width of rubbing surface on brake wheel, inches; V = velocity of point on circumference of wheel, feet per minute; K = coefficient; then

$$K = WV \div \mathrm{Hp} \tag{3}$$

Professor Flather recommends the following:

Water-cooled brake, noncompensating, $K = 400$
Water-cooled brake, compensating, $K = 750$
Noncooling brake, with or without compensating device, $K = 900$

F. A. Halsey (Ref. 2) advised a minimum brake-drum surface for water-cooled Prony brakes of 0.09 sq ft per brake horsepower to avoid the danger of the blocks taking fire.

The Alden absorption dynamometer (Ref. 3) is a friction brake capable in moderate sizes of absorbing large powers with unusual steadiness and complete regulation. A smooth cast-iron disk is keyed on the rotating shaft. This is enclosed in a cast-iron shell, formed of two disks and a ring at their circumference, which is free to revolve on the shaft. To the interior of each of the sides of the shell is fitted a copper plate, enclosing between itself and the side a watertight space. Water under pressure is admitted into each of these spaces, forcing the copper plates against the central disk. The chamber enclosing the disk is filled with oil. To the outershell is fixed a weighted arm, which resists the tendency of the shell to rotate with the shaft, caused by the friction of the plates against the central disk.

Hydraulic Brakes. The usual type of hydraulic or water brake consists of a rotor keyed to the shaft and rotating in a casing filled with water. Power absorbed heats the water, which is circulated at such a rate as to prevent boiling. Torque resistance is controlled by varying the quantity of water in the casing.

For measuring power at high rotational speeds the rotor may consist simply of a disk or series of disks. For medium and low speeds the disk usually is provided with vanes or pockets to increase the drag by promoting turbulence. For formulas and design details for this type of brake, see E. P. Culver (Ref. 4).

Reaction torque on the casing may be determined by any of the usual methods of weight measurement. Alternatively, the torque may be set by placing a predetermined weight on the lever arm and providing automatic control of water flow, as described in the reference just cited.

Hydraulic dynamometers are commercially available in sizes ranging from small capacities up to several thousand horsepower.

Fan Brakes. For relatively high-speed power measurement a fan, propeller, or "club" may be attached to the revolving shaft. Air resistance to blade motion provides the load, which is determined by measuring the torque reaction at the motor or engine. The method requires mounting the motor or engine on a torque stand which permits it to revolve freely so that the torque reaction may be determined accurately. Alternatively, a fixed mounting and a calibrated fan may be used, for approximate measurements.

Electric Dynamometers. This type of dynamometer is basically an electric generator cradle mounted so that the stator is free to turn about the axis of the shaft. Torque reaction on the stator is measured by beam scales or other convenient means. Electric power which is generated may be dissipated in load resistors consisting of metal wire rheostats or, for larger powers, water rheostats.

By operating the dynamometer as an electric motor, "motoring" tests may be run for determining friction in a machine, or the power absorbed by any type of driven machine such as a lathe. While cradle mounting permits precise, direct measurement of torque, a generator without it may be made to serve, with less accuracy, by calibrating its output kilowatts (or input when motored) in terms of input (or output) power at various speeds and loads.

Eddy Current Brake. A simple type of electrical dynamometer is the eddy current brake, which consists of a disk revolving in a magnetic field. Resisting torque is controlled by varying the field strength, as in the generator type, but there is no current output, the eddy currents being dissipated in the disk, which must be cooled by air circulation if not by other means.

TRANSMISSION DYNAMOMETERS are used where power must be measured under service conditions, as in a ship at sea or an airplane in flight or a machine tool performing its function.

Torquemeters measure the angle of twist in a calibrated length of shaft or some form of flexible coupling through which power is being transmitted. Inasmuch as the twist is usually relatively small, mechanical, optical, or electrical means are employed to magnify the indication. To simplify measurement, some torque meters measure the twist at only one instant during the shaft rotation, which is satisfactory only if the torque is unvarying.

If pulsations are present in the torque of either the driving or driven machine, a continuous torque record or an integrated reading is required. Electric strain gages (see Article 7 in this section) are being used for this purpose, among other devices, and there are many types of instruments commercially available.

Belt dynamometers provide a simple means of measuring power transmitted to a belt-driven machine. Essential characteristics are illustrated in Fig. 3. Either flat belts or double-vee belts may be employed, and the principle can also be applied to roller chain drives. Torque reaction on the cradle which carries the idler sheaves is proportional to the difference in belt tension in the tight and loose sides.

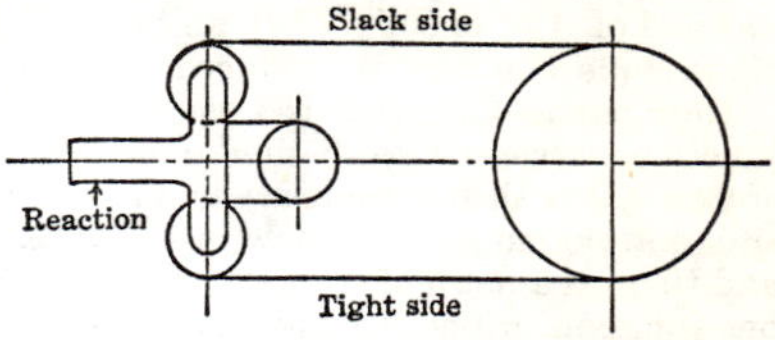

FIG. 3. Elements of belt dynamometer.

Gear Dynamometers. Any type of gear box which reduces, increases, or reverses the speed between input and output shafts produces a torque reaction equal to the algebraic difference between input and output torques. By cradle mounting the housing in anti-friction bearings and measuring the torque reaction a useful form of transmission dynamometer is obtained. Such a measuring device may be incorporated in a complete machine; for example, the stationary sun gear in many aircraft engine planetary reduction units is so mounted, the torque indications being transmitted hydraulically to the cockpit instrument. By feeding speed and torque to an integrator, horsepower may be indicated or recorded direct.

Electric Transmission. Where power is transmitted electrically, as in a diesel-electric locomotive or ship, measurement of the electrical power furnishes an indication of the power developed by the prime mover or absorbed by the driven machine. Electrical losses in the generator and motor at various speeds and loads must be known for calibration purposes.

Hydraulic Transmission. As with electrical transmission, hydraulic power transmitted between a pump and fluid motor may be determined, though less readily, by measuring rate of flow and pressure change.

REFERENCES

1. W. W. Beaumont, *Proc. Inst. Chem. Engrs.*, 1889.
2. F. A. Halsey, *Am. Machinist*, April 4, 1912.
3. G. I. Alden, *Trans. Am. Soc. Mech. Engrs.*, Vol. xi, p. 958; xii, p. 700; xiii, p. 429.
4. E. P. Culver, Investigation of a simple form of hydraulic dynamometer, *Mech. Eng.*, Vol. 59, Oct. 1937, p. 749.

THERMOMETERS AND PYROMETERS

By Robert B. Sosman

A thermometer is a device intended for the measurement of temperature. A **pyrometer** is a thermometer intended for use principally at temperatures above the first visible "red heat," which is about 500 C or 930 F.

The commonest type of thermometer is the liquid-in-glass type. When the liquid is mercury, the thermometer can be used industrially from −35 C to about 400 C (−30 F to 750 F), whereas with special glasses and a filling of nitrogen under pressure above the mercury, the range can be extended to about 500 C (930 F). Alcohol and liquid hydrocarbons serve for very low temperatures.

The expansion of a liquid in a metal container in place of a glass container gives a less fragile thermometer, though with essentially the same ranges as stated above. The change of volume is commonly registered by means of a Bourdon gage, and this device is also much used with thermometers which depend on the change of pressure of a gas in a metal bulb or on the change of vapor pressure of a liquid contained in such a bulb.

Resistance thermometry, depending on the change in electrical resistance of a wire, preferably of a pure and reproducible metal such as platinum, extends the range into the pyrometer field, to about 1000 C (1830 F), and is also usable to very low temperatures.

The thermometer of widest range is the thermoelectric, covering, with various thermocouples, all temperatures from −200 to 1650 C or from −330 to 3000 F. It stands first in versatility among the three principal types of industrial pyrometer. The other two are the radiation type (including the photoelectric) and the optical type. The radiation instruments are useful mainly, and the optical pyrometer entirely, in the pyrometric field, above 500 C.

THERMOELECTRIC PYROMETRY. The thermoelectric pyrometer consists of three parts: (1) thermocouple, (2) extension wires, and (3) apparatus for measuring the electromotive force of the thermocouple.

The Thermocouple. Take two pieces of wire, A and B, of different composition; weld one end of A to one end of B, and the other end of A to the other end of B, so that a continuous piece of wire is produced. This piece of wire is a thermocouple.

Note particularly that the piece of wire is a complete electric circuit. At every weld or junction between two dissimilar metals in such a circuit an electromotive force is generated. If both junctions are at the same temperature, the emf's will be equal and opposite, and no current will flow in the wire. But make one junction hotter than the other, and there is a net emf which is capable of causing a current to flow around the circuit. If one junction, which we may call the *cold junction* or, better, the *reference junction,* is held at a constant temperature, it is evident that the emf in the circuit can be used as a measure of the temperature of the other junction, the *hot junction* or *measuring junction.*

The Meter. The simplest method of measuring the emf is to cut the thermocouple at any point in its circuit and insert a current-measuring device such as an ammeter or a millivoltmeter (which is essentially a current measurer). The indication of the meter will then depend on the difference in temperature between the two junctions and, unfortunately, also on the total electrical resistance in the circuit, including both thermocouple and meter, in accordance with Ohm's law. Because the reading does depend on the resistance, and also because of certain deficiencies of moving-coil instruments under the severe conditions met with in industrial plants, the millivoltmeter type of instrument has been replaced in many installations by the potentiometer.

The potentiometer, in its fundamental form, balances a known and measured emf against the unknown emf in the thermocouple circuit, and its readings are consequently independent of resistance. Originally used as a hand-operated indicator only, the potentiometer has been made completely automatic in operation, and a number of types of the instrument are now available.

The earlier types of automatic potentiometer utilize a galvanometer and detect its departure from a balanced position by mechanical means. More recent types convert the unbalance of emf into some form of periodic or alternating current and utilize electron tubes for its amplification. These electronic instruments are faster than the mechanical but introduce the difficulties that are peculiar to electron tubes.

Extension Wires. We have been assuming that the thermocouple was cut at some point in order to insert the millivoltmeter or the potentiometer. Suppose the measurement is wanted at a place some distance away. A pair of copper wires leading off to the panel board where the instrument is placed would serve the purpose provided the two ends of the cut remain identical in temperature, and provided furthermore that the reference junction remains actually at a constant temperature. But the prevailing custom is to let the reference junction shift for itself, rising and falling in summer and winter, while its variations are compensated for by some device in the millivoltmeter or the potentiometer. This means that the reference junction must be placed at the instrument. Now some thermocouple materials capable of withstanding high temperature, such as platinum, are too expensive to be used in lengths of hundreds of feet in order to reach a distant instrument. This difficulty is taken care of by extension wires, consisting of a pair of wires of relatively inexpensive alloys which have approximately the same curve of emf against temperature (referred to some average air temperature for the reference junction) as the thermocouple itself. The thermocouple may now be cut open at the reference junction, and the pair of extension wires inserted in the proper direction, whereby the reference junction is transferred effectively to the far end of the extension wires where the meter is to be inserted.

Compensation for temperature change at the instrument must now be provided. In the millivoltmeter type a spring can be used whose elasticity changes with temperature in such a way as to balance the effect of the change in emf produced by the change in temperature of the reference junction. In the potentiometer, the compensation is electrical instead of mechanical. A resistance coil is inserted in the potentiometer circuit, of the proper material and resistance so that, as it warms or cools, its effect on the balancing emf is just equal and opposite to the change in emf produced by the change in reference junction temperature. For the highest accuracy, however, it is best to control the temperature of the reference junction, as by immersing it in a mixture of ice and water.

The foregoing picture of the thermocouple as a complete circuit has been adopted here in place of the customary picture of a pair of wires with free ends, with the deliberate purpose of emphasizing that no thermocouple can function without at least two junctions, one of which is just as important as the other.

The choice of instrument depends on local conditions. The principal advantages of the millivoltmeter are low cost and small bulk; in most other respects, including accuracy and dependability, the potentiometer is preferable.

The instrument may be given any or all of three functions: (1) to indicate the temperature on a scale; (2) to record the temperature either by a continuous curve or by a series of marks on a chart (the chart may be either circular or a continuous strip of paper); and (3) to control the source of heat by way of relays, valves, switches, and the like.

Thermocouple Materials. Three types of thermocouples are in current industrial use: (1) iron-constantan, (2) Chromel-Alumel, and (3) platinrhodium-platinum. (In each case the wire which is positive at the colder junction is named first in the pair.) The iron-constantan couple is used up to temperatures where the iron wire oxidizes away; at 750 C (1400 F) No. 8 iron wire will last about 1000 hr. Chromel-Alumel can be used to about 1200 C (2200 F), but oxidizes away too rapidly at higher temperatures. The platinum combination can be used at all temperatures up to the melting point of platinum at 1773 C (3223 F), although the susceptibility of platinum to contamination at high temperatures usually limits its use to about 1650 C and lower. It is the only couple available for continuous use at temperatures above 1200 C in air.

Chromel and Alumel are proprietary nickel alloys. All the Chromel and Alumel wire used in the United States is made by one manufacturer and is made to exact specifications with respect to thermal emf. The emf-temperature curve of the Chromel-Alumel couple has been standardized by the National Bureau of Standards, and all of the eight or ten instrument makers who furnish industrial pyrometers now use the same emf curve.

The platinrhodium-platinum thermocouple is made in two types. Both use pure platinum as one wire, while the platinrhodium alloy contains either 10% rhodium to 90% platinum, or 13% rhodium to 87% platinum. Standard curves for both these thermocouples have been determined by the Bureau of Standards and are now in general use.

Constantan is made in several compositions by different manufacturers. The iron of the iron-constantan couple is not the pure metal but is a low metalloid steel or "ingot iron." It likewise comes from several different sources. The iron-constantan couple has therefore not been standardized as to composition or emf-temperature curve.

The thermoelectric pyrometer is adaptable, without essential change, to any temperature below 500 C (930 F) in downward extension of its more commonly used range above 500 C. It can be used even at the temperature of liquid air. For these purposes copper-constantan can be used as the thermocouple, with a standard table available which can always be matched with commercially obtainable constantan and pure copper.

OPTICAL PYROMETRY. Although much less used than the thermoelectric pyrometer, the optical pyrometer is essential for temperatures of objects that can be seen but are not easily reached; for example, the tapping and pouring temperatures of steel. To the physicist, the optical pyrometer is a photometer, an instrument for measuring the brightness of a source of light. The brightness of a hot surface, however, is not alone dependent on its temperature but also on its emissivity, and the optical pyrometer consequently can not read temperature directly, except in the single instance of the "blackbody." The best example of a blackbody is a small opening through the wall of a furnace so uniformly heated in the interior that no details can be seen by observation through the opening. On all other objects the optical pyrometer reads a brightness temperature which must be interpreted by a knowledge of the emissivity. The brightness temperature of the flat surface of a reproducible material, such as low-carbon steel, fortunately, does bear a fixed relation to its true temperature, and the optical pyrometer is therefore a reliable instrument for the reproduction of heating, rolling, welding, and finishing temperatures. It is capable of very high precision, especially at the highest temperatures, but the readings are subjective, being dependent on the eye and the judgment of an individual observer. An impersonal check of the reading is therefore not possible.

RADIATION PYROMETRY. The total radiation pyrometer, often called simply the *radiation pyrometer*, measures energy rather than brightness. Its receiver absorbs all the energy radiated from a certain area of a hot surface, both the visible radiation and the heat radiation which at industrial furnace temperatures constitutes the greater part of the energy.

The energy is absorbed in a group of tiny thermocouples connected in series, whose emf thus becomes a measure of the temperature of the radiating source. Like the optical brightness of a surface, its energy radiation is not dependent solely on its temperature, as can easily be observed by holding the hand near a heated black surface and a polished reflecting surface at the same temperature. The polished surface has a much lower emissivity and feels correspondingly cooler, if not touched.

Although the emissivity for total radiation of a sheet of metal coated with a thin adherent film of oxide is less than the emissivity of a true blackbody, it is nevertheless reasonably constant. This fact is bringing the total radiation pyrometer into increasing use for the routine measurement of surface temperatures in the open. Sighted on metal inside a furnace, it may give false results because of reflection of heat coming from walls and roof. Sighted into a tube with closed end, it can serve as a controller of furnace temperature, and has the advantage over the thermoelectric pyrometer that there are no wires exposed to contamination and oxidation. The total radiation pyrometer is not well adapted for use as a portable instrument. The receiver is best mounted in a fixed position, and should be provided with air cooling or water cooling if the place is unusually hot.

The same types of electrical instrument (millivoltmeter, automatic potentiometer) are used with radiation and photoelectric receivers as with thermocouples, and they are adaptable to the same functions of indicating, recording, and controlling.

PHOTOELECTRIC PYROMETRY. A special type of radiation pyrometer utilizes the photoelectric cell as a receiver of energy. This type has been coming into use in industrial plants, particularly for high-speed pyrometry, as in the measurement of the brightness temperature of steel strip, rod, and tube. The photocell may be either of the electron-tube type or the self-generative (barrier-layer or Photronic) type. It is not a substitute for the optical pyrometer because it exercises no judgment as to brightness. It merely takes from a defined surface that part of the radiant energy which can produce a photoelectric current, and is therefore best classified as a partial radiation pyrometer. Since the current is extremely small, some form of amplification must be used, and care must be taken with electrical leakage. The indicating, recording, and controlling instruments, however, are otherwise no different from those used with the thermoelectric and total radiation pyrometers.

PYROMETER CALIBRATION AND MAINTENANCE. Thermocouples, pyrometer lamps, potentiometers, and similar physical apparatus are always subject to deterioration, particularly under the severe conditions of dust, high temperature, and vibration which are unavoidable in many industrial plants. An essential feature of the pyrometric equipment of the up-to-date plant is, therefore, its pyrometric laboratory for calibration and maintenance of the instruments. The plant with heavy demands on its pyrometric equipment cannot afford the frequent interruptions incident to sending pyrometers to a central calibrating station, such as the Bureau of Standards, nor can it afford the risk of spoilage incident to neglecting its pyrometric equipment for weeks at a time. The larger plants make up their own thermocouples from tested stocks of wire; calibrate the couples either by comparison with standards or by measuring the freezing points of samples of standard metals, such as tin, zinc, and copper; check the accuracy of recorders, controllers, standard cells, millivoltmeters, and milliammeters by means of portable testing sets; check optical pyrometers at frequent intervals against a standard source of brightness; and check total-radiation pyrometers by simultaneous readings on the same surface with a calibrated optical pyrometer. For small plants that cannot afford their own pyrometric laboratory, pyrometric calibrating and maintenance services are available in the larger industrial centers.

GENERAL CONSIDERATIONS. It must always be kept in mind that a thermocouple measures only the temperature of its own hot junction, and an optical, photoelectric, or total-radiation pyrometer measures the average apparent surface temperature of the objects seen by the instrument. For example, a thermocouple may be inserted into a flue with the hope that it will measure the temperature of the passing gas. What actually happens is an approach to an equilibrium between heat supplied by the gas, heat conducted away along the wires and protection tube, and heat radiated to or from the colder or hotter walls of the flue. The resulting temperature of the hot junction is seldom the temperature of the gas. In heat-treating operations on metals, it is not sufficient that the hot junction of the thermocouple is inside the furnace; it should be as near as possible to the piece of metal being treated. A study of the environment and the exercise of good judgment in placing the pyrometer are just as important as the proper selection and maintenance of the instruments.

GENERAL REFERENCES

M. F. Behar, *Handbook of Industrial Temperature and Humidity Measurement and Control*, Instruments Publishing Co., Pittsburgh, 1932.

G. K. Burgess and H. Le Châtelier, *The Measurement of High Temperatures*, John Wiley and Sons, New York, 1912. (Out of print.)

P. D. Foote, C. O. Fairchild, and T. R. Harrison, *Pyrometric Practice, Natl. Bur. Standards Technologic Paper* 170, 1921.

R. T. Glazebrook, *Dictionary of Applied Physics*, Vol. I, pp. 643–677, 693–710, 836–871, 901–921, London, 1922.
E. Griffiths, *Methods of Measuring Temperature*, Griffin and Co., London, 1926.
Pyrometry, American Institute of Mining and Metallurgical Engineers, New York, 1920. (A symposium on high-temperature measurement.)
Temperature, Its Measurement and Control in Science and Industry, American Institute of Physics, New York, 1941. (A symposium of papers touching many fields of thermometry.)
W. P. Wood and J. M. Cork, *Pyrometry*, McGraw-Hill Book Co., New York, 1927.

CALCULATING DEVICES

By W. J. Eckert

Calculating devices fall into two distinct classes, depending on the basic principle of operation, called digital and nondigital. Digital calculators (Ref. 1) employ the basic principle of counting discrete units. The nondigital devices (Ref. 2) depend on the measurement of a physical quantity such as length, angle, and electric current. In each class there is a wide range in size and complexity extending all the way from desk devices to the large central installations.

With the digital devices which depend on counting there is no limit to the precision which may be attained, whereas those which depend on measurement can be no more accurate than the measurements involved. On the other hand, a solution obtained on a digital device is a discrete solution, whereas the nondigital machines usually facilitate the investigation of a solution over a wide range of the quantities involved. Generally speaking, the nondigital devices are more specialized in their application.

8. DIGITAL CALCULATORS

Digital machines perform one or more of the common arithmetical operations, addition, subtraction, multiplication, and division, and sometimes others, such as square root. These operations are combined with the use of mathematical tables.

In the digital method the problem which has been stated in terms of mathematical formulas must be transcribed into a numerical procedure. The numerical processes for interpolation, differentiation, integration, and other mathematical operations are described in standard reference works (Refs. 3, 4, and 5). Since the processes are numerical, a given device will serve for problems from any field.

As an example of a numerical process consider the motion and the acceleration of a knife-edge follower actuated by a rotating cam, where the radius (r) of the cam in terms of the angle of rotation (θ) is given by the following expression:

$$r = f(\theta) = 1 + \sin^2 \theta + E(\theta)$$

$E(\theta)$ is an empirical quantity in the form of a table. The required velocity and acceleration are proportional to $(dr/d\theta)$ and $(d^2r/d\theta^2)$, which can be evaluated numerically from a table of values of $f(\theta)$ and the differences, δ', δ'', $\cdots$. The values of δ' are obtained from those of $f(\theta)$ by subtracting each one from the following; the values of δ'' are obtained from those of δ', etc.

θ	$E(\theta)$	$\sin\theta$	$\sin^2\theta$	$f(\theta)$	δ'	δ''	δ'''
0°	0010	0.0000	0.0000	1.0010			
					+0.0030		
2°	0028	0.0349	0.0012	1.0040		+31	
					+ 61		−3
4°	0052	0.0698	0.0049	1.0101		+28	
					+ 89		−1
6°	0081	0.1045	0.0109	1.0190		+27	
					+ 116		−1
8°	0112	0.1392	0.0194	1.0306		+26	
					+ 142		
10°	0147	0.1736	0.0301	1.0448			

The customary formulas for numerical differentiation may be written:

$$\frac{dr}{d\theta} = \frac{1}{\omega}\left\{[\delta'] - \frac{1}{6}[\delta'''] + \cdots\right\}$$

$$\frac{d^2r}{d\theta^2} = \frac{1}{\omega^2}\left\{\delta'' - \frac{1}{12}\delta'''' + \cdots\right\}$$

where ω is the tabular interval of θ; $2° = 0.0349$ radian in this case. For any given value of θ, $[\delta']$ is the average of the values of δ' above and below the horizontal line containing θ, and similarly for $[\delta''']$. δ'' and δ'''' are the values on the line. Thus at the point $\theta = 4°$ the velocity

$$\frac{dr}{d\theta} = \frac{1}{0.0349}\left\{\left(\frac{61+89}{2} - \frac{1}{6}\frac{-3-1}{2}\right) \times 10^{-4}\right\}$$

$$= \frac{1}{0.0349}\{(75 + 0.3) \times 10^{-4}\} = 0.216$$

and the acceleration

$$\frac{d^2r}{d\theta^2} = \left(\frac{1}{0.0349}\right)^2 \{0.0028\} = 2.3$$

Commercial machines are of two main types, those in which numbers are inserted by means of a keyboard and those which automatically read punched cards.

Printing adding machines are the simplest keyboard machines in general use. Numbers are inserted by means of a keyboard; totals are accumulated in a single register. The individual items and the totals are printed on a paper ribbon. These machines may be electric or hand driven. Some are limited to addition only, and others handle positive and negative numbers. Printing adding machines are also available with more than one adding register; the more complicated of them are called bookkeeping machines. Some of them have a movable carriage so that results may be printed in several columns. A programming device may be attached to the carriage so that the position of the carriage determines the arithmetical and printing operations to be performed.

Desk calculators are equipped with movable adding registers so that they may be used not only for addition and subtraction but for multiplication and division as well. These machines have a wide range of complexity. In some of them the arithmetical operations must be controlled largely by the operator; in others they are highly automatic. The results must be read from dials.

Punched card machines read and record the numbers as punched holes on cards. The machines read these holes and perform the prescribed operations automatically. These operations include rearranging the cards according to the data punched on the cards, printing data from the cards on a sheet of paper or on the cards, transcribing data from one card to another with or without rearrangement of the data, and the arithmetical operations of addition, subtraction, multiplication, or division (Ref. 6). Mathematical tables in the form of punched cards are consulted automatically with as many as 50,000 digits per hour. Punched card machines are of two kinds; those where the operation is controlled mechanically by means of linkages and those where it is controlled electrically by means of relays and plugboards.

Special Calculators. The machines described in the foregoing are all commercially available. In recent years a number of large special machines have been built. The most elaborate of them is an electronic calculator with an internal storage capacity for 400,000 digits, and programming facilities to handle automatically the most complex computational procedures.

9. NONDIGITAL CALCULATORS

The most common nondigital device is the slide rule on which lengths are added or subtracted. They also are available in circular form using angles, including the general numerical variety and also special ones for special fields such as engineering.

The planimeter is a desk device for giving the integral $\int f(x)\,dx$. The function is used in the form of a graph, and the planimeter gives the area under the curve.

Linear Algebraic Equation Solver. Somewhat more elaborate are the devices for such operations as the solution of linear algebraic equations. Examples of them have been made which depend on the measurement of mechanical, hydraulic, electric, and electronic quantities. A common type employs a resistance network where the resistances are set according to the coefficients of the equations to be solved, voltages representing trial values of the unknowns are applied, and the resulting currents represent the amounts by which the equations fail to be satisfied. The operator then adjusts the voltages by inspection until the residual currents become satisfactorily small. At this stage the applied voltages represent the values of the unknowns which satisfy the equations.

Machines of this general type have been built in several forms, depending on (1) the method of input, (2) the method of adjusting the voltage, and (3) the criterion used for adjusting the voltage; in all of them the results are read from an ammeter. Coefficients may be set up on potentiometers or on punched cards. The variables are adjusted by hand or by means of automatic devices. The criterion may be the residual current from a single equation or the sum of the squares of the residuals from all the equations (Ref. 7). An accuracy of 0.1% is reasonable.

The various gun directors and computing sights used during the war are other examples of nondigital devices.

Large central installations include differential analyzers and network analyzers. The differential analyzer is capable of handling the solution of complicated differential equations. Network analyzers are used to investigate power networks.

REFERENCES

1. Calculating machines, *Encyclopaedia Britannica*, 14th ed., 1947 printing.
2. Mathematical instruments, *Encyclopaedia Britannica*, 14th ed., 1947 printing.
3. J. B. Scarborough, *Numerical Mathematical Analysis*, Johns Hopkins Press, Baltimore, Md.
4. E. T. Whittaker, *The Calculus of Observations*, Blackie and Son, Glasgow.
5. C. Runge and H. König, *Vorlesungen über Numerisches Rechnen*, Springer, Berlin.
6. W. J. Eckert, Punched card techniques and their applications to scientific problems, *J. Chem. Educ.*, Vol. 24, Feb. 1947.
7. Francis J. Murray, *The Theory of Mathematical Machines*, King's Crown Press, Columbia University, New York.

SECTION 19

METAL MOLDING AND CASTING PROCESSES

BY

CHARLES K. DONOHO, *Chief Metallurgist, American Cast Iron Pipe Co.*

JAMES H. LANSING, *Technical and Research Director, Malleable Founders' Society*

H. M. ST. JOHN, *Crane Co.*

W. P. GOEPFERT, *Aluminum Company of America*

WILLIAM H. GROSS, *Magnesium Division, Dow Chemical Co.*

G. L. WERLEY, *Development Engineering Division, New Jersey Zinc Co. (of Pa.)*

F. S. BADGER, *Haynes Stellite Co.*

W. O. SWEENY, *Haynes Stellite Co.*

CLAIRE C. BALKE, *Director of Research, Roberts and Mander Corp.*

IRON AND STEEL CASTING

BY CHARLES K. DONOHO

MALLEABLE IRON CASTINGS

BY JAMES H. LANSING

COPPER-BASE ALLOY CASTING

BY H. M. ST. JOHN

ALUMINUM-ALLOY CASTING

BY W. P. GOEPFERT

MAGNESIUM-ALLOY CASTING

BY WILLIAM H. GROSS

ZINC-ALLOY CASTING

BY G. L. WERLEY

PRECISION INVESTMENT CASTING

BY F. S. BADGER AND W. O. SWEENY

POWDER METALLURGY

BY CLAIRE C. BALKE

IRON AND STEEL CASTING

By Charles K. Donoho

1. CAST IRON

DEFINITION AND TYPES. Cast iron is essentially an alloy of iron, carbon, and silicon in which carbon is present in excess of the amount which can be held in solid solution. The maximum amount of carbon which can be held in solid solution in iron at any temperature is usually considered to be about 1.7%. Most cast iron contains 2.0 to 4.0% of carbon. In *gray cast iron* at least a portion of the total carbon content is rejected from solution during freezing in the form of graphite flakes which are interspersed throughout the steel-like matrix. A useful concept of gray iron is that it may be considered to be "steel plus graphite." In *white cast iron* all the carbon is retained in the combined form. The massive iron carbides cause this material to be hard, brittle, wear resistant, but almost unmachinable.

Where cast iron contains specially added elements in amounts sufficient to affect the properties measurably, it is called *alloy cast iron.* Silicon, manganese, sulfur, and phosphorus as normally obtained from raw materials are not considered alloy additions.

USES. The engineering properties of gray iron which determine its uses are largely dependent on its graphite content. The precipitation of graphite from solution during freezing in the mold counteracts shrinkage and makes the metal easier to cast into sound parts than other metals with high shrinkage. As compared to other ferrous metals, the higher carbon content of cast iron enables it to be cast at lower temperatures and in molds of lower refractoriness. Cast iron may be melted in the cupola furnace, which is the most economical method of melting metals. These facts influence the choice of cast iron for a variety of castings and make it the largest tonnage foundry metal.

Where bulk and inertia are important the uses of gray iron castings are well known. Machine-tool frames and bases are cast of gray iron in intricate forms with high accuracy to finished dimensions. In such applications the graphite content increases the useful property of *damping capacity,* which limits vibration set up by moving parts. For services involving *wear,* cast iron is used in motor blocks, piston rings, brake drums, and many types of cylinders and plungers. For services where *corrosion resistance* is important, cast iron is widely used for underground pipe, and also has many applications in the chemical industries. Gray cast iron is more easily *machinable* at a given hardness level than most other metals because the graphite flakes break up the chip formation. For this reason many parts requiring high-speed production machining are cast of gray iron.

There are other large-scale uses of gray iron castings in these fields of application:

For *heat resistance:* grate bars, stove parts, radiators, firepots, and ingot molds.

For *wear and abrasion resistance:* brake shoes, car wheels, rolls, crankshafts, plowshares, plowpoints, wheels, crushers, wear plates, and sludge pumps.

For *corrosion resistance:* pipe, retorts, evaporators, valves, and refining equipment.

For *production machined parts:* automobile parts, washing machines, refrigerators.

For *architectural services:* columns, railings, ornamental work, sash weights, and stair treads.

For *municipal services:* fire plugs, manhole covers, lamp posts, mail boxes, and traffic markers.

PROPERTIES. The physical and mechanical properties of cast iron are varied widely by varying the composition, section size, freezing rate, and thermal treatment. Generally, those factors which tend to increase graphite—high carbon, high silicon, slow freezing—also progressively soften and weaken the iron. Increasing graphite content also confers the properties of easier machining, increased damping capacity, and better resistance to heat shock.

Strength Properties. Gray cast iron may be specified with tensile strengths from 20,000 psi minimum up to 60,000 psi minimum. Strengths of 25,000 to 40,000 psi encompass the majority of regular iron castings. Beam strength or bending strength is conveniently expressed as modulus of rupture. The modulus of rupture of cast iron is 1.5 to 2.5 times the tensile strength, the higher ratios occurring with the lower strength irons. Torsional strength of cast iron, expressed as modulus of rupture in torsion, varies from about 1.2

to 1.4 times the tensile strength. However, torsional and bending strengths of specimens other than circular in section will vary with the specimen shape. The compressive strength of gray cast iron is about 2 1/2 to 4 times the tensile strength, the higher ratios again occurring with the irons of lower strength. (See Fig. 1.)

Hardness. The hardness of cast iron varies from about 120 Brinell in the softest gray irons up to about 600 Brinell for chilled or white irons. Brinell or Rockwell hardness is often specified for gray iron castings, usually as a maximum limit to insure to some extent the degree of machinability. In general, softer irons are easier to machine than hard irons, although there are factors other than hardness which influence machinability to a considerable degree. There is a broad general relation between hardness and strength of gray cast iron although the degree of correlation is not so high as in steel or other more homogeneous metals. Figure 2 shows the relation of tensile strength to Brinell hardness for a large number of commercial cast irons. In general, engineering irons of highest quality fall to the top and left of the field in Fig. 2, i.e., irons with the highest strength for a given hardness. Irons with high tensile-Brinell ratio also tend to be more uniform in structure and better for wear resistance.

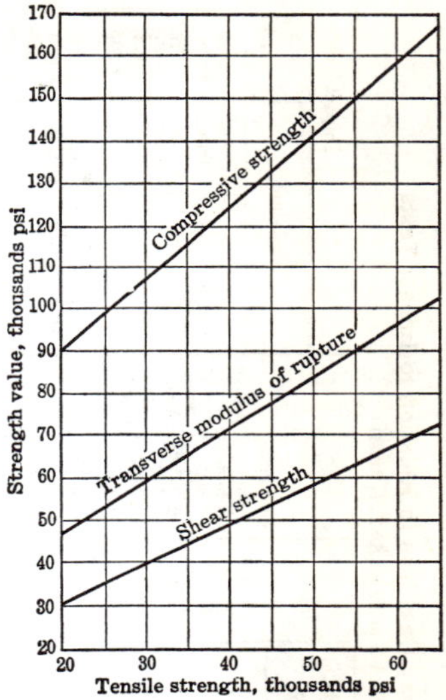

FIG. 1. Idealized chart showing approximate relation of other strength values to tensile strength for gray cast iron.

Impact Resistance. The engineering conception of gray cast iron generally is of a brittle material, incapable of shock or impact resistance. However, gray iron castings are used in many places where knowledge of impact resistance is important, as in pistons for internal-combustion motors and car wheels that are specified and purchased on impact tests.

A far-reaching investigation undertaken by ASTM Committee A-3 on the relation of impact properties to chemical and physical properties of cast iron was completed and reported in 1933. The developed data of the investigation are given in Table 1. Each test includes an investigation of three to six sets of specimens with varying chemical analyses and

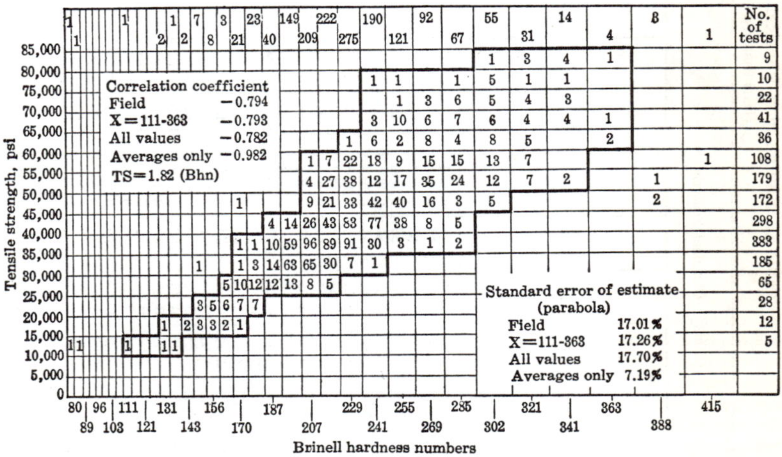

FIG. 2. Data from 1553 Brinell hardness–tensile strength tests. (Reproduced by permission from paper by MacKenzie, *Proc. ASTM*, Vol. 46, 1946)

different types of foundry conditions. Besides impact values, transverse, tensile, and compression tests were made and hardness values established.

In Table 1, summarized from data presented, average values are given. Brinell hardness is an average of values taken at the rim, center, and midway on the cross section of test

Table 1. Summary of Physical Properties of Gray Cast Iron in Relation to Impact Tests

(Data from report of ASTM Committee A-3 in *Proc. ASTM*, 1933)

Type of Furnace	Chemical Analysis									Transverse Tests, 1.20 in. Diam. x 18 in. Bars *			Tensile Tests, Specimen, 0.505 in. Diam. Load in 1000 psi	Compression Tests, Specimen, 1.20 Diam. x 3 in. Long. Load in 1000 psi	Brinell Hardness, Average, Rim to Center	Impact Tests					
																Pendulum			Falling Hammer 1.20 in. Diam. Bar		
																1.20 in. Diam. x 18 in. Span		1.20 in. Diam. x 8 in. Span	25-lb Hammer, 6-in. Span, 2-in. First Drop, 1/2-in. Increments	11-lb Hammer, 5-in. Span, 1-in. First Drop, 1-in. Increments	25-lb Hammer, 6-in. Span, 1-in. First Drop, 1-in. Increments
	T.C.	G.C.	Si	P	Mn	S	Ni	Cr	Mo	Load, lb	Deflection, in.	Resilience, ft-lb				Charpy, ft-lb	Russell, ft-lb	Russell, ft-lb			
Cupola	3.49	2.72	1.68	0.74	0.54	0.074	......	0.01	......	2457	0.284	35.4	29.47	111.0	191	47.3	46.0	29.2	10.5	20	17.3
Cupola	3.53	2.69	1.67	0.68	0.56	0.076	......	......	......	2342	.293	34.7	28.50	107.2	184	54.7	41.3	28.8	9.2	20	16.7
Cupola	3.95	3.09	1.90	0.07	0.65	0.060	0.01	......	......	1420	.352	27.1	18.32	65.2	132	42.2	39.4	24.9	7.0	15.3	12.0
Cupola	3.25	2.66	2.08	1.99	0.53	0.050	0.03	......	......	1254	.141	8.3	21.28	90.9	185	17.6	12.6	12.6	4.3	11	6.7
Cupola	3.12	2.44	2.18	0.63	0.44	0.096	......	......	......	2758	.301	42.0	32.50	120.8	195	60.9	53.4	34.7	11.2	22	20.3
Cupola	3.61	2.93	2.06	0.75	0.52	0.062	......	......	......	1857	.251	23.3	22.42	91.0	169	38.5	35.8	23.2	7.5	15.7	11.7
Cupola	3.49	2.78	2.08	0.19	0.55	0.078	1.17	0.65	......	2651	.331	44.3	29.98	108.8	188	56.6	48.9	31.9	10.8	23	17.7
Cupola	3.39	2.75	2.50	0.43	0.74	0.062	0.01	......	......	2407	.245	29.3	31.50	116.1	195	41.1	38.4	28.3	9.2	19	14.0
Cupola	3.41	2.85	2.44	0.63	0.57	0.070	0.01	0.02	......	1840	.274	25.7	20.00	87.2	159	40.3	35.6	24.2	7.0	16	11.7
Cupola	2.61	1.73	2.38	0.06	0.77	0.105	1.08	0.09	......	3489	.272	45.6	51.44	156.6	270	67.0	80.5	42.6	15.2	31	23.3
Arc. elec.†	3.14	2.30	2.69	0.18	0.82	0.094	0.25	0.09	0.07	2590	.293	38.4	35.65	115.9	195	57.6	53.4	33.4	10.2	24.7	15.3
Arc elec.‡	3.05	2.20	2.70	0.16	0.79	0.089	0.60	0.24	0.45	3083	.301	45.5	39.20	131.2	228	67.4	67.2	40.6	13.5	31.3	22.3
Cupola §	3.46	2.82	2.35	0.15	0.69	0.092	0.65	0.02	0.02	2200	.335	39.0	24.45	90.8	160	61.7	56.8	35.1	9.8	21.3	16.3
Arc elec.‖	2.53	2.03	3.26	0.02	1.01	0.035	10.59	1.55	......	2414	.471	61.1	29.73	134.2	220	139.8	158.9	76.8	13.5	30.7	16.7
Cupola	3.43	2.75	2.35	0.20	0.73	0.092	0.01	0.01	......	2218	.341	41.2	25.02	95.0	164	61.1	51.2	36.1	9.5	21	14.7
Rocking arc	2.88	2.25	1.99	0.43	0.51	0.098	0.01	0.01	......	3177	.439	78.9	40.92	119.1	189	84.2	80.3	48.5	13.8	29.3	24.0
Air	3.24	2.41	1.63	0.42	0.56	0.084	0.06	0.01	......	2905	.326	48.7	35.08	120.8	195	63.3	58.9	43.1	12.2	27	18.7
Cupola	3.38	2.59	1.18	0.14	0.82	0.077	0.95	0.01	......	2455	.267	33.1	30.65	112.0	192	61.7	50.2	36.7	10.8	25	18.0
Cupola ¶	3.49	2.16	0.53	0.33	0.42	0.133	......	0.01	......	2835	.294	45.3	34.95	123.5	212	50.7	48.8	39.8	12.6	28.3	16.5
Arc elec	3.52	2.68	2.94	0.04	0.66	0.042	0.01	0.01	......	1651	.370	32.5	19.43	72.8	150	42.4	41.8	28.3	7.8	17.3	13.7
Cupola	2.50	1.50	2.20	0.04	0.74	0.088	......	0.08	......	3472	.230	37.3	47.23	159.0	266	55.1	48.3	32.6	13.2	30.3	20.3
Cupola	3.79	2.99	1.47	0.12	0.73	0.083	0.01	0.01	......	1771	.336	31.6	18.27	77.9	142	51.2	42.5	31.3	8.2	18	12.7
Arc elec.**	3.08	2.32	2.07	0.04	1.21	0.047	16.29	1.89	......	1970	.745	94.5	16.70	86.0	121	128.7	93.3	87.8	12.3	28	24.0
Rotary coal	3.07	2.22	1.97	0.40	0.63	0.072	0.01	......	......	2776	.308	43.2	29.20	124.5	210	56.8	46.9	29.8	12.0	23	18.3
Arc elec.	2.79	1.94	2.44	0.03	0.50	0.060	0.50	0.09	......	4200	.490	114.2	56.70	135.7	234	138.8	136.0	77.9	21.4	47	28.0

* Bars cast to 1.4 in. and machined to 1.20 in. diameter. † Mo = 0.07. ‡ Mo = 0.45. § Not suitable for heavy section. ‖ Cu = 4.07; a martensitic iron not commercially used at the time of the investigation. ¶ Not suitable for light section. ** Cu = 6.12; an austenitic iron for high temperature and corrosion resistance.

specimens. The final columns include impact values established on unnotched specimens by the Charpy and Russell pendulum-type testing apparatus, the specimen being supported rigidly near the ends and struck at the center. Three of the columns are variations of the falling hammer test on a specimen 1.20 in. in diameter supported at 6 in., 5 in., and 6 in., respectively. The first of these falling hammer tests used a 25-lb weight from an elevation of 2 in., with increasing increments of $^1/_2$ in. The table gives average height from which hammer dropped when the bar broke. The second of the falling hammer tests was on specimens supported at 5 in. centers with an 11-lb weight starting at 1 in., with increasing increments of 1 in. The third summarizes average tests at 6 in. centers with 25-lb weights starting at 1 in., with increasing increments of 1 in.

The impact test, essentially a measure of energy absorbed in breaking bars, may be expressed by the work done, namely force times distance through which force moves. Results of Charpy and Russell tests are in foot-pounds. The others may be computed from the drop and weight of hammer.

Fatigue Strength. The endurance limit of gray cast iron in repeated stress varies almost linearly with tensile strength, the ratio of endurance limit in reversed bending to static tensile strength usually being between 0.40 and 0.55. An important property of cast iron in fatigue service is its relative lack of notch sensitivity. Sharp notches in cast-iron parts do not affect the endurance limit as seriously as in the more homogeneous metals.

Wear Resistance. Gray cast iron has considerable value as a wear-resistant material for bearings, automobile cylinders, piston rings, and slides on machine tools. The heterogeneous structure with components varying from soft graphite and ferrite to hard phosphides and carbides gives gray iron nongalling and nonseizing properties in the same manner that hard particles in a soft matrix confer wear resistance on bearing metals. Recent studies have shown clearly that the best gray irons for wear resistance have a normal graphite structure in a pearlitic matrix. Irons with interdentritic segregates of fine graphite with free ferrite have relatively poor wear resistance. For some important wear services a micro structure requirement is an important part of the specifications.

For other types of wear service, chilled or white irons are valuable because of the high hardness conferred by massive carbides. Chilled railway-car wheels and chilled camshafts are examples of their class. Martensitic white irons alloyed with nickel and chromium are often used for severe wear service where the very high degree of hardness is desirable.

Corrosion Resistance. Gray iron does not deteriorate rapidly under normal corrosive conditions. Cast-iron pipe for underground service, cast-iron kettles, retorts, and evaporators are examples of iron castings used for corrosion resistance. The graphite particles in cast iron are quite noble and noncorrodible, and phosphides are also corrosion resistant. When cast iron corrodes on the surface, the corrosion products tend to remain firm and intact, preventing or reducing subsequent attack on the underlying metal.

There are several valuable corrosion-resistant special cast irons. Cast iron with about 15% silicon makes a rather brittle, unmachinable white iron with remarkable resistance to most chemicals and all strong acids except hydrofluoric and hot, concentrated hydrochloric acid. Gray irons made austenitic by high percentage additions of nickel, copper, and chromium have many times greater resistance in most corrosives than unalloyed iron or steel.

Heat Resistance. Some engineering codes established more than thirty years ago limited the use of cast iron to temperatures of 450 F where pressure is also involved. According to a study by Burgess and Barlow (*American Foundryman*, May 1945), properly controlled gray iron can be successfully employed in pressure service at temperatures up to at least 750 F. Best heat resistance is found in gray irons with carbon equivalent, % C plus 0.3 (% Si + % P), in the range of 3.4 to 3.9. Unalloyed irons with a carbon equivalent of 4.0 or greater normally are not suited for high-temperature pressure or steam applications.

Most failure of gray iron castings in high-temperature service is due to *growth*, which is a permanent increase in volume under repeated heating and cooling. Growth is considered to be caused by the decomposition of the combined carbon content of the gray iron to graphite and to progressive oxidation along the graphite flakes. Irons lower in carbon and silicon are generally less susceptible to growth. Small additions of carbide stabilizing alloys such as chromium, molybdenum. and vanadium reduce growth tendency as does any alloy which increases the density of the metal. The highly alloyed austenitic irons have significantly greater heat resistance than plain or low-alloy irons as well as superior corrosion resistance.

OTHER PHYSICAL PROPERTIES OF CAST IRON. Electrical Resistivity. Cast irons vary widely in resistivity, depending on composition. Usual grades of gray irons will have specific electrical resistance of the order of 100 microhm cm but may vary from about 50 to 200 microhm cm. Increasing silicon and carbon both increase resistivity. Because of its rather high specific resistance and ability to be cast in thin and intricate sections, cast iron is used extensively in resistance grids.

Magnetic Properties. Cast iron has poorer permeability and less ability to carry high flux density than rolled mild steel and is no longer used extensively in magnetic applications. Austenitic cast irons high in nickel are of advantage where nonmagnetic properties are desired.

Thermal Conductivity. Thermal conductivity of most cast irons varies from about 0.100 to 0.140 cal per sq cm per sec per cm per °C. Although the heat conductivity is lower than mild steel, cast iron finds some application in heat exchangers, radiators, stoves, etc.

Coefficient of Thermal Expansion. The figure of 0.00001 per °C is most commonly used for cast iron. Table 2 gives data on linear contraction and shrinkage of ten different cast irons.

Solid Contraction. The commonly used figure for patternmakers' shrinkage is $^1/_8$ in. per ft for gray iron and $^1/_4$ in. per ft for white iron.

Melting and Freezing Ranges. The melting and freezing range of cast irons depends largely on carbon equivalent, % C plus 0.3 (% Si + % P); as carbon equivalent approaches 4.30% the solidification temperature is at the minimum. Most gray irons become completely molten on heating at 2100 to 2400 F. (See liquidus temperatures, Table 2.) On freezing the solidus temperature of unalloyed cast irons is always close to 2100 F.

Specific gravity varies from about 7.00 for a coarse-grained gray iron to about 7.7 for a white iron. Commonly used density figures for gray iron are 0.26 lb per cu in. or 450 lb per cu ft. Densities at several important temperatures in grams per cubic centimeter are given in Table 2.

PRODUCTION. Melting. Cast iron may be produced in the cupola, air furnace, or electric furnace. By far the greatest tonnage of cast iron is cupola-melted. The cupola furnace is a cylindrical-shaft-type furnace similar to the blast furnace but smaller and simpler in construction. In operation, air from pressure blowers is forced into the cupola at the tuyères and through a bed of white hot coke to melt and superheat the iron. The metal charges, consisting essentially of graded pig iron and selected scrap, are interspersed with layers of coke to replenish the fuel bed continually, as the metal is melted. Fluxing material, usually limestone, is charged to remove coke ash and other nonmetallics as liquid slag. The cupola is a continuous melting medium. By selection of raw materials and controlled operation most grades of cast iron may be produced most economically by cupola melting.

For details of construction and operation of the cupola furnace the *Handbook of Cupola Operation*, published by the American Foundrymen's Association, is the most complete and authoritative reference.

Molding. Molding consists of making the form into which the molten iron is cast to shape. The most common mold material is sand, in which the refractory element is silica (quartz) grains bonded with clays or other binders. In natural molding sands the silica and clay occur together in nature. Synthetic sands are compounded of a silica sand to which bonding materials are added. Sand molds are used in both the "green" or moist state and the dry state. Most iron castings are produced by green-sand molding where the molding sand is tempered with water for ramming around the pattern and the metal is poured into the moist mold. In dry-sand molding the moist-rammed mold is oven-dried before casting. Sand molds are rammed by hand or by one of several types of molding machines, the method used depending on the type of casting, the number of castings to be produced, and the facilities of the individual foundry. By the sand-casting method an unlimited number of castings may be produced from a single pattern.

Interior surfaces of castings are usually formed by sand cores. Some interior surfaces may be shaped by the pattern itself, the cores being made of green sand. Most cores, however, are made in core boxes using special core-sand mixtures. For baked cores, a drying oil binder (core oil) is usually used. The cores are baked and affixed into the mold after the pattern has been withdrawn. Baked-oil sand "cores" are sometimes used to form exterior as well as interior surfaces of castings.

Casting. The molten metal drawn from the melting furnace in ladles is poured into the molds. The rate of pouring is controlled by the size of the gate opening into the mold cavity. The metal is allowed to freeze and cool in the mold, and then the casting is removed from the sand for cleaning.

Cleaning. In cleaning castings, adhering sand is removed by brushing, tumbling, or abrasive blasting. Gates, risers, and fins are removed and roughnesses on the casting surface smoothed by chipping and by grinding. Inspection of castings is carried on during cleaning and final inspection after all finishing operations.

Table 2. Data on Volume Change * of Cast Iron

(Reprinted by permission from *Cast Metals Handbook*, American Foundrymen's Association, Chicago, 1944)

	No. VIII		No. VII		No. I		No. IV		No. IX		No. XIII		No. II		No. III		No. XIV		No. VI	
	1 1/2-in. Bar	3/4-in. Bar	1 1/2-in. Bar	3/4-in. Bar	1 1/2-in. Bar	3/4-in. Bar	1 1/2-in. Bar	3/4-in. Bar	1 1/2-in. Bar	3/4-in. Bar	1 1/2-in. Bar	3/4-in. Bar	1 1/2-in. Bar	3/4-in. Bar	1 1/2-in. Bar	3/4-in. Bar	1 1/2-in. Bar	3/4-in. Bar	1 1/2-in. Bar	3/4-in. Bar
Linear contraction from 2102 F (1150 C), %	0.67	1.08	0.83	1.08	0.93	1.13	0.95	1.11	0.97	1.08	0.98	1.09	1.06	1.11	1.08	1.12	1.12	1.17	1.31	1.77
Linear contraction from 1958 F (1070 C), %	0.90	1.17	1.04	1.14	1.13	n.d.	n.d.	1.22	1.12	1.19	1.03	1.13	1.24	1.25	1.19	n.d.	1.17	1.21	n.d.	n.d.
Shrinkage during solidification, %	0.07	1.65	0.62	1.11	n.d.	n.d.	1.94	1.73	0.82	1.10	0.07	1.31	n.d.	n.d.	1.59	1.94	0.00	0.07	5.82	4.85
Liquid contraction per 100 C (212 F), %	1.1	1.1	1.1	1.1	n.d.	n.d.	1.1	1.1	1.1	1.1	1.1	1.1	n.d.	n.d.	1.1	1.1	1.1	1.1	1.1	1.1
Density at room temperature, grams per cu cm	7.02	7.01	7.09	7.10	7.10	7.13	7.26	7.29	7.03	7.04	7.08	7.08	7.12	7.12	7.27	7.31	7.14	7.15	7.66	7.70
Density at 2102 F (1150 C), grams per cu cm	6.89	6.80	6.90	6.85	6.90	6.89	7.04	7.04	6.85	6.80	6.89	6.80	6.90	6.89	7.04	7.04	6.90	6.90	7.35	7.30
Density of liquidus, grams per cu cm	6.90	6.90	6.94	6.90	n.d.	n.d.	6.92	6.92	6.89	6.89	6.89	6.89	n.d.	n.d.	6.92	6.92	6.89	6.89	6.95	6.95
Liquidus temperature:																				
°F	2102	2102	2102	2102			2282	2282	2102	2102	2111	2111			2282	2282	2183	2183	2462	2462
°C	1150	1150	1150	1150	n.d.	n.d.	1250	1250	1150	1150	1155	1155	n.d.	n.d.	1250	1250	1195	1195	1350	1350
Total carbon, %	3.69	3.63	3.67	3.76	3.63	3.63	3.10	3.10	3.39	3.60	3.27	3.27	3.71	3.71	3.08	3.08	2.90	2.89	2.18	2.29
Graphitic carbon, %	3.53	3.42	3.26	3.35	2.97	2.99	2.31	2.27	3.20	3.40	2.88	2.90	2.99	2.96	2.18	2.23	2.68	2.65	0.56	0.42
Silicon, %	2.87	2.87	2.10	2.10	2.08	2.08	1.69	1.69	2.86	2.86	2.87	2.87	2.11	2.11	1.68	1.68	2.88	2.88	1.24	1.24
Phosphorus, %	0.68	0.68	0.46	0.46	0.44	0.44	0.35	0.35	0.67	0.67	0.59	0.59	0.46	0.46	0.35	0.35	0.66	0.66	0.27	0.27
Manganese, %	0.59	0.59	0.54	0.54	0.50	0.50	0.48	0.48	0.58	0.58	0.52	0.52	0.50	0.50	0.44	0.44	0.44	0.44	0.35	0.35
Sulfur, %	0.03	0.03	0.05	0.05	0.05	0.05	0.04	0.04	0.03	0.03	0.03	0.03	0.05	0.05	0.04	0.04	0.03	0.03	0.04	0.04

* In this table "n.d." denotes "no determination."

2. SPECIFICATIONS FOR CAST-IRON CASTINGS

GENERAL CASTINGS. Specification A-48-46 of the American Society for Testing Materials covers general gray iron castings not covered by other more specific specifications and applies to castings where strength is a consideration. Other general cast-iron specifications, such as Federal QQI-652, use the same scheme and identical requirements as ASTM A48-46. The following important details of this specification are given below, and for the complete specification consult the latest edition of *ASTM Standards.*

It is a purpose of the specification to classify irons according to the minimum strengths of tensile test bars as follows:

Class number	20	25	30	35	40	50	60
Minimum tensile strength, psi	20,000	25,000	30,000	35,000	40,000	50,000	60,000

Transverse tests are optional. When specified the breaking loads shown in Table 3 apply.

Table 3. Dimensions and Minimum Breaking Loads for Transverse Tests

As-cast Round Test Bars			Class No.						
Diam., in.	Length, in.	Distance between Supports, in.	20	25	30	35	40	50	60
			Minimum Breaking Load at Center, lb						
0.875	15	12	900	1025	1150	1275	1400	1,675	1,925
1.20	21	18	1800	2000	2200	2400	2600	3,000	3,400
2.00	27	24	6000	6800	7600	8300	9100	10,300	12,500

Test bars are cast separately from the casting. Three standard sizes of bars are available under these specifications. Tension-test specimens are machined from the round bars cast for the transverse test to the dimensions shown in Fig. 3 and Table 4. The manufacturer and the purchaser should agree on the controlling section of the casting. Then the corresponding test bar shall be used, as shown:

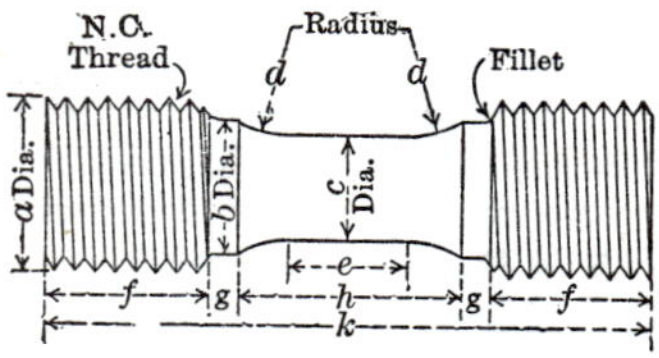

FIG. 3. Standard test specimen. See Table 4.

Table 4. Dimensions of Cast-iron Tensile Test Bars

(See Fig. 3)

Dimensions, in.	Test Bar A	Test Bar B	Test Bar C
a	3/4	1 1/8	1 7/8
b	5/8	15/16	1 7/16
c	0.505	0.750	1.25
d	1.00	1.00	2.00
e *	Shall be equal to or greater than *c* (diameter)		
f	1.00	1.00	1 3/4
g	0.25	0.25	5/16
h, minimum	1.25	1.50	2 1/4
k, minimum	3.75	4.00	6 3/8

* Length of parallel.

Controlling Section of Casting, in.	Test Bar Diameter, in.
0.50 and under	0.875
0.51 to 1.00	1.20
1.01 to 2.00	2.00
Over 2.00	2.00, or by agreement

Test bars should be made under the same sand conditions as the castings and the bars poured from ladles of iron used to pour the castings. If castings are heat treated, the test bars receive the same heat treatment.

Tension-test specimens must be loaded axially, preferably with self-aligning shackles, and at a rate of not over $^1/_8$ in. per min above 15,000-psi stress. The transverse test is made on the bar as cast with central loading between the supports. Corrections are made for sizes of round bars as shown in Table 5. In all bars the controlling dimensions are the

Table 5. Correction Factors for Transverse Cast-iron Test Bars

0.875-in. Diameter Bar			1.20-in. Diameter Bar			2.00-in. Diameter Bar		
Diameter of Test Bars, in.	Correction Factor		Diameter of Test Bars, in.	Correction Factor		Diameter of Test Bars, in.	Correction Factor	
	Load	Deflection		Load	Deflection		Load	Deflection
.825	0.838	1.061	1.10	0.770	1.091	1.90	0.857	1.053
.830	0.853	1.054	1.11	0.791	1.081	1.91	0.871	1.047
.835	0.869	1.048	1.12	0.813	1.071	1.92	0.885	1.042
.840	0.885	1.042	1.13	0.835	1.062	1.93	0.899	1.037
.845	0.900	1.036	1.14	0.857	1.053	1.94	0.913	1.032
.850	0.916	1.029	1.15	0.880	1.043	1.95	0.927	1.026
.855	0.933	1.023	1.16	0.903	1.034	1.96	0.941	1.021
.860	0.949	1.017	1.17	0.927	1.026	1.97	0.955	1.015
.865	0.966	1.012	1.18	0.951	1.017	1.98	0.970	1.010
.870	0.983	1.006	1.19	0.975	1.009	1.99	0.985	1.005
.875	1.000	1.000	1.20	1.000	1.000	2.00	1.000	1.000
.880	1.017	0.994	1.21	1.025	0.992	2.01	1.015	0.995
.885	1.034	0.989	1.22	1.051	0.984	2.02	1.030	0.990
.890	1.051	0.983	1.23	1.077	0.976	2.03	1.046	0.985
.895	1.069	0.978	1.24	1.103	0.968	2.04	1.061	0.980
.900	1.087	0.972	1.25	1.130	0.960	2.05	1.076	0.976
.905	1.106	0.967	1.26	1.158	0.952	2.06	1.092	0.972
.910	1.125	0.962	1.27	1.185	0.945	2.07	1.109	0.967
.915	1.143	0.956	1.28	1.214	0.938	2.08	1.125	0.962
.920	1.162	0.951	1.29	1.242	0.930	2.09	1.141	0.957
.925	1.181	0.946	1.30	1.271	0.923	2.10	1.158	0.952

diameter of the bar at fracture. The rate of application of load should be such that fracture is produced in not less than 15 sec for the 0.875-in. diameter bar, 20 sec for the 1.20-in. diameter bar, and 40 sec for the 2.00-in. diameter bar.

In case of a failure of a test bar to meet the requirements, a retest may be made. If the retest fails, the castings should be rejected, except that a failed transverse bar may be qualified by a satisfactory tension test on a specimen machined from the broken end of the transverse test bar.

It is the intent of these specifications to subordinate chemical composition to physical properties. The quantities of any chemical elements in the cast iron may be specified by agreement between the manufacturer and the purchaser.

As an aid to selecting the proper composition of gray cast iron, Table 6 gives suggested analyses for various section sizes to meet the strength requirements of ASTM A-48-46. The values set forth in this table are only suggestions and not a part of the specifications, but they have been computed carefully and checked against actual successful practice. The analyses shown presuppose that no alloying elements are to be used.

AUTOMOTIVE CASTINGS. A more specific specification for gray iron castings is ASTM A-159-44T for automotive gray-iron castings. The physical requirements based on the 1.20-in. diameter bar and the standard tests described in A-48-41 are given in Table 7. This specification is identical with the SAE specifications for automotive castings.

An appendix to specification A-159-44T gives uses and typical base compositions of the five grades. This is not a part of the specification and is shown for information only in Table 8.

ALLOY IRON CASTINGS. There are many applications for cast iron where the efficiency or service life of the castings may be materially increased by alloying. Table 9 shows some specific alloy compositions which have been found to be of commercial value. The applications were selected from *The Alloy Cast Iron Handbook*, published in 1944 by the American Foundrymen's Association. For specification purposes the information in Table 9 may be used as a guide to the alloy type and generally expected properties. In purchasing castings it is advisable to specify the fewest possible compositional and mechanical values which will reasonably assure the required properties. Needlessly detailed specifications work a hardship on the foundry and often significantly increase the cost of the castings.

Table 6. Suggested Analyses of Cast Iron Based on Strength and Section

(Reprinted by permission from *The Cupola Handbook*, published by the American Foundrymen's Association, 1946)

Type	Chemical Range, %					Average Carbon Equivalent	Metal Section Range, in.	Brinell Hardness	Transverse Load, lb	Strength Deflection, in.	Tensile Strength, psi
	T.C.	Si	P	S	Mn						
Class 20 Light section 0.875 test bar	3.50 to 3.80	2.40 to 2.60	0.20 to 0.80	0.08 to 0.13	0.50 to 0.70	4.56 3.45	Up to 0.50	160 to 200	900 to 1,200	0.10 to 0.15	22,000 to 26,000
Class 20 Medium section 1.2 test bar	3.40 to 3.60	2.30 to 2.50	0.20 to 0.60	0.08 to 0.80		4.34 3.49	1/2 to 1	160 to 180	1,600 to 2,200	0.20 to 0.27	18,000 to 24,000
Class 20 Heavy section 2.0 test bar	3.10 to 3.30	2.20 to 2.40	0.20 to 0.40	0.08 to 0.13	0.50 to 0.80	3.98	1 and Up	130 to 180	4,500 to 6,500		18,000 to 22,000
Class 25 Light section 0.875 test bar	3.30 to 3.50	2.20 to 2.40	0.20 to 0.50	0.08 to 0.13	0.50 to 0.80	4.20	Up to 1/2	160 to 180	950 to 1,300	0.11 to 0.16	26,000 to 29,000
Class 25 Medium section 1.2 test bar	3.20 to 3.40	2.20 to 2.40	0.15 to 0.40	0.08 to 0.12	0.50 to 0.80	4.08	1/2 to 1	172 to 207	1,800 to 2,400	0.22 to 0.28	26,000 to 29,000
Class 25 Heavy section 2.0 test bar	3.00 to 3.30	1.90 to 2.20	0.15 to 0.25	0.08 to 0.12	0.50 to 0.80	3.82	1 and Up	179 to 217	6,000 to 7,800		26,000 to 30,000
Class 30 Light section 0.875 test bar	3.20 to 3.40	2.10 to 2.30	0.15 to 0.30	0.08 to 0.12	0.50 to 0.80	4.03	1/2 to 1	179 228	1,250 to 1,500		30,000 to 34,500
Class 30 Medium section 1.2 test bar	3.10 to 3.30	2.10 to 2.30	0.15 to 0.25			3.92					
Class 30 Heavy section 2.0 test bar	2.90 to 3.20	1.70 to 2.10	0.15 to 0.25	0.08 to 0.12	0.45 to 0.70	3.68	1 and Up	207 to 228	6,500 to 8,200		30,000 to 34,500
Class 35 Light section 0.875 test bar	3.10 to 3.30	2.00 to 2.20	0.15 to 0.30	0.08 to 0.12	0.45 to 0.70	3.90	Up to 1/2	179 to 228	1,150 to 1,450		36,000 to 40,000
Class 35 Medium section 1.2 test bar	3.00 to 3.25	1.80 to 2.10	0.15 to 0.25	0.07 to 0.12	0.46 to 0.70	3.77	1/2 to 1	207 to 228	2,300 to 3,000	0.25 to 0.35	35,000 to 39,000
Class 35 Heavy section 2.0 test bar	2.80 to 3.10	1.60 to 2.00	0.10 to 0.20	0.06 to 0.12	0.45 to 0.70	3.54	1 and Up	183 to 217	7,500 to 9,000	0.32 to 0.38	35,000 to 38,000
Class 40 Light section 0.875 test bar	3.00 to 3.20	1.90 to 2.20	0.10 to 0.25	0.07 to 0.12	0.45 to 0.65	3.77	Up to 1/2	212 to 241	1,275 to 1,550		42,000 to 46,000
Class 40 Medium section 1.2 test bar	2.95 to 3.15	1.70 to 2.00	0.10 to 0.20	0.06 to 0.11	0.45 to 0.70	3.65	1/2 to 1	207 to 241	2,500 to 3,400	0.25 to 0.35	40,000 to 47,000
Class 40 Heavy section 2.0 test bar	2.75 to 3.00	1.50 to 1.90	0.07 to 0.15	0.05 to 0.12	0.50 to 0.70	3.42	1 and Up	180 to 217	8,400 to 9,800	0.30 to 0.38	41,000 to 45,000
Class 50 Light section 0.875 test bar	2.90 to 3.10	1.70 to 2.10	0.10 to 0.20	0.06 to 0.12	0.50 to 0.70	3.62 3.69	Up to 1/2	228 to 269	1,600 to 1,800		51,000 to 55,000

Table 6. Suggested Analyses of Cast Iron Based on Strength and Section—*Continued*

(Reprinted by permission from *The Cupola Handbook*, published by the American Foundrymen's Association, 1946)

Type	Chemical Range, %					Average Carbon Equivalent	Metal Section Range, in.	Brinell Hardness	Transverse Load, lb	Strength Deflection, in.	Tensile Strength, psi
	T.C.	Si	P	S	Mn						
Class 50 Medium section 1.2 test bar	2.70 to 3.00	1.70 to 2.00	0.10 to 0.20	0.06 to 0.11	0.60 to 0.80	3.45	1/2 to 1	228 to 269	3,000 to 4,000	0.28 to 0.34	50,000 to 57,000
Class 50 Heavy section 2.0 test bar	2.55 to 2.85	1.40 to 1.70	0.07 to 0.15	0.06 to 0.11	0.60 to 0.80	3.20	1 and Up	207 to 241	10,000 to 12,500	0.38 to 0.48	50,000 to 54,000
Class 60 Light section 0.875 test bar	2.70 to 3.00	1.90 to 2.20	0.10 to 0.20	0.06 to 0.12	0.50 to 0.70	3.51		228 to 272	1,750 to 2,000		60,000 to 65,000
Class 60 Medium section 1.2 test bar	2.50 to 2.85	1.90 to 2.10	0.05 to 0.15	0.05 to 0.10	0.70 to 1.00	3.37		248 to 290	3,400 to 4,500	0.25 to 0.40	60,000 to 65,000
Class 60 Heavy section 2.0 test bar	2.50 to 2.80	1.20 to 1.50	0.07 to 0.15	0.05 to 0.12	0.50 to 0.80	3.09		212 to 248	11,500 to 13,500	0.35 to 0.50	60,000 to 64,000

Table 7. Physical Requirements of Automotive Gray Iron Castings

Alloy No.	Brinell Hardness No.	Transverse Strength Minimum, lb	Deflection Minimum, in.	Tensile Strength Minimum, psi
110	187 max	1800	0.15	20,000
111	170 to 223	2200	0.20	30,000
120	187 to 241	2400	0.24	35,000
121	202 to 255	2600	0.27	40,000
122	217 to 269	2800	0.30	45,000

Table 8. Typical Base Compositions of Automotive Gray Iron Castings

Alloy No.	Typical Uses	Suggested Composition, %				
		Total Carbon	Silicon	Manganese	Sulfur	Phosphorus
110	Miscellaneous soft iron castings	3.40 to 3.70	2.80 to 2.30	0.50 to 0.80	0.12	0.25
111	Air-cooled cylinders, pistons, clutch, plates	3.25 to 3.50	2.30 to 2.00	0.60 to 0.90	0.12	0.20
120	Cylinder blocks, flywheels	3.20 to 3.40	2.20 to 1.90	0.60 to 0.90	0.12	0.15
121	Heavy blocks and beads, tractor parts	3.10 to 3.30	2.10 to 1.80	0.60 to 0.90	0.12	0.12
122	Diesel engine castings, heavy parts	3.00 to 3.20	2.10 to 1.80	0.70 to 1.00	0.10	0.10

Table 9. Some Specific Applications of Alloy Cast Iron

(Data from *The Alloy Cast Iron Handbook*, American Foundrymen's Association, 1944)

Type	Chemical Composition Range, %										
	Si	T.C.	S	P	Mn	Ni	Mo	Cr	Other	Bhn	T.S.
Automotive Castings											
Brake drums, aircraft	1.10 to 1.30	3.70 to 3.90	0.10 max	0.20 max	0.60 to 0.80	1.75 to 2.25	0.35 to 0.50	0.35 max		210 to 230	28,000 to 34,000
Brake drums, automobile	2.10 to 2.30	3.15 to 3.40	0.10 max	0.27 to 0.33	0.60 to 0.80	0.15 to 0.30		0.15 to 0.30		180 to 228	
Brake drums, automobile	2.00 to 2.20	3.60 to 3.75	0.12 max	0.30 max	0.60 to 0.80			0.20 to 0.25		180 to 210	25,000 to 28,000
Camshafts, engine	2.00 to 2.20	3.00 to 3.20	0.08 to 0.10	0.08 to 0.10	0.70 to 0.90	0.40 to 0.60	0.40 to 0.60	0.80 to 1.00		286 to 311	50,000 to 60,000
Crankshafts, engine	2.40 to 2.80	2.40 to 2.90	0.04 to 0.07	0.05 to 0.08	0.80 to 1.00	1.00 to 1.25	1.00 to 1.25	0.10 to 0.20		290 to 320	60,000 to 75,000
Cylinder blocks, engine	2.15 to 2.25	3.10 to 3.40	0.10 to 0.14	0.12 to 0.18	0.50 to 0.70	0.20 to 0.40		0.20 to 0.40		187 to 228	32,000 to 37,000
Cylinder liners, engine	1.90 to 2.20	3.30 to 3.60	0.13 max	0.25 max	0.55 to 0.75			0.50 to 0.70	Cu 1.00 to 1.50	217 to 235	34,000 min
Cylinder sleeves, diesel	1.80 to 2.00	3.30 to 3.50	0.10 max	0.20 max	0.50 to 0.70	0.25 to 0.50		0.20 to 0.30	Cu 1.00 to 1.50	241 to 269	26,000 to 33,000
Piston rings, engine	2.60 to 2.80	3.60 to 3.75	0.06 to 0.09	0.30 to 0.40	0.60 to 0.80		0.15 to 0.25	0.15 to 0.25			22,000 to 26,000
Chemical Equipment											
Kettles for * corrosive chemicals	1.20 to 1.40	2.80 to 2.90				20.0 to 22.0		1.50 to 2.50		130 to 150	24,000 to 26,000
Pump castings * for brine	2.00 to 2.20	2.80 to 3.00	0.10 max	0.20 max	1.00 to 1.10	14.0 to 16.0		2.00	Cu 5.5 to 6.5	108 to 160	25,000 to 30,000
Filter grids for brine	2.00 to 2.20	2.80 to 3.00	0.10 max	0.20 max	1.00 to 1.10	14.0 to 16.0		2.00	Cu 5.5 to 6.5	160 to 180	25,000 to 30,000
Pressure castings, steam	1.80 to 2.10	2.80 to 2.90	0.04 to 0.08	0.10 to 0.20	0.60 to 0.80	0.15 to 0.25	0.40 to 0.60	0.15 to 0.25		220 to 240	50,000 to 53,000
Crushing and Grinding											
Jaw crusher plates †	0.40 to 0.50	3.20 to 3.40	0.08 to 0.10	0.15 to 0.18	0.50 to 0.60	4.4 to 4.6		1.4 to 1.6		600 to 700	35,000 to 50,000
Roll crusher shells †	0.40 to 0.50	3.20 to 3.40	0.08 to 0.10	0.15 to 0.18	0.50 to 0.60	4.4 to 4.6		1.4 to 1.6		600 to 700	35,000 to 50,000
Pump liners dredge †	0.50 to 1.00	3.15 to 3.45	0.12 max	0.10 to 0.20	0.40 to 0.60	4.5 to 5.5		1.5 to 2.0		500 min	30,000 min
Metal Working Dies											
Drawing dies, heavy duty	1.35	2.90	0.09	0.07	0.80	1.75		0.35		400 to 450 ‡	88,240 ‡
Forging dies	1.00 to 1.50	2.80 to 3.20	0.10 to 0.12	0.10 to 0.15	0.70 to 0.90		0.70 to 0.90			241 to 277	46,000 to 50,000

Table 9. Some Specific Applications of Alloy Cast Iron—*Continued*
(Data from *The Alloy Cast Iron Handbook*, American Foundrymen's Association, 1944)

Type	Chemical Composition Range, %									
	Si	T.C.	S	P	Mn	Ni	Mo	Cr	Other	Bhn	T.S.
Oil Refinery Equipment											
Pipe and fittings for acids *	1.50 to 1.80	2.70 to 2.90	0.10 max	0.20 to 0.60	1.00 to 1.20	14.0 to 16.0		2.0 to 2.5	Cu 5.5 to 6.5	180 to 200	25,000 to 30,000
Pipe and fittings	1.40 to 1.50	3.40 to 3.50	0.06 to 0.08	0.60 to 0.70	0.50 to 0.60	0.16 to 0.20		0.28 to 0.35		200 to 250	30,000 to 40,000
Pipe and fittings, condenser coils	1.40 to 2.00	3.40 to 3.50	0.06 to 0.08	0.30 to 0.60	0.50 to 0.60				Cu 0.75 to 1.00	200 to 225	30,000 to 35,000
Miscellaneous											
Chain links, heat treating	1.00 to 2.00	1.75 to 2.00	0.10 max	0.10 max	0.30 to 0.75			22.0 to 25.0		320 to 350	80,000 to 90,000
Gears, gen. machinery	2.40 to 2.70	2.70 to 2.95	0.10 max	0.25 max	0.55 to 0.70	1.40 to 1.60	0.40 to 0.60	0.35 to 0.50		240 to 270	50,000 to 60,000
Grate bars, stoker service	1.40 to 1.60	3.20 to 3.60	0.04 to 0.10	0.15 to 0.40	0.60 to 0.80	1.25 to 1.75		0.50 to 0.75		200 to 240	32,000 to 35,000
Grate bars and stoves	1.75 to 2.25	3.00 to 3.50	0.14 max	0.40 max	0.50 to 0.80			0.85 to 1.25		200 to 300	30,000 to 45,000
Stove tops, heavy ranges *	1.50 to 2.00	2.60 to 2.80	0.12 max	0.25 max	1.00 to 1.25	14.0 to 16.0		2.00 to 3.00	Cu 6.0 to 7.0	160 to 180	20,000 to 30,000

* "Ni Resist"—austenitic iron.
† "Ni Hard"—martensitic iron.
‡ Oil-quenched and drawn.

3. FACTORS AFFECTING THE QUALITY OF IRON CASTINGS

METAL STRUCTURE. Suggested compositions for various castings are shown in the preceding article of this section under specifications, and the general effects of the normal elements in iron and steel are given in Section 2. The engineering properties of cast iron are dependent on the degree of graphitization or the relative proportion of carbon in the combined and in the free or graphitic form as well as on the total carbon content.

Carbon and silicon are the principal graphitizing elements and as the amounts of these are increased, with other factors the same, the graphite content of the iron is increased and the metal becomes progressively softer and weaker. As amounts of carbon and silicon are lowered the iron becomes progressively harder and stronger. It seems paradoxical that carbon in gray cast iron is a softener, whereas carbon in steel is always a powerful hardening agent. The explanation is, of course, that as carbon in cast iron is increased the content of soft graphite is increased, and the combined carbon, the hardening agent, is usually decreased. Other elements promoting graphitization are nickel, copper, aluminum, titanium, and zirconium. Elements which retard graphitization include manganese, sulfur, chromium, molybdenum, vanadium, and tellurium.

Other than composition, the prime factor affecting the degree of graphitization in cast iron is the cooling rate during and subsequent to freezing. Slow freezing and cooling allow graphitization to proceed to the maximum extent, and rapid cooling prevents graphitization. For sand castings, section size, by affecting freezing rate, is the principal factor affecting graphitization. To produce uniform properties of metal the composition must be balanced with respect to section size and cooling rate. Table 6 exemplifies the balancing of composition by lowering silicon and carbon to retain the same strength as section size increases. Ladle additions of graphitizing or graphite-retarding alloys are also valuable to control graphitization of iron.

The best combination of strength and toughness in cast iron is obtained when graphitization is so controlled that about 0.50 to 0.90% carbon is retained in the combined form. This content results in a substantially pearlitic matrix. Reduction of the total carbon content while retaining a pearlitic matrix reduces the graphite content and produces a stronger, tougher iron. Maximum softness and machinability in cast iron occur when all the carbon is in the free or graphitic form and when the graphite content is high. Maximum hardness and wear resistance in iron occur with all the carbon in the combined form (white iron).

Besides the proportion of combined and graphitic carbon, another factor affecting the engineering properties of cast iron is the graphite-flake type and distribution. A joint committee of AFA and ASTM has provided a standard method of rating graphite flake size and type. (ASTM A247-41T.) Fine, interdendritic segregates of graphite (Type D) confer ease of machining but impart many undesirable properties, such as poor wear resistance and a low ratio of strength to hardness. The randomly distributed normal graphite flakes (Type A) are usually preferred for engineering irons. A ladle treatment of molten iron known as "inoculation" is very effective in promoting Type A graphite and preventing Type D. Inoculation consists of adding to the molten iron some elements, such as silicon, graphite, titanium, or zirconium, either singly or in combination with other alloys. Addition of an inoculating alloy to the ladle just before casting is much more effective than melting the same alloy through the cupola. The basis of some proprietary cast irons is a standardized inoculation practice for structure control.

HEAT TREATMENT. Although iron castings are not as regularly heat treated as steel castings, many of the properties of as-cast iron may be conveniently varied by several types of heat treatment. The graphite pattern of gray iron, which is formed during freezing, is not significantly changed by heat treatment, but the matrix properties are affected as they are in the heat treatment of steel. Annealing consists of heating the iron casting to above the transformation range, which is usually at 1300 to 1350 F, and cooling slowly. This decomposes some of the combined carbon content to graphite, relieves stresses, and softens the castings. Annealing is used advantageously to increase the ease of machining, particularly for production-machined small castings in the automotive and household appliance industries.

Hardening of iron castings is accomplished by the same methods used in hardening steel, i.e., by heating to above the transformation temperature and quenching, usually in oil. Cast-iron engine-cylinder liners are often oil hardened to produce an effective wear-resistant material with a martensitic matrix but with graphite flakes to prevent galling and to act as oil reservoirs. Iron castings are also flame hardened to increase the hardness at wearing surfaces.

Stress-relief heat treatment consists of heating castings to a subcritical temperature, about 1100 to 1300 F. This treatment increases dimensional stability and is used for castings that are subsequently to be machined to close dimensional tolerances.

WELDING. The welding of iron and steel is discussed in Section 21. The most-used methods of welding cast iron are: (1) fusion welding, with oxygen and acetylene; (2) brazing or bronze welding; and (3) arc welding with a high nickel rod. Most newer specifications permit welding of iron castings to repair minor defects. With proper precautions in preheating and stress relieving such welds are entirely satisfactory.

INSPECTION. Castings inspection depends so much on the service required of the part that it is difficult to generalize. Where possible, the degree and kind of inspection required should be written into the specification or order. The material specifications given in Article 2 of this section are useful in designating the type of iron and the strength properties required. For specific services special inspection is often required. Castings subject to pressure should always be tested prior to shipment at a higher pressure than that expected in service. Castings subject to gas pressure must always be tested with gas or air, not with hydraulic pressure.

Soundness in critical areas may be insured by specifying radiographic inspection. Freedom from cracks and hot tears is best guaranteed by specifying magnetic particle inspection of critical areas. There is an increasing trend, for critical services, to specify mechanical proof testing of castings under some sort of simulated service conditions.

4. STEEL CASTINGS

DEFINITIONS. Steel is an alloy, primarily of iron and carbon, which is usefully malleable as cast. Steel is differentiated from cast iron not only by its malleability but also by the fact that steel contains neither carbide eutectic nor graphite eutectic in its structure. Steel castings are steel parts which are cast in molds to shape so that no further mechanical working is required.

TYPES. There are five classes of commercial steel castings: (1) low-carbon steels (carbon content below 0.20%); (2) medium-carbon steels (carbon content 0.20 to 0.50%); (3) high-carbon steels (carbon content above 0.50%); (4) low-alloy steels (alloy content below 8%); and (5) high-alloy steels (alloy content above 8%).

The medium-carbon steel class represents the bulk of steel-castings output and is considered the "regular grade" product. The high- and low-carbon grades are special carbon steels made to develop the special qualities imparted by high- and low-carbon contents. The distinction between carbon steels and low-alloy steel is necessarily vague, as accurate definitions of alloy steels have not been formulated. In general, if a steel casting contains intentionally added alloy in more than the following percentages it will be considered an alloyed steel casting: manganese, 1.00; silicon, 0.70; nickel, 0.50; copper, 0.50; chromium, 0.25; molybdenum, 0.10; vanadium, 0.05; tungsten, 0.05; aluminum, 0.10; titanium, 0.05; zirconium, 0.05; sulfur, 0.06; and phosphorus, 0.05.

In the low-alloy steels the elements manganese, chromium, molybdenum, and nickel are usually added either singly or in combination for the purpose of increasing hardenability. High-alloy steels, which may be considered to be those containing more than 8% of total alloys, range from the simpler heat- and stain-resisting steels up to the "alloy castings," which contain so little iron that they may be considered not steels at all.

USES. Steel in the cast form is not essentially different in engineering properties from rolled or forged steel, and in general any steel part which may be economically formed by the casting method may be specified as a steel casting. Modern use of steel castings for such critical parts as engine crankshafts, aircraft radial-engine cylinder barrels, and highly stressed gears has largely dispelled the old idea that cast steel may be unsuitable for the most severe services. As compared to other cast ferrous materials, steel castings have higher strengths, higher ductility, and greater toughness.

The following list shows by industries the more important industrial applications where the use of steel castings is well established.

Automotive Industries. Although gray iron castings predominate in castings for passenger-car parts, in heavy-duty truck, bus, and tractor parts many steel castings are used for their higher strength and dependability. Such parts as spring shackles, engine supports, clutch supports, axle parts, axle housings, side frames, tractor treads, and sprockets are often made of cast steel.

Boiler and Tank Construction. Many of the fittings used with tanks and boilers are cast steel. Cast-steel saddles and nozzles, weld necks, pipe flanges, manifolds, and cross-overs are regularly used.

Crushing and Grinding Equipment. Crusher jaws and other abrasion-resisting parts are often made of high (12 to 14%) manganese steel, which develops special resistance to battering types of service. Typical steel castings used in cement mills include riding rings or tires, feed and discharge-end heads, rollers, liners, and plates. Many heat- and corrosion-resistant alloy castings are used in these services.

Electrical Machinery. Armature parts, commutator rings, couplings, stator and rotor parts, motor frames, and housings are some of the more common types of steel castings used by the makers of electrical machinery.

Engine Castings. For many types of power units, reciprocating, air, gas, and diesel, steel castings are used for such parts as cylinders and cylinder heads, gears, flywheels, crankshafts, pistons, connecting rods, manifolds, frames, and housings.

Machine Tool Industry. For such tools as lathes, boring mills, planers, milling machines, and grinders, steel castings are used for cams, chuck bodies, clutches, levers, gears, bases, sprockets, wheels, and worms.

Oil Fields and Refineries. For the oil industry many heavy machinery castings and pressure castings are made of steel. Elevated temperature problems in refinery services require many alloy steel castings for strength at high temperatures.

Railways. Railroads were the first large users of steel castings. In locomotives, cast-steel underframes weighing 20 to 50 tons are used. Driving wheels have cast-steel centers. Couplings are often steel castings. Freight and passenger car underframes may be made of one-piece steel castings. Railroad-car truck bolsters and side frames are of one-piece cast-steel construction, including the journal boxes. About 35% of the weight of the modern locomotive and 16% of the modern freight car are cast steel.

Rolling Mills and Steel Plants. Rolling-equipment parts that are subject to high stresses are usually cast steel. The mill roll housings, pinions, couplings, etc., are steel castings. The rolls themselves are sometimes steel castings. Table rollers and guides are often made of cast alloy steels. In furnace equipment the charging boxes, annealing boxes, and ingot car bodies are steel castings.

Ship and Marine Castings. On ships the stem is a steel casting, the rudder posts and frames, hawse pipes, davits, winches, struts, and other castings are usually steel. Propul-

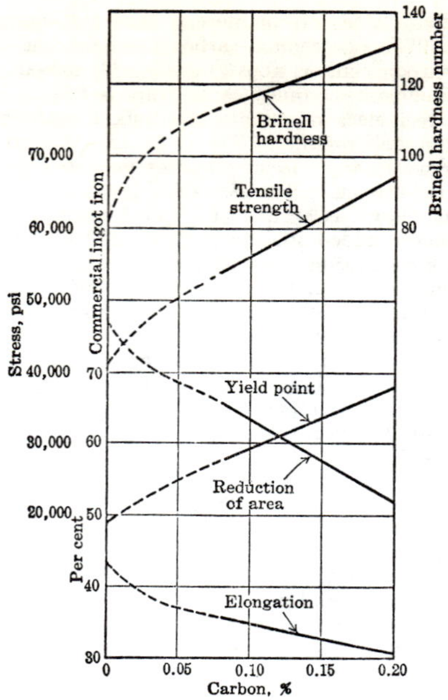

Fig. 4. Mechanical properties of full-annealed low-carbon cast steel. (Reproduced by permission from *Steel Castings Handbook*, Steel Founders Society of America)

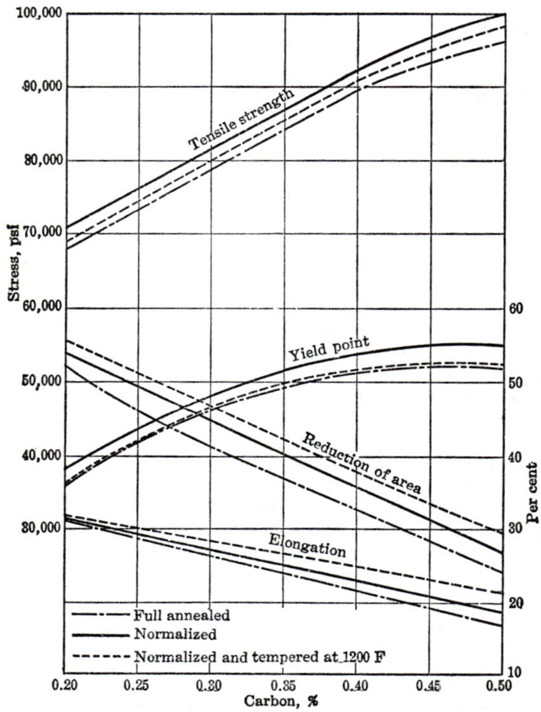

Fig. 5. Mechanical properties of heat-treated medium-carbon cast steel. (Reproduced by permission from *Steel Castings Handbook*)

sion shafting has been made of centrifugally cast steel tube sections welded together. Many cast-steel valves and pressure fittings are used in marine construction.

Many other industries such as paper, rubber, sugar, and textile use cast-steel parts.

PROPERTIES. The mechanical properties and physical constants of steel in the cast state are not essentially different from those of steel in the rolled or forged form. In the ordinary tension test for steel products cast steel will show lower values of elongation and reduction of area than comparable rolled steel tested in the direction of rolling. Rolled steel tested transversely, however, may show lower ductility than cast steel which, when properly heat-treated, does not show directional properties.

The average tensile values to be expected for cast carbon steels are very conveniently given by Figs. 4, 5, and 6, from the *Steel Castings Handbook*, published by the Steel Founders Society of America. The properties shown result from representative testing of standard round tensile bars machined from cast coupon bars. These are expected values and are not to be used as minimum specification values.

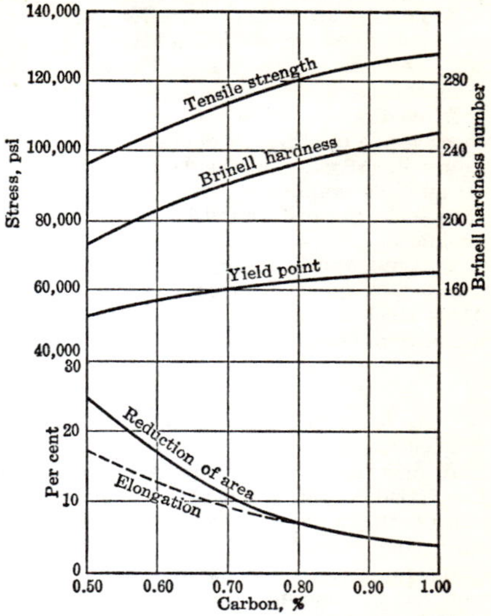

FIG. 6. Mechanical properties of full-annealed high-carbon cast steel. (Reproduced by permission from *Steel Castings Handbook*)

Quenched and tempered carbon steels and alloy steels normalized or quenched and tempered of course show much higher strength and higher ductility at a given strength level. The values of Fig. 7, also taken from the *Steel Castings Handbook*, show typical tensile values obtainable from a low-alloy cast steel quenched and tempered.

Where cast-steel parts are subject to fatigue stresses, an endurance ratio of 0.40 to 0.50 (endurance limit to tensile strength) is usual for cast steels. A recent study by C. E. Sims, reported in *The Foundry*, May 1945, showed that there was little difference in the endurance limits obtained on bars cut from forged and from centrifugally cast engine cylinder barrels.

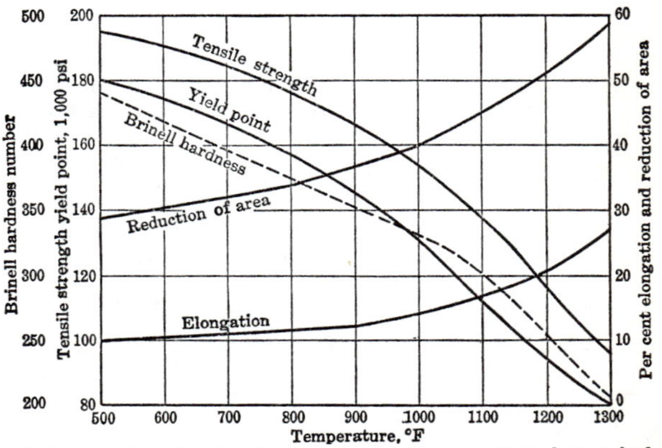

FIG. 7. Mechanical properties of nickel-chromium-molybdenum cast steel quenched in water from 1545 F and tempered at indicated temperatures. (Reproduced by permission from *Steel Castings Handbook*)

PRODUCTION. Melting. The various types of melting equipment used for the production of wrought steel products are all adaptable for producing castings. The available processes are: open-hearth (acid and basic), electric arc (acid and basic), converter, electric induction, and crucible.

Although the equipment is the same as for wrought steel, the processes may be different, since only thoroughly "killed" or deoxidized steel is suitable for castings (see Section 2,

Article 2). Good-quality steel for castings may be produced by any of the foregoing types of melting, although the crucible process is rarely used. The basic processes allow reduction of sulfur and phosphorus during the steel melting, but in the acid processes control of these elements is obtained only by selection of the raw materials. In foundries where large tonnages of steel are cast the open-hearth processes are most favored. Smaller jobbing foundries usually prefer the flexibility offered by the electric-arc furnaces.

Molding. Molding practice for steel differs from that for cast iron in several important respects. Because of the higher pouring temperatures required for steel, sands of greater refractoriness are necessary. Silica sands bonded with refractory clays are the essential components of steel molding sand. In order to reduce cost, new "facing" sand is often employed where contact is made with molten steel, and used "backing up" sand of a lower grade in other parts of the mold. No carbonaceous mold materials or washes may be used in steel foundry molding because of the tendency of the metal to dissolve carbon. In another important respect steel molding differs from gray iron molding in that all steel castings must be fed in the heavy sections if castings free from shrinkage are to be produced.

Cleaning and Finishing. Feedheads and gates on rough castings are removed by torch burning, sledging, shearing, or sawing. Torch burning followed by grinding is most used. Most steel castings are heat treated, after which the casting surfaces are cleaned by abrasive or shot blasting.

5. SPECIFICATIONS FOR STEEL CASTINGS

In drawing up specifications for steel castings to insure the desired quality there are two broad considerations: (1) the quality of the metal from which the castings are made and (2) the quality of the castings themselves. The quality of the metal is controlled by specifying chemistry, heat treatment, and physical and mechanical test values on specially cast coupon bars in which the metal is perfectly sound and in the best possible cast condition. It is sometimes argued that the physical properties of the coupon bar have little relation to the properties of metal in the actual testing. However, the coupon bar tests are properly considered to be nothing more than tests of the quality of metal as melted and cast in the best possible form. As such, these tests actually prove steel quality and melting practice and do not guarantee casting quality. The quality of castings themselves is controlled by visual inspection, by specification of some type of soundness test, or by other requirements to insure that the casting will have the desired qualifications. Metal quality is more easily generalized in the standard specifications than the actual casting quality, which often must be controlled by specific requirements for each particular type of casting.

ASTM SPECIFICATIONS. The material specifications for nearly all classes of steel castings are well covered by the ASTM. Most of these general specifications require some chemical limits, broad requirements as to heat treatment, and tensile tests on the standard 1/2-in. diameter, 2-in. gage length tensile bar. Tensile bars are machined from coupons cast either separately or attached to the casting.

Specification A-27 was completely revised in 1946 and issued as A-27-46T Carbon Steel Castings, Mild to Medium Strength, for General Application (Tentative). Table 10 gives the essential chemical and physical test requirements of this specification.

All castings of grades N-2, N-3, 60-30, 65-35, and 70-36 are required to be heat treated. Other grades may be furnished as cast. Provisions are made in this specification for radio-

Table 10. Carbon-steel Casting Materials Requirements
(ASTM A-27-46T)

Grade	Chemical Requirements, max %					Tensile Requirements, min			
	C	Mn	Si	S	P	T.S., psi	Y.P., psi	Elong., %	R.A., %
N-1	0.25	0.75	0.60	0.06	0.05	....	....	....	....
N-2	0.35	0.60	0.60	0.06	0.05	....	....	....	....
N-3	....	1.00	....	0.06	0.05	....	....	....	....
U-60-30	0.25	0.75	0.60	0.06	0.05	60,000	30,000	22	30
60-30	0.30	0.60	0.60	0.06	0.05	60,000	30,000	24	35
65-30	....	....	....	0.06	0.05	65,000	30,000	20	30
65-35	0.30	0.70	0.60	0.06	0.05	65,000	35,000	24	35
70-36	0.35	0.70	0.60	0.06	0.05	70,000	36,000	22	30

graphic inspection and magnetic particle inspection of the castings when agreed on by manufacturer and purchaser.

ASTM A-148 specification was also extensively revised in 1946 and reissued as A-148-46T, High Strength Steel Castings for Structural Purposes (Tentative). In this specification the chemistry required to meet the specified physical values is left to the discretion of the individual manufacturer except that sulfur is specified at 0.06% max and phosphorus at 0.05% max. Tensile requirements are given in Table 11.

Table 11. High-strength Steel Castings Requirements
(ASTM A-148-46T)

Grade	Tensile Requirements, min			
	T.S., psi	Y.P., psi	Elong., %	R.A., %
80-40	80,000	40,000	18	30
80-50	80,000	50,000	22	35
90-60	90,000	60,000	20	40
105-85	105,000	85,000	17	35
120-100	120,000	100,000	14	30
150-125	150,000	125,000	9	22
175-145	175,000	145,000	6	12

Heat treatment is required for all grades, but the specific type of heat treatment is left to the judgment of the manufacturer.

Radiographic and magnetic particle inspection of the castings are provided for when agreed on by manufacturer and purchaser.

Other ASTM specifications for more specific types of steel castings are:

A128-33	Austenitic Manganese-steel Castings.
A95-44	Carbon-steel Castings for Valves, Flanges, and Fittings for High-temperature Service.
A157-44	Alloy-steel Castings for Valves, Flanges, and Fittings for High-temperature Service.
A216-44T	Carbon-steel Castings Suitable for Fusion Welding for High-temperature Service (Tentative).
A217-46T	Alloy-steel Castings Suitable for Fusion Welding for High-temperature Service (Tentative).
A221-30	Chromium Alloy-steel Castings.
A198-39	20% Chromium, 9% Nickel Alloy-steel Castings.
A222-39	Chromium-nickel Alloy-steel Castings.
A223-39	Nickel-chromium Alloy-steel Castings.
A296-46T	Corrosion-resistant Iron-chromium and Iron-chromium-nickel Alloy castings for General Application (Tentative).
A297-46T	Heat-resistant Iron-chromium and Iron-chromium-nickel Alloy Castings for General Application (Tentative).

For detailed requirements of these specifications refer to the latest edition of *ASTM Standards*, Part 1-A, Ferrous Metals.

SAE SPECIFICATIONS. A *Recommended Practice* for Automotive Steel Castings has been developed by SAE to control the specific attributes of steel castings used in the automotive and allied services. Requirements cover three groups of steel castings: (1) carbon steel castings specified by chemical composition and tensile properties; (2) high-strength steel castings specified by tensile properties; and (3) high-strength steel castings specified by tensile properties and hardenability requirements.

The essential chemical and tensile requirements for the various grades of automotive steel castings are given in Table 12. The test bar for tensile properties is the standard $^1/_2$-in. diameter, 2-in. gage length bar machined from a cast coupon as in ASTM specifications.

Hardenability. Since many cast-steel automotive parts are of alloy steel used in the quenched and tempered condition, a method of specifying minimum hardenability for cast steel is provided. Requirements are based on the End-quench Hardenability Test (as described in SAE Recommended Practice, Method of Determining Hardenability, *SAE Handbook* or ASTM A-255), using the standard 1-in. around Jominy specimen. Hardenability is specified by the use of Table 13.

For example, it is desired to have a 1-in. thick casting with 105,000 psi tensile strength and of such hardenability that it will harden to the center of section in a water quench.

Table 12. Essential Chemical and Tensile Requirements of SAE Recommended Practice for Automotive Steel Castings

(Adapted by permission from the *SAE Handbank*, 1949 edition)

Grade	Description	Chemical Analyses					Minimum Tensile Properties				Bhn
		C	Mn	Si	P	S	T.S., psi	Y.P., psi	Elong., %	R.A., %	
0022	For carburizing	0.12 to 0.22	0.50 to 0.90	0.60 max	0.05 max	0.06 max	Not required				
0030	General-purpose weldable	0.30 max	0.70 * max	0.60 max	0.05 max	0.06 max	65,000	35,000	24	35	131
0050	For surface hardness	0.40 to 0.50	0.50 to 0.90	0.20 to 0.60	0.05 max	0.06 max	85,000 † 100,000 ‡	45,000 † 70,000 ‡	16 † 10 ‡	24 † 15 ‡	170 † 207 ‡
080	Carbon or alloy cast steel	Not required			0.05 max	0.06 max	80,000	40,000	18	30	163
090	Alloy cast steel	Not required			0.05 max	0.06 max	90,000	60,000	20	40	187
0105	Alloy cast steel	Not required			0.05 max	0.06 max	105,000	85,000	17	35	217
0120	Alloy cast steel	Not required			0.05 max	0.06 max	120,000	100,000	14	30	248
0150	Alloy cast steel	Not required			0.05 max	0.06 max	150,000	125,000	9	22	311
0175	Alloy cast steel	Not required			0.05 max	0.06 max	175,000	145,000	6	12	363

* For each reduction of 0.01% carbon below the maximum specified an increase of 0.04% Mn above the maximum specified will be permitted to a maximum of 1% Mn.

† Normalizing heat treatment.

‡ Quench and temper heat treatment.

Table 13. Critical Section Thickness of Casting

(Reproduced by permission from the *SAE Handbook*, 1949 edition)

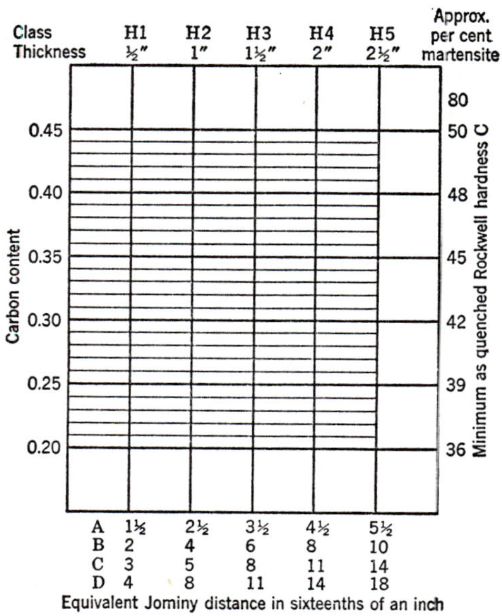

A. For desired quench hardness to a depth of one-eighth of critical section thickness—agitated water quench.

B. For desired quench hardness to a depth of one-fourth of critical section thickness—agitated water quench.

C. For desired quench hardness at center of critical section thickness—agitated water quench.

D. For desired quench hardness at center of critical section thickness—agitated oil quench.

One inch thick is class H2. The condition desired corresponds to C, which gives a Jominy distance of $^5/_{16}$ in. The specification, therefore, would read SAE-0105 H-2C, which indicates to a foundry operating with, say, 0.30% carbon that the material must have sufficient hardenability to produce a minimum hardness of 42 Rockwell C at $^5/_{16}$ in. from the quenched end of the Jominy bar.

The ASE Recommended Practice for Automotive Steel Castings also provides for radiographic inspection and magnetic particle inspection when so specified in the inquiry and purchase order and when mutually agreed upon by producer and purchaser. For full details of this specification, consult the *SAE Handbook* for 1949 or a later edition.

6. FACTORS AFFECTING THE QUALITY OF STEEL CASTINGS

SOUNDNESS. By modern steel foundry practice it is possible to produce sound steel castings except in the case of an impossible design. Gas porosity, "pinholes," etc., are prevented by proper and thorough deoxidation of the steel and by control of mold conditions. Shrinkage porosity is prevented by adequate feeding and controlled directional solidification. Where an entirely sound steel part is necessary the inspection should be so specified as to guarantee freedom from gas porosity and shrinkage.

There is no reason to tolerate gas porosity in modern steel castings, but some degree of shrinkage porosity is permissible and justifiable economically in many castings. On a plate section of uniform thickness a single riser will feed for only a limited distance, say 12 in. Therefore, to produce a perfectly sound plate casting would require risers spaced not more than 24 in. apart. If a large plate is to be cast, a number of risers must be spaced over the surface, all risers to be cut off and ground in the cleaning process. When such a plate section is insufficiently fed there is shrinkage porosity occurring at the center line of the section. In many cases some center-line shrinkage is not objectionable in the performance of the casting. Parts subject to bending and torsional stresses are little affected by some micro-shrinkage porosity at the center of the section. On the other hand, parts subject to critical tensile stresses should be perfectly sound throughout for greatest efficiency. The cast-steel railroad-truck side frame is an example of a complex plate-section casting which is not perfectly fed, yet withstands the most severe tests and has an excellent record of service. If possible, the user of steel castings should specify the degree of soundness required.

HEAT TREATMENT. Since there is no hot or cold working of the metal in cast parts, grain refinement and structure control are accomplished largely by heat treatment. Most steel castings are improved by some recrystallizing heat treatment, one exception being the very low carbon steels which are relatively unaffected by heat treatment.

Annealing, which consists of heating to well above the critical range and cooling slowly, is the simplest heat treatment as it can be done economically in large batch-type furnaces. Annealing serves the purpose of recrystallizing the metal, refining the grain, and modifying dendritic as-cast structure which has marked directional properties. Annealing usually increases the ductility and toughness of cast steel and promotes a uniform softness throughout the casting.

Normalizing heat treatment consists of air cooling from above the critical range. Normalizing accomplishes the same results as annealing, and the more rapid cooling tends to reduce segregation and promote a more uniform structure. Normalizing may be followed

Table 14. Effect of Varying Heat Treatments on Carbon Cast Steel

(Reprinted by permission from *Steel Castings Handbook*) (Melmoth)

Analysis: Carbon, 0.36%; Manganese, 0.75%; Silicon, 0.39%; Phosphorus, 0.021%; Sulfur, 0.032%

Heat Treatment, °F	Ultimate Strength, psi	Yield Point, psi	Elong. in 2 in., %	Reduction of Area, %	Brinell Hardness Number	Charpy* Impact, ft lb
As cast	88,250	42,500	15.0	15.5	187	3.2
1700 FC	85,750	49,750	25.0	38.8	179	8.4
1700 AC	96,000	55,000	24.5	39.7	196	23.1
1700 AC, 1500 AC	92,500	54,000	26.5	46.3	187	26.2
1700 AC, 1500 AC, 1200 T	83,750	51,000	29.0	46.6	170	31.1
1700 AC, 1500 WQ, 1050 T	104,500	70,000	18.0	43.9	207	30.5
1700 AC, 1500 WQ, 1200 T	94,000	65,750	25.0	56.7	196	43.5
1700 AC, 1500 WQ, 1300 T	85,250	62,000	27.5	62.3	179	44.9

FC = Furnace cooled. AC = Air cooled. T = Tempered.

* Standard keyhole notch.

by tempering at a subcritical temperature to relieve stresses and lower the hardness induced by the more rapid cooling.

Quenching heat treatment effects even more rapid cooling through the critical, with consequently greater uniformity of structure and higher hardness. Quenching is nearly always followed by tempering to stress-relieve and lower the hardness to the desired range. Often, before quenching or normalizing, a preliminary homogenizing heat treatment consisting of a soak at a temperature well above the critical is used. Homogenizing corrects by diffusion the micro segregation of constituents.

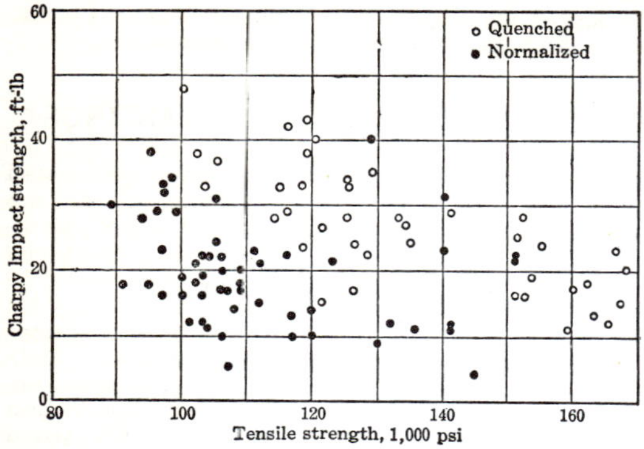

FIG. 8. Effect of heat treatment on Charpy impact strength of 0.30% carbon alloy steels cast in one-inch sections. (Reproduced by permission from paper by Hawkes, *Trans. ASM*, 1947)

Table 14 shows the effect of several heat treatments on the significant properties of a medium carbon steel. The highest ductility and impact values are obtained by quenching followed by tempering at 1200 and 1300 F.

Figure 8 shows how superior impact strength at a given tensile strength level is usually obtained by a quenching rather than a normalizing heat treatment for low-alloy cast steels.

WELDING. The welding of cast steels follows the same principles and methods as the welding of rolled steel of similar composition. All the various welding processes can be used on steel castings, although arc welding and gas welding are the principal processes used in the steel foundries. Resistance, pressure, thermit, and bronze welding are also used in fabricating with steel castings.

Repair welding of steel castings is generally accepted as a satisfactory method of reclaiming castings with minor surface defects. Where practicable, the repair welding should be done before heat treatment, or else the castings should be given a thorough stress-relief heat treatment after welding. The rules to be observed for preheating and stress relieving depend on the steel type and composition and are generally the same as for rolled products. For major repair welds by arc or gas welding the welding rod must be selected so that the weld metal will reasonably match the parent metal in chemistry and hardenability. Rules for welding are given in the ASME Boiler Construction Code and in several ASTM specifications.

INSPECTION. The types and degree of inspection given steel castings depend largely on the intended use. Use of one of the standard material specifications given in Article 5 of this section insures the quality of the metal, but the inspection must further insure the quality of the casting. Visual inspection after cleaning may reveal the surface defects, many of which may be repaired by welding.

Unsoundness due to gas porosity is often evident by visual inspection, as the "pinholes" are usually at or near the surface. Unsoundness due to shrinkage, however, is often deep seated, revealed only by cutting into representative castings or by radiography. Radiographic inspection of metal castings is described by ASTM E-15-39T. Cracks and hot tears are often detectable by visual inspection but are more positively detected and identified by magnetic particle testing. Magnetic particle testing and inspection of commercial steel castings is described in ASTM A-272-44T.

7. DESIGN FOR CASTING

Designers of iron and steel parts to be produced by casting play an important role in controlling the quality and cost of the castings obtained. For most satisfactory results the designer must at all times keep in mind the inherent behavior of molten iron during solidification and the principles of foundry practice. Repetitive castings demand the greatest care in design. The pattern equipment is often complicated and expensive.

Design errors and unnecessary complexities are multiplied in importance by quantity production. Consultation with the foundryman is very often the most effective way for the engineer to produce the design of greatest satisfaction to all concerned.

METALLURGICAL FACTORS, SOLIDIFICATION SHRINKAGE. When a mold is filled with liquid metal, solidification begins by the freezing of a thin shell of metal in contact with the mold. This shell of solid metal grows by thickening inward as heat is conducted away through the mold wall. Obviously, the thinner sections of the casting normally will be completely frozen while the thicker sections remain partially liquid. In the heavier sections, which are the last to freeze, the volume contraction during solidification is manifested by shrinkage cavities unless provision is made to feed in additional liquid metal.

The solidification contraction of medium carbon steel is on the order of 3% by volume, and if steel castings were entirely unfed they could contain up to 3% of shrinkage voids. Since shrinkage will inevitably occur in steel during solidification, the foundryman's only solution to the problem of producing sound castings is to place risers even larger in section over the heaviest sections of the casting so that the riser is the last to freeze and the shrinkage cavity will occur in the riser rather than in the useful casting.

Gray cast iron has a much lower solidification contraction or shrinkage than steel because of graphitization, and for this reason less feeding is required than for steel castings. Many soft gray iron castings of relatively uniform section are cast satisfactorily without any feeding. When the best-quality engineering castings of highest strength gray iron are required, however, the amount of shrinkage approaches that of steel and the steel foundry practice of feeding the incipient shrinkage points must be used.

Solid Contraction. When a casting is completely frozen in the mold, all the shrinkage cavities have been formed and further reduction in volume is due to solid contraction. The linear decrease in size of a casting in cooling to room temperature is compensated for by making the pattern somewhat larger than the desired finished casting. For cast iron this patternmaker's shrinkage allowance is $^1/_{12}$ to $^1/_8$ in. per ft, and for cast steel $^1/_8$ to $^1/_4$ in. per ft, depending on the size and construction of the part.

Solid contraction may cause complications more serious than mere reduction in size. When a part is designed so that it is restricted by the mold or cores from contracting freely, cracking or tearing of the hot casting may occur during cooling. In extreme cases where a long casting is securely held by the mold at both ends, the casting may be pulled in two by solid contraction. Cast steel, with its higher degree of solid contraction, is more susceptible to this effect than gray iron.

The factor of section size influences greatly many properties of the resulting castings. Radical changes in section size in the same casting give rise to localized shrinkage and also increase the dangers of cracking during solid contraction. Thick sections freeze slowly with larger crystals and coarser grain, and will generally have lower strength and ductility than thin-section castings from the same material. On the other hand, large thin-sectioned castings are difficult to run. In gray iron castings there is an additional effect of section size due to the effect of cooling rate on graphitization. An iron which will be soft and gray in a 1-in. section may cast white in a $^1/_4$-in. section. For this reason the effect of section size on hardness and strength is much greater in cast iron than in steel.

PRODUCTION FACTORS. After design, the first step in producing a casting is making the pattern. Single loose patterns require most handwork in molding and leave the gates and risers to be cut by the molder. Gated patterns standardize at least a part of the practice. Metal match-plate patterns are most useful for repetitive production of small castings. Many castings require split cope and drag patterns so that the top half and the bottom half of the mold may be rammed separately. Core boxes for ramming the cores required are also a part of the pattern equipment. Although patternmaking is too large a subject for extensive coverage here, the designer should consider the type of pattern equipment to be used as this may often influence the design to a considerable extent.

The following specific considerations in connection with pattern equipment should be kept in mind by the designer of cast parts. *Machining allowance* for surfaces to be finished depends on the metal used, the shape of the part, the size, the machining method. In general, $^3/_{32}$ in. to $^1/_4$ in. allowance for cast iron and the same or a somewhat larger allowance for steel castings are sufficient. *Draft* is the taper which must be allowed on all vertical faces of a pattern to permit its withdrawal from the sand. Interior surfaces require more draft than exterior surfaces. The amount of draft normally recommended is never less than one degree. *Core prints*, where internal cores must be used, are necessary to anchor and fix the position of the core in the mold. In general the length of a core print should equal its diameter or width. Often it is necessary for the designer to provide additional openings into an interior cavity to allow adequate core prints.

Parting lines should be made as even as possible to facilitate molding. Often a casting can be redesigned to permit a straight parting line on one plane with consequent saving in cost.

CASTING DIFFICULTIES DUE TO DESIGN. Shrinkage cavities are probably the most commonly encountered type of defect found in castings. The foundryman prevents shrinkage cavities by adequate feeding of the heavy sections, which tend to solidify last.

For casting a part of simple contour and uniform section only one riser may be necessary for a steel casting and none for a casting of soft gray iron where the slight shrinkage occurring will be so uniformly distributed that it becomes inconsequential. Many cast parts, however, must have light and heavy sections. Where light and heavy sections adjoin, risers must be used to feed into the heavy section to produce a sound casting.

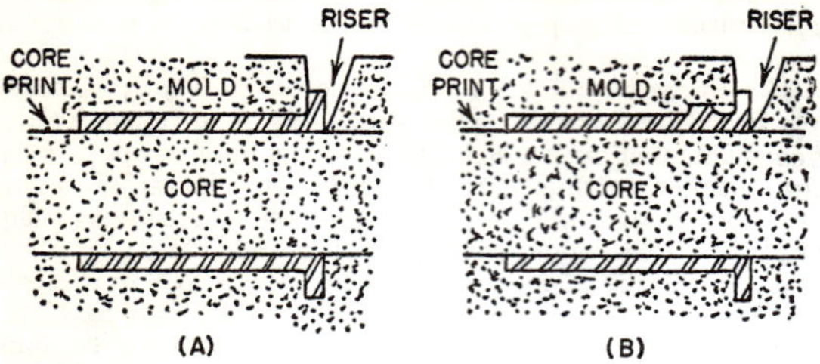

FIG. 9. Flange casting (*A*), and with tapping boss (*B*).

Figure 9*A* shows a flanged sleeve casting which is adequately fed with one riser, since freezing will progress directionally from the plain end, and the riser will be the last to freeze. Figure 9*B* shows the same casting with a tapping boss at the top near the flange. To cast this boss sound will require a second riser over the boss. If the boss is run into the flange as shown by the dotted line, however, one riser will still suffice because the isolated heavy section has been eliminated. Avoidance of isolated heavy sections which are difficult to feed simplifies the production of sound castings.

Hot tears and checks due to solid contraction are particularly amenable to treatment by design revision. Some designs will almost inevitably hot-tear when cast. If the sleeve casting in Fig. 9 had flanges at both ends it would be very susceptible to hot tearing, and the defect would usually appear just back of one of the flanges where the sharp corner of sand creates a "hot spot" which will be weak after the rest of the casting has started to contract. Generous fillet radii at the flanges are the best remedy for hot tearing in this case. Where long straight spokes of a wheel casting tend to pull out because of contraction the trouble has been cured by designing curved spokes which can give without pulling out.

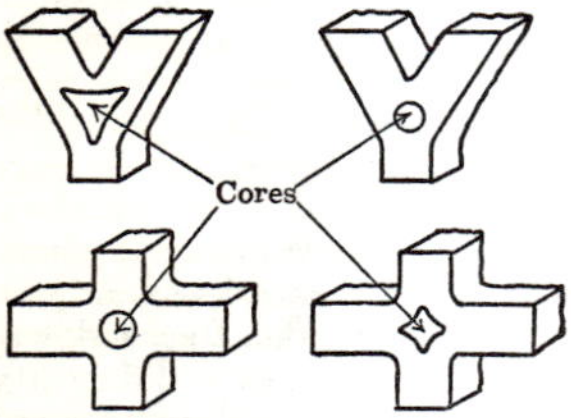

FIG. 10. Methods of insuring equalization of cooling stresses and acceleration of solidification in various types of joined members. (Reproduced by permission from *Cast Metals Handbook*, American Foundrymen's Society, Chicago)

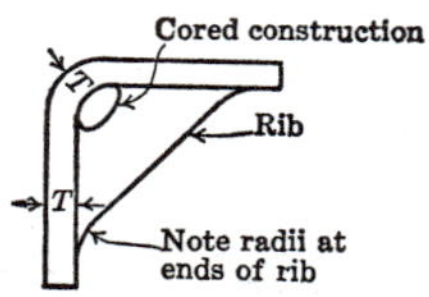

FIG. 11. Cored rib construction. Coring promotes more uniform cooling without impairing the stiffening qualities of the bracket. (Reproduced by permission from *Cast Metals Handbook*)

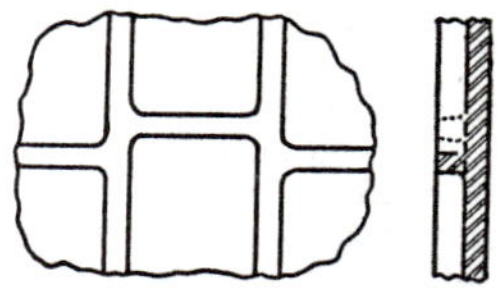

FIG. 12. Avoiding concentration of metal by staggering ribs and webs promotes more uniform cooling and is preferable to the use of cored holes in the center of rib junctions. (Reproduced by permission from *Cast Metals Handbook*)

Mis-runs occur frequently when large thin sections are cast. Core shifts are often due to insufficient openings for prints. Deep recesses and sharp internal angles in cast parts increase the difficulty of cleaning.

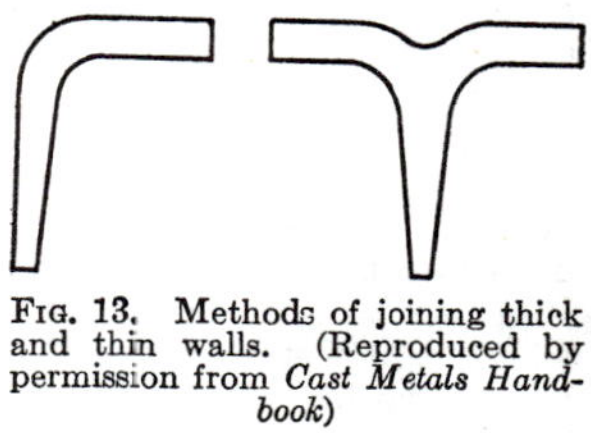

FIG. 13. Methods of joining thick and thin walls. (Reproduced by permission from *Cast Metals Handbook*)

SPECIFIC RECOMMENDATIONS FOR CASTING DESIGN.

1. Try to maintain sections as nearly uniform in thickness as possible (Figs. 10, 11, 12, 13, and 14). Abrupt changes in section are undesirable.
2. Avoid isolated heavy sections. The connecting of adjacent heavy sections simplifies feeding.
3. Indicate fillets at intersections. They should be large enough to prevent hot tearing, but not so large as to cause heavy cross sections which may not be fed through thinner walls.
4. Make castings as simple as possible. When a structure becomes very complicated it is often advisable to cast in two sections instead of one.
5. Use cores as sparingly as possible and keep them simple.

6. Avoid large areas of thin sections, which are difficult to run.
7. Allow sufficient draft to permit easy withdrawal of patterns, and avoid undercuts to permit straight drawing.
8. Design for easy division of the pattern; straight parting lines are most desirable.
9. Indicate allowances for finish-machining on individual surfaces, or note on the drawing.
10. Incorporate in the drawing pertinent information such as pressure tightness, highly stressed sections, and radiographic or magnetic particle inspection.
11. Consult with the foundryman on questionable points or where there is a choice of several possible designs.

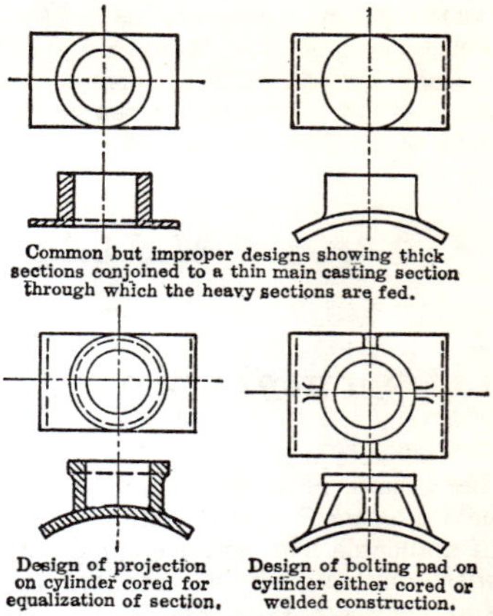

FIG. 14. Improved designs for various types of projections, bosses, lugs, and pads. (Reproduced by permission from *Cast Metals Handbook*)

8. SPECIAL CASTING METHODS

Most iron and steel castings are produced by the usual static casting methods in sand molds, but some casting methods which require special designs and techniques are also of commercial importance. Among them are (1) precision casting, (2) metal-mold casting, and (3) centrifugal casting.

PRECISION CASTING. The casting process using wax (or other meltable) patterns and poured "investment" molds is called precision casting. The process is fully described in Articles 22, 23, and 24 of this section. Not often used for regular grades of iron and steel, precision casting is of considerable importance for some of the special high-alloy steels and irons. Many of the stainless and heat-resisting steels which are difficult to forge and machine to form intricate small parts are formed to shape economically by precision casting. Very hard steels, like some tool steels, which are practically unmachinable, are best formed to accurate final dimensions by precision-casting methods.

METAL-MOLD CASTING. The principal use of metal-mold or die-casting processes is in the production of lower-melting nonferrous metals, but there are some important uses of these methods in iron and steel casting. The metal-mold material for casting ferrous metals is usually either a soft high-carbon gray iron or a ductile low-carbon alloy steel. Resistance to cracking by thermal shock is the important property for a mold material. Copper molds are sometimes used.

The largest use of a metal mold process in gray iron casting is in the production of metal-mold centrifugal cast-iron pipe. Also, several types of small automotive and household-appliance iron castings are produced by static metal-mold casting (gravity die casting). Where machinable and impact-resistant gray iron castings are required, metal-mold castings usually must be annealed to decompose excess combined carbon formed by the chilling action of the mold. Some steel casting in metal molds is practiced in conjunction with centrifugal casting.

CENTRIFUGAL CASTING. The use of centrifugal force in casting metals may confer several advantages both in economy of production and in quality of product. Several centrifugal casting methods are used in commercial production of iron and steel castings.

True centrifugal casting is used for pipe, tubes, cylinder barrels, and other generally cylindrical parts. The mold is rotated about its own axis, usually in the horizontal plane. Molten metal introduced into the spinning mold is thrown to the mold wall where it is held by centrifugal force until solidified. A perfectly cylindrical interior is thus formed without the use of a center core. Spinning speeds are such as to give a centrifugal force of the order of 50 to 100 times gravity. In addition to the advantage of eliminating center cores, the centrifugal pressure helps to eliminate gas holes and nonmetallic inclusions.

The true centrifugal process finds its largest application in the production of centrifugal cast-iron pipe. Processes using both sand molds and metal molds are in commercial use. The metal-mold pipes are annealed or normalized after casting. Centrifugal cast-iron pipes are cast in 16-, 18-, and 20-ft lengths and in diameters up to 48 in. Steel tubes and cylinders are cast by the true centrifugal process. The process is important for production of tubing in several types of high-alloy compositions which are difficult to pierce and roll into seamless tubing.

Disk- or wheel-shaped parts are centrifugally cast by what is called the semicentrifugal process. The mold is spun about the vertical axis, and metal is introduced into a downgate at the center of the mold. Centrifugal force throws the liquid metal to the periphery, and the pressure aids in feeding the rim sections. Such parts as gears and flywheels in both iron and steel are cast by this method.

Odd-shaped parts not adaptable to the other centrifugal processes are cast by the centrifuge method. Molds are spaced at the periphery of a revolving turntable, and the metal is introduced at the center axis and led into the mold cavities by radial sprues. The increased fluid pressure by reason of centrifugal force aids in filling the mold. Centrifuging is often used in conjunction with precision casting.

MALLEABLE IRON CASTINGS

By James H. Lansing

9. CHARACTERISTICS AND USES OF MALLEABLE IRON

Malleable iron is an alloy of iron and carbon, including smaller amounts of silicon, manganese, phosphorus, and sulfur, which after being cast is converted structurally by heat treatment into a matrix of ferrite containing nodules of temper carbon. The value of malleable iron as an engineering material is in its versatile castability, its ability to withstand and absorb repeated shocks, its ductility, its excellent corrosion resistance, and the remarkable ease and speed with which it can be machined. It was produced originally in Europe following Réaumur's discovery early in the eighteenth century that brittle white cast iron (iron having all its carbon in the combined form), heated to a high temperature for an extended period, will become relatively soft, tough, and ductile. It was known as "white-heart" malleable iron because malleability was acquired more by decarburization (removal of the carbon) than by a permanent conversion of its form.

Since early in the nineteenth century, malleable iron in America, and largely in Europe, too, has been of the "black-heart" type, discovered by Seth Boyden. It attains its properties of toughness and malleability, yet easy machinability, by a complete heat conversion of the carbon from the combined to the free nodular form. Europeans refer to it as black-heart malleable iron, because of its dark fracture, but in the United States it is properly called malleable iron.

Malleable iron castings are produced to physical, rather than to chemical, specifications, because their content of certain elements may be varied, depending on the amount of other elements present. In addition, the quality of the product depends largely on the heat conversion of all its carbon to the free nodular form, called temper carbon.

USES. Applications of malleable iron have progressed as have man's forms of transportation. For safety in transportation, tough and shock-resistant materials are required. Early use included saddlery hardware, then wagon and carriage parts. Agricultural implements, although not strictly forms of transportation, are akin to them and early relied on malleable castings for many of their important parts. Then, and now, plows, cultivators, mowers, reapers, binders, and spreaders incorporate malleables in their construction.

In the development of the railroad industry, malleable castings also found wide use as brake heads and fulcrums, journal boxes and lids, drawbars, brackets, door fasteners, stake pockets, tie plates, rail braces, etc. Today the largest railroad uses of malleables are in brake heads and fulcrums, brake wheels, geared brake parts, and steam and airline connectors.

With the advent of automotive transportation, however, malleable iron entered what has come to be its greatest field of usefulness. Steering-gear housings, differential cases, carriers and bearing caps, hubs, pedals and pedal brackets, motor-truck spring hangers, mountings, rear-axle housings and brake supports all must stand severe service and yet permit high-speed quantity manufacture. Malleables, being castings, can incorporate many required elements in one part and are easily and rapidly machined for their final use. Today 500,000 tons of automotive malleables are required annually.

METALLURGICAL STRUCTURE OF MALLEABLES. Malleable iron castings have a tough but relatively soft and ductile matrix of ferrite, throughout which are scattered nodules of temper carbon. Gray iron, on the other hand, has a ground mass of a hard material, pearlite, and a soft material, ferrite, with graphitic flakes distributed through it. The proportions of ferrite, pearlite, and graphite are largely governed by chemical composition, temperature, and rate of cooling. Hence a gray iron casting with both light and heavy sections may vary considerably in structure due to the difference in cooling rates of the various sections, more ferrite and graphite being present in the heavy sections and more combined carbon in the light.

The average steel casting has a ground mass of free ferrite and pearlite. The amount of pearlite and possible hardness are dependent on the carbon content. Steel castings, however, like malleable castings, have good ductility. Their structure does not vary to any considerable degree in light and heavy sections unless the difference in size is substantial. In such cases grain size may vary considerably. Steel castings are generally used in heavier applications than malleable, have better resistance to abrasive wear, but are not so easily and rapidly machined as malleable castings.

The matrix of malleable iron is ferrite; thus it somewhat resembles wrought iron, with nodules of temper carbon taking the place of the slag particles in wrought iron. Malleable iron has the greater strength due to the higher silicon content of its ferrite. Its uniform structure and fine grain size result from the heat-conversion process which is a part of its method of manufacture. Softness, ductility, and freedom from internal stress are thus assured in malleable iron castings.

10. PRODUCTION OF MALLEABLE IRON CASTINGS

CASTING. The white iron castings from which malleable ones are made are produced in green sand molds from metal melted in an air furnace of the acid-hearth reverberatory type. The metal may originally have been melted in a cupola; but in such case subsequent refinement, in an air or electric furnace, prior to casting, is essential for the proper control of the elements, so that the composition may permit conversion into malleable iron by subsequent heat treatment.

For direct pulverized-coal-fired air furnace melting, the metal charge usually consists of

Pig iron, 25 to 40%; typical amount 33%
* Sprue, 45 to 55%; typical amount 50%
Malleable scrap, 10 to 25%; typical amount 12%
Steel scrap, 0 to 15%; typical amount 5%

* Gates, runners, and any scrap castings from previous heats.

For duplex melting, original cupola melt with subsequent air furnace or electric furnace refinement, the cupola metal charge is usually

Pig iron, 5 to 20%; typical amount 12%
Sprue, 45 to 55%; typical amount 50%
Malleable scrap, 0 to 15%; typical amount 8%
Steel scrap, 15 to 45%; typical amount 30%

The decreased amount of pig iron and increased amount of steel scrap used in the cupola charge prevents the production of a metal too high in carbon content. In direct-air furnace melting more pig iron and less steel scrap may be used, as the metal is not in direct contact with the fuel, as is true in the cupola, and therefore does not absorb as much carbon.

With a proper initial silicon and carbon content of the air-furnace charge, the oxidizing action of the furnace atmosphere lowers the silicon and carbon content of the metal, both during melting and while it is being superheated, to a point where no carbon (primary graphite) will be precipitated on solidification. The castings therefore exhibit a white fracture, and are known as white or "hard iron." Duplex melting shows the same result, except that the original, and continued, charge of the air furnace consists of molten metal. Duplex melting is generally employed in conveyor-equipped plants where a continuous supply of molten metal is required.

COMPOSITION. In respect to factors other than carbon content, malleable castings may be of the same chemical composition regardless of whether medium elongation or high elongation material is being produced. Medium elongation malleable iron is covered by ASTM A-47 grade 32510, and high elongation malleable iron by ASTM A-47 grade 35018. (See Article 11 following.)

For the production of either grade, the silicon content normally ranges from 0.85 to 1.30%, most plants generally averaging slightly over 1.00%. As high silicon decreases the annealing time requirement, some plants produce castings with silicon of 1.30 to 1.80% or even higher. In such cases precautions must be taken to prevent the formation of primary graphite as the castings solidify. The precautions include the use of a largely nongraphitic charge, higher superheat of the metal, accelerated cooling of the castings, and the avoidance of production of heavy sectioned castings. High-silicon malleables, in general, meet the requirements of the 32510 grade ASTM A-47.

Sulfur content ranges from 0.05 to 0.20%, depending on whether direct air furnace melting or duplex melting equipment is employed. The sulfur content of air furnace malleable will normally run from 0.06 to 0.10%, that of duplexed malleable from 0.12 to 0.18%. The manganese content is controlled to take care of the sulfur through the formation of manganese sulfide. Otherwise the sulfur would have a stabilizing effect on the iron carbide retarding the anneal. A manganese content twice that of the sulfur plus approximately 0.15% affords a satisfactory margin of safety. Thus with an average air furnace metal of 0.08% sulfur, a manganese content of 0.30 or 0.32% will be used, whereas, with an average duplexed metal of 0.16% sulfur, a manganese content of 0.46 or 0.48% is desirable.

Phosphorus should not exceed 0.18% as it tends to reduce the impact values of the material, especially if accompanied by relatively high silicon. In the highest silicon ranges, discussed above, phosphorus content should not exceed 0.12%.

The carbon content of the metal when cast, as earlier indicated, is the most important factor in determining whether the final product will be of the medium-elongation or high-elongation type. Hence foundries operating to the 32510 specification grade normally produce metal with a carbon content of 2.40 to 2.60%. Those producing to the 35018 specification grade normally pour metal within the 2.20 to 2.40% carbon range. Those producing the extremely high silicon material must keep their carbon content in the lower range even though operating to the 32510 grade specification. Otherwise their combined silicon and carbon content would produce castings containing primary graphite even in relatively light sections. Such castings would be distinctly inferior in ultimate strength, yield point, and elongation values, and would not be true malleable iron castings.

MALLEABLE HEAT-CONVERSION PROCESS. The malleable anneal, as indicated, is really a heat-conversion process in which brittle white iron castings of suitable composition are rendered tough, ductile, and easily machinable by a complete metallurgical transformation of the metal structure. In the as-cast product all the carbon must be in the iron-combined form; in the final malleable castings none of the carbon is in the iron-combined form.

The castings may be annealed or "malleablized" in periodic furnaces which are coal, oil, or gas fired or electrically heated, or in continuous ovens similarly fired or heated. A type of periodic oven in wide use consists of a large brick chamber in which some 10 or 20 tons of castings are placed in protecting white iron boxes some 2 ft square by 1 1/2 ft deep which are stacked to a height of some 6 ft. The oven is pulverized coal or oil fired, the flame passing through the oven and out through small flues. The iron boxes, containing either sand or slag as packing material or covered by plates, protect the castings from scaling. When gas-fired radiant tubes or electric-heating elements provide the heat, or if a muffle is used, the protecting pots or boxes are not required. Continuous ovens may be of the car or tray type, the former generally being oil or gas fired and requiring the protective pots. Tray-type continuous ovens, radiant tube or electrically heated, again do not require protective pots, but may employ a gas-produced atmosphere to prevent scaling.

A typical pulverized coal-fired periodic type of oven cycle is:

	20 hr to attain 1600 F
	40 hr at 1600 F
	14 hr cooling to 1375 F
	30 hr from 1375 to 1250 F
	12 hr cooling to handle
Total	120 hours

In a radiant-tube or electrically heated periodic oven the total cycle may require but 60 hr, as the castings are much more rapidly heated and cooled. Continuous ovens further reduce the annealing time requirement as smaller batches are heated in each oven zone and the temperatures of the various zones are constantly maintained.

The decomposition of the primary "cementite and pearlite" structure of the original castings is accomplished by heating the castings and holding them, for a period, well above the critical temperature between 1330 and 1370 F. First the pearlite is transformed into austenite, the structure consisting of free cementite in a matrix of austenite. The austenite is a solid solution of iron containing varying amounts of cementite, depending on temperature and holding period. Then, aided by increased temperature, part of the cementite is dissolved in the austenite. At these elevated temperatures, 1550 to 1700 F, the austenite will dissolve more carbon as cementite than as graphite. Hence, as the castings are held longer at the high temperatures, some of the carbon which was in the cementite separates out of the solid solution in the form of temper carbon. The austenite, no longer saturated with cementite, then dissolves more cementite. The dissolving of the cementite and the precipitation of carbon by the austenite thus proceed simultaneously until all the free cementite has been dissolved and most of the carbon precipitated.

The temperature is then dropped to slightly below the 1370 to 1330 F critical range, and the castings are held there in the 1330 to 1290 F range for a sufficient number of hours to accomplish the breaking down of the austenite into temper carbon and ferrite. Some pearlite is formed at the lower temperatures, but with adequate holding time the cementite of the pearlite is spheroidized and decomposed into ferrite and temper carbon.

11. PROPERTIES AND DESIGN

By malleable casting use it is often possible to conserve metal, machine tools, and man hours. From a strength standpoint, advantage may be taken of the fact that in malleable iron the ratio of yield point to ultimate strength, 65%, is considerably higher than that of many materials of comparable elongation. Also, the notch fatigue ratio of malleable iron is high, this characteristic being further supported by its uniform rather than "directional" strength. Physical and mechanical properties are listed in Table 1.

In the design of parts to be made of malleable iron, its high fluidity and uniform strength and toughness make it possible to place metal in the exact section thickness in the exact locations desired. Interiors may be cored and many desired, or required, features incorporated in one casting with but little subsequent machining.

Table 1. Physical and Mechanical Properties of Malleable Iron

	Standard Grades 32510 and 35018
Shear strength, psi	48,000
Yield strength in shear, psi	23,000
Modulus of rigidity, million psi	10.6
Poisson's ratio	0.17
Brinell hardness number	110 to 135
Endurance limit, psi	31,600
Notch-fatigue strength, psi	About one-third ultimate tensile strength
Modulus of rupture in torsion, psi	58,000
Yield strength (torsion), psi	24,800
Angle of twist at rupture	
Round 0.9 in. diam. 5-in. long	360°
Round 1.0 in. diam. 10 in. long	790°
Charpy (keyhole and V notch) ft-lb	6.5 to 8
Izod (V notch), ft-lb	16.5
Specific gravity, average	7.32
Weight (average), lb per cu in.	0.264
Shrinkage allowance, in. per ft	1/8 to 3/16
Coefficient of thermal expansion, per °F	0.0000066
Specific heat, cal per gram per °C	0.122
Resistivity, microhms per cc	32
Coercive force, amp turns per cm	1.16 to 1.55

MACHINABILITY. It has often been stated that malleable iron is the most easily machinable and free cutting of ferrous metals of comparable or higher strength. The fact that this is so, in a metal that is both tough and possesses versatile castability, accounts for the large use of malleables in the high-speed production automotive industry.

Its high machinability is due to its uniformity of structure throughout and its nodular temper carbon which serves as a substantial lubricant on the cutting point of the tool. The formation of the chips, which generally break up without creating undue pressure on the top face of the tool, avoids the generation of high heat, with a resultant lengthening of tool life.

Illustrative of the machining properties of malleable iron is a table of relative machinability values developed by a producer of machine tools and published in *American Malleable Iron—a Handbook*. This table assigns relative machinability values to a group of the more readily machined steels and to malleable iron, with SAE 1112 steel being rated as the base value of 100. The comparison is shown in Table 2, which is abstracted from the aforementioned table.

SPECIFICATIONS FOR MALLEABLE CASTINGS. The most widely recognized standards for malleable iron castings are those of the American Society for Testing

Table 2. Relative Machinability Values

(Data from *American Malleable Iron—a Handbook*, Malleable Founders' Society, Cleveland, 1944, by permission)

		Machinability Value	Brinell Hardness Number
Standard malleable iron		120	110–145
Pearlitic malleable iron		90	180–200
Pearlitic malleable iron		80	200–240
Steels			
SAE number	AISI number		
1112	B1112	100	179–229
1113	B1113	135	179–229
1120	......	80	143–179
1118	C1118	80	143–179
1137	C1137	70	187–229
1022	C1022	70	159–192
4130	A4130	65	187–229

Materials. Specification A-47 of the ASTM provides for two grades of malleable iron castings with minimum tensile properties as follows:

	Grade 32510	Grade 35018
Tensile strength, min, psi	50,000	53,000
Yield point, min, psi	32,500	35,000
Elongation in 2 in. min, %	10	18

IMPACT RESISTANCE. Competent metallurgists disagree as to the existence of authoritative impact tests for various materials. It is believed, however, that an indication of the toughness and shock resistance of malleable iron is afforded by a wedge test that many producers employ as a shop control of their product.

The wedge and apparatus for making this test are shown in Fig. 1. The wedge is 1 in. wide, and tapers from 1/2-in. to 1/16-in. thickness in a distance of 7 in. It is placed on the anvil of the wedge-test machine, and a weight of 21 lb is allowed to fall on the tip of the wedge from a distance of 3.33 ft between the bottom of the tup and the tip of the wedge. Blows of 70 ft-lb are thus struck on the specimen which is tested as cast. It is an extremely severe test, in that the first blow falls on one-sixteenth of a square inch, and each succeeding blow on a somewhat larger section, depending on the extent of the curl of the wedge resulting from the impact of the tup. Wedges tested in this manner regularly run in excess of twenty blows and provide an excellent illustration of toughness and impact resistance.

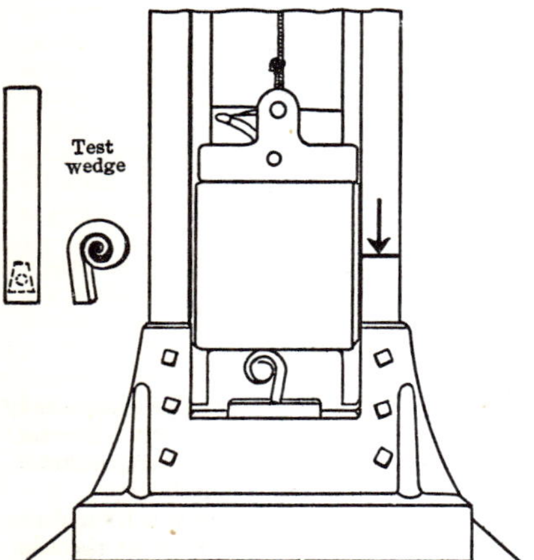

Fig. 1. Wedge test for malleable iron.

CORROSION RESISTANCE. In agricultural implement manufacture, mower and reaper knife guards and shoes, draft brackets, housings, binder needles, yokes, chain, and other castings are of malleable iron. This widespread use in machines which are exposed to the weather, in many cases at all times, attests to the excellent corrosion resistance of the material and further contributes to its large use in tractor, bulldozer, and similar farm, road, and construction machinery. Relatively unprotected railroad use is in malleable brake heads and fulcrums, box-car door hangers, hand-brake wheels and housings, tank-car dome rings and fittings.

In the electrical transmission field, an outstanding malleable application is the insulator caps used in high-voltage lines. Here also, the basic corrosion resistance of malleable iron and the facility with which it takes an excellent hot-dip galvanizing coating, influences its specification for this application.

Unprotected malleable iron castings show excellent corrosion resistance because of the formation of an adherent coating on exposure to the elements. F. L. Wolf and L. A. Meisse (*ASTM and AFA Symposium*, 1931) give the results of a corrosion test of various materials suspended in the smokejacks of a roundhouse which was in continuous use. A six-year comparison established the fact that one month's exposure in the smokejack is equivalent to two years in service. For flat specimens of standard malleable iron the loss by corrosion was only 64% of that shown by similar pieces of basic open-earth steel. Corresponding figures were 51% for malleable iron with 1% copper, and 89% for wrought iron. Round specimens exhibited similar though somewhat less pronounced differences.

SIZE OF CASTINGS. Malleable iron is cast in sections $^1/_{16}$ to 2 in. or more thick. The 2-in. section can be malleablized as completely as the $^1/_{16}$-in. section. Generally, a metal thickness of $^3/_{16}$ or $^1/_4$ in. with strengthening ribs is well adapted for small castings. For castings of medium size, sections of $^1/_2$ to $^3/_4$ in. are typical. Within reason, light and heavy sections may be incorporated in the same castings. For example, an automotive differential carrier, weighing some 20 lb, may have bearing seat sections which incorporate a 1 $^1/_4$-in. cube of metal, but some of its wall sections and ribs may be not over $^1/_4$ in. thick. Approximately 75% of malleable castings produced do not exceed 18 in. in length or 25 lb in weight. Large quantities produced weigh less than 1 lb, and this statement applies also to 5 and 10 lb classifications. Illustrative of what is considered a heavy malleable iron casting is a 300-lb automotive truck rear-axle housing. Such a casting generally has sections varying from $^1/_2$ to 2 $^1/_2$ in. thick, depending on the functions of its various parts.

UNIFORMITY OF STRENGTH AND ELONGATION. The uniform structure of malleable iron and the fact that its strength and elongation are remarkably constant, irrespective of whether the castings are machined or not, is illustrated by a test of as-cast and machined bars.

Bars of $^5/_8$-in., $^3/_4$-in., and $^7/_8$-in. diameter were cast from twenty-seven consecutive heats in one foundry. After anneal, the $^5/_8$-in. bars were tested in the as-cast condition. The bars of larger diameter were machined to the standard $^5/_8$-in. test-specimen diameter before being tested. Ultimate strengths of the as-cast and machined bars were, respectively, 54,800, 53,460, and 53,530 psi. Yield points were 36,500, 35,950, and 36,100 psi, and elongations in 2 in. were 23.4, 22.0, and 20.5%.

12. PEARLITIC AND ALLOY MALLEABLE IRON

PEARLITIC MALLEABLES. Modifications of normal malleable iron, known as pearlitic malleables, have been produced for the past few years when wearing properties, accompanied by a reasonable amount of toughness, are in demand. The product has a matrix which consists of silico-ferrite and pearlite and may contain some spheroidized cementite. The mechanical properties depend largely on the character of the matrix. Briefly, the ultimate strength varies between 60,000 and 90,000 psi, the elongation in the former averaging around 12% and in the latter not more than 3%. The type of material to be specified within the ranges indicated depends on the use to which the material is to be applied.

In general, pearlitic malleable has higher average yield point and ultimate strength, and lower elongation than normal malleable iron. It machines less readily and has a higher indentation hardness. When compared with drop forgings and steel-bar stock of identical Brinell hardness, pearlitic malleable may be machined in less time, with greater tool life, and at less cost. In comparison with forgings it also possesses the general advantages inherent in a cast part. A minimum of draft is needed, interiors may be cored out as required, machine operations are reduced, and much metal is saved. Applications include use where rigidity and wear resistance are important factors, as in gears, sprockets, air tools, and wearing pads.

The ASTM Specification for Pearlitic Malleable Iron Castings, A220-Tentative, provides for four grades conforming to the following minimum requirements for tensile properties:

	Grade 43010	Grade 48005	Grade 60003	Grade 70002
Tensile strength, min, psi	60,000	70,000	80,000	90,000
Yield point or yield strength, min, psi	43,000	48,000	60,000	70,000
Elongation in 2 in., min, %	10	5	3	2
Typical Brinell hardness number range	163–207	179–228	179–241	241–285

ALLOY MALLEABLE IRON. Some malleable iron producers also are engaged in the manufacture of the conventional type of malleable iron to which they usually add a small amount of copper and molybdenum. These alloy castings have found many applications, since they have a yield point of 40,000 to 45,000 psi and an ultimate strength of 58,000 to 65,000 psi, accompanied by an elongation of 15 to 20%.

COPPER-BASE ALLOY CASTING

By H. M. St. John

This chapter is concerned with the casting of alloys in which copper is present as 50% or more of the composition. The engineer who is planning the use of such alloys must deal with the following considerations.

1. *Selection of Alloy.* What alloy will best meet the requirements of the job in hand? If a number of alloys seem suitable, which one can be produced at least cost?

2. *Selection of Casting Method.* The method to be used, sand casting, die casting, etc., may be governed by the alloy selected. Which method will be most economical?

3. *Design of Casting.* How should the casting be designed to permit most economical production from the standpoint of foundry and machine shop?

4. *Specifications and Inspection.* To what extent and in what manner should the properties desired in the casting be insured by examination and test?

The mechanical engineer will also be asked more frequently than in the past to assist in planning foundry layout and in the choice of foundry equipment. These aspects will be discussed separately at the end of this chapter.

13. SELECTION OF ALLOY

Hundreds of copper-base alloys are in common use, but in many instances the differences among them are not significant. On p. 4-26 is a simplified list which includes all the useful alloy types with their properties and the uses for which they are suitable. For our present purpose they may be grouped as follows: (1) Red brasses and tin bronzes. (2) High-zinc alloys, including manganese bronze. (3) High-lead alloys, containing little or no zinc. (4) Aluminum bronzes. (5) Nickel bronzes. (6) Silicon bronzes. (7) Pure copper and alloys consisting of copper combined with minute percentages of hardening elements.

Of these, items 3 and 7 are used only when nothing else will serve, with no question of choice or comparison involved. The aluminum, nickel, and silicon bronzes are usually selected for special properties, particularly their corrosion resistance. The high-zinc alloys are chosen for high strength and rigidity or, sometimes, for low per-pound cost. They are also best suited for forging or permanent-mold casting. The red brasses and bronzes lend themselves best to the sand casting of intricate shapes, with a desirable combination of moderately good physical properties, good bearing properties, and excellent resistance to ordinary corrosion. They are also more readily machinable than most other alloys. The tin bronzes are the usual choice for pressure castings to be used at moderate pressures and at temperatures ranging from subzero to about 500 F.

If sand casting is to be the chosen method and either a red brass or a yellow brass seems suitable, so far as the properties of the product are concerned, red brass or bronze will usually prove to be cheaper, despite the higher per-pound cost, unless the part to be cast is of very simple design or is to be used for structural purposes only, with little or no machining. The use of the high-zinc yellow alloys imposes upon the foundryman a number of problems which substantially increase his processing costs.

14. SELECTION OF CASTING METHOD

The available casting methods are (1) ordinary sand casting; (2) gravity casting in metal molds; (3) plaster-mold casting; (4) pressure die casting; (5) centrifugal casting.

The design engineer must also consider whether the part in question might perhaps be made more economically from rod stock on a screw machine or by hot forging. The latter method is more nearly comparable to casting since it is capable of producing non-symmetrical parts which cannot be made by the screw machine.

SAND CASTING. All the alloys listed can be sand cast. Practically speaking, the red brasses and bronzes cannot be processed in any other way because of their shrinkage characteristics, their hot shortness, and the manner of their crystallization during shrinkage. An exception to this might be made in favor of centrifugal casting which, in an early stage of its development, promises eventually to be applicable to nearly all casting requirements. The nickel bronzes and most of the nearly pure copper alloys are also limited to sand casting.

Of all casting methods, sand casting requires the least investment in equipment and is most flexible with respect to changes in the design of the product. It permits the use of intricate cored cavities which can be undercut to any desired degree. Precise dimensions are not attainable; machine operations are essential if close dimensional control is required. If the number of parts to be made is not large or if the individual part weighs more than a few pounds, sand casting is the preferred method, regardless of other considerations.

METAL-MOLD GRAVITY CASTING. This method is not widely used but is of particular advantage in connection with the high-lead alloys. Since lead does not mix with the other constituents of the alloy, it tends to segregate undesirably during solidification. Poured into a metal mold the alloy chills so rapidly that the lead can be trapped in a highly dispersed condition. Molds of artificial graphite are sometimes used in place of metal and are nearly as effective. The casting and mold must be designed so as to impose no strain on the hot-short alloy as it cools. If cores are used they must be oil-sand cores or designed to be mechanically collapsible.

PLASTER-MOLD CASTING. In this process the molten metal is poured into plaster molds which have been formed on a pattern, very much as sand molds are made. The resulting castings are very close to dimension and have a smooth finish. Plaster molding is well adapted for making small castings of intricate shape, such as small gears or hardware, which in this way can be made true to finished shape and dimension but would require costly machine operations if made in sand castings. Oil-sand cores are used if cored cavities are desired.

The economical production of plaster-mold castings requires rather elaborate equipment for preparing the plaster mix and for making and drying the molds, which are poured while still hot from the oven. A portion of the broken mold material is reclaimed and processed for re-use. The process is not well adapted to short runs or to the production of any but small castings. High-zinc alloys are used exclusively.

PRESSURE DIE CASTINGS. This process is suitable for certain of the high-zinc alloys, aluminum bronze, and silicon bronze when a large number of small castings are to be made without change of design. Dies and casting equipment are costly.

In this process molten metal in a viscous condition is forced under high pressure into a steel die, which usually has several cavities connected by channels through which the metal flows almost instantaneously. Because of the speed with which the metal fills the die cavities the die designer must provide adequate vents to permit the rapid escape of air from the cavities, particularly from pockets and corners; otherwise the casting will be porous because of trapped air. The metal must be viscous rather than fluid or the metal also will squirt through the vents, and yet this very viscosity makes the trapping of air more likely. Consequently, only light sections of the casting are likely to be perfectly sound. Brass die castings having relatively heavy sections (more than $^1/_4$ in. thick, for example) will commonly be acceptable if they are used for structural purposes only, but are likely to give trouble if there is any machining to be done in the heavy sections.

The refractory alloy steels used for dies have low heat conductivity and are subject to "heat checking" after a few thousand shots, with the result that there is a network of tiny fins on the surface of the casting. If this condition is objectionable from the standpoint of appearance, the casting must be polished or the die must be redressed frequently. This difficulty is less pronounced with the aluminum and silicon bronzes.

There is also a successful patented method for casting aluminum bronze by vacuum instead of by pressure. An assembled die held together by clamps or similar means is connected to a vacuum line and dipped into molten metal, which is thus sucked into the die. Here venting is no problem since the air in the die is sucked out as the metal enters. The process is limited to small castings made in substantial volume.

CENTRIFUGAL CASTING. In this process molten metal is poured into the center of a rapidly rotating mold which may be of either metal or sand. The metal is thrown with great force to the periphery of the mold interior surfaces. The method is in use for making symmetrical brass parts such as rings or cylinders and is capable of giving a very dense, sound structure. Work done with iron and steel would seem to indicate that castings of any desired shape and lack of symmetry can be made in this way. (See Centrifugal Castings, Peter Blackwood and John Perkins, *Trans. Am. Foundrymen's Assoc.*, Vol. 52, 1944, p. 273.) There seems no reason why this method could not be applied to any alloy since the hot shortness of the red metals should offer no difficulty if sand molds are used.

15. DESIGN OF CASTINGS

Much difficulty can be avoided and costs can be reduced if the design engineer will consult the foundryman before the final design is adopted. A design unsuited to the casting method to be used may so handicap the foundryman that he cannot produce economically. Although nothing can fully take the place of such consultation it is possible to state a few basic principles for the general guidance of the designer.

Equipment, Shrinkage. First of all, the foundryman should be given a voice in the selection of pattern and core equipment. Added investment in equipment will often pay its way in reduced operating costs. In patternmaking an allowance must be made for shrinkage of the metal as it solidifies and cools. For the copper alloys this allowance varies from $^1/_8$ to $^1/_4$ in. per ft, but, depending on the design of the casting and the pattern, it will sometimes be found that certain parts of the casting show no shrinkage at all while other parts shrink more than expected.

Thickness, Tolerance, Draft. While it is important to make the casting as light as possible since the copper alloys are expensive, it is false economy to design critical parts of the casting so light as to make foundry production difficult and costly. Minimum

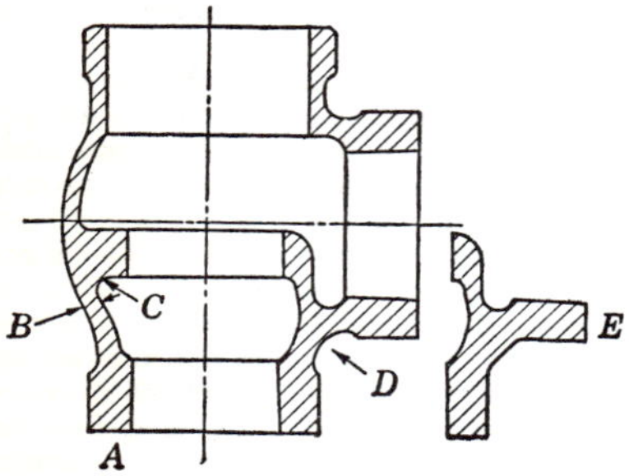

Fig. 1. Addition of rib on detail shown at right permits adequate flow of metal to heavy seat section of casting.

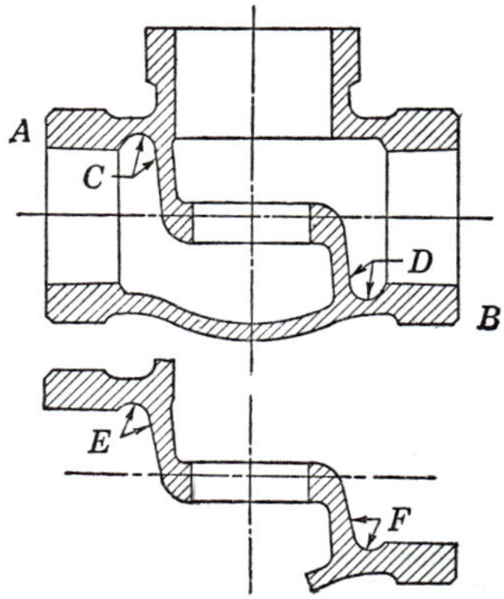

Fig. 2. Redesign shown at E and F improves flow of metal to heavy seat section.

sections in sand casting are approximately $^1/_8$ in. for the high-zinc alloys and for aluminum and silicon bronzes. The red brasses and bronzes can be poured to a minimum of $^3/_{32}$ in. In die casting, sections as thin as $^1/_{16}$ in. are permissible. Machine finishes should be held to a minimum and the foundryman consulted about the tolerance he requires. Sufficient draft must be provided for drawing the pattern or die.

Gating. Abrupt changes from a light to a heavy section are to be avoided, and provision must be made for gating the casting at an effective point. The liberal use of fillets is recommended. Figure 1 illustrates a casting so designed as to make economical production practically impossible. If the casting is gated at A—as good a place as any—the constricted section B will tend to cause a shrinkage cavity at C. By putting a rib across D at the parting line, as shown in the detail at right, the casting can be gated at E and the difficulty disappears. The casting shown in Fig. 2 can be gated at A or B, but the thin section at C and D interferes with proper feeding of the heavy seat section. By changing the design, as shown in the detail at E and F, the trouble is avoided. For a more detailed treatment of this subject refer to the *Cast Metals Handbook*, 1944 edition, American Foundrymen's Society.

Die Casting. In die casting the die cavities are filled almost instantaneously and both castings and gates solidify immediately so that no further feeding is possible. For this reason the whole structure of castings and gates must be designed as if it were a single casting. To this end the die designer must be allowed considerable latitude in determining the design of the casting.

16. SPECIFICATIONS AND INSPECTION

As a general rule the engineer should name the properties he desires in the finished casting without attempting to specify how those properties are to be attained. Visual inspection takes care of matters of appearance; dimensional tolerances can be gaged; castings designed to withstand fluid pressure can be tested for leakage. But the strength of a casting can be determined only by a destructive test. To this end specifications provide for test bars to be cast from each heat of metal as desired and for the destructive testing of these test bars as a measure of the properties to be expected from castings made from the same heat. However, it is of the greatest importance to remember that the properties of the test bar are merely a measure of the quality of the molten metal in the ladle, ready for pouring into castings. The physical properties of the casting itself are further influenced by its correctness of design, the temperature at which it is poured, and by a number of other factors that have nothing to do with the test bar. Standard test bars are castings which have been designed in such a way as to develop maximum physical properties, somewhat higher properties than can be expected in all parts of castings designed for use. Test coupons attached to the casting are not recommended.

DEFECTS. The principal defects in castings, from an engineering standpoint, are shrinkage, blow holes, sand holes, and oxide inclusions. Shrinkage holes are usually found at a point where the heavy and light sections join. Blow holes are due to a faulty molding and melting condition, and sand holes are caused by imperfect molding. Oxide inclusions may be caused by bad material but also, in some alloys, by excessive agitation during the pouring operation either inside or outside the mold.

SPECIFICATIONS. For most applications standard specifications and methods of test have been set up by the American Society for Testing Materials; they are generally accepted and in common use.

17. FOUNDRY LAYOUT AND EQUIPMENT

Modern foundries are rapidly being mechanized and are thus increasingly in need of skilled engineering. This trend results in part from a need for better working conditions since the cheap labor formerly employed in the more laborious and disagreeable phases of foundry work is no longer available. And, since labor is no longer cheap, labor-saving equipment is greatly in demand.

Formerly molders were craftsmen who used their skill and experience in the making of molds, then poured and shook them out. Now the molder is a machine operator who does nothing but make molds, with everything provided for his convenience. Molding machines in great variety are available for the most efficient production of different types of castings. In all but the smallest foundries the molds are poured and dumped with a minimum of human effort; the castings and sand are carried from one operation to the next on conveyors. In a modern sand system the sand is dried and put through a fine screen to remove metal and core butts, bentonite or clay and other additions are made, water is added to produce the proper temper, and then, after thorough mixing and mulling, the sand is transported to chutes at the molding machines. Molding sand so processed is of such quality that specially prepared facing sands are unnecessary.

MELTING EQUIPMENT. In the brass foundry, the melting operation is of first importance. Improper melting is the cause of many defective castings that are commonly attributed to bad molding or other causes. Five types of furnaces are in common use, and each of them is built in several sizes.

1. Crucible furnaces, in which the crucible is removed from the furnace and used for pouring into molds when melting is complete. In all other types, with negligible exceptions, the furnace is tilted to pour into a crucible or ladle. Furnaces of this type have been in use for centuries, with little change except in the adoption of modern fuels and more efficient combustion controls and in the improvement of crucible quality. They are still the cheapest and most convenient furnaces to be used in very small foundries or wherever operation is intermittent and a variety of alloys is handled in small quantities. Gas and oil are now the most common fuels, although coke and hard coal are in occasional use.

2. Crucible furnaces, in which the crucible is permanently installed in the furnace. Oil or gas is the fuel. More expensive to install than lift-out crucible furnaces and less flexible in operation, furnaces of this type are well adapted to production foundries of moderate size.

3. Hearth furnaces, in which metal is melted on a hearth by direct contact with flame and combustion gases. Furnaces of this type are available in a large variety of shapes and sizes, fired by oil or gas. They are commonly used in large-production foundries or in small foundries which have a steady production limited to a few alloys. Under such conditions their operating cost is low, and they lend themselves fairly well to mechanical handling of the metal and other labor-saving devices.

4. Hearth turnaces, in which metal is melted on a hearth by heat radiated from an electric arc. A motor-driven rocking mechanism thoroughly stirs the metal during melting. Furnaces of this type are of highly mechanized design and costly to install, but under favorable circumstances they show very low operating costs, and almost ideal working conditions. For best performance a dependable supply of electric energy at reasonable cost is necessary. and the foundry production must continue at a fairly high level. Intermittent operation is fatal to operating economy and to burden costs. In foundries depending principally on electric furnaces it is not uncommon to find one or more fuel-fired crucible furnaces or hearth-type furnaces to take care of peak production or to maintain production in slack periods. Hearth-type electric furnaces are particularly well suited to mechanized handling and are fairly flexible with respect to alloy changes. When necessary the furnace shell can be removed and a fresh one substituted in order to avoid contamination of the new alloy.

5. Electric induction furnaces, in which metal is melted by the flow of electric current through the metal itself, in a submerged induction channel which pumps hot molten metal into the main furnace chamber above. These furnaces also are costly to install and even more dependent than the arc type on continuity of production, since they cannot be allowed to cool and consequently require the continuous use of sufficient electric current to maintain molten metal in the induction channel during standby periods such as nights, Sundays, and holidays. Furnaces of this kind are particularly well suited to melting the high-zinc alloys such as manganese bronze. If run not less than 40 hours per week, their operating costs are reasonably low, but they are most economical if they can be operated in two or three shifts, 6 days per week. Electric induction furnaces are in satisfactory use on a wide variety of alloys including the high-copper tin bronzes and even some of the high-nickel alloys. Changing from one alloy to another is inconvenient.

OTHER FURNACE TYPES. Furnaces in less common use include cupolas, similar to those used for iron; electric hearth-type furnaces in which a radiant graphite resistor takes the place of the electric arc; crucible-type furnaces inductively heated by high-frequency electric current; and hearth-type furnaces in which flame and hot gases superheat molten metal on a hearth and then pass upward through a chimney into which the metal to be melted is charged from the top.

SELECTION OF FURNACE SIZE. With any type of furnace the melting capacity of individual units needs to be carefully matched with the foundry requirements for metal. The size of each unit should be such that, when metal is ready to be poured, it can be disposed of promptly, the furnace recharged, and melting resumed. Both economy of operation and a high standard of molten-metal quality are dependent on size of the units.

The bulk of nonferrous foundry work is in small castings in small flasks. The average weight of molten metal per mold is 12 to 20 lb. For a well-designed casting the temperature range within which satisfactory castings can be poured is about 150 F. Usually no more than 15 molds can be poured within this range from any one pot of metal before it becomes too cold. The best crucible or ladle capacity is 240 to 300 lb. Many foundries prefer 200 lb for very light work. If a 250-lb ladle is to be used, the furnace charge should be a multiple of 250. If there is only one pouring crew, the melting capacity of the furnace per heat should not greatly exceed 300 lb. For four pouring crews the proper furnace capacity is 1000 to 1200 lb. For a diversity of alloys the size of the furnace is limited to the weight of any one alloy that will be needed when the furnace is ready to pour.

Hearth-type furnaces, whether fuel-fired or electric, commonly have a molten-metal holding capacity about twice their rated melting capacity. When large castings are to be made this extra capacity can be used to advantage. For example, the molten metal in a 1000-lb furnace can be built up to 2000 lb or more by adding further increments of metal after the original charge has melted.

ALUMINUM-ALLOY CASTING

By W. P. Goepfert

Aluminum alloys suitable for casting are listed in Tables 2, 3, and 4 of Aluminum and Its Alloys, pp. 4-03, 4-04, and 4-05. The three most significant methods are sand casting, permanent-mold casting, and die casting. Die-casting alloys are listed in Table 6, p. 4-81.

18. ALUMINUM MELTING AND CASTING

MELTING. Pot furnaces of the indirectly fired gas or oil type using iron pots coated with a protective wash or nonmetallic pots of graphite or refractory are the most commonly used types for remelting for casting. The temperature of the melt should be kept as low as practicable, preferably below 1450 F, and the time a melt is held should be kept to a minimum to prevent excessive gas absorption and large grain-size development. The molten metal should be given only the minimum amount of agitation necessary to produce uniform composition. Fluxing is sometimes required to remove oxides and gas. For this purpose the agents most commonly used are aluminum chloride as solid flux, or chlorine or nitrogen as gaseous fluxes. To obtain the strongest, most ductile castings, pouring temperatures should be the lowest at which the mold can be filled without mis-runs or cold shuts. Proper arrangement of gates and risers will facilitate the use of low pouring temperatures.

Oxide entrapped in the molten alloys causes machining difficulties in the finished castings. These difficulties can be minimized by careful melting and pouring practices.

SAND CASTING. Large castings, such as crankcases for internal-combustion engines, transmission housings, engine blocks, and architectural castings, usually are made as sand castings. This process is also used frequently where the production lot is too small to warrant the expense of the steel or iron mold used in die or permanent-mold methods, or where the part to be cast requires intricate coring. The ordinary methods of molding with green-sand molds are employed, along with the good melting practices and good patterns necessary for the best results. Cores may be either green or baked sand with binders of oil, cereal flour, or synthetic resins. Ample gates and risers are necessary. Metal chills are sometimes used to avoid cracks and shrinkage and to minimize porosity in specific areas. Molding sands with high permeability are desirable to allow for the escape of mold gases.

Size Limitations. There is no inherent reason for weight limitations on sand castings. Sand castings range from an ounce to about 500 lb in weight, although castings weighing as much as 7000 lb have been produced. Although small castings having section thicknesses as thin as $^1/_8$ in. have been produced, $^3/_{16}$ in. is generally considered the practical minimum section for small and medium-sized castings.

Finish Allowance. For large castings an allowance of $^1/_4$ in. or greater should be provided for finishing. For small castings an allowance of $^1/_8$ in. is sufficient.

Tolerances. Thickness tolerances of $\pm^1/_{32}$ in. can be maintained on small castings, but greater tolerances are required for larger castings.

Sand-casting Alloys. Nominal chemical compositions, densities, and typical mechanical properties are listed in Table 2, p. 4-03.

PERMANENT-MOLD CASTING. When the quantity of an article to be produced warrants the expense of the more costly mold, castings are often made by the permanent-mold process to obtain better mechanical properties, better finish, and closer tolerance than are obtained by sand casting. Automotive and diesel pistons, cooking utensils, fuel-pump bodies, washing-machine parts, and cylinder heads are typical examples of permanent-mold castings. In this process the mold is made of dense cast iron, and the aluminum alloy is fed by gravity. Where cores are required, they are usually of alloy steel except for semi-permanent-mold casting, where sand cores are used.

Size Limitations. Permanent-mold castings as heavy as 250 lb and semi-permanent-mold castings as heavy as 500 lb have been produced. The usual size, however, ranges from a few ounces to about 20 lb. Sections as thin as $^1/_{10}$ in. have been produced on some castings. The practical minimum generally accepted, however, is $^1/_8$ in.

Finish Allowance. Depending on the size of the casting, an allowance of $^1/_{32}$ to $^1/_{16}$ in. should be provided for finishing.

Tolerances. Thickness tolerances of $\pm$0.010 can be maintained in some cases; however, a tolerance of $\pm$0.020 in. is generally applicable.

Permanent-mold-casting Alloys. Nominal chemical compositions, densities, and typical mechanical properties are contained in Table 3, p. 4-04.

PRESSURE DIE CASTINGS. Where it is desired to obtain good reproduction of detail, surface smoothness, and dimensional accuracy without subsequent machining on a large number of the same small or medium-sized casting, use of the pressure die-casting method is indicated. In this process the molten metal is forced into a steel die, set up with steel cores as required, under air or hydraulic ram pressure depending on the type of machine. The commonly called "gooseneck" machine uses air injection developing pressures as high as 750 psi, 500 psi being the more usual operating pressure; whereas in the "cold-chamber" machine a hydraulic ram is used which develops pressures as high as 50,000 psi, a range of 3000 to 15,000 psi being more usual.

Size Limitations. Castings produced by this method are usually not very large, ranging from a fraction of an ounce to 20 lb. The limitation on size is governed by the capacity of the machine. Small castings having sections as thin as 0.035 to 0.065 in., and larger castings having sections as thin as 0.085 to 0.100 in. are normally producible.

Finish Allowance. Because of the good dimensional accuracy obtained by the die-casting method, subsequent machining is not usually necessary.

Tolerances. A tolerance of $\pm$0.0015 in. per in. of dimension, $\pm$0.002 in. minimum, is required on any dimension. For each additional inch over 10 in., this tolerance can be reduced to $\pm$0.001. An additional tolerance of 0.010 in. (usually all plus) is required for dimensions across parting lines or other moving parts of the die.

Pressure Die-casting Alloys. Nominal chemical compositions, and typical mechanical properties are listed in Tables 6 and 7, pp. 4-81 and 4-82.

MAGNESIUM-ALLOY CASTING

By William H. Gross

Magnesium is most commonly cast by sand, permanent-mold, or die casting. Suitable casting alloys are listed in Tables 1 and 3, pp. 4-41 and 4-42. Other casting methods, such as precision investment (lost wax process), centrifugal, and plaster casting are also used commercially to some extent.

19. MAGNESIUM-CASTING PROCESSES

SAND CASTING. When relatively small quantities are involved, sand casting usually is the cheapest method. When parts are large or quite intricate in design, then sand casting is the only practical method. Natural and synthetic sands are both satisfactory for magnesium and should be as open as possible. The principal difference between magnesium sand foundry practice and methods used for other metals is that, for magnesium casting, protective sand addition agents are mixed into the molding sand to prevent reaction between molten magnesium and the steam generated in green-sand molds. A typical sand addition agent contains eight parts of ammonium silicofluoride and one part each of boric acid and powdered sulfur. The quantity of sand agent necessary in a green-sand mold will vary from 4 to 8% of the sand weight. Cores used for the production of magnesium castings should be resin bonded and as permeable as possible. It is desirable to have small amounts of sulfur and boric acid in the core sand; in addition, cores to be used in heavy sections should be sprayed with a solution containing a mixture of ammonium borofluoride, boric acid, and white talc in methyl alcohol. Chills are best made of cast iron, which does not alloy with magnesium and is cheap. A chill wash containing rosin and white talc in alcohol is sprayed on the chills to prevent blows. Vents and risers should be used generously in magnesium castings.

PERMANENT-MOLD CASTING. Iron molds are used for the production of castings when the quantities desired are sufficiently large. For magnesium, permanent molds can be considered when quantities approaching 1000 pieces are wanted. Both full and semi-permanent molding equipment are used. Molds are coated with a mixture (colloidal clay and water glass in a water solution) to prevent too rapid chilling. This mixture is brushed or sprayed on the mold surfaces which have been preheated to about 400 F, after which the mold is ready for use.

DIE CASTING. When small castings are required in large quantities, the die-casting process usually is considered. Quantities of 1000 or more usually justify consideration of the die-casting process. Both cold- and hot-chamber methods have been employed on magnesium. The choice of method depends largely on design of casting, end use, and cost.

MELTING PRACTICE. Several melting techniques are used on magnesium, depending on the purpose for which they are used. Most sand casting employs the crucible process in which the metal is melted in steel crucibles with a protective and purifying flux that forms a crust on the top of the molten metal. Crucibles are heated in a gas- or oil-fired brick setting or in an electric pot furnace. Flux and chunks of metal are put into the hot crucible and are allowed to melt down together. When molten, the bath is thoroughly stirred with a circular motion to refine the metal, after which the flux and dross are skimmed off the top. Heating the bath to a point between 200 and 300 F above the melting point and holding it at this temperature at least 15 min prior to pouring is called *superheating*, and is done to refine grain size of the casting. Crucibles are lifted out of the setting and poured into the mold. Permanent-mold casting employs what is known as the *open-pot process*, in which a fluid-type flux is used and castings are poured by ladling the metal out of the pot. The die-casting process employs a still different melting technique in which the molten bath is protected by a dome containing sulfur dioxide to protect the metal from burning. Refining is done by means of a heavy flux which sinks to the bottom of the molten bath.

20. MAGNESIUM-CASTING DESIGN

SAND AND PERMANENT-MOLD CASTINGS. Magnesium alloys normally exhibit a shrinkage during casting of approximately $^{11}/_{64}$ in. per ft. Where shrinkage is restrained by cores, or on large castings, the shrinkage factor may be as low as $^{1}/_{8}$ in. Wall thicknesses of sand and permanent-mold casting should not be thinner than $^{5}/_{32}$ in. unless the area is very small. These thickness limitations are about the same as for most nonferrous metals.

Probably the most important consideration in magnesium sand-casting design is the avoidance of sharp corners. Generous fillets and radii should be provided. Holes should be beaded and section changes should be very gradual. Where a relatively thick wall or a large boss adjoins a thin section the thin wall should be gradually blended into the heavier section.

DIE CASTING. In general, the suggestions for a good sand and permanent-mold-casting design apply to die castings with one primary exception. Thin walls, which should be avoided in sand-casting practice, are desirable in die castings. Stiffness of walls can be obtained by ribbed construction. Allowance for draft should be great enough to allow castings to be readily withdrawn from the die. Undercuts should be avoided because they require loose pieces in the die and thus lower production rates.

One advantage in die casting is that parts can be cast with smoother surfaces and to closer tolerances than is possible with sand castings. Dimensions on the same side of the parting line of a die casting can be held to within plus or minus 0.005 in. per in. Dimensions which cross the parting line can be held to a tolerance of plus or minus 0.010 in. per in. Wall thicknesses should preferably be held within the range of $^1/_{16}$ to $^3/_{16}$ in. since maximum mechanical properties occur within this range. Walls as thin as $^1/_{32}$ in. are permissible if the area of such thickness does not exceed 10 sq in. A wall thickness as great as $^1/_2$ in. can be used on die castings which will be lightly stressed. External threads of 16 per in. or closer can be die-cast if the axis of the threads is in the parting plane. Internal threads or threads whose axis is not in the parting plane require loose pieces in the die which will slow up production. It usually is more economical to tap or thread in such cases.

ZINC-ALLOY CASTING

By G. L. Werley

Zinc alloys can be fabricated commercially by any of the common casting practices.

21. ZINC-CASTING PROCESSES

SAND CASTING. Sand-cast stamping dies made from wooden or plaster patterns, for drop-hammer forming, are used in the fabrication of aluminum, magnesium, zinc, stainless steel, and other sheet materials. They constitute the major production of zinc-alloy sand castings.

Unalloyed zinc was first used for these dies, but, later, alloys were found to be superior. The alloys with the greatest hardness and strength were the best. ASTM Alloys XXI and XXV (see Section 4, Article 28) are the ones used for drop-hammer work, and they produce ten to fifty times as many stampings as had been obtained from unalloyed zinc.

Since many of these dies, at the end of their production runs, are stored for the manufacture of replacement parts, ASTM Alloy XXV, with its greater dimensional stability, may prove to be more desirable than XXI even though its production record is slightly inferior to the latter alloy.

SLUSH AND PERMANENT-MOLD CASTING. Slush casting is restricted to the fabrication of hollow castings. The process consists of gravity-pouring molten alloy into a split mold (generally made of bronze) until it is filled, and then, almost immediately inverting it, so as to allow escape of unfrozen metal. The time required for this casting operation is sufficient to freeze a metal shell in the mold corresponding to the shape of the cavity wall. The thickness of the casting wall is dependent on the time interval between the filling and inverting of the mold and, to a lesser extent, on the chemical and physical properties of the alloy and of the mold.

Two zinc alloys are used for slush casting. One has a nominal composition of 5.5% aluminum with the remainder Special High Grade zinc. The other contains 4.75% aluminum, 0.25% copper, and the remainder Special High Grade zinc. The former alloy is readily slush-cast with the production of very thin walls, which tends to make its use popular from an initial-cost standpoint. However, unless the impurity limits are rigidly controlled, castings made from this alloy are susceptible to subsurface network corrosion and premature failure by warping and cracking. The latter alloy produces castings with improved mechanical properties and also with longer service life because of the protective effect of the copper addition. This alloy slush-casts with greater difficulty and produces castings with thicker wall sections. Therefore, the initial cost of the castings is generally greater than that for like castings made from the copper-free alloy. It is important to keep the composition of the copper-containing alloy under close control since increases in the aluminum content make the alloy brittle and decreases make the alloy sluggish.

The copper addition must be sufficient to afford the degree of durability desired, but should not exceed an amount which seriously affects production by causing excessive weight of the castings.

Permanent-mold casting differs from slush casting in that all the metal poured into the molds is allowed to remain there until it has solidified. The resultant castings are solid instead of hollow. The same two zinc alloys which are used in slush casting are also the only alloys used in permanent-mold casting.

DIE CASTING is the art of repeatedly producing accurately dimensioned castings by forcing molten metal under pressure into metal dies. All three types of die-casting machines—direct air, plunger, and cold chamber—can be used to prepare zinc-alloy die castings, but economical practice dictates the use of the plunger machine.

In the production of zinc-alloy die castings in plunger machines the temperature of the metal should never exceed 425 C, or coarse-grained, low-shock-resistant castings will result. The preferred metal temperature range is 390 to 415 C.

The preferred range of die temperature for the production of die castings with the best surface finish, soundness, and mechanical properties is 150 to 220 C. When thermocouples are not used to indicate the die temperature it is possible to estimate the temperature from the appearance of the castings; run marks or cold shut marks indicate that the die temperature is too low and pits or blisters indicate excessive temperature.

The third major factor in the production of zinc-alloy die castings is the pressure applied on the metal. The pressure should never be less than 1500 psi. Lower pressures will result in low mechanical properties and in some unnecessary porosity in the castings. Higher pressures than 1500 psi are advantageous at times, but generally there is little to be gained by using pressures above 2500 psi. The velocity with which the metal enters the die is controlled by the pressure and the design of the die. The optimum velocity may differ for each casting, and the only way it can be determined is by trial and error.

Casting methods cannot be dismissed without some mention of the controls necessary for the maintenance of the composition of the three standard zinc alloys. Specifications for these alloys have been adopted by the American Society for Testing Materials (alloys XXI, XXIII, and XXV) and by the Society of Automotive Engineers (alloys 921, 903, and 925, respectively). (See Section 4, Article 28.) Only two of these alloys (ASTM XXIII and XXV or SAE 903 and 925) are in common use today.

In view of the limits of corrosion that must be observed for zinc-die-casting alloys, it is essential that analytical control be a part of the casting technique. This is especially true since lead and tin contamination is likely to occur at any time in a die-casting plant; aluminum is lost as a compound with iron in the skimmings when the alloy is overheated and magnesium is volatilized during fluxing.

Next in importance to composition and casting technique is employing a well-designed die. In die design, after the parting line and the general die construction have been planned, it is highly important to gate and vent the casting cavity properly. The gates and vents, as thick as can be tolerated, must be placed at their proper positions. In addition, large overflow wells should be connected to the cavity at the vent end of the die and at other places where there is a likelihood of trapping air. Since the loss of metal is small during the remelting of overflow wells, the cost of such wells is small, considering the improved quality. In most instances a saving actually results, because overflow wells accomplish a vast reduction in the percentage of rejected castings. Gates, vents, and overflow wells, considered collectively, have more influence than any other factor in the production of uniform high strength, good surface finish, and soundness in die castings.

PRECISION INVESTMENT CASTING

By F. S. Badger and W. O. Sweeny

Investment casting, or the "lost-wax" process as it is sometimes called, is a method of casting simple or intricate parts under controlled conditions so as to produce relatively accurate castings characterized by a smooth surface.

PROCESS. The process involves making a wax pattern in a die and then using this pattern to make a mold out of an investment material. The investment is a mixture of refractory particles and a liquid, which will pour and harden or set at room temperature. When the mold has set, the wax pattern is melted and burned out. Plastic patterns which vaporize when baked have been successful, but wax is the material generally used for production purposes. Parts made by this process, which vary from simple to complex designs, require a minimum of finishing and can be held to close metallurgical specifications.

MATERIALS. Practically all metals are suitable for precision-casting purposes including both ferrous and nonferrous metals, as well as highly alloyed metals. However, the process is especially suitable for expensive, hard-to-machine alloys that cannot be handled by conventional machining operations. Typical parts being manufactured in these alloys by the precision-casting process are turbosupercharger buckets, turbine blades, and turning vanes and axial-flow compressor blades for jet-propelled aircraft. Precision casting has also been used to produce such parts as reciprocating slides for cloth-cutting machines, fuel parts for aviation carburetors, pawls and claws for movie camera projectors, and also for rock drills, thread-chaser holder blocks, adjustment cams, and certain parts for diesel engines that require use of a high-strength corrosion-resistant metal.

22. METHOD OF MAKING THE MOLD

The first step in making the mold is the preparation of a die in which to produce a wax pattern. This die is usually metal, although rubber, plastic, and other materials are sometimes used. It can be produced in the same fashion as a plastic or die-casting die by direct machining from steel. However, the usual practice is to use a soft-metal die similar to the core box of ordinary foundry practice. Before the die can be produced it is necessary to make a master pattern.

THE MASTER PATTERN. The first operation is the preparation of one or more master patterns that are replicas of the parts to be produced. They are usually made about 1 1/2% oversize to compensate for wax and metal shrinkage, although this is not an exact figure and should be used only as a guide. Mastern patterns are generally made of brass, bronze, or stainless steel. In some cases, however, plaster and wood master patterns can be used, although neither of these materials is as satisfactory as a metal pattern. Where necessary, master patterns have core prints similar to those in conventional foundry practice, but it is not normally the practice to split master patterns for precision casting. Slight draft allowance is required to provide clearance from the die.

THE DIE. A two-piece die is made from the master pattern by placing the pattern on a wooden form which covers one-half of it. Plaster is then poured around the exposed half of the pattern and permitted to harden. The wooden form is removed, and a bismuth alloy is cast in a strong metal shoe over the half of the pattern previously in the board. When this is completed, the plaster support is removed; then, with the pattern still in place on half the die, the other half is cast in a similar shoe. The low-melting alloy cast has practically no shrinkage, and the die cavity produced is within 0.001 in. of the master pattern size. In order that the master pattern may be removed without harming the die, no undercuts are allowed; and a draft of about one-half degree is desirable on vertical surfaces in both halves of the die. A die is usually cast for several pieces instead of just one, since with more cavities each wax injection produces more wax patterns, thus providing faster and more economical production. When the die is completed and ready for use, it is clamped in a vise, which is water cooled to prevent any dimensional change as a result of the heat present during the wax injection.

WAX INJECTION. The wax generally used is a blend of microcrystalline waxes giving the desired properties of strength, ductility, and shrinkage. Two different blends of wax are used: one where toughness is the primary requisite; and the other, a more brittle type, for thin sections, where distortion might result if the tough ductile wax were used.

The wax is heated in a specially controlled furnace to a semi-molten condition. The temperature of the wax must be carefully controlled to within 5 F to obtain uniformity of size in the wax pattern. The actual temperatures and pressures used vary with different types of waxes, and with the shape and thickness of the part. Likewise, holding time in the die, rate of cooling, and other operating conditions will vary with the individual part.

After the wax has been brought to a molten condition, it is injected into the die. The pressure used varies with the size of the pattern, type of wax, and type of die. Obviously, a steel die will withstand higher pressures than a soft-metal die. Pressures used vary from 100 to 5,000 psi on injection machines. When the wax has solidified, usually in about one or two minutes, the die is opened and special knockout pins are operated to free the wax pattern so that it can be lifted out by hand. The die is then sprayed with a lubricant to prevent sticking and is ready for another injection.

ASSEMBLING WAX PATTERNS. Whenever possible, a group of wax patterns are joined together so that they can be placed in one mold. They are assembled by "wax welding" to the gates and risers. This is accomplished with the aid of a "hot wire welder," a short piece of resistance wire heated by an electric current to a low temperature—just hot enough to melt the wax. The gates and sprues are cast in the same manner as the wax pattern. The assembly represents the complete cavity in the finished mold.

INVESTMENT. There are two methods used in applying the refractory or investment materials.

(1) Mix and Pour Method. The assembly is surrounded by a metal box or cylinder, which is called the flask. This flask will retain the refractory mold-forming material as it solidifies around the patterns. A finely ground refractory containing some bonding agent is mixed with liquid chemicals or water and poured into the flask. Vacuum is usually employed to remove entrapped air. The mix solidifies somewhat like plaster of Paris.

(2) Dip Coat Method. This method employs a similar refractory as a thin coat on the patterns. Then this thin coat is backed up with a coarser, much cheaper material. The material cost is less, and the surface finish is generally finer, but the labor cost is more because of the two steps. The assemblage is sprayed or dipped in a fine silica powder suspended in a suitable medium, and somewhat coarser grains are screened or "stuccoed" to the dip coating. The fine material next to the wax provides a smooth finish on the casting.

The pouring mouth of the assembly is sealed to a steel plate and the stuccoed assembly thoroughly dried. A flask made of heat-resisting metal is then placed around the pattern and sealed with wax to the plate. The assembly must be liquid tight. The flask surrounding the patterns is next filled with a wet investment mixture made up of silica, sand, and crushed firebrick in a solution containing ethyl silicate, ethanol, and hydrochloric acid.

In both methods the filled flask containing this chemically hardening investment material is placed on a vibrator to pack the investment around the wax patterns and to make sure that all air is removed. This process is carried on until hydrolysis has taken place in a manner to set the investment mixture. After about one hour, the investment is gelled and the excess fines have risen to the top. The excess investment material is cut off with a saw, and the mold is allowed to age for about eight hours. At the beginning of the aging period the plate is knocked off the bottom of the mold, thereby exposing the pouring mouth into which the metal will ultimately be poured.

BAKING THE MOLD. The molds then are placed upside down in a furnace. In the preheating stage the wax is melted out, after which the mold is heated to a temperature of 1400 to 2000 F, any residue of wax being burned out as the temperature increases.

23. METHODS OF CASTING

The casting is usually done in some type of machine that will apply pressure to the metal, either centrifugally, or by pressure of a gas, usually air. For larger castings, gravity casting is used to some extent. Gravity casting is more economical than pressure casting and is just as satisfactory for simple castings. However, where complicated contours, shapes, serrations, and thin edges are involved, pressure casting is usually essential. Pressures ranging from 3 to about 30 psi, comprising the vast bulk of operations, can be applied by air. Where higher pressures are required they can be obtained by centrifugal casting.

The metal is melted by conventional means, usually in small indirect-arc or high-frequency induction furnaces. Generally, the indirect-arc furnace is used where the casting is to be made under gas pressure, and the induction furnace is used when the casting is to be made with centrifugal force.

GAS-PRESSURE METHOD. An accurately weighed charge of the desired alloy is melted under carefully controlled conditions in an indirect-arc furnace. Mold temperatures and metal pouring temperatures vary with the individual part and with the grade of alloy. Generally speaking, mold and metal temperatures are higher for thinner sections.

After the temperature has been checked, the heated mold is placed directly over the pouring spout of the furnace and is clamped in place. The assembly is rocked several times and then inverted so that the liquid metal runs from the furnace into the mold. At the time the metal is poured, air pressure is admitted into the furnace. The air pressure makes it possible to cast very thin edges and helps to produce sound and dense castings. The entire operation of bringing up the hot mold, clamping it into place, and pouring requires only a few seconds. Thus oxidation of the metal because of overcontact with air is avoided, and the flask will not cool off excessively and upset shrinkage calculations. The mold is then removed from the furnace and set aside to cool for about four hours.

VACUUM CASTING. Another casting method imposes atmospheric pressure on the casting by utilizing a vacuum on the mold. A vacuum line is attached to a chamber that is sealed to the bottom of the mold. Air and gases are drawn through the porous mold as the metal is poured in from a ladle to produce sound dense castings.

CENTRIFUGAL CASTING. The material to be cast is weighed and placed in a crucible that is then placed in a high-frequency heating coil. The charge is usually melted in 2 or 3 min, the actual time depending on the weight of charge and available power. Generally, a ratio of about 2 kw per min per lb is required. The crucible with molten metal is then transferred to the casting machine. If it is a horizontal machine, the metal is poured from the crucible or from a ladle into the central reservoir of the machine. If it is a vertical machine, the crucible itself is set in the machine; then the heated mold is inserted above the crucible and a cover plate is attached. The carrier revolves at high speed, and the centrifugal action causes the molten metal to flow to all extremities of the cavity.

FINISHING OPERATIONS. When the mold is cooled, the investment material is broken away from the casting by a pneumatic plunger. The casting is vibrated to loosen and remove the friable investment material between the nested finished pieces. Gates and feeders are removed, the gate areas are ground down, and any surface irregularities are removed by spot grinding. The ground castings are sandblasted, submitted to visual and dimensional inspection, and given x-ray and penetrative oil inspection when specified.

24. DESIGN OF INVESTMENT CASTINGS

The precision-casting process is, first, a means of producing shapes in steel or other machinable metals that cannot be economically made by other methods of manufacture, and, second, a means of producing intricate shapes in harder alloys that are difficult or impossible to forge, machine, or grind.

Parts made by the precision-investment-casting process do not compete on a price basis with parts normally made by screw-machine operations, or with forgings, coinings, stampings, die castings, or other simple castings that are readily machined.

Precision castings have been made in aluminum, magnesium, bronze, or other copper-bearing alloys, using plaster molds; but the following discussion is pertinent only to alloys with relatively high melting points. Closer tolerances, thinner edges, and other characteristics can be obtained under laboratory conditions or under circumstances where special handling is involved. The following specifications, however, reflect only what is commercially practicable at the present time.

TOLERANCES. The term *precision-investment-casting process* does not actually mean that castings will be held to fractions of a thousandth of an inch. Precision castings are held to close dimensions when compared with sand castings, but they do not compete with the tolerances obtained by precision machining or grinding. At present, ±0.005 in. per in. with a minimum tolerance of ±0.003 in. is commercially practicable. This means that to obtain a tolerance of ±0.003 in., the dimension in question must be 0.60 in. or less. These tolerances cannot be held across parting lines or on surfaces into which it is necessary to gate.

The yield of satisfactory castings is directly proportional to the dimensional tolerances allowed. The greater the tolerance, the greater the recovery, and, consequently, the lower the cost of any specific part. Since a price penalty must be put on close tolerances, all tolerances that are not critical should be widened as much as possible. For those surfaces where finishing operations are required, 0.010 to 0.030 in. of stock per dimension is satisfactory.

HOLES. Generally a hole that is small (up to 1/4 in. in diameter) and shallow (approximately 1/2-in. deep) can be held to a tolerance of ±0.003 in., both with respect to location (if dimension is 0.60 in. or less) and diameter. Larger or deeper holes require wider limits. Small blind holes cored with refractory material can be held to ±0.003 in. on diameter but not on location. Deep holes—except in hard, brittle alloys—can be successfully cored with refractory material. However, these close tolerances can be held only on holes that do not have a parting line. Since the elimination of parting lines in holes frequently necessitates loose cores, more handling is required and the cost of the part is increased.

The smooth surface common to precision-investment castings is partly the result of hand grinding to remove surface irregularities. These minor projections cannot be removed readily in small holes. Consequently, such holes frequently appear to be undersize when measured on a pin. For a hole that is to be used as cast, an allowance of a few thousandths of an inch should be made to compensate for this roughness.

Hole Finishing. If the cast hole is to be finished, this roughness will be removed at the time the hole is machined or ground to size. When the hole is cast "tap-drill size," the possibility of excessive tap wear due to the presence of small surface irregularities must be taken into consideration. With most alloys, it is advisable to plan to remove a small amount of metal by drilling or reaming so as to have parent metal exposed for the tapping operation. When a thread of 75% or less is required, the hole should be made slightly oversize to compensate for the surface irregularities.

Refractory Cores. When refractory material is used as a core, it gives a much smoother hole than if the hole is allowed to core itself with investment material. Refractory cores result in closer tolerances; however, this generally involves a slightly higher cost for the casting. The minimum-size hole that can be cored depends on the amount of surrounding metal. Thus a small hole that can be cored in a light section might not be practicable in a heavy section. The minimum-size hole that can be cored practicably is 0.050 in., even in light sections.

Special soluble wax can be used for holes when they are of intricate shape, or when the core cannot be withdrawn readily from the wax pattern because of a recess or some other structural detail. However, this procedure is fairly costly, and tolerances cannot be so close as when retractable steel cores are used.

WALLS. The minimum wall thickness practicable for production is approximately 0.50 in. for a uniform small section. But if there is sufficient taper to permit accurate feeding, it is possible to have a minimum wall thickness of 0.030 in. The tolerance on the wall thickness can generally be held to ±0.005 in. except when soluble cores are used, where a tolerance of ±0.010 in. is required.

EDGES. Approximately 0.012 to 0.015 in. is the minimum practical edge thickness, provided that everything else about the casting is favorable to the production of thin edges. To obtain the minimum thickness, it is necessary to have a gradual tapering rather than an abrupt sectional change. Moreover, it is not feasible to have a heavy section immediately adjacent to a very thin section. Thinner edges can be cast, but the recovery is generally lower and the handling problem is increased, with a resultant increase in cost. For parts subjected to high temperatures, a minimum edge of 0.025 in. is recommended. The thermal stresses created by the different rates of cooling between the thin and heavy sections are likely to crack edges that are too thin. Also, such edges are easily damaged in shop handling or in actual service.

The peak of serrations can be cast with a 0.005-in. radius, particularly with a fine pitch. The principal objection to casting serrations is the problem of eliminating small globules of metal in the root of the serration.

THREADS. Generally, it is more economical to thread parts by machining, tapping, or grinding than to cast them by the investment-casting process. Class 1 fits are the best that can be precision-cast. Unless a loose core is used, external threads will have a parting line. Internal threads do not have a parting line since the metal core chases the wax; but the cost is increased because the removal of the metal core from the wax pattern is fairly slow.

Globules of metal in threads are a serious problem, particularly in the threads of hard alloys, and it is difficult to clean them out satisfactorily by hand grinding.

In order to cast threads, the master pattern must have a pitch compensated for shrinkage. This allowance must be especially accurate, particularly on long sections, or the pitch of the thread on the casting will be incorrect.

SURFACES. Castings from investment molds generally have smoother and cleaner surfaces than those normally obtained from other types of molds. It is rather difficult, however, to obtain satisfactory castings of medium and large size without minor surface flaws. The larger and flatter the surface area to be cast, the greater the likelihood of obtaining small surface imperfections. The smooth surfaces are in part the result of hand grinding, which removes surface irregularities. Areas that cannot be readily ground will be rougher than the standard accepted precision-casting finish.

Both parting lines and draft are necessary in precision castings. The amount of draft should be as generous as permissible, yet can be as little as 1/2 degree per face. For the average piece the parting line is approximately 0.002 in. thick by 0.006 in. high.

The part should be studied so that the precision-casting manufacturer can be given the following information: (1) those surfaces which should be free from pits; (2) those surfaces into which it is not desirable to gate; (3) those surfaces on which no draft is permissible; and (4) those surfaces on which there should be no parting line.

Some applications, such as parts for gas turbines and food machinery, do not permit even minor surface pits; but, for many industrial applications, minor pits on one or more surfaces are acceptable. This information should be conveyed to the manufacturer so that the price will be based on the surface finish required.

SIZE. The size is limited primarily by the die. For cast soft-metal dies, current restrictions are a cross section of 10 in. by 5 in., with lengths of about 10 in.; but dies up to 20 in. long are feasible provided that the cross section of the part is much smaller. The die is not the limiting factor, however, when it can be machined from aluminum or other suitable metal. In such cases, the governing factors are the size of injection equipment, flasks, and other equipment.

WEIGHT of a precision casting is subject to size limitations. The maximum weight of an individual casting is dependent on the method of casting. For "pressure casting" in an indirect-arc tilting furnace, where air pressure is turned on after the furnace is inverted, the current maximum is approximately 3 1/2 lb. There is no minimum-weight limitation, provided that the casting meets the minimum requirements for thicknesses, hole diameters, and edges. A casting made by the conventional gravity-pour method has no weight limitations, but must be within the limits set by dies and molds.

FILLETS, which are frequently advantageous from a stress standpoint, are difficult to obtain by conventional means of manufacture because a square corner is normally obtained in milling or other machining operations. Generous fillets are definitely helpful in the precision-casting process, and with many parts they are almost essential. In addition, fillets are more readily cleaned by hand grinding than are square corners. At least 1/32-in. fillet is desirable, and generally a 1/16- to 1/8-in. fillet is not excessive.

INTEGRAL CASTING. Often it is possible to combine two or more parts into one integral casting, thus eliminating assembly costs and frequently improving design. For this type of casting, two or more wax patterns are joined together, a single cast part resulting.

POWDER METALLURGY

By Claire C. Balke

25. DEFINITION AND SCOPE OF POWDER METALLURGY

Powder metallurgy is the art of making objects by the heat treatment of compressed metallic powders so that none or only a part of the mass is ever melted, and the shape of the object remains substantially unaltered, even though considerable shrinkage may take place. The two distinguishing features, pressing powders to shape and sintering below the melting point, have permitted this production method to supply special metallurgical and mechanical properties unobtainable by any other production process involving one or more of the many techniques of casting, forging, stamping, screw-machine operation, or the machining of cast or wrought metal.

The advantages of the powder-metallurgy method arise from the diversity of its applications to a number of different products, which may be grouped into the following three main categories as they appeared in industry.

1. SPECIAL METALLURGICAL PROPERTIES. The refractory, or high-melting metals, such as tungsten, molybdenum, tantalum, columbium, titanium, zirconium, and their alloys with each other, must of necessity be made by pressing their powders into bar form, sintering and hammering or rolling to massive and useful metal. No refractory crucibles are known which can be used commercially for melting these metals, and, indeed, the resulting properties of some of them, if melted, would make them unfit for their present uses.

The cemented carbide industry is built on the use of the powder metallurgy technique. Part of WC-Co composition melts and causes up to 20% linear shrinkage, although the parts retain all the details of their former pressed shape.

Metals which are immiscible in the liquid state may be intimately mixed in the powdered condition. Examples include tungsten-copper for welding electrodes and silver-graphite and silver-molybdenum for electrical contact points.

2. SPECIAL MECHANICAL OR STRUCTURAL PROPERTIES. No other known production method can give controlled porosity in a piece of metal. Porosity up to 50% by volume of interconnecting pores has made possible the porous-bearing industry. Other porous parts and available materials include metallic filters made from a variety of metals, tantalum condenser elements, and porous molybdenum.

Duplex structures include thermal strips which are made by pressing one layer of powder on top of another, such as iron and nickel, and then sintering and rolling to strip. The two metals are perfectly welded together so that use can then be made of their differences in thermal expansion.

Nonmetallics may be blended with metallic powders to give a friction material for brake linings. The metallic part dissipates the heat while the nonmetallic constituent supplies the frictional quality.

Insulated metallic powders are used extensively for electronic applications. Particles of carbonyl iron powder are so treated that when pressed together they are insulated from each other. This special structure is advantageous in tuning coils and cannot be made by any other production method.

3. ECONOMIC ADVANTAGES OVER COMPETING PRODUCTION METHODS. Many structural or functional parts are made by powder metallurgy in preference to the machining of cast or wrought metal because of economies which may arise from any or all of the following considerations: (1) reduction or elimination of secondary operations; (2) the saving of metal, which may run as high as 50%; (3) a tooling time which is often more favorable; and (4) many design features that can be used which are costly to machine by conventional methods. Tolerances and production rates may be other deciding factors in choosing the powder metallurgy process for making the part under consideration.

Many parts exhibit elements of all three of the main advantages described above. Although the first two are of prime importance in supplying special products, the third or economic advantage is the one that must be carefully analyzed before powder metallurgy is chosen as the production method to be used in making a part that might be possible to obtain by other means. Since success is so dependent on the design of the part, considerable attention is given to this subject in the following articles.

26. THE POWDER METALLURGY PROCESS

Although the process has permitted the growth of completely new industries, has successfully entered into competition with established methods, and has been employed in such a wide variety of applications, it may be reduced in all applications to a series of five basic operations: (1) preparation of the powder, (2) pressing, (3) sintering, (4) coining, and (5) finishing. These operations may be altered, repeated, or even rearranged on occasion to enhance metallurgical properties or yield desired results.

PREPARATION OF THE POWDER. The great majority of elemental powders are prepared by one of two methods: electrolysis or reduction of an oxide. While other methods may be used to obtain special particle shapes or alloy powders, the great bulk of commercial powders is made by either one of these two methods. The blended powders may also include nonmetallic constituents. In most cases they are lubricated to assist in the second step of pressing them in a die. The powders are closely controlled as to purity, particle size, flowability, compacting factor, and apparent density. The last three characteristics are important factors in the use of a powder in high production presses and must be known by the die designer if the part is to be made satisfactorily.

THE PRESSING OPERATION is the most distinguishing characteristic of the powder-metallurgy process. Pressures range from only a few tons to as high as 50 tons to the square inch. Pressures up to 100 tons total capacity can be had on automatic mechanical presses operating from about eight to several hundred times a minute, depending on the size, complexity of part, and the type of powder used. Pressures above 100 and up to 3000 tons total capacity are obtained in the slower-acting hydraulic presses which may or may not be automatically fed with powder.

Dies. Hardened-steel dies are most commonly used and will yield up to 150,000 parts before wearing out of tolerance. For more abrasive powders and where parts are needed in greater quantities, it is often more economical to line the die cavity with hard carbide inserts shrunk into a steel case. Production often runs over a million parts before such a die wears out of size. It is obvious that the dies and punches must be designed to withstand the heavy pressures. They must also have sufficient capacity to hold the voluminous powder which generally has a three-to-one compacting factor, so that the final part is only about one-third the height of the powder fill. Most dies are flush filled by a powder hopper which vibrates over the die cavity and, as it sweeps aside, levels the top surface of the powder flush with the surface of the die.

Practically all metal powders, with the exception of the very soft powders such as lead, have little or no hydrostatic character even under the great loads applied. Sometimes it is possible to lubricate the die walls before each fill to reduce die-wall seizure which causes cracking when the compact is ejected. All compacts expand slightly when the pressure is released and they are removed from the die. Cracking difficulties arise more often with hard powders such as tungsten than with more ductile powders such as brass.

The size, complexity of design, and material of a given part must be carefully considered before the powder metallurgy method is attempted in its production. The main limitation lies in the pressing operation because of the cost of massive equipment needed to give high production at high unit pressures, and the comparative lack of flow in the mass of powder while under pressure.

Figure 1 illustrates the details of the pressing operation for a simple cylindrical part having a flange and a center hole. The operation is divided into four main steps, as follows: (*A*) The die cavity is flush filled by a hopper to a depth approximately three times the height of the compact. The essential punches, Fig. 1*A*, consist of an upper punch which has a hole to fit over the stationary core rod, a lower stripper punch, and a lower center punch. (*B*) The top and lower punches move toward the center of the compact to give as uniform a density as possible. (*C*) The lower punches move together to start the ejection of the part. (*D*) The stripper punch stops flush with the die as the center lower punch continues the ejection to free the part from the die, at which time it is swept aside by the hopper containing the powder, and the cycle is repeated.

It is obvious that die-wall friction must be considered in the design of the part as this directly affects the density to which the powder may be pressed. The density in turn

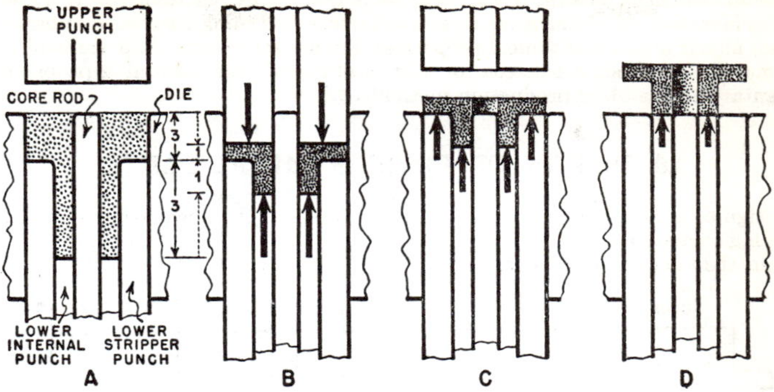

FIG. 1. Four stages in pressing a part from metal powders.

affects the properties of the part to a greater extent than any other single factor. The greater the side-wall area compared to the sectional area, the lower will be the density and the poorer will be the physical properties.

SINTERING may be done in ordinary brazing furnaces with controlled atmospheres or in highly specialized vacuum furnaces. The temperature is chosen so that either no melting takes place or only a chosen portion of the composition melts. In many instances the dimensions of the pressed part can be retained with practically no shrinkage taking place during the sintering operation. In other instances, especially where part of the composition melts, considerable shrinkage may take place and the part must be pressed oversize by a predetermined amount. Cemented carbide compacts suffer as much as 20% linear shrinkage when the binder portion melts to cement the carbide particles together. For many powders no further heat treatment is required. For others a second pressing and a second sintering are required to attain optimum physical or metallurgical properties.

COINING of a part is resorted to when tolerances must be held to close limits or the surface must be made smoother. This often relieves the necessity for too close dimensional control in the original pressing and sintering operations.

FINISHING, or secondary, operations include the use of one or more of those metallurgical or shop practices required to add features not supplied by the preceding powder metallurgy operations. They may include deburring, rolling, heat treating, or machining. Coining is sometimes classed as a secondary operation, but it is so common to the practice of the powder metallurgy method that it is here classed separately.

27. ECONOMIC ASPECTS OF POWDER METALLURGY

In those industries where powder metallurgy supplies some very special metallurgical or structural advantage, the economic aspects are of secondary importance. When the method is applied to metals and products already established, it must compete with the established procedures for making cast and machined products. While it cannot compete at every turn with all the products of precision casting and stamping methods, it does find a broad field of economic application in the manufacture of parts formerly made by these methods.

Powder metallurgy allows tolerances of 0.001 in. to be maintained. At higher costs, dimensions may be held even closer. Savings in metal are realized by the elimination, to a great extent if not completely, of all machining. Often a part will weigh only half as much as the original blank if much machining has to be performed in its manufacture. The surfaces of powder-metallurgy parts are very smooth and require no finishing although deburring may be needed, especially if high pressures are used in forming the part.

Metal powders cost more per pound than massive cast metals. However, savings resulting from the elimination of machining offset the differential somewhat, especially with nonferrous materials.

The quantity of a given part is a determining factor in the choice of the powder-metallurgy method. Small runs raise the cost out of proportion to the value of the part. Usually the quantity should exceed 50,000 pieces in order to reduce the relatively high die and pressing costs to a reasonable figure.

In contemplating the possible use of this production tool for manufacturing a given part, careful consideration should be given to all the factors, including die cost, number of pieces, tolerances, physical and mechanical properties, tooling setup time, and value of the part. Powder metallurgy is not a panacea for all metallurgical and machining problems. It is supplementary to the older production methods.

28. DESIGN OF POWDER METAL PARTS

The design of a part, even for the older applications of the powder-metallurgy method, is a deciding factor in the successful application of this method of manufacture. It has been found that density affects the properties more than any other single factor. When

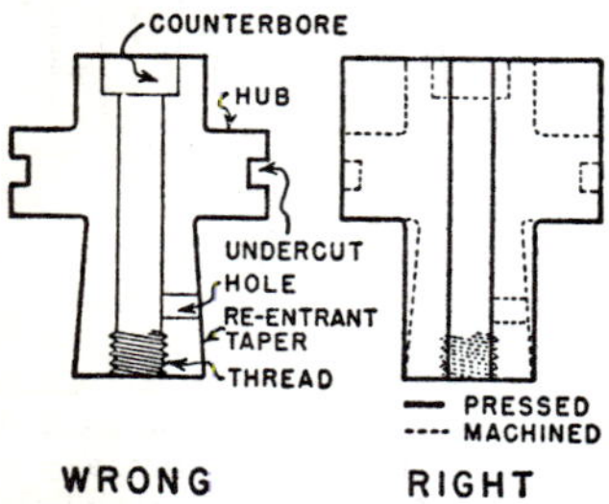

Fig. 2. Design features which require machining after sintering: holes not parallel to direction of pressing; re-entrant features like undercuts, angles, threads, some tapers, counterbores, and hubs.

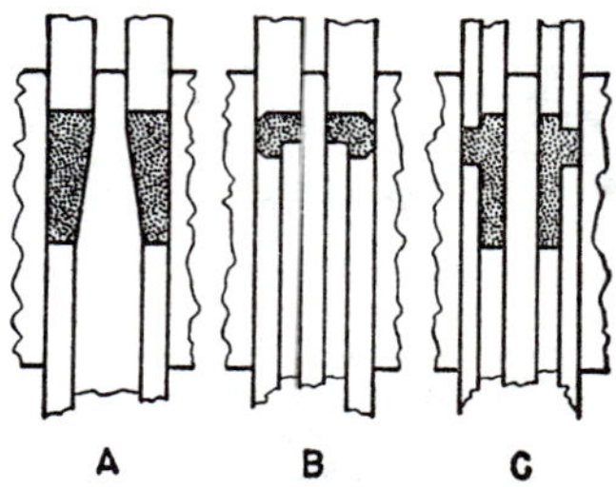

Fig. 3. Special methods or equipment can be used to make tapers (*A*) if short straight length is provided to receive upper punch; counterbores (*B*) made by lower punch and strongly tapered for ejection; and hubs (*C*) if near top of part, formed by compound upper punch.

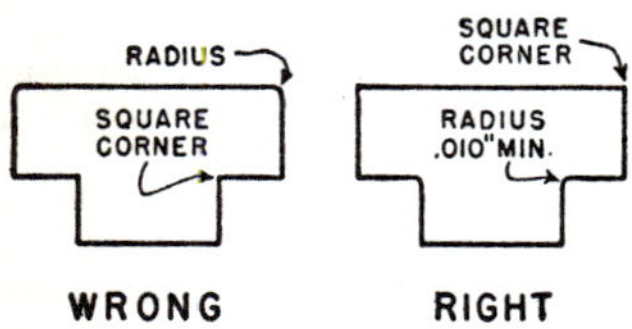

Fig. 4. Treatment of sharp corners. Upper punch should make either a square corner or a generous chamfer with a land; radius would require feather edge on punch. Corners under flange require radius to prevent wear on punch and to permit ejection without damage to compact.

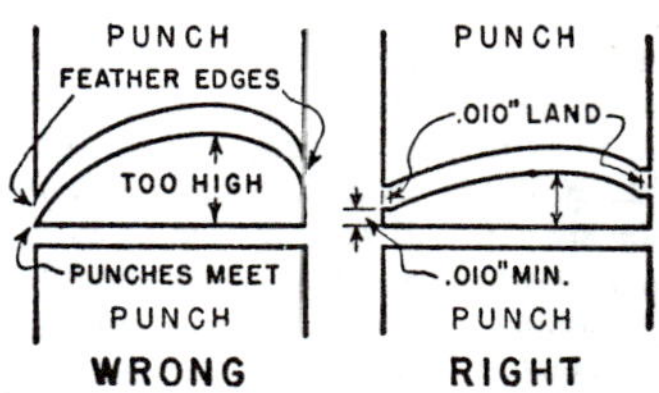

Fig. 5. Edges which require punches to meet should be avoided. Feather edges are replaced with small lands to extend life of punches. Too great a difference in vertical section should be avoided to give more uniform pressed density.

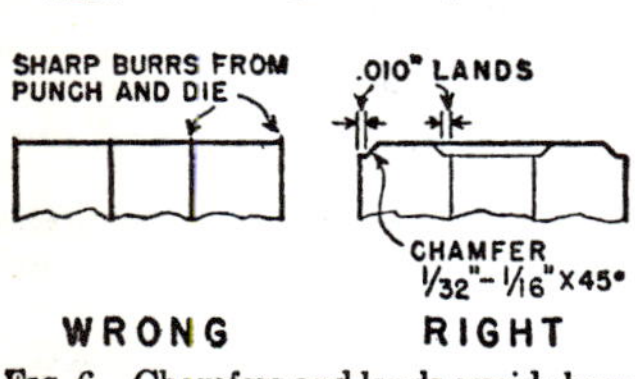

Fig. 6. Chamfers and lands avoid sharp corners and burrs which might have to be removed from the pressed piece in subsequent secondary operations.

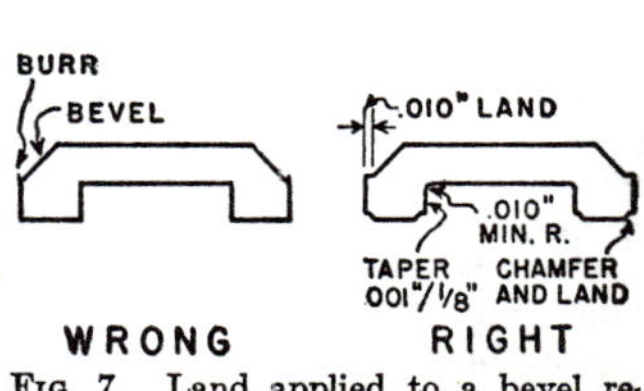

Fig. 7. Land applied to a bevel removes feather edge in punch. Radius and taper in counterbore permit easy removal of part from punch.

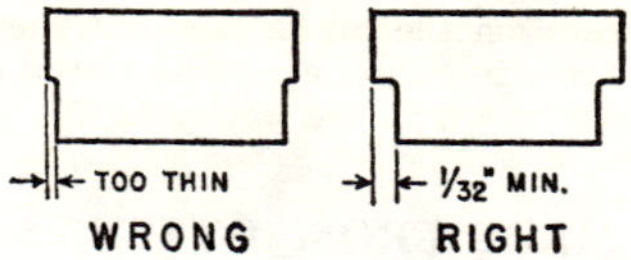

Fig. 8. Treatment of a step to avoid the use of a very thin walled punch and allow the use of a stripper punch.

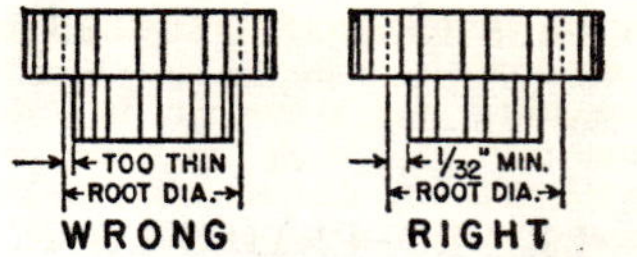

Fig. 9. Extension of the step correction of Fig. 8 to a cluster gear in which the major diameter of the smaller gear should be appreciably smaller than the root diameter of the larger gear.

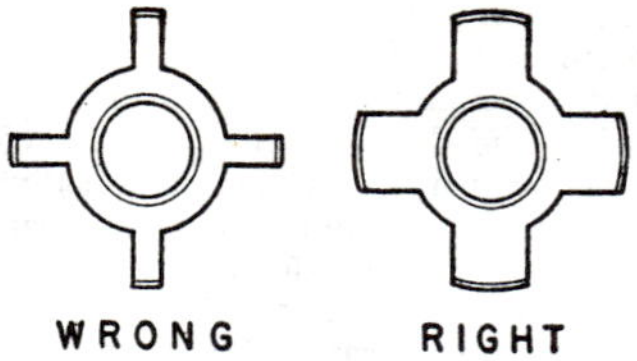

Fig. 10. Thin projections, such as splines, offer difficulty in pressing because of friction due to large die wall surface compared to cross-sectional area. More difficulty is encountered in sintering when warpage takes place.

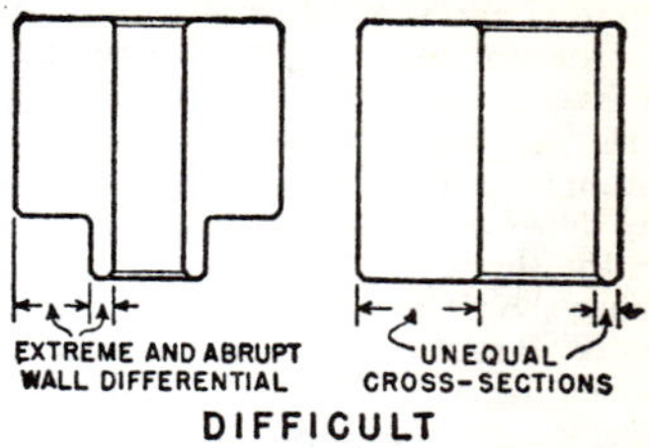

Fig. 11. Abrupt changes in section give the same warpage and shrinkage difficulties as Fig. 10.

powders are pressed in a die, they do not act hydrostatically. Because considerable friction results as the powder is pressed against the die wall, the density will be lower toward the center of the compact, with resulting lower strength. The design of the part should then take into account the ratio of wall area to the cross-sectional area against the punch, and this ratio should be kept as low as possible to obtain the highest density.

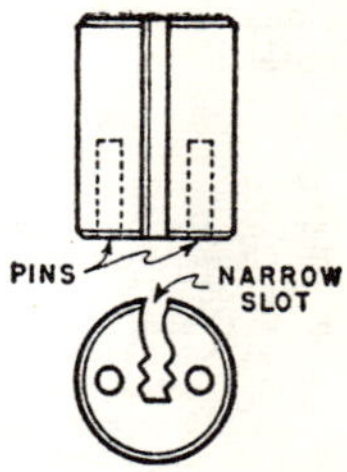

Fig. 12. Thin pins on punches to make blind holes and odd-shaped hollows which must be made by core rod or die are weak features in die design and should be avoided.

The designing of many parts by a number of manufacturers in the industry has resulted in certain practices which have become more or less standard. They are illustrated in Figs. 2 to 13. The design features may be grouped into four main types.

1. FEATURES WHICH MUST BE MACHINED into the part after the powder-metallurgy operations have been completed include holes not parallel to the direction of pressing, re-entrant angles, undercuts, threads, and certain tapers, counterbores, and hubs (Fig. 2). On the other hand, if properly used, some tapers, counterbores, and hubs can be made by powder metallurgy (Fig. 3).

2. SHARP CORNERS, RADII, FEATHER EDGES, BEVELS, LANDS, CHAMFERS, EDGES MEETING, AND STEPS must all be handled in special ways. In Fig. 4 the step should have a radius to prevent wear on the edge of the punch. Unless a chamfer and land are specified, the top punch will leave a square corner with a possible burr. A radius at this corner would require a feather edge on the punch. Figure 5 shows the corrections to be applied to punches with feather edges and the meeting of two punches. A curvature of too great proportions weakens the punch if heavy pressures are used. Figure 6 shows the application of chamfers and lands to avoid sharp corners and burrs. Figure 7 applies the land to a bevel, and a radius and severe taper to a counterbore, to permit easy removal of the part from the punch. In Figs. 8 and 9 steps should be not less than about $^1/_{32}$ in. so that a stripper punch can be used. If not used, the corner of the part tends to break off in the absence of exceptionally heavy tapers.

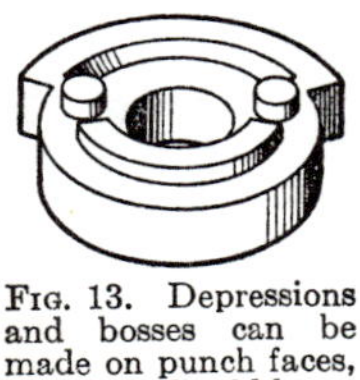

Fig. 13. Depressions and bosses can be made on punch faces, but they should be as shallow as possible and generously tapered.

3. NARROW PROJECTIONS such as splines, and abrupt changes in thickness or section, are not only difficult to press but offer their greatest trouble in warpage and non-uniform shrinkage in the sintering operation. Figure 10 shows the usual correction for thin projections. Figure 11 emphasizes the difficulty in thickness and section changes. Figure 12 shows the difficulty of thin pins in the punch or core rod as well as a thin slot which would have to be incorporated in the die or punches. These weak members of the die and punch assembly cause expensive maintenance.

4. **DEPRESSIONS AND BOSSES** can be made on the punch face but they should not be very deep and should have considerable taper to permit easy withdrawal from the punch face (Fig. 13).

29. PROPERTIES OF PARTS MADE FROM POWDERS

Powder-metallurgy parts, which are in competition with parts made by other methods, do not have the properties of wrought metal but often are superior to cast properties, depending of course on the degree of density attained in the powder-metal parts. Those powder-metal parts, which supply some special metallurgical feature such as a duplex or a porous structure, are unique and have no competing methods.

Each manufacturer of parts has developed his own techniques to achieve certain physical properties which are constantly being improved. The designer is therefore referred to them for more exact data. It must be emphasized, however, that the mechanical properties are more dependent on density than on any other factor, and that density in turn depends more on the shape and method of pressing than on any other single factor. It is advisable to compare the wall area to the pressing area of the part in question. This ratio is referred to as the W/P ratio. For a cube it would be 4 : 1. When the ratio rises (long parts pressed on end) the density decreases; when it is smaller (flat parts) the density is higher for the same operating conditions. In addition, the density of a given part varies, being highest near the face of a moving punch and lowest near a stationary punch. Comparing the density of a section taken from a part, with the physical properties plotted against density for the same material, will give a rough approximation to the lowest value of these properties in the part.

The materials generally used for powder-metallurgy parts include iron, with or without various amounts of carbon and with or without impregnation with copper, and brass, bronze, or copper, either alone or mixed with graphite flakes. The other important powders constitute only a small percentage of the total weight of powder used. They include powders for cemented carbides, contact materials, refractory metals, and electronic applications. Some use of aluminum powder has been reported, but as yet its use is not widespread.

GENERAL REFERENCES

Anon., Designing powder metal parts, *Steel*, Vol. 118, 1946, pp. 108–110, 156–159.
H. Chase, Powdered metal vs. other high production methods, *Materials and Methods*, Vol. 24, 1946, pp. 363–369.
E. V. Crane, A. G. Bureau, Presses and processes for metal powder products, *Trans. Electrochem. Soc.*, Vol. 85, 1944, pp. 63–87.
C. G. Goetzel, Pressing complicated shapes from iron powders, *Trans. Am. Inst. Mining Met. Engrs.*, Vol. 166, 1946, pp. 506–514.
F. P. Peters, Design factors for the metal forms with which powder metallurgy may compete, *Trans. Am. Inst. Mining Met. Engrs.*, Vol. 161, 1945, pp. 527–530.
P. Schwarzkopf, Engineering properties of sintered versus cast metals, *Product Eng.*, Vol. 17, 1946, pp. 122–126.
R. P. Seelig, Molding some dense machine parts, Chapter 23, pp. 264–270, *Powder Metallurgy* (edited by J. Wolff), American Society for Metals, 1942.
R. P. Seelig, The pressing of metal powders, *Metals and Alloys*, Vol. 12, 1940, pp. 744–748.
Alexander Squire, *Powder Metallurgy*, Watertown Arsenal Laboratory, Mapleton House, 1947. Report 671/15, *A Critical Survey of Powder Metallurgy and Its Application to Ordnance Design* (28 pp.).
Alexander Squire, *Ibid.* Report 671/16, *Density as a Criterion of the Mechanical Properties of Iron Powder Compacts* (28 pp.).
Alexander Squire, *Ibid.* Report 671/23, *The Effect of Dimensions upon the Density and Mechanical Properties of Iron Powder Compacts* (20 pp.).
M. T. Victor and C. A. Sorg, Design of powder metallurgy parts, *Metals and Alloys*, Vol. 19, No. 3, 1944, pp. 584–589.

SECTION 20

PLASTIC WORKING OF METALS

By

EUGENE C. CLARKE, *President, Chambersburg Engineering Co.*

WALDEMAR NAUJOKS, *Special Projects Engineer, Ladish Co.*

LESTER V. COLWELL, *Associate Professor, Department of Metal Processing, University of Michigan*

J. O. ALMEN, *Research Laboratories Division, General Motors Corp.*

FORGING

By Eugene C. Clarke and Waldemar Naujoks

STAMPING PROCESSES

By Lester V. Colwell

SHOT PEENING

By J. O. Almen

PLASTIC WORKING OF METALS

Plastic working of metal, by force of pressure, changes the form, shape, contour, or cross section of an original shape into a preconceived and desired shape. By contrast, the whittling operations of machine tools change an original shape by removal of metal. The principal processes for plastic working of metals are *forging*, *forming*, *drawing*, *shearing*, *extruding*, *rolling*, and *spinning*. Original shapes used for plastic working are *ingots*, *blooms*, *billets*, *bars*, *plates*, *sheets*, *strips*, and *extrusions*.

Plastic working of metal may be performed hot or cold. The distinction is whether the metal is to be worked at room temperature or whether it is to be heated to a high temperature preparatory to working. At room temperatures, metals are worked below their recrystallization temperatures. When heated preparatory to working, the metals are worked, usually, above their recrystallization temperatures, where slip-plane distortion through the crystal is corrected at once by what is considered to be spontaneous annealing, and the crystal structure is unstrained on completion of the work. Actually, some working strains may remain in parts hot worked or forged and are removed by suitable normalizing or annealing operations. Cold working of metals causes cumulative strain hardening, and after strain hardening in one or a series of operations the metal must be annealed to restore an unstrained crystal structure for further cold-working operations. Normal cold-working processes, *shearing* (*blanking*), *stamping*, *drawing*, and *forming* are considered in a subsequent chapter of this section. Normal hot-working processes are *forging*, *mill rolling*, and *extruding*. Plastic working of metals by forging is considered in the following chapter; mill rolling and extruding are beyond the scope of this Handbook.

FORGING

By Eugene C. Clarke and Waldemar Naujoks

Forging performs changes in shape and/or of cross section by use of powerful pressure from a hammer or press on metal which has been heated to its plastic range. The normal plastic range for steels and high-strength alloys is from about 1900 to about 2300 F; for brasses and bronzes, from about 1100 to about 1700 F; and for aluminum and magnesium alloys, from about 650 to about 900 F. At the elevated temperatures the amount of work performed on the metal is almost unlimited, except by the capacity of the part to hold temperature at the most suitable temperature working range. The forging processes are further influenced by the characteristics of the metal being forged and by physical and mechanical limitations of the equipment employed for the plastic deformation. Hammers and presses used are generally very rugged and powerful. Forging metals, iron, steels, and nonferrous alloys, are identified and described in Sections 2, 3, and 4.

Forging operations may be considered in two general groups, (1) smith die forging (flat die) and (2) impression die forging. Smith forging work includes cogging of ingots, drawing out bar stock, die block forging, weldless ring forging, and general smith forging. Impression die forging includes forgings produced by the use of impression dies where quantities of forgings are produced as commercial duplicates of each other. Impression dies are sometimes known as closed dies.

ENERGY REQUIRED. If, in mechanical forging, E represents the forging energy, V the velocity of impact of the forging element of a mechanism, and W the weight or force applied, $E = WV^2/2g$. From this it is readily appreciated that the weight or pressure applied must be greater than the affected mass of the forging and that, with the size of the forging placing a practical limitation on mechanical equipment employed, the classes of forging equipment are economically complementary rather than competitive.

The rate of plastic flow of the subject metal at forging temperature limits the amount of deformation per operation and governs the application of forging power. If deformation is accomplished beyond the rate of plastic flow, rupture of the metal occurs. The rate of plastic flow varies with the subject material, its form and density, its temperature during the forging operation, and the velocity and continuity of power application.

LIMITATIONS. This discussion is confined to consideration of actual forging applications. It, therefore, cannot include details of heating, shearing, heat treatment, etc., but a few characteristics of the metal being forged exert influences common to all procedures and are briefly mentioned.

During all forging operations the metal is cooling. Hence its fluidity for plastic flow is being reduced. Thus the duration of any forging operation is limited, the limitation varying further with the kinds of metal being forged. The forging dies are extracting heat from the forging in proportion to the time in contact with it and the amount of surface contact area with relation to the bulk.

When the forging is larger in the area to be deformed than the impinging or contacting dies, as in cogging down an ingot, the forging effect is reduced by the resistance to distortion of that part of the metal not in contact with the dies of the forging mechanism. The deformation is necessarily limited in scope to that which will not cause laps at adjoining forged areas and which does not shear the metal at those sections under the edges of the dies.

When the forging is smaller in area than the die face, a limitation on height is introduced. Generally, the greater the area on which the dies impinge, as related to the height between impinging dies, the less is the control of plastic deformation, with the limitation that the distance between the dies at the point of impact cannot safely be greater than three times the smaller sectional dimension. When the height limitation is exceeded, the forging frequently will bend, forming a lap, rather than follow the desired plastic deformation. The hammer with high impact velocity is effective on thin stock which is unworkable in a press, and the press is more effective on relatively thick sections. The difference in metal flow under the two tools can be observed at the ends of sections reduced by forging (Fig. 1). That reduced under a hammer will have the edges extending beyond the center; that reduced under a press will have the center extending beyond the edges.

FIG. 1. Metal flow—presses and hammers.

FORGING PRESSURE. The force of forging pressure may be one of three kinds: (1) impact pressure, a high-velocity pressure as imparted by the hammer; (2) push pressure, a low-velocity pressure as imparted by the hydraulic press; and (3) a combination of impact and push pressure as imparted by the modern high-speed mechanical forging press.

Impact pressure (high-velocity pressure), as given by the forging hammer, is dependent on the velocity and weight of a falling mass on the work. The maximum potential energy is available at the instant the falling weight touches the work, and as the impact pressure exerts its energy on the metal to change its shape, the weight is brought to rest, expending all its available energy. The weight is raised, and again falls for the succeeding impact blow. One, several, or many blows may be required to complete the forging. If the hammer blows are of sufficient rapidity, advantage is gained of the momentum of metal flow, and a greater rate of flow is attained than with blows at greater intervals.

Push pressure (low-velocity pressure), applied and maintained in motion hydraulically, is limited by the metal's resistance to flow, the increased density of the metal between the dies, the friction between the dies and the work, and the increasing resistance to flow caused by reduction in temperature during the forging operation.

Combination impact and push pressure is developed through eccentric crank motion in the modern high-speed mechanical forging press. On the forging stroke, the pressure increases in magnitude as the work is being deformed while the velocity of stroke decreases. At the end of the forging stroke, maximum pressure is exerted and the velocity is nil for an instant. High-speed mechanical forging presses are used almost entirely on impression die (closed die) forging work.

1. SMITH FORGING

Smith forging, or flat die forging, includes the broad field of forging work produced between flat-faced dies and possibly supplemented by stock tooling, the final shape of the forging depending largely on the skill of the smith for size and shape. This type of forging work dates back through the centuries, when anvil and hand sledges were the only means of shaping metal. The present design of hammer and press forging equipment had its inception in comparatively modern times, beginning with the Naysmith Steam Forging Hammer in 1838. Present forging practice makes use of small to large power forging units which include helve and trip hammers, steam hammers, and hydraulic forging presses. Much of the large forging equipment is used in cogging large cast ingots into blooms and billets, into drop forging die blocks, and into large shafts, cranks, blocks, plates, ship rudder frames, and large weldless rings. Medium-sized forging units work billets into bars, large square and hexagon nut blanks, wrenches, shafts, trunnions, and weldless forged rings. Small power hammer units forge hand tools, chisels, drill points, tongs, hooks, clamps, tool bits, and dress tools. Shapes are approximate in most cases, and the finished shape is obtained by partial or by overall machining.

POWER HAMMERS. All power hammers employ the same general principle of operation, a falling weight striking the blow, with the entire energy being absorbed by the work. Where further blows are necessary, the striking weight is raised for the succeeding blow. Some hammers employ only a gravity fall, the energy delivered on the work being the product of the weight of the hammer head and the distance of the fall. Other hammers increase the striking velocity of the hammer head by mechanical means. The size rating for any power hammer is given as the weight of the striking parts. The helve hammer and the trip hammer are in the small power hammer group, with their striking weights ranging from about 15 to about 500 lb. Hammer strokes range from about 400 a minute for small sizes to about 175 for large sizes. Double-acting power hammers, commonly known as steam forging hammers, though many are operated on compressed air, produce the largest amount of smith-forged products. Steam or air pressure is used to raise the striking weight and to increase the velocity on the striking stroke. Steam pressure at the hammer is usually 90 to 100 psi, and air pressure is usually 80 to 90 psi. Rated sizes of steam forging hammers range from about 800 to approximately 16,000 lb. Some

Table 1. Capacity of Pneumatic Hammers

(Courtesy of Chambersburg Engineering Co., Chambersburg, Pa.)

Rated Size Hammer, lb	Maximum Energy per Stroke, ft-lb	Square Stock Sizes			
		Occasional Production	Manufacturing Production		
		0.20 Carbon Steel, in.	0.20 Carbon Steel, in.	0.90 Carbon Steel, in.	Aircraft Type Alloy Steel, in.
200	1,180	3 1/4	2 1/4	2	1 1/2
300	1,850	4 1/4	3	2 5/8	2 1/4
500	4,050	6	4	3 1/2	3
750	6,280	7 1/2	5	4	3 1/2
1000	8,500	8 3/4	6	5 1/2	4 1/4
1500	13,700	11	7 1/2	6 3/4	5 1/4
2000	18,800	12 1/2	9	8	6 1/2
3000	28,900	15	11	9 3/4	7 3/4
5000	49,200	19	14	12 1/2	10

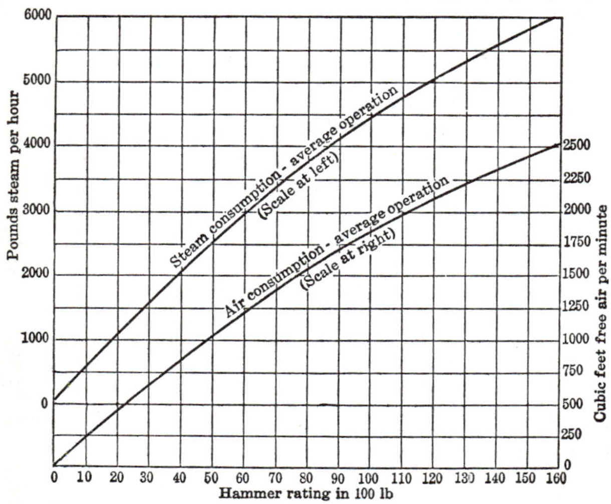

FIG. 2. Power requirements, steam- and air-driven hammers, for flat die forging.

of the newer types in the small and medium sizes are operated on compressed air, with the air compressor being built inside the hammer frame. Recommended hammer sizes for the nominal sizes of forging stock are shown by Table 1.

Power consumption of double-acting hammers varies with every job, and when several hammers are served by a single power plant the stand-by requirements are materially reduced. Figure 2 indicates the stand-by requirements of a single hammer for normal operation. For a group of six or more hammers, the stand-by requirements are frequently reduced 60%. No general data on power consumption are reliable, and the manufacturer should be consulted for specific information.

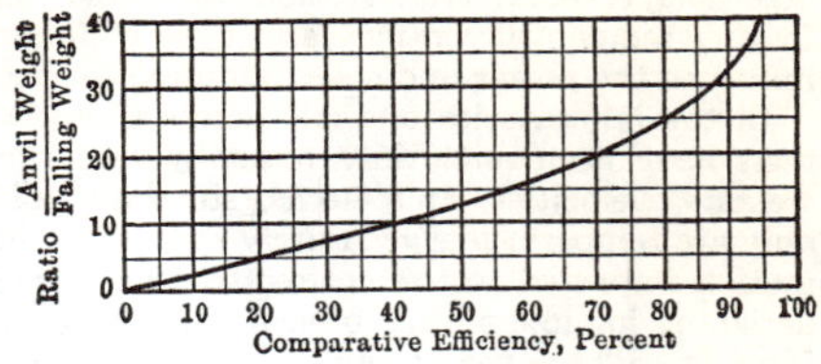

FIG. 3. Comparative efficiency of anvil ratios.

Weight of the anvil as compared with the weight of the falling parts has a marked effect on the amount of forging accomplished per blow. Figure 3 shows the relative efficiency of various ratios of anvil weight to falling weight. It has been found that a 15:1 anvil is the most economical for this type of hammer.

FORGING PRESSES for smith forging work are usually of the hydraulic type, with either steam or electric motors being the primary source of power. Intensifiers are employed to raise basic steam or pump pressure to high pressures in the operating cylinders, these being raised as high as 6000 psi. Hydraulic forging presses commonly range in sizes from about 250 tons forging pressure to about 10,000 tons forging pressure.

2. IMPRESSION DIE FORGING

Commonly known as die forging or drop forging, impression die forging makes use of cavities in alloy steel dies to produce forged shapes in quantities, the finished forgings being commercial duplicates of each other. In addition to the final or finishing shape, impression dies may contain preliminary shaping steps to permit the change from the original forging stock to the finished forging without mechanical defects. The number of forging steps necessary to produce the finished forging in the impression die is dependent on the size and shape of the part, on production quantities, and on the kind of material being forged. When impression die forgings are made on the drop hammer, they are correctly called drop forgings; products of the forging machine or upsetter are called machine or upsetter forgings; and when the forging press is used, the forgings are called press forgings. All forgings made by impression dies are considered impression die forgings. Generally, where the forging can be made by either of the several methods, the quality is comparable, and the choice is made for economic reasons.

DROP HAMMERS. Two types of drop hammers are in common use for drop forging in the United States, the gravity type usually known as the board drop hammer, and the double-acting type generally called the steam drop hammer. The gravity or board drop hammer depends on the free fall of the ram and die for its impact pressure. The ram is raised by maple boards fastened to the top of the ram, by rolls in the head of the hammer. All blows are of uniform intensity and the hammer is operated by a foot treadle. The double-acting steam drop hammer (sometimes operated with compressed air) uses steam or air pressure to raise the ram and to increase the velocity of the striking blow. The force of the blow can be regulated from a light tap to the full blow at the will of the operator. The foot treadle is used on small and medium-sized units, and hand levers are used on large units. Board drop hammer sizes range from about 200 to about 7500 lb, and the height of fall is about 30 in. for small sizes to about 60 in. for large sizes. Steam drop hammer sizes commonly range from about 1000 to about 50,000 lb. The force of the full blow of the steam drop hammer is more than two times the blow given by a board drop hammer of equivalent rating.

Hammer Foundations. Because of the shocks produced by drop hammers in operation, the importance of proper installation for the foundation cannot be overemphasized. For a given forging hammer, size and bearing area of the concrete foundation depend on the nature of the soil. Oak timbers are used for cushion between the concrete and the base of the anvil. Special pads are used sometimes to try to dampen vibration effects. A recent foundation designed to eliminate vibration is a spring suspension type, where a heavy concrete block with the drop hammer rests on a series of springs. Well-designed foundations with proper installation insure the highest production, lowest maintenance costs, and freedom from shock transmission to surrounding structures.

Power Required for Deformation. No generally accepted method or formula exists for predetermining the power required to deform metals in the forging hammer. Many tests on sample shapes under varying conditions have been made from time to time by individuals and groups, but the variables are so many and so little is known of strains beyond elastic limits that the practical application of formulas has not been developed. Size, shape, bulk, cross-section variations, material composition, plastic temperatures, forging temperature range, and type of forging hammer are some of the variables that influence the power and speed required for drop forging.

In the laboratories of the Chambersburg Engineering Company, Chambersburg, Pa., tests have been made in a specially constructed apparatus which, with varying falling weights, velocities, and anvils, simulates all conditions met in current practice. Not only are temperatures accurately controlled, but the rebound of the impacting die and the motion imparted to the entire structure also are measurable. Although useful data for designing hammers have been developed, all are empirical and incapable of reduction to formula. Three very useful points were established, however, in the research conducted to the end of year 1947. (1) Within the range of present hammer velocities there is practically no difference in work accomplished by different falling weights and varying velocities to give the same available energy. Hence, within reason, it is desirable to forge on large rather than small hammers to save time and heat by shorter strokes. (2) The speed of forging can be increased beyond current practice as hammer construction is developed to control such speeds. (3) The work done per blow is definitely related to the ratio of falling weight to anvil weight. Figure 3 shows the relative efficiency of various ratios of anvil weight to falling weight. Ratio of falling weight to anvil weight of 20 : 1 or greater is standard practice for drop hammers.

Power Consumption of Drop Hammers. Gravity drop hammers whether lifted by direct-connected motors, by belts, or by steam or air use energy only to lift the falling mass. Hence the speed of the lift and the friction within the mechanism readily indicate the power requirements. It is common to calculate 1 hp per 100 lb of rated falling weight. Manufacturers of the various types of gravity drop hammers should be consulted for specific information.

Steam and air-driven hammer power consumption is shown in Fig. 4.

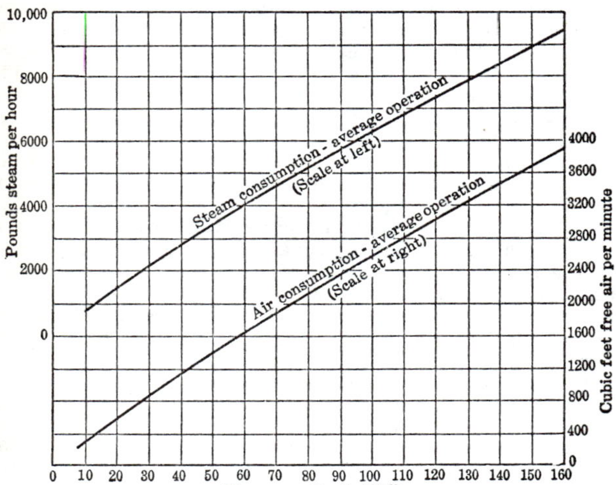

FIG. 4. Power requirements, steam- and air-driven hammers, for drop forging.

Trimming Presses. Drop hammer forgings require trimming to remove the flash. Crank-driven power presses are generally employed. For hot trimming, such presses are provided with a side shear for cutting the sprue, or stub end, from a bar.

MACHINE FORGING. The displacement of metal in closed dies, one of which is reciprocated by a crank mechanism, is plastic flow induced principally by upsetting and by piercing, or by both, for the mass production of identical shapes. Since each operation fills the die cavity under high hydraulic-like pressure, the stationary, or *grip*, dies are split to permit placing and extraction of the stock. The grip dies hold the stock, by frictional grip, against the forging force of the heading tool. Bar stock normally is used, and successive deformations to complete the shape are done in multi-impression dies. Positioning stops for each impression and a *cut-off* to shear the finished forging from the bar are provided.

The plastic flow engendered in machine forging is illustrated by Fig. 5, wherein both upsetting and piercing have contributed to the desirable internal structure. In properly designed dies, the stock is plastically flowing under compression in all impressions, resulting in a radial grain flow and refinement that is essential to a great variety of forgings, such as gear blanks and bearing races.

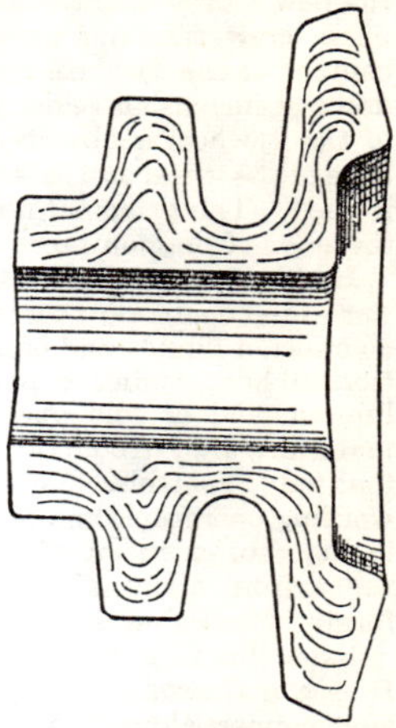

FIG. 5. Grain flow in machine forging.

DIES. E. R. Frost (Ref. 1) gives three rules as follows: Let d = diameter of bar. (1) The limit of length of unsupported stock that can be gathered or upset in one blow without injurious buckling is not more than $3d$. (2) Lengths of stock more than $3d$ can be successfully upset in one blow, provided the diameter of the upset made is not more than $1\,{}^1/_2\,d$. (3) In an upset requiring more than three diameters of stock in length, and in which the diameter of the upset is $1\,{}^1/_2\,d$, the amount of unsupported stock beyond the face of the die must not exceed $1d$.

The dies in a forging machine must resist pressure, abrasion, and heat. In automatically fed machines, the drawing of the temper of the dies generally reduces the forgings per dressing well below the number of forgings secured per dressing when hand fed. Cold-heading operations, as used in forming rivet heads, enjoy the longest die life. High chrome-nickel or vanadium steels, or other alloys, which retain their characteristics at relatively high temperatures are desirable when large radiation surfaces occur in the forming operation. Before being put in service, dies should be checked for hardness, which should conform to the requirements of the particular operation while avoiding excess hardness that may introduce brittleness. Figure 6 indicates practical selection of die steels. *Metals Handbook*, 1948 edition, should be consulted for details regarding the selection and heat treatment of dies for machine forging, cold heading, etc.

FORGING MACHINES. Requirements of the modern forging machine are: grip sufficiently tight to prevent die separation and consequent fins; maintenance of alignment in both the gripping and upsetting mechanism to insure concentricity and to avoid breakage

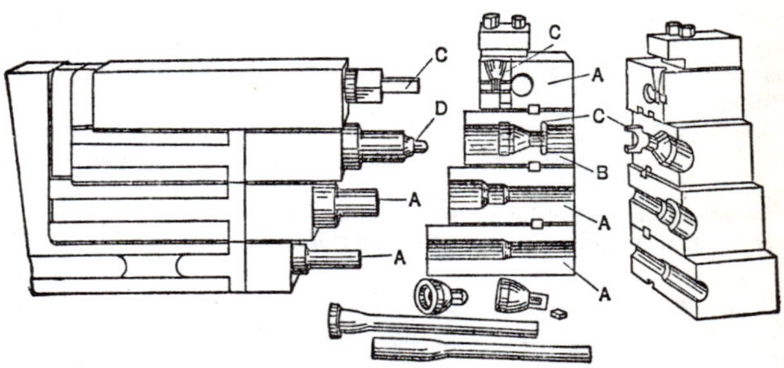

FIG. 6. Die setup for forging machine.

A Straight carbon steel
B Chrome steel
C High-tungsten steel
D Low-tungsten steel

of punches; rapid starting and stopping; ease of control by the operator. Of major importance is the rigidity of the frame enclosing the mechanism. It must be extremely massive, and currently is made of a single steel casting, heat treated and prestressed with heavy steel bars, shrunk in place. Commercial forging machines are nominally rated in inches of stock, a rating which usually requires interpretation by the builder.

GENERAL. The application of crank-driven forging machines to production of bolts, nuts, and rivets, as well as to the manufacture of mechanical parts, has encouraged the development of many special machines for which specific information cannot be given. The manufacturers of such machines should be consulted.

PRESS FORGING. Impression dies for production on forging presses are similar in basic design to drop forging dies. Forging steps can be made in separate blocks and fastened in a suitable shoe if desired. Although it has been considered that press forgings should be relatively symmetrical in design for steel and hard alloy forgings, recent advances in press construction and newer techniques produce press forgings not considered desirable a few years ago. Forging presses also find considerable use in cold and hot coining, accurate sizing at low temperature, and restriking.

Percussion presses, or power screw presses, are an intermediate development between the power press and the drop hammer. The slide is driven up and down by a large steep-pitch screw, traveling through a fixed nut in the frame to a light flywheel on top. Friction contact of the flywheel rim with a constantly revolving disk on one side drives the slide down at increasing velocity until the die comes in contact with the work and all the energy of the flywheel is absorbed. Then another revolving disk makes contact with the other edge of the wheel and lifts the slide. Foot treadle control, similar to that of a drop hammer, is used. The principal use of the percussion press is on brass and bronze impression die forgings. Common press sizes range up to about 1000 tons forging pressure.

Hydraulic forging presses have followed the forging hammer in development. They were essentially slow-moving machines because of fluid friction and the inertia of long columns of liquids and of accumulators. These characteristics tended to limit their operations to hot forming on parts heavy enough to hold heat over a considerable time and to a limited field of cold operations. Subsequent developments have increased their speed materially by introduction of variable-delivery motorized pumps, gravity fill systems, and foot valve controls. Fast-acting hydraulic presses are adapted to some production metal-working operations and have been used on some impression die forgings. Other operations include hot extrusion of shells and shapes, heavy bending, and flattening. Common sizes range from about 250 tons forging pressure to about 10,000 tons. Special hydraulic forging presses up to 18,000 tons are in operation.

Extruding is limited in its usefulness in so far as hydraulic forging of steel is concerned. It is a major operation in the production of various cross-sectional shapes of lead, tin, zinc, copper, aluminum, etc. Such shapes in the form of rods or bars are made by forcing the metal through an aperture die by high pressure, producing smooth surfaces, free from flaws, to close dimensions, and generally imparting higher physical properties to the subject. Reference 2 contains much detailed information on extrusion of nonferrous metals.

Punches and Dies. Punches must be rigid to maintain concentricity and tough to withstand abrasive action. Chrome-nickel steels generally have the longest life. Since scoring due to abrasion starts at the tip, renewable tough cast-iron tips frequently are used in order to reduce the expense of renewals. Cooling and lubrication are required between piercing operations.

Dies for piercing operations commonly are constructed in the form of a steel jacket completely enclosing a renewable cast-iron liner, bored to suit the outside diameter required of the forging. Such dies have stripper plates to strip the forging from the punch, and the larger sizes are slidably mounted on the lower platen of the press to permit mechanical handling.

The mean average life of punches and die liners for forgings over 6 in. in diameter is approximately 600 forgings. Smaller forgings may result in tool life of double or triple this amount.

Drawing dies are made of the hardest close-grained iron. It is common practice to use 20 degrees as an entrance angle and 12 degrees as the friction angle of the die. (See Fig. 7.)

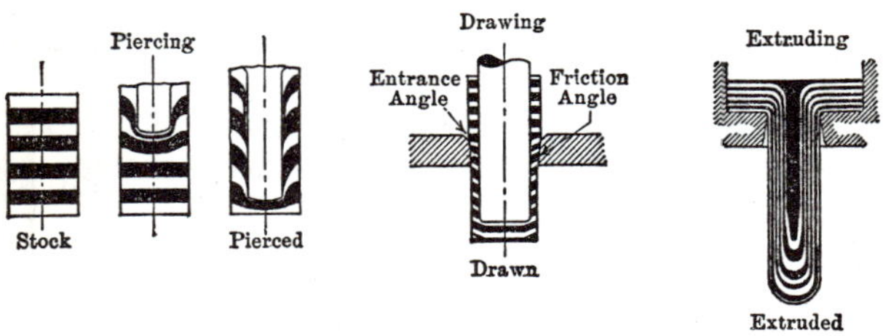

FIG. 7. Plastic flow in hydraulic press operations.

Some dies have a curved entrance. Dies may be mounted in series for multiple-drawing operations. The larger drawn forgings are produced on horizontal presses called draw benches.

Mechanical forging presses, of the eccentric crank type, have extended their activity in the impression die forging field in recent years, their rugged rigidity and high speeds being important factors. Present speeds are from about 85 to 125 strokes a minute for smaller presses to about 45 to 70 strokes a minute for larger presses. Common sizes range from about 750 tons forging pressure to about 6000 tons.

3. IMPRESSION DIES

FUNCTION OF IMPRESSION DIES. The purpose of the impression die is to change the shape of the forging stock, which is normally round, square, or flat in cross section, to the shape and dimensions of the finished forging, at the same time regulating the metal flow to maintain sound structure and unbroken flow lines. Metal, heated to its plastic state, is flowed under forging pressure through a series of forging steps contained in the impression dies. The number of forging steps is dependent on the kind of shape change necessary and the amount of change.

Simple symmetrical parts may be forged directly in the finishing impression (finishing die cavity) without preliminary shaping. The more difficult shapes may require two to ten different steps to produce the part as a finished forging. The most-used preliminary forging step is the *edger*, which serves to proportion the cross-section area along the length by flowing metal from a section being reduced to a section being enlarged. The *fullering step*, or *fuller*, reduces the cross-section area between the ends of the forging stock without appreciable change to the end sections. The *bending step*, or *bender*, forms the length of the forging stock to a shape for the finishing impression. The *blocking* impression, or *semifinishing impression*, shapes metal to an approximation of the final shape. Where possible, all the forging steps, including the finishing impression, are contained in one set of die blocks. Pieces with large surface areas, or large forgings, may necessitate the use of two or more sets of die blocks. When the drop hammer and the forging machine are used to produce a forging, two sets of dies are needed.

FORGING STEPS must be designed to make use of the law of natural metal flow, which states that plastic metal spreads and flows along the path of least resistance. Corners, sidewalls, raises, and thin, rapid-cooling sections are barriers to natural metal flow in varying degrees. Similar to a liquid, metal in plastic state will not drop suddenly to fill a pocket but flows over into the cavity in an arc. Thin sections cool rapidly and become less plastic in ratio to their cooling rate. All factors promoting and restricting movement of plastic metal are considered in design of forging steps. The ultimate aim is to proportion the shape in each step so that the metal in the finishing impression will spread to reach all sidewalls at the same time. Experience in forging die design is important.

DIE BLOCKS. Die blocks are available in a variety of compositions to meet a variety of needs. The more commonly used die blocks are made from chromium-molybdenum or nickel-chromium-molybdenum grain-controlled alloy steels, of which these are typical:

Carbon	Manganese	Chromium	Nickel	Molybdenum
0.50–0.70	0.50–0.90	0.60–1.30	1.00–2.00	....
0.50–0.70	0.60–0.90	0.75–1.00	....	0.15–0.25
0.50–0.70	0.50–0.90	0.50–1.00	1.00–2.00	0.15–0.35
0.50–0.70	0.50–0.80	0.75–1.00	1.50–2.50	0.60–0.90

Die blocks are forged on all six faces—upset to destroy all original directional fibers, an essential factor since the forging forces are exerted in all directions in a die cavity. Die block steel is never permitted to become cold from the time the ingot is cast until final tempering treatment. Blocks are normalized, quenched, and tempered. The blocks are machined and ground on the working surfaces to secure a well-prepared surface for hardness testing. Standard alloy blocks are available in four normal hardnesses for commercial use, the highest of the standard hardnesses being 56 to 60 Scleroscope hardness. Where hardnesses beyond machinable ranges are desired for use on thin light sections, annealed blocks are machined and then hardened and tempered after impression machining. Figure 8 shows a hammer die.

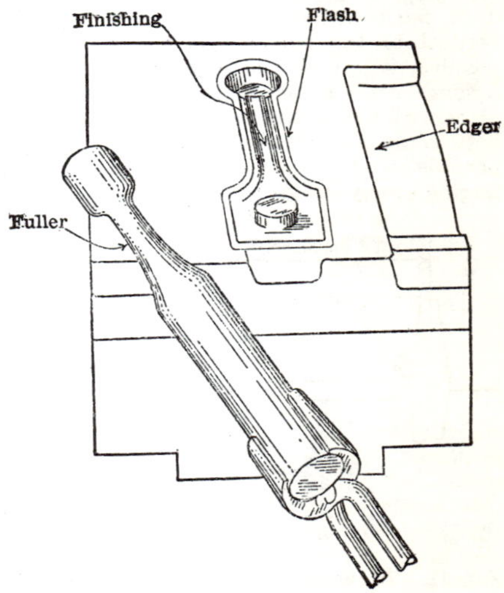

FIG. 8. Hammer die.

IMPRESSION MACHINING and preparation of the rough die blocks for use in the drop hammer, forging machine, or forging press is performed in standard machine tool equipment such as the planer, shaper, drill press, lathe, and in special vertical milling machines known as die sinking machines. Suitable benches are used for final polishing and sizing of the finishing impressions. It is important that the finishing impression be polished to a high degree of smoothness. Blocks are first squared up for impression layout. Drop forging die blocks have shanks cut on by planing or shaping, working surfaces are machined parallel with shanks, and match edges are machined, to assure alignment of cavities. Blocks for machine forging are planed or shaped to fit the machine pockets. Impressions and preliminary forging steps are machined, and impressions are polished to high smoothness.

DIE SETTING or installation of impression dies is important to produce smooth, closely dimensioned die forgings. The final blow struck on a die forging, and producing the exact reproduction from the dies, also is that blow which has the least vertical displacement and hence absorbs the greatest amount of available energy in the least time. It, therefore, is important that in the reproduction of exact forgings, the dies be set so that the displacement load of the final impression is centered under the falling mass. Since die design can only approximate the load distribution, it is necessary that adjustable yet rigid means of fastening dies in the hammer be employed. Figure 9 shows a typical dovetail fastening for dies. Shims are used on the sides of the slot for left to right adjustment, and a series of dowels with graduated offsets provides adjustment from front to back. Forging machine dies and forging press dies are fastened in place by the use of clamps and dogs.

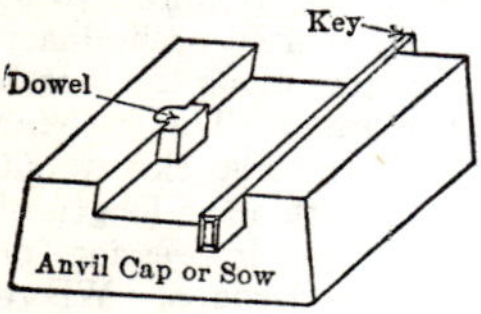

FIG. 9. Die notch.

4. DESIGN OF FORGED PARTS

Forging design as applied to impression die forgings is conventional design to completely satisfy the service requirements of a component, incorporating, as far as possible, design in keeping with good production forging practice and subsequent processing. Good original design for a particular type of production method is highly desirable where possible. All impression die forgings have similar basic requirements whereas a particular method of forging production may entail some modifications. The largest volume of impression die forgings are made as drop forgings.

DROP FORGING DESIGN must give first consideration to the die parting plane, since forging draft must be added to all sides vertical to the parting plane. If possible, it is desirable to part the forging with a single horizontal plane with half of the forging in the bottom die and half in the top die. The single horizontal parting plane (normal plane) is possible on a large portion of drop forgings, although it is not possible always to divide the forging evenly between the two dies. Some shapes make it desirable to part in two or more planes. Dies parted in two or more planes are called locked dies. Figure 10 sketches normal and locked die parting planes.

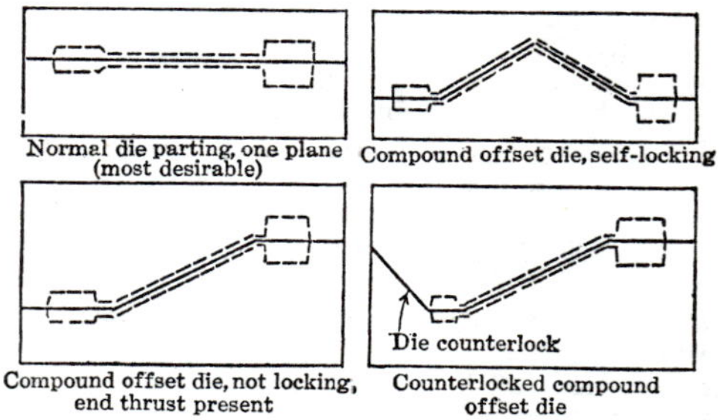

FIG. 10. Parting planes in drop forging dies.

Forging draft is taper given to all sides vertical to the parting plane of the drop forging to permit easy removal of the forging from the dies. Normal drop forging draft is 7 degrees for all outside surfaces and 10 degrees for all inside surfaces. Thin flat drop forgings may permit small reduction in draft requirements, and drop forgings with high bosses or deep pockets may require an increase in the amount of draft.

Fillets and corner radii connect and blend the several planes in the forging. They should be made as large as the design will permit, particularly on parts containing deep pockets

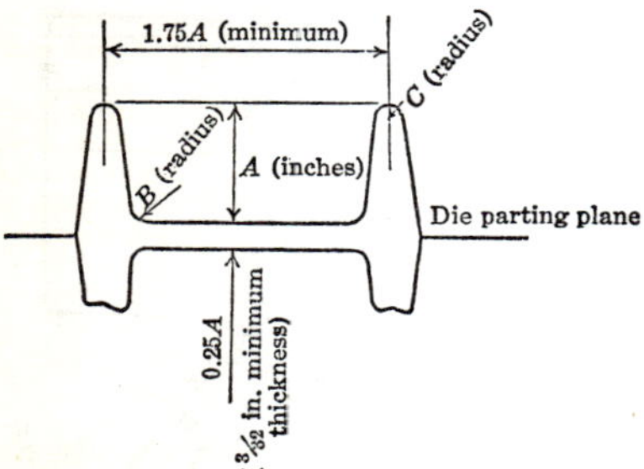

FIG. 11. Thickness, fillet, and corner radius limits. See Table 2.

or high bosses. Sharp corners restrict normal metal movement, increase die wear, and promote defects. Ample radii and fillets permit gradual changes in directional flow of plastic metal. Table 2 and Fig. 11 suggest minimum fillets and radii.

Table 2. Suggested Minimum Fillet and Radii (Fig. 11)

A	3/8	1/2	3/4	1	1 1/4	1 1/2	1 3/4	2	2 1/4	2 1/2	2 3/4	3
B	3/16	1/4	5/16	3/8	1/2	9/16	5/8	5/8	11/16	11/16	3/4	3/4
C	1/32	3/64	1/16	5/64	3/32	7/64	1/8	9/64	5/32	11/64	11/64	3/16

Ribs and Thin Sections. Ribs in forgings restrict metal flow into the narrow die cavity; both ribs and thin sections cool metal quickly, making it less plastic. Maximum possible draft angles and ample radii promote greater forging ease in forging full sections with suitable tolerances.

Bosses and stems are projections extending away from the main body of the drop forging. The addition of draft is dependent on their position with relation to the die parting plane. Vertical bosses and stems usually require addition of draft. Horizontal bosses and stems generally require only end draft. Figure 12 illustrates draft, fillets, pockets, bosses, and radii.

Die shift is a condition in the drop forging where the top die is not in complete alignment with the bottom die, for one of several reasons: improper die alignment, tendency of top die to pull off center, loose die key, or excessive running clearance in hammer ram. A small amount of die shift may be unavoidable in some types of drop forgings.

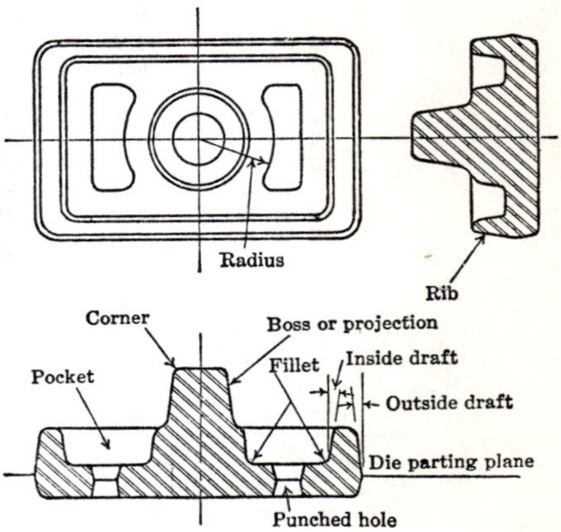

FIG. 12. Forging terms.

TOLERANCES on drop forgings should be to standard commercial practice where use of the part will permit. Drop forging tolerances, as outlined by the Drop Forging Association for drop forgings of 0.1 to 100 lb, are given here.

Tolerances shall be either "Special" or "Regular."

"Special" tolerances are those which are particularly noted in the specifications and may state any or all tolerances in any way as occasion may require. Special tolerances apply only to the particular dimension or thing noted. In all cases where Special tolerances are not specified, Regular tolerances shall apply.

"Regular" tolerances are divided into two divisions—"Commercial Standard" and "Close Standard." "Commercial Standard" tolerances are for general forging practice, but when or where extra close work is desired involving additional expense and care in the production of forgings, "Close Standard" may be specified. "Close Standard" may be specified for one or more of the following classes. When no standard is specified, "Commercial Standard" shall apply.

Classes. Regular tolerances are applicable to the following classes: (1) thickness; (2) width, including (*a*) shrinkage and die wear, (*b*) mismatching, and (*c*) trimmed size; (3) draft angle; (4) quantity; and (5) fillets and corners.

Class 1. Thickness Tolerances (Table 3). Thickness tolerances shall apply to the overall thickness of a forging. When applied to drop hammer forgings, they shall apply to the thickness in a direction perpendicular to the main or fundamental parting plane of the die. When applied to upset forgings, they shall apply to the thickness in a direction parallel to the direction of travel of the ram, but only to such dimensions as are enclosed by and actually formed by the die.

Class 2. Width and Length Tolerances. Width and length tolerances shall be alike and shall apply to the width and/or length of a forging. When applied to drop hammer forgings, they shall apply to the width or length in a direction parallel to the main or fundamental parting plane of the die, but only to such dimensions as are enclosed by and actually formed by the die. When applied to upset forgings, they shall apply to the width or length in a direction perpendicular to the direction of travel of the ram.

Width and length tolerances shall consist of three subdivisions:

Class 2. (*a*) Shrinkage and die wear tolerance.
Class 2. (*b*) Mismatching tolerance.
Class 2. (*c*) Trimmed size tolerance.

Class 2 (*a*). Shrinkage and Die Wear Tolerances. Shrinkage and die wear tolerances shall apply to that part of the forging formed by a single die block only. They shall not apply to any dimension crossing the parting plane. They shall be the sum of the shrinkage tolerances and the die wear toler-

Table 3. Thickness Tolerances

(Inches)

Net Weights up to — lb	Commercial −	Commercial +	Close −	Close +
.2	.008	.024	.004	.012
.4	.009	.027	.005	.015
.6	.010	.030	.005	.015
.8	.011	.033	.006	.018
1	.012	.036	.006	.018
2	.015	.045	.008	.024
3	.017	.051	.009	.027
4	.018	.054	.009	.027
5	.019	.057	.010	.030
10	.022	.066	.011	.033
20	.026	.078	.013	.039
30	.030	.090	.015	.045
40	.034	.102	.017	.051
50	.038	.114	.019	.057
60	.042	.126	.021	.063
70	.046	.138	.023	.069
80	.050	.150	.025	.075
90	.054	.162	.027	.081
100	.058	.174	.029	.087

Table 4. Shrinkage and Die Wear Tolerances

(Inches)

Shrinkage			Plus	Die Wear	
Lengths or Widths up to — in.	Commercial + or −	Close + or −	Net weight up to — lb	Commercial + or −	Close + or −
1	.003	.002	1	.032	.016
2	.006	.003	3	.035	.018
3	.009	.005	5	.038	.019
4	.012	.006	7	.041	.021
5	.015	.008	9	.044	.022
6	.018	.009	11	.047	.024
Each additional inch add	.003	.0015	Each additional 2 lb add	.003	0015
For example:			For example:		
12	.036	.018	21	.062	.031
18	.054	.027	31	.077	.039
24	.072	.036	41	.092	.046
36	.108	.054	51	.107	.054

ances as given in Table 4. The shrinkage tolerances and die wear tolerances shall not be applied separately, but shall only be used as the sum of the two. They shall not be so applied as to include draft or variation thereof.

Class 2 (*b*). Mismatching Tolerance (Table 5). Mismatching is the displacement of a point in that part of a forging formed by one die block of a pair, from its desired position when located from the part of the forging formed in the other die block of the pair. Mismatching does not include any displacement caused by variation in thickness of the forging, but is only the displacement in a plane parallel to the main or fundamental parting plane of the dies.

Mismatching tolerances are independent of, and in addition to, any other tolerances.

Class 2 (*c*). Trimmed Size Tolerances. The trimmed size shall not be greater or less than the limiting sizes at the parting plane imposed by the sum of the draft angle tolerances and the shrinkage and die wear tolerances.

Class 3. Draft Angle Tolerances (Table 6). Draft angle tolerances are the permissible variations from the standard or nominal angle of draft.

Class 5. Fillet and Corner Tolerances. Fillet and corner tolerances apply to all meeting surfaces even though drawings and/or models indicate sharp corners, unless such drawings and/or models

Table 5. Mismatching Tolerances

(Inches)

Net Weight up to — Pounds	Commercial	Close
1	.015	.010
7	.018	.012
13	.021	.014
19	.024	.016
Each additional 6 lb add	.003	.002
For example:		
37	.033	.022
55	.042	.028
79	.054	.036
97	.063	.042

Table 6. Draft Angle Tolerances for Drop Hammer Forgings

(Degrees)

	Nominal Angle	Commercial Limit, Maximum	Close Limit, Maximum
Outside	7	10	8
Inside holes and depressions			
Commercial limits	10	13	..
Close limits	7	..	8

Draft Angle Tolerances for Upset Forgings

(Degrees)

	Nominal Angle	Commercial Limit, Maximum	Close Limit, Maximum
Outside	3	5	4
Inside holes and depressions	5	8	7

have or indicate (even though actual dimensions are not specified) fillet and/or corner dimensions of larger radii than the following standards, in which case such actual or indicated larger dimensions shall be considered as actually specified and the tolerances shall be "Special Tolerances."

Fillet tolerances apply to inside corners and edges in all cases in which surfaces meet at an angle less than 180 degrees.

Corner tolerances apply to outside corners and edges in all cases in which surfaces meet at an angle greater than 180 degrees.

Where a corner tolerance applies on the meeting of two drafted surfaces, the tolerance shall apply to the narrow end of such meeting and the radius will increase towards the wide end. The total increase in the radius will equal the length of the drafted surface in inches, multiplied by the tangent of the nominal draft angle.

The radii of fillets and corners may be any value not greater than those given in Table 7.

Table 7. Fillet and Corner Tolerances

(Maximum radii in inches)

Net Weights up to — Pounds	Commercial	Close
.3	3/32	3/64
1	1/8	1/16
3	5/32	5/64
10	3/16	3/32
30	7/32	7/64
100	1/4	1/8

MACHINE FORGING DESIGN permits modification with respect to forging draft, inasmuch as most forging-machine forgings do not require draft. Consideration similar to drop forgings must be given machine forgings with respect to ribs and thin sections. Center holes in machine forgings can be straight and deep. In general, forging machine products are relatively symmetrical in nature.

PRESS FORGING DESIGN in steels and harder alloys is similar to drop forging design. The high-speed forging press is especially adapted for symmetrical parts in large production quantities, with most of such shapes being made with less draft than the drop forging, and some without draft since ejecting devices can be employed. Many shapes that are not symmetrical in themselves can be paired to obtain reasonable symmetry. Some forgings are preshaped by forging rolls or forging hammers and finished forged on the forging press.

Design of press forgings in impression dies on the softer metals such as brasses and bronzes is almost unlimited as to shape, and little consideration need be given draft angles or radii. Such impression die forgings must not contain undercuts. The fact that brass and bronze forging metal can be extruded as well as flowed under pressure permits the making of shapes that are not considered practical in the harder compositions such as steel. The older types of press-forging equipment have draft requirements from 1 to 3 degrees, and the newer equipment can produce brass and bronze forgings without draft. The working pressures and temperatures are less severe on the softer nonferrous metals, and dimensional tolerances can be held fairly close. Some of the high-strength forging bronzes do not work quite as readily as the forging brasses and may impose moderate limitations on difficult designs.

SUBSEQUENT PROCESSING OPERATIONS. Forgings made by the use of impression dies usually require one or more steps in the forging plant before they are ready for further processing in the heat treat or machine shop. All drop forgings and most press forgings are trimmed to remove excess metal in the form of flash. Trimming is done cold or hot in mechanical or hydraulic power presses. Other press operations may be forming, coining, restriking, sizing, straightening, and planishing. Shot blasting, pickling, or tumbling are routine operations to remove forging scale. Normalizing, annealing, quenching, and tempering may be treatment requirements.

Allowance for Machining. The amount of metal added to finish machined surfaces on the impression die forging is dependent on the size and shape of the part, metal composition, and machining operations. Average impression die forgings which require normal machining operations such as turning, boring, shaping, or milling should allow a minimum of $^1/_{32}$ in. of metal on each surface to be machined on small forgings weighing a pound or under, and $^1/_{16}$ in. is desirable where it can be permitted. Slight variations in size, small scale pits, surface blemishes, and die shift may cause surface or width variations from a few thousandths of an inch to $^1/_{32}$ in. Average forgings from 1 to 20 lb should allow a minimum of $^1/_{16}$ in. for each machined surface; forgings from 20 to 50 lb, a minimum of $^3/_{32}$ in; forgings from 50 to 100 lb., a minimum of $^1/_8$ in.; and forgings over 100 lb, a minimum of $^1/_4$ in. Modern heavy-duty machine tools involve less difficulty and expense in the removal of a little extra surface metal than are entailed in the design of jigs and fixtures for exact positioning of the forging.

Smith forgings require a greater allowance for machining than is necessary for impression die forgings. Weldless forged smith rings up to 8 in. OD require $^3/_8$ in. added to OD and thickness and $^3/_8$ in. deducted from ID. Rings of 8 in. OD to 15 in. add $^1/_2$ in. to OD and thickness and deduct $^1/_2$ in. from ID. Rings 18 in. OD to 30 in. add $^5/_8$ in. to OD and thickness and deduct $^5/_8$ in. from ID. Rings 30 in. OD to 50 in. add 1 in. to OD and thickness and deduct 1 in. from ID. General smith shapes require a minimum of $^1/_4$ in. added for each surface to be machined.

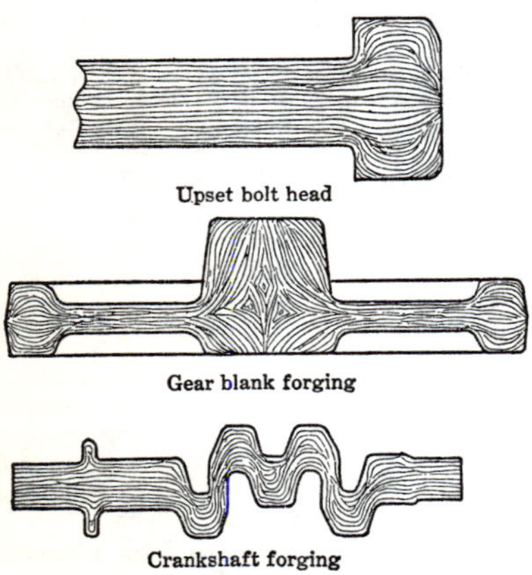

FIG. 13. Flow lines in forgings.

Machining heat-treated forgings is increasing in use with newer steels, machine tools, and improved techniques. Any greater difficulty in machining the forging in the heat-treated state is compensated for by minimizing or elimination of problems in warpage, size changes, scaled or discolored surfaces, spoiled tolerances, and heat treating cracks due to sharp machined corners.

FLOW LINES. Flow lines, generally known as *grain flow* or *fiber structure*, are present in wrought metals, the result of mechanical working given to the original cast ingots. The minute impurities, inclusions, and oxides are lengthened in the direction of working by mill rolling or hammer forging and develop a grain in the wrought metal somewhat like the grain in fine hard wood. The importance of flow-line position or direction in the forging is the great increase in resistance strength to dynamic stresses of impact, shock, vibration, and fatigue by correct positioning.

Results of tensile tests and Izod notch impact tests on standard specimens indicate that although there is little or no practical change in tensile strength and yield strength on specimens taken in the longitudinal and transverse directions, the percentage of elongation and the reduction in area drop sharply for specimens taken in the transverse direction. Values in foot-pounds obtained from notch impact specimens in the longitudinal direction were nearly three times the values obtained from specimens in the transverse direction. Interpreted broadly, flow-line position is of little importance in a component where stresses are static, occasional, and not severe. Flow-line position is important on all parts that encounter variable and dynamic stresses, such as variable and reversible loads, intermittent shock, impact, vibration, wear, and fatigue. Positioning of flow lines is especially important in aircraft forgings where maximum strength with minimum weight is mandatory and where "unpredictable stresses" can be encountered. Figure 13 illustrates flow lines in forgings.

REFERENCES

1. E. R. Frost, *Laws Governing Forging Machine Die Design*, paper read before the American Drop Forge Association.
2. D. K. Crampton, Extrusion of metals, *Trans. Am. Soc. Metals*, Vol. xxv, 1937.

GENERAL REFERENCES

Metallurgy

C. L. Clark, Evaluating steel forgeability, *Iron Age*, March 16, 1944.
E. W. Nelson, Flow lines in forged steel, *Heat Treating and Forging*, Vol. 18, 1932.
H. B. Pulsifer, *Inspection of Metals*, American Society for Metals, 1941.
D. D. Stone, Fatigue life of parts as affected by grain flow, *Product Eng.*, July 1945.
R. W. Thompson, Grain flow in precision drop forgings, *Metal Progress*, Vol. 28, 1935.

Techniques and Processes

R. W. Bolz, *Production Processes—Their Influence on Design*, Vol. 1, Penton Publishing Co., Cleveland, 1949.
Dyrkacz and Fonda, Recent developments in forging practice, *Steel*, Jan. 22 and 29, Feb. 5 and 12, 1945.
Improved forging techniques, *Steel*, 1942.
Waldemar Naujoks, Forging stainless metals, *Steel Processing*, Dec. 1947.

Dies

John Mueller, Forging die design, *Steel Processing*, 1946 and 1947.
Waldemar Naujoks, Die forged parts, *Product Eng.*, June 1938.
Waldemar Naujoks, Dies and tools for forging processes, *Ind. Heating*, July and Aug. 1941.

General Subjects

Metal Quality, Drop Forging Association, 1944.
Waldemar Naujoks, Fundamental forging practice, *Steel*, 1948.
Naujoks and Fabel, *Forging Handbook*, American Society for Metals, 1939.
Safety code for forging and hot metal stamping, *Bull. U. S. Bur. Labor Statistics* 451.

STAMPING PROCESSES

By Lester V. Colwell

The purpose of this chapter is to aid the mechanical or process engineer in making decisions about the selection of material, process, and the equipment for metal stamping. These selections are necessarily preliminary to the final design of a component part or product. A product must be designed to meet service requirements, but the cost of production may be unnecessarily high unless some consideration is given to the limitations of the methods of production before the design of the product is frozen. In addition, slight alteration in design of the product often makes it possible to use a process with much lower production costs. Thus it is the purpose of this chapter to present the production advantages offered by the many operations that can be performed by metal stamping, and in particular to emphasize the limitations of those operations.

In general if a part can be made as a metal stamping it can be produced in large quantities at a lower cost than by any other process. A wide range of sizes and quantities can be produced economically. A very broad range of properties is available in the metal ranging from low strength nonferrous alloys to high-strength stainless steels. Any metal which can be prefabricated as sheet or strip can be processed further by at least some of the metal stamping operations.

The steps in planning production by stamping generally are taken in this sequence: (1) Preliminary design of the component. (2) Selection of material for the component. (3) Selection of stamping operations. (4) Grouping of stamping operations. (5) Final design of component. (6) Selection of presses and auxiliaries. (7) Selection of heat-treating processes and equipment. (8) Selection of cleaning processes and equipment. (9) Selection of finishing processes and equipment. (10) Design of stamping dies.

5. ANALYSIS OF STAMPING OPERATIONS

The limitations of the stamping process are not very well established quantitatively because of the apparent complexity of the causes. Consequently they are not well understood, and often are misinterpreted. There is a considerable quantity of published as well as unpublished data on metal stamping and die design. Some of these data are empirical, but the vast majority are completely arbitrary. Therefore a general discussion of the theoretical aspects should aid in simplifying as well as minimizing errors in applying them.

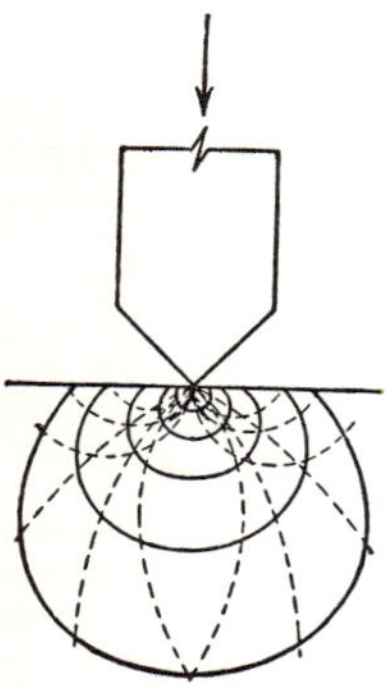

FIG. 1. Typical internal stress pattern set up by a sharp-edged tool applying a concentrated load at the surface. Dashed lines represent surfaces of potential flow in shear.

All metal stamping operations can be grouped into either of two classifications, *cutting* and *forming*. A few operations, such as burring and lancing, apparently belong to both classes, but actually the cutting phase of the operation either precedes or follows the forming phase so that two or more distinct operations are performed in sequence by the same tool.

CUTTING operations are distinguished by the obvious fact that the material is completely sheared or cut by stressing it beyond its ultimate strength. These operations seldom give rise to serious operating difficulties. Therefore an understanding of a relatively few characteristics of the cutting process makes it possible to apply all the operations of this class with assurance of success.

Figure 1 gives an indication of the mechanism of metal cutting or shearing. It illustrates the stress set up in metal by a sharp-edged tool like a cold chisel. Each of the solid curved lines represents a region of uniform compression or principal stress. The shear stress, which is the first to reach yield condition, is a maximum on surfaces at 45 degrees to the surfaces of uniform compression. The dashed lines represent possible surfaces of propagation of shear strain which ultimately become surfaces of rupture such as the two designated by the heavy dashed lines. The stress obviously is greatest at the edge of the tool, and shearing rupture begins at this point. Shearing may begin several times in different directions before yielding, and consequent relaxation of stress finally permits the shape of the chisel or cutting edge to impose a fixed pattern of stress, thus propagating rupture to appreciable depth below the surface.

Figure 2 shows a typical flat-end tool like a piercing punch being forced into the surface of the metal. The stress pattern is similar to that shown in Fig. 1 but it is influenced by the shape of the punch and the amount of penetration of the punch in the metal before the shearing is completed. It is apparent that the cut surface is not cut in the same sense as a turned or a milled surface. The cut surface is created by a true rupture in shear propagated ahead of the punch. Therefore a hole pierced under ideal conditions increases in cross section as it approaches the bottom of the sheet.

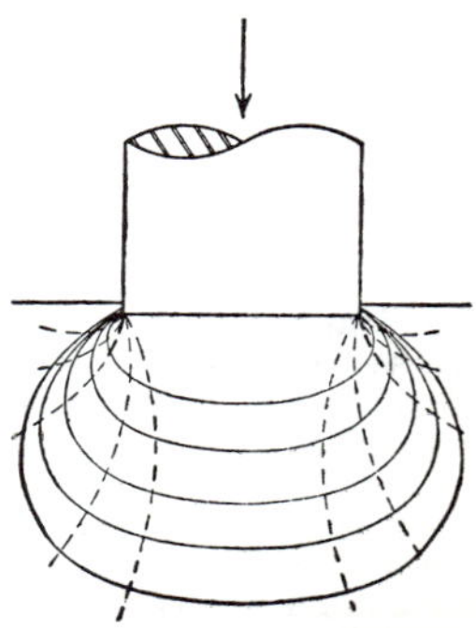

FIG. 2. The internal stress pattern set up by a punch similar to those used for cutting sheet metal.

Figure 3 exaggerates the conditions of shearing in a hypothetical blanking operation wherein the relatively thin sheet metal bends elastically prior to the completion of the shearing and exerts an additional influence on the direction of propagation of the sheared surface.

It has been claimed that the shear surfaces start almost simultaneously from the upper and lower cutting edges, and that ideal clearance between punch and die opening will permit the two sheared surfaces to meet in a common surface near the middle of the sheet. This is highly improbable since the necessary conditions are perfect homogeneity of the sheet and identical stress conditions at both cutting edges. This latter condition is possible only with identically sharp cutting edges.

At low cutting speeds it is probable that cutting begins at the sharpest cutting edge and propagates very rapidly through the sheet almost to the opposite cutting edge before any appreciable cutting takes place at that point. At higher cutting speeds the inertia of the sheet causes a greater force on the cutting edge in motion relative to the sheet. Thus it can be expected that cutting will be initiated by the moving cutting edge in dies which have been recently reground or in dies where the rate of dulling is the same for both cutting edges.

The nature of cutting or shearing in stamping dies is such that the best job of cutting still produces a cut surface not perpendicular to the surfaces of the sheet, but inclined from one cutting edge to the opposite cutting edge. This idealized condition is illustrated in Fig. 3.

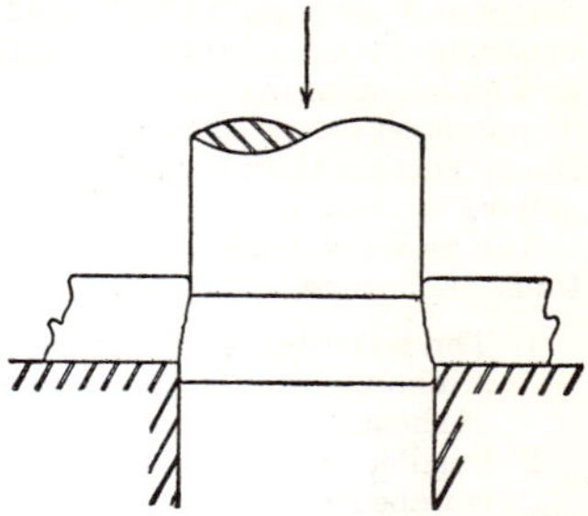

FIG. 3. With ideal conditions the sheet metal is cut by the propagation of shear surface angularly from the sharp cutting edges until the surfaces meet in a continuous surface. Shearing is completed before the punch penetrates the entire thickness of the sheet metal. The degree of angularity is a function of the sheet metal properties and determines the proper clearance between punch and die for a clean cut.

In practice it is difficult and uneconomical to maintain perfectly sharp cutting edges, uniform clearance between the cutting edges, and uniform properties of the sheet metal. Dull cutting edges cause less stress concentration than sharp cutting edges; consequently, greater penetration is required before shearing is effected. Greater penetration inclines the sheared surface at a greater angle from the line of motion of the punch, thus creating the effect of too little clearance. Soft ductile metals require greater penetration to effect shearing than the harder, less ductile metals. Therefore the proper clearance between cutting edges increases with increase in ductility or per cent elongation of the sheet metal. Figure 4*a* illustrates the proper clearance condition for both hard and soft metals.

Figure 4*b* shows the effect of dull cutting edges and too little clearance. Too little clearance or dull cutting edges usually cause a burr to be left on the die side of the stock surrounding the punch. This burr may be accompanied by a slight undercut at the same point when cutting relatively thick stock. The same defects, i.e., burr and undercut, will be present on the punch side of the material cut out of the stock. Too much clearance gives opposite results as shown in Fig. 4*c*. The burr and undercut occur near the punch side of the stock surrounding the punch and near the die side of the material cut out of the stock. It will be noted in Figs. 4*b* and 4*c* that the burr created by too little clearance increases the effective thickness of the material whereas the burr created by too much clearance is turned away from the surface of the stock and does not affect the thickness of either the part or the stock from which it was cut. Consequently, too little clearance is more serious than too much clearance where the proper clearance cannot be maintained.

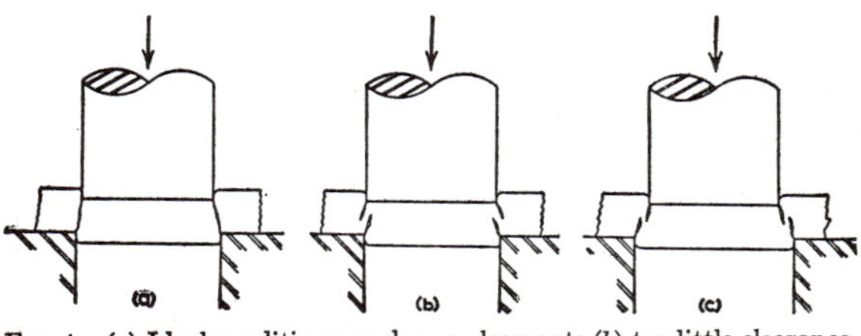

FIG. 4. (*a*) Ideal conditions produce a clean cut; (*b*) too little clearance, dull cutting edges, or increased ductility cause condition shown; (*c*) too much clearance or decreased ductility cause this condition.

The most common occurrence is nonuniform clearance combined with directional differences in the ductility of sheet metal. The most effective ways to control the frequency and size of burrs are: (1) Use only rigid, accurately built dies. (2) Set up and maintain an adequate regrinding schedule. (3) Maintain uniformity of stock through acceptance inspection.

These are rules for maintaining satisfactory production, once it has been set up. However, the foregoing discussion of the mechanism of the cutting operations is important to the selection of stamping as a means of processing a part. It is obvious that even ideal cutting conditions will not maintain dimensions within limits as close as is possible with machining operations. The tolerance on the size and shape of a pierced hole or of a sheared blank need not exceed the proper total clearance.

The proper clearance is determined by the thickness, strength, and ductility of the sheet metal. It is usually expressed as a percentage of the metal thickness varying from 2 to as much as 15%. The ductile metals, which are usually low in hardness, require considerable clearance and relatively large tolerances. Dead soft copper, 2-SO aluminum, and Armco iron are typical metals possessing high ductility. An increase in hardness or a decrease in strength accompanied by a decrease in ductility reduces the penetration and consequently the clearance for a clean fracture. High silicon steel which requires as little as 2 to 3% clearance is typical of the other extreme. The tolerances on relative location of component details such as center distances are dependent on the type of die construction. This problem is discussed in Article 7 under the heading Characteristics and Applications of Dies.

The essential characteristics of the cutting group of stamping operations impose limitations summarized as follows:

1. The selection of ductile metals such as are required for forming operations requires minimum component tolerances of 10 to 15% of stock thickness for the size and shape of sheared details.
2. Smaller tolerances than those designated in (1) above can be obtained on parts stamped from ductile metals by subsequent machining operations, such as shaving, broaching, boring, and reaming.
3. Relatively small tolerances can be maintained down to about 0.003 in. by the selection of a material too hard for forming.
4. Smaller tolerances can be maintained by selecting sheet or strip which has been rolled in two directions rather than continuous strip or coil stock which can be rolled in only one direction. A straight-line cut, parallel to the direction of rolling, requires less clearance than a similar cut across the direction of rolling.
5. Tolerances on location of detail cannot be expected to be smaller than those for size and shape and often must be larger when simple, progressive, or transfer dies are used.

FORMING operations are used to produce a desired shape from the sheet metal. This is accomplished by straining it beyond the yield point so that it will take a permanent set and retain the new shape. The operation must be controlled to such an extent that all regions of the part which have assumed a new shape will have been stressed at least beyond the yield point, but not beyond the ultimate strength where rupture would take place. It is usually difficult to confine the stresses to a sufficiently small range within these critical stresses. Most operating difficulties stem from stresses too close to either the yield point or the ultimate strength.

Most of the operations in the forming group can be classified in one of two subgroups based on the predominant type of stress set up during the operation. One subgroup is distinguished by the fact that the metal flow produced by the operation sets up both compressive stresses and tensile stresses. The other subgroup is characterized by a predominance of tensile stresses. The latter are frequently called stretching operations. Those operations which set up both compression and tension usually will permit more drastic changes in the shape of the formed part.

The group of forming operations which involves both compressive and tensile stresses of considerable magnitude can be further subdivided on the basis of the type of metal flow. One class of operations, of which conventional drawing is typical, produces the change in shape with no substantial change in the thickness of the sheet metal. Consequently, the area of the part remains almost constant during several stages of processing. In the other class of operations, of which ironing and swaging are typical, the change in shape and size is accomplished primarily through an increase or a decrease in the thickness of the metal. The metal flow is much more severe than in other forming operations.

The degree of difficulty in producing a part by forming is determined largely by the stress pattern, the influence of work hardening, the speed of metal flow, the grain size of the metal, and the friction between the part and the elements of the die. A satisfactory part has good finish, holds its shape against residual stresses, and obviously has not been torn or ruptured during forming. The greatest ease of metal flow occurs when the principal stresses are both tension and compression of equal magnitude, and well beyond the elastic limit but safely below the ultimate strength. Metals differ greatly in their capacity for and rate of work hardening. Obviously a metal with low work-hardening capacity and a high rate of work hardening can be deformed relatively little before the stresses exceed the ultimate strength and the part is spoiled by tearing.

Metals show considerable differences in sensitivity to forming speed. The factors influencing the sensitivity to speed are not completely understood, and considerable research may be necessary before a working knowledge of all variables is obtained. However, it is significant that the degree of success in forming sheet metal may depend on the magnitude of the velocity and the sensitivity of the metals to changes of velocity. The grain size of the metal influences both the ultimate extent of forming and the surface finish. The most desired grain size is often a compromise where both finish and metal flow are critical. Lubrication is especially important in the so-called stretching group of operations. Good lubrication will reduce the friction between the work and the die elements. This reduction of friction will reduce stresses and prevent seizure or welding of the work to the die. A good lubricant is frequently the difference between uninterrupted production on the one hand, and scored or torn parts and short die life.

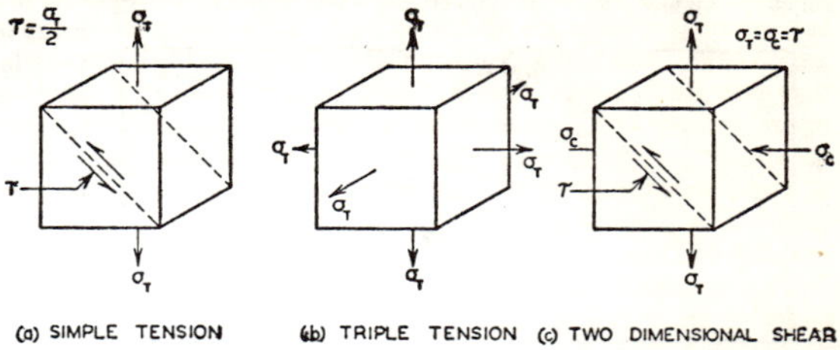

FIG. 5. Basic theoretical stress conditions which can be approached in stamping operations.

Stress Conditions. All metal flow is ultimately reduced to relative movement between adjacent parallel planes and is known as shear; the associated stress is known as shear stress. No permanent set or change in shape can take place unless the shear stress has passed the yield point. However, the loads or motions applied to a metal in an attempt to form it give rise directly to tensile or compressive stresses. Therefore, it is well to review the relationship of the shear stresses to tension and compression. Figure 5 illustrates three basic stress conditions. In simple tension, the maximum shear occurs on a surface at 45 degrees to the direction of the applied external load representing tension (Fig. 5*a*). It can be demonstrated mathematically that the shear stress in pounds per square inch is equal to one-half the tensile stress expressed in the same units. Consequently, the tensile stress required to cause yielding and permanent set is twice the yield point in shear.

If tension is applied equally in all three directions, the net result is no shear. This condition is shown in Fig. 5*b*. Since all metal flow is in a shearing direction, there can be

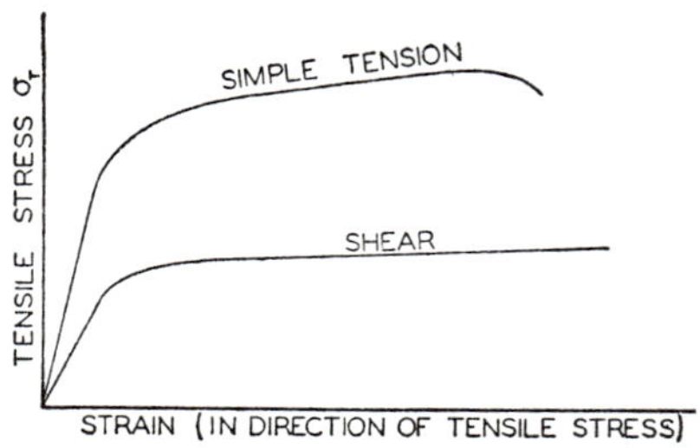

FIG. 6. Forming operations can be more extensive or severe when the punch motion produces two dimensional shear instead of simple tension since metal flow takes place at a lower tensile stress.

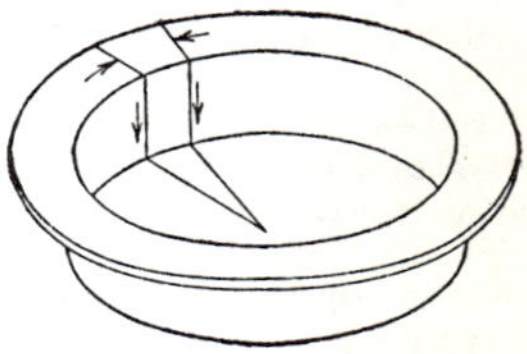

FIG. 7. In conventional drawing of a cup the applied force causes elastic tensile stresses longitudinally in the side wall and radially in the flange; this is resisted by circumferential compressive stresses in the flange which must yield and flow for the drawing to proceed.

no flow in this case until the ultimate strength is reached, at which condition the metal will rupture. This idealized stress condition has been verified experimentally. The same result, i.e., no shear stress, is obtained if the applied stress is uniform compression in all three directions. Obviously this type of stress pattern should be avoided in metal-forming operations.

Figure 5*c* shows a third basic stress pattern in which one principal or applied stress is tension whereas the other is compression. The shear stress in this case is equal to one-half the sum of the principal stresses. The shear stress will be equal to the tension and compression when these stresses are equal to each other. This represents an ideal in that metal forming can be accomplished with a minimum of applied force when the induced tension and compression are equal. Figure 6 shows load-deflection diagrams for simple tension and two-dimensional shear wherein the tensile and the compressive stresses are equal.

The advantages of a stress pattern including both tension and compression are demonstrated clearly.

Figure 7 shows the type of metal flow which takes place in conventional drawing. It will be noted that the metal is pulled toward the center of the part by a tensile load or stress. As the metal moves toward the center, however, it must become shorter in a circumferential direction, giving rise to tensile stresses in a radial direction and compressive stresses in a circumferential direction. The load applied by the punch is relatively low compared with the stretching group of operations, where tension is the only significant stress produced by the punch.

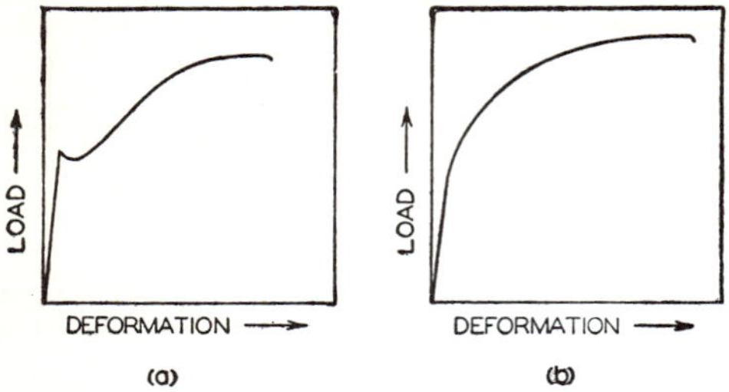

Fig. 8. (*a*) Typical load-deformation curve for annealed low-carbon steel; (*b*) slightly cold-worked low-carbon steel showing absence of yield point elongation.

Stress-strain Characteristics. It has been pointed out that sheet metal possessing a high rate of work hardening and a low capacity of work hardening is not suitable for forming operations. Nevertheless it is true that work hardening has some beneficial effects on the surface finish of the formed parts. A rather common defect in the surface of the formed sheet metal parts is the presence of so-called stretcher strains. In industry this condition is referred to variously as stretcher strains, alligator skin, or orange peel; scientific references usually designate stretcher strains as Lüder's lines, the Piobert effect, Hartmann lines, etc. These stretcher strains are a manifestation of incomplete yield point elongation, which can occur in any metal possessing a load-deformation curve similar to that shown in Fig. 8*a* for an annealed low-carbon steel.

If the steel is even slightly cold worked, the load-deformation curve is similar to that shown in Fig. 8*b*, where there is no yield point elongation. A sheet metal possessing this property, i.e., no yield point elongation, will not produce stretcher strains which might mar the surface of the formed part. The necessary amount of cold working or work hardening to produce this desirable effect can be obtained by a small amount of reduction in thickness of the sheet metal during the cold-rolling process; this is sometimes known as a temper roll or temper pass. Figure 9 contains a series of stress-strain curves illustrating the effect of progressively increasing amounts of cold reduction, and consequently an increasing amount of cold working. It will be noted that the stress-strain curve for the annealed sheet of mild steel possesses distinct yield point elongation, whereas as little as 5% reduction in thickness by cold rolling is sufficient to remove the undesirable yield point elongation which produces stretcher strains. It will also be noted that any further work hardening succeeds only in reducing the amount of elongation required to produce rupture of the material. Therefore, in summary, a small amount of work hardening obtained by cold rolling will aid in producing desirable surface finish of the part, although a greater amount of work hardening simply reduces the extent to which the shape of the part can be changed during processing.

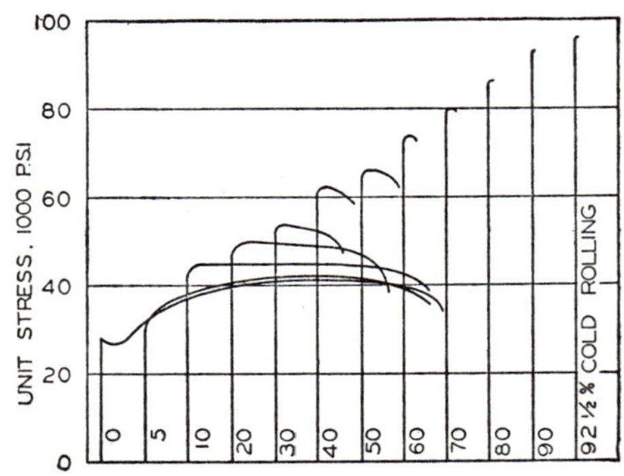

Fig. 9. Effect of cold reduction on the shape of the stress-strain curve of mild steel sheets. (Hayes and Burns, *Trans. Am. Soc. Metals*, Vol. 25, 1937, p. 129)

The velocity at which a metal is made to flow during forming operations also has a very influential effect. There is a need, however, for considerable more research on this problem. The research which has been carried out, although inadequate, has emphasized two things: first, there is a considerable difference between metals as to their sensitivity to velocity; and second, increases in the velocity of straining metals exert a very large influence on physical properties. Figures 10 to 13, inclusive, show typical results of a study made by Winlock and Leiter.* This study involved tests of five samples of a low-

* J. Winlock and R. W. E. Leiter, Some factors affecting the plastic deformation of sheet and strip steel and their relation to the deep drawing process, *Trans. Am. Soc. Metals* Vol. 25, 1937, p. 163.

carbon steel representing grain sizes at 100 magnification of 20 grains per sq. in. to 200 grains per sq. in. The tests were conducted with standard tensile test specimens for sheet metal with speeds of load applications varying from as low as 0.0046 to as high as 10.1 in. per min.

It will be noted in Fig. 10 that there is a substantial increase both in the yield point

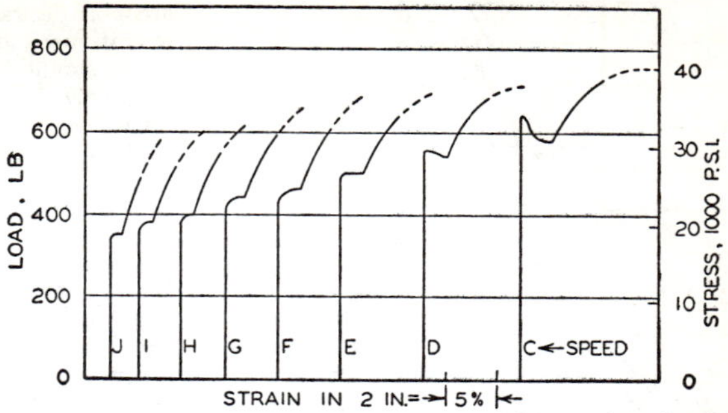

FIG. 10. Load-deformation curves at different constant rates of deformation; $J = 0.0046$ in. per min, $C = 10.1$ in. per min. Low-carbon steel with 0.05% carbon; grain size equals 20 grains per sq in. at 100×. (Winlock and Leiter, *Trans. Am. Soc. Metals*, Vol. 25, 1937, p. 163)

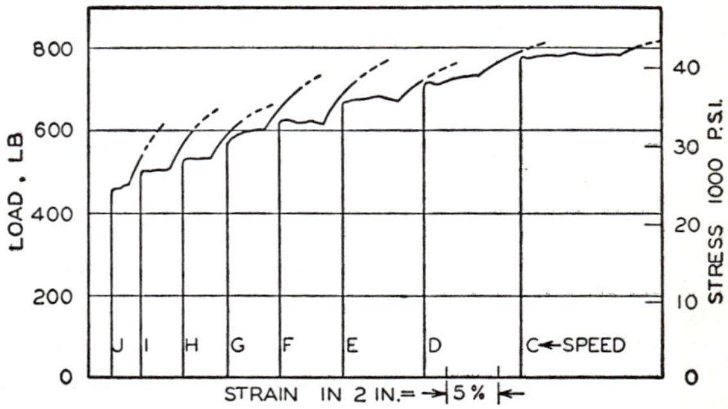

FIG. 11. Load-deformation curves at different constant rates of deformation; $J = 0.0046$ in. per min, $C = 10.1$ in. per min. Low-carbon steel with 0.04% carbon; grain size equals 200 grains per sq in. at 100×. (Winlock and Leiter, *ibid.*)

and in the yield point elongation. A similar influence of speed or velocity is shown in Fig. 11, which gives the results obtained from the 200-grain steel. Figure 12 summarizes the influence of grain size on yield point elongation at four different strain velocities.

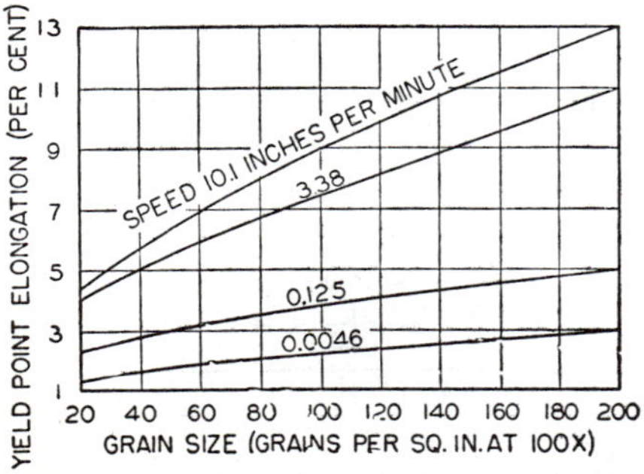

FIG. 12. Yield point elongation versus grain size for different constant rates of deformation. (Winlock and Leiter, *ibid.*)

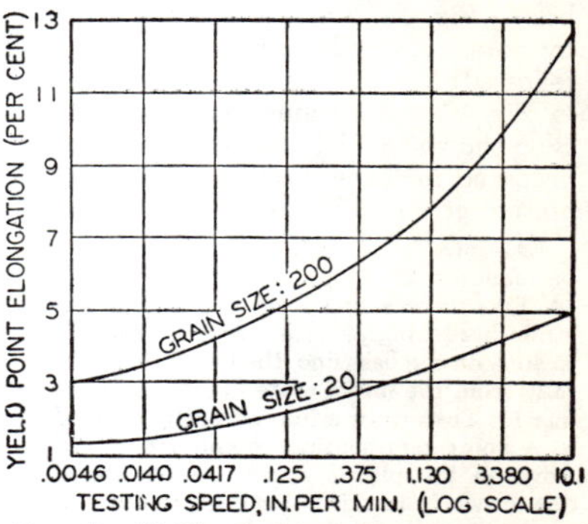

FIG. 13. Yield point elongation versus rate of deformation for steels of different grain size. (Winlock and Leiter, *ibid.*)

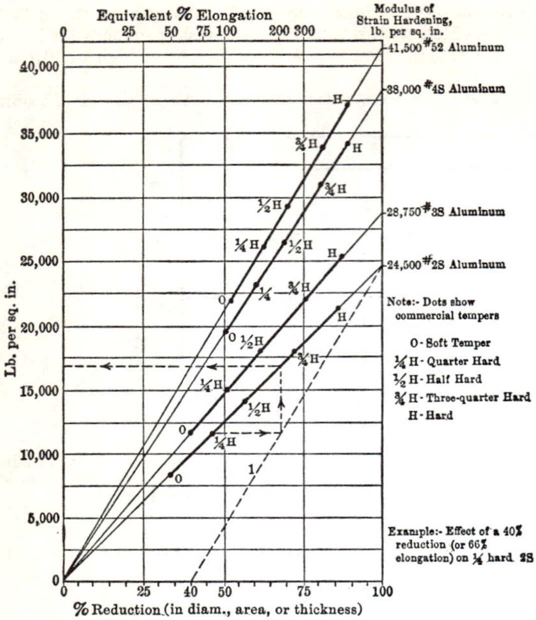

FIG. 14. Plastic range chart for four grades of aluminum at room temperature.

Figure 13 summarizes the influence of testing speed on yield point elongation for both the 20- and 200-grain steels. These results seem to indicate that fine grain steels are more desirable for forming operations, and that the operations should be performed at relatively high velocity. It must be recalled, however, that the strain velocities represented in these tests are substantially below those in general use in industry. Consequently, there remains a broad range of strain velocities which need to be investigated as to influence on physical properties, and relation to composition and microstructure of the metal.

If the factors which influence the formability of metals are summarized, it must be concluded for the present that insufficient information is available on the influence of the speed of forming and the grain size. The composition, microstructure, and grain size of the metals to be used for forming operations are subject to constant revision and improvement, and can be specified most reliably by the metals suppliers. A more useful type of information is available on the effect of strain hardening, as exemplified in Figs. 14 and 15. Figure 14 permits adding the effect of a series of cold-working operations, judging the need for annealing, and finding the yield point or resistance developed in an operation for determining equipment required. The plastic range, or useful working range for each metal, is indicated by heavy lines; and the standard aluminum tempers (listed beside Fig. 14) are indicated in order by dots. A similar procedure can be followed for the information shown in Fig. 15. Additional curves of the type shown in Figs. 14 and 15 can be prepared for other metals from information given in Table 1.

EXAMPLE 1. A reduction of 40% is to be made on 2S aluminum 1/4H. On Fig. 14, first draw a line from the modulus of strain hardening of 2S aluminum (24,500), to 40% on the base line, then a line horizontally from the initial state (1/4H) across to line 1. Then draw a line vertically upward to a point representing completion of the work on the plastic curve. A horizontal line drawn from this point back to the unit stress line shows about 16,500 psi yield point reduction.

EXAMPLE 2. Determine the cumulative

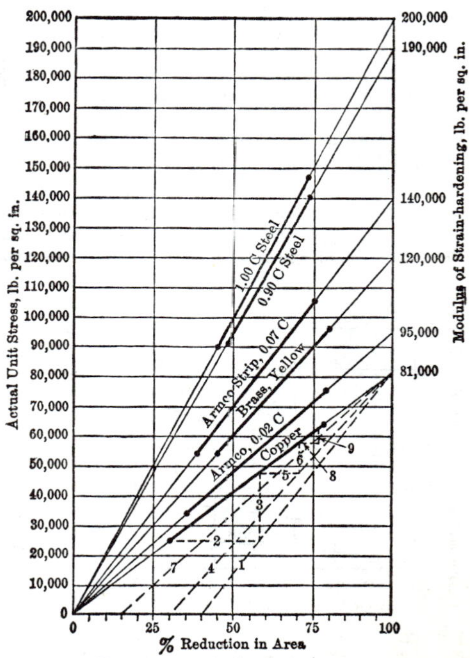

FIG. 15. Strain-hardening curves.

Table 1. Plastic Physical Properties of Metals

Material	Theoretical Yield, Point, psi		Reduction in Area, Max, Annealed, %	Shearing Resistance and Penetration for Clean Fracture				Tensile Strength, (Nominal) Annealed, psi	Elongation in 2 in., Average, Annealed, %	Young's Modulus of Elasticity, psi	Modulus of Strain Hardening, Tentative, psi	Bulk Modulus, (Volume), psi	Annealing Temperature, Approx., Commercial, °F	Forging Temperature, Approx., °F	Thermal Coefficient of Expansion, in. per in. per °F
	Minimum, Commercially Annealed	Maximum, Severely Cold-worked		Annealed to Soft or "O" Temper		Cold-worked to Hard Temper									
				psi	%	psi	%								
Aluminum, 2S, comm'l, pure	(4,000) *–8,000	H- 21,000	80	8,000	60	H-13,000	30	13,000	35–45	10,300,000	25,000	10,200,000	650		0.000 010
" 3S, manganese	11,500	H- 25,000	80	11,000		H-16,000		16,000	30–40	10,300,000	29,000		750		
" 17S, for heat-treat	30,000												630–650	400	
" 52S, chromium	21,000	H- 37,000	80	14,000		H-36,000		29,000	25–30	10,300,000	38,000		650		
Brass, yellow, for cold working	(10,000) *–40,000	95,000	75	32,000	50	52,000	20	50,000	65	13,400,000	120,000	8,800,000	1100		
Brass B, for forging														1200–1400	
Bronze, alloy X	25,000	120,000	53	36,000	25	1/4H-42,000	17	60,000	46		250,000				
Copper	(10,000) *–25,000	62,000	65	22,000	55			33,000	68	14,500,000	81,000	17,400,000		1600–1800	0.000 009
Gold										11,600,000		23,200,000	600–800		0.000 008
Iron, gray cast	8,000 † 30,000 ‡							23,000		12,000,000		13,900,000			0.000 0055
" high test	15,000 † 40,000 ‡							45,000		20,000,000					
Lead	(1,000) *–3,000	4,000		3,500	50					2,470,000		1,100,000	§		0.000 017
Monel metal	28,000	H-112,000	63					70,000	30–50				1100–1200		
Nickel	21,000	H-115,000	75	35,000	55			70,000	47	32,000,000		24,600,000	1200–1300		0.000 007
Silver										10,900,000		14,500,000	600–800		0.000 010
Steel, 0.03 C Armco	(20,000) *–35,000	70,000	76	34,000	60			42–48,000	48	30,000,000	100,000		1300–1400		
" .15 C	(32,000) *–55,000	110,000	60	48,000	38	61,000	25	57,000	35	30,000,000	150,000	23,200,000			0.000 009–0.000 016
" .50 C	(55,000) *–70,000		45	71,000	24	90,000	14	85,000	25	30,000,000			1300–1400	1800–2100	
" 1.00 C	90,000	145,000	20–40	115,000	10	150,000	2	120,000	10	30,000,000	200,000				
" Silicon high				65,000	30										
" Stainless Enduro AA	50,000	120,000	60					80,000	25	30,000,000					
" Low carbon (18-8)	(35,000) *–50,000	165,000	71	57,000	39			95,000	57				1300–1900	2100	
Tin	(2,000) *–4,000	4,500		5,000	40					7,260,000		7,260,000	§		0.000 015
Zinc	(12,000) *–28,000		43	14,000	50	19,000	25	30–37,000	27	13,100,000		5,100,000	‖	200–300	0.000 0145

* Figures in parentheses are misleadingly low test values. † Tensile. ‡ Compression. § Below room temperature. ‖ Room temperature.

strain-hardening effect of a series of operations on copper, and check approximately whether or not an annealing operation is required. It is assumed that copper is received in the soft temper or commercially annealed state and that the operations to be performed involve successive reductions of 40, 30, and 50%. (See Fig. 15.) The construction lines, numbered in order, indicate successive yield points as follows: Before the first operation, about 25,000 psi; after the first operation, about 42,000 psi; after the second operation, about 51,000 psi; after the third operation, about 62,000 psi. The final point is still within the plastic range of the material so that an annealing operation has not yet been required. It is obvious, however, that the material is now severely strain hardened and must be annealed before any further working is possible.

Annealing and lubrication are important auxiliary functions in the processing of sheet metal even though they are not closely integrated with the theory of metal cutting and forming. Heat-treating practice and equipment are metallurgical subjects and are not discussed in this chapter. It appears desirable, however, to call attention to one heat-treating problem which is the most common source of difficulty with sheet metals. Most metals possess the property of abnormal grain growth at particular combinations of stress-relieving temperature and amount of prior work hardening. Parts formed from sheet metal are usually strained different amounts in the various regions of the part. Thus, if a partially formed part is stress relieved at a particular temperature, it may contain a region in the form of a band or ring within which the grains have grown abnormally. This region will be much weaker than the other regions with normal grain size, and the part may be ruptured at this point in subsequent forming operations. The temperature at which abnormal grain growth occurs for a particular amount of strain hardening is known as the germination temperature. In order to eliminate abnormal grain growth during the stress-relieving or so-called annealing operations, it is necessary to heat very rapidly to a temperature substantially above all the germination temperatures peculiar to any one group of metal-forming operations. This is a metallurgical problem and must be solved experimentally during the setting up of operations.

Lubricants. Table 2 gives summary recommendations of lubricants for use in metal-forming operations. The recommendations made are typical of broad general use in industry. The so-called lubricating materials included in Table 2 are segregated both as to the metals for which they are recommended and the relative severity of forming designated as mild, medium, or severe. All lubricants should be avoided for metal-forming

Table 2. Lubricant Recommendations

STEEL

Mild

Most automobile body forming and shallow draws on low-carbon steel.

1. Mineral oil of medium-heavy to heavy viscosity.
2. Soap solutions (0.03 to 2.0%, high titer soap)
3. Fat, fatty oil, or fatty and mineral-oil emulsions in soap-base emulsions.
4. Lard oil or other fatty oil blends (10 to 30% fatty oil).

Medium

Deep draws on low-carbon sheet steel and drawing wire of low-carbon steel.

1. Fat or oil in soap-base emulsions containing finely divided fillers such as whiting or lithopone.
2. Fat or oil in soap-base emulsions containing sulfurized oils.
3. Fat or oil in soap-base emulsions with fillers and sulfurized oils.
4. Dissimilar metals deposited on steel plus emulsion lubricant or soap solution.
5. Rust or phosphate deposits plus emulsion lubricants or soap solution.
6. Dried soap film.

Severe

Wire drawing medium- to high-carbon steel, seamless-tube drawing, and cartridge-case drawing.

1. Dried soap or wax film, with light rust, phosphate, or dissimilar metal coatings.
2. Sulfide or phosphate coatings plus emulsions with finely divided fillers and sometimes sulfurized oils.
3. Emulsion lubricants containing flowers of sulfur as combination filler and sulfide former.
4. Oil-base sulfurized blends containing finely divided fillers.

STAINLESS STEEL

Mild

1. Corn oil or castor oil.
2. Castor oil plus emulsified soap.
3. Waxed or oiled paper.

Medium

1. Powdered graphite, suspension dried on work before operation (to be removed before annealing)
2. Filler-bearing emulsion lubricant at heavy concentration.
3. Solid wax films.

Severe

1. Lithopone and boiled linseed oil.
2. White lead and linseed oil to a consistency of 600W oil.

Table 2. Lubricant Recommendations—*Continued*

BRASS

Mild

1. Soap solution (0.03 to 2% high titer soap).
2. Fat or oil emulsions with soap emulsifier.
3. Lard-oil blends (10 to 20% lard oil in mineral oil).

Medium

1. Soap solution. Soap should be high titer (39 to 42%), fatty acids and free alkali should be less than 0.07%. Solution should be low concentration (0.3 to 1.0%), and lubricant should contact work at least 30 sec.
2. Fairly rich fat or fatty-oil emulsions with soap emulsifiers. Free fatty acid in the paste base, diluted with 8 parts water, should be at least 2%.
3. Lard-oil blends (25 to 100% lard oil in mineral oil). Free fatty-acid content 1.5 to 5%. Addition of about 1% melted tallow, 0.25% stearic acid to 1 to 2% soap solution is desirable for medium to severe draws.

Severe

1. Soap solution of 0.3 to 1% or one of 1 to 2%, containing 1 to 2% tallow or 0.25% stearic acid. Lubricant and work should be in contact longer than with less-severe draws.
2. Rich lard-oil blends (75 to 100%).
3. Dried soap properly added.

In most instances, brass recommendations apply to copper, although copper welds more rapidly and requires more efficient lubrication than brass.

MAGNESIUM

Mild

1. Graphite (usually colloidal) in mineral oil, high flash point.
2. Beeswax or paraffin and tallow.

Medium

Flake or colloidal graphite in a volatile solvent such as carbon tetrachloride, naphtha or alcohol spread on the work and solvent-evaporated. Add 20% graphite in tallow on the die.

ALUMINUM

Mild

1. Mineral oil, increasing in viscosity as severity of operation increases.
2. Fatty oil blends in mineral oil (10 to 20% fatty oil) or petroleum jelly.

Medium

1. Tallow and paraffin.
2. Sulfurized fatty-oil blends (10 to 15%), preferably enriched with 10% fatty oil.

Severe

1. Dried soap film or wax films.
2. Mineral-oil or fatty-oil blends or sulfurized-oil blends, plus finely divided fillers.
3. Fat emulsions in soap water plus finely divided fillers.

operations wherever possible, since they necessitate relatively expensive cleaning operations.

In some instances, where high production continues for an extended period of time, it may prove desirable to reduce friction between the sheet metal and the forming die elements either by chromium-plating the die elements or by plating the sheet metal with tin or copper, particularly where the sheet metal is to be steel. The latter method is particularly useful where soldering or brazing operations are to be employed in fabricating several components into a complete product. The use of sintered carbide for die elements not only reduces friction between dies and sheet metal but also greatly extends the life of the die. The use of sintered carbide for small and medium-sized dies is increasing very rapidly as a result of the development of much more suitable grades of carbides for this purpose.

6. APPLICATIONS AND LIMITATIONS

The following discussion concerns the applications and limitations of most of the metal stamping operations. Because the nomenclature has not been standardized and there are regional differences in the names applied to different operations, an attempt is made either to define or to describe the operations in each instance. The cutting operations used most frequently are discussed first; the forming operations are arranged in a sequence representing increasing difficulty or severity of metal flow.

BLANKING is the act of cutting out a piece of metal of the desired shape and size for the final part (Fig. 16). It is most frequently the first stamping operation and is usually followed by cutting or forming operations or both. When blanking is to be followed by cutting operations only, the blank may be cut to the exact final shape and size of the part. When forming operations are to be performed, the shape and size of the blank must be such as to provide some excess of stock in the blank to allow for unpredictable variations of metal flow. The name blanking is sometimes applied somewhat erroneously to the cutting out or removal of a partially completed or even a completed part being processed in a progressive die; the usage is correct to the extent that the operation does establish the perimeter or outline of the part. Blanking is limited only by the ability of the cutting tools to withstand the forces and by ability to machine the punches and dies for complex shapes. With complex shapes it is often advisable to simplify the blanking operation by splitting it up into several punching, slitting, and notching operations. It is usually uneconomical to attempt to maintain a tolerance on the size of a blanked part closer than 0.003 in., and it is often necessary to permit variations as great as 15% of the stock thickness for the softer more ductile metals.

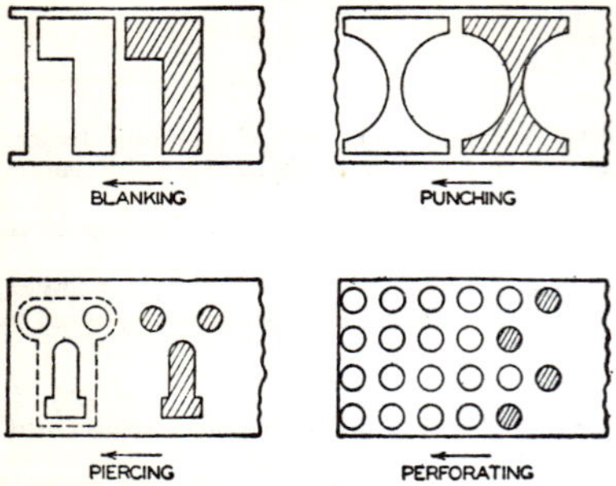

FIG. 16. Typical blanking, punching, piercing, and perforating operations performed in a progressive die.

PUNCHING could be defined as partial blanking since it is often used in progressive dies to cut a major portion of the perimeter of the blank, leaving a small connection with the stock for the purpose of transferring the part between work stages. Like blanking, punching involves a closed cut, i.e., the cutting edge extends entirely around the perimeter of the punch. The edges cut by punching operations seldom constitute any portion of the stamped part since it is used prior to forming operations which require trimming for the purpose of creating the desired perimeter of the part. Therefore, punching imposes no limitations on the design of the stamped part, and the use of the punching operation is limited only by ability to machine and harden the punch and die details.

PIERCING is the hole-cutting operation of the stamping process. It is used in all types of stamping dies for cutting holes of any size and shape for which the punch and die can be machined. The minimum cross-sectional dimension of any hole to be pierced should be at least equal to, and preferably greater than, the thickness of the stock. Figure 16 shows a typical application of a piercing operation in which two round and one irregular hole are pierced in a progressive die prior to the final blanking or cutting out of the finished part.

PERFORATING can be defined as multiple piercing. It is used to cut both round and irregularly shaped holes in making filter sheets for the chemical industry, sieves, and similar products. This type of operation often requires relatively long slender punches crowded close together. For this reason the name perforating is frequently used to describe the piercing of two or more holes in close proximity to each other. Figure 16 shows a typical perforating operation as it might be performed in a progressive die, thus permitting dispersal of the several punches and simplifying die construction. As in all progressive dies, however, the relative location of the holes is subject to ability to feed the stock accurately between the strokes of the press.

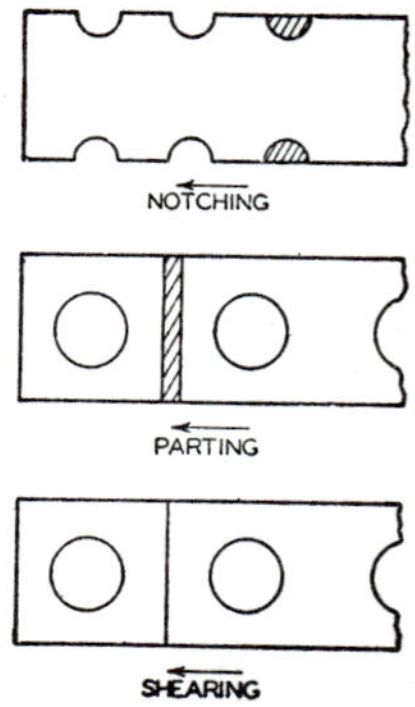

FIG. 17. Typical notching, parting, and slitting operations performed in a progressive die.

NOTCHING is the operation of cutting notches in an edge which may be either curved or straight (Fig. 17). Unlike the four previous operations notching involves cutting along a discontinuous cutting edge. This is accompanied by an unbalance of lateral forces tending to crowd the work away from the punch, thereby making it difficult to control dimensions as accurately as can be done with a continuous cutting edge. Notching is sometimes used as a follow-up operation to blanking, punching, or piercing with the objective of simplifying the contour of the punch and die elements for these operations. Notching is limited only by ability to construct sufficiently rigid tools.

PARTING and **SHEARING** are two similar operations in that their purpose is to separate a part or blank from the stock or from scrap material. Figure 17 shows typical applications of parting and shearing in progressive dies. The main distinction between the two operations is that parting produces some scrap whereas shearing does not. Parting usually involves balanced lateral forces whereas shearing does not. Parting, therefore, can usually be performed with somewhat closer control of dimensions than can be obtained with shearing.

SLITTING is similar to shearing in that cutting progresses along a single discontinuous edge. They differ in that slitting does not extend all the way across a strip of metal and, therefore, does not separate it completely from the adjacent metal. Slitting has many applications, one of which is illustrated in Fig. 18. In this application the slitting is usually followed by a bending operation to create a tab on the component part. Slitting may also be used to free the metal for flow in forming operations in which the metal to be formed is surrounded by large areas of material which would provide too much restraint. A rather minor limitation on the use of slitting is the fact that the metal on adjacent sides of the cut must be displaced relatively, making it necessary to follow up with a flattening operation where this is objectionable.

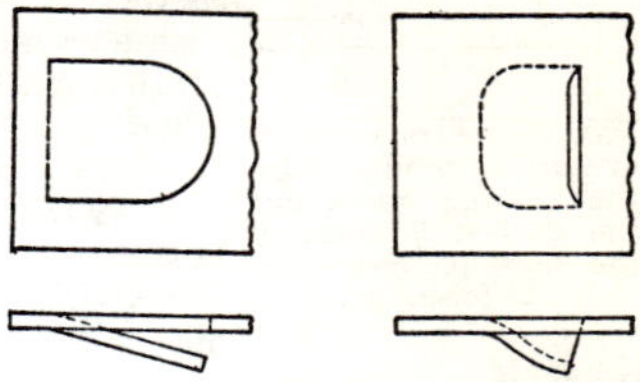

FIG. 18. An example of slitting preliminary to bending a tab, and an example of lancing to produce either a ventilating louver or a locating projection.

LANCING is a combined cutting and forming operation with very limited application. It is intended to cut and form the metal so as to produce a louver or similar shape, as shown in Fig. 18. The cut is usually along a straight line, and if the forming is restricted to a displacement approximately equal to the thickness of the stock it creates a projection which can be used to locate one part relative to another where two or more stamped parts are combined to produce component assemblies.

TRIMMING is the operation of cutting off the excess stock provided for forming operations where it is difficult to predict the exact nature and extent of the metal flow. There are three variations of this operation known as conventional trimming, pinch trimming, and lateral or cam trimming. Conventional trimming is used after drawing or other forming operations which leave a flange on the perimeter of the part. Conventional trimming also leaves a flange but cuts off the excess metal so as to create the desired size and shape of the perimeter. Pinch trimming is used to cut off all the flange remaining on the formed or drawn parts, but it leaves a sharp ragged edge. Where it is necessary to remove the flange a better job can be done by using conventional trimming, leaving a flange equal to three or four times the stock thickness in width and following up with a drawing operation. When the latter procedure is either inadvisable or not feasible, then lateral or cam trimming may be used. In this instance the part is usually mounted on a suitable locating punch and then driven laterally by cams for the cutting phase of the operation. This usually calls for relatively expensive dies and is used only as a last resort for continuing large quantity production. Conventional trimming and lateral or cam trimming can be performed with the same degree of accuracy as blanking. Pinch trimming usually creates greater variations in the length or diameter of the part; it is more pronounced with the softer and thicker metals.

SHAVING is a follow-up operation for increasing the accuracy of surfaces or edges which have been cut by blanking, punching, notching, shearing, and other so-called first operations. Shaving is usually performed as a single operation, although it may be divided into several steps where extreme accuracy is required. Where careful control is exercised it is possible to control dimensions within 0.0002 in. by shaving. A typical shaving cut on the perimeter of a part removes a chip equal in thickness to about 15% of the stock thickness. The punches and dies are usually constructed with zero clearance, and the parts to be shaved must be located as accurately as possible. Shaving cuts less than 0.005 in. thick should not be attempted on stainless steel and other metals which work-harden severely.

BROACHING can be defined as internal shaving since its purpose as an operation is to improve the accuracy of pierced holes. Broaching cutters often have more than one cutting edge and can be used in stamping presses to produce the final size of holes in castings and forgings as well as in stampings. If the broaching tool is constructed with more than one cutting edge the pitch should be such as to maintain at least one cutting edge in contact with the work piece at all times during the cut.

DINKING is an operation similar to blanking and shearing used to cut novelty items from paper, leather, plastics, and similar materials. The lower half of a dinking die usually consists of a plain flat plate or board, and the upper half consists of a plate on which strip tool steel is mounted on edge with the opposite edge ground to a knife edge. A typical application of this operation is in the making of jigsaw puzzles from lithographed cardboard.

BURRING, like lancing, is a combined operation involving both cutting and forming. In its simplest form a bullet-nosed punch is forced into the sheet metal causing it to rupture and bend out the opposite side. The purpose of this operation is to raise a boss on the sheet metal to provide greater bearing area for a screw or a rivet used to fasten the part to another part. There are several other variations of the burring operation, one of which is illustrated in Fig. 19. In this method a hole is first pierced from one side and then a larger-diameter bullet-nosed punch is forced through from the other side of the sheet, thus raising a crude boss. The height of a boss which can be raised by this method is usually limited to about two times the stock thickness. Bosses of greater height can be created by first drawing a shallow cup and then piercing out the bottom of the cup.

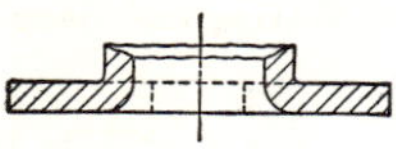

FIG. 19. A raised boss or burr as produced by the burring operation; the dashed lines show the hole pierced prior to forming.

BENDING. The usual connotation of bending applies to the operation as it is performed in the stamping process. It always involves bending the metal about a straight-line axis rather than a curved axis. The extent of a bend is designated by the included angle after bending, and the sharpness is designated by the radius at the inner corner. Some grades of aluminum and brass can be bent to a zero radius. Other materials possess low ductility at bending speeds and require substantial radii at the inner corner to prevent rupture of the metal. For example, a hard rolled magnesium alloy requires a minimum radius of seventeen times the thickness of the stock for a 90-degree bend. Most metals can be bent successfully with a minimum radius of three times the stock thickness. The design of a part to be produced by stamping should include as large bending radii as possible. When bend radii must be held to a minimum it is advisable to consider selecting material suitable for this purpose. With strip material which is rolled only in the longitudinal direction, a bend can be more severe when the axis is across the direction of rolling than when it is parallel to the direction of rolling. When a part contains critical bends with axes at 90 degrees to each other it may be necessary to use sheet stock which has been rolled in both directions.

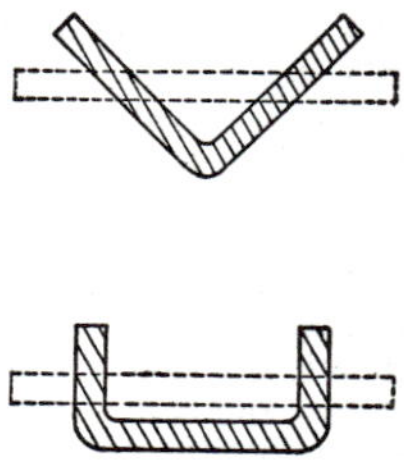

FIG. 20. Bending in a V-die permits striking in the corner to overcome springback; making a U or channel bend usually requires over-bending to offset the effects of springback.

It is more economical to specify a generous tolerance on the angle of bend in view of the spring-back inherent in the bending process. The amount of spring-back is greater with large bend radii. It is possible to minimize the amount of spring-back for a given bend through proper design of the bending die. Figure 20 shows two simple bends, the first a V bend and the second a U or channel bend. In the first instance it is possible to strike the metal in the corner at the bottom of the stroke, thus extending more of the stressed fibers through the yield point, creating more permanent set, and thereby reducing the amount of spring-back. This cannot be done readily with the channel type of bend unless the bending punch can be made to approach the corner at a substantial angle from both sides. Therefore, where spring-back must be controlled, it is advisable to design the part to permit bending in a V die.

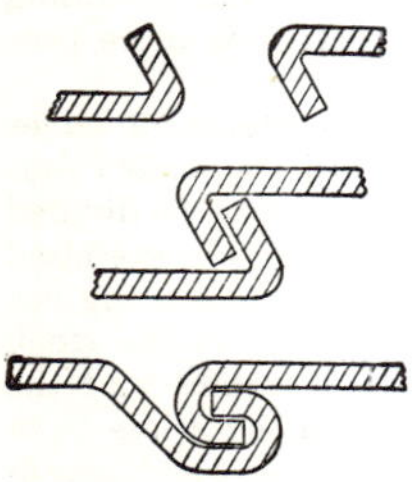

FIG. 21. A series of steps for producing a locked joint by seaming.

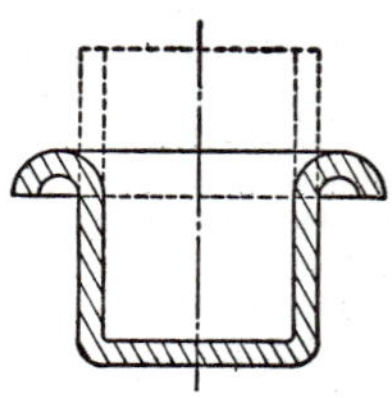

FIG. 22. A curling operation used to produce a rim at the open end of a drawn part; the rim can also be curled inward.

SEAMING is in reality a bending operation with a unique application. It is used for joining the ends of a strip of sheet metal in the making of cans, containers, etc. Three stages of a simple seaming operation are illustrated in Fig. 21. The same limitations peculiar to bending are also applicable to the seaming operations.

CURLING is the operation of rolling over the sharp edge of a drawn part. It differs from bending in that the axis of the bend is a curved line. Figure 22 shows a typical curling operation in the forming of a rim around the top of a drawn cup. The upper edge or rim of the cup could be curled inward as well as outward. The general limitations applying to bend radii and bending also apply in curling; however, there are additional limitations arising from the curved bend axis. For example, in curling the rim of the cup illustrated in Fig. 22, the metal is not only bent but it is also stretched in a peripheral direction. The tensile stresses created by the stretching are superimposed upon the tensile stresses set up in bending; therefore, the minimum bend radii are larger than for the corresponding bending operations with the same material. Most materials which can be deep drawn can be curled easily with a minimum radius of five to six times the stock thickness.

STAMPING is a shallow forming operation involving no substantial changes in the thickness of the metal. It is used to produce decorative configurations, lettering, and beads intended to increase rigidity of otherwise plain areas by increasing the section inertia. The word stamping is now used to describe all operations performed on the sheet metal in mechanical and hydraulic presses. In the early stages of development of sheet metal operations drop hammers were used to produce lettering and decorative configurations; that is why the name stamping is still used in connection with these operations. At present, most stamping operations are performed in mechanical crank-type presses or in hydraulic presses. The name impact stamping is applied to all operations performed on sheet metal by dies operated in either drop hammers or air hammers. Impact stamping is discussed in Article 9. Figure 23 shows a typical arrangement of the tools and work for a stamping operation. The metal is stretched to produce the bead; there is some flow of metal from the adjacent regions into the bead. The height of a bead or letter is usually restricted to no more than ten times the metal thickness, and it is considered good practice to make the width of the bead at least equal to its height.

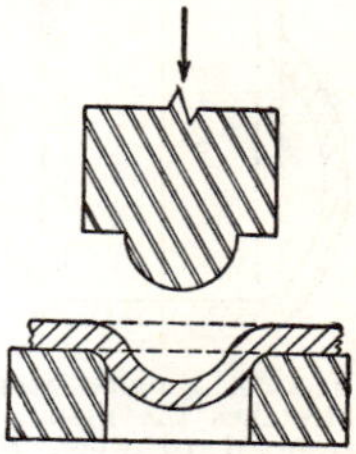

FIG. 23. The forming of a bead, letter, or decorative configuration by stamping or shallow drawing.

CONVENTIONAL DRAWING is illustrated in Fig. 24. It is called conventional drawing because it is the method used most frequently for producing cups in small and medium sizes. The cups can be drawn with or without a flange and in almost any desired cross section. The metal blank is permitted to flow toward the die cavity without any constraint other than against buckling. Internal resistance to flow of the metal toward the punch and die sets up compressive stresses which, combined with tensile stresses created by the motion of the punch, give two-dimensional shear. This is a very desirable internal-stress condition which requires a minimum of applied force to form the metal.

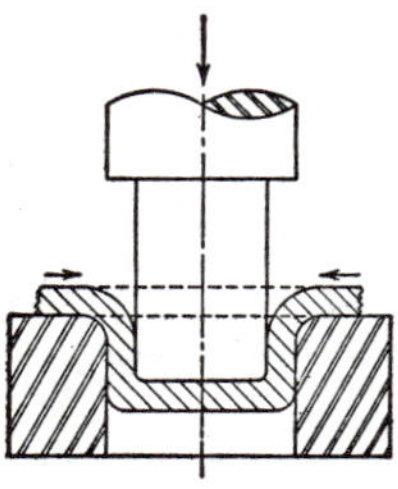

FIG. 24. An example of conventional drawing.

In the simple case of drawing a cylindrical cup from a circular blank the operation is limited theoretically to a reduction of 50% in diameter from the blank to the cup. If a greater reduction is attempted the tensile stresses will be greater than the compressive stresses at the start of the draw, and the bottom of the cup will be torn out by the punch. Further consideration of work hardening, friction, and superimposed bending stresses makes it advisable to attempt substantially less than a 50% reduction in diameter by this process. A conservative rule to follow is one which limits successive reductions in diameter or overall dimension to two-thirds of the per cent elongation of the material as obtained from standard tensile tests. The radii at the corner of the punch and die should be at least as great as the minimum established for bending the same material. The procedure for determining the number of successive draws and redraws that can be obtained prior to annealing is demonstrated in Examples 1 and 2 in Article 5.

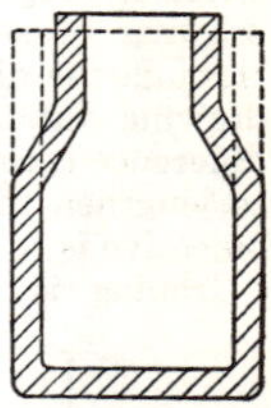

FIG. 25. An example of reducing to produce a neck on a container. The operation is similar to the drawing of seamless tubing except that the part is pushed into rather than pulled through the drawing die.

REDUCING is illustrated in Fig. 25. It is used most often in the production of containers where it is necessary to produce a neck. Consequently, the operation also is referred to as necking. The metal in a previously drawn part is made to flow in compression, thus reducing the diameter locally. Since the sheet metal is compressed, it is subject to buckling or wrinkling in a circumferential direction; this makes it necessary to limit the amount of attempted reduction in a single stage. The amount of reduction that can be accomplished without wrinkling is a function of the thickness of the metal, the friction between the die and the metal, and the included angle of the conical portion connecting the neck with the main body of the container. The commercial range of reduction by this operation varies from 5 to 20%; the lower values should be used for relatively thin-walled material in the work-hardened state and when the included angle of the conical portion approaches 90%. The higher values can be obtained when the wall thickness is relatively great, when the metal work-hardens slowly, and when the included angle of the conical portion is relatively small. When the included angle of the conical portion exceeds 90 degrees it is advisable to bulge the part rather than to attempt necking or reducing.

BULGING is illustrated in Fig. 26. This operation is usually applied to previously drawn cups. It can be accomplished in mechanical dies with either segmental steel punches or with rubber when the punch can be made to collapse sufficiently to permit entry and removal through the neck. More extensive bulging can be accomplished with hydraulic dies wherein a fluid medium such as water is compressed by a punch entering the neck, thereby stretching the metal out against the surfaces of the constraining die. The height of the cup will be decreased during the bulging operation because of unpredictable variations of friction and flow characteristics of the metal. It is therefore necessary to provide some excess height in the drawn cup and to obtain final height of the bulged part by trimming, curling, or some similar operation.

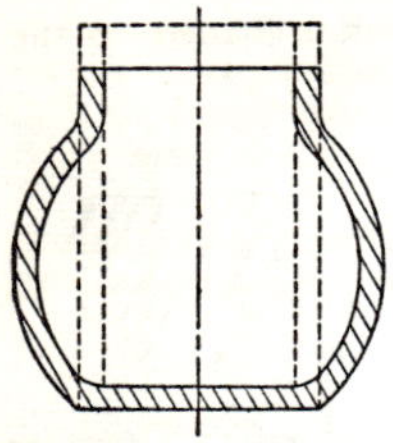

Fig. 26. Bulging, which stretches the metal, is resorted to when the inverse operation reducing either is too expensive or cannot be performed because of the danger of buckling.

STRETCH DRAWING is the name applied to the forming of such parts as automobile doors, body panels, and tops from large sheets of metal. The metal is relatively thin in proportion to the total area involved and would wrinkle badly if drawn in a conventional manner. Therefore the edges of the sheet or blank are constrained in varying degrees to induce a predominance of tensile stress, thereby preventing buckling or wrinkling. Friction plays a very prominent part in this operation and practically controls the severity of stress in any particular region. Consequently, details of the die as well as the shape of the part must be developed by cut-and-try methods. It is not unusual in the automotive industry to allow as much as a year for the design, construction, and try-out of large dies for stretch drawing.

An oversimplified arrangement of the material and tools for stretch drawing is illustrated in Fig. 27. The pressure ring surrounding the punch is used to constrain the sheet metal around its perimeter. A series of beads and depressions on the surface of the die and within the pressure ring increase the grip on the perimeter of the stock. By varying the height and number of the beads, it is possible to control the amount of flow of the metal toward the punch, thus permitting sufficient flow to prevent tearing and yet restraining the flow sufficiently to prevent wrinkling. Since friction plays such a prominent role, lubrication and surface roughness of the punch and die elements are very important and require competent consideration.

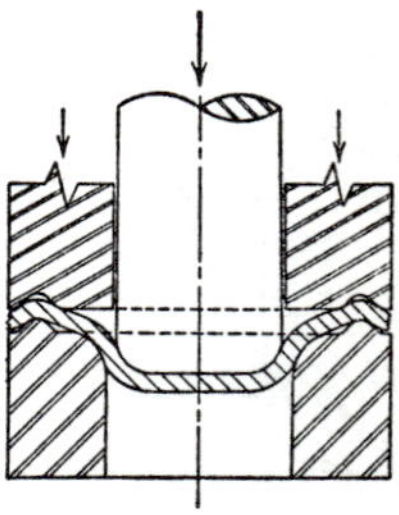

Fig. 27. Stretch drawing requires a firm grip on the perimeter of the blank to retard the inward flow of the relatively thin metal; unrestrained flow would result in serious wrinkling. This method is used to produce automotive body panels and similar large stampings.

IRONING is an operation used to increase the depth of previously drawn cups by thinning out the side wall as illustrated in Fig. 28. This operation is usually divided into several steps wherein the diameter of the punch is maintained constant and is used with a series of progressively smaller die rings. It is very similar to the drawing of seamless tubing in that the cup is literally pulled through the die. The same operating conditions that are used for drawing tubing must also be used for ironing. The principal difference is related to the fact that the main objective of ironing is to lengthen the part by thinning out the walls, whereas in tube drawing the principal objective is to reduce diameter.

Ironing is particularly adapted to the manufacture of cartridge cases from previously drawn cups. Brass is most suitable for this operation although steel has been used successfully in recent years. Application of the process is limited only by economic considerations. The long stroke required for deep cups calls for large, expensive, and slow operating presses, and frequent annealing adds considerably to the cost of processing. The maximum reduction in wall thickness which should be attempted in the first stage with a 70 : 30 brass is 45%. This should be tapered off to about 30% in the last stage, provided it is to be followed by a reducing or necking operation. The procedure outlined in Fig. 15 can be used to determine the extent of reduction of wall thickness prior to annealing.

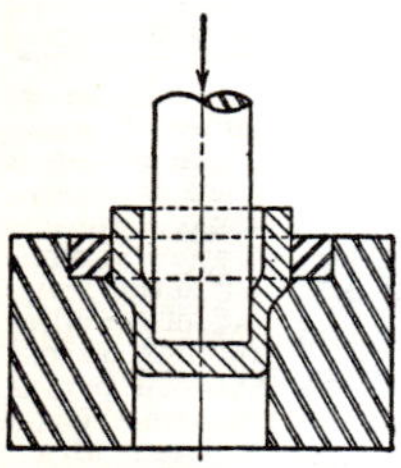

Fig. 28. Ironing is used to lengthen tubular parts by thinning out the side wall.

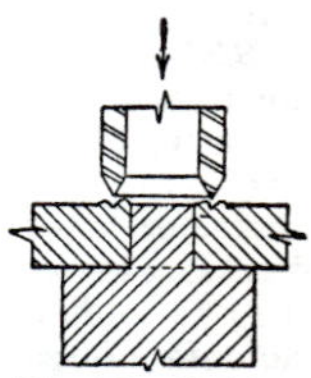

Fig. 29. Staking, the inverse of riveting, is used for fastening one part to another by deforming or peening the metal surrounding the stud or rivet.

STAKING is a press operation used for assembly of component parts. It is somewhat the inverse of a riveting operation as illustrated in Fig. 29. A hollow punch, the rim of which is ground to a sharp edge like that on a cold chisel, is forced into the metal adjacent to the hole, thus peening it against the shaft or stud placed in the hole. The staking punch creates compressive stresses generally near the surface of the work while at the same time causing very high tensile stresses locally at the sharp edge; this causes permanent set in the region of the tensile stress, thus leaving residual compression when the punch is removed. The stud or shaft can be serrated or knurled to increase its torque capacity. This method is used extensively for fastening brass, aluminum, and zinc parts to other parts which can be made of any metal.

SIZING is literally a cold-forging operation used to produce the desired size of a forging or sheet metal assembly by squeezing it between constraining punch and die elements. It is necessary that the metal be stressed thoroughly in compression so that it will hold its shape through permanent set. Application of the operation is restricted to relatively simple shapes for which rigid die elements can be made to apply considerable force over the entire areas concerned. A typical application of this operation is in the production of the desired contour of an automotive brake shoe which has been assembled by riveting and spot welding from several sheet metal components. This operation is occasionally referred to erroneously as "coining," in which the metal is constrained against flow in all directions.

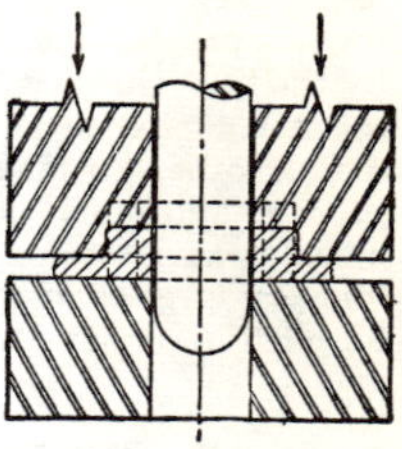

FIG. 30. An example of cold forging a part by swaging in an open die; the dashed lines indicate the original shape of the prepared blank. The perimeter of the swaged part must be trimmed to final shape and size.

SWAGING begins with a relatively thick blank which is squeezed in compression between a punch and die causing the metal to flow outward through the diminishing orifice between the punch and die. Figure 30 shows a typical arrangement of tools for producing a small wheel or gear blank with an integral hub by this method. Almost any metal can be swaged although it is obvious that the extent of the change in shape can be greater with metal possessing a high ductility. Brass, aluminum, or zinc are usually selected for this purpose. The metal will not flow out through the orifice in a uniform wave front; therefore it is necessary to use an excess of material so that trimming or trimming and shaving can produce a final size and shape of the perimeter. The shape and size of the blank should be such that the metal on the perimeter of the part will be stretched no more than the corresponding per cent elongation obtained from standard tensile tests, and the reduction in thickness should be kept within the working range of the material as given in a chart similar to Fig. 14 or Fig. 15.

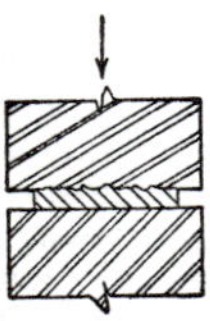

FIG. 31. Schematic sketch of an embossing operation wherein the desired configuration is produced without bending; the metal is free to flow outwardly between the two halves of the die.

EMBOSSING is the act of producing a desired configuration on one side of a piece of sheet metal without any corresponding change in the surface of the other side. This operation is illustrated schematically in Fig. 31, wherein the desired configuration is machined into the upper die block while the lower die remains flat. The metal is not constrained on the perimeter of the part being embossed. This operation is used to produce lettering and decorative configurations on the surfaces of such soft ductile metals as brass and aluminum. The amount of metal movement is slight, but the sharpness of reproduced detail and the life of the die are increased by the selection of a relatively soft ductile metal.

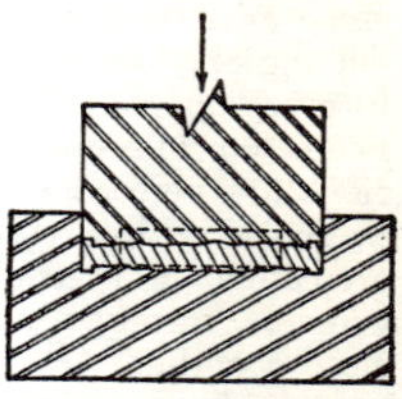

FIG. 32. Coining is similar to embossing except that the metal is constrained within a positive die which determines the outside dimensions and may be used to produce serrations on the perimeter.

COINING is very similar to embossing in that it is used to produce shallow configurations, but in addition the metal is constrained against lateral flow through the use of a positive punch and die arrangement as illustrated in Fig. 32. Thus serrations can be produced on the perimeter of the coin. Only relatively soft materials or alloys are selected for processing by this method; such alloys are gold, silver, copper, aluminum, brass, etc.

EXTRUDING produces the most severe metal flow of all the stamping operations. It involves squeezing the metal through an orifice by the application of pressure with a punch. There are two variations of this process in general use. The first variation, which might be called conventional extruding, begins with placing a blank of the material to be extruded in a closed die and applying pressure with a punch. The metal is extruded through the clearance space between the punch and die cavity and moves upward around the shank of the punch. This is the method used for making toothpaste tubes and other containers of this type. The bottom of the die cavity can be given almost any shape that may be desired on the end of the extruded tube. The metal does not extrude with a uniform wave front, and therefore it is necessary to use an excess of metal, making it possible to trim the extruded part to the desired length.

The other method of extruding is known as the Hooker process. In this process the die opening extends down through the die block, and the extruding orifice is created by the clearance space between the punch and the wall of the die. Pressure is applied by a shoulder on the punch which forces the metal to be extruded downward. The extruded part remains in the die when the punch is removed, and it is necessary to follow up with another punch to pinch out the excess material and push the completed part the remainder of the distance through the die. Products made by this process are frequently referred to as Hooker tubing. Various preliminary steps may be required in preparing the blank for this process, but it is relatively fast and the necessary trimming has been accomplished by the time the part leaves the die. Lead, tin, and soft grades of aluminum are most often selected for parts to be made by extrusion in stamping presses. Copper, brass, and zinc may also be extruded successfully. A large variety of containers is manufactured by this process, and much development work has been done by the materials suppliers who are usually eager to pass on the benefits of their development work. The process is used only for very large-quantity continuing production since some development is usually required for each new shape and size of part.

7. CHARACTERISTICS AND APPLICATIONS OF DIES

Most of the stamping or press operations can be grouped two or more to a single set of tools or stamping dies. Consequently, several operations can be performed simultaneously or in sequence within the same stamping press. The various ways in which the operations can be grouped impart unique characteristics and possess certain advantages and limitations which should be taken into consideration during the earlier steps of production planning. Most stamping dies can be included in one or more of the following classifications: (1) simple, (2) compound, (3) combination, (4) progressive, (5) transfer, and (6) special.

SIMPLE DIES are those intended to perform but one operation. That operation may be either the cutting type or the forming type. As the name implies, this type of die is usually of the simplest possible arrangement and construction consistent with the intended quality of product and life of the die. These dies vary in quality, complexity of design, and cost from very simple cheap dies for small production runs to very elaborate well-engineered expensive dies for large-quantity continuing production.

COMPOUND DIES are single-stage dies for performing two or more operations of the same general type. For example, a compound blanking and piercing die wherein two cutting operations are performed or a compound drawing and embossing die wherein two forming operations are performed. The several operations may be performed simultaneously or in sequence. The most important characteristic of this die is that all the operations are performed on the part in one location. Consequently, stamped parts can be produced within relatively close limits since the dimensions between component elements are controlled by the degree of precision achieved during the construction of the die. The principal disadvantage of this type of die is that the die elements are often crowded very close together. As a result of the crowding, they may be relatively delicate and will require slower operating speeds and greater maintenance.

COMBINATION DIES are single-stage stamping dies performing operations from both the cutting and the forming groups. These dies possess the same general characteristics as the compound dies discussed previously.

PROGRESSIVE DIES are multiple-stage dies in which two or more operations are performed during the same stroke of the press, but at several work stations. The work is fed progressively through several stages, and one or more operations may be performed at each stage. A progressive die actually may consist of several simple dies mounted on common supporting elements. Some progressive dies incorporate combination and compound as well as simple stages. However, the primary objective of the progressive-die arrangement is to achieve greater rigidity and simplicity as compared with compound and combination dies, consequently permitting increased operating speeds in addition to reducing maintenance and construction costs.

Progressive dies are normally used to process continuous strip or coil stock which is mounted on reels and fed by automatic feeding equipment during a predetermined interval of the press stroke. This permits the use of automatic presses for continuous operation. The part or blank is left integral with the strip until the last stage, where it is removed. Since the strip is used to transfer the part from one stage to the next the tolerances on location of part details must be at least as large as the expected error in feeding the strip. Consequently, parts made in progressive dies may not be as accurate as those made in simple dies, compound dies, or combination dies.

TRANSFER DIES possess many of the characteristics of progressive dies since they also contain multiple stages. Some parts, because of complexity or large size, cannot be processed readily in a progressive die because the strip cannot be used to feed the part from one stage to the next. Transfer dies can be used for these parts because they incorporate a transfer mechanism for moving the part. The transfer mechanism may be radically different for each new part; therefore, it is usually considered as a part of the die rather than as standardized press equipment.

Transfer dies are usually more expensive than progressive dies because of their greater size and the additional cost of the transfer mechanism. They are not as common as progressive dies because large and complex parts are not made as often or in as large quantities as the smaller and simpler parts. Transfer dies are used only for large-quantity continuing production. Compared with simple, compound, and combination dies, they will operate at higher production rates with lower maintenance cost and longer life. However, they may require wider tolerances on the relative location of component elements.

SPECIAL DIES are those whose design and functions are so radical that they possess none of the dominant characteristics of the more general classes of dies. Typical of dies which can be classified in this group are lateral trimming dies, hydraulic dies, universal dies, and rubber dies.

Lateral Trimming Dies. Occasionally a part, like a cylindrical cup without a flange, must be trimmed to length. The trimming can be done in a stamping press with an unorthodox die incorporating a sequence of lateral motions. A patented variation of this type of die is known as a Brehm die.

Hydraulic Dies. A hydraulic medium like oil or water can be used with a die for bulging or expanding parts which can be made to contain a fluid during the operation. This type of die is particularly useful in the manufacture of musical instruments and several types of containers.

Standard or universal dies are usually very simple in design and possess certain universal characteristics which permit them to be used for a range of sizes or different types of operations. These dies are basically simple dies intended to perform single operations, but they are designed so that punch and die elements can be quickly adjusted to new locations or replaced with different sizes. This makes it economical to use stamping for producing as few as thirty to fifty parts. There are many different design variations in universal dies, and some of the more useful types can be purchased as standard catalog items.

Rubber dies, as used extensively by the aircraft industry, are similar to universal dies in that they are well adapted to the production of relatively few parts. In addition they possess certain advantages for the forming of aluminum and magnesium alloys. These dies and their application are discussed in greater detail in Article 9 under the heading of Unique Process Variations.

8. SELECTION OF PRESSES

The selection of a press for the performance of stamping operations is based on the type of press and the peculiar characteristics which make it adaptable for certain purposes; the size of the press as to platen and bed area, bed opening, size of bolster plate, length of stroke, etc.; the force or tonnage capacity both at midstroke and near the bottom of the stroke; the work or energy capacity; and, finally, the electric motor for driving the press. The catalogs and engineering bulletins of press manufacturers should be consulted whenever available. The Automotive Joint Industry Conference in cooperation with press manufacturers and other suppliers has developed standards on presses, bolster plates, and other pressroom equipment. When they are adopted more generally, these standards will do much to simplify the problems of equipment selection.

The three basic types of machines used for metal stamping operations are hammers, hydraulic presses, and mechanical presses.

HAMMERS were used rather extensively during the early history of the development of metal stamping. Board-lift and steam hammers are used almost exclusively for forging operations. Rope-lift hammers have been used to some extent in the aircraft industry, although the faster and more versatile air hammers are finding increasing favor in this field.

HYDRAULIC PRESSES are built in sizes varying from small $^3/_4$ ton bench-type presses up to huge 15,000-ton four-post presses. Unlike mechanical presses, the rated force or tonnage capacity is available over the entire length of stroke. The available operating stroke of hydraulic presses is also substantially longer than the corresponding size of mechanical press. Most hydraulic presses are essentially slow-moving machines because of the limitations of pumping equipment in handling the large quantities of oil necessary to fill relatively large cylinders. However, the relatively recent development of rapid-traverse circuits has expanded the economical applications of hydraulic presses to the point where they are being used very successfully on many competitive high-production, deep-drawing jobs. They have also found extensive use in the aircraft industry in connection with rubber dies. (See Article 9.) Small and medium-sized hydraulic presses are usually built with square platens varying in dimension from about 8 in. square on up to several feet on a side. The larger presses are usually custom built to the customer's specifications. A high-speed hydraulic stamping press has recently become available on the market. This press with a maximum capacity of 50 tons can operate with a 4-in. stroke up to 100 strokes per minute.

MECHANICAL PRESSES may be classified in several ways on the basis of the construction of the frame, the mechanism for providing motion to the ram, and whether or not it is provided with the auxiliaries required by an automatic press. On the basis of frame design, the presses are classified as gap or C frame, arch type, straight side, and pillar presses. Most of the presses in the small and medium sizes may be mounted in a vertical position or in a tilted or even horizontal position to facilitate stock removal. Motion to the ram may be provided by cranks, eccentrics, cams, toggles, screws, knuckle joints, and in one instance by a Scotch crosshead. Mechanical presses are also classified as single-action, double-action, or triple-action presses in which case reference is made to the number of moving slides or rams. Double-action presses consist of an inner ram contained within an outer ram; while triple-action presses have the added facility of a third slide operating from the bottom of the press.

Gap or C-frame presses are constructed in a wide range of sizes from very small bench presses up through the medium-sized range. They are available in many design variations, making them adaptable to a very broad range of application. They may be equipped with automatic indexing dial feeds for assembly work and for performing a number of operations in sequence on simple parts. They may also be obtained with the crankshaft running from front to back and with or without a solid steel arm or horn extending from the frame for folding or seaming operations. This type of press is usually available only as a single-action press, but it can be used for every type of stamping operation except those which require long stroke, large bed area, or exceptional force capacity.

The press brake, a special variation of the gap-type press, has been developed particularly for making long bends. Relatively complex shapes can be bent from long stock with relatively simple dies. These presses are particularly useful in aircraft, bus, railroad, and similar fields where long structural parts must be fabricated from sheet metal.

The arch type of frame is not used very extensively except in percussion presses. These are power screw presses which occupy an intermediate position between hammers and the relatively high-speed mechanical presses. The slide is driven by a large steep-pitched screw, operating through a fixed nut in the frame and a light flywheel on top. Friction contact at the flywheel rim with the constantly revolving disk on one side drives the slide at increasing velocity until the die comes in contact with the work and all the energy stored in the flywheel is absorbed. Then another revolving disk makes contact with the other edge of the wheel and lifts the slide. Foot treadle control, similar to that of a drop hammer, is used. This press is very well adapted to slow-speed squeezing operations where it is desirable for the operator to exercise some control over the process by varying the speed of application and the magnitude of the applied force. The press is useful for hot press forging and for such critical operations as the pressing of sintered carbide wire drawing dies into alloy steel holders.

Straight-side presses along with gap presses represent the major proportion of all mechanical presses in use. They are built in medium sizes as well as in the very large sizes. The single-action presses of this type are available with force applications to the ram at one, two, or four points, depending on whether one, two, or four cranks or eccentrics are used. The number of points of force application to the ram increases with the size of the press. Single-action presses of this type are built in sizes large enough for stretch drawing of automotive turret tops and floor panels. In this case the punch is mounted on the bolster plate, and the pressure or holding ring contacts a series of air cushions by means of pins through holes in the bolster plate. Single-action straight-side presses are being used more and more for stretch drawing of very large parts.

Double-action presses are usually of the straight-side type. They are intended for drawing operations wherein the outer ram can be used for blank holding while the inner ram carries the punch. The outer ram must precede the punch to the work and grip the blank firmly, holding it during the entire working stroke of the drawing punch. The inner ram which carries the forming punch is operated by one or more cranks in all double-action presses. The outer ram is usually operated by cams in the smaller double-action presses. A toggle linkage is used to actuate the ram in the larger sizes. These linkages are usually compounded and are driven by a crank on the end of the crankshaft through long pitmans and crossheads which permit the entire linkage to act like a huge spring, thus maintaining the pressure ring in contact with the blank during a large fraction of the revolution of the crankshaft. These so-called toggle presses are used extensively for stretch drawing of large parts.

Knuckle-joint presses are in reality toggle presses since a single toggle linkage is used to actuate the ram. These presses combine long stroke with considerable force capacity and are used for ironing, swaging, embossing, coining, and similar operations of the squeezing group which require very high force capacity.

Automatic presses are available in many types. These presses are intended for completely automatic and continuous operation rather than for intermittent operation by the press operator. They are usually equipped with automatic feeding devices, scrap cutters, safety limit switches, and similar auxiliaries. Some gap-type presses are so equipped although most automatic presses are basically of the straight-side design.

There are some special automatic presses which differ radically in design from those discussed previously. One of these marketed with the name Super-Speed Press uses the principle of the Scotch crosshead to balance out the reciprocating forces. This makes it possible to operate the press at exceptional speeds and to perform piercing and blanking operations at very high production rates.

Another type of press marketed under the name U. S. Multi-Slide Press features several independent slides or rams actuated by cams whose phases can be adjusted to produce any desired cycle in sequence. The first slide of this press will mount a progressive die of orthodox design so that the stock can feed through the progressive die into an area approached by additional slides or rams from three or more sides, thus making it possible to perform many complicated piercing, bending, and forming operations which otherwise must be accomplished by relatively expensive cam or toggle actuated slides built into the progressive die. This machine is equipped also with an infinitely variable-speed transmission, making it possible to operate the press at optimum speed. Variable-speed transmissions can be used to advantage on other types of stamping presses, particularly for forming operations wherein the velocity of metal flow exerts considerable influence on the forming characteristics.

The Henry and Wright dieing machine is another design of automatic press which has found broad acceptance in the stamping field. It is a single-crank four-post or pillar type of press in which the crank and crankshaft are mounted within the housing beneath the bed. The ram or upper platen is reciprocated by the four posts extending through long bearings in the bed to the crosshead within the housing. This machine can be obtained with automatic roll feeds at both ends as well as with a scrap cutter. The feeding rolls are driven through a unique mechanism featuring a Vernier distribution of pawls acting on ratchet wheels. This makes possible the adjustment of the amount of feed in small differentials over a wide range. Thus it is possible to operate a wide variety of progressive dies at relatively high speed.

SIZE OF A STAMPING PRESS involves several more or less independent specifications. The first of these, the platen or bed area, refers to the amount of clear space available for mounting a die. It is designated by the available distance front to back and left to right in inches. The bed of the press usually has an opening either to permit stamped parts to fall through or to contain air cushions, rubber bumpers, and similar blank-holding equipment for drawing operations. The next dimension of importance is the vertical opening available for the die. This is the distance between the ram and the bed or bolster plate with the ram in its lower position. Since the ram is adjustable over a certain distance, the vertical die opening must be specified with maximum and minimum values. The length of stroke is fixed on all mechanical presses but is subject to adjustment or presetting on hydraulic presses.

Tonnage capacity of a mechanical press of a crank or eccentric type is variable, depending on that part of the stroke at which the maximum force is required. A relatively small short-stroke single-crank press suitable for blanking may have the same tonnage capacity as a much larger toggle press used for drawing operations where the maximum force requirement may come at midstroke. When a single tonnage capacity is specified, as is usually the case with small crank presses, it is applicable only near the bottom of the stroke and is determined for the most part by the diameters of the crankpin and main bearings. On larger presses intended for drawing operations the tonnage capacity should be specified both at midstroke and at the bottom. Tonnage ratings vary considerably with details of design and the conservativeness of the builder.

The energy capacity available for doing the work required in the stamping operation comes from both the drive motor and the flywheel. The amount of energy which is to be supplied by the drive motor during a short working stroke is negligible, and the flywheel must be relied upon for practically all the energy needed at the instant of doing work. However, when some of the larger presses are used for performing drawing operations, the working stroke may be as long as half the total stroke and the drive motor can supply significant amounts of the required energy. These presses are rarely operated continuously, and therefore there is usually adequate idle time during which the motor can store up sufficient energy in the flywheel.

Flywheel slowdown is arbitrarily figured up to 10% for continuous running equipment; up to 20% for intermittent operating conditions; and occasionally up to 40%, which approaches stalling. The working energy W_e, inch-tons, delivered by the average design of flywheel at 10% slow-down may be taken as

$$W_e = \frac{N^2 \times D_f^2 \times W_f}{5{,}250{,}000{,}000} \tag{1}$$

where N = revolutions per minute; D_f = diameter of flywheel, inches; and W_f = weight of flywheel, pounds. For 20% slowdown, multiply W_e by 1.9.

Motor sizes specified for punch presses usually are blanket recommendations to cover the wide range of work which may be put in the all-purpose press. They frequently are high for blanking, forging, or coining operations, and low for deep-drawing operations. In checking, the following approximations may be used:

$$\text{Hp} = 0.005 \times W_e \times n \tag{2}$$

where Hp = horsepower of motor; W_e = working energy, inch-tons; n = number of strokes per minute. Equation 2 holds good in cases where the flywheel has ample energy to meet the total requirements. Otherwise a check should be made with eq. 3 and the higher value used in selecting the motor.

$$\text{Hp} = \frac{1.82 \times n \times (W_{e1} - W_{e2})}{\theta} \tag{3}$$

in which $(W_{e1} - W_{e2})$ is the difference between the job requirements and the available flywheel energy. θ is the working portion of the cycle in degrees.

In eq. 2 the motor may use the entire cycle to build up flywheel energy. In eq. 3 it is selected to carry a portion of the working load. The motor lags some at the beginning of the demand, but then builds up to something over its rating to compensate. An allowance must always be made for friction losses and starting inertia losses, though these are small compared with average working loads.

Punch press motors must be capable of taking a slowdown up to about 10% on continuous running jobs, and up to about 20% (and higher on some jobs) for intermittently operating presses. Squirrel-cage motors are suitable for most work, but high-slip motors often are required on deep draws. Instantaneous peak loads may amount to 100% overload.

FORCE REQUIRED for performing stamping operations can usually be estimated very simply and reliably. The force required to perform all cutting operations can be obtained simply by adding the lengths of all cutting edges in operation at any instant and multiplying the result by the thickness and shearing strength of the stock. The shearing strengths of a large variety of materials are given in Table 1.

All forming operations can be divided into two classes for the purpose of estimating required force. The first group or class of operations is characterized by the fact that application of the punch to the sheet metal sets up tensile stresses directly. Failure or tearing of the metal in tension would reduce or eliminate the force on the punch; therefore a conservatively high estimate of required force will be obtained by multiplying the cross-sectional area of the metal stressed in tension by the ultimate strength of the material in tension. This procedure will provide adequate protection for the press for all types of forming operations, where it is not possible for the punch to bottom against the die. This procedure may give an unnecessarily high value of force requirement for shallow and stretch drawing operations. In such cases the drawn part seldom has vertical walls, a fact which may be taken into consideration by using the cosine of the minimum angle from the vertical as a multiplier. For example, in extremely shallow draws, where the sides of the part are never inclined more than 30 degrees from the horizontal, the vertical component of force which must be furnished by the punch is only half the total force required to rupture the metal. Some authorities recommend using the elastic limit of the metal as well as the angular deviation from the vertical in obtaining a lower estimate of the required drawing force.

In some bending operations and in swaging, embossing, and coining, the punch is permitted to bottom against the die. In these cases it is necessary to multiply the total area which can come into contact by the compressive strength of the material to obtain a reliable estimate of the required force. Even then, it is advisable to add a substantial safety factor as a protection against careless setting-up of the die in the press.

Where pressure pads or draw rings are used for drawing operations this force must also be taken into consideration in selecting the press rating. When stretch drawing is performed in single-action presses, the air cushions limit the load on the ram to that required to hold the blank; the cushions usually are attached to a 100-psi supply. In toggle presses it is necessary to assume that the force required for blank holding will be as much as two times the force required by the punch.

The energy required for cutting operations is estimated by assuming that the cutting force remains constant during the working stroke, which is equal to the per cent penetration required for cutting. The latter value, that is the per cent penetration, can be obtained from Table 1. The energy required by all forming operations, except those of the squeeze group, is estimated by taking 80% of the value obtained by assuming that the force remains constant throughout the working stroke. Energy required for the squeeze operations is relatively small despite the fact that the forces may be very high. This is because the known working strokes are invariably relatively short. In any case, the required energy is assumed to be equal to the product of the compressive force and the length of the working stroke.

9. UNIQUE PROCESS VARIATIONS

There are some unique variations of the stamping process which involve radical differences in design of the dies and equipment as compared with orthodox procedures. Typical of these process variations are impact stamping, the Whistler hydrodynamic process, and the rubber die process. These processes have limited but unique applications as well as unique designs and procedures.

IMPACT STAMPING is the plastic forming of metallic shapes by impact or hammer-like blows. The principal difference between impact stamping on the one hand and conventional and stretch drawing on the other is that the former squeezes or hammers the material between the punch and the die over the entire area, thus stressing more of the material beyond the elastic limit, as is done in striking the corner of a bend. The relatively high velocity of impact stamping is necessary to provide the very high localized forces required to give a set to the metal over its entire area.

For many years silversmiths, cornice makers, and other sheet metal workers have employed rope-lift gravity-drop hammers in their work. Airplane manufacturers adapted impact stamping as metal displaced wood and cloth in airplane construction. World War II accelerated the development of impact stamping and generated the demand for more productive equipment. Impact stamps or hammers, utilizing compressed air for lifting and for accelerating the hammer blow, became available. These not only can form by impact but also have limited pressing capacity. Current application is in bus and truck body building and other shaping where the quantities being produced do not justify steel dies and mechanical presses. They also may be used because of the impact feature which can be used to minimize spring back common to those metals possessing a high ratio of yield point to ultimate strength.

Nonferrous dies are generally used for this process since they can be cast to almost final shape in a very short period of time, thus greatly reducing the die cost. The construction of steel dies to contact the sheet metal over its entire area on both sides would cost even more than the corresponding dies for stretch drawing. The procedure for making the nonferrous dies involves making plaster patterns either from mock-ups or templates. These patterns are used to prepare sand molds for the casting of Kirksite dies and lead punches. After removal from the molds the Kirksite dies are ground and polished. The lead punches being soft require no grinding since they can be made to flow to the contour of the die after mounting in the press.

Air-actuated hammers for performing impact stamping make it possible for the press or hammer operator to control the magnitude of the hammer blow over a very wide range. Some parts can be formed from some materials in a single blow whereas others may require several strokes in a series of either diminishing or increasing blows. It is impossible to exercise this type of control in the orthodox stamping press.

THE WHISTLER HYDRODYNAMIC PROCESS.* The hydrodynamic process is a patented development available for use under license agreement. It is a special variation

* Covered by United States Patents 2,066,085; 2,399,775; 2,156,889. And by Canadian Patent 383,472.

of a hydraulic die that is particularly well adapted to stamping and stretch drawing operations. The process uses a die consisting of two halves mounted respectively on the upper and lower platens of a hydraulic press. The upper die contains a cavity having the desired shape of the formed part. In operation a previously cut blank is placed on a pressure pad located in the lower half of the die; then the lower platen of the hydraulic press is raised until the sheet metal blank is gripped firmly between the pressure pad and the flange area surrounding the cavity in the upper die. Then water is pumped through the center of the pressure pad, causing the sheet metal to be stretched by a mode of least resistance until it conforms to the shape of the upper die cavity. Many parts which would require several steps for forming in orthodox dies can be formed in one step by this process. A favorable stress distribution makes it possible to form some of the heat-treated alloys of aluminum as well as the highly alloyed, high-strength stainless materials.

THE RUBBER DIE PROCESS. Blanking, piercing, bending, and shallow forming operations can be performed with rubber dies. The rubber, usually in the form of one or more relatively thick blankets, is used to apply a relatively uniform pressure over the entire area of the part or the sheet metal from which the part is being made. Thus with proper application of the pressure the sheet metal can be either cut or formed.

For the cutting operations a hardened tool steel template or punch is mounted on a suitable support at some predetermined distance above the bed of the press; one or more rubber blankets are fastened to the ram of the press and the sheet metal is placed over the tool steel template. The ram is lowered, applying pressure through the rubber to the sheet metal until sufficient stress concentration is set up at the edges of the template to shear the metal. It is necessary that there be enough excess metal beyond the perimeter of the template to permit cutting rather than folding of the metal. The resulting sheared edge is relatively ragged compared to that obtained with orthodox blanking and piercing dies. The final product, however, is good enough even for some load carrying structural parts of air frames. The dies can be made up very quickly so that where small quantities are required it is possible to design and construct the dies and complete the total production run in less time than would be required to build conventional steel dies. Rubber cutting dies can be operated in hydraulic presses, mechanical presses, and even in drop hammers.

A very satisfactory job of forming can also be done with rubber dies. As in rubber cutting dies, the only rubber involved is in the rubber blankets which are used to apply force to the sheet metal. In rubber die forming the sheet metal can be forced either into a die or around a punch, depending on the nature of the part. The solid punch or die is cast from either low-melting metals or from ceramic aggregates. In one variation of this process the rubber blankets are mounted on the lower platen or bed of the press and the sheet metal is placed on top of them; the formed punch is then brought down in contact with the sheet metal, building up sufficient force to form the metal around the punch. In another arrangement the die cavity is mounted on the bed of the press and the rubber blankets are mounted on the movable ram.

The amount of scrap which must be trimmed from parts formed by rubber dies may be considerably greater than that which must be trimmed from the corresponding parts drawn by orthodox procedures. However, it will be recalled that this method is intended primarily for small and medium quantities. Forming with rubber dies can be performed in hydraulic presses and mechanical presses as well as in drop hammers.

GENERAL REFERENCES

M. Asimow, An investigation of the plastic flow processes involved in drawing cylindrical shells from flat cylindrical blank, *University of California Publications*, Vol. 3, 1936, p. 235.

C. S. Atchinson and L. B. Tuckerman, The pack method of compressive tests of thin specimens of materials used in thin walled structure, United States National Advisory Committee for Aeronautics, *Report* 649, 1939.

W. M. Baldwin, Jr., and Thomas S. Howald, Folding in the cupping operation, *Trans. Am. Soc. Metals*, Vol. 38, 1947, p. 757.

A. J. Becker, The strength and stiffness of steel under biaxial loading, *Univ. Ill. Bull.* 85, Engineering Experiment Station, 1916.

F. K. Bloom, G. N. Goller, and P. G. Mabus, The cold work hardening properties of stainless steel in compression, *Trans. Am. Soc. Metals*, Vol. 39, 1947, p. 843.

G. A. Brewer and R. B. Glassco, Determination of strain distribution by photo grid process, *J. Aeronautical Sciences*, Vol. 9, No. 1, pp. 1–7, Nov. 1941.

G. A. Brewer and M. M. Rockwell, Measurement of the drawing properties of aluminum alloys by the wedge draw test, *Metal Progress*, July 1942.

G. A. Brewer and M. M. Rockwell, Mechanics of deep drawing sheet metal parts (Lockheed Aircraft Corporation Research Memorandum 1083), *Aero Digest*, Vol. 40, No. 2, pp. 126–135, Feb. 1942

P. W. Bridgeman, Effect of high shearing stress combined with high hydrostatic pressure, *Phys. Rev.*, Vol. 48, 1935, p. 825.

L. A. Carapella and W. E. Shaw, Plastic deformational analyses on pure magnesium, *Trans. Am. Soc. Metals*, Vol. 38, 1947, p. 729.

E. V. Crane, *Plastic Working of Metals and Non-metallic Materials in Presses*, 2nd ed., John Wiley and Sons, New York, 1939; 3rd ed., 1944.

R. L. Dietrich and G. Ansel, Calculation of press forging pressures and application to magnesium forgings, *Trans. Am. Soc. Metals*, Vol. 38, 1947, p. 709.

J. E. Dorn and E. G. Thomsen, The ductility of metals under general conditions of stress and strain, *Trans. Am. Soc. Metals*, Vol. 39, 1947, p. 741.

C. L. Eksergian, Plastic behavior of metal in drawing, *Trans. Am. Soc. Mech. Engrs.*, Vol. 48, 1926, p. 609; *Metal Ind.*, Vol. 30, 1927, pp. 405, 433, 459, 483.

D. M. Finch, S. P. Wilson, J. E. Dorn, Deep drawing aluminum alloys at elevated temperatures, Part I, Deep drawing cylindrical cups, *Trans. Am. Soc. Metals*, Vol. 36, 1946, p. 254. Part II, Deep drawing boxes, p. 290.

M. Gensamer, Strength and ductility, *Trans. Am. Soc. Metals*, Vol. 36, 1946, p. 30.

M. Gensamer, Yield point of mild steel, *Metals Tech.*, Vol. 5, Feb. 1938, TP-894.

N. P. Goss and Wm. Brenner, Jr., The effect of microstructure upon the work hardening characteristics of a 0.74 per cent carbon strip steel, *Trans. Am. Soc. Metals*, Vol. 30, No. 4, Dec. 1942, p. 936.

H. Hencky, New theory of plasticity, strain hardening, creep and the testing of inelastic behavior of metals, *Trans. Am. Soc. Mech. Engrs.*, 1933, APM55-18.

J. H. Hollomon, L. D. Jaffe, D. E. McCarthy, and M. R. Norton, The effects of microstructure on the mechanical properties of steel, *Trans. Am. Soc. Metals*, Vol. 38, 1947, p. 807.

J. H. Hollomon and C. Zener, High speed testing of mild steel, *Trans. Am. Soc. Metals*, Vol. 32, 1944, p. 111.

J. E. Howard, Notes on the cubic compression and cold flow and crushing strength of iron, steel and other metals, *Iron Age*, Vol. 46, 1890, p. 692.

S. L. Hoyt, Plastic deformation in metals, *Metal Progress*, Vol. 38, 1940, p. 659.

Z. Jeffries, Plasticity of metals, *Sci. Monthly*, 1931, Vol. 33, p. 235; *Mech. Eng.*, 1931, Vol. 53, p. 262.

J. D. Jevons, *The Metallurgy of Deep Drawing and Pressing*, 2nd ed., John Wiley and Sons, New York, 1942.

P. G. Jones and H. E. Moore, Investigation of the effect of rate of strain on the results of tension tests of metals, *Proc. Am. Soc. Testing Materials*, Vol. 40, 1940, p. 610.

R. L. Kenyon and R. S. Burns, Autographic stress strain curves of deep drawing steel, *Trans. Am. Soc. Steel Treating*, Vol. 21, 1933, p. 577.

J. R. Low, Jr., *Mechanical Tests for Deep Drawing Sheets*, Modern Industrial Press, Nov. 1940.

D. J. McAdam, Jr., Fracture of metals under combined stresses, *Trans. Am. Soc. Metals*, Vol. 37, 1946, p. 538.

D. J. McAdam, Jr., G. W. Geil, and R. W. Mebs, Effects of combined stresses and low temperatures on the mechanical properties of some nonferrous metals, *Trans. Am. Soc. Metals*, Vol. 37, 1946, p. 497.

D. J. McAdam, Jr., G. W. Geil, and D. H. Woodard, Influence of the strain rate and the stress system on the mechanical properties of copper, *Trans. Am. Soc. Metals*, Vol. 38, 1947, p. 551.

C. W. MacGregor, The tension test, *Proc. Am. Soc. Testing Materials*, Vol. 40, 1940, p. 508.

M. Manjoine and A. Nadai, High speed tension tests at elevated temperatures, *Proc. Am. Soc. Testing Materials*, Vol. 40, 1940, p. 822.

C. H. Marshall, Report on the usefulness and limitations of various tests applied to sheet metals, *Proc. Am. Soc. Testing Materials*, Vol. 37, 1937 (I), p. 518.

C. H. Mathewson, C. S. Trewin, and W. H. Finkeldy, Some properties and application of rolled zinc strip and drawn zinc rod, *Trans. Am. Inst. Mining Met. Engrs.*, Vol. 64, 1920, p. 305.

W. S. Montgomery and E. R. Enders, Some attempts to measure the drawing properties of metals, *Mech. Eng.*, Vol. 48, 1926, p. 119.

A. Nadai, Phenomena of slip in plastic materials, *Proc. Am. Soc. Testing Materials*, Vol. 31, 1931, p. 11.

A. Nadai, *Plasticity*, United Engineering Trustees, McGraw-Hill Book Co., New York, 1931.

A Nadai, Review of recent research work in plasticity, *Trans. Am. Soc. Mech. Engs.*, Vol. 58, 1936, A-104.

A. Nadai, The forces required for rolling steel strip under tension, *Trans. Am. Soc. Mech. Engrs.*, Vol. 61, 1939, A54.

A. Nadai, Theories of strength, *Trans. Am. Soc. Mech. Engrs.*, 1933, APM55-15.

A. Nadai and E. A. Davis, Plastic behavior of metals in the strain hardening range, Parts I and II, *J. Applied Phys.*, Vol. 8, 1937, p. 205.

J. T. Nichols, E. S. Taylerson, and J. C. Whetzel, Tension test specimens of sheet metal, *Proc. Am. Soc. Testing Materials*, Vol. 27, 1927, p. 259.

T. Y. Olsen, Ductility testing machines, *Proc. Am. Soc. Testing Materials*, Vol. 20, 1920 (II), p. 398.

E. R. Parker, Intercrystalline cohesion of metals, *Trans. Am. Soc. Metals*, Vol. 33, 1944, p. 150.

Report on study of cup testing of sheet steel, Report of Committee E1, *Proc. Am. Soc. Testing Materials*, Vol 38 (I), 1938, p. 490.

G. Sachs, "Deep Drawing and Related Processes," *Modern Industrial Press*, March 1940.

G. Sachs and G. Espey, Forming of the aluminum alloy 75S, *Trans. Am. Soc. Metals*, Vol. 37, 1946, p. 468.

G. Sachs, Flow of Metals, *Iron Age*, Feb. 1 and 8, 1940, p. 36.

R. Saxton, Lubrication in bar and tube drawing, *Wire Industry*, Vol. 4, 1937, p. 584.

F. B. Seeley and W. J. Putnam, Relation between elastic strength of steel in tension, compression and shear, *Univ. of Ill. Bull.* 115, Engineering Experiment Station, 1919.

F. R. Shanley, Elastic theory in sheet metal forming problems, *J. Aeronautical Sciences*, Vol. 9, No. 9, p. 313.

Sam Spring, Drawing compounds improve press potentials, *Am. Machinist*, Feb. 13, 1947, p. 117.

R. G. Sturm and B. J. Fletcher, Determining spring-back, *Product Eng.*, Oct. 1941, pp. 526–528; November 1941, pp. 590–594.

J. M. Taub, Stretching characteristics of aluminum alloy sheet, *Trans. Am. Soc. Metals*, Vol. 40, 1948, p. 180.

R. L. Templin, Ductility testing of aluminum and aluminum alloys, *Proc. Am. Soc. Testing Materials*, Vol. 36 (II), 1936, p. 239.

R. L. Templin, Effect of cold working on the physical properties of metals, *Trans. Am. Inst. Mining Met. Engrs.*, Vol. 89, 1930, p. 466.

R. L. Templin, Methods for determining tensile properties of thin sheet metal, *Proc. Am. Soc. Testing Materials*, Vol. 27, 1927, p. 235.

R. L. Templin and R. G. Sturm, Some stress-strain studies of metals, *J. Aeronautical Sciences*, Vol. 7, No. 5, pp. 189–198, March 1940.

J. A. Van den Brock, The effects of cold working on the elastic properties of steel, *Iron Steel Inst. London, Carnegie Schol. Mem.*, Vol. 9, 1918, p. 125.

L. J. Weber and J. T. Weinzierl, Drawing of aluminum, *Trans. Am. Soc. Metals*, Vol. 27, No. 4, Dec. 1939, p. 1052.
L. J. Weber and J. T. Weinzierl, Drawing of aluminum, *Trans. Am. Soc. Metals*, Vol. 27, 1939, pp. 1052–1071.
H. M. Westergaard, On the resistance of ductile materials to combined stresses in two or three directions perpendicular to one another, *J. Franklin Inst.*, Vol. 189, 1920, p. 627.
J. Winlock and G. L. Kelley, Sheet steel and strip steel for automobile bodies, *Trans. Am. Soc. Steel Treating*, Vol. 18, 1930, p. 241.
J. Winlock and G. L. Kelley, Testing automobile body sheet steel, *Trans. Am. Soc. Steel Treating*, Vol. 12, 1927, p. 635.
J. Winlock and R. W. E. Leiter, Some factors affecting the plastic deformation of sheet and strip steel and their relation to the deep drawing process, *Trans. Am. Soc. Metals*, Vol. 25, 1937, p. 163.
J. Winlock and R. W. E. Leiter, Some observations on the yield point of low carbon steel, *Trans. Am. Soc. Engrs.*, Vol. 61, 1939, p. 581.
K. J. B. Wolfe, Relationship between the deep drawing properties of autobody sheet, its austenitic grain size and McQuaid-Ehn carburizing characteristics, *Trans. Am. Soc. Metals*, Vol. 37, 1946, p. 1.
J. K. Wood, Jr., Pole figures of the effect of some cold rolling mill variables on low carbon steel, *Trans. Am. Soc. Metals*, Vol. 39, 1947, p. 725.
C. Zener and J. H. Hollomon, Plastic flow and rupture of metals, *Trans. Am. Soc. Metals*, Vol. 33, 1944, p. 163.

SHOT PEENING

By J. O. Almen

The purpose of shot peening is to permit more efficient use of structural metals. Just as earlier generations of designers were enabled to reduce the weight and cost of their structures by the substitution of steel for wrought iron, by the introduction of heat treatment, and by the development of modern alloy steels, so prestressing by several processes, including shot peening, increases the net work that may be obtained per unit weight of machine parts.

10. APPLICATION OF SHOT PEENING

Shot peening is coming into extensive use as the most versatile of the several prestressing processes for strengthening metal machine parts against cohesive failure. Cohesive failures that are characterized by brittle fracture are caused only by tensile stresses and result from (1) impact loads on brittle metals in which the tensile stress acting at the surface is greatly magnified by various forms of stress raisers, (2) stress corrosion in which cracks (usually intercrystalline) propagate through metal that is stressed in tension, and (3) fatigue failures in which, because of the magnification of surface tensile stresses, fractures occur from repeated loads that stress the metals less than their nominal yield stress.

Shot peening is effective against these forms of cohesive failure because it induces residual compressive stress in the surface of the metal being peened. The surface tensile stresses from externally applied loads will be reduced by the amount of the residual compressive stress. The quality of the peening is determined by (1) the magnitude of the induced residual stress, which is a function of the yield stress of the peened metal and the state of strain in the metal at the time of peening, (2) the depth of the induced residual stress, which is a function of the hardness and ductility of the peened metal, the state of strain in the metal at the time of peening, and the characteristics of the shot and the shot stream, and (3) the degree of coverage, by which is meant the number of shot impacts per unit of area.

PROCESS. In the process of shot peening, relatively hard particles—usually spherical chilled cast iron or steel shot—are projected against the metal being peened with sufficient velocity to indent the surface. The indentations at each point of impact result from local plastic yielding of the peened metal. The deformed areas tend to expand elastically, but they are restrained by adjacent, deeper metal that was not plastically deformed by the shot impact. Since the plastically deformed surface layer seeks to occupy more space, it is compressively strained; that is, it is residually stressed in compression.

STRESS-CORROSION CRACKING. Since stress-corrosion cracking occurs only in metal that is stressed in tension, it is necessary to protect the surface by only a thin layer of compressively stressed metal to prevent corrosive attack, even though the underlying material is still highly stressed in tension. Thus welded or cold-formed vessels that are residually stressed in tension from the forming operation, or from the localized heat of welding, and/or from the working loads, may be protected against stress corrosion or season cracking by the corrosion-impervious layer produced by shot peening. Surface corrosion will still occur, and may actually be increased by the increased roughness, but the intercrystalline corrosion characteristic of stress-corrosion cracking will be stopped so long as the surface stress remains compressive. Shot peening will become ineffective against stress-corrosion cracking if the residual compressive stress is lost, as, for example, by applying external tensile loads that exceed the yield strength of the material. When tensile plastic extension has occurred throughout the section, the residual stress induced by peening will have been largely dissipated. Some materials may lose much of the residual stress by slow plastic yielding, usually known as "creep," without the aid of external loads. In such materials, the protection afforded by shot peening will become less effective with time.

IMPACT STRENGTH. The resistance to impact fracture of many brittle materials, such as hard steel, may be increased by shot peening to produce residual compressive stress in a relatively thin surface layer. The hardest steel is still sufficiently plastic to develop residual stress, although the indentations from shot impact may be too shallow to be detected without sensitive instrumentation.

The susceptibility to impact fracture is increased by surface imperfections and discontinuities (often submicroscopic), such as micro-stresses, scratches, minute notches, corroded areas, and similar stress raisers. These stress raisers cannot be eliminated by polishing, but they are rendered ineffective when the surface metal is compressively stressed to a depth equal to the depth of the discontinuities.

The impact strength is increased by an amount somewhat less than the residual compressive stress acting on the "notched" areas. If the effective depth of the notches is greater than the depth of the induced stress, the peening operation will be ineffective. Shot peening will also be ineffective for increasing the resistance to impact failure in metals that are sufficiently plastic to yield under the applied loads. When appreciable plastic yielding occurs, the residual stress induced by the peening operation will be dissipated.

FATIGUE STRENGTH. Fatigue failures, like impact failures, result from tensile stress only. Under conditions of repeated load applications, cohesive failures will occur from stresses considerably less than the nominal yield strength of the material. Since there is no apparent yielding, the fracture will be brittle in character, even though the material is quite ductile as measured by the conventional static tensile and bending tests.

Under ordinary conditions of specimen preparation, fatigue fractures will originate very near the surface, because the surface tensile stress is always augmented by micro-stresses and various kinds of stress raisers, or perhaps because a surface is, itself, a discontinuity. The stress increment that is added by surface stress raisers increases as the ductility of the material decreases until, in hard steel, the effective tensile stress may be several times as great as the nominal stress. As the hardness of steel increases, the effective "notch" stress increases at a greater rate than the potential strength of the steel, and there is, therefore, a well-recognized limit of hardness for best fatigue durability. Thus the upper limit of hardness of ordinary springs lies in the range of Rockwell C 40-45 because of surface vulnerability.

However, this hardness limitation does not apply when the hardness is obtained by carburizing or by nitriding, as, for example, in carburized gear teeth, carburized wrist pins, and nitrided crankshafts. The fatigue strength of machine parts that are surface hardened by these processes generally increases as the hardness increases. This deviation from the rule of brittleness is due to the residual compressive stress that is developed in the surface by the volume increase that occurs in the hardened case. It follows that the surface vulnerability (notch sensitivity) of the hardened surface of the steel cannot assert itself until the initial compressive stress is overcome by tensile stresses from external loads. Since, in most nitrided specimens and in many carburized specimens, the magnitude of the residual compressive stress is of the order of 100,000 to 150,000 psi, the tensile stresses from externally applied loads are often not sufficient to overcome the residual compressive stresses in the hardened case. The surfaces are, therefore, never stressed in tension, and the fatigue failures will originate in the core.

SHOT-PEENING STRESS. The vulnerability to fatigue failure of machine parts and of laboratory specimens is reduced by shot peening because of the layer of uniform residual compressive stress that is induced when the peening process is properly performed. It must be recognized, however, that there are important differences between the protection against fatigue failure that may be gained by the surface-hardening processes previously described and by shot peening. Prestressing by shot peening alone is not so effective as prestressing by several other processes, but it is the most versatile, because it may be applied at relatively low cost to machine parts of all shapes and to metals of all kinds. The magnitude of the residual compressive stress obtainable by shot peening is equal to approximately one half the nominal yield stress of the metal when the peening coverage is adequate and when the peening is performed on the specimen while in an unstrained state. This compressive stress of 50% yield strength is negative to the dangerous tensile stress. It is available to offset surface fatigue vulnerability when, for example, the peening is applied to the tension-stressed side of a beam that is subjected to repeated bending loads in one direction. However, much or all of the protective residual surface stress may be lost from a shot-peened beam that is subjected to repeated reversed-bending loads.

REVERSED-BENDING LOADS. In reversed loading, one side of the beam will be compressively stressed at each load application. The compressive stress on that side will, then, be the residual stress that was induced by shot peening, plus the compressive stress from the external bending load. If this total exceeds the compressive yield strength of the material, the residual compressive stress will be reduced, and a corresponding part of the protection against surface vulnerability will be lost. The actual yield stress under repeated loading is considerably less than the nominal yield stress as measured by static loading, and it is therefore possible to lose all the residual stress induced by shot peening. If all or most of the residual compressive stress is lost, the fatigue strength of the part may actually be less than before peening, because of the increased notch effects in the roughened surface. Such loss of residual compressive stress under repeated reversed-bending loads does not readily occur in carburized and in nitrided parts because of the greatly increased yield strength of the hardened surfaces.

ONE-WAY BENDING LOADS. The residual compressive stress from shot peening is not lost from the tension side of parts subjected to bending loads that act in one direction only. In such parts, the peening need be applied only to the tension-stressed side, and since this side is not compressively stressed by external loads, the total compressive stress remains that which is induced by peening.

Any plastic yielding that may occur on the tension side will increase the residual compressive stress and will therefore tend to increase the fatigue strength of the part. In fact, such yielding actually does occur in nonpeened, as well as in peened specimens, which is one reason—perhaps the principal reason—for the greater fatigue strength of specimens subjected to one-direction loading. The effect usually referred to as "understressing" is also explained by such yielding.

Plastic yielding also occurs on the side of fatigue specimens that are loaded in compression. Such compressive yielding results in residual tensile stresses, which occasionally become great enough to cause fatigue failures to originate on the side nominally stressed in compression. The maximum tensile stress occurs, of course, when the external load is removed. From such occurrences arises the erroneous belief that fatigue failures can be caused by compressive stress.

Because of the greater protection against surface vulnerability that is given to fatigue specimens subjected to one-way bending loads, it is often desirable to increase the compressive stress induced by shot peening. This may be accomplished by performing the peening on the tension side while the part is statically loaded, in the direction of the normal service load, to approximately one-half the yield stress of the material. The peening will then superimpose upon the static tensile stress, compressive stress equal to about 50% yield stress, which will be increased to approximately 100% yield stress when the static load is removed. The static load to be applied during peening should not exceed 50% yield stress because, upon relaxing the external load, the compressive stress will exceed the yield strength, and plastic deformation will occur. This plastic deformation will be in the opposite sense to the plastic deformation that occurred during shot peening, and the surface layer might be damaged by being excessively cold worked.

FATIGUE TESTS MAY BE MISLEADING. From the foregoing, it will be seen that it is not possible to evaluate the effect of shot peening or to measure the fatigue strength of parts that, in normal use, are to be loaded in one-direction bending, by subjecting them to reversed-bending fatigue tests. This is also true of specimens in which favorable residual stresses have been developed by understressing and by such processes as surface rolling, overloading to cause yielding, and quenching from tempering temperature.

MEMBERS LOADED IN TORSION. It is usually convenient to consider members that are loaded in torsion to be stressed in shear. However, to understand the effects of prestressing, it is necessary to apply a stress pattern that may be expressed in terms having positive and negative values. Therefore, instead of shear, the tensile and compressive stress components of the shear stress are used. These stresses act at angles of 45 degrees to the shear plane and at right angles to one another; that is, when the tensile stress acts in the direction of a right-hand helix, the compressive stress will act in the direction of a left-hand helix.

In repeated reversed-torsional loading, the tensile and compressive stresses are reversed at each load application, and therefore the residual compressive stress induced by shot peening may be lost, wholly or in part, in the same manner as in specimens subjected to reversed-bending loads.

For torsional loads that are applied in one direction only, the residual compressive stress acting in the direction of the tensile stress component remains fully effective. Much or all of the residual compressive stress acting in the direction of the compressive stress component may be lost by plastic yielding, but this loss does no harm unless the yielding is so extensive as to induce dangerous residual tensile stress when the external load is at its minimum.

For greater protection against fatigue in one-way torsional loading, the magnitude of the residual compressive stress induced by shot peening may be increased by performing the peening while the specimen is statically loaded, in the normal direction of twist, to approximately half the yield stress of the material. The surface stress prior to peening will then be one-half the tension yield stress in the direction of the tensile stress component and one-half the compressive yield stress in the direction of the compressive stress component. After peening, while still statically loaded, the surface stress in both directions will be compressive and equal to one-half the yield stress of the metal. The peening will not have altered the surface stress in the direction of the compressive component, but in the direction of the tension component the surface stress will have been changed from one-half the yield stress in tension to one-half the yield stress in compression, a plastic flow representing 100% of the yield stress of the metal.

Upon release of the external load, the bar will untwist to nearly its original angle of zero stress. In so doing, however, the strain in the direction of the tensile component will increase the compressive stress and the strain in the direction of the compressive component will reduce the compressive stress. The residual compressive stress will then be, roughly, 100% of the compressive yield stress in the direction of the tensile component and zero stress in the direction of the compressive component. Since, under ordinary test load range from zero to maximum, slow yielding will occur under repeated loading, the residual compressive stress in the direction of the tensile component may decrease slightly to equal the dynamic compressive yield stress. In the direction of the compressive component, tensile residual stress will be induced by slow yielding. Extensive plastic yielding by creep, added to the plastic deformation that occurred during peening, may damage the material sufficiently to cause fatigue failure from the induced tensile residual stress, which acts in the direction of the nominal compressive stress. Being residual, the maximum tensile stress, which may cause fatigue fracture, occurs when the external load is removed. Since the fatigue fracture will originate on a line at right angles to the tensile stress causing the failure, it is possible, by observing the direction of the initial fracture, to determine whether the failure was caused by the tensile component of the external load or by induced residual tensile stress acting in the direction of the compressive stress component. Fracture from induced tensile stress cannot occur from a test in which the load is varied from the same maximum, but to a lesser minimum; for example, in a test in which the load range is plus 25% to plus 100%. This again demonstrates that the effect of shot peening cannot be evaluated by an arbitrary test procedure based on nominal stress or nominal stress range.

AXIAL-LOADED SPECIMENS. A few machine parts, and occasional fatigue specimens, are loaded in tension or in combinations of tension and compression. Shot peening will increase the fatigue strength of specimens repeatedly loaded in tension by reducing the subsequent tensile stress in the vulnerable surface layer. However, since the residual compressive stress in the surface layer must be balanced by tensile stress in the core to maintain force equilibrium, the total tensile stress in the core will be the tensile stress from the external load, plus the residual tensile stress. If the area of the specimen is small in relation to the depth of the peened layer, the residual tensile stress may become so great as to require only a small external tensile load to cause subsurface fatigue failure. Slight peening may be very beneficial from the standpoint of fatigue, and heavy peening may be very harmful.

Under push-pull fatigue loading, the residual compressive stress from shot peening may be lost, in whole or in part, depending on the magnitude of the external compressive load. When the residual compressive stress is lost, the specimen will have been damaged by shot peening because of the increased stress raisers created by the peening.

CARBURIZED AND SHOT-PEENED METALS. The residual stress that results from carburizing can be made much more effective by shot peening as a final operation. The residual stress from carburizing extends to the depth of the hardened case, and the compressive stress is of the order of 100,000 psi. However, the maximum stress occurs at some depth, with considerable loss of stress near the surface. Shot peening effectively increases the maximum compressive stress, and, since the compressive yield stress in the hardened case is great, there will be little loss of residual compressive stress during fatigue testing or in normal service. The fatigue fracture will often originate in the core in carburized and shot-peened specimens, for the reason that the surface stress does not become tensile. For a given specimen, a close approach to the best possible treatment will have been attained when fatigue fractures in successive specimens originate with equal frequency at the surface and in the core. This must not be taken as "best," however, because the relative areas of core and case may be such as to stress the core in tension unduly.

An effect similar to carburizing may be obtained by other case-hardening processes. In steel sections of adequate diameter, low-hardenability steel, such as severely quenched SAE 1040-1050, will harden only in a relatively thin case. This hardened case will be residually stressed in compression because of the volume increase upon hardening, just as occurs in a carburized case. The fatigue strength of such specimens in bending and in torsion may be further increased by shot peening.

TREATMENT FOR BEST FATIGUE STRENGTH. For repeated reversed loading in bending and in torsion, the best fatigue strength from prestressing of steel is obtained by case hardening followed by shot peening. The deep, and relatively great, residual compressive stress reduces the tensile stress from external loads, and the high yield strength of the case prevents serious loss of the protective stress. Shot peening is not required for nitrided parts, because the residual stress is sufficiently great as heat treated.

For one-way bending or one-way torsion, the best fatigue strength may be obtained from parts that are deep hardened. The material may be in the hardness range Rockwell C 50-55; it should be overloaded to produce plastic yielding in the same direction as the normal load; and, finally, it should be shot peened while statically loaded to approximately half the yield strain.

Because of the suggested hardness of the specimen, the operation of prestressing by overloading is hazardous, particularly since the plastic deformation should be extensive enough to extend to approximately 15 to 20% of the specimen thickness. Unless suitable precautions are observed, this extensive cold setting may damage the material and may even cause immediate fracture. The overloading should therefore be applied at an elevated temperature to reduce the yield stress temporarily and to increase the ductility of the steel. Shot peening at one-half yield strain prior to presetting will also aid in preventing fracture or other damage during the presetting operation. Finally, the shot-peening operation should be applied while the part is statically loaded in the direction of the normal load to approximately half the yield strain.

FIG. 1. Almen shot-peening test strip. Specifications are: Hardness, Rockwell C 44-50; flatness, ±0.0015 in. on Almen gage; *A* strip thickness, 0.051 ± 0.001 in.; *C* strip thickness, 0.094 ± 0.001 in. Ends only may be shear finish, burrs must be removed and corners broken.

11. SHOT-PEENING INSTRUMENTATION

To specify that a machine part is to be shot peened without stating how much, in measurable quantities, is equivalent to specifying "heat treat" without mentioning temperature or time. To meet this need, instrumentation is available that is sufficient for the purpose in accuracy, cost, and simplicity.

The principal instrument that is used for measuring, and thereby controlling, shot peening consists of a standard steel strip (Fig. 1) that is shot peened on one side while attached by screws to a heavy hardened steel base (Fig. 2). When the test strip is removed from the heavy base after peening, the residual compressive stress that was induced causes convex curvature of the strip on the peened side. The curvature is measured on the unpeened side in terms of arc height over a standard chord by a special indicator (Fig. 3).

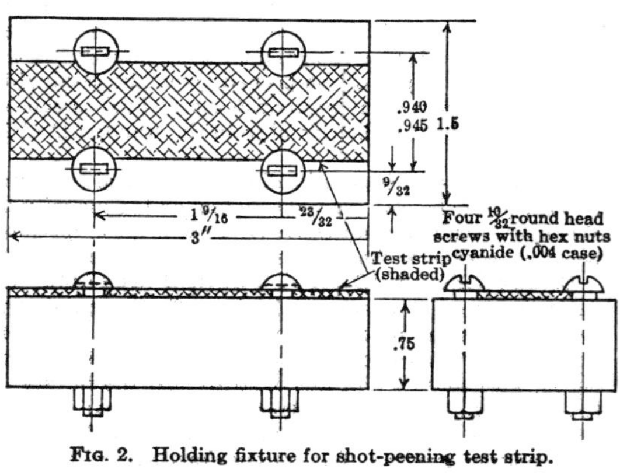

FIG. 2. Holding fixture for shot-peening test strip.

The amount of curvature, that is, the measured arc height of the strip, is called *peening intensity*. The peening intensity is the overall effect of the magnitude of the induced stress, the depth of the stressed layer, and the uniformity of the stress, in magnitude and in depth, over the surface of the strip—hereafter called the *coverage*. Since the peening intensity measurement is made on a steel strip of standard dimensions and standard physical properties, the magnitude and depth of the induced stress are those of the test strip and not of the metal being peened.

Two thicknesses of test strips have been standardized to meet the needs of sensitivity for light and for severe peening. The dimensions of the two strips, designated as *A* strips and C strips, are given in Fig. 1. A chart correlating peening intensity readings from A strips and C strips is shown in Fig. 4.

In production use, the hardened holding fixture (Fig. 2), with attached test strip, is passed through the peening machine in the same manner as the parts being peened. The surface of the holding fixture, to which the test strip is attached, is located in such a way as to represent a critical area of the work, in order that the peening received by the test strip will be substantially the same as is received by the work. Usually, the holding fixture is welded or bolted to a dummy machine part that has been prepared for the purpose. Complex machine parts having many angles and planes may require a number of holding blocks, each representing a different plane or critical area of the work. Peening intensity measurements are made several times each working day. New test strips are used for each location every time the measurements are made.

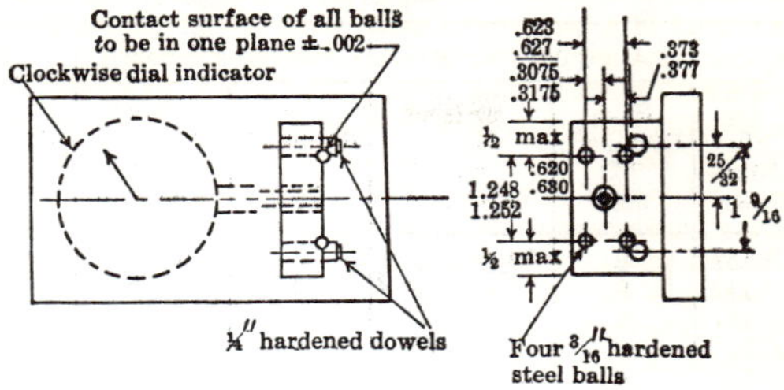

FIG. 3. Almen gage.

Peening intensity is a measure of the quality of the shot stream which may be likened to the measurement of temperature in a heat-treating operation. The temperature of the metal being heat treated is usually arrived at indirectly by measuring the temperature

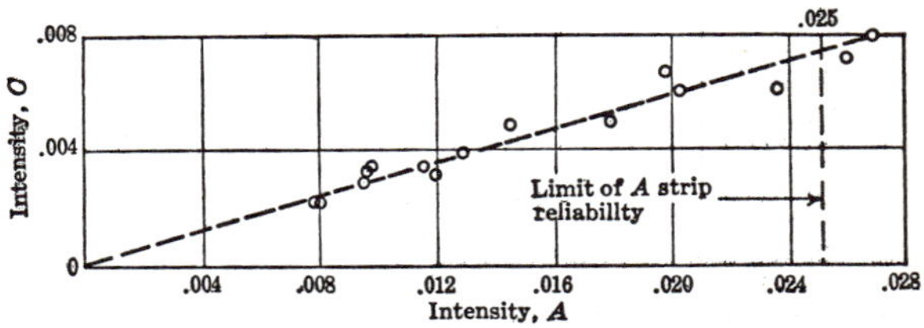

FIG. 4. Correlation of gage readings for Almen A and C strips.

of the furnace in the vicinity of the work and exposing the work to that temperature for a sufficient time to saturate the metal. Unless both temperature and time are correctly specified, and unless the specifications are followed, the heat-treating operation cannot be successful.

12. PEENING SPECIFICATIONS

A shot-peening specification consists of the indicator reading for the test strip, which is the arc height in inches, followed by the letter designating the strip size.

Shot-peening specifications must take into account the dimensions and physical properties of the peening shot, the characteristics of the shot stream, and the physical properties and manner of loading of the machine part being peened. The factors affecting peening intensity are shot size, shot velocity, coverage (quantity of shot per unit area), shot cleanliness, direction of shot relative to work, and shot quality (hardness, mass, shape, etc.).

SHOT SIZE. A system of peening shot sizes has been adopted by the Society of Automotive Engineers and accepted by shot manufacturers. This standard designates shot size by a number which is the nominal shot diameter in mills, preceded by the letter "P" to distinguish peening shot from shot intended for other purposes. Table 1 is a reproduction of the shot-size specifications included in the SAE Recommended Practice.

The selection of shot size depends on (1) the radii of fillets and of surface irregularities such as indentations and cutter marks, (2) the section thickness of the work, (3) the appearance desired, (4) the abrasion desired, and (5) the manner of loading the peened part.

Radii of Fillets. The radius of the shot should be approximately half, or less than half, the radius of the smallest fillet or surface irregularity in the critical area. This rule does not apply to shallow cutter marks or scratches, the depth of which is small with respect to the depth of the compressively stressed layer. If the radii of the shot are large with respect to the radii of deep cutter marks or of fillets, the compressive stress induced in these regions will be too small to afford effective protection.

Section Thickness. When the thickness of the work being peened is small compared to the size of the shot, there is danger that the induced compressive stress will extend too

Table 1. Peening Shot Numbers and Screening Tolerances

Per cent of total sample by weight retained on each pertinent screen

(Reprinted by permission from the *SAE Handbook*)

Peening Shot Number	High-limit Screen		Oversize Screen		Nominal Screen		Low-limit Screen	
	Shot Retained, % max	Screen Number and Aperture	Shot Retained, % max	Screen Number and Aperture	Shot Retained, % min	Screen Number and Aperture	Shot Retained, % max	Screen Number and Aperture
P 157	0	3 1/2(.223)	2	4(.187)	90	5(.157)	5	6(.132)
P 132	0	4(.187)	2	5(.157)	90	6(.132)	5	7(.111)
P 111	0	5(.157)	2	6(.132)	90	7(.111)	5	8(.0937)
P 93	0	6(.132)	2	7(.111)	90	8(.0937)	5	10(.0787)
P 78	0	7(.111)	2	8(.0937)	90	10(.0787)	5	12(.0661)
P 66	0	8(.0937)	2	10(.0787)	90	12(.0661)	5	14(.0555)
P 55	0	10(.0787)	5	12(.0661)	85	14(.0555)	7	16(.0469)
P 46	0	12(.0661)	5	14(.0555)	85	16(.0469)	7	18(.0394)
P 39	0	14(.0555)	5	16(.0469)	85	18(.0394)	7	20(.0331)
P 33	0	16(.0469)	5	18(.0394)	85	20(.0331)	7	25(.0280)
P 28	0	18(.0394)	5	20(.0331)	80	25(.0280)	12	30(.0232)
P 23	0	20(.0331)	5	25(.0280)	80	30(.0232)	12	35(.0197)
P 19	0	25(.0280)	5	30(.0232)	80	35(.0197)	12	40(.0165)
P 16	0	30(.0232)	5	35(.0197)	80	40(.0165)	12	45(.0138)
P 13	0	35(.0197)	5	40(.0165)	75	45(.0138)	17	50(.0117)
P 11	0	40(.0165)	5	45(.0138)	75	50(.0117)	17	60(.0098)
P 9	0	45(.0138)	5	50(.0117)	70	60(.0098)	22	70(.0083)
P 8	0	50(.0117)	5	60(.0098)	70	70(.0083)	22	80(.0070)
P 7	0	60(.0098)	5	70(.0083)	70	80(.0070)	22	100(.0059)
P 6	0	70(.0083)	5	80(.0070)	70	100(.0059)	22	120(.0049)

Maximum "fines" allowed through smallest screen, 3%.
The shot number is roughly the size of the shot pellets in thousandths of an inch.
Percentages given are on the basis of weight.

deeply into the work instead of affecting a relatively thin layer, as is desired. The diameter of the shot should be less than the thickness of the work, since the depth of the compressively stressed layer increases as the shot diameter increases.

Appearance. The section thickness of the work also influences the shot size from the standpoint of appearance of the peened surface. Coarse peening is less offensive to the eye on thick sections than on thin sections.

Abrasion Desired. The amount of material that is abraded from the surface of the work, for any peening intensity, is greater when small shot is used than when large shot is used.

Loading of Part. For any shot velocity, the thickness of the compressively stressed layer is roughly proportional to the shot diameter. The desired thickness of the compressively stressed layer will depend on the manner in which the peened part is loaded in service, as well as on the section thickness. For members loaded in tension and for surfaces where peening has been superimposed on a deep prestress obtained by other means such as carburizing or overstressing, the compressively stressed layer should be thin. For members loaded in bending or in torsion, which are not otherwise prestressed, the compressively stressed layer should be relatively thick.

SHOT VELOCITY. The shot velocity that will be required for any specified peening intensity will vary with (1) the size of the shot, (2) the hardness of the shot, (3) the mass of the shot, and (4) the direction of the shot relative to the work.

Size of Shot. The size of the shot affects the shot velocity to be used for any specified peening intensity, because the depth of the compressively stressed layer is proportional to the shot size.

Hardness of Shot. If the shot is so soft as to deform on striking the work, some of the shot energy will be lost. The shot will also present greater areas of contact, and therefore the unit pressure and the peening intensity will be reduced unless velocity is increased.

Mass of Shot. The energy of the shot will vary with its mass and with its velocity. If the mass of shot is changed, it is necessary to change the velocity to maintain constant peening intensity.

Direction of Shot. The energy of the shot that is absorbed by the work will vary approximately as the sine of the angle between the plane of the work and the line of motion of the shot. To maintain constant peening intensity of the work, it is therefore necessary to change the velocity of the shot as the angle of impact is changed.

COVERAGE. By *coverage* is meant uniform distribution of peening stress over the surface of the work. Uniform stress means that the points of impact are so numerous that the stress induced by each impact overlaps the adjacent stressed area to produce essentially uniform stress.

The same instruments that are used to measure peening intensity are used to determine the time of exposure of the work to the shot stream to obtain adequate coverage. Since the total residual stress will continue to increase until the peened surface is uniformly stressed, the curvature of the test strip will continue to increase until saturation is attained.

Figure 5 is a chart of peening intensity plotted against time of exposure. The upper curve shows a condition in which the specified peening intensity reading is attained at A, but since the curve continues to rise with increased time of exposure, it is evident that the peening stress at point A was not uniform; that is, the coverage was inadequate.

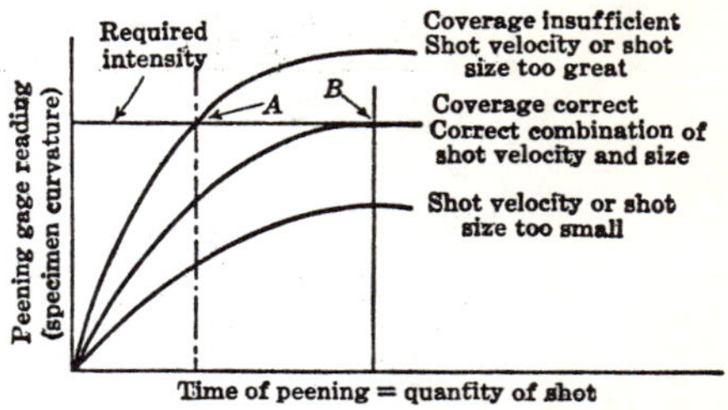

FIG. 5. Peening intensity versus time of exposure.

The second curve shows correct combinations of shot size and velocity so that the required peening intensity is attained at B after a time of exposure sufficient to saturate the surface.

The lower curve indicates conditions under which the desired peening intensity cannot be attained because of insufficient depth of the induced stress.

To determine the time of exposure required to obtain adequate coverage, successive test strips are passed through the shot stream, varying only the time of exposure. The arc height reading of each strip is plotted against the time of exposure, as in Fig. 5. Alterations are then made in shot velocity or in shot size until the required peening intensity is obtained as the limit of arc height is approached.

SHOT CLEANLINESS. Good peening requires that only reasonably round shot of uniform size be used in the machine. Since a considerable percentage of normal commercial shot will eventually fracture or wear as a result of repeated impact against the work or against the wall of the cabinet, means have to be provided for discarding split or other fractured shot, undersized shot, and dust.

Undersized shot, split shot, and smaller fragments are objectionable because they (1) reduce peening intensity, (2) increase abrasion, (3) cut, rather than peen, (4) contaminate the surface of the work, and (5) increase the cost of peening.

1. Peening intensity is reduced because the small particles strike the work with less energy than the original-sized particles.
2. Abrasion is increased because the broken, sharp-edged particles wear away the surface much more readily than the round shot.
3. Many of the broken shot cut into the work like a knife, and may therefore reduce the fatigue strength of the work, since such cuts may extend deeply into the compressively stressed layer.
5. To maintain the specified peening intensity, it is necessary to increase the velocity as the shot breaks down. This increased velocity accelerates the breakdown of the remaining whole shot as well as the added new shot, thus requiring more replacement than if only clean, round material of the specified size were used. As the shot size is decreased, not only must the velocity be increased, but often the time of exposure must also be increased.

SHOT QUALITY. Under shot quality may be listed such properties as (1) hardness, (2) impact fatigue strength, (3) shape, (4) soundness, (5) size range, and (6) mass.

Hardness. Residual compressive stress can be induced by shot softer than the material being peened, but, generally, shot should be as hard as, or harder than, the work. If the shot is too soft, it will lose its energy in deformation, and the induced stress in the work will be too small to produce the desired compressive stress unless the velocity is increased.

Impact Fatigue Strength. The rate of fracture of sound shot, used in peening, is measured by the impact fatigue strength. The impact fatigue strength of commercial chilled iron shot is low—hence the importance of good devices for separating fractured material from whole shot.

Table 2

Name of Part	Material	Hardness (R_c or Equivalent)	Thickness of Critical Section, in.	Surface Finish before Peening	Manner of Stressing	Kind of Shot Used	Size of Shot	Peening Intensity
Crankshaft	Steel	10–21	0.75	Polished	Reversed torsion	Chilled iron	P-28	0.014–0.015A
Connecting rod	Steel	37–44	0.16	Polished	Reversed torsion	Chilled iron	P-19	0.015A
Valve rocker arm	Steel	26–30	0.5	Polished and forged	One-way bending	Chilled iron	P-28	0.022A
Valve rocker arm	Malleable iron	3–16	0.2	As cast	One-way bending	Chilled iron	P-28	0.014A
Quill shaft	Steel	34–38	0.5	Polished	Reversed torsion	Chilled iron	P-19	0.14A
Tank track pin	Steel	24–47	0.9	Rough machined	One-way bending	Chilled iron	P-66	0.017C
Bolt head	Steel	33–35	0.25	Polished	Tension tension	Chilled iron	P-19	0.017–0.022A
Transmission gear	Steel *	60–62	0.3	As machined	One-way bending	Chilled iron	P-28	0.020A
Steering knuckle	Steel	26–34	1.5	Ground	Reversed bending	Chilled iron	P-28	0.018–0.026A
Truck axle shaft	Steel	24–32	1.75	As machined	Reversed torsion	Chilled iron	P-39	0.006C
Splined shaft	Steel	45–52	1.25	As machined	Reversed torsion	Chilled iron	P-23	0.028A
Universal joint	Steel *	60	0.8	Ground	Reversed bending	Chilled iron	P-23	0.008C
Coil spring	Steel	52–53	0.3	As rolled	One-way torsion	Chilled iron	P-28	0.017A
Torsion bar spring	Steel	48	2.2	Ground	One-way torsion	Chilled iron	P-60	0.015C
Coil spring	Steel	57–59	0.15	As drawn	One-way torsion	Chilled iron	P-28	0.012A
Flat spring	Steel	44–48	0.020	Zinc plated	One-way bending	Chilled iron	P-28	0.005A
Flat spring	Phosphor bronze		0.008–0.010	As rolled	Reversed bending	Malleableized iron	P-17	Too low to measure
Specimen	Magnesium		0.063	As rolled	Reversed bending	Malleableized iron	P-39	0.005A
Supercharger part	Magnesium		0.150	As machined	Service	Malleableized iron	P-17	0.008–0.010A
Specimen	Malleable iron	131 BHN	0.9	As machined	One-way bending	Chilled iron	P-93	0.018–0.025C
Specimen	Alclad aluminum	105 BHN	0.04	As rolled	Tension tension	Malleableized iron	P-66	0.002A
Specimen	Arma steel	20	0.9	As machined	One-way bending	Chilled iron	P-93	0.024C
Specimen	Cast iron	18–20	0.9	As machined	One-way bending	Chilled iron	P-93	0.019–0.024C
Specimen	Nickel	21–22	0.3	Polished	Reversed bending	Malleableized iron	P-66	0.022A
Welded fan blades	Steel		0.4	As welded	Reversed bending	Chilled iron	P-19	0.015–0.018A

* Carburized.

Shape. The shape of the shot is important from the standpoint of uniformity of peening. However, it is not commercially possible to obtain only true spherical shot.

Soundness. A considerable percentage of normal chilled-iron shot is hollow. This usually fractures more readily than sound shot and is quickly eliminated when an adequate separator is used.

Size Range. The range of size of commercial screened shot has been specified by the Society of Automotive Engineers.

Mass. Shot for peening purposes may be made from material other than chilled iron or steel. For example, aluminum may be peened with glass shot, granules of plastic material, or walnut shell granules. These alternate materials avoid electrolytic corrosion, especially in magnesium, and the walnut shell and plastic granules also avoid abrasive contamination. The softer peening materials are not capable of inducing residual stress in steel; hence the standard peening intensity strip cannot be used to measure peening intensity. Aluminum or other soft metal strips, not yet standardized, will serve, however.

It is not possible to generalize with regard to peening intensity specifications because of the wide variety of machine parts made from many kinds of metals variously loaded and heat treated. As in specifying heat treatments, each case presents problems peculiar to itself. It will perhaps be helpful to indicate peening intensity specifications that have been reasonably successful when applied to typical machine parts. Such data are presented in Table 2.

SECTION 21

JOINING OF MATERIALS

By

C. W. OBERT, *Consulting Engineer, New York*

SAMUEL DAMON, *Westinghouse Electric Corporation*

THOMAS D. PERRY, *Engineer in Wood Work, Moorestown, N. J.*

WELDING OF METALS

By C. W. Obert

BRAZING

By Samuel Damon

ADHESIVES AND GLUES

By Thomas D. Perry

WELDING OF METALS

By C. W. Obert

1. TYPES OF WELDS

Recent developments and advances have given the welding art a dominant place in the metal industries. It has become common practice to shape and prepare metal parts and then join them by welds to produce engineering structures. It is the purpose of this chapter to classify the processes by which welds of various types are applied and to outline their limiting conditions as well as the effects that result therefrom; to refer to the important engineering materials that can be welded, the physical properties of the resulting welds, and the design requirements thereby influenced. Reference will also be made to the allied processes of surfacing and hard facing. Metal cutting by flame is discussed in Section 23.

A weld is defined as a localized coalescence of metals wherein coalescence is produced by heating to suitable temperature, with or without the application of pressure and with or without the use of filler metal. The welding heat may be derived from a number of different sources, and the processes of applying the heat are divided into two general groups: (1) methods in which the joint surfaces are heated to the temperature range in which the metal becomes sufficiently plastic and pressure is then applied; and (2) methods in which the joint surfaces are heated above the melting temperature of the base metal and fusion obtained without application of pressure (fusion welding). These two main divisions are further classified according to the source of the welding heat; for example, the plastic or pressure processes are subdivided into forge welding, oxyacetylene-pressure welding, electric-resistance welding, and thermit-pressure welding; the fusion welding processes are subdivided into gas welding, electric-arc welding, and thermit welding.

FORGE WELDING is the oldest form of welding. The parts to be joined are heated locally to the proper temperature range and then subjected to pressure by hammering or rolling, either by hand or by machine. This process is used for a wide variety of applications, from the miscellaneous work of the blacksmith to the joining of steel plates up to approximately 2 in. thick in the fabrication of boiler drums and other pressure vessels. Large tonnages of lap-welded and butt-welded tubes have been produced by passing heated formed skelp through pressure rolls. Forge welding has been largely superseded by more recent welding processes.

OXYACETYLENE-PRESSURE WELDING. There are two methods by which gas welding utilizes pressure to make the welding. In the first, carefully machined and cleaned ends of pieces are butted tightly together under pressure; then heat is applied. This application of pressure welding is characterized by the fact that the bond is formed by diffusion and recrystallization at a temperature lower than the melting point of the materials being joined. Closed-face pressure welding is the only process effecting the bond completely in the solid phase. The second method involves heating with flames impinging directly on the welding faces, which have been spaced a short distance apart. These faces have previously been made substantially square. When the ends of the pieces have been brought to the fusion point, the parts are rapidly joined under pressure to effect the weld. In both processes an upset is created by the shortening of the parts. No filler metal is added in either process.

In oxyacetylene-pressure welding, the strength of the joint can be made equal to that of the annealed base metal. The finished weld is homogeneous, and there is no evidence of the juncture except where steels of different alloy content have been joined. Dissimilar metals react differently to an etching reagent, and the area of the joint can be detected although no fine line of demarcation exists under microscopic inspection.

RESISTANCE WELDING. The various electric-resistance welding processes are in principle similar to forge welding in that the weld is made by application of pressure while the surfaces being joined are heated to plastic temperature or even to the melting point. All resistance-welding processes involve local heating of the parts to be joined by the passage of a heavy current across the joint. Heat is generated locally at the joint by the

---

Revision of material originally prepared by J. C. Dodge and R. E. Hiller.

greater electrical resistance of the circuit at the joint contact surfaces. Either direct or alternating current may be used, but generally it is most convenient to use alternating current, since the desired high amperage at low voltage is readily supplied through suitable transformers. Because of the high current required and the necessary "push-up" forces, the surfaces which may be joined by this method are limited. For a given cross section to be welded, the power requirements vary widely with the time interval of current application to produce the weld. For the usual resistance butt or flash weld, with low-carbon steel, requiring a heating period of several seconds, 30 kva or more are required per square inch of weld surface. As the time interval is decreased the power requirement increases; where welds are made in a fraction of a second, 1500 to 2000 kva per sq in. are not uncommon. The necessary pressure or push-up force applied after the joint surfaces have been heated to plastic temperature is approximately 5000 psi of weld surface.

Resistance-welding processes are particularly applicable to mass production because special equipment is required for each type of welding job.

Electrodes for introducing the welding current into the work are generally faced with a hard copper alloy to avoid excessive replacement, which would be necessary if soft copper alone were used.

Resistance Butt Welding. In resistance butt welding the parts to be joined are pressed into contact, and electric current is passed through the grips (electrodes) and across the junction of the two parts. When the temperature of the contact surfaces has reached the plastic range, pressure is applied, and upsetting of the plastic metal to a greater or less degree accompanies the making of the weld.

Electric-resistance flash welding is a modification of resistance butt welding and is carried out in much the same manner. Contact edges are slightly separated, however, and arcing or flashing occurs from edge to edge of joint. This arcing may burn or flash off a considerable amount of metal at the joint before the ends become sufficiently hot to form the weld by pressure. During the flashing period this loss in metal is compensated by feeding the two edges together to provide a uniform distance between them. Pressure is applied after flashing has heated up the contact surfaces to welding temperature. Flash welding is applicable to the welding of larger joint areas than resistance butt welding. This method also requires less care in the preparation of joint surfaces and smaller current densities for satisfactory welding than resistance butt welding. Often a hydrogen atmosphere is maintained at the weld to prevent oxidation of the joint surfaces.

Both butt and flash resistance welding processes are widely used for joining ends of wires, bars, and tubes. They have been more recently developed for producing tubing from flat strips by a continuous longitudinal butt weld as the formed strip passes through electrical contact rolls. In such tube welding processes, very high currents are utilized, the weld being effected at point of contact in a fraction of a second.

After completion of the weld, it may be heat-treated in the resistance welding machine, the current being used to bring the portion between the grips or electrodes to the desired heat-treating temperature.

Resistance spot welding is applied to the welding of overlapped metal sheets or other comparatively thin sections. The overlapped parts are held between electrodes, which apply pressure while electric current is maintained through them; the current is cut off before the pressure is released. The area welded is approximately the same size as the electrode tip, usually $^1/_8$ to $^1/_4$ in. in diameter. Spot welding is particularly applicable to fabrication of metal sheets up to $^1/_4$ in. thick, but it may also be applied to heavier sections. Automatic equipment makes possible the production of many welds per minute.

Shot welding is spot welding in which an extremely high welding current is applied for a short period of time. This method is particularly valuable for the welding of certain alloy steels such as nickel-chromium steels where time at welding temperature is harmful to the metal. In spite of the large welding currents employed, the total heat required is relatively small because of the small time interval of current application.

Resistance seam welding may be considered as a series of overlapping spot welds, the electrodes generally being in the form of rolls which also transmit pressure as the overlapping sheets pass between them. Automatic equipment for this type of welding generally utilizes a current interrupter which so times the current impulses with relation to the speed of travel of the sheets that a series of overlapping spot welds actually results.

Percussive welding processes are forms of resistance welding in which electrical energy is discharged suddenly at the junction, accompanied by the application of pressure. Closing of the discharge circuit often accompanies the bringing of the two parts into contact with pressure. The energy discharge may be obtained from a condenser, or by the transformation of the energy stored in a magnetic field by the collapse of the field.

THERMIT-PRESSURE WELDING. A modification of the thermit process is the thermit-pressure method of butt welding. This is applicable to joining such parts as rails, tubes, or bars which may be conveniently pressed together by end pressure. The surfaces to be joined are butted together, and the thermit slag and metal poured into the mold around the joint. The alumina slag produced in the thermit reaction is poured into the mold first and covers the surfaces of the parts. The superheated metal does not come into direct contact with the parts but heats them to the plastic temperature range. When the ends are heated, as judged by a suitable time interval, pressure is applied and produces what is essentially a forge weld. The thermit metal and slag may be removed later.

NONPRESSURE FUSION WELDING PROCESSES. These processes do not involve the union of plastic metals but utilize fusion of the metal to effect the desired union. The joining is accomplished by fusion of the surfaces of the parts. The fused metal is generally obtained by deposit of suitable filler metal which fuses with the base metal, although in some applications abutting or overlapping edges are merely fused together to form the joint.

GAS WELDING. In this fusion welding process the welding heat is obtained from a gas flame, oxygen and acetylene being generally the gases used. Fuel gases other than acetylene may be used in combination with oxygen. With the oxygen-acetylene combination a flame temperature of 5000 to 6000 F is obtained. By varying the proportions of oxygen and acetylene, a neutral, reducing, or oxidizing atmosphere may be obtained. For most purposes a neutral or slightly reducing atmosphere is used. The process is applied successfully to the welding of practically all metals. It is of particular advantage in making welds of any shape or position on comparatively thin sections, especially at sharp edges and corners. It is not, however, limited to thin sections, but may be satisfactorily used to produce welds in heavy metal parts. For joining thin parts, a filler metal may or may not be required, but for thicker sections a filler rod of suitable composition must be used. The reducing flame protects the molten metal from oxidation, but in the welding of some materials a flux is necessary to help remove oxides.

The fuel gas and the oxygen are mixed for combustion in a suitable torch which has regulating valves that allow a flame having the desired characteristics to be produced. The oxygen is generally supplied to the torch through hose connections from compressed gas cylinders, in which it is charged at pressures up to 2000 psi, so that pressure-reducing regulators are needed to control the pressure at the torch. The acetylene may be obtained from transportation cylinders or, if large quantities are required, it may be derived from generators that produce acetylene by reaction of calcium carbide and water.

ARC WELDING. In arc welding the welding heat is obtained from an electric arc formed between the base metal and an electrode, or between two electrodes. Welding current may be either alternating or direct. Depending on the character of the electrode, conventional arc welding is designated as either *carbon arc welding* or *metal arc welding*. In both processes the heat is highly localized and the temperature of the molten metal is considerably higher than in gas welding. Arc voltages vary from 15 to 45 volts, and the arc current from 20 to 600 amp, although some processes use current values as high as 2000 amp. The magnitude of the current that can be successfully used depends on the size of the electrodes, thickness of parts being welded, and rate of weld travel.

In arc welding a molten pool forms on the work at the arc location. By manipulation of the electrode the molten pool is made to travel along the joint as desired. Additional metal is introduced by a filler rod with the carbon arc process. In the metal arc process, the molten metal is supplied by melting from the tip of the electrode. Metal arc welding is more readily adaptable to a wider variety of applications than carbon arc welding.

The carbon arc method is always used with direct current, and for most purposes with the electrode negative in order to avoid excessive consumption of electrode and absorption of carbon in the weld metal. The procedure, essentially a puddling one, is not adapted, to welding in the vertical or overhead positions. Its best application is in automatic welding in the flat position, in which it is general practice to shield the arc and molten metal from the atmosphere by introducing a combustible material and fluxing agents into the arc. The method has had wide application for high-speed welding of thin sections in the fabrication of steel barrels, tanks, and light tubular products. It has also been used for welding heavy sections in pipe and pressure vessels.

The metal arc method carries metal across the arc from the electrode to the work in very small globules and deposits them in the molten pool on the work. The electrode is consumed, and the metal is deposited at a rate dependent on the current used. By proper choice of electrodes, either direct or alternating current may be used. The procedure is adaptable for welding of sections $^1/_{16}$ in. thick or more and may be carried out in flat, vertical, or overhead position. Best-quality welding is obtained by protecting metal in the arc from oxygen and nitrogen of the atmosphere by covering the electrodes with suitable fluxing materials and combustible agents. These exclude oxygen and nitrogen from the arc and give a slag covering to the pool of molten metal. A wide variety of electrodes are available for particular applications. The method is the most versatile and most widely used of the various fusion processes.

ARC-WELDING EQUIPMENT. Either alternating or direct current may be used for arc welding, depending on the type of electrodes. Carbon electrodes, bare wire, and covered electrodes with covering principally of combustible materials generally require direct current. Covered electrodes with covering principally of minerals are, in general, satisfactory for direct or alternating current. Alternating current has the advantage of minimizing arc disturbances created by magnetic fields, and is conveniently supplied from power lines through transformers.

Motor-generator Sets. The most satisfactory source of direct current is a motor-generator set. The most common type is a portable unit, supplying one arc. The generator is of the variable-voltage type, which adjusts its output voltage to the arc demands. A less common unit is a larger set, suitable for supplying several arcs. This type usually has a constant-voltage type generator to supply 70 to 100 volts. The voltage for individual arcs is adjusted by resistances in each line.

Alternating current is supplied from single-phase transformers, either with variable-voltage taps or with a variable inductance in each arc circuit to obtain desired current values. To obtain the necessary constant-current characteristics, inductance always is employed, although sometimes it is inherent in the transformer design. The magnitude of the constant current may be adjusted by varying the inductance.

The amount of current required for each size of electrode will vary considerably with the type of coating and the mass of metal being welded. Approximate values are listed in Table 1. These values are for usual manual or automatic welding procedures. Some

Table 1. Approximate Current for Various Electrodes for Welding Steel

Electrode Diameter, in.	Current Required, Amperes		
	Carbon	Bare Wire	Heavy-coated Electrodes
3/32	....	50–80	60–100
1/8	....	75–125	80–150
5/32	40	100–175	125–200
3/16	75	150–225	150–250
1/4	100–200	200–300	225–400
5/16	200–400	....	275–500
3/8	300–500	....	350–700
1/2	400–700	....	....

special methods use currents as high as 1500 to 2000 amp with metal electrodes of 1/2 in. diameter or larger. One such method uses a bare electrode and keeps the arc completely buried in powdered flux.

Electrodes. All important metal arc welds now are made with covered electrodes because of the greater strength and ductility of the weld metal as compared to that obtained from bare wire. A large variety of materials is used for covering electrodes, but coverings may be classed as those consisting principally of combustible material, those consisting principally of mineral substances, and those using a combination of combustible and mineral materials. The combustible type gives protection to the arc largely by producing a gaseous envelope about the arc which excludes oxygen and nitrogen. The mineral-coated types give protection by forming a slag coat on the individual globules of metal as they are transferred through the arc and providing a heavy slag layer over the weld metal pool. The action of this slag may be considered similar to that of slag in steel-melting furnaces. The coverings of electrodes also aid in stabilizing the arc and, to a very large degree, determine the welding characteristics of the electrode. Electrode coverings also may be used to some extent for introducing alloying elements into the weld metal.

ATOMIC-HYDROGEN WELDING is actually a combination of both arc welding and gas welding. A stream of hydrogen is made to pass through an electric arc established between two tungsten electrodes. The heat of the arc causes disassociation of the molecular hydrogen to the atomic form. After passing through the arc the atomic hydrogen recombines into the molecular form, liberating the heat obtained originally from the arc. The atomic-hydrogen flame is used in fundamentally the same manner as the oxyacetylene flame. This process is of particular advantage for welding along edges and corners of thin sections. Its uses include welding stainless steels. The hydrogen flame gives excellent protection to the readily oxidizable chromium in these steels. Filler metal of suitable composition may be melted in the atomic-hydrogen flame as required.

INERT GAS-SHIELDED ARC WELDING. This is an electric-arc welding process that employs a blanket of monatomic inert gas to shield the weld area. Highly concentrated heat is produced by an arc drawn between a single, virtually nonconsumable tungsten electrode and the work piece. The inert gas (argon or helium) is fed through a nozzle surrounding the electrode in the head of the torch and flows out to blanket the electrode completely, as well as the arc and weld puddle, for the total exclusion of the oxidizing atmosphere.

The process is used successfully for welding magnesium, aluminum, stainless steel, high-alloy steel, nickel and nickel alloys, brass, silicon bronze, pure silver, and other metals. The outstanding characteristic is that all these metals can be welded without the use of flux. Because no flux is needed, corrosion from entrapment of flux in the weld is eliminated.

The necessary equipment consists of a torch used to hold the electrode and to provide a means of getting the helium or argon around the weld; a tungsten electrode; a supply of inert gas, governed by a regulator and measured by a low-pressure flowmeter; and a source of electric current.

Direct current of either polarity or alternating current may be used; the choice depends somewhat on the alloy and the thickness. Reverse-polarity direct current (electrode positive) has been used for welding magnesium, but many fabricators prefer alternating current because of its greater penetration, speed, and range of thickness capacity. Likewise, alternating current is almost always used for welding aluminum since straight-polarity direct current does not produce a satisfactory weld. For stainless-steel alloy steels and some copper alloys, either alternating current or direct current may be used. Alternating current is generally preferred for light material because of better arc stability. Direct current is favored for heavier material because of the deeper and narrower weld beads.

SUBMERGED-ARC WELDING PROCESS. This is a method of welding electrically beneath a protective layer of granular and molten mineral material. The process is used for welding virtually all steels, as well as other ferrous and nonferrous metals and alloys. The granular material is usually laid down directly ahead of the welding zone. In the welding zone, the granular material fuses and protects the weld from the effects of the atmosphere. Since the atmosphere is excluded from the weld, bare welding wire can be fed directly into the welding zone. There is no flash, glare, or spatter. The molten slag also removes impurities from the weld, and as it cools it solidifies, forming a protective cover over the weld. When cool, the fused material cracks off or can be removed easily. The welds produced are dense and uniform, and are as strong as the parent metal, or stronger.

The use of the process is limited only by the capacity of the equipment. With present equipment, material ranging from 22-gage sheet to 3-in. plate can be welded in a single pass. Material of any thickness can be welded by a suitable number of passes.

Speeds of about 200 in. per min can be used on thin material. The currents required vary with the material and are usually about six to ten times those used with the manual electric-arc welding methods. A maximum of about 4000 amp can be applied with the present equipment. This would be approximately the current required to weld plates 3 in. thick in a single pass.

Because of the nature of the process, welds must be made in a horizontal or nearly horizontal position. Such welds can be made without special plate-edge preparation in material less than $^1/_2$ in. thick. Thicker material is usually prepared with a single or double V. Single-pass butt welds are usually backed up by a copper strip or a layer of granular material pressed against the back of the weld. With this process, butt, fillet (positioned and nonpositioned), and plug welds can be made.

THERMIT WELDING. In the thermit welding process, superheated steel is obtained by the exothermic reaction which results when a mixture of iron oxide and aluminum is ignited. The iron oxide is reduced by the aluminum to iron. The resulting superheated steel is poured into a mold around the joint to be made, mold and joint having been preheated. The superheated steel resulting from the iron oxide–aluminum reaction has a temperature far in excess of the melting point of steel, and hence there is sufficient heat to produce fusion with the surfaces of the parts to be joined. The composition of the metal may be varied by the introduction of alloying elements. The process has been applied to all types of wrought or cast steel and iron. It is particularly applicable to repair of heavy sections and for making butt welds in rails.

2. WELDING OF VARIOUS MATERIALS

STEELS. General. The difficulty which may be encountered in the welding of various steels is, in general, in proportion to the hardening developed in the heat-affected zone along the weld. With high hardness, underbead cracking may result from welding stresses.

As the carbon content of steel increases, the potential hardness increases, and the addition of alloying elements in general emphasizes the hardening effect.

Excessive hardening can be somewhat controlled by regulating the cooling rate from the welding temperature. With thermit welding and thermit-pressure welding, a slow cooling rate can be obtained by the insulating effect of the mold surrounding the entire weld. With gas welding, particularly on thin sections, the torch can be used to heat the base materials and keep the zone adjacent to the weld sufficiently hot to retard cooling. With arc welding the area of high temperature is very localized, and preheating of the parts is sometimes necessary to obtain slow cooling and to reduce welding stresses. With a composition giving high hardness, more difficulty is likely to occur with heavy sections, and preheating will be necessary in whatever welding method is used. Heating to 400 to 500 F is effective, and, depending on the steel and the particular structure, this temperature may need to be maintained until the weld has been post heat treated.

Variation in the quality of steel may result in welding difficulties even for grades normally considered weldable, such as plain carbon steels below 0.30% carbon. Fortunately, such difficulties are exceptional. No definite information on the reasons for such difficulties is available, although it is generally conceded that trouble may occur because of excessive segregation in the steel or because of the presence of abnormally large amounts of residual elements such as arsenic and copper. Laminations in rolled steel products will, of course, produce welding difficulties, especially if the lamination extends along the joint edge.

CASTINGS. Castings may be satisfactorily welded, provided rolled or forged steel parts of the same composition can be satisfactorily welded. This assumes that the casting adjacent to the weld region is sound and free from shrinkage cracks which otherwise would open under the contraction stresses during cooling from welding.

IRON AND CARBON STEELS UP TO 0.35% CARBON. Iron and plain carbon steels up to 0.35% carbon are readily welded by all processes, and completed welds can be obtained with strength equivalent to the base materials. Preheating is not necessary, except to reduce cracking in extremely rigid joints. In fusion welding, with gas or electric arc, filler metal of approximately 0.15% carbon is recommended. With shielded-arc methods (carbon or metal arc), ductile welds with strength of 60,000 to 70,000 psi are obtained. This strength can be increased to 70,000 to 80,000 psi, without loss of ductility, by small additions of alloying elements. Molybdenum in amounts of 0.2 to 0.4% is very effective.

Where stress relieving is required, heating at 1100 to 1200 F for a period of one hour per inch of thickness is recommended.

MEDIUM-CARBON STEELS OF 0.35 TO 0.60% CARBON. As the carbon content of steel increases above 0.35%, difficulty in welding is likely to develop. With thin sections and a flexible design, gas or arc welding may be used without preheating, but in joints over $^3/_8$ to $^1/_2$ in. thick, preheating is necessary. In oxygen-acetylene gas welding, better welds are obtained with a flame having excess acetylene. To develop equivalent strength in the weld, a low-alloy filler metal should be used for both gas and arc welding methods. A rod containing chromium and molybdenum is recommended. These steels are readily welded by the thermit process.

Welds should be stress relieved at 1100 to 1200 F and, where preheating has been necessary or with the higher carbon ranges, this stress relieving should be carried out before permitting the weld to cool below 300 F.

HIGH-CARBON STEELS OF 0.60 TO 1.50% CARBON. Steels above 0.60% carbon may be welded quite satisfactorily by the thermit process, a method that has been used considerably for butt welds in rails. Gas and arc welding can be done only with considerable difficulty, and the resulting welds are only fairly satisfactory. Gas welding is more easily performed. A joint can be made by arc welding by depositing low-carbon metal over the surface of each part and then making the arc weld between these low-carbon surfaces. The resulting weld, of course, has the properties of a low-carbon weld.

LOW-ALLOY HIGH-YIELD-STRENGTH STRUCTURAL STEELS. There are a considerable number of low-alloy steels especially designed to produce yield-point values of 50,000 to 70,000 psi and with ultimate strengths of 70,000 to 100,000 psi. Most of these steels have been produced with a low carbon content, and no particular difficulty is encountered in welding them. In general, where carbon contents are 0.20% or less, gas or arc welding can be performed without preheating. With higher carbon contents, preheating will often be necessary. To develop weld strengths equivalent to the base material, it is necessary to use filler metal with some alloy content. Low-carbon filler metal with alloy composition similar to the base materials usually can be used. Stress relieving should be carried out at 1100 to 1200 F.

SAE ALLOY STEELS. For welding of the SAE alloy steels, the same general procedure may be used as for the alloy structural steels. The ease of welding will depend on the hardness developed in the heated zone. Where high hardness occurs, preheating to 300 to 500 F will be necessary. Filler metal of similar alloy composition to the base material usually can be used.

CHROMIUM STEELS OF 4 TO 6% CHROMIUM are definitely air hardening, even with low carbon content, and preheating at 300 to 500 F is essential. Welds generally must be kept above 300 F until annealed or stress-relieved. Heat treatment after welding should be a full anneal or long-time stress relieving at 1300 to 1350 F. The same procedure should be used with steels of 2 1/2% chromium. With such procedure, welds in these steels with carbon less than 0.20% are very ductile and equal to the base metal in every respect.

CORROSION- AND HEAT-RESISTING STEELS have high percentages of chromium or chromium and nickel, with low carbon content. To weld steels with 12 to 30% chromium requires preheating to above 200 F and an annealing treatment after welding in order to obtain ductility. The lower chromium ranges, 12 to 16% chromium, may be welded without great difficulty if preheating precautions are taken. As the chromium content is increased, the welds are much less ductile, and the very high chromium range becomes very difficult to weld. Heat treatment should consist of annealing at 1450 to 1550 F. To maintain equivalent chromium content in the welds, filler metal should have a slightly higher chromium content than the base metal. Because of rapid carbon absorption in high-chromium steels, the carbon arc method is not suitable.

The stainless steels containing high percentages of both chromium and nickel are not subject to hardening from heat treatment and may be welded without difficulty. Typical steels of this group are: 18% Cr–8% Ni, 25% Cr–12% Ni, 25% Cr–20% Ni; there are many other variations of analysis. Welds in all these materials are soft and ductile, and no heat treatment after welding is required, except as may be needed to improve corrosion resistance. In this respect the 18% Cr–8% Ni type is subject to serious loss of corrosion resistance, due to precipitation of carbides in the temperature range of 900 to 1500 F. For this reason, very low carbon contents (less than 0.07%) are desirable for these steels if they are to be welded. The carbides may be stabilized by addition of titanium or columbium, and steels so stabilized are recommended for welding. Filler metal also should be of the stabilized analysis. Titanium should be used in amounts at least six times the carbon content, and columbium in amounts at least ten times the carbon content. If the 18% Cr–8% Ni steels are not of the stabilized type, heat treatment after welding should consist of heating to 1900 to 2100 F, followed by rapid cooling. In the higher alloy steels, carbide precipitation is not serious.

Because of higher electrical resistance and lower heat conductivity, electric welding of these steels requires about 10% lower current than carbon steels. Also, the high coefficient of expansion requires additional care to prevent excessive distortion.

MANGANESE AUSTENITIC STEEL with 12 to 14% manganese and approximately 1 to 1 1/2% carbon is quite brittle in the as-cast condition, because of large carbide areas, but upon heating to 1900 F and quenching, the carbides are retained in solid solution and a tough ductile material results. On reheating to 1100 to 1400 F, carbides are again precipitated and the material becomes hard and brittle. In welding this steel it is necessary to deposit weld metal which is ductile and to avoid embrittlement of the base material by excessive heating.

Welding may be accomplished by either the gas or arc methods. A filler metal of 12 to 14% manganese and 3 to 5% nickel is recommended to obtain a good-quality weld. Care should be taken to avoid excessive heating of the base metal.

CAST IRON. Cast irons, either gray, white, or malleable, are not considered adaptable to welding, but may be welded with proper care and procedure. White cast iron is the most easily welded. The process may be the gas or metal arc, using a white cast-iron filler rod. Resistance welding may be used where shape permits. Gray or malleable castings may be fusion welded with proper precautions, and the same general procedure can be applied for both types of castings. Whatever method is used, because of the low tensile strength and poor ductility of these materials, it is essential to avoid excessive variations in temperature throughout the casting to prevent cracking from thermal stresses.

Gas welding is best performed by preheating the casting to approximately 1000 F and completing the weld while it is kept at that heat. After welding, the casting must be cooled slowly. A suitable flux such as borax should be used, and the filler metal should be of a high-silicon cast-iron composition. Welds made in this manner will be machinable.

Carbon arc welding can be performed in a manner similar to that used for gas welding. Use of cast-iron filler rod and borax flux is recommended.

Metallic arc welding may be performed using either coated cast-iron composition electrodes or coated steel electrodes. Preheating is necessary in using the cast-iron electrodes, and this method will produce a machinable weld. The method is best adapted for welding in the flat position. Using coated steel electrodes, welds may be made without preheating the casting, but a hard unmachinable zone will be present. Small-diameter electrodes ($^1/_8$ in. maximum) should be used, and welding should progress slowly with intermittent peening. This will keep thermal stresses in the casting, and contraction stresses in the weld metal, at a minimum. Welds made with steel electrodes will be considerably stronger than the cast-iron base metal. Welds not so strong, but very satisfactory and machinable, may be made, using an alloy electrode of high nickel composition such as Monel metal (68% Ni, 29% Cu). Similarly, welds may be made with copper or copper alloy filler metal.

ALUMINUM ALLOYS. Gas, elecric-arc, or electric-resistance welding may be used for joining aluminum alloys. Alloys suitable for welding are discussed in Section 4, Article 7.

MAGNESIUM ALLOYS. Details of welding techniques for magnesium alloys are discussed in Section 4, Article 15.

3. MECHANICAL PROPERTIES OF FUSION WELDS IN STEEL

CONDITIONS FOR SOUND WELDS. The physical properties of fusion welds depend, of course, on the relative freedom of the weld metal from such defects as porosity, entrapped slag, cracks, and incomplete fusion. Assuming that a sound weld has been obtained, physical properties depend on the degree of protection which has been given to the molten metal from the gases of the atmosphere. Avoidance of contamination by nitrogen and oxygen is especially important in metal arc welding. Welds deposited with bare wire electrodes with no arc protection will contain approximately 0.10 to 0.15% nitrogen and approximately 0.3% oxygen. With a satisfactorily designed covered electrode, the deposited weld metal will contain less than 0.03% nitrogen, with oxygen correspondingly low.

The tensile strength of steel welds may be considerably increased by addition of suitable alloying elements, but in all cases, for a given composition, superior properties are obtained in metal arc welding by using covered electrodes.

SOUNDNESS OF FUSION WELDS. Producing welds by the gas or arc fusion processes introduces many variables of technique and operator skill which may cause mechanical defects in the weld. These may be gas pockets, entrapped slag from the electrode covering, zones of incomplete fusion between weld metal and base material, or cracks resulting from contraction stresses.

Several methods have been utilized for inspection of welds to detect such defects. The common medical stethoscope has been used to a limited extent, by detecting variation in sound issued at positions along the welds when struck by a hammer. This method is not very sensitive and is limited to welds up to about 1 in. thick. Magnetic methods have been used quite successfully in magnetic materials, to detect cracks at or near the surface of welds. They have not proved adaptable to detect defects far below the surface. The simplest of the magnetic methods consists of establishing a magnetic field approximately perpendicular to suspected defects and dusting a magnetic powder over the surface. The powder tends to concentrate at locations of fissures or cracks. The magnetic method may be used to advantage in some cases to supplement radiographic methods for the detection of small surface cracks which cannot be disclosed readily by the x-ray method. The magnetic method also may be used to advantage for examining welds which cannot for one reason or another be x-rayed.

The most thorough and reliable method of inspection is x-ray examination of the complete weld. The x-ray tube is placed on one side of the weld and photographic film on the opposite side. Radiographs of excellent quality can be obtained which register the presence of all types of defects and give a clear picture of their size, location, and distribution. This method may be used for examination of steel sections 4 $^1/_2$ to 5 in. thick, and will detect defective areas that are in thickness equal to 2% or more of the steel section. Gamma rays emanating from radioactive materials may be used in the same manner as x-rays. They are particularly applicable to heavy sections and to use in locations where x-ray equipment cannot be placed. Use of both x-rays and gamma rays requires suitable precautions to avoid excessive exposure of personnel to the radiation. See X-ray protection (*U. S. Bureau of Standards Handbook* HB20) for recommended practice.

STRESSES IN WELDS. The thermal contraction in a welded structure produces residual stresses in the welds. If the structure is sufficiently rigid, the stresses may become triaxial and of high magnitude. The residual welding stresses are somewhat reduced by preheating and may be virtually eliminated by postheating the entire structure to 1100 to 1200 F (if steel) for one hour per inch of thickness. If it is possible to heat only part of the structure at a time, the temperature gradient between heated and unheated portions should not exceed 200 F per foot.

The postheat treatment not only reduces the residual stresses but also produces desirable metallurgical changes, particularly in the heat-affected areas, which improve the mechanical performance of the welded structure.

Tendency to crack, particularly when welding a highly restrained structure, can sometimes be reduced by preheating or peening the intermediate passes.

Shrinkage of the welded structure may produce distortion. By control of procedure, distortion may be reduced considerably. Peening the weld metal also is of assistance in controlling distortion.

4. WELDING DESIGN

ADVANTAGES OF WELDED CONSTRUCTION. The design and assembly of a structure with welded connections offer a number of advantages as compared to mechanical joints, such as riveting. The connections are simplified in that the joint generally is made direct, and no third member, as a butt strap, is required. Welded joints permit placing the added metal exactly where needed, and produce an integral structure which, with mild steel, can develop 100% joint efficiency on any thickness of section. Elimination of extra material results in a weight saving of up to 25%. This saving in weight is an important consideration in transportation fields.

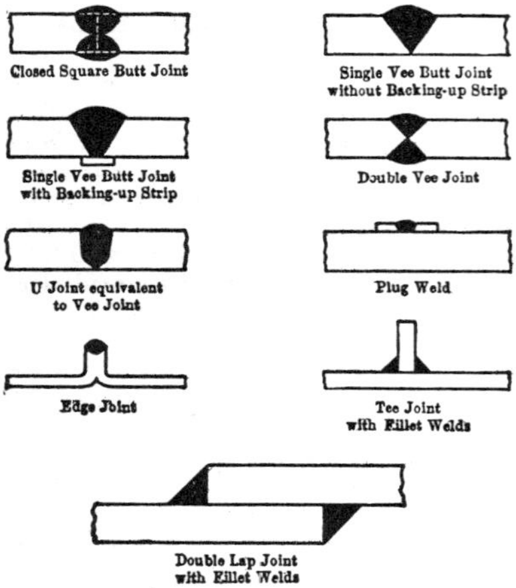

FIG. 1. Forms of joints for fusion welding.

FORMS OF WELDED JOINTS. Many forms of joints are used in fusion welding processes. Some of the more widely used are shown in Fig. 1. Drawings of welded construction should contain full information on weld details, type of joint, dimensions of welds, etc. The American Welding Society has adopted welding symbols which may be conveniently used on drawings specifying welding. In designing a welded structure, consideration also should be given to the following points:

(1) The joint should be made as accessible as possible to the welding operator. This means that wherever possible welding should be done in the downhand position rather than taxing the skill of the operator by welding in an overhead position.

(2) The dimensions of the joint should be such as to permit complete fusion through the thickness of the sections being welded. If incomplete fusion is obtained, an undesirable stress concentration will result. (See Fig. 2.)

(3) Butt joints are preferable from the viewpoint of uniform stress distribution across the weld. In the case of fillet welds a 30–60-degree weld provides a more uniform transfer of stress through the weld than a 45–45-degree weld.

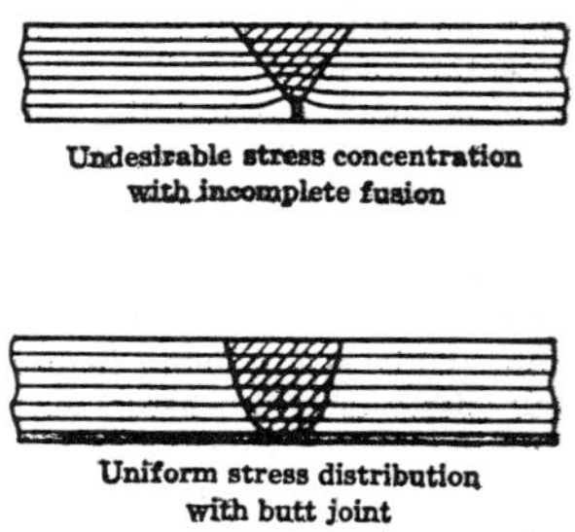

FIG. 2. Stress distribution in welds with incomplete and complete fusion.

(4) Where the structure previously has been a casting or built by riveting, and where the structure is being redesigned for welded construction, it should be appreciated that welding cannot be used interchangeably with the older riveted or casting design, and the designer should adapt his structure as far as possible to utilize the full advantages of welded construction.

(5) Where distortion of the welded structure by reason of the contraction of the weld metal is likely to occur, and where the distortion is objectionable, the welds should, if possible, be so positioned that the contraction of the welds will tend to balance each other on each side of some central plane through the structure.

The design stresses will, of course, vary, depending on the quality of the welds and the conditions of service of the structure. For example, pressure vessels may be built on the basis of factors of safety of 4 or 5, corresponding to an allowable working stress of about 10,000 to 12,000 psi (55,000 psi tensile strength steel), while an open storage tank could reasonably employ much higher stresses. The allowable working stresses for welds of various types have been set by engineering codes governing different types of construction.

A list of the commonly used construction codes is given below:

ASME Boiler Construction Code, Section I, Power Boilers.
ASME Boiler Construction Code, Section III, Locomotive Boilers.
ASME Boiler Construction Code, Section IV, Heating Boilers.
ASME Boiler Construction Code, Section V, Miniature Boilers.
ASME Boiler Construction Code, Section VIII, Unfired Pressure Vessels.
ASME Boiler Construction Code, Section IX, Qualification of Welding Operators.
API-ASME Code for Unfired Pressure Vessels.
Office of Merchant Marine Safety, U. S. Coast Guard.
Bureau of Ships, U. S. Navy Department.
ASA Pressure Piping Code.
National Board Rules for Repairs by Fusion Welding.

Some of the constructional codes issued by the American Welding Society are: Code for Fusion Welding and Gas Cutting in Building Construction (*J. Am. Welding Soc.*, March 1935); Specification for the Design, Construction, Alteration and Repair of Railway and Highway Bridges; Tentative Code for Fusion Welding and Flame Cutting in Machinery Construction; Marine Code for Welding and Gas Cutting, Part D, Rules for the Fusion Welding of Hulls and Hull Parts (*J. Am. Welding Soc.*, Feb. 1935).

5. FACING OR SURFACING BY WELDING

Facing or hard-surfacing is a process of welding onto a wearing part a layer, edge, or point of a metal that is resistant to abrasion or corrosion. Hard-facing has proved to be an easy and economical means of keeping equipment in good operating condition. There are also soft-surfacing operations—for example, the application of a bronze bearing metal.

HARD-FACING. Hard-facing, applied originally to metal-working dies, oil-well drilling tools, and excavating equipment has numerous other applications. No single grade of hard-facing material is suitable for all applications. To secure hardness, toughness, resistance to shock, resistance to corrosion, and other desired qualities, alloys of widely different compositions have been developed. They may be classified in five groups.

(1) Alloy Steels That Contain Less Than 20% of Alloying Elements. Alloying elements used are mainly chromium, tungsten, manganese, silicon, and carbon. Although not so hard as higher-priced hard-facing metals, alloy steels of this group possess greater toughness and resistance to shock. They have much greater resistance to wear than low-carbon steels, and some compare favorably with Hadfield's high-manganese steel in this respect.

(2) Ferrous Alloys That Contain More Than 20% of Alloying Elements. Chromium, tungsten, manganese, silicon, carbon, and sometimes cobalt, nickel, and other elements are used in these alloys. High hardness and greater resistance to abrasion insure longer service for alloys of this group. As a result, they are not so tough or resistant to shock as alloys of Group 1, but are particularly well adapted for use as the final wear-resisting surface, after the worn part has been built up with high-strength welding rod.

(3) Nonferrous Alloys. Cobalt-chromium-tungsten alloys are available in several grades, all highly resistant to wear and corrosion. Many of them retain their original hardness virtually unimpaired at elevated temperatures.

A newer development is the use of nickel-base alloys (Hastelloys) for the facing of parts subjected to severe chemical corrosion. This type of alloy combines outstanding resistance to corrosion with excellent mechanical properties, some grades being comparable to high-strength steels in tensile properties. Some grades will also take a high polish and retain it even when subjected to severe atmospheric and chemical corrosion.

(4) So-called Diamond Substitutes. These are the hardest and most wear resistant of all hard-facing materials. Some are almost pure tungsten carbide; others contain 90 to 95% tungsten carbide with the remainder cobalt, nickel, iron, or similar elements. Some materials contain boron compounds or tantalum carbide. The alloying elements are added to impart toughness and impact strength to the tungsten carbides; the hardness, however, is slightly reduced as a result.

The diamond substitutes, furnished in the form of small castings, are embedded in the cutting and wearing surfaces of oil-well drilling bits and similar tools. Tungsten carbide inserts are not melted or damaged to any appreciable extent in the usual method of application by the oxyacetylene process. The process of applying tungsten carbide shapes to wearing surfaces is known as *hard-setting*. After hard-setting, the entire wearing surface usually is hard-faced, which produces a composite material that has excellent resistance to abrasion with maximum cutting qualities at edges.

(5) Crushed Tungsten Carbides of Various Screen Sizes. These are furnished packed in steel tubes, fused to steel strips, or bonded with steel as a composite rod. Both tubes and strips come in short lengths, which are applied like ordinary welding rods. Grains of crushed tungsten carbide are also supplied loose for application with a Haystellite powder dispenser.

HARD-FACING PROCEDURE. Although hard-facing materials can be applied by any standard fusion-welding method, certain factors govern the choice of method for any specific applications: (1) base material to be faced, (2) hard-facing material selected, (3) size, shape, and thickness of base material and deposited facing materials, (4) initial cost of equipment and facing material, (5) smoothness of deposits, (6) overall cost, and (7) quality and appearance of deposits.

Use of the oxyacetylene blowpipe allows close control of the operation, produces a smooth deposit, insures easy elimination of particles of scale and foreign matter and easy forming of edges and corners. Although arc welding is not so flexible as gas welding, the resultant deposits are of good quality. The metal arc welding process is particularly suitable for use on thick base metals where thin, smooth deposits are not required. Another advantage is that only a small amount of distortion is produced in hard-facing large steel castings and similar heavy parts because the heat from the arc is localized.

Loose scale, dirt, rust, or other foreign substances should be removed from the area to be hard-faced, preferably by grinding or machining. When the hard-facing material is to be applied in a groove, corner, or recess, it is desirable to have all corners well rounded. If only a small area is to be hard-faced, a suitable shallow recess or indentation may be prepared for receiving the hard-facing metal.

THICKNESS OF HARD-FACING DEPOSIT. The proper thickness of deposit depends on the specific application, but usually ranges from $^1/_{16}$ to $^1/_4$ in.; the average is approximately $^1/_8$ in. Usually, however, a part that requires a deposit thicker than $^1/_4$ in. should be rebuilt before hard-facing to within $^1/_{16}$ to $^1/_4$ in. of the finished size. For this purpose an alloy of Group 1 should be used, or material similar to the base metal. The final hard-facing deposit is applied thick enough for grinding to finished size.

BRAZING

By Samuel Damon

Brazing is the joining of similar or dissimilar metals with a filler metal or alloy of lower melting point than that of the metals joined. Sometimes referred to as hard soldering, brazing involves joining or filler metal having a melting point in excess of 1100 F. Soft soldering involves filler metal which melts below 600 F (see p. 4-84).

Essential requirements of a brazing metal or alloy are high fluidity and ability to "wet" or alloy with the metals to be joined when at the brazing temperature. In respect to wetting, both the metals joined and the molten brazing material must be free from oxides upon attaining the brazing temperature. The accomplishment of this requires (1) a reducing atmosphere; (2) the addition of flux to the joint; or (3) the exclusion of oxygen during heating by immersion in a molten bath.

Copper brazing, a comparative newcomer to the field, is limited largely to furnaces containing controlled reducing atmospheres. However, recent developments in high-temperature salt baths have made available a new heating medium; in some cases induction heating has been applied using either a flux or a reducing atmosphere to prevent oxidation.

Silver brazing, in which all types of heating are applicable, is widely utilized. Flux is employed in every case unless a protective atmosphere, such as is used in furnace brazing, makes possible the complete prevention of oxide formation.

GENERAL CONSIDERATIONS. Brazing has many advantages over other joining methods for certain applications. High strength, high resistance to vibration and impact, and a strong bond in high-temperature service are among these advantages. This is particularly true of copper-brazed assemblies. Because of the slight clearances used and the deep penetration into the joint obtained with these alloys, the production of pressure-tight joints is assured. Usually a mere visual examination after brazing is sufficient to establish this.

Concomitant properties of such joints are high thermal and electrical conductivity. Distortion and residual stress build-up due to the high temperatures obtained locally, as occurs in welding, is not a serious drawback in torch or induction brazing because of the comparatively low temperatures involved. Where methods such as furnace heating are utilized, with attendant uniform heating and cooling, distortion and stress concentrations are practically eliminated.

Owing to the feasibility of building up brazed shapes by proper design, many new possibilities in fabrication are opened. This is especially true in furnace brazing where a number of joints can be brazed simultaneously.

A further advantage which may be gained from using furnace or salt-bath brazing is the bonding and simultaneous hardening of steels, especially those of the air-hardening type. This becomes particularly important if the treatment is carried out in an atmosphere or bath designed to prevent decarburization, as the part can then be completely machined and assembled before brazing and can emerge after brazing as a completely finished and hardened tool ready for use, with only tempering required. An additional factor favoring the use of protective atmospheres is the clean, bright surface resulting. Much saving in cleaning and pickling expense is thus assured.

Perhaps the most distinct benefit of the brazing technique, however, is the high production rate inherently possible. Where mass production is a necessity, brazing, especially by furnace or induction heating, has become an essential tool of industry.

6. BRAZING METHODS

TORCH BRAZING. Hand or torch brazing, the best-known method, is widely employed in industry. The work is assembled after cleaning and fluxing and, if necessary, is supported in a fixture. Heating is accomplished by playing a torch flame, which may be oxyacetylene, oxyhydrogen, or even air-acetylene or air-gas, maintained on the reducing side, onto the assembly at a point near the joint to be bonded and, where possible, on the more massive member or the material of higher heat conductivity in the assembly. The torch flame is moved constantly to prevent local overheating and is not allowed to impinge too long upon the flux or silver solder. Usually insulating brick is used on the work table to retain the heat around the work, thus enabling more rapid heating. When the joint area is at temperature as judged by eye, the brazing alloy is fed into the joint. The alloy melts and is drawn into the opening by capillary attraction until the joint itself is completely filled, after which the molten alloy forms a fillet on the outside of the joint.

This method requires skilled operators, and although slow as compared to other brazing methods it is widely used because of the low cost of the equipment and its versatility.

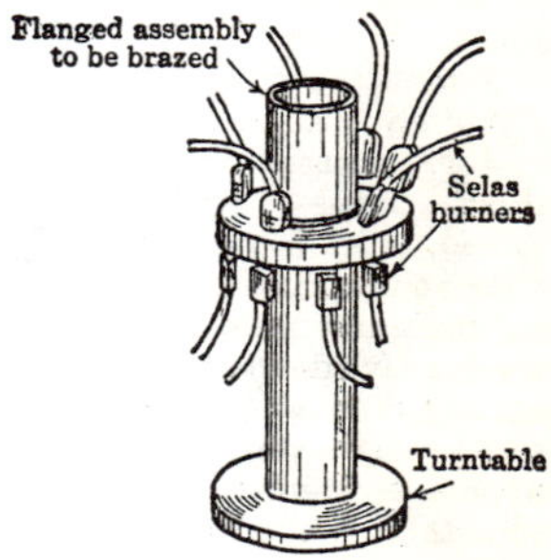

FIG. 1. Typical setup for brazing with gas burners.

GAS BURNER BRAZING. Arrangements for automatic torch brazing have been introduced which utilize gas burners (Fig. 1) set in suitable relationship to the work. The work to be brazed is held on a conveyor and carried into and held within the range of the flames from the burners for a predetermined interval of time. The work then moves away from the hot burners and the next set-up moves into the brazing position. For circular joints the fixture may be one that rotates the work while it is in the hot zone until brazing is accomplished and then moves it out to cool. This technique enables high production since the rate of heating with these gas burners is high, and, once the positioning of the burners and the time cycle have been determined, the brazing operation can be performed automatically with unskilled personnel handling the set-up of the assemblies.

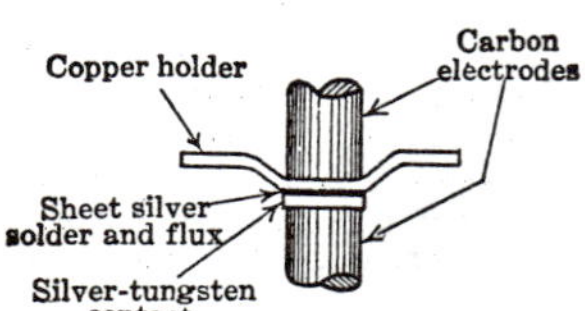

FIG. 2. Method used for silver brazing contacts by resistance heating.

RESISTANCE BRAZING. As indicated by the name, direct electrical resistance heating such as obtained in a spot or butt welder is used to join the parts. The members are placed between the carbon or molybdenum electrodes of the welding machine with brazing alloy and flux in the joint. The current is then applied until the filler metal melts and flows. (See Fig. 2.)

This method can be very rapid but is limited to assemblies which lend themselves to mounting in such a machine.

CARBON RESISTANCE BRAZING is similar to the electrical resistance brazing method except that carbon electrodes are heated by the current, and the heat is transferred to the work by direct conduction and some radiation. The work is similarly set up, with the brazing material usually in foil form inside the joint. The current is then applied until the filler metal flows freely. In this and the resistance brazing method, pressure as well as heat is applied to the joint to make for good cohesion.

SALT-BATH BRAZING. Molten salt baths, usually neutral in character (although fluxing agents can be included to dissolve oxides), are well suited to brazing of assemblies which are of appropriate dimensions and shape. The part is fluxed, if necessary, to prevent oxidation, and the silver alloy is placed at or in the joints. After the flux is completely dry, to avoid the danger of an explosion if wet flux is introduced into the pot, the assembly in a suitable jig or hanger is immersed in the molten salt far enough to cover entirely the joint region. The salt bath is maintained at a temperature somewhat higher than the top of the melting range of the brazing material, and, therefore, the operator is assured that the alloy has fused and flowed into the joint when the assembly arrives at the bath temperature. (See Fig. 3.)

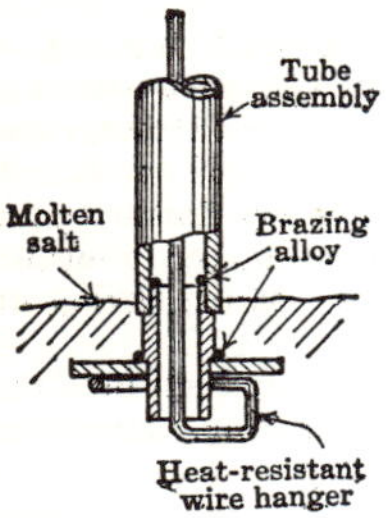

FIG. 3. Brazing in a molten salt bath.

This method has added advantages in that temperature is controlled and carburization or decarburization as well as oxidation can be completely eliminated, especially in steel pieces.

ATMOSPHERE FURNACE BRAZING. The most versatile of brazing methods, this fashion of joining metals is also ideally suited to mass production of brazed articles. The process is most widely applied in copper brazing and is also extensively utilized in silver solder brazing.

The equipment in the pusher-type furnace consists of a gastight furnace, with a cooling chamber, also gastight, separated from the heated region by an externally operated door, and with close-fitting doors at the front entrance into the hot zone and at the rear exit from the cooling chamber. (See Fig. 4.) In the conveyor-type furnaces, purging chambers at the front and rear are used, and doors are adjusted to allow just enough room for entrance of the particular assembly being joined.

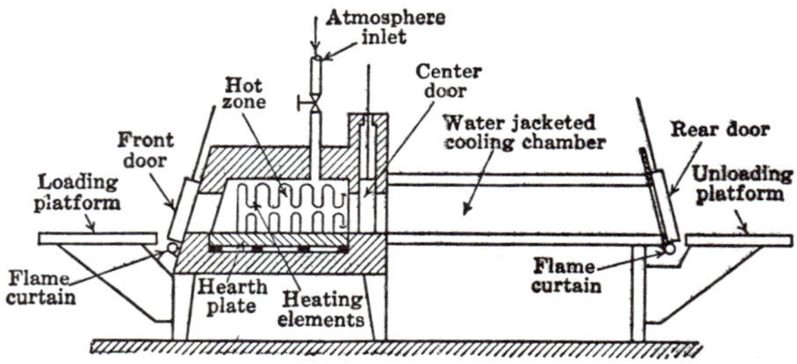

FIG. 4. Section through a pusher type furnace.

A reducing atmosphere produced in a special generator, or consisting of hydrogen or dissociated ammonia, is fed into the furnace and cooling chamber to maintain a constant pressure inside the furnace against the ingress of contaminating air or water vapor. The formation of oxides is inhibited and oxides present upon entering the hot zone are reduced, thus permitting the molten brazing metal to flow smoothly through the joint and uniformly wet the surfaces to be joined.

The same general technique previously described for the set-up of the articles to be brazed is pertinent. In copper brazing no flux is used unless the steels being joined happen to be high in alloys such as chromium, aluminum, or manganese. In silver brazing, flux is generally applied unless the components of the article consist of materials, such as copper, silver, or nickel, whose oxides are easily reduced in the furnace atmosphere, and the brazing alloy contains no zinc or cadmium.

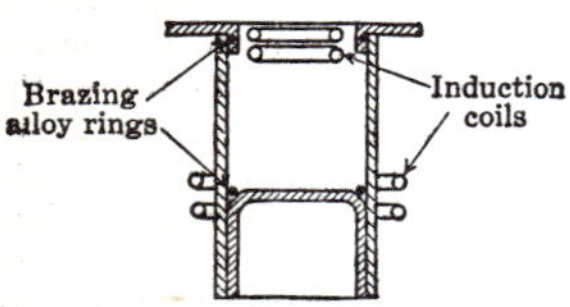

FIG. 5. Typical setup using internal and external coils in the induction brazing of a tube assembly.

INDUCTION BRAZING. Heat generated as a product of the resistance of metals to eddy currents induced within them by high-frequency currents passing through adjacent coils has gained wide acceptance as a means of treating metals. This induction brazing is likewise becoming an important tool in the high-production brazing of assemblies. (See Fig. 5.)

Where shape and size are favorable, the use of indexing tables or conveyer belts and suitable coils gives extremely rapid joining. At present, however, this mode of brazing is limited mainly to silver brazing jobs because of the objectionable oxidation and the difficulty of obtaining satisfactory bonds in copper brazing. This is solely a result of the longer time required to heat to the higher temperature required to melt copper; in some specific applications the use of chambers filled with a reducing atmosphere has overcome this difficulty.

Since the heat is generated within a shallow skin at the surface of the joint, extremely rapid heating and flow of the filler metal to bond the joint is obtained, with little change in the parent metal.

7. APPLICATION OF BRAZING

PLAIN CARBON STEELS. All ordinary steels of the plain carbon or low-alloy type are readily brazed with either copper or silver alloys. Furnace copper brazing is most commonly applied here, as the atmospheres used are reducing to any oxides present on the surface of the steel or in the copper brazing material itself, thus eliminating flux as a requirement. Steel is readily wet by molten copper in the absence of any oxides, and the joints in the assembly are quickly penetrated. Usually the time in the furnace is limited to that required to heat the assembly to at least the melting point of the brazing material, after which the assembly is introduced into the cooling chamber where also a protective atmosphere is maintained. After sufficient cooling time is allowed, the parts can be removed into the air and are found to have an extremely bright, clean surface.

ALLOY STEELS. As the alloy content of the steel increases, particularly in chromium, silicon, or aluminum, the ability to copper-braze without flux is lessened unless atmospheres of high purity such as dry hydrogen are used. Since the purity of the atmosphere in the furnace is the criterion of its value, it becomes necessary to use muffle-type furnaces when the high-alloy steels are to be brazed, to obtain the necessary freedom from contaminating gases. Usually the added problem of prevention of decarburization, and sometimes carburization, must be overcome with these steels. However, these difficulties are not met in other forms of heating, as by torch or induction, and frequently such methods are employed with great success in the brazing of nitriding or stainless steels. The brazing material used is one of the silver solders, and flux is always necessary.

Attention must be given to the need for further heat treatment which may exist, and, if such is the case, brazing alloys of sufficiently high melting ranges should be employed to forestall the occurrence of melting of the joint material during the subsequent heat treatment.

SURFACE PREPARATION. Some thought must be given also to the surface most desirable for purposes of wetting and bonding. A smooth, highly polished surface is not conducive to creeping of the molten brazing material because of the failure of the surface attraction of the metal to overcome the surface tension of the liquid copper or silver alloy. On the other hand, a clean, slightly roughened surface such as that obtained by shot blasting, pickling, or machining will result in rapid and uniform flow of the brazing metal over the surface of the steels. Judicious use of this type of surface inhibits excessive flow of the joining material where this is of importance.

CONTROL OF FLOW. Other methods for preventing undue flow of the material involve the use of grooves or ledges at appropriate points, the embedding of the brazing alloy in grooves concealed in the joint, a decrease of the hydrogen content of the atmosphere, or the use of the minimum amount of brazing material compatible with the good quality of the brazed joint.

Another method employed with success is the painting on of refractory powders such as chromium oxide, magnesium oxide, or aluminum oxide in the form of thin paste to act as barriers to the undesirable spread of the liquid braze metal. Where localized heating such as induction, torch, dip, or resistance can be used, the problem of unwanted flow away from the joint is less troublesome if due care is exercised in the application of heat to the joint.

Since a forged surface is highly conducive to superfluous creep and such surfaces are frequently involved as parts of assemblies, it is sometimes found advisable to postpone the steps to control the amount of excess brazing alloy until after the brazing cycle. The system used then is removal of the excess material by grinding or scraping or by chemical or electrolytic stripping. Where copper is involved, either sodium cyanide or chromic acid solutions are effective stripping agents.

CAST IRON. The most common practice in brazing cast irons is the application of heat by torch or induction to the metal adjacent to the joint after proper preparation of the surfaces and fluxing. When the assembly has reached a temperature at or above the lower melting point of the brazing alloy, the alloy, usually in wire form, is fed into the joint until a fillet of metal is formed around the joint. Cast iron is not as easily wet as ordinary steels, and fluxes containing fluorides are usually required, especially when the iron contains alloys such as silicon, chromium, or manganese in percentages higher than those normally found in gray cast iron.

It is possible and in some cases is the manufacturing procedure to copper-braze in reducing atmospheres assemblies made up completely or in part of gray or malleable iron.

This is done in the complete absence of flux since the atmosphere maintained in the furnace acts to suppress the formation of oxides, and it is necessary only to assure good contact and clean mating surfaces at assembly. Where poor wetting is encountered in either copper or silver brazing of this material due to high alloy content it is sometimes possible to pretreat the surface by some such method as the Kolene bath treatment to reduce the alloy content at the surface, thus implementing the action of the atmosphere or flux and achieving satisfactory brazing.

Caution must be exercised in copper-brazing of cast irons because of their low melting point and the attendant danger of injury to assemblies by exposure to high temperatures due to a lack of accurate control. Furthermore, there is a definite tendency for growth in irons when treated at brazing temperatures, although allowance can be made for this in the design of joints if it is not otherwise detrimental.

COPPER-BASE ALLOYS. Copper and its alloys, such as brass and bronze, are confined to the silver brazing category because of the limits imposed by their melting points. Since, however, copper and brass are readily wet and bond well with either the silver solders or the phosphorus-copper brazing alloys, extensive application is made of this joining method. Practically any means of heating can be employed but the most common are torch, furnace, or induction heating. Flux is required for either brass or copper when the process used does not include a protective atmosphere. By using a reducing atmosphere, such as in furnace brazing, it becomes possible to use a brazing technique which does away with the need for flux. This involves the use of an alloy brazing material which contains neither zinc nor cadmium; copper plating of the brass or bronze components permits the fluxless brazing of these metals, inasmuch as copper will not oxidize in a reducing atmosphere and will wet freely with the filler metal, while the easily oxidizable offenders (zinc and cadmium) are not present.

Ordinarily the aluminum bronzes are brazed only with fluxes high in fluoride content, owing to the refractory character of the aluminum oxide formed during heating. However, these bronzes can be brazed with much greater ease with ordinary flux in torch brazing or even without flux in atmosphere-furnace brazing by the method outlined above. This method of brazing is also open to the use of the phosphorus-copper brazing alloys with equal facility. A precaution, however, is necessary with these alloys. They cannot be used in the brazing of assemblies made up in part of ferrous materials unless there is no possibility of the molten phosphorus-copper alloy contacting the ferrous portion of the structure. This care is required because of the alloying effect which takes place between iron and phosphorus, resulting in a weak, brittle iron phosphide compound which detracts seriously from the strength of the bond.

The application of atmospheres to furnace brazing of copper itself is confined to those containing a minimum of hydrogen and carbon monoxide, for unless the copper is oxygen-free some hydrogen embrittlement will occur from brazing treatments in atmospheres wherein these reducing gases are present. This is not, however, a great drawback where the mass of the copper assembly is such as to obviate the danger of bending failure.

NICKEL ALLOYS. When the assemblies are made up of parts with melting points well above that of copper (such as nickel or its alloys with copper, chromium, etc.), copper brazing furnishes consistently satisfactory brazed joints. With nickel or Monel no flux is necessary if a sufficiently reducing atmosphere is employed. The nickel-chromium alloys require flux unless special steps are taken to provide a pure dry hydrogen or dissociated ammonia atmosphere. These steps would include the use of gastight muffle furnaces and close control of the atmosphere to assure maintenance of an adequately low dew point. Silver solder is applicable to the brazing of the nickel-chromium alloys as long as sufficiently active fluxes are used. In the silver brazing of Monel, use of any of the solders melting above 1400 F is not recommended because of the occurrence of intergranular attack by the silver above this temperature. However, those melting below 1400 F are very successfully employed to obtain high-strength bonds in Monel. The phosphorus-copper brazing alloys are here again ruled out because of an embrittling effect ascribed to the phosphorus.

HARD INSERTS. Extended use is being made of the brazing process for joining tungsten carbide, stellite, etc., to steel in the manufacture of cutting tools, dies, punches, etc. Both silver and copper brazing are employed, and all the means of heating such as furnaces, torch, and induction have been found useful in the varied applications which arise. The method of simultaneous brazing and heat treatment is here also brought into play where the steel parts brazed to the insert consist of air-hardening tool steels. Flux is called for unless sufficiently reducing atmospheres are within reach.

ALUMINUM ALLOYS. With the advent of specially developed fluxes and filler materials, the brazing of aluminum is a commercial process. The filler material consists of a silicon-aluminum alloy slightly lower in melting point than the aluminum parts to be joined. The methods adopted in production include torch, furnace, and flux bath treatments. Very careful temperature control is required because of the slight difference in the melting points of the filler metal and the aluminum assembly. The joint obtained is fully as strong as similar joints which have been welded, and appearance is good because of the similarity in color of the brazing material and the aluminum. (See also Section 4, Article 7.)

STRENGTH. For low-carbon low-alloy steel, when the parts to be brazed are sufficient in mass to prevent changes in clearance during the brazing cycle, it is found that the shear strength of the joint decreases as the clearance increases. Figure 6 gives average

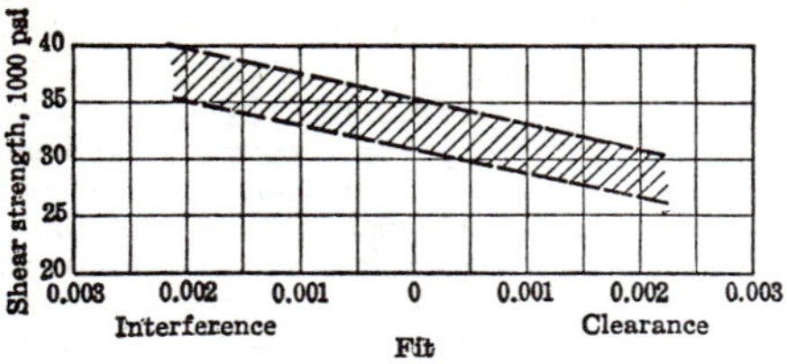

FIG. 6. Shear strength versus clearance for copper-brazed joints.

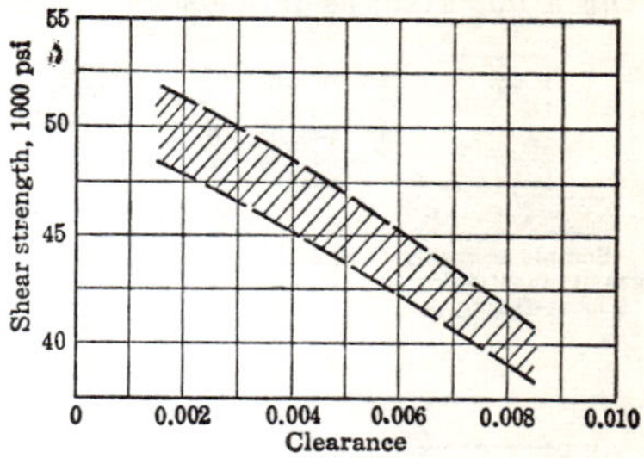

FIG. 7. Shear strength versus clearance for silver-brazed AISI E 4337 steel.

values to be expected for fits usually employed. These values are for copper-brazed joints, and it is obvious that the recommended fit would be one of no greater than 0.001 in. clearance to 0.002 in. interference. It must be noted, however, that where higher strength steels are employed and where these steels have been hardened and tempered, the values are much higher, averaging 55,000 psi in shear when clearance is within the limits specified.

Because of the lower fluidity of silver alloys at the brazing temperature. interference fits are not recommended. However, even with the prescribed clearance of 0.001 to 0.003 in., the shear strength values of silver brazed joints are quite high, being between 24,000 and 40,000 psi with the higher value coinciding with the lesser clearance. (See Fig. 7.)

Generally it can be said that the tensile strength of a brazed joint is approximately twice the shear strength. Figure 8 gives average values to show what tensile strengths can be expected. Tensile strength also is found to increase as the strength of the parent material is increased and to vary inversely as the clearance.

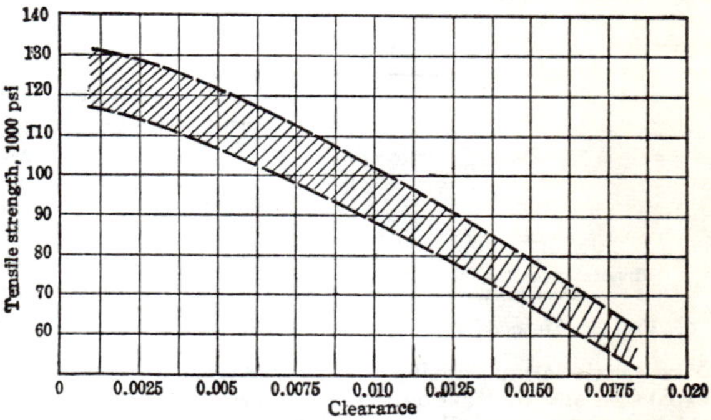

FIG. 8. Tensile strength versus clearance for butt joints in stainless steel of 160,000 psi tensile strength.

8. JOINING REQUIREMENTS AND JOINT DESIGN

In all brazing methods two factors, cleanliness and proper positioning of the individual parts of an assembly, must be given prime consideration.

SURFACE PREPARATION. Prior removal of oxides, dirt, grease, and oil is essential in assuring complete cohesion and penetration of the brazing material throughout the joint. If flux is used as a protection against oxidation during heating, its action is then more effectively applied to the work of preventing access of oxidizing constituents of the atmosphere to the surfaces of the joint. Conversely, if the protective medium is a reducing atmosphere, the work required of the atmosphere in curtailing oxidation of the joint area is lessened, and its effectiveness in reducing oxides already present is enhanced.

Mechanical elimination of oxides and tarnish can be carried out prior to brazing by filing, grinding, tumbling, shot blasting, sand blasting, or wire brushing. The last two are not highly recommended. In sand-blasting there is a tendency toward surface retention of particles of sand which later interfere with the flow of brazing material; in wire brushing there is the possibility of smearing metal over the oxide, rather than clean removal. A second well known method, pickling, is at times employed when, for some reason, a clean machined surface is not present.

Removal of grease and oil is carried out by thorough degreasing in an efficient solvent. This is important as during the rapid heating of the joint area the grease or oil is not usually burned entirely away but leaves a carbonaceous residue which may interfere seriously with the flow of the brazing alloy.

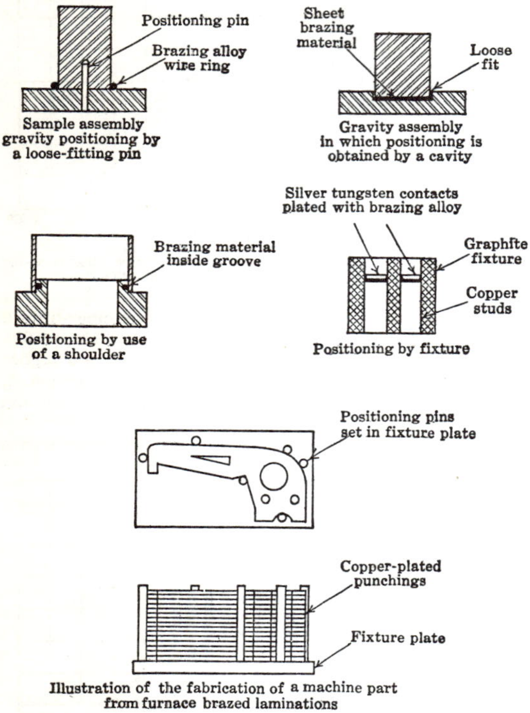

FIG. 9. Methods of positioning assemblies for brazing.

POSITIONING. In order to preserve correct alignment and relationship between the individual parts of an assembly during bonding, many devices are called into play. Among the fits which are self-aligning, the simplest might be considered the one maintained by gravity. Positioning is here attained by the use of protuberances or grooves in one or all of the parts or by utilizing loose pins or even by specially designed fixtures. (See Fig. 9.)

In respect to fixturing it must be remembered that if parts of the fixture are placed so as to be in direct line with the flame in torch brazing, or if the jig must be heated with the work as in furnace or possibly molten bath brazing, the fixture material must be heat resistant and must also resist wetting by the brazing alloys involved. Furthermore, it may be necessary to select a material with a rate of expansion and contraction suitable for the assembly so as to obtain dimensional accuracy. Many materials are available such as carbon or the stainless steels, and various rates of expansion and contraction can be selected from the alloys of chromium. If a particular thickness of brazing alloy is desired in a gravity assembly, it can be managed by the use of shims of the desired thickness which themselves should consist of a material readily wet by the brazing alloy but not dissolved by it.

In adding the brazing materials, it may be convenient to use brazing alloy in sheet form preplaced inside the joint or in wire form laid along the outer edge of the joint. If torch brazing is involved the alloy may be fed into the joint after the work attains the brazing temperature. At times, it will be found convenient to bury the filler material in cavities within the joint so that when the assembly is at brazing temperature the molten alloy will be drawn from this reservoir into the joint by capillary attraction. If the cavity method is used, particularly in furnace brazing, care must be exercised to provide vent holes to prevent entrapment of air within the reservoir.

A much-employed manner of joining is what might be called the pin or tube and hole type. (See Fig. 10, *a*, *b*, *c*, *d*.) Usually, simply making the parts a slight press fit preserves the positioning. If desired, it is quite feasible to apply a straight knurl to the male piece, thus avoiding the maintenance of close tolerances which might be objectionable from a cost or material standpoint. Other available means are peening of the hole or the pin and staking, swaging, or hot or cold riveting of the pin.

Of course, any of these methods can be adapted to positioning of tubes or bushings, and, in addition, such assemblies might profitably be spun, crimped, or expanded into place to accomplish the necessary dimensional control. Quite often it is simpler to design the male member of this type of assembly with a shoulder, thus automatically removing the positioning difficulty. A minimum fillet at the base of the shoulder and no countersink on the hole are then design requirements to prevent stoppage of the molten filler metal.

Oftentimes tack welding may be called into use or, if the thickness of the material is within limits, spot welding may be employed. The older methods of joining such as screwing, pinning, or riveting, at one time used alone, are also occasionally used where conditions so indicate.

In brazing of sheet metal assemblies, lap or interlocking joints are frequently employed, and usually the type of joint provides all the positioning necessary. If this is not the case, simply spot welding, peening, or sometimes screwing may be applied. (See Fig. 10*e*.)

Ordinary means such as described above for preserving alignment are, practically speaking, sufficient to carry out the joining of most assemblies. Nevertheless, it sometimes occurs in furnace brazing, for instance, that the most satisfactory method available may consist of simply holding the positioning of the parts by use of some material such as cellophane tape, wood, or perhaps lacquer until they are placed in the hot zone of the furnace, and then, after sufficient time has been allowed for brazing the assembly, cooling with the furnace to below the freezing point of the brazing material before introducing the assembly into the cooling chamber. Another means of resolving assembly difficulties may be the brazing of subassemblies with the higher-melting-point filler metals followed by a joining of these brazed subassemblies into a final assembly using one of the lower-melting-point brazing alloys.

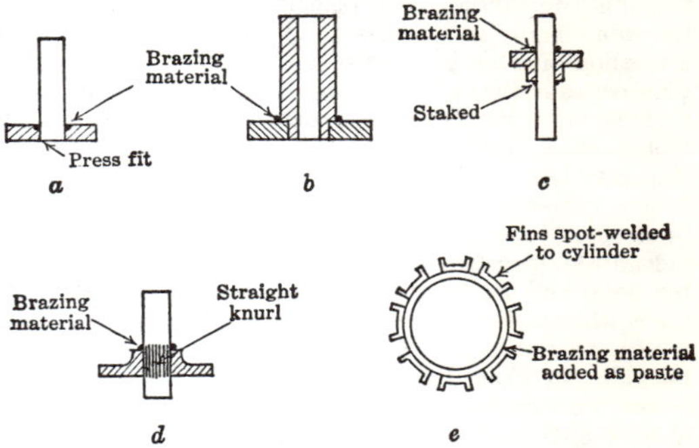

FIG. 10. Examples of positioning by simple fit.

9. BRAZING-ALLOY SELECTION

A point which demands consideration is the type of alloy to be used as a filler metal as well as its form and placement in the assembly. The choice of alloy depends on factors such as its melting point compared to that of the assembly components and the absence of any materials in the alloy which might have a harmful effect on the members of the assembly; i.e., phosphorus combining with iron to form phosphide and the intergranular penetration of silver into some nickel alloys creating an embrittling effect. Table 1 gives the analysis and melting range for some commonly employed brazing materials. Other alloys are listed in Table 2, page 4-85.

With these points in mind, the selection of the alloy is carried out. The form in which it is to be used and whether it shall be preplaced or fed into the joint are determined by such considerations as (1) whether the part is to be torch-brazed where feeding is possible or furnace-brazed where it is not; (2) whether counterbores or concealed grooves or cavities are permissible in the assembly, in which case wire, sheet, or slugs of brazing material can be used; (3) whether the flow of the alloy into the joint will be aided by gravity or whether capillary attraction is solely depended upon. If it is difficult to place the alloy where wanted, applying pastes made from the powdered brazing metal, or plating with the brazing metal, or even spraying the metal onto the desired location offer alternative methods.

Table 1. Commonly Used Brazing Alloys

Designation	Composition, % Cu	Ag	Zn	Cd	P	Melting Range
Copper	99+					1083 C (1982 F)
Cu-Ag alloy	50	50				778–850 C (1432–1562 F)
Easy Flo	15.5	50	16.5	18		627–734 C (1160–1173 F)
Phos-Copper	93				7	704–800 C (1300–1472 F)
Eutectic	28	72				778 C (1432 F)
ETX	34	50	16			695–775 C (1284–1427 F)

ADHESIVES AND GLUES

By Thomas D. Perry

The general terms *glues* and *adhesives* have similar definitions and are used somewhat interchangeably by most writers. Within the industry, glue more commonly refers to the older types used chiefly for wood and paper products, while the word adhesive is usually applied to the more recently developed resins and their chemical combinations for cementing together a wide variety of like or unlike materials.

PRINCIPLES OF ADHESION. While adhesion is the technique of sticking materials together, its fundamental principles have been rather inadequately investigated. It is a far more complex process than first appears. The earlier concepts represented it as a method of adhering two pieces together, chiefly by an interlocking of the prongs of the glue or adhesive, which flowed as a viscous fluid or was forced into the cavities of the surfaces to be bonded, and hardened. This is *mechanical adhesion*, and is still an important factor where porous materials are to be joined together. Later investigations developed the fact that *specific adhesion* is the more significant feature, since many relatively nonporous materials can be made to adhere in a satisfactory manner. Specific adhesion can be divided into several phases. (1) The *polarity* of the materials to be joined must be recognized, since polar solids bond readily with polar liquids, and nonpolar with nonpolar; but polar and nonpolar materials do not adhere satisfactorily. For example, water, alcohol, phenol-formaldehyde resins, and wood are polar; conventional glues (animal, vegetable, casein, soybean, and the like) have rather complicated chemical structures, but tend to be polar in character; metals, rubber, and most oils are nonpolar. (2) The *chemical* reaction of the surface of the solid to the constituents of the adhesive may indicate favorable or unfavorable combinations. (3) The pH of the glue line is important, since *high acidity* may weaken the molecular structure of the solid.

Other factors involved are the *wettability* of the surfaces to be joined with regard to the liquid glue employed; the *porosity* of the surface and its *smoothness;* the *viscosity* of the adhesive solution and the resulting *thickness* of the adhesive film; and the *evaporation* or *diffusion* of the volatile solvents.

CLASSIFICATION OF ADHESIVES. From an industrial standpoint, the volume of adhesives and glues used for *wood products* exceeds that of all other "glueable" materials. Since glued wood is frequently used for furniture, radio cabinets, and structural purposes, it is important that the adhesives produce strong and durable bonds.

Glues and pastes used for *paper* and *paper products* probably rank second in commercial importance. However, strength and durability are less essential here than in wood, since paper is relatively weak, and normal exposures are seldom exacting.

The gluing together of *metals* is a relatively new development, applying largely to thin metals where welding, brazing, or soldering are not practical. Metal-to-wood and metal-to-metal bonds are both increasing in importance.

There are numerous other adhesive applications of lesser magnitude, such as *safety glass*, *plastic facings* to various basic materials, *textiles*, *felts*, *leather*, *linoleum*, *ceramics*, and *rubber*, which will be briefly mentioned. Suppliers should also be consulted.

In addition to the problem of joining metal and wood, many problems are involved in uniting unlike materials such as paper and cloth to wood; plastics to wood and metal; and abrasives and granular materials to paper, cloth, and felt backings, for discussion of which the reader is referred to the literature of the industry and to adhesive manufacturers.

Primary glues and adhesives are those used for large areas, such as plywood, paper, corrugated boards, plymetal (wood to metal sheets), and plastic overlays. *Secondary glues* and adhesives are employed for assembly work, or, in the case of primary glued products, for subsequent attachment of parts. The line of demarcation is not rigid, and some of the glues and adhesives may be used for both purposes.

10. ADHESIVES FOR WOOD

FACTORS COMMON TO MOST WOOD GLUES AND ADHESIVES. Wood is a relatively porous material, with cavities that are appreciably penetrable by liquid adhesives. Hence mechanical adhesion is an important factor, supplemented by specific adhesion, as related to the physical and chemical characteristics of the adjoining solids and the intervening adhesives. Certain glues may be preferable for certain species of wood or constructions, since there is no all-purpose adhesive or glue that meets all requirements.

Wettability of the wood surface by the volatile solvents in which the glue is dispersed or dissolved is an important factor. Hence all surfaces should be clean and free from oil, dust, or other impurities. Glues of high solvent ratios, such as animal and vegetable glues, require wood of extreme dryness—3 to 5% m.c., whereas those of low solvent ratios, like resin adhesives, are best made of moisture content (m.c.) ratios of 6 to 12% (in the wood), and for exposed service up to 18%. The preceding range is for room-temperature hardening. For hot press work, moisture contents should not exceed 10% for veneer layers, or be above 6% for lumber layers.

Double spreading, on both faces of a single joint, although not usually employed because of cost, often equalizes the wetting of the two surfaces that form the glue bond. *Single spreading* is customary with most of the plywood glues.

Smoothness of the wood surface is important, and fresh, sharply machined surfaces are preferable to surfaces exposed to air for periods of 2 or 3 days and up. Tooth-planed surfaces were once regarded as necessary, but this practice has fallen into disuse. Burnished surfaces, resulting from dull cutters or excessive cutting pressures, are distinctly unfavorable for gluing. Sanded surfaces may be desirable to remove gloss or polish, but are not otherwise recommended in raw wood, where sanding may cause unevenness by cutting away the softer areas. Sawed surfaces can be glued together successfully, if of accurately uniform thickness, but require higher pressures to secure close contact at the glue line.

For the best quality glue joints, all surfaces should be in *close contact*, whether flat or curved. Experience with most glues and adhesives indicates that glue line thicknesses of 0.005 to 0.010 in. give the strongest and most durable joints, although some types tolerate thicker joints.

Working life of most glues is limited, and prepared batches should come within this range, from 2 hours to a working day. Over-age batches that have passed their usefulness should be discarded. Fresh and old batches should never be combined, and receptacles should be cleaned between batches.

Various glues and adhesives *set, harden, or cure* in different ways: by (1) evaporation of the solvent, as in animal and soybean glues; by (2) polymerization or condensation, or a combination of them, under heat or catalytic action, as in resin adhesives (thermosetting); others by (3) a process of cooling, as in thermoplastics; and (4) blood albumins by coagulation. In the *thermosetting* types, heat may be required and is usually beneficial in shortening the bonding cycle. In the *thermoplastics*, heat provides a necessary flow period as an intermediate step between the application and the final curing. In glues of the solvent type moderate heat may accelerate the evaporation of the solvents, but high temperatures may weaken the bond. On the other hand, few adhesives or glues harden or cure at temperatures appreciably below 70 F, especially resin adhesives that are prepared for room temperature work. Casein is an exception, as explained under that heading.

Adequate pressure is important with most glues and adhesives, although there are a few so-called contact resins, substantially tacky in effect. Pressure should be enough to produce a thin glue line, and should be maintained uninterruptedly until the surfaces are firmly gripped together. In room temperature work, pressure should approach, but not exceed, those that will compress the wood; from 75 psi for low density woods to 125 psi for high densities. In hot pressing, since wood softens under heat, pressures that compress the wood slightly, 5 to 10%, are usually preferable; a range of 125 to 150 psi for low densities and 175 to 250 psi for high densities.

Time allowance required to obtain a firm grip between surfaces varies widely, according to temperatures and character of the glues. An overall rule might be 4 hr and up for all room temperature work, while hot press cycles may be as short as 5 min. Induction (radio frequency) heat may have as short cycles as 1 min or less, since heat is applied more directly on the glue line and is not retarded by the insulating quality of the wood to the same extent as external heat applied on the surface.

Further specific details within these general details will be found under the descriptions of the various glues and adhesives. See also references to the literature of the art, Refs. 1–13 at the end of this chapter (p. 21-30).

ANIMAL GLUES. Animal glue, the oldest known type of wood glue, was used in ancient Egypt. It approaches an all-purpose glue, but weakens rapidly when wet. It is a derivative of hide and bones, and is supplied in eighteen standard grades (see p. 21-24), varying in viscosity and jelly strength. *Hide* glue is stronger and more costly than *bone* glue. Blends are often made to meet grade specifications. Refined grades find industrial uses as gelatins; extra-refined grades become edible gelatin. *Fish* glues come under the same general classification.

The animal glue of commerce is a hard bone-like substance, tan or brown in color, supplied in flake or granular form. When soaked in cold water it swells appreciably. The swollen glue is melted with moderate stirring at temperatures up to 140 F, and is kept at approximately that temperature while in use. Since animal glue tends to harden when chilled, wood is often warmed slightly before application. Both brush and roller application are used. Average clamping time is 4 hr, although some low water ratios can be unclamped in 2 hr, both at room temperature. Heat should not be used to shorten the setting time.

Utility Range. Animal glue is widely used for assembly work, as its cost bars it from most plywood work, except occasional special pieces. As a bench glue for general purposes it is unsurpassed. Many traditional cabinet makers prefer it to all other glues. As a dilute size, or as an ingredient in various combinations, it may be used to fill the pores of the wood and increase their cohesion, as in strengthening fragile face veneers.

Durability. Animal glue joints, when not exposed to moisture, have retained their strength for many years, even for centuries. Animal glue is not suited to products exposed to the weather, to boat work, to high humidities, or to high temperatures. Properly made animal glue joints not unduly exposed are stronger than the wood itself.

BLOOD ALBUMIN GLUES. These glues, more correctly designated as *dried-blood glues*, are divided into *soluble* and *insoluble* types. Soluble grades were used for aircraft in World War I, before resin adhesives were available, when few hot presses had been installed.

Commercial dried blood of the soluble type is supplied in powder form and is dissolved in water, usually with some toxic ingredients to inhibit mold and fungus growths. Insoluble types require the addition of caustic soda and other chemicals to convert them to adhesive uses. Roller spreading is employed for plywood work, single spread on either of the two mating surfaces. If roller spreaders are not available, brushes may be used. Blood glues harden by the coagulation of the albumin, at hot press temperatures of 120 to 180 F. Except with special formulas, seldom used, blood glues do not set at room temperatures. The irreversibility of the coagulation process, under heat, marked an early step in waterproof plywood.

Utility Range. While soluble blood glues were used for hot-pressed plywood with minor additional ingredients from 1910 to 1925, they were gradually displaced by casein and resin adhesives, which were more uniform in grade and more reliable to use. Albumin glues are not suitable for assembly or bench work. Both soluble and insoluble types of blood glue are currently used largely as modifying and extending agents in resin adhesives. In such combinations the blood additives regulate the spread and flow and shorten the hot press cycles, as well as reduce the average cost of the adhesives.

Durability. Plywood made with straight blood glues has excellent durability under severe exposure, but deteriorates slowly in storage, and is adversely affected by mold, fungus, and insect pests, except where proper inhibiting reagents have been added. Bonds in which blood is combined with other adhesives partake largely of the durability of the other adhesives. Blood and resin combinations are used in Douglas fir plywood on the Pacific coast, but do not meet the exacting requirements of standard exterior grades.

CASEIN GLUES. Primitive glues of this type, made of sour milk and quick lime, date back several hundred years, but modern commercial adhesive uses began in the early 1900's. Casein is the dry powder from curds of milk, natural- or acid-soured, and an important by-product of the dairy industry. After the removal of whey, the curds (crude casein) are washed, pressed dry, and the cake is broken and ground into a powder of the proper degree of fineness. Raw caseins vary widely in quality, and suitable adhesive types must be selected and segregated.

Preparation. Casein glues for woodwork consist chiefly of a suitable grade of casein, hydrated lime up to 15% (of casein) and caustic soda with other minor ingredients including, sometimes, certain chemical preservatives. Prepared powder combinations are usually employed, although the caustic may be added at the time of mixing. Casein for plywood work is applied by corrugated (12 per in.) metal rollers. For assembly and bench work, brush spreading is employed.

Clamping requirements for plywood at room temperatures are 4 hr and up, although most factory practice is to allow the plywood to remain overnight in clamps. This hardening is both by evaporation and by the chemical formation of calcium caseinate. It is an irreversible process of the order of concrete. Setting may be accelerated by heat. In field operations in housing, casein cures at temperatures of 50 F, but at a retarded rate.

Utility Range. Casein glues are used for water-resistant plywood as well as for assembly work. When hardened, casein joints have a somewhat abrasive effect on tools. Casein is used only to a limited extent in furniture and interior equipment because of cost. It is used extensively in laminated timbers and has many structural applications where exposure is not extreme. Casein is frequently combined with dried soluble blood, soybean, and silicate of soda glues, and is a useful component in such combinations.

Durability. Plywood made with the better grades of casein is highly water resistant and finds many fields of utilization where extreme durability is not required. It has excellent resistance to heat, below that which chars wood, but delaminates on prolonged soaking or exposure to high humidities. It is considerably below urea resins in water resistance, and superior to ureas in heat resistance.

SILICATE OF SODA GLUES. Soluble silicates are made by melting sand and soda ash in an open hearth regenerative furnace at temperatures of 1300 to 1500 C. The melt is leached of water-soluble constituents. The grades used for wood adhesives are:

Soda-silicate ratio, 1 : 3.22; Baumé rating, 41° to 42°; solid form
Soda-silicate ratio, 1 : 2.9; Baumé rating 47°; liquid form

The former is preferable for open porous woods, the latter for hardwood plywood.

Most formulas provide for combinations of silicate of soda with lime and casein, to reduce the overall cost of casein. Formulas are somewhat complex and suppliers should be consulted. Spreading is by corrugated (10 per in.) metal rollers, single spread. The hardening action is by solvent evaporation, and the joint has a slightly abrasive action on edge tools. Heat is not employed to accelerate the setting.

Utility Range. Silicate of soda glues have been extensively used for plywood box shooks, but their use has substantially declined because of their somewhat complicated formulation. This type of glue finds wider use in corrugated fiber and paper packages. It is also an important low-cost extender for other glues, such as soybean glue.

Durability. Bonds made with silicate of soda are reasonably strong, are adequate for box shooks, are highly resistant to mold and fungus growths, but have low water resistance.

SOYBEAN GLUES. Soybean is a relatively new adhesive for plywood, developed on the Pacific coast just prior to 1930, especially for the Douglas fir industry. The oil is first pressed out of the soybeans, and the resulting cake or meal, or such part of it as has desirable adhesive qualities, is dried and ground to a powder. Peanut meal flour, similarly extracted, is also used as an adhesive but is not as highly regarded.

For room temperature work single spreading is employed, corrugated (10 per in.) metal rollers being used. For hot pressing, rubber rollers, with 16 to 20 corrugations per inch, are used. Soybean glues will harden at room temperatures in 4 hr and up; by one-stage hot pressing requiring 4 1/2 min for 1/2 in. 5-ply, when platens are at approximately 225 F; or by a two-stage process under the same conditions in 2 1/2 min.

Utility Range. Soybean glues are widely used for Douglas fir plywood and with associated species. They are also used to a considerable extent for hardwood plywood box shooks. They are seldom used for furniture, except for drawer bottoms, dust backs, etc. Soybean glues are low in cost, but their high alkalinity is somewhat of a handicap.

Durability. Soybean glues produce bonds of substantial moisture resistance, for all grades of Douglas fir except exterior and concrete form types.

SYNTHETIC-RESIN ADHESIVES. PHENOL FORMALDEHYDE. Synthetic-resin adhesives began to attract attention in the early 1930's. The earliest commercial type was made from phenol, a coal tar derivative, combined with formaldehyde, partly condensed or polymerized by the supplier and sold to the consumer as a thermosetting adhesive for wood. Cresol was sometimes used instead of phenol. The polymerization is completed in contact with the wood, under heat and pressure, resulting in a bond of maximum durability.

Types. The best-known phenol type is Tego film, which is supplied on a paper carrier ready to use, furnishing a uniform and adequate amount of adhesive per joint. Its principal use is in hardwood plywood. Phenolic adhesives are also available in liquid and powder form, which in the latter form require preparation by the user. Liquid phenolics, 40% solids, are widely used for Douglas fir plywood. Most phenolics cure at platen temperatures of 275 to 325 F, but modified phenolics cure at temperatures of 150 to 200 F.

Durability. Phenolic-resin adhesives are used chiefly for exterior and marine grades of plywood, both hardwood and softwood, but are not used for assembly work. Exposure tests of many years' duration have demonstrated that a properly cured phenolic bond is more durable than the wood itself.

Impregnation with Phenolic Resins. Phenol-formaldehyde resins, less advanced than those used for adhesives, are found suitable for impregnation, the preliminary step to making Compreg, Pregwood, etc., under high pressures. Such resins, of small molecular size, penetrate into the minute cavities of the cell walls of the wood. Thorough impregnation requires veneers of 1/8 in. and thinner. The impregnating treatment may be by soaking for 1 to 2 days, by several vacuum-pressure cycles, or by pressure cycles alone; the last two methods require 2 to 3 hr for a resin pick-up of 25 to 30%. Pick-up is the percentage of dry cured resin impregnant by weight, based on the raw dry wood as 100%.

Pressures required for Compreg are 1500 to 3000 psi, with a resulting specific gravity of 1.3 to 1.4. Hardwoods, like maple and birch, are compressed under this process to approximately two-thirds of their original thickness. Less dense species such as poplar may be compressed to one-half the original thickness. Impreg is similarly impregnated with resin, but cured at lower pressures, 200 to 300 psi.

Other impregnating processes include Tegowood, made by similarly curing Tego interlayers between veneers approximately $^1/_{30}$ in. and thinner. Another process is by double-spreading a liquid phenolic on veneers $^1/_{16}$ in. thick, followed by the cure cycle outlined above.

SYNTHETIC-RESIN ADHESIVES. UREA FORMALDEHYDE. Urea-resin adhesives are a later development than the phenolics, appearing in the late 1930's. Urea is a synthetic product, based on a reaction between carbon dioxide and ammonia. When combined with formaldehyde, its thermosetting adhesive characteristics resemble those of the phenolics, but with shorter curing cycles, lower temperature requirements, and reduced ultimate durability. Urea-resin adhesives have a wide range of versatility, and when properly cured exhibit durability considerably higher than the earlier conventional glues.

Urea-resin adhesives are available in powder and liquid form, are soluble in water, may be catalyzed for room temperature cure at 70 F, or hot pressed at 225 to 275 F, may be extended with wheat flour in varying ratios, and may be fortified by additional ingredients to produce bonds approaching those of the phenolics for durability. Single spreading may be by grooved (16–20 per in.) rubber rollers, brush, or spray. Plywood bonds in panels $^1/_4$ in. thick can be cured in 6 to 7 min with hot press platens at 250 F. This time in the hot press can be shortened by completing the bond under residual heat in a subsequent hot stack for 4 hr immediately following removal from the hot press. Assembly bonds are usually cured at room temperatures of 70 F and up, with 4 to 6 hr and up under full clamp pressure. Properly catalyzed urea adhesives are extensively used under electronic heating. Each manufacturer should be consulted for bonding schedules on his resin adhesives.

Durability. Ureas are by far the most-used resin adhesive, with excellent records of durability under normal protected exposures. Extreme moisture or exposure to heat above 150 F is distinctly unfavorable, and phenolics should be employed.

OTHER SYNTHETIC-RESIN ADHESIVES. Resorcinol formaldehyde is a phenolic type in liquid form, requiring a separate catalyst. Cures are at room temperatures of 75 F and up. Durability of bonds for marine and weather exposure is improved by curing bonds at 180 to 220 F, in heated chambers with high humidity to prevent wood shrinkage and pressure losses. Resorcinols are especially adapted to laminated lumber and assembly joints subject to severe exposures. For massive and bulky assemblies where heat applications are difficult and slow, resorcinols give bonds equal to standard phenolics in durability. Resorcinols are in high cost ranges and hence impractical for plywood work.

Melamine formaldehyde is a urea type, for hot pressing only, with better durability than the standard ureas, but in higher cost brackets. When properly applied, melamines approach the durability of the phenolics and resorcinols. Their background of development and their experience in durability tests are both quite limited in comparison with the ureas and phenolics.

Thermoplastic Resin Adhesives. Polyvinyl resins of several types have distinct adhesive qualities and have had a number of successful applications in wood and like products. Because they were developed no longer ago than 1946 they should be used cautiously.

VEGETABLE GLUES have a starch base, and wood glues consist principally of cassava or tapioca flour, made from the fibers of the cassava roots and imported from the East Indies. Corn and potato starches are also used, especially for low-grade plywood, as in box shooks. Vegetable glues appeared about 1910, and were widely used for plywood until the advent of extended urea-resin adhesives about 1940.

Utility Range. When low cost is important and water resistance is not a factor, vegetable glue may be used. It has been generally employed in medium- and low-grade furniture. Its use declined severely during 1940–1945, when imports of cassava were drastically curtailed, and it is not clear whether it will regain its former importance as a wood adhesive.

Durability. While vegetable glue bonds have a long life when protected from moisture, they are distinctly deteriorated by high humidity and moisture. They can be recommended only for locations well protected from adverse conditions.

GLUE TESTING. Animal glue standards adopted by the National Association of Glue Manufacturers are not applicable to any other glue or adhesive. The eighteen standard grades are based on jelly strength and viscosity. Grades 9 to 12 are recommended for high-grade joint glues in woodworking. A wider range may be used for less exacting wood joints, i.e., 7 to 15. Little hide glue is found in grades below 7; lower grades are usually blends of bone and hide glue, and the lowest grades are chiefly bone glue.

Jelly strength test is the weight in grams required to embed a plunger 12.7 mm in diameter to a depth of 4 mm in a standard mixture. *Viscosity* is a measure of the time in seconds required for a solution to flow between two points on a capillary-type pipette that has been properly calibrated in millipoises.

TESTING GLUE JOINTS. The testing of glue itself is limited to grades of animal glue, since there are so many variables in other glues and adhesives, due to heat and chemical reactions, that no practical standards have been found that are reasonably reproducible under normal conditions. However, standards of glue joint strength applicable to most glues and adhesives have been developed by the Forest Products Laboratory and have an adequate background of experience for evaluating glue joint strength under normal and exposure conditions. These tests show *tensile shear* strength in plywood (Fig. 1) and *compression shear* strength in solid wood blocks (Fig. 2).

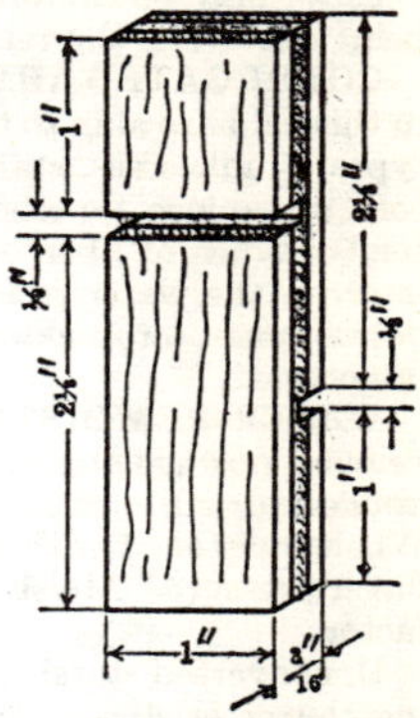

FIG. 1. Test specimen for tensile shear strength of glue joint in plywood.

Tensile Shear Strength in Plywood. Sample test strips of $^3/_{16}$-in. birch plywood, 1 in. wide by 3 $^1/_4$ in. long (Fig. 1), with single saw slots on opposite sides extending two-thirds through the central cross layer, are cut from selected or random locations in the plywood sheet. These samples are tested to destruction, usually breaking in the central cross layer. Thus the test is *across* the grain of the central layer.

Birch is a sufficiently strong species to give an adequate demonstration of adhesive strength, and the usual requirements for breaking tests are approximately 400 psi *dry* and 300 psi *wet*, after cycles of wet and dry exposure. In less dense woods the requirements are approximately 230 psi *dry*, and 130 psi *wet*. Reference is made to specifications AN-G-8 (1942) and AN-P-69a (1947) for further details as to government requirements for dry and alternate cycle tests of this type. The commercial requirements are given in CS-35-47 and CS-45-47. Because revisions are frequent, the latest editions should be obtained from the Bureau of Standards.

Compression Shear Strength in Solid Wood Blocks. Sample test blocks are glued together, sawed apart into specimens 2 in. wide by 2 in. long by approximately 1 $^1/_2$ in. thick, stepped down to give a test area of 3 sq in. *along* the grain, as shown in Fig. 2. Maple is chosen as standard, being a wood difficult to glue and of adequate strength for comparative purposes. Single test specimens can be made of strips of dense maple, glued together and cut apart as shown in Fig. 2, or they can be cut from multilaminated pieces.

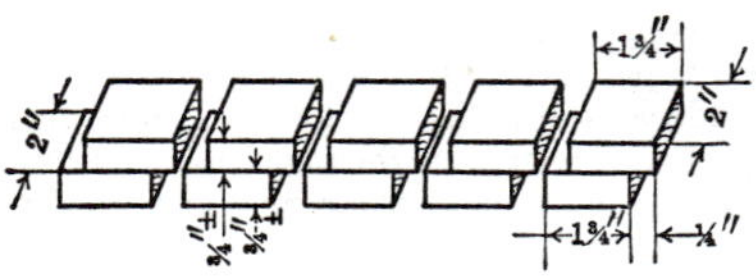

FIG. 2. Test specimen for compression shear strength of glue joints in solid wood blocks.

The requirement for dense hard maple is 2800 psi, for white oak 1800 psi, and for Douglas fir and other less dense species lower strengths. Cyclic wet and dry tests are employed for joints to be subject to marine and weather exposure.

Reference is made to specification AN-G-8 (1942) and JAN-A-397 (1946) for further details of government requirements for dry and cycle tests of this type. No commercial requirements have been established.

11. GLUES AND ADHESIVES FOR PAPER PRODUCTS

Such adhesives function largely on porous materials such as paper and fiber boards. Included also, for lack of a more logical classification, are a few relatively nonpenetrable materials, such as transparent adhesive tapes and packaging closures of the cellulose types. Paper-reinforced metallic foils are of a similar character. The ramifications of these adhesive problems and the range of combinations of basic adhesive materials and supplementary ingredients are very great. The most practical approach, for a brief survey, is to describe the groups of materials to be adhered and to indicate the adhesives that have proved generally useful. For further information the reader is referred to adhesive suppliers.

LAMINATED PAPER AND TEXTILES. Many types of standard papers are combined, somewhat as in plywood, to provide attractive or printable surfaces on center or backing layers of strong and relatively thick fiber boards of low cost. For such laminated products, adhesives are chiefly of raw starch and partially (20 to 30%) dextrinized starch. While cassava starch (tapioca) is noted for the cohesiveness of its paste, corn and potato starches are widely used.

Silicate of soda may be combined with the starches, either to reduce the time of gelatinization or to decrease cost. Neither starch nor sodium silicate glues have appreciable water resistance, but this is seldom important in laminated papers.

Cloth may be combined with paper as a reinforcing layer, the above glue combinations being used, but the relative shrinkage ratios of the materials need careful consideration.

CORRUGATED AND SOLID LAMINATED FIBER BOARDS. The adhesive problems in this field are akin to those of paper, but the glues must be in minimum cost ranges. Two types of adhesives, starch and silicate of soda, are widely used. The starches are of the corn or tapioca types, whichever is cheapest at the time, and are modified to meet particular types of fiber board. Starches combined with urea-resin adhesives substantially improve the water resistance of the bond. Silicate of soda is widely used, since a rapid development of tackiness is essential and the required moisture resistance is seldom important.

WATER-REMOISTENABLE APPLICATIONS. Adhesives for gummed labels, envelopes, postage stamps, etc., are known as water-remoistenable types, since they must be applied, dried, and stored in reasonable locations, without causing premature adhesion. When moistened with water or saliva, they must develop prompt tack and adhere with finger pressure. Moisture resistance, after they are stuck together, is seldom an important factor.

Unconverted starch gives the best bonding qualities, but exhibits the poorest remoistening characteristics. Partially converted dextrine (50 to 75%) is more soluble but gives hygroscopic bonds. Animal glues and dextrines are used in various combinations to impart greater strength and to reduce crystallization.

For postage stamps, either dextrine or gum arabic is employed, and curling is retarded by passing the dry gummed sheets between metal rollers that tend to break the glue film by slight corrugations. Gummed labels to adhere to glass bottles and paper surfaces present special problems of adhesion. Animal glues are frequently added to the dextrines. Decalcomania labels are usually coated with animal glue or glycerol.

GUMMED AND ADHESIVE TAPES. Tapes fall into two general types: (1) those that are *gummed* and *remoistenable*, as in the case of gummed labels; and (2) *adhesive* tapes that do not require remoistening. Gummed tapes for splicing veneers usually employ modified animal glues; but for package closures, the starches, dextrines, and animal glues are used. Nonblocking to the ungummed face is important for storage purposes, and surface sizings are frequently used. Gummed tapes may be made of cloth or perforated paper, the latter being employed for inner layers of veneer in plywood constructions.

The other principal type is the so-called *adhesive* tape, which is pressure sensitive and requires no moistening during application. A considerable number of chemical combinations have been developed for this purpose, many of which have rather complicated formulas. Most of them contain substantial proportions of various types of rubber and of a number of resin adhesives.

Masking types of tapes, for spray lacquering, are usually on crepe paper with a protective coating on the nonadhesive side. Such masking tapes do not require particular strength but, when removed, must leave no deposit or residue for the application of another color of lacquer.

Adhesive tapes for sealing packages are made with a crepe paper or cloth backing, requiring strong and durable adhesives on a substantial background. Their removability is unimportant.

BOOKBINDING ADHESIVES. Bookbinders customarily use two types of glue, one for the flexible parts, and another for the rigid parts, such as "board" covers. Selected and modified grades of animal glue have long been used for the back and flexible covers, both of which are continually subject to bending stresses. Certain types of resin adhesives are coming into wider use because of the shorter setting cycles, better flexibility under aging, and less susceptibility to moisture.

Starch pastes are usually used for "casing-in" or attaching the book to the cover, since these pastes are in the low-cost ranges and flexibility is not a requirement.

ABRASIVE MATERIALS. Adhesives perform an important function in sandpaper, emery cloth, and other sheeted abrasives. Formerly high grades of animal glue were employed. With the demand for better moisture resistance, as in wet rubbing, and for heat resistance, as under heavy cutting feeds, phenol-formaldehyde resin adhesives have come into use. These are thermosetting and have greatly increased the all-round durability of abrasive papers and cloths.

LIQUID ADHESIVES. For package closures, starch adhesives are prepared in liquid form with a wide range of ingredients for innumerable applications. Silicate of soda is also used, but its preparation is more complicated. Suppliers should be consulted for further details.

12. ADHESIVES FOR METALS AND DENSE MATERIALS

PROBLEMS OF METAL ADHESION. In the cementing together of metal, specific adhesion becomes important, since the penetration of adhesives into the surface is insignificant. Adhesives for metals were developed recently, under the urge of military necessity. It appears quite certain that their importance will increase as they are better understood and evaluated.

Surface Preparation. An important phase of any metal-to-metal gluing is the proper and adequate degreasing of the metal surfaces to give the adhesives a strong "bite." The surfaces of the metal should be cleansed thoroughly of any oil, grease, or other foreign materials. This may be done by the solvent wipe, vapor degrease, or by immersion in a heated acid bath, all as explained in British specification D.T.D. 915. After such treatment, the metal surfaces should be washed and dried. An alternate method is abrasion by steel wool, sandpaper, emery paper, or sandblasting, after which the surfaces should be washed, wiped, and dried to remove all dust particles.

A slightly roughened, rather than a smooth or polished, metal surface is desirable for strong durable joints. A simple test for adequately cleaned metal surfaces is to flow water over them, where it should form a continuous and unbroken film.

It is important that there be no unnecessary delay between cleaning the surfaces of the metal and applying the adhesive. Intervals of one to three hours may not be serious, but overnight delays may reduce the quality of the bond. After a week or ten days surfaces should be recleaned before bonding.

Design Requirements. Metal-to-metal joints, in lapped constructions, have attracted especial attention, since riveting has distinct limitations in thin lightweight metals, and welding, spot or seam, presents other types of difficulties, such as metallurgical changes in alloys at welding temperatures.

Any lap joint in tensile shear tends to distort, as shown in Fig. 3, thus combining a major factor of true shear along the joint with a minor effect of true tension across the joint. The extent of distortion, or eccentricity, is materially influenced by the thickness of the metal and the length of the overlap. The simplest form of lap is the plain, or square-end, overlap as shown in Fig. 3. It might be expected that a lap 1 in. long would be twice as strong in shear, as a lap $^1/_2$ in. long, but such is not the case. The disturbing influence is the elasticity of the metal, together with the distortion noted above, which causes greater stresses at the extremities of the lap than at its center. If the overlap is beveled down, the relation between the 1 in. and $^1/_2$ in. laps is greatly improved. A comparison of such lap constructions is shown in Fig. 4, where the "expectation" of normal strength increase is a straight line at the left of the chart, while the actual shear strength of plain lap joints is shown in the lower line, which approaches the horizontal. The shear strength of the beveled lap joint falls between the two lines, and does not diverge greatly from the expectation line. However, such beveled joints are likely to present difficult production problems and their strength, while recognized, has not been evaluated to any important extent.

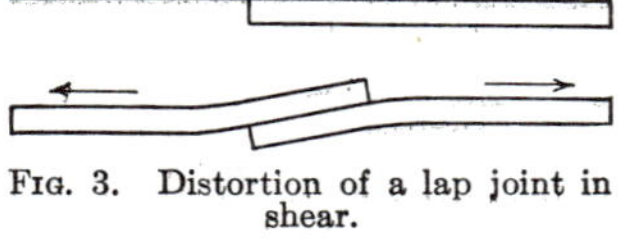

FIG. 3. Distortion of a lap joint in shear.

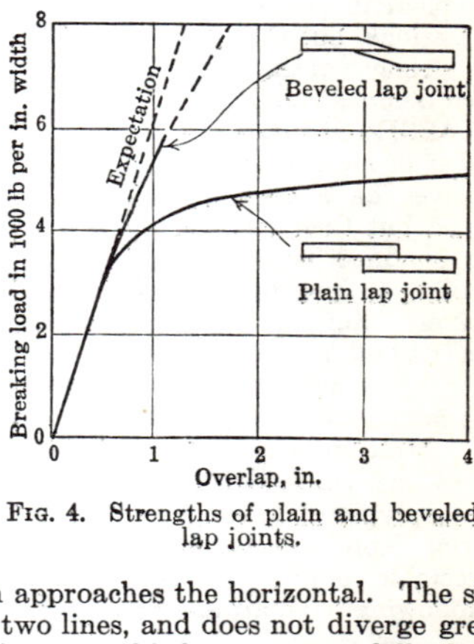

FIG. 4. Strengths of plain and beveled lap joints.

Studies have been made by Dr. N. A. de Bruyne of Aero Research, Limited (Ref. 14), on plain lap joints cemented together by Redux which are believed to be typical of other metal-to-metal adhesives of comparable strength. This research work has determined that the strength of the metal-to-metal lap joint varies directly as the width, while the effect of the metal thickness and overlap is expressed by the equation:

$$\text{Joint stress factor} = \frac{\sqrt{\text{Metal thickness}}}{\text{Overlap of joint}} \text{ (expressed in inches)}$$

A nomograph was developed expressing these relations, and from this nomograph, Figs. 5 and 6 have been derived. These curves apply to duralumin and steel.

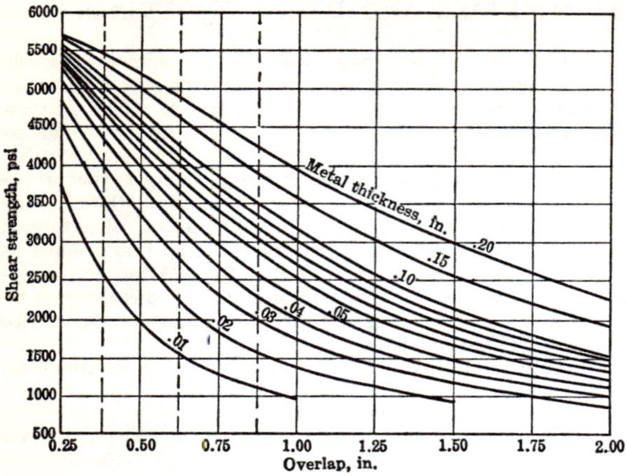

FIG. 5. Ultimate shear strength, psi, of metal-to-metal Redux lap joints.

It is to be noted (Fig. 5) that, although the maximum shear strength may be between 5000 and 6000 psi, the actual obtainable unit shear strength on most constructions is distinctly less, with definite decreases as the joints become longer and increases as the metal becomes thicker. The data given in these curves are in *pounds per square inch* of actual contact area. If the anticipated loading and metal thickness are known, the required overlap can be read from the chart.

The data given in Fig. 5 have been converted into the total strength of lap joints 1 in. wide, and have been recharted in Fig. 6. These curves indicate the length of overlap for a given metal thickness that will produce *maximum joint strength per inch of width*. Thus with two known variables, among three, the third can be determined from Figs. 5 and 6. These fundamentals of attainable strength are important in design problems.

Exhaustive tests by the U. S. Navy have shown that a combination of rivets and Redux adhesives gives several times the tensile strength of rivets alone. The economic phases of such a combination joint may be somewhat of a problem in commercial applications, but the ultimate strength of the joints is well demonstrated.

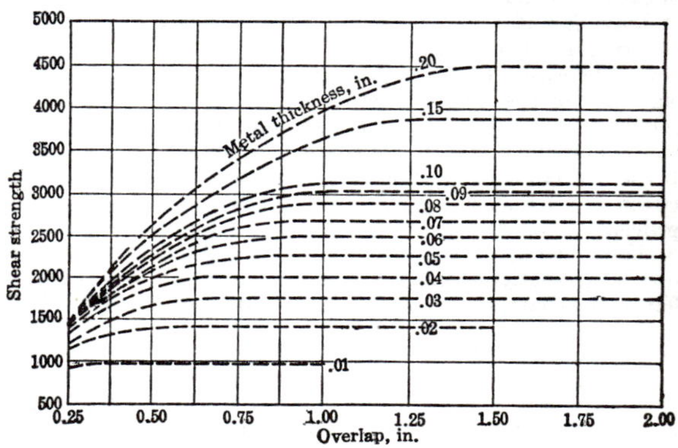

FIG. 6. Ultimate shear strength of metal-to-metal Redux lap joints one inch wide and of indicated overlap.

CLASSIFICATION. A number of metal-to-metal adhesives have been developed but their use has been confined largely to lap joints or spot joints of relatively limited area. Laminated all-metal sheets have not yet become commercially important. Metal-to-wood adhesives, as employed in making plymetal, have industrial applications of appreciable volume, and are discussed in this article. Among the metal-to-metal adhesives, the following have attained recognition:

Redux, Rohm and Haas Co., Resinous Products Division, Philadelphia, Pa.
Cycleweld, Chrysler Corporation, Detroit, Mich.
Plastilock, B. F. Goodrich Chemical Company, Akron, Ohio.
Reanite, U. S. Stoneware Company, Akron, Ohio.
Metlbond, Consolidated Vultee Aircraft Corporation, San Diego, Calif.
Araldit, Ciba Products Corporation, Kimberton, Pa.
Robertson Bonded Metal, H. H. Robertson Company, Ambridge, Pa.

The processing requirements for the first two are outlined, briefly, herein. For further details on these, and for data on the others, the literature of the suppliers should be consulted.

METAL-TO-METAL ADHESIVES—REDUX. This is a two-component adhesive, used successively and not combined, a brownish liquid (phenolic) and a white granular powder, developed originally in England. It is chiefly thermosetting, with some thermoplastic characteristics, and cannot be catalyzed for room temperature work. The liquid should be stored in tight receptacles at approximately 40 F. If separation occurs before use, it can often be redispersed by stirring. If a smooth mixture is not obtainable, the material should be discarded. When properly applied and cured, it produces lap joints with tensile strengths exceeding 5000 psi on joints of favorable design. Tensile shear tests are made for evaluation, as shown in Fig. 1.

Application. Metal surfaces should be properly cleansed, as previously mentioned. As soon as practicable after cleaning, the liquid should be brush-applied on the joint area, at the rate of 20 to 25 lb per 1000 sq ft per surface, or 40 to 50 lb per joint. As much powder as will stick is immediately dusted on the liquid, and any surplus is tapped off. Solvent evaporation should follow, from 1 to 24 hr at room temperature, sometimes longer. Forced drying at 125 to 150 F may be used to accelerate this drying. The pieces are assembled, taped to prevent slippage, and cured between metal platens heated to 275 to 325 F, at pressures of 300 to 500 psi for approximately 15 min. Pressures should be adequate to reduce glue-line thickness to 0.003 in. to 0.006 in. Under favorable conditions, with close-fitting surfaces, pressures may be as low as 100 psi. Pressure may be released, usually, while assemblies are hot, except in the case of unlike metals, in which rates of heat expansion vary significantly.

Metal Types to Be Bonded. Most Redux tests have been made on aluminum alloys, but satisfactory results also have been obtained on stainless steel, magnesium alloys, cold-rolled steel, and galvanized steel. Results with nickel and its alloys have not been satisfactory. Lead and copper, with their derivatives, do not bond well. Joints of significant area between metals of unlike rates of heat expansion are seldom satisfactory. An example of this would be a cast-aluminum handle against a steel blade in a plasterer's trowel.

Durability. Redux bonds in aluminum have excellent durability under room temperatures and at low temperatures down to −125 F. At elevated temperatures of 175 F, Redux joints may exhibit 60 to 75% of normal strength, but weakening progresses significantly above that point. Continuous exposures to salt-water spray for 10 days show little loss of strength. Strengths are well maintained on exposure to liquid fuels (45% aromatic), engine oil, and hydraulic fluids, such as oils, glycol, and water.

METAL-TO-METAL ADHESIVES—CYCLEWELD. Cycleweld cements come in several types for different purposes, so that suggestions as to applications must be in general terms, and further details should be obtained from the suppliers. Most types of this adhesive are supplied as dark-colored liquids of limited stability, containing a base of natural or synthetic rubber. In general those that gel in storage can be reconditioned by heat and stirring. When properly applied and cured, they produce lap joints with tensile shear strengths of 4000 psi where designs are favorable for maximum strength.

Application. Metal surfaces should be properly cleansed, as previously mentioned. As soon as practicable after cleaning, the liquid should be applied, preferably by spray gun, but sometimes by brushing or roller coating, usually with 3 to 5 coats on each surface of the joint, with intervening flash periods of 5 to 10 min between coatings. These combined applications, after drying 15 to 48 hr at room temperatures, or accelerated in an oven at 180 F or less, are ready for assembly and should be taped in position to prevent slippage. Hot-press curing is at temperatures of approximately 325 F, for 15 to 20 min, at pressures of 200 psi. Some types can be cured at lower temperatures, as can well-fitted joints at lower pressures.

Metal Types to Be Bonded. Most Cycleweld tests have been made on aluminum and its alloys, but different types of this adhesive are available for most of the common metals, as well as for cast, molded, and laminated plastics and rubbers. Special types are available for materials that deteriorate under normal temperature and pressure requirements. Joints of significant area between materials of different expansion factors require special consideration.

Durability. Cycleweld bonds exhibit excellent durability at normal room temperatures for extended periods, as well as at low temperatures down to −70 F. At elevated temperatures some types withstand 300 F, others 150 F, and suppliers should be consulted. Oil resistance is excellent, but in general salt spray has a deteriorating effect unless the joint is adequately protected. With diverse nonmetallic materials, durability to exposure is usually that of the nonmetallic material.

METAL-TO-WOOD ADHESIVES. Plymetal is a combined wood-metal product, in which metal sheets are used with wood layers, either for facings or interlayers, adding greatly to stiffness, strength, and surface hardness. Since the expansion, under heat, between metal and wood creates undue stresses, most metal-to-wood bonding is done with room temperature adhesives. Exceptions are facings of very thin metals, almost in the foil range, where the expansive force of the metal is not dominant.

Casein latex glues have been used for many years in plymetal constructions with excellent results. Such joints have good strength properties and satisfactory durability under moderate exposures, but are not suitable for marine or extreme weather exposure. The application procedure follows closely that of the wood ahesives, except that the metal facing is usually a separate bonding operation after the plywood has been dimensioned to size and planed or sanded to accurate thickness. Metal surfaces, usually aluminum or steel, must be thoroughly cleansed from oil, grease, etc., before gluing. Interlayers may be assembled in single or double gluing steps.

Coated Metal Sheets. A more recent development is the sizing, priming, or precoating of the metal sheets with Redux, Cycleweld, or other metal adhesive, which is cured in situ by oven or platen heat, against a nonadhering caul. Such precoatings must be thin, not over 0.007 in., and of uniform application without lumps or irregularities. A second gluing operation is one that combines these precoated sheets with layers of veneer or wood, utilizing room temperature resin adhesives of compatible types, under usual plywood pressures with required clamping cycles. This method avoids undue stresses in the gluing process.

A unique development in this class is Robertson bonded metal, a process which solders a textile fabric to a thin metal sheet with a low-temperature solder, quenched quickly to avoid burning the fabric seriously. Layers of veneer can be glued normally to the cloth side.

Wood-to-metal Assembly Joints. Assembly joints of any wood-to-metal area can be made by the foregoing precoating process. Where assembly joints are of limited area, and can be adequately heated, a metal adhesive can be employed on the metal side, with a compatible wood adhesive on the wood side, to be cured simultaneously by heat. This process has been used in constructing wood-metal aircraft in England.

13. ADHESIVES FOR RUBBER, GLASS, LEATHER, ETC.

Adhesives have been developed for a wide range of materials other than those described for wood, paper, and metal.

RUBBER. Many rubber cements have been developed from natural and synthetic rubber ingredients and latices. Blood albumin, casein, and the phenols are some of the additional ingredients. Other chemical derivatives have been used, butadiene being an outstanding example. All the larger rubber companies have been active in this field; their literature should be consulted for further details.

GLASS. The adhesives used for optical purposes are usually Canadian balsam. Several of the polyester resins have given satisfactory results and are approved for certain types of lenses and prisms. Sheets of safety glass are usually glued together with polyvinyl butyral resins, but a number of other similar adhesive resins have also been used successfully. The suppliers should be consulted.

LEATHER. Animal glues have long been popular in this field, as have other chemical compounds, most of which are tacky and set quickly at room temperatures. Electronic heat with resin adhesives is used in some cases for short progressive bonding cycles. Many types of latex are used alone, or in combinations, since flexibility is an important consideration in leather products. The literature of the industry should be consulted. There are many other adhesive applications in diverse fields for which data can be obtained in Ref. 12.

REFERENCES

1. B. C. Boulton, *Use and Manufacture of Plywood and Glue*, Sir Isaac Pitman & Son, London, 1920.
2. R. H. Bogue, *Chemistry and Technology of Gelatin and Glue*, McGraw-Hill Book Co., New York, 1922.
3. C. H. Teesdale, *Modern Glues and Glue Testing*, Periodical Publishing Co., Grand Rapids, Mich., 1922.
4. Piper and Morse, *Soybeans*, McGraw-Hill Book Co., New York, 1923.
5. T. R. Truax, *Gluing of Wood*, Forest Products Laboratory, No. 1500, 1924.
6. J. G. Vail, *Soluble Silicates in Industry*, Reinhold Publishing Corp., New York, 1928.
7. Sutermeister and Browne, *Casein and Its Industrial Applications*, Reinhold Publishing Corp., New York, 1939.
8. H. Braude, *Adhesives*, Chemical Publishing Co., New York, 1943.
9. P. I. Smith, *Synthetic Adhesives*, Chemical Publishing Co., New York, 1943.
10. T. D. Perry, *Modern Wood Adhesives*, Pitman Publishing Corp., New York, 1944.
11. Kline, Reinhart, Rinker, and DeLollis, Effect of catalysts and *pH* on strength of resin-bonded-plywood, *J. Research Natl. Bur. Standards*, Nov. 1946, and *NACA Tech. Note* 1161, 1947.
12. J. Delmonte, *Technology of Adhesives*, Reinhold Publishing Corp., 1947.
13. Rinker and Kline, *Survey of Adhesives and Adhesion*, *NACA Tech. Note* 989, 1945.
14. N. A. de Bruyne, *Aircraft Eng.* (London), Vol. 16, April 1944, pp. 115–118.

SECTION 22

INDUSTRIAL HEATING PROCESSES

BY

W. TRINKS, *Professor Emeritus of Mechanical Engineering, Carnegie Institute of Technology*

R. M. CHERRY, *Industrial Heating Divisions, General Electric Company*

INDUSTRIAL FURNACES

BY W. TRINKS

INDUSTRIAL ELECTRIC HEATING

BY R. M. CHERRY

INDUSTRIAL FURNACES

By W. Trinks

The purpose of industrial furnaces is to heat solid, processed material so that the product, after cooling, has no flaws or defects which can be attributed to incorrect heating. In general, *ovens* heat material to less than 1000 F, and furnaces heat material to more than 1000 F. In *kilns*, water or carbon dioxide is first driven off the charge, which is then heated further. Materials to be heated are, as a rule, metals, but ceramic materials are also heated in industrial furnaces and plastics are heated in ovens.

The temperatures to which iron and steel are heated for various operations are given in Table 1. For other metals, temperatures are kept well below the melting points. For the latter values, see p. 1-04, or Fig. 1.

Table 1. Furnace Temperature Range for Industrial Processes on Iron or Steel

Process	Temperature Range, °F	Process	Temperature Range, °F
Hammer or press welding	2350 to 2550	Hardening carbon steel (quenching)	1450 to 1550
Hardening high-speed steel *	2200 to 2400	Annealing cold-rolled steel	1250 to 1400
Drop forging	2250 to 2370	Drawing	1150 to 1200
Rolling (low-carbon steel)	2150 to 2275	Nitriding	950
Bending and rough forming	1900 to 2100	Tempering	500 to 800
Normalizing, also for carburizing	1650 to 1750	Bluing †	500 to 650
Rough annealing	1500 to 1650		

* Special steels require various temperatures for hardening, usually ranging from 2400 F for greatest hardness to 1850 F for greatest toughness (for tungsten steel).

† Temperatures for drying and japanning are still lower.

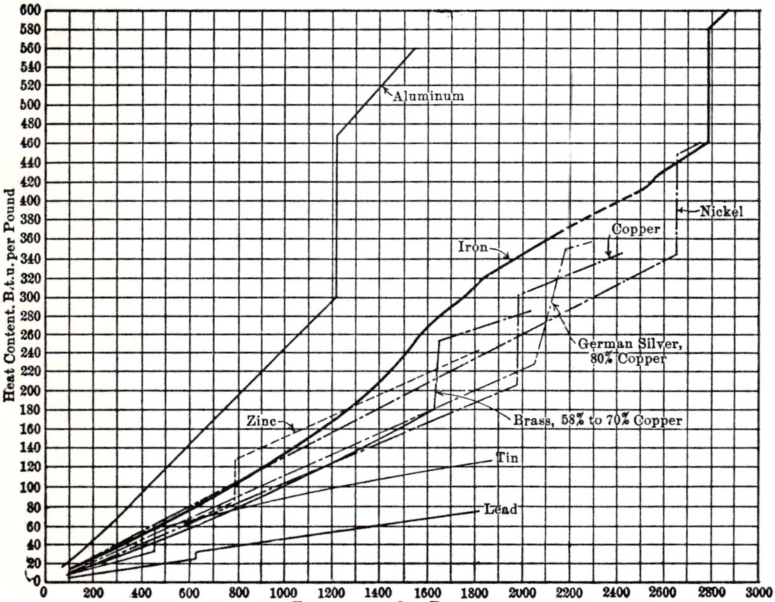

FIG. 1. Heat content of metals at various temperatures.

Design and construction of industrial furnaces have reached a high degree of perfection in the United States. It pays to install a furnace that was designed and built by a reputable furnace builder.

1. FURNACE CHARACTERISTICS

CLASSIFICATION OF FURNACES. Furnaces are classified:

According to use, such as reheating, forge, annealing, bright-annealing, heat-treating and normalizing furnaces; and soaking pits.

According to the material or shape which they heat, as rivet, bolt-heading, plate, and angle furnaces; and brick kilns.

According to the movement of the charge in the furnace, as batch (also called in-and-out type), continuous-pusher, automatic or continuous conveyor, rocker-bar or intermittent-conveyor, and car type furnaces; and furnaces with a rotating hearth or with revolving drum.

According to the fuel or other source of heat energy, as combustion, gas-fired, oil-fired, electric-resistor, and induction furnaces.

According to the method of fuel saving, as regenerative, recuperative, and preheating-chamber types.

In muffle furnaces the charge is heated in a gastight enclosure and is separated by it from the source of heat. Pot furnaces are muffle furnaces.

ECONOMY OF HEATING must be based on unit weight of perfect product. The term "perfect product" involves such features as freedom from internal cracks caused by too rapid heating, uniformly high quality of heat-treated products, uniform annealing, uniform enamel of first quality, proper distribution of hardness, freedom from scale and ashes, freedom from burning of high-carbon steels, etc. In certain operations, such as plate bending, railroad spike forming, bolt heating, and tool hardening, uniformity of temperature throughout the heated stock is not necessary.

True heating economy is inversely proportional to total cost per unit of product. It can be expressed by

$$E = W \div (I + M + L + F)$$

where E = true heating economy; I = interest on investment; M = cost of maintenance; L = cost of labor, including overhead; F = cost of fuel; and W = unit weight of perfect product.

The amount of capital that profitably can be invested in a furnace installation depends on regularity and quantity of output of heated material. Cost of maintenance, M, depends on furnace temperature, furnace design, selection of refractories, care used in operation, and cost per unit of labor and materials used in repairs. Cost of operating labor, L, depends on regularity of production with regard to continuous output, size and shape of stock, and the provision of labor-saving equipment conjointly with unit cost of labor per hour. Cost of fuel or energy, F, depends on heating temperature, heating process, unit cost of fuel or energy, furnace design, and care used in operation. Frequently the importance of fuel cost is overemphasized at the expense of the other items (including quality of finished product), because it is more readily amenable to calculation and measurement.

DISTRIBUTION OF HEAT. The potential heat energy of the fuel (or other sources of heat) is distributed as follows:

Useful Heat in the Charge. (See Fig. 1.) If ceramic materials are the charge, their mean specific heat from room temperature to furnace temperature is used, including driving off the water of crystallization. Chemical changes, like expelling carbon dioxide, are also employed.

Heat Unavailably Stored in Furnace Walls. In furnaces that operate only during the day and are cooled down overnight much heat is lost by storage in (and, later, by dissipation from) the firebrick walls. Table 2 roughly conveys this loss.

Table 2. Heat Stored in Firebrick Walls and Conducted through Walls during Heating-up Period in Btu per Square Foot of Hearth

Type of Furnace	Heat Loss, Btu/sq ft
Continuous-pusher type	125,000
Small forge furnace	60,000
In-and-out, 2500 F after 12-hr shutdown	360,000
Same, after 50-hr shutdown	500,000

This loss is one of the reasons why furnace builders cannot guarantee fuel economy unless all operating conditions are known. For intermittent heating, the use of insulating refractories is advisable. Furnace temperature is reached in 30 to 35% of the time needed with firebrick walls, and heat loss is reduced correspondingly. In furnaces that are shut down overnight, insulating refractories save almost one-third of the total fuel.

Heat Losses through Solid Walls. Figure 2 contains information on this loss. In practice, the loss is up to 20% greater, because cracks open up between the bricks, and the thermal conductivity of the bricks changes with time. For heat losses through hearths and to water-cooled parts, see the references.

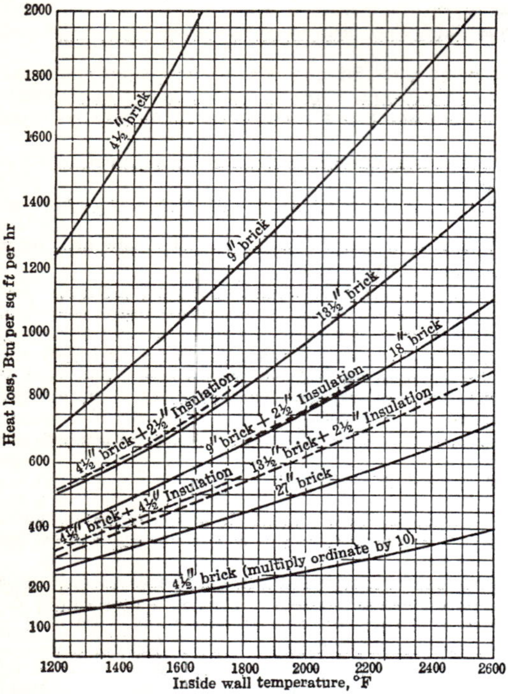

Fig. 2. Heat loss through furnace walls. (Reproduced by permission from *Industrial Furnaces*, by W. Trinks, Vol. 1, John Wiley and Sons, 1934)

Heat loss from containers and conveyors passing through the furnace equals weight times temperature rise, times mean specific heat.

Heat radiated through openings in the furnace walls equals black body radiation (for which see *Power* Volume of Kent's *Mechanical Engineers' Handbook*) through the opening multiplied by a factor which is a function of the ratio (least width of opening) divided by thickness of wall at opening. Average values are:

Ratio	0.25	1.0	4.0
Factor	0.3	0.6	0.86

Unconsumed Combustible in Refuse. This loss occurs only in coal-fired furnaces. Coal-fired industrial furnaces are very little used in the United States.

Unconsumed Combustible in Stack Gases. With automatic control of fuel-to-air ratio this loss is small, seldom exceeding 4% of the lower heating value of oil or gas. The loss is greater if coal is burned on the grate. The loss is very great in coal-fired brick kilns of the periodic type.

Sensible Heat in Stack Gases. This loss varies with furnace temperature, type of fuel, and amount of excess air. Average values (including a small amount of unconsumed combustible) are given in Table 3. The products of combustion of lean fuels (blast furnace or producer gas) carry away an even greater fraction of heat than is indicated by Table 3.

Table 3. Sensible Heat (Fraction of Lower Heating Value) in Products of Combustion

Temperature of gases, °F	1000	1200	1400	1600	1800	2000	2300
Percentage of lower heating value	26	31	36	42	47	53	62

Heat conducted through resistor terminals in electric furnaces or through part of the charge projecting into the open becomes a sizable loss at furnace temperatures of 2200 F and above. For its calculation, see the references at the end of this chapter.

The consumption of fuel per unit weight of material heated can be computed by detailed calculation of each of the eight losses. Although this method is followed for new, large furnaces, the fuel consumption of most furnaces is determined by comparison with that of existing similar furnaces. Table 4 contains average values of fuel consumption. On account of the influence which the eight losses have and because of the human element, the fuel consumption may be as low as 60% of the values in the table, or it may exceed them by 100%.

CAPACITY OF FURNACES. The number of furnaces required to heat a given weight of metal to a given temperature in a given time and the size of each furnace depend on a number of variables, such as: (1) thermal conductivity and specific heat of metal; (2) requirements for uniformity of temperature in charge; (3) rate at which heat may be applied without injury to stock; (4) ratio of surface of charge to mass; (5) ratio of surface of furnace

Table 4. Average Fuel Consumption of Industrial Furnaces

(Material heated is steel excepted where otherwise noted.)

Type of Furnace and Operation	Highest Temperature to Which Stock Is Heated, °F	Btu in Fuel per Pound of Material Heated	Thermal Efficiency, %
Drop-forging or bolt-heading furnace, oil fuel	2300	3200	11
Bolt-heading furnace, coal fired	2100	6000	6
Ingot-forging furnace, in-and-out type, oil fired	2200	2500	14
Ingot-forging furnace, in-and-out type, regenerative, gas fired	2200	1900	18
Mill-type continuous furnace	2200	1100	32
Mill-type continuous furnace, recuperative	2200	900	39
Car-type annealing furnace for castings	1600	1200	21
Annealing furnace for brass	750	750	10
Case-hardening (carburizing) furnace	1700	2500	11
Heat-treating (quenching) furnace	1400	1150	19
Heat-treating (tempering) furnace	1150	800	20
Enameling furnace (muffle)	....	....	15
Electric heat-treating furnace	1500	2500	80; * 10 †

* Based on electrical energy. † Based on fuel.

interior to exposed surface of charge; (6) state of subdivision of charge; (7) circulation of furnace gases; (8) method of heat liberation and size of heat-liberating devices; (9) cycle of operation of heating process; (10) arrangement of charge in furnace; and (11) radiation of furnace gases. In determining number and size of furnaces, total required hearth area is calculated either from the above given variables or from the following rules. It is then subdivided into standard-size units to suit plant conditions.

In the commonly used method of industrial heating, furnace temperature exceeds final temperature of charge by 50 to 125 F. Under that condition, heat "penetrates" into iron or steel at the rate of $^1/_8$ in. in 5 min. This means that a plate $^1/_4$ in. thick, standing upright in a furnace, is heated with sufficient uniformity in 5 min. This rule was derived from forging practice and is sufficiently accurate as long as furnace temperature is not much higher or lower than forging temperature. For brass, temperatures are equalized twice as fast, for aluminum six times as fast, and for copper nine times as fast as for steel; but, in spite of this fact, the rate at which these metals can be heated, per square foot of furnace hearth area per hour, usually is less than for steel which is being heated to forging temperature. Reasons are:

1. The nonferrous metals are heated to temperatures much below 2200 F. At low temperatures, radiation from furnace gases and from electrical resistors is weak compared to radiation at high temperatures. If steel is heated to 1400 F, the rate of heating is about one-half of the 2200 F rate. The absorption of radiant heat is reduced by the smooth, bright surface of most nonferrous metals. Black copper surfaces, however, absorb heat almost as rapidly as oxidized steel surfaces.

2. Forging ingots are heated at the rate of 16 to 20 min per in. diameter, if of soft steel, and at the rate of 30 to 40 min per in. of diameter, if of "tender" steel. Certain steels are very sensitive to temperature changes at "blue brittleness" (around 500 to 800 F). Above dull red heat, they cannot be injured by rapid heating. When a new alloy is to be heated, experimenting is necessary in order to find the safe rate of heating.

3. In ordinary furnaces, steel can be heated at the rate of 25 to 35 lb per sq ft of hearth area per hr for heat-treating; at the rate of 60 lb for annealing; 80 lb for uniform-temperature forging and rolling; and 110 to 130 lb for work in which 40 F temperature difference per in. of thickness of stock is allowable.

4. Steel bars for drop forging can be heated at the rate of 5 to 10 min per in. of diameter.

5. The time for heating plus soaking grows more rapidly than the diameter of the piece being heated because the soaking time is roughly proportional to the square of the diameter. Very large forging ingots (in excess of 100 in. in diameter) require up to 80 min per in. of diameter.

The heating time is shortened materially if the cold charge is heated just as rapidly as possible without injury to the charge until the surface temperature of the charge slightly exceeds the desired final temperature. The furnace temperature is then dropped slightly below the desired final temperature of the charge. Equalization of temperature in the charge, if it is solid and without voids. takes place quickly. The effect of high furnace temperatures upon the rate of heating square steel bars (between 1 and 5 in. in diameter) is illustrated by Fig. 3. Depending on the material being heated, heating plus soaking can be accomplished in 30 to 70% of the time required with constant furnace temperature. In continuous furnaces, this heating method is accomplished by a very hot heating compartment and a cooler soaking zone. When a new alloy or a ceramic material is to be heated by this method, experimentation is necessary to find the highest rate of heat input that the material can endure without injury.

6. Coiled or piled material requires from 80 to 200 min per in. of thickness, if heat is applied from both sides. The wide variation is caused by the number of air spaces per inch in the charge, by the number of contact points, and by the heat flowing in at the edges of sheets.

Heat Liberation. For limits of heat liberation in electrically heated furnaces, see the chapter on industrial electric heating, which follows this one.

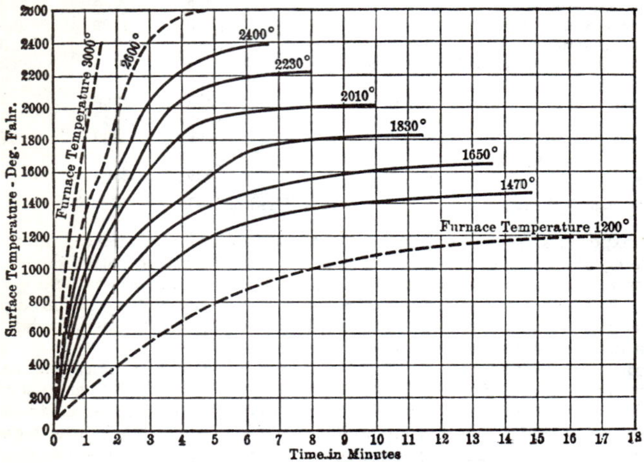

Fig. 3. Heating time per inch of diameter, square steel bars, at various furnace temperatures.

In furnaces of the combustion type, from 1.5 to 100 Btu are liberated per cubic foot of combustion space per second. The low value applies to very poor mixing of fuel and air, very imperfect utilization of volume of combustion chamber by flame or gas jets, coarse fuel, and cold air. The high value is attained by thorough mixing of fuel and air, perfect utilization of combustion space, very fine atomization or powdering of fuel, and preheating of air to 1000 F. Even higher rates are attained in modern cup burners in which temperatures in the neighborhood of 3000 F are carried.

FLUES AND PORTS. If furnaces are properly proportioned, the following port sizes are approximately correct, based on steel as stock to be heated, and for 100 lb per sq ft of hearth per hr. If a smaller quantity is to be heated, the outlet flues should be partly covered by bricks.

Temperature of furnace, °F	1400	1800	2200
Port area, sq in. per sq ft of hearth	2.0	3.5	7.3

These figures hold for batch-type furnaces of medium size without recuperation or regeneration. For large and important furnaces, the ports and flues must be calculated on the basis of flue gas quantity in unit time and pressure drop.

MATERIALS OF CONSTRUCTION FOR FURNACES. The most commonly used material is firebrick (see also Section 5), either solid or porous. The porous form is called insulating firebrick and is weaker than solid firebrick but lighter and, therefore, carries less superimposed load. High-alumina bricks withstand higher temperatures than ordinary firebrick.

If furnaces heating steel for forging or rolling are driven hard, the scale softens and forms ferrosilicate (slag) with firebrick or sand. Formation of slag is avoided, if hearths are made of chromite or magnesite. If the furnace temperature does not exceed 2300 F, no slag is formed, even with firebrick hearths. Silicon carbide serves for muffles (enclosing the charge or the flame), for baffles, and for hearth rollers. Cast iron serves for doors, for door frames, and for hearth plates of furnaces carrying not more than 1600 F. Plain carbon steel is used for binding and occasionally for door frames. Heat-resisting alloys serve for hearth chains, hearth rollers, muffles, containers, electrical resistors, and other parts subjected to furnace heat. The higher the temperature to which they are subjected, the higher must be the content of chromium, nickel, and other alloying elements. At high temperatures even the best alloys creep under stress and have a limited life.

2. FUELS

SELECTION OF FUEL OR ENERGY. The cost of fuel (or other sources of heat energy) is, in many instances, an extremely small fraction of the total cost of the manufactured product, and a minor fraction of the heating cost. In such cases, the effect of the form of heat energy on the finished product is most important, and the unit cost of heat energy is not a matter of consideration. In other cases, the fuel cost is a large item in the total cost, and cheapness of fuel is the deciding factor. A skilled and careful operator can do good heating with almost any fuel. The following fuels are commonly used (see also Eshbach's *Handbook of Engineering Fundamentals* and *Power* Volume of Kent's *Mechanical Engineers' Handbook*).

Natural gas is an excellent fuel for all heating operations. In many localities it is available at low cost. Wherever uninterrupted furnace operation in cold weather is required, a standby installation is needed for burning oil, butane, casinghead gasoline, or some other fuel which can be stored because, in many industrial sections, the domestic consumer (who pays a higher rate) is given preference in cold weather.

City gas is a mixture of water gas and retort gas. Since the cost of this excellent fuel is high, its use is commonly restricted to small and medium-sized furnaces.

Water gas is a clean but very poisonous fuel which burns with a nonluminous flame. It is made in an elaborate gas generator with waste-heat boiler and automatic valve operation. A gas storage tank is also needed. This gas is useful for large installations only. The impending availability of cheap oxygen will greatly simplify the production of water gas.

Raw producer gas is the standard fuel for rough heating in large furnaces, where the higher grade fuels are not available, or where low fuel cost is an important item. Gas producers are located close to the furnaces, so as not to lose the tarry vapors by condensation in the ducts.

Cold, clean producer gas was, at one time, the standard fuel for a group of scattered factory furnaces. The steady increase in the efficiency of electric power generation has caused the more general adoption of electrically heated furnaces and has caused cold, clean producer gas to be all but abandoned.

By-product coke oven gas and **refinery gas** are seldom available at some distance from the works where they are generated. Both gases are very good fuels.

Liquefied petroleum gases, such as butane, propane, and casinghead gasoline are useful fuels wherever an uninterrupted supply is obtainable. In most localities, the cost is rather high.

Heavy Fuel Oil (Nos. 5 and 6 on the commercial scale). Number 6 is also known as Bunker C oil. Both grades are the residuum left from distillation or cracking. These oils must be preheated and must flow continuously past the burners in a ring-type piping system. They are very good for rough heating.

Light Fuel Oil (No. 3 on the commercial scale; there is no No. 4). This oil is a distillate and needs no preheating. It is also known as gas oil or as diesel oil. It costs more than heavy fuel oil. Number 2 oil is also used occasionally.

Bituminous coal is seldom burned on grates or stokers of industrial furnaces in the United States, except in so-called air furnaces, which lie outside the scope of this section. Coke is occasionally burned under core ovens in foundries. Solid fuels are burned in industrial furnaces in some European countries where cheap labor as well as willing and attentive craftsmanship is available.

Powdered coal is useful in large furnaces only, because satisfactory small individual pulverizers have not yet been developed for powdering coal (at reasonable cost) to sufficient fineness for combustion in industrial furnaces of average size. If powdered coal is burned in small furnaces, unburned coke particles and ash are discharged into the workshop and into the stack.

Coal tar is available in steel works only. It is similar to No. 6 oil but is more difficult to handle. Topped tar is coal tar from which the tarry acids have been removed.

Electrical energy takes the place of fuel in many heating operations. The ever-increasing size of turbines and generators and the utilization of high steam pressures and temperatures have reduced the cost of labor and fuel to such an extent that electrical energy can compete with fuel in industrial furnaces up to 1800 F. Moreover, electrically heated furnaces can be placed in the production line without discomfort to the workers in the shop.

3. HEAT SALVAGE

Heat salvage, although very desirable from the standpoint of conservation of national resources, always leads to complications and is, for that reason, practiced on large furnaces only.

Wall losses, in a continuously operated furnace, can be considerably reduced by thick walls and insulation, provided the wall material does not sag or melt at furnace temperatures. The loss to water-cooled parts is reduced by covering the parts with refractory material as much as possible. This loss is also reduced by circulating distilled water through the parts and cooling it by raw water in a separate cooler. In that manner, water temperatures up to 190 F can be used without deposits of lime in the tubes.

Regenerators are piles of brick which are alternately heated by outgoing products of combustion and cooled by incoming combustion air, whereby the latter is preheated. As a rule, direction of flame in the furnace also alternates. Furnaces with regenerators become part of the building. Regenerators are in use in comparatively few furnaces, mostly under those in which heavy pieces of steel are heated. As a rule, the shape and weight of these pieces changes from time to time.

Recuperators are heat exchangers in which heat flows in a steady stream, through a partition, from hot products of combustion to cooler, incoming combustion air. Many industrial furnaces do not readily lend themselves to heat salvage by recuperators. For forging and rolling temperatures (2200 F), this dilemma exists: Refractory recuperators eventually spring leaks, but most metallic recuperators burn out. The latter misfortune is avoided by mixing cold air with the outgoing products of combustion under control of a thermostat.

Heat in chain conveyors is salvaged by keeping the chain in the furnace during the return travel. In tunnel kilns, the heat in both the outgoing products and in products of combustion is largely recovered by preheating the incoming products and also the combustion air.

4. COMBUSTION DEVICES

BURNERS for gas and oil have been developed to a high state of perfection. Their selection may well be left to the furnace builder, because the burner must be adapted to the furnace with which it is to be used. Burners must be so designed and so located in the furnace walls that the charge is uniformly heated at all points of the hearth. Burners must permit adjustment of furnace atmosphere from excess air to deficiency of air.

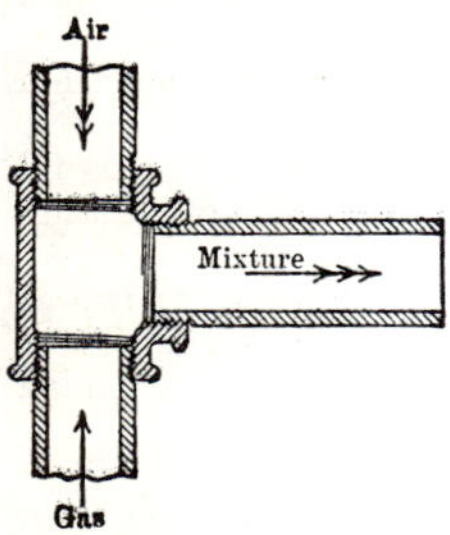

FIG. 4. Tee-mixing burner for gas.

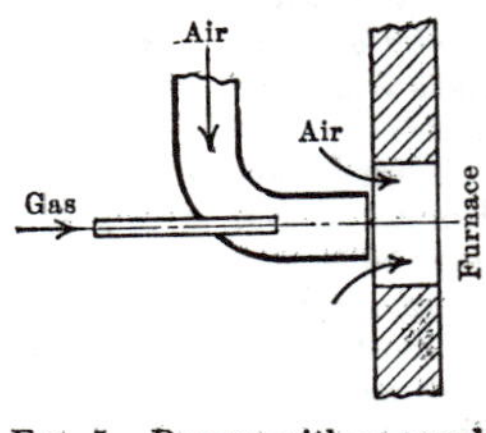

FIG. 5. Burner with gas and part of air under pressure.

Gas burners are mixers of gas and air. The mixing is obtained by gas and air impinging upon each other (Fig. 4), by turbulent flow in a pipe (Fig. 5), or by inspiration (Fig. 6). Burners of the inspirator type are very common. However, their turndown is limited, not only because the air-to-gas ratio changes, but also because the flame creeps back and causes backfires. Flames carrying unconsumed carbon particles are luminous and transmit more heat by radiation than nonluminous gases of the same temperature. For that reason, luminous-flame burners are much used. Such a burner is illustrated in Fig. 7. As a rule, raw producer gas is admitted to the furnace through ports of refractory material, which alternate with ports through which air is admitted. Such ports do not permit adjustment of flame length. Burners are on the market which, when using raw producer gas, permit adjustment of flame length and of furnace atmosphere.

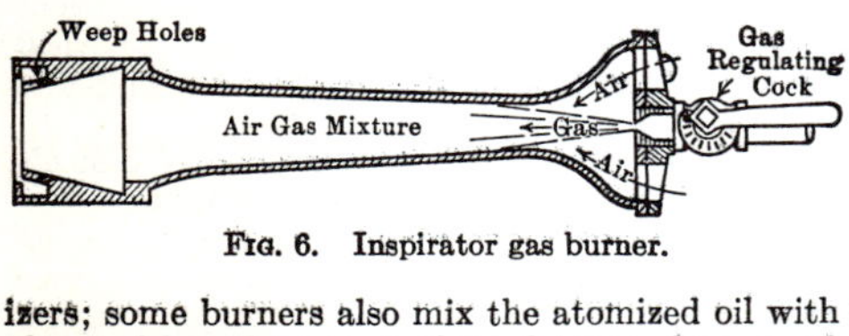

FIG. 6. Inspirator gas burner.

Oil burners are mainly oil atomizers; some burners also mix the atomized oil with combustion air. Atomization is never obtained by oil pressure alone, because fineness of atomization suffers when the burner is turned down. Three methods of atomization are in use:

1. By steam or air at high pressure. This method is preferred when a long flame must be given direction. A burner of this type is shown in Fig. 8.

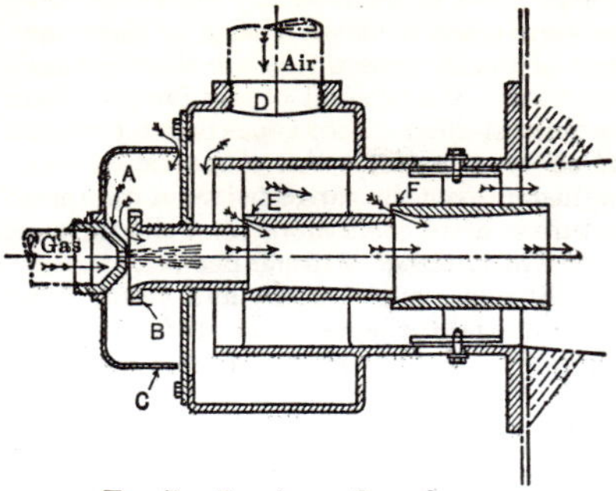

Fig. 7. Luminous-flame burner.

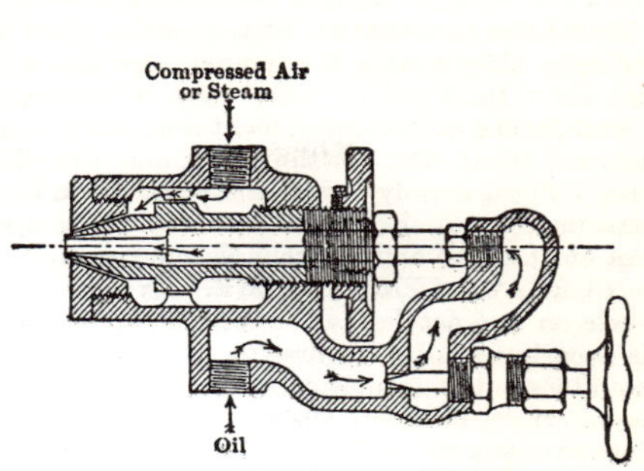

Fig. 8. High-pressure oil burner.

2. By air of 1 to 5 psi pressure and by oil pressure of 30 to 90 psi, the high oil pressures going with low air pressures. These burners are called low-pressure burners. Examples are shown in Fig. 9.

3. By fan blast of 1/4 to 1 psi and oil pressure of 30 to 150 psi, the high oil pressure again going with the low air pressure. This method of atomization is shown diagrammatically in Fig. 10. The rotary burners that are frequently applied to steam boilers and water heaters fail, if applied to industrial furnaces, because of the high-temperature heat which is radiated from the furnace interior to the burner.

Burners for powdered coal consist of two pipes, one for coal and one for air. The characteristics of the powdered coal flame depend mainly on the size of the coal particles and on turbulence in the furnace.

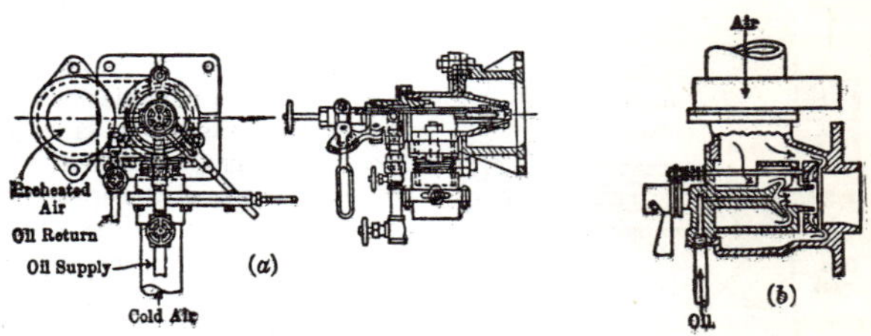

Fig. 9. Low-pressure oil burners.

Grates and stokers serving industrial furnaces do not differ from those used for other purposes, with the exception that air and steam are blown under the grate into the closed ashpit whereby the grate or stoker becomes a gas producer and some pressure can be maintained on the furnace hearth.

LABOR-SAVING EQUIPMENT. In many factories, rising costs of labor make the cost of fuel (or electrical energy) a secondary consideration and put labor saving in the first place. Labor is saved by charging (and discharging) machines, by continuous furnaces of the pusher type, by furnaces with rotating hearth, and by so-called automatic furnaces, through which the material to be heated is conveyed by mechanical means. The suitability of the various conveying devices depends largely on the furnace temperature. Up to 1200 F (with careful design, up to 1400 F), the chain conveyor with solid links or woven wire mesh is satisfactory. Up to 1800 F, rocker bars (walking beams) and driven rollers of heat-resisting metal are suitable. Up to 2300 F (but not above), rollers of silicon carbide have given a good account of themselves. For still higher temperatures, the pusher-type continuous furnace, the rotating-hearth furnace, and the batch-type furnace with charging machine serve well. Materials handling equipment may be employed for bringing pieces to the furnace and conveying them after discharge from the furnace.

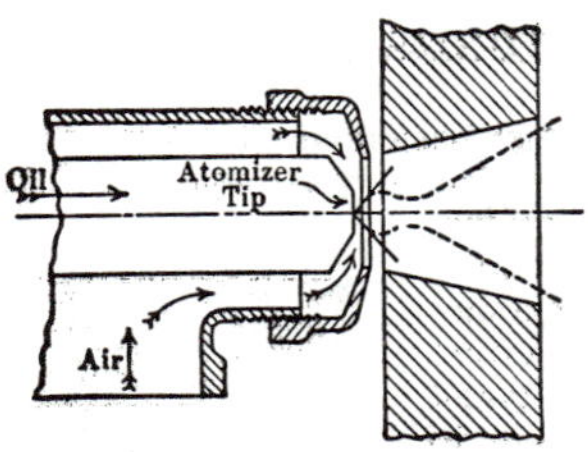

Fig. 10. Fan-blast oil burner.

TEMPERATURE CONTROL. Automatic devices are often installed for controlling the furnace temperature, usually for keeping it constant, but occasionally for varying it with time according to a set program. The measuring device is a thermocouple which actuates a millivoltmeter. Deviation from the correct position is amplified either mechanically by the depressor-bar method, or electrically by electronic tubes. The energy set free by the relay interrupts or alters the supply of electrical energy in electrically heated furnaces or adjusts the opening of fuel valves and of air valves in fuel-fired furnaces. Temperature controllers have been developed to a high state of perfection. However, they control the temperature of the furnace, not the temperature of the charge. More specifically, they control the temperature of that spot of the furnace where the thermocouple is located (or which is "seen" by a radiation pyrometer). On account of this fact, the best temperature control instrument must fail if improperly applied. Two types of temperature control are in use: (1) the all-on and almost all-off type, and (2) the proportional type, in which the supply of fuel and air is gradually adjusted over the range between minimum and maximum. For small furnaces, the temperature control costs more than the furnace. An inexpensive and yet reliable temperature control instrument is much needed.

FURNACE ATMOSPHERE. Oxygen, carbon dioxide, water vapor, and sulfur form scale on the metals that are ordinarily heated in industrial furnaces. The surface of steel is decarburized by the first three of the foregoing gases. Carbon monoxide and hydrogen counteract the formation of scale. Rate of scale formation is a function of chemical equilibrium between charge and furnace atmosphere. That equilibrium varies with furnace temperature. If fuel is burned with a deficiency of air, scale formation is reduced, but not eliminated, unless the deficiency of air is so great that an uneconomically great portion of the fuel passes out into the stack.

Natural atmospheres are controlled by maintaining a fixed but adjustable ratio between fuel and air. Devices serving this purpose are on the market for gaseous and for liquid fuels.

If the charge is to be heated without formation of any scale, artificial or "prepared" atmospheres surround the charge. Such prepared atmospheres consist of hydrogen, nitrogen, carbon monoxide, or a mixture of these gases. Lithium vapor is also used. A common method for producing artificial atmosphere is to burn natural gas or city gas in insufficient air, and to freeze water out of the products of incomplete combustion. Another method is based on the thermal decomposition of ammonia. Generators embodying these principles are expensive and for that reason are applied mainly to large furnaces. In fuel-fired furnaces, the charge rests in (or moves through) a muffle which is filled with prepared atmosphere. In electrically heated furnaces, muffles are, as a rule, not needed. In small electric furnaces for tool hardening and similar work, where the door must remain partly open most of the time, a nonoxidizing atmosphere is maintained in the furnace by providing inside the door opening a gas-flame curtain, as shown in Fig. 11. For bright annealing of copper or brass, an atmosphere of steam is frequently used. A regularly renewed supply of charcoal or petroleum coke helps to prevent oxidation and decarburization. For nitriding, the atmosphere consists of ammonia. For gas-carburizing, a carbon-rich gas such as propane is introduced into the furnace muffle.

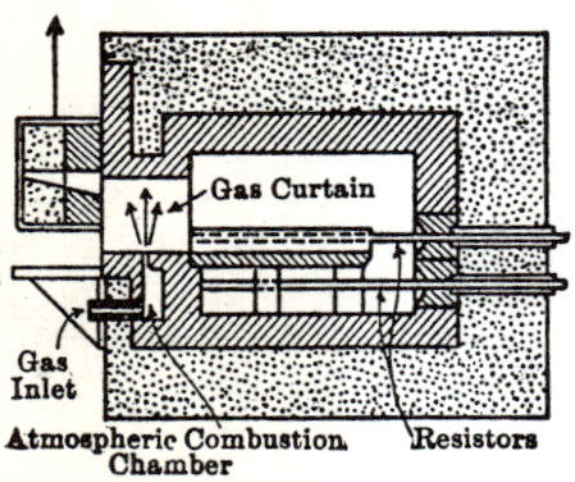

FIG. 11. Tool-hardening furnace with gas curtain.

All salt baths are prepared atmospheres, in the extended sense of that word. For a description of the large number of different salts which are used in industrial heating, see the references at the end of this chapter.

5. TYPES AND SELECTION OF FURNACES

Furnaces purchased for the purpose of heating one kind of part or material are often used later on to heat parts of another kind or are operated at another temperature. Furnaces designed for fuel of one kind are often fired, later on, by fuel of another kind. Furnaces intended for continuous operation throughout the week may be operated only eight hours each day in times of depression. For these reasons, extreme specialization in furnace types is advisable only where a long-continued demand for goods of the same material, weight, and shape is likely to exist.

FIG. 12

There are between 90 and 100 different types of furnaces in use. Only the most important types are mentioned here. Although the blacksmith's coal fire is still found occasionally, gas- or oil-heated furnaces, as shown in Figs. 12, 13, and 14 are preferred. For more specialized work, the types shown in Figs. 15 or 16 are common. Coal-fired furnaces (with steam and air blown into the ashpit) for general forge work are built as in Fig. 17, with an end slot or else with a side door. Heat-treating furnaces are often underfired, for the sake of having a hot hearth, as in Fig. 18. For pack hardening, the hearths of underfired furnaces are often perforated, for greater uniformity of temperature. Tool-

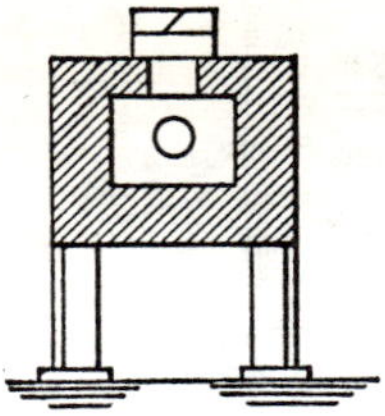

FIG. 13

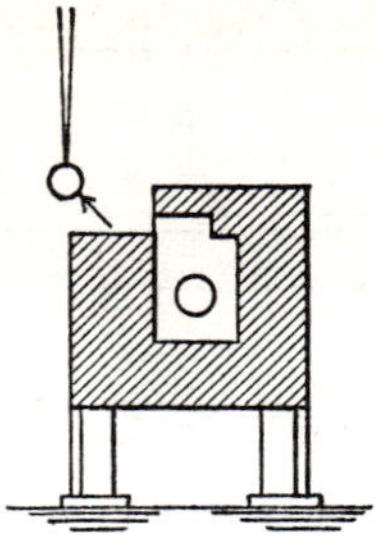

FIG. 14

FIG. 15

hardening furnaces usually have a preheating chamber above the main heating chamber (Fig. 19). Furnaces for heavy or bulky material are of the car type. If oil serves as fuel,

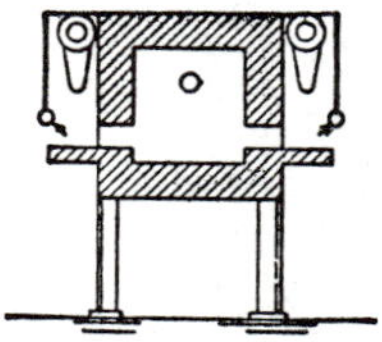

FIG. 16

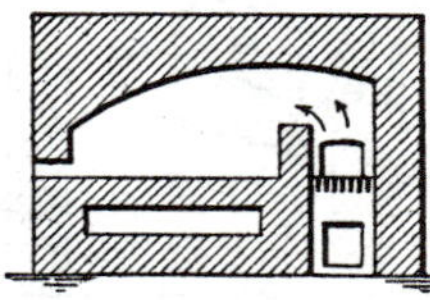

FIG. 17. Coal-fired furnace.

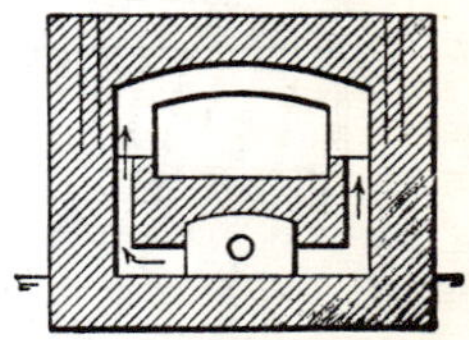

FIG. 18. Underfired furnace.

they are either sidefired, as in Fig. 20, or overfired, as in Fig. 21. With gas-firing, the burners either point directly into the heating chamber or else fire behind muffle walls. The car top is cold; hence, the charge must be blocked up above the hearth. Automatic-conveyor furnaces, with the conveyor returning either below the furnace, as in Fig. 22, or else in the furnace, are used for quantity heating of comparatively light material. Drum furnaces (Fig. 23) are used for small and light pieces. They also serve for gas carburizing. Furnaces with roller conveyors (Fig. 24) are practical for flat or long pieces that cannot fall between the rollers. Rocker-bar (intermittent-conveyor) furnaces (Fig. 25) are much used for heat treating wide material, such as automobile axles. Furnaces with rotating hearths (Fig. 26) have grown enormously in popularity and have been built with hearth diameters in excess of 60 ft.

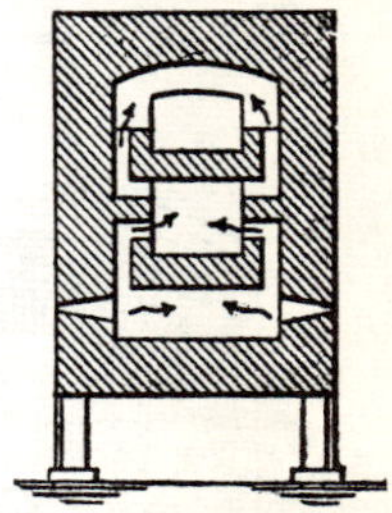

FIG. 19. Furnace with preheating chamber above main heating chamber.

Nonferrous metals are frequently bright annealed in steam or in a reducing atmosphere of products of combustion, as shown in Fig. 27. Any furnace of the muffle type can be used for bright-annealing steel, if the muffle is filled with a prepared, protective atmosphere (see p. 22-10). In electrically heated furnaces, bright-annealing is done without a muffle. A much-used furnace for bright-annealing flat sheets is illustrated in Fig. 28.

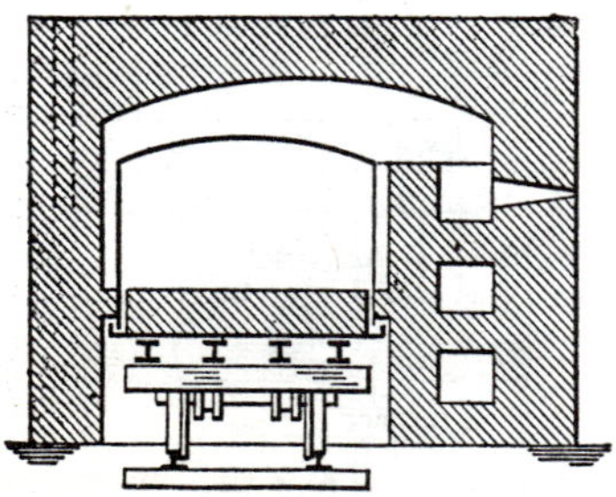

FIG. 20. Sidefired furnace.

FIG. 21. Overfired furnace.

Gas is burned in radiant tubes which are muffles for the flames while sheets are piled in an inner cover or annealing box filled with prepared, deoxidizing atmosphere (abbreviated "deox"). If direct firing is employed instead of radiant tubes, the outside of the annealing box is given a protective coating against rapid oxidation. The same principle is employed for bright-annealing coils of strip steel. They are stacked, and a cylindrical annealing box covers them. Deoxidizing gas is circulated by a fan up through the hole of the coils and down along the outer circumference. This furnace is built either with radiant tubes or with direct firing. Pot furnaces (Fig. 29) likewise conserve a bright finish on the charge. Filled with lead or one of many different salts, they are much used in the tool industry. Cyanide pot furnaces for carburizing (Fig. 30) are always equipped with a hood and a connection to the stack, for the purpose of carrying poisonous fumes off into the open.

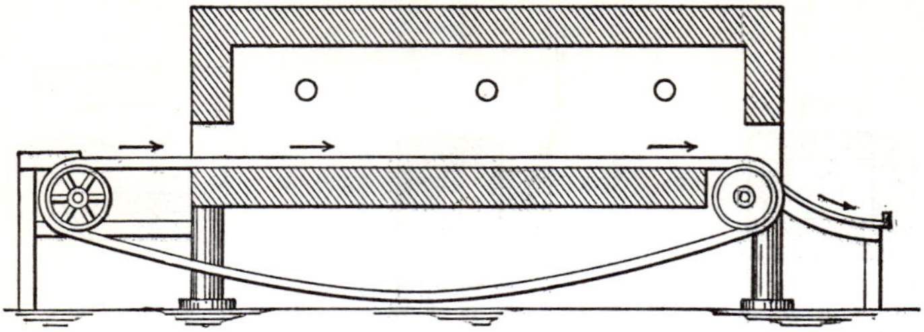

Fig. 22. Automatic conveyor furnace.

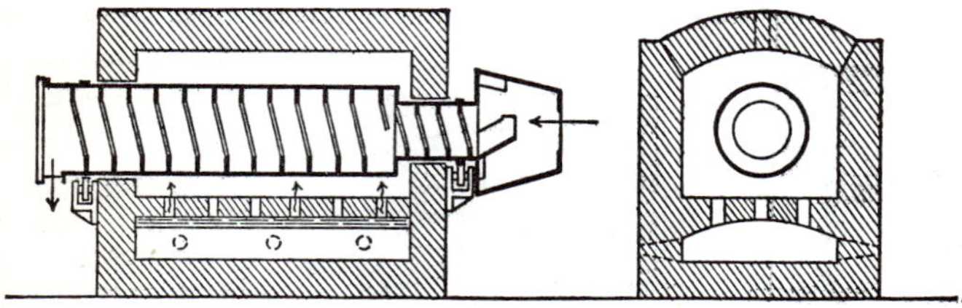

Fig. 23. Rotary cylindrical furnace.

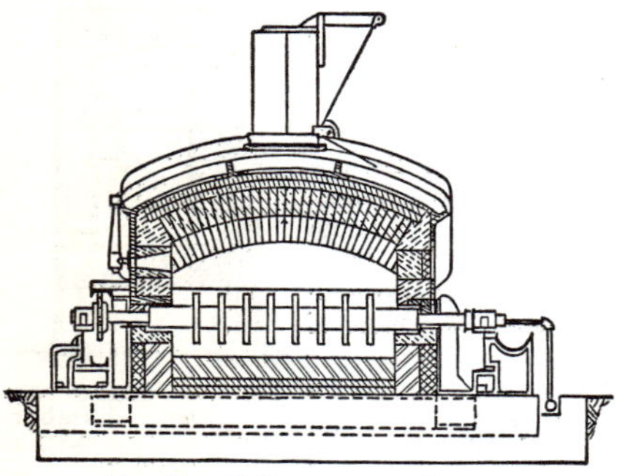

Fig. 24. Roller conveyor furnace.

A typical furnace with recirculation of furnace gases is illustrated in Fig. 31. It serves for maintaining a very uniform temperature in the charge at low furnace temperatures.

In steel works and forges for heavy material, pit furnaces are used for reheating ingots, and continuous furnaces of the pusher type serve for heating heavy blooms and billets. (See Fig. 32.)

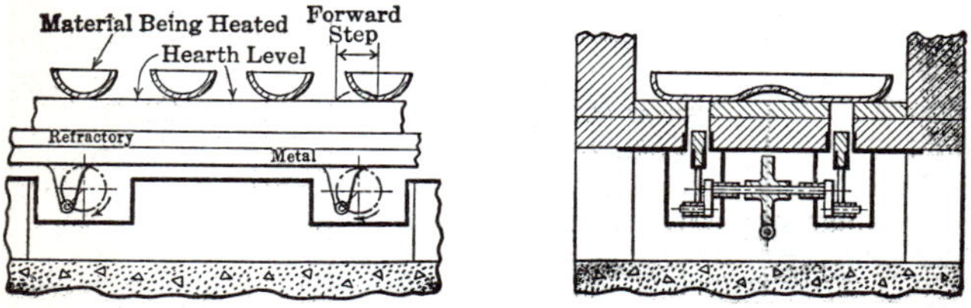

Fig. 25. Intermittent conveyor furnace.

Although electrically heated furnaces are the subject of the following chapter, they are mentioned here for the sake of completeness in the listing of furnace types. Although

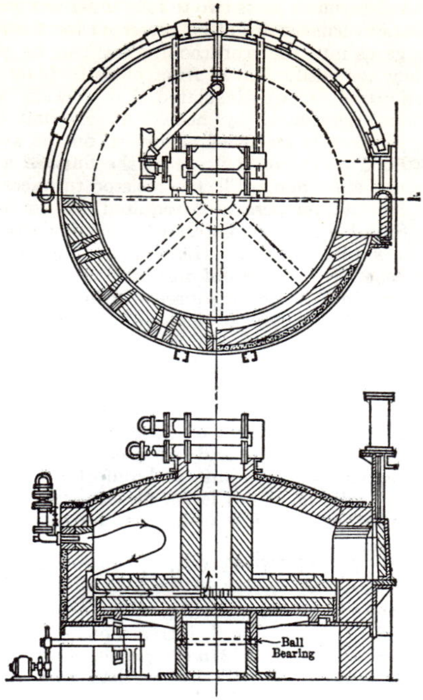

FIG. 26. Rotating hearth furnace.

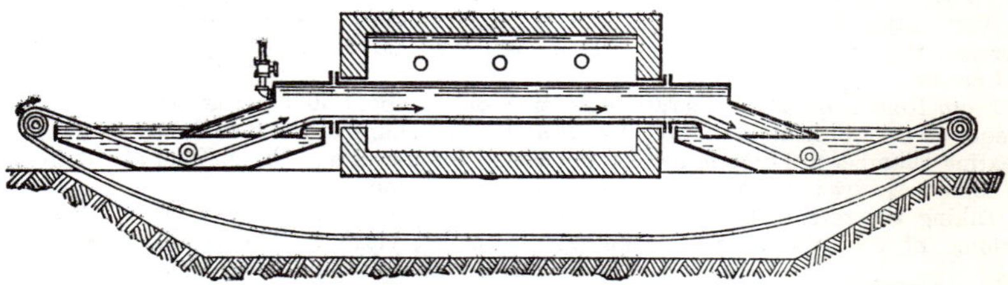

FIG. 27. Bright annealing furnace.

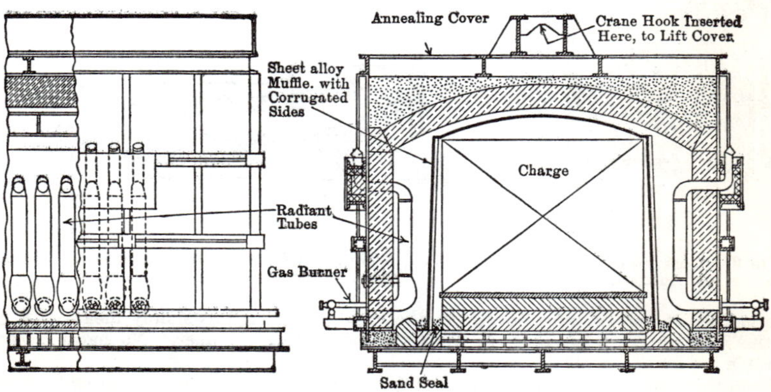

FIG. 28. Portable annealing furnace.

the source of heat in electric furnaces costs two to four times as much as it does in fuel-fired furnaces, electrically heated furnaces are often preferred for the following reasons: They discharge no obnoxious gases into the workroom; they can be placed anywhere in the production line; and they are easily moved from place to place. In order to conserve expensive heat, they are usually so well insulated that workers near the furnace are not inconvenienced by heat. At one-half cent per kilowatthour, the cost of electrical energy is, on the average, in the neighborhood of 1% of the cost of the finished article and is, therefore, not a burden. Electric-resistor furnaces are not recommended for the temperatures required for forging or rolling of steel. Suitable resistor material is not available, and resistor terminals are troublesome. Heating by high-frequency induction has been very successful for bars up to 4 in. in diameter or square. Although the efficiency (heat imparted to steel divided by heat equivalent of kilowatthours) is only 50%, the heating time is so short that practically no scale is formed. The heating process is a function of automatically controlled time, and all pieces are heated absolutely alike and uniformly. For solid metals of larger dimensions, induction heating offers no advantages.

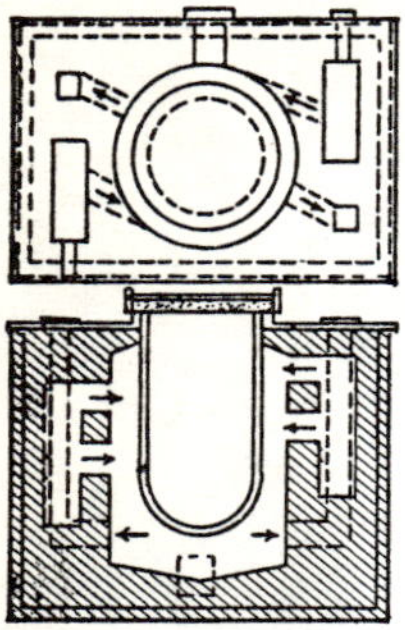

FIG. 29. Deep pot furnace, sidefired on two levels. (Reproduced by permission from Trinks, *op. cit.*)

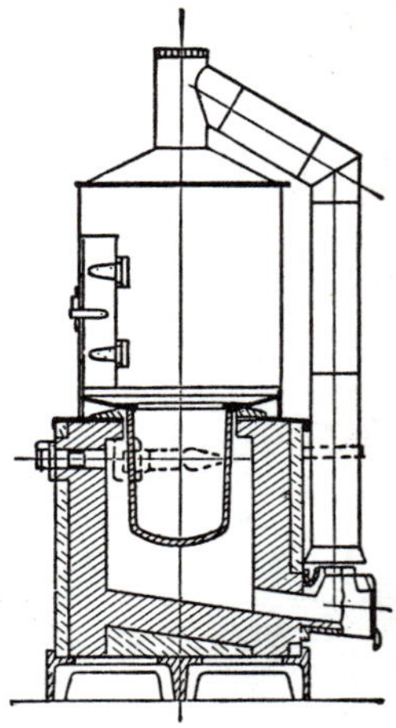

FIG. 30. Pot furnace for salt bath. (Reproduced by permission from Trinks, *op. cit.*)

EXAMPLE OF FURNACE CALCULATION. Furnace is required for heat-treating small steel parts, at the rate of 500 lb per hr. Furnace is of the underfired type (Fig. 18). Temperature of furnace = 1450 F. From Fig. 1, increase of heat content from 60 to 1450 F = (222 − 6) = 216 Btu per lb. For steel, the heat content is actually slightly higher than this value, which applies to pure iron. From Table 4, the thermal efficiency of the given type of furnace under average conditions = 19%. Heat required in fuel per pound of steel = 216/0.19 = 1140 Btu. If the fuel to be used is oil of 140,000 Btu per gal, the fuel consumption is (500 × 1140)/140,000 = 4.1 gal per hr.

From Rule 3, p. 22-05, the rate of heating will be about 30 lb per hr per sq ft of hearth. Required area of hearth = 500/30 = 16.7 ft. The shape of the hearth and height of the heating chamber are determined by the size of the pieces and the method of charging, or in some cases by the nearest standard sizes of the furnace builders.

Taking the rate of heat liberation by combustion as 12 Btu per cu ft per sec, the volume of combustion space required is (500 × 1140) ÷ (3600 × 12) = 13.2 cu ft. The height of chamber required to give this volume of combustion space depends on the shape and size of hearth supports, but would approximate 15 in. The rate of 12 Btu per cu ft per sec is somewhat high for the combustion of fuel oil with burners of the usual type, but this simply means that part of the combustion takes place in the upper or heating chamber. No harm is caused thereby.

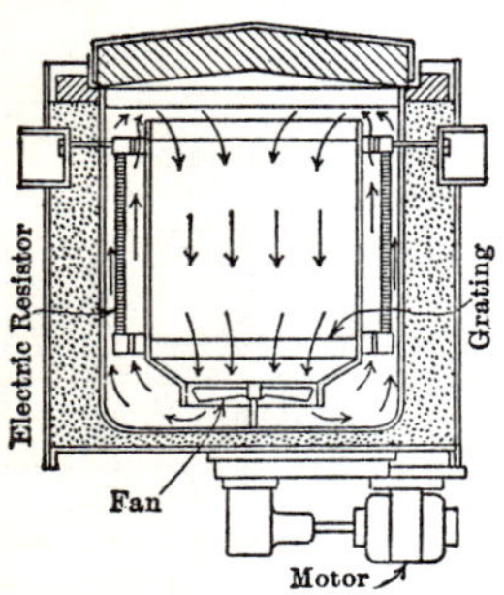

FIG. 31. Convection type electric furnace.

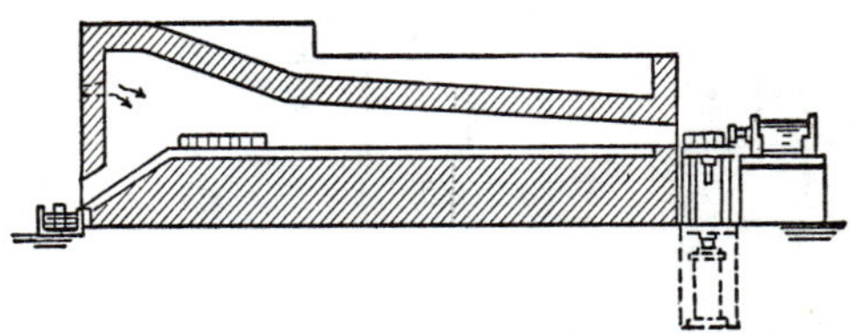

FIG. 32. Continuous furnace with end discharge. (Reproduced by permission from Trinks, *op. cit.*)

From the table of flue sizes (see p. 22-06), the area of flue openings required for furnace temperature of 1400 F = 2.0 sq in. per sq ft of hearth; for 1450 F furnace temperature, it would be somewhat more, or about 2.5 sq in. Hence, $2.5 \times 16.7 = 42$ sq in., total area of flues. The flue dimensions are made multiples of 2 $^1/_2$ in. and 4 $^1/_2$ in.

The size of doors is determined by the size and shape of heating stock and the method of charging.

GENERAL REFERENCES

H. Etherington, *Modern Furnace Technology*, J. B. Lippincott, Philadelphia, 1938.
Industrial Heating, Pittsburgh, monthly publication.
W. Trinks, *Industrial Furnaces*, John Wiley and Sons, New York, Vol. 1, 1934, Vol. 2, 1942.

INDUSTRIAL ELECTRIC HEATING

By R. M. Cherry

Electric heat may replace fuel and steam advantageously in many industrial heating processes because the heat generated by electricity usually can be applied at a higher efficiency than is realized with fuel. Its application for heating often results in increased production, a better quality of work, and in many cases lower overall costs.

Whether or not electric heating is used for a given application depends on the complete cost analysis, including such factors as cost of heat, quality of product, labor of attendants, subsequent operations, working conditions, complaints, rejects, etc.

Industrial electric heating gives perfect heat distribution and permits great flexibility in temperature control. Automatic control is recommended for practically all applications, as it not only gives better control of temperature and heat distribution, but also eliminates the labor of attendants, removes the danger of spoilage, and insures that parts subjected to heat receive exactly the same treatment, thereby producing uniform results.

In general, industrial electric heating is classified as:

1. Industrial heaters and devices for such applications as small ovens, heating of watchmen's houses, crane cabs, soft-metal melting pots, liquid heating, glue pots, process machinery, etc.
2. Ovens for baking, drying, etc.
3. Resistance furnaces for heat-treating metals, porcelain enameling, etc.
4. Melting furnaces for melting steel, copper, and brass.
5. Induction heating.
6. Dielectric heating.

6. INDUSTRIAL HEATERS AND DEVICES

Manufacturers' catalogs contain fairly complete information on small heating units and devices, method of application, and determination of type, size, and rating of the units required for a particular service.

Industrial heaters and devices are, in general, classified as air heating, liquid heating, and contact heating. For these applications, there are various types, sizes, and ratings of standard heaters, generally known as strip heaters, tubular-type heaters, and cartridge heaters.

STRIP HEATERS are used principally for air heating and contact heating. They consist of an outer metal sheath with the coiled resistance wire electrically insulated from the outer sheath and provided with suitable terminals. These heaters are usually approximately 1 $^1/_2$ in. wide and $^1/_2$ in. thick or less, and they are available in standard lengths from approximately 7 in. up to 4 ft. They are used principally for heating watchmen's houses, crane cabs, valve houses, isolated rooms, small ovens, pump houses, ticket booths, process machinery, etc.

TUBULAR-TYPE HEATERS are used principally for liquid heating and contact heating. They comprise a tubular outer metal sheath in which the resistance wire, electrically insulated from the outer sheath, is coiled. The outer sheath material may be steel, nickel, copper, stainless steel, Monel metal, etc., depending on the application of the unit. This type of unit can be formed into almost any desired shape. Also, it may be cast into gray iron, alloy iron, aluminum, or other metals. Such cast-in units provide a large area in contact with the liquid to be heated and form a mechanically strong unit.

Tubular units, either of the plain sheath type or cast-in type, are used principally for heating such liquids as water, oils, paraffin, alkaline baths, nickel-, copper-, or chrome-plating bath solutions, and mild sulfuric acid solutions up to 10% concentration, if no other acids are present; for melting soft metals, as tin, lead, solder, babbitt, type metal, and similar alloys; and for heating salt baths to temperatures up to approximately 1100 F. Locating the heating units directly in the liquid to be heated gives a maximum efficiency

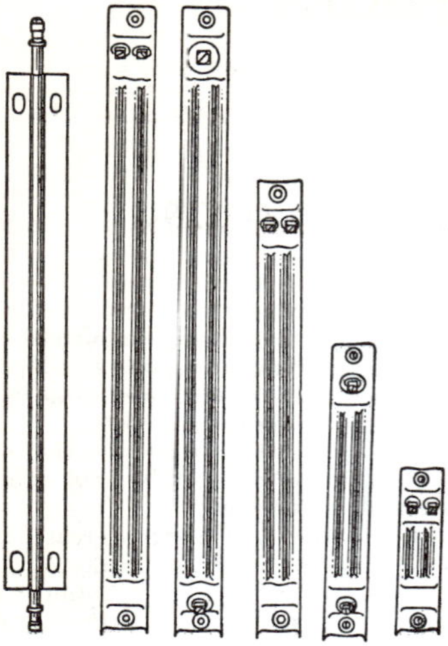

FIG. 1. Strip heaters.

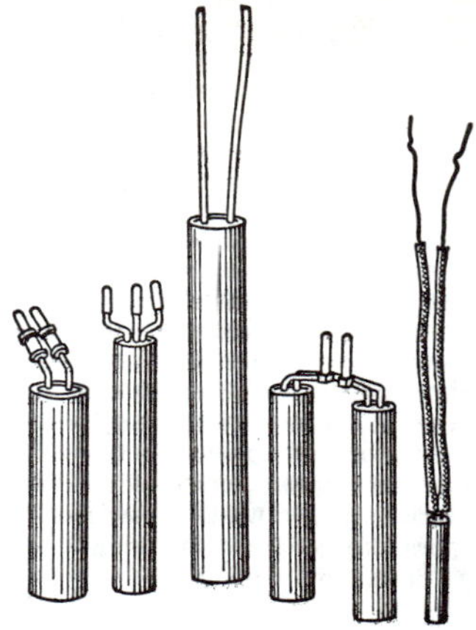

FIG. 2. Cartridge heaters.

of heating in that all the heat is generated in the liquid itself; also, this method of applying heat provides for the best possible temperature control.

The tubular-type heating unit is made in sizes varying from approximately $^1/_4$ to $^3/_4$ in. in diameter and from 2 to 20 ft long. The size and rating depend on the application.

CARTRIDGE HEATERS. Cartridge heating units are used principally for contact heating. The construction is similar to that of the tubular unit in that it consists of an outer shell of metal and coiled resistance wire suitably insulated from the outer shell. The two terminals of the heating unit are brought out at one end. Cartridge heaters are used where small quantities of heat are required at a definite location. They are applied usually by drilling a hole in the part to be heated and inserting the heating unit.

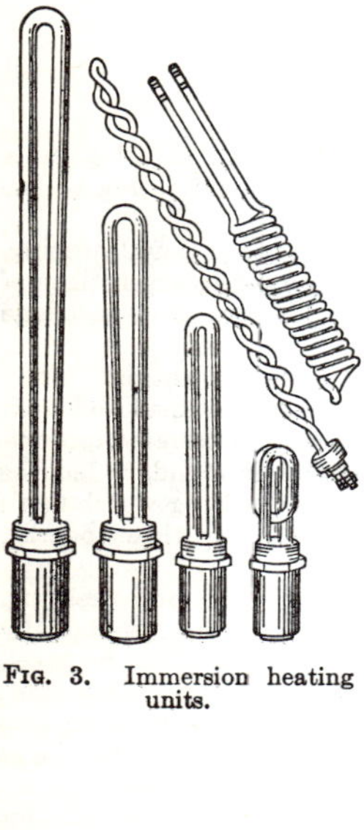

FIG. 3. Immersion heating units.

Cartridge heating units are available in sizes from $^3/_8$ in. in outside diameter by 1 $^1/_2$ in. long to approximately 1 $^1/_4$ in. in outside diameter by 8 $^1/_2$ in. long. Ratings range from 50 watts in the smaller size up to approximately 1200 watts in the larger size. These ratings are based on a sheath material of brass. Cartridge heating units also are available with an alloy steel sheath, suitable for a maximum sheath temperature of 1200 F. The electrical rating for these units is approximately 50% higher than those listed above.

Typical heaters are shown in Figs. 1, 2, and 3. Figure 1 shows several forms of strip heaters, and Fig. 2 various arrangements of cartridge heaters. Figure 3 illustrates a number of varieties of liquid heating units, technically known as immersion units. As shown, the units may terminate in a pipe fitting which may be screwed into the wall of the chamber containing the fluid to be heated. Typical methods of application of the strip, tubular, and cartridge units are shown in Figs. 4, 5, and 6.

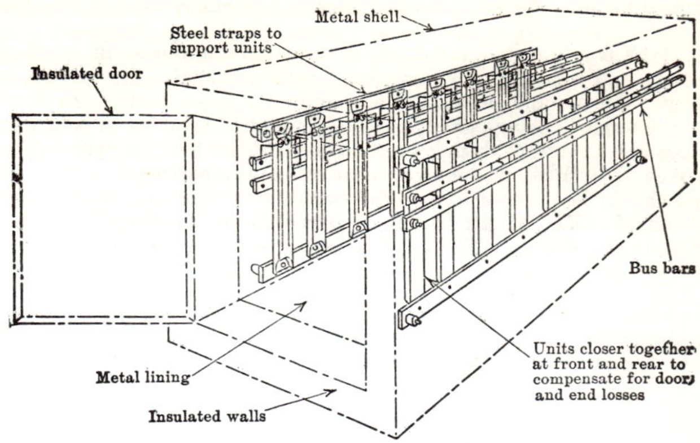

FIG. 4. Strip heaters mounted in a small oven.

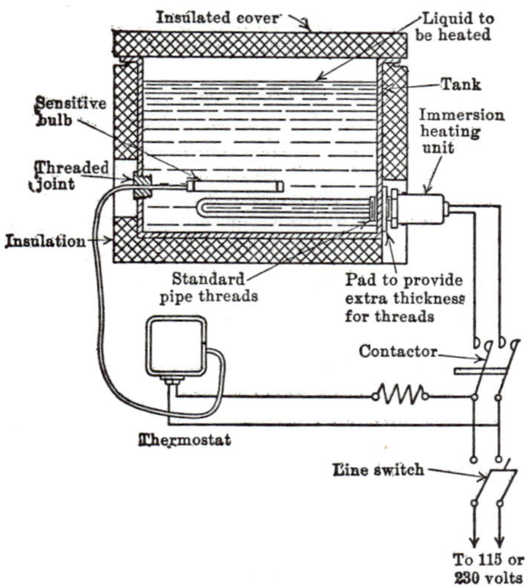

FIG. 5. Immersion heater applied to a tank, including automatic control equipment.

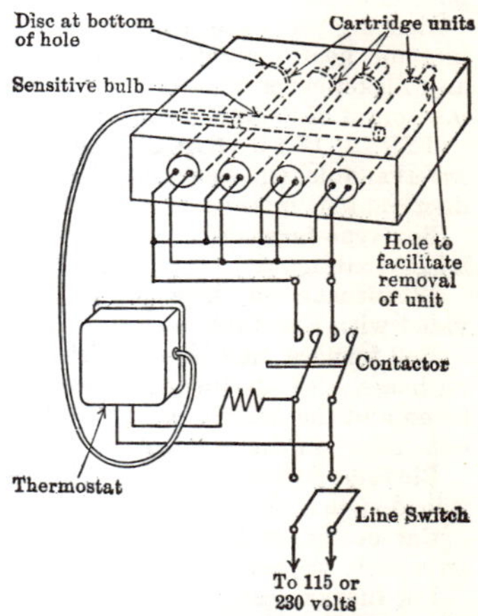

FIG. 6. Application of cartridge units with automatic temperature control.

7. OVENS FOR BAKING AND DRYING

Electric ovens are widely used for many industrial applications. Examples are core baking, paint drying, japanning, bread and cake baking, and baking of such electrical parts as armatures and coils.

Batch-type ovens are used principally for small production. A typical batch oven is the box type, with a door at one or both ends. The material to be baked is loaded on trays, trucks, or an overhead intermittently operated conveyor.

Continuous ovens are used for quantity production. Typical continuous ovens are the straight-through conveyor type, the straight-through conveyor oven with ventilating air passing counterflow to the work, or the continuous counterflow oven where outgoing hot material preheats incoming cold work. The electrical equipment for such ovens consists of standard types of oven heaters and suitable automatic control equipment. The heaters may be mounted on the oven walls or in a compartment outside the oven. When located inside the oven, the heater usually consists of ribbon resistors mounted on insulators which in turn are supported in a metal frame. When the outside heater compartment is used with a recirculating fan, the fin tubular heater or the edgewound ribbon-resistor heater is used.

VENTILATION is an important factor for many oven applications, such as those used for paint drying and japanning, where considerable quantities of volatiles are given off during heating. The amount of ventilation required depends on the size of oven, the size and surface area of parts being baked, and the amount of volatiles to be driven off. Most of the volatiles are driven off during the first part of the heating cycle. Therefore, in a continuous oven they must be removed near the entrance to the oven. All dipped or sprayed parts should be allowed to drip thoroughly before being placed in the oven.

ENERGY REQUIRED for the baking of particular parts depends on the production and the size and type of oven. The approximate energy required for various applications is as follows:

Application	Japanning		Core baking		Bread baking
Type of oven	Batch	Continuous	Batch	Continuous	Continuous
Approx. lb/kw-hr	15–20	25–45	10–12	12–15	10–12

8. RESISTANCE FURNACES

The resistance furnace consists of a heating chamber, containing heating units, which may be mounted on the sidewalls, roof, floor, or doors, depending on the application. Heat is transmitted to the work by radiation and convection. Resistance furnaces are used for such heating applications as annealing, normalizing, hardening, drawing, carburizing, porcelain enameling, and brazing.

Heating units are principally the metallic type in rod, ribbon, or cast form, or a non-metallic heating unit, as silicon carbide, usually in round-bar form. Nickel-chromium and nickel-chromium-iron alloys are universally used for operating temperatures up to a maximum of 2100 F, which covers the majority of industrial heating applications.

TYPES OF FURNACE. (See also pp. 22-10 to 22-14.) Many standard types of furnace are available, the selection of type and size of furnace depending on the application, production, and method of handling desired. The more common types of furnaces are:

Box-type furnace is essentially a box, with a door at one or both ends. The parts to be treated are handled by hand, either individually or in containers.

Pit furnace may be either circular or rectangular, with the opening at the top, and provided with a suitable cover.

Bell furnace may be either circular or rectangular. The material to be treated is loaded on bases, and the furnace is lifted and moved from base to base; or the furnace may be lifted and the bases moved. Bell furnaces are used either with or without an inner retort, depending on the application.

Elevator furnace may be either rectangular or circular, usually rectangular. The work is loaded on a car, which is pushed under the furnace and elevated into the furnace chamber.

Car-bottom furnace is similar to the box-type furnace, except that the hearth is mounted on wheels to permit it to be removed for loading and unloading.

Pot furnace consists of an inner metallic pot, with the heating units located outside of the pot. Pot furnaces are used principally for lead hardening, cyanide or salt hardening, and hot dip galvanizing.

Pusher furnace, in which the work is pushed through the furnace, one piece against the other where the parts can be so handled; or the parts are loaded on stools or trays, which are pushed through the furnace, one position at a time.

Rotary-hearth furnace, in which a rotating table, provided with sand seals between the rotating table and stationary part of the furnace, carries the parts to be treated. Loading and removal of work is through the same or adjacent doors, thereby providing for loading and unloading from one central point.

Roller-hearth furnace has a hearth of driven rolls, on which the material being treated is carried through the furnace. The spacing between rolls depends on the parts or trays to be handled. The rolls extend through the walls of the furnace to bearings mounted on the outside.

Conveyor furnaces are of many types, depending on the type of conveyor used, as a mesh belt, cast-link belt, monorail conveyor, etc. For such applications as hardening of steel, the conveyor should be arranged to remain in the furnace at all times to conserve heat. When the furnace is used for annealing, the parts being cooled in the furnace, the conveyor usually returns underneath it.

Catenary furnace, a continuous furnace for annealing steel strip in a protective atmosphere, consists of heating and cooling chambers. The strip is supported on rolls located outside the heating chamber so that the strip hangs in a catenary without any contact with supports in the heating chamber.

Tower furnace consists of one or more vertical heating chambers and may be equipped with one or more vertical cooling chambers. Each chamber is provided with supporting drums top and bottom. This type of furnace is used with a protective atmosphere for continuous annealing of strip.

ATMOSPHERE EQUIPMENT FOR RESISTOR FURNACES. Many of the applications of heating of metal parts require a protective atmosphere in the furnace heating and cooling chambers to protect the surface of the parts being treated. Practically all these applications require the prevention of the oxidation of the metal. In addition there may be other requirements such as preventing the removal of carbon, the elimination of sulfur discoloration, or the addition of carbon.

Typical applications requiring a protective atmosphere are bright-annealing of copper and steel, scale-free hardening, scale-free annealing of castings, electric furnace brazing, sintering, etc. A suitable protective atmosphere is provided by equipment which processes the available gas, such as city gas, natural gas, and propane.

Standard atmosphere equipments are available for providing the proper atmosphere for the many metal-treating processes.

One type of standard equipment consists of visual flowmeters to indicate the amount of air and gas used, a motor-driven air-gas mixer and compressor, a combustion chamber with burner, and a surface cooler for removing excess moisture. In operation, air and gas are fed into the combustion chamber at a definite ratio, which depends on the gas used and the required analysis of the output gas. In the combustion chamber, the gas is burned with a deficiency of air in the presence of a catalyst under controlled conditions. The output gas contains principally nitrogen, hydrogen, carbon monoxide, carbon dioxide, and water vapor. The relative proportions of the constituent gases depend on the original air-gas ratio. The resultant atmosphere is suitable for many of the furnace applications.

For some applications, the output gas must be further purified. For example, most city gases contain a small amount of sulfur which must be removed if material such as copper and nickel are to be bright annealed. In other applications, such as annealing or heat-treating of high-carbon steel, the carbon dioxide and water vapor must be removed to provide a nondecarburizing atmosphere. Standard equipments are available to provide such purification of the gases.

The cost of a protective atmosphere is relatively low. For instance, in bright-annealing of copper with available city gas of 530 Btu content per cu ft, the air-gas ratio used is approximately 4 : 1. Therefore, a gas output of 1000 cu ft would require approximately 250 cu ft of city gas. For scale-free hardening, the air-gas ratio may be approximately $2\,{}^{1}/_{2}$: 1, and there would be required approximately 400 cu ft of city gas of 530 Btu per 1000 cu ft of output gas.

When a higher reducing gas than that provided by processing available gases is desired, an ammonia dissociator may be used. The equipment is essentially a tube or coil containing a catalyst through which the ammonia gas passes, inserted in an electric furnace. The ammonia gas dissociates when heated to the required temperature, the resulting gas containing 75% hydrogen and 25% nitrogen. Many sizes of standard atmosphere equipments are available for a wide range of outputs and applications. Makers should be consulted in regard to equipment for special conditions.

The quantity of atmosphere required depends on the type and size of furnace, and the application. In general, it is a relatively small item in the total cost. Such protective atmosphere easily can be supplied to practically all types of furnaces, if they are properly designed for use with protective atmosphere.

The protective atmosphere eliminates oxidation of most materials. A considerable saving is effected thereby, by the elimination of the subsequent cleaning operation, and the quality and appearance of the product is improved. The cost of fabrication, such as brazing fabricated metal parts in the electric furnace, is lowered.

SELECTION OF TYPE. ENERGY REQUIRED. Table 1 gives, in general, the types of furnaces used for various applications. The type and size of furnace to be selected depend on the size, shape, weight, and material of parts to be treated, as well as the production and the desired heating and cooling cycle. The energy required, usually stated in net pounds per kilowatthour, based on continuous operation, depends on the size of furnace, production, and the heating cycle. The data in Table 1 are approximate and are to be used only as a rough guide.

ENERGY CONSUMPTION AND ELECTRICAL RATING. An estimate of the energy required for a given application should include the heat absorption of the materials being heated, heat absorption of auxiliaries for carrying the work, as trays, conveyor,

Table 1. Typical Applications of Electric Resistance Furnaces

Process	Material	Type of Furnace	Typical Applications	Approx. Energy Required, lb/kwhr
Annealing	Copper and brass	Box Pusher tray Conveyor Roller hearth Bell	Copper wire on reels Copper tubing Brass wire or strip in coils	25–35 12–20 15–20
	Malleable iron	Elevator Pusher tray Roller hearth	Malleable castings	5– 7
	Steel bar Stock and tubes	Car bottom Pit Elevator Roller hearth		6– 8 7–10 6– 8 7–11
	Steel sheets Coiled strip	Bell Elevator	Steel strip in coils	11–14
	Steel strip, continuous	Roller hearth Catenary Tower		8–12
	Steel wire in coils	Bell		10–13
	Steel castings and forgings	Car bottom Elevator Pusher tray Roller hearth		7–10
	Stainless steels	Mesh belt Roller hearth Catenary Tower		4– 7 6–10 6–10 6–10
Annealing Heat-treating	Aluminum Aluminum alloys	Box Pit Elevator Roller hearth Salt Baths		Aluminum 8–14 Aluminum alloys 6–12
Brazing (copper-brass and silver solder)	Steel Copper Aluminum	Box Mesh belt Roller hearth		For copper brazing 4– 6 3– 5 5– 7
Carburizing	Steel	Box Elevator Pusher tray Pit	Pack method Gas carburizing	2– 4 3– 6
Drawing	Steel	Box Pit Oil bath Salt bath Conveyor	At approx. 400 F At approx. 800 F At approx. 1200 F	40–60 20–25 10–15
Galvanizing	Steel	Pot	Hot-dip galvanizing	12–20
Hardening	Steel	Box Pot Pit Conveyor Pusher Rotary hearth Roller hearth	At approx. 1500 F	7–11
Porcelain enameling	Steel Cast-iron	Box Continuous-conveyor	Sheet steel, single coat In box furnace In continuous furnace Cast iron, wet process, single coat In box furnace In continuous furnace	 5– 6 10–15 6–10 12–15
Sintering	Pressed metal powders	Bell Box Conveyor Roller hearth		

etc., heat absorption of the ventilating air if used, heat absorption of the protective gas atmosphere if used, heat losses through the walls, doors, etc.

The heat absorption of the material to be heated, including the net work, conveyor, trays, etc., can be calculated by

$$\text{Kilowatthours} = \frac{w \times h_s \times (t_2 - t_1)}{3412}$$

where w = weight of material being heated, h_s = average specific heat, t_1 and t_2 = respectively, initial and final temperatures of the material in degrees Fahrenheit. Heat losses for many applications may be estimated by means of Figs. 7–11. The heat loss data in

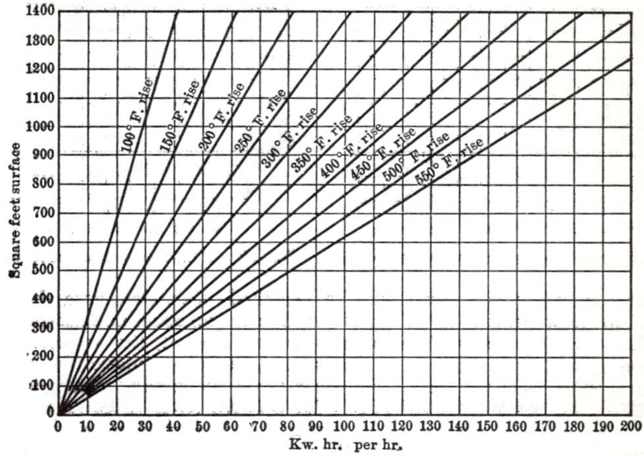

Fig. 7. Heat loss through 1-in. thick insulation of 85% magnesia or similar insulator. Heat loss varies inversely as thickness.

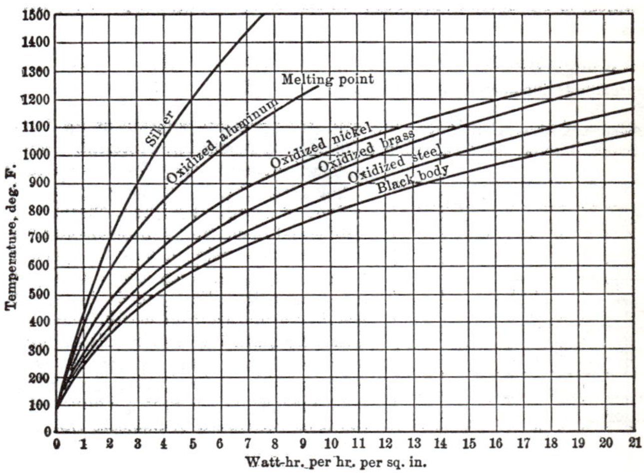

Fig. 8. Heat losses from various metals.

the series of curves are principally for low-temperature applications. For the higher-temperature resistance-type furnaces, operating up to approximately 1850 F, a good rule-of-thumb method for estimating the heat loss in kilowatthours is

$$\text{Kilowatthours} = \frac{0.8A \times T}{10{,}000}$$

where A = inside wall area of heating chamber in square feet and T = operating temperature in degrees Fahrenheit. The application of this rule will give the approximate heat losses through the insulated walls and closed doors, based on a well-insulated furnace.

The selection of the right type and size of furnace for a given application requires experience and, in some cases, tests. When there is any question as to size and type of furnace,

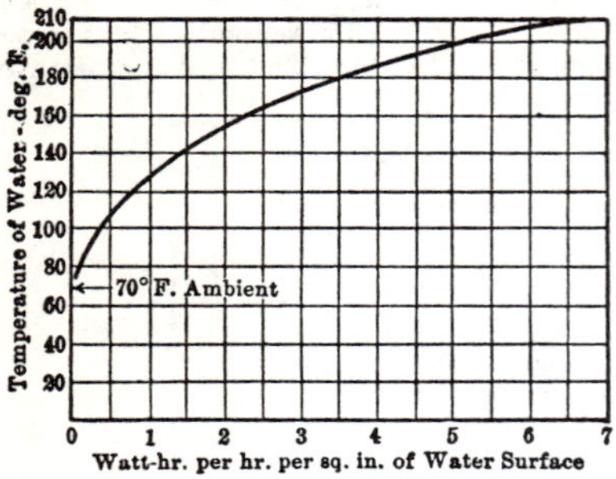

FIG. 9. Heat losses from open water surfaces.

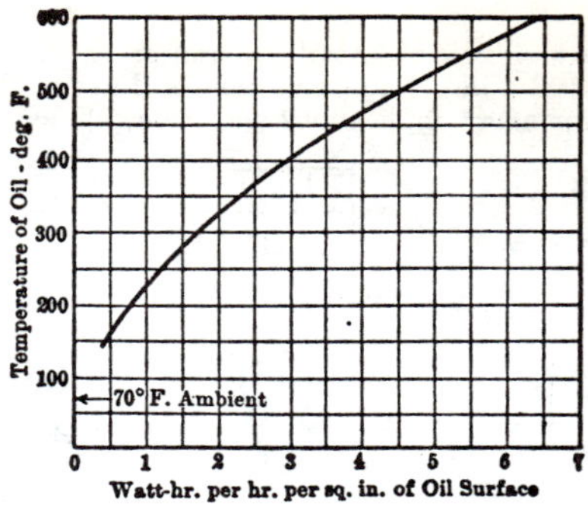

FIG. 10. Heat losses from surfaces of oil baths.

manufacturers of such equipment should be consulted. Complete data should be given, i.e., voltage, phase, and frequency of the power circuit; floor space and headroom available; the size, shape, weight, and material of parts to be treated; the complete heating and cooling cycle desired; the efficiency of loading, i.e., pounds of material to be loaded per square foot or cubic foot; production per given time; and hours to be operated per day.

REFRACTORIES AND INSULATION FOR OVENS AND RESISTOR FURNACES. The usual construction of ovens, for temperatures up to approximately 600 F, comprises a metal inner liner and an outer shell, the space between being filled with insulating material, as 85% magnesia, rock wool, mineral wool, etc., care being used to have little or no through metal.

The thickness of insulation used depends on the operating temperature and, in general, should be at least as follows:

Oven temperature, °F	200	400	600	800
Approximate thickness of insulation, in.	2	4	5	6

For electric-resistance furnaces, the wall construction consists of an inner refractory lining of brick, backed by a suitable thickness of insulating material. Several types of lightweight refractory are available in standard brick sizes. These bricks are of less than half the weight of regular firebrick, thereby providing for additional insulation and absorbing less heat while the furnace is being heated to operating temperature. For the average-size heating chamber, the inner refractory lining of 4 1/2-in. standard lightweight refractory brick is sufficient. For high walls and long spans for arches, this thickness should be increased. For data on refractory brick see pp. 5-46 to 5-56 of this book. For additional data on heat insulation see *Power* Volume of Kent's *Mechanical Engineers' Handbook*, pp. 3-34 to 3-49.

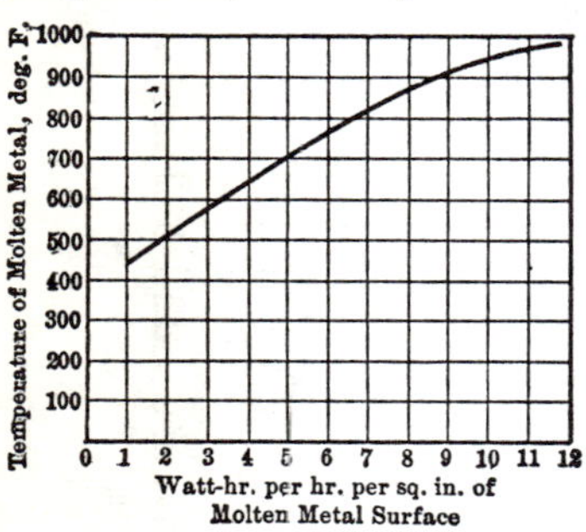

FIG. 11. Heat losses from surface of molten lead, tin, type metal, babbitt, etc.

The thickness of insulation depends on the operating temperature and the complete time-temperature cycle. A furnace to be held at operating temperature for long periods should be well insulated. For a cycle requiring cooling of the charge in the furnace, the thickness of the insulation must be in accordance with the required heating and cooling cycle.

In furnaces held at the operating temperature, the thickness of the insulation, in addition to the refractory lining, should be approximately as follows:

Operating temperature, °F	800–1100	1100–1400	1400–1800	1800–2100
Approximate thickness of insulation, in.	5	7 1/2	10	12 1/2

9. MELTING FURNACES

The three types of electric melting furnaces are, in general, the resistance furnace, the arc furnace, and the induction furnace.

RESISTANCE FURNACES, usually of the pot type, are used for melting soft metals, as tin, babbitt, lead, solder, type metal, etc. The heating units may be immersed directly in the metal to be melted or placed outside of the melting crucible, depending on the application.

ARC FURNACES, in general, are of two types, the direct arc and the indirect arc.

The direct arc furnace is used principally for melting and refining steel. The three-phase arc furnace, generally used, has three electrodes. The electric current passes between the electrodes and the bath of metal. Each electrode is provided with a motor-driven mechanism which, by suitable automatic control, raises and lowers the electrodes to maintain a constant power input.

The indirect arc furnace usually has two electrodes. The current passes between the electrodes and above the bath, which is heated by radiant heat from the electric arc, as well as by conduction and convection. This type of furnace is used frequently for melting of copper and brass.

Electrical equipment of an electric arc furnace consists of a stepdown transformer, primary and secondary control, motors for raising and lowering the electrodes, and a tilting motor if the furnace is to be of the tilting type. The primary control consists of an oil circuit breaker on the high-voltage side of the stepdown transformer, suitable means for changing taps on the transformer, and metering equipment. The secondary control consists of control equipment for controlling the movement of the electrodes to maintain constant power input. Voltages used on the arc vary, depending on the character of the application. The higher voltages are used normally for melting down, and the lower voltages for refining.

INDUCTION FURNACES are of two types, the core type, used on standard frequencies, and the coreless or high-frequency furnace.

Core-type furnaces consist essentially of an electrical transformer circuit with a primary coil, a laminated iron core, and the bath arranged for completion of the secondary circuit. Heating is accomplished by the induced current in the secondary. This type of induction furnace is used principally for melting brass and copper.

Coreless or high-frequency furnaces consist of a crucible to hold the material to be melted, around which is placed a water-cooled copper coil to which the high-frequency source is connected. Heat is generated by the electric current induced in the charge. The rate of heating is limited practically only by the capacity of the high-frequency source. This type of furnace provides for rapid heating, accurate control of heating, and agitation of the bath, and is used principally for the melting of alloy steel.

ENERGY REQUIRED FOR MELTING in induction furnaces depends on the size of furnace and the material to be melted. In general, the approximate electric energy consumption for melting for various applications is as follows:

Brass, yellow (core induction furnace)	220–350 kwhr per ton (2000 lb)
Brass, red (core induction furnace)	250–400 kwhr per ton (2000 lb)
Copper (indirect arc furnace)	250–400 kwhr per ton (2000 lb)
Steel (three-phase arc furnace melting only)	450–600 kwhr per ton (2000 lb)
Steel (coreless induction furnace)	500–800 kwhr per ton (2000 lb)

10. INDUCTION HEATING

Induction heating has been widely applied for the hardening of steel parts and for heating for forging.

The physical requirements for induction heating are not complicated. Besides the metal parts to be heated, the heating coil or inductor block and a source of alternating current are required. The high-frequency current passing through the heating coil produces a fluctuating magnetic field. This in turn induces a current to flow in the metal part, with resultant heating due to the I^2R loss in the portion of the metal in which the current flows.

The advancement of induction heating is closely related to the development of electrical equipment for converting 60-cycle frequencies to the higher frequencies required. Today, three types of commercial equipments covering a wide range of power output and frequency ratings are available. The three sources of high-frequency power (high frequency covering a range of 1000 to 1,000,000,000 cps) are:

1. Motor generators, for frequencies 1000, 3000, or 10,000 cps and output ratings up to 1200 kw.

2. Spark-gap oscillators, providing frequencies of 100 to 400 kc per second at power output ratings up to 25 kw.

3. Vacuum-tube oscillators, providing alternating current with frequencies of 100 to 500 kc, and power output rating in the lower range of frequencies up to 200 kw.

All three types of equipment are in general use. Suitability of each for any particular job depends on several factors, including size of part, power requirement, metallurgical results desired, and relative cost.

Induction heating is particularly suited for surface-hardening or localized hardening of steel parts.

An example of surface-hardening of a part is illustrated in Fig. 12, which shows the induction-hardening fixture for surface-hardening of a cluster gear. The gear is placed on a holding device which slowly rotates it inside the heating coil which is connected to the source of high frequency. Control of the heating is by automatic timing. After the proper temperature is reached, the high-frequency power is shut off and the gear is quenched by a water spray issuing from the numerous small holes in the surrounding jacket.

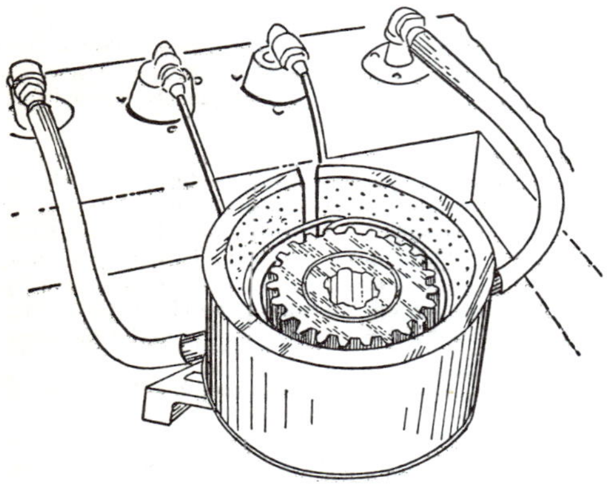

FIG. 12. Gear being heated in induction-hardening fixture.

For the heating of large parts for forging, such as 3-in. diameter round solid bars, where it is desired to heat all the way through, a frequency of 1000 or 3000 cps is usually employed. Where it is desired to surface-harden the wearing surface of a fairly heavy part such as a crankshaft, a frequency of 3000 or 10,000 cps is used. For the heating of small parts or irregularly shaped parts, and where shallow depth of heating is desired, frequencies up to approximately 500 kc are used.

In addition, high frequency is used for annealing, brazing, and soldering operations where localized heating is desired.

Normally, the high-frequency equipment is provided with coils, fixtures, and conveyors designed to handle the particular part.

Some outstanding and typical examples of the use of high frequency for surface hardening are: bearing surfaces of crankshafts, cam shafts, cutting edges of mowing-machine sections, wearing surfaces of the ways of lathe beds, wearing surfaces of gears, and cutting edges of pliers, axes, and hatchets.

Some of the outstanding advantages in the use of high frequency in heating for forging are: improved working conditions, quick heating, less oxidation of the steel (giving longer die life), less loss of metal, and better control for forging temperature.

Other excellent applications of induction heating using frequencies obtained from motor-generator equipment are the heating of steel strip continuously for annealing and the reheating of tinplate strip for tin reflowing.

11. HEATING OF ELECTRICAL NONCONDUCTORS

Parts to be heated are placed between plate electrodes to which the high-frequency voltage is supplied. The heating is the result of dielectric loss within the material.

For the heating of these electrical nonconductors, frequencies of 3 to 100 megacycles are normally used, depending on the material to be heated.

Some of the typical examples of commercial applications in use are: heating of plywood and other laminated wood structures for bonding; preheating of plastic preforms prior to molding; heating for controlled drying, as in heating paper, textiles, etc.

It must be kept in mind that dielectric heating equipment is not economical for mass heating or for evaporation of large quantities of water where other forms of heating can be successfully applied.

One of the outstanding applications of dielectric heating has been the heating of plastic preforms before molding. With dielectric preheating, the heat is generated within the preforms by the high-frequency electric field. A uniform temperature distribution is obtained throughout the charge. The older method of preheating depended on conduction, convection, or radiation. The heating was slow, and the high temperature on the outside of the part initiated a chemical reaction while the interior temperature was still low. The benefits obtained from a uniform, quick heating by dielectric preheating are: improved quality and reduction of scrap; reduction of mold repairs; increased production from a given mold; reduced cost of finishing molded parts, because of thinner flash on the mold parting line; increased range of designs which can be molded.

FIG. 13. Dielectric heater for plastic preforms.

Figure 13 shows a typical dielectric heater for plastic preforms, rating 5 kw output at 40 megacycles.

Where it is desired to obtain a uniformity of drying or a greater degree of drying than is possible or practical with any other method of heating, dielectric heating may be applicable. The equipment may be provided with conveyorized ovens that provide for passing the material to be dried between electrodes in the oven.

When dielectric heating is being considered, the problems should be referred to the manufacturer of such equipment. Tests on the product are frequently required to determine the practicability of the heating of the particular material and the frequency required.

SECTION 23

METAL-CUTTING PROCESSES

BY

ROBERT C. DEALE, *The Sylvester Co.*

E. L. MURRAY, *The Warner and Swasey Co.*

JOHN M. WALTER, *Vice President in Charge of Engineering, The G. A. Gray Co.*

A. G. BAUMGARTNER, *The Cincinnati Shaper Co.*

MARIO MARTELLOTTI, *Research Engineer, The Cincinnati Milling Machine Co.*

HANS ERNST, *Director of Research, The Cincinnati Milling Machine Co.*

C. J. OXFORD, *Chief Engineer, National Twist Drill and Tool Co.*

R. E. W. HARRISON, *Consultant, Management Engineering, Chambersburg, Pa.*

E. T. LARSON, *Norton Co.*

E. L. HEMINGWAY, *Research Engineer, Gisholt Machine Co.*

ROGER W. BOLZ, *Mechanical Engineer, Cleveland, and Associate Editor, Machine Design*

W. T. GLOOR, *The Motch and Merryweather Machinery Co.*

A. H. CANDEE, *Gleason Works*

C. W. OBERT, *Consulting Engineer, New York*

SINGLE-POINT METAL-CUTTING TOOLS

BY ROBERT C. DEALE

LATHE WORK

BY E. L. MURRAY

PLANER WORK

BY JOHN M. WALTER

METAL-WORKING SHAPERS

BY A. G. BAUMGARTNER

MILLING MACHINE PRACTICE

BY MARIO MARTELLOTTI AND HANS ERNST

DRILLS AND DRILLING

By C. J. Oxford

ABRASIVE PROCESSES

Revised by R. E. W. Harrison

BROACHING

By Roger W. Bolz

METAL SAWING

By W. T. Gloor

GEAR-CUTTING PROCESSES

By Allan H. Candee

FLAME CUTTING

By C. W. Obert

SINGLE-POINT METAL-CUTTING TOOLS

By Robert C. Deale

1. DEFINITIONS

The following definitions for the various elements of a single-point metal-cutting tool, which have been adopted for general use in the United States, are covered by Standard B5.13-1939, Terminology and Definitions for Single-point Cutting Tools, of the American Standards Association.

COMPONENTS. The shank is that part of the tool on one end of which the point is formed or the tip or bit supported.

The size of a tool of square or rectangular section is expressed by giving in the order named the width of shank, the height of shank, and the total tool length in inches, i.e., $^3/_4$ by 1 $^1/_2$ by 12 in.

The point or nose is all that part of a tool which is shaped to produce the cutting edges and face. (See Figs. 1 and 2.)

The neck is an extension of the shank but of reduced sectional area. (See Fig. 3.)

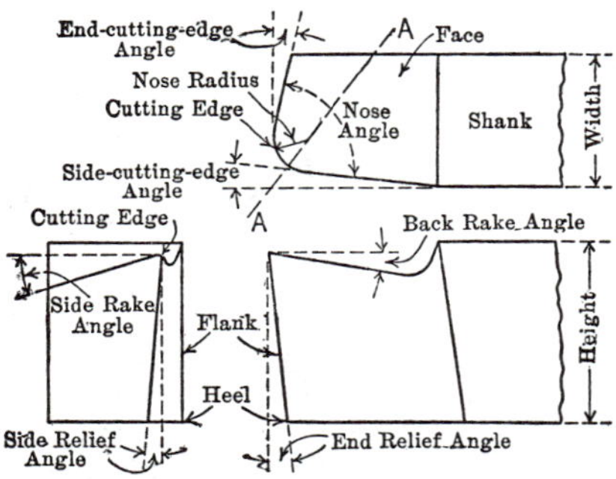

FIG. 1. Right-cut, straight-shank single-point ground tool.

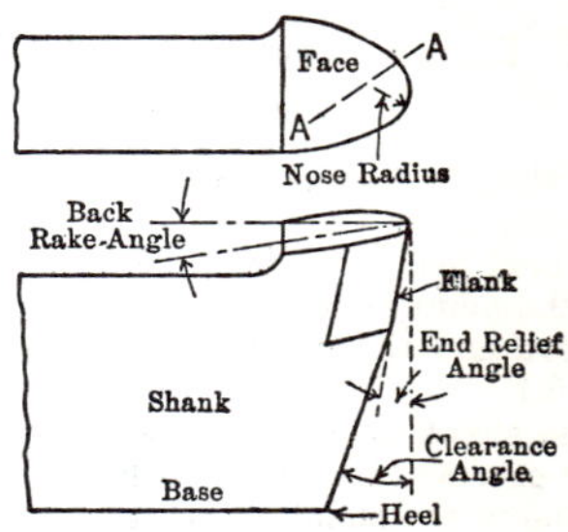

FIG. 2. Right-cut, straight-shank, single-point raised face forged tool.

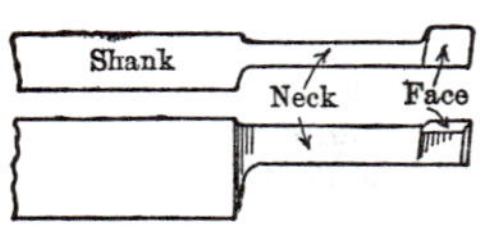

FIG. 3. Necked tool, forged.

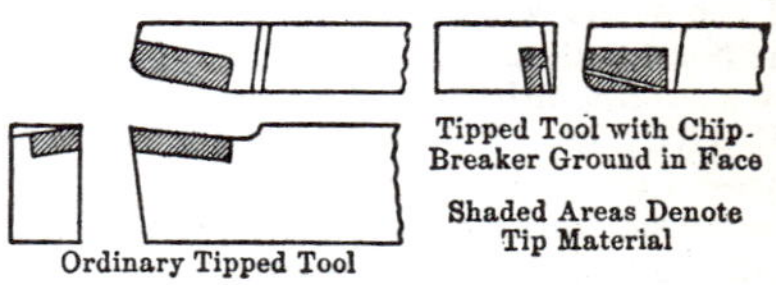

FIG. 4. Tipped tools.

The base is that side of the shank which bears against the support taking the tangential pressure of the cut.

The face is that surface on which the chip impinges as it is cut from the work.

The heel is the end of the base immediately below and supporting the face.

The cutting edge, or lip, is that portion of the face edge along which the chip is separated from the work. It consists usually of the side cutting edge, the nose radius, and the end cutting edge. (See Fig. 1.)

The nose is the curve formed by joining the side and end cutting edges.

The profile is a plan view when looking at the face from a point at right angles to the base. It is formed by joining the side and end cutting edges with the nose.

The flat is a straight portion of the cutting edge intended to produce a smooth machined surface.

The flank of the tool is the surface below the cutting edge.

TYPES OF TOOLS. Single-point tools may be made up as follows:

A ground tool is one in which a point is formed on the end of a bar (shank) of tool steel by grinding. (See Fig. 1.)

A forged tool is one in which the point is forged roughly to shape on the end of a bar (shank) of tool steel and subsequently hardened and ground. (See Fig. 2.)

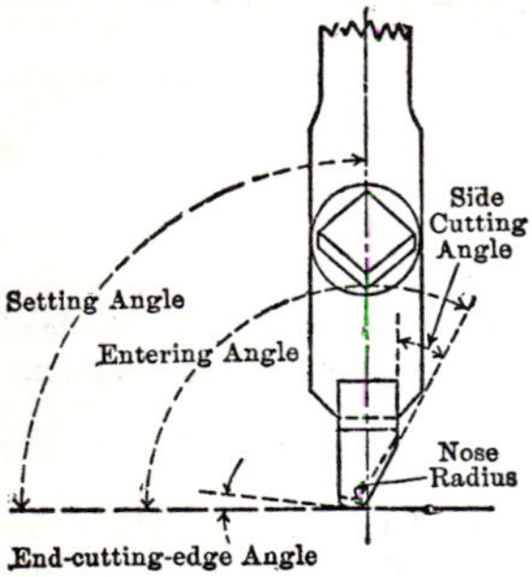

A tipped tool is one with a tip of cutting material permanently attached to a shank of noncutting metal. (See Fig. 4.)

A bit tool is one in which a small part of tool material of square, rectangular, or other section, or forged to a special shape, is held in the end of a holder or shank. (See Fig. 5.)

A tool bit is a relatively small piece of tool material clamped in a tool holder in such a way that it can be readily removed and replaced.

A right-cut single-point tool is one in which the cutting edge is on the right side when looking at the point end of the tool with the face upward and the shank pointing away.

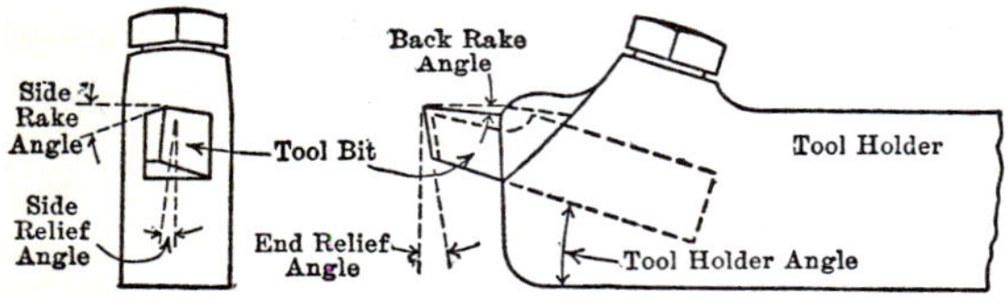

FIG. 5. Tool bit and holder.

A left-cut single-point tool is one in which the cutting edge is on the left side when looking at the point end with the face upward and the shank pointing away.

An end cut tool is one having its principal cutting edge on the end.

A straight tool is one in which the point is on the forward edge of a straight shank. (See Fig. 1.)

A bent tool is one in which the point is bent to the right or the left. A right bent tool has the point bent to the right when looking at the tool from the point end with the face upward and the shank pointing away. (See Fig. 6.)

The shank angle is the angle by which the point of a bent tool deviates from the straight portion of the shank. (See Fig. 6.)

An offset tool is one in which the point is at either side of, but parallel to, the shank. In a right offset tool, the point is offset to the right of the shank when looking at the tool from the point end with the face upward and with the shank pointing away. (See Fig. 7.)

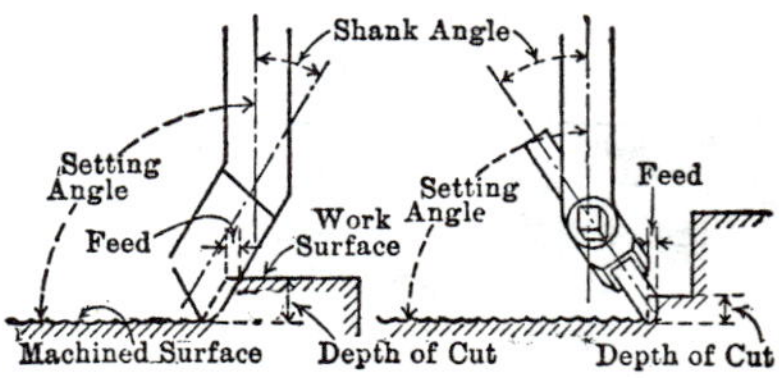

FIG. 6. Bent-shank, single-point tool.

TOOL ANGLES. The angles defined below are angles formed between various portions or surfaces of the tool. They are angles formed in forging or grinding the tool, and should not be confused with the working angles which mainly are angles formed between some surface of the tool and of the work.

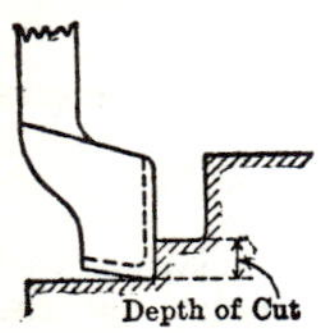

FIG. 7. Right-cut, right-offset, single-point tool.

Back-rake angle is the angle between the face of a tool and the shank base, measured in a plane at right angles to the base and parallel to the center line of the nose. The angle is positive if the face slopes downward from the nose toward the shank, and negative if the face slopes upward toward the shank.

Side-rake angle is the angle between the face of the tool and the shank base, measured in a plane at right angles to the base and at right angles to the center line of the nose.

Relief angle is the angle between a plane perpendicular to the base of a tool or tool holder and that portion of the flank immediately adjacent to the cutting edge.

Side relief angle is measured in the plane of the side-rake angle to the center line of the tool nose.

The end relief angle is measured in the plane of the back-rake angle.

The clearance angle is an angle between a perpendicular to the base of a tool and that portion of the flank immediately adjacent to the base. The side clearance is measured in the plane of the side-rake angle. The end clearance is measured in the plane of the back-rake angle.

The side-cutting-edge angle is the angle between the straight cutting edge on the side of the tool and the side of the tool shank.

The end-cutting-edge angle is the angle between the cutting edge on the end of the tool and a line at right angles to the side edge of the tool shank.

Nose angle is the angle included between the side cutting edge and the end cutting edge. (See Fig. 1.)

WORKING ANGLES are those angles between the tool and work which depend not only on the shape of the tool but also on its position with respect to the work.

Setting angle is the angle made by the shank of a tool with a line at right angles to the finished surface of the work. (See Figs. 5 and 6.)

The entering angle is the angle which the side cutting edge of the tool makes with the finished surface of the work. (See Fig. 5.)

True rake angle (or top rake) is the actual slope of the tool face from the active cutting edge in the direction of chip flow. It is a combination of front and side-rake angles and varies with the setting of the tool and with the feed and depth of cut.

Lip angle is the included angle of the tool material between the face and the ground flank, measured in a plane at right angles to the cutting edge. When measured in a plane perpendicular to the cutting edge at the end of the tool it is called the end lip angle. When measured in the direction of chip flow, it is called the true lip angle. (See Fig. 8.)

The working relief angle is the angle between the ground flank of the tool and a line, tangent to the cut surface, passing through the active cutting edge.

Cutting angle is equal to the lip angle plus the working relief angle at the point of action.

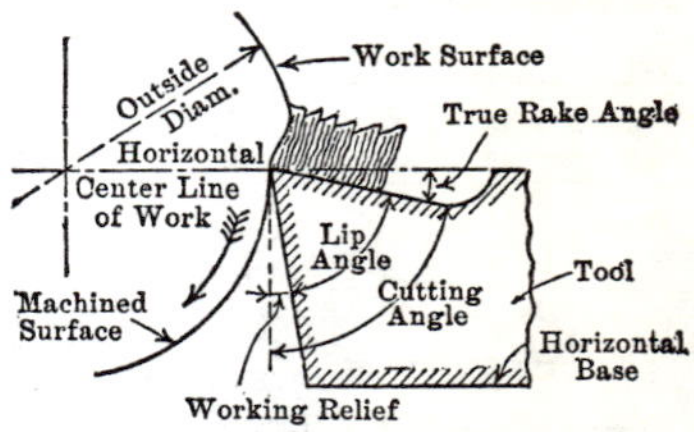

FIG. 8. Section through plane of chip flow, indicated as *AA* in Figs. 1 and 2.

The working end-cutting-edge angle is the angle between the straight end cutting edge and a plane tangent to the machined surface at the point of cutting.

CUTTING SPEED is the peripheral or surface speed of the work with respect to the tool. In turning, it usually is measured on the uncut or work surface of the work ahead of the tool.

DEPTH OF CUT is the distance between the bottom of the cut and the uncut surface, measured in a direction at right angles to the finished surface of the work.

FEED is the relative amount of motion of the tool into the work for each revolution, stroke, or unit of time.

MACHINED SURFACE is the surface left by the cutting tool. (See Figs. 6 and 8.)

WORK SURFACE is the surface that is to be machined. (See Figs. 6 and 8.)

2. TOOL CONTOUR AND SIZE

For complete data on the use of cutting tools, including feed and speed data for many varieties of alloy steel, see Ref. 1.

CHIP PROPORTIONS. The proportions of a chip, as affected by the nose contour of a tool or by the relation between depth of cut and feed, have great effect on the cutting speed which may be used. A change in chip proportions caused by a change in nose contour of the tool has the same effect on cutting speed as a similar change caused by a change in depth of cut and feed. When removing a chip of given cross-sectional area, changes in the tool contour may cause variations in the cutting speed of as much as 60%, while changes of as much as 400% may be caused by changes in the relation between depth of cut and feed.

TOOL SIZES. The number of sizes of tools used in a given shop should be as small as possible. The following sizes should meet all ordinary shop requirements: *Rectangular Tools:* $^3/_8$ by $^3/_4$ in.; $^1/_2$ by 1 in.; $^5/_8$ by 1 $^1/_4$ in.; $^3/_4$ by 1 $^1/_2$ in.; 1 by 2 in.; 1 $^1/_2$ by 3 in. *Square Tool Bits:* $^3/_{16}$, $^1/_4$, $^3/_8$, $^1/_2$, $^5/_8$, $^3/_4$, 1, 1 $^1/_4$, and 1 $^1/_2$ in. *Rectangular Tool Bits:* $^3/_{32}$ by $^1/_2$ in.; $^1/_8$ by $^3/_4$ in.; $^3/_{16}$ by 1 $^1/_8$ in.; $^1/_4$ by 1 $^3/_8$ in. A relatively small number of the foregoing sizes should be ample for all except the largest machine shops.

Tool bits forged to the desired form are available. Such bits must be used in special holders, usually made by the company furnishing the bits. Such bits are at times very convenient for shop use, but are much more expensive per pound than bar stock. When such tools and holders are adopted, both the stiffness of support to the cutting edge and the percentage of the new tool which may be used before it is discarded should be considered.

Effect of Tool Size. The size of a tool has no effect other than to give strength to the tool, to reduce deflection, or to increase the tool width available for a forming cut. A properly supported small tool may be run at the same cutting speed as a large one having the same shape, for the same depth of cut and feed. Lack of rigidity of a tool tends to increase the possibility of chatter.

TOOL FORM. When heavy roughing cuts are to be taken, and it is desired to use heavy forged tools, there are no better shapes of tool than those developed by Carl Barth. They are shown in Fig. 9. Dimensions are given in Table 1.

Table 1. Dimensions of Round-nose Forged Single-point Tools

(Left-cut straight tool with raised face. Barth standardization. See Fig. 9.)

Tool size, in., W	1/2	5/8	3/4	7/8	1	1 1/4	1 1/2	1 3/4	2
Blunt Tool for Hard Steel and Cast Iron									
Nose radius, in., R	0.101	0.169	0.236	0.303	0.371	0.506	0.641	0.776	0.911
Flank radius, in., F	0.820	0.888	0.955	1.022	1.090	1.225	1.360	1.495	1.630
Sharp Tool for Medium and Soft Steel									
Nose radius, in., R	0.072	0.126	0.180	0.235	0.289	0.397	0.506	0.614	0.723
Flank radius, in., F	0.791	0.845	0.899	0.954	1.008	1.116	1.225	1.333	1.442

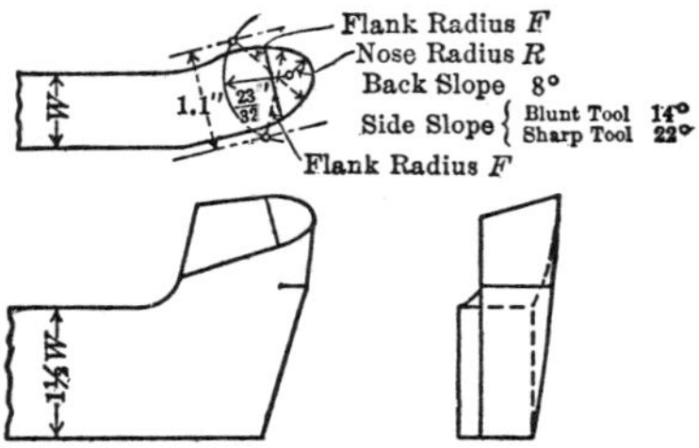

FIG. 9. Left-cut, straight-shank, round-nose tool with raised face.

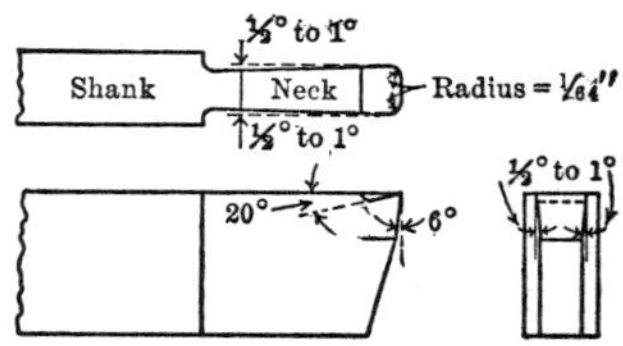

FIG. 10. Recommended shape of parting tool.

Tools of the form shown in Fig. 10 are used for parting cuts. They usually are 1/2 to 3/8 in. wide at the nose, except when the diameter of work to be parted is large, when they may be wider. To obtain the necessary stiffness, the cutting section of such a tool usually is forged down from a shank of comparatively large section. Note that the sides of the neck of this parting tool converge at a slight angle, usually 1/2 degree to 2 degrees, to avoid interference with the side of the cut. The side relief angle is small, usually not over 2 degrees. At times a blade having a height of five to eight times its width, held in a special holder, is used as a parting tool.

The contours shown in Fig. 11 are suitable and recommended for right-cut tool bits, or small solid tools ground directly from the bar. A similar set of tools, with the cutting edge on the right-hand side of the tool, would be necessary for cutting from left to right.

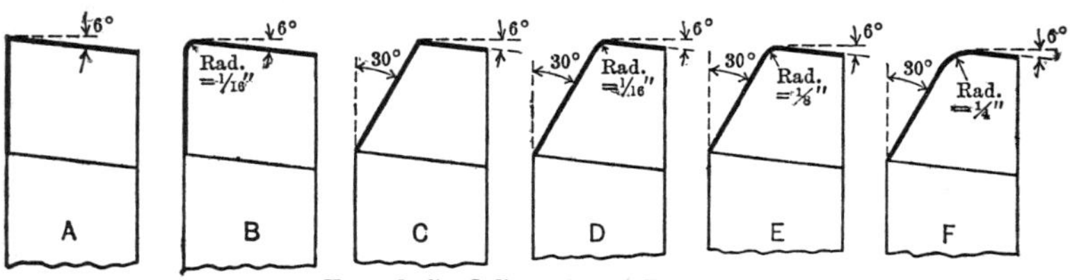

Heavy Outline Indicates Cutting Edge

FIG. 11. Recommended contour, right-cut roughing tool.

For facing operations the active cutting edge is on the end of the tool, preferably with a back rake equal to the side rake used on ordinary roughing tools, and with a side slope of approximately 8 degrees. The right-cut contours shown in Fig. 12 should be ample for all ordinary operations. A similar set, with the active cutting edge on the opposite side of the tool, would be required for left-cut tools.

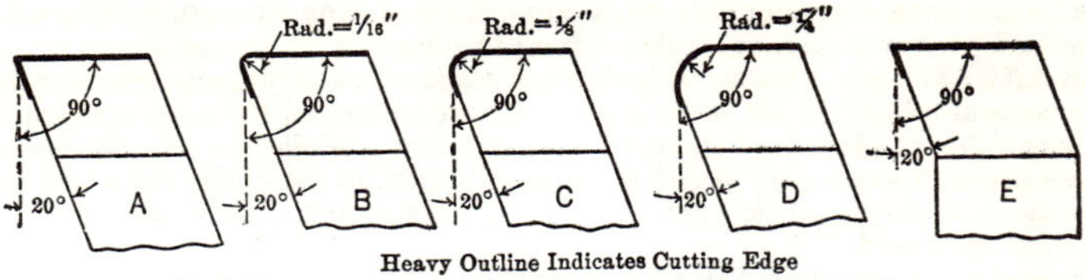

Heavy Outline Indicates Cutting Edge

FIG. 12. Recommended contour, right-cut facing tool.

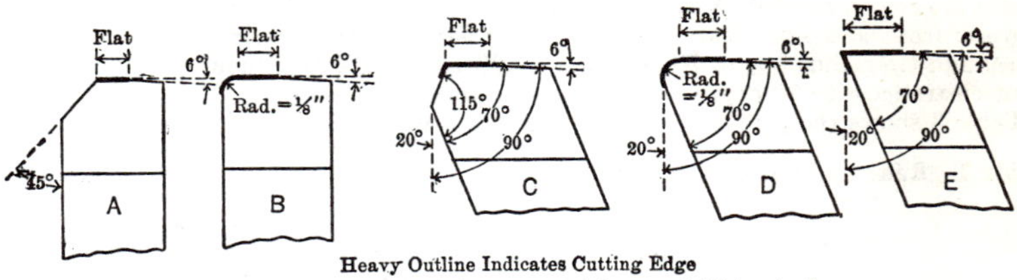

Heavy Outline Indicates Cutting Edge

FIG. 13. Recommended contour, right-cut finishing tool.

Table 2. Rake and Relief Angles for High-speed Steel Tools Cutting Various Metals

Metal Cut	Operation	Back Rake, degrees	Side Rake, degrees	Side Relief, degrees	End Relief, degrees
Steel, soft	Roughing	6–10	14–22	5–9	5–9
	Finishing	14–22	0	0	5–9
	Parting	14–22	0	1/2–2	5–9
	Forming	14–22	0	0	3–5
Steel, hard	Roughing	4–8	10–14	5–9	5–9
	Finishing	8–14	0	0	5–9
	Parting	8–14	0	1/2–2	5–9
	Forming	4–8	0	0	3–5
Steel, very hard	Roughing	3–7	5–10	5–9	5–9
	Finishing	5–10	0	0	5–9
	Parting	5–10	0	1/2–2	5–9
	Forming	0–5	0	0	3–5
Cast iron	Roughing	8	14	5–9	5–9
	Finishing	6–10	0	0	5–9
	Parting	4–8	0	1/2–2	5–9
	Forming	4–8	0	0	3–5
Brass and bronze	Roughing	6–8	4–10	5–9	5–9
	Finishing	14–22	0	0	5–9
	Parting	6–12	0	1/2–2	5–9
	Forming	12–20	0	0	3–5
Copper	Roughing	8–12	16–28	5–9	5–9
	Finishing	8	16–28	0	5–9
	Parting	6–12	0	1/2–2	5–9
	Forming	12–20	0	0	3–5
Monel metal	Roughing	4–8	10–14	5–9	5–9
	Finishing	14–22	0	0	5–9
	Parting	4–8	0	1/2–2	5–9
	Forming	4–8	0	0	3–5
Aluminum	Roughing	8	16–22	5–9	5–9
	Finishing	8	16–22	0	5–9
	Parting	6–12	0	1/2–2	5–9
	Forming	12–20	0	0	3–5
Magnesium alloys	Roughing	5–8	3–5	6–10	6–10
	Finishing	10–15	0	0	6–10
	Parting	10–15	0	6–10	6–10
	Forming	5–10	0	0	6–10

Tools for finishing cuts should have an appreciable flat on the end, substantially wider than the feed per revolution or stroke. (See Fig. 13.)

RAKE ANGLE. There seems to be little necessity for making any considerable variation in rake and relief. While cutting pressures are decreased by an increase in rake, the cutting speed is not greatly affected by such changes. While the chip is removed with less expenditure of energy by the tool having the greater rake, the reduction in cutting angle tends to weaken the tool, and so causes failure sooner than would be the case if a smaller rake were used.

The values of rake given in Table 2 are based on the practice of a leading manufacturing company, for use with high-speed steel tools. For cutting ingot iron the following tool angles are recommended by a leading producer: back rake 15 degrees; side rake 28 to 32 degrees; front and side clearance 8 to 10 degrees; side and end cutting edge each 15 degrees; and tip radius $1/32$ to $1/16$ in. For cutoff tools back rake should be 20 to 30 degrees, front clearance 8 to 10 degrees, and side clearance 2 degrees.

Table 3 shows the practice recommended for use with cemented carbide tools.

Table 3. Rake and Relief Angles for Cemented Carbide Tools Cutting Various Materials

Material to Be Cut	Rake, degrees		Relief, degrees	
	Side	Back	Front	Side
Aluminum alloys	10–20	10–20	8–12	6–10
Aluminum castings	10–20	10–20	6	6
Brass-yellow	2–6	0–5	4–8	4–8
Bronze and bronze alloys	0–10	0–5	4–6	4–6
Cast iron, soft	4–6	0–4	3–6	3–6
hard	3–6	0	3–5	3–4
chilled	3–6	0	3–5	3–4
malleable	4–8	0–4	3–5	3–5
Copper	8–15	0–10	6–8	6–8
Die castings	6–12	0–10	6–10	6–10
Fiber	6–12	0–10	6–10	6–10
Plastics	6–15	0–10	6–10	6–10
Rubber, hard	8–15	5–15	6–10	6–10
soft	8–20	10–20	8–15	8–15
Steel, to 200 Brinell	4–15	0–10	6–8	6–8
to 200–275 Brinell	4–10	0–8	5–8	5–8
to 275–350 Brinell	4–8	0–5	5–8	5–8
to 350–425 Brinell	0–6	0–3	4–6	4–6
to 425 and up	0	0	4–6	4–6
Stainless steel	8–16	4–6	4–6	4–6
Manganese steel	0–4	0	3–4	3–4
Nitralloy	4–15	0–10	4–6	4–6

3. CUTTING SPEEDS AND FEEDS

ECONOMIC TOOL LIFE. Under a given set of conditions, it may be shown that there is a particular tool life which gives lowest overall costs. This tool life, and the accompanying cutting speed, may be calculated if all data are available. The values given in Table 4, however, may be used without great loss in efficiency.

Table 4. Economic Tool Life under Cut

(High-speed steel tools, cutting steel, and cast iron)

Machine	Tool Setting	Economic Tool Life, min
Engine lathe or planer	Ordinary	40
	Accurate	120
Drill press	Ordinary	30
Hand screw machine	Ordinary	120
Automatic screw machine	Ordinary	400
Automatic milling machine, simple cutters	....	600
Milling machine, ordinary cutters	....	250
Milling machine, large forming cutters	....	2000

Speeds and feeds for cemented carbides are given in Table 26. In general, feeds are approximately the same as those used with high-speed steel tools, up to a maximum of approximately 0.063 in. per revolution, with speeds five to six times those used with high-speed steel tools. Tool life under cut should be at least 60 min.

Table 5 shows the effect of variations in cutting speed on tool life. This table is based on a value of unity for a tool life of 60 min. If the proper cutting speed is known for a cut

Table 5. Effect on Tool Life of Variations in Cutting Speed

Tool Life, min	Carbon Steel Tools			High-speed Steel Tools						Stellite Tools			Cemented Carbide Tools		
	Metal Cut														
	Steel		Cast Iron	Steel		Cast Iron	Bronze		Brass	Steel		Cast Iron	Steel		Cast Iron
	Rough	Finish	Rough, Finish	Rough	Finish	Rough, Finish	Rough	Finish	Rough, Finish	Rough	Finish	Rough, Finish	Rough	Finish	Rough, Finish
	Tool Life Factor F														
15	1.32	1.11	1.11	1.19	1.15	1.15	1.34	1.26	1.26	1.29	1.22	1.22	1.25	1.20	1.20
30	1.15	1.05	1.05	1.09	1.07	1.07	1.16	1.12	1.12	1.13	1.11	1.11	1.12	1.09	1.09
45	1.06	1.02	1.02	1.04	1.05	1.03	1.06	1.05	1.05	1.05	1.04	1.04	1.05	1.04	1.04
60	1.00	1.00	1.00	1.00	1.00	1.00	1.00	1.00	1.00	1.00	1.00	1.00	1.00	1.00	1.00
90	0.92	0.98	0.98	0.95	0.96	0.96	0.94	0.92	0.92	0.93	0.94	0.94	0.94	0.95	0.95
120	0.87	0.95	0.95	0.92	0.93	0.93	0.86	0.89	0.89	0.88	0.91	0.91	0.90	0.91	0.91
180	0.80	0.92	0.92	0.87	0.89	0.89	0.79	0.83	0.83	0.82	0.86	0.86	0.84	0.87	0.87
240	0.76	0.90	0.90	0.84	0.87	0.87	0.75	0.79	0.79	0.78	0.82	0.82	0.80	0.84	0.84
300	0.72	0.89	0.89	0.82	0.85	0.85	0.71	0.77	0.77	0.75	0.79	0.79	0.77	0.81	0.81
360	0.70	0.87	0.87	0.80	0.84	0.84	0.69	0.74	0.74	0.72	0.77	0.77	0.75	0.79	0.79
400	0.68	0.86	0.86	0.78	0.82	0.82	0.66	0.72	0.72	0.70	0.75	0.75	0.73	0.78	0.78
480	0.66	0.86	0.86	0.77	0.81	0.81	0.65	0.71	0.71	0.69	0.74	0.74	0.72	0.76	0.76

A of given duration, the speed for a cut B of different duration may be obtained by the formula

$$S_b = S_a \times \frac{F_b}{F_a}$$

where S_b, S_a = respectively cutting speeds of A and B, and F_a, F_b = respectively tool life factors for A and B from Table 5.

Example. A cutting speed of 95 ft per min in rough turning steel with high-speed steel tools gives a tool life of 45 min. Required the cutting speed to give a tool life of 120 min.

Solution. $A = 45$; $B = 120$; $S_a = 95$. From Table 5, tool life factor $F_a = 1.04$, and $F_b = 0.92$. Then $S_b = 95 \times (0.92/1.04) = 84$ ft per min.

ECONOMIC PRODUCTION. To keep costs at a minimum in rough machining operations, the maximum volume of chips per minute must be removed. This involves: (1) The maximum depth of cut that can be taken without distortion of the part machined, up to the point where all surplus stock is removed in a single roughing cut. (2) A relatively large feed, which gives thick chips, preferably above 0.030 in. (3) The use of a tool having a contour which gives the smallest possible chip thickness for a given feed without setting up injurious chatter. Increase in the nose radius or the side-cutting-edge angle decreases chip thickness under constant conditions of feed and depth of cut. (4) A cutting speed such that the life of the tool approximates the economic life.

It must be realized that the power available for driving the machine tool may limit the rate of stock removal.

MATERIALS CUT. Tables 6, 7, and 9 give the approximate relative machinability, based on the speed with which a given cut may be taken, of a large number of common materials. These tables may be used to determine the correct cutting speed and horsepower requirements for the various materials under a variety of working conditions, by use of the multipliers for speed and horsepower in connection with the basic data given in Tables 10 to 17. Table 6 is based on C1030 hot-rolled steel (see Tables 10 to 13) and Table 7 is based on cast iron of 177 Brinell hardness and tensile strength of 30,000 psi (see Tables 14 to 17).

Example. Required the correct cutting speed and the horsepower for turning 2330 steel, normalized, using a round-nose tool set at an angle of 30 degrees to the work, with a 1/8 in. depth of cut and a 1/32 in. feed.

Solution. From Table 12, the desirable cutting speed for C1030 hot-rolled steel, under the conditions given, is 80 ft per min. The horsepower required at the nose of the tool is 2.5. From Table 6, the multiplier for speed for 2330 normalized steel is 0.68 and the multiplier for horsepower is 0.65. Cutting speed for 2330 normalized steel equals 80 × 0.68 = 54.4 ft per min. Horsepower required at nose of tool = 2.5 × 0.65 = 1.63.

(Continued on p. 23-13)

Table 6. Machinability Constants of Steel

(For analysis of the various steels listed, see pp. 2-25 to 2-28.)

Steel, AISI No.	Condition*	Heat Treatment			Physical Properties					Chip Pressure Constant, K_1	Multipliers for C1030 HR Values (See Tables 10 to 13)		Relative Machinability
		Heat to, °F	Quench in	Draw at, °F	Tensile Strength, psi	Yield Point, psi	Elong. in 2 in., %	Reduction of Area, %	Brinell Hardness No.		Speed	Hp	
C1010	HR	......	......	......	55,000	35,000	35	55	110	114,000	0.53	0.47	2.5
C1010	CD	......	......	......	66,000	60,000	23	51	141	125,000	0.53	0.52	2.5
C1015	HR	......	......	......	57,000	37,000	33	55	143	116,000	0.62	0.56	2.9
C1015	CD	......	......	......	73,000	62,000	22	51	150	127,000	0.61	0.60	2.9
C1020	HR	......	......	......	66,000	32,000	35	52	127	122,000	1.05	0.99	4.9
C1020	CD	......	......	......	78,000	65,000	21	50	160	131,000	1.03	0.95	4.6
C1019	HR	......	......	......	61,000	43,000	31	61	126	119,000	1.05	1.06	4.9
C1019	CD	......	......	......	86,000	80,000	26	57	156	138,000	0.99	1.06	4.6
C1025	HR	......	......	......	66,000	39,000	29	54	160	122,000	0.90	0.87	4.1
C1025	CD	......	......	......	82,000	67,000	20	50	168	135,000	0.84	0.88	3.9
C1025	HT	1600	Water	900	80,000	56,000	28	59	168	133,000	0.93	0.95	4.3
C1030	HR	......	......	......	75,000	47,000	26	50	168	129,000	1.00	1.00	4.7
C1030	CD	......	......	......	85,000	70,000	19	50	180	137,000	0.79	0.83	3.7
C1030	HT	1600	Water	900	89,000	63,000	26	56	185	140,000	0.81	0.87	3.7
C1035	HR	......	......	......	82,000	49,000	25	51	174	135,000	0.88	0.93	4.1
C1035	HT	1550	Water	1300	85,000	50,000	34	68	174	137,000	0.87	0.92	4.1
C1035	HT	1550	Water	860	105,000	75,000	23	52	212	152,000	0.62	0.72	2.9
C1040	HR	......	......	......	90,000	53,000	24	41	187	141,000	0.78	0.84	3.6
C1040	CD	......	......	......	92,000	75,000	18	46	195	142,000	0.72	0.78	3.3
C1040	HT	1550	Water	1300	92,000	55,000	31	64	186	142,000	0.76	0.83	3.5
C1040	HT	1550	Water	800	113,000	83,000	19	49	230	158,000	0.55	0.66	2.5
C1045	HR	......	......	......	93,000	56,000	26	55	187	143,000	0.77	0.83	3.5
C1045	N	......	......	......	93,000	......	......	......	185	143,000	0.78	0.86	3.6
C1045	HT	1500	Water	1300	98,000	60,000	28	56	197	147,000	0.79	0.89	3.7
C1045	HT	1500	Water	800	121,000	91,000	15	46	248	165,000	0.49	0.61	2.2
C1050	HR	......	......	......	103,000	60,000	18	41	201	150,000	0.63	0.72	2.9
C1050	HT	1500	Water	1300	104,000	65,000	25	......	208	151,000	0.59	0.69	2.7
C1050	HT	1500	Water	800	129,000	99,000	11	......	266	171,000	0.42	0.55	1.9
C1060	HR	......	......	......	117,000	71,000	16	26	217	162,000	0.54	0.66	2.5
C1060	N	......	......	......	110,000	......	......	......	225	156,000	0.59	0.71	2.7
C1060	HT	1500	Water	1300	116,000	74,000	19	......	230	161,000	0.49	0.60	2.2
C1060	HT	1500	Water	800	144,000	114,000	......	......	315	182,000	0.28	0.39	1.3
C1095	HR	......	......	......	147,000	74,000	11	18	280	184,000	0.38	0.54	1.8
C1095	N	1500	Furnace cooled		93,000	45,000	6	7	207	143,000	0.64	0.71	2.9
B1112	CD	......	......	......	88,000	75,000	20	45	167	84,000	1.76	1.13	8.2
B1113	CD	......	......	......	88,000	75,000	20	45	183	84,000	1.98	1.28	9.3
C1117	HR	......	......	......	80,000	68,000	23	52	167	79,000	1.03	0.63	4.6
C1117	CD	......	......	......	80,000	62,000	20	50	162	95,000	1.23	0.91	5.7
C1118	HR	......	......	......	67,000	43,000	42	64	120	70,000	1.31	0.71	6.1
C1118	CD	......	......	......	90,000	75,000	20	45	161	85,000	1.05	0.69	4.9
C1332	HR	......	......	......	83,000	42,000	20	40	165	81,000	1.21	0.76	5.7
C1332	CD	......	......	......	97,000	73,000	14	32	200	91,000	0.86	0.69	4.0
C1137	HR	......	......	......	92,000	57,000	35	41	197	87,000	0.66	0.45	3.0
C1137	CD	......	......	......	90,000	75,000	20	45	197	86,000	0.66	0.44	3.0
C1137	HT	1525	Oil	1000	114,000	88,000	21	57	229	101,000	0.53	0.42	2.5
1330	HR	......	......	......	111,000	73,000	22	57	212	157,000	0.91	1.09	4.2
1330	HT	1550	Oil	1000	139,000	121,000	20	53	277	178,000	0.60	0.81	2.7
1335	HT	1550	Oil	1000	129,000	103,000	17	52	262	171,000	0.64	0.84	3.0
1340	HR	......	......	......	115,000	75,000	20	54	217	160,000	0.87	1.08	4.1
1340	HT	1525	Oil	1000	135,000	106,000	15	51	277	176,000	0.60	0.80	2.7
2317	HR	......	......	......	76,000	55,000	31	65	153	121,000	1.13	1.06	5.3
2317	N	......	......	......	75,000	52,000	34	67	......	121,000	1.05	0.98	4.9
2317	CD	......	......	......	87,000	78,000	17	61	192	123,000	0.74	0.70	3.4
2330	A	......	......	......	100,000	92,000	18	48	197	123,000	0.71	0.68	3.2
2330	HR	......	......	......	105,000	68,000	21	49	207	126,000	0.64	0.63	3.0

* A = annealed; CD = cold drawn; HR = hot rolled; HT = heat treated; N = normalized.

Table 6. Machinability Constants of Steel—*Continued*

(For analysis of the various steels listed, see pp. 2-25 to 2-28.)

Steel, AISI No.	Condition	Heat Treatment			Physical Properties					Chip Pressure Constant, K_1	Multipliers for C1030 HR Values (See Tables 10 to 13)		Relative Machinability
		Heat to, °F	Quench in	Draw at, °F	Tensile Strength, psi	Yield Point, psi	Elong. in 2 in., %	Reduction of Area, %	Brinell Hardness No.		Speed	Hp	
2330	N	......	......	......	102,000	67,000	24	50	......	124,000	0.68	0.65	3.2
2330	CD	......	......	......	124,000	119,000	12	43	223	136,000	0.46	0.48	2.1
2330	HT	1450	Oil	1000	129,000	106,000	13	36	271	139,000	0.51	0.55	2.4
2330	HT	1450	Oil	800	160,000	128,000	18	48	324	172,000	0.42	0.55	1.9
2335	A	......	......	......	108,000	95,000	18	45	211	125,000	0.67	0.65	3.1
2335	HR	......	......	......	108,000	63,000	23	56	195	127,000	0.72	0.70	3.3
2335	N	......	......	......	115,000	77,000	23	47	......	129,000	0.60	0.59	2.8
2335	HT	1450	Oil	1000	132,000	108,000	22	60	275	142,000	0.53	0.58	2.4
2335	HT	1450	Oil	800	170,000	147,000	17	54	335	186,000	0.40	0.57	1.8
2340	HR	......	......	......	114,000	78,000	19	42	240	129,000	0.58	0.59	2.7
2340	N	......	......	......	106,000	74,000	22	47	223	134,000	0.64	0.66	3.0
2340	HT	1425	Oil	1000	136,000	120,000	21	60	300	147,000	0.49	0.55	2.3
2340	HT	1425	Oil	800	180,000	165,000	17	53	376	200,000	0.38	0.59	1.8
2345	A	......	......	......	118,000	100,000	13	40	220	127,000	0.64	0.62	2.9
2345	HR	......	......	......	129,000	94,000	16	35	260	140,000	0.54	0.58	2.5
2345	N	......	......	......	126,000	82,000	20	39	......	137,000	0.53	0.55	2.4
2345	HT	1425	Oil	1100	125,000	113,000	22	62	273	137,000	0.55	0.58	2.5
2345	HT	1425	Oil	900	160,000	150,000	18	55	340	172,000	0.40	0.53	1.8
2512	A	......	......	......	115,000	93,000	15	42	214	124,000	0.66	0.64	3.0
2512	HR	......	......	......	105,000	73,000	27	52	210	126,000	0.64	0.63	3.0
2512	N	......	......	......	84,000	58,000	32	64	......	122,000	0.90	0.84	4.2
2512	CD	......	......	......	140,000	......	......	......	......	130,000	0.68	0.67	3.1
2515	HR	......	......	......	91,000	68,000	28	69	179	123,000	0.81	0.76	3.7
2515	N	......	......	......	91,000	69,000	29	69	179	123,000	0.81	0.76	3.7
2515	CD	......	......	......	115,000	75,000	28	60	226	130,000	0.62	0.63	2.9
3115	HR	......	......	......	71,000	46,000	32	69	134	90,000	1.00	0.70	4.6
3115	N	......	......	......	72,000	59,000	23	69	128	90,000	1.00	0.70	4.7
3115	CD	......	......	......	77,000	68,000	25	68	163	92,000	0.85	0.60	3.9
3120	HR	......	......	......	88,000	56,000	29	57	149	95,000	0.73	0.53	3.4
3120	N	......	......	......	85,000	63,000	21	65	144	94,000	0.83	0.61	3.9
3120	CD	......	......	......	98,000	80,000	20	63	174	119,000	0.60	0.55	2.8
3125	HR	......	......	......	91,000	57,000	27	62	164	99,000	0.69	0.53	3.2
3125	N	......	......	......	90,000	67,000	29	62	160	96,000	0.71	0.53	3.3
3125	CD	......	......	......	111,000	92,000	16	58	187	127,000	0.50	0.49	2.3
3130	A	......	......	......	98,000	88,000	18	55	188	120,000	0.65	0.60	3.0
3130	HR	......	......	......	99,000	63,000	25	60	178	122,000	0.62	0.58	2.9
3130	N	......	......	......	97,000	71,000	26	58	176	116,000	0.64	0.58	3.0
3130	CD	......	......	......	118,000	113,000	14	55	210	132,000	0.45	0.46	2.1
3130	HT	1525	Oil	1000	131,000	102,000	20	62	262	141,000	0.45	0.48	2.1
3130	HT	1525	Oil	800	166,000	138,000	14	54	335	180,000	0.33	0.46	1.5
3135	A	......	......	......	98,000	88,000	18	50	188	120,000	0.65	0.61	3.0
3140	A	......	......	......	103,000	90,000	15	48	212	121,000	0.58	0.55	2.6
3140	HR	......	......	......	112,000	74,000	20	50	228	128,000	0.53	0.52	2.4
3140	N	......	......	......	110,000	79,000	21	50	207	127,000	0.54	0.49	2.5
3140	CD	......	......	......	125,000	119,000	13	45	241	137,000	0.42	0.45	1.9
3140	HT	1500	Oil	1000	143,000	118,000	18	57	285	155,000	0.41	0.48	1.9
3140	HT	1500	Oil	800	175,000	157,000	13	48	370	195,000	0.31	0.47	1.4
3141	HR	......	......	......	112,000	66,000	20	53	210	128,000	0.60	0.59	2.8
3141	A	......	......	......	102,000	68,000	26	55	207	125,000	0.62	0.60	2.9
3141	HT	1500	Oil	1100	130,000	106,000	18	56	255	140,000	0.49	0.52	2.2
3141	HT	1500	Oil	840	168,000	143,000	13	47	219	241,000	0.36	0.67	1.6
3145	HR	......	......	......	116,000	79,000	17	48	210	137,000	0.49	0.51	2.2
3145	N	......	......	......	117,000	84,000	19	48	223	131,000	0.49	0.49	2.3
3145	HT	1475	Oil	1100	128,000	108,000	19	58	259	139,000	0.45	0.48	2.1
3145	HT	1475	Oil	900	170,000	143,000	15	52	332	186,000	0.33	0.47	1.5
3150	HR	......	......	......	120,000	85,000	15	42	280	133,000	0.45	0.46	2.1
3150	N	......	......	......	120,000	88,000	16	43	239	133,000	0.46	0.47	2.1
3150	A	......	......	......	97,000	52,000	27	42	179	91,000	0.65	0.46	3.0
3150	HT	1450	Oil	1100	132,000	112,000	18	57	262	142,000	0.44	0.47	2.0

(*Table continued on p. 23-12*)

Table 6. Machinability Constants of Steel—*Continued*

(For analysis of the various steels listed, see pp. 2-25 to 2-28.)

Steel, AISI No.	Con-dition	Heat Treatment			Physical Properties					Chip Pressure Constant, K_1	Multipliers for C1030 HR Values (See Tables 10 to 13)		Relative Machinability
		Heat to, °F	Quench in	Draw at, °F	Tensile Strength, psi	Yield Point, psi	Elong. in 2 in., %	Reduction of Area, %	Brinell Hardness No.		Speed	Hp	
3150	HT	1450	Oil	900	173,000	150,000	14	50	342	191,000	0.33	0.48	1.5
E3310	HR	......	......	......	133,000	100,000	17	50	296	144,000	0.41	0.46	1.9
E3310	N	......	......	......	160,000	120,000	17	50	321	172,000	0.36	0.47	1.6
E3310	A	......	......	......	105,000	85,000	26	72	212	126,000	0.60	0.58	2.8
E3310	HT	1400	Oil	1100	111,000	99,000	25	73	228	128,000	0.55	0.54	2.5
4017	HR	......	......	......	89,000	56,000	31	67	167	96,000	0.81	0.59	3.7
4017	N	1625	......	......	87,000	59,000	32	68	158	95,000	0.86	0.63	4.0
4130	HR	......	......	......	95,000	72,000	22	56	192	111,000	0.65	0.55	3.0
4130	HT	1500	Water	1060	141,000	134,000	18	63	293	118,000	0.67	0.60	3.1
E4135	HR	......	......	......	131,000	86,000	16	48	272	142,000	0.44	0.48	2.0
E4135	N	......	......	......	114,000	......	18	53	......	127,000	0.49	0.48	2.3
E4135	HT	1500	Oil	1000	183,000	163,000	15	55	364	185,000	0.32	0.46	1.5
E4135	HT	1550	Oil	800	219,000	198,000	11	48	444	225,000	0.24	0.42	1.1
4140	A	......	......	......	108,000	95,000	20	48	212	125,000	0.58	0.56	2.6
4140	HR	......	......	......	134,000	92,000	20	45	269	142,000	0.44	0.48	2.0
4140	N	......	......	......	137,000	81,000	19	42	277	145,000	0.42	0.48	2.0
4140	HT	1550	Oil	1000	187,000	171,000	11	46	363	191,000	0.29	0.42	1.3
4140	HT	1550	Oil	800	216,000	198,000	10	35	444	222,000	0.24	0.42	1.1
4150	HR	......	......	......	144,000	108,000	16	47	287	156,000	0.40	0.47	1.8
4150	N	......	......	......	153,000	......	12	28	300	158,000	0.36	0.45	1.7
4340	HR	......	......	......	198,000	159,000	9	22	401	202,000	0.26	0.40	1.2
4340	N	......	......	......	203,000	147,000	12	20	415	207,000	0.27	0.42	1.2
4340	HT	1500	Oil	1200	150,000	129,000	20	57	302	155,000	0.40	0.47	1.8
4340	HT	1500	Oil	1000	184,000	165,000	16	48	375	186,000	0.31	0.45	1.4
4340	HT	1500	Oil	800	212,000	194,000	12	41	429	217,000	0.26	0.43	1.2
4615	HR	......	......	......	82,000	62,000	27	66	183	109,000	0.74	0.63	3.5
4615	N	......	......	......	83,000	60,000	30	65	170	110,000	0.74	0.64	3.5
4615	CD	......	......	......	107,000	102,000	14	58	212	125,000	0.54	0.51	2.5
4615	HT	1575	Oil	1000	115,000	89,000	27	70	229	128,000	0.54	0.53	2.5
4615	HT	1575	Oil	800	129,000	92,000	23	66	251	138,000	0.47	0.49	2.2
4615	HT	1575	Oil	600	141,000	99,000	20	60	271	147,000	0.42	0.48	2.0
4620	N	......	......	......	85,000	61,000	31	68	179	111,000	0.74	0.63	3.4
4620	HT	1475	Oil	1000	120,000	97,000	24	67	246	131,000	0.49	0.49	2.3
4620	HT	1475	Oil	800	139,000	107,000	21	63	280	147,000	0.42	0.48	2.0
4620	HT	1475	Oil	600	178,000	156,000	15	60	345	182,000	0.32	0.45	1.5
4640	HR	......	......	......	119,000	93,000	13	32	241	130,000	0.46	0.46	2.1
4640	N	......	......	......	122,000	97,000	25	62	248	132,000	0.49	0.49	2.1
4640	HT	1500	Oil	1000	145,000	131,000	18	57	320	151,000	0.30	0.45	1.8
4640	HT	1500	Oil	800	182,000	169,000	14	42	392	182,000	0.29	0.42	1.4
4815	N	......	......	......	94,000	68,000	28	62	187	118,000	0.67	0.60	3.1
4820	HT	1550	Oil	600	199,000	180,000	15	61	390	203,000	0.29	0.45	1.3
5120	HR	......	......	......	82,000	48,000	28	61	......	110,000	0.78	0.66	3.6
5120	N	......	......	......	73,000	50,000	33	65	149	105,000	0.86	0.69	4.0
5120	CD	......	......	......	98,000	88,000	16	55	183	120,000	0.70	0.65	3.2
5120	HT	1560	Water	842	155,000	122,000	12	46	335	160,000	0.35	0.42	1.6
5120	HT	1600	Water	930	98,000	70,000	21	67	210	122,000	0.59	0.55	2.7
5130	HR	......	......	......	105,000	66,000	24	63	207	125,000	0.60	0.58	2.8
5130	N	......	......	......	104,000	67,000	24	63	201	125,000	0.60	0.58	2.8
5130	HT	1550	Oil	1200	120,000	99,000	22	67	235	136,000	0.53	0.55	2.5
5130	HT	1550	Oil	1000	148,000	130,000	18	60	302	154,000	0.40	0.47	1.8
5130	HT	1550	Oil	800	193,000	168,000	12	49	375	196,000	0.29	0.43	1.3
5135	A	......	......	......	90,000	58,000	25	37	207	116,000	0.69	0.61	3.2
5140	A	......	......	......	102,000	93,000	15	40	207	124,000	0.62	0.60	2.8
5140	HR	......	......	......	114,000	75,000	20	58	223	131,000	0.53	0.54	2.5
5140	N	......	......	......	110,000	70,000	22	61	212	128,000	0.55	0.55	2.6
5140	HT	1550	Oil	1200	124,000	100,000	22	63	241	138,000	0.49	0.52	2.3
5140	HT	1550	Oil	1000	152,000	130,000	18	55	302	157,000	0.38	0.47	1.8
5140	HT	1550	Oil	800	180,000	160,000	13	42	363	182,000	0.32	0.45	1.5

Table 6. Machinability Constants of Steel—*Continued*

(For analysis of the various steels listed, see pp. 2-25 to 2-28.)

Steel, AISI No.	Condition	Heat Treatment			Physical Properties					Chip Pressure Constant, K_1	Multipliers for C1030 HR Values (See Tables 10 to 13)		Relative Machinability
		Heat to, °F	Quench in	Draw at, °F	Tensile Strength, psi	Yield Point, psi	Elong. in 2 in., %	Reduction of Area, %	Brinell Hardness No.		Speed	Hp	
5150	A	......	....	......	108,000	95,000	20	40	208	125,000	0.62	0.60	2.8
5150	HR	......	....	......	135,000	101,000	15	50	277	145,000	0.42	0.48	2.0
5150	N	......	....	......	135,000	98,000	15	50	285	145,000	0.41	0.46	1.9
5150	HT	1550	Oil	1200	133,000	114,000	21	57	269	144,000	0.46	0.51	2.1
5150	HT	1550	Oil	1000	168,000	150,000	17	45	352	171,000	0.35	0.45	1.6
5150	HT	1550	Oil	800	220,000	202,000	9	30	461	226,000	0.23	0.41	1.1
52100	A	......	....	......	95,000	58,000	30	61	187	125,000	0.66	0.64	3.0
6120	HR	......	....	......	114,000	75,000	23	64	255	131,000	0.50	0.51	2.3
6120	N	1600	....	......	80,000	46,000	33	73	164	109,000	0.81	0.69	3.8
6120	HT	1625	Water	1150	112,000	95,000	24	72	220	130,000	0.55	0.55	2.6
6120	HT	1625	Water	800	150,000	130,000	17	63	310	155,000	0.38	0.46	1.8
6145	HR	......	....	......	140,000	81,000	10	26	302	173,000	0.38	0.52	1.8
6145	HT	1525	Oil	900	183,000	166,000	15	50	401	280,000	0.32	0.69	1.5
6145	HT	1565	Oil	800	231,000	206,000	11	44	444	370,000	0.26	0.74	1.2
6150	A	......	....	......	108,000	98,000	15	40	212	128,000	0.62	0.62	2.8
6150	CD	......	....	......	....	....	......	......	......	124,000	0.72	0.69	3.3
6150	HR	......	....	......	141,000	107,000	18	44	285	177,000	0.44	0.60	2.0
6150	N	......	....	......	136,000	100,000	20	47	277	159,000	0.46	0.57	2.1
6150	HT	1575	Oil	1200	133,000	110,000	19	55	262	158,000	0.47	0.58	2.2
6150	HT	1575	Oil	1000	171,000	150,000	12	45	341	265,000	0.35	0.70	1.6
6150	HT	1575	Oil	800	228,000	210,000	10	38	461	300,000	0.24	0.55	1.1
302	HT	1950	Water	......	91,000	55,000	71	80	187	126,000	0.36	0.35	1.6
410	A	......	....	......	75,000	33,000	57	63	142	122,000	0.45	0.42	2.1
410	HT	1290	Air cooled		93,000	75,000	26	74	198	118,000	0.67	0.60	3.1

TOOL PRESSURE. F. W. Taylor (Ref. 2) found that no definite relation exists between the cutting speed of tools and the pressure, tangential to the finished surface of the work, exerted on the tool by the chip.

The value of this tangential component, which is of interest in the design of machine tools, is

$$P = K_1L^{1.1}T^{0.78}, \text{ for cutting steel}$$

and

$$P = K_2LT^{0.65}, \text{ for cutting cast iron}$$

where K_1, K_2 = constants, given in Tables 6 and 7; L = total width of chip, inches, which for the average tool is equal to the total length of cutting edge of the tool in contact with the work (see Fig. 14); T = average thickness of the chip, inches. This usually equals (Feed × Depth)/L. (See Ref. 1, Chap. 22.)

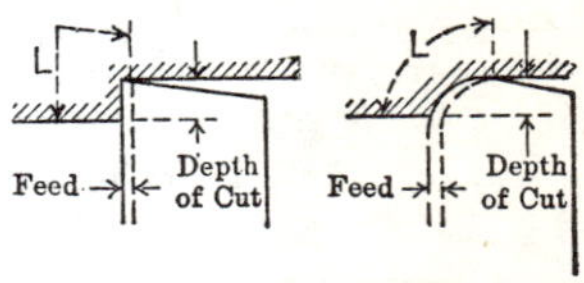

FIG. 14. Determination of factor L in formula for tool pressure.

METAL CUTTING TESTS, to be on a thoroughly scientific basis, require the consumption of much more time and material than usually are available in the average shop. Standard B5.19-1946, of the American Standards Association, outlines a procedure for the making of such tests. For further discussion of the subject, see Refs. 1, 3, 4, and 5.

CUTTING SPEED TABLES. Table 10 gives cutting speeds and horsepower at the point of the tool for cutting C1030 hot-rolled steel with high-carbon steel tools, and Table 14 gives corresponding data for cutting a medium cast iron with carbon steel tools.

Tables 11, 12, and 13 give cutting speeds and net horsepower at the point of the tool for cutting C1030 hot-rolled steel with high-speed steel tools of 18-4-1, 14-4-2, or molybdenum tungsten grades. Tables 15, 16, and 17 give corresponding data for cutting a medium cast iron (177 Brinell) with high-speed steel tools.

These tables are based on a 60-min tool life. To obtain cutting speeds that permit obtaining some other tool life, multiply the speeds given in the speed and feed tables by the factor corresponding to the desired tool life, as obtained from Table 5.

The horsepower given is that developed at the nose of the tool and does not include any losses in the machine tool. To allow for the efficiency of the average machine tool, it is necessary to increase the powers given in the tables by approximately 50% to obtain the power input to the machine required to drive a particular cut. In addition, there is an idle load, ranging from one-half to 5 horsepower which must be added to the gross horsepower obtained above to secure the required motor input.

Table 7. Machinability Constants of Cast Iron

Chemical Composition				Physical Properties			Chip Pressure Constant, K_2	Multipliers for Cast Iron of 177 Brinell (See Tables 14 to 17)		Relative Machinability
Total Carbon	Combined Carbon	Silicon	Manganese	Tensile Strength, psi	Compressive Strength, psi	Brinell Hardness No.		Speed	Hp	
2.80	0.90	2.85	0.41	50,000	159,000	224	111,000	0.59	1.11	2.4
2.90	....	1.12	0.71	44,000	152,000	218	104,000	0.66	1.16	2.7
3.30	1.05	0.88	0.94	36,000	111,000	199	66,000	0.70	0.78	2.8
3.04	....	1.70	0.69	40,000	129,000	198	82,000	0.71	0.98	2.9
3.16	....	2.50	0.79	37,000	118,000	194	74,000	0.71	0.88	2.9
2.65	....	2.34	1.05	38,000	130,000	205	84,000	0.71	1.01	2.9
2.70	0.65	2.00	0.60	38,000	133,000	204	87,000	0.76	1.12	3.1
2.92	0.88	1.57	0.36	41,000	139,000	195	92,000	0.80	1.24	3.3
3.13	....	1.51	1.06	33,000	111,000	183	66,000	0.91	1.02	3.7
2.88	0.32	1.47	0.60	30,000	102,000	177	59,000	1.00	1.00	4.1
3.03	....	1.68	0.98	33,000	120,000	181	74,000	1.03	1.28	4.2
3.10	....	1.89	0.80	36,000	116,000	182	71,000	1.04	1.25	4.2
3.20	....	2.01	0.79	33,000	122,000	182	76,000	1.05	1.35	4.3
3.39	0.80	1.60	0.52	30,000	102,000	161	59,000	1.10	1.10	4.5
2.98	0.24	2.41	0.14	....	....	144	36,000	1.20	0.74	4.9
3.37	0.18	2.34	0.50	18,500	81,000	156	42,000	1.32	0.97	5.4
3.51 *	0.10	2.31	0.69	33,000	121,000	168	76,000	1.48	0.91	6.1 *
3.71	....	1.80	0.91	17,000	66,000	127	31,000	1.67	0.88	6.8
2.46 †	....	1.00	0.26	....	....	137	70,000	1.10	1.31	4.5 †
3.43	....	1.13	0.53Mo 1.44Ni	28,550	....	165	42,000	1.29	1.25	5.3
3.52	....	1.07	0.56Mo 2.26Ni 0.49Cr	34,000	....	206	71,500	0.81	0.98	3.3
3.54	....	1.07	0.61Mo 0.61Ni 0.80Cr	33,200	....	249	92,500	0.46	0.72	1.9
3.38	0.35	1.99	0.36Ni 0.28Cr	44,520	....	205	92,000	0.81	1.43	3.3
3.40	0.20	1.97	2.00Ni 0.51Cr	44,720	....	215	100,000	0.72	1.54	3.0
3.25	0.20	1.92	0.86Mo 0.30Ni 0.40Cr	55,590	....	270	120,000	0.49	1.00	2.0

* Centrifugally cast. † Malleable iron.
Machinability given for cutting clean metal. For cutting scale, speeds must be reduced 10 to 20%.

Table 8. Cutting Speeds for Wrought Monel Metal

(High-speed Steel Tools)

Feed, in. per rev	Depth of Cut, in.								
	1/32	1/16	1/8	1/4	3/8	1/2	5/8	3/4	1
	Cutting Speed, ft per min								
.002	262	223	205	192	190	186	185	183	182
.004	209	179	163	153	149	147	147	145	144
.008	168	144	138	124	121	118	116	114	113
1/64	139	121	110	103	98	95	93	92	90
1/32	118	104	68	57	52	49	47	45	43
1/16	104	60	48	39	34	31	29	26	25
1/8	69	48	34	29	24	21	18	17	14

MONEL METAL. Table 8, giving cutting speeds for Monel metal, with high-speed steel tools, is based on experiments by the International Nickel Company, and by Melvin Matson at Montana State College. Speeds given should permit securing a tool life of 60 min between grinds.

Tools for turning Monel metal should be of much the same shape as those used for turning medium steel. Because of the toughness of the chip it is particularly desirable to use a chip breaker.

(*Continued on p. 23-18*)

Table 9. Relative Machinability of Nonferrous Metals

(Based on cutting speed)

Material	Condition	Chemical Composition					Relative Machin-ability
		Copper	Tin	Lead	Zinc	Nickel	
Muntz metal	Wrought	....	....	....	....	....	15.2
Muntz metal	Cast	....	....	....	....	....	11.8
Soft brass	Wrought	....	....	....	....	....	13.6
Soft brass	Cast	....	....	....	....	....	10.5
Phosphor bronze	Wrought	....	....	....	....	....	11.5
Phosphor bronze	Cast	....	....	....	....	....	7.1
Navy rolled brass	Wrought	....	....	....	....	....	11.5
Navy rolled brass	Cast	....	....	....	....	....	7.1
Navy "G"	Wrought	88.0	8.0	....	4.0	....	11.5
Navy "G"	Cast	88.0	8.0	....	4.0	....	7.1
Hard brass	Wrought	....	....	....	....	....	12.1
Hard brass	Cast	....	....	....	....	....	9.4
Bronze	Wrought	90.0	10.0	....	....	....	12.1
Bronze	Cast	90.0	10.0	....	....	....	9.4
Bronze	Wrought	85.0	6–11	0–1	....	....	9.2
Bronze	Cast	85.0	6–11	0–1	....	....	7.1
Bronze	Wrought	84.0	13.0	....	....	3.0	9.2
Bronze	Cast	84.0	13.0	....	....	3.0	7.1
Marine bronze	Wrought	....	....	....	....	....	11.5
Marine bronze	Cast	....	....	....	....	....	7.1
Tobin bronze	Wrought	....	....	....	....	....	11.5
Tobin bronze	Cast	....	....	....	....	....	7.1
Bronze	Wrought	88.0	11.0	1.0	....	....	11.5
Bronze	Cast	88.0	11.0	1.0	....	....	7.1
Monel metal	Forged	....	....	....	....	....	4.1
Monel metal	Cast	....	....	....	....	....	3.5
Government babbitt	Cast	....	....	....	....	....	13.9
Aluminum	Wrought	....	....	....	....	....	40.0
Aluminum	Wrought	....	....	....	....	....	34.0
Copper	Cast	....	....	....	....	....	9.5
Hard rubber	....	....	....	....	....	....	12.9
Bakelite	....	....	....	....	....	....	5.0

Table 10. Cutting Speed and Horsepower Required for Cutting Steel with Carbon Steel Round-nose Tools (Tool life, 1 hr)

Steel: C1030, hot rolled. Tool: 1.15 carbon steel; nose radius, 1/8 in.; side-cutting-edge angle, 0; front-cutting-edge angle, 6 degrees; side rake, 14 degrees; back rake, 8 degrees; clearance, 6 degrees. See Fig. 11*F*.

Feed, in. per rev	Depth of Cut, in.																	
	1/32		1/16		1/8		1/4		3/8		1/2		5/8		3/4		1	
	Ft per min	Hp	Ft per min	Hp	Ft per min	Hp	Ft per min	Hp	Ft per min	Hp	Ft per min	Hp	Ft per min	Hp	Ft per min	Hp	Ft per min	Hp
.002	103	.1	71	.1	62	.2	56	0.4	54	0.6	52	0.8	52	1.0	51	1.2	51	1.7
.004	66	.1	47	.2	40	.3	36	0.5	34	0.7	33	0.9	33	1.1	32	1.3	31	1.8
.008	42	.1	32	.2	26	.3	23	0.5	22	0.7	21	1.0	21	1.2	20	1.4	20	1.9
1/64	28	.1	21	.2	18	.3	15	0.6	15	0.8	14	1.1	13	1.3	13	1.5	13	2.1
1/32	20	.2	15	.2	12	.4	11	0.7	10	1.0	9	1.2	8	1.5	8	1.7	8	2.2
1/16	15	.2	12	.3	10	.5	8	0.9	7	1.2	6	1.4	6	1.6	5	1.9	5	2.3
1/8	13	.4	10	.5	8	.8	6	1.2	5	1.6	4	1.8	4	1.9	3	2.1	3	2.3

Table 11. Cutting Speed and Horsepower Required for Cutting Steel with High-speed Steel Side-cutting Tools (Tool life, 1 hr)

Steel: C1030, hot rolled. Tool:18-4-1 high-speed steel; nose radius, 0; side-cutting-edge angle, 0; side rake, 14 degrees; back rake, 8 degrees. See Fig. 11*A*.

Feed, in. per rev	Depth of Cut, in. 1/32		1/16		1/8		1/4		3/8		1/2		5/8		3/4		1	
	Ft per min	Hp	Ft per min	Hp	Ft per min	Hp	Ft per min	Hp	Ft per min	Hp	Ft per min	Hp	Ft per min	Hp	Ft per min	Hp	Ft per min	Hp
.002	328	0.2	327	0.5	327	1.1	325	2.2	325	3.4	323	4.7	323	5.9	323	7.2	323	9.9
.004	209	0.3	212	0.6	206	1.1	206	2.4	205	3.7	205	5.1	203	6.5	203	7.9	203	10.8
.008	136	0.4	135	0.7	134	1.3	130	2.7	130	4.1	130	5.7	129	7.1	129	8.6	128	11.7
1/64	92	0.5	91	0.8	87	1.5	86	3.0	85	4.5	84	6.1	83	7.7	83	9.3	82	12.5
1/32	64	0.8	61	1.2	59	2.0	56	3.7	55	5.4	54	7.1	53	8.7	52	10.4	51	13.7
1/16	65	1.2	44	1.9	41	2.9	39	4.8	36	6.6	35	8.5	34	10.2	33	11.9	32	15.4
1/8	66	1.9	42	2.9	31	5.0	27	7.0	24	8.8	23	10.7	22	12.5	21	14.1	20	17.5

Table 12. Cutting Speed and Horsepower Required for Cutting Steel with High-speed Steel Round-nose Cutting Tools (Tool life, 1 hr)

Steel: C1030, hot rolled. Tool: 18-4-1 high-speed steel; nose radius, 1/4 in.; side-cutting-edge angle, 30 degrees; side rake, 14 degrees; back rake, 8 degrees. See Fig. 11*F*.

Feed, in. per rev	Depth of Cut, in. 1/32		1/16		1/8		1/4		3/8		1/2		5/8		3/4		1	
	Ft per min	Hp	Ft per min	Hp	Ft per min	Hp	Ft per min	Hp	Ft per min	Hp	Ft per min	Hp	Ft per min	Hp	Ft per min	Hp	Ft per min	Hp
.002	737	0.7	515	1.0	447	1.6	408	3.0	390	4.5	382	5.9	379	7.4	371	8.9	368	12.1
.004	475	0.8	342	1.1	285	1.8	254	3.3	245	4.8	241	6.7	236	8.0	234	9.7	231	13.0
.008	302	0.9	220	1.2	197	2.1	164	3.6	157	5.3	151	7.5	149	8.9	147	10.4	146	14.1
1/64	199	1.0	144	1.3	120	2.2	106	3.9	102	5.7	98	7.6	96	9.4	95	11.3	93	15.1
1/32	134	1.2	95	1.6	80	2.5	70	4.5	65	6.4	63	8.4	61	10.1	60	12.2	59	16.3
1/16	94	1.5	68	2.0	55	3.1	47	5.3	43	7.3	41	9.4	40	11.6	39	13.6	36	17.6
1/8	81	2.2	55	2.9	42	4.2	34	6.8	30	9.0	28	11.1	27	13.5	27	16.3	23	19.6

Table 13. Cutting Speed and Horsepower Required for Cutting Steel with High-speed Steel Parting Tools (Tool life, 1 hr)

Steel: C1030 hot rolled. Tool: 18-4-1 high-speed steel; corner radius, 1/64 in.; setting angle, 90 degrees; back rake, 20 degrees; relief, 6 degrees. See Fig. 10.

Feed, in. per rev	Width of Cut, in. 1/16		1/8		1/4		3/8		1/2		3/4	
	Ft per min	Hp	Ft per min	Hp	Ft per min	Hp	Ft per min	Hp	Ft per min	Hp	Ft per min	Hp
.002	383	0.7	378	1.3	378	2.6	378	3.9	378	5.3	378	8.1
.004	240	0.8	240	1.5	239	2.8	239	4.3	236	5.8	236	8.8
.006	187	1.0	187	1.6	186	3.1	183	4.6	183	6.2	183	9.4
.008	156	1.0	154	1.7	154	3.3	151	4.8	151	6.5	150	9.7
.010	136	1.1	134	1.9	133	3.4	133	5.1	130	6.7	130	10.1
.012	122	1.2	120	1.9	118	3.5	116	5.1	116	6.9	115	10.3

Table 14. Cutting Speed and Horsepower Required for Cutting Cast Iron with Carbon Steel Round-nose Cutting Tools (Tool life, 1 hr)

Cast iron: Brinell hardness, 177; total carbon, 2.88%; combined carbon, 0.32%; silicon, 1.47%; manganese, 0.60%. Tool: 1.15 carbon steel; nose radius 1/8 in.; side-cutting-edge angle, 0; front-cutting-edge angle, 6 degrees; side rake, 14 degrees; back rake, 8 degrees; clearance, 6 degrees. See Fig. 11*F*.

Feed, in. per rev	Depth of Cut, in. 1/32		1/16		1/8		1/4		3/8		1/2		5/8		3/4		1	
	Ft per min	Hp	Ft per min	Hp	Ft per min	Hp	Ft per min	Hp	Ft per min	Hp	Ft per min	Hp	Ft per min	Hp	Ft per min	Hp	Ft per min	Hp
.002	95	.1	77	.2	69	0.3	64	0.5	63	0.7	62	0.9	61	1.1	60	1.3	60	1.7
.004	71	.1	58	.2	52	0.3	48	0.6	47	0.8	46	0.9	45	1.3	45	1.5	44	1.9
.008	54	.2	45	.2	42	0.4	37	0.6	35	0.9	34	1.2	34	1.5	33	1.7	33	2.3
1/64	49	.2	36	.3	32	0.5	29	0.8	27	1.1	26	1.4	25	1.7	25	2.1	24	2.6
1/32	33	.3	29	.4	26	0.6	24	1.1	22	1.4	20	1.7	20	2.1	19	2.4	18	3.0
1/16	29	.4	26	.6	23	0.9	20	1.4	18	1.9	16	2.2	15	2.5	14	2.8	13	3.4
1/8	27	.6	24	.9	22	1.4	18	2.1	15	2.6	13	2.9	12	2.1	11	3.4	9	3.7

Table 15. Cutting Speed and Horsepower Required for Cutting Cast Iron with High-speed Steel Side-cutting Tools (Tool life, 1 hr)

Cast iron: Brinell hardness, 177; total carbon, 2.88%; combined carbon, 0.32%; silicon, 1.47%; manganese, 0.60% Tool: 18-4-1 high-speed steel; nose radius, 0; side-cutting-edge angle, 0; side rake, 14 degrees; back rake, 8 degrees. See Fig. 11*A*.

Feed, in. per rev	Depth of Cut, in. 1/32		1/16		1/8		1/4		3/8		1/2		5/8		3/4		1	
	Ft per min	Hp	Ft per min	Hp	Ft per min	Hp	Ft per min	Hp	Ft per min	Hp	Ft per min	Hp	Ft per min	Hp	Ft per min	Hp	Ft per min	Hp
.002	169	0.2	168	0.3	167	0.6	166	1.1	166	1.7	166	2.3	166	2.8	166	3.4	166	4.5
.004	127	0.2	127	0.4	126	0.7	126	1.4	125	2.0	124	2.7	124	3.3	124	4.0	123	5.3
.008	97	0.3	97	0.4	96	0.9	94	1.6	93	2.4	93	3.2	93	4.0	92	4.7	92	6.2
1/64	77	0.4	76	0.6	74	1.1	73	2.0	71	2.9	71	3.8	70	4.7	69	5.5	68	7.2
1/32	63	0.6	61	0.9	58	1.5	56	2.5	54	3.6	53	4.6	52	5.6	52	6.6	51	9.0
1/16	62	0.9	52	1.6	48	2.3	45	3.6	43	4.7	41	5.9	40	7.0	39	8.1	37	11.3
1/8	62	1.4	49	1.7	41	4.0	37	4.6	34	5.1	32	8.1	31	9.3	29	10.2	28	12.4

Table 16. Cutting Speed and Horsepower Required for Cutting Cast Iron with High-speed Steel Round-nose Cutting Tools (Tool life, 1 hr)

Cast iron: Brinell hardness, 177; total carbon, 2.88%; combined carbon, 0.32%; silicon, 1.47%; manganese, 0.60%. Tool: 18-4-1 high-speed steel; nose radius, 1/8 in.; side-cutting-edge angle, 30 degrees; side rake, 14 degrees; back rake, 8 degrees. See Fig. 11*E*.

Feed, in. per rev	Depth of Cut, in. 1/32		1/16		1/8		1/4		3/8		1/2		5/8		3/4		1	
	Ft per min	Hp	Ft per min	Hp	Ft per min	Hp	Ft per min	Hp	Ft per min	Hp	Ft per min	Hp	Ft per min	Hp	Ft per min	Hp	Ft per min	Hp
.002	284	0.4	230	0.5	207	0.8	192	1.5	187	2.1	185	2.7	182	3.4	181	4.0	180	5.3
.004	212	0.4	172	0.6	154	1.0	144	1.7	140	2.5	137	3.2	136	4.0	135	4.7	134	6.2
.008	160	0.5	130	0.7	122	1.2	107	2.0	104	2.9	103	3.8	101	4.6	100	5.5	99	7.1
1/64	122	0.6	100	0.9	89	1.4	82	2.4	79	3.4	77	4.4	77	5.4	76	6.3	74	8.2
1/32	96	0.8	78	1.1	69	1.7	63	2.9	60	4.1	58	5.2	58	6.4	56	7.5	55	9.6
1/16	77	1.0	63	1.4	56	2.2	50	3.7	47	5.0	45	6.4	43	7.6	42	8.9	41	11.2
1/8	69	1.5	56	2.1	47	3.1	41	5.0	37	6.5	35	7.9	34	9.3	33	11.1	31	13.1

Table 17. Cutting Speed and Horsepower Required for Cutting Cast Iron with High-speed Steel Parting Tools (Tool life, 1 hr)

Cast iron: Brinell hardness, 177; total carbon, 2.88%; combined carbon, 0.32%; silicon, 1.47%; manganese, 0.62%. Tool: 18-4-1 high-speed steel; corner radius, 1/64 in.; setting angle, 90 degrees; back rake, 20 degrees; relief, 6 degrees. See Fig. 10.

Feed, in. per rev	Width of Cut, in.											
	1/16		1/8		1/4		3/8		1/2		3/4	
	Ft per min	Hp	Ft per min	Hp	Ft per min	Hp	Ft per min	Hp	Ft per min	Hp	Ft per min	Hp
.003	94	.3	94	0.6	94	1.1	94	1.5	94	2.0	93	2.9
.005	76	.4	76	0.7	75	1.2	75	1.7	74	2.2	74	3.3
.007	67	.5	67	0.7	66	1.3	66	1.9	65	2.5	65	3.6
.010	58	.5	58	0.8	57	1.5	56	2.1	56	2.7	56	4.0
.015	50	.6	49	1.0	49	1.7	48	2.4	47	3.1	47	4.4
.020	45	.8	44	1.1	43	1.9	43	2.6	42	3.3	41	4.7

ALUMINUM. Information on cutting tools, speeds, and feeds for aluminum and its alloys is found in Section 4, Article 6.

MAGNESIUM. Details concerning machining practices and cutting tools for magnesium and its alloys are given in Section 4, Article 14.

INGOT IRON. For maximum metal removal and best finish, relatively fine feeds and deep cuts should be used with ingot iron and other ductile, low-carbon ferrous materials.

4. CUTTING FLUIDS

The various fluids used in metal-cutting operations serve these purposes: (1) Cooling the tool. (2) Cooling the work. (3) Improving the quality of the surface produced. (4) Lubricating the surfaces in contact between the tool and work, or preventing welding of chip or work to the tool, thus aiding in reducing the temperature of the tool. (5) Flushing away the chips. (6) Reducing the pressure of the chip on the tool.

For cooling tool and work and flushing away chips, the quantity of fluid is more important than the kind. A thin soluble oil solution, approximately 40–1, or a solution of soda or borax in water is very satisfactory. A pure oil or a rich mixture of soluble oil does not appear to be so efficient a coolant as a fluid which is mainly water.

An oil containing a large percentage of sulfur appears to reduce the tendency of a steel chip to weld to the tool. Such oil is advantageous when a good quality of machined finish is desired. For finishing operations, an oil with a high degree of oiliness generally is used.

There is a difference of opinion as to the desirability of using cutting fluids with cemented carbide tools. There seems to be a definite advantage in such a use, provided only that the work and tool are thoroughly flooded at all times that the tool is engaged with the work. A small or intermittent flow of cutting fluid on a carbide tool is dangerous, since the tool may be cracked by the fluid hitting the overheated tool. Rates of flow up to 40 and 50 gal per min are desirable on multiple-tool operations. When cutting fluid is used at the speeds desirable for cutting with carbides, it is necessary to shield the tool and work very thoroughly to prevent splashing of the fluid on the operator.

Cooling mechanisms have been introduced into the cutting-fluid circulating system to hold the fluid at a constant temperature. Such cooling, when adjusted to hold the work at room temperature during the cutting operation, tends to facilitate the accurate holding of size of the work piece, as well as to permit the use of increased cutting speeds.

QUALITY OF SURFACE. There are few data available as to the effect of cutting fluids on the quality of the surface produced. Where high quality of finished steel surface is required, a sulfurized mineral or a sulfurized mineral lard oil is generally used, except that water or soluble oil is generally used to obtain a high grade of finish with a spring tool on a lathe.

EFFECT OF QUANTITY OF CUTTING FLUID. When a cutting fluid is used it is particularly important that it be applied steadily; intermittent application may tend to crack the tool. When surfaces which are not easily accessible are being machined, special means should be provided for supplying the cutting fluid at the cutting edge of the tool; for example, oil holes through the boring bar used for boring deep holes.

Table 18, based on the work of Georg Schlesinger as reported in Ref. 6, gives an indication of the effect of variation in the quantity of fluid used on the rate of metal removal. Frederick W. Taylor had recognized the importance of an adequate quantity of cutting fluid and recommended the use of at least 3 gal per min on roughing cuts. It will be seen from Table 18 that the use of 3 gal per min of water results in a condition close to the maximum for the experiments on which this table is based.

Table 18. Effect of Quantity of Cooling Water on Permissible Cutting Speed

Rough turning chrome-nickel steel with high-speed steel tools; 0.079 in. depth of cut; 0.039 in. feed. Tool with 45-degree side-cutting-edge angle, 30-degree front-cutting-edge angle, sharp nose, 14-degree rake.

Water per min, gallons	0	0.2	1.1	2.2	3.3	4.4
Cutting speed, ft per min *	67	71	81	84	85	85
Relative rate of metal removal	1.00 †	1.05	1.21	1.25	1.26	1.27

* For 20-min tool life. † The dry condition is taken as unity.

For light cuts on steel, where the chip thickness is less than 0.015 in., use of a copious supply of cutting fluid permits use of cutting speeds approximately 16% greater than could otherwise be used. The effect of the cutting fluid appears to vary with the thickness of the chip from 16% for thin chips to approximately 50% for chips approximately 0.090 in. thick.

GENERAL RECOMMENDATIONS. Table 19 is based on experiment and on the recommendations of the suppliers of cutting fluid. Where possible, it is advisable to obtain the recommendation of a reliable manufacturer concerning the fluid to be used for a particular application.

Table 19. Recommended Practice for the Use of Cutting Fluid for Metals

Material Cut	Rough Turning and Planing	Finishing, Forming, and Turning and Planing
Cast iron	Dry Air blast Soluble oil, 50 to 1 Water, with 1 1/2% borax or soda	Dry Air blast Soluble oil, 50 to 1 Water, with 1 1/2% borax or soda
Malleable iron	Dry Air blast Soluble oil, 50 to 1 Water, with 1 1/2% borax or soda	Dry Air blast Soluble oil, 50 to 1 Water, with 1 1/2% borax or soda
Wrought iron or low-carbon steel	Soluble oil, 40 to 1 Sulfurized mineral oil Water, with 1 1/2% borax or soda	Soluble oil, 40 to 1 Sulfurized mineral oil Water, with 1 1/2% borax or soda
Medium-carbon steel	Soluble oil, 30 to 1 Sulfurized mineral oil Water, with 1 1/2% borax or soda	Soluble oil, 30 to 1 Sulfurized mineral oil Mineral lard oil Sulfurized mineral lard oil Water, with 1 1/2% borax or soda
High-carbon steel	Soluble oil, 20 to 1 Mineral oil Water, with 1 1/2% borax or soda	Soluble oil, 20 to 1 Mineral oil Water, with 1 1/2% borax or soda
Alloy steel Monel metal Nickel	Soluble oil, 20 to 1 Sulfurized mineral oil Water, with 1 1/2% borax or soda	Soluble oil, 20 to 1 Sulfurized mineral oil Water, with 1 1/2% borax or soda Mineral lard oil
Copper	Soluble oil, 50 to 1 Light mineral oil Sulfurized mineral oil Water, with 1 1/2% borax or soda	Light mineral oil Water, with 1 1/2% borax or soda
Brass or bronze	Soluble oil, 50 to 1	Light mineral oil
Aluminum, duralumin, pewter, zinc	Light mineral oil	Soluble oil, 50 to 1 Very light mineral oil Water, with 1 1/2% borax or soda

5. FACTORS AFFECTING RATE OF METAL REMOVAL

SET-UP MATERIAL. Provision of an adequate quantity of well-chosen and well-arranged set-up material saves much time in making ready the average job. More time is lost on nonrepetitive work through failure to provide an adequate equipment of set-up material, with racks or cabinets for keeping it in an orderly manner, than through any other failure on the part of management.

A supply of the frequently used equipment should be kept at each machine tool. The cost is low, when compared with the reduction in set-up time made possible by a carefully engineered equipment of make-ready material, and should prove to be a very profitable investment.

EFFECT OF LENGTH OF TOOL EXTENSION. Tools should be so clamped in the tool post as to be supported, as closely as possible, directly beneath the active cutting edge. Excessive overhang may cause chatter and make necessary an appreciable reduction of speed, feed, or both. The following figures, based on the shop practice of a leading manufacturer of machine tools in the use of large forged planer tools, give an indication of this effect:

Tool extension, in	0–8	8–12	12–20	20–34	34 and over
Reduction in cut and feed, %	0	25	50	60	65

EFFECT OF WORK SET-UP ON RATE OF METAL REMOVAL. The size of chip removed and the cutting speed used depend on the stiffness of the work, the amount it overhangs the supporting surface, the amount of unbalance, and the rigidity with which it is clamped. Experience has shown that at times, when the part is weak, and cannot be solidly held, it is necessary to reduce the rate of removing metal as much as 75% over that possible with stiff, well-held work.

EFFECT OF ATTENUATION OF WORK ON RATE OF METAL REMOVAL. When long, slim work must be turned on centers, the ratio between length and diameter of the work piece has an appreciable effect on the size of chip which may be taken, and on the cutting speed that may be used. Figure 15 shows the effect on production rates of variations in the ratio of unsupported length to the diameter of the work, when cutting SAE 1035 hot-rolled steel, with a tool having a 1/16-in. nose radius, 90-degree entering angle, 6-degree nose drag angle, 14-degree side rake, 8-degree back rake, 6-degree relief angle, when taking a cut 1/16 in. deep, with a feed of 0.033 in., used for the roughing cut. The finishing cut was taken with a similar tool having a 1/16 in. flat, with a depth of cut of 0.015 in. and a feed of 0.033 in. The curves of Fig. 15 have been used successfully, in the shop in which they were developed, for setting production rates on a variety of long slim shafts approximately 1 to 2 in. in diameter, supported on centers and made in lots of 1 to 5. The extent to which they are applicable to conditions different from those for which they were developed is uncertain. They, at least, may be used as an indication of the reduction in the rate of chip production which becomes necessary as work on centers decreases in stiffness.

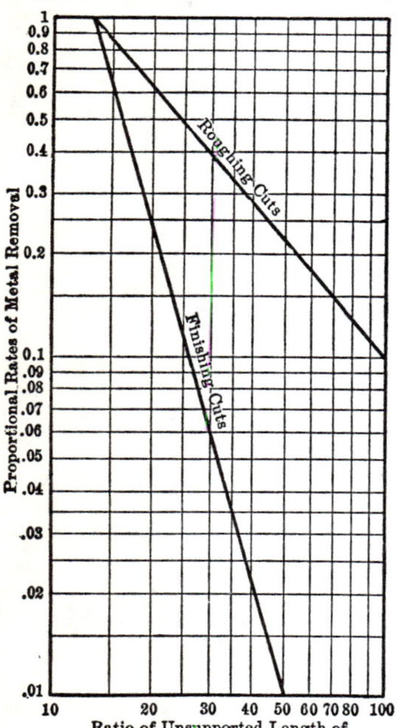

FIG. 15. Effect on metal removal of ratio of unsupported length and diameter of work. Steel work piece supported on centers.

CHATTER may be affected by several variables, of which these probably are the most important: (1) material cut; (2) chip proportions, as affected by *a*, depth of cut, *b*, feed, *c*, tool contour; (3) cutting speed; (4) rigidity of work; (5) stiffness of tool; (6) stiffness of work support; (7) stiffness of tool support; (8) vibrations caused by nature of design of machine tool, such as the gear ratios and tooth forms used; and (9) setting of tool with respect to work.

Many of these variables are fixed by design of the part to be machined and the machine tool available for doing the work. It appears that the best method of controlling chatter under a given set of conditions is to vary the chip proportions, cutting speed, the rigidity of the work support, or the setting of the tool with respect to the work.

Chip Proportions. Study of experimental data indicates that a cut taken with a relatively thick chip has less tendency to chatter than one taken with a thin chip. It would seem that it is desirable, in order to reduce chatter, to increase the feed, or to so change the shape of tool as to increase the thickness of chip with respect to its width. A tool having a nose radius greater than half the depth of cut is very likely to cause chatter unless a heavy feed is used.

Cutting Speed. A reduction in cutting speed may reduce or eliminate chatter. Some experimenters have reported that an *increase* in cutting speed may reduce chatter.

Rigidity of Work Support. Use of a center or follower rest frequently eliminates chatter. The use of opposed tools, to balance the normal components of the cutting pressure, is a very efficient method of reducing chatter.

Setting of Tool. It has been found that a change in the setting of the nose of a tool with respect to the lathe center may affect the amount of chatter. Carl G. Barth has told the author that in an aggravated case, setting of the tool nose considerably above the work center eliminated chatter. It is believed that this elimination resulted from the steadying effect of the contact between tool flank and work.

TOOL GRINDING. Improper grinding of metal-cutting tools leads to appreciable losses in production. Failure to use the most effective tool contours and angles requires

Table 20. Recommended Grinding Wheel Practice for Grinding Single-point Carbon Steel, High-speed Steel, and Stellite Tools

Grinding				Wheel					
								Grade	
Machine	Grind on	Wet or Dry	Rough or Finish	Diam., in.	Abrasive	Bond	Grain	Norton Co. Classification	Carborundum Co. Classification
Off-hand	Per.*	Dry	R	6–12	Al. oxide	Vit.† or Sil.‡	24–40	N–Q	H–K
Off-hand	Per.*	Dry	F	6–12	Al. oxide	Vit.† or Sil.‡	46–80	L–N	J–M
Off-hand	Per.*	Wet	R & F	12–24	Al. oxide	Vit.† or Sil.‡	24–36	O–P	I–K
Off-hand	Per.*	Wet	R & F	24–36	Al. oxide	Vit.† or Sil.‡	24–36	O–P	H–J
Off-hand	Per.*	Wet	R & F	36–48	Al. oxide	Sil.‡	24–30	O–P	H–J
Off-hand	Face §	Wet	R	....	Al. oxide	Vit.† or Sil.‡	30–36	K–L	M–N
Off-hand	Face §	Wet	F	....	Al. oxide	Vit.† or Sil.‡	46–60	J–K	N–O
Sellers semi-precision	Per.*	Wet	R & F	15	Al. oxide	Vit.† or Sil.‡	36	J–N	K–P
Sellers semi-precision	Per.*	Wet	R & F	24	Al. oxide	Vit.† or Sil.‡	24	N–Q	H–K
Gisholt semi-precision	Face §	Wet	R & F	....	Al. oxide	Vit.† or Sil.‡	24	L–M	K–P

* Periphery of wheel. † Vitreous. ‡ Silicate. § Cup wheel.

Table 21. Recommended Grinding Wheel Practice for Grinding Cemented Carbide Tools

Grinding			Wheel				
						Grade	
Machine	Grind on	Rough or Finish	Abrasive	Bond	Grain	Norton Co. Classification	Carborundum Co. Classification
Off-hand	Side	R	Silicon carbide	Vit.*	60–80	H–I	P–S
Off-hand	Side	F	Silicon carbide	Vit.*	100–120	H–I	R–T
Off-hand	Side	R & F	Silicon carbide	Vit.*	60–120	H–I	R–T
Off-hand	Per.†	R	Silicon carbide	Vit.*	60–80	H–J	O–R
Off-hand	Per.†	F	Silicon carbide	Vit.*	100–120	H–J	O–R
Sellers semi-precision	Per.†	R & F	Silicon carbide	Vit.*	60–100	H–I	P–T
Gisholt semi-presion	Face ‡	R & F	Silicon carbide	Vit.*	60–100	H–I	P–T
Lapping	Face cylindrical lap	....	Diamond, Silicon carbide, Boron carbide	......	280–320	....	....

* Vitreous. † Periphery of wheel. ‡ Cup wheel.

a reduction in the cutting speed. Grinding of tools by the individual machine operator causes a loss in production, as the machine tool must stand idle while tools are being ground. There is a growing tendency toward the grinding of all tools in a central tool room.

The type of grinding wheel to be used for tool grinding depends on the tool material, the area of contact between tool and wheel, whether the grinding operation is roughing or finishing, the surface of the wheel on which the grinding takes place, whether the tool is rigidly or loosely held with respect to the grinding wheel, and whether or not water is used.

Tables 20 and 21 summarize the recommendations of the two largest manufacturers of grinding wheels concerning the wheel that should be used under different conditions.

6. TOOL MATERIALS

The materials used for making tools for the cutting of metals may be classified in the following groups: (1) carbon tool steel; (2) low-alloy tool steels; (3) high-speed steel; (4) nonferrous alloys; (5) cemented carbides; (6) diamond.

CARBON TOOL STEEL. A tool steel containing little or no alloying metals, and with a carbon content of 1.10 to 1.30%, is used for cutting operations where a hard, keen cutting edge is required, at low cutting speeds, particularly when a high quality of finished surface is desired. It is used at times for forming tools, particularly where it is desirable to form the tool before hardening. It also is used at times where the size of the tool or work, coupled with speeds of available shop equipment, limit cutting speeds.

LOW-ALLOY TOOL STEELS include the fast finishing steels and some steels used for twist drills.

HIGH-SPEED STEEL is a steel containing at least 12% tungsten, 7% molybdenum, or an equivalent combination of the two, together with appreciable percentages of chromium and vanadium and, at times, cobalt. Analyses of the steels in most common use are given in Table 22. High-speed steel tools are much more generally used than those of

Table 22. Materials Used for Metal Cutting Tools

Type	Relative Cutting Ability	Carbon	Tungsten	Molybdenum	Chromium	Vanadium	Cobalt
High-speed Steels							
18-4-1	1.00	0.70–0.75	18.0	....	4.0	1.0	....
18-4-2	1.03	0.80–0.85	18.0	0.75	4.0	2.0	....
18-4-3	1.06	0.85–1.10	18.0	....	4.0	3.0	....
14-4-2	1.00	0.70–0.75	14.0	....	4.0	2.0	....
Moly-tungsten	1.00	0.80–0.85	1.4	9.0	4.0	1.0	....
Molybdenum	1.00	0.80–0.85	....	10.0	4.0	1.0	....
Cobalt Steels							
18-4-1 + 4% cobalt	1.07	0.70–0.75	18.0	0.5	4.0	1.0	4.0
18-4-2 + 7% cobalt	1.11	0.75–0.80	18.0	0.75	4.0	2.0	7.0
20-2 1/2-1 1/2 + 12% cobalt	1.12	0.80–0.85	20.0	1.00	4.5	1.5	12.0
14-4-2 + 4% cobalt	1.07	0.75–0.80	14.0	0.25	4.0	2.0	4.0
Moly-tungsten + 7% cobalt	1.07	0.80–0.85	1.4	9.0	4.0	1.0	7.0

Note. The relative cutting ability is based on relative cutting speed possible under uniform conditions of cut, tool life, and material cut.

other tool materials. They combine superior cutting qualities, strength, toughness, and relatively low first cost for a very wide range of general machining operations. Cutting qualities vary with composition. For history of the development of high-speed steel, see Ref. 2.

In general, the cobalt steels may be run at higher speeds, particularly on heavy roughing cuts. They are more brittle and weaker, when hardened to secure maximum cutting speeds, and are appreciably more expensive than the standard high-speed steels. They must be hardened at temperatures approximately 50 F higher than the equivalent standard steels. Particular care, therefore, is necessary in hardening to prevent surface burning and decarburization. The use of cobalt steel is largely confined to tools in which the contour is formed by grinding after hardening, such as lathe and planer roughing and finishing tools, tool bits, and inserted blades for milling cutters.

NONFERROUS ALLOYS. There are several well-known tool materials in this classification, including Stellite, Tantung, and Rexalloy. These materials, which occupy a position between the high-speed steel tools and cemented carbides, can be used at cutting speeds 30 to 60% higher than those which can be used with high-speed steel, particularly when rough-machining cast iron, brass, and bronze. For use with old, and underpowered, machines, these tool materials may be all that is required to operate a machine tool at its full potential output.

CEMENTED CARBIDE TOOLS are made of various proportions of tungsten, tantalum, or titanium carbides, and a double carbide of tungsten and titanium, with cobalt matrix, formed into the desired shape by powder metallurgy and sintering, after which they can be formed only by grinding. The various carbides are characterized by extreme hardness, high transverse rupture strength, high resistance to compression, high modulus of elasticity, and considerable tendency toward brittleness. Their rate of thermal expansion is approximately half that of steel.

These materials are made in a number of grades, in which hardness, strength, and other characteristics are varied to make them particularly suitable for specific applications. Table 23, which shows the approximate comparison of the grades manufactured by seven leading producers of carbide tools, is in use in the plant of one of the leading automotive

Table 23. Carbide Comparison Chart Standard Grades

(Courtesy of Buick Motor Division of General Motors Corp.)

	Carboloy Grade	Firthite Grade	Vascoloy-Ramet Grade	Kennametal Grade	Willey Grade	Carmet Grade	Wesson Grade
1	44A	HC	2A68	K-6	E-8	CA 3	GS
2	883	HA	2A5	K-6	E-6	CA 4	GI
3	905	HE	2A7	X-201	E-5	CA 7	GA
4	999	HF	2A9	X-201	E-4	CA 8	GF
5	78C	T-04	EE	K2S	710	CA 5	WS
6	78B	T-A	EM	KM	710	CA 1	WM
7	78	T-16	E	K3H	606	CA 2	WH
8	831	T-31		X-50	509	CA 6	
9		T-41	2A5	KWH	E-12		

1. Roughing cuts on cast iron, nonferrous materials, cutter blades; also used for wear purposes.
2. General purpose—cast iron. Nonferrous materials, stainless steels, also reamer and cutter blades.
3. Light finishing—cast iron and nonferrous materials.
4. Precision boring—cast iron and nonferrous materials.
5. Roughing cuts—steel.
6. General purpose—steel.
7. Finishing cuts—steel.
8. Precision boring and light finishing—steel.
9. Wear materials—centers rest blades—gages.

manufacturers. At present the manufacture of cemented carbide tools has not reached the degree of standardization obtained in the manufacture of high-speed tool steel, so that it is advisable to supplement the data given here by specific recommendations from the manufacturer whose tools are to be bought.

A large percentage of the carbide tools used in the past have been in the form of tips brazed to a high-carbon or alloy-steel shank. This method of construction was apparently used at the outset because of the high initial cost of cemented carbide material. For the sake of economy, the smallest tip that seemed at all practicable was used.

The marked difference in expansion rates between the carbide tips and shank material has caused an increasing use of tips held in place mechanically by clamps or otherwise. The reduced cost of the carbide tips, in turn, has tended toward the use of much heavier tips than those originally used. It is claimed that such mechanically held tips not only give appreciably better performance than the brazed tips, but also make possible certain types of cutting which would otherwise be impracticable, notably high-speed planer operation and wide form cutting, as on chilled iron rolls. It seems probable that such tips are particularly desirable whenever the length of contact between tip and shank is at all large. One of the recent developments makes use of small rods of solid carbide material held in a holder in such a manner that the cutting surface is on the end of the tip. Such a tool makes possible the use of several cutting edges per grind, through revolution of the tip in the holder.

When tool tips are clamped on, the holder should be of hardened high-carbon steel, with both tip seat and tip base accurately ground to insure a seating of the tip throughout its entire area, and the avoidance of localized bending strains, which will tend to break the tip.

The high moment of inertia has been found to give added stiffness to narrow grooving or cut-off tools, thus reducing the tendency to deflect in either a vertical or horizontal plane.

The size of holders used with cemented carbide tools should be sufficient to give a considerable degree of rigidity, with a depth of shank under the tool blank of at least twice, and preferably three times, the initial tip thickness.

The usual carbide tool has a side-cutting-edge angle of 0 to 20 degrees, an end-cutting-edge angle of 5 to 25 degrees, and a small nose radius, usually $^1/_{64}$ to $^1/_{16}$ in., or approximately 1 $^1/_2$ times the feed per revolution.

There has been some tendency toward the use of negative-rake angles with cemented carbide tools. It is believed that the principal value of negative rake is for use when taking interrupted cuts. It has been found that a carbide cut-off tool operated satisfactorily with a positive back rake angle of 8 degrees and a chip breaker, whereas a negative back-rake angle of 5 degrees caused chipping of the cutting edge.

Chip breakers are usually employed with cemented carbide tools when cutting steel and other materials tending to give a continuous chip, as a means of chip control, although

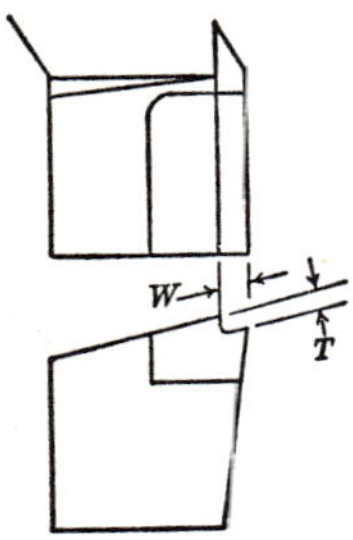

FIG. 16. Parallel type chip breaker. See Table 24.

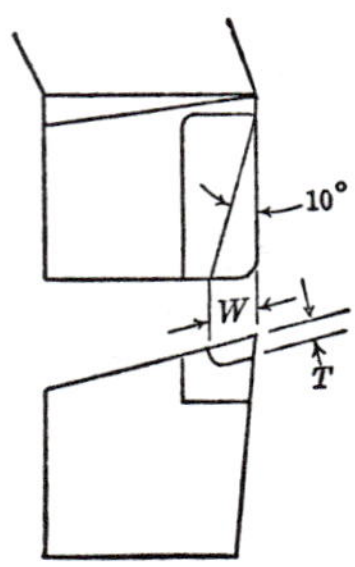

FIG. 17. Angular type chip breaker. See Table 24.

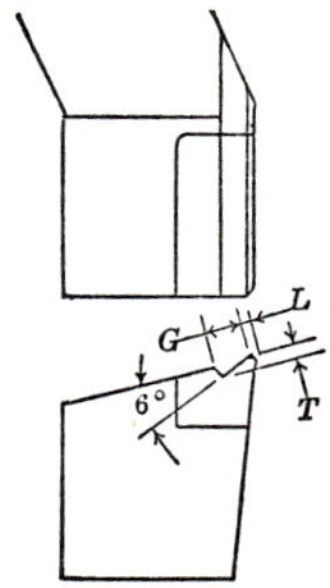

FIG. 18. Groove type chip breaker. See Table 25.

it is known that such a chip breaker reduces the cutting speed which may be used. There are several types of chip breakers, including those ground in the face of the tool tip and those formed by the use of an additional piece of carbide clamped in a suitable position. There are three usual forms of ground chip breakers: the parallel (shown in Fig. 16), the angular (shown in Fig. 17), and the groove types (shown in Fig. 18). See also Tables 24 and 25. The angular type is usually employed for light cuts, particularly finishing and semi-finishing cuts; the parallel type for light feeds, 0.005 to 0.018 in.; and the groove type for general purposes, where depth of cut varies and feeds from 0.012 to 0.065 in. are used.

Table 24. Chip Breaker Dimensions for Parallel and Angular Types (Figs. 16 and 17)

Depth of Cut	Width, *W*, for Feeds of			
	0.006–0.012	0.013–0.017	0.018–0.027	0.028–0.035
$^1/_{64}$–$^3/_{64}$	$^1/_{16}$	$^5/_{64}$	$^7/_{64}$	$^1/_8$
$^1/_{16}$–$^1/_4$	$^3/_{32}$	$^1/_8$	$^5/_{32}$	$^3/_{16}$
$^5/_{16}$–$^1/_2$	$^1/_8$	$^5/_{32}$	$^3/_{16}$	$^3/_{16}$
$^9/_{16}$–$^3/_4$	$^5/_{32}$	$^3/_{16}$	$^3/_{16}$	$^3/_{16}$
Thickness, *T*	0.006–0.012	0.013–0.017	0.018–0.027	0.030

Table 25. Chip Breaker Dimensions for Groove Type (Fig. 18)

G	Groove width	To be 3 to 4 times the feed used.
L	Land	Parallel to top surface. To be 1 to 1 $^1/_2$ times the feed used.
T	Thickness or depth of groove	To be not over 0.010 in.

Table 26. Feeds and Speeds for Use with Cemented Carbide Tools

(Courtesy of Vascoloy-Ramet Corp.)

CUTTING STEEL

Depth of cut		1/2–3/8	3/8–3/16	3/16–3/32	3/32–1/64	1/64–0.004
Feed, in. per rev		0.062–0.032	0.032–0.015	0.025–0.010	0.015–0.006	0.012–0.004
AISI No.	Condition	Cutting Speed, ft per min				
C1020	CD	200–275	225–350	275–400	275–400	400–700
C1035	CD	175–250	200–275	250–300	250–300	275–500
C1045	N	160–225	180–200	200–250	250–300	275–400
B1112	CD	200–300	250–300	275–325	300–425	400–1000
C1117	CD	250–300	275–350	300–375	350–425	400–1000
2512	CD	250–300	275–300	300–320	370–400	400–500
3120	CD	100–150	200–275	200–300	250–350	275–375
3140	A	125–225	150–225	175–250	200–275	225–300
4130	A	150–175	175–200	200–225	225–250	250–275
4150	A	140–160	160–180	180–200	200–225	225–275
302	A	150–200	175–225	200–250	225–250	275–400

CD = Cold drawn. N = Normalized. A = Annealed.

Table 27. Feeds and Speeds for Use with Cemented Carbide Tools

(Courtesy of Vascoloy-Ramet Corp.)

CUTTING CAST IRON AND CAST STEEL

Depth of cut		3/4–3/8	3/8–3/16	3/16–3/32	3/32–1/64	1/64–0.004
Feed, in. per rev		0.062–0.025	0.032–0.015	0.025–0.010	0.015–0.006	0.012–0.004
Material	Brinell	Cutting Speed, ft per min				
Cast iron	to 160	200	225	250	275	275–400
	to 200	175	200	225	250	250–300
Alloy cast iron	to 220	200	225	250	275	300
	to 300	150	175	200	220	246
Meehanite		100–150	150–200	200–250	250–300	300–350
Steel castings	to 175	175	200	250	275	325
	to 235	125	150	200	225	275
Alloy steel castings	to 200	150	170	190	210	225
	to 250	125	145	165	180	200
	to 300	100	125	145	165	180

Table 28. Feeds and Speeds for Use with Cemented Carbide Tools

(Courtesy of Vascoloy-Ramet Corp.)

CUTTING NONFERROUS MATERIALS

Depth of cut	3/8–3/16	5/32–5/64	1/16–1/64	0.012–0.004
Feed, in. per rev	0.025–0.015	0.020–0.010	0.015–0.006	0.008–0.002
Material	Cutting Speed, ft per min			
Aluminum	200–500	300–600	400–1000	500–1500
Brass, hard	125–275	150–350	200–400	300–600
Brass, soft	175–300	200–350	300–500	400–1000
Bronze, hard	125–250	150–300	200–300	300–500
Bronze, soft	135–250	175–300	225–350	350–750

Feeds and speeds for use with cemented carbide tools cutting various materials are given in Tables 26, 27, and 28.

DIAMONDS. Diamonds cut and polished to the desired contour, and set in an appropriate form of holder, are used as cutting tools, particularly when it is desired to turn or bore a part with extreme accuracy; for example, the piston rod bushings of a gasoline engine. They usually are used for cutting relatively soft materials, as bronze, hard rubber, or molded synthetic resins, which may have considerable abrasive action on the tool.

7. HEAT TREATMENT OF TOOL STEELS

CARBON STEEL used for cutting tools should be hardened as follows:

Anneal. Heat to 375 to 1425 F and cool slowly, either in furnace or buried in an insulating material, as lime, ashes, or infusorial earth. Tools should be annealed after each hot or cold working, to reduce hardness and prevent the formation of a coarse crystalline structure.

Normalize. Heat tool uniformly to 1575 to 1650 F, until there is complete penetration of the heat and refinement of the grain, and cool in still air.

Preheat. Preheating of carbon steel tools is unnecessary unless the tool is exceptionally thick or is extremely irregular in shape. If preheating is believed to be necessary, the tool may be placed in a cold furnace which is brought slowly to hardening temperature; or it may be given a preheat of approximately 900 F before being placed in the hardening furnace.

Hardening. Carbon tool steel should be hardened at temperatures which vary with the carbon content of the steel:

Carbon content, %	0.65–0.80	0.80–0.95	0.95–1.10	1.10 and over
Hardening temperature, °F	1450–1550	1410–1460	1390–1430	1380–1420

Quenching. Carbon steel tools usually are quenched in a brine, consisting of sodium or calcium chloride in water, at about 70 F. They may be removed from the brine and placed in oil while still retaining a considerable portion of their heat.

Tempering. The hardening of steel tends to set up internal stresses which cause brittleness and a tendency for the cutting edge to break under cut. These conditions may be relieved by tempering as follows: To relieve strains, temper in oil at 300 to 375 F; to relieve strains and reduce brittleness, temper in oil at 375 to 500 F; to relieve strains and toughen, temper in lead at 500 to 650 F. The strength of an untempered carbon steel tool, as measured by the torsion impact test, is approximately one-fourth its strength when tempered to develop maximum strength.

HIGH-SPEED STEEL. The method employed and the temperatures used in the heat treatment of high-speed steel tools are very important and may easily have a greater effect on the cutting qualities of a tool than the particular grade and make of steel used. There is a tendency toward the use of lower quenching temperatures than those which will secure maximum cutting qualities of a tool, either to avoid brittleness or to prevent burning of the tool through the use of a furnace which does not afford full protection to the tool during the heating cycle. There is a growing tendency toward the use of salt baths, both for heating a tool to the required high heat and as a quenching medium. Such a heat treatment affords full protection to a tool against localized overheating, surface decarburization, and burning. It is quite possible to harden a tool, even when following what is generally considered good practice, in a manner to obtain a cutting capacity, measured in possible cutting speed, as much as 20% below the capacity inherent in the tool steel.

Annealing. All high-speed steel should be annealed before machining, after forging, or before rehardening. Pack bars or tools in annealing boxes, allowing about 1 1/2 in. between tools and sides of box, using lime, mica, infusorial earth, ashes or, at times, charcoal, as the packing medium. Each box should be sealed with fireclay, raised slowly to 1550 to 1650 F in the furnace, and allowed to remain at the annealing temperature long enough to anneal thoroughly. This time varies with the size of the tools and the weight of the charge; it will range from 5 to 10 hr. Boxes and work should be allowed to cool slowly in the furnace.

Preheat. Because of the low thermal conductivity of high-speed steel, it usually is brought to one or more intermediate temperatures in other furnaces, before it is placed in the hardening furnace. Tools usually are brought to a temperature of 1550 F in a single stage of preheat. Large, intricate tools may be heated to approximately 1000 F in the first preheat furnace and to 1550 F in the second, before hardening.

Hardening. High-speed steels must be heated to a point close to their fusion point, in an atmosphere as nearly neutral as possible, to develop maximum efficiency as metal cutting tools. This may be obtained by manual control of the fuel and air supply; by enclosing the tool in a container lined with a material which gives off carbon monoxide when heated; by heating in a closed chamber, the atmosphere of which is controlled by an auxiliary gas burner at the door; or by an inert gas, such as hydrogen, in a closed furnace chamber. There is a tendency toward the use of furnaces having some method of atmospheric control.

To obtain maximum practical cutting speeds, high-speed steel should be hardened at approximately the following temperatures:

	18-4-1	18-4-2	14-4-2	18-4-1+Co	14-4-2+Co	W-Mo	W-Mo+Co
Hardening temp., furnace, °F	2400	2400	2350	2450	2400	2250	2300
Hardening temp., bath, °F	2350	2350	2300	2400	2350	2200	2250

High-speed tools should remain in the hardening furnace sufficiently long to make certain that they have reached a uniform temperature and that the thermal reactions have taken place completely. For a modern furnace, in which there is little tendency to burn the surface and edges of the tool, the following times are approximately correct:

Tool thickness, in.	1/4	1/2	3/4	1	1 1/2	2	3	4	5	6	8	10
Heating time in furnace, min	2.0	3.0	4.0	5.0	7.0	8.5	12	15	18	20	25	30

These times are based on complete penetration of the heat. It frequently is undesirable or impracticable to use hardening times as long as those required for the larger thicknesses. For very thick tools, the practical procedure is to harden to a depth sufficient to give a satisfactory cutting edge, leaving the interior of the tool relatively soft.

Quenching. High-speed steel usually is quenched in oil, or an air blast. If large tools are cooled too rapidly, there is danger of causing shrinkage cracks, and it may be desirable first to quench in a lead bath at 900 to 1000 F, and to complete the quenching in oil.

Straightening When Quenching. A tool that warps during quenching may be straightened within the few minutes immediately following quenching. The tool cannot be straightened later, as it becomes hard within a very short time after quenching.

Tempering. After the hardened tool has been reduced to atmospheric temperature, it should be reheated to approximately 1000 to 1100 F, held at that temperature for 15 to 240 min, and cooled in air, in order to obtain maximum cutting efficiency and maximum transverse strength. This second heat treatment both increases the cutting qualities and the strength of the tool and reduces brittleness. The maximum strength under impact, as measured by the torsion impact test, is obtained when the tool is given a secondary heat treatment at 900 F.

Heat Treatment in Small Shops. Where it is impracticable to install a modern type of heat-treating furnace in a shop, tools should be purchased in the hardened condition, preferably in the form of bits. It is unwise to attempt to harden high-speed steel in the blacksmith's forge because of the uneven results obtained.

8. FORM TOOLS

There are four types of form tools in general use: (1) Flat tools, such as are used in engine lathes, having no back rake. (2) Flat form tools having a positive back rake. (3) Circular form tools, such as are commonly used on automatic lathes. (4) Dovetail form tools, which are frequently used on automatic lathes.

FLAT FORM TOOLS WITHOUT RAKE are similar in general appearance to the turning tools in common use. They are usually set to cut on center, and present no design difficulties, since the contour of the tool is identical with that of the finished piece. Such tools are commonly employed for producing small quantities of parts, because of the ease of making such a tool. No tool drawing is necessary, since the part drawing may be used by the tool room in making the tool. Such a tool does not cut freely, and must be operated at a much reduced speed to secure a good grade of finish; it is consequently undesirable for use with long production runs.

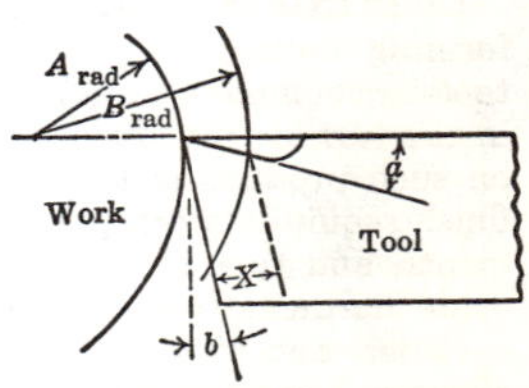

Fig. 19. Flat form tool with back rake, point of tool set on center.

FLAT FORM TOOLS WITH A BACK RAKE must have the contour corrected to permit turning the drawing contours. The following equations give the necessary basis for designing such a tool to cut a contour of any desired shape. Figure 19 shows the conditions under which such a tool cuts, the nomenclature being: X = distance on tool, normal to front flank, to cut a difference in radii on work equal to $B - A$; A = minimum radius of surface cut; B = maximum radius of surface cut; a = back-rake angle of tool; and b = front-relief angle (also setting angle of dovetail tool).

When the point of the tool is set on center, as in Fig. 19, the distance X is given by the equation

$$X = [\sqrt{B^2 - (A \sin a)^2} - A \cos a] \cos (a + b)$$

When the point of the tool is set above center by a distance h at the minimum radius of the surface cut, the distance X becomes

$$X = [\sqrt{B^2 - \cos^2 a(2h\sqrt{A^2 - h^2} - A^2)} - \sqrt{A^2 - \cos^2 a(2h\sqrt{A^2 - h^2} - A^2)}] \cos (a + b)$$

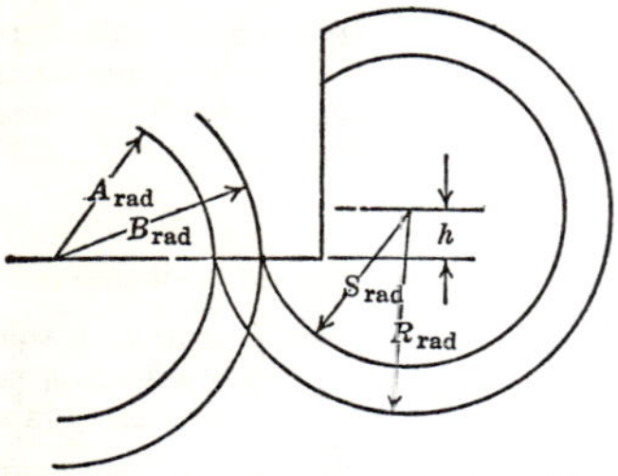

FIG. 20. Circular form tool, cutting edge of tool set on center of work.

For circular form tools, with the cutting edge on the center of the work and the center of the form tool above the center of the work by a distance h (Fig. 20), the radius S is given by the equation

$$S = \sqrt{R^2 - 2AB\sqrt{R^2 - h^2} + (B - A)^2}$$

Dovetail form tools, which are frequently used because of their inherent stiffness, may be calculated by the method given for flat form tools with positive back rake. When this is done, the setting angle is taken as the relief angle b of the formula for X.

REFERENCES

1. *Manual of Metal-Cutting*, ASME, 1939.
2. F. W. Taylor, *Trans. Am. Soc. Mech. Engrs.*, xxviii, 1908.
3. Test code for high speed steel for turning tools, *Mech. Eng.*, March 1931.
4. Standardization of testing conditions for metal cutting tools, *Mech. Eng.*, September 1932.
5. A metal cutting test method for machine shop use, *Iron Age*, May 13, 1937.
6. *Stahl u. Eisen*, Vol. 48, p. 341.
7. Standard B5.13–1939, *Terminology and Definitions for Single-point Cutting Tools*, American Standards Association, New York, 1939.

LATHE WORK

By E. L. Murray

9. THE TURNING PROCESS

The turning process is a method of metal removal whereby a work piece is gripped in a suitable holding device and rotated under power against a single-point cutting medium which is fed radially or longitudinally to the axis of the work piece. The process is essentially one of surface generation, and thus true cylindrical, conical, spherical, or planar forms with coincident axes are reproduced in the work piece, depending on the direction of cutter travel.

COMMON OPERATIONS. Literal interpretation of the turning process is extended to include cutting operations such as turning, boring, facing, forming, parting off, drilling, reaming, tapping, and threading.

CUTTING MEDIA. Single-point shank-type tools are used for turning, boring, facing, forming, cutting off, and some types of threading operations, whereas multi-flute cutting tools are commonly used for drilling, reaming, tapping, and threading. Cutting tools used in the turning process may be made from various materials, the choice of which depends on such production factors as power available and condition of the machine tool, work finish required, work piece material, speed of machining as related to production requirements, and type of cutting operation. In the turning process, these cutter materials include hardened carbon and high-speed tool steel, the cast nonferrous alloys, cemented carbides, and to a limited extent diamonds. For further information, see the preceding chapter, beginning on page 23-22.

CUTTING FLUIDS. Factors are present in the relationship between work piece and cutting medium which may require the use of cutting fluids for optimum results. Machine tools used in the turning process are therefore usually equipped with suitable systems to provide coolants or lubricants in required quantities at the cutting area as well as properly designed guards to protect both machine and operator from deleterious effects of these fluids. For types and functions of cutting fluids, see Article 4 in this section.

FORMULAS. The following formulas give the relations between the diameter of a rotating work piece, number of revolutions, and cutting speeds. Let d = diameter of rotating work piece, inches; n = revolutions of work piece per minute; S = speed of work circumference, feet per minute. Then

$$S = \frac{\pi dn}{12} = 0.2618dn \tag{1}$$

or

$$n = \frac{S}{0.2618d} = \frac{3.82S}{d} \tag{2}$$

or

$$d = \frac{3.82S}{N} \tag{3}$$

Time Required for Any Given Cut. Let T = time, minutes, total length of cut; f = feed per revolution of work, thousandths of an inch; r = revolutions of work per minute; L = total length of cut, inches. Then

$$T = \frac{L}{fr} \tag{4}$$

Horsepower required in a machine tool for cutting metal: Let Hp = horsepower required; t = torque required, inch-pounds; n = revolutions of work per minute. Then

$$\text{Hp} = \frac{tn}{63{,}000} \tag{5}$$

or

$$\text{Hp} = \frac{2\pi Tn}{33{,}000} \tag{6}$$

where T = torque, foot-pounds.

When horsepower must be calculated as above and cutting torque is not known, the following formula applies. Let Hp = horsepower required; d = depth of cut; f = tool feed per revolution in thousandths of an inch; S = work speed in surface feet per minute; R_c = cutting resistance constant of material. Then

$$\text{Hp} = \frac{dfSR_c}{33{,}000} \tag{7}$$

Cutting Resistance Constants, R_c:

Material	R_c	Material	R_c
Bronze (80-10-10)	95,000	SAE-4140 (Bhn 207)	275,000
Cast iron	175,000	SAE-4340 (Bhn 207-302)	275,000
Screw stock	225,000	SAE-52100 (Bhn 187)	315,000
SAE-1040 (Bhn 180)	250,000	SAE-1095 (Bhn 190)	315,000
SAE-1020 (Bhn 137)	275,000	SAE-4140 (Bhn 429)	345,000

A horsepower check list for material versus speed versus feed versus depth of cut is given in Table 1. These data assume use of single-point cemented carbide cutters and are based on extensive metal-cutting research conducted to determine power required to turn with carbide cutters at high speeds.

Tapping torque:

$$T = KrA \times 10^4$$

where T = torque, inch-pounds; K = material constant; r = radius from center of tap to center of gravity of cross section of area of metal being removed in one thread, inches; A = area of cross section of metal being removed by one thread of tap, square inches; rA = torque factor which may be calculated from Fig. 1.

Material Constants, K

Material	K
No. 72 F.M. Brass	$K = 17.3$
17ST Aluminum	$K = 22.9$
No. 73 Naval Brass	$K = 33.1$
No. 3 Cast Iron	$K = 41.0$
X-1315	$K = 41.0$
X-1112	$K = 41.0$
X-1117	$K = 41.0$
A-4140	$K = 55.0$

Constants, K, for other materials may be estimated by their relative machinability to those listed above. Calculated torque should be doubled for dull cutting taps and dies.

Table 1. Horsepower Check List for Material versus Speed versus Feed versus Depth

(Reprinted by permission from Job Planning on the Turret Lathe, by E. L. Murray, *The Tool Engineer*)

The tabulated data provide a quick check on power requirements for various combinations of materials, depths of cut, and speeds and feeds. Common sizes of turret lathe drive motors are listed in the extreme left-hand column. The next column corrects the power for headstocks having an efficiency of 85%. For example, a 0.375 in. depth of cut in bronze, if at 600 surface ft per min and at 0.032 in. per revolution feed, would require 21.25 hp at the spindle. If the efficiency of this lathe headstock, considering all factors, is 85%, a drive motor rated at 25 hp would be required.

Hp at Motor		Hp at Spindle	Bronze (80-10-10)	Cast Iron	Screw Stock	SAE-1040 Bh-180	SAE-1020 Bh-137 SAE-4140 Bh-207 SAE-4340 Bh-207-302	52100 Bh-187 SAE-1095 Bh-190	SAE-4140 Bh-429
Mean surface ft per min			600	250	350	300	250	200	200
5	Depth	4.25	.187	.187	.187	.187	.187	.187	.027
	Feed		.013	.019	.010	.011	.011	.012	.012
7 1/2	Depth	6.37	.187	.187	.187	.187	.187	.187	.187
	Feed		.019	.029	.014	.016	.017	.019	.017
10	Depth	8.5	.250	.250	.250	.250	.250	.250	.250
	Feed		.019	.028	.015	.016	.017	.019	.017
15	Depth	12.75	.250	.250	.250	.250	.250	.250	.250
	Feed		.028	.042	.020	.023	.025	.027	.025
20	Depth	17.00	.375	.375	.375	.375	.375	.375	.375
	Feed		.025	.037	.019	.020	.022	.024	.022
25	Depth	21.25	.375	.375	.375	.375	.375	.375	.375
	Feed		.032	.047	.023	.026	.027	.030	.028
30	Depth	25.50	.500	.500	.500	.500	.500	.500	.500
	Feed		.028	.041	.021	.022	.024	.027	.025
40	Depth	34.00	.500	.500	.500	.500	.500	.500	.500
	Feed		.037	.054	.027	.030	.033	.035	.033
50	Depth	42.5	.750	.750	.750	.750	.750	.750	.750
	Feed		.031	.044	.023	.025	.027	.029	.027

Torque for Taper Pipe Threads. Compute torque for a straight thread equivalent in pitch and diameter to the large end of the taper pipe thread. Multiply by 3 to compensate for 100% engagement of taper pipe tap or die.

EXAMPLE. What torque will be required to tap a 2 1/2–8 pitch straight thread in cast iron, assuming an 80% thread?

From Fig. 1, A for 8 pitch at 80% is approximately 36×10^{-4}. From Fig. 1, x for 8 pitch at 80% is approximately $4.6 \times 10^{-2} = 0.046$. Therefore

$$r = \frac{D}{2} - x$$

$$= \frac{2.5}{2} - 0.046$$

$$r = 1.204$$

Therefore

$$rA = 1.204 \times 36 \times 10^{-4}$$
$$= 43.34 \times 10^{-4}$$

and

$$T = KrA \times 10^4$$
$$= 41 \times 43.34$$
$$= 1777 \text{ in.-lb}$$

$$\text{Double torque for dull tap, } T = 1777 \times 2$$
$$= 3554 \text{ in.-lb required}$$

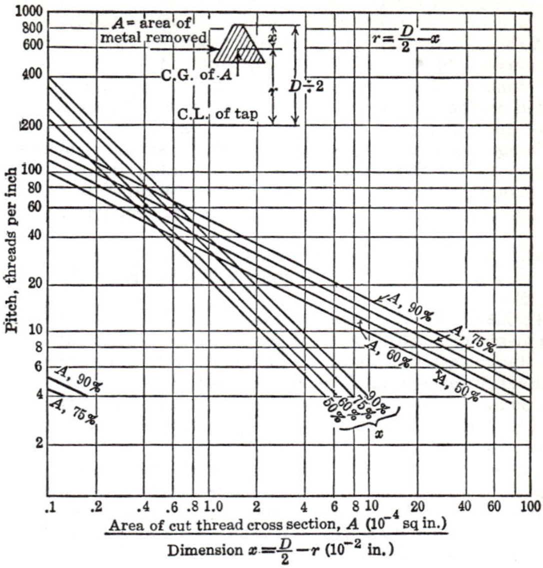

FIG. 1. Torque factor for American National form gun taps, at various percentages of full thread, is rA, where r is calculated as shown at top of cut, and A is obtained from chart.

RELATION TO OTHER MACHINING PROCESSES. Most machined parts at some point in their total routing incorporate a turning operation. For a very large percentage of these parts, the turning operations appear early in the routing and establish starting surfaces for subsequent gear cutting, grinding, honing, milling, lapping, etc., operations. The turning operation thus assumes a significant relation to the total machining of any given work part. It is understood that many machining operations such as milling and gear cutting are impractical on most types of turning equipment and then only under emergency or very low production conditions. However, it is frequently possible to duplicate on turning equipment certain finishes and dimensional control more generally expected from specific machine tools built for grinding, lapping, honing, tapping and threading, and like operations. For example, whether an engine lathe or thread miller is used to produce a certain thread involves factors of production, flexibility, and cost; and selection of the proper machine must be made on that basis.

Tolerances of 0.001 in. and less and finishes of 20 or less microinches, rms, are commonly held on turning equipment. Care should be taken to establish proper compatibility between the machine tool selected for the job and the blueprint limits, lot sizes, shop flexibility, and cost factors.

10. TURNING EQUIPMENT

Present designs in turning equipment are keyed to the omnipresent demand for lower cost production. Rapid strides in the cutting-tool industry have resulted in higher speed ranges, more power, engineered construction giving greater strength without sacrifice in accuracy, and improved maneuverability of machine elements.

Drives. Except for isolated applications, current designs of turning equipment provide for self-contained motor drives rather than overhead line and countershaft powering from a central motor source. Higher unit-power requirements per machine, machine portability, general improvement in shop lighting and cleanliness, operating efficiency, operator safety, and less cost of installation and maintenance are generally responsible for this change-over. Multiple unit V belts are in general use except for some instances where headstock speeds are obtained by shifting flat belts on cone pulleys rather than shifting gears in trains.

Constant-speed alternating-current drive motors are in wide use where gear trains and cone pulleys are used to determine spindle speeds. Direct-current motors are frequently used for some machine tools using variable-speed electronically controlled drives.

Headstocks. All-geared headstocks or boxes with gear trains, preferably arranged to give speed steps in geometric progression, are in common use today. Speed changing may be accomplished through shifting double or triple gears or by connecting and disconnecting fixed gear trains by clutches or by a combination of both. Gear shafts including spindles are generally mounted in anti-friction bearings, and teeth are usually ground on the form and rounded on the engaging face. Current headstocks are usually lubricated with a light engine oil through a spray or mist or splash system.

Bedways. These machine surfaces guide movable machine elements such as carriages and tailstocks and may be flat form, V form, or a combination of both according to the dictates of the manufacturer. Surfaces may be hand-scraped medium-hard nickel iron, ground surface flame or induction-hardened iron, or hardened and ground steel strips or blocks.

Spindle Noses. A trend toward uniformity of spindle noses has occurred in recent years to permit easy interchange of work-holding devices (see Ref. 1).

Tool Posts. (See Ref. 2.)

THE ENGINE LATHE. The engine lathe is a machine tool built to provide an extremely flexible means of machining a wide variety of work in small lots with a minimum investment in attachments and tooling. Engine-lathe practice requires the resetting of cutters for identical surfaces from piece to piece; hence for repetitive work, the hand turret lathe or automatic machine using preset cutters eliminates the costly repetition of resetting cutters to size. Conversely, when only a few pieces are required as in job shop work, repair shop work, or where production shop requirements are low, the engine lathe produces less expensively than the turret lathe or automatic principally because of the difference in set-up time and tooling cost. Because of improved designs and production methods, less difference in ability to reproduce accurate sizes exists today between engine lathes, turret lathes, or automatics, so that it is almost eliminated as a prime factor governing a choice between these machines for any given job.

Since the chargeable shop overhead is normally less on an engine lathe, it is frequently an additional advantage to perform simple-purpose cuts as secondary operations on an engine lathe for the purpose of controlling close limits of concentricity, parallelism, and size on work pieces.

Types of Engine Lathes. Engine lathes are variously classed as bench type and floor leg type, either of which may be equipped as a standard-change gear lathe or quick-change gear lathe for screw thread cutting and establishing power feeds.

Selection of Lathe Change Gears. For screw cutting with simple geared lathes, let G_s = number of teeth on stud gear; G_d = number of teeth on lead-screw gear; T_s = number of threads per inch on lead screw; and T_c = number of threads per inch to be cut. Select any gear for G_s. Then

$$G_d = \frac{G_s \times T_s}{T_c} \tag{8}$$

If the result is a fractional number, select another gear for G_s, and repeat the calculation. The equation assumes that the stud rotates at the same speed as the spindle. If these speeds are different, divide by the ratio of stud speed to spindle speed.

For screw cutting with compound-geared lathes, let G_{1r}, $G_{2r} \cdots G_{nr}$ = number of teeth on driving gears; G_{1d}, $G_{2d} \cdots G_{nd}$ = number of teeth on driven gear; T_s = number of threads per inch on lead screws; and T_c = number of threads per inch to be cut. Select at random all the driving gears and also all the driven gears except the last one. Then

$$G_r = \frac{T_c \times G_{1r} \times G_{2r} \times \cdots G_{nr}}{T_s \times G_{1d} \times G_{2d} \times \cdots G(n-1)d} \tag{9}$$

Transposing gears are used when it is desired to cut both English and metric screw threads on the same standard-change gear lathe. English transposing gears are used for cutting English screw threads on lathes having metric lead screws. Metric transposing gears are used for cutting metric screw threads on lathes having English lead screws.

For cutting multiple screw threads on lathes it is necessary to revolve the work in relation to its original position with respect to the lathe spindle as many parts of a turn as there are thread starts after the first thread has been cut.

Engine Lathe Attachments. Several attachments are available that increase the flexibility of the engine lathe. Among them are the collet chuck for work machined from bar stock, fixed and follower steady rests, four-position indexing turret for cross slide, six-position indexing turret for tailstock, milling and keyway cutting attachments, gear-cutting attachment, and grinding attachment.

Turning Tapers in a Lathe. Tapers can be turned in a lathe either by means of a taper attachment, which is supplied by many lathe manufacturers, or by setting the tailstock center a certain distance out of line with the headstock center. No exact rule for setting over the center can be given because the centers enter the work an indefinite distance. An approximate rule is: Divide the difference in diameters of the large and small ends of the taper by 2. Multiply the quotient by the ratio of the total length of the shaft to the length of the taper portion.

Miscellaneous Classes of Work. In addition to the many types of standard cutting operations covered by the *turning process*, the engine lathe may be used for certain additional special classes of work. They are knurling, filing and polishing, lapping, coil winding, spring winding, turning nonmetallic substances such as wood, fiber, and plastics; automotive service shop work such as armature truing, valve refacing, piston finishing, crankshaft turning and testing, and truing crankshaft bearings.

THE TURRET LATHE. The turret lathe is primarily distinguished from the tool room engine lathe by an arrangement of multiple-station tool holders or turrets, in place of the lathe compound rest and tailstock. These turrets permit the presetting of the total number of cutters required for the job and allow multiple and combined cuts from both turrets to operate on the work piece. This reduces cutting and handling time on repetitive work, and the function of the turret lathe in the production scheme is thus to produce lot sizes of work between those considered economical for the engine lathe and multiple-spindle automatics.

Types of Turret Lathes. Turret lathes are broadly classified as horizontal and vertical. Each type is available as a hand or semi-automatically controlled machine. Common usage limits the term *automatic turret lathe* to machines designed for chucking work only. See The Automatic Machine, p. 23-34, for details of the horizontal and vertical semi-automatic turret lathe.

The Hand-operated Vertical Turret Lathe. These machines are equipped with horizontal work tables designed with plain T slots or with built-in three-jaw combination or four-jaw independent chucks. Advantages of the horizontal table are immediately apparent for (1) large heavy work, (2) ungainly work with appended arms requiring a large swing out of proportion to small cutting diameters, or (3) overhanging work with lengthy distances between chucking and working or cutting areas. Such work may be laid down on the horizontal table and gripped with simple chuck jaws or clamps more easily and less expensively than is possible by gripping the same work against the vertical face plate or chuck used on the larger sizes of horizontal turret lathes. The capacity of the vertical turret lathe is normally designated by the diameter of the horizontal work table, and in this respect one manufacturer offers machines with tables 30 to 74 in. in diameter. Actual swing capacities run somewhat larger than the table diameter, depending on the presence of and working position of the side head.

Attachments. Thread cutting, drum scoring, angle turning, and scroll cutting may be accomplished on the vertical turret lathe by the use of suitable attachments applicable to either the main head or the side head. Turret and ram saddles may be swiveled on the cross rail within limits for taper turning and boring operations.

The Hand-operated Horizontal Turret Lathe. The horizontal turret lathe, although somewhat limited with respect to work types outlined in (1), (2) and (3) above, obtains its advantage on work of a more compact nature, the application and flexibility of taper and thread-cutting attachments and power-holding devices, and to the fact that many sizes of horizontal turret lathes are built with overlapping capacities, thus permitting a most efficient correlation of work range to machine size or mass.

The basic fact that a horizontal turret lathe is so constructed permits free interchange of chucking and bar stock jobs on the machine with the use of proper tooling. Horizontal turret lathes are subclassified as bar and chucking machines. The basic machine may be equipped with suitable tooling and work-holding devices for either or both types of work.

The horizontal bar or chucking machine may be further classified as ram type or saddle type. The ram type is built in the smaller capacities handling bar sizes up to approximately 2 1/2 in. in diameter for round stock and with chucking swings up to about 20 in. in diameter. The ram type is so named from its hex turret, slide, and saddle construction.

That is, the indexing hex turret is mounted on a ram slide guided through its stroke by a saddle which, though adjustable on the bed, is clamped in one position for any particular job. The indexing stroke of the ram is hand operated through a turnstile, rack and pinion, and power stroke by clutch engagement to the power feed train. Because of this ram construction its principal limitation is the length of the work stroke which varies with the size machine from about 4 to 15 in. Its principal advantage lies in the close correlation possible between machine and work size. Ram-type turret lathes may be powered with drive motors ranging between 3 and 15 hp.

The saddle-type turret lathe is designed with the indexing hex turret mounted directly to the saddle, without the use of an intermediate ram, the saddle in turn sliding directly on the bedways. This, in effect, greatly increases its length of work stroke compared to the ram machine, the only limitation being in the given length of the bed. Saddle machines are available with bar chuck capacities up to 6 in. round, and chucking swings up to approximately 30 in. in diameter. Machines of this type are powered by drive motors of 15 to 50 hp.

Turret Lathe Attachments. The flexibility of turret lathes is greatly enhanced by a variety of attachments. Discounting the wide variety of standard work-holding devices and standard tool holders, which in themselves make the turret lathe a versatile producing unit, several types of threading and taper attachments are available.

Screw Cutting. Threads in production may be cut from the hexagon turret by taps and dies in one pass or they may be generated by single-point tools with suitable lead control from the cross-slide carriage on ram-type machines and from the hex turret of a saddle-type machine, provided the latter is equipped with a cross-sliding hex turret. There are four types of threading attachments: (1) leading-on attachment for the ram-type turret lathes, (2) leader and follower attachment for the cross slide of ram-type machines and for both the cross slide and hex turret of saddle-type machines, (3) change-gear full-length lead screw for the cross slide of ram machines and the cross slide and hex turret of saddle-type machines, and (4) full-length lead screw with quick-change gear box for the cross slide of ram machines and the cross slide and hex turret of saddle machines. Changes in pitch for different threads are obtained with (1) and (2) by changing leaders and half nuts in the threading attachment. Separate replaceable change gears in the head end gear box are used to establish pitch for (3), and different pitches are determined by (4) through changing the position of the tumbler lever in the quick-change gear box.

Taper Attachments. Tapered surfaces may be produced on turret lathes by taper attachments or compounds. Simple contours may be produced with the taper attachment by substituting a profile guide plate for the straight plate or by the use of special contouring devices.

Steady Rests. Fixed-position steady rests may be used on turret lathes for the same work-holding requirements as found in the case of engine lathes. They are usually built to special specifications and may be (1) of the closed type, where the carriage may work on either but not both sides of the rest, or (2) of the open type, which permits free travel of the carriage on either side of the steady rest when cuts must be taken on both ends of the work piece during the course of one chucking.

THE AUTOMATIC MACHINE. This machine is profitable for producing larger lots of work parts where longer set-up times and frequently greater tool cost can be prorated over a large enough number of pieces so as not to endanger the cost savings per piece accruing from automatic cycling. Secondary advantages of automatic machinery are (1) the possibility of spreading unit labor costs over more than one machine and (2) removing all or most of the human element from the time cycle.

Types of Automatics. These machines may be broadly classified as: (1) full-automatic, where the entire production cycle is automatic including feeding of stock material, such as in bar work or hopper and magazine-fed set-ups, and (2) semi-automatic, where the cutting and index cycles are automatic, except that work parts are manually loaded and unloaded as in the case of chucking jobs.

Machines in either of these classes may occur as single- or multiple-spindle construction. Single-spindle semi-automatics are built in horizontal and vertical designs. Horizontal single-spindle semi-automatics are subclassified as general-purpose tool slide lathes, special-purpose tool slide lathes, and turret lathes.

The Single-spindle Full Automatic. This type of machine is invariably used for producing parts from bar stock in lot sizes between the hand turret lathe and multiple-spindle full automatic. The principle of combined cuts for time savings is effectively realized on this machine although it is not considered standard practice to split up cuts between tool stations as in the case of multiple-spindle machines. No consistent saving in time per piece is ordinarily realized over identical jobs done on hand turret lathes. The principal advantages of the machine are the elimination of the human element and its adaptability to multiple-machine operation by a minimum of operators and set-up men, through which factors lower costs per piece are usually realized over the similarly tooled job on the hand-operated bar turret lathe.

Single-spindle full-automatic bar machines are available in a variety of capacities. The unispindle design permits the speed of the spindle to be changed in simple ratios for greatly different types of cuts as turning and tapping, or forming and facing, without sacrificing tool feed or cutter life in favor of an abnormally high cutting speed set by other cutting operations on the same job, as may be true in multiple-spindle setups.

Speeds and feeds are set and controlled variously by clutches, speed and feed change gears, standard and special strip and disk cams, and electrical rheostat controls, depending on the manufacturer.

Cuts taken with circular or dovetail forming tools, whereby part or all of the exterior work piece contours may be finish machined easily from the cross slides, are more common on this type of machine than on the turret lathe bar machine.

Attachments. Auxiliary attachments are available for single-spindle bar automatics for performing added work on either end of the work piece. A pick-off arm allows index drilling, cross slotting, rear end drilling, threading, burring, etc., after the work has been parted off from the bar. Similar operations can be performed on the work prior to parting off by special attachments on the turret or cross slides.

Scope. Types of work ideally suited to this machine are all sizes and forms of studs, pins, washers, nuts, shafts, etc.

The Multiple-spindle Full Automatic. For production of bar stock work in large lots this machine is unique. It combines the previously cited advantages of automatic cycling with substantial reduction of time per piece over the hand turret lathe and single-spindle machine.

There are several basic principles of operation on a multiple-spindle automatic. (1) A number of tools are made to cut simultaneously. In this manner, during the operation of the tool with the longest cut, all other tools complete their operation. Therefore, the time it takes to perform the longest cut plus the constant idle time of the machine for indexing, traversing, etc., is the machining time for the component. Since there are always n components, n being the number of spindles in the machine, partly but successively machined, a finished component is obtained with each index of the spindle carrier. Therefore, it is standard practice on machines of this type to split up individual tool operations to reduce slide travel and shorten total production time. (2) Each machine set up has a particular operation which governs the machine spindle speed. All other speeds (in terms of surface feet per minute) for all other operations are directly proportionate to cutting diameters. (3) Each machine setup includes a particular operation which governs the feed gear selection. Feeds for all other operations are in direct proportion to the length of cut. (4) Each machine setup has only one spindle speed, and high differentials in required cutting speeds for tools within the cycle are obtained by attachments.

To achieve the most economical performance, it is important that every tool has only the feed stroke required for its operation during the whole cycle time. Feed strokes on currently available multiple-spindle automatics are adjusted by (1) standard or special replaceable cams and (2) quadrant arm mechanisms which are infinitely adjustable within the limits of the stroke.

Tool feeds on the principal makes of multiple-spindle automatics are obtained by (1) changing feed gears and (2) quick-change selective gear box.

Spindle speeds are selected by means of change gears in multiple-spindle full automatics. Because of the inherent designs of multiple-spindle carriers, it is mandatory that all spindles rotate at one speed. This speed is selected from a range of available speeds as being most efficient for the types of cutting operations in the cycle.

ATTACHMENTS. Since automatics of this type may have four, five, six, or eight spindles over which the work part may be successively machined, several types of attachments have been developed to speed up production and also perform miscellaneous operations otherwise done on additional machines. These attachments permit reaming, tapping and threading, recessing, high-speed drilling, slotting, cross milling, and knurling, and include pick-off devices for finishing back or cut-off ends after parting.

SCOPE. Because of the large number of work stations available on the multiple-spindle automatic, the scope of jobs handled is greatly expanded. Main limitations controlling scope are (1) collet capacity, (2) maximum length of feed stroke, (3) depth of holes drilled, and (4) types of fits required for tapping and threading operations.

The Multiple-spindle Semi-automatic. This type of machine may be vertical or horizontal. If horizontal, the collet chucks and bar feeds of the full automatic are replaced by two- or three-jaw power-operated chucks, and the work parts must be loaded and unloaded manually.

The horizontal semi-automatics are generally available in four, five, six, and eight spindles, and the capacity of each machine is expressed by the diameter of mountable chucks. Although the semi- and full automatic horizontal machines are generally identical except for chucking devices and are available in similar spindle combinations, they are not offered to handle bar and chucking work interchangeably as are hand turret lathes.

The vertical multiple-spindle semi-automatic also operates on the theory of machining a work part progressively from cutter station to station. They are available in four, six, eight, twelve, and sixteen spindle combinations with chucking capacities ranging from 10 in. to 34 in. in diameter, depending on the spindle combination. From this it may be assumed that for large work of the proper complexity and rated production requirements, this type of automatic may be profitably tooled to secure the advantages of automatic cycling not offered by any other type of turning equipment large enough to handle this range.

The Single-spindle Semi-automatic. Like the multiple-spindle semi-automatic, this machine is available as a vertical or a horizontal machine.

The vertical single-spindle machine has all the design features and applications of the hand-operated vertical turret lathe, previously described, except that control units are applied to the turret, ram, and side heads which may be preset to eliminate all manual operations except loading or unloading. The presence of these automatic-control units does not eliminate the possibility of switching the machine to manual control at will.

Horizontal single-spindle semi-automatics are available in two types, (1) tool slide lathes and (2) automatic turret lathes. Tool slide lathes are a form of engine lathe without screw-cutting characteristics but with one or more power-fed automatic tool slides. Work may be chucked or mounted between centers and gang-turning or facing cuts taken on a work piece separately or simultaneously as required. Some slides may be set at angles for taper work. These machines may be used profitably for medium-size production lots on simplified work which would not require the full complement of tool stations available on the hand or automatic horizontal turret lathe. Automatic turret lathes retain all function of the hand-operated turret lathe in addition to the usual advantages of automatic cycling.

11. COST ANALYSES

Certain fundamental relationships exist between the relative productive capacities of machine tools in the turning field and the investment required to obtain these potential capacities. Many factors enter into a production engineer's decision to tool a certain job on a given machine, not the least of which are (1) quantity of planned production of work piece, (2) expected production rate on the given machine, and (3) relation of cost savings to required investment.

The following examples of typical situations are offered as a guide to factors governing such analyses.

EXAMPLE 1. An identical job setup on a ram-type turret lathe could use either (*A*) predominantly standard tools or (*B*) special tools. The following analysis is for the purpose of determining the quantity of work for which the cost per piece would be equal.

	Method A	*Method B*
1. Cost of standard tools	$1,234.75	$ 59.50
2. Consider depreciating standard tools with machine at 15-year rate—yearly depreciation	82.32	3.97
3. Consider 2000 hours of yearly depreciation—hourly depreciation	0.041	0.002
4. Considering standard overhead rate @ $2.50 per hour, excluding depreciation; then total overhead = line 3 plus $2.50	2.541	2.502
5. Assume rate for direct labor @	1.50	1.50
6. Total labor and overhead per hour = line 4 plus line 5	4.041	4.002
7. Total labor and overhead per minute	0.0673	0.0667
8. Cost of special items all depreciated in one lot	$ 197.00	$2,616.70
9. Production time per piece	5.2 min	2.2 min
10. Cost per piece (labor and overhead): line 7 times line 9	$ 0.34996	$ 0.14674
11. Difference in cost per piece: 0.34996 less 0.14674	0.20322	
12. Difference in cost of special items: $2,616.70 less $197.00		2,419.70

13. Then $2,419.70 divided by $0.20322 = 11,900 pieces, which is the point in production where the cost per piece is the same for both Method A and Method B. Method B would be justified only if the quantity exceeds 11,900 pieces.

EXAMPLE 2. Method of determining quantity of work, where equal cost per piece exists, of an identical job setup on (A) a single-spindle semi-automatic chucking machine and (B) a multiple-spindle semi-automatic chucking machine.

	Method A	*Method B*
1. Cost of machine	$10,750.00	$17,325.00
2. Cost of standard tools	750.00	750.00
3. Total cost of standard items	$11,500.00	$18,075.00
4. Consider depreciating of standard items at 15-yr rate—yearly depreciation	$ 766.00	$ 1,205.00
5. Consider 4000 hr of yearly operation—hourly depreciation	0.19	0.30
6. Considering standard overhead rate @ $2.50, excluding depreciation, total overhead = line 5 plus $2.50	2.69	2.80
7. Direct labor	1.50	1.50
8. Total labor and overhead per hour	4.19	4.30
9. Total labor and overhead per minute: line 8 divided by 60	$ 0.07	$ 0.072
10. Setup time	1.5 hr	4 hr
11. Setup cost: line 8 times line 10	$ 6.29	$ 17.20
12. Cost of all special items depreciated in one lot	350.00	700.00
13. Total of special item and set-up cost: line 11 plus line 12	$ 356.29	$ 717.20
14. Production time per piece	0.9 min	0.33 min
15. Cost per piece (labor and overhead): line 9 times line 14	$ 0.063	$ 0.024
16. Difference in cost per piece: 0.063 less 0.024	0.039	
17. Difference in cost of special items: $717.20 less $356.29		360.91

18. Then 360.91/0.039 equals 9216 or the number of pieces made on each machine at equal cost

Proving above: 0.063 (cost per piece, labor and overhead) times 9216 plus $356.29 (special items and setup) equals $936.89; and 0.024 (cost per piece, labor and overhead) times 9216 plus $717.20 (special items and setup) equals $938.38

The cost per lot of 9216 pieces is virtually equal on both machines. Obviously, the cost per piece will be less on the multiple-spindle machine as the quantities are increased, and the lower cost per piece will be attained on the single-spindle machine as the quantities decrease.

EXAMPLE 3. Method of determining amount of production required to justify a faster but costly producer A instead of a less-expensive but slower producer B.

Fast Producer A

				Depreciation per year	Depreciation per minute
Machine A cost	$13,877.00	@	6.6% per yr =	$ 915.77	$0.0076
Tools cost	2,895.00	@	100% per yr =	2,895.00	0.0240
Total equipment	$16,772.00			$3,810.77	
Labor @ $1.50 per hr and overhead @ $2.50 per hr excluding depreciation = $4.00 per hr or					0.0660 per min
				Total charge	$0.0976

Machine A produces one piece in 2.8 min flat time or at rate of 3.5 min (48 min hour) steady production.

Therefore,

$$\text{Cost per piece} = 3.5 \times 0.0976 = 0.3423$$

Slow Producer B

				Depreciation per year	Depreciation per minute
Machine B cost	$7,274.00	@	6.6% per yr =	$ 480.08	$0.004
Tools cost	1,547.00	@	100% per yr =	1,547.00	0.013
Total equipment	$8,821.00			$2,027.08	
Labor @ $1.50 per hr and overhead @ $2.50 per hr, excluding depreciation = $4.00 per hr or					0.066 per min
				Total charge	$0.083

Machine B produces one piece in 4.5 min flat time or at rate of 5.63 min (48 min hour) steady production.

Therefore,

$$\text{Cost per piece} = 5.63 \times 0.083 = 0.4674$$

Savings per piece when made on A machine 0.4674 less 0.3423 = $0.1251

A yearly depreciation charges	$3,810.77
B yearly depreciation charges	2,027.08
Difference	$1,783.69

This difference must be paid for out of the savings made by machine A to warrant its purchase.

Saving per piece when made on A machine is \$0.1251. A machine makes one piece in 3.5 min or 136 pieces per 8-hr day; hence savings per day are 0.1251 × 136 = \$17.01. Therefore, A machine must run 105 days to pay for \$1,783.69 difference in annual depreciation. The manufacturer must make at least \$1,783.69 divided by 0.1251, or 14,260 pieces per year to warrant purchase of A machine.

EXAMPLE 4. A typical replacement statement outlining the conditions under which an obsolete model machine may be replaced by a modern machine of the same type. Time to produce one piece on obsolete machine is 93 min; on modern machine 31 min.

Assume direct labor	\$1.50 per hr
Assume factory overhead, excluding allowances for depreciation	2.50 per hr
Total	\$4.00 per hr
or	\$0.066 per min

Average cost per piece on obsolete machine: 93 min × 0.066	\$6.14
Average cost per piece on modern machine: 31 min × 0.066	2.05
Savings per piece	\$4.09

Production per month with modern machine:

$$\frac{60 \text{ min (hr)} \times 8 \text{ hr (per day)} \times 23 \text{ (working days)}}{31 \text{ min per piece (new equipment)}} = 356 \text{ pieces}$$

Total savings per month will then be:

356 (pieces per month) × 4.09 (savings per piece)	\$ 1,456.04
If the new equipment is kept busy 80% of the time, deduct 20% of \$1,456.04 (savings per month)	291.21
Value of usable savings per month	\$ 1,164.83
Investment in new equipment (machine and tools)	16,772.00
Less resale value of old equipment	1,500.00
Net cash investment	\$15,272.00

The new equipment will pay for itself in

$$\frac{\$15{,}272.00 \text{ (net cash investment)}}{\$1{,}164.83 \text{ (usable savings per month)}} = 13.1 \text{ months}$$

The net profit return per year will be:

\$1,164.83 (usable savings per month) × 12 months =	\$13,977.96
Less 6.6% depreciation on \$15,272.00 (net cash investment)	1,007.95
Net profit per year after depreciation	\$12,970.01

The rate of net profit will be

$$\frac{\$12{,}970.01 \text{ (the net profit)}}{\$15{,}272.00 \text{ (the net cash investment)}} = 84.9\%$$

REFERENCES

1. *Bull. Am. Standards Assoc.*, B5.9, Nov. 1936.
2. *Bull. Am. Standards Assoc.*, B5.2, 1943.

PLANER WORK

By John M. Walter

As a general rule, planers are used for generating flat surfaces. These may be in vertical, horizontal, or angular planes. Form planing of flat or curved surfaces is done with special tools. By the use of cam followers or tracer devices, surfaces which are curved transversely or longitudinally, or a combination of both, may be planed.

The extreme simplicity of operation, and of the cutting tools used and their application, makes the planer one of the most versatile of all machine tools. With very simple cutting tools and work-holding accessories, a wide variety of work can readily be planed. Particularly is this so on larger work pieces where the surfaces to be machined are frequently located in different planes and at angles to one another. In some cases the machined surfaces are located deep down in pockets and other hard-to-reach places. Such work can be machined on the planer with inexpensive tools. The heads can be quickly swiveled to the required setting for doing angular surfaces. On many modern planers power swiveling of the heads is built into the machine.

The initial investment for the machine and the cost and maintenance of the cutting tools are far less for the planer than for other equipment capable of machining surfaces on fairly large pieces.

A high degree of accuracy is obtained in planing because of the simple generating principle involved, the ease with which the cutting edge on the finishing tool is maintained, and the fact that the tool can be positioned readily and held in proper relation to the work. In the construction of machine tools, planing is used almost exclusively for finish-machining large bed and table ways, dovetails, slides, and other guiding surfaces.

12. TYPES AND SIZES OF PLANERS

THE DOUBLE HOUSING PLANER is the most common type. It has two uprights for supporting the cross rail. Four heads are usually used, two on the cross rail and one on each upright.

THE OPEN-SIDE PLANER has an upright or column on the right-hand side only. It can accommodate work of varying widths, as the work piece can extend out over the table on the left side. Where the work is extremely wide, it is sometimes supported on an auxiliary rolling support placed alongside the planer. Open-side planers usually have two heads on the cross rail and a side head on the column.

CONVERTIBLE OPEN-SIDE PLANERS have a removable upright on the left side. This upright may be bolted to the bed or may be movable transversely on a floor plate or runway. A side head can be used on the upright. By removing the upright the planer is converted into an open-side type.

MILLING PLANERS, which are combination machines, are used for planing, milling, boring, and drilling. Various combinations of heads can be used. Table speeds range from milling feed rates to fast planing and return speeds. On these machines various types of operations can be performed on a work piece in a given setting, resulting in greater accuracy and economy. Milling planers may be of the double housing, open-side, or convertible type.

SIZE OR CAPACITY of planers is designated by the maximum size of work that can be planed on the machine. Thus a 60 in. by 48 in. by 20 ft planer is one that can plane a piece 60 in. wide, 48 in. high, and 20 ft long.

HYDRAULIC PLANERS. The table on this type of planer is reciprocated by the use of one or more hydraulic cylinders secured to the bed, the piston rods being secured to the table. Sometimes the compressibility of the oil tends to cause jerky action of the table. A column of oil will, under ordinary conditions, compress approximately $1/2\%$ of its length when subjected to a pressure of 1000 psi.

POWER REQUIRED FOR PLANING. From the standpoint of metal removal in cubic inches per minute, per horsepower, planing is more efficient than any other type of metal cutting. Analysis of the chip formation will reveal a minimum of distortion and internal working of the material, compared to turning (except at very large diameters), milling, and drilling.

The power required for planing can be expressed approximately as horsepower $= DFSK$, where D is the depth of the cut, F is the feed or thickness of the cut, S is the speed in feet per minute, and K is a constant. For planing medium-hard cast iron, K equals 3, approximately; for free-cutting steel, 6; and for soft bronze, 1.5. The horsepower derived from this equation represents the net horsepower at the tool. The power required to drive the planer idle should be considered, particularly for large planers operating at high speeds, to obtain the gross horsepower.

PLANER MOTORS. Variable-voltage reversing motors are used almost exclusively in gear-driven planers. A motor generator set energizes the reversing drive motor. The speed of the motor generally is adjustable from 40 to 1200 rpm and is independently adjustable on the cut and return strokes. In the most commonly used types of drive, the motor develops its rated capacity at 250 rpm, and the torque is constant at speeds of 40 rpm upward. The speed where the torque begins to drop off depends on the capacity of the motor generator set. In many installations its capacity is double the rated horsepower of the drive.

The size of the motor required for a given size planer depends on the kind of work to be done, number of cuts taken simultaneously, size and length of the planer, operating speeds, etc., and when feasible all factors should be considered in determining the horsepower of the motor. For general-purpose work the following may be used as a guide. For 30-in.

planers 15 to 25 hp, 48-in. planers 35 to 50 hp, 72-in. planers 50 to 75 hp., and for larger planers 75 to 125 hp. It must be borne in mind that these values represent the rated horsepower at 250 rpm of the drive motor. At higher speeds, the horsepower depends on the size of the motor driving the generator, which in many cases is double the rated horsepower.

LOAD-CARRYING CAPACITY OF PLANERS. Damaged ways sometimes occur because of overload. Safe loading of cast-iron ways should not exceed 20 psi of projected way bearing area. Even then the planer ways must be accurately level, and a good lubricating oil, preferably an extreme pressure (EP) oil must be used. This value includes the weight of the table, which in itself produces a pressure of 4 to 10 psi of projected way bearing area, depending on the size and type of planer. Heavy work, as for example the rams of large drop hammers, must sometimes be supported on heavy longitudinal bars to avoid highly concentrated loading of the ways. When the work overhangs the table, as on the left side of open-side planers, high concentrated loads sometimes result, even though the work piece is not very heavy. When unusually heavy or overhung loads are encountered, careful analysis should be made to avoid overloading.

Nonmetallic table bearings recently developed by the G. A. Gray Company, will safely carry 50 psi, without danger of scoring.

13. CUTTING TOOLS

CARBIDE TOOLS are being used extensively for planing, there being several methods of applying the carbide tip.

The most common type of carbide tool has the tip brazed on the shank, the shank being recessed for the tip. Tools of this kind are readily purchased in various sizes and shapes for different kinds of cutting operations. New carbide tips are available for replacement of broken tips. They can easily be brazed on to the shank after the cracked or broken tip is removed by heating the tip and shank. In regrinding the tip on tools of this kind, the shank must be ground back to obtain clearance. For this reason the recess for the tip on the shank has to be remilled after some usage.

Tools having removable bits are widely used. The bits are held in the shank or holder by serrations and wedges. They can be removed from the holder while the holder is left in the tool block. It is apparent that the operator exerts much less physical effort in removing the small bit than in removing the complete tool. Planer tools, as a rule quite heavy, are held in place by four large studs which require the use of a heavy wrench. The removable bit may be made of high-speed steel, cast alloy, or carbide tipped.

Tools having a carbide disk clamped on a shank are frequently used. The disk can be rotated after unclamping for presenting a new cutting edge. It is clamped by one screw in the center of the disk.

Tools having clamped-on carbide tips are widely used and have the advantage that the tip can be removed without removing the shank or holder. The carbide tips are ground to the desired shape and angles in a fixture and are then ready for mounting in the shank. A backing-up screw in the shank is adjustable for advancing the carbide tip as it is ground back at the cutting edge. The shank is provided with a hardened steel insert for the contacting surface of the carbide tip. Both the insert and the tip are ground flat at the interfaces. The tip is held in place by a small clamp having one or two small screws. This clamp is not for the purpose of holding the tip against cutting forces, but only to hold the tip in place, as the tool is designed so that the cutting forces act against the restraining surfaces and against the adjusting screw.

When carbide tools are used on planers it is very important to lift the tools clear of the work on the return stroke to avoid chipping the cutting edge. Modern planers are equipped with automatic tool lifters for this purpose. It is important that the tool-lifter mechanism be accurately synchronized with the table movement to avoid dragging the tool back over the work piece for even a small distance. It must also reseat without shock; otherwise the accuracy of slide setting will be affected.

Forged steel tools are sometimes used, usually being made of high-speed steel. Sometimes a high-speed steel end is welded to the shank or the shank is faced with a piece of high-speed steel.

Double-cutting tools, for cutting in both directions of table travel, are used successfully in some very special applications. One such application is in planing the T part of T slots. A two-edged cutting tool is used, being fast on a trunnion which is caused to oscillate a small amount at each reversal of the table. After one side of the T slot is planed, the trunnion is rotated 180 degrees for planing the opposite side. With this type of tool, several slots can be planed simultaneously in considerably less time than was formerly required to plane one slot. Operators who formerly lifted the tools by hand, high out of the slot on the return stroke, express their appreciation of the new double-cutting method.

Gang tools can be used to advantage on many jobs. Such tools have three or more tools set in a holder, one behind the other, so that each tool takes its proportional share of the total feed. One type of gang tool employs carbide disks which can be rotated to present a new cutting edge, after the disks are unclamped.

TOOL SHARPENING. The planer operator should not sharpen the tools. It is much more efficient to have the tools ground in the tool room to fixed specifications. Once the tool angles for best results have been definitely established under actual operating conditions, it is easy to maintain these angles in the tool-grinding department.

14. SPEEDS AND FEEDS

The cutting speed depends principally on the material being cut and the cutting tool. The feed may sometimes be limited by the ability of the work piece to support the cutting forces. When three or four heads are used on the planer simultaneously, the power of the motor may limit the amount of feed that can be used. Table 1 shows approximate speeds

Table 1. Cutting Speeds in Feet per Minute

	For High-speed Steel Tools				For Cast Alloy Tools				For Carbide Tools			
Depth of cut	1/8	1/4	1/2	1	1/8	1/4	1/2	1	1/8	1/4	1/2	3/4
Feed	1/32	1/16	3/32	1/8	1/32	1/16	3/32	1/8	1/32	1/16	3/32	3/32
Cast iron, soft	95	75	60	50	160	135	110	95	300	235	190	165
Cast iron, medium	70	55	45	35	125	105	90	75	220	175	140	125
Cast iron, hard	45	35	25	.....	95	80	65	.....	140	110	.....	.....
Steel, free cutting	90	70	55	40	140	105	85	65	330	260	200	.....
Steel, average	70	55	40	30	105	80	60	45	280	220	170	.....
Bronze, soft	150	150	125	.....	*	*	*	*	*	*	*	*
Aluminum, free cutting	200	200	150	.....	*	*	*	*	*	*	*	*

* Maximum table speed.

and feeds for various materials and kinds of cutting tools. The most efficient cutting speed and feed can be determined only by actual trial on the job.

FINISHING CUTS are made with broad tools, a coarse feed being used when cast iron, bronze, aluminum, and nonmetallic materials are planed. The tools are 1 to 1 $1/2$ in. wide, and feeds $1/4$ to 1 in. are used. For the best quality of finish, the depth of the cut is 0.001 to 0.002 in. The finishing tool must be set so that its cutting edge is exactly parallel to the surface being planed.

For finish planing steel, the tool is frequently formed so that the cutting edge will be at an angle to the line of travel of the work piece. This angle, which may vary from 20 to 45 degrees, is best determined by trial. Covering the surface of the steel to be finished with heavy cutting oil minimizes tearing and improves the finish.

In general, finish planing is done at greatly reduced speeds in order to preserve the accuracy of the cutting edge of the tool.

SETTING UP WORK. Since cutting time has been so greatly reduced through the use of higher speeds it is important to also reduce the setup time to obtain maximum output from the planer.

When the work is repetitive, fixtures for holding the parts should be used. The fixtures must be carefully designed to clamp the piece rapidly without springing it. The locating surfaces must also be arranged to minimize the time for cleaning off the chips.

Setup plates can sometimes be used to reduce the standing time of the planer. Such plates are arranged for rapid mounting and removal from the planer table. Two or more plates must be used in order to remove and set up work on one or more plates while one plate is in use on the planer.

Sometimes duplex tables are used to reduce the standing time. Such tables are divided at midlength so that one-half can be unloaded and reloaded while work on the other half of the table is being planed. The standing half of the table can readily be positioned at the end of the bed through the use of a quick-acting latch, and valves in the ends of the tables where they join together automatically establish oil circuits for proper functioning of the lubricating system. The two tables can be joined together and operated as one piece for long stroke work.

Magnetic chucks are used for rapid setting up when planing thin plates of ferrous materials. These chucks are frequently used for planing taper gibs by placing the chuck on a wedge-shaped piece made to the proper angle.

In such cases where setup fixtures are not used and the work pieces are not of simple form, the surfaces to be planed should first be laid off. This work should be done on a layout plate before the pieces are brought to the planer. The layoff lines enable the planer operator to set the job up in less time than would otherwise be required.

TOOL SETTING. For many jobs a tool-setting gage can be used to good advantage, particularly for the rough and semifinish cuts. Such gages usually consist of a plate fastened to the planer table, at the rear of the work, and in line with it. The cutting tools can be set to the gage, thereby saving time and greatly minimizing the chance of errors occurring in the dimensions. Where such gages are to be used for a long period of time, the gaging surfaces should be of hardened steel. Tool-setting gage surfaces can sometimes be built into work-holding fixtures.

TABLE SPEEDS. The speed at which the table of a planer runs depends on the speed of the driving motor and the gear ratio of the driving gears. This gear ratio is expressed as the number of revolutions through which the motor must turn to move the table one foot. On high-speed planers the gear ratio is usually 4, and heavy-duty frog and switch planers operating at low speed have ratios of approximately 12. It will be apparent that for a given motor torque the table pull, or cutting force, will vary in direct proportion to the gear ratio.

TIME REQUIRED FOR PLANING. The time required to take one cut over a given surface depends on the length of the stroke, width of surface, feed per stroke, cutting speed, and return speed. The approximate time may be expressed as

$$T = \frac{W}{F}\left(\frac{L}{C} + \frac{L}{R} + A\right)$$

where T = time, minutes; W = width of surface to be planed, inches; F = feed, inches; L = length of stroke, feet; C = cutting speed, feet per minute; R = return speed, feet per minute; and A = added time required for reversals.

Under normal operating conditions, the value of A varies from 0.015 min for low motor speeds to 0.040 min for high motor speeds.

Table 2 may be used as a guide for estimating planing time for short strokes. The values given are approximate maximums for a medium-size planer having a gear ratio of 5.

Table 2

Length of Stroke, in.	Number of Strokes per Minute
8	40 to 52
12	32 to 42
18	24 to 32
27	17 to 22
40	12 to 16
60	10 to 13

15. DEVELOPMENTS IN PLANERS

A very large majority of the planers in use cannot be operated at speeds above 100 ft per min, and therefore are not suitable for using carbide tools. With modern planers, cutting speeds up to 250 ft per min and return speeds up to 400 ft per min are being used. Compared with its predecessor of 15 years ago, the modern planer makes 3 1/2 or more times as many strokes in a given time, and in most instances the metal removal is in the same proportion. Rigidity of the machines has been very greatly increased, the housings, cross rails, heads, etc., being larger and heavier.

The increased speed and acceleration of the table have necessitated new developments in the other components of the machine. The feed mechanisms used on older planers are not practicable for rapid starting and stopping of the feed as required for high-speed operation, because of their high inertia factor. The new planer feed employs a different principle, eliminating the destructive inertia effects of the former feed devices. The feed mechanism of a planer plays an important part in the operation of the machine, and it can can truly be said "A planer is no better than its feed."

The ways of a planer must be in good condition to stand up under the severe service of high-speed operation. In this connection the nonmetallic ways previously mentioned have proved to be entirely satisfactory in filling a long-felt need. Scoring of the bedways is eliminated, and distortion of the table from heat generated by friction is greatly reduced. It may be in order to explain at this point that, with the usual way construction, during high-speed operation the frictional heat causes the lower side of the table to elongate with reference to the top side, thereby causing the table to curl up at the ends. With the nonmetallic way, this curling is greatly reduced because of the heat-insulating qualities of the table-bearing material.

Better lubrication of the ways through a closed circuit, under pressure, floats the table.

For safety in operation, a table-stopping device is necessary when high speeds are used. Failure of the electrical control or equipment may permit the table to run away at greater than maximum operating speeds. At such speeds the table has sufficient stored energy to carry it farther than normal shop clearances permit. The stopping device acts as a brake immediately when the table leaves the bull gear, and normally stops the table in approximately one foot or less of travel.

Owing to the use of much longer drive motors, the old arrangement of placing the motor perpendicular to the planer has given way to an angular drive that places the motor behind the right-hand upright. In many installations much floor space is saved through the use of this drive as the planer can be placed closer to the wall. Hypoid or spiral bevel gears are used for the initial reduction.

METAL-WORKING SHAPERS

By A. G. Baumgartner

The shaper is a machine tool used primarily for producing flat surfaces in horizontal, vertical, or angular planes. The shaper also is used for machining curved surfaces, helixes, serrations, and special work involving odd and irregular shapes.

The metal-working shaper, like the planer, develops cutting action from straight-line reciprocating motion between the tool and the work. On the shaper, the tool is reciprocated and the work is fed sidewise. The horizontal shaper with the ram movement in the horizontal plane is the most common form of shaper. Vertical shapers, with ram movement in a vertical plane, also come in the metal-working shaper group. Vertical shapers are also known as slotters.

Types of shaper work often may be done on a milling machine or a planer. The shaper, however, excels in handling tool-room work, die-shop work, and small-lot manufacturing, owing to its great versatility and its flexibility and speed of setup. Setup costs and tool costs are comparatively small. Planing, contouring, internal shaping, and high-speed shaping are accomplished rapidly and economically with little need for special fixtures and with simple cutting tools. The metal-working shaper's unusual flexibility and its power to perform easily work which might be difficult or impossible on other machines make it a basic machine tool. The shaper is especially advantageous in machining tough die steels of great hardness and alloyed high-strength cast iron. Such steels are used in the forging field, while alloyed cast iron is widely used in glass molds.

16. TYPES AND SIZES OF SHAPERS

HORIZONTAL SHAPERS include the push-cut and draw-cut design. Almost all manufacturers of horizontal shapers build the push-cut, crank-operated type, which are built with cutting strokes 6 to 7 in. long in bench models on up through 36 in. long in heavy-duty models.

A typical push-cut shaper employs a rocker arm reciprocated by a crankpin mounted on the crankgear and adjustable radially, the crank mechanism being an application of the Whitworth quick-return linkage. The return speed is higher than the cutting speed for increased efficiency, the ratio being varied according to the design.

Horizontal push-cut shapers are also operated hydraulically, using a piston and cylinder to operate the ram. The cost of a proper hydraulic unit, the comparatively low efficiency of the hydraulic drive, and the problems in accuracy have influenced almost all domestic shaper builders to use the crank drive.

Horizontal shapers are furnished with either plain or universal tables. The universal table adds to the flexibility and versatility of the shaper and, in conjunction with the swivel vise, provides rotation on all three axes. Graduations on all three movements speed angular setups.

VERTICAL SHAPERS OR SLOTTERS. The typical vertical shaper, or slotter, has an operating mechanism similar in principle to that used on crank shapers. Vertical shapers are also operated hydraulically. The vertical shaper differs from the horizontal in that the ram is vertical and the table is rotary. The ram can be inclined slightly from vertical for angular work.

The table on a vertical shaper can be rotated about the vertical axis and also fed longitudinally and transversely. The flexibility of the rotary motion, together with the two-way horizontal feeding of the table on the vertical shaper, makes it a valuable tool for keywaying, slotting, and internal work. Vertical shapers are commonly built in maximum stroke length from approximately 6 to 36 in. Strokes of 48 in. and longer can be furnished.

HORSEPOWER REQUIREMENTS for fairly heavy cutting on shapers are given in Table 1.

Table 1. Horsepower Requirements for Shapers

Horizontal Shapers		Vertical Shapers	
Stroke, in.	Hp	Stroke, in	Hp
12	2–3	6	3
16	3–5	12	3–5
20	5	18	5
24	7 1/2–10	24	7 1/2–10
28	10	30	10
32	10–15	36	10–15
36	15		

Larger motors can be used for extra-heavy cutting, but they are rarely required because the demands of average cutting on shapers are well within the ratings given.

17. CUTS AND FEEDS

Because of the great variety of materials, cutting tools, and character of work, only general comments can be made on cuts and feeds. Horizontal shapers are commonly provided with a horizontal power feed range of 0.010 in. through approximately 0.200 in. and vertical power feeds on the head of 0.005 in. through approximately 0.100 in.

Vertical shapers have longitudinal and transverse power feeds with ranges approximately from 0.002 to 0.100 in. for small stroke machines, up to 0.150 in. maximum on 36-in. stroke machines. The rotary feeds on vertical shapers range approximately from 0.004 to 0.175 in. on a 20-in. circle. These figures are general and vary with particular machines.

The depth of cut is normally governed by the amount of material to be removed, how much the work will stand with regard to breaking out, deflection under the cut, and distortion from heat. The cut also depends on the character of material being cut and the limitations of the cutting tools.

Time required for shaping a flat horizontal surface is the result of WT/FS, where W represents the width of the work, F the amount of cross feed, S the strokes per minute,

Table 2. Cutting Speeds of Shapers

Strokes per min	Length of Stroke, in.																	
	2	4	6	8	10	12	14	16	18	20	22	24	26	28	30	32	34	36
	Average Cutting Speed, ft per min																	
10	5	7	10	13	16	18	21	24	27	30	32	33	35	37	39	41	43	45
15	6	10	15	19	23	28	33	37	40	43	46	50	54	58	61	64	67	70
20	7	13	20	26	32	38	44	50	55	60	65	70	74	79	83	87	91	95
35	12	23	34	45	55	65	74	83	92	101	110	118	125	133	140	147	153	159
50	17	32	48	64	78	92	106	119	131	145	157	168	179	190				
75	25	49	72	95	116	136	156	176	196	216								
100	33	65	97	129	160	190	220	250										
150	48	95	140	190	235	280												
200	65	127	190	255														
300	99	195	298															

and T the number of trips across the work to obtain the finish required. F may be different for each trip. T commonly falls within 1 to 3.

Cutting speeds are of consequence especially on horizontal shapers. Average cutting speeds for different lengths of stroke and strokes per minute on shapers with a typical return ratio are given in Table 2.

18. SHAPER WORK

The versatility of shapers should be emphasized again. Applications of the horizontal shaper include the following major functions.

SHAPING OR PLANING OF FLAT SURFACES. The majority of this work is held in a vise. The standard vise has a graduated base and can be set with the stroke, or at right angles to the stroke, or at any other angle required. Vises are furnished in a variety of types. The single-screw vise and the double-screw vise (Fig. 1) are the most popular. The shaper vise on the tilting top of the universal table gives great flexibility in angular shaping.

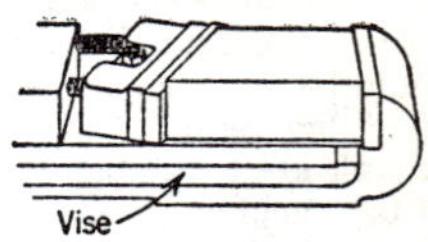

FIG. 1. Tapered workpiece in double-screw vise.

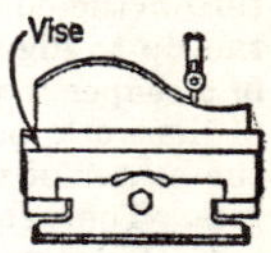

FIG. 2. Shaping contours.

Work can also be fastened directly to the shaper table. The surface gage and the indicator are two of the tools essential in setting up and checking work.

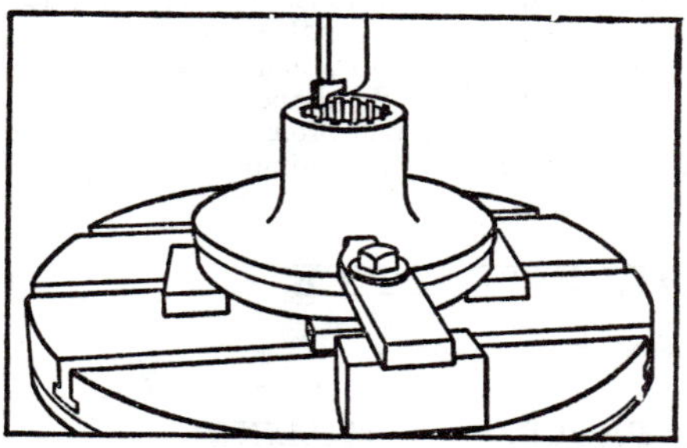

FIG. 3. Cutting internal splines on a vertical shaper.

CONTOURING, or machining irregular shapes, is another function of the shaper (Fig. 2). This operation can be done by operating the vertical feed on the head by hand while the cross feed is operated by power. For more difficult work, the cross feed is also frequently operated by hand in conjunction with the vertical feed on the head. For such operation, an additional or auxiliary control at the front of the table for cross feeding is helpful to the operator.

AUTOMATIC DUPLICATING equipment is available on shapers for developing contours and irregular surfaces. Automatic duplicating equipment can be had in elaborate forms for developing compound surfaces, such as the blade of a ship propeller, or in simpler form for curved surfaces that can be produced with straight-line cutting.

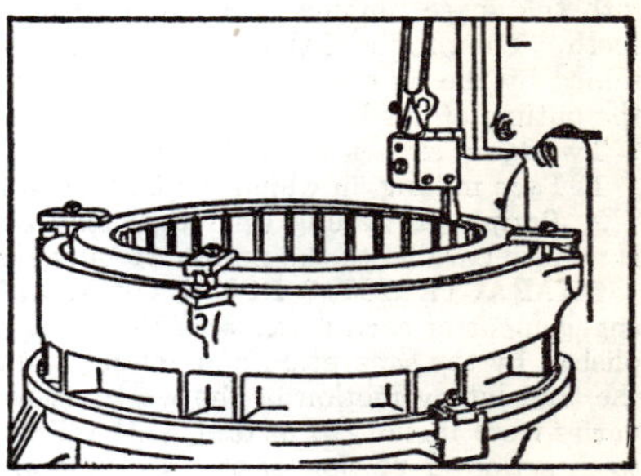

FIG. 4. Slotting on a vertical shaper.

KEYWAYING AND SLOTTING are usually done on a vertical shaper or slotter (Figs. 3 and 4). The horizontal shaper, however, is also used for this purpose (Fig. 5).

Internal shaping can include keywaying and slotting. Internal shaping, however, covers much more. The internal guides of valves, for example, are machined on horizontal shapers on a manufacturing basis.

CUTTING TOOLS. A great deal more attention can be given to cutting tools by the shaper user; incorrect and inadequate tools seem to be the rule rather than the exception. The usefulness and productivity of a shaper can be greatly increased by adequate cutting tools. Shapers, in general, can also be operated at higher speeds in most shops.

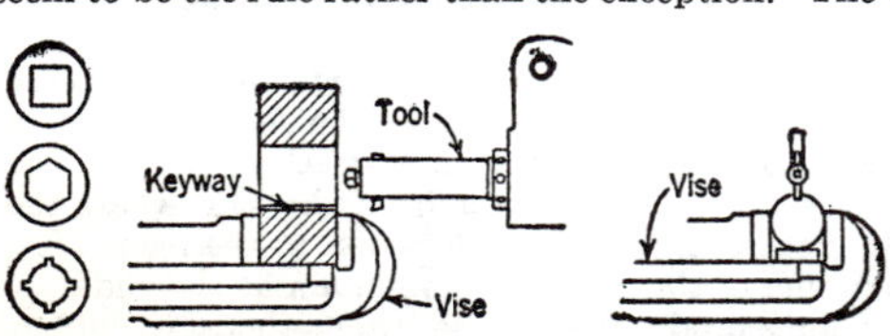

FIG. 5. Keywaying and slotting on a horizontal shaper.

The grinding of cutting tools is an enormous subject. The actual angles can be as varied as the classes of materials and the character of work. High-speed steel is commonly used on shapers today. The use of a round-nose tool is a common error, especially in cutting tough materials such as B die blocks. A tool with a very small radius, approximately $1/_{32}$ in., and with a straight cutting edge, gives the best performance. Such a tool should also be used on jobs that require excessive overhang of work or tools, and where there is a tendency for chatter. The sharp nose and the straight cutting edge, in general, are to be recommended. Negative front rake can be used with benefit on high-speed steel as well as carbide tools.

Tools and tool holders specially manufactured for both horizontal and vertical shapers are available from tool manufacturers. Shaper tools are also furnished in sets and for general purposes are quite satisfactory.

The use of the ordinary lathe tool on the horizontal shaper is another common error that should be avoided. The lathe toolholder supports the cutting bit at an angle and throws the cutting tip ahead of the shank of the holder. The deflection of the tool causes the tip to dig into the work, and the result is chatter and inaccuracy. The cutting point of a shaper tool should be near the center line of the shank so that the point will swing out of the work instead of digging in. Carbide holders which hold the bit parallel to and near the center of the shank are available. These holders should also be used for high-speed bits. Tools should be machine ground, not hand ground.

Carbides would seem to have limited application on shapers because they are usually associated with high production. Actually, the use of carbides is increasing rapidly because the modern shaper with its high speeds, fully automatic feeds, improved controls, and increased power is taxing the best high-speed steels. This is particularly true on die blocks, high-carbon, high-chrome steels, and special alloyed cast irons. Shockproof-grade carbides are already in regular use in machining these materials with complete satisfaction. A dependable fully automatic tool lifter that can operate at high speeds for many millions of operations is essential to the success of carbides on the shaper.

Like other machine tools, the shaper continues to develop and promises to enlarge its field of operation with the aid of better cutting tools and continually improved designs.

MILLING MACHINE PRACTICE

By Mario Martellotti and Hans Ernst

19. ANALYSIS OF THE PROCESS OF MILLING

FUNDAMENTAL CONCEPT. The milling process may be considered as a method of metal removal in which the portion of the work piece that is to be removed is projected into the space dominated by a rotating body carrying one or more individual cutting teeth. The profile of the finished surface produced on the work piece, therefore, is determined by the intersection of a plane, perpendicular to the motion of the work piece, and the outline of the body of revolution generated by the rotating cutter.

Two main classes of milling operations are:

1. **Face milling,** in which the finished surface is mainly parallel to the face of the cutter.
2. **Peripheral milling** (including slab and form milling), in which the finished surface is mainly parallel to the periphery of the cutter.

CHARACTERISTIC FORM OF MILLING CHIP. Because of the limited period of engagement of each tooth with the work piece, the removal of metal in milling is accomplished by the separation of small individual chips. The rotary motion of the cutter and the translatory motion of the work piece combine to produce a chip whose cross section varies from instant to instant as the direction of motion of a tooth changes with respect to direction of motion of the work piece. Chip thickness is a maximum at the instant that these motions are perpendicular to each other, and substantially zero when they are parallel.

SURFACE GENERATED IN MILLING. In peripheral milling, two distinct methods are used:

Up-milling. The work piece advances toward the cutter from the side where the teeth are moving upward. The cutter rotates against the direction of the feed. The chip thickness is a minimum at the beginning of the cut and a maximum at the end.

Down-milling. The work piece advances toward the cutter from the side where the teeth are moving downward. The cutter rotates in the same direction as the feed. The chip thickness is a maximum at the beginning of the cut and a minimum at the end.

Figure 1 compares the shape of milling chips produced by these two methods, the difference in shape being exaggerated by the use of a large feed per tooth. In both methods the path of a tooth through the work material is a curtate trochoid. Its actual shape varies with changes in rate of feed, diameter of cutter, depth of cut, and cutting speed. Figure 1 shows that the actual rake and clearance angles change continually as the tooth forms a chip. In up-milling, the actual clearance angle θ is greatest at the beginning of the tooth contact with the work, decreasing to a minimum value of θ_1 when the tooth leaves the work. The actual rake angle, however, has a minimum value α at the beginning of the cut, increasing to maximum value α_1 at the point of leaving the work. In down-milling, the actual clearance angle θ is least at the beginning of the cut and greatest, θ_1, at the end. The actual rake angle α is greatest at the beginning of the cut, and least, α_1, at the end.

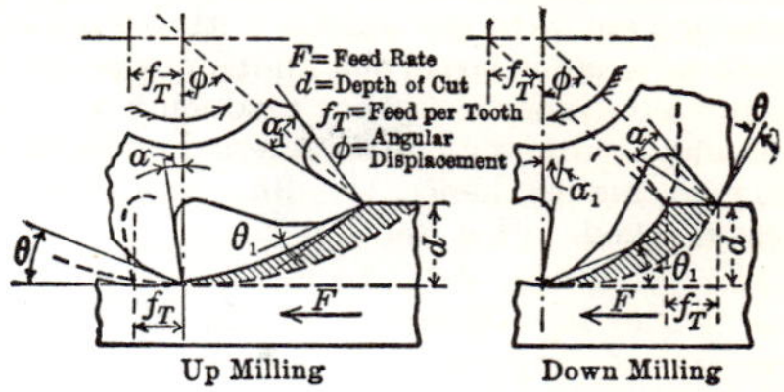

FIG. 1. Comparison of up milling and down milling.

In this respect, milling differs radically from other machining processes, like turning, planing, etc., in which clearance and rake angles remain constant throughout the cut.

CHIP FORMATION. Up-milling and down-milling produce surfaces of entirely different character, the reason lying in the nature of the process of chip formation. When a cutting tool engages a work piece, the material immediately ahead of the cutting edge is at first compressed and escapes by flowing in a direction parallel to the face of the tool. As the tool continues to advance, however, escape in this direction becomes increasingly difficult; the pressure on the tool rises; the stress in the material farther ahead of the tool increases, finally reaching a value great enough to cause rupture or flow in a plane roughly perpendicular to the face of the tool. If the material is relatively brittle, or the thickness of chip is great, a definite rupture occurs in this plane. The entire portion of severed material then moves out of the path of the advancing tool and permits the engagement of a fresh portion. The chip is not a continuous ribbon, but a series of individual segments more or less closely joined. This type of chip is known as a discontinuous chip (Fig. 2).

FIG. 2. Discontinuous chip.

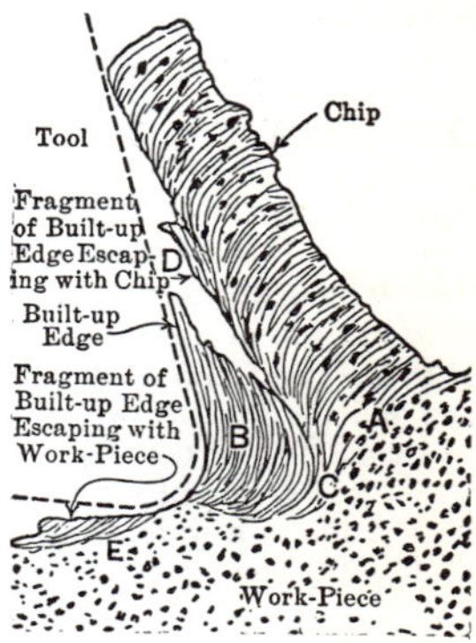

FIG. 3. Continuous chip.

If the material is relatively ductile, definite successive ruptures do not take place; instead, the material ahead of the tool yields by plastic flow over a broad region. As this material (A, Fig. 3) is forced out of the path of the advancing tool, it separates from the initially compressed portion B, which is held by frictional contact with the tool face. The body of the chip passes off as a substantially solid ribbon, with flow lines perpendicular to the tool face. The wedge-shaped portion B, with flow lines parallel to the tool face, remains to form the so-called *built-up edge*. This type of chip is generally known as a continuous chip. The built-up edge is added to continuously by material compressed at point C, but is periodically reduced by fragments, like D and E, torn off the chip and the work piece, respectively.

Roughness of a machined surface is due solely to the presence of fragments like E; consequently, for good finish, their magnitude should be kept to a minimum by so arrangeing conditions that the stable size of the built-up edge itself is a minimum. Such arrangements include: (1) decrease in chip thickness; (2) increase in rake angle; (3) increase in tool sharpness; (4) increase of cutting speed; (5) use of cutting fluid; (6) use of work material of lower ductility. (See Ref. 1.)

In *up-milling* the portion of the finished surface produced by a tooth is formed at the beginning of its contact with the work piece, when the chip thickness is extremely small. Hence, the built-up edge itself is very small, and the resultant surface is smooth and shiny. In *down-milling* an element of the finished surface is produced at the end of the tooth engagement with the work. With a ductile material such as a low-carbon steel, where the built-up edge is large, the built-up edge persists to the extreme end of the chip; thus fragments continue to escape, producing a somewhat torn finish similar to that obtained by planing or shaping. With a less ductile material, such as a high-carbon steel, the built-up edge is smaller; hence, the difference in finish between up-and-down methods of milling is less marked. (See Ref. 2.)

20. MILLING CUTTERS

Cutting materials generally used for milling cutters are of these types: (1) (*a*) carbon tool steel and (*b*) alloy tool steel (high-speed steel); (2) cast nonferrous alloys; and (3) sintered carbides.

CARBON AND HIGH-SPEED STEEL-CUTTING MATERIALS. Carbon tool steel and high-speed steel milling cutters are generally made in accordance with the standards approved by the American Standards Association. See Ref. 3 for discussions of the most commonly used types.

The classifications adopted are listed below. Although carbon tool steel is still used to some extent for screw slotting and slitting cutters, and other light-duty work, practically all heavy-duty milling cutters of these types are made of high-speed steel.

For most classes of work, high-speed steel cutters can be run 2 to 2 $^1/_2$ times as fast as carbon steel cutters.

A. CLASSIFICATION BASED ON RELIEF OF TEETH. 1. **Profile cutters** include all forms of cutters which are sharpened by grinding on the periphery of the teeth, the clearance angle (or relief) being obtained by grinding a narrow land back of the cutting edge. Where the cutting edges are of curved or irregular shape, they are designated *shaped profile cutters.*

2. **Formed cutters** include all cutters where the eccentric relief back of the cutting edge is of the same contour as the cutting edge itself. These cutters are sharpened by grinding the face of the teeth.

B. CLASSIFICATION BASED ON METHOD OF MOUNTING. 1. **Arbor Cutters.** A cutter with hole for mounting on arbor.

2. **Shank Cutters.** All cutters having either a straight or taper shank integral with the cutter.

3. **Face Milling Cutters.** All cutters designed to be attached directly to the spindle nose, or to the end of a short arbor known as a stub arbor.

C. CLASSIFICATION BASED ON TYPE AND USES. The commonly used types of cutters and the work to which they are adapted are discussed below.

The plain milling cutter (Fig. 4) is a cylinder with teeth on the circumferential surface only. It is used to produce a flat surface parallel to the axis of the cutter. Plain milling cutters are made in a wide variety of diameters and widths, for the various requirements of slab milling. They generally have helical teeth. The helical form enables each tooth to take a cut gradually, thus reducing shock and minimizing tendency to chatter. Cutters with a helix angle of 25 to 45 degrees, commonly, but incorrectly, are termed *spiral mills;* if helix angle is more than 45 degrees the cutter generally is known as a *helical mill.*

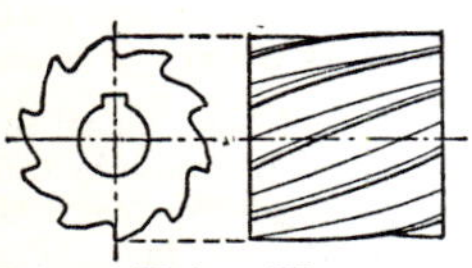

FIG. 4. Plain milling cutter.

The helical mill, which may be either of the shank or hole type, is free cutting and particularly desirable for light cuts on soft steel or brass. It can be used to advantage for profiling work, such as cam milling, and for milling intermittent surfaces. The shank type, with pilot end, commonly is used for elongating slots. For general slab milling, the helical mill is not as efficient as the common "spiral" mill.

The side milling cutter is a plain milling cutter of cylindrical form with teeth on the circumferential surface and on both sides. The side teeth extend a portion of the distance from the circumference to the axis. Half side and interlocking side milling cutters also are made. Side milling cutters are used in a large variety of work. Two or more of them often are placed on the same arbor with a space between them; they are then known as straddle mills.

Straddle mills are used to advantage where the work is to be milled on two parallel sides, e.g., bolt heads and tongues.

Interlocking cutters are used to mill slots to a standard width. They are maintained at a constant width by thin shims or collars between the inner hubs.

Staggered tooth milling cutters (Fig. 5) are cylindrical cutters with the cutting teeth on the circumferential surface only. Alternate teeth are of opposite helix angle. (See Fig. 5.) The side teeth extending from the circumference a short distance towards the axis are for chip clearance only. They are not ground for cutting purposes. This type of cutter, used to obtain exact width of slots, is the most efficient type for milling slots where depth exceeds width. Because of the alternate right- and left-hand helix angle of the teeth, with considerable undercut, these cutters can remove large amounts of metal without vibration or chatter. The free-cutting action makes possible an increased feed and speed, without detriment to the production of smooth, accurate work.

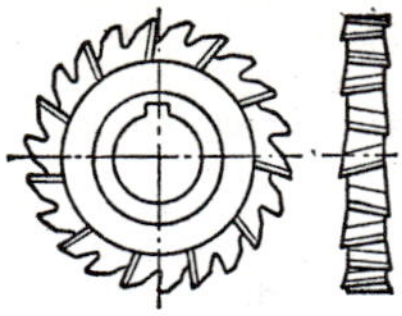

FIG. 5. Staggered tooth cutter.

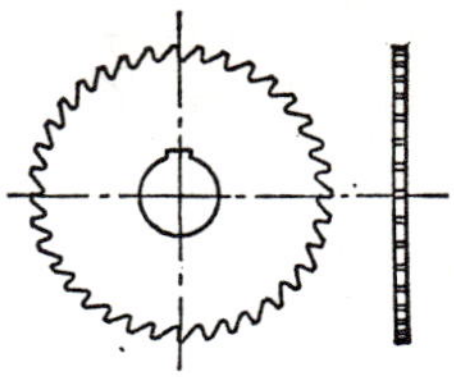

FIG. 6. Slitting saw.

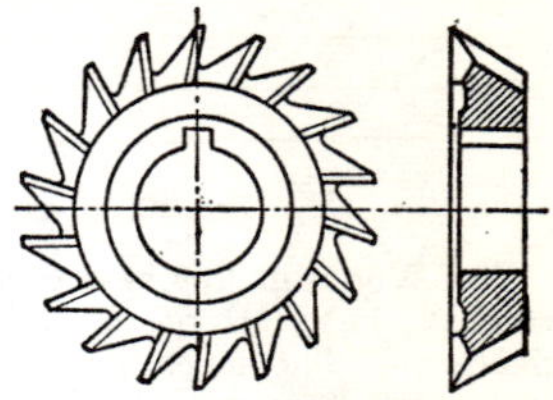

FIG. 7. Single-angle cutter.

Metal slitting saws (Fig. 6) are plain milling cutters with sides relieved or "dished" to give side clearance. They generally are made $^3/_{16}$ in. thick or less, and usually have more teeth for a given diameter than a plain milling cutter. They are used to cut off work, or to mill very narrow slots. Slitting saws also are made with side teeth similar to side milling cutters, and with staggered teeth, similar to staggered tooth milling cutters. The metal slitting saw with staggered teeth usually is made $^3/_8$ to $^3/_{16}$ in. thick, and is used for heavy sawing in steel.

Angle milling cutters are made both **single angle** and **double angle,** with teeth on the conical surfaces. (See Figs. 7 and 8.) Single-angle cutters are made both with and without teeth on one or both of the flat sides. Angle cutters are used for various fluting operations, or for milling the edge of a piece to a given angle.

Formed cutters usually have a curved tooth outline and are used in milling contours of various shapes. They may be divided into two classes: (*a*) the shaped or formed profile cutters and (*b*) the form or cam relieved cutters. They can also be classified as profile-ground and face-ground formed cutters, respectively, according to the method of sharpening.

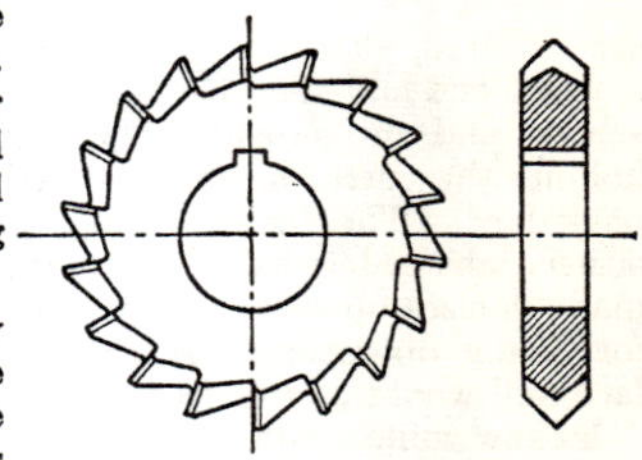

FIG. 8. Double-angle cutter.

(*a*) *Shaped or formed profile cutters* are sharpened by grinding a small land back of the cutting edge, as in the case of plain milling cutters, and the contour of the cutter must be reproduced every time the cutter is resharpened.

(*b*) *Formed or cam relieved cutters* are so termed in contradistinction to shaped profile cutters because the clearance back of the cutting edge is produced by a form or master tool in a cam relieving machine, giving all the teeth the contour of the master tool. They are sharpened by grinding the face of the teeth.

So long as the face of a tooth is maintained in its original plane with respect to the axis of rotation, the contour of the tooth will remain unchanged.

This type of cutter includes gear cutters, multiple-gear cutters, sprocket cutters, convex cutters, concave cutters, corner rounding cutters, spline cutters, etc.

Formed and shaped cutters are used for the accurate duplication of varying outlines or shapes and make possible economical milling of complicated contours.

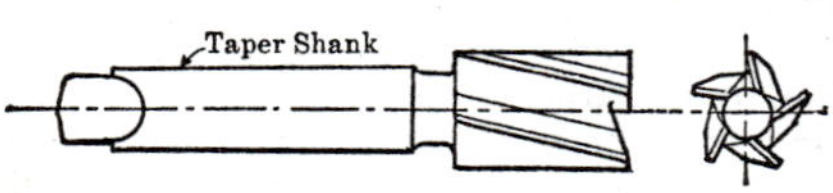

FIG. 9. End mill.

End mills (Fig. 9) are cutters with teeth on the circumferential surface and on one end. The teeth may be either parallel to the axis of rotation or helical, and either right or left hand. End mills of moderate angle commonly are called spiral end mills. End mills are made of five general types: solid end mill; two-lip end mill or slotting mill; shell end mill; hollow mill; helical end mill.

Solid end mills are used for light milling operations, as the milling of slots, profiling, and facing narrow surfaces. The teeth are integral with the shank, which may be either straight or tapered.

Two-lip end mills comprise a shank cutter with two cutting teeth on the circumferential surface and end teeth cut to the center. Flutes are either straight or helical. The cutter can be sunk into the material and then fed longitudinally. A depth of cut equal to one-half the diameter of the mill usually can be taken from the solid stock.

T-slot Cutter. A shank cutter designed for milling T slots. (See Fig. 10.) The teeth are on the circumferential surface and both sides. In making a T slot, a groove first is cut with an ordinary side milling cutter or a two-lipped end mill. The wide groove at the bottom then is cut with the T-slot cutter.

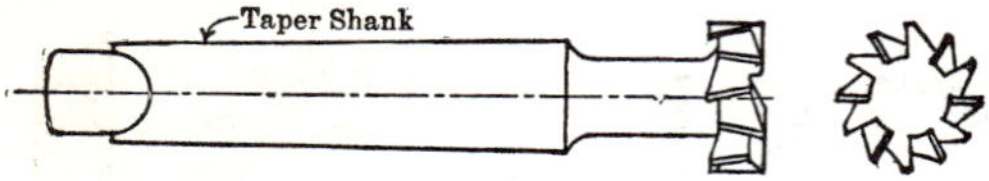

FIG. 10. T-slot cutter.

The Woodruff key seat cutter is made in the *shank* and *arbor* types. The shank type generally has teeth on the circumferential surface only, with the sides slightly relieved for clearance. The arbor type is generally used in sizes larger than 2 in. in diameter. These cutters also are made with staggered teeth. They are made with teeth on both sides which are ground for clearance only and not for cutting. Both types of cutters are used for the specific purpose of milling semi-cylindrical keyways in shafts, to permit the use of Woodruff keys.

A fly cutter consists of a single cutting tool, held and rotated by an arbor. It is used in the experimental room or tool room. The fly cutter can be formed exactly to any desired shape.

Face mill has a cylindrical body with slots on its periphery. Blades or teeth of high-speed steel are inserted in the slots. The teeth are ground to cut both on the circumferential surface and the side, as in the *side mill.* These cutters are adapted to be attached directly to the spindle nose or to the end of a stub arbor. They are made in *light-duty* and *heavy-duty* types. The *light-duty* has a lighter body and a large number of teeth, being used principally for finishing cuts. The *heavy-duty* face mill has a stronger, heavier body and fewer blades and is employed for heavy-duty roughing cuts.

Face mills are used to mill large surfaces and are highly efficient in the removal of metal.

Like face mills, almost all types of milling cutters may be made with inserted teeth. The inserted tooth cutter is most economical and useful when cutters of large dimensions are required, although they are obtainable also in diameters as low as 3 in.

Shell-end mill (Fig. 11), a face mill of the solid type with teeth cut on the circumferential surface and on one end. The tooth end is recessed to receive a nut or a screw head for holding the cutter on the stub arbor. It usually is driven from the key slot across the back face. The teeth usually are helical and either right or left hand. This type of cutter is made in sizes up to 6 in. in diameter and is used for facing operations, where a small diameter face mill would be required.

Hollow mill, a cutter of tubular construction with teeth on one end, and having an internal clearance. The internal clearance sometimes is obtained by a plain tapered hole with back taper and sometimes by internal cleared flutes. The hollow mill generally is used for producing bosses or cylindrical projections from solid bodies.

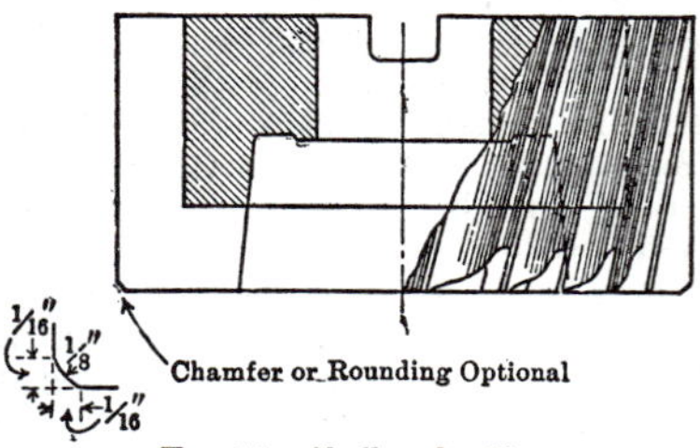

FIG. 11. Shell end mill.

CAST TOOL MATERIALS are produced by fusing various alloys at high temperature.

Cast High-Speed Steel. Milling cutters are also made of cast high-speed steel. Its characteristics lie between those of forged high-speed steel and sintered carbide. The proper physical characteristics are obtained by heat treatment.

Its hardness varies between 63 and 70 Rockwell C, depending on heat treatment. Because it has greater red hardness than forged high-speed steel, this material can be run at higher cutting speeds and is suitable for milling diversified materials.

Cast nonferrous tool materials are cast to desired shape, are fully hard after cooling at room temperature, and are finished to size by grinding. They are known under such various names as Stellite, Tantung, and Crobalt.

These materials are used as inserted blades in a variety of milling cutters, especially those of the face mill type. They are used for milling a variety of materials and can be run at higher cutting speeds than forged high-speed steel.

SINTERED CARBIDE TOOL MATERIAL is extremely hard at both room temperature and the temperature attained by the cutting edge during cutting. Cutting speeds are three to ten times as fast as with high-speed steel cutters with a comparable increase in tool life between resharpenings.

Sintered carbide is used in the form of tips, brazed to inserted blades, directly to the cutter body or bits mechanically clamped.

Made in different grades to suit operating conditions and material cut, sintered carbide consists of a mixture of powdered carbide of tungsten or of tungsten, tantalum, and titanium carbides in various combinations, with cobalt or nickel as low-melting-point cementing or binding material.

As the binder content is increased, the material becomes tougher and vice versa. Greater toughness, however, is obtained at the expense of hardness. Hardness of various grades may vary between 80 and 93 Rockwell A. Cutting fluids are not used with carbide materials.

Basic types—tungsten carbide is suitable for machining cast iron, mild steel, bronze, aluminum, magnesium, and alloys of these metals, as well as nonmetallic materials such as workable plastics, Bakelite and hard rubber. A generally used grade has 87% tungsten carbide with 13% cobalt binder. Rockwell A hardness varies between 87 and 87.5.

Tantalum carbide is used for machining carbon steels and alloy steels up to 250 Bhn. A typical composition consists of 75% tantalum carbide, 13% tungsten metal, and the remainder binder. It has a Rockwell A hardness of approximately 88.

Tungsten—titanium carbide, suitable for machining steels and alloy steels up to 500 Bhn, is predominantly tungsten carbide, with titanium carbide and 13% binder material. The Rockwell A hardness is approximately 90.

The harder grades of carbides are more wear resistant, but also display more of a tendency to chip under impact.

DIRECTION OF ROTATION. A cutter is *right-hand* if it rotates counterclockwise, and *left-hand* if it rotates clockwise, when viewed from the front end as mounted on the spindle.

NUMBER OF TEETH. Milling cutter teeth are equally spaced and project from the body of the cutter either radially as in helical mills, or both radially and axially as in face mills. The teeth may either be an integral part of the cutter or be inserted in the body in the form of inserts or blades. No standard rule exists for the number of teeth in milling cutters.

High-speed steel cutters generally are of the "coarse-tooth" type. The coarse-tooth cutter provides a great chip space between successive teeth, which permits a free flow of the chip and prevents clogging of the teeth. The efficiency of the coarse-tooth cutter is considerably higher than that of the old-style fine-tooth cutter. (See Refs. 4 and 5.) In metal-slitting saws, the number of teeth varies according to the type of operation. Coarse-tooth saws having side teeth for chip clearance, as in side mills, are used when cutting deep slots. Saws with finely pitched teeth are used for cutting thin material. Special formed saws, with teeth ground as in formed relieved cutters, are used for slitting copper and other soft nonferrous materials. Formed relieved cutters are of the coarse-tooth type to provide sufficient material for a large number of resharpenings before the cutter is used up.

Cast nonferrous tool materials cutters have approximately the same number of teeth as cutters made of high-speed steel.

Sintered Carbide-tipped Cutters. The number of teeth can be calculated for given values of feed per tooth, depth of cut, and width of cut, the object being maximum power utilization without overloading the driving motor, by means of the following formula:

$$T = \frac{KHp}{F_t N d w}$$

where T = number of teeth; K = constant, which may be taken equal to 0.650 for steel, 1.5 for cast iron, and 2.5 for aluminum. (These are conservative values which take into consideration the dulling of the cutter in service.) Hp = horsepower available at the cutter; F_t = feed per tooth, inches; N = cutter revolutions per minute; d = depth of cut, inches; and w = width of cut, inches.

In general, the number of teeth in a milling cutter should be selected on the basis of all-round good performance. The following rule has been found to give satisfactory cutter performance, strong tooth section, ample chip space, economy in cost and maintenance, and good cutting efficiency, and yet assure a continuous cutting action:

> Provide the cutter with a number of teeth so that no more than two teeth are engaged in the cut at any time.

On this basis, the following formula can be used for obtaining the number of teeth in face milling cutters:

$$T = \frac{\pi D n}{w}$$

where T = number of teeth; D = cutter diameter, inches; n = number of teeth in contact; and w = width of cut, inches.

For peripheral milling cutter:

$$T = \frac{4\pi D}{D + 4d} \cos a$$

where T = number of teeth; D = cutter diameter, inches; d = depth of cut; and a = helix angle, degrees.

21. OPERATION OF MILLING CUTTERS

CUTTING SPEED. Because of the countless variations in conditions encountered in milling practice, fixed rules for cutting speeds are impossible. The most effective speeds are governed by the structure and hardness of the part to be machined, the amount of metal to be removed, the type of cutting material, etc. In general, cutting speed is selected to give the best compromise between maximum production and maximum life of the cutter.

Approximate values of the cutting speed for different cutting materials, based on the Brinell hardness number or tensile strength of the material being cut, are obtained from Figs. 12, 13, and 14.

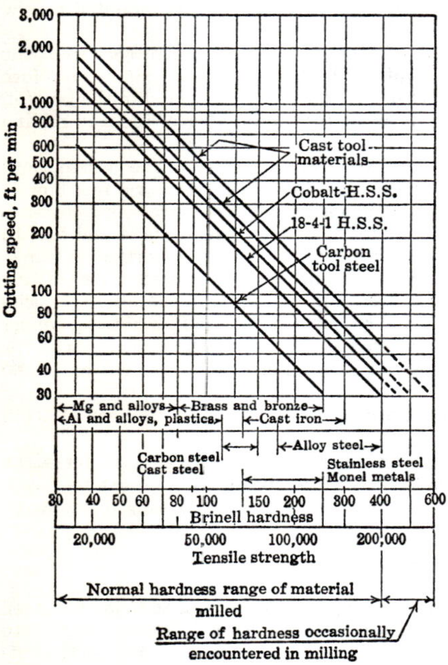

FIG. 12. Cutting speeds for different cutting tool and work materials.

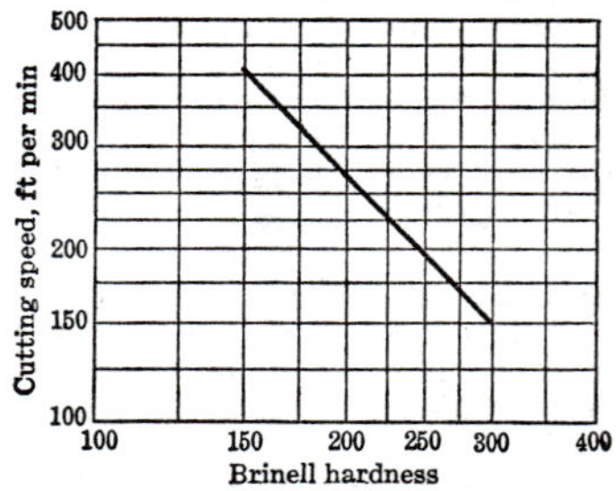

FIG. 13. Cutting speeds in milling cast iron with carbide-tipped cutters.

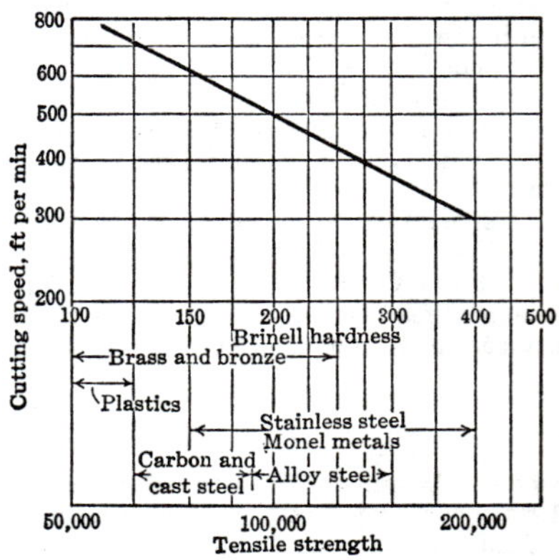

FIG. 14. Cutting speeds for milling steel with carbide-tipped cutters.

In milling with sintered carbide-tipped cutters, there is considerable variation in the cutting speed according to the grade of carbide used, the cutter design, the material being cut, and the nature of the operation.

The selected cutting speed may have to be adjusted in the actual operation to provide satisfactory cutter life, good finish of the milled surface, and operation of the machine within the power available in the driving motor.

Magnesium and aluminum alloys are milled with sintered carbide-tipped cutters within a wide range of cutting speeds. Cutting speeds of 3500 to 12000 ft per min and higher have been used when milling these materials. In general, the selection of the cutting speed is determined by practical consideration, such as the nature of the work, the setup of the machine; and the spindle bearing design and the power available in the machine.

The cutting speed for finishing milling often is 40 to 80% higher than the roughing speed.

FEED RATE. The rate of production in milling depends directly on the rate at which the work passes the cutter. This value in inches per minute may be computed from

$$F = F_t T N \tag{1}$$

where F_t is the feed per tooth, T the number of teeth, and N the rpm of the cutter. Selection of the proper feed rate for milling a given work piece involves several factors, among them the limitations of machine, cutter, work piece, and fixture; the finish desired, etc.

Estimates of feed rates usually are based on the assumption of a certain value of F_t, the feed per tooth.

Table 1. Suggested Feed per Tooth for High-speed Steel Milling Cutters

(The Cincinnati Milling Machine Co.)

Material	Face Mills	Helical Mills	Slotting and Side Mills	End Mills	Form Relieved Cutters	Circular Saws
Plastics	.013	.010	.008	.007	.004	.003
Magnesium and alloys	.022	.018	.013	.011	.007	.005
Aluminum and alloys	.022	.018	.013	.011	.007	.005
Free-cutting brasses and bronzes	.022	.018	.013	.011	.007	.005
Medium brasses and bronzes	.014	.011	.008	.007	.004	.003
Hard brasses and bronzes	.009	.007	.006	.005	.003	.002
Copper	.012	.010	.007	.006	.004	.003
Cast-iron, soft (150–180 Bh)	.016	.013	.009	.008	.005	.004
Cast iron, medium (180–220 Bh)	.013	.010	.007	.007	.004	.003
Cast iron, hard (220–300 Bh)	.011	.008	.006	.006	.003	.003
Malleable iron	.012	.010	.007	.006	.004	.003
Cast steel	.012	.010	.007	.006	.004	.003
Low-carbon steel, free machining	.012	.010	.007	.006	.004	.003
Low-carbon steel	.010	.008	.006	.005	.003	.003
Medium-carbon steel	.010	.008	.006	.005	.003	.003
Alloy steel, annealed (180–220 Bh)	.008	.007	.005	.004	.003	.002
Alloy steel, tough (220–300 Bh)	.006	.005	.004	.003	.002	.002
Alloy steel, hard (300–400 Bh)	.004	.003	.003	.002	.002	.001
Stainless steels, free machining	.010	.008	.006	.005	.003	.002
Stainless steels	.006	.005	.004	.003	.002	.002
Monel metals	.008	.007	.005	.004	.003	.002

Table 2. Suggested Feed per Tooth for Sintered Carbide-tipped Cutters

(The Cincinnati Milling Machine Co.)

Material	Face Mills	Helical Mills	Slotting and Side Mills	End Mills	Form Relieved Cutters	Circular Saws
Plastics	.015	.012	.009	.007	.005	.004
Magnesium and alloys	.020	.016	.012	.010	.006	.005
Aluminum and alloys	.020	.016	.012	.010	.006	.005
Free-cutting brasses and bronzes	.020	.016	.012	.010	.006	.005
Medium brasses and bronzes	.012	.010	.007	.006	.004	.003
Hard brasses and bronzes	.010	.008	.006	.005	.003	.003
Copper	.012	.009	.007	.006	.004	.003
Cast iron, soft (150–180 Bh)	.020	.016	.012	.010	.006	.005
Cast iron, medium (180–220 Bh)	.016	.013	.010	.008	.005	.004
Cast iron, hard (220–300 Bh)	.012	.010	.007	.006	.004	.003
Malleable iron	.014	.011	.008	.007	.004	.004
Cast steel	.014	.011	.008	.007	.005	.004
Low-carbon steel, free machining	.016	.013	.009	.008	.005	.004
Low-carbon steel	.014	.011	.008	.007	.004	.004
Medium-carbon steel	.014	.011	.008	.007	.004	.004
Alloy steel, annealed (180–220 Bh)	.014	.011	.008	.007	.004	.004
Alloy steel, tough (220–300 Bh)	.012	.010	.007	.006	.004	.003
Alloy steel, hard (300–400 Bh)	.010	.008	.006	.005	.003	.003
Stainless steels, free machining	.014	.011	.008	.007	.004	.004
Stainless steels	.010	.008	.006	.005	.003	.003
Monel metals	.010	.008	.006	.005	.003	.003

In practice the feed per tooth is made as large as possible, consistent with these limitations and the permissible tooth load. Tables 1 and 2 give average values of permissible feed per tooth for various combinations of work material and types of milling cutters for high-speed steel and sintered carbide-cutting materials. For cast nonferrous tool materials the values of the feed per tooth are the same as those for sintered carbide. These values must be reduced when cutter, or work, is of a relatively fragile nature.

DIAMETER OF CUTTER usually is determined by the limitations of the work piece, either in regard to the size of the surface to be milled or interference with adjacent portions of work or fixture. In the case of cutters of small diameters, the limitation frequently is the size of the arbor required to support it properly.

CLEARANCE ANGLE (Relief). The clearance angle is that angle included between the land on the back or flank by a milling cutter tooth, and the tangent to the periphery of the cutter at the cutting edge. Its sole function is to prevent undue interference between the land and the surface being generated on the work piece by the cutting edge.

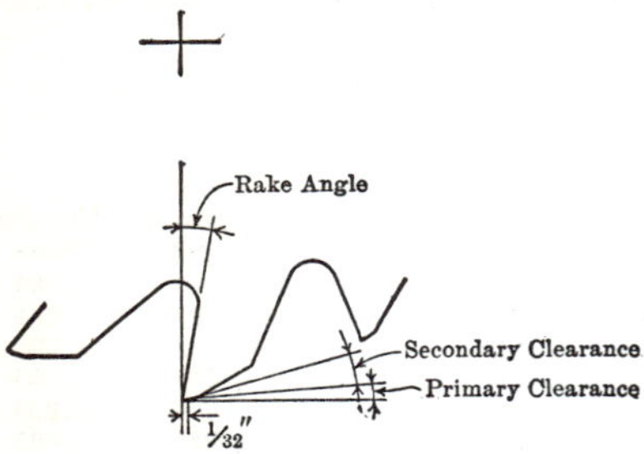

FIG. 15. Clearance and rake angles, and land.

As shown in Fig. 15, the clearance angle is sometimes referred to as the *primary clearance*, since a *secondary clearance* is usually provided when sharpening cutters to restore the keenness of the cutting edge.

The clearance angle should be carefully selected in all types of cutters. The value of the clearance angle plays an important role in obtaining good cutter performance, high cutting efficiency, good finish, and long cutter life. Any clearance greater than the minimum amount required increases the likelihood of chatter and reduces the life of the tool.

For general-purpose work, milling cutters $^1/_8$ to 3 in. in diameter have clearance angles of 13 to 5 degrees, respectively, decreasing proportionately as the diameter increases. In cutters over 3 in. in diameter, cutter manufacturers usually provide a clearance of 5 degrees.

The land is usually about $^1/_{64}$, $^1/_{32}$, and $^1/_{16}$ in. wide in cutters of small, medium, and large diameter, respectively.

The value of the clearance angle is also changed according to the material of the work piece. Generally used values are given in Table 3.

Table 3. Milling Cutter Clearance Angles for Various Work Materials

(The Cincinnati Milling Machine Co.)

Material to Be Cut	Clearance Angle, degrees
Low-carbon steel	3–5
High-carbon and alloy steels	3–5
Cast iron	4–7
Medium and hard bronze	4–7
Brass and soft bronze	10–12
Aluminum, magnesium, plastics, etc.	10–12

RAKE ANGLE. In *peripheral milling cutters*, the rake angle or "undercut" is generally defined as the angle in degrees measuring the deviation of the tooth face from a radial line to the cutting edge. This is the radial rake angle (Fig. 15).

In *face milling cutters*, the teeth are inclined with respect to both the radial and axial lines (Fig. 16). These angles are called *radial* and *axial rake*, respectively.

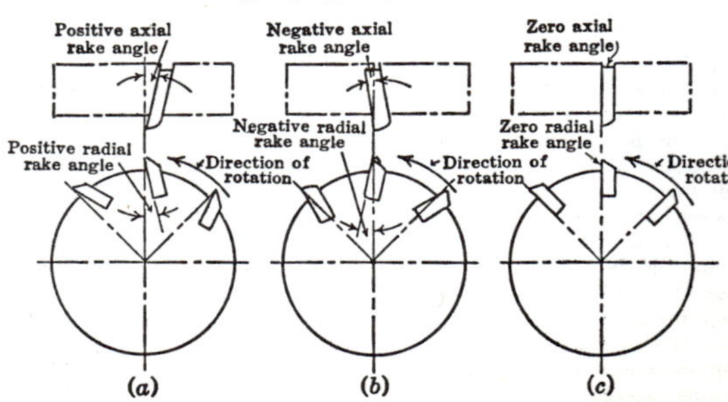

FIG. 16. Radial and axial rake angles in face milling cutters.

The rake angles may be *positive*, *zero*, or *negative*.

Selection of Rake Angles. *High-speed Steel.* Positive rake angles should be used in all types of high-speed steel milling cutters. Angles of 10 to 15 degrees are used for milling steels of various compositions. For softer materials such as aluminum and magnesium, the rake angle may be increased to 25 degrees or more.

In special milling cutters, the rake angle is selected to suit the particular operation.

Metal slitting saws have rake angle of 5 to 10 degrees. Zero and sometimes slightly negative rake angles are used in saws operating on copper or other soft nonferrous metals to counteract the tendency to "hog in."

In standard *formed relieved* cutters, the rake angle may be 10, 5, or zero degrees.

Face milling cutters usually have 10-degree positive radial and axial rake angles, for milling cast iron, steels, and nonferrous metals such as bronze and brass. It may be increased to 25 degrees or more when milling aluminum and magnesium. Rake angles for face milling cutters are given in Table 4.

Table 4. Face Milling Cutter Rake Angles for Various Cutting Materials and Work Materials

(The Cincinnati Milling Machine Co.)

Material Cut	Cutter Material			
	High-speed Steel		Cast Nonferrous Tool Material	
	Radial Rake, degrees	Axial Rake, degrees	Radial Rake, degrees	Axial Rake, degrees
Soft cast iron	10–15	10–15	6–8	6–8
Mild steel	10–15	10–15	3–6	3–6
Hard cast iron	10	10	3–6	3–6
Hard alloy steel	10	10	0–3	0–3
Aluminum and alloys	20–35	20–35	10–15	10–15
Magnesium and alloys	20–35	20–35	15–25	15–25
Yellow brass and bronze	10	10	5	5

CAST NONFERROUS TOOL MATERIALS. Positive rake angles are generally provided on milling cutters using cast nonferrous tool materials. Negative rakes may be employed for some operations where shock loads are encountered or when it is necessary to eliminate vibration when milling thin sections. Values of rake angles for face milling cutters are given in Table 4.

Sintered Carbide. *Face Mills.* Recent developments in the design of carbide-tipped face mills have indicated five principal angles which control cutting performance: (1) axial rake, (2) radial rake, (3) true rake, (4) inclination, and (5) corner angle. These angles are shown in Fig. 17. (See Ref. 6.)

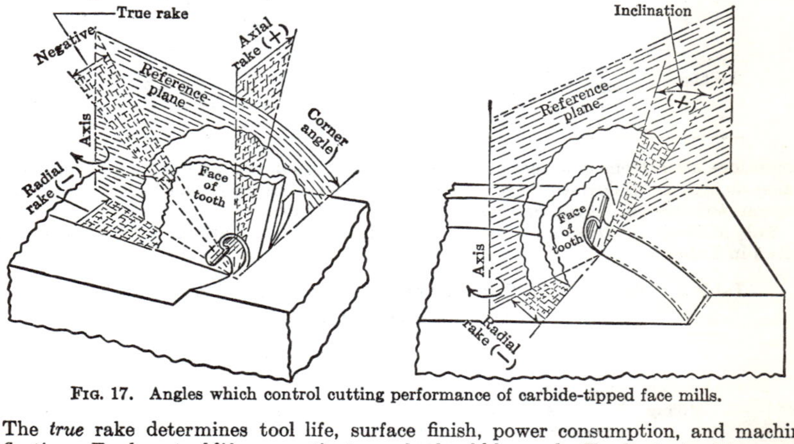

Fig. 17. Angles which control cutting performance of carbide-tipped face mills.

The *true* rake determines tool life, surface finish, power consumption, and machine deflection. For long tool life, a *negative true rake* should be used. To eliminate chip crowding by directing them away from the cutter body, a *positive inclination* should be used. A number of rake-angle combinations can be used to obtain desirable results (see Fig. 18).

A desirable chip flow is obtained by using *positive axial rakes* in combination with *negative radial rakes* and *large corner angles*.

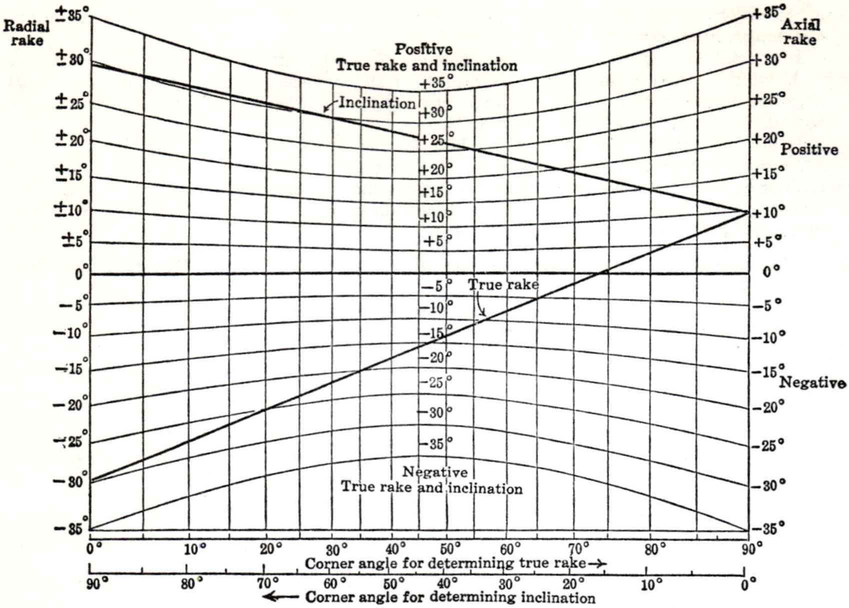

FIG. 18. True rake and inclination of face milling cutter teeth. On left-hand scale use upper sign for true rake, lower sign for inclination.

Milling Cutter Angles for General-purpose Milling of Steel. *Face mills* should have a *negative true rake*, a positive inclination, and a corner angle of at least 45 degrees. The value of the negative true rake should increase with the hardness of the steel. Combinations which have proved satisfactory in practice are given in Table 5. Where the cutting

Table 5. Face Mills

Recommended Rake Angles for Steel

(The Cincinnati Milling Machine Co.)

		Axial Rake	Radial Rake	Corner	True Rake	Inclination
Face Milling	Low-carbon and Annealed Alloy Steels	+5	−12	45	−5	+12
	Hardened Alloy Steels	+5	−18	45	−10	+16

conditions are extremely severe, because of hard inclusion or chip crowding, true negative rakes as high as 25 degrees may be used. When high feed rates can be used, it is advantageous to increase the corner angle to as high as 75 degrees.

Slab Milling, Slotting, and Sawing. Satisfactory combinations of tooth angles are given in Table 6.

Table 6. Tooth Angles Combinations for Slab Milling, Slotting, and Sawing

(The Cincinnati Milling Machine Co.)

		Axial Rake	Radial Rake	True Rake
Slab Milling	Low-carbon and Annealed Alloy Steels	5	−5	−5
	Hardened Alloy Steels	10	−10	−10
Slotting and Sawing	Low-carbon and Annealed Alloy Steels	0	−5	−5
	Hardened Alloy Steels	0	−10	−10

A general grade of sintered carbide for milling steel is represented by Carboloy 78-B, Kennametal KM, Vas-Ramet EM, and Firthite TA.

Softer and harder grades can be obtained. The harder grades are more wear-resistant but also have a tendency to chip under impact or vibration.

Milling Cutter Angles for General-Purpose Milling of Cast Iron. Face mill tooth angles vary over a wide range. Values of true rakes for any combination of axial rake, radial rake, and corner angle may be obtained from Fig. 18. For average conditions, the true positive rake should vary from 4 to 10 degrees. Corner angles, from 30 to 45 degrees. A face mill which allows high feed rates and minimizes break out of work piece at the edge where cutter leaves the work has corner angles of 60 to 75 degrees. One combination of angles that has proved satisfactory is: Axial rake $+15°$; Radial rake $-30°$; corner angle 75°; true rake 6°; inclination $+32°$.

A general grade of sintered carbide for milling cast iron is Carboloy 44A, Kennametal K6, Vas-Ramet 2468, and Firthite H.

Milling Cutter Angles for Milling Light Alloys. For aluminum and magnesium alloys, positive rake angles are generally used, ranging from 10 to 20 degrees for aluminum and 15 to 25 degrees for magnesium.

Flywheels. In all carbide applications it is now the practice to use flywheels mounted as close as possible to the milling cutter to avoid tip breakage due to vibrations.

CUTTING FLUIDS play an important role in the majority of metal-cutting operations. They are almost invariably used in milling operations on steel, alloy steel, brass, and bronzes with high-speed steel cutters. Aluminum and magnesium alloys and cast iron are mostly milled dry, although cutting fluids are used in some particular cases.

With sintered carbide-tipped cutters, cutting fluids are generally not used.

Compressed air or a rotating wire brush is useful in removing the chips off the cutter. In milling aluminum and magnesium an abundant supply of cutting fluid is used to wash away the chips.

In addition to reducing the temperature of cutting, cutting fluids prevent adhesion between the chip and the cutter teeth and improve the quality of finish.

The cutting fluids most generally used may be grouped as follows:

1. Emulsions of water and other elements such as waxes, mineral oil containing soap or sulfur, and pastes, with added antirust compounds.
2. Cutting oils, which may be straight mineral oils, animal and vegetable oils, or compounded oils having a sulfurized or sulfurized-chlorinated base.
3. Air supplied in the form of a jet.

In order to be of practical value in the shop, a cutting fluid should have certain desirable characteristics:

1. It should have a high cooling capacity to reduce cutting temperature, increase cutter life and production, and improve dimensional accuracy.
2. It should prevent adhesion between the material of the chip and the cutter teeth to improve chip formation and surface finish.
3. It should be suitable for a variety of milling operations on different work materials to be cut, thus reducing the number of cutting fluids used in the shop.
4. It should have long life and should not emit obnoxious odors or vapors harmful to the operator.
5. It should not cause rust or corrosion.
6. It should have low viscosity, to permit gravity separation of impurities and chips collected by the fluid as it is circulated in the cooling system.
7. It should not oxidize or form gummy deposits to clog the circulating system and thus halt production while the system is cleaned.
8. It should permit a clear view of the work, which is especially desirable in precision work.
9. It should be safe, particularly in regard to fire and accident hazards.
10. It should not cause skin irritation.

Of the variety of cutting fluids available, water emulsions may be prepared to possess most of the desirable properties that a cutting fluid should have, and accordingly are used in milling operations.

COMPARISON OF UP-MILLING AND DOWN-MILLING. At various times during the past 50 years down-milling (or milling with the feed) has been advocated for particular classes of work, in view of certain advantages inherent in this method. In recent years this method has been used with special types of feed systems, designed to prevent an overrunning of the milling machine table due to the action of the cutter on the work.

When thin metal slitting cutters are used, the likelihood of producing crooked slots is less when milling down than when milling up. In milling the flutes in twist drills, and other similar operations, a longer tool life and more desirable finish have been obtained when milling down, and the holding of the work piece during the milling operation is greatly simplified. In carbide applications, the down-milling method should be used wherever possible. The greatest practical objection to the down-milling method for heavy milling work is the danger of a serious accident if the work piece or fixture should be insecurely fastened to the table.

From the standpoint of power required, there is little to choose between the two methods of milling. Careful tests have shown in most cases that there is very little difference between the power actually used by the cutter when removing chips by either up-milling or down-milling.

POWER REQUIRED FOR MILLING is a function of many factors, including the cutting speed, the feed, the material being cut, the rake angle, the sharpness of the cutter, the depth of cut, the smoothness of the face of the teeth, and the type of cutting fluid used.

An empirical formula that can be used for an approximate determination of the power required in milling is

$$\text{Hp} = aV^{3/4} \tag{2}$$

where Hp = horsepower actually used in cutting; a = a constant which is different for different materials, values being given in Table 7; V = rate of stock removal, cubic inches per minute, equal to

$$V = FWD \tag{3}$$

where F = feed rate; W = width of cut; and D = depth of cut. The power required can also be evaluated from Fig. 19.

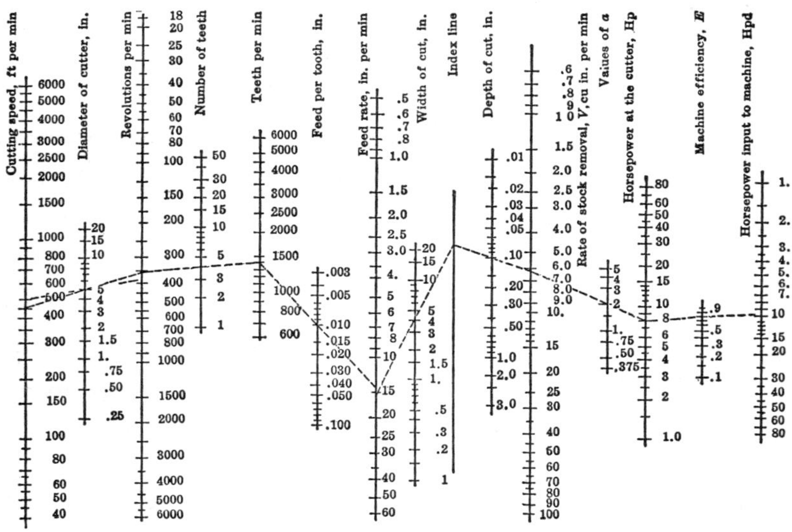

FIG. 19. Alignment chart for calculating power required in milling.

In order to obtain the total actual power required to perform a milling operation in any given type of milling machine, it is necessary to add to the result of calculation by eq. 2, the idle power required to drive the machine. Idle power varies in relation to the type of milling machine used. However, it generally is found that the idle power is approximately constant within the range of loads encountered in ordinary milling machine operation.

TYPICAL MILLING JOBS. SPEEDS. FEEDS. The following are representative examples of good commercial practice:

Automotive Practice. (1) Rough mill cylinder blocks, top and bottom, manifold, bearing cap slots, and valve cover pads, using Stellite tools: Cutting speed, 110 ft per min; feed, 10.5 in. per min; depth of cut, $1/8$ in.; production, 13.6 pieces per hr; pieces per grind, 200.

(2) Mill clutch housing on cylinder blocks, using cemented carbide tools: Cutting speed, 250 ft per min; feed, 12 $1/2$ in. per min; depth of cut, $1/32$ in.; production, 60 pieces per hr; pieces per grind, 6720.

Table 7. Range of Values of *a* Obtained in Various Milling Operations with High-speed Steel and Sintered Carbide Cutters

(The Cincinnati Milling Machine Co.)

Type of Milling Cutter	Cutting Material	Material Cut		Values of *a*		
		Kind	Brinell Hardness Number	Minimum	Maximum	Average
1. Face mills; shell end mills Plain mills; 50° and 70° Helical mills	High-speed steel	SAE 4345 Steel SAE 4150 Steel NE 8949 Steel SAE 3145 Steel	330 190 240 180	0.90 2.00	1.45 2.50	1.25 2.25
2. Face mills	High-speed steel	Cast iron	170	0.51	1.01	0.76
3. Plain mills	High-speed steel	Cast iron	170	1.15	1.45	1.30
4. Form relieved cutters	High-speed steel	SAE 1020 steel	160–170	1.48	2.45	1.96
5. Plain mills	High-speed steel	SAE 1020 steel	160–170	2.10	2.90	2.50
6. Face mills	Sintered carbide	NE 8630 Steel	200	1.50	2.50	2.00
7. Face mills	Sintered carbide	Cast iron	190	0.50	1.00	0.75
8. Peripheral mills	Sintered carbide	14 S-T aluminum alloy	....	0.50	1.00	0.75

Table 8. Milling Operations Performed on Cylinder Blocks and Heads of a Twelve-cylinder Motor

Operation	Pieces per Hour	Cutters				Rpm	Cutting Speed, ft per min	Feed in. per min	Feed per Tooth, in.	Stock Removed, in.
		No.	Type	Diam., in.	Material					
CYLINDER BLOCKS										
Mill oil-pan face, slot for bearing caps (Fig. 20)	16	2	Inserted-blade face mills	6 1/2	Tungsten carbide	107	183	19	0.0089	1/8
		1	Inserted-blade end mill	6 3/4	J-Stellite	64	113	10	0.0098	1/8 to 3/16
Rough-mill cylinder head seats and finish-mill valve cover faces (Fig. 21)	18	2	Inserted-blade face mills	7	Tungsten carbide	111	204	12 3/4	0.0058	1/8
		1	Inserted-blade face mills	5 1/2	Tungsten carbide	155	223	12 3/4	0.0052	1/8
Finish-mill cylinder head seats and mill top and water-jacket pad (Fig. 22)	18	2	Inserted-blade cutters	7	Tungsten carbide	111	204	14 5/16	0.0064	1/32
		2	Inserted-blade cutters	4	Tungsten carbide	189	199	14 5/16	0.0076	1/8
Mill cylinder block ends (Fig. 23)	13	2	Inserted-blade face mills	18	J-Stellite	31	146	7	0.0057	1/8
Straddle-mill main bearing sides (Fig. 24)	52	6	Half side mills	6	High-speed steel	49	77	5	0.0085	1/8
CYLINDER HEADS										
Mill cylinder head bottom (Fig. 25)	24	1	Inserted-blade face mill	7	Tungsten carbide	110	202	15	0.0069	3/32
Mill manifold, valve cover face and hold-down bosses (Fig. 26)	14	1	Inserted-blade shell end mill	4 1/2	Stellite	111	131	10 1/2	0.0086	1/8
		1	Inserted-blade formed mill	6	Stellite	86	120	10 1/2	0.0100	1/8
Mill ends and seal pad (Fig. 27)	57	2	Solid shell end mills	4	Stellite	111	117	8	0.0036	1/8
		1	Shell end mill	2 1/4	High-speed steel	111	66	8	0.0072	1/8

(3) Finish-mill rear end of cylinder blocks, using cemented carbide tools: Cutting speed 250 ft per min; feed, 16 in. per min; depth of cut, 0.015 in.; production, 44 pieces per hr; pieces per grind, 3168.

(4) Mill mounting pads of master cylinder body, using Stellite cutters: Cutting speed, 140 ft per min; feed, 11 $^1/_4$ in. per min; depth of cut, $^1/_8$ in.; production, 130 pieces per hr.

(5) Mill bosses on shock absorber body, using Stellite cutters: Cutting speed, 128 ft per min; feed, 25 in. per min; depth of cut, $^3/_{16}$ in.

Table 8 gives data concerning the milling operations on a 12-cylinder automobile motor block.

General Practice. (1) Bottle mold halves; mill parting faces and tongue, using tungsten carbide and high-speed steel cutters: Material, cast iron; cutting speed, tungsten carbide cutters, 170 ft per min; high-speed steel cutters, 53 ft per min; feed rate, tungsten carbide cutters, 9 $^7/_8$ in. per min; high-speed steel cutters, 3 $^5/_8$ in.; depth of cut, $^1/_{16}$ in.

(2) Roller-spider; mill three cams using high-speed form cutter: Material, steel forging; cutting speed, 78 ft per min; feed variable, average, 5 $^1/_2$ in. per min; depth of cut, $^1/_{16}$ in.; time per piece, 1.22 min; production, 42 pieces per hr.

(3) Transmission case; mill face, using high-speed steel cutters: Material, cast iron; cutting speed, 60 ft per min; feed, 5 in. per min; production, 8 pieces per hr.

(4) Compressor cylinder: Material, SAE 1112 steel; high-speed steel slotting cutter: cutting speed, 64 ft per min; feed, 5 $^3/_8$ in. per min; depth of cut, 0.025 in.; production, 229 pieces per hr.

(5) Master cylinder body; mill mounting pads, using Stellite face mill cutter: Material, cast iron; cutting speed, 140 ft per min; table feed, 11 $^1/_4$ in. per min; production, 130 pieces per hr.

(6) Block plunger feed trip; mill slot and angle using a gang of high-speed steel cutters: Material, SAE 1112 steel; cutting speed, 65 ft per min; feed rate, 2.4 in. per min; depth of cut, 1 $^1/_2$ in.; production, 12 pieces per hr.

(7) Fork; mill inside and outside of fork; mill radius and end, form-mill other radius and end. Material, steel casting; 1st and 2nd operation: Use gang of four high-speed steel cutters; cutting speed, 55 ft per min; feed rate, 5 $^3/_8$ in. per min; depth of cut, $^1/_8$ in. 3rd operation: Use high-speed steel form cutter; cutting speed, 50 ft per min; feed rate, 3 in. per min; depth of cut, $^1/_8$ in.; production, 6 pieces per hr.

(8) Bracket; mill bottom and inside edges using high-speed steel face mill cutter and high-speed steel end mill: Material, cast iron; cutting speed, roughing, 80 ft per min; finishing, 120 ft per min; feed rate, roughing, 11 $^3/_4$ in. per min; finishing, 9 $^1/_4$ in. per min; depth of cut, $^3/_{16}$ in.; production, 2 pieces per hr.

22. MILLING AND MILLING MACHINES

MILLING. Application of milling, in relation to the number of parts being manufactured, may be broadly classified as: *toolroom milling* and *production milling.*

Toolroom milling refers specifically to milling operations performed on one or a few parts only, such as may be needed for experimental or development work and for repairs. Jobbing shop work is also included in this category. The parts to be milled are located and held on the machine table in temporary setups, utilizing standard attachments, commercial or home-made clamping, and locating devices because changes in setup are frequent. For this reason, milling equipment must be flexible to permit quickly changing the setup to accommodate different parts.

Production milling is characterized as small lots or large-quantity production, depending on the number of identical parts processed through a shop at any given time. The objectives are low production cost and interchangeability of parts.

Parts are usually located and held in fixtures clamped to the machine table. The fixtures may be especially designed for locating and clamping a specific part, or, by means of removable units, may be adapted to locate and clamp a series of different parts. The milling equipment, once set up, will operate on the same job for long periods of time and, for this reason, is not as flexible as that used in toolroom milling.

Fully automatic operation of the equipment may be necessary to achieve a high production rate per hour. The production rate is determined as follows:

$$P = \frac{60E}{t}$$

where P = production per hour, number of pieces; t = time per piece, minutes; E = efficiency of use of the equipment, usually ranging from 80 to 85%, depending upon shop conditions.

For efficient production milling a thorough study must be made of the requirements of the job as to rate of production, accuracy and finish of the milled parts, the material to be milled, the method of holding and locating the parts in the fixture, the type and material of the milling cutters, the type of fixture, the type of milling machine, whether standard, special or semispecial, and the method of milling, that is, the method of arranging the parts on the machine to achieve maximum reduction in time per piece and, consequently, a maximum gain in production rate.

Methods. Typical methods of milling are:

Single piece. One piece or part is milled per machine cycle. Machine is idle to allow operator to load and unload the fixture. This method is generally used when milling medium-large-size parts.

String milling. A number of parts are placed, like beads on a string, along the machine table. Part of the loading and unloading time is saved. This method is used when milling small-size parts.

Reciprocal milling. The parts are arranged on opposite sides of the cutter. Loading and unloading take place on one end of the table, while the parts on the other end of the table are being milled, thus saving loading and unloading time. This method is used for small to medium-size parts.

Abreast milling. This is a method in which two or more parts are milled simultaneously. They are located side by side along the table width. Milling time is reduced in relation to the number of parts abreast. This method is also used for milling medium and small-size parts.

Transfer base milling. Parts are located on an index base having two, four, or more stations. The parts in each station are milled in sequence. Loading and unloading of parts in the other station take place while milling the parts in one station. This method is similar to reciprocal milling and avoids the operator's having to walk from one end of machine to the other. It is used for medium and small-size parts.

Rotary milling. The parts are arranged on a continuously rotating table, which continuously presents parts to be milled to the cutter, while the operator loads and unloads the fixture not in use. This method provides maximum utilization of the equipment, since the time per cycle is milling time.

The foregoing methods are usually combined, depending upon the application.

MACHINE CLASSIFICATION AND USES. The types and sizes of milling machines are numerous, and their designs merge into one another. It is, therefore, difficult to classify them. The following classification is based on general appearance or design.

(1) **Knee and column type** is so called because the spindle is fixed in the column or main body; the table is capable of manual adjustment and power feed and rapid traverse rates of movement longitudinally, transversely, and vertically. By set trip dogs the table can be operated through an automatic cycle of feed and rapid traverse in both directions of movement. Machines of this type are of rather universal application and generally are selected for that class of production milling that involves medium and small parts, processed in limited quantities, with frequent setups. They are especially used in toolroom, jobbing shops, and experimental laboratory work.

These machines can be equipped with fixtures and multispindle attachments to permit their use as production units. Such procedure pays in many shops, where certain jobs are manufactured in sufficient quantities to justify a knee and column type machine for high production, with the idea that it can easily be stripped of its high-production elements and put back on regular run of work.

Knee and column machines are subdivided as follows: (1) *Plain horizontal*, with spindle located horizontally, using facing and peripheral milling cutters; (2) *universal*, equipped with a swiveling table, for the cutting of helical grooves; and (3) *vertical*, with spindle located vertically, using chiefly face milling cutters.

(2) **Hand millers** usually are small machines of the knee and column type, in which the feeding movement of the table is controlled by hand. They are used in the manufacture of comparatively small quantities of light parts. Hand millers are subdivided into (*a*) floor type and (*b*) bench type. The latter are used for the machining operations of very light work.

(3) **Fixed bed type** machines are intended for use in the manufacture of parts in large quantities, and for this reason they are more rigidly built and simpler than the knee and column type millers. These machines are built on the plan of standard unit construction, which permits the make-up, from standard and complementary units, of a great variety of spindle combinations and feed cycles to meet the specific requirements of manufacture of a given product. (See Figs. 20 to 27.)

Fixed bed machines, with proper fixtures to clamp the work, operate on definite automatic cycles, such as rapid advance of work to cutter, cutting feed, rapid return to the starting position, and stop. These machines are subdivided into plain (single spindle); duplex; multiple spindle; rail; bridge; special types.

(4) **Rotary and drum type millers** are provided with fixtures and are used for mass-production work in which the cutting is continuous, with no idle time. The parts are loaded on one side of the machine. They pass first the roughing, and then the finishing cutters, and return to the loading position, where they are replaced. Drum type machines derive their name from the fact that the parts to be milled are mounted on a drum-like table that rotates continuously.

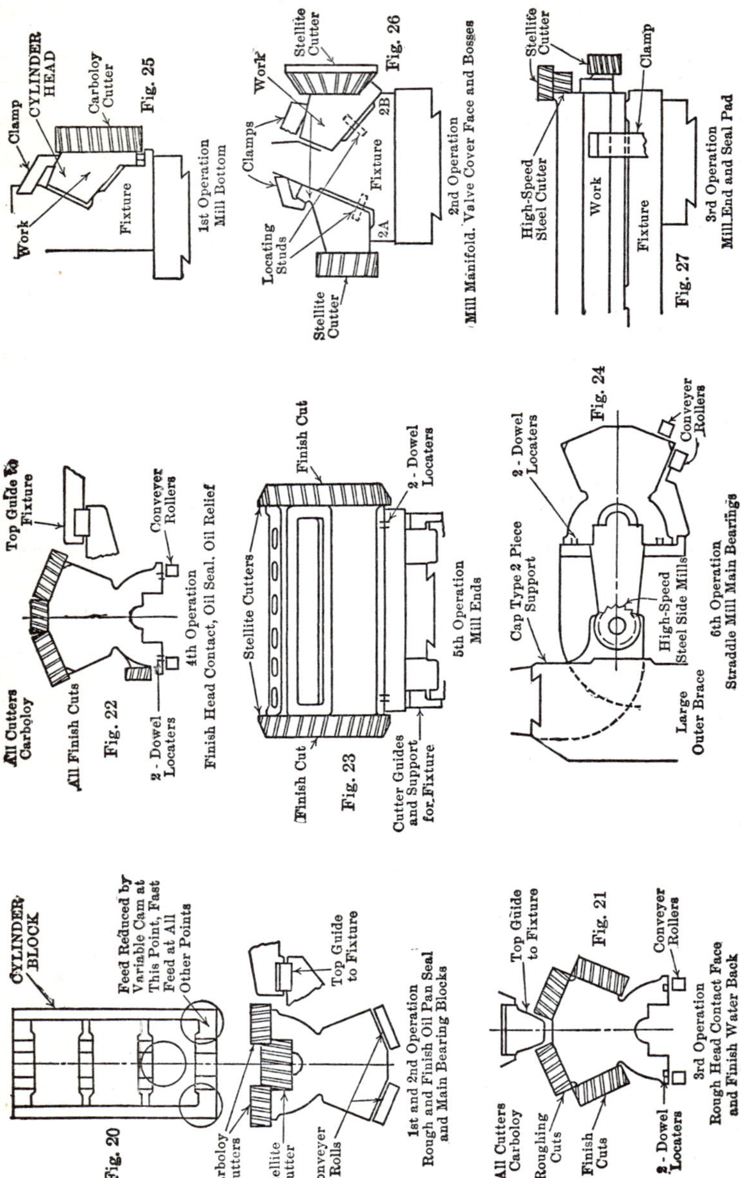

FIGURES 20 to 27. Milling operations on automotive cylinder blocks and heads. See Table 8.

(5) **Planetary milling machines** are used to mill plain or formed internal surfaces. The cutter is inserted in the bore to be milled and, while rotating, is fed to depth radially, makes a single sweeping cut about the bore.

(6) **Thread milling machines** are designed to cut threads and worms, using properly designed milling cutters.

(7) **Die-sinking and profiling machines** are designed to reproduce accurately complicated forms and contours from a master by means of a tracer, contact with which is maintained by the master form throughout the operation, thus controlling the position of work and cutter. These machines are subdivided into: (*a*) Hand controlled, in which the tracer is guided by hand over the master form. (*b*) Automatic, where the tracer, once the cycle is started, follows automatically the master form without requiring the assistance of the operator.

MILLING MACHINE ATTACHMENTS. The range of work that a milling machine can do is greatly increased by the use of attachments, of an almost endless variety to satisfy special requirements. Many of these attachments have now become standard units. Among these the most important are (1) universal dividing head; (2) index head; (3) heavy vertical milling attachment; (4) circular milling attachment; (5) universal spiral milling attachment; (6) slotting attachment; (7) rack milling attachment; (8) transfer or index base; (9) cam milling attachment; (10) toolmaker's vise; (11) vertical milling attachment with automatic depth control, for die sinking.

REFERENCES

1. H. Ernst and M. Martellotti, Metal cutting, *Mech. Eng.*, Aug. 1935.
2. Discussion of paper by O. W. Boston and C. E. Kraus, The elements of milling, *Trans. Am. Soc. Mech. Engrs.*, RP 54-4, 1932, p. 96.
3. *Bull. Am. Standards Assoc.*, 5C-1930 and 5C1-1947.
4. A. L. De Leeuw, Milling cutters and their efficiency, *Trans. Am. Soc. Mech. Engrs.*, Vol. 33, 1911, p. 245.
5. J. A. Hall and B. P. Graves, Effect of variation in design of milling cutters on power requirements and capacity, *Trans. Am. Soc. Mech. Engrs.*, Vol. 65, 1923, p. 165.
6. M. Kronenberg, Tool angles govern cutting efficiency, *Am. Machinist*, Jan. 15, 1948.

DRILLS AND DRILLING

By C. J. Oxford

23. TWIST DRILLS

DEFINITION. Twist drills are end cutting tools having one or more cutting edges, and having helical grooves adjacent thereto for the passage of chips and for admitting coolant to the cutting edges. Twist drills are used either for originating holes or for enlarging existing holes.

NOMENCLATURE. The nomenclature shown in Fig. 1 has been adopted by the American Standards Association.

Twist drills are manufactured with either straight or taper shanks as indicated in Fig. 1. Drills smaller than 1/8 in. in diameter are made with straight shank only. Drills 1/8 in. in

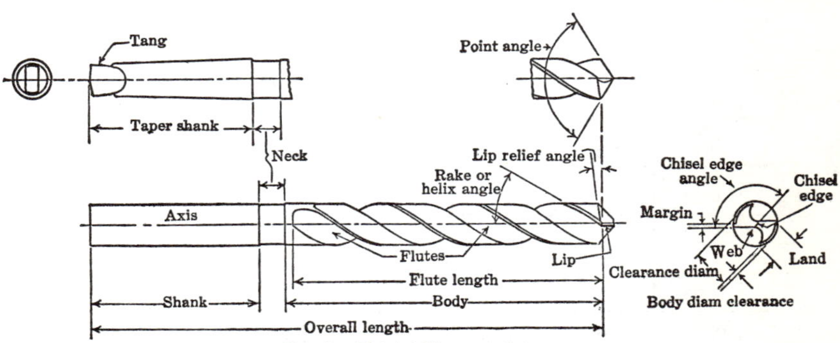

Fig. 1. Twist drill nomenclature.

diameter and larger are made with either straight or taper shanks in the fractional series (see next paragraph). Taper shank drills are made with Morse taper. For dimensions of Morse tapers, see p. 18-19.

TWIST-DRILL PROPORTIONS. Twist drills are manufactured and listed by the makers in three separate series of diameters. They are the fractional series, the wire gage series, and the letter size series. Table 1 combines all these series in order of drill diameters.

Table 1. Standard Sizes of Twist Drills

Diameters			Lengths			
			Short Series		Long Series	
Decimal Equivalent	Fractional	Number or Letter	Overall	Flute	Overall	Flute
.0135		80	3/4	3/16		
.0145		79	3/4	3/16		
.0156	1/64		3/4	3/16		
.0160		78	7/8	3/16		
.0180		77	7/8	3/16		
.0200		76	7/8	3/16		
.0210		75	1	1/4		
.0225		74	1	1/4		
.0240		73	1 1/8	5/16		
.0250		72	1 1/8	5/16		
.0260		71	1 1/4	3/8		
.0280		70	1 1/4	3/8		
.0292		69	1 3/8	1/2		
.0310		68	1 3/8	1/2		
.0313	1/32		1 3/8	1/2		
.0320		67	1 3/8	1/2		
.0330		66	1 3/8	1/2		
.0350		65	1 1/2	5/8		
.0360		64	1 1/2	5/8		
.0370		63	1 1/2	5/8		
.0380		62	1 1/2	5/8		
.0390		61	1 5/8	11/16		
.0400		60	1 5/8	11/16		
.0410		59	1 5/8	11/16		
.0420		58	1 5/8	11/16		
.0430		57	1 3/4	3/4		
.0465		56	1 3/4	3/4		
.0469	3/64		1 3/4	3/4		
.0520		55	1 7/8	7/8		
.0550		54	1 7/8	7/8		
.0595		53	1 7/8	7/8		
.0625	1/16		1 7/8	7/8		
.0635		52	1 7/8	7/8		
.0670		51	2	1		
.0700		50	2	1		
.0730		49	2	1		
.0760		48	2	1		
.0781	5/64		2	1		
.0785		47	2	1		
.0810		46	2 1/8	1 1/8		
.0820		45	2 1/8	1 1/8		
.0860		44	2 1/8	1 1/8		
.0890		43	2 1/4	1 1/4		
.0935		42	2 1/4	1 1/4		
.0937	3/32		2 1/4	1 1/4		
.0960		41	2 3/8	1 3/8		

Table 1. Standard Sizes of Twist Drills—*Continued*

Diameters			Lengths			
			Short Series		Long Series	
Decimal Equivalent	Fractional	Number or Letter	Overall	Flute	Overall	Flute
.0980		40	2 3/8	1 3/8		
.0995		39	2 3/8	1 3/8		
.1015		38	2 1/2	1 7/16		
.1040		37	2 1/2	1 7/16		
.1065		36	2 1/2	1 7/16		
.1094	7/64		2 5/8	1 1/2		
.1100		35	2 5/8	1 1/2		
.1110		34	2 5/8	1 1/2		
.1130		33	2 5/8	1 1/2		
.1160		32	2 3/4	1 5/8		
.1200		31	2 3/4	1 5/8		
.1250	1/8		2 3/4	1 5/8	5 1/8	2 3/4
.1285		30	2 3/4	1 5/8		
.1360		29	2 7/8	1 3/4		
.1405		28	2 7/8	1 3/4		
.1406	9/64		2 7/8	1 3/4	5 3/8	3
.1440		27	3	1 7/8		
.1470		26	3	1 7/8		
.1495		25	3	1 7/8		
.1520		24	3 1/8	2		
.1540		23	3 1/8	2		
.1562	5/32		3 1/8	2	5 3/8	3
.1570		22	3 1/8	2		
.1590		21	3 1/4	2 1/8		
.1610		20	3 1/4	2 1/8		
.1660		19	3 1/4	2 1/8		
.1695		18	3 1/4	2 1/8		
.1719	11/64		3 1/4	2 1/8	5 3/4	3 3/8
.1730		17	3 3/8	2 3/16		
.1770		16	3 3/8	2 3/16		
.1800		15	3 3/8	2 3/16		
.1820		14	3 3/8	2 3/16		
.1850		13	3 1/2	2 5/16		
.1875	3/16		3 1/2	2 5/16	5 3/4	3 3/8
.1890		12	3 1/2	2 5/16		
.1910		11	3 1/2	2 5/16		
.1930		10	3 5/8	2 7/16		
.1960		9	3 5/8	2 7/16		
.1990		8	3 5/8	2 7/16		
.2010		7	3 5/8	2 7/16		
.2031	13/64		3 5/8	2 7/16	6	3 5/8
.2040		6	3 3/4	2 1/2		
.2055		5	3 3/4	2 1/2		
.2090		4	3 3/4	2 1/2		
.2130		3	3 3/4	2 1/2		
.2187	7/32		3 3/4	2 1/2	6	3 5/8
.2210		2	3 7/8	2 5/8		
.2280		1	3 7/8	2 5/8		
.2340		A	3 7/8	2 5/8		
.2344	15/64		3 7/8	2 5/8	6 1/8	3 3/4
.2380		B	4	2 3/4		
.2420		C	4	2 3/4		

(*Table continued on p. 23-66*)

Table 1. Standard Sizes of Twist Drills—*Continued*

Diameters			Lengths			
			Short Series		Long Series	
Decimal Equivalent	Fractional	Number or Letter	Overall	Flute	Overall	Flute
.2460		D	4	2 3/4		
.2500	1/4		4	2 3/4	6 1/8	3 3/4
.2500		E	4	2 3/4		
.2570		F	4 1/8	2 7/8		
.2610		G	4 1/8	2 7/8		
.2656	17/64		4 1/8	2 7/8	6 1/4	3 7/8
.2660		H	4 1/8	2 7/8		
.2720		I	4 1/8	2 7/8		
.2770		J	4 1/8	2 7/8		
.2810		K	4 1/4	2 15/16		
.2812	9/32		4 1/4	2 15/16	6 1/4	3 7/8
.2900		L	4 1/4	2 15/16		
.2950		M	4 3/8	3 1/16		
.2969	19/64		4 3/8	3 1/16	6 3/8	4
.3020		N	4 3/8	3 1/16		
.3125	5/16		4 1/2	3 3/16	6 3/8	4
.3160		O	4 1/2	3 3/16		
.3230		P	4 5/8	3 5/16		
.3281	21/64		4 5/8	3 5/16	6 1/2	4 1/8
.3320		Q	4 3/4	3 7/16		
.3390		R	4 3/4	3 7/16		
.3437	11/32		4 3/4	3 7/16	6 1/2	4 1/8
.3480		S	4 7/8	3 1/2		
.3580		T	4 7/8	3 1/2		
.3594	23/64		4 7/8	3 1/2	6 3/4	4 1/4
.3680		U	5	3 5/8		
.3750	3/8		5	3 5/8	6 3/4	4 1/4
.3770		V	5	3 5/8		
.3860		W	5 1/8	3 3/4		
.3906	25/64		5 1/8	3 3/4	7	4 3/8
.3970		X	5 1/8	3 3/4		
.4040		Y	5 1/4	3 7/8		
.4062	13/32		5 1/4	3 7/8	7	4 3/8
.4130		Z	5 1/4	3 7/8		
.4219	27/64		5 3/8	3 15/16	7 1/4	4 5/8
.4375	7/16		5 1/2	4 1/16	7 1/4	4 5/8
.4531	29/64		5 5/8	4 3/16	7 1/2	4 3/4
.4687	15/32		5 3/4	4 5/16	7 1/2	4 3/4
.4844	31/64		5 7/8	4 3/8	7 3/4	4 3/4
.5000	1/2		6	4 1/2	7 3/	4 3/4

The table also shows the respective overall and flute lengths of the various series. Fractional series drills are made in two separate lengths for drills 1/8 in. in diameter and larger, as shown in the table.

SPECIAL DRILLS. In addition to the general-purpose drills described in the foregoing, many special designs of drills are used for definite applications and for special materials to be drilled. Some are available from manufacturers' stocks in the listed diameters and lengths. Others must be designed and manufactured to suit a particular application or material. Figure 2 shows a drill designed for use when drilling deep holes in

FIG. 2. Drill for use in aluminum, magnesium, slate, etc.

aluminum and magnesium alloys and for drilling holes in slate and some plastic materials used for electrical switchboard panels.

Figure 3 shows a heavy-duty type of drill used extensively for drilling interrupted holes in hard and tough steels. Drilling of cross holes for cotter pins is a good example of the use of this type of drill.

FIG. 3. Heavy-duty drill for interrupted holes in hard, tough steel.

Figure 4 shows a low helix drill much used for drilling brass, and for horizontal drilling of brass and soft steel in automatic screw machines.

FIG. 4. Low-helix drill for brass, and for horizontal drilling in automatic screw machines.

Figure 5 is a drill design used for drilling molded plastics, hard rubber, wood, and some types of aluminum and magnesium alloys.

FIG. 5. Drill suitable for molded plastics, hard rubber, wood, and some aluminum and magnesium alloys.

For drilling exceptionally deep holes in steel, such as oil holes in automotive crankshafts and connecting rods, the type of drill shown in Fig. 6 finds wide use. A large helix angle

FIG. 6. Drill for deep holes in steel.

and very heavy web are characteristic of this type of drill. Special sharpening is necessary to make such drills perform to best advantage. The split point or so-called "crankshaft point" illustrated in Fig. 7 is found most satisfactory for this purpose.

FIG. 7. Split point for drill in Fig. 6.

FIG. 8. Combined drill and countersink.

Other types of drills that find wide use in industry are three-flute and four-flute drills used for enlarging existing holes, and the combined drill and countersink shown in Fig. 8. The last is used extensively for producing center holes in lathe work.

24. DRILLING PRACTICE

SPEEDS FOR DRILLING. The varying conditions under which drills are called upon to operate, as well as the wide variety of materials, having different characteristics, to be drilled, make it impracticable to lay down any hard and fast rules for drill speeds. In general it may be stated that permissible speeds depend on the hardness and the toughness of the material to be drilled. The harder or tougher the material, the lower the speed should be. Conversely, the softer the material, the higher the speed.

When using high-speed steel drills the figures given in Table 2 may be used as a convenient starting guide. Observation of the behavior of drills when so operated indicates if these speeds can be increased or if they must be decreased. If drills operate easily and without undue signs of heating, higher speeds should be tried. If, on the other hand, drills show signs of overheating, or if corners wear away rapidly, it is an indication that speeds are too high.

Carbon-steel drills should be operated at approximately 40 to 50% of the speeds given in Table 2.

Table 2. Recommended Drilling Speeds for Various Materials with High-speed Steel Drills

(Courtesy National Twist Drill & Tool Co.)

Material	Recommended Speed Surface ft per min
Aluminum and its alloys	200–300
Bakelite	100–150
Brass and bronze, soft	200–300
Bronze, high tensile	70–100
Cast iron, soft	100–150
Cast iron, hard	70–100
Magnesium and its alloys	250–400
Malleable iron	80–90
Nickel and Monel metal	40–60
Slate, marble, and stone	15–25
Steel, forgings	50–60
Steel, manganese (15% Mn)	15–25
Steel, soft	80–110
Steel, stainless	30–40
Steel, tool	50–60
Wrought iron	50–60
Wood	300–400

Drill speeds are commonly designated as peripheral or surface speed in feet per minute (sfpm). Convenient formulas for converting revolutions per minute (rpm) into sfpm or vice versa are

$$\text{sfpm} = \frac{\text{rpm} \times d \times 3.1416}{12} \quad \text{or} \quad \text{rpm} = \frac{\text{sfpm} \times 12}{d \times 3.1416}$$

where d is the diameter of drill in inches.

FEEDS FOR DRILLING. Feeds per revolution of drills do not bear the same relation to hardness or toughness of materials to be drilled as speeds. For a great majority of drilling operations it is desirable to keep the feed per revolution for a given diameter of drill approximately constant, regardless of the material to be drilled. Table 3 shows the recommended feed per revolution for various diameters of drills.

Table 3. Recommended Feeds for Drills of Various Diameters

(Courtesy National Twist Drill & Tool Co.)

Diameter of Drill, in.	Feed, in. per rev
Under 1/8	0.001 to 0.002
1/8 to 1/4	0.002 to 0.004
1/4 to 1/2	0.004 to 0.007
1/2 to 1	0.007 to 0.015
1 in. and over	0.015 to 0.025

The formula for computing rate of penetration in inches per minute (ipm) for a given feed per revolution (fr) is

$$\text{ipm} = \text{rpm} \times \text{fr}$$

CUTTING FLUIDS FOR DRILLING. Cutting fluids serve a two-fold purpose in drilling: (1) as a coolant; (2) as a lubricant.

Table 4. Recommended Cutting Fluids for Various Materials

(Courtesy National Twist Drill and Tool Co.)

Material	Cutting fluid
Aluminum and its alloys	Kerosene, kerosene and lard oil, soluble oil
Bakelite	Dry
Brass, soft	Dry, soluble oil, kerosene and lard oil
Bronze, high tensile	Soluble oil, lard oil, mineral oil, dry
Cast iron	Dry, air jet, soluble oil
Copper	Soluble oil, dry, mineral lard oil, kerosene
Magnesium	Low-viscosity neutral oils
Malleable iron	Dry, soda water
Monel metal	Lard oil, soluble oil
Slate	Dry
Steel, forging	Soluble oil, sulfurized oil, mineral lard oil
Steel, manganese	Soluble oil, sulfurized oil, mineral lard oil
Steel, soft	Soluble oil, mineral lard oil, sulfurized oil, lard oil
Steel, stainless	Sulfurized mineral oil, soluble oil
Steel, tool	Soluble oil, mineral lard oil, sulfurized oil
Wrought iron	Soluble oil, mineral lard oil, sulfurized oil

In production work cooling of both drill and work is usually the most important factor to consider. The amount of heat generated is so great that tool life depends on the ability of the cutting fluid to carry away the heat at approximately the same rate it is generated. For that reason lubrication must often be subordinated to cooling.

Some lubrication is desirable, particularly when drilling steel, as it will help to prevent chips from sticking at the cutting edges of the drill and so cause clogging in the flutes. Cutting fluids should be selected on the basis of their heat-dissipating or lubricating qualities as conditions demand.

The recommendations in Table 4 for cutting fluids and other media, when drilling various materials, have been found to give generally good results.

DRILL SHARPENING. It has been estimated that 90% of drilling troubles can be traced to improper sharpening. Therefore, it is highly important that care be taken when resharpening drills. A good drill point will have: (1) Both lips at the same angle with the axis of the drill. (2) Both lips of the same length. (3) Correct lip clearance angles for the material to be drilled. (4) Correct web thickness.

For drilling hard or tough steels, the point angle *C* (Fig. 9) should be 135 degrees. For drilling soft materials and for general purposes, this angle should be 118 degrees. Drills for cast iron are often sharpened with a point angle of 60 to 90 degrees. Plastics are drilled successfully with drills sharpened with a 60-degree point angle.

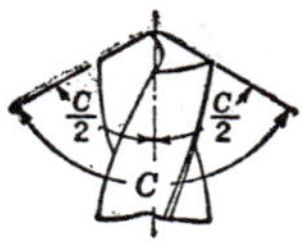

FIG. 9. Drill point angle.

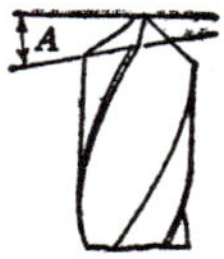

FIG. 10. Lip clearance angle.

FIG. 11. Chisel-edge angle.

The lip-clearance angle *A* (Fig. 10) should be as follows. When drilling steel this angle should be 6 to 9 degrees. When drilling soft cast iron and other soft materials, *A* should be made from 12 to 15 degrees at the periphery of the drill. These clearance angles should increase toward the center of the drill so as to produce the proper chisel-edge angle *B* (Fig. 11).

The proper combination of lip-clearance angles and chisel-edge angles is as follows. When angle *A* is made 12 to 15 degrees for softer materials, angle *B* should be made approximately 135 degrees. When angle *A* is made 6 to 9 degrees for harder materials, angle *B* should be made 115 to 125 degrees.

CAUSES OF FAILURES OF DRILLS. Manufacturers of twist drills have issued to users the following instructions to aid them in analyzing troubles that may develop when drilling.

Causes of Failures of Twist Drills

Indications	Causes
Outer corners break down	Cutting speed too high. Hard spots in material. No cutting compound at drill point.
Cutting edges chip	Too much feed. Lip clearance too great.
Checks or cracks in cutting edges	Overheated or too quickly cooled while sharpening or drilling.
Margin chips	Oversize jig bushing.
Drill breaks	Point improperly ground. Feed too heavy. Spring or back lash in drill press, fixture, or work. Drill is dull. Flutes clogged with chips.
Tang breaks	Imperfect fit between taper shank and socket caused by dirt or chips, or burred or badly worn sockets.
Drill breaks when drilling brass or wood	Flutes clogged with chips. Improper type drill.
Drill splits up center	Lip clearance too small. Too much feed.

(Table continued on p. 23-70)

Causes of Failures of Twist Drills—*Continued*

Indications	Causes
Drill will not enter work	Drill is dull. Lip clearance too small. Too heavy a web.
Hole rough	Point improperly ground or dull. No cutting compound at drill point. Feed too great. Fixture not rigid.
Hole oversize	Drill point improperly ground.
Chip shape changes while drilling	Drill becomes dull or cutting edges chipped.
Large chip coming out of one flute, small chip out of other flute.	Point improperly ground, one lip doing all the cutting.

THRUST, TORQUE, AND POWER FOR DRILLING. The design of drilling machines must be based on the forces involved when such machines are to be operated at ordinary ranges of feeds and speeds and in materials of low machinability.

Thrust and torque for various sizes of drills have been investigated at various times and by different investigators. (See Refs. 1, 2, and 3.) The results have varied somewhat because of differences in designs of drills used and variations in materials and testing equipment. It has been established, however, that torque and thrust for drilling can be expressed by the following equations:

$$\text{Torque } T = Cf^m d^n$$

where T = torque, foot-pounds; f = feed, inches per revolution; d = diameter of drill, inches; C = a constant depending on the material to be drilled, the design of drill, and the cutting fluid. The exponents m and n vary with the material to be drilled.

$$\text{Thrust } B = Kf^p d^r \text{ for aluminum and brass}$$

where B = thrust, pounds; and K = a constant corresponding to C in torque equation. The values of exponents p and r will vary with the material to be drilled.

$$\text{Thrust } B = Kf^p \left(\frac{d}{5} + \frac{w}{d}\right) \quad \text{for iron and steel}$$

where w = web thickness of drill, inches.

Tables of values for constants K and C as well as for exponents m, n, p, and r are to be found in the references cited.

As a practical reference, Table 5 gives values for thrust and horsepower as applied to drills operating at normal rates of feed.

Table 5. Thrust and Horsepower of Twist Drills at Indicated Rates of Feed

(National Twist Drill and Tool Co.)

Diameter of Drill	Feed per Revolution	Thrust, lb		Horsepower			
		Cast Iron	Steel	Cast Iron		Steel	
				100 rpm	100 sfpm	100 rpm	60 sfpm
1/8	.004		130			0.011	0.200
5/32	.0045		185			0.018	0.260
3/16	.005		245			0.028	0.340
1/4	.006		380			0.053	0.490
3/8	.008		550			0.210	1.308
1/2	.010	200	750	0.26	1.60	0.375	1.717
5/8	.011	275	1000	0.40	2.30	0.580	2.133
3/4	.012	325	1275	0.59	2.90	0.855	2.620
7/8	.013	400	1590	0.80	3.44	1.180	3.078
1	.014	500	1910	1.06	4.02	1.550	3.550
1 1/4	.016	700	2650	1.83	5.58	2.680	4.918
1 1/2	.016	850	3200	2.44	6.22	3.740	5.718

Thrust or feeding pressures should not be confused with horsepower required to drive a drill. It can be shown that only about 1% of the total power applied in drilling is used up to overcome the resistance due to thrust.

Both thrust and horsepower values in these tables are based on materials of average machinability. The tables for horsepower also make an allowance of 20% for friction losses in the machine.

All values are for sharp drills. Extra allowances for increases in thrust and horsepower due to dulling should be

Thrust 50 to 75%
Horsepower 40 to 60%

Horsepower requirements for a drill operating at any given speed in rpm may be found by multiplying the figures shown in columns under 100 rpm by the speed at which the drill is operating, and dividing by 100.

EXAMPLE. If a drill 1/2 in. in diameter is operated at 500 rpm in steel, the power required would be $0.375 \times 500/100 = 1.875$ hp.

REFERENCES

1. Dempster Smith and A. Poliakoff, abstracted in *Am. Machinist*, May 1909.
2. O. W. Boston and C. J. Oxford, *Trans. Am. Soc. Mech. Engrs.*, RP55-1, 1933, and RP58-2, 1936.
3. O. W. Boston and C. J. Oxford, *SAE Journal*, March 1931.

ABRASIVE PROCESSES

By R. E. W. Harrison *

25. ABRASIVE WHEELS †

Every grinding wheel has two components: *abrasive*, which does the actual cutting, and *bond*, which holds the abrasive grains while they cut. The retaining force with which the abrasive grains are held in place by the bonding material determines the *grade*, while the actual spacing of the abrasive grits themselves provides the definite characteristics known as *structure*.

ABRASIVES. There are two common types of manufactured abrasives used for grinding wheels and other abrasive products, *aluminum oxide* and *silicon carbide*. Both are products of the electric furnace and are marketed under various trade names, as Alundum, Aloxite, Borolon, Carborundum, Crystolon, et cetera. Emery and corundum, natural forms of aluminum oxide, contain impurities and are unsuitable for uniform quality production grinding wheels.

A typical aluminum oxide abrasive, Norton Alundum, is made by fusing bauxite in an electric furnace. Crystolon, a typical silicon carbide abrasive also a product of the Norton Company, is produced by electric furnace reaction of pure silica sand and coke. The aluminum oxide abrasives generally are used for grinding materials of high tensile strength, like carbon steel, alloy steel, high-speed steel, annealed malleable iron, tough bronzes, and other materials that are hard, tough, and strong. Silicon carbide abrasives are hard and sharp, but not as tough as the aluminum oxide abrasives. They are used for grinding materials of low tensile strength like gray iron, chilled iron, brass, soft bronze, aluminum, copper, stone, marble, rubber, cemented carbides, and other materials that are hard and brittle.

GRAIN SIZE. Numbers used to designate grain size represent the approximate number of openings per linear inch in the screen through which the grits will pass. Standard grain sizes from coarse to fine are 4, 6, 8, 10, 12, 14, 16, 20, 24, 30, 36, 46, 60, 70, 80, 90, 100, 120, 150, 180, 220, and 240. The fine flours are classified hydraulically to the sizes 280, 320, 400, 500, and 600.

BOND. There are six principal types of bonds used in grinding wheels.

1. **Vitrified bond** consists of a mixture of clays, subjected to a sufficiently high temperature to fuse to a molten, glass-like condition. The porosity and strength of this bond promote high stock removal, and its inelasticity helps in the attainment of high precision. It is not affected by water, acid, oils, or ordinary temperature variations.

2. **Silicate bond** consists basically of clays and silicate of soda, which at comparatively low temperatures form a hard cement. The silicate bond releases the abrasive grains more readily so that the wheels are milder acting and, therefore, are particularly suited for grinding edged tools and other work where heat generated in grinding must be kept at a minimum. Silicate bond is also used in making very large solid wheels.

3. **Shellac bond,** an organic bond consisting of shellac and other gums, has no porosity, high tensile strength, and good elasticity. Wheels are capable of producing highly refined finishes on camshafts and steel-mill rolls; are used also for grinding cutlery and for cut-off work requiring thin wheels.

---

* Revision of material originally prepared by R. T. Kent.

† Revised by E. T. Larson, Norton Co.

4. **Rubber bond,** consisting of rubber, either natural or synthetic, mixed with certain minerals in definite proportions, on vulcanization produces a desired hardness and toughness. Rubber-bonded wheels are used for most centerless control wheels and for precision ball race grinding; also for high-speed snagging where finish is important and for cut-off wheels where burr and burn must be held to a minimum.

5. **Resinoid bond** consists of synthetic resins. It is used for high-speed wheels in foundry cleaning rooms and billet shops, for sharpening and gumming saws, and for thin cut-off wheels. These wheels are designed to operate at 9500 sfpm; cut-off wheels at 16,000 sfpm.

6. **Magnesium oxychloride,** a cold set cement used as a bond in wheels and segments, chiefly for disk grinding.

GRADE is a term indicating the relative strength (holding power) of the bond which holds the abrasive grains in place. It does not express the hardness of the abrasive itself. A grinding wheel may be made of a very hard abrasive and still be a soft wheel with respect to cutting action. In general, with a given type of bond, it is the *amount* of bond which determines the hardness of grade. If the abrasive grains are easily broken out of the bond, the wheel is said to be soft. A wheel is of the proper grade when the cutting grains, as they become dulled, are released or fractured and either entirely new, sharp grains or new, sharp-cutting edges and points are presented to the work. Wheels that are too hard will glaze and must be resharpened by dressing.

While the ideal way to measure grade is to grind on each wheel, the best-known practicable method is to measure the resistance offered by the wheel to the penetration of a steel screw driver-like blade. Under a uniform pressure and twisting force, the blade will penetrate deeper into a softer wheel than into a hard one. Grades are designated by letters of the alphabet, so that there is a gradual change in hardness from an extremely soft wheel to an extremely hard one.

In the standard marking system adopted in 1944 by all the grinding-wheel manufacturers, the grade letters range from A to Z in the order of increasing "hardness." In other words, wheels in such grades as F, G, H, I near the beginning of the alphabet are relatively soft. Wheels around grade L, M, N are considered medium, and wheels in grades Q, R, S, etc., are hard. However, it should be borne in mind that this new marking system is a uniform system of marking only, and not of grinding action. In other words, wheels that happen to bear the same standard markings, if made by different wheel manufacturers, may not, and probably will not, produce exactly the same grinding action.

STRUCTURE is determined by the arrangement or spacing of the abrasive grits in the wheel. The same total volume of voids in two different wheels may be made up of a relatively small number of voids of large volume, or of a large number of voids each of small volume. The cutting action of two wheels of different structures will be different, although they may be of the same grain and grade.

Structure in a grinding wheel is denoted by a number. When the abrasive grains are close together relative to their grain size, the wheels have low structure numbers such as 0, 1, 2, 3. Wider spacing relative to the grain size is designated by higher structure numbers, such as 8, 9, 10, 11, 12. Some wheels, commonly referred to as being of the "open structure" or "porous type," have a special structural make-up that gives them a distinctive porous appearance. This type of wheel in soft grades has been found effective for broad-contact grinding, tool and cutter grinding, surface grinding, and internal grinding of large diameter holes.

WHEEL MARKING. Grinding wheels are marked with a combination of numbers and letters, called the *wheel marking*, which completely describes and identifies a wheel. In the new standard marking system adopted by all the grinding-wheel manufacturers a few years ago, there are six positions indicating a particular characteristic of the wheel:

Position No.	1	2	3	4	5	6
Designates	Abrasive	Grain size	Grade	Structure	Bond	Bond modification
Example	32A	46	J	5	V	BE

Position 1 denotes the type of abrasive material, A for aluminum oxide and C for silicon carbide. Numbers preceding the letter may be added by the manufacturer to indicate the brand of abrasive. The symbols in positions 2, 3, and 4 are explained in the preceding paragraphs under Grain Size, Grade, and Structure. Position 5 designates the bond material: V for vitrified, S for silicate, B for resinoid, R for rubber, E for shellac, and O for magnesite (oxychloride). Position 6 indicates any special manufacturing process and is a symbol chosen by the manufacturer. In the example, the complete marking is 32A46-J5VBE, which describes a wheel of Norton 32 brand Alundum abrasive, grain size 46, grade J, structure 5, vitrified bond of Norton BE type.

DIAMOND WHEELS. Wheels in which the abrasive surface consists of crushed and sized natural bort diamonds are used extensively for sharpening cemented carbide tools and cutters and for cutting carbide blanks. These wheels are made in a variety of sizes and shapes for use on carbide tool grinders, tool and cutter grinders, surface grinders, and other precision-type grinding machines. Diamond wheels are available in three different bond types, resinoid, metal, and vitrified. The resinoid-bonded are the fastest cutting, but the least wear resistant. The metal-bonded cut somewhat slower, but are the most durable. The vitrified-bonded approach the resinoid-bonded in rate of cut and the metal-bonded in resistance to wear. Each bond type has specific applications. The grit sizes most commonly used for carbide grinding are 100 for roughing, 220 and 320 for finishing, and 120 and 150 for both rough and finish grinding in a single wheel. Diamond wheels are available in several different grades of hardness, in three different diamond concentrations, namely, 25 (low), 50 (medium), 100 (high), and three common depths of diamond section, namely, $^1/_{16}$, $^1/_8$, and $^1/_4$ in.

Diamond wheels are operated at the same speeds as ordinary tool grinding wheels. For best results, they should be used wet, preferably with ordinary water-soluble grinding solution. If diamond wheels must be used dry, the depth of cut per pass should not exceed 0.0005 in.

26. FACTORS AFFECTING GRINDING-WHEEL SELECTION *

Selection of a grinding wheel for a given operation involves consideration of both the fixed conditions surrounding the job and the factors that may vary.

FIXED CONDITIONS. 1. Material to Be Ground. The physical properties of the material to be ground govern the type of abrasive to be used. Materials of high tensile strength, such as steel and the tougher varieties of bronze, as well as other materials that are hard, yet tough and strong, can best be ground with aluminum oxide abrasive wheels. The problem of different degrees of hardness, toughness, and sensitiveness to burning in metals is solved by using aluminum oxide abrasives of varying degrees of purity, strength, and fracture characteristics.

Materials of low tensile strength, which are penetrated easily, such as the soft bronzes, cast and chilled iron, aluminum, granite and marble, and glass, are ordinarily ground with silicon carbide abrasive wheels; also, the cemented carbides and other materials which are both exceptionally hard and brittle. The regular form of gray silicon carbide abrasive is used for general grinding, heavy-duty snagging, cylindrical, centerless, and internal grinding. The other form of silicon carbide abrasive has a characteristic light green color, and is generally preferred for grinding cemented carbide tools.

In general, hard dense materials require a relatively soft wheel. A large number of abrasive grains or cutting points in the wheel is necessary, since the penetration of each point is slight. Exceptionally hard materials, such as the cemented carbides, are best ground with open-structure wheels; likewise, soft, tough, and ductile materials.

2. **Amount of Material to Be Removed.** For relatively heavy stock removal, coarse-grain wheels with open structure should be used, particularly where maximum efficiency is desired and for rough work such as off-hand snagging of castings. Finish depends largely on grain size, except in precision grinding. Here, proper truing facilities and skillful grinding technique make possible an acceptable finish with relatively coarse-grain wheels.

It is a common mistake among production executives to think of grinding costs solely in terms of wheel cost. They overlook the more important factor of cutting speed. There are three factors to consider in computing grinding costs: (1) rate of stock removal; (2) wheel life; and (3) cost per unit of production. This includes direct labor cost, factory overhead, and wheel cost.

A very fast-cutting wheel may have a relatively short life and yet show a low *total cost* for the amount of material removed because both the direct labor and overhead costs are low for the amount of stock removed. Furthermore, the rate of production is high.

On the other hand, a hard slow-cutting wheel may last twice as long as a softer fast-cutting wheel. For the same amount of stock removal, the labor and overhead costs for such a wheel might be much higher, so that the total grinding costs would be much higher than in the preceding example. To be sure, if the wheel is too soft for fast cutting, the wheel has to be changed more frequently and work spoilage may result, so that some care must be used in selecting the right wheel for the job.

3. **Area of Contact.** The area of contact influences the selection of grain size, grade, and structure number. In general, fine grain is employed where the area of contact is small and a coarse grain where the area of contact is large. Cylindrical grinding is an example of relatively high unit pressure and small area of contact. Such conditions call for a wheel of medium to fine grain, medium structure, and medium grade.

---

* Revised by E. T. Larson, Norton Co.

Surface grinding, with the rim of a cup, cylinder, or segmental type wheel in contact with the work, is an example of high total pressure of wheel on the work, but low unit pressure. The wheel characteristics under such conditions would be a coarse grain, with wide grain spacing (open structure) to distribute the pressure per unit of area over a small number of cutting points, each subjected to sufficient pressure to cut effectively and also to provide ample chip clearance. The grade should be soft, in order that the dull grains may be released promptly, inasmuch as the forces tending to fracture them or tear them from the wheel are light.

4. Type of Grinding Machine. The grinding machine usually is already installed and imposes conditions that must be considered in wheel selection. Machines that are light and flimsy, or of the flexible type, require harder wheels than heavy, rigid machines. The combination of speeds and feeds on some precision machines may affect the grade of wheel desirable for best results. Surface-grinding machines employing cup wheels require softer wheels of more open structure than similar machines using straight wheels.

VARIABLE FACTORS. Wheel Speed. Too slow a speed consumes abrasive without corresponding useful work. Excessive speed may result in hard grinding action and danger of breakage. It is best to operate a grinding wheel at or as near as possible to the speed recommended by the maker. The grit sizes, grades, and bond types recommended for given grinding operations are based on the assumption that the wheels will be operated at the recommended speeds. If these speeds cannot be used, compensation should be made in the grade selection. Table 1 gives the peripheral speeds recommended for various classes of work.

Table 1. Recommended Wheel Speeds for Various Classes of Grinding

Type of Grinding	Wheel Speed, surface feet per minute
Cylindrical grinding (vitrified)	5500–6500
Cylindrical grinding (resinoid)	6500–9500 *
Surface grinding	4000–6000
Internal grinding	2000–6000
Internal grinding (diamond wheels)	5000–6500
Cutter and tool grinding	4500–6000
Cutting-off (rubber, shellac and resinoid)	9000–16000 *
Snagging, off-hand grinding (vitrified wheels)	5000–6500
Snagging (rubber and resinoid wheels)	7000–9500

* Higher speed recommended only if bearings, protection devices, and machine rigidity are adequate.

Work Speed. In machine grinding, high work speed tends to wear the wheel faster than slow work speed. Wheel wear depends on the ratio of wheel speed to work speed, both in surface feet per minute. The higher the ratio, the less work the wheel is required to do in a given time, and the wheel wears at a lower rate. A decreased ratio, resulting from increased work speed, requires the wheel to do more work in a given time and to wear faster.

In general, the longer the arc or the larger the area of contact in precision grinding operations, the faster should be the work speed, in order to have the wheel cut properly. In off-hand grinding the pressure employed in forcing the work against the wheel, or the method of applying the work, influences the grade selection. The more severe the pressure and application conditions, the harder should be the wheel in order to give economical wheel life. This relationship also exists in semi-automatic and high-production precision grinding, such as crankshaft and camshaft grinding. Work speed will also vary somewhat according to the material being ground, approximately as shown in Table 2.

Table 2. Work Speeds in Grinding

Material	Speed at Time of Cutting,* surface feet per minute
Soft steel	30–50
Hard steel	70–100
Cast iron	200–400
Aluminum	100–200

* Lower side of range used for roughing cuts and higher for finishing.

Skill of Workmen. Grinding costs on off-hand grinding may vary as much as 100% on some work with the same kind of machine in the same shop, because of the difference in the method of applying the work to the wheel, and the variation in the skill and knowledge of the different workmen. Even in machine grinding, the operator can vary the results by his method of handling the machine.

The foregoing considerations are summarized in Table 3. Table 4 gives the complete marking of wheels recommended for the most common materials that may have to be ground. The recommendations presuppose that operating conditions comply with good grinding practice.

Table 3. Factors Influencing the Selection of Abrasive Wheels

Selection of	Influential Factors	Kind of Wheel	Material Ground, Result Desired, or Condition to Be Met
Abrasive	Physical properties of material ground	Aluminum oxide	Carbon, alloy, or high-speed steel; annealed malleable iron; wrought iron; tough bronzes; tungsten, etc.
		Silicon carbide	Gray iron; chilled iron; brass; soft bronze; aluminum; copper; marble; granite; pearl; rubber; leather; very hard alloys as tungsten carbide, etc.
Grain Size	Amount of material to be removed	Coarse grain	Fast removal of stock
	Finish desired	Fine grain	Fine finish
	Physical properties of material to be ground	Coarse grain	Soft, ductile materials
		Fine grain	Dense, hard, or brittle material
Grade	Physical properties of material to be ground	Hard wheel	Soft material
		Soft wheel	Hard material
	Area of contact	....	Hardness increases as area of contact decreases
	Wheel speed	Hard wheel	Low wheel speed
		Soft wheel	High wheel speed
	Work speed	Hard wheel	High work speed
		Soft wheel	Low work speed
	Condition of grinding machine	Hard wheel	Machine in poor condition
		Soft wheel	Machine in good condition
	Skill of operator	Harder wheels	Unskilled operator; piece work
		Softer wheels	Skilled operator; day work; economical production
Structure	Physical properties of material to be ground	Wide spacing	Soft, tough, ductile materials
		Close spacing	Hard, brittle materials (except cemented carbides)
	Finish desired	Wide spacing	Medium or coarse finish
		Close spacing	Fine finish
	Nature of operation	Wide spacing	Snagging; operations with flexible application of pressure; surfacing
		Medium spacing	Cylindrical, centerless, tool and cutter grinding
		Close spacing	Heavy pressures tending to deform wheel face
Bond	Wheel dimensions	Vitrified	For general use; rapid removal of stock
		Silicate	Solid wheels of very large diameter
		Resinoid Shellac Rubber	Thin cutting-off wheels; wheels subject to deflection
	Operating speed	Vitrified	Wheel speeds up to 6500 surface ft per min
		Resinoid Shellac Rubber	Wheel speeds above 6500 surface ft per min
	Finish desired	Resinoid Shellac Rubber	High finish where stock removal is not a factor
		Silicate	To replace sandstone on cutlery, etc.

Table 4. Grinding Wheel Recommendations

(Courtesy of Norton Co.)

Work—Material—Operation	Abrasive	Grain Size	Grade	Structure	Bond	Bonding Variation
Agate						
Offhand roughing	37C	80	M	8	V	
Offhand finishing	37C	180	L	8	V	
Cutting-off	D	60	L	25	M	
Gem edging—roughing	D	100	N	50	M	
Gem edging—finishing	D	800	N	25	M	
Alnico						
Offhand	57A	36	N	5	V	BE
Cylindrical	57A	54	L	5	V	BE
Surfacing	38A	60	H	8	V	BE
Cutting-off (wet) up to 7500 sfpm	A	80	K	0	R	30
Aluminum						
Cylindrical	37C	364	K	5	V	
Surfacing (cups and cylinders)	37C #8 Treated	24	J	8	V	
Surfacing (disks)	37C	20	K	4	B	
Internal	37C	36	K	5	V	
Floor stands, 5000–6500 sfpm	37C #12 Treated	24	O	5	V	
Floor stands, 7000–9500 sfpm	37C #12 Treated	30	O	4	B	5
Floor stands, 7000–9500 sfpm	37C	30	O	5	B	5
Portable grinders, 5000–6500 sfpm	37C #12 Treated	24	O	5	V	
Portable grinders, 7000–9500 sfpm	37C #12 Treated	30	O	4	B	
Cutting-off (dry), 9000–16000 sfpm	A #12 Treated	245	Q	8	B	
Cutting-off (wet), 7500–12000 sfpm	A	46	V	8	R	30
Asbestos						
Cutting-off	37C	24	M	8	B	
Boron Carbide (Norbide)						
Cylindrical						
Roughing	D	100	L	100	V	
Semifinishing	D	220	J	100	B	
Finishing	D	320	J	100	B	
	D	400	J	100	B	
Cutting-off	D	100	N	100	B	
Internal grinding						
Roughing	D	100S	L	100	V	
Finishing	D	320	N	100	V	
Offhand grinding						
Roughing	D	100	J	100	B	
Finishing	D	320	J	100	B	
Brake Lining (Woven)						
Surfacing—(disks)	37C	20	J	4	B	L
Brake Lining (Molded)						
Surfacing—(disks) roughing	37C	16	M	4	B	
Surfacing—(disks) finishing	37C	46	L	5	B	
Cutting-off (dry)	37C	30	O	8	B	
Brass						
Centerless	37C	36	N	5	V	
Cylindrical	37C	36	K	5	V	
Internal	37C	36	J	5	V	
Surfacing (cups and cylinders)	37C	24	H	8	V	
Surfacing—(disks)	37C	20	M	4	B	
Snagging (floor stands), 5000–6500 sfpm	37C #12 Treated	24	P	5	V	
Snagging (floor stands), 7000–9500 sfpm	37C #12 Treated	24	P	4	B	5

Table 4. Grinding Wheel Recommendations—*Continued*

(Courtesy of Norton Co.)

Work—Material—Operation	Abrasive	Grain Size	Grade	Structure	Bond	Bonding Variation
Brass—*Continued*						
Cutting-off, gates and risers (dry, Tabor mach., 16,000 sfpm)	A	214	R	14	B	N
Cutting-off rod (dry) 9000–16,000 sfpm	A	30	U	7	B	
Cutting-off rod (wet) 7500–12,000 sfpm	A	60	V	8	R	30
Cutting-off tubing (wet) 9000–16,000 sfpm	A	120	V	8	R	30
Cutting-off tubing (dry) 7500–12,000 sfpm	A	80	V	8	R	30
Brick (Vitrified)						
Cutting-off (coping)	{ 37C	16	P		E	3
	{ 37C	24	Q		B	4
Cutting-off (portable saw)	37C	24	O	8	B	
Floor stands, 5000–6500 sfpm	37C	16	P	5	V	
Surfacing (cups and cylinders)	37C	20	M	5	V	
Surfacing (disks), hard	37C	16	K	4	B	
Surfacing (disks), refractory	37C	14	M	4	B	
Bronze (Soft)						
Use same wheels as for brass						
Bronze (Hard)						
Centerless	A	60	M	5	V	BE
Cylindrical	A	46	L	5	V	BE
Internal	A	60	L	5	V	BE
Snagging (floor stands), 5000–6500 sfpm	A	20	Q	8	V	BE
Snagging (floor stands), 7000–9500 sfpm	57A	20	O	4	B	5
Cutting-off (dry), 9000–16,000 sfpm	A	30	O	7	B	
Cutting-off, gates and risers (dry, Tabor mach., 16,000 sfpm)	37C	245	T	8	R	29
Surfacing (straight wheels)	{ 38A	60	J		V	
	{ 57A	60	J	5	V	BE
Surfacing—(disks)	37C	16	L	4	B	L
Carboloy (*See* Cemented Carbide)						
Carbon						
Surfacing	37C	16	J	8	V	
Surfacing (disks)	37C	36	M	5	B	
Carbon (Soft)						
Cutting-off (dry)	37C	36	N	8	B	K
					D	Sides
Cutting-off (wet)	37C	54	L	8	B	K
					D	Sides
Cutting-off	D	46	N	25	M	
Carbon (Hard)						
Centerless	37C	36	N	5	V	
Cutting-off (dry)	37C	46	N	8	B	K
					D	Sides
Cutting-off (wet)	37C	54	K	8	B	K
Cutting-off	D	60	L	50	M	
Carbon (Hard-plate)						
Stripping	37C	30	O	5	B	
Carbon (Metallic)						
Cutting-off	37C	36	K	8	B	
					D	Sides
Carmet (*See* Cemented Carbide)						
Cast Iron						
Centerless	37C	46	L	5	V	
Cutting-off (dry), 9000–16,000 sfpm	A	36	P	8	B	
Cylindrical	37C	36	J	5	V	

(*Table continued on p. 23-78*)

Table 4. Grinding Wheel Recommendations—*Continued*

(Courtesy of Norton Co.)

Work—Material—Operation	Abrasive	Grain Size	Grade	Structure	Bond	Bonding Variation
Cast Iron—*Continued*						
Internal	37C	46	J	5	V	
	37C	46	I	12	V	P
Surfacing (cups and cylinders)	37C	16	H	8	V	
	38A	601	F	12	V	BEP
Surfacing (segments)	38A	601	E+	12	V	BEP
Surfacing (disks)	37C	16	M	5	B	5
	37C	30	K	5	V	
Surfacing (straight wheels)	37C	30	J	8	V	
Snagging (floor stands), 5000–6500 sfpm	37C	20	R	5	V	
Snagging (floor stands), 7000–9500 sfpm	37C	16	P	2	B	
Snagging (swing frame), 5000–6500 sfpm	37C	16	S	5	V	
Snagging (swing frame), 7000–9500 sfpm	37C	14	Q	2	B	7
Snagging (portable grinder), 5000–6500 sfpm	37C	24	S	5	V	
Snagging (portable grinder), 7000–9500 sfpm	37C	20	Q	4	B	7
Catalin						
Cutting-off (dry)	37C	90	P		B	
Cutting-off (wet)	37C	100	R	0	R	30
Cylindrical	37C	46	J	5	V	
Cylindrical (form grinding)	37C	100	M	6	V	
Surfacing	37C	36	H	8	V	
Cemented Carbides						
Single-point tools, offhand						
Cup or plate mounted wheels						
Roughing (dry), 10″ and 14″ diam	39C	60	F	12	V	P
Less than 10″ diam	39C	60	I	7	V	
Roughing (wet), 10″ and 14″ diam.	39C	60	G	12	V	P
	D	100	L	100	V	
Semi-finishing (dry)	39C	100	H	7	V	
Semi-finishing (wet)	39C	100	I	7	V	
Finishing (wet)	D	320	P	100	V	
Combination R. & F. (wet)	D	150	P	100	V	
Straight wheels						
Roughing (dry)	39C	60	I	7	V	
Roughing (wet)	39C	60	J	7	V	
Semi-finishing (dry)	39C	100	H	7	V	
Semi-finishing (wet)	39C	100	I	7	V	
Single-point tools, machine grinding						
Straight wheels						
Roughing and finishing (wet)	39C	80	J	7	V	
Chip breaker grinding						
Straight wheels						
Less than 1/4″ thick	D	100S	N	100	BX	
1/4″ and thicker	D	150	N	100	V	
Milling cutters, reamers, etc.						
Backing-off, cup wheel						
Roughing (type D11B)	39C	60	I	7	V	
	D	100	L	100	B	
Finishing (type D11B)	D	150	L	100	B	
Surface grinding straight wheels						
Roughing (dry)	39C	60	I	7	V	
Roughing (wet)	39C	60	K	7	V	
	D	100	L	100	B	
Finishing (dry)	39C	100	J	7	V	
Finishing (wet)	39C	100	I	7	V	
	D	220	J	100	B	
Cylindrical grinding						
Roughing (wet)	39C	60	K	7	V	
	D	100	L	100	B	
Finishing (wet)	39C	100	J	7	V	
	D	220	J	100	B	
Internal grinding	D	150	J	100	B	

Table 4. Grinding Wheel Recommendations—*Continued*

(Courtesy of Norton Co.)

Work—Material—Operation	Abrasive	Grain Size	Grade	Structure	Bond	Bonding Variation
Cemented Carbides—*Continued*						
Hand honing or stoning						
Rectangular hone	D	320			V	
	37C	280	N		V	
Lapping						
Cup wheel (wet)	D	400	J	50	B	
Abrasive grain and cast iron disk		320	F			
Cutting-off	D	150	N	100	B	
Chilled iron						
Snagging (floor stands), 5000–6500 sfpm	37C	16	R	5	V	
Snagging (floor stands), 7000–9500 sfpm	37C	16	P	2	B	5
Surfacing (cups and cylinders)	37C	24	H	8	V	
Surfacing (straight wheels)	37C	36	I	8	V	
Cylindrical (*see* Rolls)						
Chromium Plating (Cylindrical)						
Commercial finish	A	100	J	7	V	BE
Good commercial finish	A	150	K	5	E	
High finish	A	500	I	9	E	
Clay (Blocks)						
Cutting-off	D	24	N	50	M	
Surfacing (cups and cylinders)	37C	14	P	4	B	
	37C	14	N	8	V	
Surfacing (segments)	37C	16	L	8	V	
Concrete						
Cutting-off solid wheel	37C	20	Q	8	B	
Cutting-off steel center wheel	37C	16	Q		B	
Surfacing						
hand	37C	24	R		V	
machine	37C	24	Q		V	
cup wheels, 6000 sfpm	37C	3022	N		V	
Copper						
Cylindrical (*also see* Rolls)	37C	100	I	9	V	
Surfacing (cups and cylinders)	37C	14	I	8	V	
Surfacing (disks)	37C	36	M	5	B	
Cutting-off						
Rod (dry)	A	60	V	8	R	30
Rod (wet)	A	80	V	8	R	30
Tubing (dry)	A	80	V	8	R	30
Tubing (wet)	A	120	V	8	R	30
Gates and risers (dry, Tabor mach., 16,000 sfpm)	A	241	R	14	B	N
Cork	38A	60	G	8	V	BE
	38A	80	F	12	V	BEP
Ebonite						
Cutting-off (dry)	37C	30	M	8	B	
Fiberglas Insulating Board						
Cutting to size, 11,000 sfpm	37C	20	Q	8	B	
Fiber						
Surfacing—Disks	37C	14	J	4	B	
Cutting-off (board)	37C	36	L	8	B	K
Firthite (*see* Cemented Carbide)						
Glass (Circular Mirrors)						
Rough beveling (automatic)	37C	80	H	5	V	
Rough and final beveling (automatic)	A	220	L	8	V	
Glass (Globes)						
Grinding edges	37C	80	L	8	V	

(*Table continued on p. 23-80*)

Table 4. Grinding Wheel Recommendations—*Continued*

(Courtesy of Norton Co.)

Work—Material—Operation	Abrasive	Grain Size	Grade	Structure	Bond	Bonding Variation
Glass (Lenses)						
Edging—machine	A	220	O	8	V	
	D	220	N	100	M	
Surfacing						
Roughing	D	180	M	100	M	
Finishing	D	400	J	00	B	
Glass (Plate)						
Edging	D	80	N	25	M	
Core drilling	D	100	N	100	M	
Glass Prisms						
Surfacing						
Roughing	D	120	N	50	M	
Finishing	D	400	L	100	B	
Glass (Tubing)						
Cutting-off (low speed wet), up to 7500 sfpm	37C	120	L	0	R	41
Cutting-off (wet)	D	180	L	25	M	
Cylindrical	37C	46	J	5	V	
Internal	37C	90	H	8	V	
Glass (Tumblers)						
Edging—lead glass	37C	120	P	8	V	
Edging—lime glass	A	150	J	8	V	
Glass (Windshield)						
Seaming—hand—side of wheel	37C	90	R	8	V	
Round edges—machine	A	80	N		V	
Semi-polish edge—hand	38A	150	O	8	V	
Edging—disks	37C	100	Q	8	V	
Cutting (wet—low speed), up to 7500 sfpm	37C	120	L	0	R	41
Granite						
Coping	D	24	N	25	M	
	37C	24	Q	4	B	4
Surfacing (planer wheels)	37C	14	N	2	B	
Molding, 5000–6500 sfpm	37C	36	K	5	V	
7000–9500 sfpm roughing	37C	12	O	4	B	2
finishing	37C	301	N	5	B	2
Surfacing (portable)	37C	16	H	4	B	
Cutting-off	37C	30	L	5	B	
Gypsum						
Cutting-off	37C	30	L	8	B	
Kennametal (*see* Cemented Carbide)						
Leather						
Misc. grinding	37C	30	J	5	V	
Suede finishing	37C	60	M	5	V	
Limestone						
Sawing, steel centered	37C	24	P		B	4
Sawing, inserted teeth	37C	30	U	8	B	2
Molding	37C	364	K	5	V	
Surfacing (drum rubber)	37C	24	K	5	V	
Cutting-off	37C	30	L	8	B	
Lucite						
Cutting-off (dry)	37C	46	N	8	B	
Cutting-off (wet)	37C	90	P	0	R	30
Magnesium						
Snagging						
5000–6500 sfpm	37C	24	N	5	V	
7000–9500 sfpm	37C	20	N	5	B	5

Table 4. Grinding Wheel Recommendations—*Continued*

(Courtesy of Norton Co.)

Work—Material—Operation	Abrasive	Grain Size	Grade	Structure	Bond	Bonding Variation
Malleable Castings (Annealed)						
Floor stands, 5000–6500 sfpm	15A	16	R	5	V	BE
Floor stands, 7000–9500 sfpm	{ 57A	16	P	4	B	5
	{ A	16	T	6	R	23
Swing frames, 5000–6500 sfpm	15A	14	R	5	V	BE
Swing frames, 7000–9500 sfpm	A	14	P	4	B	5
Portable grinders, 5000–6500 sfpm	A	20	Q	5	V	BE
Portable grinders, 7000–9500 sfpm	A	20	Q	4	B	H
Malleable Castings (Unannealed)						
Floor stands, 5000–6500 sfpm	37C	20	R	5	V	
Floor stands, 7000–9500 sfpm	{ 37C	20	P	2	B	5
	{ A	20	T	6	R	23
Surfacing—disks	A	20	L	4	B	5
Marble						
Balusters (rough turning)	37C	10	M		V	
Balusters (finish turning)	37C	364	K	5	V	
Coping (inserted teeth)	37C	30	P	3	B	4
Sawing	{ 37C	24	Q		E	3
	{ D	24	N	50	M	
Surfacing (segments)						
1st gritting	{ D	60	N	100	M	
	{ 37C	60	K		V	
2nd gritting	37C	120	K		V	
3rd gritting	A	220	I	9	V	
Surfacing (wheel)						
Roughing	37C	10	M		V	
Finishing	37C	364	K	5	V	
Molding, 5000–6500 sfpm						
Roughing	37C	10	M	5	V	
Finishing	37C	364	K	5	V	
Molding, 7000–9500 sfpm						
Roughing	37C	12	M	4	B	
Finishing	37C	36	N	5	B	
Cutting-off	37C	30	P	8	B	
Metallized (*see* Sprayed Metal)						
Metallographic Specimens						
Offhand (1st operation)	57A	46	L	5	V	BE
Offhand (2nd operation)	37C	120	M	6	V	
Machine surfacing	38A	46	J	5	V	BE
Cutting-off (wet)	A	60	M	6	R	30
Minerals—Lapidary (Offhand)						
Moh hardness 7 or less						
Roughing	37C	60	L	5	V	
Finishing	37C	180	K	8	V	
Cutting (wet)	{ 37C	60	L	0	R	41
	{ D	60			M	
Moh hardness over 7						
Roughing	37C	60	J	5	V	
Finishing	37C	180	K	8	V	
Cutting (wet)	{ D	100	L	25	M	
	{ D	80			M	
Monel Metal						
Cutting-off (dry)	A	36	N	8	B	
Cutting-off (wet)	A	60	T	6	R	1
Floor stands, 5000–6500 sfpm	A	24	Q	5	V	BE
Floor stands, 7000–9500 sfpm	57A	20	P	4	B	5
Cylindrical	37C	60	J	8	V	
Nickel						
Cutting-off (dry)	A	46	Q	8	B	
Cutting-off (wet)	A	60	T	6	R	1

(Table continued on p. 23-82)

Table 4. Grinding Wheel Recommendations—*Continued*

(Courtesy of Norton Co.)

Work—Material—Operation	Abrasive	Grain Size	Grade	Structure	Bond	Bonding Variation
Nitralloy						
Before nitriding	38A	60	I	12	V	BEP
	A	46	K	5	V	BE
After nitriding	38A	60	H	8	V	BE
	37C	100	I	8	V	
Plastics						
Cutting-off (dry)	37C	60	N	8	B	
Cutting-off (wet)	37C	90	P	0	R	30
Porcelain						
Cutting-off (wet), fired	D	100	L	100	M	
	37C	120	L	0	R	41
Cutting-off (low speed) (dry), prefired	37C	46	J	8	B	
Cylindrical	37C	46	J	5	V	
Removing imperfections (bricks)	38A	80	M		V	BE
Centerless	37C	36	K	5	V	
Rubber (Soft)						
Cylindrical	37C	24	J	12	V	P
	37C	24	K	5	B	
	37C	24	K	2	E	
Rubber (Hard)						
Cutting-off	37C	30	M	8	B	
Cylindrical	37C	30	K	5	B	
Sand Cores						
Cutting-off (dry)	37C	24	N	8	B	
Slate						
Cutting-off (portable)	37C	24	O	8	B	
Coping	37C	24	Q		E	3
Molding	37C	46	K	5	V	
Sprayed Metal						
Cylindrical	37C	60	J	5	V	
Cutting-off (wet), 9000–12,000 sfpm	A	46	U	8	R	30
Steel Castings (Low Carbon)						
Swing frames, 5000–6500 sfpm	15A	14	R	5	V	BE
Swing frames, 7000–9500 sfpm	A	14	P	4	B	5
Floor stands, 5000–6500 sfpm	15A	16	P	5	V	BE
	.0115 Treated					
Floor stands, 7000–9500 sfpm	A	16	P	4	B	5
	.0115 Treated					
Portable grinders, 5000–6500 sfpm	A	20	Q	5	V	BE
Portable grinders, 7000–9500 sfpm	A	20	Q	4	B	H
	A	20	T	6	R	30
Steel Castings (Manganese)						
Swing frames, 5000–6500 sfpm	A	20	Q	5	V	BE
Swing frames, 7000–9500 sfpm	A	14	Q	4	B	5
Floor stands, 5000–6500 sfpm	A	20	P	5	V	BE
Floor stands, 7000–9500 sfpm	A	16	P	4	B	5
Portable grinders, 5000–6500 sfpm	A	24	Q	5	V	BE
Portable grinders, 7000–9500 sfpm	A	20	Q	4	B	5
Surfacing—planer type, 5000–6500 sfpm	57A	14	P	5	V	BE
Surfacing—planer type, 7000–9500 sfpm	A	16	O	4	B	
Surfacing—lathes and boring mills, 5000–6500 sfpm	A	16	P	8	V	BE
Surfacing—lathes and boring mills, 7000–9500 sfpm	A	20	O	4	B	
Internal—rough grinding, 5000–6500 sfpm	A	20	O	5	V	BE
Internal—rough grinding, 7000–9500 sfpm	A	24	Q	4	B	H

Table 4. Grinding Wheel Recommendations—*Continued*

(Courtesy of Norton Co.)

Work—Material—Operation	Abrasive	Grain Size	Grade	Structure	Bond	Bonding Variation
Steel (Hardened)						
Centerless	57A	60	M	5	V	BE
Centerless (fine finish)	A	80	T	8	R	23
	A	80	R	8	R	22
Cylindrical	A	60	L	5	V	BE
Surfacing (segments)	32A	30	E	12	V	BEP
Surfacing (cups and cylinders)	38A	30	G	8	V	BE
Surfacing—(disks), roughing	A	16	J	4	B	
Surfacing—(disks), finishing	A	46	J	5	B	
Surfacing (straight wheels)	32A	46	H	8	V	BE
	32A	60	G	12	V	BEP
Internal	32A	60	J	5	V	BE
Cutting-off (wet), 9000-12,000 sfpm	A	60	P	0	R	30
Cutting-off (dry), 9000–12,000 sfpm	A	46	M	8	B	
Steel (Soft)						
Centerless	57A	60	N	5	V	BE
Cylindrical	A	54	N	5	V	BE
Surfacing (segments)	38A	30	E	12	V	BEP
Surfacing (cups and cylinders)	38A	24	J	8	V	BE
Surfacing (disks) roughing	A	16	L	4	B	
Surfacing (disks) finishing	A	46	K	5	B	
Surfacing (straight wheels) narrow	38A	46	J	5	V	BE
Surfacing (straight wheels) wide	38A	36	L	8	V	BE
Internal grinding	A	46	M	5	V	BE
Cutting-off (dry), 12,000–16,000 sfpm	A	245	Q	8	B	
Cutting-off (wet) 9000–12,000 sfpm	A	46	V	8	R	30
Steel (High Speed)						
Centerless	57A	60	L	5	V	BE
Cylindrical	38A	60	K	5	V	BE
Surfacing (segments)	32A	30	E	12	V	BEP
Surfacing (cups and cylinders)	38A	46	G	8	V	BE
Surfacing (straight wheels)	32A	46	H	8	V	BE
	32A	60	F	12	V	BEP
Cutting-off (soft, dry), 12,000–16,000 sfpm	A	24	Q	8	B	
Cutting-off (soft, wet) 9000–12,000 sfpm	A	46	V	8	R	30
Cutting-off (hard, dry), 9000–12,000 sfpm	A	46	M	8	B	
Cutting-off (hard, wet), 9000–12,000 sfpm	A	60	P	0	R	30
Internal	32A	46	J	5	V	BE
Billet Grinding						
Annealed Billets 5000–6500 sfpm	57A	14	Q	5	V	BE
Annealed Billets, 7000–9500 sfpm	A	14	Q	4	B	H
Unannealed Billets, 5000–6500 sfpm	A	16	P	5	V	BE
Unannealed Billets, 7000–9500 sfpm	A	16	O	4	B	
Steel (Stainless, Free Machining)						
Centerless	37C	46	M	5	V	
Cylindrical	37C	46	M	5	V	
Surfacing (cups and cylinders)	38A	36	G	8	V	BE
Surfacing (segments)	32A	601	E+	12	V	BEP
Surfacing—straight wheels	32A	46	H	8	V	BE
Cutting-off (dry), 12,000–16,000 sfpm	A	46	Q	8	B	
Cutting-off (wet), 9000–12,000 sfpm	A	46	V	8	R	30
Tubing—Cutting-off						
Dry 12,000–16,000 sfpm	A	80	V	10	R	29
Wet 9000–12,000 sfpm	A	80	U	10	R	29
Billets and slabs						
Swing frames, 7000–9500 sfpm	A	123	R	4	B	H
	A	16	V	8	R	30
Portable Grinders, 7000–9500 sfpm	A	20	Q	4	B	H
	A	24	T	6	R	30
Steel (Stainless, Hardened)						
Centerless	57A	60	L	5	V	BE
Cylindrical	A	60	L	5	V	BE

(*Table continued on p. 23-84*)

Table 4. Grinding Wheel Recommendations—*Continued*

(Courtesy of Norton Co.)

Work—Material—Operation	Abrasive	Grain Size	Grade	Structure	Bond	Bonding Variation
Steel (Stainless, Hardened)—*Continued*						
Cutting-off (dry)	A	46	M	8	B	
Cutting-off (wet), 9000–12,000 sfpm	A	60	P	0	R	30
Hemming and Klotz						
Cutlery, small						
Roughing	A	60	L	7	B	4
Finishing	A	100	P	7	E	3
Surfacing (cylinders and cups)	38A	24	G	8	V	BE
(Segments)	38A	301	E	12	V	BEP
(Straight wheels)	38A	46	H	8	V	BE
Internal	A	60	L	5	V	BE
Stellite (also Rexalloy, Tantung)						
Cylindrical	A	46	M	5	V	BE
Cutter grinding	32A	46	J	5	V	BE
Surfacing (cups and cylinders)	38A	46	G	5	S	
Surfacing (disks)	A	36	L	5	V	BE
Surfacing (straight wheels)	32A	46	H	8	V	BE
Cutting-off (thin wheel 1/32 in. thick)	A	60	O	8	B	
Tools						
Offhand	A	46	N	5	V	BE
Machine	38A	46	L	5	V	BE
Stone (Artificial)						
Surfacing	37C	364	K	5	V	
Coping	37C	30	P	4	B	4
Stoodite						
Offhand grinding	A	30	P	5	V	BE
Tantalum Carbide (*see* Cemented Carbide)						
Tile						
Cutting-off	D	46	L	25	M	
Cutting-off (dry), 9000–12,000 sfpm	37C	24	O	8	B	
Cutting-off (wet—low speed), up to 7500 sfpm	37C	36	N	8	B	
Titanium Carbide (*see* Cemented Carbide)						
Transite Board						
Cutting to size (dry)	37C	24	L	8	B	
Tungsten Carbide (*see* Cemented Carbide)						
Tungsten Rods						
Cutting-off	A	150	U	10	R	36
Centerless	37C	60	J	8	V	
Vascaloy-Ramet (*see* Cemented Carbide)						
Welds						
Carbon alloy steels						
Portable grinders, 5000–6500 sfpm	A	24	Q	5	V	BE
Portable grinders, 7000–9500 sfpm	A	16	R	4	B	7
Stainless steel						
Portable grinders, 7000–9500 sfpm	A	24	T	6	R	30
Wrought Iron						
Floor stands, 5000–6000 sfpm	A	16	R	7	V	BE

27. WHEEL TYPES, SPEEDS

WHEEL SHAPES. While a tremendous variety of wheel shapes are in use and can be had, industry has standardized on wheel shapes which can be obtained from stock. Nonstandard wheels, however, can be obtained on order. The standard shapes of wheels and wheel faces are shown in Figs. 1 and 2. Wherever possible, the design of parts which are to be ground should contemplate use of the simplest possible standard wheel shape for maximum economy.

OPERATING SPEED. Not all types of wheels possess equal strength; consequently it is not safe to operate them all at equal speeds. As a general rule, vitrified and silicate-bonded wheels are classified in one group and the organic-bonded (shellac, rubber, and resinoid) in another. For vitrified and silicate wheels, which have relatively low elastic modulus, a maximum operating speed of 6500 sfpm has been established. Wheels in the softer grades and of relatively coarse grain size should not be operated as fast as this. The organic-bonded wheels are stronger and can be operated at higher speeds with safety. A maximum operating speed of 9500 sfpm has been established for this group, with the exception of thin cut-off wheels, which may be operated up to 16,000 sfpm under certain conditions. In the organic-bonded group, as in the vitrified and silicate, the operating speeds for the softer and coarser wheels should be lower.

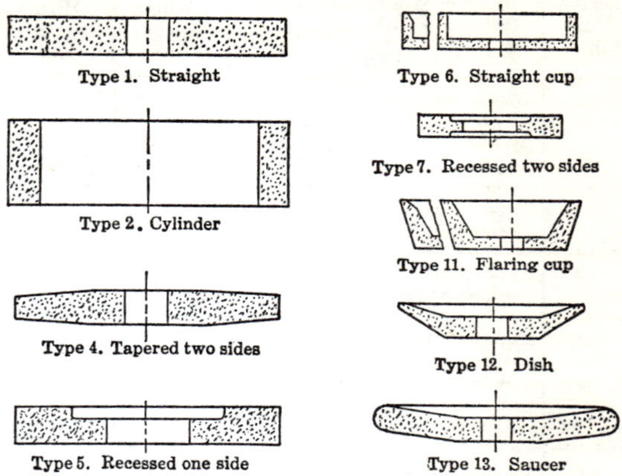

FIG. 1. Standard grinding wheel shapes.

It is general practice for the wheel manufacturer to show the maximum operating speed on the blotters, or tag, accompanying the wheel. It is also standard practice to test all wheels 6 in. in diameter, and larger, at speeds sufficiently higher to insure a reasonable factor of safety. Table 5 shows maximum peripheral speed per minute permitted by the ASA Safety Code for various types of wheels and bonds.

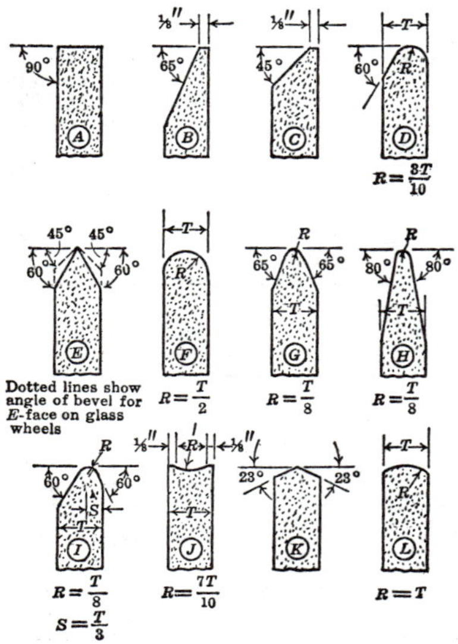

FIG. 2. Standard shapes of grinding wheel faces.

INTERNAL GRINDING OF SMALL AND MEDIUM-SIZE HOLES. C. R. Alden (Ref. 1) describes the grinding of holes as small as $1/16$ in. in diameter. An air turbine was used to give a spindle speed of 65,000 rpm. The wheel was a mounted point, that is, an abrasive tip cemented on a steel shank, substantially less in diameter than the tip, which, in turn, was smaller than the hole to be ground. An important consideration in the internal grinding of small holes is the type of grinding spindle and the bearings used. A solid projection-type of spindle, in which the wheel is mounted on an integral projection of the spindle, was found to be best. The projection should be of the largest possible diameter and shortest possible length consistent with the diameter of the hole to be ground. This will limit the amount of wheel wear that can be permitted, since allowance of excessive wheel wear will necessitate a spindle and bearings that will lack rigidity. It is recommended that wheels used for grinding 2-in. holes be allowed $1/2$ in. reduction of diameter by wheel wear, and that they then be transferred to the grinding of 1 $1/2$ in. holes, with a further allowance of $3/8$ in. for wheel wear, after which they could be used for the grinding of 1 $1/8$ in. or 1-in. holes. The wheel thus can be used up, and at the same time the conditions under which it works will be those of maximum efficiency. Figure 3 shows the spindle speeds recommended for the grinding of small holes. It should be noted that high-frequency motor-driven spindles have been developed and are expected to prove satisfactory along with the air turbine type of spindle for the speeds in excess of those conveniently obtainable through belt drives.

Table 5. Maximum Permissible Speeds for Grinding Wheels

Classification Number	Types of Wheels (See also Fig. 1)	Vitrified and Silicate Bonds			Organic Bonds		
		Low Strength, fpm	Medium Strength, fpm	High Strength, fpm	Low Strength, fpm	Medium Strength, fpm	High Strength, fpm
1	Type 1—Straight wheels (Including plate mounted and inserted nut wheels) Type 4—Taper wheels	5500	6000	6500	6500	8000	9500
2	Types 5 and 7—Recessed wheels	5500	6000	6500	6500	8000	9500
3	Type 2—Cylinder wheels (Including plate mounted and inserted nut wheels)	4500	5500	6000	6000	8000	9500
4	Dovetail wheels Types 11 and 12—Dish and flaring cup wheels Type 13—Saucer wheels	4500	5500	6000	6000	8000	9500
5	Type 6—Deep recessed cup wheels	4500	5000	5500	6000	7500	9000
6	Cutting wheels larger than 16 in. in diameter						7500 to 14,000
7	Cutting wheels 16 in. and smaller						10,000 to 16,000
8	Thread grinding wheels	5500 to 8000	6000 to 10,000	6500 to 12,000			9500 to 12,000
9	Automotive and aircraft crank grinding	5500	6000 to 7300	6500 to 8500			
10	Automotive and aircraft cam grinding	5500	6000 to 8000	6500 to 8500			
11	Diamond wheels	Cutting wheels: Metal bonded with steel center Metal bonded with metallic compound center Resin bonded with resin or metallic compound center All other types					 14,000 7500 7500 6500

satisfactory along with the air turbine type of spindle for the speeds in excess of those conveniently obtainable through belt drives.

GRAIN DEPTH OF CUT. An analysis of the action of the wheel when in operation shows how theoretical considerations bear out the truth of the empirical rules for the use of grinding wheels in machine grinding. A paper by George I. Alden (Ref. 2) gives the essential distinction between the radial or real depth at which the wheel cuts and the depth which the abrasive grain in the wheel cuts into the material being ground. The latter depth is called the *grain depth of cut.* This grain depth of cut is the controlling factor in securing the correct working of the wheel. A formula is deduced for computing the grain depth of cut, the application of the analysis is explained, and these conclusions reached by Professor Alden: (1) Other factors remaining constant, increase of work speed increases grain depth of cut, and makes a wheel appear softer. (2) A decrease of wheel speed increases grain depth of cut. (3) Diminishing the diameter of the wheel increases the grain depth of cut; increasing the diameter of the wheel decreases the grain depth of cut. (4) Decreasing the diameter of work increases the grain depth of cut; conversely, increasing the diameter of work decreases the grain depth of cut. A table of arcs of contact of wheel and work for a limited range of diameters is given, also a table of values of one of the factors in the formula for grain depth of cut.

W. H. Chapman (Ref. 3), basing his study on Alden's paper set forth above, gives a mathematical analysis of the laws involved in cylindrical grinding and an analysis for grinding action for both draw-in and traverse cuts, deriving formulas and from them drawing practical conclusions as to the best methods to follow for maximum grinding efficiency.

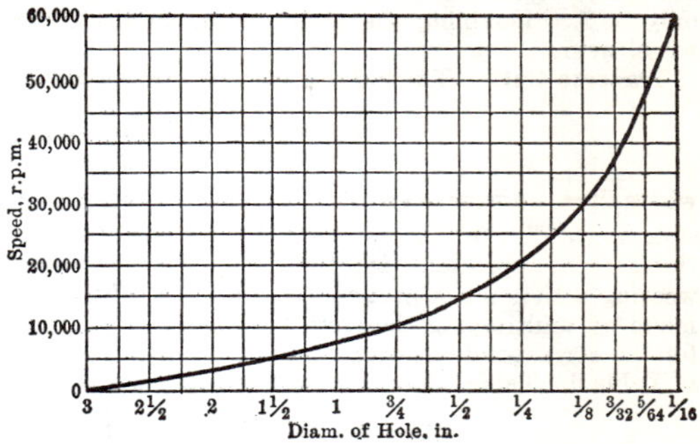

Fig. 3. Wheel speeds for grinding small holes.

Draw-in Cuts. For draw-in cuts, the maximum depth of chip, in inches, corresponding to Alden's grain depth of cut, is given by

$$G = \frac{\mu u}{V}\sqrt{\frac{2f}{r}} \text{ approximately} \quad (1)$$

where G = maximum depth of chip, inches; μ = intervals between grains of wheel in fractions of an inch; u = work speed, feet per minute; V = surface speed of wheel, feet per minute; f = depth of cut, inches, on work radius; r = radius of work, inches.

The index of the wheel-wearing action, W_1, due to grinding, is

$$W_1 = \frac{\mu u}{V^2}\sqrt{\frac{2f}{r}} \quad (2)$$

The actual wheel-wear can be computed by multiplying the value of W_1 by a constant derived by experiment, which may be taken as (1800×10^4). The amount of metal removed per minute may be determined by

$$P = 2\pi r f w \times \text{rpm} \quad (3)$$

where P = cubic inches of metal removed per minute; w = width of wheel, inches, and rpm = speed of rotation of the work, other notation as before.

Table 6 compares wheel wear and quantity of metal removed at different work speeds in a given machine, as calculated from the above formulas. Note that wheel-wear values increase rapidly as work speed increases. The value of 0.0166 for work 1 in. diameter at 53 rpm becomes 0.1040 for work 4 in. diameter at 167 rpm. A wheel sufficiently soft to cut freely in the first case would wear so fast as to be nearly useless in the second case.

Table 6. Comparison between Wheel Wear and Quantity of Metal Removed at Different Work Speeds

(Wheel diameter, 18 in.; surface speed, 6000 ft per min; diametral feed per work revolution (automatic feed), 0.001 in.; grain of wheel, 46 Alundum; material ground, mild machinery steel.)

Work Speed, rpm		Work Diameters			
		1 in.	2 in.	3 in.	4 in.
53	Production *	.0835	.1670	.2500	0.3340
	Wheel wear	.0166	.0236	.0290	0.0334
72.5	Production *	.1140	.2280	.3420	0.4560
	Wheel wear	.0226	.0319	.0390	0.0450
87	Production *	.1370	.2740	.4110	0.5480
	Wheel wear	.0273	.0387	.0475	0.0546
101	Production *	.1586	.3172	.4758	0.6344
	Wheel wear	.0316	.0448	.0540	0.0635
138	Production *	.2160	.4320	.6480	0.8640
	Wheel wear	.0433	.0612	.0645	0.0865
167	Production *	.2620	.5240	.7860	1.0480
	Wheel wear	.0522	.0740	.0910	0.1040

* In cubic inches of metal removed per minute.

Where the wheel-wear values are nearly the same, however, the same wheel will act properly in both cases, as, for instance, on 1-in. work at 138 rpm and 4-in. work at 72.5 rpm. From Table 6, and by experimental tests with various wheels, certain ranges of wheel-wear values may be found to correspond to different grades of suitable grain size and abrasive for the material to be ground, and wheel-wear indices may be computed for any set of conditions to enable the proper wheel to be selected.

Production can be increased with a minimum increase in wheel wear by an increase in the feed, up to the limit of the grain penetration. An example in the paper shows that doubling the feed approximately doubled the production but increased the wheel wear only 35%. Doubling the work speed, on the contrary, doubled both production and wheel wear.

Traverse Cuts. The index of wheel wear for traverse cuts is

$$W_t = \frac{\mu u}{V^2} \frac{u}{\sqrt{T^2 + u^2}} \sqrt{\frac{2f}{r}} \tag{4}$$

where T = traverse speed, feet per minute, other notation being as before. The expression $(u/(\sqrt{T^2 + u^2}))$ is known as the traverse factor. Where T is large with respect to u the factor is appreciably less than unity, but as the speed u increases the reduction in wheel wearing action is less pronounced. A table of wheel-wear values for traverse cuts can be made by multiplying the traverse factors determined as above by the wheel-wear value for the corresponding work speed and work diameter for draw-in cuts, such as are given in Table 6. Table 7 is such a table compiled for the same machine as Table 6. By assigning certain wheels to specified ranges of wheel-wearing action, the proper wheel for traverse cuts may be selected from Table 7. The wheel used, however, would be somewhat softer in the traverse cut for the same index figure, because the actual depth of cut does not follow the indicated depth, as shown by the feeding mechanism, as closely for traverse work as for draw-in work. The following wheels are found to be satisfactory for traverse cuts on mild steel under the assumed conditions:

Table 7. Wheel-wear Values for Traversed Cuts

(For same conditions as given for Table 6)

Traverse, ft per min	Work, rpm	Lead, in.	Work Diameters 1 in.	2 in.	3 in.	4 in.
10.02	53	2.3	.01340	.02240	.02820	.03280
14.00	53	3.1	.01170	.02080	.02740	.03230
	72.5	2.3	.01820	.02450	.03775	.04390
17.00	53	3.8	.01050	.02005	.02685	.03185
	72.5	2.8	.01650	.02900	.03720	.04370
	87	2.3	.02190	.03620	.04620	.05390
19.7	53	4.4	.00955	.01930	.02620	.03142
	72.5	3.2	.01560	.02820	.03675	.04350
	87	2.7	.02060	.03540	.04570	.05355
	101	2.3	.02540	.04185	.05245	.06260
26.9	53	5.9	.00763	.01695	.02430	.03000
	72.5	4.4	.01290	.02600	.03520	.04240
	87	3.6	.01760	.03320	.04420	.05235
	101	3.1	.02220	.03980	.05125	.06185
	138	2.3	.03470	.05130	.06195	.08500
32.6	53	7.35	.00660	.01545	.02290	.02885
	72.5	5.3	.01140	.02420	.03380	.04130
	87	4.4	.01580	.03150	.04315	.05160
	101	3.8	.02000	.03820	.05000	.06100
	138	2.8	.03230	.05580	.06095	.08450
	167	2.3	.04210	.06940	.08800	.10230

Wheel designation	36 or 46I	36 or 46J	36 or 46K	36 or 46L	36 or 46M
Wheel-wear values, in Table 6	0.0250	0.0250–0.0400	0.0400–0.0500	0.0500–0.0600	0.0600

Practical Conclusions. These practical conclusions are drawn from the foregoing mathematical analysis: The power drive must maintain the speed constant except as changed by the operator. The wheel must be in good running balance and in absolute truth, and held in position relative to the work within the closest possible limits. The work must be accurately held with respect to the wheel and must be uniformly rotated. The relative traverse between work and wheel must be uniform. The work must be rigidly supported over its entire length and no vibration allowed to occur between centers. Feeding must be at such a rate that the feed increment never exceeds the maximum grain penetration. Work must be kept at a uniform temperature and local heating prevented. A copious supply of grinding compound should be directed to the arc of contact at all times. In truing a wheel with a diamond, the cooling medium is vital to accuracy. Never feed a diamond over 0.001 in. per traverse. Do not try to use a single wheel for a variety of work sizes or materials.

FACTOR OF SAFETY. Since the centrifugal force tending to break a wheel increases as the square of the speed, it is, at 9000 sfpm 2.3 times greater than at 6000 sfpm, and 3.3 times greater than at 5000 sfpm. These values, 2.3 and 3.3, are called the established factor of safety for the operating speeds mentioned.

$$\text{Established factor of safety} = \text{testing speed}^2 \div \text{operating speed}^2$$

$$\text{True factor of safety} = \text{breaking speed}^2 \div \text{operating speed}^2$$

The true factor of safety cannot be calculated unless the breaking speed actually is measured. Experience shows the true factor of safety to be, usually, many times greater than the established factor.

The forces tending to break grinding wheels are due to other causes besides speed of rotation. Among them are poor mountings, wheel flanges of unequal diameter, unrelieved flanges, expansion from excessive and rapid heating of one periphery, work catching between wheel and work rest, improper grinding on side of wheel, and excessive out-of-truth or out-of-balance conditions. These conditions are cumulative and increase the breaking stresses due to rotation. Since these conditions exist, the established factor of safety is kept large. If the grinding conditions are correct, the wheel which has been subjected to the speed test should not give trouble, as it has been established that it is sufficiently strong to withstand the stress due to ordinary grinding conditions. A thorough investigation of a grinding wheel broken in operation usually will show the breakage to have been due to faulty operating conditions and not to the wheel itself.

28. GRINDING MACHINES

Grinding machines may be classified as cylindrical, internal, centerless, surface, and special grinders, based on the general character of the work. Each of these classes may be further subdivided according to special characteristics either of work or machine. The following paragraphs cover the types of grinding machines most commonly used, together with their fields and limitations.

CYLINDRICAL GRINDING MACHINES. Plain Grinders. Cylindrical grinding machines for general-purpose plain grinding of cylindrical surfaces such as spindles and axles, range from 3 to 36 in. swing. Standard sizes are 3 by 12 in.; 4 by 18 in.; 6 by 18 in.; 6 by 30 in.; 10 by 18 in.; 10 by 36 in.; 10 by 48 in.; 10 by 72 in.; 10 by 96 in.; 10 by 120 in.; 14 by 36 in.; 14 by 48 in.; 14 by 72 in.; 14 by 96 in.; 14 by 120 in.; 14 by 144 in.; 14 by 168 in.; 16 by 36 in.; 16 by 48 in.; 16 by 72 in.; 16 by 96 in.; 16 by 120 in.; 16 by 144 in.; and 16 by 168 in. Machines 20 in. and larger generally start at 96 in. and run up to 192 in. in 2-ft increments.

Machines are built with hydraulic, mechanical, or hand traverse, or with all three, of both table and grinding wheel head. Grinding wheel and work drives usually are from separate motors through V belts. Work tables of the machines have power headstocks and sliding tailstocks. Mounted on the longitudinally traveling main table of the machine, the work table can be swiveled up to about ten degrees relative to the main table travel, adjustment varying with different size machines.

Cylindrical grinders can be built to function automatically or semi-automatically, through a complete cycle, producing work ground within predetermined limits. They also may be fitted with attachments for cam, roll, or crankshaft grinding, although for regular production of this character special machines are advisable. For grinding faces of shoulders and the adjacent cylindrical surfaces, cylindrical grinders may be built with the wheel slide set at an angle of 45 degrees to the axis of the work, a V-shaped wheel being used. (See Fig. 4.)

Universal Grinders. Second most widely used grinding machine for general-purpose work, the universal machine, was originally a tool-room grinder but is now a useful production unit. In addition to features offered by plain grinders, universal machines are provided with a swiveling headstock and a swiveling wheel head. Completely flexible, these machines are well suited for production of small- or large-quantity lots of parts requiring face as well as cylindrical and taper grinding. Standard machines range in size from 12 to 18 in. swing and from 24 to 72 in. between centers.

INTERNAL GRINDERS comprise a reciprocating wheel head carrying a grinding wheel on a spindle long enough to reach to the bottom of the hole, and a work head in which the work can be rotated while in contact with the wheel, both carried on a suitable bed or frame, with provision for adjustment of both wheel and work. The field of the internal grinder is the finishing of holes in gears, bushings, bearings, etc., and the finishing of the inside surface of cylinders, as automotive cylinders. The cylinder grinder is a special form of internal grinder, and in one type the wheel head gives a planetary motion to the wheel spindle. H. L. Blood (Ref. 4) describes an automatic internal grinder which rough-grinds, trues the wheel, finish-grinds the hole to size, runs the wheel out of the hole, and brings all parts to the rest position. The range of capacity of internal grinders is holes from 0.040-in. to 50-in. in diameter up to 16 in. deep.

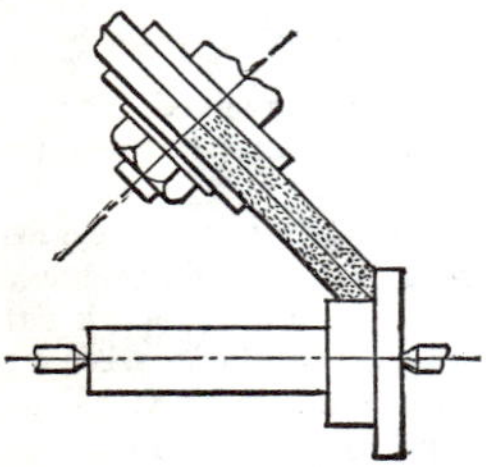

FIG. 4. Arrangement for grinding shoulders.

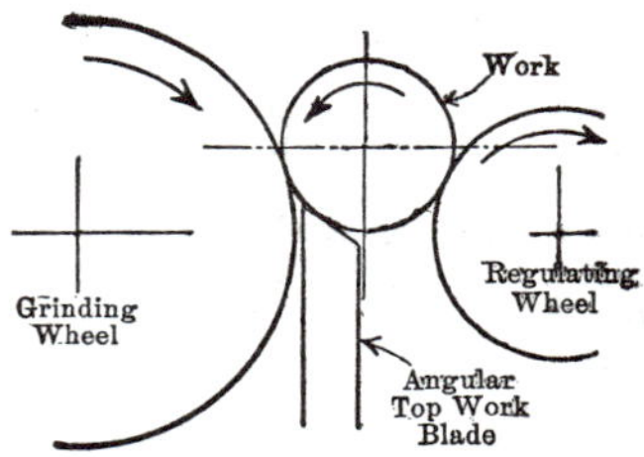

FIG. 5. Principle of centerless grinding.

CENTERLESS GRINDERS are a special form of cylindrical grinder. Instead of the work being supported on centers, it is supported on a work rest and held against a regulating wheel. (See Fig. 5.) By setting the regulating wheel at an angle to the grinding wheel, the former will act as a feeding device and will traverse the work past the grinding wheel. The regulating wheel is positively driven, and its speed determines the rotational speed of the work. The centerless grinder operates on a diametral dimension instead of on a radial dimension of the work, as in grinders where the work is held on centers, and therefore under certain conditions it will remove double the amount of metal in a given time. On slender work, much higher grinding pressure can be used in the centerless machine, as the work is rigidly supported by the work rest and regulating wheel, whereas when supported on centers it tends to spring away from the wheel under heavy pressure.

External centerless grinders can be either of the through-feed or in-feed type. With through-feed, machines available handle parts from about 0.020 in. to about 7 in. in diameter and lengths up to 20 ft or more. Internal centerless machines handle parts with bores from 1/4 to 24 in. Belt centerless external grinders handle diameters up to 9 in. by practically any length. Centerless thread grinding machines grind from the solid threaded parts or shafts with 8- to 80- pitch threads, 1/8 to 36 in. in length and up to 5 in. in diameter.

SURFACE GRINDING MACHINES. Surface grinders are divided into several distinct classes, such as (1) horizontal spindle, reciprocating table, using periphery of wheels; (2) horizontal spindle face grinder with reciprocating table, using segments; (3) horizontal spindle, rotary table; (4) vertical spindle, rotary table; and (5) vertical spindle, reciprocating table. Surface grinding machines vary all the way from small tool-room machines, using 7 by 1/2 in. wheels, to massive face grinding machines, using 48 and even 60 in. diameter segmental wheel chucks.

Reciprocating tables usually are traversed hydraulically. In certain machines the rotating table can be set at an angle to permit grinding of concave or convex surfaces. Both direct motor drive and belt drive of grinding wheels are available. The capacity of the standard reciprocating table grinders ranges from work 6 by 18 in. to 30 by 192 in. Rotating-table surface grinders are built with tables up to 60 in. in diameter. Work usually is held on the table by magnetic chucks.

Disk grinders are a form of surface grinder using abrasive disks mounted on a steel backing. The work is ground against the side of the disk. Single-spindle machines, with a disk on each end of the spindle, are built to use disks up to 40 in. in diameter. The work is supported on tables, equipped with either hand or power feed. The spindles are driven either direct from the motor or through V belts.

The double-spindle disk grinder is a machine for grinding two parallel surfaces. It comprises, virtually, two single-end disk grinders, mounted on a common base, the wheel spindles being in line, with the wheels facing each other. The work, supported by a table, is fed in between the wheels, which are spaced the correct distance apart. Wheel spindles are driven direct from the motor or through V belts. These machines are built in various sizes, with disks ranging from 15 to 72 in. in diameter and with maximum opening between wheels ranging from 20 to 30 in.

Face grinders are a form of surface grinder in which the work travels past the side of a wheel, carried on a horizontal spindle mounted at right angles to the line of travel of the work. The wheel grinds the vertical surface of the work at right angles to the surface of the table. The table is traversed hydraulically, and the wheel is driven by silent chain or V belt. Work can be held by bolting to the table or to angle plates or by magnetic chucks. These machines are used for surface grinding and for the grinding of machine knives, shear blades, etc. Tables range from 22 to 27 in. wide and from 60 to 168 in. long.

SPECIAL GRINDING MACHINES. Tool and cutter grinders are designed primarily for grinding of reamers, milling cutters, etc. In their simplest form they comprise a motor-driven grinding wheel and a pair of centers upon which the work is mounted, carried on a table that can be traversed past the wheel by hand. More elaborate machines comprise a power-operated swiveling table, a power-driven headstock to permit the machine to be used for cylindrical grinding, and attachments that permit of internal grinding and various forms of surface grinding. The capacity of tool grinders ranges usually up to 12 in. diameter swing by 20 to 27 in. between centers.

Roll Grinders. Roll grinding machines are cylindrical grinders of 20, 24, 28, 32, and 36 in. swings by 96, 120, 144, 168, or 192 in. lengths between centers. These are standard sizes. Special machines have been built up to 60 in. swing by 26 ft long between centers. Roll grinding machines are built with the work traversing past a grinding wheel in a fixed position or with the wheel traversing past the work. Rolls are crowned or concaved by moving the wheel, by appropriate mechanism, away from, or toward, the wheel as the wheel traverses it.

Cam and shape grinders are a special form of cylindrical grinder. In these machines, the work is rotated in a swinging member and is moved toward and away from the grinding wheel by means of a master cam in contact with a roller. The master cam is on the swinging member, the roller being fixed on the machine. Where the cams are integral with the camshaft, they are carried on centers in the machine, one pair being in the master cam spindle and the other in a footstock mounted on the swinging member. Single cams are mounted in fixtures or on arbors carried by the master cam spindle. For high production, as automotive camshafts, completely automatic machines are used. One type is a hydraulically operated machine that successively grinds each cam to size, indexing after each grinding operation. When the last cam is ground, the table reverses, returns to the starting position, and the work stops rotating. The wheel is trued automatically by a diamond, and its position changed to compensate for the truing. The table then stops in the loading position. This machine, made by the Norton Co., has a swing over the rocking bar of 5 in. Length of work between centers is 30, 36, or 42 in.

29. STOCK ALLOWANCE AND GRINDING TOLERANCES

ALLOWANCE FOR GRINDING. The amount of stock to be left for grinding varies with character of the work and type of machine. In cylindrical grinding the practical minimum is about 0.005 in., but because of variables such as warpage it may run up to 0.025 or 0.030 in. For economy the stock allowance should never exceed $^1/_{16}$ in. In internal grinding the stock might vary from as little as 0.004 in. for holes around $^1/_8$ in. diameter up to 0.035 in. for holes around 8 in. diameter. On centerless work, normal allowance for rough grinding may run to 0.010 in. and for finish grinding 0.001 to 0.002 in. above the high limit.

HORSEPOWER REQUIRED FOR GRINDING varies directly with the quantity of metal removed. R. E. W. Harrison gives (Ref. 5) the relation between horsepower and metal removal in centerless or controlled cylindrical grinders as follows:

Horsepower at grinding wheel spindle	5	10	15	20	25	30	35	40
Metal removal capacity, cubic inches per minute	$^1/_2$	1	$1\,^1/_2$	2	$2\,^1/_2$	3	$3\,^1/_2$	4

GRINDING TOLERANCES have decreased with improvements in grinding machines, grinding wheels, and gaging devices. On cylindrical grinding, tolerances in diameter of ±0.00001 in. are possible of attainment when an automatic sizing device similar to a Nortonizer (Norton Co.) is used, if the machine and temperature conditions approach ideal. In out-of-roundness and taper, tolerances as small as 0.00005 in. are practicable. Readily held tolerances, however, range around ±0.0001 to 0.0005 in., with surface finish as fine as 4 to 5 microinches rms.

30. HONING, LAPPING, AND SUPERFINISHING

HONING is a term applied to the refined grinding of cylindrical surfaces. It is used for the accurate finishing of automotive cylinders, compressor cylinders, cylinders in hydraulic and refrigeration units, bushings, bearings, rods, etc. It consists essentially in the passage of a honing stone or stones in an axial direction through the bore, the hone being rotated at the same time. J. W. Hindes and J. G. Young (Ref. 6) state that a ratio of about 1 to 3 of reciprocation to rotation speed gives the most satisfactory results (65 to 70 surface ft per min reciprocation, 200 ft per min rotation). If the hone is hard and tends to glaze, this condition can be corrected by increasing the ratio of reciprocating to rotating speed.

The hone usually is a form of expanding segmental wheel, whose surface is formed by a series of abrasive stones carried in a holder. Means are provided to expand the stones to the wall of the cylinder being ground, and to hold them against it under pressure, while the hone is reciprocated and rotated. Wear of the stones is compensated by floating cones in the holders which feed the stones outward and maintain them in parallelism.

J. A. Carlin (Ref. 7) describes a method of honing blind end bores, by shifting the center of pressure on the abrasive stones toward the blind end to compensate for the loss of abrasive effect resulting from inability of the hone to overrun the end of the cylinder. J. E. Kline (Ref. 8) illustrates a fixture used for precision finishing by honing connecting rod crankpin bores.

Honing is primarily a surface-refining method intended to correct internal or external cylindrical geometry of parts. Although as much as 0.090 in. can be removed from a bore in honing, stock removal should be as small as possible for best economy. Honing will remove out-of-roundness, high spots, and taper, but will not correct position or alignment of a hole, bore, or diameter. Honing generates surface finish entirely independent of stock removal.

Accuracy in modern honing operations can be held within 0.0005 to 0.001 in. total variation for roundness and straightness with diameters over 4 in.; 0.0003 to 0.0005 in. on diameters from 1 to 4 in.; and on bores below 1 in. tolerances are usually within 0.0001 to 0.0003 in. In general, tolerances of ±0.0001 to 0.00025 in. are common, and particularly fussy jobs have been held to a total variation of 0.000025 in. but at a cost in slower production. Straight-line honed rods can be finished within 0.0001 to 0.0002 in. on roundness and taper.

Surface finish in hard steel may run as low as 1 microinch, rms; in cast iron, soft steel or bronze may run to as low as 3 microinches; and with aluminum a finish quality as fine as 15 microinches can be produced. Finest finish can be attained economically, as a rule, only when metal removal is small.

Honing, in general, should not be considered for diameters under 1/4 in. Vertical honers will handle work up to 30 in. in diameter and 6 to 7 ft long. Horizontal machines will handle bores in diameters up to 50 in. and lengths to 70 ft. External work can be honed in diameters up to 6 in. and lengths from 2 to 12 ft.

Lubricant. Kerosene or mineral seal oil is generally used as a coolant and lubricant in honing cast iron and other metals. Turpentine gives a freer cutting action on bronze and brass. Water is unsatisfactory.

LAPPING produces exceedingly smooth and highly accurate surfaces, either flat, cylindrical, tapered, shouldered, or contoured. A lap, usually of cast iron, charged with fine abrasive, is rubbed against the surface of the work either by hand or by machine. Since about 1930, a lap consisting of fine abrasives bonded together has been used in many instances to replace the metal lap charged with loose abrasive. Abrasive-coated paper laps also are used. Laps are used with kerosene and lard oil lubricants.

Formerly limited in use to the production of precision surfaces in gages and measuring instruments, lapping has since 1930 been generally adopted as a production process in many industries, particularly the automotive. In this industry, it is used for the finishing of crankpins, bearings, piston rings, ball-bearing races, valve tappets, and other wearing parts. Lapping tolerances of 0.00005 in. in diameter and 0.000025 in. in straightness and roundness are possible of attainment.

Sydney Player describes (Ref. 9) the lapping of plug gages 0.875 in. in diameter ±0.00002 in. Two lapping operations were required. The grinding limit was +0.00015 to +0.0003 in. Rough lapping with M-304 American Optical emery brought the gages to within 0.00005 in. of finished size. For finish lapping, the laps were hand scraped and worked together with levigated alumina until the surfaces of the laps were mirror finished. The author also describes lapping, in lapping machines, of piston pins to limits of 0.0001 in. for straightness and roundness, the production being 500 pins per hour.

Stock allowance in lapping should never exceed 0.0005 in., preferably not over 0.0002 in. for rough lapping or 0.0001 in. for finish lapping from a rough lapped surface. Preliminary operations should be grinding or honing and should be as accurate as possible relative to geometrical and location accuracy because lapping provides little correction for these factors.

Honing machines using charged laps can be used to lap parts that fall within the size ranges mentioned under honing. Hand-operated centerless lappers using a charged lapping roll and a control roll, both of which are driven, can be employed to lap parts from 0.030 to 10 in. in diameter.

Vertical-spindle lapping machines consist essentially of a pair of laps which may be rotated independently of each other and in opposite directions if desired. The work to be lapped, which may be either cylindrical or flat, is carried in work holders, which position the work between the two laps and at the same time move it between the lap faces with a planetary motion in a constantly varying line of travel relative to the lap surfaces. For cylindrical work, the pieces are carried in the work holder at an angle to radial lines. An eccentric motion imparted to the work holder, in conjunction with the nonradial position of the work, causes all portions of the surface to come in contact with the laps and a true lapped surface is produced. The capacity of lapping machines ranges up to 3 in. in thickness or diameter by 6 in. long.

The centerless lap-grinding or lapping machine is similar in its principle of operation and in general construction to the centerless grinding machine. The lapping is done by wheels of finely divided abrasive suitably bonded. The lapping wheel is driven at a speed of 150, 195, or 225 rpm. Through-feed work from 1/2 to 3 in. in diameter and of any length can be lapped in the centerless machine. By means of special equipment, work which cannot be passed transversely through the machine can be lapped in diameters of 1/4 to 6 in. and 15 in. long.

SUPERFINISHING * complements the dimension-creating efficiency of the grinding operation by removing defective surface conditions. Ground surfaces have two general types of defects, geometrical and metallurgical. In addition to the visible spiral ridge, there may be chatter and feed marks. In plunge grinding, parallel ridges are characteristic. The heights of all these defects are tremendous as compared to the size of an oil molecule. They readily penetrate an oil film to result in true metallic contact and rapid wear.

When a hardened surface is ground, the objective is always defeated to some extent because the heat of grinding alters the applied heat treatment. A large amount of investigation of this condition by the writer has shown that the "burning" of the average hardened and ground surface extends slightly deeper than the last vestiges of the grinding scratch.

Another result of the inefficient removal of stock by the negative angles of the abrasive grit is the distortion of the crystalline structure to an appreciable depth. Furthermore, there is a production of partially loosened splinters of metal covering the surface which usually may be observed in the oil oozing from a newly started bearing.

In Superfinishing, the spurious layer of metal resulting from the foregoing conditions is removed by the mild scrubbing action of a bonded abrasive applied to the rotating surface under low pressure and a flood of light mineral oil lubricant (Fig. 6). An appropriate motion is always imparted to the abrasive or "stone" to prevent repetition of the grit paths and to cause each of them to pass over several of the various heights of the grinding defects. More efficient cutting is thus secured together with a much greater tendency to generate a true geometry of surface.

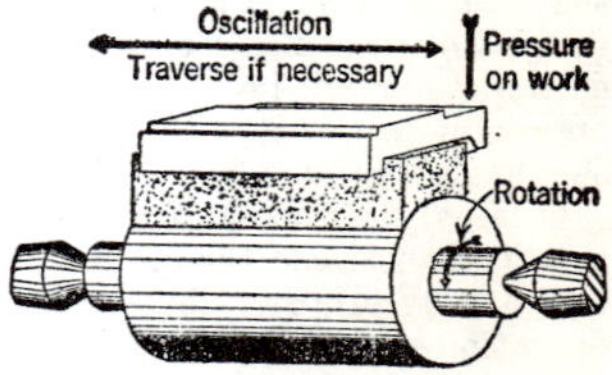

FIG. 6. Superfinishing a cylindrical surface. (Courtesy of Gisholt Machine Co.)

Surface shapes that can be superfinished are somewhat limited by their degrees of symmetry. Cylinders, cones, flat circles, and spherical sections are examples of the most easily performed operations. There are numerous borderline applications; for example,

* Contributed by E. L. Hemingway.

most reamers are readily Superfinished because their lands are contained in one cylindrical plane (Fig. 6); and flat surfaces can deviate somewhat from a true circle, but a square is usually difficult because the stone will not be equally supported by the corners and the narrow dimension during rotation of the part. Flanges or shoulders seldom interfere with work on a cylinder, but the flange surface itself may offer a problem. Any diameter or length of cylinder can be done except that adequate technique and equipment have not yet been developed for diameters as small as perhaps $^3/_{32}$ in. Likewise, there appears to be no practical limit to the size of flat circle or ring that could be Superfinished.

Shapes of Abrasives. Two shapes of abrasives are employed: square or rectangular sticks for cylindrical or cone-shaped surfaces and cup wheels for flat and spherical surfaces. Just as in grinding, the bond hardness must be chosen to accord with that of the metal. Vitrified bond is employed in much the larger percentage of cases, although a resinoid bond is sometimes used on very small diameters and in a few second operations where the best finish is desired on certain very difficult metals. For bright surfaces of 3 microinches or less rms (see Section 24, Article 9) 600 grit stones are usually employed. Where a turned surface is to be Superfinished or where a rougher finish is acceptable, stones of 120 to 500 grit are used. However, it is often surprising how good a finish can be obtained with a 320 or 400 grit stone especially in flat and spherical work.

Pressure on the stone may be applied either through spring or hydraulic means. It may vary from 10 to 40 psi of stone contact area with an average of 20, although certain high-production special machines working from turned surfaces, such as the brake drum and pressure plate Superfinishers, do operate at somewhat higher pressures and speeds than normal. Surface footage of the work is usually 30 to 90 ft per min, with an average around 55. Under these conditions of low speed and pressure, there is no perceptible creation of heat, and no "charging" with abrasive has ever been discovered.

Superfinishing Mechanism. *Cylindrical Parts.* In the Superfinishing of cylindrical parts, the essential mechanism is an arm containing a spring-loaded quill, on the end of which is mounted the stone (Fig. 7). This assembly is mounted on a base having a mech-

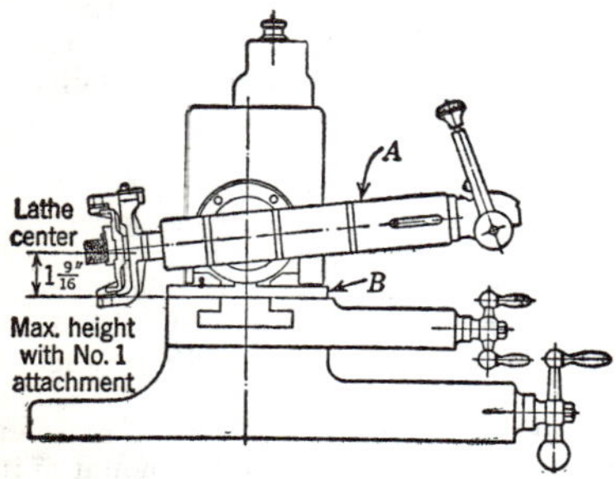

FIG. 7. Superfinishing attachment for lathe. (Courtesy of Gisholt Machine Co.)

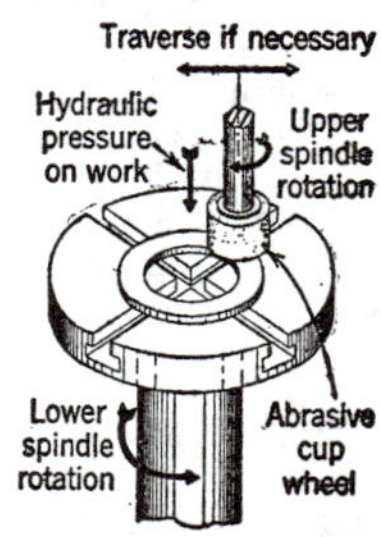

FIG. 8. Principle of flat Superfinishing. (Courtesy of Gisholt Machine Co.)

anism by which the arm and stone are oscillated, parallel to the length of the part, by a distance usually of $^3/_{16}$ in. and at a rate of 425 complete cycles per minute. Arm and base can be mounted on the cross slide of an engine lathe, as a means of rotating the work, or they can be incorporated into a machine especially designed for this purpose. If the surface is not too long (1 to 4 in., depending on the diameter), traverse will not be necessary as the stone will be the same length as the surface. The width of the stone should be approximately 75% of the part diameter. Thus, a surface 1 in. in diameter by 2 in. long should be Superfinished with a stone $^3/_4$ by $^3/_4$ by 2 in.

Original dressing-in of the stone is done either by drawing a strip of emery cloth back and forth between the stone face and the work or by means of a special diamond boring bar. Once properly dressed-in, the stone seldom needs any further attention. As the operation proceeds, the stone is slowly worn to the average radius of the part, which is a true and smooth curve, and it then acts as a master surface to reproduce its true geometry on the work.

Flat and spherical Superfinishing are performed with a machine containing an upper and a lower rotating spindle (Fig. 8). The upper spindle has a spring-loaded quill, on the lower end of which the cup wheel is located. The upper end of the lower spindle carries the work. (If it appears desirable, the positions of cup and work can be reversed.) As the lower spindle and work revolve, the end face of the stone in the rotating upper spindle is brought into contact with the work. Some degree of offset is given the cup with relation to the work position so that the path of any single grit is not often repeated.

When the two spindles are exactly parallel, the combined effect of the rotations of both stone and work results in the generation of a very smooth and flat surface. If the upper spindle is adjusted to some angularity with the fixed lower, a spherical shape is generated. Just as in cylindrical work, the stone automatically wears to the most desirable shape.

Finish Control. In either cylindrical or flat work, the cutting speed and finish secured are controllable in several ways. As cutting speed is increased, the quality of finish is likely to decrease. These are the means of control, in the order of their importance:

1. Decreasing the bond hardness of the stone increases cutting speed.
2. Decreasing the surface footage of the work increases the speed of cutting.
3. Decreasing the viscosity of the stone lubricant increases the speed of cutting.
4. Increasing the rate of oscillation of the stick-type stone or of the rotation of the cup wheel increases the rate of cutting.

If the quality of grinding is not rougher than 25 microinches rms, small parts are usually Superfinished in 30 seconds or less. One approximating 1 1/2 in. round by 2 in. long will be refined to 3 microinches in 1 min. Surfaces containing in excess of 20 sq in., usually requiring traverse, can be completed at a rate of 10 sq in. per min. Flat and spherical surfaces are Superfinished in somewhat shorter times than cylindrical. Tappet heads, for example, are corrected in geometry and smoothened to less than 3 microinches in no more than 30 sec.

The stock removal in Superfinishing depends on the quality of grinding also. A fairly accurate rule is that one-tenth of a thousandth of an inch will be removed from a diameter for each 10 microinches Profilometer reading of the ground surface.

Standard machines available for Superfinishing are lathe attachments for cylindrical work; several sizes of self-contained machines for cylindrical work up to 4 in. in diameter and 36 in. in length, some of which can be provided with flat attachments; a series of multispindle automatic machines for cylindrical, flat, spherical, and special parts such as brake drums and clutch plates; and a series of crankshaft machines.

31. BARREL FINISHING *

Barrel finishing, rotofinishing, or precision tumbling is basically a mechanical grinding and honing process which depends primarily on the free-mass abrasive action of special abrasive compounds, quartzite, or manufactured abrasive chips, stones, or pebbles under agitated rotary actuation. Processing may be carried out either dry or flooded with water. Grinding, deburring, removal of metal fragments left from previous machining operations, producing of predetermined radii on edges, polishing, honing, and even coloring can be readily accomplished with closely controlled results in mass production of parts.

Parts made of practically any material can be easily finished either in the hardened or the soft condition. Brass castings, forgings, stampings, spinnings, precision castings, or accurately finished machined pieces such as cams, gears, threaded units, gun mechanisms, housings, bearing rollers and cages, and rolls are among the parts normally handled. Design of parts is practically unlimited, and size may range from a few ounces to around 75 lb. Not only can removal of burrs or sharp projections be accomplished at extremely low cost per piece, but tremendous improvement in surface quality can be produced, often making it unnecessary to specify the most exacting finish possible in the primary processing operations.

Continuous wet barrel finishing is recommended and is best suited for grinding, deburring, and finishing the majority of parts. Grinding utilizes the largest chips or stones for fast cutting in eliminating burrs, fragmented metals, sharp or rough corners, and edges. Grinding also removes spinning, draw and die marks, tool marks on machined surfaces, etc. Honing action is obtained by the use of the smaller stones or chips and, naturally, results in a much finer surface finish, approaching a polish. So-called coloring, an extremely fine finishing operation using the smallest chips or in some cases highly polished hard-steel balls, is especially suited to the processing of small irregularly shaped parts.

Dry barrel finishing is somewhat limited and is used mainly for the final honing of die castings and brass castings after a preliminary fast wet grinding operation with large stones. It is also used to some extent for finishing small light-gage stampings of all metals as well as brass stampings, etc., prior to electroplating.

A wide variety of tumbling barrels are available for continuous processing. They are generally divided into five types: (1) tilting, (2) compartment horizontals, (3) tapered horizontals, (4) cylindrical horizontals, and (5) octagonal horizontals. Tilted barrels with capacities of 1 to 14 cu ft are usually lined with seasoned hardwood, and the tilting feature

* Contributed by Roger W. Bolz.

makes them highly economical of handling time. Compartment barrels are usually octagonal in shape and may have one to five compartments, wood lined. Capacities may run around 3 to 20 cu ft per compartment. Cast-iron tapered octagonal horizontal barrels, with capacities up to around 5 cu ft are primarily for water-rolling and sand-tumbling metal parts whose rough surfaces are to be cut down and smoothed. Cylindrical horizontal steel barrels with capacities up to about 30 cu ft are primarily for removing fins and flanges from large, heavy castings. Octagonal horizontal barrels, usually of wood with one or more screened faces, are for dry tumbling of plastics.

Time required for processing parts varies considerably with the material, design of the part, and finish desired. Time for various metals is roughly: for machined brass, 15 min to 2 hr; for machined aluminum, 15 min to 3 hr; for cast brass, 2 to 48 hr; and for deburring steel, 1 to 8 hr. With steels, the harder the metals, the brighter the surfaces obtainable, but the longer the processing time needed. Production of simple bushings may range from 12,000 to 15,000 per hr while with more complicated parts, such as the larger ones which require a compartmented barrel, finishing time may run to several hours per part.

Commonly, the practical range of radii which can be generated on parts is 0.002 to 0.020 in., but radii up to 0.050 in. and greater have been produced. Normally, the grinding operation can be expected to produce a finish of approximately 25 microinches rms on unhardened parts, the honing operation will bring this down to about 10 microinches, and the wet coloring operation to about 3 microinches. Combinations of tumbling media can be used to radius edges and at the same time hone surface to a high quality. Finished tolerances are affected by prior processing operations, and actual cases have shown that microinch readings on parts can be reduced from 15 to about 3, from 60 to about 15, or from about 500 to 80. Surface finishes obtained in current production on aircraft engine parts such as gears, gear and spline shafts, bearing retainers, special bolts, and housings range around 6 to 8 microinches, and corner radii obtained range from 0.008 to 0.010 in consistently.

Often, the use of barrel finishing can produce results comparable to grinding and at a much lower cost. Broached parts, for instance, originally hardened and finish ground have been tumbled once after broaching and once after hardening with extremely satisfactory finish results of 8 to 16 microinches along with smoothly finished and rounded corners. With extremely complicated machine parts, where high production rates must be met, barrel finishing serves a dual purpose in improving finish and deburring. Hand methods in such cases would present an almost unsurmountable problem.

32. POLISHING AND BUFFING

POLISHING is the removal of metal by means of a flexible wheel coated with abrasive. The wheels are made of various materials as described below. The abrasive may be any of the many forms of manufactured abrasives or natural emery. It is held to the face of the wheel by means of an adhesive, usually glue.

POLISHING WHEELS. The earliest form of polishing wheel was a wood disk with a strap of leather fastened to its circumference. This form of wheel has practically disappeared. Polishing wheels in common use are made of buff sections sewed or glued together to make a wheel of the desired face, of disks of canvas sewed or glued together, or of disks of leather, sheepskin, walrus hide, or similar material. The compress wheel is made of blocks of canvas, leather, sheepskin, paper, and other materials, compressed to the required density. The field of application and the characteristic of the several types of wheel are described below.

Clothflex wheels are made both of full disk buffs and sewed piece buffs. Both types are made with cloth of different weaves, and each type and weave have their special fields of service. Manufacturers of polishing wheels should be consulted on the selection of the grade and construction of wheel best adapted to a particular class of work.

Polishing wheels made of sewed piece buffs are used on plumbing supplies, such as faucets, tubular plumbing goods, and on brass, iron, and copper castings with contoured surfaces, and on steel stampings. They may be used on work where the corners are rounded and where it is not necessary to maintain sharp edges. They are not adapted to work where the edges of openings must be held square or where corners must not be rounded over.

The full-disk polishing wheel, i.e., made of full disk buff sections, is adapted to use on automatic machines and to work where perfect wheel balance must be maintained. Any desired density of full disk clothflex wheel may be obtained by building it of packed buff sections, that is, sections with disks of smaller diameter in between the buff disks.

Disk canvas wheels are made of disks of canvas sewed or glued together, either singly or in sections of two or three disks. Density can be varied in the same manner as with clothflex wheels by variations in sewing and gluing. The disk canvas wheel, which is adapted to heavier and rougher work than the clothflex wheel, is widely used for polishing automobile bumpers, mold boards of plows, harrow disks, etc. It is especially adapted to breaking through scale and to removing deep pits or irregularities in the surface of the work. The disk canvas wheel often is used for roughing with the coarser abrasive grains, finishing being done by a clothflex or a compress wheel coated with finer grain.

In the **compress canvas wheel,** seven densities are regularly manufactured ranging from super-soft to super-hard. The softer wheels are used for contour work or work where the surface may vary more or less from established dimensions, and a yielding wheel is necessary to adapt itself to the surface. In general, the softer densities are used in the cutlery trade, on double-header machines, for finishing large knives and on other work where the wheel must flex to cover the surface. Soft wheels should not be used under heavy pressure, as the resulting flexure of the fabric will generate sufficient heat to char the fabric.

The medium-density compress canvas wheel is a general all-round wheel largely used as a finishing wheel on such work as automobile bumper bars, sewing machine parts, hardware, automobile accessories, and similar work. It frequently is made with formed faces to fit contoured work. The harder densities of wheels are widely used in the cutlery trade for glazing cutlery and table knives. The super-hard density is used for the honing of razor blades. The selection of the proper density is highly important, a difference of even one-half grade in density often being the cause of success or failure of a wheel on a given job.

The **compress leather wheel** is used where edges and corners are to be held square and where holes are present in the work. These wheels are largely used to polish flat-iron sole plates, both for roughing and finishing, excepting the final operation which is performed with a compress canvas wheel. Compress leather wheels also are used in the cutlery trade for the finishing of knife blades.

Compress walrus and compress sheepskin wheels are used in the cutlery trade, the latter wheel giving a fine surface and good color.

POLISHING WHEEL SPEEDS. A surface speed of 7500 ft per min has commonly been used for polishing wheels working on steel, iron, or brass, and from long usage has been regarded as the most efficient speed. Some recent research, however, has demonstrated that the commonly accepted surface speed of 7500 ft per min is not necessarily the most efficient speed. A speed of 6700 ft per min in polishing medium steel showed an improvement of about 15% from the standpoint both of metal removal and time consumed. Similar results were obtained with a speed of about 7000 ft when polishing cast iron. Investigation has not been sufficiently detailed to determine whether or not these two speeds are the most efficient. It, therefore, is advisable for manufacturers polishing large quantities of metal to conduct experiments to determine the most efficient speeds for their particular conditions.

ABRASIVES. Type and size of abrasive used on polishing wheels depend on the material polished and the finish desired. Turkish emery formerly was used almost exclusively. It has been displaced by the manufactured abrasives, except as a finishing abrasive where high color is desired. For steel and brass the aluminum oxide abrasives give the highest cutting efficiency and longest wheel life. For cast iron, a silicon carbide abrasive should be used.

The size of abrasive grain to be used depends on the initial character of the surface and the finish to be produced. The first wheel should be coated with abrasive that will cut down to the bottom of the deepest pits and break through scale. The second wheel should be coated with a finer abrasive that will cut out the grain scratches made by the coarser wheel. The next wheel should have still finer grain, and so on until the desired surface has been produced. The final wheels should be greased with oil or tallow to modify the cut and to produce a more lustrous surface.

A common sequence of grain sizes for the hand lathe polishing steel is: first wheel, No. 60 or 90; second wheel, No. 90 or 120; third wheel, No. 120 or 150. If deep pits are present in the work, it may be necessary to use a coarser abrasive on the first wheel, say No. 46. It should be remembered, however, that each wheel must remove the grain scratches of the preceding wheel, and it may be more economical to use a finer wheel for a longer time to remove pits than to spend time removing coarse grain scratches. Highly polished steel is finished with a sequence of grain sizes of Nos. 60, 90, 120, 180, and 220. All the wheels finer than No. 90 should be oiled or greased. The sequence of grains may be extended even farther to produce mirror-like, scratchless surfaces. Rolls for copper and brass rolling mills have been finished with a final wheel coated with levigated alumina, which is about equivalent to a grain size of No. 640.

POLISHING WITH COATED ABRASIVE BELTS (Contributed by Behr-Manning Corp., Troy, N. Y.). As under any other method of polishing, the purpose is to remove a certain amount of material to reduce any roughness resulting from machining or previous grinding. This may be accomplished dry or with oil or grease as the situation requires.

Depending on the purpose of the operation and the selection of the proper materials, finishes on steel and many other metals may be obtained as fine as 1 or 1 $^1/_2$ rms.

Equipment. The backstand is a device which makes possible the use of coated abrasive belts in much the same way that a setup wheel would be used. It is an idler pulley mounted on a floor column or a bench base, depending on the purpose. In any case it has an attachment for tracking the belt to the center of the contact wheel and a spring or weight arrangement to maintain tension on the belt and to compensate for stretch.

The floor type is used with any standard polishing lathe or jack by mounting behind the spindle and approximately 4 ft between the center of the idler pulley and the center of the lathe spindle. This dimension is not important and may be increased or decreased according to the space available. Abrasive belts are made endless in all lengths.

The bench type may be mounted on any bench behind a bench grinder in the same manner, but the grinder should have not less than a $^1/_2$ hp motor. There are also complete units put out by several manufacturers, including motor, backstand, and contact wheel complete. These are made single and double, similar to any single or double spindle lathe, and operate one or two belts.

To complete this assembly, there must be a contact or back-up wheel mounted on the lathe spindle or bench grinder. It is over the idler pulley and the contact wheel that the abrasive belt is run, and the contact wheel provides the support for grinding.

Wheels that are acceptable for this purpose are made in a great variety but should be selected for density or flexibility, size, and construction, according to the type of work. It is obvious that any flat-faced wheel turned true and capable of supporting an abrasive belt can be used as a contact or supporting wheel, but a careful selection is necessary to obtain maximum belt life and proper flexibility.

Among the more popular wheels for belt polishing are the compressed canvas and clothflex previously described in this article. These are preferred for flat surfaces. Of the two, the compressed canvas is more expensive, but it lasts longer, holds its shape better, and offers a more controllable variety of densities. There are nine densities in all, listed from Super Hard to Super Soft and in a full range of widths and diameters.

For surfaces which diverge from the flat, that is, those which have convex curves, all such curves can be reached with a flexible abrasive belt running over a soft wheel similar to the Divine Beltflex or the Jackson Airway. These will form the abrasive belt to almost any convex curve without leaving flats or distortions.

Abrasive Belts. Abrasive cloth belts are made in two weights, "X" and "J." The X weight is used for all flat polishing and in a limited way for contour work, that is, where the surface is slightly curved, either convex or concave. For work requiring a more severe flexing, use "J" weight. This is a different type of cloth, somewhat lighter and especially flexed for use with the soft Beltflex or Jackson Airway type of wheel.

Both the foregoing types of backings are coated with either aluminum oxide or silicon carbide grain. These are considered best for all metal polishing. Aluminum oxide is preferred in all cases except for aluminum, white metal, and zinc. The standard grit numbers available in aluminum oxide are 24, 30, 36, 40, 50, 60, 80, 100, 120, 150, 180, 220, 240, 280, 320, and 400, and, except for 320 and 400, these are duplicated in the silicon carbide.

As lubricants practically any type of vegetable, mineral, or animal oil or grease may be used with coated abrasive belts.

BUFFING consists in leveling off minute projections on the surface of metal, and in producing a sheen or luster on the surface so prepared. The first operation is termed *cutting down* and the second *coloring*. The work is performed with a buffing wheel which is composed of a number of circular disks of cloth mounted on an arbor and rotated at high speed. Buffs or buffing wheels are made in several forms, each of which has its particular field of application.

Buffing wheels are made by assembling on an arbor a sufficient number of *buff sections* to form a wheel of the requisite width of face. The various forms of buff that are used are: (1) Full disk buffs, which may be either *loose* or *sewed.* (2) Sewed piece buffs, which always are sewed. (3) Some form of patented folded buff.

Buff material usually, but not always, is cotton sheeting, bleached or unbleached. The most widely used grade of sheeting has a warp of 64 threads per inch and a filling of 68 threads. It commonly is known as 64 by 68 cloth. Another widely used fabric is woven 84 by 92. Buffs made of this fabric are harder and will stand rougher usage than those made of 64 by 68 fabric. They generally are used for the cutting-down operation, the 64 by 68 buff generally being used for coloring. Buffs made of bleached material are harder than those of unbleached fabric.

Full disk buffs, made of complete disks of the fabric, are assembled in *sections*, usually of 20 plies of fabric. The loose buff is sewed with three or four closely spaced rows of sewing around the hole to hold the disks together. The sewed buff in addition to the sewing around the center hole is sewed with concentric or spiral rows of sewing, spaced at any desired interval, from the center hole to the periphery. The wider the spacing of this sewing, the more flexible and yielding is the buff.

The 64 by 68 loose buff is used for coloring soft nickel, brass, and cast iron. When sewed it may be used for cutting down copper, soft brass, zinc, and similar metals on hand lathes. The 84 by 92 buff both in sewed and loose form is used to cut down copper, brass, German silver, and heavy nickel plate on bumper bars, builders' hardware, etc. For chromium plate, it should be sewed.

Sewed Piece Buffs. The sewed piece buff is built up of more or less irregularly shaped pieces held between two full disks of fabric and sewed from the center hole to the periphery with spiral sewing. The same grades of fabric are used in sewed piece buffs of the better grades as in the full disk buffs. Many of the cheaper buffs of this sort contain much coarser and flimsier fabric. Where sewed piece buffs are purchased in large quantities, it is wise to procure them only from manufacturers who have established reputations for quality. The sewed piece buff costs less than the full disk, and for many buffing operations is equally satisfactory.

Sewed piece buffs are largely used for cutting down cast plumbing goods, brass stampings, brass tubular goods, and aluminum castings.

Folded Buffs. Folded buffs consist of full disks folded and assembled in various ways to form pockets, which then are assembled to form a complete circle and are sewed either as a loose buff or as a spirally sewed buff. These buffs are used both for cutting down and for coloring. Their action is somewhat different from that of the loose or the sewed piece buff, in that the edge of the fabric passes across the work at an angle to the plane of the buff wheel. Buff marks on the work which are produced by each separate layer of fabric in the other types of buff thus are avoided. Another advantage of the folded buff is that the pockets tend to trap air, which has a cooling effect on the buff surface. Folded buffs are relatively more expensive than any other type because of the large amount of hand labor required in their manufacture.

Buff Ventilation. Buffs continuously operating under heavy pressure must be ventilated in order to dissipate the heat generated. Without adequate ventilation the buff may char or even be set afire. Folded buffs have provision for ventilation inherent in their construction. Full disk buffs may be ventilated by the process known as *packing*. This consists in assembling one or two disks of smaller diameter than the buff disk between each pair of buff disks, or at such intervals as may be necessary, depending on the severity of the work. Sewed piece buffs can be ventilated by inserting a few disks of fabric or washers of felt or leather between each section. Packed buffs should be oscillated axially in relation to the work in order to avoid distinct streaks at the points where the disks or sections are separated by the packing.

Speed of Buffs. The surface speed at which a buff should operate depends on the type of metal being buffed. Table 8 shows the speeds commonly used with various metals and various grades of buffs. In general, higher speeds are used for cutting down than for coloring with the copper bearing metals.

Table 8. Buffing Practice for Various Metals

Material	Cutting Down					Coloring				
	Buff			Composition		Buff			Composition	
	Weave	Sewing	Surface Speed, ft per min	Kind *	Grade	Weave	Sewing	Surface Speed, ft per min	Kind *	Grade
Aluminum	84 x 92	0	8500	T	X67	64 x 68	0	7500	WD	XXX
Brass (cast)	84 x 92	0	8000	T	X67	64 x 68	0	7500	WD	XXX
	Pieced	3/8	8000	T	X67					
Bronze (cast)	84 x 92	0	8000	T	X67	64 x 68	0	7500	WD	XXX
Copper	84 x 92	0	9000	T	X67	64 x 68	0	7500	T	XX
Chromium	84 x 92	0	8000	CO	....	....	....	....	....	......
Monel metal	84 x 92	0	7500	C	....	64 x 68	0	7500	L	......
Nickel	64 x 68	0	7500	L	....	64 x 68	0	7500	L	......
Stainless steel	84 x 92	0	8500	CO	.....	84 x 92	0	8500	CO	......
	Pieced	3/8	8500	CO	.....					
White metal (Die cast)	Pieced	3/8	6000	T	X67	64 x 68	0	6000	WD	XXX

* C = Crocus. CO = Chromium oxide. L = Lime. T = Tripoli. WD = White diamond.

Since a buff wears down in use, its peripheral speed constantly decreases. Provision should be made to maintain the peripheral speed fairly constant either by the provision of a variable-speed drive or by changing the buff to a machine of higher spindle speed for each decrease of 2 in. diameter.

Pressure and Horsepower for Buffing. The effectiveness of buffing depends largely on the pressure of the wheel on the work. Increasing the pressure increases the rapidity of cutting, but at the same time increases the liability of cutting through the plating on plated articles. Definite rules cannot be given since the pressure that can be safely applied depends on the hardness of the plating and its thickness. A heavy coat of hard nickel can withstand much greater pressure than a thin coat of soft nickel. The horsepower required for buffing is a function of the pressure of the wheel on the work. In general, for buffing nickel plate on steel on hand lathes 3 to 5 hp at the buff is the usual practice. For heavy work, such as automobile bumper bars, hand lathes usually are equipped with 10- to 15-hp motors.

Brass requires considerably more horsepower for buffing than steel or nickel. Experiments indicated that a pressure of 15 lb per in. of width is required for cutting down sheet brass, and 10 lb per in. of width is required for coloring.

Buffing Compositions. The following compositions are adapted to the different varieties of work: for cutting down brass, bronze and Britannia metal preparatory to plating, tripoli; for smooth surfaces on nickel and brass, crocus; for coloring brass, copper, nickel, bronze, German silver, etc., either in solid or plated metal, White Diamond XXX composition; for chased or embossed parts, or for cutting down silver-plated pieces which are afterward to be colored with rouge and alcohol, White Diamond XXXX composition is used; for nickel-plated pieces with a high luster, Vienna lime. Where rapid, sharp, even cutting is desired, emery cake is used. Chandelier rouge is used to produce a deep color on brass and bronze parts. Chromium oxide is used for chromium-plated articles and stainless steel.

POLISHING AND BUFFING MACHINERY. For irregularly shaped pieces, or where work is performed in relatively small lots, the hand lathe is almost universally used both for polishing and buffing. For long runs of similar work, automatic or semi-automatic polishing and buffing machines are advisable. They are economical of labor, have high capacity, and produce uniform work. In one instance, automatic equipment reduced the number employed on a given line of work from 31 to 5 operators, increasing production at the same time. Automatic machines for automobile bumpers have been built with an output of upwards of 300 sets of bumpers per hour. Automatic machines may be divided generally into three classes, straight-line machines, multiple-head machines, and contour machines.

Straight-line machines comprise essentially a conveyor on which the work is carried in appropriate chucks or holders underneath a series of polishing wheels, successively coated with the various sizes of grain necessary to obtain the desired finish. The wheels are arranged to float, either on an arm oscillating vertically on a horizontal shaft or between vertical guides. Each wheel is driven by an independent motor, direct-connected to the wheel spindle or driving the wheel spindle through V belts. The conveyor is driven by an independent motor. The more common form of conveyor is a belt, although various other forms embodying roller chain, or other construction, have been built. The conveyor usually is driven through a speed changing device at speeds ranging from 15 to 35 ft per min.

Multiple-head Machines. Multiple-head machines are used principally for the finishing of circular work. In their simplest form they consist of a head carrying two spindles on which chucks are mounted to hold the work. The head is indexed by hand to bring the work on each spindle in contact with the polishing or buffing wheel. The spindle in the working position is rotated by a small motor mounted on the machine. The idle spindle is unloaded and reloaded during the operating time of the working spindle.

Contour machines for polishing or buffing articles of rectangular outline, or forms not truly circular, are essentially a rotating spindle which carries the work, combined with a mechanical movement which causes the work to approach or recede from the wheel in conformity with the contour, and maintains uniform contact of the work and wheel at all times. The mechanical movement may be a cam of appropriate shape or a geared chuck.

Special machines have been built for many different purposes including the finishing of ice-skate blades, windshield wipers, blank book backs, sewing thimbles, tubing, sheet and strip metal, heads of screws and bolts, and various forms of cutlery. Some of these machines employ polishing wheels in pairs, the work passing between the wheels in order to finish two sides at one operation.

Principles for Polishing and Buffing Machine Design. The following principles should be embodied in polishing machines. (1) *Variable Wheel Speed.* Where a machine is to be used with wheels of a variety of diameters, or where the diameter decreases in service, as with a buffing wheel, provision should be made for a variable-speed wheel spindle. Wheel speed then may be maintained constant at the most efficient point. (2) *Variable Work Speed.* Provision should be made to vary the speed at which the work moves past the wheel. The time required for a wheel to do its work varies with different materials, and provision should be made to change the work speed as necessary. Work speeds range from 200 ft per min with strip zinc, to 15 ft per min with stainless steel. Hardness of material and character of finish desired also influence work speed. (3) *Variable Wheel Pressure.* Excessive pressure of the wheel on the work may damage the wheel and work, change the character of the finish, and increase the cost of polishing or buffing. Hence all machines should have provision to regulate wheel pressure. (4) *Wheel oscillation* in an axial direction is desirable to avoid deep scratches and wheel marks. (5) *Vibration* should be eliminated by adequate foundations, sufficient weight in machine bases, shafts of adequate diameter, ample bearings, and dynamic balance of rotating parts. Machine vibration will spoil finished surfaces and increase costs. (6) *Work-holding fixtures* of such design as to properly present work to the wheel and to facilitate rapid loading and unloading, greatly reduce the cost of machine operation.

For many data on the design and operation of polishing and buffing machines, see Ref. 10.

REFERENCES

1. C. R. Alden, *Trans. Am. Soc. Mech. Engrs.*, MSP-51-15, 1929.
2. George I. Alden, *Trans. Am. Soc. Mech. Engrs.*, Vol. 36, 1914, p. 451.
3. W. H. Chapman, *Trans. Am. Soc. Mech. Engrs.*, Vol. 42, 1920, p. 595.
4. H. L. Blood, *Trans. Am. Soc. Mech. Engrs.*, MSP-54-1, 1932.
5. R. E. W. Harrison, *Trans. Am. Soc. Mech. Engrs.*, MSP-51-18, 1929.
6. J. W. Hindes and J. G. Young, *Trans. Am. Soc. Mech. Engrs.*, MSP 53-13, 1931.
7. J. A. Carlin, *Automotive Inds.*, Dec. 9, 1933.
8. J. E. Kline, *Iron Age*, April 5, 1934.
9. Sydney Player, *Trans. Am. Soc. Mech. Engrs.*, MSP 52-14, 1930.
10. Robert T. Kent, Fundamentals of machine polishing, *Trans. Am. Soc. Mech. Engrs.*, MSP-53-9, 1931.

GENERAL REFERENCES

American Machinist, Grinding Book.
A Handbook on Abrasives and Grinding Wheels, Norton Company.
Grits and Grinds, Norton Company.
Points about Grinding Wheels and Their Selection, Brown & Sharpe Manufacturing Company.
Treatise on Tool Room Grinding, Carborundum Company.
Table of Causes of Grinding Wheel Accidents, Independence Inspection Bureau.
Safety Code for Use, Care, and Protection of Abrasive Wheels, American Standards Association.
Grinding Wheels and Their Uses, The Grinding Wheel Manufacturers' Association.

BROACHING

By Roger W. Bolz

During the last quarter century broaching has risen from a relatively unknown production process to one of major importance in present-day mass-production manufacturing. The rise of broaching into industrial prominence is due primarily to its two salient characteristics—extremely high speed of production, which makes possible minimum cost per piece of parts suited to broaching, and high repetitive accuracy.

Although broaching is recognized principally for its value as a production method suitable for manufacture at extremely high rates of output, it is now being applied to considerable advantage in even medium-volume production work where rather large lots of 2500 to 5000 pieces can be run. Where multiple jobs can be tooled up to run in succession to make up such quantity lots, relatively smaller quantities of certain parts can thus be designed to obtain the benefits of broaching. Cases are on record where as many as twelve different setups are used in this manner.

33. THE BROACHING TOOL

Developed from so-called drifts almost a century ago, broaches were first used to cut keyways and were driven through the part by means of a hammer. Round holes followed later, and by the turn of the century surface broaching was in limited use. Today, holes of any cross section from plain round or oval to inverted or helical spline can be readily produced to exacting tolerances. Over and above a wide variety of flat, related flat, or flat and curved surfaces, compound shapes such as gear teeth, chain sprocket teeth, sinuous slots, and turbine blades comprise an ever-broadening field of possibilities for economical production. The principal limitation of the process, however, must be observed in judging the advantages of broaching or in designing specific parts for such production. Owing to the necessity for each tooth gullet of a broach to retain the chip created until the entire pass across the part is completed, the process is not particularly suited for removing large amounts of material. Stock removed per tooth and length of broach must be economically balanced to achieve optimum results.

THE BROACH. Unlike turning, milling, shaping, planing, or similar tools on which the most favorable operating conditions can be secured by simple adjustments in feed or speed, a broach has a predetermined fixed feed or cut represented by the rise in height of each tooth (Fig. 1). This, of course, is usually established by the broaching tool designer

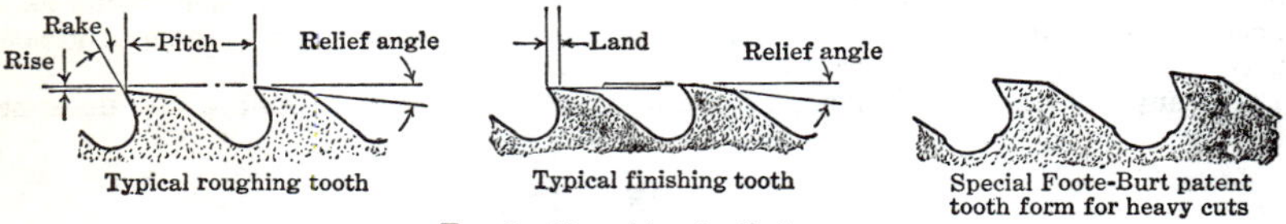

FIG. 1. Broaching tooth form.

but also must be recognized in designing parts to be broached in order to assure the most practical and economical manufacture.

Because each tooth gullet of a broach must retain the chip created until the tooth has finished the cut, obvious limitations are present in broaching tool design (Fig. 2). Depth of cut or rise per tooth is thus broadly restricted by the practical size of gullet allowable between teeth. Size of tooth gullet, of course, is directly related to tooth size or pitch, tooth depth usually ranging around one-third the pitch or slightly greater. As a rule, a minimum of two teeth should be cutting at all times, and preferably three; a minimum of three teeth per inch is recommended for parts having a length of 1 in. or less.

Desirable tooth pitch, and therefore chip thickness, is further restricted or controlled by machine power available and tooth strength. Internal or hole broaches of the pull type are limited by the strength of the tool itself, and this can be calculated from the following formula normally employed:

FIG. 2. Correct chip formation in broaching.

$$P = \frac{AT}{S} \tag{1}$$

where P = allowable pull, pounds per square inch; A = area of the minimum section which occurs at the root of the first roughing tooth or at the pull end, square inches; T = tensile strength of the broach material, pounds per square inch; and S = factor of safety to prevent broach damage because of sudden overloads due to hard spots in material, etc. Minimum tooth pitch, as can readily be seen, will help attain maximum broach strength and life.

In push broaches length is a critical factor, and the tendency is for the tool to buckle in the center, usually the weakest point. When L/D is greater than 25, a push broach is considered as a long column, and strength in this case is computed as

$$P = 13{,}500{,}000 \times D^4/SL^2 \tag{2}$$

where P = allowable push, pounds per square inch; D = root diameter of broach at $^1/_2 L$, square inches; L = length of broach from push end to first cutting tooth; and S = factor of safety. When ratio of L/D is less than 25, a broach is considered to act as a short column which resists straight compression loads only, and, as a rule, broaching loads are not critical for a tool of these proportions.

Length of broach which can be used for internal pull broaching depends on the machine stroke available. Economy, however, dictates the shortest possible tool for the job. Reasonable length-to-diameter ratio also dictates the shortest possible tool to assure adequate stiffness in broaching and sharpening. Standard machines available have strokes up to 66 in. and are capable of handling broaches up to 62 in. in length. Larger machines are in use, however, and some standard broaches available run up to as much as 74 in. in length; for special jobs such as rifling, even much longer tools are used.

MATERIAL REMOVAL. Tool strength in surface broaching is not the critical factor. Machine capacity, however, is, and again relates back to tooth pitch and rise inasmuch as volume of material removed is limited by both stroke and power. The total amount of material which can be removed for each available ton of ram pressure per foot of stroke is 0.06 cu in. for cast iron, 0.04 cu in. for medium-hard steel, and 0.03 cu in. for tough alloy steels having Rockwell hardness over C28. In Table 1, the relationship between tooth

Table 1. Relationship of Tooth Pitch and Chip Area per Tooth *

(Courtesy of Cincinnati Milling Machine Co.)

Pitch of Broach Teeth, in.	Chip Area for Steel, sq in.	Chip Area for Cast Iron, sq in.
0.2500	.0014	.0028
0.3125	.0022	.0044
0.3750	.0031	.0062
0.4375	.0042	.0084
0.5000	.0055	.0110
0.5625	.0070	.0140
0.6250	.0086	.0172
0.6875	.0104	.0208
0.7500	.0124	.0248
0.8125	.0145	.0290
0.8750	.0168	.0336
0.9375	.0193	.0386
1.0000	.0220	.0440
1.2500	.0330	.0660
1.5000	.0495	.0950
1.7500	.0673	.1346
2.0000	.0880	.1760

* Chip thickness times length

pitch and chip area is correlated for the various commonly used pitches. Since these chip area figures represent the product of chip thickness and lengths of cut, the maximum stock removal for any size part can be approximated for any machine available.

Material removal in single-pass broaching thus is usually limited to a maximum of about $1/2$ in. or less in thickness, depending to a large extent on the capacity of the machine to be used. Standard machines, as noted previously, allow for up to 62 in. of broach; continuous machines run up to 90 in. standard and specials have been made carrying up to 120 in. Normally, surface broaches are made in sections of 6 to 15 in. in length primarily for economy, but special sections such as for gear teeth are made longer. Where amounts exceeding the removal capacity or stroke of the machine are necessary, repetitive or multiple-stroke cuts must be made. Rise or cut per tooth can be made heavy if desirable by ingenious broach design. Staggering of the cutting portions of each tooth so as to achieve complete overlap over the entire surface or hole often can be employed to keep the total of the material being removed at any instant well within machine horsepower. This method is widely used in surface broaching as the "progressive" broach (Fig. 3).

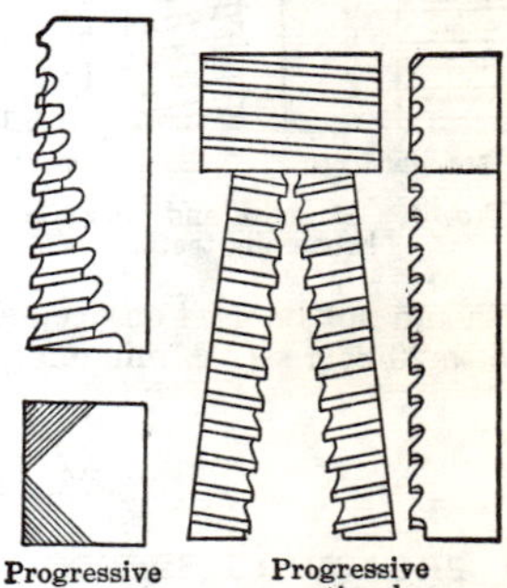

FIG. 3. Progressive broaches.

DATA NEEDED FOR BROACH DESIGN. General information requested by broaching tool manufacturers in order to design a successful broach is as follows: (1) Drawing of part giving tolerances. (2) Material specifications including prior heat treatments, hardness, and whether cast, forged, cold rolled, etc. (3) Prior machining operations, including machined locating surfaces if provided. (4) Subsequent heat treatment which may affect accuracy of broached dimensions through shrinkage, warpage, etc. (5) Production quantities and production rate. (6) Finish quality desired. (7) Broaching machine to be used, including maximum and minimum stroke, power capacity in pounds of pull or push, and details of broach support fixtures, shank type required with length to first tooth, and type of puller. In addition, for hole broaching: (8) Dimensions and shape of hole before broaching including length of hole and how prepared—whether drilled, reamed, hot or cold punched, forged, or cored. (9) If recesses or counterbores are used, state whether machined before or after broaching. (10) Whether more than one hole is to be broached at one time. In relation to splines: (11) If hole is to be ground after broaching, give stock allowance for grinding. (12) Radius or chamfer desired at ID and OD. (13) Whether broach is to size ID as well as cut splines and concentricity of splines with ID. (14) Spline fit and how controlled. (15) All gear elements for involute forms and lead and helix angle if required. For keyways: (16) Guide to be used. (17) Relation of keyway to other dimensions. For surface broaches: (18) Method of holding part. (19) Length and width of cut. (20) Maximum stock removal. (21) Total number of surfaces or parts to be completed at once. (22) Design of broach holder.

BROACHING TOOTH DESIGN. **Pitch** is determined primarily by length of cut and thickness of chip removed per tooth, consideration being given to type and condition of material to be broached. Formula sometimes used to estimate approximate tooth spacing for round broaches is $P = 0.35L$, where P = tooth pitch, inches; and L = length of cut to be made. Actual pitch used must be modified to suit necessary chip gullet capacity (Table 1).

Gullet. Area of gullet space should be ten times the cross-sectional area of the chip for steels, and five times for noncontinuous chip materials such as cast iron (Table 1).

Rise per Tooth. In free-machining steel this is usually 0.0015 to 0.003 in. for round broaching; 0.004 to 0.006 in. for spline; 0.001 to 0.006 in. for keyway and surface broaching. Practical minimum rise is about 0.00025 in. Rise per tooth as high as 0.015 in. can be used on narrow trimming broaches, where the cut is across the piece and the chip need not be contained in the tooth gullet. Slotting broaches should not exceed 0.00275 in. rise to avoid jamming of heavier chips against the sides of the slots. Preferable rise is 0.0025 in. Values can often be doubled for cast iron, malleable iron, and brass. For burnishing or sizing, rise per tooth is not over 0.001 in. and often 0.0001 in. or less.

Rake Angle. Rake angle, sometimes termed face angle, varies with the material and its condition. It is usually 6 to 8 degrees positive for cast iron; 8 to 12 degrees positive for hard steel; 15 to 20 degrees positive for soft steel; approximately 10 degrees positive for aluminum; and 5 degrees positive to 5 degrees negative for brittle brasses.

Relief Angle. Normally this angle falls between $1/2$ and 3 degrees for internal broaches and is up to 3 $1/2$ degrees for surface broaches. For steel, the relief angle may vary from $1/2$ to 2 degrees; for cast iron, from 2 to 5 degrees; and for brasses and bronzes, from $1/4$ to $1/2$ degree. Relief may vary from maximum at one end to minimum at the finishing end to minimize loss of size in resharpening.

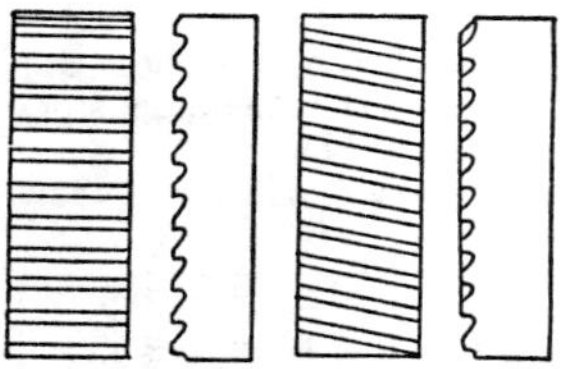

FIG. 4. Straight and shear cut broaching teeth.

Land. Varies from zero to $1/16$ in. in width and is usually graduated in width from the first finishing tooth to the last on internal broaches. Often calculated as 0.2 of the pitch. The land varies from zero to 0.005 in. for surface broaches and is usually backed off about $1/4$ degree.

Shear Angle. Shear cutting teeth (Fig. 4) are used wherever possible to obviate chatter and improve finish. Should not be used on deep slotting. Angle of teeth relative to direction of cut is usually 5 to 20 degrees. Minimum angle with surface broaching is usually 15 degrees.

34. THE BROACHING PROCESS

BROACHING SPEEDS. Cutting speed or travel of a broach across the work is usually fifteen to twenty-five times faster than in milling. Steel castings and forgings are normally broached at speeds of 20 to 33 ft per min whereas cast iron, brass, or aluminum are often broached at speeds up to 40 ft per min. While brass and aluminum can be satisfactorily cut at speeds even as high as 80 ft per min, most modern broaching machines have a much lower maximum ram travel, the fastest of which is about 43 ft per min. Faster speeds normally are found impracticable owing to lack of sufficient time for fixture loading and unloading. In the high-production manufacture of locks, however, broaching speeds as high as 59 ft per min are used to cut the narrow sinuous key slots in brass.

BROACHING MACHINES. There are two broad classes of broaching machines available for performing almost any variety of broaching operations, the vertical and the horizontal, either of which may have one or more rams. Although most modern machines are hydraulically driven, a few are operated mechanically. Plain horizontal machines are usually of the pull type, but vertical machines are available as pull-up, pull-down, or push-down types. Most of these machines can be arranged for efficient production of the quantities required and where necessary semi-automatic operation can be employed. Where extremely large quantities are in view as in some automotive and similar work, continuous broaching is a possibility. In these machines the broaches are stationary, a continuous chain conveyor with fixtures carrying the work pieces past the broaches. Other special machines such as for broaching radial surfaces are also in use.

ROUND PULL BROACHING. This method is used for producing holes of high accuracy and fine finish (Fig. 5). Holes may be finish broached from drilled, reamed, bored, cored, punched, or hot-pierced holes. Starting hole size should be $^1/_{32}$ in. smaller than the finished hole for small diameters, increasing to as much as $^1/_{16}$ in. smaller on large diameters. To assure clean-up and good finish, the allowance on diameter should never be less than $^1/_{64}$ in. Where cored holes are planned, draft angle and size variations must be taken into consideration to assure entry of the pilot and good clean-up. If extremely smooth finish is necessary, burnishing teeth or buttons can be used at the end of the tool (Fig. 5). However, these are recommended only for cast iron and nonferrous materials. Long holes should be avoided or chambered to reduce the length of broach required.

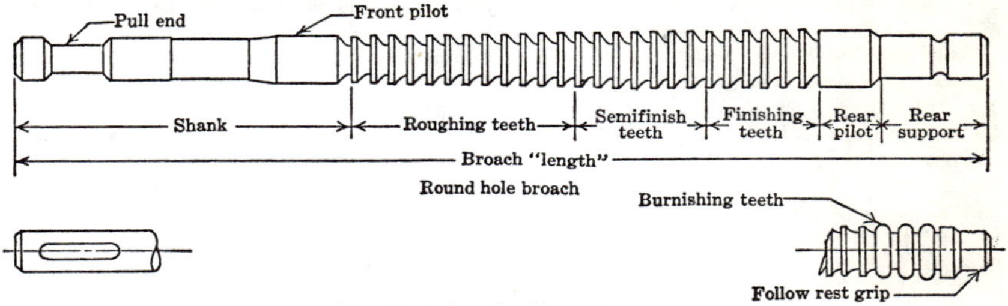

FIG. 5. Round pull broach design.

ROUND PUSH BROACHING. Employed to produce the same results as pull broaching, this method usually is restricted to parts requiring a relatively small amount of stock removal, since the tool must be comparatively short in order to have adequate stiffness to withstand the force exerted by the press ram. Push broaches are used for sizing holes after drilling, reaming or pull broaching round or splined holes, and for smoothing of die-cast and similarly produced holes. Special sizing broaches can be used for removing scale and correcting distortion in the holes of hardened gears, correcting errors in spacing of splines, correcting distortion in splines, etc. Push broaches are widely used for forming or sizing "blind" holes through which the tool cannot be completely passed. To prevent the cutting teeth from dragging on the return stroke, burnishing buttons are sometimes used to open the hole slightly.

BURNISHING. A cold-working operation, burnishing removes no metal but only compresses and smoothes out fine surface irregularities. Contraction usually follows the expansion of a hole produced by burnishing. Amount of contraction varies with the material, size of the hole, and construction of the part and can be determined most easily by experiment. Only soft, ductile materials are well suited to the burnishing operation. Burnishers are usually short and consequently are pushed through the hole.

REGULAR AND IRREGULAR-SHAPED HOLE BROACHING. This phase of the process is similar to round push and pull broaching. Symmetrical internal shapes usually are started from drilled, punched, or cored holes. The method of generating a square hole is illustrated in Fig. 6. However, where possible, an oversize starting hole should be used to make possible a shorter, more economical tool and much faster production. Because a broach is a form-generating tool, irregular holes of all varieties may be produced within the previously mentioned metal removal capacity.

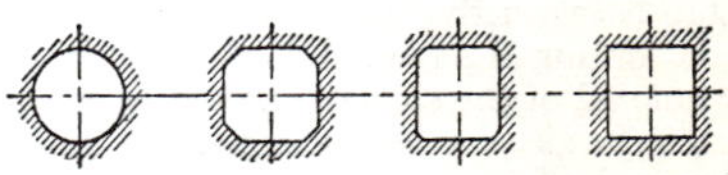

FIG. 6. Generation of a square hole by broaching a round hole.

KEYWAY, SLOTTING, AND CUTOFF BROACHING. Broaches for this type of work may be push or pull variety and can be used for producing single plain keyways, multiple plain splines by means of indexing fixtures, or for finishing round holes simultaneously with a plain keyway. A keyway-type broach is also adaptable for sawing or cutoff work and also is useful in producing various slot designs, especially where very narrow or sinuous key slots are required to exacting dimensions. Effectiveness of coolant action with this type of tool obviates heat distortion troubles otherwise present in such thin-bladed tools. Rise per tooth usually is restricted to a maximum of 0.00275 in. to avoid marking or scoring of the slot sides by the spreading action of chips formed with heavier cuts. This limits the depth of slot to about $^7/_{16}$ in. on a 66-in. capacity standard machine. Slots $^5/_8$ to $^3/_4$ in. deep are possible on 120-in. machines.

MULTIPLE-SPLINE SERRATION AND GEAR BROACHING. Splines of any required shape can be broached and are usually generated rather than cut directly. Both straight and spiral splines are produced having straight, involute, angular-sided, radius-lobe, and inverted shapes, Fig. 7. Broaches are usually of the pull type owing to length of tool necessary for removing the large quantity of material. As a rule the work is first drilled, bored, broached, or reamed to proper minor diameter and then broached, cutting the splines only. Where high concentricity is desired between the splines and the inside diameter, the hole can be prepared as for round broaching and a combination round-and-spline broach employed to finish the inside diameter as well as the splines. This type of tool is particularly useful on high-production jobs requiring extreme accuracy. Present practice is to size the inside diameter with the broaching tool when producing involute splines.

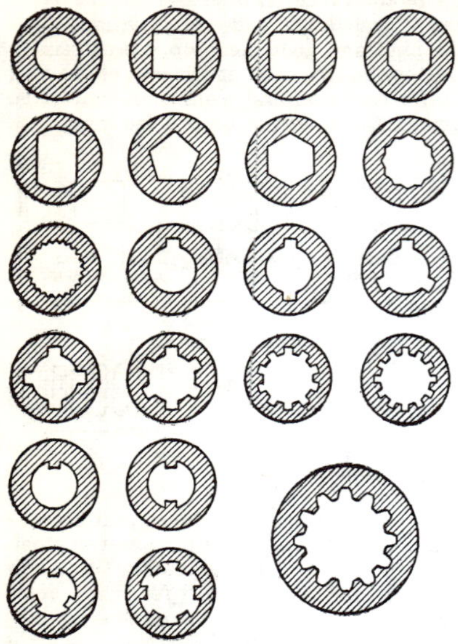

FIG. 7. Typical internal broaching. (Courtesy of Illinois Tool Co.)

Spiral-spline broaches generate tooth shape in much the same manner as straight-spline broaches. Where the helix angle of the spiral is less than 15 degrees and preferably under 8 degrees, it is possible to pull the tool through the part with a ball-bearing puller, the broach itself revolving to generate the proper helix. If the angle is larger than 15 degrees, a lead screw is necessary to assure positive and accurate generation. It is not usually practical to broach splines with a helix angle greater than 60 degrees.

Involute spline or **internal-gear** broaches, which can be made to cut any number of teeth, are used for producing spur or helical gears with teeth as small as 48 diametral pitch. In these gears the involute form is generated and checked from the base diameter. For extremely accurate parts or those heat treated after the first cut, a sizing or finishing operation should be used.

Serration broaches, similar to spline broaches, are usually made to SAE standards although, of course, special designs can be made to suit. Sizes range from $^1/_8$- to 3-in. nominal diameters with 36 to 48 serrations and may be tapered or straight. Serration drifts are used to produce taper serrations after broaching with a straight serration broach. A combination push broach and taper drift can be employed in some cases.

SURFACE BROACHING. A fast, accurate method for producing flat, curved, or irregular surfaces, this method of broaching is unexcelled for making a relatively smooth, even finish at low cost. One of the many combinations of tools available usually can be adapted to suit the job on hand. Gear segments, racks, serrations of any variety, slots, teeth, curved guides, etc., can be machined at one pass with exceptionally accurate relationship between the various surfaces. Typical procedure for surface broaching a connecting-rod forging is shown in Fig. 8. Removal of the surface skin on such forgings relieves certain internal stresses which in turn cause the bore to open up slightly. A boring operation is necessary after broaching such a cap and rod to assure a perfectly circular bearing seat. For best economy, surface areas should be kept to a bare minimum.

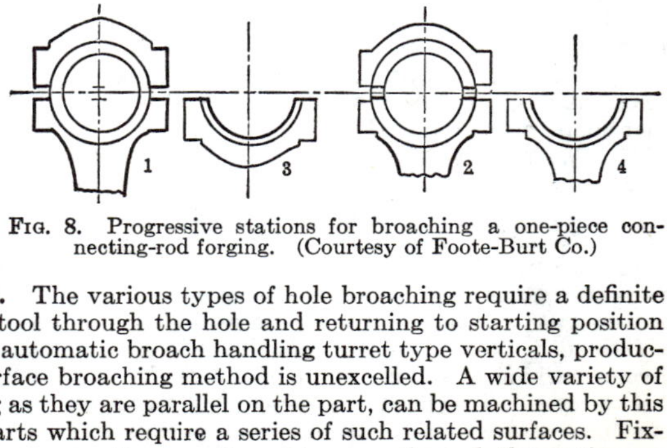

FIG. 8. Progressive stations for broaching a one-piece connecting-rod forging. (Courtesy of Foote-Burt Co.)

CONTINUOUS BROACHING. The various types of hole broaching require a definite cycle for starting the broaching tool through the hole and returning to starting position after a cut; and even considering automatic broach handling turret type verticals, production output of the continuous surface broaching method is unexcelled. A wide variety of plain or irregular surfaces, so long as they are parallel on the part, can be machined by this method and particularly those parts which require a series of such related surfaces. Fixtures can be designed to index one or more times as the part passes through the stationary broaches to produce squares, hexagons, slots, bevel gears, and similar forms on a single pass through the machine. Table 2 gives actual figures concerning current practice on the various parts shown in Fig. 9, produced on both regular and continuous machines.

Table 2. Production Data on Typical Broached Parts

(Courtesy of Foote-Burt Co.)

No. *	Part	Operation	Material	Broaching Machine	Pieces per hr	Pieces per grind
1	Connecting rod and cap	Broach sides, back of bolt bosses, partial splitting, final splitting, joint faces, and half round rod and cap	SAE 1035 steel forging	15-ton, 66-in. stroke duplex vertical hydraulic surface	120 sets	20,000 to 25,000
2	Fuse body	Broach tongue, top of tongue, wrench slots, and chamfer tongue	SAE 1112 steel	No. 15 continuous horizontal surface	1200	45,000 to 60,000
3	Drive end expander	Broach 0.230 x 0.375 deep slot and radius edges	SAE 1112 steel	No. 10 continuous horizontal surface	2200	35,000 to 40,000
4	Fuel pump rocker arm	Broach 0.317 x 0.812 deep slot	SAE 1115 steel forging	No. 15 continuous horizontal surface	1200	......
5	Connecting rod and cap	Broach sides, back of bolt bosses, split, joint faces, and half round rod and cap	Cast iron	No. 5 and No. 10 continuous horizontal surface	600 sets	......

*See Fig. 9.

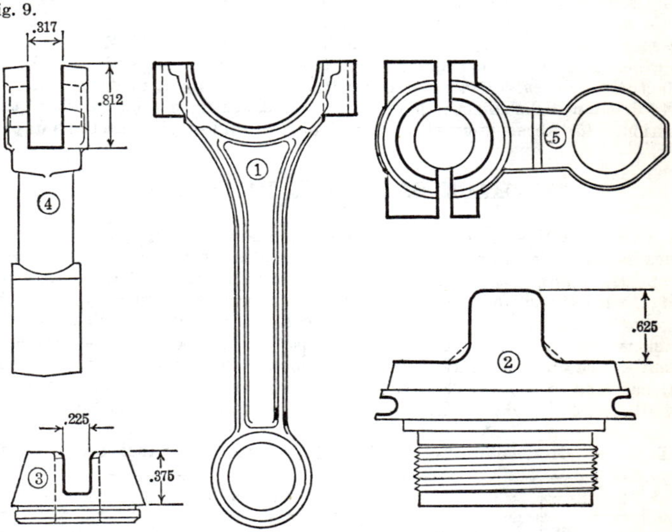

FIG. 9. Typical surface-broached machine parts. Production data relating to each part are given in Table 2. (Courtesy of Foote-Burt Co.)

MATERIALS. Any material that can be machined can be broached, but for the most practical and economical results hardness should range between Rockwell C25 to C35. Harder materials involve additional wear, and softer ones tend to tear. In any case, hardness should be rather carefully controlled to prevent excessive variation because such variation often results in poor surface finish. Internal stresses due to heat treatment should be avoided in parts which must be accurate because the relief of such stresses in broaching results in warpage.

TOLERANCES. Close accuracy and uniformity in surface broaching depend to a large extent on heavy, rigid fixtures, and a solid machine slide having a minimum of deflection. Internal broaching accuracy, however, is dependent more on uniformity of material. Holes can be held to within ± 0.0005 to 0.001 in. in production. Plain splined holes can be held to $+0.001$ in., -0.000 in. on the major and minor diameters with width of splines within ± 0.001 in. Outside diameter of serrations and diameter measured over wires can also be held to $+0.001$-inch, -0.000 in. Closer limits can be held if cost warrants and standard involute splines are normally held to $+0.0007$ in., -0.0000 in. on major diameter and $+0.0015$ in., -0.0000 in. on minor diameter.

In surface work, straddle-broached faces can be held to ±0.001 in. but, where design demands, plus or minus one quarter-thousandth can be obtained. Slots can be held to ±0.0002 in., but ±0.002 in. are the normal economical limits. Spacings of lugs, straddle-broached, can be held to 0.0003-in. accuracy. External gears and racks, broached one tooth at a time, can be held to 0.0005 in. on tooth spacing and form. On high-production runs where the entire segment is completed in one pass, even closer tolerances can be held. On large chain sprockets, tooth form and spacing can be held to ±0.005 in.

Manufacturers. Broaching machines are built by American Broach & Machine Co., Ann Arbor, Mich.; Cincinnati Milling Machine Co., Cincinnati, Ohio; Foote-Burt Co., Cleveland, Ohio; LaPointe Machine Tool Co., Hudson, Mass.; Colonial Broach Co., Detroit, Mich.; Oilgear Co., Milwaukee, Wis.; Acme Broach Co., Lexington, Ky.; Zagar Tool Co., Cleveland, Ohio.

METAL SAWING

By W. T. Gloor

The generally used methods of sawing metals are: (1) **Hack saws,** of which there are two types: (*a*) *hand hack saws* for occasional use at a bench, on assembly work, and in the field; (*b*) *power hack saws* for cut-off of light, medium, and large sections in production work. (2) **Band saws,** of which there are three types: (*a*) *contour saws* for sawing irregular designs; (*b*) *metal cutting band saws* for production cut-off of light and medium sections; (*c*) *friction saws* for sawing light sections and harder materials. (3) **Metal-cutting circular saws,** of which there are two types: (*a*) *milling saws* for slitting and slotting operations in manufacturing; (*b*) *cut-off saws* for production work on light, medium, and heavy sections. (4) **Friction disks** for cut-off of medium and heavy sections in either the hot or cold condition. (5) **Abrasive disks** for cut-off of light and medium sections in production work.

35. HACK SAWS AND BAND SAWS

HACK SAW BLADES are generally made of high-speed steel, though semihigh-speed steel blades find a moderate use and carbon steel blades a very limited use in present-day production. *Hand hack saw blades* are available in full hard, soft back, spring back, and double edge, with the spring back blades finding increased use because of low breakage and yet good cutting qualities. *Power hack saw blades* are either of the full hard or the spring back types. The saw teeth have various tooth shapes and are arranged on the blade to cut in one direction only. The teeth generally have no rake angle and are "set" in several patterns to provide clearance for the chips removed by the cutting action as well as the back of the blade. Table 1 gives data on standard high-speed steel blades.

Table 1. Dimensions and Teeth of High-speed Steel Hack Saw Blades

	Hand Saws		Power Saws											
Length, in.	10	12	12		14			17			18		21	24
Width, in.	9/16	9/16	1	1	1	1	1 1/4	1	1	1 1/4	1	1 1/4	1 1/2	1 1/2
Thickness, in.	.025	.025	.049	.065	.049	.065	.065	.049	.065	.065	.065	.065	.065	.065
Teeth per in.														
Regular	18	18	…	…	…	…	4	…	…	4	…	4	4	…
Medium	24	24	…	6	…	6	6	…	6	6	…	6	6	6
Fine	32	32	…	10	…	10	10	…	10	10	10	…	10	10
Tubing	…	…	14	…	14	…	…	14	…	…	…	…	…	…

These are the fields of the several classes of teeth: *Regular*—soft steel, cast iron, bronze. *Medium*—annealed high-carbon steel, high-speed steel. *Fine*—solid brass, iron pipe, heavy tubing. *Tubing*—thin tubing, sheet metals. It is advisable to use finer teeth for the sawing of thin sections or hard materials, whereas coarser teeth are more advantageously used for sawing thick sections or soft materials. To avoid possible stripping of the teeth from hack saw blades, it is recommended that two or more teeth be in contact with the work wherever possible.

The approximate sawing speeds in feet per minute for high-speed steel saw blades are shown in Table 2.

Coolant aids in decreasing sawing time and increasing blade life for ferrous materials as well as nonferrous materials. No coolant is required in the sawing of cast iron, brass, magnesium, and a few other exceptions.

Table 2. Cutting Speeds for High-speed Steel Hack Saw Blades

Material	Range of Speed, sfpm
Brass	160–200
Cast iron, gray	60–80
Steel, mild	140–180
Steel, tool, annealed	100–120
Steel, stainless	100–120
Rails	60–100
Steel, nickel	80–120
Phosphor-bronze	160–200

HACK SAW MACHINES, or power hack saws, range in capacity from stock 4 in. square up to stock 18 in. square. One type, built with a gap frame, can cut stock 16 by 26 in. by turning the work over when sawed halfway through. The stroke of the saw is 4 in. in the smaller machines and 9 in. in the larger machines, though this varies with manufacturers. All machines lift the saw clear of the cut on the return stroke, thus eliminating tooth drag and wear. Some machines feed by gravity, the saw frame being provided with weights that can be shifted to give greater or less pressure on the saw as required. Other machines have a positive feed which ranges from 0.001 to 0.025 in. per stroke, depending on the class of material being cut. When hard spots are encountered the automatic feed is stopped or shortened to decrease the cutting pressure on the saw until the hard spot has been cut through. Hydraulically operated machines in general provide variable cutting pressures and greater versatility. Automatic stock feeds have been adapted to a number of machines where numbers of pieces are to be cut off. Hack saws for making miter cuts are available in both the small and large capacities. Hack saws are powered either by belt or by a motor within the base. Generally three stroke speeds are supplied, and the amount of stroke is adjustable to suit the class of work.

METAL CUTTING BAND SAWS are used in production for cutting all types of materials up to a 12-in. square. Essential to the efficient cutting of a band saw is the proper adjustment of the pulley for alignment and tension. The saw blade is made in widths of 1/2 to 3 in., depending on the class of work to be cut. The generally recommended sawing speeds, in feet per minute are: Aluminum—1000–1500. Brass—soft castings 500–700, hard castings 150–300, sheets 750–1500, tubes 750–1500. Cast iron—100–150. High-speed steel—100–125. Cold-rolled steel—150–200. Malleable iron—150–200.

Machines manufactured for this use are of either vertical or horizontal type. Generally the vertical type is used in the cutting of sheets of material, and the work is fed into the blade. The horizontal type is used for cut-off of stock, and in this case the band saw blade is fed through the rigidly clamped work, usually by a power feed. Band saws are built with a variety of speeds to allow a range of material to be cut. They are also used in removing sprues, risers, and gates from castings.

CONTOUR BAND SAWS are similar to the general metal cutting band saws in method, but differ in the width of blade used. Contour saws vary from 1/16 to 1/2 in. in width up to 1 in. and are generally of an alloy steel. Table 3 shows the proper width of saw blade to be used in cutting radii.

Table 3. Minimum Radii for Contour Saws

Contour Saw Width, in.	Minimum Radii Heavy Set Saws
1/16	90 degrees
3/32	1/16 in.
1/8	1/8 in.
3/16	5/16 in.
1/4	5/8 in.
5/16	1 in.
3/8	1 7/16 in.
1/2	2 1/2 in.

Sawing speeds are somewhat dependent on the class of work to be cut, which determines the width of the contour saw. The thickness of contour saws falls in the range of 0.025 to 0.040 in. For saw widths of 1/16 to 1/4 in. the following cutting speeds in feet per minute are recommended. Aluminum—forged or dural, 1000–1500. Brass—cast 500–1000, naval (rolled) 250–350. Carbon steel—200–300. Steels—chromium 120–150, stainless 75–100, high speed 120–150. Malleable iron 180–220. For wider contour saws the cutting speeds are reduced. *Machines* used in contour sawing are of the vertical type with a deep throat and a speed range to cover a variety of work. A resistance welding unit is built into the machine to rejoin the saw when a central portion is being cut out of a piece.

FRICTION BAND SAWS are applicable to cutting single sheets or shapes. Sawing speeds used are in the range of 3000 to 15,000 surface feet per minute. A burr is generally left on the under side of the work. Feed rates of 25 to 60 in. per minute are recommended for material less than $^1/_4$ in. thick, with a range of 5 to 25 in. per minute for $^1/_2$-in. thick material. Heavier materials require greater care. The method is recommended for ferrous materials such as stainless, armor plate, etc., and for tool salvage. However, the method is not applicable to materials of low melting point or to those having a high thermal conductivity, e.g., aluminum, magnesium, brass, copper.

36. CIRCULAR SAWS AND FRICTION DISKS

Circular saws fall into two general classes: (1) Saws with teeth whose action is similar to that of a milling cutter. (2) Friction and abrasive disks.

METAL CUTTING SAWS are made in three varieties: (1) Teeth cut in the periphery of the disk. (2) Teeth inserted individually into the periphery of the disk. (3) Teeth cut in a segment of steel fastened to the periphery of the disk. Saws of diameters up to and including 8 in. generally are termed *metal slitting saws*. Saws of diameters 9 to 72 in. or more are generally referred to as *cold saws*, as they cut metal in the cold condition.

METAL SLITTING SAWS are made with a variety of tooth forms for sawing different materials. These saws are generally obtainable with side clearance and teeth having positive rake angles of 0 to 20 degrees. Saws without side clearance are used for slotting in order to hold a definite width of cut. Slitting saws with side milling teeth are used for heavier work and generally are made with alternate teeth having side clearance on opposite sides. For cutting cast iron, teeth are more widely spaced than for cutting steel, and rake angles of 3 to 5 degrees are commonly used. Nonferrous materials use widely spaced teeth and rake angles of 10 to 15 degrees, whereas steels utilize rake angles of 10 to 20 degrees with 5 to 10 degrees in the harder types. Saws with side clearance should be used wherever possible to reduce the side drag. Generally, the side clearance in slitting saws is machined into the disk either to the hole or to a hub of the same thickness as the new saw. Slitting saws are usually made of high-speed steel for production use.

Speeds and feeds using high-speed steel vary with the class of the material to be cut: Cast iron—speed of 60 to 90 ft per min, feed of 3 to 6 in. per min. Mild steel—speed of 100 to 125 ft per min, feed of 5 to 8 in. per min. Alloy steel—speed of 40 to 60 ft per min, feed of $^1/_2$ to 3 in. per min. Brass—speed of 150 to 250 ft per min, feed of 8 to 12 in. per min. Copper—150 to 350 ft per min, feed of 10 to 20 in. per min. Aluminum—speed of 500 to 1000 ft per min, feed of 10 to 20 in. per min. In general, higher speeds and feeds are used for the softer varieties of materials in each class.

INSERTED TOOTH SAWS were developed to allow a superior material to be used in the cutting action of larger saws and a lower quality steel to act as the disk and driving means. Inserted tooth saws are restricted to wider spacing of teeth because of strength requirements in transmitting the power from the disk to the tooth. The tooth is usually of a high-speed steel machined to fit into a socket provided in the periphery of the disk. The inserted tooth is machined with side clearance and is wider than the disk. Inserted tooth saws are used in metal cutting where wide tooth spacings can be used, depending on the type of material and the length of cut. Carbide-tipped inserted tooth saws are used in the cutting of nonferrous materials.

SEGMENTAL SAWS were developed to replace inserted tooth saws where heavy strains were to be absorbed by the saw due to cutting, without destroying the tooth mounting. The segments are generally of high-speed steel and are machined with side clearance and fastened by a suitable means to the thinner disk. The teeth are cut into the periphery of the segment. The segmental form allows a number of teeth to be cut on the segment with the result of narrower tooth spacing. These saws are adaptable to cutting all sizes of materials either ferrous or nonferrous.

SOLID CIRCULAR SAWS have teeth cut into the periphery and are generally made of semihigh-speed or high-speed steels with the side clearance cut into the sides of the disk to a hub. These are limited in use to smaller-diameter saws, usually under 24 in. for semihigh-speed and under 16 in. for high-speed steels.

Sharpening of metal cutting saws is generally done on either automatic grinders or hand-operated grinders. Various tooth forms are used to distribute the cutting load between successive teeth by breaking up the chip width. Most popular of the tooth grinds are the alternately high and low teeth in which the high tooth is beveled on both sides, leaving a cutting edge of one-third to one-half the width of the tooth in the center. The high tooth is often referred to as the roughing tooth, and the low tooth as the finishing tooth. The high tooth removes part of the chip width and leads the low tooth usually about 0.015 in., while the low tooth removes the remaining corners of the chip width and finishes the cut.

For longer saw life and faster cutting it is recommended that coolants be used wherever possible. Richer mixtures are generally used in circular sawing than in other types of machining, with mixtures running as high as 4 to 1 in soluble oils. For best results, as in all types of sawing, circular saws must be ground correctly. For mild steel cutting, a rake angle of 15 to 20 degrees is recommended, with a top clearance angle of 9 to 14 degrees. The top clearance angle should always be greater than the feed dictates lest the back of the tooth bind on the material being cut. Table 4 illustrates the rake angles, cutting speeds and feeds generally used in circular-sawing various metals.

Table 4. Cutting Speeds, Feeds, and Rake Angles for Circular Saws Having High Speed Steel Teeth

Material	Cutting Speed, ft per min	Saw Feed, in. per min	Tooth Rake, degrees
Aluminum—soft	2000–4000	25–40	20–30
Aluminum—hard	500–1500	15–30	10–20
Brass—soft	400–600	15–30	15–20
Brass—hard	300–450	8–15	10–15
Bronze	250–400	8–30	15–20
Cast iron	35–60	3–9	5–10
Copper	800–2000	20–40	20–25
Steels—40–65,000 psi tensile	80–100	5–8	15–20
Steels—70–85,000 psi tensile	60–80	5–8	15–20
Steels—85–100,000 psi tensile	50–65	4–7	10–15
Steels—rails—soft	55–65	3–6	5–10
Steels—rails—hard	40–55	1–3	5–10
Steels—cast—hard	40–55	3–6	5–10
Steels—tubes—thick wall	80–100	5–8	15–20
Steels—tubes—thin wall	120–160	6–15	15–20

FRICTION DISKS are relatively thin disks of alloy steel which are rotated at peripheral speeds of 20,000 to 25,000 ft per min. Although the trade often refers to the method as friction sawing, actually the cut is made by heating the work by friction of the disk to the molten or plastic state, and then removing it by notches or grooves in the disk. Generally there are two types of applications, cutting work in the hot condition and in the cold condition, which the trade calls "Hot Saws" and "Cold Saws," respectively. The *hot* disks are usually notched with V-type teeth, and the points swaged for side clearance. The *cold* disks are notched at regular intervals, usually by a chisel, to provide side clearance and roughen the friction surface. A flash of metal is generally left by the cutting action on the side where the disk leaves the work. Friction disks find application in mills and other places where quality of cut is not important. Friction disks are mounted on motor-driven machines and are usually water cooled. Rapid cutting times are obtained by this method, e.g., 8 to 12 in. I beams in 5 to 15 sec and 40 in. I Beams in 1 $^1/_2$ min.

ABRASIVE CUT-OFF DISKS are very thin abrasive wheels used in cut-off machines mainly for cutting smaller sections and bars. The method is applicable to cutting mild and hard materials of ferrous and nonferrous types generally using a coolant. The cutting speed of the disk is usually in the range of 10,000 to 16,000 ft per min.

CIRCULAR-SAWING MACHINES are built for cutting ferrous or nonferrous stock. Standard capacities will cut a range of sizes up to 14 in. by 24 in. or 16 $^1/_2$ in. diameter with one pass of the saw. Usually the work is cut singly, but in certain instances multiple set-ups are advantageous. In the sawing operation the work is clamped solidly and the saw fed through. Clamping and feeding are accomplished either mechanically or hydraulically. General-purpose sawing machines require manual loading of stock and operation of clamp and feed controls. Automatically operated circular-sawing machines are used where production quantities must be cut to a given length. Length tolerances of less than plus or minus 0.010 in. are obtained in sawing smaller size stock. Circular sawing features fast cutting with little or no burr left on the piece cut.

Circular-sawing machines are built with a range of feeds and speeds depending on the type of material to be cut. Saw speeds and feeds of ferrous and nonferrous metals generally fall within the range given in Table 4. The saw is usually driven by the spindle on which it is mounted. Power is transmitted to the saw spindle through a gear train from a motor drive integrally mounted on the machine. Another method of driving the saw employs a bevel drive gear which meshes with slots cut in the saw disk.

Several manufacturers utilize an *up-cutting* method on sawing machines designed for heavier work. The saw rotation in this method is such that the cutting load holds the front of the saw head against the ways. Chip-removers are used wherever possible in conjunction with the saw to remove unexpelled chips from the saw teeth. Such chip-removers generally mesh with the saw teeth, and are recommended particularly where ferrous alloys are being cut. A large variety of sawing machines has been built for such special production operations as cutting off sprues, gates, and risers on both large and small castings; for mitering large and small work; for cutting plates; and for performing machining operations.

GEAR-CUTTING PROCESSES

37. SPUR, HELICAL, AND WORM GEARS

Machining of gear teeth is performed in a variety of ways, all of which are based on planing, milling, or shaping. The processes generally are classified as (1) forming (milling), (2) template (planing), and (3) generating. It is not considered good practice to operate gears cut by the generating processes with gears cut by any of the others, and vice versa. Hobbed gears can be interchanged with shaper-generated gears provided proper tooling is used.

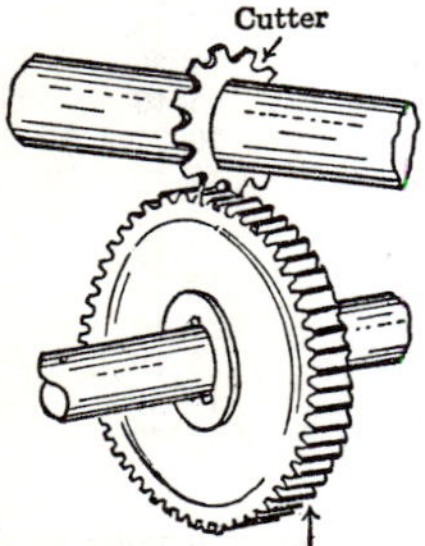

FIG. 1. Circular form cutter milling the spaces between the teeth of a spur gear. (Courtesy of Brown and Sharpe Manufacturing Co.)

FORMING PROCESSES utilize a tool having the shape of the space between the teeth. The cutter may be a planer or shaper-type tool, which is reciprocated while being fed into the blank. One tooth space is cut at a time, after which the blank is indexed through an angle corresponding to the circular pitch of the gear. The method is slow, and for this reason is little used.

Circular form cutters are milling cutters having profiles corresponding to the shape of the space between the teeth of the gear (Fig. 1). They find their widest application in job shops and for short production runs. Standard cutters are available for cutting teeth with a composite profile based on the 14 1/2-degree full-depth tooth with cycloidal modifications (see Article 3, Section 14). Theoretically, a different cutter is needed for each pitch and each gear diameter, inasmuch as the shape of the space is a function of base circle diameter as well as pitch and pressure angle. In practice a selection of 8 cutters is usually considered sufficient to cover the range from 12-tooth gears up to a rack. These are indicated by the whole numbers in Table 1. For greater accuracy in between these steps, additional cutters may be used. These, with their ranges, also are shown in Table 1. Special cutters may be obtained for specific numbers of teeth, as well as for special tooth forms.

Table 1. Cutters for Standard 14 1/2-degree Composite Gear Tooth Systems

	Range of Numbers of Teeth Cut	
Cutter No.	Using Set of 8	Using Set of 15
1	135–rack	135–rack
1 1/2		80–134
2	55–134	55–79
2 1/2		42–54
3	35–54	35–41
3 1/2		30–34
4	26–34	26–29
4 1/2		23–25
5	21–25	21–22
5 1/2		19–20
6	17–20	17–18
6 1/2		15–16
7	14–16	14
7 1/2		13
8	12–13	12

Helical teeth may be cut by circular-form cutters with the aid of a dividing head which imparts rotation to the blank corresponding to the lead of the helix. Selection of the cutter is based on normal pitch, number of teeth, and helix angle. Inasmuch as a section through the blank parallel to the plane of rotation of the cutter shows the pitch and base circles to be ellipses, the shape of the tooth space corresponds to a gear of radius equal to the radius of curvature of the ellipse at the end of the minor axis. An additional correction is required to avoid undercutting the flanks and relieving the faces of the tooth profiles, resulting in the following formula for formative number of teeth, N_f, to be used in selecting the cutter from Table 1:

$$N_f = \frac{N}{\cos^3 \phi} + p_n d \tan^2 \phi$$

where N = actual number of teeth in gear; ϕ = helix angle; p_n = diametral pitch of cutter = normal diametral pitch of gear; and d = pitch diameter of the cutter.

End milling cutters having profiles ground to the shape of the space between the teeth offer an alternative forming method (Fig. 2).

Broaching is used to cut the teeth of internal and external (Fig. 3) gears where the quantity justifies the cost of the tool. (See Article 33 in this section.)

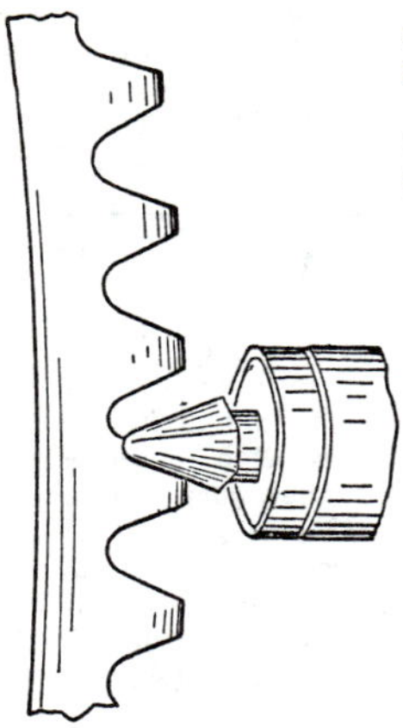

Fig. 2. Forming the space between teeth of a large spur gear by means of an end-milling cutter. (Courtesy of United Engineering and Foundry Co.)

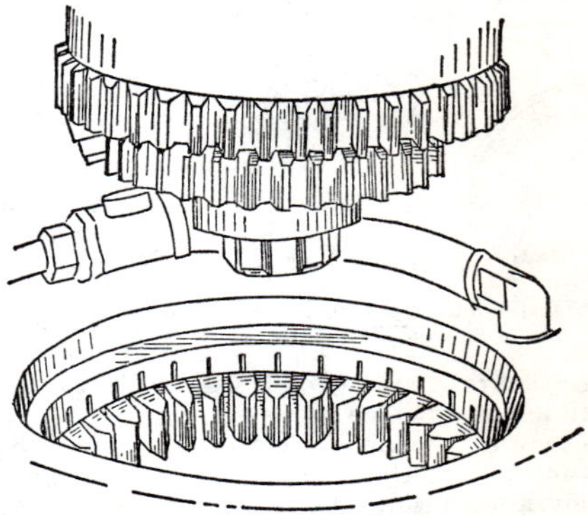

Fig. 3. Combination broaching and shaving a spur gear. (Courtesy of Michigan Tool Co.)

TEMPLATE PROCESSES employ a planing tool which is guided by a tracer mechanism, the tracer point moving over a template having the shape of the tooth profile. The method is used to make gears from forged blanks in which holes are first drilled in the area of the tooth space, the remainder of the metal being removed by the planer which produces the correct profile. It is also used for finishing teeth which already have been cast to shape approximately, and for cutting bevel gears. (See Article 38 of this section.)

GENERATING PROCESSES. Except for the broaching of internal gears, generation of gear tooth profiles represents the only method adapted to high production. Spur, helical, and worm gear generating processes may be classified as shaper generating and hobbing.

Gear shapers employ a cutting tool in the form of a rack or a pinion having the same pitch as the gear to be cut. In cutting a spur gear, the cutter and the blank are mounted on parallel axes and the cutter reciprocated axially while being fed into the blank until the pitch circles of cutter and blank are tangent (Fig. 4). Cutting of the gear is completed by rotating the cutter and the gear as if they were two gears in mesh, while the cutter continues to reciprocate. Helical and herringbone gears can be cut with equal facility by imparting proper oscillation to the work. Pinion-shaped cutters are adapted to all types of gears, including internal gears. Inasmuch as the involute rack is straight sided, a rack-type tool may be produced easily to consider-

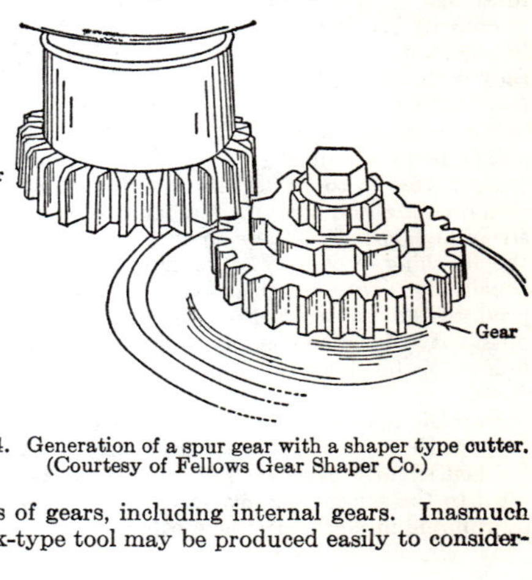

Fig. 4. Generation of a spur gear with a shaper type cutter. (Courtesy of Fellows Gear Shaper Co.)

able accuracy. The gear shaper-generating machine is not limited in scope to the production of gears but is adaptable to the making of a considerable variety of shaped parts.

Hobbing employs a rotating cutter known as a hob, which is, in effect, a worm gear having gashed teeth so relieved as to form cutting edges. The hob and blank are mounted on axes so oriented that the pitch helix of the hob is tangent to the pitch elements of the gear teeth (Fig. 5). For a helical gear, the angle between the axis of the hob and the plane of rotation of the gear, if both are right-handed or left-handed, is equal to $(\phi - \lambda)$, where ϕ is the helix angle of the gear and λ is the lead angle of the hob. If gear and hob are opposite handed, the angle would be $(\phi + \lambda)$. The speed ratio of the hob to the gear must be

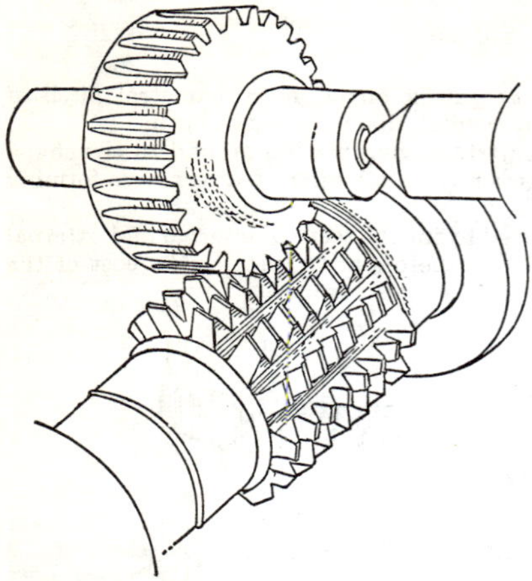

FIG. 5. Gear hobbing. (Courtesy of Michigan Tool Co.)

$$\frac{\omega_h}{\omega_g} = \frac{N_g}{N_h\left(1 \pm \dfrac{f \tan \phi}{\pi D_g}\right)}$$

where ω_h and ω_g denote speed of hob and gear, respectively; N_g = number of teeth in gear; N_h = number of threads in hob; ϕ = helix angle of gear; D_g = pitch diameter of gear; and f = feed of the hob parallel to the axis of the blank for N_g/N_h revolutions of the hob. The plus sign is used when hob and gear are of the same hand, the minus sign being used where they are to opposite hand. While hob and blank are rotated the hob is fed parallel to the axis of the blank, thus generating all the teeth in one complete pass.

Hobs may be straight, for cutting spur and helical gears up to 35-degree helix angle, tapered for higher helix angles. Formed hobs are made for special work. One straight hob of given pitch can be used to cut any size of spur or helical gear of either hand up to 30-degree helix angle; above this angle the hob should be of the same hand as the gear. For greatest accuracy ground single-thread hobs are recommended, but multiple-thread hobs give higher production.

GEAR FINISHING. Shaving, burnishing, lapping, or grinding, or a combination of these improve surface finish and accuracy over and above that possible by the foregoing methods. In particular, distortion during heat treatment necessitates subsequent refinishing.

Shaving involves the use of a cutting tool in the form of a hardened helical pinion or rack with teeth serrated at points along the face. The cutter and gear are run together in mesh at high speed (Fig. 6), the cutting action being obtained either by running on crossed axes (5 to 10 degrees usually) or by reciprocating the shaving tool at comparatively high speed. Where shaving is used as the finishing process, the rough cutting is usually done with a special tooth form which produces an undercut, thus eliminating the possibility of gouging the root fillet, and confining the finishing operation to the tooth faces.

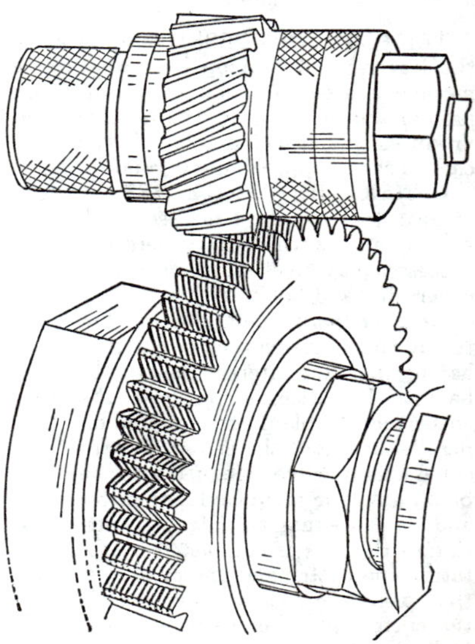

FIG. 6. Rotary-gear shaver. (Courtesy of Michigan Tool Co.)

Burnishing consists of running the gear in mesh with three hardened burnishing gears finished slightly oversize, pressure being applied to the burnishing gears. Mating gears, particularly if hardened, can be burnished by running in mesh with each other.

Grinding machines for gears employ either the forming process or the generating principle.

In form grinding the abrasive wheel has the shape of the space between the teeth and is used in a similar manner to form cutters. The diamonds which true the wheel are controlled by a suitable mechanism based on the use of templates.

Generating gear grinders employ a narrow straight-sided abrasive wheel, the flat side of which corresponds to the profile of an imaginary rack tooth. The machine imparts relative motion to the gear and the wheel similar to that of the gear meshing with a rack.

When gears are to be finished by grinding, special roughing cutters are generally employed in the preliminary stages of manufacture. These cutters are designed to provide a variable allowance for stock removal along the tooth profiles, tapering from a maximum at the tips to zero in the region of the fillet.

Lapping usually is done by rotating the production gear in mesh with a lapping gear in the presence of an abrasive compound. The gear and lap are so mounted that an endwise sliding motion occurs between the teeth. Sometimes production gears are lapped with each other.

Laps are sometimes made with high or low pressure angles with respect to the piece to be lapped, so as to remove metal in specific contact zones.

38. BEVEL-GEAR CUTTING

By Allan H. Candee

The tooth profiles of bevel gears are generated by tools carried on a rotating cradle and representing teeth of a crown gear or another bevel gear. Machines differ principally in the type of tool used. Straight teeth are cut by tools reciprocating on straight slides, and curved (spiral) teeth usually by rotating cutters.

STRAIGHT BEVEL GEARS. Method 1. Generation. Two reciprocating tools represent teeth of an imaginary generating crown gear, and work on opposite sides of the tooth being generated. After one tooth is finished the blank is indexed for the next tooth. Tools with straight cutting edges generate curved tooth profiles as the result of relative motion, obtained in some machines directly by rolling segments and in other machines by gear trains. (See Figs. 7 and 8.) This is the method most widely used. Machines of latest type generate slight lengthwise crowning so as to ease off the contact at the ends of the teeth. Straight bevel-gear generators range in capacity up to approximately 30-in. gear diameter or 15-in. cone distance.

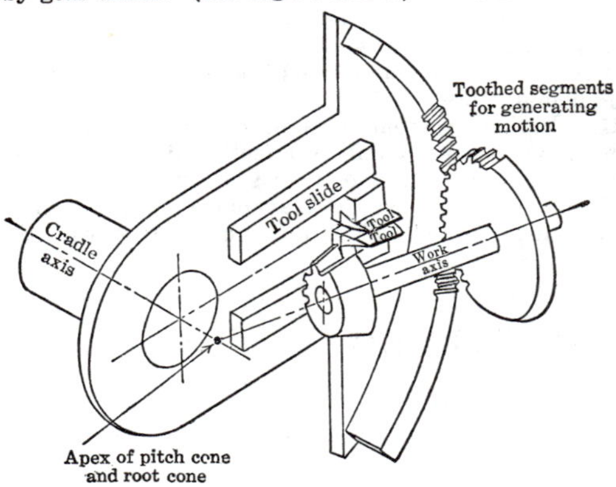

FIG. 7. Principle of bevel-gear generator.

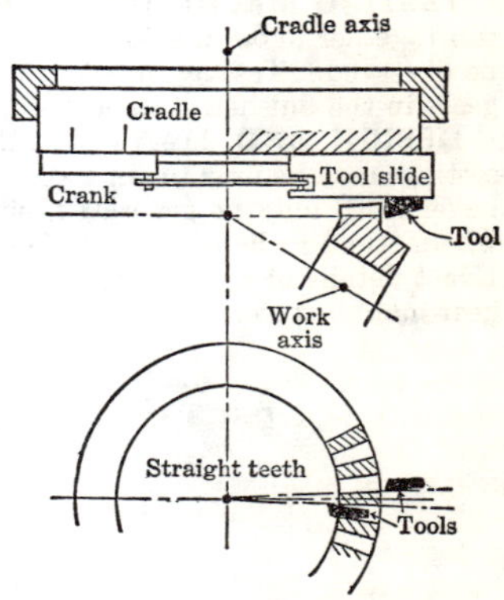

FIG. 8. Straight bevel-gear generator.

Method 2. Generating-milling. For mass production of gears for automobile differentials and similar applications, the so-called Revacycle process was introduced in 1938. A continuously rotating disk cutter feeding from end to end of the teeth rough-cuts and finishes a tooth space during each revolution. The teeth of the cutter vary in height, thickness, and form in order to produce properly tapering bevel gear teeth. The method is three to four times faster than any other. A special cutter must be calculated and manufactured for each job, which makes the process uneconomical for small quantities.

Method 3. Form-planing. In bevel-gear planers a single reciprocating tool with rounded point is controlled by a former to produce the profile shape of the tooth. Different formers are required for different-pitch angles. This is the oldest type of correct bevel-gear-cutting. Gear planers have been built to cut gears up to 20 ft in diameter. Planer-cut gears are not equal to generated gears in smoothness of action.

SPIRAL BEVEL GEARS. Method 1. (See Fig. 9.) Rotating cutters of face-mill type are used for most spiral bevel gears up to about 30-in. diameter. A relative motion between the work spindle and the cradle generates conjugate tooth profiles. Most "flat" gears for tooth ratios of 3 to 1 or higher are cut without generation and are known as Formate gears. All other gears and pinions are generated. Machines can be set to cut any spiral angle from 0 degrees up.

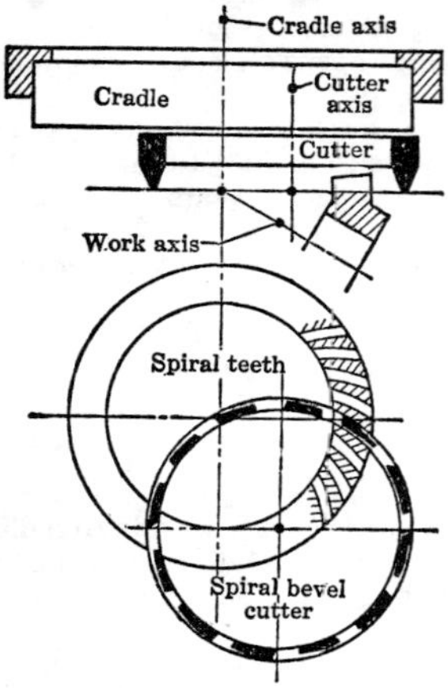

FIG. 9. Spiral bevel-gear generator.

Method 2. Larger gears, up to about 100 in. in diameter or 50 in. cone distance, are cut on planing generators by single reciprocating tools with side-cutting edges and with continuous rotation of the blank.

Method 3. In Germany a method of generating spiral bevel gears by means of taper (conical) hobs is used to some extent. Although invented in the United States, this method has not been adopted here.

HYPOID GEARS are so closely related to bevel gears that the same machines are used for both. It is necessary, however, that the workhead be provided with vertical adjustment. For generating a hypoid pinion the work spindle must be offset *below center* or *above center*.

TOOTH GRINDING. Spiral teeth of the type produced by the rotating spiral-bevel cutter can be ground after hardening. A grinding wheel in the general form of the cutter is used. The gears may be bevel or hypoid, and the teeth may be generated or Formate. Gear sizes are limited by machine capacities.

SUPPLEMENTARY MACHINES. Roughing machines are customarily used for gears of large and intermediate pitches. Burnishing and lapping machines are used in production shops.

TESTING MACHINES. On these machines the gears are set in correct position and run together under a brake load. Smoothness of action, sound, and tooth bearing can all be observed. Testing in this way is essential for demonstrating the correctness of bevel gears in the finished state and during manufacture.

DESIGN AND MANUFACTURE. Various combinations of cutters and machine settings may be used to cut a gear and pinion of given numbers of teeth and pitch. Spiral-bevel teeth may be generated, either one side at a time or both sides in one operation. The method to be used often depends on cutters and machines available in a given shop. Exact details of tooth design therefore should be determined after consultation with the gear manufacturer.

FLAME CUTTING

By C. W. Obert

39. GAS CUTTING

Gas cutting of ferrous metals is a process of preheating material to its kindling or ignition temperature and then oxidizing it rapidly by means of a closely regulated jet or stream of oxygen issuing from a special *cutting blowpipe* or *torch.* Hence the process is primarily chemical, based on the chemical affinity of oxygen for ferrous metals that have been heated to the kindling temperature or higher.

The chemical reaction that results when commercially pure oxygen is brought into contact with most steels or iron at the kindling point is very active. For this carbon steel must be preheated to a bright red color, a condition reached between approximately 1400 and 1600 F. For accurate control, both the gases supplied and the rate of progression of the cutting should be closely regulated.

Only the metal within the direct path of the oxygen jet is acted upon. In linear cutting, a *kerf* is formed having uniformly smooth and parallel walls. Under skilled workmanship, using mechanically guided cutting torches or blowpipes, cutting tolerances, with respect to squareness and longitudinal alignment of the cut surface, may be held very closely.

The faces or walls of the kerf, in the commonly used carbon steels, are not injured by the cutting operation. On the contrary, the strength and toughness of the faces are slightly increased. The effect of heat on the metal, as a result of normal cutting operations, rarely penetrates farther than approximately 0.1 in. below the cut surface, except in the so-called stainless steels, where precipitation of carbide may extend as far as 0.375 in.

In addition to linear gas cutting or severing, other types of controlled oxidation are in use, employing the free-burning reaction of oxygen with ferrous metals that have been heated to kindling temperature, in order to sever, shape, and work such materials. One application, known generally as *oxygen-lance cutting*, utilizes a relatively long section of small-diameter iron pipe. A pipe with a stream of oxygen flowing through it is ignited and kept aflame, thus forming a lancelike cutting tool that furnishes preheat as well as free oxygen for cutting and is used primarily for emergency and heavy-duty cutting, such as opening furnace tap holes.

Another application of gas cutting is known as *flame machining*. This method is used primarily to surface ferrous metals in much the same way as a planer or milling machine. The cut does not penetrate the work as in severing or linear gas cutting.

In actual gas cutting, the iron or steel removed from the kerf is, as already noted, not entirely oxidized or consumed by the oxygen. Because of the eroding effect of the cutting oxygen jet, however, about 30 to 40% of the metal is washed out of the cut as unconsumed or metallic iron. As a result of this fortunate circumstance, it is not uncommon to obtain satisfactory cuts with a consumption of cutting oxygen less than one cubic foot of oxygen per cubic inch of material removed from the kerf.

FUEL GASES. A number of fuel gases of high flame temperature have been employed for cutting ferrous metals, with varying degrees of success. The more important of these are acetylene, hydrogen, city gas, natural gases, coke-oven gas, ethylene, and the lighter petroleum distillates, including methane, ethane, propane, and butane.

Aside from hydrogen, all these fuel gases are hydrocarbons and give off carbon dioxide (CO_2) and water vapor (H_2O) as products of complete combustion. They may be burned in different manners but, when used for gas cutting, the combustion is made to take place in the way that will develop maximum flame temperatures. For quick starting and rapid progress of a cut, hot preheating flames are a prime essential. This requirement is fulfilled by designing cutting torches or blowpipes that burn the fuel gases according to the Bunsen principle and premix commercially pure oxygen instead of air with the gases before combustion. The substitution of oxygen for air, in the primary combustion, results in a notable increase of flame temperature of the fuel gas, because the nitrogen, constituting nearly four-fifths of the air, does not need to be heated by the flame when oxygen is used. Acetylene produces a hot, short flame, with an easily distinguished, bright inner cone; the hottest point is at the tip of the inner cone.

FLAME CHARACTERISTICS. Depending on the ratio of oxygen to fuel gas in the mixture, the characteristic of the flame produced may be carburizing, neutral, or oxidizing. The neutral flame results when just sufficient oxygen is supplied to complete the primary combustion in the inner cone of flame, yielding carbon monoxide (CO) and hydrogen (H_2) as products. Neither of these two gases carburizes or oxidizes molten metal; in fact, both are neutral to it. Regardless of the amount of primary oxygen supplied, complete combustion in the inner cone of flame is impossible since, at such extremely high temperatures, carbon dioxide and water vapor are unstable and dissociate or revert in part to carbon monoxide, hydrogen, and oxygen. This dissociation limits maximum temperatures attainable in a flame of oxy-fuel gas; otherwise these temperatures might go much higher.

Maximum flame temperatures are obtained with a neutral flame. With strongly oxidizing flames, however, burning of the metal itself occurs, and this type is generally used for gas cutting. Oxidizing flames preserve the general appearance of neutral flames, but their inner cone is shorter, "necked in" on the sides, and more sharply defined. Because of dissociation, the products of the primary combustion in the inner cone of excess oxygen flames consist of carbon monoxide, carbon dioxide, hydrogen, water vapor, and free oxygen, in proportions that depend on the amount of excess oxygen premixed with the fuel gas.

Secondary combustion of the fuel gas takes place in the outer or envelope flame. Here any carbon monoxide and hydrogen contained in the primary combustion products complete their combustion to carbon dioxide and water vapor, respectively, since the lower temperature is insufficient to cause dissociation of these latter compounds. The necessary oxygen required for completing the secondary combustion in the envelope flame is taken from the surrounding air by diffusion into the zone.

Most hydrocarbons are exothermic, generating heat during formation and absorbing the same amount of heat during combustion. But acetylene is highly endothermic, absorbing heat during formation and liberating it during burning or decomposing. The endothermic heat liberated by acetylene is in addition to that produced by the combustion of its component elements.

Acetylene contains the highest percentage of carbon (92.3% by weight) of any common gaseous hydrocarbon compound. This, coupled with its exceptionally large endothermic heat, gives it the highest flame temperature of any known fuel gas, by a wide margin.

EFFECT ON METAL STRUCTURE. In gas cutting, the temperature of a slight depth of metal immediately adjacent to the cut face is raised considerably above the transformation range. As the cut progresses, the steel cools again through this range. The speed of cooling depends on the heat conductivity of the surrounding metal, radiation losses, and the fact that the heat is applied only momentarily at any given point, since the cutting torch is kept in constant motion while cutting. When the steel being cut is at room temperature, the rapidity of the cooling is sufficient to have an actual chilling or quenching effect on the cut edges, particularly in heavier cuts involving large masses of cold metal.

When etched specimens are examined under the microscope, the material is found to be altered when compared with the original metal. However, the change is largely a physical one, concerned with crystalline grain structure. The normal pearlitic steel is transformed into one or more of several structures ranging from spheroidized carbides in ferrite to the harder constituents, depending on the particular chemical composition of the steel cut and the rate of cooling after the cutting operation, as well as the carbon content and the alloying elements present. Also the heat developed by the cutting makes the grains at the edge of the cut either larger or smaller than the unaffected grains behind the cut edge.

LOW-CARBON AND MILD-CARBON STEELS. At normal temperatures, gas cutting of carbon steels that contain less than 0.30 or 0.35% carbon is not detrimental unless alloying elements are present. Gas-cut edges of such steels are suitable for any uses where edges are employed that are prepared by any of the mechanical methods of severing, such as planing, milling, shearing, friction sawing, and grinding. Gas-cut edges are adapted to applications involving stresses, such as mechanical calking and holes in which tubes are to be expanded or rivets driven. Clean, smooth gas-cut edges of low-carbon or medium-carbon steel plates or forgings have physical properties that render them actually superior to edges that are sheared, planed, or otherwise prepared mechanically.

Where edges are gas beveled to form a V for welding, and later are welded together by either oxyacetylene blowpipe or electric arc, the heat of the welding operation acts to dissolve all marginal structures previously formed and operates to anneal the metal for an appreciable depth below the cut surface. Any change in structure that the preceding cutting may have produced in and adjacent to the cut surface is therefore obliterated.

Tests of flame-cut edges of low-carbon and medium-carbon steels to determine their physical properties (such as hardness, ductility, microstructure, and resistance to impact) have shown that they are satisfactory for almost any type of mechanical treatment or character of service.

HIGHER-CARBON AND ALLOY STEELS. Gas cutting of higher-carbon and alloy steels, particularly at room temperature, may produce a thin layer of hard material on the cut surfaces, which may not be sufficiently ductile to withstand the stresses set up in cooling, without cracking, or the stresses imposed by the subsequent use of the part cut. The torch heats the steel at the cut edges to a high temperature momentarily, but normal thermal expansion is prevented by the adjoining cold metal. The action called *upsetting* results. Later, when cooling occurs, contraction of the upset metal may develop severe stresses in tension along the cut edges. With these steels, the ductility may not be sufficient to prevent incipient cracks and checks that can prove dangerous in service, since they are likely to develop into large fractures.

Experience has shown that if these higher-carbon and alloy steels are preheated and, in certain instances, annealed afterward, no cracks will occur on the cut edges. Naturally, with the adjoining metal preheated, the difference in expansion is less between it and the metal heated during cutting, and upsetting is greatly reduced or eliminated.

When equipment is used that delivers large volumes of oxygen through large tips at low pressures, large billets and rounds are successfully cut at rolling temperatures (1700 to 1800 F). Occasionally very high cutting speeds are obtained. Evidently, at temperatures above the transformation range, steel reacts more vigorously with the cutting oxygen, producing slags with great amounts of iron and relatively little Fe_3O_4.

To obtain the most efficient use of the oxygen and, at the same time, leave the steel in the best condition, a preheat temperature of 500 to 600 F should ordinarily suffice for carbon steels. The additional expense that accompanies preheating is compensated for by the more rapid rate at which preheated steel can be cut.

Annealing completely restores the original pearlitic structure of the steel at the cut edges, wherever this may be desired, and removes any residual stresses set up in the metal. The majority of steels do not require annealing if they have been preheated before the flame cutting.

GAS-CUTTING EQUIPMENT consists ordinarily of the cutting torch or blowpipe with separate cutting tips of different sizes, oxygen and fuel-gas hose with connections, oxygen and fuel-gas pressure regulators, and gas cylinders. The pressure regulators are usually equipped with cylinder and working-pressure gages. Regulators for the cutting oxygen are designed especially to deliver oxygen in large volumes and at higher pressures and, hence, are usually fitted with gages that measure working pressures as high as 400 psi.

Where large quantities of oxygen are required, oxygen cylinders are often manifolded. Acetylene generators, rather than cylinders, are used where the consumption justifies it. In such installations, distribution pipe lines are frequently employed to carry both gases to the various points of consumption throughout the premises. For gas cutting by machine, the torch is held and operated mechanically, but its essential features and the accessory equipment remain unchanged. To insure safety and efficient results, the manufacturer's instructions for setting up and operating the equipment should be followed closely.

DRAG. In flame cutting, the jet of cutting oxygen usually enters the cut in line with the axis of the cutting oxygen orifice. After traversing about one-half the thickness of the cut, however, the jet may curve backward in a direction opposite to the travel. When this occurs, the amount by which the oxygen jet falls behind the perpendicular in passing through the material is known as the "drag" of the cut. Faint markings or ridges called "drag lines," left on the face of the cut, enable the drag to be measured. In heavy cuts, these drag lines are more pronounced. Cuts with short or negligible drags are known as *high-quality* cuts; those with long drags are termed *low-quality* cuts.

Cutting studies have shown these facts: (1) As the drag is decreased, the oxygen consumption is increased. For very short drags, a slight decrease in drag length can be procured at the expense of increased oxygen consumption. (2) Except for low cutting speeds, a certain minimum length of drag is characteristic of a given tip and speed regardless of the operating pressure (or the oxygen consumption). (3) An increased drag may result if an attempt is made to obtain a drag shorter than the minimum by increasing the operating pressure beyond that at which the minimum drag is obtained.

In considering the costs of flame cutting, therefore, it is essential to determine how short the drag must be to fulfill the requirements. If it is too long, a final corner will remain uncut, and this is usually not permissible. Straight-line cuts can frequently be made with considerable drag, without a particularly detrimental effect, since the drag lines lie within the plane of the cut and do not prevent both top and bottom edges of the cut from being left clean and sharp. This procedure, which saves considerable oxygen, is also applied in cutting scrap, where cut quality is nonessential as long as severing is accomplished. However, in cutting shapes, when rounding curves or corners, less drag is permissible, because the bottom contour of the cut will tend to come out different from the top, with the edges not square. Such cuts must be made with short drag. They require a good deal more oxygen than long-drag cuts that barely penetrate the material. Precision-machine flame cuts are necessarily of high quality.

PRESSURE. In heavy cutting, greatest care must be used in selecting the proper operating pressure. Not only is the range of suitable pressures considerably narrower for the heavier cuts, but the pressure also decreases at the hose-line connections, and a larger torch should be used with the increasingly greater flows of oxygen required for heavy cutting. The large volumes of oxygen needed are better supplied through the use of lower pressures and torches with large cutting orifices. Extreme preheating is necessary. Attempts to obtain the quantity of oxygen required by the use of high pressures in common machine cutting torches result in gouges and pockets in the kerf. Variations in the quality of steel are more important in heavy cutting.

Considerable skill is sometimes required for starting a heavy cut, in order to sever the initial lower corner completely and to avoid pocketing the stream of cutting oxygen in the lower portion of the cut. It is essential that the front edge of the cut be very clean, particularly the lower portion of the front face. To align the axis of the tip with the front edge of the material, the stream of cutting oxygen should be turned on with preheating flames burning to render it visible. In starting the cut, the movement of the torch should be retarded just long enough to allow the cutting stream to attack the material actively throughout almost the entire depth. Then the torch is permitted to proceed normally. If the piece is very heavy, the lower portion of the front face is sometimes heated with a large welding flame. In certain instances, starting is facilitated by first undercutting the forward edge of the material with a hand-cutting torch. However, unless the operator is experienced, the resulting cuts may be poor and ragged.

Care is required also in completing the final corner of heavy cuts. A slight retardation of the forward movement of the torch is desirable here as well.

THICKNESS. When the standard cutting torches are not assisted by oxygen lances. present limits of penetration are about 36 in. for manual cuts and 54 in. for machine cuts. Particularly for freehand flame cutting, the upper limit depends largely on the skill and steadiness of the operator. With oxygen lances it is possible to penetrate virtually any thickness of metal.

MATERIALS. Almost any ferrous material can be cut by gas; however, certain steels contain sufficient alloying elements to offer considerable resistance to the process. The uncombined carbon in cast iron has a similar effect. For such materials, special flame-cutting techniques must be applied, as described later. Oxygen lances permit penetration of virtually any metal, ferrous or nonferrous.

CROSS-CUT TOLERANCES. Where the torch is held firmly and advanced at uniform speed, as in machine flame cutting, cross-cut tolerances can be kept within narrow limits. In ordinary steel 6 in. thick, cut surfaces can be held to cross-sectional squareness accurate within $^1/_{32}$ in. Cuts in thinner sections can be held to proportionately smaller limits. The degree of longitudinal precision of a machine cut depends on the trueness of the guide rails of the cutting machine, the clearance of the operating mechanism, and the regularity of the propelling unit. Machine cuts can be made so smooth and square and can have such sharp edges that they require no further finishing of any kind.

BEVELS. Angular or bevel flame cuts are accomplished by inclining the torch and tip sideways to the desired angle or by using a bent tip. Such cuts may be either by hand or by machine, in straight or irregular lines, and with the same range of depth as in normal cutting at right angles. The usual angular limits of bevel and straight-faced fillet cutting range from 10 to 60 degrees.

SHAPES. Straight lines, circles, or contours of the utmost complexity are flame-cut with equal ease by either mechanical or hand-guided torches. Flame cutting, in sharp contrast with most machine-tool processes, has almost unlimited flexibility in this respect.

MULTIPLE-TORCH CUTTING. In many applications of machine flame cutting, as many as twenty torches are used to cut a like number of identical shapes simultaneously. Concentric circles for flanges and the like are often cut in one operation, with two or more torches.

When the sheets are thin and level enough to be clamped economically into a tight pile, stack cutting is indicated for multiple production. When the plate is heavy or wavy, multiple cutting is used—sometimes along with stack cutting.

STACK FLAME CUTTING. The operation known as *stack flame cutting* offers a means of cutting several thicknesses of metal simultaneously with a single torch, instead of employing multiple-torch cutting.

The plates in the stack should be clean and flat, with edges in alignment where the cut is started. The number varies with the thickness of individual plates, to keep the thickness of the stack within reasonable limits. The plates should be clamped together firmly, as nearly as possible along the line of cut. When thin plates are to be flame-cut in stacks, a *waster plate* that has sufficient rigidity to lie flat while the cut is being made is usually clamped on to prevent the top thin plate from buckling with the heat. A waster plate serves also to maintain ignition in the cut by its own combustion, thereby increasing the speed of cutting in this kind of work.

UNDERWATER CUTTING. Equipment for underwater cutting has been highly developed. Although oxyacetylene is used for shallow work, the pressure of acetylene needed soon rises to more than 15 psi because of the hydrostatic head. As a result oxy-hydrogen torches are used at depths greater than 25 ft. Operation under water involves the creation of an artificial atmosphere or *bubble*, in which the preheating flames burn and in which the actual cutting is done. Some torches create such a bubble by the use of a skirt around the tip. The skirt is filled with the combustion gases in order to keep the water away from the flame. The most successful equipment for deep-water operation requires a supply of compressed air to form the bubble in the skirt.

Operating pressures are determined largely by the hydrostatic head to be overcome. The chilling effect of the water on the work requires heavier preheats. Because of the difficulties under which the diver must work—such as clumsy dress, darkness, insecure footing, cold, and high pressure—only severing cuts are made, and oxygen and acetylene consumption figures are meaningless, although the specific cutting speed may be higher than in the open air.

Another form of underwater cutting equipment uses the electric arc for preheating.

40. ELECTRIC-ARC CUTTING

The process of electric-arc cutting consists in applying the heat energy of the electric arc to melt the metal along the desired line of cut. Because of the extremely high temperature developed in the electric arc, it can be used to fuse almost any material that conducts electricity, and in this way it is employed for cutting metals. Arc cutting is not a true burning process, since the metal is merely melted away. Arc cutting does not equal gas cutting in smoothness, quality, or accuracy of cut surfaces, and hence is not used where high-quality, accurate cuts are required. However, it is useful where smooth, uniform cut surfaces are not essential, and mere severing is the main object, as in scrapping operations. Arc cutting is used also for metals not easily cut in any other manner, such as cast iron, manganese steel, and large sections of nonferrous metals. It is applied particularly to the cutting of risers from large manganese steel castings. Arc cutting is also used for the cutting of rivets and piercing of holes.

CARBON-ARC CUTTING consists essentially of drawing an arc between a carbon or graphite electrode and the material to be cut. Direct current is generally used, with the carbon electrode negative (straight polarity), although alternating current can be used also. Graphite electrodes are preferred to carbon electrodes because they have a much longer service life and retain a point longer. A point is essential for clean, smooth cuts. The heavy currents used for arc cutting make ordinary electrode holders unsuitable except for light work. Special holders are recommended that have a long shank, a special mechanism for gripping, and a shield for protecting the hands.

MANIPULATION of the cutting arc, when cutting plates $^1/_2$ in. thick or less, consists simply in advancing along the line of cut at a rate equal to that at which the metal is being melted away. The most smooth and uniform cuts are made when the plate is in a horizontal plane, but the operation may be performed when the plate is in an inclined or a vertical position. For horizontal cutting, the electrode is held vertical or pointed slightly forward. For inclined or vertical cutting, the electrode is pointed downward. When materials heavier than $^1/_2$ in. are being cut, the vertical face of the cut is melted away progressively from top to bottom. Thus the bottom of the cut is advanced slightly ahead of the top of the cut, producing an undercut that permits the molten metal to run out more easily. Skilled operators can make narrower cuts. The width and speed of the cut increase with the current and the diameter of the electrode.

ARC CUTTING WITH COATED ELECTRODES. The application of heavily coated electrodes to cutting is not always an economical process, but it has certain merits, particularly where no oxyacetylene cutting torch or graphite electrode is available. Electrodes with either a gaseous or slagging type of coating can be used, but the slagging type gives better results because it will handle higher current densities. Plates $^3/_8$ in. thick or less can be cut simply by advancing the electrode along the line of cut. On heavier plates the electrode is pointed forward and moved from the top to the bottom of the cut, so as to keep the bottom slightly ahead. Electrodes smaller than $^5/_{32}$ in. in diameter are not suitable for arc cutting. The width of the cut is approximately constant for all electrodes $^1/_4$ in. in diameter or less. Alloy steels and many nonferrous metals can be cut readily by the coated-electrode cutting process.

OXYGEN-ARC CUTTING WITH HOLLOW ELECTRODES. Using the equipment and technique described for underwater cutting with an oxygen arc, rough cuts at high speeds can be made in air if the work piece is moderate in thickness. However, the electrode must be changed at frequent intervals, and the speed decreases when the sections to be cut are heavy.

SECTION 24

DIMENSIONAL CONTROL

By

HAROLD S. SIZER, *Assistant Director of Design, Brown and Sharpe Manufacturing Co.*

ERMAND L. WATELET, *Charge of Design, Small Tool Division, Brown and Sharpe Manufacturing Co.*

JAMES A. BROADSTON, *Aerophysics Field Laboratory, North American Aviation, Inc., and Chief Engineer, Surface Checking Gage Co.*

CHARLES D. FERRIS, *General Electric Co., Bridgeport, Conn.*

SPECIFICATION OF DIMENSIONS AND TOLERANCES

By Harold S. Sizer

DIMENSIONAL MEASUREMENTS

By Ermand L. Watelet

MEASUREMENT AND DESIGNATION OF SURFACE QUALITY

By James A. Broadston

STATISTICAL QUALITY CONTROL

By Charles D. Ferris

SPECIFICATION OF DIMENSIONS AND TOLERANCES

By Harold S. Sizer

1. DIMENSIONING

There are two basic purposes to be accomplished in dimensioning a drawing: *First:* The dimensions must define the piece exactly, giving workmen and inspectors a positive understanding of the form desired by the designer. *Second:* The dimensions must recognize how the piece is to be manufactured and should give the workman producing the piece as much help as possible.

STANDARDS. Some good help in fulfilling these basic purposes is given by two standards:

(1) American Standards Association Z14.1-1946, *Drawing and Drafting Room Practice.*

(2) Society of Automotive Engineers, *Aeronautical Drafting Manual.*

Much of the material in the following paragraphs has been suggested by or taken from these standards. Do not, however, hesitate to deviate from these standards when such deviation will help a drawing to fulfill its basic purposes. Recognize, also, that drawings made in different plants for the same piece can be dissimilar in dimensioning and still comply rationally with the basic requirements, for methods of manufacturing have a strong influence on the specification of dimensions and tolerances.

USE INCHES. If your work falls in the class where lengths are mostly less than 100 in., use inch values exclusively. No unit need be given with the dimension values; and,

FIG. 1. Dimension in inches.

FIG. 2. Dimension in feet and inches.

wherever a figure appears without units specified, it is understood to be inches (Fig. 1).

In structural work, long lengths are specified in feet and inches. Here a unit (′) is given to identify feet, but inches are understood (Fig. 2). A hyphen should separate the numerals for feet and inches.

HOW WILL PIECE BE MEASURED? Try to specify a dimension in the form in which it will be measured, for every time an operator has to transpose drawing figures to working figures there is a possibility of error.

If the lengths between shoulders on a shaft will be turned to scale measurements, dimension these lengths with fractional figures which can be read on the common sixty-

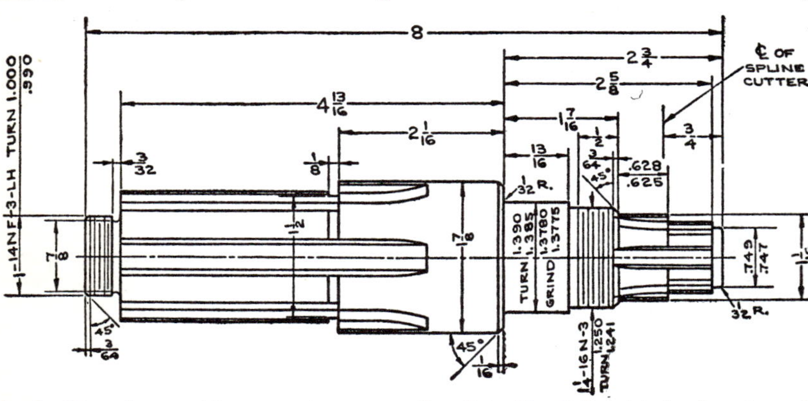

FIG. 3. Dimensions requiring no greater accuracy than that with which a sixty-fourth scale can be read are given in fractions. Where greater accuracy is needed, decimal limits are specified.

fourth scale. No limits need be placed on fractional dimensions because their use indicates that the practical accuracy with which a sixty-fourth scale can be read (± 0.010) is acceptable as the tolerance. Some companies extend the tolerance to $\pm ^1/_{64}$ (Fig. 3).

If the ground diameter of a sleeve will be checked with a micrometer, dimension it with decimal figures that can be read directly on a micrometer (Fig. 4).

If a ground and lapped roller (Fig. 5) will be checked for diameter and taper in a comparator, dimension it with plus and minus limits that may be read on the zero center scale of a comparator.

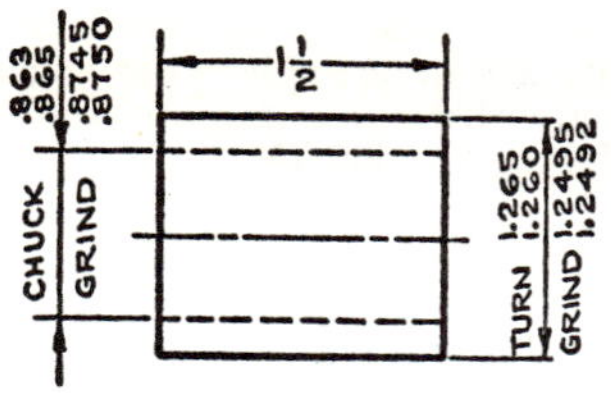

FIG. 4. Use of dimensions that can be read on a micrometer.

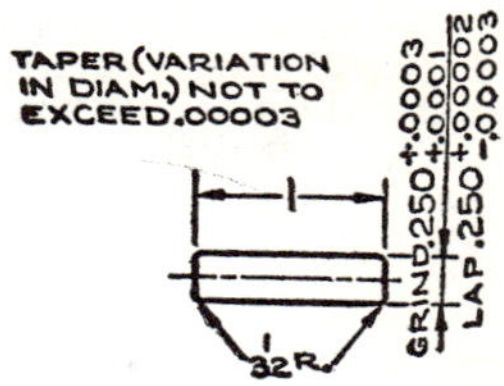

FIG. 5. Dimensions that may be read on a comparator.

If a screw thread is to be made to NF-3 gages, dimension it as an NF-3 thread and avoid the confusing details of pitch-line dimensions which are best checked by the gages.

If a first piece will not justify tools and must be laid out before machining, dimension the piece using arcs, radii, and angles so that a machinist with his square, center punch, and dividers can readily mark out and then machine the part.

POSITION OF LIMITS. As a workman reads a drawing, the upper or top figure on a dimension line is most likely to catch his eye. Make this figure the extra stock or first approached limit, which leaves him the tolerance to go if another or corrective cut is necessary. On an outside diameter this calls for the high limit on top whereas, for a hole, positions would be reversed (see Fig. 4). The width of a land would have the largest figure uppermost; for a slot the smallest figure would be given the top position (Fig. 6).

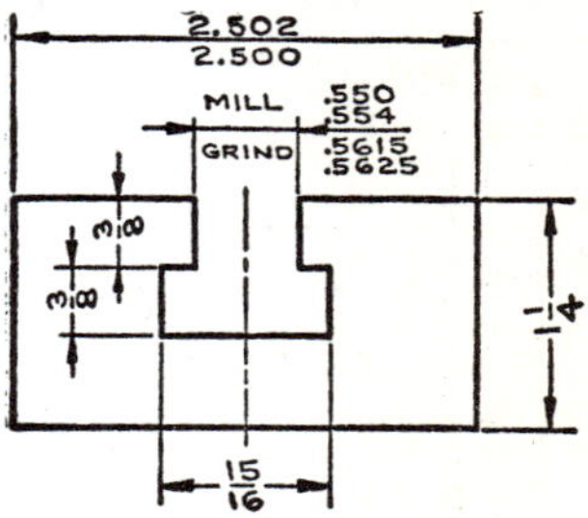

FIG. 6. Dimensioning so that top figure of limits corresponds to extra stock.

REFERENCE LINES. The reference-line method of dimensioning is a simple way to control the surfaces of a part. A few important surfaces, which will be established in the first machining operations, are taken as reference lines or planes, and all dimensions are given from these reference lines. Figure 7 is such a drawing. Each point, axis or surface, has an absolute position within the limits of its own tolerances and is not affected by the tolerances used in machining intermediate surfaces. Tolerances are not cumulative as they are in step-to-step dimensioning, and an operator does not have the responsibility of balancing plus and minus deviations incurred in successive operations.

Reference-line drawings are particularly helpful in tool work or work to be done on jig borers. The work tables on these machines can be positioned with micrometer accuracy, and settings are readily made when drawings give dimensions from reference lines.

A shaft dimensioned from a reference line established at its largest shoulder is shown in Fig. 3. Note that there is one close relationship of shoulders which is dimensioned with a small tolerance and does not conform to the strict rules of reference-line dimensioning. To maintain the close tolerance length by reference-line dimensioning, it would be necessary to hold two shoulders to very close tolerances in order that the difference between them would be within the desired limits. It is a less costly operation to machine a single shoulder to get a desired length between shoulders than it is to machine two shoulders accurately to obtain the same between shoulder length. Departures from reference-line dimensions are approved where they will simplify shop work. Figure 8 shows a gear center distance dimensioned independently.

SCREW THREADS. Most screw threads are made to the American Standard, which is described in American Standards Association B1.1-1949, *Screw Threads*, or in Screw-thread Standards for Federal Services, *National Bureau of Standards Handbook* H28 (1944). See also Section 10, Article 2, of this volume.

In dimensioning, it is necessary only to give the OD of the screw and to identify the type, size, and class of the screw thread (see Fig. 9). It is assumed that the thread will be cut to, and inspected with, thread gages and that details of pitch diameter, included angle, etc., are not necessary on the drawing. All threads are right hand unless LH (left hand) appears on the drawing. If full threads must approach close to a shoulder or to the bottom of a hole, this should be specified with a dimension on the drawing.

Large machine screws and special screws should be specified in complete detail (Fig. 10). Operators doing the turning, milling, and grinding should find all necessary figures on the drawing and should not have to hunt up values for wire measurements or lead angle.

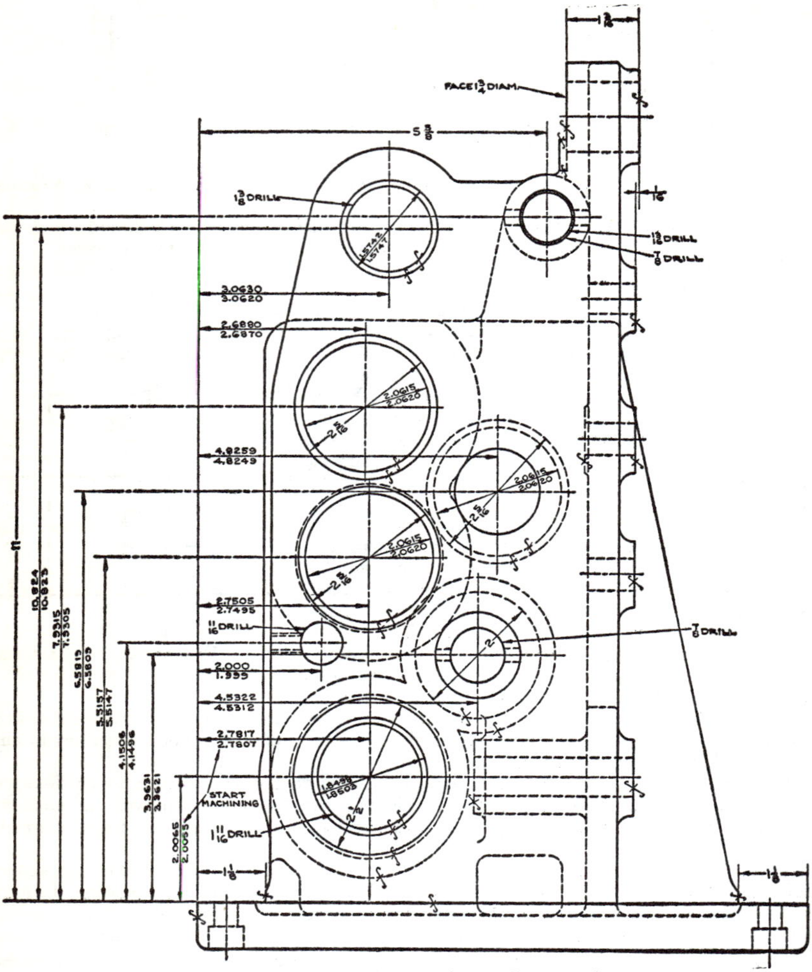

FIG. 7. Dimensioning from a reference line or surface.

HOLES. Holes to be drilled are dimensioned with a leader line and a note such as:

$^3/_8$ DRILL—6 HOLES

The total number of similar holes is specified to reduce congestion on the drawing. It is understood that the tolerance on a drilled hole is plus or minus 0.010. No limits are placed on the drawing (Fig. 8).

Counterbored holes are treated in the same manner as drilled holes. It is, however, often convenient to add the depth of counterbore to the leader note (Fig. 8).

Reamed, bored, or ground holes should be dimensioned with high and low limits.

POINTS FREQUENTLY OVERLOOKED. Do not expect a workman to do your arithmetic. Every time an operator has to recompute dimensions there is a chance for error.

Do not dimension from imaginary lines such as center lines or pitch lines; you cannot butt a scale against a center line or set the anvil of a micrometer on a pitch line.

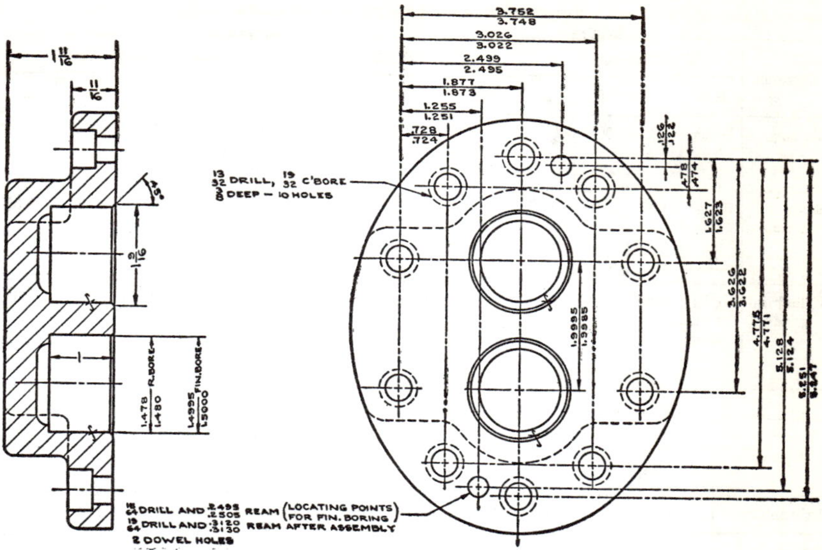

FIG. 8. Dimensioning from reference lines, but with gear center distance independently dimensioned.

Do not hide behind "good practice" and assume that tolerances are understood. If run-out should be less than 0.0005, put this on the drawing; and if taper should not exceed 0.00025, say so.

Do not specify a dimension in two ways and thus produce a conflict of tolerances. If you put tolerances on all the component lengths, do not put a tolerance on the total

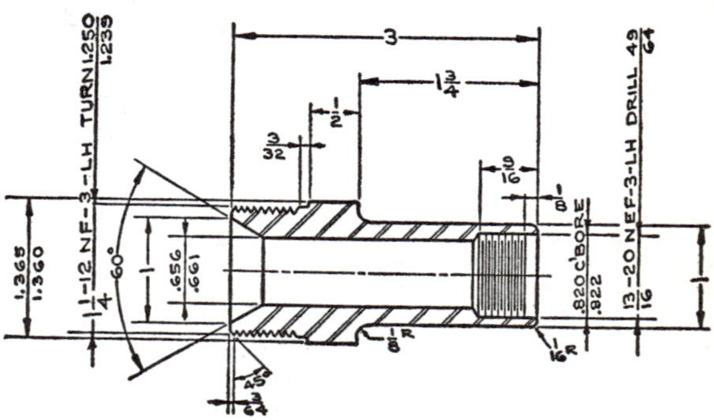

FIG. 9. Screw-thread identification.

length. If you want a tolerance on overall length, leave off the tolerance on some one of the steps.

Be sure that the surface from which you are dimensioning actually exists at the time the dimensioned operation is being performed.

Do not present a workman with an upside-down view of his work, especially if it contains unsymmetrically located holes. As he looks down on the work piece in the vise or on the machine table, he should see the same plan as that presented on the drawing.

PRACTICAL TOLERANCES. The accompanying table suggests the tolerances which can be maintained in various machine tool operations. See Art. 2 for definition of tolerance. There are a thousand exceptions to this table and the values given must be modified to suit the peculiarities of each part, the abilities of the operators in the plant, and the condition of its machine tools.

Suggestions on Specifying Tolerances

Operation	Diameter				Length				Face or Shoulder Runout			
	Size	*C*	*B*	*A*	Size	*C*	*B*	*A*	Size	*C*	*B*	*A*
Turning	Under 6 in. diam.	.005	.002	.001	Under 12 in. length	1/32	.010	.005	Runout per inch of diam.	.001	.0005	.00025
Screw machines	Under 1.5 in. diam.	.003	.001	.0005	Under 2 in. length	.005	.002	.00075	Under 1.5 in. diam.	.002	.0005	.0002
Cylindrical grinding	Under 6 in. diam.	.0005	.0002	.0001	Under 12 in. length	.005	.002	.001	2 in. shaft shoulder	.0005	.00025	.0001
									6 in. disk on arbor	.006	.003	.0015

Operation	Diameter				Taper in Length of Hole		
	Size	*C*	*B*	*A*	*C*	*B*	*A*
Drill and ream	Under 3 in. diam.	.002	.001	.0005	Same as tolerance on hole diameter	1/2 the tolerance on hole diameter	1/4 the tolerance on hole diameter
Boring	Under 3 in. diam.	.001	.0005	.00025			
Hole grinding	Under 3 in. diam.	.001	.0005	.00025			

Operation	Length				Flatness				
	Size	*C*	*B*	*A*	Size		*C*	*B*	*A*
Milling	Slots and splines under 2 in.	.005	.001	.0005	Bed type machine Double this for knee type	12	.003	.0015	.001
						24	.004	.0025	.0015
						36	.006	.004	.002
Planing	Ways and channels under 12 in.	.005	.002	.001	Length	24	.003	.002	.001
						72	.005	.003	.0015
Surface grinding	Thickness under 6 in. — Area 4 x 4	.001	.0003	.0001	Length	6	.001	.0003	.0001
	Thickness under 6 in. — Area 18 x 18	.003	.001	.0005		24	.003	.001	.0005
	Thickness under 6 in. — Area 2 x 60	.003	.001	.0005		72	.003	.001	.0005
Scraping	Any length that can be measured	.0005	.0002	.0001	Values per foot of length		.0002	.0001	.00005

The three classes of tolerances may be described as:

C. Practical tolerances which can readily be maintained by an average operator pushing for high production on an average machine. In general time cannot be saved or costs reduced by using more liberal tolerances for these operations.

B. A good operator using a good machine can maintain these limits in production. Tools must be sharpened and wheels trued more often than in (A), and added passes or finishing cuts must be made. Costs will be 15 to 50% higher than for "A" tolerances.

A. These tolerances are about the smallest that should be placed on drawings. This work requires continuous vigilance. Machines must be kept in first-class condition, and fine gages and measuring tools must be available. Before "A" tolerances are specified, need for such precision and willingness to pay the costs should be ascertained.

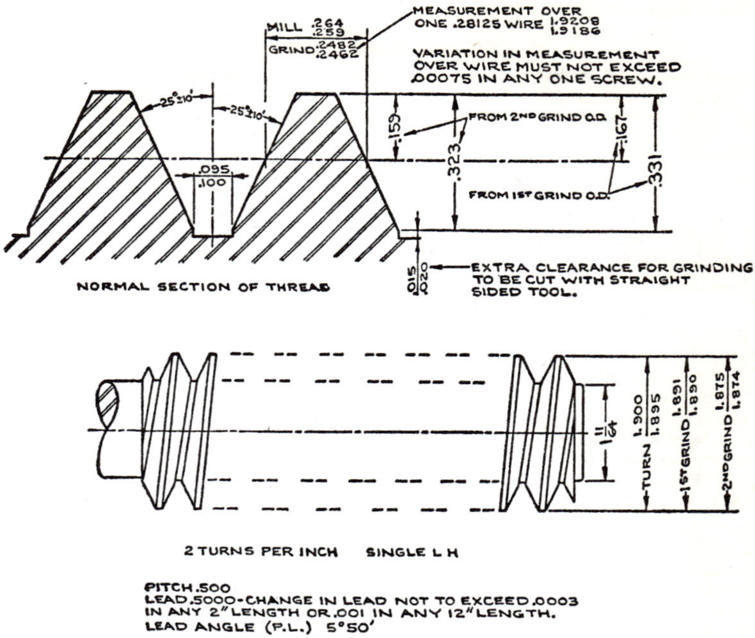

FIG 10. Detailed specification of thread dimensions for large or special screws.

2. FITS

The degree of tightness or looseness between two parts that are intended to act together is known as the fit of the parts. The character of the fit depends on the use to which it is to be put. Thus the fit between members that move or rotate relative to each other, as a journal rotating in a bearing, is considerably different from the fit which is designed to prevent any relative motion of the fitted members, as, for example, a tire fitted to a locomotive driving-wheel center.

Manufacturers and engineering associations have adopted in the past many different standards for fits to meet their particular needs. For details of these practices, see earlier editions of this book and also Ref. 1. In 1925 the American Standards Association published a standard (B4a-1925) called Tolerances, Allowances and Gages for Metal Fits. In 1947 additional recommendations were made in ASA B4.1-1947, which is Part 1 on Limits and Fits. These two standards are the source of most of the material which follows. For fits in journal bearings, see pp. 12-50 and 12-51, and for fits for ball and roller bearings see pp. 12-85 to 12-92.

TERMS RELATING TO FITS (B4.1-1947).

Nominal Size. The size used for purposes of general identification.

Limits of Size. The maximum and minimum sizes permissible for a specific dimension.

Tolerance. The total permissible variation in size. The tolerance is the difference between the limits of size.

Allowance. A prescribed difference in dimensions of mating parts. It is a minimum clearance or a maximum interference between mating parts.

Clearance Fit. A fit between mating parts having limits of size so prescribed that a clearance always results in assembly.

Interference Fit. A fit between mating parts having limits of size so prescribed that an interference always results in assembly.

Transition Fit. A fit between mating parts having limits of size so prescribed as partially or wholly to overlap, so that either a clearance or an interference may result in assembly.

PREFERRED SIZES (B4.1-1947). To conserve the number of tools and gages required and to limit bar stock to an orderly series of sizes, the following tables of preferred values have been prepared:

Preferred Nominal Sizes

.0100	.0313	.0938	3/16	7/16	11/16	1 1/8	1 5/8	2 1/8	2 5/8	3 1/4
.0125	.0400	.1000	1/4	1/2	3/4	1 1/4	1 3/4	2 1/4	2 3/4	3 1/2
.0156	.0500	1/8	5/16	9/16	7/8	1 3/8	1 7/8	2 3/8	2 7/8	3 3/4
.0200	.0625	5/32	3/8	5/8	1	1 1/2	2	2 1/2	3	4
.0250	.0800									

Preferred Tolerances and Allowances

.0001	.00025	**.0005**	**.0010**	.0015	.0025	.0040	.0060	**.0100**	.0150	.0250
.00015	.0003	.0006	.0012	**.0020**	.0030	**.0050**	.0080	.0120	**.0200**	**.0300**
.0002	.0004	.0008								

Bold-faced values are preferred values.

In using the tables of fits given in the following pages, if possible select nominal sizes which are listed in the preferred nominal sizes table. Do not, for example, call for a 15/16 diameter if 1 can be used; also, design for 3 or 3 1/4 instead of 3 1/8. Tolerances taken from the table of fits may be changed to conform to the preferred tolerances and allowances values. If, for example, "go"-"not-go" gages are available for holes ground standard to plus 0.0005, then the limits for a class 4, 1 diameter hole which are standard to +0.0006 can be reduced to standard to +0.0005.

CLASSIFICATION OF FITS. Fits are classified as follows:

Class 1, Loose Fit, Large Allowance. This fit provides for considerable freedom and embraces certain fits where accuracy is not essential. Examples of its use are the machined fits on agricultural and mining machinery, control apparatus for marine work, machinery in the textile, rubber, and canning industries, and general machinery of similar grades. It is used for strictly interchangeable assembly. Tolerances and allowances are determined by the following formulas, in which d = diameter of fit, inches:

$$\text{Allowance} = 0.0025\sqrt[3]{d^2};\ \text{hole tolerance and shaft tolerance} = 0.0025\sqrt[3]{d}$$

Dimensions are given in Table 1.

Class 2, Free Fit, Liberal Allowance. This fit is designed as a running fit for speeds of 600 rpm or over, and journal pressures of 600 psi or more. Examples of its use are electric generators, engines, machine tool parts, and automotive parts. It is used for strictly interchangeable assembly. Allowances and tolerances are determined by the following formulas:

$$\text{Allowance} = 0.0014\sqrt[3]{d^2};\ \text{hole and shaft tolerance} = 0.0013\sqrt[3]{d}$$

Dimensions are given in Table 1.

Class 3, Medium Fit, Medium Allowance. This fit is designed for running fits under 600 rpm and with journal pressures of less than 600 psi. It also is used for sliding fits, and the more accurate machine tool and automotive parts. It is designed for strictly interchangeable assembly. Allowances and hole tolerances are determined by the following formulas:

$$\text{Allowance} = 0.0009\sqrt[3]{d^2};\ \text{hole tolerance and shaft tolerance} = 0.0008\sqrt[3]{d}$$

Dimensions are given in Table 2.

Class 4, Snug Fit, Zero Allowance. A snug fit is the closest fit that can be assembled by hand. It necessitates work of considerable precision. It is used where no perceptible shake is permitted, and where moving parts are not intended to move freely under load. An example of its use is the assembly of the body and jaws of a drill chuck. It is designed for strictly interchangeable assembly. Allowance = 0.0000. Hole and shaft tolerances are determined by the following formulas:

$$\text{Hole tolerance} = 0.0006\sqrt[3]{d};\ \text{shaft tolerance} = 0.0004\sqrt[3]{d}$$

Dimensions are given in Table 2.

Class 5, Wringing Fit, Zero to Negative Allowance. This fit also is known as a *tunking fit.* It is practically a metal-to-metal fit. Assembly usually is selective and not interchangeable. Examples of its use are locating keys, tongues, or dowels. The selected

Table 1. American Standard Loose and Free Fits

A.S.A. B4a–1925. All dimensions in inches

Size			Class 1. Loose Fit, Large Allowance, Interchangeable					Class 2. Free Fit, Liberal Allowance, Interchangeable				
			Limits			Tightest Fit, Allowance	Loosest Fit, Allowance +Tolerances	Limits			Tightest Fit, Allowance	Loosest Fit, Allowance +Tolerances
From	Up to and Including	Mean	Hole or External Member *	Shaft or Internal Member				Hole or External Member *	Shaft or Internal Member			
			Plus	Minus	Minus	Plus †	Plus †	Plus	Minus	Minus	Plus †	Plus †
0	3/16	1/8	0.001	0.001	0.002	0.001	0.003	0.0007	0.0004	0.0011	0.0004	0.0018
3/16	5/16	1/4	.002	.001	.003	.001	.005	.0008	.0006	.0014	.0006	.0022
5/16	7/16	3/8	.002	.001	.003	.001	.005	.0009	.0007	.0016	.0007	.0025
7/16	9/16	1/2	.002	.002	.004	.002	.006	.0010	.0009	.0019	.0009	.0029
9/16	11/16	5/8	.002	.002	.004	.002	.006	.0011	.0010	.0021	.0010	.0032
11/16	13/16	3/4	.002	.002	.004	.002	.006	.0012	.0012	.0024	.0012	.0036
13/16	15/16	7/8	.002	.002	.004	.002	.006	.0012	.0013	.0025	.0013	.0037
15/16	1 1/16	1	.003	.003	.006	.003	.009	.0013	.0014	.0027	.0014	.0040
1 1/16	1 3/16	1 1/8	.003	.003	.006	.003	.009	.0014	.0015	.0029	.0015	.0043
1 3/16	1 3/8	1 1/4	.003	.003	.006	.003	.009	.0014	.0016	.0030	.0016	.0044
1 3/8	1 5/8	1 1/2	.003	.003	.006	.003	.009	.0015	.0018	.0033	.0018	.0048
1 5/8	1 7/8	1 3/4	.003	.004	.007	.004	.010	.0016	.0020	.0036	.0020	.0052
1 7/8	2 1/8	2	.003	.004	.007	.004	.010	.0016	.0022	.0038	.0022	.0054
2 1/8	2 3/8	2 1/4	.003	.004	.007	.004	.010	.0017	.0024	.0041	.0024	.0058
2 3/8	2 3/4	2 1/2	.003	.005	.008	.005	.011	.0018	.0026	.0044	.0026	.0062
2 3/4	3 1/4	3	.004	.005	.009	.005	.013	.0019	.0029	.0048	.0029	.0067
3 1/4	3 3/4	3 1/2	.004	.006	.010	.006	.014	.0020	.0032	.0052	.0032	.0072
3 3/4	4 1/4	4	.004	.006	.010	.006	.014	.0021	.0035	.0056	.0035	.0077
4 1/4	4 3/4	4 1/2	.004	.007	.011	.007	.015	.0021	.0038	.0059	.0038	.0080
4 3/4	5 1/2	5	.004	.007	.011	.007	.015	.0022	.0041	.0063	.0041	.0085
5 1/2	6 1/2	6	.005	.008	.013	.008	.018	.0024	.0046	.0070	.0046	.0094
6 1/2	7 1/2	7	.005	.009	.014	.009	.019	.0025	.0051	.0076	.0051	.0101
7 1/2	8 1/2	8	.005	.010	.015	.010	.020	.0026	.0056	.0082	.0056	.0108

Table 2. American Standard Medium and Snug Fits

A.S.A. B4a–1925. All dimensions in inches

Size			Class 3. Medium Fit, Medium Allowance, Interchangeable					Class 4. Snug Fit, Zero Allowance, Interchangeable		
			Limits			Tightest Fit, Allowance	Loosest Fit, Allowance + Tolerances	Limits		Loosest Fit, Allowance + Tolerances §
From	Up to and Including	Mean	Hole or External Member *	Shaft or Internal Member				Hole or External Member *	Shaft or Internal Member ‡	
			Plus	Minus	Minus	Plus †	Plus †	Plus †	Minus	Plus †
0	3/16	1/8	0.0004	0.0002	0.0006	0.0002	0.0010	0.0003	0.0002	0.0005
3/16	5/16	1/4	.0005	.0004	.0009	.0004	.0014	.0004	.0003	.0007
5/16	7/16	3/8	.0006	.0005	.0011	.0005	.0017	.0004	.0003	.0007
7/16	9/16	1/2	.0006	.0006	.0012	.0006	.0018	.0005	.0003	.0008
9/16	11/16	5/8	.0007	.0007	.0014	.0007	.0021	.0005	.0003	.0008
11/16	13/16	3/4	.0007	.0007	.0014	.0007	.0021	.0005	.0004	.0009
13/16	15/16	7/8	.0008	.0008	.0016	.0008	.0024	.0006	.0004	.0010
15/16	1 1/16	1	.0008	.0009	.0017	.0009	.0025	.0006	.0004	.0010
1 1/16	1 3/16	1 1/8	.0008	.0010	.0018	.0010	.0026	.0006	.0004	.0010
1 3/16	1 3/8	1 1/4	.0009	.0010	.0019	.0010	.0028	.0006	.0004	.0010
1 3/8	1 5/8	1 1/2	.0009	.0012	.0021	.0012	.0030	.0007	.0005	.0012
1 5/8	1 7/8	1 3/4	.0010	.0013	.0023	.0013	.0033	.0007	.0005	.0012
1 7/8	2 1/8	2	.0010	.0014	.0024	.0014	.0034	.0008	.0005	.0013
2 1/8	2 3/8	2 1/4	.0010	.0015	.0025	.0015	.0035	.0008	.0005	.0013
2 3/8	2 3/4	2 1/2	.0011	.0017	.0028	.0017	.0039	.0008	.0005	.0013
2 3/4	3 1/4	3	.0012	.0019	.0031	.0019	.0043	.0009	.0006	.0015
3 1/4	3 3/4	3 1/2	.0012	.0021	.0033	.0021	.0045	.0009	.0006	.0015
3 3/4	4 1/4	4	.0013	.0023	.0036	.0023	.0049	.0010	.0006	.0016
4 1/4	4 3/4	4 1/2	.0013	.0025	.0038	.0025	.0051	.0010	.0007	.0017
4 3/4	5 1/2	5	.0014	.0026	.0040	.0026	.0054	.0010	.0007	.0017
5 1/2	6 1/2	6	.0015	.0030	.0045	.0030	.0060	.0011	.0007	.0018
6 1/2	7 1/2	7	.0015	.0033	.0048	.0033	.0063	.0011	.0008	.0019
7 1/2	8 1/2	8	.0016	.0036	.0052	.0036	.0068	.0012	.0008	.0020

* Minimum limit = 0.0000 in. for all sizes. † Plus here denotes clearance or amount of looseness.
‡ Maximum limit = 0.0000 in. for all sizes. § Tightest fit allowance = 0.0000 for all sizes.

average interference of metal is 0.0000 in. Hole and shaft tolerances are determined by these formulas:

$$\text{Hole tolerance} = 0.0006\sqrt[3]{d};\ \text{shaft tolerance} = 0.0004\sqrt[3]{d}$$

Dimensions are given in Table 3.

Class 6, Tight Fit, Slight Negative Allowance. These fits are used where parts are to be more or less permanently assembled, and light pressure is required in the assembly operation. They are used for drive fits in thin sections, or very long fits in other sections, and also for shrink fits on very light sections. Examples of the use of this fit are jig bushings, and fixed ends of studs for gears, pulleys, rocker arms, in automotive, ordnance,

Table 3. American Standard Wringing Fits and Tight Fits

A.S.A. B4a–1925. All dimensions in inches

Size			Class 5. Wringing Fit, Zero to Negative Allowance. Selective Assembly				Class 6. Tight Fit, Slight Negative Allowance Selective Assembly					
			Limits				Limits					
From	Up to and Including	Mean	Hole or External Member *	Shaft or Internal Member *	Tightest Fit, Allowance	Loosest Fit, Allowance + Tolerances	Hole or External Member *	Shaft or Internal Member		Tightest Fit, Allowance	Loosest Fit, Allowance + Tolerances	Selected Fit, Average Interference of Metal
			Plus	Plus	Minus †	Plus ‡	Plus	Plus	Plus	Minus †	† ‡	Minus †
0	3/16	1/8	0.0003	0.0002	0.0002	0.0003	0.0003	0.0003	0.0000	0.0003	+0.0003	0.0000
3/16	5/16	1/4	.0004	.0003	.0003	.0004	.0004	.0005	.0001	.0005	+ .0003	.0001
5/16	7/16	3/8	.0004	.0003	.0003	.0004	.0004	.0005	.0001	.0005	+ .0003	.0001
7/16	9/16	1/2	.0005	.0003	.0003	.0005	.0005	.0006	.0001	.0006	+ .0004	.0001
9/16	11/16	5/8	.0005	.0003	.0003	.0005	.0005	.0007	.0002	.0007	+ .0003	.0002
11/16	13/16	3/4	.0005	.0004	.0004	.0005	.0005	.0007	.0002	.0007	+ .0003	.0002
13/16	15/16	7/8	.0006	.0004	.0004	.0006	.0006	.0008	.0002	.0008	+ .0004	.0002
15/16	1 1/16	1	.0006	.0004	.0004	.0006	.0006	.0009	.0003	.0009	+ .0003	.0003
1 1/16	1 3/16	1 1/8	.0006	.0004	.0004	.0006	.0006	.0009	.0003	.0009	+ .0003	.0003
1 3/16	1 3/8	1 1/4	.0006	.0004	.0004	.0006	.0006	.0009	.0003	.0009	+ .0003	.0003
1 3/8	1 5/8	1 1/2	.0007	.0005	.0005	.0007	.0007	.0011	.0004	.0011	+ .0003	.0004
1 5/8	1 7/8	1 3/4	.0007	.0005	.0005	.0007	.0007	.0011	.0004	.0011	+ .0003	.0004
1 7/8	2 1/8	2	.0008	.0005	.0005	.0008	.0008	.0013	.0005	.0013	+ .0003	.0005
2 1/8	2 3/8	2 1/4	.0008	.0005	.0005	.0008	.0008	.0014	.0006	.0014	+ .0002	.0006
2 3/8	2 3/4	2 1/2	.0008	.0005	.0005	.0008	.0008	.0014	.0006	.0014	+ .0002	.0006
2 3/4	3 1/4	3	.0009	.0006	.0006	.0009	.0009	.0017	.0008	.0017	+ .0001	.0008
3 1/4	3 3/4	3 1/2	.0009	.0006	.0006	.0009	.0009	.0018	.0009	.0018	− .0000	.0009
3 3/4	4 1/4	4	.0010	.0006	.0006	.0010	.0010	.0020	.0010	.0020	− .0000	.0010
4 1/4	4 3/4	4 1/2	.0010	.0007	.0007	.0010	.0010	.0021	.0011	.0021	− .0001	.0011
4 3/4	5 1/2	5	.0010	.0007	.0007	.0010	.0010	.0023	.0013	.0023	− .0003	.0013
5 1/2	6 1/2	6	.0011	.0007	.0007	.0011	.0011	.0026	.0015	.0026	− .0004	.0015
6 1/2	7 1/2	7	.0011	.0008	.0008	.0011	.0011	.0029	.0018	.0029	− .0007	.0018
7 1/2	8 1/2	8	.0012	.0008	.0008	.0012	.0012	.0032	.0020	.0032	− .0008	.0020

* Minimum limit = 0.0000 in. for all sizes.
† Minus here denotes interference of metal or negative allowance.
‡ Plus here denotes clearance or amount of looseness.

and general machine manufacturing. In this fit, the average interference of metal is the desired condition. It must be obtained by selective assembly, i.e., by mating large shafts in large holes, and small shafts in small holes. In this fit, selected average interference of metal = $0.00025d$. Hole tolerances and shaft tolerances are given by the formula

$$\text{Tolerance} = 0.0006\sqrt[3]{d}$$

Dimensions are given in Table 3. The stresses set up in the members, and the forces required for making these fits, are given in Table 4. The values of the greatest hole stress in Table 4 are for mean diameters only. For other sizes in the same group, the interference per inch of diameter, and hence the greatest hole stress, varies from the values given by less than 10%. If greater accuracy for intermediate steps is required, the interference per inch of diameter should be computed by the formula given above. The values given in Table 4 are true only when hub diameter equals twice the hole diameter.

Class 7, Medium Force Fit, Negative Allowance. These fits are used where the parts are to be permanently assembled. Considerable pressure is required for assembly. They are the tightest fits recommended for cast-iron holes or external members, since they will stress cast iron to its elastic limit. Examples of the use of this class of fit are the attachment to their axles or shafts of locomotive and car wheels, generator and motor armatures, and engine crank disks. This class of fit also is used as a shrink fit on medium sections or long fits. These fits are used for selective assembly only. Selected average interference of metal = $0.0005d$. The shaft and hole tolerances are determined by the formula

$$\text{Tolerance} = 0.0006 \sqrt[3]{d}$$

Dimensions are given in Table 5. Stresses set up in the parts, and the pressure required to make the fits, are given in Table 6. The values given in Table 6 for stress and force are true only when the hub diameter equals twice the hole diameter. The values of

Table 4. Stresses and Forces for Tight Fits, Class 6

A.S.A. B4a–1925

Force values are for fits 1 in. long. For longer fits multiply tabular values by length of fit in inches

Mean size, in.	Interference of Metal per inch of Mean Size*: Tightest Fit, in.	Interference of Metal per inch of Mean Size*: Loosest Fit, in.	Greatest Hole Stress, Steel Shaft in Steel Hole†: Tightest Fit, lb. per sq. in.	Steel Hole†: Loosest Fit, lb. per sq. in.	Cast-iron Hole‡: Tightest Fit, lb. per sq. in.	Cast-iron Hole‡: Loosest Fit, lb. per sq. in.	Force Required to Press Steel Shaft into Steel Hole: Tightest Fit, tons	Steel Hole: Loosest Fit, tons	Steel Hole: Selected Fit, tons	Cast-iron Hole: Tightest Fit, tons	Cast-iron Hole: Loosest Fit, tons	Cast-iron Hole: Selected Fit, tons
1/8	0.00240	No values, because of smallest shaft member in largest hole member.	69,600	No values, because of smallest shaft member in largest hole member.	25,037	No values, because of smallest shaft member in largest hole member.	0.389	No values, because of smallest shaft member in largest hole member.	0.000	0.224	No values, because of smallest shaft member in largest hole member.	0.000
1/4	.00200		58,000		20,864		.649		.130	.374		.075
3/8	.00133		38,700		13,909		.649		.130	.374		.075
1/2	.00120		34,800		12,518		.779		.130	.448		.075
5/8	.00112		32,480		11,684		.909		.260	.523		.149
3/4	.00093		27,100		9,737		.909		.260	.523		.149
7/8	.00091		26,500		9,538		1.038		.260	.598		.149
1	.00090		26,100		9,389		1.168		.389	.672		.224
1 1/8	.00080		23,200		8,346		1.168		.389	.672		.224
1 1/4	.00072		20,880		7,511		1.168		.389	.672		.224
1 1/2	.00073		21,300		7,650		1.428		.519	.822		.299
1 3/4	.00063		18,240		6,557		1.428		.519	.822		.299
2	.00065		18,850		6,781		1.687		.649	.971		.374
2 1/4	.00062		18,040		6,491		1.817		.779	1.046		.448
2 1/2	.00056		16,200		5,842		1.817		.779	1.046		.448
3	.00057		16,433		5,911		2.207		1.038	1.270		.598
3 1/2	.00051		14,910		5,365		2.336		1.168	1.345		.672
4	.00050		14,500		5,216		2.596		1.298	1.494		.747
4 1/2	.00047	0.00002	13,530	644	4,868	232	2.726	0.130	1.428	1.569	0.075	.822
5	.00046	.00006	13,340	1740	4,799	625	2.985	.389	1.687	1.719	.224	.971
6	.00043	.00007	12,570	1933	4,521	695	3.375	.519	1.947	1.942	.299	1.121
7	.00041	.00010	12,010	2900	4,322	1043	3.764	.909	2.336	2.167	.523	1.345
8	.00040	.00010	11,600	2900	4,173	1043	4.154	1.038	2.596	2.391	.598	1.494
			Stress = $29{,}000{,}000\,A/d$ §		Stress = $10{,}432{,}000\,A/d$ §		Force = $1298\,A$ §			Force = $747.2\,A$ §		

* Interference of metal on selected fits, all sizes, = 0.00025 in.
† Greatest stress in hole on selected fits, all sizes, = 7250 lb. per sq. in.
‡ Greatest stress in hole on selected fits, all sizes, = 2608 lb. per sq. in.
§ A = allowance; d = mean size.

greatest hole stress are for mean diameter sizes only. For other sizes in the group, the interference per inch of diameter, and hence the greatest hole stress, will vary from the values given by less than 10%. If greater accuracy for intermediate steps is required, the interference per inch of diameter should be computed by the formula given above.

Class 8, Heavy Force and Shrink Fit, Considerable Negative Allowance. These fits are used for steel holes where the metal can be highly stressed without exceeding its elastic limit. They cause excessive stress in cast-iron holes. Shrink fits are used where heavy force fits are impracticable, as on locomotive wheel tires and heavy crank disks. They are for selective assembly only. Selected average interference of metal = $0.001d$. Hole and shaft tolerances are given by the formula

$$\text{Tolerance} = 0.0006 \sqrt[3]{d}$$

Dimensions are given in Table 7. Stresses set up by, and the force required to make these fits, are given in Table 8. The values given in Table 8 for stress and force are true only when the hub diameter equals twice the hole diameter. The values of the greatest hole stress are for mean diameter sizes only. For other sizes in the group, the interference per inch of diameter, and hence the greatest hole stress, varies from the values given by less than 10%. If greater accuracy for intermediate steps is required, the interference per inch of diameter should be computed by the formula given above.

STRESSES RESULTING FROM FORCE FITS. In any force or shrink fit, the metal of the fitted member is deformed by an amount equal to the interference of metal between the external and internal members. Discussions of the stress set up by this deformation usually are based on the assumption that the materials are elastic and follow Hooke's law. A. Lewis Jenkins (*Am. Mach.*, March 4, 1915) shows that for the usual combination of a cast-iron hub shrunk on a steel shaft, Hooke's law does not apply to the cast-iron hub. From a study of a wide variety of force and shrink fits, he deduced the following equations:

For cast-iron hubs on steel shafts:

$$p = \frac{102{,}900{,}000(e/r_2)}{7.16 + (r_3/r_2)}$$

Table 5. American Standard Medium Force Fit, Negative Allowance, Selective Assembly

A.S.A. B4a–1925. All dimensions in inches

Size			Limits				Tightest Fit Allowance	Loosest Fit, Allowance + Tolerances	Selected Fit, Average Interference of Metal
From	Up to and Including	Mean	Hole or External Member		Shaft or Internal Member				
			Plus		Plus	Plus	Minus*	*†	Minus*
0	3/16	1/8	0.0003	0.0000	0.0004	0.0001	0.0004	+0.0002	0.0001
3/16	5/16	1/4	.0004	.0000	.0005	.0001	.0005	+ .0003	.0001
5/16	7/16	3/8	.0004	.0000	.0006	.0002	.0006	+ .0002	.0002
7/16	9/16	1/2	.0005	.0000	.0008	.0003	.0008	+ .0002	.0003
9/16	11/16	5/8	.0005	.0000	.0008	.0003	.0008	+ .0002	.0003
11/16	13/16	3/4	.0005	.0000	.0009	.0004	.0009	+ .0001	.0004
13/16	15/16	7/8	.0006	.0000	.0010	.0004	.0010	+ .0002	.0004
15/16	1 1/16	1	.0006	.0000	.0011	.0005	.0011	+ .0001	.0005
1 1/16	1 3/16	1 1/8	.0006	.0000	.0012	.0006	.0012	.0000	.0006
1 3/16	1 3/8	1 1/4	.0006	.0000	.0012	.0006	.0012	.0000	.0006
1 3/8	1 5/8	1 1/2	0007	.0000	.0015	.0008	.0015	− .0001	.0008
1 5/8	1 7/8	1 3/4	.0007	.0000	.0016	.0009	.0016	− .0002	.0009
1 7/8	2 1/8	2	.0008	.0000	.0018	.0010	.0018	− .0002	.0010
2 1/8	2 3/8	2 1/4	.0008	.0000	.0019	.0011	.0019	− .0003	.0011
2 3/8	2 3/4	2 1/2	.0008	.0000	.0021	.0013	.0021	− .0005	.0013
2 3/4	3 1/4	3	.0009	.0000	.0024	.0015	.0024	− .0006	.0015
3 1/4	3 3/4	3 1/2	.0009	.0000	.0027	.0018	.0027	− .0009	.0018
3 3/4	4 1/4	4	.0010	.0000	.0030	.0020	.0030	− .0010	.0020
4 1/4	4 3/4	4 1/2	.0010	.0000	.0033	.0023	.0033	− .0013	.0023
4 3/4	5 1/2	5	.0010	.0000	.0035	.0025	.0035	− .0015	.0025
5 1/2	6 1/2	6	.0011	.0000	.0041	.0030	.0041	− .0019	.0030
6 1/2	7 1/2	7	.0011	.0000	.0046	.0035	.0046	− .0024	.0035
7 1/2	8 1/2	8	.0012	.0000	.0052	.0040	.0052	− .0028	.0040
8 1/2	10	9 1/4	.0012	.0000	.0057	.0045	.0057	− .0033	.0045
10	11 1/2	10 3/4	.0013	.0000	.0063	.0050	.0063	− .0037	.0050
11 1/2	13	12 1/4	.0014	.0000	.0074	.0060	.0074	− .0046	.0060
13	15	14	.0014	.0000	.0084	.0070	.0084	− .0056	.0070
15	17	16	.0015	.0000	.0095	.0080	.0095	− .0065	.0080
17	19	18	.0016	.0000	.0106	.0090	.0106	− .0074	.0090
19	22	20	.0016	.0000	.0116	.0100	.0116	− .0084	.0100
22	26	24	.0017	.0000	.0137	.0120	.0137	− .0103	.0120
26	30	28	.0018	.0000	.0158	.0140	.0158	− .0122	.0140
30	34	32	.0019	.0000	.0179	.0160	.0179	− .0141	.0160
34	38	36	.0020	.0000	.0200	.0180	.0200	− .0160	.0180
38	44	40	.0021	.0000	.0221	.0200	.0221	− .0179	.0200
44	52	48	.0022	.0000	.0262	.0240	.0262	− .0218	.0240
52	60	56	.0023	.0000	.0303	.0280	.0303	− .0257	.0280
60	68	64	.0024	.0000	.0344	.0320	.0344	− .0296	.0320
68	76	72	.0025	.0000	.0385	.0360	.0385	− .0335	.0360
76	88	80	.0026	.0000	.0426	.0400	.0426	− .0374	.0400
88	104	96	.0027	.0000	.0507	.0480	.0507	− .0453	.0480
104	120	112	.0029	.0000	.0589	.0560	.0589	− .0531	.0560
120	136	128	.0030	.0000	.0670	.0640	.0670	− .0610	.0640

* Minus here denotes interference of metal or negative allowance.
† Plus here denotes clearance or amount of looseness.

For steel hubs on steel shafts:

$$p = \frac{30{,}000{,}000(e/r_2)[(r_3/r_2)^2 - 1]}{2(r_3/r_2)^2}$$

where p = the unit stress between shaft and hub; r_2 = common radius of fitted shaft and hub; r_3 = external radius of hub; e = total deformation of shaft and hub in pressing $= r_s - r_h$; r_s, r_h = respectively, external radius of shaft and internal radius of hub before pressing.

Table 6. Stresses and Forces for Medium Force Fits, Class 7

A.S.A. B4a–1925

Force values are for fits 1 in. long. For longer fits multiply tabular values by length of fit in inches

Mean Size, in.	Interference of Metal per Inch of Mean Size *		Greatest Hole Stress, Steel Shaft, in				Force Required to Press Steel Shaft into					
			Steel Hole †		Cast-iron Hole ‡		Steel Hole			Cast-iron Hole		
	Tightest Fit, in.	Loosest Fit, in.	Tightest Fit, in.	Loosest Fit, in.	Tightest Fit, in.	Loosest Fit, in.	Tightest Fit, tons	Loosest Fit, tons	Selected Fit, tons	Tightest Fit, tons	Loosest Fit, tons	Selected Fit, tons
1/8	0.00320		92,800		33,382		0.519		0.130	0.299		0.075
1/4	.00200		58,000		20,864		.649		.130	.374		.075
3/8	.00160		46,400		16,691		.779		.260	.448		.149
1/2	.00160		46,400		16,691		1.038		.389	.598		.224
5/8	.00128	See	37,120	See	13,353	See	1.038	See	.389	.598	See	.224
3/4	.00120	Note 1	34,800	Note 1	12,513	Note 1	1.168	Note 1	.519	.672	Note 1	.299
7/8	.00114		33,140		11,922		1.298		.519	.747		.299
1	.00110		31,900		11,475		1.428		.649	.822		.374
1 1/8	.00107		30,930		11,127		1.558		.779	.897		.448
1 1/4	.00096		27,830		10,015		1.558		.779	.897		.448
1 1/2	.00100	0.00007	29,000	1,934	10,432	695	1.947	0.130	1.038	1.121	0.075	.598
1 3/4	.00091	.00011	26,510	3,314	9,538	1192	2.077	.260	1.168	1.196	.149	.672
2	.00090	.00010	26,100	2,900	9,389	1043	2.336	.260	1.298	1.345	.149	.747
2 1/4	.00084	.00013	24,490	3,866	8,809	1391	2.466	.389	1.428	1.420	.224	.822
2 1/2	.00084	.00020	24,340	5,800	8,763	2086	2.726	.649	1.687	1.569	.373	.971
3	.00080	.00020	23,200	5,800	8,346	2086	3.115	.779	1.947	1.793	.448	1.121
3 1/2	.00077	.00026	22,370	7,457	8,048	2682	3.505	1.168	2.336	2.017	.672	1.345
4	.00075	.00025	21,750	7,250	7,824	2608	3.894	1.298	2.596	2.242	.747	1.494
4 1/2	.00073	.00029	21,270	8,377	7,652	3014	4.283	1.687	2.985	2.466	.971	1.719
5	.00070	.00030	20,300	8,700	7,302	3130	4.543	1.947	3.245	2.615	1.121	1.868
6	.00068	.00032	19,810	9,183	7,129	3303	5.322	2.466	3.894	3.064	1.420	2.242
7	.00066	.00034	19,060	9,942	6,855	3577	5.971	3.115	4.543	3.437	1.793	2.615
8	.00065	.00035	18,850	10,150	6,781	3651	6.750	3.634	5.192	3.885	2.092	2.989
9	.00063	.00037	18,370	10,630	6,607	3875	7.399	4.283	5.841	4.259	2.466	3.362
10	.00063	.00037	18,270	10,730	6,572	3860	8.177	4.803	6.490	4.707	2.765	3.736
12	.00062	.00038	17,890	11,120	6,433	3999	9.605	5.971	7.788	5.529	3.437	4.483
14	.00060	.00040	17,400	11,600	6,259	4173	10.903	7.269	9.068	6.276	4.184	5.230
16	.00059	.00041	17,220	11,780	6,194	4238	12.331	8.437	10.384	7.098	4.857	5.978
18	.00059	.00041	17,080	11,920	6,143	4289	13.759	9.605	11.682	7.920	5.529	6.725
20	.00058	.00042	16,820	12,180	6,051	4381	15.057	10.903	12.980	8.668	6.276	7.472
24	.00057	.00043	16,550	12,450	5,955	4477	17.783	13.369	15.576	10.236	7.696	8.966
28	.00056	.00044	16,360	12,640	5,887	4173	20.508	14.538	18.172	11.806	8.369	10.461
32	.00056	.00044	16,220	12,780	5,835	4597	23.234	18.302	20.768	13.375	10.536	11.955
36	.00056	.00044	16,110	12,890	5,796	4636	25.960	20.768	23.364	14.944	11.955	13.450
40	.00055	.00045	16,020	12,980	5,764	4668	28.686	23.234	25.960	16.513	13.375	14.944
48	.00055	.00045	15,830	13,170	5,694	4738	34.008	28.296	31.152	19.577	16.289	17.933
56	.00054	.00046	15,690	13,310	5,644	4788	39.329	33.359	36.344	22.640	19.203	20.922
64	.00054	.00046	15,570	13,410	5,607	4825	44.651	38.421	41.536	25.704	22.117	23.910
72	.00053	.00047	15,500	13,490	5,578	4854	49.973	43.483	46.728	28.767	25.031	26.899
80	.00053	.00047	15,440	13,560	5,555	4877	55.295	48.545	51.920	31.831	27.945	29.888
96	.00053	.00047	15,320	13,680	5,509	4923	65.809	58.799	62.304	37.883	33.848	35.866
112	.00053	.00047	15,250	13,750	5,486	4946	76.452	68.924	72.688	44.010	39.676	41.843
128	.00052	.00048	15,180	13,820	5,461	4972	86.966	79.178	83.072	50.062	45.579	47.821
			Stress = $29{,}000{,}000A/d$§		Stress = $10{,}432{,}000A/d$§		Force = 1298 A §			Force = 747.2A §		

NOTE 1. No values are given for sizes 1/8 in. to 1 1/4 in. inclusive, because of the absence of interference of metal, due to the smallest shaft member being smaller than the largest hole member.

* Interference on selected fits, all sizes, = 0.0005 in.

† Greatest hole stress on selected fits, all sizes, = 14,500 lb. per sq. in.

‡ Greatest hole stress on selected fits, all sizes, = 5216 lb. per sq. in.

§ A = allowance; d = mean size

Table 7: American Standard Heavy Force and Shrink Fit, Considerable Negative Allowance, Selective Assembly, Class 8

A.S.A., B4a–1925. All dimensions in inches

Size			Limits				Tightest Fit, Allowance	Loosest Fit, Allowance + Tolerances	Selected Fit, Average Interference of Metal
From	Up to and Including	Mean	Hole or External Member		Shaft or Internal Member				
			Plus		Plus	Plus	Minus*	*†	Minus*
0	3/16	1/8	0.0003	0.0000	0.0004	0.0001	0.0004	+0.0002	0.0001
3/16	5/16	1/4	.0004	.0000	.0007	.0003	.0007	+ .0001	.0003
5/16	7/16	3/8	.0004	.0000	.0008	.0004	.0008	.0000	.0004
7/16	9/16	1/2	.0005	.0000	.0010	.0005	.0010	.0000	.0005
9/16	11/16	5/8	.0005	.0000	.0011	.0006	.0011	− .0001	.0006
11/16	13/16	3/4	.0005	.0000	.0013	.0008	.0013	− .0003	.0008
13/16	15/16	7/8	.0006	.0000	.0015	.0009	.0015	− .0003	.0009
15/16	1 1/16	1	.0006	.0000	.0016	.0010	.0016	− .0004	.0010
1 1/16	1 3/16	1 1/8	.0006	.0000	.0017	.0011	.0017	− .0005	.0011
1 3/16	1 3/8	1 1/4	.0006	.0000	.0019	.0013	.0019	− .0007	.0013
1 3/8	1 5/8	1 1/2	.0007	.0000	.0022	.0015	.0022	− .0008	.0015
1 5/8	1 7/8	1 3/4	.0007	.0000	.0025	.0018	.0025	− .0011	.0018
1 7/8	2 1/8	2	.0008	.0000	.0028	.0020	.0028	− .0012	.0020
2 1/8	2 3/8	2 1/4	.0008	.0000	.0031	.0023	.0031	− .0015	.0023
2 3/8	2 3/4	2 1/2	.0008	.0000	.0033	.0025	.0033	− .0017	.0025
2 3/4	3 1/4	3	.0009	.0000	.0039	.0030	.0039	− .0021	.0030
3 1/4	3 3/4	3 1/2	.0009	.0000	.0044	.0035	.0044	− .0026	.0035
3 3/4	4 1/4	4	.0010	.0000	.0050	.0040	.0050	− .0030	.0040
4 1/4	4 3/4	4 1/2	.0010	.0000	.0055	.0045	.0055	− .0035	.0045
4 3/4	5 1/2	5	.0010	.0000	.0060	.0050	.0060	− .0040	.0050
5 1/2	6 1/2	6	.0011	.0000	.0071	.0060	.0071	− .0049	.0060
6 1/2	7 1/2	7	.0011	.0000	.0081	.0070	.0081	− .0059	.0070
7 1/2	8 1/2	8	.0012	.0000	.0092	.0080	.0092	− .0068	.0080
8 1/2	10	9 1/4	.0012	.0000	.0102	.0090	.0102	− .0078	.0090
10	11 1/2	10 3/4	.0013	.0000	.0113	.0100	.0113	− .0087	.0100
11 1/2	13	12 1/4	.0014	.0000	.0134	.0120	.0134	− .0106	.0120
13	15	14	.0014	.0000	.0154	.0140	.0154	− .0126	.0140
15	17	16	.0015	.0000	.0175	.0160	.0175	− .0145	.0160
17	19	18	.0016	.0000	.0196	.0180	.0196	− .0164	.0180
19	22	20	.0016	.0000	.0216	.0200	.0216	− .0184	.0200
22	26	24	.0017	.0000	.0257	.0240	.0257	− .0223	.0240
26	30	28	.0018	.0000	.0298	.0280	.0298	− .0262	.0280
30	34	32	.0019	.0000	.0339	.0320	.0339	− .0301	.0320
34	38	36	.0020	.0000	.0380	.0360	.0380	− .0340	.0360
38	44	40	.0021	.0000	.0421	.0400	.0421	− .0379	.0400
44	52	48	.0022	.0000	.0502	.0480	.0502	− .0458	.0480
52	60	56	.0023	.0000	.0583	.0560	.0583	− .0537	.0560
60	68	64	.0024	.0000	.0664	.0640	.0664	− .0616	.0640
68	76	72	.0025	.0000	.0745	.0720	.0745	− .0695	.0720
76	88	80	.0026	.0000	.0826	.0800	.0826	− .0774	.0800
88	104	96	.0027	.0000	.0987	.0960	.0987	− .0933	.0960
104	120	112	.0029	.0000	.1149	.1120	.1149	− .1091	.1120
120	136	128	.0030	.0000	.1310	.1280	.1310	− .1250	.1280

* Minus here denotes interference of metal or negative allowance.
† Plus here denotes clearance or amount of looseness.

In designing force and shrink fits, the interference e usually is assumed. The resulting stress then should be calculated to make sure that the elastic limit of the material is not exceeded.

DESIGN DATA. The design of press and shrink fit assemblies is well summarized in the design data sheets prepared by O. J. Horger and C. W. Nelson (Ref. 2).

TEMPERATURES FOR SHRINKAGE FITS. The temperature to which the outer member in a shrinkage fit must be heated in order to assemble the parts depends upon the coefficient of linear expansion of the metal. For wrought iron and steel this is about 0.0000065 for each °F and for cast iron it is 0.0000062. Low redness corresponds to about 900 F and hence causes an expansion of the bore of about 0.005 in. per in. of diameter.

EXPANSION FITS. It is sometimes more convenient to contract the shaft than to expand the ring when assembling members for an interference fit. Dry ice (solid CO_2)

Table 8. Stresses and Forces for Heavy Force and Shrink Fits, Class 8

A.S.A. B4a–1925

Force values are for fits 1 in. long. For longer fits, multiply tabular values by length of fit in inches

Mean Size, in.	Interference of Metal per Inch of Mean Size			Greatest Hole Stress, Steel Shaft in Steel Hole			Force Required to Press Steel Shaft into Steel Hole		
	Tightest Fit, in.	Loosest Fit, in.	Selected Fit, in.	Tightest Fit, lb. per sq. in.	Loosest Fit, lb. per sq. in.	Selected Fit, lb. per sq. in.	Tightest Fit, tons	Loosest Fit, tons	Selected Fit, tons
1/8	0.00320		0.001	92,800		29,000	0.519		0.130
1/4	.00280	See	.001	81,200	See	29,000	.909	See	.389
3/8	.00213	Note 1	.001	61,870	Note 1	29,000	1.038	Note 1	.519
1/2	.00200		.001	58,000		29,000	1.298		.649
5/8	.00176	0.00016	.001	51,040	4,640	29,000	1.428	0.130	.779
3/4	.00173	.00040	.001	42,270	11,600	29,000	1.687	.389	1.038
7/8	.00171	.00034	.001	49,710	9,943	29,000	1.947	.389	1.168
1	.00160	.00040	.001	46,400	11,600	29,000	2.077	.519	1.298
1 1/8	.00151	.00044	.001	43,840	12,890	29,000	2.207	.649	1.428
1 1/4	.00152	.00056	.001	44,080	16,230	29,000	2.466	.909	1.687
1 1/2	.00147	.00053	.001	42,530	15,470	29,000	2.856	1.038	1.947
1 3/4	.00143	.00062	.001	41,430	18,230	29,000	3.245	1.428	2.336
2	.00140	.00060	.001	40,600	17,400	29,000	3.634	1.558	2.596
2 1/4	.00138	.00066	.001	39,960	17,910	29,000	4.024	1.947	2.985
2 1/2	.00132	.00068	.001	38,280	19,720	29,000	4.283	2.207	3.245
3	.00130	.00070	.001	37,700	20,300	29,000	5.062	2.726	3.894
3 1/2	.00126	.00074	.001	36,460	21,540	29,000	5.711	3.375	4.543
4	.00125	.00075	.001	36,250	21,750	29,000	6.490	3.894	5.192
4 1/2	.00122	.00078	.001	35,440	22,560	29,000	7.139	4.543	5.841
5	.00120	.00080	.001	34,800	23,200	29,000	7.788	5.192	6.490
6	.00118	.00082	.001	34,317	23,680	29,000	9.216	6.360	7.788
7	.00116	.00084	.001	33,570	24,440	29,000	10.514	7.658	9.086
8	.00115	.00085	.001	33,350	24,650	29,000	11.942	8.826	10.384
9	.00113	.00087	.001	32,870	25,130	29,000	13.240	10.124	11.682
10	.00113	.00087	.001	32,770	25,233	29,000	14.667	11.293	12.980
12	.00112	.00088	.001	32,390	25,610	29,000	17.393	13.759	15.576
14	.00110	.00090	.001	31,900	26,100	29,000	19.989	16.355	18.172
16	.00109	.00091	.001	31,720	26,290	29,000	22.715	18.821	20.768
18	.00108	.00091	.001	31,580	26,420	29,000	25.441	21.287	23.364
20	.00108	.00092	.001	31,370	26,680	29,000	28.037	23.883	25.960
24	.00107	.00093	.001	31,050	26,940	29,000	33.359	28.945	31.152
28	.00106	.00094	.001	30,860	27,130	29,000	38.680	34.008	36.344
32	.00106	.00094	.001	30,710	27,280	29,000	44.002	39.070	41.536
36	.00106	.00094	.001	30,600	27,390	29,000	49.324	44.132	46.728
40	.00105	.00095	.001	30,510	27,480	29,000	54.646	49.194	51.920
48	.00105	.00095	.001	30,330	27,680	29,000	65.160	59.448	62.304
56	.00104	.00096	.001	30,190	27,810	29,000	75.673	69.703	72.688
64	.00104	.00096	.001	30,080	27,900	29,000	86.187	79.957	83.072
72	.00103	.00097	.001	30,000	27,990	29,000	96.701	90.211	93.456
80	.00103	.00097	.001	29,940	28,060	29,000	107.113	100.465	103.840
96	.00103	.00097	.001	29,810	28,180	29,000	128.115	121.103	124.608
112	.00103	.00097	.001	29,750	28,250	29,000	149.140	141.612	145.736
128	.00102	.00098	.001	29,690	28,320	29,000	170.038	162.250	166.144
				Stress = 29,000,000 A/d *			Force = 1298A *		

* A = allowance; d = mean size.

Note 1. No values are given for sizes 1/8 in. to 1/2 in. inclusive because of the absence of intererence of metal, due to the smallest shaft member being smaller than the largest hole member.

in alcohol, or liquid air, may be used for the cooling medium. Dry ice has a temperature of −110 F, and from a 70 F room temperature can contract steel about 0.001 in. per in. of diameter. Liquid air boils at −310 F, and from a 70 F room can contract steel better than 0.002 in. per in. of diameter. (See Ref. 3.)

BASIC SHAFT SYSTEM. The fits just tabulated have been set up using a basic hole or one that always has the basic size as the minimum limit of the hole. The allowance for various fits is obtained by varying the dimensions of the shaft. Now the same fit or the same degree of clearance or interference can be obtained using a basic shaft system in which the maximum limit of the shaft is the basic size and where the allowance for various fits is obtained by varying the hole size. To use the basic shaft system, adopt the allowance or tightest fit figures from the tables of fits and work out limits to give these allowances.

REFERENCES

1. *Machinery's Handbook*, 9th ed., published by The Industrial Press, New York.
2. O. J. Horger and C. W. Nelson, *Trans. Am. Soc. Mech. Engrs.*, 1937 (A-183) and 1938 (A-32).
3. E. V. David and W. S. Farr, Making expansion fits with liquid air, *Machinery*, Nov. 1931.

DIMENSIONAL MEASUREMENTS

By Ermand L. Watelet

Industrial dimensional measurements may be divided into three classifications: nonprecision, precision, and high precision. The range of precision in each group, however, is quite wide and depends on the inherent accuracy of the device plus the expertness of the user. In the precision and high-precision groups additional factors of temperature, measuring pressure, friction, and cleanliness affect the amount of precision obtainable. In general, nonprecision measurements are made with an accuracy of 0.015 to 0.001 in., precision measurements within an accuracy of 0.001 to 0.0001 in., and high precision 0.0001 in. or less.

3. NONPRECISION MEASURING DEVICES

THE STEEL RULE or scale, ordinarily made in lengths of 6 in. or multiples thereof up to 6 ft, is perhaps the most commonly used device for direct measurements in the shop. It is made with graduations 0.005 to 0.010 in. wide, etched in polished tempered steel and is divided into inches subdivided into 8ths, 16ths, 32nds or 64ths. Decimal rules are divided into inches subdivided into 10ths, 20ths, 50ths or 100ths. Few rules are divided into 100ths because such closely spaced graduations are difficult to read. Although rules of good quality are inherently accurate beyond the precision to which they are used, the average person can use them only to an accuracy of about 0.010 in. Experts, with the aid of an eyeglass, can estimate to as close as 0.002 or 0.003 in.

THE HOOK RULE is a plain rule to which has been added an adjustable member flush with one end of the rule and projecting beyond the measuring edge for the purpose of conveniently making a measurement from a shoulder or edge without the necessity of setting the first end by eye. It is particularly useful in measuring from a hidden edge.

THE RULE DEPTH GAGE is a narrow rule equipped with a sliding head having a working surface at right angles to the rule. The head is provided with clamping means for locking it in position when a measurement has been made. This device is used for measuring the depth of holes and slots or determining the distance from a shoulder to another point.

THE CALIPER RULE has a fixed jaw at one end of the scale and a sliding jaw movable along the scale. The measuring faces of the jaws are parallel to each other and square with the edge of the rule. For the measurement of outside diameters and thicknesses, it affords a greater degree of accuracy than a plain rule because graduations are more easily sighted from the edge of the movable jaw. The ends or nibs of the jaws are rounded and held to a fixed width for the purpose of making internal measurements. An index line is marked on the movable jaw for the purpose of indicating internal measurements. The movable jaw is provided with locking means for preserving a specific setting.

THE SHRINK RULE. In the making of metal castings the metal contracts or shrinks as it cools. Suitable allowance must therefore be made in the dimensions of the pattern used to make the mold in which the molten metal is poured. The shrinkage value for a specific metal is fairly constant, and it is possible therefore to graduate shrink rules to give shrinkage allowances directly. Thus the pattern maker can set his pattern sizes directly from the dimensions on the drawing by the use of a shrink rule without the necessity of computing shrinkage allowance. The graduations on a shrink rule, while indicated in inches and fractions, are actually greater. Thus, 1 in. on a shrink rule for cast iron which has a shrinkage of $^1/_8$ in. per ft is actually 1.0105 in. long.

CALIPERS. It is often necessary to make nonprecision measurements which cannot be made directly with a rule. For instance, it would not be possible to determine the diameter of a piece being turned in a lathe with a rule. For applications of this type a caliper, comprising a pair of legs hinged at one end and made adjustable to measure at the free ends, is frequently used. Since calipers are not graduated, the actual measurement is determined by referring the setting or space between the ends of the legs to a graduated rule.

Outside spring calipers have the measuring ends of the legs curving inwardly so that no part of the legs except the ends ever contact the part being measured.

Inside spring calipers are similar except that the measuring ends of the legs curve outwardly and are used to measure inside dimensions such as the diameter of a hole. Spring calipers are limited in size to about 8 in.

Firm joint calipers, both outside and inside types, are similar in construction to the spring type except that the legs are held in a set position by means of a friction joint. Firm joint calipers are particularly convenient in making large measurements and are therefore made in sizes up to 24 in. or larger.

Hermaphrodite calipers have one inside or outside leg and one leg terminating in a divider point. They are used for such layout work as scribing a line parallel to a shoulder or edge, and may be either of the spring or firm joint type.

Transfer calipers are used for making measurements of recesses or diameters which are larger than the opening through which the calipers must be inserted. They are provided with an auxiliary arm against which one leg bears, thereby allowing collapsing of the leg for removal from the hole while preserving the measurement.

DIVIDERS are calipers with both legs terminating in sharp points, used for scribing arcs and circles as well as transferring and making measurements. They may be either of the spring or the firm joint types.

BEAM TRAMMELS, which are used in sizes beyond the range of dividers, consist of a long bar on which two arms or trammels slide. The arms terminate in one of a variety of points and are arranged to be clamped in the desired position. A fine adjustment may be provided with one of the legs for easy setting.

4. PRECISION MEASURING DEVICES

THE VERNIER (Fig. 1), consists of a fixed scale and a vernier scale, which is movable in relation to the former. The vernier scale makes possible the accurate reading of a measurement to a fractional part of the smallest division on the fixed scale and therefore converts a scale into a precision device. Although a vernier scale can theoretically be designed to indicate to very small fractions of length, it is limited practically to about 0.001 in. when applied to simple scales.

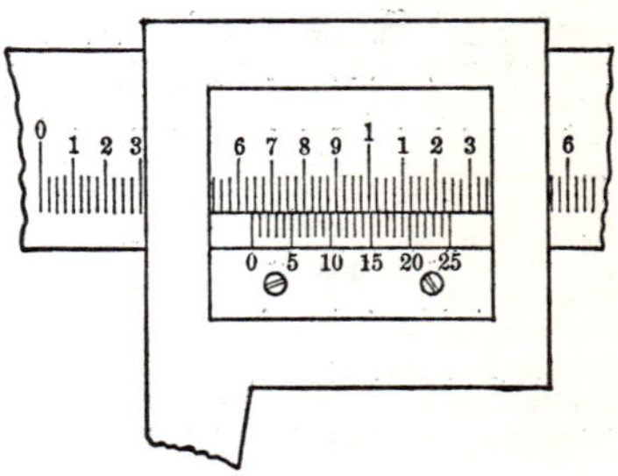

FIG. 1. Vernier.

EXAMPLE. The number of graduations on the vernier scale is equal to the number of parts into which the smallest graduations of the fixed scale may be subdivided. The total length of these graduations is made equal to the length of an equivalent number on the fixed scale minus one. For instance, a typical vernier (Fig. 1) has a fixed scale graduated into 0.025 in. or $^{25}/_{1000}$ in. It is desired to read directly to 0.001 in., which is $^{1}/_{25}$ of one graduation. Therefore, the vernier scale is graduated into 25 divisions, the total length of which is equal to 24 divisions on the fixed scale. Thus, the difference between the length of one graduation on the fixed scale and that of one division on the vernier scale is $^{1}/_{25}$ of 0.025 in. or 0.001 in. The first or zero graduation on the vernier scale is the index line, i.e., the point at which the measurement is observed. If the zero line on the vernier scale coincides with a graduation on the fixed scale, the measurement is to an even 0.025 in. If the zero line is between graduations on the fixed scale, the remainder of the measurement is determined by noting the number of the vernier scale graduation that coincides with a graduation on the fixed scale. This number is the number of $^{1}/_{1000}$ in. which must be added to the reading on the fixed scale.

In Fig. 1, the reading on the fixed scale is 0.625+ in. (Each small graduation on the scale = 0.025 in.) Graduation 19 (representing $^{19}/_{1000}$ or 0.019) on the vernier scale coincides with a graduation on the fixed scale. Thus the reading is 0.625 + 0.019 = 0.644 in.

THE VERNIER CALIPER is similar in principle to the caliper rule described before except that a vernier scale is added for making precision measurements. In addition, much greater refinements are made in the quality of the device with respect to the accuracy of the graduations, the squareness and parallelism of the jaws, and the fit of the movable jaw on the bar. The width of graduations on vernier devices is smaller than those on rules, being of the order of 0.003 or 0.004 in. It will be evident that in order to use the vernier caliper to the observed accuracy some degree of skill and judgment is necessary. The vernier caliper is used for making inside as well as outside measurements. The ends of the jaws are provided with nibs that are ground on the outside to a radius less than the smallest inside diameter which may be measured.

Separate vernier plates and graduations are provided for inside and outside measurements.

THE VERNIER HEIGHT GAGE is used to make measurements of height to 0.001 in. above a plane surface. In general, it is used on a commercial surface plate which has the required accurate flat surface. The vernier height gage consists of a vertical graduated bar, a base, and a movable jaw. The bottom of the base is lapped flat and square with the bar. The movable jaw has its upper and lower surfaces finished parallel to the bottom of the base for making measurements over or under a part. It is attached to a slide which carries a vernier plate. The slide is provided with a clamp screw and an additional adjusting slide for making fine adjustments. A combination marker and extension is furnished for clamping to the jaw and is used for scribing lines on work or for making measurements which cannot be reached by the jaw.

THE VERNIER DEPTH GAGE is similar to the rule depth gage except for the addition of a vernier scale for reading to 0.001 in., better accuracy of graduations, and refinement of construction.

THE MICROMETER is probably the most commonly used precision measuring device in industry for obtaining measurements of the order of 0.001 to 0.0001 in. Its measuring accuracy is based on an accurate screw and a fixed nut. The micrometer (Fig. 2) is composed of a U-shaped frame, the end of one leg of which carries a fixed anvil, the other leg carrying a tubular member or barrel. The micrometer nut is fixed in the barrel and is slotted and provided with a tapered thread and nut at its outer end to allow adjustment for wear. The measuring spindle and screw are integral, being supported on the nut at the threaded end and in an accurately fitted hole in the frame at the spindle end. To the outer end of the spindle is fastened a shell or thimble which extends over and turns around the barrel. The measurement is made between the faces of the anvil and spindle.

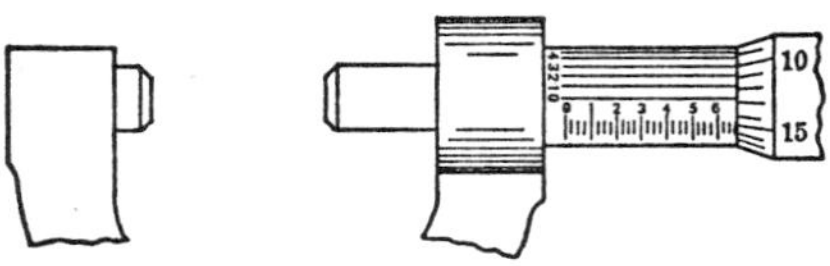

FIG. 2. Vernier micrometer.

EXAMPLE. The screw in the standard micrometer has 40 threads per inch and thus advances 1/40 in. per turn or 0.025 in. The beveled edge of the thimble is graduated into 25 equal divisions to allow measurements of 1/25 of a turn or 0.001 in. The barrel is graduated along its length into divisions of 0.1 in. and subdivided to 0.025. Thus, one turn of the thimble moves one division on the barrel. A longitudinal line on the barrel serves as an index for the thimble graduations. A vernier scale (Fig. 2) may be added to the barrel in conjunction with the thimble graduations. Ten graduations, equal to nine on the thimble, enable subdivision of the thimble graduations to 0.0001 in. The total measurement is equal to the reading on the barrel plus the reading on the thimble plus the reading of the vernier.

The thimble of the micrometer in Fig. 2 has uncovered 6 main divisions and 3 subdivisions on the barrel. The scale on the thimble shows that it has been turned through 12+ divisions past the index line on the barrel. The third line on the vernier scale coincides with a line on the thimble scale. The reading of the micrometer then is:

Scale on barrel	$0.6 + (3 \times 0.025) =$	0.675
Scale on thimble	$12 \times 0.001 =$	0.012
Scale on vernier	$0.0003 =$	0.0003
Micrometer reading		0.6873

The Ratchet Stop. In the use of the micrometer the measuring pressure is dependent on the operator's sense of "feel." Errors of 0.0001 to 0.0003 in. may be introduced in a measurement as a result of this factor. In order to overcome this source of error, a device known as a ratchet stop is provided for use with the micrometer. Attached to the end of the thimble, it serves as a constant-torque turning means. The outer knurled member, when turned, exerts a turning force on the thimble through ratchet teeth under spring pressure. When the spindle and anvil contact the part to be measured and the correct pressure has been reached, the ratchet teeth will slip, preventing further pressure from being applied. Ratchet stops are set to give a measuring pressure of approximately 25 lb per sq in. of spindle face area.

INSIDE MICROMETERS are used for measuring the diameters of holes and other inside dimensions. Two general types are in use. The first, an inside micrometer caliper for measuring small inside dimensions of 0.2 to 1 in., is provided with two jaws, one fastened to the end of the spindle, the other to the barrel of the micrometer. The spindle and screw are separate parts of this device.

The spindle moves longitudinally with the screw but is restrained from turning since the jaws must remain parallel at all measuring positions. The measurement is made on the outside of nibs provided at the ends of the jaws.

For measuring holes and inside dimensions larger than 1 in. and up to several feet, inside micrometers are provided that measure from anvils placed at the ends of thimble and barrel. By means of interchangeable measuring rods a wide range of measurements is possible. The micrometer head has a range of 1 in. whereas the rods are made in multiples of 1 in. so that measurements are possible from minimum to maximum size by 0.001-in. intervals.

THE MICROMETER DEPTH GAGE is used for measuring the depth of holes and slots or the distance from one surface to another. It is similar in general construction to the rule depth gage and vernier depth gage except that a micrometer head is used for measuring. A base whose lower surface is at right angles to the micrometer axis is fastened to the micrometer barrel and is used as the reference point. A measuring rod is encased in the hollow screw and is clamped by means of a chuck at the end of the thimble at one of several scribed graduations on the rod. The distance between the graduations is equal to the micrometer screw movement so that a range of measurements from zero to several inches by intervals of 0.001 in. is possible by clamping the rod progressively at the separate scribed graduations. Each graduation on the rod is a V groove into which fingers in the chuck fit for accurately locating the rod.

Other types of micrometer depth gages are made with several interchangeable measuring rods of different lengths in contrast to a single rod with spaced grooves.

THE DIAL INDICATOR (Fig. 3) is an instrument in which the movement of a spring-loaded measuring spindle is magnified by means of a rack and train of gears which actuate a pointer on a graduated dial. Typical scale graduations are 0.001, 0.0005, and 0.0001 in. for various types and sizes while the magnification varies from about 50 to one to 1250 to one. The rim of the dial or bezel is rotatable so that for a certain setting the zero on the scale may be brought opposite the position of the pointer for convenient reading. Although the dial indicator is a direct-measuring type of device, it is commonly used as a comparator for determining the variations of a measurement from a standard figure.

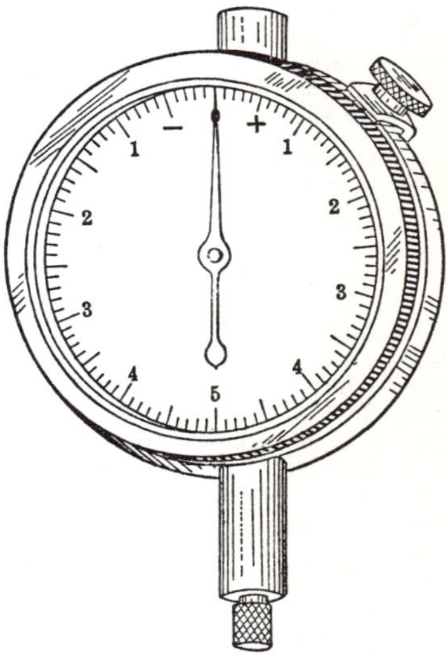

Fig. 3. Dial indicator. (Courtesy of Federal Products Corp.)

The dial indicator has many common uses. It may be used as a means of checking runout or out-of-roundness of a piece rotated between centers or to check the trueness of the rotating part of a machine such as the revolving center of a lathe. In these applications the indicator is fixed to a stationary part of the machine either by a clamp or by means of a heavy base to which it is attached. In a similar manner it may be used to set or check the parallelism of one surface with respect to another. When mounted to a base and upright commonly known as a *test set*, it may be used on a surface plate to test for size, straightness, and parallelism of parts. Built-in dial indicators are frequently used in such devices as calipers, snap gages, bore gages, thickness gages, depth gages, height gages, and bench comparators for determining variations in size from a master or standard.

5. HIGH-PRECISION MEASUREMENTS

MEASURING MACHINES are used for making measurements where the accuracy must be maintained to better than 0.0001 in. or of the order of 0.00001 in. Such machines are very rarely used in production work, but are used for measuring master gages and standards, which are in turn used to check production gages such as master plug, taper plug, and thread gages. In making measurements of the order of 0.0001 to 0.00001 in., it is of utmost importance that control of the measuring pressure be made, since slight variations in pressure will cause errors in the readings too large to allow the accuracy required. For this reason, measuring machines must be provided with pressure control and high-magnification indicators which determine when the exact pressure has been reached. Furthermore, since the dimension of a part varies with the temperature in accordance with its temperature coefficient, it is necessary to measure parts at a specific temperature in order that all measurements shall be consistent. For this reason, 68 F has been established as the standard temperature at which fine measurements must be made. Measuring machines should be used where the temperature is maintained constant at the above value.

A TYPICAL MACHINE, made by the Pratt & Whitney Company (Fig. 4), consists of a heavy bed *A* designed for stability and elimination of distortion, a master bar *B* fastened to the bed and containing fine, accurately spaced lines one inch apart on polished steel plugs, a measuring head *C* with a 1-in. range micrometer screw graduated to 0.0001 in., having a vernier making possible readings to 0.00001 in., a microscope *D* fixed to the measuring head and focused on the master bar for setting to even inches, and a tailstock *E* with adjustable measuring pressure and a high magnification indicator for determining the exact measuring pressure.

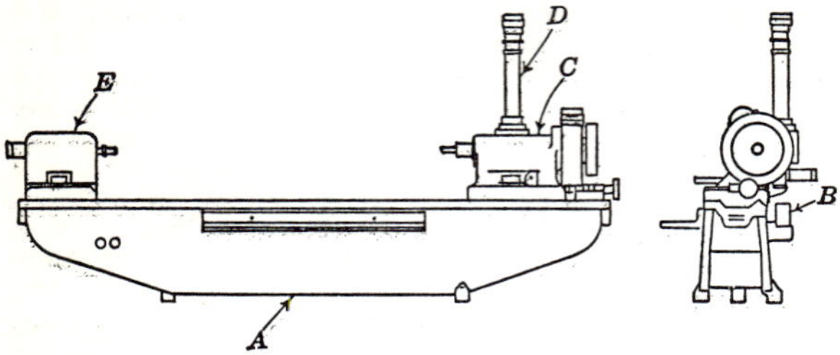

FIG. 4. Measuring machine. (Courtesy of Pratt and Whitney Division, Niles Bement Pond)

In use, the machine is first set to zero. This is done by moving the measuring head until the microscope is focused on the zero line on the master bar and setting the micrometer dial to zero. The tailstock is moved and clamped when its measuring anvil contacts the anvil on the measuring head. The micrometer dial is moved slightly until the pressure indicator is on zero. Finally, the vernier on the micrometer dial is set to zero. On making a measurement the headstock is set so that the microscope is focused on the nearest higher inch line. The part to be measured is supported between the anvils and the micrometer dial turned until the anvils contact the part and the pressure indicator is on zero. The measurement to the nearest 0.00001 in. is then read on the micrometer dial.

THE INTERFEROMETER is an apparatus for making measurements by the use of light waves. The wavelength of monochromatic light emitted by the incandescent gas of an element is the best known natural standard of length. Under constant atmospheric conditions the length of the wave of such a light is always identical for a given substance. If the light from the incandescent gas of an element is observed in the spectrograph it will be found to consist of colors of several frequencies. These colors, which appear as narrow lines on the spectrum, are characteristic of the element. Actually each color is not truly monochromatic, that is, it does not have a single frequency but is composed of several slightly different frequencies. One of the spectrum lines which is most nearly monochromatic is the red radiation of the element cadmium, whose wavelength has been determined in terms of the standard meter and used to make precision measurements. The wavelength of the red radiation of cadmium is 0.000638+ mm.

DETAILS OF THE METHOD by which the wavelength of light is used to make measurements with the interferometer is beyond the scope of this chapter, but the principle may be briefly explained. Light is an electromagnetic disturbance the characteristics of which may be graphically represented by a sine wave whose wavelength λ and frequency are characteristic of the color of the light. The speed of light in air, 186,000 miles per second, is the same for all colors, but is reduced when the light passes through another medium such as glass. If a ray of light passes through two points and it were possible to count the number of wavelengths occupying the distance between them, a measure of the distance could be obtained. This cannot be done directly, since the only effect of light upon the human eye is the sensation of color and intensity. However, by means of the phenomenon of light interference, the interferometer can be used to determine the number of wavelengths of light between two points and, consequently, the measurement. If two rays of light of equal wavelength arrive at a given point in phase, that is, with either the crests or the troughs of the waves in unison, the intensities of the rays are additive, and brilliance results. If, however, the rays are out of phase and the crest of one wave coincides with a trough of another wave, interference occurs, the two intensities cancel each other, and darkness results. In the former case the difference in path lengths of the two rays is equal to a whole number of wavelengths; in the latter the difference is an odd number of half wavelengths.

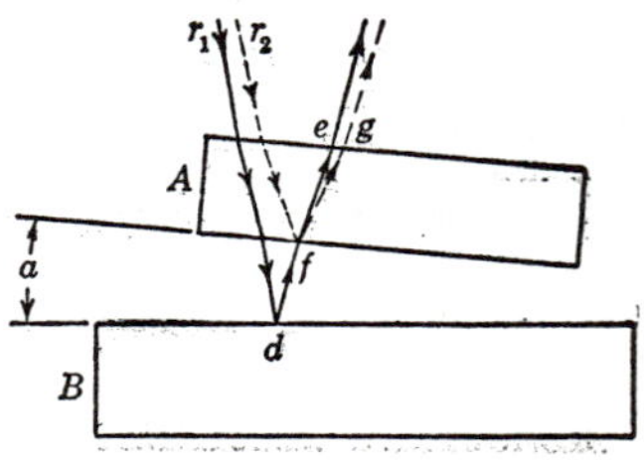

FIG. 5. Principle of the interferometer.

In Fig. 5, *A* is a transparent glass member known as an optical flat whose lower surface is truly flat. *B* is a member under test whose upper surface is also truly flat. r_1 and r_2 are rays from a source of light striking the glass at an angle. A portion of the light represented by r_1 passes through the glass and is reflected from the upper surface of *B* at the point *d*, the path of the reflected light being *dfe.* Another portion of the light represented by r_2 is reflected from the lower surface of *A* at point *f*, the path of the reflected light being *fg*. At the point *f*, therefore, portions of each ray coincide; and since a portion of r_1 traveled an additional distance through the air from the lower surface of *A* to the upper surface of *B* and back to *A*, it differs from that portion of r_2 reflected at the lower surface of *A* by a whole or fractional number of wavelengths. Assuming that the difference is an odd number of half wavelengths, interference will occur at *f*. In practice a large source of light would cause an infinite number of parallel rays to strike the optical flat. If the surfaces of *A* and *B* are perfectly parallel, the result of interference is a solid dark color at the lower surface of *A* when viewed from above. If *A* and *B* are at a slight inclination *a* to each other, a series of straight alternate light and dark bands is seen, as shown in Fig. 6. These are known as interference fringes. It should be noted that the change from light to dark bands is not as sharp as indicated in Fig. 6 but occurs gradually. The distance *d* between two dark bands represents a change in thickness of the wedge of air between *A* and *B* of one-half wavelength of light. The wavelength of light as furnished by commercial monochromatic light devices is 0.0000232 in. Viewed in such a light, the change in thickness would be 0.0000116 in. By counting the whole and fractional number of bands appearing and multiplying by 0.0000116, the total difference in thickness of the wedge of air is obtained. This will indicate to the reader a method of comparing the length of slightly different specimens. If the member *B* is moved in a direction perpendicular to its upper surface, the bands will be seen to move across the field of view. The number of bands passing a given reference point multiplied by one-half wavelength of light is equivalent to the movement. If the surface of *B* is not flat, the interference bands will be irregular instead of straight, will describe points of equal elevation much as contour lines on a topographical map. In this manner the exact deviation from flatness of a piece may be determined.

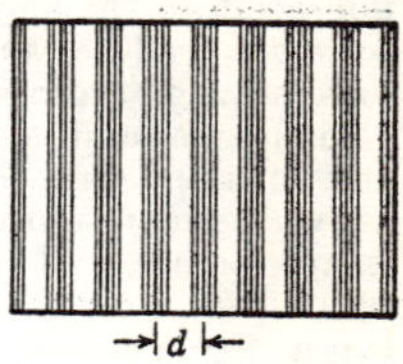

FIG. 6. Interpretation of interferometer bands.

Thus by use of the foregoing principles, the interferometer, using a very accurate optical system of lenses and mirrors, can make absolute measurements of length, comparisons of length, and deviations from flatness.

The simplest and most common use of light interference measurements is with the well-known optical flats. These are commercial glass disks made accurately flat on the sides. Together with the use of a monochromatic light source and gage blocks, the optical flat may be used to compare lengths in the manner described above. They are commonly used also for determination of flatness. It is to be noted that the surfaces of the parts being measured must be highly polished or lapped in order for interference fringes to appear.

For further information on the interferometer, see Refs. 1–4.

PRECISION GAGE BLOCKS are the most accurate commercial standards of length made. They consist of hardened steel blocks of rectangular or square cross section whose two opposite measuring faces are lapped flat and parallel to each other, and are held to a dimensional accuracy of a few millionths of an inch for lengths up to one inch and of an equivalent amount per inch for lengths over one inch. This accuracy requires the use of alloy tool steel of exceptional stability, careful heat treatment, and proper seasoning. Gage blocks are made in lengths of 0.010 in. to 20 in. or more and are furnished in sets having a variety of sizes. By using the blocks singly or in combinations, a large variety of sizes is possible. With sets having sufficient specific sizes, it is possible to make combination sizes over a range in increments of 0.0001 in. For example, a typical set of 81 blocks is divided into the following:

1 Block—0.050 in.
10 Blocks—0.100 in. to 0.1009 in. by 0.0001-in. steps
50 Blocks—0.101 in. to 0.150 in. by 0.001-in. steps
16 Blocks—0.200 in. to 0.950 in. by 0.050-in. steps
4 Blocks—1.000 in. to 4.000 in. by 1-in. steps

With such a set it is possible to obtain any dimension from 0.100 in. to 0.200 in. in increments of 0.001 in., and any dimension from 0.200 in. to over 10 in. in increments of 0.0001 in.

The utility of gage blocks in combinations is greatly enhanced by their ability to adhere to each other when suitably wrung together. This quality is due to the intimacy of contact of two highly finished and flat surfaces. The force of this adherence is sufficient to more than support the weight of several blocks thus held together. The errors due to joining several blocks together are so small as to be negligible, being in the order of a fraction of one millionth inch per joint.

Gage blocks are used extensively for setting measurements of reference and master gages, for calibration of measuring instruments, as masters for comparator instruments, and for establishing any high-precision measurement. With the use of suitable attachments and holders, gage blocks may be built up into outside and inside caliper combinations.

COMPARATORS. The comparator is a device which measures a dimension to an accuracy of 0.0001 in. to better than 0.00001 in., by comparison with a known standard as against the direct measuring types described heretofore. It is used for rapid inspection of parts in production as well as for making a variety of precision measurements. Its main element is a high-magnification low-range measuring head or indicator having a constant measuring pressure. The magnification is the ratio of the actual size of scale indication to the movement of a measuring spindle. For instance, a comparator whose scale has graduations 0.10 in. apart representing a spindle movement of 0.0001 in. would have a magnification of 0.10/0.0001 = 1000. In consequence of the limited space available on a scale, it is obvious that the range of the measuring head is, of necessity, small. It is for this reason that the device can be used only to compare the size of a part with a known standard of the same nominal size, its scale indicating by how many units the part is large or small. The comparing range, however, may be made as large as desired simply by moving the measuring head mechanically with respect to a reference anvil. Several types of commercial comparators are made. Mechanically, they all have the same basic elements but differ in refinements and in the principle used to obtain magnification.

Mechanical comparators use a measuring head or indicator which obtains its sensitivity from some form of mechanical multiplying system such as a gear train, levers, or other means. The conventional dial indicator belongs in this category. Because of practical limitations in mechanical movements, magnifications in excess of about 1000 are obtained by other methods.

Optical comparators employ the multiplying effect of an optical mirror and lens system. In one such scheme a measuring point is fastened to a mirror in such a manner that movement of the point causes tilting of the mirror. Light from an illuminated accurate glass scale is reflected from the mirror through an objective lens and is projected on a screen where a magnified image appears. Movement of the measuring point causes the scale to move past a fixed reference line on the screen, and thus indicates the amount of movement. A comparator made by the Sheffield Corporation uses part mechanical and part optical means for magnification. In this device movement of the measuring point causes the movement of an arm through an amplifying reed mechanism. The arm carries a target which interrupts a light beam projected through a lens system and casts a shadow on a scale. Movement of the target is amplified by the reed mechanism, and movement of its shadow further magnified by the optical system. Magnifications up to 20,000 are obtained by this method.

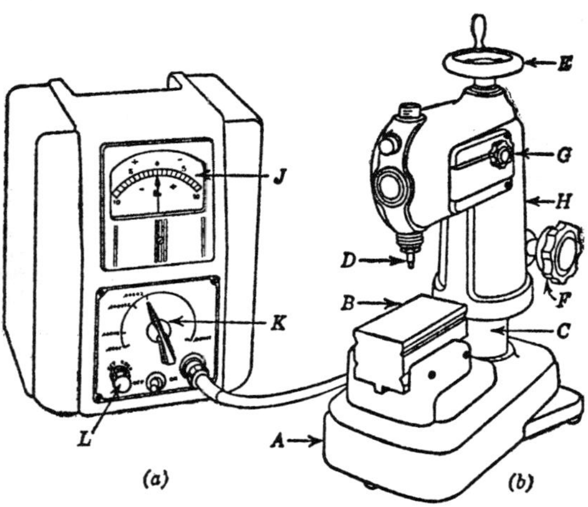

Fig. 7. Electronic comparator. (Courtesy of Brown and Sharpe Manufacturing Co.)

The electronic comparator, shown in Fig. 7, made by the Brown & Sharpe Manufacturing Company, obtains magnification through electrical and electronic means. It is made in two units, the comparator *b*, and the electronic amplifier and indicator *a*. The comparator consists of a base *A* to which is fastened an anvil *B* and an upright column *C*. The upper surface of the anvil is accurately lapped to a high degree of flatness. The head *H* is movable vertically on the column *C* and carries the measuring spindle *D* and the gaging mechanism within. Movement of the measuring spindle causes the deflection of a sensitive element employing four bonded type resistance wire strain gages which are all active and connected in a Wheatstone bridge circuit. The bridge circuit is fed with a low-voltage alternating current, and its output signal is determined by the amount of unbalance of the bridge as determined by the spindle movement. The output signal is then amplified through the electronic circuit of amplifier *a* and impressed upon a sensitive galvanometer whose scale *J* is calibrated to indicate the spindle movement. By means of selector *K* the graduations of the scale may be made to equal any value from 0.0001 to 0.00001 in.

In the use of the comparator, a master is placed on the anvil *B* and the head lowered by means of the handwheel E until the spindle D contacts the master. The head is then locked by means of clamp F. Further adjustment to bring the pointer of the scale near zero is made by knob G on the comparator, and final zero adjustment by knob L on the amplifier. The master is removed, and the instrument is ready for use in checking the work. Instruments of this type may be adapted with the use of additional equipment for automatically or semi-automatically sorting parts into categories of several sizes, such as small, large, and within tolerance.

Air comparators or air gages have been developed and are used mainly for the measurement of internal diameters. The measuring member is a cylindrical plug having a central air duct which terminates in one or more jets on the outside diameter of the plug. A narrow circular land surrounds the jet opening; a circular groove surrounds the land. Suitable clearance cuts on the periphery of the plug serve to allow free escape of air once it has passed between the jet land and the wall of the hole. If air is fed to the jets under constant pressure, the rate of air escape is proportional to the clearance between the jets and the sides of the hole being checked. In other words, the rate of flow of air is proportional to the size of hole. The relationship, however, is true only for small clearances between the plug jets and the hole. For this reason, the measuring range is seldom made over 0.004 to 0.005 in.

Two types of air circuits are used to convert the rate of air escape to an indicated measurement. The first type measures the back pressure caused by limiting the discharge of air through the jets. This comparator receives compressed air from a source and passes it through a filter and pressure regulator. From the regulator the air passes through a restricting orifice to a pressure indicator and the gaging plug. The scale of the pressure indicator is calibrated in graduations representing parts of an inch.

The second type of air circuit uses the *rate of flow principle.* Figure 8 shows an air comparator made by the Sheffield Corporation, which incorporates this principle. Compressed air enters the gage through a filter and pressure regulator and passes directly through the vertical transparent indicator tube A to the gaging plug B. The indicator tube hole is tapered slightly, the small end being at the bottom. An indicator float is free to move up and down in the tube and will adjust itself at some position corresponding to the velocity, or rate of flow of air through the tube and gaging plug. The gage is set to maximum and minimum rings representing the limits of tolerance of a hole, and adjustable markers are set at the points at which the float comes to rest. For all parts within tolerance the float will come to rest at some point between the two markers.

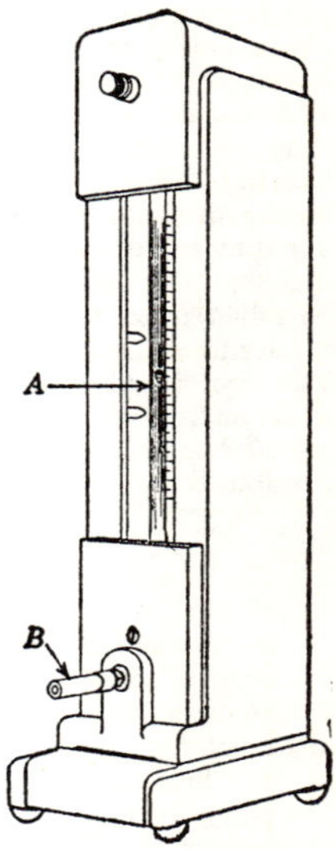

FIG. 8. Air comparator. (Courtesy of Sheffield Corp.)

Air gages are convenient for checking taper, bell mouth, and out-of-round as well as size of holes. Since the gaging members or jets do not contact the part, there is no need for critical positioning of the gaging plug.

The optical projection comparator is used for measuring curves, contours, radii, short angles, thread forms, gear tooth forms, etc., which are either impossible or impractical to measure with linear measuring devices. A beam of light is directed against that portion of a part being checked, the resulting shadow of which is magnified and projected upon a ground glass screen. The shadow is brought into coincidence with a master magnified outline of the particular form engraved or otherwise impressed upon a transparent sheet placed over the screen. Deviations of the part from standard can thus easily be seen. Common magnifications used are 20 : 1 and 62.5 : 1, the latter value being convenient in that 0.001 in. appears as $^1/_{16}$ in. on the screen. A measurement from one point to another of a part may be made by moving the table on which the part is held with a micrometer screw so that the image on the screen moves past a fixed reference line on the screen, and the amount of movement is noted on the micrometer. The device is arranged so that both lateral and vertical measurements are possible.

6. MEASUREMENT OF ANGLES

THE UNIVERSAL BEVEL PROTRACTOR (Fig. 9) is commonly used for measuring and setting angles. It consists of a body graduated through a complete circle in quadrants of 0 to 90 degrees. A dial carrying a vernier scale can be rotated about the center of the body. The main scale and vernier are so graduated as to enable readings to 5 min of arc. The angular measurement is made between a blade clamped to the dial and one of two projecting members of the body. This type of protractor can be used to measure inside as well as outside angles. In using the instrument care must be taken that it is held at right angles to the planes of the angle being measured.

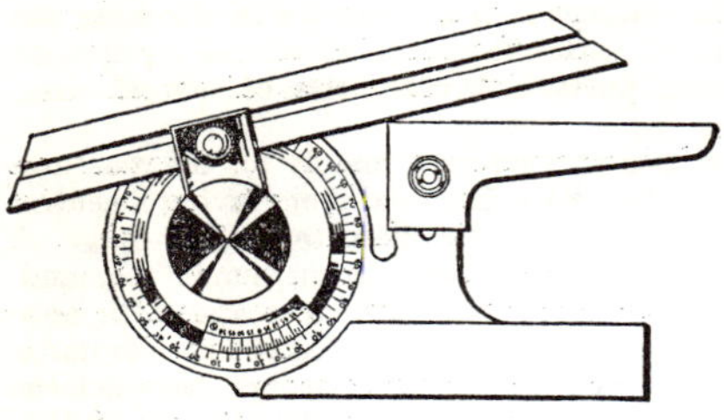

FIG. 9. Universal bevel protractor.

THE SINE BAR is a device used for accurately setting and checking angles by linear measurements. A typical arrangement is shown in Fig. 10. A bar *B*, whose upper surface is ground and lapped straight and flat, is machined to receive two rolls *R*, as shown. The diameters of the rolls and their distances from the top surface of the bar are made as nearly alike as possible. The distance *L* is made to equal either 5 or 10 in. The bar is set to a desired angle a by placing it with one roll on the plane surface *S* and the second roll on a gage block *G* having a specific length. Since the diameters of the rolls are equal, the length of *G* is equal to *H*, and $H = L \sin a$. The necessary length of *G* is determined, therefore, by multiplying the sine of the desired angle by *L*. A 5-in. sine bar whose various dimensions are accurate to 0.0001 in., together with gage blocks, can be relied upon to set small angles to an accuracy of a few seconds of arc. The error rapidly increases, however, when settings of large angles approaching 90 degrees are made. For this reason, when large angles are required, it is best if possible to set to the complement of the angle.

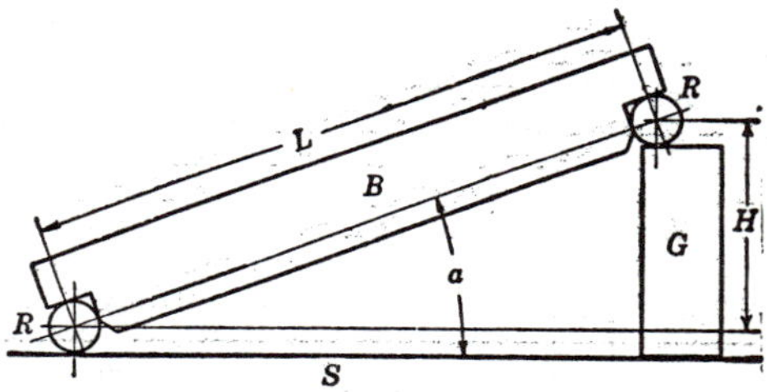

FIG. 10. Sine bar.

7. SCREW-THREAD MEASUREMENTS

The important measurements of screw threads are those made of the pitch diameter, the angle of thread, and the lead. The pitch diameter is the diameter at which the thread thickness is equal to the space between the threads. The angle of thread is the included angle between the sides of adjacent threads measured in the same plane as the axis of the screw. The lead is the distance advanced along the screw axis in one turn of the thread. For single-thread screws, the lead is equivalent to the pitch, or distance between centers of adjacent threads.

THREAD MICROMETER. A common and simple method of measuring pitch diameter is by means of the screw-thread micrometer. The micrometer is equipped with a pointed spindle and a V anvil. The angle of the point and V are made equal to the angle of the thread to be measured. The anvil is made rotatable about its axis so that in use it will conform to the helix angle. The micrometer measures from the theoretical sharp point of the thread crest to the theoretical sharp point of the root, this being equivalent to the pitch diameter. Actually, the end of the spindle and bottom of the V have clearance so that they will not rest on the bottom or top of the thread. The screw-thread micrometer is more accurate when set to a thread standard and used as a comparator. When set to zero with the faces closed, errors in measurement will occur due to the effect of the helix angle. However, for low helix angles the thread micrometer may be used to an accuracy of about 0.00025 in.

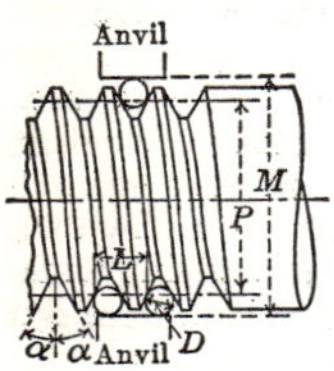

FIG. 11. Thread measurement by three-wire method.

THREE-WIRE SYSTEM. The most accurate method of measuring pitch diameter is by the three-wire system. Two wires are placed in adjacent thread spaces and held against a fixed anvil on one side of the screw; a third wire is placed between a thread space and a movable anvil on the opposite side, as shown in Fig. 11. The requisites of thread-measuring wires are extreme accuracy of size, roundness, and straightness. The diameter is held within 25 millionths of an inch of exact size. The material must be hard and wear resistant. The size of wire used in relation to pitch of the thread is such that contact will be made with the sides of the thread at the pitch diameter. This particular size is used because the resulting pitch diameter measurement is least affected by any error in the thread angle. The recommended sizes of wire for measuring 60-degree threads are given in Table 1. The pitch diameter of

Table 1. Recommended Wire Sizes for Measuring 60-degree Threads

(All dimensions in inches)

Threads per in.	Wire Size	Threads per in.	Wire Size	Threads per in.	Wire Size	Threads per in.	Wire Size	Threads per in.	Wire Size	Threads per in.	Wire Size
80	0.00722	44	0.01312	27	0.02138	14	0.04124	10	0.05773	5	0.11547
72	.00802	40	.01443	24	.02406	13	.04441	9	.06415	4 1/2	.12830
64	.00902	36	.01604	20	.02887	12	.04811	8	.07217	4	.14434
56	.01031	32	.01804	18	.03208	11 1/2	.05020	7	.08248	3 1/2	.16496
48	.01203	28	.02062	16	.03608	11	.05249	6	.09623	3	.19245

a thread whose sides are symmetrical about a line through the vertex perpendicular to the axis of the thread is

$$E = M + \frac{\cot \alpha}{2n} - w(1 + \operatorname{cosec} \alpha) \tag{1}$$

For a 60-degree thread with a helix angle smaller than 6 degrees, eq. 1 becomes

$$E = M + \frac{0.86603}{n} - 3w \tag{2}$$

For a helix of 6 degrees or more, the equation becomes

$$E = M + \frac{\operatorname{cotan} \alpha}{2n} - w\left(1 + \operatorname{cosec} \alpha + \frac{T^2}{2} \cos \alpha \cot \alpha\right) \tag{3}$$

In these equations, E = pitch diameter; M = measurement over the wires; α = 1/2 included angle of thread; n = threads per inch; w = diameter of the wires; T = tangent of helix angle = $(L/\pi)P$, where L = lead of thread.

The included angle of a thread may be determined by measuring over wires in a manner similar to the foregoing. In this case two sizes of wire are used. First, a measurement is taken over three wires having a diameter equal to or somewhat smaller than the recommended size for a given pitch. A second measurement is then made over three wires having a larger diameter, preferably near or equal to the maximum size which will fit the thread. The formula for determining the angle of thread is

$$\operatorname{cosec} \alpha = \frac{M_1 - M_2}{w_1 - w_2} - 1 \tag{4}$$

or

$$\frac{M_1 - M_2}{w_1 - w_2} = \operatorname{cosec} \alpha + 1 \tag{5}$$

where M_1 = the measurement over large wires of diameter w_1; M_2 = the measurement over small wires of diameter w_2; and α = 1/2 included angle of thread.

Measurements over wires may be made with micrometers to an accuracy of 0.0001 in. For greater accuracies, comparators and measuring machines are used.

The accurate measurement of the lead of a screw requires more elaborate equipment than is used for pitch diameter determination. However, the chance of variation in lead of screws is not so great as in other dimensions because the accuracy of lead is determined by the more or less fixed accuracy of the machine producing it. Accurate lead measurements, therefore, are usually made only on long screws, such as machine lead screws, or on very accurate short screws such as micrometer screws and thread gages. For this purpose, lead comparators are used. This type of device measures the movement of a stylus from one thread to another in relation to either a calibrated line standard or gage blocks, and the variation from standard lead is noted by means of a microscope or a high-magnification indicator. The radius of the point of the stylus is such as to contact the thread at the pitch line. It is essential that the measurement be made along the true axis of the screw.

8. LIMIT GAGES

Limit gages, as differentiated from measuring devices and comparators, do not indicate the exact measurement of a part but only show whether or not the part is within the allowed variation. In the mass production of parts for interchangeable manufacture it is not possible to make the dimensions exact; therefore a permissible variation in the nominal size is allowed. The total variation is known as the tolerance; the limiting dimensions of the tolerance are known as the limits. In practice, the amount of tolerance prescribed for a dimension is determined by its function in an assembly. In the interests of economy, the tolerance is made as large as possible consistent with the proper functioning of the part. A limit gage is used to determine whether a part is below size, within tolerance, or oversize. In general, the gage has incorporated in it two gaging members, one set to the low limit and one set to the high limit. The gaging member that will pass over or enter the part being measured, depending on whether it is an external or an internal gage, is known as the "go" member. The member that should not pass over or enter the part is known as the "not go."

SNAP-GAGES are used for gaging external dimensions. Figure 12 shows a fixed-type limit snap gage having two sets of jaws, one for the acceptable or "go" dimension, the other for the rejecting or "not go" dimension. These two sets of jaws are made to the upper and lower limits of the dimension, respectively. This type of snap gage has advantages of simplicity and cheapness. Its measuring dimensions are not easily changed by tampering. However, it is susceptible to wear, and for that reason another type is made, known as an adjustable snap gage. A commercial make of this gage is shown in Fig. 13. The anvil *A* is solid and fixed in position. Anvils *B* and *C* are adjustable, the former being set to the "go" and the latter to the "not go" dimension. Adjustment is made by differential screws as shown, gage blocks or master gages being used between the fixed and movable anvils. When properly adjusted the anvils are locked in position by clamp screws *d*. If the work being gaged is within the prescribed limits, anvil *B* will pass it, but anvil *C* will not.

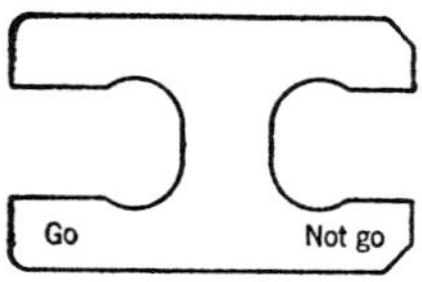

FIG. 12. Snap gage.

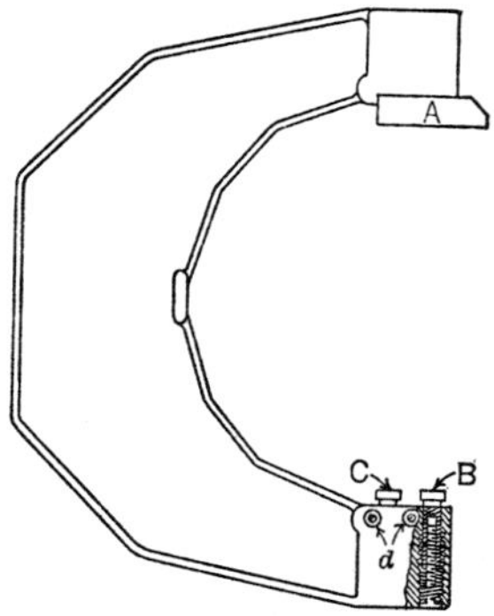

FIG. 13. Adjustable snap gage.

PLUG GAGES are cylindrical gages used for checking the size of holes. The gaging member is a simple cylinder ground and lapped to the required dimension and fastened to a hexagonal handle. When used as a limit gage, two separate cylinders are used, each finished to the limiting dimensions and fastened to the opposite ends of the handle, as in Fig. 14. The design of plug gages has been standardized by the American Gage Design Committee whose specifications have generally been followed by the gage industry. In sizes up to approximately 1.5 in. gaging members and handle are provided with tapered shanks and holes, respectively, for fastening. This type is known as the Taperlock. In sizes above 1.5 in., the member is made with a hole counterbored at each end and fastened to the handle by means of a screw. For resistance against turning of the member in the handle, three locking prongs project from the end of the handle and fit into corresponding grooves in the member. This is known as the Trilock type. The "go" and "not go" members are distinguished by the difference in length, the "go" member being long and the "not go" member short.

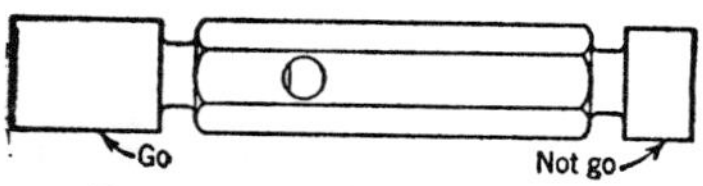

FIG. 14. Double-end plug gage.

A single-end type of limit gage is also made in which the gaging member is divided into two diameters. The diameter at the front of the member is made to the "go" dimension; the second diameter is made to the "not go" dimension. This type is known as a progressive plug gage.

RING GAGES are used for gaging diameters. The gage consists of a cylindrical member having a ground and lapped hole made to the required dimensions. Separate members are used for limit gaging, one for the "go" dimension and one for the "not go." The outside diameter of the members are knurled for easy handling, and the "not go" gage is provided with an annular groove for identification.

THREAD PLUG GAGES are used for checking internal threads. Except for the fact that the gaging members are threaded, these gages are similar in general design to plug gages described above. Thread gages are made in the single-end type to be used as reference gages or setting gages for thread ring gages, and in the double-end type for use as limit gages.

The thread angle and lead of the screw are made to the basic dimensions while the pitch diameter is made to the limiting dimension, the "go" gage being made to the lower limit of pitch diameter while the "not go" is made to the upper limit. The root of the thread is provided with suitable clearance while the crest of the thread is flatted to prescribed dimensions to avoid interference beyond normal thread contact.

THREAD-RING GAGES are used for gaging external threads. They are cylindrical members provided with very accurate central threaded holes. Since internal threads may not be measured as easily as external threads, thread-ring gages are made to the correct size by setting to a reference thread plug. The ring is provided with three equally spaced radial slots; two of them terminate in drilled holes, and the third passes entirely through the wall of the gage to allow expansion or contraction of the threaded hole. The gage is adjusted and locked to size by means of adjusting screw, locking screw, and sleeve associated with the through slot. For limit gaging, separate "go" and "not go" thread-ring gages are used.

THREAD-ROLL SNAP GAGES are combination "go" and "not go" limit gages for checking external threads rapidly.

A frame, similar to that of a plain snap gage, carries two sets of threaded rolls, each free to rotate. The first set of rolls is set to check the "go" or upper tolerance limit; the second, the "not go" or lower limit. If a threaded part passes through the first set of rolls, it is evidence that the pitch diameter is below the maximum limit. If the part does not pass the second or "not go" rolls, it is evidence that the pitch diameter is over the minimum limit, and therefore the part is correctly made within tolerance.

REFERENCES

1. *Iron Age*, May 22, 1919, p. 1345.
2. *Engineering*, Oct. 3, 1924, p. 491.
3. *Machinery*, July 1928, p. 809.
4. *Product Eng.*, Dec. 1932, p. 496.

MEASUREMENT AND DESIGNATION OF SURFACE QUALITY

By James A. Broadston

Skilled craftsmen have worked from drawings without tolerances or finish designations for many years. As it was their practice to make and assemble all the parts of a given machine, they knew from experience which surfaces of each part required special fitting or finishing. In recent years improved manufacturing methods and mass production have made it necessary for machine operators, inspectors, and others to depend on drawings and specifications for sufficient information to permit satisfactory parts to be made, as they seldom know from their own experience the function of the part or are able to select optimum finishes. Responsibility for the control of surface quality and its proper designation on drawings now rests with the designer since he is usually the only one who knows the exact function of each surface of a part and can specify its proper finish.

9. FINISH SYMBOLS AND TERMINOLOGY

EARLY FINISH SYMBOLS. The finish symbol f was originally adopted to show both machinist and pattern maker which surfaces of a casting were to be machined. Advancing techniques soon required that a coded symbol, such as f_3, be used on drawings to indicate that, for example, a "smooth machine finish" was required. Soon these brief descriptions, based strictly on personal opinion rather than on actual measurement and therefore subject to wide variations in interpretation, were supplanted by similarly coded symbols designating that certain processes, such as *smooth grind*, be used as it was known that some processes produced smoother finishes than others. Even with this improved control, parts were often produced having unsatisfactory finishes because their acceptance was still a matter of opinion rather than measurement. Now that machine finishes whose perfection approaches that of a fine lens may be rapidly produced, within dimensional and surface-finish tolerances unbelievable a decade ago, mechanisms are designed that depend on the precise control of these factors for their successful operation. The old surface-finish designations and the loose descriptive controls they connote have therefore become inadequate.

SURFACE-ROUGHNESS MEASUREMENT AND DESIGNATION STANDARDS. In 1932 the ASA, recognizing the need for standards for the designation of surface qualities, organized a committee to study the problem under the sponsorship of the ASME and the SAE. After many years of work with industry they published, in Feb. 1947, an approved American Standard (ASA B46.1) titled Surface Roughness, Waviness and Lay.

The new standard established an approved series of values expressed in microinches (millionths of an inch, 0.000001) for designating the size of the minute geometrical irregularities existing in the surfaces of solid materials. These values, which divide the normal range of the irregularities of machined surfaces into convenient steps, serve as a measurement index to their size when used with the approved new symbol.

The standard does not specify the finishes suitable for any specific purpose or the means by which they may be produced or measured. Neither is it concerned with other surface qualities such as luster, appearance, color, hardness, microstructure, or corrosion and wear resistance, any of which may be governing considerations in specific applications.

SURFACE IRREGULARITIES COMPLEX IN CHARACTER. The precise definition and measurement of surface quality are relatively impossible because surface irregularities are usually of very complex shape and do not lend themselves to direct measurement. Surfaces produced by various methods exhibit distinct differences in character such as those that make ground, turned, milled, honed, lapped, or polished surfaces relatively easy to identify. In general, all surfaces of solid materials are composed of a series of more or less microscopic peaks and valleys. These irregularities influence such vital design factors as wear, strength, dependability, and performance of parts, and must, therefore, be considered for proper design. Although both shape and length of the irregularities may affect the overall performance of a given surface, the standard deals only with their height, width, and direction. Fortunately, surfaces normally produced by a given process have irregularities whose shapes are similar; hence it is possible through control of only their average size also to control the expected performance of a given surface within very close limits. Since this is all the designer is expected to do, adoption of the new nomenclature and the controls made possible by using measured roughness values, rather than vague descriptions based only on opinions, have made possible a great forward step in design procedures.

APPROVED SURFACE-ROUGHNESS TERMINOLOGY. Roughness refers only to relatively finely spaced surface irregularities such as those produced on machined surfaces

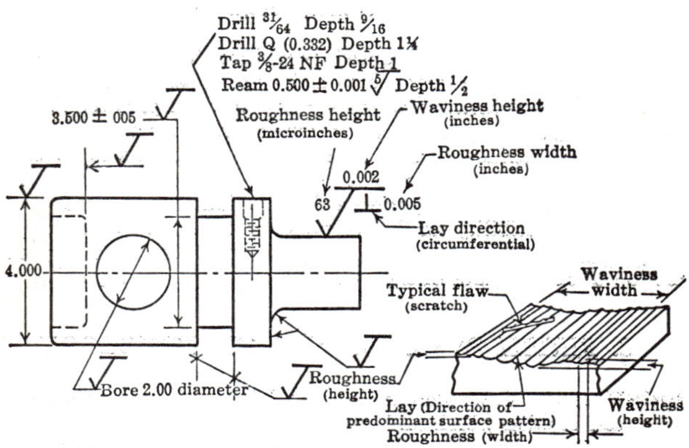

FIG. 1. Typical applications of surface roughness symbol.

by the cutting action of tool edges or abrasive grains. Roughness includes those minute irregularities that cover a surface completely and may be said to be characteristic of it. Measurement of roughness is the most important factor considered by the approved standard, and its designation, in most cases, permits ample control of surface quality even though the standard also provides for the designation of other values. The maximum allowable roughness height expressed in microinches is normally specified by the use of the approved symbol (Fig. 1) and one value from the approved standard series of Table 1.

Table 1. Standard Roughness Height Values (Microinches)

1/4	**4**	13	40	**125**	400
1/2	5	**16**	50	160	**500**
1	6	20	**63**	200	600
2	**8**	25	80	**250**	800
3	10	**32**	100	320	1000

In inspecting a part for conformance with a given standard roughness value, any lesser height than the one designated shall be satisfactory except that when two values are given, e.g., 32–8, they indicate maximum and minimum values. It is recommended that the *bold-faced* values be given preference because standard sets of flat surface roughness specimens for fingernail tactual comparison conforming to these roughnesses are available, making conformance with specified values easier to check.

The standard also permits the maximum allowable roughness width (repetitive units of the dominant surface pattern due to feed) to be specified in inches (adjacent to the lay symbol), as shown in Fig. 1. Roughness height specifications shall be based on irregularities having widths up to and including the maximum specified or, when not specified, up to and including the irregularities due to machine feed.

THEORY OF ROUGHNESS MEASUREMENT. Because surface irregularities are so minute and so complex it is impossible to make a simple direct measurement of roughness height. The value specified is therefore only an index value. The American Standard recognizes that it may be one of several approved indices that will serve to indicate the approximate size or magnitude of the surface irregularities.

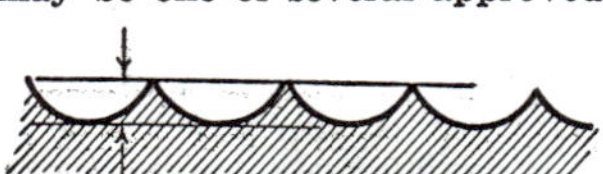

FIG. 2. Ideal regular profile of surface.

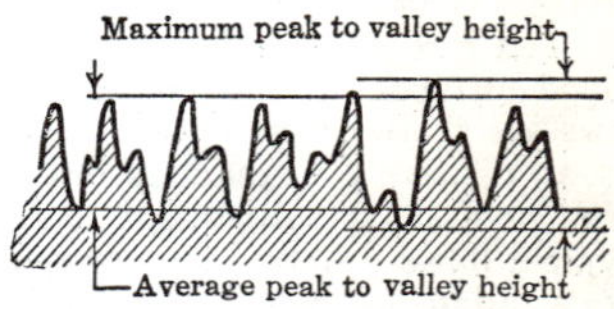

FIG. 3. Irregular profile of finished surface.

If a cross section of a given surface were highly magnified, the contour or profile of the surface irregularities might look like Fig. 2. For such a simple, regular profile, where all the peaks and valleys occur at regular elevations, the *peak-to-valley height* is simply the distance between these elevations, expressed in microinches. For a highly irregular surface profile, as shown in Fig. 3, it is customary to use one of two interpretations: (1) the *maximum peak-to-valley height*, which is the vertical distance between the elevation of the highest peak and that of the lowest valley, or (2) the *average peak-to-valley height*, which requires a personal factor of interpretation to determine the average elevation of the peaks and the average elevations of the valleys and then the vertical distance between them. These considerations make it plainly evident that these measurements depend either on human ability to find the highest peak and the lowest valley or to estimate their average heights, hence can hardly be classed as specific or duplicable measurements. Furthermore, they make no allowance for the shape of the irregularities, a vital factor in performance.

Height ratings may also be expressed as an *average deviation from the mean surface.* The mean surface is a surface located in such a way that the volume of the peaks above it is equal to the volume of the valleys below. It is the perfect surface that would result if all the peaks were leveled off and the material used to fill the valleys. The *arithmetic average* deviation of the surface of the irregularities from the mean surface is defined as the result of taking a great many uniformly spaced measurements in microinches, such as *a*, *b*, *c*, *d*, and *e*, as shown in Fig. 4, and averaging them. Should each value be squared and the square root of the average square be taken, the result represents the *root-mean-square* average. These two averages are not strictly mathematically equivalent, but general agreement exists that the difference between them is

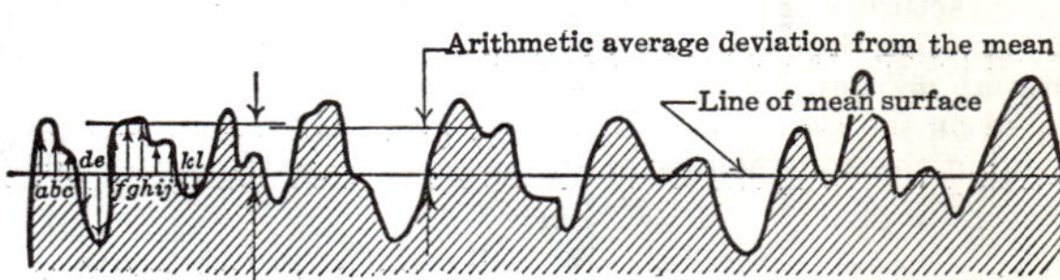

FIG. 4. Arithmetic and root-mean-square average deviations.

negligible for surface-roughness measurements. Because these averages make some allowance for the shape of the irregularities as well as their height, they have been selected as suitable indices for surface-roughness measurement.

It appears (1949) that the *arithmetic average deviation from the mean surface* should be used in preference to other values since commercial roughness-measuring instruments are calibrated for a similar value, and as future national standards may prefer it because it is simple to define and explain.

Even at best the approved standard values serve only as approximate indices to the performance of a given surface since it is easily possible that widely differing surfaces may have the same average roughness. However, if two surfaces made by the same process exhibit the same average roughness, their performance will probably be similar.

APPROVED TERMINOLOGY FOR OTHER VALUES. *Waviness* refers to those irregularities which are of greater spacing than roughness irregularities. On machined surfaces such irregularities may result from machine or work deflections or may be caused by warping as a result of the release of internal stresses, temperature effects, or other causes. Waviness heights are usually specified as maximum allowable peak-to-valley heights and with the approved standard values given in Table 2.

Table 2. Standard Waviness Height Values (Inches)

0.00002	0.00008	0.0003	0.001	0.005	0.015
0.00003	0.0001	0.0005	0.002	0.008	0.020
0.00005	0.0002	0.0008	0.003	0.010	0.025

Waviness is usually measured by using a sensitive dial indicator. Waviness widths may be specified directly in inches, but no provision for such specification is included in the approved surface-roughness symbol.

Flaws are those irregularities not characteristic of a surface that occur at one or more places at relatively infrequent intervals and are caused by lack of uniformity of the material or damage subsequent to processing. Scratches, cracks, pits, dents, and holes are typical flaws. Acceptance or rejection of a part having flaws is a matter of the inspector's judgment, as flaws are not considered in the general surface finish designation.

Lay refers to the direction of the predominant visible surface marks. Certain approved symbols are used in connection with the approved surface roughness symbol to indicate the lay direction with respect to the boundary line of the surface indicated by the symbol, i.e., = equals parallel; ⊥ equals perpendicular; × equals angular in both directions. M indicates multidirectional, C indicates approximately circular, and R indicates approximately radial relative to the center of the surface indicated by the symbol.

10. SPECIFICATION OF SURFACE QUALITY

APPLICATION OF STANDARD SURFACE-ROUGHNESS SYMBOL. The approved symbol for designating surface irregularities consists of a check mark and horizontal extension, as shown in Fig. 1. It is preferably made in the position shown. The point of the symbol may be on the line indicating the surface, on the witness line, or on an arrow pointing to the surface. In some cases it may be used in conjunction with a dimension and will then indicate surface roughness values of surfaces defined by the dimension. Values must be located exactly as shown, but it is not necessary that they all be given.

The ASA B46.1 standard requires that a general note be included in all specifications and on drawings to indicate which type of roughness height rating is intended. Many manufacturers prefer to use the NAS-30 (National Aircraft Standards) Surface Roughness Designation Standard as a drawing and specification reference, as it indicates automatically that an *average deviation from the mean* surface height rating is to be used. The NAS-30, based on the B46.1 standard, offers other distinct advantages such as provisions for notes limiting the roughnesses of surfaces not individually specified, as in Fig. 5, and for the

All machined surfaces except as noted.

FIG. 5.

optional use of a symbol having a flag, as shown, which is more distinctive and conspicuous, hence is less likely to be overlooked on the drawing.

ACTUAL ROUGHNESS MEASUREMENT. The closest commercial approach to actual roughness measurement (1949) is made by tracer point instruments such as the Profilometer and Brush Surface Analyzer. These precision instruments interpret the up and down motion of a finely pointed diamond as it is drawn slowly over the irregularities of the surface whose roughness is being measured. The diamond is flexibly suspended in a suitable tracer head, and its minute irregular motion moves a coil through a magnetic field (or deforms a piezoelectric crystal), thereby generating a variable voltage which is suitably amplified and used to indicate or record the roughness of the surface. The Profilometer, which indicates average deviation from the mean surface values directly on a meter, provides an excellent method for making rapid roughness comparisons of finely finished critical surfaces. Special Profilometers and numerous accessories to improve their versatility are available. The Brush Surface Analyzer makes a greatly magnified record on a moving paper tape of the irregular motion of the diamond tracer point. This record, which depicts an approximate contour of the irregularities being traced, provides a permament record and is not subject to the human error incident to reading a roughness meter having a rapidly oscillating needle. From such a tape peak-to-valley heights may be estimated and average roughness values determined by calculation. Surface performance may be more accurately evaluated from a knowledge of the shape of the surface irregularities, as indicated by the tape record, than by a meter reading alone. The Brush Surface Analyzer can also be equipped with a direct reading roughness meter.

All roughness-measuring instruments have inaccuracies and inherent limitations. Readings, therefore, must not always be taken at face value but should receive the benefit of intelligent interpretation based on a study of the surface configuration and performance and a knowledge of the characteristics of the instrument used.

ROUGHNESS COMPARISON. While measurement of surface roughness is a complicated procedure requiring precision instruments, the comparison of most machined surfaces with similarly finished standards of known roughness is quite simple. If the fingernail is held in scratching position and drawn lightly over a standard surface and then over the work, it is easy for even inexperienced personnel to determine which is the rougher unless their difference is so little as to be negligible. The surface quality of most machined parts can be controlled by the extensive distribution of accepted roughness standards which permit the direct comparison of prepared surfaces with the limiting values of roughness designated on the drawing by the designer. For very fine surfaces only instruments are sensitive enough for accurate comparison.

Roughness standards are commercially available in many forms for direct fingernail comparison. Some are original machined surfaces, some are plated replicas of machined surfaces, and some are molded of a hard black phenolic plastic. Since roughness comparison is essentially a tactual rather than visual procedure, the idea that a brighter surface is necessarily the smoother being false, the material of which the standard roughness specimen is made need not be the same as the material being processed, as long as the specimen is durable and corrosion resistant.

Finish comparison of finely finished surfaces can be made by using commercial devices that illuminate the surface in a manner similar to that used in the light field photomicrography of metals. These devices use a light-sensitive photoelectric cell and a meter to measure the percentage of light reflection—a very sensitive index to the surface roughness of fine surfaces. It is also possible to record and compare roughnesses by making transparent film replicas and projecting the image onto a screen. Many other methods of roughness comparison have been used that have not proved of commercial importance.

Table 3. Typical Roughnesses of Natural and Machine Finishes

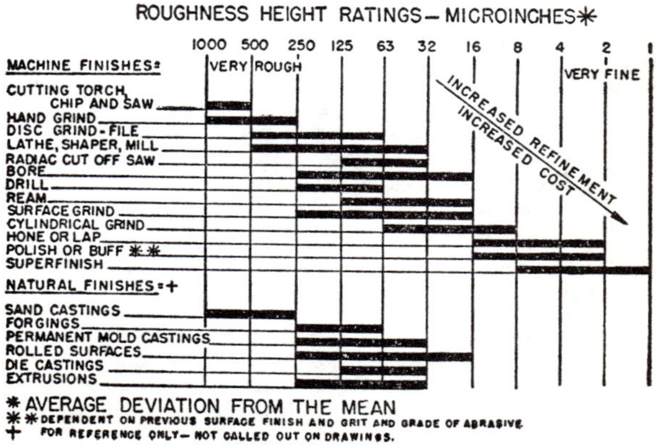

SELECTING THE PROPER SURFACE ROUGHNESS. Although handbooks now specify tolerances and dimensional clearances that are optimum for many specific applications, the state of the art (1949) is such that suitable surface roughness values based on the new standards cannot yet be specified. Designers can contribute to future progress by adopting and using the standard designations, determining the roughnesses of surfaces in their respective fields that have given satisfactory performance, and making this information available to others. A knowledge of the practical roughness range of normal machine work, together with the roughness height ratings given in Table 3, will provide a basis upon which designers can estimate surface-roughness values suitable for a given application. In general, typical surface-roughness applications are:

500√ A very rough, low-grade surface resulting from heavy cuts and coarse feeds. Suitable only for clearance surfaces that are not subject to vibration, fatigue, or stress concentration and do not make contact with other parts.

250√ This surface definitely shows tool marks from rapid feeds in producing a medium machine finish or may be produced by a very coarse surface grind, rough file, coarse disk grind, etc. It is used as a rough finish for soft alloys and for steel and hard alloys where moderate notch sensitivity exists, but a rough finish is otherwise tolerable. It is also suitable for parts where an average machine finish is acceptable and on deep bores that do not require special finishing.

125√ This is a fine machine finish resulting from high-grade machine work using high speeds and fine feeds taking light cuts with sharp cutters. It may be produced by all methods of direct machining under proper conditions, including coarse surface and cylindrical grinding, average disk grind, and ordinary hand filing. It is the coarsest finish suitable for rough bearing surfaces where loads are light and infrequent. It may also be used on moderately stressed machined parts which require moderately close fits.

63√ This is a smooth machine finish suitable for ordinary bearings and ordinary machine parts where fairly close dimensional tolerances must be held. It may be used for highly stressed parts that are not subject to severe stress reversals. It is just about as smooth a finish as can be economically produced by turning and milling without subsequent operations. On flat parts such a surface can be produced on a grinding machine.

32√ This surface corresponds to an extremely fine machine finish, a commercial carbide or diamond bore, a medium surface grind, a coarse cylindrical grind, reaming, and similar operations. In the case of turned parts, this finish may be produced by subsequent hand work with emery cloth. In using this finish and finer ones, careful consideration should be given to increased shop costs. May be used on parts subject to stress concentration and vibration. While difficult to produce on lathe, mill, or other direct machining operations, it is relatively easy to produce as centerless, cylindrical, or surface grind and is most suitable finish for hardened steel parts. May be used for bearings where motion is continuous but loads are light, provided lay is in direction of motion.

16√ Fine cylindrical grind, very smooth ream, fine surface grind, smooth emery buff, coarse hone, coarse lap, etc. Seldom used except where surface finish is of primary importance for proper functioning of part. Typical applications are for rapidly rotating shaft bearings, heavily loaded bearings, extreme tension members. It is also required for static sealing rings, bottom of sealing ring grooves, etc.

8√ Very, very fine cylindrical grind, microhone, hone, lap, buff, etc. For use only when coarser finishes are proved or known to be inadequate.

Note. 8 and finer finishes may have either a dull or bright appearance, depending on method used to produce them. Appearance of a surface shall not be considered in judging its quality. Either "feel" or roughness-measuring instruments should be used.

4√ Hone, lap, superfinish, very fine buff, or bright polish. This finish is not to be called for on drawings without special approval. It is required only in cases where packings and rings must slide across the direction of the finish grain, as in the interior honed surfaces of hydraulic struts and on buffed chrome-plated piston rods and similar operations.

GENERAL REFERENCES

Surface Roughness Designation—NAS-30, Aircraft Industries Association of America, Washington 5, D.C., approved Dec. 1948.
Surface Roughness, Waviness, and Lay, Part I, ASA B46.1, 1947, American Standards Association, New York, approved Feb. 1947.
James A. Broadston, Measurement and designation of surface quality, *Iron Age,* Oct. 1944.
James A. Broadston, Surface finish measurement instrumentation, *Instruments,* April 1947.
James A. Broadston, *Control of Surface Quality,* Surface Checking Gage Co., Hollywood, Calif., August 1949.
Ben C. Brosheer, How Smooth Is Smooth?, *American Machinist,* Sept. 9 and 23, 1948.
H. R. Nelson, Surface contour by taper sectioning, *Am. Machinist,* Aug. 6, 1941.
Mary R. Norton, Development of standards for Army Ordnance finishes, *Mech. Eng.,* Oct. 1942.
George Schlesinger, *Surface Finish,* ASME, New York, 1942. (A report of the Research Department of the Institution of Production Engineers of Great Britain.)
Gustav Schmalz, *Technische Oberflachenkunde* (*Evaluation and Measurement of Surface Quality*), Julius Springer, Berlin, 1936. (This basic reference work has been translated into English by J. Ehrlich for the General Motors Corporation Research Laboratories Section, Detroit, Mich., and was reviewed in *Mech. Eng.,* Nov. 1937.)
Arthur M. Swigert, Jr., *The Story of Superfinish,* Lynn Publishing Co., Detroit, Mich., 1940.

STATISTICAL QUALITY CONTROL

By Charles D. Ferris

OBJECTIVES. Statistical quality control techniques are used to detect, measure, and control variability in product quality. This variability may exist in the hardness of steel gears, in the diameter of a machined shaft, in the finish of a ball bearing, in the life of a switch, to mention but a few examples. Fluctuations such as these may be responsible for reworked and scrapped parts, high inspection costs, uneven production, and, ultimately, consumer dissatisfaction if the defective work is not caught before it leaves the factory. On the other hand, the variations in the part or assembly characteristics may be well within the specification limits set by the engineer and cause no trouble in the plant or in the hands of the customer. Statistical quality control has the task of determining which is the case as economically as possible. When goods of poor quality are being produced, action must immediately be taken to determine and eliminate the cause.

The basic techniques had their beginnings over twenty years ago in the Bell Telephone Laboratories. From here their use first spread to some of the other large concerns engaged in mass production. Later the methods were adopted more generally as interest grew in academic, industrial, and military circles. Real impetus was given to their application, however, by World War II because of the urgency and magnitude of quality problems and nationwide courses offered by the Office of Production Research and Development of the War Production Board. Since the war, statistical quality control has had ever-growing acceptance in firms large and small throughout the country. At present those using its techniques are represented by the American Society for Quality Control.

In the articles that follow, the specific techniques and basic principles are considered as they apply to the fundamental phases of product quality. References to important books and articles are provided for detailed treatment.

11. INCOMING QUALITY

Maintenance of a high level of quality in incoming material is of great importance in industry. Acceptance of poor parts or subassemblies leads to difficulties in processing in the shop and assembly on the line. This, in turn, is almost certain to bring about rework and spoilage costs, production delays, and expensive inspection. The answer to these problems seems simple: subject all incoming material to a rigid incoming inspection. But such a procedure can be decidedly expensive if there are many parts. Hence, there is the practical problem of deciding when, where, and how much inspection is absolutely necessary. Several tools of statistical quality control may be employed by the incoming inspection department to strike the right balance between economy and protection. All these discussed, however, fall under the general heading of sampling inspection.

FREQUENCY DISTRIBUTION. Problem 1. Machined gears are purchased from a gear company. A critical dimension of the gear is its thickness, which must be held between 0.8270 and 0.8350 in. A lot of 2000 gears has just been received from the vendor. An inch micrometer is available for measurement. How shall this dimension be inspected?

Since it is known that successive machined gears coming off the grinding machine may vary appreciably in thickness because of wear, hardness of material, etc., we certainly want to check more than one gear. On the other hand, it does not seem reasonable to check all the gears—at least not until a reliable preliminary estimate of lot quality is obtained. To obtain such an estimate, select a sample of 25 to 50 pieces *at random*, i.e., so that each piece in the lot has an equal chance of being selected, and measure the diameters. From these data a diagram of the type shown in Fig. 1 can be constructed. Such a graph is known as a *frequency distribution.* It shows at a glance where the diameters in the sample are located. Three have a diameter between 0.8320 and 0.8329 in., seven between 0.8330 and 0.8339 in., and so on.

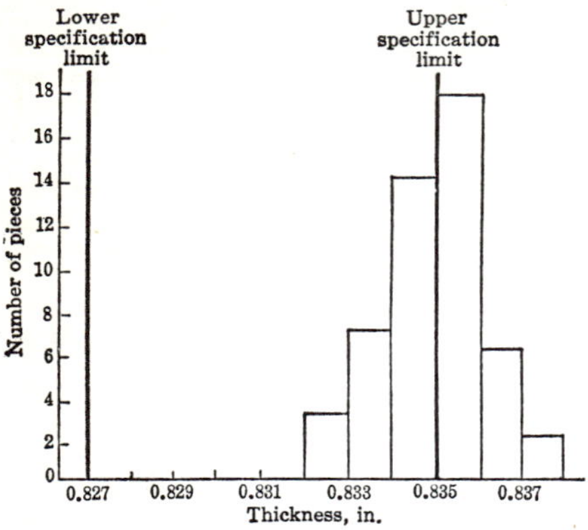

FIG. 1. Frequency distribution of thicknesses of 50 gears.

This chart, then, depicts the sample thickness distribution. But theory tells us it also provides some information about the position of diameters in the remainder of the lot; exactly how much information will be discussed later. Here the frequency distribution certainly indicates that a high percentage of the pieces in the lot have thicknesses outside the upper specification limit. In other cases there might be two distinct groupings indicating two sets of manufacturing conditions. Or the distribution might have too large a spread and extend beyond both specification limits. As a working rule, visual inspection of the frequency distribution constructed from a sample of 50 will usually indicate the true situation. If the lot represented by Fig. 1 is to be used, it must obviously be sorted 100%.

For detailed information on the construction of frequency distributions and calculation of means and standard deviations, where required, consult Ref. 1.

There are some situations, mostly borderline, when it is desirable to make a more precise statement about the position of the lot distribution. In these instances select the largest value in the sample and the smallest value. The accompanying table then lists the sample size required to estimate that at least a certain percentage, b, of the lot falls between these two values while running a risk that a times in 100 actually less than b% of the lot lies between them. Such a statement does not depend on the shape of the frequency distribution of the entire lot and is based on the assumption that the lot size is very large compared to the sample size. If the latter is not the case, all risks will be less than those tabulated.

Sample Size Required

	$b = 99$	$b = 95$	$b = 90$
$a = 1$	661	130	64
$a = 5$	473	93	46
$a = 10$	388	76	37

For example, the table tells us that we can predict from 50 samples that at least 90% of the lot falls between 0.8320 and 0.8380 in. However, roughly 4% of the time we make such an estimate there will actually be less than 90% between these two end points. Observe that a sample of 661 is required to predict that at least 99% of the lot lies between the sample extremes, with a 1% chance of being wrong. For the necessary formulas, consult Refs. 2 and 3. Also of use in such problems is Ref. 4, Chapter 2.

But for the most part, visual inspection of the frequency distribution indicates the nature of incoming quality and suggests corrective action (e.g., in Fig. 1 the entire distribution should be shifted to the left). Once the reliability of incoming lots is established, the size of the sample may be substantially reduced. A sample of only 4 or 5 pieces is needed to reveal whether the lot is acceptable (see Article 12, Variable Control Charts, for details). Advantage is taken of the confidence in the supplier's performance.

ACCEPTANCE SAMPLING INSPECTION. Problem 2. Lots of a molded plastic part are being received from a vendor. A standard has been drawn up on the finish of this part. The quality of finish may vary considerably within the lot. What type of inspection should be used to protect the customer against acceptance lots of poor quality?

This problem clearly falls into a different category from Problem 1. The quality characteristic cannot be measured but only observed; the piece is either good or bad. Such inspection is known as *attribute inspection,* whereas the use of the frequency distribution is an example of *variable inspection.*

Since any per cent defective may occur in the lot, it will not do merely to inspect one or two pieces. Nor would 100% inspection seem justifiable until some preliminary investigation had proved it necessary. In other words, sampling inspection is again needed.

Before the sampling plan can be set up, the protection desired by the plan must be specified. This protection can take two forms. The first is *lot protection.* We must determine that lot per cent defective which is definitely nonacceptable (*lot tolerance per cent defective*). The second is *average outgoing quality* protection. We must set down a per cent defective which shall rarely be exceeded in the long run in outgoing lots, i.e., *average outgoing quality limit* (AOQL). Such protection assumes that all rejected lots are sorted 100%.

Finally, we must prescribe a per cent defective which is acceptable in incoming lots, i.e., *acceptable quality level* (AQL). It may be very low, say one-fourth of 1%; however, if sampling inspection is to be used, there must always be some lot per cent defective which will be accepted practically all the time by the sampling plan.

Assume that in this problem we definitely want to reject any lot in which 4% of the pieces have defective finish (the lot tolerance per cent defective). However, because of the price of the part, commercial practice, and the fact that the appearance of the molded part is not too serious a quality characteristic, we are willing to accept lots with no more than 1% (AQL) defective pieces. An attribute sampling plan can be constructed that treats lots of 1% and 4%, as specified above, practically all the time. Lots with intermediate per cent defective will be accepted varying percentages of the time, when such lots are submitted. This characteristic of a sampling plan is illustrated in Fig. 2 by a curve known as an *operating characteristic.* Such a curve traces the probability of accepting lots over the entire range of per cent defective.

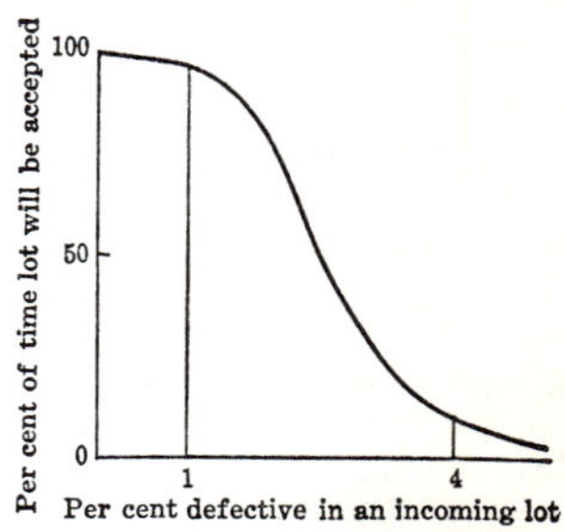

FIG. 2. Operating characteristic.

Sampling plans which provide this type of protection and their operating characteristics are contained in Ref. 5. In this reference also is the single sampling plan called for by these tables which gives protection similar to that portrayed in Fig. 2, for lot sizes greater than ten times the indicated sample size: Select a random sample of 300 pieces from the lot. If 7 or less defective pieces are found, accept the lot. Otherwise, reject.

The AOQL protection corresponding to each sampling plan and operating characteristic are listed in the tables of Ref. 5. Double and multiple sampling plans are also furnished. These sampling schemes give the same protection for less inspection but with the disadvantage of more complicated inspection procedures. For a discussion of the underlying principles of sampling inspection, see Ref. 6, Chapters 13–15.

Thus, it is possible through sampling methods to discriminate between good and bad lots, even though the type of defect does not permit precise measurement. Such a system of inspection could, of course, be applied to our first problem if we regard all gears with thicknesses outside specification limits as defective and all those inside as nondefective. Instead of using a micrometer, a snap gage of the go, not-go type could furnish necessary data.

In planning such inspection due account must always be taken of past history, confidence in the vendor, and nature of the incoming quality. If no trouble is ever experienced with a quality characteristic, inspection can certainly be slashed and, if necessary, greater risks can be taken. If the part being inspected is a molded piece, it probably will suffice to check a few pieces to make certain that the mold cavity is according to print.

ESTIMATION OF LOT PER CENT DEFECTIVE. Problem 3. In a sample of 300 pieces selected at random from a large incoming lot, 30 (10%) are found defective. What valid prediction can be made of the true percentage of defective pieces in the lot?

There are many times when we are not content merely to reject a lot as in the previous section, but wish to make an estimate of exactly how bad the lot really is. When the sample size is greater than 100, the following simple rule applies: If in a sample of n units selected at random from a large lot the fraction of defective units is designated by p, the fraction defective in the lot will lie between $p \pm 3\sqrt{p(1-p)/n}$, conservatively, 99 times in 100. In this problem $p = 0.10$, $n = 300$; hence we can predict that the fraction defective will fall between $0.10 \pm 3\sqrt{[(0.10)(0.90)]/300}$ or 0.152 and 0.048, with small risk that the true fraction defective will fall outside these values. Consult Ref. 4, Chapter 10, for an exact solution to this problem.

12. PROCESS QUALITY

Once the incoming material is accepted, two things may happen to it. First, it may be immediately used for assembly into the end product. Second, and more likely, it must undergo a certain amount of fabrication before it is ready for assembly. This may take the form of stamping, punching, plating, machining, etc. Let us look first at some applications of statistical quality control techniques in the machine shop.

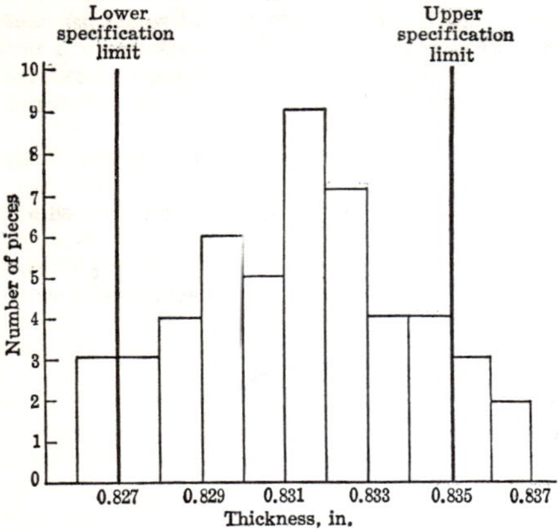

FIG. 3. Machine accuracy study.

MACHINE ACCURACY STUDY. Problem 4. The plant which had been receiving gears in an earlier example decided to do all the machining on the raw castings. What is the first step that should be taken after a machine has been set up for a production run?

Once the machine is set up, a machine accuracy study should be made. Select, say, 50 pieces at random from a short trial run, measure the thicknesses, and plot a frequency distribution as described in Article 11. By comparing the spread of the sample values with the width of the tolerance band, an idea can be obtained of how well the process will meet the specification limits during a short interval. Figure 3 is a frequency distribution of 50 measurements taken in such a study. It is clear that no matter how the process is shifted, there still will be a substantial portion of the material outside specification limits.

VARIABLE CONTROL CHARTS. Problem 5. How can the variability of gear thickness in the foregoing problem be reduced?

The secret to the solution of this problem is to determine how this process fluctuates in time. The first step is to select small samples of four or five from production at regular frequent intervals. Calculate the mean and range (largest value minus the smallest value) of each sample and plot in the order that the parts are produced. One of two situations can appear. *Situation I:* At any given moment in the grinding of the gears the thicknesses will vary about some constant thickness value as widely as the process illustrated in Fig. 3. This is an example of a state of *statistical control*—the process distribution remains constant at every point in time. *Situation II:* At any given moment in the grinding of the gears the spread is much less than that in Fig. 3, but the overall spread is increased by assignable causes which pull all the diameters being machined at that time toward either specification limit. Here we have *lack of statistical control*.

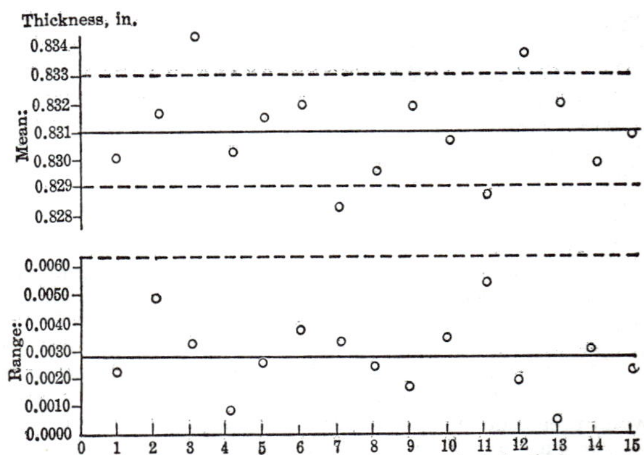

FIG. 4. Variable control chart.

A method of distinguishing between Situation I and Situation II on a practical basis was developed by Walter A. Shewhart of the Bell Telephone Laboratories more than twenty years ago and appeared in his classic treatise on statistical quality control (Ref. 7). Through simple analysis of the variability within each sample, horizontal lines are calculated and placed around the average of the small sample means. Figure 4 shows what was found when the procedure was applied to the thickness data by using samples of four.

The fact that sample means fall above the upper dotted line and below the lower dotted line indicates that one or more assignable causes of variation are present. Typical causes might be excessive hardness of the steel being used or some instability in the grinder. In this particular case the chart reveals that it is possible for the process to produce parts that meet the specification limits, provided the assignable causes can be ferreted out and eliminated. If all the sample means had fallen within the dotted limits, it would have indicated Situation I, and a fundamental methods change would be required to reduce variability. The fact that all sample ranges fall beneath the upper dotted limit signifies that the spread of thicknesses at any given moment is approximately constant.

The mean and range plots in Fig. 4 are known as *control charts;* the limits, as *control limits.* Their calculation, interpretation, and use are clearly described in Refs. 1, 6 (Chapters 4–8), and 8.

Problem 6. What should be done in the future to control the variability of gear thickness in Problem 4?

To solve, continue to select small samples from successive segments of production. Such segments should be selected on the basis that each is a potential period of change in basic manufacturing conditions. Changes in operators, steel, shifts, and tools affect the selection. These segments are known as *rational subgroups.* Plot the results on the chart, and take action whenever assignable cause of variation is indicated. This is real process control. Detailed instructions for process control procedures using variable control charts are to be found in Ref. 9 and also in Refs. 1, 6, and 8.

The techniques used in solving this problem must be flexible. If the specification limits are wide compared to the variability in thickness, and the thickness is expected to shift because of wear, *warning limits* may be set in from each specification limit (see Ref. 6, Chapter 8). The moment a mean falls outside these limits the cause must be investigated and the machine reset if excessive wear is the reason. Often a mere plot of sample points on the chart shows when the dimension is coming too close to the specification limit. Resetting will effectively prevent defective material from being produced and accepted. Regardless of the statistical technique, it is important that small samples be taken regularly from rational subgroups of production, measurements made and noted, and action be taken according to a predetermined criterion.

PER CENT DEFECTIVE CONTROL CHARTS. Problem 7. At a certain point in an assembly line, a 100% visual check is made of the finish of a critical bearing surface. Rejects have been averaging 15% of production. Production is approximately 100 pieces per hour. What can be done with these figures that will help reduce the percentage of defects?

First, calculate the fraction of rejects for each hourly, daily, or weekly interval. The unit of time depends on the least interval of time between changing factors affecting finish.

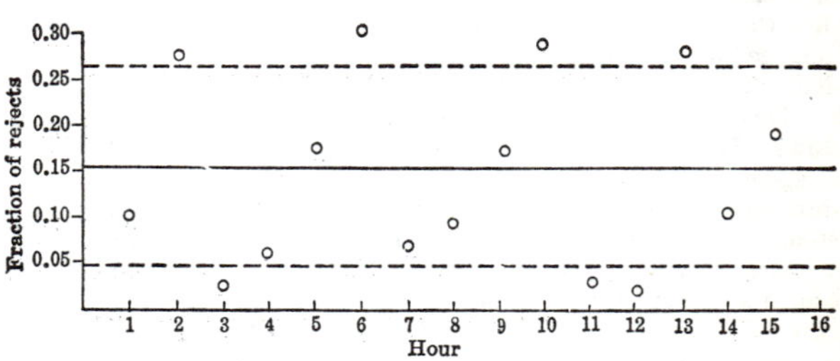

Fig. 5. Per cent defective control chart.

Since it was thought that the condition of the cutting tool from hour to hour was the responsible factor, successive hourly rejections expressed as a fraction of hourly production were computed. The points in Fig. 5 are a record of this hourly performance.

Common sense tells us that if the process were producing 15% rejections steadily, we should not necessarily expect to find 15 rejects in every subgroup of 100 inspected. Just as the per cent defective in the sample varied from that in the lot in the estimate of incoming quality in Problem 3, so we may expect these hourly percentages to fluctuate around 15%. Replacing p by 0.15, the fraction of rejects, and n by 100 in the formula cited earlier, we find that the sample percentage fluctuates between 0.04 and 0.26. These values are the *control limits* (dotted lines) on Fig. 5. The fact that hourly percentages fall above the upper control limit and below the lower control limit indicates that at times the finishing process is producing bearings that have more than 15% rejects; points below the lower limit represent a process that is producing less. Such a fact is significant because, if we can eliminate the causes of high rejections and benefit from the practice followed while producing low rejections, the average rejection level can be dropped. Actual investigation disclosed that the boring machine making the bearings was not properly adjusted during the second, tenth, and sixteenth hours, and that the cutting tool required sharpening during the sixth and thirteenth hours. After these corrections had been made it was possible to run the process "in control" at a level of 6%, further reduction being impossible until basic changes could be made in the machine.

This is an example of one of the most useful applications of the per cent defective control chart in the process. Points are plotted as the data become available, and assignable causes of variation are sought out whenever a point goes outside control limits on the chart. Of course such charts can be used on other than 100% data. They have a very important application in incoming inspection in keeping track of the quality of the vendor's product. Here the per cent defective in the sample selected from incoming lots is treated in exactly the same fashion as the 100% data above. Finally, the quality of any fabrication or assembly operation can be sampled, results recorded on an attribute basis, charts kept, and action taken. For example, the strength of a crimped connection or the presence of burrs in a reamed hole could be checked by such methods on a sampling basis. By tying down faulty performance to men, machines, materials, or methods we can put our finger on the source of the trouble and initiate corrective action.

For details on the techniques and interpretation of the attribute control chart, see Refs. 8 and 9; also Ref. 6, Chapter 10.

ACCEPTANCE SAMPLING INSPECTION. Problem 8. A washing-machine line receives 1000 gears daily from the plant machine shop. Of particular concern is the eccentricity of the gear teeth produced by six hobbing machines. What method of acceptance should be used to make certain the gears are meeting the eccentricity specification limit satisfactorily?

The vendor-customer relationship discussed in Article 11 has its parallel here inside the plant. Department A in effect buys the product of Department B. The production line accepts parts from the machine shop for purposes of assembly. In these circumstances, positive protection against acceptance of poor material is desirable. If attribute inspection is to be used, the principles discussed under Acceptance Sampling Inspection, p. 24-34, apply, with one exception. Both the cost of sorting defective material and the cost of sampling inspection are borne by the same company. Hence if the machine shop is producing gears that are 1% defective (the process average) and the line wishes lot tolerance protection at 4%, a sampling plan should be used which minimizes the total amount of inspection at the 1% level. Single and double sampling plans constructed on this principle are published in Ref. 10. The plans are in terms of AOQL and lot tolerance protection. In this instance the single sample required by the tables for a lot of 1000 pieces is 165. Accept the lot if there are 3 or less defectives in the sample; otherwise, reject.

These are not the only sampling plans that can be used in the process. When sorting costs are not a large factor or little history has been built up, do not hesitate to use sampling plans of the Navy type discussed under Problem 2.

Problem 9. Gears are coming off a boring machine and proceeding directly to the hobbing machine. How can we be sure that no more than 2% of the outgoing flow will be outside the specification limits?

Here it is impossible to accumulate a lot, and so none of the lot acceptance plans can be applied. Instead, a type of inspection known as continuous sampling must be adopted (see Ref. 11). Such a plan might provide that 10% of the pieces be checked, that sorting be instituted once a single piece is found outside specification limits, and that sorting be discontinued once 55 pieces are found free of defects. Under the usual production conditions no more than 2% of outgoing production should be defective.

To summarize, the tools of statistical quality control that can be used in the process fall into two categories. One group *samples the process* and acts to indicate the presence of assignable causes of variation. The other group *samples the product* and acts to give positive assurance that excessively defective material will not be fabricated or assembled at subsequent stages. All of them operate to reduce expensive spoilage and rework and to cut down inspection costs by sampling. Inspection has become truly preventive.

13. OUTGOING QUALITY

All the controls discussed in the previous articles operate to produce a reliable end product at minimum cost. By keeping our finger on the pulse of quality throughout the process, by trying to build quality into the product rather than cull the bad from the good at the end of the line, a healthy and economic environment for quality is created. However, there remains the necessity of checking the final unit to determine whether it conforms to all the specifications and whether the in-line controls have been properly exercised. The problem is: To what extent can sampling inspection be used at final inspection?

On some very critical performance characteristics sampling can be used only as a check on the efficiency of 100% inspection. Acceptance sampling inspection plans play a definite role in this area since inspection fatigue is a well-known phenomenon and automatic testing devices are not infallible.

In the case of less critical characteristics, acceptance sampling plans may well replace 100% inspection. Naturally the sampling plan cannot guarantee a 100% perfect product in the field, but it can assure an almost perfect product and distinguish between good and bad lots.

In addition to the usual inspection, an exhaustive quality audit and life test should be made periodically on the finished product. Appropriate action should be taken, when necessary, to change quality standards or strengthen controls.

Finally, management is interested in field results. Is poor material reaching the field? What is the expected life of a unit in the customer's hand? Statistics can help management evaluate such data properly. Action should be taken in the factory to prevent the recurrence of customer complaints.

14. QUALITY OF DESIGN

All the applications discussed so far have dealt with the treatment of manufacturing and production quality problems. Statistical techniques, however, can also assist the mechanical engineer in designing the product. Statistical principles play an important role in laying out experiments to determine the relationship among variables. Analysis of variance point out those variables that should be controlled (see Ref. 12). Proper analysis of pilot lot runs, manufactured under production conditions, permit a valid prediction of how well the product will meet specification limits once production is under way.

Problem 10 is a design problem of practical importance to the mechanical engineer.

Problem 10. A shaft will pass through a bushing in assembly. The specification limits on the shaft are 0.6008 and 0.6012 in. It is desired that the clearance between the shaft and the bushing be no less than 0.0002 in. and no more than 0.0005 in. for a satisfactory fit. Within what limits may the inside diameter of the bushing vary if this assembly specification is to be met?

The answer to this question must depend on the shape and position of the shaft and bushing diameters. It will be assumed that the diameters can be produced under a state of statistical control in the center of their respective specification bands. The diameter distribution should be bell shaped. Assembly should be completely random. The proportion of these distributions falling outside their limits should be small if the clearance specification is to be met most of the time.

To determine the bushing specification limits, B_1 and B_2, we use the formula

$$B_1 - B_2 = \sqrt{(0.6012 - 0.6008)^2 + (0.0005 - 0.0002)^2}$$

Then

$$B_1 - B_2 = \sqrt{(0.0004)^2 + (0.0003)^2} = 0.0005 \text{ in.}$$

But if the average shaft diameter is 0.6010 in. and the average clearance is 0.00035 in., the average bushing diameter is 0.6010 + 0.00035 or 0.60135 in. Hence the specification limits for the bushing are 0.60135 ± 0.00025 or 0.6016 in. and 0.6011 in.

The foregoing formula describes the manner in which variability adds up statistically, provided the assumptions hold. Although it is possible to select a bushing with a smaller diameter than the shaft, it is not at all probable. The theory can be extended to any number of pieces. See Ref. 6, Chapter 12, for a detailed discussion of the theory, with examples.

REFERENCES

1. *ASTM Manual on Presentation of Data*, American Society for Testing Materials, Philadelphia, 1945.
2. S. S. Wilks, *Mathematical Statistics*, p. 93, Princeton University Press, 1943.
3. H. Scheffe and J. W. Tukey, A formula for sample size for population tolerance limits, *Annals Math. Statistics*, Vol. 15, 1944, p. 217.
4. Statistical Research Group (C. Eisenhart *et al.*, editors), *Techniques of Statistical Analysis*, McGraw-Hill Book Co., New York, 1947.

5. *General Specifications for Inspection of Material*, Appendix X, U. S. Navy Department. (Available at U. S. Government Printing Office, Washington, D. C.)
6. E. L. Grant, *Statistical Quality Control*, McGraw-Hill Book Co., New York, 1946.
7. W. A. Shewhart, *Economic Control of Quality of Manufactured Product*, D. Van Nostrand Co., New York, 1931.
8. *Guide for Quality Control and Control Chart Method of Analyzing Data*, American Standards Association, New York, 1941.
9. *Control Chart Method of Controlling Quality during Production*, American Standards Association, New York, 1942.
10. H. F. Dodge and H. G. Romig, *Sampling Inspection Tables*, John Wiley and Sons, New York, 1944.
11. H. F. Dodge, A sampling plan for continuous production, *Annals of Math. Statistics*, Vol. 14, 1943.
12. K. A. Brownlee, *Industrial Experimentation*, Chemical Publishing Co., Brooklyn, 1947.

SECTION 25

WORKING OF NONMETALLIC MATERIALS

BY

THOMAS D. PERRY, *Engineer in Woodwork, Moorestown, N. J.*

J. E. FALOON, *Plastics Division, Chemical Department, General Electric Company, Pittsfield, Mass.*

WOODWORKING

BY THOMAS D. PERRY

PLASTICS MOLDING

BY J. E. FALOON

WOODWORKING

By Thomas D. Perry

1. LOGS AND LOGGING

RIVER LOGGING formerly was practiced in all forest areas where cold weather facilitated operations by freezing the ground, making roads passable, and where snow aided transportation of logs on sleds to the river side. Logs were decked high at convenient waterways, awaiting spring freshets that would float the logs to the sawmills. Winter logging still prevails in sparsely wooded areas, often with second-growth timber, but is little used in the far West and deep South.

TRACTOR LOGGING was developed because the immense size of trees on the Pacific Coast necessitated mechanical handling of logs, especially as logs up to 40- and 60-ft lengths were available and desirable for large timbers. The usual practice is to "snake" or "skid" logs by cables. A track cable is attached at one end to a high tree stump, and at the other to a steam or diesel-driven "skidder." The skidder cable is reeved through a sheave that travels on the track cable, and is attached to one end of the log. The winding drum on the skidder then lifts and drags the logs to a convenient central point, where they are loaded by derrick or small skidders on railroad cars or tractor trucks. Skidders can operate through a radius of 1000 ft.

Logging railroads usually are temporary, with little grading, and are pulled up and relocated as stumpage is exhausted. Because of excessive grades and curvature, geared locomotives are used. Corduroy roads often are required for truck and trailer operation. Trucks and trailers are of several types, trailers with regular or track-laying tractors, and trucks of the "straddle" type, where logs are hoisted and suspended between the wheels, being common.

SWAMP LOGGING often is imperative in the warmer districts where roadways do not freeze. Corduroy roads usually are necessary, with smaller mechanical equipment than on the Pacific Coast, as long logs are few in number, and most logs can be loaded by man power. In the larger swamps, like the Dismal Swamp on the Virginia-North Carolina line, canals have been dug to simplify log transportation, and many mills in such areas are supplied with logs by barge and tug. Logs also are handled by skidders to logging railroads built on crib-work or fills.

SELECTIVE LOGGING is in contrast to the earlier procedure, where timber was felled without regard to constructive forest perpetuation. This destructive logging left in its wake abandoned timber areas that were a serious fire hazard and did not develop new growth of commercial value; it wiped out industrial communities dependent on the sawmill; and it nearly obliterated all real estate values in the vicinity. Selective logging is now obligatory in removing timber from government-owned and government-controlled areas. Only merchantable trees can be removed, and the remainder are protected for future growth.

SUSTAINED YIELD, which is in accord with the modern and the economic policy, has been adopted by many large timber owners who are operating in privately held timber tracts. Not only are the trees selectively logged as they mature, but promising young growths are preserved and protected. In many cases this younger growth is supplemented by new plantings. As these protected forests develop, the smaller trees, suitable for pulp and other purposes, are removed, and thus the better trees are encouraged to grow into mature sawmill and veneer-mill material. If the brush is kept clear and fires prevented, a 20-year stand will yield pulp wood, and a 40-year growth in softwoods will begin to be available for saw logs. Hardwoods require 60 to 100 years to become desirable as lumber and veneer logs. The modern technique is to maintain and supplement young growth, to protect against fire and insect depredation, and progressively to harvest trees suitable for various purposes. It is the modern recognition of timber as a crop, in contrast to the destructive logging, which was a counterpart of the exhausted and abandoned mine.

LOG SCALES resulted from the attempts of the early timberman to state the amount of lumber that reasonably could be expected from the average log. He endeavored to eliminate the slabs, edgings and saw kerf (see p. 25-34), and to allow for the oversize of board thickness provided for shrinkage during drying. These scales were applied to the small end of the log, and the figures usually stamped thereon. The taper of the log was a factor in some scales, so that a 24-ft log does not necessarily scale twice that of a 12-ft log.

From this unscientific and rule-of-thumb beginning, numerous log scales developed, many of them of only local importance. A few which have received definite and wide-spread recognition are given in Table 1 (for a log 16 ft long). Other sizes, under these

Table 1. Comparison of Log Scales, 16-ft Lengths

(Cylindrical content = $\frac{1}{4}\pi D^2 \times (16/12) = \frac{1}{3}\pi D^2$ board feet)

Diam., Small End, in.	Doyle Scale	Scribner Scale	Doyle-Scribner Scale	Scribner Decimal Scale	Spaulding Scale	Cylindrical Content	Diam., Small End, in.	Doyle Scale	Scribner Scale	Doyle-Scribner Scale	Scribner Decimal Scale	Spaulding Scale	Cylindrical Content
	Board Feet							Board Feet					
6	4	18	4	20	....	37.699	28	576	582	576	580	569	821.003
7	9	28	9	30	....	51.313	29	625	609	609	610	612	880.693
8	16	32	16	30	....	67.021	30	676	657	657	660	656	942.478
9	25	40	25	40	....	84.823	31	729	710	710	710	701	1006.36
10	36	50	36	60	50	104.720	32	784	736	736	740	748	1072.33
11	49	65	49	70	63	126.711	33	841	784	784	780	796	1140.40
12	64	79	64	80	77	150.796	34	900	800	800	800	845	1210.56
13	81	97	81	100	94	176.976	35	961	876	876	880	896	1282.82
14	100	114	100	110	114	205.250	36	1024	923	923	920	950	1357.17
15	121	142	121	140	137	235.620	37	1089	1029	1029	1030	1006	1433.60
16	144	159	144	160	161	268.083	38	1156	1068	1068	1070	1064	1512.15
17	169	185	169	180	188	302.640	39	1225	1120	1120	1120	1124	1592.80
18	196	213	196	210	216	339.292	40	1296	1204	1204	1200	1185	1675.52
19	225	240	225	240	245	378.038	41	1369	1272	1272	1270	1248	1760.40
20	256	280	256	280	276	418.879	42	1444	1343	1343	1340	1312	1847.26
21	289	304	289	300	308	461.814	43	1521	1396	1396	1400	1377	1936.26
22	324	334	324	330	341	506.844	44	1600	1480	1480	1480	1448	2027.38
23	359	377	359	380	376	553.968	45	......	1518	1518	1520	1512	2120.53
24	400	404	400	400	412	603.186	46	1764	1587	1587	1590	1581	2215.87
25	441	459	441	460	449	654.499	47	......	1656	1656	1660	1652	2312.20
26	484	500	484	500	488	707.906	48	1936	1728	1728	1730	1724	2412.74
27	530	549	530	550	528	763.407	......	......	......	......	......	....	....

scales, and other scales can be found in several published woodsmen's handbooks. Figure 1 is a graphical comparison of the several log scales applied both to sawed lumber and to veneer; a log 18 in. in diameter is used as an example.

Table 2. Comparison of Yield of Veneer and Sawed Lumber by Various Log Scales

(See Fig. 1)

	Veneer		Sawed Lumber	
	Fig. 1*a*	Fig. 1*b*	Fig. 1*c*	Fig. 1*d*
Cylindrical Measure:				
Total content, board feet	339	339	339	339
Core or waste, board feet	51	26	131	114
Core or waste, %	15	8	39	34
Net product, board feet	288	313	208	225
Net product, %	85	92	61	66
Net Product by Log Scales:				
Doyle, board feet	196	196	196	196
Doyle, %	147	160	106	115
Scribner, board feet	213	213	213	213
Scribner, %	135	147	98	105
Scribner Decimal, board feet	210	210	210	210
Scribner Decimal, %	137	149	99	107
Spaulding, board feet	216	216	216	216
Spaulding, %	133	145	96	104

Log scales are used both for the purchase of logs and to estimate timber stands for valuation purposes. An experienced timber cruiser should be employed for the second function. The Spaulding and Doyle scales and their combinations are extensively used in the Midwest and the South; the Spaulding and Scribner Decimal scales are more usual on the Pacific Coast. In general, log scales favor the purchaser, rather than the seller of logs. Logs of a good grade should produce the scale footage of lumber and substantial over-runs of veneer.

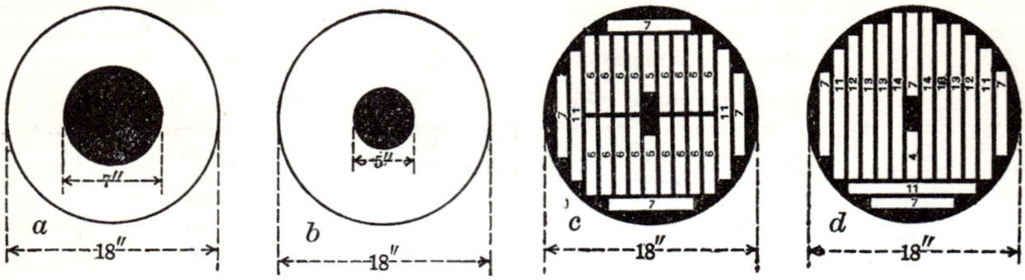

FIG. 1. Comparison of yield from 18-in. log. (See Table 2.)

2. LUMBER

SAW LOGS, suitable for conversion into lumber, usually are those between the selected peelers required for veneer and the smaller logs used for poles, posts, mine props, etc. Small blocks and down-fall are dimensioned into shingle blocks or prepared for gang sawing into lath. One-inch boards usually are cut about 1 $^1/_4$ in. thick to allow for shrinkage-thickness variations, and saw marks. Corresponding allowances are made on other thicknesses. One-inch boards should *dress clean* to $^{13}/_{16}$ in., 2-in. stock to 1 $^5/_8$ in., and 4-in. stock to 3 $^1/_2$ in.

Lumber usually is marketed as *air-dried,* containing between 15 and 25% moisture, and as *kiln-dried,* approximating 10% moisture. While considerable lumber is machined or dressed green, allowance must be made for shrinkage, which may run considerably over 10%. The shrinkage and distortion vary with the position in the log from which a piece is cut. (See Fig. 2.) Tangential shrinkage is about twice the radial shrinkage. For data on the various species of wood, see p. 5-73.

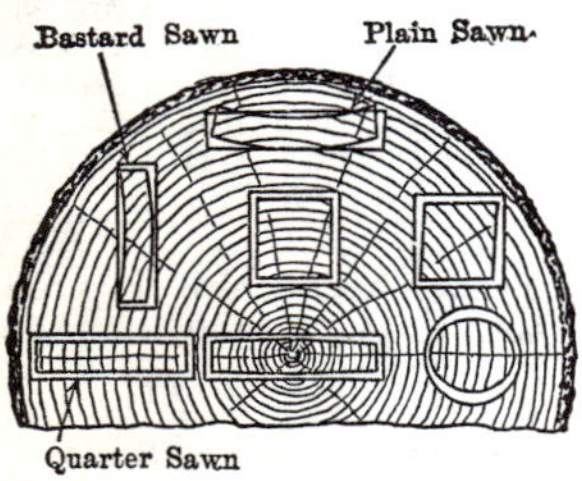

FIG. 2. Effect of direction of annual rings on shrinkage.

TYPES OF SAWING. Logs are sawed in various ways as shown in Fig. 2.

Plain-sawed or flat grain lumber customarily is cut right through the log and nominally varies from a tangent on the growth ring to a 45-degree angle.

Quarter-sawed, or vertical or edge grain, theoretically is cut on a radial line across the growth rings. In practice it may extend to 45 degrees from the radial line. In general, quarter-sawed is more carefully selected, and its angular range should be considerably less than for plain-sawed.

Bastard-sawed is a term usually applied to lumber cut between 30 and 60 degrees to the radial or tangent line. Certain species like white oak, when quarter-sawed, show distinct flake markings from the wood rays (radial), which are somewhat less conspicuous in hickory, elm, sycamore, etc.

CLASSES OF WOOD. The following terms are commonly used to describe different types of wood.

Sapwood is a name given to the several layers next to the bark, which usually are lighter in color, less dense, and more easily dried than the *heartwood* that forms the center of the tree. The life processes of growth are more active near the bark, and decrease toward the heart. Heartwood is sometimes more valuable for color, as in walnut and red gum. Its density makes it more durable for use as flooring, and, in general, its greater solid content makes it more resistant to decay than sapwood. In some trees the line of color demarcation between heart and sap is not easy to detect.

Softwood, in general, means lumber from coniferous trees, usually evergreens.

Hardwood refers to lumber from broad-leaved trees. The terms have no reference to the hardness of the wood; e.g., yellow pine, a softwood, is much harder than basswood, which technically is a hardwood.

Spring wood is that portion of the growth ring that is formed the early part of the seasonal growth. It usually is less dense than the *summer wood,* formed during the latter part of the yearly growth. Summer wood usually is mechanically stronger than spring wood. A large number of growth rings per inch is desirable from a strength standpoint.

Factory, Shop, and Yard Lumber. In general, *factory* and *shop* lumber are intended to be cut up for use in further manufacture. *Yard* lumber is of moderate sizes and intended for general building purposes. *Structural timbers* are relatively large in size, and are

Table 3. Board Measure Content of Dimension Stock

Thickness, in.	Width, in.	Length in inches										Length in feet						
		4	6	8	10	12	14	16	18	24	30	4	6	8	10	12	14	16
		Feet, Board Measure										Feet, Board Measure						
4/4	1	0.03	0.04	0.06	0.07	0.08	0.10	0.11	0.13	0.17	0.21	0.33	0.50	0.67	0.83	1.00	1.17	1.33
	2	.06	.08	.11	.14	.17	.19	.22	.25	.33	.42	.67	1.00	1.33	1.67	2.00	2.33	2.67
	3	.08	.13	.17	.21	.25	.29	.33	.38	.50	.63	1.00	1.50	2.00	2.50	3.00	3.50	4.00
	4	.11	.17	.22	.28	.33	.39	.44	.50	.67	.83	1.33	2.00	2.67	3.33	4.00	4.67	5.33
	5	.14	.21	.28	.35	.42	.49	.56	.63	.83	1.04	1.67	2.50	3.33	4.17	5.00	5.83	6.67
	6	.17	.25	.33	.42	.50	.58	.67	.75	1.00	1.25	2.00	3.00	4.00	5.00	6.00	7.00	8.00
	7	.19	.29	.39	.49	.58	.68	.78	.88	1.17	1.46	2.33	3.50	4.67	5.83	7.00	8.17	9.33
	8	.22	.33	.44	.56	.67	.78	.89	1.00	1.33	1.67	2.67	4.00	5.33	6.67	8.00	9.33	10.67
	9	.25	.38	.50	.63	.75	.88	1.00	1.13	1.50	1.88	3.00	4.50	6.00	7.50	9.00	10.50	12.00
	10	.28	.42	.56	.69	.83	.97	1.11	1.25	1.67	2.08	3.33	5.00	6.67	8.33	10.00	11.67	13.33
	11	.31	.46	.61	.76	.92	1.07	1.22	1.38	1.83	2.29	3.67	5.50	7.33	9.17	11.00	12.83	14.67
	12	.33	.50	.67	.83	1.00	1.17	1.33	1.50	2.00	2.50	4.00	6.00	8.00	10.00	12.00	14.00	16.00
5/4	1	.04	.05	.07	.09	0.10	0.12	0.14	0.16	0.21	0.26	0.42	0.63	0.83	1.04	1.25	1.46	1.67
	2	.07	.10	.14	.17	.21	.24	.28	.31	.42	.52	.83	1.25	1.67	2.08	2.50	2.92	3.33
	3	.10	.16	.21	.26	.31	.37	.42	.47	.63	.78	1.25	1.88	2.50	3.13	3.75	4.38	5.00
	4	.14	.21	.28	.35	.42	.49	.56	.63	.83	1.04	1.67	2.50	3.33	4.17	5.00	5.83	6.67
	5	.17	.26	.35	.43	.52	.61	.69	.78	1.04	1.30	2.08	3.13	4.17	5.21	6.25	7.29	8.33
	6	.21	.31	.42	.52	.63	.73	.83	.94	1.25	1.56	2.50	3.75	5.00	6.25	7.50	8.75	10.00
	7	.24	.37	.49	.61	.73	.85	.97	1.09	1.46	1.82	2.92	4.37	5.83	7.29	8.75	10.21	11.67
	8	.28	.42	.56	.69	.83	.97	1.11	1.25	1.67	2.08	3.33	5.00	6.67	8.33	10.00	11.67	13.33
	9	.31	.47	.63	.78	.94	1.09	1.25	1.41	1.88	2.34	3.75	5.63	7.50	9.38	11.25	13.13	15.00
	10	.35	.52	.69	.87	1.04	1.22	1.39	1.56	2.08	2.60	4.17	6.25	8.33	10.42	12.50	14.58	16.67
	11	.38	.57	.76	.96	1.15	1.34	1.53	1.72	2.29	2.86	4.58	6 88	9.17	11.46	13.75	16.04	18.33
	12	.42	.63	.83	1.04	1.25	1.46	1.67	1.88	2.50	3.13	5.00	7.50	10.00	12.50	15.00	17.50	20.00
6/4	1	.04	.06	.08	0.10	0.13	0.15	0.17	0.19	0.25	0.31	0.50	0.75	1.00	1.25	1.50	1.75	2.00
	2	.08	.13	.17	.21	.25	.29	.33	.38	.50	.63	1.00	1.50	2.00	2.50	3.00	3.50	4.00
	3	.13	.19	.25	.31	.38	.44	.50	.56	.75	.94	1.50	2.25	3.00	3.75	4.50	5.25	6.00
	4	.17	.25	.33	.42	.50	.58	.57	.75	1.00	1.25	2.00	3.00	4.00	5.00	6.00	7.00	8.00
	5	.21	.31	.42	.52	.63	.73	.83	.94	1.25	1.56	2.50	3.75	5.00	6.25	7.50	8.75	10.00
	6	.25	.38	.50	.63	.75	.88	1.00	1.13	1.50	1.88	3.00	4.50	6.00	7.50	9.00	10.50	12.00
	7	.29	.44	.58	.73	.88	1.02	1.17	1.31	1.75	2.19	3.50	5.25	7.00	8.75	10.50	12.25	14.00
	8	.33	.50	.67	.83	1.00	1.17	1.33	1.50	2.00	2.50	4.00	6.00	8.00	10.00	12.00	14.00	16.00
	9	.38	.56	.75	.94	1.13	1.31	1.50	1.69	2.25	2.81	4.50	6.75	9.00	11.25	13.50	15.75	18.00
	10	.42	.63	.83	1.04	1.25	1.46	1.67	1.88	2.50	3.13	5.00	7.50	10.00	12.50	15.00	17.50	20.00
	11	.46	.69	.92	1.15	1.38	1.60	1.83	2.06	2.75	3.44	5.50	8.25	11.00	13.75	16.50	19.25	22.00
	12	.50	.75	1.00	1.25	1.50	1.75	2.00	2.25	3.00	3.75	6.00	9.00	12.00	15.00	18.00	21.00	24.00
7/4	1	.05	.07	0.10	0.12	0.15	0.17	0.19	0.22	0.29	0.37	0.58	0.88	1.17	1.46	1.75	2.04	2.33
	2	.10	.15	.19	.24	.29	.34	.39	.44	.58	.73	1.17	1.75	2.33	2.92	3.50	4.08	4.67
	3	.15	.22	.29	.37	.44	.51	.58	.66	.88	1.09	1.75	2.63	3.50	4.38	5.25	6.13	7.00
	4	.19	.29	.39	.49	.58	.68	.78	.88	1.17	1.46	2.33	3.50	4.67	5.83	7.00	8.17	9.33
	5	.24	.37	.49	.61	.73	.85	.97	1.09	1.46	1.82	2.92	4.38	5.83	7.29	8.75	10.21	11.67
	6	.29	.44	.58	.73	.88	1.02	1.17	1.31	1.75	2.19	3.50	5.25	7.00	8.75	10.50	12.25	14.00
	7	.34	.51	.68	.85	1.02	1.19	1.36	1.53	2.04	2.55	4.08	6.13	8.17	10.21	12.25	14.29	16.33
	8	.39	.58	.78	.97	1.17	1.36	1.56	1.75	2.33	2.92	4.67	7.00	9.33	11.67	14.00	16.33	18.67
	9	.44	.66	.88	1.09	1.31	1.53	1.75	1.97	2.63	3.28	5.25	7.88	10.50	13.13	15.75	18.38	21.00
	10	.49	.73	.97	1.22	1.46	1.70	1.94	2.19	2.92	3.65	5.83	8.75	11.67	14.58	17.50	20.42	23.33
	11	.54	.80	1.07	1.34	1.60	1.87	2.14	2.41	3.21	4.01	6.42	9.63	12.83	16.04	19.25	22.46	25.67
	12	.58	.88	1.17	1.46	1.75	2.04	2.33	2.63	3.50	4.38	7.00	10.50	13.00	17.50	21.00	24.50	28.00
8/4	1	.06	.08	0.11	0.14	0.17	0.19	0.22	0.25	0.33	0.42	0.67	1.00	1.33	1.67	2.00	2.33	2.67
	2	.11	.17	.22	.28	.33	.39	.44	.50	.67	.83	1.33	2.00	2.67	3.33	4.00	4.67	5.33
	3	.17	.25	.33	.42	.50	.58	.67	.75	1.00	1.25	2.00	3.00	4.00	5.00	6.00	7.00	8.00
	4	.22	.33	.44	.56	.67	.78	.89	1.00	1.33	1.67	2.67	4.00	5.33	6.67	8.00	9.33	10.67
	5	.28	.42	.56	.69	.83	.97	1.11	1.25	1.67	2.08	3.33	5.00	6.67	8.33	10.00	11.67	13.33
	6	.33	.50	.67	.83	1.00	1.17	1.33	1.50	2.00	2.50	4.00	6.00	8.00	10.00	12.00	14.00	16.00
	7	.39	.58	.78	.97	1.17	1.36	1.56	1.75	2.33	2.92	4.67	7.00	9.33	11.67	14.00	16.33	18.67
	8	.44	.67	.89	1.11	1.33	1.56	1.78	2.00	2.67	3.33	5.33	8.00	10.67	13.33	16.00	18.67	21.33
	9	.50	.75	1.00	1.25	1.50	1.75	2.00	2.25	3.00	3.75	6.00	9.00	12.00	15.00	18.00	21.00	24.00
	10	.56	.83	1.11	1.39	1.67	1.94	2.22	2.50	3.33	4.12	6.67	10.00	13.33	16.67	20.00	23.33	26.67
	11	.61	.92	1.22	1.53	1.83	2.14	2.44	2.75	3.67	4.58	7.33	11.00	14.67	18.33	22.00	25.67	29.33
	12	.67	1.00	1.33	1.67	2.00	2.33	2.67	3.00	4.00	5.00	8.00	12.00	16.00	20.00	24.00	28.00	32.00

selected and graded with reference to strength factors. Table 3 gives the board feet content of various sizes of dimension stock. Trade usage designates 1-in. thickness as $^4/_4$ lumber, 1 $^1/_4$-in. as $^5/_4$ lumber, etc.

AMERICAN LUMBER STANDARDS embody provisions for softwood lumber dealing with recognized classifications, nomenclature, basic grades, seasoning standards, sizes, uniform workings, description, measurement, tally, shipping provisions, grade markings, tally cards, and inspection of lumber. These standards have become well integrated in the grading rules and regulations of the leading lumber manufacturers' associations. They are widely recognized and have become well established. These standards have, on a national scale, superseded practices in various limited areas, in behalf of certain species, and in most industrial applications. They are now sponsored and maintained by the National Bureau of Standards in Washington and are kept abreast of progressive developments.

Lumber Grading Rules are promulgated by the various lumber manufacturing associations. They usually are revised from year to year, so that it is important to use the latest edition which may be obtained from the listing below. Most of the hardwood lumbermen have combined their rules in the publications of the National Hardwood Lumber Association. They also make available an inspection service, under the same rules, for all controversial points of interpretation. The softwood lumbermen are grouped in several associations, usually according to species and locality of trees. The leading lumber association is the National Lumber Manufacturers' Association at Washington, D. C., supported by and affiliated with most of the national, regional, and species associations, of which the following are the more important:

* *American Walnut Manufacturers Association*, Chicago; American walnut.
Appalachian Hardwood Manufacturers, Cincinnati; ash, basswood, beech, birch, butternut, chestnut, cherry, elm, hickory, maple, yellow poplar, red oak, white oak, walnut.
California Redwood Association, redwood.
* *Mahogany Association, Inc.*, Chicago; mahogany.
* *Maple Flooring Manufacturers Association*, Chicago; maple, birch and beech flooring.
* *National Hardwood Lumber Association*, Chicago; all hardwood species of lumber and veneer. Has grade inspection service available in all principal centers.
Northeastern Lumber Manufacturers Association, New York; northern white pine, Norway pine, eastern spruce, balsam fir, northern hardwoods.
Northern Hemlock and Hardwood Manufacturers Association, Oshkosh; hemlock, birch, maple, basswood, elm, ash, beech, tamarack, northern white pine.
Northern Pine Manufacturers Association, Minneapolis; northern white pine, Norway pine, eastern spruce, tamarack.
Red Cedar Shingle Bureau, Seattle; western red cedar.
Southern Cypress Manufacturers Association, Jacksonville; southern cypress.
Southern Hardwood Producers, Inc., Memphis; ash, basswood, beech, yellow cypress, cottonwood, elm, black gum, red gum, hackberry, hickory, red maple, magnolia, white oak, red oak, poplar, pecan, persimmon, sycamore, tupelo, willow.
Southern Pine Association, New Orleans; longleaf and shortleaf pine.
West Coast Lumberman's Association, Seattle; Douglas fir, west coast hemlock, Sitka spruce, western red cedar, Port Orford cedar.
Western Pine Association, Portland; ponderosa pine, Idaho white pine, sugar pine, larch, Douglas fir, white fir, Englemann spruce, western red cedar, incense cedar.

3. VENEER OR THIN LUMBER

VENEERS are thin sheets of wood, which are, excepting a minor amount of sawed veneer, knife-cut by various processes. (See below.) There is no saw kerf waste in knife cutting, rotary veneer being cut continuously from the log (see Fig. 3), and sheets of sliced veneer being cut from flitches, i.e., log sections, in widths as they come, without edging (see Fig. 5). Hence the conversion of logs into veneer utilizes much more of the log than does sawing into lumber. (See Fig. 1 and Table 2.) The assembly of veneers into plywood distributes wood strength more effectively than does solid lumber. Hence thinner veneers (in plywood form) provide equivalent strength factors as compared with thicker solid lumber. The four principal types of veneer, distinguished by different cutting processes are rotary veneer, sliced veneer, half-round veneer, and sawed veneer.

Rotary veneer, cut on lathes, accounts for approximately 90% of all veneer. As very little face veneer is made by this process, with such exceptions as bird's-eye and quilted

* **National in scope, usually including Canada.**

maple, rotary-cut veneer usually is called commercial veneer. It is used for cross bands, cores, and faces where strength is more important than appearance. Rotary veneer may be as thin as 0.01 in. It is difficult to cut thicker than $^1/_4$ in., although $^5/_{16}$-in. and $^3/_8$-veneer sometimes is cut from softer woods like basswood and poplar. Few veneer lathes are over 8 ft long (some are 10 ft) because of the difficulty of cutting accurately across the unsupported distance between lathe centers. Figure 3 shows rotary cutting, the veneer being unrolled over a knife, much as paper is unrolled. The convex side is the *tight* side and often is marked with crayon during lathe cutting.

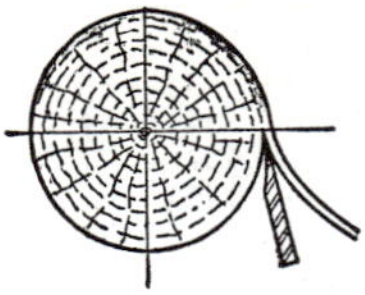

FIG. 3. Rotary veneer cut on a lathe.

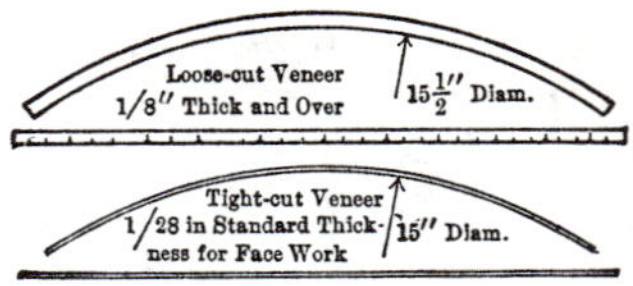

FIG. 4. Tight and loose cutting of veneer.

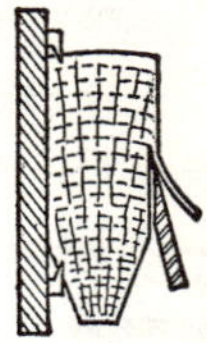

FIG. 5. Sliced veneer cut quartered from a segment flitch.

Figure 4 shows tight- and loose-cut veneer. The cutting checks on the loose side may result from inadequate pressure. They cannot be eliminated entirely and frequently cause the veneer to flatten better. In some stock, however, they sometimes are objectionable. The process of flattening veneer from its original curved shape need not result in checks or splits on the concave side in veneer under $^1/_8$ in. thick. Proper pressure produces tight-cut veneer and avoids cutting checks. Both sides of the veneer then can be used, as in matched faces. See Fig. 20 for book matching.

Sliced Veneer. Figure 5 shows the method customarily used in cutting valuable face veneers. Much of the most attractive face figure, such as mahogany and walnut, is developed from slicing on the "quarter" as shown. Adjacent sheets can be "book-matched" by opening a flitch like the pages of a book, to produce balanced symmetrical figure. Sliced veneer does not have the cutting checks characteristic of rotary veneer. It may be as thin as $^1/_{100}$ in. but seldom can be thicker than $^1/_8$ in. It is regularly cut up to 16 ft long.

Half-round veneer cutting is an intermediate process between rotary and sliced. Two distinct types are cut, *regular and back cut.* (See Fig. 6.) Cutting is done in a lathe in which the log or flitch is swung eccentrically between centers on a *stay log,* the radius of knife cut being much greater than in ordinary rotary cutting.

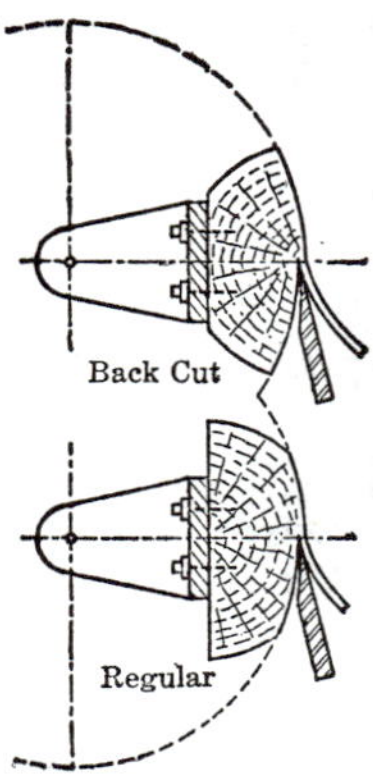

FIG. 6. Half-round veneer.

Skillful cutting of these three types of veneer depends principally on proper adjustment and manipulation of the relations between the knife and the pressure bar. (See Fig. 7.)

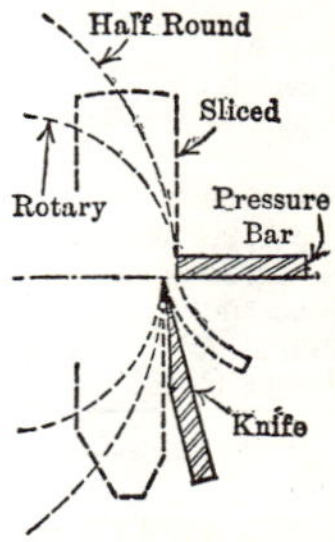

FIG. 7. Positions of pressure bar and knife in veneer cutting.

Sawed Veneer. A few face veneers are sawed on a large diameter segment saw, particularly to reveal the flake of woods like quartered white oak. The minimum thickness for sawed veneer is $^1/_{20}$ in., when the saw kerf waste is equal to the veneer; the maximum may be $^3/_8$ in. or more.

Weights of Veneer. Weights of single-ply veneer, representing the more common species and the thicknesses usually employed in veneer and plywood work, are given in Table 4. Since wood, within a species, varies considerably as between heartwood and sapwood, between river bottom and mountain growth, and between butt logs and upper logs, reasonable tolerances must be observed.

Standard thicknesses for veneer and cutting tolerances have been established (Army Navy–Aeronautical Specifications AN-P-69a, February 1946) for government aircraft products, but have not been adopted, generally, for commercial work. Commercial standards (National Bureau of Standards) have been established for both softwood and hardwood plywood, including the suitable types and grades of veneer to be employed. For a description of grade rules and face figure nomenclature, consult the publications of the veneer associations, the more prominent of which are American Walnut Manufacturers Association, Chicago; Mahogany Association, Chicago; National Hardwood Lumber

Association, Chicago; Pacific Veneer Package Association, Tacoma, Wash.; Rotary Cut Lumber Association, New Orleans; Veneer Association, Chicago; Veneer Package Association, Washington.

Table 4. Weights of Veneer

In ounces per square foot of single-ply, oven-dry.

The weight of wood varies within a species, so that although the table gives average weights, substantial variations may sometimes be found.

Species	S.G. (1)	M.C. (2)	Thickness in Inches												
			1/100	1/80	1/60	1/40	1/32	1/28	1/24	1/20	1/16	1/10	1/8	3/16	1/4
Ash, black	.50	10.4	.42	.52	.69	1.04	1.30	1.49	1.74	2.08	2.60	4.16	5.20	7.81	10.41
Ash, commercial white	.58	8.9	.48	.60	.80	1.21	1.51	1.72	2.01	2.41	3.02	4.82	6.04	9.05	12.06
Basswood	.38	8.4	.32	.40	.53	0.79	0.99	1.13	1.32	1.58	1.98	3.16	3.96	5.94	7.92
Beech	.63	11.2	.52	.66	.87	1.31	1.64	1.87	2.19	2.62	3.28	5.24	6.56	9.84	13.12
Birch, yellow	.63	9.6	.52	.66	.87	1.31	1.64	1.87	2.19	2.62	3.28	5.24	6.56	9.84	13.12
Butternut	.39	7.6	.32	.41	.54	0.81	1.02	1.16	1.35	1.62	2.03	3.25	4.06	6.09	8.12
Cedar, Spanish	.37	7.3	.31	.38	.51	0.77	0.96	1.10	1.28	1.54	1.92	3.08	3.85	5.78	7.70
Cherry, black	.51	9.2	.42	.53	.71	1.06	1.33	1.52	1.77	2.12	2.65	4.25	5.31	7.96	10.62
Chestnut	.44	8.6	.37	.46	.61	0.92	1.14	1.31	1.53	1.83	2.29	3.66	4.58	6.87	9.16
Cottonwood	.43	4.7	.36	.45	.60	0.90	1.12	1.28	1.49	1.79	2.24	3.58	4.47	6.71	8.96
Cypress, bald	.44	9.0	.37	.46	.61	0.92	1.14	1.31	1.53	1.83	2.29	3.66	4.58	6.87	9.16
Douglas fir (coast)	.51	6.2	.42	.53	.71	1.06	1.33	1.52	1.77	2.12	2.65	4.25	5.31	7.96	10.62
Douglas fir (mountain)	.44	9.4	.37	.46	.61	0.92	1.14	1.31	1.53	1.83	2.29	3.66	4.58	6.87	9.16
Elm, white	.51	8.8	.42	.53	.71	1.06	1.33	1.52	1.77	2.12	2.65	4.25	5.31	7.96	10.62
Gum, black	.52	7.2	.43	.54	.72	1.08	1.35	1.55	1.80	2.17	2.71	4.35	5.42	8.12	10.82
Gum, cotton	.52	6.1	.43	.54	.72	1.08	1.35	1.55	1.80	2.17	2.71	4.35	5.42	8.12	10.82
Gum, red	.49	11.3	.41	.51	.68	1.02	1.28	1.46	1.70	2.04	2.55	4.08	5.10	7.66	10.20
Hackberry	.54	9.2	.45	.56	.75	1.12	1.40	1.61	1.87	2.25	2.81	4.49	5.62	8.43	11.24
Hemlock, western	.42	8.6	.35	.44	.58	0.87	1.09	1.25	1.46	1.75	2.18	3.50	4.37	6.56	8.74
Magnolia (evergreen)	.51	8.8	.42	.53	.71	1.06	1.33	1.52	1.77	2.12	2.65	4.25	5.31	7.96	10.62
Mahogany (C. Amer.)	.49	7.9	.41	.51	.68	1.02	1.28	1.46	1.70	2.04	2.55	4.08	5.10	7.66	10.20
Mahogany (African)	.46	8.0	.38	.48	.64	0.96	1.19	1.37	1.59	1.91	2.39	3.83	4.78	7.17	9.57
Maple (silver)	.48	8.2	.40	.50	.67	1.00	1.25	1.43	1.67	2.00	2.50	4.00	5.00	7.50	7.00
Maple (sugar)	.62	10.5	.52	.65	.86	1.29	1.61	1.85	2.15	2.58	3.23	5.16	6.46	9.69	12.91
Oak, commercial red	.64	10.7	.53	.67	.89	1.33	1.66	1.90	2.22	2.66	3.33	5.32	6.66	9.99	13.32
Oak, commercial white	.68	11.0	.57	.71	.94	1.41	1.77	2.02	2.36	2.83	3.54	5.66	7.08	10.61	14.16
Pine, long-leaf	.66	9.2	.55	.69	.92	1.37	1.72	1.96	2.29	2.75	3.44	5.50	6.88	10.32	13.75
Pine, sugar	.37	11.4	.31	.38	.51	0.77	0.96	1.10	1.28	1.54	1.92	3.08	3.85	5.78	7.70
Pine, short-leaf	.54	11.0	.45	.56	.75	1.12	1.40	1.61	1.87	2.25	2.81	4.49	5.62	8.43	11.24
Pine, western yellow	.41	10.8	.34	.43	.57	0.85	1.07	1.22	1.42	1.71	2.13	3.41	4.27	6.40	8.54
Pine, white	.39	9.9	.32	.41	.54	0.81	1.02	1.16	1.35	1.62	2.03	3.25	4.06	6.09	8.12
Poplar, yellow	.41	6.1	.34	.43	.57	0.85	1.07	1.22	1.42	1.71	2.13	3.41	4.27	6.40	8.54
Spruce, Sitka	.38	8.9	.32	.40	.53	0.79	0.99	1.13	1.32	1.58	1.98	3.16	3.96	5.94	7.92
Sycamore	.50	9.2	.42	.52	.69	1.04	1.30	1.49	1.74	2.08	2.60	4.16	5.20	7.81	10.41
Tanguile (3)	.54	11.8	.45	.56	.75	1.12	1.40	1.61	1.87	2.25	2.81	4.49	5.62	8.43	11.24
Walnut, black	.57	4.8	.47	.59	.79	1.19	1.48	1.70	1.98	2.37	2.97	4.75	5.94	8.91	11.87

(1) Specific gravity, oven-dry based on volume air-dry.
(2) Air-dry moisture content in percentage.
(3) So-called Philippine mahogany.

4. SAWMILLS

Sawmills are either permanent or portable. The former are used to saw the timber from large tracts, where the life of the operation will be 10 years or more. The latter are used in small tracts. They are so constructed that they can be moved easily from place to place, and are driven by steam, gasoline, or diesel engine. Both types of mill use either band saws or circular saws. The former are higher in first cost, but the waste in saw kerf is much less.

ARRANGEMENT OF SAWMILLS. Sawmills may be either single or double, i.e., the equipment of a double mill is largely in duplicate, and arranged symmetrically on either side of the center line of the mill. Modern practice tends to the use of band saws;

a mill with one such saw is known as a *single-band mill;* one with two saws is a *double-band mill.*

A typical double-band mill can be designed to cut 100,000 board feet of lumber per day, with a steam gang saw for sawing balks of timber into boards. In many mills a horizontal or vertical band resaw replaces the gang saw. The machinery floor of the mill is elevated 15 to 20 ft above ground level. Line shafts or other drives are located below the floor.

Logs are drawn from the long pond or other storage and elevated to the mill floor by a *log slip,* i.e., an inclined chain conveyor, which may be single or in duplicate. At the head of the log slip, the logs are cut into convenient lengths by a deck saw. This is a horizontal or vertical swing saw, 6 ft or more in diameter, belt or motor driven, moved to and fed through the log by steam or air cylinders. Steam-actuated *kickers,* i.e., steam cylinders whose piston rods terminate in large forked ends, kick the sawed logs from the log slip to the *log deck,* whence they are transferred to the saw carriages. Steam *niggers* transfer the logs from the log deck to the saw carriages, and also turn over the partially sawed logs to present new surfaces to the saw.

The carriages carry the logs back and forth past the band saws. An operator riding on the carriage, by means of appropriate *set works,* advances the log toward the saw after each cut to obtain the desired thickness of lumber.

The lumber as it comes from the saw is moved by the live roll conveyors and chain transfers to any one of a series of operations. Lumber sawed to correct widths moves on the first roller conveyor to the transfer chain to the trimming saws, which cut it to proper length. Lumber of irregular width is moved by chain transfers to the double edgers which cut it to width, whence it passes to the trimming saws. Large balks of timber are moved by the chain transfers to pass through the steam gang saw which saws the balk into planks; the planks then move to the trimming saws.

If a band resaw is used instead of a gang saw, the balks travel on the first live roll bed to the far end of the mill. A chain transfer then moves them to the middle live roll bed, which returns them in the reverse direction to the band resaw, where they are sawed into boards. The chain transfer moves these boards again to the first live roll bed, which delivers them to the trimming saw.

Refuse from the various operations is disposed of in several ways. Sawdust from each operation drops into a conveyor which moves it either to a fuel storage or direct to the boiler plant. Trimmings, edgings, and other pieces that are large enough are placed on conveyors to transport them to the lath and shingle mills for conversion into marketable product. Other waste is diverted to the slasher which saws it into pieces short enough to be used as fuel.

Finished lumber from the trimming saws is placed on the sorting chain. This is a two- or three-strand chain conveyor 400 to 800 ft long, on which the lumber is graded and sorted into convenient stacks for transfer to the lumber yard or dry kilns.

SAWMILL MACHINERY. Band Mills. A band mill comprises a heavy cast-iron frame and column carrying an upper and lower band wheel on which runs a band saw. The shaft and bearings of the upper wheel are movable, vertically, for the purpose of putting proper tension on the saw. The mill is driven by a belt pulley on the shaft of the lower wheel. Band mills are commonly made with band wheels 6 to 10 ft in diameter, and the diameter of the wheel designates the size of the mill. The saws range in dimensions from 10 in. wide, No. 16 gage, on 6-ft mills, up to 18 in., No. 12 gage, on 10-ft mills.

Saw carriages are massive timber frames carried on 8 to 12 wheels, riding on T rails. The headblocks, which carry movable knees, are mounted on the carriage. The knees each have two sets of dogs with teeth pointing upward and downward, respectively. These dogs grip the round log or *balk* of partially sawed timber, and hold it firmly against the knee while the carriage moves past the saw. After each saw cut, the log or balk is advanced the desired amount for the next cut, by means of *set works* which move the knees forward on the carriage. Set works usually are designed to advance the knees by multiples of $^1/_{32}$ in. The operator rides on the carriage and manipulates the set works in accordance with directions from the sawyer.

Two general methods are in use for moving the carriage past the saw: (1) *Steam* or "*shot-gun*" feed, consisting of a steam cylinder, in line with the carriage, and of a stroke equal to the travel of the carriage. (2) *Twin-engine* feed, in which a pair of opposed engines operate an endless cable from a drum to the carriage.

Gang saws are reciprocating saws, mounted in gangs, to saw large balks into a number of planks in one operation. Twenty-five to 54 saws are mounted parallel to each other, the requisite distance apart to saw planks of the desired thickness, and either belt or engine driven. The saws are carried in a steel sash, which is reciprocated vertically by a pitman driven by a crankshaft below the sash. Work is fed to the saws by fluted feed rolls, driven by a friction feed, whose speed can be varied by the operator. Pressure rolls actuated by steam cylinders hold the work firmly in place during sawing.

Band resaws are of both horizontal and vertical types. The vertical type is, in general, similar to the band mill, with adjustable feed rolls on either side to provide for resawing various thicknesses of stock. The horizontal band resaw has a horizontal blade, the upper run of the saw operating across one or more adjustable feed tables, which control the thickness. Pressure rolls, independently adjustable on each table, permit the simultaneous sawing of different thicknesses.

Edgers. Double edgers consist of a series of circular saws that can be moved in an axial direction on their arbor. The position of the saws is adjusted by the operator to the desired width of stock. Double edgers are used to remove rough or wany edges from slabs to form a single plank, or to saw a wide slab into narrower planks. Saws range from 20 to 24 in. in diameter, and usually are direct motor driven, although belt drive is also used. Edgers are provided with pressure and feed rolls both before and behind the saws.

Slashers. A slasher consists of a series of circular saws spaced at equal intervals on an arbor above the feed chains conveying the waste material from the live rolls to the slasher saws. Saws are set for cutting suitable lengths for lath, shingles, or waste, and a conveyor transfers the sawed slashing to its proper destination.

Trimming Saws. Trimming saws comprise a series of swing saws suspended over transfer chains, somewhat similar to those used at the slasher. The saws are placed so that any two of them will saw boards to commercial lengths, as may be determined by the operator for each successive board or plank.

Niggers and log loaders are mechanical devices, located below the log deck, for turning or moving the logs and balks toward or against the knees of the saw carriage.

Manufacturers. Sawmill machinery is built by Allis-Chalmers Co., Milwaukee, Wis.; American Saw Mill Machinery Co., Hackettstown, N. J.; F. L. Estes Saw and Machinery Co., Nashville, Tenn.; Filer and Stowell Co., Milwaukee, Wis.; Gainesville Iron Works, Gainesville, Ga.; Standard Saw Mill Machinery Co., Erie, Pa.; Taylor Iron Works and Supply Co., Macon, Ga.; Wheland Co., Chattanooga, Tenn.; J. A. Vance Co., Winston-Salem, N. C.

SAWMILL DRIVES. Sawmills are, in general, driven either from one or more line shafts extending the length of the mill, underneath the machinery floor, or by electric motors driving machines individually or in groups. In line-shaft drives, the main shaft is belted to a steam engine or electric motor. Sometimes a rope drive is used to transmit power from the engine. Where the driven machine is at an angle to the line shaft, mule pulley drives (see p. 15-52) or large bevel gears with wood teeth or bevel friction gears (see p. 15-83) are used to change the direction of the power supply. Final drive to the individual machines usually is by belt. Line-shaft speeds are from 150 to 250 rpm. Line shafts usually are carried in rigid post bearings. In the more modern mills, all bearings are force lubricated from a central point by an oil pump, reservoir, and oil filter.

Steam engines used for line-shaft drive are of various types. Steam economy seldom is given consideration in sawmill practice, as wood waste is used for fuel, and usually there is more available than can be burned for steam making. However, it should be remembered that wasteful engines require a larger boiler plant, with consequently greater first cost and overhead and operating charges. Many mill engines exhaust direct to atmosphere and use live steam in the dry kilns, but this practice also requires greater investment in boilers than where exhaust steam is used in the kilns. A few mills which have ample water available operate condensing, and use live steam in the kilns.

Power Requirements. The power required to drive a sawmill with line shafts varies widely with the type of equipment, the character of maintenance, and other factors. A mill of the type described usually has installed engine capacity of about 1000 hp. Indicator cards from engines driving several mills of this type showed a maximum demand of 850 to 900 hp.

Where the mill is to be motor driven, manufacturers of equipment will recommend the size of the motors to be installed. For line-shaft drive, the power must be estimated from previous performances of similar mills. The following figures were used in estimating the requirements of a double-band mill described above. They are given as a general guide only, and should be checked against local conditions. The figures are indicated horsepower except as otherwise stated. Deck saw, 30 hp; 8-ft band saw, 225 hp; 48-in. gang saw, 225 hp; 72-in. double edger, 150 hp; slasher, 50 hp; trimming saw, 40 hp; chain transfer to gang saw, 5 hp; chain transfer to trimming saw, 15 hp; nigger, 40 boiler hp; shot-gun feed for carriage, 14 in., 120 boiler hp, 12 in., 80 boiler hp. The load factor generally is taken as 80%.

5. VENEER LATHES AND SLICERS

The accuracy requirements for veneer cutting demand more careful log preparation than for lumber sawing. Logs for rotary cutting are cut to length on a *drag saw*, blade, circular, or band. The short lengths, or *bolts*, are sorted by length and species, and placed in vats for cooking or steaming. To produce sliced, half-round, and sawed veneer, logs are sawed into flitches, partly to obtain face figure and partly to facilitate clamping on the machine bed. Every veneer mill producing these latter types requires a small band (band-saw) mill both for flitching and to salvage lumber from unsuitable and defective logs that will not make veneer. Flitching technique varies widely according to log shape and veneer character. Figure 8 shows a suitable method of flitching oak logs, 24 to 30 in. in diameter, for quartering. In crotches, stumps, and burls, only experience and judgment can solve the widely varying individual problems of flitching. Flitches, except for the sawed veneer, go to the vats.

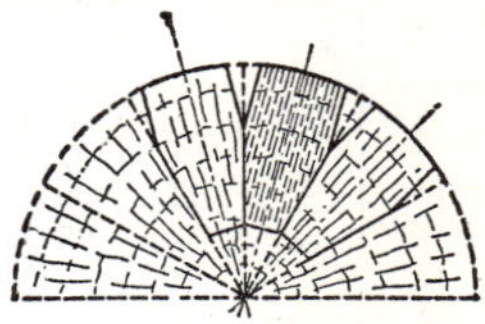

Fig. 8. Flitching for quarter-sawed veneer.

COOKING OR STEAMING. A few species of veneer woods, as basswood, poplar, and similar woods, are best cut "cold," without cooking. All logs are improved for veneer cutting by soaking in a log pond. The best authorities advocate cooking logs in hot water. Temperature and duration of cooking vary widely with the species, size, texture, and freshness from the woods. A minimum is 4 to 6 hr at 200 F up for coniferous woods, extending to 15 to 20 hr at 150 to 175 F for hardwoods. Flitches of fragile figure are best soaked for longer periods at 100 to 125 F. Too much cooking produces "fuzzy" veneer. Logs are barked before mounting and should be not more than 30 min from vat to machine. Some operators prefer steaming, a shorter and more intense process that is too harsh for fragile woods and does not properly penetrate others.

A cooking vat should be built of concrete liberally reinforced, especially at corners, to withstand severe temperature strains. It usually is made about 6 ft deep by 10 ft wide; the length is varied to suit capacity. Covers, preferably of steel plates with central rings, are handled by the crane used for bolts and flitches.

VENEER LATHES. Figure 3 shows the principle of operation of a rotary veneer lathe. Details of the cutting mechanism are shown in Fig. 9. Logs are mounted in the lathe from the back by crane. Standard operating speed, traditionally, is 25 to 30 rpm to accommodate 30-in. logs. This is slow for small logs. Spindle chucks or centers usually are 5 or 6 in. in diameter for 60- to 100-in. bolts, leaving a 7-in. core that is defective in many species but sometimes can be cut into crating. Lathes usually are equipped with 50- to 75-hp motors for 60- to 100-in. lengths and logs up to 30-in. in diameter. For larger logs, as on the Pacific Coast, power demand is greater. An efficient setup would be mechanical speed control permitting speeds to be changed without stopping lathes, giving 48, 36, 30, and 24 rpm, and permitting linear veneer delivery speeds of 100 (for thick veneer) to 200 ft per min. Electrical speed controls are to be avoided, as motors incline to race under wide load variations, caused by thick and thin cuts, and large and small logs. Table 5 shows the resulting delivery speeds in such a setup. Rotary box veneer often is cut at 60 rpm, but quality requirements are low. Figure 10 shows a typical lathe layout, with delivery tables, clipper, and crew location. In species with sound cores, as birch, a supplementary high-speed (50 rpm) lathe sometimes is provided with 3-in. chucks to cut $^1/_{20}$-in. veneer from 7- and 8-in. cores, 42 in. long or less.

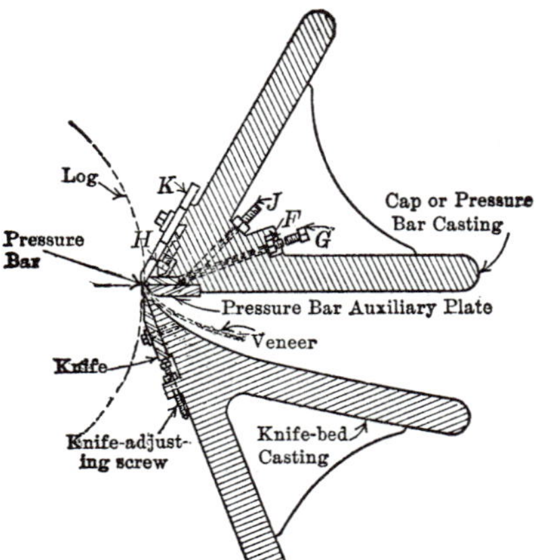

Fig. 9. Details of veneer lathe cutting mechanism.

The veneer comes off the lathe in long ribbons and frequently goes direct to the clipper on a conveyor, where it is dimensioned for width, singly or in piles. In most cases either the clipper or the lathe is slowed down by this direct connection, or the quality of the veneer is sacrificed by rapid clipping in piles. An alternate procedure, which has been

Table 5. Actual Veneer Delivery Speeds, Feet per Minute

Log Diameter, in	Rpm of Spindle				Log Diameter, in.	Rpm of Spindle				Log Diameter, in.	Rpm of Spindle	
	24	30	36	48		24	30	36	48		24	30
	Lineal Feet per Minute of Lathe Delivery					Lineal Feet per Minute of Lathe Delivery					Lineal Feet per Minute of Lathe Delivery	
8	....	....	75.4	100.5	19	119.4	149.2	179.1	238.8	30	188.5	235.6
9	....	....	84.8	113.1	20	125.7	157.1	188.5	251.3	31	194.8	243.5
10	....	78.5	94.2	125.7	21	131.9	164.9	197.9	....	32	201.1	251.3
11	....	86.4	103.7	138.2	22	138.2	172.8	207.3	....	33	207.3	259.2
12	75.4	94.2	113.1	150.8	23	144.5	180.6	216.7	....	34	213.6	....
13	81.7	102.1	122.5	163.4	24	150.8	188.5	226.2	....	35	219.9	....
14	88.0	110.0	131.9	175.9	25	157.1	196.4	235.6	....	36	226.2	....
15	94.2	117.8	141.4	188.5	26	163.4	204.2	245.0	....	37	232.5	....
16	100.5	125.7	150.8	201.1	27	169.6	212.1	254.5	....	38	238.8	....
17	106.8	133.5	160.2	213.6	28	175.9	219.9	....	....	39	245.0	....
18	113.1	141.4	169.6	226.2	29	182.2	227.8	....	....	40	251.3	....

used in Europe for some time and has been adopted by several American veneer mills, is to reel the green veneer on drums direct from the lathe, store the drums on racks under a conveyor, and transfer them to the clipper table at a rate that permits careful quality cutting. This method conserves quality and increases yield in the better grades of veneer.

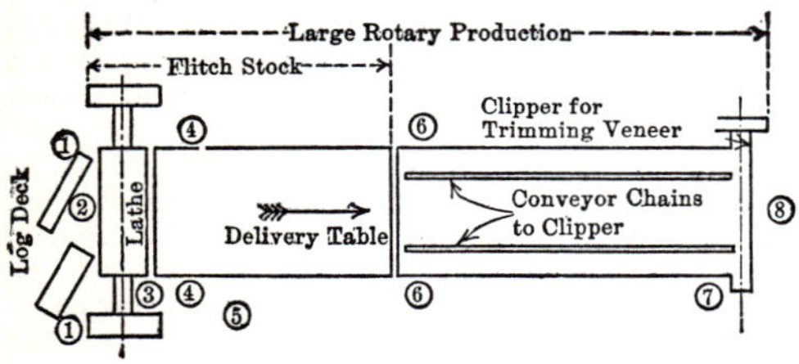

Fig. 10. Standard veneer layout and crew. Standard crew is: (1) 2 log peelers, (2) 1 log hanger, (3) 1 lathe operator, (4) 2 off bearers or breakers, (5) 1 waste collector, (6) 2 off bearers for large production, (7) 1 clipper operator, (8) 1 clipper off bearer.

VENEER SLICERS. Figure 5 shows the operating principle of a veneer slicer. Flitches are mounted on the table or bed by dogs, the table moving up and down on angling ways to produce a shear cut. The typical cutting stroke is about 3 ft. The draw or lateral movement of the knife is about 1 $^1/_2$ ft. Cutting is done on the down stroke, and the log is fed forward on the return stroke of the knife. Horizontal slicers are used more in Europe than in the United States and afford greater latitude in range of angle cuts, often desirable in highly figured flitches. Veneer delivery equipment for slicers is simpler than for lathes. Flitch-cut veneer always is kept in sequence to permit matching.

HALF-ROUND VENEER is an intermediate method between rotary and sliced veneer. (See Fig. 6.) The log is flitched or sawed, as in slicing, and cooked. It is cut on a lathe provided with a *stay-log* swung from the lathe centers, with radial adjustment and provided with a table or bed to which flitches are clamped with bolts. Half-round veneer has more regular grain than rotary veneer.

CUTTING MECHANISM (Fig. 9) is essentially the same in both lathes and slicer. Its proper adjustment and manipulation are the critical factors in good veneer cutting. The feed mechanism in both lathes and slicers is mounted on two screws, moving the lathe carriage with the cutting mechanism into the log or flitch a distance equal to the veneer thickness at each revolution or stroke.

Rotary veneers usually are cut to widths providing whole, half, third, and quarter parts of the desired sizes. Flitch veneers are kept together after numbering and taking several sample swatches from top, center, and bottom of each flitch.

VENEER DRYERS comprise rollers, traveling plates, or mesh belts, on which veneer passes progressively through drying chambers. Controlled temperature and humidity establish the necessary moisture deficit to dry various thicknesses at a single pass, the speed for thicker veneers being reduced. Veneer usually is dried to 10% moisture content. It must be dried within a few hours after cutting to prevent discoloration.

Manufacturers. Veneer machinery is built by Merritt Engineering and Sales Co., Lockport, N. Y.; Capitol Machine Co., Indianapolis, Ind.; Coe Manufacturing Co., Painesville, Ohio; Proctor and Schwartz Co., Philadelphia, Pa.

6. LUMBER DRY KILNS

EQUILIBRIUM MOISTURE CONTENT. Wood products, in service and storage, should have, as nearly as possible, an equilibrium moisture content in the wood parts to keep at a minimum dimensional changes (shrinkage and swelling) due to moisture changes. Under ordinary room conditions this is about 6 to 8% moisture content; in normal outdoor weather 10 to 15%. Lumber is stronger and stiffer when dry than when wet. Dry wood (to the extent mentioned above) will retain paint and finishes, whereas wet lumber will not hold surface coatings. Table 6 gives conditions for maintaining moisture content equilibrium in lumber. Correct atmospheric conditions are particularly important in raw lumber storage rooms.

Table 6. Variation of Equilibrium Moisture Content of Wood

Wet-bulb Depression, °F	Dry-bulb Temperature, °F										
	50	55	60	65	70	75	80	85	90	95	100
	Equilibrium Moisture Content, %										
4	15.7	16.2	17.0	17.3	17.6	17.9	18.1	18.1	18.2	18.2	18.3
5	13.0	14.5	15.0	15.3	15.6	15.9	16.2	16.4	16.6	16.8	17.0
6	12.5	13.0	13.5	13.0	14.2	14.5	14.8	15.0	15.1	15.3	15.5
7	11.3	11.9	12.4	12.8	13.1	13.4	13.7	13.9	14.1	14.3	14.4
8	10.2	10.8	11.4	11.8	12.2	12.5	12.8	13.1	13.3	13.5	13.7
9	9.2	9.9	10.4	10.9	11.3	11.7	12.0	12.2	12.4	12.6	12.7
10	8.3	9.9	9.6	10.1	10.6	10.9	11.2	11.5	11.8	12.0	12.2
11	7.3	8.1	8.7	9.3	9.9	10.3	10.6	10.8	11.0	11.2	11.3
12	6.4	7.3	8.1	8.6	9.1	9.6	10.0	10.2	10.4	10.6	10.8
13	5.5	6.4	7.2	7.8	8.4	8.8	9.2	9.6	9.9	10.1	10.2
14	4.6	5.7	6.6	7.3	7.9	8.3	8.7	9.1	9.4	9.7	9.9
15	....	5.0	5.8	6.6	7.3	7.8	8.2	8.5	8.8	9.1	9.3
16	....	....	5.2	6.0	6.7	7.2	7.7	8.0	8.3	8.6	8.8
17	....	....	....	5.4	6.1	6.7	7.2	7.5	7.8	8.1	8.3
18	....	....	....	4.8	5.6	6.2	6.7	7.1	7.4	7.7	7.9
19	....	....	....	....	5.0	5.6	6.2	6.6	7.0	7.3	7.5
20	....	....	....	....	....	5.2	5.7	6.2	6.6	6.9	7.1
21	....	....	....	....	....	....	5.3	5.8	6.2	6.5	6.7
22	....	....	....	....	....	....	....	5.3	5.8	6.1	6.3
23	....	....	....	....	....	....	....	....	5.4	5.7	6.0
24	....	....	....	....	....	....	....	....	5.1	5.4	5.7
25	....	....	....	....	....	....	....	....	....	5.1	5.5

LUMBER DRY KILNS are of two principal types. In the single-charge (*box* or *compartment*) kiln, the lumber charge remains stationary during the drying cycle; in the *progressive* kiln the lumber moves in the kiln through the various stages in the drying cycle, from the initial low-temperature-high-humidity condition to the final high-temperature-low-humidity stage. Figure 11 compares the two types.

The method of piling lumber in the dry kilns may be *crosswise*, as shown in Fig. 11, in which the length of the boards is at right angles to its movement into, through and out of the kiln; or *endwise*, in which the boards are parallel to the movement of the lumber, as shown in the left-hand sketch of Fig. 12. Either method is suitable for ventilated or condenser kilns, but the endwise method is preferred for fan kilns.

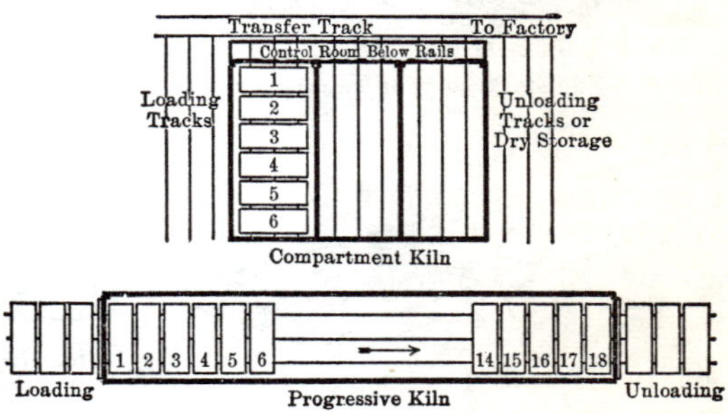

Fig. 11. Dry kilns with equal capacity.

In ventilated or condenser kilns (Fig. 12) the boards in each layer are placed 2 or 3 in. apart, since the air moves both horizontally and vertically, so that only 60 to 75% of the space in each layer can be utilized. In fan kilns (Fig. 12) the boards are laid close together edgewise, since the air travels only horizontally. Hence the kiln capacity is considerably greater. In all cases the intervening stickers, between the layers of boards, should be of uniform height, and laid in accurate vertical alignment, so that all boards will be held straight and true during the drying process.

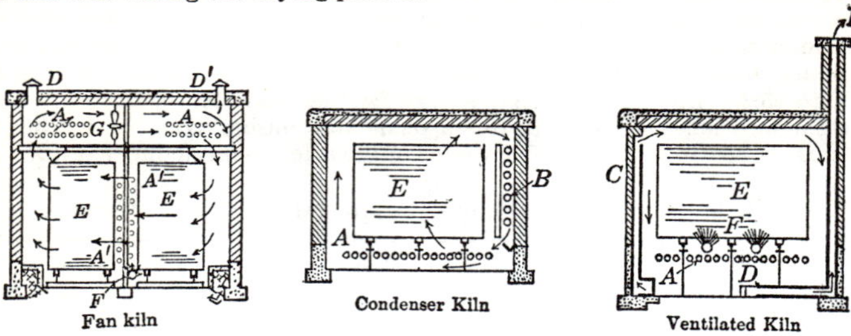

FIG. 12. Types of kilns. A, A', heating unit. B, condensing unit. C, fresh air inlet. D, D', used air outlets. E, lumber charge. F, humidifying unit. G, reversible fans.

DRYING TECHNIQUE is of three principal forms, as shown in Fig. 12. (1) Forced circulation, or *fan kilns*, in which the air is forced mechanically through between the layers of lumber, and the fan circulation reverses at stated intervals to equalize the drying effect. Supplementary or booster heating coils are frequently placed between the loads of lumber as shown. (2) *Condenser kilns*, where moisture-laden air passes over condensing coils (usually circulating cold water pipes), is reheated, and again passes through the lumber, and so on for a series of cycles. (3) Natural circulation, or *ventilated kilns*, where the air movement is accomplished by gravity. Heating pipes are often installed in the outlet ducts to accelerate the ventilation. In this case adequate air inlets and outlets are of prime importance.

The present tendency is toward fan kilns, since their capacity is greater by eliminating the spaces between the boards; and their steam consumption is less, due to the recirculating feature that conserves the atmospheric heat within the kiln. Condenser kilns are seldom used, and many ventilated kilns are being replaced with fan types.

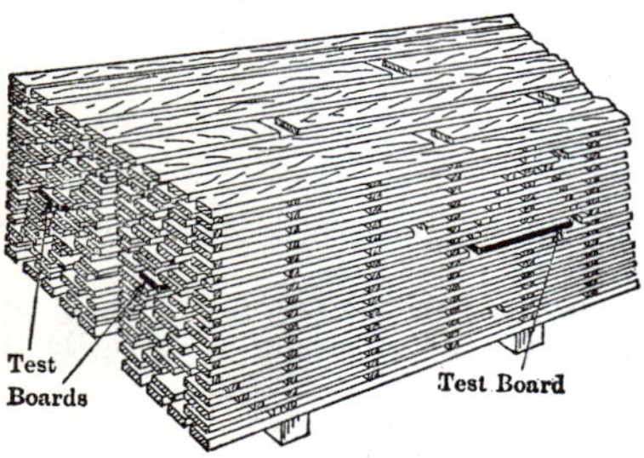

FIG. 13. Kiln stack, showing locations of test boards.

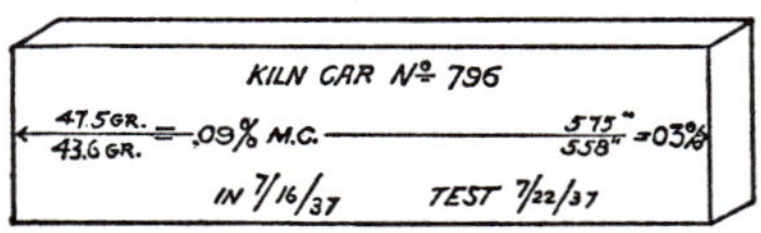

FIG. 15. Method of marking test samples.

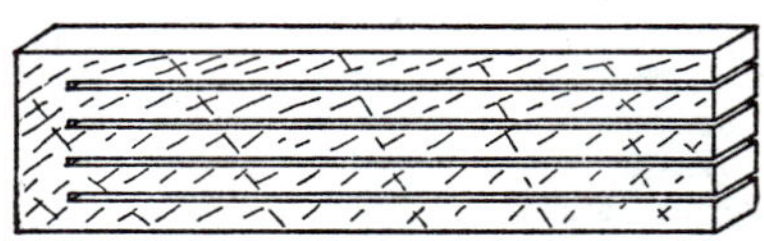

FIG. 16. Test sample for case-hardening.

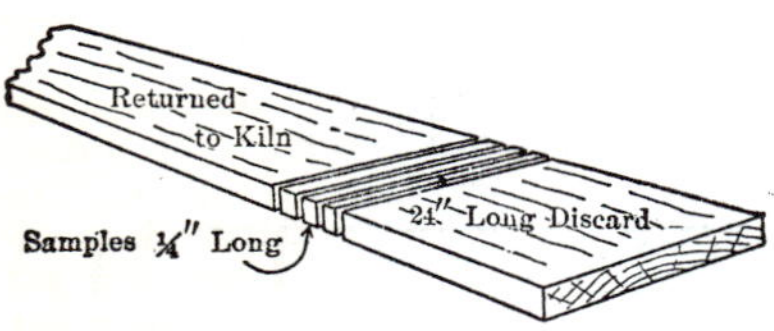

FIG. 14. Location of sample in test boards.

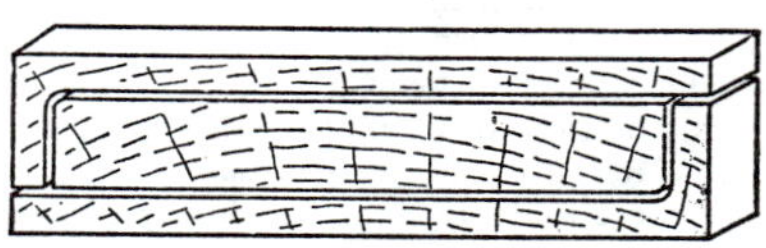

FIG. 17. Test sample to determine dryness between center and outside.

Most kilns have steam or water sprays to maintain the humidity. Temperature and humidity controls, usually with recording attachments, are essential to maintain schedule requirements. In kiln-drying programs, *humidity* refers to air conditions, and *moisture content* means the moisture contained in the lumber. The latter is computed by weight, weight of dry wood being divided into the weight of moisture removed:

$$MC = \frac{W_i - W_d}{W_d} \times 100$$

where MC = moisture content, per cent; W_i = initial weight; and W_d = oven-dry weight.

In general, temperatures of 200 F can be used safely on 1-in. (4/4) softwoods; for hardwoods the temperature range is 150 to 160 F. With thicker and greener lumber, these temperatures must be reduced.

TIME REQUIRED FOR KILN DRYING of pine and similar softwoods usually is 3 to 5 days, from the original tree condition of greenness to about 10% moisture content, usually considered shipping dry. Hardwoods require varying times, according to thickness, original and desired moisture content, and types of kilns available. A general survey of these drying times is given for internal fan kilns in Table 7.

Table 7. Time Required to Dry Hardwoods to 5% Moisture Content with Internal Cross Circulation Fan Kilns

(By permission, Standard Dry Kiln Co.)

Wood	4/4 in. Thick		8/4 in. Thick	
	Air Dried, 25% M.C. or less	Green, 50–80% M.C.	Air Dried, 25% M.C. or less	Green, 50–80% M.C.
	Days of 24 hr Continuous Operation			
Hardwoods:				
Ash	4–5	8–10	8–10	18–20
Beech	6–7	10–12	12–14	26–28
Birch	5–6	10–12	10–12	24–26
Cherry	6–7	10–12	12–14	26–28
Chestnut	4–6	8–10	10–12	16–20
Elm, soft	5–6	10–11	10–12	20–22
Elm, rock	6–7	12–14	12–14	22–24
Gum, red	7–8	12–14	13–15	26–28
Gum, sap	5–6	8–9	8–10	16–18
Gum, tupelo	4–5	7–8	8–9	15–16
Hickory	7–8	14–16	15–16	30–32
Mahogany	5–6	11–12	12–14	22–24
Maple, hard	5–6	8–10	10–12	18–20
Maple, soft	4–5	7–8	8–10	16–18
Oak, red	4–6	16–18	10–12	34–36
Oak, white	5–7	18–21	13–15	40–42
Sycamore	4–5	8–10	9–10	16–18
Walnut	6–8	12–14	12–14	25–27
Semihardwoods:				
Basswood	3–4	5–6	6–7	10–12
Cottonwood	3–4	5–6	6–7	10–12
Poplar	3–4	5–6	6–7	10–12
Willow	3–4	5–6	6–7	10–12

It is highly important that lumber be uniformly kiln dried to the required moisture content, and be maintained at that point during the entire product manufacturing procedure. In fact, most kiln authorities advocate kiln drying slightly below required conditions, and allowing reabsorption to equilibrium point in dry storage.

Tests for moisture content usually are made by weighing samples. When tests are required, samples are cut from test boards whose locations in the kiln load are shown in Fig. 13. The locations of the test samples in the test boards are shown in Fig. 14. The board ends, which probably are overdried, are discarded. Figure 15 shows the best procedure for marking weighed samples for filing, the figures on the left being moisture content and those on the right shrinkage. In both cases the upper figures are values immediately after cutting test samples; the lower figures are values after oven drying to minimum obtainable conditions, i.e., bone dry. Figure 16 illustrates a test sample prepared to show case hardening. If the prongs flare out, the sample is case hardened but dry; if they pinch, it was case hardened before drying, and the stresses were not relieved during drying. A further test for exterior and interior moisture content is shown in Fig. 17. The test sample is band sawed, and outer and inner portions weighed before and after oven drying to determine the moisture content of each. A normal moisture content gradient usually will show an excess of 1 to 2% for the inner section.

Kiln Accessories. For kilns and adjoining storage tracks, 35-lb rail generally is used, supported at 6-ft intervals. Kiln bunks or trucks should have roller bearings, frequently lubricated because of the heat. Kiln transfer cars will facilitate lumber distribution from railroad unloading siding direct to kilns, and from kilns to dry storage and cutting rooms. They are usually about 12 in. high, rail top to rail top, and preferably are motor driven. Lumber lifts elevated by motor-driven revolving screws, in a pit below, are essential both for loading and unloading kiln cars, as well as for keeping the top of the lumber load at convenient height for serving rough planers or cut-off saws. Some mills use locomotive cranes with 40-ft booms to transfer lumber in packets to and from kilns and lumber piles in the yard.

Manufacturers. Lumber dry kilns are supplied by Standard Dry Kiln Co., Indianapolis, Ind.; Moore Dry Kiln Co., Jacksonville, Fla., and Portland, Ore.

7. MACHINE OPERATIONS AND LAYOUTS

In wood products factories, machine, and department arrangement are substantially similar in major items, with considerable divergence in minor details, because of difference in character of products and degree of premanufacture in raw materials. A typical example of standard procedure is given in Table 8.

Table 8. Standard Sequence of Machine Operations in Woodworking

Lumber yards
↓
Dry kilns
↓
Dry storage and tempering
↓
Transfers and lumber lifts
↓
Rough planing (sometimes omitted)
↓
Cut-off or stock saws
↓
Rip saws
↓
Edge jointers
↓
Edge gluing (for solid lumber parts or plywood cores)
↓
Finish planing to net thickness
↓

(Plywood parts) *	(Solid parts)
Veneer preparation	Dimensioning, i.e., machining to size
Glueroom	Molding
Plywood redrying	Shaping
Dimensioning	Turning
Bending and curving	Routing
Routing	Special machine work
Molding edges	

↓
Boring and drilling
↓
Corner construction
↓
Sanding
↓
Stockroom for machined parts
↓
Cabinet assembly
↓
Polish sanding
↓
Finishing (stain, filler, lacquer, and rubbing)
↓
Trimming with hardware and glass
↓
Packing and crating

Inspection stations according to product requirements

* Omit plywood operations if plywood is purchased.

For further details of plywood and finishing, see Articles 8 and 9. For kiln drying, see Article 6. For details of operation of the other types of woodworking machines, see Articles 13 and 14.

8. PLYWOOD AND GLUING

FLAT PLYWOOD. Standard plywood consists of an odd number of layers of veneer, with opposed grain in alternate layers, rigidly glued together. The along-the-grain strength reinforces the across-the-grain weakness, and distributes wood strength in both directions. The central layer (core), of veneer or lumber, has the lengthwise grain in the direction requiring strength. In 5-ply, face and core grains are parallel; in 3-ply, they are opposed. Fundamentally the layers should be balanced on the core to avoid uneven stresses. (See Figs. 18 and 19.) The face wood is chosen for appearance; the other layers

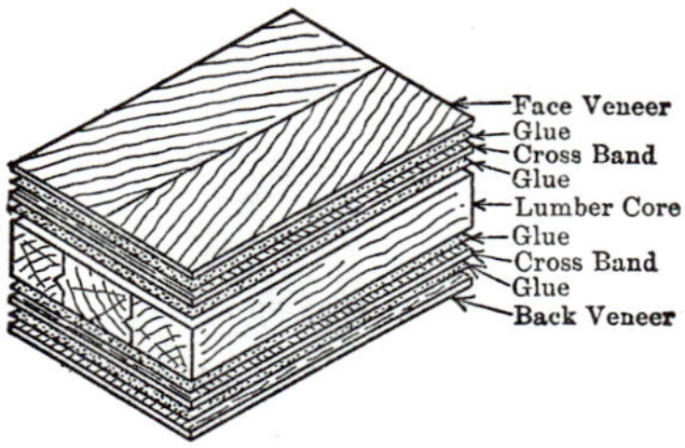

Fig. 18. 5-ply plywood.

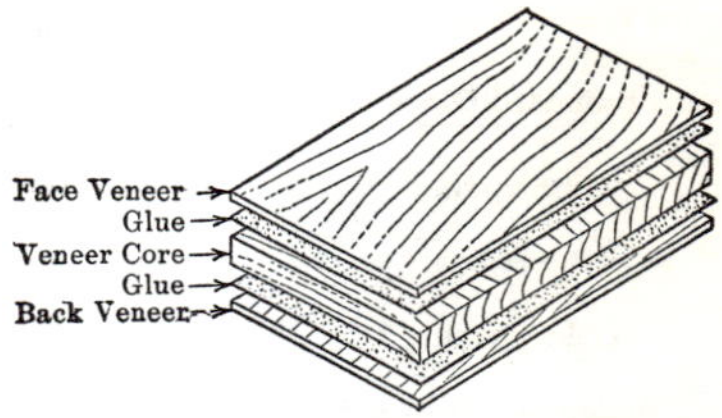

Fig. 19. 3-ply plywood.

supply the strength. For most purposes $^1/_4$-in. 5-ply plywood is stronger than $^{13}/_{16}$-in. solid lumber. Plywood is unsplittable, and has more nail- and screw-holding power than solid wood.

The large bulk of plywood is made flat as described. Recent techniques have been developed either to bend flat plywood into simple curvatures or to produce simple and compound curved plywood as it is glued together from veneers. For these techniques, see p. 25-22.

ARRANGEMENT OF EQUIPMENT. Plywood manufacturing falls in two general classes: (1) Factories assembling plywood as a semi-raw material for others. (2) Factories making the ultimate product, as furniture, with an integral plywood department, in which plywood operations are seldom complete. Plywood factories are located preferably with reference to the supply of the bulkiest material, i.e., rotary commercial veneer and core lumber, the face veneer supply being far less important. Propinquity to a consumer furniture area is a doubtful advantage because of the higher wage levels.

Table 9 shows the general sequence of plywood operations. The face veneer preparation, commercial veneer preparation, and lumber core stock converge to a central tempered-air stockroom, where the gluing allocation for each day must be ready not later than the night before. The glueroom is invariably the "bottleneck," has a higher man-hour-ratio per machine than any other plywood operation, and must produce at maximum efficiency.

FACE VENEER. Sliced, half-round, and sawed face veneers are received in flitches, and their suitability for the job in question demands experienced judgment. Unused portions of flitches (lay backs) are a serious problem, often because of wrong size or figure. In general, whole flitches should be charged to a job, leaving lay backs as a by-product; otherwise costs may be seriously distorted.

Highly figured face veneers, if "crinkly," must be flattened before dimensioning. For less fragile species, platen veneer driers are used; it is advisable to dampen fragile veneers and apply gradual pressure with heated boards. In extreme cases, as feathered crotches, veneers are dipped in a dilute glue size to make them more leathery, and are flattened even more gradually. Moisture content should be 5 to 6% after flattening.

Longwood veneers, longer flitches in moderate figure, usually are cross cut on saws in packages of 25 sheets and edge clipped (sheared) with reference to figure matching. Clipped edges are *jointed* (see Veneer Jointers, p. 25-43), planed straight and square, to produce tight joints without cracks. *Stumps* (butts), *crotches*, and *burls* and other highly figured veneers usually are cut both ways in packages on a paper-cutting machine, with an angling shear cut under pressure. Such veneers cannot be jointed as the edges will tear out.

Book-matched veneer comprises two or four adjacent sheets of a flitch, with uniformly (as to grain figure) trimmed edges, laid together to balance the figure as in Fig. 20. It resembles an unfolded 2- or 4-sheet booklet.

Table 9. General Arrangement of Plywood Operations

FACE VENEERS	BACKS, BANDS, AND ROTARY CORES	LUMBER CORES
Flattening ↓	Flattening ↓	Cross and rip sawing ↓
Dimensioning ↓	Dimensioning ↓	Jointing edges ↓
Jointing edges ↓	Taping or splicing ↓	Matching ↓
Matching ↓		Edge gluing ↓
Taping or splicing ↓		Banding edges ↓
Gluing edges ↓		Planing or surfacing ↓

↓

Tempered storage for all parts ready to glue

↓

Standard Glues	Resin Adhesives	Film Adhesives
↓ Glue mixing ↓	↓ Preparation ↓	↓ Film cutting ↓
Spreading ↓	Spreading ↓	Interlaying veneer with film ↓
Cold pressing * ↓	Hot pressing * ↓	Hot pressing ↓
Baling and storage ↓		Sponging and tape removal ↓
Redrying on stickers ↓		

↓

Plywood dimensioning

↓

Sanding

Inspection (subinspection as required)

* Cold pressed resin adhesives, with special catalysts, are finding substantial use, followed by baking and storage but eliminating redrying on stickers.

Dimensioned veneers (longwood or book-matched) are *taped* (see Fig. 20) with gummed tape from edge to edge on one surface to hold the pieces in position during gluing. The tape is sanded off later. Taping machines have angling rollers that draw the edges together and press the moistened gummed tape over the adjoining edges. Intricate *combination faces*, of vari-colored species, with or without angling grains, give remarkable effects. Figure 21 shows general arrangement of machine operations in a face-veneer department.

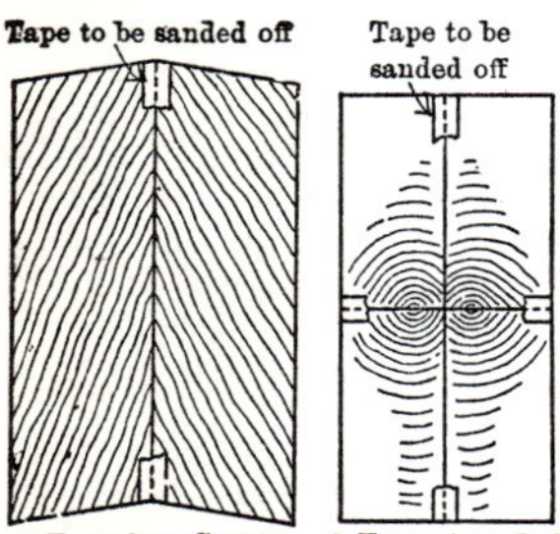

a Two piece Center Matched *b* Four piece End and Side Matched

FIG. 20. Matched veneer.

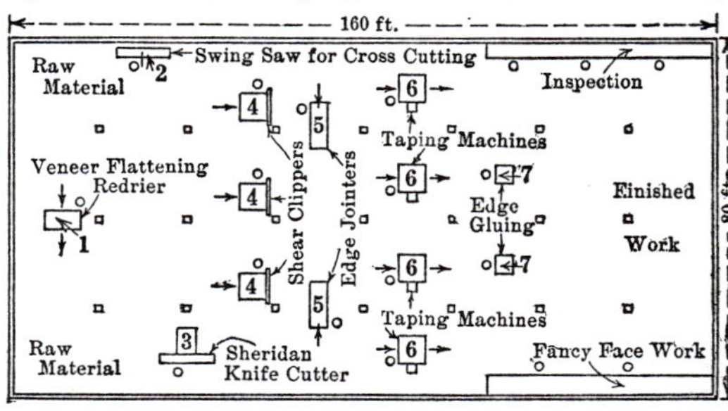

FIG. 21. Arrangement of face-veneer department. ○ indicates position of operator. Numerals show sequence of operations.

Taping machines, applying gummed tape over each veneer joint, are still widely used. They are being displaced by *tapeless splicers*, which push the jointed edges of the veneer together by converging feed chains under heater bars, which cure the resin adhesive in the joint. Such splicing is superior to taped joints, since tape is eliminated from all inner layers and does not have to be removed from face and back layers. Tapeless splicers are not suitable for highly figured face veneers. The tapeless splicers (Fig. 21) take the place of the taping machines (6) and edge gluing (7).

CROSS BANDS, ROTARY CORES, AND BACKS, unless two faces are required, are of commercial rotary veneer, flattened, edge-clipped, sometimes jointed if 3-ply, and taped or spliced as in face veneer. Cross bands under faces, as well as cores in 3-ply plywood, should be one-piece or spliced, but should not be taped, as tape will show through. Back cross-bands, backs and 5-ply cores may be taped or spliced. Commercial veneers, for long runs of standard sizes, often are dimensioned at the veneer mill into whole, half, one-third, two-third, one-quarter, and three-quarter sizes for taping. Moisture content must be kept close to 5%, except that certain layers in film bonding should be 8 to 12%. A commercial veneer department arrangement would be similar to Fig. 21, omitting the swing saw, Sheridan cutter, and edge gluing.

LUMBER CORES generally are used for plywood 1/2 in. and thicker. Five-ply construction is the more stable; face and back veneers in 3-ply usually are thin and do not effectively balance the lumber core. Core lumber is preferably of poplar, chestnut, gum, or other available species. Only one species should be used in a core. Chestnut does not give a sound edge for molding or shaping.

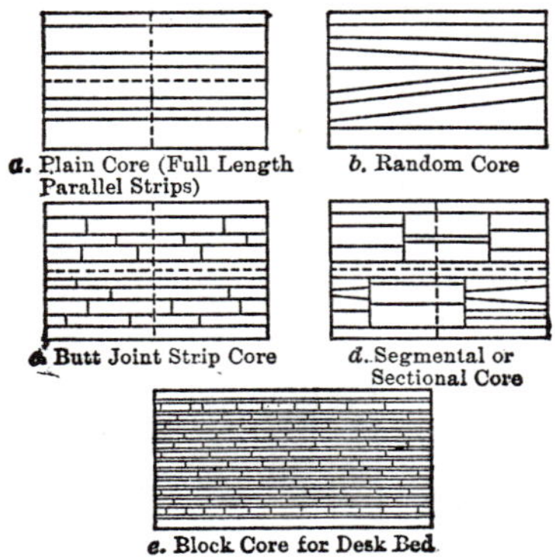

FIG. 22. Types of cores.

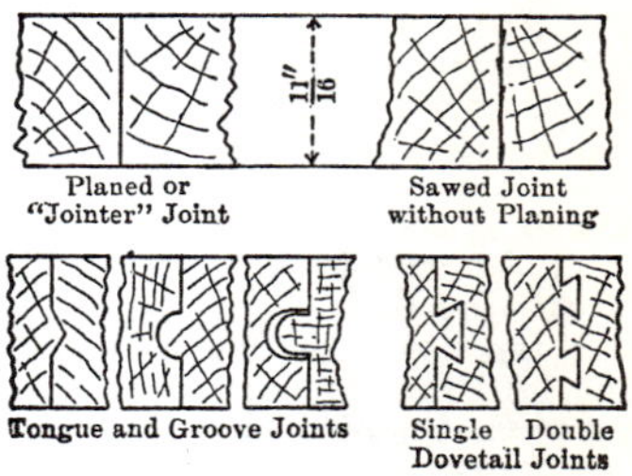

FIG. 23. Types of edge joints in plywood cores.

To avoid warp, cores preferably should be made of strips 2 to 3 in. wide. End joints are not objectionable and tend to eliminate twisting stresses. Figure 22 shows various core constructions, and Fig. 23 various core joints. The dotted lines in Fig. 22 indicate multiple cores to be cut apart after planing or gluing, to avoid showing endwise splices. The veneer strips used as glue carriers in core *e* are patented.

Assembled core strips may be glued together by one of several methods. (1) An endless chain *revolving glue clamp*, with a cycle of 30 to 60 min. When the glued core returns to the loading point, it is removed, and a fresh assembly inserted. The application of edge glue swells the lumber, increasing the thickness at the joint, and the glued cores should be seasoned, or equalized in moisture content, before planing. (2) A continuous chain feed *lumber splicer*, in which core strips are placed successively on a rubber-faced chain, pushed between steam-heated platens, and automatically sawed apart into required widths as they emerge from the heated platens. In an alternate feed method corrugated rollers replace rubber-faced chain. (3) A batch or continuous-feed electronic *lumber splicer*, in which the assembled core strips, under edge and face pressure, are passed through a high-frequency electrostatic field to cure the resin adhesive. These core units, in methods 2 and 3, can be planed as soon as the adhesive has attained its grip.

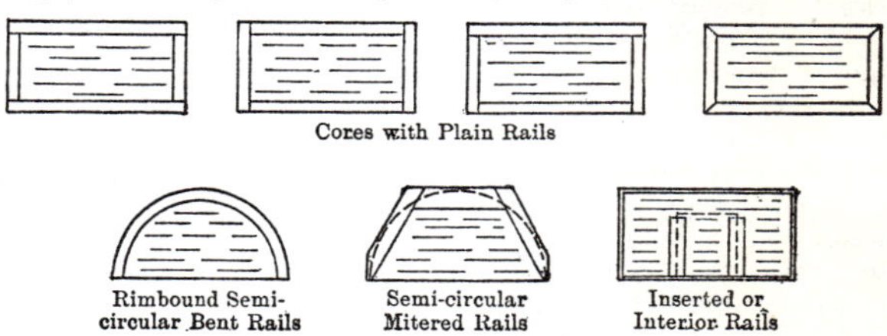

FIG. 24. Banded or railed cores.

Banded or railed cores (Fig. 24) have a band or rail, of the same species as the face veneer, on one or more edges, to give molded edges the same appearance as the upper surface. These bands are additional core strips, glued on edges or ends, after the ends are trimmed.

All plywood elements, faces, bands, backs, and cores for an order should meet in an air-conditioned central stockroom to avoid subsequent glueroom delays.

Figure 25 shows the arrangement of a core department. Lumber is most economically handled on kiln cars from dry storage and placed on a lumber lift; top of load then always is at machine table height. If, before cross cutting, both sides are rough planed to reveal defects, a 2-side rough planer is placed adjacent to the lift, with the swing cut-off saws next in line. The stock then goes from cross cut to rip saws (see pp. 25-34 and 25-39) for trimming edges and removing wane (tapered and bark edges). If specifications are not too exacting, ripped edges may be glued together without further finishing operations, but quite frequently edges must be jointed true, straight, and smooth for gluing. The sliding matching table (operation 5) is used for the purpose of arranging core strips to distribute stresses, and to rip the last strip to net core width.

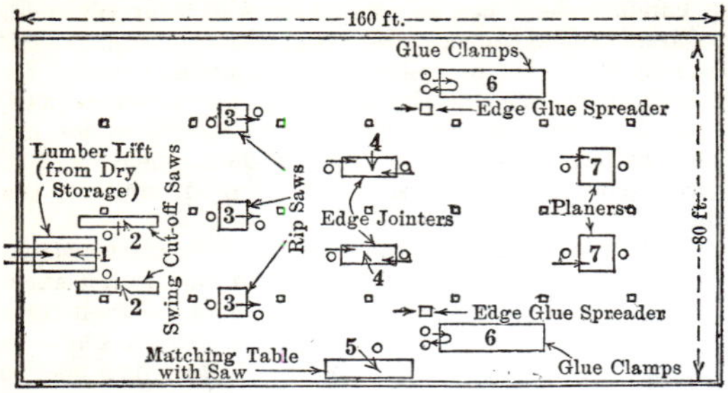

FIG. 25. Arrangement of core department. ○ indicates position of operator. Numerals show sequence of operations.

PLYWOOD GLUING DEPARTMENTS (for flat plywood) vary considerably in their arrangement; three types are shown in Fig. 26. In each type a single line is shown, but in multiple lines, where more than one working shift is employed, the ratio of mixers, spreaders, and lay-up tables may be varied. The single lines shown have an approximate average capacity of 5000 sq ft of thin plywood, and a like area of lumber core plywood per 8-hr shift, total 10,000 surface feet.

These types may be described briefly. (*a*) Using standard liquid glues, *casein*, *vegetable* (cassava), *soybean*, *silicate of soda*, etc. Pressing is done at room temperature. The clamped bundles or bales are removed from the press for overnight storage while the glue hardens; they are then stacked up for redrying to remove surplus moisture, and are subsequently unstacked. Operating cycle 2 to 3 days. (*b*) Using liquid resin adhesives, *ureas*, *phenolics*, *melamines*, and also *albumin* glues, where hot pressing is employed, with multiple steam-heated platen presses. Operating cycle, 1/2 to 1 day, including equalizing period for cooling and moisture diffusion. Alternately, certain types of urea resins can be used in room-temperature work, for moderate production requirements, clamped and stored overnight. Redrying is not required. (*c*) Using *phenolic film* adhesives, Tego, hot pressed, and hot plywood usually dipped quickly in water to permit tape removal and to shorten the time of cooling and moisture equalization. Operating cycle as in type *b*. The type of equipment and order of processing steps are shown in Fig. 26.

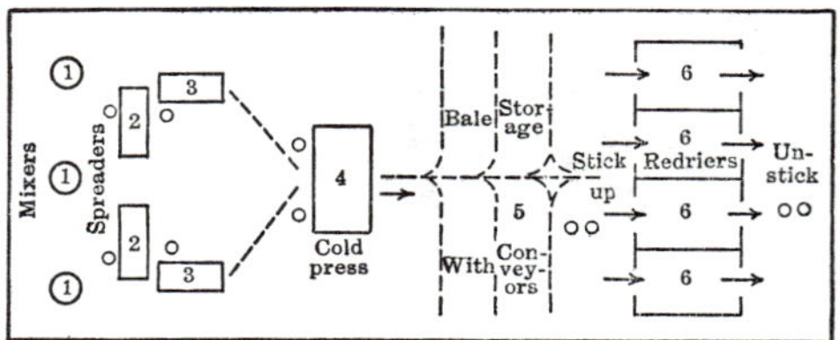

(*a*) Cold press (room temperature) plywood department

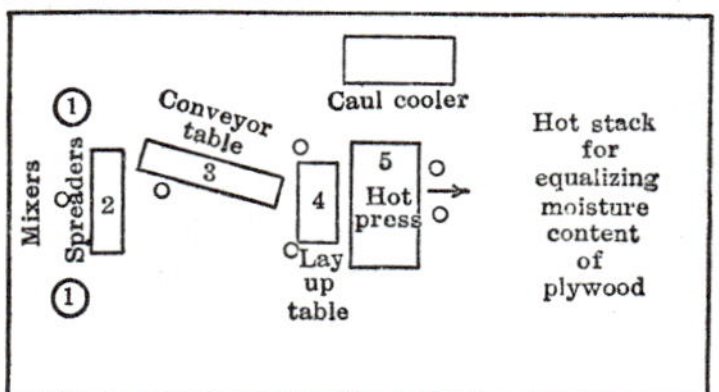

(*b*) Hot press plywood department for liquid resins

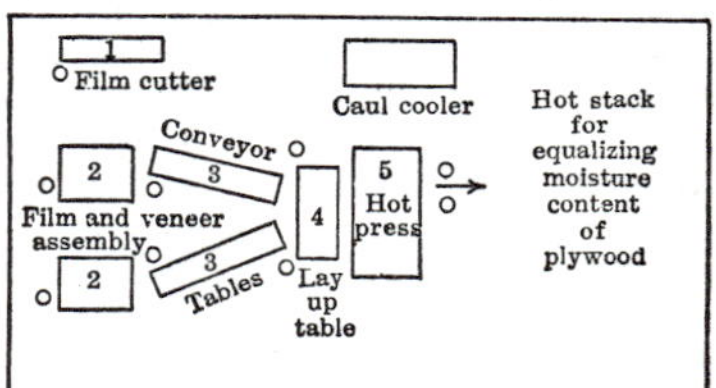

(*c*) Hot press plywood department for film resins

FIG. 26. Gluing department layouts for flat plywood. ○ indicates position of operator. Numerals show sequence of operations.

Following are the various machines and items of equipment.

Glue mixers are steam and water-jacketed kettles, and have revolving double paddles, usually on a vertical shaft. Horizontal shaft mixers are used to a limited extent. Discharge openings from mixers must be adequate for viscous mixtures and easily accessible for frequent cleaning. For details of preparing liquid glues and their processing requirements, see Section 21, Article 10.

Glue Spreaders. Double-roller spreaders are the rule, applying glue to both sides of the veneer or lumber layer; corrugated metal rollers for casein, vegetable, soybean, and silicate glues; grooved (20 per in.) rubber rollers for all resin adhesives and albumin. Spreading ratios vary from 60 to 90 lb of liquid glue per 1000 sq ft of joint for standard glues and 30 to 50 lb for resin adhesives.

Lay ups, or assemblies, from bottom to top in 5-ply constructions, are back layer without glue, cross-band glue coated on both sides, core without glue, cross band with glue on both sides, and face without glue. Adjacent assemblies are often reversed, with a waxed plywood or metal caul between the two faces to prevent damage. In room-temperature work, bundles are built up, 24 to 30 in. high, on thick head blocks, with another head block on top, clamped with I beams and turnbuckle rods in the press, removed from the press, and stored overnight for hardening of the glue. In hot pressing, either liquid or film, a lay-up table is provided in front of the hot press, where assembled units are placed between aluminum cauls for loading into the hot press.

Pressures for room-temperature work, often called cold pressing, are from 100 psi for softwoods to 150 lb and up for dense hardwoods. For hot-press work, where thickness compression may be of the order of 5 to 10%, pressures range from 150 to 250 lb, respectively.

Pressing time for room-temperature work is 4 to 6 hr, usually overnight; for hot pressing, 5 to 30 min, depending on thickness of the assembly. After removal from the hot press, plywood is stacked solid to cool and equalize.

Redrying is required for most standard glues where the solids content of the mixture averages 30 to 40% and plywood after pressing may have 15 to 20% moisture content. Redrying is not required for resin adhesives with 50 to 60% solids content and pressed at room temperatures. Hot-pressed plywood tends to be overdry and needs time to reach equilibrium moisture content. Phenolic bonded plywood is often dipped in water to shorten the cooling and equalizing period, and to reduce warping in thin plywood. Such plywood should be dipped quickly when hot and piled solid.

While steam heat is usually employed for hot-pressed plywood, as it is easily available in woodworking factories, some installations use high frequency or resistance heat from electrical sources. For details, consult manufacturers of such electrical equipment.

DIMENSIONING for flat plywood is done preferably on double cross cut and double rip saws of the double-end tenoner type (p. 25-46), in which a pair of endless chains, with lugs, carry the plywood through the saws. One saw head is fixed, and the other adjustable to the size required, usually up to 96 in. Many other types of saws and saw tables may be used for this work. Ripping usually precedes cross cutting. For center matched plywood faces, pointers on the machine are used to insure proper centering of face veneer joints.

Sanding. Plywood is sanded to remove the veneer tape and to prepare surfaces for filling, staining, and lacquering. Usual operations are: (1) Preliminary sanding of the back side of the plywood on a roll or endless belt sander, to clean it and remove most of the tape; other back sanding seldom is necessary. (2) Similar sanding on the face, taking care that high spots do not sand through. (3) Finish sanding on belt or stroke sanders, with very fine sandpaper. Some authorities advocate doing all sanding of plywood with highly figured face on belt or stroke sanders, to prevent the waste of sanding-through by careful operation. Figure 27 shows a typical arrangement of a dimensioning and sanding department.

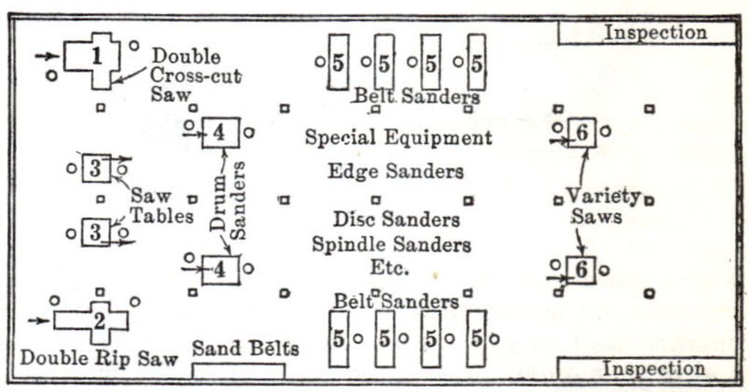

FIG. 27. Arrangement of dimensioning and sanding department. O indicates position of operator. Numerals show sequence of operations.

Redimensioning sometimes is practiced when plywood is glued and sanded in multiples, as with small parts, where separate handling would greatly increase costs.

Inspection. The chief inspector usually is stationed between the sanding and crating departments, with subinspectors only in the face veneer and lumber core departments.

CURVED PLYWOOD. Flat plywood can be bent into several types of simple curvature more cheaply than it can be curved during the gluing process. Alternately veneers can be bent and made into simple and compound curvatures during the gluing process. The principal methods are these:

1. Flat plywood bent into *simple* curvature, with or without heat.
   a. Without recessed backs on concave side.
   b. With recessed backs on concave side.
2. Veneers bent into curves during the process of gluing into plywood.
   a. *Simple* curvature, with or without heat.
   b. Moderate *compound* curvature, formed plywood, between pairs of heated dies with unidirectional pressure.
   c. Wide range of *compound* curvature, molded plywood, using one rigid die and omnidirectional pressure from flexible bags, usually with heat.

Combinations of these methods are often advantageous, as suggested in the following descriptions.

CURVING FLAT PLYWOOD. Flat plywood, up to $^1/_4$ in. and sometimes up to $^3/_8$ in., can be bent into gradual curvature with relatively large radii, without steaming or other-

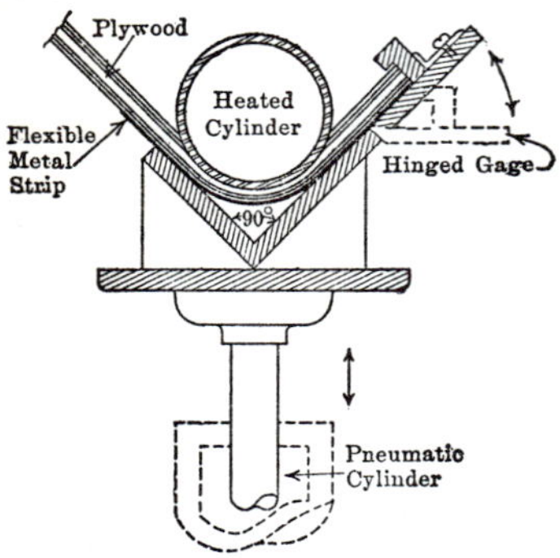

FIG. 28. Plywood bending press.

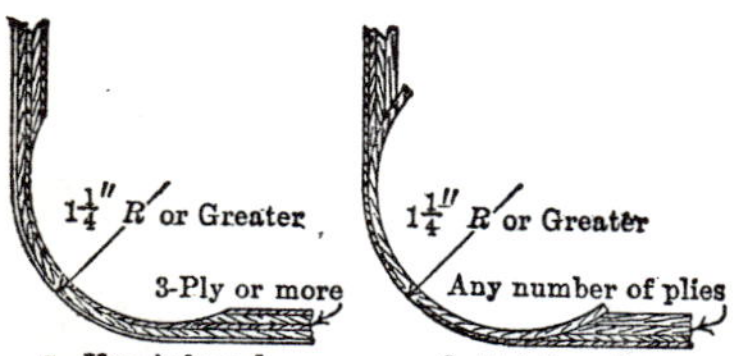

a. Unreinforced *b.* Reinforced

FIG. 29. Recessed backs for bending thin plywood.

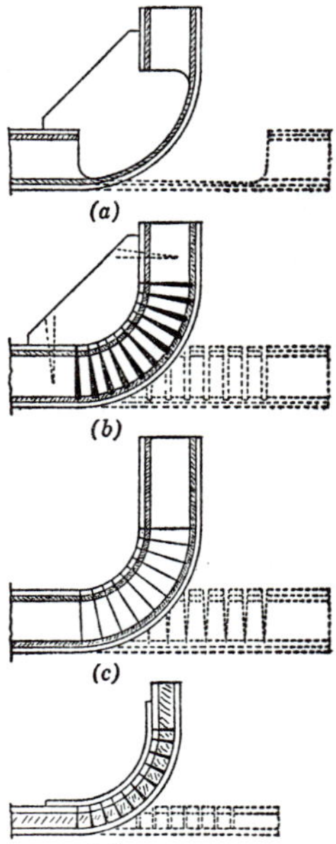

FIG. 30. Recessed backs for bending thick plywood.

wise heating, but such curves must be held in place by framework or grooves. Much depends on the grain direction of the veneer layers and their relative thickness. There is a distinct limit to which such simple curvature can be made without preliminary steaming and drying under pressure in some type of curved forms. When heat is employed, the bending press shown in Fig. 28, or the heated endless bed mechanism in Fig. 43, may be substituted for the curved forms. Under such heat bending, plywood curved from flat sheets is not dimensionally stable and tends to open or close somewhat, requiring bracing, but not to the same extent as plywood bent without heat.

Another technique is to recess the plywood on the concave side, with either a wide groove behind the entire curved area or a series of saw slots, both as shown in Figs. 29 and 30. Curved plywood with such recessed backs is quite limber and requires bracing by a supporting framework, by reinforcing strips of veneer glued in place, or by supporting blocks screwed or nailed in position. Moderate steaming is often used to soften the convex veneer layers over the recess and reduce danger of wood fracture. Where veneer strips are glued for such reinforcing, heat curing resin adhesives may be employed in the bending press shown in Fig. 28. When either reinforcing veneers or mechanically fastened strips are employed, the curves will be dimensionally stable and will not require supporting grooves or framework.

Where plywood is to be steamed for bending, or reinforcing veneers are to be attached with a heat curing resin adhesive, such plywood must be selected for heat and moisture resistance, preferably made with a phenolic or melamine resin adhesive (see Section 21, Article 10), although urea resins can be used under limited steaming or heating. In some cases, where heat only is employed, casein glued plywood may be used.

CURVING PLYWOOD DURING GLUING ASSEMBLY. As a practical manufacturing problem, only such curves can be economically bent in plywood assembly operations as can be formed in hydraulic presses with vertical pressure. There is a definite limit to the extent of curved segments that can be glued effectively. Figure 31 shows that, with a given direct pressure A, a 90-degree segment will have a pressure of approximately $^2/_3\ A$ at its extremity B, while a 120-degree segment will have a pressure of $^1/_2\ A$ at the extremity C. It is possible to bend a 120-degree segment successfully, and it will include enough more than a quadrant to produce grooved and well-fitted corners. A 180-degree segment would have no vertical pressure component at D; pressure would be derived only from the snugness of the forms. The limit of 120-degree segmental angle applies to both simple and compound curved surfaces.

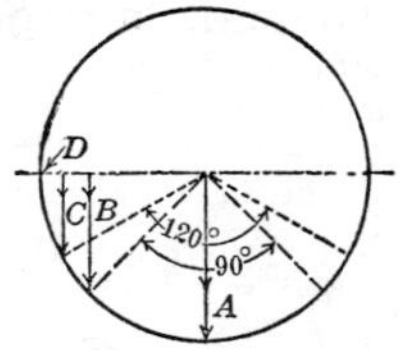

FIG. 31. Diagram showing how edge pressure (at B and C) diminishes with angle subtended, in bending plywood.

Figures 32 and 33 show typical curved segment plywood constructions. The thickest layers of veneer are bent with the grain, and thin sheets only are bent against the grain. Increasing the number of plies (Fig. 32) often facilitates the bending operation, and reduces the thickness of the layers.

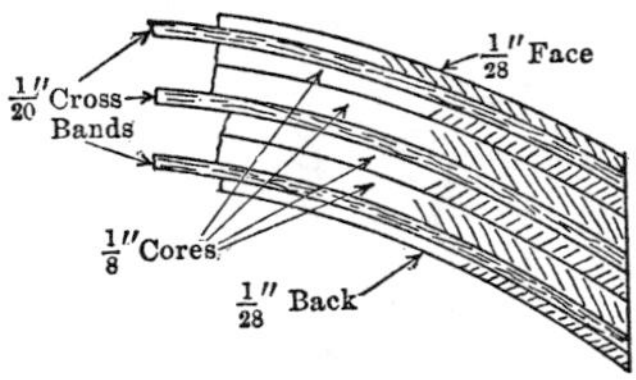

FIG. 32. Typical curved construction, thick plywood.

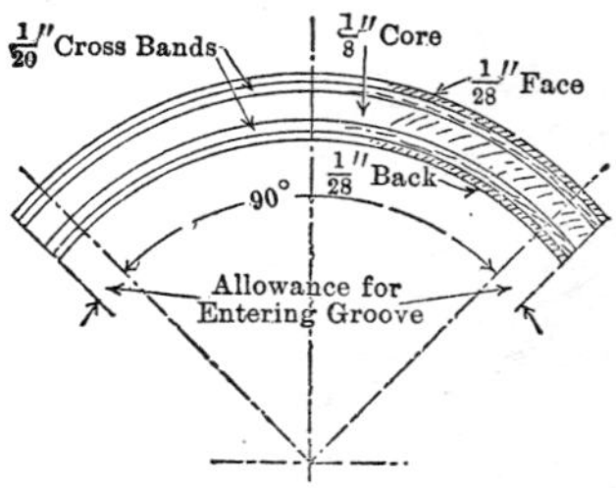

FIG. 33. Typical curved construction, thin plywood.

When curved plywood with lumber cores is required, such cores can be band sawed from laminated lumber blocks of suitable size, as shown in Fig. 34. In the left sketch (*a*) the lumber core units are of uniform thickness, on a radial line, and shaded crescent-shaped fillers are sawed out and employed for pressure purposes. The right sketch (*b*) sacrifices the feature of uniform thickness and eliminates the filler, as is practical for many designs. In either case it is important that the saw kerf be wide enough to accommodate the several layers of veneer that are inserted for curved faces and backs. Crossings are used, are coated on both sides with glue, and the assembly laid up in the usual manner. The top and bottom segments are used against the head blocks, and the bale is hydraulically pressed and clamped as in flat plywood.

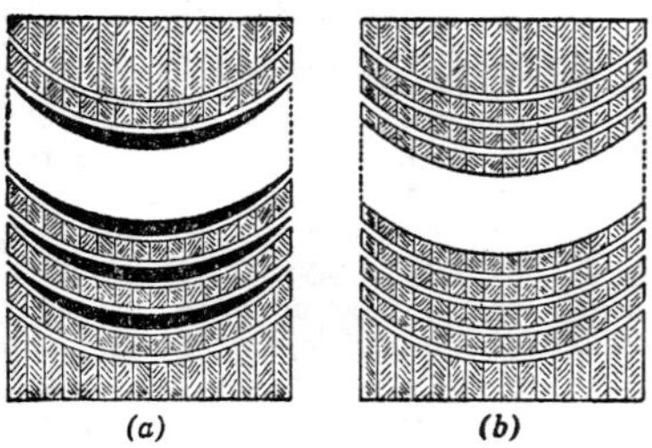

FIG. 34. Band-sawed lumber cores for curved plywood.

Adhesives for Bonding Plywood Curves at Room Temperature. Plywood bending operations can be carried out with a wide range of liquid glues or adhesives, vegetable (cassava), casein, or any of the room-temperature resin adhesives. A quick-setting glue, with special accelerator, is desirable to permit the re-use of forms and dies. Standard hydraulic presses, I beams, and clamping bolts are employed, as all forms should have square end blocks for firm clamping (see Figs. 34 and 35). After removal from gluing forms, the curved plywood should be racked in curved skeleton forms and weighted during redrying.

Hot Press Adhesives for Curved Plywood. Curved plywood can be made with much shorter operating cycles in hot presses, where pairs of heated metal dies are used. Such dies must be accurately shaped and can be used only for a given thickness of plywood. These dies may have heating units incorporated, steam or electric, or may take their heat from the platens of a hot press, to which they can be firmly attached. The same limits control the extent of simple plywood curvature from a unidirectional pressure piston as outlined on p. 25-22. Synthetic resin adhesives are preferred for this type of work (see p. 21-23) and permit operating cycles of 5 to 30 min.

FORMED PLYWOOD refers to plywood pressed to shape between pairs of heated metal dies during the gluing operation, but having only a moderate degree of compound curvature. Wood, when heated, acquires a certain degree of ductility, and can be pressed into such shapes as Toastmaster trays, in which the center is depressed below the surrounding rim; or a barrel stave, with both body and bilge curves combined; or a crowned or

(a)

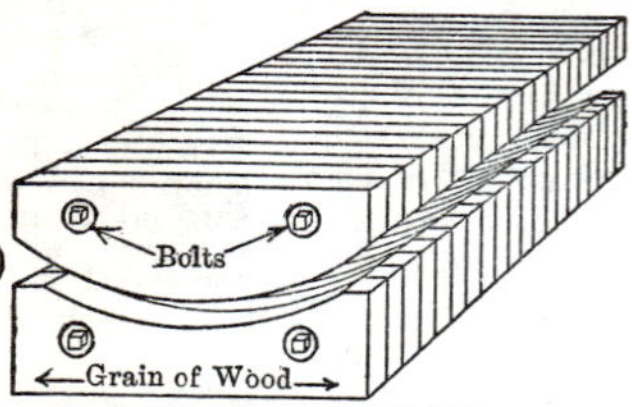

Single Wood Forms for Cold Pressing
(For Hot Pressing made integral of Aluminum)

(b)

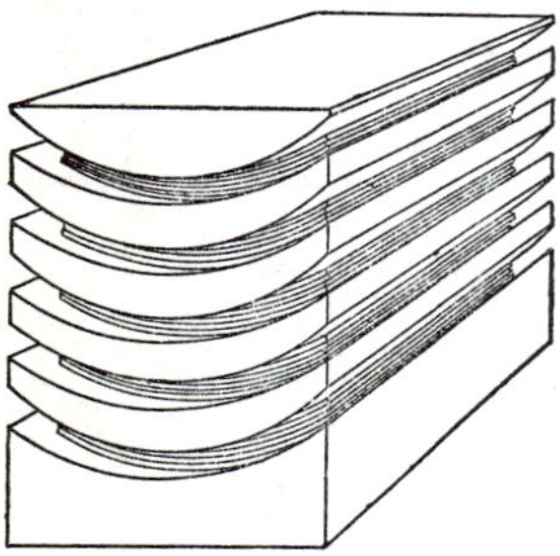

Multiple Aluminum Forms for Cold Pressing

Fig. 35. Platens for bending plywood.

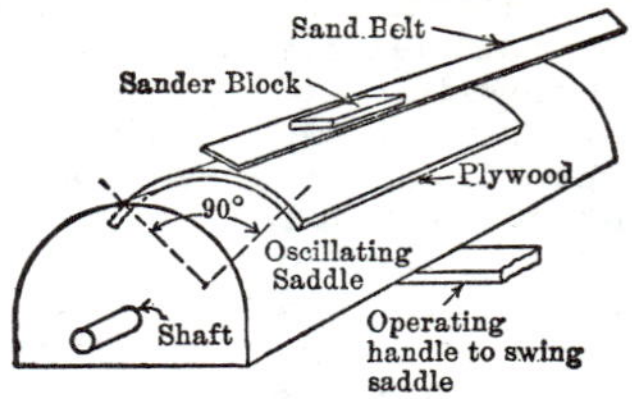

Fig. 36. Method of sanding convex plywood.

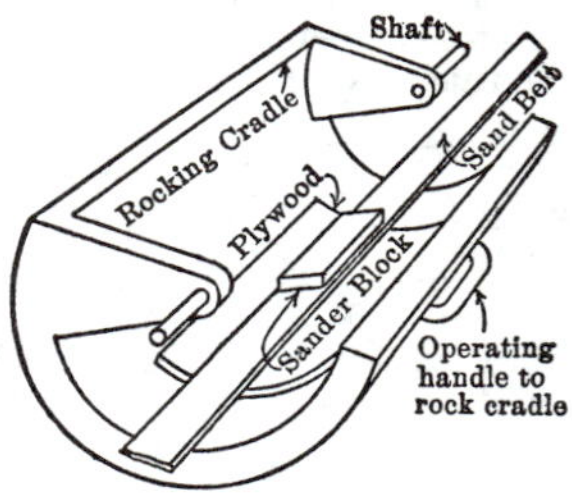

Fig. 37. Method of sanding co-cave plywood.

dished head, as for a barrel or wardrobe trunk. Such formed plywood can be compared with molded plywood, described below, in which compound curvature is much more pronounced.

FLEXIBLE PLYWOOD may be made 3-ply with a very thin core, and thick loosely cut outsides. It is suitable only for rough work that is to be painted, and is not adaptable to figured face woods.

SANDING OF CURVED SHAPES presents some unusual problems. Good results may be obtained by attaching either the saddle or cradle (Figs. 36 and 37) to a belt sander to facilitate sanding either the inside or outside face of curved plywood.

MOLDED PLYWOOD. The flexible bag pressure process, developed during the 1930's, employs a rigid member as half of the mold (matrix, form, or die) and a nonrigid or flexible bag as the other half of the mold. By various methods of inflating or deflating, the pressure can be applied in all directions, in net effect, perpendicular to tangents on all curved plywood surfaces. This gives great versatility to the resulting plywood and permits the production of intricate plywood shapes of various degrees of compound curvature, such as boat hulls, airplane fuselages and wings, reflectors, domes, hollow floats, and stream-lined fuel tanks. Since pressure is wholly flexible, the thickness of the plywood can be increased for strength or decreased for weight reduction. The average pressure requirements are less than for flat plywood.

The veneer strips, of which the various layers are composed, are tailored to shape; for example, strips like small sections of an orange peel for a hemisphere, or tapering strips for a boat hull. These strips are coated on both sides with a suitable resin adhesive; this coating is dried before cutting to shape. The coated strips are properly assembled, laid up, on a male form or dummy, and temporarily fastened in place by tacks, tape, or staples, before pressure is applied. During the pressure application, the veneer strips are forced into place, and as the pressure continues, are firmly bonded by the resin adhesive at the proper temperature. A resin adhesive for this purpose must have a relatively long flow period at around 180 F, considerably below the curing range, to permit the veneers to become properly nested before the cure takes place, usually around 250 F. The bags are of synthetic or natural rubber that can be clamped tight against leakage. The pressure medium is usually steam, sometimes supplemented by compressed air, although this combination deteriorates the bag more rapidly than steam alone.

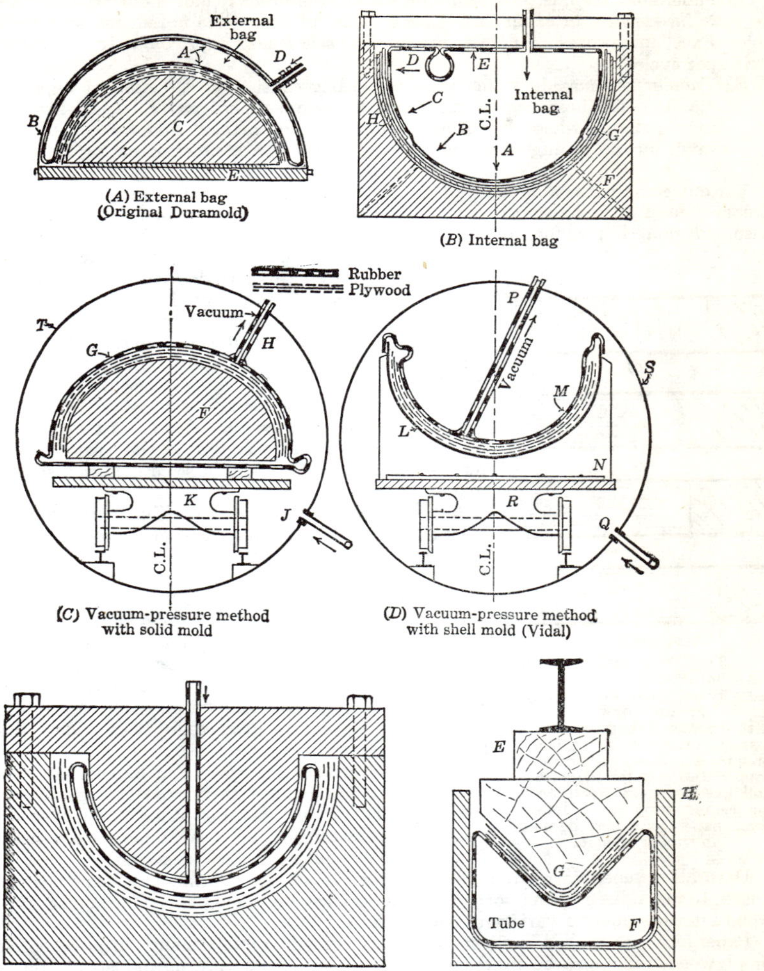

(A) External bag (Original Duramold)

(B) Internal bag

(C) Vacuum-pressure method with solid mold

(D) Vacuum-pressure method with shell mold (Vidal)

(E) Pressure cushion bag

(F) Tubular pressure, multidirectional

Fig. 38. Omnidirectional pressure applications in making compound curved plywood by flexible bag (fluid) pressure.

There are many ways of applying this omnidirectional pressure, some of the more important being illustrated in Fig. 38, and briefly outlined as follows.

A. *External bag, inflated* against an enclosing outer shell. Mold of male type, solid and may be heated. Assembly laid up directly on mold.

B. *Internal bag, inflated* against a closed mold, which may be heated. Mold of female type, assembly usually laid up on a male dummy and inverted into the mold.

C. *Autoclave process,* plywood assembled on the male mold, enclosed within bag, which is deflated by vacuum before being placed in autoclave. Autoclave requires mechanically manipulated doors, for quick opening and closing and tight sealing. Autoclave pressure supplements the vacuum within the bag. Efficient method for large complicated curvature, but with an operating cycle of 2 to 3 hr.

D. Autoclave process, as in *C*, except that plywood assembly is laid up flat as a blanket, or upon a male dummy. Plywood assembly transferred to the female mold, usually a braced steel shell. Vacuum-pressure process as in *C*.

E. A modification of *B*, an *internal inflated* bag against a closed mold, in which the upper male mold extends to within 2 or 3 in. of the lower female mold. This upper mold is faced with an expandable bag so that inflation time is reduced to a minimum. Excellent process for fast production schedules, approximately 10 to 15 min per cycle.

F. Tubular pressure, in which the flexible bag consists of a section of 2 1/2-in. inside diameter fire hose, which is inflated by steam or hot water. Ideal for relatively simple shapes where curvature is chiefly in one plane. Adaptable and simple for small moldings, angles, coves, pilasters, etc., with a short operating cycle as in *E*.

Various combinations of these methods as well as other flexible pressure combinations are also used to a lesser extent. These various methods of making molded plywood assumed great importance in World War II (1942–1945), when thin light metals were scarce and plywood was found to be an acceptable alternate. They continue to give decided promise of extensive industrial use for civilian plywood products.

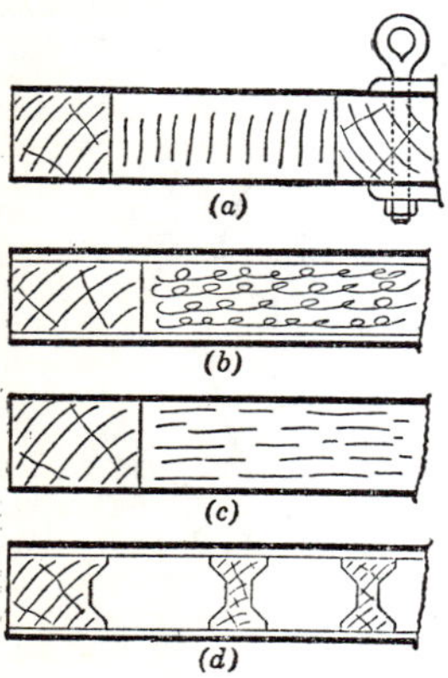

FIG. 39. Sandwich constructions: (*a*) metal faces, end grain balsa cores, solid wood inserts for bolts; (*b*) plywood faces with sponge rubber or other foamed core material; (*c*) Compreg faces with side grain wood cores; (*d*) Plymetal faces with hollow cores, using wood strips for center supports. Honeycomb paper cores may be used in similar fashion.

MODIFIED PLYWOOD. Standard and conventional plywood consists wholly of wood layers, with the wood grain of alternate layers crossed. There are many comparable constructions, in which other materials may be used for some or all of the layers, in which case less attention is paid to the cross-laid effects. Adhesives must be properly selected and adapted to the various materials, for which information should be secured from resin adhesive manufacturers.

Another technique, developed chiefly during the war emergency, is to impregnate one or more layers of wood or other material with a synthetic resin to facilitate densification, as well as to increase hardness and strength. Under normal pressures, up to 500 psi, this densified plywood product is called Impreg; under pressures of 1000 to 2000 psi, Compreg, Pregwood, etc. When layers are of paper or like products, the name Papreg is used.

Metal layers, with special metal-to-wood adhesives (see Section 21, Article 12) may be one-side, two-side, or internal, and are called Plymetal. It is widely used for transportation equipment.

Sandwich constructions (see Fig. 39) consist of a low-density or hollow core, with normal or high-density plywood exteriors or with metal or other high-tensile-strength material as coverings. Solid wood blocks are inserted in the core for fastening purposes as shown.

Decorative veneer faces are applied to many types of insulating fiber boards, to asbestos board, to metal fireproofing partitions, and to plaster surfaces, all to harmonize with other wood interiors under a variety of trade names.

Paper faces are applied to plywood, of various grades and for many purposes; on Douglas fir plywood, for improved protection and painting properties; on industrial plywood, to permit die punching; on plywood box shooks, for strength, tightness, and appearance.

Laminated resin products, paper, cloth, glass cloth, etc., are used as plywood faces for kitchen, hospital, and store equipment.

In general, these modified plywood products are made flat, usually by hot pressing, employing suitable resin adhesives, in much the same way as standard plywood. However, the supplementary methods described of bending and curving plywood apply to certain of these combination laminated materials.

Manufacturers of Plywood Equipment. The following are some of the leading makers: Merritt Engineering and Sales Co., Lockport, N. Y.; Chas. E. Francis Co., Huntington, Ind.; Williams White and Co., Moline, Ill.; Black Brothers Co., Mendota, Ill.; Plycor Co., Chicago, Ill.; Columbia Machinery and Engineering Corp., Hamilton, Ohio; Union Tool Corp., Warsaw, Ind.; G. M. Diehl Machine Works, Wabash, Ind.; Handy Manufacturing Co., Chicago, Ill.

Bibliography for Plywood

Knight and Wulpi, *Veneers and Plywood,* Ronald Press Co., New York, 1926.
A. Mora, Plywood, *Timber & Plywood,* London, 1932.
C. B. Norris, *Technique of Plywood,* I. F. Laucks, Inc., Seattle, 1942.
T. D. Perry, *Modern Plywood,* 2nd ed., Pitman Publishing Corp., New York, 1948.
Wood and Linn, *Plywoods,* W. & A. K. Johnston, Ltd., London, 1942; American Edition, Chemical Publishing Co., Brooklyn, N. Y., 1943.

9. WOOD FINISHING

The term *finishing* generally includes *polish sanding, filling, staining, lacquering,* and *waxing.* The preliminary sanding operations (see p. 25-16) precede the cabinet assembly stage. Final *polish sanding,* either by hand or by mechanical devices, after cabinet assembly, should leave a very smooth surface which will not place an undue burden on the filler. Many factories sponge before polish sanding to raise the grain, and cause the slender sander shreds to become more or less erect for removal in sanding. Since sponging tends to clean out the wood pores, they must be filled even with the surface, especially in open-grain woods like oak.

Filler. Standard fillers usually consist of silex and other ingredients, in a paste form, diluted with naphtha. They are brushed on, allowed to dry partially, and rubbed into the pores with a circular motion. The surplus is wiped off, without removing any filler below the finished surface. Defective filling is indicated by pockmarks, when too much filler has been wiped off, or the depressions are too open for the filler to stick. Correct results should follow the first filling; deficiencies may be remedied by repeating the process. In modern finishing technique, filler replaces the many coats of varnish formerly used.

Stain. After filling, stain is applied (this sequence is sometimes reversed), usually with a spray gun, to obtain the desired color. The color may be that to which the species will normally mellow from air exposure or may be a purely artificial color. Stain and filler make end wood appear darker than side wood, because of the volume of penetration. Care must be taken to limit the amount of stain on colored areas. Stains are distinguished by the solvent as: (1) *Water stains,* erroneously called acid stains. They are the most permanent in color, but darken a trifle as the wood ages. They raise the grain more than other stains. (2) *Spirit stains,* dissolved in alcohol. They dry more quickly, fade easily, but do not raise the grain. (3) *Oil stains,* which are not extensively used in high-grade furniture because they "bleed" with the solvents used in later lacquer applications. In cheaper finishes, filler and stain are combined and applied in one operation; results are inferior. Water stain and aniline dyes, made from coal tar, are preferred for best all-round results.

Varnish, formerly a widely used final finish, has largely given place to lacquers, which have much shorter drying cycles.

Lacquer. Many clear lacquers, of similar characteristics, are gradually developing into high-solid, quick-drying coatings. They perform more satisfactorily than varnish in drying ovens, and can be varied in constituents to meet the conditions of exacting service. A large number of them are made from a synthetic resin base, and many are formulated to withstand, without discoloration or blister, water, heat, and alcoholic solutions. The number of coats applied varies, in general, with the degree of permanent finish desired. Consult the literature of the well-known makers of finishing materials for further details.

Force dry finish is a recently developed technique consisting of a combination of synthetic resins and other chemical ingredients that can be dried rapidly at elevated temperatures, 140 F $\pm$ in 60 to 90 min. It permits the use of high solids content, 40 to 50%, and thus requires fewer coats for a given weight of finish. The shortened time schedule facilitates conveyorized finishing systems, thus reducing the time of handling. This force dry finish method may be applied over a lacquer sealer or used as a self-sealer; in the latter case a subsequent top coat and a second baking are required. It may be used for gloss, semigloss, or as a conventional rub and polish finish in the regular manner. Consult the literature of the well-known makers of finishing materials for further details.

Wax is one of the oldest finishes known. It is applied as a paste over about two coats of shellac. The shellac is applied on properly stained wood, which then usually is sanded. After the wax is dry, it is rubbed to a dull polish. If rewaxed occasionally an excellent and easily maintained finish is obtained.

Other finishing materials are *colored lacquers* and *enamels* and various trade-marked products, some of which are of excellent quality. Their application is similar to the process outlined above. Usually a less expensive undercoater is applied, followed by the finish coats. Care must be taken to prevent the solvents used in the successive coats from bleeding the preceding coats, thus producing muddy colors and sticky effects.

ARRANGEMENT OF FINISHING EQUIPMENT. Figure 40 shows a typical finishing department. Finishing rooms must be dustfree, and thoroughly fireproof, since many solvent fumes are highly explosive. All fan motors must be sparkproof. Conveyor chains and other moving parts should be guarded against friction and sparks. Fully air-conditioned finishing rooms permit standardization of drying conditions and make them independent of climate and weather.

Many finishing rooms have overhead conveyors, from which semifinished units are suspended, progressing successively from station to station in the finishing cycle.

Some cheaper products are finished, in whole or in part, by dipping in tanks. Dipping applies especially to stains and sealers. Provision then must be made to drain and drip, to avoid excess material forming beads in objectionable locations.

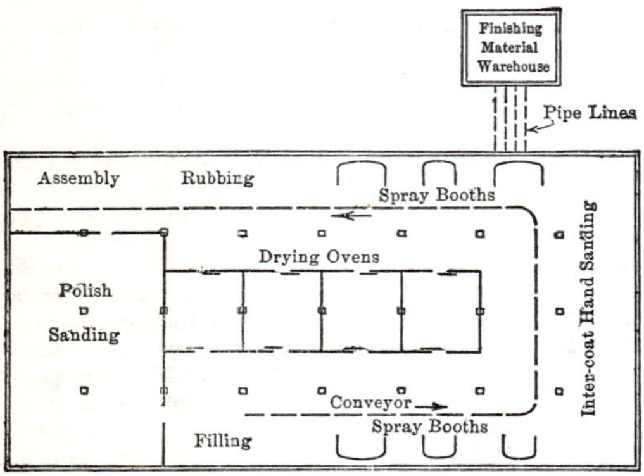

FIG. 40. Arrangement of finishing department.

The polish sanding department, immediately preceding the application of finishing materials, should be separated from the spray booths by airtight partitions to reduce dust hazards. Drying ovens should be centrally located, with several separately controlled, well-ventilated rooms to provide conditions required for drying filler, stain, undercoater, and lacquer. Finishing materials are best stored in a separate firesafe building, where all standard mixing can be done. All material should be piped and pumped to individual booth locations.

Hand sanding often is necessary between coats to remove small pimples that form around small particles of dust and undissolved solids. Many better grades of products require *rubbing*, to produce a dull nonlustrous surface. On irregular shapes, rubbing may be by hand or by holding the work against a cushioned revolving spindle. For table tops and similar products, mechanical stroke rubbers, with reciprocating cushion pads, are used. Rubbing down a good grade of lacquer gives a much better finish than semigloss lacquer, which lacks durability. The usual rubbing compounds are powdered pumice and oil, mineral oil, and diatomaceous earth.

Spray booths are almost universally used in finishing rooms. They are closely regulated by local ordinances, requiring direct fan connections to outdoors, to preserve health of operators and to reduce fire hazard. Spray booths generally are equipped with automatic revolving tables, with pneumatic vertical travel, to give access to legs, backs, and under parts.

Spray guns are atomizers, operated by compressed air at 75 to 100 psi. Guns giving smooth, glasslike fan sprays are available that cover up to 60 in. width. Air volume may vary from $^1/_2$ to 50 cu ft per min. Dual, triple, and quadruple heads are available for attachment on spray guns for special products. Consult manufacturers' literature for details of spray booths and spray guns.

10. WOOD PRESERVATION

Preservatives that lengthen the life of wood exposed to decay, insects, marine borers, and fire are: (1) preservatives of an oily nature injected into wood as such; (2) chemicals injected into wood in water solutions; (3) petroleum oils fortified with toxic chemicals; (4) fire-retardant chemicals injected into wood in water solutions. Intelligent selection for a particular service greatly prolongs usefulness of wood, with excellent ultimate economy.

PRESERVATIVE OILS. Coal-tar Creosote. As used in wood preserving, creosote is a distillate of coal tar produced by high-temperature carbonization of bituminous coal. It is the most generally adaptable preservative for use with wood in contact with ground or situated under water. Its principal advantages are its general widespread use and availability and its highly permanent quality in service. Its principal disadvantages are its sometimes objectionable odor and appearance and the fact that it is difficult to work with, especially in the hot sun. It cannot be painted successfully. The odor is not strictly objectionable except where it may be transferred to stored food products. Freshly creosoted wood is inflammable, but after it has seasoned it is no more inflammable than untreated wood. Specifications for creosote and its blends for both pressure and non-pressure processes are listed under *Specifications* at the end of this article.

Creosote and Coal-tar Solutions. Occasionally creosote is diluted with coal tar for treating items, except poles, that are generally creosoted. The blends most commonly used contain 20 to 30% coal tar.

Creosote-petroleum Solutions. Creosote is also blended with petroleum quite extensively for the treatment of crossties used in the arid regions of the West. On the average, blends contain 50% petroleum, and although the toxicity is reduced, the treated products are cheaper and possess a measurable resistance to checking.

Water-gas Tar Creosote and Wood-tar Creosote. These are somewhat less effective preservatives and used only occasionally.

CHEMICALS IN WATER SOLUTION. Chemicals injected into wood in water solutions have the distinction of leaving the wood clean, and although its color might be changed, it is readily paintable after it has dried. Their principal use is for sawed lumber in structures which must be painted and for general indoor and outdoor construction where clean treatment is necessary. Such uses include wet-process mill roofs, piers and boardwalks, bridges and highway structures, mine timbers, refrigerated buildings, sanitary works and water supply, and stadiums.

Zinc Chloride. This chemical was used quite widely several years ago, but it has dropped off in favor of improved preservatives. It was used in a 3 to 5% solution which left about 1 lb of the dry chemical per cubic foot of wood. It is still covered in some specifications. See under *Specifications* at end of this article.

Chromated Zinc Chloride. The chromated zinc chloride, which has replaced straight zinc chloride, contains not less than 17.5% sodium dichromate. The American Wood-Preservers' Association requires a net retention of 3/4 lb of dry chemical per cubic foot of wood in their specifications. See also under *Specifications* at the end of this article.

Wolman Wood Preservative Salts. The Tanalith compound is the most widely used of the Wolman wood preservatives. It contains 37 1/2% sodium chromate, 25% disodium hydrogen arsenate, 25% sodium fluoride, and 12 1/2% dinitrophenol. It is noted for its anticorrosive qualities. Lumber treated with Tanalith can be successfully glued with phenol or resorcinol glues. It makes no difference whether the wood is treated first, then glued, or whether it is treated after being glued; a satisfactory bond will result. The American Wood-Preservers' Association, in its specifications, requires a net retention of 0.35 lb of dry chemical per cubic foot of wood. See also under *Specifications* at the end of this article.

Zinc Meta Arsenite. This preservative has been used somewhat in the past, but at present commercial treatments are not available. The American Wood-Preservers' Association specify a 0.35 lb per cu ft minimum net retention for this preservative. See also under *Specifications* at the end of this article.

Greensalt. This preservative, also known by the trade name of Erdalith, is a mixture of 56% sodium or potassium dichromate, 33% copper sulfate, and 11% arsenic acid. It becomes highly permanent in the wood and is especially effective against fungi which attack wood in ground contact. It is a clean, paintable preservative used for the same purposes as creosote. See under *Specifications* at the end of this article.

Miscellaneous. There are other water-borne preservatives in use which are not listed in the American Wood-Preservers' Association specifications and deserve mention. They are (1) Chemonite, which is copper-arsenate with a small excess of copper hydroxide in a strongly ammoniacal solution; (2) Celcure, which is a slightly acid aqueous solution of a soluble chromate and a soluble copper salt; (3) Osmose, a fluoride-phenol-chromate mixture. Other preservatives, which have practically disappeared from use, are straight sodium fluoride, mercuric chloride, copper sulfate, and arsenic salts.

PETROLEUM OILS FORTIFIED WITH TOXIC CHEMICALS. The chemicals pentachlorophenol, copper naphthenate, and zinc naphthenate are used in petroleum solutions of various types. For products such as millwork, the petroleum is of the volatile type, so that the wood will eventually take paint. When used for poles, crossarms, etc., the petroleum is on the non-volatile type. See under *Specifications* at the end of this article.

FIRE-RETARDANT CHEMICALS IN WATER SOLUTIONS. Fire-retardant chemicals are injected into wood in chemical retentions five to ten times as great as the toxic salt preservative chemicals. A heavy retention, 3 to 6 lb per cu ft, is needed to delay ignition and check the spread of fire over wood surfaces. The important fire-retardants, as listed in the Navy Department Specifications, are as follows:

Type I. Consists of a mixture of 3% sodium dichromate, 78% ammonium sulfate, and 19% ammonium phosphate (either monobasic or dibasic).

Type II. Consists of a mixture of 10% diammonium phosphate, 60% ammonium sulfate, 10% anhydrous sodium tetraborate (borax), and 20% boric acid. This formula is used under the trade name Minalith.

Type III. Consists of a mixture of 60% anhydrous sodium tetraborate and 40% boric acid.

Type IV. Straight zinc chloride or chromated zinc chloride.

Type V. Any undisclosed formula as may be acceptable by performance tests.

PREPARATION OF WOOD FOR TREATMENT. Before adequate treatment can be obtained by any of the commercial methods in common use, freshly cut wood must be properly prepared for processing by the following methods.

Peeling. As bark is impermeable to liquids, it must be peeled from round timbers to facilitate impregnation.

Air Seasoning or Kiln Drying. Because green wood is harder to treat than dry wood, it should be dried before treatment.

Preframing. Where practicable, all cutting, framing, boring, etc., should be done before treatment.

Incising is an improved method of increasing penetration in species that are hard to treat. The timbers are passed through a machine, in which teeth, on horizontal and vertical rollers, slit into the wood $^1/_2$ to $^3/_4$ in.

Steam Conditioning. Where it is either impracticable or undesirable to air-season or kiln-dry lumber before treatment, a steaming-and-vacuum process is commonly employed. This involves steaming the lumber in the pressure-treating cylinder for several hours, at 7 to 20 psi gage pressure, after which a vacuum is drawn for an hour or more. The effect of the vacuum is to lower the boiling temperature in the cylinder and facilitate evaporation of moisture from the wood.

Boiling under Vacuum. Another substitute for air-seasoning or kiln-drying before creosote treatment is boiling the wood in creosote under a vacuum. This process is called Boultonizing. The standard cylinder charge is heated in the preservative oil at 180 to 210 F under vacuum, which causes the water in the wood to vaporize off.

WOOD PRESERVING PROCESSES. Nonpressure Processes. Methods of treating that do not use pressure do not result in as thorough impregnation as pressure processes, but they are the only possible means of treating where pressure equipment is not available or where pressure treating is impracticable. The results of nonpressure processes, with the exception of the hot and cold bath, are only fair at best, but are no doubt better than no treatment at all. Nonpressure processes include the following: (1) Hot and cold bath, which is the most effective nonpressure process. The wood is heated in preservative in an open tank for several hours, and then cooled in preservative for a like period. In the first bath, air is greatly expanded by heating, and in the cold bath this air contracts and draws the preservative into the wood. This is an important process in butt treating cedar poles. (2) Brush and spray treatments. (3) Dipping. (4) Steeping. (5) Charring. (6) Diffusion process.

Pressure Processes. These are the most effective methods of treatment. The lumber, or round timbers, separated to expose all surfaces, are placed on steel tram cars and run into a long steel cylinder, which then is closed and filled with preservative under pressure. The preservative pressure may range from 85 to 175 psi, and the temperature may range from 100 to 200 F. The full-cell and empty-cell are the two important pressure processes.

1. *Full-cell Process.* This process is preferable for both oil and water solutions, when the maximum absorption of preservative is desired. The steps in the process are: (1) charging the cylinder; (2) applying a preliminary vacuum (at least 22 in. of mercury) to remove all possible air from the cylinder and wood; (3) admission of preservative to the cylinder without breaking the vacuum; (4) application of pressure until required absorption is obtained; (5) withdrawal of preservative; (6) a short final vacuum or standing period to free the charge from dripping preservative.

2. *Empty-cell Process.* This process is preferable when the maximum penetration is desired with only a moderate amount of preservative and is practicable only with easily treated lumber. It is generally called the Rueping process. It differs from the full-cell process in that instead of an initial vacuum being applied to the charge, initial air pressure of 25 to 100 psi is applied. The preservative is then introduced while the charge is still under air pressure. After the charge has been treated and pressure released, the trapped air in the wood will force part of the preservative out. The charge, therefore, will be deeply penetrated, but will have only a moderate amount of preservative, which reduces the cost of treating.

FIELD TREATMENT. Insofar as practicable, lumber and timbers should be framed, bored, and cut to length before treatment, and field work of this character should be limited to a minimum. The use of cant hooks, pickaroons, peavies, and other pointed tools, except end hooks, should not be allowed in handling treated wood. The greatest care should be exercised in unloading and handling so as not to expose any untreated wood. All timbers which must be cut or bored after treatment should have the surfaces so exposed covered with hot preservative solution. Where cut, the surfaces should be painted thoroughly with two coats of hot preservative. Where holes are bored, they should be poured full of hot preservative. Horizontal holes may be filled by pouring the preservative in them through a bent funnel. The use of a special instrument for applying the preservative to the holes under pressure will result in much better protection.

DRYING AFTER TREATMENT. When wood is treated with water solutions, it is desirable to air-season or kiln-dry it after treatment and before it is installed, to insure against shrinkage and other bad effects of wet wood. When the lumber is to be painted, it is imperative that it be dry when the paint is applied.

SPECIFICATIONS FOR WOOD-PRESERVATION TREATMENTS. A summary of the various specifications for the preservative treatment of wood is given in the following paragraphs.

American Wood-Preservers' Association. Coal tar creosote, P1-49. Creosote-coal tar solutions, P2-48. Creosote-petroleum solutions, P3-48. Zinc chloride, P5-49. Chromated zinc chloride, P5-49. Wolman wood preservative (Tanalith) P5-49. Zinc meta arsenate, P5-49. Greensalt, P5-49. Pentachlorophenol, P8-49. Copper naphthenate, P8-49.

Treating Methods. Lumber and timbers, T1-49 and T2-49. Piles, T1-49, and T3-49. Poles, T1-49, and T4-49. Posts, T1-49, and T5-49. Crossties, T1-49 and T6-49. Wood blocks, T1-49 and T7-49 Pole butt, T9-49.

American Railway Engineering Association. Committee 17. Coal tar creosote (1946). Creosote-coal tar solutions (1946). Creosote-petroleum solutions (1945). Zinc chloride (1926). Chromated zinc chloride (1948). Wolman wood preservative (Tanalith) (1948).

American Association of State Highway Officials. Coal tar creosote, M-133-42. Creosote-coal tar solutions, M-133-42. Creosote-petroleum solutions, M-133-42. Zinc chloride, M-133-42. Chromated zinc chloride, M-133-42. Wolman wood preservative (Tanalith), M-133-42. Zinc meta arsenate, M-133-42.

Federal Standard Stock Catalog. Coal tar creosote, TT-W-556a. Creosote-coal tar solutions, TT-W-566a. Creosote-petroleum solutions, TT-W-568. Zinc chloride, TT-W-576. Chromated zinc chloride, TT-W-551. Wolman wood preservative (Tanalith), TT-W-573. Zinc meta arsenate, TT-W-581. Celcure, TT-W-546. Chemonite, TT-W-549. Pentachlorophenol, TT-W-570.

Treating Methods. TT-W-5716.

American Society for Testing Materials. Coal tar creosote, ASTM designation: D 391.

American Telephone and Telegraph Company. Greensalt, AT-6993.

U. S. Department of Commerce. Zinc naphthenate, Commercial Standard, TS-4895.

Navy Department Fireproofing Specifications. Types I-V, 51 C 38, and 51 C 40.

Treating Methods. Federal Specification TT-W-571b.

Bibliography for Wood Preservation

Federal Standard Stock Catalog. For sale by Superintendent of Documents, Washington, D. C. Revised periodically.
H. L. Henderson, *Air Seasoning and Kiln Drying of Wood,* J. B. Lyon, Albany, N. Y., 1936 and 1946.
Hunt and Garratt, *Wood Preservation,* McGraw-Hill Book Co., New York, 1936.
Manual of American Railway Engineering Association, Chicago, Ill. Revised annually.
Manual of Recommended Practice, American Wood Preservers' Association, Washington, D. C. Revised annually.
Manual of American Association of State Highway Officials. Revised annually.
Weiss, *Preservation of Structural Timber,* McGraw-Hill Book Co., New York, 1916.
Wood Handbook, Forest Products Laboratory, U. S. Department of Agriculture, 1940.

11. BENDING SOLID WOOD

Several processes of bending are in use. The technique differs between solid wood and plywood (see p. 25-22) as the fiber strain usually is much greater and more unbalanced in solid wood. Fundamentally, in bending wood strips the individual wood fibers tend to slide by each other slightly; outside fibers are stretched under tension, and inside fibers are compressed.

WET BENDING IN SOLID WOOD. Any wood part is more easily bent when soaked or steamed to a point of thorough saturation. Consequently, bending is better and breakage less in steamed or cooked air-dried lumber than in kiln-dried lumber, which is brash. Pieces to be bent should be dressed as nearly as possible to final sizes before bending.

The following process is applicable to pieces of approximately square cross section; for wider or flat pieces, see alternate method below. Three factors are important in this first type of wet bending: (1) a smooth rigid form without abrupt changes in curvature on the concave side of the curve; (2) a thin flexible metal strip, on the convex side, drawn tight, to reduce breakage; (3) sufficient end pressure to eliminate danger of shear on any fiber line. (See Fig. 41.) The cycle of bending and drying such rectangular members is much shortened if forming is done in a press under heat, as described below.

In another method of wet bending in solid wood members which are required in substantial quantities, such as chair backs of oblong cross section, pairs of metal forms are used. A series of steamed wet flats is placed between these forms, pressed in a hydraulic press, and the pressure maintained by bolts, as shown in Fig. 42. These clamped pairs of forms can be placed in a kiln at 150 to 180 F, the metal forms transmitting the heat rapidly to the wood members, and the pressure cycle reduced to a couple of hours, during which the bent members are dried to hold their shape with reasonable uniformity. Steam-heated forms, without supplementary bolt pressure, are also used, in which case higher temperatures may be employed, but a number of hydraulic presses are required. In general, wood curves bent by this process tend to open up somewhat on pressure release, and the design of forms should anticipate this factor. The literature of the Forest Products Laboratory at Madison, Wisconsin, should be consulted.

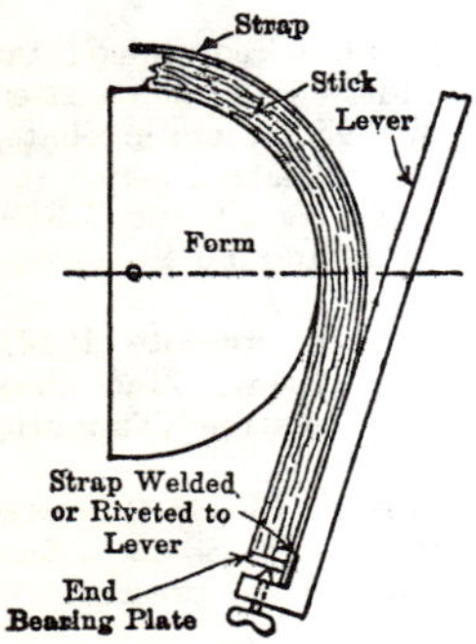

FIG. 41. Fixture for wet bending.

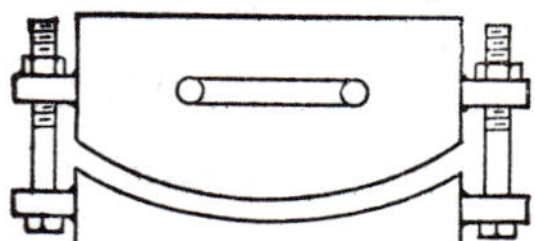
FIG. 42. Pair of metal forms for bending chair parts of solid wood.

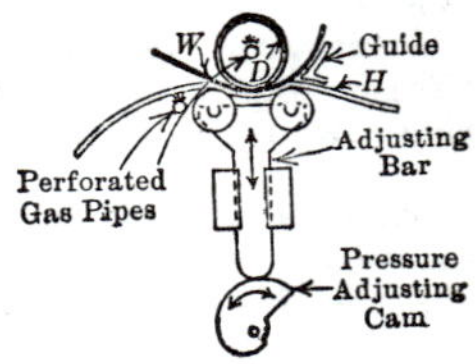

FIG. 43. Arrangement for dry bending solid wood.

DRY BENDING IN SOLID WOOD. A preferable method of bending wider and thinner pieces of solid wood is a series of three rollers, as in Fig. 43. Heat is applied to the upper roller *D*, and below the endless steel flexible conveyor belt *H*. The wood *W* is rocked back and forth until the desired degree of curvature is obtained. Temperatures may be as high as possible without charring wood surfaces. This method can be used for such items as curved slats for chair backs, and for plywood up to $^5/_{16}$ in. thick, if the adhesive is heat resistant (resin or casein). The bent pieces, both of solid wood or plywood, should be placed in skeleton curved forms until thoroughly cooled and normalized as to moisture content.

12. LAMINATED LUMBER

Solid timbers in long lengths and of substantial cross section have always been difficult to obtain since long straight logs of adequate size, without defects, are scarce. When timbers are cut from such logs they usually contain the heart section of the log, which is of questionable quality. Drying such large timbers is difficult, requiring many months, and usually results in splits, twists, and other deformations.

ADVANTAGES OF LAMINATED LUMBER. Laminated lumber fabricated from standard boards, glued together into timber sizes, while somewhat more costly, overcomes most of the difficulties enumerated above. Lumber in board form can be kiln dried without serious deterioration, glued up to any size, and such laminated products made available in a reasonable length of time and in any required quantity. End joints, adapted to anticipated loadings, can be staggered in the different layers and permit the use of relatively short boards. Narrow boards can be glued together edgewise as required. Large timbers, laminated, can be made from small trees.

Adhesives are available, stronger than the wood itself, and of maximum durability under severe weather and marine exposure. In laminated lumber constructions selected material can be used in the outer layers, where maximum stresses are encountered, and material of lower grade in the inner layers. Laminated lumber assemblies can be bent, during the gluing process, into curves for bow string trusses, cantilever arches, and other structural requirements. At points of greatest stress, such as the elbow in a cantilever arch, extra layers can be added for reinforcement. The versatility of laminated lumber is much greater than that of solid timbers.

END AND EDGE JOINTS. For column loading, square butt, or 30- to 60-degree-angle, joints may be used, properly staggered in adjacent layers. For beams subjected to bending stresses, end joints should be scarfed and lapped on a slope not steeper than twelve times the thickness of the layer. Scarfs should be cut true, closely fitted, glued with suitable adhesives, and anchored firmly to prevent slipping during pressure application. Best results are obtained when individual layers are scarfed, glued, and surfaced to uniform thickness before assembly. With adequate anchoring provisions, scarfed joints and laminae can be assembled in one operation. Scarf joints, properly made, give 80 to 90% the strength of solid lumber, and when properly staggered in the different layers, the overall strength compares favorably with solid timbers.

Edge joints are square and may or may not require pregluing and surfacing before assembly, depending on the direction of the loading.

THICKNESS TOLERANCES. All surfaces should be freshly planed before gluing, within an hour or two if possible. Thickness tolerances for softwoods are of the order of $^1/_{64}$ in., for hardwoods, $^1/_{100}$ in.

ADHESIVES. The most durable adhesive for laminated lumber is resorcinol (see Section 21, Article 10), which can be cured normally at room temperatures of 75 F and up, giving a bond as strong as the wood. Adhesives should be spread on both the mating surfaces in a joint, 5 to 7 $^1/_2$ lb of liquid resin per 100 sq ft of joint. Layers should be assembled immediately after being spread with glue, and pressure should be applied within 30 to 60 min. For maximum durability, especially with oak for ship construction, clamped assemblies should be cured in a kiln, with controlled humidity to prevent shrinking, until the most remote glue line (from the source of heat) reaches 180 to 210 F. Casein glue can be used for timbers protected from the weather, but not for marine work. Urea resin adhesives, unextended, are also used where exposure is not severe. Manufacturers of adhesives and glue should be consulted for further details.

PRESSURE. Pressure is usually applied in screw clamps (see Fig. 44), tightened by torque wrenches, indexed in foot-pounds. Wrenches should be calibrated for a standard

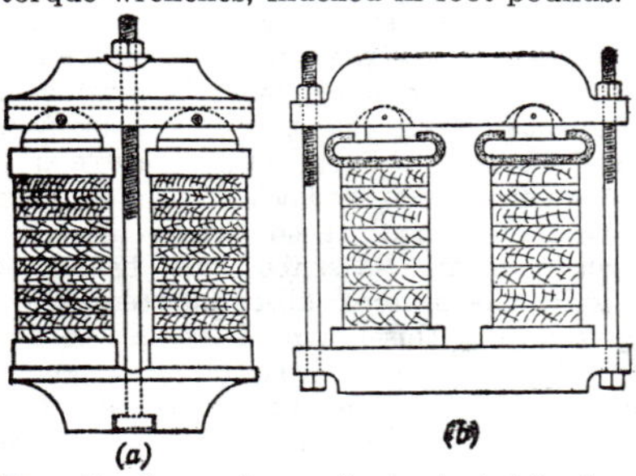

FIG. 44. Screw clamps for laminated lumber: (*a*) double C open clamp with single screw and equalizer; (*b*) closed clamp with twin screws and equalizer.

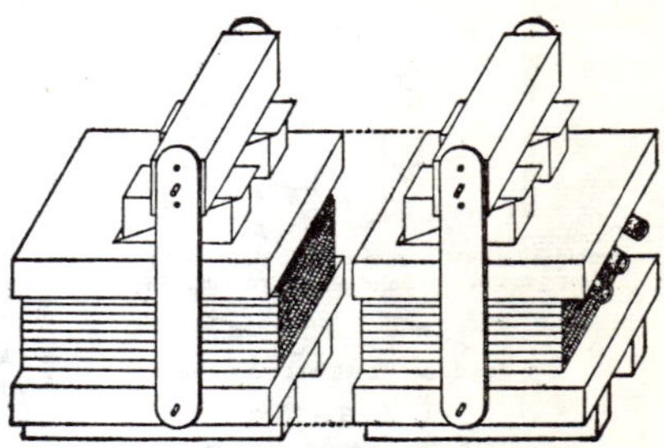

FIG. 45. Method of applying pressure for laminated lumber by flexible tubing. Wedges are used for take-up before inflating hose.

size and type of screw in a given clamp, with compressometers, and with pressure on the lumber assembly calculated to be no less than 100 psi. The closed type of clamp with double screws (Fig. 44*b*) is preferable to the open type with one screw. Equalizers or eveners should be provided, as shown, to insure uniform pressure. Cushion pads of rubber, or other resilient material, aid in such equalization. Where headblocks 2-in. thick are used, clamps should be spaced on 15-in. centers; with $^3/_4$-in. outside layers without headblocks, the spacing should be 9 in.

With straight members, pressure should be applied progressively from one end of the assembly to the other, or from the center simultaneously toward both ends. For curved members, the pressure should first be applied at the point of maximum curvature, and then progressively from this point toward each end. Retightening of bolts is often advisable, in 30 to 60 min, to compensate for the initial yielding of the wood under pressure.

In another method of pressure application flexible tubing, usually 2 $^1/_2$-in. (inside diameter) fire hose, is expanded; when deflated it is 4 in. wide. The technique of this method is shown in Fig. 45. The wedges are used to flatten the tube before inflation, since it is important to have a maximum of contact area after inflation. Pressure requirements within the hose increase as this contact area decreases. Alternatively, screw clamps can be used instead of wedges for flattening the tubes. This tubular pressure method has not been used extensively, but it holds considerable promise for quick and controllable clamping conditions.

CUSTOM VERSUS STOCK LAMINATING. Custom-made laminated lumber units, to a special design, have been made with entirely satisfactory results for some years. Bridge trusses and arches have given excellent service in Europe, and have been found sound after 25 to 30 years of use.

Plans are being developed by the lumber industry to establish standard sizes and grades of laminated items which can be carried as stock in lumber yards. Grade rules and tests are in process of formulation, and fabrication manuals are being prepared under the supervision of the Central Committee of Lumber Standards, in cooperation with the National Bureau of Standards.

13. WOODWORKING TOOLS

Small tools for woodworking can, in general, be used for working the plastic and phenolic materials, and also rubber. As the density of the product increases, carbide-tipped cutting edges become more and more essential for long service and smooth cutting.

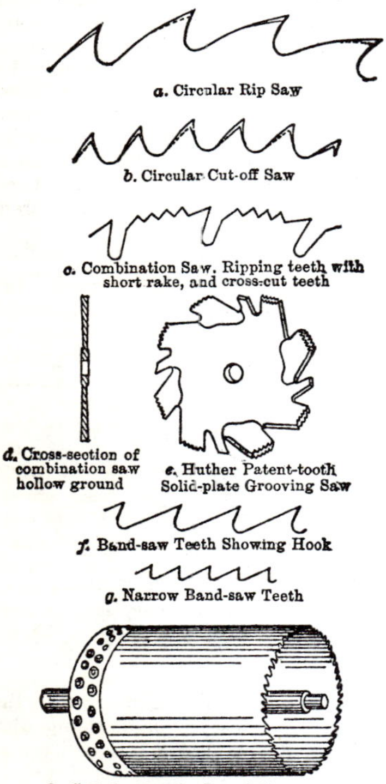

FIG. 46. Forms of saw teeth.

THEORY OF CUTTING WOOD. Wood is made up of parallel fibers. Cutting lengthwise is simply peeling a layer of fibers off from those below, as with a splitting axe or carpenter's plane. In cutting across the fibers, every fiber must be severed. A knife or chisel driven forcibly across grain is the simplest form of cross cutting. Both examples represent "line" cutting without waste.

Saws divide the fibers, lengthwise or crosswise, by cutting out a swath or *kerf*. In a cross-cut kerf the fiber is first severed at both sides of the swath; the short cut fiber is then removed by a lengthwise cutting at the bottom of the kerf. In ripping a kerf with the grain, the fiber in the swath to be removed is simultaneously cross cut and peeled loose from the sides of the kerf. Cross cutting requires a knife point cut at each side of the kerf. This is made by teeth alternately high at the sides and pointed and beveled on the front. *Raker* teeth may be used to peel or split the severed fiber and remove it. Ripping requires chisel-pointed teeth, set or swaged to the width of the kerf, and with enough hook to sever the fiber with an adzing cut. It is simultaneously split loose at both sides and removed.

Saws are set, or hollow ground, to reduce friction. Surface tension or case hardening of the lumber may cause boards to pinch or close the kerf during a cut. This is one of the principal reasons for wide clearance and wide kerfs. (See Fig. 46*d*.)

With revolving *cutter heads* having knives or bits, the cut generally is lengthwise. It is not exactly a splitting cut, since only at the finish line of straight grain woods is the cut parallel with the grain. At other parts of the cut, the knives are chiseling, or adzing, and cross-cutting the grain.

SAWS have two general classifications. (1) Saws for the *primary conversion* of logs into lumber and timber. Because they are used on live or green logs, there is no danger of overheating and damaging saw temper. Such saws should be designed to cut to approximate dimensions, as subsequent shrinkage necessitates machining to final size and shape. The cutting of tough green wood requires a different tooth from the one used in the cutting of stiffer and stronger, but more brittle, dry wood. (2) Saws for the *secondary conversion* of the board after seasoning, into intermediate or final products. Saws may overheat and lose temper in dry wood, and may scorch or discolor it. Accurate, smooth cuts must be made, to produce final dimensions with maximum smoothness and minimum waste. Teeth must be shaped to cut rapidly without heating, and accurately, with minimum power requirements. The thickness of saws is designated by Stubs gage. (See p. 6-05.)

Large saws, chiefly for sawmills in primary production, can be provided with *inserted teeth* of special hard steel, mechanically locked in position, that are interchangeable and can be replaced to compensate for excessive wear due to grinding.

Carbide-tipped teeth are also used on smaller saws in secondary production, brazed electronically to the standard steel blade. Carbide-tipped teeth give long cutting service, on densified wood as well as on casein and resin glue joints, where abrasive characteristics may be encountered. Carbide-tipped teeth have rather blunt cutting edges, cannot be sharpened in the regular way, and require high speeds for smooth cutting.

The two general classifications of primary and secondary conversion may be still further subdivided as below. Figure 46 shows the various forms of saw teeth.

Saws for Primary Production (converting green logs into timber and lumber):

- Circular saws.
  - Solid disks.
    - Parallel-to-grain or rip cut (Fig. 46*a*).
    - Perpendicular-to-grain or cross cut (Fig. 46*b*).
  - Inserted teeth.
  - Shingle saws, thin edges with thick centers.
- Band saws, continuous or endless (Fig. 46*f*).
  - Parallel-to-grain for cutting logs into lumber and timber.
  - Perpendicular-to-grain, drag saws for log lengths.
- Band saws, reciprocating, mostly gang resaws.
- Special-purpose saws.
  - Cylindrical, for cooperage stock (Fig. 46*h*).
  - Concave.

Saws for Secondary Production (converting dry boards into diversified wood products):

- Circular saws.
  - Ripping (Fig. 46*a*).
  - Cross-cutting (Fig. 46*b*).
  - Combination (Fig. 46*c*).
  - Special-purpose (Fig. 46*e*).
- Band saws, continuous or endless.
  - Ripping or resawing, wide (Fig. 46*f*).
  - Curved and special shape, narrow (Fig. 46*g*).
- Band saws, reciprocating.
  - Jig and scroll saws.
- Special-purpose saws.
  - Dadoing (Fig. 46*e*).
  - Tenoning and mortising.

KNIVES AND CUTTERS. The function of a knife or cutter on a planer, molder, or shaper is to reduce the surface or edge of a part to the desired shape and dimensions without marks or scratches, and with a smoothness that can be finished. A roughly sawed surface or edge requires more subsequent machining than one smoothly sawed. Close correlation between the roughness of the saw cut and the depth of the planer cut is necessary. Often, as in jointing plywood cores, a saw can be made to cut smooth enough to eliminate a separate jointing operation. A slightly rough cut is considered better for gluing than one with a very smooth edge.

Following are the basic machines using knives and cutters. (1) Planers or jointers using straight cutting edges to produce plane or level surfaces. They cut continuously along a straight face, not on the side, and parallel or nearly parallel to the grain. (2) Molders or stickers, which are, in effect, multiple planers, usually designed to cut straight or irregular surfaces on all four sides of a piece simultaneously. Cutting may be with either face or side of the knife, but substantially parallel with the grain. (3) Shapers, which are particularly designed to produce irregular and curved shapes or edges both in vertical and horizontal planes. They use a variety of solid, sectional, and laminated cutters, often cutting at an abrupt angle to the grain.

On relatively large diameter wheels, the method of attachment of knives and cutters for planers, molders, and shapers determines the classification. This involves both the provision made for knife attachment, and the corresponding design of head to which knives are attached. Rigid mounting is imperative because of high operating speeds, strong centrifugal forces, and the deep cuts often required.

Knives are distinguished, in general, by thickness. (1) *Thick knives*, usually comprising a tempered steel face and cutting edge, welded to a softer steel back used for clamping. Narrow knives of the chisel type may be entirely of tool steel, tempered principally at the cutting edge. (2) *Thin knives* which require clamp bars and plates, or attachments. Such tool-steel knives seldom are recessed for attachment. They require stiffener bars to provide rigidity and firmness. A knife that chatters will not cut smoothly.

Cutter heads, knives, and router bits can be provided with *tungsten carbide* cutting edges that give long service against very hard species, impregnated, fireproofed, and densified woods, and also resin bonded plywood. The cutting edges of these carbide tips, brazed electronically to a standard steel backing, are rather blunt and cannot be readily sharpened or ground to shape. High linear cutting speeds are necessary for smooth cutting. Solid tungsten carbide cutter heads or knives are seldom practicable, since this alloy lacks the toughness and strength of the standard steels under severe torsional and tension stresses.

Knives and cutters may be classified by machines and by thickness, as follows:

Planers and Jointers (for cutting straight surfaces, single and double, parallel to wood grain):
Thick rigid knives, parallel to axis of head (Fig. 47*a*).
- Square heads.
  - Two-knife (Fig. 48*a*).
  - Four-knife (Fig. 48*c*).
- Face clamping.
  - Oval holes in knife with bolts adjustable in slotted head (Figs. 47*b*, 48*c*).
  - Slotted holes in knife, with tapped holes in head (Figs. 47*a*, 48*a*).
  - Ribbed-surface knives for rigid clamping (Fig. 47*c*).
- Round heads.
  - Clamped with countersunk bolt heads.
  - Internal bolt clamping devices.

Thin Knives, semiflexible, requiring support and clamp bars (Fig. 47*d*).
- Knives parallel to axis of head.
  - External clamp bars and bolts.
    - Square heads (Fig. 48*d*).
    - Round heads (Fig. 48*g*).
  - Internal clamp bars and bolts (Fig. 48*e*).
  - Wedge and taper clamps (Fig. 48*f*).
- Knives spiral to axis of head.
  - External clamp bars and bolts.
  - Internal clamp bars and bolts.

Molders and Stickers (for cutting multiple straight and irregular surfaces parallel to wood grain [Figs. 50*a*, 50*b*, 50*c*]):
Thick rigid knives (straight, not spiral).
- Angular heads.
  - Two-knife (Fig. 48*a*).
  - Three-knife (Fig. 48*b*).
  - Four-knife (Fig. 48*c*).
- Face clamping.
  - Tapped holes in head (Fig. 48*a*).
  - Bolt slots in head (Figs. 48*b*, 48*c*).
  - Ribbed-surface knives for rigid clamping (Fig. 47*c*).
- Edge clamping.
  - External bolts (Figs. 49*a*, 49*b*).
  - Vise or jaw clamps (Figs. 49*c*, 49*d*).
  - Slotted bolts with adjustable cutting angles (Fig. 49*c*).
- Round heads.
  - Clamped with countersunk bolt heads.
  - Internal bolt clamping device.
  - Wedge and taper clamps (Fig. 49*f*).

Thin knives, semiflexible; requiring support and clamp bars.
- Knives, parallel to axis of head.
  - External clamp plates and bolts (Fig. 48*d*).
  - Internal clamp plates and bolts (Fig. 48*e*).
  - Wedge and taper clamps (Fig. 48*f*).
- Knives spiral to axis of head.
  - External clamp plates and bolts.
  - Internal clamp plates and bolts.

Shapers (for cutting irregular surfaces in two planes and frequently at abrupt angle to wood grain):
Knife blades clamped to head.
- Face clamped to head.
  - Slotted hole, tapped hole in head (Fig. 48*a*)
  - Slotted knife and bolt in head (Figs. 48*b*, 48*c*).
- Edge clamped.
  - External bolts (Figs. 49*a*, 49*b*).
  - Vise or jaw clamps (Figs. 49*d*, 49*e*).
  - Slotted bolts with adjustable cutting angles (Fig. 49*c*).
- Clamped with pressure bars or plates (thin knives).
  - External clamped with countersunk bolt heads.
  - Internal clamped.
  - Wedge and taper clamps.
- Thin knives of slightly irregular shapes (Figs. 48*a*, 48*b*, 48*c*).

Solid knife heads mounted on arbor.
- Thick knives of a wide range of shapes (Figs. 48*d*, 48*e*).

BORING TOOLS. The principle underlying a boring tool or bit is similar to that of a shaper tool, except that the boring tool advances into the wood. Mechanical means must be provided for this movement, as well as for the removal of chips from the enclosing hole along the spiral shank of the bit. The two basic types of boring tools are these: (1) Boring into *side wood*, where the bit requires side spurs to sever the wood fibers on the circumference of the hole, before the cutting and lifting edge of the bit separates the chips along the fiber line. (2) Boring into *end wood*, where the bit shears off the fibers as it advances, the cutting of the circumference being merely a separation of the fibers. (See Fig. 51.)

Bits. Figure 52 shows four main types of bits. The *double-spur* bit is a general utility tool, usually made with a single-thread screw point. The *extension lip* bit usually has a double-thread point. It is used to produce smoother holes, with less liability to split hard wood. Both types are used to bore holes in side wood. The *taper head drill* is designed for boring end wood. It may be used for side wood if a rough hole is not objectionable. The *spur machine drill* is a modification of the taper head drill. It has spurs and a brad point, to adapt it to better boring service in side wood. It is used in gang boring where holes may be at various angles to the grain.

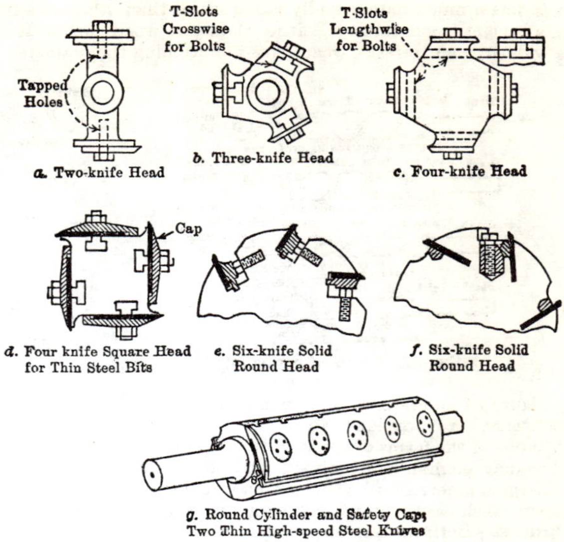

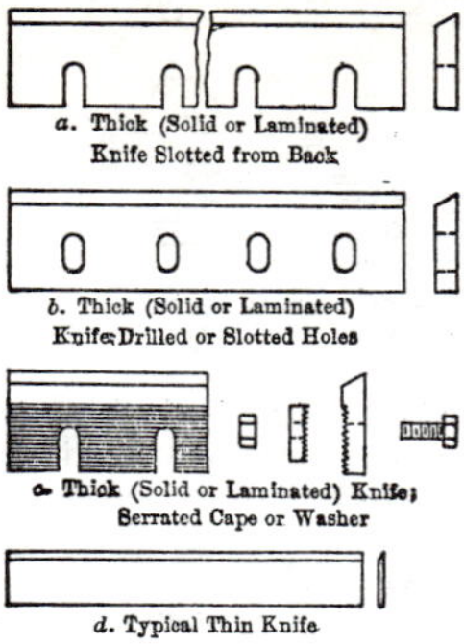

FIG. 47. Forms of cutter knives.

FIG. 48. Forms of cutter heads.

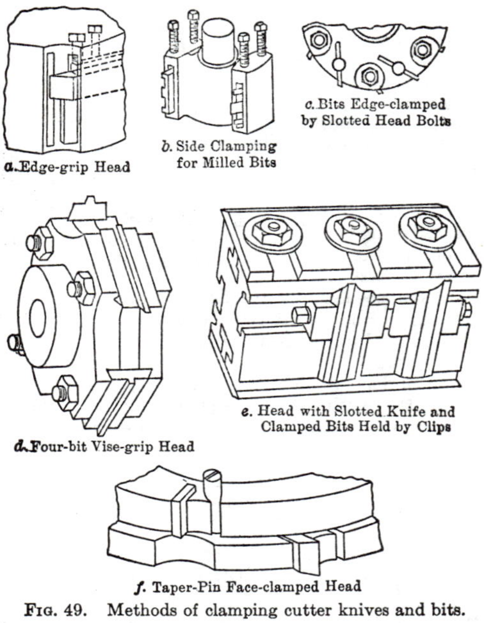

FIG. 49. Methods of clamping cutter knives and bits.

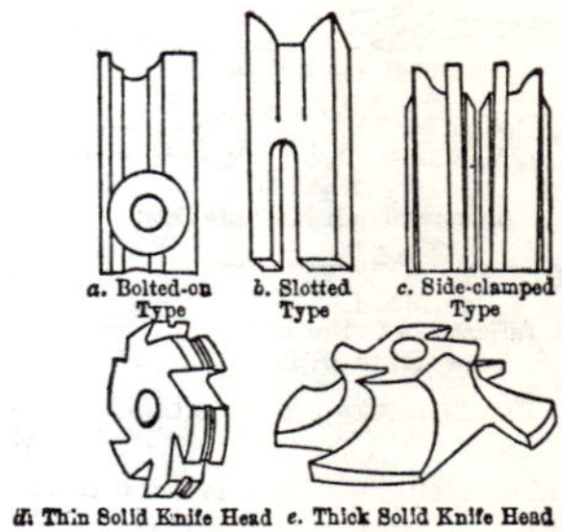

FIG. 50. Bits and solid knife heads.

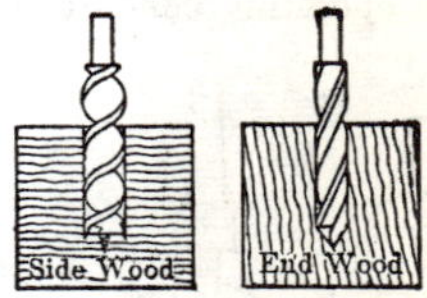

FIG. 51. Boring into side and end wood.

The kind of spiral groove or twist for chip removal depends on bit size, depth of hole, and direction of grain. Figure 53 shows the more important types. The *double twists* are usual for medium and large regular boring; left-hand is standard for a bit turning clockwise. Some gang boring machines require right-hand twist bits, rotating counterclockwise. The *solid center twist*, with a round stem or shank in the center, is used for small bits. The *ship auger twist* has the largest body of metal on the edges, the center being used for more adequate chip removal. It is most suitable for deep boring in large timbers. The *drill twist* has a much more rapidly rising spiral than others, its function being the removal of smaller end wood cuttings, rather than side wood chips. It is ground so that the margin is wide enough to resist wear, thus maintaining the diameter.

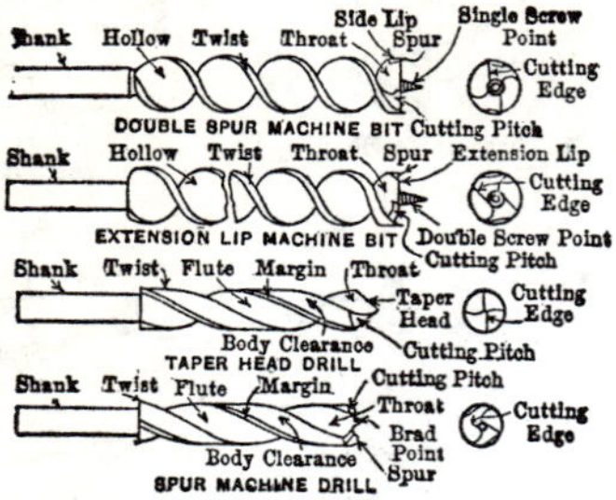

FIG. 52. Types of boring bits.

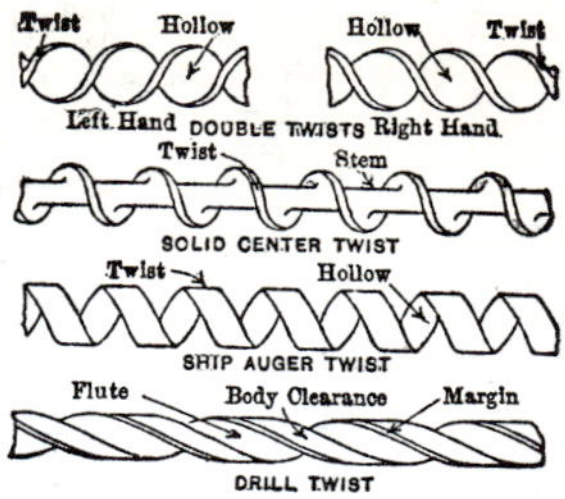

FIG. 53. Types of chip removal grooves.

Figure 54 shows bits for boring concentric holes, sometimes called counterbores. The countersink *a* is for flat-head screws, while recessing, as for washers or bolt heads, is done with one of the forms of counterbore of which *b* and *c* are examples. Many other combinations may be made for both side and end wood.

Large-diameter shallow holes do not require bits with spiral shanks, as chips seldom clog. Figure 55 shows two types, *a* for solid wood and *b*, with saw teeth on the circumference, for plywood. Both have brad points.

Figure 56 shows *plug cutters* with ejector plungers, one of which has inserted high-speed steel teeth for abrasive plywood adhesives, as casein.

Router bits (Fig. 57), while similar in appearance to boring tools, have different functions. A router is chiefly used to cut recessed grooves or to cut out, by means of grooves, regular or irregular outlines. A router bit has a cutting lip throughout its length to produce smooth-edge grooves. It rotates at much greater speeds than a boring bit. The function of side cutting is much more important than end cutting. The advance cut of a spur is not needed, and the fluted clearance slot, for chip removal, may be straight instead of spiral. Router bits may be designed for various shapes, like T slots, ogee grooves, and the like. In this form they may resemble shaper knives and heads.

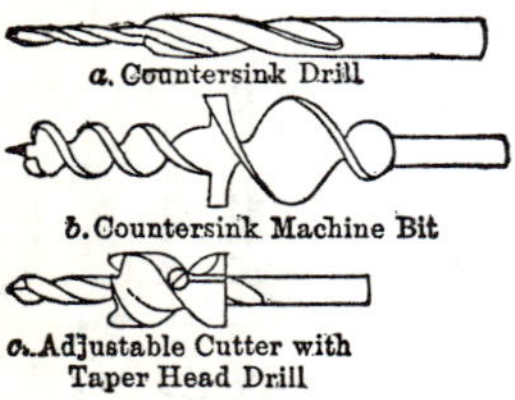

FIG. 54. Counterbore bits.

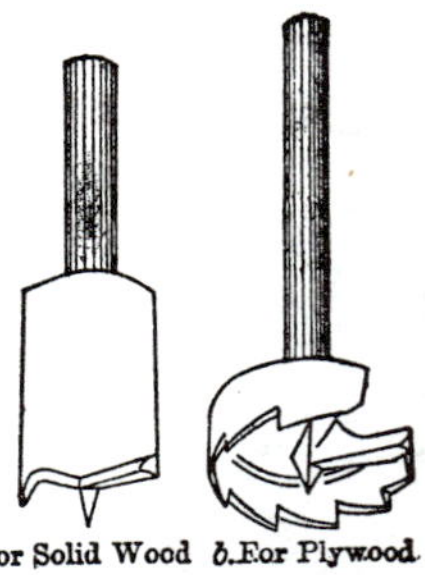

FIG. 55. Bits for large diameter shallow holes.

Router bits, for densified types of wood and plywood, are frequently provided with *carbide-tipped* cutting edges, as shown in Fig. 58. Such router bits give long service under severe operating conditions, but must be carefully designed with adequate reinforcing backings of tough standard steels to withstand excessive torsional stresses. Carbide-tipped cutting edges are relatively blunt, require high speeds for smooth cutting, and are difficult to sharpen.

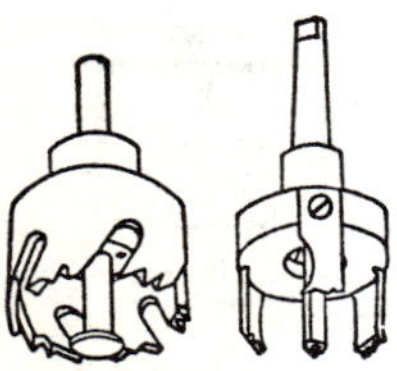

FIG. 56. Plug cutter.

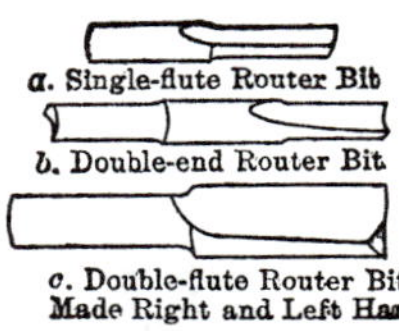

FIG. 57. Router bits.

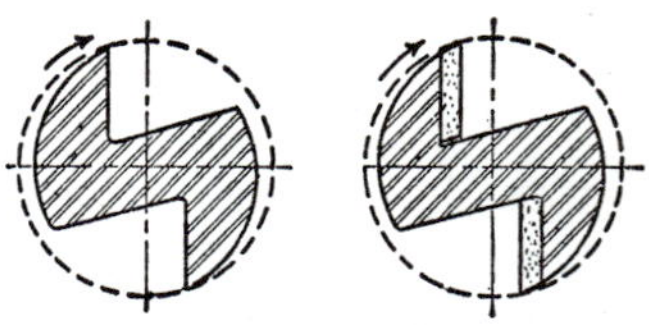

FIG. 58. Sections through router bits without and with carbide-tipped cutting edges on each flute.

MORTISING TOOLS (Fig. 59) are duplex boring and chisel cutting tools. A central standard bit bores a round hole, and, surrounding the bit, a square four-side chisel removes or broaches out the round fillet corners. Chips from both cuts are lifted out by the spiral shank of the bit and discharged through clearance clots in the square chisel shank (one for soft wood, two for hardwood). The slots are located above the wood surface at maximum cutting depth. Chisel points are longer at the corner than at the center, to insure a shearing cut in removing corner fillets. Oblong holes, up to 1 1/2 by 1 dimension ratios, can be made with a rectangular oblong chisel, as in Fig. 59*b*. If a 2 by 1 ratio is required, and the shape of hole is concealed by shoulders on the tenon, the result can be obtained at a single stroke by the hollow chisel (Fig. 59*c*). The cutting edges of the chisel are arranged for shear cuts on all four sides. Otherwise, oblong holes are made by multiple strokes of a square mortising tool. Proper clearances must be provided between internal bits and external chisels to avoid friction, heating and burning the wood. In setting mortising tools, the chisel is set first, and then the bit, with the bit cutting lips extending a little beyond the extreme cutting tips of the chisel. Mortising tools are not suited for cutting end wood.

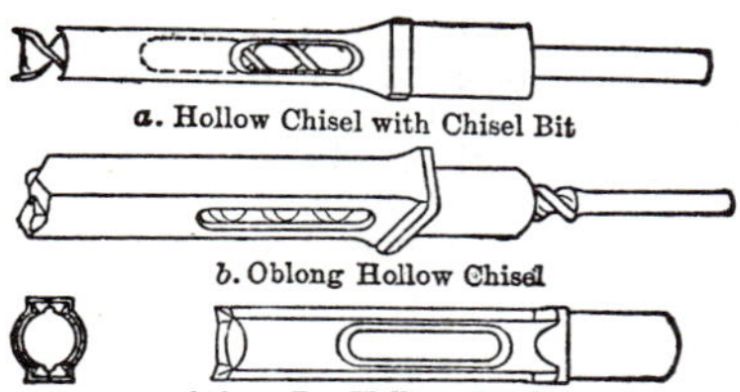

a. Hollow Chisel with Chisel Bit

b. Oblong Hollow Chisel

c. Oblong Bar Hollow Chisel

Fig. 59. Mortising tools.

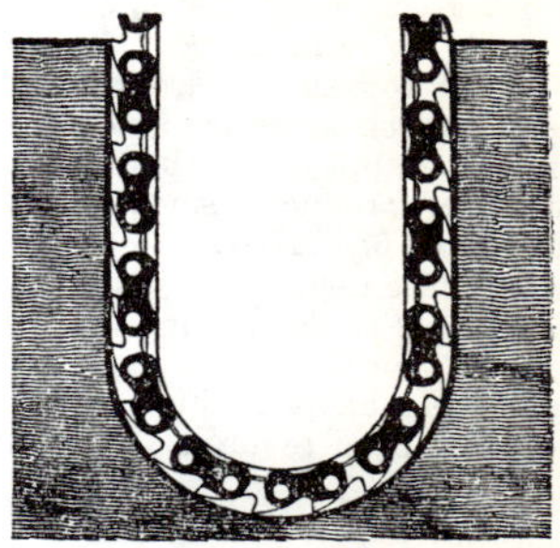

Fig. 60. Mortising chain with saw teeth on outer periphery.

Mortising chains are available for oblong holes, being a series of cutters attached to an endless chain. They are chain saws in principle (Fig. 60), not boring tools, but produce a similar result. The hole has a rough edge, but may be satisfactory for such products as mortise-and-tenon doors, although the joint cannot be very snug.

14. MACHINE TOOLS FOR WOOD

SAW MACHINES. The five principal types of saw machines are: (1) *rip-saws*, circular cutting; (2) *cross-cut saws*, circular cutting; (3) *variety saw tables*, circular cutting; (4) *band saws*, continuous band cutting; (5) *jig* and *scroll saws*, reciprocating band cutting. In many machines, rip, cross-cut, or combination-tooth saw disks may be used interchangeably, while some are single-purpose machines. See p. 25-34 for saw disk types and teeth.

Saw cuts are required on a variety of machines, where the saw cut is auxiliary to the general purpose of the machine. Examples are double-end tenoners, dado heads, or grooving cuts on molders, trim cuts on veneer jointers, or a cut-off on a dowel machine. In general, the best cutting speed for either a circular saw or band saw is approximately 10,000 ft per min.

Rip Saws. Machines designed exclusively for ripping are of two main classes. (1) Single or multiple ripping of full length boards, often the preliminary dimensioning operation in a planing mill, and sometimes in a furniture factory. In gang setups, to avoid a winding cut, saws must be of equal diameter and approximately the same degree of set and sharpness. Saws usually are mounted on the lower arbor, with pressure or feed rollers above. (2) Single ripping of cross-cut stock (often called straight-line rip saws). The saw usually is mounted above, with an endless feed chain below. Such saws often give a sufficiently straight and true edge for gluing. Arbors are sometimes below the work, the saws extending upward between a divided feed chain, above which are mounted idling pressure or feed rollers.

Cross-cutting saws (also called stock-cutting saws) may be of these types. (1) Swing saw (an older type) hung from the ceiling and swinging into the work on an arc. (2) Clip saw, swinging from below, with the motor drive pulley concentric with the pivot. (3) Saws on sliding ways, or other mechanical device, to move them horizontally into the work and out. Cross cutting usually is the first dimensioning operation in a furniture factory.

Variety saw tables may be equipped interchangeably with rip, cross-cut, or combination (rip and cross-cut) saw disks, depending on the work. For grooving or tenoning, dado heads or *wobble saws* may be used. Wobble saws are angle mounted, with wedge-shaped collars, between two perpendicular saws for slotting or rabbeting. Variety saw tables generally are equipped with side gages, back gages, and angle adjustments for miter sawing. For angle cutting, the table may tilt or the arbor may be pivoted. In size, they range from small bench saws up to floor machines. Speed of floor units usually is 3400 to 3600 rpm, requiring 5 hp.

Since all circular saws cut against the feed, the operator should take great care to avoid dangerous "kick backs." Most modern saws have mechanical devices to protect against both kick backs and hand hazards from the saw disks.

Band Saws. Resaws for such operations as ripping special thicknesses usually are about 6 in. wide, with heavy corrugated feed rollers on vertical shafts. Regular band saws may be equipped with a variety of blades, seldom wider than 1 in. for straight work, and down to $^1/_4$ in. for curved work. Band saw wheels range from 30 to 36 in. in diameter, with direct-connected motor on the lower wheel. Band saw throats take work up to 36 in. wide by 18 in. thick. Motors are 3 to 7 $^1/_2$ hp; wheel speeds are 600 to 1200 rpm. Small bench models also are made.

Jig and scroll saws are used chiefly for cut-outs that cannot be made on band saws, and are much less costly than band saws for small shops. The development of the high-speed router, producing smoother edges, has greatly lessened the utility of the jig saw.

PLANING MACHINES. The underlying principles of cutting wood surfaces in the planer also are used in the jointer, molder (sticker), and flooring machine. The following explanation of these principles in the planer applies also to these other types of machines. See p. 25-37 for the forms, shapes, and methods of attachment of knives.

Cutting Mechanism. Figure 61 shows a section of a typical planer cutting mechanism. The cutting head revolves against the infeed of the stock, and the grooved *infeed rolls*

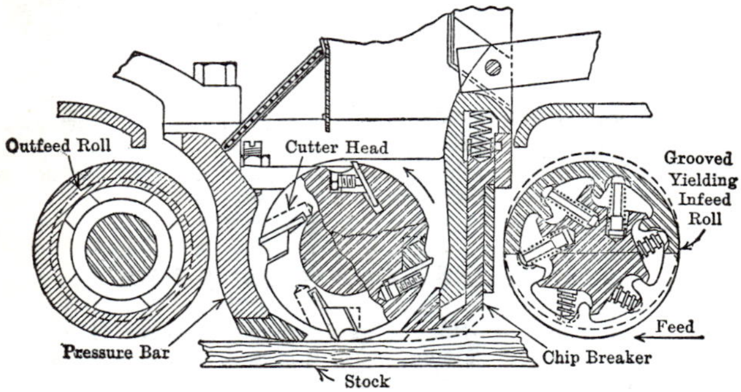

FIG. 61. Planer cutting mechanism.

should be powerful enough to withstand the double load of feeding stock and holding it against the action of the cutters. Infeed rolls often are made in sections to bear more uniformly on thick and thin stock. Large single surfacers and all double surfacers have two pairs of infeed rolls. Pressure of the feed rolls tends to straighten crooked lumber, but the ordinary planer cannot be relied upon to remove wind and warp entirely. Special types are used for this purpose. (See p. 25-42.) The *chip breaker* prevents lifting and tearing of the grain, and reduces all chips to a size that will not clog the cutting head or be difficult to move as shavings in the dust collector system. The steel-tipped chip-breaker shoes ride the top surface of the incoming lumber, and are substantially concentric with the cutter head. They are suspended from a sturdy cross bar, as a heavy cut places a severe strain on chip-breaker mechanism. The smooth *outfeed roll* and the *pressure bar* require exact adjustment to the final planed thickness. They must be rigid enough to prevent any chatter or end clipping in the board during the planer cut.

The *cutter head* in Fig. 61 is of the round type with four thin knives, clamped in slots. While it is possible to set the several knives in approximate alignment, it is necessary to joint or grind all knives, at operating speed, to a true cutting line, after they have been set and clamped in the head. This is done preferably with a special motor-driven traveling grinder mounted over the main bearings of the planer. In some cases (as an enclosed lower knife head), the head is partially or wholly removed, with the knives clamped in place and ground in that position.

Planer cutting circles usually are about 6 in. in diameter but vary with the different makers and shaft lengths. Since the standard speed of planer shafts usually is 3600 rpm (7200 when a frequency changer is used), it is important that they be in perfect mechanical and static balance, requiring careful attention to the individual weights of knives and clamping devices. One of the most difficult problems of planer design is to overcome shaft whip or deflection at high speeds, especially in the wider machines, tending to produce uneven thicknesses and rough planing.

Proper *cutting angles* are important. They vary with the species and dryness of the wood. Table 10 gives their dimensions.

Table 10. Cutting Angles and Knife Cuts per Inch for Planers and Similar Machines

(Data courtesy of S. A. Woods Machine Co., Boston)

Kind of Wood	Dry		Air Dry		Wet		Green	
	Cutting Angle	Knife Cuts per in.	Cutting Angle	Knife Cuts per in.	Cutting Angle	Knife Cuts per in.	Cutting Angle	Knife Cuts per in.
Cedar	20–25	8–12	25–28	7–10	28–35	7–10	28–35	7–10
Redwood	20–25	8–12	25–28	7–10	28–35	7–10	28–35	7–10
Cypress	20–25	8–12	25–28	7–10	28–35	7–10	28–35	7–10
White pine	15–20	9–13	20–25	9–12	28–35	8–12	28–35	8–12
Spruce	15–20	9–12	20–25	9–12	28–35	8–10	28–35	8–10
Norway pine	9–18	9–12	18–23	9–12	23–25	8–10	25–28	8–10
Hemlock	9–18	9–12	18–23	9–12	23–25	8–10	25–28	8–10
Douglas fir	9–18	9–12	18–23	9–12	23–25	8–10	25–28	8–10
Larch	9–18	9–12	18–23	9–12	23–25	8–10	25–28	8–10
Poplar	9–18	9–12	18–23	9–12	23–25	8–10	25–28	8–10
Yellow pine	9–13	9–12	13–18	8–11	18–23	7–10	23–28	7–10
Gum	9–13	10–16	9–18	10–16	18–25	9–12	25–28	9–12
Oak	9–13	10–14	13–18	10–14	25–28	10–12	25–28	10–12
Ash	9–13	10–14	13–18	10–14	25–28	10–13	25–28	10–12
Mahogany	4–9	12–14	9–13	12–14	13–18	10–13	18–25	10–13
Beech	4–9	12–14	9–13	12–14	13–18	10–13	13–18	10–13
Birch	4–9	12–14	9–13	12–14	13–18	10–13	13–18	10–13
Maple	4–9	12–15	9–13	12–14	13–18	12–14	13–18	12–14

Above figures apply to stock 1/2 in. thick or thicker. For stock less than 1/2 in. thick, increase the knife cuts per inch 10–50%.

To prevent tearing out grain on dry stock, decrease the cutting angle or increase the knife cuts per inch.

To prevent roughing up or raising the grain on wet or green stock, increase the cutting angle and decrease the knife cuts per inch.

Knife Cuts per Inch. A definite relation exists between the rate of feed, the peripheral speed of the cutter and the number of knives in the shaft. At best, any planer cut is a series of miniature waves in the planed surface, the degree of smoothness depending on the shortness of the individual waves. This relation is clearly indicated in Table 11.

Table 11. Knife Cuts per Inch with 3500 Rpm Spindle Speed

(Rpm × Number of knives)/(Ft per min × 12) = Cuts per inch

Feed of Work, ft per min	Number of Knives per Head								Feed of Work, ft per min	Number of Knives per Head				
	1	2	3	4	6	8	10	12		4	6	8	10	12
15	19.4	.....	.....	.....	.....	.....	.....	.....	160	7.3	10.9	14.6	18.2	21.8
20	14.6	29.2	.....	.....	.....	.....	.....	.....	170	6.9	10.3	13.8	17.1	20.6
25	11.7	23.4	.....	.....	.....	.....	.....	.....	180	6.5	9.7	13.0	16.2	19.4
30	9.7	19.4	29.2	.....	.....	.....	.....	.....	190	6.1	9.2	12.2	15.3	18.2
35	8.3	16.6	25.0	.....	.....	.....	.....	.....	200	5.8	8.7	11.6	14.6	17.4
40	7.3	14.6	22.0	29.2	.....	.....	.....	.....	220	....	7.9	10.6	13.2	15.8
45	6.5	13.0	19.4	26.0	.....	.....	.....	.....	240	....	7.3	9.7	12.1	14.6
50	5.8	11.6	17.5	23.2	.....	.....	.....	.....	260	....	6.7	9.0	11.2	13.4
60	.....	9.7	14.6	19.4	29.2	.....	.....	.....	280	....	6.2	8.3	10.4	12.4
70	.....	8.3	12.5	16.6	25.0	.....	.....	.....	300	....	5.8	7.8	9.7	11.7
80	.....	7.3	10.9	14.6	22.0	29.2	.....	.....	325	....	.....	7.2	9.0	10.8
90	.....	6.5	9.7	13.0	19.4	26.0	.....	.....	350	....	.....	6.7	8.3	10.0
100	.....	5.8	8.7	11.6	17.5	23.2	29.2	.....	375	....	.....	6.2	7.8	9.3
110	.....	.....	7.9	10.6	15.9	21.2	26.5	.....	400	....	.....	5.8	7.3	8.7
120	.....	.....	7.3	9.7	14.6	19.4	24.3	29.2	425	....	.....	.....	6.8	8.2
130	.....	.....	6.7	9.0	13.5	18.0	22.4	27.0	450	....	.....	.....	6.5	7.8
140	.....	.....	6.2	8.3	12.5	16.6	20.8	25.0	475	....	.....	.....	6.1	7.4
150	.....	.....	5.8	7.8	11.7	15.6	19.4	23.4	500	....	.....	.....	5.8	7.0

Types of Planers. The principal types of planing machines are (1) general-purpose planers; (2) cabinet planers, for high quality work; (3) mill-type planers, for surfacing full length boards.

General-purpose planers, ordinarily used in so-called planing mills and job shops, have a bed (thickness) adjustment in vertical ways by means of two geared screws. With this construction a slight rocking motion is unavoidable, when stock is entering the infeed or leaving the outfeed rolls, which may be unimportant in many rough planing operations that do not require precision. Since the average cost of this type is less than that of the cabinet planer, it usually lacks the better attachments and more accurate adjustments of the latter.

Cabinet planers owe much of their precision to bed adjustment by motor-actuated sliding wedges, giving both accuracy and rigidity. The range of adjustment is less than that of the general-purpose planers, but tendency to rock is entirely eliminated. In cabinet planer design, the point of the chip breaker is placed as close as possible to the actual point of cutting, to reduce the size of chips and lessen the tendency to tear out ahead of the cut. The chip breaker usually is of sectional or flexible construction, to ride equally well on thick, thin, and rough spots in the stock. High-grade planer work demands a large number of knife cuts per inch, which is obtainable either by increasing the number of knives (up to six, rarely more) in the head or stepping up the speed of the shaft, a point on which opinion is divided. Straight knives generally are used, but spiral knives occasionally are used to distribute the load and reduce pull-outs at the back end of the stock.

Motor equipment for planers varies with the service. Standard power ratings for single surface cabinet planers are:

Planer width, in.	26	30	36	40	44	50	60	64
Horsepower	10–15	15–20	20–25	20–25	20–25	20–25	25–30	25–30

Feed mechanism may be driven direct from the motor on the cutter shaft by belt to a constant diameter pulley, or by a variable-speed V-belt drive (between expandable flanged pulleys), thus permitting an adjustment of the feed with relation to the speed of the cutter head. When frequency changers are used on the cutter heads it is an advantage to have separate multiple-speed constant torque motors on the feeding mechanisms.

Speeds usually are up to 60 ft per min, occasionally 100 ft per min, with considerably slower speeds for exacting work.

Production cabinet planers often are fitted with *hopper feed*, the infeed bed or table being equipped with an endless rubber-faced conveyor chain, with slats or side rollers for random lengths. Hopper feeds usually are set on an angle to feed pieces diagonally.

Cabinet planers are made for double surfacing, the lower cutter shaft then being arranged to be pulled out for adjustment and jointing. Double surfacing planers usually are built from 30 to 44 in. wide with larger power requirements than single surfaces.

Special feeding mechanisms are required when the warp and wind in the lumber are beyond the capacity of the infeed rolls to control sufficiently by pressure. These mechanisms, on a single surfacer, called a *facing planer*, have an overhead flexible fingered infeed chain, and produce a true surface on the underside of the board against the cutter head. Two or more surfacings may be required for especially crooked lumber. The rough side is surfaced in a subsequent operation, in the same facing planer or in a standard cabinet planer.

A combination double surfacer, called the *Straitoplane*, has this flexible fingered infeed chain for the lower cutter head to produce an initial true surface on the underside of the board, followed by a standard upper planer head, rigid bed, and outfeed roll for the subsequent facing of the upper side of the board. The second cut can be adjusted for accurate thickness. This machine operates efficiently on stock as short as 10 in., and is particularly adapted to furniture work where most pieces are relatively small. Each cutter head is driven by a 15-hp motor at 3500 rpm. A 5-hp, 2-speed motor drives the feed. The machine is standard at 36 in. wide. Feeds up to 60 ft per min are practicable.

Mill planers are designed primarily for use with full-length lumber. They have been developed, and are used principally, on the Pacific Coast. Feed mechanism comprises endless chains with cross slats of ribbed cast iron, giving a more powerful and positive feed than is possible with rolls, but lacking somewhat in accuracy of cut. The cutter head is above, with the adjustments of a cabinet planer. Bed adjustment is of the sliding wedge type. Capacity is stock 8 in. thick, 30 to 36 in. wide. Feeds of 100 ft per min are practicable.

Jointers are classified as:

Standard Jointers. The cutting mechanism of the jointer is quite similar to that of the planer with a cutter head below the bed, but the feed mechanism is omitted, since edged jointing is the predominant function of the jointer. Jointers are provided with a high *fence* (side gage) to permit them both to square and to true the edge. Portable bench jointers are made and extensively used in preference to the hand plane. Jointers seldom are made over 16 in. wide, and sometimes are hand fed to take the wind out of dimension stock. This is particularly important in producing a true bed surface for molder or sticker work.

Lumber Jointer. The continuous feed lumber jointer, often called a *glue jointer*, has an automatic return feed for edge jointing strips for plywood core stock. This is a duplex or return cycle machine, with an endless feed chain in the center, mounted on vertical shafts. The two cutter heads revolve in opposite directions. A board to be jointed on both edges is placed edgewise at the infeed end against the feed chain, jointed on the lower edge, passed over a tipping mechanism that turns the unjointed edge down and into the reverse feed track. The board travels back to the infeed end, moved by the opposite side of the feed chain, passing over the second cutter head. This machine is suitable only for use on ripped stock, as the cut is insufficient for rough or wany-edge stock, and it will not produce uniform width.

Veneer jointers are highly specialized jointing machines. They operate on a flat *package* of sheets of veneer, aggregating 2 in. thick, which must be clamped firmly during the entire cutting period. The cutter head, usually mounted on a vertical shaft, carries as many as twelve knives to give the smooth edge necessary for matching high-grade veneer faces. In some machines the clamped package travels past the revolving knife head. In others, the clamp is stationary, and the head travels.

Molders or stickers are essentially four-sided planers, with two horizontal and two vertical shaft cutter heads, usually square four-side heads (see Fig. 48*c*). The feed may be by pressure rolls against a rigid bed or by endless feed chain below. Cuts frequently made are irregular and curved faces on moldings. Cutter heads, so far as possible, have chip breakers and the other auxiliary devices of the planer. Since all cutter shafts are short, knives and cutter heads are bolted on ends of shafts, and it is possible to use high speeds with frequency changers, up to 7200 rpm. Fewer knives are required at high speeds to give the same number of knife cuts per inch. This is a distinct advantage in grinding special knives as most molder heads are four sided, and the uniform setting of four irregularly shaped knives is most difficult. It is usual to operate a molder head at high speed with two opposed knives of the same shape and a correspondingly greater rate of feed. For long runs on standard shapes, solid cutter heads are often used, with knives jointed in a profiling machine.

Modern molders have brackets for *cope heads*, which make re-entrant cuts that are not possible with regular heads. They also often supplement the regular heads by making deep cuts that would require impractical or very costly knives. Molders are close coupled and compact, usually with individual motors for each head and for the feed. A popular size takes stock up to 4 by 4 in., but they are made up to 16 by 6 in. for heavy timbers.

Flooring machines are specially designed molders, the two vertical shafts making the tongue and groove, while the lower horizontal head makes the flooring face, and the upper head cuts the hollow back. A steel lettering wheel often grades and trademarks the back as it leaves the machine. Flooring machines are designed to operate on full-length lumber and at high speeds.

THE SHAPER is one of the oldest types of machines. It still is important in woodworking, although many of its functions have been absorbed by more specialized machines. It is a versatile machine for small shops, as it is more easily set up for short runs than many of the specialized machines that are more efficient for long runs.

The shaper consists essentially of a large flat metal table with a vertical spindle, driven from below, extending up through the center of the table. Revolving cutter heads, mounted on the end of the spindle, accommodate a variety of knives for a wide range of cuts. (See Shaper Knives, p. 25-36.) Solid heads may be mounted for long runs. A job to which a shaper is best adapted is the shaping of the edges of table tops, especially those of other than rectangular form. The table top blank is firmly mounted, by two or more points, on a jig of the desired shape. A spindle collar is mounted at the machine table level, above which the knives are set and clamped for the required shape, ogee or otherwise. The operator revolves the combined jig and blank, holding it snug against the collar for one complete revolution of the jig, and thus producing the ogee edge around the periphery. Shaper knives cut with the grain rather than against it, as far as possible. Slow feeds are necessary when cutting at right angles to the grain. Blanks may be reversed on the jig to help this condition.

The shaper is especially useful in re-entrant curves, octagons, and other odd contours not suited to machines designed for rectilinear or circular cutting. Small molding, flutings, both straight and curved, and many edge and end cuts can be made on a shaper.

Shapers frequently are equipped with two spindles, usually rotating in opposite directions. All cuts around an irregular table top can then be made in a direction favorable to the grain, and no cuts have to "dig into the grain," leaving a rough and torn edge. The jig, with mounted work, is moved from spindle to spindle according to grain requirements. Automatic shapers are equipped with mechanically operated jigs or revolving tables to which blanks are held with pneumatic clamps. These are suited only to long production runs.

THE ROUTER is a relatively modern machine, achieving its effectiveness with the development of high spindle speeds. It is somewhat the reverse of the shaper, and usually is designed for considerably lighter cuts. The cutting head is above the table, which is provided with an upwardly projecting, jig-guiding round plug of the same diameter as, and directly under, the cutter. The cutter head carries a router bit (see p. 25-38) rotating at speeds of 20,000 rpm or higher. The table may be moved up to the bit, or the bit lowered into the work. The router is admirably adapted to cutting out (with grooved cuts) grills or irregular shapes. It works much better and more quickly than jig saws. Router bits may have shoulders for shaping the edges of cut-outs, or T heads for slotting. Straight cuts on narrow strips can be made by clamping suitable gages to the table. A straight gage, through or near the bit center, permits shaping of straight-edge table tops and similar work.

Router motors, both electric and pneumatic, must be very accurately balanced for excessively high speeds. Brakes are provided to slow down the bits, which otherwise would require a long time to come to rest. Blowers, directed to the exact point of cut, are standard equipment to prevent chips and shaving from clogging the cut and heating the bits. Some routers are equipped with tilting tables. In these the bit is lowered into the work. Motors are 1 to 5 hp for heavy duty. Spindle speeds often are stepped up by V belts.

Portable hand routers are built to operate electrically and pneumatically. Electric machines operate at 10,000 rpm maximum; pneumatic, at 60,000 rpm maximum.

CARVING MACHINES are of two main types, single and multiple.

Single carvers may be rigidly mounted on a floor standard, the work being moved free-hand by the operator. The single carver also is available in the balanced suspension type, hung much like a dentist's drill. The work is held stationary and the drill moved over the pattern by the operator. In both types the mechanism consists of a small high-speed motor with a small spur drill mounted directly on the motor shaft. Single carvers are used for a certain amount of undercutting. Final retouching usually is by hand.

Multiple carvers may carry as many as twenty-four blanks, all to be roughed out uniformly from a master pattern or model, usually of metal. The spur cutters are moved in unison by link levers that are fixed in their relation to each other. Cutters usually are belt driven from one or more motors. Individual motors can be used for extra heavy cuts. Such machines can only rough out the carvings, but their efficiency consists in removing all surplus wood, even on deeply recessed carvings, leaving a minimum of hand carving of finer lines and undercuts. This type of machine has greatly extended the use of carving on classical furniture at very reasonable cost. But one operator is required, who guides the blank cutter over the pattern, carefully watching the direction of the cut to avoid tearing the wood. The hand finishing work reproduces the character of hand carving. This process is displacing pressed and overlaid imitation carved work.

BORING MACHINES range over a wide variety of woodworking machines, from a single bit held in clutch jaws on the end of a motor shaft, to gang or multiple machines in various arrangements, sometimes equipped for as many as thirty bits. Some bore only vertical holes, others bore horizontally, and still others bore at various angles, depending on the complexity of the work. The maximum number of individual bits seldom exceeds thirty, but special machines have been made for a large number of boring problems. The literature of the industry should be consulted for more specific information.

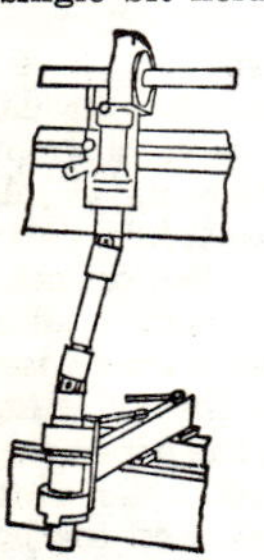

FIG. 62. Boring mill spindle with universal joint.

Vertical multiple borers carry a series of fixed or adjustably positioned spindles, usually arranged in approximately single or double vertical ranks, with a clamp-equipped work table moving up the bits. Spindles may be fixed or *rigid*, i.e., maintain a vertical alignment with the splined drive shaft above, or *adjustable* to points at quite a horizontal distance from the plane of the drive shaft, by a telescopic universal joint (see Fig. 62).

Horizontal multiple borers differ from vertical borers principally in boring a series of horizontal holes, having a work table that moves in and out. They are used mostly for end boring. Universal-joint spindles are not generally adaptable to horizontal boring. However, adjustable multiple or cluster boring heads, to which bits are fixed in position in the head, are available.

Cluster vertical borers (not made in horizontal types) have universal-joint spindles, adjustable over rectangular or circular areas up to 40 in. in diameter, with many combinations of bits that may be attached. Motors are above, and operate the bits through gear trains which terminate in fixed sockets for spindle connections. The universal-joint spindles permit a wide range of adjustment of bits in any horizontal direction within operating limit.

Multiple boring heads are principally (1) fixed cluster (Fig. 63*a*); (2) adjustable multi-spindle (Fig. 63*b*). Many other types of heads are made, for use in both vertical and horizontal boring machines.

Standard spindle speeds range from about 1000 to 2000 rpm, varying principally with size of hole, hardness of the wood, and rate of feed.

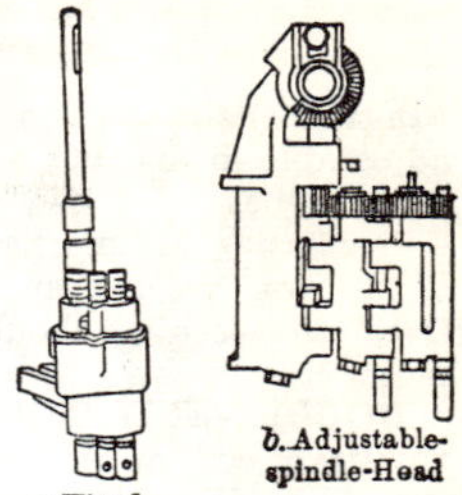

a. Fixed Center Cluster

b. Adjustable-spindle-Head

FIG. 63. Multiple boring heads.

LATHES differ widely in basic principles, developing by degrees from the plain hand lathe, which is one of the oldest woodworking machines and suited to a wide range of individual work. The principle of the high-speed cutter head has been borrowed from the shaper and adapted to many lathe cuts, especially for duplicating work. Patterns and cams have been adapted from such machines as the carver, and other devices, for various other types of cut. Only the principal types are described below.

The copying lathe works entirely from a model or pattern of the exact shape required. It can be used for gun stocks, axe handles, shoe lasts, parts of period furniture, etc. The blank to be turned and the pattern are mounted on slowly rotating gear-driven centers. The cutting member is either a high-speed revolving cutter head or occasionally a heavy saw. Its movement is controlled by a tracing pin, or roller, which traverses all portions of the pattern, producing an exact duplicate. Usually the cutter moves along the length of the blank, but in axe handle lathes the frame, with centers, blank and pattern, moves past the cutter head.

Back knife lathes first turn the blank to a rough round by a ring or die with inside cutting edges, supported on a traveling carriage. The die is moved out of the way, over the dead center, for the turning operation. The long back knife is so set, milled to such a shape, that when in an angling cutting position against the blank, it gradually cuts in, and is stopped when the final turning is of correct size and contour. The back knife never is the exact reverse of the turning, but a mechanically developed extension of it.

The shaping lathe is an exceedingly versatile machine, turning rounds and polygonal sections in almost every conceivable combination. Both lathe centers are "alive," and oscillate slowly into and out of the cutting zone, under the operator's control. Cutter knives, similar to a series of shaper heads on a single shaft, are mounted on a cutter spindle or arbor in the rear, and revolve at high speed. All knives are mounted on the arbor, whether for square, round, or hexagon cuts. As a general rule, where squares and rounds are on one turning, flat surfaces are cut first, the lathe centers turning through a small arc, and keeping the work tangential to the revolving cutters by a cam. The piece to be turned is withdrawn from the cutters, turned through approximately a 90-degree angle (for a square), and the next flat surface is cut. The angle of movement controls the number of sides for squares, hexagons, octagons, etc. After the flats are completed, the rounds are turned, with lathe centers slowly and uniformly rotating. Various modifications may be made in this procedure for complicated types of turning or shaping pieces other than turnings. The setup of the cutter head may be so complex as to demand long production runs for good economy.

Many different types of lathes are built in addition to the major types described above. The literature of the industry should be consulted for detailed information.

MORTISING MACHINES may be classified as (1) hollow chisel mortisers; (2) chain saw mortisers; (3) oscillating bit mortisers.

Hollow chisel mortisers usually provide a vertical movement of the tool into the work. Horizontal types also are available. The tool (see p. 25-39) is attached directly to the motor, the rotating bit being motor driven. The surrounding nonturning chisel is mounted in the chisel carriage on the motor frame, and both are forced together into the work. The tables are equipped with clamps, adjustments, and multiple end-gages, as several mortises may be made in the same piece. Tool strokes may be arranged automatically to repeat cuts for oblong holes, up to about 50 per minute. Horizontal movement of the table also may be automatic for a series of mortises. Motor sizes range up to 4 hp at 3600 rpm. Multiple hollow chisel mortisers always are horizontal stroke, limited to six heads, with a minimum of 4 in. center spacing, and a maximum spacing of 72 in. between outer heads.

Chain saw mortisers operate with cutting teeth mounted on the outside periphery of a link chain. (See Fig. 60.) The minimum size of hole is $^3/_{16}$ by $^3/_4$ in.; the maximum may be 1 $^1/_2$ by 4 in. Longer mortises may be cut at one stroke by using a double roll bar, or by successive cuts. The table moves the work up into the chain cutters. Sprocket motors usually are 3600 rpm, giving the following approximate chain-cutting speeds:

No. of sprocket teeth	5	7	9
Chain speed, ft per min	1330	1930	2400

Multiple-chain saw mortisers are available up to 8 heads, which may be stationary, or adjustable in and out as well as sideways. Chain saw mortisers are not made in horizontal types.

Oscillating bit mortisers make round-end holes, whose depth is limited by the size of the router-type bit. The bit oscillates between stops set for the desired lengths. Oscillations seldom should exceed 200 per min; bit speeds are up to 7200 rpm. Oscillating motor usually is 1 hp; bit motors may be 4 hp. Air clamps hold stock on the tables.

DOUBLE-END TENONERS are highly developed machines that accomplish a wide variety of cuts at one passage of the work. Their functions have been broadened and increased by individual motors on each cutting head, which are more flexible than the belt-driven heads of earlier machines. This machine originally was intended to cut tenons simultaneously on both ends of a piece of stock. Cutting heads of various types were added gradually for a diversity of end and intermediate cuts.

In essential principles, the double-end tenoner consists of a sole (or bed) plate carrying two pedestals. One pedestal is stationary, and carries all operating switches. The other may be moved along the sole plate to accommodate the work, limited in standard machines to about 90 in. Between the two pedestals, and below the work, a pair of endless feed chains, with adjustable lugs, carry the work into the saws and other cutting units. The feed chains are kept relatively close to the pedestals but are adjustable along a spline feed shaft between them. They are fixed in the horizontal plane, all vertical adjustments being made with the cutter heads. Two idling pressure chains hold the work firmly, operating directly above the feed chains.

For double cut-off and double rip cuts, as in plywood, only one saw is used in each pedestal. This cut is made from above. On modern double-end tenoners, the main cutter head can be tilted to make angle cuts. One or more additional horizontal (sometimes tilting) cutter heads are located on each pedestal to make end tenons, grooved, shoulder or shaper cuts on either or both sides of the stock, and at one or both ends. Each pedestal also carries a vertical cutter head. An arbor can be placed between pedestals for intermediate saws or cutters. Cope units, dado heads, relishing cutters, etc., if required, may be attached in various ways. Jump dado cams can be used to enter and withdraw dado cutters near the edge of the stock. Dovetailing attachments may be used for either male or female cuts. Methods of adjustment, attachment of various cutter heads, and other operating features vary considerably in the product of the several machine builders. The variety of work is limited only by the available cutter heads and the extent to which production runs justify elaborate setup.

Single-end tenoners, with corresponding attachments, are available for short runs. Stock is fed on a roller feed carriage, with the cutting units at the left of the operator.

SANDERS. Power-feed drum-sanders are of two general types. (1) Work is forwarded by a rubber-cushioned endless-belt bed. (2) Work is moved ahead by pairs of pressure rollers. Widths range from 31 to 103 in. Generally, the roll-feed type sands one side only, giving thinner and more accurate surface removal as in figured faced veneers on plywood. Two-side types are available, if there is plenty of surface to be removed, and accurate centering between surfaces is unimportant. Endless-bed sanders sand one side only.

Endless-bed Sanders. In a typical machine of this type, the sandpaper drums, usually three, are on top. They are approximately 12 in. in diameter, covered with felt cushion and sandpaper, sometimes wound spirally, but more often attached by clamps in slots. The feed belt, with rubber cushions, is below. Surface speed of the first two drums usually is 3700 ft per min (1200 rpm); that of the third drum is 5600 ft per min (1800 rpm). Drums revolve against, rather than with the work. The work is fed at speeds ranging from 12 to 24 ft per min. The coarsest sandpaper is on the first roll, and the finest on the third roll. Each roller and the feed mechanism are driven by an individual motor. Motor sizes, depending on machine width, are: first and second rollers, 5 to 10 hp; third roller, 7 1/2 to 15 hp; feed mechanism, usually, 3 hp. A brush roller is frequently installed at the delivery end to remove sandpaper dust. Standard drums oscillate slightly endwise (75 to 100 per min) to simulate the side movement of hand sanding.

Single-surfacing Roll-feed Sanders. The drum arrangement is similar to that of endless-bed sanders, except that drums are below the work, and sand through slots in the feed bed. There is an idler-pressure roller above the work. Pairs of smooth feed rollers are located at the entering end, on both sides of the center drum and at the discharge end.

The roll type is not well suited to small dimensioned stock. The sandpaper drums are not so conveniently located for paper changing as in the endless-bed type.

Two-side Roll-feed Sanders. The upper and lower drums are directly over each other in pairs, with the same arrangement of pairs of feed rollers as in the single-surface type.

Belt sanders have a 6- to 8-in. endless sandpaper belt running clockwise on two 18-in. pulleys, 8 to 12 ft apart. In the overhead type the work table, moving on ways toward and away from the operator, is below the lower belt, and the smooth side of the paper is against the pulley. In the underslung type, the table is between the belts, with the sand side of the paper against the pulley. In both types, the work is held on the table by stops at the left and an edge clamp at the right. The operator bears down on the cushion pad, hand operated or mounted on two-way sliding levers. The pad is above and applies controlled pressure through the moving sand belt to sand the stock. Belt sanders are preferred to drum sanders for work that requires a thin cut, as on plywood faces.

Stroke sanders usually are used for polish and finish sanding. They are somewhat similar to belt sanders, except that the pad is mechanically reciprocated under weights. The same type of stroke machine is used for rubbing finish, the sandpaper then being omitted, and a rubbing compound used.

Automatic belt sanders for sanding moldings consist of driven rubber feed rolls which carry the stock under a sandpaper belt, and a flexible shoe shaped to the reverse of the molding contour. Feed is 32 to 50 per min.

Specialized types of sanding machines are belts on vertical pulleys for edge sanding; sand disks on rotating horizontal or vertical tables for drawers and boxes; sand drums of large diameter for curved work; sand arbors for chair legs and irregular pieces, etc. A small portable motor-driven belt sander is available for bench use.

Manufacturers of woodworking machinery include: E. C. Atkins & Co., Indianapolis, Ind.; Babcock Machinery Co., New York, N. Y.; Bell Machine Co., Oshkosh, Wis.; Buss Machine Works, Holland, Mich.; G. M. Diehl Machine Works, Wabash, Ind.; Henry Disston & Sons, Inc., Philadelphia, Pa.; Ekstrom, Carlson & Co., Rockford, Ill.; Greenlee Brothers & Co., Rockford, Ill.; Hermance Machine Co., Williamsport, Pa.; Jenkins Machinery Co., Sheboygan Falls, Wis.; Leitelt Iron Works, Grand Rapids, Mich.; Mattison Machine Works, Rockford, Ill.; Mereen-Johnson Machine Co., Minneapolis, Minn.; J. M. Nash Co., Milwaukee, Wis.; Oliver Machinery Co., Grand Rapids, Mich.; Onsrud Machine Works, Inc., Chicago; C. O. Porter Machinery Co., Grand Rapids, Mich.; B. M. Root Co., York, Pa.; Tannewitz Works, Grand Rapids, Mich.; Vonnegut Moulder Corp., Indianapolis, Ind.; Baxter D. Whitney & Son, Winchendon, Mass.; Wilkin-Challoner Co., Oshkosh, Wis.; S. A. Woods Machine Co., Boston; Yates-American Machine Co., Beloit, Wis.

GENERAL REFERENCES

N. C. Brown, *Timber Products and Industries*, John Wiley and Sons, New York, 1937.
N. C. Brown, *Lumber*, John Wiley and Sons, New York, 1947.
N. C. Brown, *Logging*, John Wiley and Sons, New York, 1949.
G. F. Cosgrove, *Modern Woodworking Machinery*, Grand Rapids, Mich., 1936.
H. S. Graves and E. A. Ziegler, *Woodsman's Handbook*, U. S. Government Printing Office, Washington, D. C., 1912.
H. L. Henderson, *Air Seasoning and Kiln Drying of Wood*, J. B. Lyon, Albany, N. Y., 1936.
Knight and Wulpi, *Veneers and Plywood*, Ronald Press Co., New York, 1926.
Proceedings of Wood Industries Division, *Trans. Am. Soc. Mech. Engrs.*, 1921–1949.
Wood Handbook, Forest Products Laboratory, Madison, Wis., 1940.

PLASTICS MOLDING

By J. E. Faloon

The economical production of quality molded plastics products is a reflection of careful choice of both the plastic material and the process to produce this material in the desired shape and size. Normally the prospective user of the plastics part submits the proposed design to the molder for cost estimate. At this time the molder proposes such design changes as may be necessary or desirable to facilitate mold construction or improve part design.

Although noteworthy exceptions exist, thermosetting materials (such as phenol-formaldehyde, urea-formaldehyde, and melamine-formaldehyde) are usually molded by the action of heat and pressure in compression molds or transfer molds. On the other hand, the thermoplastic materials (such as cellulose acetate, polystyrene, and methyl-methacrylate) are generally molded by the injection process wherein the material is heated to desired plasticity and injected into the relatively cool mold to solidify. Most frequently this entire cycle takes less than a minute, although heavy plastics sections or certain other limiting features may require a much longer cycle.

15. COMPRESSION AND TRANSFER MOLDING

COMPRESSION MOLDING is that process wherein the molding material is compressed between mating mold halves known as cavity and plunger (Fig. 1). Flow of the material is accomplished by the interaction of heat and pressure. Mold temperatures range from 290 to 350 F, depending on types of material being molded. Pressures required range between 1 and 10 tons per sq in. of projected area of molded part. Molding materials with coarse fillers require the heavier tonnages. Sources of mold heat are influenced by local costs and available supply; steam, high-pressure hot water, electricity, and gas are commonly used. In selecting type of heat supply, serious consideration should be given to the problem of temperature control as some materials and processes may require sensitive temperature adjustment to achieve efficient operation.

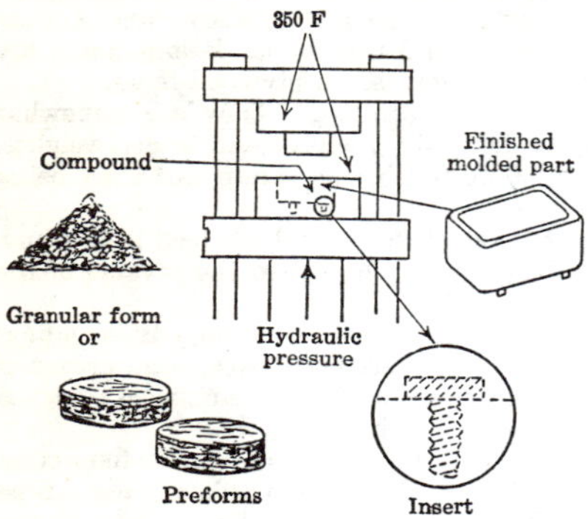

FIG. 1. Compression molding.

TRANSFER MOLDING is a process for molding thermosetting materials in compression-type press equipment wherein molding compound is loaded into a specially provided loading chamber, then, by the action of heat and plunger pressure, caused to plasticize and flow through channels to the molding cavity where final curing occurs (Fig. 2).

Advantages of transfer molding center about the fact that the molding material is in a semiliquid condition before it is transferred into molding cavities which are already clamped together at their parting line. The semiliquid nature of the molding compound allows it to flow freely around rather delicate mold pins or inserts and thereby has allowed molding of many parts from molds which could not stand up in compression molding. With mold parting line closed before transfer of compound, built-up tolerances can be held closer than in compression molding. This characteristic facilitates flash burr removal and is especially worthwhile on parts molded from compounds that contain coarse fillers.

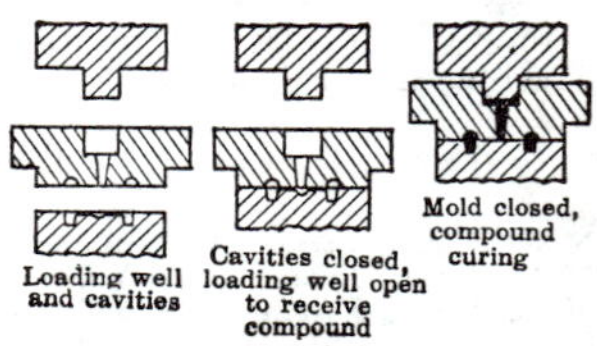

FIG. 2. Conventional transfer molding.

Along with the numerous advantages of transfer molding, consideration must also be given to certain matters of expense inherent in the process. One of them is the item of compound waste since a slug of material remains in the loading well with each molding cycle. Since this is thermosetting material it cannot be reused. Another problem that must be investigated is location and removal of the molded gates where compound is introduced into the molded part. The optimum condition would be in the use of a small gate located at some inconspicuous spot on the molded product. Since optimum conditions seldom exist, it is frequently necessary to provide for a grinding or sawing operation to remove the gates and, if appearance is paramount, it may be necessary to buff the gate area.

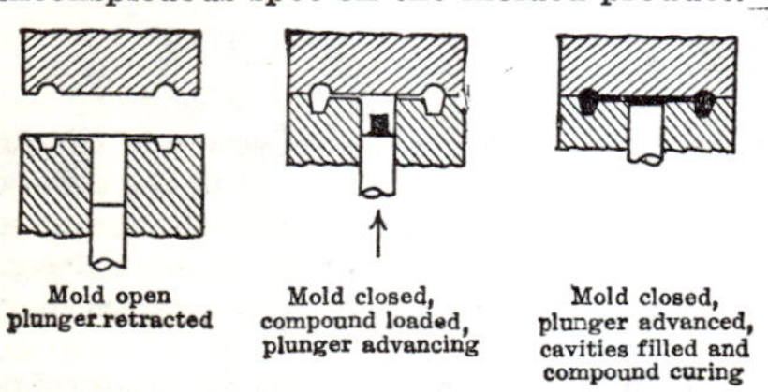

FIG. 3. Plunger-type transfer molding.

Although there are several variations of transfer molding, two basic types predominate. The first of these types is the three-piece mold, as illustrated in Fig. 2, which consists of a loading cavity and plunger connected through sprue and runners to the molding cavity and plunger. All these members are appropriately contained in mold frame components; commonly the loading cavity and molding plunger are fastened in the same center frame member.

The other popular variety of transfer molding consists fundamentally of a two-piece mold clamped together by the normal action of the molding press. This variety, sometimes known as plunger-type transfer molding, is illustrated in Fig. 3. A separately actuated plunger supplies the pressure for transferring material to the cavities. Frequently this type of operation is dependent on electronic preheating of compound to achieve maximum refinement of operation. Outstanding rates of production can be attained with use of plunger-type transfer molding. Special equipment to the extent of a separately actuated plunger mechanism to be run in conjunction with standard presses is the only extra requirement. Once obtained, this plunger mechanism is a useful piece of molding shop equipment which might be considered of a standard nature.

CLASSES OF MOLDS. Compression molds may be designed to operate as hand molds, semi-automatic molds, or completely automatic molds.

Hand molds, representing the simplest form of mold construction, are designed to be removed from the press with each molding cycle for loading material and unloading the molded part. Mold heat is supplied by means of hot plates permanently mounted in the press. Molds are opened and parts removed by hand tools or simple auxiliary fixtures in a work table, which should be located adjacent to the press. The cycle is then repeated by loading molding compound in cavity, reassembling mold, and relocating the mold between heated press platens. The advantages of this design are low cost and easy setup. However the slow resultant production rate and high labor costs of hand molds seldom justify their use except for experimental purposes or very low production requirements.

Semi-automatic molds differ from hand molds to the extent that the mold frames include (1) members to eject the molded part by knockout pins, (2) provision for continuous heating of mold, and (3) provision for fixed mounting of mold in press. With this design the normal action of the press is also utilized to open the mold and eject the molded part from the cavity. The press operator has only to remove the molded part to the tote pan, clean the excess flash from the mold, reload the cavity, and operate the press-control levers. Semi-automatic operation is the most widely used compression molding method since the additional expense for frame construction can easily be justified in increased production rates. Although the setup of the semi-automatic mold is not as simple as the hand mold, standardization of frame and press platens design can reduce the average setup to some fraction of an hour.

Automatic molds are designed to operate in conjunction with special presses to supplant the need for manual loading and unloading of the molds. In design, the automatic mold is very similar to the semi-automatic type, but it is mounted in a press which provides hopper feed of the molding compound, suitable timing devices adjusted to suit the part to be molded, equipment to remove all excess mold flash, and provision for fully automatic ejection of the molded part.

PREPARATION OF MOLDING COMPOUND. Although it is possible to use loose molding powder in loading compression molds, it is most often advantageous to compress the loose powder before using. The nomenclature applied to this operation varies among different molding shops but is perhaps known most commonly as *pilling, tableting,* and *preforming.* Although attempts are sometimes made to define these terms as being distinctly different, their usage overlaps in practice, and all may be considered as molding powders compressed but not chemically changed since no heat is used in the operation. The benefits from such treatment are numerous. They include (1) ease of handling by press operator and frequent elimination of need for weighing the compound charge since weight of compressed material can usually be held closely; (2) reduction of bulk factor of compound—ratio of volume of a given amount of material before and after molding—thereby lessening the need for building molds with deep cavities and long plungers to accommodate bulky materials; (3) ease of distribution of molding material to sections of the mold that are difficult to fill out with normal placement of compound; and (4) improved transmission of heat throughout the compound (this is very important where materials are preheated before molding).

PREFORMING EQUIPMENT. Three principal forms of preforming equipment are commonly used. They are (1) single-stroke machine, (2) rotary machine, and (3) hand-loaded press. In the single-stroke machine a die equipped with an upper and lower punch is volume loaded with the free flowing molding powder, compressed, then ejected from the machine to ready it for a repetition of the cycle. Normally the single-stroke machine can produce ten to sixty preforms per minute—more in instances where multiple-cavity dies can be utilized. Single-stroke machines may be likened to a motor-driven punch press in fundamental operation.

Preforms may be produced more rapidly on the rotary machine, wherein a series of identical die holes are located on the periphery of a revolving horizontal table. Each die hole is equipped with both top and bottom punches which revolve with the table but are located in the vertical direction by means of a cam track. Through this relation the individual die hole is first volume loaded from a hopper feed, the molding powder is then compressed as both punches simultaneously pass between pressure rollers, and preform is finally ejected before the continuous cycle is repeated. The rotary machines commonly produce 350 preforms per minute, and models are available to produce as many as 1000 per minute, on smaller preforms. In general the rotary machines are used to produce circular preforms while the single stroke can produce any outline compatible with tool-strength limitations on dies and punches. Both single stroke and rotary are normally limited to molding powders which are granular in nature and pour easily through the hopper feed arrangements.

Preforms can also be produced in standard hydraulic or mechanical presses by pressing molding powder in an unheated special mold. This process is frequently used to handle molding powders containing long-fibered fillers which will not flow through hopper feed mechanisms on standard preforming machines. Preforms can also be produced by this method from nonfibrous compounds, but use for these materials is generally restricted to preforms too large to be formed on automatic machines. The formation of this type of preform is much slower than the automatic machines mentioned, inasmuch as the powder must be hand loaded into the cavity. If close control of weights is required, the charge must be weighed before loading.

PREHEATING OF THERMOSETTING MATERIALS. It is common practice to heat the charge of molding compound before loading it into the mold. Probably the greatest single benefit gained is reduction in cure time. Since any reduction in cure time leads to improved production rate with greater return on tool and equipment investment, that benefit is obvious. Other significant benefits include improved flow of compound within the mold, improved insulating properties imparted to plastic by driving off moisture content, improved density of plastic product, and improved product appearance under certain conditions.

The basic problem of preheating is getting the maximum amount of heat into the molding material without causing material to precure and thereby lose its properties of plastic flow. Too much preheat may be evidenced in either of two ways: (1) a total precuring of the compound with a general stiffening or lack of flow and (2) localized burnt spots wherein certain particles may be overheated although the total charge may still retain its flow characteristics.

Hot Plates. Probably the simplest form of preheating equipment consists of a steam or electric hot plate located close to the molding press. Its use requires reasonable caution to avoid localized burning. Steam or electrically heated ovens represent some improvement although they may be awkward to use under many circumstances. Rotating ovens for granular materials are one special form of oven equipment which minimizes tendency of localized overheating.

Infrared heat lamps are widely used for preheating, inasmuch as they produce effective results with almost a minimum investment in equipment. A number of ingenious arrangements can be utilized in locating the infrared lamps for optimum preheat benefits.

Live Steam. Interesting developments are in progress toward use of live steam on molding compounds for preheating purposes. This method is in use in some plants with satisfactory results.

Dielectric preheating is a recent development that has already proved to be one of the milestones in plastics production equipment. These electronic devices are available in a wide range of portable units designed to handle either small- or large-capacity molding charges.

By means of this type of preheater, the molding compound is subjected to rapid changes in direction of application of electrical energy (in the magnitude of several million cycles per second), with an ensuing creation of frictional molecular heat within the compound. Accordingly, the molding compound is uniformly heated throughout in a matter of seconds. The benefits from electronic preheating are worthy of serious consideration from all molders of thermosetting materials.

FINISHING OPERATIONS. In the plastics industry, the term *finishing operations* is rather loosely applied to all process operations which follow the actual pressing operations. Accordingly, the finishing of compression or transfer molded parts may include any or all such problems as removal of flash burrs which occur at mold parting lines, punching holes or slots, drilling, tapping, cleaning flash from metallic inserts, sanding, grinding, buffing, and polishing. The foregoing list is by no means complete but is indicative of the variety involved.

The important place of finishing operations in the overall economy of plastic molding cannot be overemphasized. As an example, one can readily visualize a reduction of pressing price by building multiple-cavity molds. However, if the product of that mold does not lend itself to mass production finishing operation on each part, the savings in pressing may become insignificant.

Rotating tumbling barrels provide an economical means of removing flash from molded parts. Barrels may be built of wood or metal and frequently have flat sides to provide agitation. Some form of speed control is desirable to take care of different sizes and quantities of parts; speeds of rotations generally vary between 10 and 50 rpm. If molded parts are relatively sturdy, tumbling is usually performed by simply rotating a sizable quantity of parts in the barrel. If parts are more delicate, wooden balls or pegs included in the barrel serve as a sort of cushion during rotation. Holes and slots can sometimes be cleaned by including shot pellets, nails, metal jacks, or like articles with the mass being tumbled. If these assisting media are included, a sorting operation is necessary after tumbling; frequently a simple screening operation gives the desired result.

Hand filing is most often used on parts that cannot be tumbled. Basically this is a rather slow operation which subjects the part to hazards of visible file marks. Cooperative effort between part designer and molder can often accomplish much in simplifying the parting line and masking its appearance. Scraping by knife is usually effective, but the operation is more expensive than filing.

Belt sanders can be used to remove flash burr from items such as open-back radio cases and clock cases. The sander can also be used to correct slight warpages which may exist between mounting pads. Since the sanding does not leave a smooth finish, its use is restricted to areas which are concealed in ultimate assembly.

Spindling. Round molded parts are often adaptable to a spindling operation. By mounting the part on a simple rotating fixture driven by a bench motor or lathe, the flash burr is removed by pressing a file against the cut-off edge.

Punching. Multiple holes, odd-shaped slots, etc., can generally be cleaned by use of punching dies. Basic construction may vary from small simple units that are hand powered to semi-automatic power-driven varieties requiring of the operator only the simplest of loading operations. The degree of refinement used in the punch construction depends, of course, on quantities involved and overall economies of the individual job.

Drilling. In drilling plastics parts, care must be taken to avoid overheating the drill since the plastic material is not a good conductor of heat. A jet of compressed air directed on the work acts as the most satisfactory coolant and avoids the need for any expensive washing operation. In drilling deep holes it is good practice to back the drill out of the hole frequently. Overheated drills can result in premature dulling of the tool and may cause cracking, burning, or extreme roughness on the molded part. The user of drills for plastics should investigate special types that are available. Most of these special types are designed to provide wider polished flutes to give plenty of chip clearance to relieve drill heating. Carboloy-tipped drills are useful to minimize necessary resharpening of drills.

Tapping operations require somewhat the same precautions noted for drilling. Two or three fluted taps are commonly used instead of four, as in metal tapping. Chrome-plated taps usually extend tap life. To obtain full thread size, it may be necessary to provide special taps that are 0.005 oversize to compensate for the slight contraction which may occur immediately after the tapping operation.

Polishing. Although the appearance of the molded product is generally dependent on the degree of polish imparted to the mold, it is sometimes necessary to enhance the finish with a buffing or polishing operation. If it is desired to remove completely a mold parting line or blemish, the area can be corrected by pressing it against a rotating cloth belt to which an abrasive compound has been applied. Care must be taken in these operations to avoid exposing the filler in the plastic compound.

There is a wide variety of commercially available polishing waxes for plastics parts (including white waxes for light-colored work). The technique for using them varies with the particular part being buffed and the nature of the wax. In all instances the polishing operation should immediately precede the actual packing and shipping of the product since further handling may disturb the lustrous finish.

Economical finishing methods are a challenge to the ingenuity of those responsible for planning the operations. By clever fixture design it is often possible to combine such operations as drilling and punching, drilling and tapping, and punching and filing.

16. INJECTION MOLDING

In injection molding, thermoplastic molding compounds are plasticized in an appropriate heating cylinder, then forced by plunger action through one or more orifices into relatively cool mold cavities where the material solidifies to the desired shape (Fig. 4). The injection molding press is usually a self-contained unit consisting of the various hydraulic, electrical, and mechanical units which are adjustable to various molding requirements. The granules of thermoplastic material are loaded into a feed hopper from which a controlled quantity is advanced into the heating cylinder with each complete cycle. The heating cylinder is surrounded by electric heating coils or hot oil. Since plastic materials are inherently low conductors of heat, all designs include some form of spreader located in the center of the plasticizing chamber to force the material close to the cylinder walls and insure uniform heating. Enough material for several molding cycles is present in the heating cylinder at any given time. The newly introduced material is as yet unplasticized, the degree of plasticity becoming more advanced until full plasticity is present adjacent to the injection nozzle.

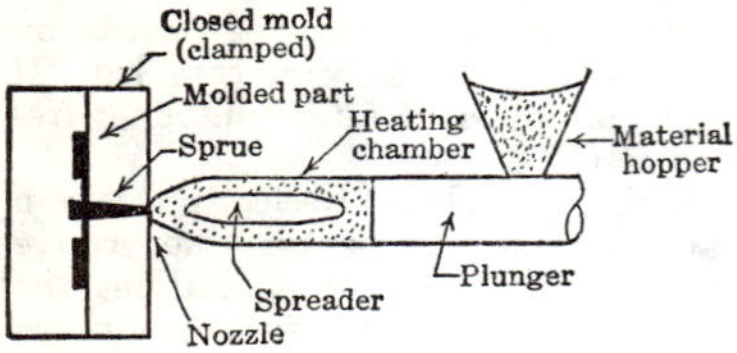

FIG. 4. Injection molding.

Temperatures of molding compounds vary from 250 to 500 F, depending on characteristics of material being used and also conditions peculiar to the size and shape of part being produced. Injection pressures vary from 10,000 to 30,000 psi.

To confine the material properly in the mold cavities during the application of these high pressures, it is essential that adequate clamping force be maintained against the mold. Most injection presses provide this force either through a hydraulic ram or through a combination of hydraulic ram and toggle action.

Injection presses are commonly rated on basis of total weight of material which can be molded per shot. This rating necessarily includes not only the net weight of desired molded parts but also that material necessary for sprues and runners. If compounds with low specific gravity are to be used (such as polystyrene or polyethylene), the rating in ounces per shot is reduced in proportion to the lower specific gravity. Machines for injection molding can be set to operate either fully automatically or semi-automatically at speeds up to about five shots per minute. Since production rate is largely dependent on time necessary to force material from the heating cylinder into the mold plus time required to chill this material in the cavities, few parts can be molded at such rapid cycles.

FINISHING OPERATIONS ON INJECTION MOLDED PARTS. One of the basic economies of injection molding can be found in its ability to produce parts without flash along the mold parting line. Accordingly it is frequently necessary only to remove the sprues or gates to provide the finished part. This removal is sometimes accomplished with such standard shop tools as knives, shears, snips, saws, or grinding belt. Frequently, however, it is expedient to provide special milling fixtures, shear fixtures, or other devices to suit the particular job.

Where grinding operations are required it may be necessary to use a water-lubricated belt or wheel to prevent overheating and possible distortion of the molded part. The same principles may apply to drilling, tapping, and other machining operations, to prevent overheating by use of coolant, reduced speeds, and possibly special chip relief on the cutting edges. If coolants other than water are used, their compatibility with the particular type of plastic must be considered.

Thermoplastic parts can be polished on fabric wheels. Very light pressures should be exerted to avoid softening through generation of heat. A number of successful polishing compounds, commercially available, are prepared to supply desired lustrous effects on specific plastics materials.

17. COLD MOLDING

Cold molding is a term applied broadly to a varied family of plastics which are pressure molded (with little or no heat) and cured by a post-molding bake that may require several hours or days. Cold-molding materials fall into two categories—those of refractory and nonrefractory binder content.

The nonrefractory type includes binders of either bituminous or phenolic origin. Refractory types are of a variety of clay, cement, or silicate combinations, and logically their right to be classed strictly as plastics may be challenged. However, their processing technique is so closely allied to the plastics industry that they are generally accepted as such.

The usual molding process consists of loading the necessary volume of molding compound into the molding cavity and applying pressure through the action of a quick-operating press. Intricate molded parts may require special placement of the molding compound to insure proper density since very little actual flow of the compound occurs. Also trial and error may require that the pressure be applied several times in rapid succession on each load to pack the compound properly.

Curing of cold-molded parts requires controlled temperature ventilated ovens which are capable of progressively raising the temperature to desired levels. Required temperatures vary with specific compounds and part design. However, baking is generally started at a low level to drive off volatile content and gradually raised to avoid distortion and blistering. Ultimate baking temperature may be as high as 500 F, and overall time may be as much as a week or more.

Cold-molding presses may be similar to types commonly used for compression molding, namely, the hydraulic cylinder vertical action. However, fast-acting mechanical types which exert their force through a crank arm motion are well adapted to the coid-molding operation.

18. EXTRUSION MOLDING

Extrusion molding is that process wherein articles of uniform cross section are formed by forcing the softened material through a die opening which is substantially the same shape as the cross section of the article. Typical extruded parts are rods, tubes, and decorative strips. Variations of this basic process include processes for continuous coating of insulated wire, clothes line, and similar articles.

Extrusion molding is generally limited to the thermoplastic materials, although some examples of thermosetting products have been in production.

The usual construction of an extrusion machine consists of a hopper-feed heating cylinder through which the material advances by means of a screw feed mechanism. Heat may be supplied by hot oil, steam, or electric units. At the forward end of the heating cylinder the plasticized compound is forced through the die opening to a continuous conveyor or *take-off* belt.

The successful operation of the extrusion machine requires a careful balance between the temperature of the heating cylinder, rate of screw feed, and speed of take-off belt. Because the raw material must be free of excess moisture, it often is necessary to preheat it before putting it into the hopper. Rates of production vary with material being extruded, capacity of machine, and particular shape being formed. Increased product strength can sometimes be achieved by stretching the product slightly as it is extruded (by advancing speed of take-off belt) and thereby orienting the structure.

19. LAMINATED PLASTICS

Sheets of materials united by a resinous plastic comprise the family of laminated plastic materials. Materials united by this process include cloth, paper, wood, asbestos cloth, glass mat, and woven glass. Laminated materials formed under pressures in the magnitude of 1500 psi, or more, are known as high-pressure laminates. Low-pressure laminates, a more recent development, are formed with pressures of 300 psi, or less. Temperatures in the range of 300 F are commonly utilized in laminating.

Resins most frequently used in laminating are of the phenol-formaldehyde type suspended in a suitable solvent and supplied as a varnish. Urea-formaldehyde resins are available for decorative applications. Melamine formaldehyde may be specified for special electrical properties, low water absorption, and decorative properties. Filler material can be impregnated by a continuous process wherein the rolls of paper (for example) are unwound, passed through a varnish bath, then passed through squeeze rolls which remove the excess varnish. The impregnated material continues through a drying oven where solvent is removed. With proper apparatus, much of this solvent can be recovered.

An alternate impregnating process involves use of a pick-up roll which rotates in the pool of varnish and transfers the varnish to the surface of the material sheet. This method is usually reserved for certain rod and tube applications which are built up from relatively thin stock.

LAMINATED SHEETS. In the production of sheets, the impregnated filler is first cut to size, then stacked into the necessary number of laminations to produce the desired finished sheet thickness. Several sheets of laminated stock can be produced simultaneously in one press opening by stacking multiple "build-ups" separated by a metal plate. The surface finish of the laminated sheet is determined largely by the finish on these metal sheets; a high polish, satin finish, or a number of other variations, can be provided.

Multi-platen presses are used for high production of laminated sheets. A typical press of this description would have perhaps fifteen openings which open vertically in an accordion-like fashion to permit horizontal placement of the multiple build-up in each opening. Platens are individually piped for steam with flexible pipe joints to follow the opening and closing of the press. Curing of the sheets requires temperatures of about 300 F (necessarily less for urea-formaldehyde resins). Cure times range from 5 min on very thin sheets to several hours on thick plates. After necessary cure (while material is still under pressure), cold water may be circulated through the platens to eliminate blisters and warpage and provide best surface finish. Edges which overhang the separating plates must be trimmed. Thicknesses ranging from 0.010 to 12 in. are produced from this basic process.

DECORATIVE AND GRAPHIC LAMINATES. By taking advantage of the built-up nature of the laminate, many useful combinations can be produced by incorporating layers of different types of sheets to provide an effect unlike the individual components.

By sandwiching layers of white urea-formaldehyde-treated sheets between thicknesses of black phenol-formaldehyde sheets, engraving stock can be produced. Items such as name plates can then be made by cutting desired letters or numbers through the outer black layer to expose the white core.

Graphic laminates are produced by embedding the printed sheet on or just below the surface. Techniques employed vary under different conditions. In some instances the sheet containing the printed copy or design is impregnated with resin and inserted with a very thin impregnated translucent overlay sheet on top of the stack of built-up layers, then pressed in the usual manner. These other sheets can be of regular economical core stock and of sufficient number to give desired thickness.

A variation of the aforementioned process involves use of a similar translucent overlay sheet with the copy or design printed (in reverse) on the underside. The economies of this method are apparent; however, its use is normally restricted to high-production operations.

In some instances it is undesirable or impossible to submit the printed sheet to an impregnating operation. This can be avoided by printing the design on a very thin sheet and stacking it (without impregnation) between the impregnated overlay sheet and an impregnated background sheet.

A sheet of metal foil can be included within the laminated build-up to make the finished product resistant to cigarette burns. The high rate of thermal conductivity of the metal foil prevents burning by distributing the concentrated heat over a wide area.

MOLDED LAMINATING. Blanks cut from impregnated laminating stock are stacked and subjected to heat and pressure in compression-type molds to provide molded laminated products (Fig. 5). Unlike compression-molded plastics, very little flow occurs during the molding process. Hence product designs should be free of sharp corners and intricate ribs. Uniform cross sections are ordinarily desirable, although nonuniform sections can be molded by stacking extra thicknesses of impregnated stock at desired locations.

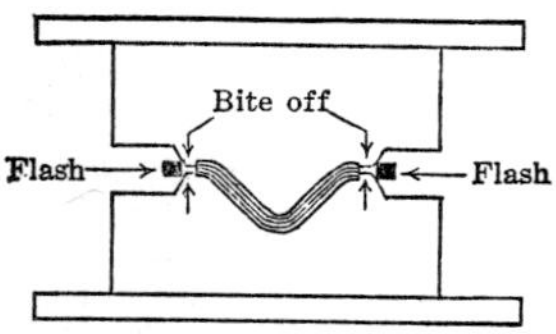

Bite off type mold

FIG. 5. Molded laminating.

A variation of molded laminated technique that can be used for rayon spinning buckets and similar applications involves use of coil-wound impregnated material. The coil must be wound with a center hole to fit over mold core pin or insert. Likewise the outside diameter must be held to fit into the mold cavity.

LAMINATED RODS AND TUBES. Two basic production methods are available for manufacture of laminated tubing. These products are known as rolled tubing and molded tubing. For rolled tubing the impregnated paper is wound under tension over heated steel mandrels. During this operation pressure and heat are applied through pressure rolls. When desired size is reached, the mandrel and tubular wrapping are removed intact to an oven, where curing is completed. With completion of cure, the mandrel is stripped out and the completed tube remains except for trimming to length.

Molded tubing is first wound on a mandrel similar to the rolled product. However, the mandrel and wrapping are next transferred to a heated compression-type mold composed of semicircular mating mold halves. Usually each mold contains several cavities so that several wrapped mandrels can be molded simultaneously. The molded tubing provides a product of superior strength but requires an additional problem of removing the flash excess which "bites off" at the mold parting line. This is commonly removed by centerless grinding.

Molded rods are formed like molded tubing except that the impregnated material is wound on a small diameter mandrel, which in turn is withdrawn before the compression molding operation. With the action of heat and pressure in molding, the space originally occupied by the mandrel disappears, and a thoroughly bonded rod results. As in the case of molded tubing, the flash line must be removed by grinding.

LOW-PRESSURE LAMINATING is a means of molding resin-coated or resin-impregnated plies of paper, cloth, woven glass, or wood by means of heat and relatively low pressures. Although no fixed line of demarcation exists, pressures used are seldom more than 300 psi with the majority of applications under 100 psi. Although there is considerable variation in techniques, the fundamental process usually involves use of a rigid mold half and also a nonrigid half (such as a flexible rubber diaphragm), through which molding pressure is transferred (Fig. 6). A typical low-pressure laminated construction of a curved panel is as described in the next two paragraphs.

The mold half conforming to the outside contour of the panel is made from some rigid material which can economically be formed to the desired shape. Inasmuch as low pressures are involved, this mold half may be made from such materials as wood, plaster, concrete, metal, or certain plastics. The required number of plies of impregnated laminating stock are clamped or stapled to the mold form in such fashion as to conform as nearly as is practicable to the mold contour. If size permits, both the mold and the built-up plies are placed inside a flexible bag with a vacuum connection which causes the bag to follow the mold contour. Care must be taken at this stage to avoid wrinkles between the rubber bag and the built-up plies since that condition would result in a similar wrinkle on the molded part.

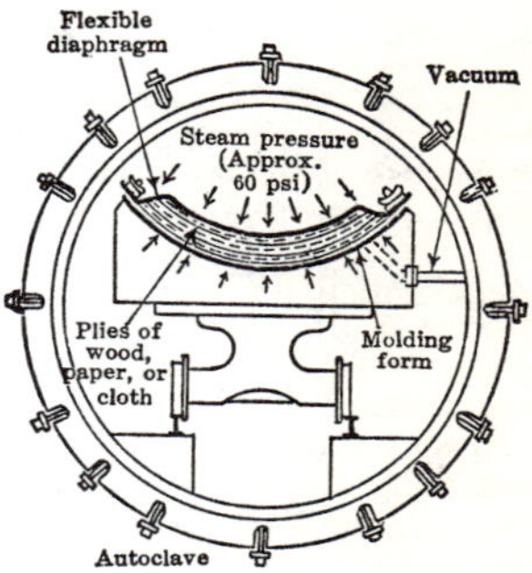

FIG. 6. Low-pressure laminating.

The rubber bag and its contents are then placed in an autoclave, sealed, and subjected to steam pressure (Fig. 6). Under these conditions of heat and pressure, the laminated plies are molded to the desired shape.

The foregoing example utilizes but one of many approaches to low-pressure laminating techniques. Variations in product design may dictate whether the inside or outside contour should be formed in the rigid mold member. Likewise, low-pressure laminating is not restricted to the flexible diaphragm technique. Such articles as cooling-tower fan blades 6 ft in length are successfully produced at low pressures between rigid mold halves which are channeled for steam heating.

While the basic resins used in the low-pressure field are similar to those utilized in high-pressure work, they are necessarily modified to provide for the special molding problems to be encountered. Lower curing temperatures must be provided through inclusion of suitable catalysts since the flexible diaphragms will not withstand high temperatures such as are used in high-pressure laminating. Likewise, since the plies must be formed to approximate final shape during the preliminary build-up operation, it is necessary to have a highly flexible material at this process stage. To achieve this quality, the drying which immediately follows the impregnation is not permitted to advance to the same point used for high-pressure work. With this higher moisture content, a more readily formed material results.

The unique features of low-pressure laminating techniques have broadened the field of plastics applications. Molded parts of almost unlimited area can be produced, such as boats and airplane fuselage and wing sections. The wide variety of structural materials offered provide lightweight products which can give high strength, electrical insulation, corrosion resistance, and excellent appearance if desired. Because the process can be used with low-cost mold equipment it can be a useful development process to prepare the way for more substantial investments in high-pressure mold equipment.

20. MOLD DESIGN

Basically a mold consists of a plunger and a cavity which form the actual molded part. Supplementing these are the frame components which provide support and guidance and also the operating members which simplify removal of parts from mold. A typical mold for a plastic cup (Fig. 7) illustrates the fundamental components normally required in a semi-automatic compression mold.

MATERIALS. The mold plunger and cavity may be made from alloy steels or tool steels. For general-purpose applications a reliable degree of success has been reported in the use of SAE 3310 alloy steel. When properly carburized this chrome nickel steel provides uniformly hard case (Rockwell 56-60C), an exceedingly tough core, and a minimum of hardening distortion.

Cavities formed by the hobbing process (wherein a hardened steel master is sunk under pressure into a soft steel blank) require a milder steel receptive to this means of forming but capable of hardening to withstand molding service conditions. Another chrome nickel steel, of the SAE 3115 type, has met these requirements well. The lower chrome and nickel content represents some sacrifice in core toughness by comparison with 3310, but the ease in hob sinking places it much in demand for this type of construction.

Advocates of tool steel as a mold material claim that such steel, because of its deep hardening properties, gives longer service life and is not as subject to crushing under high molding pressures. In general this class of steels is subject to a greater hardening distortion. A wide variety of suitable tool steels are commercially available. One formulation of chrome silicon tool steel that finds considerable usage is of the following approximate analysis (in percentages): carbon, 0.42; chromium, 1.45; vanadium, 0.25; manganese, 0.30; and silicon, 1.45.

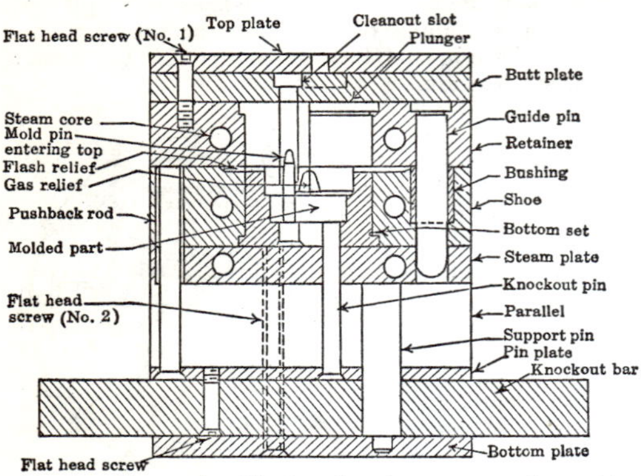

FIG. 7. Conventional bottom-knockout compression mold.

Where possible, mold pins are commonly made from commercial drill rod or music wire. Drill rod (hardened to Rockwell 48-50C) is usually preferred. Music wire (without further hardening) is used for long slender pins which are subjected to some flexing action in service. Since music wire tends to score where friction with a mating mold member is encountered, its use may be somewhat restricted.

Plate steel is usually utilized for many of the miscellaneous frame components. This low-carbon steel, produced by the open-hearth process, is both adequate and economical because cleanliness of steel is not of primary importance here. Retainer shoes, steam plates, parallels, and knock-out bars are examples of this application. Although these members normally do not require hardening, as much as $^1/_{16}$-in. depth of case can be obtained by carburization although considerable correction of distortion may then be required.

INJECTION MOLD. The fundamental construction of an injection mold follows along the same lines noted for the compression mold except that provision must be made for a sprue bushing through the top retainer to convey the plasticized compound from the injection machine nozzle to the mold cavities. A representative injection mold is illustrated in Fig. 8.

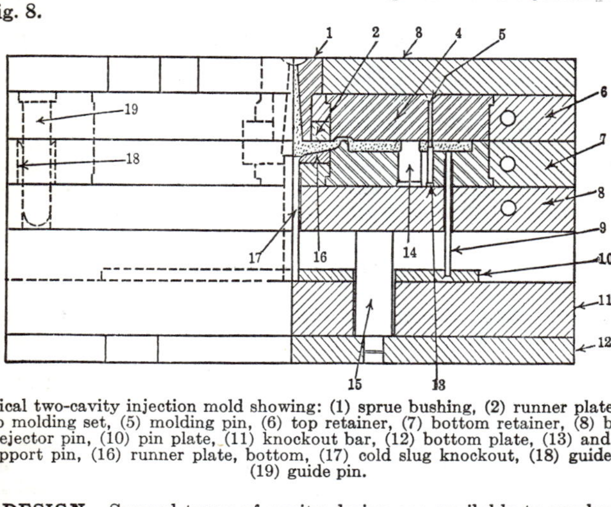

FIG. 8. Typical two-cavity injection mold showing: (1) sprue bushing, (2) runner plate, top, (3) top plate, (4) top molding set, (5) molding pin, (6) top retainer, (7) bottom retainer, (8) butt plate, (9) knockout or ejector pin, (10) pin plate, (11) knockout bar, (12) bottom plate, (13) and (14) molding pins, (15) support pin, (16) runner plate, bottom, (17) cold slug knockout, (18) guide pin bushing, (19) guide pin.

CAVITY DESIGN. Several types of cavity design are available to produce any given molded part. In order to compare these types, the molding of the same plastic cup will be compared with the different cavity designs applied.

Flash Type (Fig. 9). Usually one of the simplest plunger and cavity relationships is obtained from the flash-type design. It will be noted that the cavity is machined only as deep as the desired molded part and the excess molding compound overflows only in a horizontal plane. While this construction may be the most economical from standpoint of original tooling, overall advantages and disadvantages should be considered before selection of this (or any other) plunger and cavity design.

Inasmuch as any excess molding material escapes quite readily over edges of cavity, the charge of material does not demand precise weight control to provide moldings with a thin flash edge. Conversely, however, this same characteristic may contribute to excessive waste of expensive molding powders. This lack of positive entrapment of the molding compound also makes it difficult to achieve high density moldings on intricate part designs since the material may escape from the cavity before the intricate section is filled. High bulk factor compounds present another problem since it may not be possible to include the necessary weight of unmolded compound in the open cavity. This condition can be alleviated by preforming the bulky material or double loading. Double loading amounts to loading the open cavity with the maximum amount of compound, quickly compressing the mold, reopening, and loading in the additional compound. While necessity may sometimes demand this method, the momentary action of heat and pressure on the partial load may result in a line of lamination within the product, so that this practice should not be encouraged.

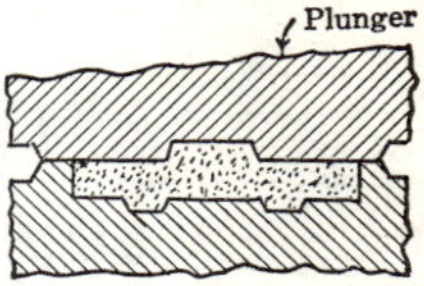

FIG. 9. Flash-type cavity design.

A characteristic of the flash-type design which contributes to low maintenance costs is the absence of any vertical rubbing action between plunger and cavity members. This means, however, that the mating members depend entirely on mold guide pins or built-in interlocking guides. It is essential to maintain the guide pins or interlocking guides diligently if reasonably close dimensional relation between elements molded on plunger and cavity are required.

Within limitations noted, flash-type design is used for compression-type molding. Of perhaps greater significance, however, is the almost exclusive use of flash-type cavity design for transfer and injection molds. In these instances the choice is entirely a matter of tool economy. Since the molding compound is introduced independently into the closed cavity in a semiliquid condition, the problem of loading space is nonexistent, and good molded-part density is achieved by adequate pressure on the material as it is introduced into the cavities.

Positive Type (Fig. 10). The action of the truly positive design may be likened to that of a piston trapping the molding compound and forcing it into the cavity. Inasmuch as there is no land surrounding the cavity, the full molding pressure is exerted against the compound, and the resultant density of the molded part is uniformly good. Uniformity of flash burr is another characteristic of this type of mold since the only flash escape occurs as a vertical burr between the fit of the plunger and cavity walls. Excellent alignment between plunger and cavity may be expected from this design.

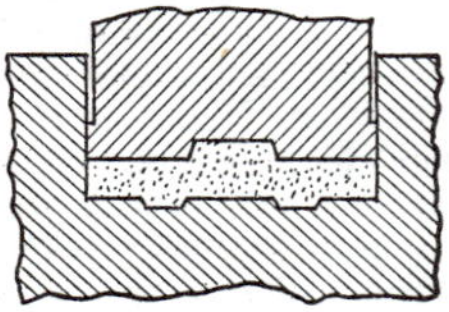

FIG. 10. Positive-type cavity design.

One of the critical features of this type of mold is the necessity of accurate weighing of the molding compound. Any excess material simply tends to increase the thickness of the part, and it is therefore difficult to maintain close tolerances on vertical dimensions.

Another feature which may discourage selection of this design is the item of mold wear due to the rubbing action that usually develops between plunger and walls of the cavity.

Some hand molds of the truly positive variety for molding simple parts such as disks have been constructed without use of guide pins or butt pads. Such molds may be intended for molding disks of various thicknesses, vertical dimensions being controlled solely by weight of compound loaded in mold. The conventional truly positive mold, however, utilizes guide pins to assist in mold assembly and installation and also utilizes external butt pads to prevent plunger from striking bottom of cavity if mold is closed without containing molding compound.

Semipositive Type (Fig. 11). The semipositive design illustrated is a variation of the basic truly positive type. This design behaves somewhat as a flash-type mold until the last fraction of an inch of closure, when the short positive portion of the plunger enters the cavity depression. This design permits overflow of slight excesses of compound charge while retaining good alignment between plunger and cavity. The uniform burr characteristic is especially beneficial for finishing operations, particularly if belt sanding can be utilized.

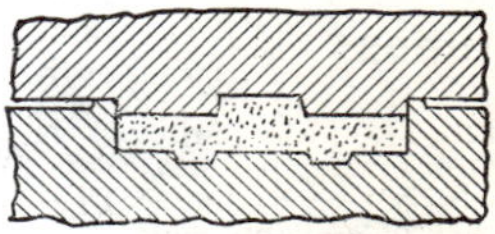

FIG. 11. Semipositive-type cavity design.

Landed Plunger Type (Fig. 12). This design incorporates many of the desirable features of other types previously described. Loading space is attained by allowing necessary space in cavity block above cavity cut-off line. The entering plunger provides excellent alignment between elements molded above and below the parting line. Wear which may occur between plunger and cavity walls ordinarily does not affect the molded part since these lines of friction are outside the outline of the molded part. Reasonable control of compound weights is required although the situation is far less critical than in the truly positive design.

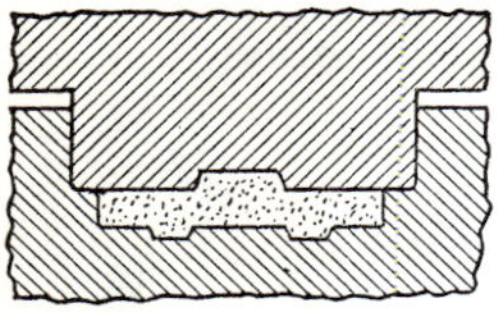

FIG. 12. Landed plunger-type cavity design.

In estimating mold construction expense it must be remembered that the cavity of the landed plunger design must be machined at a level which is depressed because of the built-in loading space. The deep machining required may be somewhat more expensive than that of a flash type for instance. Because of the many obvious advantages of the landed plunger type, the added initial expense is usually rewarded with excellent performance. This is verified by the widespread employment of this design for compression molds.

Subcavity Type (Fig. 13). This is a hybrid variety which may be adapted to high-production multiple-cavity molds for relatively small simple parts. A horizontal line of flash will be noted between the various individual cavities, but the extremities of the common cavity area are of a landed plunger type. The operating economy of this type of mold may be realized in considering the ease of loading compound into such a common cavity. Rather than having to load each cavity with an individual preform, the entire lot may be loaded with one large preform. Rather high molding pressure should be planned to assure minimum flash thickness between the individual molded parts. With careful consideration of the foregoing characteristics and qualifications, this design may contribute to very economical production of selected molded parts.

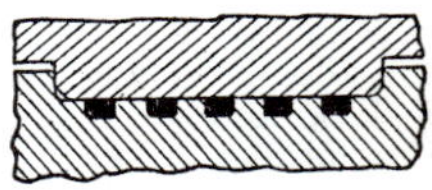

FIG. 13. Subcavity design.

21. MOLDED PART DESIGN

In addition to providing necessary functional requirements, the optimum molded part design should give consideration to factors involving mold making (economy of construction and strength of mold members) and simplification of processing problems. The latter includes simplification of pressing operations, finishing operations, handling and shipping. All the foregoing involve savings reflected either in the form of reduced costs or improved delivery on the molded product. Although it may not be possible to fulfill all these considerations, the economies involved should be carefully explored.

RADII. In general a radius of $^1/_{32}$ in. or more is preferred to sharp corners throughout the cavity. (The mold cut-off line is the outstanding exception since a radius at that line requires expensive matching of upper and lower mold members. Another exception exists where it is necessary to build up sections of steel to form the cavity.) The corner radii may be formed by normal milling cutters. A sharp corner usually requires laborious hand filing and chipping of the mold steel.

Round corners provide better flow of compound and generally improve the strength and appearance of the part. Square corners may create "dead spots" in the mold since flow direction of compound is changed abruptly and may result in poor knitting of compound in adjoining areas. Certain materials such as urea formaldehydes are especially susceptible to this condition, and in them age cracking may result.

SLOTS. Deep slots must be formed by a correspondingly long blade of steel. If these are made integral with the main mold member they may crack during the hardening process or under service conditions. In this circumstance, some insurance can be allowed by setting in this blade with a sturdy embedded shank. The outline of this shank will show on the face of the molded product. If design permits, added mold pin strength can be gained by rounding out the center portion as shown in Fig. 14.

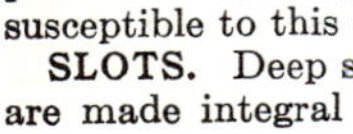

FIG. 14. Mold pin to form rectangular slot, left, can be strengthened by rounding out center portion as shown at right.

TAPER on vertical molding surfaces aids in ejection of part from the mold member. Specific recommended tapers vary with the particular part design and type of plastics material being utilized. If design permits, it is desirable to allow a 1-degree taper per side, although deep barriers or bosses may require 3 to 5 degrees. Many plastics molders work to as little as a quarter degree per side on most vertical surfaces.

RIBS are commonly used to add mechanical strength or increase electrical creepage distances. They also may improve moldability of a part by improving flow of material to remote areas of the mold. In general, ribbing may improve warpage control although it may be desirable to stagger the rib pattern to interrupt shrinkage tendencies over large areas. Rather thin plain surfaces may evidence a shadow effect from ribs on the opposite face (particularly true in translucent materials). The sudden increase in cross section may also contribute to sink spots. Either the shadow effect or the sink spots can usually be rendered unnoticeable if the plain surfaces can be interrupted with a decorative pattern. Well-proportioned rib design is illustrated in Fig. 15.

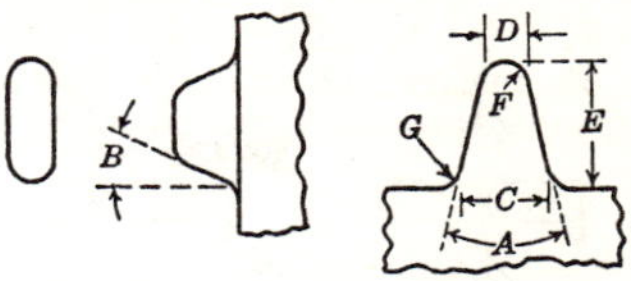

Recommended proportions for ribs							
	A	B	C	D	E	F	G
Hot molded	10°	5°	n	$n/2$	$3n$	$n/4$	$n/4$

Fig. 15. Rib design.

HOLES. Since molded holes are formed by steel mold pins which must withstand bending or breaking action of the molding compound, the designer must avoid exceeding certain proven limits of slenderness. For compression molding the depth of blind holes is generally limited to twice the diameter; if the diameter is $^1/_{16}$ in. or less, the depth should not exceed the diameter. If deeper holes are required, it may be necessary to mold a drill spot to maximum permissible depth and finish drill as an added operation (Fig. 16*b*).

If thickness of molded part does not exceed foregoing ratios, through holes can be molded by butting the pin against opposite mold member (Fig. 16*a*). The horizontal flash can be removed by a subsequent punching operation.

Through holes can also be molded by butt pins which extend from both top and bottom mold member (Fig. 16*d*). Since the individual pin can be twice the diameter, an overall hole length of approximately four times diameter can be achieved by this method. To insure a through hole, one of the pin diameters should be 0.015 in. greater than the other to allow for possible misalignment between upper and lower mold halves.

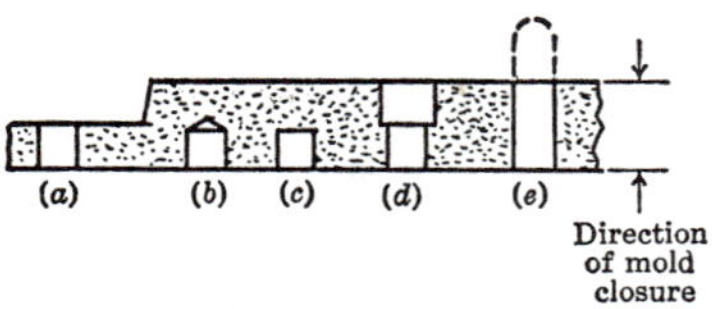

Fig. 16. Alternative pin designs: (*a*) through hole formed by pin abutting on opposite mold half, (*b*) blind hole, with drill point for finish drilling, (*c*) blind hole, (*d*) hole formed by butt pins, one diameter intentionally oversize to allow for mold misalignment, (*e*) through hole formed by entering mold pin.

Transfer and injection molds (wherein the compound is somewhat liquefied before entering actual molding cavity) allow considerable increase of slenderness ratio for mold-pin lengths. It is quite common to allow ratios of three or four to one, and successful moldings of six to one are sometimes practicable.

Through holes can also be made by mold pins which enter a hole on the opposite half of the mold (Fig. 16*e*). This gives the effect of supporting the pin (against sideward stresses) from both halves of the mold. With this added support, slenderness ratios of five to one have been obtained on compression molds and twelve to one on transfer and injection molds. For normal practice it is desirable to stay well under these extreme limits.

Entering pins do not require any flash cleaning around the molded hole as long as a close fit has been maintained between pin and mating member. After some production service a vertical burr is usually present around one end of the hole. It is usually quite light and can be removed by a light burring operation.

The designer must also avoid locating holes too close together or too close to an edge of the molded part. The reasons for this precaution are twofold: (1) weakness of molded part with tendency toward cracking and (2) weakness of the mold member. By remembering that holes are usually formed by pins set into a larger block of steel it can readily be visualized that cracks may occur in the hardened block if there is insufficient space between holes. Holes $^3/_{32}$ in. in diameter or less should be at least one diameter from wall of the part. Larger holes may have a lower permissible ratio of wall thickness. In general the distance between holes should be at least $^3/_{32}$ in.

Holes should be located parallel to direction of mold travel for simplified construction and minimum operating costs. However, if product design cannot be altered to this extent, the molder has a choice of alternate mold designs to provide holes at right angles to direction of mold travel. The necessary pins can be located on a removable mold member which ejects with the molded part and subsequently is stripped from the product manually or with the aid of fixtures. Before reloading the mold for the next cycle, the loose member is relocated in the cavity opening.

Side holes can also be formed from pins which extend out through the retaining shoe. Obviously such pins must be withdrawn from the molded part before any action of the knockout pins has occurred. Such pins must be adequately reinforced to overcome the internal pressure of the molding compound. This reinforcement may be applied through mechanical locking devices or hydraulic cylinders.

By mechanical cam operation it is frequently possible to translate the action of press opening into a side movement to actuate pins forming side holes. The side molding pins are attached to a sturdy cam block which is moved through an offset in the cam pin mounted on the other mold half. As the press opens, the offset on the cam pin moves the cam block outward so that the molding pin is free of wall of the molded part (Fig. 17). During the succeeding cycle, the same offset in the cam-pin positions the cam block for proper molding through the closing action of the press. In general, this type of cam action is reserved for transfer or injection molding.

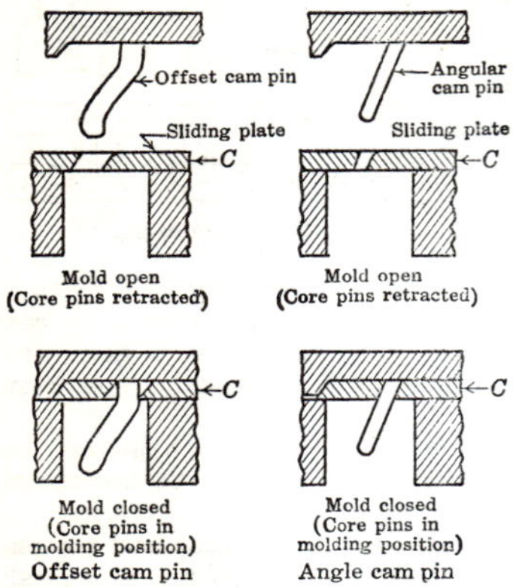

FIG. 17. Means of operating side core pins by opening and closing action of press. Core pins are mounted at C on sliding plate.

INSERTS may be included in molded parts to facilitate subsequent assembly, to provide added strength in specific areas and a conducting path through the insulating material, or perhaps to enhance the insulating properties in localized areas. A variety of insert materials may be used such as brass, steel, silver, Mycalex, glass fiber, and paper, depending on ultimate properties desired.

Brass is the material chosen for the vast majority of insert applications for assembly or conduction properties. Reasons for popularity are threefold: (1) ease of machining, (2) good electrical conduction, and (3) inherent softness which does not damage hardened mold surfaces (particularly in instances where an insert is inadvertently misplaced in cavity and subsequently crushed).

Strips of laminated material, fiber board, or wire mesh may be used to gain mechanical strength in areas of molded parts. In considering such a selection, caution should be observed to avoid serious reduction of some other property of the plastic such as dielectric strength.

A typical example of using inserts to enhance electrical properties is the inclusion of a Mycalex insert in an area subject to electrical arc. The superior arc resistance of this material often suffices for the area in the arc path, whereas the remainder of the part is made from phenolic material that is adequate for the general electrical and mechanical properties.

Anchorage. Allowance must be made to provide adequate anchorage in the molded compound. If round stock is used, a coarse knurl suffices. Hexagonal stock may be grooved to resist pull out; the hexagonal shape resists any twisting action. If stamped, punched, or cast inserts are used, anchorage can be achieved by inclusion of a series of through holes or serrations on the embedded portion.

Locating Inserts. Careful consideration must be given to the method of locating and holding inserts in the mold during molding cycle. Female inserts (Fig. 18) are usually located over mold pins. For best results, inside diameter or minor diameter should be held to ± 0.001 in. to permit accurate location, discourage tipping, and minimize compound seepage into the threads. Where inserts must be located on upper half of the mold, a slight taper on the entering portion of the top pins (properly selected with relation to closely held minor diameter) can hold inserts in place if insert is lightly driven over the taper. Knurl on shank of insert should terminate about $1/32$ in. from open end to avoid distortion of rim on open face. It is usually good practice to counterbore the mold 0.005 to 0.010 in. deep around the mold pin to allow insert shoulder to project above adjoining surface of molded part. This shoulder effect tends to prevent compound from flashing over the exposed insert face. From an assembly standpoint the raised insert shoulder absorbs the stress from objects tightened against it, thus avoiding any tendency to pull the embedded insert from the compound.

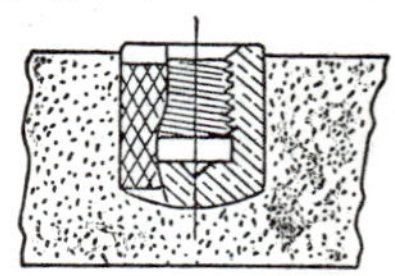

FIG. 18. Threaded female insert.

Rod-type inserts must be located in a hole in the mold, thereby presenting certain molding problems which differ from the female type. If design requirements dictate that the insert be free of flash, very careful consideration must be given to a means of sealing off compound flow lest an expensive and perhaps unsatisfactory finishing operation result. Figure 19 shows three variations in male insert design. The double shoulder provides excellent compound seal and gives added strength against lateral pressures during molding cycle if exposed diameter is small. The single-shoulder type is widely used with reasonably good results. Lacking any shoulder, the third design provides no compound seal and requires an added finishing operation with resultant increase in product cost. This design also may make it difficult to control height of exposed portion of insert. Inclusion of a

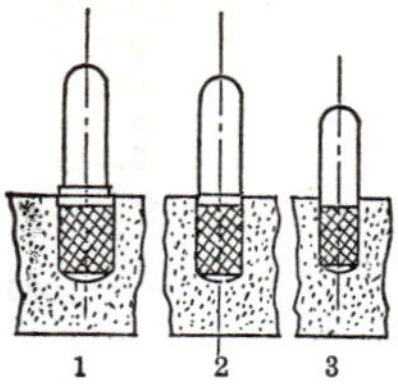

FIG. 19. Rod-type inserts: (1) double shoulder, (2) single shoulder, (3) no shoulder.

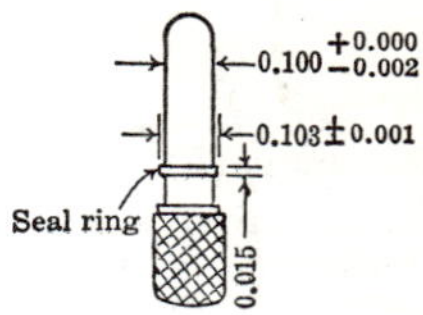

FIG. 20. Seal ring to hold insert in upper half of mold. Also prevents seepage of molding compound beyond level of seal ring.

"seal-ring," as illustrated in Fig. 20, will provide a means of trapping the molding compound below the level of the ring. This ring requires a light pressure to locate properly the insert in the hole in the mold. It is also an effective means of holding inserts in a hole on the top plunger if some design feature prevents locating the insert in a downward position in the cavity.

Stamped, punched, or cast inserts are less frequently used than the machined type but may be selected if they perform some unique function or contribute to overall economy of the design. If chosen, close tolerances are required at dimensions where they must fit the mold. As previously noted, anchorage may be obtained by inclusion of through holes, serrations, or fins in the embedded portion.

22. TOLERANCES FOR MOLDED PARTS

Tolerances are necessary in molded part construction because of unintentional variations in tools, materials, or processing techniques. In the construction of a mold, for example, it is not practicable for the toolmaker to hold every one (of perhaps a hundred dimensions) to precisely the tool drawing dimension. This is particularly true when molds are hardened, polished, and plated after the majority of machining operations have been completed. Molding compounds vary somewhat from batch to batch insofar as shrinkage is concerned. The mold designer must specify all mold dimensions correspondingly larger than the desired molded part, but precision results must be modified by any variation from the nominal shrinkage that may be present in the particular batch of material being used. The final result is also subject to normal variations such as length of cure, amount of preheat, mold temperature, and prevailing temperature and humidity.

TYPES OF TOLERANCES. The average molded part has three different conditions of tolerance because parts are formed between mating mold halves. Dimensions like those indicated by A on Fig. 21 may be considered as fixed mold dimensions since they are

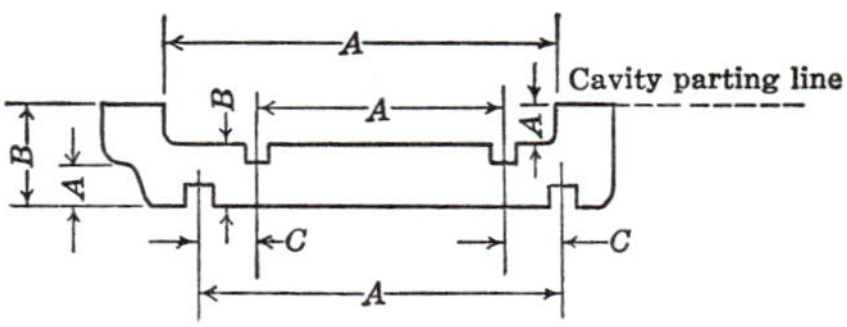

FIG. 21. Conditions affecting tolerance.

independent of the interrelation of the two mold halves. Dimensions like those indicated by B are known as build-up dimensions and are influenced by the degree of mold closure. Excess molding compound trapped between the mold halves increases the basic dimensions. A third set of conditions (designated by C) exists where points being measured are on different halves of the mold. Any misalignment between upper and lower mold halves is reflected in such dimensions.

The designer should not specify close tolerances unless absolutely essential. Violation of this common-sense principle invariably results in unnecessarily high tool cost, increased product cost, annoying production delays, and higher percentage of rejects.

TOLERANCE VALUES, as suggested below, are intended as a guide for designing molded phenolic parts.

Fixed Mold Dimensions (*A*)

Normal Dimension, in.	Preferable Tolerance, in.	Close Tolerance, in.
1/2	±0.005	±0.002
1	±0.008	±0.003
2	±0.012	±0.005
4	±0.016	±0.010
6	±0.020	±0.015

Build-up Tolerances (*B*)

Small parts, transfer molded	+0.005 in.
Compression molded, wood flour, filled, small parts	+0.010 in.
Compression molded, larger parts, wood flour filled	+0.015 in.
Compression molded, small rag filled	+0.015 in.
Multiple cavity, wood flour filled	+0.015 in.
Large rag filled	+0.025 in.

Note: If depth of molded part exceeds 1 in. it may be necessary to allow fixed mold tolerances in addition to above.

Tolerance for Mold Misalignment (*C*). Additional tolerance of ±0.006 in. (for average molds) should be allowed where points being measured are formed in different halves of the mold. More allowance should be made on very large molded parts.

WARPAGE. A good working tolerance for warpage on most molded plastic parts is ±0.003 in. per in., measured from an average plane through the warped surface.

ADDITIONAL CONSIDERATIONS ON TOLERANCES.

1. Grinding or machining finished part is sometimes the most economical procedure for obtaining extremely close dimensional accuracy.

2. All molded parts are subject to some after-shrinkage upon aging, particularly if subjected to hot, dry service conditions. This shrinkage is extremely slight on mineral-filled phenolics and is most noticeable on ureas and certain thermoplastics. Accordingly dissimilar plastics parts should not be assembled too rigidly together. Assembly of plastics part to metal frame should also be governed by this consideration.

3. Cold-molded parts require a tolerance of about double the preferable tolerances suggested for phenolic moldings.

4. Tolerances on injection-molded parts usually should be somewhat larger than those required on phenolic parts. Because of wide variety of injection materials available, the subject cannot be generalized. Molders or material suppliers should be consulted before such tolerances are established if critical assembly problems are anticipated.

SECTION 26

MATERIALS HANDLING

By

MATTHEW W. POTTS, *Materials Handling Consultant, New Rochelle, N. Y.*

Revision of material originally prepared by Frederic V. Hetzel

HOISTING

1. MANILA-ROPE HOISTING

MANILA-ROPE HOISTING IN PRACTICE. New manila rope is stiffer than used rope, friction losses due to bending are greater, and hoisting efficiency is less. To determine load capacity of a hoist, note that manila rope limber enough to show higher efficiency than new rope has lost some of its strength and will not be as safe. *Safety in a hoist is more important than efficiency.* For instance, a 1000-lb pull on a lead line might be rated to lift 3400 lb in an eight-part hoist with new rope and 4400 lb with used rope, but the latter would not be so safe.

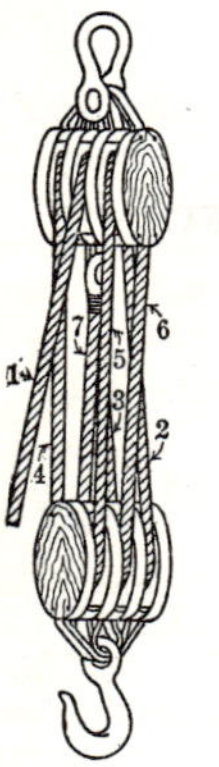

FIG. 1. Triple tackle block.

Manila rope hoists are seldom used with more than six or eight parts of line. Figure 1 shows a six-part hoist with rope properly reeved. The lower block is at right angles to the upper, lead line 1 comes off a middle sheave in one block, and becket line 7 comes off a middle sheave in the other. A six-part hoist has 6 parts of rope at the *movable* block.

MANILA-ROPE BLOCKS are made in a variety of patterns for different services, as occasional or steady, slow or rapid hoisting. Table 1 gives working loads for regular-pattern manila-rope blocks used in pairs. These loads are what the blocks can continue to lift safely in daily service until sheave, pin, and shell are worn out. Bronze self-lubricating bushings prolong the life of sheave and block. If sheaves for hoisting rope can be hung or mounted independently of the block, it is often wise to make them larger than block sheaves. The rope will last longer.

LOADS SUDDENLY APPLIED. Calculations of rope hoists assume the load to be uniformly divided among all parts of rope. With many parts of rope on multiple-sheave blocks, the distribution may be unequal if load is applied suddenly or hoist started at full speed. In such cases, a rope may break at an apparently light load.

Table 1. Working Loads (Pounds) for a Pair of Manila-rope Blocks

(Courtesy of Western Block Co.)

Diam. of Rope, in.	Size of Block	Diam. of Sheave	Singles	Doubles	Triples	Diam. of Rope, in.	Size of Block	Diam. of Sheave	Singles	Doubles	Triples
3/8	3	1 3/4	150	250	350	1	9	5 1/2	2000	2700	3400
1/2	4	2 1/4	300	450	600	1 1/8	10	6 1/4	2400	3100	3800
5/8	5	3	500	750	1000	1 1/4	12	8	2800	3700	4600
3/4	6	3 1/2	800	1200	1600	1 3/8	14	9 1/2	3800	4900	6000
7/8	7	4 1/4	1200	1800	2400	1 5/8	16	11	4400	5700	7000
1	8	4 3/4	1600	2300	3000						

Extra heavy wood blocks and metal frame blocks are good for loads listed for blocks one size larger.

2. WIRE-ROPE HOISTING

EFFICIENCY OF WIRE-ROPE HOISTING. Table 2 shows percentage ratio of theoretical to actual pull (called efficiency). Values in Table 2 probably represent the best that can be obtained in wire-rope hoisting. Loads in the American Bridge Company

Table 2. Percentage Efficiencies of Wire Rope

(American Bridge Co.)

Number of Parts	2	4	6	8	10	15	20
Rope, 5/8 in. diameter	95	90	87	83	80	73	67
Rope, 3/4 in. diameter	93	89	86	82	79	71	65
Rope, 7/8 in. diameter	93	88	85	82	78	70	63
Rope, 1 in. diameter	92	88	84	80	76	68	61

tests were 20 to 60 tons, riggers were skilled, and operating conditions good. Makers' tests of tackle blocks with lighter loads show lower efficiency. Table 3 gives test results that represent average practice in wire-rope hoisting.

Table 3. Percentage Efficiencies with Standard Wire-rope Block

(Courtesy of Western Block Co., Lockport, N. Y.)

Sum of No. of sheaves, both blocks	1	2	3	4	5	6	7	8	9	10	11	12
Efficiency, %	98	96	91	88	84	82	80	76	74	73	70	68

Combinations of rope diameter and sheave diameter are those listed in Table 4.

Table 3 refers to 6 by 19 wire rope used with manufacturers' standard blocks carrying sheaves of sizes shown in Table 4 as related to rope diameters, fitted with self-lubricated bronze bushings.

Table 4. Commercial Sizes of Wire-rope Sheaves

Diameter of rope, in.	3/8	1/2	5/8	3/4	7/8	1
Diameter of sheave, in.	6	8	10, 12	12, 14	14, 16	16, 18, 20

If only life of rope is considered, sheaves of commercial tackle blocks are of too small diameter; practically, it is inconvenient to make them larger. Some single blocks for whip or rapid cargo hoisting have construction refinements, making the blocks safer, more durable, and less troublesome, and add life to the rope. Sheaves are relatively large, of manganese steel, with ball or roller bearings. Shells are cast steel with heat-treated alloy steel pins.

WIRE-ROPE BLOCKS are made as regular, heavy, and extra heavy, with hook (lower block, Fig. 1) or shackle (upper block, Fig. 1) connections. Shackles are used for permanently attached blocks, and hooks for detachable blocks. Hooks fail by straightening, thus giving notice of overloading before failure. Shackles fail by shearing the pin at loads about 5 times those at which round iron hooks straighten, size for size. Hooks flattened to give more depth of cross section at the back are 10 to 15% stronger than hooks bent from round bars. Drop-forged hooks are more generally used and are even stronger. Table 5 gives recommended working loads for a pair of wire-rope blocks. Recommended working loads refer both to ultimate strength of block and its parts, and to durability and safety of block in use.

Table 5. Working Loads (Tons) for a Pair of Wire-rope Blocks

(Courtesy of Western Block Co.)

Regular Pattern							Extra-heavy Pattern						
Diameter of Sheave, in.	With Hooks			With Shackles			Diameter of Shave, in.	With Hooks			With Shackles		
	Singles	Doubles	Triples	Singles	Doubles	Triples		Singles	Doubles	Triples	Singles	Doubles	Triples
6	1 1/2	2	2 1/2	2	3	4	8	2 1/2	3 1/2	4 1/2	4	6	7
8	2	2 1/2	3	3	5	6	10	3 1/2	4 1/2	5	5	7	9
10	2 1/2	3	3 1/2	4	6	7	12	4 1/2	5	6	7	10	12
12	3	4 1/2	5 1/2	5	7	8	14	5	6	8	8	12	14
14	4 1/2	5 1/2	6 1/2	7	9	10	16	6 1/2	8	10	12	15	20
16	5 1/2	6 1/2	8	8	10	11	18	7 1/2	9	11	15	20	25

DRUMS AND SHEAVES FOR WIRE ROPE. A wire rope may be used for hoisting haulage or power transmission in various ways.

1. By winding on a drum, grooved or smooth, in one or more layers. The drum may be cylindrical, a frustum of a cone or a combination of the two. Winding on a drum generally is used in hoists and in haulage back and forth.

2. By winding on a winch head. Rope leads on a large diameter, makes several wraps, and pays off on a smaller diameter. Tension is applied to the slack rope. Under load, a side slip tends to wear rope and winch head. This method is used on capstans, car pullers, and for intermittent hoisting and hauling.

3. By winding several wraps of an endless rope on a concave-faced drum or sheave shaped like a winch head. Tight side of rope leads on a high diameter; slack rope leads off a low diameter to a tail sheave to which tension is applied. The side slip of the rope on the drum is objectionable, and this method is not used widely.

4. By winding and rewinding an endless rope around a pair of grooved drums, and from drums to a tail sheave to which tension is applied. This method is used in endless

rope hauling, in traction drives for high-speed elevators, etc. (See Fig. 2.) The smaller this angle, the larger the friction, but the wear on rope and groove from slip, creep, and motion at engagement and disengagement may be destructive if the load is great or the hours of service long.

5. By tractive effort of a sheave with grooves lined with wood, leather, or rubber friction blocks. This method is used for power transmission with high-speed wire ropes. Such large diameter sheaves are used with counterbalanced cages in the Koepe system of hoisting. See Peele, *Mining Engineers' Handbook* (John Wiley and Sons) and catalogs of wire-rope makers.

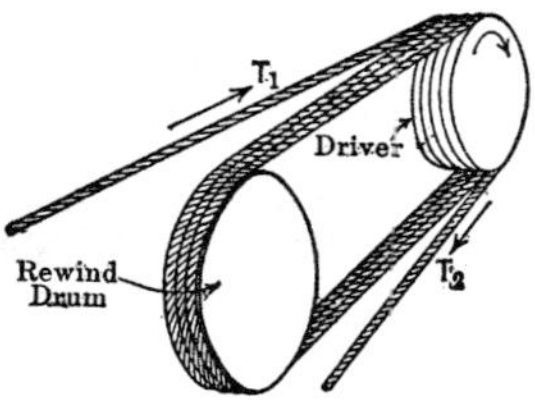

Fig. 2. Winding machine drive.

6. By the use of a grip sheave. Within the rim of the sheave pairs of jaws hinged together are spaced. As they yield to the pressure of the rope they grip it. Springs open the jaws at point of disengagement. This method is used for endless rope carhauls for rope tramways. Older forms of grip sheaves have been known to cut and break ropes; for an improved form (American Steel and Wire Company, Worcester, Mass.) see Ref. 1.

SHEAVE AND DRUM SIZES. The larger the diameter over which rope runs, the longer it will last, but the cost and trouble of renewing rope usually are less than the difficulty of accommodating larger drums and sheaves and the greater weight and cost of heavier machinery. Tackle blocks, small hoists, shop cranes, derricks, dredges, excavators, etc., are examples of such machinery.

Ropes on small sheaves and drums are subjected both to a larger stress due to direct bending (see Bending Stresses) and to pinching of wires and strands on each other, preventing free movement; also, they may be nicked and broken by the pressure. When a rope runs over a sheave, the tightening of the outer wires tends to twist the rope. The smaller the sheave, the greater the twist, and also the greater the unit pressure of the rope on the sheave tread. This twist under pressure causes the rope to cut the sheave, producing sharp edges which wear away outer wires and shorten rope life. To avoid this trouble, cast-steel sheaves should be used. Manganese-steel sheaves, ground true in the groove, are still better. C. D. Meals (Ref. 2) gives a table of allowable pressures for various kinds of rope on cast-iron, cast-steel, or manganese-steel sheaves. One way to destroy a new rope is to run it on sheaves worn and corrugated by an old one.

Table 6 gives, in the form of ratios of sheave or drum diameter to rope diameter, the following: (1) Minimum economical diameters; these should be chosen where sizes of

Table 6. Diameter Ratios of Sheaves and Drums for Wire Rope

	6 x 7	6 x 8	6 x 19	6 x 25	6 x 30	6 x 37	8 x 19	18 x 7
Minimum ratios	42	42	30	30	30	18	21	34
Desirable ratios	72	72	45	45	45	27	31	51

sheaves and drums are not limited by considerations of available space or weight of parts, and where economy in the use of rope is important. (2) Desirable diameters are recommended for heavy work in important installations, as mine hoists, skip hoists, heavy cranes, hot-metal cranes, and in places where safety is important, and where shutdown from rope breakage cannot be tolerated.

PROPER GROOVE CONTOUR. Groove diameter as measured by template or gage should be slightly larger than rope diameter. New ropes are slightly oversize, and for a proper fit groove diameter as measured by template should be larger than nominal rope diameter as follows: $^5/_{16}$-in. rope and smaller, $^1/_{64}$ to $^1/_{32}$ in.; $^3/_8$- to $^3/_4$-in. rope, $^1/_{32}$ to $^1/_{16}$ in.; $^{13}/_{16}$- to 1 $^1/_8$-in. rope, $^3/_{64}$ to $^3/_{32}$ in.; 1 $^3/_{16}$- to 1 $^1/_2$-in. rope, $^1/_{16}$ to $^1/_8$ in.; 1 $^9/_{16}$- to 2 $^1/_4$-in. rope, $^3/_{32}$ to $^5/_{32}$ in.; 2 $^5/_{16}$-in. rope and larger, $^1/_8$ to $^3/_{16}$ in. (See Fig. 3.)

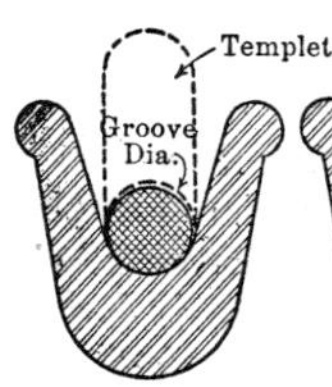

Fig. 3. Correct design of rope groove.

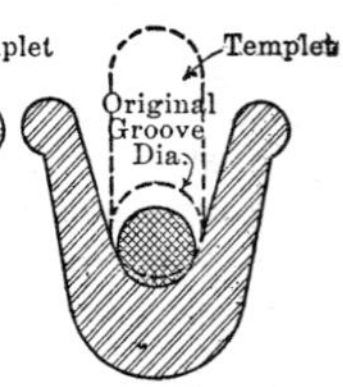

Fig. 4. Correct design with rope worn.

As metal wears away in the tread or bottom of groove, and as rope wears smaller in diameter, there is a new condition of fit, with groove diameter equal to that of the worn rope. (See Fig. 4.) If a new rope runs over worn sheaves, it will not fit the worn groove with proper clearance but will be jammed into it, preventing free movement of the rope, and concentrating abrasion on two lines. Also, there will be a tendency to distort the shape of the rope and cause nicking and breaking of wires. Rope diameter is the largest diameter outside of strands.

DETERIORATION OF WIRE ROPE. Ropes deteriorate by (1) frictional wear; (2) breakage of wires due to defects of materials; (3) breakage due to bending stresses; (4) corrosion, inside and outside; (5) fatigue in steel in the wires. All defects finally show as broken wires, and their appearance guides decisions when to discard a rope. After a new rope has been put in service, the crowns of outside wires wear to an extent necessary to form a wearing surface or contact area to support the rope on sheaves and drums. The reduction in strength of rope is slight and need not cause concern if wear is slow. The appearance of broken wires does not necessarily indicate rapid deterioration. If the rope has started with the proper factor of safety, early abrasion and breakage of wires may be considered normal deterioration within safe limits, but when broken wires continue to appear regularly, removal of the rope should be considered.

Safety of operation depends on strength of the rope as it nears the time of removal. The factor of safety at this time is hard to determine; it depends both on wear and on breakage of visible outside wires, and on corrosion and breakage of invisible inside wires. These combine to cause a reduction in diameter of the rope, which, when greater than wear on outside wires would account for, may indicate internal corrosion. *Coal Age* (Ref. 3) reports a 1 $^3/_8$ in., 6 by 19 mine hoist rope regularly inspected and reported in good condition, which suddenly broke at the part in contact with the head sheave, with cage at bottom of the shaft. Examination of the broken ends showed internal corrosion due to entrance of water; the hemp core was crushed to a pulp 10 ft on each side of break and rope diameter was reduced from 1 $^3/_8$ to 1 $^9/_{32}$ in.

If a rope has not been corroded, its remaining strength may be estimated by counting broken wires in one rope lay in the worst part of the rope, and measuring amount of metal removed by abrasion of outside wires. In one rope lay all outside wires in that length of rope can be seen. Most of the strength of the rope is in these wires, ranging from 83% for 6 by 7 rope to 57% for 6 by 19 Warrington rope. John A. Roebling's Sons Company publishes charts which show, from observed facts of breakage and abrasion, the remaining strength of a used rope, provided there is no internal corrosion. American Cable Company publishes tables showing when ropes in various kinds of services should be discarded. These tables may be summarized as follows. Rope should be discarded when in one rope lay 8 to 20% of visible wires are broken; or when 15 to 25% of wires in one strand are broken, especially if they are adjacent wires; or when outside wires are worn to one-half or two-thirds of original diameter. For high-speed work, heavy loads, and when risk to life or property must be kept low, the lower percentages are recommended. If breakage of a rope merely means a machine stoppage, ropes can be run until outside wires are worn off or until a strand breaks.

Rules and tables for discarding ropes are suggestive only. For instance, it is not possible to group all shop cranes in one class. It is better to treat each installation separately, examine rope systematically and regularly, and note and record breakage and wear of wires. A minimum life based on observed facts can be set up as a standard; if tensile tests of parts of discarded rope show it to have more than requisite strength, considering a proper factor of safety, a new basis can be set up. After a few ropes have been discarded and tested, a standard for that installation can be established that will allow rope to be used with safety and economy.

EARLY FAILURE OF WIRE ROPE. Faults which appear early in the use of a rope should not be laid always to poor manufacture. Careless handling and reeving of the rope often are responsible. The life of a rope often can be increased by putting an obligation of responsibility and competition on the men whose duty it is to use it (Ref. 4).

ATTACHING WIRE ROPE (Ref. 5). The rope may be attached to its load by a socket (Fig. 5) or it may be bent back on itself to form an eye and the loose end clamped to the rope with rope clips (Fig. 6). Properly done, the first method develops the full strength of the rope but has the following disadvantages: poor workmanship is concealed; it cannot be readily inspected; the rope is rigidly held and the effect of bending is concentrated just above the socket; it is more difficult to make. Fastening by clips develops not over 85% of the strength of the rope. This fastening bruises and crushes the rope where the clips are applied, and is likely to slip, but it can be inspected readily and requires little skill in making. Rope makers recommend the socket method of fastening, which, if made according to instructions, never fails. After either a socket or a clip attachment is made it should be tested by making several trips under a heavy load before men are allowed to ride.

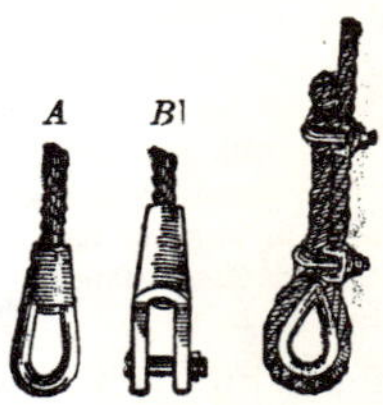

FIG. 5. Socket. FIG. 6. U-bolt and clip.

The clip attachment is made as follows. The rope is bent back over a thimble of the proper size and the loose end clamped against the long end of the rope with the proper number of clips. The Crosby type of U-bolt-and-drop-forging clip has proved most satisfactory. The clips should be spaced a distance apart of six times the rope diameter, the drop forging being against the long end and the U bolt against the loose end of the rope. The number of clips required to develop approximately 80% efficiency of the rope and the proper length of wrench to tighten the bolts are given in Table 7. Never put the drop forging of the clip against the loose end of the rope.

Table 7. Number of Crosby-type Rope Clips Required to Develop 80% of Strength of 6 x 19 Plow-steel Rope

Diameter of rope, in.	3/4	7/8	1	1 1/8	1 1/4
Number of clips	5	5	5	5	6
Length of wrench, in.	18	18	24	24	24
Efficiency, %	77	79	77	80	82

The clips must be carefully inspected occasionally and tightened if they show any signs of loosening through the reduction in diameter which comes from stretching of the rope and compression of the hemp core.

Thimbles (Fig. 7) may be held in the bend of the rope by clips or they may be spliced in. A spliced end, properly made, will not pull out, but it will not develop the full strength for 1/2-in. rope, and not over 60% for 2 1/2-in. rope. Thimbles and sockets form the means of attachment of hooks and shackles and eyes of various kinds. (See Figs. 10, 11, and 12.) For high-tensile ropes (plow steel, etc.), it is best to use the special heavy thimbles and sockets listed by wire rope makers.

FIG. 7. Thimble.

Swaged or compressed fittings for wire rope are being extensively used on wire rope. Examples of such fittings are illustrated and discussed on p. 6-20.

FASTENING WIRE ROPES TO DRUMS. Figure 8 (Link-Belt Company) shows a method of readily attaching or detaching the rope from the circumference of a drum on an electric hoist. The last groove sweeps with a large radius into a depression which allows the socket on the end of the rope to be slipped into the inside of the drum; a pipe plug closes the cavity and locks the socket in place. By unscrewing the plug, the socket can be pulled out and the rope removed.

Where the end of the drum or its hub is readily accessible the rope can be secured by a few wraps around the hub with the end clipped to itself or to one of the arms or spokes of the drum.

The hole in the rim through which the rope passes should be well rounded to avoid a sharp kink in the rope. Wherever possible the drum should be cast with a ledge or lug to lead the rope tangent to the hub over a radius of 6 or 8 diameters of the rope (similar

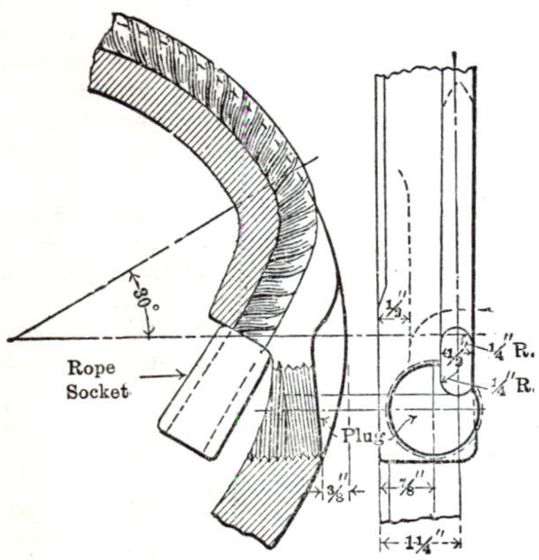

FIG. 8. Method of fastening wire rope to drum.

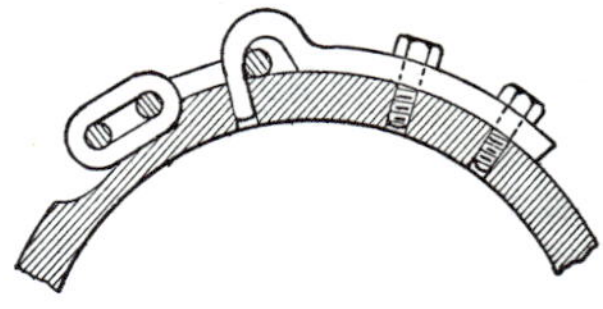

FIG. 9. Method of fastening chain to drum.

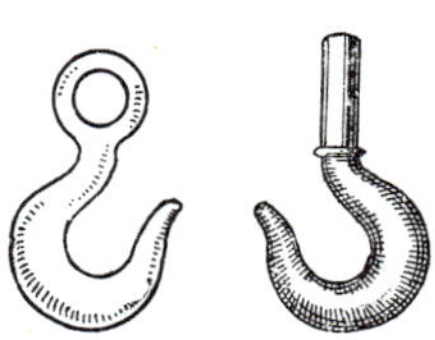

FIGS. 10 and 11. Drop-forged hoist hooks.

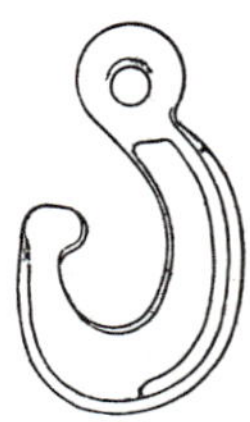

FIG. 12. Bemis safety hook.

to Fig. 8). To fasten the end link of a chain to a drum, it is customary to use a forged clamp, tap-bolted to the rim of the drum, and provided with a finger equal in diameter to the size of the chain (Fig. 9).

DROP-FORGED HOIST HOOKS of steel without welds are made in sizes up to 25 tons capacity. They are heat treated after forging and tested to 50% beyond rated working load. (See Figs. 10 and 11 and Table 8.) Drop-forge companies have stock

Table 8. Drop-forged Hoist Hooks

(Courtesy of J. H. Williams & Co., Brooklyn, N. Y.)

Shank Hook				Hook with Eye				Capacity, Net Tons		
Diam. of Shank, in.	Length of Shank, in.	Extreme Length, in.	Extreme Width, in.	Inside Diam. of Eye, in.	Outside Diam. of Eye, in.	Extreme Length, in.	Extreme Width, in.	Safe Working Load	Aver. Load at Elastic Limit	Approximate Load Required to Straighten Out
1/2	2	5	2 7/8	3/4	1 1/2	4 3/8	2 7/8	0.5	0.9	1.9
9/16	2 1/4	5 1/2	3 1/8	7/8	1 3/4	4 7/8	3 1/8	0.6	1.2	2.3
5/8	2 1/2	6	3 1/2	1	2	5 3/8	3 1/2	0.7	1.5	3.0
3/4	2 3/4	6 3/4	3 7/8	1 1/8	2 1/4	6 3/16	3 7/8	1.2	2.5	5.7
7/8	3	7 1/2	4 3/8	1 1/4	2 1/2	6 7/8	4 3/8	1.7	3.5	7.0
1	3 1/4	8 1/4	4 7/8	1 3/8	2 3/4	7 5/8	4 7/8	2.1	4.2	8.5
1 1/8	3 1/2	9 1/4	5 5/8	1 1/2	3	8 9/16	5 5/8	2.5	5.4	10
1 1/4	3 3/4	10 1/4	6 3/8	1 5/8	3 1/4	9 9/16	6 3/8	3.0	6.2	13
1 3/8	4	11	6 7/8	1 3/4	3 1/2	10 3/8	6 7/8	4.0	8	17
1 1/2	4 1/4	12	7 1/2	2	4	11 1/2	7 1/2	4.7	9	19
1 5/8	4 1/2	13	8 1/4	2 3/8	4 5/8	13	8 1/4	5.5	11	26
1 3/4	5	14 1/2	9 1/4	2 3/4	5 1/4	14 3/4	9 1/4	6.8	13	32
2	5 1/2	16	11	3 1/8	6 1/8	16 3/4	11	8	17	35
2 3/8	6	17 7/8	13 1/2	3 1/2	7	19 1/8	13 1/2	11	21	48
2 1/2	6 1/2	19 3/4	14 1/2	....	....	....	....	15	25	55
2 3/4	7	21 1/2	15	4	8 1/2	22 3/4	15	20	40	80
2 3/4	7	21 1/2	15	4	8 1/2	22 3/4	15	25	50	....

dies for most sizes and furnish hooks in moderate quantities without making charge for dies. Drop-forged hooks of dimensions in Table 8 are good for loads of 3 1/2 to 5 times rating, if sling or fixture rests in the bottom of the jaw. If the load improperly comes on the point of the hook, the carrying capacity is much less. In no event is the hook likely to break suddenly; the jaw will open out first, and signify overloading or improper engagement of the hook.

The Bemis patent safety hook (Fig. 12) (American Locomotive Company, New York) has rated capacities up to 14,000 lb. It is made in four sizes. Hooks fitted with movable guards are not favored by stevedores. They are likely to get out of order, and time is lost in putting slings on or off.

Figure 13 shows various forms of forged connections used with hoists and slings. *A* is a slip hook used on end of rope or chain sling passed around a load to tighten on the load when hoisted. *B* is a grab hook used on chain slings which are not to tighten on the load; the throat of the hook allows the chain to enter but not to slip. *C* is a bunk hook used on tackle to take several rope slings at one time. *D* is a finger hook to avoid the danger of the hook's catching undeı the edges of hatches in ships. *E* is a cargo hook for the same purpose. *F* is a grab link used instead of a grab hook; one end of the chain sling is fastened to the large end of ring; the chain is passed through and when the loop is of proper size the small end of the link is pulled between two links of chain. *G* is a double grab link used to shorten a chain or join two quickly and temporarily. *H* is a swivel used to prevent twisting of load by twisting of the hoisting rope or chain. *K* is a cold shut for repairing a chain or connecting two chains. *L* is a stone hoist hook, used in quarry work; the points of the hooks engage the sides of the stone blocks.

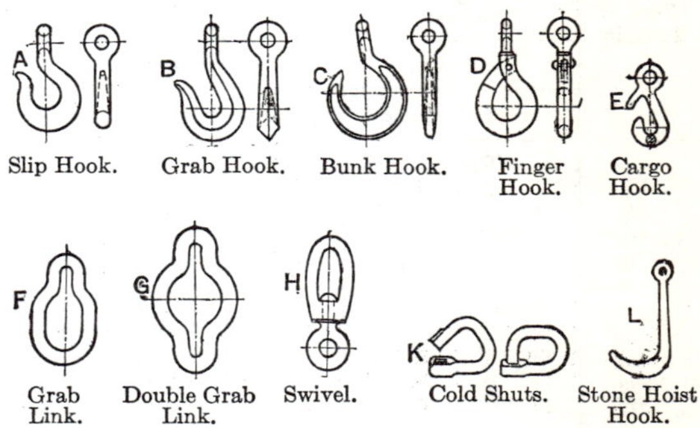

FIG. 13. Forms of forged connections used with hoists and slings.

Steel for hooks, rings, and hoisting fixtures should be of known analysis. Test specimens should be subjected to bending to insure absence of pipes, seams, or other imperfections in the bar. Hot-rolled steel of forging quality, SAE 1010, is used by some makers.

In one important plant all chains, slings, hooks, and lifting fixtures are inspected once a month; parts that can be benefited by annealing are so treated once a year (Ref. 6).

WIRE-ROPE SLINGS, as compared with chain slings, are lighter and, in general, will not fail without warning. Breakage of outer wires is the visible warning, but the inner wires alone will carry loads at which the slings are rated by the makers.

Plain slings are made of standard 6 by 19 or 6 by 37 rope spliced endless, or fitted with sockets, thimbles, or eye ends. In the Atlas patent slings each member consists of one right and one left lay rope braided around a hemp core back and forth so that the sling member has a strength of eight or more strands of the rope. (See Fig. 14.) The ends are

FIG. 14. Atlas patent sling.

loops that can be used as eyes, or fitted to sockets or thimbles. In the Drew sling an endless wire rope is braided into a band about 2 rope diameters thick, 6 or more diameters wide. John A. Roebling's Sons Company, Trenton, N. J., makes slings whose members are of *grommet* construction, i.e., one strand braided back on itself to form a six- or seven-strand rope, endless except for a splice in the one strand. The ends preferably are held by patented thimbles, which are large, permit some slip to equalize the load, and in the form shown in Fig. 15, one thimble can be passed through the other to form a loop or anchor

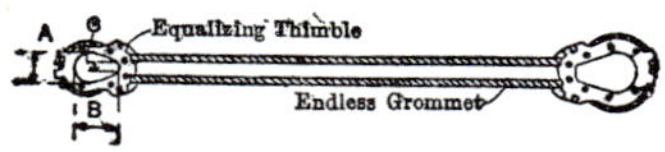

FIG. 15. Grommet sling.

sling. The grommet link (Fig. 16) avoids using a forged steel ring to connect the two or three legs of a sling. All tension comes on a wire-rope ring spliced endless as described above.

Table 9. Improved Plow Steel Wire Rope Slings

(Courtesy of John A. Roebling's Sons Co., Trenton, N. J.)

The safe load for any number of parts may be obtained from values given below by direct multiplication. Slings 3/8 to 1 1/8 in. in diameter inclusive, 6 x 19 rope; slings 1 1/4 in. in diameter and larger, 6 x 37 rope. All ropes have independent wire rope core.

Diam. of Rope, in.	Safe Load Values When Used in Single Part: Vertical		60°		45°		30°	
	Connection		Connection		Connection		Connection	
	Socket	Splice	Socket	Splice	Socket	Splice	Socket	Splice
3/8	1,000	800	900	700	700	600	500	400
1/2	2,000	1,800	1,700	1,600	1,400	1,300	1,000	900
5/8	4,000	3,600	3,400	3,100	2,800	2,500	2,000	1,800
3/4	6,000	5,200	5,100	4,500	4,200	3,700	3,000	2,600
7/8	8,000	7,000	6,800	6,100	5,600	5,000	4,000	3,500
1	10,000	9,000	8,700	7,800	7,000	6,400	5,000	4,500
1 1/8	12,500	11,400	11,000	10,000	9,000	8,000	6,200	5,700
1 1/4	14,500	13,000	12,500	11,300	10,000	9,200	7,200	6,500
1 3/8	17,000	16,000	14,500	14,000	12,000	11,300	8,500	8,000
1 1/2	20,000	19,000	17,000	16,500	14,000	13,400	10,000	9,500
1 5/8	23,000	22,000	20,000	19,000	16,000	15,500	11,500	11,000
1 3/4	27,000	25,000	23,000	22,000	19,000	18,000	13,500	12,500
1 7/8	30,000	29,000	26,000	25,000	21,000	20,000	15,000	14,500
2	34,000	33,000	30,000	29,000	24,000	23,000	17,000	16,500

Figure 17 shows a sling with clamp hooks for lifting plate. It has an equalizing thimble, and carries plate clamps with a positive grip. Structural-steel slings may be fitted with a toggle and releasing pin for detaching the sling, when slack, by pulling a hand line from a distance.

The rated carrying capacity of a sling depends on the kind of fittings, the way the sling engages the load, and other factors obtained by experience. Table 9 gives carrying capacities of improved plow-steel slings which bind against the load. If the sling is not in contact with the load or if it passes around a large diameter, so that stresses are not concentrated at corners, safe loads may be increased somewhat above the tabular figures.

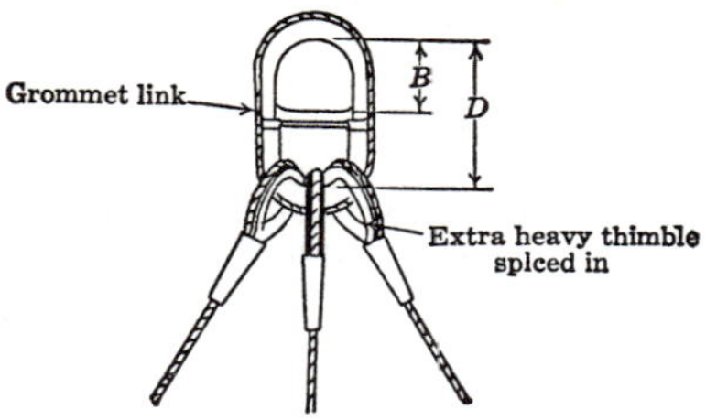

FIG. 16. Grommet link.

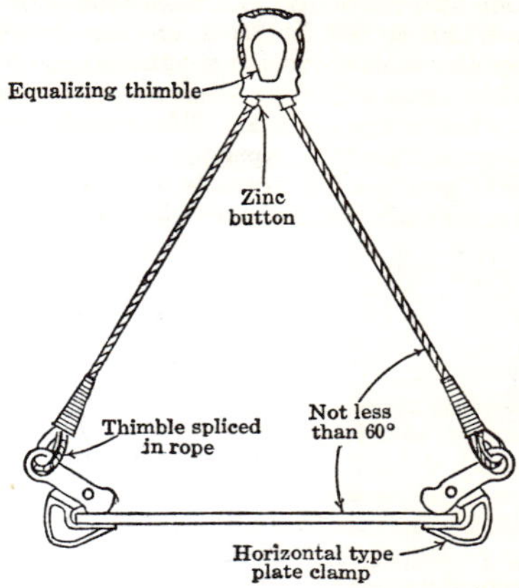

FIG. 17. Sling with clamp hooks for lifting plate.

REFERENCES

1. *Trans. Am. Soc. Mech. Engrs.*, 1931, MH-53-1.
2. C. D. Meals, *Civil Eng.*, 1930.
3. *Coal Age*, Nov. 1924.
4. *Proc. Am. Soc. Civil Engrs.*, May 1932.
5. Bureau of Mines, *Tech. Paper* 237.
6. *Iron Age*, May 22, 1930.

HANDLING MATERIAL IN BULK

3. WIRE-ROPE HAULAGE

WIRE-ROPE TRAMWAYS, or simply **ropeways,** are transporting systems in which moving carriers or containers hang from a wire rope carried on a series of spaced supports, and are moved along by a rope, generally from one terminal to another. Ropeways that both hoist the load and carry it are cableways.

For distances over a few hundred feet a ropeway often is the best, cheapest, and sometimes only means of transport. Under favorable conditions it is low in first cost, cheap to operate, and maintenance and depreciation are low. Ropeways are durable and serviceable. According to one European builder over 90% of those built by him in 30 years are in use; of those built within 50 years, 75%. A ropeway 18 miles long carrying 36,000 tons a year at one-third the cost of transportation by truck is described in Ref. 1. Tramways are single rope and double rope.

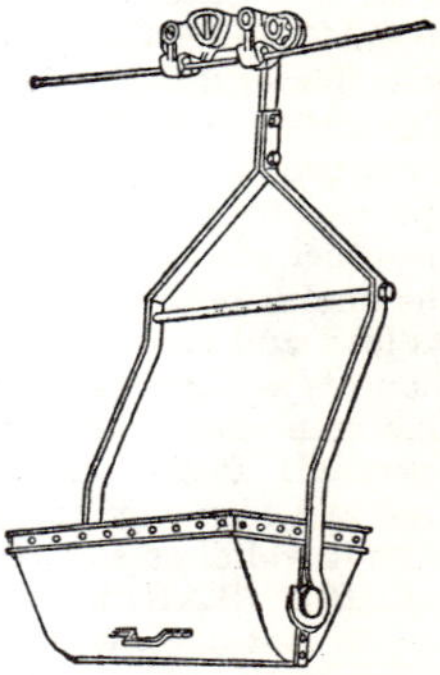

FIG. 1. Single-rope tramway carrier

Single-rope tramways comprise one endless moving rope which both supports and moves the carriers (Fig. 1). At the driving terminal is a large grip sheave (see p. 26-04) or a multiple-groove sheave working with a rewind sheave (see p. 26-04). At the other end of the line, an idler sheave with a pull-back weight puts tension in the rope. Towers on the run carry sheaves to support rope and carriers. The carrier has a pair of clamps to grip the rope. In one construction (Pohlig) the grip permits some axial rotation of the rope, to lessen wear on grip and rope. Grips are automatically engaged or disengaged at terminals, and the carrier runs on its own rollers over fixed track to be loaded or unloaded. Ropeways are best built in a straight line, but angle stations can be provided. Single-rope tramways are used for relatively light loads, which can be spaced with some regularity, and where grades on the line are not over 30%.

Double-rope tramways are those most used in the United States; they are suited to a wider variety of loads, grades, and spans than single-rope tramways. They have two track cables, one for loads, one for empties, usually of locked coil construction anchored at one end, with tension weights at the other. The carriers (Fig. 2) have rollers running on the track cables and are pulled along by a hauling rope which they grip. The hauling rope is driven, tensioned, and supported as is the rope of a single-rope tramway. The grip is worked by two jaws on a threaded pin to which a lever is attached. At the terminal a stop throws the lever over and tightens or loosens the clamp.

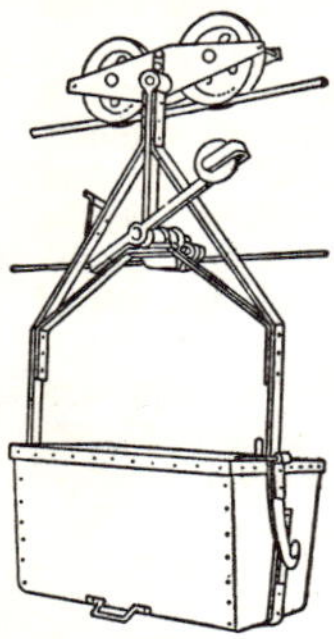

Fig. 2. Double-rope tramway carrier

For details of American ropeways, see Ref. 2 and makers' catalogs. For European practice, see Ref. 3. Costs of handling are compared with costs by other methods in Ref. 4. See also Ref. 5 on recent improvements in automatic dumps, etc.

SUSPENSION CABLEWAYS OR CABLE HOIST CONVEYORS are systems of hoisting and transporting materials consisting of one or two track cables stretched in one span between terminals, a bucket, container, or grab suspended from a carriage running on the track cable, ropes for hoisting the suspended load and moving the carriage, and winding drums for the ropes. In quarrying, rock cutting, stripping, piling, dam building, and many other operations where it is necessary to hoist and convey large individual loads economically it frequently happens that the application of a system of derricks is impracticable, by reason of the limited area of their efficiency and the room which they occupy. To meet such conditions cable hoist conveyors are used, as they can be operated in clear spans up to 1500 ft, and in lifting individual loads up to 15 tons. Two types are made—one in which the hoisting and conveying are done by separate running ropes, and the other applicable only to inclines in which the carriage descends by gravity, and but one running rope is required. In the former the carriage is moved by means of an endless rope. These conveyors are commonly known as *endless-rope* hoist conveyors to distinguish them from the other type, known as *inclined* hoist conveyors.

The general arrangement of the endless-rope hoist conveyors consists of a main cable passing over towers, A frames, or masts, as may be most convenient, and anchored to the ground at each end. Cable tension is maintained by a turnbuckle at one anchorage.

The carriage travels on this cable. It is moved back and forth by means of the endless rope. The hoisting is done by a separate rope, both ropes being operated by an engine specially designed for the purpose, which may be located at either end of the line, and is constructed in such a way that the hoisting rope is coiled up or payed out automatically as the carriage is moved in and out. Loads may be picked up or discharged at any point along the line. Where sufficient inclination can be obtained in the main cable for the carriage to descend by gravity, and the loading and unloading are done at fixed points, the endless rope can be eliminated. The carriage, which is similar in construction to the carriage used in the endless-rope cableways, is arrested in its descent by a stop block, which may be clamped to the main cable at any desired point, the speed of the descending carriage being controlled by a brake on the engine drum.

SLACKLINE CABLEWAY Following is the procedure in the digging, conveying, and elevating of material by a dragline bucket (see p. 26-13) used with a slackline cableway. The bucket (Fig. 6) travels on an inclined track cable, one end of which is anchored. The other end is so rigged to the top of the mast that it can be tightened or slackened by the operator to vary its sag. One end of the hauling rope (load cable) leads to the bucket; the other end winds on a drum. The track cable spans the area served. When it is slack the bucket rests on the material to be dug, the hauling rope is wound in, the bucket digs its load and fills. The track cable is tightened, the bucket raised, and, according as the dump is at the high or low end, the carriage is pulled to or allowed to run to the dump. One man controls all operations. This type of plant is cheaply built and economically operated. It digs sand and gravel from pits and river beds, strips overburden, and makes cuts and fills. By changing the position of the track cable anchorage and turning the mast it will dig a semicircular pit.

CABLE CRANES are similar to suspension cableways, but by having one or both end towers movable they serve a greatly increased area. They are used in building canals, dams, locks, and to serve large storage areas. In American practice, work of this kind usually is done by power shovels (locomotive or crawler cranes) or slackline cableways, or combinations of them, with drag scrapers.

POWER DRAG SCRAPER SYSTEM OF STORAGE. A typical application of this system is a plant receiving coal by rail. The coal-handling system crushes, elevates, and delivers coal to the boiler house, or forms a small pile near the railroad track. Two ropes run across the storage from a double-drum winding machine through tail blocks fastened to back posts. They are hitched to the front and back of an open-end steel box, the scraper. Faced one way the scraper drags coal from the starting pile into storage; faced the other way it drags coal back to the track hopper. If the coal is frozen it goes to the crusher; otherwise the crusher is by-passed.

In small plants, tail blocks are hooked to tail posts, which are short steel columns in concrete piers. They may be hitched to a bridle cable stretched between adjacent posts. In large plants, back posts may be mounted on a car on a standard-gage track, and counterweighted to resist pull of the scraper rope. For high piles the tail block may be carried by a steel tower, 20 or 30 ft high, on a broad flat car on wide-gage tracks. In one plant (Beaumont) the tail sheave is mounted on a caterpillar tractor which can travel anywhere on the storage lot and form a pile of any size or shape.

Advantages claimed for this system are these. (1) Storage area can be any shape or size up to 900 ft scraper travel. (2) Simple machinery, low first cost, cheap repairs, one-man operation. (3) Low fire risk in stored coal; ability to reach fire easily should it occur. (4) Wide range of economical handling capacities up to 1000 tons per hr.

Dragline scraper systems are built in smallest size, $^{1}/_{3}$ cu yd; largest size, 12 cu yd. Small drag scrapers are made for mines, tunnels, quarries.

A combination slackline cableway and dragline scraper system was successfully used in digging a canal in Russia and disposing of excavated material. All three towers are mounted on wheeled trucks running parallel to the canal. This combination has a higher capacity than could be obtained with a cable crane (see p. 26-10) spanning the whole width of the plant.

Drag Scrapers in Cargo Boats. A number of self-unloading boats on the Great Lakes and in coastwise service carrying bulk cargoes have a belt conveyor on a boom which can be slung to either side in order to discharge cargo ashore. The conveyor on the boom is fed from an inclined conveyor from the hold of the boat. There are several ways to take material from the hold and deliver it to the inclined conveyor. Some boats have two belt conveyors, others two apron conveyors, in tunnels extending the length of the cargo hold. Another method, patented by L. D. Smith, Sturgeon Bay, Wis., uses a pair of drag scrapers, each operated by its own double-drum winding machine, in two tunnels with smooth steel floors. Material flows into the tunnels through side openings. Along the top of each tunnel is an I-beam trolley track. The haulage cable is hitched to the open end of the scraper; the pull-back cable for each of the scrapers runs through a small tunnel, around end sheaves, over a trolley sheave hung from the I beam, thence to the back of the scraper. The scrapers drop material into a hopper over the inclined conveyor. On the back pull the scraper is lifted up to the trolley, passes over material which has flowed into the tunnel, back to the starting point. Unloading capacities are 750 to 2000 tons per hr. Small scrapers are 4 cu yd capacity; large ones, 12 cu yd. This system has been applied to tow barges with electric motors for operating the winding drums, but without generating plant. Power cables are run from the dock for unloading.

4. LOCOMOTIVE CRANES

LOCOMOTIVE CRANES usually consist of a lower frame fitted with 4 or 8 wheels for use on track, and a rotating upper frame carrying the power plant and operating machinery. A boom, hinged to the upper frame, can be raised or lowered to handle loads at various radii. The crane will act as a portable hoisting engine, a swinging derrick, a grab bucket unloader, or a switching engine. It will handle a lifting magnet, scoop shovel, dragline scraper, or pile driver. Locomotive cranes with (usually) 8 wheels find widest application around industrial plants served by tracks. Besides lifting loads by hook block, they handle pig iron and scrap with a magnet, and shift cars.

Steam-operated cranes require a boiler, fuel tank, and water tank. Usually a licensed engineer is required to run the boiler. Many later cranes have gasoline or diesel engines, which require less space, can be started or shut down instantly, and have lower stand-by expense.

Crawler cranes, instead of having wheels, have a pair of treads made of steel plates linked together and driven by sprocket wheels. The weight is carried by a row of rollers on the lower frame. The width of tread plates and length of assembly make the unit bearing pressure so low that the crane can travel on soft, rough, or uneven ground, and it requires no track or mats. Crawler cranes as made by several companies are fitted with digging shovels, dragline buckets, trench hoes, skimmer buckets, etc., and can be converted easily from one to the other.

The power shovel is for general excavating, digging from the bank, handling blasted rock, and for grading, including shallow cuts below grade. Its particular field is making cuts into and through banks, loading blasted rock on cars or trucks, where the machine stands in the hole or at base of the excavation. It can be used on loose piled materials, but if reach is long or lift high a grab bucket often works better.

The dragline crawler is for removing overburden, digging under water, cutting wide or deep trenches, and where it is necessary to work above material handled. Dragline excavators work outside the hole, and when fitted with long booms can cover a large area, pile material high, and load cars some distance away. The dragline bucket is not so well suited to side cutting as the trench hoe.

The trench hoe is used to dig trenches up to 12 ft wide and 20 ft deep in types of earth that can be cut and that will not cave in. It is faster than the dragline, can be more accurately controlled, and will cut a greater length of trench in a given time.

The skimmer bucket is a form of shovel comprising an open-end box with teeth. It travels back and forth on the underside of the crane boom and is used for grading, leveling, tearing up old pavements, etc.

STORAGE SYSTEMS USING LOCOMOTIVE CRANES. Figure 3 shows a system of storing coal in semicircular piles. Drop-bottom cars discharge to a central pit around which as a center are broad gage tracks for a long-boom crane. In large plants, working radius is 100 ft. The grab bucket has a capacity of 1 to 4 tons of coal which it digs from the pit and drops on the pile; in reclaiming, it digs from the pile and drops into cars. With coal piled 25 ft deep, a crane of 100-ft radius will store 70,000 tons in a full circle with crane tracks covered. By continuing the tracks into more loops, successive storage piles can be formed by one crane. The handling capacity, depending on size of bucket, is 100 to 250 tons per hr.

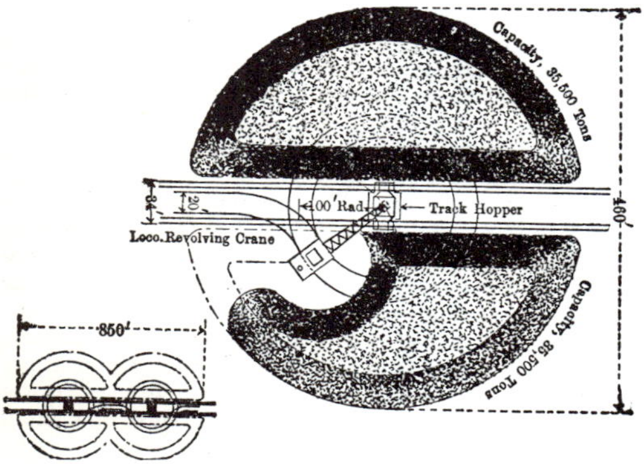

FIG. 3. Semicircular pile system of coal storage.

Locomotive shovel cranes and crawler shovels are used with belt conveyors to remove excavated material or to build levees. Dragline cranes, feeding a system of belt conveyors in tandem, are used in building levees along the Mississippi River. In 1929, over 6,000,000 cu yd were dug from a hill in the middle of Seattle, Wash., by crawler shovels, dumped into feed hoppers, thence to belt conveyors, some of them portable, and carried 3000 ft for disposal. For the system used in building the Grand Coulee dam, see Ref. 6.

5. GRAB BUCKETS

GRAB BUCKETS. Most rope-operated grab buckets, clam-shell buckets, or orange-peel buckets work on the same principle, viz., an upper head, and movable, relative to it, a lower head. The holding line, attached to the upper head, sustains the weight of the bucket when it opens. The lower head is connected to the bowl of the bucket and the closing line, so that when this line slacks the bucket opens. When it is pulled the bucket closes. Various styles of buckets differ in the manner of reeving closing line through sheaves or around a drum (power wheel) or of attaching it to arms or a crosshead so that pulling the closing line will cause the digging edges of the bowl to dig material and fill the bucket. (See Fig. 4.) Very heavy buckets may be operated with two holding and two closing lines to reduce size of rope, sheaves, and drum. Figure 5 is a diagram of two holding lines run from drum H and clipped to an equalizer lever at the top of the bucket. The closing line runs from drum C, around sheaves U and L in the bucket, thence to an anchorage or back to the drum. This is the *fleeting through* method of reeving. If the crane handles a hook block or buckets of different sizes, the bucket is

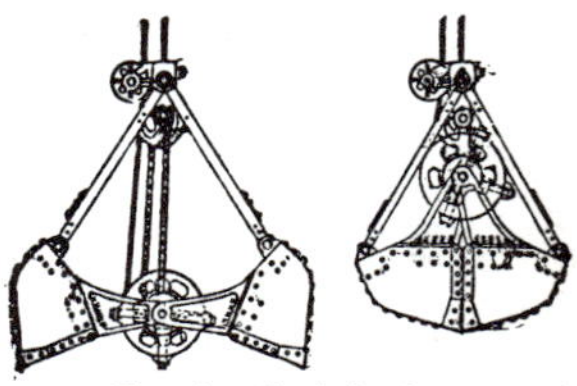

FIG. 4. Grab bucket.

readily detachable if the two closing lines in the bucket are dead-ended, instead of being in one length.

The following is the cycle of operation of any rope-operated bucket. Assume bucket to be in upper position (Fig. 5), empty and closed; the weight is on the closing line. Drum *C* turns to lower bucket; drum *H* pays out holding line at the same rate of travel. As the bucket approaches material to be dug, *H* is braked, *C* pays out closing line, the weight of the parts *L* pushes hinge of the bucket down, the halves open, and the bucket is ready to bite. Both lines then are let go, and the bucket drops on the material. With holding line slack, the weight of the bucket resting on its digging lips, the closing line is pulled in; upper head *B* sinks, bucket closes, digs, and fills. To raise bucket, *C* turns to pull in closing line, which now takes the weight; *H* pulls in the holding line. If the loaded bucket hangs from a trolley or carriage both drums are braked while the carriage moves. To lower the loaded bucket, both drums, under brake control, pay out line. To open the bucket, *H* is braked or stopped, *C* pays out closing line, lower head *A* sinks, and bucket opens. In its traverse back to the loading point, both drums are braked and the bucket stays open.

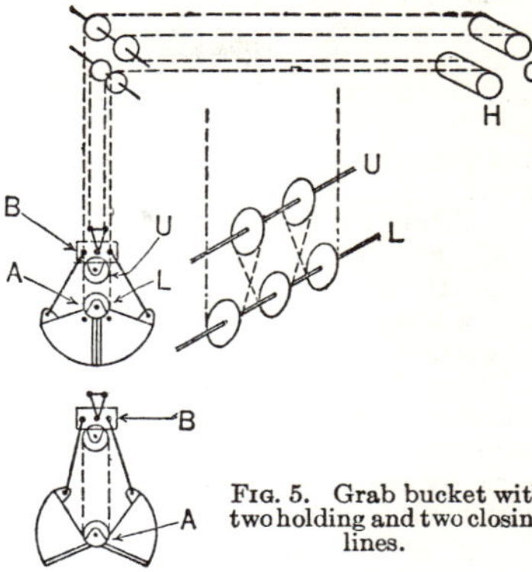

FIG. 5. Grab bucket with two holding and two closing lines.

The choice of a grab bucket depends chiefly on the material to be dug. Buckets with half-round bowls (Fig. 4) will dig loose free-flowing material and coal that is not too lumpy. To dig material that is lumpy or packed hard, the open bucket should have its digging edges in position to penetrate the material and not skid over it.

Orange-peel buckets have a hemispherical bowl divided into sections fitted with digging points. Holding, opening, and closing are the same as in clam-shell buckets operated by two lines. They are made in sizes up to 1 $^1/_2$ cu yd; also in very small sizes that can be used in digging test holes and holes for concrete piles.

Single-line buckets are made in the smaller sizes to rehandle loose material from cars or stock piles or for light excavating. They are hung on a single rope which runs from the hoist drum of a crane or derrick, through sheaves on the bucket, to a closing arm which pulls the halves of the bucket together. A latched arm holds bucket closed; it is tripped open by a hand line. It is not possible to lower a single-line bucket to a pile and then dump or discharge a part of its contents.

Electric Motor Buckets. The Hayward electric motor bucket differs from all others. It uses an electric motor with a compact geared hoist in the upper head. Winding up a chain raises the lower head and closes the bucket; reversing the motor opens the bucket. A slip clutch prevents overload on the motor if the current is not cut off when the bucket is closed. Current is supplied through a flexible cable, which, when the travel is considerable, is automatically wound and unwound from a drum operated by an internal coiled spring, and with collector rings on the end of its axle. An electric controller, hand-operated, starts, stops, or reverses the motor to close, hold, or open the bucket. This bucket can be hung on a single hoisting rope, or from the hook of a rope tackle used with a single drum, or from the hook of an ordinary hoist or crane.

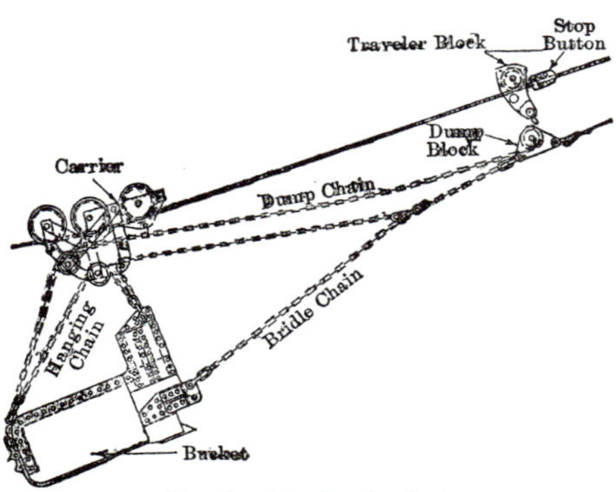

FIG. 6. Dragline bucket.

DRAGLINE BUCKETS (Fig. 6) are used with locomotive cranes or slack-line cableways. The bucket is an open-end box with a digging lip, with or without teeth. The dragline is hitched by a pair of chain bridles with clevis connections to front of bucket. When used with a locomotive crane, the bucket is pulled into dumping position by a line from one of the drums over the end of the boom, and hitched to the back end of bucket. In the slack cable system, the bucket, hung from the carriage, runs up the track cable with a traveler block, hits a stop button;

further motion of the hauling rope pulls on the dump chain hitched to back of bucket. (See Fig. 6.)

6. LIFTING MAGNETS

LIFTING MAGNETS are used with bridge cranes, monorail carriers, locomotive cranes, and derricks to handle iron and steel of all kinds, sizes, and shapes, hot or cold.

In general the only labor needed is that of the crane man, who operates both crane motor and the magnet controllers. A magnet, in combination with a bridge crane, can pile steel higher and more compactly than any other means; cars can be unloaded more quickly and demurrage avoided; magnets will handle steel hot from the roll and will work under water.

A lifting magnet consists of a circular or rectangular cast-steel case enclosing copper coils, inner and outer pole shoes to form contact with the material lifted, and a non-magnetic shield to protect the coils from the material. Rectangular magnets have less lifting power than circular magnets. They lift only what they actually touch, but round magnets will operate through an air gap of 5 to 15 in. according to their size.

Figure 7 is a cross section of a round magnet, and Table 1 gives capacity per lift for different kinds of service for average 10 hr steady work. If service is not continuous, the coils run cooler and the lifting capacity is greater. Figure 8 shows three rectangular magnets hung from a spreader beam for lifting long plates.

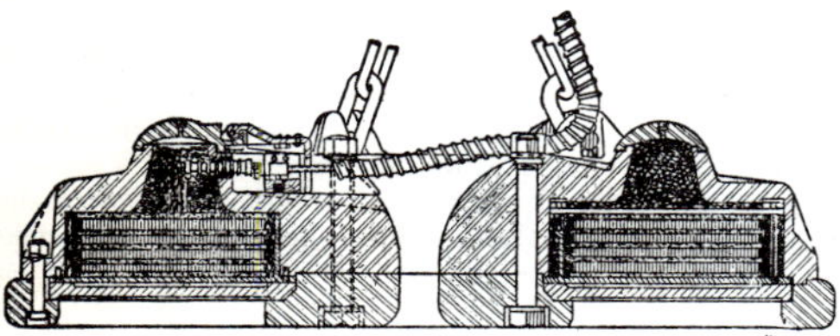

FIG. 7. Lifting magnet.

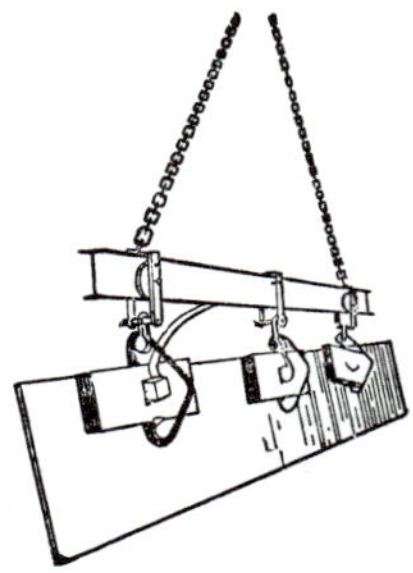

FIG. 8. Lifting magnet, handling long plates.

To provide direct current needed for a magnet, locomotive cranes are fitted with a generating set in the cab. On a-c bridge cranes, a motor-generator set on a platform over the operator's cage supplies direct current. The trolley has a cable winding drum for taking up or paying out electric cable to the magnet.

Table 1. Average Lift, Pounds, for All-day Service of Lifting Magnets

(Courtesy of Cutler-Hammer, Milwaukee, Wis.)

Size of magnet, in. diam.*	65	55	45	39	24
Skull cracker balls	40,000	30,000	24,000	18,000	6000
Slabs	60,000	50,000	35,000	20,000	6500
Machine-cast pig from cars	2,800	2,280	1,275	750	250
Sand-cast pig from cars	2,650	2,170	1,150	680	225
Broken scrap	1,800	1,300	1,000	600	200
Cast-iron borings	2,300	1,500	1,100	600	200
Steel turnings	1,100	750	450	300	100
Principal uses	Large capacity; heavy duty	Open-hearth plants; steel mills	General work; larger capacity with locomotive cranes	Pig iron; scrap; medium capacity with locomotive cranes	Light-finished parts; pipe; castings

* Round magnets 18 in., 12 in., 7 in., 5 in. in diameter also are made.

Round magnets often are hung over a belt conveyor to remove stray iron from coal, foundry sand, etc., carried on the belt. The diameter of the magnet should be one-third larger than the belt width, and it should not be higher above the belt than one-fourth the belt width. Belt speed should not exceed 450 ft per min.

REFERENCES

1. *Trans. Am. Soc. Mech. Engrs.*, 1929, MH-51-2.
2. R. Peele and J. A. Church, *Mining Engineers' Handbook*, 3rd ed., John Wiley & Sons, New York, 1941.
3. P. Stephan, *Die Drahtseilbahnen*, Julius Springer, Berlin.
4. *Trans. Am. Soc. Mech. Engrs.*, 1930, MH-52-4.
5. E. J. Lloyd, *Trans. Am. Soc. Mech. Engrs.*, 1931, MH-53-4.
6. *Eng. News-Record*, Aug. 1, 1935.

CRANES AND DERRICKS

A hoist is a machine for raising and lowering loads. A crane is a hoist or number of hoists so mounted as to move loads both vertically and in other directions.

SAFETY STANDARDS FOR CRANES, DERRICKS, AND HOISTS. The safety code sponsored by the ASME and ASA defines terms, states requirements of design and construction, and gives rules to promote safety in operating.

Cranes generally are tested by their makers with a load 25% greater than the load for which they are rated.

7. TYPES OF CRANES

Cranes which lift and swing the load are pillar cranes, jib cranes in various forms, derricks, rotary bridge cranes, locomotive cranes. Cranes which lift the load and travel with it comprise many forms of bridge and gantry cranes.

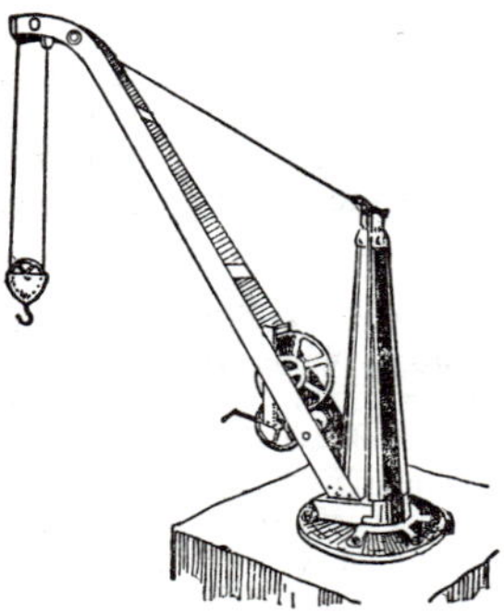

FIG. 1. Pillar crane.

A PILLAR CRANE consists of a stiff self-supported column, turning on a vertical axis and fitted with an inclined boom, the end of which carries the load. The boom swings in a full circle on ball or roller bearings. Hoisting is by hand-power winch on the column. Figure 1 shows a type made in sizes of 2 to 20 tons; working radius 12, 15, or 20 ft.

Pillar jib cranes have a trolley traveling on a horizontal boom to carry the load. The hoist is hand or power; the trolley usually is moved by hand. The jib may be two channels or a single I beam for chain, electric, or air hoist. (See Fig. 2.)

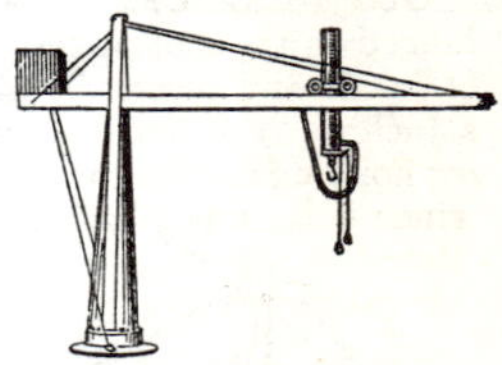

FIG 2. Pillar jib crane.

TOP-SUPPORTED JIB CRANES require the top of the column to be braced to a wall or ceiling, or by guy ropes or struts. They include yard derricks, foundry cranes, and sidewall cranes for machine shops. Figure 3 shows one with separate hand winches for hoist and traverse. Jib cranes sometimes are made with single I-beam jib for hoists hung from four-wheel trolleys as auxiliaries to traveling bridge cranes.

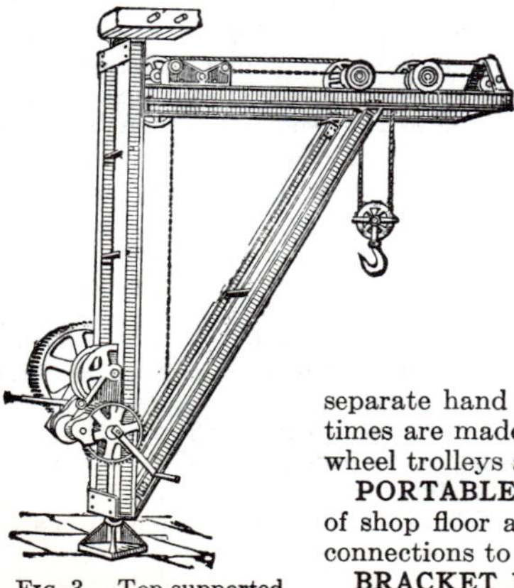

FIG. 3. Top supported jib crane.

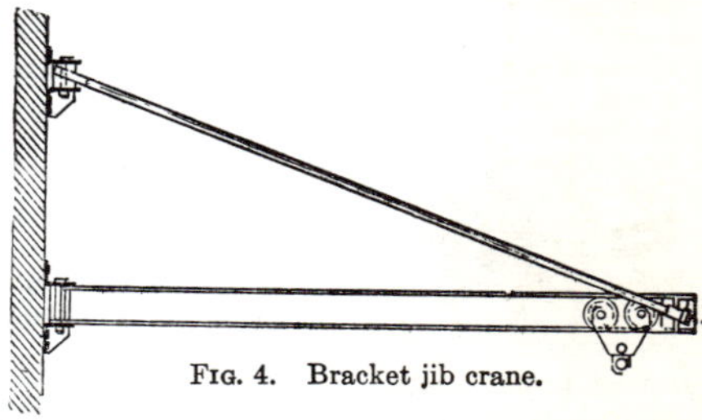

FIG. 4. Bracket jib crane.

PORTABLE JIB CRANES can be made to serve different areas of shop floor as needed, by fitting building columns with socket connections to fit loose-pin connections on the crane mast.

BRACKET JIB CRANES have a horizontal jib hinged to a wall bracket by a vertical pin, and stayed by a rod hinged to an upper bracket. The form shown in Fig. 4 has an I-beam jib for a trolley, from which hangs a hand or power hoist. Bracket jib cranes also are made with built-up girder rail jibs for monorail. (See p. 26-28.) The ends of the special rail on the jib can be fitted with interlocks to engage with fixed tramrail. Trolleys on the jib thus also may run on tramrails radiating from the bracket crane, and serve a large area. To serve trolley tracks running through a doorway, bracket jib cranes are made hinged to lie flat against the wall and allow the door to close.

WALKING CRANES and **traveling wall cranes** are jib cranes mounted for longitudinal travel; the first runs on a track on the shop floor with an upper rail on the wall or building columns to take the thrust at the top of the mast. The wall crane (Fig. 5) has an upper

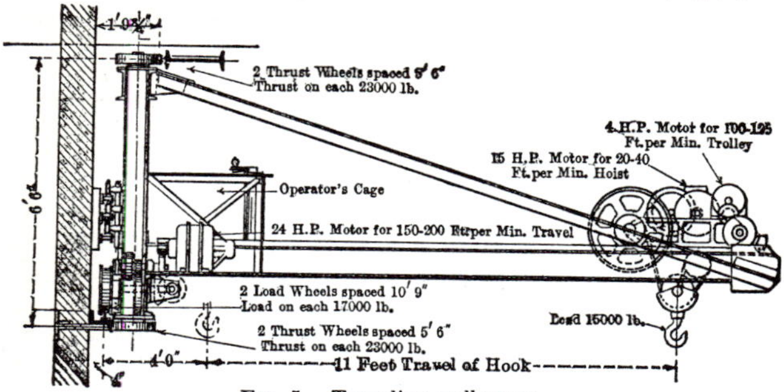

Fig. 5. Traveling wall crane.

and a lower thrust track on which run flat-faced rollers, while the vertical load is taken on a separate rail mounted on wall brackets. The thrust load is much greater than the dead load when the trolley is at the extreme end of the crane. Such cranes act as auxiliaries to shop bridge cranes in serving long lines of machine tools or in handling riveters on long structural steelwork.

GUY DERRICKS have a mast, with a step and vertical pivot pin, at the base, a boom hinged to the foot of the mast, whose inclination can be varied by rope tackle joining tops of boom and mast. A pivot pin on top of the mast fits a cap to which guy ropes are attached. For hand operation, a double-drum winch is mounted at the base of the mast for hoisting and luffing respectively. If angle of boom is not often changed, a single-drum winch will suffice, luffing being by a hand line hitched to the mast. Figure 6 shows a guy derrick used in quarries and by contractors in bridge and building construction. It is operated by steam or power hoist with two hoisting drums for load and luffing, and a pair of swinging drums. Ropes from the latter lead to opposite sides of a bullwheel, on the foot of the mast, to swing mast and boom. The luffing rope runs to foot of mast, to masthead, thence to tackle between masthead and top of boom. The hoisting rope follows a similar path, and runs through sheaves on the boom to the suspended load. Stress in boom-hoist tackle increases as boom is lowered. Hoisting rope is generally 6 by 19 steel wire rope; boom hoist line is the same, but with more parts in the tackle. Stresses in guy lines can be calculated from their arrangement around the mast with respect to the position of the load and also with respect to the angle they make with the horizontal. For other points on derrick design, see Ref. 1. Contractors' derricks often are made of steel for safety, and in sections for ease in handling.

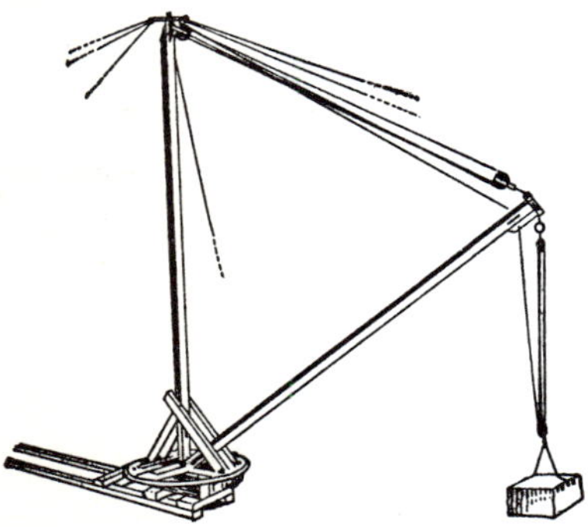

Fig. 6. Guy derrick.

STIFF-LEG DERRICKS are used where it is not feasible to run guy ropes, or where derrick must be shifted often. (See Fig. 7.) Struts from masthead, inclined at 45 degrees, are framed, or hinged, to two sills set on the ground at 90 degrees, meeting under the step of the mast. The boom can swing 270 degrees. Sills can be bolted to foundations or held down by weights.

JINNIWINK DERRICKS are similar to stiff-leg derricks. An A frame replaces the mast, and with its back strut is mounted on a T-shaped frame acting as a base. The boom, hinged to the base frame, is adjustable for angle. These derricks, light and easily set up, are used by building contractors.

TOWER DERRICKS consist of a square tower of wood or steel with footings for booms on one or more corner posts. They are used in building construction, booms being shifted higher on the posts as the building goes up. A steel tower derrick with a revolving pivoted counterweighted 50-ft boom is used in building construction in Germany. It will lift 1 ton with boom horizontal; maximum height of lift with boom luffed, 100 ft. The tower rests on a wheeled platform, and has three small electric winches, controlled by a man on the platform.

The concrete distributing apparatus (Fig. 8) is a modified mast and tower derrick. The mast hoist has a framed steel mast, about 20 in. square, made in 20-ft sections with bolted end connections and guy attachments. It has guides for a dump bucket which discharges into a receiving hopper adjustable in height on the mast, from which chutes distribute concrete. The upper end of the chute is a boom, guyed from the mast head. Extensions of the chute hang from trolleys on a chute-extension cable which spans the area over which concrete is distributed. Figure 8 shows a 60-ft mast with a 40-ft boom chute, served by a 1-yd hoist bucket. Larger plants have steel or wood guyed towers, with one or two hoist buckets. Masts and towers for this system of distributing concrete have been designed and standardized for rapid and easy erection.

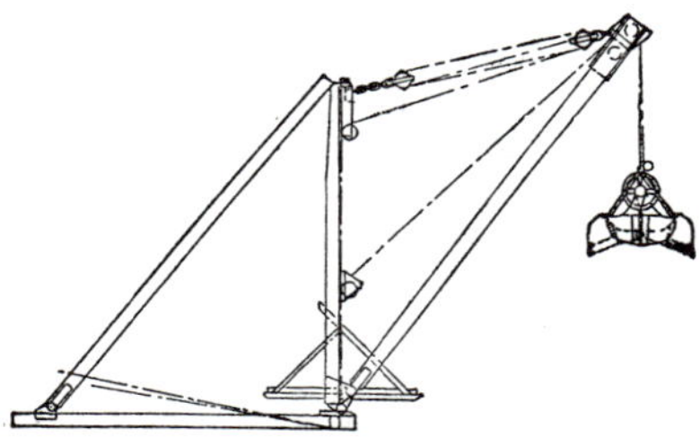

Fig. 7. Stiff-leg derrick.

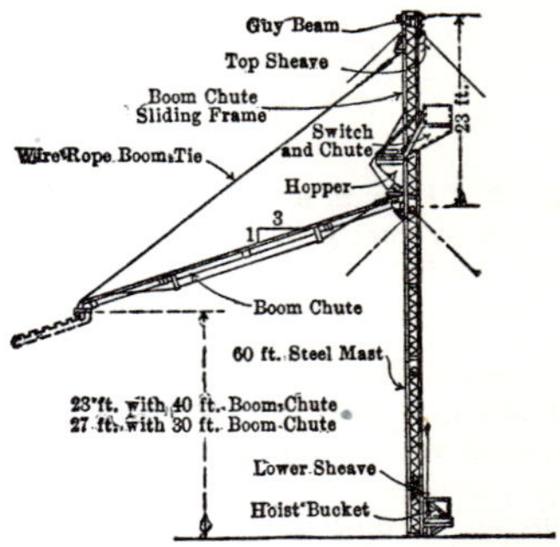

Fig. 8. Tower derrick and boom chute. (Insley Manufacturing Co., Indianapolis, Ind.)

TRAVELING BRIDGE CRANES. Makers of tramrail equipment (see p. 26-28) list and illustrate in their bulletins a wide variety of light cranes for 2 tons or less load. Table 1 classifies Cleveland tramrail bridge cranes; similar cranes are made for American mono-

Table 1. Cleveland Tramrail Bridge Cranes

(Courtesy of Cleveland Crane and Engineering Co., Wickliffe, Ohio)

Bridges Propelled by Hand (Pull or Push Load)

Type CH: For any Cleveland carrier or hoist, hand or electric up to 2 tons capacity; 2 runways for bridge lengths up to 30 ft.; 30 to 50 ft. 3 runways.

Type CHL: For any Cleveland carrier or hoist, hand or electric, up to 2 tons capacity; 2 runways for bridge lengths up to 30 ft.; low headroom.

Type DGH: Double girder bridge; trolley hoist on top. Capacities up to 1 ton; 2 runways for bridge lengths up to 30 ft.; low headroom.

Bridges Propelled by 2 or 3 Motors

Type UCE: For any Cleveland carrier or hoist, hand or electric up to 2 tons capacity, 2 runways for bridge length up to 30 ft.; for 30 to 50 ft. 3 runways.

Type DGE: Double girder bridge; hand pushed trolley; capacities up to 2 tons; 2 runways for bridge lengths up to 30 ft.; low headroom.

Type DGEE: Same as DGE except that trolley is motor-driven.

Type DGC: Like DGEE, but operator controls all motions from cab mounted on bridge.

rail. These light cranes are notable for several reasons. (1) End trucks have equalizer levers and swivel wheels; wheel flanges do not bind on the rail. (2) Structure is light; wheels have ball bearings; crane is easy to move. (3) The light construction of bridge, flexible mounting of the wheels, and absence of a squaring shaft permit cranes to be hung from three runways if bridge is longer than 30 or 35 ft. By using three runways weight of the bridge can be kept so low that a 50-ton crane can be moved by pushing or pulling the load, and there is no danger of pinching or binding the runway. (4) Most types can be fitted with interlocks to register bridge rails with fixed tramrail, so that hoists, carriers, and trailers can travel on or off crane bridge. (5) Electrical conductors and mountings, control, and accessories are well standardized. (See Fig. 9.)

Other bridge cranes, for loads up to 5 tons and for spans under 30 ft, are made of single I-beam spans with trolley hoists hung from lower flanges. The bridge travel of these cranes is generally by hand chain hung from a chain wheel on a squaring shaft that drives truck wheels at each end of the span. For loads over 3 tons, it is better to have a gear reduction between chain gear and squaring shafts.

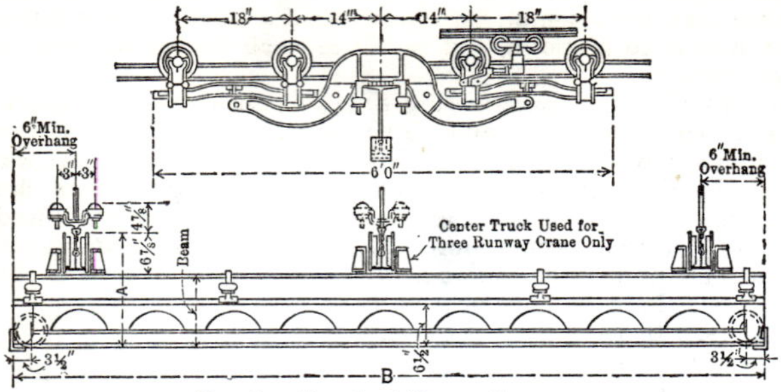

FIG. 9. Cleveland Tramrail crane.

Double I-beam cranes are used for loads up to 20 tons and for spans of 20 to 50 ft. The trolley runs on top of the girders. The hoist rope is wound on a drum which may be geared to hoist light loads at faster speed than heavy loads. With double I-beams, the load hook can go between the beams, and for a given lift, the building need not be so high as it would be if trolleys were hung below the beam. Some cranes have two trolleys for handling long or heavy pieces.

The hand-power cranes mentioned are for what crane builders call class D, i.e., very intermittent, service. If hand operation is too slow, I-beam cranes can be fitted with one or two motors (class C service), or with three motors for class B service, which means the ordinary work of a crane in shops, foundries, yards, etc., where the crane is busy 8 or 10 hr and adjustments, inspections, and repairs can be made in idle time.

A three-motor bridge crane for class B service is shown in Fig. 10. The bridge is made of two I-beams to carry the load, with two light lattice-girders to stiffen the frame and

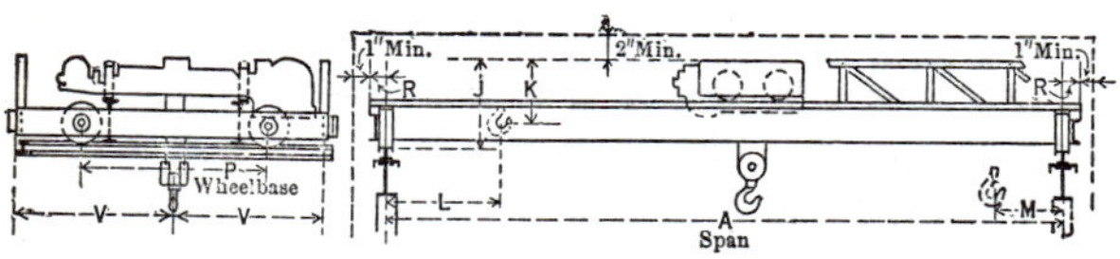

FIG. 10. Light-duty bridge crane.

to help carry the travel machinery. The bridge often can be shipped and erected as a unit. The trolley frame is a one-piece steel casting; the double-flanged trolley wheels run on flat bars welded to top flanges of the girder beam.

Cranes for class A service work 24 hr a day in steel mills, cement mills, and foundries, or on heavy hoist and grab bucket or magnet work, where interruption of service is serious. (See Fig. 11.) Speeds of hoists, trolley, and travel often are higher than for cranes in

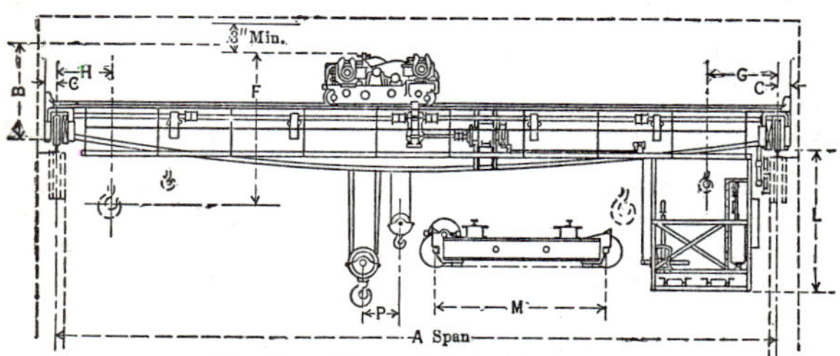

FIG. 11. Heavy-duty bridge crane.

lighter service; motors are larger; parts are designed with a higher factor of safety and with more regard to access to working parts and certainty of lubrication. Cranes made by various makers do not differ much in essential dimensions.

AUXILIARY HOISTS FOR CRANES. The dimensions *P* and *G* in Fig. 11 for cranes over 10 tons refer to position of hook of the auxiliary hoist, which is used for rapid hoisting of loads lighter than those for which the crane is designed. For this duty the trolley carries two motors, each with its own set of drums and gearing. One set is for the main hoist of the rated capacity of the crane; the other is for 3, 5, or 10 tons auxiliary capacity.

CRANES FOR OPERATING GRAB BUCKETS for handling coal at power plants, raw material and clinker at cement plants, unloading bulk material from cars, etc., have a trolley with two drums, one for closing line and one for holding line. The drums are mounted parallel to the bridge girders; each has its own motor, solenoid brake, and reduction gearing. If the crane does miscellaneous hoisting and a bucket is used only occasionally, the trolley can be of the ordinary hoisting type, and the bucket of the single-line or one of the motor-driven types, to be put on the hook block when needed.

Grab bucket cranes are made by most builders of heavy-type bridge or gantry cranes. Dimensions of trolley vary with weight and type of grab bucket and height of lift. Parts generally are of heavy construction, and precautions are taken to keep dust out of gears. All motions of crane and bucket are electrically controlled from the operator's cab, usually mounted on the trolley.

TRENDS IN CRANE DESIGN (Ref. 2). Modern cranes are characterized by less dead weight, by rolled beams instead of box girders, lighter motors, alloy steel parts, ball or roller bearings, heat-treated gears, gears enclosed for protection and running in oil baths, grease cups, and intermittent oiling being avoided. Bridge cranes with structural parts of aluminum alloy are described in Ref. 3. Advantages claimed are reduction of weight, less power for travel motions, and no necessity of painting.

GANTRY CRANES. For loads up to 2 tons Cleveland tramrail can be fitted to two gantry legs. (See Fig. 12.) All wheels have ball bearings, and the structure, with load, can be moved by one man at either end pushing or pulling. For rapid or long travel the crane can be electrically propelled. These cranes can be used in shops as auxiliaries to traveling bridge cranes. Cleveland gantry cranes can be interlocked to register bridge rails with fixed tramrail so that hoists, carriers, and trailers can move on and off the gantry bridge.

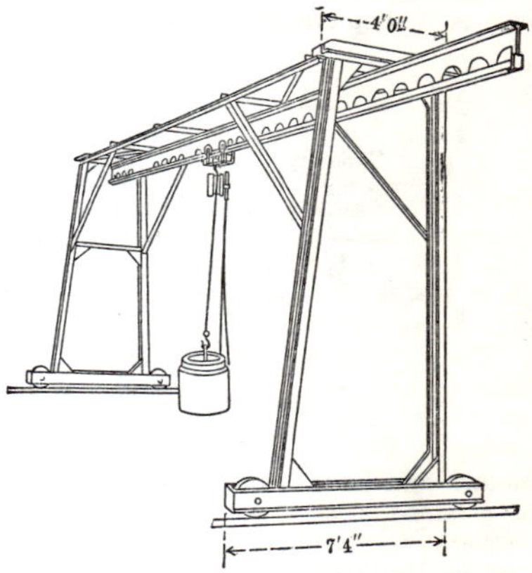

Fig. 12. Tramrail gantry crane.

Gantry cranes of heavier type range in size from 5 to 250 tons capacity, of any span. In construction they are like bridge cranes built for hoist, bucket, or magnet work, but have better protection from weather for operator and machinery. For travel motion, power is transmitted from a motor on the bridge through horizontal and vertical shafts to geared truck wheels at ground level. Truck should have brakes to prevent high winds moving the bridge. Such mechanical and electrical brakes are described in Ref. 4. If the crane serves a yard alongside a building, it is sometimes convenient and economical to support one crane track on the building wall. The crane is then a *single-leg gantry.*

The advantages of gantry cranes are that there is no expense for overhead runways; tracks can be extended or changed in position; and there is no obstruction of yards by overhead track supports.

Portal cranes (double-leg) or **semi-portal** cranes, single-leg, are similar to gantry cranes. They generally have a rotating crane or a crane with luffing motion mounted on top of the structure to transfer freight from ship to wharf or vice versa.

SPECIAL CRANES FOR STEEL WORK. Ladle cranes are made up to 300 ton capacity to control all the motions of filling, transporting, and pouring hot-metal ladles in open-hearth plants. They have two independent trolleys with a pair of girders for each. The main trolley travels on the outer girders and hoists the ladle; the auxiliary trolley travels on the inner girder and has two hoists to tilt the ladle one way or the other. To insure safety, the two hoist motors and two sets of hoisting machinery are geared together to prevent the ladle from tipping sidewise and spilling metal if one motor or set of gears should fail.

Ingot-stripping cranes are of several types. In one, a vertical arm suspended from a trolley carries jaws which seize the hooks of the ingot mold, while a screw-actuated plunger pushes the ingot from the mold. Some cranes carry auxiliary trolleys to handle stripped ingots. In rope-operated cranes, the hoist block carries tongs which lift the mold while a stationary ram pushes the ingot out.

Soaking-pit cranes do all the work of handling ingots into and out of soaking pits. In one type, a trolley carries a guided vertical ram, gear-operated, with gripper jaws on the lower end. These have an automatic closing motion, suitable to ingots of different widths. The ram rotates on a vertical axis. This type requires high headroom. Rope-operated types require less headroom, have a shorter ram, and the guides are below the trolley.

Crane-type Charging Machines. A vertical frame moves with the trolley and is braced against the underside of the bridge girders. It carries an arm that holds the charging box and turns it over when it has been pushed into the open-hearth furnace.

ELECTRICAL EQUIPMENT FOR OVERHEAD CRANES. Alternating current generally is used for most cranes. More delicate control can be obtained than with direct current, especially for light loads. If both alternating and direct current are available, choose alternating; if only direct current, do not convert it to alternating, but use it.

OPERATING SPEEDS OF CRANES. Table 2 shows motor sizes and operating speeds of shop cranes in intermittent service. For heavy-duty and 24-hr service, heavier

Table 2. Light-duty Crane Ratings

Capacity, tons	Full Load Speeds, ft per min			Horsepower of Motors			Diameter of Rope, in.
	Hoist	Trolley	Bridge	Hoist	Trolley	Bridge	
1	42	110	175	3	3/4	3	1/4
2	35	110	175	5	3/4	3	1/4
3	33	110	175	7 1/2	1	3	3/8
3 1/2	28	110	175	7 1/2	1	3	3/8
5	26	110	175	10	1	3	3/8
7 1/2	24	110	175	15	2	5	9/16
10	18	110	175	15	2	5	9/16

motors and higher speeds generally are used. High speeds of hoist and travel cannot be attained in short movements, because much of the time is used in accelerating and decelerating moving parts. (See p. 26-21.) In estimating capacity of a crane to hoist and transfer loads, allowance must be made for time thus lost. For short travel, as in transferring loads from car to adjacent storage, the allowance may be 25% or more.

CRANE SUPPORTS. Running rails are furnished by the user; size is recommended by the maker for the particular service. They must be laid straight and at exact center-to-center distance. Track supports should be strong enough to carry maximum load close to one end of bridge, and stiff enough laterally to take the shock of sudden stops of the loaded crane trolley. Column connections and columns must be designed on a basis of maximum loads. It is well to consult crane builders before design of the building has gone too far.

HOISTING ROPES are cast-steel wire rope, usually 6 by 37, but 6 by 19 and 8 by 19 also are used. Sheave and drum diameter should be (30 × rope diameter). For use of wire rope, see p. 26-02.

8. POWER REQUIRED TO DRIVE CRANES

POWER REQUIRED TO DRIVE CRANES.* The current for hoisting is expended partly in lifting the load and partly in losses in the motor, the solenoid brake, the hoisting gearing and bearings, and the friction and bending of the ropes on the drums and sheaves. As a rough average for shop cranes, these current expenditures may be 70% useful work and the losses 14, 1, 12, and 3%, respectively, of the total current furnished. The size of the motor is determined by the torque required to lift the load plus the torque necessary to bring the load and all the rotating mechanism up to speed in the time allowed by the designer for the hoist to attain its full speed. The overload torque required for acceleration to a speed v in t seconds varies as $1/t$, and that factor, rather than the torque for direct lift, determines the size of the motor in all hoists except very slow ones. In a typical crane hoist (Fig. 13) the effect of the acceleration of all the moving parts on the torque of the motor can be obtained by calculating the weight and radius of gyration of the moving parts at the drum shaft and on all other rotating shafts, and then finding an equivalent

* Adapted from *Electric Crane Construction*, by C. W. Hill, C Griffin and Co., Ltd., London, 1911.

weight which, at the center of gyration of the motor armature, will produce the same energy effect.

PRACTICAL RULES. Builders of traveling electric cranes used in shops assume a hoisting efficiency of about 70%, which corresponds to 23,000 ft-lb per min per hp of motor rating. For cranes used in steel mills, grab-bucket hoisting, etc., subjected to occasional overload and rough handling 20,000 ft-lb per min per hp is used. The margin of 30 or 40% generally is considered sufficient to allow for acceleration and for electrical and mechanical losses, but for high hoisting speeds and quick acceleration, the forces due to acceleration should be calculated and allowance made for them.

For travel speeds not over 250 ft per min a rule used in practice is

$$\text{Motor horsepower} = 2\left[\left(W \times \frac{d}{D} \times 0.07 \times 1.07^{n} \times S\right) \div 33{,}000\right] \tag{1}$$

where W = weight of loaded crane, pounds; d = diameter of truck axles, inches; D = diameter of truck wheels, inches; 0.07 = coefficient of axle friction, and also is the loss at each cut gear reduction, including journal friction of countershaft; n = number of gear reductions; S = speed, feet per minute. The factor 2 covers overload due to acceleration, bad condition of crane track, "notching-up" controller too rapidly, etc. For greater safety against these adverse conditions, the factor sometimes is made 2 1/2. For speeds over 250 ft per min and for quick acceleration, the forces due to acceleration should be calculated and allowed for.

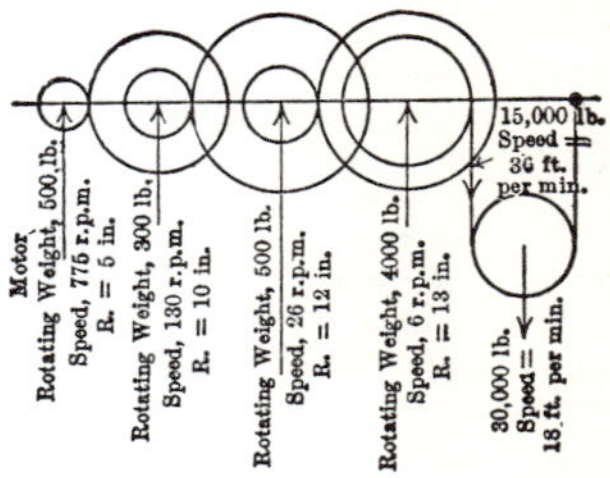

FIG. 13. Diagram of crane hoist speeds.

The following rules are given in Ref. 5: Horsepower of hoist motor = (Load in tons × Lifting speed, feet per minute)/11. This is equivalent to assuming an overall efficiency of 66%. Horsepower of trolley motor or bridge motor = $(WS/132)(d/D)$, if bearings are bronze bushed; or $SW[30 + (S/20)]/33{,}000$, if axles run in roller bearings. In these equations, W = total weight moved, tons; S = travel speed, feet per minute; d = diameter of axle, inches; D = diameter of wheel, inches.

9. LOADING AND UNLOADING EQUIPMENT

ORE BRIDGES AND COAL-STORAGE BRIDGES are gantry cranes with a grab bucket to take ore or coal from boats, cars, or a dump to storage, and to reclaim it from storage to cars or boats. As used at Great Lakes ports, these often are bridges spanning 300 ft or more over the storage area, with cantilever extensions over the water at one end and over railroad tracks at the other. In modern bridges up to 10-ton buckets have all drums for control of bucket mounted, together with operator's cab, on the trolley.

In some storage bridges a revolving crane, with a grab bucket on a horizontal boom, runs on the top chords of the bridge. Coal from boats is delivered to a belt conveyor on the bridge and tripped off into storage. In reclaiming from storage, the crane boom swings sideways from the bridge and the grab bucket drops the coal into a hopper delivering to the belt conveyor. Coal can be reclaimed at the rate of 300 tons per hr.

BOAT LOADING. At the upper end of the Great Lakes, over 80 million tons of iron ore are loaded into steamers each year. The ore is dug by steam shovel and loaded into large bottom-dump cars, which at the shipping pier run on high trestles. When the bottom doors open, the ore drops and and runs freely through long chutes into the boat.

On the Atlantic coast, coal is shipped from Norfolk, Baltimore, Philadelphia, and other ports the year round. In winter, it sometimes is necessary to thaw the railroad cars in a closed thawing shed before dumping. At some piers the cars run into a rotary dumper, are turned over, and the coal runs into transfer cars. The transfer cars may be lifted to the top deck of the pier to dump their coal into hoppers and ship-loading chutes, or they may travel along the pier and discharge to a conveyor on a boom extending over the boat. In a plant at Baltimore the turn-over motion of the Wellman rotary dump is so controlled that the coal slides out, without much breakage, to an apron conveyor which in turn discharges to the conveyor on the boom. At the end of the boom a vertical chute with a control gate, to ease the fall, delivers coal to the boat. Boats loading coal at Atlantic ports are of all sorts; hatches generally are small, and chutes must be arranged to suit them and to clear ship's rigging. At a coal pier at Pensacola, Fla., a Link-Belt car dumper feeds two bucket elevators, of 300 tons per hr capacity each. They can discharge at heights variable by 45 ft to suit a great variety of boats and conditions of loading. Loading chutes have telescopic ends and a horizontal swing of 180 degrees.

The large boats bringing ore from the head of the Great Lakes and taking coal back are designed for the purpose. At coal shipping ports, mostly in Ohio, boats are loaded by coal dumpers, generally of the lift type. An example is the Link-Belt rolling dumper which pours the contents of a 40-ft car into a barge. When car is empty the cradle runs down an inclined track. For history and description of car dumpers, see Ref. 6.

BOAT UNLOADING. Clam-shell buckets are in general use for unloading boats. For intermittent service and small capacities, a mast and gaff with a clam-shell bucket and 2-drum hoist is good and cheap. An inclined boom (gaff) swings around the mast, the bucket being hung from the top end of the gaff. Guide sheaves for holding and closing lines (see p. 26-13) are spread apart at the heel of the gaff, so that the swing of the gaff toward and from a belt hopper can be controlled by putting more tension on one line than on another by means of clutches and brakes at the hoisting machine. Such hoists are used for boats and barges having no rigging to interfere with the swing of the gaff; for rigged boats, a tower with a boom hinged to swing vertically is necessary.

Hoisting towers, with separate hoists to lift the bucket and trolley it on the boom, permit rapid work, as the two motions are simultaneous part of the time. They often work in combination with cable railways or conveying systems to handle coal at storage pockets and power houses on water fronts. To transfer coal or ore from boat to cars or to storage piles, the hinged boom may be a cantilever extension of a bridge tramway.

The Hulett unloader (Wellman Engineering Company) comprises a large gantry frame spanning a number of railroad tracks and carrying a trolley. To the trolley is pivoted a pair of arms carrying a vertical bucket leg. An operator in the leg controls all motions of bucket, trolley, and gantry. A second operator on a traveling scale lorry, hung under the gantry, receives the discharge of the bucket through a hopper holding three bucket loads. After weighing, the larry discharges to railroad cars or to the reserve storage space under the cantilever end of the gantry. Large Hulett buckets spread open to 24 ft pick up 17 tons of iron ore, and make the cycle of load, traverse, and dump in about 50 sec.

FREIGHT HANDLING TO AND FROM SHIP. In the United States ships generally are loaded and unloaded by the ship's own winches and rigging. In Europe most of the work is done by wharf cranes, as much of the freight comes and goes in barges without hoisting machinery. To handle freight to and from ships faster and cheaper than by use of the ship's winches, cargo cranes on the wharf must be kept to that one duty and not be burdened with distribution on the wharf. Monorail hoists and tramrail carriers have been used on a few wharves, but electric or gasoline tractors with trains of trailers are more flexible and useful.

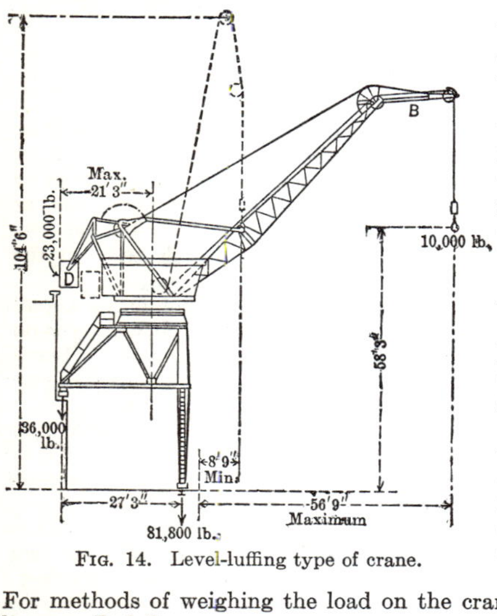

Fig. 14. Level-luffing type of crane.

Most wharf cranes are traveling gantries with a revolving crane on top of the bridge. Crane booms often are rigged with a "level-luffing" device. Figure 14 shows one type. Outrigger *B* and counterweight *D* are mounted so that with hoist drum braked, boom can be moved in a vertical plane without raising or lowering the load. The advantages are that several cranes can work in one hatch without interfering with each other or with the ship's rigging; time is saved; and the boom hoist motor may not be so large. See Bernard Dunell (Ref. 7). For methods of weighing the load on the crane according to the tension in the hoisting line, see Ref. 8.

Cargo cranes generally have a lifting capacity of 10 tons or less, but important European harbors always have some wharves with large cranes to handle heavier loads. A typical hammer-head crane is one that will lift 150 tons from a large ship at a distance of 65 ft from the wharf front, and 80 or 90 ft above the wharf. Heavy lifting also is done by pontoon cranes; they often work from the harbor side of the ship while the wharf cranes work on the other side. A typical 150-ton capacity crane of this kind has a diesel engine power plant for hoists, lighting, and driving the screws which propel the pontoon at a speed of 4 knots. See articles in Refs. 9 and 10.

REFERENCES

1. American Cable Co., *Wire Rope Users' Handbook.*
2. *Trans. Am. Soc. Mech. Engrs.*, 1928, MH-50-1.
3. *Eng. News-Record*, Oct. 8, 1931.
4. *Trans. Am. Soc. Mech. Engrs.*, 1930, IS-52-4.
5. *Crane Engineering*, Whiting Corp., Harvey, Ill.
6. A. F. Case, *Trans. Am. Soc. Mech. Engrs.*, 1928, MH-50-6.
7. Bernard Dunell, Cargo cranes, *Trans. Am. Soc. Mech. Engrs.*, 1928, MH-50-5.
8. *Fördertechnik*, Jan. 13, 1933.
9. *Demag Nachrichten*, Feb. 1932.
10. *Fördertechnik*, Feb. 12, 1932.

HOISTS

10. HAND-OPERATED HOISTS

HAND POWER HOISTS (or chain blocks) are made in three types: (1) Differential hoists contain two chain wheels above, cast as one piece, and a single wheel below. An endless chain passes over one upper sheave, down around the bottom wheel that carries the load hook, then up over the other wheel. For theory of this device, see p. 7-17. (2) Screw-geared hoists have a worm gear that holds the suspended load by friction of the thread of the worm. (3) Spur-geared hoists have a train of spur gears with a friction-disk brake of the Weston type to hold the load. (See Fig. 1.) Type 3, when fitted with ball-bearings, will deliver 85% of the work of the operator to the load hook. Its mechanical efficiency is about twice that of a screw-geared hoist and nearly three times that of a differential hoist. (See Table 1.)

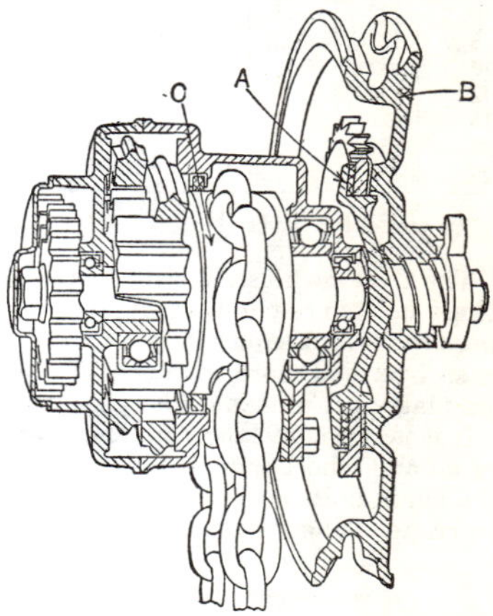

Fig. 1. Spur-geared hoist with friction brake.

Table 2 shows hoisting speeds and pulls on hand chain for the three types of hoist as stated by one maker. Other makers list the same sizes; their dimensions and ratings are nearly the same. The speed of a chain hoist depends on the pull required on the hand chain and number of feet of hand chain overhauled to lift load 1 ft. Speeds given in Table 2 are for short lifts with men accustomed to the work. For easy continuous lifting, use two-thirds of the values given for hoisting speeds. The spur-geared hoist has less internal friction, lifts and lowers load more rapidly than the others. The screw-geared hoist is lighter, takes up less headroom, will work in any position (as for horizontal pulling), and is well suited to work of riggers and repair gangs. The differential hoist is still lighter and serves well for occasional use.

Table 1. Comparison of Chain Block Hoists

	Differential	Screw-geared	Spur-geared
Mechanical efficiency	35%	40%	85%
Relative durability	20%	80%	100%
Relative cost	40%	70%	100%

Figure 1 shows the Cyclone hoist made by Chisholm-Moore. Gearing is compact and entirely enclosed in a grease-filled case. Brake wheel *A* is keyed to the shaft that carries hand chain wheel *B*, whose bore is threaded to fit a thread of hub on *A*. The pull of the load chain on wheel *C* acts through the gears to move *B* to the left, squeezing the leather and steel friction rings against *A*. A pawl engages ratchet teeth on the rim of *A*, prevents the latter from turning, and holds load suspended while the friction surfaces are squeezed together. To lift load, *B* is pulled in direction in which ratchet is free, the friction surfaces being engaged. To lower, load *B* is pulled in the opposite direction; it moves on threaded hub of *A* just enough to overcome friction imposed by the load; the pawl engages the ratchet and holds brake wheel when the motion of hand chain stops. This description applies generally to all screw-type brakes (known also as Weston brakes) used on hand-driven or power-driven hoists.

Table 2. Hand-power Chain Hoists

(Courtesy of Yale & Towne Manufacturing Co., Philadelphia)

Capacity, tons (2000 lb)	Pull on Hand Chain to Lift Full Loads, lb			Feet of Chain Overhauled to Lift Load 1 ft			Hoisting Speeds Attainable, ft per Min; Number of Men Needed to Lift Stated Loads, without Pulling over 76 lb on Hand Chain							
							Spur-geared				Screw-geared		Differential	
	Spur-geared	Screw-geared	Differential	Spur-geared	Screw-geared	Differential	Full Load	Half Load	Quarter Load	No. of Men	Full Load	No. of Men	Full Load	No. of Men
1/4	47	....	72	12 1/2	....	18	17.0	30.0	45.0	1	......	.....	6.0	1
1/2	57	68	110	21	40	24	8.5	17.0	26.0	1	4.0	1	6.0	2
1	76	87	190	31	59	30	4.2	8.5	13.0	1	2.0	1	3.7	3
1 1/2	103	94	225	35	80	36	5.1	10.2	15.3	2	2.4	2	2.5	3
2	112	115	300	42	93	42	3.9	7.8	11.7	2	1.8	2	2.3	4
3	107	132	....	70	126	....	2.5	5.0	7.5	2	1.1	2	.....	.....
4	116	142	....	84	155	....	1.9	3.8	5.7	2	0.8	2	.....	.....
5	102	145	....	126	195	....	1.4	2.8	4.2	2	0.65	2	.....	.....
6	122	....	....	126	....	....	1.2	2.4	3.6	2	.....	.....	.....	.....
8	127	....	....	168	....	....	0.85	1.7	2.6	2	.....	.....	.....	.....
10	132	....	....	210	....	....	0.65	1.3	2.0	2	.....	.....	.....	.....
12	122 *	....	....	126 *	....	....	1.2	2.4	3.6	4	.....	.....	.....	.....
16	127 *	....	....	168 *	....	....	0.85	1.7	2.6	4	.....	.....	.....	.....
20	132 *	....	....	210 *	....	....	0.65	1.3	2.0	4	.....	.....	.....	.....
25	165 *	....	....	210 *	....	....	0.50	1.0	1.5	4	.....	.....	.....	.....

* On each of the two hand chains.

The Cyclone hoist, in sizes up to 2 tons capacity, also is made with aluminum alloy parts. These hoists are one-third lighter than regular hoists, and can be put up and taken down more easily. A safety device throws the hand chain out of gear if the hoists are subjected to an overload more than the 50% excess over normal rating to which they are tested by the maker. Constants for speed and pull are about the same as those given in Table 2.

If it is not convenient for the operator to stand under the hoist, a fixed winch operated by a hand wheel or by a hand chain may be used. (See Fig. 2.) The mechanism is that of a chain hoist enclosed in a box that can be bolted to a wall, building post, or mast of a jib crane. These hoists, up to 2 tons capacity, are listed by several makers.

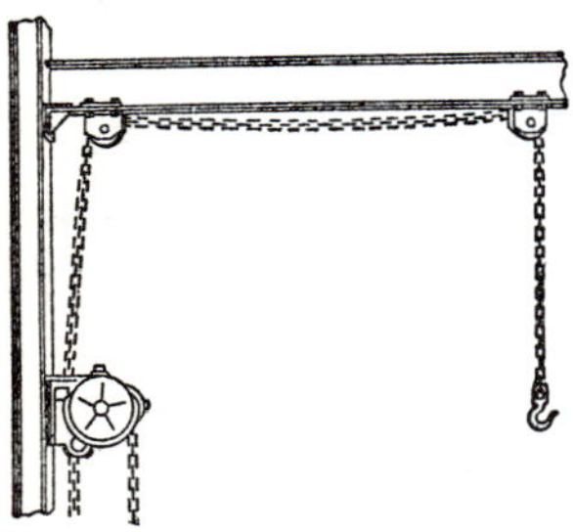

FIG. 2. Spur-geared hand winch.

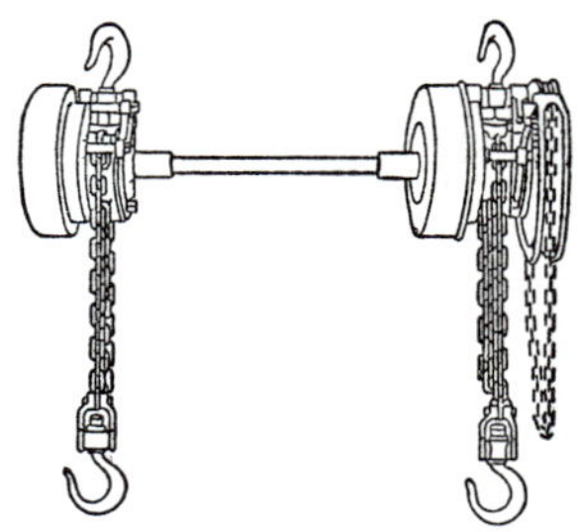

FIG. 3. Twin hook chain block.

Hoist makers furnish double hoists (Fig. 3) for lifting lumber, tanks, platforms, truck bodies, etc. They are made in sizes up to 10 tons, and with hoists spaced to suit dimensions of pieces handled. Four- or eight-wheel carriers with double hoists hooked on, or built in, are listed by some builders of tramrail systems.

The usefulness of a hand hoist can be increased if it is hung from a trolley on an overhead track. If the track forms the bridge of a traveling crane or is attached to a jib crane, a greater area can be covered. These trolleys are moved by moving the load or are geared and propelled by a hand chain. The Ace geared trolley (Robbins & Myers) is shown in Fig. 4. It has cast-steel side frames with adjustable spacing for track beams of different widths. The load wheels have no flanges; guiding is done by small rollers which bear against lower flanges of the beam. Similar geared trolleys are furnished by other makers of hoists. Trolleys with four or eight ball-bearing wheels, not geared, are made to take standard makes of chain hoists up to 2 tons capacity, to run on American monorail or Cleveland tramrail tracks. (See Fig. 5.)

The so-called Army type of chain hoist is a combination of a trolley, either geared or plain, with the hoist mechanism built in. The advantage is that the load hook can come closer to the bottom of the I-beam track. The disadvantage is that the hoist does not work so well if the pull of the load is much off to the side of the track. Figure 6 shows a hoist of this type, the "Rail Hugger" (Yale & Towne). The hoist is spur geared with ball bearings; the trolley frame is steel plate; the chilled wheels have roller bearings.

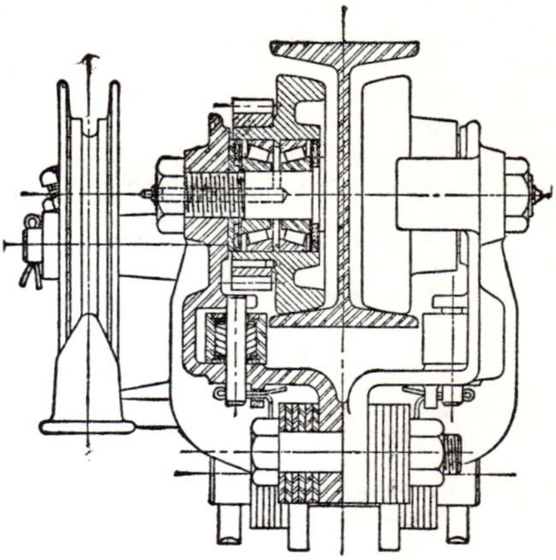

FIG. 4. Geared trolley for hand hoist.

TRACKS FOR HOISTS. Standard I beams are in general use to carry trolleys of hoists, especially for heavy loads. The beams should be straight; where they join, the bottom flanges should be supported by a plate, otherwise the corners may be bent down by the trolley wheels. It is not usual to splice webs. Supports or hangers are fastened to top flange in various ways to leave room for trolleys. Under frequent passage of heavy loads, lower flanges tend to bend down from peening action of the wheels. This can be prevented by riveting strips to lower flanges to act as a track. The double-rail construction in Fig. 7 serves the same purpose. The I-beam track made by Richards-Wilcox Company, Aurora, Ill., consists of straightened beams; Fig. 8 shows a joint and Fig. 9 an I-beam switch. Pulling the hand chain releases a latch pin, then draws the movable beam over to the next position. A guard plate keeps the trolley from running off the open end of tracks not in register.

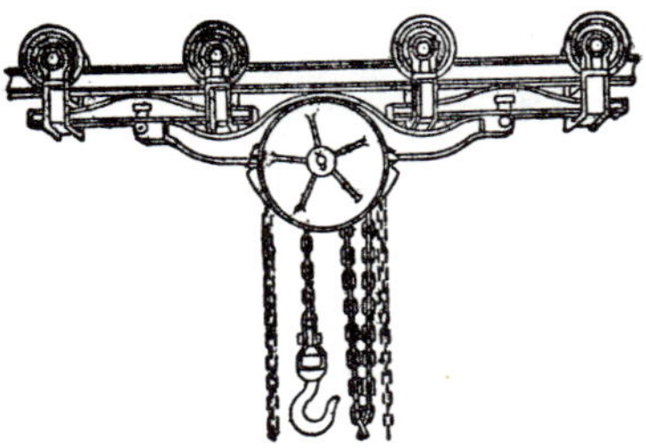

FIG. 5. Tramrail chain hoist carrier.

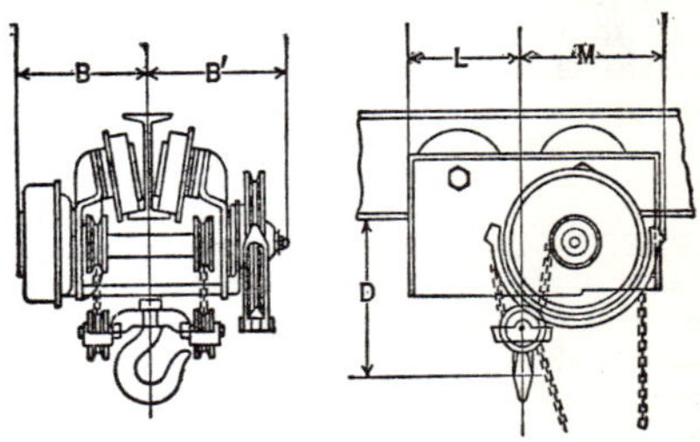

FIG. 6. Army-type chain hoist.

Similar track work is made by Philadelphia Tramrail Company, Philadelphia, and some builders of hoists. See also Tramrail Systems, p. 26-28.

PORTABLE HOISTS of the type shown in Fig. 10 are made for loads up to 3 tons. They serve where it is not convenient to hang up a chain hoist. Lifting is by a winch built in the frame, or a chain hoist hung from top of the column.

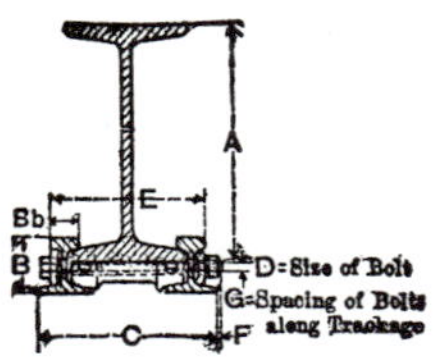

FIG. 7. Double rail hoist track.

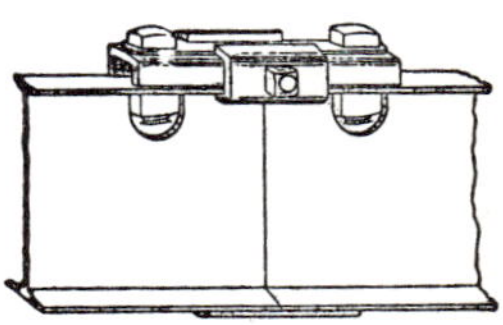

FIG. 8. I-beam track joint.

SPUR-GEAR HOISTS in capacities up to 1 ton, operated by hand rope or hand wheel, and fitted with automatic stops and safety devices, are available. They are fitted with platforms or containers for hoisting or lowering from floor to floor or from sidewalk to basement. These machines are also made with electric motor drive.

Small hand-operated counterbalanced hoists for carrying parcels, tools, and books rapidly from floor to floor are available. In some cases, installation of these machines is subject to State Elevator Codes and local regulations in regard to safety and fire.

CRABS OR HAND WINCHES are made as self-contained units for use with derricks, with block and tackle for riggers, or for lifting through heights beyond the range of chain blocks. In the portable form, hand winches consist of two frames carrying a drum with, usually, two sets of gears on shafts with squared ends for the cranks. Sometimes drum shafts also are squared, to allow slack rope to be taken up quickly by using cranks directly on the drum. One pinion is made to shift in and out of gear, so that cranks can be used on either pinion shaft for different hoist speeds. The load is held and controlled for lowering either by a ratchet and pawl, by a strap brake, or by a friction disk brake of the Weston

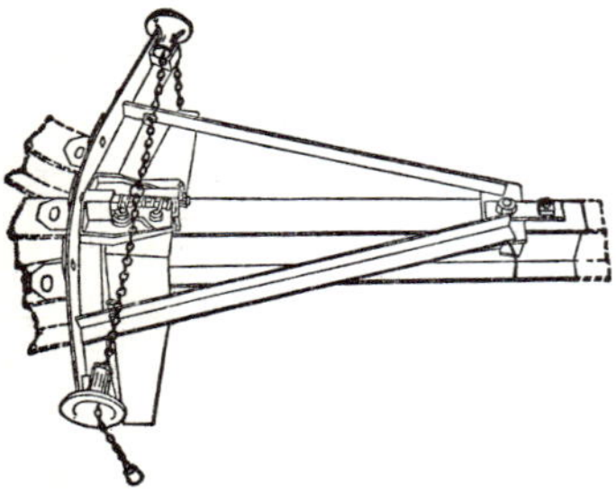

Fig. 9. I-beam switch.

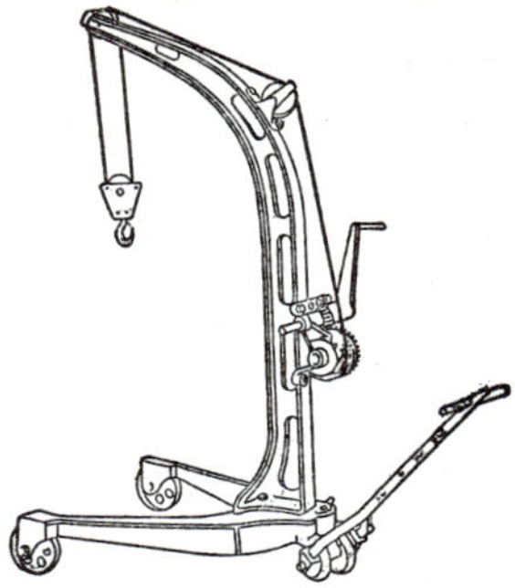

Fig. 10. Portable hoist.

type. The friction brake will prevent the cranks from accidentally flying back and injuring workmen, as often happens with crabs equipped only with ratchet and pawl. The principle of the Weston brake is shown in Fig. 1. The same mechanism frequently is fitted to flat frames for use on masts of jib cranes and derricks. Simple worm-geared winches of a capacity of 1 ton or less are made by nearly all makers of conveying machinery.

11. ELECTRIC HOISTS

ELECTRIC HOISTS generally use ball and roller bearings, alloy steels, and heat-treated parts. Pressed steel has been substituted for castings in some designs. Welding has replaced bolting in many cases. Motors usually are enclosed, ball bearing, and specially designed for hoist work; high-torque squirrel-cage induction motors are more general than d-c motors. The latter are useful for close control of load movement. Both rope and push-button control are used. Push buttons give quicker, safer, and more convenient control. Electric hoists range in size from $^1/_8$ to 6 tons, with a wide range of hoist speeds. They are made in floor-operated and cab-operated types. For single-speed hoisting, push-button control is the most convenient; one hand is free to steady or guide the load, operation is faster, and the load can be placed more accurately. Cab-controlled hoists are used to transport loads over a considerable distance, where the operator must observe the load from above, or where rapid transfer is necessary.

Worm-driven electric hoists hold the load by friction on the worm thread. Spur-geared hoists use a screw brake (Weston) to hold the load. (See Fig. 1.) All have a limit switch, to prevent overtravel of hoist hook at top of the lift, and a motor brake to stop the rotor if current fails, and to prevent drifting when controller is in "off" position.

In electric chain hoists a chain drive from the motor operates gearing like that in a chain hoist. The load is lifted by a chain driven by a pocket wheel; slack chain is paid out as load ascends. For high lift and for pulling at loads not directly underneath, a chain hoist often is preferable to a rope hoist. Chain hoists are made in sizes of $^1/_8$ to 2 tons.

SPUR-GEARED HOISTS. Figure 11 shows a hoist built in sizes up to 2 tons. The size of main frame determines distance from overhead track to upper position of hook. It is of steel plate with welded joints, and holds drum, drum shaft, and motor brake. Motor and controller are bolted to one side, and an oiltight case with spur gears and brake to the other. All shafts have roller bearings, and all load-carrying parts are steel.

WORM GEAR HOISTS are made in sizes up to 6 tons. In this type a hardened steel worm works in a bronze worm wheel. The thread angle is low enough to make the gears self-locking, thus avoiding a mechanical brake. Some hoists have, beside the worm gears, a pair of spur gears to get greater speed reduction between motor and hoist drum. For foundry work, the worm gear hoist can be fitted with special control. A nonspring-return controller used with an oversized motor gives a speed range of $^1/_{10}$ to full speed, which is useful in handling flasks and molds. A jointed control rod worked by one hand permits operator to use the other hand to guide the load.

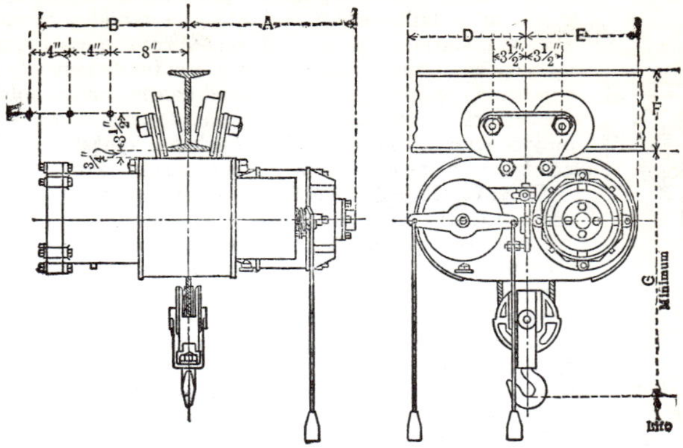

FIG. 11. 2-ton spur-geared hoist.

HOIST CARRIERS. Hoists with top hook or clevis may be used as fixed hoists or attached to a commercial trolley. More headroom is required than when trolley and hoist are combined, and wiring is less neat. Most electric hoists can be made with top connections to fasten to an eight-wheel push-type trolley, which is strong, light, and can pass curves of small radius. Figure 5 is typical; overall length is 66 in.; distance bottom of rail to hook, 21 in.

Plain push-type trolleys for I-beam tracks are useful for moderate distances. If load or hook hangs out of reach, or must be spotted accurately, it is better to use a hand-chain geared trolley. If the hoist must travel much, or if the loads are hot, heavy, or bulky, the trolley can be electrically motor driven. The control cords may be kept away from the hoist by carrying the controller on a separate trolley when the hoist operates through a wide hatchway, from vats or beyond the edge of a platform.

Electric hoists up to 2 tons capacity especially for use with Cleveland tramrail are mounted on four-wheel hand-propelled carriers for 500-lb hoists, and eight-wheel carriers for 1- or 2-ton hoists. An electrically propelled carrier mounted with an electric hoist coupled to a trailing operator's platform has dimensions: Maximum overall length, with single motor drive, 12 ft 3 in.; operator's platform to bottom of rail, 6 ft 11 in. minimum; hook to bottom of rail, 29 in. Carriers also are made for pendant cord control, and also with enclosed operating cab. Travel speeds are up to 600 ft per min, and the carrier can climb grades, run through switches, and pass short radius curves. Most builders supply cab-operated machines to travel in single I-beam tracks. For heavy loads and high speeds, the double T-rail track, shown in Fig. 7, is preferable to the lower flange of the I beam.

TELPHER HOISTS. In Europe, the track rail for outdoor work usually is a T rail on a longitudinal supporting beam. The traveling frame, with cab and hoist or grab bucket, hangs from the trolleys by hook-shaped hangers which permit some side swing. In England, this is known as the Telpher system. A German monorail carrier of this type has no operator but runs instead under an automatic block-control system. Cab control, however, permits higher traveling speed, requires fewer cars, and has less apparatus to maintain.

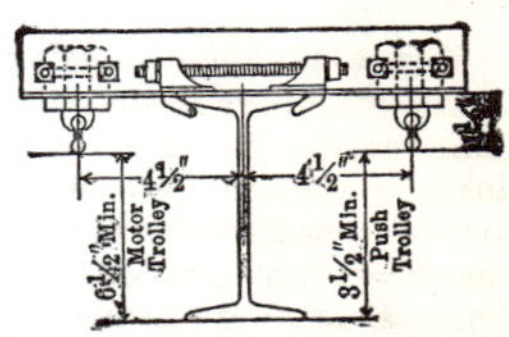

FIG. 12. 8-shaped conductors and I-beam track

ELECTRIC CONDUCTORS AND COLLECTORS. If travel of hoist is short, a flexible cable can be supported so as to allow the necessary movement. If cable drags or interferes, a spring-actuated reel can pay out and take up cable automatically. For long travel, shoe or wheel collectors bear against conductors. These may be copper wire, angle iron, flat bar on edge, or an 8-shaped bar. Figure 12 shows the latter with an I-beam track, and Fig. 13 a special track and conductor support made for Cleveland tramrail.

The American monorail (see below) uses small channel conductors on ebonite blocks mounted on the hanger bolts.

To use two conductors with three-phase current, the primary current is brought through a special transformer. One phase of the secondary current is grounded and connected to the track which is separately grounded. The conductor bars transmit the other two phases. The Cleveland Tramrail Company in one of its bulletins states that this method of transmitting current to tramrail carriers does not improve operating conditions of a-c tramrail motors. Such motors are fundamentally unsuited for conditions requiring delicate speed control and should not be used for extremely severe service. This reference to a-c motors applies only to tramrail service, where loads are light and lift is short, and not to hoist and crane service in general.

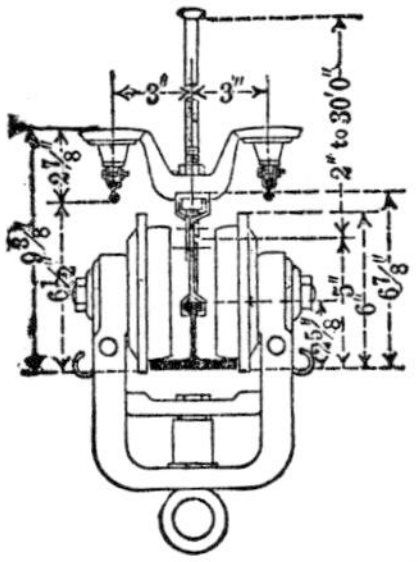

FIG. 13. Tramrail and electric conductor support.

TRAMRAIL SYSTEMS. The advantages of tramrail systems are: (1) Little machinery required, cost and maintenance low. (2) No floor space required. (3) Switches permit many tracks to be served by few carriers. (4) Overhead tracks for storage of suspended loads avoid the use of floor space. (5) Standard carriers are available for all kinds of bulk and package material.

Louden Machinery Company, Fairfield, Iowa, makes a standard track of I section, with wide flanges to carry trolley wheels, and for heavier loads a track similar to a 6-in. I beam with strengthened flanges, one flange wide, the other narrow. They are 6 to 16 ft long, carried by hangers spaced to suit the load. Trolley wheels are drop forged, ball bearing, to carry loads up to 5000 lb. Trolleys can be driven by motor or operated by hand.

Richards-Wilcox Company makes an I-beam track (Figs. 8 and 12) for heavy work, and a pressed sheet steel track (Fig. 14) for lighter work, in 10- or 12-ft lengths of No. 10, 13 or 14 gage steel. Trolley wheels for the size shown are 3 in. in diameter, ball or roller bearing, and are assembled in groups of 4 to 16 with equalizer bars to carry loads up to 3000 lb. The track is made with switches, crossovers, turntables, drop sections, hangers, and brackets to suit different conditions.

Figure 15 shows American Monorail track, made by the American Monorail Company, Cleveland, Ohio, in lengths up to 16 ft, and clamped to drop-forged suspenders. The track is made with staggered joints. It assembles 2 in. wide, 2 1/2 in. deep, and will carry loads up to 4000 lb, depending on spacing of hangers and number of trolley wheels. For spans in the open, or where hanger bolts cannot be used, the halves of the rail are welded to the web of a T beam. Trolley wheels are drop forged, 4 in. in diameter, ball bearing, and are assembled in groups of 4- or 8-in. carriers with drop-forged wheel yokes, equalizers, and load bars.

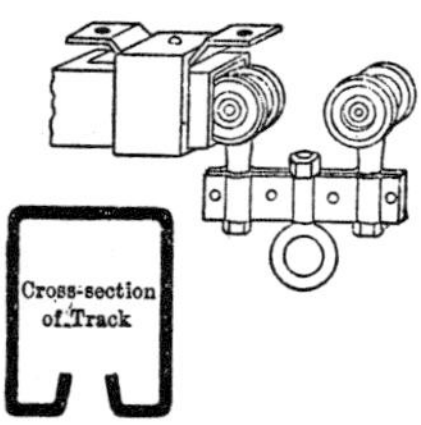

FIG. 14. Pressed steel tramrail track.

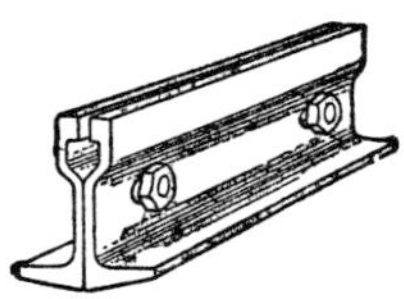

FIG. 15. American Monorail track.

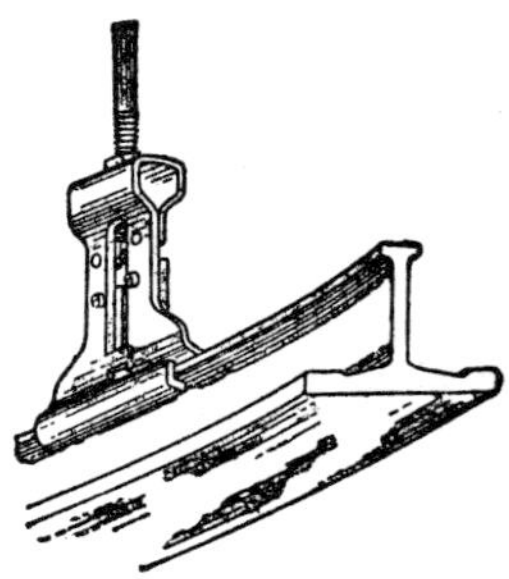

FIG. 16. Cleveland Tramrail track support.

Figure 16 shows Cleveland track. It is about 3 by 3 in., 7 lb per ft. Hanger rods, spaced 3 to 10 ft, support the rail, a clamp on the lower end gripping the rail. A ball-and-socket joint at the upper end permits some free movement, allowing track to adjust itself to load, giving trolley wheels a fair bearing on tread of the rail. On curves, where the load tends to swing outward, the rail is free to tilt slightly, and wheels on the inside of the curve are not overloaded. Trolley wheels are 3, 4, or 5 in. in diameter, ball bearing. Curves, switches, turntables, transfers, and fire-door arrangements have been standardized. This system uses hand- or motor-propelled carriers, with or without hoists, to carry loads of 500 to 4000 lb. Carriers are made for suspended tiering machines and furnace chargers. Foundry ladles for one-man operation have hoist control buttons on the tilting handle of the ladle. Power-propelled carriers with suspension frames for ladles have controllers for tilt, travel, and hoist in the operator's cab or platform. This system also provides automatic dispatch with distance control and special devices for high-speed or continuous travel. With any tramrail system it is possible to arrange track turn-outs to register with various positions of transfer bridges or overhead cranes, enabling carriers to cover a large area.

12. POWER-OPERATED HOISTS

POWER-OPERATED HOISTS. Steam, gasoline and electric hoists for mine, quarries, building erection, logging, and grab-bucket work are available. Single-drum hoists are used for cargo hoisting and building construction, double-drum hoists for general hoisting, and grab-bucket work with straight booms. To operate grab buckets on derricks, three drums generally are used, for the holding line, closing line, and for luffing the boom, respectively. For operation of grab buckets, see p. 26-12. Swing boom derricks require an extra pair of drums to swing the boom. (See p. 26-16.) On small rigs, the bucket-holding line can be wound on a drum, separate from the main hoist, operated by a counterweight rope on the same drum and controlled by a brake. With one of these, a hoist with a single drum for the closing line will work a grab bucket.

Small winches, winding drums, car pullers, and drag scrapers, made to operate by compressed air, are furnished in sizes from 750 to 2000 lb. In the Sullivan Turbinair portable hoist the air motor is located in the hollow of the drum. A 3-hp hoist will wind 350 ft of $^{5}/_{16}$-in. rope and exert a pull of 1000 lb at 130 ft per min. It weighs 175 lb, and is 20 by 15 by 12 in. overall. The handles control clutch and brake. These machines are made with single or double drums or winch heads. They can be bolted to wood or steel plate bases, or clamped to pipe posts or to the masts of derricks. Air-driven hoists mounted on trailer trucks with gasoline engine, compressors, and air tanks also are available.

Many makers of electric hoists supply winches or pulling machines for bolting to floor or ceiling, or mounted on a truck or skids to form a portable winch. American Engineering Company, Philadelphia, makes single-drum pullers of this type in three sizes. The smallest is 21 by 23 in., 15 in. high, has a 2-hp motor, will wind 250 ft of $^{3}/_{8}$-in. rope, and exert a pull of 1000 lb.

CAR PULLERS OR SPOTTERS for moving or spotting freight cars are made by various manufacturers. The car spotter is a compact machine of the vertical capstan type, with gearing enclosed and running in oil. The operator throws a few wraps of rope around the capstan and maintains a hand pull on the slack rope as it pays off.

AIR HOISTS FOR SHOP USE. Direct lift hoists (Fig. 17) have cylinders in diameters up to 19 in. and lengths to suit the work, with lifting capacities up to 10 tons at 80 psi air pressure. They are made single acting for simple rapid lifting, or as balanced-pressure cylinders, in which air acts at full pressure on one side of the piston but is varied on the other by a hand-operated valve. The piston moves under a difference of pressure, and speed and travel are accurately controlled. Air cylinders can be hung from trolleys if the travel is not too great.

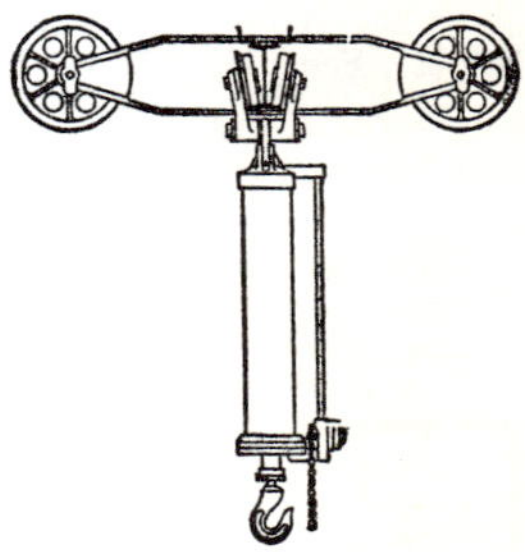

Fig. 17. Direct lift air hoist.

Advantages claimed for direct acting air hoists are: (1) They will work under conditions of dust, moisture, fumes, etc., injurious to other hoists. (2) Few parts to get out of order; repairs can be made by any mechanic. (3) Accuracy of control. (4) Low first and operating cost, including maintenance and depreciation.

GEARED HOISTS operated by compressed air are available. A balanced 2-, 3-, or 4-cylinder engine drives a wire rope drum or chain pocket wheel, through gears enclosed in a tight lubricant-filled case. Sizes range from 500 lb to 10 tons capacity. They weigh two or three times as much as hand-chain hoists but are lighter than the ordinary electric hoists, especially in smaller sizes. They are useful in shops and foundries where compressed air is available and where the hoist can be hung from a fixed point or a trolley with travel not too long for the air hose. They are not easily affected by heat and dust in foundries, and since the valve control is accurate and delicate, patterns, cores, and flasks can be handled with certainty and safety.

GENERAL CONVEYING AND ELEVATING

13. CARRIERS AND ELEVATORS

SKIP HOISTS comprise a single-drum winding machine, a wire rope, a skip or bucket which takes its load below, and dumps it above, together with the necessary lead sheaves, skip filling device, and electrical-control apparatus for the winding machine. Skip hoists are simple in construction and relatively cheap to maintain and operate. Carrying capacities can be made to suit the smallest or largest boiler houses. The path of the skip can be vertical, inclined, or a combination of both. Carrying capacities depend on size of skip, speed of travel, method of loading, whether or not the skip is counterbalanced, and whether operation is full or semi-automatic. Makers' catalogs give tables of expected capacity under various conditions. Skip hoists have been well standardized by manufacturers of conveying and hoisting machinery. Their catalogs give detailed information about track hoppers, loaders, skips, steel structures, winding machines and electrical control apparatus.

Skip hoist with traverse motion is a form of skip hoist, first used to load anthracite or coke to silo bins in retail coal yards. A latched-bottom bucket takes its load of 1 or 2 tons from a track hopper with an automatic feed chute. When hoist is started the bucket rises, comes in contact with a trolley carriage, moves horizontally to a chosen point, then lowers. When bottom of the bucket touches material in the bin, the doors open and contents slide out. The bucket then returns for a new load. Some of these hoists have full-automatic push button control. See Refs. 1 and 2.

PIVOTED BUCKET CARRIERS consist of round-bottom buckets, pivoted between two long-pitch roller chains, arranged to dump on a horizontal run by tipping the bucket. They can be run vertically, horizontally, or in an inclined path. Suitably combined with crusher and feeder, they fit many arrangements of railroad track, boiler house, and storage bin. Where lift, travel, and capacity required are not too great, a pivoted bucket carrier is economical and convenient to install and operate.

Most carriers of this type have overlapping buckets which can be fed continuously from a chute. The chains have hardened pins and bushings, self-oiling rollers, and other refinements. The difference between different makes is chiefly in the suspension of the buckets, and making them lap properly at corner turns. Sizes of buckets, capacities, clearances, power required are practically the same for all such carriers made in the United States. See Tables 1 and 2.

Table 1. Standard Sizes of Pivoted Bucket Carriers

(Courtesy of Link-Belt Co., Philadelphia)

Size of Bucket, in., Length × Width	Pitch of Chain, in.	Carrying Capacity of Bucket, cu ft	Coal Capacity, tons per hour	Speed, ft per min
18 × 15	18	C.68	15– 20	30–40
18 × 21	18	0.94	20– 30	30–40
24 × 18	24	1.68	40– 50	40–50
24 × 24	24	2.24	55– 70	40–50
24 × 30	24	2.80	75–100	40–50
24 × 36	24	3.36	90–120	40–50
30 × 24	30	3.50	95–120	45–60
30 × 30	30	4.37	110–160	45–60
30 × 36	30	5.25	140–190	45–60
36 × 36	36	8.50	210–330	50–80

The horsepower required at the head shaft of a pivoted bucket carrier is

$$\text{Horsepower} = AL + BR \tag{1}$$

where L = total vertical lift, feet, R = total horizontal run, feet (see cut in Table 2), A and B are factors whose values are given in Table 2. Add 10% for each cast gear drive, and 5% for each drive through belt, chain, or cut gears.

Pivoted bucket carriers are quiet in operation, convey and elevate at the same time and thus avoid transfers and deep pits. They require little power. They cannot be fed on a vertical run nor under a turn wheel. The dump must always be on a horizontal run. Catalogs of manufacturers give dimensions of drive groups, corner turns, and clearances.

Table 2. Horsepower Factors of Pivoted Bucket Carriers

(Courtesy of Jeffrey Manufacturing Co., Columbus, Ohio)

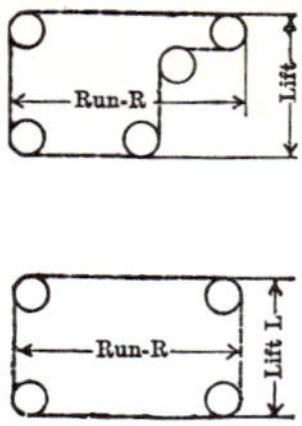

Size of Bucket		Loaded on One Vertical and Both Horizontal Runs			
		Power Factors			
		Material 50 lb per cu ft		Material 100 lb per cu ft	
Length, in.	Width, in.	*A* For Lift *L*, ft	*B* For Run *R*, ft	*A* For Lift *L*, ft	*B* For Run *R*, ft
18	18	0.044	0.017	0.088	0.023
24	18	.074	.039	.148	.051
24	24	.097	.045	.194	.060
24	30	.120	.051	.240	.069
30	30	.198	.061	.396	.085
30	36	.238	.070	.476	.100

CENTRIFUGAL DISCHARGE ELEVATORS. The simplest form of elevator consists of malleable-iron buckets fastened at intervals to a single chain. It is called a "centrifugal discharge" machine because the contents of the buckets are thrown out at the head wheel by centrifugal force. A full, clean discharge into the head chute depends chiefly on the size of the head wheel, its speed, and the bucket spacing. Table 3 gives the result of practical experience with vertical centrifugal discharge coal elevators. Belts with buckets attached are used for fine, dry, or gritty materials, but chain is better for the rough work of elevating coal. Figure 1 shows style A malleable-iron buckets, used in coal elevators. The buckets will not discharge clean if set too closely together. For analysis of conditions of pick-up and discharge of centrifugal bucket elevators, see Ref. 3.

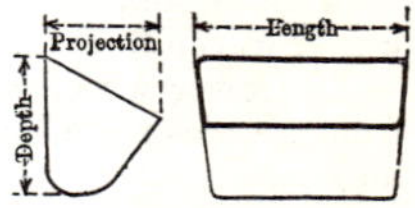

FIG. 1. Malleable-iron elevator bucket.

Table 4 gives the closest possible spacing for the speeds given in Table 3. If spacing does not match pitch of chain, increase it to a multiple of that pitch. If the coal is lumpy,

Table 3. Speeds for Vertical Centrifugal-discharge Coal Elevators

Diameter, head wheel, inches	18	21	24	27	30	36	42
Revolutions per minute	46	43	41	39	37	34	32
Speed of chain, feet per minute	217	237	257	276	290	320	352

Note. Permissible variation not over a few rpm either way.

Table 4. Minimum Spacing Style A Buckets for Clean Discharge

Projection of bucket, inches	4	5	6	7	8
Closest spacing, inches	10	12	14	17	20

increase the spacing. To determine the size of chain, working strengths alone, as given in makers' catalogs, are not a suitable guide. A link should be chosen which bears a proper proportion to the dimensions of the bucket, and which has a bucket attachment large enough for the bucket. K1 attachments have two bolt holes; K2 attachments have four bolt holes and are advisable for buckets 8 by 5 in. and larger. The greater the projection of the bucket, the greater should be the pitch of the link to give the bucket a proper support in picking up its load. Table 5 represents good practice.

Table 5. Pitch of Chain for Style A Malleable-iron Buckets

Size of bucket, in., length × projection	6 × 4	7 × 4 1/2	8 × 5	10 × 6	12 × 7	14 × 8	16 × 8
Pitch of malleable-iron chain, in.	2.6	2.6	3.1	3.1	4.7	6	6

Table 6 gives weights and capacities of buckets used in simple coal elevators. The diameter of the foot wheel of a centrifugal discharge coal elevator should be at least four

Table 6. Weights and Capacities, Style A, Malleable-iron Buckets

Size of bucket, in., length × projection	6 × 4	7 × 4 1/2	8 × 5	10 × 6	12 × 7	14 × 8	16 × 8
Weight, lb	2.2	3	4	6.4	9.6	13.2	21
Coal capacity, lb (*C* in formulas)	1.5	2.2	3	5.2	8.3	13.9	18.9

times the pitch of chain used. If the wheel is too small, buckets will not fill properly. The following formulas are applicable to bucket elevators:

Notation. C = pounds of coal per bucket (see Table 6); D = diameter of foot wheel, inches; hp = horsepower to be delivered to the elevators; L = height of lift, feet; P = pull in chain, pounds, due to pick-up of coal at foot of elevator; S = spacing of buckets, inches; T = travel of chain, feet per minute; U = unbalanced pull in chain due to weight of coal in buckets; W = capacity of elevator, pounds of coal per minute; X = theoretical horsepower required.

$$P = \frac{12DC}{S} \tag{2}$$

$$U = \frac{12LC}{S} \tag{3}$$

Combining eqs. 2 and 3 gives

$$X = 12\frac{C}{S}(L + D)\frac{T}{33{,}000} \tag{4}$$

Allowing 30% for losses in power transmission,

$$\text{Hp} = 0.0005\frac{CT}{S}(L + D) \tag{5}$$

$$W = \frac{12CT}{S} \tag{6}$$

The head chute of a centrifugal discharge coal elevator should have a slope of at least 35 degrees, and its upper end should not be above a line drawn at 45 degrees through the center of the head wheel.

The lower wheel of a centrifugal discharge coal elevator generally is enclosed in a cast-iron or steel plate box, called a boot, which serves as a support for shaft bearings and also confines material to the path of the buckets. Boots with fixed bearings generally are not used for coal elevators, those with adjustable bearings being convenient. For sizes and dimensions, see makers' catalogs.

Advantages and Limitations. The centrifugal discharge elevator is cheap and simple, takes up little space, does not require deep or large pits, and can be encased easily. It will give good service when worked for a few hours each day, but for lumpy material, large capacities, and long hours of service, it is better to use a heavier machine at slower speed. Such machines comprise continuous bucket elevators, perfect-discharge elevators, and V-bucket elevators.

CONTINUOUS BUCKET ELEVATORS consist of single or double chains with an endless series of steel buckets so shaped that in passing over the head wheel each bucket acts as a deflector or chute for the discharge of the contents of the following bucket. Single-strand machines use K2 attachments fastened to the backs of the buckets, the depth of the bucket being 2, or rarely 3, pitches of the chain. Angles usually are fastened to the bucket ends to guide the vertical runs. Long buckets for large capacities require two chains, fastened to the ends. The chains usually are of steel, with a pitch equal to the depth of the bucket. In this position the chains guide the vertical runs. For inclined runs, they can be fitted with rollers that run up and down on suitable tracks. Buckets for vertical elevators handling crushed coal (Fig. 2) are made with the angle B in the bottom equal to 55 degrees, thus giving a chute or discharge angle of 35 degrees, which is steep enough for crushed coal. If the coal is all fine or wet, B should be 45 degrees; angle A then is 45 degrees, giving a better delivery into the head chute. When the elevator is inclined 25 or 30 degrees from the vertical, as in Fig. 3, B can be increased to 70 degrees to give more bucket capacity, and for ordinary crushed coal the discharge will be clean. This is one advantage of inclining such an elevator; another is that material can be delivered directly to the buckets by a chute without spill. These buckets will pick up soft material like crushed coal from a boot, but not hard lumpy material like stone. Stone and ores should be handled in inclined elevators, with the feed directly from a chute set 2 or 3 buckets above the foot wheel, and the foot of the elevator must be free and open to allow spilled material to fall clear of the buckets and be shoveled up. The head chute never

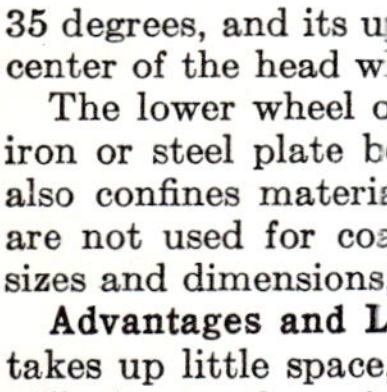

FIG. 2. Continuous bucket elevator.

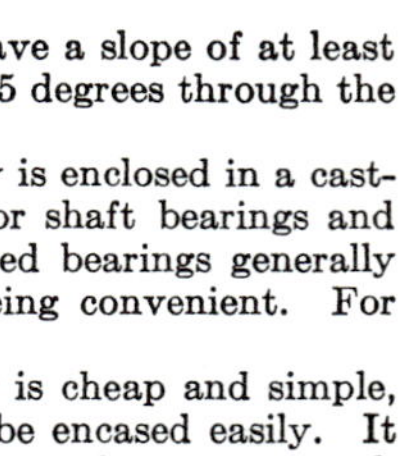

FIG. 3. Inclined continuous bucket elevator.

must be above a line drawn at 45 degrees through the center of the head wheel. Speeds range from 90 to 200 ft per min in handling coal.

For horsepower required by continuous bucket elevators, use eq. 5; for capacities, use eq. 6; for diameter of head shaft, see p. 26-35.

The continuous-bucket elevator is especially adapted for handling stone, but in elevating crushed coal for boiler houses it has the merit of giving high capacities at speeds much lower than those of centrifugal discharge elevators. For handling run-of-mine or large lump coal, the V-bucket elevator is generally better.

PERFECT-DISCHARGE ELEVATORS have two strands of chain fastened to the ends of the buckets. The buckets are not continuous, and the speed is slow as compared with centrifugal discharge elevators. In order to effect a clean discharge, the chains are deflected backward under the head wheels by a pair of supplementary wheels in such a way that the contents of each bucket will fall clear of the bucket ahead. Figure 4 shows the location of the deflector wheels with reference to the head wheels. Table 7 is based

Table 7. Data for Perfect-discharge Elevators, Style A Buckets

Projection of bucket, in.	5	6	7	8	9	10
Center of chain from lip, in. (*L*, Fig. 4)	3	3 3/8	4 1/2	5 3/8	5 7/8	6 1/4
Diameter of head-wheel, in. (Fig. 4)	22	24	26	28	30	32
Diameter of deflector wheel, in. (Fig. 4)	14	15	16	17 1/2	19	20
Revolutions per minute of head wheel	19	19	19	19	19	19
Speed of chain, ft per min	109	119	130	140	149	158
Spacing of buckets, in.	17	19	21	22	23	25
Buckets per min	77	75	74	76	77	76
Distance *A* (Fig. 4), in.	18	19	21	23	24	26
Distance *B* (Fig. 4), in.	7	7 1/2	8	8 3/4	9 1/2	10

on good practice and gives dimensions for elevators using malleable-iron style A buckets and with wheels arranged according to Fig. 4. These elevators often are built with sheet steel buckets for larger capacities of coal, at speeds between 100 and 150 ft per min.

Measure *A* down from center of head wheel. Draw line 50 degrees from vertical. Measure *B* from intersection to center of deflector.

For horsepower, use eq. 5; for capacities, use eq. 6; for diameter of head shaft, see p. 26-35.

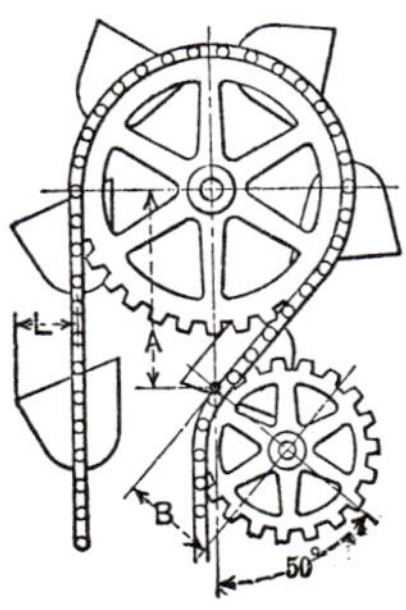

FIG. 4. Perfect-discharge elevator.

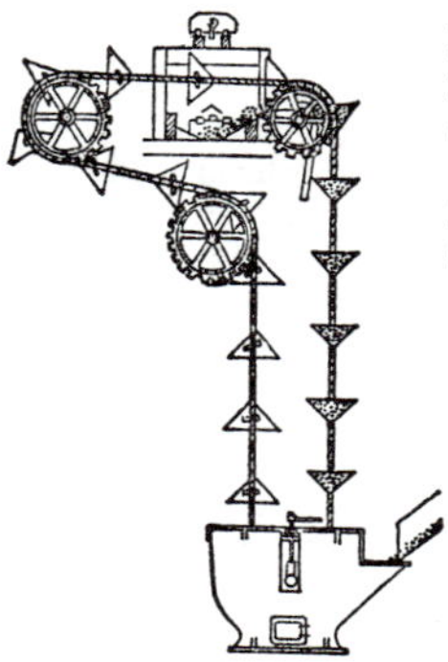

FIG. 5. V-bucket carrier.

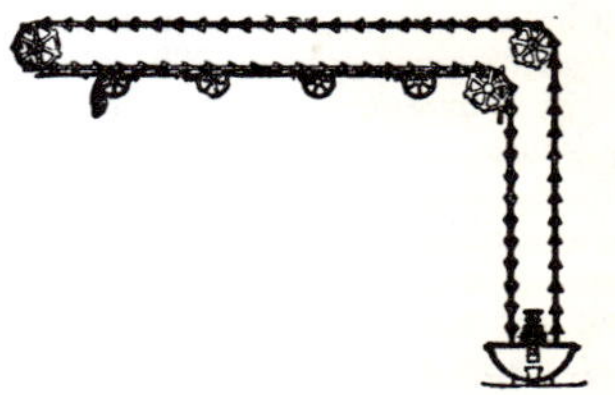

FIG. 6. V-bucket carrier with boot.

Advantages of the Perfect-discharge Elevator. (1) It has two chains, and as usually designed, has a large margin of safety between the working strain and the breaking strength of the chains. (2) The pick-up of material is easy. (3) On account of the slow speed, the chains, buckets, wheels, gears, and bearings last longer and require less attention and repair. As an offset, the machine costs more than a single-strand elevator.

THE V-BUCKET CARRIER GRAVITY DISCHARGE MACHINE is used extensively for handling coal, either lump or crushed. It consists of two roller chains fastened to the ends of V-shaped buckets which carry the coal on vertical or inclined runs and push it along a trough on the horizontal run. Gates in the trough permit the coal to be discharged where it is wanted. When arranged as in Fig. 5 or Fig. 6, it will pick up from a boot; with the path shown in Fig. 7, it will pick up at any point on the bottom run as well as at the corner turn.

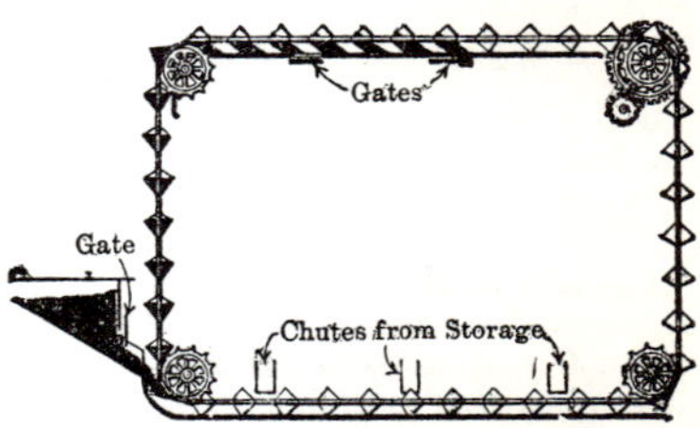

FIG. 7. V-bucket carrier in boiler house.

The 12-in. and 16-in. buckets sometimes are carried on malleable-iron roller chains, but steel bar chains generally are used for the larger sizes, and are better for the small sizes also. Take-ups or adjustable bearings always should be placed at a corner where the buckets pass empty, not at a corner around which coal travels. Vertical runs always should be encased to confine the spill and to avoid exposure to rain and snow.

Advantages of V-bucket Carrier. (1) Strong and rugged; will carry any coal up to run-of-mine size, at speeds up to 120 ft per min. (2) Will receive on any run or from a boot. (3) Handles coal without breaking it. (4) Gives larger capacities than pivoted bucket carriers.

Disadvantages. (1) Requires more power on horizontal runs than other types. (2) Not suitable for ashes or hard and gritty material. (3) As compared with other elevators, it cannot discharge over a head wheel, but always must discharge on a run between two sets of wheels.

HORIZONTAL DISTRIBUTION BY BELT OR FLIGHT CONVEYOR. Inclined conveyors sometimes are used for elevating coal when the boiler house is at a distance from the track. If the house is not too long, the inclined conveyor may be carried over a hump into the horizontal; but if the house is long, or if its axis is not in line with the conveyor from the track, a separate conveyor must be used for distribution in the house. A similar condition exists in boiler houses where the ash disposal is by car or hoist. In such cases, where a pivoted-bucket or a V-bucket carrier cannot well be used for distributing, a simple coal elevator or a hoist is used for lifting the coal.

Whether to use a belt conveyor or a flight conveyor depends partly on the structures and operating conditions, and partly on the personal preferences of engineers and owners. Considering only physical conditions, belt conveyors may be regarded as best (1) where the bin is long, say, 150 ft or more; (2) where the capacity is more than 100 tons an hour; (3) where noiseless operation is essential.

A flight conveyor may be best (1) where there is no room to admit a belt tripper; (2) where the first discharge is less than 20 ft from loading point of the conveyor, a distance too short to admit a tripper; (3) when the conveyor is shorter than 50 ft center to center; (4) where separate spaced bins are used, instead of one continuous bin; (5) where the conveyor is exposed to rain, snow, or freezing.

For either machine, these limits are not hard and fast, and between them there are chances to exercise personal preference or judgment. Both machines have been on the market for many years and have been standardized. There is no "universal" conveyor; for handling heavy or gritty ores, as in the Western smelters, or in conveying heavy tonnages of coal over long distances, as in by-product coke plants, a flight conveyor is not to be considered; but in many boiler houses it makes the best distributor, even though it requires more power. The older forms of flight conveyor were noisy, but modern roller-flight conveyors, or double-strand roller-chain conveyors, are by comparison almost noiseless. They are strong and rugged, easy to load and discharge, and will work under conditions and lack of care that would be harmful to belts. Some engineers have a prejudice against scraping the coal in a trough. It does consume more power, but it does not harm even anthracite. As to wear on the trough, the item of replacement is not an important one.

CHAIN PULLS AND HORSEPOWERS. The general formula for any ordinary flight conveyor, apron conveyor, bucket elevator, or bucket carrier is

$$\text{Pull} = P = X + Y + Z \tag{7}$$

where X = pull to move the weight of material, coal, etc., on the loaded run; Y = pull to move the conveyor parts on the loaded run; Z = pull to move the conveyor parts on the empty run.

$$X = WL(F \cos \alpha + \sin \alpha);\ Y = W_1L(f \cos \alpha + \sin \alpha);\ Z = W_1L(f \cos \alpha - \sin \alpha),$$

where W = weight of coal, etc., pounds per foot; W_1 = weight of conveyor parts, pounds per foot; L = length of conveyor, feet; α = angle of inclination from the horizontal; F = coefficient of friction for the material (coal, etc.); f = coefficient of friction of moving conveyor parts.

For anthracite, $F = 0.33$; for bituminous coal, $F = 0.50$. For conveyor parts, sliding in a trough or on tracks, $f = 0.33$; for conveyor parts, rolling on tracks, $f = Nd \div D$, where D = diameter of chain roller, inches; d = size of hole in roller, inches; N = factor determined by the kind of hole and the grade of lubrication. For a smooth hole, not lubricated, $N = 0.5$; for a smooth hole, greased, $N = 0.33$; for a bored hole, greased, $N = 0.25$; for a bored hole, well oiled, $N = 0.20$. For vertical lifts, $\cos \alpha = 0$; for horizontal runs, $\sin \alpha = 0$. For ordinary bucket carriers, apron conveyors, etc., where the material is carried, and not scraped, $F = f$.

If material is carried down hill or lowered, sin α becomes negative and then

$$X = WL(F \cos \alpha - \sin \alpha);\ Y = W_1L(f \cos \alpha - \sin \alpha);\ Z = W_1L(f \cos \alpha + \sin \alpha).$$

For combined elevator conveyors of complicated path, make a diagram of the path, calculate the pull for each run separately, note it on the diagram with a plus sign if positive and with a minus sign if the pull is in the direction of motion of the machine. Add the several values, allowing at each turn, where pull is transmitted, 5% of the accumulated pull up to that point. Note P, the maximum pull where the chain enters the driving sprocket wheel, and P_1, the pull on the chain leaving the wheel. Then P determines the stress in the chain, and $(P - P_1)$ is the torsional pull on the driving wheel. From this

$$\text{Hp} = \frac{T(P - P_1)}{33{,}000} \tag{8}$$

where T = chain speed, feet per minute. For the size of the motor, increase horsepower by 10% for each pair of cast-tooth gears, and by 5% for each drive through belt, chain, or cut gears.

SIZES OF SHAFTING FOR ELEVATORS AND CONVEYORS OF ALL KINDS. Head shafts and important turn shafts should be calculated for torsion and bending. From eq. 8, $(P - P_1)$ is the torsional pull; this multiplied by the radius of the sprocket wheel gives the torsional moment in inch-pounds. The load which produces bending is the resultant of two forces. One of these is $(P - P_1)$; the other is the weight of the shaft, wheels, and gear. In a vertical machine, the resultant is the algebraic sum of the forces. For other machines the weight of shaft, etc., is ignored, if it is proportionately small; otherwise the resultant is found by means of a diagram of forces. The shaft then may be considered as a beam supported by its bearings, and the bending moment in inch-pounds determined. By observing where the bending stress and the torsion stress act together on the same cross section of the shaft, the shaft diameter may be obtained from Article 2, Section 15.

CHAINS FOR FLIGHT CONVEYORS. Malleable-iron hook-joint chains are used for simple flight conveyors without rollers when the pull is not over 1500 lb. For roller flights, malleable-iron, closed-joint chains, or chains of the steel bar type are common. For double-strand conveyors with flights 20 by 10 in. or smaller, and of moderate length, malleable-iron roller chains are used; for larger flights and for heavy duty, steel bar chains are better. They are made in pitches 9 to 24 in. and of a strength to meet any service.

HANDLING ASHES IN BOILER HOUSES. For pneumatic system, see p. 26-43; for hydraulic methods, see p. 26-45; for skip hoists, see p. 26-30; for pivoted bucket carriers, see p. 26-30.

In small plants, ashes often are collected by hand car and taken to a waste dump, or discharged into a skip or elevator for storage in a bin. Ashes elevators which work only an hour or two per day are of the centrifugal discharge type (see p. 26-31) with a chain strong enough to withstand grit and corrosion. If the machine must work longer, it is better to use a slower speed elevator or a skip hoist. For handling large clinkers, old brick, soot, and other boiler house refuse, a car and skip system is better than any bucket elevator. For high lifts a skip hoist is better under any conditions.

A simple ashes conveyor used in some small boiler houses consists of a wide malleable-iron chain which runs at about 5 ft per min in a heavy cast-iron trough and partly drags, partly carries the ashes to the foot of an elevator. It is cheap, requires no lubrication, and lasts long enough to pay for itself in some places.

14. ELEVATORS FOR MATERIALS OTHER THAN COAL

It is necessary to consider whether material to be handled is fine or in large pieces, abrasive or nonabrasive, friable or nonfriable, wet or dry, whether or not it must be handled at slow speed. It is not possible to give exact rules, but for most cases the following directions are useful.

(1) Ashes, crushed ores, small coal, foundry sand, dry clay, gypsum, acid phosphate, and similar materials. Use centrifugal discharge elevators with malleable-iron buckets on single-strand closed-joint, or steel-bushed malleable-iron, chain.

(2) Grains, feeds, and other free-flowing materials. Use centrifugal discharge elevators with chains lighter than those in (1) or rubber belts, with malleable-iron or sheet-steel buckets.

(3) Cement, crushed coal, coke, slag, ores, glass house cullet, and other material, not too large, which should be handled at relatively slow speeds. Use single-strand continuous-

bucket elevators with steel buckets on closed-joint or steel-bushed malleable-iron chains. These elevators are vertical, or nearly so, and have a boot and casing.

(4) For materials in (3), handled in large quantities or to high lifts, use double-strand continuous-bucket elevators with steel buckets on steel chains.

(5) Heavy coarse material, ores, crushed stone, gravel, and rock which cannot be picked up in a boot. Use inclined continuous-bucket elevators with one or two strands of chain or a rubber belt.

Elevators for handling coal in boiler houses and for many other materials have been classified and standardized by manufacturers of conveying machinery, whose catalogs give dimensions, capacities, speeds, power required, and details of casings, boots, etc. For belt elevators, see handbooks issued by rubber belt makers: Goodyear Rubber Company, Akron, O.; Goodrich Rubber Company, Akron, O.; Manhattan Rubber Company, Passaic, N. J., and others.

PORTABLE ELEVATORS OR LOADERS are bucket elevators with a motor or gasoline engine mounted on wheels or caterpillar treads, and arranged to pick up their load at ground level and deliver it to wagons or trucks through a chute, screen, or short belt conveyor. Some loaders for hard coal, coke, or excavated gravel have shaking screens. In coal yards, they save time and labor and reduce cost of handling by keeping trucks at work. A man will shovel a ton of coal from ground to wagon in 20 min; a loader in 3 min or less. Four men load a 5-ton truck by hand in 30 min; one man with a loader in 5 min. The system of using loaders in connection with ground storage is preferred by many to storage in overhead bins with chutes to wagons. First cost and depreciation are less and there is less machinery to maintain. For large capacities, or where a machine handles sand, stone, or gravel, a self-propelled machine (Fig. 8) is an advantage. The loader backs into the pile by its own power, which is quicker and easier than breaking down the pile by hand. Loaders that work on roads or in yards with soft floors should have track-laying treads.

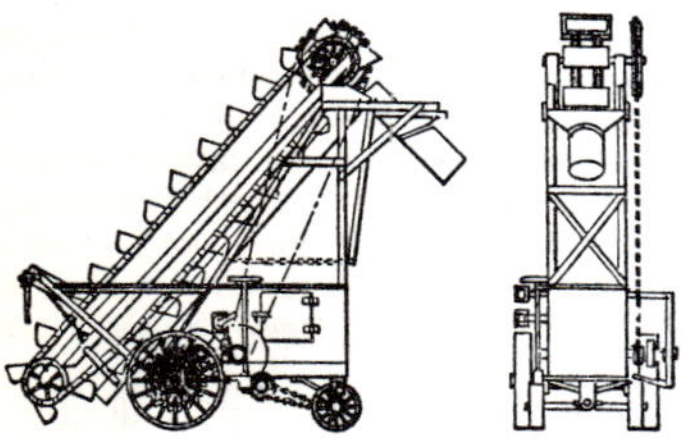

Fig. 8. Self-propelled loading machine.

Portable elevators, in modified form, excavate earth, dig trenches for pipe lines, remove snow from city streets. A development by Barber-Breene is a portable road-building machine that travels under its own power, picks up sand and gravel, mixes it with bituminous binder, spreads, tamps, and levels it to make a finished road at the rate of 1 mile per hour.

VIBRATING CONVEYORS, mounted on roller races for use in mines, are described below. Similar conveyors with the trough mounted on spring legs are used for sugar and other materials not easily handled by screw, flight, or apron conveyors.

Vibrating pipe conveyors made by Traylor Vibrator Company, Denver, Colo., are used to convey roasted ores, sintered concentrates, and other hot, dusty, and gritty materials. The pipes are 10 to 26 in. in diameter, hung by flexible suspension from above. Vibrating units are clamped to the pipe at intervals, each consisting of an armature and a stator supplied with alternating current. At 60 cycles per second the armature is attracted 3600 times per minute against the flexure of powerful bar springs. The inertia of heavy weights attached to the frame causes the pipe and armature to vibrate in unison. The plane of vibration is inclined to the axis of the pipe, and the material receives a component of motion, which moves it through the pipe. See Ref. 4.

The Redler conveyor (Stephens-Adamson Manufacturing Company) consists of a flexible chain or wire rope carrying pusher bars or links at short intervals. These move in a closed box, embedded in the material conveyed. By a proper arrangement of turns, horizontal and vertical runs can be combined, and the machine acts both as elevator and conveyor. Moving parts do not completely fill the cross section of the box, and the conveyor is said to be free from danger of overloading.

15. COAL CONVEYING IN THE MINE

COAL CONVEYING BY MINE CAR HAULAGE. Robert Peele (Ref. 5) says that locomotive haulage is justified at mines with easy track grades and large output. Compared with haulage by animals, the high speed, large tonnage per trip, and few men required may balance the larger interest and depreciation charges. Compared with rope haulage, especially if there are branches or many curves, it is more flexible and easier to extend or maintain. It requires less headroom, heavier rails, and easier curves than animal haulage, but cannot compete with it on thick seams in daily output of less than about 175 net ton-miles on grades below 1.25%, or with rope haulage on grades averaging 4% against or 5% with loads.

Gasoline locomotives have the flexibility of steam locomotives in radius of travel, and for mining work are geared for speeds of 3 to 12 miles per hour. They have no open fire, consume no fuel when idle, and are less dangerous than steam locomotives in gaseous mines. Exhaust gases and cost of gasoline are factors to consider.

Compressed-air locomotives are flexible in operation within the radius of travel from charging stations for which they are designed, require no fire, do not spark or emit noxious exhaust, but aid ventilation and are the safest type for gaseous, dry-timbered or poorly ventilated workings.

Electric storage-battery locomotives are flexible, cheaper to install than trolley or air locomotives, simple to operate, and safer than any other type except compressed air. They are suitable for short hauls, in temporary workings, as in gathering, or for infrequent long hauls where trolley lines are not warranted or where low-roofed wet workings make them dangerous. Grades must be easy.

Electric trolley locomotives are the prevalent types in mines. They are compact, simple to operate at high efficiency, and usually have lowest power cost of all types. Disadvantages are cost and danger of trolley circuit; danger of sparking at commutator or trolley wheel; necessity of wired and bonded track.

Rope haulage conditions are large tonnage and grades which make locomotive haulage inefficient. Straight haulways and uniform grades are desirable but not necessary. See Rope Haulage, p. 26-09.

UNDERGROUND CONVEYING BY BELT CONVEYOR. A belt conveyor system may handle large quantities of materials in mines at a lower cost than any system of haulage. At the Colonial mines of H. C. Frick Coke Company, near Brownsville, Pa., trains of cars in the mine are turned over in a rotary dumper; the coal falls into a bin and is fed by feeders to a 60-in. belt which delivers to a series of nineteen 48-in. belts run at 500 ft per min, and of lengths of 321 to 2439 ft. The total run is 4 1/2 miles; the total lift is 357 ft.

The system was planned and installed after the estimate had shown a probable saving of 3 cents per ton, 10,000 tons a day, based on the following comparisons between the belt system and haulage by electric locomotive. (1) First cost, belts 25% less. (2) Depreciation, belts 15% more. (3) Cost of power, belts 100% more. (4) Cost of labor, belts, 50% less. After eight years in service, the system had carried over 25,000,000 tons of coal, and eight of the original belts were still in use. Records and tests have shown the cost of power for the belt system to be less than half of that estimated, the saving in labor to be greater than estimated, and saving in cost of conveying to be over 3 cents per ton. For a description of this system, see Ref. 6 and Ref. 3, pp. 273–274. A belt conveyor similar to the above is used at the Palmer mines of H. C. Frick Coke Company. It has twelve belts with a total length of travel of over 3 miles.

RETARDING CONVEYORS. In parts of West Virginia and Kentucky, coal veins are on the hills and railroad tracks in the valley. An old way to get coal down is in coupled mine cars lowered down the slope by a tail rope. A broken car coupling or rope may cause a runaway resulting in great damage or loss of life. Coal sometimes is lowered in a large skip (monitor or gunboat), but unless it is shipped as run-of-mine, breakage due to filling and emptying the skip is objectionable. Wire-rope tramways have been used for capacities of a few hundred tons per day (see p. 26-09). For larger capacities, retarding conveyors are used.

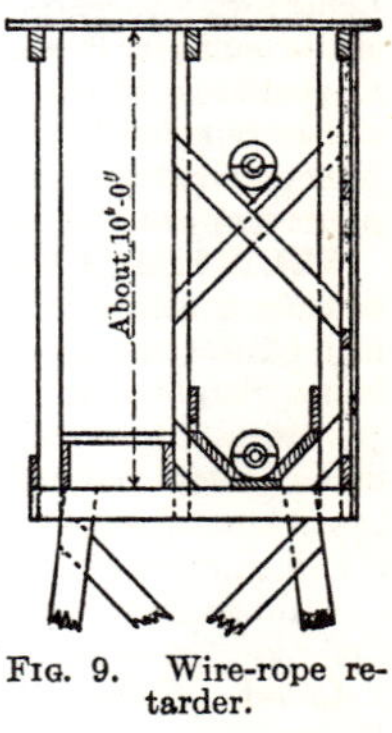

Fig. 9. Wire-rope retarder.

The wire-rope retarder, Fig. 9 (Jeffrey Manufacturing Company, Columbus, Ohio) comprises an endless wire rope to which 12 in. diameter cast-iron disks are clamped at intervals of 3 or 4 ft. On the carrying (lower) run the disks travel in a V-shaped trough with rounded bottom. On the return they run in a small guide trough. The rope, 7/8 to 1 1/8 in., generally is made with six flattened strands (see p. 6-14) arranged alternately, regular lay and lang lay so that the rope does not twist or rotate. The ends are joined in a split clamp disk. Rope retarders can be used on any slope up to 35 degrees and up to 1000 ft long. The best slope for soft coal is about 26 degrees. For slopes over 30 degrees, hinged baffle boards over the trough stop lumps if they start to avalanche. A straight run is preferable, but if slope must change from steep to flat, a large radius curve (1000 ft) should be made in the trough, or one conveyor must be made to discharge into another. As the wire rope is submerged in coal, wheels or tracks cannot be used to hold the rope down. At terminals, the rope runs over toothed wheels. On all but flat slopes, brakes prevent the conveyor drifting when power is off. On steep slopes a hand-operated brake also is used. Rope retarders have been used for 50 years or more. When properly designed and built, they are good and relatively inexpensive. Carrying capacity is 100 tons to 200 tons per hour.

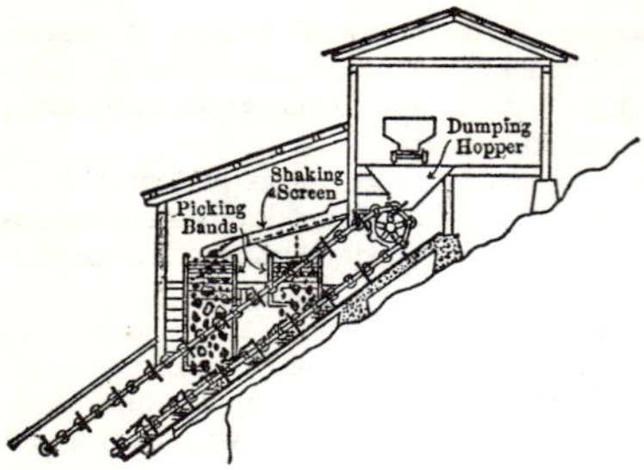

FIG. 10. Retarding conveyor loading end.

Chain and flight-retarding conveyors consist of two strands of steel roller chain attached to ends of steel flights, running in a steel trough. The chains roll on tracks, and the path of the conveyor can be made to suit the slope and contour of the ground. A 36 by 12 in. flight will carry 360 tons per hour at 100 ft per min. Some retarders have 48 by 12 in. or 60 by 16 in. flights for large carrying capacity. Figure 10 shows the upper end of a retarding conveyor where a shaking screen separates the fines before the coal goes to the picking bands. With such an arrangement, refuse is thrown out on the slope of the hill.

16. BELT CONVEYORS

For fundamentals of design and operation, see Ref. 3. In its simplest form a belt conveyor comprises two end pulleys, a moving belt, and idlers to support the belt. The lower idlers have flat-faced rolls; the upper or carrying idlers generally are arranged to curve or trough the belt to make it carry more and avoid risk of spill. There is not much difference between 3- and 5-roll troughing idlers in carrying capacity. Three-roll idlers cost less, have fewer bearings, lubrication is more certain. The best troughing idlers have ball or roller bearings; the saving of power, reduced cost of maintenance, and longer life of belt often pay for the extra cost of the improved idlers. On grain conveyor belts, the best practice is to run the belt nearly flat with concentrator or "tuck-up" idlers at intervals to give a slight troughing.

Troughing idlers are spaced 4 to 5 $^1/_2$ ft apart on narrow belts, according to weight of material carried; on wide belts, 3 to 4 ft apart. Return idlers are spaced 9 or 10 ft apart.

For detailed information about belt idlers and other parts, see catalogs of the Link-Belt Company, Chicago; Robins Conveying Belt Company, New York; Stephens-Adamson Manufacturing Company, Aurora, Ill., and other manufacturers of conveying machinery.

THE CONVEYOR BELT may be of solid woven fabric, or of cotton duck with the plies stitched together. Both kinds are impregnated with oils or compounds to resist moisture. Rubber belts, however, are in more general use. They are built up with layers of cotton duck cemented together by rubber compounds, and then vulcanized. Lengthwise threads of the cotton duck are heavier than crosswise to give tensile strength with flexibility for troughing. An extra thickness or cover of rubber on carrying side of the belt resists abrasion and increases its life. For crushed coal, the cover may be $^1/_{16}$ in. thick, but for run-of-mine coal and abrasive ores a thicker cover is required to avoid puncture by sharp pieces, and to resist wear occasioned by making material at the loading point acquire the belt velocity.

BELT THICKNESS. In general, for durability and long life, tension in a rubber or cotton belt should not exceed 24 lb per in. of width per ply of thickness, if duck is the usual 28-oz duck. In a belt conveyor, this 24 lb cannot all be used to move the loaded belt. Some of it must be used to supply enough tension to prevent slip on the drive pulley. This tension is applied by the take-up. The smaller the arc of wrap on the drive pulley and the lower the coefficient of friction between belt and pulley, the greater must be the tension. By increasing angle of wrap and lagging drive pulley, more of the strength of the belt can be usefully employed.

Rubber belts are made in different grades, differing as to weight of duck, quality of "friction" used to cement plies together, and thickness and quality of rubber covers. For detailed information, see Ref. 3, pp. 52ff. Reference 7 gives considerable detailed information concerning belts, tandem drives, stresses in inclined conveyors, etc. See also Ref. 8.

DIAMETER OF BELT PULLEYS. The larger the pulley, the less the stretching of the rubber "friction," or of the stitching which holds the plies of the belt together, and the longer the belt will last. The diameter of driving pulleys in inches should be at least five times the number of plies in the belt. The diameter of foot pulleys and snub pulleys should be four times the number of plies, and of tandem pulleys, geared together, driving thick belts with heavy rubber covers, at least six times the number of plies.

LOADING BELTS. To reduce the wear on the belt surface, the material should be fed by a chute of such length and slope as will give to the material the velocity of the belt, if possible. To prevent scatter, and to pile the load properly, skirt boards 5 to 7 ft long are used. They must be carefully designed. If not well supported, they may sag and cut the belt.

DISCHARGE FROM BELT CONVEYORS. The simplest belt conveyor discharges over the head pulley, as in Fig. 11. For intermediate discharge at one or more points, fixed trippers may be used, as in Fig. 12, but they double or triple the wear on the belt surface whenever they feed material back on the belt, and by repeated reverse bending over pulleys they tend to separate the plies of the belt. For most conditions of intermediate discharge, a traveling or movable tripper (Fig. 13) is preferable. Some materials,

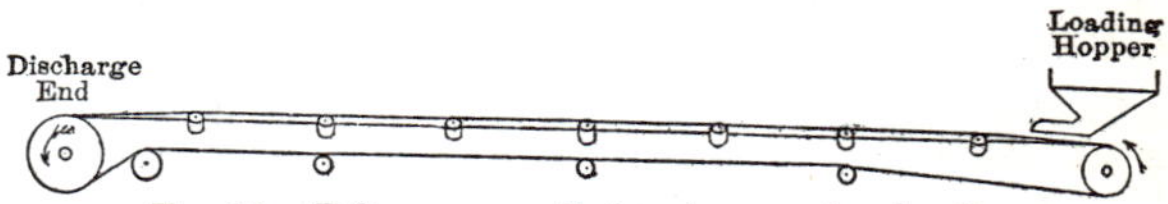

FIG. 11. Belt conveyor discharging over head pulley.

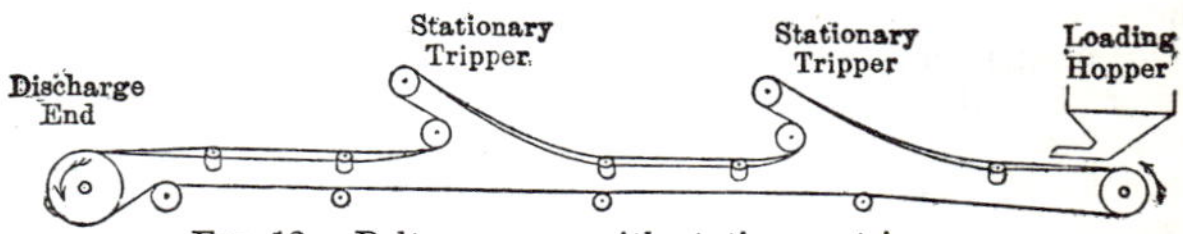

FIG. 12. Belt conveyor with stationary trippers.

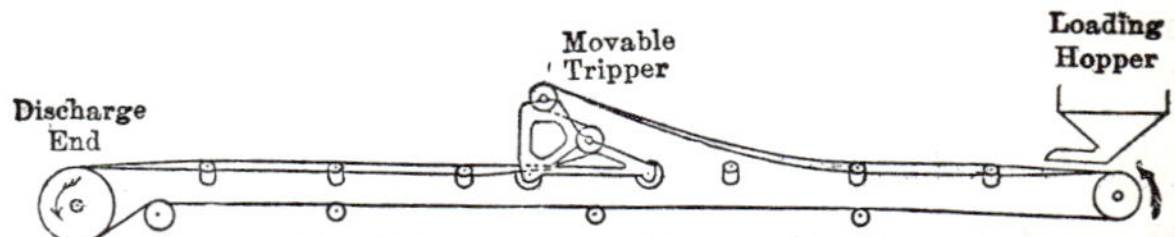

FIG. 13. Belt conveyor with movable trippers.

like foundry sand or wood chips, can be carried on flat belts, and at slow speeds the discharge can be divided at several intermediate points by "strike-off" boards set at various positions above the belt, or all can be thrown off at any point if the boards are hinged so as to swing into and out of position. This method is not good with lumpy material.

INCLINED BELTS. With uniform feed, fine coal may be carried on a belt at an angle from the horizontal of 22 degrees; but when mixed with lumps, at an angle of not over 20 degrees. Run-of-mine coal, coal, and coke can be carried at an angle of not over 18 degrees, crushed stone and ore, at not over 18 degrees, and tempered foundry sand at not over 24 degrees. On the other hand, numerous substances under certain conditions can be carried at not over 15 degrees.

ADVANTAGES OF BELT CONVEYOR. (1) Low power consumption. (2) Noiseless operation. (3) Adaptable to all sorts of material. (4) Suitable for high capacities and long distances beyond the range of other conveyors. (5) Simplicity of construction.

LIMITATIONS. The belt is expensive and must be protected from these sources of injury: (1) poor arrangement of feed; (2) chute or skirt-board sagging on belt; (3) lumps falling on return belt and jamming between foot pulley and belt; (4) belt running crooked and fraying the edges; (5) oil or grease getting on rubber belt; and (6) continued exposure to great heat or cold, or direct sunlight.

When belts are inclined, especially at or near the maximum angle, there is a shift or slip of material in passing over each troughing idler. This adds to the wear on the carrying surface, and with gritty materials it may be serious. Short belts, steeply inclined, are likely to be short lived under such conditions, because the wear from the pick-up at the loading point is already large, due to the fact that each foot of belt passes the loading point frequently. In long belts the wear from pick-up is distributed over a greater amount of belt surface, and the bending over pulleys does not occur so often. Hence long belts last proportionately longer than short belts.

STANDARD BELT CONVEYORS, 16 to 48 in. wide for various lengths and capacities, are shown in catalogs of manufacturers of conveying machinery. For many conditions of ordinary service, time and expense can be saved by buying a conveyor from a catalog rather than designing one.

Sectional belt conveyors 18 and 24 in. wide (Barber-Greene Company) consist of standard drive and take-up groups with machinery parts mounted on steel frames. Conveyor length is made up of light steel frames of various standard lengths, which, with belts, idlers, and guards, weigh not more than 60 lb per ft. They will span 25 to 50 ft in the clear.

Portable conveyors of the same light construction, listed by the same makers, are used to receive concrete from mixers and deliver it to forms in building construction. By having one conveyor discharge to the next, concrete can be distributed over large areas. In some cases, 15 conveyors have been used for a distance of 500 ft or more. The 24-in. belt runs over 2-roll idlers to form a deep trough. A flexible scraper under each head pulley removes concrete from the belt. Frames are made up to 50 ft long and are mounted on large caster wheels. For handling concrete on belt conveyors, see Ref. 9.

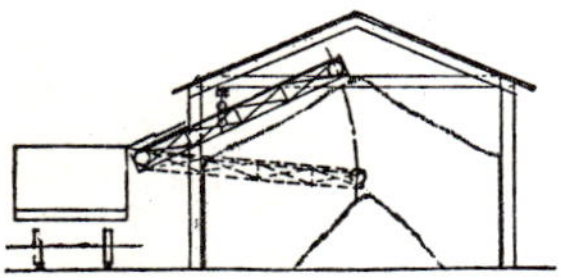

Fig. 14. Portable conveyor for car unloading.

Other portable belt conveyors are made. Figure 14 shows one which stores in a shed, material shoveled over side of a car. The frame is hung from a chain block and trolley on an overhead track that runs the length of the shed. Motor- or gasoline-driven portable conveyors on wheeled frames, with 12-, 14- or 16-in. belts on frames 20 to 40 ft long, will receive material from bottom-dump or side-dump cars and form storage piles or deliver to bins or wagons. Several machines can be placed in tandem to deliver to a distance, or a high pile can be formed by arranging machines as in Fig. 15. These machines are light, a 16 in. by 30 ft size weighing less than a ton. Heavy machines of this type are made by Barber-Greene Company, with 18- or 24-in. belts on frames up to 60 ft long, with a hand-power hoist for adjusting height of boom.

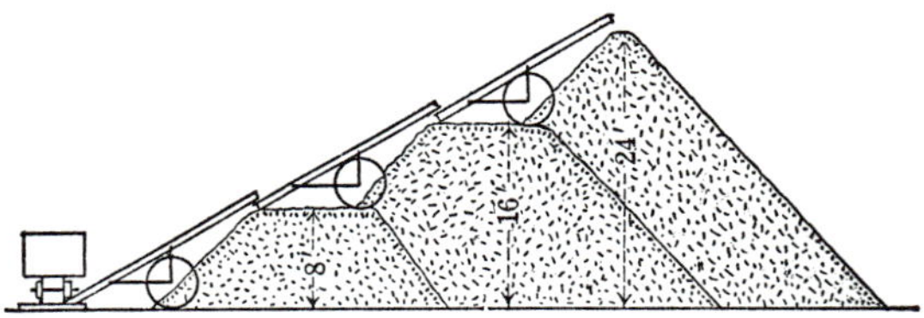

Fig. 15. Portable conveyors arranged for high piling.

CAR LOADER BELTS. Short belt conveyors mounted on wheeled frames easily handled by one man, and run at 1200 ft per min or more, will throw fine material a distance of 15 to 20 ft. The machine is placed in position at door of a box car, receives material from a chute, throws it into a pile at each end of the car, and distributes it for proper loading. It saves the labor of several men.

CAR UNLOADER BELTS (Barber-Greene) are mounted on a shallow steel frame, 4 or 5 in. deep, that can be laid across a track above the rails, or in a shallow pit below them, to receive discharge from a drop-bottom car. They are used for free-flowing materials without lumps. The service is severe, but the car can be unloaded anywhere on the track and a deep pit is unnecessary.

WEIGHING ON BELT CONVEYORS. The Merrick Weightometer (Merrick Scale Manufacturing Company, Passaic, N. J.) for weighing material in transit consists of (1) a part of the upper run of the belt suspended from counterbalanced weighing levers and (2) a totalizer driven by the moving belt which continuously multiplies weight by travel to give tonnage passing over the conveyor. This is recorded on a counting machine or chart recorder.

STEEL BELTS. The Sandvik belt (Sandvik Steel, Inc., New York) is a tempered, polished, cold-rolled steel band, about 1/32 in. thick, made up to 32 in. wide and up to 375 ft long. It runs flat over rollers, spaced 6 to 12 ft, or slides on wooden strips. These belts handle clinker, phosphate rock, sugar, clay, foundry sand, and similar wet, sticky mixtures that do not discharge clean from fabric or rubber belts. They discharge either over the head pulley or by scrapers over the belts. The scraper is simple, cheap, and works well with materials that cannot be discharged cleanly by a belt tripper. The belts are light, run quietly, and when supported on rollers are said to take less power than rubber belts on troughing idlers. Belts wider than 32 in. can be made by a method of longitudinal jointing.

STEEL MESH BELTS are used for conveying through acid or alkaline baths, chemical solutions, and scalding tanks, and in processes of drying, baking, annealing, spraying, etc., for which fabric belts would be unsuited. They are made in the United States by Cambridge Wire Cloth Company, Cambridge, Md., and Wickwire-Spencer Steel Company, New York. They consist of woven wire mesh, with or without edge reinforcement, or of coils of wire interlocked and flattened. Sometimes the coils are joined over wire rods which may be bent at the ends to form a protective edge or border. These belts can be made of copper, brass, or the various alloys which impart strength or resist heat, corrosion, or the action of chemicals. They can be made with steel plate selvages or partitions that will bend around driving drums. Belts can be woven up to 12 ft wide. Those made for lehrs and annealing ovens will work regularly in temperatures over 1000 F.

LONG-DISTANCE CONVEYING BY BELT. See pp. 26-12 and 26-37. For an installation (Jeffrey Manufacturing Company) of long, wide belts of large carrying capacity, see Ref. 10.

17. SCREW CONVEYORS

Screw conveyors consist of a helical blade of metal fastened to a central pipe or shaft, and set in a trough or box. The blade generally is of steel plate, but, depending on service and nature of material handled, it can be cast iron, cast steel, or an alloy that resists corrosion and oxidation. The trough generally is U- or V-shaped with a rounded bottom. Some materials can be conveyed through a round pipe if the length is short enough to permit the screw conveyor bearings to be kept outside the pipe.

Bearings are of various kinds, babbitt, bronze, chilled cast iron, hardwood. Success with screw conveyors often depends on using the proper kind of hanger bearings and the right kind of trough. It is advisable to consult the makers on these points.

CONVEYING CAPACITIES. Figure 16 shows curves developed by Link-Belt Company, Chicago, for capacities of screw conveyors. Diagonals 3, 4, 6, etc., are diameters of screw, in inches. Curves *A*, *B*, *C*, etc., are advisable speed limits, rpm, but rating should be multiplied by a capacity factor *X*, values of which follow.

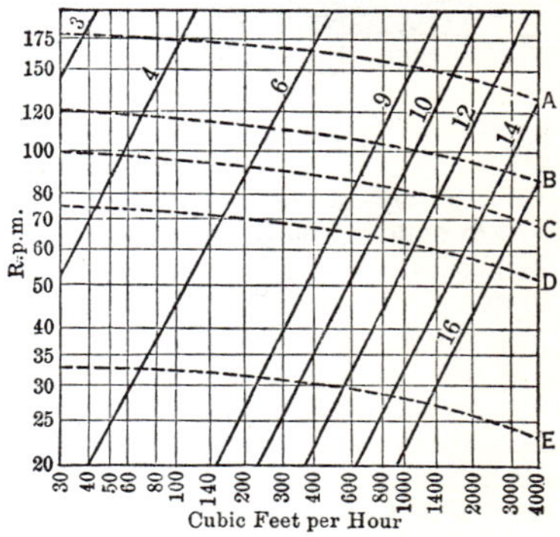

Fig. 16. Capacities of screw conveyors.

Class 1. Dry, free-flowing, not gritty; as grain, malt, flour, meal, pulverized coal; advisable speeds not beyond curve *A*; $X = 1.00$.

Class 2. Shelled corn, clean cottonseed, fine coal, crushed ice, pulverized minerals that are not abrasive; advisable speeds not beyond curve *B*; $X = 0.75$.

Class 3. Wet malt, starch, sawdust, ground cork, carbon black, dry paper pulp stock, nonabrasive minerals, with occasional lumps (not over 25%) of size not over $1/8$ diameter of screw; advisable speeds not beyond curve *C*; $X = 0.66$ to 0.75.

Class 4. Damp, trashy cottonseed, dry clay, raw sugar, foundry sand, fertilizers, crushed gypsum, lime, and other abrasive materials with lumps (not over 50%) of size not over $1/6$ diameter of screw; advisable speeds not beyond curve *D*; $X = 0.50$ to 0.66.

Class 5. Very gritty materials, as ashes, wet or dry, silica sand, pulverized flint or quartz, flue dust, fly ash; advisable speeds not beyond curve *E*; $X = 0.30$ or less.

The capacity factor *X* refers to screw conveyors with intermediate hanger bearings. If conveyor is short, has no hanger bearings, and is properly fed, capacity may be 50% greater than use of the factor would indicate.

Screw conveyors at slopes up to 10 or 15 degrees have about the same capacity as level conveyors. At greater slopes, capacity tends to fall off, but for such conditions, screws with short-pitch flights can be used. Screws at steeper angles should be referred to the makers. Vertical screws, run at the proper speed and properly fed, will elevate ground cork and some other light materials in closed pipes. In "batching" plants that serve concrete mixers, Link-Belt vertical screws have been used to elevate cement 20 or 30 ft. They require more power than a bucket elevator and are likely to get out of order from cement working into the step bearing. The advantage is that they are light and portable.

HORSEPOWER OF SCREW CONVEYORS. If a screw conveyor is crooked and out of line, if speed is excessive, if hangers and bearings are not suited to the work, if discharge outlets choke, the power consumed cannot be estimated. For screws working under proper conditions, Link-Belt Company gives the following general formula:

$$\text{Hp} = 1 + \frac{ALN + CWLF}{1{,}000{,}000} \qquad (1)$$

where L = length of conveyor, feet; N = rpm of screw; C = cubic feet conveyed per hour;

W = weight of material, pounds per cubic foot; A and F are factors. For conveyors with lubricated bearings, values of A are:

Diam. of screw, in.	3	4	6	9	10	12	14	16
A	15	21	33	54	66	96	135	186

For bearings not lubricated, multiply A by 1.75; for chilled iron bearings, multiply A by 2.5.

Factor F varies with kind of material (see classification under Conveying Capacities), and within a given classification with the resistance material offers to moving in the trough and past hanger bearings. Values of F are:

Material class	1	2	3	4	5
F	0.4–0.6	0.4–0.9	0.6–1.2	1.4–2.0	2.0–5.0

For specific values of F, see Link-Belt Company catalog.

SCREW CONVEYOR FEEDERS. For withdrawing material from a bin at a definite and regular rate, a screw has these advantages over other feeders: (1) compact, leak-proof, dust-tight; (2) relatively cheap in first cost and operation; (3) can be fitted to bin bottoms of any shape; (4) acts as a shut-off valve as well as a feeder.

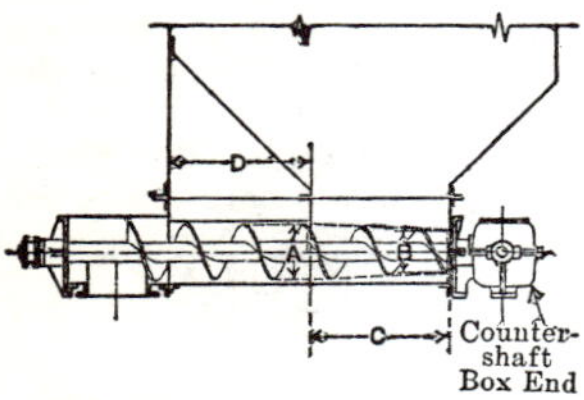

FIG. 17. Screw conveyor feeder.

Figure 17 shows a short feeder of diameter A in a closed pipe for the length D. The pipe prevents material from flooding out of the bin and acts as a shut-off when screw is idle. Screw is tapered over length C, which is attached to the bin, and the trough is flared. Diameter B is one size smaller than A. The screws are made with heavy steel blades on heavy pipe. Table 8 gives dimensions of the three sizes used for feeders. The carrying capacity for fine material, or material containing not more than 25% of lumps, is about 0.02 cu ft per revolution for 9-in.; 0.04 cu ft for 12-in.; and 1.1 cu ft for 16-in. screws.

Table 8. Dimensions of Screw Conveyor Feeder

Size of Conveyor, in.	Max. Size of Lumps, in.	Dimensions, in. (See Fig. 17)			
		A	B	C	D
9XX	1 1/2	9	6	24	24
12XXX	2	12	9	36	36
16XXX	2 1/2	16	12	48	48

For longer feed screws, where an intermediate bearing must be used, sizes are either 9 in. or 12 in. A U-shaped trough is used beyond the length D (Fig. 17). It contains the hanger bearing and a screw of full diameter. Clearance between screw and inside of trough should be 3 to 4 in. Length $D + C$ has a screw of a reduced diameter B. The rated capacity is the same as for short feed screws, but the lumps may be 50% of total volume.

For cement, pulverized coal, and materials which at times flow like fluids, screw feeders are inclined upward at 20 or 30 degrees, with outlet of the trough higher than the intake point. Screws generally have double flights.

THE THRUST IN SCREW CONVEYORS should be taken on end castings. These are of various kinds, with ball or roller thrust bearings for heavy loads. Driving ends with bevel gears are general. The Link-Belt Company places the motor worm gearing and oiltight gear box on a base plate; gear box is bolted to end of the trough.

TROUGHS are well standardized. For cement and similar dusty materials no attempt is made to bolt on a dust-tight cover. The cover plate is flanged downward all around and rests in grooves formed by Z-bars on top edge of trough. Dust collecting in the grooves seals cover against further dust emission.

Screw conveyors originally were used for conveying grain, but their use has been greatly extended. With properly designed screws, troughs, bearings, and drive ends, they handle a wide variety of materials, some of which cannot be conveyed by any other kind of conveyor. For details of screw conveyors and accessories, see catalogs of makers of conveying machinery.

18. PNEUMATIC AND HYDRAULIC CONVEYING

PNEUMATIC TUBE CONVEYORS are used for transmitting small carriers between departments of shops, newspaper offices, industrial plants, etc. The conveyor tube is round or oval, the carriers are short, of a cross section to fit the tube, and are propelled at 20 or 30 ft per sec by vacuum or air pressure. The round tubes are 1 $^1/_4$ in. in diameter or larger; the oval tubes are 3 by 6 or 4 by 7 in. Short lengths of small tube, up to 200 ft, can be served by a foot-power pump, but usually a motor-driven blower is used. For a history and description of this system, see Ref. 11.

PNEUMATIC CONVEYING FOR BULK MATERIAL was first used in England, developed chiefly in Germany, and now finds extensive use in the United States. The suction or vacuum system (Fig. 18) comprises an exhauster or pump to produce a partial vacuum, a pipe line through which mingled air and material move, a receiver or cyclone to separate the material from the air, and usually a bag filter to clear the air and keep dust from the exhauster, and prevent its discharging to atmosphere. In a pressure system, the pump delivers air to a pipe line. Material is fed from a supply bin through an airtight rotary valve into the pipe, moves with the air to a cyclone or separator. The air is discharged to a filter or to atmosphere. The material settles to the bottom of the cyclone.

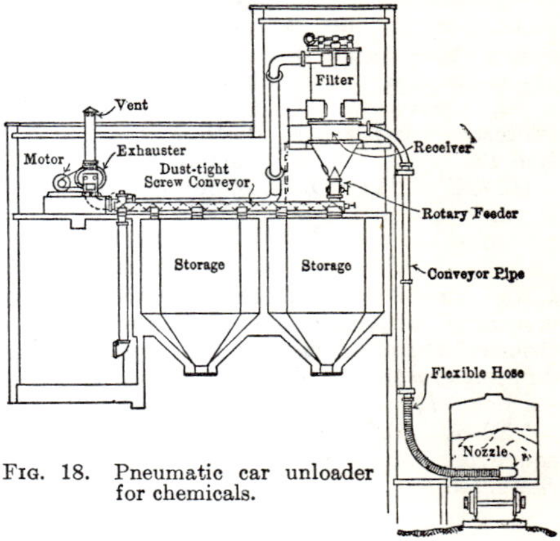

FIG. 18. Pneumatic car unloader for chemicals.

Carrying capacity depends on amount of air pumped and difference in air pressure between the two ends of the system. Theoretically, almost any material can be conveyed, but feasibility and cost are determined by practical considerations. (1) Material too large requires large-diameter pipes and a large air pump. (2) Material too dusty may require too large a filter. (3) Material too fragile may be reduced in value by breakage at pipe turns, or dust produced by breakage may require a costly filter. (4) If air pressure is kept low enough to avoid damage to material or to avoid excessive wear at pipe turns, carrying capacity may be too low to be economical.

The cost of a pneumatic system may be higher than that of a mechanical system. Power cost may be five or ten times as much, but a pneumatic system often is justified by other considerations. (1) In boats, a flexible suction hose is more easily handled than an elevator leg, takes up less room, cleans up better, requires less shoveling, can work under closed hatches in wet weather. (2) In unloading railroad box cars, time often is saved, labor costs reduced, delays and demurrage charges avoided. Avoidance of dust sometimes is important. (3) In handling bulk materials in manufacturing plants, labor often is saved, less dust is produced, and less space occupied. Changes in the pipe system to reach new points of intake or discharge are not difficult, and, what often is most important, conditions of safety and health for workmen are better.

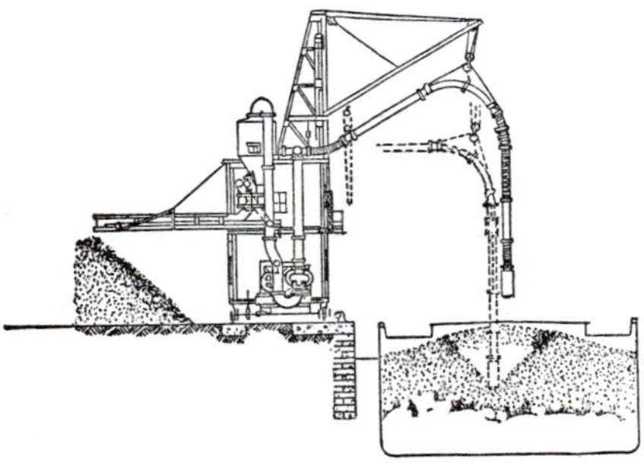

FIG. 19. Fuller pneumatic conveying system.

A combination of pneumatic handling with mechanical conveying often is desirable. In Fig. 18 material is drawn from the cyclone separator to a distributing screw conveyor. At the New York State Barge Canal terminal in Brooklyn, grain is lifted 80 ft vertically from barges, and discharged to a distributing belt conveyor.

Figure 19 shows a Fuller conveyor system on a movable tower for unloading small-size soft coal at a cement plant. Coal, at rate of 100 tons per hour, passes through the cyclone separator, is discharged to a short belt conveyor, then dropped on the ground to be distributed later over the storage area by a drag scraper. The tail pulley of the drag line is attached to the rear of the tower and moves with it. See Drag Scraper System, p. 26-10. The boat also carries cement in bulk from the cement plant. At destination, the cement is unloaded by a Fuller-Kinyon pump system on the boat. See below.

Pneumatic systems for conveying bulk material are made by Dust Recovery and Conveying Company, Detroit; Brady Conveyors Corporation, Chicago; Federal Pneumatic Systems, Inc., Chicago; Holly Pneumatic Systems, Inc., New York; United Conveyor Corporation, Chicago; Fuller Company, Catasauqua, Pa.

SUCTION SYSTEMS for collecting and conveying chips and sawdust from woodworking machines and dust from grinders are made by B. F. Sturtevant Company, Boston; Allington and Curtis Manufacturing Company, Saginaw, Mich.; Kirk and Blum Manufacturing Company, Cincinnati, and also by makers of blowers and exhausters. A Kirk and Blum installation in Camden, N. J., handles over 1,000,000 cu ft of air per min, collects 150 tons of sawdust daily, and has filters with an area of 200,000 sq ft of cloth surface. Similar systems for taking ground coal, rock, etc., from pulverizing mills are made by several makers of grinding mills.

For precautions against fires in blower systems, see publications of National Fire Protection Association, Boston. For safety codes for preventing dust explosions, see Ref. 12.

GRAIN UNLOADING. Suction apparatus for unloading grain from boats is used at many ports in Europe. The whole system, including power plant and separators, generally is mounted on a barge or float that works on the harbor side of the ship. It transfers grain to river barges while the ship discharges other cargo on the wharf. In the United States, one of the largest suction systems for unloading grain is the one in Brooklyn mentioned above; it lifts 6500 bushels per hour. At the public grain elevator, New Orleans, where it was necessary to provide for a 21-ft range of water level, and for boats differing greatly in width of beam and depth of hold, a pneumatic unloader of 6000 bushels per hour capacity has been used since 1917.

THE FULLER-KINYON CONVEYING SYSTEM (Fig. 20) is not strictly a pneumatic conveyor. At one end of a pipe line a short impeller screw 1 is driven by motor 2. Material is fed through hopper 3. At discharge end of cylinder in which the screw revolves,

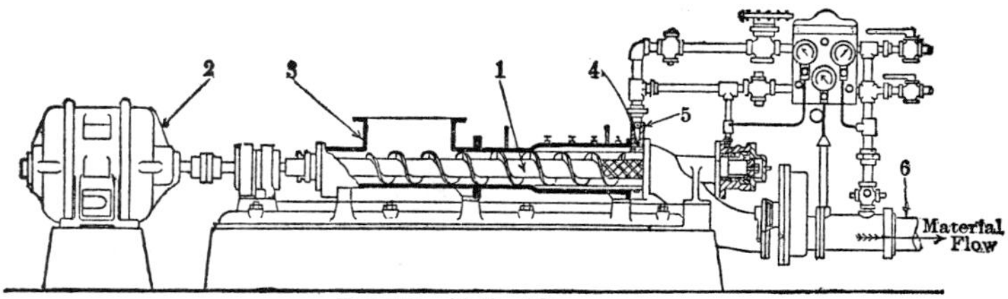

FIG. 20. Fuller-Kinyon pump.

compressed air enters through rings 4 and 5. The material when mixed with air acts as a fluid. In the fluent condition it is forced through the pipe line 6 by the displacement of material fed in by the screw and also by the expansion of the compressed air used to create the fluent condition.

Practically any dry pulverized material in quantities up to 300 tons per hour can be transported by this system over distances of 3000 ft and elevations of 150 ft or more. The system is widely used in cement mills, and for other pulverized materials as lime, soda ash, starch, kaolin, bauxite, etc. The pipe line may range from 2 to 8 in. in diameter according to amount of material. Branch pipes are connected to the main line by two- or three-way valves, usually operated by hand, but sometimes by automatic and remote control. Bins can be filled in desired sequence; the system will continue to work automatically so long as any bin has room to receive material. Deliveries under automatic control have been made to 48 bins in one system. To mix and blend raw materials in cement making, the discharge from several pipe systems may be controlled automatically according to predetermined time intervals. Hot materials up to 1000 F are successfully conveyed. Pipe lines may be water jacketed for cooling cement in process of manufacture. Steam-jacketed lines also are in use.

Figure 21 shows a Fuller-Kinyon portable pump, used chiefly on construction jobs to unload bulk cement from box cars or barges, to convey it to mixer bins or storage, and to reclaim it from storage to the bins. The pump is a motor-driven self-contained unit. Clutches control the power drive to the traction wheels. Cement is discharged through flexible hose which may be connected to any of the steel pipes leading to delivery point.

Several boats on the Great Lakes are equipped with the L. D. Smith system of drag scrapers (see p. 26-11) and a Fuller-Kinyon pump. Cement is pumped ashore at rate of 1400 barrels per hour through a flexible hose to the pipe lines which distribute cement to silos or pack houses.

The Fuller-Kinyon system often is cheaper and more convenient than any combination of mechanical elevators and conveyors. For short distances power consumption may be more, but for average or long distances it is likely to be less. Its advantages are that pipe can be run anywhere over, under, or around obstacles; and that material conveyed is always enclosed so that risk of fire, explosion, or injury to workmen is low. For temporary jobs of road making or building construction, its erection, dismantling, and moving are relatively cheap.

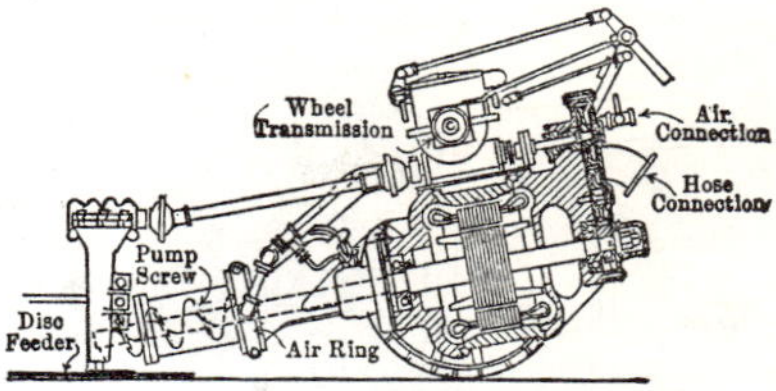

Fig. 21. Portable Fuller-Kinyon pump.

TRANSFER OF MATERIALS UNDER DIRECT PRESSURE. The Fluxo pump (F. L. Smidth & Company, New York) is used in some cement plants to transfer cement or dry pulverized raw materials to and from storage bins. A carriage runs on a track at one side of a row of silos or storage bins, and connects with a pipe in the base of the silo by flexible hose. On the carriage are two tanks, a vacuum pump with an air filter, an air compressor, and an electropneumatic control apparatus. When the current is being drawn into one tank by suction, it is being expelled from the other by air pressure. The operation is reversed by the control apparatus, the tanks alternately filling and emptying. The flow, from silo to transport line to which the machine is connected, is continuous.

The Kennedy transport system consists of a single tank, to which cement or other pulverized materials flow, suitable valves, air under pressure, and a transport pipe line. At a predetermined weight of material in the tank, springs on which the tank rests deflect, the supply is shut off, the bottom valve opens, and air admitted to top of the tank forces the material out through the pipe line. The cycle automatically repeats.

STEAM-EJECTOR SYSTEM FOR BOILER-HOUSE ASHES. The Steamatic ash-conveyor system has in many cases replaced the steam-jet system of handling ashes. In the Steamatic system, air flow is produced by a ring-nozzle ejector and a venturi tube placed at the discharge end. These claims are made: (1) Low steam consumption. (2) Relatively little wear on pipe lines and pipe turns. (3) Since ashes are in contact with steam for only 3 ft of travel, they enter the bin dry, and do not pack, or freeze in winter. (4) Less noise. (5) No danger of explosion in storage bin. This system requires steam at 70 psi or higher, will carry up to 200 tons of ashes per day up to 300 ft distance, and lift through 100 ft. Pipes are 4, 5, 6, 8 in. in diameter. The 8-in. size handles large clinkers and is most economical of steam per ton of ashes carried. Small pipes may be used as branch lines for fly ash and soot.

The Nuveyor system is similar in principle to the Steamatic. The pipe line is the same, but to produce suction a positive air exhauster of the Roots blower type or a multijet exhauster is used. In both cases ashes pass to a cyclone separator; then by an automatic timing device, which acts every 90 seconds, a release valve opens, suction stops for 5 seconds, and ashes in the cyclone drop into the storage bin. Figure 22 shows such a system with an air washer between cyclone and exhauster. No dust enters the latter, and discharge to atmosphere is clean. The air washer acts also as a silencer. The Nuveyor will handle 30 tons of ashes per hour, is quiet in operation, and ashes entering the storage bin are dry. The system is operated by remote-control buttons at points where ashes are fed to pipe line.

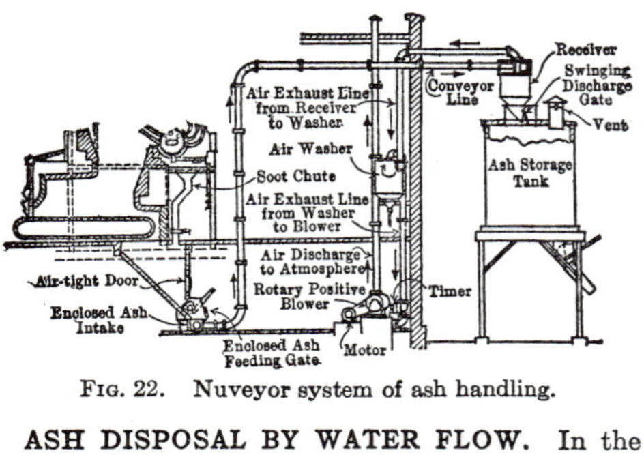

Fig. 22. Nuveyor system of ash handling.

ASH DISPOSAL BY WATER FLOW. In the Hydrojet system ashes accumulated under boiler furnaces are dislodged by a water jet at 100 psi and moved through a trench below the basement floor by similar jets. Figure 23 shows the system applied to the ash pit of a chain grate stoker. The whole apparatus is enclosed. The jet breaks down the pile of ashes; the mixture of water and ashes flows into the trench under the floor, lined with U-shaped cast-iron plates; nozzles at intervals deliver jets in the direction of flow

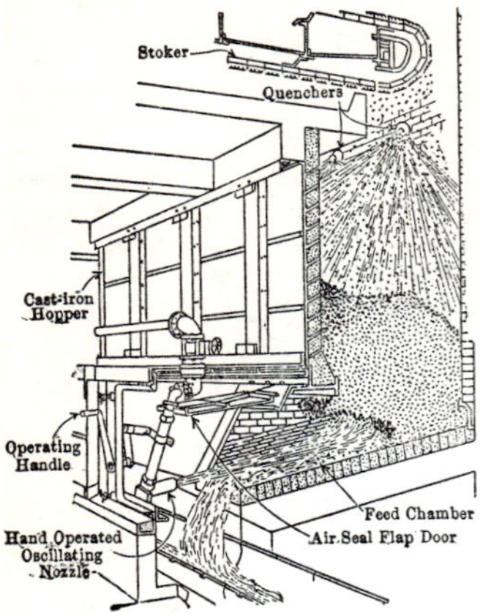

Fig. 23. Hydrojet system of ash removal.

Outside the boiler house the discharge can flow to waste land or to a sump from which ashes are removed by grab bucket or pumped to a drainage bin for removal by car or truck. Advantages claimed for this system are: (1) No escape of dust, gas, or steam into basement. (2) Men cannot be gassed or burned. (3) No leakage of air from basement into furnace to disturb combustion. (4) Operation is intermittent; time is available for repairs when necessary. (5) Overhead ash hoppers and cars are not used; boiler house can be built 10 ft lower.

The Hydrovac dust-handling system for boiler houses removes siftings under stokers, fly dust from boiler settings, soot and floor sweepings, by means of air flow under vacuum produced by high pressure water jets in a venturi tube. The siftings can be blown into a furnace, refuse can be collected in a tank for removal or discharged into the water trench of the Hydrojet system.

For mechanical methods of ashes handling, see p. 26-35. For hydraulic ash handling systems, see Ref. 13.

19. PACKAGE HANDLING

TYPES OF PACKAGE ELEVATORS. (1) For boxes, barrels, bales, and sacks, the rigid-arm elevator (Fig. 24) will discharge over head wheel only. (2) The automatic-arm elevator (Fig. 25) has carriers arranged to lift and lower; lifting side is so hinged that it will dump a barrel or sack forward when a lever on the arm contacts with an adjustable cam on the lifting side of elevator. On lowering side, the arm is rigid. Packages going down are deposited on skids at unloading point. (3) Suspended tray elevators (Fig. 26)

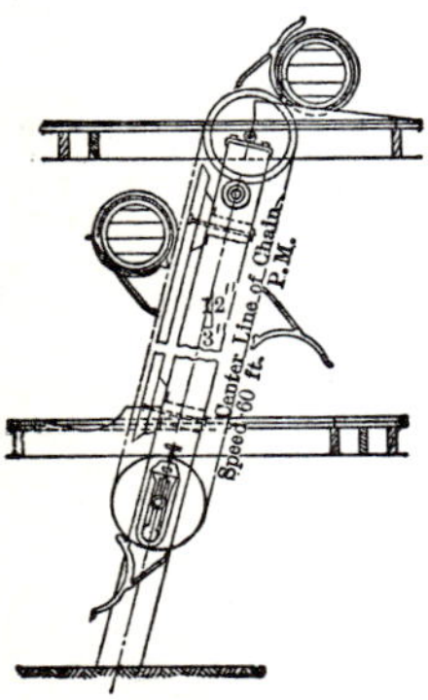

Fig. 24. Rigid-arm elevator.

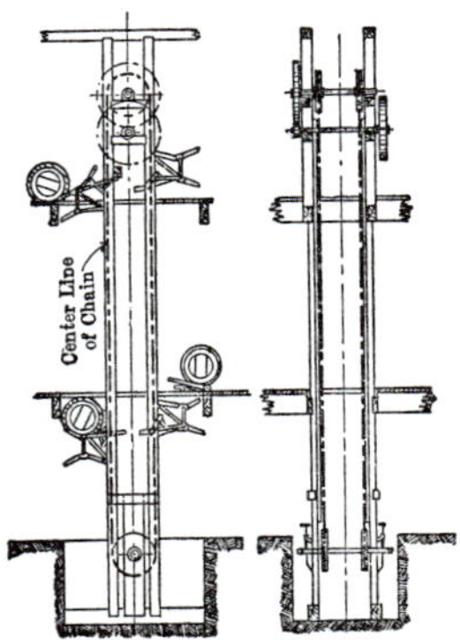

Fig. 25. Automatic-arm elevator.

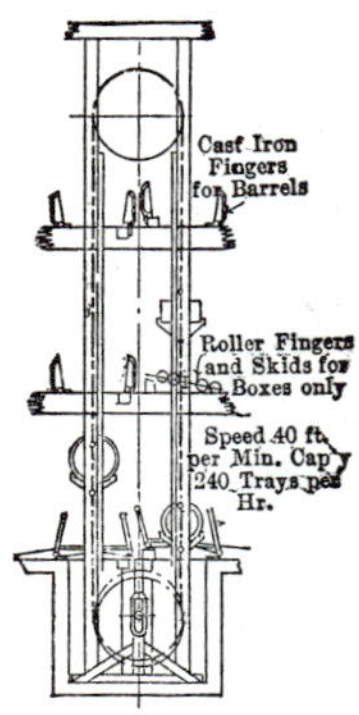

Fig. 26. Suspended tray elevator.

have trays so hung on pivots between two chains that center of gravity of loaded tray is well below points of suspension. These trays, when made with finger bottoms, will pick up packages from skids on the up run and deposit them on skids on the down run. Solid-bottom trays require loads to be put on or taken off by hand. Suspended tray elevators run at not over 30 ft per min. (4) Elevators with corner-hung trays (Fig. 27) are similar to (3) but the chains are not directly opposite each other and pivots are not in line. This makes tray more stable in passing around turns and on straight runs, but wheels are out of line and the drive is not so simple. (5) Tray elevators with selective discharge: Style A, long trays with fingered bottoms with one loading compartment for each discharge point; receiving fingers at each floor are set to register with com-

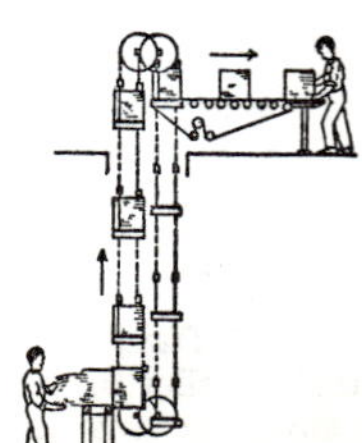

Fig. 27. Corner-hung tray elevator.

partment for that floor. Style B, fingered trays dumping forward on up run; tray guides have a throw switch at each floor to tilt the tray; switches being controlled by latches on tray which are set by an indicator at loading point.

Package elevators in these various styles are made by nearly all builders of conveying machinery. Much ingenuity has been spent on special forms. Elevators are made for trays of dishes, to deliver to or to take from roller conveyors, and for selective discharge.

PACKAGE CONVEYORS. (1) Packages of uniform size, not too heavy, can be moved in a trough by pushers which are attached to a single moving chain. (2) Packages can be carried on slats which slide on both runs. (3) Slat conveyors moving on roller chains will handle packages of any size or weight. Discharge is generally over the end, but movable plows can be used to deflect boxes sideways for intermediate discharge. (4) Belt conveyors are used in shops, assembly departments, stores, post offices, etc., for handling stacks of light package goods of all kinds. For this work the conveyor has straight tubular idlers, and the edges of the belt are supported to prevent sagging between idlers. Figure 28*A* shows the construction with steel guards used in department stores, the goods being delivered by chute from above. Figure 28*B* shows the construction where packages are tossed on the belt from below. For light work the belt may slide on a hard wood plank. (5) Packages of uniform size can be carried long distances and at any angle of slope by wheeled carriages attached to a single moving chain. (6) Inclined conveyors generally are used for handling and piling bags and bales. They have a belt to which cleats are attached, or else push bars mounted on two strands of chain.

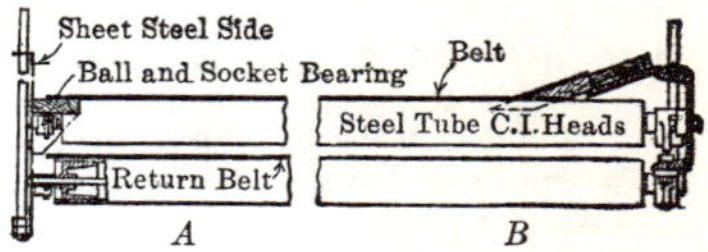

FIG. 28. Belt conveyor for packages.

Package conveyors are built by all makers of conveying machinery. For information concerning them and special conveyors to work with gravity roller systems, see catalogs of the several makers. Selective discharge from belt or roller conveyors is described by W. O. Hildreth, Ref. 14. For selective discharge actuated by photoelectric relay, see Ref. 15. The electrical apparatus is made by General Electric Company.

A novel form of portable flexible conveyor for goods in bales and sacks is the Twinveyor (Clark Tructractor Company, Battle Creek, Mich.). It comprises several 8-ft sections coupled end to end by flexible joints. Each section is made of two parallel tubes 4 in. in diameter geared together and driven by a 5-hp motor, with a blunt screw thread on the outside. As the screws revolve, the bag or bale resting on them acts as a nut and travels at 90 ft per min. The original installation is described by J. T. Buzzo, Ref. 16.

THE MONORAIL CHAIN CONVEYOR consists of a power-driven chain suspended from trolleys on a continuous track, usually overhead. The chain joint pins are in a vertical plane. It bends around sprocket wheels or nests of rollers, or it can bend around tracks curved in a vertical plane if the proper chain is used. Loads are carried on hooks, racks, trays, slings, etc., hung from the trolleys. Speed generally is less than 75 ft per min. The chain may be in one plane or several. It may serve several floors of one building or connect several buildings. Work can be done on suspended pieces as they travel along to dipping vats, baking ovens, etc. Monorail chain conveyors do work for which floor trucks and freight elevators otherwise would be required. They occupy no floor space, require little power, and provide a steady movement of parts from one operation to another and to final assembly, inspection, and shipment

Calculation of Monorail Chain Conveyor.* (1) Make a layout, and scale off amount of chain required. (2) Assume a chain speed and a spacing of loads, as determined by time required to travel a certain distance, or the number of pieces to be delivered per minute. (3) Sketch tray or carrier and make the horizontal and vertical turn so that carriers clear each other even if they swing. (See Fig. 29.) (4) From catalogs of makers select a trolley suitable for the load. These have been standardized. (5) As a first approximation, assume horsepower pull for a single-plane conveyor to be 5% of total moving weight (chains, trolleys, carriers, loads), about 7% for a conveyor with a few easy vertical curves, and 10% for one with many curves. (6) Select a chain strong enough for (5), and suited to the trolley. (7) Locate drive where sprocket wheel has at least a 90-degree arc of contact with chain; 180 degrees is better. The take-up should be where chain is under light load. Vertical turns work best if chain passing them is not under heavy tension. (8) For more accurate determination of chain tension and horsepower, calculate chain pull

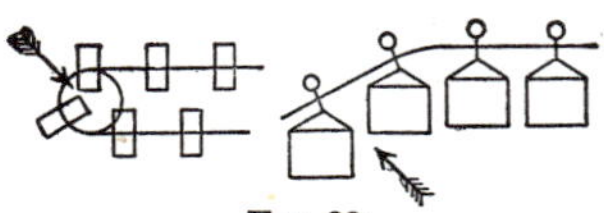

FIG. 29.

* Adapted from Link-Belt *Data Book*.

for each run and each turn. Thus, $C_H = WLX$; $C_C = YC$, where W = weight of moving parts, pounds per foot, empty or loaded as the case may be; L = length of run, feet; $X = 0.03$ for ball- or roller-bearing trolleys; $= 0.05$ for bronze-bushed trolleys; C_H = pull for each straight horizontal run, pounds; C_C = pull for each corner, take-up idler or roller turn in horizontal plane, pounds; C = chain pull at turn, pounds; $Y = 0.03$ for sprocket wheel or roller turn; $= 0.02$ for a traction wheel turn. The pull for a double curve in passing from one level to another is

$$C_V = WSX + CZ + HW$$

where W and X are as above; S = horizontal length of double curve, feet; H = height of double curve, feet, taken as (+) when travel is up, (−) when travel is down; C = chain pull at entrance to the curve, pounds; Z = friction factor due to radial pressure of trolley on curve; for ball or roller bearings $Z = 0.001$ per degree of inclination from the horizontal of tangent between curves; for bronze-bushed trolley wheels, $Z = 0.002$. (9) To algebraic sum $C_H + C_C + C_V$, add 3% for friction at drive wheel. Check this against strength of chain, assumed in (6). (10) The radial pressure on each trolley is $P = BC/R$, where B = trolley spacing, inches; R = radius, inches, to center of chain on curve; C = maximum chain pull at any curve, pounds. Check this with permissible trolley load. If too great, space trolleys closer or make radius longer. The longer the radius, the better the operating results.

Where conditions make it difficult or impossible to drive a chain by a sprocket wheel, an auxiliary drive acting on the straight run of chain may be used. The auxiliary chain is short and has pushers which engage links of the conveyor chain. The Webb system of using a number of these devices synchronized, and with automatic speed control, is described by H. C. Rundle, Ref. 17.

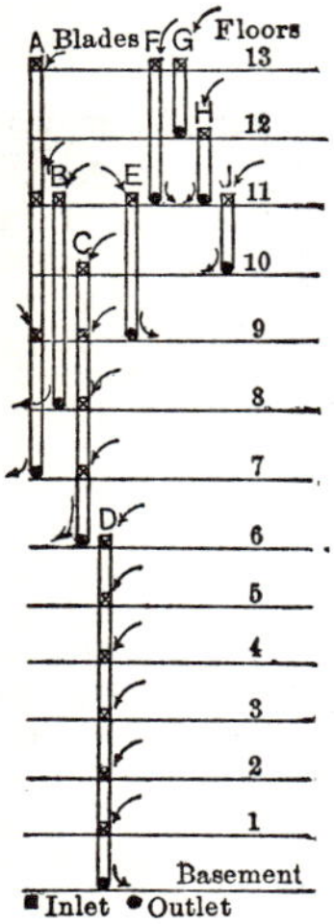

FIG. 30. Scheme of spiral conveyor system.

ROLLER CONVEYORS will readily convey down slopes of 1 1/2 to 5%, boxes, packages, cans, bottles, lumber, steel products, almost anything that has a smooth face or side. The goods travel over rollers of steel tubing fitted with ball bearings. Capacities range from a few pounds up to 4 tons per roller. Some conveyors have spaced wheels instead of rollers. Rollers are assembled in straight sections, curves, switches, Y-assemblies, and roller spirals. They can be set on floors supported from walls or hung from ceilings. Roller conveyors have been highly developed and have a wide field of usefulness. For a description of an elaborate system of Logan Company conveyors with a new form of switch, see C. K. Pevear, Ref. 18.

SPIRAL CHUTES are made by manufacturers of roller conveyors. Some are made with roller beds, others have steel plate bottoms. They carry boxes, packages, and even some kinds of unwrapped goods from upper to lower floors. Some are made with two or more spirals. They provide means for delivering different classes of goods to separate unloading stations at the lowest floor of a building, or goods can be passed from one floor to another on one spiral without interfering with goods on other spirals. Figure 30 shows diagrammatically a chute with six spirals serving thirteen floors of a clothing factory. The letters A, B, C, D, E, F indicate separate spirals, and the arrows show inlets and outlets. G, H, and J are straight chutes in the system, not part of the spiral chutes, but all combine to provide a gravity flow for raw materials from the thirteenth and eleventh floors to manufacturing floors below, and finally as finished goods to the shipping room in the basement.

PROCESSING OVENS. Monorail and other forms of chain conveyors are used to carry cores, lithograph plates, and enameled metal parts through drying and baking ovens, and to carry metal parts through ovens in which they are dipped and drained. Other forms of processing ovens are equipped with roller beds for handling cores and other parts on pallets. Some carry the work on trays suspended from two chains.

The conveyor tunnel kiln conveys tile, refractory brick, and similar ceramic materials through a firing kiln by means of rocking frames which support the materials. In other forms of the kiln, castings, steel plate, or other heavy materials are annealed while they move through the kiln on a train of cars propelled by a ram.

REFERENCES

1. F. D. Campbell, *Trans. Am. Soc. Mech. Engrs.*, 1931, MH-53-1.
2. W. W. Sayers, *Trans. Am. Soc. Mech. Engrs.*, 1931, MH-53-8.
3. F. V. Hetzel and R. K. Albright, *Belt Conveyors and Belt Elevators*, 3rd ed., John Wiley and Sons, 1941.
4. A. F. Beasley, *Eng. Mining J.*, March 24, 1930.

5. R. Peele and J. A. Church, *Mining Engineers' Handbook*, 3rd ed., John Wiley and Sons, 1941.
6. *Coal Age*, Dec. 25, 1924.
7. Goodyear's 1930 *Handbook.*
8. Goodrich's *Manual of Belt Practice*, No. 2800.
9. *Trans. Am. Soc. Mech. Engrs.*, 1930, MH-52-10.
10. *Eng. News-Record*, Aug. 1, 1935.
11. *Trans. Am. Soc. Mech. Engrs.*, 1929, MH-51-10.
12. *U. S. Dept. Labor Bull.* 433. 1927.
13. *Trans. Am. Soc. Mech. Engrs.*, 1928, MH-50-12.
14. W. O. Hildreth, *Trans. Am. Soc. Mech. Engrs.*, 1929, MH-51-8.
15. *Materials Handling and Distribution*, May 1931, and other issues in 1931 and 1932.
16. J. T. Buzzo, *Trans. Am. Soc. Mech. Engrs.*, 1928, MH-50-2.
17. H. C. Rundle, *Trans. Am. Soc. Mech. Engrs.*, 1931, MH-53-7.
18. C. K. Pevear, *Trans. Am. Soc. Mech. Engrs.*, 1930, MH-52-6.

VEHICLES

20. TIERING MACHINES, LIFT TRUCKS, SKIDS, AND PALLETS

TIERING MACHINES (portable elevators, stackers) for piling case goods, barrels, bales, rolls of paper, etc., are made in a variety of styles and sizes. These machines also are used to handle dies to the level of press beds, put up and take down overhead motors and shafting, build riggings and platforms for painting, cleaning, and repair work, etc. For light service and occasional use, they are operated by hand crank, fitted with a safety brake to prevent it from flying back and injuring the operator. For regular service, an electric-driven machine is more useful.

Figure 1 shows a hand-powered stacker for lifts from 4 to 12 ft. The frame is hinged so that the machine can pass through a doorway. Smaller hand-powered and electric machines are made with telescoping frames for 2000-lb loads and 12-ft lifts. Most of the electric machines are driven by extension cables to a power source, but models are also available which are self-contained power units, driven from electric storage batteries (Fig. 2). These have advantages over the extension cable type for portability and maneuverability.

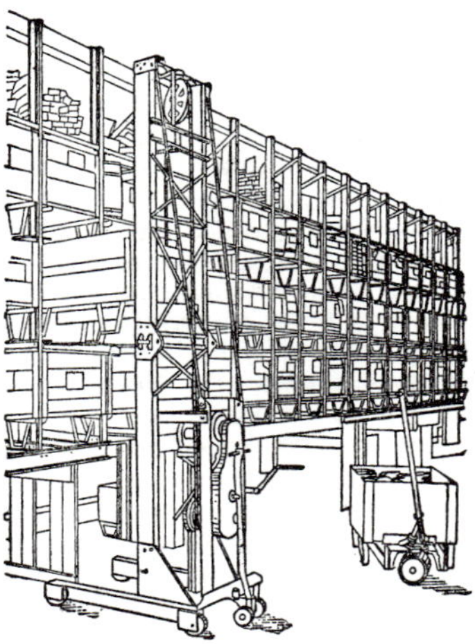

FIG. 1. Hand-powered stacker.

FIG. 2. Yale battery-powered industrial tilting-fork portable elevator.

One manufacturer builds these units with a turn-table in the base (Fig. 3), which permits the platform or uprights to be turned in the complete 360-degree circle. This type of unit has certain advantages in tight locations in narrow aisles.

The hinge type, or telescoping lift type with single uprights, is generally limited to a lift of about 18 ft, and 2000-lb capacity; but other units have been built with multiple uprights, the platform riding between end supports similar to a freight elevator.

Units have also been built in what is known as the box type, that is, one box frame telescoping inside the other like a set of toy blocks; and these units have been built to elevate to a height of 96 ft for window cleaning, building maintenance, repairs on aircraft motors, etc.

SKIDS AND PALLETS are related in principle but are of different construction. That is, they are accessories for lift truck operations, and involve the principle of a movable floor, as they provide sectional areas to pile materials on so that these materials can be handled in unit loads.

Skids. The skid platform, or skid (Fig. 4), is a flat rectangular platform, made of wood or metal, and elevated a short distance above the floor, resting on longitudinal members, skid legs or casters, or a combination of legs and casters. There are several types of skid platforms, built in standard widths from 24 to 54 in., and in standard lengths from 24 to 96 in., both in light- and heavy-duty types. The most common type has steel legs and a hardwood surface. Skid platforms are manufactured in two standard clearance heights, from floor level to inside of skid platform, namely 8 and 12 in. This type of skid can be handled either by hand lift truck or power lift truck. The minimum horizontal width between the inside of the runner legs or other supports is 29 in. when the skid is to be handled by power lift trucks. Platform skids are generally classified under the following types:

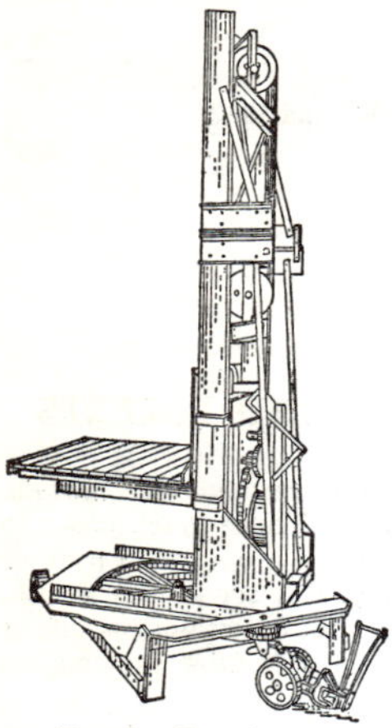

Fig. 3. Revolvator.

1. Dead skid, which means a skid with fixed legs or runners.
2. Semi-live skid, which means a skid with two fixed legs at one end, and two rigid casters or wheels at the other end (Fig. 5).
3. Live skid, which means a skid equipped either with four rigid casters, one at each corner, or with two swivel casters at one end, and two rigid casters at the other.

The construction of the skid depends on the nature of the load. The capacity of the skid depends on the strength and stiffness of the top platform and the type of supporting members. If the loads are evenly distributed over the entire surface of the platform, the skid has more capacity than if the weight is centrally concentrated. Too concentrated a load may cause the platform to sag. Excessively long or wide loads require additional support for the skids on the sides or the ends. Skids must be of the proper capacity for the loads to be handled. An undercapacity skid platform is unsatisfactory in service, is frequently in the repair shop, and has a short life.

Since the underside of the platform skid must be left clear for the entry of the lift truck, the legs or support members must be located near the outer ends, and must also be close to the ends to prevent tipping. When not supported by the lift truck, the load is carried on the top of the platform skid, the side angles and the supporting legs, casters or other members. For that reason, the capacity of a very long or very wide platform is much less than that of a short or narrow one, even though the angles and top members are of the same thickness. Thus the capacity of a skid platform varies inversely with its length and width. Also, because of the construction of the platform, uneven or springy floors require stronger skids, as the weight of the load must at times be carried on two or three legs. This feature makes it difficult to stack skid loads one on top of the other, unless they are placed in racks or with sufficient dunnage to spread the load over the top of the load beneath. Many varieties of special platform superstructures are built by manufacturers to meet specific operating conditions. This includes such features as superimposed removable boxes, fixed shelves and racks, removable vertical posts at sides or ends, and special attachments for particular commodities.

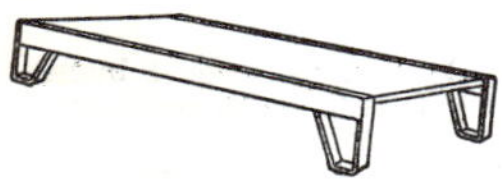

Fig. 4. Steel frame skid.

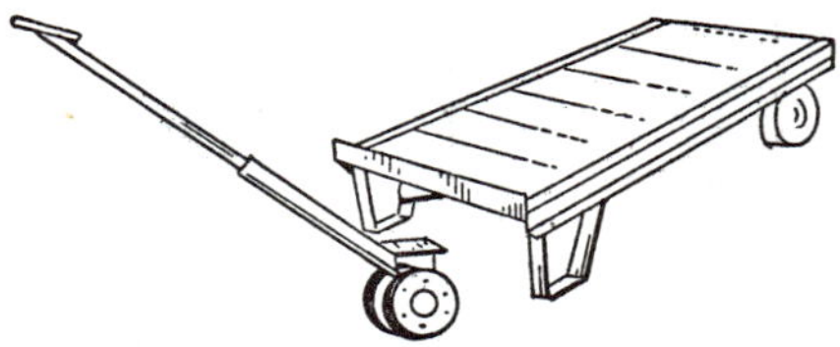

Fig. 5. Semi-live skid.

Skids can also be built that do not require a lift truck. These are generally constructed with two wheels in the back and two legs in the front. A separate handle fitted with wheels is thrust under the leg end of the platform (Fig. 5) and either engages a pin or a plate mounted on the front end of the platform. When the handle is pushed down, the front of the platform is lifted. The weight is taken off the legs, and the platform is ready to roll. However, in the majority of installations, some type of lift truck is used for handling skids.

Pallets are different from skids and can be constructed of a variety of materials, such as metal, paper, soft and hard woods, and are either open faced or double faced. The term *face* means the surfaces, top and bottom.

The most common type is of hardwood construction, with three stringers. The top and bottom boards are full lengths, and spaces are provided in the lower deck so that the pallet can be handled by pallet trucks as well as power fork trucks.

Pallets have been made with steel stringers and wooden decks, and also of all steel construction, especially for handling hot materials. They are classified under three general types, regardless of the material used in their construction:

1. Open-faced pallet with stringers mounted to an upper deck only, and with no lower deck.
2. Double-faced pallets (Fig. 6), constructed with stringers between an upper and a lower deck, with openings in the lower deck to provide space for the drop wheels of the hand pallet truck; generally approximately 3 1/2 to 3 3/4 in. is required for clearance between the upper and lower deck for operation with hand pallet trucks or power fork trucks.
3. Box pallets, constructed of a framework with one or more open sides and mounted on standard pallets so as to provide for handling of small, odd-size, or crushable items; the box frame is to provide supports for multiple stacking.

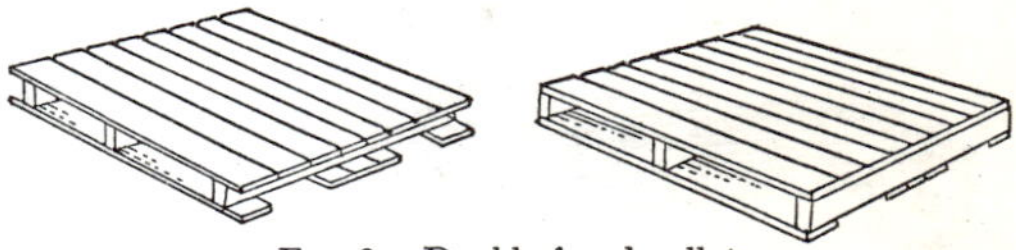

FIG. 6. Double-faced pallets.

Pallets are made in a wide variety of sizes. In stating the measurements, the length of the pallet is always the dimension in the direction of the stringers or the distance required for the length of the forks or trucks, even though this dimension might be shorter than a width dimension, which is the length in the direction across or at right angles to the truck center line.

In building pallets a number of handling operations should be taken into consideration so as to locate the stringers properly, as these can be placed at the extreme edge of the top and bottom deck. If, however, the pallets are to be handled by slings from cranes, ships' gear, or other similar handling, the stringers should be set inboard approximately 3 1/2 in. from each side of the end of the top and bottom deck.

The double-faced pallet (Fig. 6) has been found to be the best unit to use for most commodities when they are being placed in storage, because it provides a good load distribution on the lower tiers.

Various attachments for standard pallets and special pallets can be built, as for skid platforms.

The advantages of handling goods on skid platforms or pallets include saving in time and labor, avoidance of loss or damage from piecemeal handling, and facility in handling and storing in large units without breaking bulk. Indirect advantages are simplified cost and inventory accounting, and greater mobility of stores, supplies and materials.

The paper and other industries are now shipping goods on skids or pallets, in what are known as unit loads. Advantages of unit load shipments include all the advantages of intra-plant unit load handling, plus savings in packing material, labor cost of handling and unloading, and elimination of breaking bulk in rehandling. Savings actually may range from 25 to 90%.

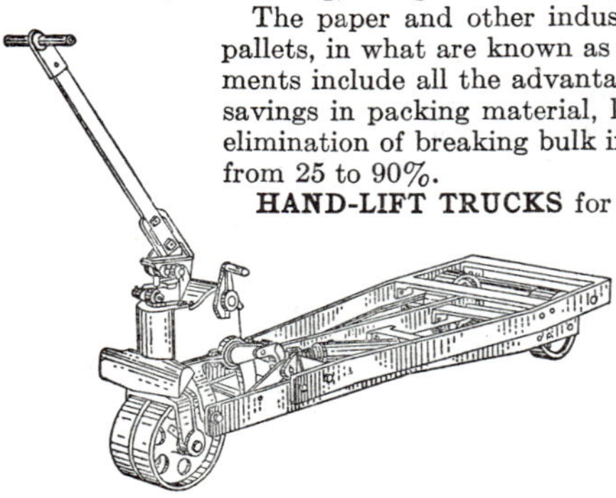

FIG. 7. Hydraulic-lift truck.

HAND-LIFT TRUCKS for handling skids are made in three or four styles, depending on the manufacturer. Some of these units have hydraulic lifts operated by a pull handle, which raise the skid platform to clear the floor 3 in. or more (Fig. 7). In some instances, this is done with step-by-step ratchet lift operated by the handle (Fig. 8); some are known as foot-lift models, worked by a foot treadle.

The principle of the truck is the same, but they differ chiefly in lift and release mechanisms. These trucks are generally constructed with forward wheels, connected to a steering handle, which is attached to a main frame, in which the rear wheels are mounted. A second framework is imposed upon the bottom framework, and, through a system of linkage, it is possible to raise the skid and its load from the floor.

When fully elevated, this lifting frame assumes a fixed position and remains rigidly locked in place until released by the operator. In order to avoid shock, the lowering of the load is generally accomplished through a powerful hydraulic check.

HAND-PALLET TRUCKS operate on the same principle as the hand-lift truck, being constructed with large forward wheels connected to a steering handle, and attached to a main frame, but the rear wheels are small and are depressed in the main frame (Fig. 9). At the end of the main frame, near the back, additional small wheels are mounted which assist the operator in propelling the fork end of the truck into the pallet, in the opening between the stringers.

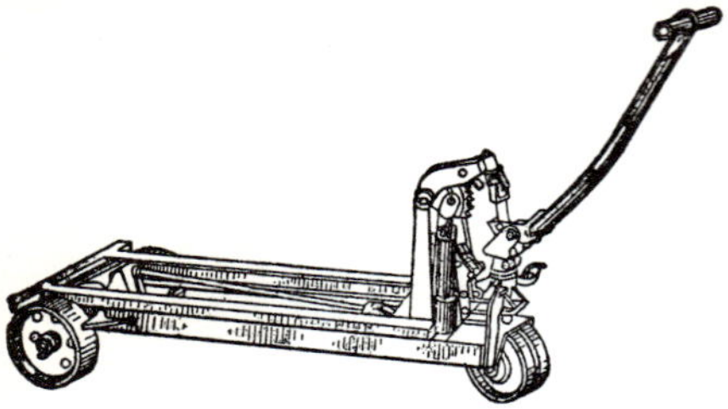

FIG. 8. Jack-lift truck.

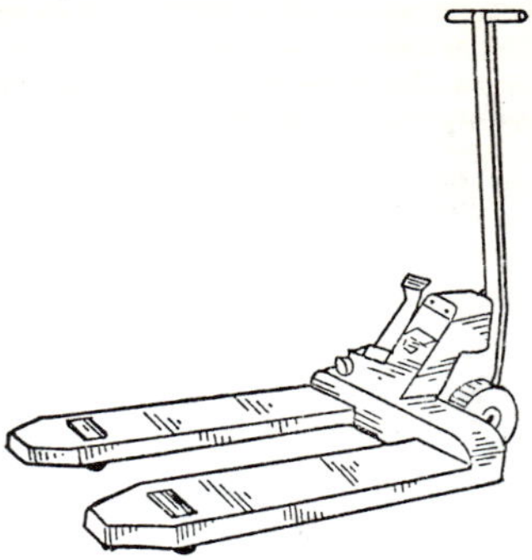

FIG. 9. Hand-pallet truck with hydraulic lift.

With double-faced pallets, great care must be taken to space the openings in the bottom deck so that these small wheels can pass through these openings, in order to raise the pallet off the floor for transportation. This lifting is accomplished as in the hand-pallet truck, and the lowering of the load is also accomplished in the same way.

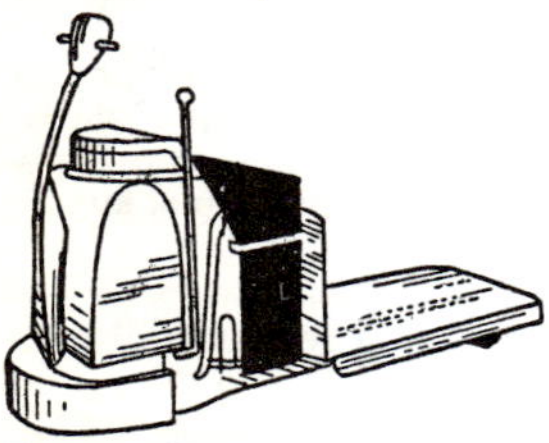

FIG. 10. Hand-controlled power-operated lift truck.

In addition to hand-propelled and hand-operated lift trucks and pallet trucks, there are now trucks of this same type with electric power to assist in the transportation operation, and in some instances for elevating the load (Fig. 10). These trucks are constructed with practically the same type of main frame as the hand-pallet trucks, but the front wheels are driven by an electric motor which receives its current from a storage battery located directly back of the steering handle. By using push-button control type electric contacts, mounted in the steering handle, the operator is able to control the forward and backward movement of the truck by applying power as desired.

21. INDUSTRIAL TRUCKS AND TRACTORS

TRUCKS WITH FIXED PLATFORMS are generally called *burden-carrying trucks.* They are power operated and can be built with either a high or a low platform (Fig. 11). They load and unload manually or by auxiliary equipment. They are built in different capacities. Some are constructed with drives at both ends, and with a low-slung platform in the middle section such as is used in baggage service in railroad stations; others are built in the high-type platform, with the platform 20 to 24 in. above the ground, with the wheels arranged either for two-wheel steer or four-wheel steer. In the majority of cases, they are electric storage battery operated.

FIG. 11. Electric platform truck.

Because they have a fixed platform, it is necessary to load and unload these trucks either manually or with auxiliary equipment, such as a crane mounted on the truck itself, or operating beside it. They are not as efficient for general handling as the elevating platform truck or a fork truck.

TRUCKS WITH LIFT PLATFORMS, generally known as *elevating platform trucks,* are of two sorts: low-lift elevating platform truck and high-lift elevating platform truck. They are made in capacities of 2000 to 20,000 lb.

In low-lift trucks (Fig. 12) a cam or link motion in the base frame raises the platform to give floor clearance, only sufficiently to raise the skid from the floor for transportation

In high-lift trucks (Fig. 13) the platform is raised by cables or chain, by a screw working in a movable nut, or by hydraulic cylinders. These trucks are designed with platforms to handle standard skids as well as special or higher skids.

The wheels under the platform of these low-lift and high-lift trucks are seldom over 10 1/2 in. in diameter, and roughness and unevenness of floor affect them more than trucks with larger front wheels.

In the high-lift elevating type of platform, the front wheels or supporting wheels sometimes offer interference when an attempt is made to place the raised load on to platforms,

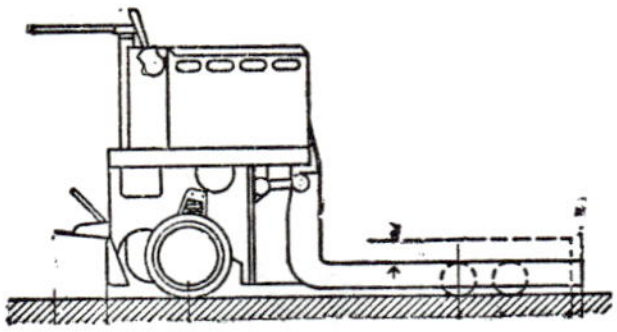

FIG. 12. 2-ton electric elevating platform lift truck.

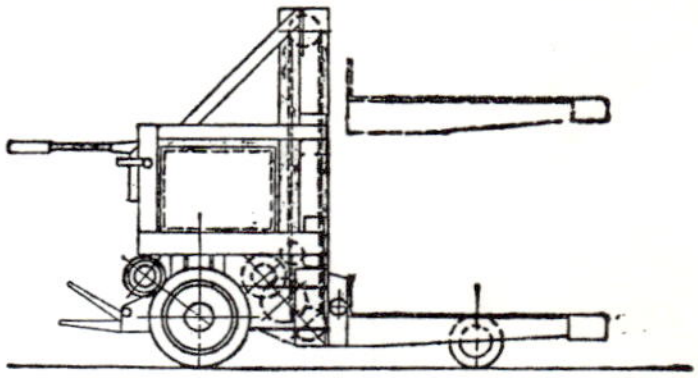

FIG. 13. 3-ton electric high-lift elevating platform truck.

trucks, trailers, etc. However, these front wheels make it possible to have a short truck, as little or no counterweight is required, since the load on the uprights is partially supported by these front wheels.

FORK-LIFT TRUCKS are a modification of the low-lift and high-lift elevating platform truck, except that these trucks are built on a cantilever principle, and therefore require considerable counterweight back of the drive wheels, in order to compensate for the load cantilevered in front of the uprights (Fig. 14).

Fork-lift trucks are of either the telescoping or nontelescoping type, are generally provided with a forward and backward tilt, and are built in capacities of 2000 to 30,000 lb. The load is supported in front of the vertical uprights, and the tines or forks project either into pallets which support the loads, or directly under unit loads for stacking or handling.

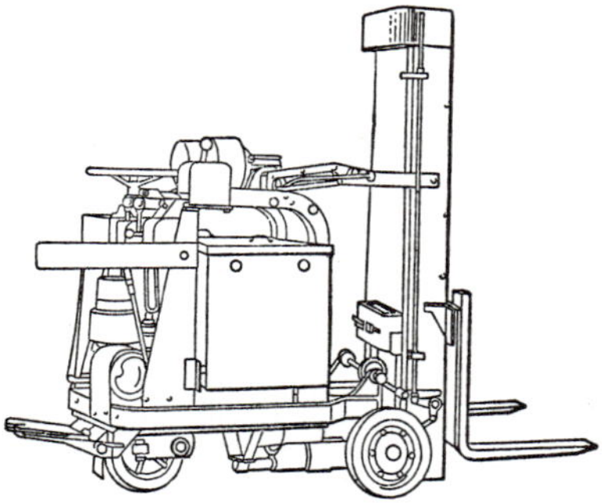

FIG. 14. Fork type lift truck.

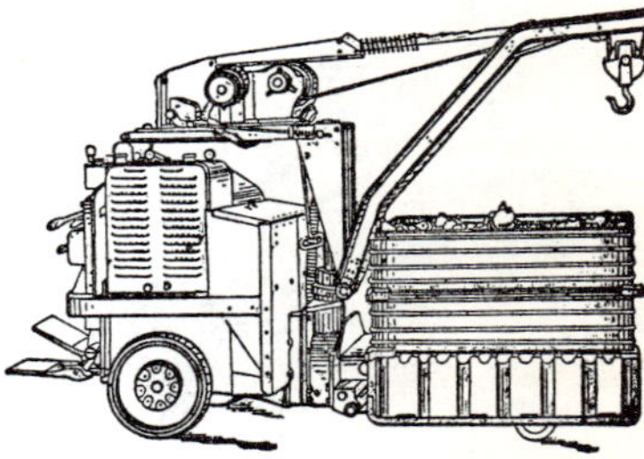

FIG. 15. Electric lift truck with crane.

Fork trucks are sometimes called chisel trucks, if they are equipped with very thin forks or steel plates for chiseling under a load, without the need of supporting the load on a pallet.

They can also be equipped with rams for picking up cylindrical materials, and with a wide variety of attachments for clamping the load, for turning it over, for tipping it forward, etc., in which cases they are sometimes given trade names or called by the name of the accessory, such as a *ram truck*. The proper term for this type of truck is a cantilever truck.

OTHER TYPES OF INDUSTRIAL TRUCKS are rigid boom, industrial cranes, swinging boom, etc. These are made up as special units or sometimes mounted on low-lift industrial trucks or burden-carrying trucks (Fig. 15). It is better to use a distinct unit for this type of operation rather than a combination unit.

The rigid boom industrial crane is small and compact, mounted on a tractor or a specially designed chassis with the boom suspended in front of the power unit and with sufficient counterweight to keep the unit from tipping (Fig. 16).

In some models, the boom can be topped to different angles into fixed positions, but not while the load is being carried. In other models, the boom is power topped and can be raised or lowered with the load suspended from the hook.

The swinging boom industrial type of crane is constructed in the same way, and in some models the boom is mounted on pedestals above the power unit, and counterbalanced with weights so that it can swing in a full 360 degrees.

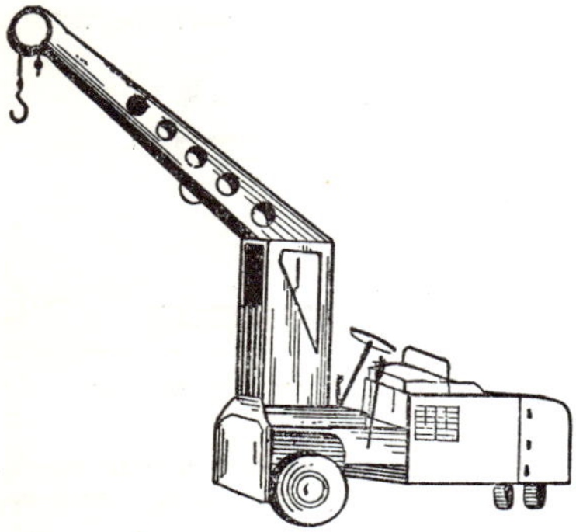

FIG. 16. Rigid boom industrial crane. (Courtesy, Hyster Co.)

FIG. 17. Swinging boom industrial crane.

In other models, the boom is mounted over the drive wheels, and is attached to a vertical mast, so that it can swing up to 300 degrees (Fig. 17). Suitable counterweights are provided on the side of the tractor units to prevent tipping.

The various types of industrial trucks that are power operated are generally powered by electric storage batteries, or gasoline-electric generator units, or straight gasoline power.

Some manufacturers mount all their models on solid rubber tires; other manufacturers mount all their models on pneumatic tires. Some manufacturers have a combination of pneumatic tires and solid rubber tires. The pneumatics have distinct advantages when used over rough ground, uneven surfaces, and to eliminate vibration on the equipment and the load.

FIG. 18. Track type tractor. (Courtesy, Caterpillar Tractor Co.)

INDUSTRIAL TRACTORS are built in a variety of sizes and with different wheel combinations. The capacity is figured by drawbar pull. Some industrial tractors are built with crawler-type treads. On outside work, the usual cleat type crawler is acceptable, but for use on finished roads and floors, crawlers equipped with rubber pads to prevent damage to the surface are required.

Tractors can be broken down into three classifications:

1. Crawler tractors, which are tractors without wheels and supported on a self-laying crawler type of tread (Fig. 18).
2. Three-point contact tractors, which have three points of contact with the road surface, the drive being through the two wheels at the rear and the steering by the front wheels (Fig. 19). The front wheels can be either a single or a dual wheel, placed close together.
3. Four-point contact tractors, which are driven by the rear wheels, and have two steering wheels, spaced wide apart, at the front, similar to an automobile (Fig. 20).

The size and type of tire and the weight of the unit have a direct bearing on the tractive effort, and the drawbar pull or capacity of the tractor.

These tractors can be operated either by electric-storage battery, electric-generator sets, or straight gasoline. Tractors are recommended where the quantity of materials handled is enough to warrant the use of trains of trailers, and where movements of trains can be arranged according to a time schedule or a route schedule.

The selection of the size of tractor will depend on the trailers to be hauled, the length of the train, the grades to be negotiated, the type of service on which they must operate, and the speed of operation.

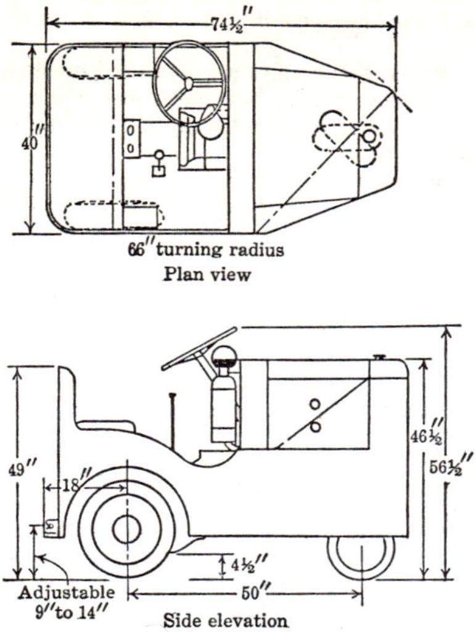

FIG. 19. Three-wheel industrial tractor (Courtesy, W. F. Hebard and Co.)

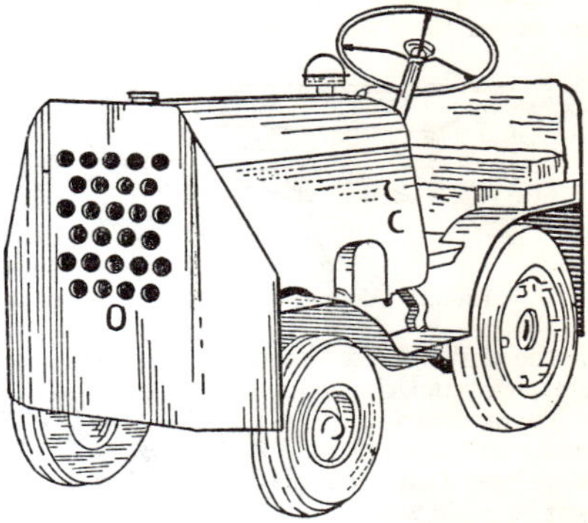

FIG. 20. Four-wheel industrial tractor. (Courtesy, W. F. Hebard and Co.)

22. INDUSTRIAL TRAILERS

Industrial trailers, which are used in combination with industrial tractors, are made in four types: four-wheel steer; caster; fifth wheel (wagon); and balance type. Table 1 can be used for selecting trailers for particular operations.

Table 1. Cars and Trailers for Different Services

Moved by	Conditions and Loads	Style of Car (*See* Note) Choice 1st	2d	3d
Power	Accurate trailing	A	B	C
Power	Accurate trailing and reverse	A	D	....
Power	Good or rough runways, heavy loads on 2 or 3 trailers	C	A	D
Power	Good runways, heavy loads on 4 or more trailers	B	A	C
Power	Rough runways or dirt floors, heavy loads on 4 or more trailers	C	A	....
Power	Rough runways or dirt floors, light loads	A	C	....
Power	In conjunction with considerable hand trucking	B	C	D
Hand	Good runways, light loads	B	D	C
Hand	Poor runways, dirt floors, heavy loads	C	....	....

Note. A—4-wheel-steer trailer; B—Caster trailer or truck; C—Fifth-wheel trailer or truck (wagon-type); D—Balanced trailer or truck. On right-angle turns in 7 ft aisles, caster trailers will creep outward 3 or 4 in.; fifth-wheel, 7 to 10 in.; balanced trailers with cross-chain connections, 7 to 10 in.; 4-wheel steer trailers will not creep. The larger the radius of turn, the less the creep.

Four-wheel steer trailer (Fig. 21) is a unit with all wheels of the same type, with front and rear axles mounted on fifth wheels, or the wheels mounted on knuckle steers, and all four wheels coupled together through cross reaches or linkage so that they steer simultaneously. On these trailers, the drawbar is removable so that it can be connected to either end of the trailer.

A four-wheel steer trailer is different from a reversible trailer, because on a four-wheel steer trailer, the front and rear axles are linked together at all times, and all four wheels steer simultaneously. This four-wheel steer makes for accurate trailing, but requires a larger radius for turning than the caster type.

When the trailer-train is pulled away from a platform or curb, the train must be steered gradually so as not to bind the wheels against the platform or curb obstruction. This difficulty is encountered because of the tendency for the second trailer, etc., to steer sideways from the trailer in front.

A reversing trailer has four wheels of the same size and the axles are mounted on fifth wheels at each end of the trailer. The drawbars are removable, but the wheels are not connected together; and, in operation, one end of the trailer is locked in position to make that pair of wheels rigid.

Caster trailers (Fig. 22) are constructed with the rear wheels on rigid axles, and these are larger than the steering wheels, which are mounted in swivel casters at the front end

Fig. 21. Four-wheel steer trailer.

Fig. 22. Caster-type trailer with automatic coupler.

of the trailer. On this type of trailer, the drawbar is mounted at the front end of the platform, and the couplers on the rear end of the platform.

The average caster-type industrial trailer is built in capacities of 2 to 3 tons, and generally with a platform 3 ft wide by 6 ft long, of wood and steel construction, although some have been built with platforms 4 ft wide and 8 ft long, of all steel plate construction. The rear or rigid axle is generally located about one-third forward of the platform, and thus the large rear wheels carry about two-thirds of the load. This relieves the weight on the caster wheels, which are located near the front of the trailer, one on each side, making it easier for the casters to swivel under the load. This type of trailer moves easily, steers well, can be maneuvered in narrow aisles, and can be moved by hand if necessary. They are generally equipped with automatic couplers but can be equipped with hook and eye type or cross chain type, if desired.

Fifth-wheel or wagon-type trailers (Fig. 23) are used for heavy loads on rough surfaces, and they are not reversible. They are generally constructed with all four wheels of the same size, although in some instances, the rear wheels are slightly larger than the front or steering wheels. The rear wheels are mounted on rigid axles, and the front wheels are mounted on an axle attached to a fifth wheel steering arrangement, with the drawbar attached directly to the front axle and the fifth wheel. The steering unit is mounted on a single king pin with a large bearing plate, and in some instances, these are made with an oscillating arrangement to allow the front wheels to conform to the unevenness of road surfaces.

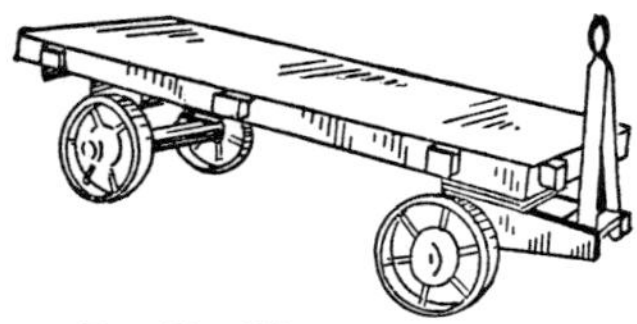

Fig. 23. Wagon-type trailer.

Balance-type trailers are like warehouse hand trucks, and carry the load on two large wheels mounted in the middle of the trailer; they have single or double casters at each end. This type of trailer turns easily by hand but does not trail well at high speeds.

On smooth new floors, trucks and trailers with steel wheels start easier and roll easier than rubber-tired wheels; but on most floors, on runways, ramps, and paved yards, the rubber-tired, canvas cushion wheels or pneumatic tired wheels have the advantage. The pull is less, goods are not shaken or damaged, wear and tear on the truck is less, noise is reduced, and damage to floor is eliminated.

In every case, trailers should be equipped with anti-friction roller bearings, and where dust or dirt is prevalent, these bearings should be protected with dust seals and should be fully enclosed in order to reduce wear, tear, and maintenance.

It is important to consider the height of the platform from the roadway so as to eliminate excessive lifting, but care should be exercised to see that the wheels are not too small for the service required.

Trailers, like pallets and skids, can be equipped with dump bodies, removable bodies, special rack bodies, etc., and these will be found to have advantages for handling specific materials.

CONDITIONS OF FLOORS AND RUNWAYS. The tractive resistance of trailers is affected by the type of floors. Table 2 gives the resistance in pounds per ton for various types of road surfaces. For each 1% of upgrade, add 20 lb per ton.

Table 2. Tractive Resistance of Various Surfaces

(*Reprinted by permission from Materials Handling Cyclopedia, Simmons-Boardman Publishing Corp.*, 1921, p. 545)

Type of Road Surface	Resistance, lb per ton
Smooth brick	30 to 50
Concrete	28 to 40
Poor concrete	45 to 65
Granite blocks	50 to 60
Wood blocks	30 to 50
Gravel road, good condition	75 to 85
Clay	200 to 400
Wood planking	35 to 50
Wood planking, sticky surface	50 to 60

Tests made on trailers fitted with roller bearings and solid rubber-tired wheels show that power required to keep them moving varies greatly. Incidental disadvantages of poor floors and runways are:

1. Tractor operator will run slowly to prevent goods from being jolted off trailers.
2. Cost of repair and maintenance of equipment is increased.
3. Labor cost due to delays and stoppage is increased.

On rough floors, the advantages of anti-friction bearings tend to disappear, especially at speeds over 3 miles per hour.

It has been found on good floors that the rolling resistance of 8-in. wheels fitted with roller bearings is four times that of 20-in. wheels.

TYPE OF POWER FOR INDUSTRIAL TRUCKS AND TRACTORS. There has been considerable discussion of the type of power to be used for industrial trucks and tractors, without any general agreements. In arriving at a formula for estimating these costs for a specific installation, it is necessary to take into consideration the following factors.

For electric trucks and tractors, consider:

A. Depreciation of the electric truck.
B. Depreciation of storage batteries for that truck.
C. Depreciation of charging equipment.
D. Interest, maintenance, and storage charges.
E. Electric current for charging batteries.
F. Labor and cost of space, for charging batteries.
G. Maintenance and repairs of truck.

On gasoline units, consider:

A. Depreciation of the gasoline truck.
B. Interest, maintenance, and storage charges.
C. Maintenance and repairs of truck.
D. Gasoline and oil.

These costs will vary considerably, depending on the number of hours per day that the equipment is required to operate. The factor of depreciation can be made to vary because electric trucks can be shown to last 25 years, whereas gasoline trucks will have a life of approximately 10 years. However, it is best to figure the truck life on the basis of 5 years minimum and 10 years maximum, in both cases, as these trucks depreciate rapidly by obsolescence due to change in design and improvements in operational function.

It is claimed for electric trucks that:

1. They are quieter in operation.
2. They do not give off fumes.
3. They are therefore more advantageous for working in close and confined quarters.

It is claimed for the gasoline trucks and tractors that:

1. They can work 24 hr a day and take a fresh supply of fuel in 5 min or less.
2. The torque of the engine and the use of the clutch exerts a strong pull to start heavy loads or get over bad places in floors or paving.
3. They can go almost anywhere, are self-contained, and independent of power lines and charging stations.
4. They can travel over streets, public roads, bad floors, rough paving.

The ready-power gas-electric unit is a self-contained power plant for use on tractors and trucks designed for storage-battery power, where time cannot be spared for recharging. It is an automobile-type gasoline engine with automatic governor and generator armature instead of a flywheel. It will deliver 120 amp at 36 volts, or 90 amp at 48 volts. It weighs approximately 1000 lb and fits in the space required for a storage battery.

PROBABLE SAVINGS BY USE OF INDUSTRIAL TRUCK OR TRACTOR SYSTEMS. It is difficult to determine mathematically whether or not the use of any system of mechanical handling will be profitable. The Coes formula (Ref. 1) gives various debit and credit items which should be considered. If proper values are assigned, the justifiable investment and the probable profit based on the assumptions will be given by the formula. It may happen, however, that indirect savings will determine whether or not a materials-handling system will pay a profit. C. B. Crockett and H. J. Payne (Ref. 2) give explicit data on operation of electric industrial trucks and tractors in twenty plants. All show saving of time and money. In some plants, indirect and incidental advantages are most important. For examples of the use of the Coes formula, see Ref. 3. See also Ref. 4. *Materials Handling and Distribution* in March 1931, described a case where eleven tractors and twenty-three trailers, costing $27,000, saved $48,000 per year. The plant covered 40 acres, and all haulage was done on a time schedule. For other cases, see a list of fifty installations in the same magazine, Jan. 1930.

INDUSTRIAL RAILWAYS. For industrial plant railways, the usual gage is 24 in., although 21 1/2-in. gage is used in many places. Rails range from 12 to 25 lb per yard. Mill railroad is usually the standard 56-in. gage (4 ft 8 1/2 in.). This type of railroad, with fixed tracks, switches, and turntables, has been replaced in a number of industrial plants by industrial trailers, but they are still used extensively in mines, quarries, and on specific industrial operations.

In some instances, portable rail sections are used; they are generally made up in 15-ft sections, 24-in. gages.

Various types of locomotives are available, some using overhead wires, some third rails, others storage batteries. For mine work, the first are in most general use. They are low, four- or six-wheel structures, weighing 5 to 35 tons, for track gages as narrow as 18 in., and with one or two trolley poles. Two are used when rails are not bonded for the return circuit. Twenty-five-volt series motors generally are used. Haulage capacity, rated as drawbar pull, usually is taken as one-fifth of the weight of the locomotive, without excessive heating in the motor. The rating may be reduced to one-half the standard rating for steady work on long hauls or steep grades. The pull for mine cars on straight level track may be taken at 30 lb per ton; for grades, at 20 lb per ton for each 1% of grade. Usually speeds are 6 to 8 miles per hour.

These are the advantages of using storage-battery locomotives: It is unnecessary to bond rails or remove rock to make legal clearances for trolley wires. They are safe in gaseous mines, can place cars at the working face, remove loaded cars, and go wherever a track can be laid. If batteries are recharged at night, the day load of the power plant is decreased. For mine service, they are made in sizes up to 10 tons; for shifting railroad cars, size may be 50 tons.

Combination locomotives, fitted with both storage batteries and current collectors, also are built. Heavy-duty locomotives with diesel engine, generator plants, storage batteries, and current collectors for railroad switching works have been constructed.

Some operators prefer straight gasoline or diesel locomotives for certain types of haulage, as they are independent of power house, have large radius of travel, and as compared with steam locomotives, require less attention and use fuel only when working. (See Fig. 24.)

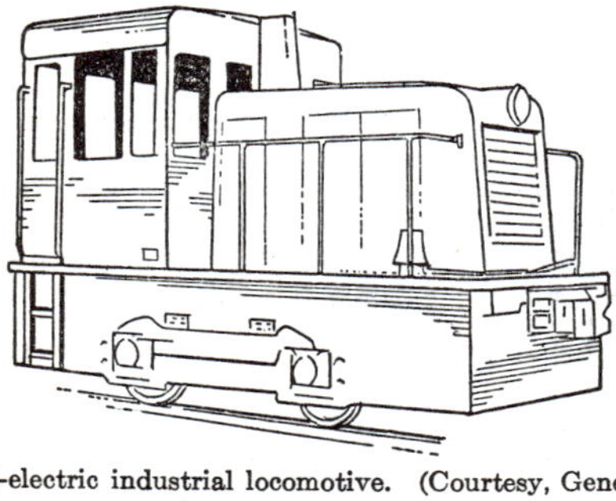

Fig. 24. Diesel-electric industrial locomotive. (Courtesy, General Electric Co.)

DUMPSTERS. This equipment, generally called a Dempster Dumpster, consists of demountable containers of various sizes and designs for operating on a special hoisting and dumping device, mounted on a motor truck chassis (Fig. 25), and recently installed on a tractor unit. The containers can be left in convenient locations for loading and unloading. It is a one-man operation to pick up the container, load it on to the truck, and transport it to the next operation. They have been used extensively in industrial plants for handling refuse, for specific handling operations, and in connection with construction work and city garbage disposal.

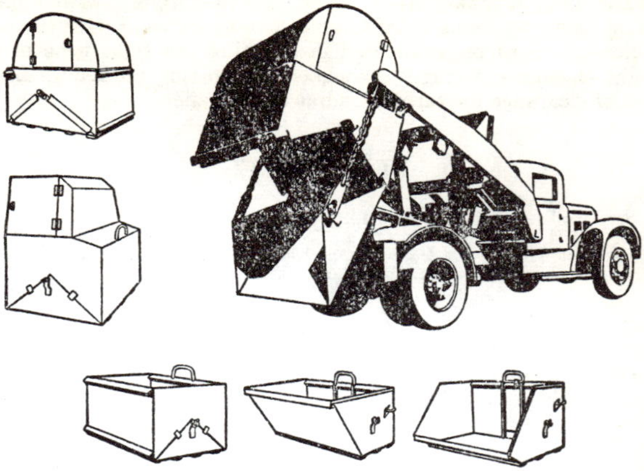

FIG. 25. Dempster Dumpster with various containers. (Courtesy, Dempster Brothers, Inc.)

OVER-THE-ROAD TRACTORS AND TRAILERS. For intraplant, intrastate, and interstate over-the-road handling, a two-wheel trailer, having its axles and wheels under the rear portion, and the frame supported in front on the fifth wheel of the vehicle that tows it, has many advantages, as it permits loading and unloading of the trailer chassis without tying up the expensive tractor unit. By using a motor transport trailer converter, which is a demountable fifth-wheel unit, for converting the semitrailers into full trailers, the same trailer chassis can be handled by either the "over-the-road" tractor unit, or any other vehicle capable of pulling the load.

MOTOR-TRUCK ACCESSORIES. Motor trucks are now a definite part of materials handling, both within industrial plants and between industrial plants. Therefore, it is essential to equip these units with the best possible loading devices.

The elevating end gate, which is a power-operated mechanism, replaces the tail gate of motor trucks, and is used for raising or lowering loads from the ground to the level of the truck platform or higher (Fig. 26).

FIG. 26. Elevating end gate.

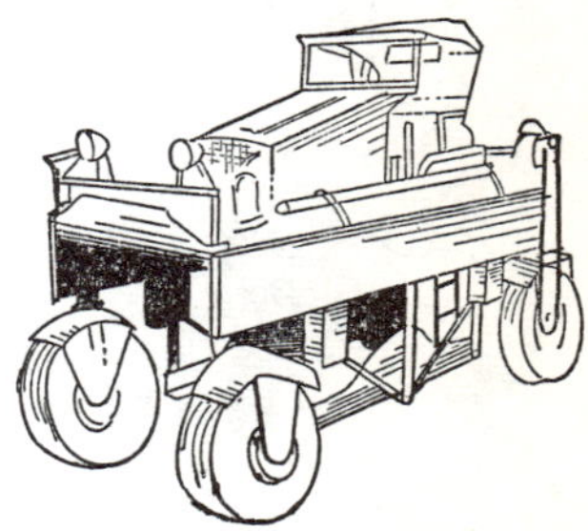

FIG. 27. Straddle truck. (Courtesy, Hyster Co.)

Power winches for motor trucks, operating from the truck motor, facilitate the pulling on and lowering of loads too heavy to be handled by auxiliary equipment.

Demountable bodies for regular truck chassis have been found advantageous in a number of operations. They are supplied in standard units, and they can be designed to fit specific operations.

THE STRADDLE TRUCK is an automotive type of truck, with a four-wheel steer constructed on an inverted U-shaped framework (Fig. 27). The motor and the operator's position are on top of the framework, and the load is straddled, lifted, and carried suspended on hoisting shoes supported by the framework.

The straddle truck generally operates with bolsters under loads, such as lumber, steel, and pipe. They are also used in steel plants and shipyards, with special racks for steel plate from the storage yard to the fabricating shops; and with special bolster-type flat skid bodies, they have been used in industrial plants for handling a wide variety of materials.

These units are made in capacities of 10,000 to 30,000 lb, and with a speed of 30 to 50 miles per hour, so they can be used for internal plant service or over the highway.

Important dimensions to be considered in selecting the right type of equipment are width and height of load and length of bolsters supporting the load. Length of load is purely a matter of clearance for turning radius and balance.

REFERENCES

1. *Trans. Am. Soc. Mech. Engrs.*, Vol. xliv, 1923.
2. C. B. Crockett and H. J. Payne, *Trans. Am. Soc. Mech. Engrs.*, 1928, MH-50-3.
3. S. J. Koshkin, *Modern Materials Handling*, p. 194, John Wiley and Sons, 1932. (Out of print.)
4. *Materials Handling and Distribution*. July, Aug. 1931.

SECTION 27

PRODUCTION PLANT CONSTRUCTION AND EQUIPMENT

By

THEODORE CRANE, *Consulting Engineer, Professor of Architectural Engineering, Yale University*

SAMUEL G. HIBBEN, *Director of Applied Lighting, Westinghouse Electric Corp.*

SIDNEY J. WILLIAMS, *Assistant to the President, National Safety Council*

A. M. BALTZER, *Assistant Manager, Industrial Department, National Safety Council*

R. B. FOLEY, *Sales Engineer, American Blower Corp.*

BUILDING CONSTRUCTION

By Theodore Crane

ILLUMINATION

By Samuel G. Hibben

SAFETY ENGINEERING

By Sidney J. Williams and A. M. Baltzer

DUST COLLECTION

By R. B. Foley

BUILDING CONSTRUCTION

By Theodore Crane

The purpose of this chapter is to enable the mechanical engineer to make a reasonably accurate selection of the structural elements comprising a building, prior to the quantitative design, which would be developed later by a structural engineer.

1. BUILDING CODES: DESIGN STANDARDS

AUTHORITIES PROMULGATING BUILDING CODES. There is no national building code in the United States. Buildings erected for a department or agency of the federal government are designed in accordance with the standards of that department. Local building codes do not apply to structures erected on federal property.

There are only a few complete state building codes, but many states have passed statutes which apply to the design of buildings erected for particular uses. These state regulations affect particularly the design of factories, or other buildings where labor is employed, places of public assembly such as theatres, and multifamily dwellings. Some state labor laws and industrial codes specify minimum requirements for ventilation and sanitary facilities, as well as standards to safeguard employees from danger due to manufacturing processes or fires.

Many cities have building codes which apply to alterations or additions to existing buildings, and to all new construction. Often these codes are supplemented by separate ordinances promulgated by city departments, such as those for fire and health, which may also affect building design.

There are only a very few county building codes in the United States, but many townships have building ordinances which cover, much more briefly, the same subjects as the municipal codes. These township codes have very little significance, provided the designer follows standards of good practice, except in the matter of zoning.

ANALYZING BUILDING CODES. As there is a wide and unjustifiable variation in the approximately 2000 building codes now in effect in the United States, it is impossible to present any average of their specific requirements. There is, however, sufficient uniformity in the subjects which they control to permit a topical analysis. A primary necessity is that buildings be structurally adequate to meet the conditions of locality and occupancy. The design must provide a degree of protection from danger incident to fire, which is appropriate to size, location, and use of the building. Minimum standards are established for the health and safety of the occupants. Adding the subject of zoning, we have a substantially inclusive résumé of the topics covered by building codes, so far as they affect the architectural or structural design.

In the following paragraphs the contents of the more complete municipal ordinances are outlined. The outline is arranged in the form of a procedure to facilitate the checking of code regulations for any specific project.

Occupancy and Construction Classifications. Most city codes have five to seven classifications, based upon the type of occupancy, such as "Public Building," "Business Building," and "Residence Building." The identification of the classification is used in determining the allowable type of construction and in relation to zoning laws. Most city codes have about the same number of classifications based upon the type of construction. This classification is determined by type of occupancy, location in reference to fire limits, or zoning laws, and height and ground area of the structure. Regulations covering these matters appear in all building codes. Identification of the construction classification, such as fireproof and semi-fireproof, provides a minimum standard of fire resistance to which the design of the building must conform.

Zoning Laws. These laws may prohibit either extremely hazardous or otherwise objectionable types of occupancy, even within industrial districts. They also restrict the type of occupancy permitted in certain zones. In the more congested municipal areas, they may limit the perpendicular height of a building as a function of the width of the street or streets upon which it fronts. Some cities have definite height limitations for even fireproof structures.

Height and Area Limitations. Many building codes place a definite height limit for buildings of other than fireproof construction, particularly within "fire limits." Ground areas may also be limited for structures of the more combustible classifications when located within congested areas. Recommended limitations are given in Tables 1 and 2.

Table 1. Allowable Heights of Buildings

(From the *Building Code Recommended by the National Board of Fire Underwriters*)

Classes of Occupancy	Classes of Construction						
	Fire-proof	Semi-fireproof	Heavy Timber	Ordinary	Light Noncom-bustible	Frame	Unpro-tected Metal
						Not permitted within Fire Limits	
Public * buildings	No limit	75 ft	Three stories nor more than 35 ft Churches: two stories nor more than 45 ft Schools: two stories Theatres †			Two stories nor more than 30 ft	One story
Institutional buildings	No limit	75 ft	Two stories nor more than 35 ft †			One story nor more than 35 ft †	One story
Residence buildings	No limit	75 ft	75 ft	Three stories nor more than 45 ft ‡	Three stories nor more than 45 ft	Two stories nor more than 35 ft Dwellings not to exceed three stories	One story
Business * buildings	No limit	75 ft	75 ft	Four stories nor more than 50 ft		25 ft	One story
Storage * buildings	No limit	50 ft §	35 ft §	35 ft §		One story nor more than 25 ft	One story

* Outside the fire limits public buildings, business buildings, or storage buildings may be erected to greater heights in the discretion of a board.

† Special occupancies: asylums, detention buildings, hospitals, penal institutions, orphanages, nurseries, and theatres hereafter erected shall be of fireproof construction or semi-fireproof construction; except that hospitals where occupants are not involuntarily detained or bedridden (sanatoriums) may be of nonfireproof construction, provided such buildings do not exceed two stories in height and the floors and partitions have fire-resistance ratings of not less than 1 hr, with fire-retardant ceilings under roofs. One-story theatres not over 30 ft in height may have wood or steel trusses and wood, steel, or gypsum roof decks if protected by fire-retardant ceilings.

‡ With a 2-hr first floor, height may be four stories or 55 ft; in multifamily houses with a 2-hr first floor and other floors 1-hr, with fire partitions subdividing floors into areas not exceeding 2500 sq ft, the height may be five stories or 65 ft.

§ Buildings which are sprinklered of semi-fireproof or heavy-timber construction may be 75 ft, and ordinary construction and light noncombustible may be 50 ft.

Author's note: Sprinklers are required by this code for certain types of occupancy.

Means of Egress. Except for single and two-family dwellings, the following matters are usually controlled by code: minimum hallway widths; location and design of stairs when serving as a required exit; location and design of fire towers and fire escapes, slope of ramps; number, location, size, construction, and operation of doors serving as required exits. In places of public assembly, width of aisles and arrangement of seats are definitely controlled.

Light and Ventilation. Minimum requirements often apply to the areas of windows and skylights; the areas of vent shafts and the widths of courts; the design of alcoves and the amount of mechanical ventilation.

Illumination, Wiring, and Electrical Equipment. Adequate lighting for safety, particularly in connection with exit facilities, is a general requirement of municipal codes. The light levels required upon working surfaces may be specified in state codes or state legislation applying to the employment of labor. The standards of the National Electrical Code and the National Board of Fire Underwriters are usually incorporated in building ordinances and should always be followed even if not mandatory.

Table 2. Allowable Areas of Buildings

(From the *Building Code Recommended by the National Board of Fire Underwriters*)

Maximum areas in square feet permitted in new buildings not protected by automatic sprinklers and without suitable subdivisions by fire walls for heavy timber, ordinary, light noncombustible, and frame construction or by fire partitions in fireproof and semi-fireproof construction.

Classes of Occupancy	Classes of Construction							
	Fireproof	Semi-fireproof	Street fronts *	Heavy Timber	Ordinary and Light Noncombustible		Frame	Unprotected Metal
					One story	More than one story	Not Permitted within Fire Limits	
Not a factor	Unlimited except as restricted by combustible occupancy		One street	6,500	6,000	5,000	5,000	Not restricted
			Two streets	8,000	7,500	6,000		
			Three streets	10,000	9,000	7,500		
			Four streets	10,000	10,500	9,000		

Area Modification.

The limiting areas may be increased by 100% when the building is sprinklered, and by 200% when the building is sprinklered and does not exceed one story or an average of 25 ft in height to the roof, or to a fire-retardant ceiling through which there are no openings unless they open into shafts or ducts to the roof, the enclosing walls of which have the same construction as the ceiling.

Outside the congested areas of the city, when a hazardous condition is not created thereby, the area of a public building, a business building, or a storage building, not over two stories high, may be increased in excess of the areas fixed by this section, in the discretion of a board consisting of the chief executive officer of the municipality, the building official, and the chief of the fire department; provided that a building of combustible occupancy, or involving considerable combustible material in its structural parts, shall be sprinklered.

* Under this section a street shall be deemed to include any avenue, boulevard, street, alley, or lane, 20 ft or greater in width, or any court, parking space, or yard, with direct connection to a street, and not less than 20 ft wide. Such court, parking space, or yard shall be the property of the owner of the building and shall not be enclosed or roofed over.

Author's note: Sprinklers are required and further area restrictions established by this code for certain types of occupancy.

Plumbing, Heating, and Gas Piping. Minimum sanitary requirements applying to factories and to multifamily dwellings are often contained in state codes or state statutes. Municipal sanitary codes cover all types of occupancy. Subjects that should be particularly checked are number and type of fixtures; location and size of toilets, latrines or bathrooms, water tanks, and standpipes; regulations regarding sprinkler systems; fire-fighting equipment and automatic fire-alarm systems for high buildings. The more complete codes cover heat installations and heat appliances of all kinds, in so far as they are related to danger incident to fire. Safety provisions for the use of gas are customary.

Elevators, Escalators, and Dumbwaiters. The requirements for such installations cover design, construction, inspection, operation, and maintenance, but only a few matters apply to building design. Most important are regulations for elevator shafts and machinery compartments; size of elevator pits and penthouses; equipment loads and maximum slope of escalators.

Quality of Materials. These regulations often reflect local practice, but most large cities follow the standards of our national organizations, such as the American Society for Testing Materials. The recommendations of the National Board of Fire Underwriters are standard for all materials and appliances in so far as they are related to the fire risk.

Weights of Materials. All building codes contain a list of weights applying to various materials such as concrete, stone, and brick masonry. These weights are used in computing the dead loads required for design purposes.

Live loads and soil loads comprise floor, roof, sidewalk, and courtyard loads. Included also are maximum soil loads for different types of soil and rock; allowance for wind pressure; permitted reductions in live loads, and requirements for moving loads.

Working Stresses and Structural Limitations. Allowable, safe working stresses, applying to all structural materials, are given in all municipal building codes. Often they are supplemented by specific limitations which apply irrespective of stress requirements. Such might demand a minimum thickness for a wall or partition of a certain height, or a minimum thickness for a structural floor slab of a particular design.

Table 3. Minimum Design Dead Loads

(From *Minimum Design Loads in Buildings and Other Structures*, sponsored by the National Bureau of Standards, approved and published by the American Standards Association, 1945)

Walls	Pounds per Square Foot
4-in. clay brick, medium absorption	39
4-in. sand-lime brick	38
4-in. concrete brick, heavy aggregate	46
4-in. concrete brick, light aggregate	33
8-in. clay brick, medium absorption	79
8-in. sand-lime brick	74
8-in. concrete brick, heavy aggregate	89
8-in. concrete brick, light aggregate	68
12 1/2-in. clay brick, medium absorption	115
12 1/2-in. sand-lime brick	105
12 1/2-in. concrete brick, heavy aggregate	130
12 1/2-in. concrete brick, light aggregate	98
17-in. clay brick, medium absorption	155
4-in. brick, 4-in. load-bearing structural-clay-tile backing	60
4-in. brick, 8-in. load-bearing structural-clay-tile backing	75
8-in. load-bearing structural clay tile	42
12-in. load-bearing structural clay tile	58
8-in. concrete block, heavy aggregate	55
12-in. concrete block, heavy aggregate	85
8-in. concrete block, light aggregate	35
12-in. concrete block, light aggregate	55
2-in. furring tile, one side of masonry wall, add to above figures	12
Partitions	
3-in. clay tile	17
4-in. clay tile	18
6-in. clay tile	28
8-in. clay tile	34
2-in. gypsum block	9 1/2
3-in. gypsum block	10 1/2
4-in. gypsum block	12 1/2
6-in. gypsum block	18 1/2
2-in. solid plaster	20
4-in. solid plaster	32
4-in. hollow plaster	22
4-in. concrete block, heavy aggregate	30
6-in. concrete block, heavy aggregate	42
8-in. concrete block, heavy aggregate	55
12-in. concrete block, heavy aggregate	85
4-in. concrete block, light aggregate	20
6-in. concrete block, light aggregate	28
8-in. concrete block, light aggregate	38
12-in. concrete block, light aggregate	55
Wood studs 2 x 4, unplastered	4
Wood studs 2 x 4, plastered 1 side	12
Wood studs 2 x 4, plastered 2 sides	20
Glass-block Masonry	
4-in. glass-block walls and partitions	18
Split Furring Tile	
1 1/2-in.	8
2-in.	8 1/2
Concrete Slabs	
Concrete, reinforced-stone, per inch	12 1/2
Concrete, reinforced-cinder, per inch	9 1/4
Concrete, reinforced, lightweight, per inch	9
Concrete, plain stone, per inch	12
Concrete, plain cinder, per inch	9
Concrete, plain, lightweight, per inch	8 1/2

Floor Finish and Fill	Finish Floor to Top of Slab, Inches	Pounds per Square Foot
Double 7/8 wood on sleepers, light-concrete fill	4	19
Single 7/8 wood on sleepers, light-concrete fill	4	23
Floor Finish	**Thickness, Inches**	
1 1/2-in. asphalt mastic flooring	1 1/2	18
3-in. wood block on 1/2-in. mortar base	3 1/2	16
Solid flat clay tile on 1-in. mortar base	2	23
2-in. asphalt block, 1/2-in. mortar	2 1/2	30
1-in. terrazzo, 2-in. stone concrete	3	38
Floor Fill		
Cinder concrete, per inch		9
Lightweight concrete, per inch		7
Ceilings		
Plaster on tile or concrete		5
Suspended metal lath and gypsum plaster		10
Suspended metal lath and cement plaster		15
Plaster on wooden lath		8
Roof Coverings		
Asbestos shingles		4
Asphalt shingles		6
4-ply felt and gravel composition		5 1/2
Sheathing, per 1-in. thickness		3
Slate, 3/16-in.		7
Slate, 1/4-in.		10
Wood shingles		3

Weights of Materials	Pounds per Cubic Foot
Cast-stone masonry (cement, stone, sand)	144
Cinder fill	57
Concrete, plain	
Stone (including gravel)	144
Slag	132
Cinder	108
Haydite (burned-clay aggregate)	90
Expanded-slag aggregate	100
Concrete, reinforced	
Stone (including gravel)	150
Slag	138
Cinder	111
Masonry, brick	
Hard (low absorption)	130
Medium (medium absorption)	115
Soft (high absorption)	100
Masonry, ashlar	
Granite	165
Limestone, crystalline	165
Limestone, oolitic	135
Marble	173
Sandstone	144
Timber, seasoned	
Cypress, southern	32
Fir, Douglas, coast region	34
Oak, commercial reds and whites	45
Pine, southern yellow	39
Redwood	28
Spruce, red, white, and Sitka	28

Weights of masonry include mortar but not plaster. For plaster, add 5 lb per sq ft for each face plastered. Values given represent averages. In some cases there is a considerable range of weight for the same construction. These variations are given in the authority from which this table is abstracted.

Methods of Computation. Although most modern building codes follow the more recent standards recommended for structural design by our national engineering associations, it is often necessary to check the formulas and computation methods applying to structural design because variations still exist.

Construction Details and Standards of Workmanship. To insure structural adequacy and fire resistance, most building codes contain articles that cover such matters as the

Table 4. Minimum Live Loads of Representative Codes

(From T. Crane's *Architectural Construction*, John Wiley and Sons, 1947)

Type of Occupancy	Minimum Live Loads per Square Foot of Floor Area					
	Department of Commerce, 1925	Pacific Coast Building Officials Conference, 1946	National Building Code of Canada, 1941	National Board of Fire Underwriters' Recommendations, 1943	Philadelphia Code, 1941	American Standards Association, 1945
Dwellings	40	40	40	40	40	40
Hotels * Lodging houses, tenements Apartments, hospitals (private rooms and wards)	40	40	40	40	40	40
Office buildings: Offices and private corridors Lobbies and public spaces †	 50 100	 50 100	 50 100	 50 100	 60 100	 80 100
Schools: Classrooms	50	40	50	50	50	40
Buildings for public assembly: With fixed seats ‡ Without fixed seats Aisles and corridors	 50 100 100	 50 100 100	 60 100 100	 75 100 100	 60 100 100	 60 100 100 §
Stores: Wholesale Retail	 100 75	 100 75	 125 100	100	 110 110	125
Garages: All types of vehicles Passenger cars only	 100 80	100 ‖	 150 ¶ 75 ‡‡	175	 100 ** 100	100–200 ††
Warehouses §§	100–250	125–250	125	120	150	125–250
Manufacturing: Heavy §§ Light §§	 100 75	 125 75	125	120	 200 120	125
Stairways	100	100	100	100	100	100 ‖‖
Sidewalks	250	250	250	300	120	250 ¶¶

* Does not apply to hotel lobbies, stairways, or any portion used for public assembly. Such spaces same as for office buildings.

† This requirement is often applied to the entire first floor irrespective of expected use.

‡ This requirement is usually applied to aisles between rows of seats.

§ Corridors, upper floors.

‖ Additional provision required for loaded trucks.

¶ Limited to 20,000 lb gross weight; minimum concentrated load, 8000 lb on an area 2 ft 6 in. square.

** Minimum concentrated load "at any point" 8000 lb.

†† One hundred for cars with loads less than 3 tons; 150 for trucks with loads 3 to 10 tons; 200 for trucks with loads above 10 tons; each classification also subject to maximum wheel concentrations.

‡‡ Any type of vehicle not exceeding 6000 lb gross weight; minimum concentrated load 4000 lb on an area 2 ft 6 in. square.

§§ These figures have significance only as minimums. Such structures should be designed for the actual loads due to the type of materials stored or manufacturing equipment.

‖‖ Or a series of concentrated loads of 300 lb each, spaced 3 ft 0 in. center to center, each occupying an area 1 ft wide by the depth of the tread, whichever will produce the greater stresses.

¶¶ Or a concentrated load of 8000 lb on an area 2 1/2 ft square placed in any position, whichever will produce the greater stresses.

bonding of masonry walls and riveting or welding structural-steel assemblies. To lessen the fire risk, there are generally regulations covering the minimum thickness of fire walls, partitions, and shaft enclosures; limitations in regard to the use of combustible materials in fireproof or semi-fireproof construction; thickness of concrete or other fireproofing material surrounding structural steel, or concrete reinforcement. "Fire stopping" is also specified.

LOADS CARRIED BY BUILDINGS comprise dead loads, live loads, wind loads, pressure due to earth or water, and in districts subject to earthquakes, lateral loads assumed to equal the probable effect of seismic disturbances.

Dead load identifies the weight of walls, columns, partitions, floors, roofs, and all other permanent construction. In building codes, the dead load is controlled by stating the weights of materials which must be used in design. For example, 150 lb per cu ft for stone concrete and 120 lb per cu ft for brick masonry. Minimum design dead loads, abstracted from *Minimum Design Loads in Buildings and Other Structures*, sponsored by the National Bureau of Standards and issued by the American Standards Association as A58.1-1945, are given in Table 3.

Live load identifies all vertical loads except dead load but, in many building codes, it includes lateral loads such as are caused by wind. The live-load allowance represents the weight of the occupants, furniture, or other equipment, including stored materials or movable partitions. As applied to a roof surface it includes the allowance for snow. All municipal building codes designate specific live loads for the various types of occupancy. Table 4 gives minimum live loads of representative codes. These loads do not include any permanent parts of the structure.

It should be clearly understood that the loads shown in Table 4 are merely minimum live-load requirements. Their use does not relieve the designer of the responsibility of providing adequate structural strength to support safely the actual loads for which the structure is intended. This warning is particularly important for manufacturing and storage buildings. It will be seen that Table 4 makes no mention of a provision for concentrated loads except for garages. Such requirements have been omitted for the sake of clarity but appear in most codes as supplementing the uniformly distributed loads. For example, the floor of an office or storage building might be required to support a concentrated weight of 2000 lb, without other live load, on any area 2 1/2 or 3 ft square. Such regulations affect the design only if the stresses produced by such concentrations are greater than those due to the uniform load.

If roofs are intended for occupancy, they are designed to sustain the same loads as corresponding floor areas. If not intended for occupancy, an allowance varying from 20 to 40 lb per sq ft of horizontal projection is customary for flat roofs and those of moderate slope. Steeper roofs are designed for dead load, wind load, and a snow load appropriate for their slope and locality.

Live-load Reductions. Certain reductions are permitted in computing the total live loads carried by supporting members such as columns, but in no case should the total reduction be less than 50%. The following standard (Table 5) is abstracted from the *Building Code Recommended by the National Board of Fire Underwriters.*

Table 5. Maximum Permissible Live-load Reductions

	% Reduction
Carrying the roof	0
Carrying one floor and roof	0
Carrying two floors and roof	10
Carrying three floors and roof	20
Carrying four floors and roof	30
Carrying five floors and roof	40
Carrying six floors (or more) and roof	50

These percentages apply to live loads carried by columns, girders, trusses, walls, piers, or foundations. The Underwriters Code allows these reductions for all buildings except those intended for storage purposes, for which no reduction is permitted. Some codes permit lesser reductions for storage and manufacturing buildings.

Moving Loads. The dynamic effect of all moving loads should be considered. For example, the static loads on beams and girders supporting elevator machinery may be increased or the working stresses reduced.

Wind Loads. A minimum pressure in pounds per square foot is specified by all building codes as an allowance for the effect of wind pressure against exposed surfaces of buildings. Table 6, abstracted from *Minimum Design Loads in Buildings and Other Structures*, shows wind pressures for zones of various heights. The figures given are recommended in this report as minimums. These requirements do not provide for tornadoes because they are based on a design wind velocity of 75 mph, corresponding roughly to a 5-min average of 50 mph (indicated 63 by a 4-cup anemometer) at 30 ft from the ground.

Special provision for wind bracing is required for narrow buildings of even medium height and for all high buildings. In towerlike structures wind bracing should be planned in the early stages of design. The overturning moment due to wind pressure should not exceed two-thirds of the moment of stability, disregarding live loads, except where foundations are securely anchored. The possibility of a structure sliding on its foundation bed

Table 6. Design Wind Pressures for Various Height Zones of Buildings or Other Structures

(From *Minimum Design Loads in Buildings and Other Structures*, American Standards Association A58.1-1945)

Height Zone, ft	Wind Pressure, lb per sq ft
Less than 50	20
50 to 99	24
100 to 199	28
200 to 299	30
300 to 399	32
400 to 499	33
500 to 599	34
600 to 799	35
800 to 999	36
1000 to 1199	37
1200 to 1399	38
1400 to 1599	39
1600 and over	40

should also be considered. In computing the wind load no allowance should be made for the shielding effect of other buildings. All roofs should be designed to resist wind suction as well as wind pressure (Ref. 1).

Earthquake Loads. In districts that have experienced earthquakes of an intensity of VII or over, in the Ross-Forel scale, adequate provision should be made in the design to resist the lateral forces caused by seismic disturbances. For an appropriate design procedure see *Minimum Design Loads in Buildings and Other Structures*, previously identified.

Soil Loads. The figures given in building codes for the allowable bearing values of various soils and rocks are merely somewhat arbitrary maximums and should be checked by subsurface investigation, often supplemented by a load test. As a guide in selecting reasonable soil-bearing values, Table 7 is abstracted from the Boston Building Code. Refer to Article 8 in this section, under Subsurface Materials, for further identification of these classifications (see p. 27-54).

Table 7. Allowable Bearing Values for Various Soils and Rocks

(Boston Building Code 1944)

Class	Material	Allowable Bearing Value, tons per square foot
1	Massive bedrock without laminations, such as granite, diorite, and other granitic rocks; gneiss, trap rock, felsite, and thoroughly cemented conglomerates, all in sound condition (minor cracks allowed)	100
2	Laminated rocks such as slate and schist, in sound condition (minor cracks allowed)	35
3	Shale in sound condition (minor cracks allowed)	10
4	Residual deposits of shattered or broken bedrock of any kind except shale	10
5	Hardpan	10
6	Gravel, sand-gravel mixtures, compact	5
7	Gravel, sand-gravel mixtures, loose; sand, coarse, compact	4
8	Sand, coarse, loose; sand, fine, compact	3
9	Sand, fine, loose	1
10	Hard clay	6
11	Medium clay	4
12	Soft clay	1
13	Rock flour, shattered shale, or any natural deposit of unusual character not provided for herein (value to be fixed by the commissioner)	

Table 8. Rated Fire-resistive Periods for Various Walls and Partitions

(From Uniform Building Code, Pacific Coast Building Officials Conference, 1946)

Material	Construction			Minimum Finished Thickness Face to Face (Including Plaster Where Mentioned), in. 4 hr	3 hr	2 hr	1 hr
Brick of clay, shale, sand-lime or concrete and plain concrete	Solid unplastered			8			4 *
	Solid plastered			9		5 *	
	Hollow (rowlock) unplastered			12	10	8	
	Hollow (rowlock) plastered			9			
Hollow clay tile wall	End or side construction. One cell in wall thickness. Plastered.						3 *
	End or side construction. Two cells in 8-in. or less thickness. Unplastered.			16	12		6 * or 8
	End or side construction. Two cells in 8-in. or less thickness. Plastered.			13	9	7 *	
Hollow clay tile ASTM load bearing	End or side construction. Two cells in wall thickness. Unplastered.					6	
	End or side construction. Two cells in wall thickness. Plastered.					5 *	
	End or side construction. Three cells in 8-in. or less thickness. Unplastered.			12			
	End or side construction. Three cells in 8-in. or less thickness. Plastered one side.				8 1/2		
	End or side construction. Three cells in 8-in. or less thickness. Plastered.			9			
Combination of brick and ASTM load-bearing tile, or hollow concrete block or tile	4-in. brick and 4-in. tile plastered one side (tile side)			9			
Concrete block or tile	Aggregate—expanded slag, burned clay or shale, cinders	1 1/4-in. face shells	Unplastered			8	4 †
			Plastered one side		8 1/2	4 1/2 †	
		1 1/2-in. face shells	Unplastered	12 ‡	8		
			Plastered each side	9			
		2 1/4-in. face shells	Unplastered	8			
	Other aggregates	1 3/8-in. face shells	Unplastered			8	4 †
			Plastered each side		9	5 †	
		1 3/4-in. face shells	Unplastered	12	8		
			Plastered each side	9			
		2 1/4-in. face shells	Unplastered	8			

(*Table continued on p. 27-10*)

Table 8. Rated Fire-resistive Periods for Various Walls and Partitions—*Continued*

(From Uniform Building Code, Pacific Coast Building Officials Conference, 1946)

Material	Construction	Minimum Finished Thickness Face to Face (Including Plaster Where Mentioned), in.			
		4 hr	3 hr	2 hr	1 hr
Solid concrete	Reinforcement not less than 0.2% in each direction.	6	5	4	2 *
Solid pneumatically placed concrete	Reinforcement not less than 0.2% in each direction	5 * or 6	4	3	2 *
Hollow gypsum blocks	Unplastered	6 *	5 *	4 *	3 *
	Plastered each side	5 *	4 *	4 *	3 *
Hollow wall of reinforced pneumatically placed concrete	Outer shell 2-in. thick for 10-in. wall and 1 1/2-in. thick for 8-in. wall.	10 *	8 *		
Solid gypsum or portland cement plaster	Incombustible studding with metal or wire lath				2 *
	Incombustible studding with metal or wire lath, neat wood fiber gypsum plaster			2 *	
Hollow stud partition with gypsum or portland cement plaster or pneumatically placed plaster on each side	Incombustible studding with metal or wire lath, 3/4-in. plaster on each side				3
	Incombustible studding with metal or wire lath, 1-in. plaster on each side			4 1/2	
	Wood studs with metal or wire lath. Fire-stopped. 3/4-in. plaster on each side.				3 * or 5
	Wood studs with metal or wire lath. Fire-stopped. 1-in. neat wood fiber plaster each side.			5 *	
	Wood studs with 3/8-in. perforated gypsum lath. Fire-stopped. 1/2-in. gypsum plaster each side.				3 * or 5

* Shall be used for nonbearing purposes only.
† 7/8-in. face shells.
‡ 8 in. for expanded slag.

RECOMMENDED WORKING STRESSES. During recent years, standardization of stress requirements has progressed considerably, but many building codes are out of date. The following references represent good practice. They have been widely accepted in current revisions of many building ordinances and are recommended as standards except when more restrictive requirements are mandatory.

Stresses for Structural Steel. *Specification for the Design, Fabrication and Erection of Structural Steel for Buildings,* contained in the manual entitled *Steel Construction,* published by the American Institute of Steel Construction.

Stresses for Reinforced Concrete. *Building Regulations for Reinforced Concrete,* adopted as a standard of the American Concrete Institute in 1941 and revised in 1947 (Ref. 2).

Design and Specification of Structural Lumber. *National Design Specification for Stress-grade Lumber and Its Fastenings,* recommended by the National Lumber Manufacturers Association.

Design and Specification of Masonry. *American Standard Building Code Requirements for Masonry,* identified as Miscellaneous Publication M174, National Bureau of Standards. It may be obtained from the Superintendent of Documents, U. S. Government Printing Office, Washington, D. C.

THE STANDARD FIRE TEST. The following procedure, given in more detail, appears in Appendix B of the Building Code Recommended by the National Board of Fire Underwriters. It prescribes a standard of exposure to fire of controlled extent and severity; in some cases the sample or assembly is immediately subjected to a fire-hose stream. Performance is defined as the period of resistance to such exposure elapsing before the first critical point in behavior is observed. If this period is 2 hr, the material or assembly, such as a partition or floor system, is given a "2-hr rating." These ratings vary from 1 to 4 hr, and they appear as requirements for the more fire-resistant construction classifications. It is consequently often necessary to limit the choice of structural assemblies to those having a rating consistent with the desired building classification. Typical regulations are shown in Table 8.

FIREPROOFING REQUIREMENTS. As the capacity of steel to support load is considerably reduced at such temperatures as may easily occur in any serious fire, adequate protection is essential for both structural steel and concrete reinforcement. Typical requirements, abstracted from the New York City Building Code, are given in Table 9.

Table 9. Thickness of Fireproofing to Protect Structural Steel Requirements for Various Fire-resistance Ratings

(From the Building Code of New York City, Amended to 1942)

Fire-resistive Materials	Minimum Thickness, in. 4 hr	3 hr	2 hr	1 hr
Brick	3 3/4	3 3/4	2 1/4	2 1/4
Brick	3 3/4	3 3/4	2 1/4	2 1/4
Concrete brick, block or tile, except cinder concrete units	3 3/4	3 3/4	2 1/4	2 1/4
Hollow or solid cinder concrete block and tile having a compressive strength of at least 700 psi of gross area	2 1/2	2	2	1 1/2
Solid gypsum block (to obtain 40 hr rating must be plastered with 1/2-in. gypsum plaster)	2	2	1 1/2	1
Gypsum poured-in-place and reinforced	2	1 1/2	1 1/2	1
Hollow or solid burned clay tile or combinations of tile and concrete	2 1/2	2	2	1 1/2
Metal lath and gypsum plaster	2 1/2	2	1 1/2	7/8
Cement concrete, Grade I	2	2	1 1/2	1
Cement concrete, Grade II	4	3	2	1 1/2
Cement concrete, Grade II, with wire mesh	3	2	2	1 1/2
Hollow gypsum block (to obtain 4-hr rating must be plastered with 1/2-in. gypsum plaster on outer side)	3	3	3	3

Grade I concrete is defined in this code as that in which the aggregate consists of limestone, traprock, blast-furnace slag, cinders, or calcareous gravel.

Grade II concrete is defined as that in which the aggregate consists of granite or siliceous gravel.

The fireproofing of reinforced concrete is obtained by specifying a minimum thickness outside the metal reinforcement. Table 10, showing typical requirements, is taken from the *Report of the Joint Committee on Standard Specifications for Concrete and Reinforced Concrete*, 1940 sponsored by the American Concrete Institute, American Society of Civil Engineers and affiliated associations (Ref. 3).

Table 10. Thickness of Concrete Fireproofing to Protect Metal Reinforcement Requirements for Various Fire-resistance Ratings

(From the *Report of the Joint Committee on Recommended Practice and Standard Specifications for Concrete and Reinforced Concrete*, 1940, published by the American Society of Civil Engineers)

Structural Member	Minimum Thickness of Concrete, in. 4 hr	3 hr	2 hr	1 hr
Columns, beams, girders, and unprotected ribbed slabs				
Group 1, aggregate	1 1/2	1 1/2	1 1/2	1
Group 2, aggregate	2	1 1/2	1 1/2	1
Solid slabs				
Group 1, aggregate	3/4	3/4	3/4	3/4
Group 2, aggregate	1	3/4	3/4	3/4

Group 1 aggregates are defined in this standard as: blast-furnace slag, limestone, calcareous gravel, trap rock, burnt clay or shale, cinders containing not more than 25% of combustible material and not more than 5% of volatile material, and other aggregates meeting the Committee's specifications and containing not more than 30% of quartz, chert, flint, and similar materials.

Group 2 aggregates are defined in this standard as: granite, quartzite, siliceous gravel, sandstone, gneiss, cinders containing more than 25%, but not more than 40% of combustible material and not more than 5% of volatile material, and other aggregates meeting the Committee's specifications and containing more than 30% of quartz, chert, flint, and similar materials.

SUPPLEMENTARY DESIGN STANDARDS. Besides the various standards mentioned in the preceding articles, the following references should be particularly noted.

The municipal building codes of New York, Boston, Chicago, and Los Angeles are useful as indicating current practice. The *Uniform Building Code*, adopted by the Pacific Coast Building Officials Conference, and the *Southern Standard Building Code*, prepared by the Southern Building Code Congress in 1946, illustrate a movement to standardize code requirements. The state codes of Wisconsin and Ohio, supplemented by the *Multiple Dwelling Law* of New York State and the *State Housing Act* of California are of interest from the viewpoint of state control.

Much valuable information on structural design will be found in the series of reports entitled *Building Materials and Structures*, sponsored by the U. S. Department of Commerce and published by the National Bureau of Standards. They are obtainable from the Superintendent of Documents, U. S. Government Printing Office, Washington, D. C. The Federal Emergency Administration of Public Works, Housing Division, has sponsored publications applying to particular aspects of building. The standards of the Federal Housing Authority furnish helpful criteria applying to residential work. They are mandatory when mortgages are to be insured by the Authority.

Many trade associations such as the Portland Cement Association, the National Lumber Manufacturers Association, and the Structural Clay Products Institute have made easily available the results of research in their respective fields.

2. DESIGN PROCEDURE

BASIS OF THE STRUCTURAL DESIGN. Prior to the design of a building, an analysis should be made of all mandatory regulations affecting either the architectural or structural work. In the absence of legal requirements, the designer should choose his own standards. These subjects have been discussed in the preceding section.

After conferences with the owners, preliminary sketches and a memorandum specification are developed. These drawings, made to an appropriate scale, should show in plan all office, manufacturing, or storage spaces with elevations or sections indicating clear story heights. They form the bases for the structural design which should be commenced at this stage of the project and thoroughly coordinated with the spatial requirements. For industrial work, the utilitarian demands of the particular industry for which the building is to be used are far more important than artistic considerations.

INFLUENCE OF LOCALITY. All characteristics of the site or geographical location, which may directly or indirectly affect structure, should be studied at this time and reduced to the form of simple notes. Subsurface conditions are most important. Methods of investigation are described briefly in Article 8 of this chapter. In localities subject to earthquakes, the probable severity of seismic disturbance would be noted. In districts where exceedingly strong winds may be expected, the probable wind loads should be compared with the minimums given in Article 1 or required by local building codes.

Owing to the expansion and contraction of building materials caused by temperature variation, joints to provide for such movement are often necessary. Such are particularly important for long concrete buildings erected in the colder climates as mentioned in Article 3. Extreme temperatures obviously require more thorough insulation of the building envelope. Acoustical considerations may, in some cases, influence the foundations and require insulating the structural frame from a source of external vibration.

ECONOMIC CONSIDERATIONS. Even in this preliminary stage of design, the relative prices for various building materials, labor rates, labor efficiency, and union restrictions for the different trades should be recorded. They may have an influence on the choice of materials and structural assemblies. There is also the very important need of checking the availability of materials.

SPECIAL STRUCTURAL REQUIREMENTS. Any special equipment loads, a future horizontal extension of the building or a provision for future, additional stories, should be recorded at this time. The type of foundations should be chosen prior to the actual design of the foundation assemblies, in order to eliminate later changes in column locations. For example, in steel-mill buildings, longer spans would be favored when columns are supported by piles, so as to reduce the number of pile clusters. If there is a probability of unequal settlement, the more rigid types of framing would be inadvisable.

The amount of day lighting or the sort of artificial light demanded by the particular type of industry may have an important effect upon the exterior wall design. If skylights are required, the roof construction is obviously affected.

MECHANICAL AND ELECTRICAL EQUIPMENT. These matters are extremely important for all modern buildings, particularly for factories. The designer should note the requirements which follow:

(1) Requirements for number, size, location, and height of rise for elevators, with particular attention to penthouse dimensions and equipment loads.

(2) General requirements for plumbing with particular attention to the location of soil stacks, standpipes, mains, pumps, water-storage tanks, and sprinkler systems.

(3) If steam is to be produced within the building, requirements of the boiler room and accessories, such as fuel storage. The probable location of steam mains and ducts and their approximate sizes in order to avoid interference with the structural members or other utilities.

(4) Typical lighting demands with particular attention to ceiling outlets as their proper location may influence the framing of the building, and the necessary space required for running the electric conduit often affects the floor design.

(5) For industrial buildings, all specific demands of the manufacturing process, such as special mechanical and electrical equipment or interior clearances, should be identified.

CONCLUSION. When these preliminary investigations have been completed and the structural significance of the subjects mentioned have been reduced to the form of definite design requirements, the choice of the structural assemblies, comprising building frame, floors, roof, walls, and foundations, can be determined. The next six articles describe the more important systems or assemblies employed in this country at the present time with recommendations concerning their appropriate applications. This selection of structural elements constitutes a qualitative design that not only is a vital prerequisite for all quantitative design but will also permit the mechanical engineer to develop the concept of the building to fit the requirements of its particular use, long before the final architectural and structural drawings are completed.

3. PLANNING THE BUILDING FRAME

CHOICE OF THE FRAMING MATERIAL. Structural steel and reinforced concrete, in skeleton construction, are the two normal options for other than comparatively low structures. Bearing-wall designs are seldom economical in this country for buildings more than three stories high. Combined with heavy timber interior framing, they were formerly widely used for factories in "slow-burning" or "mill-construction." Since about 1917, however, such designs have been largely supplanted by structural steel or reinforced concrete. Exterior bearing walls, often combined with interior masonry bearing partitions, are still popular for low residential and commercial structures, but skeleton designs have been found economical for even two-story multifamily housing. Wall-bearing structures usually have an interior framing of structural steel, wood, or lightweight cold-formed steel sections. Reinforced concrete is seldom as appropriate. Complete wood framing is popular for smaller residential and commercial buildings in rural and suburban districts. Structural timber, in the form of trusses, is widely used for industrial work. Timber is also appropriate for rigid frames and arches in buildings as diversified as airplane hangars and auditoriums.

In the general field of building construction, skeleton frames of structural steel or reinforced concrete compete on an economic basis for all buildings up to and including those of moderate height. Above twenty stories, structural steel is the usual choice, and it is often more economical when the height exceeds fifteen stories. Height, however, is not the only consideration. A structural-steel frame can be erected much more rapidly than one of reinforced concrete and is particularly adapted to the one-story industrial building and to the comparatively light floor loads of the multistory city office building or apartment house. On the other hand, the delay in filling an order for structural steel may cancel the advantage of a speedier erection, and either a steel or reinforced-concrete frame may be chosen for the purpose of using a particularly economical type of floor construction, inappropriate for the other material. Such a choice is illustrated by the Master Printers' Building in New York City. This structure is twenty stories high, but the heavy floor loads and the requirement for a comparatively uniform clear story height favored the use of a "flat-slab" (see p. 27-18) concrete design. Because this floor system could not be combined with a structural steel frame, the entire building was constructed of reinforced concrete.

The choice between structural-steel and reinforced-concrete framing may, in some cases, be affected by local building ordinances. An example is the building code of San Francisco which limits reinforced-concrete buildings to a height of 135 ft. The desired architectural treatment of exterior surfaces, as well as the form of structural members dictated by the architectural design, may also play an important part. Structural steel can provide an adequate frame, but if concrete surfaces and forms appropriate to concrete are desired, it is logical to employ concrete for framing purposes.

In the industrial field, structural steel frames are used for textile mills only a few stories high, but for types of industry requiring heavier loads a reinforced-concrete frame with a flat-slab floor system is more economical unless local conditions, or specific design requirements, are unfavorable to concrete. Concrete is particularly appropriate for heavy floor loads, ranging from 150 to 300 lb per sq ft. As the single-story industrial building can be built more cheaply of steel than of concrete, the selection of the framing material often depends on the option between a one-story or a multistory building.

Comparing two alternative designs, both structures have the same gross floor area, 51,000 sq ft. The one-story design has a definite advantage for many types of industry but no valid generalization can be made because each project demands individual study. On the basis of gross floor area, the single-story building generally costs less. There is also a very appreciable saving in the loss between gross area and usable area. The single-story structure is more flexible from the viewpoint of future changes, more easily daylighted, and, except, where gravity systems of distribution can be employed advantageously, is often better suited to manufacturing operations.

On the other hand, a high land value obviously favors a multistory design. The heating load in cold climates and the heat gain in hot climates are also less for multistory buildings. Maintenance of the roof areas again favors this type. Furthermore, in any comparison between steel and concrete framing for the roofs of industrial buildings, the designer should consider the required fire resistance because the cost of fireproofing steel trusses might well be an important factor. This has been one of the reasons for employing concrete rigid frames and shell-roof designs for one-story industrial buildings. They are described in Article 5.

Structural steel framing is the usual choice for pitched roofs supported by either a skeleton design or bearing walls. Steel trusses are employed in building construction for spans varying from 40 to 200 ft. There are many types of steel arches and rigid frames which are suitable for spans up to several hundred feet. All of these wide-span roof assemblies are described in Article 5.

Within recent years, the use of wood as a framing material has come into competition with lightweight steel designs composed usually of cold-formed rather than hot-rolled sections. Complete steel frames of this kind are appropriate for buildings not over three stories high, provided the type of occupancy does not demand heavy floor loads. Where standardization is practicable, the first cost may closely approach that of wood-framed construction with the advantage of using an incombustible material and the elimination of shrinkage. Such designs have proved very economical for machine shops, warehouses, hutments, service buildings, and other single-story structures required for the Armed Forces during World War II. Any kind of exterior or interior finish may be applied over steel with practically the same facility as over wood. Combined with insulated, cellular-steel wall panels, lightweight steel frames are suitable for some types of industrial buildings; with porcelain enameled steel sheets they are used for commercial work. Many of these lightweight steel designs are more or less prefabricated. In most cases the framing members are cut to length in the shop and trusses are shop-welded. For a description of the more or less prefabricated designs suitable for industrial buildings, see p. 27-21.

ARRANGEMENT OF THE FRAMING. The structural design should obviously satisfy the spatial requirements and the demands identified in the preceding sections of this chapter. It is equally important that component parts of the framing should be chosen and combined in a manner to effect economy in field construction as well as in stress analysis. This means that the design should permit simple, direct framing and spans within economical limits except where a prime requirement justifies more expensive construction. It is exemplified by the functional use of appropriate building materials and structural forms.

STRUCTURAL STEEL OR REINFORCED CONCRETE. Column Locations. The position of each column, or tier of columns, should be marked on the floor plans. For the multistory building of skeleton design, the best column spacing usually results from dividing the supported area into squares or rectangular areas of approximately equal size. For example, a building 150 ft long and 65 ft wide would suggest a longitudinal spacing of columns at 25 ft and a spacing of about 22 ft across the building. A column spacing of less than 18 or 20 ft is seldom economical. If reinforced-concrete girder framing is used, a maximum span of about 25 ft, and for rolled-steel girders a maximum of about 40 ft, is desirable. This limit for concrete may be extended to about 40 ft, but it would generally be uneconomical. The heaviest of the standard wide-flange steel girders, 36 in. deep, may occasionally be used to advantage on spans up to as much as 60 ft.

Exterior steel columns are placed a clear distance of 5 or 9 in. back from the building line. The larger dimension allows sufficient depth to accommodate 8-in. bond stones or a nominal 4-in. thickness of stone facing with one course of brick backing. Exterior reinforced-concrete columns, if covered by masonry or other architectural surfacing, are set back the distance that the surface treatment requires. Where not so covered, their outer surfaces may coincide with the building line as the reinforcement is adequately protected by the required concrete fireproofing.

In wall-bearing construction, the location of interior columns follows the same principles, except that masonry partitions may serve as bearing partitions, or as interior bearing walls, when their position is such as to provide support along lines that divide the floor area into spans suitable for the floor system. Bearing partitions, or bearing walls, should be shown upon the structural floor plans.

Column Types. Structural steel columns are usually of standard H section. Other types, e.g., pipe, and angle, are occasionally employed to meet special conditions. Cover plates are often used to increase the thickness of the flanges for heavily loaded columns. When sizes are computed and weights of structural steel columns are estimated, it should be remembered that fireproofing is an essential for all fire-resistant construction.

The flexibility of reinforced concrete permits changing the concrete section, the concrete mixture, and the steel reinforcement at any floor level. Standardization of the form work by avoiding all changes not absolutely justified by a reduction in material quantities, or other economic advantage, is a basic principle of concrete design. On the other hand, oversized columns waste floor area and those of minimum size, requiring maximum percentages of reinforcement, are often uneconomical. The form material should also be considered. Sheet-steel cylinders, commonly used to form round columns, so widely employed for industrial plants, are easily altered to suit varying column dimensions whereas a change in a wood form may require a round trip to the job form-building shop.

Reinforced concrete columns may be of practically any cross section provided design requirements are met. The only other restrictions are the practicability of constructing, filling, and stripping the forms and the appropriateness of the column section to receive the beams, girders, or other structural elements which they support. The common sections are square, oblong, round, oval, or as shown in Fig. 1. The types of columns ap-

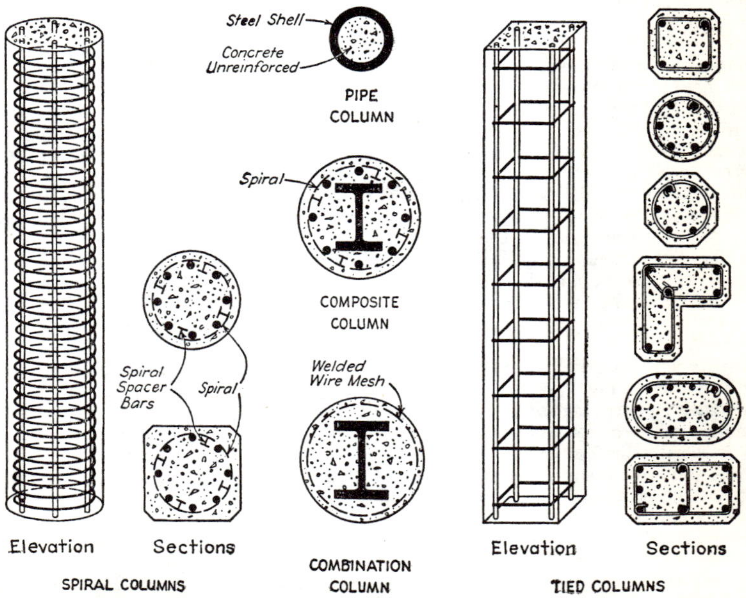

FIG. 1. Reinforced-concrete column sections. (Reproduced by permission from Crane, *Architectural Construction*, John Wiley and Sons, 1947)

propriate for reinforced-concrete designs are described in the following paragraphs based on the classification given in the *Report of the Joint Committee on Standard Specifications for Concrete and Reinforced Concrete* (Ref. 3).

(1) *The tied column* is reinforced with longitudinal bars or rods and separate lateral steel ties. This is the simplest design and widely used but, as its carrying capacity for a given cross section is less than any of the following designs, it would be discarded in favor of another type when the required size became too large for either economy or appearance.

(2) *The spirally reinforced column* is reinforced with longitudinal bars or rods enclosed by a closely spaced steel spiral. This design is particularly applicable to heavily loaded buildings such as factories, warehouses, and garages. In such structures a tied column is often appropriate for the top story, below which the spiraled design would be used.

(3) *The composite column* consists of a structural steel, or cast-iron section, encased in concrete reinforced with both longitudinal and spiral steel. A hollow-metal core can be used if filled with concrete. Loads are transferred to the metal core by bearing members such as billets, brackets, or other positive connections; they are placed at the top of the metal core and at intermediate floor levels where required. This design permits a much smaller cross section for any given load than even the spiraled column. Structural steel H sections are the usual choice for the column cores. They can be economically erected in one piece through a height of about six normal stories and are often helpful in reducing the size of heavily loaded columns through the lower stories of a comparatively high concrete-framed building.

(4) *The combination column* is a structural steel section encased in concrete 2 1/2 in. thick over all metal (except rivet heads), reinforced with a welded-wire mesh. Special brackets are used to receive the entire floor load at each floor level. This design has the same application as the composite column but is not as widely used.

(5) *The pipe column* is a steel pipe section filled with concrete. This type is similar to the Lally column. Its particular application is in low buildings and where fire-resistive standards are not severe. Often it may take the place of the interior wood column in wall-bearing construction and serve as a support for the lighter types of structural work.

Beam and Girder Locations. The designer should first decide between a "short-span" or a "long-span" system for the floor construction. The term short-span identifies a group of structural floor designs that are used on spans varying from about 6 to 10 ft and consequently require one or more intermediate beams between each line of columns.

Long-span floor systems are those capable of spanning, under various loads within the usual range, from 20 to about 30 ft, thus eliminating the need for intermediate beams. As the decision between these two different types of floor construction affects the design of the framing, it should be made at this time, prior to the choice of the particular type of floor system.

Long-span designs are particularly suitable for light and medium floor loads where there are no heavy concentrations. Avoiding beams except along column axes, particularly where columns and connecting beams can be concealed by partitions, improves the general appearance of a room and facilitates ceiling lighting. To obtain a flat ceiling with a short-span design, by the use of a hung ceiling, is usually more expensive than to choose a long-span design, but there may be structural reasons which favor the shorter slab spans.

The direction in which to run the girders or deeper beams may be decided by any one of several factors. If windows are to extend as high as possible, girders should run across the building, permitting spandrel beams of lesser depth. This arrangement also favors structural rigidity, particularly desirable for high, narrow buildings. Another consideration is the desirability of placing the girders, or heavier beams, on the shorter spans for the sake of conserving headroom or improving the appearance of an open floor area. The designer will usually find that the special conditions of his particular building result in a definite preference.

In structural steel design, the beams and girders may be framed with their top surfaces flush, as shown in Fig. 2, or the beams may be dropped sufficiently to avoid the expense of coping their upper flanges, as shown in Fig. 3. If precast-concrete slabs, gypsum plank, or similar units are used as a structural floor or roof deck, a flush surface is required. The same is usually preferable for long-span designs as the principal beams frame into the columns and are not supported by girders. Either type of assembly may be preferable for short-span designs with poured concrete slabs. If panels are square or nearly square, the girders determine the clear story height, and it may be advantageous to drop the beams.

Flat roofs, identified as those that are horizontal, or only slightly sloped in one or more directions for the purpose of drainage, are usually framed with the same materials and often with the same arrangement of framing members as used for the floors below. For multistory factories, the same system should usually be retained unless special features of the roof, or obvious economy, justifies a change. Pitched roofs are framed of steel or wood because of the difficulty of pouring concrete on steep slopes. The various designs that have been developed for precast-concrete framing have not, up to the present time, been very widely used.

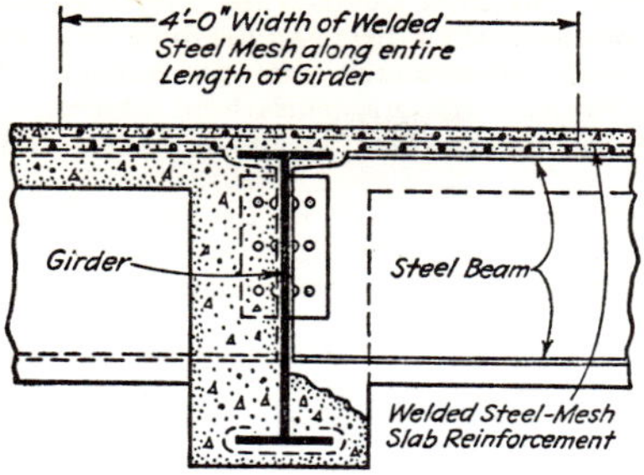

FIG. 2. Beam-and-girder connection with upper surfaces flush. (Reproduced by permission from Crane, *op. cit.*)

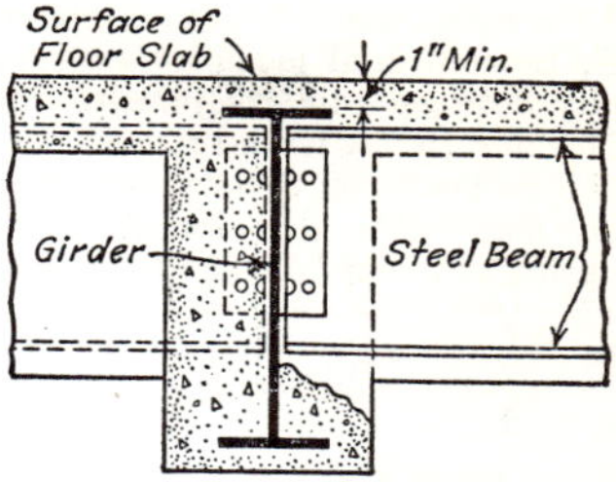

FIG. 3. Beam-and-girder connection with beams dropped. (Reproduced by permission from Crane, *op. cit.*)

Beam and Girder Types. A variety of structural steel shapes and built-up sections have been used for framing buildings. Present practice favors the wide-flange H sections for the principal beams and girders, unless the span or load requirements demand built-up units such as plate girders or trusses. Around openings, such as stair wells, the steel section is chosen to facilitate the stair erection, and various lightweight types have been developed for roof construction. Typical stair arrangements are shown in Fig. 4.

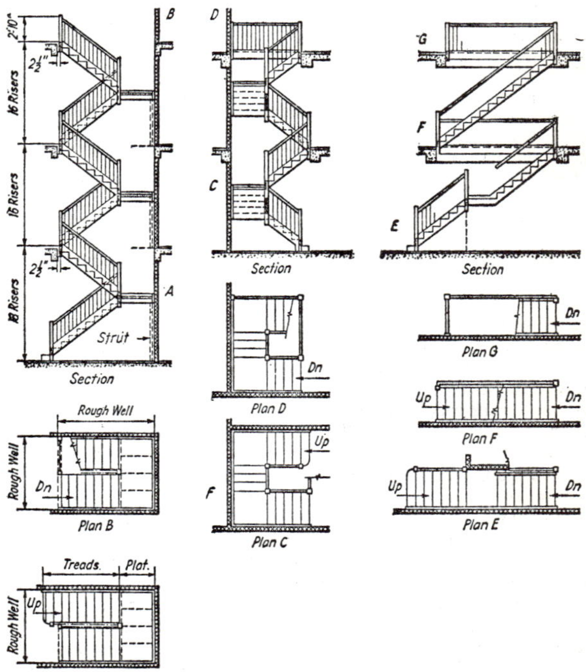

FIG. 4. Steel stair designs. (Courtesy of the National Association of Ornamental Metal Manufacturers)

The usual reinforced-concrete beam or girder is of T section, the flange of the T being supplied by the contiguous floor construction which is placed monolithically with the beam or girder. Around openings, such as stair wells, the section takes the form of an inverted L. In the case of an exterior beam spanning between wall columns, along an open well, the section becomes rectangular. Figure 5 shows the usual sections, including shallow beams, which have the advantage of facilitating ceiling installations but require more reinforcement because of their reduced effective depth.

Concrete Flat-slabs Designed with Specified Moment Coefficients. For multistory factories this system often offers more advantages than beam-and-slab or beam-and-girder construction. There are obvious exceptions where many openings, changes in floor level, concentrated loads, offset columns, or other structural variations interrupt continuity between adjacent panels. Furthermore, in some cases, beams or girders may be helpful in resisting lateral forces.

The conventional flat-slab assembly is shown in Fig. 6. It is particularly suitable for heavy floor loads. If columns can be placed without reference to interior requirements, a spacing of about 20 to 22 ft is economical for superimposed floor loads up to 300 lb per

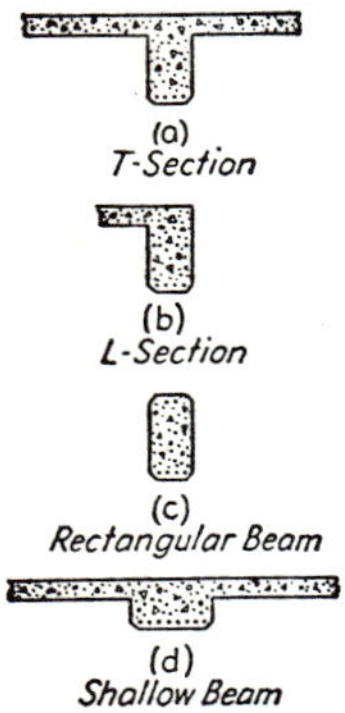

Fig. 5. Types of reinforced-concrete beams and framing plans. (Reproduced by permission from Crane, *op. cit.*)

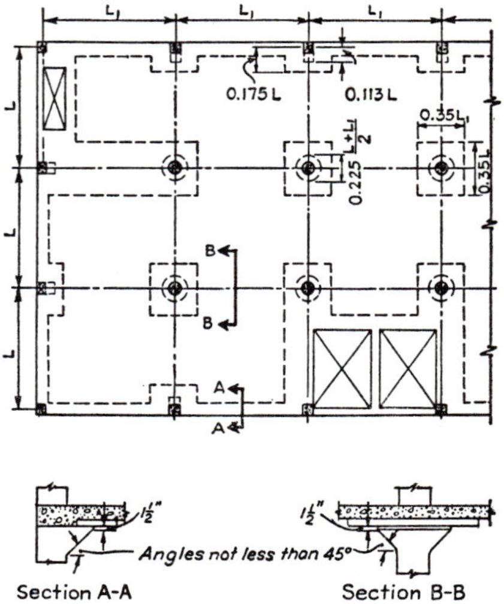

Fig. 6. Typical reinforced-concrete flat-slab design. (Reproduced by permission from Crane, *op. cit.*)

sq ft. Garages are often designed with a column spacing of about 26 by 22 ft. A 30-ft spacing approaches the practicable maximum for a uniformly distributed superimposed load of 200 lb per sq ft, unless the dead load of the construction is decreased by fillers or voids. In Fig. 6, wide, shallow sections of the same depth as the drop panels are used along all discontinuous edges of the slab. Narrower and deeper beams are often desirable around stair and elevator wells above partitions. Openings of any size may be cut through a flat slab if adequate structural provision is made for the stress requirements.

The designs employing specified moment coefficients, such as are given in the *Building Regulations for Reinforced Concrete of the American Concrete Institute*, are limited in their application. The ratio of length to width of panels must not exceed 1.33; there must be at least three continuous panels in each direction. Successive span lengths in each direction must not differ by more than 20% of the shorter span.* Interior columns have flared capitals; brackets are used for exterior columns. Drop panels, as shown, are usually placed over columns as they are structurally economical but can be omitted if the slab is made thick enough to satisfy stress requirements. When drop panels are used over interior columns, half drops are used over exterior columns. Interior columns are usually of circular cross section with capitals in the form of an inverted frustum of a cone. Exterior columns are usually square or rectangular, windows often extending from column to column and the spacing chosen to fit stock sash, or the dimensions of other materials. Spandrel beams may be placed either above or below the slab with which they are cast monolithically. They may be comparatively wide and shallow or of any adequate section to suit the exterior wall construction.

Concrete Flat Slabs Designed by Elastic Analysis. For structures having less than three consecutive panels in each direction, or where the size or shape of panels does not

* This is the requirement of the American Concrete Institute. The *Report of the Joint Committee on Standard Specifications for Concrete and Reinforced Concrete*, 1940, allows only 15% variation.

fall within the specified limits, or where the designer desires to eliminate column capitals, specified moment coefficients are inapplicable. An appropriate procedure, implementing the method of elastic analysis, is given in *Building Regulations for Reinforced Concrete of the American Concrete Institute*, 1947. It should be noted that the use of specified moment coefficients for flat-slab designs which meet the requirements mentioned above is merely a special case for which it is possible, within these limits, to use a definite moment distribution and greatly simplify the computations. This practice corresponds to the use of specified moment coefficients in the design of a series of continuous concrete beams or girders, having spans of approximately the same length.

The broader application of continuity in concrete design has been widely accepted during recent years, as illustrated by the omission of column capitals and the use of flat slabs having a uniform thickness, or combined with "slab bands." Designs of this type are particularly appropriate for the comparatively light loads and moderate spans of apartment houses but have also been used for industrial buildings. Any such system, comprising a solid concrete slab, provides facility for embedment of electric conduit or other pipes, and the smooth surfaces of floor and ceiling can be utilized to save expense on fill or finishes.

Expansion or Contraction Joints. All commonly used building materials expand or contract as their temperature is increased or lowered. In the case of masonry assemblies, the resulting deformations, except under very extreme exposures, are more or less uniformly distributed throughout the joints and seldom require special provision. Structural-steel frames, used in building construction, do not usually require special provision for expansion or contraction joints except in the case of roof trusses of considerable length, as mentioned in Article 5. In reinforced-concrete design, however, joining is required for many types of structures. The following paragraphs on this subject are quoted from the *Report of the Joint Committee on Standard Specifications for Concrete and Reinforced Concrete.*

Purpose and Location. (*a*) If a reinforced-concrete structure is free to expand and contract with variations in temperature or moisture conditions (uniform over the section), no stresses of any importance will be developed. When the movement of the structure is restrained, however, or when temperature or moisture variations are not uniform over the section, stresses are introduced which should be provided for in the design.

(*b*) Points requiring special consideration are located where large changes in cross section occur or at corners in long members where expansions or contractions may result in rupture of the side members. No amount of reinforcement will prevent the formation of cracks in a restrained structure in which the required change in length exceeds the extensibility of the concrete, but reinforcement properly designed will serve to distribute the cracks.

(*c*) Joints providing a complete separation in the structure, when properly located, can relieve the restraint and prevent or greatly minimize the development of cracks. Such joints should be carefully designed with respect to the type of structure and should provide for the full expected movements without introducing local stresses severe enough to rupture the concrete. They should be located at predetermined points and fully detailed on the drawings. They may or may not be designed to carry load or distribute stress across the joint. Frequently this type of joint requires some form of protection to keep out extraneous material.

Expansion Joints in Long Buildings. (*a*) Expansion joints are expensive and in some cases difficult to maintain. They are, therefore, to be avoided if possible. In relatively short buildings, expansion and contraction can be provided for by additional reinforcement. No arbitrary spacing for joints in long buildings can be generally applicable. In heated buildings joints can be spaced farther apart than in unheated buildings. Also, where the outside walls are of brick or of stone ashlar backed with brick, or where otherwise insulated, the joints can be farther apart than with exterior walls of lower insulating value.

(*b*) In localities with large temperature ranges, the spacing of joints for the most severe conditions of exposure (uninsulated walls and unheated buildings) should not exceed 200 ft. Under favorable conditions buildings 400 to 500 ft long have been built without joints even in localities with large temperature ranges.

(*c*) In localities with small temperature ranges, the spacing of joints for unheated buildings or with uninsulated walls should not exceed 300 ft. In such localities buildings up to 700 ft long have been successfully built without joints where other conditions were favorable.

(*d*) In roof construction, provision for expansion is an important factor. The joints in the roof may be required at more frequent intervals than in the other portions of the building because of more severe exposure. In some cases expansion joints spaced 100 ft apart have been provided in roofs and not in walls or floors.*

(*e*) Joints should be located at junctions in L-, T-, or U-shaped buildings and at points where the building is weakened by large openings in the floor construction, such as at light wells, stairs, or elevators. Joints should provide for a complete separation from the top of the footings to the roof, preferably by separate columns and girders.

---

* In regard to paragraph (*d*), it is the author's opinion that if a joint across a roof slab is expected to function, the slab should not be rigidly connected to the supporting structure, unless corresponding joints are provided in the latter.

Joints of this type provide for contraction or expansion of the concrete by an actual separation of structural members and should not be confused with "construction joints" which are merely horizontal or vertical planes along which it is necessary to interrupt the continuous placing of concrete as the work progresses from day to day.

Control Joints. These are a modification of the expansion or contraction joint which produce a "weakened plane" through a concrete section for the purpose of controlling the location of potential cracks due to shrinkage occurring either during the early stages of hydration or subsequently by reason of drying or chilling. Such should be seriously considered where a change in the concrete cross section, or extensive unbroken areas of walls or floors, present the probability of developing secondary stresses due to volumetric change. Other critical locations are where wall panels join pilasters, where steel columns are cast into concrete walls, and where openings occur adjacent to long or thick wall sections.

WOOD FRAMING AND PREFABRICATION. Masonry Bearing Walls with Interior Wood Framing. This type of construction is often identified in building codes as "Ordinary Construction." Interior bearing walls are frequently used, particularly in basements and cellars. Steel or iron columns and structural steel girders are occasionally combined with wood-joist floor construction and wood-framed roofs. Most building codes have specific regulations applying to the manner in which structural members are assembled and supported.

If wood posts are used in the basement they should rest upon cast-iron bases for protection against moisture. Pipe columns are also supported upon cast-iron or steel-plate bases; they are generally preferable to wood posts. Any type of post or column, bearing wall or bearing partition, should rest upon concrete footings rather than merely upon a concrete cellar floor unless the latter is laid directly over rock. The lighter structural-steel sections are appropriate for girders supporting wood-joist floors and are greatly preferable to wood girders as an aid in diminishing differential settlement between the interior and exterior of the building. In this connection it should be noted that when exterior masonry bearing walls are used with interior wood framing there is the obvious tendency for the shrinkage of the wood to cause settlement of the interior assembly. Timber shrinks only slightly in direction parallel to the grain, but the cross-grain shrinkage may be considerable. For this reason, the framing should be arranged to eliminate such effects as far as practicable. The studs of wood-bearing partitions should rest directly upon the plate below when supported by a partition and directly upon the wood girder where such is used. If carried by a steel girder, this is also desirable unless a sole plate is considered necessary to permit nailing the lower ends of the studs.

Wood girders supported by a masonry wall are usually either held in place by a metal wall hanger or built into a recess of the wall and supported by a metal plate. The recess is made large enough to provide ventilation against dry rot. Wood joists should have a 4-in. bearing on exterior walls and should be secured by metal anchors placed on at least one joist in every 6 ft. The ends of joists are also beveled to permit release in case of fire without dislodging the masonry. Where supported by interior masonry walls, joists in small dwelling-house construction are merely lapped across the wall and spiked together, but special methods of support may be required for larger structures. Where supported by wood partitions which, in two-story dwellings, would normally require 2 by 4-in. studs, the floor joists are similarly lapped across the partition plate and spiked. Solid bridging is used at midheight of all wood-stud partitions. Where lateral stiffness is essential, partition bridging is inclined. Studs are doubled at sides of openings, with wood lintels or trusses across the heads.

Typical floor construction is of wood joists, usually 2 by 10 in. or 2 by 12 in., nominal sizes, spaced 12 to 16 in. on centers. A 3-in. minimum nominal thickness may be demanded by a municipal code within fire limits. Bridging is placed at a maximum spacing of 8 ft. The subflooring may be of $^7/_8$-in. matched boards, laid diagonally over the wood joists, or of plywood. Metal lath, occasionally wood lath, or gypsum lath serve as a base for plaster. Well-nailed metal lath, lapped and wired where sheets intersect, affords the best insurance against plaster cracks. Gypsum lath, often called "rock lath," requires less plaster and gives slightly better heat insulation. Many types of fiber and composition boards, finished to present a suitable ceiling surface or adapted to the application of paint, are available. If the roof surface is comparatively flat, construction is similar to that of the floors. If pitched, the rafters and ridge, forming the skeleton-like frame, carry shingle lath, solid boarding, or other desired material serving as a base for the roof surface and providing rigidity for the entire roof assembly. Where attic spaces are heated, an appropriate type of insulation is used between the roof rafters instead of in the attic floor.

Heavy Timber or Mill Construction. This term identifies buildings with exterior masonry walls and in which the interior structural framing is of heavy timbers designed to resist the effect of fire. For this reason it is sometimes called "slow-burning construction." Interior bearing walls of masonry are often used, particularly in structures of large area, for which they may be designed as fire walls. Wood posts are generally connected by steel or iron caps, pintles, and base plates, or by timber splice blocks of special design. Columns are not permitted to rest on floor timbers, and either stone, cast-iron, or steel bases are used to transmit the load to masonry foundations. Where beams or girders rest on masonry walls, boxes of self-releasing type, or an appropriate hanger, are required. Most building codes give specific regulations for connections between beams or girders and the supported columns or walls, and for the support of beams carried by girders.

In typical mill construction, wood girders span from an exterior wall to an interior column and then from column to column in one direction only. The girder spans between columns vary from 12 to 18 ft. No beams are used at right angles to the girders. The distance between the girders, varying from 7 to 10 ft, is spanned by heavy plank flooring. The smooth under surface of the floor plank lessens the flammability of the construction. Lath and plaster are not used on typical designs. Wood columns are of a minimum 8-in. nominal thickness in any dimension; corners are rounded or chamfered. Occasionally steel or cast-iron columns are used; they are generally permitted for this classification by building codes if protected to furnish a 1-hr fire-resistance rating. Wood beams and girders supporting floors have a minimum nominal thickness of 6 in. and a minimum nominal depth of 10 in., irrespective of stress requirements. Floors are usually composed of wood plank, tongued and grooved or preferably splined, not less than 3-in. nominal thickness and laid across the timber framing. Planks set on edge and spiked together are also used to form a laminated floor; the minimum nominal thickness of this design is 4 in. Over the structural wood floor, a 1-in. finished wood floor is laid to take wear and improve the appearance.

As the complete wood-framed structure, so widely used in this country for dwellings and small, commercial buildings, erected in rural districts, has practically no application for industrial buildings, the subject is omitted in this book.

Prefabrication. In the building industry this term identifies a degree of factory assembly beyond that customary in conventional practice. Several of the wall designs shown in Article 7 are types of shop-fabricated units similar in principle to those used in prefabricated structures. Many of the newer designs in laminated wood, used for both beams and trusses, are characteristic of present-day trends which gain economy by shop manufacture. Prefabrication is consequently a relative rather than specific designation, implying merely a method of construction by which economy is gained through the standardized, high-mechanized shop manufacture of building elements, such as wall, floor, and roof panels. Up to the present time the more complete types of prefabrication have been largely confined to single-family dwellings, military hospitals, barracks, and small service structures.

These designs are usually of steel or wood-panel construction. Although there have been many fairly complete prefabricated concrete systems patented during the last generation, most of them were dropped after the construction of a few test houses. Others have been applied, more or less successfully, over a period of years, but have little significance for industrial work. Precast structural members of reinforced concrete have, however, interesting possibilities for factories and warehouses. During World War II precast concrete columns, girders, and slabs were erected, and two large single-story warehouses were built for the Navy of thin-shell precast concrete. The framing comprises a system of bents supporting precast roof panels. These bents are of hollow section, and reinforcement is welded to obtain continuity where adjacent segments join. For a detailed description of these buildings see Refs. 4 and 5. In 1947 four dormitories were erected for the University of Vermont, at Burlington, Vt., using precast concrete slabs 8 in. thick for the structural portion of the exterior walls and precast concrete slabs 6 in. thick for the interior partitions. These elements supported concrete floors and roofs which were poured-in-place. The complete exterior wall section comprised a 4-in. exterior surface of brick, a 2-in. air space, and the 8 in. of precast concrete slab. No plaster was applied to the interior surfaces of either walls or partitions. Although these uses of precast concrete slabs were for residential work, systems of this type have particular significance for warehouses and garages. The following year eight large dormitories of a similar design were erected for the University of Connecticut, at Storrs, Conn.

In another noteworthy design the exterior walls are composed of Latisteel sections with 2 in. of concrete. Panels are cast in a horizontal position. After a 24-hr curing period, they are erected by crane and bolted in place. The roof is of conventional design, having a steel deck of welded sections supported by steel trusses. It is surfaced with insulation and waterproofing. This system has particular application to industrial work.

4. CHOOSING THE FLOOR CONSTRUCTION

Structural floor systems can be divided into two groups, the long-span and the short-span designs. A choice between these two types should be made, as mentioned in Article 3, prior to the planning of the building frame. The selection of a particular design for the structural floor follows logically as the next step in determining the structural assembly for a building. In this book *system* identifies the structural assembly spanning between supports, such as means or girders, or bearing walls. The following paragraphs describe the more popular types of floor construction and indicate the particular application of each type to conditions which will be known by the designer at this stage of his work. The maximum spans and limiting dimensions noted for the various designs conform to structural standards generally accepted, but they should be checked against the requirements of local building codes.

The term *long-span* identifies designs which ordinarily require no beams except along the center lines of columns. As a group, they are particularly suited to light and medium, uniformly distributed superimposed loads, 40 to 100 lb per sq ft, and to spans 15 to 30 ft. They have a wide application except for the heavier industrial loads. The *short-span* designs require structural supports, usually in the form of beams, spaced 6 to 10 ft apart. These designs are used for all types of occupancy.

LONG-SPAN DESIGNS. Metal-pan System. This is one of the most popular of the reinforced-concrete ribbed designs. A comparatively thin, lightly reinforced-concrete slab, normally 2 to 3 in. thick, is poured over voids formed by steel forms as shown in Fig. 7. This slab, called the "topping," is supported by and cast monolithically with reinforced-concrete ribs, or joists, spaced 12 to 30 in. apart, face to face. Supplemented by the slab, these ribs form a series of T beams, spanning between the principal framing members of the floor assembly and furnishing the structural support for the system. The width of the ribs is normally 5 in. They should be straight and of a depth not over three times their average width. The slab thickness should be at least one-twelfth of the clear distance between ribs. Reinforced-concrete bridging, usually 4 in. wide and of the same depth as the ribs, is required on long spans. The width of ribs supporting masonry partitions is increased as required to carry the additional load. Rigid, electric conduit is sometimes placed in the open spaces beneath the pans, or the thickness of the topping is increased to accommodate such installations. In many cases a fill, about 3 in. deep, is placed over the structural floor. This provides space for the conduit or other pipes.

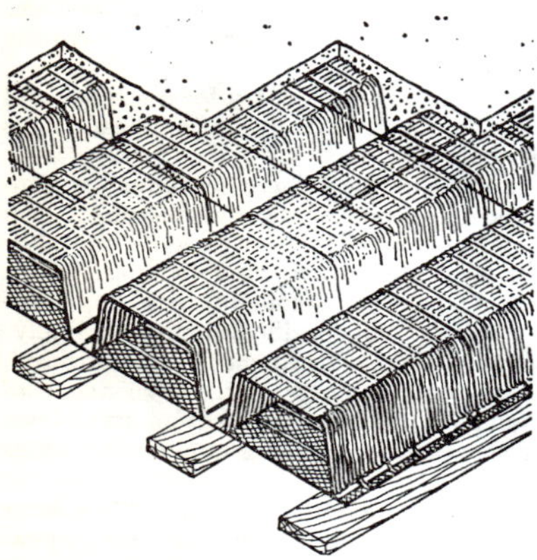

Fig. 7. Metal-pan system. (Courtesy of Truscon Steel Co.)

Metal pans are manufactured in standard widths of 20 and 30 in.; the latter is a generally accepted maximum. Smaller widths are available for filling odd spaces. Standard depths vary from 6 to 14 in. by 2-in. increments. Metal pans are assembled by lapping the ends. Appropriate lengths are supplied by manufacturers to suit any length of floor panel. The end units are furnished with tapered sides or with both side and top tapered, so as to enlarge the cross section of the concrete rib, or of both rib and slab, adjacent to supports. This system can be used with either steel or reinforced-concrete framing or with bearing walls. For additional information on this subject, refer to the Concrete-reinforcing Steel Institute.*

Structural Clay Tile System. This design, shown in Fig. 8, is similar to the metal-pan system except that structural clay-tile fillers are used in place of pans. The tile may be 12 by 12 in., with depths varying from 3 to 12 in.; 12 by 16 in., or 16 by 16 in., with depths from 3 to 10 in. A 12- by 20-in. tile is also manufactured with depths varying from 4 to 10 in. The larger sizes of tile usually furnish the more economical design. Oblong tile are assembled with the longer dimensions perpendicular to the ribs. As a 5-in. rib would normally be used with a 16-in. tile, the center-to-center spacing of the ribs is 21 in.

* Wood or plywood forms are occasionally employed for the same purpose, but metal is generally more economical. Except for the material used to enclose the void between ribs and details in regard to the forming, all characteristics are identical.

A 6-in. width of rib is appropriate for a 20-in. tile, resulting in a 26-in. center-to-center spacing. A 1 1/2-in. minimum thickness of concrete topping is generally required for the 12- by 12-in. tile and a 2-in. minimum thickness for the larger sizes.

The lower faces of the tile provide an excellent surface for plaster. Soffit blocks of the same material are often used, as shown in Fig. 8. This practice is recommended. The ends of each line of tile, where in contact with the concrete, are generally closed by sheets of metal or other appropriate means. Bridging is not required because of the lateral support furnished by the tile. The thickness of ribs is increased beneath partitions as noted for the metal-pan system. Rigid electrical conduit is generally placed in the concrete slab over the tile or in a fill laid over the structural floor. This system can be used with steel or reinforced concrete framing or with bearing walls.

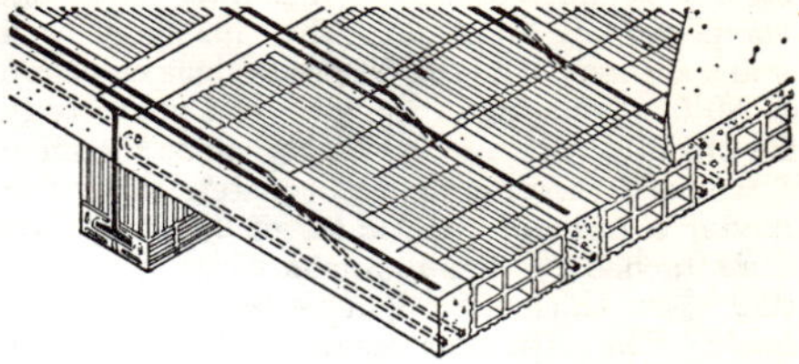

FIG. 8. Standard clay-tile system. (Courtesy of National Fireproofing Co.)

Concrete-block Systems. The essential characteristics of these designs are similar to clay-tile construction except that the fillers between the reinforced-concrete ribs are hollow block composed of portland cement combined with sand, cinders, slag, Haydite, Waylite, Super-Rock, or other lightweight aggregate. Haydite is composed of burned shale ground to the required size; Waylite and Super-Rock are made from expanded blast-furnace slag. Although these materials are produced in many parts of the United States, their availability at a competitive price should be checked for any particular locality. Figure 9 shows a typical design, and Fig. 10 a variation of the standard concrete-block system which provides a uniform surface for plastering without the use of soffit blocks. These systems can be used with steel or reinforced-concrete framing or with bearing walls.

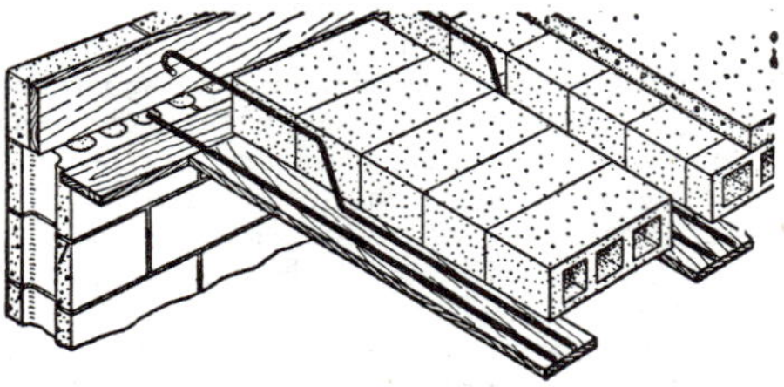

FIG. 9. Standard concrete-block system. (Courtesy of Portland Cement Assoc.)

Various other types of filler blocks, such as gypsum and compressed fiber units, are less frequently used and have little application to industrial buildings.

Two-way Systems. If the framing of a floor assembly results in panels that are square, or approximately square, it may be economical to use any one of the concrete-ribbed systems in a two-way design. This results in forming a type of flat plate which requires a supporting beam, or bearing wall, on each of the four sides of the panel. The particular application of such designs is for buildings which require a number of large, square, or approximately square panels, 20 to 25 ft between column center lines, and preferably for floor areas large enough to permit several continuous panels of about the same size in each direction. Two-way designs are similar to the corresponding one-way systems except that the metal-pan or filler block must present a solid surface on all four sides, or one which will at least prevent the entrance of any appreciable amount of concrete into the interior of the void or block.

In the metal-pan system, this requirement is met by the use of steel domes identified as the "Grid System." The Grid System steel domes have their particular application for industrial buildings. Domes are available in an 8-in. depth, 19 by 19 in., and in a 4-in. depth, 18 by 18 in. The width of the rib is controlled by the flange width of the dome; it is 5 in. for the deeper type and 4 in. for the shallower. The 8-in. depth, with 6 in. of concrete topping over the metal, has been used on panels 36 by 38 ft to support a uniformly distributed superimposed load of 125 lb per sq ft. The 4-in. depth has been used on 20 by 20 ft panels with a 2-in. concrete topping for a similar load of 100 lb.

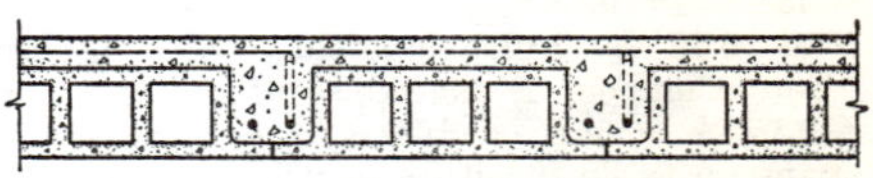

FIG. 10. Concrete-block system, with special block forming soffits of concrete ribs. (Courtesy of Portland Cement Assoc.)

The clay-tile system may also be used in a two-way design. Closures are generally placed across the open ends of the tile to eliminate the entrance of concrete into the cells. Specially designed tile, however, are available for two-way systems, such as the gypsum unit, manufactured with four solid faces, but they have little application to industrial buildings.

Precast Concrete Joist Construction. The joists may be cast of stone, cinder, Haydite, or other lightweight-aggregate concrete. They are manufactured in standard depths of 6 to 14 in., varying by 2-in. increments. Special joists are available up to 22 in. in depth with flanges up to 16 in. wide. These precast concrete joists are particularly appropriate for wall-bearing designs and comparatively low buildings. They may be used with either steel or reinforced-concrete framing. Although identified ordinarily with light loads and comparatively long spans, they may also be found satisfactory for commercial buildings and warehouses, as the deeper designs can support comparatively heavy loads on spans up to 30 ft. Bridging in the form of a concrete rib, cast-in-place, is used on spans over 16 ft for the 8-in. joist and on longer spans for the deeper units. Joists are usually placed 2 to 5 ft apart, carrying precast concrete slabs 2 to 3 in. thick, or cast-in-place concrete slabs having a minimum thickness of 2 1/2 in. Concrete plank are often appropriate. Plank are attached by clips to anchors cast in the joists. They can be sawed, nailed, or drilled. A 2 3/4-in. thickness might be used on a joist spacing of 5 ft for light and medium floor loads. The particular economy of this system occurs where it is permissible to leave the joists exposed on the underside of the floor.

Precast Concrete Cored Designs. The Flexicore precast system comprises structural units 12 in. or more in width and of a length and depth required for the particular span and load. These units are available in many localities and can be assembled with bearing walls, steel or reinforced-concrete framing. Round cores are obtained by expanding elastic tubes to form the inner void around which stone, gravel, or lightweight-aggregate concrete is poured in steel molds. With adequate steel reinforcement, prestressed in shop manufacture, a 6-in. total depth can be used to support light loads on spans up to about 22 ft. For short-span designs, up to 10 ft, superimposed loads of several hundred pounds per square foot may be supported. Because of the slight variations in top surfaces, characteristic of all such unit construction, a mortar topping about 1/2 in. thick should be applied as a base for linoleum, asphalt tile, or similar surface finishes. Ceiling surfaces are reasonably smooth, and electric conduit can be run within the structural thickness of the slab. These units are particularly appropriate for low residential and commercial structures. They are used with steel framing or supported by bearing walls.

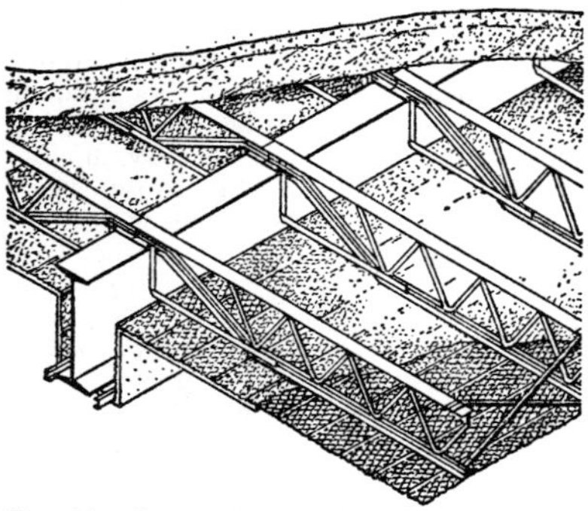

FIG. 11. Open-truss steel joists. (Courtesy of Truscon Steel Co.)

Open-truss Steel-joist Construction. The system shown in Fig. 11 is particularly adapted to buildings with comparatively light floor loads, and although occasionally used for high structures is more appropriate for low buildings. It competes in the same field as the precast-concrete joist assembly but is not recommended for use with a reinforced-concrete frame. Although inappropriate for heavy industrial loads and as a support for machinery which might induce vibration, open-truss steel joists might well find their place in the more lightly loaded buildings of an industrial group. The spacing of joists is limited to 24 in. for floors, 30 in. for roofs having a poured concrete slab, and 7 ft for roofs having a steel or wood deck. The structural slab, supported by the joists, is usually of concrete 2 to 2 1/2 in. thick, poured in place over ribbed-wire fabric of close mesh or welded-wire fabric backed by heavy paper. Joists are supplied in depths varying from 8 to 16 in. by 2-in. increments. Steel bridging, in the form of rods or fabricated from 3/4-in. channels, is placed at a maximum spacing of 7 ft. A 16-in. joist supporting a 2 1/2-in. concrete slab, can carry a uniformly distributed superimposed load of 75 lb on spans up to 30 ft with a joist spacing of 21 in. This system is used with steel framing or supported by bearing walls. For further information on this subject refer to the *Steel Joist Institute Specification.*

Stran-steel Joists. These lightweight, cold-formed steel sections are used as floor joists for the lighter load requirements and moderate spans. They are manufactured of 16 to 11 gage copper-bearing steel and in depths of 6 to 9 in. A typical assembly comprises a 2-in. reinforced-concrete slab, supported by corrugated metal laid over the joists, which are usually spaced 12 to 32 in. on centers. With a spacing of 18 in., a 9-in., 13-gage joist, with a 2-in. concrete slab, can support a uniformly distributed load of approximately 60 lb per sq ft on an 18-ft span. Twenty feet is about the limit for light loads. These joists have their particular application for low residential buildings. They would seldom be appropriate for the requirements of even light manufacturing, but might be an economical choice for floors of an industrial housing development. They are used with steel framing or bearing walls.

Concrete Flat Slabs. These designs, particularly appropriate for industrial buildings, were discussed in Article 3. They are types of construction rather than floor systems.

SHORT-SPAN DESIGNS. Solid Reinforced-concrete Slabs. In combination with steel framing, either stone or lightweight aggregate may be used. With reinforced-concrete framing, stone-concrete slabs represent a standard design. However, it is not practicable to use a lightweight aggregate with concrete framing except in the rather unusual case where the supporting structural members, as well as the floor slabs, are cast of the same material. This has occasionally been done, however, where a prime requirement was to lighten the dead load of the entire assembly. Concrete slabs used as structural floors normally have a minimum thickness of 4 in.; a 3 1/2-in. thickness is permitted for roofs. Figure 12 shows a typical section with structural-steel framing which would appear the same for either stone or lightweight aggregate.

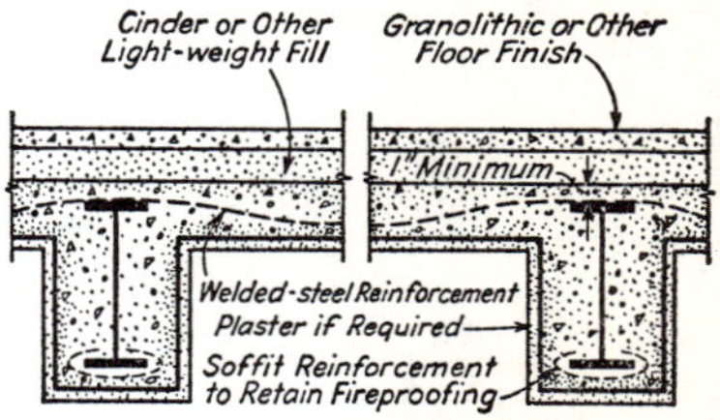

FIG. 12. Short-span, reinforced-concrete slabs supported by structural-steel framing. (Reproduced by permission from Crane, *op. cit.*)

In short-span assemblies of this type, the beams may be connected to the girders so that the top surfaces of the two are at the same level, or the beams may be dropped sufficiently to avoid coping their upper flanges. In such designs, the structural-concrete slab is often finished flush with the upper surface of the structural steel, but it is better practice to provide a minimum thickness of 1 in. of concrete over the upper flanges. This concrete serves as a protection for the welded-wire reinforcement of the slabs where it passes over the beams and permits the introduction of a 4-ft width of the same steel mesh along the girders, placed with the heavier wires at right angles to the girder flanges. The purpose of this additional reinforcement (shown in Fig. 2) is to minimize the probability of cracks developing parallel with the sides of the girders.

The so-called cinder-concrete arch, actually a short-span slab, as shown in Fig. 12, has been a standard type of construction in New York City for many years. The more general use of this design is limited to clear spans between steel beams of not over 8 ft. The reinforcement, in the form of welded-wire mesh, draped over the supporting steel beams, is designed for the span and loading conditions. Haydite, Waylite, and various lightweight cement-aggregate mixtures, such as Aerocrete, Porete, and Gritcrete, are also used for both short-span floor and roof slabs.

Precast cement-aggregate slabs are occasionally appropriate for short-span floor systems supported by structural-steel framing, especially where it is unnecessary to fireproof the steel, but their special application is for roofs. Slabs are cast of small-size stone aggregate, Haydite or other lightweight material, and reinforced with galvanized, welded-steel mesh. They are anchored to the girders by bolts. Such systems have been used successfully for many types of occupancy varying from industrial buildings to schools. When lightly reinforced, a web thickness of 2 1/4 in. with channels 4 3/4 in. deep is suitable for a uniformly distributed superimposed load of 75 lb per sq ft on spans up to about 10 ft. Considerably heavier loads can be carried by increasing the amount of reinforcement. The channel sections rest upon either steel or concrete beams to which they may be bolted if required. Joints are grouted with cement mortar and troweled level. A 1/2-in. mortar surfacing is applied beneath linoleum or asphalt tile. If wood floors are required, metal clips are inserted in the grout between the slabs for the retention of wood sleepers.

Lightweight concrete is also employed in the form of precast channel sections and plank. The former are appropriate for light or medium, uniformly distributed floor loads on spans up to about 8 ft; the 2-in. plank, which is used over steel beams, open-truss steel joists, or wood joists, can support similar loads on spans up to 4 ft. Although suitable for floor construction, when the surface is properly protected, plank of this type have their particular application for the roofs of industrial buildings.

Field-poured Gypsum Systems. Gypsum mixed with wood shavings, usually in the proportion of 12 1/2% by weight of fiber and 87 1/2% of gypsum, is used for short-span floor and roof construction in the form of slabs, for field pouring. Plywood forms are hung from the steel framing, which can be successfully fireproofed with the same material, or with precast gypsum units manufactured to fit a wide range of structural shapes. Such a design would not be used with reinforced-concrete framing. Slabs poured of the same mixture, or with less wood fiber for the purpose of producing a greater compressive strength, may also be used above open-truss steel joists, spaced not over 36 in. on centers. This design does not require bulb tees. The gypsum slabs 2 to 3 in. thick are always reinforced with welded-steel mesh. For additional information on this subject see *Reinforced Gypsum Concrete*, American Standards Association, A59.1 (1945).

Brick and Structural-clay Tile Arches. These designs are seldom used in modern buildings. In most districts stone concrete, or lightweight-aggregate concrete is far more economical for short-span designs.

Reinforced-brick Slabs. Brick laid on edge with small rods embedded in the mortar joints have been used for short-span structural floor slabs. This system, however, although employed to a limited extent in England, India, and other parts of Asia, is usually uneconomical in the United States.

Steel-plate and Sheet-steel Cellular Floor Systems. Several designs in this group are appropriate for use with structural-steel framing. Although some systems are capable of supporting medium floor loads on spans up to 20 ft, they are most suitable for short-span designs. If the type of occupancy permits the use of a steel ceiling on the soffit of the structural floor, they have an added economic advantage. Short-span designs may be adapted to either light or heavy floor loads. A system providing space for the installation of electric wiring within the thickness of the structural floor is illustrated in Fig. 13. This

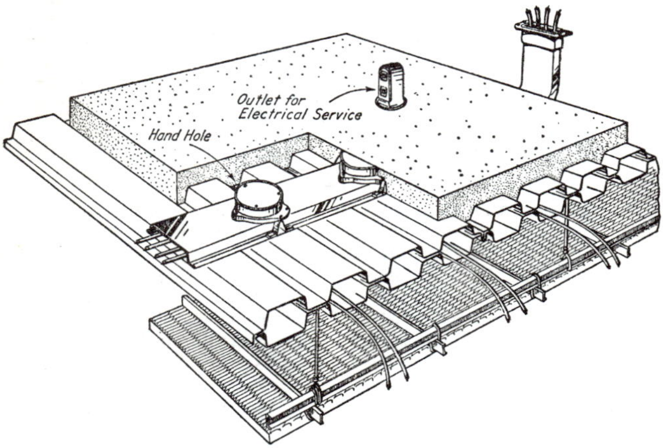

FIG. 13. Sheet-steel cellular floor construction. (Courtesy of H. H. Robertson Co.)

particular system, which has a broad application for many types of occupancy, has recently been used in a number of large structures, including office buildings, government buildings, and factories.

Wood Plank Floors. For industrial buildings such as textile mills, where the weight of the machines, except for looms, is comparatively light, heavy plank may occasionally be used to advantage for structural floors spanning between steel beams if the fire risk can be accepted. Such a floor has the advantage of facilitating the bolting down of machines in any desired location. An automatic sprinkler system would often be installed beneath floors of this type.

FLOOR SYSTEMS FOR TIMBER FRAMING. If wood framing is used with exterior masonry bearing walls, an ordinary wood-joist design is customary except for industrial building or warehouses, where "slow-burning" or mill construction might be used. For structures where timber forms the entire framework, including that of the exterior walls, wood-joist floor construction is generally employed. Such designs, however, are omitted as they have practically no application to industrial buildings.

FLOOR CONSTRUCTION ON OR NEAR THE GROUND. First-story Floors. If the floor system is self-supporting, even if there is no cellar beneath, any appropriate structural system previously described may be used. When framed in wood, it is imperative that the space beneath be ventilated. Ventilation is obtained by openings in the foundation walls or between foundation piers. If continuous walls are used for the foundation, openings of at least 1 sq ft in area should be placed just above grade on all four sides of even small buildings. All openings should be covered by grills. If part of the cellar is fully excavated and contains a heating unit, the heat from this section can be allowed to pass entirely beneath the floor. Such an arrangement obviates the need for openings in the foundation walls and minimizes the need for insulating the floor.

If the floor is laid upon the ground, a concrete slab with a granolithic, or other desired surface finish, is often appropriate. A bituminous compound may be added to the concrete composing the floor slab or incorporated in the finish, to increase resilience. Figure 14, sketches *a* and *b*, show typical details for various conditions. Sketches *b* and *d* show wood flooring supported by sleepers. For these may be substituted a 2-in. thickness of lightweight concrete composed of portland cement and a fine mineral aggregate, forming a

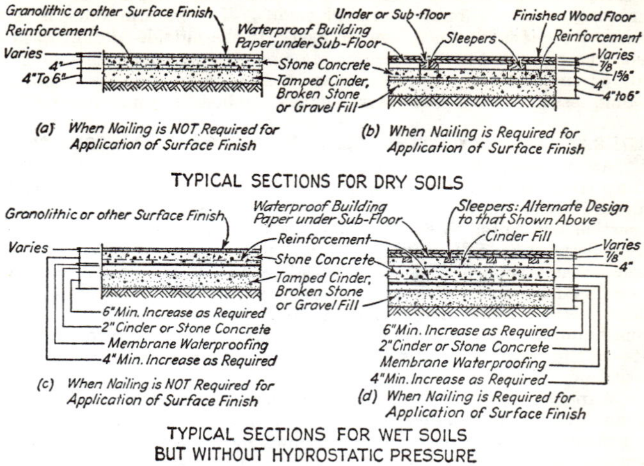

FIG. 14. Floor construction on soil. (Reproduced by permission from Crane, *op. cit.*)

base to which a wood floor may be nailed. Wood flooring may also be bedded in mastic, applied directly to the concrete slab. If wood sleepers are used they should be treated with a preservative. Because the elimination of moisture is a primary necessity, a porous fill, composed of cinders or gravel, is recommended even over "dry" soils. Even if floors are laid slightly above the level of the surrounding ground, some of the provisions shown in sketches *c* and *d* of Fig. 14 might be desirable. These two designs, however, have their particular application to basement and cellar floors. Roof water should be effectively diverted from the building by shallow gutters or drains. It is common practice to raise the grade of the ground about 6 in. adjacent to building walls. The surface of the floor slab should be a foot or more above the adjacent ground level, and a greater height may be necessary if the building is located at the bottom of a natural drainage area.

Basement and Cellar Floors. When concrete floors are laid over soil known to remain comparatively dry at all seasons, 4 to 6 in. of porous fill, cinders, gravel, or crushed stone, as in the case of floors slightly above the natural grade, should be sufficient as a base for

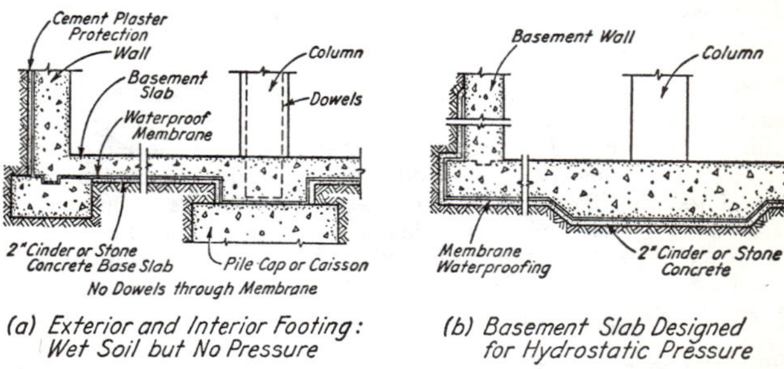

FIG. 15. Arrangement of waterproof membrane for deep basements. (Reproduced by permission from Crane, *op. cit.*)

concrete slabs. Even then, however, it is desirable to provide land tile drains around the building at the footing level. Such a design is particularly appropriate for comparatively shallow basements.

When concrete floors are laid over soil which is actually wet, or likely to become wet, at certain seasons of the year, the thickness of the fill should be increased to perhaps 12 in.

In extreme cases, parallel lines of land tile are laid beneath the entire cellar area at a spacing of 3 to 4 ft. All land-tile lines should be connected with tight-jointed clay sewer pipe leading to a sewer, storm-water drain, dry well, or, preferably, if grades permit, to the surface of the ground.

In deep cellars, and often in shallow cellars where grades are unfavorable, it may be necessary to install either a sump pit with an automatic electric pump or a complete system of waterproofing. Figure 15 shows typical details. The waterproofing is carried up the exterior walls above any potential water level. If the outside of the basement walls is inaccessible, the same type of waterproofing may be placed on the inner surface and retained by either a masonry or concrete seal. If conditions do not demand membrane waterproofing, less expensive treatments may be applied to either the exterior or interior surfaces.

CONSIDERATIONS AFFECTING THE CHOICE OF THE STRUCTURAL FLOOR SYSTEM. At this stage of the design procedure it will be known whether structural steel, reinforced concrete, or wood is to be used for the structural framing. This fact is one of several guides in selecting the floor system as it has been shown that some systems are not only particularly suited to steel or concrete but also cannot be used except with one or the other. For example, if a building were framed in steel, the floor construction could not be designed as a typical "flat slab" or rigid frame, but any other system mentioned in this book could be used. If framed in reinforced concrete, steel-plate and sheet-steel cellular floor systems would not be appropriate. The selection of the floor system for wall-bearing buildings follows the type of interior framing. For wood-framed structures, wood joists are the normal choice, supported by wood framing or unprotected steel girders and pipe columns.

The size, shape, and continuity of floor panels may influence the choice of the system as the dead load of the heavier designs becomes excessive for long spans. The ratio of length to width of panel usually determines the advantage of a one-way as compared to a two-way system. This same ratio of length to width of panel and the requirements for continuity between at least three consecutive panels may preclude the use of the standard flat-slab design with specified moment coefficients.

The need for structural stability is particularly important for types of industrial occupancy where manufacturing processes cause vibration and would preclude the lighter systems such as the open-truss steel joist. Where the effect of lateral forces, due to wind or earthquakes, is a factor in the design, preference is given to systems that have sufficient strength to act as stiffening elements of the building assembly.

As there is a wide range in the relative fire resistance of the structural floor systems, the designer should choose one that not only meets the demands of the local code, or good practice, for the locality and type of occupancy, but is also consistent with the rest of the construction. This should be done to obtain a favorable insurance rate on both building and contents.

The nature of the under surfaces of the various floor systems should be considered in relation to the ceiling finish. For example, structural clay tile, gypsum, or Haydite block provide an excellent base for two-coat plaster, whereas the metal-pan and steel-joist systems require metal lath to carry the plaster. Solid concrete slabs should not be plastered. If the concrete is poured over well-built forms, two coats of paint, applied by sprayer, can give an excellent surface.

The nature of the floor finish may also influence the choice of a structural floor system and, in particular, the need for concealing electrical conduit or other pipes within the thickness of the floor. Some systems provide space within the structural depth; others require a lightweight fill, 3 or 4 in. deep, between the structural slab and the surface finish. Conduit and small pipes can very easily be embedded in any solid concrete slab. They can be easily accommodated within the depth of open-truss steel joists. The same is true of the precast concrete-joist system unless the plastered ceiling is omitted. It is not generally practicable to place even electric conduit within the ribs of the concrete-ribbed designs, such as the metal-pan and cinder-block assemblies. For these it is usually necessary to increase the thickness of the topping over the pan or block, or to place a fill above the topping.

The weight of the floor system is a minor matter for buildings of only a few stories, but obviously important for higher structures. The thickness should also be considered for buildings of many stories as every inch added to the floor thickness, from the face of the ceiling to the surface of the finish, adds an additional inch to the height of every vertical element in the entire building. Although the relative heat transmission of competing designs is usually negligible for floors, it should be considered if there are extensive areas beneath unheated spaces. The acoustical requirements may occasionally affect the choice of a floor system but would seldom be important except for residential buildings.

A careful analysis of the various possible designs for the structural floor will usually result in reducing the practicable alternates to two or three types. The final step is to make preliminary designs of the proposed assemblies, including the supporting members of the structural frame as well as the actual floor panel. When accompanied by the necessary general information concerning building dimensions and plot conditions, these sketches should enable any competent builder, familiar with local conditions, to make a fairly accurate estimate of cost for each system. From this estimate the designer can make a final decision.

5. SELECTING THE TYPE OF ROOF FRAMING

Except for reinforced concrete used in the form of rigid frames or bents, arches or domes, all pitched and curved roofs are framed in steel or wood. Short spans, particularly over wall-bearing designs, may be supported by steel or timber rafters or wood trusses, with or without a system of purlins. For longer spans, steel or wood trusses, rigid frames and arches of steel, concrete or wood, supplemented by domical forms, may furnish the support for the roof deck. Trusses and rigid frames are particularly appropriate for the wide spans required in factories and hangars.

The framing for flat roofs of multistory buildings often follows the same arrangement as that of the floors below, the final design reflecting the difference in live load. This choice is particularly desirable for multistory concrete structures, in order that the same forms may be used with a minimum of changes. Only where the structural requirements of the roof design are materially different is economy obtained by changing to a lighter-weight design.

The framing for flat roofs of single-story industrial buildings usually comprises some type of parallel chord truss, in wood or steel, with or without purlins between the trusses. Rolled-steel beams of heavy section offer a possible alternate for moderate spans. Reinforced-concrete bents have occasionally been used for long spans. Wood girders are seldom appropriate except in "heavy-timber construction" (see p. 27-20) or where exposed and finished to present an attractive appearance.

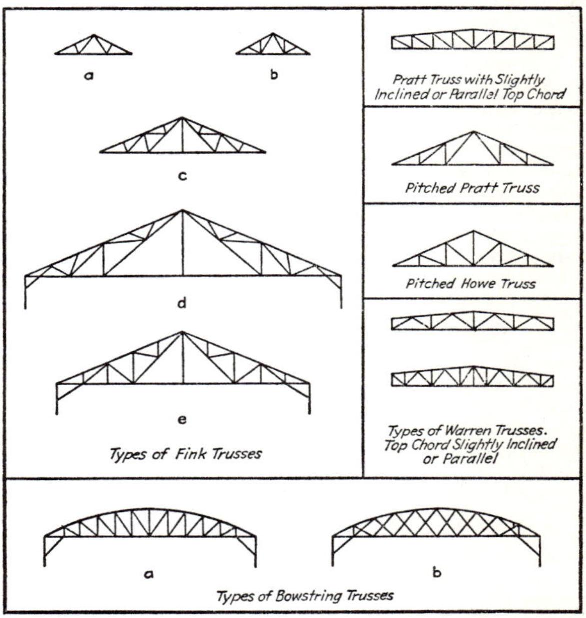

FIG. 16. Steel roof trusses. (Reproduced by permission from Crane, *op. cit.*)

STEEL TRUSSES. The more common types of steel trusses are widely used for the roofs of one-story industrial buildings. Figure 16 shows popular designs. It is difficult to make any specific recommendations which would have significance for a particular project as the choice depends on many factors. From a structural viewpoint, however, the following comments may be of help.

The Warren truss, with slightly sloped top chord, is often an appropriate choice for spans of 30 to 200 ft, although most economical for spans up to 125 ft. The top chords are sloped from $^3/_4$ to $1\ ^1/_4$ in. per ft, a suitable incline for built-up waterproofing.

The Pratt truss, with slightly sloped top chord, may be considered an alternate for the Warren truss, but the latter is often preferred. The Howe truss, with either parallel or slightly sloped top chord, is usually not as appropriate as a Warren or Pratt. Most designers prefer a slight pitch for drainage of roof surface, as the top chords of all these trusses normally serve as direct support for the roof construction. Consequently, those having absolutely parallel chords are not extensively used on spans over 50 ft.

The Fink truss is widely accepted for all types of occupancy wherever a pitched design is desired. It has a particular application for steel-framed mill buildings as a support for steel purlins carrying corrugated roofing or other roof deck. It is particularly economical for spans up to 125 ft. A popular slope is 5 in. vertical in 1 ft horizontal.

Both the Pratt and the Howe trusses, with steeply sloped top chords, are also used for pitched roofs, but they are usually more costly than the Fink. The Pratt would probably be cheaper than the Howe. Where there is no objection to a curved upper chord, the steel bowstring truss is often suitable for moderate spans and has an economic application up to about 125 ft.

The depth of roof trusses, with slightly inclined top chords, supporting only roof loads on a 100-ft span, would be about 11 ft at the center and 6 ft at the ends for both the Warren and Pratt designs. If the chords were parallel, both types would have a depth of about one-twelfth the span. For fairly heavy loads, in addition to the weight of the roof construction, the depth of the parallel chord design might be about one-tenth the span instead of one-twelfth. When supporting only roof loads, the Fink truss will have a depth of about 21 ft, and the bowstring a depth of about 10 ft, for a span of 100 ft.

The spacing of roof trusses is governed by the requirements of the roof construction, the length of span, the cost of foundations, and the loads that the trusses are required to support in addition to that of the roof. In one-story mill-building construction, the length of the typical bay, which determines the spacing, usually varies between 20 and 28 ft. Costly foundations, such as the need for piling, would favor the longer limit. It is also important, in planning manufacturing buildings, to minimize the number of columns wherever this may favor the present or probable future use of the structure. A column once installed is a fixture which requires expensive alterations to remove. Long spans, providing extensive clear floor spaces, are often desirable, notwithstanding a higher first cost.

Provision for expansion or contraction due to temperature changes is important for long, exposed steel trusses resting on concrete or masonry abutments. When framing into steel columns, or resting on top of them, trusses spanning up to 125 ft normally need no special provision for end movement. Where support is furnished by masonry walls or piers, the matter should be considered for spans over 50 ft. There are many instances where much longer trusses have been placed over masonry bearings without causing trouble, but experience dictates caution, particularly in colder climates. In mill-building construction, featuring multiple aisles serviced by cranes, either double columns or a longitudinal expansion joint is often used if the building is over 500 or 600 ft wide. Where it is apparent that movement of a magnitude greater than can be safely absorbed by the supporting assembly may occur, sliding plates or another similar device is required at one end of the truss. The type chosen deserves particular attention, as many designs used in the past have failed to function.

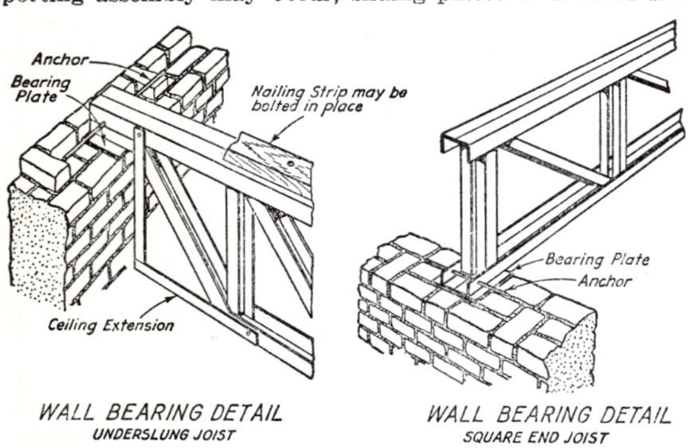

Fig. 17. Open-truss steel joists for wide-span designs. (Courtesy of Ceco Steel Products Corp.)

Supplementing the open-truss steel joists described on p. 27-29, a heavier design is manufactured capable of economically supporting roof loads on spans up to about 60 ft. These units have upper and lower chords formed of hot-rolled structural-steel shapes welded to form a chord section. They vary in depth from 18 to 32 in. and are appropriate for lightly loaded floors and roofs on spans up to fifteen times their depth for the 18-in. size and twenty-four times their depth for the 32-in. size. Typical designs and details of end supports are shown in Fig. 17. These trusses are manufactured with either square or underslung ends; top chords may be either pitched or flat. If pitched, the slope may be toward one end or in each direction from the center. Bridging is essential and is normally placed at approximately 10 ft on centers.

STEEL RIGID FRAMES AND ARCHES. Rigid frames of structural steel are particularly appropriate for spans of about 60 to 200 ft. A common form is illustrated by the familiar aircraft hangar. Types providing exceptionally clear height, to meet the requirements of a particular industry, can be designed with longer legs. Rigid frames of structural steel have a broad application for university field houses, gymnasiums, armories, and amphitheaters, as well as for hangars and industrial plants.

A patented application sponsored by the American Bridge Company is known as the beam-arch construction, although this type is also a rigid frame. The name is derived from the fact that the ends of the beams comprising the principal elements of the assembly are split longitudinally through the webs, and the inner flanges bent to predetermined curves, providing the necessary increase in depth at points of maximum moment. Plates of a thickness conforming to the webs are then fitted into the resulting V-shaped openings. As these are securely welded, or otherwise spliced to the beam webs, they function as an integral part of the expanded beam section.

Field erection is facilitated by shop fabrication in the largest sections that can be transported. Single-span rigid frames are usually designed and erected as two-hinged assemblies. If it is desired to assume a three-hinged condition for dead loads, this may be maintained until the steel work is completely erected by using at the peak of the arch a temporary third pin joint, which is later replaced by riveting or welding. Particularly appropriate for buildings designed for public assembly, the typical form provides an open floor space with unobstructed vision and ample head room. It is structurally applicable to a wide range of roof slopes and to spans up to 200 ft.

Wide-span steel arches are also built of trussed design. They are particularly suitable for hangars, where considerable height is desired. The steel lamella arch, although used for gymnasiums, auditoriums, and restaurants, has little application for industrial work and is less widely used than its counterpart in wood.

CONCRETE RIGID FRAMES. These designs have been used throughout the United States for the roof construction of auditoriums, gymnasiums, hangars, armories, markets, and factories. The structural deck is usually formed of either precast-concrete slabs or concrete poured-in-place, but sometimes the decking is made of wood. Although many rigid frames have been built of greater span, 100 ft is about the economical limit for solid reinforced-concrete sections, as the dead load usually makes one of greater span more costly than a corresponding design in material of a lighter weight.

When comparing the rigid frame in concrete with those meeting the same structural requirements in steel or wood, the choice may be influenced, if not controlled, by the type of occupancy. If either wood or steel must be protected from fire or deleterious gases, the additional cost may be an important item.

Rigid frames of concrete have also been built with hollow sections for both girders and columns. Those employed in one design have clear spans of 77 ft and are spaced 18 ft apart. Precast-concrete joists serve as purlins carrying the roof deck.

CONCRETE SHELL ARCHES. These designs have been widely used for factories, hangars, auditoriums, markets, and sewage-disposal plants. Spans up to approximately 300 ft have been constructed, and longer ones are contemplated. These designs comprise thin, reinforced-concrete curved plates, supported and stiffened when necessary by structural steel or reinforced-concrete framing or masonry walls. Shells may be of single, or multiple span, or various combinations.

The reinforced-concrete columns supporting shell roofs of the multiple barrel and butterfly types usually provide sufficient flexibility to make hinges unnecessary. Stiffness against buckling increases with the curvature of the roof. A strong curvature, however, steepens the section of the shell for some distance above the springing line, and this may result in additional expense for double forms. In some designs, where it is necessary to use a comparatively flat roof with a long radius of curvature, stiffness is obtained either by thickening the shell section or by adding ribs.

For designs of this type, the scaffolds supporting the forms represent a large proportion of the construction cost. Thus buildings of little length are comparatively more expensive than long structures for which the forms may be used a considerable number of times. Both barrel shells and the butterfly design, which is a variation particularly suitable for garages, have been used for industrial plants such as the single-story engineering shop erected at Wright Field during World War II. This building is 360 ft wide and 420 ft long. The roof is composed of nine concrete barrel shells, each of 40-ft span and 3 1/2 in. thick, except at the haunches, which are 5 in. thick. The columns are spaced longitudinally 56 ft on centers, and the skylight openings are glazed with sandblasted glass. After the shop was built, a Signal Corps hangar was built with the same forms, but the longitudinal spacing of the columns was increased to 120 ft.

Several large hangars, armories, auditoriums, and industrial plants built during recent years illustrate the use of reinforced-concrete arch rings combined with shell construction. The concrete arches should be designed so that the shells comprising the structural slab are supported from the intrados. This arrangement eliminates any projection of structural members below the under surface of the roof and facilitates the movement of the forms as the structure is poured section by section. Shell-roof designs have also been used for domical structures as important as the Hayden Planetarium in New York City, but such applications have little significance for industrial buildings.

The asphalt-mixing plant, erected a few years ago in New York City, comprises four arch rings spaced 22 ft on centers. They have a clear span of 90 ft and a height to the intrados of approximately 85 ft. Concrete barrel arches 8 in. thick were poured between the arch rings, which vary in depth from 6 ft 6 in. at the springing line to 5 ft at the crown. The width of the arches is 2 ft 4 in. at the extrados and 3 ft at the intrados. The change in thickness occurs along the inner face of the barrels and provides a shelf for their support. Structural steel shapes were used to reinforce the arch rings in place of rods or bars. This development, which is comparatively new, eliminated all false work within the structure as the steel served to support the forms. Although not a shell roof, this plant presents an interesting example of designing an enclosed area to meet the requirements of its particular use.

WOOD BEAMS AND GIRDERS. Glued, laminated beams and girders of rectangular section are used for spans up to about 60 ft, but for spans less than 30 ft, solid timbers are usually cheaper except for heavy loads. When the cost is estimated merely on a structural basis, timber girders of considerable span are generally more expensive than steel serving the same purpose. Often, however, they may be economical when employed as a finished product and exposed as part of an architectural treatment. Because laminated beams are manufactured with surfaces machine planed and sanded, they present an attractive appearance. Standard widths vary from 3 $^5/_{16}$ to 11 in. with depths as required.

The approximate depth of such units may be roughly estimated from the fact that on a span of 40 ft, with spacing of 14 ft on centers, a uniformly distributed superimposed load of 50 lb per sq ft could be carried by a laminated section 25 in. deep by 9 $^1/_2$ in. wide. If the span were 60 ft, and the spacing of the beams reduced to 12 ft on centers, the same superimposed load would require a laminated section about 32 in. deep and 11 $^1/_2$ in. wide. These figures are based on stresses generally accepted for such designs. Beams are usually cambered as the deflection is high.

WOOD TRUSSES. Wood trusses may be supported by steel, concrete, or wood columns or by masonry piers or walls. Solid timbers are widely used except for curved members, such as the upper chord of a bowstring truss. Steel rods are occasionally used for tensile members. Figure 18 shows the more common designs. For spans of 40 to 150 ft, the bowstring type is generally the most economical choice. It has very wide use for roofing industrial structures and is particularly suitable for spans between 80 and 120 ft. In a typical design laminated sections are used for both upper and lower chords; lower chords are often made of solid section.

When supporting a uniformly distributed, superimposed load of 40 lb per sq ft of horizontal protection, bowstring trusses may be appropriately spaced about 20 ft on centers for spans over 60 ft. For spans of lesser length, it may sometimes be economical to use a closer spacing, eliminating the roof joists or purlins, and applying the roof sheathing directly to the top chords of the trusses. On the other hand, for longer spans, trusses are often spaced somewhat farther apart and purlin trusses used to carry the roof deck. Comparative estimates of cost are often necessary to determine which design is more economical.

Bowstring trusses used to support only roof loads on a 100-ft span would have an overall height of about 13 ft. For similar service on a 40-ft span, the height is about 7 ft. For industrial work erected during World War II bowstring trusses were frequently used on spans between 150 and 200 ft, but in some cases the choice was dictated by the need to conserve structural steel. Although one of the more important aspects of modern design is the use of longer spans for both floors and roofs, it should be realized that, from the structural viewpoint, the first cost is definitely greater for long than for moderate spans. If we consider the cost of wooden bowstring trusses on a square-foot basis of roof surface, one well-known manufacturer states that a design suitable for a span of 75 ft might cost 25% more than for a span of 50 ft. Similarly, for a span of 100 ft, the cost of the trusses might be about 50% more than for a 50-ft span.

Used for spans up to about 90 ft but seldom an economical choice beyond 60 ft, Fink trusses are particularly appropriate for pitches greater than 25 degrees. The pitched Pratt and Belgian trusses are used on the same spans as the Fink and are preferable to the latter when the roof pitch is appreciably less than 25 degrees. The pitched Howe truss has been practically replaced by these types in the Tecco connector system of timber construction.

The Pratt truss, with slightly inclined or parallel top chord, shown in Fig. 18*i*, is often economical for spans up to about 120 ft. During World War II spans of 150 ft were used over an aircraft assembly plant. These trusses were 18 ft deep. All these types of wooden trusses can be designed to carry loads resulting from monorails or other factory

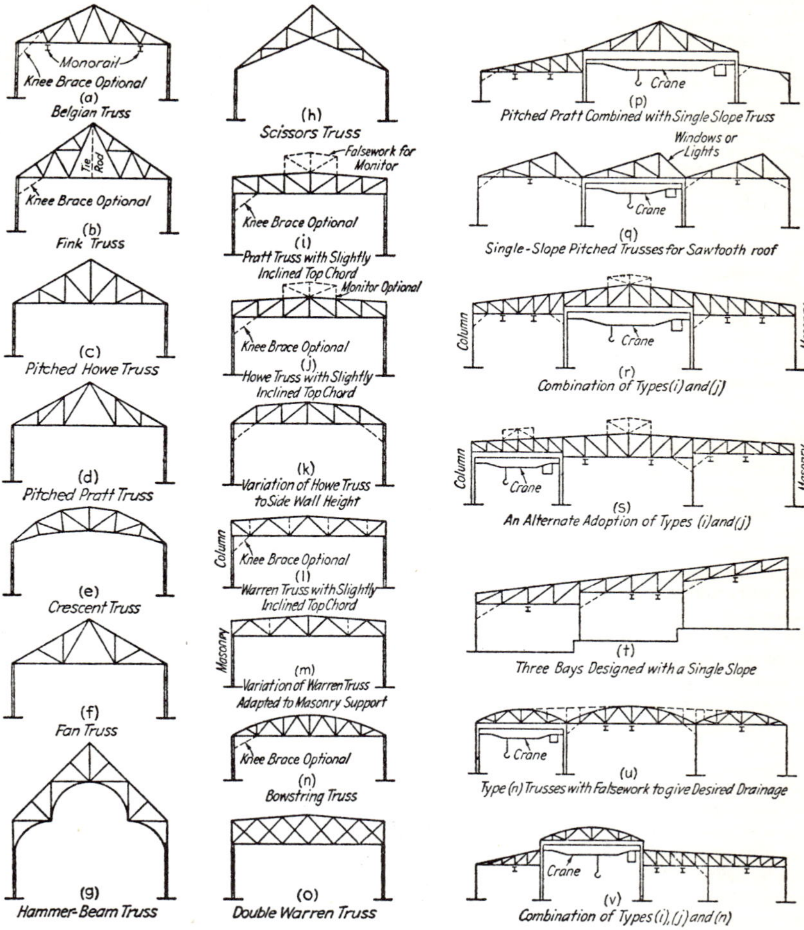

FIG. 18. Popular types of timber trusses and typical combinations. (Courtesy of Timber Engineering Co.)

equipment. The roof deck, composed of solid boards, boarding carried by wood purlins, or any of the appropriate materials described in Article 6, helps to brace the upper chords. Lateral bracing is used between trusses as required by the stress analysis.

The crescent truss, employing glued, laminated chords, is not widely used but may be economical for spans under 80 ft where increased interior clearance is essential. It was employed in hangar construction for a span up to about 160 ft. In this construction the crescent design was chosen instead of the bowstring as the former permitted the side walls of the hangars to be greatly reduced in height.

The hammer-beam truss is illustrated in Fig. 18 merely for the purpose of identification as ornamental designs of this type have no use in industry. The various combinations of wooden trusses shown in this same illustration have their particular application for one-story factory construction.

For roof spans of 25 to 40 ft, a design identified as the "trussed rafter," may be economical for stores, barns, and some housing projects. It may also have an application for shorter span requirements in industry. These trusses are placed up to 2 ft on centers and surfaced with $^7/_8$-in. boards or other suitable material.

WOOD RIGID FRAMES AND ARCHES. The rigid-frame types of laminated wood sections are particularly appropriate for spans between 30 and 100 ft.

Frames may be spaced 6 to 10 ft on centers and spanned by 2-in. plank forming the roof deck. If this spacing is too close, heavier units may be used and spaced up to about 20 ft on centers. In this case wood purlins, or a type of deck capable of spanning such a distance, would be used between the frames. These designs have a broad application for gymnasiums and dance pavilions as well as for hangars and garages. The usual forms for rigid frames and arches, designed of laminated sections, are shown in Fig. 19. For long spans combined with high rise, trussed frames and arches constructed of short-length lumber, joined with bolts and metallic connectors, have been built during recent years for army and navy hangars.

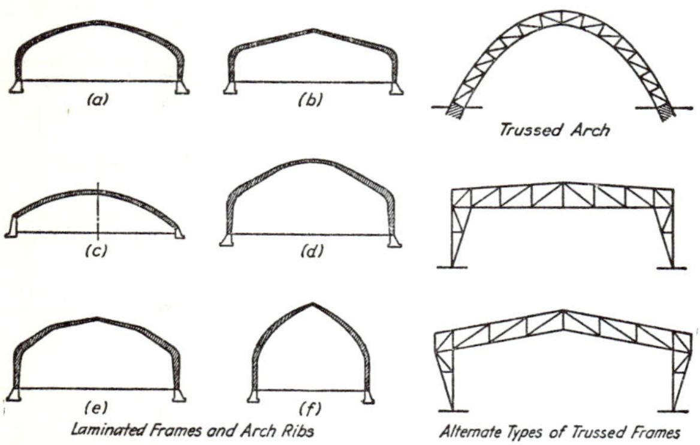

Fig. 19. Types of wood frames and arches. (Courtesy of Unit Structures, Inc.)

The various arch-rib designs, many of them appropriate for small and medium-sized hangars, are supplied by the manufacturers in two pieces, each half continuous from foundations to roof ridge. For the lighter types they are usually placed on 2-ft centers, and the roof deck is composed of $^7/_8$-in. boards nailed directly to the arch ribs. A design identified as the "tied rafter" is also used on a close spacing for spans up to about 50 ft, or obtained in a heavier design suitable for a 10-ft spacing on spans up to 100 ft. These rafters have their particular application for markets, dance pavilions, and garages, but they might well find their place in industry. The "segment rafter" is another design similarly employed. The lamella arch in steel was mentioned on p. 27-31, but it has been more widely used in wood. Its particular application is for auditoriums and garages. An arrangement of short units, placed in a diamond pattern, is characteristic.

6. CHOOSING THE ROOF DECK

The various types of structural roof systems, or "roof decks" as they are often called, that are most widely used in this country at the present time are described in this article. The term *system* identifies the structural assembly spanning between the framing members such as beams, rafters, or principal purlins. The maximum spans and limiting dimensions conform with standards generally accepted for the various systems, but they should be checked against the requirements of local building codes.

When selecting the floor system, as described in Article 4, the designer was primarily concerned with the material used as structural framing, nature and amount of superimposed load, clear span, degree of fire resistance demanded by building code or good practice, and the general fitness of the design to meet economically the requirements of the entire building. When choosing the deck for a flat roof, we have practically the same considerations except in regard to the live load, which is seldom over 40 lb per sq ft unless designed for occupancy. As used in this book, the term *flat* identifies a horizontal surface or one only slightly sloped, in one or more directions, for the purpose of drainage.

In selecting an appropriate deck for a pitched or curved roof, the same general considerations apply as for flat roofs, but there is often the additional requirement of providing a surface to which the materials comprising the finish, such as shingles or slate, can be easily attached. Precast cement-aggregate slabs and channel sections, corrugated cement-asbestos board, corrugated wire glass, and many sheet-steel designs have a wide application to industrial work. Wood plank of a nominal 2-in. thickness is sometimes used. Many of these types require a waterproof surfacing, whereas others are designed to shed water.

Where a roof surfacing such as slate, asbestos shingles, or clay tile is to be applied, or a wide salvage type of roll roofing, the roof deck must be of a material to permit easy nailing and good nail retention.

Flat roofs, or those sloped only as may be required for drainage, are usually surfaced with a built-up waterproofing and finished with slag or gravel. If the roof is intended for occupancy, the slag or gravel is omitted and the desired surface finish laid over the waterproofing. The structural deck may be formed with gypsum plank, concrete plank, precast cement-aggregate slabs or channel sections, corrugated cement-asbestos board, or some type of sheet-steel assembly. For flat roofs there is also the possibility of using any one of the structural floor systems. If a lightweight design is desired, however, it is usually best to avoid cast-in-place materials. For wood-framed structures, wood-joist construction is often used on short spans and parallel-chord wood trusses for longer spans. Precast cement-aggregate units are occasionally laid over wood framing, but wood boarding, plywood, composition board, or gypsum plank are more appropriate.

The minimum desirable incline for wood, asbestos, or asphalt shingles, slate, and the ordinary types of roofing tile, such as the Spanish Imperial and Mission, is 6 in. vertical to 1 ft horizontal. Wood shingles, however, are often laid on a pitch of 4 in. vertical to 1 ft horizontal. Slate may also be laid to a pitch of somewhat less than 6 in. horizontal to 1 ft vertical, if the "head lap" is increased from 3 to 4 in. For comparatively slight slopes, the slate is laid in bitumen.

"Built-up" roofing, comprising usually either three or four layers of roofing felt, cemented together with coal-tar pitch or asphalt, and surfaced with slag or gravel, is used principally for flat roofs but may be laid on any slope up to 4 in. to 1 ft if "nailability" is provided. This is ordinarily accomplished by using a type of roof deck that has good nail penetration and retention, but the roofing may be nailed into a suitable insulation if the insulation is cemented with bitumen to the roof deck. If the slope is less than 2 in. to 1 ft, nailing is not required. This is the usual condition for flat-roof designs. Roofs sloped more than 4 in. to 1 ft may be surfaced with a wide-selvage type of roll roofing or with a built-up roofing laid in either hot bitumen or cold plastic cement, but without slag or gravel. In both cases nailing is required. Roll roofing has a mineralized surface available in a number of colors. It has been used for roofing concrete arches, for instance, in the field house at Swarthmore College, Pa.

Thin-sheet metals like copper and aluminum can be used on a slope of $1/4$ in. per ft or greater, provided the proper type of seam is specified, but they are more appropriate for steeper surfaces. Sheets of corrugated metal and corrugated cement-asbestos board, serving both as a structural roof deck and surface finish, should generally have a minimum slope of 4 in. to 1 ft. Interlocking steel-plate designs, when occasionally used in the same manner without a membrane covering, are difficult to make tight on slopes less than about 3 in. to 1 ft.

As roofs form one of the enclosing surfaces of a building, the heat transmission is important where interior temperatures, appreciably different from exterior temperatures, are to be maintained for extended periods. A low transmission coefficient would obviously reduce the cost for additional insulation. The appearance of the under surface, which varies considerably for different designs, or the probable cost of applying a desired finish is also important for some types of occupancy. The general subject of flashing is discussed on p. 27-51. It does not influence the choice of the structural roof system and should be carefully planned in relation to wall construction. In districts subject to earthquakes, those designs having the stronger connections between deck and framing elements would be preferred. In comparing the advantages of competing designs, the comparative cost of maintenance should be included.

PITCHED-ROOF DESIGNS. Precast Cement-aggregate Slabs, Channel Sections, and Plank. These lightweight units, composed of portland cement and aggregates such as Haydite, cinders, slag, or slag products, are widely employed for structural roof decks over steel framing. Slabs of uniform thickness, channel, or rib sections are used. They are generally placed on a purlin spacing up to 12 ft, but the heavier units can carry normal roof loads on spans up to 20 ft. All are thoroughly reinforced. When nailing is required, the slabs are cast with a special material $1/4$ in. thick applied monolithically to the top surface. Some designs furnish a layer of insulating or acoustical material cast integrally with the concrete on the underside. Tile are also manufactured with glass panels.

Precast slabs of chemically expanded concrete are also manufactured in both plain rectangular section and channel design. For example, one manufacturer supplies plain slabs 24 in. wide and 2 to 3 $1/2$ in. thick which are suitable for normal roof loads on spans 5 to 6 ft between purlins or other supports. Channel sections are of the same width and 3 $1/2$ in. deep for spans carrying only light loads up to 10 ft. All are thoroughly reinforced.

These designs compete with those employing Haydite or other lightweight aggregate without an aerating compound or special process.

Lightweight concrete plank, described on p. 27-50, are also suitable for pitched roofs. As the material is "nailable," any desired surface finish may be applied.

Corrugated Cement-asbestos Board. This material is usually composed of about 85% portland cement and 15% by weight of asbestos fiber. It is fire resistant, and it has low heat transmission and good surface reflection. The average thickness is $^5/_8$ in.; sheets are usually 3 ft 6 in. wide with lengths varying from 3 to 10 ft. Weight is about 4 lb per sq ft. The maximum purlin spacing for standard types is 4 ft 6 in. Sheets are lapped one corrugation at sides and 6 in. at ends. They can be used without other covering for roofs having a minimum slope of about 3 in. vertical to 12 in. horizontal, but a steeper pitch is preferable. When used over steel framing they are attached to purlins by means of clips and bolts. Units of the same material are furnished for ridges, eaves, and other special requirements. Although this type of roof deck has been used for gymnasiums, auditoriums, and even small residential buildings, it is particularly appropriate for industrial structures.

Corrugated Wire Glass. Two types of glass are available: the white and the actinic, which has an amber color. The latter diminishes the effect of the radiant heat from direct sunlight. Both types are furnished in widths of about 2 ft 4 in. and of a suitable length to span between purlins spaced up to 5 ft on centers. Corrugated glass should not be used on slopes less than 4 in. vertical to 12 in. horizontal. At least this pitch is essential to permit the condensation gutters to function. When properly installed, the surfaces are practically self-cleansing. The corrugated glass panels are not lapped at the sides but placed $^1/_2$ in. apart with a metal cover cap over the joint. Caps may be of galvanized steel, copper, aluminum, Monel metal, or stainless steel. The same metals are employed for ridge caps and other special requirements. The sheets are lapped 3 in. at the ends and are attached to the purlins either by clips or by bolts. This type of roof deck has a general application wherever daylighting is essential. It is widely used for large skylights and sawtooth construction in industrial buildings and is also appropriate for museums and art galleries.

Corrugated metal sheets, supported by steel purlins or steel or wood rafters, are often used for roof decks in cheaper types of industrial work. Corrugated galvanized steel or iron is generally available in standard-sized sheets. Zinc alloys furnish a more durable surface which does not require painting. Cold-rolled hard copper has also been used, but a copper-clad corrugated steel sheet, with a layer of asbestos felt cemented with asphalt between the two metals, is more popular. The asbestos felt is helpful as insulation. Other types comprise corrugated sheets surfaced with a bitumen and felt, factory-applied sealing coat.

A purlin spacing of 2 ft 6 in. is required for plain corrugated copper sheets of 16-oz weight, but a purlin spacing of 5 ft is appropriate for the standard-weight steel and zinc alloys. The side connections between sheets, which vary in width from 2 to 3 ft, are generally made by lapping the corrugations. At the ends, sheets are lapped 4 to 6 in. Sheets may usually be obtained of any desired length up to 10 ft. The built-up designs, comprising corrugated steel, insulation, and surface protection are superior to those of bare steel in permanence and heat transmission. For these types, the purlin spacing varies with the gage of the metal; 10 ft would be appropriate for 18 gage and 7 ft for 24 gage. Some of the more modern steel-deck designs are appropriate for the pitched roofs of industrial buildings. As their chief application is for flat roofs, however, they are described on p. 27-37. The thin sheet metals, such as copper and aluminum, except the corrugated designs mentioned above, are used for surface finishes and not for structural roof decks.

Designs for Wood Framing. Wood plank of a 2-in. nominal thickness is sometimes used for steel-framed, pitched roof decks with a purlin spacing up to 6 or 8 ft, but is more appropriate when the support is furnished by wood trusses, arches, or rigid frames of wood. For buildings framed entirely of wood and for wall-bearing construction where wood is used for the interior support, the same material is appropriate for the roof framing. Structural designs of this type are well standardized and appear in many reference books. The roof deck, supported by wood rafters, may be of solid boarding, plywood, shingle lath, or various composition boards such as laminated Sheetrock. The roof boards are usually matched, either tongue-and-groove or shiplap. A $^7/_8$-in. thickness is standard.

When slate, tile, or asbestos shingles are used as a surfacing material, over the roof deck, a layer of roofing felt and sometimes insulation is placed between the two. If Sheetrock is used, it should be protected from rain by either built-up waterproofing or a type of roll roofing. Corrugated sheets of steel, asbestos-protected steel, copper, zinc, corrugated asbestos board, or corrugated glass can also be used over wood rafters, but they are usually supported by structural steel framing.

FLAT-ROOF DESIGNS. Precast Gypsum Tile and Plank. The most popular gypsum unit for flat roofs is the plank. Standard sizes are 2 in. by 15 in. by 10 ft and 2 in.

by 15 in. by 8 ft. The plank are reinforced with welded, galvanized steel mesh. The edges are steel bound and of tongue-and-groove design. Assembly is by clips to the supporting purlins, which are spaced 30 in. to about 6 ft, as required by the superimposed load. The plank weigh 12 lb per sq ft. After erection the plank are surfaced with a built-up waterproofing. Insulation is placed between if required. Light weight and ease in erection make them very popular for industrial use, but the under surfaces are vulnerable to moisture. Although much less widely used for industrial roofs than the gypsum plank, any type of precast gypsum unit that is suitable for pitched roofs can also be used on flat roofs. Nailing facilities are obviously unnecessary.

Precast Cement-aggregate Units. Many types of precast cement-aggregate units described on p. 27-35 as suitable for pitched roofs can also be used on flat roofs. The Flexicore slab already discussed is an example. Built-up waterproofing is required over the roof deck. As nailing is not necessary, stone-aggregate slabs may be used, but those composed of Haydite and similar lightweight aggregates are more popular.

A lightweight, tongue-and-groove plank, cast with cinders or blast-furnace slag, is widely used. These units, which are 2 or 2 $^3/_4$ in. thick, may be obtained in appropriate lengths conforming with a desired purlin or rafter spacing. The weight of the 2-in. plank is 13 lb per sq ft and it may be placed on a purlin spacing of 7 ft for superimposed loads up to about 50 lb per sq ft.

Slabs composed of mineralized wood fiber, combined with portland cement to form a porous and fire-resistant unit, are identified by the manufacturers as Porex. They are useful for both heat insulation and sound absorption. For slabs 1 ft 10 in. wide and 1 $^3/_4$ in. thick, a subpurlin spacing of 22 in. would be appropriate. The total weight of this design, including $^1/_4$-in. cement-mortar finish and subpurlins, spanning up to 8 ft, is about 9 lb per sq ft of roof surface. Products of this type have been shipped to many sections of the country, but as all are manufactured in New Jersey, the designer should check their availability before planning to use them.

Cemesto board is composed of Celotex cane fiber surfaced on both sides with $^1/_8$-in. cement-asbestos board. The standard width is 4 ft, and lengths vary from 4 to 12 ft. The 1 $^9/_{16}$-in. and 2-in. thicknesses are used for roof decks. Corresponding weights are 4.8 and 5.3 lb per sq ft. A 4-ft purlin spacing is often used. Support may be furnished by rolled-steel sections or open-truss steel joists.

Corrugated Cement-asbestos Board. This material is appropriate for flat roof decks when one or more layers of rigid insulation are laid over the corrugations to provide a smooth surface for the application of built-up waterproofing. Additional insulation is often obtained by filling the corrugations with an insulating material.

Steel Roof Decks. Designs composed of interlocking plates have a particularly broad application for one-story industrial buildings. They are economical in first cost, speedily erected, and some types may be combined with any required degree of insulation or acoustical treatment applied to the under surface. Built-up waterproofing is required.

The units comprising a steel deck vary considerably in the different designs. Usual widths are between 12 and 16 in., depths of ribs 1 $^1/_2$ to 2 in. for the shallower or lighter types and 3 to 5 $^1/_2$ in. for those intended for longer spans. Manufactured of copper-bearing steel or iron, the different styles vary in weight from about 200 to over 600 lb per 100 sq ft. Thicknesses vary from 22 to 16 gage. The 18- and 20-gage thicknesses are particularly popular. Their respective weights are about 320 and 240 lb per 100 sq ft of roof surface. Sheets are welded to steel beams or purlins. The lighter designs are used on spans 4 to 5 ft, and the heavier types on spans up to 25 ft to support a superimposed load of 50 lb per sq ft. Plates usually span over several purlins and are lapped or otherwise joined to form a tight, continuous surface. Special plates are provided for ridges and valleys, monitors, and saw-tooth designs. The lighter types may be bent to form a curved surface where the radius is not less than 60 ft.

Installing with ribs up provides a smooth steel ceiling on the soffit but requires sheathing on top which may be in the form of steel or iron sheets, wood, or preferably a rigid insulation laid over the tops of the ribs to serve as a base for the built-up waterproofing. If the units are assembled with ribs down, an advantage is that a smooth top surface is available to support the waterproofing. If the appearance or sound reflection of the ribbed ceiling is objectionable, various combinations of acoustical materials or additional insulation may be applied to the lower surface. A design employing Fiberglas has been used for industrial buildings to provide low heat transmission, excellent sound absorption, a surface which greatly assists in shadowless lighting, and one which can be made both fire and shatter resistant. With slight modifications to meet special conditions, this assembly has been used successfully in a number of manufacturing plants.

Designs for Wood Framing. Roof decks of flat-roofed buildings framed in wood are ordinarily supported by wood-joist construction similar to that used for the floors. The structural surface is sloped to provide drainage. Wood board, plywood, various composition boards, and gypsum plank may be used for the roof deck. Built-up waterproofing would be laid over any of these surfaces. The lighter precast units of cement-aggregate composition can also be placed over wood joists but are less appropriate than for steel framing.

HEAT TRANSMISSION AND INSULATION. When comparing the characteristics of competing roof systems it is often desirable to investigate their relative resistance to the passage of heat. For this reason a résumé of the subject, together with a brief discussion of the principal types of insulating materials, is given in this article. No tables of heat-transmission coefficients are included as they may be found in the current edition of the *Guide*, yearly publication of the American Society of Heating and Ventilating Engineers, and in the *Power* Volume of Kent's *Mechanical Engineers' Handbook*, pp. 12–02 to 12–12.

Heat Losses and Heat Gain. If interior temperatures are appreciably higher or lower than exterior temperatures, insulation is often necessary to reduce the heat loss in cold weather or the heat gain in hot weather. The principal means by which heat passes through the envelope of a building are:

1. The infiltration of air through cracks due to faulty construction or the normal leakage along contact faces of operating sash and doors, obviously minimized by careful fitting of all elements comprising the building envelope.

2. By transmission through the enclosing surfaces of the building. For example, if the structural roof were composed of concrete, with a hung ceiling below, heat would pass through the plaster of the ceiling by conduction. It would then cross the air space between ceiling and slab by radiation, conduction, and convection. Passing through the concrete slab by conduction, it would leave the surface by radiation, conduction, and convection.

For homogeneous materials conductivity varies with their density and the amount of moisture contained. In general, it is higher for the more dense materials, but for some of fibrous composition there are optimum densities for lowest conductivities. The amount transferred from a surface by radiation depends upon the emissivity of the surface as well as the temperature difference between the boundaries of the air space across which radiation takes place. Both conduction and convection are influenced by the width and shape of the air space, the amount of air movement within the space, and the roughness of the boundary surfaces.

A large proportion of the total heat transferred across air spaces, such as exist in furred walls or between the joists of roof construction, may be by radiation. This fact indicates the value of metallic surfaces such as aluminum foil and specially coated sheets of steel having capacity to reflect heat rays. Although the transfer or radiant heat is not greatly influenced by width of such a space, it should be at least $^3/_4$ in. wide in order to limit the total conductance. In referring to heat-reflective materials it is undesirable to use the term *brightness;* the important characteristics are emissivity and reflectivity.

Types of Insulating Materials. There are two main groups. The first comprises those which offer resistance to the passage of heat by conduction as they have low density and comparatively low conductivity. The second group comprises metallic surfaces used as boundaries of air spaces within the construction of walls, ceilings, or roofs. The first group is divided into three classes: rigid insulation, flexible types, and insulating fills. The second group has no subdivision, but various metals are used to accomplish the same purpose, which is the reflection of radiant heat waves.

Rigid Insulation. This is generally in the shape of boards $^1/_2$ to 1 in. thick or blocks 1 to 4 in. thick. One variety is composed of wood, cane, and other vegetable fibers pressed into sheets sufficiently rigid to be nailed upon joists or studding, or laid over a roof deck beneath built-up waterproofing. Pressed cork, Foamglas, and mineral wool are also available in blocks varying in size from 12 by 18 in. to 12 by 36 in. and in thicknesses from 2 to 6 in. Cork has a wide application to low-temperature installations. When laid upon a roof deck these blocks are set in coal-tar pitch or asphalt.

The thinner types of rigid insulation, particularly the fiber boards, may be used as a base for plaster on ceilings or exterior walls, and some are attractively finished on one face for use as exposed interior surfaces. Others are coated with bitumen or enclosed in waterproof paper to prevent the passage of water vapor. At least one type is furnished with aluminum foil applied to one surface in order to combine the function of low heat conduction with the reflection of radiant heat.

Flexible Insulation. Varieties in this group may be composed of organic materials, such as wood fiber and cotton, treated in a manner to render them very light and fire resistant. Another type is derived from inorganic materials and identified as rock or glass wool. Mineral wool is composed principally of silicates of calcium and aluminum.

It has the appearance of loose wool or cotton. Rock wool is made from natural rock or a combination of limestone and shale. Slag wool is derived from iron, copper, or lead blast-furnace slag. Glass wool is made from silica sand, soda ash, and limestone.

These materials are furnished in the form of batts, blankets, or quilts as may be required to fit the spaces where insulation is to be placed. The thicknesses vary from 2 to 4 in. Waterproofed paper is often applied to one or both surfaces for the purpose of excluding moisture. Insulation of this type is not sufficiently rigid to be used beneath built-up waterproofing but has a wide application between wood studs in exterior wall construction and between the joists of flat roofs and rafters of pitched roofs. Although most widely used with wood-framed construction, it is also appropriate between furring strips on the inside of masonry walls and between floor or roof joists in steel-framed assemblies. In the steel-framed assemblies definite provision should be made for attachment. When installing either the rigid or flexible type of insulation, it is important to avoid any cracks between the sheets, batts, or blankets through which heat transfer can occur.

Insulating Fills. Over concrete roof decks various types of granular materials, often the size of small pellets, are used for the purpose of insulation. Several of the newer varieties are derived from expanded mica and diatomaceous earth. Pumice is used for the same purpose. When applied over a roof deck, these materials are usually mixed with portland cement. However, they also have a wide application as merely loose fills in locations where it would be difficult to install other types of insulation.

Precast concrete roof slabs are manufactured containing a vermiculite composition which provides considerable insulation. Although most widely used as a fireproofing material, vermiculite plaster has a high heat insulating value and may be employed for this purpose where plaster applications are appropriate. The word vermiculite identifies a group of minerals derived from certain varieties of mica.

Reflective Insulation. Several types of metals are used for this purpose. Aluminum foil mounted upon a strong paper backing is an example. Thin sheets of steel, finished to a dull luster, are also effective and have an application for low-temperature installations. These materials are installed as one of the boundaries of an air space which should be at least $^3/_4$ in. wide. The better types maintain their effectiveness under service conditions.

CONDENSATION AND VAPOR SEAL. Two possibilities for condensation exist under normal conditions of occupancy. (1) Moisture may form on the interior surfaces of roofs, ceilings, or walls. It is referred to as surface condensation. (2) Moisture, in the form of water vapor contained in the interior atmosphere of the building, may penetrate the enclosing surfaces and condense within the roof or wall assembly. It is referred to as interstitial condensation.

Surface Condensation. Surface condensation may present a serious problem in factories where the manufacturing process produces high humidities. It can be eliminated by either reducing the relative humidity of the interior or maintaining the interior surfaces of the enclosing elements above dew-point temperature. If the second alternative is necessary, the chart (Fig. 20) will indicate the permissible relative humidities for various transmission coefficients and outside temperatures ranging from minus 30 to 40 F. The dotted line indicates the permissible relative humidity (64%), if surface condensation must be avoided, for an exterior wall or roof having a transmission coefficient of 0.26, when the exterior temperature is minus 10 F.

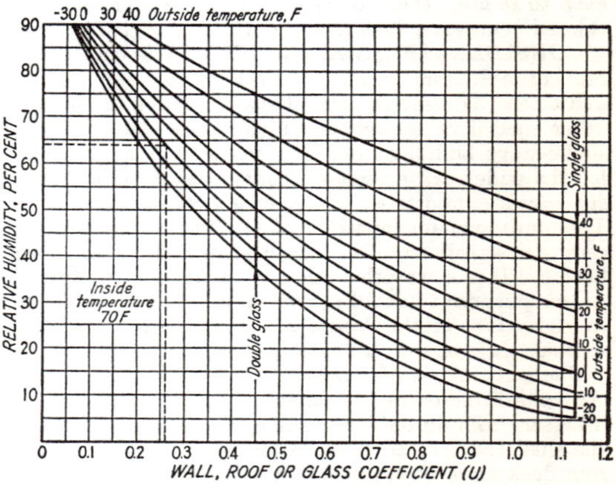

FIG. 20. Permissible relative humidities for inside temperature 70 F and various transmission coefficients. (Reproduced by permission from *Heating, Ventilating, Air Conditioning Guide*, 1946, Chapter 6)

Interstitial Condensation and Vapor Seal. Even with interior temperatures such as are normally maintained within a building during the heating season, if combined with a moderately moist atmosphere, such as 30% relative humidity, there may be a higher vapor pressure within the building than outside. As water vapor passes very easily through most building materials, it may enter the surfaces of plastered ceilings or walls. On its way through the roof or wall construction, a colder zone is reached and the temperature of the moisture-laden air is lowered to the dew-point, resulting in condensation. Condensation has often caused not only the discoloration of interior surfaces but also serious deterioration of insulating materials.

The lowering of the relative humidity of the interior atmosphere would lessen the probability of interstitial as well as surface condensation, but moderately high humidities are often desired and, even where not provided intentionally, may be caused by conditions of occupancy. As it is usually impracticable to permit the free passage of water vapor through walls and roofs, because of the need of minimizing heat losses, some definite provision must be made to obviate this condition. Ventilation between roof joists, or in the open spaces occupied by the trusses supporting the roof deck, is the most effective means. It is particularly recommended in sealed-off spaces beneath steel roof decks. If ventilation is impracticable, the most effective means of guarding against condensation, within either roofs or walls, is to introduce a barrier formed of a material offering a high resistance to the passage of water vapor.

Such a barrier should be placed as near as practicable to the interior surface. It is imperative that it be installed on the warmer side of the insulation. If roofs are framed in open-joist construction, wood or steel, the seal may be in the form of a glossy-surfaced asphalt-impregnated and surface-coated sheathing paper, weighing at least 50 lb per 500 sq ft. Such paper should be applied directly to the underside of the joists or rafters immediately above the plaster or other interior finish. Sheets should be well lapped with mastic applied at all joints. The same type of seal may be used above a hung ceiling. Somewhat less effective are various brush or spray applications of a vapor-resistant finish applied to the interior surface of a ceiling or wall.

For the typical flat-roof designs of fire-resistant buildings, where a reinforced-concrete slab forms the structural roof deck, a vapor seal should be placed on top of the slab beneath the insulation. It may be formed with two plies of impregnated roofing felt, laid in hot asphalt or coal-tar pitch. As some moisture may pass through this seal and be prevented from escaping by the four-ply built-up waterproofing, which would normally form the exterior surface of such designs, a type of insulating material should be chosen which will not absorb moisture or deteriorate when exposed to dampness. A critical condition has occurred in many cases where promenade tile have been used as a wearing surface over the built-up waterproofing. In some instances sufficient deterioration has taken place to require a complete resurfacing of the roof. It has occasionally been considered desirable even to remove the promenade tile and cover the roof with ordinary slag or gravel, or a colored aggregate, such as ground slate, to give a more attractive surface finish.

CONSIDERATIONS AFFECTING THE CHOICE OF THE ROOF DECK. For flat roofs of multistory buildings, it is generally appropriate to use the same framing materials and often the same structural system as for the floors, unless there are large skylights or other special requirements which make such a choice impracticable. The flat roofs of single-story buildings and all pitched roofs are normally framed in steel or wood, except for the wide-span designs in concrete such as the arch or rigid frame. In all cases, the spacing of supports should provide economical spans for the selected type of roof deck. The limiting spans for which the different designs are suitable have been given in the preceding descriptions. The longer span systems obviously reduce the number of purlins or beams. If a sawtooth design is required, a system should be chosen that is adaptable to this purpose to avoid introducing dissimilar types of construction. In districts subject to high winds or the danger of earthquakes, the strength of the connections between the various parts of the roof assembly should be studied.

For some types of industry, the cost of providing adequate fire resistance may be a controlling factor in the choice of the roof assembly. This is particularly true if it is necessary to fireproof steel trusses and may lead to the consideration of designs comprising the shell arch or other reinforced-concrete forms. The weights of the different types of roof deck have been given in the previous descriptions. They may favor one system at the expense of another, as reflected in the steel tonnage or cost of other supporting material. For certain classes of occupancy appearance may influence the choice. Some assemblies are more attractive than others, and some offer superior facilities for applying ceiling finishes.

There is considerable difference in the heat transmission of various roof assemblies. The lightweight porous materials are generally superior in this respect to those that are more dense, such as stone concrete. Some designs comprising especially processed lightweight cement-aggregate units have considerable insulating value. The unprotected sheet metals, on the other hand, are particularly undesirable. The steel deck requires considerable insulation for most classes of occupancy. In comparing competitive designs the relative heat loss should be roughly checked as the cost of insulation may be an important factor. The extent to which a roof surface is capable of reflecting, rather than absorbing, the radiant heat from the sun is also worthy of consideration for systems that serve both as structural deck and surface finish. In buildings where the manufacturing process either requires or produces a relatively high humidity, the designer should beware of choosing materials susceptible to moisture. As previously mentioned, it is always desirable to use an insulating material that will not deteriorate in the presence of moisture or show any appreciable degree of absorption.

The selected roof assembly should provide for the application of the desired surface finish. Because of the extensive use of built-up waterproofing for flat roofs, this feature has little importance except for pitched roofs. Fortunately, the various plank, slab, and channel designs of gypsum or cement-aggregate composition are available with surfaces prepared for nailing to which slate, clay tile, and similar materials may be attached. Some designs, however, offer better facilities than others. Durability has its particular application to roof surfacing materials rather than to the structural roof deck. It should be considered for systems that serve both purposes such as the red-tile and corrugated cement-asbestos board, and where manufacturing processes cause critical exposures. The cost of maintenance is often an important item for extensive roof areas and should be estimated for competing designs. This item has particular significance, as previously noted, when comparing the advantages of a one-story and multistory industrial building.

An analysis of the various possible systems that may be employed for the structural roof deck will usually result in reducing the practicable alternates to two or three designs at most. Comparative estimates should then be made as in floor construction.

7. CHOOSING THE WALL CONSTRUCTION

There are many types of wall and partition construction which differ widely in architectural appearance, structural characteristics, fire resistance, and heat and sound transmission. Their relative advantages for a specific project may be compared by making a functional analysis. For this purpose any type of wall from the primitive adobe of the Southwest to the complex assemblies including exterior facing, masonry backing, furring, and plaster, typical of our larger cities, may be analyzed (see Fig. 21). The following characteristics are always essential.

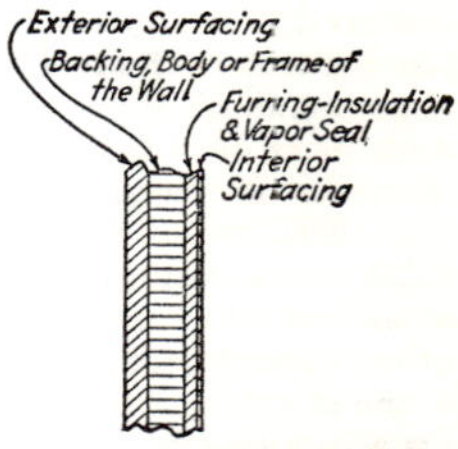

Fig. 21. Diagrammatic section of an exterior wall. (Reproduced by permission from Crane, *op. cit.*)

1. Exterior surface should be reasonably resistant to atmospheric conditions and have an acceptable appearance; in combination with the other elements of the assembly, it must be rain- and windproof.

2. Body, backing, or frame of the wall, in combination with the other elements of the assembly, must be structurally adequate to support its own weight, resist all applied forces, often transmitting them to the other elements of the structural frame, and provide the necessary fire resistance.

3. Insulation from cold or heat and adequate provision against condensation of water vapor, either on the inner surface or inside the wall, appropriate for the locality and type of occupancy, should be furnished. It may also be desirable to compare the sound transmission of competing designs.

4. Interior surfaces should meet the utilitarian demands of occupancy, including, where necessary, light reflection and suitable sound absorption. They should also have an acceptable appearance.

5. The wall assembly which satisfies these requirements at the lowest first cost and expense for maintenance will generally be the best choice.

In the conventional residential and commercial designs, different materials are used to meet these demands, but for some types of structures a far simpler wall section may serve the same purposes. For example, an 8-in. reinforced-concrete panel wall may often be used for industrial buildings. Good concrete is reasonably resistant to atmospheric conditions. If cast in well-constructed forms, the exterior surface may be satisfactory without finish. If well designed and well built, such a wall is durable and rain- and windproof. An 8-in. thickness should meet all structural and fire-resistance requirements. For many industrial buildings no special insulation is necessary. If the interior concrete surface is acceptable for industrial occupancy, this simple concrete section may meet every requirement demanded of the more complicated wall designs.

DEFINITIONS. The following definitions are abstracted from *A Glossary of Housing Terms*, Building Materials and Structures Report BMS91, National Bureau of Standards, 1942.

Bearing wall is a wall that supports any vertical load in addition to its own weight.

Common wall is a wall owned by one party but jointly used by two parties, one or both of whom is entitled to such use under the provision of a lease.

Curtain wall is a nonbearing wall between columns or piers that is not supported by girders or beams.

Enclosure wall is an exterior nonbearing wall in skeleton construction anchored to columns, piers, or floors, but not necessarily built between columns or piers.

Exterior wall is any outside wall or vertical enclosure of a building other than a party or common wall.

Fire-division wall is a wall that subdivides a building to restrict the spread of fire, but is not necessarily continuous through all stories or extended through the roof.

Fire wall is a wall that subdivides a building to restrict the spread of fire. It starts at the foundation and extends continuously through all stories to and above the roof.

Foundation wall is a bearing wall or pier below the first-floor construction.

Nonbearing wall is a wall which supports no vertical load other than its own weight.

Panel wall is a nonbearing wall in skeleton construction built between columns or piers and wholly supported at each story.

Parapet wall is that part of an exterior, party, or fire wall extending above the roof line.

Partition is a wall that subdivides spaces within any story of a building. If designed to restrict the spread of fire, it is called a **fire partition.**

Party wall is a wall used jointly by two parties under easement agreement and erected at or upon a line separating two parcels of land that may be held under different ownership.

Retaining wall. (1) Any wall subjected to lateral pressure other than wind pressure. (2) A wall built to support a bank of earth.

Spandrel or spandril is the panel between the top of a window or door at one story and sill of a window or door at the story above.

MASONRY WALLS. All materials used in masonry wall construction should meet the requirements for strength and durability demanded by good practice or mandatory code. As mentioned on p. 27-10, the *American Standard Building Code Requirements for Masonry*, 1944, furnishes an excellent standard for such designs. Because in this country masonry bearing walls are used only for comparatively low buildings, the minimum thicknesses demanded by building codes or common practice usually result in adequate structural strength. Consequently, the most important requirement for exterior walls above grade is often impermeability to wind-driven rain. This fact has not been sufficiently emphasized in either the specification or the supervision of masonry.

Specification of Materials. The quality of materials should conform with the appropriate standard of the American Society for Testing Materials. As constant changes are being made in the specifications no revision would have more than temporary significance. The designer has the option of obtaining the current designation from the publications of the Society or merely stating that the latest revision shall apply.

The recommendations in the following articles have a general application to all masonry assemblies when used in exterior wall construction. It is important that the materials not only should conform to the mandatory or accepted standards but also should be chosen for the purpose of obtaining a wall section impervious to wind-driven rain. For example, in specifying mortar, it is highly desirable to choose a type of lime which, with the selected proportions, will produce a satisfactorily high water retentivity. This requirement favors the better grades of pulverized quick lime made into putty or type S mason's hydrate, rather than the standard variety of hydrated lime identified as type N. Either choice would meet any requirement of building code or common practice, but the first definitely facilitates laying a water-tight wall.

Although it is necessary to forbid the use of bricks which do not conform to the requirements for grade SW (ASTM C62-44) in locations where they may be frozen when saturated, those having an absorption as high as 17 to 20% * have been used successfully in exposed walls in both Chicago and New York City. It is essential, however, that their suction rate be reduced to a satisfactorily low limit before they are laid. The suction † of the bricks when placed in the wall, rather than their total absorption when dry, affects the permeability of the masonry.

In some sections of the country a white powdery deposit, generally referred to as efflorescence, occurs on the outer face of exterior brick walls. It is caused by the action of water penetrating the wall and dissolving salts contained in the mortar or masonry units. In order for efflorescence to occur, soluble salts must be present in the wall materials, and the construction must permit water to enter the wall, dissolve the salts, and then carry them to the surface, where evaporation results in their deposit. Most clay and shale brick manufactured in this country do not have this efflorescence, but in some localities efflorescence presents a serious problem. Any masonry unit may be easily tested by partial

---

* Percentage by weight after 5-hr boiling.

† *Brick suction* is defined as the amount of water in grams or ounces absorbed by a brick (30 sq in.) placed on the flat side in water to a depth of 1/8 in. for 1 min.

immersion for a period of one week in distilled water. If there is likelihood of efflorescence after incorporation in walls, it will show on the surface of the sample. Efflorescence may be removed from the face of a wall with a solution of one part commercial muriatic acid mixed with ten parts of water, applied by means of a fiber brush. Walls should be thoroughly washed with water both before and after scrubbing with this solution. Such a treatment is seldom of more than temporary value.

As most varieties of light-colored stones commonly used for wall facing or veneer become discolored or otherwise injured when brought in contact with standard portland cement, such masonry units should always be set in a mortar containing a white or nonstaining cement. The lime should also be of a type incapable of causing injury to the stone. In order to break the contact between the facing of a masonry wall and the backing (the backing is normally laid in mortar containing standard portland cement), the backsides of the facing units, including bond stones, should be coated to a depth of $^1/_2$ to $^3/_4$ in. with a layer of the same nonstaining mortar used for the facing. The operation is called *parging*. (See p. 27-44.)

Proportions for Mortars. Provided that the ingredients of the mortar and the proportions of the mixture meet building code requirements or accepted standards for structural strength, the designer is primarily concerned with obtaining a "workable" mortar that will facilitate the laying of the masonry units. There is no nationally accepted test for workability, but it is closely related to the water retentivity of the mixture, which can be checked by reference to the Federal Specification for Masonry Cement, SS-C-181. In the following recommendations all proportions are by volume, and sand or other aggregate is measured in a damp and loose condition which conforms to usual practice on the job.

If the essential requirement for the masonry is structural strength or durability to resist alternate freezing and thawing in the presence of moisture, a mortar containing a relatively high proportion of portland cement is desirable. A good mixture is: 1 part portland cement : $^1/_4$ part lime putty : 3 parts sand. This mortar has a somewhat lower workability than would be chosen if less strength were demanded, but it is still workable and is appropriate for foundation walls composed of hollow masonry units, for the thinner cavity wall designs, and where strengths comparable to that shown in Table 12 for type A mortar is demanded by the design. The cost is generally more than for a mortar containing a higher proportion of lime and sand, which is preferable where high strength is not necessary.

For masonry walls above grade, where the essential requirement is impermeability to wind-driven rain, a mortar having greater workability is desirable. A good mixture is: 1 part portland cement : 1 part lime putty : 5 to 6 parts sand. Such a mortar is suitable for all general purposes except where one of higher strength is definitely required. It is an appropriate choice for foundation walls laid with solid masonry units, load-bearing walls of solid units above grade, and often for isolated piers above grade. It may also be used for hollow masonry units, hollow walls of solid masonry units, and cavity walls more than 10 in. in nominal thickness. It would not be appropriate for use in districts where the earthquake hazard demands that masonry have a higher structural strength than required elsewhere. It would not be appropriate for gypsum blocks that are laid in gypsum mortar, and it should be remembered that special mortar mixtures are used with refractory cements.

Except for the specially processed hydrates, lime should be used in the form of putty. This is, of course, necessary for the quick limes, but it is also desirable to do it when the ordinary hydrates are used. The resulting mortar is more workable.

As an alternate to this last specification, the designer has the option of choosing a masonry cement such as that identified by Federal Specification SS-C-181b, Type II, or a natural cement conforming to ASTM Specification C-10-37. As the natural cements may be included in the general classification of "masonry cements," they are not specifically mentioned in Table 11. The usual mixture is: 1 part cement : 3 parts sand. Such mortars

Table 11. Volumetric Proportions for Masonry Mortars

(From *American Standard Building Code Requirements for Masonry*)

Mortar Type	Cement	Hydrated Lime or Lime Putty, Allowable Range	Aggregate, Measured in a Damp and Loose Condition
A	1 (portland)	0 to $^1/_4$	Not more than 3
B	1 (portland)	1 to 1 $^1/_4$	Not more than 6
B	1 (masonry FS type II)	....	Not more than 3
C	1 (portland)	2 to 2 $^1/_2$	Not more than 9
C	1 (masonry FS type I)	....	Not more than 3
D	0 to $^1/_2$ (portland)	1 to 1 $^1/_4$	Not more than 3 parts for each part of cementitious material

have been used successfully for many years. In making a comparison with the portland cement and lime and sand mixtures they should be evaluated on the basis of compressive strength and workability.

Table 11 gives six different sets of proportions for mortars. The first three correspond to the mixtures mentioned in the foregoing discussion. The two types identified by capital C are not recommended in locations where mortar is in contact with soil. Their use may be permitted in exterior and interior walls of solid masonry, in interior nonloadbearing walls or partitions of hollow units, and in other locations for which types A or B are not specifically required.*

Mortar of type D hardens very slowly and may soften if kept wet. It is not recommended for parapet walls, rubble-stone walls, or for uses where it will be in contact with soil. Where this type is used, lower limits should be set for height and spacing of lateral supports than for walls built with stronger mortars (see p. 27-48).

The only advantage of using either type C or type D would be economy, and the designer should carefully check the requirements of his building, as well as local code restrictions before making such a decision. In districts subject to earthquakes, proportions of the mortar ingredients are often definitely controlled. For example, many West Coast cities require that cement-lime motar be mixed in the volumetric proportions of 1 part portland cement : 1/2 part lime putty : not more than 4 1/2 parts sand. Where a so-called cement mortar, corresponding to type A in the table, is required many codes in various parts of the country limit the permitted amount of lime to less than the maximum allowed by this standard. Such limitation is seldom justified but may be mandatory.

Table 12 gives the minimum compressive strengths which the mortars mixed in the corresponding proportions may be assumed to develop at an age of 28 days when tested in 2-in. cubes. Such tests should be performed as prescribed in the *Federal Specification for Masonry Cements* SS-C-181b (except for type D).

Table 12. Classification of Masonry Mortars on a Basis of Strength

(From *American Standard Building Code Requirements for Masonry*)

Type	Minimum Compressive Strength of 2-in. Cubes at 28 Days, psi
A	2500
B	600
C	200
D	75

Mortar Colors and Admixtures. Inorganic compounds that are free from acids or soluble salts, and of a type that will not react with calcium hydroxide, are used for coloring mortars. With the exception of carbon black and other colors of specific gravity less than 3.0, a quantity up to about 10% by weight of the cement may be used without harmful result.

Experience in the field and laboratory tests indicate that some mortars may be improved by incorporating proprietary compounds. Some of these diminish the absorption of surface water, others accelerate hydration, lower initial shrinkage of the mortar, or increase workability. No nationally accepted standards have as yet been promulgated for gaging the value of these materials. The designer will be guided in their selection by tests furnished by the manufacturers and by the history of their performance. Anti-freeze compounds of a proprietary type are not recommended. The use of salt for lowering the freezing point of mortar should be prohibited.

Solid Brick Masonry. Except for rare cases where porous brick have been used, the improper filling of the joints, particularly the vertical joints, is the usual cause of leakage. All brick should be laid upon an adequate thickness of mortar evenly spread over the horizontal bed. In order to make certain that the vertical joints are filled, brick should be laid with a full "shove-joint," or plenty of mortar should be "thrown" on the end of the brick to be placed, or on the end of the adjoining brick already laid, and the brick pushed into place so that the mortar oozes out at the top of the joint. The common practice of "slushing" vertical joints is not recommended. All exposed joints should be thoroughly compacted. The application of a 1/2-in. mortar coat to the inner surface of an exterior vertical layer of brick, or to the outer surface of the backing material, forms a helpful water barrier. Mortar used in this manner and the operation of applying it are called "parging."

---

* This and the following paragraph have been adapted but not directly quoted from *American Standard Building Code Requirements for Masonry*, ASA A41-1-1944.

The allowable working stresses in solid brick masonry, corresponding to various mortar types and the strength of the brick, are given in Table 13.

Table 13. Compressive Stresses in Solid Brick Masonry

(From *American Standard Building Code Requirements for Masonry*)

Brick and other solid units of clay or shale, sand lime, or concrete brick.

Average Compressive Strength of Units Tested in the Position Taken in the Masonry, psi	Allowable Compressive Stresses Corresponding to Mortar Types, psi			
	A	B	C	D
8000 plus	400	300	200	100
4500 to 8000	250	200	150	100
2500 to 4500	175	140	110	75
1500 to 2500	125	100	75	50

The thickness and bonding of solid brick masonry walls should be checked upon requirements of the local building code, particularly in districts subject to earthquakes. As a guide to the designer, the following recommendations are abstracted from *American Standard Building Code Requirements for Masonry.*

Thickness: Solid Masonry Bearing Walls. (*a*) The thickness of solid masonry bearing walls shall be sufficient at all points to keep the combined stresses due to live, dead, and other loads for which the building is designed within the limits given in the preceding table.

(*b*) Except as otherwise provided in this section, the minimum nominal thickness of solid masonry bearing walls shall be 12 in. for the uppermost 35 ft of their height and shall be increased 4 in. for each successive 35 ft, or fraction thereof, measured downward from the top of the wall.

(*c*) Where solid masonry bearing walls are stiffened at distances not greater than 12 ft apart by cross walls, or by internal or external offsets or returns at least 2 ft deep, or by reinforced concrete floors, they may be of 12-in. nominal thickness for the uppermost 70 ft, measured downward from the top of the wall, and shall be increased 4 in. in thickness for each successive 70 ft or fraction thereof.

(*d*) In buildings not more than three stories in height, solid masonry bearing walls of the top story may be of 8-in. nominal thickness when the total height of the wall does not exceed 35 ft, provided that such 8-in. walls do not exceed 12 ft in height and that the roof beams are horizontal.

(*e*) In residential buildings not more than three stories in height, solid masonry bearing walls may be of 8-in. nominal thickness when not over 35 ft in height. Such walls in one-story single-family dwellings and one-story private garages may be of 6-in. nominal thickness when not over 9 ft in height, provided that when gable construction is used an additional 6 ft is permitted to the peak of the gable.

(*f*) Solid masonry walls above roof level, 12 ft or less in height, enclosing stairways, elevator shafts, penthouses, or bulkheads may be of 8-in. nominal thickness and may be considered as neither increasing the height nor requiring any increase in the thickness of the wall below, provided that the requirements for allowable stresses are met. Requirements for lateral support are as given on p. 27-48.

Thickness: Solid Masonry Nonbearing Walls. Nonbearing walls of solid masonry may be 4 in. less in thickness than required for bearing walls, but the nominal thickness shall be not less than 8 in., except where 6-in. walls are specifically permitted.

Bond. The facing and backing of solid masonry walls shall be bonded, either with at least one full header course in each seven courses, or with at least one full-length header in each 1.5 sq ft of wall surface. This distance between adjacent full-length headers shall not exceed 20 in. either vertically or horizontally. In solid brick walls of more than 8-in. nominal thickness, the inner joints of header courses shall be covered with another header course which shall break joints with the course below.

The more common methods of finishing the exposed surfaces of mortar joints are shown in Fig 22. The concave or rodded joint is the best. The second choice is the weathered joint. For standard work a 1/2-in. joint thickness is usually satisfactory. If the principles of modular design are applied, the joint thickness will be fixed.

Solid Stone Masonry. The allowable working stresses in solid stone masonry, corresponding to various mortar types and various stones, are given in Table 14.

Table 14. Compressive Stresses in Solid Stone Masonry

(From *American Standard Building Code Requirements for Masonry*)

Material	Allowable Compressive Stresses, Corresponding to Mortar Types, psi			
	A	B	C	D
Granite, ashlar	800	640	500	400
Limestone, ashlar	500	400	325	250
Marble, ashlar	500	400	325	250
Sandstone, ashlar	400	320	250	160
Cast stone	400	320	250	160
Rubble stone	140	100	80	...

The recommended minimum thickness and bonding of solid masonry are given in the following abstract from *American Standard Building Code Requirements for Masonry.*

Thickness. (*a*) The thickness of stone walls shall be sufficient at all points to keep the combined stresses due to live, dead, and other loads for which the building is designed within the prescribed limits.

(*b*) The minimum thickness of walls of stone ashlar shall be not less than that required for solid masonry walls on p. 27-45.

(*c*) Rubble-stone walls shall be 4 in. thicker than is required for solid masonry walls of the same respective heights, but in no case less than 16 in. thick. Requirements for lateral support are as given on p. 27-48.

Bond. (*a*) In ashlar masonry bond stones uniformly distributed shall be provided to the extent of not less than 10 per cent of the area.

(*b*) Rubble-stone masonry 24 in. or less in thickness shall have bond stones with a maximum spacing of 3 ft vertically and horizontally, and if the masonry is of greater thickness than 24 in., shall have one bond stone for each 6 sq ft of wall surface on both sides.

In ashlar masonry, the thickness of both vertical and horizontal joints usually varies between 1/4 and 3/16 in. When the stone is laid the mortar is either kept back from the face or raked out to a depth of 3/4 in. At some later period, the exterior joints are filled and finished to the required surface appearance. The more usual types are the same as those shown in Fig. 22.

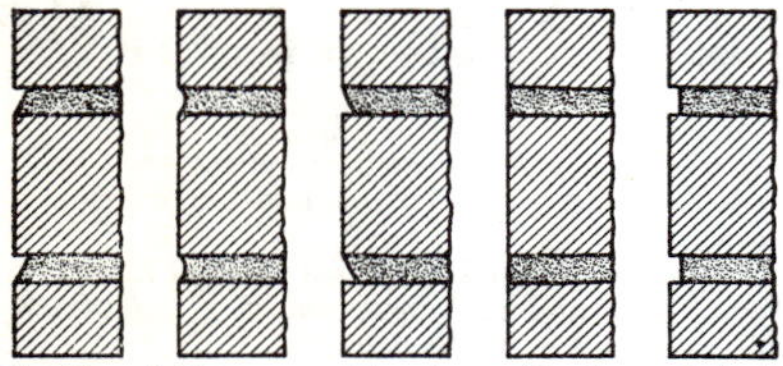

Fig. 22. Common types of exposed joints. (Reproduced by permission from Crane, *op. cit.*)

Walls of Hollow Masonry Units. This classification applies to walls built of structural-clay tile or hollow concrete-masonry units. These materials are widely used for back-up purposes with stone or brick facings. The joint thickness is usually 3/8 in. Portland cement-stucco is normally applied when structural-clay or concrete-masonry units are used without other surfacing. An alternate finish applicable to the concrete-masonry is two coats of cement paint. Special concrete-masonry units are manufactured to serve as a wall-surfacing material and are laid in various patterns.

The allowable compressive stresses in pounds per square inch of gross cross-sectional area, for masonry of hollow units of structural-clay tile or of hollow concrete masonry, should not exceed 85 psi when laid in type A mortar or 70 psi when laid in type B mortar.* The following recommendations are abstracted from *American Standard Building Code Requirements for Masonry.*

Thickness and Height. (*a*) The thickness of walls of structural-clay tile or hollow concrete-masonry units shall be sufficient at all points to keep the combined stresses due to live, dead, and other loads for which the building is designed within the allowable stresses.

(*b*) The minimum thickness ot walls of structural-clay tile or hollow concrete-masonry units shall be not less than that required for solid masonry walls, as given on p. 27-45.

(*c*) Walls of structural-clay tile or hollow concrete-masonry units shall not exceed 50 ft in height above the support of such walls. Requirements for lateral support are as given on p. 27-48.

Bond. (*a*) Hollow masonry units shall have full mortar coverage of the face shells in both the horizontal and vertical joints.

(*b*) Where two or more hollow units are used to make up the thickness of a wall, the stretcher courses shall be bonded at vertical intervals not exceeding 34 in. by lapping at least 3 3/4 in. over the unit below or by lapping with units at least 50 per cent greater in thickness than the units below at vertical intervals not exceeding 17 in.

(*c*) Where walls of hollow masonry units are decreased in thickness, a course of solid masonry shall be interposed between the wall below and the thinner wall above.

Structural Glass Block. The usual thickness is 3 7/8 in. with sizes 5 3/4 by 5 3/4, 7 3/4 by 7 3/4, and 11 3/4 by 11 3/4 in. Mortar bearing surfaces are often corrugated and coated with an alkali and moisture-resistant aggregate material to insure proper bond. Noncorrodible wire ties are placed in the joints. If panels are not set into chases they are anchored to adjoining walls with perforated metal strips. Panels greater than 144 sq ft in area, over 25 ft long, or 20 ft high should be divided horizontally by mullions or vertically by shelf angles. Glass block are usually laid in mortar of volumetric proportions; 1 part white portland cement : 1 part lime putty : 4 parts fine sand. Joints vary from 3/16 to 3/8 in. Expansion strips are used at jambs and heads of panels.

Faced and Veneered Walls.† Ashlar masonry of natural or cast stone, face brick, architectural terra cotta, structural glass, and occasionally other materials are used as a

* Also quoted from *American Standard Building Code Requirements for Masonry,* ASA A41.1-1944.

† The requirements applying to masonry for faced walls have been adapted but not directly quoted from *American Standard Building Code Requirements for Masonry,* ASA A41.1-1944.

facing or as a veneer over a backing of cheaper masonry units. Figure 23 shows typical examples of brick facing over clay tile and concrete block. A "faced wall" is one in which the masonry facing and backing are bonded to act together under load; a veneered wall is one in which the masonry facing is not so bonded, being merely attached to the backing.

In conventional wall assemblies, the minimum thickness of the facing should be 2 $^1/_4$ in. or not less than one-eighth the height of the unit. The full cross section of both the facing and the backing is then included when computing the unit stresses, which should not exceed those allowed for the weaker of the two materials. The total structural thickness of faced walls should not be less than that required for masonry walls of either of the types forming the facing and backing. Brick facing should be bonded to the backing, as noted on p. 27-44 for solid brick masonry. Ashlar facing of either natural or cast stone should have at least 20% of the superficial wall area extending not less than 3 $^3/_4$ in. into the backing to form bond stones, which should be uniformly distributed throughout the wall. Substantial noncorrodible metal anchors, at least 0.2 sq. in. in cross section, should be required for every projecting stone, and, except when alternate courses are full bond courses,

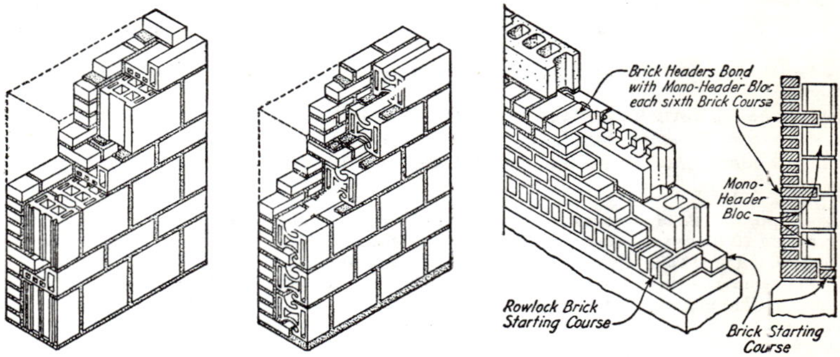

FIG. 23. Brick facing over tile and block backing. (Reproduced by permission from Crane, *op. cit.*)

every stone not a bond stone should be similarly anchored. Each stone more than 2 ft in length and 30 sq ft in superficial area should have two anchors. Facing stones not over 12 sq ft in area should have at least one anchor to each 4 sq ft of superficial face area. The bond required between a backing of structural-clay tile or hollow concrete-masonry and a facing of hollow units is the same as described for walls of hollow masonry units.

Precast concrete panels 2 to 3 in. thick and ranging up to 8 by 10 ft in area may be used over concrete backing, poured-in-place, or over masonry. For concrete backing, the panels serve as exterior forms.

Precast concrete wall facings have a wide application to many classes of occupancy. They have been used for commercial and even monumental buildings (Ref. 6).

Concrete Walls Above Grade. Cast-in-place concrete may often be used to advantage for low structures. Such structures may be of panel, curtain, or bearing wall design. The minimum thickness for panel walls should be 4 in., increased as required by local building code or structural demands. They should be reinforced with small rods representing about $^2/_{10}$% of the cross-sectional area of the wall both horizontally and vertically, and recessed into the adjoining elements of the structural frame to obtain stability against wind pressure and a weathertight joint. When serving as curtain walls or bearing walls, additional reinforcement is often required. In these cases, the minimum thickness is usually 8 in. It is important that additional reinforcement be placed around all openings.

Concrete walls may be veneered with practically any surfacing material, and either precast ornament or designs poured monolithically may be used as embellishment. For industrial buildings a satisfactory finish may often be obtained by wetting the wall surface with a thin cement mortar and lightly grinding with silicon carbide or other abrasive. This treatment is appropriate for conditions that prevent removal of forms prior to hardening of the concrete. If the forms can be promptly removed, as soon as security permits, a somewhat better finish may be obtained by wetting and rubbing the surface with fine sand beneath a wood float. In any case, all honeycomb, holes left by form ties, or other imperfections should be remedied before any surface treatment is applied. Standard references on the subject of concrete are *Report of the Joint Committee on Standard Specifications for Concrete and Reinforced Concrete*, obtainable from the Portland Cement Association, and Report of American Concrete Institute Committee 318, *Journal of the American Concrete Institute*, Sept. 1947.

Furring Masonry Walls. Between the body or backing of a masonry wall and the interior finish, an air space is often provided. This may be formed by tile furring block, a free-standing tile partition, or any of the other means shown in Fig. 24. In damp locations corrodible metal furring is obviously inferior to masonry furring. Gypsum tile is appropriate only where there is no probability of moisture. The purpose of furring is to provide a more or less dead air space that assists in preventing the direct passage of wind-driven rain and increases the resistance of the wall to the transmission of heat. A clear air space at least $^3/_4$ in. wide is desirable.

Exterior walls of solid masonry, including concrete, should ordinarily be furred for residential occupancy and often for commercial structures. In garages and factories, where

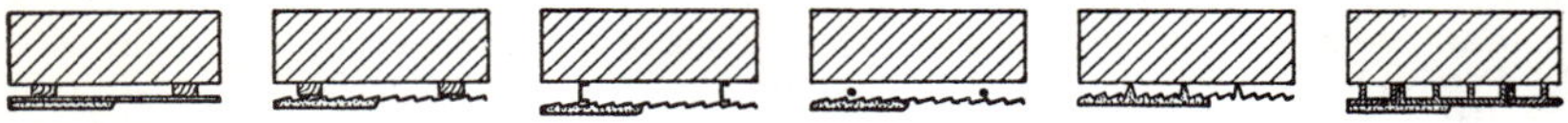

FIG. 24. Exterior masonry-wall furring. (Reproduced by permission from Crane, *op. cit.*)

there is plenty of ventilation and the heat loss is not an important factor, furring is often omitted.

Miscellaneous Structural Requirements. The following recommendations are quoted from *American Standard Building Code Requirements for Masonry.* They conform with the most modern practice in this field but are more liberal in some respects than those of many building codes.

Lateral Support. Solid masonry walls shall be supported at right angles to the wall face at intervals not exceeding 20 times the nominal wall thickness if laid in type A, B, or C mortar and not exceeding 12 times the nominal wall thickness if laid in type D mortar.

Walls of structural-clay tile or hollow concrete-masonry units and hollow walls of masonry shall be supported at right angles to the wall face at intervals not exceeding 18 times the nominal wall thickness.

Cavity walls shall be supported at right angles to the wall face at intervals not exceeding 14 times the nominal wall thickness.

Lateral support may be obtained by cross-walls, piers, or buttresses, when the limiting distance is measured horizontally, or by floors and roofs, when the limiting distance is measured vertically. Sufficient bonding or anchorage shall be provided between the walls and the supports to resist the assumed wind force, acting either inward or outward. Piers or buttresses relied upon for lateral support shall have sufficient strength and stability to transfer the wind force, acting in either direction to the ground. When walls are dependent upon floors or roofs for their lateral support, provision shall be made in the building to transfer the lateral forces to the ground.

Except for window-panel backs and permissible chases and recesses, walls shall not vary in thickness between their lateral supports. When a change in thickness, due to minimum thickness requirements, occurs between floor levels, the greater thickness shall be carried up to the higher floor level.

Bonding of Walls at Intersections. Masonry walls shall be securely anchored or bonded at points where they intersect and where they abut or adjoin the frame of a skeleton frame building. When two bearing walls meet or intersect and the courses are built up together, the intersections shall be bonded by laying in a true bond at least 50 per cent of the units at the intersection.

When the courses of meeting or intersecting bearing walls are carried up separately, the perpendicular joint shall be regularly toothed or blocked with 8-in. maximum offsets, and the joints provided with metal anchors having a minimum section of $1/4$ by $1\ 1/2$ in. with ends bent up at least 2 in. or with cross-pins to form anchorage. Such anchors shall be at least 2 ft long, and the maximum spacing shall be 4 ft. Meeting or intersecting nonbearing walls shall be bonded or anchored to each other in an approved manner.

Anchoring of Walls. Masonry walls shall be securely anchored to each tier of wood joists or wood beams bearing on them at maximum intervals of 6 ft in one- and two-family dwellings, and 4 ft in other buildings, by metal anchors having a minimum cross-section of $1/4$ by $1\ 1/4$ in. and at least 16 in. long, securely fastened to the joists or beams and provided with split and upset ends or other approved means for building into masonry. Girders shall be similarly anchored at their bearings. Anchors shall be attached in a manner to be self-releasing.

Masonry walls parallel to wood joists or wood beams shall be provided with similar anchors at maximum intervals of 8 ft in one- and two-family dwellings, and 6 ft in other buildings, engaging three joists or beams. Upset and T-ends on anchors shall develop the full strength of the anchor strap.

Cast-in-place concrete slabs bearing on masonry walls shall be considered as sufficient anchorage for the supporting walls.

Chases and Recesses. There shall be no chases in walls of less than 12-in. nominal thickness or within the required area of any pier, and no chase in any wall shall be deeper than one-third the wall thickness, except that in dwellings not over two stories in height vertical chases may be built in 8-in. walls under the following limitations: In 8-in. bearing walls the chases shall not exceed 4 in. in depth, 30 in. in width, and 2 ft in height, and shall not extend below the level of joist bearing, provided that where such chases occur below window sills the width may be not in excess of the width of the window opening above. In any case, not less than 4 in. of masonry shall remain between the back of chase and exterior surface of wall, and the backs and sides of all such chases in exterior walls shall be waterproofed and insulated. Masonry directly over chases wider than 12 in. shall be supported on lintels. Chases permitted in 8-in. walls shall not be cut but shall be built in as construction progresses.

No horizontal chase shall exceed 4 ft in length, nor shall the horizontal projection of any diagonal chase exceed 4 ft in length. There shall be at least 7 3/4 in. of masonry between chases and the jambs of openings.

Recesses for stairways or elevators may be left in walls, but in no case shall the walls at such points be reduced to less than 12 in. unless reinforced by additional piers or by columns or girders of steel, reinforced masonry, or concrete, securely anchored to the walls on each side of such recesses. Recesses for alcoves and similar purposes shall have not less than 8 in. of material at the back. Such recesses shall be not more than 8 ft in width and shall be arched over or spanned with lintels.

The aggregate area of recesses and chases in any wall shall not exceed one-quarter of the whole area of the face of the wall in any story. Chases and recesses shall not be cut in hollow walls, cavity walls, or walls of hollow masonry units, but when permitted may be built in.

Lintels and Arches. The masonry above openings shall be supported by arches or lintels of metal or masonry, plain or reinforced, which shall bear on the wall at each end for not less than 4 in. Stone or other nonreinforced-masonry lintels shall not be used unless supplemented on the inside of the wall with iron or steel lintels or with suitable masonry arches or reinforced-masonry lintels carrying the masonry backing.

Steel or reinforced-masonry lintels shall be of sufficient strength to carry the superimposed load without deflection of more than 1/360 of the clear span. Masonry arches shall have at least 1-in. rise for each foot of span and shall be designed to carry the superimposed load. Proper provision shall be made for resisting lateral thrust.

Separation of Combustible Structural Members. No wall of 8-in. nominal thickness shall be broken into, subsequent to building, for the insertion of structural members. A separation of at least 4 in. of solid masonry shall be provided between combustible members which may enter walls from opposite sides.

When unprotected steel or combustible structural members frame into hollow walls of thickness not greater than 12 in., they shall project not more than 4 in. into the wall and shall be spaced so that the distance between embedded ends is not less than 4 in. The space above, below, and between such members shall be filled solidly with burned-clay materials, mortar, concrete, or equivalent fire-resistive material to a depth of not less than 4 in. on all sides of the members.

All open cells in tiles or blocks occurring at wall ends shall be filled solidly with concrete for a depth of at least 6 in., or closure tiles set in the opposite direction shall be used.

Beam Supports. Beams, joists, girders, or other concentrated loads, supported by a wall or pier, shall have bearing at least 3 in. in length upon solid masonry not less than 4 in. thick or on a metal bearing plate of adequate design and dimensions to distribute safely the loads on the wall or pier.

Parapet Walls. Parapet walls shall be at least 8 in. in nominal thickness. They shall be not higher than 4 times their thickness unless laterally supported, provided that, when reinforced both horizontally and vertically with not less than 1/4-in. rods spaced not more than 2 ft on centers, the height shall be not more than six times the thickness. All parapet walls shall have a coping of incombustible material.

Piers. The unsupported height of piers shall not exceed 10 times their least dimension. When structural-clay tile or hollow concrete-masonry units are used for isolated piers to support beams and girders, the cellular spaces shall be filled solidly with concrete of type A mortar, and when so constructed the allowable stresses may be increased 25 per cent, provided that unfilled hollow piers may be used if their unsupported height is not more than 4 times their least dimension.

WALLS SURFACED WITH SHEET METAL. Specially formed sections of steel sheets have come into general use for the exterior wall surfaces of industrial buildings. These supplement corrugated iron, corrugated zinc, and sheet copper. An interesting design appropriate for industrial work is employed in a so-called windowless factory or, more correctly expressed, a controlled-conditions plant. Light and interior atmosphere are regulated to meet the requirements of the occupants. Objectionable sound is largely eliminated. The structural support of the wall and roof is steel. The masonry wall below sill level is of brick backed by Fiberglas board with an interior finish of clay tile. Above the sill, steel panels are used for the exterior finish. These are 30 in. wide, fabricated from 18-gage steel and ribbed 10 in. on centers. They are supported by girts or other framing on spans up to 12 ft. The shape of the panels, backed by the Fiberglas insulation board, results in the formation of vertical flues, which provide air circulation within the wall. A vapor seal is applied to the inner face of the board. Four-inch Fiberglas bats, compressed to 3 in. and covered by a 7-mm Fiberglas mat, retained in place with expanded metal lath, serve as insulation and interior finish. This particular combination of materials was designed to obtain low heat transmission, good sound absorption, and light reflection.

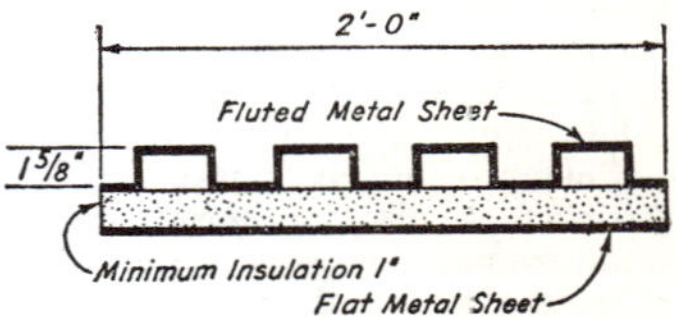

FIG. 25. Insulated, fluted, sheet-metal wall panels. (Courtesy of H. H. Robertson Co.)

A design employing fluted panels on the exterior, backed by insulation and surfaced on the inside with flat metal sheets, is shown in Fig. 25. If desired, the flat surface can be used on the exterior. The more recent installations made by the sponsors of this system employ aluminum sheets separated by an insulation of Fiberglas. The advantage of dry-wall construction, combined with speed in erection resulted in the wide acceptance of these metallic designs for industrial buildings during World War II. In addition to these purely utilitarian types, several of the more expensive metals are now furnished in practicable shapes to serve as exterior surfaces. Both aluminum and stainless steel are included in this category supplemented by various designs in porcelain-enameled iron or steel. In considering the appropriateness of any proposed design, the fire rating should be checked where it is important, as some assemblies unbacked by masonry are classed merely as incombustible.

GUNITE AND STUCCO SURFACES. Gunite. This is the trade name of a portland-cement-sand mixture combined with water when discharged in the form of mortar from the nozzle of a cement gun. The process has a broad application for remedial treatments applied over deteriorated concrete or masonry surfaces, but it may also be used advantageously as a means of constructing thin walls and partitions.

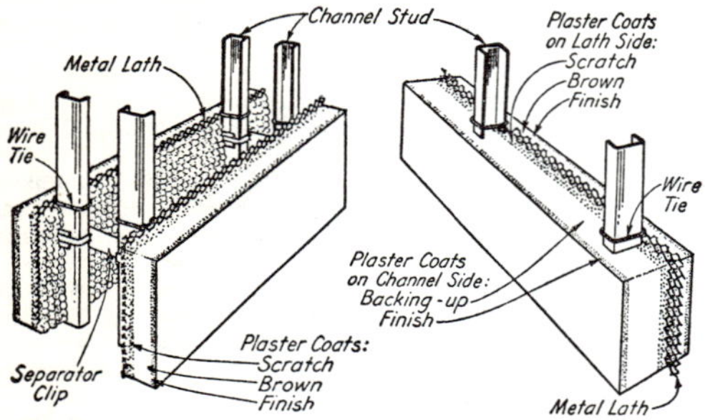

Fig. 26. Hollow and solid channel-stud partitions. (Courtesy of Metal Lath Manufacturing Assoc.)

Gunite may be shot against any solid backing such as a temporary wood panel or insulating board serving as a permanent part of the wall assembly. When insulation is not desired, the mortar may be shot against a backing of heavy paper with reinforcing mesh attached to a steel or wood frame. The same treatment may be used to form both an exterior and an interior surface, resulting in a hollow wall. The thickness of a solid section usually varies from 1 1/2 to 2 in., but may be made thicker when desired. The finished surface is of uniform, slightly rough texture.

Stucco Surfaces. The term *stucco* generally signifies cement-aggregate mortar applied by hand to form an exterior surface. Coloring is usually incorporated in the final coat. Plasticizing and waterproofing agents may be used. The material is applied directly to a masonry base or to welded-wire mesh attached to framed construction of steel or wood. When properly applied, portland-cement stucco is durable and impervious to wind-driven rain, but it is absolutely essential that good practice be followed in every detail and that differential movement be avoided between the exterior stucco and the structural assembly supporting it. Stucco is particularly vulnerable to even slight settlement of the supporting wall, steel or wood frame. If appreciable differential volumetric change occurs between the base over which the stucco is applied and the enclosing envelope of stucco, the surface will crack. The best recommendations now available are given in *Standard Specification for Portland Cement Stucco*, approved by the American Standards Association in 1946.

WALLS SURFACED WITH CEMENT-AGGREGATE PLANK AND COMPOSITION BOARDS. Cement-aggregate Plank. Lightweight concrete planking composed of portland cement combined with slag, slag products, hard-coal cinders, or similar materials is used as an exterior wall surfacing for industrial buildings. Thoroughly reinforced with welded, galvanized wire, tongued and grooved, it may offer an economical choice for the lighter types of wall construction. Planks are usually bolted or clipped to T bars or other appropriate structural shapes, spaced 6 to 8 ft on centers vertically or horizontally.

Cement-asbestos Boards. The corrugated cement-asbestos boards mentioned in Article 6 are also appropriate for wall surfacing over structural steel and occasionally over wood framing for certain types of industrial buildings. Sheets may be either plain or corrugated. A typical size is 3 ft 6 in. by 8 ft. Supports are required at a maximum spacing between 5 and 6 ft on centers. They are usually in the form of steel girts. One particular assembly is composed of 3/8-in. corrugated sheets backed by 1-in. insulating boards, to which the interior surface is veneered. The design furnishes a lightweight section with good insulation. The better materials in this group are highly resistant to the attack of most acid fumes or gases, extreme temperature variations, and alternate wetting and drying. They are also fire resistant. Of a light gray color, cement-asbestos boards are usually left unpainted.

Cemesto Board. Identified on p. 27-37, these sheets, as used for wall construction, vary in thickness from 1 1/8 to 2 in. They have an application for some types of industrial work and are used in more or less prefabricated designs for small dwellings.

PARTITIONS. Clay tile, concrete-masonry units, and gypsum tile have been widely used in fire-resistant buildings. The 2-in. solid metal-lath and plaster partition has a particular application for large housing projects for which the reduced thickness has special significance in increasing the available floor area. The solid plaster partition may also be useful for enclosing office spaces within a manufacturing plant. Figure 26 shows the construction of steel-stud partitions, both of hollow and of solid type. In connection with these designs it should be noted that the horizontal elements placed at top and bottom of the partition assembly can be easily attached to a concrete floor or ceiling by means of steel nails. Such installations have been highly developed by the New York City Housing Authority but have an application for other types of work. Figure 27 shows a comparison of weights for the more popular partition types. Table 15 shows the minimum thicknesses recommended for nonbearing masonry partitions.

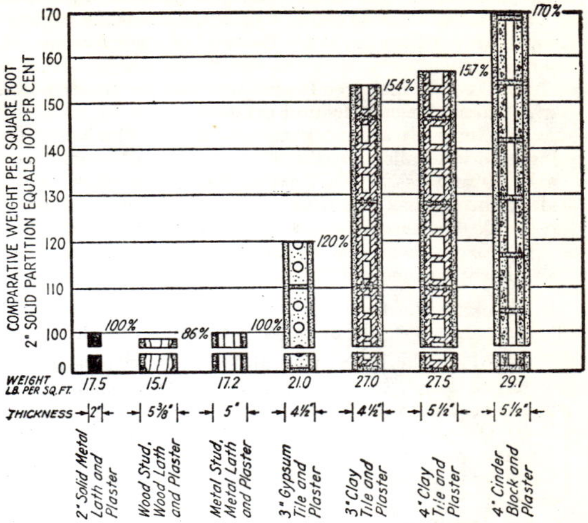

FIG. 27. Weights of various types of partitions. (Courtesy of Metal Lath Manufacturing Assoc.)

Table 15. Minimum Thickness of Nonbearing Masonry Partitions

(From *American Standard Building Code Requirements for Masonry*)

Maximum Unsupported Height, ft	Thickness Exclusive of Plaster, in.	Maximum Unsupported Height, ft	Thickness Exclusive of Plaster, in.
9 *	2	20	6
12	3	25	8
15	4		

* Not over 6 ft in length.

HEAT INSULATION AND VAPOR SEAL. These subjects as applying to roofs were discussed in Article 6. They are also important in relation to exterior walls for which insulation is often desirable to reduce the heat loss in cold weather or, more rarely, the heat gain in hot weather. Provision to avoid condensation on an interior surface of an exterior wall might require insulation, even when the heat loss could be accepted. Prevention of condensation of water vapor within an exterior wall assembly may require either a vapor seal on the warm side of the insulation or ventilation within the wall. This is particularly true if interior humidities are high.

Valuable information on this subject will be found in Accumulation of Moisture in Walls of Frame Construction during Winter Exposure, by Weber and Reichel, *Building Materials and Structures Report* BMS 93, National Bureau of Standards, 1942; Moisture Condensation in Building Walls, by Harold W. Wooley, also a Bureau of Standards publication, BMS 63, 1940; Comparative Resistance to Vapor Transmission of Various Building Materials, by L. V. Teesdale, *Journal of the American Society of Heating and Ventilating Engineers*, 1943; *Condensation in Walls and Attics*, by L. V. Teesdale, Forest Products Laboratory, Forest Service, United States Department of Agriculture, 1937.

FLASHING AND CAULKING. Locations Where Flashings Are Required. Flashings are generally necessary in the following locations:

(*a*) Wherever a parapet, chimney, or wall intersects a roof surface.

(*b*) Directly under the copings of parapets, chimneys, moat walls, and other masonry courses which form the top surface of exposed walls or piers, unless a continuous metallic coping is installed. Parapets, other than low parapets, should have an additional through-flashing, at the level of the cap flashing, or just above the flashing block.

(*c*) Either directly beneath or over masonry cornices and belt courses, for the purpose of preventing leakage through the vertical joints separating the masonry units forming the projection. The same statement applies to recesses.

(*d*) Usually across the top of basement walls to prevent moisture from traveling upward by capillary action. If metal is used for this purpose, it may be designed as a termite shield.

(*e*) Overall spandrel beams in skeleton construction at every floor and roof level, with particular care around columns.

(*f*) Generally at both heads and sills of all window, door, and other openings penetrating exterior walls. If sills are impervious to moisture and made in one piece, flashing may be unnecessary. If exterior stucco surfaces are used, flashings may be necessary along the sides as well as at heads and sills of openings. If stucco is placed over flashings, it should be carried by metal reinforcement thoroughly protected from corrosion.

(*g*) Except where eaves have a considerable overhang, between the top of the wall and lower surface of the roof.

In checking the flashings for a building, the designer would also give attention to the following locations, although they do not affect the structural work.

(*h*) Along the boundaries of flat roofs without parapets when surfaced with built-up waterproofing. In this case, the gravel stop may often be combined with a flashing which protects the eaves.

(*i*) On pitched roofs, all valleys are flashed. Hips and ridges are designed in conformity with common practice applying to the particular roof-surfacing material. Flashing is also necessary along horizontal lines where a change occurs in the slope of a pitched roof.

(*j*) Around all elements penetrating a roof surface, such as pipes or flues, in addition to the various wall locations previously mentioned. On comparatively flat roofs, the gutters and drainage connections are flashed into the surrounding waterproofing.

(*k*) Built-in gutters which require particular attention.

Excellent details illustrating good practice in the design of flashings may be obtained from the Copper and Brass Research Association.

The Choice of Flashing. The flashing material should be of a type suited to its particular exposure. As a guide to making a suitable choice, flashings may be divided, from the viewpoint of their appropriate uses, into the following groups.

Metals with Smooth Metallic Surfaces. Except where there is danger of staining with copper wash, soft-rolled 16- to 24-oz sheet copper is generally satisfactory. At least a 20-oz weight is recommended where permanence is a prime requisite. Zinc is also an acceptable flashing material but is more vulnerable than copper to certain types of atmospheric pollution. Both the 11-gage and 13-gage weights have been used. However, because of the particular characteristics of this metal, designs contemplating its use should follow the recommendations of the American Zinc Institute. When zinc is placed in contact with mortar containing lime, a heavy coating of paint or a bituminous compound free from acid is essential.

The lead and antimony alloy, generally called *hard lead*, can be used for flashings in weights of 2 to 3 lb per sq ft. Because hard lead has a high coefficient of thermal expansion and low elasticity, large sheets of it tend to "creep" and acquire a permanent set after wrinkles have developed. Designs should be checked with the Lead Industries Association. An asphaltum coating is recommended if lead is placed in contact with fresh mortar or concrete. Lead-coated copper is widely used for high-grade work in locations where copper staining would be objectionable or the color of copper inappropriate. There is also the general belief that the lead coating increases the permanence of the copper. This is generally true up to the point that the lead is worn through in spots. Beyond this stage, corrosion may be accelerated by electrolytic action. Aluminum also has possibilities as a flashing material but is not used to any extent for this purpose. As the coefficient of expansion due to temperature changes is very high aluminum should be used with caution. If brought in contact with mortar or concrete, it should be protected by a coating of lacquer to prevent corrosion.

As a class, the metallic flashings are appropriate for standard cap-and-base designs and for through-flashings. Galvanized steel sheets and the cheaper grades of tin plate should not be used except where their probable life will equal that of the roof-surfacing material. The choice should be based upon authentic records of performance under similar exposure or accelerated tests of a conclusive nature.

Thin Copper Sheets Surfaced with Bitumen-saturated Fabric. These sheets are very flexible as the copper is only 2- to 3-oz weight. They are particularly appropriate for spandrel waterproofing.

Membrane and Composition Flashings. Fabrics saturated with bitumen and laid in a bituminous waterproofing compound are appropriate for a dampcheck across the top of masonry walls and for spandrel waterproofing as an alternate to the material described in the preceding paragraphs. For this purpose two layers of fabric are laid with three coats of a cold-troweled mastic. Bituminous materials also serve as a cheaper but inferior substitute for the metal flashings of window sills and lintels. They are the usual choice for the junction of flat roofs surfaced with built-up waterproofing and parapets or penthouse walls.

Caulking. This term identifies the use of a plastic compound resembling putty instead of mortar for filling joints between masonry or other building units and for sealing the cracks around window frames and other wood or metal elements built into masonry walls. The material may be applied by extruding from a caulking gun, thereby forming a bead along the joint, or by the use of a trowel or knife. Consistencies suitable for either type of application are furnished by the manufacturers. Various colors are available. As certain types of caulking compounds have been found to stain porous materials, and considerable difference exists in both consistency and lasting quality, care should be exercised in choosing an appropriate material. Additional information on this subject will be found in Plastic Caulking Materials, by Tregoning, Milliken, Hockman, Sligh, and Kessler, *Building Materials and Structures Report* BMS 33, National Bureau of Standards, 1940.

DESIGN PROCEDURE. The problem is to select a wall construction which fulfills the desired functions at the least ultimate cost. The exterior finish should meet the utilitarian requirements of durability without costly upkeep. In some cases the appearance of the exterior and interior surfaces will be an important consideration. Interior surfaces should be suitable for the type of occupancy. The requirement that the wall be rain- and windproof is very important, but these qualities depend rather upon careful detailing, specification, and supervision than choice of the assembly.

In specifying the structural requirements of an exterior wall or partition, it is essential that the standards of a mandatory code or those dictated by good practice be constantly borne in mind. This is particularly important in districts subject to earthquakes.

Well-standardized methods have long been established for accommodating electric conduit, plumbing, and heating pipes within the outer surfaces of conventional wall assemblies, but when the newer prefabricated or semi-prefabricated designs are employed, the concealing or "building-in" of building elements should be carefully considered.

The outdoor-temperature range, solar intensity, and relative humidity for the various seasons may dictate the amount of heat insulation or favor one type of wall surface over another, as having a higher reflection of radiant heat or a greater time lag in heat transmission. When a particular manufacturing process causes or requires a high interior humidity, its possible effect should be considered.

The weights per square foot of the different wall assemblies are seldom important for low buildings, but should be considered for comparatively high structures, or in localities where a low soil-bearing capacity demands an appreciable increase in foundation cost to carry an increased dead load. The thickness of walls obviously affects the usable floor area. This condition is quite important in selecting partitions for multifamily dwellings, but it has little significance for industrial projects.

After these requirements are checked for each of the probable designs, preliminary estimates should be obtained for comparison. Only by using the material prices and labor rates applying to the time and place of construction, and with the local standard of man-hour production, can a valid estimate of cost be obtained. Unit prices in cents per square foot of wall area may sometimes serve as a general guide but are always unreliable and usually valueless for border-line cases.

8. CHOOSING THE TYPE OF FOUNDATION

The foundation of a building should transmit the weight of the superstructure, and any other loads, to an underlying stratum of soil or rock which is capable of furnishing support without objectionable settlement. The term *objectionable settlement* means an amount of settlement resulting in injury to the structure from either a utilitarian or an aesthetic viewpoint. All foundations, except those supported by sound rock, will settle somewhat. Slight, uniform settlement is of no importance for most buildings. Unequal settlement of the different parts of a structure, however, may cause serious stresses in the frame and cracks in floors and walls. A critical condition occurs where a portion of a building is supported on rock and the remainder on soil. This situation should be avoided, if practicable, by the use of concrete piers or piles carried down through the soil to a rock bearing. Where there is a probability of unequal settlement between two portions of a structure, vertical joints should be provided from top to bottom of the building.

Building records indicate that objectionable settlement is usually due to an incorrect evaluation of soil conditions rather than error in structural design. This condition indicates a need for adequate subsurface investigation. Information obtained in this manner may be supplemented by settlement records from other buildings, erected over similar soil conditions, soil loading tests made upon the site of the proposed building, and laboratory investigation of soil samples.

Foundations represent an appreciable portion of the building cost, and injury caused by settlement is expensive to repair. Although important advances have been made, particularly during the last generation, in the study of soil mechanics, this subject is at present far from an exact science. Because of the vastly different combinations of soils which may underlie a building foundation, both experience and good judgment, as well as a thorough knowledge of the rather extensive literature relating to this field, are necessary for important decisions regarding them. This section is consequently written to assist the building designer in correlating his procedure with the services of a foundation engineer, rather than to present a definite procedure for his own use.

SUBSURFACE MATERIALS. Types of Soil. Soil is the unconsolidated material formed by the disintegration of rock or organic substances. Sand and clay are the two principal classes. Sand is composed of comparatively large grains and clay of very small grains. Sand is often referred to as a cohesionless soil; clay as a cohesive soil. Sand is highly permeable to water; clay is comparatively impermeable. Sand is inelastic. Clay may have considerable elasticity, that is, ability to regain its original form and dimensions when an applied load, which has caused deformation, is released. Sand is not plastic, whereas clay has this characteristic, which means that it may change in shape without appreciable change in volume when subjected to continued loading.

The Boston Building Code gives the following definitions corresponding to the designations in Table 7.

Shale. A laminated, fine-textured soft rock composed of consolidated clay or silt, which cannot be moulded without the addition of water, but which can be reduced to a plastic condition by moderate grinding and mixing with water.

Slate. A dense, very fine-textured, soft rock which is readily split along cleavage planes into thin sheets and which cannot be reduced to a plastic condition by moderate grinding and mixing with water.

Schist. A fine-textured, laminated rock with a more or less wavy cleavage, containing mica or other flaky minerals.

Gravel. An uncemented mixture of mineral grains 1/4 in. or more in diameter.

Sand. A type of soil possessing practically no cohesion when dry, and consisting of mineral grains smaller than 1/4 in. in diameter.

Coarse Sand. A sand consisting chiefly of grains which will be retained on a 65-mesh sieve.

Fine Sand. A sand consisting chiefly of grains which will pass a 65-mesh sieve.

Compact Gravel, Compact Sand. Deposits requiring picking for removal and offering high resistance to penetration by excavating tools.

Loose Gravel, Loose Sand. Deposits readily removable by shoveling only.

Hardpan. A thoroughly compact mixture of clay, sand, gravel, and boulders, for example, boulder clay; or a cemented mixture of sand or of sand and gravel, with or without boulders, and difficult to remoke by picking.

Clay. A fine-grained, inorganic soil possessing sufficient cohesion when dry to form hard lumps which cannot readily be pulverized by the fingers.

Hard Clay. A clay requiring picking for removal, a fresh sample of which cannot be moulded in the fingers.

Medium Clay. A clay which can be removed by spading, a fresh sample of which can be moulded by a substantial pressure of the fingers.

Soft Clay. A clay which, when freshly sampled, can be moulded under relatively slight pressure of the fingers.

Rock Flour (Inorganic Silt). A fine-grained, inorganic soil consisting chiefly of grains which will pass a 200-mesh sieve, and possessing sufficient cohesion when dry to form lumps which can readily be pulverized with the fingers.

Ground Water. Below a certain level all soil is saturated with water. This level is often called the *water table*. It follows, approximately, the surface contours, but may be lowered by dry seams, by continued pumping, or by covering the ground surface with buildings or pavements. The ground water level has significance as affecting the bearing capacity of some soils, particularly fine sand and clay; as determining the level of "cut-off" for wood piles; and as causing increased cost of excavation and eventual provision for either waterproofing basements or the relief of water pressure by pumping.

SUBSURFACE EXPLORATION. Subsurface investigations should be made well in advance of the structural design. When there is a choice between two or more building sites, the result may determine the selection. Even where rock underlies the entire site of a future building, an investigation of its nature, by means of core borings, may be necessary. An illustration of a log of borings is published in *Bulletin* 208, Graduate School of Engineering, Harvard University.

For all important work "dry-sample" borings (see below) should be made as a preliminary step toward determining the most economical type of foundation assembly and as a guide in choosing the construction methods. In most cases this type of soil sampling gives sufficient data for design purposes, but over plastic soils such as clay, where the structural characteristics of the material, as well as the depth of the strata, may be in doubt, laboratory tests performed upon "undisturbed samples" (see p. 27-56) may be desirable.

The number of borings, or other type of soil exploration, obviously depends upon site conditions, the amount of supplementary information available, and the degree to which it seems necessary to develop a subsurface profile. In the general case, at least one boring should be made at points corresponding to the corners of the future structure. If soil conditions are variable, as they are particularly within glaciated regions, additional borings may be desirable. There is also a greater probability of encountering peat, organic or inorganic silt, and soft clay within glaciated regions. This area lies north of a line extending through Long Island, northern New Jersey, and Pennsylvania, along the Ohio and Missouri rivers into Kansas, thence northwesterly roughly parallel to and a little west of the Missouri River to the Rocky Mountains, thence westerly south of the Canadian border to the Pacific Coast.

At least one, and preferably several, of the borings should be carried down to a definitely determined stratum of rock or to a depth of not less than 1 1/2 times the width of the structure. For very wide buildings, where this depth appears excessive, judgment based on available information may warrant a lesser depth. If heavy loads are involved, the investigation of rock by means of core borings is often desirable, but the crushing strength of most rocks equals or exceeds that of the materials used for the substructure.

The records of foundation experience during construction and the subsequent settlement of adjacent structures, known to be supported by similar types of soil, often furnish valuable information. Geological advice is occasionally helpful. All available records should be checked to identify underground obstacles such as sewers, water and gas pipes, old foundations, and pier heads buried under fill.

METHODS EMPLOYED FOR SUBSURFACE EXPLORATION. **Test pits** are occasionally used where necessary to examine the soil in its natural state, but deep pits are costly as they generally require shoring and lining. Test pits, however, may offer an inexpensive means of making comparatively shallow investigations for small structures. For example, a pit may be dug to the level of the future footings, and a sample of soil obtained by means of an auger, or post-hole digger, from a depth 10 to 15 ft below that level.

Sounding rods are usually solid steel rods, or pipe sections, provided with points at the lower end. They are driven by hand or by a small gasoline-operated hammer. Provided that boulders or other obstacles are not mistaken for ledge rock, sounding rods may have a useful but limited application for determining the depth to rock, or other satisfactory bearings, such as gravel.

Auger borings are seldom used in building construction except for preliminary survey. The auger is rotated by hand or power. As the object is to raise samples of soil, its effective use is limited to cohesive materials, or noncohesive materials above ground-water level.

Wash-sample borings are obtained by driving a steel pipe casing, usually 2 or 2 1/2 in. in diameter, in 5-ft lengths by means of a hand-powered or power-driven drop hammer. Inside the casing, a 1-in. wash pipe, usually fitted at the lower end with a chopping bit, is churned up and down and rotated by hand to advance the hole.

A hand- or power-driven pump circulates the wash water down through the wash pipe and back to the surface through the annular space between the pipe and the casing. The water is then drained into a tub from which it is again recirculated. In its upward flow the water carries the loosened soil in suspension. As it is emptied into the comparatively still water of the tub, this material is partially deposited in the form of sediment. Samples of soil are then obtained by scraping a receptacle along the bottom of the tub. Such samples are referred to as "wash samples." As any individual sample may contain material from every stratum penetrated up to that time, and the finer particles of soil remain in suspension in the wash water, the results are of so little value that this method is not recommended.

Dry-sample Borings. This method employs the same procedure for drilling the hole as for wash-sample boring, but as soon as the foreman detects a change in the nature of the soil, as he turns the wash pipe at the end of the stroke, he raises the pipe off the bottom of the hole and continues pumping until the water flowing from the casing shows a substantial reduction in suspended soil.

The next step is to withdraw the wash pipe from the casing and substitute an open piece of 1-in. pipe, or a sampling spoon, for the chopping bit. The wash pipe is then returned to the hole and driven into the soil to obtain a sample. After the pipe is again withdrawn, the sample is removed and placed in a bottle. All samples are labeled with the number of the boring and the depth at which they were taken. After the chopping bit has been replaced on the end of the wash pipe, the operation is continued until another change of soil is encountered. As these samples have been removed ahead of the washing, they are identified as "dry samples" even if taken below ground-water level. This method of sampling has a broad application and in many cases furnishes adequate information for foundation design.

Undisturbed Sampling. In this method the same general procedure is followed up to the point that the soil specimens are taken. The object is to obtain a sample of soil with as little change in structure as possible. This type of investigation applies only to plastic soils, as the grains of granular soils suffer displacement in any known method of sampling. It is much more costly than dry sampling, but is used to procure specimens of soil for laboratory analysis. The usual tests that are performed are clearly described in Ref. 7.

Core borings are samples in the form of a more or less continuous core cut from rock. They are obtained by sinking a casing to refusal with earth-boring tools and tightly sealing it to the rock. Within the casing, an appropriate type of rotary drilling tool, such as a diamond or shot drill, is used to cut a core. It is not very large, often about 1 3/8 in. in diameter.

SPREAD FOUNDATIONS. The various types may be classified as wall footings, isolated or independent column and pier footings, combined footings, cantilever footings, continuous footings, and raft or mat and rigid-frame foundations. The surface or soil or rock upon which the foundation directly rests is called the *foundation bed.*

Wall Footings. Figure 28 shows the common designs employed for wall footings. Concrete is the most appropriate material wherever suitable aggregates can be obtained at a reasonable cost. A single rectangular section like the one shown in Fig. 28 (sketch *a*) is often sufficient to support basement walls of low, lightly loaded buildings. If plain concrete is used, the depth is made twice the projection from the face of the wall to prevent stresses that may rupture an unreinforced section. Even in footings of this type some longitudinal steel, such as the two rods shown in sketch *a*, may be desirable if there is any variation in the nature of the soil forming the foundation bed. If such is used, it should be placed 4 in. from the bottom and run continuously, with adequate splices, entirely around the building.

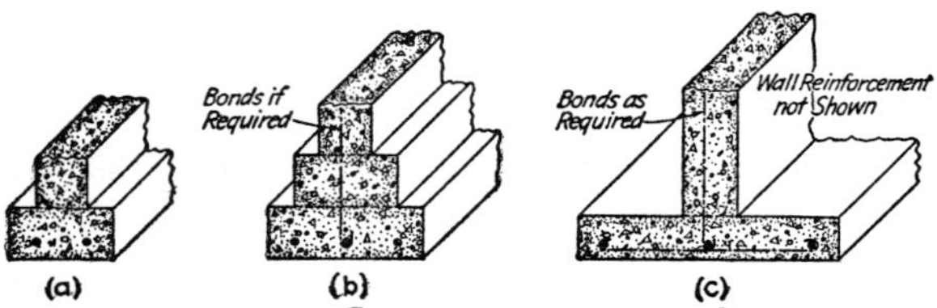

FIG. 28. Wall footings. (Reproduced by permission from Crane, *op. cit.*)

If the load per linear foot of footing, when divided by the allowable bearing value of the soil per square foot, results in a required projection of over 6 to 8 in., the single rectangular block shown in sketch *a* becomes too bulky. In such cases, another step may be introduced as shown in sketch *b*. This procedure of adding additional steps, each one of which has a depth of twice the projection from the wall, or superimposed block, could be carried on indefinitely. It is usually cheaper, however, if the total required width is more than about 3 ft, to use a laterally reinforced-concrete slab as shown in sketch *c*. When the stepped footing is used, adequate bond between the blocks should be obtained either by pouring the entire depth in a continuous operation, or otherwise effectively bonding the slabs comprising the footing.

Isolated or Independent Column or Pier Footings. Figure 29 shows various typical designs. The simplest type is of unreinforced concrete, as shown in Fig. 29, sketch *a*. This is appropriate beneath a wood post, lightly loaded pipe column, or similar compression member. A stepped footing may also be used under the same conditions for somewhat heavier loads. Both are also employed for concrete or masonry piers.

For other than very light loads, the designs shown in Fig. 29, sketches *b*, *c*, and *d*, are appropriate over either soil or piles. A square, rather than oblong plan, is desirable if space limitations permit. The choice between a single slab, sketch *b*, and a stepped design, sketch *c*, is merely one of economy. The single slab is wasteful of concrete; the stepped design is more costly to form. The area of the footing slab is located concentrically with the supported column or pier. A two-way reinforcement, placed 3 or 4 in. from the bottom of the slab, is customary; the ends of the rods should be hooked and protected by 3 or 4 in. of concrete. Pedestals are often used between the top of a footing and the base of the first tier of columns.

Beneath structural steel columns, a steel slab is normally employed, as shown in sketch *d*, to distribute the column loads. Beneath very heavily loaded columns, both a steel slab and a grillage, composed of I beams placed between the bottom of the steel slab and the top of the concrete footing, may be necessary for load distribution. Steel grillages enclosed in concrete are also used in place of reinforced-concrete slabs over a rock foundation bed. They are brought to level by means of grouting. In the past, such designs were employed over soil, but they are seldom found in present-day American practice. Various timber assemblies are hardly appropriate except for temporary structures, but they have been used for wood-framed buildings when placed beneath permanent ground-water level.

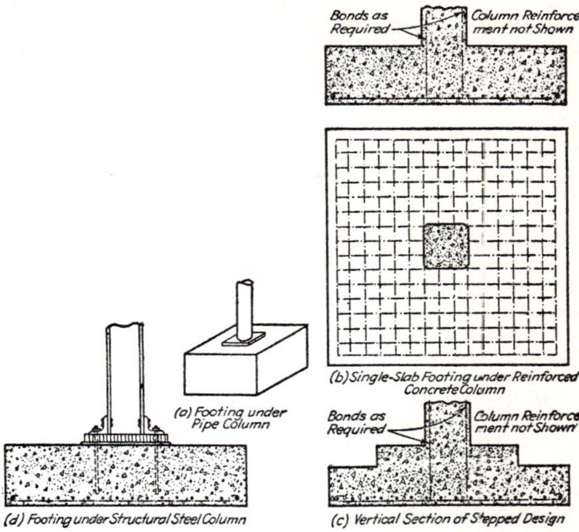

FIG. 29. Independent column footings. (Reproduced by permission from Crane, *op. cit.*)

Combined Footings. Exterior columns are often placed adjacent to property lines where space is not available for a concentric footing. Some method of overcoming the resulting eccentricity is necessary. The common solution is to combine the footing of the exterior column with that of the adjacent interior column. The shape of the combined footing is then determined so that the centroid of the area corresponds to the line of action of the resultant of the column loads. Figure 30, sketches *a*, *b*, and *c*, show typical designs. Sketch *a* is the more desirable plan as it is easier to design and construct than the plan shown in sketch *b*. However, space limitations may make the latter plan necessary. Sketch *c* shows a typical plan for a combined footing supporting four columns, a design that is often necessary at the corners of buildings where little or no projection is allowed beyond the faces of the columns on adjacent sides.

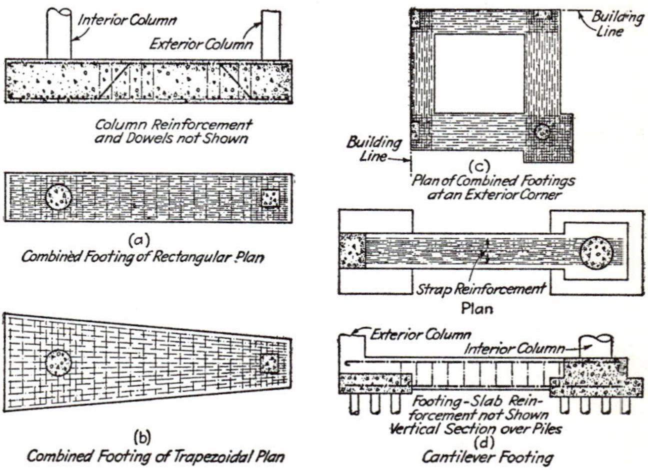

FIG. 30. Combined and cantilever footings. (Reproduced by permission from Crane, *op. cit.*)

Cantilever Footings. The design shown in Fig. 30, sketch *d*, is used to meet the same conditions as sketches *a* and *b*. Although this alternate is occasionally less expensive, it is not recommended for use except over piles adequately supported by point bearing.

Continuous Footings. A typical design is shown in Fig. 31. Although continuous footings may be used for the support of interior columns, they have their particular application for exterior columns when a foot or two is available between the exterior face of the column and the building line. Such a limited distance is generally insufficient to accommodate an isolated concentric footing, but may permit a continuous footing to be designed concentrically with the columns. Any of these designs may be used over either soil or piles with the exception previously noted for the cantilever footing.

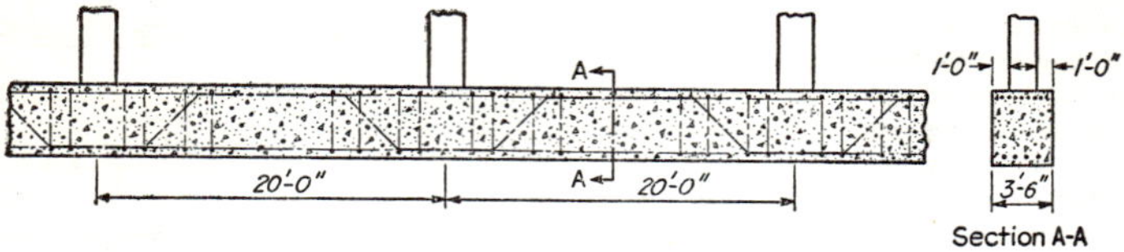

FIG. 31. Continuous footing. (Reproduced by permission from Crane, *op. cit.*)

Raft or Mat and Rigid-frame Foundations. In their simplest form, the raft or mat foundation is an inverted slab of reinforced concrete, often several feet thick, which covers the entire basement area of a building. Such designs have apparently been an economical solution in some cases, but there is always the danger of excessive soil pressures occurring beneath heavily loaded columns. Furthermore, even if the design is made sufficiently rigid to distribute the loads over the soil between columns, a mat foundation may be uneconomical. This is because the load from individual column footings, at normal spacing, will often be spread over the entire basement area at a depth of less than 20 ft below the level of the foundation bed, and the increased area of the contact surface, consequently, may have no significance.

Rigid-frame designs are also occasionally used below grade to furnish a foundation assembly of sufficient rigidity to distribute the concentrated loads beneath columns. Such have their application over soils of low bearing capacity extending to a depth so great as to make the use of piles impracticable. One such rigid-frame design was used for the New England Mutual Life Insurance Company's building in Boston. Recent reports indicate that the design was a good solution for this particular project over the soil conditions existing at this site. Its success should not, however, lead to the unconsidered acceptance of such a design as normally safe and economical over all deep strata of low bearing capacity. Every building presents a special problem. Where conditions permit a foundation to be carried to rock bearing, it is structurally preferable to do so. This statement applies particularly to buildings of considerable height and to those of tower design.

BEARING CAPACITY OF SPREAD FOUNDATIONS. Table 7, p. 27-08, will serve as a general guide in cases where a local code does not enforce more restrictive requirements. Such bearing values, however, should be considered merely as maximums. Wherever there is any question of the quality of the material beneath the foundation bed, or its ability to support the proposed design without objectionable settlement, it should be checked by subsurface investigation, often supplemented by a soil-loading test.

Assumed Pressure Distribution beneath Footings. The soil-bearing values given in all building codes are commonly applied to the foundation bed which is the surface of the soil or rock immediately underlying the lower surface of the footing, or spread foundation. The transfer of load to underlying strata, even when the strata are of lower bearing value, is often neglected. This practice has so often resulted in excessive settlement that it is specifically mentioned in the following paragraph, also abstracted from the Boston Building Code.

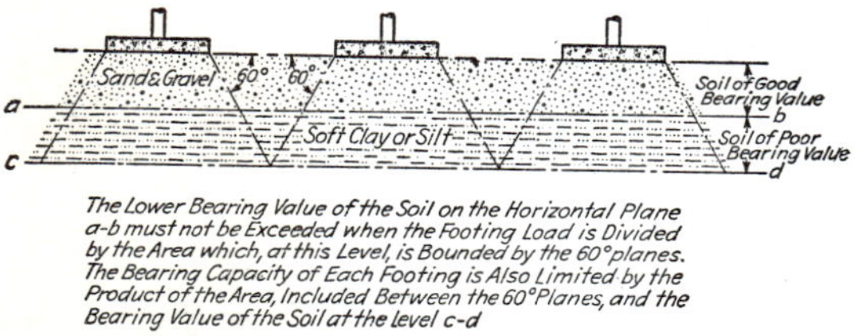

FIG. 32. An assumption of pressure distribution beneath footings. (Building Code of Boston, Mass., 1944)

Where the bearing materials directly under a foundation overlie a stratum having smaller allowable bearing values, these smaller values shall not be exceeded at the level of such stratum. Computation of the vertical pressure in the bearing materials at any depth below a foundation shall be made on the assumption that the load is spread uniformly at an angle of 60° with the horizontal, but the area considered as supporting the load shall not extend beyond the intersection of 60° planes of adjacent foundations.

Figure 32 shows the condition described in the preceding paragraph. Although the assumed distribution of pressure is only a crude approximation, and varies considerably for different types of soil, it should serve to direct the designer's attention to the danger of exceeding the safe bearing capacity of underlying strata that may be a considerable distance below an apparently satisfactory surface at the footing level.

Soil-loading Tests. There seems to be a common belief that an allowable soil-bearing value may be obtained by reducing the failure load, derived from small-scale loading tests, by a liberal factor of safety. This, however, is not a safe practice because the depth of pressure effect within the soil increases rapidly with the size of the loaded area. The settlement record derived from a small-scale test depends on the properties of a comparatively small volume of soil beneath the loading block or platform. The much deeper distribution of pressure beneath an entire group of footings, representing the complete foundation of a building, affects a much larger volume of soil. If the material is not uniform and homogeneous for a considerable depth, which, under some conditions, might approximate 1 $^1/_2$ times the width of the structure, little similarity may exist between the record of a loading test and the future settlement of a building. (See Refs. 8, 9, 10).

Furthermore, even if it can be established that soil conditions are homogeneous beneath the entire foundation bed, the pressure effect from adjacent footings will often overlap, thereby causing an excessive load upon some critical subsurface stratum. Increasing the size of the footings will obviously lessen the contact pressures on the foundation bed but may be of no value in relieving the pressure on a lower plane if, at that level, the load is already spread fairly uniformly over the entire available area. As pointed out by Prentis and White in their very practical book (Ref. 11), a foundation consisting of isolated spread footings would, under these conditions, be as efficient as a much more expensive mat or raft foundation placed over the entire basement area. This statement suggests choosing the footing areas, for any required column spacing, of just sufficient size to spread the load over the entire available subsurface stratum, which is considered to be the critical bearing medium.

It is suggested that bearing values derived from loading tests be studied in relation to all other available data such as the records of dry-sample borings, settlement records of other buildings erected over similar coil conditions and, as a last resort, in the case of cohesive soils, laboratory tests upon undisturbed samples.

The following requirements for a soil loading test are also taken from the Boston Building Code, 1944:

(*a*) For bearing materials of classes 1 to 5 [see Table 7, p. 27-08], inclusive, the loaded area shall be at least 1 sq ft and for other classes at least 4 sq ft. For materials of classes 6 to 13 [see Table 7], inclusive, the loaded area shall be the full size of the pit and at such depth that the ratio of the width of the loaded area to its depth below the immediately adjacent ground surface is the same as the larger of the following two values:

1. Ratio of the width of any footing to its depth below the immediately adjacent ground surface.
2. Ratio of the width of the entire foundation or group of footings to its depth below the average surrounding ground surface.

(*b*) When loading tests are made on bearing materials of classes 10 to 13, inclusive, suitable methods shall be used to prevent evaporation from the materials being tested.

(*c*) A test load shall be applied which will produce a unit pressure equal to that for which the proposed foundations are designed. This load shall be allowed to remain undisturbed until no measurable settlement occurs during a period of 24 hours. The load shall then be doubled in increments not exceeding 25 per cent of the design load. At least 4 hours shall elapse between the application of successive increments. The total load shall be allowed to remain undisturbed until no measurable settlement occurs during a period of 24 hours.

(*d*) Measurements of settlement shall be accurate to $^1/_{32}$ in. and shall be taken and recorded every hour during the first 6 hours after the application of each increment, and at least once every 12 hours thereafter.

(*e*) When the design load upon bearing materials of classes 1 to 10, inclusive, causes settlement of less than $^3/_8$ in. and twice the design load causes settlement of less than 1 in., the design load shall be allowed; but if medium or soft clay underlies these materials, the vertical pressure in such clay shall not exceed that allowed in the table.

(*f*) Whenever the proposed foundation rests on or is underlain by bearing materials of classes 11 to 13, inclusive, the results of loading tests must be interpreted in conjunction with accurate soil profiles showing magnitude and variation of the thickness of these strata. If this information, in the opinion of the commissioner, is not sufficient to determine whether the design load will cause excessive settlement, as might occur due to a thick stratum of clay, or dangerous differential settlement, as might occur when the underlying clay stratum varies considerably in thickness, the commissioner may require an analysis to be made of the probable magnitude, rate, and distribution of settlement of the proposed structure. Such analysis may be based upon:

1. A study of settlement records of near-by structures having essentially the same foundation conditions.
2. Consolidation tests and other investigations of undisturbed samples of the compressible materials.

TYPES OF PILES AND PILE DRIVING. Bearing piles are of the following types: wood piles; concrete piles, both precast and cast-in-place; structural-steel piles; metal-pipe piles; and composite piles, composed of wood or steel superimposed by concrete. All types are ordinarily driven in groups called clusters; a staggered arrangement of two or more parallel lines may be appropriate beneath bearing walls or continuous footings. Piles may also be distributed over an entire basement area beneath a reinforced concrete mat. The superimposed footings, known as pile caps, are similar in shape to those employed over soil.

Wood Piles. These are usually tree trunks driven with the small end down. In comparison with other types of piles, they are more appropriate for light than for heavy loads. Practically any kind of sound timber may be used.* For difficult driving a metal point may be employed. Lengths are determined to suit soil requirements. The diameter of the tip usually varies from 6 to 8 in. The diameter at the butt is seldom less than 12 in.; the taper should be uniform. For building foundations, untreated wood piles should be used only below permanent ground-water level. Except for temporary structures, even treated wood piles should be used with due consideration of the possibility of future replacement, when not entirely below ground-water level. This requirement alone may involve deep excavation with consequent expense for sheathing and pumping. In comparing the cost of wood piles with other types these items may be important.

If entirely embedded in soil and cut off below the lowest present or future ground-water level, or entirely submerged in fresh water, untreated timber piles may be considered permanent.† When exposed to unfavorable conditions, durability can be greatly increased by the use of coal-tar creosote impregnated by standard methods (Ref. 12). Either a treated or untreated pile, particularly the latter, may be injured by ground water containing alkali or acids. However, the principal danger to untreated timber piles is an unexpected lowering of the ground-water level, which causes the upper part of the pile to decay. Where there is even a remote possibility of this condition occurring in future years, the design should allow for such a contingency.

Concrete piles may be divided into two classifications, precast and cast-in-place. As they are immune to attack by biological organisms, they may be used where untreated wood piles would be inappropriate. If entirely embedded in soil, concrete piles may be considered permanent, whether above or below ground-water level, unless destructive acids, alkalies, or chemical salts are present.‡ Concrete piles extended above ground are subject to weathering and deleterious atmospheres to the same extent as concrete columns. Each type of concrete pile has its particular advantages and disadvantages in relation to soil conditions and the duty to be performed.

Precast Concrete Piles. Typical designs are square and octagonal. Circular sections are occasionally used. Although tapered designs are appropriate for the shorter lengths, up to about 40 ft, parallel sides and pointed ends are more usual. The parallel-side type has been cast with diameters up to 30 in. and in lengths over 100 ft.

The particular application of precast concrete piles is for conditions where the upper portion of the pile extends above the ground or water level, acting as a column for the superstructure. They are always reinforced internally to resist stresses caused by handling and driving. When serving as a column, or required to resist horizontal forces, additional reinforcement may be required. Pipes are occasionally placed in precast concrete piles to permit jetting while being driven. They are usually cast at the site of the future building.

Cast-in-place concrete piles are of two types: those in which a steel shell is left in the ground, and those which are shell-less or unencased. In the first type, the shell is usually of thin steel with a boot at the bottom. A heavy steel core, called a mandrel, is placed within the shell and carries it down when driven. The shell is uniformly tapered. When sufficient resistance has been obtained, the core is withdrawn, and the shell inspected by means of an electric light and filled with concrete. A standard design for the type just described has an 8-in. point diameter, the thickness increasing at the rate of 0.4 in. for each foot of pile length. As the practical limit of the uniformly tapered type is about

---

* Specifications of the American Society for Testing Materials and those of the American Railway Engineering Association.

† The statements in this article regarding the permanence of various types of piles follow the opinions expressed in Pile Foundations and Pile Structures, *Manual of Engineering Practice* 27, American Society of Civil Engineers, 1946. This manual, an excellent reference on the entire subject identified by the title, has been quoted in several portions of this chapter.

‡ This subject is treated very practically, with particular reference to concrete piles, in Pile Foundation, by A. E. Cummings, *Proceedings of the Purdue Conference on Soil Mechanics and Its Applications*, sponsored by the Society for the Promotion of Engineering Education, Purdue University, Lafayette, Indiana, 1940. Reprints are available from the Raymond Concrete Pile Company.

37 ft, a "step-taper" design has been developed which is driven in lengths up to nearly 100 ft. For greater lengths, a pipe-step-taper pile is available in lengths up to about 150 ft.

For the unencased type, a heavy steel shell or tube is used in combination with a steel core. After sufficient driving resistance has been obtained, the core is withdrawn and the tube filled with concrete. In this case, however, the core is then placed on top of the newly deposited concrete, at the upper end of the tube, and held in this position as the tube is pulled out of the ground. A modification of this procedure results in what is called a "pedestal pile." The pedestal is formed by ramming a portion of the concrete out of the lower end of the tube prior to complete withdrawal.

The cast-in-place pile, rather than the precast type, is the usual choice for ordinary building foundations. Installation is more rapid as no time is lost either in driving test piles, for the purpose of determining pile lengths, or in waiting for the piles to be cast and cured. Cast-in-place piles are not generally reinforced except where resistance to bending is required. Where soil conditions are favorable, the shell-less design may be more economical, but in many soils there is danger that subsurface pressures, or the driving of adjacent piles, will reduce or otherwise injure the concrete section.

Structural-steel piles are generally of rolled-steel H sections, designed with wide flanges. Their greater strength makes them appropriate for driving through dense gravel, or similar types of soil which would be difficult to penetrate with either wood or precast concrete piles. Their smaller cross section causes less compaction of the soil and, when driven in cohesive soils, less swelling of the ground. They are particularly suitable for deep driving where it is necessary to extend the foundation to rock. Lengths up to 120 ft have been successfully driven beneath the John Hancock Mutual Life Insurance Company's building in Boston, Massachusetts (Ref. 13).

If entirely embedded in comparatively impervious soil, steel piles may be considered permanent. When extended above the ground, there is the probability of rusting at the surface and for a short distance below grade. To avoid this danger the steel piles may be encased in concrete or coated with coal tar, extending from 6 in. above grade to 2 ft or more below the surface. If corrosive materials are present in the soil or ground water, such protection should be carried to a greater depth. When there are coal pockets, alkali soil, cinder fills of comparatively recent deposit, or waste from certain manufacturing plants in the immediate locality, the soil should be analyzed.

Metal Pipe Piles. Heavy, open-end cylindrical shells are the common type. Diameters usually range from 10 to 18 in. Shell thickness normally varies from $^1/_4$ to $^1/_2$ in.; a $^7/_8$-in. thickness has been used for 12-in. pipe where it was necessary to drive through obstructions. As in the case of the structural-steel piles, this type is normally driven to rock and can be designed to carry up to 150 tons. Lengths up to 100 ft have been used. When driven with a hammer, which is the usual procedure except when working beneath a structure, sections are sunk in lengths of about 20 ft. After driving, the soil contained within the pipe is blown out by means of compressed air and the pipe filled with concrete.

Composite piles may be of two types. In both cases the upper portion is of concrete superimposed upon either wood or steel. The wood and concrete combination is more common. The wood portion is designed of such length as to be always below ground-water level. The concrete portion is usually cast-in-place but may be precast. A typical composite pile comprises a cast-in-place step-taper pile above a wood pile. The advantage of such a combination is the comparatively low cost of the wood pile and the elimination of deep excavation which would be required if only wood were used and the cut-off elevation dropped below permanent water level.

Means by Which Piles Support Load. The two terms ordinarily used to identify the load that can be sustained by a pile foundation, without objectionable settlement, are *allowable safe load* and *bearing capacity.* A very important consideration is the bearing capacity of the subsoil strata underlying the points of the piles, as these strata must support the load placed upon the piles, in addition to the weight of the soil above this level, supplemented by any other imposed loads.

Piles may transfer their loads by point bearing to an underlying stratum, or by skin friction to the mass of soil in which they are embedded. Piles frequently furnish support by a combination of these two means, but all types may be roughly classified as either *end-bearing* or *friction* piles, depending on the means by which they obtain their chief resistance. In either case, however, the load is eventually imposed upon the material around and beneath the points of the piles.

Sketch *a* of Fig. 33 illustrates an end-bearing pile where practically all load is transmitted by point bearing to rock or other very hard stratum upon which it rests. The pile acts as a column with an indefinite amount of lateral support furnished by the material through which it is driven. Such piles must be strong enough to support the design load without injury to the tip. Sketch *b* illustrates a pile embedded in a firm stratum, such as dense sand or gravel, without clay or other soft strata below. Such a pile supports its load by friction through the lower portion, aided by end bearing. As the load in this case is transferred to the sand at the level of the pile points, the bearing capacity of such a stratum would limit the bearing capacity of the pile for any particular spacing.

Sketch *c* illustrates a similar condition, except that a stratum of soft clay, or other material of a low bearing capacity, is assumed to exist beneath the stratum of sand or gravel. In this case, the surface of the clay is the critical bearing area, and such a condition warrants very careful investigation. The load from a single pile, or even that of one cluster, may be distributed over a sufficient area to cause only slight settlement when a loading test is applied. When the entire foundation is in place, however, the overlapping of pressures from adjacent clusters may result in placing an excessive load upon the soft clay and cause serious settlement. Piles transmitting their load to any deposit of compressible material should be spaced so that the load-carrying capacity of the selected bearing

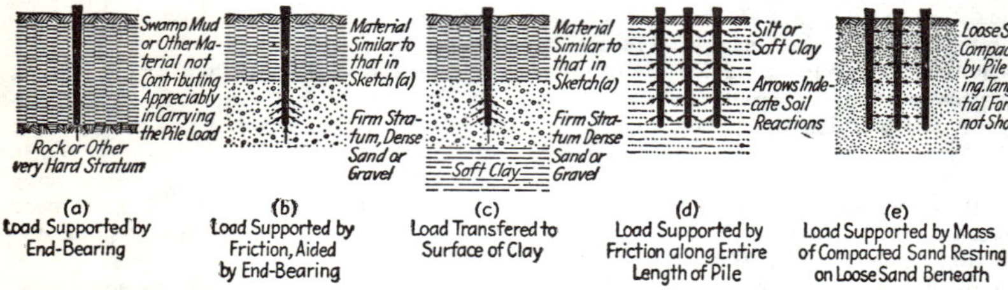

FIG. 33. Means by which piles support load. (Reproduced by permission from Crane, *op. cit.*)

stratum, and any underlying strata to a depth of not less than 1 1/2 times the width of the entire foundation, is not exceeded. This consideration may require limiting the load upon critical subsurface strata to that imposed by the original overburden.

Sketch *d* illustrates the case where piles are supported by friction throughout their full length. When embedded in silt, or soft clay, such an assembly is often referred to as a *floating* pile foundation. The segment of soil surrounding a pile, to which the load is transmitted, is known as the *bearing body*. For an entire pile cluster the same term is applied to the mass of soil lying within the perimeter of the group and closely surrounding it. As in the preceding case, the load-carrying capacity may be limited by the bearing value of the undisturbed soil beneath the level of the pile points. Where a compressible stratum lies below, a common type of failure is the general downward displacement of the bodies of soil around the pile clusters.

The bond between soil and pile, generally referred to as *skin friction*, is also important as a gage of the pile's resistance against slipping through the surrounding material. The value of the ultimate shearing stress in the soil is a measure of the tangential resistance which the earth can transmit. Sketch *e* shows piles driven their entire length into a stratum of loose sand and having the primary function of consolidating the material by vibration and lateral displacement. When such piles are driven in sufficient number and at proper spacing, the result is to form a more or less solid block bounded approximately by the perimeter of the pile cluster. Provided that there is no underlying stratum of lower bearing value, the bearing capacity of such a design is limited by the adequacy of the loose sand below the pile points to support the entire bearing body of consolidated sand.

Pile Driving. The selection of suitable pile-driving equipment is extremely important. It should be selected by a competent engineer who is familiar with the soil conditions and probable pile loads.

A drop hammer is essentially a heavy metal weight, raised by a hoisting line and allowed to fall on top of the pile by gravity. A single-acting steam hammer is a freely falling weight with steam pressure applied only to the underside of the piston for the purpose of raising the ram. In the double-acting steam hammer, the ram is likewise raised by steam. At the top of the stroke, however, steam is admitted to the upper side of the piston, and its pressure, added to the weight of the ram, increases the driving force. Piles are occasionally sunk by means of hydraulic jacks. Placed on top of a short pile, they force it down with resistance against the superstructure of a building, a temporary rig, or previously driven piles.

A water jet is often used when necessary to penetrate a considerable depth of sand or gravel. Jetting, however, should be used with discretion as there is danger of loosening adjoining piles; and if support is furnished by skin friction, the bearing value may be lessened by the soil's not settling back into place after driving. When piles are driven through soft clay, or silt, to a firmer material, such as coarse sand or gravel, it is important to obtain sufficient penetration in the firmer soil, and driving should often be continued even after adequate resistance has apparently been reached. The compression of the soil due to driving may raise piles already in place to such an extent that they no longer bear properly. If this condition is suspected, the levels of all pile heads should be checked after each cluster is complete. Those that have risen appreciably should be redriven.

Formulas for Computing Bearing Capacity. Although the limitation of data obtained solely by driving records is clearly apparent, dynamic formulas have been used for many years and with comparative success when they have been applied with discretion. For additional information on this subject see Pile Foundations and Pile Structures, *Manual of Engineering Practice* 27, American Society of Civil Engineers, 1946.

Engineering News *—drop hammers: $R = \dfrac{2Wh}{s + 1.0}$

Engineering News—single-acting steam hammers: $R = \dfrac{2Wh}{s + 0.1}$

Modified Engineering News formula: $R = \dfrac{2Wh}{s + 0.1P/W}$

Hiley formula (simplified) for drop hammers: $R = \dfrac{3Wh}{s + k} \times \dfrac{W}{W + P}$

Hiley formula (simplified) for single-acting steam hammers; $R = \dfrac{3.6Wh}{s + k} \times \dfrac{W}{W + P}$

where R = allowable safe static load on pile, pounds; W = weight of a drop hammer or striking parts of a single-acting steam or air hammer, pounds; P = weight of pile as driven, pounds; h = height of fall of a drop hammer or length of stroke of a steam or air hammer, feet; s = penetration of pile per blow, inches (either the last blow or the average of the last few blows); k = half the total rebound of pile hammer used in Hiley formula and expressed in inches.

These references will be found particularly helpful in connection with the application of pile-driving formulas:

Pile-Driving Handbook, by Robert D. Chellis, Pitman Publishing Corporation, 1944.

Dynamic Pile-Driving Formulas, by A. E. Cummings, *Journal of the Boston Society of Civil Engineers*, January 1940.

CONCRETE PIERS. Concrete piers are used to carry column loads through a material of low bearing value down to rock, hardpan, or gravel. Provided that the bearing capacity of the rock or soil is adequate, the design of a plain or reinforced-concrete pier does not present the many uncertain factors that enter into the design of spread foundations and piles. During the last thirty or forty years, methods of field construction have been highly developed, but their detailed description is not within the scope of this book.

The first problem involved is to choose the most economical method of sinking a shaft of the desired dimensions to the chosen bearing stratum. In building construction, this is often accomplished by means of open wells unless a considerable depth must be penetrated below ground-water level where soil conditions necessitate using pneumatic caissons. The system used for either full-lot or open-well excavation depends on the nature of the soil, the estimated amount of ground water, the required depth, and the necessity of avoiding injury to adjacent structures, as well as other considerations pertaining to the individual operation. For both full-lot excavation and individual pier foundations, steel sheet piling has a broad application. It may often be used in connection with perforated pipes, called *well points*. Such installations have permitted lowering the ground-water level as much as 30 ft.

A system known as the Chicago method, because of its particular use in that city, is applicable to open-well excavation where soil conditions are favorable. The open wells are dug beneath each column location. After the earth has been excavated to a depth of 3 to 6 ft, depending on the stability of the soil, the pits are sheathed with boards held in place by metal rings. The procedure of excavation and subsequent sheathing is then repeated, increasing the depth of the well, until the required level is reached. Piling of considerable length is used when it is necessary to excavate through water-bearing strata. In such cases, the tips of the piles, or other sheathing, may be sunk into clay, thereby sealing the bottom of the shaft and permitting practically dry excavation through a water-bearing stratum. In some types of soil both wood cylinders and open steel shells have been sunk by jacking.

* For a discussion of these and other pile formulas, see *Proc. Am. Soc. Civil Engrs.*, May 1941 and Sept. 1941

The Gow caisson, composed of telescoping steel cylinders, power-driven successively and withdrawn in reverse order after filling with concrete, has its particular application for concrete piers supported by rock or soil of suitable bearing capacity, lying at moderate depths below grade or below an open excavation.

REFERENCES

1. Wind bracing in steel buildings, *Trans. Am. Soc. Civil Engrs.*, 1940, p. 1713.
2. Report of ACI Committee 318, *J. Am. Concrete Inst.*, Sept. 1947.
3. Report of the Joint Committee on Standard Specifications for Concrete and Reinforced Concrete, published by the American Society of Civil Engineers in *Proceedings*, Part II, June 1940. (Available in pamphlet form.)
4. Arshom Amirikian, Precast concrete storehouses, *J. Am. Concrete Inst.*, June 1947.
5. Louis P. Corbetta, Precast concrete warehouse construction, *J. Am. Concrete Inst.*, June 1947.
6. Stewart Wagner, Research Laboratory for Hoffman-LaRoche, Inc., *Architectural Concrete*, Vol. 12, No. 1, 1946.
7. Henry S. Jacoby and Donald P. Davis, *Foundations of Bridges and Buildings*, 3rd Ed., McGraw-Hill Book Co., New York, 1941.
8. Karl Terzaghi, *Theoretical Soil Mechanics*, John Wiley and Sons, New York, 1943.
9. A. Casagrande and R. E. Fadum, Application of soil mechanics in designing building foundations. *Trans. Am. Soc. Civil Engrs.*, Vol. 109, 1944.
10. Karl Terzaghi, *Settlement of Structures*, Opening discussion of the International Conference on Soil Mechanics and Foundation Engineering, Graduate School of Engineering, Harvard University, 1936.
11. Edmund Astley Prentis, Jr., and Lazarus White, *Underpinning*, Columbia University Press, New York, 1931.
12. Timber piles and construction timbers, *Manual of Engineering Practice* 17, American Society of Civil Engineers.
13. Welded steel piles support 26-story skyscraper, *Civil Eng.*, March 1947.

For more detailed treatment of the subjects covered in the foregoing chapter, the reader is referred to *Architectural Construction* by Theodore Crane, John Wiley and Sons, 1947.

ILLUMINATION

By Samuel G. Hibben

9. LIGHT—CHARACTERISTICS AND MEASUREMENTS

THE RADIATION SPECTRUM. Light is that form of radiant energy which is capable of producing the sensation of sight. Visible energy is an exceedingly small portion of the *electromagnetic spectrum*, a tremendous range of radiant energy which travels through

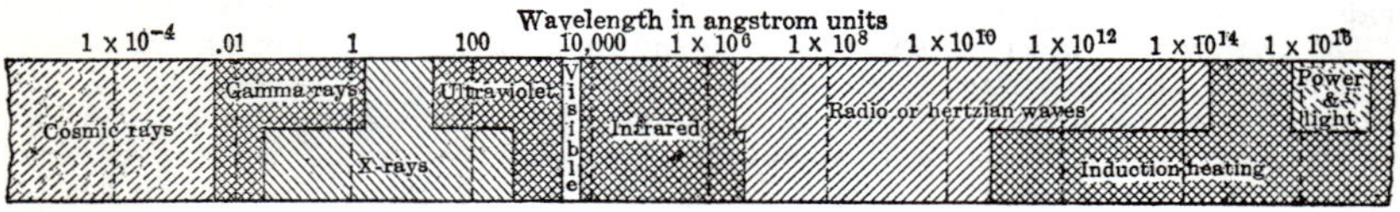

FIG. 1. Electromagnetic spectrum.

space in the form of electromagnetic waves. All these radiations are similar in nature and in the speed at which they travel (186,300 miles per second). They differ only in wavelength and frequency, and in the ways in which they manifest themselves. The present known spectrum ranges from cosmic rays with a wavelength of 1×10^{-15} cm, and a frequency of 3×10^{25} cycles per second, to 60-cycle a-c waves 3100 miles long (see Fig. 1).

(*Continued on p. 27-70*)

Table 1. Color Temperatures, °K

(Approximate Values)

Blue sky	10,000 to 30,000
Overcast sky	7000
Daylight fluorescent lamp	6500
Noon sunlight	5250
500-Watt daylight incandescent lamp	4000
White fluorescent lamp	3500
Photoflood lamp	3415
General-service incandescent lamps	2500 to 3050
Candle flame	1800

Table 2. Terminology and Measurement of Light and Illumination

Quantity	Symbol	Unit	Definition	Fundamental Equations	Method of Measurement	Principal Use
Luminous Intensity (Candlepower) Light density in a specified direction	I One "Hefner" equals 0.9 cp	Candle (c) The luminous intensity of a source expressed in candles is its candlepower (cp)	The standard unit of luminous intensity in a given direction is the *international candle.* An ordinary wax candle has a luminous intensity in a horizontal direction of approximately one candle. The international candle is the basic quantity in all measurements of light. Candlepower is always a property of a source of light, and gives information regarding luminous flux at its origin.	CP = Foot-candles $\times D^2$ (D = Distance in feet from source to illuminated surface). See Illumination. $\text{MSCP} = \frac{\text{Lumens}}{12.57}$ (Mean spherical candlepower is the average candlepower of a source in all directions.)	Candlepower measurements are primarily a laboratory procedure requiring special instruments. Rough estimates of the candlepower of a source or fixture can be made in the field by (1) holding a light meter at a distance of at least five times the greatest dimension of the source; (2) aiming the cell of the meter directly at the source; and (3) multiplying the foot-candle reading by the square of the distance in feet. (See Fundamental Equations.) There must of course be no other light in the room, and it may be necessary to make allowance for light reflected from walls and ceilings.	Candlepower is used not only to indicate the luminous intensity of a source in one particular direction; candlepower measurements are often taken at various angles around a source or a fixture, and the results plotted to give a *candlepower distribution curve.* Such a curve shows luminous intensity in any direction, and from it illumination calculations can be made.
Luminous Flux Time rate of flow of light Light is actually a form of radiant energy in motion. In common practice, however, the time element is neglected, and luminous flux is considered as a definite quantity.	F	Lumen (lm)	A lumen is the light flux falling on a surface one square foot in area, every point on which is one foot from a uniform point source of one candle. (Such a surface is a one-foot-square section of a sphere of one-foot radius, with a one-candle source at its center.) The lumen differs from the candle in that it is a measure of light flux *irrespective of direction.*	Lumens incident on a surface = Foot-candles $\times$ Area (sq ft) Lumens emitted or reflected by a surface = Foot-Lamberts $\times$ Area (sq ft) Lumens = MSCP $\times$ 12.57 (Since a sphere of one-foot radius has a surface area of 4π (12.57) square feet, a uniform point source of one candle must produce 12.57 lumens. The same relationship exists between the mean spherical candlepower of any source and its total lumen output.)	Lumen measurements of light sources are a laboratory procedure requiring special equipment. The lumens falling upon a surface may, however, be estimated with the aid of an ordinary light meter. First obtain foot-candle readings at various points on the surface in order to arrive at an average value; then multiply the average foot-candles by the area of the surface in square feet. (See Basic Formulas in Article 13.)	The lumen is used primarily to express the total output of a light source. It can also be used to indicate amount of light absorbed, transmitted, or reflected. The Lumen Method of calculating illumination provides average foot-candle values by the use of relatively simple formulas.

(*Table continued on p. 27-66*)

Table 2. Terminology and Measurement of Light and Illumination—*Continued*

Quantity	Symbol	Unit	Definition	Fundamental Equations	Method of Measurement	Principal Use
Illumination Density of luminous flux on a surface Luminous flux may be called the *cause*, and illumination the *effect* or result. Usually the target visibility is the result of the light it reflects, rather than being measured by what falls upon it.	E	Foot-candle (ft-c) The unit, lux, is one lumen per square meter, so one foot-candle = 10.76 lux.	A foot-candle is the illumination at a point (A) on a surface that is one foot from and perpendicular to a uniform point source of one candle. 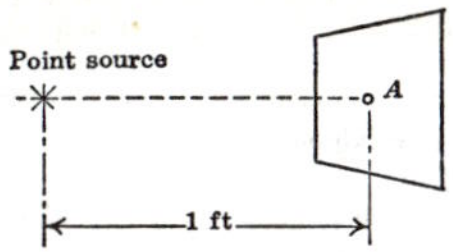 From the definition of a lumen it is obvious that one lumen uniformly distributed over one square foot of surface produces an illumination of one foot-candle.	Inverse Square Law Illumination decreases inversely as the square of the distance. 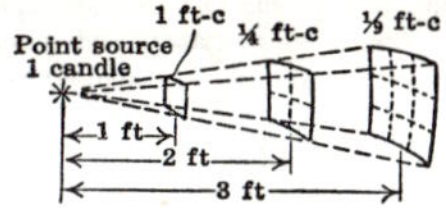 When light rays are perpendicular to the surface, the inverse square law as stated above applies: 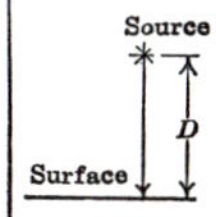 $E = \frac{I}{D^2}$ E = Foot-candles I = Candlepower D = Distance in feet When light rays are not perpendicular to the surface: 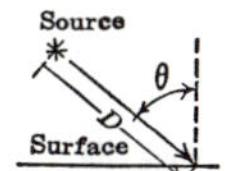 $E_h = \frac{I \times \cos\theta}{D^2}$ 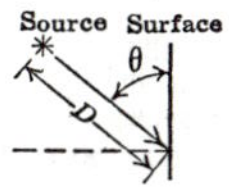 $E_v = \frac{I \times \sin\theta}{D^2}$ Foot-candles incident on a surface = $\frac{\text{Lumens}}{\text{Area (sq ft)}}$	Various models of direct-reading light-sensitive cell foot-candle meters and visual photometers are available. A discussion of these instruments and their use is found in the technical literature. The principle is to generate small electric currents through the action of light on selenium-copper oxide surfaces.	Foot-candle readings are used to indicate the illumination at a specific point, or the average illumination on a surface. The inverse square law is the basis of calculation in the Point-By-Point Method of lighting design. The inverse square law applies strictly only to a *point source.* With most types of interior-lighting fixtures, however, it is safe to assume that the law operates with sufficient accuracy for all practical purposes if the distance at which the measurements are taken is at least five times the greatest dimension of the light source. For special considerations involving linear sources note that the estimates of useful lumens are more useful than calculated foot-candles.

Brightness Luminous intensity in a given direction per unit of (projected) area A surface or an object has *brightness* by reason of light emitted, reflected, or transmitted. Brightness is ordinarily independent of distance of observation.	B or B^1	Candle per square inch ($c/in.^2$) or Foot-Lambert (ft-L)	Brightness is expressed in two ways: in *candles* per unit area or in *lumens* per unit area. A surface emitting or reflecting light in a given direction at the rate of one candle per square inch of projected area has a brightness in that direction of *one candle per square inch.* A surface which has a brightness equal to the uniform brightness of a perfectly diffusing surface emitting or reflecting one lumen per square foot has a brightness of *one foot-Lambert.* The foot-Lambert is also the average brightness of any surface emitting or reflecting light at the rate of one lumen per square foot. A *lambert* is the brightness of a surface emitting or reflecting one lumen per square centimeter; a *millilambert* is one-thousandth of a lambert.	Ft-L = Foot-candles × Reflection factor of surface $\text{Ft-L} = \dfrac{\left\{\begin{matrix}\text{Lumens (incident)} \times \\ \text{Reflection factor}\end{matrix}\right\}}{\text{Area (sq ft) of surface}}$ 1 candle per sq in. = 452 foot-Lamberts If the surface under consideration departs widely from the properties of a perfect diffuser, the lumens emitted or reflected cannot safely be calculated on the basis of a single brightness reading taken from any one angle. 1 lambert = 929 foot-Lamberts = 2.054 candles per sq in. 1 millilambert = 0.929 foot-Lambert = 0.002 candle per sq in.	Methods of making brightness measurements and the meters used for the purpose are described in Table 3. Real or apparent brightness of a source is candlepower divided by the projected area in that direction, i.e., the intrinsic brightness.	Relatively high brightnesses, such as those of light sources, are usually expressed in terms of candles per square inch. Since the average brightness of a surface in foot-Lamberts can be calculated by multiplying the illumination in foot-candles by the reflection factor, the foot-Lambert is a very convenient unit in which to express the brightnesses of illuminated surfaces.

Table 3. Reflection, Transmission, Refraction, and Polarization

Type of Control	Illustration	Unit	Method of Measurement	Instrument	Materials	Use
Reflection When a ray of light striking a surface is turned back, it is said to be *reflected.* Reflection may be of several types, the most common of which are *specular* (regular), *diffuse, spread,* and *mixed.*	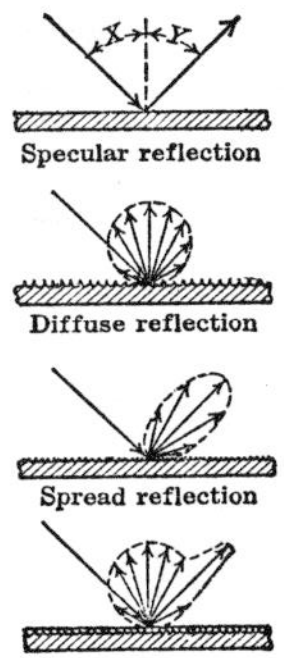 Specular reflection Diffuse reflection Spread reflection Mixed reflection	Reflection Factor The ratio of the light reflected from a surface to that incident upon it. The reflection factor of a given surface may vary considerably according to the direction and nature of the incident light. Specular reflection increases with angle of incidence, almost total reflection being obtainable at grazing angles. With colored surfaces the reflection factor may be quite different for different colors of light.	Place light meter cell against surface. Withdraw meter from surface slowly until constant reading is obtained (2 to 6 in.). (*A*) Place meter against surface with cell facing out (*B*) and note reading. Reflection factor $= \frac{\text{Reading } (A)}{\text{Reading } (B)}$	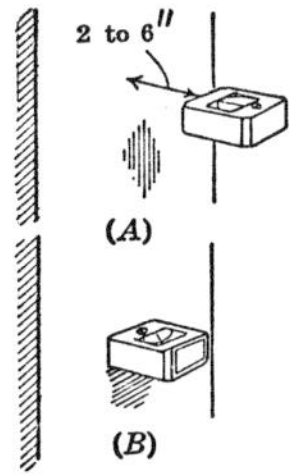 	Reflecting Surface — Per Cent Light Reflected Magnesium Carbonate — 97–98 Polished silver — 90 White plaster — 85–90 Mirrored glass — 80–90 Alzak aluminum — 80–85 White blotting paper — 80–85 Porcelain enamel — 77–82 White paint — 75–85 Polished chromium — 65 Polished aluminum — 60–70 Polished nickel — 65 Aluminum paint — 50–75 Black paint — 3–5	In *specular,* or *regular,* reflection (mirrors, highly polished metals) the angle of incidence is equal to the angle of reflection (see Illustration: Angle X = Angle Y). In *diffuse* reflection (matte surfaces like white blotting paper, fresh snow) the maximum intensity is perpendicular to the surface, regardless of the angle of the incident beam. *Spread* reflection, as in sanded glass, is intermediate between specular and diffuse. Diffusing surfaces with a glazed superficial coat, like porcelain enamel, exhibit *mixed* reflection, a combination of specular and diffuse.
Transmission Light rays passing through transparent or translucent materials are said to be *transmitted.* The degree of diffusion of the transmitted light depends on the type and density of the material.	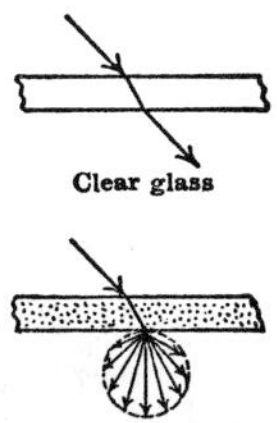 Clear glass Diffusing glass	Transmission Factor The ratio of the light transmitted by a material to that incident upon it. Transmission depends to some extent upon the direction and quality of the light.	Place material to be tested over cell of light meter. Note reading (*A*). Remove material. Note reading (*B*). Transmission factor $= \frac{\text{Reading } (A)}{\text{Reading } (B)}$	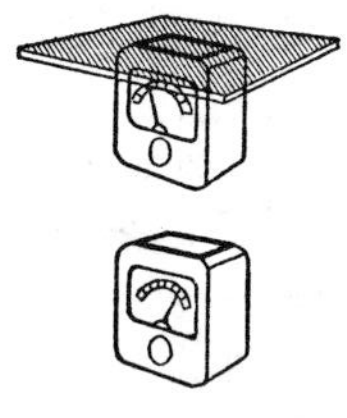	Type of Glass — Per Cent Light Transmitted Clear — 80–90 Frosted — 75–85 Figured or Ribbed Crystal — 70–85 Light-Density Diffusing — 55–70 Heavy-Density Diffusing — 10–45	In *regular* transmission (clear glass and plastics) the direction of the incident light is not changed. Diffusing media, such as dense opal glass, scatter the transmitted light so that its maximum intensity is normal to the surface. As in reflection, between the two extremes of regular transmission and perfectly diffuse transmission are to be found all degrees of diffusion.

Refraction A light ray bent by passing obliquely from one transparent medium to another in which its velocity is different (as from air into glass) is said to be *refracted.*	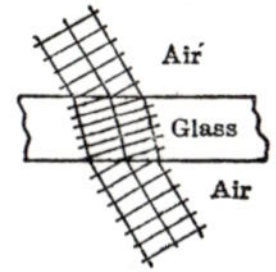	Index of Refraction The ratio of the speed of light in free space to the speed of light in the medium in question. $V = \frac{\text{vel. (air)}}{\text{vel. (medium)}}$	By special laboratory apparatus only.	Index of refraction is the ratio of the speed of light in free space, or a vacuum, compared to its speed in the substance.	Index of Refraction for Various Materials Water 1.33 Alcohol 1.36 Turpentine 1.47 Crown Glass 1.53 Flint Glass 1.67 Diamond 2.47	The principle of *refraction* is utilized to control the direction of light by means of prismatic or ribbed glass plates, or in lens systems. It has wide application in certain types of general lighting systems, as well as in signal lighting and street lighting.
Polarization Light in which the waves vibrate in one plane only is said to be *polarized.* The vibrations which make the wave motion in a ray of light are at right angles to the direction in which the light is traveling, and in a beam of ordinary light these vibrations take place in all possible directions in that plane. By passing light through a material with a crystalline structure such that it transmits only waves vibrating in a certain direction, it is possible to produce polarized light, all of whose vibrations are parallel. 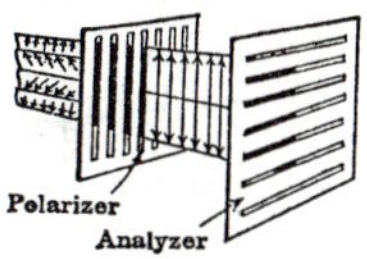 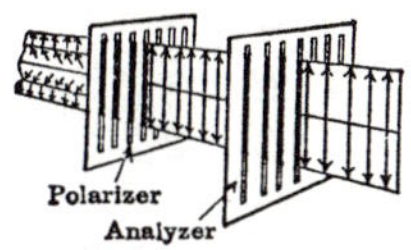				Two polarizing screens are ordinarily used in a system that involves polarization. The first, called the *polarizer*, produces the polarization, and the second, called the *analyzer*, selects or rejects the polarized light, according to the position in which it is placed.	Crystals of Iceland spar, calcite, and tourmaline; Polaroid, (a cellophane-like material available commercially). Reflection from specular or polished surfaces partially polarizes light. Ordinary radiant energy, with waves in a plurality of planes, is not polarized.	The principle of *polarization* is used in certain kinds of laboratory equipment, and in testing for stress and strain in transparent materials; in producing third-dimension effects in motion pictures; in sun glasses and automobile visors to reduce reflected glare from road surfaces and water; in photographic filters. Experimental work on the control of automobile headlight glare by means of polarizing material is under way.

The human eye is tuned to respond only to the energy within the limits of the *visible spectrum,* a narrow band of wavelengths between 3800 Angstroms and 7600 Angstroms.* Energy in this region, evaluated in terms of the eye sensitivity curve is *light.*

Color of light is determined by its wavelength. Light at the short-wave end of the visible spectrum, from 3800 to about 4500 Angstroms, produces the sensation of violet; the longest visible waves, from about 6400 to 7600 Angstroms, appear as red. Between these limits lie the wavelengths which the eye distinguishes as blue (4500–4900 Angstroms), green (4900–5600), yellow (5600–5900), and orange (5900–6400)—the colors of the rainbow.

The spectrum of a light source may be *continuous,* including all the visible wavelengths, or it may be a *line* or a *band* spectrum, containing only one or a few separated groups of wavelengths. A tungsten filament has a continuous spectrum, a mercury arc a line spectrum. An *equal-energy* spectrum—that is, all the visible wavelengths in equal quantities—produces the sensation of white light. Noon sunlight approximates an equal-energy spectrum.

Color temperature is a term sometimes used to describe the color of the light from a source by comparing it with the color of a *blackbody,* a theoretical "complete radiator" which absorbs all radiation that falls on it, and in turn radiates a maximum amount of energy in all parts of the spectrum. A blackbody, like any other incandescent body, changes color as its temperature is raised. The light from a white fluorescent lamp is similar in color to the light from a blackbody at a temperature of approximately 3500 Kelvin,† and the lamp is accordingly said to have a color temperature of 3500 K. The light from a daylight fluorescent lamp is bluer, and the blackbody must be raised to 6500 K to match it. Hence the daylight lamp has a color temperature of 6500 K.

Color temperature is not a measure of the *actual temperature* of an object. It defines *color only.* Some light sources, such as a sodium vapor lamp, or a green or pink fluorescent lamp, will not match the color of a blackbody at any temperature, and therefore no color temperatures can be assigned to them. (See Table 1, p. 27-64.)

TERMINOLOGY AND MEASUREMENTS. Table 2 gives definitions for lighting intensity and other information concerning the quantities used in the measurement of lighting intensity. Table 3 provides data on the modification of light by reflection, transmission, refraction, and polarization.

In general, the following facts regarding light should be kept closely in mind when dealing with illumination problems:

Light travels in straight lines, unless it is modified or re-directed by means of a reflecting, refracting, or diffusing medium.

Light waves pass through one another without alteration of either: for example, a beam of red light will pass directly through a beam of blue light unchanged in direction or color.

Light is invisible in passing through space unless some medium (such as dust) scatters it in the direction of the eye.

10. THE FACTORS OF SEEING

SIZE of the object is the most generally recognized and accepted factor in seeing. The larger an object in terms of visual angle (the angle subtended by the object at the eye) the more readily it can be seen. The familiar eye test chart illustrates this principle (Fig. 2). The person who brings a small object close to his eye in order to see it more clearly is unconsciously making use of the size factor by increasing the visual angle. Under any conditions an object must subtend about a one-minute angle to be seen clearly.

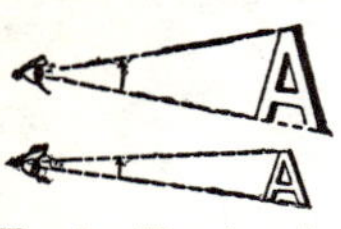

FIG. 2. Visual angle.

Visual acuity, expressed as the reciprocal of the visual angle in minutes, is a measure of the smallest detail that can be seen. Since visual acuity increases markedly with increase in illumination, light is sometimes said to act as a "magnifier," making visible small details that could not be seen with less light.

BRIGHTNESS is one of the primary factors in visibility. The brightness of an object depends both on the intensity of the light striking it and the proportion of that light reflected in the direction of the eye. A white surface will have a much higher brightness than a black surface receiving the same illumination. However, by adding enough light to a dark surface it is possible to make it as bright as a white one. The darker an object or a visual task, the greater the illumination necessary for equal brightness and, under like circumstances, for equal visibility.

---

* The Angstrom is a wavelength unit equal to one ten-millionth of a millimeter, or approximately four-billionths of an inch. The *micron,* equal to one-millionth of a meter, is another unit often used to express wavelength. One micron is 1×10^4 Angstroms.

† Kelvin is a temperature scale which has its zero point at −273 C.

CONTRAST in brightness or color between the visual object and its immediate background is as important for seeing as general brightness level. The difference in the visual effort required to read the two halves of Fig. 3 is a simple demonstration of the effectiveness of contrast. High levels of illumination partly compensate for low brightness contrasts and are of great assistance where poor contrast conditions cannot be avoided.

TIME. Seeing requires time. The dark-adapted eye can see under low levels of illumination, if sufficient time is allowed. "Snapshot" eye pictures require more light energy. The time factor is particularly important where the visual object is in motion.

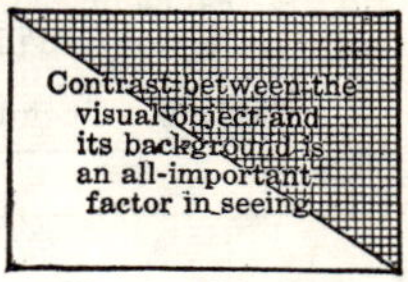

FIG. 3. Contrast.

11. LIGHT SOURCES

The primary purpose of a light source is the production of light, and the efficiency with which a lamp fulfills this purpose is expressed in terms of lumens emitted per watt of power consumed. If a source could be developed that would radiate all the energy it received as monochromatic yellow-green light in the region of maximum sensitivity of the eye, 5550 Angstroms, it would produce approximately 650 lumens for each watt of power consumed. A theoretical source of white light of maximum efficiency, emitting only visible energy without any infrared or ultraviolet, would produce about 200 lumens per watt. Compared with these figures, the best available lamp efficiencies today seem disappointingly low (see Table 4). However, reference to earlier products of lamp manufacture shows the

Table 4. Efficiencies of Various Light Sources

	Approximate Lumens per Watt
Candle (luminous efficiency equivalent)	0.1
Oil lamp (luminous efficiency equivalent)	0.3
Original incandescent lamp (1879)	1.4
60-watt carbon filament lamp (1905)	4.0
60-watt coiled coil tungsten filament lamp (1947)	13.9
1000-watt general-service lamp (1947)	21.5
No. 1 photoflood lamp (1947)	34.6
400-watt A-H1 mercury vapor lamp (lamp only) (1947)	40.0
40-watt white fluorescent lamp (lamp only) (1947)	58.0

progress that has been made. The efficiency of the 60-watt lamp, for example, has been increased almost 350% during the last 40 years by changing from carbon to tungsten as a filament material, from vacuum to gas-filled construction, and from straight filament wire to coiled and then coiled coil filaments.

However, the filament lamp has certain characteristics which make it inherently inefficient as a source of light, and, although it is probable that efficiencies will still be raised slightly by further refinements in manufacturing processes, the maximum possible values have already been approached. For any considerable increase in efficiency it is now necessary to turn to another type of light source, the electric discharge lamp. Present mercury vapor lamps have efficiencies of 40 to 65 lumens per watt, and sodium vapor lamps 55 lumens per watt, whereas fluorescent lamps have already reached figures as high as 65 lumens per watt, with every likelihood of further increases.

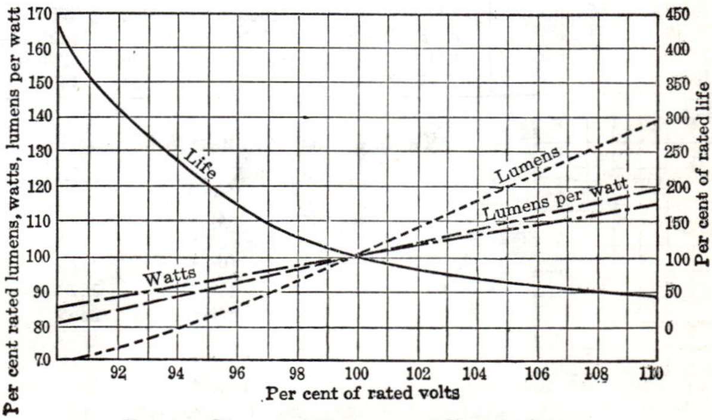

FIG. 4. Characteristic curves of filament lamps.

FILAMENT LAMPS. The typical tungsten filament lamp, in clear or frosted bulb, when burned on a multiple circuit, exhibits the characteristics shown by the curves in Fig. 4. Efficiencies are plotted against rated watts in Fig. 5, and data on lamps of various sizes are given in Table 5. Approximate filament temperatures are indicated in Table 6.

Table 5. Technical Data on Representative 120-volt Standard Filament Lamps

General Lighting Service Lamps

Watts	Bulb	Base	Filament Construction	Rated Average Life, hours	Approx. Initial Lumens	Lumens per Watt	
						Initial	70% Life
25	A-19	Medium	C-9	1000	260	10.5	9.0
40	A-19	Medium	C-9	1000	470	11.7	10.7
50	A-19	Medium	CC-6	1000	665	13.3	12.5
60	A-19	Medium	CC-6	1000	835	13.9	13.2
75	A-21	Medium	C-9	750	1120	14.9	13.5
100	A-21	Medium	CC-6	750	1630	16.3	15.3
100	G-30	3-Contact Mogul	2 C-2R	1000	1410	14.1	....
200					3350	16.7	....
300					4760	15.8	13.3
150	PS-25	Medium	C-9	750	2600	17.3	15.6
200	PS-30	Medium	C-9	750	3700	18.5	16.4
300	PS-30	Medium	C-9	750	5900	19.7	17.3
300	PS-35	Mogul	C-9	1000	5650	18.8	16.7
500	PS-40	Mogul	C-7A	1000	9950	19.9	17.0
750	PS-52	Mogul	C-7A	1000	15500	20.6	17.6
1000	PS-52	Mogul	C-7A	1000	21500	21.5	17.4
1500	PS-52	Mogul	C-7A	1000	33000	22.0	15.6

Lumiline Lamps

Watts	Bulb	Length, inches	Base	Filament Construction	Rated Average Life, hours	Approx. Initial Lumens	Initial Lumens per Watt
30	T-8	17 3/4	Disk	C-8	1500	245	8.2
40	T-8	11 3/4	Disk	C-8	1500	345	8.7
60	T-8	17 3/4	Disk	C-8	1500	550	9.2

Projector and Reflector Lamps

Watts	Bulb	Base	Filament Construction	Rated Average Life, hours	Initial Max. Beam Candlepower	Approx. Initial Zone Lumens	Description
Projector							
150	PAR-38	Med. Skt.	CC-6	1000	12500	1050 (0–15°)	Spot
150	PAR-38	Med. Skt.	CC-6	1000	2800	1300 (0–30°)	Flood
Reflector							
150	R-40	Medium	C-11	1000	7000	700 (0–15°)	Spot
300	R-40	Medium	CC-2V Horiz.	1000	12500	1270 (0–15°)	Spot
150	R-40	Medium	C-11	1000	1200	700 (0–30°)	Flood
300	R-40	Medium	CC-2V Horiz.	1000	2350	1400 (0–30°)	Flood

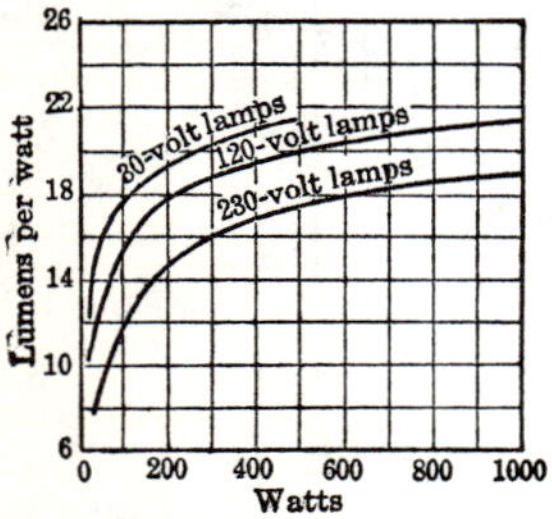

Fig. 5. Efficiencies of 1000-hr. lighting service lamps

Table 6. Approximate Filament Temperatures

Watts	°C
40 *	2475
60 *	2500
100 *	2575
200 *	2620
300 *	2665
500 *	2670
1000 *	2720
1500 *	2765
No. 1 Photoflood	3160

* Standard 120-volt general lighting service lamps.

Shape of a lamp bulb is designated mnemonically by letters (Fig. 6), and the maximum diameter, in eighths of an inch, by figures. For example, a PS-52 bulb is pear-shaped, measuring 6 1/2 in. in diameter. There may be added the bulb glass finish, the base designation, the design volts and watts, and the filament shape, "C" for coiled; "CC" for double or coiled-coil, etc.

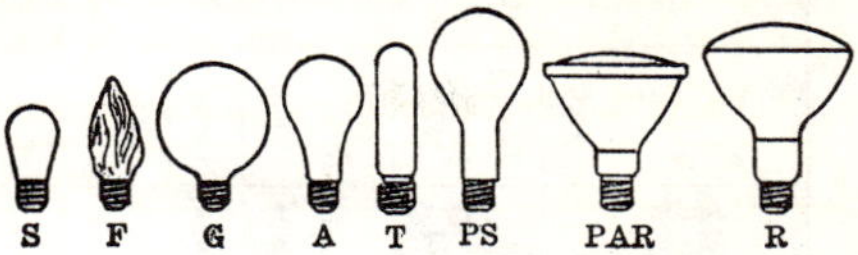

FIG. 6. Lamp bulb shape designations.

FLUORESCENT LAMPS. The fluorescent lamp is essentially an electric discharge source. It consists of a tubular bulb having electrodes sealed into each end and containing mercury vapor at low pressure with a small amount of argon (for starting), Fig. 7. The inner walls of the bulb are coated

(*Continued on p. 27-76*)

Table 7. Reference Data on Fluorescent Lamps

Lamp [1]	Base	Approx. Operating		Open-Circuit Volts [2]	Preheat Amperes [3]	Rated Initial Lumens [4]		
		Amperes	Volts			White [5]	4500 White	Daylight
8-W 9" T-5	Min. Bipin	0.165	55	118	0.21	330	310	295
14-W 15" T-12	Med. Bipin	0.370	41	118	0.55	495	475	420
15-W 18" T-8	Med. Bipin	0.300	56	118	0.55	650	620	550
15-W 18" T-12 [6]	Med. Bipin	0.330	48	118	0.55	610	580	520
20-W 24" T-12 [6]	Med. Bipin	0.360	60	118	0.55	960	915	815
25-W 33" T-12	Med. Bipin	0.480	59	118	0.75	1430	1370	1210
30-W 36" T-8 [6]	Med. Bipin	0.340	103	200	0.53	1600	1530	1360
40-W 48" T-12 [6,7]	Med. Bipin	0.420	106	200	0.65	2480	2370	2100
85-W 60" T-17	Mog. Bipin	[10]	......	....	....	4600	4400	3900
100-W 60" T-17	Mog. Bipin	1.450	72	150	1.78	4600	4400	3900
Instant-Start								
40-W 48" T-12	Med. Bipin [8]	0.420	106	450	....	2480	2370	2100
40-W 60" T-17	Mog. Bipin [8]	0.400	110	450	....	....	2370	......
Slimline [9]								
18-W 42" T-6	Single-pin	0.120	175	450	....	1070	1020	900
25-W 42" T-6	Single-pin	0.200	150	450	....	1425	1360	1200
33-W 42" T-6	Single-pin	0.300	130	450	....	1800	1720	1530
27.5-W 64" T-6	Single-pin	0.120	270	600	....	1690	1610	1430
39-W 64" T-6	Single-pin	0.200	230	600	....	2300	2200	1950
51-W 64" T-6	Single-pin	0.300	200	600	....	2900	2760	2460
26-W 72" T-8	Single-pin	0.120	240	600	....	1710	1640	1460
38-W 72" T-8	Single-pin	0.200	220	600	....	2420	2310	2060
51-W 72" T-8	Single-pin	0.300	200	600	....	3170	3030	2700
34-W 96" T-8	Single-pin	0.120	320	750	....	2280	2180	1940
51-W 96" T-8	Single-pin	0.200	295	750	....	3300	3150	2800
69-W 96" T-8	Single-pin	0.300	265	750	....	4350	4150	3700
59-W 72" T-12	Single-pin	0.425	152	525	....	3660	3500	3100
72-W 72" T-12	Single-pin	0.600	137	525	....	4300	4100	3660
75-W 96" T-12	Single-pin	0.425	195	625	....	4800	4575	4080
96-W 96" T-12	Single-pin	0.600	175	625	....	5800	5540	4925
Circlarc								
18-W 12" diam. T-8	Med. Bipin	0.380	54.5	118	0.55	710	....	......
Circline								
32-W 12" diam. T-10	Four-pin	0.430	84	150	0.65	1600	....	......

[1] Lamp watts, nominal overall length (one lamp plus two standard lampholders), and bulb designation (T for tubular; number indicates tube diameter in eighths of an inch).

[2] Minimum open-circuit volts at rated line voltage of ballast.

[3] Center value.

[4] Lumens when measured at 80 F ambient, and under specified test conditions. The light output of a new lamp will appreciably exceed the above rated initial values, which apply when the lamps have burned 100 hr.

[5] The majority of lamp types are available also in warm white and soft white.

[6] Regularly available also in blue, green, gold, pink, and red.

[7] A low-temperature 40-watt T-12 lamp, marked with the letters "LT," gives reliable starting down to 0 F.

[8] Base pins shorted inside base.

[9] Slimline lamps may be operated at any intermediate value within their design current range. The light output increases with the current loading.

[10] Operates on standard 100 w ballast.

Table 8. Mercury Vapor Lamps

Type	A-H4	B-H4	C-H4 (Sp) E-H4 (Fl)	S-4	A-H5	C-H5	RS	A-H1 B-H1	D-H1	E-H1	F-H1	A-H6 [1]	A-H12	B-H12	A-H9	B-H9
Watts (lamp only) [2]	100	100	100	100	250	250	275	400	400	400	400	1000	1000	1000	3000	3000
Outer bulb	T-10	T-16	PAR-38	A-21	T-14	T-14	R-40	T-16	T-20	T-20	T-16	T-2	T-28	T-28	T-9 1/2	T-9 1/2
Outer bulb glass	772	5872	772	721	774	172	776	772	774	172	772	Quartz	172	774	172	973
Outer bulb finish	Clear	Nat. Red Purple	Alum. Reflect.	Clear	Clear	Clear	I.F. Alum. Reflect.	Clear	Clear	Clear	Clear	Clear	Clear	Clear	Clear	Clear
Arc tube material	Quartz	Quartz	Quartz	Quartz	Quartz	Quartz	Quartz	172	Quartz	Quartz	172	Quartz [3]	Quartz	Quartz	172 [3]	973 [3]
Arc length, inches	1	1	1	1	1 5/8	1 5/8	1 5/8	6	2 3/8	2 3/8	6	1	5	5	48	50
Base	Admed. Sc.	Admed. Sc.	Admed. Sc. Skt.	Admed. Sc.	Mogul Sc.	Mogul Sc.	Med. Sc.	Mogul Sc.	Mogul Sc.	Mogul Sc.	Mog.Sc. Mech.	3/16″ Sleeve	Mog.Sc. Mech.	Mog.Sc. Mech.	S/C Term.	S/C Term.
Initial lumens (at 100 hr)	3300	....	....	3000	11,250	11,250	....	16,000	21,000	20,000	16,000	65,000	60,000	60,000	120,000	120,000
Rated average life:																
5 hours per start	1000	1000	1000	1000 [4]	1000	....	1000 [4]	....	1000 [5]	....	4000	75	....	2000	....	3000
10 hours per start	....	....	....	....	....	4000	....	6000	[5]	4000	6000	....	3000	....	6000	4500
Maximum overall length, inches	5 5/8	5 1/2	5 7/16	5 1/4	8	8	6 7/8	13	11	11	13	3 1/4	14	14	54 7/8	56 7/8
Light center length, inches	3 7/16	3 7/16	....	3 7/16	5	5	....	7 3/4	7	7	7 3/4	....	9	9	....	....
Burning position	Any	Any	Any	Any	Any	Any	Any	[6]	Any	Any	[6]	Horiz.	Any	Any	Any	Any
Electrode type	Thor.	Thor.	Thor.	Thor.	Thor.	Thor.	Thor.	Oxide	Thor.	Thor.	Oxide	Mercury	Thor.	Thor.	Oxide	Thor.
Vapor pressure, atmospheres	8	8	8	8	4.5	4.5	1.1	1.2	2.5	2.5	1.2	110	1.5	1.5	0.4	0.4
Open-circuit volts	250	250	250	250	250	250	....	220	220 [7]	220 [7]	220	1200	220 [7]	220 [7]	850	850
Operating volts	130	130	130	130	135	135	110–125 [8]	135	135	135	135	840	135	135	535	535

Starting current, amperes	1.3	1.3	1.3	1.3	2.9	2.9	3.2	4.7	4.7	4.7	4.7	2.6	12	12	9.3	9.3
Operating current, amperes	0.9	0.9	0.9	0.9	2.1	2.1	2.4	3.2	3.2	3.2	3.2	1.4	8.2	8.2	6.1	6.1
Starting time, minutes	3	3	3	3	4	4	2	7	4	4	7	2 Sec.	4	4	7	7
RECOMMENDED APPLICATIONS																
General lighting	X	....	....	....	....	X	....	X	....	X	....	....	X	....	X	....
Floodlighting	....	....	X	....	....	X	....	X	....	X	....	....	X	....	....	....
Street lighting	....	....	....	....	....	X	....	X	....	X	X	....	X	....	....	....
Black light	X	X	X	....	X	....	....	....	X	....	....	X	....	X	....	X
Photochemical	X	....	X	X	X	....	....	....	X	....	....	X	....	X	....	X
Blueprint and photography	X	....	....	....	X	....	....	....	X	....	....	X	....	X	....	X
Sun lamp service	....	....	....	X	....	....	X	....	....	....	....	....	....	....	....	....
Searchlight and projection	X	....	....	....	....	X	....	....	....	X	....	X	X	....	....	....

1 The A-H6 is water cooled and requires an outer water jacket, generally of quartz or 774 heat-resisting glass. A similar lamp for air jet cooling is the B-H6 rated at 900 watts.
2 For total wattage add transformer watts which range from 8 to 25% of lamp watts, depending chiefly on the type of lamp and transformer used.
3 The A-H6, A-H9, and B-H9 are single-bulb lamps. The outer bulb is the arc tube.
4 Life of the S-4 and RS sun lamps is rated at 1000 applications in normal sun lamp use in the home, or 1000 hours at 5 hr per start.
5 Rated average life at 5 hr per start for D-H1 in any position is 1000 hr. At 10 hr per start the rated average life is 3000 hr in the vertical position only.
6 The A-H1 and F-H1 are designed for base up burning, the B-H1 for base down burning. These types must be operated within 10° of vertical.
7 For normal indoor use. Higher open-circuit voltages are desirable for dependable starting at lower temperatures.
8 For 50-60 cycle a-c operation only. Operates directly from regular lighting circuit, no transformer required.

with fluorescent powders which give off light when activated by ultraviolet energy. When the proper voltage is impressed on the electrodes, a flow of electrons is driven from one electrode and attracted or pulled to the other. As these electrons speed through the tube they collide with the mercury atoms, causing a state of excitation which produces short-wave ultraviolet radiation (2537 Angstroms). The fluorescent powders, commonly known as phosphors, absorb this invisible energy and radiate visible light. Table 7 gives data on fluorescent lamps and Table 8 on mercury vapor lamps.

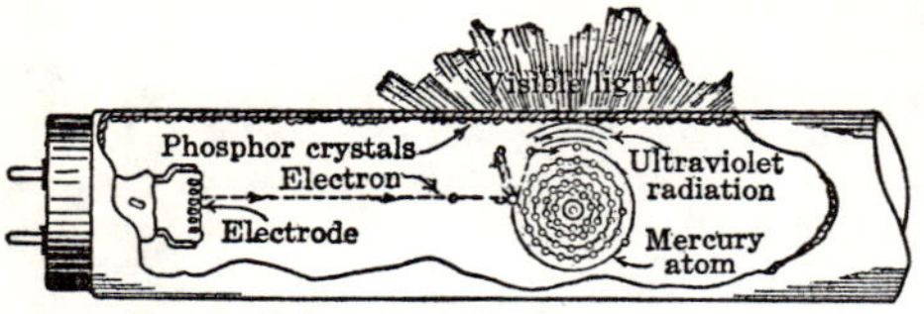

FIG. 7. Principle of the fluorescent lamp.

12. ILLUMINATION LEVELS

The high levels of illumination produced by natural daylight are desirable for many seeing tasks. Before the development of modern light sources many severe visual tasks had to be performed only in daylight hours because of the weakness and expense of illumination provided by the artificial illuminants then in use. The illumination intensities which were considered to be standard in past years were largely determined by the light sources then available. Present-day practice recognizes that the foot-candle levels now recommended do not represent the ultimate in see-ability. They do, however, represent practical values which may readily be obtained by means of existing equipment.

The relationship between foot-candle levels and ability to see is geometric rather than arithmetic, and the chart in Fig. 8 lists approximately equal steps in foot-candle effectiveness.

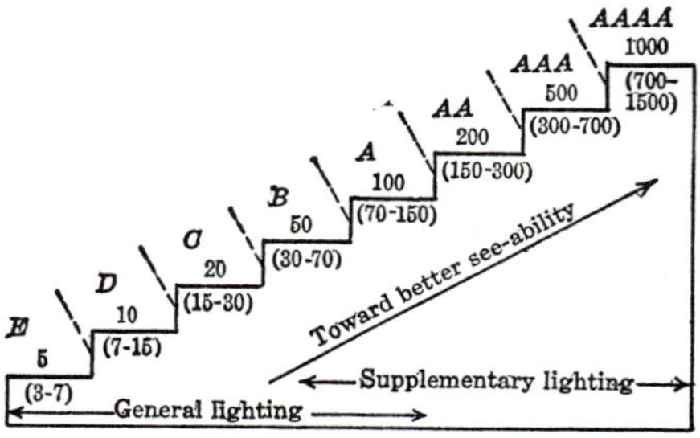

FIG. 8. Levels of illumination.

The quantitative requirement for good illumination of a specific seeing task is primarily a function of the difficulty of the task in terms of fineness of detail, brightness or color contrast, and the speed demanded. Other factors such as the length of time for which the task is to be performed, the surrounding conditions, and the physiological state of the eyes which have to do the work are also involved.

GENERAL LIGHTING. The following classification of foot-candles assumes substantially white light illumination on the work or target, not in the eyes. Usually the values are those maintained in service, or the going average; and if a work plane is not specified, it is assumed to be the horizontal surface 30 in. above the floor.

Level A—100 Foot-candles. For very exacting and prolonged seeing tasks, such as fine bench and machine work, extra fine hand painting and finishing, and for the discrimination of fine detail of low contrast. For showcases, wall cases and open counter displays in stores and reading medium to fine print, or sewing. (Under various circumstances the foot-candles may range from 30 below to 40% above the value given.)

Level B—50 Foot-candles. For severe and extensive seeing tasks, such as medium and fine machine work, medium fine assembly and inspection; fine sanding and finishing of woodwork, drafting, and proofreading. For general merchandising areas in stores. (Range —minus 30 to plus 40%.)

Level C—20 Foot-candles. For moderately critical and lengthy seeing tasks, such as rough bench and machine work, medium assembly and inspection, hand painting and finishing, for weaving light woolens and reading large type. For circulation areas in stores. (Range—minus 30 to plus 40%.)

Level D—10 Foot-candles. For visually controlled work in which seeing is important, but more or less interrupted or casual, and does not involve discrimination of fine details or low contrasts. For rough manufacturing processes, such as molding clay products and cements, glass-blowing machines, billet, blooming and bar mills in steel manufacturing. For stockrooms and active storage areas for a variety of small articles such as for merchandise stocks.

Level E—5 Foot-candles. For exteriors where crude manual tasks are intermittently carried on, such as required for grinding clay products and cements, stone crushing, hand furnaces and boiling tanks in chemical plants, stockrooms and active storage areas for medium-size materials; for the safe assembly or movement of people in auditoriums, through corridors and stairways. For active work areas out-of-doors—loading docks.

SUPPLEMENTARY LIGHTING. Level AA—200 Foot-candles. For extra fine inspection, such as required in making jewelry and precision instruments. For featured items and merchandise displays in show windows in secondary business areas of large cities. (Range—minus 30 to plus 40%.)

Level AAA—500 Foot-candles. For feature merchandise displays in show windows in main business areas of large cities. For color identification in industry. (Range—minus 30 to plus 40%.)

Level AAAA—1000 Foot-candles. Outdoor levels of illumination. For photography, hospital operating rooms, and the daytime illumination of show windows. (Range—minus 30 to plus 40%.)

STREET LIGHTING. Average foot-candles or lumens per square foot on roadway, based on medium reflectance of surface on the order of 10%, and for urban streets are:

	Vehicular Traffic Classification			
Pedestrian Traffic Likely to Cross	Very Light (Under 150)	Light (150–500)	Medium (500–1200)	Heavy (1200 up)
Heavy	...	0.8	1.0	1.2
Medium	...	0.6	0.8	1.0
Light	0.2	0.4	0.6	0.8

13. INTERIOR LIGHTING DESIGN

BASIC FORMULAS for interior lighting design are:

$$\text{Total lumens} = \frac{\text{Foot-candles} \times \text{area}}{\text{Coefficient of utilization} \times \text{maintenance factor}}$$

$$\text{Number of lamps} = \frac{\text{Foot-candles} \times \text{area}}{\left\{\begin{array}{l}\text{Lumens per lamp} \times \text{coefficient of utili-}\\ \quad\text{zation} \times \text{maintenance factor}\end{array}\right\}}$$

$$\text{Luminaires} = \frac{\text{Foot-candles} \times \text{area}}{\left\{\begin{array}{l}\text{Lamps per luminaire} \times \text{lumens per lamp} \times \text{co-}\\ \quad\text{efficient of utilization} \times \text{maintenance factor}\end{array}\right\}}$$

$$\text{Foot-candles} = \frac{\left\{\begin{array}{l}\text{Lamp lumens} \times \text{coefficient of utilization} \times\\ \quad\text{maintenance factor}\end{array}\right\}}{\text{Area}}$$

$$\text{Area per luminaire} = \frac{\left\{\begin{array}{l}\text{Lamps per luminaire} \times \text{lumens per}\\ \quad\text{lamp} \times \text{coefficient of utilization} \times\\ \quad\text{maintenance factor}\end{array}\right\}}{\text{Foot-candles}}$$

Total (generated) lumens are determined by the published initial lumen figures given for each type or size of lamp, multiplied by the number of lamps involved.

Foot-candles are either measured by a light meter, or pre-chosen from published tables of desirable intensities.

Areas are normally the square footages of the floors or of a large area target.

Maintenance factor is the combined allowance for dirt on the fixture or immediate surroundings and the normal reduction in lumen output of the lamps themselves. This factor, for medium maintenance and average (70% life point) condition of lamp, is roughly 55%. Poor cleaning might reduce this to 40%; good maintenance raise it to 70%.

Coefficient of utilization expresses the percentage of generated (lamp) lumens that reach the work area or target. It takes into account the lumens lost in the fixtures or reflector, plus those absorbed by floors, ceilings, and walls that are not the work areas. The larger the room, the less light is lost on walls. Fifteen to 35% of the generated lumens is lost through fixture absorption, so that for an average office installation of fluorescent luminaires, this coefficient may be taken as roughly 50%; for large factories, with open-bottom industrial reflectors, about 70%.

Example. To illuminate a large, clean general factory space, 60 by 200 ft, for medium rough machine operations, the advisable foot-candles would be chosen to average, say, 30. If using fluorescent (100 watt) lamps in open reflectors, the coefficient of utilization would be about 70%. With good maintenance, the maintenance factor would be roughly 65%. Therefore

$$\text{Total lumens} = \frac{30 \times 12{,}000}{0.70 \times 0.65} = 792{,}000$$

Each 100-watt (white) fluorescent lamp is rated at 4600 lumens, so 172 lamps are needed.

Table 9. Allowable Current-carrying Capacities of Conductors in Amperes

Not more than three conductors in raceway or cable

(Based on room temperature of 30 C, 86 F)

Size AWG MCM	Rubber Type R Type RW Type RU (14–6) / Thermoplastic Type T (14–4/0) Type TW (14–4/0)	Rubber Type RH	Paper / Thermoplastic Asbestos Type TA / Var-Cam Type V / Asbestos Var-Cam Type AVB	Asbestos Var-Cam Type AVA Type AVL	Impregnated Asbestos Type AI (14–8) Type AIA	Asbestos Type A (14–8) Type AA
14	15	15	25	30	30	30
12	20	20	30	35	40	40
10	30	30	40	45	50	55
8	40	45	50	60	65	70
6	55	65	70	80	85	95
4	70	85	90	105	115	120
3	80	100	105	120	130	145
2	95	115	120	135	145	165
1	110	130	140	160	170	190
0	125	150	155	190	200	225
00	145	175	185	215	230	250
000	165	200	210	245	265	285
0000	195	230	235	275	310	340
250	215	255	270	315	335	....
300	240	285	300	345	380	....
350	260	310	325	390	420	....
400	280	335	360	420	450	....
500	320	380	405	470	500	....

Table 10. Branch Circuit and Feeder Wire Sizes for Various Lengths of Run

On short runs, where voltage drop does not affect wire size, the table shows minimum permissible commercial size Brown & Sharpe gage rubber-covered copper wire according to the National Electrical Code.

Based on { 2% loss in voltage on 115-volt, 2-wire circuits.
1% loss in voltage on 220-volt, 2-wire circuits.
1% loss in voltage on 115/230-volt, 3-wire circuits.

Wattage Load, 115-volt Circuit	Ampere Load	Length of Run, ft																	
		30	40	50	60	70	80	90	100	120	140	160	180	200	240	280	320	360	400
575	5	14	14	14	14	14	14	12	12	12	10	10	10	10	8	8	8	6	6
690	6	14	14	14	14	14	12	12	12	10	10	10	8	8	8	8	6	6	6
805	7	14	14	14	14	12	12	12	10	10	10	8	8	8	6	6	6	6	4
920	8	14	14	14	12	12	12	10	10	10	8	8	8	8	6	6	6	4	4
1035	9	14	14	12	12	12	10	10	10	8	8	8	8	6	6	6	4	4	4
1150	10	14	14	12	12	10	10	10	10	8	8	8	6	6	6	4	4	4	4
1380	12	14	12	12	10	10	10	8	8	8	8	6	6	6	4	4	4	4	2
1610	14	14	12	10	10	10	8	8	8	6	6	6	6	4	4	4	2	2	2
1840	16	12	12	10	10	8	8	8	8	6	6	6	4	4	4	2	2	2	2
2070	18	12	10	10	8	8	8	8	6	6	6	4	4	4	4	2	2	2	1
2300	20	12	10	10	8	8	8	6	6	6	4	4	4	4	2	2	2	1	1
2875	25	10	10	8	8	6	6	6	6	4	4	4	2	2	2	1	1	0	0
3450	30	8	8	8	6	6	6	6	4	4	4	2	2	2	1	1	0	0	2/0

INTERIOR WIRING FOR LIGHTING. After determining the foot-candles needed for a general-purpose interior, the required power in watts per square foot of floor area may be roughly estimated from the chart in Fig. 9. This assumes good maintenance, a light color of interior surfaces, and medium-sized rooms.

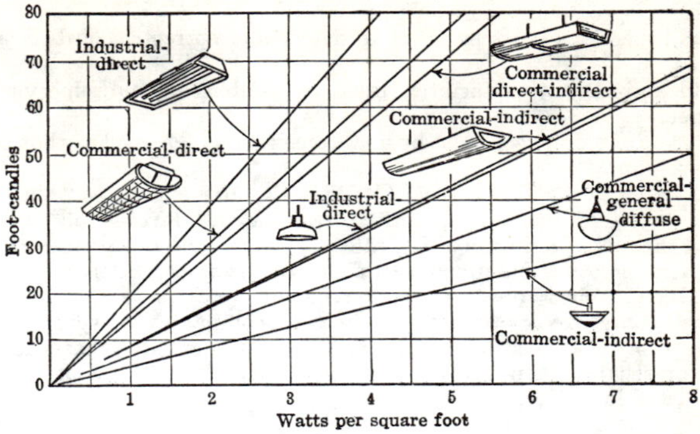

Fig. 9. Power required for commercial and industrial lighting.

Manufacturing areas should have at least one convenience outlet for every 20 linear ft of wall space; private commercial offices, one outlet for each 10 ft of baseboard.

The wire sizes for various ampere loadings and the voltage losses for different lengths of run (of circuit, not of single wire) are given in Tables 9 and 10.

SAFETY ENGINEERING

By Sidney J. Williams and A. M. Baltzer

14. BASES OF ACCIDENT PREVENTION

The guarding of dangerous machinery and dangerous places is required by the laws of most states. High accident rates are penalized, also, by an increase in the rate charged for workmen's compensation insurance.

Most industrial concerns endeavor to eliminate or guard all danger points and to correct all unsafe practices, even where not demanded by state or insurance requirements because they have found that: (1) Accidents cause, or indicate, an irregularity in production that results in lost time and inefficiency. (2) Accidents cause the temporary or permanent loss of experienced and valuable workmen. (3) Careful attention to safety, by both mechanical and educational means, raises the morale of the working force. Such companies often require that equipment which they purchase be guarded in accordance with certain specifications.

A generation of accident-prevention experience in American industry has proved that safety engineering is indispensable to efficient production and that any organization, large or small, can protect itself against unnecessary loss. In the 19-year period of 1930–1948 the industrial injury frequency rate of thousands of companies reporting to the National Safety Council dropped an average of over 38% whereas the severity rate went down over 43%. It has been established that practically all accidents are preventable and that most of them involve both personal and mechanical causes.

PRINCIPLES which guide industry's effort are:

1. Management must be sincerely interested and must give continuing and active support to the program.
2. Supervisors must accept clearly defined responsibilities for the removal of hazards within their departments.
3. The accident-prevention program must be directed by a capable person on a full or part-time basis.
4. The nature of the problem must be known through an analysis of accident records and an inventory of physical hazards.

5. Unsafe conditions must be made safe, machines must be guarded, protective equipment must be provided, housekeeping must be improved, and physical hazards must be removed.
6. Supervisors and employees must be trained to do their jobs safely.
7. Simple but informative accident records must be kept so that trends may be known and accident-producing causes eliminated.
8. Regular inspections must be made to detect and correct hazards *before* they cause accidents.
9. First-aid and sanitation facilities must be established and supervised by trained personnel.
10. The interest of employees must be maintained on a high level so that they cooperate in the program.

ADMINISTRATION OF THE SAFETY PROGRAM is entrusted to a safety director who, in larger companies, has had specialized training and devotes full time to this work. In smaller and medium-sized companies, safety is combined with personnel, industrial relations, insurance, or other functions. In all cases the safety director analyzes accident causes, and effects their removal through recommendations to top management. He exerts no "line" control over production. In some companies, however, the safety engineer is given authority, if need be, to stop operations immediately hazardous.

In starting or rehabilitating a safety program, help should be sought early from the organizations listed in Article 23.

15. PLANT DESIGN AND EQUIPMENT

Safety features built into the plant reduce the need for costly alterations later.

RAILROAD SIDING clearances are subject to control by state public utility commissions, but in general they require 4-ft clearance between the nearest rail and the sides of buildings or other obstructions. Overhead clearance must be at least 22 ft from rail to the under side of structures or building entrances and at least 35 ft from high-tension power lines.

AISLES should be not less than 2 ft wider than the widest vehicle loaded. For two-way traffic, the aisle should be not less than 3 ft wider than twice the width of the widest vehicle loaded. Aisle areas should be clearly marked with stripes of a contrasting color painted on the floor. The speed of trucks is generally limited to 6 or 8 miles per hour.

EXITS must be adequate for any emergency. At least two, remote from each other, should be provided for each room or working area in conformance with the Building Exits Code of the National Fire Protection Association.

STAIRS should preferably be at an angle of 30 to 35 degrees from the horizontal with treads, exclusive of the nosing, at least 9 1/2 in. wide and risers not over 8 in. high. For heavy traffic the stair treads, particularly the nosing, should be surfaced with a nonslip material to prevent falls.

HAND RAILS 30 to 34 in. in height above the stair nosing are specified by the American Standard Code for Railings and Toe Boards. Stairs less than 44 in. wide, with both sides enclosed, should have at least one hand rail on the right side descending. Stairways between 44 and 88 in. require hand rails on each side and over 88 in. an additional hand rail in the center.

FLOOR OPENINGS require enclosure with a 42-in. high standard railing, a 4- to 6-in. toeboard at the floor level, and an intermediate rail approximately 18 in. above the floor.

FLOORS must be strong enough to support machine and human loads, loaded power trucks, and the heaviest material that might be stored on them. Floors should be as level as possible and smooth without being excessively slippery.

ACCESS should be provided, by runways and stairs or permanent ladders, to all points which must be visited for maintenance, cleaning, oiling, or other purposes.

PIPES and other equipment should not encroach on safe headroom (at least 6 ft 6 in.) over stairs and passageways.

ELECTRICAL HAZARDS. High voltage equipment should be available only to specially trained employees authorized to work with it. Transformers, switchboards, and other exposed parts can be protected by grounded enclosures or standard railings and by grounding noncurrent-carrying metal parts as recommended by the National Electrical Code of the National Bureau of Standards. Warning signs and devices for locking out equipment under repair provide further protection.

Plant wiring should also conform to the National Electrical Code and to local codes. Makeshift wiring should be eliminated because of the shock and fire hazard. Extension cords and other equipment should be of a type listed by Underwriters Laboratories.

STEAM BOILERS. Special safety measures include provision for tagging and locking valves while boilers are under repair, reinforced gage glass guards, enclosure of stoker drives, and safe, convenient access to all overhead work areas. Refer also to ASME Power Boiler Code.

UNFIRED PRESSURE VESSELS. Compressed air tanks, steam-jacketed kettles, digesters, vulcanizers, and similar equipment should be designed and constructed in accordance with the ASME Code for Unfired Pressure Vessels. The hazard of corrosion, external damage, or collapse due to vacuum is aggravated if such equipment is operated by untrained persons. Regular inspection and maintenance of safety valves, rupture disks, fusible plugs, and relief valves are of great importance.

GOOD LIGHTING is conducive to operating efficiency and the prevention of accidents. Good general lighting requires at least 10 foot-candles at the work level and avoidance of glare and strong shadows.

GENERAL VENTILATION prevents the accumulation of dust, fumes, and vapors without excessive draft. However, general ventilation is not an adequate substitute for local exhaust, whereby harmful materials are removed close to their source.

OVERHEAD EXHAUST HOODS are suitable only where the material is heated or is lighter than air and then should be designed so that the harmful materials are not drawn past the breathing level. Down-draft exhausts are suitable for the collection of dust, such as in a foundry shakeout, and for the collection of solvent vapors which are generally heavier than air.

LADDER shoes of abrasive material, rubber, cork, or sharp spikes will prevent slipping, but lashing or otherwise holding the ladder is recommended. For regular or continuous work, safety platform ladders, equipped with guard rail, are recommended.

SCAFFOLDS 10 ft or more above the ground or floor should be equipped with guard rails and standard toeboard. The principal causes of scaffold accidents are poor or defective materials, poor construction, overloading, and the falling of men or materials due to inadequate protection.

PIPING IDENTIFICATION systems help prevent confusion which may cause injury. The following classifications are recommended by the National Safety Council and the American Standards Association:

Fire protection materials and equipment (F)—red.
Dangerous materials (D)—yellow (or orange).
Safe materials (S)—green.
Protective material (P)—bright blue.
Valuable materials (V)—deep purple.

More specific identification can be accomplished by stenciling a painted legend on each pipe or by adding one or more color stripes in conformance with the standard code.

WARNING SIGNS and tags are particularly useful in locking out valves and boilers and chemical or petroleum pipelines, for warning employees against the starting of electrical equipment while repairs are being made, for preventing smoking in hazardous areas, for identifying high-voltage electrical equipment, and for instructing employees on the use of personal protective equipment or machine guards.

FIRST-AID FACILITIES. Large companies generally provide professional medical personnel with at least one or more registered nurses in attendance on all shifts. Smaller plants train competent first-aid attendants for treating very minor injuries and for rendering emergency care for serious injuries. The advice of insurance companies and private physicians should be sought in setting up first-aid facilities adequate for the exposure.

Minimum first-aid room equipment includes two chairs, a small table, running water, wash basin, soap, paper towels, waste basket, and a dustproof, well-equipped cabinet of needed materials. It is particularly desirable to provide cots made private with screens.

16. MACHINE GUARDING

Machinery must be reasonably well guarded before management can expect supervisors and employees to cooperate wholeheartedly in the safety program.

ESSENTIALS OF GUARDING. Built-in guards are always preferable to built-on guards. Temporary, makeshift, and flimsy guards not only give a false sense of security but also imply that management is not sincere in its efforts to protect the employee.

Excellent commercial guards are available from machine manufacturers and from distributors of safety equipment. Guards should be constructed in conformance with the codes of the American Standards Association and pamphlets of the National Safety Council. State codes generally represent only minimum requirements, and positive protection often requires more thorough treatment. A good guard: (1) Provides positive protection and prevents all access to the danger zone during operations. (2) Does not interfere with the operation of the machines; permits convenient and safe lubrication, adjustment, and repairs. (3) Is strong and durable, fire and corrosion resistant. (4) Does not have splinters or sharp edges and projections which might cause injury. (5) Is designed to prevent tampering and yet permit authorized removal and adjustment. (6) Does not interfere with the operator or cause eye strain or other discomfort.

GUARD MATERIALS. Sheet metal, heavy wire mesh, expanded metal, pipe, and structural steel are most commonly used (Fig. 1). Cast iron should not be used where

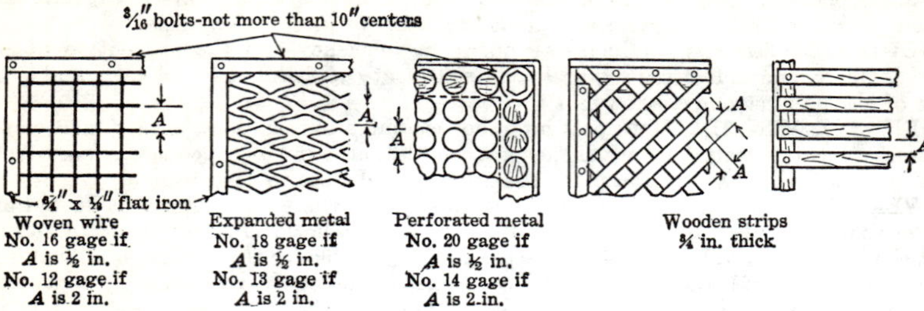

Solid sheet metal guards should be No. 22 gage or heavier. Supports should be 1 in. or 1½ in. angle iron or ¾ in. to 1½ in. ID pipe.

If the distance from moving part to guard is less than 4 in., opening *A* should not exceed ½ in. For distances more than 4 in. and less than 15 in., opening *A* should not exceed 2 in. except wooden slot guards where *A* should not exceed 1 in.

FIG. 1. Guard materials. (Reproduced by permission from *Accident Prevention Manual for Industrial Operations*, National Safety Council)

impact resistance is required; wood guards are permissible in exposures which might corrode metals; shatterproof glass and transparent plastic are used where visibility rather than strength is required.

Where resistance to rust or possible damage to tools is an important factor, guards of aluminum or other alloy are often used. Guards exposed to trucking operations and the movement of material are often protected by extra bumpers.

PRIME MOVERS AND TRANSMISSION MACHINERY GUARDS. Flywheels and other heavy machine parts which cannot be completely enclosed or otherwise guarded should be protected by standard 42-in. high railings placed not less than 6 or more than 20 in. from the flywheel or other moving part, (Fig. 2). Standard 6-in. toeboards are

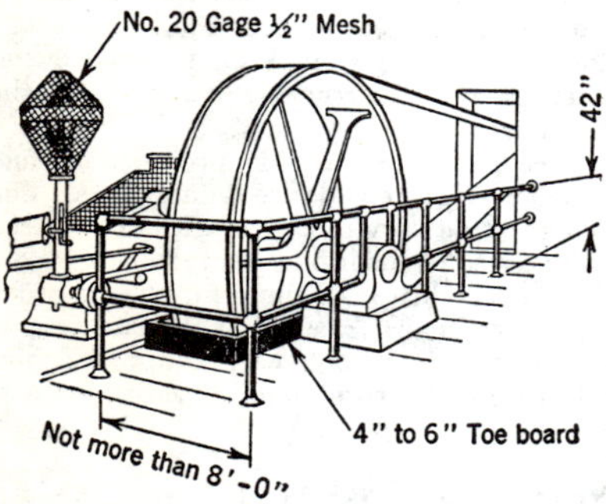

FIG. 2. Standard railing and toeboard, guarding flywheel and belt too large to be enclosed. Note governor enclosure. (National Safety Council, *op. cit.*)

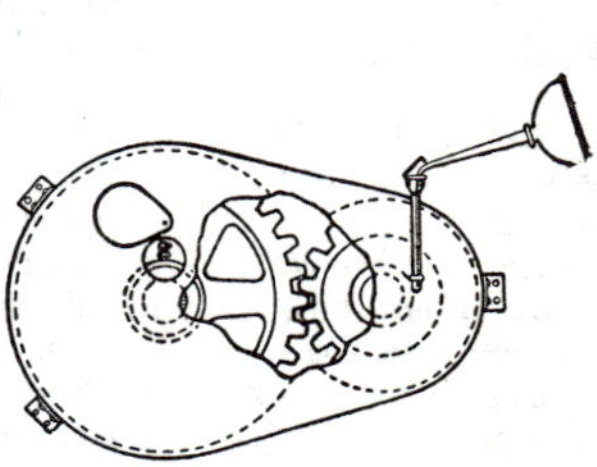

FIG. 3. Solid sheet-metal gear guard, illustrating arrangements for safe oiling. The swinging opening at the left provides access for a grease gun; the pipe through the guard at the right permits use of an oil can with safety. (National Safety Council, *op. cit.*)

necessary where there is a floor opening or pit. Governor balls, couplings, belts, and pulleys 6 ft. or less from the floor or working platform should be enclosed with a substantial guard.

Where machinery must run during oiling and when automatic lubrication is not feasible, remote-control devices provide the best protection. Oil holes in guards can be designed so as to permit access for the end of the oil can spout but not the hands (Fig. 3). Central oiling systems are becoming more widely used as they also speed the oiling.

BELTS AND BELT GUARDS. Standard codes call for the guarding of V-belts, round belts more than $1/2$ in. in diameter, and flat belts more than 2 in. wide, running at 250 ft per min or more. (See Figs. 4 and 5.) Larger belts are likely to cause serious injury if they break and whip about, and belts of all sizes with metal lacings may cause scratches and cuts, and even catch unprotected hair.

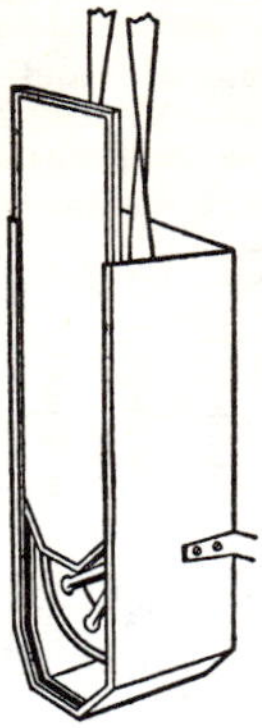

Fig. 4. Sheet-metal guard for vertical belt, with one side removable for access to moving parts. Note movable side falls automatically into place. (National Safety Council, *op. cit.)*

If belt dressing must be applied while the belt is in motion an opening may be built in the guard so that the dressing can be applied near the point where the belt *leaves* the pulley.

Leather and rubber belts tend to create static sparks which may ignite flammable liquids, dusts, or gases. Conductive dressings can be applied, but conductive rubber belts are better. Static brushes or collectors can also be used to drain away excessive static.

Horizontal belts should be totally enclosed if both the upper and lower runs are less than 7 ft from the floor or working area; the lower run should be enclosed if it is less than 7 ft above the working floor.

All belts traveling more than 1800 ft per min or located directly over aisles or work areas should have the lower run and pulley enclosed on the sides and bottom.

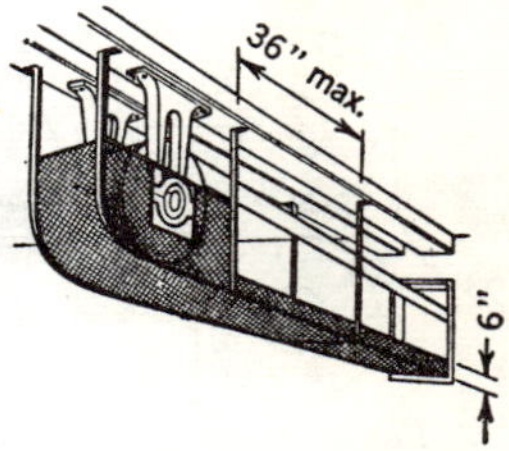

Fig. 5. Typical horizontal belt and pulley guard with removable sides. (National Safety Council, *op. cit.*)

Vertical and inclined belts should be enclosed so that no portion of the exposed belt is less than 6 ft 6 in. from the floor (Fig. 6). Pulleys lower than 7 ft above the floor should be enclosed up to that point.

To facilitate cleaning the floor, guards should be attached to the machine rather than supported from the floor.

Belt shifters with positive locks prevent accidental starting. Belt shipper poles for placing belts on moving pulleys obviate the need to shift belts by hand, and the latter practice should be prohibited. Belts should always be unshipped by throwing them off the *driving* pulley. "Do not use" tags should be used on belt shifters to prevent unauthorized starting of the machine while someone is working on it.

GEARS, SPROCKETS AND CHAINS. Power-driven gears and sprockets should be guarded by complete standard enclosures with no openings larger than $1/2$ in. if the guard is within 4 in. of any moving part, or not larger than 2 in. if the guard is between 4 and 15 in. from the moving part. Guards which enclose only the teeth of gears or which shield only the in-running nip point are accepted by some state codes. They are not so safe as complete enclosures, however, because they do not provide positive protection against persons falling or slipping into them.

Chains require the same guarding as belts.

SHAFTING, COUPLINGS, AND KEYS. Many serious accidents have occurred when hair, wiping rags, or loose clothing has caught on smooth shafting. Horizontal or vertical shafting within 7 ft of the floor should be guarded by fixed barriers or three-sided trough guards, the open end of which extends at least 2 in. beyond the nearest side of the shaft. Pulleys, shafting, and other moving parts located less than 15 ft above aisles or passages used by power-driven vehicles should be similarly enclosed.

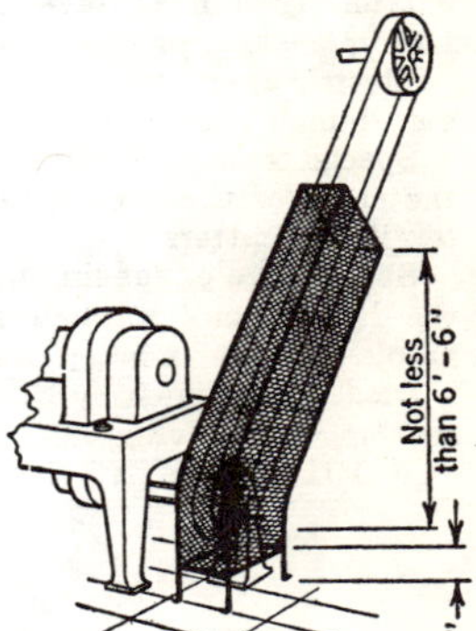

Fig. 6. Inclined belt totally enclosed to a height of 6 ft, 6 in. above floor. Note 6-in. clearance to permit cleaning floor. (National Safety Council, *op. cit.*)

Shaft ends which project beyond pulleys or bearings should be enclosed, and open keyways should be filled or enclosed. Projecting keys should always be cut off flush with end of shaft.

Bolts, nuts, and set screws on couplings should be of the countersunk flush type, or the coupling should be completely enclosed with a stationary guard.

POINT OF OPERATION GUARDS should be as nearly foolproof as possible. They may be of the following types.

Automatic or semi-automatic feed with enclosure so that no part of the body can reach the moving part. Example—punch presses, wood planers.

Two-hand trip, either electrical or mechanical, which require simultaneous pressure of two hands on separated buttons while the dangerous part is in motion. Example—punch presses, paper cutters, dough mixers.

Complete fixed enclosure which admits only the stock being fed. Example—metal square shears, punch presses.

Adjustable barrier which admits only the stock being fed (Fig. 7). Example—candy-cutting rolls.

Warning enclosures which admit stock thicker than the hand and, therefore, also the hands, but tend to deflect the hand or remind the operator that he is entering the danger zone. Example—circular saw and jointer guards.

Barrier with electric contact or mechanical stop actuating a mechanical or electric brake (Fig. 8). Approach of any part of the body to the danger zone stops the machine. Example—punch presses, rubber mills, paper calenders.

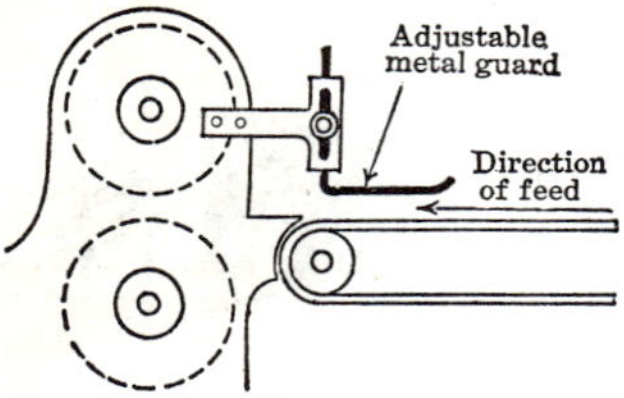

FIG. 7. An adjustable barrier, projecting horizontally, permits a feed opening large enough to admit stock but not the arm of the operator. (National Safety Council, *op. cit.*)

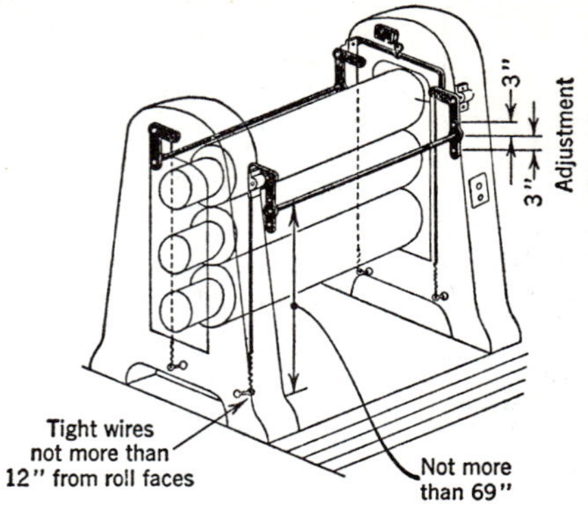

FIG. 8. A combination tight wire and rigid bar device, when touched, trips a contact and stops rolls before the operator is caught by them. (National Safety Council, *op. cit.*)

Enclosure with electrical or mechanical interlock. Opening of guard stops the machine and starting the machine closes the guard. Example—laundry extractors, foundry tumblers, punch presses.

Hand-removal guards of the sweep or hand-pullaway type connected to the machine to remove the hand from the danger zone. Example—hand-fed platen presses, punch presses, leather-embossing presses.

Limiting plunger travel to $^3/_8$ in. or less so that fingers cannot enter the danger point. Example—kick presses, punch presses.

Electric eye; interruption of the beam actuates a brake which stops the machine. Example—punch presses, embossing presses.

Special tools or handles on dies; long-handled tongs, vacuum lifters, or holders eliminate the need for the operator to place his hand in the danger zone. Example—leather and textile die cutters.

Special jigs or feeding devices which keep the operator's hand at a safe distance from the danger point through use of a fixed or movable enclosure built to fit the stock. Example—circular saws, jointers, punch presses.

Enclosures which by their complex shape or size of opening prevent the operator from reaching the moving part.

WOOD-WORKING MACHINERY. Circular table saws, jointers, band saws, swing saws, and sliding yoke circular saws are commonly used in most industries and cause a great many disabling injuries.

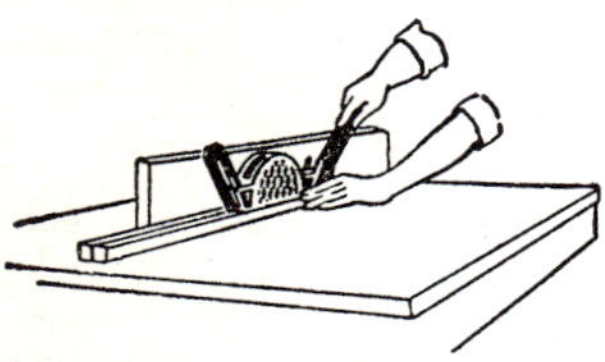

FIG. 9. Rip-saw guard with splitter and built-in anti-kickback device. For extra safety, the operator uses a push stick. (National Safety Council, *op. cit.*)

Circular rip saws require a lightweight hood enclosure of soft metal, a splitter to prevent the work from binding on the saw, and an anti-kickback device (Fig. 9). All parts of circular saws should be completely enclosed above and below the work table. Sometimes an exhaust system for the removal of chips serves to enclose the lower part of the saw. Swing saws or sliding yoke saws require a device which will limit the length of the saw travel and prevent bouncing. Saw guards should provide a view of the cut. The operator should stand to one side of the line of the cut, the floor surface should be free of obstructions and not slippery, and a push stick should be used for feeding short pieces of stock.

Jointers and planers. All jointers should have cylindrical safety heads with only enough clearance between the head and the table for the knife to clear. A fixed or adjustable guard should cover the unused end of the head, and an adjustable guard should completely cover the operating side of the head (Fig. 10). These guards may be of the swinging type or of the type that rises and rides on top of the work, as well as swinging in a horizontal plane. The latter type of guard gives better protection for surface joining.

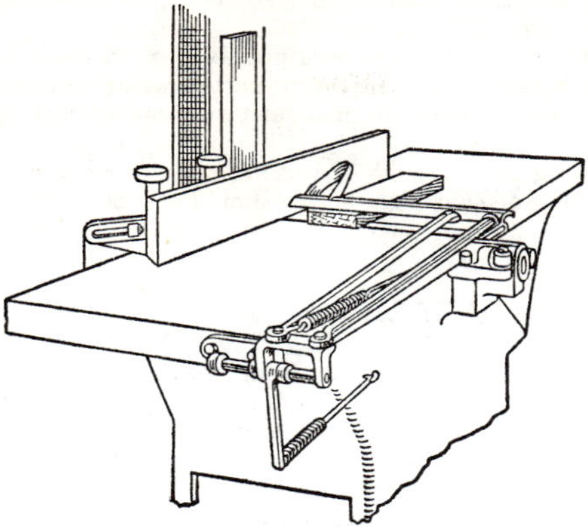

FIG. 10. This jointer guard swings and also rises over the work to permit safer surfacing. The knife on the other side of the guide is fully protected by an automatic guard. (National Safety Council, *op. cit.*)

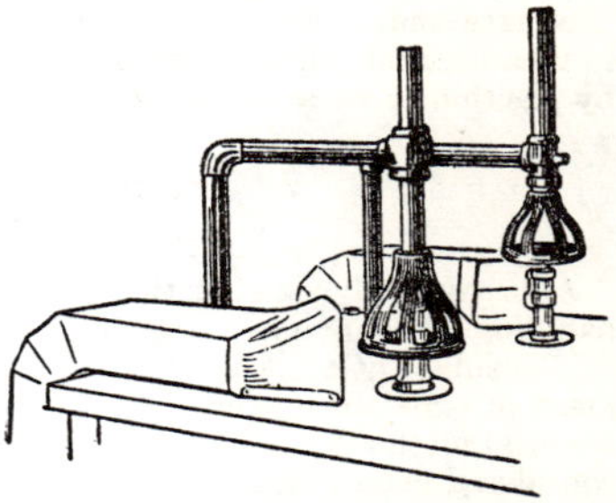

FIG. 11. Adjustable shaper head guards, with exhaust pipes to remove chips. (National Safety Council, *op. cit.*)

A small push block provides an additional margin of safety when feeding short stock. The hands should be kept well back on the work—never over the end of the stock as it is fed into the knife.

Band saw wheels should be completely enclosed. The only portion of the blade exposed should be that necessary for the thickness of stock being cut. An adjustable guard set close to the work will minimize the length of saw blade exposed.

Shapers require an adjustable ring or bell-shaped guard (Fig. 11). The extreme hazard of shaper knives being thrown can be reduced by the use of solid cutters, perfectly balanced knives, and safety collars which tend to retain the knife.

METAL-WORKING MACHINERY. Power Presses. Most of the methods of guarding points of operation listed previously can be applied to power presses. Length of run and type of job often determine the type of guard that can be used—an expensive, automatic feeding device or a simple barrier enclosure. With all points of operation guarded, power press nonrepeat devices provide an additional factor of safety.

Many plants have adopted the policy of building a guard for or into each die. This assures maximum production efficiency and gives the operator no excuse for neglecting to use the guard.

Foot treadles can be protected against accidental tripping through the use of sheet metal hoods (Fig. 12).

A regular system of inspection and maintenance for power-press guards is imperative, to insure safe condition and proper adjustment. The action of the guard should be tested before each run is started and, in the case of the two-hand pull back guard, for the individual operator's reach.

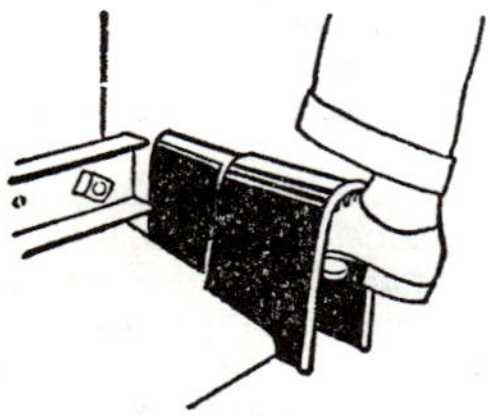

FIG. 12. Substantial hood guard over foot treadle prevents accidental tripping of power presses and similarly controlled machines. (National Safety Council, *op. cit.*)

Grinders and abrasive wheels require substantial metal hood guards and safety flanges. (See Fig. 13, p. 27-101.) The tool rest should be set within $^1/_8$ in. of the wheel to prevent fingers or stock being caught between the rest and the wheel. (See Fig. 13.)

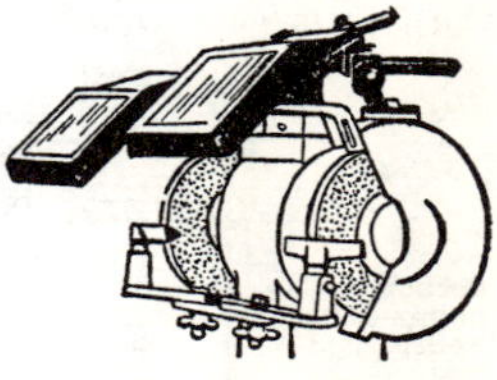

FIG. 13. A well-guarded grinder with safety glass shields and tool rests adjusted to within ⅛ in. of the abrasive wheels. (National Safety Council, *op. cit.*)

Safety-glass shields with built-in light to illuminate the point of operation reduce the danger of flying objects, but safety goggles or plastic eye shields should also be worn where there is danger of heavy impact. Gloves should not be worn on abrasive wheels or other moving machinery of similar type.

Abrasive wheels should be provided with exhaust hoods, particularly if the work involves grinding or polishing of material which gives off toxic dusts. See the American Standards Association Safety Code for the Use, Care, and Protection of Abrasive Wheels.

Lathes. All gears should be completely enclosed and, where possible, speed changes should be handled mechanically. Safety dogs without projecting parts should be used, and a guard over the chuck is advisable. Splash shields and drip pans will prevent cutting oil from reaching the floor and causing falls.

Square shears are comparatively easy to guard through the use of simple barriers of iron bars, expanded or slotted metal. The take-off side of the shears should also be guarded by position or by a barrier.

17. MATERIALS HANDLING

Approximately one-fourth of all compensated injuries result from unsafe methods of handling and moving materials. Of this number, about half are due to improper lifting.

The substitution of mechanical handling does not necessarily bring about an improvement in experience even though the exposure has apparently been reduced. Since mechanical methods are more subject to engineering and management control, however, it is considered good practice to revise procedures to take advantage of mechanical aids.

MANUAL HANDLING. Lifting injuries such as back strains may be caused by improperly lifting even a light object. The U. S. Department of Labor sets up the following limits for continuous or repetitive lifting operations: male workers—50 lb in compact form; female workers—25 lb in compact form.

Employees should be trained and reminded to lift according to the following principles: (1) Bend the knees and lift with the legs, not the back. (2) Keep the load close to the body and avoid lifting in an awkward or twisted position. (3) Get good footing and a solid grip. (4) Get help or use mechanical equipment for heavy or awkward loads. (5) Don't jerk the load or attempt to throw it.

MECHANICAL HANDLING. Two-wheel and four-wheel hand trucks are responsible for a great many serious injuries. Truck handles should not project into aisles, and truck handles should have metal guards to protect the hands and fingers.

In general, four-wheel trucks should be pushed; two-wheel trucks should be pulled up an incline and pushed down an incline. On level and smooth floors a decision to push or pull a two-wheel hand truck depends on the amount of traffic, visibility, size of the load, and convenience in unloading or loading the truck.

Electric or gasoline power trucks require the establishment of traffic rules which should be rigidly enforced. Riding of unauthorized persons on trucks must be strictly prohibited. The operator of high-lift trucks should be protected from falling objects by a strong overhead guard. Material to be piled safely must be cross tied, kept within safe floor load limits, and should not project into aisles or obstruct overhead sprinkler heads.

18. HAND TOOLS

Facilities for inspecting and maintaining hand tools should be provided by management. Employees must be constantly reminded (1) to use the right tool for the job, (2) to keep the tool in good condition and turn in defective tools, (3) to use tools correctly, and (4) to store and carry tools in a container, kit, or other safe place.

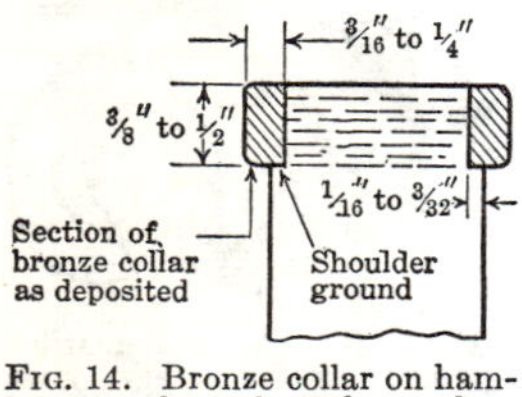

FIG. 14. Bronze collar on hammer-struck tools reduces dangerous mushrooming of tool heads. A bronze-ended tool also outlasts three ordinary tools without cracking. (National Safety Council, *op. cit.*)

Hammerstruck tools, such as chisels and punches, are known for their tendency to develop mushroom heads which frequently chip off, causing the loss of an eye or other serious injury. At the first sign of mushrooming, tool heads should be either dressed or safe-ended by the addition of a bronze collar, (Fig. 14). A regular schedule for dressing or bronze tipping is advisable.

PORTABLE ELECTRIC TOOLS. Potential accident hazards are electric shock, burns, cuts, and injuries from flying particles or dropping tools. The preventives are selection of the right tool, regular inspection and maintenance, and training in proper use.

Grounding of noncurrent-carrying metal parts is particularly important. The best ground consists of a third wire within the power cord, one end connected to the metal body of the tool and the other to a three-prong socket which is automatically grounded when the circuit is completed (Fig. 15).

Extension cords should be checked for breaks in the insulation, for loose plugs, and for other hazards which might cause shock. For work in extremely wet areas or inside steel tanks it is advisable to use 6-volt tools and a portable transformer.

Moving parts of portable power saws, wire brushes, and abrasive wheels should be enclosed in standard guards. Eye protection is important where there is danger of sparks and flying objects.

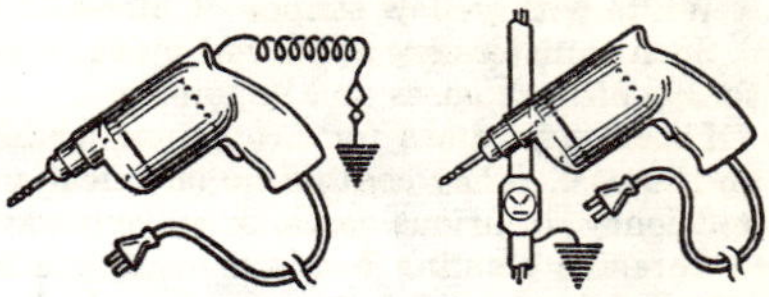

FIG. 15. Methods of grounding portable power tools. At left, a separate ground wire to metal casing; at right, the ground wire is built into the extension cord and a three-prong plug automatically grounds the tool. (National Safety Council, *op. cit.*)

19. PERSONAL PROTECTIVE EQUIPMENT

Personal protective equipment should not be regarded as a "cure-all" or substitute for built-in safety. However, it does have its place as secondary protection, particularly where no other protection is possible or where the operation is of such short duration that other methods are not practicable.

EYE PROTECTION. Hardened glass goggles, plastic eye shields, and plastic face shields provide protection against sparks, dust, flying objects, glare, and splashes of corrosive chemicals or molten metal. See the American Standards Association Code for Protection of Heads, Eyes, and Respiratory Organs, Z-2, for detailed specifications and classes of equipment suitable for each exposure.

Lenses for protection against impact are required to withstand without shattering the impact of a $^7/_8$-in. steel ball dropped from a height of 50 in. The research report, *Plastic Eye Protectors*, released by the National Safety Council, indicates that most plastics had an impact resistance comparable to that of hardened glass. However, there is a possibility of optical distortion due to surface scratches on the comparatively soft plastic.

Most industrial companies furnish their employees, without charge, plano goggles of the spectacle or eye-shield type. A few companies provide for complete eye examinations and furnish prescription-ground hardened glass goggles, but the majority of companies assume only a portion of the extra cost of prescription goggles.

Some companies report that a rule requiring *all* employees to wear eye protection is a good investment; others limit the use of goggles to certain hazardous areas or to specific operations.

Plastic face shields give an excellent sense of security, particularly when worn over regular eye protectors. They are used on spot welding operations, babbitting, and in handling corrosive chemicals where full protection of the face is desired.

RESPIRATORS. Where local exhaust cannot be employed to provide respiratory protection, gas masks, hose masks, and respirators protect against harmful fumes, mists, gas, vapors, and dusts. Respirators should be of a type approved by the U. S. Bureau of Mines. These are the types of respiratory equipment:

Air-supplied suits for complete protection of the body in atmospheres that are extremely corrosive to the skin.

Hose masks with complete face piece, used where there is an inadequate supply of oxygen or a dangerous concentration of dust, vapor, or gas. The large-diameter, flexible hose permits breathing even though the hand- or motor-operated, low-pressure pump is not functioning.

Oxygen-breathing apparatus, which provides a self-contained oxygen or compressed air tank, or oxygen-generating mechanism, for work in atmospheres deficient in oxygen where mobility is required.

Air-line respirators consisting of a small mouth and nosepiece connected to a small airline which supplies air for work in atmospheres that are not immediately hazardous to life.

Gas masks and cartridge respirators, for protection against solvent vapors and injurious gases provided there is an adequate supply of oxygen. The chemical canister or cartridge is identified by different colors for different hazards:

Black—Organic vapors.
White—Acid gases.
Yellow—Organic vapors and acid gases.
Green—Ammonia.

Brown—Organic vapors, acids, gases, and ammonia.

Red—All industrial gases, including carbon monoxide, smoke, and fumes.

White with green stripes—Hydrocyanic acid gas.

White with yellow stripes—Chlorine.

Such canisters are approved by the U. S. Bureau of Mines for use in concentrations up to 2% of both gases and vapors.

Filter respirators for protection against nontoxic dusts, including silica, toxic dusts, and fumes. They contain no chemical and therefore are of no value if there is an oxygen deficiency, injurious gases, or solvent vapors.

Abrasive blasting hoods or complete helmets and hoods provided with a window and supplied by air at low pressure. Such equipment is commonly used by sand blasters.

HEAD AND EYE PROTECTION. Hard hats, commonly used in the construction, mining, shipbuilding, quarrying, and meat-packing industries are built of aluminum or lightweight composition which withstands a considerable impact without possibility of injury to the wearer.

Hair protection for women is provided better by safety caps with a high crown and stiff peak than by bandanas, turbans, and hair nets.

Acid hoods of rubber or synthetic resin protect the entire head and shoulders against splashing and dripping of acids and caustics. In such exposures, rubber gloves, aprons, and boots are also called for.

FOOT AND LEG PROTECTION. Work shoes with a built-in, steel toe cap capable of supporting a static load of 2500 lb or the impact of a 50-lb weight dropped 1 ft are readily available at most retail stores at prices comparable to good-grade work shoes. Either such shoes or steel toe caps which fit over ordinary shoes should be worn when handling heavy objects.

Leggings of chrome leather, asbestos, or flame-resistant canvas are necessary for foundry workers and others exposed to splashes of molten metal and sparks.

HAND PROTECTION. Hand leathers, pads, or gloves protect the worker against cuts, abrasions, and puncture wounds when handling rough lumber, sheet metal, castings, and other material which might injure the hands. Rubber gloves must be worn when handling corrosive chemicals and for work on high-voltage electrical equipment.

20. FIRE PREVENTION AND PROTECTION

MATCHES, improperly used, and smoking are outstanding fire causes according to the National Fire Protection Association. Smoking and the use of open flames should, of course, be absolutely prohibited in hazardous areas such as woodworking shops, textile mills, certain chemical plants, and wherever flammable liquids, vapors, or gases may be present. If smoking is prohibited in a plant or in certain departments, employees should be granted permission to smoke in designated fire-safe areas.

OTHER FIRE CAUSES are overheated bearings, belt friction, insufficient insulation for stoves, heaters, and electrical appliances, spontaneous ignition, static electricity, and defective electric wiring.

GOOD PLANT HOUSEKEEPING is essential to fire prevention as well as accident prevention. A regular system of rubbish and waste disposal should be set up, and approved type containers should be provided for the storage and disposal of all combustible material and rubbish.

FIRE-SAFE CONSTRUCTION. A building constructed of "fireproof" materials does not necessarily protect the occupants. Adequate fire exits should be provided, should not be blocked, and should be identified by fully visible exit signs and lights. Fire-alarm systems, adequate fire protection equipment, and a carefully prepared plan for evacuating occupants in an emergency are necessary adjuncts to safe construction.

Both commercial and home-made solutions are available for flameproofing fabrics. Wood may be made more fire resistant by coating or impregnating it with chemical solutions.

AUTOMATIC WATER SPRINKLING SYSTEMS are recognized as the best means of extinguishing fires. Systems are the wet, dry, thermostatically operated, and open. The wet system must be protected against freezing and also must be carefully inspected and maintained to insure operation in an emergency. Sprinkler heads are commonly rated at 155 to 165 degrees, with others rated up to 500 F.

FOAM AND CARBON DIOXIDE systems may be installed where water sprinklers cannot be used. Water spray and fog nozzles provide good protection against flammable liquids and may even be used safely by experts on live electrical equipment such as transformer banks.

PORTABLE FIRE EXTINGUISHERS. The right type of portable extinguishers should be available for each of the three classes of fires.

Class A (wood, paper, rubbish, etc.). Soda-acid, water pump, water and gas cartridge, foam.

Class B (oil, gasoline, paints, etc.). Foam, loaded stream, carbon dioxide, vaporizing liquids, dry chemical.

Class C (live electrical equipment). Carbon dioxide, vaporizing liquid, dry chemical.

Water base extinguishers must be protected against freezing and cannot be used on electrical equipment because of the shock hazard.

21. ACCIDENT REPORTING

These types of disabling injury are included in the computation of accident rates: death, permanent total disability, permanent partial disability, temporary total disability (commonly called a *lost-time accident*).

THE ACCIDENT FREQUENCY RATE is the number of disabling injuries per million man-hours worked. The average frequency rate for all industries in recent years has been between 10 and 15. Certain industries, formerly considered hazardous, have established frequency rates of considerably less than 10.

THE ACCIDENT SEVERITY RATE is based on the number of days of disability plus standard time charge for deaths, permanent total, and permanent partial disabilities. The severity rate is stated as the number of days of disability, including such time charges, per thousand man-hours worked. The average severity rate for all industries is approximately one.

22. TRAINING AND SUPERVISION

SELECTION AND TRAINING of a new employee requires an accurate knowledge of the employee's physical and mental abilities and of the job. Physical examinations should be for correct job placement, rather than for "weeding out." Many large companies also give aptitude tests, to place new employees better.

Safety training starts with the interview of the applicant and continues until he leaves the company. It is handled as a part of production training, with the foreman constantly impressing the employee with the importance of working safely.

Successful training includes: (1) Planning and preparing for a safe job. (2) Telling and showing the employee how to do the job safely. (3) Supervising and encouraging the employee while he does the job. (4) Following up frequently to check unsafe practices.

THE FOREMAN is the key man in accident prevention. He is in direct contact with the causes of accidents at the job level. He can discover and remove unsafe conditions before they cause injury, and he represents top management in carrying out company policies.

The principal safety duties of the foreman are: (1) To make safety an integral part of each job and to instruct new and old employees in the right way to do the job. (2) To detect and correct both unsafe conditions and unsafe acts in his department. (3) To insist on prompt reporting of injuries and to furnish accident reports to management. (4) To set a good example for his employees.

The foreman, who often must first be convinced of the need for safety, should receive special training in the more technical phases of machine guarding, personal protective equipment, and health hazards, and in the technique of training his employees. Special aids for training the foremen and for his use in training employees are available from many sources.

INSPECTION SYSTEMS vary but always require planning to avoid wasted time and effort. A thorough and very complete appraisal of physical hazards is necessary at the start of the program. Often very complete inspections are made annually to furnish the management with an accurate picture of the progress made during the year so that long-range plans may be made for the removal of hazards. Such inspections may be for fire hazards, fire equipment, housekeeping, machine guarding, and the like. Less detailed inspections and reports are often made monthly.

Inspections also can serve to educate employees and supervisors, particularly if made by a rotating committee. Then the forms or check lists are of the "reminder" type and are valuable for checking progress over daily, weekly, or monthly intervals.

Inspection forms must be appropriate for the type of inspection desired and should be arranged so that progress on recommendations can be easily checked. It is extremely important that action be taken promptly on all recommendations and that necessary delays in compliance be explained.

The forms may give a running or continuous report of plant changes or furnish information on new processes or equipment. The common fault of reporting unsafe physical conditions and overlooking unsafe acts should be guarded against.

Inspections may be made by specially trained men such as boiler or elevator insurance inspectors, by the company safety engineer, by plant maintenance men, and by special committees.

23. SOURCES OF INFORMATION

NATIONAL SAFETY COUNCIL, 20 North Wacker Drive, Chicago 6, Ill., is a nonprofit, nonpolitical association, comprising thousands of industrial and other members who work cooperatively for the solution of their common accident problems. The Council has over twenty separate industrial sections and engages also in traffic, farm, school, and home safety work. It provides statistical, consultation, and library service as well as "Safe Practices" pamphlets, magazines, posters, and other publications; it conducts training courses, research, and has numerous local chapters and affiliated local safety councils.

AMERICAN SOCIETY OF SAFETY ENGINEERS, 20 North Wacker Drive, Chicago 6, Ill., is composed of several thousand professional safety engineers with chapters located in every section of the United States. It sponsors safety codes and research projects.

LOCAL SAFETY COUNCILS are affiliated with the National Safety Council and offer a localized consultation service, facilities for exchange of information, and joint educational programs.

AMERICAN STANDARDS ASSOCIATION, 70 E. 45 St., New York 17, N. Y., is a federation of national associations and government departments dealing with standardization. Through its member bodies it formulates standard codes which are often used as the basis for city and state health and safety codes.

NATIONAL FIRE PROTECTION ASSOCIATION, 60 Batterymarch St., Boston 10, Mass., is the national authority on fire prevention and protection. It publishes the Crosby-Fiske-Forster *Handbook of Fire Protection*, the *Quarterly*, technical reports, numerous posters, and other material.

NATIONAL BOARD OF FIRE UNDERWRITERS, 85 John St., New York 7, N. Y., maintained by the stock fire-insurance companies, provides engineering, statistical, and educational information, aids in the promotion of safe building construction, and publishes numerous technical pamphlets.

ASSOCIATION OF CASUALTY AND SURETY COMPANIES, 60 John St., New York 7, N. Y., represents the stock casualty companies. It publishes numerous safety booklets and serves as a clearing house on accident prevention information for its members.

AMERICAN MUTUAL ALLIANCE, 919 North Michigan Ave., Chicago, Ill., represents the mutual casualty companies and provides similar service.

OTHER SOURCES OF INFORMATION. Insurance carriers, state and federal agencies, consulting safety engineers, and safety-equipment manufacturers are in a position to furnish helpful materials and consultation service.

DUST COLLECTION

By R. B. Foley

24. CHARACTERISTICS OF DUSTS

CLASSIFICATION. Solid particulate matter in air is generally classified as dusts, fumes, or smoke.

Dusts are mechanical dispersoids formed by the reduction of larger masses into fine particles by such processes as grinding, crushing, and drilling.

Fumes are condensed dispersoids formed by condensation or sublimation from the vapor phase.

Smoke is the term generally applied to the special form of fumes produced by combustion.

Size of such particles is usually expressed in terms of the particle diameter in microns. A micron equals $^1/_{1000}$ mm or approximately $^1/_{25,000}$ in.

Dusts are usually particles 1 micron and larger in diameter, fumes range from 0.10 to 1.0 micron in size, and smoke particles vary from 0.001 to 0.25 micron. Fumes very often agglomerate into larger particles which can be considered dusts in their behavior and treatment.

PHYSICAL CHARACTERISTICS OF DUSTS. The important physical properties of dust particles which affect their behavior in air or any other fluid are size, shape, and weight. Although it has been a common practice to express the behavior of dust particles in an air stream as a function of their size or diameter, the other two properties are of equal importance. The weight which must be considered is the true density, or specific gravity, of the particle; not the bulk density of the material.

MEASUREMENT OF DUST CHARACTERISTICS. Screens. The size of dust particles is most commonly measured by grading with standard screens or sieves. These screens are made of woven wire cloth with openings of definite and uniform sizes. The mesh size equals the number of wires per linear inch, so the number of openings per square inch equals the square of the mesh number. The Tyler Standard Screen Scale consists of screens in which the size of the openings vary in the fixed ratio of the square root of 2 or 1.414. Table 1 illustrates the proportions of common screens. The finest standard screen, however, has 400 meshes per inch, and many industrial dusts consist principally of particles which will pass through this screen.

Table 1. Properties of Common Screens

Screen mesh	20	40	60	80	100	150	200	325	400
Width of mesh opening, inches	0.0328	0.0150	0.0097	0.0069	0.0058	0.0041	0.0029	0.0017	0.001425
Approximate microns across net opening	833	380	246	175	147	104	74	43	36

Microscopic Examination. The measurement of particle size by means of microscopic examination is one of the simplest methods for measuring subsieve-sized particles. It also provides a means for determining the shape of dust particles. Since most dust particles are very irregular in shape, the determination of the average linear dimension is rather difficult. This method is also limited by the difficulties involved in dispersing a truly typical sample within the small field which can be studied under a high-powered microscope.

Elutriation and Sedimentation. The two methods now most frequently used for analyzing subsieve dusts depend upon their aerodynamic characteristics. They are the elutriation and sedimentation methods. Since both these methods involve the determination of the terminal velocities of the particle, they measure the combined effect of all three of their physical properties upon their behavior in air or any fluid. The terminal velocity of a dust particle is the ultimate velocity which it will obtain when falling through a given quiescent fluid. That is, it is the velocity at which the resistance of the fluid to the motion of the particle equals the gravitation force on it, so that acceleration ceases and the velocity becomes constant.

In the sedimentation methods, the terminal velocity, or rate of fall, is measured in a quiescent fluid, usually a liquid. The analysis is made by determining the percentage, by weight, of the sample which falls through a certain distance in various time intervals.

Elutriation methods are the reverse of sedimentation but are based on the same laws and thus measure the same characteristics. The upward speed of a fluid column which will just sustain a particle, so that it will neither rise nor fall, is equal to the terminal velocity which the particle would obtain in free fall in the same fluid. In elutriation, the analysis is made by carrying upward, in a fluid stream, those particles which have a terminal velocity less than the upward velocity of the fluid. Air and water are the fluids most commonly used. Regardless of the fluid used for the analysis, terminal velocity is generally expressed in terms of the particle velocity in standard air, since most dusts are contained or handled in air.

Although it has been quite common practice to express the results of analyses made by the sedimentation or elutriation methods in terms of particle diameters in microns, the transition from terminal velocity is usually based on the assumption that the particles are spheres of uniform density. Since most dusts are very irregular in shape, and a heterogeneous mixture of materials of different specific gravities, these assumptions are not usually accurate. The relationship between terminal velocity and particle size is expressed by Stokes' law:

$$V_m = \frac{\alpha_v}{K} g d^2 \left(\frac{\rho - \rho_0}{\mu} \right) \tag{1}$$

when V_m = terminal velocity of the particle; d = diameter of the particle; ρ = density of the particle; ρ_0 = density of the fluid medium; μ = viscosity of the fluid medium; g = acceleration due to gravity; α_v = a constant, depending on the shape of the particle, and such that $\alpha_v d^3$ measures the volume of the particle; and K = a constant which varies as the particle shape and is 3π for spheres.

For solid spherical particles the particle diameter in microns when cgs units are used is:

$$d = \sqrt{\frac{18\mu V_m \times 10^8}{g(\rho - \rho_0)}} \tag{2}$$

Figure 1 is a chart giving this relationship for solid spherical particles of various specific gravities.

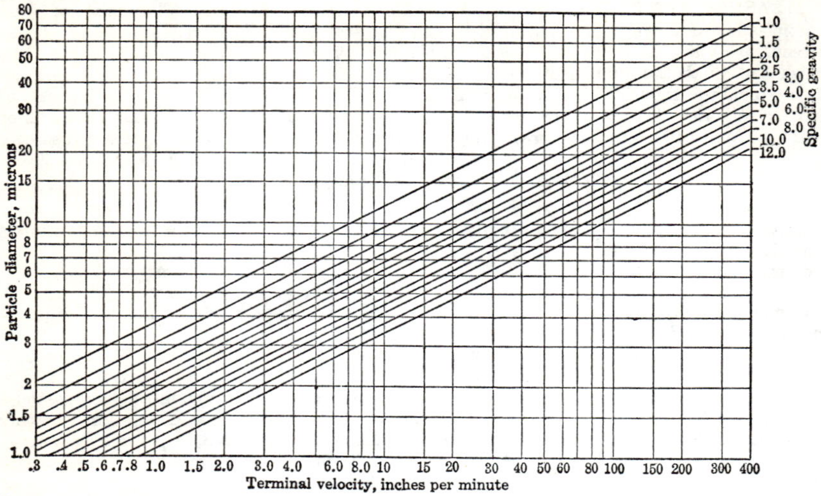

FIG. 1. Relationship between particle size and terminal velocity for solid spherical particles, based on Stokes' law.

Within the range generally encountered in industrial dusts, the cumulative analysis in terms of terminal velocity is usually a straight line when plotted on log-probability scales. Figure 2 illustrates such plots of several common dusts.

Miscellaneous Methods. There are several other methods used to measure average particle size, most of them dependent on the surface area of the particles. Those falling in this category are the photoelectric or turbidimetric methods, the permeability method, and the adsorption methods. Although the results obtained by these methods are of value in determining properties that affect chemical reactions, they do not measure the physical characteristics that affect their behavior in an air stream.

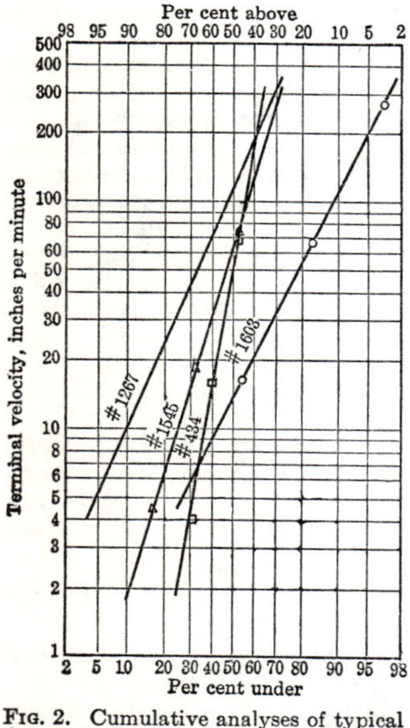

FIG. 2. Cumulative analyses of typical dusts, plotted in terms of terminal velocity on log-probability scales.

TYPES OF DUSTS. Dusts may also be classified in groups determined by the types of material from which they are produced.

Toxic Dusts. In addition to dusts produced from poisonous materials, these include also dusts that are injurious to health in their finely divided state, although the material from which they are formed is not harmful. Dusts that contain a large percentage of free silica are of this type since they produce silicosis. Particles less than 10 microns in size appear to be most dangerous in these types of dusts because they are not filtered out in the nasal or throat passages.

Explosive Dusts. All fine inflammable dusts are explosive when present in air at certain concentrations. At one time, only dusts produced from carbonaceous materials such as coal, grain, starch, sugar, and wood were considered explosive. Recent experience and laboratory tests, however, have shown that many metallic and plastic dusts may form explosive mixtures with air. Zirconium, magnesium, aluminum, and titanium are among the most explosive of the metal powders, whereas many of the dusts from synthetic resins, especially those of the phenolic type, must also be included in this group. Although generally dust concentrations below 10 grains per cubic foot of air are not explosive, tests have proved that very fine dust of the more inflammable types will form explosive mixtures at much lower concentrations. When dust of this type is handled it is very important to eliminate all possible sources of ignition. All metal parts should be grounded, and the rotating parts of all machinery, such as fans, should be made of sparkproof material.

Nuisance Dust. A large number of industrial dusts fall in this group. Although they are not directly injurious to health or safety, their control and collection result in improved working conditions, lower maintenance, longer equipment life, and improved public relations. In many cases the dust collected is of no value. However, research has led to the discovery of uses for some of these dusts which pay for at least a portion of the cost of collecting and handling them.

Products. In many industries the final product, or some of its component parts, are handled in a powdered form and their separation from air streams is an important part of the process. Recent developments in powder metallurgy and in spray-drying processes have added many materials to this group.

By-products. Many valuable by-products are salvaged in the form of dusts or powders, and, in some cases, dusts which were formerly considered a nuisance have become as valuable as the original product. Powdered catalysts might also be included in this group, since although not a part of the final product their control is essential to its processing.

25. TYPES OF DUST COLLECTORS

GRAVITATIONAL. Settling chambers, which are the oldest and simplest form of dust separators, depend on gravitational forces. The velocity of the air or gas is reduced to the point where the dust particles will settle out of the air stream. Figure 1 illustrates the low velocities necessary to settle out the finer dust particles. Turbulence and eddy currents also tend to suspend particles that would theoretically settle out, and the size of a unit that would be required to give velocities in the streamlined flow region would be prohibitive.

To be effective, not only must the cross-sectional area be large enough to give a low velocity, but the length must be great enough to prevent the particles from being carried to the outlet by their inertial force. Although relatively inefficient, settling chambers are sometimes sufficient when the dust is coarse and heavy. Their principal use, however, is as preliminary separators, when the concentration is high, to reduce the dust loading before the air or gas passes on to a more efficient type of dust collector. They are also sometimes used to obtain a rough preliminary grading of the material.

INERTIAL. Baffle types of inertial separators are a modification of the settling chamber in which a baffle or a series of baffles is employed to cause a sudden change of direction in the dust-laden air stream and thus throw out the dust particles due to their greater inertia. Although they can be designed to occupy somewhat less space than settling chambers without baffles, they are generally used for similar applications. Care must be taken in the design and location of the baffles so that the dust is not re-entrained in the air-stream because of the eddies and turbulence set up by the baffles.

CENTRIFUGAL. In centrifugal-type collectors, radial acceleration or centrifugal force is utilized to separate the dust particles from the air or gas stream. The centrifugal settling velocity, which is the outward or radial velocity of a dust particle in a centrifugal collector, can be expressed by the following equation for particles within the Stokes' law range:

$$V_c = \frac{\alpha_v}{K} d^2 \left(\frac{\rho - \rho_0}{\mu}\right) \frac{V_t^2}{R} \tag{3}$$

when V_c = centrifugal settling velocity, V_t = tangential velocity of the particle, and R = radius of the circular path of the particle.

A comparison of this equation with eq. 1 for the terminal velocity, V_m, shows that $V_c = V_m \times V_t^2/Rg$. The factor V_t^2/Rg, called the "separation factor," is a measure of the increase in separation velocity obtained by centrifugal action as against gravitational action.

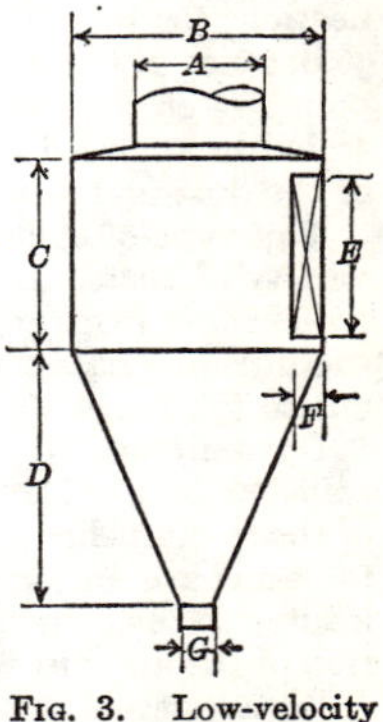

FIG. 3. Low-velocity cyclone.

Low-velocity Cyclones. The most familiar type of centrifugal collector is the common or low-velocity cyclone shown in Fig. 3. In this type of cyclone, the dust-laden air enters the cylindrical body of the collector tangentially at a high velocity and is thus given a rotary direction. The clean air forms a vortex and travels up and out through a central discharge pipe at a low velocity, while the dust moves down and is removed at the lower end of the cone. Since the velocity is reduced after it once enters the collector, the dust is separated from the air by a combination of centrifugal and gravitational forces.

This type of cyclone is generally used for collecting shavings, chips, and sawdust from wood-working equipment, grains, and coarse heavy particles. There are a great number of designs and modifications of the low-velocity cyclone, but generally the proportions are within the limits shown in Table 2. The inlet velocity should be at least 3500 ft per

Table 2. Proportions of Low-velocity Cyclones

Dimension (See Fig. 3)	*A*	*B*	C	*D*	*E*	*F*	*G*
Ratio to outlet diameter	1	2	1 to 1.5	1.75 to 3	1 to 1.5	0.2 to 0.3	0.2 to 0.3

Outlet area = 3 to 4 times the inlet area.

min and the outlet velocity not more than 1000 ft per min. The resistance through this type of collector is usually between one and three velocity heads, based on the inlet velocity. Although these collectors may be used on either the exhaust or pressure side of the fan, most of them are operated under pressure. When operated under suction it is important that the dust outlet be sealed against any inward leak of air.

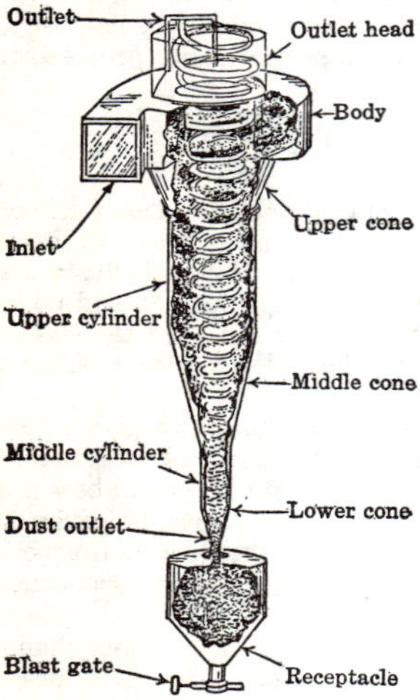

FIG. 4. High-velocity vortex collector. (Courtesy of American Blower Corp.)

High-velocity Cyclones. There are several patented designs of centrifugal cyclones which, although somewhat similar in principle to the low-velocity cyclone, are much more efficient for the separation of fine dusts, as they develop considerably greater centrifugal force. The dust-laden air or gas stream usually enters the collector at a lower velocity, but the velocity is continually increased as it rotates down through the unit until it reaches the vortex, similar to the vortex of a natural cyclone, near the dust outlet where the cleaned air starts rotating up in the inner air column. Since at this point the tangential velocity is high and the radius of the path small, a great centrifugal force is developed to separate the dust from the air.

The swirl or rotation of the gases may be started by an involute scroll, such as in the collector shown in Fig. 4, by spiral vanes, or by one or more tangential slots or inlets. Since the air velocities are high, it is important that the internal turbulence and eddies be eliminated to prevent re-entrainment of the dust particles. The resistance through the collector is usually higher than that for the low-velocity type of cyclone, and, within the limits of proper design, better collection of fine dust is obtained with the designs that have the higher resistance to air flow.

Since the collection efficiency decreases as the size of the unit increases, two or more collectors are often used in parallel when large volumes of air or gas are to be handled. Many high-velocity cyclones are made in multiple units with common inlet and outlet connections for this purpose.

Because of the vortex that is formed near the dust outlet, there is a high suction at this point. For this reason, it is very important that the dust outlet be sealed. It can be sealed by a closed receptacle of adequate storage capacity or, when continuous discharge of the dust is desired, by an air-tight dust seal.

This type of collector is widely used in industrial application and is suitable for a great variety of dusts. Even though the collection efficiency is high, it is not recommended that the exhaust from any centrifugal collector be recirculated into an occupied building, since the ultrafine particles which are not separated from the air are usually those most injurious to the lungs.

Dynamic fan-type or rotary-type dust collectors consist of a combination fan and dust collector unit. The principle of operation of this type is illustrated in Fig. 5. The blades of the fan impeller are designed so that the dust particles are forced outward by centrifugal force and are concentrated at the tip of the blade, where they are ejected through an opening leading to the dust chamber, while the cleaned air is blown out the fan outlet. A small portion of the air conveys the dust into dust chamber and then returns to the impeller circuit.

Water sprays are sometimes added to this type of unit when the dust can be handled as a sludge. The wetted impingement surface, thus provided by the blades of the impeller, results in an increased efficiency with some types of dusts. Since the fan and collector are combined, minimum space is required. All the dust is carried over the surface of the impeller blades at high velocities, and wear is a factor when handling abrasive dusts.

Filters. Investigation of the theory of air filtration has shown that contact and adhesion of the dust particles to the surface of the filtering media are probably more important factors than the straining or sieving action. Filters are thus of two types. (1) Those in which the air is passed through a closely woven or slightly porous material are the type most commonly used in industrial applications. (2) The impingement type, which have a more open but tortuous path for the air flow, are generally used in air-conditioning and ventilation systems where the dust concentration is low.

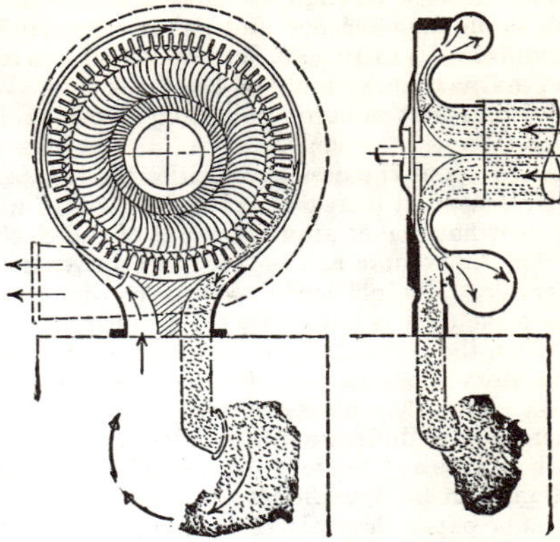

Fig. 5. Dynamic fan collector. (Courtesy of American Blower Corp.)

Cloth is the filtering medium used most extensively in the industrial units. The type of media used limits the maximum gas temperature for which this type of collector is suitable. Cotton fabrics can be used up to a temperature of 180 F and wool up to 220 F. Asbestos and glass cloth have been used for temperatures as high as 450 F, but their practical application is limited by their mechanical strength. One type of filter suitable for small volumes of air, when relatively high pressures are available, uses porous stainless steel as the filtering medium.

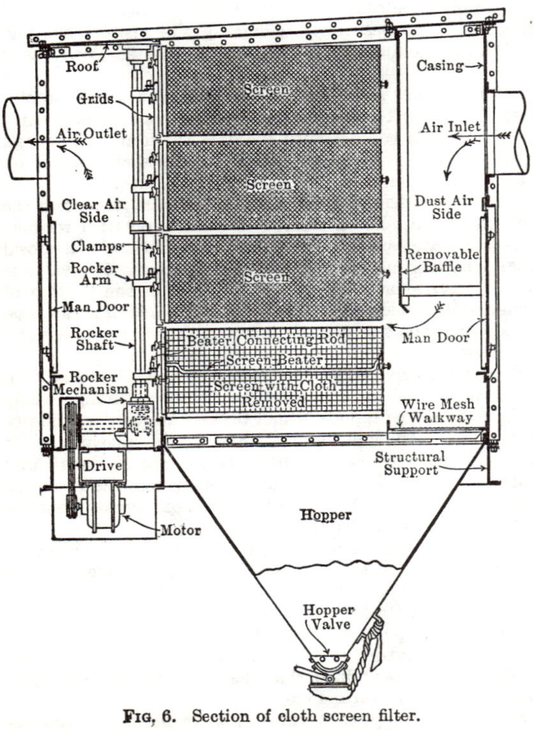

Fig. 6. Section of cloth screen filter.

Bag-type filters consist of cloth socks or tubes suspended vertically in a chamber and connected to an inlet header so that the air passes through the cloth to reach the outlet of the chamber. As the dust fills the pores and collects on the surface of the cloth, it is necessary to provide some means for removing the collected dust. It is usually removed by a mechanical means of shaking, or flexing, the bags so that the dust shakes loose and drops down into the hopper below. This is generally done when the unit is shut down, or the air flow is dampered off from the unit being cleaned. This shaking can be either a manual or motor-driven operation, and, in some cases, it is done automatically when the filter resistance has reached a certain value. In one type of bag collector, this cleaning is done by a moving nozzle which blows a small volume of air in the reverse direction through the cloth.

Cloth screen filters are a modification of the bag type in which cloth panels are supported by wire or metal frames. The frames are connected to a shaking or rapping device for dislodging the collected dust from the cloth. With this design more cloth area can usually be obtained in a given sized unit. The coarser heavier dust is often settled out in a chamber ahead of the screens. Figure 6 shows a typical cross section of this type of filter.

The velocity through the cloth depends on the type and concentration of the dust, as well as on the efficiency and power requirements. Although the capacity of cloth filters varies from 1/2 to 10 cu ft per min per square foot of cloth area, for most industrial application the maximum should be 4 cu ft per min. With most materials, lower velocities through the cloth result in better collection efficiency, lower maintenance and operating costs, and longer life.

Although at the design velocities the resistance through the cloth is quite low when the cloth is clean, it increases as the dust builds up on the surface. The efficiency of collection is somewhat higher after the pores of the cloth have filled with dust; however, the reduction in air volume as the resistance increases necessitates cleaning or shaking at regular intervals. The efficiency of properly designed cloth filters is generally very good, and, where the dust is not toxic, the cleaned air can sometimes be returned to the building.

When the concentrations are high, centrifugal-type collectors are often used as primary separators ahead of cloth filter, to reduce the frequency of cleaning and increase the life of the cloth. In addition to the limitation of gas temperature, cloth filters are not suitable when there is moisture in the gases, or with some types of dust which tend to matt on the cloth surface and prevent the normal flow of air.

Impingement-type filters are made in a great many forms but they usually consist of metal or paper channels, or loosely packed fibers such as steel or glass wool, coated with a viscous material. The air is forced to take a tortuous path through the filter, and the dust is impinged against the surfaces where it is retained by the viscous coating. As the dust builds up on the surfaces, the resistance to flow increases and the filter must be replaced or cleaned.

Since this type of filter is suitable only for low concentrations of dusts, its industrial application is usually as a final cleaner following one of the other types of dust collectors, or when only normally air-borne dust must be removed from the air. There are several automatic types that are self-cleaning. In these oil is generally the viscous coating. Either the oil is flushed over the filtering medium, or the medium is rotated slowly as it passes through a bath of oil.

WASHERS. Many dusts, which are either soluble or wettable in water, can be efficiently washed from the air stream. In order to be effective cleaners, washers must provide a large area of water surface in intimate contact with the dust-laden air. Although there are many designs of air or gas washers, they all provide this surface area either by the use of sprays, wetted impingement surfaces, or a combination of the two.

The wettability of the dust, desirability of handling the dust in a sludge form, and the available water supply are factors which must be considered in connection with washer-type collectors. Unless an inexpensive water supply and adequate facilities for disposal of the contaminated water are available, settling tanks are usually used and the water is recirculated through the collector. A conveyor is sometimes installed in the bottom of the settling tank for automatic removal of the sludge. When finely atomizing spray nozzles are used, however, it is important that the water be clean enough to prevent plugging of the nozzles.

Dusts that are highly inflammable or explosive when dry, such as magnesium dust, can be handled most safely in wet washers. Some dusts and gases, however, although easily cleaned in this type of collector, form corrosive liquids which necessitate construction of special materials to resist the corrosion. Although many of these washers clean the air sufficiently to allow its recirculation in the building, the intimate contact of the air and water necessary for maximum cleaning efficiency also produces nearly complete saturation so that the effect of increased humidity must always be considered.

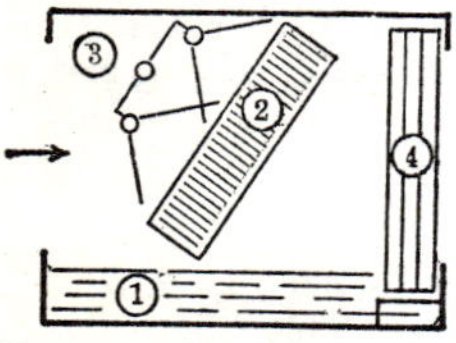

FIG. 7. Section of a capillary air washer, showing (1) tank, (2) cell of fine glass fibers, (3) low head distributing sprays, and (4) metal moisture eliminators.

Spray-type Washers. The simplest of this type is the common air washer, which consists of a bank of spray nozzles located above a tank and followed by a series of baffles or eliminator plates that remove entrained moisture. Washers of this type use 3 to 15 gal of water per 1000 cu ft of air, depending on the type of dust and temperature of the air. The air velocity is low, usually about 500 ft per min so that the resistance to flow is generally less than 1/2 in. water gage.

The simple spray type is suitable only for light dust concentrations. In a newer design of air washer the water is sprayed against the face of a cell filled with fine glass fibers, oriented to form a multitude of minute wetted air passages. Since they provide a large area of wetted surface, fine atomization is not necessary, and coarse low-head spray nozzles can be used

Three to 8 gal of water per 1000 cu ft per min is normally used, and, with an air velocity of 400 ft per min, the resistance is 0.4 to 0.5 in. water gage. This type of unit is quite efficient in the elimination of wettable dust present in moderate concentrations. Figure 7 illustrates the arrangement of its component parts.

Scrubbers are washers in which the dust is impinged against a large area of wetted scrubbing surface. In its simplest form the water flows over a distributor plate and down over a series of impingement plates countercurrent to the flow of dust-laden air or gas. In some units of this type, flowing water or sprays are used in combination with centrifugal action in the gases to throw both the dust and water against wetted surfaces. In the packed scrubber tower, the tower is filled with solid hurdles such as grids, terra cotta shapes, or coarse coke. The water quantities used depend on the type of dust, dust concentration, and gas temperature. Velocities and resistance to flow are generally considerably higher than in the spray type of units.

In several popular types of wet scrubbers, such as the one shown in Fig. 8, a fine spray or water curtain is produced, without the use of nozzles and a pump, by impinging the air at a high velocity against the surface of the water in the tank. The air passes through the water curtain, and the spray and air are usually impinged against baffle plates. Although the resistance to flow is higher than in units that use nozzle sprays, the only make-up water required is that lost by evaporation. In another type a high-speed rotor carries the water from the tank into the air stream.

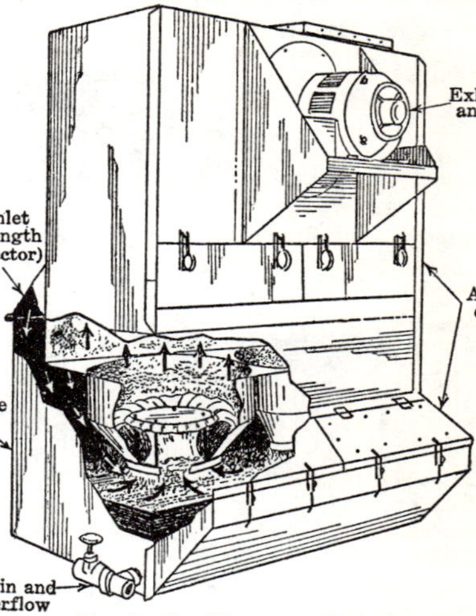

FIG. 8. Scrubber type air washer.

Froth-type collectors are generally a combination type of washer. After the air has been sprayed or scrubbed by one of the foregoing methods the partially saturated air is forced through some type of nozzles or orifices into a liquid bath. The liquid bath is usually water to which some wetting agent, such as flotation oil, has been added to reduce the surface tension and produce a finer froth. After leaving the surface of the bath the air and entrained liquid impinges against baffle plates. Although the cleaning efficiency of these froth-type collectors is very good, the inherent resistance through the collector is high; and the velocity through the liquid bath must not be too great or the pressure drop will be excessive.

Disintegrators. In the disintegrator type of gas washer the liquid is dispersed, and the air and liquid intimately mixed, mechanically. This is accomplished by a series of rotating cages, which are usually turned by the same shaft that drives the fan impeller that moves the gas through the unit. In some designs the rotating cages are alternated between similar stationary cages. Disintegrators are effective even on very fine particles. However, because of their high power requirements, they are generally used only on hot gases, such as blast furnace gas, where both a cooling and a thorough cleaning are desired.

ELECTROSTATIC PRECIPITATORS. In electrostatic precipitators the air or gas is passed through an electrostatic field produced by a high-voltage d-c discharge between two sets of electrodes. The dust particles become ionized and are attracted to the collecting electrodes.

Electric precipitators are capable of very high collection efficiencies with very fine dusts when properly designed and operated for the particular application. Since the collection efficiency is a function of the time that the gas remains in the active field, precipitators designed for maximum efficiency must be large in both cross-sectional area and length. In any case the velocity of the gases must be low enough so that the precipitated dust is not redispersed in the gas stream. Gas velocities usually range from 300 to 600 ft per min, and consequently the resistance to gas flow is low.

Uniform gas distribution is essential to proper operation, and straightening vanes or perforated distributor plates are often used at the inlet of the precipitator. The surface conductivity of the dust particles is an important factor in their precipitation, and with some dry nonconductive dusts the gas and dust are conditioned by introducing some conductive wetting agent, such as water or steam. Since electrical precipitators are most efficient with low dust concentrations, and are effective on even the finest particles, they are often used in conjunction with the centrifugal type of collectors for the final cleaning.

The two methods of electrical precipitation are represented by the single-stage and double-stage types.

Single-stage Type. The single-stage type, commonly known as the Cottrell system of electrical precipitation, is used most generally for industrial applications. The particles are charged and collected between the same set of electrodes, and the discharge is maintained throughout the unit.

In order to obtain the corona discharge, the discharge electrodes must either be small or have sharp edges. Either fine wires or twisted square steel rods are generally used. There are several designs of collecting electrodes, and the type used depends on the application. In the earlier designs they consisted of vertical pipes, 3 to 12 in. in diameter and 5 to 15 ft long. These are now most commonly used in the wet-type precipitators, whereas plate-type collecting electrodes are generally used in the dry precipitators.

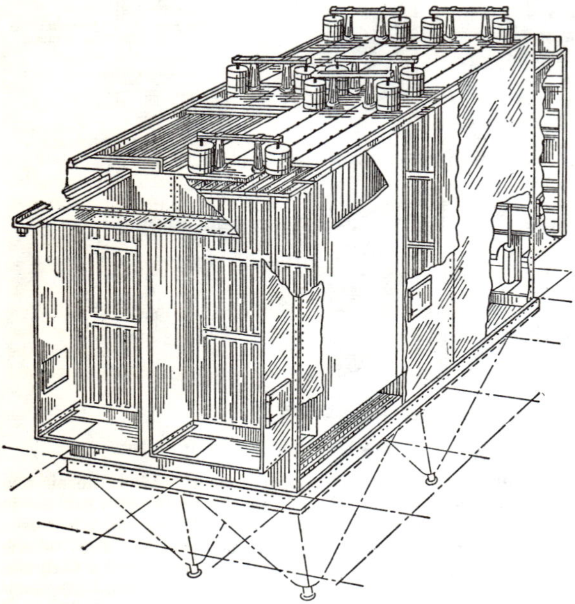

FIG. 9. Single-voltage plate type electrostatic precipitator.

The plate type consists of a series of parallel plates between which the discharge electrodes are suspended. There are several modifications of the plate type, such as steel plates that are sometimes corrugated, rod curtains that consist of a number of small rods or pipes hung close together, concrete plates with the conductor rods embedded in them, perforated or dual plate type consisting of perforated plates welded up in the form of narrow boxes with perforated sides, and pocket electrodes in which the precipitated dust is trapped in V-shaped pockets. When the plate-type electrodes are used the spacing between the opposing electrodes is ordinarily 3 to 6 in.

The electrodes are usually assembled in cells or sections and one or more cells are used in series, depending on the degree of cleaning desired. Figure 9 shows a single-stage plate-type unit.

Since the dust particles are deposited on the collecting electrodes, some method must be provided for removing them from the surface. In the wet precipitator the gas flow is upward, and water flows downward over the collecting surface to carry off the precipitated particles. In the dry type of units the gas flow may be either vertical or horizontal, and the dust is removed from the plates by scraping or rapping them. Compressed air or motor-driven rappers are most commonly used, and the frequency and duration of the rapping periods depends on the type and concentration of the dust.

The electrical equipment necessary to produce the high-voltage direct current consists of a transformer and either a mechanical or an electronic rectifier. The voltages most commonly used range from 60,000 to 100,000 volts. The mechanical rectifier is usually driven by a synchronous motor, whereas the electronic type consists of a group of electron tubes arranged to give full-wave rectification. Either type of rectifier gives equally satisfactory precipitation, and, although the electronic type requires less space, the control equipment required for its operation is somewhat more complicated. The power consumed is about the same for either type, and, although the voltage is high, the current is very low so that the power consumption is only 0.5 to 1 kwhr per 100,000 cu ft of gas cleaned.

Double-stage type electric precipitators consist of units in which a pre-ionization stage is used ahead of the collecting stage. Figure 10 illustrates the principle of operation of this type of precipitator. The corona discharge takes place in the first or ionizing stage, and the dust is precipitated on alternating, oppositely charged, plates in the second or collecting stage.

Lower voltages are used, usually 10,000 to 14,000 volts in the ionizing section and 5000 to 7000 volts on the collecting plates. With these lower voltages the plate spacing must be considerably closer, usually $^5/_{16}$ to $^1/_2$ in. As a result of this close spacing, this type of precipitator is generally used only when the dust concentration is low. High concentrations would require too frequent cleaning to prevent the dust from building up on the plates to a point where the current would arc-over.

The collecting plates, which are usually coated with a viscous material to hold the dust, are contained in cells that can easily be removed for cleaning by washing off the plates. Each cell handles 600 to 750 cu ft per min.

Provision is made in some units for automatically washing and recoating the plates, without removing the cells, by the use of movable spray nozzles. The cells being cleaned, however, must be disconnected electrically and dampered off from the gas flow. In one type of unit the cleaning plates are continuously rotated through a bath of viscous fluid for automatic cleaning.

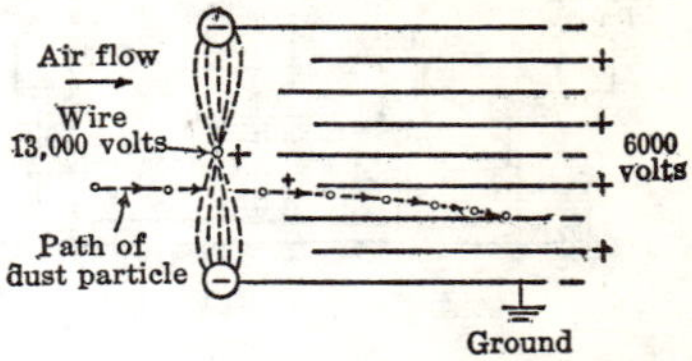

Fig. 10. Principle of operation of double-voltage type electrostatic precipitator.

Electronic rectifiers are used with this type of precipitator, and the power requirements are of the order of 0.01 to 0.02 kw per 1000 cu ft per min. The gas velocity is 300 to 400 ft per min so that the resistance to flow is only a few tenths of an inch water gage. Since the voltages are much lower than in the single-stage type, the ozone produced is much less. Although the double-stage precipitator is suitable only for use with low dust concentrations, it is most satisfactory for cleaning air to be used for ventilation in an occupied space since its cleaning efficiency is very high even for the finest particles, and the ozone produced is very low.

26. DUST-COLLECTOR SYSTEMS

FACTORS INFLUENCING DESIGN AND REQUIREMENTS. The most important factors that influence the design and requirements of a dust-collection system are those that determine the need for the system.

Improvement of working conditions is probably one of the most common reasons for the installation of a dust-collector system, and its importance should not be underestimated. In any process where dust is produced, the elimination or control of the dust pays dividends in improved efficiency of both workmen and machines. The control of the dust at its source is the important requirement, and the best design is the one that will accomplish this control with minimum amount of air.

Health of workers is the most important reason for the control of dusts, and, in the design of systems where health conditions are involved, adequate design must take precedence over economic considerations.

State laws, which usually govern the requirements of such systems, should be consulted when the design is being considered. For example, the maximum allowable concentration of such poisonous dusts as lead dust in some states is 0.15 mg per cu meter of air. Table 3, which gives the probable safe limits for several harmful types of dusts in millions of particles per cubic foot of air, is based on the light field microscope counting technique.

Table 3. Probable Safe Limits for Several Harmful Dusts in Million Particles per Cubic Foot

Type of Dust	
Asbestos dust	5
High silica dust (100 to 50% free silica)	5
Medium silica dust (50 to 5%)	20
Low silica dust (5 to 0%)	50
Marble	20
Granite	10
Silver	12

Nuisance Elimination. In addition to removing the dust from the working area, dust-collection systems are needed to prevent dust's being a nuisance to the surrounding neighborhood and general public. The conditions necessary to eliminate this type of nuisance often govern the degree of cleaning and type of dust collector required.

Recovery of material, which is either the product or a by-product of the dust-producing process, is the purpose of many dust-collector systems. The form in which the material can be utilized influences the design and type of dust collector, and the value of the material governs the requirements regarding collection efficiency.

TYPES OF SYSTEMS. Central exhaust systems are the type most commonly used in industry. In this type of system the hoods, or enclosures, at the dust sources are connected by branch pipes to the main duct or header, which conveys the air and dust to the dust collector and fan. The design of such systems is often specified by state laws and codes, and the general design data given here should be checked to be sure they conform with local regulations. In general, the requirements of the central exhaust system are positive control of the dust at its source, adequate conveying to a safe and convenient point for collection and disposal, and a collection efficiency sufficient to eliminate any hazard or nuisance. The planing mill exhaust system shown in Fig. 11 is a typical example of this type of system.

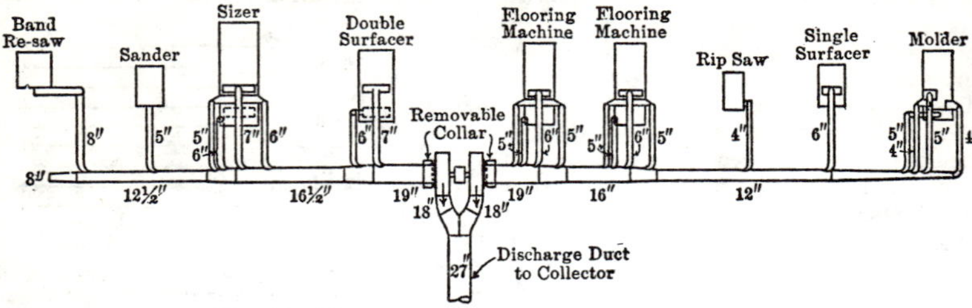

FIG. 11. Plan of typical planing mill exhaust system.

Hood design is of primary importance since it is here that the dust is brought under control and removed from the working area. The first rule in hood design is to enclose the dust source as completely as possible without interfering with the work. This precaution not only makes it easier to control the dust but also reduces the quantity of air required. The design of the hood is usually governed by the type of work or the machine that is producing the dust and by the surrounding physical conditions. Figure 12 shows typical hoods for woodworking machines, and Fig. 13 illustrates the types used for grinding and buffing wheels. The hoods should be located so that the air movement is in the same direction as the particles are discharged by the machine. If any air currents are set up by the machine or process, they should also be in the same direction as the flow produced by the hood.

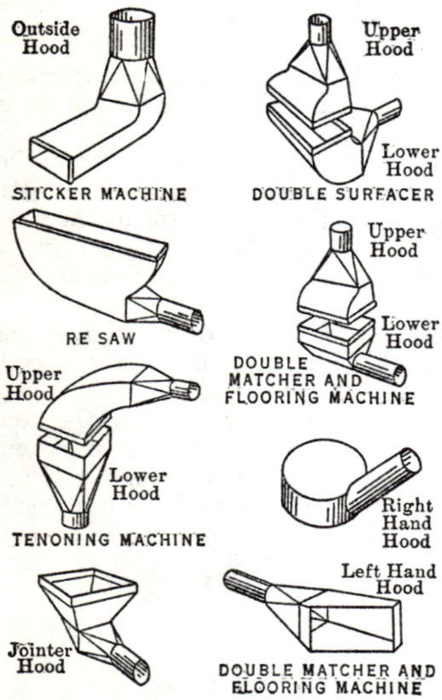

FIG. 12. Typical hoods for woodworking machines.

The velocity necessary at the face of the hood, or enclosure, depends on the type of dust being handled and the surrounding conditions. The face of the hood should be as close as possible to the dust source since, in general, the air velocity varies inversely with the square of the distance from the hood. When drafts, or opposing air currents, cannot be eliminated in the vicinity of the hood, the hood velocity must be enough higher to overcome them.

Generally, the air velocity should be high enough to pick up only the float dust which would escape or remain in suspension in the room air. When coarse heavy dust particles must be handled in the hood, provision should be made for settling out such particles as near the hood as possible so that they will not have to be conveyed in the piping system. Conveying of these coarser, heavy particles results in more rapid wear in the pipes and greater power requirements due to the higher conveying velocities that are necessary. Table 4 gives the air velocities required at the source to control various common types of dusts.

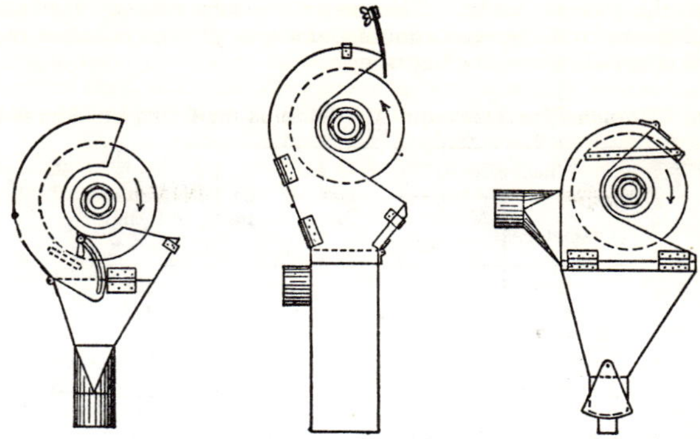

FIG. 13. Typical hoods for grinding and buffing wheels.

Table 4. Air Velocities for Controlling Various Types of Dust

Type of Dust or Fume	Process or Operation	Required Air Velocity, ft per min	
		At Dust Source	At Face of Hood
Granite	Hand pneumatic tool	200	....
	Surfacing machine	1500	....
	All tools	....	1500
Grain	Elevator boot and head garner	....	500
Grinder	Swing grinder	....	200
Fine Powders	Bagging	200–400	....
	Belt conveyors (at transfer points)	250	....
Sand	Bagging	400	....
Lead	Metal spraying	200	....
Zinc	Metal spraying	125	....
Welding	Electric welding	200	....

Piping Design. The size of the branch pipe and the velocity of the air in the pipes must be adequate to convey the material picked up in the hood. Table 5 shows the size of the hood connections required for various types of woodworking machines.

Table 5. Sizes of Hood Connections for Woodworking Machinery

Type of Machine	Diameter of Pipe, in.	Type of Machine	Diameter of Pipe, in.
Matcher heads, each	5	Rip saws and resaws	
Sash and cabinet shaper, each head	4	Up to 14 in. diameter	4
Door tenoner	5	14 to 18 in. diameter	4 1/2
Sash tenoner	5	18 to 24 in. diameter	5
Door and sash sticker, each head	5	Cut-off saws not over 16 in. diameter	4
Blind slat sticker	5	Band saws	
Blind rail router	4	Up to 1 1/2 in. wide	4
Panel raiser, each head	4	1 1/2 to 3 in. wide	5
Sand drums		3 to 4 in. wide	6
24 in. long	5	4 to 6 in. wide	7
30 in. long	6	Buzz planers, moulders, planers, and all other machines having knives	
36 in. long	7		
48 in. long	8	Up to 6 in. long	4
60 in. long	10	6 in. to 8 in. long	5
Belt sander		8 in. to 12 in. long	6
Up to 4 in. wide	4	18 in. to 24 in. long	8
8 to 12 in. wide	6	24 in. to 30 in. long	9
Disk sander		Floor sweep-up	6
Up to 24 in. diameter	5		
24 to 36 in. diameter	6		

Although some state laws still specify the suction that should be maintained in the branch pipes from the hoods on grinding and buffing wheels, most of these laws and codes are being revised to give the requirements in terms of the volume of air that should be handled through the hood as this is a more correct measure of hood effectiveness. Table 6 gives the air volume in cubic feet per minute and size of pipe connections usually required for various sizes of grinding and buffing wheels.

Table 6. Branch Pipe Sizes and Air Volumes for Grinding Wheel Hoods

Wheel Size, in.		Branch Pipe Diameter, in.	Air Volume, cu ft/min
Diameter	Maximum Width		
Up to 9	1	3	225
9 to 18	3	4	400
10 to 24	4	5	600
24 to 30	5	6	900
30 to 36	6	7	1200

Table 7 gives the air velocities needed to convey various types of materials. Some laws and codes, however, set minimum velocities regardless of the type of dust handled.

Table 7. Air Velocities for Conveying Various Types of Materials

Material	Air Velocity, ft per min
Ashes or ground clinkers	6000–9000
Buffing lint, dry	2500–3500
Buffing lint, sticky	3000–4000
Cement	6000–9000
Coal, powdered	4000–5000
Cotton lint	1500–2000
Cotton seed	4000–6000
Flour	3500–6000
Fly ash	3000–4000
Foundry shakeout dust	3500–4500
Grain	4000–5000
Grain dust	2000–3000
Grinding dust	3500–4500
Lime	5000–7000
Metal turnings	4000–5000
Pulp chips, dried	4000–6000
Rags	4000–6000
Salt	6000–7500
Sand	6000–9000
Sand blast dust	3500–4000
Sawdust and shavings, dry	2000–3000
Sawdust and shavings, wet or green	3000–4000
Sugar	5000–7000
Wood chips	4000–5000
Wood flour, sander dust	2000–4000

It is worth while to give careful consideration to the design of the entry connection from the hoods to the branch pipes, and all points where there are turns or changes in section. Proper design at these points not only will prevent settling of the dust in the pipes but will also result in power savings from the lower resistance to flow. Table 8 gives the entry losses for the various designs of entry conditions for round pipes. Similar designs and entry losses from plenums or large hoods are given in Table 9.

In order to have a balanced system with the desired volume being drawn through each hood, it is necessary that the system be designed so that the total pressure in each branch is equal to the total pressure in the main header at the point where the branch enters the main. This means that the resistance of the system from the hood to the fan must be the same for all the connections on the same system. Usually the resistance of the system from the longest equivalent run to the fan is calculated, then the other branches are designed and sized for the same resistance. The main duct should be increased in area at each branch by an amount at least equal to the area of the branch. Some codes recommend an increase in area of 10 to 20% more than the branch pipe area, but when this method is used care must be taken that the velocity does not become too low for conveying the dust.

The piping is generally made of black or galvanized sheet steel. Round pipes are preferable to rectangular or square pipes. All joints should be made as smooth as possible,

Table 8. Entry Losses for Round Pipes in Percentage of Velocity Pressure at Diameter D

Type of Entry	Entry Loss, % of Velocity Pressure
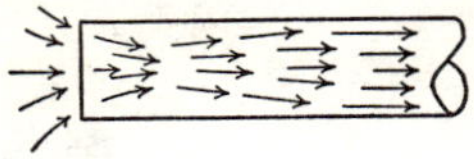 *a*	87
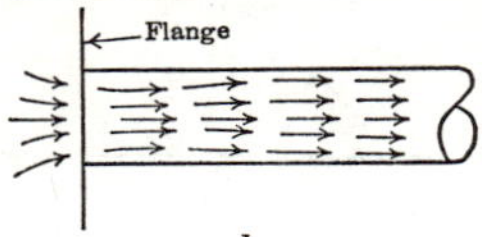 *b*	48
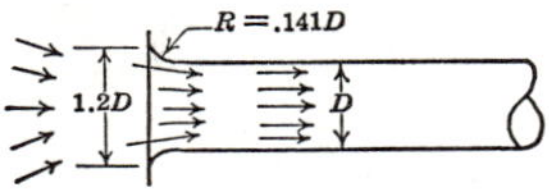 *c*	15
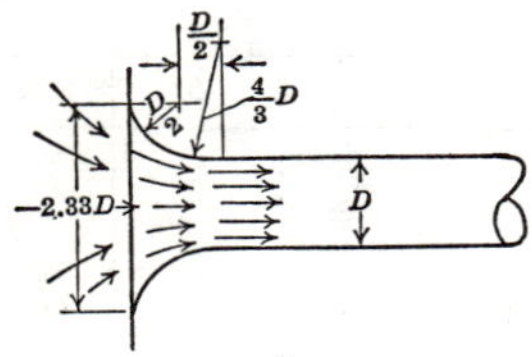 *d*	2.5
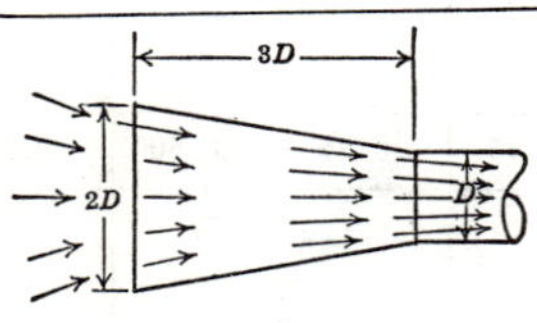 *e*	12
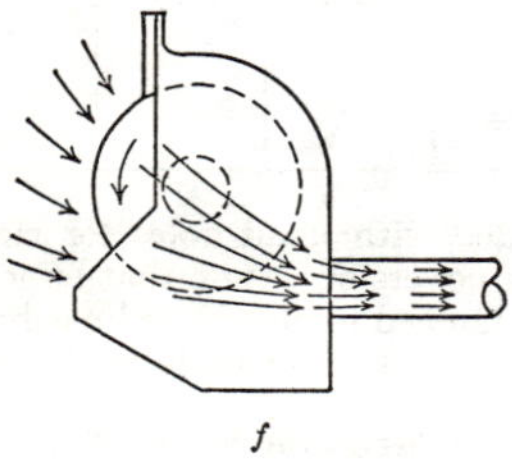 *f*	65

and lapped joints should be lapped in the direction of air flow. All parts of the system should be made as free as possible from air leakage. The gage thickness of the piping material should be adequate for structural strength, and, when abrasive dusts are handled, heavier gages are required because of the abrasion on the pipes. Table 10 gives the minimum thickness that should be used for round pipes. Rectangular pipes should usually be made heavier for equal strength.

Elbows should be at least two gages heavier than that required for the pipes since they are subject to the greatest abrasion. Long radius elbows, with an inside or throat radius of at least two pipe diameters, are preferable and result in minimum resistance losses. The branch pipes should enter the main header at an angle of not more than 45 degrees, and an included angle of 30 degrees is preferable.

Table 9. Entry Losses from Plenums or Large Hoods

Type of Entry	Entry Loss, % of Velocity Pressure					
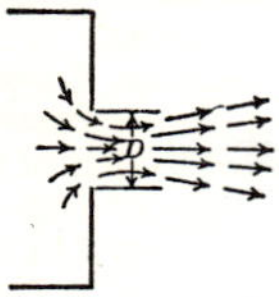 *a*	256					
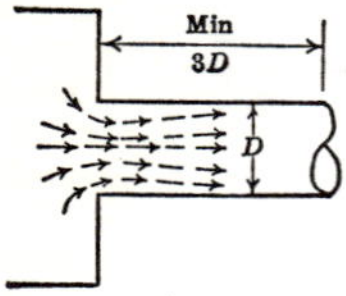 *b*	47					
∠α	5°	10°	15°	20°	25°	30°
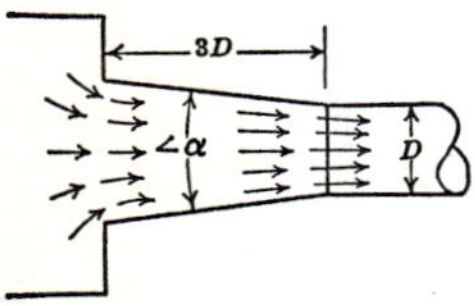 *c*	12.0	11.3	11.3	11.7	12.0	25.0
R/D	0.15	0.25	0.50	0.75	1.00	1.25
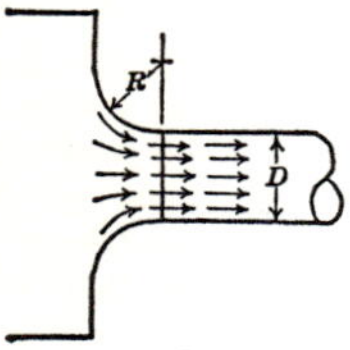 *d*	13.4	12.4	11.5	11.0	10.7	10.6

All horizontal pipes should be provided with hand holes for cleaning and inspection. These openings should have tight-fitting covers of a type that offer no obstruction on the inside of the pipe. With welded piping, flanged joints are ordinarily used every 10 or 15 ft to make installation and repair easy. The end of the main header should be capped with

Table 10. Gages of Metal for Round Pipes

Diameter of Pipe	U. S. Gage Number: For Abrasive Material	U. S. Gage Number: For Other Material
Up to 6 in.	20	24
6 to 8 in.	20	22
8 to 18 in.	18	20
18 to 32 in.	16	18
Over 32 in.	14	16

a tight-fitting but removable cap. Dampers or gates are sometimes provided to aid in balancing the system, but in a system handling dust they are unnecessary obstructions if the system is carefully designed.

Dust-collector selection for the central exhaust system will depend upon the types of dust being handled and the degree of cleaning desired. Low-velocity cyclones are most commonly used for systems exhausting from woodworking machinery. High-velocity cyclones or cloth filters are generally used for exhausts from grinding operations and for foundry exhaust systems. If the dust is abrasive, the dust collector is usually placed ahead of the fan. However, in many systems handling such material as chips and sawdust, the material is handled through the fan and blown into the collector.

Fan Requirements. The volume of air and the total resistance of the system are the values which determine the capacity of the fan required for the dust-collector system. The resistance of the dust collector and its connections must be added to that of the hoods, branch pipes, and main header to determine the total resistance of the system. A fan of the desired capacity can then be selected from the capacity tables published by fan manufacturers. Centrifugal fans are usually used, and, when the fan is placed ahead of the dust collector, an impeller of the radial blade type is used since the material must be carried through the fan wheel.

Unit exhaust systems consist of combination fan and dust-collector units located at or near the dust source. They are frequently used at locations isolated from the central exhaust system or when there are only a few scattered sources of dust. Unit systems are also desirable when the dust is of an inflammable or explosive type, which should not be conveyed through a long piping system or mixed with other dusts in a central system. The dust collector and fan may be supported separately or they may be assembled as a unit in a single casing.

The combination type of unit, which is somewhat more portable, is used extensively with machines that may be moved frequently. Most of the combination units are designed to recirculate the cleaned air in the working space, a fact which will often determine the type of dust collector required or the advisability of using this type of unit.

There are many different designs of unit dust collectors incorporating various types of dust collectors. Most designs are manufactured in a range of standard sizes. The smaller units, designed for use with grinding wheels or small individual machines, often consist of a fan which either draws or blows the air through some type of filter. Either dry or viscous filters may be used. In some units of this design the coarser heavy particles are removed ahead of the filter by a settling chamber, baffle type, or centrifugal type collector. This feature is very desirable unless the dust concentration is quite low. The dynamic fan type of dust collector, such as is illustrated in Fig. 5, is often built into a combination unit with filters on the fan outlet. The filter area should be adequate for the design air volume and suitable for the sort of dust to be handled. Also, the fan should be capable of maintaining the rated capacity throughout normal operation without too frequent cleaning or replacement of the filters.

Wet scrubber-type collectors, such as the one shown in Fig. 8, are commonly assembled in combination units. They are especially adaptable when the dust is of an inflammable or explosive nature. In most of these units a settling tank is built into the common casing and the water is recirculated so that no drain connection is needed. Provision must be made, however, for addition of make-up water. When the dust is of the nuisance type, its removal in the form of a sludge also eliminates the possibility of contamination of the air when the dust is being removed from the unit.

Electric precipitators of the low-voltage, double-stage type are frequently combined with a fan into a unit suitable for low concentrations of fine dusts, fumes, or oil mists. Since with this type of unit it is important that the dust does not build up on the plates, a regular and frequent cleaning schedule should be established.

PNEUMATIC CONVEYING. Systems for conveying bulk materials in air can be divided into two types: (1) low-pressure systems, in which the air-to-material ratio is high and the velocities are fairly low so that the pressure loss in the system is low, and (2) high-pressure systems, which use a lower air-to-material ratio requiring higher velocities and thus have higher pressure losses. The limiting pressure for the low-pressure system is generally the maximum pressure that can economically be obtained with a centrifugal fan. High-pressure systems are thus those that require a vacuum pump, positive pressure blower, or steam jet to produce the necessary pressure. This discussion will be limited to the low-pressure type of system since the high-pressure types are discussed elsewhere in this book. (See p. 26-43.)

The nature of the material to be conveyed is generally the most important factor to consider in pneumatic conveying. Free-flowing granular materials can be very easily conveyed in air, but this method of conveying is not satisfactory for materials which tend to cake or pack. Fragile materials should not be handled in this system if reduction of the particle size is undesirable. Materials containing hazardous dust can be conveyed safely, and many materials can be transported with less contamination pneumatically than by mechanical conveyors. With proper dust-collecting equipment, the loss of material can be kept at a minimum when the quantity of material to be handled is not large. The advantages of pneumatic conveying include cleanliness, flexibility, and safety. Such systems are not so compact as mechanical conveying systems because of the space needed for the dust-separation equipment, and they have the added disadvantage of requiring more power for operation. Although the low-pressure systems require less power than the high-pressure systems, the initial cost is generally higher because of the larger air volumes handled.

Conveying Velocities. Theoretically the critical velocity for conveying a particle in air is its terminal velocity. Figures 1 and 2 illustrate the terminal velocities for fine dusts within the Stokes' law range. Although this law can be used for the fine dusts handled in exhaust systems, many of the materials handled in conveying systems consist mainly of particles larger than those for which this law applies. Stokes' law defines the terminal velocity of particles between 1 and 100 microns. The limits vary, however, depending on the density and shape of the particle. The fall of particles larger than these follows Newton's law, from which the terminal velocity can be expressed in this way:

$$V_m = K\sqrt{\frac{\rho - \rho_0}{\rho_0} d} \tag{4}$$

where K is a constant, d = diameter of the particle, ρ = density of the particle, and ρ_0 = density of the conveying fluid.

In air ρ_0 is so small compared to ρ that it can be neglected in the numerator of the fraction, and the equation shows that the velocity for a given-sized particle is directly proportional to the square root of density of the material, and inversely proportional to the square root of density of the air.

In conveying the particles must be accelerated, instead of just balanced, and the velocity must also be great enough to overcome the friction between the particle and pipe and the other particles. For these reasons, and because the velocity in the pipe is not uniform, the conveying velocity must be several times the terminal or critical velocity of the largest particle.

Pneumatic conveyor systems are similar in principle to exhaust systems, except that in the conveying systems the material concentration is considerably greater. Table 7 gives the air velocities needed to convey various types of material. The lower velocity would be satisfactory in an exhaust system; the upper velocity is required at the concentrations used in a low-pressure conveying system. These values are for standard air with a density of 0.075 lb per cu ft and, as shown by eq. 4, should be corrected for air of different density.

Carrying Capacity. The velocities given in Table 7 can be used when 50 cu ft, or 3.75 lb of standard air at 70° F, are used per pound of material to be conveyed. The weight of material that can be transported per pound of air varies directly as the square of the air velocity. Therefore, the carrying capacity of a given size of pipe increases directly as the cube of the velocity.

Since the fan horsepower required to move the air also varies as the cube of the velocity, the horsepower is proportional to the weight of material moved through a given pipe. However, since the frictional resistance of the pipe for a given velocity increases as the size of the pipe decreases, it requires more power to use high velocities and small pipes. Also smaller pipes tend to clog more easily. A pipe 10 in. in diameter is usually considered about the smallest that should be used for a pneumatic conveying system of this sort.

The same recommendations given for the central exhaust system regarding pipe smoothness, elbow design, etc., should be followed in the design of a pneumatic conveying system. The conveying pipes should, however, be made of considerably heavier gage metal because of the higher pressures and higher material concentrations. Because abrasion increases very rapidly with increases in velocity, very abrasive materials are usually handled at higher air ratios and lower velocities.

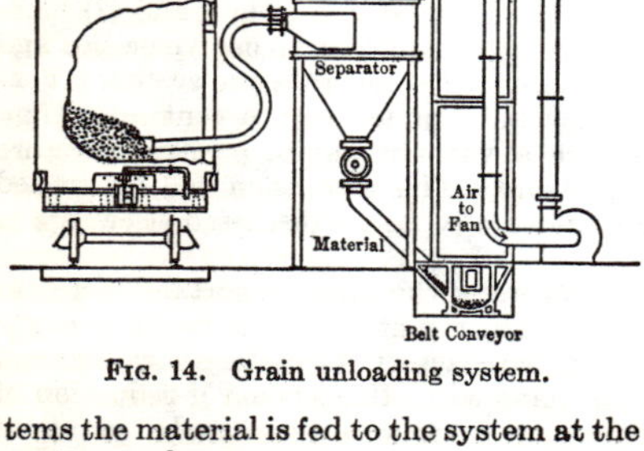

Fig. 14. Grain unloading system.

Material Feeders. In most suction systems, such as the grain-unloading system shown in Fig. 14, the material is drawn into the system through a nozzle. In this type the fan is usually placed on the clean air side of the dust separator. In some pressure systems the material is fed to the system at the fan inlet and then blown through the conveying system under pressure.

Material subject to damage by the fan impeller, or of a type harmful to the fan, may be fed into the pipe beyond the fan with a feeder like the one shown in Fig. 15. It is sometimes advantageous to use a closed system and return the air and dust not caught by the separator to the inlet side of the fan. When this is done, it is necessary to provide a small vent, which should be located as near the null point in the system as possible in order to minimize the dust loss.

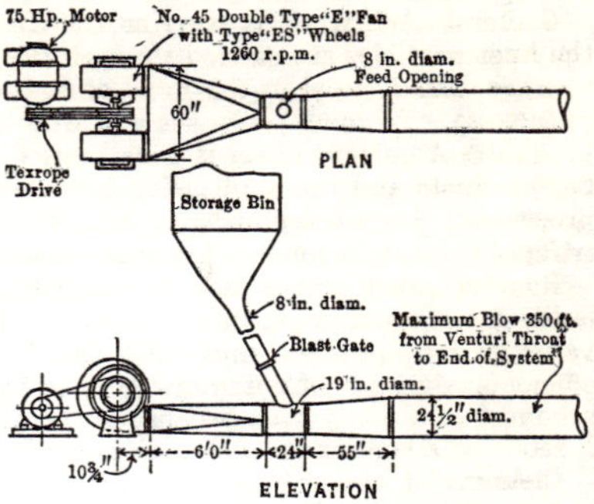

FIG. 15. Material fed beyond fan.

Material Separators. The type of separator or dust collector used to separate the material from the air stream will depend on the quantity being handled and the nature and value of the dust. When a large quantity of fairly coarse material is handled it is often blown into a large separating tank or settling chamber. Sometimes a centrifugal or cloth filter collector is used on the exhaust of the tank to catch the fine dust which does not settle out.

If the material is somewhat finer the centrifugal-cyclone type of units are used either alone or as the primary cleaner. If the value of the material is high, the high-velocity type should be used to obtain the maximum efficiency. In most pneumatic conveyor systems the concentrations are too high for the use of a cloth filter without some sort of primary cleaner ahead of it.

PRODUCT AND BY-PRODUCT RECOVERY SYSTEMS. In many industrial processes dust is formed that is valuable either as a part of the product or a by-product, and the dust-collector system is an important part of the process equipment. The collection efficiency required is usually governed by the value of the product.

Grinders, crushers, and pulverizers are typical of the type of process equipment that produces dust of this nature. In some equipment of this type all the resulting product is carried in the air stream, and the system becomes a pneumatic conveying system. More often, however, vents are provided to handle only the fine dust which is produced. The fine dust may be either mixed with the product or utilized as a by-product. The air is usually at normal room temperature, and, when the material is to be handled dry, cloth filters or centrifugal collectors are generally used. If the process is such that the material can be used wet, washer or scrubbers are often used.

Kilns. In the processing of many organic materials the product is formed by burning the raw material in a vertical or rotary kiln. The fine dust carried with the exhaust gases usually has a value either as a part of the product or as a by-product. In some cases, however, the value of the dust would not justify the cost of the recovery system and equipment required to handle the collected material. When this is true, the collection efficiency required is that necessary to eliminate a nuisance from the discharge of gases. The gases are usually handled at a high temperature and thus centrifugal collectors are most commonly used. When the temperatures are above 1000 F special stainless steel alloys are usually used in the collector construction. Single-stage electrostatic precipitators are often used when the maximum recovery is desired.

Rotary Dryers. The dusts carried by the drying gases in rotary dryers are often a valuable part of the product. Centrifugal dust collectors are generally used because of the temperature of the gases. When maximum recovery is desired cloth filters are sometimes used by bleeding in enough air to cool the gases to a safe temperature for the cloth. This latter method is limited, however, by the type of material and the amount of moisture in gases, since some dusts will cake on the cloth, and the cooling required may lower the temperature of the gases below their dew point.

Spray dryers are generally of two types. In one type a vertical drying tower is used and the velocity of the drying gases is low enough so that the majority of the dried particles fall to the bottom of the tower. This type is commonly used when the fine particles are not desired in the finished product, and the dust carried up with the gases is collected and returned to the raw material or spray liquid. Wet-type collectors are sometimes adaptable for the recovery of the dust since it can often be reused in the wet condition. Centrifugal collectors are also frequently used.

In the second type of spray drying system the gas velocity is high enough so that most of the dried material is carried with the drying gases and the material collected by the recovery system is the actual product. In this type of system the material must be collected dry, and, since a high collection efficiency is usually required, high-velocity centrifugal collectors are generally used. When the dried material is a food or drug product, the collectors can be built of the special metals required to prevent contamination of the product. Cloth filter units however, will give maximum recovery when the nature of the material and the temperature and dew point of the gases permit their use.

Coolers. When air is used as the direct cooling medium for cooling bulk materials, the finer particles are carried through the cooler with the air. Centrifugal or cloth collectors are usually used in this type of dust-collecting system.

Dusting. In some processes a powdered coating must be applied to the surface of the product. A current of air is usually used to remove the surplus powder. Dusting of rubber products and the application of bronze powder in lithographing are examples of such processes. Since the powder is usually quite valuable, cloth filters or high-efficiency centrifugal-type collectors are generally used.

Roasters, such as are used in the roasting and treatment of mineral ores, produce dust-laden gases which must be handled in a dust-collector system. The dust is sometimes of value as a by-product, and when this is the case its value will determine the collection efficiency desired. Centrifugal-type collectors or single-stage electrostatic precipitators are most commonly used because of the high gas temperature. When maximum efficiency is required a combination of both types is sometimes used.

Cleaning of ores or other powdered materials is another process in which the dust sometimes has a by-product value. In the cleaning of castings, core and molding sands are separated from the cleaning air for reuse. Almost any of the various types of dust collectors can generally be used in such systems, and the collection efficiency required will determine the type selected.

Bagging. In the bagging or packaging of powdered materials the dust produced is usually removed by ventilation air, collected and returned to the product. Cloth filters are ordinarily used unless the material is of such a nature that it plugs the cloth.

FACTORS INFLUENCING THE CHOICE OF THE TYPE OF DUST COLLECTOR. These factors will influence the type of dust collector selected for the system: (1) nature of the dust; (2) value of the dust; (3) dust concentration; (4) method of handling the collected dust; (5) collection efficiency required; (6) relative capital, operating, and maintenance costs.

MAINTENANCE AND OPERATION OF DUST-COLLECTOR SYSTEMS. Even the best designed dust-collector system can become a liability rather than an asset if not properly operated or maintained. Regular operating and inspection schedules should be set up and maintained. The entire system must be kept free of air leaks and accumulated material. Lack of attention to these requirements may not only reduce efficiency but also decrease the capacity of the system to the point where it is practically ineffective in the performance of its required functions.

Dust Handling. When receptacles or storage bins are used they should be emptied at scheduled intervals frequent enough to prevent their being filled. The method of emptying should be such that the dust is not redispersed in the air. When the dust is handled by a continuous discharge system the dust seals or valves should be maintained so that they are free of leaks.

Filters. Filter-type collectors should be cleaned or replaced before their resistance to air flow increases to the point where the capacity of the system is reduced below its rated design. Manometers should be installed to indicate when this point has been reached.

Washers. In washer-type collectors, the sprays and water piping should be kept clean to prevent plugging. Pumps should be properly lubricated and packed, and the sludge tanks should be cleaned at regular intervals.

Electrostatic precipitators should be cleaned at frequent intervals to prevent arcing and redistribution of the dust in the air or gas stream. The transformers and rectifying equipment should be maintained to operate at the design voltage.

Pipes and Fans. The pipes and air ducts should be regularly cleaned and inspected for leaks. Fans and rotating equipment must be adequately lubricated and maintained so that they are in proper balance. When there is a possibility of liquids collecting in the fan, a drain should be provided at the lowest point in the fan casing.

RECIRCULATION OF AIR FROM DUST-COLLECTOR SYSTEMS. When very large quantities of air are exhausted from a room or building, an equal amount of air must be supplied either by infiltration or by some positive source of supply such as a supply fan. In fact, some exhaust systems fail to operate properly because adequate provision has not been made to supply the quantity of air that should be exhausted. During the heating season this supply air must be heated, and, when large exhaust systems are used, the additional heating load becomes a major item in the heating cost. When air conditioning is used, the cost of conditioning all the make-up air may be quite excessive. It is apparent that serious consideration must sometimes be given to recirculation of all or part of the air after it has been thoroughly cleaned. When the portable combination units are used, all the air is generally recirculated.

Some state laws and codes have prohibited recirculation under most conditions, but the present trend is to allow it under conditions approved by the enforcement agency. The Michigan Bureau of Industrial Health does not approve installations with provision for recirculating any of the exhaust air if the contaminant is a material which may possibly impair health, regardless of the type of dust collector used. The bureau cites these reasons for disapproval: *

A. Many types of air cleaners do not collect toxic contaminants efficiently enough to remove the health hazard.

B. Poor maintenance of the air cleaner would result in the deliberate return of highly contaminated air to the breathing zone of the workers. Not being production equipment, air cleaners are too often poorly maintained.

C. Non-operation of the cleaner, through mechanical failure or through ignorance or neglect on the part of the operators, would also result in the return of highly contaminated air.

The Michigan bureau allows the recirculation of air from systems handling nuisance dust under certain conditions. In general a maximum of 10 million particles per cubic foot, on a dust count basis, is allowed in the recirculated air if certain other provisions are met. The exhaust system should be so arranged that at least 30 cu ft per min per worker is exhausted from the building. Except for portable unit-type dust collectors, the system should be constructed to provide no recirculation in the summer months if air conditioning is not used, and to prevent recirculation in case of a shutdown or failure of the air-cleaning system.

Some codes allow a recirculation of 80% of the exhaust air provided it is properly cleaned. Others allow recirculation of air equal to a certain percentage of the room or building volume.

The Division of Industrial Hygiene and Safety Standards, of the New York State Department of Labor, has tested quite a number of different types of unit dust collectors and has found that many of them do not meet their standards regarding recirculation of air from systems handling nontoxic dusts. They have set up a suggested set of standards for this type of unit: †

1. Maximum permissible recirculation for units having effluent dust count no higher than ambient without recirculation may be 1 cfm for every 10 cu ft of room volume.
2. Maximum permissible test dust count above which no recirculation would be permitted should be 3 mppcf in excess of ambient without recirculation when the unit is tested at a dust loading of 1.5 grains per cu ft of a dust having a mean size of 3 microns on a particle count basis.
3. Maximum permissible recirculation for units having effluent dust count of 3 mppcf in excess of ambient air without recirculation under above test conditions should be 1 cfm for every 100 cu ft of room volume.
4. In the intermediate range an additional 10 cu ft per cfm could be allowed for each step of 1/3 mppcf above ambient without recirculation between 0 and 3 mppcf.

The exhaust from inertial or centrifugal-type collectors, even though the efficiency is very high by weight, should never be recirculated since these dust collectors do not catch the ultrafine particles that are most injurious to the lungs. Generally, when any air is recirculated, frequent dust count surveys should be made to insure that the system is operating satisfactorily and that other conditions have not added to the dust count in the breathing zone.

---

* Bureau of Industrial Health, Michigan Department of Health, Publication, May 22, 1947.

† A. C. Stern, J. Baliff, A. E. Perina, R. Crowley, B. Feiner, and A. A. Urbano, *Heating Piping and Air Conditioning*, ASHVE Journal Section, May 1946.

SECTION 28

MATHEMATICAL TABLES

COMMON LOGARITHMS

Logarithms (abbreviation *log*).—The log of a number is the exponent of the power to which it is necessary to raise a fixed number to produce the given number. The fixed number is called the *base*. Thus if the base is 10, the log of 1000 is 3, for $10^3 = 1000$. There are two systems of logs in general use, the *common*, in which the base is 10, and the Napierian, or *hyperbolic*, in which the base is 2.718281828. . . . The Napierian base is commonly denoted by e, as in the equation $e^y = x$, in which y is the Napierian log of x. The abbreviation $\log_e$ is commonly used to denote the Napierian log.

In any system of logs, the log of 1 is 0; the log of the base, taken in that system, is 1. In any system the base of which is greater than 1, the logs of all numbers greater than 1 are positive and the logs of all numbers less than 1 are negative.

The modulus of any system is equal to the reciprocal of the Napierian log of the base of that system. The modulus of the Napierian system is 1, that of the common system is 0.4342945. The log of a number in any system equals the modulus of that system × Napierian log of the number. The *hyperbolic* or *Napierian* log of any number equals the common log × 2.3025851.

Every log consists of two parts, an integral part called the *characteristic*, or index, and the decimal part, or *mantissa*. The mantissa only is given in the usual tables of common logs, with the decimal point omitted. The characteristic is one less than the number of figures to the left of the decimal point in the number whose log is to be found. The characteristic of numbers from 1 to 9.99+ is 0, from 10 to 99.99+ is 1, from 100 to 999+ is 2, from 0.1 to 0.99+ is −1, from 0.01 to 0.099+ is −2, etc. Thus,

log of 2000 is 3.30103;	log of 0.2 is −1.30103,	or	9.30103 − 10
" " 200 " 2.30103;	" " 0.02 " −2.30103,	"	8.30103 − 10
" " 20 " 1.30103;	" " 0.002 " −3.30103,	"	7.30103 − 10
" " 2 " 0.30103;	" " 0.0002 " −4.30103,	"	6.30103 − 10

The minus sign is frequently written above the characteristic thus: log 0.002 = $\bar{3}.30103$. The characteristic only is negative, the decimal part, or mantissa, being always positive.

When a log consists of a negative index and a positive mantissa, it is usual to write the negative sign over the index, or to add 10 to the index, and indicate the subtraction of 10 from the resulting logarithm. Thus log 0.2 = $\bar{1}.30103$, may be written 9.30103−10. The difference between a logarithm and 10 is its *arithmetical complement* or **cologarithm.**

In tables of logarithmic sines, etc., the −10 is generally omitted, as being understood.

RULES FOR USE OF THE TABLE OF COMMON LOGARITHMS.—To Find the Log of a Decimal Fraction or of a Whole Number and a Decimal.—First find the log of the quantity as if there were no decimal point, then prefix the index according to rule; the index is one less than the number of figures to the left of the decimal point.

Example. log of 3.14159.	log of 3.141	= 0.497068.	Diff. = 138.
From proportional parts	5 =	690	
" " "	09 =	1242	
	log 3.14159	0.4971494	

If the number is a decimal less than unity, the index is negative and is one more than the number of zeros to the right of the decimal point. Log of 0.0682 = $\bar{2}.833784$ = 8.833784 − 10.

To Find the Number Corresponding to a Given Log.—Find in the table the log nearest to the decimal part of the given log and take the first four digits of the required number from the column N and the top of the column containing the log which is the next less than the given log. To find the 5th and 6th digits subtract the log in the table from the given log, and multiply the difference by 100, and divide by the figure in the Diff. column opposite the log; annex the quotient to the four digits already found, and place the decimal point according to the rule; the number of figures to the left of the decimal point is one greater than the index. The number corresponding to a log is called the anti-logarithm.

Find the anti-log of.............................. 0.497150
Next lowest log in table corresponds to 3141........ 0.497068 Diff. = 82.
Tabular diff. = 138; 82 ÷ 138 = 0.59+

The index being 0, the number is therefore 3.14159+.

Multiplication by Means of Logarithms.—Add together the logs of the two numbers to be multiplied. The sum is the log of the product.

a. Where both factors are greater than unity.

EXAMPLE. $31 \times 1274 = 39{,}494$.

Solution. $\log 31 + \log 1274 = 1.491362 + 3.105169 = 4.596531 = \log 39{,}494$.

b. Where one or more factors are less than unity, the logs with a negative characteristic can be handled most conveniently by adding and subtracting 10.

EXAMPLE. $.000028961 \times .084507 = .0000024474$.

Solution. $\log .000028961 + \log .084507 = \bar{5}.461814 + \bar{2}.926893$.

$$= (5.461814 - 10) + (8.926893 - 10) = 14.388707 - 20 = \bar{6}.388707$$
$$= \log .0000024474.$$

Division by Means of Logarithms.—Subtract the log of the divisor from the log of the dividend. The remainder is the log of the quotient.

a. When the divisor is smaller than the dividend.

EXAMPLE. $2987 \div 63 = 47.41284$.

Solution. $\log 2987 - \log 63 = 3.475235 - 1.799341 = 1.675894 = \log 47.41284$.

b. When the divisor is larger than the dividend, add and subtract as many tens as may be necessary to the log of the dividend and proceed as before.

EXAMPLE. $.000672 \div 263 = .00000255513$.

Solution. $\log .000672 - \log 263 = \bar{4}.827369 - 2.419956$
$$= 6.827369 - 2.419956 - 10 = \bar{6}.407413 = \log .00000255513.$$

c. The log of a fraction is obtained by subtracting the log of the denominator from the log of the numerator. Thus,

$$\log \frac{a}{b} = \log a - \log b.$$

To Raise a Number to Any Given Power.—Multiply the log of the number by the exponent of the number, and find the number whose log is the product.

a. Where the exponent consists of one figure.

EXAMPLE. $16.2^3 = 4251.528$.

Solution. $3 \times \log 16.2 = 3 \times 1.209515 = 3.628545 = \log 4251.528$.

b. Where the exponent consists of two or more figures, it is best to multiply the characteristic and mantissa separately.

EXAMPLE. $.005624^{.37} = .147067$.

Solution. $.37 \times \log .005624 = .37 \times \bar{3}.750045 = .37 \times (-3) + .37 \times .750045$
$$= -1.11 + .277517 = \bar{1}.167517 = \log .147067.$$

c. Where the number is a fraction, first find the log of the fraction and then multiply it by the exponent.

EXAMPLE. $\left(\frac{276}{.032}\right)^{.72} = 681.9396$.

Solution. $.72\,(\log 276 - \log .032) = .72\{\,2.440909 - (8.505150 - 10)\,\}$
$$= .72 \times 3.935759 = 2.833746 = \log 681.9396.$$

To Extract Any Root of a Number.—Divide the log of the number by the index of the root, and find the number whose log is the quotient.

To extract the root of a decimal: *a.* When the root index is positive and evenly divisible into the negative characteristic of the log of the number, the division may be performed with the negative characteristic written in its usual place.

EXAMPLE. $\sqrt[4]{.0006954} = .16239$.

Solution. $\log .0006954 \div 4 = \bar{4}.842235 \div 4 = \bar{1}.210559 = \log .16239$.

b. When the root index is positive and not even divisible into the negative characteristic, add to the log of the number, and indicate the subtraction from it, the smallest integral multiple of the root which will eliminate the negative characteristic. Divide the result by the root index and ascertain the number whose log corresponds to the quotient.

EXAMPLE. $\sqrt[3.1]{.00002785} = .03393$.

Solution. $\log .00002785 = \bar{5}.444825;\ \bar{5}.444825 \div 3.1 = \{(2 \times 3.1) + \bar{5}.444825 - 2 \times 3.1\} \div 3.1$
$$= \{1.644825 - 6.2\} \div 3.1 = \bar{2}.530589 = \log .03393.$$

c. When the root index is negative, determine the excess of the negative characteristic over the positive mantissa. Divide the result by the root index and ascertain the number whose log corresponds to the quotient.

EXAMPLE. $\sqrt[-4]{.000003976} = 22.394$.

Solution. $\log .000003976 = \bar{6}.599446 = .599446 - 6 = -5.400554;$
$$-5.400554 \div (-4) = 1.350138 = \log 22.394.$$

Solution of Exponential Equations.—In an exponential equation, the unknown quantity is the exponent; thus $a^x = b$. This may be transformed to $\log a^x = \log b$, or $x \log a = \log b$, whence $x = \log b \div \log a$.

a. When the base is greater than unity, put the equation in the form $x = \log b \div \log a$. Then $\log x = \log (\log b) - \log (\log a)$.

EXAMPLE. $32.6^x = 14.632$.

Solution. $\log x = \log (\log 14.632) - \log (\log 32.6) = \log 1.165303 - \log 1.513218$
$= .066439 - .179901 = \bar{1}.886538;\ x = .77008.$

b. When both the known quantities are decimal, put the equation in the form $x = \log b \div \log a$. Subtract the positive mantissa from the negative characteristic in both divisor and dividend, obtaining negative remainders. Change the signs of divisor and dividend and proceed as in Case *a*.

EXAMPLE. $.0729^x = .2693$.

Solution. $x = \log .2693 \div \log .0729 = \bar{1}.430236 \div \bar{2}.862728$
$= (-.569764) \div (-1.137272) = .569764 \div 1.137272;$
$\log x = \log .569764 - \log 1.137272 = \bar{1}.755695 - .055864 = \bar{1}.699831$
$x = .50099.$

c. When only one of the known quantities is a decimal, put the equation in the form $x = \dfrac{\log b}{\log a}$. Subtract the positive mantissa from the negative characteristic of the numerator or denominator as the case may be, and rewrite the fraction with the remainder so obtained as the new numerator or denominator. Make both numerator and denominator positive, but write a minus sign in front of the fraction, to signify that the result will be a negative quantity. Solve the fraction by logarithms and write a minus sign in front of the result.

EXAMPLE. $.726^x = 802.7$.

Solution. $x = \dfrac{\log 802.7}{\log .726} = \dfrac{2.904553}{\bar{1}.860937} = \dfrac{2.904553}{-.139063} = -\dfrac{2.904553}{.139063}.$

$\log x = -(\log 2.904553 - \log .139063) = (.463079 - \bar{1}.143211) = -(1.319868)$
$x = -20.8867$

d. When the exponent is negative and one of the known quantities is less than unity, put the equation in the form $(-x) = \dfrac{\log b}{\log a}$. Subtract the positive mantissa from the negative characteristic, as in Case *c*, and multiply both sides of the resulting equation by (-1). Find the value of x as in Case *c*.

EXAMPLE. $10.78^{-x} = .09431$.

Solution. $x = \dfrac{\log .09431}{\log 10.78} = \dfrac{\bar{2}.974558}{1.032619};\ x = \dfrac{1.025442}{1.032619}.$

$\log x = .010911 - .013940 = \bar{1}.996971.\quad x = .99305.$

Table 1. Logarithms of Numbers from 1 to 100

N	Log	N	Log	N	Log	N	Log	N	Log
1	0.000000	21	1.322219	41	1.612784	61	1.785330	81	1.908485
2	0.301030	22	1.342423	42	1.623249	62	1.792392	82	1.913814
3	0.477121	23	1.361728	43	1.633468	63	1.799341	83	1.919078
4	0.602060	24	1.380211	44	1.643453	64	1.806180	84	1.924279
5	0.698970	25	1.397940	45	1.653213	65	1.812913	85	1.929419
6	0.778151	26	1.414973	46	1.662758	66	1.819544	86	1.934498
7	0.845098	27	1.431364	47	1.672098	67	1.826075	87	1.939519
8	0.903090	28	1.447158	48	1.681241	68	1.832509	88	1.944483
9	0.954243	29	1.462398	49	1.690196	69	1.838849	89	1.949390
10	1.000000	30	1.477121	50	1.698970	70	1.845098	90	1.954243
11	1.041393	31	1.491362	51	1.707570	71	1.851258	91	1.959041
12	1.079181	32	1.505150	52	1.716003	72	1.857332	92	1.963788
13	1.113943	33	1.518514	53	1.724276	73	1.863323	93	1.968483
14	1.146128	34	1.531479	54	1.732394	74	1.869232	94	1.973128
15	1.176091	35	1.544068	55	1.740363	75	1.875061	95	1.977724
16	1.204120	36	1.556303	56	1.748188	76	1.880814	96	1.982271
17	1.230449	37	1.568202	57	1.755875	77	1.886491	97	1 986772
18	1.255273	38	1.579784	58	1.763428	78	1.892095	98	1.991226
19	1.278754	39	1.591065	59	1.770852	79	1.897627	99	1.995635
20	1.301030	40	1.602060	60	1.778151	80	1.903090	100	2.000000

See pp. 28–05 to 28–22 for a complete table of six-place logarithms.

Table 2. Common Logarithms of Numbers

N	0	1	2	3	4	5	6	7	8	9	Diff.
100	**000000**	**000434**	**000868**	**001301**	**001734**	**002166**	**002598**	**003029**	**003461**	**003891**	432
1	004321	004751	005181	005609	006038	006466	006894	007321	007748	008174	428
2	008600	009026	009451	009876	010300	010724	011147	011570	011993	012415	424
3	012837	013259	013680	014100	014521	014940	015360	015779	016197	016616	420
4	017033	017451	017868	018284	018700	019116	019532	019947	020361	020775	416
5	021189	021603	022016	022428	022841	023252	023664	024075	024486	024896	412
6	025306	025715	026125	026533	026942	027350	027757	028164	028571	028978	408
7	029384	029789	030195	030600	031004	031408	031812	032216	032619	033021	404
8	033424	033826	034227	034628	035029	035430	035830	036230	036629	037028	400
9	037426	037825	038223	038620	039017	039414	039811	040207	040602	040998	397
110	**041393**	**041787**	**042182**	**042576**	**042969**	**043362**	**043755**	**044148**	**044540**	**044932**	393
1	045323	045714	046105	046495	046885	047275	047664	048053	048442	048830	390
2	049218	049606	049993	050380	050766	051153	051538	051924	052309	052694	386
3	053078	053463	053846	054230	054613	054996	055378	055760	056142	056524	383
4	056905	057286	057666	058046	058426	058805	059185	059563	059942	060320	379
5	060698	061075	061452	061829	062206	062582	062958	063333	063709	064083	376
6	064458	064832	065206	065580	065953	066326	066699	067071	067443	067815	373
7	068186	068557	068928	069298	069668	070038	070407	070776	071145	071514	370
8	071882	072250	072617	072985	073352	073718	074085	074451	074816	075182	366
9	075547	075912	076276	076640	077004	077368	077731	078094	078457	078819	363
120	**079181**	**079543**	**079904**	**080266**	**080626**	**080987**	**081347**	**081707**	**082067**	**082426**	360
1	082785	083144	083503	083861	084219	084576	084934	085291	085647	086004	357
2	086360	086716	087071	087426	087781	088136	088490	088845	089198	089552	355
3	089905	090258	090611	090963	091315	091667	092018	092370	092721	093071	352
4	093422	093772	094122	094471	094820	095169	095518	095866	096215	096562	349
5	096910	097257	097604	097951	098298	098644	098990	099335	099681	100026	346
6	100371	100715	101059	101403	101747	102091	102434	102777	103119	103462	343
7	103804	104146	104487	104828	105169	105510	105851	106191	106531	106871	341
8	107210	107549	107888	108227	108565	108903	109241	109579	109916	110253	338
9	110590	110926	111263	111599	111934	112270	112605	112940	113275	113609	335

PROPORTIONAL PARTS

Diff.	1	2	3	4	5	6	7	8	9
434	43.4	86.8	130.2	173.6	217.0	260.4	303.8	347.2	390.6
432	43.2	86.4	129.6	172.8	216.0	259.2	302.4	345.6	388.8
430	43.0	86.0	129.0	172.0	215.0	258.0	301.0	344.0	387.0
428	42.8	85.6	128.4	171.2	214.0	256.8	299.6	342.4	385.2
426	42.6	85.2	127.8	170.4	213.0	255.6	298.2	340.8	383.4
424	42.4	84.8	127.2	169.6	212.0	254.4	296.8	339.2	381.6
422	42.2	84.4	126.6	168.8	211.0	253.2	295.4	337.6	379.8
420	42.0	84.0	126.0	168.0	210.0	252.0	294.0	336.0	378.0
418	41.8	83.6	125.4	167.2	209.0	250.8	292.6	334.4	376.2
416	41.6	83.2	124.8	166.4	208.0	249.6	291.2	332.8	374.4
414	41.4	82.8	124.2	165.6	207.0	248.4	289.8	331.2	372.6
412	41.2	82.4	123.6	164.8	206.0	247.2	288.4	329.6	370.8
410	41.0	82.0	123.0	164.0	205.0	246.0	287.0	328.0	369.0
408	40.8	81.6	122.4	163.2	204.0	244.8	285.6	326.4	367.2
406	40.6	81.2	121.8	162.4	203.0	243.6	284.2	324.8	365.4
404	40.4	80.8	121.2	161.6	202.0	242.4	282.8	323.2	363.6
402	40.2	80.4	120.6	160.8	201.0	241.2	281.4	321.6	361.8
400	40.0	80.0	120.0	160.0	200.0	240.0	280.0	320.0	360.0
398	39.8	79.6	119.4	159.2	199.0	238.8	278.6	318.4	358.2
396	39.6	79.2	118.8	158.4	198.0	237.6	277.2	316.8	356.4
394	39.4	78.8	118.2	157.6	197.0	236.4	275.8	315.2	354.6
392	39.2	78.4	117.6	156.8	196.0	235.2	274.4	313.6	352.8
390	39.0	78.0	117.0	156.0	195.0	234.0	273.0	312.0	351.0
388	38.8	77.6	116.4	155.2	194.0	232.8	271.6	310.4	349.2
386	38.6	77.2	115.8	154.4	193.0	231.6	270.2	308.8	347.4
384	38.4	76.8	115.2	153.6	192.0	230.4	268.8	307.2	345.6
382	38.2	76.4	114.6	152.8	191.0	229.2	267.4	305.6	343.8
380	38.0	76.0	114.0	152.0	190.0	228.0	266.0	304.0	342.0
378	37.8	75.6	113.4	151.2	189.0	226.8	264.6	302.4	340.2
376	37.6	75.2	112.8	150.4	188.0	225.6	263.2	300.8	338.4
374	37.4	74.8	112.2	149.6	187.0	224.4	261.8	299.2	336.6
372	37.2	74.4	111.6	148.8	186.0	223.2	260.4	297.6	334.8
370	37.0	74.0	111.0	148.0	185.0	222.0	259.0	296.0	333.0
368	36.8	73.6	110.4	147.2	184.0	220.8	257.6	294.4	331.2
366	36.6	73.2	109.8	146.4	183.0	219.6	256.2	292.8	329.4
364	36.4	72.8	109.2	145.6	182.0	218.4	254.8	291.2	327.6
362	36.2	72.4	108.6	144.8	181.0	217.2	253.4	289.6	325.8
360	36.0	72.0	108.0	144.0	180.0	216.0	252.0	288.0	324.0

N	0	1	2	3	4	5	6	7	8	9	Diff.
130	**113943**	**114277**	**114611**	**114944**	**115278**	**115611**	**115943**	**116276**	**116608**	**116940**	333
1	117271	117603	117934	118265	118595	118926	119256	119586	119915	120245	330
2	120574	120903	121231	121560	121888	122216	122544	122871	123198	123525	328
3	123852	124178	124504	124830	125156	125481	125806	126131	126456	126781	325
4	127105	127429	127753	128076	128399	128722	129045	129368	129690	130012	323
5	130334	130655	130977	131298	131619	131939	132260	132580	132900	133219	321
6	133539	133858	134177	134496	134814	135133	135451	135769	136086	136403	318
7	136721	137037	137354	137671	137987	138303	138618	138934	139249	139564	316
8	139879	140194	140508	140822	141136	141450	141763	142076	142389	142702	314
9	143015	143327	143639	143951	144263	144574	144885	145196	145507	145818	311
140	**146128**	**146438**	**146748**	**147058**	**147367**	**147676**	**147985**	**148294**	**148603**	**148911**	309
1	149219	149527	149835	150142	150449	150756	151063	151370	151676	151982	307
2	152288	152594	152900	153205	153510	153815	154120	154424	154728	155032	305
3	155336	155640	155943	156246	156549	156852	157154	157457	157759	158061	303
4	158362	158664	158965	159266	159567	159868	160168	160469	160769	161068	301
5	161368	161667	161967	162266	162564	162863	163161	163460	163758	164055	299
6	164353	164650	164947	165244	165541	165838	166134	166430	166726	167022	297
7	167317	167613	167908	168203	168497	168792	169086	169380	169674	169968	295
8	170262	170555	170848	171141	171434	171726	172019	172311	172603	172895	293
9	173186	173478	173769	174060	174351	174641	174932	175222	175512	175802	291
150	**176091**	**176381**	**176670**	**176959**	**177248**	**177536**	**177825**	**178113**	**178401**	**178689**	289
1	178977	179264	179552	179839	180126	180413	180699	180986	181272	181558	287
2	181844	182129	182415	182700	182985	183270	183555	183839	184123	184407	285
3	184691	184975	185259	185542	185825	186108	186391	186674	186956	187239	283
4	187521	187803	188084	188366	188647	188928	189209	189490	189771	190051	281
5	190332	190612	190892	191171	191451	191730	192010	192289	192567	192846	279
6	193125	193403	193681	193959	194237	194514	194792	195069	195346	195623	278
7	195900	196176	196453	196729	197005	197281	197556	197832	198107	198382	276
8	198657	198932	199206	199481	199755	200029	200303	200577	200850	201124	274
9	201397	201670	201943	202216	202488	202761	203033	203305	203577	203848	272

PROPORTIONAL PARTS

Diff.	1	2	3	4	5	6	7	8	9
358	35.8	71.6	107.4	143.2	179.0	214.8	250.6	286.4	322.2
356	35.6	71.2	106.8	142.4	178.0	213.6	249.2	284.8	320.4
354	35.4	70.8	106.2	141.6	177.0	212.4	247.8	283.2	318.6
352	35.2	70.4	105.6	140.8	176.0	211.2	246.4	281.6	316.8
350	35.0	70.0	105.0	140.0	175.0	210.0	245.0	280.0	315.0
348	34.8	69.6	104.4	139.2	174.0	208.8	243.6	278.4	313.2
346	34.6	69.2	103.8	138.4	173.0	207.6	242.2	276.8	311.4
344	34.4	68.8	103.2	137.6	172.0	206.4	240.8	275.2	309.6
342	34.2	68.4	102.6	136.8	171.0	205.2	239.4	273.6	307.8
340	34.0	68.0	102.0	136.0	170.0	204.0	238.0	272.0	306.0
338	33.8	67.6	101.4	135.2	169.0	202.8	236.6	270.4	304.2
336	33.6	67.2	100.8	134.4	168.0	201.6	235.2	268.8	302.4
334	33.4	66.8	100.2	133.6	167.0	200.4	233.8	267.2	300.6
332	33.2	66.4	99.6	132.8	166.0	199.2	232.4	265.6	298.8
330	33.0	66.0	99.0	132.0	165.0	198.0	231.0	264.0	297.0
328	32.8	65.6	98.4	131.2	164.0	196.8	229.6	262.4	295.2
326	32.6	65.2	97.8	130.4	163.0	195.6	228.2	260.8	293.4
324	32.4	64.8	97.2	129.6	162.0	194.4	226.8	259.2	291.6
322	32.2	64.4	96.6	128.8	161.0	193.2	225.4	257.6	289.8
320	32.0	64.0	96.0	128.0	160.0	192.0	224.0	256.0	288.0
318	31.8	63.6	95.4	127.2	159.0	190.8	222.6	254.4	286.2
316	31.6	63.2	94.8	126.4	158.0	189.6	221.2	252.8	284.4
314	31.4	62.8	94.2	125.6	157.0	188.4	219.8	251.2	282.6
312	31.2	62.4	93.6	124.8	156.0	187.2	218.4	249.6	280.8
310	31.0	62.0	93.0	124.0	155.0	186.0	217.0	248.0	279.0
308	30.8	61.6	92.4	123.2	154.0	184.8	215.6	246.4	277.2
306	30.6	61.2	91.8	122.4	153.0	183.6	214.2	244.8	275.4
304	30.4	60.8	91.2	121.6	152.0	182.4	212.8	243.2	273.6
302	30.2	60.4	90.6	120.8	151.0	181.2	211.4	241.6	271.8
300	30.0	60.0	90.0	120.0	150.0	180.0	210.0	240.0	270.0
298	29.8	59.6	89.4	119.2	149.0	178.8	208.6	238.4	268.2
296	29.6	59.2	88.8	118.4	148.0	177.6	207.2	236.8	266.4
294	29.4	58.8	88.2	117.6	147.0	176.4	205.8	235.2	264.6
292	29.2	58.4	87.6	116.8	146.0	175.2	204.4	233.6	262.8
290	29.0	58.0	87.0	116.0	145.0	174.0	203.0	232.0	261.0
288	28.8	57.6	86.4	115.2	144.0	172.8	201.6	230.4	259.2
286	28.6	57.2	85.8	114.4	143.0	171.6	200.2	228.8	257.4
284	28.4	56.8	85.2	113.6	142.0	170.4	198.8	227.2	255.6
282	28.2	56.4	84.6	112.8	141.0	169.2	197.4	225.6	253.8
280	28.0	56.0	84.0	112.0	140.0	168.0	196.0	224.0	252.0

N	0	1	2	3	4	5	6	7	8	9	Diff.
160	**204120**	**204391**	**204663**	**204934**	**205204**	**205475**	**205746**	**206016**	**206286**	**206556**	271
1	206826	207096	207365	207634	207904	208173	208441	208710	208979	209247	269
2	209515	209783	210051	210319	210586	210853	211121	211388	211654	211921	267
3	212188	212454	212720	212986	213252	213518	213783	214049	214314	214579	266
4	214844	215109	215373	215638	215902	216166	216430	216694	216957	217221	264
5	217484	217747	218010	218273	218536	218798	219060	219323	219585	219846	262
6	220108	220370	220631	220892	221153	221414	221675	221936	222196	222456	261
7	222716	222976	223236	223496	223755	224015	224274	224533	224792	225051	259
8	225309	225568	225826	226084	226342	226600	226858	227115	227372	227630	258
9	227887	228144	228400	228657	228913	229170	229426	229682	229938	230193	256
170	**230449**	**230704**	**230960**	**231215**	**231470**	**231724**	**231979**	**232234**	**232488**	**232742**	255
1	232996	233250	233504	233757	234011	234264	234517	234770	235023	235276	253
2	235528	235781	236033	236285	236537	236789	237041	237292	237544	237795	252
3	238046	238297	238548	238799	239049	239299	239550	239800	240050	240300	250
4	240549	240799	241048	241297	241546	241795	242044	242293	242541	242790	249
5	243038	243286	243534	243782	244030	244277	244525	244772	245019	245266	248
6	245513	245759	246006	246252	246499	246745	246991	247237	247482	247728	246
7	247973	248219	248464	248709	248954	249198	249443	249687	249932	250176	245
8	250420	250664	250908	251151	251395	251638	251881	252125	252368	252610	243
9	252853	253096	253338	253580	253822	254064	254306	254548	254790	255031	242
180	**255273**	**255514**	**255755**	**255996**	**256237**	**256477**	**256718**	**256958**	**257198**	**257439**	241
1	257679	257918	258158	258398	258637	258877	259116	259355	259594	259833	239
2	260071	260310	260548	260787	261025	261263	261501	261739	261976	262214	238
3	262451	262688	262925	263162	263399	263636	263873	264109	264346	264582	237
4	264818	265054	265290	265525	265761	265996	266232	266467	266702	266937	235
5	267172	267406	267641	267875	268110	268344	268578	268812	269046	269279	234
6	269513	269746	269980	270213	270446	270679	270912	271144	271377	271609	233
7	271842	272074	272306	272538	272770	273001	273233	273464	273696	273927	232
8	274158	274389	274620	274850	275081	275311	275542	275772	276002	276232	230
9	276462	276692	276921	277151	277380	277609	277838	278067	278296	278525	229
190	**278754**	**278982**	**279211**	**279439**	**279667**	**279895**	**280123**	**280351**	**280578**	**280806**	228
1	281033	281261	281488	281715	281942	282169	282396	282622	282849	283075	227
2	283301	283527	283753	283979	284205	284431	284656	284882	285107	285332	226
3	285557	285782	286007	286232	286456	286681	286905	287130	287354	287578	225
4	287802	288026	288249	288473	288696	288920	289143	289366	289589	289812	223
5	290035	290257	290480	290702	290925	291147	291369	291591	291813	292034	222
6	292256	292478	292699	292920	293141	293363	293584	293804	294025	294246	221
7	294466	294687	294907	295127	295347	295567	295787	296007	296226	296446	220
8	296665	296884	297104	297323	297542	297761	297979	298198	298416	298635	219
9	298853	299071	299289	299507	299725	299943	300161	300378	300595	300813	218

PROPORTIONAL PARTS

Diff.	1	2	3	4	5	6	7	8	9
278	27.8	55.6	83.4	111.2	139.0	166.8	194.6	222.4	250.2
276	27.6	55.2	82.8	110.4	138.0	165.6	193.2	220.8	248.4
274	27.4	54.8	82.2	109.6	137.0	164.4	191.8	219.2	246.6
272	27.2	54.4	81.6	108.8	136.0	163.2	190.4	217.6	244.8
270	27.0	54.0	81.0	108.0	135.0	162.0	189.0	216.0	243.0
268	26.8	53.6	80.4	107.2	134.0	160.8	187.6	214.4	241.2
266	26.6	53.2	79.8	106.4	133.0	159.6	186.2	212.8	239.4
264	26.4	52.8	79.2	105.6	132.0	158.4	184.8	211.2	237.6
262	26.2	52.4	78.6	104.8	131.0	157.2	183.4	209.6	235.8
260	26.0	52.0	78.0	104.0	130.0	156.0	182.0	208.0	234.0
258	25.8	51.6	77.4	103.2	129.0	154.8	180.6	206.4	232.2
256	25.6	51.2	76.8	102.4	128.0	153.6	179.2	204.8	230.4
254	25.4	50.8	76.2	101.6	127.0	152.4	177.8	203.2	228.6
252	25.2	50.4	75.6	100.8	126.0	151.2	176.4	201.6	226.8
250	25.0	50.0	75.0	100.0	125.0	150.0	175.0	200.0	225.0
248	24.8	49.6	74.4	99.2	124.0	148.8	173.6	198.4	223.2
246	24.6	49.2	73.8	98.4	123.0	147.6	172.2	196.8	221.4
244	24.4	48.8	73.2	97.6	122.0	146.4	170.8	195.2	219.6
242	24.2	48.4	72.6	96.8	121.0	145.2	169.4	193.6	217.8
240	24.0	48.0	72.0	96.0	120.0	144.0	168.0	192.0	216.0
238	23.8	47.6	71.4	95.2	119.0	142.8	166.6	190.4	214.2
236	23.6	47.2	70.8	94.4	118.0	141.6	165.2	188.8	212.4
234	23.4	46.8	70.2	93.6	117.0	140.4	163.8	187.2	210.6
232	23.2	46.4	69.6	92.8	116.0	139.2	162.4	185.6	208.8
230	23.0	46.0	69.0	92.0	115.0	138.0	161.0	184.0	207.0

N	0	1	2	3	4	5	6	7	8	9	Diff.
200	**301030**	**301247**	**301464**	**301681**	**301898**	**302114**	**302331**	**302547**	**302764**	**302980**	217
1	303196	303412	303628	303844	304059	304275	304491	304706	304921	305136	216
2	305351	305566	305781	305996	306211	306425	306639	306854	307068	307282	215
3	307496	307710	307924	308137	308351	308564	308778	308991	309204	309417	213
4	309630	309843	310056	310268	310481	310693	310906	311118	311330	311542	212
5	311754	311966	312177	312389	312600	312812	313023	313234	313445	313656	211
6	313867	314078	314289	314499	314710	314920	315130	315340	315551	315760	210
7	315970	316180	316390	316599	316809	317018	317227	317436	317646	317854	209
8	318063	318272	318481	318689	318898	319106	319314	319522	319730	319938	208
9	320146	320354	320562	320769	320977	321184	321391	321598	321805	322012	207
210	**322219**	**322426**	**322633**	**322839**	**323046**	**323252**	**323458**	**323665**	**323871**	**324077**	206
1	324282	324488	324694	324899	325105	325310	325516	325721	325926	326131	205
2	326336	326541	326745	326950	327155	327359	327563	327767	327972	328176	204
3	328380	328583	328787	328991	329194	329398	329601	329805	330008	330211	203
4	330414	330617	330819	331022	331225	331427	331630	331832	332034	332236	202
5	332438	332640	332842	333044	333246	333447	333649	333850	334051	334253	
6	334454	334655	334856	335057	335257	335458	335658	335859	336059	336260	201
7	336460	336660	336860	337060	337260	337459	337659	337858	338058	338257	200
8	338456	338656	338855	339054	339253	339451	339650	339849	340047	340246	199
9	340444	340642	340841	341039	341237	341435	341632	341830	342028	342225	198
220	**342423**	**342620**	**342817**	**343014**	**343212**	**343409**	**343606**	**343802**	**343999**	**344196**	197
1	344392	344589	344785	344981	345178	345374	345570	345766	345962	346157	196
2	346353	346549	346744	346939	347135	347330	347525	347720	347915	348110	195
3	348305	348500	348694	348889	349083	349278	349472	349666	349860	350054	194
4	350248	350442	350636	350829	351023	351216	351410	351603	351796	351989	193
5	352183	352375	352568	352761	352954	353147	353339	353532	353724	353916	
6	354108	354301	354493	354685	354876	355068	355260	355452	355643	355834	192
7	356026	356217	356408	356599	356790	356981	357172	357363	357554	357744	191
8	357935	358125	358316	358506	358696	358886	359076	359266	359456	359646	190
9	359835	360025	360215	360404	360593	360783	360972	361161	361350	361539	189
230	**361728**	**361917**	**362105**	**362294**	**362482**	**362671**	**362859**	**363048**	**363236**	**363424**	188
1	363612	363800	363988	364176	364363	364551	364739	364926	365113	365301	
2	365488	365675	365862	366049	366236	366423	366610	366796	366983	367169	187
3	367356	367542	367729	367915	368101	368287	368473	368659	368845	369030	186
4	369216	369401	369587	369772	369958	370143	370328	370513	370698	370883	185
5	371068	371253	371437	371622	371806	371991	372175	372360	372544	372728	184
6	372912	373096	373280	373464	373647	373831	374015	374198	374382	374565	
7	374748	374932	375115	375298	375481	375664	375846	376029	376212	376394	183
8	376577	376759	376942	377124	377306	377488	377670	377852	378034	378216	182
9	378398	378580	378761	378943	379124	379306	379487	379668	379849	380030	181

PROPORTIONAL PARTS

Diff.	1	2	3	4	5	6	7	8	9
228	22.8	45.6	68.4	91.2	114.0	136.8	159.6	182.4	205.2
226	22.6	45.2	67.8	90.4	113.0	135.6	158.2	180.8	203.4
224	22.4	44.8	67.2	89.6	112.0	134.4	156.8	179.2	201.6
222	22.2	44.4	66.6	88.8	111.0	133.2	155.4	177.6	199.8
220	22.0	44.0	66.0	88.0	110.0	132.0	154.0	176.0	198.0
218	21.8	43.6	65.4	87.2	109.0	130.8	152.6	174.4	196.2
216	21.6	43.2	64.8	86.4	108.0	129.6	151.2	172.8	194.4
214	21.4	42.8	64.2	85.6	107.0	128.4	149.8	171.2	192.6
212	21.2	42.4	63.6	84.8	106.0	127.2	148.4	169.6	190.8
210	21.0	42.0	63.0	84.0	105.0	126.0	147.0	168.0	189.0
208	20.8	41.6	62.4	83.2	104.0	124.8	145.6	166.4	187.2
206	20.6	41.2	61.8	82.4	103.0	123.6	144.2	164.8	185.4
204	20.4	40.8	61.2	81.6	102.0	122.4	142.8	163.2	183.6
202	20.2	40.4	60.6	80.8	101.0	121.2	141.4	161.6	181.8
200	20.0	40.0	60.0	80.0	100.0	120.0	140.0	160.0	180.0
198	19.8	39.6	59.4	79.2	99.0	118.8	138.6	158.4	178.2
196	19.6	39.2	58.8	78.4	98.0	117.6	137.2	156.8	176.4
194	19.4	38.8	58.2	77.6	97.0	116.4	135.8	155.2	174.6
192	19.2	38.4	57.6	76.8	96.0	115.2	134.4	153.6	172.8
190	19.0	38.0	57.0	76.0	95.0	114.0	133.0	152.0	171.0
188	18.8	37.6	56.4	75.2	94.0	112.8	131.6	150.4	169.2
186	18.6	37.2	55.8	74.4	93.0	111.6	130.2	148.8	167.4
184	18.4	36.8	55.2	73.6	92.0	110.4	128.8	147.2	165.6
182	18.2	36.4	54.6	72.8	91.0	109.2	127.4	145.6	163.8
180	18.0	36.0	54.0	72.0	90.0	108.0	126.0	144.0	162.0

N	0	1	2	3	4	5	6	7	8	9	Diff.
240	**380211**	**380392**	**380573**	**380754**	**380934**	**381115**	**381296**	**381476**	**381656**	**381837**	181
1	382017	382197	382377	382557	382737	382917	383097	383277	383456	383636	180
2	383815	383995	384174	384353	384533	384712	384891	385070	385249	385428	179
3	385606	385785	385964	386142	386321	386499	386677	386856	387034	387212	178
4	387390	387568	387746	387924	388101	388279	388456	388634	388811	388989	
5	389166	389343	389520	389698	389875	390051	390228	390405	390582	390759	177
6	390935	391112	391288	391464	391641	391817	391993	392169	392345	392521	176
7	392697	392873	393048	393224	393400	393575	393751	393926	394101	394277	
8	394452	394627	394802	394977	395152	395326	395501	395676	395850	396025	175
9	396199	396374	396548	396722	396896	397071	397245	397419	397592	397766	174
250	**397940**	**398114**	**398287**	**398461**	**398634**	**398808**	**398981**	**399154**	**399328**	**399501**	173
1	399674	399847	400020	400192	400365	400538	400711	400883	401056	401228	
2	401401	401573	401745	401917	402089	402261	402433	402605	402777	402949	172
3	403121	403292	403464	403635	403807	403978	404149	404320	404492	404663	171
4	404834	405005	405176	405346	405517	405688	405858	406029	406199	406370	
5	406540	406710	406881	407051	407221	407391	407561	407731	407901	408070	170
6	408240	408410	408579	408749	408918	409087	409257	409426	409595	409764	169
7	409933	410102	410271	410440	410609	410777	410946	411114	411283	411451	
8	411620	411788	411956	412124	412293	412461	412629	412796	412964	413132	168
9	413300	413467	413635	413803	413970	414137	414305	414472	414639	414806	167
260	**414973**	**415140**	**415307**	**415474**	**415641**	**415808**	**415974**	**416141**	**416308**	**416474**	
1	416641	416807	416973	417139	417306	417472	417638	417804	417970	418135	166
2	418301	418467	418633	418798	418964	419129	419295	419460	419625	419791	165
3	419956	420121	420286	420451	420616	420781	420945	421110	421275	421439	
4	421604	421768	421933	422097	422261	422426	422590	422754	422918	423082	164
5	423246	423410	423574	423737	423901	424065	424228	424392	424555	424718	
6	424882	425045	425208	425371	425534	425697	425860	426023	426186	426349	163
7	426511	426674	426836	426999	427161	427324	427486	427648	427811	427973	162
8	428135	428297	428459	428621	428783	428944	429106	429268	429429	429591	
9	429752	429914	430075	430236	430398	430559	430720	430881	431042	431203	161
270	**431364**	**431525**	**431685**	**431846**	**432007**	**432167**	**432328**	**432488**	**432649**	**432809**	
1	432969	433130	433290	433450	433610	433770	433930	434090	434249	434409	160
2	434569	434729	434888	435048	435207	435367	435526	435685	435844	436004	159
3	436163	436322	436481	436640	436799	436957	437116	437275	437433	437592	
4	437751	437909	438067	438226	438384	438542	438701	438859	439017	439175	158
5	439333	439491	439648	439806	439964	440122	440279	440437	440594	440752	
6	440909	441066	441224	441381	441538	441695	441852	442009	442166	442323	157
7	442480	442637	442793	442950	443106	443263	443419	443576	443732	443889	
8	444045	444201	444357	444513	444669	444825	444981	445137	445293	445449	156
9	445604	445760	445915	446071	446226	446382	446537	446692	446848	447003	155
280	**447158**	**447313**	**447468**	**447623**	**447778**	**447933**	**448088**	**448242**	**448397**	**448552**	
1	448706	448861	449015	449170	449324	449478	449633	449787	449941	450095	154
2	450249	450403	450557	450711	450865	451018	451172	451326	451479	451633	
3	451786	451940	452093	452247	452400	452553	452706	452859	453012	453165	153
4	453318	453471	453624	453777	453930	454082	454235	454387	454540	454692	
5	454845	454997	455150	455302	455454	455606	455758	455910	456062	456214	152
6	456366	456518	456670	456821	456973	457125	457276	457428	457579	457731	
7	457882	458033	458184	458336	458487	458638	458789	458940	459091	459242	151
8	459392	459543	459694	459845	459995	460146	460296	460447	460597	460748	
9	460898	461048	461198	461348	461499	461649	461799	461948	462098	462248	150

Proportional Parts

Diff.	1	2	3	4	5	6	7	8	9
182	18.2	36.4	54.6	72.8	91.0	109.2	127.4	145.6	163.8
180	18.0	36.0	54.0	72.0	90.0	108.0	126.0	144.0	162.0
178	17.8	35.6	53.4	71.2	89.0	106.8	124.6	142.4	160.2
176	17.6	35.2	52.8	70.4	88.0	105.6	123.2	140.8	158.4
174	17.4	34.8	52.2	69.6	87.0	104.4	121.8	139.2	156.6
172	17.2	34.4	51.6	68.8	86.0	103.2	120.4	137.6	154.8
170	17.0	34.0	51.0	68.0	85.0	102.0	119.0	136.0	153.0
168	16.8	33.6	50.4	67.2	84.0	100.8	117.6	134.4	151.2
166	16.6	33.2	49.8	66.4	83.0	99.6	116.2	132.8	149.4
164	16.4	32.8	49.2	65.6	82.0	98.4	114.8	131.2	147.6
162	16.2	32.4	48.6	64.8	81.0	97.2	113.4	129.6	145.8
160	16.0	32.0	48.0	64.0	80.0	96.0	112.0	128.0	144.0
158	15.8	31.6	47.4	63.2	79.0	94.8	110.6	126.4	142.2
156	15.6	31.2	46.8	62.4	78.0	93.6	109.2	124.8	140.4

N	0	1	2	3	4	5	6	7	8	9	Diff.
290	**462398**	**462548**	**462697**	**462847**	**462997**	**463146**	**463296**	**463445**	**463594**	**463744**	
1	463893	464042	464191	464340	464490	464639	464788	464936	465085	465234	149
2	465383	465532	465680	465829	465977	466126	466274	466423	466571	466719	
3	466868	467016	467164	467312	467460	467608	467756	467904	468052	468200	148
4	468347	468495	468643	468790	468938	469085	469233	469380	469527	469675	
5	469822	469969	470116	470263	470410	470557	470704	470851	470998	471145	147
6	471292	471438	471585	471732	471878	472025	472171	472318	472464	472610	146
7	472756	472903	473049	473195	473341	473487	473633	473779	473925	474071	
8	474216	474362	474508	474653	474799	474944	475090	475235	475381	475526	
9	475671	475816	475962	476107	476252	476397	476542	476687	476832	476976	145
300	**477121**	**477266**	**477411**	**477555**	**477700**	**477844**	**477989**	**478133**	**478278**	**478422**	
1	478566	478711	478855	478999	479143	479287	479431	479575	479719	479863	144
2	480007	480151	480294	480438	480582	480725	480869	481012	481156	481299	
3	481443	481586	481729	481872	482016	482159	482302	482445	482588	482731	143
4	482874	483016	483159	483302	483445	483587	483730	483872	484015	484157	
5	484300	484442	484585	484727	484869	485011	485153	485295	485437	485579	142
6	485721	485863	486005	486147	486289	486430	486572	486714	486855	486997	
7	487138	487280	487421	487563	487704	487845	487986	488127	488269	488410	141
8	488551	488692	488833	488974	489114	489255	489396	489537	489677	489818	
9	489958	490099	490239	490380	490520	490661	490801	490941	491081	491222	140
310	**491362**	**491502**	**491642**	**491782**	**491922**	**492062**	**492201**	**492341**	**492481**	**492621**	
1	492760	492900	493040	493179	493319	493458	493597	493737	493876	494015	139
2	494155	494294	494433	494572	494711	494850	494989	495128	495267	495406	
3	495544	495683	495822	495960	496099	496238	496376	496515	496653	496791	
4	496930	497068	497206	497344	497483	497621	497759	497897	498035	498173	138
5	498311	498448	498586	498724	498862	498999	499137	499275	499412	499550	
6	499687	499824	499962	500099	500236	500374	500511	500648	500785	500922	137
7	501059	501196	501333	501470	501607	501744	501880	502017	502154	502291	
8	502427	502564	502700	502837	502973	503109	503246	503382	503518	503655	136
9	503791	503927	504063	504199	504335	504471	504607	504743	504878	505014	
320	**505150**	**505286**	**505421**	**505557**	**505693**	**505828**	**505964**	**506099**	**506234**	**506370**	
1	506505	506640	506776	506911	507046	507181	507316	507451	507586	507721	135
2	507856	507991	508126	508260	508395	508530	508664	508799	508934	509068	
3	509203	509337	509471	509606	509740	509874	510009	510143	510277	510411	134
4	510545	510679	510813	510947	511081	511215	511349	511482	511616	511750	
5	511883	512017	512151	512284	512418	512551	512684	512818	512951	513084	133
6	513218	513351	513484	513617	513750	513883	514016	514149	514282	514415	
7	514548	514681	514813	514946	515079	515211	515344	515476	515609	515741	
8	515874	516006	516139	516271	516403	516535	516668	516800	516932	517064	132
9	517196	517328	517460	517592	517724	517855	517987	518119	518251	518382	
330	**518514**	**518646**	**518777**	**518909**	**519040**	**519171**	**519303**	**519434**	**519566**	**519697**	131
1	519828	519959	520090	520221	520353	520484	520615	520745	520876	521007	
2	521138	521269	521400	521530	521661	521792	521922	522053	522183	522314	
3	522444	522575	522705	522835	522966	523096	523226	523356	523486	523616	130
4	523746	523876	524006	524136	524266	524396	524526	524656	524785	524915	
5	525045	525174	525304	525434	525563	525693	525822	525951	526081	526210	129
6	526339	526469	526598	526727	526856	526985	527114	527243	527372	527501	
7	527630	527759	527888	528016	528145	528274	528402	528531	528660	528788	
8	528917	529045	529174	529302	529430	529559	529687	529815	529943	530072	128
9	530200	530328	530456	530584	530712	530840	530968	531096	531223	531351	

PROPORTIONAL PARTS

Diff.	1	2	3	4	5	6	7	8	9
154	15.4	30.8	46.2	61.6	77.0	92.4	107.8	123.2	138.6
152	15.2	30.4	45.6	60.8	76.0	91.2	106.4	121.6	136.8
150	15.0	30.0	45.0	60.0	75.0	90.0	105.0	120.0	135.0
148	14.8	29.6	44.4	59.2	74.0	88.8	103.6	118.4	133.2
146	14.6	29.2	43.8	58.4	73.0	87.6	102.2	116.8	131.4
144	14.4	28.8	43.2	57.6	72.0	86.4	100.8	115.2	129.6
142	14.2	28.4	42.6	56.8	71.0	85.2	99.4	113.6	127.8
140	14.0	28.0	42.0	56.0	70.0	84.0	98.0	112.0	126.0
138	13.8	27.6	41.4	55.2	69.0	82.8	96.6	110.4	124.2
136	13.6	27.2	40.8	54.4	68.0	81.6	95.2	108.8	122.4
134	13.4	26.8	40.2	53.6	67.0	80.4	93.8	107.2	120.6
132	13.2	26.4	39.6	52.8	66.0	79.2	92.4	105.6	118.8
130	13.0	26.0	39.0	52.0	65.0	78.0	91.0	104.0	117.0
128	12.8	25.6	38.4	51.2	64.0	76.8	89.6	102.4	115.2

N	0	1	2	3	4	5	6	7	8	9	Diff.
340	**531479**	**531607**	**531734**	**531862**	**531990**	**532117**	**532245**	**532372**	**532500**	**532627**	
1	532754	532882	533009	533136	533264	533391	533518	533645	533772	533899	127
2	534026	534153	534280	534407	534534	534661	534787	534914	535041	535167	
3	535294	535421	535547	535674	535800	535927	536053	536180	536306	536432	126
4	536558	536685	536811	536937	537063	537189	537315	537441	537567	537693	
5	537819	537945	538071	538197	538322	538448	538574	538699	538825	538951	
6	539076	539202	539327	539452	539578	539703	539829	539954	540079	540204	125
7	540329	540455	540580	540705	540830	540955	541080	541205	541330	541454	
8	541579	541704	541829	541953	542078	542203	542327	542452	542576	542701	
9	542825	542950	543074	543199	543323	543447	543571	543696	543820	543944	124
350	**544068**	**544192**	**544316**	**544440**	**544564**	**544688**	**544812**	**544936**	**545060**	**545183**	
1	545307	545431	545555	545678	545802	545925	546049	546172	546296	546419	
2	546543	546666	546789	546913	547036	547159	547282	547405	547529	547652	123
3	547775	547898	548021	548144	548267	548389	548512	548635	548758	548881	
4	549003	549126	549249	549371	549494	549616	549739	549861	549984	550106	
5	550228	550351	550473	550595	550717	550840	550962	551084	551206	551328	122
6	551450	551572	551694	551816	551938	552060	552181	552303	552425	552547	
7	552668	552790	552911	553033	553155	553276	553398	553519	553640	553762	121
8	553883	554004	554126	554247	554368	554489	554610	554731	554852	554973	
9	555094	555215	555336	555457	555578	555699	555820	555940	556061	556182	
360	**556303**	**556423**	**556544**	**556664**	**556785**	**556905**	**557026**	**557146**	**557267**	**557387**	120
1	557507	557627	557748	557868	557988	558108	558228	558349	558469	558589	
2	558709	558829	558948	559068	559188	559308	559428	559548	559667	559787	
3	559907	560026	560146	560265	560385	560504	560624	560743	560863	560982	119
4	561101	561221	561340	561459	561578	561698	561817	561936	562055	562174	
5	562293	562412	562531	562650	562769	562887	563006	563125	563244	563362	
6	563481	563600	563718	563837	563955	564074	564192	564311	564429	564548	
7	564666	564784	564903	565021	565139	565257	565376	565494	565612	565730	118
8	565848	565966	566084	566202	566320	566437	566555	566673	566791	566909	
9	567026	567144	567262	567379	567497	567614	567732	567849	567967	568084	
370	**568202**	**568319**	**568436**	**568554**	**568671**	**568788**	**568905**	**569023**	**569140**	**569257**	117
1	569374	569491	569608	569725	569842	569959	570076	570193	570309	570426	
2	570543	570660	570776	570893	571010	571126	571243	571359	571476	571592	
3	571709	571825	571942	572058	572174	572291	572407	572523	572639	572755	116
4	572872	572988	573104	573220	573336	573452	573568	573684	573800	573915	
5	574031	574147	574263	574379	574494	574610	574726	574841	574957	575072	
6	575188	575303	575419	575534	575650	575765	575880	575996	576111	576226	115
7	576341	576457	576572	576687	576802	576917	577032	577147	577262	577377	
8	577492	577607	577722	577836	577951	578066	578181	578295	578410	578525	
9	578639	578754	578868	578983	579097	579212	579326	579441	579555	579669	114
380	**579784**	**579898**	**580012**	**580126**	**580241**	**580355**	**580469**	**580583**	**580697**	**580811**	
1	580925	581039	581153	581267	581381	581495	581608	581722	581836	581950	
2	582063	582177	582291	582404	582518	582631	582745	582858	582972	583085	
3	583199	583312	583426	583539	583652	583765	583879	583992	584105	584218	
4	584331	584444	584557	584670	584783	584896	585009	585122	585235	585348	113
5	585461	585574	585686	585799	585912	586024	586137	586250	586362	586475	
6	586587	586700	586812	586925	587037	587149	587262	587374	587486	587599	
7	587711	587823	587935	588047	588160	588272	588384	588496	588608	588720	112
8	588832	588944	589056	589167	589279	589391	589503	589615	589726	589838	
9	589950	590061	590173	590284	590396	590507	590619	590730	590842	590953	

PROPORTIONAL PARTS

Diff.	1	2	3	4	5	6	7	8	9
128	12.8	25.6	38.4	51.2	64.0	76.8	89.6	102.4	115.2
126	12.6	25.2	37.8	50.4	63.0	75.6	88.2	100.8	113.4
124	12.4	24.8	37.2	49.6	62.0	74.4	86.8	99.2	111.6
122	12.2	24.4	36.6	48.8	61.0	73.2	85.4	97.6	109.8
120	12.0	24.0	36.0	48.0	60.0	72.0	84.0	96.0	108.0
118	11.8	23.6	35.4	47.2	59.0	70.8	82.6	94.4	106.2
116	11.6	23.2	34.8	46.4	58.0	69.6	81.2	92.8	104.4
114	11.4	22.8	34.2	45.6	57.0	68.4	79.8	91.2	102.6

N	0	1	2	3	4	5	6	7	8	9	Diff.
390	**591065**	**591176**	**591287**	**591399**	**591510**	**591621**	**591732**	**591843**	**591955**	**592066**	
1	592177	592288	592399	592510	592621	592732	592843	592954	593064	593175	111
2	593286	593397	593508	593618	593729	593840	593950	594061	594171	594282	
3	594393	594503	594614	594724	594834	594945	595055	595165	595276	595386	
4	595496	595606	595717	595827	595937	596047	596157	596267	596377	596487	
5	596597	596707	596817	596927	597037	597146	597256	597366	597476	597586	110
6	597695	597805	597914	598024	598134	598243	598353	598462	598572	598681	
7	598791	598900	599009	599119	599228	599337	599446	599556	599665	599774	
8	599883	599992	600101	600210	600319	600428	600537	600646	600755	600864	109
9	600973	601082	601191	601299	601408	601517	601625	601734	601843	601951	
400	**602060**	**602169**	**602277**	**602386**	**602494**	**602603**	**602711**	**602819**	**602928**	**603036**	
1	603144	603253	603361	603469	603577	603686	603794	603902	604010	604118	108
2	604226	604334	604442	604550	604658	604766	604874	604982	605089	605197	
3	605305	605413	605521	605628	605736	605844	605951	606059	606166	606274	
4	606381	606489	606596	606704	606811	606919	607026	607133	607241	607348	
5	607455	607562	607669	607777	607884	607991	608098	608205	608312	608419	107
6	608526	608633	608740	608847	608954	609061	609167	609274	609381	609488	
7	609594	609701	609808	609914	610021	610128	610234	610341	610447	610554	
8	610660	610767	610873	610979	611086	611192	611298	611405	611511	611617	
9	611723	611829	611936	612042	612148	612254	612360	612466	612572	612678	106
410	**612784**	**612890**	**612996**	**613102**	**613207**	**613313**	**613419**	**613525**	**613630**	**613736**	
1	613842	613947	614053	614159	614264	614370	614475	614581	614686	614792	
2	614897	615003	615108	615213	615319	615424	615529	615634	615740	615845	
3	615950	616055	616160	616265	616370	616476	616581	616686	616790	616895	105
4	617000	617105	617210	617315	617420	617525	617629	617734	617839	617943	
5	618048	618153	618257	618362	618466	618571	618676	618780	618884	618989	
6	619093	619198	619302	619406	619511	619615	619719	619824	619928	620032	
7	620136	620240	620344	620448	620552	620656	620760	620864	620968	621072	104
8	621176	621280	621384	621488	621592	621695	621799	621903	622007	622110	
9	622214	622318	622421	622525	622628	622732	622835	622939	623042	623146	
420	**623249**	**623353**	**623456**	**623559**	**623663**	**623766**	**623869**	**623973**	**624076**	**624179**	
1	624282	624385	624488	624591	624695	624798	624901	625004	625107	625210	103
2	625312	625415	625518	625621	625724	625827	625929	626032	626135	626238	
3	626340	626443	626546	626648	626751	626853	626956	627058	627161	627263	
4	627366	627468	627571	627673	627775	627878	627980	628082	628185	628287	
5	628389	628491	628593	628695	628797	628900	629002	629104	629206	629308	102
6	629410	629512	629613	629715	629817	629919	630021	630123	630224	630326	
7	630428	630530	630631	630733	630835	630936	631038	631139	631241	631342	
8	631444	631545	631647	631748	631849	631951	632052	632153	632255	632356	101
9	632457	632559	632660	632761	632862	632963	633064	633165	633266	633367	
430	**633468**	**633569**	**633670**	**633771**	**633872**	**633973**	**634074**	**634175**	**634276**	**634376**	
1	634477	634578	634679	634779	634880	634981	635081	635182	635283	635383	
2	635484	635584	635685	635785	635886	635986	636087	636187	636287	636388	
3	636488	636588	636688	636789	636889	636989	637089	637189	637290	637390	
4	637490	637590	637690	637790	637890	637990	638090	638190	638290	638389	100
5	638489	638589	638689	638789	638888	638988	639088	639188	639287	639387	
6	639486	639586	639686	639785	639885	639984	640084	640183	640283	640382	
7	640481	640581	640680	640779	640879	640978	641077	641177	641276	641375	
8	641474	641573	641672	641771	641871	641970	642069	642168	642267	642366	
9	642465	642563	642662	642761	642860	642959	643058	643156	643255	643354	99

PROPORTIONAL PARTS

Diff.	1	2	3	4	5	6	7	8	9
112	11.2	22.4	33.6	44.8	56.0	67.2	78.4	89.6	100.8
110	11.0	22.0	33.0	44.0	55.0	66.0	77.0	88.0	99.0
108	10.8	21.6	32.4	43.2	54.0	64.8	75.6	86.4	97.2
106	10.6	21.2	31.8	42.4	53.0	63.6	74.2	84.8	95.4
104	10.4	20.8	31.2	41.6	52.0	62.4	72.8	83.2	93.6
102	10.2	20.4	30.6	40.8	51.0	61.2	71.4	81.6	91.8
100	10.0	20.0	30.0	40.0	50.0	60.0	70.0	80.0	90.0
98	9.8	19.6	29.4	39.2	49.0	58.8	68.6	78.4	88.2

N	0	1	2	3	4	5	6	7	8	9	Diff.
440	**643453**	**643551**	**643650**	**643749**	**643847**	**643946**	**644044**	**644143**	**644242**	**644340**	
1	644439	644537	644636	644734	644832	644931	645029	645127	645226	645324	
2	645422	645521	645619	645717	645815	645913	646011	646110	646208	646306	
3	646404	646502	646600	646698	646796	646894	646992	647089	647187	647285	98
4	647383	647481	647579	647676	647774	647872	647969	648067	648165	648262	
5	648360	648458	648555	648653	648750	648848	648945	649043	649140	649237	
6	649335	649432	649530	649627	649724	649821	649919	650016	650113	650210	
7	650308	650405	650502	650599	650696	650793	650890	650987	651084	651181	
8	651278	651375	651472	651569	651666	651762	651859	651956	652053	652150	97
9	652246	652343	652440	652536	652633	652730	652826	652923	653019	653116	
450	**653213**	**653309**	**653405**	**653502**	**653598**	**653695**	**653791**	**653888**	**653984**	**654080**	
1	654177	654273	654369	654465	654562	654658	654754	654850	654946	655042	
2	655138	655235	655331	655427	655523	655619	655715	655810	655906	656002	
3	656098	656194	656290	656386	656482	656577	656673	656769	656864	656960	96
4	657056	657152	657247	657343	657438	657534	657629	657725	657820	657916	
5	658011	658107	658202	658298	658393	658488	658584	658679	658774	658870	
6	658965	659060	659155	659250	659346	659441	659536	659631	659726	659821	
7	659916	660011	660106	660201	660296	660391	660486	660581	660676	660771	95
8	660865	660960	661055	661150	661245	661339	661434	661529	661623	661718	
9	661813	661907	662002	662096	662191	662286	662380	662475	662569	662663	
460	**662758**	**662852**	**662947**	**663041**	**663135**	**663230**	**663324**	**663418**	**663512**	**663607**	
1	663701	663795	663889	663983	664078	664172	664266	664360	664454	664548	
2	664642	664736	664830	664924	665018	665112	665206	665299	665393	665487	94
3	665581	665675	665769	665862	665956	666050	666143	666237	666331	666424	
4	666518	666612	666705	666799	666892	666986	667079	667173	667266	667360	
5	667453	667546	667640	667733	667826	667920	668013	668106	668199	668293	
6	668386	668479	668572	668665	668759	668852	668945	669038	669131	669224	
7	669317	669410	669503	669596	669689	669782	669875	669967	670060	670153	93
8	670246	670339	670431	670524	670617	670710	670802	670895	670988	671080	
9	671173	671265	671358	671451	671543	671636	671728	671821	671913	672005	
470	**672098**	**672190**	**672283**	**672375**	**672467**	**672560**	**672652**	**672744**	**672836**	**672929**	
1	673021	673113	673205	673297	673390	673482	673574	673666	673758	673850	
2	673942	674034	674126	674218	674310	674402	674494	674586	674677	674769	92
3	674861	674953	675045	675137	675228	675320	675412	675503	675595	675687	
4	675778	675870	675962	676053	676145	676236	676328	676419	676511	676602	
5	676694	676785	676876	676968	677059	677151	677242	677333	677424	677516	
6	677607	677698	677789	677881	677972	678063	678154	678245	678336	678427	
7	678518	678609	678700	678791	678882	678973	679064	679155	679246	679337	91
8	679428	679519	679610	679700	679791	679882	679973	680063	680154	680245	
9	680336	680426	680517	680607	680698	680789	680879	680970	681060	681151	
480	**681241**	**681332**	**681422**	**681513**	**681603**	**681693**	**681784**	**681874**	**681964**	**682055**	
1	682145	682235	682326	682416	682506	682596	682686	682777	682867	682957	
2	683047	683137	683227	683317	683407	683497	683587	683677	683767	683857	90
3	683947	684037	684127	684217	684307	684396	684486	684576	684666	684756	
4	684845	684935	685025	685114	685204	685294	685383	685473	685563	685652	
5	685742	685831	685921	686010	686100	686189	686279	686368	686458	686547	
6	686636	686726	686815	686904	686994	687083	687172	687261	687351	687440	
7	687529	687618	687707	687796	687886	687975	688064	688153	688242	688331	
8	688420	688509	688598	688687	688776	688865	688953	689042	689131	689220	89
9	689309	689398	689486	689575	689664	689753	689841	689930	690019	690107	
490	**690196**	**690285**	**690373**	**690462**	**690550**	**690639**	**690728**	**690816**	**690905**	**690993**	
1	691081	691170	691258	691347	691435	691524	691612	691700	691789	691877	
2	691965	692053	692142	692230	692318	692406	692494	692583	692671	692759	
3	692847	692935	693023	693111	693199	693287	693375	693463	693551	693639	88
4	693727	693815	693903	693991	694078	694166	694254	694342	694430	694517	
5	694605	694693	694781	694868	694956	695044	695131	695219	695307	695394	
6	695482	695569	695657	695744	695832	695919	696007	696094	696182	696269	
7	696356	696444	696531	696618	696706	696793	696880	696968	697055	697142	
8	697229	697317	697404	697491	697578	697665	697752	697839	697926	698014	87
9	698100	698188	698275	698362	698449	698535	698622	698709	698796	698883	

PROPORTIONAL PARTS

Diff.	1	2	3	4	5	6	7	8	9
98	9.8	19.6	29.4	39.2	49.0	58.8	68.6	78.4	88.2
96	9.6	19.2	28.8	38.4	48.0	57.6	67.2	76.8	86.4
94	9.4	18.8	28.2	37.6	47.0	56.4	65.8	75.2	84.6
92	9.2	18.4	27.6	36.8	46.0	55.2	64.4	73.6	82.8
90	9.0	18.0	27.0	36.0	45.0	54.0	63.0	72.0	81.0
88	8.8	17.6	26.4	35.2	44.0	52.8	61.6	70.4	79.2

N	0	1	2	3	4	5	6	7	8	9	Diff.
500	**698970**	**699057**	**699144**	**699231**	**699317**	**699404**	**699491**	**699578**	**699664**	**699751**	
1	699838	699924	700011	700098	700184	700271	700358	700444	700531	700617	
2	700704	700790	700877	700963	701050	701136	701222	701309	701395	701482	
3	701568	701654	701741	701827	701913	701999	702086	702172	702258	702344	
4	702431	702517	702603	702689	702775	702861	702947	703033	703119	703205	
5	703291	703377	703463	703549	703635	703721	703807	703893	703979	704065	86
6	704151	704236	704322	704408	704494	704579	704665	704751	704837	704922	
7	705008	705094	705179	705265	705350	705436	705522	705607	705693	705778	
8	705864	705949	706035	706120	706206	706291	706376	706462	706547	706632	
9	706718	706803	706888	706974	707059	707144	707229	707315	707400	707485	
510	**707570**	**707655**	**707740**	**707826**	**707911**	**707996**	**708081**	**708166**	**708251**	**708336**	
1	708421	708506	708591	708676	708761	708846	708931	709015	709100	709185	85
2	709270	709355	709440	709524	709609	709694	709779	709863	709948	710033	
3	710117	710202	710287	710371	710456	710540	710625	710710	710794	710879	
4	710963	711048	711132	711217	711301	711385	711470	711554	711639	711723	
5	711807	711892	711976	712060	712144	712229	712313	712397	712481	712566	
6	712650	712734	712818	712902	712986	713070	713154	713238	713323	713407	
7	713491	713575	713659	713742	713826	713910	713994	714078	714162	714246	84
8	714330	714414	714497	714581	714665	714749	714833	714916	715000	715084	
9	715167	715251	715335	715418	715502	715586	715669	715753	715836	715920	
520	**716003**	**716087**	**716170**	**716254**	**716337**	**716421**	**716504**	**716588**	**716671**	**716754**	
1	716838	716921	717004	717088	717171	717254	717338	717421	717504	717587	
2	717671	717754	717837	717920	718003	718086	718169	718253	718336	718419	83
3	718502	718585	718668	718751	718834	718917	719000	719083	719165	719248	
4	719331	719414	719497	719580	719663	719745	719828	719911	719994	720077	
5	720159	720242	720325	720407	720490	720573	720655	720738	720821	720903	
6	720986	721068	721151	721233	721316	721398	721481	721563	721646	721728	
7	721811	721893	721975	722058	722140	722222	722305	722387	722469	722552	
8	722634	722716	722798	722881	722963	723045	723127	723209	723291	723374	
9	723456	723538	723620	723702	723784	723866	723948	724030	724112	724194	82
530	**724276**	**724358**	**724440**	**724522**	**724604**	**724685**	**724767**	**724849**	**724931**	**725013**	
1	725095	725176	725258	725340	725422	725503	725585	725667	725748	725830	
2	725912	725993	726075	726156	726238	726320	726401	726483	726564	726646	
3	726727	726809	726890	726972	727053	727134	727216	727297	727379	727460	
4	727541	727623	727704	727785	727866	727948	728029	728110	728191	728273	
5	728354	728435	728516	728597	728678	728759	728841	728922	729003	729084	
6	729165	729246	729327	729408	729489	729570	729651	729732	729813	729893	81
7	729974	730055	730136	730217	730298	730378	730459	730540	730621	730702	
8	730782	730863	730944	731024	731105	731186	731266	731347	731428	731508	
9	731589	731669	731750	731830	731911	731991	732072	732152	732233	732313	
540	**732394**	**732474**	**732555**	**732635**	**732715**	**732796**	**732876**	**732956**	**733037**	**733117**	
1	733197	733278	733358	733438	733518	733598	733679	733759	733839	733919	
2	733999	734079	734160	734240	734320	734400	734480	734560	734640	734720	80
3	734800	734880	734960	735040	735120	735200	735279	735359	735439	735519	
4	735599	735679	735759	735838	735918	735998	736078	736157	736237	736317	
5	736397	736476	736556	736635	736715	736795	736874	736954	737034	737113	
6	737193	737272	737352	737431	737511	737590	737670	737749	737829	737908	
7	737987	738067	738146	738225	738305	738384	738463	738543	738622	738701	
8	738781	738860	738939	739018	739097	739177	739256	739335	739414	739493	
9	739572	739651	739731	739810	739889	739968	740047	740126	740205	740284	79
550	**740363**	**740442**	**740521**	**740600**	**740678**	**740757**	**740836**	**740915**	**740994**	**741073**	
1	741152	741230	741309	741388	741467	741546	741624	741703	741782	741860	
2	741939	742018	742096	742175	742254	742332	742411	742489	742568	742647	
3	742725	742804	742882	742961	743039	743118	743196	743275	743353	743431	
4	743510	743588	743667	743745	743823	743902	743980	744058	744136	744215	
5	744293	744371	744449	744528	744606	744684	744762	744840	744919	744997	
6	745075	745153	745231	745309	745387	745465	745543	745621	745699	745777	78
7	745855	745933	746011	746089	746167	746245	746323	746401	746479	746556	
8	746634	746712	746790	746868	746945	747023	747101	747179	747256	747334	
9	747412	747489	747567	747645	747722	747800	747878	747955	748033	748110	

Proportional Parts

Diff.	1	2	3	4	5	6	7	8	9
86	8.6	17.2	25.8	34.4	43.0	51.6	60.2	68.8	77.4
84	8.4	16.8	25.2	33.6	42.0	50.4	58.8	67.2	75.6
82	8.2	16.4	24.6	32.8	41.0	49.2	57.4	65.6	73.8
80	8.0	16.0	24.0	32.0	40.0	48.0	56.0	64.0	72.0
78	7.8	15.6	23.4	31.2	39.0	46.8	54.6	62.4	70.2

N	0	1	2	3	4	5	6	7	8	9	Diff.
560	**748188**	**748266**	**748343**	**748421**	**748498**	**748576**	**748653**	**748731**	**748808**	**748885**	
1	748963	749040	749118	749195	749272	749350	749427	749504	749582	749659	
2	749736	749814	749891	749968	750045	750123	750200	750277	750354	750431	
3	750508	750586	750663	750740	750817	750894	750971	751048	751125	751202	
4	751279	751356	751433	751510	751587	751664	751741	751818	751895	751972	**77**
5	752048	752125	752202	752279	752356	752433	752509	752586	752663	752740	
6	752816	752893	752970	753047	753123	753200	753277	753353	753430	753506	
7	753583	753660	753736	753813	753889	753966	754042	754119	754195	754272	
8	754348	754425	754501	754578	754654	754730	754807	754883	754960	755036	
9	755112	755189	755265	755341	755417	755494	755570	755646	755722	755799	
570	**755875**	**755951**	**756027**	**756103**	**756180**	**756256**	**756332**	**756408**	**756484**	**756560**	
1	756636	756712	756788	756864	756940	757016	757092	757168	757244	757320	**76**
2	757396	757472	757548	757624	757700	757775	757851	757927	758003	758079	
3	758155	758230	758306	758382	758458	758533	758609	758685	758761	758836	
4	758912	758988	759063	759139	759214	759290	759366	759441	759517	759592	
5	759668	759743	759819	759894	759970	760045	760121	760196	760272	760347	
6	760422	760498	760573	760649	760724	760799	760875	760950	761025	761101	
7	761176	761251	761326	761402	761477	761552	761627	761702	761778	761853	
8	761928	762003	762078	762153	762228	762303	762378	762453	762529	762604	**75**
9	762679	762754	762829	762904	762978	763053	763128	763203	763278	763353	
580	**763428**	**763503**	**763578**	**763653**	**763727**	**763802**	**763877**	**763952**	**764027**	**764101**	
1	764176	764251	764326	764400	764475	764550	764624	764699	764774	764848	
2	764923	764998	765072	765147	765221	765296	765370	765445	765520	765594	
3	765669	765743	765818	765892	765966	766041	766115	766190	766264	766338	
4	766413	766487	766562	766636	766710	766785	766859	766933	767007	767082	
5	767156	767230	767304	767379	767453	767527	767601	767675	767749	767823	
6	767898	767972	768046	768120	768194	768268	768342	768416	768490	768564	**74**
7	768638	768712	768786	768860	768934	769008	769082	769156	769230	769303	
8	769377	769451	769525	769599	769673	769746	769820	769894	769968	770042	
9	770115	770189	770263	770336	770410	770484	770557	770631	770705	770778	
590	**770852**	**770926**	**770999**	**771073**	**771146**	**771220**	**771293**	**771367**	**771440**	**771514**	
1	771587	771661	771734	771808	771881	771955	772028	772102	772175	772248	
2	772322	772395	772468	772542	772615	772688	772762	772835	772908	772981	
3	773055	773128	773201	773274	773348	773421	773494	773567	773640	773713	
4	773786	773860	773933	774006	774079	774152	774225	774298	774371	774444	**73**
5	774517	774590	774663	774736	774809	774882	774955	775028	775100	775173	
6	775246	775319	775392	775465	775538	775610	775683	775756	775829	775902	
7	775974	776047	776120	776193	776265	776338	776411	776483	776556	776629	
8	776701	776774	776846	776919	776992	777064	777137	777209	777282	777354	
9	777427	777499	777572	777644	777717	777789	777862	777934	778006	778079	
600	**778151**	**778224**	**778296**	**778368**	**778441**	**778513**	**778585**	**778658**	**778730**	**778802**	
1	778874	778947	779019	779091	779163	779236	779308	779380	779452	779524	
2	779596	779669	779741	779813	779885	779957	780029	780101	780173	780245	
3	780317	780389	780461	780533	780605	780677	780749	780821	780893	780965	**72**
4	781037	781109	781181	781253	781324	781396	781468	781540	781612	781684	
5	781755	781827	781899	781971	782042	782114	782186	782258	782329	782401	
6	782473	782544	782616	782688	782759	782831	782902	782974	783046	783117	
7	783189	783260	783332	783403	783475	783546	783618	783689	783761	783832	
8	783904	783975	784046	784118	784189	784261	784332	784403	784475	784546	
9	784617	784689	784760	784831	784902	784974	785045	785116	785187	785259	
610	**785330**	**785401**	**785472**	**785543**	**785615**	**785686**	**785757**	**785828**	**785899**	**785970**	
1	786041	786112	786183	786254	786325	786396	786467	786538	786609	786680	**71**
2	786751	786822	786893	786964	787035	787106	787177	787248	787319	787390	
3	787460	787531	787602	787673	787744	787815	787885	787956	788027	788098	
4	788168	788239	788310	788381	788451	788522	788593	788663	788734	788804	
5	788875	788946	789016	789087	789157	789228	789299	789369	789440	789510	
6	789581	789651	789722	789792	789863	789933	790004	790074	790144	790215	
7	790285	790356	790426	790496	790567	790637	790707	790778	790848	790918	
8	790988	791059	791129	791199	791269	791340	791410	791480	791550	791620	
9	791691	791761	791831	791901	791971	792041	792111	792181	792252	792322	

PROPORTIONAL PARTS

Diff.	1	2	3	4	5	6	7	8	9
78	7.8	15.6	23.4	31.2	39.0	46.8	54.6	62.4	70.2
76	7.6	15.2	22.8	30.4	38.0	45.6	53.2	60.8	68.4
74	7.4	14.8	22.2	29.6	37.0	44.4	51.8	59.2	66.6
72	7.2	14.4	21.6	28.8	36.0	43.2	50.4	57.6	64.8
70	7.0	14.0	21.0	28.0	35.0	42.0	49.0	56.0	63.0

N	0	1	2	3	4	5	6	7	8	9	Diff.
620	**792392**	**792462**	**792532**	**792602**	**792672**	**792742**	**792812**	**792882**	**792952**	**793022**	**70**
1	793092	793162	793231	793301	793371	793441	793511	793581	793651	793721	
2	793790	793860	793930	794000	794070	794139	794209	794279	794349	794418	
3	794488	794558	794627	794697	794767	794836	794906	794976	795045	795115	
4	795185	795254	795324	795393	795463	795532	795602	795672	795741	795811	
5	795880	795949	796019	796088	796158	796227	796297	796366	796436	796505	
6	796574	796644	796713	796782	796852	796921	796990	797060	797129	797198	
7	797268	797337	797406	797475	797545	797614	797683	797752	797821	797890	
8	797960	798029	798098	798167	798236	798305	798374	798443	798513	798582	
9	798651	798720	798789	798858	798927	798996	799065	799134	799203	799272	**69**
630	**799341**	**799409**	**799478**	**799547**	**799616**	**799685**	**799754**	**799823**	**799892**	**799961**	
1	800029	800098	800167	800236	800305	800373	800442	800511	800580	800648	
2	800717	800786	800854	800923	800992	801061	801129	801198	801266	801335	
3	801404	801472	801541	801609	801678	801747	801815	801884	801952	802021	
4	802089	802158	802226	802295	802363	802432	802500	802568	802637	802705	
5	802774	802842	802910	802979	803047	803116	803184	803252	803321	803389	
6	803457	803525	803594	803662	803730	803798	803867	803935	804003	804071	
7	804139	804208	804276	804344	804412	804480	804548	804616	804685	804753	
8	804821	804889	804957	805025	805093	805161	805229	805297	805365	805433	**68**
9	805501	805569	805637	805705	805773	805841	805908	805976	806044	806112	
640	**806180**	**806248**	**806316**	**806384**	**806451**	**806519**	**806587**	**806655**	**806723**	**806790**	
1	806858	806926	806994	807061	807129	807197	807264	807332	807400	807467	
2	807535	807603	807670	807738	807806	807873	807941	808008	808076	808143	
3	808211	808279	808346	808414	808481	808549	808616	808684	808751	808818	
4	808886	808953	809021	809088	809156	809223	809290	809358	809425	809492	
5	809560	809627	809694	809762	809829	809896	809964	810031	810098	810165	
6	810233	810300	810367	810434	810501	810569	810636	810703	810770	810837	
7	810904	810971	811039	811106	811173	811240	811307	811374	811441	811508	**67**
8	811575	811642	811709	811776	811843	811910	811977	812044	812111	812178	
9	812245	812312	812379	812445	812512	812579	812646	812713	812780	812847	
650	**812913**	**812980**	**813047**	**813114**	**813181**	**813247**	**813314**	**813381**	**813448**	**813514**	
1	813581	813648	813714	813781	813848	813914	813981	814048	814114	814181	
2	814248	814314	814381	814447	814514	814581	814647	814714	814780	814847	
3	814913	814980	815046	815113	815179	815246	815312	815378	815445	815511	
4	815578	815644	815711	815777	815843	815910	815976	816042	816109	816175	
5	816241	816308	816374	816440	816506	816573	816639	816705	816771	816838	
6	816904	816970	817036	817102	817169	817235	817301	817367	817433	817499	
7	817565	817631	817698	817764	817830	817896	817962	818028	818094	818160	
8	818226	818292	818358	818424	818490	818556	818622	818688	818754	818820	**66**
9	818885	818951	819017	819083	819149	819215	819281	819346	819412	819478	
660	**819544**	**819610**	**819676**	**819741**	**819807**	**819873**	**819939**	**820004**	**820070**	**820136**	
1	820201	820267	820333	820399	820464	820530	820595	820661	820727	820792	
2	820858	820924	820989	821055	821120	821186	821251	821317	821382	821448	
3	821514	821579	821645	821710	821775	821841	821906	821972	822037	822103	
4	822168	822233	822299	822364	822430	822495	822560	822626	822691	822756	
5	822822	822887	822952	823018	823083	823148	823213	823279	823344	823409	
6	823474	823539	823605	823670	823735	823800	823865	823930	823996	824061	
7	824126	824191	824256	824321	824386	824451	824516	824581	824646	824711	
8	824776	824841	824906	824971	825036	825101	825166	825231	825296	825361	**65**
9	825426	825491	825556	825621	825686	825751	825815	825880	825945	826010	
670	**826075**	**826140**	**826204**	**826269**	**826334**	**826399**	**826464**	**826528**	**826593**	**826658**	
1	826723	826787	826852	826917	826981	827046	827111	827175	827240	827305	
2	827369	827434	827499	827563	827628	827692	827757	827821	827886	827951	
3	828015	828080	828144	828209	828273	828338	828402	828467	828531	828595	
4	828660	828724	828789	828853	828918	828982	829046	829111	829175	829239	
5	829304	829368	829432	829497	829561	829625	829690	829754	829818	829882	
6	829947	830011	830075	830139	830204	830268	830332	830396	830460	830525	
7	830589	830653	830717	830781	830845	830909	830973	831037	831102	831166	
8	831230	831294	831358	831422	831486	831550	831614	831678	831742	831806	**64**
9	831870	831934	831998	832062	832126	832189	832253	832317	832381	832445	

PROPORTIONAL PARTS

Diff.	1	2	3	4	5	6	7	8	9
70	7.0	14.0	21.0	28.0	35.0	42.0	49.0	56.0	63.0
68	6.8	13.6	20.4	27.2	34.0	40.8	47.6	54.4	61.2
66	6.6	13.2	19.8	26.4	33.0	39.6	46.2	52.8	59.4
64	6.4	12.8	19.2	25.6	32.0	38.4	44.8	51.2	57.6
62	6.2	12.4	18.6	24.8	31.0	37.2	43.4	49.6	55.8

N	0	1	2	3	4	5	6	7	8	9	Diff.
680	**832509**	**832573**	**832637**	**832700**	**832764**	**832828**	**832892**	**832956**	**833020**	**833083**	
1	833147	833211	833275	833338	833402	833466	833530	833593	833657	833721	
2	833784	833848	833912	833975	834039	834103	834166	834230	834294	834357	
3	834421	834484	834548	834611	834675	834739	834802	834866	834929	834993	
4	835056	835120	835183	835247	835310	835373	835437	835500	835564	835627	
5	835691	835754	835817	835881	835944	836007	836071	836134	836197	836261	
6	836324	836387	836451	836514	836577	836641	836704	836767	836830	836894	
7	836957	837020	837083	837146	837210	837273	837336	837399	837462	837525	
8	837588	837652	837715	837778	837841	837904	837967	838030	838093	838156	63
9	838219	838282	838345	838408	838471	838534	838597	838660	838723	838786	
690	**838849**	**838912**	**838975**	**839038**	**839101**	**839164**	**839227**	**839289**	**839352**	**839415**	
1	839478	839541	839604	839667	839729	839792	839855	839918	839981	840043	
2	840106	840169	840232	840294	840357	840420	840482	840545	840608	840671	
3	840733	840796	840859	840921	840984	841046	841109	841172	841234	841297	
4	841359	841422	841485	841547	841610	841672	841735	841797	841860	841922	
5	841985	842047	842110	842172	842235	842297	842360	842422	842484	842547	
6	842609	842672	842734	842796	842859	842921	842983	843046	843108	843170	
7	843233	843295	843357	843420	843482	843544	843606	843669	843731	843793	
8	843855	843918	843980	844042	844104	844166	844229	844291	844353	844415	
9	844477	844539	844601	844664	844726	844788	844850	844912	844974	845036	
700	**845098**	**845160**	**845222**	**845284**	**845346**	**845408**	**845470**	**845532**	**845594**	**845656**	62
1	845718	845780	845842	845904	845966	846028	846090	846151	846213	846275	
2	846337	846399	846461	846523	846585	846646	846708	846770	846832	846894	
3	846955	847017	847079	847141	847202	847264	847326	847388	847449	847511	
4	847573	847634	847696	847758	847819	847881	847943	848004	848066	848128	
5	848189	848251	848312	848374	848435	848497	848559	848620	848682	848743	
6	848805	848866	848928	848989	849051	849112	849174	849235	849297	849358	
7	849419	849481	849542	849604	849665	849726	849788	849849	849911	849972	
8	850033	850095	850156	850217	850279	850340	850401	850462	850524	850585	
9	850646	850707	850769	850830	850891	850952	851014	851075	851136	851197	
710	**851258**	**851320**	**851381**	**851442**	**851503**	**851564**	**851625**	**851686**	**851747**	**851809**	
1	851870	851931	851992	852053	852114	852175	852236	852297	852358	852419	61
2	852480	852541	852602	852663	852724	852785	852846	852907	852968	853029	
3	853090	853150	853211	853272	853333	853394	853455	853516	853577	853637	
4	853698	853759	853820	853881	853941	854002	854063	854124	854185	854245	
5	854306	854367	854428	854488	854549	854610	854670	854731	854792	854852	
6	854913	854974	855034	855095	855156	855216	855277	855337	855398	855459	
7	855519	855580	855640	855701	855761	855822	855882	855943	856003	856064	
8	856124	856185	856245	856306	856366	856427	856487	856548	856608	856668	
9	856729	856789	856850	856910	856970	857031	857091	857152	857212	857272	
720	**857332**	**857393**	**857453**	**857513**	**857574**	**857634**	**857694**	**857755**	**857815**	**857875**	
1	857935	857995	858056	858116	858176	858236	858297	858357	858417	858477	
2	858537	858597	858657	858718	858778	858838	858898	858958	859018	859078	
3	859138	859198	859258	859318	859379	859439	859499	859559	859619	859679	60
4	859739	859799	859859	859918	859978	860038	860098	860158	860218	860278	
5	860338	860398	860458	860518	860578	860637	860697	860757	860817	860877	
6	860937	860996	861056	861116	861176	861236	861295	861355	861415	861475	
7	861534	861594	861654	861714	861773	861833	861893	861952	862012	862072	
8	862131	862191	862251	862310	862370	862430	862489	862549	862608	862668	
9	862728	862787	862847	862906	862966	863025	863085	863144	863204	863263	
730	**863323**	**863382**	**863442**	**863501**	**863561**	**863620**	**863680**	**863739**	**863799**	**863858**	
1	863917	863977	864036	864096	864155	864214	864274	864333	864392	864452	
2	864511	864570	864630	864689	864748	864808	864867	864926	864985	865045	
3	865104	865163	865222	865282	865341	865400	865459	865519	865578	865637	
4	865696	865755	865814	865874	865933	865992	866051	866110	866169	866228	
5	866287	866346	866405	866465	866524	866583	866642	866701	866760	866819	
6	866878	866937	866996	867055	867114	867173	867232	867291	867350	867409	59
7	867467	867526	867585	867644	867703	867762	867821	867880	867939	867998	
8	868056	868115	868174	868233	868292	868350	868409	868468	868527	868586	
9	868644	868703	868762	868821	868879	868938	868997	869056	869114	869173	

PROPORTIONAL PARTS

Diff.	1	2	3	4	5	6	7	8	9
64	6.4	12.8	19.2	25.6	32.0	38.4	44.8	51.2	57.6
62	6.2	12.4	18.6	24.8	31.0	37.2	43.4	49.6	55.8
60	6.0	12.0	18.0	24.0	30.0	36.0	42.0	48.0	54.0
.58	5.8	11.6	17.4	23.2	29.0	34.8	40.6	46.4	52.2

N	0	1	2	3	4	5	6	7	8	9	Diff.
740	**869232**	**869290**	**869349**	**869408**	**869466**	**869525**	**869584**	**869642**	**869701**	**869760**	
1	869818	869877	869935	869994	870053	870111	870170	870228	870287	870345	
2	870404	870462	870521	870579	870638	870696	870755	870813	870872	870930	
3	870989	871047	871106	871164	871223	871281	871339	871398	871456	871515	
4	871573	871631	871690	871748	871806	871865	871923	871981	872040	872098	
5	872156	872215	872273	872331	872389	872448	872506	872564	872622	872681	
6	872739	872797	872855	872913	872972	873030	873088	873146	873204	873262	
7	873321	873379	873437	873495	873553	873611	873669	873727	873785	873844	
8	873902	873960	874018	874076	874134	874192	874250	874308	874366	874424	58
9	874482	874540	874598	874656	874714	874772	874830	874888	874945	875003	
750	**875061**	**875119**	**875177**	**875235**	**875293**	**875351**	**875409**	**875466**	**875524**	**875582**	
1	875640	875698	875756	875813	875871	875929	875987	876045	876102	876160	
2	876218	876276	876333	876391	876449	876507	876564	876622	876680	876737	
3	876795	876853	876910	876968	877026	877083	877141	877199	877256	877314	
4	877371	877429	877487	877544	877602	877659	877717	877774	877832	877889	
5	877947	878004	878062	878119	878177	878234	878292	878349	878407	878464	
6	878522	878579	878637	878694	878752	878809	878866	878924	878981	879039	
7	879096	879153	879211	879268	879325	879383	879440	879497	879555	879612	
8	879669	879726	879784	879841	879898	879956	880013	880070	880127	880185	
9	880242	880299	880356	880413	880471	880528	880585	880642	880699	880756	
760	**880814**	**880871**	**880928**	**880985**	**881042**	**881099**	**881156**	**881213**	**881271**	**881328**	
1	881385	881442	881499	881556	881613	881670	881727	881784	881841	881898	
2	881955	882012	882069	882126	882183	882240	882297	882354	882411	882468	57
3	882525	882581	882638	882695	882752	882809	882866	882923	882980	883037	
4	883093	883150	883207	883264	883321	883377	883434	883491	883548	883605	
5	883661	883718	883775	883832	883888	883945	884002	884059	884115	884172	
6	884229	884285	884342	884399	884455	884512	884569	884625	884682	884739	
7	884795	884852	884909	884965	885022	885078	885135	885192	885248	885305	
8	885361	885418	885474	885531	885587	885644	885700	885757	885813	885870	
9	885926	885983	886039	886096	886152	886209	886265	886321	886378	886434	
770	**886491**	**886547**	**886604**	**886660**	**886716**	**886773**	**886829**	**886885**	**886942**	**886998**	
1	887054	887111	887167	887223	887280	887336	887392	887449	887505	887561	
2	887617	887674	887730	887786	887842	887898	887955	888011	888067	888123	
3	888179	888236	888292	888348	888404	888460	888516	888573	888629	888685	
4	888741	888797	888853	888909	888965	889021	889077	889134	889190	889246	
5	889302	889358	889414	889470	889526	889582	889638	889694	889750	889806	56
6	889862	889918	889974	890030	890086	890141	890197	890253	890309	890365	
7	890421	890477	890533	890589	890645	890700	890756	890812	890868	890924	
8	890980	891035	891091	891147	891203	891259	891314	891370	891426	891482	
9	891537	891593	891649	891705	891760	891816	891872	891928	891983	892039	
780	**892095**	**892150**	**892206**	**892262**	**892317**	**892373**	**892429**	**892484**	**892540**	**892595**	
1	892651	892707	892762	892818	892873	892929	892985	893040	893096	893151	
2	893207	893262	893318	893373	893429	893484	893540	893595	893651	893706	
3	893762	893817	893873	893928	893984	894039	894094	894150	894205	894261	
4	894316	894371	894427	894482	894538	894593	894648	894704	894759	894814	
5	894870	894925	894980	895036	895091	895146	895201	895257	895312	895367	
6	895423	895478	895533	895588	895644	895699	895754	895809	895864	895920	
7	895975	896030	896085	896140	896195	896251	896306	896361	896416	896471	
8	896526	896581	896636	896692	896747	896802	896857	896912	896967	897022	
9	897077	897132	897187	897242	897297	897352	897407	897462	897517	897572	
790	**897627**	**897682**	**897737**	**897792**	**897847**	**897902**	**897957**	**898012**	**898067**	**898122**	55
1	898176	898231	898286	898341	898396	898451	898506	898561	898615	898670	
2	898725	898780	898835	898890	898944	898999	899054	899109	899164	899218	
3	899273	899328	899383	899437	899492	899547	899602	899656	899711	899766	
4	899821	899875	899930	899985	900039	900094	900149	900203	900258	900312	
5	900367	900422	900476	900531	900586	900640	900695	900749	900804	900859	
6	900913	900968	901022	901077	901131	901186	901240	901295	901349	901404	
7	901458	901513	901567	901622	901676	901731	901785	901840	901894	901948	
8	902003	902057	902112	902166	902221	902275	902329	902384	902438	902492	
9	902547	902601	902655	902710	902764	902818	902873	902927	902981	903036	

PROPORTIONAL PARTS

Diff.	1	2	3	4	5	6	7	8	9
60	6.0	12.0	18.0	24.0	30.0	36.0	42.0	48.0	54.0
58	5.8	11.6	17.4	23.2	29.0	34.8	40.6	46.4	52.2
56	5.6	11.2	16.8	22.4	28.0	33.6	39.2	44.8	50.4
54	5.4	10.8	16.2	21.6	27.0	32.4	37.8	43.2	48.6

N	0	1	2	3	4	5	6	7	8	9	Diff.
800	**903090**	**903144**	**903199**	**903253**	**903307**	**903361**	**903416**	**903470**	**903524**	**903578**	
1	903633	903687	903741	903795	903849	903904	903958	904012	904066	904120	
2	904174	904229	904283	904337	904391	904445	904499	904553	904607	904661	
3	904716	904770	904824	904878	904932	904986	905040	905094	905148	905202	
4	905256	905310	905364	905418	905472	905526	905580	905634	905688	905742	54
5	905796	905850	905904	905958	906012	906066	906119	906173	906227	906281	
6	906335	906389	906443	906497	906551	906604	906658	906712	906766	906820	
7	906874	906927	906981	907035	907089	907143	907196	907250	907304	907358	
8	907411	907465	907519	907573	907626	907680	907734	907787	907841	907895	
9	907949	908002	908056	908110	908163	908217	908270	908324	908378	908431	
810	**908485**	**908539**	**908592**	**908646**	**908699**	**908753**	**908807**	**908860**	**908914**	**908967**	
1	909021	909074	909128	909181	909235	909289	909342	909396	909449	909503	
2	909556	909610	909663	909716	909770	909823	909877	909930	909984	910037	
3	910091	910144	910197	910251	910304	910358	910411	910464	910518	910571	
4	910624	910678	910731	910784	910838	910891	910944	910998	911051	911104	
5	911158	911211	911264	911317	911371	911424	911477	911530	911584	911637	
6	911690	911743	911797	911850	911903	911956	912009	912063	912116	912169	
7	912222	912275	912328	912381	912435	912488	912541	912594	912647	912700	
8	912753	912806	912859	912913	912966	913019	913072	913125	913178	913231	
9	913284	913337	913390	913443	913496	913549	913602	913655	913708	913761	53
820	**913814**	**913867**	**913920**	**913973**	**914026**	**914079**	**914132**	**914184**	**914237**	**914290**	
1	914343	914396	914449	914502	914555	914608	914660	914713	914766	914819	
2	914872	914925	914977	915030	915083	915136	915189	915241	915294	915347	
3	915400	915453	915505	915558	915611	915664	915716	915769	915822	915875	
4	915927	915980	916033	916085	916138	916191	916243	916296	916349	916401	
5	916454	916507	916559	916612	916664	916717	916770	916822	916875	916927	
6	916980	917033	917085	917138	917190	917243	917295	917348	917400	917453	
7	917506	917558	917611	917663	917716	917768	917820	917873	917925	917978	
8	918030	918083	918135	918188	918240	918293	918345	918397	918450	918502	
9	918555	918607	918659	918712	918764	918816	918869	918921	918973	919026	
830	**919078**	**919130**	**919183**	**919235**	**919287**	**919340**	**919392**	**919444**	**919496**	**919549**	
1	919601	919653	919706	919758	919810	919862	919914	919967	920019	920071	
2	920123	920176	920228	920280	920332	920384	920436	920489	920541	920593	
3	920645	920697	920749	920801	920853	920906	920958	921010	921062	921114	52
4	921166	921218	921270	921322	921374	921426	921478	921530	921582	921634	
5	921686	921738	921790	921842	921894	921946	921998	922050	922102	922154	
6	922206	922258	922310	922362	922414	922466	922518	922570	922622	922674	
7	922725	922777	922829	922881	922933	922985	923037	923089	923140	923192	
8	923244	923296	923348	923399	923451	923503	923555	923607	923658	923710	
9	923762	923814	923865	923917	923969	924021	924072	924124	924176	924228	
840	**924279**	**924331**	**924383**	**924434**	**924486**	**924538**	**924589**	**924641**	**924693**	**924744**	
1	924796	924848	924899	924951	925003	925054	925106	925157	925209	925261	
2	925312	925364	925415	925467	925518	925570	925621	925673	925725	925776	
3	925828	925879	925931	925982	926034	926085	926137	926188	926240	926291	
4	926342	926394	926445	926497	926548	926600	926651	926702	926754	926805	
5	926857	926908	926959	927011	927062	927114	927165	927216	927268	927319	
6	927370	927422	927473	927524	927576	927627	927678	927730	927781	927832	
7	927883	927935	927986	928037	928088	928140	928191	928242	928293	928345	
8	928396	928447	928498	928549	928601	928652	928703	928754	928805	928857	
9	928908	928959	929010	929061	929112	929163	929215	929266	929317	929368	
850	**929419**	**929470**	**929521**	**929572**	**929623**	**929674**	**929725**	**929776**	**929827**	**929879**	
1	929930	929981	930032	930083	930134	930185	930236	930287	930338	930389	51
2	930440	930491	930542	930592	930643	930694	930745	930796	930847	930898	
3	930949	931000	931051	931102	931153	931204	931254	931305	931356	931407	
4	931458	931509	931560	931610	931661	931712	931763	931814	931865	931915	
5	931966	932017	932068	932118	932169	932220	932271	932322	932372	932423	
6	932474	932524	932575	932626	932677	932727	932778	932829	932879	932930	
7	932981	933031	933082	933133	933183	933234	933285	933335	933386	933437	
8	933487	933538	933589	933639	933690	933740	933791	933841	933892	933943	
9	933993	934044	934094	934145	934195	934246	934296	934347	934397	934448	

PROPORTIONAL PARTS

Diff.	1	2	3	4	5	6	7	8	9
56	5.6	11.2	16.8	22.4	28.0	33.6	39.2	44.8	50.4
54	5.4	10.8	16.2	21.6	27.0	32.4	37.8	43.2	48.6
52	5.2	10.4	15.6	20.8	26.0	31.2	36.4	41.6	46.8

N	0	1	2	3	4	5	6	7	8	9	Diff.
860	**934498**	**934549**	**934599**	**934650**	**934700**	**934751**	**934801**	**934852**	**934902**	**934953**	
1	935003	935054	935104	935154	935205	935255	935306	935356	935406	935457	
2	935507	935558	935608	935658	935709	935759	935809	935860	935910	935960	
3	936011	936061	936111	936162	936212	936262	936313	936363	936413	936463	
4	936514	936564	936614	936665	936715	936765	936815	936865	936916	936966	
5	937016	937066	937116	937167	937217	937267	937317	937367	937418	937468	
6	937518	937568	937618	937668	937718	937769	937819	937869	937919	937969	
7	938019	938069	938119	938169	938219	938269	938320	938370	938420	938470	**50**
8	938520	938570	938620	938670	938720	938770	938820	938870	938920	938970	
9	939020	939070	939120	939170	939220	939270	939320	939369	939419	939469	
870	**939519**	**939569**	**939619**	**939669**	**939719**	**939769**	**939819**	**939869**	**939918**	**939968**	
1	940018	940068	940118	940168	940218	940267	940317	940367	940417	940467	
2	940516	940566	940616	940666	940716	940765	940815	940865	940915	940964	
3	941014	941064	941114	941163	941213	941263	941313	941362	941412	941462	
4	941511	941561	941611	941660	941710	941760	941809	941859	941909	941958	
5	942008	942058	942107	942157	942207	942256	942306	942355	942405	942455	
6	942504	942554	942603	942653	942702	942752	942801	942851	942901	942950	
7	943000	943049	943099	943148	943198	943247	943297	943346	943396	943445	
8	943495	943544	943593	943643	943692	943742	943791	943841	943890	943939	
9	943989	944038	944088	944137	944186	944236	944285	944335	944384	944433	
880	**944483**	**944532**	**944581**	**944631**	**944680**	**944729**	**944779**	**944828**	**944877**	**944927**	
1	944976	945025	945074	945124	945173	945222	945272	945321	945370	945419	
2	945469	945518	945567	945616	945665	945715	945764	945813	945862	945912	
3	945961	946010	946059	946108	946157	946207	946256	946305	946354	946403	
4	946452	946501	946551	946600	946649	946698	946747	946796	946845	946894	
5	946943	946992	947041	947090	947140	947189	947238	947287	947336	947385	
6	947434	947483	947532	947581	947630	947679	947728	947777	947826	947875	**49**
7	947924	947973	948022	948070	948119	948168	948217	948266	948315	948364	
8	948413	948462	948511	948560	948608	948657	948706	948755	948804	948853	
9	948902	948951	948999	949048	949097	949146	949195	949244	949292	949341	
890	**949390**	**949439**	**949488**	**949536**	**949585**	**949634**	**949683**	**949731**	**949780**	**949829**	
1	949878	949926	949975	950024	950073	950121	950170	950219	950267	950316	
2	950365	950414	950462	950511	950560	950608	950657	950706	950754	950803	
3	950851	950900	950949	950997	951046	951095	951143	951192	951240	951289	
4	951338	951386	951435	951483	951532	951580	951629	951677	951726	951775	
5	951823	951872	951920	951969	952017	952066	952114	952163	952211	952260	
6	952308	952356	952405	952453	952502	952550	952599	952647	952696	952744	
7	952792	952841	952889	952938	952986	953034	953083	953131	953180	953228	
8	953276	953325	953373	953421	953470	953518	953566	953615	953663	953711	
9	953760	953808	953856	953905	953953	954001	954049	954098	954146	954194	
900	**954243**	**954291**	**954339**	**954387**	**954435**	**954484**	**954532**	**954580**	**954628**	**954677**	
1	954725	954773	954821	954869	954918	954966	955014	955062	955110	955158	
2	955207	955255	955303	955351	955399	955447	955495	955543	955592	955640	
3	955688	955736	955784	955832	955880	955928	955976	956024	956072	956120	
4	956168	956216	956265	956313	956361	956409	956457	956505	956553	956601	
5	956649	956697	956745	956793	956840	956888	956936	956984	957032	957080	**48**
6	957128	957176	957224	957272	957320	957368	957416	957464	957512	957559	
7	957607	957655	957703	957751	957799	957847	957894	957942	957990	958038	
8	958086	958134	958181	958229	958277	958325	958373	958421	958468	958516	
9	958564	958612	958659	958707	958755	958803	958850	958898	958946	958994	
910	**959041**	**959089**	**959137**	**959185**	**959232**	**959280**	**959328**	**959375**	**959423**	**959471**	
1	959518	959566	959614	959661	959709	959757	959804	959852	959900	959947	
2	959995	960042	960090	960138	960185	960233	960280	960328	960376	960423	
3	960471	960518	960566	960613	960661	960709	960756	960804	960851	960899	
4	960946	960994	961041	961089	961136	961184	961231	961279	961326	961374	
5	961421	961469	961516	961563	961611	961658	961706	961753	961801	961848	
6	961895	961943	961990	962038	962085	962132	962180	962227	962275	962322	
7	962369	962417	962464	962511	962559	962606	962653	962701	962748	962795	
8	962843	962890	962937	962985	963032	963079	963126	963174	963221	963268	
9	963316	963363	963410	963457	963504	963552	963599	963646	963693	963741	

PROPORTIONAL PARTS

Diff.	1	2	3	4	5	6	7	8	9
52	5.2	10.4	15.6	20.8	26.0	31.2	36.4	41.6	46.8
50	5.0	10.0	15.0	20.0	25.0	30.0	35.0	40.0	45.0
48	4.8	9.6	14.4	19.2	24.0	28.8	33.6	38.4	43.2

N	0	1	2	3	4	5	6	7	8	9	Diff.
920	**963788**	**963835**	**963882**	**963929**	**963977**	**964024**	**964071**	**964118**	**964165**	**964212**	
1	964260	964307	964354	964401	964448	964495	964542	964590	964637	964684	
2	964731	964778	964825	964872	964919	964966	965013	965061	965108	965155	
3	965202	965249	965296	965343	965390	965437	965484	965531	965578	965625	
4	965672	965719	965766	965813	965860	965907	965954	966001	966048	966095	
5	966142	966189	966236	966283	966329	966376	966423	966470	966517	966564	
6	966611	966658	966705	966752	966799	966845	966892	966939	966986	967033	
7	967080	967127	967173	967220	967267	967314	967361	967408	967454	967501	
8	967548	967595	967642	967688	967735	967782	967829	967875	967922	967969	
9	968016	968062	968109	968156	968203	968249	968296	968343	968390	968436	
930	**968483**	**968530**	**968576**	**968623**	**968670**	**968716**	**968763**	**968810**	**968856**	**968903**	
1	968950	968996	969043	969090	969136	969183	969229	969276	969323	969369	
2	969416	969463	969509	969556	969602	969649	969695	969742	969789	969835	
3	969882	969928	969975	970021	970068	970114	970161	970207	970254	970300	
4	970347	970393	970440	970486	970533	970579	970626	970672	970719	970765	
5	970812	970858	970904	970951	970997	971044	971090	971137	971183	971229	
6	971276	971322	971369	971415	971461	971508	971554	971601	971647	971693	
7	971740	971786	971832	971879	971925	971971	972018	972064	972110	972157	
8	972203	972249	972295	972342	972388	972434	972481	972527	972573	972619	
9	972666	972712	972758	972804	972851	972897	972943	972989	973035	973082	
940	**973128**	**973174**	**973220**	**973266**	**973313**	**973359**	**973405**	**973451**	**973497**	**973543**	
1	973590	973636	973682	973728	973774	973820	973866	973913	973959	974005	
2	974051	974097	974143	974189	974235	974281	974327	974374	974420	974466	
3	974512	974558	974604	974650	974696	974742	974788	974834	974880	974926	
4	974972	975018	975064	975110	975156	975202	975248	975294	975340	975386	**46**
5	975432	975478	975524	975570	975616	975662	975707	975753	975799	975845	
6	975891	975937	975983	976029	976075	976121	976167	976212	976258	976304	
7	976350	976396	976442	976488	976533	976579	976625	976671	976717	976763	
8	976808	976854	976900	976946	976992	977037	977083	977129	977175	977220	
9	977266	977312	977358	977403	977449	977495	977541	977586	977632	977678	
950	**977724**	**977769**	**977815**	**977861**	**977906**	**977952**	**977998**	**978043**	**978089**	**978135**	
1	978181	978226	978272	978317	978363	978409	978454	978500	978546	978591	
2	978637	978683	978728	978774	978819	978865	978911	978956	979002	979047	
3	979093	979138	979184	979230	979275	979321	979366	979412	979457	979503	
4	979548	979594	979639	979685	979730	979776	979821	979867	979912	979958	
5	980003	980049	980094	980140	980185	980231	980276	980322	980367	980412	
6	980458	980503	980549	980594	980640	980685	980730	980776	980821	980867	
7	980912	980957	981003	981048	981093	981139	981184	981229	981275	981320	
8	981366	981411	981456	981501	981547	981592	981637	981683	981728	981773	
9	981819	981864	981909	981954	982000	982045	982090	982135	982181	982226	
960	**982271**	**982316**	**982362**	**982407**	**982452**	**982497**	**982543**	**982588**	**982633**	**982678**	
1	982723	982769	982814	982859	982904	982949	982994	983040	983085	983130	
2	983175	983220	983265	983310	983356	983401	983446	983491	983536	983581	
3	983626	983671	983716	983762	983807	983852	983897	983942	983987	984032	
4	984077	984122	984167	984212	984257	984302	984347	984392	984437	984482	
5	984527	984572	984617	984662	984707	984752	984797	984842	984887	984932	**45**
6	984977	985022	985067	985112	985157	985202	985247	985292	985337	985382	
7	985426	985471	985516	985561	985606	985651	985696	985741	985786	985830	
8	985875	985920	985965	986010	986055	986100	986144	986189	986234	986279	
9	986324	986369	986413	986458	986503	986548	986593	986637	986682	986727	
970	**986772**	**986817**	**986861**	**986906**	**986951**	**986996**	**987040**	**987085**	**987130**	**987175**	
1	987219	987264	987309	987353	987398	987443	987488	987532	987577	987622	
2	987666	987711	987756	987800	987845	987890	987934	987979	988024	988068	
3	988113	988157	988202	988247	988291	988336	988381	988425	988470	988514	
4	988559	988604	988648	988693	988737	988782	988826	988871	988916	988960	
5	989005	989049	989094	989138	989183	989227	989272	989316	989361	989405	
6	989450	989494	989539	989583	989628	989672	989717	989761	989806	989850	
7	989895	989939	989983	990028	990072	990117	990161	990206	990250	990294	
8	990339	990383	990428	990472	990516	990561	990605	990650	990694	990738	
9	990783	990827	990871	990916	990960	991004	991049	991093	991137	991182	

PROPORTIONAL PARTS

Diff.	1	2	3	4	5	6	7	8	9
48	4.8	9.6	14.4	19.2	24.0	28.8	33.6	38.4	43.2
46	4.6	9.2	13.8	18.4	23.0	27.6	32.2	36.8	41.4
44	4.4	8.8	13.2	17.6	22.0	26.4	30.8	35.2	39.6
42	4.2	8.4	12.6	16.8	21.0	25.2	29.4	33.6	37.8

N	0	1	2	3	4	5	6	7	8	9	Diff.
980	**991226**	**991270**	**991315**	**991359**	**991403**	**991448**	**991492**	**991536**	**991580**	**991625**	
1	991669	991713	991758	991802	991846	991890	991935	991979	992023	992067	
2	992111	992156	992200	992244	992288	992333	992377	992421	992465	992509	
3	992554	992598	992642	992686	992730	992774	992819	992863	992907	992951	
4	992995	993039	993083	993127	993172	993216	993260	993304	993348	993392	
5	993436	993480	993524	993568	993613	993657	993701	993745	993789	993833	
6	993877	993921	993965	994009	994053	994097	994141	994185	994229	994273	
7	994317	994361	994405	994449	994493	994537	994581	994625	994669	994713	44
8	994757	994801	994845	994889	994933	994977	995021	995065	995108	995152	
9	995196	995240	995284	995328	995372	995416	995460	995504	995547	995591	
990	**995635**	**995679**	**995723**	**995767**	**995811**	**995854**	**995898**	**995942**	**995986**	**996030**	
1	996074	996117	996161	996205	996249	996293	996337	996380	996424	996468	
2	996512	996555	996599	996643	996687	996731	996774	996818	996862	996906	
3	996949	996993	997037	997080	997124	997168	997212	997255	997299	997343	
4	997386	997430	997474	997517	997561	997605	997648	997692	997736	997779	
5	997823	997867	997910	997954	997998	998041	998085	998129	998172	998216	
6	998259	998303	998347	998390	998434	998477	998521	998564	998608	998652	
7	998695	998739	998782	998826	998869	998913	998956	999000	999043	999087	
8	999131	999174	999218	999261	999305	999348	999392	999435	999479	999522	
9	999565	999609	999652	999696	999739	999783	999826	999870	999913	999957	
1000	**000000**	**000043**	**000087**	**000130**	**000174**	**000217**	**000260**	**000304**	**000347**	**000391**	43

Table 3. Natural (Napierian or Hyperbolic) Logarithms of Numbers

Table gives natural logarithms of numbers from 1.0 to 9.99 directly. To find logarithms of numbers outside that range add or subtract natural logarithm of the powers of 10 as follows:

$\log_e 10 = 2.302585$	$\log_e 10^4 = 9.210340$	$\log_e 10^7 = 16.118096$
$\log_e 10^2 = 4.605170$	$\log_e 10^5 = 11.512925$	$\log_e 10^8 = 18.420681$
$\log_e 10^3 = 6.907755$	$\log_e 10^6 = 13.815511$	$\log_e 10^9 = 20.723266$

EXAMPLE.—$\log_e 679. = \log_e 6.79 + \log_e 10^2 = 1.9155 + 4.6052 = 6.5207$

$\log_e .0679 = \log_e 6.79 - \log_e 10^2 = 1.9155 - 4.6052 = -2.6897$

The base of natural logarithms is $e = 2.7182818$. Therefore $\log_e N = \log_e 10 \times \log_{10} N = 2.302585 \log_{10} N$; and $\log_{10} N = \log_{10} e \times \log_e N = 0.434294 \log_e N$.

N	0	1	2	3	4	5	6	7	8	9
1.0	**0.0000**	**0.0100**	**0.0198**	**0.0296**	**0.0392**	**0.0488**	**0.0583**	**0.0677**	**0.0770**	**0.0862**
1.1	0.0953	0.1044	0.1133	0.1222	0.1310	0.1398	0.1484	0.1570	0.1655	0.1740
1.2	0.1823	0.1906	0.1989	0.2070	0.2151	0.2231	0.2311	0.2390	0.2469	0.2546
1.3	0.2624	0.2700	0.2776	0.2852	0.2927	0.3001	0.3075	0.3148	0.3221	0.3293
1.4	0.3365	0.3436	0.3507	0.3577	0.3646	0.3716	0.3784	0.3853	0.3920	0.3988
1.5	0.4055	0.4121	0.4187	0.4253	0.4318	0.4383	0.4447	0.4511	0.4574	0.4637
1.6	0.4700	0.4762	0.4824	0.4886	0.4947	0.5008	0.5068	0.5128	0.5188	0.5247
1.7	0.5306	0.5365	0.5423	0.5481	0.5539	0.5596	0.5653	0.5710	0.5766	0.5822
1.8	0.5878	0.5933	0.5988	0.6043	0.6098	0.6152	0.6206	0.6259	0.6313	0.6366
1.9	0.6419	0.6471	0.6523	0.6575	0.6627	0.6678	0.6729	0.6780	0.6831	0.6881
2.0	**0.6931**	**0.6981**	**0.7031**	**0.7080**	**0.7129**	**0.7178**	**0.7227**	**0.7275**	**0.7324**	**0.7372**
2.1	0.7419	0.7467	0.7514	0.7561	0.7608	0.7655	0.7701	0.7747	0.7793	0.7839
2.2	0.7885	0.7930	0.7975	0.8020	0.8065	0.8109	0.8154	0.8198	0.8242	0.8286
2.3	0.8329	0.8372	0.8416	0.8459	0.8502	0.8544	0.8587	0.8629	0.8671	0.8713
2.4	0.8755	0.8796	0.8838	0.8879	0.8920	0.8961	0.9002	0.9042	0.9083	0.9123
2.5	0.9163	0.9203	0.9243	0.9282	0.9322	0.9361	0.9400	0.9439	0.9478	0.9517
2.6	0.9555	0.9594	0.9632	0.9670	0.9708	0.9746	0.9783	0.9821	0.9858	0.9895
2.7	0.9933	0.9969	1.0006	1.0043	1.0080	1.0116	1.0152	1.0188	1.0225	1.0260
2.8	1.0296	1.0332	1.0367	1.0403	1.0438	1.0473	1.0508	1.0543	1.0578	1.0613
2.9	1.0647	1.0682	1.0716	1.0750	1.0784	1.0818	1.0852	1.0886	1.0919	1.0953
3.0	**1.0986**	**1.1019**	**1.1053**	**1.1086**	**1.1119**	**1.1151**	**1.1184**	**1.1217**	**1.1249**	**1.1282**
3.1	1.1314	1.1346	1.1378	1.1410	1.1442	1.1474	1.1506	1.1537	1.1569	1.1600
3.2	1.1632	1.1663	1.1694	1.1725	1.1756	1.1787	1.1817	1.1848	1.1878	1.1909
3.3	1.1939	1.1969	1.2000	1.2030	1.2060	1.2090	1.2119	1.2149	1.2179	1.2208
3.4	1.2238	1.2267	1.2296	1.2326	1.2355	1.2384	1.2413	1.2442	1.2470	1.2499
3.5	1.2528	1.2556	1.2585	1.2613	1.2641	1.2669	1.2698	1.2726	1.2754	1.2782
3.6	1.2809	1.2837	1.2865	1.2892	1.2920	1.2947	1.2975	1.3002	1.3029	1.3056
3.7	1.3083	1.3110	1.3137	1.3164	1.3191	1.3218	1.3244	1.3271	1.3297	1.3324
3.8	1.3350	1.3376	1.3403	1.3429	1.3455	1.3481	1.3507	1.3533	1.3558	1.3584
3.9	1.3610	1.3635	1.3661	1.3686	1.3712	1.3737	1.3762	1.3788	1.3813	1.3838

Table 3. Natural (Napierian or Hyperbolic) Logarithms of Numbers—*Continued*

N	0	1	2	3	4	5	6	7	8	9
4.0	**1.3863**	**1.3888**	**1.3913**	**1.3938**	**1.3962**	**1.3987**	**1.4012**	**1.4036**	**1.4061**	**1.4085**
4.1	1.4110	1.4134	1.4159	1.4183	1.4207	1.4231	1.4255	1.4279	1.4303	1.4327
4.2	1.4351	1.4375	1.4398	1.4422	1.4446	1.4469	1.4493	1.4516	1.4540	1.4563
4.3	1.4586	1.4609	1.4633	1.4656	1.4679	1.4702	1.4725	1.4748	1.4770	1.4793
4.4	1.4816	1.4839	1.4861	1.4884	1.4907	1.4929	1.4951	1.4974	1.4996	1.5019
4.5	1.5041	1.5063	1.5085	1.5107	1.5129	1.5151	1.5173	1.5195	1.5217	1.5239
4.6	1.5261	1.5282	1.5304	1.5326	1.5347	1.5369	1.5390	1.5412	1.5433	1.5454
4.7	1.5476	1.5497	1.5518	1.5539	1.5560	1.5581	1.5602	1.5623	1.5644	1.5665
4.8	1.5686	1.5707	1.5728	1.5748	1.5769	1.5790	1.5810	1.5831	1.5851	1.5872
4.9	1.5892	1.5913	1.5933	1.5953	1.5974	1.5994	1.6014	1.6034	1.6054	1.6074
5.0	**1.6094**	**1.6114**	**1.6134**	**1.6154**	**1.6174**	**1.6194**	**1.6214**	**1.6233**	**1.6253**	**1.6273**
5.1	1.6292	1.6312	1.6332	1.6351	1.6371	1.6390	1.6409	1.6429	1.6448	1.6467
5.2	1.6487	1.6506	1.6525	1.6544	1.6563	1.6582	1.6601	1.6620	1.6639	1.6658
5.3	1.6677	1.6696	1.6715	1.6734	1.6752	1.6771	1.6790	1.6808	1.6827	1.6845
5.4	1.6864	1.6882	1.6901	1.6919	1.6938	1.6956	1.6974	1.6993	1.7011	1.7029
5.5	1.7047	1.7066	1.7084	1.7102	1.7120	1.7138	1.7156	1.7174	1.7192	1.7210
5.6	1.7228	1.7246	1.7263	1.7281	1.7299	1.7317	1.7334	1.7352	1.7370	1.7387
5.7	1.7405	1.7422	1.7440	1.7457	1.7475	1.7492	1.7509	1.7527	1.7544	1.7561
5.8	1.7579	1.7596	1.7613	1.7630	1.7647	1.7664	1.7681	1.7699	1.7716	1.7733
5.9	1.7750	1.7766	1.7783	1.7800	1.7817	1.7834	1.7851	1.7867	1.7884	1.7901
6.0	**1.7918**	**1.7934**	**1.7951**	**1.7967**	**1.7984**	**1.8001**	**1.8017**	**1.8034**	**1.8050**	**1.8066**
6.1	1.8083	1.8099	1.8116	1.8132	1.8148	1.8165	1.8181	1.8197	1.8213	1.8229
6.2	1.8245	1.8262	1.8278	1.8294	1.8310	1.8326	1.8342	1.8358	1.8374	1.8390
6.3	1.8405	1.8421	1.8437	1.8453	1.8469	1.8485	1.8500	1.8516	1.8532	1.8547
6.4	1.8563	1.8579	1.8594	1.8610	1.8625	1.8641	1.8656	1.8672	1.8687	1.8703
6.5	1.8718	1.8733	1.8749	1.8764	1.8779	1.8795	1.8810	1.8825	1.8840	1.8856
6.6	1.8871	1.8886	1.8901	1.8916	1.8931	1.8946	1.8961	1.8976	1.8991	1.9006
6.7	1.9021	1.9036	1.9051	1.9066	1.9081	1.9095	1.9110	1.9125	1.9140	1.9155
6.8	1.9169	1.9184	1.9199	1.9213	1.9228	1.9242	1.9257	1.9272	1.9286	1.9301
6.9	1.9315	1.9330	1.9344	1.9359	1.9373	1.9387	1.9402	1.9416	1.9430	1.9445
7.0	**1.9459**	**1.9473**	**1.9488**	**1.9502**	**1.9516**	**1.9530**	**1.9544**	**1.9559**	**1.9573**	**1.9587**
7.1	1.9601	1.9615	1.9629	1.9643	1.9657	1.9671	1.9685	1.9699	1.9713	1.9727
7.2	1.9741	1.9755	1.9769	1.9782	1.9796	1.9810	1.9824	1.9838	1.9851	1.9865
7.3	1.9879	1.9892	1.9906	1.9920	1.9933	1.9947	1.9961	1.9974	1.9988	2.0001
7.4	2.0015	2.0028	2.0042	2.0055	2.0069	2.0082	2.0096	2.0109	2.0122	2.0136
7.5	2.0149	2.0162	2.0176	2.0189	2.0202	2.0215	2.0229	2.0242	2.0255	2.0268
7.6	2.0281	2.0295	2.0308	2.0321	2.0334	2.0347	2.0360	2.0373	2.0386	2.0399
7.7	2.0412	2.0425	2.0438	2.0451	2.0464	2.0477	2.0490	2.0503	2.0516	2.0528
7.8	2.0541	2.0554	2.0567	2.0580	2.0592	2.0605	2.0618	2.0631	2.0643	2.0656
7.9	2.0669	2.0681	2.0694	2.0707	2.0719	2.0732	2.0744	2.0757	2.0769	2.0782
8.0	**2.0794**	**2.0807**	**2.0819**	**2.0832**	**2.0844**	**2.0857**	**2.0869**	**2.0882**	**2.0894**	**2.0906**
8.1	2.0919	2.0931	2.0943	2.0956	2.0968	2.0980	2.0992	2.1005	2.1017	2.1029
8.2	2.1041	2.1054	2.1066	2.1078	2.1090	2.1102	2.1114	2.1126	2.1138	2.1150
8.3	2.1163	2.1175	2.1187	2.1199	2.1211	2.1223	2.1235	2.1247	2.1258	2.1270
8.4	2.1282	2.1294	2.1306	2.1318	2.1330	2.1342	2.1353	2.1365	2.1377	2.1389
8.5	2.1401	2.1412	2.1424	2.1436	2.1448	2.1459	2.1471	2.1483	2.1494	2.1506
8.6	2.1518	2.1529	2.1541	2.1552	2.1564	2.1576	2.1587	2.1599	2.1610	2.1622
8.7	2.1633	2.1645	2.1656	2.1668	2.1679	2.1691	2.1702	2.1713	2.1725	2.1736
8.8	2.1748	2.1759	2.1770	2.1782	2.1793	2.1804	2.1815	2.1827	2.1838	2.1849
8.9	2.1861	2.1872	2.1883	2.1894	2.1905	2.1917	2.1928	2.1939	2.1950	2.1961
9.0	**2.1972**	**2.1983**	**2.1994**	**2.2006**	**2.2017**	**2.2028**	**2.2039**	**2.2050**	**2.2061**	**2.2072**
9.1	2.2083	2.2094	2.2105	2.2116	2.2127	2.2138	2.2148	2.2159	2.2170	2.2181
9.2	2.2192	2.2203	2.2214	2.2225	2.2235	2.2246	2.2257	2.2268	2.2279	2.2289
9.3	2.2300	2.2311	2.2322	2.2332	2.2343	2.2354	2.2364	2.2375	2.2386	2.2396
9.4	2.2407	2.2418	2.2428	2.2439	2.2450	2.2460	2.2471	2.2481	2.2492	2.2502
9.5	2.2513	2.2523	2.2534	2.2544	2.2555	2.2565	2.2576	2.2586	2.2597	2.2607
9.6	2.2618	2.2628	2.2638	2.2649	2.2659	2.2670	2.2680	2.2690	2.2701	2.2711
9.7	2.2721	2.2732	2.2742	2.2752	2.2762	2.2773	2.2783	2.2793	2.2803	2.2814
9.8	2.2824	2.2834	2.2844	2.2854	2.2865	2.2875	2.2885	2.2895	2.2905	2.2915
9.9	2.2925	2.2935	2.2946	2.2956	2.2966	2.2976	2.2986	2.2996	2.3006	2.3016

Table 4. Values and Logarithms of Exponential and Hyperbolic Functions

x	Values					Common Logarithms			
	e^x	e^{-x}	Sinh x	Cosh x	Tanh x	e^x	Sinh x	Cosh x	Tanh x
0.00	1.0000	1.0000	0.0000	1.0000	.00000	0.00000	$-\infty$	0.00000	$-\infty$
0.10	1.1052	.90484	0.1002	1.0050	.09967	0.04343	$\bar{1}$.00072	0.00217	$\bar{2}$.99856
0.20	1.2214	.81873	0.2013	1.0201	.19738	0.08686	$\bar{1}$.30392	0.00863	$\bar{1}$.29529
0.30	1.3499	.74082	0.3045	1.0453	.29131	0.13029	1.48362	0.10926	$\bar{1}$.46436
0.40	1.4918	.67032	0.4108	1.0811	.37995	0.17372	$\bar{1}$.61358	0.03385	$\bar{1}$.57973
0.50	1.6487	.60653	0.5211	1.1276	.46212	0.21715	1.71692	0.05217	1.66475
0.60	1.8221	.54881	0.6367	1.1855	.53705	0.26058	$\bar{1}$.80390	0.07389	$\bar{1}$.73001
0.70	2.0138	.49659	0.7586	1.2552	.60437	0.30401	$\bar{1}$.88000	0.09870	$\bar{1}$.78130
0.80	2.2255	.44933	0.8881	1.3374	.66404	0.34744	$\bar{1}$.94846	0.12627	1.82219
0.90	2.4596	.40657	1.0265	1.4331	.71630	0.39087	0.01137	0.15627	$\bar{1}$.85509
1.00	2.7183	.36788	1.1752	1.5431	.76159	0.43439	0.07011	0.18839	$\bar{1}$.88172
1.10	3.0042	.33287	1.3356	1.6685	.80050	0.47772	0.12569	0.22233	$\bar{1}$.90336
1.20	3.3201	.30119	1.5095	1.8107	.83365	0.52115	0.17882	0.25784	$\bar{1}$.92099
1.30	3.6693	.27253	1.6984	1.9709	.86172	0.56458	0.23004	0.29467	$\bar{1}$.93537
1.40	4.0552	.24660	1.9043	2.1509	.88535	0.60801	0.27974	0.33262	$\bar{1}$.94712
1.50	4.4817	.22313	2.1293	2.3524	.90515	0.65144	0.32823	0.37151	$\bar{1}$.95672
1.60	4.9530	.20190	2.3756	2.5775	.92167	0.69487	0.37577	0.41119	$\bar{1}$.96457
1.70	5.4739	.18268	2.6456	2.8283	.93541	0.73830	0.42253	0.45153	$\bar{1}$.97100
1.80	6.0496	.16530	2.9422	3.1075	.94681	0.78173	0.46867	0.49241	$\bar{1}$.97626
1.90	6.6859	.14957	3.2682	3.4177	.95624	0.82516	0.51430	0.53374	$\bar{1}$.98057
2.00	7.3891	.13534	3.6269	3.7622	.96403	0.86859	0.55953	0.57544	$\bar{1}$.98409
2.10	8.1662	.12246	4.0219	4.1443	.97045	0.91202	0.60443	0.61745	$\bar{1}$.98697
2.20	9.0250	.11080	4.4571	4.5679	.97574	0.95545	0.64905	0.65972	$\bar{1}$.98934
2.30	9.9742	.10026	4.9370	5.0372	.98010	0.99888	0.69346	0.70219	$\bar{1}$.99127
2.40	11.023	.09072	5.4662	5.5569	.98367	1.04231	0.73769	0.74484	$\bar{1}$.99285
2.50	12.182	.08208	6.0502	6.1323	.98661	1.08574	0.78177	0.78762	$\bar{1}$.99415
2.60	13.464	.07427	6.6947	6.7690	.98903	1.12917	0.82573	0.83052	$\bar{1}$.99521
2.70	14.880	.06721	7.4063	7.4735	.99101	1.17260	0.86960	0.87352	$\bar{1}$.99608
2.80	16.445	.06081	8.1919	8.2527	.99263	1.21602	0.91339	0.91660	$\bar{1}$.99679
2.90	18.174	.05502	9.0596	9.1146	.99396	1.25945	0.95711	0.95974	$\bar{1}$.99737
3.00	20.086	.04979	10.018	10.068	.99505	1.30288	1.00078	1.00293	$\bar{1}$.99785
3.10	22.198	.04505	11.077	11.122	.99595	1.34631	1.04440	1.04616	$\bar{1}$.99824
3.20	24.533	.04076	12.246	12.287	.99668	1.38974	1.08799	1.08943	$\bar{1}$.99856
3.30	27.113	.03688	13.538	13.575	.99728	1.43317	1.13155	1.13273	$\bar{1}$.99882
3.40	29.964	.03337	14.965	14.999	.99777	1.47660	1.17509	1.17605	$\bar{1}$.99903
3.50	33.115	.03020	16.543	16.573	.99818	1.52003	1.21860	1.21940	$\bar{1}$.99921
3.60	36.598	.02732	18.285	18.313	.99851	1.56346	1.26211	1.26275	$\bar{1}$.99935
3.70	40.447	.02472	20.211	20.236	.99878	1.60689	1.30559	1.30612	$\bar{1}$.99947
3.80	44.701	.02237	22.339	22.362	.99900	1.65032	1.34907	1.34951	$\bar{1}$.99957
3.90	49.402	.02024	24.691	24.711	.99918	1.69375	1.39254	1.39290	$\bar{1}$.99964
4.00	54.598	.01832	27.290	27.308	.99933	1.73718	1.43600	1.43629	$\bar{1}$.99971
4.10	60.340	.01657	30.162	30.178	.99945	1.78061	1.47946	1.47970	$\bar{1}$.99976
4.20	66.686	.01500	33.336	33.351	.99955	1.82404	1.52291	1.52310	$\bar{1}$.99980
4.30	73.700	.01357	36.843	36.857	.99963	1.86747	1.56636	1.56652	$\bar{1}$.99984
4.40	81.451	.01228	40.719	40.732	.99970	1.91090	1.60980	1.60993	$\bar{1}$.99987
4.50	90.017	.01111	45.003	45.014	.99975	1.95433	1.65324	1.65335	$\bar{1}$.99989
4.60	99.484	.01005	49.737	49.747	.99980	1.99775	1.69668	1.69677	$\bar{1}$.99991
4.70	109.95	.00910	54.969	54.978	.99983	2.04118	1.74012	1.74019	$\bar{1}$.99993
4.80	121.51	.00823	60.751	60.759	.99986	2.08461	1.78355	1.78361	$\bar{1}$.99994
4.90	134.29	.00745	67.141	67.149	.99989	2.12804	1.82699	1.82704	$\bar{1}$.99995
5.00	148.41	.00674	74.203	74.210	.99991	2.17147	1.87042	1.87046	$\bar{1}$.99996
5.10	164.02	.00610	82.008	82.014	.99993	2.21490	1.91386	1.91389	$\bar{1}$.99997
5.20	181.27	.00552	90.633	90.639	.99994	2.25833	1.95729	1.95731	$\bar{1}$.99997
5.30	200.34	.00499	100.17	100.17	.99995	2.30176	2.00072	2.00074	$\bar{1}$.99998
5.40	221.41	.00452	110.70	110.71	.99996	2.34519	2.04415	2.04417	$\bar{1}$.99998
5.50	244.69	.00409	122.34	122.35	.99997	2.38862	2.08758	2.08760	$\bar{1}$.99999
5.60	270.43	.00370	135.21	135.22	.99997	2.43205	2.13101	2.13103	$\bar{1}$.99999
5.70	298.87	.00335	149.43	149.44	.99998	2.47548	2.17444	2.17445	$\bar{1}$.99999
5.80	330.30	.00303	165.15	165.15	.99998	2.51891	2.21787	2.21788	$\bar{1}$.99999
5.90	365.04	.00274	182.52	182.52	.99998	2.56234	2.26130	2.26131	$\bar{1}$.99999
6.00	403.43	.00248	201.71	201.72	.99999	2.60577	2.30473	2.30474	$\bar{1}$.99999

Table 4 on the preceding page gives values of e^x, e^{-x}, sinh x, cosh x, and tanh x for values of x from 0.00 to 6.00 in intervals of 0.10.

To facilitate computations involving multiplication, the common logarithms of e^x, sinh x, cosh x, and tanh x are also given.

For values of x greater than 6: e^x may be computed from the relationship $e^x = \log^{-1}(x \log_{10} e) = \log^{-1} 0.43429x$; e^{-x} approaches zero; sinh x and cosh x are approximately equal and become 0.5 e^x; and tanh x and coth x have values approximately equal to unity.

Where more accurate values of the exponentials and functions are required they may be computed from the following relationships.

$$e = 2.71828\ 18285 \qquad \frac{1}{e} = 0.36787\ 94412$$

$$M = \log_{10} e = 0.43429\ 44819 \qquad \frac{1}{M} = \log_e 10 = 2.30258\ 50930$$

$$e^x = \log^{-1} Mx \qquad e^{-x} = \log^{-1} - Mx$$

$$\sinh x = \frac{e^x - e^{-x}}{2} \qquad \cosh x = \frac{e^x + e^{-x}}{2} \qquad \tanh x = \frac{e^x - e^{-x}}{e^x + e^{-x}}$$

$$\operatorname{csch} x = \frac{1}{\sinh x} \qquad \operatorname{sech} x = \frac{1}{\cosh x} \qquad \coth x = \frac{1}{\tanh x}$$

Table 5. Certain Constants Containing e and π

$e = 2.7182818285$ $\quad M = \log_{10} e = 0.4342944819$

$\pi = 3.1415926536$ $\quad M^{-1} = \log_e 10 = 2.3025850930$

Powers of e			Multiples of π			Fractions of π		
e^n	Value	Logarithm	$n\pi$	Value	Logarithm	π/n	Value	Logarithm
e	2.718282	0.434294	π	3.141593	0.497150	$\pi/2$	1.570780	0.196120
e^{-1}	0.367879	$\bar{1}$.565706	2π	6.283185	0.798180	$\pi/3$	1.047198	0.020029
e^2	7.389057	0.868589	3π	9.424778	0.974271	$\pi/4$	0.785398	$\bar{1}$.895090
e^{-2}	0.135335	$\bar{1}$.131411	4π	12.566371	1.099210	$\pi/180$	0.017453*	$\bar{2}$.241877
$e^{1/2}$	1.648721	0.217147	5π	15.707963	1.196120			

Reciprocals of π			Powers of π			Roots of π		
n/π	Value	Logarithm	$\pi^{\pm n}$	Value	Logarithm	$\pi^{\pm 1/n}$	Value	Logarithm
$1/\pi$	0.318310	$\bar{1}$.502850	π^2	9.869604	0.994300	$\sqrt{\pi}$	1.772454	0.248575
$2/\pi$	0.636620	$\bar{1}$.803880	$1/\pi^2$	0.101321	$\bar{1}$.005700	$1/\sqrt{\pi}$	0.564190	$\bar{1}$.751425
$3/\pi$	0.954930	$\bar{1}$.979971	π^3	31.006277	1.491450	$\sqrt[3]{\pi}$	1.464592	0.165717
$180/\pi$	57.295780†	1.758123	$1/\pi^3$	0.032252	$\bar{2}$.508550	$1/\sqrt[3]{\pi}$	0.681784	$\bar{1}$.834283

* Number of radians per degree. † Number of degrees per radian.

Table 6. Factorials

n	$n! = 1\cdot2\cdot3\ldots n$	$1/n!$	n	$n! = 1\cdot2\cdot3\ldots n$	$1/n!$
1	1	1.	11	$399,168 \times 10^2$	0.250521×10^{-7}
2	2	0.5	12	$479,002 \times 10^3$	$.208768 \times 10^{-8}$
3	6	.166667	13	$622,702 \times 10^4$	$.160590 \times 10^{-9}$
4	24	$.416667 \times 10^{-1}$	14	$871,783 \times 10^5$	$.114707 \times 10^{-10}$
5	120	$.833333 \times 10^{-2}$	15	$130,767 \times 10^7$	$.764716 \times 10^{-12}$
6	720	$.138889 \times 10^{-2}$	16	$209,228 \times 10^8$	$.477948 \times 10^{-13}$
7	5,040	$.198413 \times 10^{-3}$	17	$355,687 \times 10^9$	$.281146 \times 10^{-14}$
8	40,320	$.248016 \times 10^{-4}$	18	$640,237 \times 10^{10}$	$.156192 \times 10^{-15}$
9	362,880	$.275573 \times 10^{-5}$	19	$121,645 \times 10^{12}$	$.822064 \times 10^{-17}$
10	3,628,800	$.275573 \times 10^{-6}$	20	$243,290 \times 10^{13}$	$.411032 \times 10^{-19}$

Table 7. Properties of Numbers

Decimal Equivalents, Squares, Cubes, Three-halves Powers, Square Roots, Cube Roots, Fifth Roots,
Reciprocals, Circumference and Area of Circles

Number, N		N^2	N^3	$\sqrt{N}$	$\sqrt[3]{N}$	$N^{3/2}$	$\sqrt[5]{N}$	$\frac{1}{N}$	Circle (N=Diam.)	
Fraction	Decimal								Circum.	Area
1/64	.015625	0.000244	$.381\times10^{-5}$	.1250	.2500	.00195	.4353	64.0	0.04909	.00019
1/32	.03125	.000977	$.305\times10^{-4}$	.1768	.3150	.00552	.5000	32.0	.09818	.00077
3/64	.046875	.002197	$.103\times10^{-3}$	.2165	.3606	.01015	.5422	21.3333	.14726	.00173
1/16	**.0625**	**.003906**	$\mathbf{.244\times10^{-3}}$	**.2500**	**.3969**	**.01563**	**.5744**	**16.0**	**.19635**	**.00307**
5/64	.078125	.006104	$.477\times10^{-3}$	.2795	.4275	.02184	.6006	12.80	.24544	.00479
3/32	.09375	.008789	$.824\times10^{-3}$	.3062	.4543	.02871	.6229	10.6667	.29452	.00690
	.10	.010	.00100	.3162	.4642	.03162	.6310	10.0	.31416	.00785
7/64	.109375	.01196	.001308	.3307	.4782	.03617	.6424	9.1429	.34361	.00939
1/8	**.125**	**.01563**	**.001953**	**.3536**	**.5000**	**.04419**	**.6598**	**8.0**	**.39270**	**.01227**
9/64	.140625	.01978	.002782	.3750	.5200	.05273	.6755	7.1111	.44179	.01554
5/32	.15625	.02441	.003814	.3953	.5386	.06176	.6899	6.40	.49087	.01917
11/64	.171875	.02954	.005077	.4146	.5560	.07126	.7031	5.8182	.53996	.02320
3/16	**.1875**	**.03516**	**.006592**	**.4330**	**.5724**	**.08119**	**.7155**	**5.3333**	**.58905**	**.02761**
	.20	.040	.0080	.4472	.5848	.08944	.7248	5.0	.62832	.03142
13/64	.203125	.04126	.008381	.4507	.5878	.09155	.7270	4.9231	.63814	.03241
7/32	.21875	.04785	.01047	.4677	.6025	.10231	.7379	4.5714	.68722	.03758
15/64	.234375	.05493	.01287	.4841	.6166	.11347	.7481	4.2667	.73631	.04314
1/4	**.250**	**.0625**	**.01563**	**.5000**	**.6300**	**.12500**	**.7579**	**4.0**	**.78540**	**.04909**
17/64	.265625	.07056	.01874	.5154	.6428	.13690	.7671	3.7647	.83448	.05542
9/32	.28125	.07910	.02225	.5303	.6552	.14916	.7759	3.5556	.88357	.06213
19/64	.296875	.08813	.02616	.5449	.6671	.16176	.7844	3.3684	.93266	.06922
	.30	.090	.0270	.5477	.6694	.16432	.7860	3.3333	.94248	.07069
5/16	**.3125**	**.09766**	**.03052**	**.5590**	**.6786**	**.17469**	**.7925**	**3.2000**	**.98175**	**.07670**
21/64	.328125	.10767	.03533	.5728	.6897	.18796	.8002	3.0476	1.0308	.08456
11/32	.34375	.11816	.04062	.5863	.7005	.20154	.8077	2.9091	1.0799	.09281
23/64	.359375	.12915	.04641	.5995	.7110	.21544	.8149	2.7826	1.1290	.10143
3/8	**.375**	**.14063**	**.05273**	**.6124**	**.7211**	**.22964**	**.8219**	**2.6667**	**1.1781**	**.11045**
25/64	.390625	.15259	.05961	.6250	.7310	.24414	.8286	2.5600	1.2272	.11984
	.40	.16	.0640	.6325	.7368	.25298	.8326	2.50	1.2566	.12566
13/32	.40625	.16504	.06705	.6374	.7406	.25894	.8351	2.4615	1.2763	.12962
27/64	.421875	.17798	.07508	.6495	.7500	.27402	.8415	2.3704	1.3254	.13979
7/16	**.4375**	**.19141**	**.08374**	**.6614**	**.7592**	**.28938**	**.8476**	**2.2857**	**1.3744**	**.15033**
29/64	.453125	.20532	.09304	.6732	.7681	.30502	.8536	2.2069	1.4235	.16126
15/32	.46875	.21973	.10300	.6847	.7768	.32093	.8594	2.1333	1.4726	.17257
31/64	.484375	.23462	.11364	.6960	.7854	.33711	.8650	2.0645	1.5217	.18427
1/2	**.50**	**.2500**	**.12500**	**.7071**	**.7937**	**.35355**	**.8706**	**2.0**	**1.5708**	**.19635**
33/64	.515625	.26587	.13709	.7181	.8019	.37025	.8759	1.9394	1.6199	.20881
17/32	.53125	.28223	.14993	.7289	.8099	.38721	.8812	1.8824	1.6690	.22166
35/64	.546875	.29907	.16355	.7395	.8178	.40442	.8863	1.8286	1.7181	.23489
9/16	**.5625**	**.31641**	**.17798**	**.7500**	**.8255**	**.42188**	**.8913**	**1.7778**	**1.7671**	**.24850**
37/64	.578125	.33423	.19323	.7604	.8331	.43957	.8962	1.7297	1.8162	.26250
19/32	.59375	.35254	.20932	.7706	.8405	.45751	.9010	1.6842	1.8653	.27688
	.60	.3600	.21600	.7746	.8434	.46476	.9029	1.6667	1.8850	.28274
39/64	.609375	.37134	.22628	.7806	.8478	.47569	.9057	1.6410	1.9144	.29165
5/8	**.625**	**.39063**	**.24414**	**.7906**	**.8550**	**.49410**	**.9103**	**1.6000**	**1.9635**	**.30680**
41/64	.640625	.41040	.26291	.8004	.8621	.51275	.9148	1.5610	2.0126	.32233
21/32	.65625	.43066	.28262	.8101	.8690	.53162	.9192	1.5238	2.0617	.33824
43/64	.671875	.45142	.30330	.8197	.8759	.55072	.9235	1.4884	2.1108	.35454
11/16	**.6875**	**.47266**	**.32495**	**.8297**	**.8826**	**.57005**	**.9278**	**1.4545**	**2.1598**	**.37122**
	.70	.4900	.34300	.8367	.8879	.58566	.9312	1.4286	2.1991	.38485
45/64	.703125	.49438	.34761	.8385	.8892	.58959	.9320	1.4222	2.2089	.38829
23/32	.71875	.51660	.37131	.8478	.8958	.60935	.9361	1.3913	2.2580	.40574
47/64	.734375	.53931	.39605	.8570	.9022	.62933	.9401	1.3617	2.3071	.42357
3/4	**.750**	**.56250**	**.42188**	**.8660**	**.9086**	**.64952**	**.9441**	**1.3333**	**2.3562**	**.44179**
49/64	.765625	.58618	.44879	.8750	.9148	.66992	.9480	1.3061	2.4053	.46038
25/32	.78125	.61035	.47684	.8839	.9210	.69053	.9518	1.2800	2.4544	.47937
51/64	.796875	.63501	.50602	.8927	.9271	.71135	.9556	1.2549	2.5035	.49874
	.80	.6400	.51200	.8944	.9283	.71554	.9564	1.2500	2.5133	.50265
13/16	**.8125**	**.66016**	**.53638**	**.9014**	**.9331**	**.73238**	**.9593**	**1.2308**	**2.5525**	**.51849**
53/64	.828125	.68579	.56792	.9100	.9391	.75361	.9630	1.2075	2.6016	.53862
27/32	.84375	.71191	.60067	.9186	.9449	.77503	.9666	1.1852	2.6507	.55914
55/64	.859375	.73853	.63467	.9270	.9507	.79666	.9702	1.1636	2.6998	.58004
7/8	**.875**	**.76563**	**.66992**	**.9354**	**.9565**	**.81849**	**.9737**	**1.1429**	**2.7489**	**.60132**
57/64	.890625	.79321	.70645	.9437	.9621	.84051	.9771	1.1228	2.7980	.62299
	.90	.81000	.72900	.9487	.9655	.85435	.9792	1.1111	2.8274	.63617
29/32	.90625	.82129	.74429	.9520	.9677	.86272	.9805	1.1034	2.8471	.64504
59/64	.921875	.84985	.78346	.9601	.9733	.88513	.9839	1.0847	2.8962	.66747
15/16	**.9375**	**.87891**	**.82398**	**.9683**	**.9787**	**.90773**	**.9872**	**1.0667**	**2.9452**	**.69029**
61/64	.953125	.90845	.86587	.9763	.9841	.93053	.9905	1.0492	2.9943	.71349
31/32	.96875	.93848	.90915	.9843	.9895	.95349	.9937	1.0323	3.0434	.73708
63/64	.984375	.96899	.95385	.9922	.9948	.97666	.9969	1.0159	3.0925	.76104

Table 7. Properties of Numbers—*Continued*

N	N^2	N^3	$\sqrt{N}$	$\sqrt[3]{N}$	$N^{3/2}$	$\sqrt[5]{N}$	$\frac{1}{N}$	Circle (N = Diam.) Circum.	Area
1.	**1.0000**	**1.0000**	**1.0000**	**1.0000**	**1.0000**	**1.0000**	**1.0000000**	**3.1416**	**0.7854**
1.125	1.2656	1.4238	1.0606	1.0400	1.1932	1.0238	.8888888	3.5343	.9940
1.25	1.5625	1.9531	1.1180	1.0772	1.3975	1.0456	.80000000	3.9270	1.2272
1.375	1.8906	2.5996	1.1726	1.1120	1.6123	1.0658	.72727272	4.3197	1.4849
1.5	2.25	3.3750	1.2247	1.1447	1.8371	1.0845	.66666666	4.7124	1.7671
1.625	2.6406	4.2910	1.2748	1.1757	2.0715	1.1020	.61538462	5.1051	2.0739
1.75	3.0625	5.3594	1.3229	1.2051	2.3150	1.1186	.57142857	5.4978	2.4053
1.875	3.5156	6.5918	1.3693	1.2331	2.5675	1.1340	.53333333	5.8905	2.7612
2.	**4.0000**	**8.0000**	**1.4142**	**1.2599**	**2.8284**	**1.1487**	**.50000000**	**6.2832**	**3.1416**
2.125	4.5156	9.5957	1.4577	1.2856	3.0977	1.1627	.47058823	6.6759	3.5466
2.25	5.0625	11.3906	1.5000	1.3104	3.3750	1.1761	.44444444	7.0686	3.9761
2.375	5.6406	13.3965	1.5411	1.3342	3.6601	1.1889	.42105263	7.4613	4.4301
2.5	6.2500	15.6250	1.5811	1.3572	3.9529	1.2011	.40000000	7.8540	4.9087
2.625	6.8906	18.0879	1.6202	1.3795	4.2530	1.2129	.38095231	8.2467	5.4119
2.75	7.5625	20.7969	1.6583	1.4011	4.5604	1.2242	.36363636	8.6394	5.9396
2.875	8.2656	23.7637	1.6956	1.4219	4.8748	1.2352	.34782609	9.0321	6.4918
3.	**9.0000**	**27.0000**	**1.7321**	**1.4422**	**5.1962**	**1.2457**	**.33333333**	**9.4248**	**7.0686**
3.125	9.7656	30.5176	1.7678	1.4620	5.5243	1.2559	.32000000	9.8175	7.6699
3.25	10.5625	34.3281	1.8028	1.4813	5.8590	1.2658	.30769231	10.2102	8.2958
3.375	11.3906	38.4434	1.8371	1.5000	6.2003	1.2754	.29629629	10.6029	8.9462
3.5	12.2500	42.8750	1.8708	1.5183	6.5479	1.2847	.28571429	10.9956	9.6211
3.625	13.1406	47.6348	1.9039	1.5362	6.9018	1.2938	.27586207	11.3883	10.3206
3.75	14.0625	52.7344	1.9365	1.5536	7.2619	1.3026	.26666666	11.7810	11.0447
3.875	15.0156	58.1856	1.9685	1.5707	7.6279	1.3112	.25806452	12.1737	11.7932
4.	**16.0000**	**64.0000**	**2.0000**	**1.5874**	**8.0000**	**1.3195**	**.25000000**	**12.5664**	**12.5664**
4.125	17.0156	70.1895	2.0310	1.6038	8.3779	1.3277	.24242424	12.9591	13.3640
4.25	18.0625	76.7656	2.0616	1.6198	8.7616	1.3356	.23529412	13.3518	14.1863
4.375	19.1406	83.7402	2.0916	1.6355	9.1510	1.3434	.22857143	13.7445	15.0330
4.5	20.2500	91.1250	2.1213	1.6510	9.5460	1.3510	.22222222	14.1372	15.9043
4.625	21.3906	98.9317	2.1506	1.6661	9.9465	1.3584	.21621622	14.5299	16.8001
4.75	22.5625	107.1719	2.1795	1.6810	10.3524	1.3656	.21052632	14.9226	17.7205
4.875	23.7656	115.8574	2.2079	1.6956	10.7637	1.3728	.20512821	15.3153	18.6655
5.	**25.0000**	**125.0000**	**2.2361**	**1.7100**	**11.1803**	**1.3799**	**.20000000**	**15.7080**	**19.6350**
5.125	26.2656	134.6113	2.2638	1.7241	11.6022	1.3866	.19512195	16.1006	20.6289
5.25	27.5625	144.7031	2.2913	1.7380	12.0293	1.3933	.19047619	16.4933	21.6475
5.375	28.8906	155.2871	2.3184	1.7517	12.4614	1.3998	.18604651	16.8860	22.6906
5.5	30.2500	166.3750	2.3452	1.7652	12.8987	1.4063	.18181818	17.2787	23.7583
5.625	31.6406	177.9785	2.3727	1.7784	13.3409	1.4126	.17777777	17.6714	24.8505
5.75	33.0625	190.1094	2.3979	1.7915	13.7880	1.4188	.17391304	18.0641	25.9672
5.875	34.5156	202.7793	2.4238	1.8044	14.2400	1.4250	.17021277	18.4568	27.1085
6.	**36.0000**	**216.0000**	**2.4495**	**1.8171**	**14.6969**	**1.4310**	**.16666666**	**18.8495**	**28.2743**
6.125	37.5156	229.7832	2.4749	1.8297	15.1586	1.4369	.16326531	19.2422	29.4647
6.25	39.0625	244.1406	2.5000	1.8420	15.6250	1.4427	.16000000	19.6349	30.6796
6.375	40.6406	259.0840	2.5249	1.8542	16.0961	1.4484	.15686275	20.0276	31.9190
6.5	42.2500	274.6250	2.5495	1.8663	16.5718	1.4542	.15384615	20.4203	33.1831
6.625	43.8906	290.7754	2.5739	1.8781	17.0522	1.4596	.15094339	20.8130	34.4716
6.75	45.5625	307.5469	2.5981	1.8899	17.5370	1.4651	.14814815	21.2057	35.7847
6.875	47.2656	324.9512	2.6220	1.9015	18.0264	1.4705	.14545454	21.5984	37.1223
7.	**49.0000**	**343.0000**	**2.6458**	**1.9129**	**18.5203**	**1.4758**	**.14285714**	**21.9911**	**38.4845**
7.125	50.7656	361.7051	2.6693	1.9243	19.0186	1.4810	.14035088	22.3838	39.8712
7.25	52.5625	381.0781	2.6926	1.9354	19.5212	1.4862	.13793103	22.7765	41.2825
7.375	54.3906	401.1309	2.7157	1.9465	20.0283	1.4913	.13559322	23.1692	42.7183
7.5	56.2500	421.8750	2.7386	1.9574	20.5396	1.4963	.13333333	23.5619	44.1786
7.625	58.1406	443.3223	2.7613	1.9683	21.0552	1.5012	.13114754	23.9546	45.6635
7.75	60.0625	465.4844	2.7839	1.9789	21.5751	1.5061	.12903226	24.3473	47.1730
7.875	62.0156	488.3731	2.8063	1.9895	22.0992	1.5110	.12698413	24.7400	48.7069
8.	**64.0000**	**512.0000**	**2.8284**	**2.0000**	**22.6274**	**1.5157**	**.12500000**	**25.1327**	**50.2655**
8.125	66.0156	536.3770	2.8504	2.0104	23.1598	1.5204	.12307692	25.5254	51.8485
8.25	68.0625	561.5156	2.8723	2.0206	23.6963	1.5251	.12121212	25.9181	53.4562
8.375	70.1406	587.4278	2.8940	2.0308	24.2369	1.5297	.11940298	26.3108	55.0883
8.5	72.2500	614.1250	2.9155	2.0408	24.7816	1.5342	.11764706	26.7035	56.7450
8.625	74.3906	641.6192	2.9368	2.0508	25.3301	1.5387	.11594203	27.0962	58.4262
8.75	76.5625	669.9219	2.9580	2.0606	25.8828	1.5431	.11428571	27.4889	60.1320
8.875	78.7656	699.0450	2.9791	2.0704	26.4394	1.5475	.11267605	27.8816	61.8623
9.	**81.0000**	**729.0000**	**3.0000**	**2.0801**	**27.0000**	**1.5518**	**.11111111**	**28.2743**	**63.6172**
9.125	83.2656	759.7989	3.0207	2.0897	27.5645	1.5561	.10958904	28.6670	65.3966
9.25	85.5625	791.4531	3.0414	2.0992	28.1328	1.5604	.10810811	29.0597	67.2006
9.375	87.8906	823.9746	3.0619	2.1086	28.7050	1.5646	.10666666	29.4524	69.0291
9.5	90.2500	857.3750	3.0822	2.1179	29.2810	1.5687	.10526316	29.8451	70.8822
9.625	92.6406	891.6660	3.1024	2.1272	29.8608	1.5728	.10389610	30.2378	72.7597
9.75	95.0625	926.8594	3.1225	2.1363	30.4444	1.5769	.10256410	30.6305	74.6619
9.875	97.5156	962.9668	3.1425	2.1454	31.0317	1.5809	.10126582	31.0232	76.5886

Table 7. Properties of Numbers—*Continued*

N	N^2	N^3	$\sqrt{N}$	$\sqrt[3]{N}$	$N^{3/2}$	$\sqrt[5]{N}$	$\frac{1}{N}$	Circle (N = Diam.)	
								Circum.	Area
10	**100**	**1000**	**3.1623**	**2.1544**	**31.623**	**1.5849**	**.10000000**	**31.4159**	**78.5398**
11	121	1331	3.3166	2.2240	36.483	1.6154	.09090909	34.5575	95.0332
12	144	1728	3.4641	2.2894	41.569	1.6438	.08333333	37.6991	113.0973
13	169	2197	3.6056	2.3513	46.873	1.6703	.07692308	40.8407	132.7323
14	196	2744	3.7417	2.4101	52.384	1.6953	.07142857	43.9823	153.9380
15	225	3375	3.8730	2.4662	58.095	1.7188	.06666667	47.1239	176.7146
16	256	4096	4.0000	2.5198	64.000	1.7411	.06250000	50.2654	201.0619
17	289	4913	4.1231	2.5713	70.093	1.7623	.05882353	53.4070	226.9801
18	324	5832	4.2426	2.6207	76.367	1.7826	.05555556	56.5486	254.4690
19	361	6859	4.3589	2.6684	82.819	1.8020	.05263158	59.6902	283.5287
20	**400**	**8000**	**4.4721**	**2.7144**	**89.442**	**1.8206**	**.05000000**	**62.8318**	**314.1593**
21	441	9261	4.5826	2.7589	96.235	1.8384	.04761905	65.9734	346.3606
22	484	10648	4.6904	2.8020	103.19	1.8556	.04545455	69.1150	380.1327
23	529	12167	4.7958	2.8439	110.30	1.8722	.04347826	72.2566	415.4756
24	576	13824	4.8990	2.8845	117.58	1.8882	.04166667	75.3982	452.3893
25	625	15625	5.0000	2.9240	125.00	1.9037	.04000000	78.5398	490.8739
26	676	17576	5.0990	2.9625	132.57	1.9186	.03846154	81.6813	530.9292
27	729	19683	5.1962	3.0000	140.30	1.9332	.03703704	84.8229	572.5553
28	784	21952	5.2915	3.0366	148.16	1.9473	.03571429	87.9645	615.7522
29	841	24389	5.3852	3.0723	156.17	1.9610	.03448276	91.1061	660.5198
30	**900**	**27000**	**5.4772**	**3.1072**	**164.32**	**1.9744**	**.03333333**	**94.2477**	**706.8583**
31	961	29791	5.5678	3.1414	172.60	1.9873	.03225806	97.3893	754.7676
32	1024	32768	5.6569	3.1748	181.02	2.0000	.03125000	100.5309	804.2477
33	1089	35937	5.7446	3.2075	189.57	2.0123	.03030303	103.6725	855.2986
34	1156	39304	5.8310	3.2396	198.25	2.0244	.02941176	106.8141	907.9203
35	1225	42875	5.9161	3.2711	207.06	2.0362	.02857143	109.9557	962.1127
36	1296	46656	6.0000	3.3019	216.00	2.0477	.02777778	113.0972	1017.8760
37	1369	50653	6.0828	3.3322	225.06	2.0589	.02702703	116.2388	1075.2101
38	1444	54872	6.1644	3.3620	234.25	2.0699	.02631579	119.3804	1134.1149
39	1521	59319	6.2450	3.3912	243.56	2.0807	.02564103	122.5220	1194.5906
40	**1600**	**64000**	**6.3246**	**3.4200**	**252.98**	**2.0913**	**.02500000**	**125.6636**	**1256.6371**
41	1681	68921	6.4031	3.4482	262.53	2.1016	.02439024	128.8052	1320.2543
42	1764	74088	6.4807	3.4760	272.19	2.1118	.02380952	131.9468	1385.4424
43	1849	79507	6.5574	3.5034	281.97	2.1218	.02325581	135.0884	1452.2012
44	1936	85184	6.6332	3.5303	291.86	2.1315	.02272727	138.2300	1520.5308
45	2025	91125	6.7082	3.5569	301.87	2.1411	.02222222	141.3716	1590.4313
46	2116	97336	6.7823	3.5830	311.99	2.1506	.02173913	144.5131	1661.9025
47	2209	103823	6.8557	3.6088	322.22	2.1598	.02127660	147.6547	1734.9445
48	2304	110592	6.9282	3.6342	332.55	2.1689	.02083333	150.7963	1809.5574
49	2401	117649	7.0000	3.6593	343.00	2.1779	.02040816	153.9379	1885.7410
50	**2500**	**125000**	**7.0711**	**3.6840**	**353.55**	**2.1867**	**.02000000**	**157.0795**	**1963.500**
51	2601	132651	7.1414	3.7084	364.21	2.1954	.01960784	160.2211	2042.820
52	2704	140608	7.2111	3.7325	374.98	2.2039	.01923077	163.3627	2123.716
53	2809	148877	7.2801	3.7563	385.85	2.2124	.01886792	166.5043	2206.183
54	2916	157464	7.3485	3.7798	396.82	2.2206	.01851852	169.6459	2290.221
55	3025	166375	7.4162	3.8030	407.89	2.2288	.01818182	172.7875	2375.829
56	3136	175616	7.4833	3.8259	419.07	2.2369	.01785714	175.9290	2463.008
57	3249	185193	7.5498	3.8485	430.35	2.2448	.01754386	179.0706	2551.758
58	3364	195112	7.6158	3.8709	441.72	2.2526	.01724138	182.2122	2642.079
59	3481	205379	7.6811	3.8930	453.19	2.2603	.01694915	185.3538	2733.970
60	**3600**	**216000**	**7.7460**	**3.9149**	**464.76**	**2.2679**	**.01666667**	**188.4954**	**2827.433**
61	3721	226981	7.8102	3.9365	476.43	2.2755	.01639344	191.6370	2922.466
62	3844	238328	7.8740	3.9579	488.19	2.2829	.01612903	194.7786	3019.070
63	3969	250047	7.9373	3.9791	500.05	2.2902	.01587302	197.9202	3117.245
64	4096	262144	8.0000	4.0000	512.00	2.2974	.01562500	201.0618	3216.990
65	4225	274625	8.0623	4.0207	524.05	2.3045	.01538462	204.2034	3318.307
66	4356	287496	8.1240	4.0412	536.19	2.3116	.01515152	207.3449	3421.194
67	4489	300763	8.1854	4.0615	548.42	2.3186	.01492537	210.4865	3525.652
68	4624	314432	8.2462	4.0817	560.74	2.3254	.01470588	213.6281	3631.680
69	4761	328509	8.3066	4.1016	573.16	2.3322	.01449275	216.7697	3739.280
70	**4900**	**343000**	**8.3666**	**4.1213**	**585.66**	**2.3389**	**.01428571**	**219.9113**	**3848.450**
71	5041	357911	8.4261	4.1408	598.26	2.3456	.01408451	223.0529	3959.191
72	5184	373248	8.4853	4.1602	610.94	2.3522	.01388889	226.1945	4071.503
73	5329	389017	8.5440	4.1793	623.71	2.3587	.01369863	229.3361	4185.386
74	5476	405224	8.6023	4.1983	636.57	2.3651	.01351351	232.4777	4300.839
75	5625	421875	8.6603	4.2172	649.52	2.3714	.01333333	235.6193	4417.864
76	5776	438976	8.7178	4.2358	662.55	2.3777	.01315789	238.7608	4536.459
77	5929	456533	8.7750	4.2543	675.68	2.3840	.01298701	241.9024	4656.625
78	6084	474552	8.8318	4.2727	688.88	2.3901	.01282051	245.0440	4778.361
79	6241	493039	8.8882	4.2908	702.17	2.3962	.01265823	248.1856	4901.669

Table 7. Properties of Numbers—*Continued*

N	N^2	N^3	$\sqrt{N}$	$\sqrt[3]{N}$	$N^{3/2}$	$\sqrt[5]{N}$	$\frac{1}{N}$	Circle (N = Diam.) Circum.	Area
80	**6400**	**512000**	**8.9443**	**4.3089**	**715.54**	**2.4022**	**.01250000**	**251.327**	**5026.547**
81	6561	531441	9.0000	4.3267	729.00	2.4082	.01234568	254.469	5152.998
82	6724	551368	9.0554	4.3445	742.54	2.4141	.01219512	257.610	5281.016
83	6889	571787	9.1104	4.3621	756.17	2.4200	.01204819	260.752	5410.607
84	7056	592704	9.1652	4.3795	769.88	2.4258	.01190476	263.894	5541.770
85	7225	614125	9.2195	4.3968	783.66	2.4315	.01176471	267.035	5674.501
86	7396	636056	9.2736	4.4140	797.53	2.4372	.01162791	270.177	5808.805
87	7569	658503	9.3274	4.4310	811.49	2.4429	.01149425	273.318	5944.679
88	7744	681472	9.3808	4.4480	825.52	2.4485	.01136364	276.460	6082.124
89	7921	704969	9.4340	4.4647	839.63	2.4540	.01123596	279.602	6221.138
90	**8100**	**729000**	**9.4868**	**4.4814**	**853.82**	**2.4595**	**.01111111**	**282.743**	**6361.725**
91	8281	753571	9.5394	4.4979	868.09	2.4650	.01098901	285.885	6503.882
92	8464	778688	9.5917	4.5144	882.44	2.4705	.01086957	289.026	6647.610
93	8649	804357	9.6437	4.5307	896.86	2.4758	.01075269	292.168	6792.909
94	8836	830584	9.6954	4.5468	911.36	2.4810	.01063830	295.309	6939.778
95	9025	857375	9.7468	4.5629	925.95	2.4863	.01052632	298.451	7088.219
96	9216	884736	9.7980	4.5789	940.61	2.4915	.01041667	301.593	7238.230
97	9409	912673	9.8489	4.5947	955.34	2.4966	.01030928	304.734	7389.812
98	9604	941192	9.8995	4.6104	970.15	2.5018	.01020408	307.876	7542.962
99	9801	970299	9.9499	4.6261	985.04	2.5069	.01010101	311.017	7697.688
100	**10000**	**1000000**	**10.0000**	**4.6416**	**1000.0**	**2.5119**	**.01000000**	**314.159**	**7853.982**
101	10201	1030301	10.0499	4.6570	1015.0	2.5169	.00990099	317.301	8011.85
102	10404	1061208	10.0995	4.6723	1030.1	2.5219	.00980392	320.442	8171.28
103	10609	1092727	10.1489	4.6875	1045.3	2.5268	.00970874	323.584	8332.29
104	10816	1124864	10.1980	4.7027	1060.6	2.5317	.00961538	326.725	8494.87
105	11025	1157625	10.2470	4.7177	1075.9	2.5365	.00952381	329.867	8659.01
106	11236	1191016	10.2956	4.7326	1091.3	2.5413	.00943396	333.009	8824.73
107	11449	1225043	10.3441	4.7475	1106.8	2.5461	.00934579	336.150	8992.02
108	11664	1259712	10.3923	4.7622	1122.4	2.5509	.00925926	339.292	9160.88
109	11881	1295029	10.4403	4.7769	1138.0	2.5556	.00917431	342.433	9331.32
110	**12100**	**1331000**	**10.4881**	**4.7914**	**1153.7**	**2.5602**	**.00909091**	**345.575**	**9503.32**
111	12321	1367631	10.5357	4.8059	1169.5	2.5649	.00900901	348.716	9676.89
112	12544	1404928	10.5830	4.8203	1185.3	2.5695	.00892857	351.858	9852.03
113	12769	1442897	10.6301	4.8346	1201.2	2.5740	.00884956	355.000	10028.75
114	12996	1481544	10.6771	4.8488	1217.2	2.5786	.00877193	358.141	10207.03
115	13225	1520875	10.7238	4.8629	1233.2	2.5831	.00869565	361.283	10386.89
116	13456	1560896	10.7703	4.8770	1249.4	2.5876	.00862069	364.424	10568.32
117	13689	1601613	10.8167	4.8910	1265.5	2.5920	.00854701	367.566	10751.31
118	13924	1643032	10.8628	4.9049	1281.8	2.5964	.00847458	370.708	10935.88
119	14161	1685159	10.9087	4.9187	1298.1	2.6008	.00840336	373.849	11122.02
120	**14400**	**1728000**	**10.9545**	**4.9324**	**1314.5**	**2.6052**	**.00833333**	**376.991**	**11309.73**
121	14641	1771561	11.0000	4.9461	1331.0	2.6095	.00826446	380.132	11499.01
122	14884	1815848	11.0454	4.9597	1347.5	2.6138	.00819672	383.274	11689.86
123	15129	1860867	11.0905	4.9732	1364.1	2.6181	.00813008	386.416	11882.29
124	15376	1906624	11.1355	4.9866	1380.8	2.6223	.00806452	389.557	12076.28
125	15625	1953125	11.1803	5.0000	1397.5	2.6265	.00800000	392.699	12271.84
126	15876	2000376	11.2250	5.0133	1414.4	2.6307	.00793651	395.840	12468.98
127	16129	2048383	11.2694	5.0265	1431.2	2.6349	.00787402	398.982	12667.68
128	16384	2097152	11.3137	5.0397	1448.2	2.6390	.00781250	402.124	12867.96
129	16641	2146689	11.3578	5.0528	1465.2	2.6431	.00775194	405.265	13069.81
130	**16900**	**2197000**	**11.4018**	**5.0658**	**1482.2**	**2.6472**	**.00769231**	**408.407**	**13273.23**
131	17161	2248091	11.4455	5.0788	1499.4	2.6513	.00763359	411.548	13478.22
132	17424	2299968	11.4891	5.0916	1516.6	2.6553	.00757576	414.690	13684.77
133	17689	2352637	11.5326	5.1045	1533.8	2.6593	.00751880	417.831	13892.91
134	17956	2406104	11.5758	5.1172	1551.2	2.6633	.00746269	420.973	14102.61
135	18225	2460375	11.6190	5.1299	1568.6	2.6673	.00740741	424.115	14313.88
136	18496	2515456	11.6619	5.1426	1586.0	2.6712	.00735294	427.256	14526.72
137	18769	2571353	11.7047	5.1551	1603.6	2.6751	.00729927	430.398	14741.14
138	19044	2628072	11.7473	5.1676	1621.1	2.6790	.00724638	433.539	14957.12
139	19321	2685619	11.7898	5.1801	1638.8	2.6829	.00719424	436.681	15174.67
140	**19600**	**2744000**	**11.8322**	**5.1925**	**1656.5**	**2.6867**	**.00714286**	**439.823**	**15393.80**
141	19881	2803221	11.8743	5.2048	1674.3	2.6906	.00709220	442.964	15614.50
142	20164	2863288	11.9164	5.2171	1692.1	2.6944	.00704225	446.106	15836.77
143	20449	2924207	11.9583	5.2293	1710.0	2.6981	.00699301	449.247	16060.60
144	20736	2985984	12.0000	5.2415	1728.0	2.7019	.00694444	452.389	16286.01
145	21025	3048625	12.0416	5.2536	1746.0	2.7057	.00689655	455.531	16512.99
146	21316	3112136	12.0830	5.2656	1764.1	2.7094	.00684932	458.672	16741.54
147	21609	3176523	12.1244	5.2776	1782.2	2.7131	.00680272	461.814	16971.67
148	21904	3241792	12.1655	5.2896	1800.5	2.7168	.00675676	464.955	17203.36
149	22201	3307949	12.2066	5.3015	1818.8	2.7204	.00671141	468.097	17436.62

Table 7. Properties of Numbers—*Continued*

N	N^2	N^3	$\sqrt{N}$	$\sqrt[3]{N}$	$N^{3/2}$	$\sqrt[5]{N}$	$\frac{1}{N}$	Circle (N = Diam.) Circum.	Area
150	**22500**	**3375000**	**12.2474**	**5.3133**	**1837.1**	**2.7241**	**.00666667**	**471.239**	**17671.46**
151	22801	3442951	12.2882	5.3251	1855.5	2.7277	.00662252	474.380	17907.86
152	23104	3511808	12.3288	5.3368	1874.0	2.7314	.00657895	477.522	18145.84
153	23409	3581577	12.3693	5.3485	1892.5	2.7349	.00653595	480.663	18385.38
154	23716	3652264	12.4097	5.3601	1911.1	2.7385	.00649351	483.805	18626.50
155	24025	3723875	12.4499	5.3717	1929.7	2.7420	.00645161	486.946	18869.19
156	24336	3796416	12.4900	5.3832	1948.4	2.7455	.00641026	490.088	19113.45
157	24649	3869893	12.5300	5.3947	1967.2	2.7490	.00636943	493.230	19359.28
158	24964	3944312	12.5698	5.4061	1986.0	2.7525	.00632911	496.371	19606.68
159	25281	4019679	12.6095	5.4175	2004.9	2.7560	.00628931	499.513	19855.65
160	**25600**	**4096000**	**12.6491**	**5.4288**	**2023.9**	**2.7595**	**.00625000**	**502.654**	**20106.19**
161	25921	4173281	12.6886	5.4401	2042.9	2.7629	.00621118	505.796	20358.30
162	26244	4251528	12.7279	5.4514	2061.9	2.7663	.00617284	508.938	20611.99
163	26569	4330747	12.7671	5.4626	2081.0	2.7697	.00613497	512.079	20867.24
164	26896	4410944	12.8062	5.4737	2100.2	2.7731	.00609756	515.221	21124.06
165	27225	4492125	12.8452	5.4848	2119.5	2.7765	.00606061	518.362	21382.46
166	27556	4574296	12.8841	5.4959	2138.8	2.7799	.00602410	521.504	21642.43
167	27889	4657463	12.9228	5.5069	2158.1	2.7832	.00598802	524.646	21903.96
168	28224	4741632	12.9615	5.5178	2177.5	2.7865	.00595238	527.787	22167.07
169	28561	4826809	13.0000	5.5288	2197.0	2.7898	.00591716	530.929	22431.75
170	**28900**	**4913000**	**13.0384**	**5.5397**	**2216.5**	**2.7931**	**.00588235**	**534.070**	**22698.00**
171	29241	5000211	13.0767	5.5505	2236.1	2.7964	.00584795	537.212	22965.82
172	29584	5088448	13.1149	5.5613	2255.8	2.7997	.00581395	540.353	23235.21
173	29929	5177717	13.1529	5.5721	2275.5	2.8029	.00578035	543.495	23506.18
174	30276	5268024	13.1909	5.5828	2295.2	2.8061	.00574713	546.637	23778.71
175	30625	5359375	13.2288	5.5934	2315.0	2.8094	.00571429	549.778	24052.81
176	30976	5451776	13.2665	5.6041	2334.9	2.8126	.00568182	552.920	24328.49
177	31329	5545233	13.3041	5.6147	2354.8	2.8158	.00564972	556.061	24605.73
178	31684	5639752	13.3417	5.6252	2374.8	2.8189	.00561798	559.203	24884.55
179	32041	5735339	13.3791	5.6357	2394.9	2.8221	.00558659	562.345	25164.94
180	**32400**	**5832000**	**13.4164**	**5.6462**	**2415.0**	**2.8252**	**.00555556**	**565.486**	**25446.90**
181	32761	5929741	13.4536	5.6567	2435.1	2.8284	.00552486	568.628	25730.42
182	33124	6028568	13.4907	5.6671	2455.3	2.8315	.00549451	571.769	26015.52
183	33489	6128487	13.5277	5.6774	2475.6	2.8346	.00546448	574.911	26302.19
184	33856	6229504	13.5647	5.6877	2495.9	2.8377	.00543478	578.053	26590.43
185	34225	6331625	13.6015	5.6980	2516.3	2.8408	.00540541	581.194	26880.25
186	34596	6434856	13.6382	5.7083	2536.7	2.8438	.00537634	584.336	27171.63
187	34969	6539203	13.6748	5.7185	2557.2	2.8469	.00534759	587.477	27464.58
188	35344	6644672	13.7113	5.7287	2577.7	2.8499	.00531915	590.619	27759.11
189	35721	6751269	13.7477	5.7388	2598.3	2.8529	.00529101	593.761	28055.20
190	**36100**	**6859000**	**13.7840**	**5.7489**	**2619.0**	**2.8560**	**.00526316**	**596.902**	**28352.87**
191	36481	6967871	13.8203	5.7590	2639.7	2.8590	.00523560	600.044	28652.10
192	36864	7077888	13.8564	5.7690	2660.4	2.8619	.00520833	603.185	28952.91
193	37249	7189057	13.8924	5.7790	2681.2	2.8649	.00518135	606.327	29255.29
194	37636	7301384	13.9284	5.7890	2702.1	2.8679	.00515464	609.468	29559.24
195	38025	7414875	13.9642	5.7989	2723.0	2.8708	.00512821	612.610	29864.76
196	38416	7529536	14.0000	5.8088	2744.0	2.8738	.00510204	615.752	30171.85
197	38809	7645373	14.0357	5.8186	2765.0	2.8767	.00507614	618.893	30480.51
198	39204	7762392	14.0712	5.8285	2786.1	2.8796	.00505051	622.035	30790.74
199	39601	7880599	14.1067	5.8383	2807.2	2.8825	.00502513	625.176	31102.55
200	**40000**	**8000000**	**14.1421**	**5.8480**	**2828.4**	**2.8854**	**.00500000**	**628.318**	**31415.93**
201	40401	8120601	14.1774	5.8578	2849.7	2.8883	.00497512	631.460	31730.87
202	40804	8242408	14.2127	5.8675	2871.0	2.8911	.00495050	634.601	32047.39
203	41209	8365427	14.2478	5.8771	2892.3	2.8940	.00492611	637.743	32365.47
204	41616	8489664	14.2829	5.8868	2913.7	2.8968	.00490196	640.884	32685.13
205	42025	8615125	14.3178	5.8964	2935.2	2.8997	.00487805	644.026	33006.36
206	42436	8741816	14.3527	5.9059	2956.7	2.9025	.00485437	647.168	33329.16
207	42849	8869743	14.3875	5.9155	2978.2	2.9053	.00483092	650.309	33653.53
208	43264	8998912	14.4222	5.9250	2999.8	2.9081	.00480769	653.451	33979.47
209	43681	9129329	14.4568	5.9345	3021.5	2.9109	.00478469	656.592	34306.98
210	**44100**	**9261000**	**14.4914**	**5.9439**	**3043.2**	**2.9137**	**.00476190**	**659.734**	**34636.06**
211	44521	9393931	14.5258	5.9533	3065.0	2.9165	.00473934	662.875	34966.71
212	44944	9528128	14.5602	5.9627	3086.8	2.9192	.00471698	666.017	35298.94
213	45369	9663597	14.5945	5.9721	3108.7	2.9220	.00469484	669.159	35632.73
214	45796	9800344	14.6287	5.9814	3130.6	2.9247	.00467290	672.300	35968.09
215	46225	9938375	14.6629	5.9907	3152.5	2.9274	.00465116	675.442	36305.03
216	46656	10077696	14.6969	6.0000	3174.5	2.9302	.00462963	678.583	36643.54
217	47089	10218313	14.7309	6.0092	3196.6	2.9329	.00460829	681.725	36983.61
218	47524	10360232	14.7648	6.0185	3218.7	2.9356	.00458716	684.867	37325.26
219	47961	10503459	14.7986	6.0277	3240.9	2.9383	.00456621	688.008	37668.48

Table 7. Properties of Numbers—*Continued*

N	N^2	N^3	$\sqrt{N}$	$\sqrt[3]{N}$	$N^{3/2}$	$\sqrt[5]{N}$	$\frac{1}{N}$	Circle (N = Diam.)	
								Circum.	Area
220	**48400**	**10648000**	**14.8324**	**6.0368**	**3263.1**	**2.9409**	**.00454545**	**691.150**	**38013.27**
221	48841	10793861	14.8661	6.0459	3285.4	2.9436	.00452489	694.291	38359.63
222	49284	10941048	14.8997	6.0550	3307.7	2.9463	.00450450	697.433	38707.56
223	49729	11089567	14.9332	6.0641	3330.1	2.9489	.00448430	700.575	39057.07
224	50176	11239424	14.9666	6.0732	3352.5	2.9516	.00446429	703.716	39408.14
225	50625	11390625	15.0000	6.0822	3375.0	2.9542	.00444444	706.858	39760.78
226	51076	11543176	15.0333	6.0912	3397.5	2.9568	.00442478	709.999	40115.00
227	51529	11697083	15.0665	6.1002	3420.1	2.9594	.00440529	713.141	40470.78
228	51984	11852352	15.0997	6.1091	3442.7	2.9620	.00438596	716.283	40828.14
229	52441	12008989	15.1327	6.1180	3465.4	2.9646	.00436681	719.424	41187.07
230	**52900**	**12167000**	**15.1658**	**6.1269**	**3488.1**	**2.9672**	**.00434783**	**722.566**	**41547.56**
231	53361	12326391	15.1987	6.1358	3510.9	2.9698	.00432900	725.707	41909.63
232	53824	12487168	15.2315	6.1446	3533.7	2.9723	.00431034	728.849	42273.27
233	54289	12649337	15.2643	6.1534	3556.6	2.9749	.00429185	731.990	42638.48
234	54756	12812904	15.2971	6.1622	3579.5	2.9774	.00427350	735.132	43005.26
235	55225	12977875	15.3297	6.1710	3602.5	2.9800	.00425532	738.274	43373.61
236	55696	13144256	15.3623	6.1797	3625.5	2.9825	.00423729	741.415	43743.54
237	56169	13312053	15.3948	6.1885	3648.6	2.9850	.00421941	744.557	44115.03
238	56644	13481272	15.4272	6.1972	3671.7	2.9875	.00420168	747.698	44488.09
239	57121	13651919	15.4596	6.2058	3694.8	2.9900	.00418410	750.840	44862.73
240	**57600**	**13824000**	**15.4919**	**6.2145**	**3718.0**	**2.9925**	**.00416667**	**753.982**	**45238.93**
241	58081	13997521	15.5242	6.2231	3741.3	2.9950	.00414938	757.123	45616.71
242	58564	14172488	15.5563	6.2317	3764.6	2.9975	.00413223	760.265	45996.06
243	59049	14348907	15.5885	6.2403	3788.0	3.0000	.00411523	763.406	46376.98
244	59536	14526784	15.6205	6.2488	3811.4	3.0025	.00409836	766.548	46759.47
245	60025	14706125	15.6525	6.2573	3834.9	3.0049	.00408163	769.690	47143.52
246	60516	14886936	15.6844	6.2658	3858.4	3.0074	.00406504	772.831	47529.16
247	61009	15069223	15.7162	6.2743	3881.9	3.0098	.00404858	775.973	47916.36
248	61504	15252992	15.7480	6.2828	3905.5	3.0122	.00403226	779.114	48305.13
249	62001	15438249	15.7797	6.2912	3929.2	3.0147	.00401606	782.256	48695.47
250	**62500**	**15625000**	**15.8114**	**6.2996**	**3952.9**	**3.0171**	**.00400000**	**785.398**	**49087.39**
251	63001	15813251	15.8430	6.3080	3976.6	3.0195	.00398406	788.539	49480.87
252	63504	16003008	15.8745	6.3164	4000.4	3.0219	.00396825	791.681	49875.92
253	64009	16194277	15.9060	6.3247	4024.2	3.0243	.00395257	794.822	50272.55
254	64516	16387064	15.9374	6.3330	4048.1	3.0267	.00393701	797.964	50670.75
255	65025	16581375	15.9687	6.3413	4072.0	3.0291	.00392157	801.105	51070.52
256	65536	16777216	16.0000	6.3496	4096.0	3.0314	.00390625	804.247	51471.85
257	66049	16974593	16.0312	6.3579	4120.0	3.0338	.00389105	807.389	51874.76
258	66564	17173512	16.0624	6.3661	4144.1	3.0362	.00387597	810.530	52279.24
259	67081	17373979	16.0935	6.3743	4168.2	3.0385	.00386100	813.672	52685.29
260	**67600**	**17576000**	**16.1245**	**6.3825**	**4192.4**	**3.0418**	**.00384615**	**816.813**	**53092.92**
261	68121	17779581	16.1555	6.3907	4216.6	3.0432	.00383142	819.955	53502.11
262	68644	17984728	16.1864	6.3988	4240.8	3.0455	.00381679	823.097	53912.87
263	69169	18191447	16.2173	6.4070	4265.1	3.0478	.00380228	826.238	54325.21
264	69696	18399744	16.2481	6.4151	4289.5	3.0501	.00378788	829.380	54739.11
265	70225	18609625	16.2788	6.4232	4313.9	3.0524	.00377358	832.521	55154.59
266	70756	18821096	16.3095	6.4312	4338.3	3.0547	.00375940	835.663	55571.63
267	71289	19034163	16.3401	6.4393	4362.8	3.0570	.00374532	838.805	55990.25
268	71824	19248832	16.3707	6.4473	4387.3	3.0593	.00373134	841.946	56410.44
269	72361	19465109	16.4012	6.4553	4411.9	3.0616	.00371747	845.088	56832.20
270	**72900**	**19683000**	**16.4317**	**6.4633**	**4436.5**	**3.0639**	**.00370370**	**848.229**	**57255.53**
271	73441	19902511	16.4621	6.4713	4461.2	3.0662	.00369004	851.371	57680.43
272	73984	20123648	16.4924	6.4792	4485.9	3.0684	.00367647	854.512	58106.90
273	74529	20346417	16.5227	6.4872	4510.7	3.0707	.00366300	857.654	58534.94
274	75076	20570824	16.5529	6.4951	4535.5	3.0729	.00364964	860.796	58964.55
275	75625	20796875	16.5831	6.5030	4560.4	3.0752	.00363636	863.937	59395.74
276	76176	21024576	16.6132	6.5108	4585.3	3.0774	.00362319	867.079	59828.49
277	76729	21253933	16.6433	6.5187	4610.2	3.0796	.00361011	870.220	60262.82
278	77284	21484952	16.6733	6.5265	4635.2	3.0818	.00359712	873.362	60698.71
279	77841	21717639	16.7033	6.5343	4660.2	3.0840	.00358423	876.504	61136.18
280	**78400**	**21952000**	**16.7332**	**6.5421**	**4685.3**	**3.0863**	**.00357143**	**879.645**	**61575.22**
281	78961	22188041	16.7631	6.5499	4710.4	3.0885	.00355872	882.787	62015.82
282	79524	22425768	16.7929	6.5577	4735.6	3.0907	.00354610	885.928	62458.00
283	80089	22665187	16.8226	6.5654	4760.8	3.0928	.00353357	889.070	62901.75
284	80656	22906304	16.8523	6.5731	4786.0	3.0950	.00352113	892.212	63347.07
285	81225	23149125	16.8819	6.5808	4811.3	3.0972	.00350877	895.353	63793.97
286	81796	23393656	16.9115	6.5885	4836.7	3.0994	.00349650	898.495	64242.43
287	82369	23639903	16.9411	6.5962	4862.1	3.1015	.00348432	901.636	64692.46
288	82944	23887872	16.9706	6.6039	4887.5	3.1037	.00347222	904.778	65144.07
289	83521	24137569	17.0000	6.6115	4913.0	3.1058	.00346021	907.920	65597.24

Table 7. Properties of Numbers—*Continued*

N	N^2	N^3	$\sqrt{N}$	$\sqrt[3]{N}$	$N^{3/2}$	$\sqrt[5]{N}$	$\frac{1}{N}$	Circle (N = Diam.)	
								Circum.	Area
290	**84100**	**24389000**	**17.0294**	**6.6191**	**4938.5**	**3.1080**	**.00344828**	**911.061**	**66051.99**
291	84681	24642171	17.0587	6.6267	4964.1	3.1101	.00343643	914.203	66508.30
292	85264	24897088	17.0880	6.6343	4989.7	3.1123	.00342466	917.344	66966.19
293	85849	25153757	17.1172	6.6419	5015.4	3.1144	.00341297	920.486	67425.65
294	86436	25412184	17.1464	6.6494	5041.1	3.1165	.00340136	923.627	67886.68
295	87025	25672375	17.1756	6.6569	5066.8	3.1186	.00338983	926.769	68349.28
296	87616	25934336	17.2047	6.6644	5092.6	3.1207	.00337838	929.911	68813.45
297	88209	26198073	17.2337	6.6719	5118.4	3.1228	.00336700	933.052	69279.19
298	88804	26463592	17.2627	6.6794	5144.3	3.1249	.00335570	936.194	69746.50
299	89401	26730899	17.2916	6.6869	5170.2	3.1270	.00334448	939.335	70215.38
300	**90000**	**27000000**	**17.3205**	**6.6943**	**5196.2**	**3.1291**	**.00333333**	**942.477**	**70685.83**
301	90601	27270901	17.3494	6.7018	5222.2	3.1312	.00332226	945.619	71157.86
302	91204	27543608	17.3781	6.7092	5248.2	3.1333	.00331126	948.760	71631.45
303	91809	27818127	17.4069	6.7166	5274.3	3.1354	.00330033	951.902	72106.62
304	92416	28094464	17.4356	6.7240	5300.4	3.1374	.00328947	955.043	72583.36
305	93025	28372625	17.4642	6.7313	5326.6	3.1395	.00327869	958.185	73061.66
306	93636	28652616	17.4929	6.7387	5352.8	3.1416	.00326797	961.327	73541.54
307	94249	28934443	17.5214	6.7460	5379.1	3.1436	.00325733	964.468	74022.99
308	94864	29218112	17.5499	6.7533	5405.4	3.1456	.00324675	967.610	74506.01
309	95481	29503629	17.5784	6.7606	5431.7	3.1477	.00323625	970.751	74990.60
310	**96100**	**29791000**	**17.6068**	**6.7679**	**5458.1**	**3.1497**	**.00322581**	**973.893**	**75476.76**
311	96721	30080231	17.6352	6.7752	5484.5	3.1518	.00321543	977.034	75964.50
312	97344	30371328	17.6635	6.7824	5511.0	3.1538	.00320513	980.176	76453.80
313	97969	30664297	17.6918	6.7897	5537.5	3.1558	.00319489	983.318	76944.67
314	98596	30959144	17.7200	6.7969	5564.1	3.1578	.00318471	986.459	77437.12
315	99225	31255875	17.7482	6.8041	5590.7	3.1598	.00317460	989.601	77931.13
316	99856	31554496	17.7764	6.8113	5617.3	3.1618	.00316456	992.742	78426.72
317	100489	31855013	17.8045	6.8185	5644.0	3.1638	.00315457	995.884	78923.88
318	101124	32157432	17.8326	6.8256	5670.7	3.1658	.00314465	999.026	79422.60
319	101761	32461759	17.8606	6.8328	5697.5	3.1678	.00313480	1002.167	79922.90
320	**102400**	**32768000**	**17.8885**	**6.8399**	**5724.3**	**3.1698**	**.00312500**	**1005.309**	**80424.77**
321	103041	33076161	17.9165	6.8470	5751.2	3.1718	.00311526	1008.450	80928.21
322	103684	33386248	17.9444	6.8541	5778.1	3.1737	.00310559	1011.592	81433.22
323	104329	33698267	17.9722	6.8612	5805.0	3.1757	.00309598	1014.734	81939.80
324	104976	34012224	18.0000	6.8683	5832.0	3.1777	.00308642	1017.875	82447.96
325	105625	34328125	18.0278	6.8753	5859.0	3.1796	.00307692	1021.017	82957.68
326	106276	34645976	18.0555	6.8824	5886.1	3.1816	.00306748	1024.158	83468.97
327	106929	34965783	18.0831	6.8894	5913.2	3.1835	.00305810	1027.300	83981.84
328	107584	35287552	18.1108	6.8964	5940.3	3.1855	.00304878	1030.442	84496.28
329	108241	35611289	18.1384	6.9034	5967.5	3.1874	.00303951	1033.583	85012.28
330	**108900**	**35937000**	**18.1659**	**6.9104**	**5994.7**	**3.1894**	**.00303030**	**1036.725**	**85529.86**
331	109561	36264691	18.1934	6.9174	6022.0	3.1913	.00302115	1039.866	86049.01
332	110224	36594368	18.2209	6.9244	6049.3	3.1932	.00301205	1043.008	86569.73
333	110889	36926037	18.2483	6.9313	6076.7	3.1951	.00300300	1046.149	87092.02
334	111556	37259704	18.2757	6.9382	6104.1	3.1970	.00299401	1049.291	87615.88
335	112225	37595375	18.3030	6.9451	6131.5	3.1989	.00298507	1052.433	88141.31
336	112896	37933056	18.3303	6.9521	6159.0	3.2009	.00297619	1055.574	88668.31
337	113569	38272753	18.3576	6.9589	6186.5	3.2028	.00296736	1058.716	89196.88
338	114244	38614472	18.3848	6.9658	6214.1	3.2047	.00295858	1061.857	89727.03
339	114921	38958219	18.4120	6.9727	6241.7	3.2066	.00294985	1064.999	90258.74
340	**115600**	**39304000**	**18.4391**	**6.9795**	**6269.3**	**3.2085**	**.00294118**	**1068.141**	**90792.03**
341	116281	39651821	18.4662	6.9864	6297.0	3.2103	.00293255	1071.282	91326.88
342	116964	40001688	18.4932	6.9932	6324.7	3.2122	.00292398	1074.424	91863.31
343	117649	40353607	18.5203	7.0000	6352.4	3.2141	.00291545	1077.565	92401.31
344	118336	40707584	18.5472	7.0068	6380.2	3.2160	.00290698	1080.707	92940.88
345	119025	41063625	18.5742	7.0136	6408.1	3.2178	.00289855	1083.849	93482.02
346	119716	41421736	18.6011	7.0203	6436.0	3.2197	.00289017	1086.990	94024.73
347	120409	41781923	18.6279	7.0271	6463.9	3.2216	.00288184	1090.132	94569.01
348	121104	42144192	18.6548	7.0338	6491.9	3.2234	.00287356	1093.273	95114.86
349	121801	42508549	18.6815	7.0406	6519.9	3.2253	.00286533	1096.415	95662.28
350	**122500**	**42875000**	**18.7083**	**7.0473**	**6547.9**	**3.2271**	**.00285714**	**1099.557**	**96211.28**
351	123201	43243551	18.7350	7.0540	6576.0	3.2289	.00284900	1102.698	96761.84
352	123904	43614208	18.7617	7.0607	6604.1	3.2308	.00284091	1105.840	97313.97
353	124609	43986977	18.7883	7.0674	6632.3	3.2326	.00283286	1108.981	97867.68
354	125316	44361864	18.8149	7.0740	6660.5	3.2345	.00282486	1112.123	98422.96
355	126025	44738875	18.8414	7.0807	6688.7	3.2363	.00281690	1115.264	98979.80
356	126736	45118016	18.8680	7.0873	6717.0	3.2381	.00280899	1118.406	99538.22
357	127449	45499293	18.8944	7.0940	6745.3	3.2399	.00280112	1121.548	100098.21
358	128164	45882712	18.9209	7.1006	6773.7	3.2417	.00279330	1124.689	100659.77
359	128881	46268279	18.9473	7.1072	6802.1	3.2435	.00278552	1127.831	101222.90

Table 7. Properties of Numbers—*Continued*

N	N^2	N^3	$\sqrt{N}$	$\sqrt[3]{N}$	$N^{3/2}$	$\sqrt[5]{N}$	$\frac{1}{N}$	(Circle N = Diam.) Circum.	Area
360	129600	46656000	18.9737	7.1138	6830.5	3.2453	.00277778	1130.972	101787.60
361	130321	47045881	19.0000	7.1204	6859.0	3.2471	.00277008	1134.114	102353.87
362	131044	47437928	19.0263	7.1269	6887.5	3.2489	.00276243	1137.256	102921.72
363	131769	47832147	19.0526	7.1335	6916.1	3.2507	.00275482	1140.397	103491.13
364	132496	48228544	19.0788	7.1400	6944.7	3.2525	.00274725	1143.539	104062.12
365	133225	48627125	19.1050	7.1466	6973.3	3.2543	.00273973	1146.680	104634.67
366	133956	49027896	19.1311	7.1531	7002.0	3.2561	.00273224	1149.822	105208.80
367	134689	49430863	19.1572	7.1596	7030.7	3.2579	.00272480	1152.964	105784.49
368	135424	49836032	19.1833	7.1661	7059.5	3.2597	.00271739	1156.105	106361.76
369	136161	50243409	19.2094	7.1726	7088.3	3.2614	.00271003	1159.247	106940.60
370	136900	50653000	19.2354	7.1791	7117.1	3.2632	.00270270	1162.388	107521.01
371	137641	51064811	19.2614	7.1855	7146.0	3.2650	.00269542	1165.530	108102.99
372	138384	51478848	19.2873	7.1920	7174.9	3.2668	.00268817	1168.671	108686.54
373	139129	51895117	19.3132	7.1984	7203.9	3.2685	.00268097	1171.813	109271.66
374	139876	52313624	19.3391	7.2048	7232.8	3.2702	.00267380	1174.955	109858.35
375	140625	52734375	19.3649	7.2112	7261.8	3.2719	.00266667	1178.096	110446.62
376	141376	53157376	19.3907	7.2177	7290.9	3.2737	.00265957	1181.238	111036.45
377	142129	53582633	19.4165	7.2240	7320.0	3.2754	.00265252	1184.379	111627.86
378	142884	54010152	19.4422	7.2304	7349.2	3.2772	.00264550	1187.521	112220.83
379	143641	54439939	19.4679	7.2368	7378.4	3.2789	.00263852	1190.663	112815.38
380	144400	54872000	19.4936	7.2432	7407.6	3.2807	.00263158	1193.804	113411.49
381	145161	55306341	19.5192	7.2495	7436.8	3.2824	.00262467	1196.946	114009.18
382	145924	55742968	19.5448	7.2558	7466.1	3.2841	.00261780	1200.087	114608.44
383	146689	56181887	19.5704	7.2622	7495.4	3.2858	.00261097	1203.229	115209.27
384	147456	56623104	19.5959	7.2685	7524.8	3.2875	.00260417	1206.371	115811.67
385	148225	57066625	19.6214	7.2748	7554.2	3.2892	.00259740	1209.512	116415.64
386	148996	57512456	19.6469	7.2811	7583.7	3.2909	.00259067	1212.654	117021.18
387	149769	57960603	19.6723	7.2874	7613.2	3.2926	.00258398	1215.795	117628.30
388	150544	58411072	19.6977	7.2936	7642.7	3.2943	.00257732	1218.937	118236.98
389	151321	58863869	19.7231	7.2999	7672.3	3.2960	.00257069	1222.079	118847.24
390	152100	59319000	19.7484	7.3061	7701.9	3.2977	.00256410	1225.220	119459.06
391	152881	59776471	19.7737	7.3124	7731.5	3.2994	.00255754	1228.362	120072.46
392	153664	60236288	19.7990	7.3186	7761.2	3.3011	.00255102	1231.503	120687.42
393	154449	60698457	19.8242	7.3248	7790.9	3.3028	.00254453	1234.645	121303.96
394	155236	61162984	19.8494	7.3310	7820.7	3.3045	.00253807	1237.786	121922.07
395	156025	61629875	19.8746	7.3372	7850.5	3.3061	.00253165	1240.928	122541.75
396	156816	62099136	19.8997	7.3434	7880.3	3.3078	.00252525	1244.070	123163.00
397	157609	62570773	19.9249	7.3496	7910.2	3.3095	.00251889	1247.211	123785.82
398	158404	63044792	19.9499	7.3558	7940.1	3.3111	.00251256	1250.353	124410.21
399	159201	63521199	19.9750	7.3619	7970.0	3.3128	.00250627	1253.494	125036.17
400	160000	64000000	20.0000	7.3681	8000.0	3.3145	.00250000	1256.636	125663.71
401	160801	64481201	20.0250	7.3742	8030.0	3.3161	.00249377	1259.778	126292.81
402	161604	64964808	20.0499	7.3803	8061.1	3.3178	.00248756	1262.919	126923.48
403	162409	65450827	20.0749	7.3864	8090.2	3.3194	.00248139	1266.061	127555.73
404	163216	65939264	20.0998	7.3925	8120.3	3.3211	.00247525	1269.202	128189.55
405	164025	66430125	20.1246	7.3986	8150.5	3.3227	.00246914	1272.344	128824.93
406	164836	66923416	20.1494	7.4047	8180.7	3.3243	.00246305	1275.486	129461.89
407	165649	67419143	20.1742	7.4108	8210.9	3.3260	.00245700	1278.627	130100.42
408	166464	67917312	20.1990	7.4169	8241.2	3.3276	.00245098	1281.769	130740.52
409	167281	68417929	20.2237	7.4229	8271.5	3.3292	.00244499	1284.910	131382.19
410	168100	68921000	20.2485	7.4290	8301.9	3.3308	.00243902	1288.052	132025.43
411	168921	69426531	20.2731	7.4350	8332.3	3.3325	.00243309	1291.193	132670.24
412	169744	69934528	20.2978	7.4410	8362.7	3.3341	.00242718	1294.335	133316.63
413	170569	70444997	20.3224	7.4470	8393.2	3.3357	.00242131	1297.477	133964.58
414	171396	70957944	20.3470	7.4530	8423.7	3.3373	.00241546	1300.618	134614.10
415	172225	71473375	20.3715	7.4590	8454.2	3.3390	.00240964	1303.760	135265.20
416	173056	71991296	20.3961	7.4650	8484.8	3.3406	.00240385	1306.901	135917.86
417	173889	72511713	20.4206	7.4710	8515.4	3.3422	.00239808	1310.043	136572.10
418	174724	73034632	20.4450	7.4770	8546.0	3.3438	.00239234	1313.185	137227.91
419	175561	73560059	20.4695	7.4829	8576.7	3.3454	.00238663	1316.326	137885.29
420	176400	74088000	20.4939	7.4889	8607.4	3.3470	.00238095	1319.468	138544.24
421	177241	74618461	20.5183	7.4948	8638.2	3.3485	.00237530	1322.609	139204.76
422	178084	75151448	20.5426	7.5007	8669.0	3.3501	.00236967	1325.751	139866.85
423	178929	75686967	20.5670	7.5067	8699.8	3.3517	.00236407	1328.893	140530.51
424	179776	76225024	20.5913	7.5126	8730.7	3.3533	.00235849	1332.034	141195.74
425	180625	76765625	20.6155	7.5185	8761.6	3.3559	.00235294	1335.176	141862.54
426	181476	77308776	20.6398	7.5244	8792.5	3.3564	.00234742	1338.317	142530.92
427	182329	77854483	20.6640	7.5302	8823.5	3.3580	.00234192	1341.459	143200.86
428	183184	78402752	20.6882	7.5361	8854.5	3.3596	.00233645	1344.601	143872.38
429	184041	78953589	20.7123	7.5420	8885.6	3.3612	.00233100	1347.742	144545.46

Table 7. Properties of Numbers—*Continued*

N	N^2	N^3	$\sqrt{N}$	$\sqrt[3]{N}$	$N^{3/2}$	$\sqrt[5]{N}$	$\frac{1}{N}$	Circle (N = Diam.) Circum.	Area
430	**184900**	**79507000**	**20.7364**	**7.5478**	**8916.7**	**3.3627**	**.00232558**	**1350.884**	**145220.12**
431	185761	80062991	20.7605	7.5537	8947.8	3.3643	.00232019	1354.025	145896.35
432	186624	80621568	20.7846	7.5595	8979.0	3.3659	.00231481	1357.167	146574.15
433	187489	81182737	20.8087	7.5654	9010.1	3.3674	.00230947	1360.308	147253.52
434	188356	81746504	20.8327	7.5712	9041.4	3.3690	.00230415	1363.450	147934.46
435	189225	82312875	20.8567	7.5770	9072.7	3.3705	.00229885	1366.592	148616.97
436	190096	82881856	20.8806	7.5828	9104.0	3.3720	.00229358	1369.733	149301.05
437	190969	83453453	20.9045	7.5886	9135.3	3.3736	.00228833	1372.875	149986.70
438	191844	84027672	20.9284	7.5944	9166.7	3.3752	.00228311	1376.016	150673.92
439	192721	84604519	20.9523	7.6001	9198.1	3.3767	.00227790	1379.158	151362.72
440	**193600**	**85184000**	**20.9762**	**7.6059**	**9229.5**	**3.3783**	**.00227273**	**1382.300**	**152053.08**
441	194481	85766121	21.0000	7.6117	9261.0	3.3798	.00226757	1385.441	152745.02
442	195364	86350888	21.0238	7.6174	9292.5	3.3813	.00226244	1388.583	153438.53
443	196249	86938307	21.0476	7.6232	9324.1	3.3828	.00225734	1391.724	154133.60
444	197136	87528384	21.0713	7.6289	9355.7	3.3844	.00225225	1394.866	154830.25
445	198025	88121125	21.0950	7.6346	9387.3	3.3859	.00224719	1398.008	155528.47
446	198916	88716536	21.1187	7.6403	9419.0	3.3874	.00224215	1401.149	156228.26
447	199809	89314623	21.1424	7.6460	9450.7	3.3889	.00223714	1404.291	156929.62
448	200704	89915392	21.1660	7.6517	9482.4	3.3904	.00223214	1407.432	157632.55
449	201601	90518849	21.1896	7.6574	9514.2	3.3919	.00222717	1410.574	158337.05
450	**202500**	**91125000**	**21.2132**	**7.6631**	**9546.0**	**3.3935**	**.00222222**	**1413.716**	**159043.13**
451	203401	91733851	21.2368	7.6688	9577.8	3.3950	.00221729	1416.857	159750.77
452	204304	92345408	21.2603	7.6744	9609.6	3.3965	.00221239	1419.999	160459.99
453	205209	92959677	21.2838	7.6801	9641.5	3.3980	.00220751	1423.140	161170.77
454	206116	93576664	21.3073	7.6857	9673.5	3.3995	.00220264	1426.282	161883.13
455	207025	94196375	21.3307	7.6914	9705.5	3.4010	.00219780	1429.423	162597.05
456	207936	94818816	21.3542	7.6970	9737.5	3.4025	.00219298	1432.565	163312.55
457	208849	95443993	21.3776	7.7026	9769.5	3.4039	.00218818	1435.707	164029.62
458	209764	96071912	21.4009	7.7082	9801.6	3.4054	.00218341	1438.848	164748.26
459	210681	96702579	21.4243	7.7138	9833.8	3.4069	.00217865	1441.990	165468.47
460	**211600**	**97336000**	**21.4476**	**7.7194**	**9865.9**	**3.4084**	**.00217391**	**1445.131**	**166190.25**
461	212521	97972181	21.4709	7.7250	9898.1	3.4199	.00216920	1448.273	166913.60
462	213444	98611128	21.4942	7.7306	9930.3	3.4113	.00216450	1451.415	167638.52
463	214369	99252847	21.5174	7.7362	9962.6	3.4128	.00215983	1454.556	168365.02
464	215296	99897344	21.5407	7.7418	9994.8	3.4143	.00215517	1457.698	169093.08
465	216225	100544625	21.5639	7.7473	10027.	3.4158	.00215054	1460.839	169822.72
466	217156	101194696	21.5870	7.7529	10060.	3.4173	.00214592	1463.981	170553.92
467	218089	101847563	21.6102	7.7584	10092.	3.4187	.00214133	1467.123	171286.70
468	219024	102503232	21.6333	7.7639	10124.	3.4202	.00213675	1470.264	172021.05
469	219961	103161709	21.6564	7.7695	10157.	3.4217	.00213220	1473.406	172756.96
470	**220900**	**103823000**	**21.6795**	**7.7750**	**10189.**	**3.4231**	**.00212766**	**1476.547**	**173494.45**
471	221841	104487111	21.7025	7.7805	10222.	3.4246	.00212314	1479.689	174233.51
472	222784	105154048	21.7256	7.7860	10255.	3.4260	.00211864	1482.830	174974.14
473	223729	105823817	21.7486	7.7915	10287.	3.4275	.00211416	1485.972	175716.34
474	224676	106496424	21.7715	7.7970	10320.	3.4289	.00210970	1489.114	176460.12
475	225625	107171875	21.7945	7.8025	10352.	3.4304	.00210526	1492.255	177205.46
476	226576	107850176	21.8174	7.8079	10385.	3.4318	.00210084	1495.397	177952.37
477	227529	108531333	21.8403	7.8134	10418.	3.4332	.00209644	1498.538	178700.86
478	228484	109215352	21.8632	7.8188	10450.	3.4347	.00209205	1501.680	179450.91
479	229441	109902239	21.8861	7.8243	10483.	3.4361	.00208768	1504.822	180202.54
480	**230400**	**110592000**	**21.9089**	**7.8297**	**10516.**	**3.4375**	**.00208333**	**1507.963**	**180955.74**
481	231361	111284641	21.9317	7.8352	10549.	3.4390	.00207900	1511.105	181710.50
482	232324	111980168	21.9545	7.8406	10582.	3.4404	.00207469	1514.246	182466.84
483	233289	112678587	21.9773	7.8460	10615.	3.4418	.00207039	1517.388	183224.75
484	234256	113379904	22.0000	7.8514	10648.	3.4433	.00206612	1520.530	183984.23
485	235225	114084125	22.0227	7.8568	10681.	3.4447	.00206186	1523.671	184745.28
486	236196	114791256	22.0454	7.8622	10714.	3.4461	.00205761	1526.813	185507.90
487	237169	115501303	22.0681	7.8676	10747.	3.4475	.00205339	1529.954	186272.10
488	238144	116214272	22.0907	7.8730	10780.	3.4489	.00204918	1533.096	187037.86
489	239121	116930169	22.1133	7.8784	10813.	3.4504	.00204499	1536.238	187805.19
490	**240100**	**117649000**	**22.1359**	**7.8837**	**10847.**	**3.4518**	**.00204082**	**1539.379**	**188574.10**
491	241081	118370771	22.1585	7.8891	10880.	3.4532	.00203666	1542.521	189344.57
492	242064	119095488	22.1811	7.8944	10913.	3.4546	.00203252	1545.662	190116.62
493	243049	119823157	22.2036	7.8998	10946.	3.4560	.00202840	1548.804	190890.24
494	244036	120553784	22.2261	7.9051	10980.	3.4574	.00202429	1551.945	191665.43
495	245025	121287375	22.2486	7.9105	11013.	3.4588	.00202020	1555.087	192442.18
496	246016	122023936	22.2711	7.9158	11046.	3.4602	.00201613	1558.229	193220.51
497	247009	122763473	22.2935	7.9211	11080.	3.4616	.00201207	1561.370	194000.41
498	248004	123505992	22.3159	7.9264	11113.	3.4630	.00200803	1564.512	194781.89
499	249001	124251499	22.3383	7.9317	11147.	3.4643	.00200401	1567.653	195564.93

Table 7. Properties of Numbers—*Continued*

N	N^2	N^3	$\sqrt{N}$	$\sqrt[3]{N}$	$N^{3/2}$	$\sqrt[5]{N}$	$\frac{1}{N}$	Circle (N = Diam.)	
								Circum.	Area
500	**250000**	**125000000**	**22.3607**	**7.9370**	**11180**	**3.4657**	**.00200000**	**1570.795**	**196349.54**
501	251001	125751501	22.3830	7.9423	11214	3.4671	.00199601	1573.937	197135.72
502	252004	126506008	22.4054	7.9476	11247	3.4685	.00199203	1577.078	197923.48
503	253009	127263527	22.4277	7.9528	11281	3.4699	.00198807	1580.220	198712.80
504	254016	128024064	22.4499	7.9581	11315	3.4713	.00198413	1583.361	199503.70
505	255025	128787625	22.4722	7.9634	11348	3.4726	.00198020	1586.503	200296.17
506	256036	129554216	22.4944	7.9686	11382	3.4740	.00197628	1589.645	201090.20
507	257049	130323843	22.5167	7.9739	11416	3.4754	.00197239	1592.786	201885.81
508	258064	131096512	22.5389	7.9791	11450	3.4768	.00196850	1595.928	202682.99
509	259081	131872229	22.5610	7.9843	11484	3.4781	.00196464	1599.069	203481.74
510	**260100**	**132651000**	**22.5832**	**7.9896**	**11517**	**3.4795**	**.00196078**	**1602.211**	**204282.06**
511	261121	133432831	22.6053	7.9948	11551	3.4808	.00195695	1605.352	205083.95
512	262144	134217728	22.6274	8.0000	11585	3.4822	.00195313	1608.494	205887.42
513	263169	135005697	22.6495	8.0052	11619	3.4836	.00194932	1611.636	206692.45
514	264196	135796744	22.6716	8.0104	11653	3.4849	.00194553	1614.777	207499.05
515	265225	136590875	22.6936	8.0156	11687	3.4863	.00194175	1617.919	208307.23
516	266256	137388096	22.7156	8.0208	11721	3.4876	.00193798	1621.060	209116.97
517	267289	138188413	22.7376	8.0260	11755	3.4890	.00193424	1624.202	209928.29
518	268324	138991832	22.7596	8.0311	11789	3.4904	.00193050	1627.344	210741.18
519	269361	139798359	22.7816	8.0363	11824	3.4917	.00192678	1630.485	211555.63
520	**270400**	**140608000**	**22.8035**	**8.0415**	**11858**	**3.4930**	**.00192308**	**1633.627**	**212371.66**
521	271441	141420761	22.8254	8.0466	11892	3.4944	.00191939	1636.768	213189.26
522	272484	142236648	22.8473	8.0517	11926	3.4957	.00191571	1639.910	214008.43
523	273529	143055667	22.8692	8.0569	11960	3.4970	.00191205	1643.052	214829.17
524	274576	143877824	22.8910	8.0620	11995	3.4984	.00190840	1646.193	215651.49
525	275625	144703125	22.9129	8.0671	12029	3.4997	.00190476	1649.335	216475.37
526	276676	145531576	22.9347	8.0723	12064	3.5010	.00190114	1652.476	217300.82
527	277729	146363183	22.9565	8.0774	12098	3.5024	.00189753	1655.618	218127.85
528	278784	147197952	22.9783	8.0825	12133	3.5037	.00189394	1658.760	218956.44
529	279841	148035889	23.0000	8.0876	12167	3.5050	.00189036	1661.901	219786.61
530	**280900**	**148877000**	**23.0217**	**8.0927**	**12202**	**3.5064**	**.00188679**	**1665.043**	**220618.34**
531	281961	149721291	23.0434	8.0978	12236	3.5077	.00188324	1668.184	221451.65
532	283024	150568768	23.0651	8.1028	12271	3.5090	.00187970	1671.326	222286.53
533	284089	151419437	23.0868	8.1079	12305	3.5103	.00187617	1674.467	223122.98
534	285156	152273304	23.1084	8.1130	12340	3.5116	.00187266	1677.609	223961.00
535	286225	153130375	23.1301	8.1180	12375	3.5130	.00186916	1680.751	224800.59
536	287296	153990656	23.1517	8.1231	12410	3.5143	.00186567	1683.892	225641.75
537	288369	154854153	23.1733	8.1281	12444	3.5156	.00186220	1687.034	226484.48
538	289444	155720872	23.1948	8.1332	12479	3.5169	.00185874	1690.175	227328.79
539	290521	156590819	23.2164	8.1382	12514	3.5182	.00185529	1693.317	228174.66
540	**291600**	**157464000**	**23.2379**	**8.1433**	**12549**	**3.5195**	**.00185185**	**1696.459**	**229022.10**
541	292681	158340421	23.2594	8.1483	12583	3.5208	.00184843	1699.600	229871.12
542	293764	159220088	23.2809	8.1533	12618	3.5221	.00184502	1702.742	230721.71
543	294849	160103007	23.3024	8.1583	12653	3.5234	.00184162	1705.883	231573.86
544	295936	160989184	23.3238	8.1633	12688	3.5247	.00183824	1709.025	232427.59
545	297025	161878625	23.3452	8.1683	12723	3.5260	.00183486	1712.167	233282.89
546	298116	162771336	23.3666	8.1733	12758	3.5273	.00183150	1715.308	234139.76
547	299209	163667323	23.3880	8.1783	12793	3.5286	.00182815	1718.450	234998.20
548	300304	164566592	23.4094	8.1833	12828	3.5299	.00182482	1721.591	235858.21
549	301401	165469149	23.4307	8.1882	12863	3.5311	.00182149	1724.733	236719.79
550	**302500**	**166375000**	**23.4521**	**8.1932**	**12899**	**3.5324**	**.00181818**	**1727.875**	**237582.94**
551	303601	167284151	23.4734	8.1982	12934	3.5337	.00181488	1731.016	238447.67
552	304704	168196608	23.4947	8.2031	12969	3.5350	.00181159	1734.158	239313.96
553	305809	169112377	23.5160	8.2081	13004	3.5363	.00180832	1737.299	240181.83
554	306916	170031464	23.5372	8.2130	13040	3.5376	.00180505	1740.441	241051.26
555	308025	170953875	23.5584	8.2180	13075	3.5388	.00180180	1743.582	241922.27
556	309136	171879616	23.5797	8.2229	13110	3.5401	.00179856	1746.724	242794.85
557	310249	172808693	23.6008	8.2278	13146	3.5414	.00179533	1749.866	243668.99
558	311364	173741112	23.6220	8.2327	13181	3.5426	.00179211	1753.007	244544.71
559	312481	174676879	23.6432	8.2377	13217	3.5439	.00178891	1756.149	245422.00
560	**313600**	**175616000**	**23.6643**	**8.2426**	**13252**	**3.5451**	**.00178571**	**1759.290**	**246300.86**
561	314721	176558481	23.6854	8.2475	13288	3.5464	.00178253	1762.432	247181.30
562	315844	177504328	23.7065	8.2524	13323	3.5477	.00177936	1765.574	248063.30
563	316969	178453547	23.7276	8.2573	13359	3.5490	.00177620	1768.715	248946.87
564	318096	179406144	23.7487	8.2621	13394	3.5502	.00177305	1771.857	249832.01
565	319225	180362125	23.7697	8.2670	13430	3.5515	.00176991	1774.998	250718.73
566	320356	181321496	23.7908	8.2719	13466	3.5527	.00176678	1778.140	251607.01
567	321489	182284263	23.8118	8.2768	13501	3.5540	.00176367	1781.282	252496.87
568	322624	183250432	23.8328	8.2816	13537	3.5553	.00176056	1784.423	253388.30
569	323761	184220009	23.8537	8.2865	13573	3.5565	.00175747	1787.565	254281.29

Table 7. Properties of Numbers—*Continued*

N	N^2	N^3	$\sqrt{N}$	$\sqrt[3]{N}$	$N^{3/2}$	$\sqrt[5]{N}$	$\frac{1}{N}$	Circle (N = Diam.)	
								Circum.	Area
570	**324900**	**185193000**	**23.8747**	**8.2913**	**13609**	**3.5577**	**.00175439**	**1790.706**	**255175.86**
571	326041	186169411	23.8956	8.2962	13644	3.5590	.00175131	1793.848	256072.00
572	327184	187149248	23.9165	8.3010	13680	3.5602	.00174825	1796.989	256969.71
573	328329	188132517	23.9374	8.3059	13716	3.5615	.00174520	1800.131	257868.99
574	329476	189119224	23.9583	8.3107	13752	3.5627	.00174216	1803.273	258769.85
575	330625	190109375	23.9792	8.3155	13788	3.5640	.00173913	1806.414	259672.27
576	331776	191102976	24.0000	8.3203	13824	3.5652	.00173611	1809.556	260576.26
577	332929	192100033	24.0208	8.3251	13860	3.5664	.00173310	1812.697	261481.83
578	334084	193100552	24.0416	8.3300	13896	3.5677	.00173010	1815.839	262388.96
579	335241	194104539	24.0624	8.3348	13932	3.5689	.00172712	1818.981	263297.67
580	**336400**	**195112000**	**24.0832**	**8.3396**	**13968**	**3.5702**	**.00172414**	**1822.122**	**264207.94**
581	337561	196122941	24.1039	8.3443	14004	3.5714	.00172117	1825.264	265119.79
582	338724	197137368	24.1247	8.3491	14040	3.5726	.00171821	1828.405	266033.21
583	339889	198155287	24.1454	8.3539	14077	3.5738	.00171527	1831.547	266948.20
584	341056	199176704	24.1661	8.3587	14113	3.5751	.00171233	1834.689	267864.76
585	342225	200201625	24.1868	8.3634	14149	3.5763	.00170940	1837.830	268782.89
586	343396	201230056	24.2074	8.3682	14186	3.5775	.00170648	1840.972	269702.59
587	344569	202262003	24.2281	8.3730	14222	3.5787	.00170358	1844.113	270623.86
588	345744	203297472	24.2487	8.3777	14258	3.5799	.00170068	1847.255	271546.70
589	346921	204336469	24.2693	8.3825	14295	3.5812	.00169779	1850.397	272471.12
590	**348100**	**205379000**	**24.2899**	**8.3872**	**14331**	**3.5824**	**.00169492**	**1853.538**	**273397.10**
591	349281	206425071	24.3105	8.3919	14368	3.5886	.00169205	1856.680	274324.66
592	350464	207474688	24.3311	8.3967	14404	3.5848	.00168919	1859.821	275253.78
593	351649	208527857	24.3516	8.4014	14440	3.5860	.00168634	1862.963	276184.48
594	352836	209584584	24.3721	8.4061	14477	3.5872	.00168350	1866.104	277116.75
595	354025	210644875	24.3926	8.4108	14514	3.5884	.00168067	1869.246	278050.58
596	355216	211708736	24.4131	8.4155	14550	3.5896	.00167785	1872.388	278985.99
597	356409	212776173	24.4336	8.4202	14587	3.5908	.00167504	1875.529	279922.97
598	357604	213847192	24.4540	8.4249	14624	3.5920	.00167224	1878.671	280861.52
599	358801	214921799	24.4745	8.4296	14660	3.5932	.00166945	1881.812	281801.65
600	**360000**	**216000000**	**24.4949**	**8.4343**	**14697**	**3.5944**	**.00166667**	**1884.954**	**282743.34**
601	361201	217081801	24.5153	8.4390	14734	3.5956	.00166389	1888.096	283686.60
602	362404	218167208	24.5357	8.4437	14770	3.5958	.00166113	1891.237	284631.44
603	363609	219256227	24.5561	8.4484	14807	3.5980	.00165837	1894.379	285577.84
604	364816	220348864	24.5764	8.4530	14844	3.5992	.00165563	1897.520	286525.82
605	366025	221445125	24.5967	8.4577	14881	3.6004	.00165289	1900.662	287475.36
606	367236	222545016	24.6171	8.4623	14918	3.6016	.00165017	1903.804	288426.48
607	368449	223648543	24.6374	8.4670	14955	3.6028	.00164745	1906.945	289379.17
608	369664	224755712	24.6577	8.4716	14992	3.6040	.00164474	1910.087	290333.43
609	370881	225866529	24.6779	8.4763	15029	3.6052	.00164204	1913.228	291289.26
610	**372100**	**226981000**	**24.6982**	**8.4809**	**15066**	**3.6063**	**.00163934**	**1916.370**	**292246.66**
611	373321	228099131	24.7184	8.4856	15103	3.6075	.00163666	1919.511	293205.63
612	374544	229220928	24.7386	8.4902	15140	3.6087	.00163399	1922.653	294166.17
613	375769	230346397	24.7588	8.4948	15177	3.6099	.00163132	1925.795	295128.28
614	376996	231475544	24.7790	8.4994	15214	3.6111	.00162866	1928.936	296091.97
615	378225	232608375	24.7992	8.5040	15252	3.6122	.00162602	1932.078	297057.22
616	379456	233744896	24.8193	8.5086	15289	3.6134	.00162338	1935.219	298024.05
617	380689	234885113	24.8395	8.5132	15326	3.6146	.00162075	1938.361	298992.44
618	381924	236029032	24.8596	8.5178	15363	3.6158	.00161812	1941.503	299962.41
619	383161	237176659	24.8797	8.5224	15400	3.6169	.00161551	1944.644	300933.95
620	**384400**	**238328000**	**24.8998**	**8.5270**	**15437**	**3.6181**	**.00161290**	**1947.786**	**301907.05**
621	385641	239483061	24.9199	8.5316	15475	3.6192	.00161031	1950.927	302881.73
622	386884	240641848	24.9399	8.5362	15513	3.6204	.00160772	1954.069	303857.98
623	388129	241804367	24.9600	8.5408	15550	3.6216	.00160514	1957.211	304835.80
624	389376	242970624	24.9800	8.5453	15588	3.6227	.00160256	1960.352	305815.20
625	390625	244140625	25.0000	8.5499	15625	3.6239	.00160000	1963.494	306796.16
626	391876	245314376	25.0200	8.5544	15663	3.6250	.00159744	1966.635	307778.69
627	393129	246491883	25.0400	8.5590	15700	3.6262	.00159490	1969.777	308762.79
628	394384	247673152	25.0599	8.5635	15738	3.6274	.00159236	1972.919	309748.47
629	395641	248858189	25.0799	8.5681	15775	3.6285	.00158983	1976.060	310735.71
630	**396900**	**250047000**	**25.0998**	**8.5726**	**15813**	**3.6297**	**.00158730**	**1979.202**	**311724.53**
631	398161	251239591	25.1197	8.5772	15850	3.6309	.00158479	1982.343	312714.92
632	399424	252435968	25.1396	8.5817	15888	3.6320	.00158228	1985.485	313706.88
633	400689	253636137	25.1595	8.5862	15926	3.6331	.00157978	1988.626	314700.40
634	401956	254840104	25.1794	8.5907	15964	3.6343	.00157729	1991.768	315695.50
635	403225	256047875	25.1992	8.5952	16002	3.6354	.00157480	1994.910	316692.17
636	404496	257259456	25.2190	8.5997	16040	3.6366	.00157233	1998.051	317690.42
637	405769	258474853	25.2389	8.6043	16077	3.6377	.00156986	2001.193	318690.23
638	407044	259694072	25.2587	8.6088	16115	3.6389	.00156740	2004.334	319691.61
639	408321	260917119	25.2784	8.6132	16153	3.6400	.00156495	2007.476	320694.56

Table 7. Properties of Numbers—*Continued*

N	N^2	N^3	$\sqrt{N}$	$\sqrt[3]{N}$	$N^{3/2}$	$\sqrt[5]{N}$	$\frac{1}{N}$	Circle (N = Diam.)	
								Circum.	Area
640	**409600**	**262144000**	**25.2982**	**8.6177**	**16191**	**3.6411**	**.00156250**	**2010.618**	**321699.09**
641	410881	263374721	25.3180	8.6222	16229	3.6423	.00156006	2013.759	322705.18
642	412164	264609288	25.3377	8.6267	16267	3.6435	.00155763	2016.901	323712.85
643	413449	265847707	25.3574	8.6312	16305	3.6446	.00155521	2020.042	324722.09
644	414736	267089984	25.3772	8.6357	16343	3.6457	.00155280	2023.184	325732.89
645	416025	268336125	25.3969	8.6401	16381	3.6468	.00155039	2026.326	326745.27
646	417316	269586136	25.4165	8.6446	16419	3.6479	.00154799	2029.467	327759.22
647	418609	270840023	25.4362	8.6490	16457	3.6499	.00154560	2032.609	328774.74
648	419904	272097792	25.4558	8.6535	16495	3.6502	.00154321	2035.750	329791.83
649	421201	273359449	25.4755	8.6579	16534	3.6513	.00154083	2038.892	330810.49
650	**422500**	**274625000**	**25.4951**	**8.6624**	**16572**	**3.6524**	**.00153846**	**2042.034**	**331830.72**
651	423801	275894451	25.5147	8.6668	16610	3.6536	.00153610	2045.175	332852.53
652	425104	277167808	25.5343	8.6713	16648	3.6547	.00153374	2048.317	333875.90
653	426409	278445077	25.5539	8.6757	16687	3.6558	.00153139	2051.458	334900.85
654	427716	279726264	25.5734	8.6801	16725	3.6569	.00152905	2054.600	335927.36
655	429025	281011375	25.5930	8.6845	16764	3.6580	.00152672	2057.741	336955.45
656	430336	282300416	25.6125	8.6890	16802	3.6592	.00152439	2060.883	337985.10
657	431649	283593393	25.6320	8.6934	16840	3.6603	.00152207	2064.025	339016.33
658	432964	284890312	25.6515	8.6978	16879	3.6614	.00151976	2067.166	340049.13
659	434281	286191179	25.6710	8.7022	16917	3.6625	.00151745	2070.308	341083.50
660	**435600**	**287496000**	**25.6905**	**8.7066**	**16956**	**3.6636**	**.00151515**	**2073.449**	**342119.44**
661	436921	288804781	25.7099	8.7110	16994	3.6647	.00151286	2076.591	343156.95
662	438244	290117528	25.7294	8.7154	17033	3.6658	.00151057	2079.733	344196.03
663	439569	291434247	25.7488	8.7198	17071	3.6669	.00150830	2082.874	345236.69
664	440896	292754944	25.7682	8.7241	17110	3.6680	.00150602	2086.016	346278.91
665	442225	294079625	25.7876	8.7285	17149	3.6691	.00150376	2089.157	347322.70
666	443556	295408296	25.8070	8.7329	17187	3.6702	.00150150	2092.299	348368.07
667	444889	296740963	25.8263	8.7373	17226	3.6713	.00149925	2095.441	349415.00
668	446224	298077632	25.8457	8.7416	17265	3.6724	.00149701	2098.582	350463.51
669	447561	299418309	25.8650	8.7460	17304	3.6735	.00149477	2101.724	351513.59
670	**448900**	**300763000**	**25.8844**	**8.7503**	**17343**	**3.6746**	**.00149254**	**2104.865**	**352565.24**
671	450241	302111711	25.9037	8.7547	17381	3.6757	.00149031	2108.007	353618.45
672	451584	303464448	25.9230	8.7590	17420	3.6768	.00148810	2111.148	354673.24
673	452929	304821217	25.9422	8.7634	17459	3.6779	.00148588	2114.290	355729.60
674	454276	306182024	25.9615	8.7677	17498	3.6790	.00148368	2117.432	356787.54
675	455625	307546875	25.9808	8.7721	17537	3.6801	.00148148	2120.573	357847.04
676	456976	308915776	26.0000	8.7764	17576	3.6812	.00147929	2123.715	358908.11
677	458329	310288733	26.0192	8.7807	17615	3.6823	.00147710	2126.856	359970.75
678	459684	311665752	26.0384	8.7850	17654	3.6834	.00147493	2129.998	361034.97
679	461041	313046839	26.0576	8.7893	17693	3.6845	.00147275	2133.140	362100.75
680	**462400**	**314432000**	**26.0768**	**8.7937**	**17732**	**3.6856**	**.00147059**	**2136.281**	**363168.11**
681	463761	315821241	26.0960	8.7980	17771	3.6866	.00146843	2139.423	364237.04
682	465124	317214568	26.1151	8.8023	17810	3.6877	.00146628	2142.564	365307.54
683	466489	318611987	26.1343	8.8066	17850	3.6888	.00146413	2145.706	366379.60
684	467856	320013504	26.1534	8.8109	17889	3.6899	.00146199	2148.848	367453.24
685	469225	321419125	26.1725	8.8152	17928	3.6909	.00145985	2151.989	368528.45
686	470596	322828856	26.1916	8.8194	17967	3.6920	.00145773	2155.131	369605.23
687	471969	324242703	26.2107	8.8237	18007	3.6931	.00145560	2158.272	370683.59
688	473344	325660672	26.2298	8.8280	18046	3.6942	.00145349	2161.414	371763.51
689	474721	327082769	26.2488	8.8323	18085	3.6953	.00145138	2164.556	372845.00
690	**476100**	**328509000**	**26.2679**	**8.8366**	**18125**	**3.6963**	**.00144928**	**2167.697**	**373928.07**
691	477481	329939371	26.2869	8.8408	18164	3.6974	.00144718	2170.839	375012.70
692	478864	331373888	26.3059	8.8451	18204	3.6985	.00144509	2173.980	376098.91
693	480249	332812557	26.3249	8.8493	18243	3.6995	.00144300	2177.122	377186.68
694	481636	334255384	26.3439	8.8536	18283	3.7006	.00144092	2180.263	378276.03
695	483025	335702375	26.3629	8.8578	18322	3.7016	.00143885	2183.405	379366.95
696	484416	337153536	26.3818	8.8621	18362	3.7027	.00143678	2186.547	380459.44
697	485809	338608873	26.4008	8.8663	18401	3.7038	.00143472	2189.688	381553.50
698	487204	340068392	26.4197	8.8706	18441	3.7049	.00143266	2192.830	382649.13
699	488601	341532099	26.4386	8.8748	18480	3.7059	.00143062	2195.971	383746.33
700	**490000**	**343000000**	**26.4575**	**8.8790**	**18520**	**3.7070**	**.00142857**	**2199.113**	**384845.10**
701	491401	344472101	26.4764	8.8833	18560	3.7080	.00142653	2202.255	385945.44
702	492804	345948408	26.4953	8.8875	18600	3.7091	.00142450	2205.396	387047.36
703	494209	347428927	26.5141	8.8917	18640	3.7101	.00142248	2208.538	388150.84
704	495616	348913664	26.5330	8.8959	18679	3.7112	.00142045	2211.679	389255.90
705	497025	350402625	26.5518	8.9001	18719	3.7123	.00141844	2214.821	390362.52
706	498436	351895816	26.5707	8.9043	18759	3.7133	.00141643	2217.963	391470.72
707	499849	353393243	26.5895	8.9085	18799	3.7144	.00141443	2221.104	392580.49
708	501264	354894912	26.6083	8.9127	18839	3.7154	.00141243	2224.246	393691.82
709	502681	356400829	26.6271	8.9169	18879	3.7165	.00141044	2227.387	394804.73

Table 7. Properties of Numbers—*Continued*

N	N^2	N^3	$\sqrt{N}$	$\sqrt[3]{N}$	$N^{3/2}$	$\sqrt[5]{N}$	$\frac{1}{N}$	Circle (N = Diam.)	
								Circum.	Area
710	**504100**	**357911000**	**26.6458**	**8.9211**	**18919**	**3.7175**	**.00140845**	**2230.529**	**395919.21**
711	505521	359425431	26.6646	8.9253	18959	3.7185	.00140647	2233.670	397035.26
712	506944	360944128	26.6833	8.9295	18999	3.7196	.00140449	2236.812	398152.89
713	508369	362467097	26.7021	8.9337	19039	3.7206	.00140252	2239.954	399272.08
714	509796	363994344	26.7208	8.9378	19079	3.7217	.00140056	2243.095	400392.84
715	511225	365525875	26.7395	8.9420	19119	3.7227	.00139860	2246.237	401515.18
716	512656	367061696	26.7582	8.9462	19159	3.7238	.00139665	2249.378	402639.08
717	514089	368601813	26.7769	8.9503	19199	3.7248	.00139470	2252.520	403764.56
718	515524	370146232	26.7955	8.9545	19239	3.7258	.00139276	2255.662	404891.60
719	516961	371694959	26.8142	8.9587	19280	3.7269	.00139082	2258.803	406020.22
720	**518400**	**373248000**	**26.8328**	**8.9628**	**19320**	**3.7279**	**.00138889**	**2261.945**	**407150.41**
721	519841	374805361	26.8514	8.9670	19360	3.7290	.00138696	2265.086	408282.17
722	521284	376367048	26.8701	8.9711	19400	3.7300	.00138504	2268.228	409415.50
723	522729	377933067	26.8887	8.9752	19440	3.7310	.00138313	2271.370	410550.40
724	524176	379503424	26.9072	8.9794	19481	3.7321	.00138122	2274.511	411686.87
725	525625	381078125	26.9258	8.9835	19521	3.7331	.00137931	2277.653	412824.91
726	527076	382657176	26.9444	8.9876	19562	3.7341	.00137741	2280.794	413964.52
727	528529	384240583	26.9629	8.9918	19602	3.7351	.00137552	2283.936	415105.71
728	529984	385828352	26.9815	8.9959	19643	3.7362	.00137363	2287.078	416248.46
729	531441	387420489	27.0000	9.0000	19683	3.7372	.00137174	2290.219	417392.79
730	**532900**	**389017000**	**27.0185**	**9.0041**	**19724**	**3.7382**	**.00136986**	**2293.361**	**418538.68**
731	534361	390617891	27.0370	9.0082	19764	3.7392	.00136799	2296.502	419686.15
732	535824	392223168	27.0555	9.0123	19805	3.7403	.00136612	2299.644	420835.19
733	537289	393832837	27.0740	9.0164	19845	3.7413	.00136426	2302.785	421985.79
734	538756	395446904	27.0924	9.0205	19886	3.7423	.00136240	2305.927	423137.97
735	540225	397065375	27.1109	9.0246	19927	3.7433	.00136054	2309.069	424291.72
736	541696	398688256	27.1293	9.0287	19967	3.7443	.00135870	2312.210	425447.04
737	543169	400315553	27.1477	9.0328	20008	3.7454	.00135685	2315.352	426603.94
738	544644	401947272	27.1662	9.0369	20049	3.7464	.00135501	2318.493	427762.40
739	546121	403583419	27.1846	9.0410	20090	3.7474	.00135318	2321.635	428922.43
740	**547600**	**405224000**	**27.2029**	**9.0450**	**20130**	**3.7484**	**.00135135**	**2324.777**	**430084.03**
741	549081	406869021	27.2213	9.0491	20171	3.7494	.00134953	2327.918	431247.21
742	550564	408518488	27.2397	9.0532	20212	3.7504	.00134771	2331.060	432411.95
743	552049	410172407	27.2580	9.0572	20253	3.7514	.00134590	2334.201	433578.27
744	553536	411830784	27.2764	9.0613	20294	3.7524	.00134409	2337.343	434746.16
745	555025	413493625	27.2947	9.0654	20335	3.7534	.00134228	2340.485	435915.62
746	556516	415160936	27.3130	9.0694	20376	3.7545	.00134048	2343.626	437086.64
747	558009	416832723	27.3313	9.0735	20417	3.7555	.00133869	2346.768	438259.24
748	559504	418508992	27.3496	9.0775	20458	3.7565	.00133690	2349.909	439433.41
749	561001	420189749	27.3679	9.0816	20499	3.7575	.00133511	2353.051	440609.16
750	**562500**	**421875000**	**27.3861**	**9.0856**	**20540**	**3.7585**	**.00133333**	**2356.193**	**441786.47**
751	564001	423564751	27.4044	9.0896	20581	3.7595	.00133156	2359.334	442965.35
752	565504	425259008	27.4226	9.0937	20622	3.7605	.00132979	2362.476	444145.80
753	567009	426957777	27.4408	9.0977	20663	3.7615	.00132802	2365.617	445327.83
754	568516	428661064	27.4591	9.1017	20704	3.7625	.00132626	2368.759	446511.42
755	570025	430368875	27.4773	9.1057	20745	3.7635	.00132450	2371.900	447696.59
756	571536	432081216	27.4955	9.1098	20787	3.7645	.00132275	2375.042	448883.32
757	573049	433798093	27.5136	9.1138	20828	3.7655	.00132100	2378.184	450071.63
758	574564	435519512	27.5318	9.1178	20869	3.7665	.00131926	2381.325	451261.51
759	576081	437245479	27.5500	9.1218	20910	3.7675	.00131752	2384.467	452452.96
760	**577600**	**438976000**	**27.5681**	**9.1258**	**20952**	**3.7685**	**.00131579**	**2387.608**	**453645.98**
761	579121	440711081	27.5862	9.1298	20993	3.7694	.00131406	2390.750	454840.57
762	580644	442450728	27.6043	9.1338	21035	3.7704	.00131234	2393.892	456036.73
763	582169	444194947	27.6225	9.1378	21076	3.7714	.00131062	2397.033	457234.46
764	583696	445943744	27.6405	9.1418	21117	3.7724	.00130890	2400.175	458433.77
765	585225	447697125	27.6586	9.1458	21159	3.7734	.00130719	2403.316	459634.64
766	586756	449455096	27.6767	9.1498	21200	3.7744	.00130548	2406.458	460837.08
767	588289	451217663	27.6948	9.1537	21242	3.7754	.00130378	2409.600	462041.10
768	589824	452984832	27.7128	9.1577	21283	3.7764	.00130208	2412.741	463246.69
769	591361	454756609	27.7308	9.1617	21325	3.7774	.00130039	2415.883	464453.84
770	**592900**	**456533000**	**27.7489**	**9.1657**	**21367**	**3.7784**	**.00129870**	**2419.024**	**465662.57**
771	594441	458314011	27.7669	9.1696	21408	3.7793	.00129702	2422.166	466872.87
772	595984	460099648	27.7849	9.1736	21450	3.7803	.00129534	2425.307	468084.74
773	597529	461889917	27.8029	9.1775	21492	3.7813	.00129366	2428.449	469298.18
774	599076	463684824	27.8209	9.1815	21533	3.7822	.00129199	2431.591	470513.19
775	600625	465484375	27.8388	9.1855	21575	3.7832	.00129032	2434.732	471729.77
776	602176	467288576	27.8568	9.1894	21617	3.7842	.00128866	2437.874	472947.92
777	603729	469097433	27.8747	9.1933	21658	3.7852	.00128700	2441.015	474167.65
778	605284	470910952	27.8927	9.1973	21700	3.7861	.00128535	2444.157	475388.94
779	606841	472729139	27.9106	9.2012	21742	3.7871	.00128370	2447.299	476611.81

Table 7. Properties of Numbers—*Continued*

N	N^2	N^3	$\sqrt{N}$	$\sqrt[3]{N}$	$N^{3/2}$	$\sqrt[5]{N}$	$\frac{1}{N}$	Circle (N = Diam.)	
								Circum.	Area
780	**608400**	**474552000**	**27.9285**	**9.2052**	**21784**	**3.7881**	**.00128205**	**2450.440**	**477836.24**
781	609961	476379541	27.9464	9.2091	21826	3.7890	.00128041	2453.582	479062.25
782	611524	478211768	27.9643	9.2130	21868	3.7900	.00127877	2456.723	480289.83
783	613089	480048687	27.9821	9.2170	21910	3.7910	.00127714	2459.865	481518.97
784	614656	481890304	28.0000	9.2209	21952	3.7920	.00127551	2463.007	482749.69
785	616225	483736625	28.0179	9.2248	21994	3.7929	.00127389	2466.148	483981.98
786	617796	485587656	28.0357	9.2287	22036	3.7939	.00127226	2469.290	485215.84
787	619369	487443403	28.0535	9.2326	22078	3.7949	.00127065	2472.431	486451.28
788	620944	489303872	28.0713	9.2365	22120	3.7959	.00126904	2475.573	487688.28
789	622521	491169069	28.0891	9.2404	22162	3.7969	.00126743	2478.715	488926.85
790	**624100**	**493039000**	**28.1069**	**9.2443**	**22205**	**3.7978**	**.00126582**	**2481.856**	**490166.99**
791	625681	494913671	28.1247	9.2482	22247	3.7987	.00126422	2484.998	491408.71
792	627264	496793088	28.1425	9.2521	22289	3.7997	.00126263	2488.139	492651.99
793	628849	498677257	28.1603	9.2560	22331	3.8006	.00126103	2491.281	493896.85
794	630436	500566184	28.1780	9.2599	22373	3.8016	.00125945	2494.422	495143.28
795	632025	502459875	28.1957	9.2638	22416	3.8025	.00125786	2497.564	496391.27
796	633616	504358336	28.2135	9.2677	22458	3.8035	.00125628	2500.706	497640.84
797	635209	506261573	28.2312	9.2716	22500	3.8044	.00125471	2503.847	498891.98
798	636804	508169592	28.2489	9.2754	22543	3.8054	.00125313	2506.989	500144.69
799	638401	510082399	28.2666	9.2793	22585	3.8064	.00125156	2510.130	501398.97
800	**640000**	**512000000**	**28.2843**	**9.2832**	**22627**	**3.8073**	**.00125000**	**2513.272**	**502654.82**
801	641601	513922401	28.3019	9.2870	22670	3.8083	.00124844	2516.414	503912.25
802	643204	515849608	28.3196	9.2909	22712	3.8092	.00124688	2519.555	505171.24
803	644809	517781627	28.3373	9.2948	22755	3.8102	.00124533	2522.697	506431.80
804	646416	519718464	28.3549	9.2986	22797	3.8111	.00124378	2525.838	507693.94
805	648025	521660125	28.3725	9.3025	22840	3.8121	.00124224	2528.980	508957.64
806	649636	523606616	28.3901	9.3063	22883	3.8130	.00124069	2532.122	510222.92
807	651249	525557943	28.4077	9.3102	22925	3.8139	.00123916	2535.263	511489.77
808	652864	527514112	28.4253	9.3140	22968	3.8149	.00123762	2538.405	512758.19
809	654481	529475129	28.4429	9.3179	23010	3.8158	.00123609	2541.546	514028.18
810	**656100**	**531441000**	**28.4605**	**9.3217**	**23053**	**3.8168**	**.00123457**	**2544.688**	**515299.74**
811	657721	533411731	28.4781	9.3255	23096	3.8177	.00123305	2547.829	516572.87
812	659344	535387328	28.4956	9.3294	23138	3.8186	.00123153	2550.971	517847.57
813	660969	537367797	28.5132	9.3332	23181	3.8196	.00123001	2554.113	519123.84
814	662596	539353144	28.5307	9.3370	23224	3.8205	.00122850	2557.254	520401.68
815	664225	541343375	28.5482	9.3408	23267	3.8215	.00122699	2560.396	521681.10
816	665856	543338496	28.5657	9.3447	23310	3.8224	.00122549	2563.537	522962.08
817	667489	545338513	28.5832	9.3485	23352	3.8234	.00122399	2566.679	524244.63
818	669124	547343432	28.6007	9.3523	23395	3.8243	.00122249	2569.821	525528.76
819	670761	549353259	28.6182	9.3561	23438	3.8252	.00122100	2572.962	526814.46
820	**672400**	**551368000**	**28.6356**	**9.3599**	**23481**	**3.8262**	**.00121951**	**2576.104**	**528101.73**
821	674041	553387661	28.6531	9.3637	23524	3.8271	.00121803	2579.245	529390.56
822	675684	555412248	28.6705	9.3675	23567	3.8280	.00121655	2582.387	530680.97
823	677329	557441767	28.6880	9.3713	23610	3.8290	.00121507	2585.529	531972.95
824	678976	559476224	28.7054	9.3751	23653	3.8299	.00121359	2588.670	533266.50
825	680625	561515625	28.7228	9.3789	23696	3.8308	.00121212	2591.812	534561.62
826	682276	563559976	28.7402	9.3827	23740	3.8317	.00121065	2594.953	535858.32
827	683929	565609283	28.7576	9.3865	23783	3.8327	.00120919	2598.095	537156.58
828	685584	567663552	28.7750	9.3902	23826	3.8336	.00120773	2601.237	538456.41
829	687241	569722789	28.7924	9.3940	23869	3.8345	.00120627	2604.378	539757.82
830	**688900**	**571787000**	**28.8097**	**9.3978**	**23912**	**3.8355**	**.00120482**	**2607.520**	**541060.79**
831	690561	573856191	28.8271	9.4016	23955	3.8364	.00120337	2610.661	542365.34
832	692224	575930368	28.8444	9.4053	23999	3.8373	.00120192	2613.803	543671.46
833	693889	578009537	28.8617	9.4091	24042	3.8382	.00120048	2616.944	544979.15
834	695556	580093704	28.8791	9.4129	24085	3.8391	.00119904	2620.086	546288.40
835	697225	582182875	28.8964	9.4166	24128	3.8401	.00119760	2623.228	547599.23
836	698896	584277056	28.9137	9.4204	24172	3.8410	.00119617	2626.369	548911.63
837	700569	586376253	28.9310	9.4241	24215	3.8419	.00119474	2629.511	550225.61
838	702244	588480472	28.9482	9.4279	24259	3.8428	.00119332	2632.652	551541.15
839	703921	590589719	28.9655	9.4316	24302	3.8437	.00119190	2635.794	552858.26
840	**705600**	**592704000**	**28.9828**	**9.4354**	**24346**	**3.8446**	**.00119048**	**2638.936**	**554176.94**
841	707281	594823321	29.0000	9.4391	24389	3.8456	.00118906	2642.077	555497.20
842	708964	596947688	29.0172	9.4429	24432	3.8465	.00118765	2645.219	556819.02
843	710649	599077107	29.0345	9.4466	24476	3.8474	.00118624	2648.360	558142.42
844	712336	601211584	29.0517	9.4503	24520	3.8483	.00118483	2651.502	559467.39
845	714025	603351125	29.0689	9.4541	24563	3.8492	.00118343	2654.644	560793.92
846	715716	605495736	29.0861	9.4578	24607	3.8501	.00118203	2657.785	562122.03
847	717409	607645423	29.1033	9.4615	24650	3.8510	.00118064	2660.927	563451.71
848	719104	609800192	29.1204	9.4652	24694	3.8519	.00117925	2664.068	564782.96
849	720801	611960049	29.1376	9.4690	24738	3.8528	.00117786	2667.210	566115.78

Table 7. Properties of Numbers—*Continued*

N	N^2	N^3	$\sqrt{N}$	$\sqrt[3]{N}$	$N^{3/2}$	$\sqrt[5]{N}$	$\frac{1}{N}$	Circle (N = Diam.)	
								Circum.	Area
850	**722500**	**614125000**	**29.1548**	**9.4727**	**24782**	**3.8538**	**.00117647**	**2670.352**	**567450.17**
851	724201	616295051	29.1719	9.4764	24825	3.8547	.00117509	2673.493	568786.14
852	725904	618470208	29.1890	9.4801	24869	3.8556	.00117371	2676.635	570123.67
853	727609	620650477	29.2062	9.4838	24913	3.8565	.00117233	2679.776	571462.77
854	729316	622835864	29.2233	9.4875	24957	3.8574	.00117096	2682.918	572803.45
855	731025	625026375	29.2404	9.4912	25000	3.8582	.00116959	2686.059	574145.69
856	732736	627222016	29.2575	9.4949	25044	3.8592	.00116822	2689.201	575489.51
857	734449	629422793	29.2746	9.4986	25088	3.8601	.00116686	2692.343	576834.90
858	736164	631628712	29.2916	9.5023	25132	3.8610	.00116550	2695.484	578181.85
859	737881	633839779	29.3087	9.5060	25176	3.8619	.00116414	2698.626	579530.38
860	**739600**	**636056000**	**29.3258**	**9.5097**	**25220**	**3.8628**	**.00116279**	**2701.767**	**580880.48**
861	741321	638277381	29.3428	9.5134	25264	3.8637	.00116144	2704.909	582232.15
862	743044	640503928	29.3598	9.5171	25308	3.8646	.00116009	2708.051	583585.39
863	744769	642735647	29.3769	9.5207	25352	3.8655	.00115875	2711.192	584940.20
864	746496	644972544	29.3939	9.5244	25396	3.8664	.00115741	2714.334	586296.59
865	748225	647214625	29.4109	9.5281	25440	3.8673	.00115607	2717.475	587654.54
866	749956	649461896	29.4279	9.5317	25485	3.8682	.00115473	2720.617	589014.07
867	751689	651714363	29.4449	9.5354	25529	3.8691	.00115340	2723.759	590375.16
868	753424	653972032	29.4618	9.5391	25573	3.8700	.00115207	2726.900	591737.83
869	755161	656234909	29.4788	9.5427	25617	3.8708	.00115075	2730.042	593102.06
870	**756900**	**658503000**	**29.4958**	**9.5464**	**25661**	**3.8717**	**.00114943**	**2733.183**	**594467.87**
871	758641	660776311	29.5127	9.5501	25706	3.8726	.00114811	2736.325	595835.25
872	760384	663054848	29.5296	9.5537	25750	3.8735	.00114679	2739.466	597204.20
873	762129	665338617	29.5466	9.5574	25794	3.8744	.00114548	2742.608	598574.72
874	763876	667627624	29.5635	9.5610	25839	3.8753	.00114416	2745.750	599946.81
875	765625	669921875	29.5804	9.5647	25883	3.8762	.00114286	2748.891	601320.47
876	767376	672221376	29.5973	9.5683	25927	3.8771	.00114155	2752.033	602695.70
877	769129	674526133	29.6142	9.5719	25972	3.8780	.00114025	2755.174	604072.50
878	770884	676836152	29.6311	9.5756	26016	3.8789	.00113895	2758.316	605450.88
879	772641	679151439	29.6479	9.5792	26061	3.8797	.00113766	2761.458	606830.82
880	**774400**	**681472000**	**29.6648**	**9.5828**	**26105**	**3.8806**	**.00113636**	**2764.599**	**608212.34**
881	776161	683797841	29.6816	9.5865	26150	3.8815	.00113507	2767.741	609595.42
882	777924	686128968	29.6985	9.5901	26194	3.8823	.00113379	2770.882	610980.08
883	779689	688465387	29.7153	9.5937	26239	3.8832	.00113250	2774.024	612366.31
884	781456	690807104	29.7321	9.5973	26283	3.8841	.00113122	2777.166	613754.11
885	783225	693154125	29.7489	9.6010	26328	3.8850	.00112994	2780.307	615143.48
886	784996	695506456	29.7658	9.6046	26373	3.8859	.00112867	2783.449	616534.42
887	786769	697864103	29.7825	9.6082	26417	3.8868	.00112740	2786.590	617926.93
888	788544	700227072	29.7993	9.6118	26462	3.8877	.00112613	2789.732	619321.01
889	790321	702595369	29.8161	9.6154	26507	3.8885	.00112486	2792.874	620716.66
890	**792100**	**704969000**	**29.8329**	**9.6190**	**26551**	**3.8894**	**.00112360**	**2796.015**	**622113.89**
891	793881	707347971	29.8496	9.6226	26596	3.8902	.00112233	2799.157	623512.68
892	795664	709732288	29.8664	9.6262	26641	3.8911	.00112108	2802.298	624913.04
893	797449	712121957	29.8831	9.6298	26686	3.8920	.00111982	2805.440	626314.98
894	799236	714516984	29.8998	9.6334	26730	3.8929	.00111857	2808.581	627718.49
895	801025	716917375	29.9166	9.6370	26775	3.8937	.00111732	2811.723	629123.56
896	802816	719323136	29.9333	9.6406	26820	3.8946	.00111607	2814.865	630530.21
897	804609	721734273	29.9500	9.6442	26865	3.8955	.00111483	2818.006	631938.43
898	806404	724150792	29.9666	9.6477	26910	3.8963	.00111359	2821.148	633348.22
899	808201	726572699	29.9833	9.6513	26955	3.8972	.00111235	2824.289	634759.58
900	**810000**	**729000000**	**30.0000**	**9.6549**	**27000**	**3.8981**	**.00111111**	**2827.431**	**636172.51**
901	811801	731432701	30.0167	9.6585	27045	3.8989	.00110988	2830.573	637587.01
902	813604	733870808	30.0333	9.6620	27090	3.8998	.00110865	2833.714	639003.09
903	815409	736314327	30.0500	9.6656	27135	3.9007	.00110742	2836.856	640420.73
904	817216	738763264	30.0666	9.6692	27180	3.9015	.00110619	2839.997	641839.95
905	819025	741217625	30.0832	9.6727	27225	3.9024	.00110497	2843.139	643260.73
906	820836	743677416	30.0998	9.6763	27270	3.9032	.00110375	2846.281	644683.09
907	822649	746142643	30.1164	9.6799	27316	3.9041	.00110254	2849.422	646107.01
908	824464	748613312	30.1330	9.6834	27361	3.9050	.00110132	2852.564	647532.51
909	826281	751089429	30.1496	9.6870	27406	3.9059	.00110011	2855.705	648959.58
910	**828100**	**753571000**	**30.1662**	**9.6905**	**27451**	**3.9067**	**.00109890**	**2858.847**	**650388.22**
911	829921	756058031	30.1828	9.6941	27497	3.9076	.00109769	2861.988	651818.43
912	831744	758550528	30.1993	9.6976	27542	3.9084	.00109649	2865.130	653250.21
913	833569	761048497	30.2159	9.7012	27587	3.9093	.00109529	2868.272	654683.56
914	835396	763551944	30.2324	9.7047	27632	3.9101	.00109409	2871.413	656118.48
915	837225	766060875	30.2490	9.7082	27678	3.9110	.00109290	2874.555	657554.98
916	839056	768575296	30.2655	9.7118	27723	3.9118	.00109170	2877.696	658993.04
917	840889	771095213	30.2820	9.7153	27769	3.9127	.00109051	2880.838	660432.68
918	842724	773620632	30.2985	9.7188	27814	3.9135	.00108932	2883.980	661873.88
919	844561	776151559	30.3150	9.7224	27859	3.9144	.00108814	2887.121	663316.66

Table 7. **Properties of Numbers**—*Continued*

N	N^2	N^3	$\sqrt{N}$	$\sqrt[3]{N}$	$N^{3/2}$	$\sqrt[5]{N}$	$\frac{1}{N}$	Circle (N = Diam.)	
								Circum.	Area
920	**846400**	**778688000**	**30.3315**	**9.7259**	**27905**	**3.9153**	**.00108696**	**2890.263**	**664761.01**
921	848241	781229961	30.3480	9.7294	27950	3.9161	.00108578	2893.404	666206.92
922	850084	783777448	30.3645	9.7329	27996	3.9169	.00108460	2896.546	667654.41
923	851929	786330467	30.3809	9.7364	28042	3.9178	.00108342	2899.688	669103.47
924	853776	788889024	30.3974	9.7400	28087	3.9186	.00108225	2902.829	670554.10
925	855625	791453125	30.4138	9.7435	28133	3.9194	.00108108	2905.971	672006.30
926	857476	794022776	30.4302	9.7470	28179	3.9203	.00107991	2909.112	673460.08
927	859329	796597983	30.4467	9.7505	28224	3.9212	.00107875	2912.254	674915.42
928	861184	799178752	30.4631	9.7540	28270	3.9220	.00107759	2915.396	676372.33
929	863041	801765089	30.4795	9.7575	28315	3.9229	.00107643	2918.537	677830.82
930	**864900**	**804357000**	**30.4959**	**9.7610**	**28361**	**3.9237**	**.00107527**	**2921.679**	**679290.87**
931	866761	806954491	30.5123	9.7645	28407	3.9246	.00107411	2924.820	680752.50
932	868624	809557568	30.5287	9.7680	28453	3.9254	.00107296	2927.962	682215.69
933	870489	812166237	30.5450	9.7715	28499	3.9262	.00107181	2931.103	683680.46
934	872356	814780504	30.5614	9.7750	28544	3.9271	.00107066	2934.245	685146.80
935	874225	817400375	30.5778	9.7785	28590	3.9279	.00106952	2937.387	686614.71
936	876096	820025856	30.5941	9.7819	28636	3.9288	.00106838	2940.528	688084.19
937	877969	822656953	30.6105	9.7854	28682	3.9296	.00106724	2943.670	689555.24
938	879844	825293672	30.6268	9.7889	28728	3.9304	.00106610	2946.811	691027.86
939	881721	827936019	30.6431	9.7924	28774	3.9313	.00106496	2949.953	692502.05
940	**883600**	**830584000**	**30.6594**	**9.7959**	**28820**	**3.9321**	**.00106383**	**2953.095**	**693977.82**
941	885481	833237621	30.6757	9.7993	28866	3.9329	.00106270	2956.236	695455.15
942	887364	835896888	30.6920	9.8028	28912	3.9338	.00106157	2959.378	696934.06
943	889249	838561807	30.7083	9.8063	28958	3.9346	.00106045	2962.519	698414.53
944	891136	841232384	30.7246	9.8097	29004	3.9354	.00105932	2965.661	699896.58
945	893025	843908625	30.7409	9.8132	29050	3.9363	.00105820	2968.803	701380.19
946	894916	846590536	30.7571	9.8167	29096	3.9371	.00105708	2971.944	702865.38
947	896809	849278123	30.7734	9.8201	29142	3.9379	.00105597	2975.086	704352.14
948	898704	851971392	30.7896	9.8236	29189	3.9388	.00105485	2978.227	705840.47
949	900601	854670349	30.8058	9.8270	29235	3.9396	.00105374	2981.369	707330.37
950	**902500**	**857375000**	**30.8221**	**9.8305**	**29281**	**3.9404**	**.00105263**	**2984.511**	**708821.84**
951	904401	860085351	30.8383	9.8339	29327	3.9413	.00105152	2987.652	710314.88
952	906304	862801408	30.8545	9.8374	29374	3.9421	.00105042	2990.794	711809.50
953	908209	865523177	30.8707	9.8408	29420	3.9429	.00104932	2993.935	713305.68
954	910116	868250664	30.8869	9.8443	29466	3.9438	.00104822	2997.077	714803.43
955	912025	870983875	30.9031	9.8477	29513	3.9446	.00104712	3000.218	716302.76
956	913936	873722816	30.9192	9.8511	29559	3.9454	.00104603	3003.360	717803.66
957	915849	876467493	30.9354	9.8546	29605	3.9462	.00104493	3006.502	719306.12
958	917764	879217912	30.9516	9.8580	29652	3.9471	.00104384	3009.643	720810.16
959	919681	881974079	30.9677	9.8614	29698	3.9479	.00104275	3012.785	722315.77
960	**921600**	**884736000**	**30.9839**	**9.8648**	**29745**	**3.9487**	**.00104167**	**3015.926**	**723822.95**
961	923521	887503681	31.0000	9.8683	29791	3.9495	.00104058	3019.068	725331.70
962	925444	890277128	31.0161	9.8717	29838	3.9503	.00103950	3022.210	726842.02
963	927369	893056347	31.0322	9.8751	29884	3.9512	.00103842	3025.351	728353.91
964	929296	895841344	31.0483	9.8785	29931	3.9520	.00103734	3028.493	729867.37
965	931225	898632125	31.0644	9.8819	29977	3.9528	.00103627	3031.634	731382.40
966	933156	901428696	31.0805	9.8854	30024	3.9536	.00103520	3034.776	732899.01
967	935089	904231063	31.0966	9.8888	30070	3.9544	.00103413	3037.918	734417.18
968	937024	907039232	31.1127	9.8922	30117	3.9553	.00103306	3041.059	735936.93
969	938961	909853209	31.1288	9.8956	30164	3.9561	.00103199	3044.201	737458.24
970	**940900**	**912673000**	**31.1448**	**9.8990**	**30210**	**3.9569**	**.00103093**	**3047.342**	**738981.13**
971	942841	915498611	31.1609	9.9024	30257	3.9577	.00102987	3050.484	740505.59
972	944784	918330048	31.1769	9.9058	30304	3.9585	.00102881	3053.625	742031.62
973	946729	921167317	31.1929	9.9092	30351	3.9593	.00102775	3056.767	743559.22
974	948676	924010424	31.2090	9.9126	30398	3.9602	.00102669	3059.909	745088 39
975	950625	926859375	31.2250	9.9160	30444	3.9610	.00102564	3063.050	746619 13
976	952576	929714176	31.2410	9.9194	30491	3.9618	.00102459	3066.192	748151.44
977	954529	932574833	31.2570	9.9227	30538	3.9626	.00102354	3069.333	749685.32
978	956484	935441352	31.2730	9.9261	30585	3.9634	.00102249	3072.475	751220.78
979	958441	938313739	31.2890	9.9295	30632	3.9642	.00102145	3075.617	752757.80
980	**960400**	**941192000**	**31.3050**	**9.9329**	**30679**	**3.9650**	**.00102041**	**3078.758**	**754296.40**
981	962361	944076141	31.3209	9.9363	30726	3.9658	.00101937	3081.900	755836.59
982	964324	946966168	31.3369	9.9396	30773	3.9666	.00101833	3085.041	757378.30
983	966289	949862087	31.3528	9.9430	30820	3.9674	.00101729	3088.183	758921.61
984	968256	952763904	31.3688	9.9464	30867	3.9682	.00101626	3091.325	760466 48
985	970225	955671625	31.3847	9.9497	30914	3.9691	.00101523	3094.466	762012 93
986	972196	958585256	31.4006	9.9531	30961	3.9699	.00101420	3097.608	763560 95
987	974169	961504803	31.4166	9.9565	31008	3.9707	.00101317	3100.749	765110 54
988	976144	964430272	31.4325	9.9598	31055	3.9715	.00101215	3103.891	766661 70
989	978121	967361669	31.4484	9.9632	31102	3.9723	.00101112	3107.033	768214.44

Table 7. Properties of Numbers—*Continued*

N	N^2	N^3	$\sqrt{N}$	$\sqrt[3]{N}$	$N^{3/2}$	$\sqrt[5]{N}$	$\frac{1}{N}$	Circle ($N = D$) Circum.	Circle ($N = D$) Area
990	**980100**	**970299000**	**31.4643**	**9.9666**	**31150**	**3.9731**	**.00101010**	**3110.174**	**769768.74**
991	982081	973242271	31.4802	9.9699	31197	3.9739	.00100908	3113.316	771324.61
992	984064	976191488	31.4960	9.9733	31244	3.9747	.00100806	3116.457	772882.06
993	986049	979146657	31.5119	9.9766	31291	3.9755	.00100705	3119.599	774441.07
994	988036	982107784	31.5278	9.9800	31339	3.9763	.00100604	3122.740	776001.66
995	990025	985074875	31.5436	9.9833	31386	3.9771	.00100503	3125.882	777563.82
996	992016	988047936	31.5595	9.9866	31433	3.9779	.00100402	3129.024	779127.54
997	994009	991026973	31.5753	9.9900	31480	3.9787	.00100301	3132.165	780692.84
998	996004	994011992	31.5911	9.9933	31528	3.9795	.00100200	3135.307	782259.71
999	998001	997002999	31.6070	9.9967	31575	3.9803	.00100100	3138.448	783828.15
1000	**1000000**	**1000000000**	**31.6228**	**10.0000**	**31623**	**3.9811**	**.00100000**	**3141.593**	**785398.16**

CIRCULAR ARCS, CHORDS, SEGMENTS AND SECTORS

Notation.—A = central angle, degrees; C = chord of arc; c = chord of 1/2 arc; D = diameter; G = area of segment; H = rise of arc or height of segment; L = length of arc; R = radius; S = area of sector.

L (exact) $= 2\pi RN/360 = 0.017453RN$

L (approx.) $= (8c - C)/3 = \{(2c \times 10H)/(60D - 27A)\} + 2c$

$= \{(\sqrt{C^2 + 4H^2} \times 10H^2)/(15C^2 + 33H^2)\} + 2c$

$C = 2\sqrt{c^2 - H^2} = \sqrt{D^2 - (D - 2H)^2} = (8c - 3L) = 2\sqrt{R^2 - (R - H)^2} = 2\sqrt{(D - H) \times H}$

$C/2 = \sqrt{H \times (D - H)}$; $c = 1/2\sqrt{C^2 - 4H^2} = \sqrt{D \times H} = (3L + C)/8$

$D = c^2/H = (1/4\, C^2 + H^2)/H$

$H = c^2/D = 1/2\,(D - \sqrt{D^2 - C^2})$ when $H < R$; $= 1/2\,(D + \sqrt{D^2 - C^2}) = \sqrt{c^2 - 1/4\, C^2}$ when $H > R$

$G = S - 1/2\, C\sqrt{R^2 - 1/4\, C^2}$ when $G <$ semi-circle; $= S + 1/2\, C\sqrt{R^2 - 1/4\, C^2}$ when $S >$ semi-circle

$= S - (C/2)(R - H) = 1/2\,\{(L - C)(D/2) + CH\}$

$S = 1/2\,LR = \pi AR^2/360 = (\pi R^2/360) \times 2[\sin^{-1}\{(C/2)/R\}]$; $R = (C^2 + 4H^2)/8H$

Table 8. Circular Arcs, Chords, and Segments. Radius = 1

Central Angle in Degrees	Arc Length L	Arc Rise H	Chord, Length C	$\frac{H}{C}$	Area of Segment* G	Central Angle in Degrees	Arc Length L	Arc Rise H	Chord, Length C	$\frac{H}{C}$	Area of Segment* G
1	0.0175	0.0000	0.0175	0.0022	0.00000	31	0.5411	0.0364	0.5345	0.0680	0.01301
2	.0349	.0002	.0349	.0044	.00000	32	.5585	.0387	.5513	.0703	.01429
3	.0524	.0003	.0524	.0066	.00001	33	.5760	.0412	.5680	.0725	.01566
4	.0698	.0006	.0698	.0087	.00003	34	.5934	.0437	.5847	.0747	.01711
5	.0873	.0010	.0872	.0109	.00006	35	.6109	.0463	.6014	.0770	.01864
6	.1047	.0014	.1047	.0131	.00010	36	.6283	.0489	.6180	.0792	.02027
7	.1222	.0019	.1221	.0153	.00015	37	.6458	.0517	.6346	.0814	.02198
8	.1396	.0024	.1395	.0175	.00023	38	.6632	.0545	.6511	.0837	.02378
9	.1571	.0031	.1569	.0196	.00032	39	.6807	.0574	.6676	.0859	.02568
10	**.1745**	**.0038**	**.1743**	**.0218**	**.00044**	**40**	**.6981**	**.0603**	**.6840**	**.0882**	**.02767**
11	.1920	.0046	.1917	.0240	.00059	41	.7156	.0633	.7004	.0904	.02976
12	.2094	.0055	.2091	.0262	.00076	42	.7330	.0664	.7167	.0927	.03195
13	.2269	.0064	.2264	.0284	.00097	43	.7505	.0696	.7330	.0949	.03425
14	.2443	.0075	.2437	.0306	.00121	44	.7679	.0728	.7492	.0972	.03664
15	.2618	.0086	.2611	.0328	.00149	45	.7854	.0761	.7654	.0995	.03915
16	.2793	.0097	.2783	.0350	.00181	46	.8029	.0795	.7815	.1017	.04176
17	.2967	.0110	.2956	.0372	.00217	47	.8203	.0829	.7975	.1040	.04448
18	.3142	.0123	.3129	.0394	.00257	48	.8378	.0865	.8135	.1063	.04731
19	.3316	.0137	.3301	.0415	.00302	49	.8552	.0900	.8294	.1086	.05025
20	**.3491**	**.0152**	**.3473**	**.0437**	**.00352**	**50**	**8727**	**.0937**	**.8452**	**.1108**	**.05331**
21	.3665	.0167	.3645	.0459	.00408	51	.8901	.0974	.8610	.1131	.05649
22	.3840	.0184	.3816	.0481	.00468	52	.9076	.1012	.8767	.1154	.05978
23	.4014	.0201	.3987	.0503	.00535	53	.9250	.1051	.8924	.1177	.06319
24	.4189	.0219	.4158	.0526	.00607	54	.9425	.1090	.9080	.1200	.06673
25	.4363	.0237	.4329	.0548	.00686	55	.9599	.1130	.9235	.1223	.07039
26	.4538	.0256	.4499	.0570	.00771	56	.9774	.1171	.9389	.1247	.07417
27	.4712	.0276	.4669	.0592	.00862	57	.9848	.1212	.9543	.1270	.07808
28	.4887	.0297	.4838	.0614	.00961	58	1.0123	.1254	.9696	.1293	.08212
29	.5061	.0319	.5008	.0636	.01067	59	1.0297	.1296	.9848	.1316	.08629
30	**.5236**	**.0341**	**.5176**	**.0658**	**.01180**	**60**	**1.0472**	**.1340**	**1.0000**	**.1340**	**.09059**

*Area of segment of any radius = factor G × square of radius

Table 8. Circular Arcs, Chords, and Segments—*Continued*

Central Angle in Degrees	Arc		Chord, Length	$\frac{H}{C}$	Area of Segment *	Central Angle in Degrees	Arc		Chord, Length	$\frac{H}{C}$	Area of Segment *
	Length L	Rise H	C		G		Length L	Rise H	C		G
61	1.0647	0.1384	1.015	0.1363	0.09502	121	2.1118	0.5076	1.741	0.2916	0.62734
62	1.0821	.1428	1.030	.1387	.09958	122	2.1293	.5152	1.749	.2945	.64063
63	1.0996	.1474	1.045	.1410	.10428	123	2.1468	.5228	1.758	.2975	.65404
64	1.1170	.1520	1.060	.1434	.10911	124	2.1642	.5305	1.766	.3004	.66759
65	1.1345	.1566	1.075	.1457	.11408	125	2.1817	.5383	1.774	.3034	.68125
66	1.1519	.1613	1.089	.1481	.11919	126	2.1991	.5460	1.782	.3064	.69505
67	1.1694	.1661	1.104	.1505	.12443	127	2.2166	.5538	1.790	.3094	.70897
68	1.1868	.1710	1.118	.1529	.12982	128	2.2340	.5616	1.798	.3124	.72301
69	1.2043	.1759	1.133	.1553	.13535	129	2.2515	.5695	1.805	.3155	.73716
70	**1.2217**	**.1808**	**1.147**	**.1576**	**.14102**	**130**	**2.2689**	**.5774**	**1.813**	**.3185**	**.75144**
71	1.2392	.1859	1.161	.1601	.14683	131	2.2864	.5853	1.820	.3216	.76584
72	1.2566	.1910	1.176	.1625	.15279	132	2.3038	.5933	1.827	.3247	.78034
73	1.2741	.1961	1.190	.1649	.15889	133	2.3213	.6013	1.834	.3278	.79497
74	1.2915	.2014	1.204	.1673	.16514	134	2.3387	.6093	1.841	.3309	.80970
75	1.3090	.2066	1.218	.1697	.17154	135	2.3562	.6173	1.848	.3341	.82454
76	1.3265	.2120	1.231	.1722	.17808	136	2.3736	.6254	1.854	.3373	.83949
77	1.3439	.2174	1.245	.1746	.18477	137	2.3911	.6335	1.861	.3404	.85455
78	1.3614	.2229	1.259	.1771	.19160	138	2.4086	.6416	1.867	.3436	.86971
79	1.3788	2284	1.272	.1795	.19859	139	2.4260	.6498	1.873	.3469	.88497
80	**1.3963**	**.2340**	**1.286**	**.1820**	**.20573**	**140**	**2.4435**	**.6580**	**1.879**	**.3501**	**.90034**
81	1.4137	.2396	1.299	.1845	.21301	141	2.4609	.6662	1.885	.3534	.91580
82	1.4312	.2453	1.312	.1869	.22045	142	2.4784	.6744	1.891	.3566	.93135
83	1.4486	.2510	1.325	.1894	.22804	143	2.4958	.6827	1.897	.3599	.94700
84	1.4661	.2569	1.338	.1919	.23578	144	2.5133	.6910	1.902	.3633	.96274
85	1.4835	.2627	1.351	.1944	.24367	145	2.5307	.6993	1.907	.3666	.97858
86	1.5010	.2686	1.364	.1970	.25171	146	2.5482	.7076	1.913	.3700	.99449
87	1.5184	.2746	1.377	.1995	.25990	147	2.5656	.7160	1.918	.3734	1.0105
88	1.5359	.2807	1.389	.2020	.26825	148	2.5831	.7244	1.923	.3768	1.0266
89	1.5533	.2867	1.402	.2046	.27675	149	2.6005	.7328	1.927	.3802	1.0428
90	**1.5708**	**.2929**	**1.414**	**.2071**	**.28540**	**150**	**2.6180**	**.7412**	**1.932**	**.3837**	**1.0590**
91	1.5882	.2991	1.427	.2097	.29420	151	2.6354	.7496	1.936	.3871	1.0753
92	1.6057	.3053	1.439	.2122	.30316	152	2.6529	.7581	1.941	.3906	1.0917
93	1.6232	.3116	1.451	.2148	.31226	153	2.6704	.7666	1.945	.3942	1.1082
94	1.6406	.3180	1.463	.2174	.32152	154	2.6878	.7750	1.949	.3977	1.1247
95	1.6581	.3244	1.475	.2200	.33093	155	2.7053	.7836	1.953	.4013	1.1413
96	1.6755	.3309	1.486	.2226	.34050	156	2.7227	.7921	1.956	.4049	1.1580
97	1.6930	.3374	1.498	.2252	.35021	157	2.7402	.8006	1.960	.4085	1.1747
98	1.7104	.3439	1.509	.2279	.36008	158	2.7576	.8092	1.963	.4122	1.1915
99	1.7279	.3506	1.521	.2305	.37009	159	2.7751	.8178	1.967	.4158	1.2084
100	**1.7453**	**.3572**	**1.532**	**.2332**	**.38026**	**160**	**2.7925**	**.8264**	**1.970**	**.4195**	**1.2253**
101	1.7628	.3639	1.543	.2358	.39058	161	2.8100	.8350	1.973	.4233	1.2422
102	1.7802	.3707	1.554	.2385	.40104	162	2.8274	.8436	1.975	.4270	1.2592
103	1.7977	.3775	1.565	.2412	.41166	163	2.8449	.8522	1.978	.4308	1.2763
104	1.8151	.3843	1.576	.2439	.42242	164	2.8623	.8608	1.981	.4346	1.2934
105	1.8326	.3912	1.587	.2466	.43333	165	2.8798	.8695	1.983	.4385	1.3105
106	1.8500	.3982	1.597	.2493	.44439	166	2.8972	.8781	1.985	.4424	1.3277
107	1.8675	.4052	1.608	.2520	.45560	167	2.9147	.8868	1.987	.4463	1.3449
108	1.8850	.4122	1.618	.2548	.46695	168	2.9322	.8955	1.989	.4502	1.3621
109	1.9024	.4193	1.628	.2575	.47844	169	2.9496	.9042	1.991	.4542	1.3794
110	**1.9199**	**.4264**	**1.638**	**.2603**	**.49008**	**170**	**2.9671**	**.9128**	**1.992**	**.4582**	**1.3967**
111	1.9373	.4336	1.648	.2631	.50187	171	2.9845	.9215	1.994	.4622	1.4140
112	1.9548	.4408	1.658	.2659	.51379	172	3.0020	.9302	1.995	.4663	1.4314
113	1.9722	.4481	1.668	.2687	.52586	173	3.0194	.9390	1.996	.4704	1.4488
114	1.9897	.4554	1.677	.2715	.53806	174	3.0369	.9477	1.997	.4745	1.4662
115	2.0071	.4627	1.687	.2743	.55041	175	3.0543	.9564	1.998	.4786	1.4836
116	2.0246	.4701	1.696	.2772	.56289	176	3.0718	.9651	1.9988	.4828	1.5010
117	2.0420	.4775	1.705	.2800	.57551	177	3.0892	.9738	1.9993	.4871	1.5184
118	2.0595	.4850	1.714	.2829	.58827	178	3.1067	.9825	1.9997	.4913	1.5359
119	2.0769	.4925	1.723	.2858	.60116	179	3.1241	.9913	1.9999	.4957	1.5533
120	**2.0944**	**.5000**	**1.732**	**.2887**	**.61419**	**180**	**3.1416**	**1.0000**	**2.000**	**.5000**	**1.5708**

* Area of segment of any radius = factor G × square of radius

Table 9. Areas of Segments of a Circle

Area = $F \times D^2$ where segment < semicircle; = $\frac{1}{4}\pi D^2$ − (area of segment of rise h). H = rise or height of segment; D = diam. of circle; F = factor corresponding to H/D. $h = (D - H)$

H/D	F	H/D	F	H/D	F	H/D	F	H/D	F	H/D	F
0.000	0.00000	0.07	0.02417	0.14	0.06683	0.21	0.11990	0.28	0.18002	0.35	0.24498
.001	.00004	.071	.02468	.141	.06753	.211	.12071	.281	.18092	.351	.24593
.002	.00012	.072	.02520	.142	.06822	.212	.12153	.282	.18182	.352	.24689
.003	.00022	.073	.02571	.143	.06892	.213	.12235	.283	.18272	.353	.24784
.004	.00034	.074	.02624	.144	.06963	.214	.12317	.284	.18362	.354	.24880
.005	.00047	.075	.02676	.145	.07033	.215	.12399	.285	.18452	.355	.24976
.006	.00062	.076	.02729	.146	.07103	.216	.12481	.286	.18542	.256	.25071
.007	.00078	.077	.02782	.147	.07174	.217	.12563	.287	.18633	.357	.25167
.008	.00095	.078	.02836	.148	.07245	.218	.12646	.288	.18723	.358	.25263
.009	.00113	.079	.02889	.149	.07316	.219	.12729	.289	.18814	.359	.25359
.01	.00133	.08	.02943	.15	.07387	.22	.12811	.29	.18905	.36	.25455
.011	.00153	.081	.02998	.151	.07459	.221	.12894	.291	.18996	.361	.25551
.012	.00175	.082	.03053	.152	.07531	.222	.12977	.292	.19086	.362	.25647
.013	.00197	.083	.03108	.153	.07603	.223	.13060	.293	.19177	.363	.25743
.014	.00220	.084	.03163	.154	.07675	.224	.13144	.294	.19268	.364	.25839
.015	.00244	.085	.03219	.155	.07747	.225	.13227	.295	.19360	.365	.25936
.016	.00268	.086	.03275	.156	.07819	.226	.13311	.296	.19451	.366	.26032
.017	.00294	.087	.03331	.157	.07892	.227	.13395	.297	.19542	.367	.26128
.018	.00320	.088	.03387	.158	.07965	.228	.13478	.298	.19634	.368	.26225
.019	.00347	.089	.03444	.159	.08038	.229	.13562	.299	.19725	.369	.26321
.02	.00375	.09	.03501	.16	.08111	.23	.13646	.3	.19817	.37	.26418
.021	.00403	.091	.03559	.161	.08185	.231	.13731	.301	.19908	.371	.26514
.022	.00432	.092	.03616	.162	.08258	.232	.13815	.302	.20000	.372	.26611
.023	.00462	.093	.03674	.163	.08332	.233	.13900	.303	.20092	.373	.26708
.024	.00492	.094	.03732	.164	.08406	.234	.13984	.304	.20184	.374	.26805
.025	.00523	.095	.03791	.165	.08480	.235	.14069	.305	.20276	.375	.26901
.026	.00555	.096	.03850	.166	.08554	.236	.14154	.306	.20368	.376	.26998
.027	.00587	.097	.03909	.167	.08629	.237	.14239	.307	.20460	.377	.27095
.028	.00619	.098	.03968	.168	.08704	.238	.14324	.308	.20553	.378	.27192
.029	.00653	.099	.04028	.169	.08779	.239	.14409	.309	.20645	.379	.27289
.03	.00687	.1	.04087	.17	.08854	.24	.14494	.31	.20738	.38	.27386
.031	.00721	.101	.04148	.171	.08929	.241	.14580	.311	.20830	.381	.27483
.032	.00756	.102	.04208	.172	.09004	.242	.14666	.312	.20923	.382	.27580
.033	.00791	.103	.04269	.173	.09080	.243	.14751	.313	.21015	.383	.27678
.034	.00827	.104	.04330	.174	.09155	.244	.14837	.314	.21108	.384	.27775
.035	.00864	.105	.04391	.175	.09231	.245	.14923	.315	.21201	.385	.27872
.036	.00901	.106	.04452	.176	.09307	.246	.15009	.316	.21294	.386	.27969
.037	.00938	.107	.04514	.177	.09384	.247	.15095	.317	.21387	.387	.28067
.038	.00976	.108	.04576	.178	.09460	.248	.15182	.318	.21480	.388	.28164
.039	.01015	.109	.04638	.179	.09537	.249	.15268	.319	.21573	.389	.28262
.04	.01054	.11	.04701	.18	.09613	.25	.15355	.32	.21667	.39	.28359
.041	.01093	.111	.04763	.181	.09690	.251	.15441	.321	.21760	.391	.28457
.042	.01133	.112	.04826	.182	.09767	.252	.15528	.322	.21853	.392	.28554
.043	.01173	.113	.04889	.183	.09845	.253	.15615	.323	.21947	.393	.28652
.044	.01214	.114	.04953	.184	.09922	.254	.15702	.324	.22040	.394	.28750
.045	.01255	.115	.05016	.185	.10000	.255	.15789	.325	.22134	.395	.28848
.046	.01297	.116	.05080	.186	.10077	.256	.15876	.326	.22228	.396	.28945
.047	.01339	.117	.05145	.187	.10155	.257	.15964	.327	.22322	.397	.29043
.048	.01382	.118	.05209	.188	.10233	.258	.16051	.328	.22415	.398	.29141
.049	.01425	.119	.05274	.189	.10312	.259	.16139	.329	.22509	.399	.29239
.05	.01468	.12	.05338	.19	.10390	.26	.16226	.33	.22603	.4	.29337
.051	.01512	.121	.05404	.191	.10469	.261	.16314	.331	.22697	.401	.29435
.052	.01556	.122	.05469	.192	.10547	.262	.16402	.332	.22792	.402	.29533
.053	.01601	.123	.05535	.193	.10626	.263	.16490	.333	.22886	.403	.29631
.054	.01646	.124	.05600	.194	.10705	.264	.16578	.334	.22980	.404	.29729
.055	.01691	.125	.05666	.195	.10784	.265	.16666	.335	.23074	.405	.29827
.056	.01737	.126	.05733	.196	.10864	.266	.16755	.336	.23169	.406	.29926
.057	.01783	.127	.05799	.197	.10943	.267	.16843	.337	.23263	.407	.30024
.058	.01830	.128	.05866	.198	.11023	.268	.16932	.338	.23358	.408	.30122
.059	.01877	.129	.05933	.199	.11102	.269	.17020	.339	.23453	.409	.30220
.06	.01924	.13	.06000	.2	.11182	.27	.17109	.34	.23547	.41	.30319
.061	.01972	.131	.06067	.201	.11262	.271	.17198	.341	.23642	.411	.30417
.062	.02020	.132	.06135	.202	.11343	.272	.17287	.342	.23737	.412	.30516
.063	.02068	.133	.06203	.203	.11423	.273	.17376	.343	.23832	.413	.30614
.064	.02117	.134	.06271	.204	.11504	.274	.17465	.344	.23927	.414	.30712
.065	.02166	.135	.06339	.205	.11584	.275	.17554	.345	.24022	.415	.30811
.066	.02215	.136	.06407	.206	.11665	.276	.17644	.346	.24117	.416	.30910
.067	.02265	.137	.06476	.207	.11746	.277	.17733	.347	.24212	.417	.31008
.068	.02315	.138	.06545	.208	.11827	.278	.17823	.348	.24307	.418	.31107
.069	.02366	.139	.06614	.209	.11908	.279	.17912	.349	.24403	.419	.31205

Table 9. Areas of Segments of a Circle—*Continued*

H/D	*F*	*H/D*	*F*	*H/D*	*F*	*H/D*	*F*	*H/D*	*F*	*H/D*	*F*
0.42	0.31304	0.435	0.32788	0.45	0.34278	0.465	0.35773	0.48	0.37270	0.495	0.38770
.421	.31403	.436	.32887	.451	.34378	.466	.35873	.481	.37370	.496	.38870
.422	.31502	.437	.32987	.452	.34477	.467	.35972	.482	.37470	.497	.38970
.423	.31600	.438	.33086	.453	.34577	.468	.36072	.483	.37570	.498	.39070
.424	.31699	.439	.33185	.454	.34676	.469	.36172	.484	.37670	.499	.39170
.425	.31798	.44	.33284	.455	.34776	.47	.36272	.485	.37770	.500	.39270
.426	.31897	.441	.33384	.456	.34876	.471	.36372	.486	.37870		
.427	.31996	.442	.33483	.457	.34975	.472	.36471	.487	.37970		
.428	.32095	.443	.33582	.458	.35075	.473	.36571	.488	.38070		
.429	.32194	.444	.33682	.459	.35175	.474	.36671	.489	.38170		
.43	.32293	.445	.33781	.46	.35274	.475	.36771	.49	.38270		
.431	.32392	.446	.33880	.461	.35374	.476	.36871	.491	.38370		
.432	.32491	.447	.33980	.462	.35474	.477	.36971	.492	.38470		
.433	.32590	.448	.34079	.463	.35573	.478	.37071	.493	.38570		
.434	.32689	.449	.34179	.464	.35673	.479	.37171	.494	.38670		

Table 10. Values of Degrees and Minutes in Radians

Deg	Radians	Deg	Radians	Deg	Radians	Deg	Radians	Deg	Radians	Deg	Radians	Min	Radians	Min	Radians
1	0.01745	31	0.54105	61	1.06465	91	1.58825	121	2.11185	151	2.63545	1	0.00029	31	0.00902
2	.03491	32	.55851	62	1.08210	92	1.60570	122	2.12930	152	2.65290	2	.00058	32	.00931
3	.05236	33	.57596	63	1.09956	93	1.62316	123	2.14676	153	2.67035	3	.00087	33	.00960
4	.06981	34	.59341	64	1.11701	94	1.64061	124	2.16421	154	2.68781	4	.00116	34	.00989
5	.08727	35	.61087	65	1.13446	95	1.65806	125	2.18166	155	2.70526	5	.00145	35	.01018
6	.10472	36	.62832	66	1.15192	96	1.67552	126	2.19912	156	2.72271	6	.00175	36	.01047
7	.12217	37	.64577	67	1.16937	97	1.69297	127	2.21657	157	2.74017	7	.00204	37	.01076
8	.13963	38	.66323	68	1.18682	98	1.71042	128	2.23402	158	2.75762	8	.00233	38	.01105
9	.15708	39	.68068	69	1.20428	99	1.72788	129	2.25148	159	2.77507	9	.00262	39	.01134
10	.17453	40	.69813	70	1.22173	100	1.74533	130	2.26893	160	2.79253	10	.00291	40	.01164
11	.19199	41	.71559	71	1.23918	101	1.76278	131	2.28638	161	2.80998	11	.00320	41	.01193
12	.20944	42	.73304	72	1.25664	102	1.78024	132	2.30384	162	2.82743	12	.00349	42	.01222
13	.22689	43	.75049	73	1.27409	103	1.79769	133	2.32129	163	2.84489	13	.00378	43	.01251
14	.24435	44	.76795	74	1.29154	104	1.81514	134	2.33874	164	2.86234	14	.00407	44	.01280
15	.26180	45	.78540	75	1.30900	105	1.83260	135	2.35620	165	2.87979	15	.00436	45	.01309
16	.27925	46	.80285	76	1.32645	106	1.85005	136	2.37365	166	2.89725	16	.00465	46	.01338
17	.29671	47	.82031	77	1.34390	107	1.86750	137	2.39110	167	2.91470	17	.00495	47	.01367
18	.31416	48	.83776	78	1.36136	108	1.88496	138	2.40856	168	2.93215	18	.00524	48	.01396
19	.33161	49	.85521	79	1.37881	109	1.90241	139	2.42601	169	2.94961	19	.00553	49	.01425
20	.34907	50	.87267	80	1.39626	110	1.91986	140	2.44346	170	2.96706	20	.00582	50	.01454
21	.36652	51	.89012	81	1.41372	111	1.93732	141	2.46092	171	2.98451	21	.00611	51	.01484
22	.38397	52	.90757	82	1.43117	112	1.95477	142	2.47837	172	3.00197	22	.00640	52	.01513
23	.40143	53	.92502	83	1.44862	113	1.97222	143	2.49582	173	3.01942	23	.00669	53	.01542
24	.41888	54	.94248	84	1.46608	114	1.98968	144	2.51328	174	3.03687	24	.00698	54	.01571
25	.43633	55	.95993	85	1.48353	115	2.00713	145	2.53073	175	3.05433	25	.00727	55	.01600
26	.45379	56	.97738	86	1.50098	116	2.02458	146	2.54818	176	3.07178	26	.00756	56	.01629
27	.47124	57	.99484	87	1.51844	117	2.04204	147	2.56564	177	3.08923	27	.00785	57	.01658
28	.48869	58	1.01299	88	1.53589	118	2.05949	148	2.58309	178	3.10669	28	.00814	58	.01687
29	.50615	59	1.02974	89	1.55334	119	2.07694	149	2.60054	179	3.12414	29	.00844	59	.01716
30	.52360	60	1.04720	90	1.57080	120	2.09440	150	2.61800	180	3.14159	30	.00873	60	.01745

Table 11. Values of Radians in Degrees

Rad.	.00	.01	.02	.03	.04	.05	.06	.07	.08	.09
	Deg.	Deg.	Deg.	Deg.	Deg.	Deg.	Deg.	Deg.	Deg.	Deg.
0.0	0.0000	0.5730	1.1459	1.7189	2.2918	2.8648	3.4377	4.0107	4.5837	5.1566
.1	5.7296	6.3025	6.8755	7.4485	8.0214	8.5944	9.1673	9.7403	10.3132	10.8862
.2	11.4591	12.0321	12.6051	13.1780	13.7510	14.3239	14.8969	15.4699	16.0428	16.6158
.3	17.1887	17.7617	18.3346	18.9076	19.4806	20.0535	20.6265	21.1994	21.7724	22.3454
.4	22.9183	23.4913	24.0642	24.6372	25.2101	25.7831	26.3561	26.9290	27.5020	28.0749
.5	28.6479	29.2208	29.7938	30.3668	30.9397	31.1527	32.0856	32.6586	33.2316	33.8045
.6	34.3775	34.9504	35.5234	36.0963	36.6693	37.2423	37.8152	38.3882	38.9611	39.5341
.7	40.1070	40.6800	41.2530	41.8259	42.3989	42.9718	43.5448	44.1178	44.6907	45.2637
.8	45.8366	46.4096	46.9825	47.5555	48.1285	48.7014	49.2744	49.8473	50.4203	50.9932
.9	51.5662	52.1392	52.7121	53.2851	53.8580	54.4310	55.0039	55.5769	56.1499	56.7228

1 Radian = 57.29578 deg. | 2 Radians = 114.59156 deg. | 3 Radians = 171.88734 deg.

Table 12. Decimals of a Degree in Minutes and Seconds

Decimal	.00		.01		.02		.03		.04		.05		.06		.07		.08		.09	
	Min.	Sec.	Min.	Sec.	Min.	Sec.	Min.	Sec.	Min.	Sec.	Min.	Sec.	Min.	Sec.	Min.	Sec.	Min.	Sec.	Min.	Sec.
0.0	0	0	0	36	1	12	1	48	2	24	3	0	3	36	4	12	4	48	5	24
.1	6	0	6	36	7	12	7	48	8	24	9	0	9	36	10	12	10	48	11	24
.2	12	0	12	36	13	12	13	48	14	24	15	0	15	36	16	12	16	48	17	24
.3	18	0	18	36	19	12	19	48	20	24	21	0	21	36	22	12	22	48	23	24
.4	24	0	24	36	25	12	25	48	26	24	27	0	27	36	28	12	28	48	29	24
.5	30	0	30	36	31	12	31	48	32	24	33	0	33	36	34	12	24	48	35	24
.6	36	0	36	36	37	12	37	48	38	24	39	0	39	36	40	12	40	48	41	24
.7	42	0	42	36	43	12	43	48	44	24	45	0	45	36	46	12	46	48	47	24
.8	48	0	48	36	49	12	49	48	50	24	51	0	51	36	52	12	52	48	53	24
.9	54	0	54	36	55	12	55	48	56	24	57	0	57	36	58	12	58	48	59	24

Table 13. Circumferences and Areas of Circles—Diameters in Feet and Inches *

Circumferences in feet; areas in square feet

Circumference

Diam., feet	0 in.	1	2	3	4	5	6	7	8	9	10	11
0	0.0000	0.2618	0.5236	0.7854	1.0472	1.3090	1.5708	1.8326	2.0944	2.3562	2.6180	2.8798
1	3.1416	3.4034	3.6653	3.9270	4.1877	4.4507	4.7124	4.9741	5.2361	5.4978	5.7595	6.0215
2	6.2832	6.5450	6.8069	7.0686	7.3293	7.5923	7.8540	8.1157	8.3777	8.6394	8.9011	9.1631
3	9.4248	9.6866	9.9485	10.210	10.471	10.734	10.996	11.257	11.519	11.781	12.043	12.305
4	12.566	12.828	13.090	13.352	13.613	13.875	14.137	14.399	14.661	14.923	15.184	15.446
5	15.708	15.970	16.232	16.493	16.754	17.017	17.279	17.540	17.802	18.064	18.326	18.588
6	18.850	19.111	19.373	19.635	19.896	20.159	20.420	20.682	20.944	21.206	21.467	21.729
7	21.991	22.253	22.515	22.777	23.037	23.300	23.562	23.824	24.086	24.347	24.609	24.871
8	25.133	25.395	25.656	25.918	26.179	26.442	26.704	26.965	27.227	27.489	27.751	28.013
9	28.274	28.536	28.798	29.060	29.320	29.583	29.845	30.107	30.369	30.631	30.892	31.154
10	31.416	31.678	31.940	32.201	32.462	32.725	32.987	33.248	33.510	33.772	34.034	34.296
11	34.557	34.819	35.081	35.343	35.604	35.867	36.128	36.390	36.652	36.914	37.175	37.437
12	37.699	37.961	38.223	38.484	38.745	39.008	39.270	39.532	39.794	40.055	40.317	40.579
13	40.841	41.102	41.364	41.626	41.887	42.150	42.411	42.673	42.935	43.197	43.459	43.721
14	43.982	44.244	44.506	44.768	45.028	45.291	45.553	45.815	46.077	46.338	46.600	46.862
15	47.124	47.386	47.648	47.909	48.170	48.433	48.695	48.956	49.218	49.480	49.742	50.004
16	50.265	50.527	50.789	51.051	51.312	51.575	51.836	52.098	52.360	52.622	52.883	53.145
17	53.407	53.669	53.931	54.192	54.453	54.716	54.978	55.240	55.502	55.763	56.025	56.287
18	56.549	56.810	57.072	57.334	57.595	57.858	58.119	58.381	58.643	58.905	59.167	59.429
19	59.690	59.952	60.214	60.476	60.736	61.000	61.261	61.523	61.785	62.046	62.308	62.570
20	62.832	63.094	63.356	63.617	63.878	64.141	64.403	64.664	64.926	65.188	65.450	65.712
21	65.973	66.235	66.497	66.759	67.020	67.282	67.544	67.806	68.068	68.330	68.591	68.853
22	69.115	69.377	69.639	69.900	70.161	70.424	70.686	70.947	71.209	71.471	71.733	71.995
23	72.257	72.518	72.780	73.042	73.303	73.566	73.827	74.089	74.351	74.513	74.874	75.136
24	75.398	75.660	75.922	76.184	76.444	76.707	76.969	77.231	77.493	77.754	78.016	78.278
25	78.540	78.802	79.063	79.325	79.586	79.849	80.111	80.372	80.634	80.896	81.158	81.420
26	81.681	81.943	82.205	82.467	82.727	82.990	83.252	83.514	83.776	84.038	84.299	84.561
27	84.823	85.085	85.347	85.608	85.869	86.132	86.394	86.655	86.917	87.179	87.441	87.703
28	87.965	88.226	88.488	88.749	89.011	89.274	89.535	89.797	90.059	90.321	90.582	90.844
29	91.106	91.367	91.630	91.892	92.152	92.415	92.677	92.939	93.201	93.462	93.724	93.986
30	94.248	94.509	94.771	95.033	95.294	95.557	95.818	96.080	96.342	96.604	96.866	97.128

Area

Diam., feet	0 in.	1	2	3	4	5	6	7	8	9	10	11
0	0.0000	0.0055	0.0218	0.0491	0.0873	0.1364	0.1963	0.2673	0.3491	0.4418	0.5454	0.6600
1	0.7854	0.9218	1.069	1.227	1.396	1.576	1.767	1.969	2.182	2.405	2.640	2.885
2	3.142	3.409	3.687	3.976	4.276	4.587	4.909	5.241	5.585	5.940	6.305	6.681
3	7.069	7.467	7.876	8.296	8.727	9.168	9.621	10.08	10.56	11.04	11.54	12.05
4	12.57	13.10	13.64	14.19	14.75	15.32	15.90	16.50	17.10	17.72	18.35	18.99
5	19.63	20.29	20.97	21.65	22.34	23.04	23.76	24.48	25.22	25.97	26.73	27.49
6	28.27	29.07	29.87	30.68	31.50	32.34	33.18	34.04	34.91	35.78	36.67	37.57
7	38.48	39.41	40.34	41.28	42.24	43.20	44.18	45.17	46.16	47.17	48.19	49.22
8	50.27	51.32	52.38	53.46	54.54	55.64	56.75	57.86	58.99	60.13	61.28	62.44
9	63.62	64.80	66.00	67.20	68.42	69.64	70.88	72.13	73.39	74.66	75.94	77.24
10	78.54	79.85	81.18	82.52	83.86	85.22	86.59	87.97	89.36	90.76	92.18	93.60
11	95.03	96.48	97.93	99.40	100.9	102.4	103.9	105.4	106.9	108.4	110. 0	111.5
12	113.1	114.7	116.3	117.9	119.5	121.1	122.7	124.4	126.0	127.7	129.4	131.0
13	132.7	134.4	136.2	137.9	139.6	141.4	143.1	144.9	146.7	148.5	150.3	152.1
14	153.9	155.8	157.6	159.5	161.4	163.2	165.1	167.0	168.9	170.9	172.8	174.8
15	176.7	178.7	180.7	182.7	184.7	186.7	188.7	190.7	192.8	194.8	196.9	199.0
16	201.1	203.2	205.3	207.4	209.5	211.7	213.8	216.0	218.2	220.4	222.5	224.8
17	227.0	229.2	231.4	233.7	236.0	238.2	240.5	242.8	245.1	247.5	249.8	252.1
18	254.5	256.8	259.2	261.6	264.0	266.4	268.8	271.2	273.7	276.1	278.6	281.0
19	283.5	286.0	288.5	291.0	293.6	296.1	298.7	301.2	303.8	306.4	308.9	311.5
20	314.2	316.8	319.4	322.1	324.7	327.4	330.1	332.7	335.4	338.2	340.9	343.6
21	346.4	349.1	351.9	354.7	357.4	360.2	363.1	365.9	368.7	371.5	374.4	377.2
22	380.1	383.0	386.0	388.8	391.7	394.7	397.6	400.5	403.5	406.5	409.5	412.5
23	415.5	418.5	421.5	424.6	427.6	430.7	433.7	436.8	439.9	443.0	446.1	449.2
24	452.4	455.5	458.7	461.9	465.0	468.2	471.4	474.6	477.9	481.1	484.3	487.6
25	490.9	494.1	497.4	500.7	504.0	507.4	510.7	514.0	517.4	520.8	524.1	527.5
26	530.9	534.3	537.7	541.2	544.6	548.1	551.6	555.0	558.5	562.0	565.5	569.0
27	572.6	576.1	579.6	583.2	586.8	590.3	594.0	597.5	601.2	604.8	608.4	612.1
28	615.8	619.4	623.1	626.8	630.5	634.2	637.9	641.7	645.4	649.2	652.9	656.7
29	660.5	664.3	668.1	672.0	675.8	679.6	683.5	687.3	691.2	695.1	699.0	702.9
30	706.9	710.8	714.7	718.7	722.6	726.6	730.6	734.6	738.6	742.6	746.7	750.7

* For circumferences and areas of circles, diameters in inches, see Table 7, pp. 28–26 to 28–42.

The following tables give the values of sin x, cos x, and tan x for values of x from 0 to 90° in intervals of 0.1 degree. By making use of the periodic character of these functions, the values can be determined from these tables for all values of x to an accuracy of 0.1 degree. The following relations facilitate this determination for any quadrant:

$$\sin x = +\sin(180° - x) = -\sin(180° + x) = -\sin(360° - x)$$

$$\cos x = -\cos(180° - x) = -\cos(180° + x) = +\cos(360° - x)$$

$$\tan x = -\tan(180° - x) = +\tan(180° + x) = -\tan(360° - x)$$

$$\text{Angle in degrees} = \text{angle in radians} \times 180/\pi$$

Table 14. Trigonometric Functions **0.0°–15.9°**

Angle in Degrees	Name of Function	Value of Function for Each Tenth of a Degree									
		0.0	0.1	0.2	0.3	0.4	0.5	0.6	0.7	0.8	0.9
	sin	0.0000	0.0017	0.0035	0.0052	0.0070	0.0087	0.0105	0.0122	0.0140	0.0157
0	cos	1.0000	1.0000	1.0000	1.0000	1.0000	1.0000	0.9999	0.9999	0.9999	0.9999
	tan	0.0000	0.0017	0.0035	0.0052	0.0070	0.0087	0.0105	0.0122	0.0140	0.0157
	sin	0.0175	0.0192	0.0209	0.0227	0.0244	0.0262	0.0279	0.0297	0.0314	0.0332
1	cos	0.9998	0.9998	0.9998	0.9997	0.9997	0.9997	0.9996	0.9996	0.9995	0.9995
	tan	0.0175	0.0192	0 0209	0.0227	0.0244	0.0262	0.0279	0.0297	0.0314	0.0332
	sin	0.0349	0.0366	0.0384	0.0401	0.0419	0.0436	0.0454	0.0471	0.0488	0.0506
2	cos	0.9994	0.9993	0.9993	0.9992	0.9991	0.9990	0.9990	0.9989	0.9988	0.9987
	tan	0.0349	0.0367	0.0384	0.0402	0.0419	0.0437	0.0454	0.0472	0.0489	0.0507
	sin	0.0523	0.0541	0.0558	0.0576	0.0593	0.0610	0.0628	0.0645	0.0663	0.0680
3	cos	0.9986	0.9985	0.9984	0.9983	0.9982	0.9981	0.9980	0.9979	0.9978	0.9977
	tan	0.0524	0.0542	0.0559	0.0577	0.0594	0.0612	0.0629	0.0647	0.0664	0.0682
	sin	0.0698	0.0715	0.0732	0.0750	0.0767	0.0785	0.0802	0.0819	0.0837	0.0854
4	cos	0.9976	0.9974	0.9973	0.9972	0.9971	0.9969	0.9968	0.9966	0.9965	0.9963
	tan	0.0699	0.0717	0.0734	0.0752	0.0769	0.0787	0.0805	0.0822	0.0840	0.0857
	sin	0.0872	0.0889	0.0906	0.0924	0.0941	0.0958	0.0976	0.0993	0.1011	0.1028
5	cos	0.9962	0.9960	0.9959	0.9957	0.9956	0.9954	0.9952	0.9951	0.9949	0.9947
	tan	0.0875	0.0892	0.0910	0.0928	0.0945	0.0963	0.0981	0.0998	0.1016	0.1033
	sin	0.1045	0.1063	0.1080	0.1097	0.1115	0.1132	0.1149	0.1167	0.1184	0.1201
6	cos	0.9945	0.9943	0.9942	0.9940	0.9938	0.9936	0.9934	0.9932	0.9930	0.9928
	tan	0.1051	0.1069	0.1086	0.1104	0.1122	0.1139	0.1157	0.1175	0.1192	0.1210
	sin	0.1219	0.1236	0.1253	0.1271	0.1288	0.1305	0.1323	0.1340	0.1357	0.1374
7	cos	0.9925	0.9923	0.9921	0.9919	0.9917	0.9914	0.9912	0.9910	0.9907	0.9905
	tan	0.1228	0.1246	0.1263	0.1281	9.1299	0.1317	0.1334	0.1352	0.1370	0.1388
	sin	0.1392	0.1409	0.1426	0.1444	0.1461	0.1478	0.1495	0.1513	0.1530	0.1547
8	cos	0.9903	0.9900	0.9898	0.9895	0.9893	0.9890	0.9888	0.9885	0.9882	0.9880
	tan	0.1405	0.1423	0.1441	0.1459	0.1477	0.1495	0.1512	0.1530	0.1548	0.1566
	sin	0.1564	0.1582	0.1599	0.1616	0.1633	0.1650	0.1663	0.1685	0.1702	0.1719
9	cos	0.9877	0.9874	0.9871	0.9869	0.9866	0.9863	0.9860	0.9857	0.9854	0.9851
	tan	0.1584	0.1602	0.1620	0.1638	0.1655	0.1673	0.1691	0.1709	0.1727	0.1745
	sin	0.1736	0.1754	0.1771	0.1788	0.1805	0.1822	0.1840	0.1857	0.1874	0.1891
10	cos	0.9848	0.9845	0.9842	0.9839	0.9836	0.9833	0.9829	0.9826	0.9823	0.9820
	tan	0.1763	0.1781	0.1799	0.1817	0.1835	0.1853	0.1871	0.1890	0.1908	0.1926
	sin	0.1908	0.1925	0.1942	0.1959	0.1977	0.1994	0.2011	0.2028	0.2045	0.2062
11	cos	0.9816	0.9813	0.9810	0.9806	0.9803	0.9799	0.9796	0.9792	0.9789	0.9785
	tan	0.1944	0.1962	0.1980	0.1998	0.2016	0.2035	0.2053	0.2071	0.2089	0.2107
	sin	0.2079	0.2096	0.2113	0.2130	0.2147	0.2164	0.2181	0.2198	0.2215	0.2232
12	cos	0.9781	0.9778	0.9774	0.9770	0.9767	0.9763	0.9759	0.9755	0.9751	0.9748
	tan	0.2126	0.2144	0.2162	0.2180	0.2199	0.2217	0.2235	0.2254	0.2272	0.2290
	sin	0.2250	0.2267	0.2284	0.2300	0.2317	0.2334	0.2351	0.2368	0.2385	0.2402
13	cos	0.9744	0.9740	0.9736	0.9732	0.9728	0.9724	0.9720	0.9715	0.9711	0.9707
	tan	0.2309	0.2327	0.2345	0.2364	0.2382	0.2401	0.2419	0.2438	0.2456	0.2475
	sin	0.2419	0.2436	0.2453	0.2470	0.2487	0.2504	0.2521	0.2538	0.2554	0.2571
14	cos	0.9703	0.9699	0.9694	0.9690	0.9686	0.9681	0.9677	0.9673	0.9668	0.9664
	tan	0.2493	0.2512	0.2530	0.2549	0.2568	0.2586	0.2605	0.2623	0.2642	0.2661
	sin	0.2588	0.2605	0.2622	0.2639	0.2656	0.2672	0.2689	0.2706	0.2723	0.2740
15	cos	0.9659	0.9655	0.9650	0.9646	0.9641	0.9636	0.9632	0.9627	0.9622	0.9617
	tan	0.2679	0.2698	0.2717	0.2736	0.2754	0.2773	0.2792	0.2811	0.2830	0.2849

Table 14. Trigonometric Functions—*Continued* **16.0°–35.9°**

Angle in Degrees	Name of Function	0.0	0.1	0.2	0.3	0.4	0.5	0.6	0.7	0.8	0.9
	sin	0.2756	0.2773	0.2790	0.2807	0.2823	0.2840	0.2857	0.2874	0.2890	0.2907
16	cos	0.9613	0.9608	0.9603	0.9598	0.9593	0.9588	0.9583	0.9578	0.9573	0.9568
	tan	0.2867	0.2886	0.2905	0.2924	0.2943	0.2962	0.2981	0.3000	0.3019	0.3038
	sin	0.2924	0.2940	0.2957	0.2974	0.2990	0.3007	0.3024	0.3040	0.3057	0.3074
17	cos	0.9563	0.9558	0.9553	0.9548	0.9542	0.9537	0.9532	0.9527	0.9521	0.9516
	tan	0.3057	0.3076	0.3096	0.3115	0.3134	0.3153	0.3172	0.3191	0.3211	0.3230
	sin	0.3090	0.3107	0.3123	0.3140	0.3156	0.3173	0.3190	0.3206	0.3223	0.3239
18	cos	0.9511	0.9505	0.9500	0.9494	0.9489	0.9483	0.9478	0.9472	0.9466	0.9461
	tan	0.3249	0.3269	0.3288	0.3307	0.3327	0.3346	0.3365	0.3385	0.3404	0.3424
	sin	0.3256	0.3272	0.3289	0.3305	0.3322	0.3338	0.3355	0.3371	0.3387	0.3404
19	cos	0.9455	0.9449	0.9444	0.9438	0.9432	0.9426	0.9421	0.9415	0.9409	0.9403
	tan	0.3443	0.3463	0.3482	0.3502	0.3522	0.3541	0.3561	0.3581	0.3600	0.3620
	sin	0.3420	0.3437	0.3453	0.3469	0.3486	0.3502	0.3518	0.3535	0.3551	0.3567
20	cos	0.9397	0.9391	0.9385	0.9379	0.9373	0.9367	0.9361	0.9354	0.9348	0.9342
	tan	0.3640	0.3659	0.3679	0.3699	0.3719	0.3739	0.3759	0.3779	0.3799	0.3819
	sin	0.3584	0.3600	0.3616	0.3633	0.3649	0.3665	0.3681	0.3697	0.3714	0.3730
21	cos	0.9336	0.9330	0.9323	0.9317	0.9311	0.9304	0.9298	0.9291	0.9285	0.9278
	tan	0.3839	0.3859	0.3879	0.3899	0.3919	0.3939	0.3959	0.3979	0.4000	0.4020
	sin	0.3746	0.3762	0.3778	0.3795	0.3811	0.3827	0.3843	0.3859	0.3875	0.3891
22	cos	0.9272	0.9265	0.9259	0.9252	0.9245	0.9239	0.9232	0.9225	0.9219	0.9212
	tan	0.4040	0.4061	0.4081	0.4101	0.4122	0.4142	0.4163	0.4183	0.4204	0.4224
	sin	0.3907	0.3923	0.3939	0.3955	0.3971	0.3987	0.4003	0.4019	0.4035	0.4051
23	cos	0.9205	0.9198	0.9191	0.9184	0.9178	0.9171	0.9164	0.9157	0.9150	0.9143
	tan	0.4245	0.4265	0.4286	0.4307	0.4327	0.4348	0.4369	0.4390	0.4411	0.4431
	sin	0.4067	0.4083	0.4099	0.4115	0.4131	0.4147	0.4163	0.4179	0.4195	0.4210
24	cos	0.9135	0.9128	0.9121	0.9114	0.9107	0.9100	0.9092	0.9085	0.9078	0.9070
	tan	0.4452	0.4473	0.4494	0.4515	0.4536	0.4557	0.4578	0.4599	0.4621	0.4642
	sin	0.4226	0.4242	0.4258	0.4274	0.4289	0.4305	0.4321	0.4337	0.4352	0.4368
25	cos	0.9063	0.9056	0.9048	0.9041	0.9033	0.9026	0.9018	0.9011	0.9003	0.8996
	tan	0.4663	0.4684	0.4706	0.4727	0.4748	0.4770	0.4791	0.4813	0.4834	0.4856
	sin	0.4384	0.4399	0.4415	0.4431	0.4446	0.4462	0.4478	0.4493	0.4509	0.4524
26	cos	0.8988	0.8980	0.8973	0.8965	0.8957	0.8949	0.8942	0.8934	0.8926	0.8918
	tan	0.4877	0.4899	0.4921	0.4942	0.4964	0.4986	0.5008	0.5029	0.5051	0.5073
	sin	0.4540	0.4555	0.4571	0.4586	0.4602	0.4617	0.4633	0.4648	0.4664	0.4679
27	cos	0.8910	0.8902	0.8894	0.8886	0.8878	0.8870	0.8862	0.8854	0.8846	0.8838
	tan	0.5095	0.5117	0.5139	0.5161	0.5184	0.5206	0.5228	0.5250	0.5272	0.5295
	sin	0.4695	0.4710	0.4726	0.4741	0.4756	0.4772	0.4787	0.4802	0.4818	0.4833
28	cos	0.8829	0.8821	0.8813	0.8805	0.8796	0.8788	0.8780	0.8771	0.8763	0.8755
	tan	0.5317	0.5340	0.5362	0.5384	0.5407	0.5430	0.5452	0.5475	0.5498	0.5520
	sin	0.4848	0.4863	0.4879	0.4894	0.4909	0.4924	0.4939	0.4955	0.4970	0.4985
29	cos	0.8746	0.8738	0.8729	0.8721	0.8712	0.8704	0.8695	0.8686	0.8678	0.8669
	tan	0.5543	0.5566	0.5589	0.5612	0.5635	0.5658	0.5681	0.5704	0.5727	0.5750
	sin	0.5000	0.5015	0.5030	0.5045	0.5060	0.5075	0.5090	0.5105	0.5120	0.5135
30	cos	0.8660	0.8652	0.8643	0.8634	0.8625	0.8616	0.8607	0.8599	0.8590	0.8581
	tan	0.5774	0.5797	0.5820	0.5844	0.5867	0.5890	0.5914	0.5938	0.5961	0.5985
	sin	0.5150	0.5165	0.5180	0.5195	0.5210	0.5225	0.5240	0.5255	0.5270	0.5284
31	cos	0.8572	0.8563	0.8554	0.8545	0.8536	0.8526	0.8517	0.8508	0.8499	0.8490
	tan	0.6009	0.6032	0.6056	0.6080	0.6104	0.6128	0.6152	0.6176	0.6200	0.6224
	sin	0.5299	0.5314	0.5329	0.5344	0.5358	0.5373	0.5388	0.5402	0.5417	0.5432
32	cos	0.8480	0.8471	0.8462	0.8453	0.8443	0.8434	0.8425	0.8415	0.8406	0.8396
	tan	0.6249	0.6273	0.6297	0.6322	0.6346	0.6371	0.6395	0.6420	0.6445	0.6469
	sin	0.5446	0.5461	0.5476	0.5490	0.5505	0.5519	0.5534	0.5548	0.5563	0.5577
33	cos	0.8387	0.8377	0.8368	0.8358	0.8348	0.8339	0.8329	0.8320	0.8310	0.8300
	tan	0.6494	0.6519	0.6544	0.6569	0.6594	0.6619	0.6644	0.6669	0.6694	0.6720
	sin	0.5592	0.5606	0.5621	0.5635	0.5650	0.5664	0.5678	0.5693	0.5707	0.5721
34	cos	0.8290	0.8281	0.8271	0.8261	0.8251	0.8241	0.8231	0.8221	0.8211	0.8202
	tan	0.6745	0.6771	0.6796	0.6822	0.6847	0.6873	0.6899	0.6924	0.6950	0.6976
	sin	0.5736	0.5750	0.5764	0.5779	0.5793	0.5807	0.5821	0.5835	0.5850	0.5864
35	cos	0.8192	0.8181	0.8171	0.8161	0.8151	0.8141	0.8131	0.8121	0.8111	0.8100
	tan	0.7002	0.7028	0.7054	0.7080	0.7107	0.7133	0.7159	0.7186	0.7212	0.7239

Table 14. Trigonometric Functions—*Continued* **36.0°–55.9°**

Angle in Degrees	Name of Function	Value of Function for Each Tenth of a Degree									
		0.0	**0.1**	**0.2**	**0.3**	**0.4**	**0.5**	**0.6**	**0.7**	**0.8**	**0.9**
	sin	0.5878	0.5892	0.5906	0.5920	0.5934	0.5948	0.5962	0.5976	0.5990	0.6004
36	cos	0.8090	0.8080	0.8070	0.8059	0.8049	0.8039	0.8028	0.8018	0.8007	0.7997
	tan	0.7265	0.7292	0.7319	0.7346	0.7373	0.7400	0.7427	0.7454	0.7481	0.7508
	sin	0.6018	0.6032	0.6046	0.6060	0.6074	0.6088	0.6101	0.6115	0.6129	0.6143
37	cos	0.7986	0.7976	0.7965	0.7955	0.7944	0.7934	0.7923	0.7912	0.7902	0.7891
	tan	0.7536	0.7563	0.7590	0.7618	0.7646	0.7673	0.7701	0.7729	0.7757	0.7785
	sin	0.6157	0.6170	0.6184	0.6198	0.6211	0.6225	0.6239	0.6252	0.6266	0.6280
38	cos	0.7880	0.7869	0.7859	0.7848	0.7837	0.7826	0.7815	0.7804	0.7793	0.7782
	tan	0.7813	0.7841	0.7869	0.7898	0.7926	0.7954	0.7983	0.8012	0.8040	0.8069
	sin	0.6293	0.6307	0.6320	0.6334	0.6347	0.6361	0.6374	0.6388	0.6401	0.6414
39	cos	0.7771	0.7760	0.7749	0.7738	0.7727	0.7716	0.7705	0.7694	0.7683	0.7672
	tan	0.8098	0.8127	0.8156	0.8185	0.8214	0.8243	0.8273	0.8302	0.8332	0.8361
	sin	0.6428	0.6441	0.6455	0.6468	0.6481	0.6494	0.6508	0.6521	0.6534	0.6547
40	cos	0.7660	0.7649	0.7638	0.7627	0.7615	0.7604	0.7593	0.7581	0.7570	0.7559
	tan	0.8391	0.8421	0.8451	0.8481	0.8511	0.8541	0.8571	0.8601	0.8632	0.8662
	sin	0.6561	0.6574	0.6587	0.6600	0.6613	0.6626	0.6639	0.6653	0.6665	0.6678
41	cos	0.7547	0.7536	0.7524	0.7513	0.7501	0.7490	0.7478	0.7466	0.7455	0.7443
	tan	0.8693	0.8724	0.8754	0.8785	0.8816	0.8847	0.8878	0.8910	0.8941	0.8972
	sin	0.6691	0.6704	0.6717	0.6730	0.6743	0.6756	0.6769	0.6782	0.6794	0.6807
42	cos	0.7431	0.7420	0.7408	0.7396	0.7385	0.7373	0.7361	0.7349	0.7337	0.7325
	tan	0.9004	0.9036	0.9067	0.9099	0.9131	0.9163	0.9195	0.9228	0.9260	0.9293
	sin	0.6820	0.6833	0.6845	0.6858	0.6871	0.6884	0.6896	0.6909	0.6921	0.6934
43	cos	0.7314	0.7302	0.7290	0.7278	0.7266	0.7254	0.7242	0.7230	0.7218	0.7206
	tan	0.9325	0.9358	0.9391	0.9424	0.9457	0.9490	0.9523	0.9556	0.9590	0.9623
	sin	0.6947	0.6959	0.6972	0.6984	0.6997	0.7009	0.7022	0.7034	0.7046	0.7059
44	cos	0.7193	0.7181	0.7169	0.7157	0.7145	0.7133	0.7120	0.7108	0.7096	0.7083
	tan	0.9657	0.9691	0.9725	0.9759	0.9793	0.9827	0.9861	0.9896	0.9930	0.9965
	sin	0.7071	0.7083	0.7096	0.7108	0.7120	0.7133	0.7145	0.7157	0.7169	0.7181
45	cos	0.7071	0.7059	0.7046	0.7034	0.7022	0.7009	0.6997	0.6984	0.6972	0.6959
	tan	1.0000	1.0035	1.0070	1.0105	1.0141	1.0176	1.0212	1.0247	1.0283	1.0319
	sin	0.7193	0.7206	0.7218	0.7230	0.7242	0.7254	0.7266	0.7278	0.7290	0.7302
46	cos	0.6947	0.6934	0.6921	0.6909	0.6896	0.6884	0.6871	0.6858	0.6845	0.6833
	tan	1.0355	1.0392	1.0428	1.0464	1.0501	1.0538	1.0575	1.0612	1.0649	1.0686
	sin	0.7314	0.7325	0.7337	0.7349	0.7361	0.7373	0.7385	0.7396	0.7408	0.7420
47	cos	0.6820	0.6807	0.6794	0.6782	0.6769	0.6756	0.6743	0.6730	0.6717	0.6704
	tan	1.0724	1.0761	1.0799	1.0837	1.0875	1.0913	1.0951	1.0990	1.1028	1.1067
	sin	0.7431	0.7443	0.7455	0.7466	0.7478	0.7490	0.7501	0.7513	0.7524	0.7536
48	cos	0.6691	0.6678	0.6665	0.6652	0.6639	0.6626	0.6613	0.6600	0.6587	0.6574
	tan	1.1106	1.1145	1.1184	1.1224	1.1263	1.1303	1.1343	1.1383	1.1423	1.1463
	sin	0.7547	0.7559	0.7570	0.7581	0.7593	0.7604	0.7615	0.7627	0.7638	0.7649
49	cos	0.6561	0.6547	0.6534	0.6521	0.6508	0.6494	0.6481	0.6468	0.6455	0.6441
	tan	1.1504	1.1544	1.1585	1.1626	1.1667	1.1708	1.1750	1.1792	1.1833	1.1875
	sin	0.7660	0.7672	0.7683	0.7694	0.7705	0.7716	0.7727	0.7738	0.7749	0.7760
50	cos	0.6428	0.6414	0.6401	0.6388	0.6374	0.6361	0.6347	0.6334	0.6320	0.6307
	tan	1.1918	1.1960	1.2002	1.2045	1.2088	1.2131	1.2174	1.2218	1.2261	1.2305
	sin	0.7771	0.7782	0.7793	0.7804	0.7815	0.7826	0.7837	0.7848	0.7859	0.7869
51	cos	0.6293	0.6280	0.6266	0.6252	0.6239	0.6225	0.6211	0.6198	0.6184	0.6170
	tan	1.2349	1.2393	1.2437	1.2482	1.2527	1.2572	1.2617	1.2662	1.2708	1.2753
	sin	0.7880	0.7891	0.7902	0.7912	0.7923	0.7934	0.7944	0.7955	0.7965	0.7976
52	cos	0.6157	0.6143	0.6129	0.6115	0.6101	0.6088	0.6074	0.6060	0.6046	0.6032
	tan	1.2799	1.2846	1.2892	1.2938	1.2985	1.3032	1.3079	1.3127	1.3175	1.3222
	sin	0.7986	0.7997	0.8007	0.8018	0.8028	0.8039	0.8049	0.8059	0.8070	0.8080
53	cos	0.6018	0.6004	0.5990	0.5976	0.5962	0.5948	0.5934	0.5920	0.5906	0.5892
	tan	1.3270	1.3319	1.3367	1.3416	1.3465	1.3514	1.3564	1.3613	1.3663	1.3713
	sin	0.8090	0.8100	0.8111	0.8121	0.8131	0.8141	0.8151	0.8161	0.8171	0.8181
54	cos	0.5878	0.5864	0.5850	0.5835	0.5821	0.5807	0.5793	0.5779	0.5764	0.5750
	tan	1.3764	1.3814	1.3865	1.3916	1.3968	1.4019	1.4071	1.4124	1.4176	1.4229
	sin	0.8192	0.8202	0.8211	0.8221	0.8231	0.8241	0.8251	0.8261	0.8271	0.8281
55	cos	0.5736	0.5721	0.5707	0.5693	0.5678	0.5664	0.5650	0.5635	0.5621	0.5606
	tan	1.4281	1.4335	1.4388	1.4442	1.4496	1.4550	1.4605	1.4659	1.4715	1.4770

Table 14 Trigonometric Functions—*Continued* **56.0°–75.9°**

Angle in Degrees	Name of Function	0.0	0.1	0.2	0.3	0.4	0.5	0.6	0.7	0.8	0.9
		Value of Function for Each Tenth of a Degree									
	sin	0.8290	0.8300	0.8310	0.8320	0.8329	0.8339	0.8348	0.8358	0.8368	0.8377
56	cos	0.5592	0.5577	0.5563	0.5548	0.5534	0.5519	0.5505	0.5490	0.5476	0.5461
	tan	1.4826	1.4882	1.4938	1.4994	1.5051	1.5108	1.5166	1.5224	1.5282	1.5340
	sin	0.8387	0.8396	0.8406	0.8415	0.8425	0.8434	0.8443	0.8453	0.8462	0.8471
57	cos	0.5446	0.5432	0.5417	0.5402	0.5388	0.5373	0.5358	0.5344	0.5329	0.5314
	tan	1.5399	1.5458	1.5517	1.5577	1.5637	1.5697	1.5757	1.5818	1.5880	1.5941
	sin	0.8480	0.8490	0.8499	0.8508	0.8517	0.8526	0.8536	0.8545	0.8554	0.8563
58	cos	0.5299	0.5284	0.5270	0.5255	0.5240	0.5225	0.5210	0.5195	0.5180	0.5165
	tan	1.6003	1.6066	1.6128	1.6191	1.6255	1.6319	1.6383	1.6447	1.6512	1.6577
	sin	0.8572	0.8581	0.8590	0.8599	0.8607	0.8616	0.8625	0.8634	0.8643	0.8652
59	cos	0.5150	0.5135	0.5120	0.5105	0.5090	0.5075	0.5060	0.5045	0.5030	0.5015
	tan	1.6643	1.6709	1.6775	1.6842	1.6909	1.6977	1.7045	1.7113	1.7182	1.7251
	sin	0.8660	0.8669	0.8678	0.8686	0.8695	0.8704	0.8712	0.8721	0.8729	0.8738
60	cos	0.5000	0.4985	0.4970	0.4955	0.4939	0.4924	0.4909	0.4894	0.4879	0.4863
	tan	1.7321	1.7391	1.7461	1.7532	1.7603	1.7675	1.7747	1.7820	1.7893	1.7966
	sin	0.8746	0.8755	0.8763	0.8771	0.8780	0.8788	0.8796	0.8805	0.8813	0.8821
61	cos	0.4848	0.4833	0.4818	0.4802	0.4787	0.4772	0.4756	0.4741	0.4726	0.4710
	tan	1.8040	1.8115	1.8190	1.8265	1.8341	1.8418	1.8495	1.8572	1.8650	1.8728
	sin	0.8829	0.8838	0.8846	0.8854	0.8862	0.8870	0.8878	0.8886	0.8894	0.8902
62	cos	0.4695	0.4679	0.4664	0.4648	0.4633	0.4617	0.4602	0.4586	0.4571	0.4555
	tan	1.8807	1.8887	1.8967	1.9047	1.9128	1.9210	1.9292	1.9375	1.9458	1.9542
	sin	0.8910	0.8918	0.8926	0.8934	0.8942	0.8949	0.8957	0.8965	0.8973	0.8980
63	cos	0.4540	0.4524	0.4509	0.4493	0.4478	0.4462	0.4446	0.4431	0.4415	0.4399
	tan	1.9626	1.9711	1.9797	1.9883	1.9970	2.0057	2.0145	2.0233	2.0323	2.0413
	sin	0.8988	0.8996	0.9003	0.9011	0.9018	0.9026	0.9033	0.9041	0.9048	0.9056
64	cos	0.4384	0.4368	0.4352	0.4337	0.4321	0.4305	0.4289	0.4274	0.4258	0.4242
	tan	2.0503	2.0594	2.0686	2.0778	2.0872	2.0965	2.1060	2.1155	2.1251	2.1348
	sin	0.9063	0.9070	0.9078	0.9085	0.9092	0.9100	0.9107	0.9114	0.9121	0.9128
65	cos	0.4226	0.4210	0.4195	0.4179	0.4163	0.4147	0.4131	0.4115	0.4099	0.4083
	tan	2.1445	2.1543	2.1642	2.1742	2.1842	2.1943	2.2045	2.2148	2.2251	2.2355
	sin	0.9135	0.9143	0.9150	0.9157	0.9164	0.9171	0.9178	0.9184	0.9191	0.9198
66	cos	0.4067	0.4051	0.4035	0.4019	0.4003	0.3987	0.3971	0.3955	0.3939	0.3923
	tan	2.2460	2.2566	2.2673	2.2781	2.2889	2.2998	2.3109	2.3220	2.3332	2.3445
	sin	0.9205	0.9212	0.9219	0.9225	0.9232	0.9239	0.9245	0.9252	0.9259	0.9265
67	cos	0.3907	0.3891	0.3875	0.3859	0.3843	0.3827	0.3811	0.3795	0.3778	0.3762
	tan	2.3559	2.3673	2.3789	2.3906	2.4023	2.4142	2.4262	2.4383	2.4504	2.4627
	sin	0.9272	0.9278	0.9285	0.9291	0.9298	0.9304	0.9311	0.9317	0.9323	0.9330
68	cos	0.3746	0.3730	0.3714	0.3697	0.3681	0.3665	0.3649	0.3633	0.3616	0.3600
	tan	2.4751	2.4876	2.5002	2.5129	2.5257	2.5386	2.5517	2.5649	2.5782	2.5916
	sin	0.9336	0.9342	0.9348	0.9354	0.9361	0.9367	0.9373	0.9379	0.9385	0.9391
69	cos	0.3584	0.3567	0.3551	0.3535	0.3518	0.3502	0.3486	0.3469	0.3453	0.3437
	tan	2.6051	2.6187	2.6325	2.6464	2.6605	2.6746	2.6889	2.7034	2.7179	2.7326
	sin	0.9397	0.9403	0.9409	0.9415	0.9421	0.9426	0.9432	0.9438	0.9444	0.9449
70	cos	0.3420	0.3404	0.3387	0.3371	0.3355	0.3338	0.3322	0.3305	0.3289	0.3272
	tan	2.7475	2.7625	2.7776	2.7929	2.8083	2.8239	2.8397	2.8556	2.8716	2.8878
	sin	0.9455	0.9461	0.9466	0.9472	0.9478	0.9483	0.9489	0.9494	0.9500	0.9505
71	cos	0.3256	0.3239	0.3223	0.3206	0.3190	0.3173	0.3156	0.3140	0.3123	0.3107
	tan	2.9042	2.9208	2.9375	2.9544	2.9714	2.9887	3.0061	3.0237	3.0415	3.0595
	sin	0.9511	0.9516	0.9521	0.9527	0.9532	0.9537	0.9542	0.9548	0.9553	0.9558
72	cos	0.3090	0.3074	0.3057	0.3040	0.3024	0.3007	0.2990	0.2974	0.2957	0.2940
	tan	3.0777	3.0961	3.1146	3.1334	3.1524	3.1716	3.1910	3.2106	3.2305	3.2506
	sin	0.9563	0.9568	0.9573	0.9578	0.9583	0.9588	0.9593	0.9598	0.9603	0.9608
73	cos	0.2924	0.2907	0.2890	0.2874	0.2857	0.2840	0.2823	0.2807	0.2790	0.2773
	tan	3.2709	3.2914	3.3122	3.3332	3.3544	3.3759	3.3977	3.4197	3.4420	3.4646
	sin	0.9613	0.9617	0.9622	0.9627	0.9632	0.9636	0.9641	0.9646	0.9650	0.9655
74	cos	0.2756	0.2740	0.2723	0.2706	0.2689	0.2672	0.2656	0.2639	0.2622	0.2605
	tan	3.4874	3.5105	3.5339	3.5576	3.5816	3.6059	3.6305	3.6554	3.6806	3.7062
	sin	0.9659	0.9664	0.9668	0.9673	0.9677	0.9681	0.9686	0.9690	0.9694	0.9699
75	cos	0.2588	0.2571	0.2554	0.2538	0.2521	0.2504	0.2487	0.2470	0.2453	0.2436
	tan	3.7321	3.7583	3.7848	3.8118	3.8391	3.8667	3.8947	3.9232	3.9520	3.9812

Table 14. Trigonometric Functions—*Continued* **76.0°–89.9°**

Angle in Degrees	Name of Function	Value of Function for Each Tenth of a Degree									
		0.0	**0.1**	**0.2**	**0.3**	**0.4**	**0.5**	**0.6**	**0.7**	**0.8**	**0.9**
76	sin	0.9703	0.9707	0.9711	0.9715	0.9720	0.9724	0.9728	0.9732	0.9736	0.9740
	cos	0.2419	0.2402	0.2385	0.2368	0.2351	0.2334	0.2317	0.2300	0.2284	0.2267
	tan	4.0108	4.0408	4.0713	4.1022	4.1335	4.1653	4.1976	4.2303	4.2635	4.2972
77	sin	0.9744	0.9748	0.9751	0.9755	0.9759	0.9763	0.9767	0.9770	0.9774	0.9778
	cos	0.2250	0.2232	0.2215	0.2198	0.2181	0.2164	0.2147	0.2130	0.2113	0.2096
	tan	4.3315	4.3662	4.4015	4.4374	4.4737	4.5107	4.5483	4.5864	4.6252	4.6646
78	sin	0.9781	0.9785	0.9789	0.9792	0.9796	0.9799	0.9803	0.9806	0.9810	0.9813
	cos	0.2079	0.2062	0.2045	0.2028	0.2011	0.1994	0.1977	0.1959	0.1942	0.1925
	tan	4.7046	4.7453	4.7867	4.8288	4.8716	4.9152	4.9594	5.0045	5.0504	5.0970
79	sin	0.9816	0.9820	0.9823	0.9826	0.9829	0.9833	0.9836	0.9839	0.9842	0.9845
	cos	0.1908	0.1891	0.1874	0.1857	0.1840	0.1822	0.1805	0.1788	0.1771	0.1754
	tan	5.1446	5.1929	5.2422	5.2924	5.3435	5.3955	5.4486	5.5026	5.5578	5.6140
80	sin	0.9848	0.9851	0.9854	0.9857	0.9860	0.9863	0.9866	0.9869	0.9871	0.9874
	cos	0.1736	0.1719	0.1702	0.1685	0.1668	0.1650	0.1633	0.1616	0.1599	0.1582
	tan	5.6713	5.7297	5.7894	5.8502	5.9124	5.9758	6.0405	6.1066	6.1742	6.2432
81	sin	0.9877	0.9880	0.9882	0.9885	0.9888	0.9890	0.9893	0.9895	0.9898	0.9900
	cos	0.1564	0.1547	0.1530	0.1513	0.1495	0.1478	0.1461	0.1444	0.1426	0.1409
	tan	6.3138	6.3859	6.4596	6.5350	6.6122	6.6912	6.7720	6.8548	6.9395	7.0264
82	sin	0.9903	0.9905	0.9907	0.9910	0.9912	0.9914	0.9917	0.9919	0.9921	0.9923
	cos	0.1392	0.1374	0.1357	0.1340	0.1323	0.1305	0.1288	0.1271	0.1253	0.1236
	tan	7.1154	7.2066	7.3002	7.3962	7.4947	7.5958	7.6996	7.8062	7.9158	8.0285
83	sin	0.9925	0.9928	0.9930	0.9932	0.9934	0.9936	0.9938	0.9940	0.9942	0.9943
	cos	0.1219	0.1201	0.1184	0.1167	0.1149	0.1132	0.1115	0.1097	0.1080	0.1063
	tan	8.1443	8.2636	8.3863	8.5126	8.6427	8.7769	8.9152	9.0579	9.2052	9.3572
84	sin	0.9945	0.9947	0.9949	0.9951	0.9952	0.9954	0.9956	0.9957	0.9959	0.9960
	cos	0.1045	0.1028	0.1011	0.0993	0.0976	0.0958	0.0941	0.0924	0.0906	0.0889
	tan	9.5144	9.6768	9.8448	10.02	10.20	10.39	10.58	10.78	10.99	11.20
85	sin	0.9962	0.9963	0.9965	0.9966	0.9968	0.9969	0.9971	0.9972	0.9973	0.9974
	cos	0.0872	0.0854	0.0837	0.0819	0.0802	0.0785	0.0767	0.0750	0.0732	0.0715
	tan	11.43	11.66	11.91	12.16	12.43	12.71	13.00	13.30	13.62	13.95
86	sin	0.9976	0.9977	0.9978	0.9979	0.9980	0.9981	0.9982	0.9983	0.9984	0.9985
	cos	0.0698	0.0680	0.0663	0.0645	0.0628	0.0610	0.0593	0.0576	0.0558	0.0541
	tan	14.30	14.67	15.06	15.89	15.46	16.35	16.83	17.34	17.89	18.46
87	sin	0.9986	0.9987	0.9988	0.9989	0.9990	0.9990	0.9991	0.9992	0.9993	0.9993
	cos	0.0523	0.0506	0.0488	0.0471	0.0454	0.0436	0.0419	0.0401	0.0384	0.0366
	tan	19.08	19.74	20.45	21.20	22.02	22.90	23.86	24.90	26.03	27.27
88	sin	0.9994	0.9995	0.9995	0.9996	0.9996	0.9997	0.9997	0.9997	0.9998	0.9998
	cos	0.0349	0.0332	0.0314	0.0297	0.0279	0.0262	0.0244	0.0227	0.0209	0.0192
	tan	28.64	30.14	31.82	33.69	35.80	38.19	40.92	44.07	47.74	52.08
89	sin	0.9998	0.9999	0.9999	0.9999	0.9999	1.000	1.000	1.000	1.000	1.000
	cos	0.0175	0.0157	0.0140	0.0122	0.0105	0.0087	0.0070	0.0052	0.0035	0.0017
	tan	57.29	63.66	71.62	81.85	95.49	114.6	143.2	191.0	286.5	573.0

Table 15. Logarithmic Trigonometric Functions

Deg.	Sine	Cosec	Versin	Tangent	Cotan	Covers	Secant	Cosine	Deg
0	− ∞	+ ∞	− ∞	− ∞	+ ∞	10.00000	10.00000	10.00000	**90**
1	8.24186	11.75814	6.18271	8.24192	11.75808	9.99235	10.00007	9.99993	**89**
2	8.54282	11.45718	6.78474	8.54308	11.45692	9.98457	10.00026	9.99974	**88**
3	8.71880	11.28120	7.13687	8.71940	11.28060	9.97665	10.00060	9.99940	**87**
4	8.84358	11.15642	7.38667	8.84464	11.15536	9.96860	10.00106	9.99894	**86**
5	8.94030	11.05970	7.58039	8.94195	11.05805	9.96040	10.00166	9.99834	**85**
6	9.01923	10.98077	7.73863	9.02162	10.97838	9.95205	10.00239	9.99761	**84**
7	9.08589	10.91411	7.87238	9.08914	10.91086	9.94356	10.00325	9.99675	**83**
8	9.14356	10.85644	7.98820	9.14780	10.85220	9.93492	10.00425	9.99575	**82**
9	9.19433	10.80567	8.09032	9.19971	10.80029	9.92612	10.00538	9.99462	**81**
10	9.23967	10.76033	8.18162	9.24632	10.75368	9.91717	10.00665	9.99335	**80**
11	9.28060	10.71940	8.26418	9.28865	10.71135	9.90805	10.00805	9.99195	**79**
12	9.31788	10.68212	8.33950	9.32747	10.67253	9.89877	10.00960	9.99040	**78**
13	9.35209	10.64791	8.40875	9.36336	10.63664	9.88933	10.01128	9.98872	**77**
14	9.38368	10.61632	8.47282	9.39677	10.60323	9.87971	10.01310	9.98690	**76**
15	9.41300	10.58700	8.53243	9.42805	10.57195	9.86992	10.01506	9.98494	**75**
16	9.44034	10.55966	8.58814	9.45750	10.54250	9.85996	10.01716	9.98284	**74**
17	9.46594	10.53406	8.64043	9.48534	10.51466	9.84981	10.01940	9.98060	**73**
18	9.48998	10.51002	8.68969	9.51178	10.48822	9.83947	10.02179	9.97821	**72**
19	9.51264	10.48736	8.73625	9.53697	10.46303	9.82894	10.02433	9.97567	**71**
20	9.53405	10.46595	8.78037	9.56107	10.43893	9.81821	10.02701	9.97299	**70**
21	9.55433	10.44567	8.82230	9.58418	10.41582	9.80729	10.02985	9.97015	**69**
22	9.57358	10.42642	8.86223	9.60641	10.39359	9.79615	10.03283	9.96717	**68**
23	9.59188	10.40812	8.90034	9.62785	10.37215	9.78481	10.03597	9.96403	**67**
24	9.60931	10.39069	8.93679	9.64858	10.35142	9.77325	10.03927	9.96073	**66**
25	9.62595	10.37405	8.97170	9.66867	10.33133	9.76146	10.04272	9.95728	**65**
26	9.64184	10.35816	9.00521	9.68818	10.31182	9.74945	10.04634	9.95366	**64**
27	9.65705	10.34295	9.03740	9.70717	10.29283	9.73720	10.05012	9.94988	**63**
28	9.67161	10.32839	9.06838	9.72567	10.27433	9.72471	10.05407	9.94593	**62**
29	9.68557	10.31443	9.09823	9.74375	10.25625	9.71197	10.05818	9.94182	**61**
30	9.69897	10.30103	9.12702	9.76144	10.23856	9.69897	10.06247	9.93753	**60**
31	9.71184	10.28816	9.15483	9.77877	10.22123	9.68571	10.06693	9.93307	**59**
32	9.72421	10.27579	9.18171	9.79579	10.20421	9.67217	10.07158	9.92842	**58**
33	9.73611	10.26389	9.20771	9.81252	10.18748	9.65836	10.07641	9.92359	**57**
34	9.74756	10.25244	9.23290	9.82899	10.17101	9.64425	10.08143	9.91857	**56**
35	9.75859	10.24141	9.25731	9.84523	10.15477	9.62984	10.08664	9.91336	**55**
36	9.76922	10.23078	9.28099	9.86126	10.13874	9.61512	10.09204	9.90796	**54**
37	9.77946	10.22054	9.30398	9.87711	10.12289	9.60008	10.09765	9.90235	**53**
38	9.78934	10.21066	9.32631	9.89281	10.10719	9.58471	10.10347	9.89653	**52**
39	9.79887	10.20113	9.34802	9.90837	10.09163	9.56900	10.10950	9.89050	**51**
40	9.80807	10.19193	9.36913	9.92381	10.07619	9.55293	10.11575	9.88425	**50**
41	9.81694	10.18306	9.38968	9.93916	10.06084	9.53648	10.12222	9.87778	**49**
42	9.82551	10.17449	9.40969	9.95444	10.04556	9.51966	10.12893	9.87107	**48**
43	9.83378	10.16622	9.42918	9.96966	10.03034	9.50243	10.13587	9.86413	**47**
44	9.84177	10.15823	9.44818	9.98484	10.01516	9.48479	10.14307	9.85693	**46**
45	9.84949	10.15052	9.46671	10.00000	10.00000	9.46671	10.15052	9.84949	**45**
	Cosine	Secant	Covers	Cotan	Tangent	Versin	Cosec	Sine	

From 45° to 90° read from bottom of table upwards.

Table 16. Fractions Expressed as Decimals of an Inch and of 1 Foot

Inches	1	2	3	4	5	6	7	8	9	10	11
Decimal of 1 ft.	0.0833	0.1667	0.2500	0.3333	0.4167	0.5000	0.5833	0.6667	0.7500	0.8333	0.9167

Fractions of 1 inch

Fraction	Decimal of 1 in.	Decimal of 1 ft.	Fraction	Decimal of 1 in.	Decimal of 1 ft.	Fraction	Decimal of 1 in.	Decimal of 1 ft.	Fraction	Decimal of 1 in.	Decimal of 1 ft.
1/64	0.015625	0.0013	17/64	0.265625	0.0221	33/64	0.515625	0.0430	49/64	0.765625	0.0638
1/32	.03125	.0026	9/32	.28125	.0234	17/32	.53125	.0443	25/32	.78125	.0651
3/64	.046875	.0039	19/64	.296875	.0247	35/64	.546875	.0456	51/64	.796875	.0664
1/16	.0625	.0052	5/16	.3125	.0260	9/16	.5625	.0469	13/16	.8125	.0677
5/64	.078125	.0065	21/64	.328125	.0273	37/64	.578125	.0482	53/64	.828125	.0690
3/32	.09375	.0078	11/32	.34375	.0286	19/32	.59375	.0495	27/32	.84375	.0703
7/64	.109375	.0091	23/64	.359375	.0299	39/64	.609375	.0508	55/64	.859375	.0716
1/8	.125	.0104	3/8	.375	.0313	5/8	.625	.0521	7/8	.875	.0729
9/64	.140625	.0117	25/64	.390625	.0326	41/64	.640625	.0534	57/64	.890625	.0742
5/32	.15625	.0130	13/32	.40625	.0339	21/32	.65625	.0547	29/32	.90625	.0755
11/64	.171875	.0143	27/64	.421875	.0352	43/64	.671875	.0560	59/64	.921875	.0768
3/16	.1875	.0156	7/16	.4375	.0365	11/16	.6875	.0573	15/16	.9375	.0781
13/64	.203125	.0169	29/64	.453125	.0378	45/64	.703125	.0586	61/64	.953125	.0794
7/32	.21875	.0182	15/32	.46875	.0391	23/32	.71875	.0599	31/32	.96875	.0807
15/64	.234375	.0195	31/64	.484375	.0404	47/64	.734375	.0612	63/64	.984375	.0820
1/4	.25	.0208	1/2	.50	.0417	3/4	.75	.0625	....	....	......

EXAMPLE: 9 19/32 in. = 0.7500 + 0.0495 = 0.7995 ft.

Table 17. Temperature Conversion—Centigrade to Fahrenheit

Values for Interpolation in Table										
Deg. C....	1	2	3	4	5	6	7	8	9	10
Deg. F....	1.8	3.6	5.4	7.2	9.0	10.8	12.6	14.4	16.2	18.0

Deg.→ C.↓	0	10	20	30	40	50	60	70	80	90
	Degrees, Fahrenheit									
−200	−328	−346	−364	−382	−400	−418	−436	−454	....	....
−100	−148	−166	−184	−202	−220	−238	−256	−274	−292	−310
−0	+32	+14	−4	−22	−40	−58	−76	−94	−112	−130
+0	+32	+50	+68	+86	+104	+122	+140	+158	+176	+194
100	212	230	248	266	284	302	320	338	356	374
200	392	410	428	446	464	482	500	518	536	554
300	572	590	608	626	644	662	680	698	716	734
400	752	770	788	806	824	842	860	878	896	914
500	932	950	968	986	1004	1022	1040	1058	1076	1094
600	1112	1130	1148	1166	1184	1202	1220	1238	1256	1274
700	1292	1310	1328	1346	1364	1382	1400	1418	1436	1454
800	1472	1490	1508	1526	1544	1562	1580	1598	1616	1634
900	1652	1670	1688	1706	1724	1742	1760	1778	1796	1814
1000	1832	1850	1868	1886	1904	1922	1940	1958	1976	1994
1100	2012	2030	2048	2066	2084	2102	2120	2138	2156	2174
1200	2192	2210	2228	2246	2264	2282	2300	2318	2336	2354
1300	2372	2390	2408	2426	2444	2462	2480	2498	2516	2534
1400	2552	2570	2588	2606	2624	2642	2660	2678	2696	2714
1500	2732	2750	2768	2786	2804	2822	2840	2858	2876	2894
1600	2912	2930	2948	2966	2984	3002	3020	3038	3056	3074
1700	3092	3110	3128	3146	3164	3182	3200	3218	3236	3254
1800	3272	3290	3308	3326	3344	3362	3380	3398	3416	3434
1900	3452	3470	3488	3506	3524	3542	3560	3578	3596	3614
2000	3632	3650	3668	3686	3704	3722	3740	3758	3776	3794
2100	3812	3830	3848	3866	3884	3902	3920	3938	3956	3974
2200	3992	4010	4028	4046	4064	4082	4100	4118	4136	4154
2300	4172	4190	4208	4226	4244	4262	4280	4298	4316	4334
2400	4352	4370	4388	4406	4424	4442	4460	4478	4496	4514
2500	4532	4550	4568	4586	4604	4622	4640	4658	4676	4694
2600	4712	4730	4748	4766	4784	4802	4820	4838	4856	4874
2700	4892	4910	4928	4946	4964	4982	5000	5018	5036	5054
2800	5072	5090	5108	5126	5144	5162	5180	5198	5216	5234
2900	5252	5270	5288	5306	5324	5342	5360	5378	5396	5414
3000	5432	5450	5468	5486	5504	5522	5540	5558	5576	5594
3100	5612	5630	5648	5666	5684	5702	5720	5738	5756	5774
3200	5792	5810	5828	5846	5864	5882	5900	5918	5936	5954
3300	5972	5990	6008	6026	6044	6062	6080	6098	6116	6134
3400	6152	6170	6188	6206	6224	6242	6260	6278	6296	6314
3500	6332	6350	6368	6386	6404	6422	6440	6458	6476	6494
3600	6512	6530	6548	6566	6584	6602	6620	6638	6656	6674
3700	6692	6710	6728	6746	6764	6782	6800	6818	6836	6854
3800	6872	6890	6908	6926	6944	6962	6980	6998	7016	7034
3900	7052	7070	7088	7106	7124	7142	7160	7178	7196	7214

Following is a comparison of temperature scales at significant points:

Temperature Scale	For Water at Sea Level		Absolute Zero
	Boiling Pt.	Freezing Pt.	
Fahrenheit	212	32	−459.69
Centigrade	100	0	−273.16
Rankine (Fahrenheit Absolute)	672	492	0
Kelvin (Centigrade Absolute)	373	273	0
Réaumur	80	0	−218.53

To convert from one scale to another, use the following formulas:

Temperature Fahrenheit	$= {}^9/_5 \times$ temp. C $+ 32°$	$= {}^9/_4$ temp. R $+ 32°$
Temperature Centigrade	$= {}^5/_9$ (temp. F $- 32°$)	$= {}^5/_4$ temp. R
Temperature Réaumur	$= {}^4/_5$ temp. C	$= {}^4/_9$ (temp. F $- 32°$)

Table 18. Temperature Conversion—Fahrenheit to Centigrade

Values for Interpolation in Table

Deg. F...	1	2	3	4	5	6	7	8	9	10
Deg. C...	0.555	1.111	1.666	2.222	2.777	3.333	3.888	4.444	5.000	5.555

Deg. → F ↓	0	10	20	30	40	50	60	70	80	90
	Degrees, Centigrade									
−400	−240.00	−245.56	−251.11	−256.67	−262.22	−267.78	−277.33	......	......	......
−300	−184.44	−190.00	−195.56	−201.11	−206.67	−212.22	−217.78	−223.23	−228.89	−234.44
−200	−128.89	−134.44	−140.00	−145.56	−151.11	−156.67	−162.22	−167.78	−173.33	−178.89
−100	−73.33	−78.89	−84.44	−90.00	−95.56	−101.11	−106.67	−112.22	−117.78	−123.33
−0	−17.78	−23.33	−28.89	−34.44	−40.00	−45.56	−51.11	−56.67	−62.22	−67.78
+0	−17.78	−12.22	−6.67	−1.11	+4.44	+10.00	+15.56	+21.11	+26.67	+32.22
100	37.78	43.33	48.89	54.44	60.00	65.56	71.11	76.67	82.22	87.78
200	93.33	98.89	104.44	110.00	115.56	121.11	126.67	132.22	137.78	143.33
300	148.89	154.44	160.00	165.56	171.11	176.67	182.22	187.78	193.33	198.89
400	204.44	210.00	215.56	221.11	226.67	232.22	237.78	243.33	248.89	254.44
500	260.00	265.56	271.11	276.67	282.22	287.78	293.33	298.89	304.44	310.00
600	315.56	321.11	326.67	332.22	337.78	343.33	348.89	354.44	360.00	365.56
700	371.11	376.67	382.22	387.78	393.33	398.89	404.44	410.00	415.56	421.11
800	426.67	432.22	437.78	443.33	448.89	454.44	460.00	465.56	471.11	476.67
900	482.22	487.78	493.33	498.89	504.44	510.00	515.56	521.11	526.67	532.22
1000	537.78	543.33	548.89	554.44	560.00	565.56	571.11	576.67	582.22	587.78
1100	593.33	598.89	604.44	610.00	615.56	621.11	626.67	632.22	637.78	643.33
1200	648.89	654.44	660.00	655.56	671.11	676.67	682.22	687.78	693.33	698.89
1300	704.44	710.00	715.56	721.11	726.67	732.22	737.78	743.33	748.89	754.44
1400	760.00	765.56	771.11	776.67	782.22	787.78	793.33	798.89	804.44	810.00
1500	815.56	821.11	826.67	832.22	837.78	843.33	848.89	854.44	860.00	865.56
1600	871.11	876.67	882.22	887.78	893.33	898.89	904.44	910.00	915.56	921.11
1700	926.67	932.22	937.78	943.33	948.89	954.44	960.00	965.56	971.11	976.67
1800	982.22	987.78	993.33	998.89	1004.4	1010.0	1015.6	1021.1	1026.7	1032.2
1900	1037.8	1043.3	1048.9	1054.4	1060.0	1065.6	1071.1	1076.7	1082.2	1087.8
2000	1093.3	1098.9	1104.4	1110.0	1115.6	1121.1	1126.7	1132.2	1137.8	1143.3
2100	1148.9	1154.4	1160.0	1165.6	1171.1	1176.7	1182.2	1187.8	1193.3	1198.9
2200	1204.4	1210.0	1215.6	1221.1	1226.7	1232.2	1237.8	1243.3	1248.9	1254.4
2300	1260.0	1265.6	1271.1	1276.7	1282.2	1287.8	1293.3	1298.9	1304.4	1310.0
2400	1315.6	1321.1	1326.7	1332.2	1337.8	1343.3	1348.9	1354.4	1360.0	1365.6
2500	1371.1	1376.7	1382.2	1387.8	1393.3	1398.9	1404.4	1410.0	1415.6	1421.1
2600	1426.7	1432.2	1437.8	1443.3	1448.9	1454.4	1460.0	1465.6	1471.1	1476.7
2700	1482.2	1487.8	1493.3	1498.9	1504.4	1510.0	1515.6	1521.1	1526.7	1532.2
2800	1537.8	1543.3	1548.9	1554.4	1560.0	1565.6	1571.1	1576.7	1582.2	1587.8
2900	1593.3	1598.9	1604.4	1610.0	1615.6	1621.1	1626.7	1632.2	1637.8	1643.3
3000	1648.9	1654.4	1660.0	1665.6	1671.1	1676.7	1682.2	1687.8	1693.3	1698.9
3100	1704.4	1710.0	1715.6	1721.1	1726.7	1732.2	1737.8	1743.3	1748.9	1754.4
3200	1760.0	1765.6	1771.1	1776.7	1782.2	1787.8	1793.3	1798.9	1804.4	1810.0
3300	1815.6	1821.1	1826.7	1832.2	1837.8	1843.3	1848.9	1854.4	1860.0	1865.6
3400	1871.1	1876.7	1882.2	1887.8	1893.3	1898.9	1904.4	1910.0	1915.6	1921.1
3500	1926.7	1932.2	1937.8	1943.3	1948.9	1954.4	1960.0	1965.6	1971.1	1976.7
3600	1982.2	1987.8	1993.3	1998.9	2004.4	2010.0	2015.6	2021.1	2026.7	2032.2
3700	2037.8	2043.3	2048.9	2054.4	2060.0	2065.6	2071.1	2076.7	2082.2	2087.8
3800	2093.3	2098.9	2104.4	2110.0	2115.6	2121.1	2126.7	2132.2	2137.8	2143.3
3900	2148.9	2154.4	2160.0	2165.6	2171.1	2176.7	2182.2	2187.8	2193.3	2198.9
4000	2204.4	2210.0	2215.6	2221.1	2226.7	2232.2	2237.8	2243.3	2248.9	2254.4
4100	2260.0	2265.6	2271.1	2276.7	2282.2	2287.8	2293.3	2298.9	2304.4	2310.0
4200	2315.6	2321.1	2326.7	2332.2	2337.8	2343.3	2348.9	2354.4	2360.0	2365.6
4300	2371.1	2376.7	2382.2	2387.8	2393.3	2398.9	2404.4	2410.0	2415.6	2421.1
4400	2426.7	2432.2	2437.8	2443.3	2448.9	2454.4	2460.0	2465.6	2471.1	2476.7
4500	2482.2	2487.8	2493.3	2498.9	2504.4	2510.0	2515.6	2521.1	2526.7	2532.2
4600	2537.8	2543.3	2548.9	2554.4	2560.0	2565.6	2571.1	2576.7	2582.2	2587.8
4700	2593.3	2598.9	2604.4	2610.0	2615.6	2621.1	2626.7	2632.2	2637.8	2643.3
4800	2648.9	2654.4	2660.0	2665.6	2671.1	2676.7	2682.2	2687.8	2693.3	2698.9
4900	2704.4	2710.0	2715.6	2721.1	2726.7	2732.2	2737.8	2743.3	2748.9	2754.4
5000	2760.0	2765.6	2771.1	2776.7	2782.2	2787.8	2793.3	2798.9	2804.4	2810.0
5100	2815.6	2821.1	2826.7	2832.2	2837.8	2843.3	2848.9	2854.4	2860.0	2865.6
5200	2871.1	2876.7	2882.2	2887.8	2893.3	2898.9	2904.4	2910.0	2915.6	2921.1
5300	2926.7	2932.2	2937.8	2943.3	2948.9	2954.4	2960.0	2965.6	2971.1	2976.7
5400	2982.2	2987.8	2993.3	2998.9	3004.4	3010.0	3015.6	3021.1	3026.7	3032.2
5500	3037.8	3043.3	3048.9	3054.4	3060.0	3065.6	3071.1	3076.7	3082.2	3087.8
5600	3093.3	3098.9	3104.4	3110.0	3115.6	3121.1	3126.7	3132.2	3137.8	3143.3
5700	3148.9	3154.4	3160.0	3165.6	3171.1	3176.7	3182.2	3187.8	3193.3	3198.9
5800	3204.4	3210.0	3215.6	3221.1	3226.7	3232.2	3237.8	3243.3	3248.9	3254.4
5900	3260.0	3265.6	3271.1	3276.7	3282.2	3287.8	3293.3	3298.9	3304.4	3310.0

MEASURES AND WEIGHTS COMMONLY USED IN THE UNITED STATES AND GREAT BRITAIN

Abbreviations

bbl.	= barrel	dwt.	= pennyweight	hr.	= hour	quin.	= quintal
bu.	= bushel	fath.	= fathom	Imp.	= Imperial (bu., gal.)	qr.	= quarter
chn.	= chain	ft.	= foot	in.	= inch	qt.	= quart
circm.	= circumference	fl. drm.	= fluid drachm	lea	= league	scr.	= scruple
cir. in.	= circular inch	fl. oz.	= fluid ounce	m. pace	= military pace	sec.	= second
cir. mil.	= circular mil	furl.	= furlong	min.	= minute	sq.	= square (in., ft., etc.)
cwt.	= hundred weight	gal.	= gallon (U.S.)	naut. mi.	= nautical mile	stat. mi.	= statute mile
deg.	= degree	grn.	= grain	pnch	= puncheon	trce.	= tierce
drm.	= drachm (dram)	hghd.	= hogshead	quad.	= quadrant	yd.	= yard

Table 19. Measures of Length

Long Measure

12 in. = 1 ft.
3 ft. = 1 yd.
5 1/2 yd. = 16 1/2 ft. = 1 rod, pole or perch
40 poles = 220 yd. = 1 furl.
8 furl. = 1760 yd. = 5280 ft. = 1 mile
4 in. = 1 hand
9 in. = 1 span
2 1/2 ft. = 1 m. pace.

Land, Surveyors, or Gunter's Measure

7.92 in. = 1 link
100 links = 66 ft. = 4 rods = 1 chn.
10 chn. = 220 yd. = 1 furl.
8 furl. = 80 chn. = 1 mile.

Nautical Measure

6080.26 ft. = 1.15156 stat. mi. = 1 naut. mi.
3 naut. mi. = 1 lea.
60 naut. mi. = 1 deg. (at equator)
6 ft. = 1 fath.
120 fath. = 1 cable
1 naut. mi. per hr. = 1 knot.

Table 20. Measures of Area

Square Measure

144 sq. in. = 183.35 cir. in. = 1 sq. ft.
9 sq. ft. = 1 sq. yd.
30 1/4 sq. yd. = 272 1/4 sq. ft. = 1 sq. rod
160 sq. rods = 10 chn. } = 1 acre
4840 sq. yd. = 43560 sq. ft. } = 1 acre
640 acres = 27,878,400 sq. ft. = 1 sq. mile
1 acre = $(208.71\text{ ft.})^2$
1 cir. in. = {area of circle 1 in. diam. = 0.7854 sq. in.
1 cir. mil = {area of circle 0.001 in. diam. = 0.0_57854 sq. in.
1,000,000 cir. mil = 1 cir. in.

Board Measure

144 cu. in. = 12 × 12 × 1 in. = 1 board ft.

The above definition of a board foot applies to rough green lumber. American practice uses a yard standard of 29/32 in. thick and an industrial standard of 7/8 in. thick for rough dry lumber; for dressed lumber the corresponding thicknesses are 25/32 in. and 13/16 in. For example, a rough dry piece 12 × 12 × 29/32 in. contains 1 board ft. Board ft. in round timber = $(1/4\, c \times d \times l)/144$. c = mean circm., d = diam., l = length, all in ft.

Table 21. Measures of Weight

(The grain is the same in all systems.)

Avoirdupois Weight

16 drm. = 437.5 grn. = 1 oz.
16 oz. = 7000 grn. = 1 lb.
14 lb. = 1 stone
28 lb. = 2 stone = 1 qr.
100 lb. = 1 quin.
4 qr. = 112 lb. = 1 cwt.
20 cwt. = 2240 lb. = 1 long ton
2000 lb. = 1 short ton.

Troy Weight (for gold and silver)

24 grn. = 1 dwt.
20 dwt. = 480 grn. = 1 oz.
12 oz. = 5760 grn. = 1 lb.
1 Assay ton = 29,167 milligrams = {Troy oz. per 2000 lb. ton avoirdupois

Carat Weight (for precious stones)

1 carat = 3.086 grn. = 0.200 gram

Apothecaries' Weight

20 grn. = 1 scr. ℈
3 scr. = 60 grn. = 1 drm. ʒ
8 drm. = 480 grn. = 1 oz. ℥
12 oz. = 5760 grn. = 1 lb.

Table 22. Measures of Volume

Cubic Measure

1728 cu. in. = 1 cu. ft.
27 cu. ft. = 1 cu. yd.
1 cord of wood = {4 × 4 × 8 ft. = 128 cu. ft.
1 perch masonry = {16 1/2 × 1 1/2 × 1 ft. = 23 3/4 cu. ft.

Liquid Measure

4 gills = 1 pint
2 pints = 1 qt.
4 qt. = 231 cu. in. = 8.3356 lb. H_2O = 1 U. S. gal. = 0.8327 Imp. gal.
1 Imp. gal. = 277.420 cu. in. = 10 lb. H_2O
1.20094 U. S. gal. = 1 Imp. gal.
7.4805 U. S. gal. = 1 cu. ft.

Old Liquid Measure

31 1/2 gal. = 1 bbl.
42 gal. = 1 trce.
2 bbl. = 63 gal. = 1 hghd.
84 gal. = 2 trce. = 1 pchn.
2 hghd. = 126 gal. = 1 pipe or butt
2 pipes = 3 pchn. = 1 tun.

Apothecaries' Fluid Measure

60 minims = 1 fl. drm.
8 drm. = 1 fl. oz.
1 fl. oz. = 1/128 U. S. gal. = 1.8047 cu. in. = 456.3 grn. H_2O at 39° F.
16 fl. oz. = 1 pint
1 fl. oz. (British) = 1.732 cu. in. = 437.5 grn. H_2O at 62° F.

Dry Measure

2 pints = 1 qt.
8 qts. = 1 peck
4 pecks = 2150.42 cu. in. = 1 bu. (struck)
1 1/4 bu. (struck) = 1 bu. (heaped)
105 qts. (struck) = 7056 cu. in. = 1 bbl.
8 Imp. gal. = 2218.192 cu. in. = 1 Imp. bu.
8 Imp. bu. = 1 qr.

Shipping Measure

100 cu. ft. = 1 Register ton *
40 cu. ft. = 31.143 U. S. bu. = 1 shipping ton (U. S.) †
42 cu. ft. = 32.719 Imp. bu. = 1 shipping ton (British) †

* For measurement of entire internal capacity. † For measurement of cargo.

Miner's Inch

A variable unit, defined by statute in different states as the quantity of water that will flow through an orifice of 1 sq. in. area, under a head ranging from 4 1/2 to 6 in. U. S. Reclamation Service defines the miner's inch as 1 cu. in. per sec. The law governing the locality in which the measurement is to be made should be consulted.

Table 23. Circular Measure

60 sec. = 1 min.
60 min. = 1 deg.
90 deg. = 1 quad.
360 deg. = circm.
57.2957795 deg. = 1 radian
= 57° 17′ 44.806″
1 radian = angle of arc equal to radius.

Table 24. Time

60 sec. = 1 min. 7 days = 1 week
60 min. = 1 hr. 52 weeks = 1 year
24 hr. = 1 day

1 year (exact) = 365 days, 5 hr., 48 min., 45.6 sec.

A year divisible by 4 is a leap year and contains 366 days, except centesimal years, of which only those divisible by 400 are leap years.

365.24734 mean solar days = 1 yr.
= 366.008515 sidereal days
24 hr. mean solar time
= 23 hr. 56 min., 49.9 sec. sidereal time
1 mean solar day
= 1 mean sidereal day + 3 min., 10.1 sec.

THE METRIC SYSTEM

The fundamental standard of length in the metric system is the **International Standard Meter,** derived from the Metre des Archives, Paris. Its length is defined as the distance between two lines at 0° C. on a platinum-iridium bar deposited at the International Bureau of Weights and Measures near Paris. 1 meter = 39.37 in.

The fundamental standard of mass is the International Standard Kilogram, which is a mass of platinum-iridium, deposited at the International Bureau, whose weight *in vacuo* is the same as that of the Kilogram des Archives.

The fundamental unit of volume is the **liter,** which is defined as the volume of a kilogram of water at the temperature of maximum density, 4° C., under a pressure of 76 cm. of mercury. The liter originally was defined as a cubic decimeter, but is slightly greater. The generally accepted relation between the cubic decimeter and the liter is 1 liter = 1.000027 cu. decimeters.

The unit of force is the **dyne,** which is the force which acting for 1 second on a mass of one gram will produce a velocity of 1 centimeter per second. 1 dyne = (1/980.665) gram. 1,000,000 dynes = 1.020 kilograms force = 2.248 lb. force. 1 lb. force = 0.4536 kilogram force = 444,800 dynes.

The unit of heat is the **gram-calorie,** which is the quantity of heat required to raise 1 gram of water 1° C.

Table 25. Units of Metric Measures*

Length			Area		
Unit	Symbol	Value in Meters	Unit	Symbol	Value in Sq. Meters
Micron	μ	0.000 001			
Millimeter	mm.	0.001	Sq. millimeter	sq. mm.	0.000 001
Centimeter	cm.	0.01	Sq. centimeter	sq. cm.	0.000 1
Decimeter	dm.	0.1	Sq. decimeter	sq. dm.	0.01
Meter (unit)	m.	1.0	Sq. meter (centiare)	sq. m.	1.0
Dekameter	dkm.	10.0	Sq. dekameter (are)	sq. dkm.	100.0
Hectometer	hm.	100.0	Hectare	ha.	10,000.0
Kilometer	km.	1,000.0	Sq. kilometer	sq. km.	1,000,000.0
Myriameter	Mm.	10,000.0			
Megameter	....	1,000,000.0			

Cubic Measure			Volume		
Unit	Symbol	Value in Cubic Meters	Unit	Symbol	Value in Liters
Cubic kilometer	cu. km.	1,000,000,000	Milliliter	ml.	0.001
Cubic hectometer	cu. hm.	1,000,000	Centiliter	cl.	0.01
Cubic dekameter	cu. dkm.	1,000	Deciliter	dl.	0.1
Cubic meter	cu. m.	1	Liter (unit)	l.	1.0
Cubic decimeter	cu. dm.	0.001	Dekaliter	dkl.	10.0
Cubic centimeter	cu. cm.	0.000 001	Hectoliter	hl.	100.0
Cubic millimeter	cu. mm.	0.000 000 001	Kiloliter	kl.	1,000.0
Cubic micron	μ^3	$0.0_{17}1$			

Weight					
Unit	Symbol	Value in Grams	Unit	Symbol	Value in Grams.
Microgram	γ	0.000 001	Gram (unit)	g.	1.0
Milligram	mg.	0.001	Dekagram	dkg.	10.0
Centigram	cg.	0.01	Hectogram	hg.	100.0
Decigram	dg.	0.1	Kilogram	kg.	1,000.0
Gram (unit)	g.	1.0	Myriagram	Mg.	10,000.0
			Quintal	q.	100,000.0
			Ton	t.	1,000,000.0

* A subscript after a figure indicates the number of times it is repeated. Thus $0.0_38 = 0.0008$

Table 26. Metric Equivalents of Length *

English to Metric			Log.	Metric to English			Log.
1 inch =	25.4	mm.	1.404834	1 mm. =	0.03937	in.	$\bar{2}.595165$
=	2.54	cm.	0.404834	=	0.003281	ft.	$\bar{3}.515984$
=	0.0254	m.	$\bar{2}.404834$	=	0.001094	yd.	$\bar{3}.038863$
1 foot =	304.800	mm.	2.484015	1 cm. =	0.3937	in.	$\bar{1}.595165$
=	30.480	cm.	1.484015	=	0.03281	ft.	$\bar{2}.515984$
=	0.3048	m.	$\bar{1}.484015$	=	0.01094	yd.	$\bar{2}.038863$
1 yard =	91.4402	cm.	1.961137	1 meter =	39.37	in.	1.595162
=	0.9144	m.	$\bar{1}.961137$	=	3.2808	ft.	0.515984
=	0.0_3914	km.	$\bar{4}.961137$	=	1.0936	yd.	0.038863
1 rod =	502.9211	cm.	2.701500	=	0.1988	rd.	$\bar{1}.298500$
=	5.0292	m.	0.701500	=	0.04971	chain	$\bar{2}.696440$
=	0.005029	km.	$\bar{3}.701500$	=	0.0_36214	mi.	$\bar{4}.793350$
1 chain =	2011.68	cm.	3.303559	1 kilometer =	3280.833	ft.	3.515984
=	20.1168	m.	1.303559	=	1093.611	yd.	3.038863
=	0.02012	km.	$\bar{2}.303559$	=	198.838	rods	2.298500
1 mile =	1609.344	m.	3.206650	=	49.7095	chains	1.696440
=	1.6093	km.	0.206650	=	0.6214	mi.	$\bar{1}.793350$

Table 27. Metric Equivalents of Area *

English to Metric			Log.	Metric to English			Log.
1 cir. mil =	0.0_35067	sq. mm.	$\bar{4}.704751$	1 sq. mm. =	1,973.55	cir. mils	3.295249
1 sq. in. =	645.163	sq. mm.	2.809669	=	0.001550	sq. in.	$\bar{3}.190331$
=	6.4516	sq. cm.	0.809669	=	0.0_410764	sq. ft.	$\bar{5}.031968$
=	0.0_36452	sq. m.	$\bar{4}.809669$	=	0.0_51196	sq. yd.	$\bar{6}.077726$
1 sq. ft. =	92,903.41	sq. mm.	4.968032	1 sq. cm. =	0.1550	sq. in.	$\bar{1}.190331$
=	929.0341	sq. cm.	2.968032	=	0.001076	sq. ft.	$\bar{3}.031968$
=	0.0929	sq. m.	$\bar{2}.968032$	=	0.0_31196	sq. yd.	$\bar{4}.077726$
=	0.0_5929	hectare	$\bar{6}.968032$	1 sq. m. =	1,549.9969	sq. in.	3.190331
=	0.0_7929	sq. km.	$\bar{8}.968032$	=	10.7639	sq. ft.	1.031968
1 sq. yd. =	836,130.74	sq. mm.	5.922274	=	1.1960	sq. yd.	0.077726
=	8,361.307	sq. cm.	3.922274	=	0.002471	sq. chain	$\bar{3}.392882$
=	0.83613	sq. m.	$\bar{1}.922274$	=	0.0_32471	acre	$\bar{4}.392882$
=	0.0_4836	hectare	$\bar{5}.922274$	=	0.0_63861	sq. mi.	$\bar{7}.586700$
=	0.0_6836	sq. km.	$\bar{7}.922274$	1 hectare =	107,638.7	sq. ft.	5.031968
1 sq. chain =	404.686	sq. m.	2.607118	=	11,959.85	sq. yd.	4.077726
=	0.04047	hectare	$\bar{2}.607118$	=	24.710	sq. chain	1.392882
=	0.0_34047	sq. km.	$\bar{4}.607118$	=	2.4710	acres	0.392882
1 acre =	4,046.86	sq. m.	3.607118	=	0.003861	sq. mi.	$\bar{3}.586700$
=	0.4047	hectare	$\bar{1}.607118$	1 sq. km. =	10,763,867.36	sq. ft.	7.031968
=	0.004047	sq. km.	$\bar{3}.607118$	=	1,195,985.26	sq. yd.	6.077726
1 sq. mile =	2,589,998	sq. m.	6.413300	=	2,471.050	sq. chains	3.392882
=	259.0	hectare	2.413300	=	247.1045	acres	2.392882
=	2.590	sq. km.	0.413300	=	0.3861	sq. mi.	$\bar{1}.586700$

Table 28. Metric Equivalents of Weight or Mass *

English to Metric			Log.	Metric to English			Log.
1 grain =	64.797	mg.	1.811568	1 milligram =	0.01543	grain	$\bar{2}.188433$
=	0.0648	gm.	$\bar{2}.811568$	=	0.0_43215	oz. Troy	$\bar{5}.507191$
1 oz. Troy or				=	0.0_43527	oz. avoir.	$\bar{5}.547454$
apothecary =	31,103.5	mg.	4.492809	1 gram =	15.4324	grains	1.188433
=	31.103	gm.	1.492809	=	0.03215	oz. Troy	$\bar{2}.507191$
=	0.03110	kg.	$\bar{2}.492809$	=	0.03527	oz. avoir.	$\bar{2}.547454$
1 oz. avoirdupois =	28,349.5	mg.	4.452546	=	0.0_22679	lb. Troy	$\bar{3}.428010$
=	28.3495	gm.	1.452546	=	0.0_22205	lb. avoir.	$\bar{3}.343334$
=	0.02835	kg.	$\bar{2}.452546$	1 kilogram =	32.1508	oz. Troy	1.507191
1 pound Troy or				=	35.2740	oz. avoir.	1.547454
apothecary =	373.2417	gm.	2.571990	=	2.6792	lb. Troy	0.428010
=	0.3732	kg.	$\bar{1}.571990$	=	2.2046	lb. avoir.	0.343334
1 pound avoirdupois =	453.5925	gm.	2.656666	=	0.0_21102	short ton	$\bar{3}.043304$
=	0.4536	kg.	$\bar{1}.656666$	=	0.0_39842	long ton	$\bar{4}.993086$
=	0.0_34536	metric ton	$\bar{4}.656666$	1 metric ton =	2204.62	lb. avoir.	3.343334
1 short ton =	907.2	kg.	2.957696	=	1.1023	short tons	0.043304
=	0.9072	metric ton	$\bar{1}.957696$	=	0.9842	long ton	$\bar{1}.993086$
1 long ton =	1,016.06	kg.	3.006914				
=	1.0161	metric ton	0.006914				

* A subscript after a figure indicates the number of times it is repeated. Thus $0.0_38 = 0.0008$

Table 29. Metric Equivalents of Capacity or Volume *

English to Metric	Log.	Metric to English	Log.
1 cu. in. = 16,387.17 cu. mm.	4.214504	1 cu. mm. = 0.0_46102 cu. in.	$\bar{5}.785496$
= 16.3872 cu. cm.	1.214504	= 0.0_32705 fluid dr.	$\bar{4}.432185$
= 0.016387 l.	$\bar{2}.214504$	= 0.0_43381 fluid oz.	$\bar{5}.529094$
= 0.0_316387 hl.	$\bar{4}.214504$	1 cu. cm. = 0.06102 cu. in.	$\bar{2}.785496$
= 0.0_416387 cu. m.	$\bar{5}.214504$	= 0.0_43531 cu. ft.	$\bar{5}.547951$
1 cu. ft. = 28,317.08 cu. cm.	4.452049	= 0.0_51308 cu. yd.	$\bar{6}.116589$
= 28.3169 l.	1.452049	= 0.0_42838 bushel	$\bar{5}.452972$
= 0.2832 hl.	$\bar{1}.452049$	= 0.2705 fluid dr.	$\bar{1}.432185$
= 0.02832 cu.m.	$\bar{2}.452049$	= 0.03381 fluid oz.	$\bar{2}.529094$
1 cu. yd. = 764,559.5 cu. cm.	5.883411	= 0.001057 quart	$\bar{3}.023944$
= 764.5595 l.	2.883411	= 0.0_32642 gallon	$\bar{4}.421884$
= 7.6456 hl.	0.883411	**1 liter** = 61.02398 cu. in.	1.785496
= 0.7646 cu. m.	$\bar{1}.883411$	= 0.035313 cu. ft.	$\bar{2}.547951$
1 bushel = 35,239.3 cu. cm.	4.547027	= 0.0013079 cu. yd.	$\bar{3}.116589$
= 35.2393 l.	1.547027	= 0.028377 bushel	$\bar{2}.452972$
= 0.3524 hl.	$\bar{1}.547027$	= 1.0567 quart	0.023944
= 0.03524 cu. m.	$\bar{2}.547027$	= 0.2642 gallon	$\bar{1}.421884$
1 fluid drachm = 3,696.7 cu. mm.	3.567815	**1 hectoliter** = 6,102.398 cu. in.	3.785496
= 3.6967 cu. cm.	0.567815	= 3.5313 cu. ft.	0.547951
1 fluid oz. = 29,573.7 cu. mm.	4.470906	= 0.13079 cu. yd.	$\bar{1}.116589$
= 29.5737 cu. cm.	1.470906	= 2.8377 bushels	0.452972
1 quart = 946.359 cu. cm.	2.976056	1 cu. meter = 61,023.38 cu. in.	4.785496
= 0.9464 l.	$\bar{1}.976056$	= 35.3133 cu. ft.	1.547951
= 0.0_3946 cu. m.	$\bar{4}.976056$	= 1.3079 cu. yd.	0.116589
1 U. S. gallon = 3,785.43 cu. cm.	3.578116	= 28.3773 bushels	1.452972
= 3.7854 l.	0.578116	= 1,056.682 quarts	3.023944
= 0.003785 cu. m.	$\bar{3}.578116$	= 264.170 gallons	2.421884

Table 30. Metric Equivalents of Density *

English to Metric	Log.	Metric to English	Log.
1 lb. per cu. in. = 27.680 gm. per cu. cm.	1.442162	1 gm. per cu. cm. = 0.03613 lb. per cu. in.	$\bar{2}.557838$
= 27,679.7 kg. per cu. m.	4.442162	= 62.430 lb. per cu. ft.	1.795381
1 lb. per cu. ft. = 0.01602 gm. per cu. cm.	$\bar{2}.204619$	= 8.3454 lb. per U. S. gal.	0.921450
= 16.0183 kg. per cu. m.	1.204619	**1 kg. per cu. m.** = 0.0_43613 lb. per cu. in.	$\bar{5}.557838$
= 0.01602 metric ton per cu. m.	$\bar{2}.204619$	= 0.062430 lb. per cu. ft.	$\bar{2}.795381$
1 lb. per cu. yd. = 0.5933 kg. per cu. m.	$\bar{1}.773254$	= 1.6856 lb. per cu. yd.	0.226746
= 0.0_35933 metric tons per cu. m.	$\bar{4}.773254$	= 0.0_28345 lb. per U. S. gal.	$\bar{3}.921450$
1 lb. per U. S. gal. = 0.1198 gm. per cu. cm.	$\bar{1}.078550$	1 metric ton per cu. m = 62.4286 lb. per cu. ft.	1.795381
= 119.826 kg. per cu. m.	2.078550	= 1685.487 lb. per cu. yd.	3.226746
1 shortton per cu. yd. = 1.1865 metric tons per cu. m.	0.074285	= 0.8428 short ton per cu. yd.	$\bar{1}.925715$
1 long ton per cu. yd. = 1.3289 metric tons per cu. m.	0.123502	= 0.7525 long ton per cu. yd.	$\bar{1}.876498$

Table 31. Metric Equivalents of Velocity *

English to Metric	Log.	Metric to English	Log.
1 in. per sec., = 2.54001 cm. per sec.	0.404834	1 cm. per sec. = 0.3937 in. per sec.	$\bar{1}.595165$
= 0.0254 m. per sec.	$\bar{2}.404834$	= 0.03281 ft. per sec.	$\bar{2}.515984$
= 1.5240 m. per min.	0.182985	= 1.9685 ft. per min.	0.294136
1 ft. per sec. = 30.4801 cm. per sec.	1.484015	= 0.02237 mi. per hr.	$\bar{2}.349653$
= 0.3048 m. per sec.	$\bar{1}.484015$	= 0.01943 knot	$\bar{2}.288367$
= 18.2880 m. per min.	1.262166	**1 m.** per sec. = 39.37 in. per sec.	1.595165
= 1.0973 km. per hr.	0.040318	= 3.2808 ft. per sec.	0.515984
1 ft. per min. = 0.5080 cm. per sec.	$\bar{1}.705864$	= 196.8500 ft. per min.	2.294136
= 0.00508 m. per sec.	$\bar{3}.705864$	= 2.2369 mi. per hr.	0.349653
= 0.3048 m. per min.	$\bar{1}.484015$	= 1.9426 knots	0.288367
= 0.018288 km. per hr.	$\bar{2}.262166$	**1 m.** per min. = 0.6562 in. per sec.	$\bar{1}.817014$
1 mile per hr. = 44.704 cm. per sec.	1.650348	= 0.05468 ft. per sec.	$\bar{2}.737833$
= 0.4470 m. per sec.	$\bar{1}.650348$	= 3.2808 ft. per min.	0.515984
= 26.8222 m. per min.	1.428495	= 0.03728 mi. per hr.	$\bar{2}.571503$
= 1.6093 km. per hr.	0.206650	= 0.03238 knot	$\bar{2}.510218$
1 knot = 51.4791 cm. per sec.	1.711631	1 km. per hr. = 0.9113 ft. per sec.	$\bar{1}.959681$
= 0.5148 m. per sec.	$\bar{1}.711631$	= 54.6806 ft. per min.	1.737833
= 30.8875 m. per min.	1.489782	= 0.62138 mi. per hr.	$\bar{1}.793357$
= 1.8532 km. per hr.	0.267933	= 0.5396 knot	$\bar{1}.732072$

* A subscript after a figure indicates the number of times it is repeated. Thus $0.0_38 = 0.0008$

Table 32. Metric Equivalents of Pressure *

English to Metric	Log.	Metric to English	Log.
1 lb. per sq. in. = 0.7031 gm. per sq. mm.	$\bar{1}.846996$	1 gm. per sq. mm. = 1.4223 lb. per sq. in.	0.153004
= 70.3066 gm. per sq. cm.	2.846996	= 204.8170 lb. per sq. ft.	2.311366
= 0.07031 kg. per sq. cm.	$\bar{2}.846996$	1 gm. per sq. cm. = 0.01422 lb. per sq. in.	$\bar{2}.153004$
= 0.7031 metric ton per sq. m.	$\bar{1}.846996$	= 2.0482 lb. per sq. ft.	0.311366
= 0.07031 metric atmosphere†	$\bar{2}.846996$	1 kg. per sq. cm. = 14.2234 lb. per sq. in.	1.153004
1 lb. per sq. ft. = 0.004882 gm. per sq. mm.	$\bar{3}.688598$	= 2048.1696 lb. per sq. ft.	3.311366
= 0.4882 gm. per sq. cm.	$\bar{1}.688598$	1 metric ton per sq. m. = 1.4223 lb. per sq. in.	0.153004
= 0.0_34882 kg. per sq. cm.	$\bar{4}.688598$	= 204.8170 lb. per sq. ft.	2.311366
= 0.004882 metric ton per sq. m.	$\bar{3}.688598$	= 0.1024 ton per sq. ft.§	$\bar{1}.010342$
= 0.0_34882 metric atmosphere†	$\bar{4}.688598$	1 metric atmosphere† = 14.2234 lb. per sq. in.	1.153004
1 ton per sq. ft.† = 9.7648 metric tons per sq. m.	0.989663	= 2048.1696 lb. per sq. ft.	3.311366
= 0.9765 metric atmosphere†	$\bar{1}.989663$	= 1.0241 tons per sq. ft.	0.010342
1 atmosphere ‡ = 1.0335 metric atmospheres †	0.014310	= 0.9676 atmosphere ‡	$\bar{1}.985696$

† 1 metric atmosphere = 1 kg. per sq. cm. ‡ 1 atmosphere = 14.7 lb. per sq. in. § 1 ton = 2000 lb.

Table 33. Metric Equivalents of Force *

English to Metric	Log.	Metric to English	Log.
1 lb. = 444,800 dynes	5.648165	1 dyne = 0.0_52248 lb.	$\bar{6}.351796$
= 453.6 grams	2.656673	= 0.0_47233 poundal	$\bar{5}.859318$
= 0.04448 joule per cm.	$\bar{2}.648165$	= 0.0_21020 gram	$\bar{3}.008600$
= 0.4536 kg.	$\bar{1}.656673$	= 0.0_51020 kg.	$\bar{6}.008600$
= 32.17 poundals	1.507451	= 0.0_61 joule per cm.	$\bar{7}.000000$
1 poundal = 13,826 dynes	4.140696	1 gram = 0.0_22205 lb.	$\bar{3}.343409$
= 14.10 grams	1.149219	= 0.07093 poundal	$\bar{2}.850830$
= 0.0_213826 joule per cm.	$\bar{3}.140696$	= 980.7 dynes	2.991536
= 0.01410 kg.	$\bar{2}.149219$	= 0.001 kg.	$\bar{3}.000000$
= 0.03108 lb.	$\bar{2}.492481$	= 0.0_49807 joule per cm.	$\bar{5}.991536$
		1 joule per cm. = 22.48 lb.	1.351796
		= 723.3 poundals	2.859318
		= 10,000,000 dynes	7.000000
		= 10,200 grams	4.008600
		= 10.20 kg.	1.008600
		1 kg. = 2.205 lb.	0.343409
		= 70.93 poundals	1.850830
		= 980,700 dynes	5.991536
		= 1000 grams	3.000000
		= 0.09807 joule per cm.	2.991536

Table 34. Metric Equivalents of Power, Work, and Energy *

English to Metric	Log	Metric to English	Log
1 ft-lb = 0.138255 kg-m	$\bar{1}.140680$	1 kg-m = 7.2330 ft-lb	0.859318
= 0.0_637650 kwhr	$\bar{7}.575765$	= $0.0_5365304$ hp-hr	$\bar{6}.562655$
= 0.32379 IT cal	$\bar{1}.510263$	= 0.0_292938 Btu	$\bar{3}.968193$
1 hp = 76.042 kg-m per sec	1.881053	1 kwhr = 2.656000 ft-lb	6.424228
= 0.76042 poncelet	$\bar{1}.881053$	= 1.3414 hp-hr	0.127558
= 1.01389 cheval vapeur	0.005994	= 3412.75 Btu	3.533103
1 hp-hr = 273,750 kg-m	5.437355	1 cheval vapeur = 0.9863 hp	$\bar{1}.994009$
= 0.74548 kwhr	$\bar{1}.872435$	1 poncelet = 1.3151 hp	0.118959
= 641,100 IT cal	5.806926	1 IT cal † = 3.0884 ft-lb	0.489738
1 Poncelet = 100 kg-m per sec	2.000000	= $0.0_5155980$ hp-hr	$\bar{6}.193070$
1 cheval-vapeur = 75 kg-m per sec	1.875061	= 0.0_239683 Btu	$\bar{3}.598605$
1 Btu = 251.996 IT cal	2.401400	= 1.00037 mean calories	0.000160
= 107.599 kg-m	2.031808		
= 0.0_329302 kwhr	$\bar{4}.466898$		

* A subscript after a figure indicates the number of times it is repeated. Thus 0.0_38 = 0.0008.

† The IT cal is defined as 1/860 of the international watthour, and is nearly equal to the mean calorie; 1 IT cal = 1.00037 mean calories, or 1 Btu = 251.996 IT cal. (See *Mechanical Engineering*, Nov. 1935, p. 710.)

SYMBOLS AND ABBREVIATIONS

Greek Letters

Α	α	Alpha	Η	η	Eta	Ν	ν	Nu	Τ	τ	Tau
Β	β	Beta	Θ	ϑ θ	Theta	Ξ	ξ	Xi	Υ	υ	Upsilon
Γ	γ	Gamma	Ι	ι	Iota	Ο	ο	Omicron	Φ	φ	Phi
Δ	δ	Delta	Κ	κ	Kappa	Π	π	Pi	Χ	χ	Chi
Ε	ε	Epsilon	Λ	λ	Lambda	Ρ	ρ	Rho	Ψ	ψ	Psi
Ζ	ζ	Zeta	Μ	μ	Mu	Σ	σ ς	Sigma	Ω	ω	Omega

Mathematical Signs and Abbreviations

$+$ plus (addition).
$+$ positive.
$-$ minus (subtraction).
$-$ negative.
$\pm$ plus or minus.
$\mp$ minus or plus.
$=$ equals.
$\geqq$ equals or greater than.
$\leqq$ equals or is less than.
$\approx$ approximately equals.
$\times$ multiplied by.
ab or $a.b = a \times b$.
$\div$ divided by.
$/$ divided by.
$\frac{a}{b} = a/b = a \div b$. $15/16 = \frac{15}{16}$.
$0.2 = \frac{2}{10}$; $0.002 = \frac{2}{1000}$.
$\sqrt{\ }$ square root.
$\sqrt[3]{\ }$ cube root.
$\sqrt[4]{\ }$ 4th root.
: is to, : : so is, : to (proportion).
2 : 4 : : 3 : 6, 2 is to 4 as 3 is to 6.
: ratio; divided by.
2 : 4, ratio of 2 to 4 = 2/4.
$>$ greater than.
$<$ less than.
° degrees, arc or thermometer.
′ minutes or feet.
″ seconds or inches.
′ ″ ‴ accents to distinguish letters, as a', a'', a'''.
a_1, a_2, a_3, a_b, a_c, read a sub 1, a sub b, etc.
( ) [ ] { } —— parentheses, brackets, braces, vinculum; denoting that the numbers enclosed are to be taken together; as, $(a + b)c = \overline{4 + 3} \times 5 = 35$.
a^2, a^3, a squared, a cubed.

a^n, a raised to the nth power.
$a^{2/3} = \sqrt[3]{a^2}$, $a^{3/2} = \sqrt{a^3}$.
$a^{-1} = \frac{1}{a}$, $a^{-2} = \frac{1}{a^2}$.
10^9 = 10 to the 9th power = 1,000,000,000.
$\sin a$ = the sine of a.
$\sin^{-1} a$ = the arc whose sine is a.
$\sin a^{-1} = \frac{1}{\sin a}$.
log = logarithm.
$\log_e$ or hyp log = hyperbolic logarithm.
% per cent.
∠ angle.
△ triangle.
sin, sine.
cos, cosine.
tan, tangent.
sec, secant.
versin, versed sine.
cot, cotangent.
cosec, cosecant.
covers, co-versed sine.

In Algebra, the first letters of the alphabet, a, b, c, d, etc., are generally used to denote known quantities, and the last letters, w, x, y, z, etc., unknown quantities.

Abbreviations and Symbols commonly used.

d, differential (in calculus).
$\int$, integral (in calculus).
$\int_b^a$, integral between limits a and b.
Δ, delta, difference.
Σ, sigma, sign of summation.
π, pi, ratio of circumference of circle to diameter = 3.14159.
g, acceleration due to gravity = 32.16 ft. per second per second.

INDEX